2017 40th International Convention on Information and Communication Technology, Electronics and Microelectronics (MIPRO 2017)

Opatija, Croatia
22-26 May 2017

Pages 1-782

IEEE Catalog Number: CFP1739K-POD
ISBN: 978-1-5090-4969-1

Copyright © 2017, Croatian Society for Information and Communication Technology, Electronics and Microelectronics (MIPRO)
All Rights Reserved

*** This is a print representation of what appears in the IEEE Digital Library. Some format issues inherent in the e-media version may also appear in this print version.

IEEE Catalog Number: CFP1739K-POD
ISBN (Print-On-Demand):   978-1-5090-4969-1
ISBN (Online):   978-953-233-090-8

Additional Copies of This Publication Are Available From:

Curran Associates, Inc
57 Morehouse Lane
Red Hook, NY 12571 USA
Phone: (845) 758-0400
Fax: (845) 758-2633
E-mail: curran@proceedings.com
Web: www.proceedings.com

2017 40th International Convention on Information and Communication Technology, Electronics and Microelectronics (MIPRO)

May 22 – 26, 2017
Opatija, Croatia

Proceedings

Edited by:
Petar Biljanovic
Marko Koricic
Karolj Skala
Tihana Galinac Grbac
Marina Cicin-Sain
Vlado Sruk
Slobodan Ribaric
Stjepan Gros
Boris Vrdoljak
Mladen Mauher
Edvard Tijan
Filip Hormot

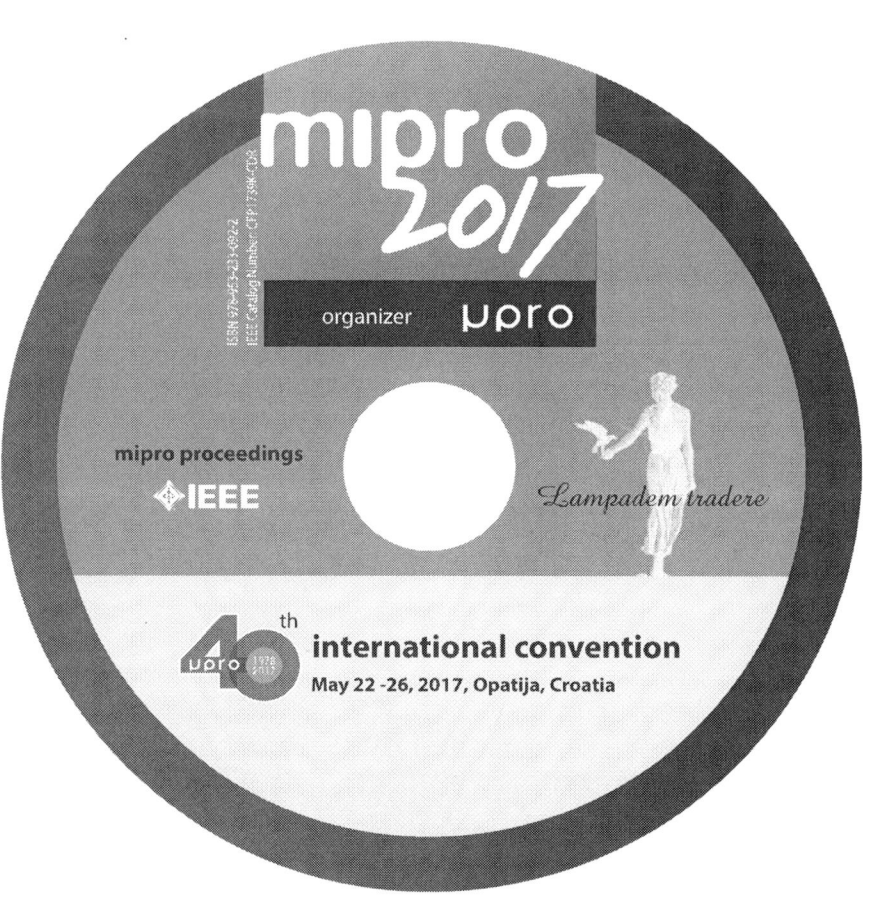

Introduction

The Conference Proceedings contain papers accepted for the 2017 40th International Convention on Information and Communication Technology, Electronics and Microelectronics (MIPRO) held from 22 to 26 May 2017 at the Grand Hotel Adriatic Congress Centre and Hotel Admiral in Opatija, organized by MIPRO Croatian Society and technically co-sponsored by IEEE Region 8.

Papers are from the following fields:

Microelectronics, Electronics and Electronic Technology /MEET
Distributed Computing, Visualization and Biomedical Engineering /DC VIS
Dew Computing /DEWCOM
Telecommunications & Information /CTI
Modeling System Behaviour/MSB
Computers in Education /CE
Computers in Technical Systems /CTS
Intelligent Systems /CIS
Information Systems Security /ISS
Business Intelligence Systems /miproBIS
Digital Economy and Government, Local Government, Public Services /DE-GLGPS
MIPRO Junior - Student Papers /SP

The authors are from industry, education, academia and public administration.

MIPRO 2017 Convention was held under the patronage of the Government of Croatia.
It was supported by many sponsors and patrons among which we single out HEP-Croatian Electricity Company Zagreb, Ericsson Nikola Tesla Zagreb, Koncar-Electrical Industries Zagreb, T-Croatian Telecom Zagreb, City of Opatija, InfoDom Zagreb, IN2 Zagreb, Transmitters and Communications Company Zagreb, King-ICT Zagreb, Hewlett Packard Croatia Zagreb, Storm Computers Zagreb, Danieli Automation Buttrio, VIPNet Zagreb, Mjerne tehnologije Zagreb, Selmet Zagreb, Institute SDT Ljubljana, Nomen Rijeka, EuroCloud Croatia, University of Zagreb, University of Rijeka, IEEE Croatia Section, IEEE Croatia Section Computer Chapter, IEEE Croatia Section Electron Devices/Solid-State Circuits Joint Chapter, IEEE Croatia Section Education Chapter, IEEE Croatia Section Communications Chapter, University of Zagreb Faculty of Electrical Engineering and Computing Zagreb, Rudjer Boskovic Institute Zagreb, University of Rijeka Faculty of Maritime Studies, Faculty of Engineering and Faculty of Economics, Faculty of Tourism and Hospitality Management, Faculty of Organization and Informatics Varazdin, University of Applied Sciences Zagreb, Croatian Regulatory Authority for Network Industries, CISEx, Kermas energija Zagreb, Business center Silos Rijeka and River Publishers.

The purpose of the Proceedings is to present the state of development and work within the ICT, electronics and microelectronics field in the world, particularly in countries of Southeast Europe – we hope we have been successful in doing that.

In Opatija/Rijeka/Zagreb, June 20th 2017

Professor Petar Biljanovic
MIPRO 2017 General Chair

The Government of the Republic of Croatia is a Patron of the convention

organized by
MIPRO Croatian Society

technical cosponsorship
IEEE Region 8

under the auspices of
Ministry of Science and Education of the Republic of Croatia
Ministry of the Sea, Transport and Infrastructure of the Republic of Croatia
Ministry of Economy, Entrepreneurship and Crafts of the Republic of Croatia
Ministry of Public Administration of the Republic of Croatia
Central State Office for the Development of Digital Society
Croatian Chamber of Economy
Primorje-Gorski Kotar County
City of Rijeka
City of Opatija
Croatian Regulatory Authority for Network Industries
Croatian Power Exchange - CROPEX

patrons
University of Zagreb, Croatia
University of Rijeka, Croatia
IEEE Croatia Section
IEEE Croatia Section Computer Chapter
IEEE Croatia Section Electron Devices/Solid-State Circuits Joint Chapter
IEEE Croatia Section Education Chapter
IEEE Croatia Section Communications Chapter
T-Croatian Telecom, Zagreb, Croatia
Ericsson Nikola Tesla, Zagreb, Croatia
Koncar - Electrical Industries, Zagreb, Croatia
HEP - Croatian Electricity Company, Zagreb, Croatia
VIPnet, Zagreb, Croatia
University of Zagreb, Faculty of Electrical Engineering and Computing, Croatia
Rudjer Boskovic Institute, Zagreb, Croatia
University of Rijeka, Faculty of Maritime Studies, Croatia
University of Rijeka, Faculty of Engineering, Croatia
University of Rijeka, Faculty of Economics, Croatia
University of Zagreb, Faculty of Organization and Informatics, Varazdin, Croatia
University of Rijeka, Faculty of Tourism and Hospitality Management, Opatija, Croatia
University of Applied Sciences, Croatia
EuroCloud Croatia
Croatian Regulatory Authority for Network Industries, Zagreb, Croatia
Selmet, Zagreb, Croatia
CISEx, Zagreb, Croatia
Kermas energija, Zagreb, Croatia
Business Center Silos, Rijeka, Croatia
River Publishers, Aalborg, Denmark

general sponsor
HEP - Croatian Electricity Company, Zagreb, Croatia

sponsors
Ericsson Nikola Tesla, Zagreb, Croatia
Koncar-Electrical Industries, Zagreb, Croatia
T-Croatian Telecom, Zagreb, Croatia
City of Opatija
InfoDom, Zagreb, Croatia
Hewlett Packard Croatia, Zagreb, Croatia
IN2, Zagreb, Croatia
King-ICT, Zagreb, Croatia
Storm Computers, Zagreb,
Croatia Transmitters and Communications Company, Zagreb, Croatia
VIPnet, Zagreb, Croatia
Danieli Automation, Buttrio, Italy
Mjerne tehnologije, Zagreb, Croatia
Selmet, Zagreb, Croatia
Institute SDT, Ljubljana, Slovenia
Nomen, Rijeka, Croatia
EuroCloud, Croatia

donor
Erste&Steiermärkische bank, Rijeka, Croatia

International Program Committee

Petar Biljanovic, General Chair, Croatia
S. Amon, Slovenia
V. Andjelic, Croatia
M.E. Auer, Austria
S. Babic, Croatia
A. Badnjevic, Bosnia and Herzegovina
M. Baranovic, Croatia
B. Bebel, Poland
L. Bellatreche, France
E. Brenner, Austria
G. Brunetti, Italy
A. Budin, Croatia
Z. Butkovic, Croatia
Z. Car, Croatia
M. Colnaric, Slovenia
A. Cuzzocrea, Italy
M. Cicin-Sain, Croatia
M. Cupic, Croatia
M. Delimar, Croatia
T. Eavis, Canada
M. Ferrari, Italy
B. Fetaji, Macedonia
R. Filjar, Croatia
T. Galinac Grbac, Croatia
P. Garza, Italy
L. Gavrilovska, Macedonia
M. Golfarelli, Italy
S. Golubic, Croatia
F. Gregoretti, Italy
S. Gros, Croatia
N. Guid, Slovenia
J. Henno, Estonia
L. Hluchy, Slovakia
V. Hudek, Croatia
Z. Hutinski, Croatia
M. Ivanda, Croatia
H. Jaakkola, Finland
L. Jelenkovic, Croatia
D. Jevtic, Croatia
R. Jones, Switzerland
P. Kacsuk, Hungary

A. Karaivanova, Bulgaria
M. Koricic, Croatia
T. Kosanovic, Croatia
M. Mauher, Croatia
I. Mekjavic, Slovenia
B. Mikac, Croatia
V. Milutinovic, Serbia
N. Miskovic, Croatia
V. Mrvos, Croatia
J.F. Novak, Croatia
J. Pardillo, Spain
N. Pavesic, Slovenia
V. Persic, Croatia
S. Ribaric, Croatia
J. Rozman, Slovenia
K. Skala, Croatia
I. Sluganovic, Croatia
M. Spremic, Croatia
V. Sruk, Croatia
S. Stafisso, Italy
U. Stanic, Slovenia
N. Stojadinovic, Serbia
M. Stupicic, Croatia
J. Sunde, Australia
A. Szabo, IEEE Croatia Section
L. Szirmay-Kalos, Hungary
D. Simunic, Croatia
Z. Simunic, Croatia
D. Skvorc, Croatia
A. Teixeira, Portugal
E. Tijan, Croatia
A.M. Tjoa, Austria
R. Trobec, Slovenia
S. Uran, Croatia
T. Vámos, Hungary
M. Varga, Croatia
M. Vidas-Bubanja, Serbia
M. Vranic, Croatia
B. Vrdoljak, Croatia
D. Zazula, Slovenia

List of paper reviewers

Aksentijevic, S.	(Croatia)	Gamberger, D.	(Croatia)
Antolic, Z.	(Croatia)	Gamulin, O.	(Croatia)
Antonic, A.	(Croatia)	Garza, P.	(Italy)
Asenbrener Katic, M.	(Croatia)	Geric, S.	(Croatia)
Avbelj, V.	(Slovenia)	Giedrimas, V.	(Lithuania)
Babic, D.	(Croatia)	Gilhespy, M.	(Netherlands)
Babic, S.	(Croatia)	Glavas, J.	(Croatia)
Bacmaga, J.	(Croatia)	Golfarelli, M.	(Italy)
Bakalar, G.	(Croatia)	Golub, M.	(Croatia)
Bako, N.	(Croatia)	Golubic, S.	(Croatia)
Bala, P.	(Poland)	Gracin, D.	(Croatia)
Banek, M.	(Croatia)	Gradisnik, V.	(Croatia)
Baotic, M.	(Croatia)	Grbac, N.	(Croatia)
Bebel, B.	(Poland)	Grd, P.	(Croatia)
Begusic, D.	(Croatia)	Grguric, A.	(Croatia)
Bellatreche, L.	(France)	Gros, S.	(Croatia)
Biondic, I.	(Croatia)	Gulic, M.	(Croatia)
Bogdan, S.	(Croatia)	Gumzej, N.	(Croatia)
Brcic, M.	(Croatia)	Henno, J.	(Estonia)
Brezany, P.	(Austria)	Holenko Dlab, M.	(Croatia)
Brkic Bakaric, M.	(Croatia)	Horvat, M.	(Croatia)
Brkic, K.	(Croatia)	Hrkac, T.	(Croatia)
Brkic, L.	(Croatia)	Hudek, V.	(Croatia)
Brodnik, A.	(Slovenia)	Huljenic, D.	(Croatia)
Brscic, D.	(Croatia)	Humski, L.	(Croatia)
Budin, A.	(Croatia)	Hyrynsalmi, S.	(Finland)
Budin, L.	(Croatia)	Inkret, R.	(Croatia)
Butkovic, Z.	(Croatia)	Ipsic, I.	(Croatia)
Candrlic, S.	(Croatia)	Ivasic-Kos, M.	(Croatia)
Capko, Z.	(Croatia)	Ivosevic, D.	(Croatia)
Cavrak, I.	(Croatia)	Jaakkola, H.	(Finland)
Cicin-Sain, M.	(Croatia)	Jakobovic, D.	(Croatia)
Costa, J.	(Portugal)	Jakopovic, Z.	(Croatia)
Cubrilo, M.	(Croatia)	Jaksic, D.	(Croatia)
Cuzzocrea, A.	(Italy)	Jakupovic, A.	(Croatia)
Delac, G.	(Croatia)	Jardas, M.	(Croatia)
Djerek, A.	(Croatia)	Jelenkovic, L.	(Croatia)
Djerek, V.	(Croatia)	Jevtic, D.	(Croatia)
Dobrijevic, O.	(Croatia)	Jezic, G.	(Croatia)
Domitrovic, A.	(Croatia)	Joler, M.	(Croatia)
Dundjer, I.	(Croatia)	Josanov, B.	(Serbia)
Dzanko, M.	(Croatia)	Jovanovic, V.	(United States)
Dzapo, H.	(Croatia)	Jovic, A.	(Croatia)
Eavis, T.	(Canada)	Jurcevic Lulic, T.	(Croatia)
Fertalj, K.	(Croatia)	Jurdana, M.	(Croatia)
Filjar, R.	(Croatia)	Juricic, V.	(Croatia)
Frankovic, D.	(Croatia)	Jurisic, D.	(Croatia)
Frid, N.	(Croatia)	Kalafatic, Z.	(Croatia)
Galinac Grbac, T.	(Croatia)	Kalpic, D.	(Croatia)

Kastelan, I.	(Serbia)	Miletic, V.	(Croatia)
Katanic, N.	(Croatia)	Milic, L.	(Croatia)
Kaucic, B.	(Slovenia)	Milicevic, M.	(Croatia)
Kazi, Z.	(Serbia)	Milicic, S.	(Croatia)
Keto, H.	(Finland)	Min Tjoa, A.	(Austria)
Kljajic Borstnar, M.	(Slovenia)	Miskovic, N.	(Croatia)
Knezevic, K.	(Croatia)	Mohorcic, M.	(Slovenia)
Konecki, M.	(Croatia)	Molnar, G.	(Croatia)
Koricic, M.	(Croatia)	Nacinovic Prskalo, L.	(Croatia)
Kovacevic, T.	(Croatia)	Nikitovic, M.	(Croatia)
Kovacic, B.	(Croatia)	Nikolovski, S.	(Croatia)
Kovacic, Z.	(Croatia)	Ocko, M.	(Croatia)
Kozak, D.	(Croatia)	Oletic, D.	(Croatia)
Kragic, D.	(Sweden)	Oreski, D.	(Croatia)
Krapac, J.	(Croatia)	Orsulic, J.	(Croatia)
Krasna, M.	(Slovenia)	Pale, P.	(Croatia)
Krivec, S.	(Croatia)	Palestri, P.	(Italy)
Krois, I.	(Croatia)	Palomaki, J.	(Finland)
Krpic, Z.	(Croatia)	Pandzic, I.	(Croatia)
Kruzic, S.	(Croatia)	Pavcevic, M.	(Croatia)
Kuhar, U.	(Slovenia)	Pecar-Ilic, J.	(Croatia)
Kurdia, A.S.	(Croatia)	Pejcinovic, B.	(United States)
Kusek, M.	(Croatia)	Pelin, D.	(Croatia)
Kuusisto, M.	(Finland)	Peric Hadzic, A.	(Croatia)
Leppaniemi, J.	(Finland)	Petkovic, T.	(Croatia)
Lerga, J.	(Croatia)	Petrovic, T.	(Croatia)
Lesic, V.	(Croatia)	Pintar, D.	(Croatia)
Lucic, D.	(Croatia)	Poljak, M.	(Croatia)
Lukac, D.	(Germany)	Poscic, P.	(Croatia)
Ljubic, S.	(Croatia)	Prokopec, G.	(Croatia)
Magdalenic, I.	(Croatia)	Pusnik, M.	(Slovenia)
Makinen, T.	(Finland)	Rantanen, P.	(Finland)
Malaric, R.	(Croatia)	Rashkovska, A.	(Slovenia)
Malekovic, M.	(Croatia)	Repnik, R.	(Slovenia)
Males, L.	(Croatia)	Ribaric, S.	(Croatia)
Mandic, T.	(Croatia)	Rolich, T.	(Croatia)
Marcetic, D.	(Croatia)	Rupnik, R.	(Slovenia)
Markovic, I.	(Croatia)	Salamon, K.	(Croatia)
Marovic, M.	(Croatia)	Sarolic, A.	(Croatia)
Marsic, D.	(Croatia)	Schatten, M.	(Croatia)
Matetic, M.	(Croatia)	Seder, M.	(Croatia)
Matic, T.	(Croatia)	Sevrovic, M.	(Croatia)
Matijasevic, M.	(Croatia)	Silic, M.	(Croatia)
Mausa, G.	(Croatia)	Sillberg, P.	(Finland)
Mekovec, R.	(Croatia)	Simunic, D.	(Croatia)
Mekterovic, I.	(Croatia)	Skala, K.	(Croatia)
Mestrovic, A.	(Croatia)	Skocir, P.	(Slovenia)
Mihajlovic, Z.	(Croatia)	Skvorc, D.	(Croatia)
Mikac, B.	(Croatia)	Slivar, I.	(Croatia)
Mikuc, M.	(Croatia)	Sluganovic, I.	(Croatia)
Milanovic, I.	(Serbia)	Sluganovic, I.	(Croatia)
Milasinovic, B.	(Croatia)	Smuc, T.	(Croatia)

Soler, J.	(Denmark)	Trobec, R.	(Slovenia)
Solic, K.	(Croatia)	Trost, A.	(Slovenia)
Sprager, S.	(Slovenia)	Trzec, K.	(Croatia)
Spremic, M.	(Croatia)	Uroda, I.	(Croatia)
Stajduhar, I.	(Croatia)	Vasic, D.	(Croatia)
Stanic, U.	(Slovenia)	Vidacek-Hains, V.	(Croatia)
Staresinic, D.	(Croatia)	Vinko, D.	(Croatia)
Stipcevic, M.	(Croatia)	Vladimir, K.	(Croatia)
Stojkovic, N.	(Croatia)	Vlahovic, N.	(Croatia)
Struc, V.	(Slovenia)	Vojkovic, G.	(Croatia)
Subasic, M.	(Croatia)	Vranic, M.	(Croatia)
Sumak, B.	(Slovenia)	Vrdoljak, B.	(Croatia)
Sunde, V.	(Croatia)	Vrhovec, S.	(Slovenia)
Supic, H.	(Bosnia and Herzegovina)	Vukadinovic, D.	(Croatia)
Susac, F.	(Croatia)	Vukovic, M.	(Croatia)
Suznjevic, M.	(Croatia)	Vukovic, M.	(Croatia)
Svelec, D.	(Croatia)	Wrembel, R.	(Poland)
Tijan, E.	(Croatia)	Yrjonkoski, K.	(Finland)
Tomic, M.	(Croatia)	Zagar, D.	(Croatia)
Topic, M.	(Slovenia)	Zilak, J.	(Croatia)
Toth, Z.	(Hungary)	Zonja, S.	(Croatia)
Trancoso, I.	(Portugal)	Zulim, I.	(Croatia)

2017 40th International Convention on Information and Communication Technology, Electronics and Microelectronics (MIPRO)

Microelectronics, Electronics and Electronic Technology

Active Learning, Labs and Maker-spaces in Microwave Circuit Design Courses — 1
B. Pejcinovic

A Learning Tool for Synthesis, Visualization, and Editing of Programming for Simple Programmable Logic Devices — 7
M. Cupic, K. Brkic, Z. Mihajlovic

Active-Learning Implementation Proposal for Course Electronics at Undergraduate Level — 13
T. Mandic, A. Baric

Decision Trees in Formative Procedural Knowledge Assessment — 17
J. Petrovic, P. Pale

Glass Based Structures Fabricated by Rf-Sputtering — 21
A. Chiasera, F. Scotognella, D. Dorosz, G. Galzerano, A. Lukowiak, D. Ristic,
G. Speranza, I. Vasilchenko, A. Vaccari, S. Valligatla, S. Varas, L. Zur, M. Ivanda,
A. Martucci, G.C. Righini, S. Taccheo, R. Ramponi, M. Ferrari

Piezoresistive Effect in Composite Films Based on Polybenzimidazole and Few-Layered Graphene — 27
V.A. Kuznetsov, B.C. Kholkhoev, A.Y. Stefanyuk, V.G. Makotchenko,
A.S. Berdinsky, A.I. Romanenko, V.F. Burdukovskii, V.E. Fedorov

Local Growth of Graphene on Cu and $Cu_{0.88}Ni_{0.12}$ Foil Substrates — 31
H.S. Funk, J. Ng, N. Kamimura, Y.-H. Xie, J. Schulze

Growth of Patterned GeSn and GePb Alloys by Pulsed Laser Induced Epitaxy — 37
J. Schlipf, J.L. Frieiro, I.A. Fischer, C. Serra, J. Schulze, S. Chiussi

Impact of Sn Segregation on $Ge_{1-x}Sn_x$ epi-Layers Growth by RP-CVD — 43
D. Weisshaupt, P. Jahandar, G. Colston, P. Allred, J. Schulze, M. Myronov

Tungsten Dichalcogenides as Possible Gas-Sensing Elements — 48
V.A. Kuznetsov, A.Y. Ledneva, S.B. Artemkina, M.N. Kozlova, G.E. Yakovleva,
A.S. Berdinsky, A.I. Romanenko, V.E. Fedorov

Flicker Noise in AlGaAs/GaAs High Electron Mobility Heterostructure Field-Effect Transistor at Cryogenic Temperature   53
S. Mouetsi, F. Zouach, D. Rechem

Device Performance Tuning of Ge Gate-All-Around Tunneling Field Effect Transistors by Means of Gesn: Potential and Challenges   57
E.G. Rolseth, A. Blech, I.A. Fischer, Y. Hashad, R. Koerner, K. Kostecki, A. Kruglov, V.S. Senthil Srinivasan, M. Weiser, T. Wendav, K. Busch, J. Schulze

Band-Structure of Ultra-Thin InGaAs Channels: Impact of Biaxial Strain and Thickness Scaling   66
S. Krivec, M. Poljak, T. Suligoj

Perimeter Effects from Interfaces in Ultra-Thin Layers Deposited on Nanometer-Deep p^+n Silicon Junctions   72
T. Knezevic, L.K. Nanver, T. Suligoj

Analysis of Hot Carrier-Induced Degradation of Horizontal Current Bipolar Transistor   77
J. Zilak, M. Koricic, T. Suligoj

Impact of the Local p-well Substrate Parameters on the Electrical Performance of the Double-Emitter Reduced-Surface-Field Horizontal Current Bipolar Transistor   83
M. Koricic, J. Zilak, T. Suligoj

Characterization of Measurement System for High-Precision Oscillator Measurements   88
I. Brezovec, M. Magerl, J. Mikulic, G. Schatzberger, A. Baric

Temperature Calibration of an On-Chip Relaxation Oscillator   93
J. Mikulic, I. Brezovec, M. Magerl, G. Schatzberger, A. Baric

Model of High-Efficiency High-Current Coupled Inductor Two-Phase Buck Converter   98
V.C. Valchev, O.P. Stanchev, G.T. Nikolov

Analog to Digital Signal Converters for BiCMOS Quaternary Digital Systems   103
D. Bundalo, Z. Bundalo, D. Pasalic, B. Cvijic

Ultra-Wideband Pulse Generator for Time-Encoding Wireless Transmission   109
L. Sneler, M. Herceg, T. Matic

Spectral-Efficient UWB Pulse Shapers Generating Gaussian and Modified Hermitian Monocycles   113
A. Milos, G. Molnar, M. Vucic

Design of Multiplierless CIC Compensators Based on Maximum Passband Deviation
G. Molnar, A. Dudarin, M. Vucic
119

Adaptive State Observer Development Using Recursive Extended Least-Squares Method
N.N. Nikolov, M.I. Alexandrova, V.C. Valchev, O.P. Stanchev
125

Inverter Current Source for Pulse-Arc Welding with Improved Parameters
V.C. Valchev, D.D. Mareva, D.D. Yudov, R.S. Stoyanov
130

Power Output Comparison of Three Phase Passive Converter Circuits for Wind Driven Generators
V.C. Valchev, P.V. Yankov, A. Van den Bossche
136

Dynamic Range Optimization and Noise Reduction by Low-Sensitivity, Fourth-Order, Band-Pass Filters Using Coupled General-Purpose Biquads
E. Emanovic, D. Jurisic
141

A Circular Economy for Photovoltaic Waste - the Vision of the European Project CABRISS
W. Brenner, N. Adamovic
146

Smart Farm Computing Systems for Animal Welfare Monitoring
M. Caria, J. Schudrowitz, A. Jukan, N. Kemper
152

Power Management Circuit for Energy Harvesting Applications with Zero-Power Charging Phase
D. Vinko
158

System for Early Condensation Detection and Prevention in Residential Buildings
M. Marcelic, R. Malaric
162

FEM Analysis and Design of a Voltage Instrument Transformer for Digital Sampling Wattmeter
M. Dadic, K. Petrovic, R. Malaric
166

Random Number Generation with LFSR Based Stream Cipher Algorithms
T. Tuncer, E. Avaroğlu
171

RAM-Based Mergers for Data Sort and Frequent Item Computation
A. Rjabov, V. Sklyarov, I. Skliarova, A. Sudnitson
176

Distributed Computing, Visualization and Biomedical Engineering

New Classes of Kochen-Specker Contextual Sets – *Invited Paper*   182
N.D. Megill, M. Pavicic

New Method for Determination Complexity Using in AD HOC Cloud Computing   188
M. Babic, B. Jerman-Blazic

Neneta: Heterogeneous Computing Complex-Valued Neural Network Framework   192
V. Lekic, Z. Babic

Cloud-Distributed Computational Experiments for Combinatorial Optimization   197
M. Brcic, N. Hlupic

Design of Digital IIR Filter Using Particle Swarm Optimization   202
F. Serbet, T. Kaya, M.T. Ozdemir

Big Data Analytics in Electricity Distribution Systems   205
S. Stoyanov, N. Kakanakov

Running HPC Applications on Many Million Cores Cloud   209
D. Tomic, Z. Car, D. Ogrizovic

DBaaS Comparison   N/A
I. Astrova, A. Koschel, C. Eickemeyer, J. Kersten, N. Offel

Modeling Heterogeneous Computational Cluster Hardware in Context of Parallel Database Processing   221
K.Y. Besedin, P.S. Kostenetskiy

Properties of Mathematical Number Model Provided Exact Computing   225
V. Golodov

Simulation of the Parallel Database Column Coprocessor   229
P.S. Kostenetskiy

Towards Flexible Open Data Management Solutions   233
B. von St. Vieth, J. Rybicki, M. Brzeźniak

Spatial Analysis of the Clustering Process   238
M. Kranjac, U. Sikimic, J. Salom, S. Tomic

Usage of Android Device in Interaction with 3D Virtual Objects   244
I. Prazina, V. Okanovic, K. Balic, S. Rizvic

Generating Virtual Guitar Strings Using Scripts 247
L. Kunic, Z. Mihajlovic

Remote Interactive Visualization for Particle-Based Simulations on Graphics Clusters 253
A. Sabou, D. Gorgan

Collaborative View-Aligned Annotations in Web-Based 3D Medical Data Visualization 259
P. Lavric, C. Bohak, M. Marolt

Feasibility of Biometric Authentication Using Wearable ECG Body Sensor Based On Higher-Order Statistics 264
S. Sprager, R. Trobec, M.B. Juric

Bio-Medical Analysis Framework 270
M. Mohorcic, M. Depolli

Finding a Signature in Dermoscopy: A Color Normalization Proposal 276
M. Machado, J. Pereira, M. Silva, R. Fonseca-Pinto

A Textured Scale-Based Approach to Melanocytic Skin Lesions in Dermoscopy 279
R. Fonseca-Pinto, M. Machado

Remarks on Visualization of Fuzziness of Cardiac Data 283
J. Opiła, T. Pełech-Pilichowski

Abdominal Fetal ECG Measured With Differential ECG Sensor 289
A. Rashkovska, V. Avbelj

Synchronization of Time in Wireles ECG Measurement 292
A. Vilhar, M. Depolli

Long-Term Follow-Up Case Study of Atrial Fibrillation after Treatment 297
M. Jan, R. Trobec

A Case Report of Long-Term Wireless Electrocardiographic Monitoring in a Dog with Dilated Cardiomyopathy 303
M. Brloznik, V. Avbelj

SaaS Solution for ECG Monitoring Expert System 308
A. Ristovski, M. Gusev

Wavelet-Based Analysis Method for Heart Rate Detection of ECG Signal Using LabVIEW 314
D. Kaya, M. Türk, T. Kaya

Parallelization of Digital Wavelet Transformation of ECG Signals   318
E. Domazet, M. Gusev

**Hilbert Transform Based Paroxysmal Tachycardia Detection
Algorithm**   324
I. Culjak, M. Cifrek

**Biomedical Time Series Preprocessing and Expert-System Based Feature
Extraction in MULTISAB Platform**   330
A. Jovic, D. Kukolja, K. Friganovic, K. Jozic, S. Car

**Evaluation of Chronic Venous Insufficiency with PPG Prototype
Instrument**   336
M. Makovec, U. Aljancic, D. Vrtacnik

**Highly Parallel Online Bioelectrical Signal Processing on GPU
Architecture**   340
Z. Juhasz

Dew Computing

The Dawn of Dew: Dew Computing for Advanced Living Environment   347
Z. Sojat, K. Skala

**Service-Oriented Application for Parallel Solving the Parametric Synthesis
Feedback Problem of Controlled Dynamic Systems**   353
G.A. Oparin, V.G. Bogdanova, S.A. Gorsky, A.A. Pashinin

**Augmented Coaching Ecosystem for Non-obtrusive Adaptive Personalized
Elderly Care on the Basis of Cloud-Fog-Dew Computing Paradigm**   359
Y. Gordienko, S. Stirenko, O. Alienin, K. Skala, Z. Sojat, A. Rojbi,
J.R. López Benito, E. Artetxe González, U. Lushchyk, L. Sajn,
A. Llorente Coto, G. Jervan

**Cloud-Dew Computing Support for Automatic Data Analysis in Life
Sciences**   365
P. Brezany, T. Ludescher, T. Feilhauer

**Distributed Database System as a Base for Multilanguage Support for
Legacy Software**   371
N. Crnko

Sag/Tension Dynamical Line Rating System Architecture   N/A
A. Pavlinic, V. Komen

**Toward a Framework for Embedded & Collaborative Data Analysis with
Heterogeneous Devices**   381
M. Goeminne, M. Boukhebouze

A Dew Computing Solution for IoT Streaming Devices   387
M. Gusev

3D-based Location Positioning Using the Dew Computing Approach for Indoor Navigation   393
D. Podbojec, B. Herynek, D. Jazbec, M. Cvetko, M. Debevc, I. Kozuh

Architecting a Hybrid Cross Layer Dew-Fog-Cloud Stack for Future Data-Driven Cyber-Physical Systems   399
M. Frincu

Telecommunications & Information

Future Applications of Optical Wireless and Combination Scenarios with RF Technology – *Invited Paper*   404
E. Leitgeb

The Golden Ratio in the Age of Communication and Information Technology   407
Y.I. Doychinov, I.S. Stoyanov, T.B. Iliev

Wireless Machine-to-Machine Communication for Intelligent Transportation Systems: Internet of Vehicles and Vehicle to Grid   411
N.R. Moloisane, R. Malekian, D. Capeska Bogatinoska

Power Control Schemes for Device-to-Device Communications in 5G Mobile Network   416
T.B. Iliev, G.Y. Mihaylov, E.P. Ivanova, I.S. Stoyanov

Brain Computer Interface Communicator : A Response to Auditory Stimuli Experiment   420
G. Madhale Jadav, L. Batistic, S. Vlahinic, M. Vrankic

Determination of Origins and Destinations for an O-D Matrix Based on Telecommunication Activity Records   424
S. Desic, M. Filic, R. Filjar

What Factors Influence the Quality of Experience for WebRTC Video Calls?   428
J. Barakovic Husic, S. Barakovic, A. Veispahic

Is There Any Impact of Human Influence Factors on Quality of Experience?   434
J. Barakovic Husic, S. Barakovic, S. Muminovic

Extended AODV Routing Protocol Based on Route Quality Approximation via Bijective Link-Quality Aggregation   N/A
V.L. Prcic, D. Kalpic, D. Simunic

Development Trends of Telecommunications Metrics 445
N. Banovic-Curguz, D. Ilisevic

Implementation and Testing of Cisco IP SLA in Smart Grid Environments 450
J. Horalek, F. Holik, V. Hurtova

Fault Management and Management Information Base (MIB) 455
O. Jukic, I. Hedji, I. Speh

Distributed Threat Removal in Software-Defined Networks 460
D. Samociuk, A. Chydzinski

Performance Analysis of Virtualized VPN Endpoints 466
D. Lackovic, M. Tomic

A Big Data Solution for Troubleshooting Mobile Network Performance Problems 472
K. Skracic, I. Bodrusic

Overlapping Blocks in Reconstruction of Sparse Images 478
I. Stankovic, M. Dakovic, I. Orovic

Sparse Signal Reconstruction Based on Random Search Procedure 482
M. Dakovic, I. Stankovic, M. Brajovic, L. Stankovic

A Fast Noise Level Estimation Algorithm Based on Adaptive Image Segmentation and Laplacian Convolution 486
E. Turajlic

Advanced Regulation Approach: Dynamic Rules for Capturing the Full Potential of Future ICT Networks 492
D. Ilisevic, N. Banovic-Curguz

LTE eNB Traffic Analysis and Key Techniques towards 5G Mobile Networks 497
T.B. Iliev, G.Y. Mihaylov, T.D. Bikov, E.P. Ivanova, I.S. Stoyanov, D.I. Radev

IoT Network Protocols Comparison for the Purpose of IoT Constrained Networks 501
I. Hedji, I. Speh, A. Sarabok

Digital Forensic Analysis through Firewall for Detection of Information Crimes in Hospital Networks 506
A. Akbal, E. Akbal

Topology Analysis for Energy-Aware Node Placement in Wireless Sensor Networks of Home Sensing Environments N/A
A. Koren, D. Simunic

Mine Safety System Using Wireless Sensor Networks 515
V. Henriques, R. Malekian, D. Capeska Bogatinoska

Replication of Virtual Network Functions: Optimizing Link Utilization and Resource Costs   521
F. Carpio, W. Bziuk, A. Jukan

Impact of Human Resources Changes on Performance and Productivity of Scrum Teams   527
D. Alic, A. Djedovic, S. Omanovic, A. Tanovic

Synergy of ITIL Methodology and Help Users Systems   533
I. Ivosic

Design of Optical Fiber Gyroscope System in Program Environment OptSim   N/A
M. Márton, Ľ. Ovseník, J. Turán, M. Spes

Value Based Service Design Elements in Business Ecosystem Architecture   N/A
D. Ramljak

Simulation Study of M-ARY QAM Modulation Techniques Using Matlab/Simulink   547
S.M. Sadinov

Modeling System Behaviour

Multiscale and Multiobjective Modelling: a Perspective for Mastering the Design and Operation Complexity of IoT Systems – *Invited Paper*   555
K. Drira

Topological Data Analysis and Applications   558
J. Pita Costa

Modelling of Pedestrian Groups and Application to Group Recognition   564
D. Brscic, F. Zanlungo, T. Kanda

eCST to Source Code Generation - an Idea and Perspectives   570
N. Rakic, G. Rakic, N. Sukur, Z. Budimac

Design and Development of Contactless Interaction with Computers Based on the Emotiv EPOC+ Device   576
B. Sumak, M. Spindler, M. Pusnik

Patterns for Improving Mobile User Experience   582
M. Pusnik, D. Ivanovski, B. Sumak

Drawing Process Recording Tool for Eye-Hand Coordination Modelling   588
V. Giedrimas, L. Vaitkevicius, A. Vaitkeviciene

Model of Calculating Indicators of Power System Reliability   N/A
A. Balota

Modelling of Variable Shunt Reactor in Transmission Power System for Simulation of Switching Transients   598
A. Zupan, B. Filipovic-Grcic, I. Uglesic

An Extended Model of a Level and Flow Control System   603
H. Siljak, J. Hivziefendic, J. Kevric

Procedure for Modelling of Soft Tissues Behavior   608
M. Franulovic, K. Markovic, S. Pilicic

Analysis of ERASMUS Staff and Student Mobility Network within a Big European Project   613
M. Savic, M. Ivanovic, Z. Putnik, K. Tütüncü, Z. Budimac, S. Smrikarova, A. Smrikarov

Computers in Education

Modern Education and Its Background in Cognitive Psychology: Automated Question Creation and Eye Movements   619
M. Höfler, G. Wesiak, P. Pürcher, C. Gütl

Learning to Program – Does it Matter Where you Sit in the Lecture Theatre?   624
A. McGowan, P. Hanna, D. Greer, J. Busch, N. Anderson

Knowledge and Skills: A Critical View   630
M. Radovan

Today is the Future of Yesterday; What is the Future of Today?   635
H. Jaakkola, J. Henno, J. Mäkelä, B. Thalheim

Interdisciplinary Utilization of IT   644
S. Neradová, S. Zitta

Maximizing Quality Class Time Using Computers for a Flipped Classroom Approach   649
C.P. Fulford, S. Paek

Quantitative Structured Literature Review of Research on e-Learning   655
L. Abazi-Bexheti, A. Kadriu, M. Apostolova

The Educators' Telescope to the Future of Technology   660
H. Jaakkola, J. Henno, B. Thalheim, J. Mäkelä

ICT Support for Promotion of Nature Park — 666
A. Zisko, A. Sorgo, M. Krasna

Experiences in Using Educational Recommender System ELARS to Support e-Learning — 672
M. Holenko Dlab

Influence of Accuracy of Simulations to the Physics Education — 678
R. Repnik, G. Nemec, M. Krasna

Competence-Oriented Model of Representation of Educational Content — 685
O.N. Ivanova, N.S. Silkina

Using CODESYS as a Tool for Programming and Simulation in Applied Education at University — N/A
D. Lukac

Developing Curiosity and Multimedia Skills with Programming Experiments — 694
J. Henno, H. Jaakkola, J. Mäkelä

Informetrics: the Development, Conditions and Perspectives — 700
A. Papic

Structuring e-Learning Multi-Criteria Decision Making Problems — 705
N. Kadoic, N. Begicevic Redjep, B. Divjak

Introducing Gamification into e-Learning University Courses — 711
A. Bernik, D. Radosevic, G. Bubas

Perceived Security and Privacy of Cloud Computing Applications Used in Educational Ecosystem — 717
T. Orehovacki, D. Etinger, S. Babic

Estimating Profile of Successful IT Student: Data Mining Approach — 723
D. Oreski, M. Konecki, L. Milic

Comparison of Game Engines for Serious Games — 728
S. Pavkov, I. Frankovic, N. Hoic-Bozic

Case Study of Online Resources and Searching for Information on Students' Academic Needs — 734
A. Cizmesija, V. Vidacek-Hains

Recursions and How to Teach Them — 740
P. Brodjanac

Learning Management System (LMS) Software Comparison: Edmodo vs Schoology — N/A
M. Filipovic Tretinjak, M. Tretinjak

Machine Learning Techniques in the Education Process of Students of Economics 750
J. Bucko, L. Kakalejcík

Moodle-Based Data Mining Potentials of MOOC Systems at the University of Szeged 755
G. Kőrösi, F. Havasi

View on Development of Information Competencies and Computer Literacy of Slovak Secondary School Graduates 761
L. Révészová

Digital Learning as a Tool to Overcome School Failure in Minority Groups 767
D. Paľová, N.M. Novak, V. Weidinger

C Based Laboratory for Teaching Digital Signal Processing to Computer Engineering Undergraduates 773
D. Bokan, M. Temerinac, Z. Lukac, S. Ocovaj

Analysis of Video Views in Online Courses 778
P. Esztelecki, F. Havasi

Education in the Field of Electronic Financial Services of Its Future Users 783
M. Vejacka

Didactic Concepts of Modern Data Analysis 789
C. Ungermanns, W. Werth

Extending the Object-Oriented Notional Machine Notation with Inheritance, Polymorphism, and GUI Events 794
M. Aglic Cuvic, J. Maras, S. Mladenovic

Mediated Transfer from Visual to High-Level Programming Language 800
D. Krpan, S. Mladenovic, G. Zaharija

Story of a 'Storyline Visualization' in High School Readings 806
K. Osmakcic, K. Kocijan

Impact of ICT on Archival Practice from the 2000s Onwards and the Necessary Changes of Archival Science Curricula 812
H. Stancic, A. Rajh, M. Jamic

Publishing of Personal Information on Facebook with Regard to Gender: Comparison of Pupils and University Students 818
P. Dzapo, M. Duic

Web Sources of Literature for Teachers and Researchers: Practices and Attitudes of Croatian Faculty toward Legal Digital Libraries and Shadow Libraries Such as Sci-Hub 824
M. Duic, B. Konjevod, L. Grzunov

Supporting Mobile Learning: Usability of Digital Collections in Croatia for Use on Mobile Devices   830
R. Vrana, D. Gascic, M. Podkonjak

The Use of ICT in the English Language Classroom   836
R. Lerga, S. Candrlic, M. Holenko Dlab

LIS Students and Plagiarism in the Networked Environment   842
I. Hebrang Grgic

A Platform for Supporting Learning Process of Visually Impaired Children   848
A. Kavcic, M. Pesek, M. Marolt

Web Application for Time Telling   N/A
A. Cobic, M. Carapina

Integration of the Future Technologies to High Schools and Colleges   858
P. Kurent

Preschool Children and Computers: Who Lives in a Meadow?   862
B. Strnad

Web Service Model for Distance Learning Using Cloud Computing Technologies   865
D. Cvetkovic, M. Mijatovic, M. Mijatovic, B. Medic

The Struggle with Academic Plagiarism: Approaches Based on Semantic Similarity   870
T. Vrbanec, A. Mestrovic

Personal Learning Environment as Support to Education   876
B. Grba, B. Kovacic

Free and Open Source Software in the Secondary Education in Bosnia and Herzegovina   882
M. Pezer, N. Lazic, M. Odak

E-Learning of Mathematics, Problem and a Possible Solution   N/A
V. Dedic, R. Cvejic, M. Andjelkovic

Research Methodology in the 21st Century   891
D. Cvetkovic, B. Medic

A Survey and Evaluation of Free and Open Source Simulators Suitable for Teaching Courses in Wireless Sensor Networks   895
D. Pesic, Z. Radivojevic, M. Cvetanovic

Excessive Internet Use among Elementary School Students in Vojvodina   901
S. Tapiska, M. Kresoja, Z. Putnik, M. Ivanovic

E-Learning on Polytechnic Nikola Tesla – Analysis and Comparison 905
B. Kovacic, A. Skendzic, K. Devcic

Determination of Time Criteria for Assessment in Learning Management Systems 910
T. Alajbeg, M. Sokele, V. Simovic

Using Moodle in English for Professional Purposes (EPP) Teaching at the University North 915
J. Lasic-Lazic, T. Ivanjko, I. Grubjesic

Air Traffic Controllers' Practical Part of Basic Training on Computer Based Simulation Device 920
M. Pavlinovic, B. Juricic, B. Antulov-Fantulin

The Perspective of Use of Digital Libraries in Era of e-Learning 926
R. Vrana

ArTeFact – Digitization of Archives of Technical Faculty from the Period 1919 – 1956 N/A
M. Tucakovic, J. Lisek

Introductory Physics Course for ICT Students: Computer-Programming Oriented Approach 937
V. Krstic

Advocacy of Born-Digital Materials: an ESP Course Example N/A
D. Pesut

Theoretical and Practical Challenges of Using Three Ammeter or Tree Voltmeter Methods in Teaching 944
V. Simovic, T. Alajbeg, J. Curkovic

National Competition of Photography as Visual Art in Croatian Primary Schools, High Schools and Schools of Applied Arts 949
Z. Prohaska, Z. Prohaska, I. Uroda

Generating Large Random Test Data Table for SQL Training 954
U. Sterle

Programming Lego Mindstorms for First Lego League Robot Game and Technical Interview 958
B. Strnad

Children Online Safety 961
J. Zufic, T. Zajgar, S. Prkic

Computers in Technical Systems

Metals Industry: Road to Digitalization – *Invited Paper*   967
A. Merluzzi, G. Brunetti

IoT Gateway for Smart Metering in Electrical Power Systems - Software Architecture   974
M. Shopov

Application Models for Ubiquitous Systems with Sporadic Communication Availability   979
I. Cavrak, I. Zagar, A. Drazic

Towards the Utilization of Crowdsourcing in Traffic Condition Reporting   985
P. Rantanen, P. Sillberg, J. Soini

Survey of Prototyping Solutions Utilizing Raspberry Pi   991
M. Saari, A. Muzaffar bin Baharudin, S. Hyrynsalmi

Evaluating Robustness of Perceptual Image Hashing Algorithms   995
A. Drmic, M. Silic, G. Delac, V. Klemo, A.S. Kurdija

Adaptive Models for Security and Data Protection in IoT with Cloud Technologies   1001
N. Kakanakov, M. Shopov

Making a Smart City Even More Intelligent Using Emergent Property Methodology   1005
A. Saric, B. Mihaljevic, K. Marasovic

Digital Chess Board Based on Array of Hall-Effect Sensors   1011
F. Susac, I. Aleksi, Z. Hocenski

Influence of Human-Computer Interface Elements on Performance of Teleoperated Mobile Robot   1015
S. Kruzic, J. Music, I. Stancic

The Challenge of Measuring Distance to Obstacles for the Purpose of Generating a 2-D Indoor Map Using an Autonomous Robot Equipped with an Ultrasonic Sensor   1021
R. Rodin, I. Stajduhar

Automated Posture Assessment for Construction Workers   1027
R.J. Dzeng, H.H. Hsueh, C.W. Ho

Software Toolbox for Analysis and Design of Nonlinear Control Systems and Its Application to Multi-AUV Path-Following Control   1032
S. Ul'yanov, N. Maksimkin

Remote Alarm Reporting System Responsive to Stoppage of Ballast Water Management Operation on Ships — 1038
G. Bakalar, M.B. Baggini, S.G. Bakalar

Parameters for Condition Assessment of the High Voltage Circuit Breakers Arcing Contacts Using Dynamic Resistance Measurement — 1044
K. Obarcanin, R. Ostojic, S. Dzuzdanovic

Measurement Noise Propagation in Distribution-System State Estimation — 1049
U. Kuhar, G. Kosec, A. Svigelj

Application of Integrated Fail-Safe Technology for Safe and Reliable Natural Gas Distribution — N/A
J. Stanic, D. Marsic, G. Malcic

Brief Review of Self-Organizing Maps — 1061
D. Miljkovic

Fault Detection for Aircraft Piston Engine Using Self-Organizing Map — 1067
D. Miljkovic

Intelligent Systems

PicoAgri. Realization of a Low-Cost, Remote Sensing Environment for Monitoring Agricultural Fields through Small Satellites and Drones — 1073
S. Marsi, A. Gregorio, M. Maris, M. Puligheddu

Architecture of an Information System for Biological Sampling of Reservoir Microhabitats — N/A
B. Lukovac, D. Simunic, N. Vuckovic

Feeding a DNN for Face Verification in Video Data Acquired by a Visually Impaired User — 1084
J. Bhattacharya, S. Marsi, S. Carrato, H. Frey, G. Ramponi

Acquiring ISAR Images Using Measurement Instruments — 1090
H.-C. Lee, S.G. Lee, S.H. Lee, C.H. Jung

Segmentation of Kidneys and Abdominal Images in Mobile Devices with the Android Operating System by Using the Connected Component Labeling Method — 1094
S. Arslan Tuncer, A. Alkan

An Overview of Action Recognition in Videos — 1098
M. Buric, M. Pobar, M. Ivasic Kos

Pothole Detection: An Efficient Vision Based Method Using RGB Color Space Image Segmentation — 1104
A. Akagic, E. Buza, S. Omanovic

Classification of the Hand-Printed and Printed Medieval Glagolitic Documents Using Differentiation in Orthography   1110
D. Brodic, A. Amelio

An Evolutionary Approach to Route the Heterogeneous Groups of Underwater Robots   1116
M.Y. Kenzin, I.V. Bychkov, N.N. Maksimkin

Low Cost Robot Arm with Visual Guided Positioning   1120
P. Djurovic, R. Grbic, R. Cupec, D. Filko

Symbolic Tensor Differentiation for Applications in Machine Learning   N/A
A. Zhabinski, S. Zhabinskii, D. Adzinets

Reasoning with Air Pollution Data in SWI-Prolog   N/A
Z. Kazi, L. Kazi, I. Berkovic, B. Radulovic

Knowledge Elicitation in Multi-Agent System for Distributed Computing Management   1138
A. Feoktistov, A. Tchernykh, S. Gorsky, R. Kostromin

Improving a Distributed Agent-Based Ant Colony Optimization for Solving Traveling Salesman Problem   1144
A. Kaplar, M. Vidakovic, N. Luburic, M. Ivanovic

Towards an Agent-Based Automated Testing Environment for Massively Multi-Player Role Playing Games   1149
M. Schatten, I. Tomicic, B. Okresa Djuric, N. Ivkovic

On the Properties of Discrete-Event Systems with Observable States   1155
N. Nagul

The Formal Description of Discrete-Event Systems Using Positively Constructed Formulas   1161
A. Davydov, A. Larionov, N. Nagul

Runtime Estimation for Enumerating All Mutually Orthogonal Diagonal Latin Squares of Order 10   1166
S. Kochemazov, O. Zaikin, A. Semenov

Improving the Effectiveness of SAT Approach in Application to Analysis of Several Discrete Models of Collective Behavior   1172
S. Kochemazov, O. Zaikin, A. Semenov

Effects of the Distribution of the Values of Condition Attribute on the Quality of Decision Rules   1178
V. Ognjenovic, E. Brtka, V. Brtka, I. Berkovic

Complexity Comparison of Integer Programming and Genetic Algorithms for Resource Constrained Scheduling Problems   1182
R. Coric, M. Djumic, D. Jakobovic

Balancing Academic Curricula by Using a Mutation-Only Genetic Algorithm 1189
K. Sylejmani, A. Halili, A. Rexhepi

A Hybrid Method for Prediction of Protein Secondary Structure Based On Multiple Artificial Neural Networks 1195
H. Hasic, E. Buza, A. Akagic

Hardware-in-the-Loop Architecture with MATLAB/Simulink and QuaRC for Rapid Prototyping of CMAC Neural Network Controller for Ball-and-Beam Plant 1201
V. Shatri, L. Kurtaj, I. Limani

Primary and Secondary Experience in Developing Adaptive Information Systems Supporting Knowledge Transfer 1207
S. Lugovic, I. Dundjer, M. Horvat

The Captology of Intelligent Systems 1211
Z. Balaz, D. Predavec

Automatic Information Behaviour Recognition 1217
S. Lugovic, I. Dundjer

Adjective Representation with the Method Nodes of Knowledge 1221
M. Pavlic, Z. Dovedan Han, A. Jakupovic, M. Asenbrener Katic, S. Candrlic

Information Systems Security

Identification of Image Source Using Serial-Number-Based Watermarking under Compressive Sensing Conditions 1227
A. Draganic, M. Maric, I. Orovic, S. Stankovic

Anti-Computer Forensics 1233
K. Hausknecht, S. Gruicic

Analysis of Mobile Phones in Digital Forensics 1241
S. Dogan, E. Akbal

Security Analysis of Open Home Automation Bus System 1245
M. Ramljak

Analysis of Credit Card Attacks Using the NFC Technology 1251
J. Jumic, M. Vukovic

Vulnerabilities of Modern Web Applications 1256
F. Holik, S. Neradová

An Experiment in Using IMUNES and Conpot to Emulate Honeypot Control Networks 1262
S. Kuman, S. Gros, M. Mikuc

A Wireless Propagation Analysis for the Frequency of the Pseudonym Changes to Support Privacy in VANETs 1269
E. Cano Pons, G. Baldini, D. Geneiatakis

Resilience of Students' Passwords against Attacks 1275
B. Brumen, T. Makari

Empirical Study on the Risky Behavior and Security Awareness among Secondary School Pupils - Validation and Preliminary Results 1280
T. Velki, K. Solic, V. Gorjanac, K. Nenadic

Interoperability and Lightweight Security for Simple IoT Devices 1285
D. Androcec, B. Tomas, T. Kisasondi

Security and Privacy Issues for an IoT Based Smart Home 1292
D. Geneiatakis, I. Kounelis, R. Neisse, I. Nai-Fovino, G. Steri, G. Baldini

Towards Overall Information Security and Privacy (IS&P) Taxonomy 1298
K. Solic, H. Ocevcic, I. Fosic, I. Horvat, M. Vukovic, T. Ramljak

Trends in IoT Security 1302
M. Radovan, B. Golub

International Cyber Security Challenges 1309
I. Duic, V. Cvrtila, T. Ivanjko

Cloud Computing Threats Classification Model Based on the Detection Feasibility of Machine Learning Algorithms 1314
Z. Masetic, K. Hajdarevic, N. Dogru

Legal Framework Issues Managing Confidential Business Information in the Republic of Croatia 1319
G. Vojkovic, M. Milenkovic

Combinatorial Optimization in Cryptography 1324
K. Knezevic

Taxonomy of DDos Attacks N/A
I. Kramaric

Business Intelligence Systems

Multidimensional Mining of Big Social Data for Supporting Advanced Big Data Analytics – *Invited Paper* 1337
A. Cuzzocrea

Accelerating Dynamic Itemset Counting on Intel Many-Core Systems 1343
M. Zymbler

ETLator – a Scripting ETL Framework 1349
M. Radonic, I. Mekterovic

**The Role of Alignment for the Impact of Business Intelligence Maturity
on Business Process Performance in Croatian
and Slovenian Companies** 1355
V. Bosilj Vuksic, M. Pejic Bach, T. Grubljesic, J. Jaklic, A.M. Stjepic

MapReduce Research on Warehousing of Big Data 1361
M. Pticek, B. Vrdoljak

**Selection of Variables for Credit Risk Data Mining Models: Preliminary
research** 1367
M. Pejic Bach, J. Zoroja, B. Jakovic, N. Sarlija

**Brand Communication in Social Media: the Use of Image Colours in
Popular Posts** 1373
L. Zailskaitė-Jakstė, A. Ostreika, A. Jakstas, E. Stanevicienė, R. Damasevicius

Recommender System Based on the Analysis of Publicly Available Data 1379
G. Antolic, L. Brkic

Alternative Business Intelligence Engines 1385
I. Kovacevic, I. Mekterovic

Insights into BPM Maturity in Croatian and Slovenian Companies 1391
V. Bosilj Vuksic, M. Indihar Stemberger, D. Susa Vugec

Efficient Social Network Analysis in Big Data Architectures 1397
I. Soric, D. Dinjar, M. Stajcer, D. Orescanin

**Integrating Evolving MDM and EDW Systems by Data Vault Based
System Catalog** 1401
D. Jaksic, V. Jovanovic, P. Poscic

Digital Economy and Government, Local Government, Public Services

Distributed Governance of Life Care Agreements via Public Databases N/A
J. Klasinc

**Using Public Private Partnership Models in Smart Cities – Proposal for
Croatia** 1412
M. Milenkovic, M. Rasic, G. Vojkovic

The Platform for the Content Exchange between Internet Music Streaming Services and Discographers — 1418
M. Sretenovic, B. Kovacic, A. Skendzic

Law and Technology in Data Processing: Risk-Based Approach in EU Data Protection Law and Implementation Challenges in Croatia — 1424
N. Gumzej

Social and Economic Effects of Investments in Primorsko-goranska County Broadband Network — 1431
S. Vojvodic, S. Cegar, D. Medved

Mobile Applications in Communication of Local Government with Citizens in Croatia — N/A
D. Bunja, G. Pavelin, F. Mlinac

The Role of Applications and their Vendors in Evolution of Software Ecosystems — 1442
S. Hyrynsalmi, P. Linna

Open Data Based Value Networks: Finnish Examples of Public Events and Agriculture — 1448
P. Linna, T. Mäkinen, K. Yrjönkoski

Financial Impact of Forensic Proceedings in ICT — 1454
S. Aksentijevic, E. Tijan, A. Jugovic

Reliability, Availability and Security of Computer Systems Supported by RFID Technology — 1459
P. Ristov, T. Miskovic, A. Mrvica, Z. Markic

Transportation and Power System Interdependency for Urban Fast Charging and Battery Swapping Stations in Croatia — 1465
I. Pavic, N. Holjevac, M. Zidar, I. Kuzle, A. Neskovic

Croatian Qualification Framework – Data Model and Software Implementation in Higher Education — 1471
Z. Kovacevic, M. Mauher, M. Slamic

Internet as a Purchasing Information Source in Children's Products Retailing in Croatia — 1476
B. Knezevic, Z. Pavlic Sipek, B. Jakovic

Valuation of Common Stocks Using the Dividend Valuation Approach and Excel — 1481
Z. Prohaska, I. Uroda, A. Radman Pesa

The Interconnection between Investment in Software and Financial Performance – The Case of Republic of Croatia — 1486
M. Boban, T. Susak

Mipro Junior – Student Papers

Data Warehouse Architecture Classification — 1491
G. Blazic, P. Poscic, D. Jaksic

Comparative Analysis of the Selected Relational Database Management Systems — 1496
R. Poljak, P. Poscic, D. Jaksic

A Tool for Simplifying Automatic Categorization of Scientific Paper Using Watson API — 1501
L. Cvetkovic, B. Milasinovic, K. Fertalj

The Role of Redundancy and Sexual Reproduction in the Conservation of the Genetic Information Tested on a Cellular Automaton — 1506
V. Kovács, V. Póser

Developing MOBA Games Using the Unity Game Engine — 1510
D. Polancec, I. Mekterovic

Comparative Analysis of Tools for Development of Native and Hybrid Mobile Applications — 1516
T. Vilcek, T. Jakopec

Exploring HTTP/2 Advantages and Performance Analysis Using Java 9 — 1522
L.M. Bach, B. Mihaljevic, A. Radovan

Software Supporting International Student Exchange Program in Higher Education — 1528
Z. Gracak, L. Brkic

Blood Vessel Segmentation Using Multiscale Hessian and Tensor Voting — 1534
A. Lukac, M. Subasic

Storytelling in Web Design: A Case Study — 1540
M. Pivac, A. Granic

Idioms in State-of-the-Art Croatian-English and English-Croatian SMT Systems — 1546
M. Manojlovic, L. Dajak, M. Brkic Bakaric

Contactless Control of Sanitary Water Flow and Temperature — 1551
D. Gecevic, T. Bjazic

PI Controller for DC Motor Speed Realized with Arduino and Simulink — 1557
M. Gavran, M. Fruk, G. Vujisic

Laboratory Model of the Elevator Controlled by ARDUINO Platform — 1562
M.A. Balug, T. Spoljaric, G. Vujisic

System for Acquisition and Processing of Pressure Data Around Body in Airflow

D. Meznaric, K. Krajcek Nikolic, D. Franjkovic

MIPRO 2017, May 22- 26, 2017, Opatija, Croatia

Active Learning, Labs and Maker-spaces in Microwave Circuit Design Courses

B. Pejcinovic[*]

[*] Portland State University, ECE dept., Portland OR, USA
pejcinb@pdx.edu

Abstract - Circuit design courses in general, and microwave circuit design courses as a subspecialty, have been taught over many decades. It is relatively recently, however, that instructors have started experimenting with more modern approaches to in-class and out-of-class instruction. In our attempt to make instruction more effective we have turned to: a) utilizing classroom interaction systems and collaborative work in class, b) studio-like approach to labs where students are encouraged to explore a problem through design, simulation, building and testing of simple structures, c) makerspaces that enable full design-build-test-redesign cycle of fairly sophisticated designs, and d) systematic literature reviews for graduate students taking the courses. We describe our experiences in designing and implementing a sequence of two courses, present assessment data, discuss obstacles to student learning, and propose additional ways to improve student learning.

I. INTRODUCTION

Microwave circuit design is a fairly well established topic for senior undergraduate and graduate courses and many good textbooks are available, e.g. [1]. Topics vary but typically include some fundamentals of electromagnetic wave propagation, transmission lines (TL), effects of matching and reflection on TLs, various passive circuits (matching, filters, couplers, etc.), linear amplifiers (usually a low-noise amplifier), mixers and power amplifiers. Active devices are also discussed at length as many parasitic effects have to be dealt with. There is no precise cut-off, but we will take "microwave" to mean anything from 1 GHz to around 30 GHz.

One of the main difficulties in setting a course in this area is expensive instrumentation that is needed for realistic measurements. For this reason, many programs have very limited measurement labs for students and rely heavily on simulations. Much of the initial design can still be done using pencil-and-paper but, like other circuit courses, simulation is indispensable for any practical designs that will be actually manufactured and tested. Interestingly, much of the practical information about microwave measurements is still presented in various application notes, although situation is slowly changing (e.g. see [2]).

A. Microwave Circuit Design at Portland State U.

We have a sequence of two courses taught back-to-back. Two quarters last 10 weeks each. The first quarter is primarily devoted to passive components, including: lumped passive components and resonators, transmission lines, matching and Smith charts, L-, T- and Π-circuit matching, 2-port devices and measurements, lumped and

microstrip filters. The tenth week is devoted to work on projects, typically a filter design using SMD and/or microstrip techniques. Weekly lab activities include: characterization and modeling of SMD components; using TDR response to determine line properties; examining reflections using TDR and VNA; designing and testing $\lambda/4$ and single-stub matching circuits; two-port measurements on VNA; deembedding of fixtures. During the term students also work on two simulation assignments which utilize Agilent ADS software.

The second quarter deals primarily with active circuits, including: passive power combiners; active devices, gain, and stability; design for gain and noise; using ADS in amplifier design; transistor biasing; nonlinear effects and their measurement; power amplifiers; mixers; complete receiver design. Weekly lab activities include: design and test of Wilkinson and quadrature-hybrid combiners; active devices and their DC and S-parameter characterization; design of amplifier matching circuits; LNA design and layout; noise figure measurement of amplifiers and other devices; using ADS to design amplifiers; on-wafer measurements using probe stations. Emphasis during this term is on producing a working prototype of an active circuit, e.g. designing, fabricating and testing an LNA using procedure given by [3]. Both courses heavily rely on hands-on, lab-based exercises, simulation assignments and a culminating team project. Typical enrollment has been 15-25 students.

II. COURSE DESIGN

Overarching goal of this course is to prepare students not only to be competent in the specialist technical area of microwave circuit design, but also to prepare students to be self-directed learners who can easily adapt to new, fast changing technologies. Specific course learning outcomes (LO-s) are:

1. Design passive & active circuits using microstrip and SMD technologies:

2. Design circuits using simulation tools.

3. Manufacture circuit prototypes in different technologies.

4. Measure circuits up to 20 GHz (PCB & on-wafer).

5. Design, build & test a microwave system.

6. Be able to read, understand and report on papers published in a trade journal (for undergraduates) and scientific journal (graduate students).

7. Write good quality reports.

Based on these LO-s and results of various research studies, we used the following guidelines, largely based on [4], for the development of specific instructional methods. Specific methods are listed below each guideline:

A. Students need to be engaged in their learning, not be just a passive listener [5]:

   o Use in-class interaction system ("clickers")

   o Follow up lectures with immediate labs

   o Encourage learning from, and teaching to, peers

B. Provide activities that engage higher cognitive functions [6]:

   o Assign readings to expand lecture material

   o Give less prescriptive lab assignments

   o Assign authentic team-based projects

C. Immediate feedback is more effective than delayed:

   o Observe students in lab and probe their understanding

   o Use clicker results to adjust lecture content and pacing

D. Provide multiple ways to retrieve learned concepts

   o Theory and designs are put to immediate use in labs

E. Prototyping and concomitant failure should be encouraged [7]

   o Quick prototyping is part of labs and projects

III. STUDENT LEARNING

Our students have very diverse backgrounds, e.g. many of our graduate students work at local high tech companies and may have been out of school for many years. International students may not be familiar at all with instrumentation our students use regularly. In addition, it is well known that students compartmentalize their knowledge and we need to bring out their previously learned knowledge by explicitly making students retrieve it. Part of this process should be some type of assessment of prior knowledge, which will be briefly discussed in the Assessment section below. Ideally, we would have reliable and valid tools that would help us with this task but there seem to be none in the open literature. All this points to the fact that some kind of scaffolding for student learning should be provided from the very beginning and is one of the reasons that we introduced active learning whenever possible.

In practical terms, the biggest changes we introduced were: a) addition of labs, b) modified role of projects, and c) repurposing of lectures. These, along with additional out-of-class assignments, are discussed below.

A. Labs

In 2010 a new lab facility was designed and built which enabled acoustic, electromagnetic and optic measurements to be made by students [8]. We acquired four TDR oscilloscopes and four 20 GHz VNAs. In combination with instrumentation available in individual faculty research labs, students now have access to outstanding instrumentation. However, this is both a blessing and a curse. Because such instrumentation is very expensive there is a tendency to keep it under "lock-and-key" where students have very limited and supervised access. This runs counter to our desire to have students experiment with construction and testing of various circuits and systems. Currently, we are attempting to work in a hybrid mode where highly motivated students are asked to take a "qualifying" exam so that they gain access to our research lab which has very similar instrumentation. The trade-off is that they have to help other students and make themselves available. This model of lab management is very similar to, but less formal than, the one used in our makerspace (LID) lab, discussed below. Actual course labs are done in the departmental EMAG lab, shown in Figure 1.

Figure 1. Electromagnetics and acoustics group (EMAG) lab at PSU.

Immediate experimentation is also enabled by the use of copper tape with conductive adhesive [9]-[11]. In combination with cheap FR4 substrates students can design and build some fairly complex circuits, e.g. branch-line couplers. Addition of soldering stations and drill machine enables addition of 0603-sized SMD components for added circuit complexity. The imprecise nature of the technology exposes students to the idea that they should make this uncertainty part of their design process. Similarly, they have to start thinking what the dominant factors influencing their designs are. This is brought out more fully during their projects.

Along the way students also start using our makerspace facility (LID lab) where they can fabricate PCB-s with much better accuracy and predictability. However, many real-life issues still remain, such as uncertainty of FR4 properties, dimensional variations etc.

Measurements at microwave frequencies are very different than at low frequencies and labs provide an opportunity to explore various issues in depth. We begin with TDR measurements on transmission lines because we have found that students grasp TDR behavior quicker than frequency domain measurements done on VNA. However, caution is needed because students can easily conflate the two so that they start bringing time-domain concepts, such as wave travel in time domain and signal delay, in inappropriate ways. Eliminating some of those misconceptions is hard [12], but they are also not easy to define and identify. Development of concept inventory for microwave circuit area would be welcome [13]. For now, our approach is to repeatedly revisit some of the key concepts such as delay vs. phase change, electrical length, difference between singe- and two-port measurements, and effects of port termination.

From early on, we also had labs dealing with circuit simulation for which we used HP/Agilent/Keysight MDS/ADS. In 2013 we realized that it made little sense to teach students how to use ADS – the company produced very nice videos and materials that enable students to learn it on their own. Instead of spending valuable face-to-face time we now require that students watch the videos and submit two labs that tie in with course material on passive component modeling and impedance matching. Students are asked to demonstrate their basic competency in using ADS by constructing and running some simple simulations during lab sessions. So far there have been no complaints from students as they seem to master the software fairly quickly.

B. Projects

The main purpose of projects is to integrate students' knowledge and move their learning up the cognitive scale of complexity towards creation, synthesis and critical thinking. Over the years, we have moved towards more project-based learning but have not given up on face-to-face component. Projects present difficulties due to conflicting criteria: they should be easy enough to be doable in relatively short time, but they also must be complex and authentic enough to help student learning and motivation. They are also logistically difficult to organize and evaluate.

To improve project experience and streamline the logistics, we have introduced three new components:

1. CATME for team formation and peer evaluation

2. Initial mini-project

3. Project management using Scrum and Trello

We have used CATME in other courses and it has proven to be very useful in team formation and collecting peer feedback [14]. During the first half of each course we run a mini-project, e.g., constructing and testing an SMA cable [15], that is meant to coalesce teams and give them an opportunity to work on smaller but relevant mini-project. Teams submit reports at the end of the mini project as well as peer evaluations. It is usually fairly easy to identify which teams are having trouble at which point we can intervene and in some rare cases disband teams. Once the final project report is done teams submit second peer evaluation on CATME and scoring is now used to allocate individual scores, i.e. students who did not perform their share of the work get lower scores and grades. We can triangulate these results with our observation during lab sessions and during the final project demo.

Currently, we are introducing Scrum approach to project management and use Trello as visual aid in planning and tracking projects. We have had good success in freshman courses and in capstone (senior) teams [16] but we do not have any firm data or observations to report on their use in microwave circuit design classes. One potential hurdle is that students have to learn yet another software tool and we have, therefore, stretched the introduction and use of Scrum and Trello over two quarters.

As is well known, motivated students can do wonderful projects and one example is shown in Figure 2. This is a full FSK receiver with antenna, LNA, divider, filters, detectors and readout circuits, as first described in [17]. The bell shaped structures are 2.4 & 2.6 GHz resonators built out of pipe fittings, separate board on the left is student designed LNA and on the right side there is a readout circuit.

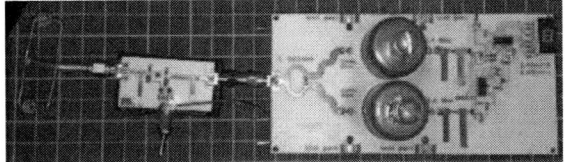

Figure 2. Student-built full FSK receiver.

C. Active Lectures

Initial experience with introduction of in-class interaction system were described in [8]. Overall, we find this instrumental in fostering in-class atmosphere where it is expected that everyone participates. One observation is that our activities are still too oriented towards individual responses and we need to expand the types of questions asked, so that teams of students can discuss them after individuals submit their responses. It is important to initially ask some very simple questions because we have been surprised many times by the kinds of misunderstandings students have. Open-ended, textual questions are best for uncovering misconceptions. One advantage of Learning Catalytics system is its ability to collect answers to graphical problems, which is very helpful in explaining tools like Smith chart.

D. Out of class assignments

Typical assignments of this type are homeworks with various calculation problems, typically assigned from a back of a textbook. We have found these to be of limited use for formative feedback because of the delay involved in grading and the kind of grading that is done. Similarly, homeworks are not very good for summative purposes because of prevalence of cheating, whether it is intentional (e.g. by copying from solutions manuals) or unintentional (e.g. due to student collaboration and sharing of results). Finally, they are time consuming to grade. We have repurposed homework assignments as part of "study guides" for preparation for quizzes and we typically hold three quizzes each term. Homeworks are not graded but are taken as part of participation score.

One item on which students spend a lot of time is writing of so-called lab reports. These are meant to answer some specific questions that come up during lab sessions and students have to present data they collected. We find that students struggle with both text and graphical presentations, both of which are key communication skills they should master. To help them develop these skills, they are allowed to re-submit their first report after they are given detailed feedback. Ideally, there would be more than one such revision opportunity but this is not possible due to time constraints. In general, we see good improvement in writing over the two courses but it is difficult to know if students transfer those skills to other courses.

To improve critical thinking and writing we are experimenting with a novel approach to writing research reports, which are required from graduate students. The idea originally came from medicine where systematic literature reviews are used for meta-analyses and guide policy decisions. It was adopted in software engineering where it was expanded into primarily educational role through development of Iterative Systematic Literature Reviews (ISLR) (see [18] and references therein). In short, teams of graduate students are asked to formulate a research question, develop search string for literature search, evaluate quality of papers and write a critical summary of the relevant literature. Given that we have two quarters we have broken this task down into two sections so that students first summarize a set of papers which gives them writing practice, and follow that up with a full ISLR projects. Defining good ISLR topic is the hardest part and so far we have used topics within Doherty power amplifier design, with mixed success. There is some resistance to this approach coming from students who do not see immediate benefit or relation to their work. In parallel, we are currently developing some metacognitive approaches that we hope will help students understand their own learning better.

One final note: one of the longstanding practices, especially in graduate education, is assignment of reading materials. We have found these to be of limited usefulness because there was no easy and efficient way to monitor student performance. This year we are experimenting with online tool Perusall [19] that allows not only monitoring of compliance, i.e., did the students open the files and read them, but also checks quality and quantity of their comments. It also encourages social interaction among readers. So far, it has done these tasks surprisingly well and we will continue to use it.

IV. MAKERSPACES

A relatively recent development in engineering education is the use of so-called makerspaces as places where students have an opportunity to develop their own ideas or work on class projects. As stated "The desire to make design and prototyping more integral to the engineering experience led to the creation of The Invention Studio, a free-to-use, 3000 ft^2 maker space and culture at the Georgia Institute of Technology" [20]. At around the same time, we started a very similar lab at PSU called "Lab for Interconnected Devices" [21]. ECE department provides a staff member to run the lab but bulk of the work is done by volunteers who are trained on all available tools before they become lab managers. Everything needed for circuit production is available, including electroplating for vias. Tools are integrated with various software to produce appropriate files for layout or 3D printing. Our students can usually get their boards and other items made within hours of coming to the lab.

LID organizes workshops on various topics and students are encouraged to participate. All of the activities centered around LID foster much needed sense of community among engineering students. While we have no data to demonstrate it, in our opinion the development of this makerspace has been a great boost to our students' motivation and learning. The sample project shown in Figure 2 was entirely built in LID and two students involved were LID managers. A portion of LID lab is shown in Figure 3 [21].

Figure 3. Laboratory for Interconnected Devices (LID) at PSU.

V. ASSESSMENT

Results of some of our initial student surveys were presented in [8]. We have continued using the same survey but eliminated a couple of questions that seemed redundant. Our most recent survey questions are given in Table I. The first part asks students about their self-efficacy, i.e. their belief in their capacity to execute behaviors necessary to produce specific performance attainments [22], such as designing a microwave circuit. Results of surveys conducted in 2012, 2013 and 2016 are summarized in Figure 4. Note that results for 2012 and 2013 were added together to have larger pool of respondents: in 2016 there were 19 responses and 31 in 2012+2013.

TABLE I.   STUDENT SURVEY QUESTIONS

I am confident that I can:	I found this technique to be:
1. Design microwave circuits 2. Build and test microwave circuits 3. Write good quality reports 4. Read and understand technical publications	5. Building and testing circuits 6. Running circuit simulations 7. Doing in-class exercises and problems 8. Solving homework problems 9. Listening to lectures in-class 10. Watching pre-recorded videos (special topics) 11. Class project
Scale: Strongly Agree, Agree, Neutral, Disagree, Strongly Disagree.	Scale: Very helpful, Somewhat helpful, Neutral, Not helpful, Waste of time.

From Figure 4 it looks like our students are reasonably confident in their abilities in design, build & test, and writing, and their confidence has changed little over time. However, reading self-efficacy fell noticeably between two data sets. One possible explanation is that we introduced SLR project for graduate students which requires extensive readings. This activity may have produced a realization that they are not that well acquainted with the literature and have difficulty reading it. Fortuitously, this year we are utilizing reading management software [19] which will enable more meaningful reading assignments and their assessment.

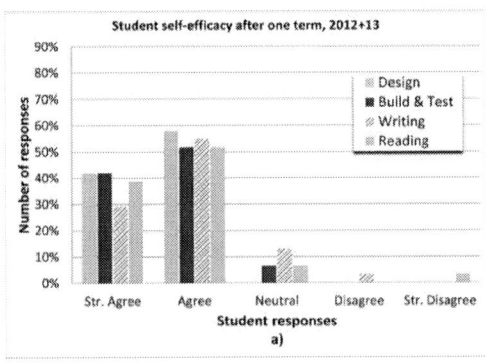

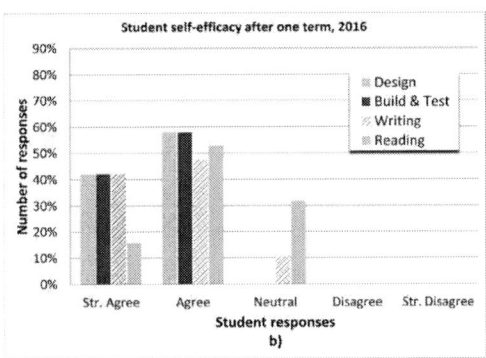

Figure 4. Student responses to survey questions about their self-efficacy. Data from a) 2012+2013 (31 responses), and b) 2016 (19 responses).

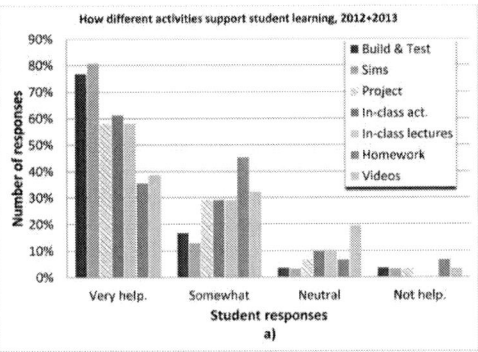

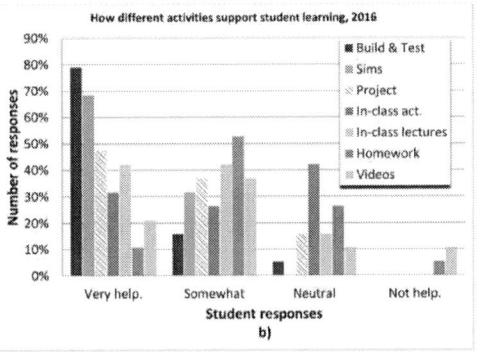

Figure 5. Student responses to survey questions regarding effectiveness of various instructional approaches. Data from a) 2012+2013 (31 responses), and b) 2016 (19 responses).

As part of the same end-of-term survey, we ask students to evaluate how helpful various instructional techniques are to their learning. Just like the previous results, these are subject to interpretation. There is also a lingering question regarding how students actually interpret various questions. Given the data in Figure 5, we can see that over the years students evaluated hands-on components: "build & test," "simulations" and "project" as the most helpful. There is a drop off in usefulness of in-class activities for which we currently do not have a good explanation. It is worth noting that students spend 2-3 hours working in the lab every week and that they are working on either the mini-project (first half of the course) or big project (second half). Therefore, time that students spend on various activities is not uniformly spread out over all activities. This is done intentionally because we believe that some activities are more effective than others, but it is possible that the time allocated affects how students perceive their usefulness. Even so, it is notable that simulation gets such high marks given that we spend no time on it in class and only ask for brief demos during labs. Homework is collected but not graded and counts only towards participation which, again, may color student perception of its usefulness. We have made some specialized videos, e.g. on deembedding, and it looks like students have not been impressed. In addition, some of our classes were recorded for later viewing but students did not get excited about those either. It is possible to make the lecture recordings more attractive and effective but it requires additional time and resources. In addition, they have to serve a well-defined instructional purpose.

We also perform other assessments, in particular with respect to writing abilities where we are still experimenting and attempting to settle on a set of rubrics that can be used across courses and assignments. One area in which there is general lack of assessment tools is the measurement of learning gains. These can be useful not only in determining efficacy of our teaching but also in finding students' misconceptions. Given that there is no concept inventory for any type of circuit design we are attempting to design a pre- and post-test that may help us in both areas of assessment. However, the results are very preliminary and will be reported at a later date.

VI. CONCLUSIONS AND FUTURE WORK

Our microwave circuit design courses have undergone several major shifts over the last 20 years. The first one was started in late 1990-s with introduction of simulation which increased complexity and realism of problems that could be addressed. Around 2010 we started the second one, which centered on introducing active learning in classroom. The third, and most recent, shift is to hands-on practice which enables true design-build-test cycle. In some cases, this also enables re-design, which has significant educational benefits. The evolution of makerspaces was the main enabler for the third shift, much as the capable and affordable software was for the second.

We have described how various components of the courses fit together and rationale for their introduction. We have also provided some data from student surveys which illustrate how well we are meeting our course learning objectives. Students seem to value hands-on activities – whether in the form of hardware or software design – and projects above other parts of the course. We will be examining potential to expand even more so in that direction but without sacrificing other important learning goals. New tools, such as Perusall and ISLR, promise to help us design proper reading and writing assignments and we have had good initial experiences in their implementation. Finally, assessment needs to be expanded to a more formalized assessment of technical performance and understanding, as well as more formalized assessment of writing.

These are times of rapid change across all levels of education. Implementation of research based instructional techniques is unavoidable for any instructor who wants to provide his/her students with the most effective learning experience. This paper provided some ideas on how this may be accomplished in microwave circuit design courses but the same techniques should be applicable in many other electronics design or similar courses.

REFERENCES

[1] D. Pozar, "Microwave Engineering," Wiley, 2011.

[2] J. P. Dunsmore, "Handbook of Microwave Component Measurements : With Advanced VNA Techniques," John Wiley & Sons, 2012.

[3] K. Payne, "Practical RF Amplifier Design Using the Available Gain Procedure and the Advanced Design System EM/Circuit Co-Simulation Capability," White Paper, Agilent Technologies, 5990-3356EN, 2008.

[4] S.A. Ambrose et al., "How Learning Works: Seven Research Based Principles for Smart Teaching," Jossey-Bass/Wiley, 2010.

[5] M. Prince, "Does Active Learning Work? A Review of the Research," J. Eng. Education, vol. 93, no. 3, pp. 223-232, 2004.

[6] R. M. Felder and R. Brent, "The Intellectual Development of Science and Engineering Students. Part 1: Models and Challenges," J. Eng. Education, vol. 93, no. 4, pp. 269–277, Oct. 2004.

[7] S. C. Zemke, "Student Learning in Multiple Prototype Cycles," in ASEE Annual Conference and Exposition, 2012.

[8] B. Pejcinovic, "Application of active learning in microwave circuit design courses," in ASEE Annual Conference and Exposition, Atlanta, GA, 2013.

[9] R.H. Caverly, "Project-Based Learning Experiences in RF and Microwave Wireless Communications Systems Components," in ASEE Annual Conference and Exposition, June 2011.

[10] R.H. Caverly, "Curriculum and Concept Module Development in RF Engineering," Proc. 2007 ASEE Annual Conf., June 2007.

[11] R. Campbell and R.H. Caverly, "RF Design in the Classroom," IEEE Microwave Magazine, pp. 74-83, June 2011.

[12] K.C. Gupta, "Concept Maps and Modules for Microwave Education," IEEE Microwave Magazine, pp. 56-63, September 2000.

[13] B. Pejcinovic and R. L. Campbell, "Active Learning , Hardware Projects and Reverse Instruction in Microwave / RF Education," in European Microwave Conference, 2013, pp. 1571–1574.

[14] M. L. Loughry, M. W. Ohland, and D. D. Moore, "Development of a Theory-Based Assessment of Team Member Effectiveness," Educational and Psychological Measurement, vol. 67, no. 3, pp. 505–524, Jun. 2007.

[15] Original idea comes from RFIC lab at Berkeley, see http://rfic.eecs.berkeley.edu/142/labs/resources/Making_SMA_Conformal_Test_Cables.pdf , last accessed March 5, 2017.

[16] R. B. Bass, B. Pejcinovic, and J. Grant, "Applying Scrum project management in ECE curriculum," in 2016 IEEE Frontiers in Education Conference (FIE), 2016, pp. 1–5.

[17] C. Furse, R. J. Woodward, and M. A. Jensen, "Laboratory Project in Wireless FSK Receiver Design," IEEE Trans. Education, vol. 47, no. 1, pp. 18–25, Feb. 2004.

[18] B. Pejcinovic, "Using Systematic Literature Reviews to Enhance Student Learning," in 2015 ASEE Annual Conference and Exhibition, Seattle, WA, 2015, p. 26.1685.1-26.1685.13.

[19] See perusall.com , last accessed March 5, 2017.

[20] C. R. Forest et al., "The Invention Studio: A University Maker Space and Culture," Advances in Engineering Education, vol. 4, no. 2, pp. 1–32, Summer 2014.

[21] Lab for Interconnected Devices – http://psu-epl.github.io/

[22] A. Bandura, "Self-efficacy: Toward a unifying theory of behavioral change," Psychological Review, vol. 84, no. 2, pp. 191–215, Mar. 1977.

MIPRO 2017, May 22- 26, 2017, Opatija, Croatia

A Learning Tool for Synthesis, Visualization, and Editing of Programming for Simple Programmable Logic Devices

Marko Čupić, Karla Brkić, Željka Mihajlović
University of Zagreb, Faculty of Electrical Engineering and Computing
Unska 3, 10 000 Zagreb, Croatia
{marko.cupic, karla.brkic, zeljka.mihajlovic}@fer.hr

Abstract—**Our experiences in teaching digital circuit design at university level indicate that students find it difficult to understand programmable logic devices (PLDs) such as PALs, PLAs, GALs and FPGAs. This is mainly due to the complexity of the topic and the lack of tools that visualize the inner workings of PLDs and enable students to modify and inspect individual components. The majority of publicly available SPLD-related tools are proprietary, platform-specific and do not expose all elements of the PLD structure to the user. In this work, we propose a learning tool that enables synthesis, visualization and editing the programming of a GAL16v8 SPLD. GAL16v8 has been chosen as it is simple enough that its physical implementation can be observed to the smallest detail, while enabling simultaneous realization of multiple Boolean functions. The tool is platform-independent and specifically tailored towards use in an educational setting, facilitating much better understanding of SPLDs through a hands-on experience.**

I. INTRODUCTION

The historical development of programmable logic devices (PLDs) can be traced from the most simple ones, such as one-time programmable read-only memories, through a bit more complicated simple programmable logic devices (SPLDs) such as PALs, PLAs and GALs, to complex programmable logic devices (CPLDs) and field programmable gate arrays (FPGAs) [1], [2]. The basic principles involving internal workings and architectures of such devices are often taught in university-level digital circuit design courses, including the Digital Logic course at our Faculty. It is our experience that students often find the topic hard to understand [3], considering the fact that the presentation of the topic typically covers both low-level details including electronical components such as FGMOS transistors and high level concepts including logical design and PLD programming.

Our previous work [4], [5], [3] indicated that the students at out Faculty readily accept e-learning tools and that such tools have a positive impact on their academic performance. In order to foster student experience with PLD technologies and improve their understanding, we have developed a multipurpose learning tool for GAL16v8. The tool can be used for direct programming, where the student can input each programmable switch programming. It can also be used for visualization of programming and generating the JEDEC file needed to program the physical GAL chip in hardware programmer. Additionally, the tool allows the students to

provide a higher level description of the desired design using logical expressions, which the synthesis module then converts into GAL programming. The generated GAL programming can be visualized in a graphical environment and modified if needed.

One of the main advantages of the tool is that it is written in Java so it can be run on many different operating systems; a fact that is not true for many commercial tools. The GAL16v8 has been chosen because it is relatively simple, so the students can understand the function of each of 2194 programmable switches and can set their programming directly. In order to better integrate the rest of the Digital Logic course topics taught at our Faculty which are based on VHDL usage, we have equipped the tool with a module that allows students to provide a logical description of the desired circuits. Using the description, the module determines the GAL programming that the student can then inspect.

II. RELATED WORK

Learning basic concepts of digital electronics represents one of the cornerstones of modern computer and electrical engineering education. These concepts are essential for understanding hardware design [6], [7], computer architectures [8], [9], systems-on-a-chip [10], etc. Without deep understanding of these concepts, the students will not be able to understand how the computer actually works, how a microprocessor is designed and implemented, how computer programs are executed and what to consider when thinking about efficient computer code. Given the importance and the ubiquity of digital electronics, there is an interest in developing better learning methods for various digital electronics concepts in the academic education community. These methods involve developing specialized learning tools that include purely software-based solutions [11], [12], [13], as well as remote hybrid software-hardware learning systems [14], [15], [16]. Improvements of the teaching process itself are also considered, e.g. through reframing the teaching to be more centered around using a hardware description language [17], or devising instructive problems that can be solved by small teams of students in a competition-based semi-professional environment [18].

One example of a specialized software learning tool for digital circuit design is Boole-Deusto, proposed by Garcia-Zubia et al. [11]. Boole-Deusto is intended for teaching the design and analysis of bit-level combinatorial and sequential circuits. Combinatorial circuits can be defined using truth tables, minterms or maxterms or Boolean logic expressions. Boole-Deusto supports Veitch-Karnaugh diagrams, finite state machines, and code generation from the designed circuits (OrCAD-PLD, VHDL, and JEDEC are supported). Similarly, Hacker and Sitte [12] propose WinLogiLab, an interactive teaching suite for combinatorial and sequential circuits. The suite is comprised of a set of increasingly complex tutorials, covering a range of topics including Boolean algebra, truth tables, Karnaugh maps, finite state machines, etc. Donzellini and Ponta [13] propose an e-learning simulation environment for digital electronics called Deeds. The environment is tailored towards teaching embedded systems, and it supports combinational and sequential circuits, finite state machines, and microcomputer interfacing and programming. Baneres et al. [19] introduce an online platform for the design and verification of digital circuits through a series of exercises, consisting of a desktop application that communicates with the course server and a web-based management system for instructors.

In hybrid software-hardware learning systems, the goal is to enable the student to interact with a real physical SPLD. For instance, El Medany [14] proposes a remote laboratory that enables the students to interactively control a physical FPGA board over the internet. Garcia-Zubia et al. [15] and Rodriguez-Gil et al. [16] propose a system that combines their previously described Boole-Deusto learning tool with a physical remote FPGA board that controls a virtual simulated water tank. Boole-Deusto is used to generate VHDL code that is synthesized and programmed into the FPGA, and the student can see the remote FPGA controlling the simulated water tank according to the developed digital circuit.

One shortcoming of Boole-Deusto, WinLogiLab and Deeds is that they run exclusively on Microsoft Windows, making them unavailable to students that use other operating systems. Furthermore, Boole-Deusto and WinLogiLab seem to be restricted to basic digital electronics concepts, offering no support for teaching SPLDs such as PALs, PLAs, GALs. Motivated by the importance of teaching basic SPLD concepts and our experience that students find it difficult to understand the inner workings of an FPGA following a classical black box vendor-specific synthesis approach [20], we have previously proposed a platform independent tool for programming, visualization and simulation of simplified FPGAs [3]. In this work, we move a step further by developing a tool that exposes the inner workings of GAL16V8 in a fully interactive manner.

III. THE LEARNING TOOL REQUIREMENTS

In the Digital Logic course at our Faculty, we cover programmable logic circuits in depth: from implementation details up to the logical model and high-level programming. The course covers implementation of permanent read-only

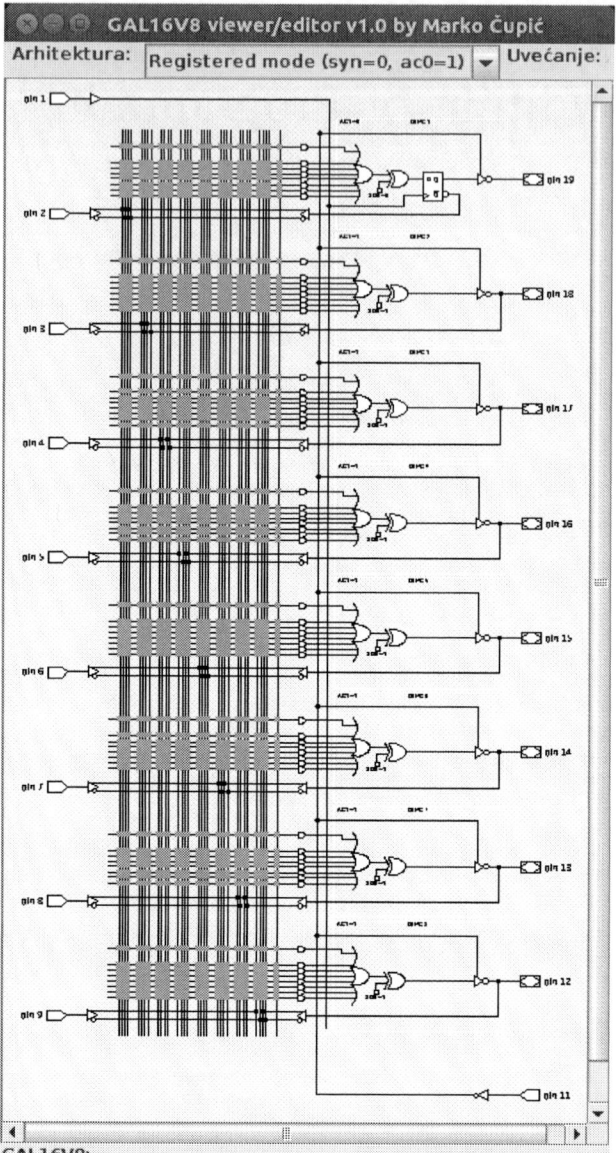

Fig. 1. Logical view of GAL16v8 in registered mode.

memories (using semiconductor diodes and then metal-oxide-semiconductor field-effect transistors, MOSFETs), principles of implementation of erasable ROMs (EPROMs) and electrically erasable ROMs (EEPROMs) which are based on floating-gate MOSFET (FGMOS), various SPLDs (such as programmable logic array - PLA, and programmable array logic - PAL) and finally complex programmable logic devices (CPLDs) and field programmable gate arrays (FPGAs).

In order to clarify the design, programming and application of SPLDs on an instructive example, we decided that our learning tool should model some existing chip which fulfils the following requirements:

1) it can still be purchased,
2) it is not too expensive,
3) it offers adequate logic programmability so that several Boolean functions can be realized simultaneously,
4) it is simple enough that its physical implementation can be observed to the smallest detail (i.e. which of the FGMOS transistors should have excess charge on its floating-gate and which not in order to realize the desired Boolean function - see Figure 2),
5) support for generating both combinatorial and sequential circuits is desired.

The fourth requirement stems from the fact that within the Digital Logic course (as taught at our Faculty) we teach students the basics of digital circuits implementation (i.e. how can various logic gates, such as NOT, AND, OR, NAND, and NOR, be implemented using bipolar transistors, using MOS-FETs and using CMOS) as well as how programmable digital circuits are implemented (i.e. programming by burning fuses, programming by blocking channel creation in MOSFETs, etc). Therefore, we required a chip which is simple enough that such level of detail can be presented.

We also wanted the selected chip to be suitable for teaching how high level SPLD programming can be done. There are two possible ways of doing SPLD programming:

1) the state for each of the programmable elements (FG-MOSFET) can be set manually, or
2) the desired functionality can be described by some language for formal specification and then the state of each of programmable element can be set by some automatic procedure.

We have chosen the GAL16v8 chip, as it fulfils all of the aforementioned requirements. It is rather cheap, it can realize eight Boolean functions and it works with a wide range of supply voltages. Additionally, it has a relatively regular structure so that each of 2194 programmable switches can be observed and its function readily understood. This chip can be configured to operate in three different architectural modes defined using two bits of information: simple mode, complex mode and registered mode. This is in itself very instructive, as it offers a great example of multiplexer usage, where

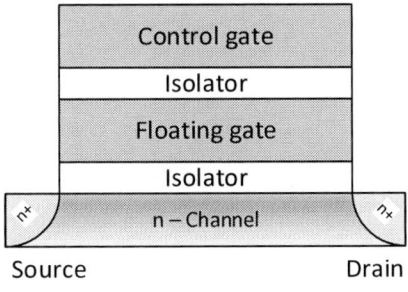

Fig. 2. Structure of FGMOSFET.

different architectures are realized by multiplexers choosing which signal is routed where.

For actual chip programming, an additional device is needed (so-called programmer) which is connected to the user's computer and in which the programmable chip is placed. On the computer, the appropriate software must be started and data based on which the programming will be performed must be provided. Today, such data is typically provided as JEDEC-formatted file, so we wanted to be able, while teaching the topic, also to explain how this file is formatted and how it is generated for selected chip.

Having selected the chip, our goal was to develop a learning tool that would satisfy the following requirements:

1) it should be portable (so that the students can use it on various operating systems),
2) it should offer graphical user interface which should show all programmable elements and allow the student to edit each FGMOS transistor state,
3) it should be able to automatically generate a JEDEC file for shown programming,
4) it should be able to read programming from a JEDEC file and adjust current programming,
5) it should support some way of formally describing Boolean functions and automatic programming based on a given formal description,
6) it should offer a command line tool for converting a direct formal circuit description into JEDEC,
7) and lastly, it must be simple enough to be adequate for education purposes (which most commercial professional tools are not).

IV. THE DEVELOPED TOOL

The learning tool that accommodates the requirements elaborated in Section III has been developed using the Java programming language. This way, we have ensured that the tool is portable across all often used desktop operating systems. When started in interactive mode, the student is presented with a graphical editor showing the logical structure of GAL16v8 (see Figure 1 for a coarse overview).

On the top of the editor window there is a combo box which allows the user to set values for two architectural bits, thus choosing the active GAL configuration. The values of these two architectural bits are connected to several multiplexers in the physical chip implementation. The developed editor, for the sake of clarity, does not show all of these details. Instead, once the user selects values of architectural bits, the editor shows the effective chip architecture.

In simple mode (see more detailed Figure 3), each macrocell can realize a Boolean function in the form of a sum of 8 products. The output of each macrocell can be configured to be permanently enabled or permanently in high-impedance state. Signals from pins 1-9, 11-14 and 17-19 are connected to the input array in the which user can program products. Each programmable element (which determines if the given signal is used in given product) is technologically realized using FGMOSFET.

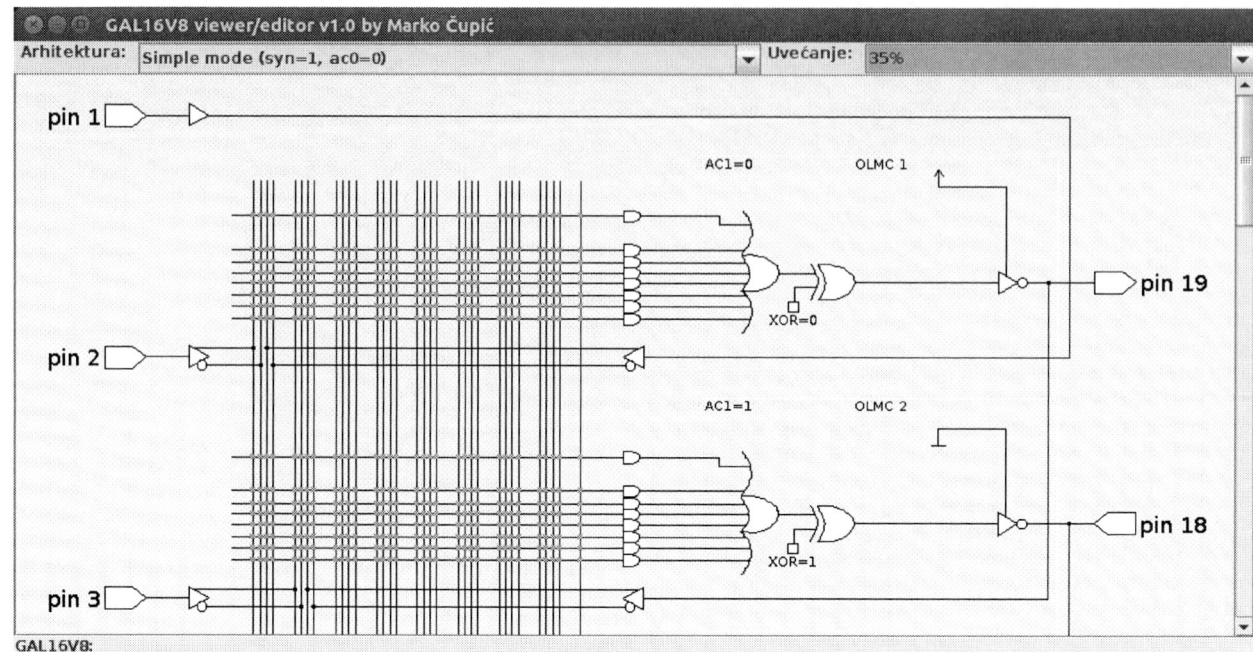

Fig. 3. A detail of the logical view of GAL16v8 in simple mode.

Programming of all the programmable elements can be performed by mouse. The state of FGMOSFET can be toggled by double-clicking on the signal-product connection, which will be visually indicated. Macrocell configuration bits can be defined by double clicking on the macrocell. In simple mode, the user can configure the logical value for lower input of XOR-gate and additionally define whether the output buffer is enabled or disabled. If a low number of Boolean functions is to be realized but of many variables, disabling some macrocells can open output pins to be used as additional inputs. A relevant part of GAL configured to calculate $f(A, B) = A \oplus B$ is shown in Figure 4. Here, A and B were connected to pins 2 and 3 while macrocell OLMC 1 (the top-most one) was used for calculation. Its output buffer was enabled and the bit for output XOR-gate was set to 1 since the output buffer actually also inverts the calculated value.

In complex mode (see smaller Figure 5) each macrocell can realize a Boolean function in the form of a sum of 7 products, while one product controls the output buffer (enabled or in high-impedance).

Finally, in registered mode (see smaller Figure 6), each macrocell can realize a Boolean function in the form of a sum of 8 products. The realized function, depending on the macrocell configuration bit, can be routed to macrocell output or it can be used as input for D-flipflop whose state is then used as macrocell output. In this mode, the output buffer is enabled/disabled for all macrocells by a common signal. In Figure 6, the top-most macrocell is configured to use a D-flipflop and the second macrocell is configured to calculate

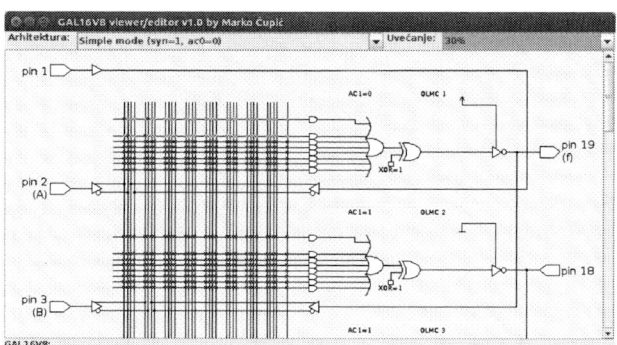

Fig. 4. A GAL configured to generate $f = A \oplus B$.

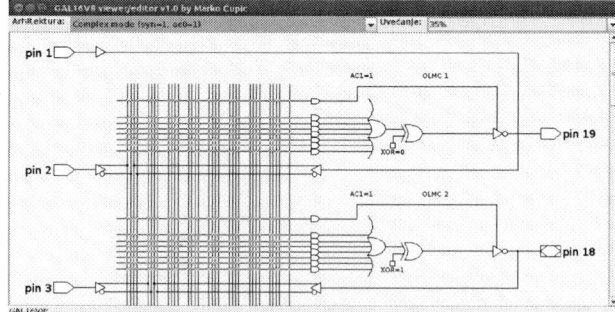

Fig. 5. A detail of the logical view of GAL16v8 in complex mode.

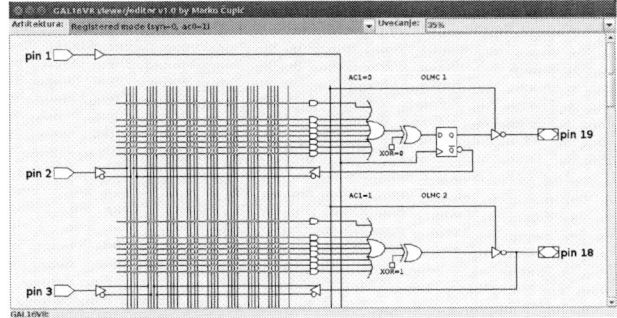

Fig. 6. A detail of the logical view of GAL16v8 in registered mode.

```
JEDEC datoteka
JEDEC datoteka za GAL16V8
U zastavicama 1 znaci visoki otpor - nema spoja
dok 0 znaci nizak otpor - postoji spoj.*
NOTE PINS A:2 B:3*
NOTE PINS f:19*
QP20*QF2194*F0*
NOTE Slijede podatci za 8 produkata makrocelije 1*
L0000
011110111111111111111111111111
101101111111111111111111111111
000000000000000000000000000000
000000000000000000000000000000
000000000000000000000000000000
000000000000000000000000000000
000000000000000000000000000000
000000000000000000000000000000*
NOTE Slijede podatci za 8 produkata makrocelije 2*
L0256
000000000000000000000000000000
000000000000000000000000000000
000000000000000000000000000000
000000000000000000000000000000
000000000000000000000000000000
000000000000000000000000000000
000000000000000000000000000000

                 Snimi u datoteku...

                       OK
```

Fig. 7. Generation of JEDEC file from GUI.

combinatorial function.

The developed tool allows students to easily generate the JEDEC file from a popup menu. JEDEC file generation for XOR example from Figure 4 is shown in Figure 7.

The tool also allows user to load configuration from JEDEC file which was generated for this GAL chip.

To support automatic programming based on a formal circuit specification, we decided to reuse a syntax used in older commercial tools, but in a simplified form. An example of such a description is shown below.

```
DEVICE GAL16V8
MODE SIMPLE
PIN 2 A
PIN ? B
PIN ? C
PIN ? f OUTPUT
PIN ? g OUTPUT

EQUATIONS

f = A*/B + C
g = A:+:B + C
```

This formal description defines two Boolean functions to be realized: $f = A \cdot \bar{B} + C$ and $g = A \oplus B + C$. The specification

also defines some constraints for the synthesiser: the input A must be attached on pin 2 while other variable-pin assignment can be arbitrary.

Descriptions such as this one can be synthesized through the GUI or the synthesis can be requested directly from the command line. In the latter case, the synthesizer will generate the a appropriate JEDEC file. If the previous description is saved in file f2.eqn, the synthesis from the command line can be started by command:

```
java -jar SimpleGAL-1.0.1.jar
     pahdl-to-jedec -in f2.eqn
     -out f2.jedec
```

which will produce the JEDEC file f2.jedec as well as additional info (including actual pin assignment). Example of such a file report is given below.

```
DEVICE: GAL16V8
MODE: SIMPLE
F SOP: A*/B+C
G SOP: /A*B+A*/B+C
INFO: Inputs: [A, B, C].
INFO: Outputs: [F, G].
INFO: Pin mapping process succeeded.
INFO: Pin 01: name=NC
INFO: Pin 02: name=A
INFO: Pin 03: name=B
INFO: Pin 04: name=C
INFO: Pin 05: name=NC
INFO: Pin 06: name=NC
INFO: Pin 07: name=NC
INFO: Pin 08: name=NC
INFO: Pin 09: name=NC
INFO: Pin 10: name=GND
INFO: Pin 11: name=NC
INFO: Pin 12: name=NC
INFO: Pin 13: name=NC
INFO: Pin 14: name=NC
INFO: Pin 15: name=NC
INFO: Pin 16: name=NC
INFO: Pin 17: name=NC
INFO: Pin 18: name=G
INFO: Pin 19: name=F
INFO: Pin 20: name=VCC
INFO: Here is the pin assignment:
                GAL16V8
        +----      ----+
        |    \__/      |
   NC |   1        20  | Vcc
    A |   2        19  | F
    B |   3        18  | G
    C |   4        17  | NC
   NC |   5        16  | NC
   NC |   6        15  | NC
   NC |   7        14  | NC
   NC |   8        13  | NC
   NC |   9        12  | NC
  GND |  10        11  | NC
        +------------+
```

INFO: Synthesis to GAL16V8 completed.

V. EXAMPLE USE CASES

Example use cases for our tool within the Digital Logic course we teach include illustrations of the topic "Programmable logical devices", where we offer interested students to complete two experiments.

Experiment 1. Elements: 2 DIP switches, 5 resistors, GAL16V8, 3 LEDs. Use 2 DIP switches and two resistors

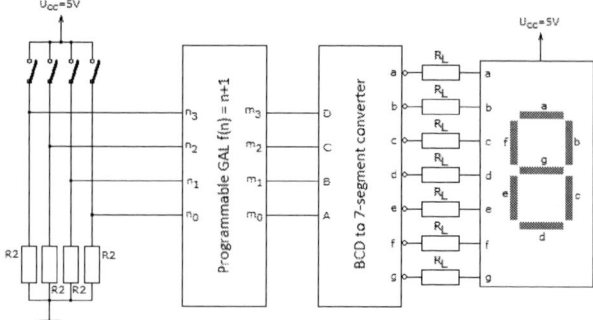

Fig. 8. A digital circuit with GAL16V8.

to create logical values for A and B. Connect values A and B to GAL16V8 inputs, and 3 LEDs to its outputs. Using our tool, program the GAL16V8 by manually clicking the programmable switches, to ensure that the first LED lights only if both A and B are high, that the second LED lights if any of A or B is high, and that the third LED lights only when one of A or B is high. Create the JEDEC file. Program the GAL and test complete circuit on the protoboard.

Experiment 2. Elements: 4 DIP switches, 11 resistors, GAL16V8, BCD to 7-segment converter, 7-segment display. Using 4 DIP switches and 4 resistors create logical values $a_3a_2a_1a_0$. Let us treat $a_3a_2a_1a_0$ as binary-encoded number. Derive by hand logical expressions for function $f(a_3a_2a_1a_0) = a_3a_2a_1a_0 + 1$ (i.e. the next number). Write a formal circuit specification and using our synthesizer create the JEDEC file. Program the GAL and test complete circuit on the protoboard (see Figure 8).

VI. Conclusion

We have presented an interactive learning tool for synthesis, visualization and programming GAL16v8 SPLD. The developed tool can be used in education for courses that cover programmable logic devices. The tool is very useful in bringing insights into the complete design process: starting with a formal description of a circuit, the tool enables learning which steps are needed to complete the programming. This is quite handy because the selected GAL16v8 is simple enough so that the programming can be traced back to each single FGMOSFET transistor.

Using this tool, students can also observe what is the end process of the synthesis - which decisions did the synthesizer make and how the programming was completed. In our experience, if the topic is clearly covered and illustrated in this way, students can more easily proceed to FPGAs, which are much more complex to grasp.

As a future work, we plan to add support for another formal language for hardware specification: a simplified version of VHDL, since the VHDL is used for FPGAs as well. We believe that consistently using a single language from the start can make the transition from simple programmable logic devices to more advanced ones easier and more natural for students.

References

[1] B. J. LaMeres, *Programmable Logic*. Cham: Springer International Publishing, 2017, pp. 371–384.

[2] V. Taraate, *Introduction to PLD*. Singapore: Springer Singapore, 2017, pp. 169–209.

[3] M. Čupić, K. Brkić, and Ž. Mihajlović, "A platform independent tool for programming, visualization and simulation of simplified FPGAs," in *2016 39th International Convention on Information and Communication Technology, Electronics and Microelectronics (MIPRO)*, May 2016, pp. 986–991.

[4] M. Čupić and Ž. Mihajlović, "Computer-based knowledge, self-assessment and training," *International Journal of Engineering Education*, vol. 26, no. 1, pp. 111–125, 2010.

[5] Ž. Mihajlović and M. Čupić, "Software environment for learning and knowledge assessment based on graphical gadgets," *International Journal of Engineering Education*, vol. 28, no. 5, pp. 1127–1140, 2012.

[6] J. Staunstrup, *A Formal Approach to Hardware Design*, ser. Kluwer international series in engineering and computer science: VLSI, computer architecture, and digital signal processing. Springer US, 1994.

[7] I. Grout, *Digital Systems Design with {FPGAs} and {CPLDs}*. Newnes, Burlington, 2008.

[8] J. L. Hennessy and D. A. Patterson, *Computer Architecture, Fifth Edition: A Quantitative Approach*, 5th ed. San Francisco, CA, USA: Morgan Kaufmann Publishers Inc., 2011.

[9] D. Patterson and J. Hennessy, *Computer Organization and Design MIPS Edition: The Hardware/Software Interface*, ser. The Morgan Kaufman Series in Computer Architecture and Design. Elsevier Science, 2013.

[10] L. H. Crockett, R. A. Elliot, M. A. Enderwitz, and R. W. Stewart, *The Zynq Book: Embedded Processing with the Arm Cortex-A9 on the Xilinx Zynq-7000 All Programmable SoC*. UK: Strathclyde Academic Media, 2014.

[11] B. S. B. Javier Garcia Zubia, Jesus Sanz Martinez, "A new approach to educational software for logic analysis and design," in *IADAT e2004, International Conference on Education*, 2004.

[12] C. Hacker and R. Sitte, "Interactive teaching of elementary digital logic design with winlogilab," *IEEE Transactions on Education*, vol. 47, no. 2, pp. 196–203, May 2004.

[13] G. Donzellini and D. Ponta, "A simulation environment for e-learning in digital design," *IEEE Transactions on Industrial Electronics*, vol. 54, no. 6, pp. 3078–3085, Dec 2007.

[14] W. El Medany, "FPGA remote laboratory for hardware e-learning courses," in *Computational Technologies in Electrical and Electronics Engineering, 2008. SIBIRCON 2008. IEEE Region 8 International Conference on*, July 2008, pp. 106–109.

[15] J. Garcia-Zubia, I. Angulo, L. Rodriguez-Gil, P. Orduna, O. Dziabenko, and M. Guenaga, "Boole-WebLab-FPGA: Creating an integrated digital electronics learning workflow through a hybrid laboratory and an educational electronics design tool," *International Journal of Online Engineering (iJOE)*, vol. 9, 2013.

[16] L. Rodriguez-Gil, P. Orduna, J. Garcia-Zubia, I. Angulo, and D. Lopez-de Ipina, "Graphic technologies for virtual, remote and hybrid laboratories: WebLab-FPGA hybrid lab," in *Remote Engineering and Virtual Instrumentation (REV), 2014 11th International Conference on*, Feb 2014, pp. 163–166.

[17] E. G. Breijo, L. G. Sanchez, and J. I. Civera, "Using hardware description languages in a basic subject of digital electronic: Adaptation to high academic performance group," in *2014 XI Tecnologias Aplicadas a la Ensenanza de la Electronica (Technologies Applied to Electronics Teaching) (TAEE)*, June 2014, pp. 1–6.

[18] R. Rengel, M. J. Martin, and B. G. Vasallo, "Supervised coursework as a way of improving motivation in the learning of digital electronics," *IEEE Transactions on Education*, vol. 55, no. 4, pp. 525–528, Nov 2012.

[19] D. Baneres, R. Clariso, J. Jorba, and M. Serra, "Experiences in digital circuit design courses: A self-study platform for learning support," *IEEE Transactions on Learning Technologies*, vol. 7, no. 4, pp. 360–374, Oct 2014.

[20] A. Sangiovanni-Vincentelli, A. El Gamal, and J. Rose, "Synthesis method for field programmable gate arrays," *Proceedings of the IEEE*, vol. 81, no. 7, pp. 1057–1083, Jul 1993.

MIPRO 2017, May 22- 26, 2017, Opatija, Croatia

Active-Learning Implementation Proposal for Course Electronics at Undergraduate Level

Tvrtko Mandic, Adrijan Baric
University of Zagreb, Faculty of Electrical Engineering and Computing
Zagreb, Croatia
Email: tvrtko.mandic@fer.hr

Abstract—The course "Electronics 1" is taught at undergraduate level and covers broad area of electronics starting from the physics of semiconductors to the complex electronic system such as operational amplifier. This course is obligatory for sophomore students enrolled in Electrical Engineering and Information Technology undergraduate program as well as in Computing undergraduate program. Due to the course comprehensiveness and recognised lack of interest, students tend to study without understanding the studied concepts and their practical application. This paper presents the proposal to increase the interest in electronics by introduction of the active learning (AL) process. This AL process is supported by the developed AL module, i.e. clicker, which provides the hands-on experience of the topics studied throughout the course. We provide questionnaire proposal that will be given to groups of students and the results will be compared to the control group to assess the usefulness of the proposed approach.

Index Terms—active-learning; electronics; hardware development; questionnaire proposal

I. INTRODUCTION

Sophomore students enrolled in Faculty of Electrical Engineering and Computing at the University of Zagreb have obligatory course "Electronics 1". This course covers wide area of electronics starting from physical properties of semiconductors to the complex electronic system such as an operational amplifier.

Every semester the students are asked to evaluate the course through a survey. According to the survey results several conclusions are drawn:

1. Although the course is well organized there is substantial lack of interest in the studied matter
2. The students find the course difficult and think of themselves as not competent enough to understand the studied matter
3. Students spend approx. 3 hours/week of additional studying
4. Laboratory exercise procedures are not clear enough

The first conclusion tells about the low level of interest in electronics. This is more pronounced for the students enrolled in Computing undergraduate program. The lack of interest has grounds in perception that their competence and prior knowledge necessary for this course is not on the required level. Their incompetence lowers motivation and hinder learning process [1]. The students find the course well organized (literature is easily available, presentation can be downloaded,

problem solving exercises are given, etc.) and necessary to spend 3 hours/week of studying to pass the exam. The last conclusion refers to lack of practical knowledge needed to successfully and with understanding perform the laboratory exercises.

In order to gain interest in the studied matter, the authors have designed active-learning (AL) module, i.e. clicker. The idea behind the clicker presented in this paper is to have students to actively contribute in the lectures by providing active feedback. In [2] clicker is replaced by smart phone which is practical but the usage of the smart phone may distract students. By using special developed clicker we believe it will improve understanding of the studied concepts through:

- gamification
- hands-on experience
- practical understanding of learned theory

The idea behind the gamification is to organize students in groups. Each group (2-4 students) will have their own clicker. During the semester they collect points according to the correct answers given. The questions asked during the lectures bring necessary breaks in order to keep their concentration at the high level. Furthermore, the work in group will support discussions among them and through explanations help better understanding of given lecture [3]. During the first two semesters of undergraduate program the students do not have opportunity to see how actually electronic components look like. By using the clicker they can bridge the gap between theory and practical application. This hand-on experience together with the actual implementation of the studied theory in the clicker itself will help them to better understand the electronics.

This paper is organized as follows. In Section II the design and features of the developed module are presented. Section III brings the explanation of the studied topics in the course Electronics and their connection to the clicker hardware. Section IV presents the proposal of the questions to be given to students in order to assess their understanding of studied concepts. Section V brings the conclusion.

II. DEVELOPED ACTIVE-LEARNING (AL) MODULE

The developed AL module i.e. clicker, is presented in Fig. 1. The clicker size is 5 cm × 5 cm. These are the main hardware features of the module:

- Supercapacitor charged via USB

- 4 tactile switches
- Two LEDs
- Microcontroller and RF module.

The module size as well as the components are selected to result in low-cost design (approx. 5 GBP @ 10 pcs.). The project is open-source and all hardware and software files are available online [4]. The students are encouraged to built their own clicker and use it as a platform for other projects. This clicker is actually small part of the functionality of the complete module. This module is developed in the frame of YAWN-Your Autonomous Wireless Node project and supports whole ecosystem of sensors, power supply solutions, communication protocols, etc. available on-board or through the expansion headers.

The students have four tactile switches denoted by A, B, C and D to provide an answer. After sending the answer to the lecturer, the clicker goes in deep-sleep mode and therefore extends the time between recharging. The charging lasts approximately 30 s and provides several days of operation without recharging. The blinking of the LED indicates proper clicker operation and according to the light intensity one can estimate the power supply voltage level.

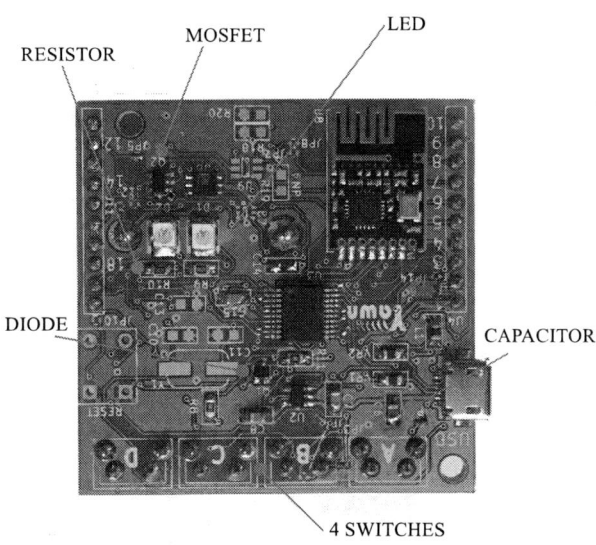

Fig. 1. Active-learning (AL) module, i.e. clicker.

The lecturer collects the student feedback by using a similar module. The lecturer's module is usually connected to the computer used for running the presentation. The small user interface is developed in open source computer programming language Processing [5] and presented in Fig. 1. Prior asking students a question, lecturer resets all previous answers by clicking the RST button. The answers together with the time tag are saved by clicking the SAVE button. The answer can be reset if a student clicks the same switch again. This allows lecturer to give the multiple-choice questions.

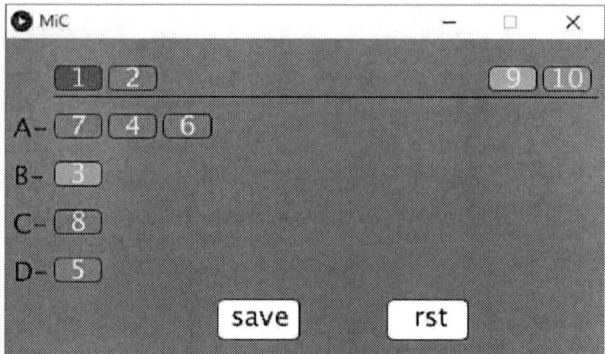

Fig. 2. Lecturer application for collecting the students answers.

III. METHODOLOGY

At the beginning of the semester students are asked to take one clicker and form the group containing 2-4 students. Each clicker is assigned with a unique number (in the example presented in Fig. 2 there are 10 clickers given to the students). The group should remember the number and use the same clicker during the semester.

The lecturer should give points according to the answers collected during the class and present the current ranking at the beginning of the following class. This gamification of the lectures, will provide additional motivation for student to actively participate in the lectures. Furthermore, the lecturer can immediately asses the students knowledge by looking at the results and if necessary give additional explanations or encourage peer discussions [6]. At the end of the semester lecturer can ask who is hiding behind the clicker with the largest number of points in order to make public acknowledgement of the winners.

Usually not more than two or three questions are needed to keep students concentrated. The questions in lecturing will bring necessary breaks. For example, if a question is asked after 15-20 minutes of lecturing, only two questions are needed during the class.

The usage of the clicker gives hands-on experience of lectures learned. Furthermore, by seeing actual implementation of the theory learned students might be able even to synthesize simple electronic circuits. Although, the complexity of the clicker is much bigger than the basics of electronics taught in the course "Electronics 1", some parts of the clicker will be analyzed through following problems.

A. RC Network

Problem : The design problem encountered while making the clicker is following. The super-capacitor C_S having the capacitance equal to 1.5 F. The input power supply is available through USB bus ($U_I = 5$ V). According to the schematic presented in Fig. 3 select the resistors to obtain the output voltage equal to 3.3 V.

Solution : To generate the output voltage equal to 3.3 V, the two resistor values have to be selected. The selection of

14

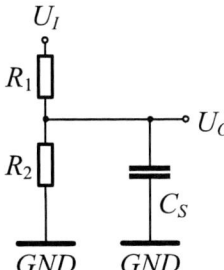

Fig. 3. Network for charging the super-capacitor.

small resistor values can lead to fast charge time but also the discharge will be fast. The power dissipation can also increase if the resistors are too small. Furthermore, the small resistor values can also lead to overloading USB bus. If the high-value resistors are selected, the discharge of capacitor will be slower as well as the charging time. Here the Thevenin's theorem should be used to calculate the time constant of the circuit.

B. Diode

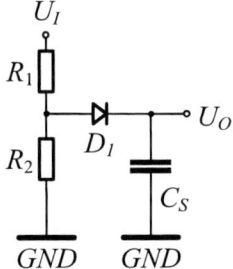

Fig. 4. Network for charging the super-capacitor with discharge prevention.

Problem : After analyzing and evaluation of the previously presented network for charging the super-capacitor the diode should be used to prevent the discharging of the super-capacitor.

Solution : The solution is presented in Fig. 4. The diode prevents discharging of the capacitor, but still the reverse current is flowing which discharges the super-capacitor. Due to the diode forward-voltage the resistors values have to be adjusted to generate output voltage equal to 3.3 V.

C. LED

Problem : For given voltage of 3.3 V it is necessary to properly bias LED.

Solution : The solution is presented in Fig. 5. The resistor is added to define the current through the LED. This problem helps better understanding of *I-V* characteristic of the diode and the role of the bias resistor. Furthermore, this is good example to relate the different forward-voltage values (due to different types of semiconductors used) to the real application.

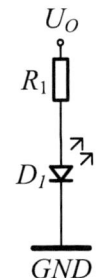

Fig. 5. LED biasing.

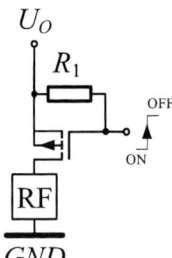

Fig. 6. MOSFET used to turn on/off the RF-module.

D. MOSFET As Switch

Problem : The RF-module power consumption is large. Use MOSFET as a switch to turn on/off the RF-module to decrease its power consumption.

Solution : The solution is presented in Fig. 6. The presented solution is based on p-MOS but the same functionality can be obtained using n-MOS. Through this problem students are taught about different current flows in p-MOS and n-MOS and different control signals.

IV. QUESTIONNAIRE PROPOSAL

This is questionnaire proposal to be given to students at the end of the semester. The idea behind is to give this proposal to three groups of students:

- regular class
- regular class + problems from Section III explained
- AL class + problems from Section III explained

A. Design problems

- Design a power supply using only two resistors. The power supply output voltage should be equal to $V_{IN} = 3.3$ V for the input voltage equal to $V_{USB} = 5$ V. The load current is negligible.

- Design a power supply using only two resistors. The power supply output voltage should be equal to $V_{IN} = 3.3$ V for the input voltage equal to $V_{USB} = 5$ V. The load current is equal to $I_L \leq 100$ mA.

15

- Design a power supply using only two resistors with the super-capacitor connected to the output. The power supply output voltage should be equal to $V_{IN} = 3.3$ V for the input voltage equal to $V_{USB} = 5$ V. Minimize the capacitor charging time.

- Design a power supply using only two resistors with the super-capacitor connected to the output. The power supply output voltage should be equal to $V_{IN} = 3.3$ V for the input voltage equal to $V_{USB} = 5$ V. Minimize the capacitor discharging current.

- Design a power supply using only two resistors with the super-capacitor connected to the output. Use diode to minimize the capacitor discharging current. The power supply output voltage should be equal to $V_{IN} = 3.3$ V for the input voltage equal to $V_{USB} = 5$ V. Minimize the capacitor discharging current.

- Design a circuit for powering one LED diode on power supply equal to $V_{IN} = 3.3$ V.

- Design a circuit for turning on and off a LED diode by a MOSFET. The power supply output voltage is equal to $V_{IN} = 3.3$ V.

- Design a circuit which will amplify input triangle-wave signal. For the 100 mV input amplitude, the output triangle-wave signal should have 5 V amplitude. The physical dimensions of the circuit should be the smallest possible.

B. Questions

- Time needed to charge super-capacitor of 1 F over the resistor equal to 1 Ω?
  a) 0.5 s
  b) 5 s
  c) 50 s

- In order to connect the LED to 5 V power supply what is additionally necessary to connect and how? Write on the line "parallel" or "serial".
  a) capacitor _____
  b) resistor _____
  c) inductor _____

- Which diode has the smallest forward voltage?
  a) LED
  b) Schottky
  c) pn-diode

- What is the value of the diode reverse current?
  a) mA
  b) uA
  c) nA

- What type of MOSFET is needed in application where a MOSFET is used to turn on an LED?
  a) nMOS

  b) pMOS
  c) both can do the job

- What these numbers represent: 0402, 0603, 0805 and 1206?
  a) value
  b) size
  c) price

- What is the approximate cost of the 1k resistor for 1000 pcs.?
  a) 0.001 GBP each
  b) 0.1 GBP each
  c) 1 GBP each

- What is the size of the passive discrete components (resistors and capacitor) usually used in electronic circuits? Write as "X mm × X mm"

  _____

- What is the DC voltage value available in USB bus?
  a) 1 V
  b) 5 V
  c) 9 V

- What is the material of printed circuit board?

  _____

V. CONCLUSION

This paper presents the proposal of active learning (AL) implementation in the course "Electronics 1". This implementation is based on the developed AL module, i.e. clicker. This clicker provides hand-on experience of the lectures begin studied and practical understanding of the theory. The gain in understanding and even ability to synthesize simple circuits will be assessed by the proposed questionnaire.

ACKNOWLEDGMENT

The authors would like to thank prof. dr. sc. Branimir Pejcinovic for boosting our motivation in the field of engineering education and for many helpful discussions.

REFERENCES

[1] R. Brent and R. Felder, "How learning works," *Chemical Engineering Education*, vol. 45, no. 4, pp. 257–258, Fall 2011.
[2] M. Jagar, J. Petrovic, and P. Pale, "AuResS: The audience response system," pp. 171–174, Sept 2012.
[3] M. K. Smith, W. B. Wood, W. K. Adams, C. Wieman, J. K. Knight, N. Guild, and T. T. Su, "Why peer discussion improves student performance on in-class concept questions," *Science*, vol. 323, no. 5910, pp. 122–124, 2009. [Online]. Available: http://science.sciencemag.org/content/323/5910/122
[4] N. Budimir, T. Mandic, and A. Baric. (2017) Yawn - your autonomous wireless node. [Online]. Available: http://github.com/tmandic/YAWN/
[5] Fry, B. and Reas, C. (2017.) A short introduction to the processing software and projects from the community. [Online]. Available: https://www.processing.org/overview/
[6] E. Mazur, "Confessions of coverted lecturer," *Lecture materials (Oporto Portugal)*, pp. 1–29, May 2007.

Decision trees in formative procedural knowledge assessment

J. Petrović and P. Pale

University of Zagreb, Faculty of Electrical Engineering and Computing
Department of Electronic Systems and Information Processing, Zagreb, Croatia
Juraj.Petrovic@FER.hr

Abstract - In this paper, an approach to automated formative assessment of procedural knowledge is described and evaluated. While assessment and representation of conceptual knowledge using visual aids such as of concept maps has often been analyzed and discussed in the literature, significantly less attention has been given to assessment of procedural knowledge. The approach described in this paper is based on automated evaluation of knowledge described by examinees in form of a decision tree against a repository of test cases. The examinee is afterwards provided with automatically generated feedback about the overall results of applying his decision tree, as well as possible changes required to correct them. The results of the pilot evaluation of a prototype system implementation are described and discussed.

I. INTRODUCTION

The importance of formative educational assessment has been emphasized many times in the published research over the last decades. A large body of research has suggested that formative assessment can be very beneficial for learning across a range of domains and settings [1]. Those benefits are the result of two main effects of formative assessments: the testing effect [2] and the benefits of formative feedback that can be provided to examinees [1]. While the testing effect results from the knowledge retrieval process [3] that inherently takes place during a knowledge assessment, in order to provide the examinees with timely feedback in a scalable manner, formative assessments have to support automated evaluation of assessment items. This is typically not a problem in situations when constrained answer item types [4] are used for assessment, for example multiple choice questions, although it requires additional labor to define feedback. On the other hand, automatically providing feedback becomes more challenging for less constrained and innovative item types, where examinees have more freedom in constructing their answers, or when their solution cannot be directly compared with a correct answer.

Concept maps are an example of a less constrained item type that has been used as a learning or assessment tool for quite a while [5] and has been praised as a knowledge representation which is structurally similar to human knowledge architecture [6]. Different algorithmic approaches have been proposed to automatically evaluate concept maps and provide their constructors with automatically generated feedback to their work [7], [8],

allowing evaluation or appropriate scoring not just based on structural equivalence.

However, while concept maps help greatly to assess conceptual or factual knowledge, less attention in the literature has been devoted to assessment of procedural knowledge and automated evaluation and feedback provision in that process. Procedural knowledge [9] can be described in a form very similar to a concept map, but usually referred to as a decision tree. While decision trees are commonly used as a machine learning technique or a method for representation of knowledge or guidelines [10], [11], their potential in knowledge assessment remains mostly unutilized and unexplored.

In this paper, a decision tree based procedural knowledge assessment approach is proposed, by evaluating a decision tree for solving a given procedural problem on a base of test cases with known solutions. The rest of this paper is organized as follows. In section 2 the general aspects of using decision trees in procedural knowledge assessment are described. In section 3 the key properties and functions of a system supporting this kind of assessment are described. The materials and methods used in the prototype evaluation are described in section 4, and the obtained results are presented in section 5. Results are discussed in section 6, and conclusions and future work are presented in the section 7.

II. USING DECISION TREES IN PROCEDURAL KNOWLEDGE ASSESSMENT

In machine learning classification problems, a decision tree usually consists of a single root node, several decision nodes, and several leaf nodes. An example of a decision tree for diagnosing problems with gasoline engine start is given in Image 1. In each of the decision nodes, depending on the corresponding attribute value, a decision in made either to investigate values of other attributes or to stop because a conclusion has been reached to assign case class or final diagnosis (a leaf node). This process also resembles a general approach to solving a diagnostic problem, for example technical failure diagnosis or clinical diagnosis. In both these processes, the diagnostician acquires information, for example a result of a medical test, and, based on its value, he either determines the final diagnosis or problem solution, or determines which information is relevant to get next. This process, described as a decision tree in terms of relevant attributes' names (decision nodes), their possible values (node links' labels), and final diagnoses (leafs), could easily be evaluated on a set of test cases for which all relevant attribute values are known, as well as final diagnoses. Test cases could be known examples from clinical practice or real examples of technical failure.

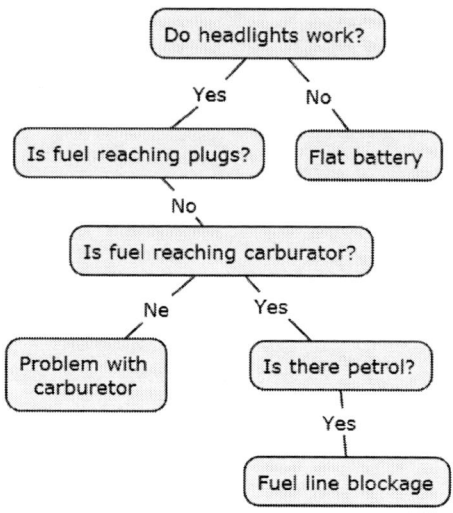

Image 1. Example of a decision tree for diagnosing problem with gasoline engine start

The main advantage of the proposed approach compared to using different items for assessing procedural knowledge is the possibility of this approach to provide its user with automatically generated evaluation and feedback. The overall evaluation of an examinee's decision tree can be performed automatically using test cases. The evaluation results can be reported as standard classification performance measures of precision and recall per diagnosis or final solution. The feedback related not to the decision tree performance, but to its included propositions can include:

- The classification path for each correctly/incorrectly diagnosed test case, as well as the node value based on which the correct diagnosis was discarded for incorrectly classified cases.

- The classification path in the decision tree for each unclassified case. Unclassified cases are the result of forgotten situations not addressed in the tree.

- The maximal correct subtree of the uploaded decision tree.

- The corrections to the decision tree, that can be automatically generated based on the test cases using information gain measure.

The feedback provided this way is a way of measuring and informing the examinee not just how well does the tree perform the classification it is supposed to do, but also why, i.e. which section of the tree is correct, which needs to be changed and which is missing completely. These components of knowledge are expected to be

relevant components of formative feedback, which is the main goal of the system, even though its usage in summative assessments can also be considered.

Also, while it would be possible to evaluate a decision tree by directly comparing it to a referent one designed by domain experts or obtained by machine learning techniques, evaluation using test cases would allow its correct evaluation when reaching justified conclusions with different attribute acquisition sequence, or even with different attributes.

The proposed approach, however, also has some disadvantages that have to be acknowledged. The main disadvantage is the need to predefine a discrete set of tree node and link labels for decision tree construction. This is necessary because the same labels have to be available in test cases to enable automatic evaluation. While defining a task template with those labels is not a technical issue, it could enable examinees to recognize instead of remember relevant parameters of the procedure they are describing. Additionally, this is also a practical limitation, since sometimes it can be difficult or impractical to split the values of a target attribute into disjoint sets and separate reactions for those sets. For example, 37°C can be considered the upper limit of a normal body temperature, but 37,01°C does not have to indicate increased body temperature.

III. IMPLEMENTATION

In order to evaluate the characteristics of the proposed knowledge assessment method and enable automated evaluation of procedural knowledge described as a decision tree, an online system to support these tasks was designed and implemented. The prototype system implementation was realized using JSF/Java, JavaScript and HTML/CSS technologies. The prototype system implementation evaluates an uploaded decision tree on a base of test cases stored in a web repository as XML files with predefined syntax. After the user's decision tree is uploaded on the system and the repository is defined using its URL, test cases from the repository are parsed and evaluated on the uploaded tree.

The prototype system implementation does not support the decision tree construction process, since there are already free online concept mapping tools like IHMC CmapTools [12] available for that purpose. The key features of IHMC CmapTools, an online free concept mapping tool include the possibility to import a template with predefined (tree) elements, a simple graphic user interface to edit a tree or a concept map, and the option to export the tree either as textual propositions or in detailed CXL file format.

The prototype system implementation supports automatically generating feedback including overall classification statistics of the uploaded decision tree, visualization of the classification path for each of the test cases, maximal correct subtree, and possible corrections of the classification path.

IV. Materials and Methods

The objective of the pilot evaluation of the prototype system implementation was to confirm two initial hypotheses related to the proposed method of assessing procedural knowledge:

- That the assessment of procedural knowledge using decision trees would be understandable to participants, not causing cognitive overload.

- That the automatically generated feedback would be relevant and understandable to participants.

The prototype system implementation was evaluated twice on a sample of 19 electrical engineering students of an 8th semester course. The participation in the research was voluntary, with the possibility to achieve an extra 5% of the course total value, based on the performance in the research.

Research participants solved two similar problems examining their procedural knowledge related to diagnosis of a technical failure in a local area network configuration. In the first research session, the participants were supposed to construct a decision tree for determining each of the seven different configuration problems based on the outcomes (success or failure) of seven *ping* commands run between different pairs of computers in the network. *Ping* commands were used as nodes in the decision tree, and success and failure were their possible values (branch labels). Possible diagnoses included, for example, incorrectly defined gateway for one of the network computers or incorrect network interface. In the second session, the participants were supposed to construct a decision tree for determining each of four different configuration problems (a subset of the previous seven), again, based on the outcomes of seven *ping* commands run between different pairs of computers in the network. IHMC CmapTools was used as the decision tree construction tool, based on the prepared template containing possible tree node labels and final diagnoses.

In each of the two sessions, the participants were instructed to:

1. Construct the decision tree using the given template and instructions.

2. Solve a questionnaire measuring their motivation and perceived complexity of the concepts and relationships between task-related concepts.

3. Upload their decision tree on the system.

4. Solve the second questionnaire, measuring how difficult was it for the participants to locate and understand feedback components.

Participants were also provided with detailed instructions for the task, containing a description of the task, all prior assumptions about it, a short explanation of decision tree meaning interpretation, and examples of solution structure. The submitted decision trees were evaluated using predefined base of 8 test cases designed by course lecturers using a design module which is a part of the prototype system implementation.

V. Results

The results of the first questionnaire were significantly different between the two sessions only on the question of how difficult did the participants find the overall topic of the tasks ($p = 0,011$, $F = 8,112$), suggesting a significant decrease in the perceived topic complexity after the first research session ($m_1 = 5,16$, $SD_1 = 2,19$, $m_2 = 4,00$, $SD_2 = 1,89$). All items were measured on a Likert scale from 1 (*Completely disagree*) to 10 (*Completely agree*). Combined results of all other first questionnaire items for both sessions are presented in the Table I.

TABLE I. Method, Topic, and Motivation Related Items

Item	Mean	SD
The task topic in general was difficult to understand.	4,58	2,10
The task-related concepts were difficult to understand.	4,03	1,85
The task-related concepts' interrelationships were difficult to understand.	4,39	1,97
The task instructions were difficult to understand.	3,05	1,84
The decision tree construction was difficult to understand.	1,76	1,02
The IHMC Cmap Tools tools was difficult to use.	2,58	1,75
The decision tree meaning interpetation was difficult to understand.	2,71	1,74
I was not motivated to participate in this research.	3,11	2,36
I find the task topic not important.	2,00	1,29
I don't care about the award points for participating in the research.	2,61	2,33
I don't care about my final course grade.	2,50	1,93

Since no statistically significant differences were found between of the two sessions data for the second survey (MANOVA results: Wilks' $\Lambda = 0,719$, $F(13, 6) = 0,847$, $p = 0,557$, $\eta^2 = 0,281$), the results for both sessions were analyzed together to increase the sample size. These results of the second questionnaire are presented in the Table II, again on a Likert scale from 1 (*Completely disagree*) to 10 (*Completely agree*).

TABLE II. Feedback Related Items

Item	Mean	SD
The total number of cases for decision tree evaluation was difficult to find.	5,11	1,29
The total number and links to correctly calssified caseses were difficult to find.	2,84	2,02
The total number and links to incorrectly calssified caseses were difficult to find.	1,63	1,44
The total number and links to uncalssified caseses were difficult to find.	0,63	0,88
The correct/incorrect sections of the uploaded decision tree were difficult to identify.	2,82	1,52
The changes required to correct the tree were difficult to understand.	3,00	1,32

Finally, all decision trees submitted by the participants were recorded on the system for later analysis. The analysis results indicated both tasks were solved completely correct by at least one participant per session (meaning all test cases were solved correctly). Detailed test case classification results are displayed in the Table III.

TABLE III. TASK SOLVING STATISTICS

Item	Session 1 (6 test cases)		Session 2 (4 test cases)	
	Mean	SD	Mean	SD
Correctly classified cases	60,50%	2,34	51,25%	1,27
Incorrectly classified cases	33,33%	1,60	31,50%	1,19
Unclassified cases	10,50%	10,2	15,75%	0,76

VI. DISCUSSION

The results presented in the section 5 are in line with the two hypotheses set in section 4. The results in Table I. suggest that the task topic in general, the related concepts or their interrelationships, with an average difficulty rating between 4,03 and 4,58 were moderately difficulty to understand, while IHMC Cmap Tools interface or the construction and interpretation of a decision tree were rated with an average rating between just 1,76 and 2,71, on a scale from 1 to 10. This result is important to show that constructing a decision tree is not difficult to the level it would be an obstacle for knowledge assessment using that method, and it is also supported by the results in the Table III. The results in the Table II are also encouraging, as they suggest the validation process and feedback are understandable to the participants. This result is in line with the second hypothesis set in the section 4, and important since the main purpose or advantage of the proposed assessment approach is formative assessment and automated feedback provision.

Even though the results of this pilot evaluation generally support the hypotheses guiding the pilot evaluation, the limitations of the study also have to be taken into account. The main limitation is the profile of the participants, since, as electrical engineering students, this was probably not their first encounter with decision trees as a knowledge representation method, which would have positively influenced their understanding of the method and its feedback.

VII. CONCLUSION

In this paper, an approach to formative assessment of procedural knowledge was described and evaluated using a prototype implementation of a system supporting the method. The proposed method is based on automatic evaluation of a decision tree constructed by an examinee, which describes which relevant information is required for solving a diagnostic problem, and how conclusions are reached based on those information.

The results obtained from a system prototype evaluation suggest that the proposed method is understandable to students and that it can provide them with valuable automatically generated feedback, which is its main value.

In the future work we hope to further elaborate the feedback provision possibilities of the system, and assess the validity of the proposed method of knowledge assessment by comparing the results obtained by using the developed tool with those obtained through other kinds of assessments.

REFERENCES

[1] F. M. V. der Kleij, R. C. W. Feskens, and T. J. H. M. Eggen, "Effects of Feedback in a Computer-Based Learning Environment on Students' Learning Outcomes A Meta-Analysis," Rev. educ. res., vol. 85, pp. 475–511, January 2015.

[2] R. P. Phelps, "The Effect of Testing on Student Achievement, 1910–2010," Int. J. Test., vol. 12, pp. 21–43, January 2012.

[3] J. R. Blunt and J. D. Karpicke, "Learning With Retrieval-Based Concept Mapping," J. Educ. Psychol., vol. 106, pp. 849–858, August 2014.

[4] K. Scalise and B. Gifford, "Computer-Based Assessment in E-Learning: A Framework for Constructing 'Intermediate Constraint' Questions and Tasks for Technology Platforms," J. Techn., Learn., Ass., vol. 4, pp. , June 2006.

[5] J. D. Novak and D. Musonda, "A Twelve-Year Longitudinal Study of Science Concept Learning," Am Educ Res J, vol. 28, , pp. 117–153, March 1991.

[6] J. D. Novak and A. J. Canas, "The theory underlying concept maps and how to construct and use them," Technical Report IHMC CmapTools, 2008.

[7] B. E. Cline, C. C. Brewster, and R. D. Fell, "A rule-based system for automatically evaluating student concept maps," J Exp. Sys. App, vol. 37, pp. 2282–2291, March 2010.

[8] A. Anohina and J. Grundspenkis, "Scoring Concept Maps: An Overview," in Proceedings of the International Conference on Computer Systems and Technologies and Workshop for PhD Students in Computing, New York, USA, 2009, pp. 78:1–78:6.

[9] D. R. Krathwohl, "A Revision of Bloom's Taxonomy: An Overview," Theory into practice, vol. 41, no. 4, Autumn 2002.

[10] T. J. Cleophas and A. H. Zwinderman, "Decision Trees," in Machine Learning in Medicine, Springer Netherlands, 2013, pp. 137–150.

[11] L. Breiman, "Random Forests," Mach. Learn., vol. 45, pp. 5–32, October 2001.

[12] IHMC Cmap Tools, http://cmap.ihmc.us/

Glass based structures fabricated by rf-sputtering

Alessandro Chiasera[1,*], Francesco Scotognella[2,3], Dominik Dorosz[4], Gianluca Galzerano[5], Anna Lukowiak[6], Davor Ristic[7], Giorgio Speranza[8,1], Iustyna Vasilchenko[1,9], Alessandro Vaccari[10], Sreeramulu Valligatla[1,9,11], Stefano Varas[1], Lidia Zur[12,1], Mile Ivanda[7], Alessandro Martucci[13], Giancarlo C. Righini[12], Stefano Taccheo[14], Roberta Ramponi[5], Maurizio Ferrari[1,12]

1 IFN - CNR CSMFO Lab. & FBK CMM, via alla Cascata 56/C Povo, 38123 Trento, Italy.
2 Politecnico di Milano, Dipartimento di Fisica and IFN-CNR, Piazza Leonardo da Vinci 32, 20133, Milano, Italy
3 Center for Nano Science and Technology@PoliMi, Istituto Italiano di Tecnologia, Via Giovanni Pascoli, 70/3, 20133, Milan, Italy
4 AGH University of Science and Technology, Faculty of Materials Science and Ceramics, Al. A. Mickiewicza 30, 30-059 Kraków, Poland
5 IFN – CNR and Politecnico di Milano, Dipartimento di Fisica, Piazza Leonardo da Vinci 32, 20133, Milano, Italy
6 Institute of Low Temperature and Structure Research, PAS, 2 Okolna St., 50-422, Wroclaw, Poland
7 Center of Excellence for Advanced Materials and Sensing Devices, Ruđer Bošković Institute, Bijenička c. 54, Zagreb, Croatia
8 FBK CMM FMPS Unit, via Sommarive 18, Povo, 38123 Trento, Italy.
9 Dipartimento di Fisica, Università di Trento, via Sommarive 14, Povo, 38123, Trento, Italy
10 FBK CMM-ARES Unit, via Sommarive 18, Povo, 38123 Trento, Italy.
11 Institute for Integrative Nanosciences, IFW Dresden, Helmholtz Straße 20, 01069 Dresden, Germany.
12 Centro di Studi e Ricerche Enrico Fermi, P.zza Viminale 1, 00184, Roma, Italy
13 Dipartimento di Ingegneria Industriale, Università di Padova, via Marzolo 9, 35122, Padova, Italy
14 College of Engineering, Swansea University, Singleton Park, Swansea, UK
alessandro.chiasera@cnr.it

Abstract - In this paper we present some results obtained by our consortium regarding rf-sputtered glass-based structures.

INTRODUCTION

Glasses activated by rare earth ions are the fundamental bricks of various photonic systems with application not only in ICT but also concerning lighting, laser, sensing, energy, environment, biological and medical sciences, and quantum optics. Recently, a remarkable increase in the experimental efforts to control and enhance emission properties of emitters by tailoring the dielectric surrounding of the source has been performed. With this aim, several approaches, using nanocomposite materials or specific geometries, such as planar interfaces, photonic crystals, solid state planar microcavities, dielectric nanospheres, and spherical microresonators, have been proposed. However, the dependence of the final product on the fabrication protocol remains an important task of the research in material science. Among the various techniques that could be employed to fabricate oxide dielectric materials rf-sputtering is a promising route to fabricate rare earth-activated waveguides and 1D photonic crystals. Here we discuss some recent results obtained by our consortium regarding: (A) 1D photonic crystals allowing Er^{3+} luminescence enhancement concerning the $^4I_{13/2} \rightarrow {}^4I_{15/2}$ transition; (B) disordered 1D photonic structures that are very interesting for the modelization and realization of broad band filters and light harvesting devices; (C) 1D microcavities, activated by a layer based on poly-laurylmethacrylate matrix containing $CdSe@Cd_{0.5}Zn_{0.5}S$ quantum dots, leading to coherent emission; (D) Er^{3+}-activated $SiO_2-P_2O_5-HfO_2-Al_2O_3-Na_2O$ planar waveguides. We show as the rf-sputtering technique is suitable for the fabrication of high quality structures based on glass matrix. The paper shows the steps used in the fabrication protocols and the optical, spectroscopic, structural and morphologic results of the samples. The examples of systems prepared with this technique are presented with the aim of show the flexibility of the rf-sputtering to realize different devices to cover several photonic applications.

A. 1D photonic crystals allowing Er^{3+} luminescence enhancement

One of the interesting features of the 1D microcavities is the possibility to enhance the luminescence, resonant with the cavity, when the defect layer is activated by a luminescent species. This is a general and significant property of photonic crystals and it has frequently been used to modulate the emission wavelength and enhance the radiative rate and intensity of luminescent objects [1]. When the cavity dimensions approach the wavelength of the emission, the density of electromagnetic states inside the cavity are strongly perturbed and can lead to significant enhancement of the luminescence quantum yield [2]. This enhancement is achieved by increasing the number of the localized modes coupled with the emitter [3-5].

Figure 1 shows the SEM image of a microcavity with an Er^{3+}-doped SiO_2 active layer inserted between two Bragg reflectors, each one constituted of ten pairs of SiO_2/TiO_2 layers. The dark regions correspond to the SiO_2 layer and the bright regions correspond to the TiO_2 layer. The substrate is located at the bottom of the images and the air on the top.

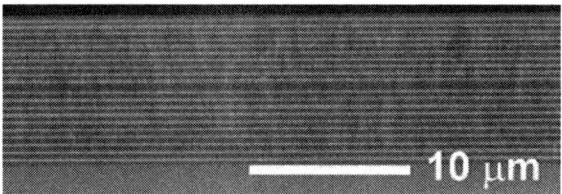

Figure 1. SEM micrograph of a 1D microcavity fabricated by RF-sputtering. The Er^{3+}-doped SiO_2 active layer is inserted between two Bragg reflectors, each one constituted of ten pairs of SiO_2/TiO_2 layers. The bright and the dark areas correspond to TiO_2 and SiO_2 layers, respectively. The substrate is located on the bottom of the images and the air on the top.

The Er^{3+} content in the active layer is about 0.6 ± 0.1 mol%. The NIR transmittance spectrum, measured at zero degree of incident angle, shows the stop band from 1490 to 1980 nm. A sharp peak in the transmittance spectrum appears at 1749 nm. It corresponds to the cavity resonance wavelength related to the half wave layer inserted between the two Bragg mirrors. The full width at half maximum of the resonance is 1.97 nm, corresponding to a quality factor of the cavity, Q, of about 890.

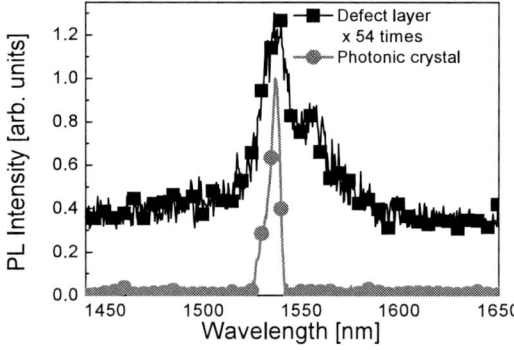

Figure 2. $^4I_{13/2} \rightarrow ^4I_{15/2}$ photoluminescence spectra of the cavity activated by Er^{3+} ion in 1D photonic crystal (●) and of the single Er^{3+}-doped SiO_2 active layer with first Bragg mirror (■). The light is recorded at 50° from the normal on the samples upon excitation at 514.5 nm.

Figure 2 shows the luminescence from the cavity and from the Er^{3+}-doped single SiO_2 layer with one Bragg reflector. To get a correct comparison, a specific procedure, assuring that the only variation is constituted by the cavity effect, was employed as detailed reported in [6]. Both the cavity and the Er^{3+}-doped single SiO_2 layer with first Bragg reflector were excited with the 514.5 nm line of an Ar^+ ion laser with an excitation power of 180 mW. The erbium emission from the reference sample is centered at 1538 nm with a FWHM of 29 nm and exhibits the characteristic shape of Er^{3+} ion in silica glass [7]. The peak luminescence intensity of Er^{3+} ions is enhanced by a factor 54, in respect to that detected for the reference at the corresponding wavelength. The Er^{3+} $^4I_{13/2} \rightarrow ^4I_{15/2}$ emission line shape is strongly narrowed by the cavity and exhibits a full width at half maximum of 5 ± 0.5 nm. The Er^{3+} emission is enhanced when the wavelength corresponds to the cavity resonant mode and weakened for the others emission wavelengths depending on the number of the localized modes coupled with the erbium ion in the defect layer.

B. Disordered 1D photonic structures

The optical properties of disordered natural and artificial systems and structures are a fascinating research topic characterized by an interdisciplinary approach. Glasses, glass ceramics, highly diffusive media, such as colloidal distributions, mixed dielectric media, and aperiodic multilayered systems are some examples of disordered photonic structures allowing photon management [8-11]. Recently, we have discussed the light transmission properties of disordered one dimensional photonic structures in which disorder is introduced by a random variation of layer thickness [12].

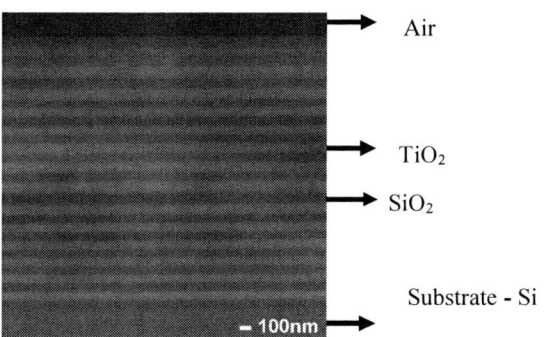

Figure. 3. SEM micrograph of a photonic structure made by alternating SiO_2 and TiO_2 layers with random thickness.

Figure 3 shows the SEM micrograph of the microcavity composed by 14 pairs of TiO_2/SiO_2 layers. To realize the disordered photonic structure, we have alternated layers of SiO_2 and TiO_2, with a thickness of $(80 + n)$ nm, where n is a random integer $0 < n < 40$. In this way, we obtain a random sequence of thicknesses, between 80 and 120 nm.

As shown in figure 4 such structures have the advantage to exhibit a broad transmission band and lower transmittance with respect to the corresponding periodic photonic crystal, opening the way to the fabrication of broad band filters. In the case of the example shown in figure 3, an average transmittance value of 0.7 was obtained for the 300 - 1200 nm transmission spectrum reported in figure 4 [12].

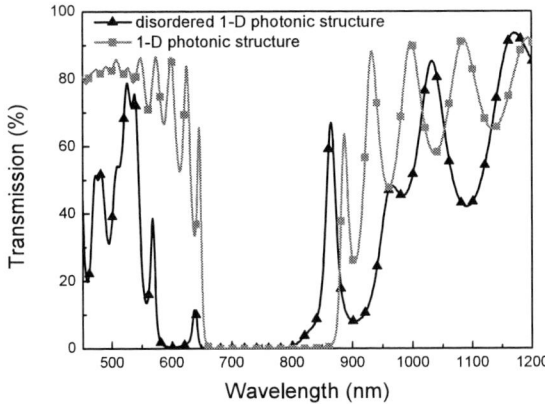

Figure 4. Transmission spectra obtained from a (■) 1D photonic crystal and a (▲) disordered 1D photonic structure (b).

Figure 5 shows the direct comparison of the reflectance properties of one dimensional photonic crystals (a) and disordered 1D photonic structures (b). The only difference is the randomness in thickness. This appealing behavior is due to the interference between waves traveling in regions with different optical paths, determined by the disordered distribution of stacked layer thicknesses.

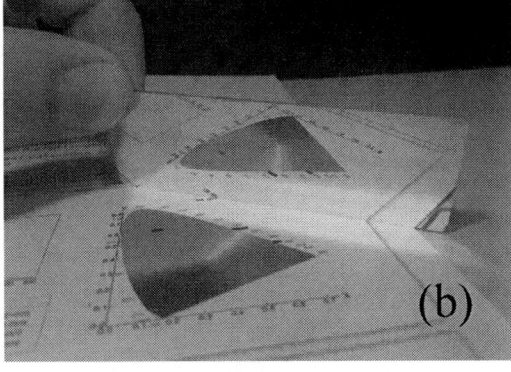

Figure 5. Reflection of the CIE 1931 diagram by a 1D photonic crystal (a) and a disordered 1D photonic structure (b).

C. 1D microcavities, activated by a layer based on poly-laurylmethacrylate matrix containing $CdSe@Cd_{0.5}Zn_{0.5}S$ quantum dots, leading to coherent emission

When the spontaneous emission of the emitter, embedded in the defect layer of a 1D photonic crystal, is strongly enhanced, the possibility of low threshold lasing could take place. In this regard, there have been considerable efforts to fabricate photonic band gap laser devices either as a distributed feedback lasing at band edge frequencies [13] or as a defect-mode lasing at localized defect mode frequencies [14]. Due to the relative simplicity of fabrication, one dimensional 1D photonic crystal laser devices have been extensively studied. The more effective devices are fabricated by organic/inorganic hybrid 1D photonic crystal and some years ago organic laser dyes as a gain medium was used to demonstrate a low threshold defect-mode lasing action [15]. We have demonstrated low threshold defect-mode lasing action in a one dimensional microcavity constituted by two Bragg reflectors, each one constituted of ten pairs of SiO_2/TiO_2 layers, with a defect layer based on poly-laurylmethacrylate matrix containing $CdSe@Cd_{0.5}Zn_{0.5}S$ quantum dots. The defect-mode laser structure is based on photophysical properties of the gain medium, and in particular to reflect the light at around 650 nm wavelength. The defect mode laser structure was optically pumped at 514.5 nm and the luminescence spectrum in the case of 2 mW of excitation power is shown in figure 6. Experimental details are given in Ref. [16]. The spectrum shows some narrow peaks in the low energy region followed in the high frequency region by a typical comb structure superimposed to the broad band assigned to the spontaneous emission. We can conclude that the spontaneous emission in the low energy region, around the localized defect modes, is enhanced by a factor proportional to the density of states at those frequencies leading to low threshold lasing.

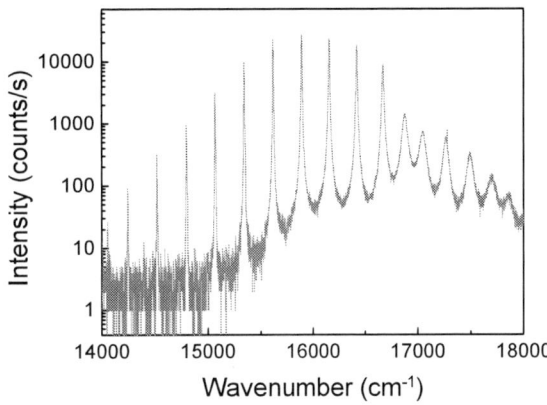

Figure 6. Luminescence spectrum obtained exciting at 514.5 nm with 2 mW, the SiO_2/TiO_2 1D microcavity, fabricated by RF sputtering, with a defect layer constituted by a poly-laurylmethacrylate matrix containing $CdSe@CdZn_{0.5}S$ quantum dots.

The measurements were taken with a 2 cm⁻¹ step and different laser power, and the intensity of the higher peak situated at 15803 cm⁻¹, was plotted as a function of the pump power; these data were then fitted via a linear function considering the points on the graph obtained at pump power above 0.5 mW. The fit line reaches the X-axis at a value of about 0.5 mW. This kind of not completely linear dependence could be due to presence of coherent emission from the sample. However, to find the proof of laser emission some more evidence had to be gathered.

As one can see in figure 7 the peak intensity can be approximately described via a linear dependence on the pump power up to 4 mW. It is important to note that in this range of pump power values, even at low power, the shape of the spectra does not change.

Above pump power of 4 mW the power density is enough high to induce modification in the polymeric active film and his refractive index is altered. As result the sample still emit light but the emitted peaks become irretrievably broader with less emitted intensity in respect to what obtained at 4mW pump power.

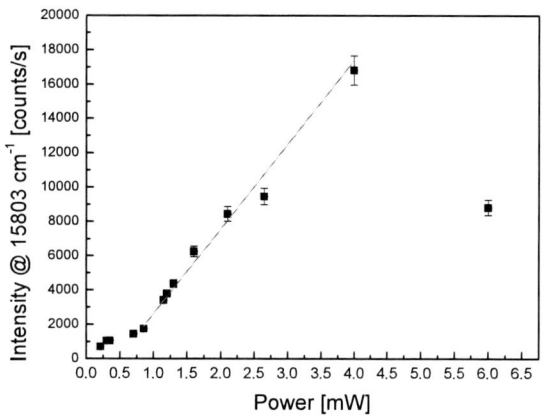

Figure 7. Intensity of the luminescence measured at 15803 cm⁻¹ at different pump power up to 6 mW focusing the excitation laser beam on the sample and detecting the luminescence at 2° with a solid angle of 7·10⁻⁴. The line is the result of the linear fit on all the points.

D. Er³⁺-activated SiO₂-P₂O₅-HfO₂-Al₂O₃-Na₂O planar waveguides

In Er-doped waveguide amplifiers (EDWAs), widely studied as active devices for integrated optical (IO) circuits, the amplification must be achieved in a length scale of few centimeters instead of some meters, as required for Er-doped fiber amplifiers, and as consequence it is required a high Er³⁺ doping level where the possibility of clustering effects of the rare earths become important [17]. A possible way to increase the amounts of rare-earth ions in the matrix avoiding or reducing clustering effects is the addition of co-doping agents, such as P_2O_5 or Al_2O_3 [18, 19]. Previously, we demonstrated that Er-doped SiO_2-HfO_2 planar

waveguides are a viable system for 1.5 µm applications [20] and, among the different fabrication techniques, we have shown rf-sputtering as a suitable technique to fabricate optical coatings and waveguides [12, 19].

Here we present the fabrication by rf-sputtering technique and the optical and spectroscopic assessment of Er³⁺-activated SiO_2-P_2O_5-HfO_2-Al_2O_3-Na_2O planar waveguides. Two glass samples labelled PSi_1Er, with composition 69P₂O₅-15SiO₂-10Al₂O₃-5Na₂O-1Er₂O₃, and PSi_0Er, with composition 70P₂O₅-15SiO₂-10Al₂O₃-5Na₂O were fabricated by melting. More information about these glasses can be found in Ref. [21].

The planar waveguide, labelled PPW, was prepared by multi target rf-sputtering technique using silicon and silica substrates with dimensions 7 × 3.5 cm². The substrates were cleaned inside the rf-sputtering deposition chamber by heating at 120 °C for 30 minutes just before the deposition procedure. Sputtering deposition of the films were performed by sputtering a 15 × 5 cm² silica target on which the PSi_1Er and 2 disks of HfO₂ (diameter 5 mm) were placed. The residual pressure, before the deposition, was about 8.0 × 10⁻⁷ mbar. During the deposition process, the substrates were not heated and the temperature of the sample holder during the deposition was 30 °C. The sputtering has occurred with an Ar gas pressure of 5.4 × 10⁻³ mbar; the applied rf power was 120 W and the reflected powers 0 W. The deposition time needed to obtain a planar waveguide that support a mode at 1.5 µm was of 70 h. As prepared samples present a non-stoichiometry structure of SiOₓ with x < 2. To achieve the correct stoichiometry, the samples were subsequently treated in air at 400 °C for 6 h [12].

The UV-Vis-NIR absorption spectrum obtained for the PSi_1Er glass is characteristic of Er³⁺-doped glasses [19,22]. These samples present a wide transparency region starting from 350 nm and confirm the potential application of these glasses for the fabrication of low losses active waveguides.

M-line technique was employed also to investigate the optical feature of the planar waveguide. The sample supports two modes at 633 nm while at 1319 nm and 1542 nm the waveguide exhibits a single mode [19].

TABLE 1. OPTICAL PARAMETERS OF THE PHOSPHATE PLANAR WAVEGUIDE MEASURED BY M-LINE APPARATUS BASED ON THE PRISM COUPLING TECHNIQUE AT 632.8 NM IN TE AND TM POLARIZATIONS AND AT 1319 AND 1542 NM IN TE/TM MODE

Refractive Index ±0.001				Attenuation coefficient (dB/cm) ±0.2		
633nm		1319nm	1542nm	633nm	1319nm	1542nm
TE	TM					
1.478	1.479	1.468	1.465	<0.2	<0.2	<0.2

The PPW waveguide presents a thickness of 3.0 ± 0.1 µm and, as reported in Table 1, supports one propagating mode at 1319 and 1542 nm. The attenuation coefficients are below the detection limit (0.2 dB/cm) of the apparatus at all three employed wavelengths. At 632.8 nm, where the waveguide supports 2 propagating modes, it is possible to obtain the refractive indexes in TE and TM polarization and it is noticeable that birefringence is negligible also for this sample. The modelling of the electric field intensity along the thickness of the film for the TE_0 mode at 1542 nm indicates that the optical parameter of the PPW waveguide appears appropriate for application in the 1.5 µm region. The ratio of the integrated field intensity inside the waveguide to the total intensity, including also the evanescent fields, is 0.84 [19].

The luminescence spectra of the PPW planar waveguide and the PSi_1Er parent glass are reported in figure 8. The two spectra are recorded with the same spectral resolution and exciting the samples with the same laser at 514.5 nm. The shape of the emission spectrum reported in figure 8 for the PPW sample in the 1.5 µm region is characteristic of the $^4I_{13/2} \rightarrow {}^4I_{15/2}$ transition of Er^{3+} ions in silicate glasses [19,22]. The spectrum for PPW exhibits a main emission peak at 1537 nm with a Full Width at Half Maximum (FWHM) of 28.5 ± 0.5nm. The spectrum obtained from the PSi_1Er glass appears different in comparison with the PPW spectrum and exhibits a FWHM of about 30.5 ± 0.5 nm.

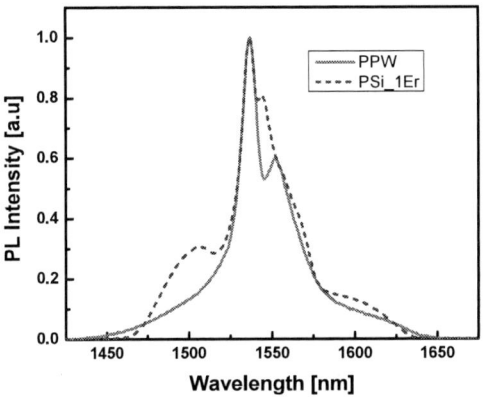

Figure 8. Room temperature photoluminescence spectra related to the $^4I_{13/2} \rightarrow {}^4I_{15/2}$ transition of Er^{3+} ions from PPW planar waveguide (continuous line ▬▬) and PSi_1Er parent glass (dash line ▬ ▬) upon excitation at 514.5nm. The spectra are normalized to the maximum.

The decay curves of the luminescence from the $^4I_{13/2}$ metastable state of Er^{3+} ions for PPW planar waveguide and PSi_1Er parent glass present a single exponential behavior with lifetime of 5.6 ms and 0.6 ms for the PPW and PSi_1Er samples respectively. One should note here, that the comparison of the spectroscopic features reported in the figure 8 put in evidence strong differences between the planar waveguide and the bulk glasses used to fabricate the active film. This variation can be explained with the different composition of the matrixes and the addition of the HfO_2 system to the planar waveguide. Zampedri et al. [23] and Gonçalves et al. [24] have, in fact, demonstrated that Er^{3+} spectroscopic properties are strongly affected by the local distortion induced by Hf^{4+} also at low content [19]. There is a possibility that the waveguide exhibits a more distorted local environment for the Er^{3+} ion in respect to the parent glasses changing

drastically the emission band shape in the 1.5 µm region. The erbium content in the active film was significantly reduced due to the presence of HfO_2 and SiO_2 in the target for the waveguide fabrication. This fact can contribute to reduction of the ion-ion interaction and increase of the 1.5 µm lifetime [24].

CONLUSION

Here we have briefly revised some recent results obtained by our consortium with 1D confined photonic structures. Low threshold lasing has been demonstrated at a single defect-mode wavelength of the 1D photonic band gap structure, activated by $CdSe@Cd_{0.5}Zn_{0.5}S$ quantum dots, resulting from the inhibited density of states of photons within the stop band and the enhanced rates of spontaneous emission at the localized resonant defect mode. On the same physical base Er^{3+} luminescence enhancement has been demonstrated. An interesting application, as broad band reflector, of disordered 1D photonic structures has been shown. In conclusion, 1D confined structures, where light can be confined over micro scale region, are a fantastic challenge for nanoscience-based photonic technologies. Finally Er^{3+}-activated SiO_2-P_2O_5-HfO_2-Al_2O_3-Na_2O planar waveguides with valuable optical and spectroscopic properties were obtained by multi-target rf-sputtering technique starting by massive Er^{3+}-activated P_2O_5-SiO_2-Al_2O_3-Na_2O glass. Manufacture of such structures has become possible due to the opportunity delivered by nanotechnology, which opens the way to the study of new functional artificial materials and structures, promising progress in miniaturization and which allow exploration of new aspects of light-matter interaction. Beyond the three examples reported here, the exploitation of their unique properties covers a range of applications possibilities and system performance that concern Lighting, Laser, Sensing, Energy, Environment, ICT, and Health.

ACKNOWLEDGMENT

This research is performed in the framework of the projects COST MP1401 "Advanced Fibre Laser and Coherent Source as tools for Society, Manufacturing and Lifescience" (2014 - 2018), PAS-CNR (2014-2016), ERANet-LAC RECOLA, "Plasmonics for a better efficiency of solar cells" bilateral project between South Africa and Italy (contributo del Ministero degli Affari Esteri e della Cooperazione Internazionale, Direzione Generale per la Promozione del Sistema Paese), MaDEleNA PAT projects.

REFERENCES

[1] W. Xu, Y. Zhu, X. Chen, J. Wang, L. Tao, S. Xu, T. Liu, H. Song, "A novel strategy for improving upconversion luminescence of NaYF4:Yb, Er nanocrystals by coupling with

hybrids of silver plasmon nanostructures and poly(methyl methacrylate) photonic crystals", Nano Research 6 (2013) pp. 795-807.

[2] G. Guida, P.N. Stavrinou, G. Parry, J.B. Pendry, "Time-reversal symmetry, microcavities and photonic crystals", Journal of Modern Optics, 48 (2001) pp. 581- 595.

[3] W. Wang, H. Song, X. Bai, Q. Liub, Y. Zhu, "Modified spontaneous emissions of europium complex in weak PMMA opals", Phys. Chem. Chem. Phys. 13 (2011) pp. 18023-18030.

[4] A. Chiasera, F. Scotognella, S. Valligatla, S. Varas, J. Jasieniak, L. Criante, A. Lukowiak, D. Ristic, R.R. Gonçalves, S. Taccheo, M. Ivanda, G.C. Righini, R. Ramponi, A. Martucci, M. Ferrari, "Glass-based 1-D dielectric microcavities", Optical Materials 61 (2016) pp. 11-14.

[5] J.D Joannopoulos, S.G. Johnson, J.N. Winn, R.D. Meade, "Photonic Crystals Molding the Flow of Light", 2nd Edition, Princeton University Press, Princeton, 2008.

[6] S. Valligatla, A. Chiasera, S. Varas, N. Bazzanella, D. Narayana Rao, G.C. Righini, M. Ferrari, "High quality factor 1-D Er3+-activated dielectric microcavity fabricated by rf-sputtering", Optics Express 20 (2012) pp. 21214-21222.

[7] A. Chiappini, A. Chiasera, C. Armellini, S. Varas, A. Carpentiero, M. Mazzola, E. Moser, S. Berneschi, G.C. Righini, and M. Ferrari, "Sol-gel-derived photonic structures: fabrication, assessment, and application," J. Sol-Gel Science & Technol. 60 (2011) pp. 408-425.

[8] Y. Lahini, A. Avidan, F. Pozzi, M. Sorel, R. Morandotti, D. N. Christodoulides and Y. Siberberg, "Anderson localization and nonlinearity in one-dimensional disordered photonic lattices" Phys. Rev. Lett. 100 (2008) pp. 013906.

[9] D. S. Wiersma, P. Bartolini, A. Lagendijk, and R. Righini, "Localization of light in a disordered medium", Nature 390 (1997) pp. 671-673.

[10] M. Bellingeri, D. Cassi, L. Criante, F. Scotognella, "Light Transmission Properties and Shannon Index in One-Dimensional Photonic Media With Disorder Introduced by Permuting the Refractive Index Layers", IEEE Photonics Journal 5 (2013) pp. 2202811.

[11] F. Scotognella, "One-dimensional photonic structure with multilayer random defect", Optical Materials 36 (2013) pp. 380-383.

[12] A. Chiasera, F. Scotognella, L. Criante, S. Varas, G. Della Valle, R. Ramponi, M. Ferrari, "Disorder in Photonic Structures Induced by Random Layer Thickness", Science of Advanced Materials, 7 (2015) pp. 1207-1212.

[13] M. Meier, A. Mekis, A. Dodabalapur, A. Timko, R.E. Slusher, J.D. Joannopoulos, O. Nalamasu, "Laser action from two dimensional distributed feedback in photonic crystals", Appl. Phys. Lett. 74 (1999) pp. 7-9.

[14] Y. Matsuhisa, R. Ozaki, K. Yoshino, M. Ozaki, "High Q defect mode and laser action in one-dimensional hybrid photonic crystal containing cholesteric liquid crystal", Appl. Phys. Lett. 89 (2006), pp. 101109-e1/3.

[15] J. Yoon, W. Lee, J.-M. Caruge, M. Bawendi, E.L. Thomas, S. Kooi, P.N. Prasad, "Defect-mode mirrorless lasing in dye-doped organic/inorganic hybrid one-dimensional photonic crystal", Appl. Phys. Lett. 88 (2006), pp. 091102-e1/3.

[16] A. Chiasera, J. Jasieniak, S. Normani, S. Valligatla, A. Lukowiak, S. Taccheo, D.N. Rao, G.C. Righini, M. Marciniak, A. Martucci, M. Ferrari, "Hybrid 1-D dielectric microcavity: Fabrication and spectroscopic assessment of glass-based sub-wavelength structures", Ceram. Int. 41 (2015) pp. 7429-7433.

[17] I. Vasilchenko, A. Carpentiero, A. Chiappini, A. Chiasera, A. Vaccari, A. Lukowiak, G.C. Righini, V. Vereshagin, M. Ferrari, "Influence of phosphorous precursors on spectroscopic properties of Er3+-activated SiO2-HfO2-P2O5 planar waveguides", Journal of Physics: Conference Series 566 (2014) pp. 012018-1/5.

[18] A. Lukowiak, R.J. Wiglusz, A. Chiappini, C. Armellini, I.K. Battisha, G.C. Righini, M. Ferrari, "Structural and spectroscopic properties of Eu3+-activated nanocrystalline tetraphosphates loaded in silica-hafnia thin film", Journal of Non-Crystalline Solids 401 (2014) pp. 32-35.

[19] A. Chiasera, I. Vasilchenko, D. Dorosz, M. Cotti, S. Varas, E. Iacob, G. Speranza, A. Vaccari, S. Valligatla, L. Zur, A. Lukowiak, G.C. Righini, M. Ferrari, "SiO2-P2O5-HfO2-Na2O glasses activated by Er3+ ions: From bulk sample to planar waveguide fabricated by rf-sputtering", Optical Materials 63 (2016) pp. 153-157.

[20] R.R. Gonçalves, G. Carturan, L. Zampedri, M. Ferrari, M. Montagna, A. Chiasera, G.C. Righini, S. Pelli, S.J.L. Ribeiro, Y. Messaddeq, "Sol-gel Er-doped SiO2-HfO2 planar waveguides: A viable system for 1.5 µm application", Applied Physics Letters 81 (2002) pp. 28-30.

[21] D. Dorosz, M. Kochanowicz, J. Zmojda, P. Miluski, M. Marciniak, A. Chiasera, A. Chiappini, I. Vasilchenko, M. Ferrari, G. Righini, "Rare-Earth Doped Materials for Optical Waveguides", Invited Paper Proceedings ICTON 2015, 17th International Conference on Transparent Optical Networks, Budapest, Hungary, July 5-9, pp. 1-5.

[22] M.P. Hehlen, N.J. Cockroft, T.R. Gosnell, Spectroscopic properties of Er3+- and Yb3+-doped soda-lime silicate and aluminosilicate glasses, Phys. Rev., vol. B 56 (1997) pp. 9302-9318.

[23] L. Zampedri, G.C. Righini, H. Portales, S. Pelli, G. Nunzi Conti, M. Montagna, M. Mattarelli, R.R. Goncalves, M. Ferrari, A. Chiasera, M. Bouazaoui, C. Armellini, "Sol-gel-derived Er-activated SiO2-HfO2 planar waveguides for 1.5 µm application", Journal of Non-Crystalline Solids 345&346 (2004) pp. 580-584.

[24] R.R. Gonçalves, G. Carturan, M. Montagna, M. Ferrari, L. Zampedri, S. Pelli, G.C. Righini, S.J.L. Ribeiro, Y. Messaddeq, "Erbium-activated HfO2-based waveguides for photonics", Optical Materials 25 (2004) pp. 131-139.

Piezoresistive Effect in Composite Films Based on Polybenzimidazole and Few-Layered Graphene

V. A. Kuznetsov[*, **, a], B. Ch. Kholkhoev[***], A. Ya. Stefanyuk[*, **], V. G. Makotchenko[*],
A. S. Berdinsky[**], A. I. Romanenko[*], V. F. Burdukovskii[***] and V. E. Fedorov[*, ****, b]

[*] Nikolaev Institute of Inorganic Chemistry, Siberian Branch of Russian Academy of Sciences, Novosibirsk, Russia
[**] Novosibirsk State Technical University / Semiconductor Devices and Microelectronics, Novosibirsk, Russia
[***] Baikal Institute of Nature Management, Siberian Branch of Russian Academy of Sciences, Ulan-Ude, Russia
[****] Novosibirsk State University, Novosibirsk, Russia
[a] e-mail address: vitalii.a.kuznetsov@gmail.com
[b] e-mail address: fed@niic.nsc.ru

Abstract – The paper reports experimental study of piezoresistive effect in composite films based on polybenzimidazole with few-layered graphene nanoparticles filler. Colloidal dispersions of few-layered graphene (FLG) were obtained by ultrasonic treatment of synthesized FLG in the solution of poly[2,2'-(p-oxydiphenylen)-5,5'-bisbenzimidazole] (OPBI) in N-methyl-2-pyrrolidone (NMP). Electroconductive films were formed from the dispersions by flow coating. To investigate dependence of electrical resistance on mechanical strain strips of the films were bonded onto beams of uniform strength (in bending) with cyanoacrylate adhesive, the beams being insulated with polymer glue. The strain gauge factors were measured for two films with filler content being 0.75 and 2.00 mass per cent. Electrical resistances were measured by two- and four-point methods, the factors being independent on the method used. The factors are the same within the error for both filler contents and equal to 21 on average.

I. INTRODUCTION

Since unique physical and chemical properties of graphene were discovered many works have been devoted to synthesis of graphene and graphene based composites and to investigation of their possible applications [1-5]. Graphite sheets as a filler material can lead to significant change in properties of composites. Carbon-based materials and graphite nanoparticles in particular in general enhance electronic, optical, thermal and mechanical properties of polymer matrixes [6-11]. To synthesize electroconductive composites graphite nanoparticles were used as filler material in many insulating polymers, for example epoxy, polystyrene, poly(methyl methacrylate), thermoplastic polyurethane and many others [8]. Such composites attracted significant attention of many researchers as possible material for strain sensing elements. Experimental and theoretical studies of piezoresistive effect in such composites have been carried out of late [12-14]. Piezoresistive effect is a change of electrical resistance of a conductor under axial strain. One of the main quantitative variables of

piezoresistive effect is strain gauge factor (SGF). SGF is relative change of electrical resistance divided to mechanical strain (relative change of the length of the conductor). Commercial strain gauges are made with metallic and semiconducting sensing elements. SGF of metallic gauges is from 2 to 6, and for semiconducting ones it is from 40 to 200 in absolute value [15-17]. Piezoresistive effect in metallic conductors is concerned with change of geometry of them. In semiconductors the effect is concerned generally with change of specific resistance of them due to change of crystal structure , the value depending on the conduction type of material and doping dose [17]. As for SGF of conducting composites based on carbon nanomaterials, there many works reporting different value of the factor depending on the type of filler and forming method of the composites. For example composites based on graphene aerogel and polydimethylsiloxane have SGF equal to 61.3 [18]; graphene platelets in epoxy matrix showed SGF equal to 56.7 [19]; amino-functionalized graphene nanoplates in epoxy exhibited SGF to be 45 [20]; carbon nanotubes based composites are reported to have SGF from several units to 25 [21-24]; SGF of CVD graphene on poly(dimethylsiloxane) substrate was reported to be 6.1 [25]; other carbon nanomaterials based composites show similar SGF. At the same time SGF can be tuned by different filler concentration, for examples in [26, 27] PECVD graphene films consisted of packed graphene nanoislands were shown to have SGF up to 300 depending on resistance of samples which simpliciter depends on weight concentration of the filler. The highest SGF reported is 10^3 for 2-6% strain and 10^6 for higher strain for films of graphene woven fabrics on/in poly(dimethylsiloxane) matrix. The value of the factor the authors explained by high density of cracks in the structure of sensing elements [28]. In the literature the nature of piezoresistive effect is generally attributed to changing of distances in networks between neighbouring conducting particles in insulating polymer matrixes when samples are stressed. Tunnelling current between the neighbouring particles and amount of conductive paths in the networks are dependent on the distances. So SGF can

---

The work was supported by the Russian Science Foundation (Grant no. 14-13-00674).

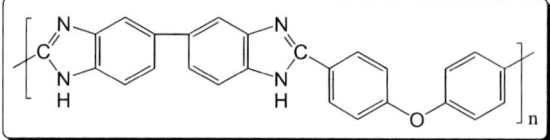

Figure 1.   Structure of poly[2,2'-(m-phenylen)-5,5'-bisbenzimidazole

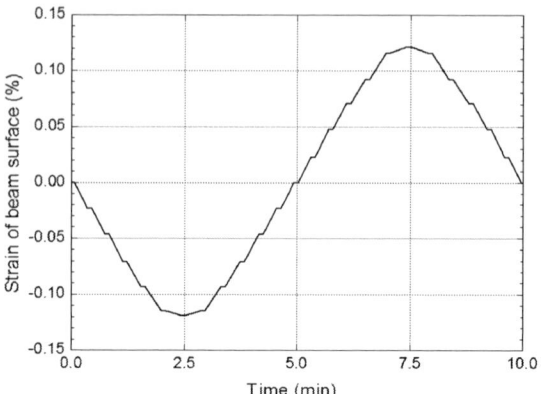

Figure 2.   Strain of the beam of uniform strength (in bending) surface at the sample bonding place

be tuned by forming composites with different filler concentration.

In practice measuring of wide range of strain is needed, temperature range being wide as well. Standard metallic gauges are used at a temperature up to 260°C, platinum alloys are used at higher temperatures [17]. Widespread polycrystalline silicon strain gauges can be used at high temperatures due to good insulating layers of silicon dioxide. Metallic gauges are used to measure strain in range from 0-3%, and annealed ones up to 10%. Semiconducting strain sensing elements are used to measure strain only up to 0.1%. It is because of different mechanical properties of the materials.

From the point of view of new functional materials for strain sensing elements electroconductive composites based on high-temperature polymers with good mechanical properties are very perspective to be investigated. One of such polymers is polybenzimidazole (PBI). There is lack of experimental data in the literature about PBI with conducting fillers.

The aim of this paper is to investigate experimentally piezoresistive effect in composite conducting films based on polybenzimidazole matrix with few-layered graphene filler.

II.   EXPERIMENTAL

A.   Synthesis of Few-Layered Graphene

Few-layered graphene (FLG) was synthesized as it had been described in our previous works [29, 30]. The precursor for FLG was polyfluorodicarbon with chlorine trifluoride with a composition $C_2F \cdot xClF_3$. The precursor was synthesized by interaction of natural graphite (ash content 0.05 mass per cent) with liquid chlorine trifluoride at room temperature by the method described in [31]. The compound was put into a quartz tube (200-300 mm in length, 30-40 mm in diameter) and the tube was placed into a pre-heated to 800°C tubular furnace for about 30 s. "Thermal shock" led to instant decomposition of the precursor. The resulted nanoparticles of few-layered graphene were 3-4 nm in thickness [30].

B.   Synthesis of Composite Films

To synthesize the composite films FLG was sonicated in 2% solution of poly[2,2'-(m-phenylen)-5,5'-bisbenzimidazole] (OPBI) in N-methyl-2-pyrrolidone (NMP) (the structure of OPBI is shown in Fig. 1). The resulted dispersion was centrifuged and decanted. Film specimen was formed by flow-coating of the dispersion on a glass substrate. The film was dried at 70-80°C for 24 h. The resulted film took off from the substrate and dried at 100°C for 24 h in vacuum and at 200°C for 2 h in air to

remove the rest solvent. That way two films were obtained with 0.75 and 2.00 mass per cent of FLG in the OPBI matrix.

C.   Experimental samples, Methods and Apparatures

The experimental samples were formed by cutting the films synthesized to strips 2 mm in width and 8 mm in length. To deform the experimental samples beams of uniform strength (in bending) were used as in [32]. The beams were covered with polymer glue for dielectric insulation. The strips were bonded onto the beams with cyanoacrylate adhesive using standard technique as for metallic strain gauges. The electrical contacts to the samples were made of silver paste and thin copper leading wires were used. To measure resistances of the samples four-point and two-point probe methods were used.

The beams with samples were loaded with the following cycles. First a beam was loaded for a sample to be compressed and tensioned for 10 min, and then the beam was kept undeformed for 10 min. Compression-tension steps are shown in Fig. 2. Every sample was stressed such cycles for 20 times.

III.   RESULTS AND DISCUSSION

The resistivities of the films synthesized are equal to 2900 and 250 Ohm·cm for 0.75 and 2.00 mass per cent of FLG filler, respectively. Time dependences of samples relative changes of resistances for first 10 loading cycles are shown in Fig. 3 and Fig. 4. One can see that there are no considerable degradations of the resistivity after the loading cycles. First three cycles should be considered as training cycles by analogy with metallic strain gauges testing [33]. Dependences of relative changes of resistances of the samples on strain are shown in Fig. 5 and Fig. 6. One can see that the dependences are not linear. It can be caused by the insulating polymer glue layer. The strain gauge factor was estimated by the formula [33]:

$$K = \frac{\Delta R / R_{\varepsilon(min)}}{\varepsilon}, \qquad (1)$$

28

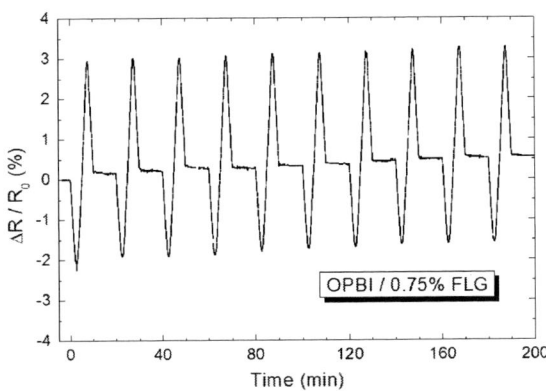

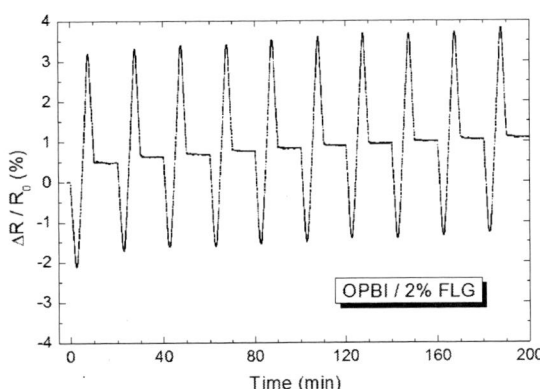

Figure 3. Time dependence of relative change of resistance of the film sample of polybenzimidazole polymer matrix with 0.75 mass per cent of few-layered graphene filler during first 10 loading cylces

Figure 4. Time dependence of relative change of resistance of the film sample of polybenzimidazole polymer matrix with 2.00 mass per cent of few-layered graphene filler during first 10 loading cylces

where $\Delta R = R_{\varepsilon(max)} - R_{\varepsilon(min)}$; $R_{\varepsilon(max)}$ and $R_{\varepsilon(min)}$ – resistance at maximum and minimum strain, respectively; $\varepsilon = \Delta l / l$ – mechanical strain.

SGF was estimated as a slope of the lines connecting two extreme values (see Fig. 5 and Fig. 6). Systematic and instrumental errors were evaluated and the resulted SGFs were equal to 20.7±1.0 and 21.2±0.9 for the films contained 0.75 and 2.00 mass per cent of few-layered graphene filler in OPBI matrix, respectively. The value of SGF does not depend on the method used, two- of four-point probe method – the result is the same within the error. So the factors are equal within the error despite the fact that the values of films resistivities are of a different order of magnitude. Piezoresistive effect in the composites concerned can be related to the factors described in the literature for similar composites. The fact that resistivities of the films concerned are rather low and the percolation threshold is far from the concentrations might lead to the identical SGFs. This fact makes the compositions to be perspective in applications, because samples with different resistivities have the same value of the strain gauge factor and deviation of the filler concentration does not result to change of SGF.

IV. CONCLUSION

Experimental study of piezoresistive effect of the composite films of polybenzimidazole with few-layered graphene nanoparticles filler was carried out. Strain gauge factors of the films with 0.75 and 2.00 mass per cent of FLG are the same within the error and equal to 21 on average. Unique properties of the polymer concerned and the fact that the strain gauge factors are relatively high and independent on the filler concentrations make such composites perspective for prospective research.

REFERENCES

[1] L. He and S. C. Tjong, "Nanostructured transparent conductive films: Fabrication, characterization and applications," Materials Science and Engineering: R: Reports, vol. 109, pp. 1-101, 2016.

[2] C. I. Idumah and A. Hassan, "Emerging trends in graphene carbon based polymer nanocomposites and applications," Reviews in Chemical Engineering, vol. 32, pp. 223–264, 2016.

[3] X. Li, J. Yu, S. Wageh, A. A. Al-Ghamdi, and J. Xie, "Graphene in Photocatalysis: A Review," Small, vol. 12, pp. 6640-6696, 2016.

[4] M. Li, D. Liu, D. Wei, X. Song, D. Wei, and A. T. S. Wee, "Controllable Synthesis of Graphene by Plasma-Enhanced Chemical Vapor Deposition and Its Related Applications," Advanced Science, vol. 3, p. 1600003, 2016.

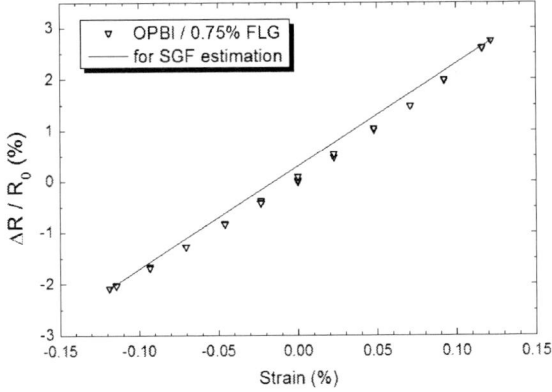

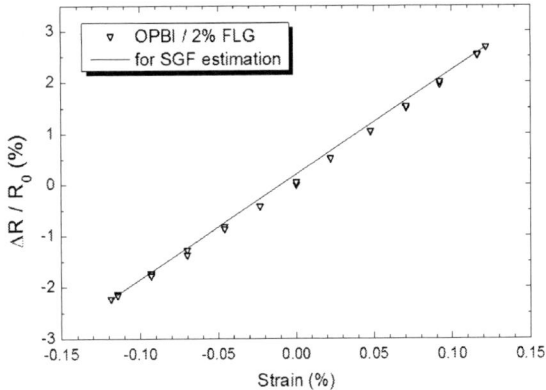

Figure 5. Relative change of resistance of the film of OPBI with 0.75 mass per cent of few-layered graphene filler. R_0 is the resistance at zero strain

Figure 6. Relative change of resistance of the film of OPBI with 2.00 mass per cent of few-layered graphene filler. R_0 is the resistance at zero strain

[5] D. N. Nguyen and H. Yoon, "Recent Advances in Nanostructured Conducting Polymers: from Synthesis to Practical Applications," Polymers, vol. 8, p. 118, 2016.

[6] R. Verdejo, M. M. Bernal, L. J. Romasanta, and M. A. Lopez-Manchado, "Graphene filled polymer nanocomposites," Journal of Materials Chemistry, vol. 21, pp. 3301-3310, 2011.

[7] X. Huang, X. Qi, F. Boey, and H. Zhang, "Graphene-based composites," Chemical Society Reviews, vol. 41, pp. 666-686, 2012.

[8] T. Kuilla, S. Bhadra, D. Yao, N. H. Kim, S. Bose, and J. H. Lee, "Recent advances in graphene based polymer composites," Progress in Polymer Science, vol. 35, pp. 1350-1375, 2010.

[9] H. Kim, A. A. Abdala, and C. W. Macosko, "Graphene/Polymer Nanocomposites," Macromolecules, vol. 43, pp. 6515-6530, 2010.

[10] A. Elmarakbi, W. Jianhua, and W. L. Azoti, "Non-linear elastic moduli of Graphene sheet-reinforced polymer composites," International Journal of Solids and Structures, vol. 81, pp. 383-392, 2016.

[11] H. Saleem, A. Edathil, T. Ncube, J. Pokhrel, S. Khoori, A. Abraham, et al., "Mechanical and Thermal Properties of Thermoset–Graphene Nanocomposites," Macromolecular Materials and Engineering, vol. 301, pp. 231-259, 2016.

[12] C. S. Boland, U. Khan, G. Ryan, S. Barwich, R. Charifou, A. Harvey, et al., "Sensitive electromechanical sensors using viscoelastic graphene-polymer nanocomposites," Science, vol. 354, pp. 1257-1260, 2016.

[13] C. Bonavolontà, C. Camerlingo, G. Carotenuto, S. D. Nicola, A. Longo, C. Meola, et al., "Characterization of piezoresistive properties of graphene-supported polymer coating for strain sensor applications," Sensors and Actuators A: Physical, vol. 252, pp. 26-34, 2016.

[14] A. Deepak, V. Ganesan, and P. Shankar, "Nondestructive evaluation of graphene-based strain sensor using Raman analysis and Raman mapping," Journal of Polymer Engineering, vol. 36, pp. 649-653, 2015.

[15] M. Elwenspoek and R. J. Wiegerink, Mechanical microsensors. Berlin: Springer-Verlag Berlin Heidelberg, 2001.

[16] N. Maluf and K. Williams, An Introduction to Microelectromechanical Systems Engineering. Boston: Artech House Inc., 2004.

[17] J. Fraden, Handbook of Modern Sensors, 5 ed.: Springer International Publishing, 2016.

[18] S. Wu, R. B. Ladani, J. Zhang, K. Ghorbani, X. Zhang, A. P. Mouritz, et al., "Strain Sensors with Adjustable Sensitivity by Tailoring the Microstructure of Graphene Aerogel/PDMS Nanocomposites," ACS Applied Materials & Interfaces, vol. 8, pp. 24853-24861, 2016.

[19] L. M. Chiacchiarelli, M. Rallini, M. Monti, D. Puglia, J. M. Kenny, and L. Torre, "The role of irreversible and reversible phenomena in the piezoresistive behavior of graphene epoxy nanocomposites applied to structural health monitoring," Composites Science and Technology, vol. 80, pp. 73-79, 2013.

[20] J.-W. Zha, B. Zhang, R. K. Y. Li, and Z.-M. Dang, "High-performance strain sensors based on functionalized graphene nanoplates for damage monitoring," Composites Science and Technology, vol. 123, pp. 32-38, 2016.

[21] N. Hu, Y. Karube, M. Arai, T. Watanabe, C. Yan, Y. Li, et al., "Investigation on sensitivity of a polymer/carbon nanotube composite strain sensor," CARBON, vol. 48, pp. 680-687, 2010.

[22] G. T. Pham, Y.-B. Park, Z. Liang, C. Zhang, and B. Wang, "Processing and modeling of conductive thermoplastic/carbon nanotube films for strain sensing," COMPOSITES PART B-ENGINEERING, vol. 39, pp. 209-216, 2008.

[23] M. K. Njuguna, C. Yan, N. Hu, J. M. Bell, and P. K. D. V. Yarlagadda, "Sandwiched carbon nanotube film as strain sensor," Composites: Part B, vol. 43, pp. 2711-2717, 2012.

[24] S. Nag-Chowdhury, H. Bellegou, I. Pillin, M. Castro, P. Longrais, and J. F. Feller, "Non-intrusive health monitoring of infused composites with embedded carbon quantum piezo-resistive sensors," Composites Science and Technology, vol. 123, pp. 286-294, 2016.

[25] Y. Lee, S. Bae, H. Jang, S. Jang, S.-E. Zhu, S. H. Sim, et al., "Wafer-Scale Synthesis and Transfer of Graphene Films," Nano Letters, vol. 10, pp. 490-493, 2010.

[26] J. Zhao, C. He, R. Yang, Z. Shi, M. Cheng, W. Yang, et al., "Ultra-sensitive strain sensors based on piezoresistive nanographene films," Applied Physics Letters, vol. 101, p. 063112, 2012.

[27] Z. Jing, Z. Guang-Yu, and S. Dong-Xia, "Review of graphene-based strain sensors," Chinese Physics B, vol. 22, p. 057701, 2013.

[28] X. Li, R. Zhang, W. Yu, K. Wang, J. Wei, D. Wu, et al., "Stretchable and highly sensitive graphene-on-polymer strain sensors," Scientific Reports, vol. 2, p. 870, 2012.

[29] V. G. Makotchenko, E. D. Grayfer, A. S. Nazarov, S.-J. Kim, and V. E. Fedorov, "The synthesis and properties of highly exfoliated graphites from fluorinated graphite intercalation compounds," Carbon, vol. 49, pp. 3233-3241, 2011.

[30] V. G. Makotchenko, E. V. Makotchenko, and D. V. Pinakov, "The ways of use of multilayered graphene in engineering ecology," Environmental Science and Pollution Research, pp. 1-10, 2016.

[31] V. G. Makotchenko and A. S. Nazarov, "Structural characteristics of dicarbon polyfluoride," Journal of Structural Chemistry, vol. 50, pp. 1088-1095, 2009.

[32] V. A. Kuznetsov, A. S. Berdinsky, A. Y. Ledneva, S. B. Artemkina, M. S. Tarasenko, and V. E. Fedorov, "Strain-sensing Element Based on Layered Sulfide Mo0.95Re0.05S2," 2015 8th International Convention on Information and Communication Technology, Electronics and Microelectronics (Mipro), pp. 15-18, 2015.

[33] F. P. u. Strukturanalyse, "Experimental structure analysis - Metallic bonded resistance strain gages - Characteristics and testing conditions," in Characteristics to be tested vol. 2635, ed. Dusseldorf: Verein Deutscher Ingenieure, 2007, pp. 1-40.

MIPRO 2017, May 22- 26, 2017, Opatija, Croatia

Local growth of graphene on Cu and $Cu_{0.88}Ni_{0.12}$ foil substrates

H. S. Funk[*], J. Ng[**], N. Kamimura[***], Y.-H. Xie[**] and J. Schulze[*]

[*] University of Stuttgart/Institute for Semiconductor Engineering, Stuttgart, Germany
[**] University of California Los Angeles/Material Sciences & Engineering, Los Angeles, United States of America
[***] University of Nagoya/Quantum Engineering Department, Nagoya, Japan
st101222@stud.uni-stuttgart.de

Abstract - A method for large single-grain graphene growth on a $Cu_{0.88}Ni_{0.12}$-alloy using a local precursor feeding setup has been reported. Using back-end of line integration devices exploiting the high mobility and good mechanical properties can be built. However, few details about the actual local feeding setup and the yield are known. A local precursor feeding setup was implemented using a conventional tube-furnace, modified to allow local precursor feeding. Ar-diluted CH_4 as the C precursor was fed through a quartz-nozzle, placed above the growth substrate. The influence of different growth-parameters was studied. Precursor flowrate, background-pressure, substrate material and nozzle-substrate distance were optimized for the experimental setup used. Local growth of poly-crystalline graphene was achieved for small substrate-nozzle distance (2 mm) near atmospheric pressure (86.5 kPa) for low precursor flowrates (5 sccm). Local growth on both Cu and $Cu_{0.88}Ni_{0.12}$ is possible for these optimized parameters. Local graphene growth yield was found to be low. A possible explanation for the dependencies based on the fluid mechanics inside the furnace was found. An implementation of the local feeding setup was presented. Important parameters were optimized to allow local growth. Dependencies were studied to gain a better understanding of the local feeding growth mechanism.

I. INTRODUCTION

As the first widely studied 2D material, graphene has attracted much attention for a wide range of applications [1]. The first method for fabricating graphene was mechanical exfoliation from natural graphite or highly oriented poly-crystalline graphite (HOPG) [2]. Since 2008/2009, chemical vapor deposition (CVD) grown poly-crystalline graphene provides an alternative to mechanically exfoliated graphene flakes [3]. With a mobility $\mu = 5500$ cm$^2 \cdot$V$^{-1} \cdot$s^{-1} [4], CVD graphene does not reach the mobility of exfoliated graphene flakes $\mu \geq 10000$ cm$^2 \cdot$V$^{-1} \cdot$s^{-1} [2]. One way to improve the mobility of CVD graphene is by reducing grain-boundary scattering. Wu et al. showed local single-crystalline growth of graphene by local precursor feeding [5]. The single-crystalline graphene from [5] shows mobility values ranging from $\mu = 10000$ cm$^2 \cdot$V$^{-1} \cdot$s^{-1} up to $\mu = 20000$ cm$^2 \cdot$V$^{-1} \cdot$s^{-1} on SiO_2. Local single-crystalline growth was achieved by creating a C precursor distribution restricting

nucleation to a small area. Details on the experimental setup to achieve the required C precursor distribution are not given in [5].

This paper introduces a simple experimental setup for local precursor feeding to achieve the required C precursor distribution. Local poly-crystalline graphene growth was observed. Parameters with an important influence on the growth pattern were studied and a parameter set allowing local growth was determined. Further understanding of the H_2 influence on the local feeding growth is gained. The deeper understanding of the process helps to optimize the setup further and can help reduce variations and improve the yield.

II. THE EXPERIMENTAL SETUP, GROWTH PROCESS AND ANALYSIS METHODS

A. Experimental setup

A tube furnace *Lindberg Blue M* (11 kW model) [6] was used for the growth experiments. The conventional gas feed was extended to allow a separate feeding of the C precursor. The conventional gas feed of the furnace was used for the Ar and H_2 making up the growth atmosphere. For the local C precursor feeding, a modified cover plate with a feedthrough was made.

A special sample holder was built out of quartz glass. Fig. 1 shows a computer generated image of the sample holder used.

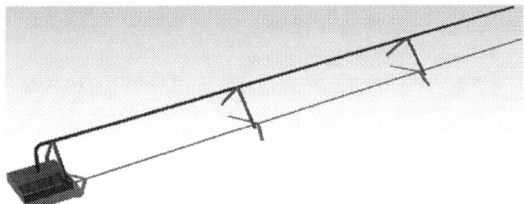

Figure 1: The modified sample holder with the tray and quartz tube to supply the C precursor

The sample holder consists of a tray for the substrate and a quartz pipe delivering the locally fed gasses to the substrate surface. Distance between the substrate and the quartz pipe outlet above the substrate (substrate-nozzle distance) is $d = 15$ mm and can be reduced by placing quartz discs under the growth substrate. The quartz pipe connects to the feedthrough in the cover plate of the furnace. The sample tray has to be placed in the center of the furnace. This requires the comparatively long horizontal pipe segment. The feedthrough and the quartz pipe must be connected gas-tight inside the furnace.

Cu foil for graphene growth (*Alfa Aesar*, No. 46365) with thickness $d_{sub} = 25\ \mu m$ was used. Square samples with a side length $L = 4$ cm were cut. The $Cu_{0.88}Ni_{0.12}$ foils were prepared by electroplating Ni and an annealing step included in the growth process to form the alloy [5]. High purity H_2 and Ar were used for the growth atmosphere. Ar-diluted CH_4 was used as the C precursor (5 % CH_4 in Ar). Initially, reduced background pressure and the resulting lower CH_4 partial pressure were intended to compensate for the higher CH_4 concentration compared to the one used in [5]. At atmospheric pressure, no single-grain growth is expected with 5 % CH_4 in Ar.

B. The growth process

The sample holder is placed inside the furnace with the foil centered under the nozzle as good as possible. Gas-tight connection between the quartz pipe and the feedthrough in the cover plate is established. The furnace is closed and pumped down. Flushing is done with Ar at a flowrate $Q_{Ar} = 1000$ sccm for a duration $t = 4$ min to remove residual air. The growth atmosphere is established afterwards. The Ar flowrate is kept at $Q_{Ar} = 1000$ sccm and H_2 is added at a flowrate $Q_{H2} = 50$ sccm. The background pressure p_{back} inside the furnace chamber is ramped up to the growth pressure. This is achieved by throttling the gas outflow with an exhaust valve. After reaching p_{back}, the furnace is heated to the growth temperature $T_{Gr} = 1050$ °C. An annealing step with a duration $t_{anneal} = 2$ h is executed before starting growth at C precursor flowrate Q_C. Characteristic growth parameters are the selected p_{back}, growth time t_{Gr} and Q_C. The furnace is cooled down after growth. The sample is taken out of the furnace after cooling time $t_{cool} = 6$ h.

C. Analysis methods employed for the grown samples

Analysis of grown samples was done using optical microscopy and Raman spectroscopy [7],[8]. The analysis was focused on the distribution of the graphene coverage to evaluate the influence of growth parameters. Raman spectroscopy was done using a *Renishaw inVia* Raman microscope with an excitation wavelength $\lambda = 514$ nm. Samples were cut into two pieces and transferred to Si/SiO₂ for analysis. Due to equipment limitations, only one dimensional scans were taken to study the lateral distribution. The Raman spectra were prepared for analysis using *Wire 3.2* software and evaluated using custom written *Matlab* scripts. Spectra taken during the linescan were analyzed for the characteristic G and 2D peaks of graphene [7]. As a simple binary criterion for graphene availability, the graphene signature was calculated. Therefore, the intensity, fit quality, signal noise ratio and peak width of G and 2D peak were compared to defined thresholds. Positive graphene signature marks a spot where graphene was found using Raman spectroscopy.

Contrast enhancement for optical microscopy was done by selective oxidation on a hotplate with a modified process from [9]. The oxidation temperature $T_{ox} = 250$ °C was selected, allowing the oxidation of uncovered Cu and the $Cu_{0.88}Ni_{0.12}$ alloy. Graphene is not damaged for $T_{ox} < 500$ °C [10]. Optical microscopy avoids problems of Raman scans such as subsampling in the linescan due to small grains compared to the sampling step-size.

III. PARAMETERS INFLUENCING THE LOCAL GROWTH

The parameters influencing the local growth of graphene are discussed in this section. As no details about the setup used by *Wu et al.* [5] are known, the characteristic parameters were studied for their influence with the setup introduced in section II A. Parameters which are assumed to be uninfluenced by the local precursor feeding were not studied experimentally. The growth temperature was not studied experimentally. Increasing the growth temperature results in lower nucleation density and faster growth [11]. The samples were grown at constant $T_{Gr} = 1050$ °C.

For this setup with no further optimization made, local growth was achieved at $p_{back} = 86.5$ kPa (sub-atmospheric, SA) with $Q_C = 5$ sccm on Cu. Fig. 2 a shows an overview over the sample. The graphene covered circular area can be seen in the center. The graphene signature result (Fig. 2 c) confirms the optical image. The Raman spectrum in Fig. 2 b was taken in the center of the covered area. The spectrum matches graphene spectra given in literature [7, 8].

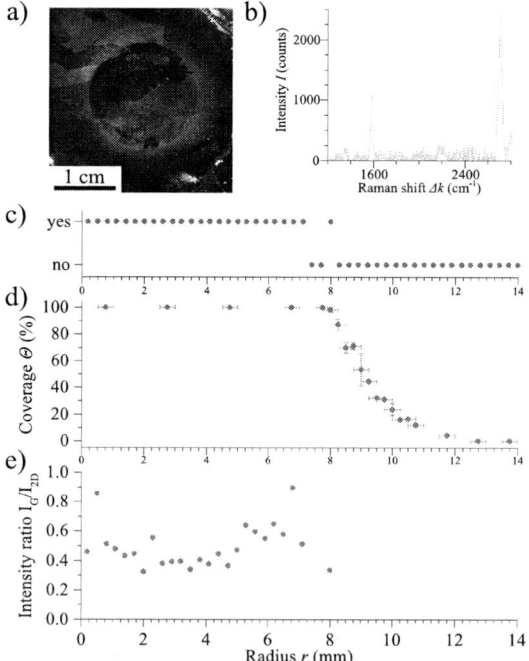

Figure 2: Locally grown graphene; optical image showing the graphene covered area in the center (a), a Raman spectrum taken in the center of the covered area (b), the graphene signature result (c), surface coverage (d) and ratio of the G and 2D peak intensity (e)

The intensity ratio of the G and 2D peaks allows to draw a conclusion about the number of graphene layers [12]. The intensity ratio of the G and 2D peaks (~ 0.47) hints to single-layer graphene. The intensity ratio of the G and 2D peaks was further calculated along the radius (Fig. 2 e). Except for some locally confined spots, the circular area along the scanline is single-layer graphene. This is as expected for graphene growth on Cu.

Influence of the characteristic growth parameters is discussed in the following subsections and compared to the results given for the reference sample in Fig. 2.

A. Influence of the substrate-nozzle distance

The substrate-nozzle distance was studied for the sample holder default $d = 15$ mm and reduced $d = 2$ mm. Cu substrate was used as growth substrate. Both samples were compared to determine the more suitable distance. No influence of the local C precursor feeding was found for $d = 15$ mm. The sample was fully covered in graphene, similar to conventionally grown graphene. Distance $d = 2$ mm was selected for all further growths.

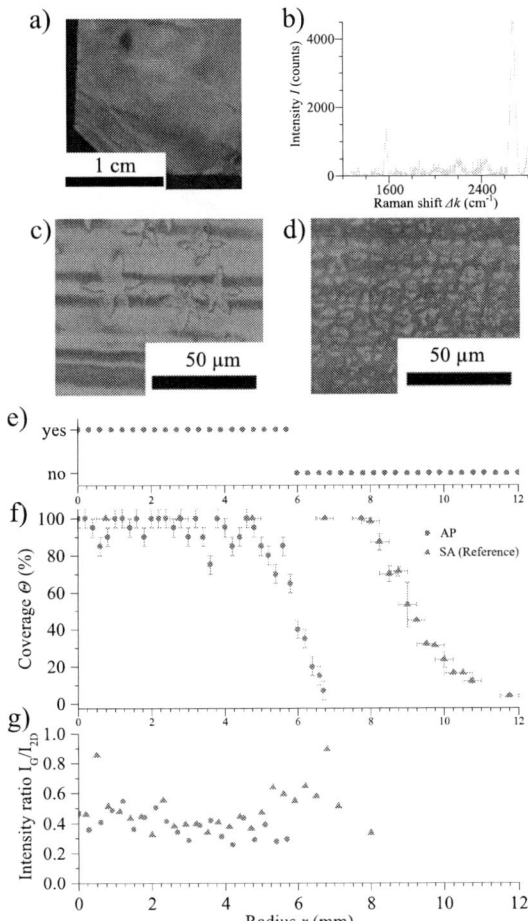

Figure 3: Sample grown at LP (a-d). Optical image with FeCl₃ contrast etch to show the transition region (a), Raman spectrum taken on a grain to confirm graphene (b), six-lobed graphene grains in the transition region (c) and outside the transition region (d). Sample grown at AP (e-f) with the graphene signature (e), surface coverage (f) and the ratio of the G and 2D peak intensities (g).

B. Influence of the background pressure

Local growth of graphene is strongly influenced by the background pressure. Graphene growth was executed for background pressures $p_{back} = 120$ Pa (low pressure, LP) and $p_{back} = 101.3$ kPa (atmospheric pressure, AP), i. e. below and above the reference sample. Precursor flowrate $Q_C = 5$ sccm and $t_{Gr} = 10$ min was set for all samples compared here. A coarse Raman scan and optical microscopy were both used to analyze the grown samples.

Usage of LP was intended to achieve the required low CH_4 partial pressure. An inverse pattern was found in the LP case, however. Fig. 3 a shows an optical image of the sample with FeCl₃ etch for contrast enhancement. Oxidation did not produce visible contrast without magnification. A transition from a bright circular area concentric to the nozzle to a darker tone can be seen. Comparison to micrographic images taken on the oxidized piece of the sample shows that the bright circle is uncovered. The darker area is partially covered. Fig. 3 c shows graphene grains in the transition region from uncovered to partially covered. The grains grow in a regular, six-lobed shape which is usually observed for growth with low H_2 availability [14]. Fig. 3 d shows the highest coverage achieved on the LP sample with irregular grain shape. The grains were confirmed to be graphene by Raman spectroscopy (See Fig. 3 d). Results from the Raman linescan are not given here as few grains were hit. The results were not representative. At AP, local graphene growth was achieved, similar to the reference sample in Fig. 2. Fig. 3 f shows the surface coverage Θ for the sample grown at AP. A concentric circular area around the nozzle is covered in polycrystalline graphene. Diameter of the area is smaller than for the SA sample. This local growth pattern is as expected from the results in [5]. Intensity ratio of the G and 2D peaks shown in Fig. 3 g show similar results for the AP and SA samples. Graphene on both samples is single-layered.

The local growth pattern on the AP and SA sample can be explained as in [5]. Highest CH_4 supersaturation in the center leads to highest generation rate of mobile C species in the center. This leads to early nucleation and high growth speed there [13]. Consequently, a circular graphene island grows from the center. The inverse growth pattern on the LP sample cannot be explained as in [5]. The lowered background pressure should lower the CH_4 partial pressure by the same factor everywhere on the sample. The pressure distribution should not be influenced. From [5] there is also no reason for the occurrence of regular and irregular grain shapes on the same sample. A possible explanation for the inverse pattern and grain shape is discussed in section IV.

C. Influence of the local feeding flowrate

The flowrate of the locally fed C precursor was studied for flowrates higher than $Q_C = 5$ sccm. Fig. 4 shows results for samples grown at $Q_C = 30$ sccm and $Q_C = 70$ sccm. Growth time $t_{Gr} = 10$ min. The graphene distribution was analyzed using optical microscopy and Raman spectroscopy. In Fig. 4 b the surface coverage as a function of the radius r is shown for the grown samples. Due to non-centric positioning of the sample, datasets have a different length.

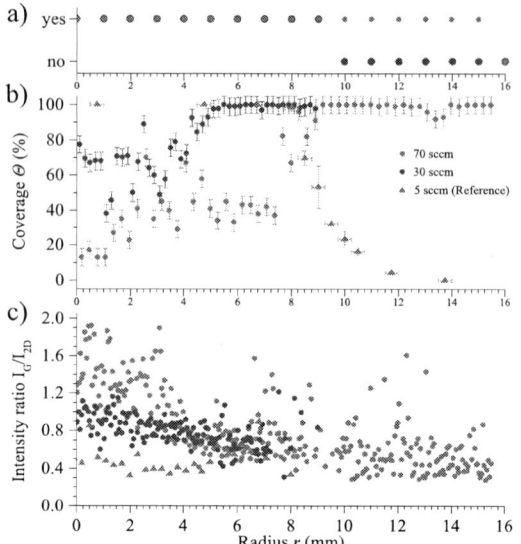

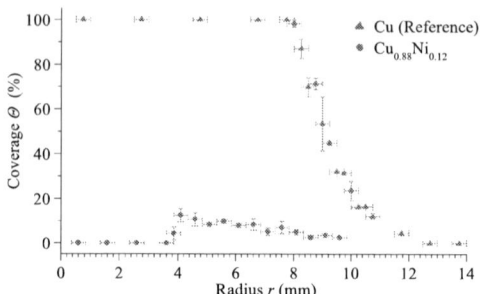

Figure 5: Surface coverage on the $Cu_{0.88}Ni_{0.12}$ alloy foil substrate compared to Cu foil substrate

The C solubility of the $Cu_{0.88}Ni_{0.12}$ foil substrate provides an alternative for the surface diffusion [15],[16]. Due to the competing processes of surface diffusion versus bulk diffusion, the concentration of C species on the surface is reduced. This increases the time until nucleation can occur. The uncovered area in the center of the sample is unexpected. The pattern is similar to the inverse pattern in III B & C. Due to the high Cu content of the $Cu_{0.88}Ni_{0.12}$ alloy the growth mechanism may be similar to pure Cu. The pattern may thus be explained as discussed in section VI.

E. Influence of the cooldown rate

For growth substrates with C solubility, precipitation driven growth of graphene must be considered besides the surface adsorbed growth [15]. The single grain growth explained in [5] is isothermal growth. This requires to suppress the cooling induced precipitation. The cooldown rate R_C has an influence on the precipitation of C from a substrate with significant C solubility [17]. The growth on $Cu_{0.88}Ni_{0.12}$ was thus studied for different cooldown rates. The cooldown rates were selected as the lowest ($R_C = 9$ K·min^{-1}) and highest ($R_C = 5$ K·s^{-1}) possible with the furnace used. The low cooldown rate was achieved by keeping the sample in the center of the furnace. High cooldown rate was achieved by pulling the sample out of the furnace center.

The samples were grown at AP without local feeding to allow pulling the sample out of the furnace center. Growth of both samples was done at $Q_C = 30$ sccm for $t_{Gr} = 15$ min. The samples were analyzed using optical microscopy and Raman spectroscopy. No precipitation grown graphene or other carbonaceous substance was found for $R_C \approx 9$ K·min^{-1}.

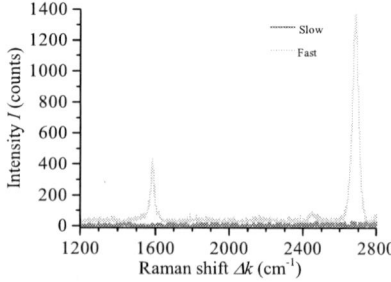

Figure 6: Raman spectrum on the samples with fast (green) and slow (red) cooldown. Fast cooldown spectrum shifted by 30 counts for clarity.

Figure 4: Graphene signature (a), coverage (b) and ratio of the G and 2D peak intensity (c) for samples grown at $Q_C = 30$ sccm (blue) and $Q_C = 70$ sccm (red)

Between $Q_C = 5$ sccm and $Q_C = 30$ sccm the growth pattern flips from the local growth pattern to an inverted pattern. The two fundamentally different patterns are clearly visible in Fig. 4 b. The sample grown at $Q_C = 70$ sccm exhibits an even stronger inverse pattern. In Fig. 4 c the intensity ratio of the G and 2D peak is shown for the samples. In the center the intensity ratio increases with flowrate. For both $Q_C = 30$ sccm and $Q_C = 70$ sccm the intensity ratio drops with increasing radius. The values in the center indicate graphene with layer number $n > 1$. Far off-center n = 1.

Growth of graphene on Cu substrate usually yields single-layer graphene. Growth with $n > 1$ on Cu was reported for high CH_4 partial pressure [14]. The CH_4 partial pressure increases with the flowrate. Increasing flow speed results in rising dynamic pressure in the sample center. The layer number therefore increases with flowrate. For the layer number to drop with the radius, the CH_4 partial pressure needs to be highest in the center, lowering with the radius.

The observed inverse pattern for $Q_C = 30$ sccm and $Q_C = 70$ sccm requires growth in the center to be slower than far from the center. This requires lowest CH_4 partial pressure in the center, rising with the radius when only considering the influence of CH_4.

This contradiction of explanations for the growth pattern and the layer count is discussed in section VI.

D. Influence of the $Cu_{0.88}Ni_{0.12}$ foil substrate

The surface coverage on the $Cu_{0.88}Ni_{0.12}$ alloy foil substrate was studied. $Cu_{0.88}Ni_{0.12}$ foil substrate was compared to reference sample on Cu. The $Cu_{0.88}Ni_{0.12}$ sample was grown at the same parameters as the reference sample. Surface coverage was determined using optical microscopy. Fig. 5 shows the surface coverage for both samples. Surface coverage is reduced by one order of magnitude when using the $Cu_{0.88}Ni_{0.12}$.

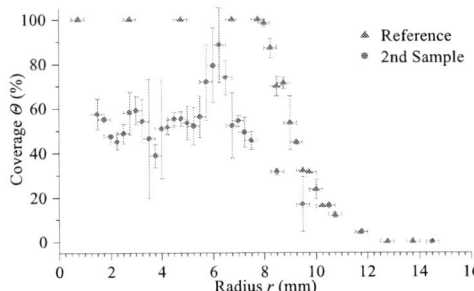

Figure 7: The surface coverage the 2nd sample grown at same parameters compared to the reference sample from Fig. 2

For $R_C \approx 5$ K·s^{-1} the sample was fully covered in graphene. Optical analysis was confirmed by Raman spectroscopy. Fig. 6 shows Raman spectra from both samples. The fast-cooled sample shows a typical single-layer graphene spectrum while the slow-cooled sample shows no significant signal in the observed range.

With the C solubility of the $Cu_{0.88}Ni_{0.12}$ alloy, C precipitation from the substrate must be considered, similar to Ni. According to [17] for slow cooldown on Ni, C continues to diffuse inside the substrate reducing the concentration immediately under the surface. Cooling induced precipitation thus does not take place. Graphene grown on samples cooled down with $R_C = 9$ K·min^{-1} is from isothermal segregation as in [5].

IV. REPEATABILITY OF THE LOCAL FEEDING GROWTH USING THE SETUP

The repeatability of the local feeding growth using the same parameters is low. A second sample was grown with $Q_C = 5$ sccm and $t_{Gr} = 10$ min and compared to the reference sample (See Fig. 7). The growth pattern for both samples is the same. Highest coverage for both samples is found in the center. Coverage of the 2nd sample occurs scaled down compared to the reference.

This behavior is contributed to uncontrolled or only weakly controlled parameters influencing the growth. The parameters determined above do not allow to fully control growth. A possible reason is discussed in section VI. Modifications to the setup to remove the variations are proposed.

V. NUCLEUS DISTRIBUTION ON $Cu_{0.88}Ni_{0.12}$

To achieve high-yield growth of single-grain graphene, the nucleation probability should only be influenced by the local precursor feeding. The distribution of nuclei was analyzed on Cu and $Cu_{0.88}Ni_{0.12}$ foil substrates. Fig. 8 shows the sample areas evaluated for the nucleus distribution on $Cu_{0.88}Ni_{0.12}$ and Cu. For nucleation without influence of the substrate, random distribution is expected. The position of individual grains in Fig. 8 was determined using *Fiji* [18]. A method described by *Clark & Evans* [19] was employed for an objective measure of the distribution. The average value of measured nearest neighbor distances $\bar{x}_M$ is compared to the average value of the expected nearest neighbor distances $\bar{x}_E$.

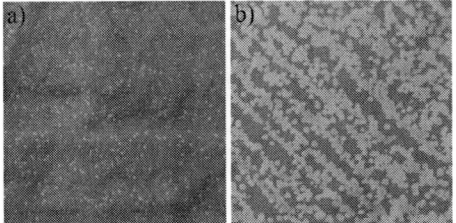

Figure 8: Optical micrograph of the nucleation distribution on Cu:Ni (a) and Cu (b) substrates. Side length of the square area $L = 125$ μm.

The ratio R between $\bar{x}_M$ and $\bar{x}_E$ according to Eq. 1 is evaluated.

$$R = \frac{\bar{x}_M}{\bar{x}_E} \quad (1)$$

The distribution is considered non-random for $R \neq 1$ For the $Cu_{0.88}Ni_{0.12}$ foil substrate $R = 1.0897$. The more regular nucleus distribution on Cu exhibits $R = 1.1875$.

The distribution on the $Cu_{0.88}Ni_{0.12}$ foil substrate is closer to random than on Cu. This coincides with the visual impression from Fig. 8. The alignment of nuclei along lines on Cu may be caused by groves on the surface from manufacturing. Electroplating and annealing leads to a smoother cleaner surface with less aligned nucleation seeds compared to the exposed Cu foil surface. It is thus less likely for the $Cu_{0.88}Ni_{0.12}$ foil substrate to adversely influence the nucleation with respect to single-grain growth.

VI. ON THE INFLUENCE OF FLUID-MECHANICS ON THE LOCAL PRECURSOR FEEDING

The experimental study showed problems which need to be solved before local growth as in [5] may be achieved. This section is intended to introduce a possible explanation for the inverse growth pattern and high variability of the growth. The discussion is given qualitatively. This is believed to be better to understand the explanations and proposed solutions.

A. Influence of H_2 on the growth pattern

In section III B an inverse growth pattern was found for the LP sample which may be not be explained by the explanation from [5]. In section III C, an inverse growth pattern was found on samples grown at high flowrates. The reduced coverage and multi-layer graphene in the center cannot be explained as in [5]. H_2 is known to serve two important functions for the growth of graphene on Cu. First, H_2 etches grown graphene. The etching of H_2 is anisotropic thereby influencing the shape of graphene grains [14]. Second, H_2 acts as a catalyst for the generation of mobile C species [14]. Mobile C species are required for graphene growth to occur. A reduction in H_2 partial pressure p_{H2} leads to reduced generation rate. This results in later nucleation [13] and slower growth [14] at constant CH_4 partial pressure p_{CH4}.

The growth behavior of graphene on Cu can be described as a function of the partial pressure ratio p_{H2}/p_{CH4}. For $p_{H2}/p_{CH4} \approx 0$, no growth occurs for small p_{CH4}. For high p_{CH4} without H_2 multilayer growth may occur. With p_{H2}/p_{CH4} increasing, growth rates increase and the growth on Cu is single-layered [14].

In the setup introduced in II A, p_{H2}/p_{CH4} is a function of the radius. The CH_4 is delivered by the nozzle and streams onto the surface. The dynamic pressure component of the total pressure leads to highest p_{CH4} under the nozzle, dropping with the radius (See Fig. 9 a). The H_2 is supplied from the growth atmosphere. It is separated from the surface by the C precursor gas streaming over the surface. To reach the surface, H_2 needs to penetrate the C precursor gas layer. This may occur by diffusion of H_2 into the moving C precursor layer. The diffusion takes the H_2 a certain time during which the gas streams from the center over the surface. H_2 partial pressure p_{H2} thus increases with the radius (See Fig. 9 a). The resulting p_{H2}/p_{CH4} is smallest in the center and increases with radius (See Fig. 9 b). Such non-constant p_{H2}/p_{CH4} may explain the inverse pattern observed in III B & C. Lack of H_2 in the center suppresses graphene growth on the LP sample ($p_{H2}/p_{CH4} \approx 0$). With increased radius, the first grain grow at low p_{H2}/p_{CH4} resulting in lobed shape. Further increased radius leads to increased p_{H2}/p_{CH4} resulting in irregular grains and higher growth speed (higher surface coverage) [14] as observed in subsection III B. On the SA samples with high flowrates, high p_{CH4} permits growth even at $p_{H2}/p_{CH4} \approx 0$. Growth speed is lower in the center, however. Surface coverage is thus lowest in the center. Small p_{H2}/p_{CH4} at high p_{CH4} also leads to multilayer growth as observed for the samples presented in subsection III C.

A fix is to feed both, H_2 and CH_4 locally. To achieve a homogeneous mixture, H_2 and CH_4 should be mixed with desired p_{H2}/p_{CH4} before injecting them into the furnace. This fixes p_{H2}/p_{CH4} to a constant value throughout the sample. This removes the issues linked to the mixing inside the furnace. A more detailed and systematic study of the influencing parameters can then be made.

B. C precursor leakage after turn-off

Even with same growth parameters set, the surface coverage shows high variation. For a more detailed study, such variations need to be reduced as far as possible.

One source of variation is the C precursor remaining inside the piping between the mass flow controller (MFC) and the quartz nozzle after MFC turn-off. This C precursor streams over the sample during cooldown. Small pressure variations inside the furnace lead to non-repeatable behavior. This stream may contribute to the growth, effectively changing the growth time. The required flexibility in the supply piping does not allow to arbitrarily reduce this uncontrolled volume. To avoid this uncontrolled leakage of the locally fed precursors onto the surface after growth, the piping can be carefully flushed after growth using Ar.

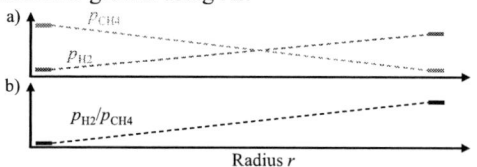

Figure 9: Relative p_{H2} and p_{CH4} (a) and resulting pressure ratio p_{H2}/p_{CH4} (b) in the center and off-center (bold lines) and dashed lines for guidance

VII. CONLUSION

A simple setup for local precursor feeding using a special sample holder was introduced. The important setup and process parameters were studied. It was shown that local growth of poly-crystalline graphene is possible with the introduced setup. An explanation for the inverse pattern and high variability based on the precursor mixing is given. An alternative experiment for local graphene growth, feeding both precursors locally, is proposed. This alternative may solve the encountered problems.

ACKNOWLEDGMENT

H. S. Funk thanks the German Academic Exchange Service (DAAD) and the University of Stuttgart for their financial support through the PROMOS scholarship.

REFERENCES

[1] J.-H. Ahn and B. H. Hong, "Things you could do with graphene," *Nat Nano*, vol. 9, no. 10, p. 737, 2014.

[2] K. S. Novoselov *et al.*, "Electric field effect in atomically thin carbon films," *Science*, vol. 306, no. 5696, pp. 666–669, 2004.

[3] X. Li *et al.*, "Large-area synthesis of high-quality and uniform graphene films on copper foils," *Science*, vol. 324, no. 5932, pp. 1312–1314, 2009.

[4] Y.-P. Hsieh, D.-R. Chen, W.-Y. Chiang, K.-J. Chen, and M. Hofmann, "Recrystallization of copper at a solid interface for improved CVD graphene growth," *RSC Adv*, vol. 7, no. 7, pp. 3736–3740, 2017.

[5] T. Wu *et al.*, "Fast growth of inch-sized single-crystalline graphene from a controlled single nucleus on Cu-Ni alloys," *Nature materials*, vol. 15, no. 1, pp. 43–47, 2016.

[6] Lindberg Blue, "3-Zone Tube furnace - Installation and Operation Manual: Models STF55346C & STF55666C," 2003.

[7] A. C. Ferrari, "Raman spectroscopy of graphene and graphite: Disorder, electron–phonon coupling, doping and nonadiabatic effects," *Solid state comm.*, vol. 143, no. 1-2, pp. 47–57, 2007.

[8] L. M. Malard, M. A. Pimenta, G. Dresselhaus, and M. S. Dresselhaus, "Raman spectroscopy in graphene," *Physics Reports*, vol. 473, no. 5, pp. 51–87, 2009.

[9] C. Jia, J. Jiang, L. Gan, and X. Guo, "Direct optical characterization of graphene growth and domains on growth substrates," (ENG), *Scientific reports*, vol. 2, p. 707, 2012.

[10] H. Y. Nan *et al.*, "The thermal stability of graphene in air investigated by Raman spectroscopy," *J. Raman Spectrosc.*, vol. 44, no. 7, pp. 1018–1021, 2013.

[11] S. Xing, W. Wu, Y. Wang, J. Bao, and S.-S. Pei, "Kinetic study of graphene growth: Temperature perspective on growth rate and film thickness by chemical vapor deposition," *Chemical Physics Letters*, vol. 580, pp. 62–66, 2013.

[12] Y. Hao *et al.*, "Probing layer number and stacking order of few-layer graphene by Raman spectroscopy," (eng), *Small*, vol. 6, no. 2, pp. 195–200, 2010.

[13] H. Kim *et al.*, "Activation energy paths for graphene nucleation and growth on Cu," *ACS nano*, vol. 6, no. 4, pp. 3614–3623, 2012.

[14] I. Vlassiouk *et al.*, "Role of hydrogen in chemical vapor deposition growth of large single-crystal graphene," *ACS nano*, vol. 5, no. 7, pp. 6069–6076, 2011.

[15] X. Li, W. Cai, L. Colombo, and R. S. Ruoff, "Evolution of graphene growth on Ni and Cu by carbon isotope labeling," *Nano letters*, vol. 9, no. 12, pp. 4268–4272, 2009.

[16] L. Baraton *et al.*, "On the mechanisms of precipitation of graphene on nickel thin films," *EPL (Europhysics Letters)*, vol. 96, no. 4, p. 46003, 2011.

[17] Q. Yu *et al.*, "Graphene segregated on Ni surfaces and transferred to insulators," *Appl. Phys. Lett.*, vol. 93, no. 11, p. 113103, 2008.

[18] J. Schindelin *et al.*, "Fiji: an open-source platform for biological-image analysis," *Nat Meth*, vol. 9, no. 7, pp. 676–682, 2012.

[19] P. J. Clark and F. C. Evans, "Distance to Nearest Neighbor as a Measure of Spatial Relationships in Populations," *Ecology*, vol. 35, no. 4, pp. 445–453, 1954.

Growth of Patterned GeSn and GePb Alloys by Pulsed Laser Induced Epitaxy

J. Schlipf* , J. L. Frieiro† , I. A. Fischer* , C. Serra‡ , J. Schulze* , S. Chiussi†

*Institut für Halbleitertechnik (IHT), Universität Stuttgart, Pfaffenwaldring 47, 70569 Stuttgart, Germany
†Dpto. Física Aplicada, Univ. de Vigo, Rua Maxwell s/n, Campus Universitario Lagoas Marcosende, Vigo, Spain
‡C.A.C.T.I., Univ. de Vigo, Rua Maxwell s/n, Campus Universitario Lagoas Marcosende, Vigo, Spain

Abstract - While modulators, waveguides and detectors have been successfully integrated in silicon devices, a laser source still remains a challenge. Unfortunately, in silicon and germanium, indirect band transitions are favored, making laser emission unlikely. Direct bandgap transitions have recently been demonstrated in germanium by introducing tensile strain with heavy n-type doping or by alloying with Sn. GePb alloys seem to be promising candidates here as well, since the required Pb concentration is predicted to be far lower than for Sn.

We examine the influence of SiO_2 hard masks on the formation of GeSn and GePb by Pulsed Laser Induced Epitaxy, a method allowing fast processing and in-situ monitoring. Main objectives are to study the spatial distribution of elements and strain as well as the possible underetching of Ge after mask removal.

Sn or Pb was deposited by thermal evaporation on an epitaxial Ge layer on Si(100), patterned with a SiO_2 hard mask. The patterns were then irradiated with ArF excimer laser pulses of 193 nm wavelength to induce melting and resolidification processes, aligned to the crystal structure of the Si(100) substrate below. Extensive characterization was performed, mainly using Atomic Force Microscopy and Raman Spectroscopy, examining the fabrication quality, thus the feasibility of future integrated laser devices.

I. INTRODUCTION

Optical interconnects could replace electrical interconnects, which have become a bottleneck in scaling integrated circuits towards smaller feature sizes. Distributing important signals optically would drastically decrease power consumption and self-heating, thus allow higher integration densities. Almost all components required for optical interconnects have already been demonstrated. The only remaining device is an integrated laser source. In silicon and germanium, which are already used in the fabrication of CMOS integrated circuits, indirect bandgap transitions dominate, making them unsuitable for laser devices.

III-V compound semiconductors are therefore predominantly used for semiconductor lasers because their band structure favors direct band transitions, which are required for stimulated emission. However, their integration with Si CMOS devices is difficult. Epitaxial growth is possible, but introduces a large amount of crystal imperfections in the III-V material due to the large mismatch in lattice constants and in thermal expansion coefficients. [1]

An alternative pure Group IV approach is the alloying of Ge or SiGe with Sn or Pb. As long as the Si or Ge percentage in these alloys is sufficiently high, their crystals will have diamond structure, enabling epitaxial growth on Ge or Si and thus their integration with Si logic devices. The band structure of Ge has already been tuned to support lasing by introducing tensile strain [2] or Sn atoms [3] into its crystal lattice. Pb in Ge is predicted to behave similarly to Sn, with lower concentrations needed to make the transition towards a direct bandgap material. [4]

Pulsed Laser Induced Epitaxy (PLIE) on the other hand allows very fast processing and out-of-equilibrium growth through short pulses of only tens of nanoseconds. ArF 193 nm excimer lasers, as used in this work, are already well-established in CMOS processes for UV photolithography and annealing, and are easily controllable and scalable. Their small emitted wavelength provides precision in lateral and vertical dimensions.

Formation of GeSn [5] and GePb [6] alloys by PLIE has already been demonstrated. This work examines the possibility of patterned epitaxy for GeSn and GePb. An SiO_2 hardmask is used to control the lateral distribution of the alloy by separating the Ge and Sn layers, thus limiting alloy formation to specific areas, as well as to improve uniformity of the alloy composition. This could allow direct patterning during the fabrication of the alloys, which would require less processing steps than lithographical patterning after full-area deposition. Additionally, three-dimensional structures could be formed by selective etching of Ge while GeSn or GePb would remain in place. Tuning possible strain fields created by patterned epitaxy would also be of interest.

II. SETUP

Samples were prepared by growing a 50 nm undoped Si buffer and 200 nm of Ge on (100)-oriented Si wafers by Molecular Beam Epitaxy (MBE). On some of the samples, 150 nm SiO_2 were added by Plasma Enhanced Chemical Vapor Deposition (PECVD). This layer was then patterned by photolithography to serve as a hardmask for the formation of the group IV alloys. The mask featured rectangular structures, ranging from 0.1 µm to 100 µm. Positive and negative structures on the mask created both windows (Fig. 1) in a flat SiO_2 surface and SiO_2 mesas, the former being studied in this work.

Additional maskless samples were processed with Ge thickness of 160 nm and 2500 nm. One sample with hardmask was examined prior to laser treatment by AFM-

Raman for evaluating position and intensity distribution of the Ge-Ge Raman peak (Fig. 2).

Fig. 1: AFM image of the hardmask.

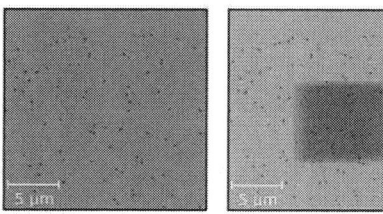

Fig. 2: 633 nm AFM-Raman maps for the untreated Ge sample visualizing the position (left) and the intensity distribution (right) of the 300.6 cm^{-1} peak (red color stands for 300.6 cm^{-1} position, blue for high, and green for low intensity values).

The position of the Ge-Ge peak is uniform with a value of (300.6 ± 0.1) nm over all of the area, indicating that there is no initial variation of strain due to the mask (Fig. 2a). The windows in the hardmask were clearly observable in the center of the image, by analyzing the intensity (difference of 4% between in and outside the window) of the Ge-Ge peak due to different reflectivity of masked and unmasked areas (Fig. 2b).

Sn or Pb was first deposited on top of the patterned and unpatterned samples by Physical Vapor Deposition (PVD) and PLIE with a pulsed 193 nm ArF Excimer laser was then used to create the group IV alloys. Beam intensity was controlled by an optical attenuator and the intensity profile shaped to a top-hat one by an optical beam homogenizer (Fig. 3), as described elsewhere [7]. Through irradiation in inert gas atmosphere (Ar), the Sn or Pb and a small volume of the underlying Ge is heated up and molten for facilitating liquid phase diffusion (LPD), thus enabling intermixing of the elements in the unmasked (window) areas. Being the substrate the thermal sink, the material closest to it solidifies, taking its crystal structure as seed. The resulting solidification front propagates towards the surface, creating a single crystal (Fig. 4). The melting/solidification process was monitored "in-situ" by time-resolved reflectivity (TRR) measurements of a diode probe laser (645 nm), since reflectivity of the material drastically changes through the transition from solid to liquid state and vice versa (Fig. 3). [7]

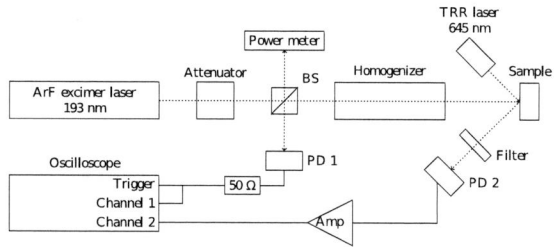

Fig.3: Setup of the excimer laser and TRR system.

III. METAL DEPOSITION

The metal layer (Sn or Pb) was deposited by PVD through thermal evaporation in high vacuum. For Sn, the thickness was 25 nm, aiming at a very high Sn percentage in the alloy. Assuming homogenous alloying through the complete Ge layer, this would yield a Sn concentration of around 11 % , thus leading to a Sn content that should be sufficient to meet the minimum of 7.3 % required for the indirect-to-direct bandgap transition, according to theoretical predictions [8]. The thickness of the Pb was 8 nm, implying final Pb concentration of around 4 %, that also should be enough to create direct-bandgap GePb. [4]

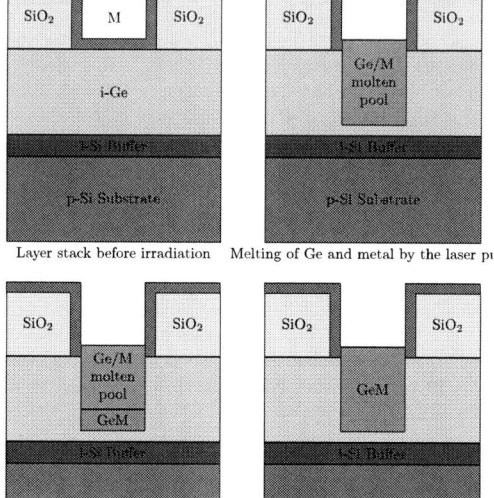

Fig. 4: Principle of the PLIE process, M stands for Sn or Pb.

The morphology of the deposited metal layers was examined by AFM. The Root Mean Square (RMS) value of the AFM probe height was calculated over a selected area to serve as an indicator for surface roughness (Tab. I).

TABLE I. COMPARISON OF SURFACE ROUGHNESS RMS FOR DIFFERENT SURFACES.

Surface Material	RMS inside / nm	RMS outside / nm
Ge/SiO$_2$	1.20 ± 0.05	1.32 ± 0.31
Sn	11.9 ± 0.6	10.8 ± 0.5
Pb	3.28 ± 0.22	3.56 ± 0.11

The SiO$_2$ hardmask appeared to have no influence on the conformality of the deposited metals. Inside and outside the windows, the RMS values were identical with overlapping uncertainties. Both Pb and Sn show a higher roughness than the original sample, as expected for PVD, the roughness of both layers being almost half the nominal thickness of the corresponding metal layer.

IV. GE IRRADIATION

a. TRR results

Ge wafers (masked and unmasked) were irradiated first, to evaluate the influence of PLIE on the crystal structure and its compatibility with the hardmask. Experiments were conducted to find the maximum power at which the mask remains intact, referred to as 'damage threshold'. For masked wafers with a 150 nm SiO_2 mask, this threshold was found to be 0.30 J/cm². This leaves only a small window for processing, since about 0.25 J/cm² are required to start heating up the Ge to its melting threshold. The results indicate that 150 nm SiO_2 caps are not adequate to mechanically resist subjacent molten Ge and should somehow be protected. TRR measurements were also used (Fig. 5) to gain insight on the PLIE triggered melting/solidification process. [7]

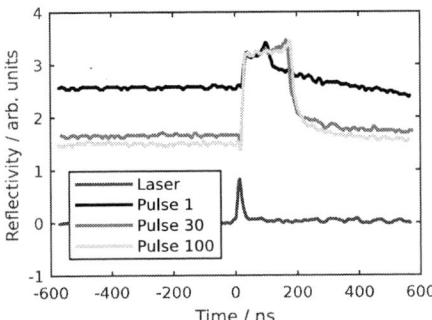

Fig. 5: Excimer laser pulse and TRR, maskless Ge sample, 0.93 J/cm².

The delay of probe laser to excimer laser signal is due to different signal paths. The full width at half maximum (FWHM) of TRR peaks was used to evaluate the timespan over which the sample surface is liquid.

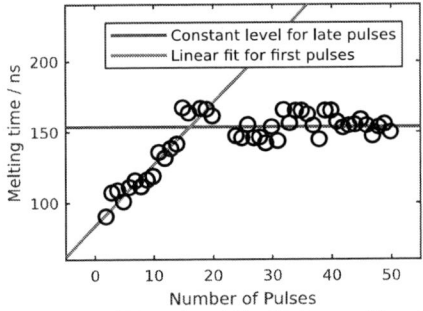

Fig. 6: Variation of TRR pulse width with time, maskless, 0.93 J/cm².

With more pulses, the zero level of the TRR decreases (Fig. 5), while melting time increases (Fig. 6) till around 15 pulses. This effects can be attributed to an increase in surface roughness. A rough solid surface reflects less light directly to the TRR detector, and absorbs more energy from the laser pulses, prolonging the duration of the melting and resolidification process.

b. Analysis of Ge samples

Maskless Ge samples (200 nm Ge) were irradiated with 10, 20, 30 and 50 laser pulses at a fluence of 0.93 J/cm². It was found that with this laser fluence, the 633 nm Raman spectrum (Inset Fig. 7) shows Ge-Ge, Ge-Si and Si-Si peaks, thus indicates the creation of a SiGe alloy. [9] The 488 nm spectrum that probes a shallower surface region on the other hand shows only the Si-Ge mode, not the Si-Si mode (Inset Fig. 7), features typical for lower Si concentrations in Ge [10]. We can therefore conclude that the molten pool was extended to the Si substrate, creating SiGe and opening the possibility for obtaining also SiGeSn

and SiGePb alloys. However, at energies that high, the Ge under the hardmask would receive too much energy, expand rapidly and destroy the hardmask, disabling pattered processing altogether. We therefore also analyzed thinner 160 nm Ge layers and found that formation of SiGe was here possible with lower fluences of 0.63 J/cm². Even thinner Ge layers and, as evident from the following figures, a sufficient number of excimer laser pulses should therefore enable SiGe, SiGeSn and SiGePb creation at even lower fluences and without mask destruction.

Detailed evaluation of the Raman spectra showed a decrease in peak position of the characteristic Ge-Ge mode with the number of pulses (Fig. 7).

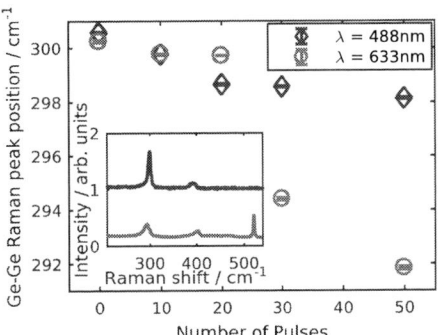

Fig. 7: Ge-Ge Raman peak position for different number of laser pulses, maskless, fluence of 0.93 J/cm². Inset: Raman spectra after 50 pulses.

Apart from alloying, the decrease in peak position can also be attributed to biaxial tensile strain in Ge, generated through cooling after each laser pulse. Since Ge has a large thermal expansion coefficient, it shrinks more than the Si substrate during cooling. The substrate prevents the shrinking of the Ge layer, forcing it into a larger lattice constant than in a relaxed state. [11] This process is prominent closer to the substrate, creating a gradient in lattice constant. Since the 633 nm laser has a larger penetration depth than the 488 nm laser due to its larger wavelength, its Raman spectra show a greater decrease in peak position due to tensile strain and alloying with Si.

An increase in peak width (Fig. 8) was also observed and can be attributed to decreasing crystal quality for the spectra with 488 nm excitation.

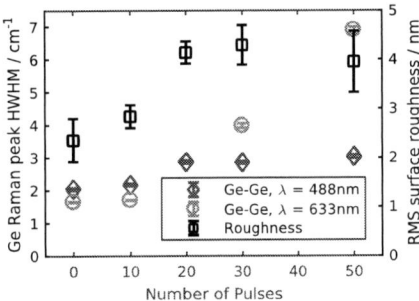

Fig. 8: Ge-Ge Raman peak width and surface roughness for different number of laser pulses, maskless, fluence of 0.93 J/cm².

The higher peak width for the 488 nm laser for the first three data points shows that the quality is worse closer to the surface, being far away from the more perfect substrate. With more pulses, the peak width of the Ge-Ge peak in the 633 nm spectrum increases, while the peak position decreases. These changes come with alloying [10]. The surface roughness RMS was found to increase as well, correlated with 488 nm peak width (Fig. 8), and the increase of melting time (Fig.6), corroborating the given explanation of surface degradation. These results suggest the possibility of introducing strain into Ge in a controlled manner by limiting the number of laser pulses.

V. GE/SN IRRADIATION

a. GeSn growth

For the creation of a GeSn alloy, 25 nm of Sn were thermally evaporated on top of the samples and TRR data were analyzed to understand the PLIE process. Higher laser fluences lead to broader TRR peaks (Fig. 9), since a higher fluence heats the material up to a higher temperature and also melts it up to a larger depth. Therefore, the samples have higher thermal energy, that takes longer to dissipate. For fluences below 0.35 J/cm², TRR pulses were shorter for unmasked samples than for masked ones. This could be attributed to Sn on the surface, which has a far lower melting point than Ge and requires less energy to melt. Since on the masked samples part of the molten Sn is on top of the SiO₂ hardmask, which has a lower thermal conductivity than the Ge layer, it takes longer to dissipate heat and solidify the Sn than on unmasked samples. For fluences exceeding 0.35 J/cm², the opposite is the case. Masked samples cool down faster. The pulse width is here dominated by the melting time of the created GeSn since a higher fluence will also cause melting of deeper regions of the sample. It has to be pointed out that the mask sustains much higher fluences (up to 0.60 J/cm²) compared to the previous ones without Sn on top. This can be attributed to the Sn on the mask, reflecting the incident laser light and preventing the Ge below the mask from melting.

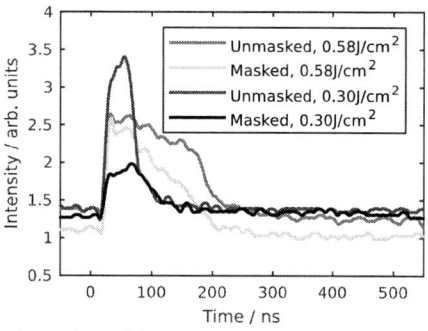

Fig. 9: Comparison of the Ge/Sn TRR pulses for masked and unmasked samples for various fluences.

b. Analysis of GeSn samples

AFM-Raman analyses of the samples clearly show the effectiveness of the hardmasks. While the areas covered by SiO₂ still show the characteristic Raman spectrum of Ge, the unmasked areas irradiated at 0.30 J/cm² yield a broader Raman peak at (281.1 ± 2.2) cm⁻¹ (Fig. 10).

The width of the peaks and the large difference in peak position compared to pure crystalline Ge let us suggest that the GeSn layer is polycrystalline. [12] At the edge of the windows, the Raman peak is broader and the center of this band shifted to the left (275 cm⁻¹), indicating amorphization.

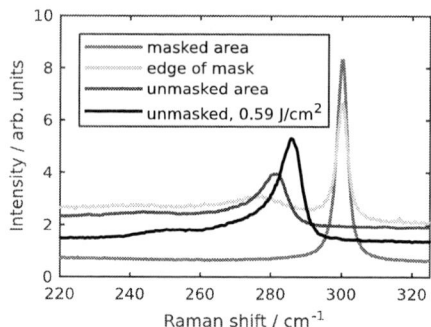

Fig. 10: Comparison of the 633 nm Raman peak of the Ge-Ge mode in masked and unmasked areas, fluence of 0.30 J/cm² (one at 0.59 J/cm²).

Additionally, the peak of the Ge-Ge mode of unchanged Ge at (300.6 ± 0.1) cm⁻¹ is visible at the window borders. This implies that the GeSn layer is thinner here than in the center of the window, so that the underlying Ge can be seen (Fig. 10).

For a sample, irradiated with 0.59 J/cm², the Ge-Ge mode was at 286 cm⁻¹, indicating a Sn concentration of at least 24 % for the case of relaxed Sn. Since the GeSn was grown on Ge, and the peak shape is sharper than those of typical amorphous material it should at least be partially strained, with a Sn concentration above this. [13]

AFM-Raman maps (Fig. 11) of the 0.30 J/cm² sample show the spatial distribution of the 281.1 cm⁻¹ and 300.6 cm⁻¹ peak. The quadratic window in the hardmask can be seen in the center of the images. Outside the windows, the 300.6 cm⁻¹ peak dominates, while there is no sign of the 281.1 cm⁻¹ peak.

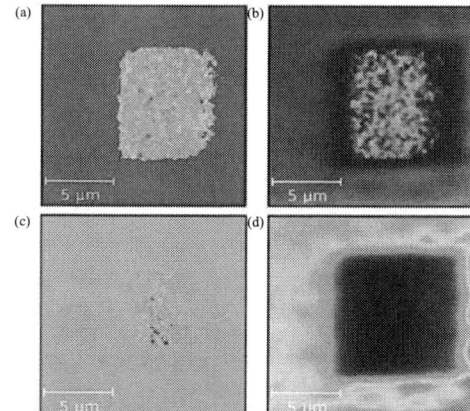

Fig. 11: 633 nm AFM-Raman maps for a masked GeSn sample, fluence of 0.30 J/cm². a) and c): Distribution of the 281.1 cm⁻¹ and 300.6 cm⁻¹ (green) position, respectively. b) and d): Intensity distribution of the 281.1 cm⁻¹ and 300.6 cm⁻¹ (red/green=high, blue=low), respectively.

This proves that the mask has successfully restricted the formation of GeSn to the window areas. The position of the 300.6 cm⁻¹ peak is uniform inside and outside the map, thus no strain fields have been created during epitaxy. The 281.1 cm⁻¹ peak is only found in the center of the window and there is no gradient in the peak position. At the edge, the 300.6 cm⁻¹ peak as well as a broad band centered at 275 cm⁻¹ was detected, with less intensity than outside the window. This corroborates the hypothesis of amorphization at the edge, or window borders, as deduced from the single spectra (Fig. 10).

The results of Time-of-flight Secondary Ion Mass Spectroscopy (TOF-SIMS) analysis (Fig. 12) are averaged over a large area, containing fragments from masked and unmasked sections alike. Since sputtering starts simultaneously at the Sn on top of the SiO_2 mask and at the 150 nm deeper GeSn windows, a combination of unmasked and masked area depth profiles are shown.

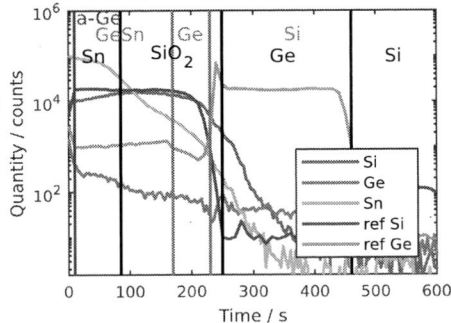

Fig. 12: TOF-SIMS depth profile results of a GeSn sample, using PLIE fluence of 0.35 J/cm^2. The results are average over the masked samples. The layer transitions at the edge (blue), outside (black) and inside (red) of the heterostructures are marked with vertical lines. Results of a sample without Sn and PLIE processing are shown as reference.

For the reference sample without metal and PLIE, one can clearly see the Si fragments from the 150 nm SiO_2 during the first 200 s of sputtering and then from the 200 nm Ge below the SiO_2 for the next 250 s. Ge fragments in the first 200 s are attributed to the Ge from the maskless window. For the GeSn sample a high amount of Ge fragments are found through the first 25 seconds, which could be caused by the faster erosion of the amorphous edges of the windows, as deduced from Raman results. On the other hand, the high Sn amount in the beginning comes from Sn residuals on the hardmask as well from the GeSn created inside the windows. Inside the GeSn layer, the Sn counts drop almost exponentially, indicating exponential distribution of Sn in the alloy. Sn is present until the drop of Ge at 230 seconds, suggesting the alloying of Ge all the way down to the Si substrate. The Si signal drops between 250 and 300 seconds, indicating the end of the hardmask, followed by a slightly increased Ge signal due to v-Ge.

VI. GE/PB IRRADIATION

Similarly to the GeSn experiments, an 8 nm Pb layer was created on the sample by PVD. As for GeSn, the TRR pulses show an increase in melting time with laser fluence (Fig. 13, left image), since higher energies introduce more heat that takes longer to dissipate. The values are similar to those for the GeSn alloys, for the tested low fluences of 0.30 – 0.40 J/cm^2 (Fig.9).

The Raman spectra (Fig. 13 right image) show, similar to the Sn samples, the Ge-Ge peak in the areas covered by the mask. The edge of the window shows that peak as well, plus a very broad and flat peak around 260 cm^{-1}, indicating partial amorphization. Inside the window, the sharper peak at 295 cm^{-1} evidences that epitaxial GePb was formed there. The spatial distribution, as visualized by AFM-Raman maps not shown here, is very similar to the results for the GeSn samples.

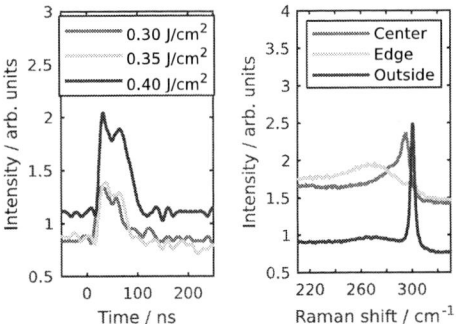

Fig. 13: Results for the irradiation of Pb on Ge. Left image: Comparison of the GePb TRR signals for various fluences. Right image: Comparison of the Raman spectra obtained with 633 nm excitation of masked samples for different positions, after PLIE with 0.35 J/cm^2.

As for Sn, TOF-SIMS shows some residual Pb on top of the hardmask (Fig. 14), as well as large initial Ge counts possibly caused by the amorphous edges of the window.

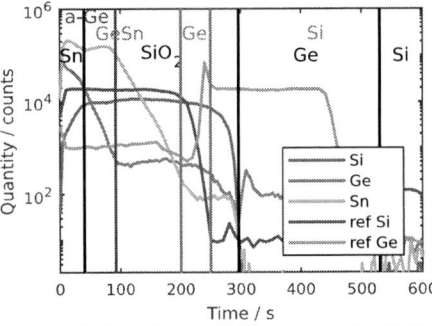

Fig. 14: TOF-SIMS results of a GePb sample, using PLIE fluence of 0.35 J/cm^2. The results are average over the masked samples. The layer transitions at the edge (blue), outside (black) and inside (red) of the heterostructure are marked with vertical lines. Results of a sample without Pb and PLIE processing are shown as reference.

However, the Pb signal does not decrease exponentially throughout the Ge inside the window, as seen for the GeSn sample, but drops around 50 seconds prior to the drop of Ge attributed to regions without hardmask. This suggest that there is still a thin Ge layer between the GePb and the Si substrate. Slopes at transitions are steeper than for GeSn, and profiles are less blurred, probably due to the lower amount of Pb used (8 nm Pb versus 25 nm Sn) and a consequently better intermixing of the elements in the molten volume.

VII. HARDMASK REMOVAL

Hardmasks were finally removed by etching three minutes in 2 % diluted HF and their topography analyzed by AFM (Fig. 15). As visible in the upper left and upper right images of Fig. 15, the features of the windows are clearly visible for both the GeSn and the GePb alloys. However, the GeSn feature seems more uniform, while the GePb seems to be slightly etched in the center of the corresponding windows. Analyzing the height profiles reveals that the GePb sample features trenches at the edge of the windows, varying in depth between 25 and 40 nanometers (Fig. 15 lower graph) and are at lower height in the center of the window with respect to the masked unalloyed areas, thus are etched. This corroborates the idea of amorphized material spotted at the edge of the window

41

by AFM-Raman. Etching of the GePb evidences its higher solubility in strongly diluted HF compared to Ge, thus the need to look for a more appropriate etching solution. The parabolic profile inside the window might be caused by isotropy of the etch.

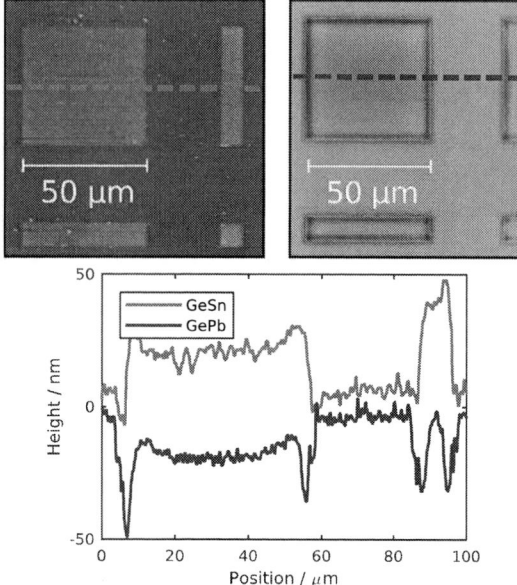

Fig. 15: AFM data for masked GeSn and GePb sample, obtained through 0.30 J/cm² PLIE and mask removal by etching. The upper left and right images represent GeSn and GePb surfaces, respectively. The lower graph shows the height profile of both samples (dashed lines in images).

For GeSn, trenches seem to exist as well, but are less visible due to the etching of the Ge in the masked areas. In contrast to the native oxide of GePb, that should have properties similar to PbO_x, which is slightly soluble in water, the native oxide of GeSn should be more akin to to SnO_x, being not significantly soluble in water. However, the smaller trenches for GeSn could also be attributed to a thinner amorphized zone close to the edges, corroborating the TOF-SIMS results with the initial Ge signal higher for Pb than for Sn.

VIII. CONCLUSION

SiO_2 hardmasks were successfully used to spatially confine the growth of GeSn and GePb alloys through PLIE. The centers and borders or edges of the windows as well as the masked areas are clearly distinguishable by their Raman spectra. Coating of the masks with a thin metal film improves their stability, allowing the use of higher laser fluences, thus better intermixing of the elements through PLIE. The adjustment of Ge thickness together with the fluence and number of laser pulses as well as masks that avoid melting of the subjacent Ge could also allow the formation of ternary alloys such as SiGeSn or SiGePb. A possible wet etching procedure to expose the GeSn patterns has been shown, while for GePb, more extensive studies are needed. If trenches due to amorphization can be evaded, PLIE patterned by hardmasks would allow the precise formation of planar, lateral heterostructures of group IV materials for photonic devices or optical waveguides with tailored band structures. Through selective etching of the patterned samples, the creation of three-dimensional structures, for example resonating cavities, is also possible and should be investigated further.

ACKNOWLEDGMENT

The authors thank Dan Buca from Forschungszentrum Jülich for providing the 160 nm and 2500 nm Ge coatings on Si(100). J. Schlipf thanks German DAAD Erasmus+ grant and J. L. Frieiro thanks the Spanish "Ministerio de Educación, Cultura de Educación, Cultura y Deporte for grant "Beca colaboración 2015/16".

REFERENCES

[1] Y. B. Bolkhovityanov, O. P. Pchelyakov, "GaAs epitaxy on Si substrates: Modern status of research and engineering", *Physics-Uspekhi*, vol. 51, no. 5, pp. 437–456, 2008.

[2] J. Liu, X. Sun, R. Camacho-Aguilera, L. C. Kimerling, and J. Michel, "Ge-on-Si laser operating at room temperature," *Optics Letters*, vol. 35, no. 5, pp. 679–681, 2010.

[3] S. Wirths, R. Geiger, N. von den Driesch, G. Mussler, T. Stoica, S. mantl, Z. Ikonic, M. Luysberg, S. Chiussi, J. M. Hartmann, H. Sigg, J. Faist, D. Buca, and D. Grützmacher, "Lasing in direct-bandgap GeSn alloy grown on Si", *Nature Photonics*, vol. 9, pp. 88-92, 2015.

[4] W. Huang, B. Cheng, C. Xue, and C. Li, "Comparative studies of clustering effect, electronic and optical properties for GePb and GeSn alloys with low Pb and Sn concentration," *Physica B*, vol. 443, pp. 43– 48, 2014.

[5] S. Stefanov, J. C. Conde, A. Benedetti, C. Serra, J. Werner, M. Oehme, J. Schulze, D. Buca, B. Holländer, S. Mantl, and S. Chiussi, "Laser synthesis of germanium tin alloys on virtual germanium," *Applied Physics Letters*, vol. 100, p. 104101, 2012.

[6] Q. Zhou, T. K. Chan, S. L. Lim, C. Zhan, T. Osipowicz, X. Gong, E. S. Tok, and Y.-C. Yeo, "Single crystalline germanium-lead alloy on germanium substrate formed by pulsed laser epitaxy," *ECS Solid State Letters*, vol. 3, no. 8, pp. 91–93, 2014.

[7] J. E. Jellison, D. H. Lowndes, D. N. Mashburn, and R. F. Wood, "Time-resolved reflectivity measurements on silicon and germanium using a pulsed excimer KrF laser heating beam," *Physical Review B*, vol. 34, no. 4, pp. 2407–2415, 1986.

[8] L. Jiang, J. D. Gallagher, C. L. Senaratne, T. Aoki, J. Mathews, J. Kouvetakis, and J. Menéndez, "Compositional dependence of the direct and indirect band gaps in $Ge_{1-y}Sn_y$ alloys from room temperature photoluminescence," *Semiconductor Science and Technology*, vol. 29, p. 115028, 2014.

[9] T. S. Perova, R. A. Moore, K. Lyutovich, M. Oehme, and E. Kasper, "Strain, composition and crystalline perfection in thin SiGe layers studied by Raman spectroscopy," *Thin Solid Films*, vol. 517, pp. 265– 268, 2008.

[10] D. Rouchon, M. Mermoux, F. Bertin, and J. M. Hartmann, "Germanium content and strain in $Si_{1-x}Ge_x$ alloys characterized by raman spectroscopy," *Journal of Crystal Growth*, vol. 392, pp. 66– 73, 2014.

[11] Y. Ishikawa, K. Wada, J. Liu, D. D. Cannon, H.-C. Luan, J. Michel, and L. C. Kimerling, "Strain-induced enhancement of near-infrared absorption in Ge epitaxial layers grown on Si substrate," *Journal of Applied Physics*, vol. 98, no. 1, p. 013501, 2005.

[12] L. Zhang, Y. Wang, N. Chen, G. Lin, C. Li, W. Huang, S. Chen, J. Xu, and J. Wang, "Raman scattering study of amorphous GeSn films and their crystallization on Si substrates," *Journal of Non-Crystalline Solids*, vol. 448, pp. 74–78, 2016.

[13] A. Gassenq, L. Milord, J. Aubin, N. Pauca, K.Guilloya, Rothman, D. Rouchon, A. Chelnokov, J. Hartmann, V. Reboud, and V. Calvo, "Raman spectral shift versus strain and composition in GeSn layers with: 6 to 15% Sn contents," 2017, submitted to *Applied Physics Letters*.

MIPRO 2017, May 22- 26, 2017, Opatija, Croatia

Impact of Sn segregation on $Ge_{1-x}Sn_x$ epi-layers growth by RP-CVD

David Weisshaupt[*], Pedram Jahandar[**], Gerard Colston[**], Phil Allred[**], Jörg Schulze[*] and Maksym Myronov[**]

[*]Universität Stuttgart, Institut für Halbleitertechnik (IHT), Pfaffenwaldring 47, 70569 Stuttgart, Germany
E-mail: David.Weisshaupt@gmx.de, Schulze@iht.uni-stuttgart.de
[**]The University of Warwick, Department of Physics, Gibbet Hill Road, CV4 7AL, Coventry, UK
E-mail: P.Jahandar@warwick.ac.uk, G.Colston@warwick.ac.uk, P.Allred@warwick.ac.uk, M.Myronov@warwick.ac.uk

Abstract - **This work investigates the impact of Sn segregation on the growth of $Ge_{1-x}Sn_x$ epi-layers using a reduced pressure chemical vapour deposition (RP-CVD) system with the common precursors Ge_2H_6 and $SnCl_4$. The investigated samples were grown on top of a 1 μm thick relaxed Ge buffer layer with different amounts of Sn incorporation, achieved by increasing the $SnCl_4$ partial pressure. The grown $Ge_{1-x}Sn_x$ epi-layers themselves are fully strained with respect to the Ge buffer underneath. A range of advanced analytical techniques have been used to characterize the material properties. The crystal structure, quality and thickness of the $Ge_{1-x}Sn_x$ epi-layers were analysed by using cross-sectional high resolution transmission electron microscopy, high resolution X-ray diffraction and fourier transform infrared spectrometry. Atomic force microscopy and Scanning electron microscopy in combination with energy dispersive X-ray spectroscopy are used for analysing the surface. It is shown that simply increasing the $SnCl_4$ partial pressure is insufficient for achieving Sn contents beyond ~8%. Above these concentrations the epitaxial growth breaks down due to the segregation of Sn resulting in the formation of dots on the epilayer surface, which consist of pure Sn.**

I. INTRODUCTION

Germanium tin ($Ge_{1-x}Sn_x$) is a semiconductor alloy with interesting applications in infrared photonics. While Germanium (Ge) is an indirect bandgap semiconductor, $Ge_{1-x}Sn_x$ undergoes a bandgap transition from indirect-to-direct for Tin (Sn) contents of over 9 % [1, 2, 3, 4]. Recently, $Ge_{1-x}Sn_x$ epilayers have been grown by the industrial standard growth technique, chemical vapour deposition (CVD), directly on relaxed Ge buffer layers on Silicon (Si) substrates; However, the highest Sn content $Ge_{1-x}Sn_x$ epilayers have been grown using molecular beam epitaxy (MBE). [1, 2, 5, 6] While MBE offers finer control over growth conditions, it is not a commercially viable process and does not exhibit excellent optical properties unlike CVD's one [6]. The growth of $Ge_{1-x}Sn_x$ by means of CVD can open a pathway for the commercialization of $Ge_{1-x}Sn_x$ devices that can be integrated into the well-established Si industry.

Due to the huge lattice mismatch of 19.48 % between Si (5.43102 Å) and α-Sn (6.4892 Å) a relaxed Ge (5.5679 Å) buffer is grown as an intermediate layer directly on the Si

substrate to reduce the lattice mismatch down to 14.69%. This lattice mismatch prevents the growth of smooth and high quality $Ge_{1-x}Sn_x$ epi-layers due to the three dimensional Stranski-Krastanov growth mode. Additionally, the solid solubility of Sn in Ge is extremely low with about 1.1 % and therefore non-equilibrium conditions are necessary to achieve higher Sn concentrations [7, 8]. However, the main issue for the growth of Sn-rich $Ge_{1-x}Sn_x$ epi-layers is the Sn segregation since Sn has a comparably low free surface energy. Segregation can take place at the surface or in the volume named precipitation. *Kasper et al.* described the Sn segregation with a linear model, assuming a constant surface energy independent from the surface coverage [1]:

$$n_s = \Delta_s \cdot n, \qquad (1)$$

with n_s the concentration of surface adatoms, Δ_s the segregation length and n the bulk concentration of Sn. To prevent segregation, the temperature must be kept as low as possible, since a low growth temperature reduces the exchange rate between the adatoms (Ge) and subsurface atoms (Sn). The time slot required for the exchange can be minimized with higher growth rates.

II. EXPERIMENTAL

$Ge_{1-x}Sn_x$ samples were epitaxially grown using an ASM Epsilon 2000 reduced pressure chemical vapour deposition (RP-CVD) system, with a total growth pressure of 100 Torr. The growth was carried out on 100 mm diameter, on-axis silicon (001) substrates. Four samples with different tin tetrachloride ($SnCl_4$) precursor partial pressures (25 mTorr, 50 mTorr, 75 mTorr, and 100 mTorr) were grown on top of a 1 μm thick Ge relaxed buffer. For $Ge_{1-x}Sn_x$ growth, digermane (Ge_2H_6) was used as the precursor gas, along with $SnCl_4$ for the Sn. All samples were grown at 270 ℃. Epitaxial layer thicknesses, crystal quality, segregation and other surface features were examined using both a JEOL 2000FX and a JEOL 2100 transmission electron microscopes (TEM) (200 kV accelerating voltage). An atomic force microscope (AFM) was used to determine the surface roughness. Measurements of lattice parameter and therefore determining strain and Sn concentration were obtained by use of a PanAnalytical high resolution X-ray diffractometer (HR-XRD). The Sn precipitation on the surface was analysed with scanning electron microscopy (SEM) and energy dispersive X-ray spectroscopy (EDS). Fourier transform infrared spectrometry (FTIR) is used to detect the thickness homogeneity across the whole wafer.

III. RESULTS

The HR-XRD rocking curves shown in Fig. 1 prove the crystalline nature of the grown layers. Full Width at Half Maximum values in a range of 0.05085 (75 mTorr) to 0.09603 (100 mTorr) are achieved, indicating a good crystal quality. Thickness fringes are used to calculate $Ge_{1-x}Sn_x$ epi-layer thickness for the sample with 25 mTorr $SnCl_4$ partial pressure (64 nm and 7.78 % Sn) and 50 mTorr $SnCl_4$ partial pressure (100 nm and 8.13 % Sn). No thickness fringes were observed on higher $SnCl_4$ grown samples as either the crystal quality was too poor or the epilayers had undergone relaxation. For the sample grown with 100 mTorr $SnCl_4$ partial pressure the $Ge_{1-x}Sn_x$ peak is shifted close to the Ge (004). The Sn concentration drops down to 1.06 %, corresponding to the natural solubility in equilibrium. The Sn concentration is determined iteratively using the elastic constant ratio and Vegard´s law with the bowing parameter 0.041 Å [9]:

$$C = 0.3738 + 0.167 \cdot x - 0.0296 \cdot x^2, \qquad (2)$$

$$a_{0,GeSn} = \frac{a_{\perp,GeSn} + 2 \cdot a_{\parallel,GeSn}}{1 + 2 \cdot C}, \qquad (3)$$

$$a_{0,GeSn} = a_{0,Ge} \cdot (1 - x) + a_{0,GeSn} \cdot x + b \cdot x \cdot (1 - x) \qquad (4)$$

with x the Sn concentration, C the elastic constant ratio and a the lattice constants for Ge and $Ge_{1-x}Sn_x$ in-plane ($\parallel$) and out-of-plain ($\perp$).

A typical tilt-corrected asymmetrical HR-XRD reciprocal space map (RSM) around the (224) Bragg peak, depicted in Fig. 2, determines the in- and out-of-plain lattice constants. The peaks lying on a straight line between Si and the origin are all fully relaxed. The $Ge_{0.92}Sn_{0.08}$ has the same in-plane lattice parameter, q_x, as the Ge, meaning it is fully strained to the buffer and has taken on the same lattice parameter as the Ge in real space. Additionally, symmetrical HR-XRD RSMs around the (004) Bragg peaks are recorded. The Bragg peaks in the (004) aligned in q_x confirm that the sample layers are not tilted (Fig. 3). The other samples with partial pressure of 25 mTorr, 75 mTorr and 100 mTorr exhibits similar (004) and (224) RSMs. The different partial pressures 25 mTorr, 50 mTorr, 75 mTorr, and 100 mTorr result in a Sn concentration of 7.78 %, 8.13 %, 7.43 %, and 1.06 %, respectively. The sample with a $SnCl_4$ partial pressure of 100 mTorr results in a segregation of Sn, since there are too many Sn molecules available in a gas phase for the given experimental conditions. Trying to increase the Sn concentration in the gas phase leads to an oversaturation of Sn near the epitaxial layer surface, which results in its segregation and the degradation of the grown $Ge_{1-x}Sn_x$ epilayer quality. Lower partial pressures result in nearly the same concentration of Sn, indicating a solubility of Sn in Ge of around 8 % for a growth temperature of 270 °C and a total growth pressure of 100 Torr.

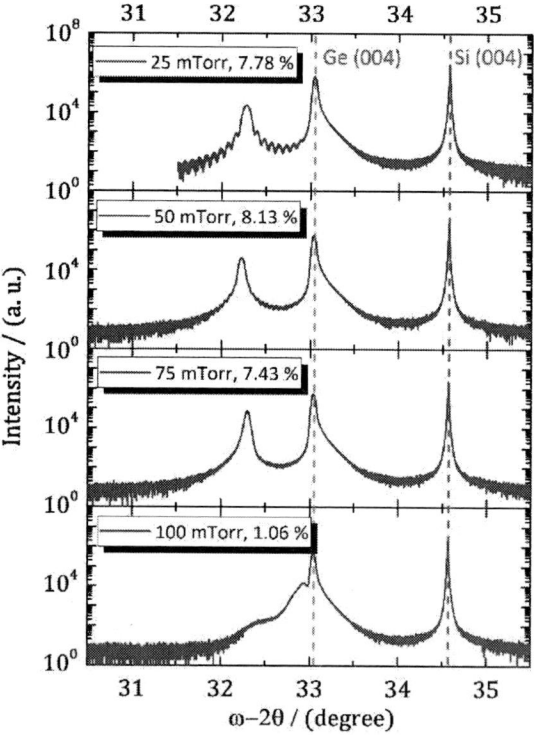

Fig. 1: HR-XRD rocking curve of the samples with different partial pressure 25 mTorr, 50 mTorr, 75mTorr and 100 mTorr.

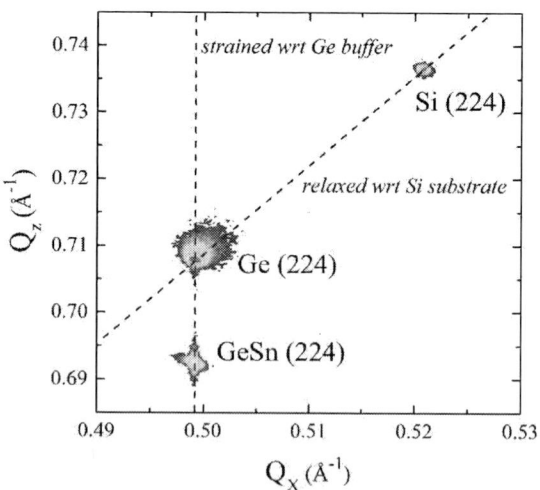

Fig. 2: HR-XRD (224) RSM of the sample with partial pressure 50 mTorr.

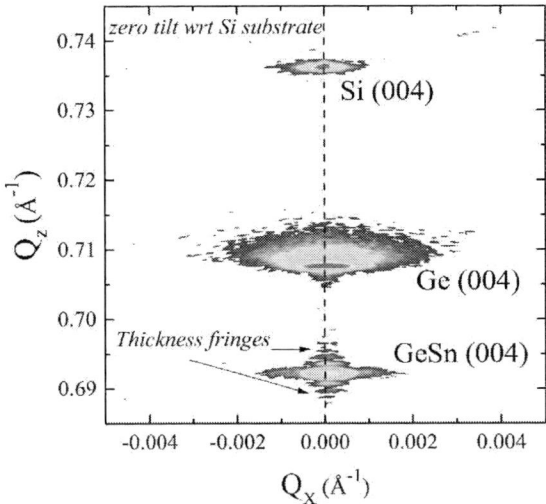

Fig. 3: HR-XRD (004) RSM of the sample with partial pressure 50 mTorr.

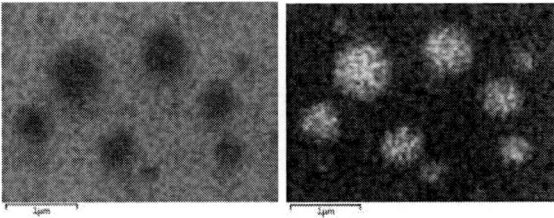

Fig. 5: EDS scan of sample containing $Ge_{0.99}Sn_{0.01}$ epi-layer for the elements Ge (red) and Sn (green), using the emission lines GeL, and SnL, respectively.

Fig. 6: EDS scan of sample containing $Ge_{0.92}Sn_{0.08}$ epi-layer for the elements Ge (red) and Sn (green), using the emission lines GeL, and SnL, respectively.

The AFM scans from sample containing $Ge_{0.99}Sn_{0.01}$ epi-layer shows dot formation on the surface (Fig. 4) and EDS proves that these dots are pure Sn without any Ge (Fig. 5). In contrast, the AFM scan from sample containing $Ge_{0.92}Sn_{0.08}$ epi-layer highlights the smooth surface before segregation occurs with a roughness of 1.23 nm. The other two samples with Sn content of 7.78 % and 7.43 % exhibits similar smooth surface. The Ge buffer has a typically roughness of 0.81 nm. Fig. 6 proves a good Sn homogeneity at the surface before Sn segregation occurs.

The cross-sectional TEM (X-TEM) image in Fig. 7 shows the epilayer surface of the segregated and non-segregated $Ge_{1-x}Sn_x$ samples. The amorphous Sn dot observed on the surface of the 100 mTorr sample has a height of 80 nm and a diameter of 380 nm. The $Ge_{0.99}Sn_{0.01}$ epi-layer thickness varies in a range from 64 nm to 98 nm. Underneath the segregated Sn dot, the $Ge_{0.99}Sn_{0.01}$ epi-layer thickness is 40 nm. The crystal quality of the grown $Ge_{0.99}Sn_{0.01}$ epi-layer turns polycrystalline or in worst case amorphous. For comparison, the X-TEM image in Fig. 8 shows the flat and crystalline $Ge_{0.92}Sn_{0.08}$ epi-layer before Sn segregation occurs.

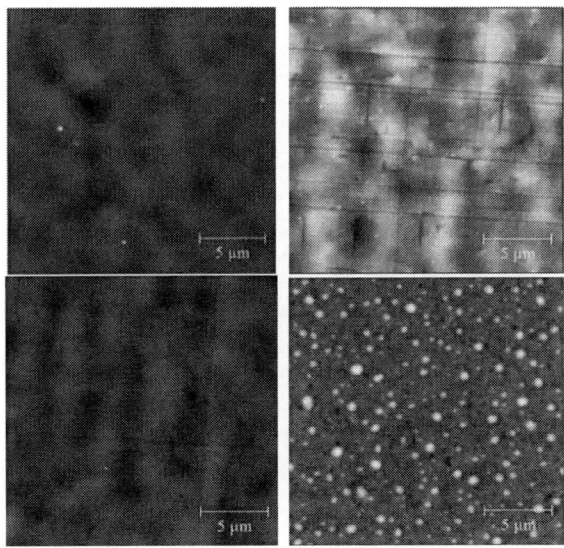

Fig. 4: (20x20) μm^2 AFM scan of the surface of samples grown with 25 mTorr (top left), 50 mTorr (top right), 75 mTorr (bottom left), and 100 mTorr (bottom right) $SnCl_4$ partial pressure.

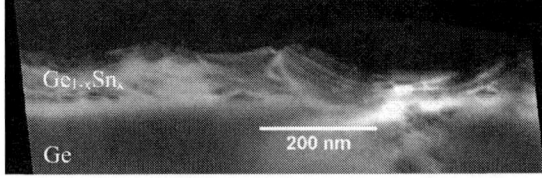

Fig. 7: Dark field (004) X-TEM image of sample containing $Ge_{0.99}Sn_{0.01}$ epi-layer.

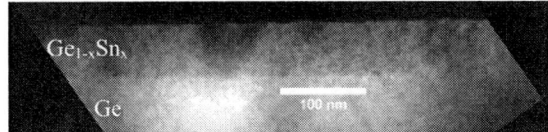

Fig. 8: Dark field (004) X-TEM image of sample containing $Ge_{0.92}Sn_{0.08}$ epi-layer.

Additionally, the X-TEM image in Fig. 9 demonstrates the perfect crystalline nature of the $Ge_{0.92}Sn_{0.08}$ epi-layer without any defects or misfit dislocations, an atomic sharp $Ge_{0.92}Sn_{0.08}/Ge$ interface and is to be used for comparison with the sample containing $Ge_{0.99}Sn_{0.01}$ epi-layer.

In good consistency with the AFM scan, the X-TEM image in Fig. 10 demonstrates the atomic flat surface of the $Ge_{0.92}Sn_{0.08}$ epi-layer. The samples containing $Ge_{0.923}Sn_{0.077}$ and $Ge_{0.928}Sn_{0.072}$ are similar.

The measured X-TEM thicknesses for the different samples, using the (004) Bragg diffraction condition, are 65 nm, 77 nm and 131 nm for the different partial pressures 25 mTorr, 50 mTorr and 75 mTorr, respectively. The X-TEM thickness for the sample with partial pressure of 25 mTorr fits well to the thickness extracted from its rocking curve. In contrary, there is a difference between the thicknesses for the sample with partial pressure of 50 mTorr. This discrepancy could occur due to the limitations of each technique to measure particular epilayer thickness. However, the most accurate technique in terms of thickness measurement should be the X-TEM. For the sample with partial pressure 50 mTorr a defect density of $4.444 \cdot 10^7$ defects/cm^2 and for a partial pressure of 75 mTorr a defect density of $4.756 \cdot 10^7$ defects/cm^2 could be determined, using the (220) Bragg diffraction condition.

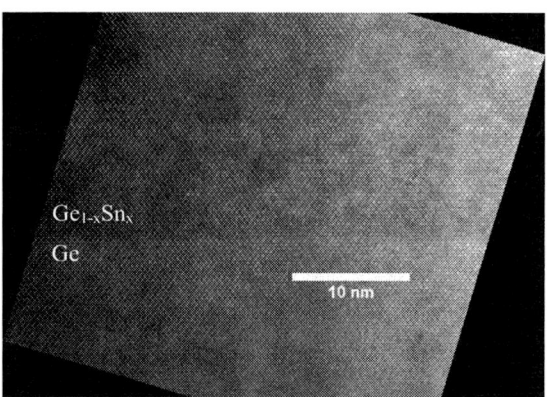

Fig. 9: Straight through X-TEM image of the $Ge_{0.92}Sn_{0.08}$ /Ge-interface of sample grown at 50 mTorr $SnCl_4$ partial pressure.

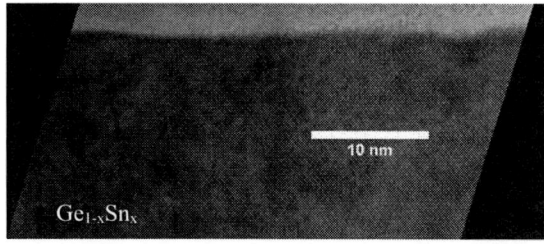

Fig. 10: Straight through X-TEM image of the $Ge_{0.92}Sn_{0.08}$ surface of sample grown at 50 mTorr $SnCl_4$ partial pressure.

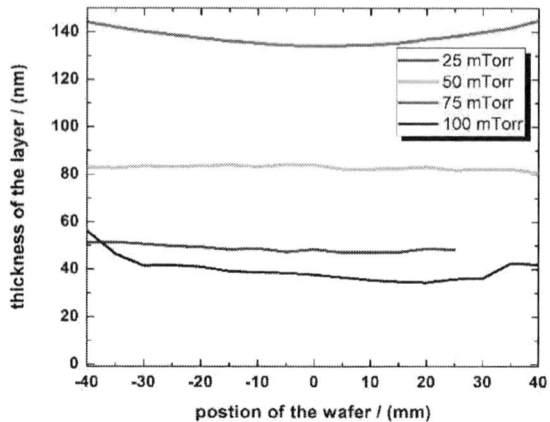

Fig. 11: FTIR thickness measurement across the wafer.

Fig. 11 depicts the measured thicknesses of the $Ge_{1-x}Sn_x$ epi-layer across the wafer over a diameter of 80 mm, using a reflectivity measurement performed on an FTIR system. As the $Ge_{1-x}Sn_x$ epilayers are thin (<150 nm) the accuracy of the FTIR measurement is uncertain, however, the variation of thickness can still be extracted from the data with good reliability. The uniformity is determined to be 4.4 %, 2.1 %, 3.8 %, and 24.0 % for the different partial pressures 25 mTorr, 50 mTorr, 75 mTorr, and 100 mTorr, respectively. It implies uniform two dimensional growth of the $Ge_{1-x}Sn_x$ epi-layer until segregation occurs. Table 1 summarizes the most important characteristics of the different samples.

TABLE I. Characteristics of growth $Ge_{1-x}Sn_x/Ge/Si$ (001) samples.

$SnCl_4$ Partial pressure / (mTorr)	Sn concentration / (%)	RMS surface roughness / (nm)	XRD FWHM / (degree)
25	7.78 ± 0.29	1.08	0.07289
50	8.13 ± 0.23	1.23	0.05734
75	7.43 ± 0.19	1.26	0.05085
100	1.06 ± 0.48	27.90	0.09603

IV. CONCLUSION

In conclusion, the impact of Sn segregation on $Ge_{1-x}Sn_x$ epi-layer growth has been demonstrated. While high levels of Sn incorporation have been demonstrated (~8%), simply increasing the $SnCl_4$ partial pressure during growth does not result in higher Sn concentrations beyond this threshold. The epitaxial growth breaks down due to the segregation of Sn and the grown $Ge_{1-x}Sn_x$ epi-layer turns polycrystalline or amorphous. The Sn concentration drops down to 1.06 %, corresponding to the natural solubility in equilibrium. Sn segregation results in a lower Sn concentration in the $Ge_{1-x}Sn_x$ epi-layer, a poorer crystal quality, an increased root mean square surface roughness, a poorer thickness uniformity and the formation of amorphous dots, which consist of pure Sn.

V. REFERENCES

[1] E. Kasper, J. Werner, M. Oehme, S. Escoubas, N. Burle, J. Schulze: Growth of silicon based germanium tin alloys. Thin Solid Film, 520 (2012) 3195-3200

[2] S. Wirths, D. Buca, S. Mantl; ScienceDirect, Progress in Crystal Growth and Characterization of Materials 62, 1-39, 2016.

[3] R. Soref, J. Kouvetakis, and J. Menendez; material Research Society Vol. 958, 2007.

[4] B. Vincent, F. Gencarelli, H. Bender, C. Marckeling, B. Douhard; Applied Physics Letters, 99, 2011.

[5] J. Kouvetakis, J. Menendez, and A.V.G Chizmeshva; Annual Review Matter Research, 36, 497-554, 2006.

[6] F. Pezzoli, A. Giorgioni, D. Patchett, and M. Myronov; ACS Photonics Letter, 10.1021, 2016.

[7] C.D. Thurmond, F.A. Trumbore, M. Kowalchik: Germanium solidus curves. In: J. Chem. Phys. 25 (1956) 799, doi:10.1063/ 1.1743083

[8] F.A. Trumbore: Solid solubilities and electrical properties of tin in germanium single crystals. In: J. Electrochem. Soc. 103 (1956) 597, doi:10.1149/1.2430167.

[9] F. Gencarelli, B. Vincent, J. Demeulemeester, A. Vantomme, A. Moussa, A. Franquet, et al.: Crystalline properties and strain relaxation mechanism of CVD grown GeSn. ECS J. Solid State Sci. Technol. 2 (2013) P134–P137, doi:10.1149/2. 011304jss

Tungsten Dichalcogenides as Possible Gas-Sensing Elements

V. A. Kuznetsov[*, **, a], A. Yu. Ledneva[*], S. B. Artemkina[*], M. N. Kozlova[*], G. E. Yakovleva[*],
A. S. Berdinsky[**], A. I. Romanenko[*] and V. E. Fedorov[*, ***, b]

[*] Nikolaev Institute of Inorganic Chemistry, Siberian Branch of Russian Academy of Sciences, Novosibirsk, Russia
[**] Novosibirsk State Technical University / Semiconductor Devices and Microelectronics, Novosibirsk, Russia
[***] Novosibirsk State University, Novosibirsk, Russia
[a] e-mail address: vitalii.a.kuznetsov@gmail.com
[b] e-mail address: fed@niic.nsc.ru

Abstract - Possible application of tungsten dichalcogenides as gas-sensing elements is discussed in this paper. The experimental results on sensitivity of pristine and niobium doped WS_2 and WSe_2 to acetone and ethanol gases are presented. Polycrystalline powder specimens were obtained by high temperature solid-state synthesis from the stoichiometric mixture of pure elements. Two types of samples were studied: 1) tablets pressed at 1.5 GPa to form bulk samples and 2) thin films prepared from 35% ethanol-water colloidal dispersions by their filtration onto membrane filters (pore diameter is 20 nm). The electrical resistances of the samples were shown to be increased in the presence of ethanol and acetone gases at room temperature, thereby revealing positive response to reducing gases.

I. INTRODUCTION

The problems of atmospheric air pollution in human environment, controlling of atmosphere composition in technological processes, in coal mines are very important subjects today. People have developed several methods with different types of functional sensing materials to solve the problem. To detect different toxic gases semiconductor, thermocatalytic, thermoconductometric, electrochemical sensors and optical/infrared spectral analysis are used [1-5].

Semiconductor gas sensors are widely used in practice today. Such sensors have resistive sensing mechanism. Adsorption of gas molecules changes conductivity of semiconducting sensing materials. If any gas is reducing agent it acts as electron donor for semiconductor, but if any gas is oxidizing agent it acts as acceptor. Despite the fact that the effect of changing semiconductor conductivity in the presence of gases has been known for long such sensors have some problems in practical applications. One of the problems is selectivity to different gases. Different gases can cause one and the same response – decreasing or increasing of resistance. Selectivity can be reached by several principles. One of them is that different gases influence the sensing material with different amplitude of resistance change and response time [5]. Another principle is use of additional surface layer on sensing element, for example layer of noble

The work was supported by the Russian Science Foundation (Grant no. 14-13-00674).

metals, which act as catalyst, thereby improving selectivity and sensibility [6-9]. One other problem is adsorption of gases on sensing element surface. In practical applications of classic semiconducting metal oxides heating of sensors is used in temperature range from 100°C to 800°C [4]. For example gas sensor based on classic semiconducting tin oxide is working at a temperature more than 250°C [1]. The temperature of highest sensitivity to different gases can be different. The other problem is time of response: the rate-of-change is very important in detecting of hazardous air pollution.

It was mentioned above that the most widely used semiconducting gas sensors were made of metal oxides, there is a wide range of the oxides, for example SnO_2, Ga_2O_3 for reducing gases and WO_3, ZnO, TiO_2 for oxidizing gases [3]. But lately layered transition metal chalcogenides (TMC) and graphene-based sensors of volatile organic compounds (VOC) have attracted many researchers attention [10-20]. The investigations are aimed at search of new gas sensing materials, which will be used without heating of sensing element and have high selectivity. For example TMC were shown to have response to reducing and oxidizing gases at room temperature with selectivity reached by noble metals on their surfaces [14, 21, 22].

Modification of TMC electronic properties can be achieved by substitution of chalcogen or metal atoms in crystal structure. For example recently we showed that substitution of Mo atoms in MoS_2 by Nb ones led to changing electronic properties – MoS_2 doped with Nb is a *p*-type semiconductor with semimetallic resistance-temperature behaviour in contrast to pristine semiconducting MoS_2 [23]. Modification of electronic properties can be achieved not only by substitution of metal or chalcogen atoms in the structure of TMC, but also by precipitation of metal nanoparticles onto TMC surface. So, since first works on dispersed TMC were published, many works have been devoted to deposition of noble metals onto 2-dimensional TMC layers and it was clearly shown that electronic properties of TMC were changed [24-26]. It was shown that chemical adsorption of noble metal nanoparticles onto surface of TMC could lead to both *n*- and *p*-doping effects; for example *n*-doping

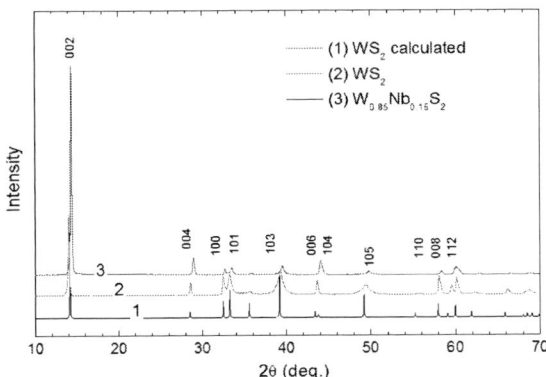

Figure 1. X-ray powder diffraction pattern of WS_2 and $W_{0.85}Nb_{0.15}S_2$ sythesized by high-temperature reaction in comparison to theoretical patten

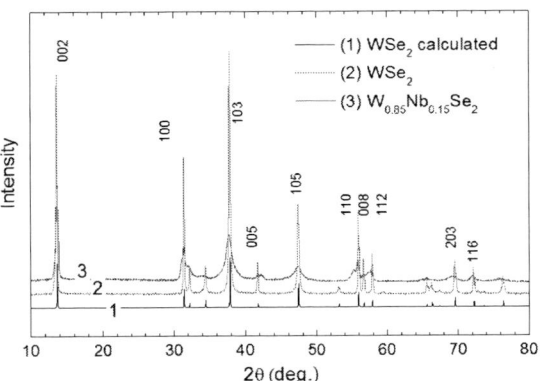

Figure 2. X-ray powder diffraction pattern of WSe_2 and $W_{0.85}Nb_{0.15}Se_2$ sythesized by high-temperature reaction in comparison to theoretical patten

effect on p-type pristine MoS_2 [13] and p-doping effect on n-type pristine MoS_2 [14] were demonstrated.

The goal of this work is to investigate experimentally the influence of reducing gases, acetone and ethanol, on electrical resistances of bulk pressed samples of pristine tungsten dichalcogenides WS_2 and WSe_2. To get different sensitivity to the gases film and bulk samples of the dichalcogenides with electronic properties modified by substitution of tungsten atoms by niobium ones were investigated as well.

II. EXPERIMENTAL

A. Synthesis of Experimental Samples

Polycrystalline samples were synthesized by high-temperature reaction. The tungsten powder was annealed in hydrogen flow at 900°C for 1 h to remove adsorbed water and traces of oxides. Stoichiometric quantities of pure elements were mixed in an agate mortar and put into a quartz ampule. The ampule was evacuated and sealed. The ampule was heated up to 400°C for 4 h, kept at this temperature for 5 h, then it was heated up to 800°C and kept for 96 h. Then the ampule was cooled and opened.

All prepared samples were investigated by X-ray powder diffraction analysis. X-ray powder diffraction patterns for the samples were collected with a Philips PW 1830/1710 automated diffractometer (Cu-Kα radiation, graphite monochromator, silicon plate as an external standard). It showed that the samples were single phase and corresponded to 2H-WS_2 ($P6_3/mmc$, № 194) (see Fig. 1 and Fig. 2). The addition of Nb atoms led to small shifts of the peaks in comparison with the unsubstituted chalcogenides. It can be explained by increasing of distance metal-chalcogen. For example the W–S distance in WS_2 is equal to 2.405 Å, and the Nb–S distance is equal to 2.421 Å. At the same time this increasing of the distances leads to increasing of lattice parameter a. These values were refined and reached 3.178 Å and 3.296 Å in $W_{0.85}Nb_{0.15}S_2$ and $W_{0.85}Nb_{0.15}Se_2$ compared to 3.153 Å and 3.282 Å in WS_2 and WSe_2, respectively [27]. Energy-dispersive X-ray spectroscopy analysis showed the presence of all the declared elements in formulae:

$W_{0.88}Nb_{0.14}S_2$ and $W_{0.83}Nb_{0.15}Se_2$ for $W_{0.85}Nb_{0.15}S_2$ and $W_{0.85}Nb_{0.15}Se_2$, respectively.

To form bulk samples 180 mg of each powder was pressed with a laboratory hydraulic press under a pressure inside tables of about 1.5 GPa. To form filtered film samples 300 mg of each powder was put into a glass flake with 60 ml of ethanol-water solution (35% ethanol / 65% water). It was sonicated for 12 h, then it was settled down for 12 h and then 2/3 of the dispersion was decanted. The resulted dispersions were filtered onto membrane filters Whatman Anodisc (pore diameter is 20 nm) with low pressure equipment. The thicknesses of the filtered films were estimated by weighting of the filters before and after the dispersions filtered, the thicknesses of the tablets were estimates from the masses of the powders and diameter of the tablets. From the estimated thicknesses, lengths, widths and electrical resistances of the samples their resistivities were estimated (see Table 1).

B. Measurement Technique

Experimental samples were formed by cutting the pressed tablets and filtered films to strips. Electrical contacts were made of graphite paste and lead thin copper wires were attached with the paste to the contacts. Four-point probe method was used to measure electrical resistance of the samples. Keithley 2000 was used to measure dc voltage on the samples and voltage on the resistance coil to control current in the electric circuit (see Fig. 3). To investigate influence of volatile organic compounds on the samples resistances the chamber with

TABLE I. RESISTIVITIES OF BULK PRESSED AND FILM FILTERED EXPERIMENTAL SAMPLES AT ROOM TEMPERATURE IN AIR

Composition of sample	Type of sample	Resistivity, Ohm·cm	Thermopower, μV/K
WS_2	pressed tablet	76	600
WSe_2	pressed tablet	21	1000
$W_{0.85}Nb_{0.15}S_2$	pressed tablet	0.012	50
$W_{0.85}Nb_{0.15}Se_2$	pressed tablet	0.005	16
$W_{0.85}Nb_{0.15}S_2$	filtered film	0.190	–
$W_{0.85}Nb_{0.15}Se_2$	filtered film	0.520	–

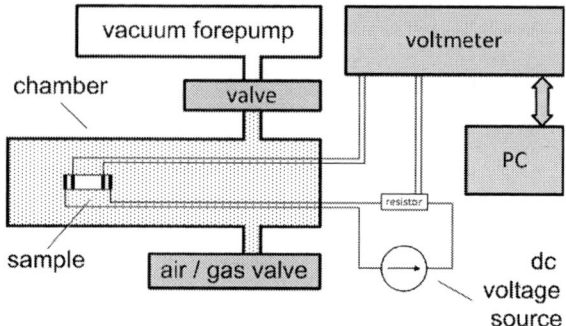

Figure 3. Schematic view of the measuring equipment for investigation of influence of gases on electrical resistance of experimental samples

vacuum forepump schematically shown in Fig. 3 was used. After the chamber having been evacuated it was filled with air or air with VOC. The volume of liquid VOC was small enough for the partial pressure to be lower than the saturation vapour pressure at room temperature. In all the experiments acetone or ethanol were used, and the partial pressure was about 4.0 and 4.7 kPa, respectively, and it corresponded to the concentrations about 10 000 ppm. After the exposure the chamber was evacuated and the experiment was repeated again.

III. RESULTS AND DISCUSSION

Relative changes of resistances of bulk WS_2 and WSe_2 in the presence of acetone and ethanol gases at room temperature are shown in Fig. 4 and Fig. 5, respectively. As one can see the rate of curves is similar to ones in the literature.

Thin polycrystalline films of n-type semiconducting WS_2 were reported to have negative response to reducing gases (reduction of resistivity), ethanol and NH_3, and positive response to oxidizing one O_2 (increasing of resistivity) [19]. It was shown that molecules of ethanol and NH_3 acted as electron donors and molecules of O_2 acted as electron acceptors. The semiconducting WS_2 was n-type and that is why increasing of electrons in WS_2 due to reducing gases led to resistivity decreasing (negative

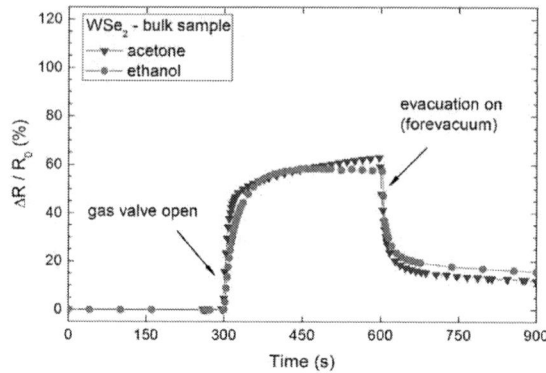

Figure 5. Relative change of electrical resistances of bulk WSe_2 under vacuum and in the presence of acetone and ethanol gases at room temperature

response in the terms we use here). Generally reducing gases act as electron donors, reducing resistivity of n-type semiconductors and vice-versa leading to increasing of p-type semiconductor resistivity. As one can see in Fig. 4 and Fig. 5 the samples concerned showed positive response. Conduction type of pristine WS_2 and WSe_2 can be either p- or n-type [28]. As conduction type of semiconductors can be defined by measuring thermopower we measured it for the bulk samples concerned. The resulted values of thermopower are positive for all bulk samples (see Table 1), it points to the fact that the samples were p-type. That may be a reason that we got positive resistance response to reducing gases, as concentration of holes was reduced by electrons donated from the gases adsorbed.

Pressed powder of $W_{0.85}Nb_{0.15}S_2$ and $W_{0.85}Nb_{0.15}Se_2$ did not show any distinguishable response to the gases. This may be because of too low resistivity of the pressed samples (see Table 1), and reducing of charge carriers concentration in near-surface layer due to the gases molecules adsorbed did not result to any distinguishable change in overall samples resistances. But at the same time the filtered films of the compositions showed positive response at the same concentration of acetone (see Fig. 6 and Fig. 7). The electrical resistance responses to the gases may be related in general to complicated

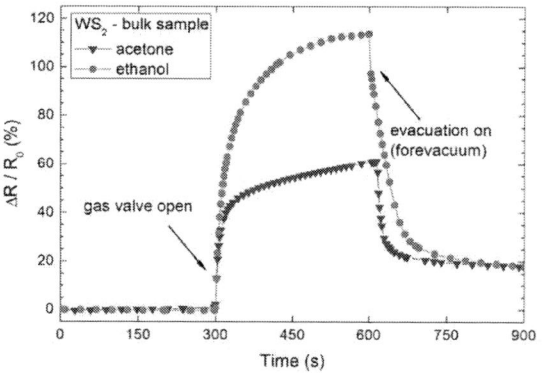

Figure 4. Relative change of electrical resistances of bulk WS_2 under vacuum and in the presence of acetone and ethanol gases at room temperature

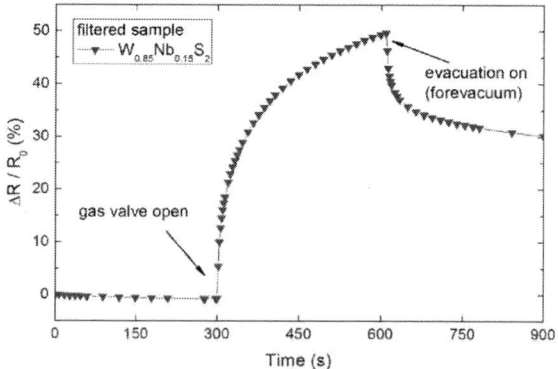

Figure 6. Relative change of electrical resistance of film filtered sample of $W_{0.85}Nb_{0.15}S_2$ under vacuum and in the presence of acetone gas at room temperature

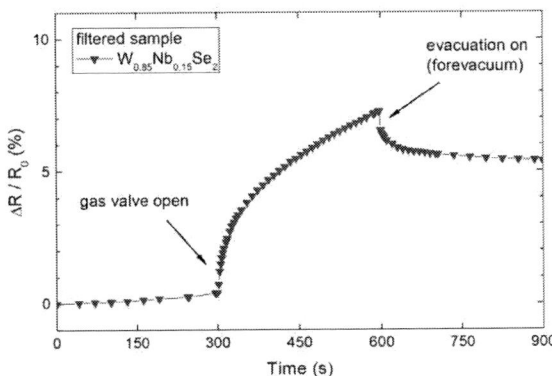

Figure 7.   Relative change of electrical resistance of film filtered sample of $W_{0.85}Nb_{0.15}Se_2$ under vacuum and in the presence of acetone gas at room temperature

mechanism of physisorption, chemisorption, and moreover capillary condensation of the gases inside pores of both pressed and filtered samples. If capillary condensation takes place it contributes to the positive responses due to wedging of conduction particles. In any case the nature of the responses should be investigated in the near future.

IV.   CONCLUSION

The study showed that tungsten chalcogenides WS_2 and WSe_2 could be used as gas sensing elements at room temperature with comparative short response time. All the semiconducting compounds concerned were p-type and exhibited positive response to the reducing gases – electrical resistances of the samples increased when exposed with the gases. This can be explained by both decreasing of charge carriers concentrations in the samples due to recombination of holes with electrons donated from gases molecules adsorbed and complicated mechanism of sorption of the gas molecules inside pores of the samples. One way or the other, the objects concerned need further consideration.

REFERENCES

[1]   Modern Sensors Handbook. London: ISTE Ltd, 2007.

[2]   S. Soloman, Sensors Handbook, Second Edition ed.: McGraw-Hill: New York, Chicago, San Francisco, Lisbon, London, Madrid, Mexico City, Milan, New Delhi, San Juan, Seoul, Singapore, Sydney, Toronto, 2010.

[3]   G. Korotcenkov, Handbook of Gas Sensor Materials, 1 ed. vol. Volume 1: Conventional Approaches: Springer-Verlag New York, 2013.

[4]   G. Korotcenkov, Handbook of Gas Sensor Materials, 1 ed. vol. Volume 2: New Trends and Technologies: Springer-Verlag New York, 2014.

[5]   J. Fraden, Handbook of Modern Sensors, 5 ed.: Springer International Publishing, 2016.

[6]   R. Huck, U. Böttger, D. Kohl, and G. Heiland, "Spillover effects in the detection of H2 and CH4 by sputtered SnO2 films with Pd and PdO deposits," Sensors and Actuators, vol. 17, pp. 355-359, 1989.

[7]   K. D. Schierbaum, J. Geiger, U. Weimar, and W. Göpel, "Specific palladium and platinum doping for SnO2-based thin film sensor arrays," Sensors and Actuators B Chemical, vol. 13, pp. 143-147, 1993.

[8]   G. Tournier, C. Pijolat, R. Lalauze, and B. Patissier, "Selective detection of CO and CH4 with gas sensors using SnO2 doped with palladium," Sensors and Actuators B: Chemical, vol. 26, pp. 24-28, 1995.

[9]   M. H. Darvishnejad, A. A. Firooz, J. Beheshtian, and A. A. Khodadadi, "Highly sensitive and selective ethanol and acetone gas sensors by adding some dopants (Mn, Fe, Co, Ni) onto hexagonal ZnO plates," RSC Advances, vol. 6, pp. 7838-7845, 2016.

[10]   K. Arshak, E. Moore, L. Cavanagh, J. Harris, B. McConigly, C. Cunniffe, et al., "Determination of the electrical behaviour of surfactant treated polymer/carbon black composite gas sensors," Composites Part A: Applied Science and Manufacturing, vol. 36, pp. 487-491, 2005.

[11]   P. Bondavalli, P. Legagneux, and D. Pribat, "Carbon nanotubes based transistors as gas sensors: State of the art and critical review," Sensors and Actuators B: Chemical, vol. 140, pp. 304-318, 2009.

[12]   W. Yang, L. Gan, H. Li, and T. Zhai, "Two-dimensional layered nanomaterials for gas-sensing applications," Inorganic Chemistry Frontiers, vol. 3, pp. 433-451, 2016.

[13]   S.-Y. Cho, H.-J. Koh, H.-W. Yoo, J.-S. Kim, and H.-T. Jung, "Tunable Volatile-Organic-Compound Sensor by Using Au Nanoparticle Incorporation on MoS2," ACS Sensors, vol. 2, pp. 183–189, 2017.

[14]   C. Kuru, C. Choi, A. Kargar, D. Choi, Y. J. Kim, C. H. Liu, et al., "MoS2 Nanosheet–Pd Nanoparticle Composite for Highly Sensitive Room Temperature Detection of Hydrogen," Advanced Science, vol. 2, p. 1500004, 2015.

[15]   A. Gaiardo, B. Fabbri, V. Guidi, P. Bellutti, A. Giberti, S. Gherardi, et al., "Metal Sulfides as Sensing Materials for Chemoresistive Gas Sensors," Sensors, vol. 16, p. 296, 2016.

[16]   Z. Feng, Y. Xie, J. Chen, Y. Yu, S. Zheng, R. Zhang, et al., "Highly sensitive MoTe2 chemical sensor with fast recovery rate through gate biasing," 2D Materials, vol. 4, p. 025018, 2017.

[17]   D. J. Late, T. Doneux, and M. Bougouma, "Single-layer MoSe2 based NH3 gas sensor," Applied Physics Letters, vol. 105, p. 233103, 2014.

[18]   B. Cho, A. R. Kim, Y. Park, J. Yoon, Y.-J. Lee, S. Lee, et al., "Bifunctional Sensing Characteristics of Chemical Vapor Deposition Synthesized Atomic-Layered MoS2," ACS Appl. Mater. Interfaces, vol. 7, pp. 2952-2959, 2015.

[19]   N. Huo, S. Yang, Z. Wei, S.-S. Li, J.-B. Xia, and J. Li, "Photoresponsive and Gas Sensing Field-Effect Transistors Based on Multilayer WS2 Nanoflakes," Scientific Reports, vol. 4, p. 5209, 2014.

[20]   F. K. Perkins, A. L. Friedman, E. Cobas, P. M. Campbell, G. G. Jernigan, and B. T. Jonker, "Chemical Vapor Sensing with Monolayer MoS2," Nano Letters, vol. 13, pp. 668-673, 2013.

[21]   D. Sarkar, X. Xie, J. Kang, H. Zhang, W. Liu, J. Navarrete, et al., "Functionalization of Transition Metal Dichalcogenides with Metallic Nanoparticles: Implications for Doping and Gas-Sensing," Nano Letters, vol. 15, pp. 2852–2862, 2015.

[22]   Q. He, Z. Zeng, Z. Yin, H. Li, S. Wu, X. Huang, et al., "Fabrication of Flexible MoS 2 Thin-Film Transistor Arrays for Practical Gas-Sensing Applications," Small, vol. 8, pp. 2994-2999, 2012.

[23]   V. E. Fedorov, N. G. Naumov, A. N. Lavrov, M. S. Tarasenko, S. B. Artemkina, and A. I. Romanenko, "Tuning electronic properties of molybdenum disulfide by a substitution in metal sublattice," in 36th International Convention on Information and Communication Technology, Electronics and Microelectronics (MIPRO), Opatija, CROATIA, 2013, pp. 11-14.

[24]   T. S. Sreeprasad, P. Nguyen, N. Kim, and V. Berry, "Controlled, Defect-Guided, Metal-Nanoparticle Incorporation onto MoS2 via Chemical and Microwave Routes: Electrical, Thermal, and Structural Properties," Nano Letters, vol. 13, pp. 4434-4441, 2013.

[25]   Y. M. Shi, J. K. Huang, L. M. Jin, Y. T. Hsu, S. F. Yu, L. J. Li, et al., "Selective Decoration of Au Nanoparticles on Monolayer MoS2 Single Crystals," Scientific Reports, vol. 3, p. 7, May 2013.

[26]   J. Zhao, Z. Zhang, S. Yang, H. Zheng, and Y. Li, "Facile synthesis of MoS2 nanosheet-silver nanoparticles composite for surface

enhanced Raman scattering and electrochemical activity," Journal of Alloys and Compounds, vol. 559, pp. 87-91, 2013.

[27] W. J. Schutte, J. L. D. Boer, and F. Jellinek, "Crystal structures of tungsten disulfide and diselenide," Journal of Solid State Chemistry, vol. 70, pp. 207-209, 1987.

[28] V. L. Kalikhman and Y. S. Umanskii, "Transition-metal chalcogenides with layer structures and features of the filling of their brillouin zones," Soviet Physics Uspekhi, vol. 15, pp. 728-741, 1973.

Flicker noise in AlGaAs/GaAs of high electron mobility heterostructure field-effect transistor at cryogenic temperature

Souheil Mouetsi[1,2], Foudil Zouach[1] and Djamil Rechem[1,2]

[1] Department of electrical engineering, faculty of sciences and applied sciences,
University Larbi Ben M'hidi,Oum El Bouaghi, Algeria.
[2] Laboratory of active components and materials, University Larbi Ben M'hidi,
Oum El Bouaghi. Algeria.

E-mail: souheil25m@gmail.com

Abstract - In this paper, Low frequency noise measurements at cryogenic temperature below 120 K in AlGaAs/GaAs of high electron mobility heterostructure field-effect transistor (HFET) was studied. In this range of temperature, the samples are biased at very low voltage to avoid velocity saturation. Assuming that the different noise sources of power noise spectral density such as flicker noise (1/f), generation–recombination (G-R) noise, thermal noise are independent. In this work, the parameters characterizing flicker noise (γ, α_H) are studied, noting that γ varies around the unity and α_H decreases with increasing temperature for the lowest temperature range from 4 to 70 K. Taking into account the effect of bidimensional electron gas (2DEG) mobility at low temperatures, the obtained results allow us to suggest that the predominant model of the flicker noise at cryogenic temperature is the fluctuation mobility model.

Keywords: flicker noise (1/f), Generation–recombination (G-R) noise, thermal noise, power spectral density (PSD), 2DEG.

I. INTRODUCTION

Some small signals received from sensors require that the used preamplifiers must be characterized by an extremely low noise conditions. Each amplifier is chosen with specific characteristics in order to overcome the difficulties imposed by the detection system. For example, in a cryogenic environment, the preamplifier must be inserted as close as possible to the detection system (a bolometer, for example). This requires that the preamplifier must be able to support very low temperatures. As a consequence, its performance comes mainly from reduced equivalent input current at low frequency noise [1].

Thus, for very low noise applications, the choice of High Electron Mobility Transistor (HEMT) is generally better than other devices. Indeed, the performances of the HEMTs improve at cryogenics temperatures [2].

Different low frequency noise contributions (thermal noise, generation – recombination (G-R) noise and flicker noise) were affected by different parameters depending on the kind of the studied devices, and were investigated to identify quality and reliability of semiconductor devices [3, 4] and to study impurity and defects in semiconductor structures [5, 6], for example, several studies on the electrical noise in semiconductor components have highlighted correlation between some extrinsic parameters of the studied devices (including electrically active defects) and the appearance of noise. The evaluation of the G-R noise can lead to extract the density of traps, their activation energy and their capture cross section [7, 8]. As a consequence, the noise measurement can be used not only to determine the minimum noise parameters of the components, but also for the identification of heterogeneity and defects induced by the growth process and manufacturing.

The present study concerns the low frequency noise in AlGaAs/GaAs of high electron mobility heterostructure field-effect transistor at cryogenic temperatures and for different applied voltages. Let us notice that our samples are similar to a HEMT but without a control gate represented by a GaAs channel with a two-dimensional electron gas (2DEG). The dimensions of the sheet resistance are given as: length L = 16 μm and the width W = 500 μm.

II. SAMPLE DESCRIPTION

The samples used in this study were grown on a semi-insulating <100> GaAs substrate by Molecular-Beam Epitaxy (MBE) [7].

Table (1) summarizes technical description of the samples:

Cap layer	n+ GaAs	
Barrier	n $Al_xGa_{1-x}As$ (x = 19.6%)	
Donor	n+ AlGaAs	15 nm, SiAs: 8×10^{12} cm^{-2}
		35 nm, Si: 10^{12} cm^{-2}
Spacer	AlGaAs	40 nm
Channel	GaAs	20 nm
	$Al_xGa_{1-x}As$	10 nm
Substrate	GaAs	45 nm

Tableau 01: Technical description of the samples [7].

The general lithographic steps were: the (Al, In) GaAs/GaAs wafer was firstly spun on a PMMA resist and baked at 170 °C; then it was exposed by the e-beam with a dose of about 400 μC/cm^2 and developed in MIBK-isopropyl alcohol. The device was realized mainly as follows: first, the active 2DEG area was defined by mesa with a chemical wet etching in H_2O_2:H_3PO_4:H_2O, then, the ohmic contacts for the source and the drain, were obtained by the Au/Ge eutectic alloy as following, the plot electrodes were made by evaporation of Ni on the GaAs layer followed by evaporation of Au/Ge eutectic. Two metallic layers made of Ni and Al were then successively deposited. Finally the samples were warmed at about 400 °C to allow Ge to diffuse through GaAs. This diffusion reduces the created depletion layer under metallic contacts.

III. EXPERIMENTS

The device under test (DUT) was inserted in the simplified experimental setup shown in figure (01) and previously described in [8].

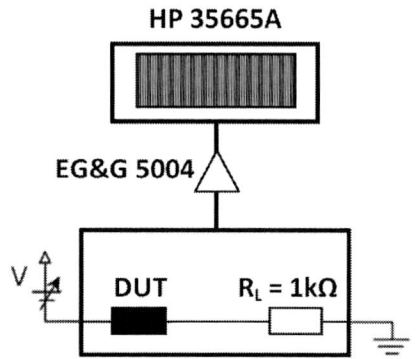

Figure 01: A simplified diagram for the low-frequency noise measurements

The measurements of spectral noise were performed at various temperatures from 300 K down to 4 K, and for different applied voltages. The voltage noise was amplified by the EG&G 5004 low-frequency noise voltage amplifier, in which the amplification was fixed to G = 10^3, equivalent noise voltage of the order is set to 0.8 nV/√Hz, and equivalent noise current tuned to 92 fA/√Hz at 1 kHz.

The noise measurements were performed using the HP 35665A spectrum analyzer in the frequency range of 1 Hz–100 kHz. The samples were mounted on a sample holder located at the end of a cryogenic cane that can be directly put in a helium reservoir. The temperature was measured by a 330 lake shore controller. The samples were maintained for a long time almost 10 min, for a given temperature, and before making each measurement in order to be sure that the thermodynamic equilibrium was reached. All measurements are made in the dark condition.

IV. RESULTS AND DISCUSSIONS

IV.1. Noise spectroscopy

In figure (02), we present the effect of the applied electric field taken at 20, 50, 70 and 100 mV on the total power spectral density (PSD) $S_v(f)$ for low frequency, and measured at 4 K.

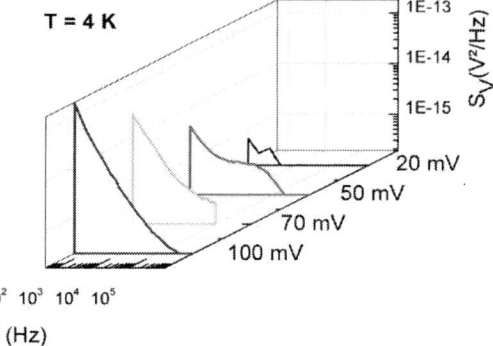

Figure 01: Typical spectral noise, measured at 4 K for different applied voltages

The PSD increases with increasing applied voltage related through ohm's law as reported in the literature [8, 9]. We also noticed that the PSD noise measured for different applied voltages in low frequency is much greater than the one measured in high frequency. During measurements, the samples were electrically polarized at 0 mV to measure the thermal noise which was negligible in most of the experiments [10].

Figure (03) gives typical total spectral noise in the low frequency range for different temperature taken at 50 mV.

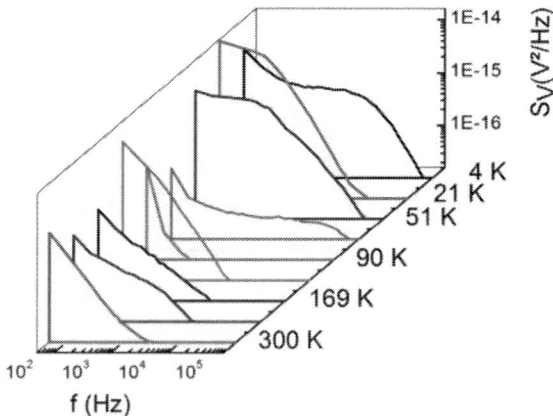

Figure 03: Total spectral noise at 50 mV versus the temperature.

We observed a pronounced dependence of noise power spectral density on the temperature, but its variation is not regular and neither clearer [7 - 12]. The above figure (03) shows inverse dependence of the PSD as a function of frequency, this is due either to the saturation of the current as shown in figure (04) or resulting from the high increase of the 2DEG mobility in figure (05).

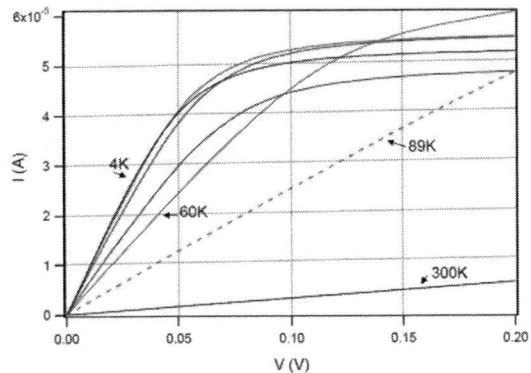

Figure 04: I-V characteristics in the temperature range of 4-300 K [7]

At cryogenic temperatures (<100 K), the *I-V* curves show a linear ohmic behavior followed by a saturation feature [13].

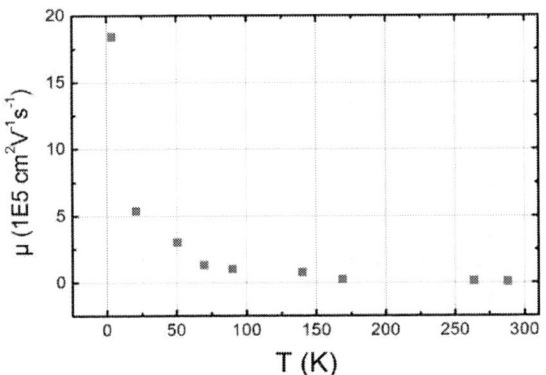

Figure 05: Mobility of load carriers measured as a function of temperature.

IV.2. Decomposition of the total spectral noise

In order to study the effect of low temperatures on the noise spectral density especially the flicker noise), we will proceed to separate between all the low frequency noise components; typical spectral are given in figure (06):

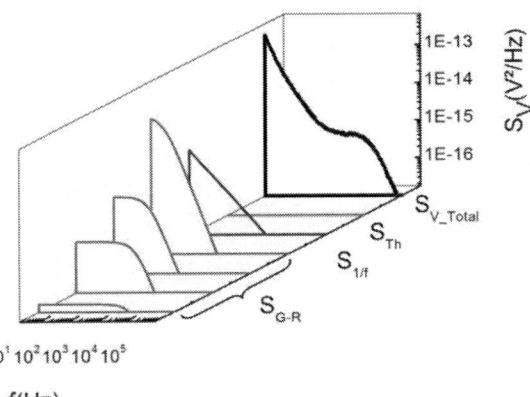

Figure 06: Decomposition of the total noise spectral at 4K and V = 50mV.

All obtained spectral noise were fitted to distinguish between the flicker noise and the others G-R noises and thermal noise (S_{Th}). We observe that thermal noise is very weak of which it can be neglected, the generation–recombination noise (S_{G-R}) contributions decrease rapidly versus frequency (f^2), which is why they disappear when the frequencies are less than 100 Hz. Therefore, we can limit our study of flicker noise for the frequency less than 1 kHz.

III. 3. Flicker noise at cryogenic temperature

As the flicker noise level decreases with the frequencies and increases with the voltage, the evolution of the flicker noise spectral density verifies the Hooge relation [14]:

$$S_{1/f}(f) = \frac{\alpha_H V^2}{N f^\gamma} \qquad (1)$$

It is convenient to determine the Hooge parameters (α_H, γ) which have been used as a reliability parameter of electronic devices [15, 16]. Usually, a lower value of α_H indicates better noise properties [17]. It is clear that systematic noise investigation is helpful in estimating the quality and applicability of the device. But usually, the source of fluctuation causing the appearance of flicker noise is expected to have some relation with the phonon scattering of the mechanism of flicker noise. That is why we must know the dependence of the Hooge parameter on temperature.

If we consider that there is no correlation in the flicker noise contributions, S_v (f = 1 Hz) can be used to calculate α_H basis of the expression given below [10]:

$$\alpha_H = S_V(1Hz) \frac{n_s(T)WL}{V^2} \qquad (2)$$

Where $n_s(T)$ is the bi-dimensional electron gas density which was measured by the Hall effect technique as a function of temperature. V is the applied voltage; W and L are the width and the length of the channel, respectively.

Figure (07) shows the evolution of α_H at low frequency for cryogenic temperatures.

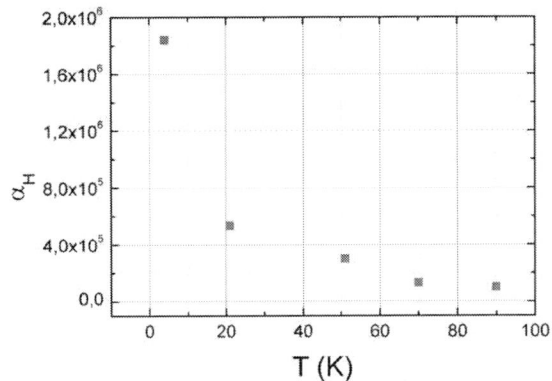

Figure 07: α_H parameter variation as a function of temperature.

The experimental value of α_H increases with decreasing temperature for the lowest temperature values even the 2DEG mobility increases rapidly in this low temperature range (figure (05)), this suggests the predominant model on the noise source is based on the fluctuation mobility model [18, 19]. Similar results have been obtained in $In_{0.52}Al_{0.48}As/In_xGa_{1-x}As$ heterostructures by Pavelka et al. at low temperature [19].

In the basis of Hooge expression, we calculate γ by the use of the following equation [10]:

$$\gamma = \frac{Ln[S_v(1Hz)/S_v(f)]}{Ln(f)} \qquad (3)$$

$S_v(1Hz)$ is the voltage noise at $f = 1$ Hz and $S_v(f)$ is the voltage noise at a given frequency $f < 1$ kHz. The equation (03) shows the linear relationship in the Ln-Ln scale of the flicker noise versus frequency. Results given above in figure (02) seem to consolidate this suggestion in low frequency.

The γ values are given in figure (08):

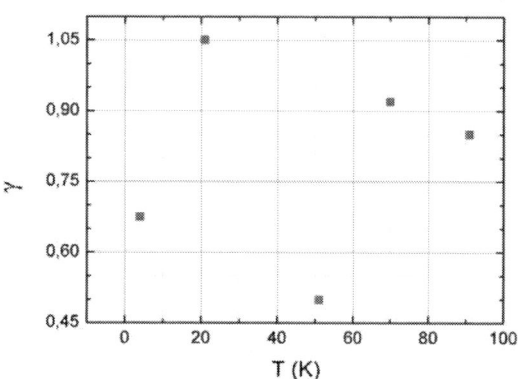

Figure 08: γ parameter variation as a function of the temperature.

The fluctuation of γ around the unit shows a regular variation with the temperature, we can conclude that the effect of temperature in low values does not influence the flicker noise rate.

V. CONCLUSION

AlGaAs/GaAs high electron mobility heterostructure field-effect transistor (HFET) has been investigated, this work is characterized by the low frequency noise method. The study was made at different applied electric fields, and at cryogenic temperatures. Different contributions (thermal noise, G-R noise and flicker noise) have been identified.

Hooge parameters (α_H, γ) were studied as a function of temperature. The variation of α_H values versus temperature are compared with theoretical ones, the obtained results confirm the predominance of the mobility fluctuation noise model at cryogenic temperature. The regular variations of γ in temperature justify the non responsibility of the temperature on the flicker noise rate.

REFERENCES

[1]. Catalano, Andrea, Alain Coulais, and Jean-Michel Lamarre. "Analytical approach to optimizing alternating current biasing of bolometers." Applied Optics 49.31 (2010): 5938-5946.

[2]. Pospieszalski, Marian W. "Extremely low-noise amplification with cryogenic FETs and HFETs: 1970-2004." IEEE Microwave Magazine 6.3 (2005): 62-75.

[3]. L. K. J. Vandamme, "Noise as a Diagnostic Tool for Quality and . 41, N° 11, pp. 2176-2187, 1994.

[4]. Jones, B. KIEE. "Electrical noise as a reliability indicator in electronic devices and components." IEE Proceedings-Circuits, Devices and Systems 149.1 (2002): 13-22.

[5]. Tartarin, Jean-Guy, et al. "Using low-frequency noise characterization of AlGaN/GaN HEMT as a tool for technology assessment and failure prediction." Second International Symposium on Fluctuations and Noise. International Society for Optics and Photonics, 2004.

[6]. Ciofi, C., and B. Neri. "Low-frequency noise measurements as a characterization tool for degradation phenomena in solid-state devices." Journal of Physics D: Applied Physics 33.21 (2000): R199.

[7]. Abdelillah El Hdiy, and Souheil Mouetsi. "Identification of traps in an epitaxied AlGaAs/GaAs/AlGaAs quantum well structure." Journal of Applied Physics 108.3 (2010): 034513.

[8]. Khlil, Rachid, Abdelillah El Hdiy, and Yong Jin. "Deep levels and low-frequency noise in Al Ga As/Ga As heterostructures." Journal of applied physics 98.9 (2005): 093709.

[9]. S. Mouetsi, A. El Hdiy, R. Khlil, Y. Jin and M. Bouchemat, "Temperature and voltage effects on the *1/f* noise in a two-dimensional electron Gas". 32nd International Convention MIPRO, Opatija, Adriatic Coast, Croatie. May 25-29, 2009.

[10]. Souheil Mouetsi, and Abdelillah El Hdiy. "Contribution to the *1/f* noise analysis in a bi-dimensional electron gas." Journal of Applied Physics 114.10 (2013): 104507.

[11]. Bora, Achyut, and A. K. Raychaudhuri. "Low-frequency resistance fluctuations in metal films under current stressing at low temperature (T< 0.3 T melting)." Physical Review B 77.7 (2008): 075423.

[12]. Camin, D. V., C. F. Colombo, and V. Grassi. "Low frequency noise versus temperature spectroscopy of Ge JFETs, Si JFETs and Si MOSFETs." Journal de Physique IV (Proceedings). Vol. 12. No. 3. EDP sciences, 2002.

[13]. Khlil, R., et al. "An unusual nonlinearity in current-voltage curves of a bidimensional electron gas at low temperatures." Journal of applied physics 98.12 (2005): 123701.

[14]. Hooge, Friits N. "1/f noise is no surface effect." Physics letters A 29.3 (1969): 139-140.

[15]. Tacano, M., et al. "Dependence of Hooge parameter of InAs heterostructure on temperature." Microelectronics Reliability 40.11 (2000): 1921-1924.

[16]. Berntgen, Jürgen, et al. "Hooge parameter of InGaAs bulk material and InGaAs 2DEG quantum well structures based on InP substrates." Microelectronics Reliability 40.11 (2000): 1911-1914.

[17]. Rumyantsev, S. L., et al. "Low-frequency noise in AlGaN/GaN heterostructure field effect transistors and metal oxide semiconductor heterostructure field effect transistors." Fluctuation and Noise Letters 1.04 (2001): L221-L226.

[18]. Garrido, J. A., et al. "Low-frequency noise and mobility fluctuations in AlGaN/GaN heterostructure field-effect transistors." Applied Physics Letters 76.23 (2000): 3442-3444.

[19]. Pavelka, Jan, et al. "1/f noise models and low frequency noise characteristics of InAlAs/InGaAs devices." physica status solidi (c) 8.2 (2011): 303-305.

MIPRO 2017, May 22- 26, 2017, Opatija, Croatia

Device performance tuning of Ge gate-all-around tunneling field effect transistors by means of GeSn: Potential and challenges

Erlend G. Rolseth[*], Andreas Blech[*], Inga A. Fischer[*], Youssef Hashad[*], Roman Koerner[*], *Student Member, IEEE*,
Konrad Kostecki[*], Aleksei Kruglov[*], V.S. Senthil Srinivasan[*], Mathias Weiser[*], Torsten Wendav[**], Kurt Busch[**†],
Joerg Schulze[*], *Senior Member, IEEE*

[*] Institute for Semiconductor engineering, Department of Electrical Engineering and Information Technology, University of Stuttgart, Stuttgart 70174, Germany
[**] Institut für Physik, AG Theoretische Optik & Photonik, Humboldt-Universität zu Berlin, Berlin 10099, Germany
[†]Max-Born-Institut, Berlin 10099, Germany
erolseth@gmail.com

Abstract— In this paper we report experimental results on the fabrication and characterization of vertical Ge gate-all-around p-channel TFETs, utilizing GeSn as a channel material. Through two sample series, the potential and challenges of implementing the low-band gap material GeSn are reviewed. It is verified that I_{ON} can be effectively enhanced by increasing the Sn-content in the GeSn-channel, due to increasing tunneling probabilities. Further it is found that when limited to a 10 nm δ-layer, $Ge_{0.96}Sn_{0.04}$ is most beneficial for I_{ON} when positioned inside the channel as opposed to in the source, with a maximum of I_{ON} = 180 µA/µm at V_{DS} = -2 V and V_G = -4 V. Enhanced leakage currents (I_{OFF}), which also degrades the subthreshold swing (SS), is a consequence of a smaller band gap and enhanced defect densities, and represent key challenges with implementing GeSn.

I. INTRODUCTION

Recent years have seen a growing interest in device concepts based on quantum mechanical tunneling. The Tunneling Field Effect Transistor (TFET) is a device that competes directly with the Metal-Oxide-Semiconductor Field Effect Transistor (MOSFET) in terms of speed, power and area [1], [2]. The drive current (I_{ON}) in a TFET is a band-to-band tunneling (BTBT) current. The subthreshold swing (SS) of this tunneling current is not constrained by the 60 mV/dec limit for MOSFETs at room temperature. This allows TFETs to potentially perform better at low supply voltages. They are hence regarded as interesting devices for low-power operation. To provide high switching speed in terms of RC time, high device I_{ON} is necessary. The international technology roadmap for semiconductors (ITRS) requirements [3] for I_{ON} have, however, proven to be a major challenge for group-IV TFET devices. The experimental results of TFETs are also currently lagging behind projections showing a disparity between the measured and simulated subthreshold characteristics [4]. A number of material and geometry modifications have been proposed in order to boost I_{ON} in group-IV TFETs: The lower-band gap material Ge has been implemented in $Si_{1-x}Ge_x$ alloy [5] or bulk Ge [6] TFETs, and device geometry modifications have been performed in order to align the tunneling direction

with the gate field in Si [7] and SiGe [8] TFETs. Those devices show improved I_{ON} compared to all-Si TFETs, but still fail to achieve the ITRS I_{ON} requirement. As a measure to further improve I_{ON}, $Ge_{1-x}Sn_x$ [9]-[11] has recently been introduced in parts of the channel region. In addition to being a CMOS-compatible alloy, GeSn has an even smaller band gap than Ge. This is favorable in terms of raising I_{ON} in TFET devices. Relaxed $Ge_{1-x}Sn_x$ is also predicted to become a direct band gap material for x > 0.11 [12]. Tunneling without the assistance of phonons would greatly increase the tunneling probability, and hence I_{ON}. Epitaxial growth of high quality $Ge_{1-x}Sn_x$ however poses many challenges. Due to the large lattice mismatch of 14.7 % between α-Sn and Ge and the low solid solubility of 1 % of Sn in Ge, a lot of experimental effort is currently directed towards the growth of $Ge_{1-x}Sn_x$ [13]-[15]. While a small band gap material raises I_{ON}, the leakage current (I_{OFF}) is inevitably also affected. In order to combine the advantages of small band gap materials (higher I_{ON}) with those of large band gap materials (lower I_{OFF}), heterostructures with small band gap materials positioned at the source-channel junction and large band gap materials at the channel-drain junction are often preferable. However, heterojunctions can also introduce parasitic effects which deteriorate device performance. Device I_{OFF} and SS in particular are influenced by crystalline quality and interface defects, as trap-assisted tunneling (TAT) and Shockley-Read-Hall (SRH) generation currents are enhanced in low band gap material and heterojunction devices [16], [17].

In this experimental study we want to assess the potential of GeSn as channel material in vertical p-channel TFET devices. This is done through two sample series, each consisting of three samples. The first series focuses on the bulk material properties of GeSn, by introducing a 200 nm GeSn-channel with 0 %, 2 % and 4 % Sn-content. In the second series the position of a 10 nm-$Ge_{0.96}Sn_{0.04}$-δ-layer at the source-channel interface is varied from completely inside the channel to completely inside the source. Its effect on the electrical characteristics of the transistor is reported. The device area

This work was funded by the Deutsche Forschungsgemeinschaft (DFG) under Grant FI 1511/2-1. T.W. and K.B. would like to acknowledge support by the Stiftung der Deutschen Wirtschaft (SDW) and by the DFG through subproject B10 within the Collaborative Research Center (CRC) 951 Hybrid Inorganic/Organic Systems for Opto-Electronics (HIOS).

and temperature dependence of the TFET devices were also studied.

II. LAYER GROWTH AND DEVICE FABRICATION

A thin tunneling barrier between the source and channel region is an important requirement for achieving high I_{ON} in TFETs. This necessitates abrupt doping profiles for the semiconductor pin diode layer structure. For the heterojunction devices reported here this was achieved by means of molecular beam epitaxy (MBE). All three samples were grown on p-doped (0.02-0.05 Ω·cm) Si (100) wafers. A detailed MBE layer sequence for the two sample series is given in Table I and Table II and a schematic of the Ge and $Ge_{0.96}Sn_{0.04}$ parts of the MBE layers of the second sample series is shown in Fig. 1. The epitaxial layer thickness of the grown GeSn-layers, is lower than the expected critical thickness for all samples [18]. We therefore expect pseudomorphic growth of GeSn, where these layers are biaxially strained with respect to the underlying Ge. Because of Fermi level pinning, Al and n-doped Ge contacts show a rectifying behavior [19]. A heavily Sb-doped Si cap layer is therefore grown as a final layer to ensure an ohmic top contact. More details on the MBE growth, in particular the virtual substrate (VS) technology necessary in order to enable the growth of high quality GeSn, and relaxed Ge on Si, is reported in [14] and [20].

After MBE growth, the vertical TFETs were fabricated using a gate-all-around (GAA) process. Details on this GAA process can be found in [9]. Al_2O_3 gate oxide was deposited by means of plasma enhanced atomic layer deposition (PEALD). A GeO_x interfacial layer was formed with an O_2 post plasma oxidation step. The formation of such a layer can reduce the interface density of states (D_{it}) with as much as one order of magnitude [21]. From CV-measurements of planar MOS-Capacitors using the same oxidation method, and ALD- and MBE systems, a D_{it} at midgap has been calculated to be $\sim$ 5E11 $cm^{-2}eV^{-1}$ for Ge. The gate stack provides an equivalent oxide thickness (EOT) of $\sim$ 7 nm for the first sample series, and an EOT $\sim$ 4.5 nm for the second sample, after reducing the number of ALD cycles after O_2 plasma post oxidation between experiments. A schematics of a finished device is shown in Fig. 2. In contrast to the devices reported in [9], the contact windows in the SiO_2 passivation were opened with photolithography for all devices investigated here.

The relatively large gate oxide thicknesses, which are unfavorable for achieving steep SSes and scaling the supply voltage, were chosen to prevent leakage current between the gate and the substrate. A complete chip consists of TFET devices with both circular and square mesa shapes and with mesa cross-sections varying from 0.79 to 100 μm^2. For this vertical geometry devices the gate width, w_g, equals the mesa perimeter, P, and is proportional to the square root of the mesa area cross-section, A, $w_g = P \propto A^{1/2}$. This can be seen in Fig. 3.

TABLE I

MBE layer sequence for the bulk GeSn channel TFETs. First sample series.

Layer	Material	Sample A Thickness (nm)	Sample B Thickness (nm)	Sample C Thickness (nm)	Doping (cm^{-3})
Source	Si	100			$N_D = 1 \cdot 10^{20}$
Source	Ge	200			$N_D = 1 \cdot 10^{20}$
Channel	Ge	200	-	-	
Channel	$Ge_{0.98}Sn_{0.02}$	-	200	-	
Channel	$Ge_{0.968}Sn_{0.04}$	-	-	200	
Drain	Ge	200			$N_A = 1 \cdot 10^{18}$
Drain	Ge (VS)	100			$N_A = 1 \cdot 10^{20}$
Drain	Si	400			$N_A = 1 \cdot 10^{20}$

TABLE II

MBE layer sequence for the GeSn-δ-layer TFETs. Second sample series.

Layer	Material	Sample A Thickness (nm)	Sample B Thickness (nm)	Sample C Thickness (nm)	Doping (cm^{-3})
Source	Si	350			$N_D = 1 \cdot 10^{20}$
Source	Ge	100	95	90	$N_D = 1 \cdot 10^{20}$
Source	$Ge_{0.96}Sn_{0.04}$	-	5	10	
Channel	$Ge_{0.96}Sn_{0.04}$	10	5	-	
Channel	Ge	40	45	50	
Drain	Ge	200			$N_A = 1 \cdot 10^{18}$
Drain	Ge (VS)	100			$N_A = 1 \cdot 10^{20}$
Drain	Si	50			$N_A = 1 \cdot 10^{20}$

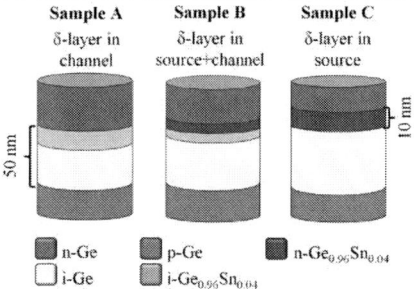

Fig. 1. Schematics of the MBE layer structures for the second sample series, where the positional dependence of a GeSn-δ-layer at the source-channel interface is investigated.

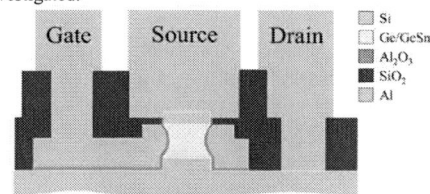

Fig. 2. Schematics of a finished vertical GeSn-TFET.

Fig. 3. Scanning electron microscope image of a vertical GeSn-TFET after gate formation. The gate width, w_g, of the transistor equals the perimeter, P, and is proportional to the square root of the mesa area, A.

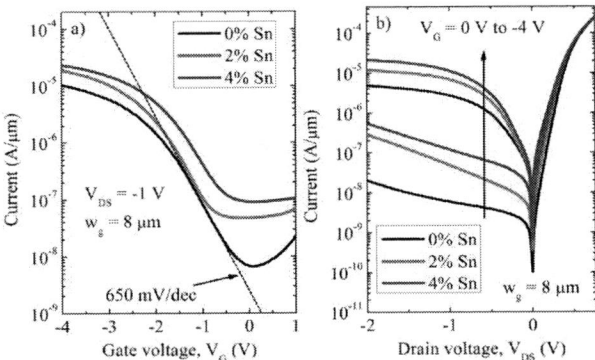

Fig. 4. a) Transfer and b) Output characteristics of Ge$_{1-x}$Sn$_x$ channel TFETs with x = 0 %, 2 % and 4 %. Increase in I$_{ON}$ as well as I$_{OFF}$ can be seen to result from increasing Sn-content in the channel.

III. MEASUREMENT RESULTS

The TFET devices were characterized through I-V measurements obtained with a Keithley 4200 Semiconductor Characterization System. The temperature dependence of the devices was studied by cooling down the measurement chuck from room temperature to 240 K. The temperature was measured with a thermocouple attached to the sample. The source (top) contact was kept at ground potential.

A. Bulk GeSn channel

Transfer and output characteristics of the first sample series are shown in Fig. 4 a) and b), respectively. A 650 mV/dec slope is drawn in Fig. 4 a) to indicate the steepness of the devices. The TFETs' I-V-characteristics can be considered to be composed of a leakage current I$_{OFF}$, and a gate controlled BTBT-current I$_{ON}$, whose steepness can be quantified by the SS. The effect of Sn-content in the GeSn channel on these three parameters will be considered in turn, and we start by examining I$_{OFF}$ which we assume can be expressed by:

$$I_{OFF}(V_{DS}) = J_A(V_{DS})\frac{A}{w_G} + J_P(V_{DS})\frac{P}{w_G} + I_G. \quad (1)$$

Here A is the device area, P the device perimeter and J_A and J_P are the area and perimeter leakage current densities, respectively. The gate leakage current, I_G, was established through measurements to be much lower than 1 pA and is negligible, compared to the two other components in (1). For our vertical TFETs the perimeter equals the gate width, $P = w_g$. A large set of transistors with A/w$_g$ ratios varying between 0.25 - 2.5 µm were measured in order to separate the J_A and J_P components. In Fig. 5, I$_{OFF}$ for V$_{DS}$ = - 1.0 V is shown as a function of A/w$_g$. The straight line fit validates (1) and J_A and J_P can be extracted from the slope and intercept, respectively. Following this approach, straight line fits were performed for all I$_{OFF}$(V$_{DS}$)-values obtained through the output characteristics. The resulting mapping of J_A and J_P as a function of V$_{DS}$ is shown in Fig. 5 b) and c), respectively.

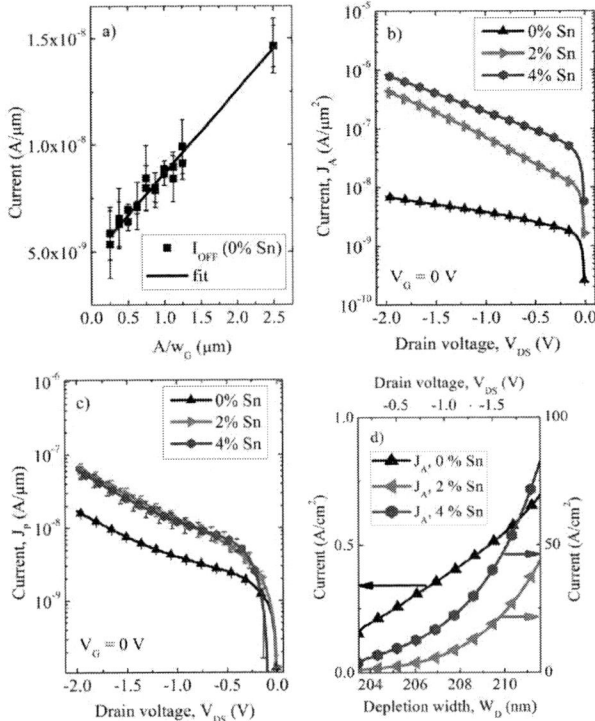

Fig. 5. a) I$_{OFF}$ for the Ge reference shown as a function of A/w$_G$ for V$_{DS}$ = -1.0 V. The area (J$_A$) and perimeter (J$_P$) leakage current densities can be determined from the slope and intersect of the fitted line, respectively. b) J$_A$ and c) J$_P$ contributions to I$_{OFF}$ as a function of V$_{DS}$. Increasing the Sn-content in the channel leads to a strong successive increase of J$_A$, whereas only a slight constant increase of J$_P$. d) J$_A$ as a function of W$_D$. The Ge reference shows a linear J$_A$-W$_D$ relationship, while the GeSn samples show an superlinear behavior due to the modification of the SRH generation rate by the electric field-enhancement factor.

J_A addresses the epitaxial quality of the MBE grown diode structure. A correlation between leakage current density and threading dislocation density has been found for SiGe pin diodes on Si-substrates [22]. For the Ge TFET we therefore expect J_A to reflect the threading dislocations resulting from the strain and growth of Ge on Si-substrates. It can be seen in Fig. 5 b) that increasing the Sn-content in the GeSn channel leads to a strong successive increase in J_A as well as a stronger V$_{DS}$ dependence. An increase in J_A is expected for GeSn compared to Ge, due to the lowering of the band gap which exponentially affects the intrinsic carrier concentration n$_i$, and SRH generation currents [23]. The magnitude of the increase and the strong V$_{DS}$ dependence of these samples, however, indicates that this is not a result of band gap lowering alone. Point defects associated with the growth of GeSn on Ge at low temperature [13] degrade the epitaxial quality. With a higher trap density, SRH generation- and TAT-currents increase. The leakage current due to these two mechanisms can be approximated by [23]-[25]:

$$J \approx q(1+\Gamma)U_{SRH}W_D. \quad (2)$$

where q is the elementary charge, W_D is the depletion width, U_{SRH} is the SRH generation rate and Γ the field enhancement factor due to TAT. For the samples presented here we approximate W_D by solving the Poisson equation for a pin diode using the depletion approximation [23]:

$$W_D = \sqrt{d^2 + \frac{2\varepsilon_s}{q}\left(\frac{N_A + N_D}{N_A N_D}\right)(V_{bi} - V_{DS})}, \qquad (3)$$

where the i-region thickness is d = 200 nm. Although containing minor errors when inserting Ge parameters for the permittivity ε_s and built-in potential V_{bi} for all samples, a qualitative comparison can in either way be made based on the J_A-$\sqrt{V_{DS}}$ relationship. In Fig. 5 d) J_A is shown as a function of W_D. For the Ge reference sample a linear J_A-W_D relationship is seen, consistent with dominating SRH generation current, i.e. $\Gamma \sim 0$, for all V_{DS}-biases. The samples with GeSn in the channel however exhibit a superlinear J_A-W_D relationship, which originates from TAT contribution to J_A. The field enhancement factor in (2) has to be considered, $\Gamma > 0$. The generation rate, U_{SRH}, can be estimated by taking the minimum of the derivative dJ_A/dW_D, found at low electric fields. When normalizing to the U_{SRH} of the Ge TFET, we get a relative increase in the generation rate of $\tilde{U}_{SRH}$ (2 % Sn) = 12 and $\tilde{U}_{SRH}$ (4 % Sn) = 48, respectively. This is a result of increased trap density and band gap reduction with increasing Sn-content. These results show how the incorporation of GeSn, leads to both higher SRH-generation currents as well as enhanced TAT leakage.

The simplest strategy to reduce the J_A component of I_{OFF} is through device dimension scaling. Extrapolating J_A gives rough estimates for the dimensions needed for I_{OFF} to equal the 10 pA/µm ITRS low power requirements (neglecting J_P contribution). For the Ge TFET reference sample a diameter of D ~ 10 nm for $V_{DS} = -1.0$ V is needed. These dimensions are in the same scale as the ITRS recommended multi-gate MOSFET body thickness [3], and also the predicted required body thickness for achieving sub-60 mV/dec SS in TFETs using Ge [16]. For the samples containing GeSn the prospects are worse, with a required D < 2 nm for $V_{DS} = -0.5$ V. At the scales discussed here, however, it is no longer possible to separate between the bulk J_A and the surface J_P currents as the gate will influence the full depth of the channel. The chip variability will also become an issue at this scale, widening the range of threshold voltages. This will in turn also affect the leakage current per unit area. Hence, other improvement strategies are needed in addition to downscaling to keep I_{OFF} manageable. MBE growth of GeSn is still a relatively new technique, which means we can expect progress as this technique matures. A hope of reducing the high leakage current of GeSn TFETs therefore lies in reducing bulk defect density through improved GeSn growth strategies. Additionally, we propose that the leakage current can be reduced through the use of a Ge/GeSn heterojunction channel. As BTBT only occurs at the channel-source interface, the GeSn can be confined within a thin layer and positioned at this interface without degrading the TFETs on-state performance

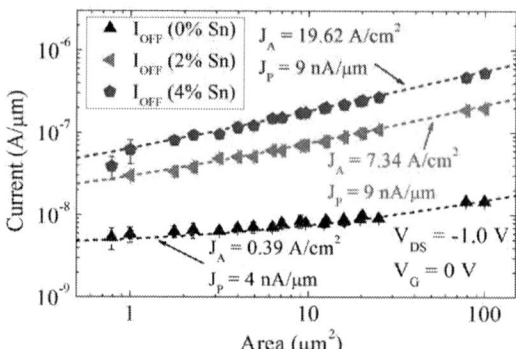

Fig. 6. I_{OFF} as a function of device area. Dashed lines show the fit of (1) to experimental data. The corresponding fit values of J_A and J_P are indicated. When scaling down the device size, the current becomes limited by J_P. Scaling is only effective to improve device performance when I_{OFF} is dominated by J_A.

with respect to all GeSn-channel TFETs. Reducing the GeSn-layer thickness would, however, reduce the total number of traps inside the depletion region. The leakage is, hence, effectively reduced and the I_{OFF} will approach the I_{OFF} of Ge TFETs. Heterojunction GeSn/Ge -channel TFETs are considered in the second sample series, which will be investigated later on.

J_P is another concern, as it represents a minimum I_{OFF} achievable by device dimension scaling for the GAA geometry. In Fig. 6 I_{OFF} together with fits of (1) are displayed as a function of device area. As the dimensions become smaller I_{OFF} will approach J_P, which was neglected in the previous discussion. For the GeSn TFETs a reduction of ~ 3 orders of magnitude in J_P is necessary to reach the ITRS low power requirement for I_{OFF}. It is hence necessary to reduce the D_{it} at the Ge(Sn)-oxide interface, which causes SRH and TAT surface leakage currents. From Fig. 5 b) and Fig. 6 we see that J_P is roughly equal for the GeSn TFETs, but increased by a factor ~ 2 compared to the Ge TFET. The increase in J_P when moving from a Ge to a GeSn system is believed to be due to the addition of interface states. If an increase of J_P was mainly due to band gap lowering, a successive increase with increasing Sn-content would be expected. The reported D_{it} for GeSn-oxide interfaces is 2E12 – 6E13 cm^{-2}eV^{-1} [26]-[28]. Current experimental work on sulfur passivation of the Ge and GeSn-surfaces is under investigation by our group and is, like reported also elsewhere [28], [29], showing promising results.

We do not expect that as much as a three order of magnitude reduction in D_{it} is necessary to reach the ITRS requirement. The devices presented here have a very large gate-drain overlap. Gate-induced drain leakage [30] together with gate-induced tunneling at the drain-channel interface (an ambipolar behavior can clearly be seen in Fig. 4. a)) are therefore also contributing to the high J_P currents. A way of reducing gate induced leakage currents could be to introduce a spacer between the buried layer and the gate electrode, reducing the gate-drain overlap. This would also enable the use of thinner gate oxides, as the leakage path between the substrate and gate is blocked.

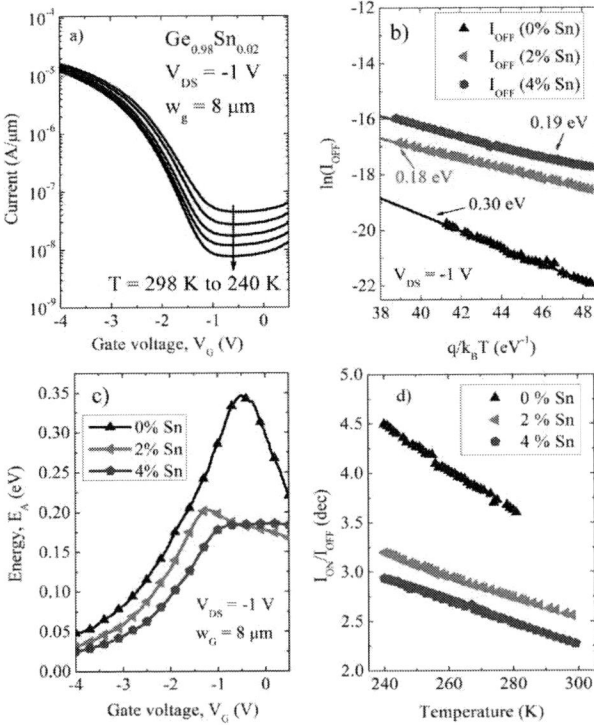

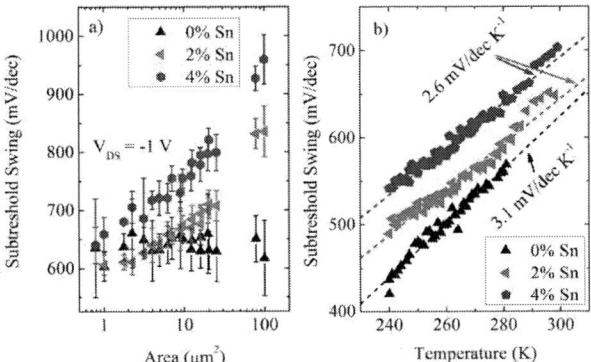

Fig. 8. a) SS as a function of device area. b) SS as a function of temperature.

Fig. 7. Transfer characteristics of a $Ge_{0.98}Sn_{0.02}$-channel TFET measured at different temperatures. b) Arrhenius plot of I_{OFF} for all three samples. Activation energies are indicated. c) Activation energies as a function of V_G. d) I_{ON}/I_{OFF}-ratio as a function of temperature

Which transport mechanisms determining I_{OFF}, was investigated further by examining the temperature dependence. Transistors with $w_g = 8~\mu m$ were measured. In Fig. 7 a) the temperature dependence of the transfer characteristics for a transistor with 2 % Sn in the GeSn channel is shown. The leakage floor and the early subthreshold region show a strong temperature dependence. Fig. 7 b) shows the Arrhenius plots of I_{OFF}, and in Fig. 7 c) fitted activation energies are shown as a function of V_G. The activation energy for I_{OFF} of the Ge reference, is close to half the band gap of Ge ($E_G/2 \sim 0.33\text{-}0.34$ eV in the temperature range investigated [31]) for low V_G-bias. This is consistent with the temperature dependence of n_i [23], and hence SRH-generation currents. The temperature dependence of I_{OFF} for the samples with GeSn in the channel relate to the bulk properties, as $I_{OFF} \approx J_A \times A/P$ for these samples. The activation energies are lower than half of the expected band gaps of GeSn with 2 % and 4 % Sn-content [32]. This is a consequence of TAT contribution and the temperature dependence of $\ln(\Gamma)$ in (2), which has an activation energy that is lower than $E_G/2$ [25]. The low temperature dependence of the BTBT current, combined with the strong temperature dependence of I_{OFF}, leads to a considerable improvement of the I_{ON}/I_{OFF}-ratio with decreasing temperature. This can be seen in Fig. 7 d).

Let us now analyze the SS, which here is defined as the steepest point in the transfer characteristics. The devices all have a high SS ~ 550-1000 mV/dec compared to the 60 mV/dec MOSFET limit. Although the SSs are similar to those of GeSn-TFETs reported elsewhere [10], they are considerably higher than the lowest SSs reported for TFETs so far (SS = 3.9 mV/dec [33]). This is on the one hand due to the thick gate oxide (EOT ~ 7 nm) necessary to eliminate I_G in (1). On the other hand a high I_{OFF} also limits the visibility of steeper SSs at low currents. This can be seen in Fig. 8 a) where the SSs of the GeSn-TFETs are improved when the device mesa area, and hence I_{OFF}, is reduced. The SS of the Ge-reference is less affected by an area reduction as I_{OFF} is already close to J_P. Although the leakage currents of GeSn TFETs are considerably higher than those of the Ge TFET, Fig. 8 a) shows that the SSs of the smallest sized transistors are roughly equal for all samples. The degradation of the SS due to bulk defects can be removed by subtracting the J_A-component of I_{OFF} from I_{DS}, $I_{SUB} = I_{DS} - J_A \times A$. The technological equivalent of this operation would be device dimension downscaling as discussed earlier on. Calculating the SS from the I_{SUB} characteristics, we see an improvement of SS with increasing Sn-content. For $V_{DS} = -1$ V we get SS = 624 ± 43 mV/dec for the Ge reference TFET, SS = 505 ± 20 mV/dec for the 2 % Sn TFET and SS = 494 ± 25 mV/dec for 4 % Sn TFET. Simulations have predicted an improvement of SS with increased Sn-content [10], due to the bandgap reduction and increased tunneling probability. The SS as a function of Sn-content is therefore determined by the tradeoff between increased tunneling probability, which makes the band pass switching mechanism more effective, and increased leakage currents. Fig. 7 c) reveals that the TFETs holds a thermal subthreshold region current, with a slow decrease of activation energy for increasing negative V_G. This indicates TAT involving SRH processes, which degrade the SS, as they reduce the energy filtering which is a precondition for steep SS in TFETs. The combination of reducing I_{OFF} and quenching the SRH processes through temperature reduction, leads to a considerable improvement of SS. This can be seen in Fig. 8 b), which shows SS as a function of temperature. A linear relationship between SS and temperature was found in the temperature range investigated. It is evident that the TAT processes involving SRH generation strongly affect SS. These results therefore highlight the difficulty of achieving sub-60mV/dec SS with materials with high defect densities. This aspect is also supported by simulations [34] and has been brought forward by others [4], [35].

Let us now discuss I_{ON}. We expect, that the main contribution of I_{ON} is indirect BTBT at the source-channel junction (often referred to as "point tunneling"), as opposed to in the inversion layer within the source layer for devices with a source-gate overlap (often referred to as "line tunneling") [36]. The devices presented here have a large source-gate overlap, but given the high dopant concentration in the source, the energy bands in the source region are largely unaffected when a voltage is applied to the gate. The BTBT transmission probability can be approximated by [23]:

$$T_{WKB} \approx \exp\left(-\frac{4\lambda\sqrt{2m^*}E_G^{3/2}}{3q\hbar(\Delta\Phi + E_G)}\right) \quad (4)$$

Here m^* is the effective mass, E_G the band gap energy and λ the spatial extent of the tunneling region. $\Delta\Phi$ is the V_G-dependent energy window of allowed tunneling transitions. The promise of using GeSn in TFETs is first and foremost through raising I_{ON}. This is achieved through lowering of E_G and reducing λ with respect to Ge in (4).

The averaged I_{ON}, defined here as $I_{DS}(V_G = -4 \text{ V})/w_g$, from all measured transistors is shown in Fig. 9 a) as a function of Sn-content in the GeSn-channel. The large gate bias needed to define I_{ON} is a consequence of the thick gate oxide needed to be certain of negligible gate oxide leakage. This gate bias is much higher than the ITRS requirement for supply voltage, which ideally should be equal to V_{DS}. Strategies for scaling the supply voltage for these TFETs would be increasing the gate oxide capacitance, through switching to higher permittivity gate oxides (e.g. HfO$_2$). Reducing the gate oxide thickness is also an option, but with the danger of increasing the gate oxide leakage. With respect to the Ge reference, I_{ON} is seen to increase by a factor ~ 2 and ~ 3 for the samples with 2 %- and 4 % Sn-content in the GeSn channel, respectively. TAT Models for TFETs have shown how traps degrade I_{OFF} and SS [16], [23]. The same models, show no influence of TAT for I_{ON}. If TAT contributed to I_{ON}, we would also expect it to be due to interface traps, as I_{ON} is a surface current. The perimeter leakage was shown to be roughly equal for the GeSn TFETs, which indicates that these samples have similar interface trap densities. TAT can therefore not explain the successive increase of I_{ON} when increasing the Sn-content from 2 % to 4 %. We therefore assume that the increase in I_{ON} can be attributed the lowering of E_G, and that the different trap densities of the samples play a minor role. It is seen that even with the incorporation of GeSn, I_{ON} is one order of magnitude below the ITRS requirement for a supply voltage of $V_{DD} = -0.82$ V (top axis of Fig. 9 a)). The trend suggests that an increase of Sn-content above 4 % would further increase

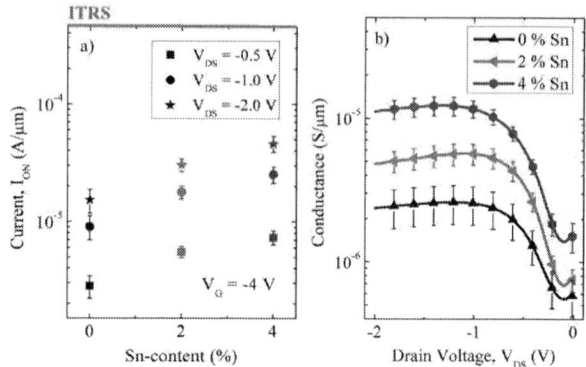

Fig. 9. a) I_{ON} as a function of Sn-content in the GeSn channel for different V_{DS}-biases. b) The averaged conductance of all measured transistors as a function of V_{DS}.

I_{ON}. However, this would seriously increase I_{OFF} which is already at an alarming level. The averaged conductance of all measured transistors as a function of V_{DS} is shown in Fig. 9 b). The conductance can be seen to increase for small V_{DS}, and saturate for $|V_{DS}| > 1$ V. When the valence band in the channel is lifted above the fermi level in the drain, holes accumulate in the channel. For a large gate bias and small V_{DS}-bias the channel potential is therefore pinned by the drain potential. The tunneling current at the source-channel junction is hence V_{DS} dependent for small V_{DS}. At high V_{DS} I_{ON} becomes limited by the saturating behavior of T_{WKB} in (4). I_{ON} can also be increased by reducing the channel length and effectively increasing the conductance and reducing the channel resistance [11].

B. GeSn-δ-layer

As reported for the first sample series, I_{ON} can be improved through incorporation of GeSn. However this comes at the expense of higher I_{OFF}. In order to benefit from higher I_{ON} by the incorporation of GeSn, but maintain manageable I_{OFF} the GeSn with 4% Sn-content was confined in a 10 nm δ-layer at the channel-source interface for the second sample series. The

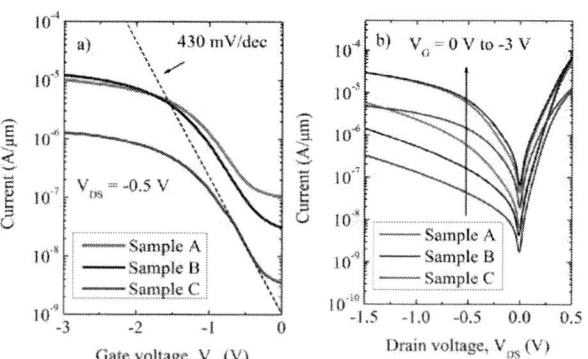

Fig. 10. a) Transfer characteristics of transistors from sample A, B and C with gate width w_g = 4 μm for V_{DS} = -0.5 V. The dashed black line indicate the minimum subthreshold slope (430 mV/dec) obtained for sample C. b) Output characteristics from transistors with gate width w_g = 8 μm.

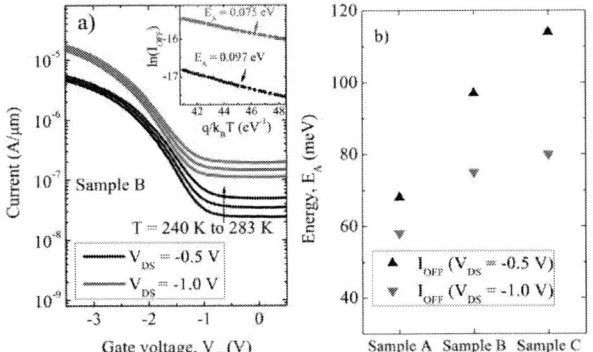

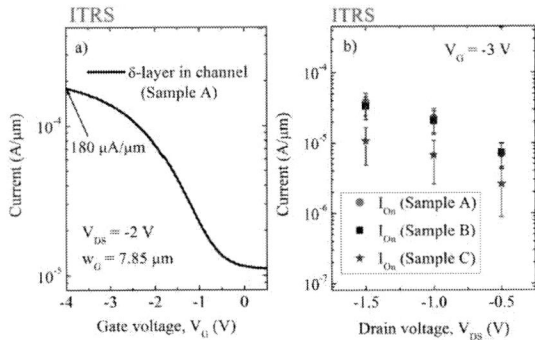

Fig. 13. a) Transfer characteristics from the transistor with the highest I_{ON}. b) I_{ON} as function of V_{DS} for the three samples. The samples with the $Ge_{0.96}Sn_{0.04}$-δ-layer partly or completely in the channel (Sample A and B) exhibit higher I_{ON} than the sample with the $Ge_{0.96}Sn_{0.04}$-δ-layer in the source (Sample C).

Fig. 11. Transfer characteristics showing the temperature dependency of a transistor from sample B. Inset shows Arrhenius plot of I_{OFF} for both V_{DS} voltages with corresponding activation energies of the fits indicated. b) The activation energy of I_{OFF} from all samples for $V_{DS} = -0.5$ V and $V_{DS} = -1$ V.

position of the δ-layer at the channel-source interface was shifted through the samples from completely inside the source to completely inside the channel, as shown in Fig. 1. In order to increase the conductance the total channel region thickness was reduced to 50 nm.

Fig. 10 a) and b) shows the transfer and output characteristics of TFETs from the three samples, respectively. The steepest subthreshold point slope ~ 430 mV/dec found for Sample C is indicated.

The temperature dependence of the transfer characteristics of a transistor from sample B is seen in Fig. 11 a), and in Fig. 11 b) the activation energies of I_{OFF} for the three samples are shown. The activation energies are lower than reported for the previous sample series. This is mostly due to the reduced channel thickness which leads to the effect often referred to as drain induced barrier thinning (DIBT) [37], and is a problem for devices with poor electrostatic control of the body as we have here. Due to DIBT the activation energy also reduces for increasing V_{DS}, as tunneling events dominate at high electrical fields.

In Fig. 12 the position of the $Ge_{0.96}Sn_{0.04}$-δ-layer can be seen to greatly affect I_{OFF}. The total thickness and Sn-content of the $Ge_{0.96}Sn_{0.04}$-δ-layer is the same for all devices, and one can expect the same total number of defects to be present in

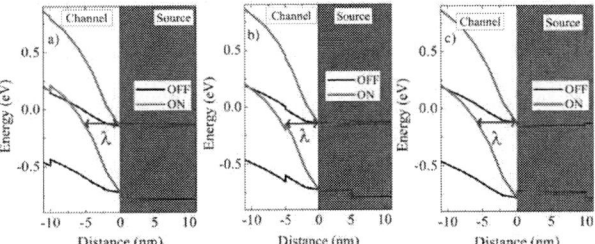

Fig. 14. Schematic band structure diagrams for the source-channel junction of the a) Sample A, b) Sample B, c) Sample C for the off-state (black) and on-state (red) of the TFET. The tunneling barrier λ is reduced when the GeSn is inside the channel.

each of the samples. Defects are, however, more likely to contribute to TAT leakage when they are positioned in the channel where the electrical field is higher. This also explains the reduced activation energy of I_{OFF} seen in Fig. 11 b) when shifting the $Ge_{0.96}Sn_{0.04}$-δ-layer from the source and into the channel.

In Fig. 13 a) the transfer characteristics of the transistor with the highest I_{ON}, $I_{ON} = 180$ μA/μm for $V_{DS} = -2$ V and $V_G = -4$ V, is shown. Although exhibiting a high I_{ON}, a high I_{OFF} results in a poor I_{ON}/I_{OFF}-ratio. In Fig. 13 b) I_{ON} is shown as a function of V_{DS} for the sample series. The effect of the position of the $Ge_{0.96}Sn_{0.04}$-δ-layer on I_{ON} can be explained by examining the band structure in each of the three cases.

Band offsets including strain dependent effects were calculated using model solid theory [38], and all model parameters, except for the band gaps, were obtained from linear interpolations of the model parameters for Ge and Sn. Quadratic interpolation according to [39] was used to calculate the band gap energies. A similar parameter set as that of [40] was used, but updated to include the newer experimental data of [39]. Band offsets between materials were approximated according to [41]. The largest calculated band offset between Ge and $Ge_{0.96}Sn_{0.04}$ is found in the valence band between the heavy hole (hh) bands, ~ 50 meV, while a conduction band offset was calculated to ~ 20 meV for the L-band, respectively. A band structure calculation of a Ge TFET

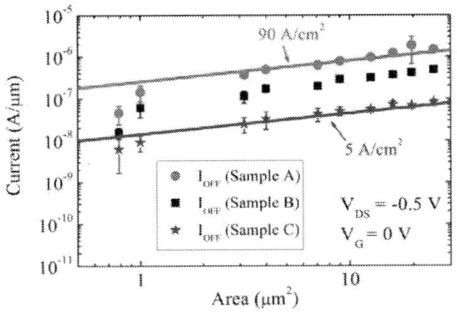

Fig. 12. I_{OFF} for $V_{DS} = -0.5$ V as a function of device area. The transistors show a device area dependent I_{OFF} indicated by the current densities (blue and red lines).

63

device was obtained using SILVACO Atlas [10] and the calculated band offsets between the hh-bands and between the L-bands of Ge and $Ge_{0.96}Sn_{0.04}$ were imposed onto these calculations for the on- and off-state of the transistor for the three different devices.

Fig. 14 shows the schematic band structure diagrams for the source-channel junction of the three device types. The behavior of I_{ON} can be understood qualitatively from those band diagrams. The main contribution to point tunneling will take place at the junction where the spatial extent of the tunneling barrier, λ, has its minimum value. Tunneling is enhanced if that region is within the low band gap material layer. Hence, I_{ON} are largely unchanged between the samples in which the $Ge_{0.96}Sn_{0.04}$-δ-layer is situated in the channel or across channel and source region, see Fig. 14. When the $Ge_{0.96}Sn_{0.04}$-δ-layer is shifted entirely into the source region I_{ON} degrades as tunneling mainly occurs within the Ge.

IV. CONCLUSION

Heterojunction GeSn TFETs could potentially be a means to realize Group-IV TFETs with high I_{ON}. However, in order to optimize overall device performance, it is important to understand how the introduction of the low-band gap material GeSn influences not only I_{ON} but also I_{OFF}. Here, we report that with increasing Sn-content in a GeSn-channel, I_{ON} can be effectively increased due to the lowering of the band gap. Furthermore it was found that, when limited to a 10 nm δ-layer, $Ge_{0.96}Sn_{0.04}$ is most beneficial for I_{ON} when positioned in the channel as opposed to in the source. This can be explained by the band offsets between Ge and $Ge_{0.96}Sn_{0.04}$, which is largest in the valence band. The highest I_{ON} are achieved in the sample with the $Ge_{0.96}Sn_{0.04}$-δ-layer completely in the channel with I_{ON} = 180 μA/μm for V_{DS} = -2.0 V and V_G = -4 V. The lowering of the band gap and the degradation of the epitaxial quality that comes with increasing Sn-content, heavily influences the I_{OFF} and SS of the devices through TAT. It is found that achieving the required I_{OFF} and SS < 60 mV/dec with GeSn is difficult. A possible strategy to boost I_{ON} consists of increasing the Sn content in the GeSn-δ-layer. However, the leakage current density has to be reduced to keep I_{OFF} manageable. This can in part be achieved by reducing the mesa volume and reducing the δ-layer thickness, but also through the investigation of alternative and optimized MBE growth strategies. Finally, our results can serve as input for calibrated device models that would enable the fine tuning of specific device and material parameters in order to boost device performance particularly at low voltages.

REFERENCES

[1] A. M. Ionescu and H. Riel, "Tunnel field-effect transistors as energy-efficient electronic switches," *Nature*, 2011.

[2] D. E. Nikonov and I. A. Young, "Overview of beyond-CMOS devices and a uniform methodology for their benchmarking," *Proceedings of the IEEE*, Vol. 101, Nr. 12, pp. 2498-2533, 2013.

[3] "International Technology Roadmap for Semiconductors," [Online]. Available: http://www.itrs.net/reports.html.

[4] H. Lu and A. Seabaugh, "Tunnel field-effect transistors: state-of-the-art," *IEEE Journal of the Electron Devices Society*, Vol. 2, Nr. 4, pp. 44 - 49, 2014.

[5] Q. T. Zhao, J. M. Hartmann and S. Mantl, "An improved Si tunnel field effect transistor with a buried strained $Si_{1-x}Ge_x$ source," *IEEE Electron Dev Let*, Vol. 32, Nr. 11, pp. 1480-1482, 2011.

[6] D. Hähnel et al., "Germanium vertical tunneling field-effect transistor," *Solid State Electronics*, Vol. 62, Nr. 1, pp. 132-137, 2011.

[7] I. Fischer et al., "Silicon tunneling field-effect transistors with tunneling in line with gate field," *IEEE Electron Device Letters*, Vol. 34, Nr. 3, pp. 154-156, 2013.

[8] M. Schmidt et al., "Line and point tunneling in scaled Si/SiGe heterostructure TFETs," *IEEE Electron Device Letters*, Vol. 35, Nr. 7, pp. 699-701, 2014.

[9] J. Schulze et al., "Vertical Ge and GeSn heterojunction gate-all-around tunneling field effect transistors," *Solid-State Electronics*, 2015.

[10] Y. Yang et al., "Germanium–Tin p-channel tunneling field-effect transistor: device design and technology demonstration," *IEEE Transactions on Electron Devices*, Vol. 60, Nr. 12, pp. 4048-4056, 2013.

[11] D. Hähnel, I. A. Fischer, A. Hornung, A. C. Köllner and J. Schulze, "Tuning the Ge(Sn) tunneling FET: influence of drain doping, short channel and Sn content," *Electron Devices, IEEE Transactions*, Vol. 62, Nr. 1, pp. 36 - 43, 2015.

[12] J. Kouvetakis, J. Menendez and A. Chizmeshya, "Tin-based group IV semiconductors: new platforms for opto- and microelectronics on silicon," *Annual Review of Materials Research*, Vol. 36, pp. 497-554, 2006.

[13] E. Kasper, J. Werner, M. Oehme, S. Escoubas, N. Burle and J. Schulze, "Growth of silicon based germanium tin alloys," *Thin Solid Films*, Vol. 520, pp. 3195-3200, 2012.

[14] K. Kostecki et al., "Virtual substrate technology for $Ge_{1-x}Sn_x$ heteroepitaxy on Si substrates," *ECS Trans.*, Vol. 64, Nr. 6, pp. 811-818, 2014.

[15] B. Vincent et al., "Undoped and in-situ B doped GeSn epitaxial growth on Ge by atmospheric pressure-chemical vapor deposition," *Applied Physics Letters*, Vol. 99, Nr. 15, p. 152103, 2011.

[16] U. Avci et al., "Study of TFET non-ideality effects for determination of geometry and defect density requirements for sub-60mV/dec Ge TFET," *IEDM 15*, pp. 34.5.1-34.5.4, 2015.

[17] C. D. Bessire, M. T. Björk, H. Schmid, A. Schenk, K. B. Reuter and H. Riel, "trap-assisted tunneling in Si-InAs nanowire heterojunction tunnel diodes," *Nano Letters*, Vol. vol. 11, pp. 4195-4199, 2011.

[18] W. Wang, Q. Zhou, Y. Dong, E. S. Tok and Y. C. Yeo, "Critical thickness for strain relaxation of Ge_{1-} xSn_x ($x\leq$ 0.17) grown by molecular beam epitaxy on Ge (001)," *Applied Physics Letters*, vol. 106, p. 232106, 2015

[19] D. R. Gajula, P. Baine, M. Modreanu, P. K. Hurley, B. M. Armstrong and D. W. McNeill, "Fermi level de-pinning of aluminium contacts to n-type germanium using thin atomic layer deposited layers," *Appl. Phys. Lett.*, no. 104, p. 012102, 2014

[20] M. Oehme, D. Buca, K. Kostecki, B. Holländer, E. Kasper and J. Schulze, "Epitaxial growth of highly compressively strained GeSn alloys up to 12.5% Sn," *Journal of Crystal Growth*, Vol. 384, p. 71–76, 2013.

[21] R. Zhang, J. Lin, M. Takenaka and S. Takagi, "Impact of plasma post oxidation temperature on interface trap density and roughness at GeOx/Ge interfaces," *Microelectronic Engineering*, no. 109, pp. 97-100, 2013.

[22] L. M. Giovane, H. C. Luan, A. M. Agarwal and L. C. Kimerling, "Correlation between leakage current density and threading dislocation density in SiGe pin diodes grown on relaxed graded buffer layers." *Applied Physics Letters*, 78(4), 541-543, 2001.

[23] S. M. Sze and K. N. Kwok, Physics of Semiconductor Devices, 3. Editon, New Jersey: John Wiley & Sons, 2007.

[24] G. A. M. Hurkx, D. B. M. Klaassen and M. P. G. Knuvers, "A new recombination model for device simulation including tunneling," *IEEE*

Transactions on Electron Devices, Vol. 39, Nr. 2, pp. 331-338, 1992.

[25] E. Simoen, F. De Stefano, G. Eneman, B. De Jaeger, C. Claeys and F. Crupi "On the Temperature and Field Dependence of Trap-Assisted Tunneling Current in Ge Junctions." *IEEE Electron Device Letters,* Vol. 30, Nr. 5, pp. 562-564, 2009.

[26] D. Lei et al., "Ge0.83Sn0.17 p-channel metal-oxide-semiconductor field-effect transistors: Impact of sulfur passivation on gate stack quality," *J. Appl. Phys.,* Vol. 119, Nr. 2, pp. 024502, 2016.

[27] C. Schulte-Braucks et al., "Low temperature deposition of high-k/metal gate stacks on high-Sn content (Si) GeSn-alloys." *ACS applied materials & interfaces,* 2016.

[28] P. Guo et al., "Ge0. 97Sn0. 03 p-channel metal-oxide-semiconductor field-effect transistors: Impact of Si surface passivation layer thickness and post metal annealing" *J. Appl. Phys.,* Vol. 114 Nr. 4, pp. 044510 2013

[29] S. Sioncke et al., "S-passivation of the Ge gate stack: Tuning the gate stack properties by changing the atomic layer deposition oxidant precursor," *J. Appl. Phys.,* Vol. 110, pp. 084907, 2011.

[30] T. Y. Chan, A. T. Wu, P. K. Ko and C. Hu, "Effects of the gate-to-drain/source overlap on MOSFET characteristics," *IEEE Electron Device Letters,* Vol. 8, Nr. 7, pp. 326-328, 1987.

[31] "Ioffe Physico Technical Institute," [Online]. Available: http://www.ioffe.ru/SVA/NSM/Semicond/Ge/bandstr.html.

[32] S. Gupta et al., "GeSn technology: extending the Ge electronics roadmap," *IEEE International Electron Devices Meeting (IEDM),* pp. 16.6.1-16.6.4 , 2011.

[33] D. Sarkar et al., "A subthermionic tunnel field-effect transistor with an atomically thin channel," *Nature,* vol. 526, pp. 91-95, 2015.

[34] M. Pala and D. Esseni, "Interface traps in InAs nanowire tunnel-FETs and MOSFETs—Part I: model description and single trap analysis in

tunnel-FETs," *IEEE Transactions on Electron Devices,* Vol. 60, Nr. 9, pp. 2795 - 2801 , 2013.

[35] S. Agarwal and E. Yablonovitch, "The low voltage TFET demands higher perfection than previously required in electronics," *73rd Annual Device Research Conference,* pp. 247 - 248 , 2015.

[36] W. G. Vandenberghe, A. S. Verhulst, G. Groeseneken, B. Soree and W. Magnus, "Analytical model for point and line tunneling in a tunnel field-effect transistor," *Proc. SISPAD ,* pp. 137-140, 2008.

[37] L. Liu, D. Mohata and S. Datta, "Scaling length theory of double-gate interband tunnel field-effect transistors" *IEEE Transactions on Electron Devices,* Vol.59, Nr. 4, pp. 902-908

[38] C. G. Van de Walle, "Band lineups and deformation potentials in the model-solid theory," *Physical Review B,* Vol. 39, Nr. 3, pp. 1871-1883, 1989.

[39] L. Jiang et al., "Compositional dependence of the direct and indirect band gaps in Ge_{1-y} Sn_y alloys from room temperature photoluminescence: implications for the indirect to direct gap crossover in intrinsic and n-type materials," *Semiconductor Science and Technology,* Vol. 29, 2014.

[40] G.-E. Chang, S.-W. Chang and C. S. L., "Strain-balanced Ge_zSn_{1-z}–$Si_xGe_ySn_{1-x-y}$ multiple-quantum-well lasers," *IEEE Journal of quantum electronics,* Vol. 46, Nr. 12, pp. 1813-1820, 2010.

[41] M. Jaros, "Simple analytic model for heterojunction band offsets," *Physical Review B,* Vol. 37, p. 7112, 1988.

[42] SILVACO ATLAS, SILVACO Inc., Santa Clara, 2012.

Band-structure of ultra-thin InGaAs channels: Impact of biaxial strain and thickness scaling

Sabina Krivec, Mirko Poljak and Tomislav Suligoj

Micro and Nano Electronics Laboratory
Department of Electronics, Microelectronics, Computer and Intelligent Systems
Faculty of Electrical Engineering and Computing, University of Zagreb, HR-10000 Zagreb, Croatia
E-mail: {sabina.krivec, mirko.poljak, tomislav.suligoj}@fer.hr

Abstract—The band-structure of ultra-thin InGaAs layers is calculated using a nearest neighbor sp3d5s* tight binding approach to assess the impact of compressive and tensile biaxial strain on effective in-plane masses, non-parabolicity factor α, and conduction band minimum (CBM) shift down to channel thicknesses of 4 nm. The reported results show that the effective mass increases with body thickness decrease, whereas it decreases with the strain increase from compressive to tensile. Furthermore, the difference between the position of pinned Fermi level and CBM increases with strain. The impact of band-structure effects on electron transport is demonstrated for the InGaAs-OI structure. The extracted band-structure parameters provide electron mobility results consistent with experiments. Our calculations make it possible to assess the electron mobility in a wide range of both compressive and tensile strain values and body thicknesses from 15 nm down to 4 nm.

Keywords—*InGaAs; tight-binding model; strain ; Fermi level pinning; interface charge; ultra-thin body; electron mobility*

I. INTRODUCTION

Due to their outstanding carrier transport properties, III-V materials are attractive as alternative channel materials for future post-silicon CMOS applications. InGaAs is considered as the most promising III-V material for *n*-channel MOSFETs owning to its high electron mobility (μ_e) and acceptable bandgap [1]. The high μ_e is attributed to the light effective mass and it is more than 10 times higher than in silicon at the comparable sheet density [1]. The short channel effects in III-V devices are alleviated by employing ultra-thin body structures and non-planar architectures [2]. One of the main challenges limiting the development of III-V MOSFETs is the high density of interface states (D_{it}) at the III-V/oxide interface, which results in a strong Fermi level pinning (FLP) and lower inversion carrier densities [3]. Fermi level pinning can be mitigated by introducing strain in InGaAs channels, i.e. the presence of tensile strain causes an increase of energy difference between the conduction band minimum (CBM) and the FLP energy, larger inversion carrier densities (N_{inv}), and higher on-current (I_{ON}) values [4]. Recent experimental data confirms the beneficial effect of tensile strain on the

electron mobility [5]. In order to perform electron mobility simulations in the presence of arbitrary strain, it is important to examine first the impact of strain on the band structure parameters. In this work, we develop an atomistic band structure simulator to obtain the effective masses (m_{eff}), band gap energies (E_G) and non-parabolicity (NP) factors (α) for unstrained InGaAs layers, as well as for structures exposed to the arbitrary in-plane biaxial strain. The objective is to examine the impact of body thickness (T_B) scaling on effective masses, band gap energies and non-parabolicity factors to obtain the electron mobilities for the strained extremely-thin InGaAs channels.

II. BAND-STRUCTURE MODELING

A. Nearest neighbor sp3d5s* tight-binding model

Band-structure calculations are done for Γ-valley electrons using nearest-neighbor (NN) sp3d5s* tight-binding (TB) approach. According to our previous results on valley occupation [6], it is sufficient to include only the Γ-valley electrons in the calculations. The tight binding Hamiltonian matrix is derived under the following assumptions: localized basis functions have atomic orbitals symmetry (atom-like orbitals); overlap of two orbitals on different atomic site is zero; overlap of two different orbitals located on the same atomic site is zero (orthogonality); maximum relative distance between atoms where the orbitals are located is limited to NN; nonzero matrix element is possible only when the potential is on one of the two atoms where the orbitals are located [7]. The inclusion of higher orbitals within the model (i.e. d and s^* orbitals) is important to correctly reproduce the band-structure dispersion, which is in turn needed for accurate calculation of subband energies and effective masses. If spin orbit coupling is not included in model, each atom is represented by 10 orbitals (s, p_x, p_y, p_z, d_{xy}, d_{yz}, d_{zx}, d_{x2-y2}, d_{z2-r2}). Since the focus of this paper is on the properties of electron mobility in III-V materials, it is not necessary to include the spin-orbit coupling as we are not interested in the features of the top of the valence band.

InGaAs has the zinc-blende crystal structure, which consists of two face-centered-cubic (FCC) lattices misplaced by a quarter of the main diagonal along the main diagonal direction. Each of these two lattices is

---

This work is sponsored by the Croatian Science Foundation under contract No. 9006 - project "High-performance Semiconductor Devices for Wireless Circuit and Optical Detection Applications" (HIPERSEMI)

composed entirely of one species of atoms which are identified as anions or cations. Therefore, the unit cell of a zinc-blende structure consists of two atoms (anion/cation), where each anion has four bonds and is connected to four nearest neighbor cations and vice versa. The matrix representing the unit cell is a 10-by-10 matrix expressed as:

$$\left[H_{unit\_cell}\right] = \begin{bmatrix} H_{aa} & H_{ac} \\ H_{ca} & H_{cc} \end{bmatrix}, \tag{1}$$

where the diagonal blocks H_{aa} and H_{cc} contain the on-site energies at the main diagonal, and the off-diagonal overlap blocks H_{ac} and H_{ca} describe the coupling between anion and cation, and cation and anion, respectively. The matrix elements between orbitals of cation and anion atoms are given by:

$$H^{ac}(\vec{k}) = \left\langle a\vec{k} \,|\, H \,|\, c\vec{k} \right\rangle = g_i(\vec{k})V^{ac}, \tag{2}$$

where V^{ac} are the overlap energies and $g_i(\vec{k})$ takes one of the following forms depending on the relative position of neighboring cations with respect to the anion and vice versa:

$$4g_0 = e^{i\vec{k}x_0} + e^{i\vec{k}x_1} + e^{i\vec{k}x_2} + e^{i\vec{k}x_3} \tag{3.1}$$

$$4g_1 = e^{i\vec{k}x_0} + e^{i\vec{k}x_1} - e^{i\vec{k}x_2} - e^{i\vec{k}x_3} \tag{3.2}$$

$$4g_2 = e^{i\vec{k}x_0} - e^{i\vec{k}x_1} + e^{i\vec{k}x_2} - e^{i\vec{k}x_3} \tag{3.3}$$

$$4g_3 = e^{i\vec{k}x_0} - e^{i\vec{k}x_1} - e^{i\vec{k}x_2} + e^{i\vec{k}x_3} \tag{3.4}$$

where x_0, x_1, x_2, x_3 are positions of cations relative to the anion and vice versa. The tight-binding Hamiltonian must be Hermitian, i.e.

$$\left[H_{ca}\right] = \left[H_{ac}\right]. \tag{4}$$

To calculate the band-structure for bulk devices, tight-binding Hamiltonian is derived by assuming the infinite lattice periodicity along all three orthogonal spatial axes. For ultra-thin InGaAs layers, the band structure Hamiltonian is discretized along the finite dimensional direction. In case of an ultra-thin InGaAs channel, we consider the [001] as a confinement direction, while [100] and [010] are the transport and width directions, respectively. Since the dimension along the transport and width directions are long enough compared to the channel thickness, infinite periodicity is assumed along these axes. The on-site orbital energies and the overlap integrals between atomic orbitals for different bond types are taken from [8], where these TB parameters are calibrated using genetic algorithm to match the bulk band structure over the entire Brillouin zone at room temperature. For a more precise calculation of E_G and m_{eff} of $In_xGa_{(1-x)}As$ as a function of x, bowing parameters have been introduced [9]. The passivation of the dangling bonds is achieved by increasing the on-site energies of surface atoms by 30 eV which leads to nearly hard-wall boundaries and negligible wave function penetration. Finally, the band dispersion can be calculated by solving

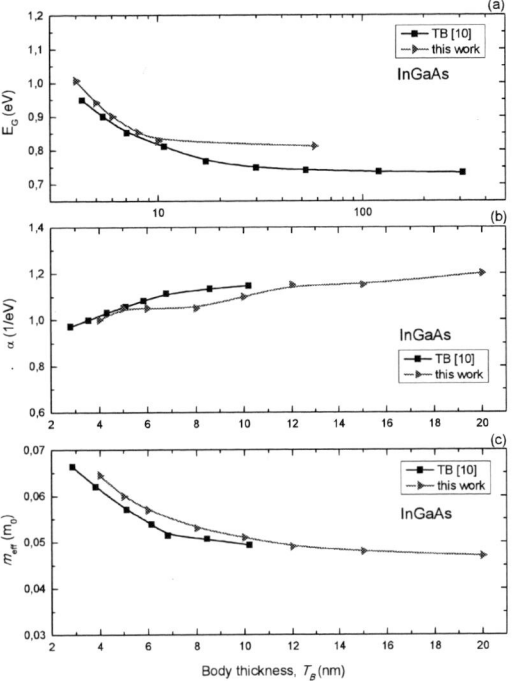

Fig. 1. (a) Band gap, **(b)** non-parabolicity factor α, and **(c)** effective in-plane mass as a function of body thickness in non-strained UTB InGaAs layers. Band structure calculations are validated by comparison to the results from [10].

the eigenvalue problem of the Hamiltonian for the k-values of interest as:

$$E = eig\left[H(k_x, k_y, k_z)\right], \tag{5}$$

From the derived band-structure dispersion, the m_{eff} and NP factor α are extracted by fitting the energy dispersion near the Γ point of the lowest subband with non-parabolic effective mass approximation energy (NP-EMA) curve: $E(k)\cdot[1 + \alpha E(k)] = \hbar^2 k^2 / 2m_{eff}$. As a result of the anisotropy of the bands, α is extracted along the in-plane transport direction [100]. The band-structure calculations for in-plane effective mass, band gap and NP factor for non-strained structures are validated by comparison to the results from [10] and shown in Fig. 1 as a function of body thickness. The effective mass increases with body thickness decrease, which is in agreement with literature [11].

B. The Tight-binding Hamiltonian with strain

For a relaxed crystal, the directional cosines for the anion-to-cation bond orientation are equal and can be written as:

$$l = \frac{1}{\sqrt{3}}, \qquad m = \frac{1}{\sqrt{3}}, \qquad n = \frac{1}{\sqrt{3}}. \tag{6}$$

However, for a strained crystal the relative position of the neighboring atoms are changed, where both bond-angle and bond-length between the nearest neighbors are modified. The new directional cosines calculated from the deformed crystal structure automatically incorporate the overlap matrix elements modification due to change in the bond angle. For the biaxial in-plane strain produced by lattice mismatch between $In_xGa_{1-x}As$ buffer layer and $In_{0.53}Ga_{0.47}As$ active layer on a (001) substrate, we define the strain tensor components:

$$\varepsilon_{xx} = \varepsilon_{yy} = \varepsilon_{\parallel} \qquad (4.1)$$

$$\varepsilon_{zz} = \frac{-2C_{12}}{C_{11}}\varepsilon_{\parallel} \qquad (7.2)$$

$$\varepsilon_{xy} = \varepsilon_{yz} = \varepsilon_{zx} = 0 \qquad (7.3)$$

where C_{12} and C_{11} are elastic stiffness constants for InGaAs taken from [12]. This strain tensor is valid for arbitrary strain and it can be decomposed into separate tensors, where the following shear tensor:

$$\varepsilon_{001} = \frac{1}{3}\begin{pmatrix} \varepsilon_{xx} - \varepsilon_{zz} & 0 & 0 \\ 0 & \varepsilon_{xx} - \varepsilon_{zz} & 0 \\ 0 & 0 & -2(\varepsilon_{xx} - \varepsilon_{zz}) \end{pmatrix} \qquad (5)$$

lowers the crystal symmetry by shortening (biaxial tensile strain) or elongating (biaxial compressive strain) the z-direction lattice spacing with respect to the x or y directions. The bond length modification in the presence of strain is introduced through modification of overlap integrals via the generalized Harisson's d^2 scaling law:

$$U = U_0 \left(\frac{d_0}{d}\right)^2 \qquad (6)$$

where d_0 and d are ideal and actual (strained) bond lengths, respectively. Fig. 2(a) and (b) show the energy dispersion for the lowest subband energy at the Γ point for compressive and tensile biaxially strained 15 nm-thick InGaAs channel, respectively. The confinement direction is along [001], while X and L direction correspond to [100] and [010], respectively. The shift of the lowest subband energy caused by compressive strain ranging from -2% to unstrained with a step of 0.5% is reported in Fig. 2(a). The arrow indicates the absolute strain value increase. The inset of Fig. 2(a) shows the enlarged area in the vicinity of the Γ point. As compressive strain increases, the CBM also increases against the CBM for unstrained channel. On the other hand, for tensile strain increase we report the CBM decrease as shown in Fig. 2(b). The CBM shift for both compressive and tensile strain values are listed in Table I. Here, the ΔCBM is difference in CBM with and without strain. When both the tensile and compressive absolute strain value increases, the CBM shift increases. which is later assessed within the mobility simulations. In Fig. 3(a) and (b), the effective mass and NP factor α are plotted as a function of strain, respectively, for $T_B = 4$ nm and $T_B = 15$ nm. For both body thicknesses, effective mass decreases with strain increase from -2% to 2%. Furthermore, as the strain value increases from compressive to tensile the non-parabolicity factor α

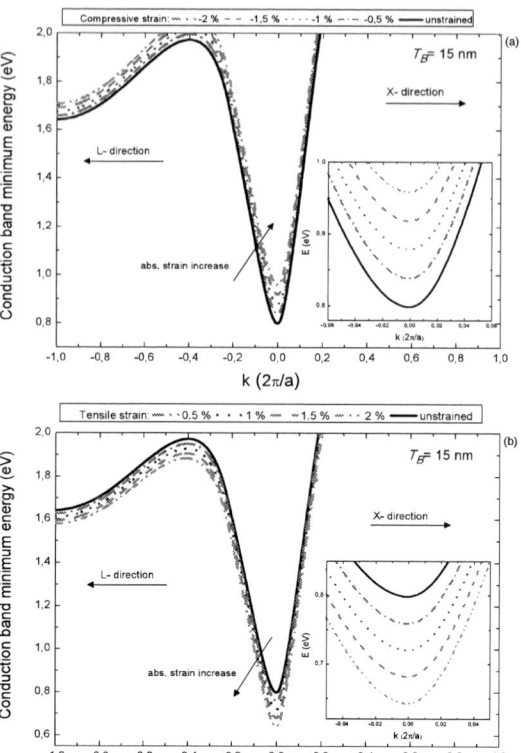

Fig. 2. Energy dispersion for the lowest subband energy near the Γ point for: **(a)** compressive and **(b)** tensile biaxially strained 15 nm-thick InGaAs channel. The strain values are ranging from absolute values of 2% to unstrained device. Quantization is along [001] direction, while X and L direction correspond to [100] and [010], respectively. Inset in figure (a) and (b) shows enlarged area near the Γ point for better understanding of conduction band minimum shift due to compressive/tensile strain.

Table I. Conduction band minimum shift due to compressive and tensile strain reported from band structure calculations in vicinity of Γ point ranging from −2 % to 2 % with step of 0.5 % for T_B= 15 nm. The ΔCBM is difference in CBM with and without strain.

	Strain	ΔCBM (meV)
Compressive	−2%	160.4
	−1.5 %	119.9
	−1 %	79.6
	−0.5 %	39.8
Tensile	0.5 %	−39.4
	1 %	−78.7
	1.5 %	−117.5
	2 %	−156.1

increases. As expected, NP factor becomes smaller when the body thickness decreases, indicating that band-structure dispersion near the Γ point is more parabolic. Due to a more parabolic and wider curve of the energy dispersion, the effective mass for thinner channel is

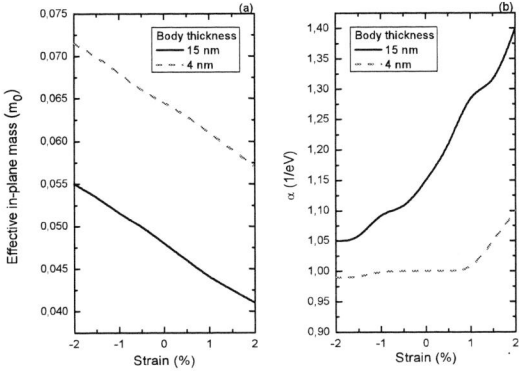

Fig. 3. (a) Effective in-plane mass and (b) non-parabolicity factor α as a function of biaxial strain (from compressive $\varepsilon = -2\%$ to tensile strain $\varepsilon = 2\%$) for $T_B = 4$ nm and $T_B = 15$ nm.

enlarged. This also applies for band-structure narrowing and widening due to compressive and tensile strain, respectively.

The extracted non-parabolicity factors and effective in-plane masses for body thicknesses from 20 nm down to 4 nm are plotted in Fig. 4, respectively, for biaxial strain values varying from compressive ($\varepsilon = -2\%$) to tensile ($\varepsilon = 2\%$) with a step of 0.5%. As strain value increases from unstrained to 2% of tensile strain, the effective mass decreases, whereas with the increase from unstrained to compressive strain of -2%, the m_{eff} increases. For each of the examined strain values, the NP factor increases with body thickness increase. The irregularities observed for the non-parabolicity factor curve are due to an approximate assessment when fitting the NP-EMA curve to the exact calculated band-structure.

III. IMPACT OF BANDSTRUCTURE EFFECTS ON ELECTRON MOBILITY IN INGAAS-OI STRUCTURE

In this section, we perform mobility calculation in order to demonstrate the impact of compressive and tensile strain on the electron mobility in ultra-thin InGaAs-OI channels. The calibration of the mobility model was carried out on experimental data for 15 nm-thick biaxially strained InGaAs-OI structure with 10 nm-thick Al_2O_3 as gate and Al_2O_3 as buried oxide [5]. Schematic figure of the examined structure is shown in Fig. 5, while the material and scattering parameters used in the simulations are listed in Table II. Detailed description of the mobility model can be found in [13, 14]. The values of strain whose impact was investigated in [5] are 0.4%, 1.2% and 1.7%, whereas the objective of our work is to examine the impact of arbitrary strain values on the electron mobility, including those not covered by the reported experimental data. Therefore, we plot total mobility enhancement due to biaxial strain ranging from compressive values of -2% to tensile strain values of 2% with a step of 0.5% in Fig. 6. Mobility calculations are done in strong inversion ($N_{inv} = 8 \times 10^{12}$ cm^{-2}). In the case of tensile strain, the

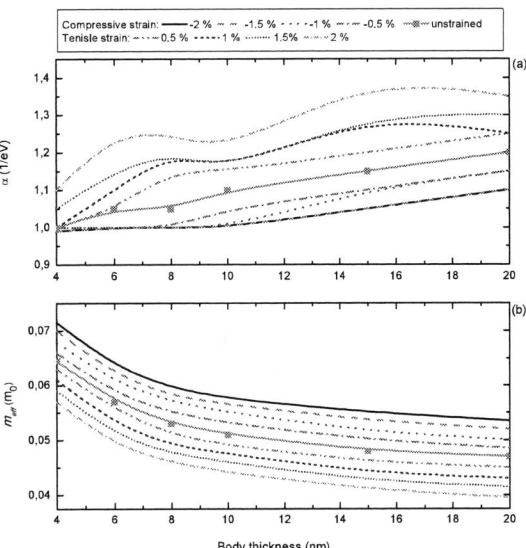

Fig. 4. (a) Non-parabolicity factor α and (b) effective in-plane mass (m_{eff}) for body thicknesses from 20 nm down to 4nm for strain values ranging from compressive (-2%) to tensile (2%) strain values. Curve for unstrained device is additionally marked with cubic symbol.

Fig. 5. Schematic illustration of the examined InGaAs-OI structure. Structure has strained 15 nm-thick InGaAs channel with Al_2O_3 as the gate and buried oxide. $T_{ox} = 10$ nm, $T_{BOX} = 30$ nm.

Table II. Material and scattering parameters used in self-consistent Schrödinger-Poisson and MRTA mobility simulations.

15 nm-thick $In_{0.53}Ga_{0.47}As$ channel			
	ε (%)	$m_l = m_t (m_0)$	NP factor α
Γ valley	0	0.048	1.15
	0.4	0.0465	1.2
	1.2	0.0435	1.3
	1.7	0.042	1.3
Bandstructure and Scattering Parameters			
$D_{AP,\Gamma} = 7$ eV $E_{OP} = 32$ meV, $D_{OP} = 10\times10^8$ eV/cm $E_{POP} = 32$ meV, $\varepsilon_0 = 13.94$, $\varepsilon_\infty = 11.64$ $v_{s,T} = 2974$ m/s, $v_{s,L} = 4253$ m/s, $\rho = 5506$ kg/m^3 $a = 5.868$ Å, $V_0 = 0.5$ eV, $N_{imp} = 3\times10^{16}$ cm^{-3} $D_{it} = D_{it}(E_0)\exp[-|(E-E_0)/E_S|])$, $D_{it}(E_0) = 10^{16}$ eV^{-1}cm^{-2}; $E_0 =$ 0.28 eV above CBM; $E_S = 20$ meV			
SOP-related parameters (Al_2O_3)			
ε_0		12.53	
ε_{int}		7.27	
ε_∞		3.20	
$\hbar\omega_{TO1}$ (meV)		48.18	
$\hbar\omega_{TO2}$ (meV)		71.41	
Al_2O_3-InGaAs Interface Parameters			
(100)	$L = 1$ nm, $\Delta = 0.275$ nm	$N_{int} = 1.35\times10^{12}$ cm^{-2}	

69

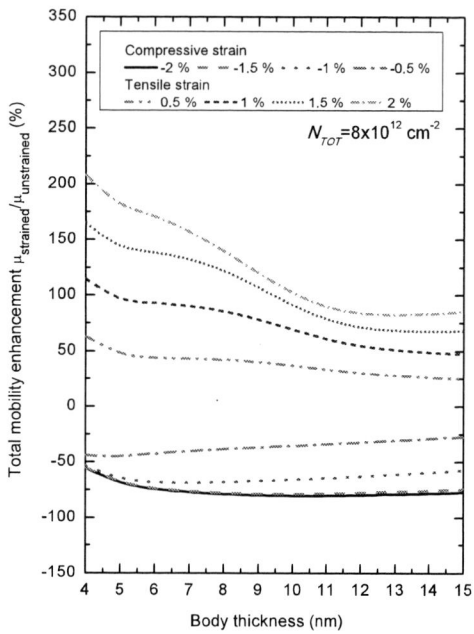

Fig. 6. Total mobility enhancement with biaxial strain as a function of body thickness for strain values from −2% (compressive) to 2% (tensile) with step of 0.5%. Strong inversion, $N_{TOT} = 8 \times 10^{12}$ cm^{-2}. $T_B = 15$ nm, $T_{ox} = 10$ nm, $T_{BOX} = 30$ nm. Al$_2$O$_3$ is GOX and BOX material.

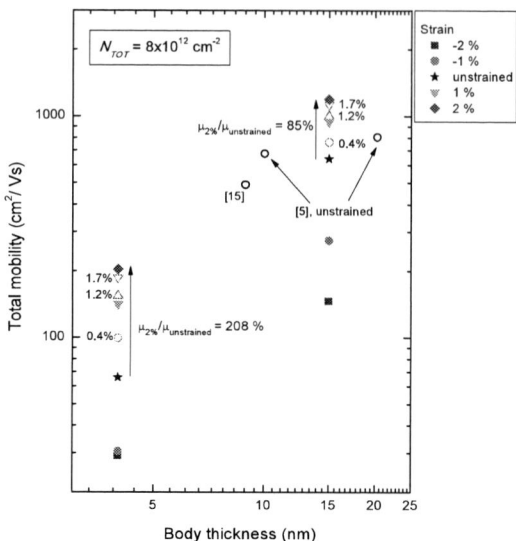

Fig. 7. Comparison of total mobility for arbitrary strain values from compressive to tensile strain values simulated in this work with experimental data. Experimental data for unstrained 10 and 20 nm-thick InGaAs channel is taken from [5]. Exp. data for 9 nm-thick channel is taken from [15], while data for 0.4%, 1.2% and 1.7% strained 15 nm-thick channel is taken from [5].

mobility enhancement increases with strain increase for all the examined T_B values. Moreover, the enhancement also increases when the body thickness is scaled down. Therefore, for the highest tensile strain value of 2% and for the thinnest body of $T_B = 4$ nm, we report the electron mobility enhancement of 208%, while the enhancement for thicker channels reaches up to 85%. When the compressive strain value increases to 2%, the mobility degrades by −77% in the case of the thickest channel ($T_B = 15$ nm). With the body thickness decrease, mobility deterioration due to compressive strain increases to −54% for $\varepsilon = -2\%$ and $T_B = 4$ nm. An improved electron mobility originates from D_{it} energy profile shift relative to the CBM in the presence of strain. According to the obtained CBM shift for tensile strain reported in Table I., the lowering of the CBM relative to the pinned Fermi energy level inside the conduction band results in an increase of energy difference between the CBM and the FLP energy, larger free inversion charge density and weaker impact of interface states charge on μ_e. On the other hand, for the compressive strain increase, the CBM is getting closer to the FLP energy, which has the opposite effect in comparison to the tensile strain. Although the m_{eff} modification due to strain does not contribute to mobility improvement as much as CBM shift does, the m_{eff} increase due to compressive strain demonstrated in Fig. 4(b) additionally reduces the electron mobility. Comparison between the experimental mobility data for unstrained and strained structures [5, 15] and total mobility calculated in this work is reported in Fig. 7. The total mobility enhancement for the

highest examined tensile strain value for $T_B = 4$ nm is larger than the enhancement obtained for $T_B = 15$ nm. The degradation of mobility in the case of thinner InGaAs channels is nearly the same for compressive strain values of −1% and −2%, whereas the mobility in thicker channels for $\varepsilon = -1\%$ is higher by nearly two orders of magnitude when compared to the case when $\varepsilon = -2\%$. As expected, the electron mobility is highest for tensile strain value $\varepsilon = 2\%$ for both body thicknesses reported in Fig. 7.

IV. CONCLUSION

We reported the impact of biaxial strain on several electronic and transport properties of ultra-thin InGaAs channels. We assessed the influence of body thickness downscaling on the effective in-plane mass, non-parabolicity factor and conduction band energy shift in the presence of strain that ranges from compressive ($\varepsilon = -2\%$) to tensile ($\varepsilon = 2\%$). The band-structure calculation were performed using the nearest neighbor sp3d5s* tight-binding approach. The results of this study showed that m_{eff} becomes larger with body thickness downscaling, whereas the effective mass decreases when strain increases from compressive to tensile, from $0.0715m_0$ to $0.057m_0$, respectively, for $T_B = 4$ nm. The conduction band minimum increases with an increasing compressive strain, while for tensile strain increase the CBM decreases with respect to the unstrained case.

Due to CBM shift relative to the Fermi level pinned energy, we report mobility enhancement and degradation in case of tensile and compressive biaxial strain, respectively. The mobility enhancement is the highest for the highest tensile strain value ($\varepsilon = 2\%$) and thinnest channel ($T_B = 4$ nm) with a mobility increase of 208%. Similarly, for the highest compressive strain value of -2% we obtain degradation up to -77% for the thickest channel ($T_B = 15$ nm). The reported electron mobility results for $\varepsilon = 2\%$ are higher for 6% and 9% than the highest ones demonstrated in experiments for both thinner ($T_B = 4$ nm) and thicker devices ($T_B = 15$ nm), respectively. This is due to larger energy difference between CBM and FLP energy level reported for higher tensile strain values. Using these band-structure results, we can assess the mobility in extremely-thin InGaAs channels with thicknesses down to 4 nm in presence of arbitrary biaxial strain.

REFERENCES

[1] S. Takagi *et al.*, "Carrier-transport-enhanced channel CMOS for improved power consumption and performance", IEEE Trans. Electron Devices 55, pp. 21, 2008.

[2] M. Yokoyama *et al.*, "Thin body III-V-semiconductor-on-insulator MOSFETs on Si fabricated using direct wafer bonding", Appl. Phys. Exp. 2, pp. 124501, 2009.

[3] N. Taoka *et al.*, "Impact of Fermi level Pinning inside conduction band on electron mobility of In$_x$Ga$_{1-x}$As MOSFETs and mobility enhancement by pinning modulation", Tech. Digest of International Electron Device Meeting (IEDM) 2011, pp. 27.2.1

[4] S. Kim *et al.*, "Strained In$_{0.53}$Ga$_{0.47}$As metal-oxide-semiconductor field-effect transistor with epitaxial based biaxial strain", Appl. Phys. Exp. 100, pp. 193510, 2012.

[5] Kim *et al.*, "Biaxially strained extremely-thin body In$_{0.53}$Ga$_{0.47}$As-on- insulator metal-oxide-semiconductor field-effect transistor on Si substrate and physical understanding on their electron mobility", J. Appl. Phys. 114, pp. 164512-1, 2013.

[6] S. Krivec *et al.*, "Electron mobility in ultra-thin InGaAs channels: Impact of surface orientation and different gate oxide materials", Solid State Electronics 115, pp.109-119, 2016.

[7] J.C. Slater and G.F. Koster, "Simplified LCAO Method for the Periodic Potential Problem", Phys. Rev. 94, pp.1498, 1954.

[8] T. B. Boykin *et al.*, "Diagonal parameter shift due to nearest-neighbor displacements in empirical tight-binding theory", Phys. Rev. B 66, pp.125207, 2002.

[9] M. Luisier *et al.*, "Investigation of InGaAs ultra thin body tunneling FETs using full band and atomistic approach", Proc SISPAD 2009., pp 1-4

[10] G. Zerveas *et al.*, "Comprehensive comparison and experimental validation of band-structure calculation methods in III-V semiconductor quantum wells", Solid State Electronics 115, pp.92-102, 2016.

[11] Y. Liu *et al.*, "Band-structure effects on the performance of III–V ultrathin-body SOI MOSFETs", IEEE Trans. Electron Devices 55, No. 5, pp. 1116–1122, 2008.

[12] Electronic archive of the Ioffe Physical-Technical Institute. http:// www.ioffe.ru/SVA/NSM/ Semicond.

[13] M. Poljak *et al.*, "Assessment of electron mobility in ultra-thin body InGaAs-on-insulator MOSFETs using physics-based modeling", IEEE Trans. Electron. Dev 59, pp.1636-1643, 2012.

[14] S. Krivec *et al.*, "Strained-induced increase of electron mobility in ultra-thin InGaAs-OI MOS transistors", Proc. EuroSOI-ULIS 2017., pp. 1-4

[15] M. Yokoyama *et al.*, "Extremely-thin-body InGaAs-on-Insulator MOSFETs on Si fabricated by direct wafer bonding", Tech. Dig. - Int. Electron devices Meet. 4, 2010.

Perimeter effects from interfaces in ultra-thin layers deposited on nanometer-deep p$^+$n silicon junctions

Tihomir Knežević[1*], Lis K. Nanver[2] and Tomislav Suligoj[1]

[1]University of Zagreb, Faculty of Electrical Engineering and Computing, Micro and Nano Electronics Laboratory, Croatia

[2]University of Twente, Faculty of Electrical Engineering Mathematics & Computer Science, Enschede, The Netherlands

tihomir.knezevic@fer.hr

Abstract - **Interface states at metal-semiconductor or semiconductor-semiconductor interfaces in ultra-thin layers deposited on nanometer-deep p$^+$nsilicon junctions that are contacted by metal, can be beneficial for suppressing the injection of majority carriers from the bulk. The effect is more pronounced as the p$^+$n junction depth becomes smaller and it dominates the electrical characteristics of ultrashallow junctions, as, for example sub-10-nm deep pure boron (PureB) diodes. The properties of the perimeter of such an interface play a critical role in the overall electrical characteristics. In this paper, a TCAD simulation study is described where nanometer-deep p$^+$n junctions have an interface hole-layer that forms an energy barrier at the semiconductor-semiconductor interface. The suppression of bulk electron injection is analyzed with respect to the barrier height and the p$^+$n junction depth. Perimeter effects are investigated by 2D simulations showing a detrimental impact on the parasitic majority carrier injection from the bulk in structures with nanometer deep p$^+$n junctions. Other than employing a guard ring, reduction of the perimeter effects by shifting the position of the metal electrode was considered.**

I. INTRODUCTION

The properties of metal-semiconductor or semiconductor-semiconductor interfaces define the electrical characteristics of many semiconductor devices. An example is given bymetal-semiconductor devices where the Schottky-barrier height is solely defined by the atomic structure of the interface [1], [2]. Bipolar transistors with a polysilicon (poly-Si) emitter also make use of the interfacial properties of the poly-Si/Si transition for suppressing the minority carrier (hole) injection from the base into the emitter[3] and increasing the current gain. In poly-Si emitters, the physical mechanisms governing the minority carrier transport are mainly affected by the formation of a thin oxide layer at the poly-Si/Si interface or the recombination of the minority carriers via interface states for devices without a deliberately grown oxide layer [4]. On the other hand, the hetero-emitter-like behavior of phosphorus doped poly-Si emitters is exploited to form a barrier both in the valence and conduction band which can suppress the hole injection from the base[5]. Apart from a band offset, in

heterostructure devices the interface states can be filled with electrons or holes thus bending the band significantly [6], [7]. Other mechanisms which impact the carrier transport and where the interface can play a role include tunneling to and from interface states and tunneling through barriers formed at the interface[8].

It can be assumed that the interface properties in the pure amorphous boron (PureB) devicesare responsible for the exceptional electrical characteristics of these devices.A potential barrier at the PureB/Si interface is formed due to an interface hole-layer [9] and suppresses the electron injection from the bulk.While being CMOS compatible [10], the PureB deposition technology allows the formation of a nanometer-deep p$^+$n junctions [11]. Such a shallow p$^+$n junction depth is expected to suffer from a large electron injection from the bulk, which would increase the saturation current density to the values larger than 10^{-14} A/μm^2, as found in Schottky-like devices [12]. However, in PureB diodes the saturation current density is measured to be lower than 10^{-19} A/μm^2[10], [12] and is comparable to the saturation current density of devices with deep-diffused pn-junctions. The presence of an effective blocking mechanism is also confirmed by the high effective emitter Gummel number measured in pnp-transistors where PureB layers are incorporated in the emitter region [10], [12]. The effective blocking mechanism of the PureB layers is a subject of ongoing research, which also yielded the wide-bandgap model of the PureB layer [13]. However, this model is made obsolete since the latest ellipsometry measurements yielded an optical bandgap lower than that of Si with values similar to the ones reported for amorphous boron layers [14], [15].

The efficient suppression of the majority carrier injection from the bulk by an interface blocking mechanism, as is the case for the PureB devices, can be deteriorated by the perimeter effects. Photodiodes fabricated with PureB layers deposited at either 400° C or 700° C without the guard ring (GR) show several orders of magnitude higher currents in the forward regime than the PureB diodes where the GR is added to the periphery [16]. This behavior is also attributed to the inherent properties of the PureB layers and the interface to Si. The PureB/Si interface is terminated at the perimeter, which

---

This work was supported by the Croatian Science Foundation under contract no. 9006.

allows a higher injection of electrons. The Al pits cannot be formed at the periphery since it is shown that the PureB servesas a diffusion barrier for pure Al deposition [17]. Moreover, the use of the Al saturated with 1-2 % of Si can completely prevent the formation of the pits.

In this paper, a pure Si test structure is proposed with the interface hole-layer which islocated in a several nanometers deep p^+ region of a p^+n-junction diode. This structure is used to analyze the impact of the perimeter on the potential barrier formed by an interface hole-layer which is responsible for suppression of the electron carrier injection. The impact of both oxide interface charge and the oxide layer thickness on the termination of the interface region at the perimeter is examined. Methods for eliminating detrimental perimeter effects are proposed.

II. SUPPRESSION OF ELECTRON INJECTION BY AN INTERFACE HOLE-LAYER

The impact of aninterface hole-layer located in the p^+ region of a p^+n-junction diode is analyzed on a pure Si test structure based on the material, geometrical and doping parameters found in PureB devices. The PureB layer can be used as an abundant source for diffusion of boron into the Si and the junction depths from sub-10 nm to several hundreds of nanometersare attainable by controlling the annealing time and temperature[10]. The thickness of the PureB layer is set by the duration of the diborane (B_2H_6) gas exposure and thegrowth rate of the PureB layers is found to be equal to 0.4 nm/min for adeposition performed at $700°$ C[11]. Depending on the application of the devices, PureB layer thickness is varied between 10 nm and 2 nm, while even the latter allows a complete coverage of the Si surface [11]. The concentration of carriers in the PureB layers is not yet measured, while the concentration of holes in amorphous boron layers fabricated using different deposition techniques is found to vary in the range between 10^{16}cm^{-3}[18]and 10^{18} cm^{-3}[19].Other properties such as bandgap[14], [15], mobility[19] and affinity of the PureB layers are neglected for simplicity and the default Si values are used.

The test structure is defined and simulated in Sentaurus Device [20] TCAD software. In simulations,the p^+region, corresponding to the as-diffused boron, is Gaussian with peak concentration, N_{p+}, at the surface of 10^{19} cm^{-3}andpn-junction depth, y_j, varied between 1 nm and 500 nm defined at a background concentration of 10^{15} cm^{-3}. The interface region is defined between the bulk-Si region and the low-doped top-Si layer which has a fixed thickness, t_{pt}, of 5 nm. The doping of the top-Si layer, N_{pt}, equals 10^{18} cm^{-3} and is set to model the hole concentration measured in amorphous boron layers[19]. The total thickness of the simulated structure is 10 μm. The cross-section of the simulated structure is depicted in Fig. 1 and the doping profile of the structure with $y_j = 10$ nm is shown for reference. The top-Si/bulk-Si interface is also indicated.

In diodes with shallow p^+n-junctions, the properties of the metal contact such as work-functionare important for the behavior of the device. In our simulations, we usedan aluminum metal contact with work-function of 4.1 eV [20]. The band alignment and the formation of the Al/Si barrier follows the Schottky model for contacts [3] while

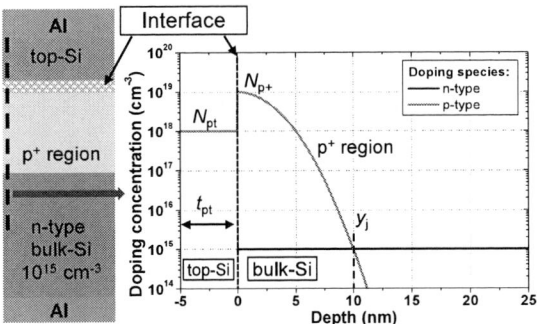

Figure 1. Cross section of a 1D test structure with simulated doping concentration profile.

the thermionic emission model is used to account for the carrier transport over the barriers [20].The simulations are also performed using the Schottky-barrier lowering model, together with the tunneling of the carriers to the anode contact[20]. Fermi-level pinning at the Schottky contact was neglected. For comparison, some of the simulations are performed using an ideal ohmic contact to Si. The electron and hole lifetimes in the Shockley-Read-Hall model equal 10^{-3} s to account for the low saturation current density characteristics typical of PureB photodiodes [10], [12].

A potential barrier for electrons is formed by a negative fixed charge at the interface with concentration N_I. The negative charge attracts holes, which then form an interface layer of holes that bends the band and forms a potential barrier at the interface. In Fig. 2, the band diagram of a device with an Al/Si contact and having a potential barrier formed by a fixed interface charge of$N_I = 5 \times 10^{12}$ cm^{-2} is compared to the band diagram of the device without this barrier. The band diagram is plotted for the device with $y_j = 10$ nm at forward diode voltage $V_D = 0$ V. For reference, the band diagram of the device with an ideal ohmic contact is also shown. An increase of the potential barrier due to the hole-layer at the interface is clearly seen to be capable of suppressing the electron injection

We analyzed the impact of the parameters of the device such as y_j and N_I on the suppression of the electron

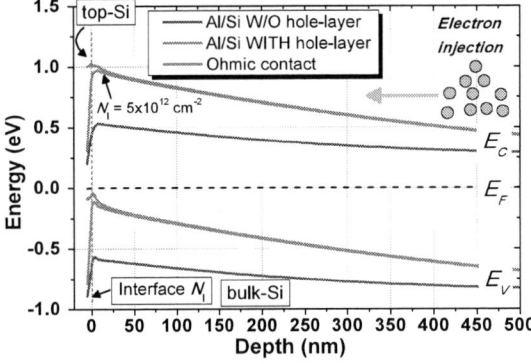

Figure 2. Simulated energy band diagram ($V_D = 0$ V and $y_j = 10$ nm) illustrating the formation of the potential barrier due to the interface hole-layer with $N_I = 5 \times 10^{12}$ cm^{-2}. Band diagram of the device with an ohmic contact is shown for reference.

(a)

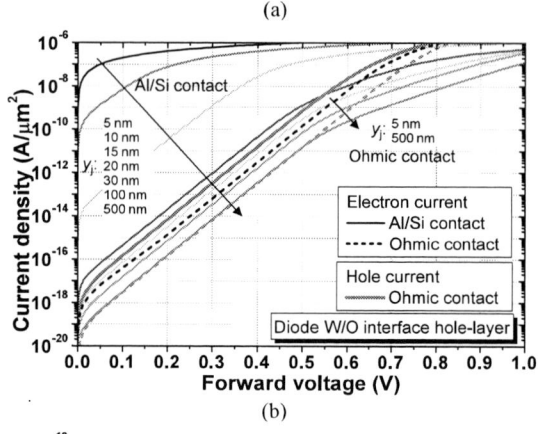

(b)

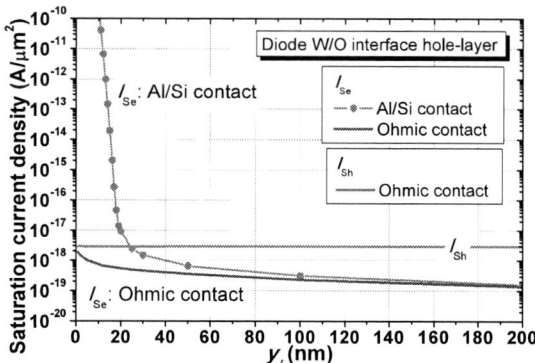

Figure 3. (a) Current-voltage characteristics of the simulated structure for y_j between 5 nm and 500 nm for the device with Al/Si contact and ideal ohmic contact to Si. (b) Extracted electron and hole saturation current density with respect to y_j.

injection by simulating the current-voltage characteristics of the diode. Simulated current-voltage characteristics of the devices with an Al/Si contact without the interface hole-layer ($N_I=0$) are shown in Fig. 3a. For $y_j < 20$ nm, the Al work-function lowers the potential barrier for electronsand a large electron current can flow, dominating the total diode current. However, for $y_j > 20$ nm a barrier is formed which can suppress the electron injection and the diode current is defined by the hole current. For the device where an ideal ohmic contact to Si is defined, there is no impact from the Al work-function and a barrier capable of suppressing the electron injection is formed even for $y_j = 5$ nm. The saturation current density of both electrons, I_{Se}, and holes, I_{Sh}, is extracted with respect to y_j and is shown in Fig. 3b. For $y_j > 150$ nm, the I_{Se} of the device with Al/Si contact is equal to the I_{Se} of the device with the ohmic contact to Si.

In the simulatedpure Si test structure, the suppression of the electron injection from thebulk is achieved by introducing a large concentration of holesat the interface. The potential barrier formed in this way can lower the I_{Se} of the diode to become comparable or lower than I_{Sh}. Electron and hole saturation current densities are extracted for devices with Al/Si contacts and anN_I interface charge that is assumed to be at the top-Si/bulk-Si interface. The results are shown in Fig. 4. For the device with $y_j = 10$ nm, an N_I larger than 6×10^{12} cm^{-2} is needed to lower I_{Se} below

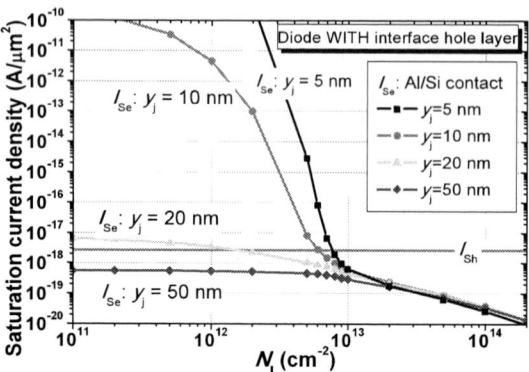

Figure 4. Electron and hole saturation current density for devices with an Al/Si contact with respect to the interface charge concentration responsible for the formation of the potential barrier. Devices with $y_j = 5$, 10, 20 and 50 nm are shown.

the I_{Sh} values, whereas for deeper pn-junctions, those values are even lower.

III. IMPACT OF PERIMETER EFFECTS ON THE POTENTIAL BARRIER AT THE TOP-SI/BULK-SI INTERFACE

In order to study the perimeter effects, we performed 2D simulations in Sentaurus Device [20]. The cross section of the 2D structure is depicted in Fig. 5. The simulated device is 2 µm wide, whereas the intrinsic diode width is set to 1 µm implying that the top-Si layer and the p$^+$ regionare equally wide. The Gaussian profile of the p$^+$ region extends laterally with a factor of 0.5. The whole perimeter is covered by oxide with thickness t_{ox}. The part of the top-Si region can also be covered by oxide which is defined by parameter d_{ox}. In this way, the aluminum contact is removed from the edge of the intrinsic diode. The default value ofd_{ox} is 0 meaning that the aluminum covers the whole intrinsic diode region. The concentration of positive oxide interface charge [21]is defined as N_{ox}.

The perimeter effects can impact the potential barrier formed at the interface due to the 2D distribution of charge. Also, if the positive oxide interface charge is located in the vicinity of the interface hole-layer or if the oxide layer is sufficiently thin, electrons are accumulated at the oxide/Si interface. These electrons compensate the charge in the hole-layer thus leading to a decrease of the potential barrier. At the same time, the accumulated electrons form a channel which can steer the electron current towards the already lowered potential barrier thus increasing the electron injection. The electron and hole

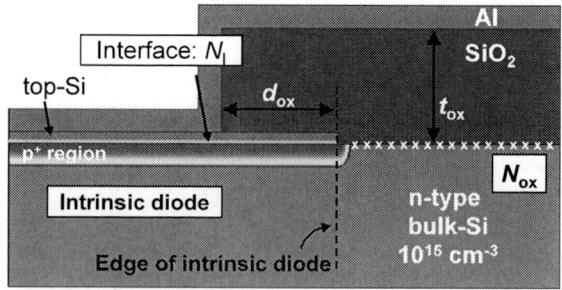

Figure 5. Cross section of a 2D test structure used to analyze the impact of perimeter effects on the potential barrier at the interface.

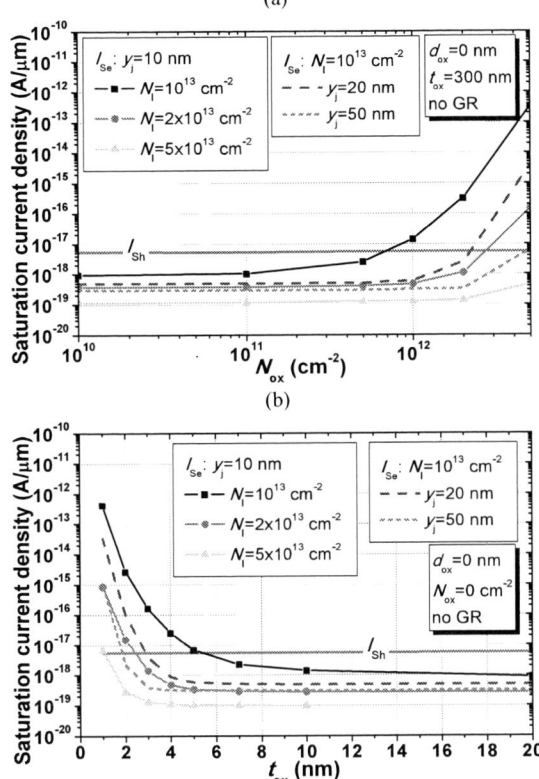

(a)

(b)

Figure 6. Electron and hole saturation current density of a 2D device for junction depths of y_j = 10 nm, 20 nm and 50 nm and N_I = 10^{13} cm^{-2}, 2×10^{13} cm^{-2} and 5×10^{13} cm^{-2} with respect to (a) oxide/Si interface charge, N_{ox} and (b) oxide thickness, t_{ox}.

saturation current density for junction depths of y_j = 10 nm, 20 nm and 50 nm and N_I = 10^{13} cm^{-2}, 2×10^{13} cm^{-2} and 5×10^{13} cm^{-2} with respect to the oxide/Si interface charge, N_{ox} is shown in Fig. 6a. The impact of the oxide thickness on I_{Se} and I_{Sh} for the same y_j and N_I is shown in Fig. 6b. For the device having y_j = 10 nm and N_I = 10^{13} cm^{-2}, the perimeter effects start to considerably increase I_{Se} for values of $N_{ox}> 7\times10^{11}$ cm^{-2} or $t_{ox}< 5$ nm which can then dominate the diode current.The lowering of the barrier due to the perimeter effects is analyzed by plotting the band diagram of the intrinsic diode and the band diagram at the

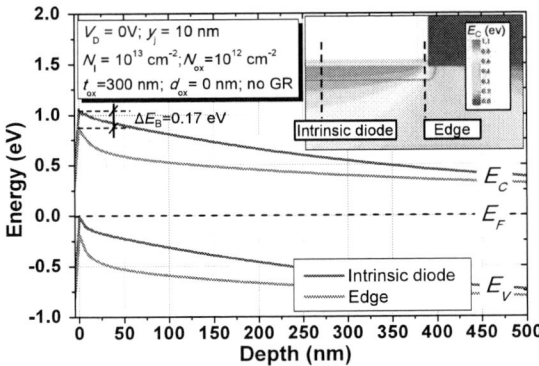

Figure 7. Band diagram of the intrinsic diode and at the edge of the intrinsic diode region indicating a lowering of the potential barrier.

edge of the intrinsic diode, which is shown in Fig. 7. The barrier lowering, ΔE_B, for the simulated device with parameters indicated in Fig. 7 is found to equal 0.17 eV. In reality, both the oxide/Si interface charge and thin oxide participate in deteriorating the barrier at the perimeter of the device, while the 3D effects found at the sharp corners of devices with rectangular layout can further decrease the barrier.

Elimination of the perimeter effects is performed by means of GR formation at the edge of the intrinsic diode or by increasing the distance between the aluminum contact and the edge of the intrinsic diode. Both methods can lower the increased electron saturation current density. In our simulations, the GR region has a Gaussian doping profile with the junction depth of y_{jGR} = 300 nm defined at a bulk doping concentration of 10^{15} cm^{-3}. The Gaussian profile extends laterally with a factor of 0.5. The peak concentration of the GR region, N_{pGR}, is located at the surface (y=0 nm). The impact of the N_{pGR} on successful elimination of the perimeter effects is analyzed and shown in Fig. 8a. The 2D device is defined having y_j = 10 nm, N_I = 10^{13} cm^{-2}, N_{ox} = 5×10^{11} cm^{-2}, t_{ox}=2 nm and d_{ox} = 0 nm. The results show that the peak surface concentration of such a GR region needs to be higher than 3×10^{18} cm^{-3} in order to efficiently decrease I_{Se} to be lower than I_{Sh}. On the other hand, increasing the distance between the aluminum contact and the edge of the intrinsic diode lowers the I_{Se} significantly. The aluminum sink electrode, which originally decreases the barrier for electrons in

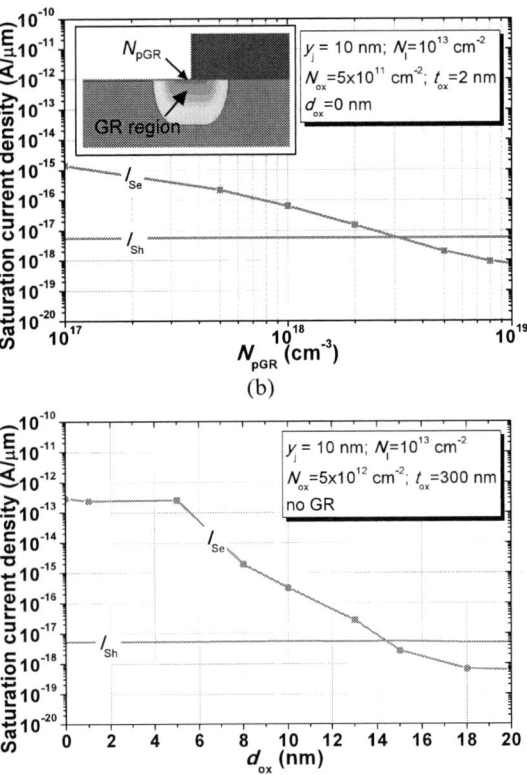

(a)

(b)

Figure 8. Electron and hole saturation current density of a 2D device with respect to the (a) peak concentration of the GR region; (b) distance between the aluminum contact and the edge of the intrinsic diode, d_{ox}.

devices with nanometer deep pn-junctions is thereafter moved away and is not subjected to the perimeter effects. The extracted I_{Se} and I_{Sh} for a device with $y_j = 10$ nm, $N_I = 10^{13}$ cm^{-2}, $N_{ox} = 5\times10^{12}$ cm^{-2}, $t_{ox}=300$ nm and no GR region are shown in Fig. 8b. The parameter d_{ox} for which I_{Se} is lower than I_{Sh} needs to be larger than ≈15 nm. This also sets the lower limit at which the perimeter effects can efficiently be suppressed.

IV. CONCLUSION

In this paper, we analyzed the impact of a layer of holes at an interfacelocated in the p$^+$ region of p$^+$n-junction diodes on the suppression of electron injection from the bulk. The layer of holes causes the formation of a potential barrier capable of reducing the otherwise large electron saturation current density found in devices with Al/Si contacts having a shallow pn-junction depth lower than 20 nm. For a device with an Al/Si contact with a junction depth of 10 nm and the thickness of the top-Si layer of 5 nm, an electron saturation current density lower than 10^{-18} A/μm^2 can be achieved for an interface hole-layer with a concentration larger than 6×10^{12} cm^{-2}.

The perimeter effects in devices employing a layer of holes can have detrimental effects on the suppression of the electron injection. Both interface oxide charge and thin oxide layers can lower the potential barrier and form a channel at the oxide/Si interface steering the electrons toward the lowered barrier at the edge of the intrinsic diode. These perimeter effects can be even more pronounced at the edges ofdevices having a circular or rectangular layout due to 3D effects. In the devices where a GR is employed, the perimeter effects can be efficiently suppressed. In this paper the peak doping concentration of the GR needed to suppress these effects for the simulated structure is found to be larger than 3×10^{18} cm^{-3}. At the same time, if part of the intrinsic diode is covered with oxide thus increasing the distance between the aluminum contact and the perimeter, the perimeter effects can be eliminated. The distance that the aluminum contact must be shifted with respect to the edge of the intrinsic diodecan be as low as 15 nm.

REFERENCES

[1] R. T. Tung, "The physics and chemistry of the Schottky barrier height," *Applied Physics Reviews*, vol. 1, no. 1, p. 011304, Mar. 2014.

[2] R. T. Tung, "Recent advances in Schottky barrier concepts," *Materials Science and Engineering: R: Reports*, vol. 35, no. 1, pp. 1–138, 2001.

[3] S. M. Sze and K. K. Ng, *Physics of Semiconductor Devices*, 3rd edition. Hoboken, N.J: Wiley-Interscience, 2006.

[4] I. R. Post, P. Ashburn, and G. R. Wolstenholme, "Polysilicon emitters for bipolar transistors: a review and re-evaluation of theory and experiment," *IEEE Transactions on Electron Devices*, vol. 39, no. 7, pp. 1717–1731, 1992.

[5] M. Kondo, T. Kobayashi, and Y. Tamaki, "Hetero-emitter-like characteristics of phosphorus doped polysilicon emitter transistors. Part I: band structure in the polysilicon emitter obtained from electrical measurements," *IEEE Transactions on Electron Devices*, vol. 42, no. 3, pp. 419–426, 1995.

[6] W. G. Oldham and A. G. Milnes, "Interface states in abrupt semiconductor heterojunctions," *Solid-State Electronics*, vol. 7, no. 2, pp. 153–165, Feb. 1964.

[7] H. Kroemer, "Heterostructure Devices: A Device Physicist Looks at Interfaces," in *Electronic Structure of Semiconductor Heterojunctions*, vol. 1, G. Margaritondo, Ed. Dordrecht: Springer Netherlands, 1988, pp. 116–149.

[8] A. G. Milnes and D. L. Feucht, *Heterojunctions and metal-semiconductor junctions.* Academic Press, 1972.

[9] L. Qi and L. K. Nanver, "Conductance Along the Interface Formed by 400 °C Pure Boron Deposition on Silicon," *IEEE Electron Device Letters*, vol. 36, no. 2, pp. 102–104, Feb. 2015.

[10] L. K. Nanver *et al.*, "Robust UV/VUV/EUV PureB Photodiode Detector Technology With High CMOS Compatibility," *IEEE Journal of Selected Topics in Quantum Electronics*, vol. 20, no. 6, pp. 306–316, Nov. 2014.

[11] F. Sarubbi, T. L. M. Scholtes, and L. K. Nanver, "Chemical Vapor Deposition of α-Boron Layers on Silicon for Controlled Nanometer-Deep p + n Junction Formation," *Journal of Electronic Materials*, vol. 39, no. 2, pp. 162–173, Feb. 2010.

[12] F. Sarubbi, L. K. Nanver, and T. L. M. Scholtes, "High Effective Gummel Number of CVD Boron Layers in Ultrashallow p+n Diode Configurations," *IEEE Transactions on Electron Devices*, vol. 57, no. 6, pp. 1269–1278, Jun. 2010.

[13] T. Knežević, T. Suligoj, A. Šakić, and L. K. Nanver, "Modelling of electrical characteristics of ultrashallow pure amorphous boron p+ n junctions," in *MIPRO, 2012 Proceedings of the 35th International Convention*, 2012, pp. 36–41.

[14] U. Kuhlmann, H. Werheit, T. Lundström, and W. Robers, "Optical properties of amorphous boron," *Journal of Physics and Chemistry of Solids*, vol. 55, no. 7, pp. 579–587, 1994.

[15] A. Hori, M. Takeda, H. Yamashita, and K. Kimura, "Absorption Edge Spectra of Boron-Rich Amorphous Films Constructed with Icosahedral Cluster," *Journal of the Physical Society of Japan*, vol. 64, no. 9, pp. 3496–3505, Sep. 1995.

[16] L. Qi, "Interface Properties of Group-III-Element Deposited-Layers Integrated in High-Sensitivity Si Photodiodes," PhD thesis, TU Delft, 2016.

[17] A. Šakić, V. Jovanović, P. Maleki, T. L. Scholtes, S. Milosavljević, and L. K. Nanver, "Characterization of amorphous boron layers as diffusion barrier for pure aluminium," in *MIPRO, 2010 Proceedings of the 33rd International Convention*, 2010, pp. 26–29.

[18] Y. Kumashiro, T. Yokoyama, and Y. Ando, "Thermoelectric properties of boron and boron phosphide CVD wafers," in *Thermoelectrics, 1998. Proceedings ICT 98. XVII International Conference on*, 1998, pp. 591–594.

[19] K. Kamimura, M. Ohkubo, T. Shinomiya, M. Nakao, and Y. Onuma, "Preparation and properties of boron thin films," *Journal of Solid State Chemistry*, vol. 133, no. 1, pp. 100–103, 1997.

[20] Synopsys, *Sentaurus Device User Guide*. Mountain View, CA, USA: Synopsys, 2016.

[21] F. Li and A. Nathan, *CCD Image Sensors in Deep-Ultraviolet: Degradation Behavior and Damage Mechanisms*, 1st edition. Berlin ; New York: Springer, 2005.

MIPRO 2017, May 22- 26, 2017, Opatija, Croatia

Analysis of Hot Carrier-Induced Degradation of Horizontal Current Bipolar Transistor (HCBT)

J. Žilak[*], M. Koričić[*] and T. Suligoj[*]

[*] University of Zagreb, Faculty of Electrical Engineering and Computing, Department of Electronics, Microelectronics, Computing and Intelligent Systems, Micro and Nano Electronics Laboratory, Zagreb, Croatia
josip.zilak@fer.hr

Abstract - The relative contribution of the hot electrons and hot holes to the reliability degradation of the Horizontal Current Bipolar Transistor (HCBT) is investigated by TCAD simulations. The base current (I_B) degradation, obtained by the reverse-bias emitter-base (EB) and mixed-mode stress measurements, is caused by a hot carrier-induced interface trap generation at silicon-oxide interfaces above and below HCBT's emitter n^+ polysilicon region. The simulation analysis is performed on the HCBT structures with different n-collector doping profiles and n-hill silicon sidewall surface treatment. The used lucky electron injection model distinguishes the hot carrier type responsible for the damage and makes it possible to predict the HCBT reliability behavior. It is shown that the majority of traps under the reverse-bias EB stress is located at the top interface and is caused by the hot holes, whereas the hot electrons produce the traps under the mixed-mode stress, located mostly at the bottom interface.

I. INTRODUCTION

The constantly growing electronic market tightens the requirements on the already highly scaled semiconductor technologies, forcing the low-power and high-speed devices to work in more restrictive operating conditions. The electric field and current density have been constantly increasing, damaging the various transistor regions and making its reliability analysis more important [1], [2]. The reliability examination of Si and SiGe bipolar transistors is performed by employing, e.g., reverse-bias emitter-base (reverse EB) [3], mixed-mode [4] or high forward current [5] stress tests. The stress accelerates the damage mechanisms and shortens, otherwise very long, duration of the reliability testing.

Under the stress conditions, a very high electric field is induced in devices, causing the impact ionization, carrier multiplication and hot carrier generation. The hot electrons and hot holes can diffuse toward the silicon-oxide interfaces and potentially damage it. If they retain sufficient energy, they can produce traps at the silicon-oxide interface by interacting with the passivated (H-terminated) Si bonds. The interface traps act as generation-recombination (G-R) centers causing the excess base current (I_B) and, consequently, the degradation of current gain (β) [6]. The 1/f noise is also increased [7], whereas the impact of the hot carrier-induced interface traps on the small-signal parameters of the SiGe HBTs and RF front-end circuits have been extensively

investigated [7]-[9]. The physics-based TCAD modeling of the stress degradation mechanisms, as an indispensable part of the reliability analysis, strives to determine safe operating area and lifetime of devices, as well as the device-to-circuit reliability interaction [10]-[12].

The Horizontal Current Bipolar Transistor (HCBT) [13], integrated with 180 nm CMOS technology, represents the flexible low-cost BiCMOS technology platform, suitable for the wireless communication circuits, (e.g., high-linearity RF mixers [14] and frequency dividers [15] recently reported). The HCBT is fabricated at the silicon hill sidewall defined by the shallow trench isolation (Fig. 1). Optimization of the doping profiles resulted in the state-of-the-art high-frequency characteristics among the implanted-base silicon BJTs (i.e., cut-off frequency $f_T = 51$ GHz, maximum frequency of oscillations $f_{max} = 61$ GHz, common-emitter breakdown voltage $BV_{CEO} = 3.4$ V and $f_T \times BV_{CEO}$ product of 173 GHzV [13]). Furthermore, high-voltage HCBTs with adjustable breakdown voltages up to 36 V [16] can be fabricated without additional cost.

In this paper, the additional HCBT reliability analysis is performed by using the TCAD device simulations based on the lucky-electron injection model and dynamic generation of traps at the interface. It is an extension of the study published in [17], where different damage locations and impact of the several HCBT technological parameters under both reverse EB and mixed-mode stresses are reported. The relative contribution of the hot carrier type (hot electrons and hot holes) to the reliability degradation of the HCBT is investigated, providing the further, in-depth analysis of damage mechanisms identified in [17].

II. MODEL DESCRIPTION

The degradation model used in TCAD simulations is based on the predictive physics-based model presented in [10]. It accounts for two degradation mechanisms: hot carrier generation and interface trap formation. Since the traps at the silicon-oxide interface act as G-R centers, the calculated interface trap concentration can be related to the base current degradation measured during the stress tests.

The concentration of traps generated at the silicon-oxide interface depends on the rate of hot carriers reaching the interface and their capability of trap creation. The trap formation rate at the given silicon-oxide interface point is given by [10]

This work was supported by the Croatian Science Foundation under contract no. 9006.

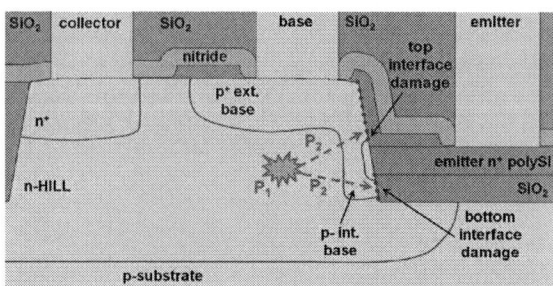

Fig. 1. Cross section of the HCBT structure examined in the reliabilty study with the highlighted top and bottom silicon-oxide interfaces damaged by the hot carriers.

$$K_F = \sum_{(x,y) \in V}^{\cdot} r_e(x,y) + \sum_{(x,y) \in V}^{\cdot} r_h(x,y), \quad (1)$$

where $r_e(x,y)$ and $r_h(x,y)$ are the hot electron and hot hole rates as function of the location within the semiconductor. The hot carrier rate at given position in semiconductor is calculated as [10]

$$r_{e,h}(x,y) = \frac{J_{n,p}(x,y)}{q} P_1(x,y) P_2(x,y) M(x,y), \quad (2)$$

where $J_{n,p}(x,y)$ is the local electron or hole current density, $P_1(x,y)$ is the probability that the carrier acquires sufficient kinetic energy and is directed toward the interface with sufficient momentum to create a trap, $P_2(x,y)$ is the probability that the hot carrier travels to the interface without losing any energy, as sketched in Fig. 1. The term $M(x,y)$ corresponds to the element area in two-dimensional mesh of the simulated device [10].

The probability P_1 of the carrier to acquire sufficient kinetic energy and to retain the appropriate momentum after redirection arises from the lucky-electron injection model [18]

$$P_1(x,y) = 0.25 \frac{\lambda F_{\text{eff}}(x,y)}{\Phi_{\text{hot}}} e^{-\Phi_{\text{hot}}/\lambda F_{\text{eff}}(x,y)}, \quad (3)$$

where $F_{\text{eff}}(x,y)$ is the effective electric field experienced by the carriers, λ is the scattering mean free path of the hot carriers and Φ_{hot} (2.3 eV from [10]) is the threshold energy required to depassivate the silicon dangling bond at the interface. The probability P_2 that the hot carrier travels to the interface without losing any energy is given by [18]

$$P_2(x,y) = e^{-d/\lambda}, \quad (4)$$

where d is the distance between a given point in device structure and a point at the interface.

Interface trap concentration (N_{it}) is calculated by using the reaction-diffusion model from [19] that includes both the hot carrier-induced depassivation of H-terminated bonds at the interface and re-passivation of the silicon dangling bonds by hydrogen. The time dependent generation of the interface traps is approximated by [19]

$$N_{it} \approx 1.16 \sqrt{\frac{K_F N_0}{K_R}} (Dt)^{\alpha}, \quad (5)$$

where K_F is the trap formation rate, K_R is the reverse rate constant (10^{-7} s in [10]), N_0 is the total areal density of dangling bonds, D is the diffusion coefficient of hydrogen

in the oxide (0.01 μm^2/s in [10]) and α is the time dependence of the trap formation (0.25 from [19]).

The part of the described model is incorporated in the Synopsys TCAD device simulator [20]. However, it is developed for the MOSFET devices and calculates the lucky electron gate current due to hot carrier injection. The model parts that calculate P_1 and P_2 and allow integration over the entire structure are the same, whereas the K_F and N_{it} equations are built in into the model. The necessary modifications are allowed by the Synopsys device simulator physical model interface (PMI).

III. REVERSE-BIAS EMITTER-BASE STRESS

Three HCBT structures are investigated under the reverse EB stress test: HCBT 1 (steep n-hill) that has optimized $f_T \times BV_{CEO}$ product, HCBT 2 (uniform n-hill) that achieves best high-current linearity in the RF mixer and HCBT 3 (CMOS n-well) that represents the low-cost HCBT version since it saves one mask in fabrication. The analyzed structures differ in the n-collector doping profiles. Furthermore, the oxide etching before the polysilicon deposition in the form of a longer HF dip is used in the case of HCBT 2. The measured electrical characteristics, doping profiles and differences between three HCBTs are described in more details in [17].

During the reverse EB stress, the hot carriers are generated in the EB depletion region due to high electric field caused by the high reverse EB bias. The stress conditions used in experiments are: constant reverse EB current of 0.5 μA, open collector, room temperature. The resulting emitter-base voltage is $V_{EB} = 3$ V. The forward Gummel characteristics of three HCBT structures are measured at the collector-base voltage $V_{CB} = 0$ V after different stress times in the range from 10 to 3000 s. The stress causes the excess non-ideal base current, as can be seen in Fig. 2, where Gummel plots of HCBT 2 are shown. Similarly, the base current increase is observed in other two HCBT structures. The I_B increase as a function of stress time is shown in Fig. 3, expressed as the ratio of the base current after and before certain stress period (i.e., $I_{B,POST}/I_{B,PRE}$). The rates are extracted at $V_{BE} = 0.9$ V, which is the bias point around peak f_T. After 3000 s of stress, HCBT 1 (steep n-hill), HCBT 2 (uniform n-hill), and HCBT 3 (CMOS n-well) exhibit the $I_{B,POST}/I_{B,PRE}$ ratio of 1.33, 1.58 and 1.31, respectively. The HCBT 1 and HCBT 3 has similar I_B increase due to the same pre-

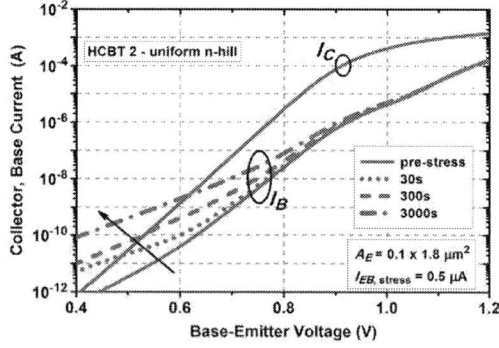

Fig. 2. Measured forward Gummel characteristics of the HCBT 2 (uniform n-hill) before and after the reverse-bias EB stres [21].

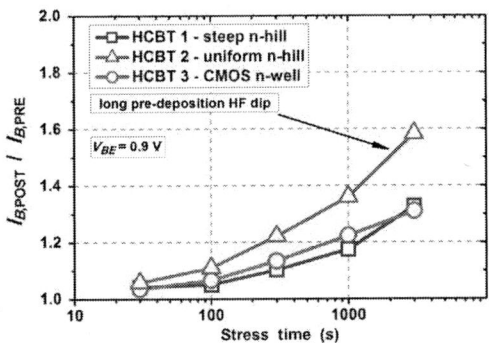

Fig. 3. Measured base current (I_B) degradation as a function of stress time of three HCBT structures. Reverse EB stress conditions: $I_{EB,stress}$ = 0.5 µA, open collector.

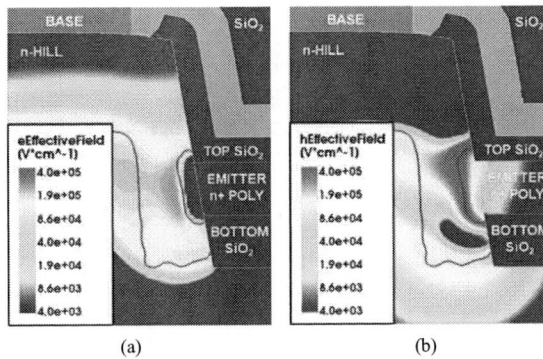

Fig. 4. Simulated electron (a) and hole (b) effective electric field under the reverse-bias EB stress (V_{EB} = 3 V, open collector) of HCBT 2 (uniform n-hill).

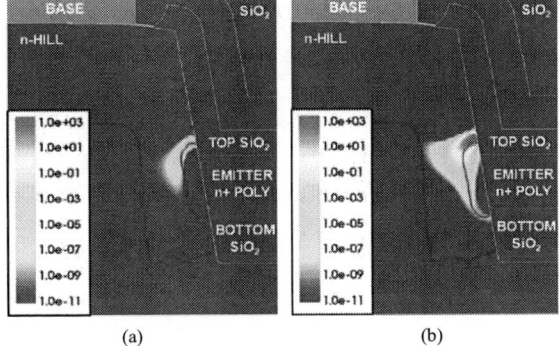

Fig. 5. Simulated rates of hot electrons (a) and hot holes (b) reaching the top silicon-oxide interface at the position just above the emitter n$^+$ polysilicon region. Rates are calculated for the HCBT 2 (uniform n-hill) under the reverse EB stress conditions.

deposition HF dip, whereas HCBT 2, where the longer pre-deposition HF dip is used, has a higher I_B increase.

The HCBT reliability behavior is investigated further by TCAD device simulations, using the model described in section II. Three structures with realistic doping profiles, obtained by the process simulator, are biased in the same stress conditions as in the measurements (i.e., V_{EB} = 3 V, open collector). The trap formation rates (K_F) and the total interface trap concentration (N_{it}) at the silicon-oxide interfaces are the outcomes of the conducted simulations.

The effective electric field (F_{eff}), as the driving force experienced by electrons and holes, is simulated by using the hydrodynamic transport model that uses the local carrier temperatures to calculate the effective electric field [20]. Hence, the F_{eff} is calculated separately for the electrons and holes and depends on the carrier temperture. The simulated electron and hole effective electric fields of the HCBT 2 (uniform n-hill) under the reverse EB stress conditions are shown in Fig. 4a and 4b. The figures show the zoomed intrinsic transistor region. Similar results are obtained in other two HCBT structures, which is expected since the fabrication of the emitter and base regions is identical in all three examined HCBTs. The hole F_{eff} component is larger than the electron component with the peak value higher than $5 \cdot 10^5$ Vcm^{-1} and the area of the strongest F_{eff} is located at the top portion of the EB junction next to the top silicon-oxide interface.

The simulated effective electric field impacts the calculated P_1 probabilities [see (3)]. The probability that the hot hole acquires the threshold energy (Φ_{hot}) and is directed toward the interface is three orders of magnitude higher than the same probability of the hot electrons. The scattering mean free path λ, used as the fitting parameter, equals 6.2 nm as in [10]. The P_1 probability distribution is very similar to the F_{eff} distribution with the peak located next to the top silicon-oxide interface. The probability P_2 [see (4)] that the hot carrier will travel to the interface without losing any energy is calculated for all positions along the interface separately. It is a function of the distance from the interface (d) and the scattering mean free path (λ). The P_2 exponentially decreases as the distance between the position at the interface and the given position within device increases.

Once the probabilities P_1 and P_2 are obtained, the rates of hot electrons [$r_e(x,y)$] and hot holes [$r_h(x,y)$] reaching the interface from the given position within the device are calculated by using (2). The rates are calculated separately for all mesh points at the interface. The rate of hot electrons and hot holes reaching the position on the interface located just above the emitter n$^+$ polysilicon region are shown in Fig. 5. The hot hole rate reaching that particular interface position is much higher than the hot electron rate.

The total hot carrier rate at the given interface position is calculated as the sum of the hot carrier rates from all points within the device (i.e., integration of data from Fig. 5) [10]. This is calculated separately for the hot electrons and the hot holes. The simulated rates of the hot carriers (r_e and r_h) reaching the silicon-oxide interface as a function of the interface position are shown in Fig. 6. The distance on the x-axis in Fig. 6 corresponds to the distance from the top to the bottom of the n-hill along the n-hill's sidewall (Fig. 1). Hence, the distance to around 0.19 µm corresponds to the top silicon-oxide interface, the distance from 0.19 µm to around 0.27 µm corresponds to the n$^+$ polysilicon emitter and the distance above 0.27 µm corresponds to the bottom silicon-oxide interface. Along the entire interface, the hot hole carrier rate is several orders of magnitude higher in comparison to the hot electron rate. Hence, the hot electron rate contribution can

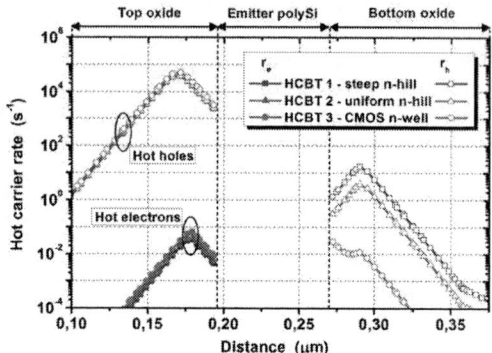

Fig. 6. Simulated rates of hot carriers (r_e and r_h) reaching the silicon-oxide interface as a function of the interface position under the reverse-bias EB stress (V_{EB} = 3 V, open collector) of all three HCBT structures.

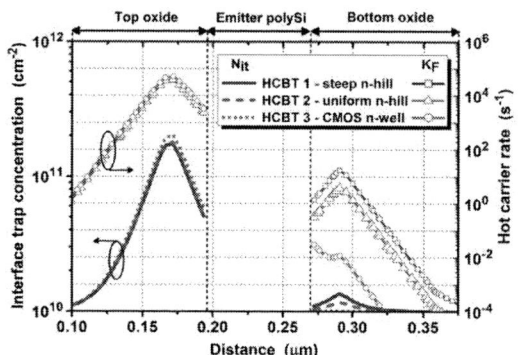

Fig. 7. Simulated interface trap concentrations (N_{it}) and trap formation rates (K_F) of all three HCBT structures as a function of the interface position. Traps are calculated after 3000 s of the reverse EB stress.

be neglected in the case of the reverse EB stress. Moreover, the hot carrier rates are much higher along the top interface than along the bottom interface. The hot carrier rates obtained in all three HCBTs are similar, coming from the similar simulated effective electric field distributions because of the identical fabrication of the emitter and base regions.

The total trap formation rate (K_F) at the given interface position is then calculated as the sum of the hot electron (r_e) and the hot hole (r_h) component obtained at the same interface position. The trap formation rate (K_F) is dominated by the hot hole component (Fig. 6). With the known K_F, the total interface trap concentration (N_{it}) is calculated by using (5). Both simulated N_{it} and K_F as a function of the interface position, after 3000 s of the reverse EB stress, are shown in Fig. 7. The majority of the interface traps, with the peak values of around $1.7 \cdot 10^{11}$ cm^{-2}, is placed along the top silicon-oxide interface. Hence, under the reverse EB stress conditions, the top interface is more damaged and it is responsible for the I_B degradation obtained during the stress tests.

Both N_{it} and K_F are similar in all three HCBT structures, which indicates that the same base current degradation is to be expected. This is in a disagreement with the stress measurement results (Fig. 3), since the higher I_B degradation is obtained in the case of HCBT 2 (uniform n-hill). This deviation is attributed to the lower

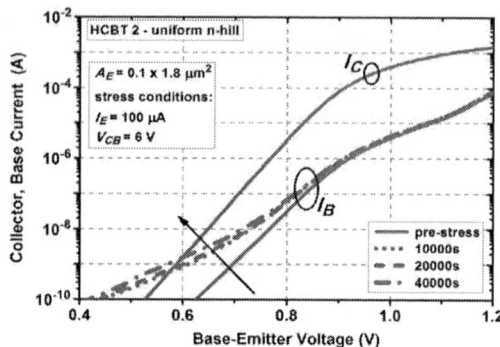

Fig. 8. Measured forward Gummel characteristics of the HCBT 2 (uniform n-hill) before and after the mixed-mode stress [21].

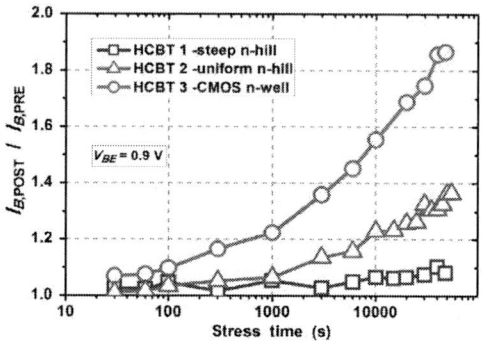

Fig. 9. Measured base current (I_B) degradation as a function of stress time of three HCBT structures. Stress conditions: $I_{E,stress}$ = 100 μA, V_{CB} = 7 V (HCBT 1), V_{CB} = 6 V (HCBT 2) and V_{CB} = 4 V (HCBT 3).

quality of the top silicon-oxide interface caused by the longer pre-deposition HF dip in HCBT 2. In the simulated structures, the top interface has the same properties explaining the similar simulated interface trap concentrations. Moreover, the differences in the n-collector region design between the HCBT structures do not impact the trap generation caused by the reverse EB stress.

IV. MIXED-MODE STRESS

During the mixed-mode stress test, the hot carriers are generated in the CB depletion region, due to simultaneously used high collector-base voltage (V_{CB}) and high collector current density (J_C). The used stress conditions are ([17]): constant emitter current I_E of 100 μA, V_{CB} = 7 V (HCBT 1), V_{CB} = 6 V (HCBT 2), V_{CB} = 4 V (HCBT 3), room temperature. The forward Gummel characteristics of three HCBT structures are measured at the V_{CB} = 0 V after different stress times in range from 100 to 50000 s. As in the reverse EB stress test, the excess non-ideal base current is observed (see Fig. 8). The I_B increase as a function of stress time (i.e., $I_{B,POST}/I_{B,PRE}$) is shown in Fig. 9. After around 50000 s of stress, HCBT 1 (steep n-hill), HCBT 2 (uniform n-hill), and HCBT 3 (CMOS n-well) exhibit the $I_{B,POST}/I_{B,PRE}$ ratio of 1.10, 1.36 and 1.86, respectively. The HCBT structures with different n-collector doping profiles exhibit different I_B increase. Despite the highest CB voltage (V_{CB} = 7 V), the smallest degradation occurs in HCBT 1. Although it is stressed with the smallest V_{CB}, HCBT 3 shows the highest degradation.

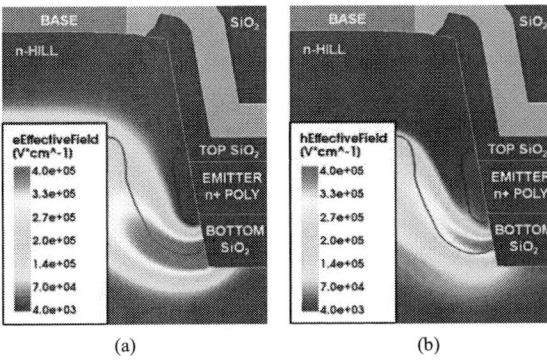

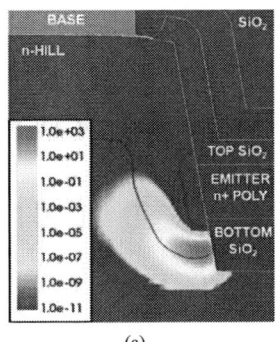
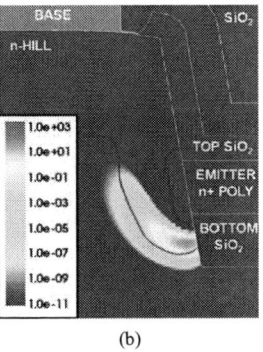

Fig. 10. Simulated electron (a) and hole (b) effective electric field under the mixed-mode stress ($I_{E,stress} = 50\ \mu A/\mu m$, $V_{CB} = 4$ V) of HCBT 3 (CMOS n-well).

Fig. 12. Simulated rates of hot electrons (a) and hot holes (b) reaching the bottom silicon-oxide interface at the CB *pn* junction position. Rates are calculated for the HCBT 3 (CMOS n-well) under the mixed-mode stress ($I_{E,stress} = 50\ \mu A/\mu m$, $V_{CB} = 4$ V).

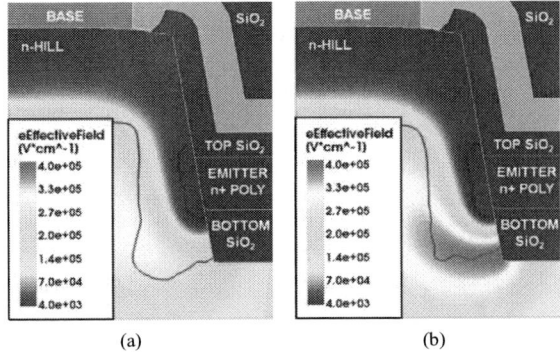

Fig. 11. Simulated electron effective electric field under the mixed-mode stress of HCBT 1 (a) and HCBT 2 (b). Stress conditions: $I_{E,stress} = 50\ \mu A/\mu m$, $V_{CB} = 7$ V (HCBT 1) and $V_{CB} = 6$ V (HCBT 2).

rates reaching the bottom silicon-oxide interface at the CB *pn* junction position are shown in Fig. 12. By integrating the data from Figs. 12(a) and 12(b), the total hot carrier rates at the particular interface position are obtained. Identical is done for all interface positions. The total hot electron (r_e) and hot hole (r_h) rates as a function of the interface position are shown in Fig. 13. The distance on the x-axis in Fig. 13 corresponds to the distance from the top to the bottom of the n-hill along the sidewall (Fig. 1). The hot electron rate is higher than the hot hole rate in all three HCBTs. Hence, the trap formation is dominated by the hot electrons. Moreover, the rate of hot carriers reaching the bottom interface is much higher than the rate of hot carriers reaching the top interface.

The total trap formation rate (K_F) is calculated as the sum of electron and hole rates, while the interface trap concentration (N_{it}) is computed by using (5). Both simulated N_{it} and K_F as a function of the interface position after 50000 s of the mixed-mode stress are shown in Fig. 14. The interface traps are placed along the bottom interface, whereas the amount of traps at the top interface is comparatively negligible. The HCBT 3 has the highest trap concentration with the peak value of $2.4 \cdot 10^{12}$ cm^{-2}, whereas the HCBT 1 has the smallest trap concentration (peak of only $6.5 \cdot 10^{10}$ cm^{-2}).

The obtained differences in the interface trap concentrations between analyzed HCBTs correspond to the base current degradation differences obtained by the

The device simulations, with included model from section II, are carried out on the same HCBT structures as in the case of the reverse EB stress. The stress conditions are equivalent to the ones used in the measurements. The simulated electron and hole effective electric fields (F_{eff}) of the HCBT 3 (CMOS n-well) under the mixed-mode stress conditions are shown in Fig. 10 and the higher electron F_{eff} can be observed [Fig. 10(a)]. The other two structures (HCBT 1 and HCBT 2) also exhibits higher electron F_{eff}, which is in contrast to the higher hole F_{eff} in the reverse EB stress. Moreover, unlike the reverse EB stress, there are differences between HCBT structures regarding the location and maximum of the F_{eff}. This can be observed if the electron F_{eff} of the HCBT 1 and HCBT 2, shown in Fig. 11, are compared with the HCBT 3 electron F_{eff} from Fig. 10. The peak electron F_{eff} of HCBT 2 and HCBT 3 is located bellow the BC junction at the bottom interface with the higher value in the case of HCBT 3 (due to higher collector concentration in the bottom part of the intrinsic transistor). The HCBT 1 (steep n-hill) exhibits smaller electron F_{eff} and the peak is pushed away from the interface due to the smaller collector concentration and the suppressed charge sharing effect [17].

The higher electron effective field causes the higher electron redirection probability P_1, and consequently, the higher hot electron carrier rates. That is mostly obvious in HCBT 3 (CMOS n-well), whose simulated hot carrier

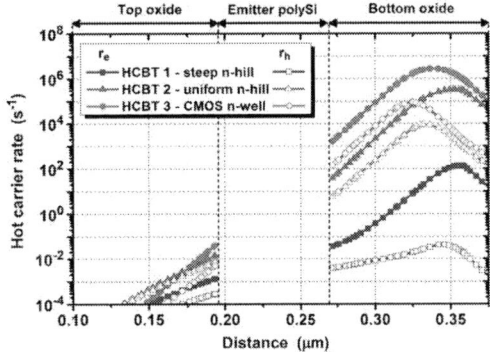

Fig. 13. Simulated rates of hot carriers (r_e and r_h) reaching the silicon-oxide interface as a function of the interface position under the mixed-mode stress of all three HCBT structures.

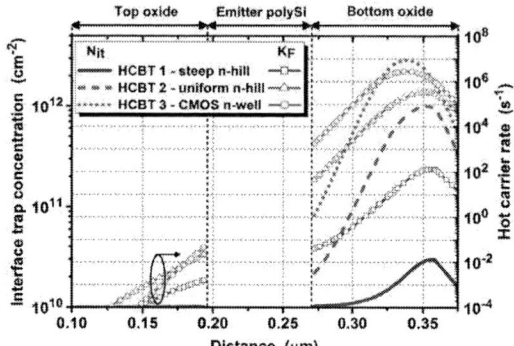

Fig. 14. Simulated interface trap concentrations (N_{it}) and trap formation rates (K_F) of all three HCBT structures as a function of the interface position. Traps correspond to 50000 s of the mixed-mode stress. Stress conditions: $I_{E,stress} = 50\ \mu A/\mu m$, $V_{CB} = 7$ V (HCBT 1), $V_{CB} = 6$ V (HCBT 2) and $V_{CB} = 4$ V (HCBT 3).

mixed-mode stress measurements. For example, the highest interface trap concentration in HCBT 3 corresponds to its highest I_B degradation.

V. CONCLUSION

The TCAD device simulations, with included lucky electron injection model, confirmed the different locations of the silicon-oxide interface damage discovered in the HCBT reliability analysis presented in [17]. The damage location reflects to the different I_B degradation rates in the various HCBT structures. The reverse-bias EB stress generates most of the damage at the top interface, whereas the mixed-mode stress introduces traps mostly at the bottom interface.

The quality of the top silicon-oxide interface in HCBT 2 structure is compromised by the longer HF dip causing more H-terminated bonds at the interface and consequently higher I_B degradation under reverse EB stress. The mixed-mode stress simulations show that the position and the peak of the effective electric field depend on the n-collector doping profile. It results in different interface trap concentration at the bottom interface between the HCBT structures, which fits different I_B degradation obtained by measurements.

The used lucky electron injection model distinguishes the hot electron and hot hole contributions to the interface trap generation. It makes possible to predict the impact of the HCBT technological parameters on the reliability behavior. The relative contribution of the hot electrons and hot holes to the reliability degradation is identified. The simulations showed that the hot holes are responsible for the damage during the reverse EB stress, while the hot electrons create most of the traps during the mixed-mode stress.

REFERENCES

[1] J. D. Cressler, "Emerging SiGe HBT reliability issues for mixed-signal circuit applications," in *IEEE Trans. Device and Mater. Rel.*, vol. 4, no. 2, pp. 222-236, June 2004.

[2] J. D. Cressler, "Device-to-circuit interactions in SiGe technology: Challenges and opportunities," in *Proc. BCTM*, Coronado, CA, 2014, pp. 45-55.

[3] G. Sasso, N. Rinaldi, G. G. Fischer and B. Heinemann, "Degradation and recovery of high-speed SiGe HBTs under very high reverse EB stress conditions," in *Proc. BCTM*, Coronado, CA, 2014, pp. 41-44.

[4] G. Zhang, J. D. Cressler, G. Niu and A. J. Joseph, "A new "mixed-mode" reliability degradation mechanism in advanced Si and SiGe bipolar transistors," in *IEEE Trans. Electron Devices*, vol. 49, no. 12, pp. 2151-2156, Dec 2002.

[5] J.-S. Rieh *et al.*, "Reliability of high-speed SiGe heterojunction bipolar transistors under very high forward current density," in *IEEE Trans. Device Mater. Rel.*, vol. 3, no. 2, pp. 31-38, June 2003.

[6] P. S. Chakraborty and J. D. Cressler, "Hot-Carrier Degradation in Silicon-Germanium Heterojunction Bipolar Transistors," in *Hot Carrier Degradation in Semiconductor Devices*, T. Grasser, Ed. Springer International Publishing Switzerland, 2015.

[7] M. Diop, S. Ighilahriz, F. Cacho and V. Huard, "250 GHz heterojunction bipolar transistor: From DC to AC reliability," *Int. Rel. Phys. Symp.*, Monterey, CA, 2011, pp. 4E.5.1-4E.5.5.

[8] P. Cheng, C. M. Grens and J. D. Cressler, "Reliability of SiGe HBTs for Power Amplifiers—Part II: Underlying Physics and Damage Modeling," in *IEEE Trans. Device Mater. Rel.*, vol. 9, no. 3, pp. 440-448, Sept. 2009.

[9] T. K. Thrivikraman, A. Madan and J. D. Cressler, "On the large-signal robustness of SiGe HBT LNAs for high-frequency wireless applications," in *Proc. SiRF.*, New Orleans, LA, 2010, pp. 156-159.

[10] K. A. Moen, P. S. Chakraborty, U. S. Raghunathan, J. D. Cressler and H. Yasuda, "Predictive Physics-Based TCAD Modeling of the Mixed-Mode Degradation Mechanism in SiGe HBTs," in *IEEE Trans. Electron Devices*, vol. 59, no. 11, pp. 2895-2901, Nov. 2012.

[11] B. R. Wier, K. Green, J. Kim, D. T. Zweidinger and J. D. Cressler, "A Physics-Based Circuit Aging Model for Mixed-Mode Degradation in SiGe HBTs," in *IEEE Trans. Electron Devices*, vol. 63, no. 8, pp. 2987-2993, Aug. 2016.

[12] U. S. Raghunathan *et al.*, "Physical Differences in Hot Carrier Degradation of Oxide Interfaces in Complementary (n-p-n+p-n-p) SiGe HBTs," in *IEEE Trans. Electron Devices*, vol. 64, no. 1, pp. 37-44, Jan. 2017.

[13] T. Suligoj, M. Koricic, H. Mochizuki, S. Morita, K. Shinomura and H. Imai, "Horizontal Current Bipolar Transistor With a Single Polysilicon Region for Improved High-Frequency Performance of BiCMOS ICs," in *IEEE Electron Device Lett.*, vol. 31, no. 6, pp. 534-536, June 2010.

[14] J. Žilak, M. Koričić, T. Suligoj, H. Mochizuki and S. Morita, "Impact of Emitter Interface Treatment on the Horizontal Current Bipolar Transistor (HCBT) Characteristics and RF circuit Performance," in *Proc. BCTM*, Boston, MA, 2015, pp. 31-34.

[15] J. Žilak, M. Koričić, T. Suligoj, H. Mochizuki and S. Morita, "A low-cost 180 nm BiCMOS technology with Horizontal Current Bipolar Transistor (HCBT) for wireless communication ICs," in *Proc. EuMIC*, London, 2016, pp. 373-376.

[16] M. Koričić, J. Žilak and T. Suligoj, "Double-Emitter Reduced-Surface-Field Horizontal Current Bipolar Transistor With 36 V Breakdown Integrated in BiCMOS at Zero Cost," in *IEEE Electron Device Lett.*, vol. 36, no. 2, pp. 90-92, Feb. 2015.

[17] J. Žilak, M. Koričić and T. Suligoj, "Reliability Degradation Mechanisms of Horizontal Current Bipolar Transistor," in *IEEE Trans. Electron Devices*, vol. 63, no. 11, pp. 4409-4415, Nov. 2016.

[18] Simon Tam, Ping-Keung Ko and Chenming Hu, "Lucky-electron model of channel hot-electron injection in MOSFET'S," in *IEEE Trans. Electron Devices*, vol. 31, no. 9, pp. 1116-1125, Sep 1984.

[19] K. O. Jeppson, C. M. Svensson, "Negative bias stress of MOS devices at high electric fields and degradation of MNOS devices, " *J. Appl. Phys.*, vol. 48, pp. 2004-2014, 1977.

[20] Synopsys, "Sentaurus Device User Guide, L-2016.03," Synopsys, Mountain View, CA, USA, 2016.

[21] J. Žilak *et al.*, "Examination of Horizontal Current Bipolar Transistor (HCBT) reliability characteristics," in *Proc. BCTM*, Coronado, CA, 2014, pp. 37-40.

MIPRO 2017, May 22- 26, 2017, Opatija, Croatia

Impact of the Local p-well Substrate Parameters on the Electrical Performance of the Double-Emitter Reduced-Surface-Field Horizontal Current Bipolar Transistor

Marko Koričić*, Josip Žilak* and Tomislav Suligoj*

* University of Zagreb, Faculty of Electrical Engineering and Computing, Department of Electronics, Microelectronics, Computing and Intelligent Systems, Micro and Nano Electronics Laboratory, Zagreb, Croatia
marko.koricic@fer.hr

Abstract - **Double-Emitter Reduced-Surface-Field Horizontal Current Bipolar Transistor is analyzed by the device simulations. Geometrical parameters of the local p-well substrate, which is used to introduce the second drift region are investigated. It is shown that the length of the p-well l_{pw}=0.5 µm is sufficient to obtain efficient electric field shielding and BV_{CEO} independent of the transistor current gain. The analysis of the distance between the extrinsic base and the p-well region (d_{pw}) shows that the optimum d_{pw} exists. The optimum structure with l_{pw}=0.5 µm and d_{pw}=0.6 µm has BV_{CEO}=30 V and f_T=7 GHz. With assumed p-well mask misalignment tolerances, a good trade-off between BV_{CEO} and f_T is achieved with transistors having the $f_T \cdot BV_{CEO}$ at the Johnson's limit.**

I. INTRODUCTION

One of the limits for the safe operating area of transistors is the value of a breakdown voltage. In bipolar transistors, the common emitter breakdown voltage (BV_{CEO}) is usually 2 to 3 times lower than the common-base breakdown voltage (BV_{CBO}) due to the physics of the breakdown mechanism which involves the common-emitter current gain (β) and the positive feedback that is closed at the onset of the breakdown. Therefore, BV_{CEO} represents a tighter constraint on the allowable voltage range in the circuit applications. High-voltage transistors are desirable components in the technology since they allow higher operating voltages which extends the application of the technology. They could be used in the input/output circuits, power amplifiers or simply in the applications where higher voltage swing is important. The most common approach to achieve higher BV_{CEO} in high-voltage bipolar transistors, is to reduce the collector doping concentration in order to reduce the maximum electric field at the collector-base junction. In that case, at least one additional lithography mask is needed for the fabrication of the high-voltage device together with high-speed device. The other approach to increase the BV_{CEO} is to change the potential distribution and the electric field profile in the base-collector depletion region by some form of a Reduced-Surface-Field (RESURF) effect [1]. In those structures, the collector is fully depleted when

transistors are operating in normal mode [2]-[4].

Horizontal current bipolar transistor (HCBT) is integrated with 180 nm CMOS with reported state-of-the art electrical performance among implanted base bipolar transistors [5]. A high-voltage double-emitter (DE) HCBT is added to the process at zero additional cost [4]. Due to the lateral orientation of the intrinsic transistor, the n-collector charge is confined in space by using the double-emitter geometry. In the normal operation mode, the collector charge is shared between two opposing intrinsic bases and the collector is fully depleted. After the full depletion, voltage drop across the intrinsic base-collector junction is limited and voltage is dropped across drift regions which are formed toward extrinsic collector. Maximum electric field at the intrinsic base-collector junction is limited and breakdown occurs when the electric field in the drift region reaches the critical value. As a result BV_{CEO} is increased from 3.5 to 12.7 V compared to the high-speed transistor. Recently, a double-emitter Reduced-Surface-Field (DE RESURF) HCBT has been reported [6]. Additional shaping of the electric field is accomplished by using the local p-well substrate, which is available in CMOS fabrication. The p-well region is used to deplete the portion of the collector and introduce the second drift region in order to limit the electric field in the abovementioned first drift region of the DE HCBT. In that case, the voltage is mainly dropped across the second drift region where the peak electric field appears and causes the transistor breakdown. BV_{CEO} as high as 36 V is reported for this structure. In this paper DE RESURF HCBT is analyzed by the device simulations with the emphasis on the influence of the local p-well substrate parameters on the electrical characteristics.

II. SIMULATION STRUCTURE

The analysis of the breakdown voltage mechanisms of the DE RESURF HCBT is given in [7], where the measurement results of fabricated transistors along with the 3D device simulations are presented. In order to have extensive examination of the structure, a large number of simulation structures have to be analyzed. 3D simulations have an order of magnitude larger number of simulation nodes compared to 2D. In order to reduce the simulation

---

This work was supported by the Croatian Science Foundation under contract no. 9006.

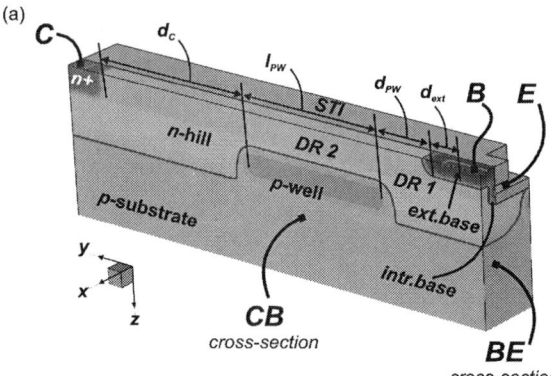

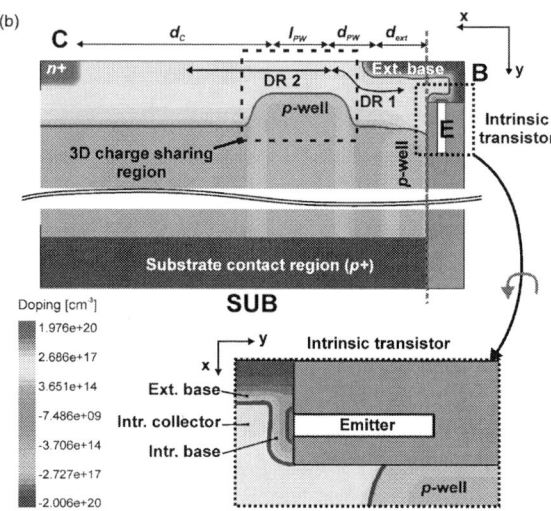

Fig. 1. DE RESURF HCBT simulation structures: a) 3D simulation structure used in [7], b) 2D simulation structure used in this paper.

time and maintain accuracy simultaneously, special attention should be given to optimization of the simulation mesh. This often leads to convergence problems especially if current boundary conditions are used as in the case of the simulations of the output characteristics of bipolar transistor. Therefore, in order to perform the analysis in a reasonable time, we have developed a 2D simulation model of the DE RESURF HCBT.

3D and 2D simulation models are presented in Figs. 1.a and 1.b, respectively. For the 3D simulations only one quarter of the device is used because structure has two lines of symmetry. Simulator's default reflective boundary condition assumes that the structure is mirrored at the lines of symmetry. The double-emitter structure uses this symmetry in order to accomplish the fully depleted collector and the structure relies on 3D effect of charge sharing. Therefore, the concept of the 2D simulation structure is not straightforward and also cannot be completely accurate. However, a good approximation can be made. In 2D simulation structure from Fig. 1.b we have only one emitter and one intrinsic base, whereas 3D simulation structure from Fig. 1.a has two emitters and intrinsic bases, because the reflective boundary condition is used. In order to accomplish the similar geometry for

the intrinsic collector charge sharing in the 2D model, the extrinsic base is folded and placed opposite to the intrinsic base. The doping profiles on the left of the red dashed line including the p-well channel stopper below the emitter and excluding the folded portion of the extrinsic base are rotated by 90° clockwise. The portion of the structure on the left of the red dashed line has the same geometry as the 3D model in Fig. 1.a in the CB-cross-section. An extrinsic base extension (d_{ext} in Fig. 1) acts as a field plate and shields the intrinsic transistor from the collector voltage after intrinsic collector is fully depleted [4]. Transition of the lateral intrinsic collector profile (to the right of the red dashed line) and the vertical extrinsic collector profile (to the left of the red dashed line) is adjusted carefully to obtain similar potential distribution and current flow through this region as in the 3D simulation structure. The last tweak to obtain sufficiently accurate 2D simulation structure is to set the doping profile of the collector above p-well region. Since the p-well extends below the shallow trench isolation (STI) oxide (see Fig. 1.a), the 3D charge sharing in the collector above the p-well takes place due to the fringing field component, which is closed through the STI between donor charge in the collector and acceptors in the p-well, as explained in [7]. In order to capture this effect an additional background acceptor doping is placed in the region marked by the dashed square in Fig. 1.b which mimics the fringing field consumption of the donor charge by the p-well acceptors. The doping is adjusted to obtain a similar potential distribution in this portion of the collector as in the 3D simulation structure. In both structures, the substrate contact is placed at the bottom with the total structure height of 5μm.

The important geometrical parameters of the transistor are marked in Fig. 1. The extrinsic base extension (d_{ext}) influence on the electrical characteristics of the double-emitter structure is given in [4] and is kept at 0.6 μm. Distance from the n+ collector contact region to the p-well (d_c) should be large enough not to limit the value of the breakdown voltage. It is kept constant at the value of 2 μm in the simulations. Since the p-well is used to introduce the second drift region (DR 2 in Fig. 1), the most important geometrical parameters are the distance between the extrinsic base and the p-well (d_{pw}) and the length of the p-well region (l_{pw}), which are investigated in this paper.

III. DEVICE SIMULATIONS

Simulated common-emitter output characteristics of the transistor with small d_{pw}=0.2 μm and small l_{pw}=0.1 μm are shown in Fig. 2.a. All transistor currents, i.e. the collector current (I_C), the emitter current (I_E) and the substrate current (I_{SUB}) are monitored in order to conclude about the breakdown mechanisms involved. Soft-breakdown is observed with the BV$_{CEO}$ around 13 V. It can also be seen that the I_C and the I_E rise at the same rate meaning that the classical BV$_{CEO}$ mechanism occurs. Avalanche holes generated in the base-collector depletion region flow to the base and are added to the constant base terminal current increasing the hole injection to the emitter. Due to the emitter efficiency, a large electron back-injection follows, which in turn increases the avalanche in the base-collector depletion region. Positive

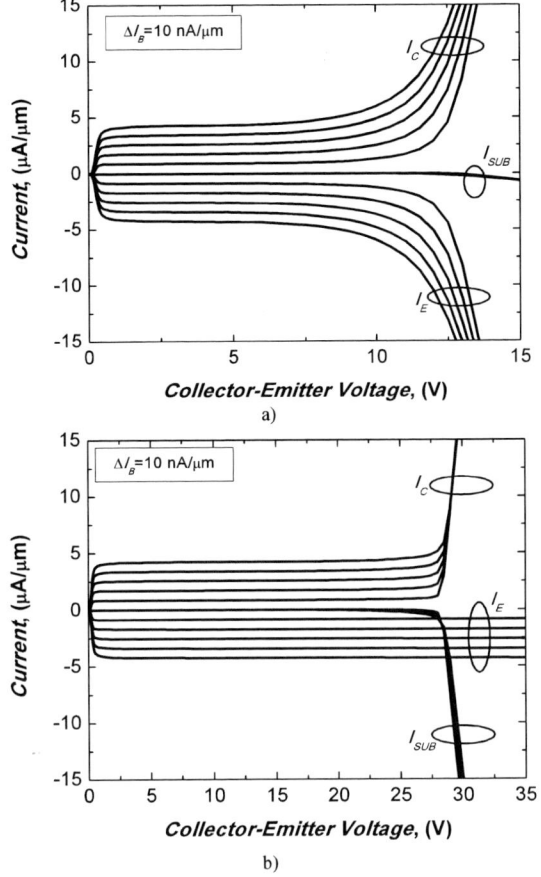

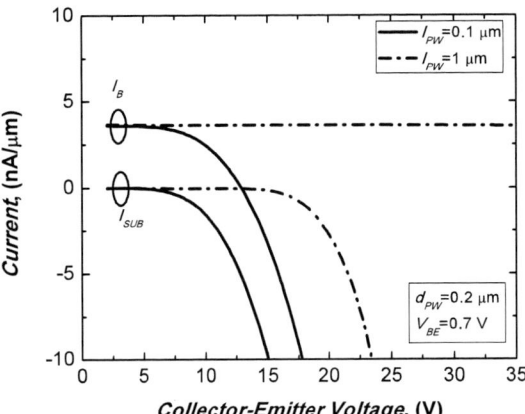

Fig. 3. Forced-V_{BE} simulations of transistors with d_{pw}=0.2 µm and diferenet l_{pw}.

the peak electric field at the end of the DR 1 is kept below the critical value and the breakdown is caused by the peak electric field at the collector-pwell junction placed at the end of DR 2.

Fig. 3 shows the simulation results of widely used forced-V_{BE} measurement of the BV$_{CEO}$. Since constant V_{BE} is set in the simulations, the hole injection to the emitter is approximately constant. The avalanche holes generated in the collector-base depletion region are then flowing to the base contact, which reduces the value of the I_B observed at the base terminal. The BV$_{CEO}$ is determined as the V_{CE} at which the I_B reverses its sign. In the case of l_{pw}=0.1 µm device, the I_B reverses its sign at V_{CE}=15 V, which is the value of simulated BV$_{CEO}$. This simulation confirms that the avalanche holes have reached the base region. We can also see that the I_{SUB} starts to change at the same V_{CE} meaning that the part of the avalanche holes flow to the substrate. In the case of long l_{pw}=1 µm transistor, the I_B remains constant even for V_{CE} above BV_{CEO} determined from the output characteristics, meaning that avalanche holes are not flowing to the base. Therefore in case of this transistor, the BV$_{CEO}$ cannot be determined by the forced-V_{BE} simulation.

Simulated output characteristics of the large d_{pw}=1 µm transistors are shown in Fig. 4. For the small l_{pw}=0.1 µm transistor we can see that the I_C and the I_E magnitude increase from around 10 V, similar as in the case of the small d_{pw} transistor. However breakdown is very soft indicating that the peak electric field responsible for avalanche has very weak dependence on the collector voltage. This means that the intrinsic part of the transistor is not properly shielded by the short p-well (i.e. small l_{pw}) and that the peak electric field at the end of the DR 1 slightly changes with the V_{CE}. However, the shielding is better than in the case of small d_{pw} transistor since the DR 1 and the DR 2 (see Fig. 1.b) are not merged and together sustain a larger voltage drop. Substantial increase in the I_C and the I_{SUB} starts around 27 V which is observed as a hard-breakdown. At this point substantial avalanche associated with the collector-pwell junction is superimposed to the soft-breakdown resulting in a slightly smaller value of the breakdown voltage compared to the

Fig. 2. Simulated common-emitter output characteristics of the structure with d_{pw}=0.2 µm and a) l_{pw}=0.1 µm, b) l_{pw}=1 µm.

feedback is closed via avalanche holes and the I_C and the I_E increase at the same rate.

Common emitter output characteristics of the transistor with small d_{pw}=0.2 µm and large l_{pw}=1 µm are shown in Fig. 2.b. BV$_{CEO}$ determined from the characteristics is increased up to 30 V compared to the small l_{pw} device. It can be observed that the I_C and the I_{SUB} increase at the onset of the breakdown, whereas the I_E remains constant indicating the different breakdown mechanism compared to the small l_{pw} structure. Since the I_E is not increased, we can conclude that the avalanche holes generated in the base-collector depletion region do not end up in the base and do not increase the hole injection to the emitter. Therefore, the positive feedback loop of the classical BV$_{CEO}$ mechanism is broken making it independent of the transistor β. The avalanche holes are collected by the local p-well substrate and the avalanche current flows between the collector and the substrate. The breakdown observed in the characteristics is a hard breakdown of the reverse polarized collector-pwell junction. Compared to the small l_{pw} device, longer p-well region acts as a more efficient field plate and shielding of the electric field in the DR 1 is more efficient. As a result,

85

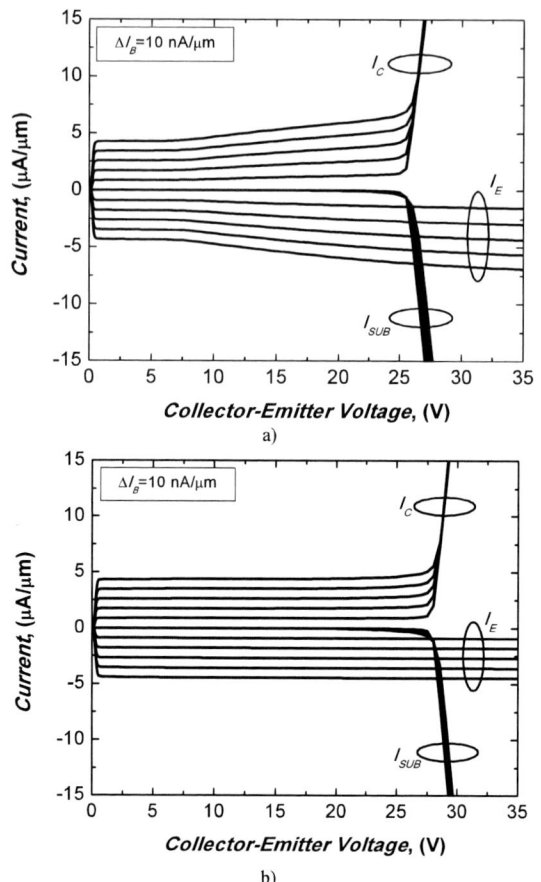

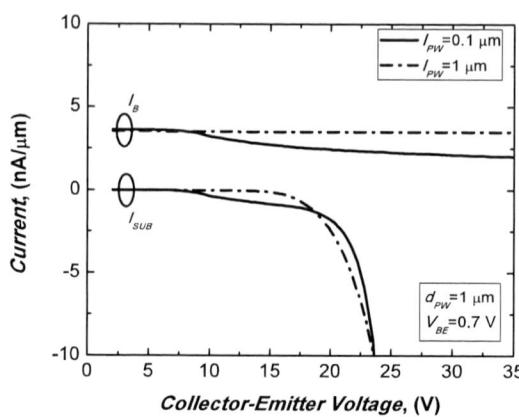

Fig. 5. Forced-V_{BE} simulation of transistors with $d_{pw}=1$ µm and differenet l_{pw}.

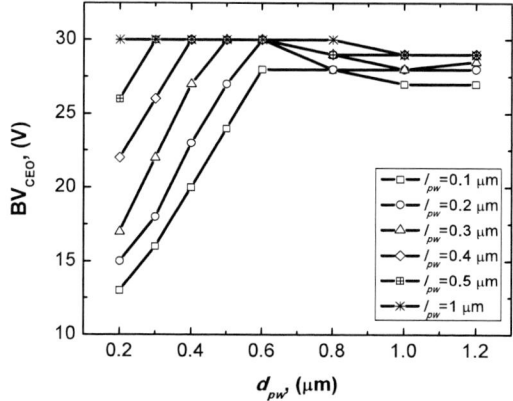

Fig. 4. Simulated common-emitter output characteristics od the structure with $d_{pw}=1$ µm and a) $l_{pw}=0.1$ µm, b) $l_{pw}=1$ µm.

Fig. 6. BV$_{CEO}$ simulated from the output characteristics of the DE RESURF HCBT structures with different d_{pw} and l_{pw}.

collector-pwell breakdown voltage alone. In the case of long $l_{pw}=1$ µm transistor, we see similar behavior as in the case of small d_{pw} transistor. BV$_{CEO}$ is determined by the collector-pwell junction breakdown and the positive feedback of classical BV$_{CEO}$ is broken resulting in the constant I_E at the onset of breakdown. Large l_{pw} presents efficient shield for the intrinsic transistor, regardless of the value of d_{pw}.

Simulated forced-V_{BE} characteristics of the large $d_{pw}=1$ µm structures are shown in Fig. 5. In the case of large l_{pw}, characteristics are the same as for the small d_{pw} device with large l_{pw}. Shielding is efficient and breakdown is caused by the collector-pwell junction breakdown. The I_B remains constant and does not change sign for V_{CE} beyond BV$_{CEO}$. In the case of small l_{pw}, we see that avalanche starts around $V_{CE}=8$ V but avalanche current slightly increases with V_{CE}, which is seen as a reduced drop in the I_B. We can also observe that the I_B is not negative at $V_{CE}=27$ V even though substantial avalanche current is generated. Since both breakdown mechanisms are involved, slightly smaller breakdown voltage is observed in the output characteristics in Fig. 4.a. compared to the long l_{pw} device characteristics in Fig. 4.b.

In order to find the optimum geometry of the DR 2, the influence of the d_{pw} and the l_{pw} on the electrical characteristics of the DE RESURF HCBT is analyzed in terms of BV$_{CEO}$ and the cutoff frequency (f_T). The l_{pw} should be chosen in order to have efficient shielding properties of the DR 2 and high value of the BV$_{CEO}$, whereas the d_{pw} should be chosen to have the best electrical performance with assumed lithography mask misalignment tolerances. Since forced-V_{BE} simulations cannot be used for the determination of the BV$_{CEO}$ it was extracted from the output characteristics. The BV$_{CEO}$ dependence on the d_{pw} with the l_{pw} as a parameter is shown in Fig. 6. For larger l_{pw}, BV$_{CEO}$ saturates at the value around 30 V, which is the value of the collector-pwell junction breakdown. Furthermore, for structures with $d_{pw}\geq0.3$ µm, BV$_{CEO}$ saturates at 30 V if $l_{pw}\geq0.5$ µm. As the d_{pw} increases, the drift regions DR 1 and DR 2 are less overlapped and the smaller l_{pw} is sufficient for the efficient shielding of the electric field in the DR 1. The f_T dependencies on the d_{pw} with the l_{pw} as a parameter are shown in Fig. 7. Since by the addition the drift regions DR 1 and DR 2, the f_T becomes dominated by the base-collector depletion region transit time, the characteristics are simulated at $V_{CE}=20$ V, which is the voltage sufficient

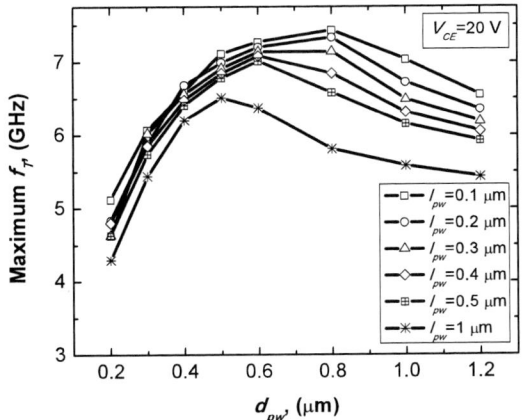

Fig. 7. Simulated f_T of the DE RESURF HCBT structures with different d_{pw} and l_{pw}. Structures with $BV_{CEO} < 20$ V are simulated at $V_{CE} = 10$ V.

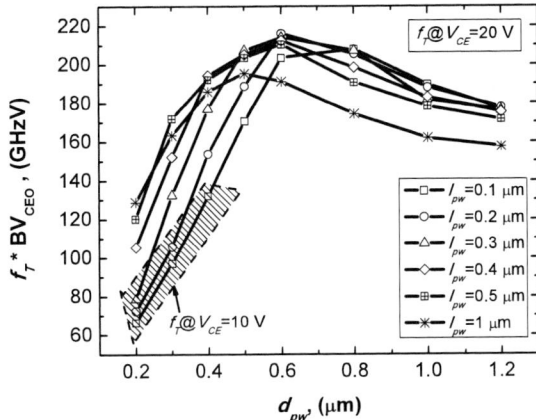

Fig. 8. Simulated $f_T \cdot BV_{CEO}$ of the DE RESURF HCBT structures with different d_{pw} and l_{pw}. Structures with $BV_{CEO} < 20$ V are simulated at $V_{CE} = 10$ V.

for the formation of both the DR 1 and the DR 2. The f_Ts for the structures with the $BV_{CEO} < 20$ V (see Fig. 6) are simulated at $V_{CE} = 10$ V. It can be observed that f_T reduces as the l_{pw} is increased. This is expected behavior since by increasing the l_{pw}, the length of the DR 2 increases resulting in the larger base-collector depletion region transit time. Interestingly, f_T is reduced for the structures with very small d_{pw}. For very small d_{pw}, the collector charge sharing between the extrinsic base and the p-well acceptors is more pronounced since there is less available donor charge between the extrinsic base and the p-well. Associated depletion regions merge at lower V_{CE} and the lateral electric field along the current path in DR 1 is lower. As a result, carriers drift through DR 1 at the velocity lower than the saturation velocity leading to the longer transit time. Furthermore, current crowding through this part of the collector is more pronounced causing the onset of the Kirk effect at lower I_C. The f_T increases for structures with d_{pw} up to 0.6 µm after which it falls off indicating that the minimum transit time through the DR 1 is achieved. For larger d_{pw}, transit time increases due to longer overall length of the DR 1 and the DR 2. The optimum performance in terms of f_T value is achieved for structures with $d_{pw} = 0.6$ µm.

Since f_T and BV_{CEO} are always in trade-off, their product is usually given as a figure of merit for transistor performance. The results of the $f_T \cdot BV_{CEO}$ are shown in Fig. 8. If we choose the d_{pw} lower limit for acceptable f_T performance (e.g., $d_{pw} \geq 0.4$ µm) we see from Fig. 6 that $l_{pw} \geq 0.4$ must be chosen in order to have the efficient shielding by the DR 2 and the highest value of the $BV_{CEO} = 30$ V. In that case, $f_T \cdot BV_{CEO}$ around Johnson's limit [8] is achieved, which is around 180 GHzV for the implanted base bipolar transistors. In order to be on the safe side, we propose that $l_{pw} = 0.5$ µm is the optimum length of the p-well region. The value of the d_{pw} should be chosen such to accommodate the p-well mask misalignment tolerance. If the optimum value with respect to the f_T is chosen (i.e. $d_{pw} = 0.6$ µm), then the mask misalignment of e.g. ±0.2 µm would give transistors with $f_T \cdot BV_{CEO}$ result around the Johnson's limit (see Fig. 8), showing a good immunity to the p-well mask misalignment tolerance.

IV. CONCLUSIONS

The impact of the geometrical parameters of the local p-well substrate on the electrical performance of the DE RESURF HCBT is investigated. Two breakdown mechanisms are identified depending on the shielding properties of the drift region which is introduced by the local p-well. The BV_{CEO} of the transistor can be made independent of the current gain by having sufficiently long p-well region. In that case, the BV_{CEO} is set by the collector-substrate junction breakdown. The optimum f_T vs. BV_{CEO} trade-off is obtained for $l_{pw} = 0.5$ µm and $d_{pw} = 0.6$ µm. Variation of the characteristics with d_{pw} shows that transistors are rather immune to the p-well mask misalignment tolerances. Transistors with high value of $f_T \cdot BV_{CEO}$ around Johnson's limit are obtained.

REFERENCES

[1] J.A. Appels, and H.M.J Vaes, "High Voltage Thin Layer Devices (RESURF Devices)," in *IEDM Tech. Dig.*, 1979, pp.238-241.

[2] H. Kondo, and Y. Yukimoto, "A New Bipolar Transistor – GAT," *IEEE Trans. Electron Devices*, vol. 27, no. 2, pp. 373-379, Feb.1980.

[3] J. Cai, M. Kumar, M. Steigenvalt, H. Ho, K. Schonenberg, K. Stein, H. Chen, K. Jenkins, Q. Ouyang, P. Oldiges, and T. Ning, "Vertical SiGe-Base Bipolar Transistors on CMOS-Compatible SOI Substrate", *BCTM 2003*, pp. 215-218.

[4] M. Koричić, T. Suligoj, H. Mochizuki, S. Morita, K. Shinomura, and H. Imai, "Double-Emitter HCBT Structure—A High-Voltage Bipolar Transistor for BiCMOS Integration,", *IEEE Trans. Electron Devices*, vol. 59, no. 12 pp. 3647 – 3650, Dec. 2012.

[5] T. Suligoj, M. Koricic, H. Mochizuki, S. Morita, K. Shinomura and H. Imai, "Horizontal Current Bipolar Transistor With a Single Polysilicon Region for Improved High-Frequency Performance of BiCMOS ICs," in *IEEE Electron Device Lett.*, vol. 31, no. 6, pp. 534-536, June 2010.

[6] M. Koричić, J. Žilak and T. Suligoj, "Double-Emitter Reduced-Surface-Field Horizontal Current Bipolar Transistor With 36 V Breakdown Integrated in BiCMOS at Zero Cost," in *IEEE Electron Device Lett.*, vol. 36, no. 2, pp. 90-92, Feb. 2015.

[7] M. Koричić, J. Žilak, T. Suligoj, "Investigation of Double-Emitter Reduced-Surface-Field Horizontal Current Bipolar Transistor Breakdown Mechanisms," in *Proc. Bipolar/BiCMOS Circuits Technol. Meeting*, New Brunswick, NJ, 2016, pp. 25-28.

[8] E. O. Johnson, "Physical limitations on frequency and power parameters of transistors," *RCA Rev.*, vol. 26, pp. 163–177, Jun. 1965.

MIPRO 2017, May 22- 26, 2017, Opatija, Croatia

Characterization of Measurement System for High-Precision Oscillator Measurements

Ivan Brezovec*, Marko Magerl*, Josip Mikulic†, Gregor Schatzberger† and Adrijan Baric*

*University of Zagreb Faculty of Electrical Engineering and Computing, Unska 3, Zagreb, Croatia
Email: ivan.brezovec@fer.hr

†ams AG, Tobelbader Strasse 30, Premstaetten 8141, Austria
Email: josip.mikulic@ams.com

Abstract—**Temperature stability of a high-precision oscillator is characterized by measurements in a temperature chamber. Two time constants are measured for a given temperature of the chamber: (i) time required for the silicon to reach the steady-state temperature obtained by measuring the time-domain voltage response of the on-chip temperature sensor; (ii) time required for the oscillator circuit to reach the steady-state frequency obtained by measuring the oscillator frequency in the time-domain. The temperature probe for measuring the chamber temperature is characterized in terms of its response to a step in temperature. The noise performance of the measurement system is characterized based on Allan deviation.**

Index Terms—**oscillator characterization, temperature calibration, on-chip temperature sensor, Allan deviation**

I. INTRODUCTION

The precise measurement of integrated circuits is not an easy task to perform, especially when various voltage and current variations as well as temperature changes have to be taken into account in order to fully test the system. Also, numerous measurements have to be repeated to prove the validity of the previous results. To speed up such processes, an automated system is shown to be a good solution.

High precision frequency measurements are important for the characterization of oscillator intellectual property (IP) integrated circuit blocks. The frequency of most oscillator circuit architectures depends on the temperature [1], therefore the new oscillator designs have to be characterized in a wide temperature range. The requirements for the precision of the measurement system increase as the precision of integrated oscillators increases into the 20 ppm/°C range [1].

In [2] a measurement system with an automatically controlled temperature chamber is presented. The authors present the control system implemented in the graphical programming language available in the National Instruments LabVIEW [3] software.

This paper presents an automated measurement system for a high-precision oscillator. The measurement system is controlled using the PyVISA module [4] in Python. The instruments and PC are connected via USB and GPIB ports. The main purpose of the presented system is the measurement of the dependence of the oscillator frequency on the ambient temperature in the range from -40°C to 150°C. This temperature range is wider than the range of the measurement system in [2] (0°C to 100°C). The measured test chip has two

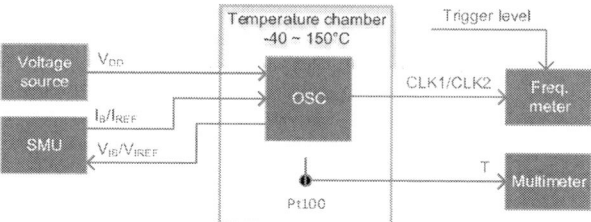

Fig. 1: Block diagram of the measurement system.

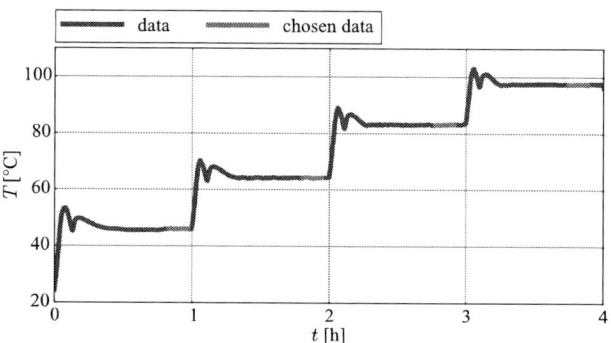

Fig. 2: Temperature sweep used for the static characteristic measurements.

temperature-dependent voltages that act as on-chip thermometers. The dependence of these voltages on the temperature is measured in the steady-state. The time-constant of these voltages is characterized by measuring the response to a step change of the ambient temperature. The noise performance of the system is characterized based on the Allan deviation [5]. The measurement methodology is verified using a commercial low-jitter signal generator and it is then applied to the measured on-chip oscillator, similarly to [6].

II. MEASUREMENT RESULTS

A. The Measurement System

The measurement system is shown in Fig. 1. It consists of the following instruments: two three-channel source and monitor (SMU) units (Keysight U2722A [7]), one dual-chan-

88

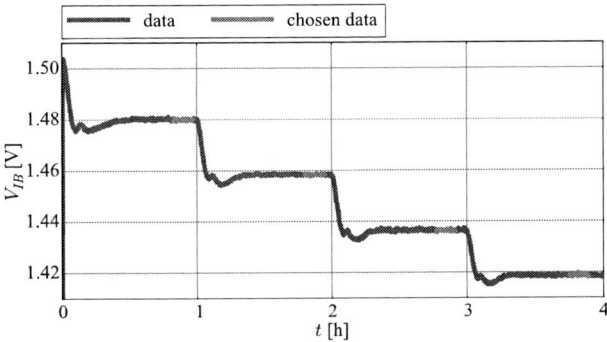

Fig. 3: Response of the voltage V_{IB} to the temperature sweep.

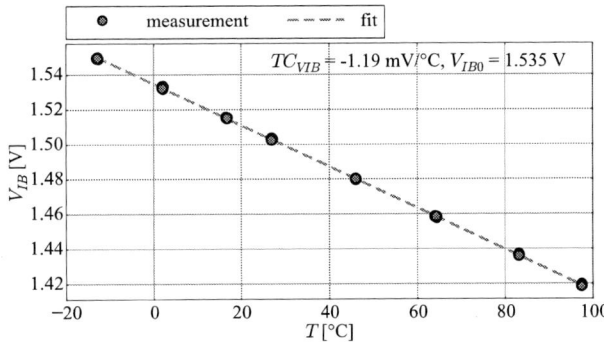

Fig. 5: Temperature coefficient of the voltage V_{IB}.

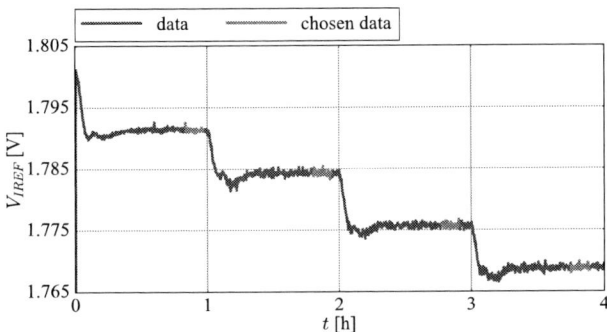

Fig. 4: Response of the voltage V_{IREF} to the temperature sweep.

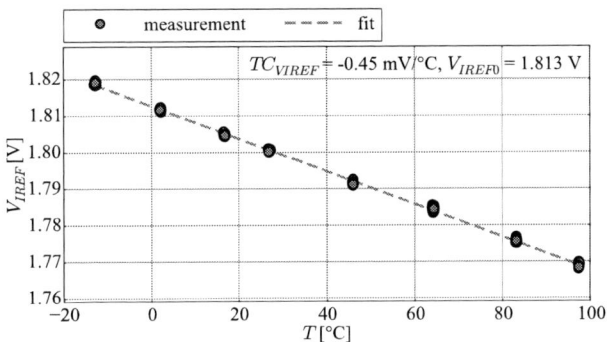

Fig. 6: Temperature coefficient of the voltage V_{IREF}.

nel voltage source (Keysight E3646A [8]), two dual-channel frequency meters (Keysight 53220A [9]) and a multimeter (Keysight 34401A [10]) set up to measure temperature using a 4-wire Pt100 RTD probe [11]. The ambient temperature is set using a temperature chamber operating from -40°C to 150°C. The SMU units and voltage source are used to set the bias voltages and currents required for the oscillator operation. The oscillator has complementary frequency outputs CLK1 and CLK2 with the nominal frequency of 1 MHz.

B. Static Temperature Characteristics

Two on-chip voltages marked as V_{IB} and V_{IREF} have linear temperature dependence by design. In order to obtain the temperature coefficients, the test chip is exposed to various ambient temperatures. Fig. 2 shows the ambient temperature profile used to obtain the steady-state voltages at the following temperatures: 25°C, 45°C, 65°C, 85°C and 95°C. The settling time of the temperature chamber is in the order of one hour. A similar measurement is also made for the temperatures below the room temperature: -30°C, -15°C, 0°C and 15°C. In order to get a more precise measurement, the averaging factor of 100 points is used for the voltage measurements.

The voltages V_{IB} and V_{IREF}, measured during the entire temperature sweep, are shown in Figs. 3, 4. The chosen steady-state points marked in red are used to plot the tem-

perature dependence of voltages V_{IB} and V_{IREF}, shown in Figs. 5, 6. The temperature coefficients are calculated using linear regression. Their values are TC_{VIB} = -1.19 mV/°C for V_{IB} and TC_{VIREF} = -0.45 mV/°C for V_{IREF}. The fitted voltage values at 0°C are marked as V_{IB0} and V_{IREF0}.

The steady-state dependence of the voltages V_{IB} and V_{IREF} on the temperature is linear. The relationship between these voltages and the temperature allows monitoring the temperature inside the package of the test chip. The measured points show that the measurement of the linear relationship is repeatable. The voltage V_{IB} is more sensitive to temperature changes and it is therefore used as the preferred of the two on-chip thermometers.

C. Dynamic Temperature Characteristics

The response of the Pt100 RTD probe and the response of the voltages V_{IB} and V_{IREF} to a step of the ambient temperature are shown in Figs. 7, 8 and 9. The expected steady-state value of the voltages V_{IB} and V_{IREF} calculated from the static temperature coefficients is also shown. The step response is achieved as follows. The chamber is heated to a the starting temperature T_0, e.g. 50°C, and the probe and test chip are left in the chamber for an hour to stabilize. The probe and the PCB with the test chip is then removed from the chamber to the room temperature. This procedure is repeated for the following starting temperatures : T_0 = -30°C, 0°C, 50°C, 75°C

89

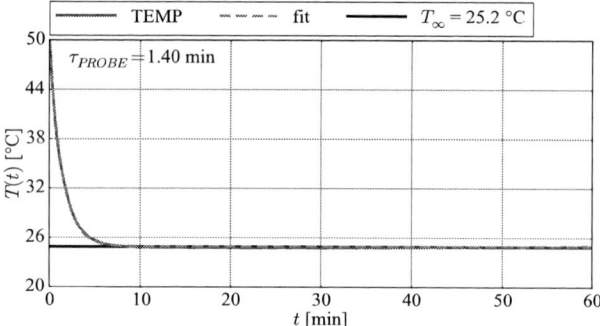

Fig. 7: Temperature response of the Pt100 RTD probe.

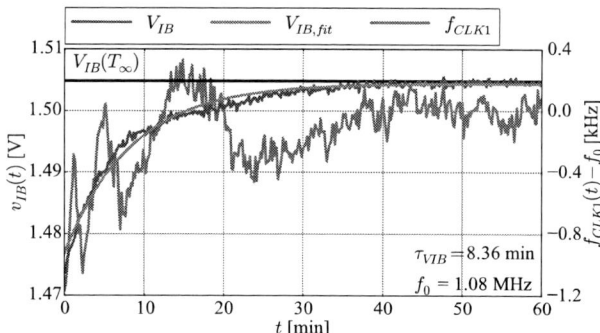

Fig. 8: Temperature response of voltage V_{IB} and oscillator frequency f_{CLK1}. The time-constant τ_{VIB} is equal to 8.357 min.

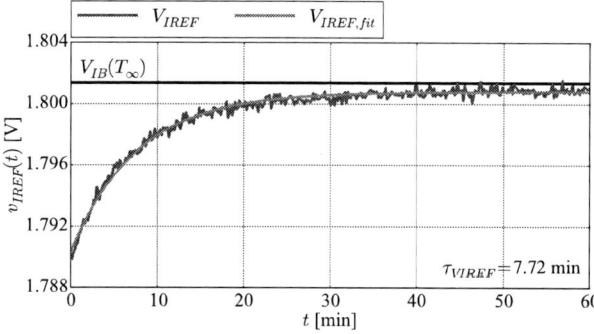

Fig. 9: Temperature response of V_{IREF}. The time-constant τ_{VIREF} is equal to 7.720 min.

and 100°C. The data is recorded every 10 seconds for 1.5 hrs after the probe and PCB are taken out of the temperature chamber.

The measured step response of the Pt100 probe is fitted to the exponential function in the form:

$$f(t) = (T_0 - T_\infty)e^{-t/\tau} + T_\infty \qquad (1)$$

The fitted coefficients τ, T_0 and T_∞ are shown in Table I. The nominal values of the starting temperatures $T_{0,nom}$ are

TABLE I: Exponential function coefficients for the probe.

$T_{0,nom}$ [°C]	τ_{PROBE} [min]	T_0 [°C]	T_∞ [°C]
-30	1.381	-30.35	25.17
0	1.435	3.35	25.58
50	1.396	49.94	24.90
75	1.640	77.64	25.40
100	1.428	104.79	25.71
average	1.456	/	25.20

TABLE II: Exponential function coefficients for V_{IB}.

$T_{0,nom}$ [°C]	τ_{VIB} [min]	V_0 [V]	V_∞ [V]
-30	8.106	1.567	1.506
0	8.804	1.530	1.505
50	8.357	1.477	1.504
75	7.388	1.449	1.503
100	6.923	1.423	1.503
average	7.915	/	1.504

TABLE III: Exponential function coefficients for V_{IREF}.

$T_{0,nom}$ [°C]	τ_{VIREF} [min]	V_0 [V]	V_∞ [V]
-30	7.180	1.826	1.801
0	8.332	1.811	1.801
50	7.720	1.790	1.801
75	7.757	1.780	1.801
100	6.900	1.771	1.801
average	7.578	/	1.801

taken between -30°C and 100°C. The average time-constant τ_{PROBE} of the Pt100 probe is equal to 1.46 min.

The step response of the Pt100 RTD probe used to measure the ambient temperature in the chamber, depicted in Fig. 7, shows that the probe requires certain time to adapt to the new ambient temperature. The probe is modeled well as a first-order system. It can be concluded from the figure that the probe needs about 10 minutes to adapt to a 25°C temperature step. This is in line with the average time-constant of 1.456 minutes shown in Table I.

The step responses of the on-chip voltages V_{IB} and V_{IREF} are also fitted to an exponential function in the form:

$$f(t) = (V_0 - V_\infty)e^{-t/\tau} + V_\infty \qquad (2)$$

The fitted coefficients τ, V_0 and V_∞ are shown in Tables II, III. The time-constants τ_{VIB} and τ_{VIREF} of the voltages V_{IB} and V_{IREF} at 50°C are equal to 7.92 min and 7.58 min, respectively.

Figs. 8, 9 show that the time-constant of the on-chip thermometers is much longer than the time-constant of the temperature probe. The time-constants of the voltages V_{IB} and V_{IREF}, given in Tables II and III, have an average of 7.91 and 7.58 minutes, respectively. The speed of the temperature profile used during the frequency measurements should be adjusted according to these time-constants. The

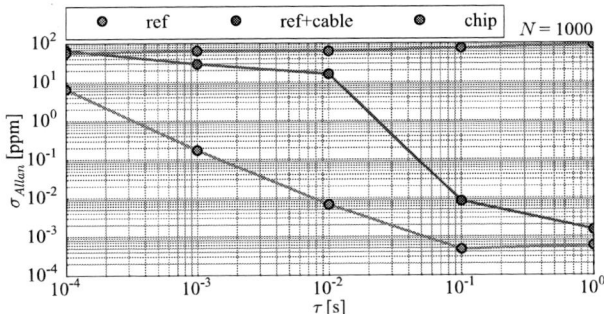

Fig. 10: Allan deviation with averaging factor $N = 1000$.

TABLE IV: Allan deviation for different averaging factors.

σ_{Allan} [ppm] $\quad N$ $\quad\tau$ [s]	ref		ref+cable		test chip	
	100	1000	100	1000	100	1000
10^{-4}	65.6	65.5	63.4	65.9	56.4	57.4
10^{-3}	1.7e-1	1.7e-1	27.8	29.2	37.8	64.2
10^{-2}	6.9e-3	6.7e-3	14.6	16.1	38.9	62.4
10^{-1}	8.2e-4	4.7e-4	6.3e-3	8.3e-3	57.1	74.8
10^{0}	6.4e-4	5.8e-4	1.1e-3	1.5e-3	82.6	91.3

slow time-constant of the temperature-dependent voltages is influenced by two factors: (i) the time required for the circuit to reach the steady-state in the new ambient temperature; (ii) the time required for the temperature change to propagate from the ambient into the package of the test chip. The oscillator chip frequency, shown in green line in Fig. 8 reaches the steady-staet at the same time as the voltages V_{IB}, V_{IREF}, however the transient shape is not exponential. This behaviour is caused by the compensation mechanism built into the oscillator circuit that becomes active during the transient phase.

Figs. 8, 9 also show the expected steady-state stable value $V(T_\infty)$ calculated from the static characteristics in Figs. 5, 6 at the stabilized ambient temperature of 25.2 °C. The error of the steady-state value is larger for V_{IREF} than for V_{IB}. This behaviour can be explained by the fact that the voltage V_{IREF} has a larger spread than the voltage V_{IB}, as well as a lower temperature coefficient.

D. Noise Performance

The noise performance of the measurement system is characterized using the Allan deviation, defined in [5] as:

$$\sigma_{allan} = \frac{\sigma}{f_{CLK}} \quad (3)$$

where σ is the standard deviation and f_{CLK} is the mean value calculated over N samples of the measured oscillator frequency. The Allan deviation is a measure used for characterizing the frequency stability and/or phase noise of circuits such as oscillators, mixers [5]. Its value is proportional to the combined noise level of the measured circuit and the

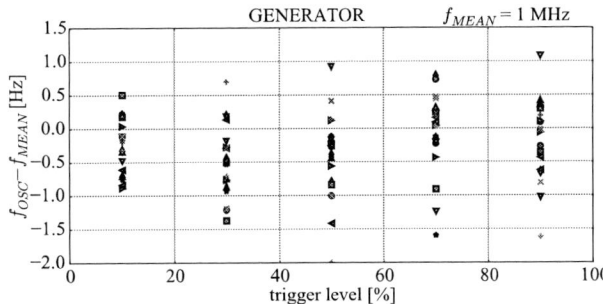

Fig. 11: Frequency measurements of the signal generator for various trigger levels.

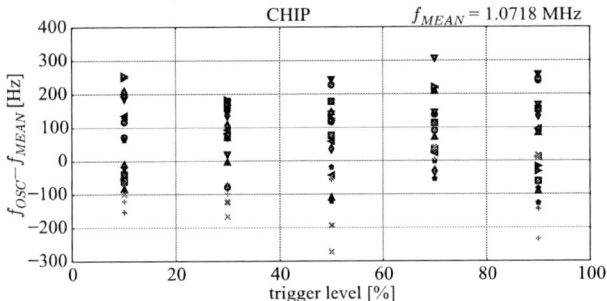

Fig. 12: Frequency measurements of the test chip the for various trigger levels.

measurement system. The number of samples N is varied between 100 and 1000. The width of the frequency-meter sampling window (i.e. the gate time) is varied from 100 μs to 1 s, with each width being 10 times larger than the previous one. In order to test the measurement system and methodology, the Allan deviation of a reference frequency generator (Keysight 33500B [12]) set to frequency $f = 1$ MHz is measured.

The Allan deviation of the reference generator connected to the frequency-meter via coaxial cables with BNC connectors as a function of the gate-time τ is shown in Fig. 10 in red. The Allan deviation is reduced with increasing gate times up to 0.1 s as the white noise component is averaged out. For higher gate times, the Allan deviation saturates at a constant level defined by the pink noise. The measurements of the test chip obtained using the same system settings are shown in green. The test chip is connected to the frequency-meter via a 2 meter long unshielded 7 core cable required to reach the chamber. Finally, the measurements for the generator are connected to the frequency-meter using the same cables that are used for the test chip measurements are shown in blue line. The Allan deviation measurements for the number of samples $N = 100$ and $N = 1000$ are summarized in Table IV.

The Allan deviation analysis in Table IV shows that in the case of the reference generator [12] (the red line), better results are obtained for higher number of samples N. In the

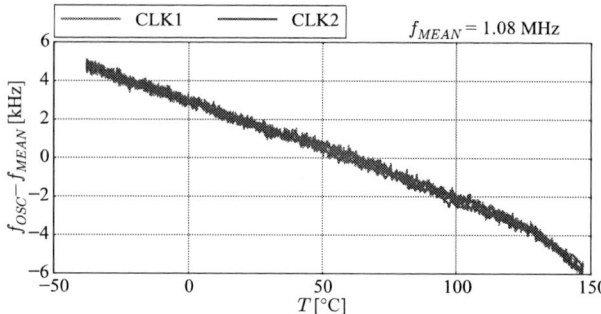

Fig. 13: TMeasured frequency of the oscillator chip as a function of temperature.

case of the generator frequency measurements with the long cables (the blue line), the measurement does not match the one obtained using the shorter coaxial cables and BNC connectors (the red line) for lower gate times τ. For higher gate times τ, the measurement with long cables becomes comparable to the reference case. It is concluded that the noise introduced by the non-shielded long cables and the used pin-header connectors is too high to measure the Allan deviation properly. Future work for improving the presented measurement system should include introducing methods that reduce the noise level, e.g. the usage of coaxial cables and BNC connectors.

E. Measurement Repeatability and Dependence on Trigger Level

The repeatability of the frequency measurements is evaluated for varying trigger levels of the frequency-meter. The trigger level of the frequency meter represents the voltage level at which the period count is incremented, i.e. it is used analogously to the trigger level on an oscilloscope. The measurements of the low-jitter signal generator are shown in Fig. 11, and of the test chip in Fig. 12. The deviation of the oscillation frequency f_{OSC} from the mean frequency value f_{MEAN} is shown. The measurement is repeated ten times for a number of trigger levels: 10%, 30%, 50%, 70% and 90%. Each point represents the mean value of 100 frequency measurements using the gate-time of $\tau = 100$ μs, and the different measurement runs are represented by different marker types. The presented figures visualize the spread of the measurement.

The obtained results show that different trigger levels give a similar spread of frequencies. It is concluded that the trigger level is not critical for the measurements, both for the reference generator, and for the test chip. The results show that the test chip has wider spread than the generator at each trigger level. The order of magnitude of the measured frequencies for the generator is ± 1.5 Hz and it is ± 300 Hz for the test chip.

The presented measurement system is used to obtain the temperature dependence of the oscillator chip frequency shown in Fig. 13. The ambient temperature is swept in the temperature range from -40°C to 150°C. The slope of the temperature

sweep is set to 0.6°C per minute. Both the rising and the falling temperature ramp are included in the presented measurement results. Since no hysteresis is observed, it can be concluded that the used temperature slope is slow enough for the oscillator frequency to follow the temperature change.

III. CONCLUSION

A measurement system for characterization of a high-precision oscillator is evaluated. The system enables the measurement of the oscillator frequency dependence on temperature. The static and dynamic characteristics of the temperature probe and two circuits used as on-chip thermometers are measured. The noise performance of the measurement system is evaluated based on the Allan deviation using a reference signal generator. The repeatability of the measurements and the dependence of the measurements on the trigger level of the frequency-meter are evaluated. The guidelines for the system improvements are given.

ACKNOWLEDGMENT

This work is funded by ams AG, Premstaetten, Austria.

REFERENCES

[1] Y. Tokunaga, S. Sakiyama, A. Matsumoto, and S. Dosho, "An On-Chip CMOS Relaxation Oscillator With Voltage Averaging Feedback," *IEEE J. Solid-State Circuits*, vol. 45, no. 6, pp. 1150–1158, June 2010.
[2] R. Szabo, A. Gontean, and I. Lie, "Temperature and climate chamber automated control," in *2011 IEEE 12th International Symposium on Computational Intelligence and Informatics (CINTI)*, Nov 2011, pp. 155–159.
[3] N. Instruments. (2016) LabVIEW System Design Software. [Online]. Available: https://www.ni.com/labview/
[4] Python. (2016) PyVISA. [Online]. Available: https://pyvisa.readthedocs.io/en/stable/
[5] F. L. Walls and D. W. Allan, "Measurements of frequency stability," *Proceedings of the IEEE*, vol. 74, no. 1, pp. 162–168, Jan 1986.
[6] A. Paidimarri, D. Griffith, A. Wang, G. Burra, and A. P. Chandrakasan, "An RC Oscillator With Comparator Offset Cancellation," *IEEE Journal of Solid-State Circuits*, vol. 51, no. 8, pp. 1866–1877, Aug 2016.
[7] K. Technologies. (2014) Keysight Technologies U2722A/U2723A USB Modular Source Measure Unit. [Online]. Available: https://literature.cdn.keysight.com/litweb/pdf/5990-7416EN.pdf?id=2028171
[8] ——. (2014) Keysight Technologies E3640A-E3649A Programmable DC Power Supplies. [Online]. Available: https://literature.cdn.keysight.com/litweb/pdf/5968-7355EN.pdf?id=1118372
[9] ——. (2016) Keysight Technologies 53200A Series RF/Universal Frequency Counter/Timers. [Online]. Available: https://literature.cdn.keysight.com/litweb/pdf/5990-6283EN.pdf?id=1942617
[10] ——. (2016) Keysight Technologies 34401A Digital Multimeter. [Online]. Available: https://literature.cdn.keysight.com/litweb/pdf/5968-0162EN.pdf?id=1000070110:epsg:dow
[11] L. Facility. (2016) General Purpose Pt100 Probe with Teflon insulated cable. [Online]. Available: https://www.farnell.com/datasheets/1523527.pdf?_ga=1.134788401.1795664537.1427902463
[12] K. Technologies. (2016) Keysight Technologies 33500B Series Trueform Waveform Generators, 20 & 30 MHz. [Online]. Available: https://literature.cdn.keysight.com/litweb/pdf/5991-0692EN.pdf?id=2202606

Temperature Calibration of an On-Chip Relaxation Oscillator

J. Mikulić*, I. Brezovec†, M. Magerl†, G. Schatzberger* and A. Barić†

*ams AG, Premstaetten, Austria
†University of Zagreb/Faculty of Electrical Engineering and Computing, Zagreb, Croatia
josip.mikulic@ams.com

Abstract – This work investigates the calibration procedure of a conventional relaxation oscillator. First, the numerical analysis is performed in MATLAB in order to evaluate the sensitivity of the procedure to the noise generated inside the chip and measurement system. Next, the theory is experimentally verified by calibrating four test chip samples designed and manufactured in 0.35-μm CMOS technology. The test chips are calibrated with two different test methods: the first method measures the output frequency in the entire temperature range from -40 to 150 °C during 12 hours; the second method measures the output frequency from 30 to 60 °C during 30 seconds. The proposed calibration methods exhibit the reduction of the frequency error by 18x and 8x, having the total post-calibration precision of ±0.1 % and ±0.22 %, respectively.

I. INTRODUCTION

Fully-integrated solutions have become mandatory for a number of industrial applications as the requirements for the size reduction and low-power consumption are taking place. Consequently, SoCs like biomedical devices, portable mobile devices and wireless sensor networks cannot utilize an external clock reference based on crystal resonator which has been a standard solution for decades [1–9]. Unfortunately, there exists no universal replacement for the quartz resonator. While MEMS and LC on-chip oscillators exhibit good performance, the excessive power consumption, large area and additional process steps often make them unacceptable from the cost and power consumption point of view [5–9]. Relaxation oscillators, on the other hand, while suited for area and power efficient design, usually have the precision in the range of several percentage points [4]. Although sufficient for most applications, in the need of more accurate reference one must rely on the temperature calibration of the oscillator. One such solution presented in [1] features the on-chip heater and the approximation of the calibration data by a polynomial function. In this manner, the average TC (temperature coefficient) is brought down from ±33.3 ppm/°C to ±1.42 ppm/°C, the precision otherwise unachievable with an on-chip RC oscillator. Nevertheless, this method has the downsides as well, namely the following:
1) The additional blocks, such as the heater and temperature sensor, consume a significant portion of area and power;
2) The post-manufacturing calibration adds to the total production cost, especially if the process is time-intensive;

3) On-chip heaters only give the opportunity to calibrate the reference at the temperatures larger than the room temperature.

Despite the mentioned drawbacks, the temperature calibrated RC oscillator is still expected to offer a better compromise between the precision and power efficiency compared to MEMS or LC oscillators. For this reason, this work further investigates the tradeoffs and the limitations of the temperature calibration. The theory developed based on numerical calculations will prove worthy during the design phase since the estimate of the post-calibration precision can be accessed when the noise parameters of the system are available upfront. For the demonstration purposes, the numerical analysis is also experimentally verified using two different test methods. In the first method, the temperature characteristic is evaluated with long test time in the entire temperature range. In this way, the limits of the calibration method are examined, since all the noise influence is virtually eliminated. Moreover, the theory presented in our earlier work [3] will have been confirmed as well. While the precision using this method will prove to be in the ppm/°C range, being time-intensive makes it manageable only for a limited number of test chip samples. For this reason, the second procedure is implemented in the limited temperature range with a minimal number of measurements, bringing the test time in the range of several seconds. Although the precision of this method is somewhat degraded compared to the previous one, a reasonable test time makes it interesting for the high volume production.

This work is organized as follows. Section II presents the conventional relaxation oscillator topology and the timing analysis. In Section III the calibration setup is explained. Section IV presents the theoretical observations and the numerical analysis. Section V exhibits the measurement results. The final conclusions are given in Section VI.

II. CONVENTIONAL RELAXATION OSCILLATOR

A. Oscillator Architecture

The topology of a conventional relaxation oscillator is shown in Fig. 1. It features two identical integrator-comparator blocks and an SR flip-flop. Each integrator-comparator block comprises a referent current source, two switches controlled by the clock signals with opposite phases, a capacitor with the capacitance value C and a

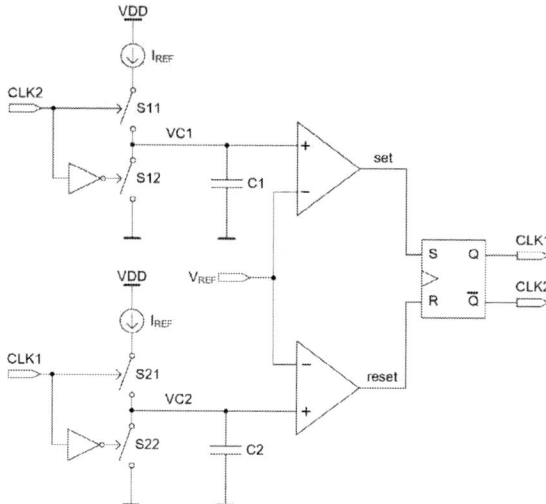

Fig. 1 – Conventional relaxation oscillator topology.

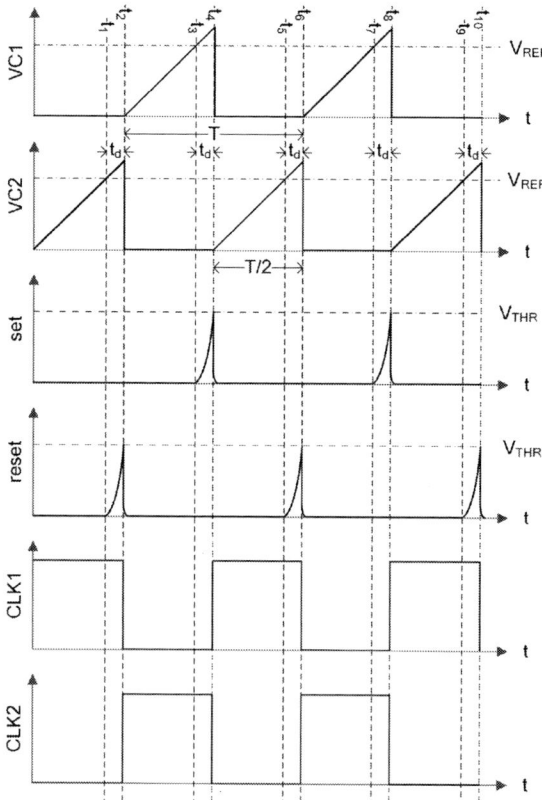

Fig. 2 – Signal waveforms for the conventional relaxation oscillator topology.

comparator. The referent current I_{REF} and the referent voltage V_{REF} are presumed to be generated within the reference generator, not shown in the schematics.

The signal waveforms of the oscillator are shown in Fig. 2. As seen in the figure, only one integrator-comparator block is active at the time. By the end of the integrating phase, the active comparator generates the *set*

or *reset* pulse, subsequently changing the state of the flip-flop. As a consequence, the integration phase of the active block ends as the integration phase in the formerly inactive integrator-comparator block is started. The oscillation cycle is permanently sustained in this manner.

B. Timing Analysis

The oscillation period is deduced from the waveforms presented in Fig. 2. First, the duration of the integrating phase is determined by the slew rate of the signals $VC1$ and $VC2$, equal to I_{REF}/C, and the value of the referent voltage V_{REF}. Next, since the comparator requires some time to generate the pulse at the output once the integrated signal has reached the referent voltage V_{REF}, the duration of the half-cycle is increased by the time delay t_d. Moreover, the comparators are featured with the input-referred offset voltage V_{OFF}, which is effectively superimposed to the referent voltage V_{REF}. Since the influence of the propagation delay of the digital gates is usually negligible, it is not considered within this analysis. Finally, with the mismatch between the blocks neglected, the expression for the oscillation period can be written as follows:

$$T_{osc} = \frac{1}{f_{osc}} = \frac{2 \cdot C \cdot (V_{REF} + V_{OFF})}{I_{REF}} + 2t_d. \quad (1)$$

As seen in (1), the sources of the temperature drift of the output frequency are various, specifically the capacitor, references, offset voltage and comparator delay. The latter one proves to be the critical factor since it is nonlinear against temperature and a large amount of power has to be invested in order to minimize it. The typical temperature drift of the oscillation period for a relaxation oscillator is simulated and shown in Fig. 3.

III. CALIBRATION METHOD

The temperature calibration, as already discussed, can resolve the clock temperature drift and bring it down to a ppm/°C range. Instead of the usual trimming at the single temperature [2], the addition of a temperature sensor and LUT (look-up-table) enables the calibration over the entire temperature range. One such setup is shown in Fig. 4 [3], where the value of the referent current is adjusted by means of DAC controlled by the temperature sensor. As seen in (1), the oscillation period can be directly altered by

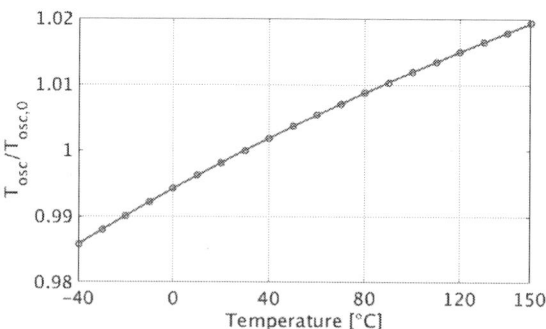

Fig. 3 – Oscillation period vs. temperature plot for a typical relaxation oscillator.

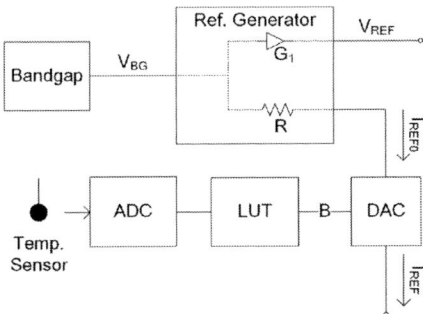

Fig. 4 – Temperature calibration setup.

changing the referent current I_{REF}. The LUT data collected during the test can be implemented either directly or as a polynomial function [1].

In this work, as the voltage and current reference are externally sourced, the calibration setup from Fig. 4 is software-emulated. For this reason, the setup is not limited by the ADC and DAC resolution, which would be the case in a silicon implementation of the setup. The temperature changes for the measurement and calibration purposes are induced by the temperature chamber, with the calibration data approximated by a polynomial function. The method used to calculate the calibration data and calibrate the oscillator is explained in the prior work [3].

IV. NUMERICAL ANALYSIS

Once the temperature behavior of the test chips is measured, the aim is to approximate it with a polynomial function. In an ideal case, given $P + 1$ points, any curve can be uniquely fitted to a polynomial of P-th order. In general, the approximation improves with the increasing degree of a polynomial. This is demonstrated in Fig. 5 where the residuals of the polynomial fitting of the curve from Fig. 3 are plotted against temperature. The maximal residual values over the temperature for 1st to 4th order polynomial are 1273 ppm, 85 ppm, 23 ppm and 6 ppm, respectively.

In practice, however, the accuracy of any measured point in the *temperature–period* (T–T_{osc}) space is compromised by several factors. To begin with, the noise generated inside the active and passive devices of the oscillator results in the uncertainty of the oscillation period (σ_y), characterized by the Allan deviation [10]. Moreover, the measurement equipment, having a finite precision as well, introduces the additional uncertainty in the temperature value (σ_x). As a result, the variation of the measured frequency at fixed temperature and supply voltage can be statistically expressed as follows:

$$\sigma^2 = \sigma_y^2 + \sigma_x^2 \cdot \left(\frac{\partial T_{osc}}{\partial T}\right)^2. \quad (2)$$

In order to analyze the expected error with respect to the total measurement deviation σ and the degree of a polynomial, numeric simulations are performed using MATLAB. First, a total of n equidistant points are assumed in the *temperature–period* space with the temperature ranging from T_a to T_b, assuming the $T_{osc}(T)$

function from Fig. 3. Next, randomly generated noise is superimposed on the data, resulting in the set of points which can be explicitly written as

$$\left\{T_i, T_{osc,i} + N(0,\sigma^2)\right\}, \begin{cases} i \in (1, 2 \ldots n) \\ T_i = T_a + (i-1) \cdot \dfrac{T_b - T_a}{n-1}, \\ T_{osc,i} = T_{osc}\left(T_i\right) \end{cases} \quad (3)$$

where $N(0, \sigma^2)$ is a normally distributed random variable, T_i is the i-th point temperature, and $T_{osc,i}$ is the i-th point oscillation period. Afterwards, the polynomial fitting is performed on the generated points (3). This procedure is repeated 10000 times for each value of σ and the maximum deviation from the ideal characteristic from Fig. 3 in the temperature range from -40 °C to 150 °C is recorded. Finally, the procedure is repeated for the different values of σ and polynomial degree in order to obtain the *error vs. σ* graph, therefore determining the upper bound of the error after the calibration caused by the measurement inaccuracy.

A. Entire Temperature Range

Fig. 6 presents the results of the numerical analysis where the 1st to 4th order polynomials are fitted with $n = \{10, 30, 100\}$ points and the temperature range $(T_a, T_b) = (-40\ °C, 150\ °C)$. From the figure one can conclude that for the low values of σ the use of a higher order polynomial is advantageous, while the lower order polynomials are more stable for high σ values. Moreover,

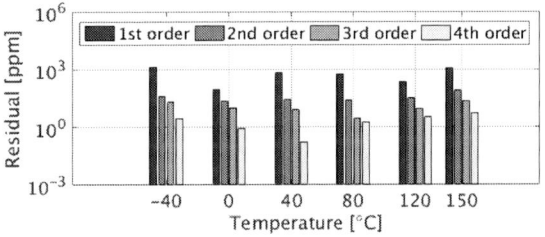

Fig. 5 – Residuals of the polynomial fitting shown for 1st to 4th order polynomial at different temperatures. The data is sampled in the entire temperature range.

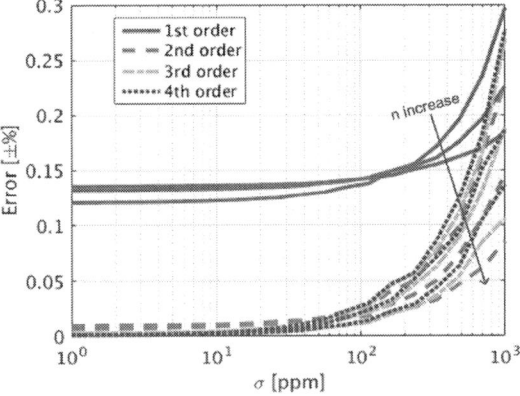

Fig. 6 – Polynomial fitting error vs. standard deviation of the measured oscillation period shown for 1st to 4th order polynomials. Simulated for $n = \{10, 30, 100\}$ and $(T_a, T_b) = (-40\ °C, 150\ °C)$.

as can be intuitively understood, the error becomes smaller as the number of points n increases.

B. Limited Temperature Range

While only the measurements in the entire temperature range can yield the post-calibration precision in the sub-100 ppm range, the excessive test time and the temperatures below the room temperature would make it unusable for any industrial application. This gives the motivation for limiting the temperature range, thereby making the calibration process cost-effective.

First, the polynomial fitting of the curve from Fig. 3 is calculated based on the data from 30 to 60 °C, and the residuals are calculated for the temperature range from -40 to 150 °C. The residuals for the 1st to 4th order polynomial are plotted against temperature in Fig. 7, with the maximal values being equal to 2214 ppm, 338 ppm, 128 ppm and 214 ppm, respectively. The statement from before that the approximation of the curve improves with the increasing number of polynomial no longer holds true. Despite no noise being added during the calculation, the 4th order approximation proves to be less stable than the 3rd order for the given graph and sampling temperature range.

Next, the numerical analysis is performed where the 1st to 3rd order polynomials are fitted with $n = \{10, 30, 100\}$ points and temperature range $(T_a, T_b) = (30 \text{ °C}, 60 \text{ °C})$. The results are presented in Fig. 8. From the figure one can observe that the sensitivity to noise increases

drastically with the increasing number of polynomial. For this reason, lower order polynomials will generally be preferred when the entire temperature range is not measured. Similar as before, the error comes down as the number of sampled points n increases.

V. MEASUREMENT RESULTS

The oscillator core from Fig. 1 is designed and manufactured in 0.35-μm CMOS process. The microphotograph of the test chip is shown in Fig. 9. The area of the oscillator core is around 0.04 mm². The typical power consumption is 150 μW with the supply voltage of $V_{DD} = 3.3$ V. The nominal frequency of the oscillator equals to $f_{osc,0} = 1$ MHz. As mentioned before, the voltage and current reference are sourced externally.

The imprecision of the used measurement system and the designed oscillator is characterized, resulting in the following parameters: $\sigma_y = 50$ ppm, $\sigma_x = 0.6$ °C and $\partial T_{osc}/\partial T = 180$ ppm/°C. The resulting σ calculated from (2) then ranges around 120 ppm. From the analysis conducted in the previous section (Fig. 6 and Fig. 8), the optimal polynomial orders are shown to be the 4th order for the entire temperature range and the 1st order for the limited temperature range measurements.

A. Entire Temperature Range

The calibration procedure analyzed in the previous section is now verified experimentally. Four test chip samples are measured in the temperature range from -40 to 150 °C in the time interval of 12 hours ($n = 5000$). High number of measurement points almost completely eliminates the noise influence. The corresponding calibration factors are calculated and fitted using the 4th order polynomial, and the calibration is performed on the test chip samples. The results are plotted in Fig. 10, showing the frequency error against temperature both for non-calibrated and calibrated test chips. While the non-calibrated samples exhibit around ±1.8 % spread

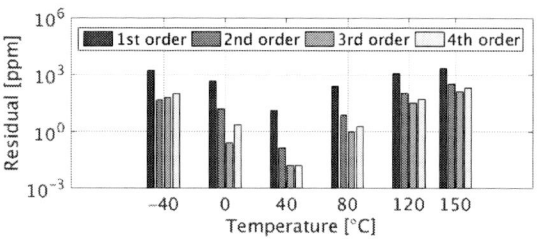

Fig. 7 – Residuals of the polynomial fitting shown for 1st to 4th order polynomial at different temperatures. The data is sampled in the temperature range from 30 to 60 °C.

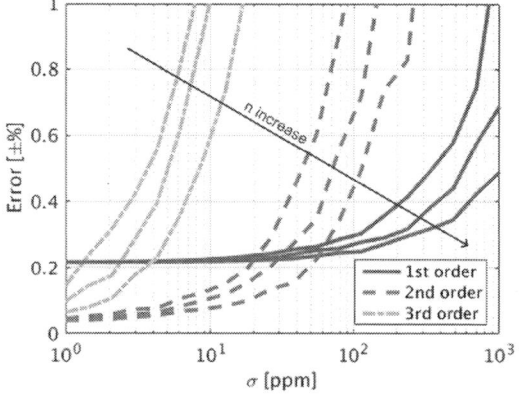

Fig. 8 – Polynomial fitting error vs. standard deviation of the measured oscillation period shown 1st to 3rd order polynomials. Simulated for $n = \{10, 30, 100\}$ and $(T_a, T_b) = (30 \text{ °C}, 60 \text{ °C})$.

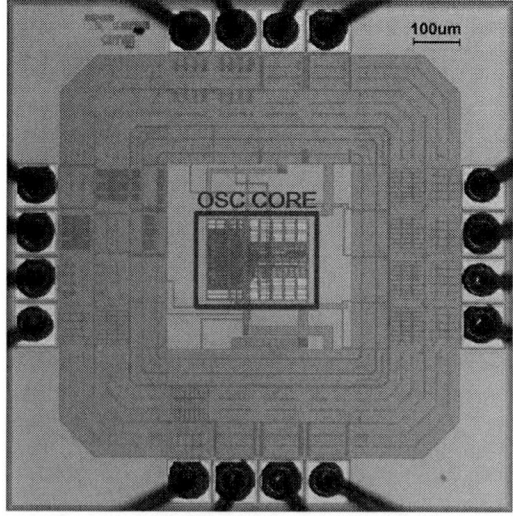

Fig. 9 – Microphotograph of the oscillator. The core area is approximately 0.04 mm².

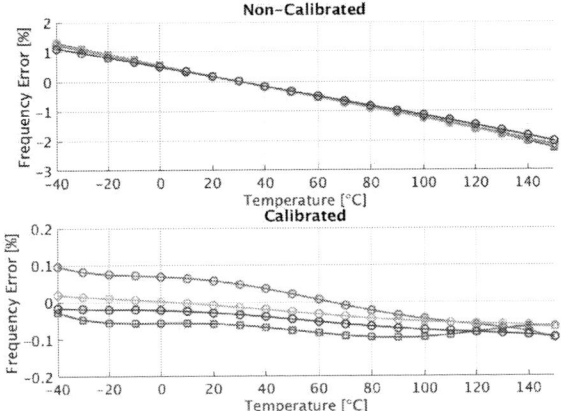

Fig. 10 – Frequency error vs. temperature shown for the non-calibrated and calibrated test chip samples. The calibration data is fitted to 4th order polynomial. The calibration data is sampled in the entire temperature range $(T_a, T_b) = (-40\ °C, 150\ °C)$ over 12 hours $(n = 5000)$.

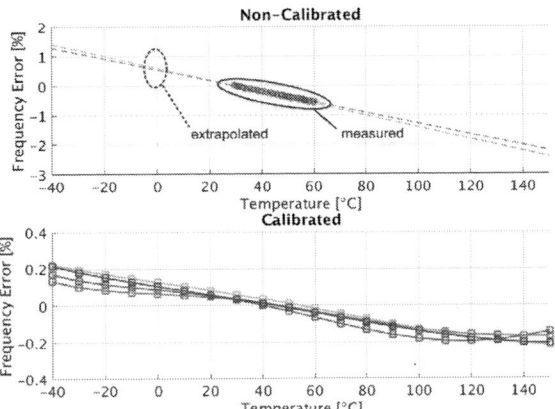

Fig. 11 – Frequency error vs. temperature shown for the non-calibrated and calibrated test chip samples. The calibration data is fitted to 1st order polynomial. The calibration data is sampled in the limited temperature range $(T_a, T_b) = (30\ °C, 60\ °C)$ over 30 seconds $(n = 30)$.

over the temperature relative to the center frequency, with calibration this is reduced down to ±0.1 %, therefore achieving the improvement of around 18x. While the graph in Fig. 6 predicts almost no error for the given σ, n and order of polynomial, the present error comes from the nonlinearity of the oscillator tuning since the single-step trimming procedure described in [3] is used.

B. Limited Temperature Range

Next, in order to cut down the test time, the test chip samples are measured in the temperature range from 30 to 60 °C in the time interval of 30 seconds $(n = 30)$. The calibration data is fitted using 1st order polynomial since higher order polynomials are expected to produce worse behavior for the given σ and n, as can be read out from Fig. 8. Fig. 11 shows the sampled data and the corresponding polynomial fits, together with the relative frequency error of the calibrated chip samples. The total frequency spread now equals ±0.22 %, which gives the improvement of around 8x compared to the precision of the non-calibrated test chips.

VI. CONLUSION

In this work the temperature calibration of a conventional relaxation oscillator is investigated. At the beginning, the numerical analysis is performed in order to quantify the influence of the noise, after which the experimental results are obtained. High correspondence between the two is observed, meaning the numerical analysis can be used in the early phases of the design, making the tradeoffs between the precision, test time, system noise and other factors. The calibration performed in the temperature range from -40 to 150 °C over 12 hours reveals the total precision of ±0.1 %. On the other hand, the fast calibration procedure executed in the temperature range from 30 to 60 °C during 30 seconds exhibits the total precision of ±0.22 % in the temperature range from -40 to 150 °C. The latter one proves to be suited for

the industrial applications, combining the reasonable test time with relatively high post-calibration precision.

REFERENCES

[1] Y. Satoh, H. Kobayashi, T. Miyaba, S. Kousai, "A 2.9mW +/- 85ppm accuracy reference clock generator based on RC oscillator with on-chip temperature calibration", in Proc. Dig. Symp. VLSI Circuits, pp. 1-2, 2014.

[2] A. Vilas Boas, A. Olmos, "A temperature compensated digitally trimmable on-chip IC oscillator with low voltage inhibit capability," in Proc. IEEE Int. Symp. Circuits and System (ISCAS), vol. 1, pp. 501-504, 2004.

[3] J. Mikulić, G. Schatzberger, A. Barić, "Relaxation oscillator calibration technique with comparator delay regulation", Information and Communication Technology Electronics and Microelectronics (MIPRO) 2016 39th International Convention on, pp. 57-61, 2016.

[4] Y. Tokunaga, S. Sakiyama, A. Matsumoto, S. Dosho, "An on-chip CMOS relaxation oscillator with voltage averaging feedback," J. Solid-State Circuits, vol. 45, no. 6, pp. 1150-1158, 2010.

[5] M.S. McCorquodale, et al., "A 25-MHz self-referenced solid-state frequency source suitable for XO-Replacement", Circuits and Systems I: Regular Papers IEEE Transactions on, vol. 56, pp. 943-956, 2008.

[6] F. Sebastiano, L.J. Breems, K. Makinwa, S. Drago, D. Leenaerts and B. Nauta, "A low-voltage mobility-based frequency reference for crystal-less ULP radios," IEEE J. Solid-State Circuits, vol. 44, no. 7, pp. 2002-2009, 2009.

[7] Y. Cao, P. Leroux, W. De Cock, M. Steyaert, "A 63000 Q-factor relaxation oscillator with switched-capacitor integrated error feedback", Solid-State Circuits Conference Digest of Technical Papers (ISSCC) 2013 IEEE International, pp. 186-187, 2013.

[8] V. De Smedt, P. De Wit, W. Vereecken, and M. Steyaert, "A 66 μW 86 ppm/°C fully-integrated 6 MHz wienbridge oscillator with a 172 dB phase noise FOM," IEEE J. Solid-State Circuits, vol. 44, no. 7, pp. 1990-2001, 2009.

[9] S. Mahdi Kashmiri, K. Souri and A.K.A. Makinwa, "A scaled thermal-diffusivity-based 16 MHz frequency reference in 0.16 μm CMOS", IEEE Journal of Solid-State Circuits, vol. 47, no. 7, pp. 1535-1545, 2012.

[10] F. Walls, D. Allan, "Measurements of frequency stability", Proc. IEEE, vol. 74, no. 1, pp. 162-168, 1986.

Model of High-Efficiency High-Current Coupled Inductor Two-Phase Buck Converter

V.C. Valchev*, O.P. Stanchev* and G.T. Nikolov**
* Technical University of Varna/Department of Electronics and Microelectronics, Varna, Bulgaria
** IDT Bulgaria Ltd., Varna, Bulgaria
vencivalchev@hotmail.com, or.stanchev@gmail.com, georgi.nikolov@idt.com

Abstract - This paper proposes a study and design considerations on a high-current high-efficiency two-phase buck converter with a coupled inductor. The converter operates close to the megahertz range. A specialized simulation model of the considered two-phase buck converter is proposed. The model allows an evaluation of current and voltage ripples at various duty cycles and inductor coupling factors. It can facilitate the design and tuning of such DC-DC converters. The model is implemented into a general-purpose simulation environment. The simulation results are presented and analyzed. The model is verified trough a dedicated experiment on a realized converter. Although coupled inductors increase the power converter efficiency, for high-current buck converter this effect is substantial at duty cycle bigger than 40 %. The advantages at lower duty cycle are related to the output voltage ripple and feedback response time. Design considerations are derived based on the analyzed simulation and experimental results of the investigated two-phase buck converter.

I. INTRODUCTION

The buck converter is well known DC-DC conversation topology when no galvanic isolation is required. The high-current multiphase buck converters that step down and stabilize voltage from 12 V to 3.3 V and less, have their well-known applications for powering microprocessor and other digital systems. Such applications require limited voltage fluctuations and fast response when the load current rises from almost zero amperes to approach maximum load while maintaining the high efficiency. The transient response is provided by the output capacitor that supplies the new demand for output current while the inductors' current changes to charge the capacitors up. For faster response a high-switching frequency like 500 kHz or more and smaller inductors are needed. Because of the two-phase topology the inductors and capacitors values can be reduced and still be able to obtain the same voltage ripple. However, the current flowing though the MOSFETs retain its peak values bringing conduction losses and high-current requirement. As a result, the buck converter suffers from limited performance for high-switching frequency high-step-down and large output current applications when the duty cycle is very small [1], [2]. An improvement can be achieved with an innovative design technique that utilizes coupled inductors and bring the following major benefits [3], [4]:

- Low leakage inductance.
- Improved transient response.
- Smaller and cheaper inductor core (Less magnetic material needed).
- Reduction of ripple currents and losses in MOSFETs.
- Improved efficiency.

The application of a unified magnetic component in two-phase buck converter reduces the inductor current ripple with more than 50 % as smaller duty ratio results in less reduction [5]. One can obtain over 80 % efficiency at more than 8 times step-down ratio, 2 MHz switching frequency and over 20 A output current [6]. In [7] a mathematical model of an electronic converter for operating energy storage elements is proposed and verified. However, the reviewed publications propose mathematical tools for analysis but do not provide any complete dedicated simulation model of two-phase buck converter that utilizes coupled inductors. This paper proposes a simulation model that can be used by engineers to analyze the limitations of the buck converter topology for high-current, high-frequency and high step-down; facilitate the design and investigate the effect of the duty ratio and coupling factor on the power converter voltage and current ripples. At such high-frequency power circuits it is very difficult to measure currents flowing through the inductor and MOSFETs, and to observe how they are affected by changing certain parameters of the components in the circuit.

II. SIMULATION MODEL STRUCTURE

The proposed simulation model is given in Fig. 1. It consists of power stage, control stage and some measurement topologies. The power stage represents the topology of a two-phase synchronous buck converter. The two subsystems "Transistor module 1" and "Transistor module 2" contain the high side and low side MOSFET for each phase and a block that models the switching losses. The transistor modules are connected to an ideal voltage source "12 V" and the model of mutual inductance "Coupled inductor". Its windings are connected out of phase in order to operate correctly and provide the previously described benefits. The output capacitor is modeled as equivalent series resistance (ESR) and capacitance "Cout". The load is represented by a subsystem "Load" that contains a controllable current source so that

---

The carried out research is realized in the frames of the project "Model based design of power electronic devices with guaranteed parameters", ДН07/06/15.12.2016 funded by Bulgarian National Scientific Fund.

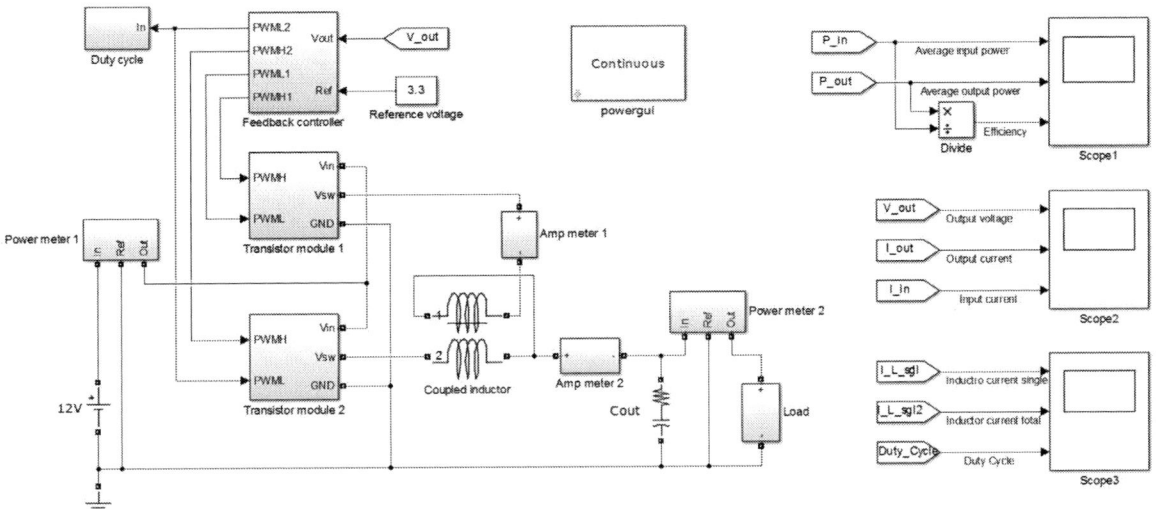

Figure 1.   Simulation model of high-frequency high-current two-phase buck converter

the load current can be precisely set. The control stage is represented by the subsystem "Feedback controller" that contains an error amplifier, PID regulator and H-bridge PWM generator. A measurement lag is implemented into the feedback so that the algebraic loops are avoided. The "Power meter 1" and Power meter 2" blocks provide voltage and current averaging and multiplication according to the buck converter operating frequency. The "Coupled inductor" block models the magnetic component behavior according to the equivalent circuit given in Fig. 2. It consists of an ideal transformer with unity turns ratio, series inductor and resistor that represent the inductor coupling. The coupled inductors' manufacturers normally provide information for coils' inductances (L1 and L2), resistances (R1 and R2), and leakage inductance (Lk). The other needed parameters have to be found. The connection between the mutual inductance M and L1, L2 and Lk is given is:

$$L_k = L_1 + L_2 - 2L_m \qquad (1)$$

The coupling factor k depends on the coils' inductance and the mutual inductance. It is given in (2):

$$k = \frac{L_m}{\sqrt{L_1 \cdot L_2}} \qquad (2)$$

When (1) is substituted in (2) the coupling factor can be directly derived in (3), using the available parameters:

$$k = \frac{L_1 + L_2 - L_k}{2\sqrt{L_1 \cdot L_2}} \qquad (3)$$

The coupled inductor mutual resistance is found by:

$$R_m = k\sqrt{R_1 \cdot R_2} \qquad (4)$$

For the selected coupled inductor with given parameters in Table I the coupling factor is k = 0.744. When using the given inductor model, if the mutual impedance is set to 0 ($L_m = 0$ H and $R_m = 0$ Ω) then the two inductors become uncoupled. This effect adds more functionality to the proposed simulation model. More precise modeling of losses could be realized using the core loss model of ferrites under square voltage waveforms [8].

The structure of "Transistor module 1" and "Transistor module 2" blocks is given in Fig. 3. The two MOSFETs perform "ideal" switching and only conduction losses are taken into account so that simulation time of the model is optimized. The switching losses are obtained using a

Figure 2.   Equivalent schematic of the model of coupled inductor

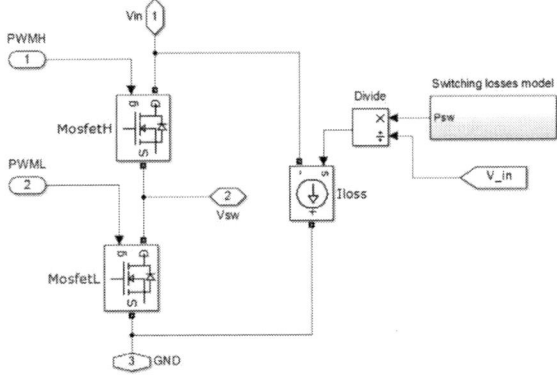

Figure 3.   Transistor module block

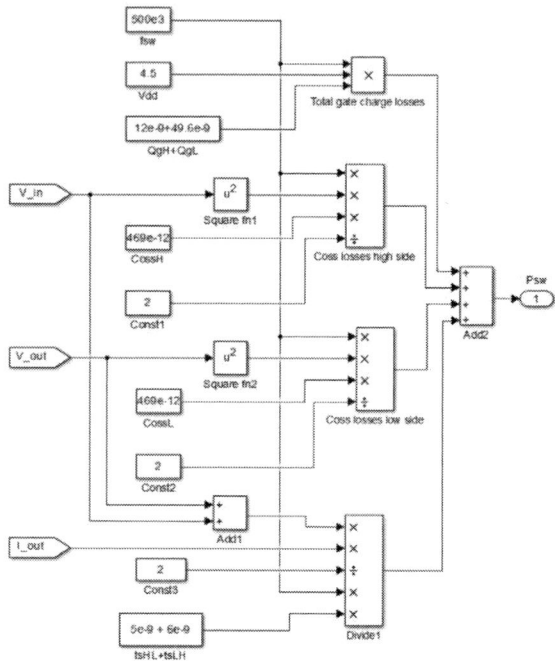

Figure 4. Switching losses model block

dedicated model represented by the "Switching losses model" block (Fig 4). It consists of blocks for mathematical operators and constants that represent the equations for high side and low side transistor switching losses generated by the switching transition times, the charging of the gates and the output capacitances according to the methodology given in [9].

III. VERIFICATION

A. Simulation Results

The proposed simulation model of two-phase buck converter is implemented in general-purpose simulation environment MATLAB – Simulink [10]. It includes the most determinant parameters of the elements of the buck converter. Thus, the model is simplified but accurate, providing optimized calculation time and iterative process of setting and tuning components' parameters. The model is set to simulate the power stage of two-phase buck converter experimental board, designed by Integrated

TABLE I. MODEL PARAMETERS

Variable	Value	Unit	Description
Coupled inductor			
L1, L2	860	nH	Winding inductace
Lm	640	nH	Mutual inductance
R1, R2	0.490	mΩ	Winding resistance
Rm	0.365	mΩ	Mutual resistance
Output capacitor			
C	2702	µF	Capacitance
ESR	0.71	mΩ	Equivalent Series Resistance
Transistor module 1 and 2			
RonH	3.65	mΩ	High side FET channel on resistance
RonL	0.85	mΩ	Low side FET channel on resistance
Switching losses model			
CossH	469	pF	High side FET output capacitance
QgH	12	nC	High side FET total gate charge
CossL	1950	pF	Low side FET output capacitance
QgL	49.6	nC	Low side FET total gate charge
tsHL	5	ns	Transistor module transition time High - Low
tsLH	6	ns	Transistor module transition time Low - High
Feedback controller			
fsw	500	kHz	Switching frequency[a]
Kp	0.1	-	Proportional factor[b]
Ki	400	-	Integral factor[b]

a. The switching frequency of each phase is half of the output voltage ripple frequency.

b. Related to the PID regulator.

Device Technology Inc. The model parameters are given in Table I. The experimental board utilizes a digital PWM controller that operates at different principle from the feedback controller of the simulation model. The focus is set on the characteristics of the power stage at fixed load where the exact PID regulator settings are not defined. The output voltage ripples at two different output voltages and maximum load current (40 A) are given in Fig. 5 and

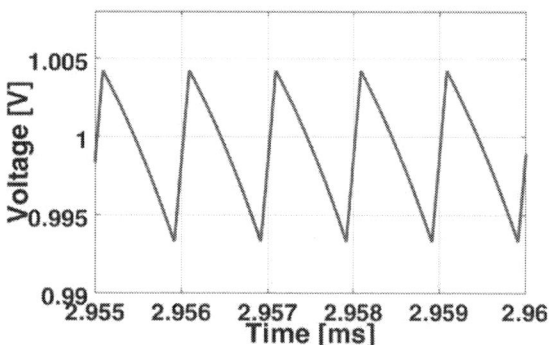

Figure 5. Output voltage ripple at V_{out} = 1 V when providing maximum load current

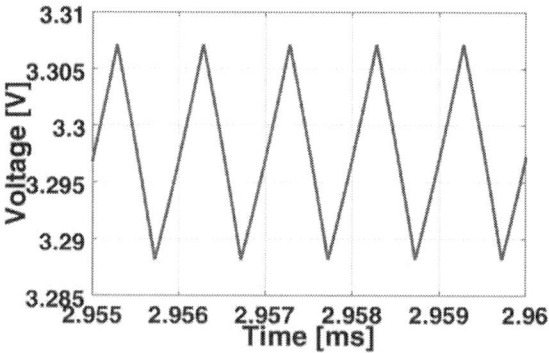

Figure 6. Output voltage ripple at V_{out} = 3.3 V when providing maximum load current

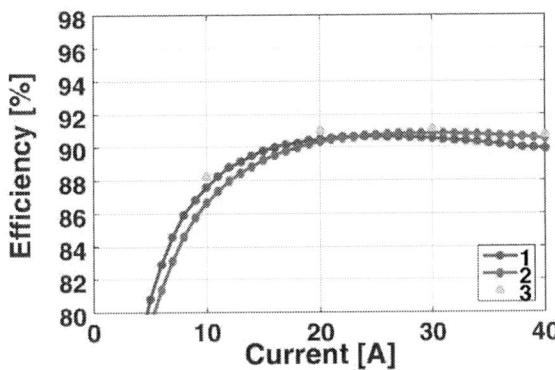

Figure 7. Buck converter efficiency dependance on current at $V_{out} = 1$ V when using: 1 - coupled inductor; 2 - separate inductors and 3 – simulation results for couplet inductor

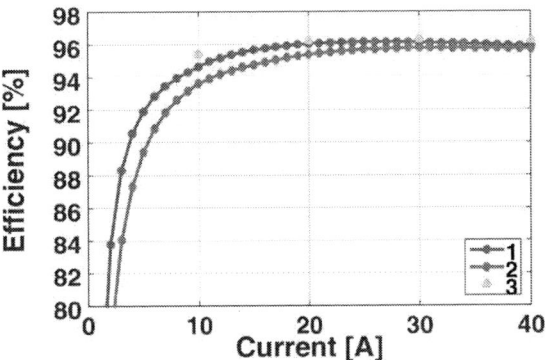

Figure 8. Buck converter efficiency dependance on current at $V_{out} = 3.3$ V when using: 1 - coupled inductor; 2 - separate inductors and 3 – simulation results for couplet inductor

Fig. 6. The output capacitor ESR directly affects the output voltage ripple. If it is set according to a real case voltage ripple, the other cases can be simulated without any physical change of components. The simulation results for efficiency of the two-phase buck converter for both output voltages are given in Fig. 7 and Fig. 8, and directly compared to the experimental results.

B. Experimental results

The simulation model is verified with a dedicated laboratory experiment. A two-phase buck converter experimental board is used, which is designed by Integrated Device Technology Inc (Fig. 9). It utilizes a dual phase state-of-art digital power controller – ZSPM1363, while the power stage of the synchronous buck converter is build using two ZSPM9060 DrMOS devices [11], [12], a shielded coupled inductor and multiple multilayer ceramic capacitors, connected in parallel, so that the ESR and respectively the output voltage ripple are reduced. The DrMOS devices consist of two high-frequency, high-current and low voltage MOSFET transistors, connected in series, their dedicated drivers and some additional components are placed in thermally enhanced ultra-compact standard package. These specialized integrated circuits are fully optimized and ultra compact with reduced switch ringing, dead times and propagation delays. A dual-phase digital PWM controller controls the power stage [13]. The experimental board is connected to a high-current power supply and a programmable load. The output voltage is measured using a high-resolution digital scope, connected to a dedicated measurement SMB connector of the board, which minimizes the noise as the AC component of the output voltage is less than 1 % from the DC one. The PCB board is designed so that either coupled inductor or two separate inductors can be used. All experiments are accomplished using 12 V input voltage and components with the values given in Table I. The high frequencies and high level of integration of the experimental board allow accurate measurement only of the input and output voltages and currents. The output voltage waveform of the buck converter at 1 V and 3.3 V and 40 A load current is given in Fig. 10 and Fig. 11. The peak-to-peak voltage ripple at 1 V mean output voltage is about 11.5 mV and at 3.3 V it is about 19 mV. Output voltage ripple is highly affected by the ESR value of the output capacitor. The waveform is

close to saw-tooth but a distortion is available at both output voltages. The voltage ripple is difficult to be precisely measured because of the highly dominating DC component. The converter total efficiency is measured for various load currents and two output voltages for the case when one coupled or two separate inductors are used (Fig. 7 and Fig. 8). The efficiency graphs prove the statement that the higher step-down ratio causes more losses. It can be noticed that when the inductors are coupled the efficiency is improved. The reason is less inductor current ripple and the cubic dependency between the conduction power losses in MOSFETs and the drain current. The high-side MOSFET has about 4 times higher resistance of the induced N-channel compared to the low-side MOSFET. When the output voltage is set at 3.3 V the high-side MOSFET is 'switched on' 20 % more time than the 1 V output voltage case. Therefore the efficiency is higher when using coupled inductors at 3.3 V than at 1 V. However, at higher step-down ratio there is not any considerable effect. This tendency is directly connected to the duty cycle of the power transistor gate drive voltage and the time interval when the inductor current is increasing. One could notice that the error when simulating efficiency is less than 1 %.

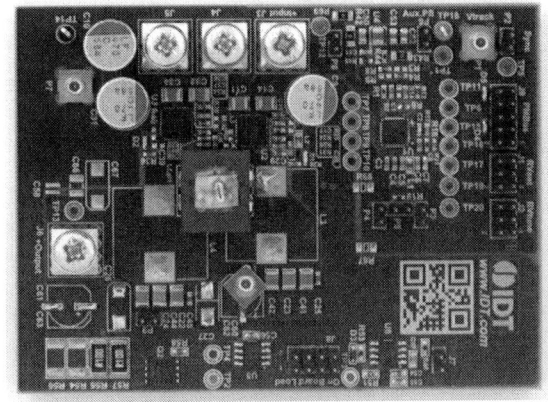

Figure 9. Two-phase buck converter experimental board, [11]

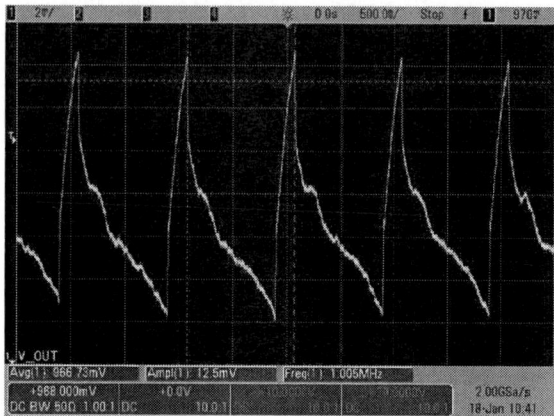

Figure 10. Output voltage ripple at $V_{out} = 1$ V and $I_{out} = 40$ A

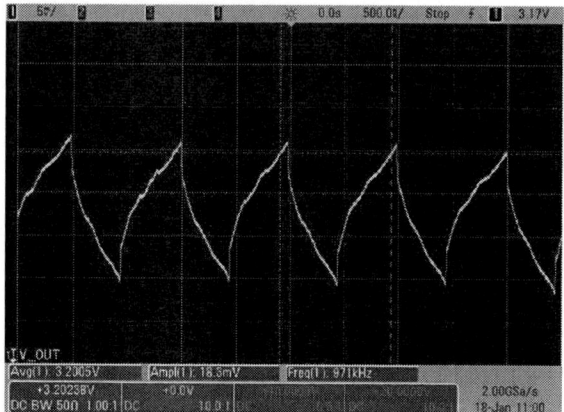

Figure 11. Output voltage ripple at $V_{out} = 3.3$ V and $I_{out} = 40$ A

IV. CONCLUSION

The proposed model of high-efficiency high-current two-phase buck converter is a powerful tool for simulation, design, analysis and precise tuning of components.

The model is verified trough comparison with experiments and measurements on a two-phase buck converter DC-DC converter. It was derived that:

- At higher step-down ratio, respectively lower duty cycle, there is not any considerable effect concerning efficiency. This tendency is directly connected to the operational time-intervals of the power switches.

- Although coupled inductors increase the power converter efficiency, for a high-current buck converter this effect is substantial at duty cycle higher than 40 %. The advantages at lower duty cycle are related to the output voltage ripple and feedback response time.

Further improvement of the model could be realized by including the dependence of core losses in the inductor on the applied voltage across the inductor. This improvement is possible by utilizing the available models of ferrite core losses under square voltage wave forms [8], [14].

Verification of the model shows that it is operational within a certain degree of accuracy.

ACKNOWLEDGMENT

The carried out research is realized in the frames of the project "Model based design of power electronic devices with guaranteed parameters", ДН07/06/15.12.2016 funded by Bulgarian National Scientific Fund.

REFERENCES

[1] Peng Xu, Jia Wei, and F.C. Lee, "Multiphase coupled-buck converter-a novel high efficient 12 V voltage regulator module," IEEE Transactions on Power Electronics, Volume: 18, Issue: 1, Jan 2003.

[2] Kaiwei Yao, Yu Meng, and F.C. Lee, "A novel winding coupled-buck converter for high-frequency, high step-down DC/DC conversion," 33rd Annual Power Electronics Specialists Conference, IEEE, 23-27 June 2002.

[3] J. Czogalla, Jieli Li, and C. R. Sullivan, "Automotive application of multi-phase coupled-inductor DC-DC converter," 38th IAS annual meeting, Conference record of the industry applications conference, 12-16 Oct. 2003.

[4] Jieli Li, and C. R. Sullivan, "Coupled inductor design optimization for fast-response low-voltage DC-DC converters," Seventeenth annual IEEE applied power electronics conference and Exposition, APEC 2002., 10-14 March 2002.

[5] J. Gallaghe, "Coupled inductors improve multiphase buck efficiency," Power Electronics Technology, Jan 2006.

[6] Kaiwei Yao, Yang Qiu, Ming Xu, and F.C. Lee, "A novel winding-coupled buck converter for high-frequency, high-step-down DC-DC conversion," IEEE Transactions on Power Electronics, Volume: 20, Issue: 5, Sept. 2005.

[7] Arnaudov D., N. Hinov, I. Nedyalkov, "Mathematical Model of an Electronic Converter for Charging of Energy Storage Elements", XXV International Scientific Conference Electronics - ET2016, Sozopol 2016, pp.215-218.

[8] Valchev V. C., A. Van den Bossche and D. Van de Sype, 'Ferrite losses of cores with square wave voltage and DC bias', IEEE Industry Electronics Society, IECON'05, 2005, North Carolina, USA, pp. 837-841.

[9] J. Klein, "AN-6005 Synchronous buck MOSFET loss calculations with Excel model," Fairchild Semiconductor, www.fairchildsemi.com, 2014.

[10] Mathworks (2016), "Matlab & Simulink - Simscape User's Guide," www.mathworks.com.

[11] Integrated Device Technology Inc, "ZSPM9060, Ultra-Compact, High-Performance, High-Frequency DrMOS Device Power Datasheet," www.idt.com, 2016.

[12] Intel Inc., "DrMOS Specifications," www.intel.com, November 2004.

[13] Integrated Device Technology Inc, "ZSPM1363, True Digital PWM Controller (Dual-Phase, Single-Rail) Datasheet," www.idt.com, 2016.

[14] Venkatachalam, K., Sullivan, C.R., Abdallah, T., Tacca, H., "Accurate prediction of ferrite core loss with nonsinusoidal waveforms using only Steinmetz parameters,", Proceedings, IEEE Workshop on Computers in Power Electronics, 3-4 June 2002, pp. 36-41.

Analog to Digital Signal Converters for BiCMOS Quaternary Digital Systems

Dušanka Bundalo*, Zlatko Bundalo**, Dražen Pašalić*** and Branimir Cvijić ****

* Faculty of Philosophy, University of Banja Luka, Banja Luka, Bosnia and Herzegovina
** Faculty of Electrical Engineering, University of Banja Luka, Banja Luka, Bosnia and Herzegovina
*** Sberbank A.D., Banja Luka, Bosnia and Herzegovina
**** Lanaco d.o.o, Banja Luka, Bosnia and Herzegovina
dusanka.bundalo@unibl.rs, zbundalo@etfbl.net, pasalic.drazen@gmail.com, brano.cvijic@gmail.com,

Abstract - Possibilities of practical development, design and implementation of analog signal to quaternary digital signal converters for application in BiCMOS quaternary digital circuits and systems are considered, proposed and described in the paper. General approaches and general structure for implementation and design of parallel analog to quaternary BiCMOS digital signal converters are proposed and described. Two digit parallel analog to quaternary BiCMOS digital signal converters are proposed and described as the illustration of proposed way for the converters design and implementation. More possibilities of converters development, design and implementation were considered and described. Two types of such converters are described in more details: the basic type converters and the modified type converters. Given solutions have been analyzed by computer simulations. All descriptions and considerations have been confirmed by computer simulations. Some of the computer simulation results are given in the paper.

I. INTRODUCTION

The binary digital systems and circuits are still dominant in practical use and practical applications. Possibilities and interest for implementation of so-called multiple-valued (MV) digital systems and circuits are increased with development of VLSI technologies [1, 2]. The greatest interest practically exists for ternary (logic basis 3) and quaternary (logic basis 4) MV circuits and systems [1-7].

The quaternary MV logic circuits and systems have many advantages comparing with the binary ones. Well known the most important advantages of MV logic circuits and systems are: reduction in the number of interconnections required to implement logic function, greater speed of logic and arithmetic operation, greater density of memorized information, better usage of transmission paths, decreasing of interconnection complexity and interconnection area, decreasing of pin number of integrated circuits and printed boards, possibilities for easier testing of digital systems [1, 2].

Constant development and progress in the monolithic integrated circuits technology and efforts to achieve and maintain good characteristics and advantages of the CMOS and TTL logic, are the main reasons that the BiCMOS technology and BiCMOS logic is increasingly used in binary digital VLSI systems. For the same reasons, there is interest for development, implementation and application of MV BiCMOS logic circuits and systems. The reasons and advantages of application of BiCMOS technology in implementation of binary digital systems and circuits are well known. All these good characteristics of BiCMOS technology should be also kept in quaternary MV logic systems and circuits.

The MV digital systems also use analog to digital signal conversion and converters, as well as digital to analog signal conversion and converters. There is need to develop appropriate converters for applications in MV digital systems [5-7]. It is well known that parallel analog to digital converters are the fastest ones. It is also needed to develop, design and implement appropriate analog to digital signal converters for applications in quaternary digital systems.

Principles and possibilities for development, design and implementation of parallel analog to digital signal converters for applications in quaternary BiCMOS digital systems are proposed and described in this paper. General principle and structure for converters implementation and design are considered and described. The concrete circuits designs for two digit BiCMOS analog to quaternary digital signal converters are proposed and presented as example of the converters design,. Two types of the converters are described: so called basic type converters and so called modified type converters. All proposed and described principles and solutions of the converters were analyzed and confirmed by computer PSpice simulations.

II. DESIGN OF ANALOG TO DIGITAL SIGNAL CONVERTERS FOR BiCMOS QUATERNARY DIGITAL SYSTEMS

General principle and general structure for development and design of parallel analog to digital signal converters for applications in BiCMOS quaternary digital systems are proposed and shown in Figure 1. There is one analog input signal (A_i) and there are m quaternary digital BiCMOS output signals (Y_i) in the design. The proposed structure consists of three logic levels and networks: CMOS binary voltage comparator network, CMOS binary encoder network and BiCMOS quaternary output network. The CMOS binary voltage comparator network and the binary CMOS encoder network are supplied by two supply voltages enabling to obtain two CMOS binary states: V_{SS} (for binary logic state 0) and V_{CC3} (for binary logic state 1). The quaternary BiCMOS output network is supplied by four supply voltages enabling to obtain four quaternary BiCMOS logic states: V_{SS} (for quaternary logic state 0), V_{CC1} (for quaternary logic state 1), V_{CC2} (for quaternary logic state 2) and V_{CC3} (for quaternary logic state 3).

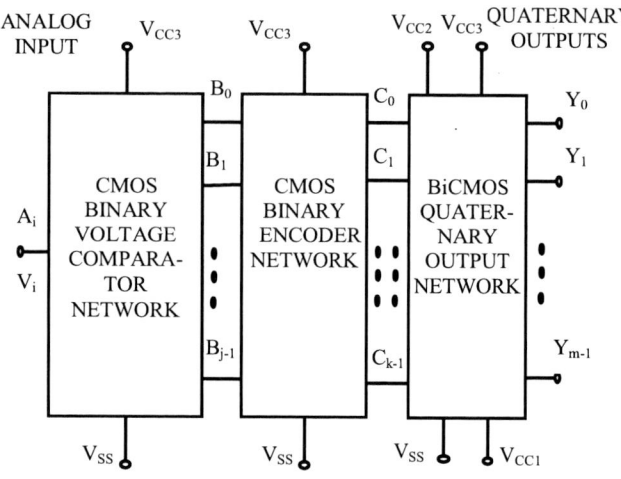

Figure 1. Proposed structure of parallel analog to digital signal converters for BiCMOS quaternary digital systems

The input CMOS binary voltage comparator network compares input analog voltage signal V_i (analog signal A_i) with appropriate threshold voltages. Since here is analog to digital quaternary signal converter, there are existing 4^m-1 threshold voltages, where m is number of quaternary digital outputs. The number of the threshold voltages depends on converter resolution and on number of quaternary digital outputs m. The threshold voltages are in the middle between successive voltage levels of quantized analog signal. This network can be realized as standard binary CMOS voltage comparator network using standard CMOS binary voltage comparator circuits. The network gives binary output signals (B_i) with voltage levels of V_{SS} and V_{CC3}.

The CMOS binary encoder network encodes the binary voltage comparator network output signals (B_i) into appropriate binary signals (C_i) for control of BiCMOS quaternary output network. This network can be realized as standard binary CMOS encoder network using standard CMOS binary logic circuits. It gives binary output signals (C_i) with voltage levels of V_{SS} and V_{CC3}.

The BiCMOS quaternary output network generates needed output quaternary signals (Y_i) depending on states at the binary inputs (C_i). It is proposed to use appropriate quaternary output BiCMOS stages in the network for every quaternary output. This network gives quaternary BiCMOS output signals (Y_i) with BiCMOS voltage levels of ($V_{SS}+V_{BE}$) for logic level 0, ($V_{CC1}+V_{BE}$) for logic level 1, ($V_{CC2}+V_{BE}$) for logic level 2 and ($V_{CC3}-V_{BE}$)

for logic level 3, where V_{BE} is voltage between base and emitter of conducting output bipolar transistor.

Complexity of design of the BiCMOS analog to quaternary digital signal converter and complexity of design of all three networks in the proposed converter design depends on the number of quaternary outputs m and increases with increasing of number m. It also depends on used ways for design of three appropriate networks in the converter structure. Practical development and design of such analog to quaternary digital signal converters includes determination of number of quaternary BiCMOS outputs, selection of used output BiCMOS quaternary stages, selection of used CMOS binary voltage comparators and design of appropriate CMOS binary encoder network.

The proposed structure shown in Figure 1 is general one and gives possibility to develop and design analog to digital signal converter for applications in BiCMOS quaternary digital systems with any number of quaternary outputs and with any resolution. Design of two digit parallel analog to digital signal converters for BiCMOS quaternary systems is proposed and shown in this paper as an example of design of such converters.

III. DESIGN OF TWO DIGIT ANALOG TO DIGITAL SIGNAL CONVERTERS FOR BiCMOS QUATERNARY DIGITAL SYSTEMS

The way for development and design of parallel analog to digital signal converters for BiCMOS quaternary digital systems with two quaternary outputs, based on the proposed structure is proposed and shown. Iit can be developed, designed and realized more different concrete solutions of such parallel analog to quaternary digital signal converters based on the proposed structure given in Figure 1. Solutions that are appropriate for some concrete applications are proposed and described. Solutions of so called basic type converters are shown and described. Way to obtain so called modified type converters is also proposed and described.

A. Basic type converters

Proposed structure and output stages of basic type two digit parallel analog to digital signal converter for BiCMOS quaternary systems is shown in Figure 2a. The circuit has one analog input (A_i) and two quaternary outputs (Y_0 and Y_1). It uses the basic type of BiCMOS quaternary output stage shown in Figure 2b.

Used CMOS voltage comparator network consists of $4^m-1 = 4^2-1 = 15$ voltage comparators (VC_i), since number of converter outputs is m=2. Each comparator has appropriate threshold voltage (V_{Ri}) for comparison with input analog signal. The threshold voltages should be equal to the voltages that are in the middle between successive voltage levels of quantized analog signal. Each comparator compares the input analog signal with its threshold voltage and gives appropriate binary output signal B_i. For implementation of this network can be used standard CMOS binary voltage comparator circuits.

The CMOS binary encoder network encodes the comparator network output signals (B_i) into appropriate binary signals (C_i). There are 15 input signals (B_i) and 8 output signals (C_i) in the encoder network. This network can be realized as binary CMOS encoder network using standard CMOS binary logic circuits.

The BiCMOS quaternary output network generates needed output quaternary BiCMOS signals (Y_i) depending on output states of CMOS encoder logic network (C_i). It is proposed here to use appropriate BiCMOS quaternary output network and appropriate quaternary output BiCMOS stages. The proposed output stages are the basic type of BiCMOS quaternary output stages (Figure 2b).

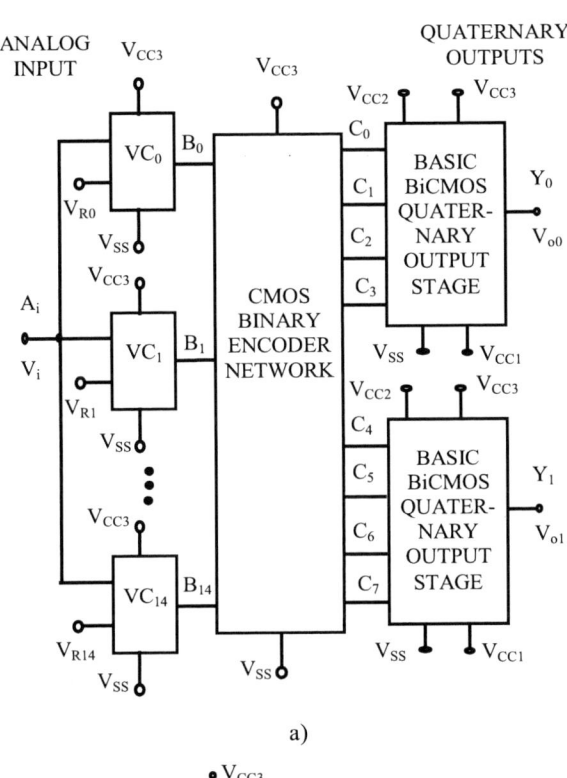

a)

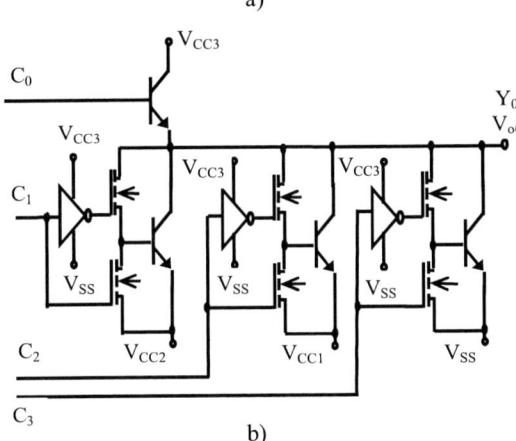

b)

Figure 2. Two digit basic type parallel analog to digital signal converter for BiCMOS quaternary digital systems (a) and basic BiCMOS quaternary output stage (b)

It can be developed and designed needed CMOS binary encoder network using appropriate logic table for description of operation way of the converter in Figure 2. Based on the logic table it can be obtained appropriate logic expressions for encoder network logic outputs (C_i) as a function of encoder network logic inputs (B_i). It can be shown that the expressions can be obtained and given in the next form:

$$C_0 = B_3\,\overline{B_2} + B_7\,\overline{B_6} + B_{11}\,\overline{B_{10}} + \overline{B_{14}}, \tag{1}$$

$$C_1 = \overline{\overline{B_2\,\overline{B_1}} + \overline{B_6\,\overline{B_5}} + \overline{B_{10}\,\overline{B_9}} + \overline{B_{14}\,\overline{B_{13}}}}, \tag{2}$$

$$C_2 = \overline{\overline{B_1\,\overline{B_0}} + \overline{B_5\,\overline{B_4}} + \overline{B_9\,\overline{B_8}} + \overline{B_{13}\,\overline{B_{12}}}}, \tag{3}$$

$$C_3 = \overline{\overline{B_0} + \overline{B_4\,\overline{B_3}} + \overline{B_8\,\overline{B_7}} + \overline{B_{12}\,\overline{B_{11}}}}, \tag{4}$$

$$C_4 = \overline{B_{11}}, \tag{5}$$

$$C_5 = \overline{B_{11}\,\overline{B_7}}, \tag{6}$$

$$C_6 = \overline{B_7\,\overline{B_3}}, \tag{7}$$

$$C_7 = \overline{B_3}. \tag{8}$$

It can be designed appropriate logic circuits and complete CMOS encoder network using the given expressions. Practically designed CMOS encoder network of the basic type two digit parallel analog to digital signal converter for BiCMOS quaternary digital systems is shown in Figure 3. Standard binary CMOS inverting logic circuits (inverters and NAND logic circuits) are used for design of the encoder network. Such is obtained analog to quaternary digital signal converter of basic type with minimized number of used transistors. All standard CMOS binary logic circuits are supplied by two supply voltages V_{SS} and V_{CC3}.

The CMOS encoder network can be also implemented using networks of NMOS and PMOS transistors for realization of logic functions given by expressions (1) to (8). In that way can be implemented more compact the CMOS encoder network using smaller number of MOS transistors. Such can be implemented the basic type converter using minimal number of transistors.

B. Modified type converters

Development and design of analog to digital signal converters for BiCMOS quaternary digital systems with smaller total average propagation delay time and smaller conversion time can be obtained if it is used different implementation of binary encoder network and different BiCMOS quaternary output stages compared with the basic type converters. It can be developed and obtained more different designs of such analog to quaternary converter depending on used design of CMOS encoder network and BiCMOS quaternary output stages

Structure of the converter and proposed output stages of modified type two digit parallel analog to digital signal converter for BiCMOS quaternary digital systems with smaller total average propagation delay time and smaller conversion time is shown in Figure 4a. The modified type converter uses modified quaternary BiCMOS output stage shown in Figure 4b. This way of design uses more control signals (C_i) for every BiCMOS output stage compared with the basic BiCMOS output stage used in the basic converter. It requires design of CMOS encoder network with more outputs (C_i).

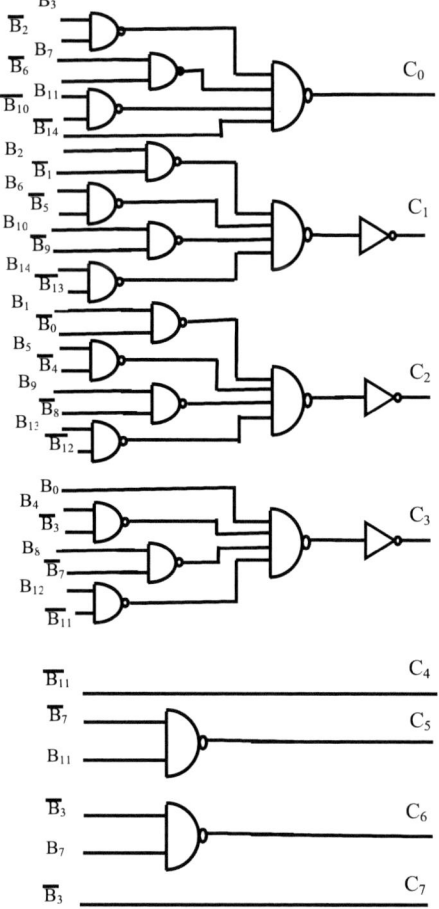

Figure 3. Encoder network of two digit basic type parallel analog to digital signal converter for BiCMOS quaternary digital systems

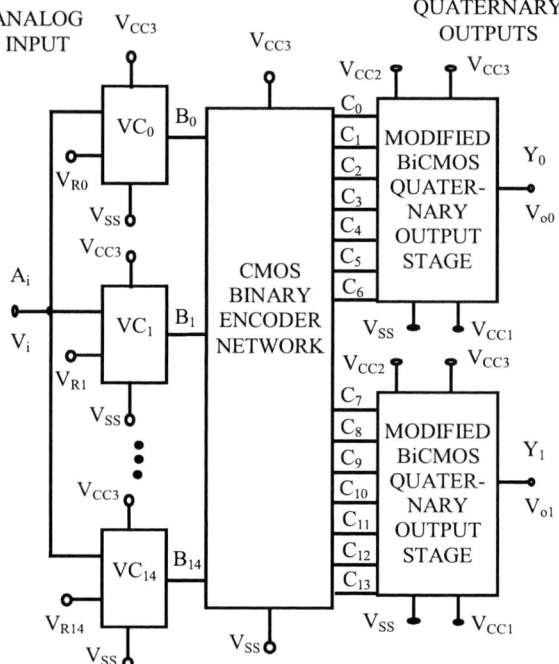

a)

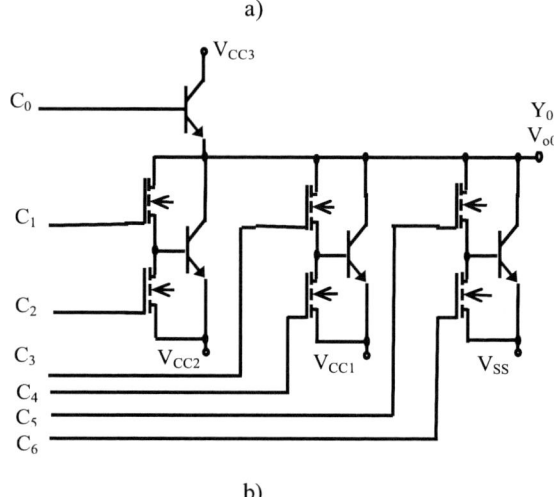

b)

Figure 4. Two digit modified type parallel analog to digital signal converter for BiCMOS quaternary digital systems (a) and modified BiCMOS quaternary output stage (b)

It can be designed appropriate CMOS binary encoder network using logic table for description of operation of the converter in Figure 4. Based on the logic table it can be obtained logic expressions for encoder network logic outputs (C_i) as a function of encoder network logic inputs (B_i). It can be shown that the expressions can be obtained and given in the next form:

$$C_0 = B_3\overline{B_2} + B_7\overline{B_6} + B_{11}\overline{B_{10}} + \overline{B_{14}}, \qquad (9)$$

$$C_1 = B_2\overline{B_1} + B_6\overline{B_5} + B_{10}\overline{B_9} + B_{14}\overline{B_{13}}, \qquad (10)$$

$$C_2 = \overline{B_2\overline{B_1} + B_6\overline{B_5} + B_{10}\overline{B_9} + B_{14}\overline{B_{13}}}, \qquad (11)$$

$$C_3 = B_1\overline{B_0} + B_5\overline{B_4} + B_9\overline{B_8} + B_{13}\overline{B_{12}}, \qquad (12)$$

$$C_4 = \overline{B_1\overline{B_0} + B_5\overline{B_4} + B_9\overline{B_8} + B_{13}\overline{B_{12}}}, \qquad (13)$$

$$C_5 = B_0 + B_4\overline{B_3} + B_8\overline{B_7} + B_{12}\overline{B_{11}}, \qquad (14)$$

$$C_6 = \overline{B_0 + B_4\overline{B_3} + B_8\overline{B_7} + B_{12}\overline{B_{11}}}, \qquad (15)$$

$$C_7 = \overline{B_{11}}, \qquad (16)$$

$$C_8 = B_{11}\overline{B_7}, \qquad (17)$$

$$C_9 = \overline{B_{11}\overline{B_7}}, \qquad (18)$$

$$C_{10} = B_7\,\overline{B_3}, \qquad (19)$$

$$C_{11} = \overline{B_7\,\overline{B_3}}, \qquad (20)$$

$$C_{12} = B_3, \qquad (21)$$

$$C_{13} = \overline{B_3}. \qquad (22)$$

Using the given expressions it can be developed and designed needed logic circuits and complete CMOS encoder network. Standard binary CMOS inverting logic circuits (inverters, NAND and NOR logic circuits) are used for design of the encoder network. All standard CMOS binary logic circuits are supplied by two supply voltages V_{SS} and V_{CC3}.

The CMOS encoder network also can be designed and realized using networks of NMOS and PMOS transistors for realization of logic functions given by expressions (9) to (22). Such can be realized more compact the CMOS encoder network using smaller number of MOS transistors. In that way can be implemented the modified type converter using minimal number of transistors.

In this way it is obtained and designed analog to digital signal converter for BiCMOS quaternary digital systems of modified type with decreased propagation delay time and decreased conversion time compared with the basic type converter. But, this converter uses increased number of MOS transistors compared with the basic type one. It is increased number of CMOS binary encoder network output control signals (C_i) for control of modified BiCMOS quaternary stages. That requires development and design of more complex CMOS binary encoder network using increased number of MOS transistors.

In design, implementation and simulation of the analog to quaternary digital signal converters were used CMOS binary encoder networks solutions based on application of standard binary CMOS inverting logic circuits (inverters, NAND and NOR logic circuits). For the implementation of the encoder network in the basic converter were used 92 MOS transistors (46 NMOS and 46 PMOS transistors). In the implementation of the encoder network in the midified converter were used 128 MOS transistors (64 NMOS and 64 PMOS transistors).

IV. RESULTS OF CONVERTERS SIMULATIONS

Operation and parameters of proposed and described analog to digital signal converters for BiCMOS quaternary digital systems have been analyzed by PSpice simulations. Technology parameters of one BiCMOS process [8] and supply voltages $V_{SS}=0V$, $V_{CC1}=5V$, $V_{CC2}=10V$ and $V_{CC3}=15V$ were used in the simulations. The CMOS voltage comparator circuits as were proposed in the paper [9], with appropriate designed threshold voltages, were used for design of CMOS voltage comparator networks of the converters. The used CMOS voltage comparator circuits are based on the circuits proposed in the paper [10]. There were simulated and analyzed the most

important static and dynamic parameters of the proposed converters.

The timing diagrams of output quaternary voltages and input analog voltage obtained by PSpice simulations for two digit basic type converter are shown in Figure 5. At the converter analog input was applied slow-changing voltage signal (V_i) for obtaining all possible quaternary states at converter outputs (V_{o0} and V_{o1}). It was also confirmed by the simulations that the same signal timing diagrams (Figure 5) are also valid for the modified type two digit analog to digital signal converter for BiCMOS quaternary digital systems.

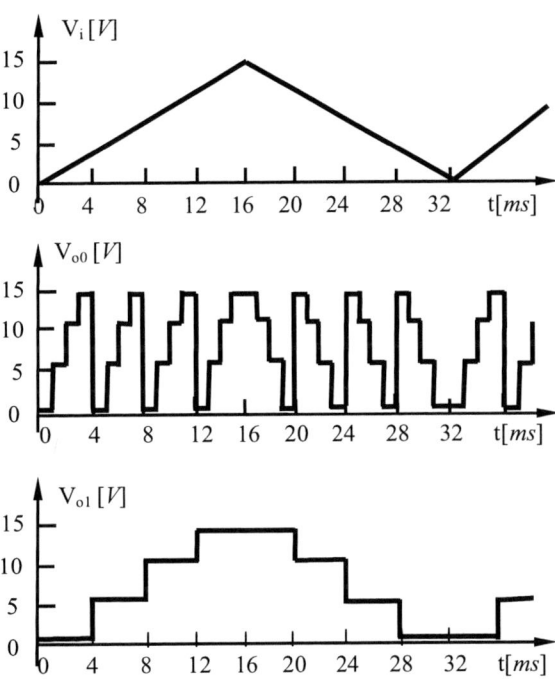

Figure 5. Timing diagrams of input analog signal and output BiCMOS quaternary signals for basic type two digit converter

Figure 6 shows conversion times (t_c) of the two digit basic type and modified type analog to digital signal converter for BiCMOS quaternary digital systems as a function of capacitive load C_L obtained by PSpice simulations. The same technology parameters and the same supply voltages as in previous simulations were used in this simulations. Obtained results for two digit basic type converter are shown by full line. Obtained results for two digit modified type converter are shown by dashed line. It can be seen from the simulation results that modified type converter has smaller conversion time for greater capacitive loads comparing with the basic type converter. That is the main advantage of the modified type converter.

It can be seen from the designs and the simulation results that the basic type analog to digital converter is simpler than the modified type converter. It uses simpler CMOS encoder network and has smaller total number of

MOS transistors. But, the basic type converter and converters obtained on such principle generally have increased conversion time for greater loads compared with the modified type converter.

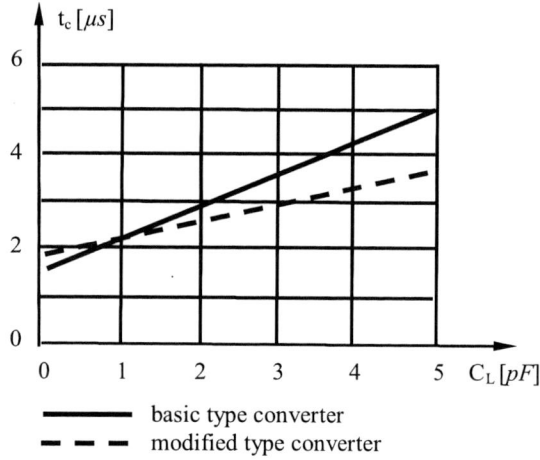

Figure 6. Conversion time of two digit analog to digital quaternary BiCMOS signal converters as a function of capacitive load

V. CONCLUSIONS

Very well known advantages of MV digital systems are the main reasons for increased interest for development, design and application of such digital systems. Practically the greatest interest exists for quaternary MV circuits and systems.

There is need to use analog to digital signal conversion and such converters in MV digital systems. Parallel analog to digital converters are the fastest ones.

The proposed and described principles for development, design and implementation of parallel analog to digital signal convertors for applications in BiCMOS quaternary digital systems are very clear and relatively simple. For implementation are used only standard MOS and bipolar transistors and the standard MOS and bipolar technology (standard BiCMOS technology). Proposed and described principles enable design of such converters with any (needed) number of quaternary BiCMOS logic outputs and with any (needed) resolution, according to converter working conditions.

The proposed and described principles and solutions enable to develop, design, implement and obtain optimal BiCMOS analog to quaternary digital signal converter depending on concrete application requirements on the place of converter application. The proposed basic type converters are simpler and use smaller number of MOS transistors. But, that type of converters have greater delay times and greater conversion times for greater loads. The proposed modified type converters are more complex and use more MOS transistors. But that type of converters have smaller delay times and smaller conversion times for greater loads. The reason for such dilay times and conversion times is that the basic type converter uses more

CMOS logic stages, compared with the modified converter. That increases delay times and conversion times when using greater converter loads. The basic type converter uses smaller total number of MOS transistors, having smaller total parasitic capaticances, compared with the modified type converter. That contributes to decresed delay time and conversion time of the converter when using smaller capacitive loads.

The proposed basic type BiCMOS analog to quaternary digital signal converters should be used in analog to quaternary digital signal conversion applications that have smaller loads and smaller needed working speeds. The proposed modified type BiCMOS converters should be used in analog to quaternary digital signal conversion applications that use greater loads and greater needed working speeds.

The parameters of one older BiCMOS technology process were used in the proposed analog to quaternary digital signal converter circuits simulations. The reason was to be able to compare results of the simulations with earlier obtained simulation results for some other similar circuits. Also, aim of the paper was proposal and description of the structure and the design, as well as confirmation of proper operation, of the proposed BiCMOS analog to quaternary digital signal convertors. The aim was not the determination of parameters and characterization of the concrete converter circuits. All it does not depend on used BiCMOS technology process in the simulations.

REFERENCES

[1] E. V. Dubova, "Multiple-valued logic in VLSI: challenges and opportunities", Proceedings of Conference NORCHIP'99, 1999, pp. 340-350.

[2] E. V. Dubrova, "Multiple - valued logic in VLSI design", International Journal on Multiple -Valued Logic, 2002, pp. 1-17.

[3] V. Patel and K. S. Gurumurthy, "Quaternary CMOS combinational logic circuits", Proceedings of International Conference on Information and Multimedia Technology, 2009., pp.538-542.

[4] V. Patel and K. S. Gurumurthy, "Quaternary sequential circuits", International Journal of Computer Science and Network Security, July 2010., pp. 110-117.

[5] T. Tanoue, M. Nagatani and T. Waho, "A ternary analog-to-digital converter system", Proceedings of the 37th International Symposium on Multiple-Valued Logic, 2007.

[6] S. Farhana, A. H. M. Zahirul Alam and S. Khan, "Development of 2-Digit analog-to-digital converter", World Applied Sciences Journal 17 (5), 2012., pp. 622-625.

[7] Z. Bundalo, D. Bundalo, F. Softić, M. Kostadinović and D. Pašalić, "Analog to quaternary digital CMOS converters", Proceedings of 22nd International Scientific Conference ERK2013, Portorose, Slovenia, September 2013., pp. 27-30.

[8] C. H. Diaz, S. Kang and Y. Leblebici, "An accurate analytical delay model for BiCMOS driver circuits", IEEE Transaction on Computer-Aided Design, vol. 10, no. 5, may 1991., pp. 577-588.

[9] Z. Bundalo, D. Bundalo, F. Softić and M. Kostadinović, "Logic circuits for interconnection of ternary and binary CMOS digital circuits and systems", Proceedings of the 55th Conference ETRAN, Banja Vrućica, Bosnia and Herzegovina, June 2011., pp. EL4.1-1-4 (in Serbian).

[10] Z. Bundalo, "CMOS and BiCMOS logic circuits for conversion from low to high logic level", Proceedings of the 51st Conference ETRAN, Herceg Novi -Igalo, Montenegro, June 2007., pp. EL2.1-1-4 (in Serbian).

MIPRO 2017, May 22- 26, 2017, Opatija, Croatia

Ultra-Wideband Pulse Generator for Time-Encoding Wireless Transmission

Leon Šneler, Marijan Herceg and Tomislav Matić
Department of Communications
Faculty of Electrical Engineering, Computer Science and Information Technology
Osijek, Croatia
tmatic@etfos.hr

Abstract—The paper presents implementation of the Ultra-Wideband (UWB) pulse generator suitable for wireless transmission of time-encoded pulse train. Application of the UWB pulse generator enables direct pulsed triggering from the output of Time Encoding Machine (TEM). The pulse generator circuit is designed in IHP 0.24 µm SG25H3 technology. The paper presents simulation results for the pulse power spectral density and time domain results for the corresponding input and output signals. The pulse generator circuit is expected to consume 189 pJ energy per transmitted pulse and covers area of 0.132 mm². Expected output pulse width is equal to 301 ns and output voltage swing 502 mV.

Keywords—Ultra-Wideband; pulse generator; Time-Encoding Machine; Integral Pulse Frequency Modulator; Power Spectral Density.

I. INTRODUCTION

Due to low energy per transmitted pulse Ultra-Wideband systems are attractive solution for low power short range wireless applications [1-6]. Besides energy efficiency constraints, increasing number of Internet of Things (IoT) connected devices requires novel solutions for wireless connectivity, due to increased spectrum congestion in lower GHz frequency bands [7, 8]. Last 10 meters connectivity is becoming crucial problem in both spectrum management and data processing. Introduction of UWB provides alternative for Bluetooth and similar competing technologies, particularly in Body Area Networks [1, 5], with high number of sensors within low distance.

The most common UWB application areas are in wireless communication systems, particularly Personal Area Networks (PAN) [1, 3], localization [9] and radar imaging [10, 11]. This work is based on analog signal processing, time encoding and asynchronous transmission. Therefore, it cannot be compared in terms of bit rate and bit error rate performances to conventional UWB communication systems. Since it aims to transmit analog signal using asynchronous pulses, its application is considered as baseband modulation technique and is limited with pulse width and pulse repetition rate. The most recent and the most relevant publication proposes similar concept for wireless sensing application, using time constant measurement and UWB transmission [12].

This work is funded in part by Croatian Science Foundation under the project UIP-2014-09-6219 "Energy Efficient Asynchronous Wireless Transmission.

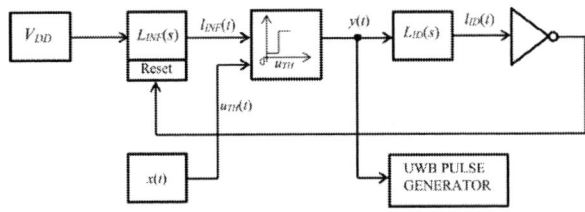

Fig. 1.   Block diagram of the UWB transmitter

Architecture of the Integral Pulse Frequency Modulator (IPFM) provides linear and energy efficient transformation of the analog input signal to time information encoded in output pulse distance. Therefore, triggering UWB pulse generator directly with TEM output signal enables extremely simple solution for wireless sensing, applicable in Internet of Things applications. To achieve multi-user analog signal acquisition, time delay-based coding is applied within IPFM feedback loop. According to that, the output pulse width depends on the user time delay ID value. Rectangular output pulse train is fed to the input of UWB generator to generate UWB pulse pairs, suitable for further wireless transmission.

Several recent papers present contributions in IC based Ultra-Wideband pulse generation and shaping [13-16]. Besides IC pulse generator implementations, the most common architectures of UWB pulse generators are based on step recovery diodes or avalanche transistors with application of micro-strip lines [17]. However, SRD and BJT implementations significantly increase power consumption of the circuit and cannot satisfy energy efficiency requirements. The proposed UWB generator, suitable for IC implementation in high frequency 0.24 µm process enables size and power efficient circuitry, suitable for wireless sensing applications.

Section II of the paper presents brief introduction to UWB transmitter architecture, followed by modulation principle explanation and UWB generator architecture. The simulation results of the UWB generator are presented in Section III, including the circuit performances analysis.

II. ULTRA-WIDEBAND TRANSMITTER

A. Integral pulse frequency modulator

The pulse generator is incorporated within IPFM wireless sensor node (Fig. 1). Feedback loop of the IPFM contains

109

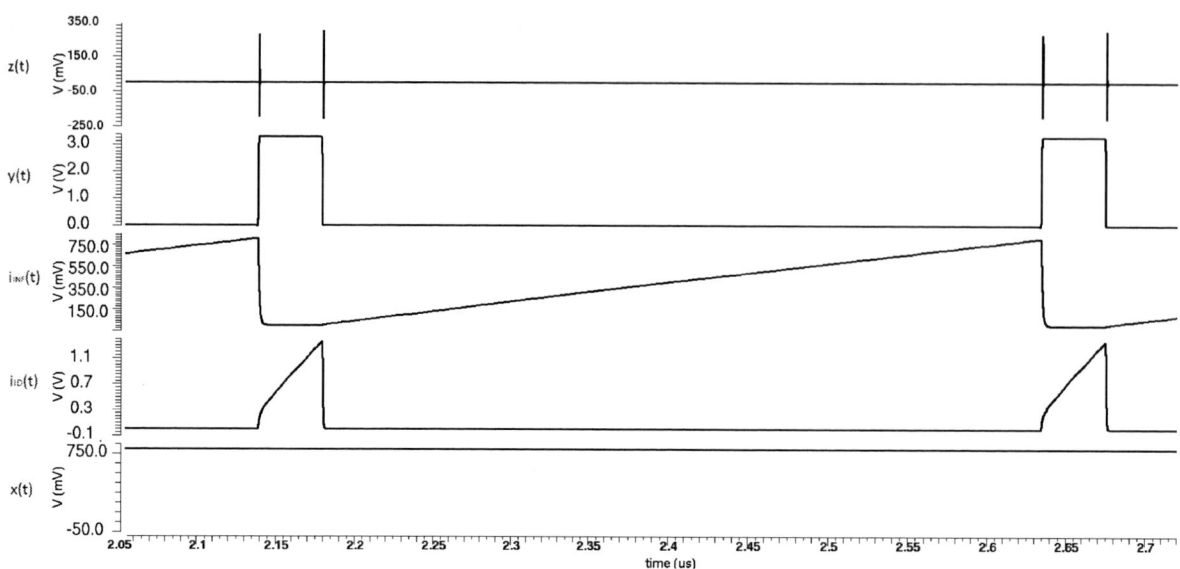

Fig. 2.   Ultra-Wideband transmitter waveforms for DC input voltage

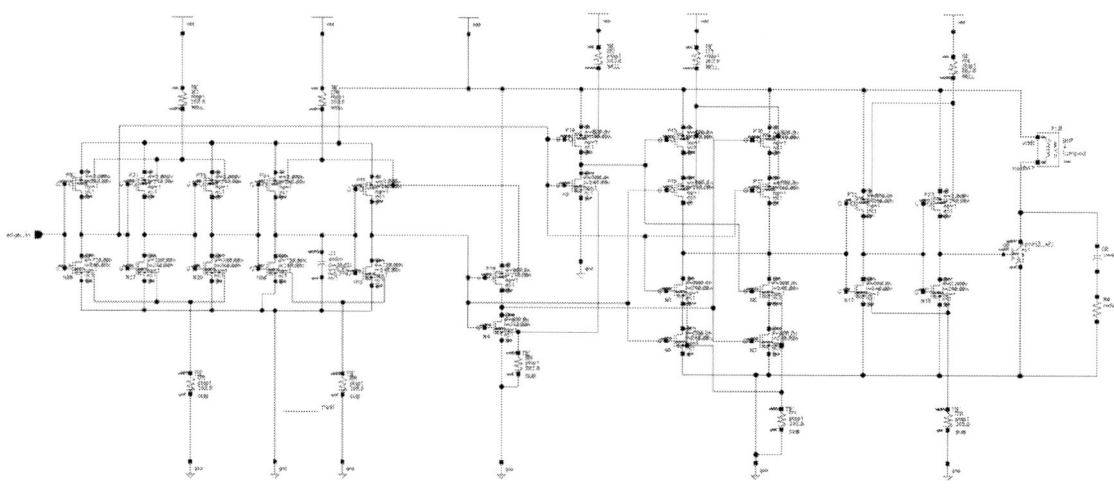

Fig. 3.   Ultra-Wideband Pulse generator circuit

unique identifying delay cell used for multi-user-coding [18, 19]. Each sensor node has different corresponding time delay value, denoted as sensor ID. In that way, each sensor node exhibits different rectangular pulse width at the output.

Corresponding waveforms, for the IPFM DC input signal $x(t) = 1$ V (bottom line), are presented in the fig. 2. Corresponding integrator waveforms present input integrator $l_{INF}(t)$ and user delay integrator $l_{ID}(t)$ signals. Output of the IPFM modulator $y(t)$ conveys information on analog input voltage $x(t)$ within pulse distance and sensor ID information within pulse duration. Following rising and falling edges of the $y(t)$ signal, UWB generator forms UWB pulses (top line).

The consecutive pulse distance is proportional to the comparator threshold $u_{TH}(t) = x(t)$ for ideal integrator $L_{INF}(s)$ with time constant τ_{INF}. The input integrator output signal in time domain can be expressed by the following equation:

$$l_{INF}(t_k) = \int_{t_{k-1}}^{t_k} \frac{V_{DD}}{\tau_{INF}} + C_{k-1}. \tag{1}$$

At the time t_k when integrator $L_{INF}(s)$ triggers comparator, under consumption that at the beginning of the integration slope t_{k-1}, the integrator output was equal to 0 ($l_{INF}(t_{k-1}) = C_{k-1} = 0$) the following equation holds:

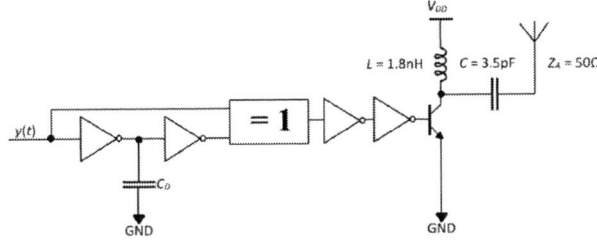

Fig. 4.   Block diagram of the UWB generator

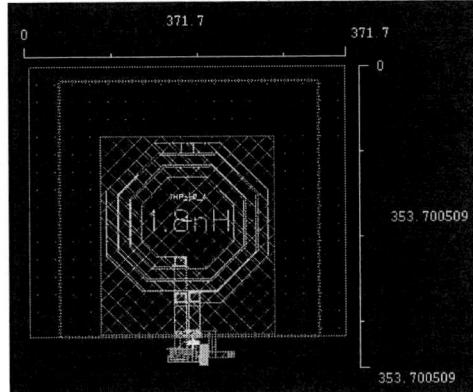

Fig. 5.   Layout of the proposed UWB pulse generator

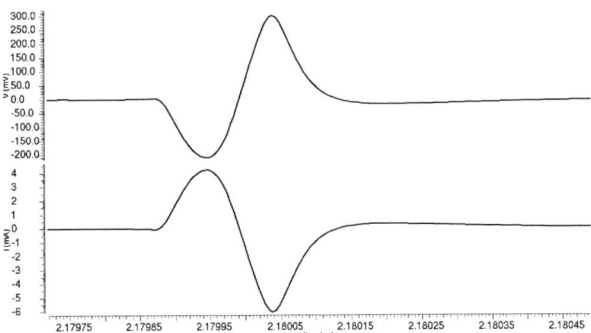

Fig. 6.   UWB pulse waveform

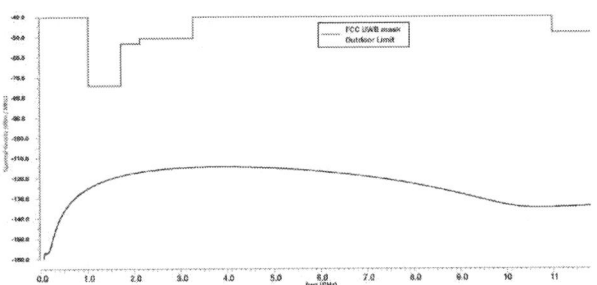

Fig.7.   Power spectral density of UWB pulse train

$$l_{INF}(t_k) = \int_{t_{k-1}}^{t_k} \frac{V_{DD}}{\tau_{INF}} dt = x(t). \qquad (2)$$

The IPFM pulse width depends on input integrator switching on and switching off times T_{ON} and T_{OFF} and sensor ID time delay T_{ID}, which is equal to:

$$T_{ID} = \frac{\tau_{ID}}{2}. \qquad (3)$$

According to (2), the distance between consecutive output pulses T_k and T_{k-1} of the signal $y(t)$, is equal to:

$$T_k - T_{k-1} = \frac{\tau_{INF}}{V_{DD}} x(t_k) + \frac{\tau_{ID}}{2} + T_{ON} + T_{OFF}. \qquad (3)$$

B. Ultra-Wideband pulse generator

To keep information on particular user ID, UWB generator (Fig. 3) produces two UWB pulses per one rectangular IPFM pulse. Input to the pulse generator contains delay cell followed by XOR gate (Fig.4). One rectangular pulse is being transformed to two pulses, one for rising and the other for falling edge. Both pulses trigger the RF part of the UWB pulse generator and at the output form two wideband pulses (top line on Fig.2).

Delay line is implemented by two CMOS inverter cells and capacitor to adjust the time delay, required for the pulse period setting. Input and output of the delay line feed the XOR gate for two consecutive pulses formation. Output of the XOR gate

is followed by the buffer which triggers the bipolar transistor coupled with capacitance and inductor for output pulse shaping. Proposed architecture enables direct application of the IPFM output pulse train to the input of the pulse generator, which is extremely important to maintain the circuit simplicity and to provide energy efficient wireless connectivity.

Resulting waveform at the output of the pulse generator is monocycle pulse in required frequency spectrum. Layout of the proposed UWB pulse generator is depicted on Fig. 5. It consumes only 0.132 mm² mostly occupied by the planar inductor.

III.   SIMULATION RESULTS

The UWB pulse generator circuit is designed and simulated in Cadence Virtuoso tools, using IHP's PDK for SG25H3 0.24 μm BiCMOS technology, providing higher RF performances. Since the pulse generator requires high slew rate inputs for triggering, the technology is highly suitable for integration of time-encoding circuitry providing satisfying rising and falling pulse times suitable to trigger the pulse generator.

The resulting UWB pulse waveform is presented on Fig. 5. Achieved simulated pulse width is 301 ps and peak-to-peak amplitude over 500 mV. The power spectral density is provided on the Fig.6. The figure represents the PSD in dBW/MHz for pulse repetition rate equal 1.2 μs. Due to low pulse repetition rate, the power spectral density is far below the allowed limits according to the FCC requirements. Advantage of asynchronous transmission at low pulse frequency provides benefits in low power spectral density, which in turn enables application of output power amplifiers application and higher output voltage swing for longer range wireless transmission.

TABLE I.   SIMULATION RESULTS COMPARISON WITH STATE OF THE ART

Reference	Area [mm^2]	Energy [pJ/pulse]	V$_{pp}$ [mV]	Pulse width [ns]
This work	0.132	0.189	502	0.301
[13]	0.123	0.54	365-583	0.28-7.5
[15]	0.182	30	400	-

IV.   CONCLUSION

The paper presents the simple implementation of the Ultra-Wideband pulse generator in 0.24 μm technology. Architecture enables application within Time-Encoding based wireless sensor node applicable for wireless sensing applications. Since wireless sensor node operates in asynchronous manner, no need for clock signal is needed for signal acquisition and processing. Multi-user coding, achieved by simple time delay cell enables microprocessor-free coding which additionally reduces power consumption. Disadvantage of the proposed system is receiver architecture, suitable for asynchronous pulse detection. It can be implemented as a sliding correlator or simple level detector. Such architecture provides increased symbol error rate, but for application in slow varying signals, pulse repetition frequency can significantly be reduced, which enables potential error correction algorithms.

REFERENCES

[1]   H. Liu *et al.*, "Performance Assessment of IR-UWB Body Area Network (BAN) Based on IEEE 802.15.6 Standard," in IEEE Antennas and Wireless Propagation Letters, vol. 15, no. , pp. 1645-1648, 2016.

[2]   H. Kassiri *et al.*, "Battery-less Tri-band-Radio Neuro-monitor and Responsive Neurostimulator for Diagnostics and Treatment of Neurological Disorders," in IEEE Journal of Solid-State Circuits, vol. 51, no. 5, pp. 1274-1289, May 2016.

[3]   B. Zhou and P. Chiang, "Short-Range Low-Data-Rate FM-UWB Transceivers: Overview, Analysis, and Design," in IEEE Transactions on Circuits and Systems I: Regular Papers, vol. 63, no. 3, pp. 423-435, March 2016.

[4]   N. S. Kim and J. M. Rabaey, "A High Data-Rate Energy-Efficient Triple-Channel UWB-Based Cognitive Radio," in IEEE Journal of Solid-State Circuits, vol. 51, no. 4, pp. 809-820, April 2016.

[5]   S. Sapienza, M. Crepaldi, P. Motto Ros, A. Bonanno and D. Demarchi, "On Integration and Validation of a Very Low Complexity ATC UWB System for Muscle Force Transmission," in IEEE Transactions on Biomedical Circuits and Systems, vol. 10, no. 2, pp. 497-506, April 2016.

[6]   C. I. Dorta-Quiñones, X. Y. Wang, R. K. Dokania, A. Gailey, M. Lindau and A. B. Apsel, "A Wireless FSCV Monitoring IC With Analog Background Subtraction and UWB Telemetry," in IEEE Transactions on Biomedical Circuits and Systems, vol. 10, no. 2, pp. 289-299, April 2016.

[7]   C. Raj and S. Suganthi, "Survey on Microwave frequency V Band: Characteristics and challenges," 2016 International Conference on Wireless Communications, Signal Processing and Networking (WiSPNET), Chennai, 2016, pp. 256-258.

[8]   Gil Reiter, "Wireless connectivity for the Internet of Things", White paper, Texas Instruments, June 2014.

N. Decarli, F. Guidi and D. Dardari, "Passive UWB RFID for Tag Localization: Architectures and Design," in IEEE Sensors Journal, vol. 16, no. 5, pp. 1385-1397, March1, 2016.

[9]   K. Mæland, K. G. Kjelgård and T. S. Lande, "CMOS distributed amplifiers for UWB radar," 2015 IEEE International Symposium on Circuits and Systems (ISCAS), Lisbon, 2015, pp. 1298-1301.

[10]   E. Bakken, T. S. Lande and S. Holm, "Real time UWB radar imaging using single chip transceivers," 2014 IEEE International Symposium on Circuits and Systems (ISCAS), Melbourne VIC, 2014, pp. 2461-2464.

[11]   J. Mao, Z. Zou and L. R. Zheng, "A UWB-Based Sensor-to-Time Transmitter for RF-Powered Sensing Applications," in IEEE Transactions on Circuits and Systems II: Express Briefs, vol. 63, no. 5, pp. 503-507, May 2016.

[12]   B. Faes, P. Reynaert and P. Leroux, "A 280 ps - 7.5 ns UWB Pulse Generator with Amplitude Compensation in 40 nm CMOS," 2015 IEEE International Conference on Ubiquitous Wireless Broadband (ICUWB), Montreal, QC, 2015, pp. 1-4.

[13]   D. Pepe, L. Aluigi and D. Zito, "Sub-100 ps monocycle pulses for 5G UWB communications," 2016 10th European Conference on Antennas and Propagation (EuCAP), Davos, 2016, pp. 1-4.

[14]   K. Na, H. Jang, H. Ma, Y. Choi and F. Bien, "A 200-Mb/s Data Rate 3.1–4.8-GHz IR-UWB All-Digital Pulse Generator With DB-BPSK Modulation," in IEEE Transactions on Circuits and Systems II: Express Briefs, vol. 62, no. 12, pp. 1184-1188, Dec. 2015.

[15]   Yi He, Xiaole Cui, Qiang Si, Chung Len Lee and Dongmei Xue, "A new programmable delay cell with good symmetry for the digital IR-UWB pulse generator," Electron Devices and Solid-State Circuits (EDSSC), 2014 IEEE International Conference on, Chengdu, 2014, pp. 1-2.

[16]   Y.W. Yeap, "Ultra Wideband Signal Generation", Microwave Journal, September 2006.

[17]   T. Matić, M. Herceg, J. Job and L. Šneler, "Ultra-wideband transmitter based on integral pulse frequency modulator," 2016 39th International Convention on Information and Communication Technology, Electronics and Microelectronics (MIPRO), Opatija, 2016, pp. 80-83.

[18]   T. Matić, M. Herceg, J. Job and L. Šneler, "The receiver circuit for ultra-wideband integral pulse frequency modulated wireless sensor," 2016 5th International Conference on Modern Circuits and Systems Technologies (MOCAST), Thessaloniki, 2016, pp. 1-4.

[19]   T. Matic, M. Herceg, J. Job, "Energy-efficient system for distant measurement of analogue signals", WO/2014/195739, 11.12.2014.

[20]   T. Matic, M. Herceg, J. Job, "Energy-efficient system for distant measurement of analogue signals", WO/2014/195744, 11.12.2014.

MIPRO 2017, May 22- 26, 2017, Opatija, Croatia

Spectral-Efficient UWB Pulse Shapers Generating Gaussian and Modified Hermitian Monocycles

Ante Milos[1], Goran Molnar[1], and Mladen Vucic[2]

[1]Ericsson Nikola Tesla d. d. Research and Development Centre
Radio Development Unit
Krapinska 45, 10000 Zagreb, Croatia

[2]University of Zagreb Faculty of Electrical Engineering and Computing
Department of Electronic Systems and Information Processing
Unska 3, 10000 Zagreb, Croatia
E-mails: ante.milos@ericsson.com, goran.molnar@ericsson.com, mladen.vucic@fer.hr

Abstract—Ultra-wideband (UWB) impulse radio uses very short pulses whose spectra are regularized by Federal Communications Commission (FCC). One technique to obtain these pulses is shaping. The shaping is realized with a bandpass filter called pulse shaper. Its impulse response approximates an FCC-compliant waveform. In this paper, we propose spectral-efficient shapers whose responses approximate high-order Gaussian and modified Hermitian monocycles. To obtain the optimum shapers, we use least-squares error criterion. Furthermore, for various orders of monocycles, we provide the transfer functions that exhibit high spectral efficiency. Within the FCC UWB passband, the transfer functions of the Gaussian shapers ensure the spectral efficiency up to 89%, whereas the functions of the modified-Hermitian shapers provide the efficiency up to 62%.

I. INTRODUCTION

Ultra-wideband (UWB) impulse radio uses very short pulses to transmit data, resulting in high rate transmission. The UWB technology was regularized 2002 in North America by Federal Communications Commission (FCC), which provides spectral masks for UWB radio [1]. Later, other regulatory organizations allocated particular frequency bands in Europe and Japan.

The UWB impulse radio employs spectral efficient pulse-forming networks called UWB pulse generators. The design of a UWB pulse generator usually starts with the choice of an ideal pulse that fits the FCC mask. The common ideal pulses utilize modified Hermite polynomials [2], prolate spheroidal wave functions [3], and Gaussian monocycles [4], [5]. In [6], the unified structure of these pulses is described. The design and implementation of the pulse shapers is described in [5], whereas their comparison with respect to the FCC compatibility and energy efficiency are elaborated in [7] and [8].

FCC-compliant pulse waveforms that maximize the bandwidth are proposed in [9]. Considering the spectral efficiency, various waveforms are developed, resulting in sinc-based [10], sinusoidal-like [11], [12], square-like [9],

[13], and wavelet-based [14] monocycles. To obtain high spectral efficiency, linear combinations of monocycles are proposed in [5], [11], [15], and [16].

The design of pulse shaper is based on the approximation of the given monocycle. Several design methods have been developed, considering Padé [17], elliptic [18], and least squares [19], [20] approximation. In addition, the design of shapers incorporating the response of UWB antennas is considered in [21] and [22].

In this paper, we propose spectral-efficient analog pulse shapers whose impulse responses approximate high-order Gaussian and modified-Hermitian monocycles in the least-squares sense. To obtain the optimum shapers, we use the time-domain synthesis proposed in [19]. Furthermore, we provide the transfer functions of the shapers generating the monocycles with very high spectral efficiency within the FCC UWB passband.

The paper is organized as follows. Section II describes approaches to the design of UWB pulse generators. The method for the design of pulse shapers is described in Section III. Section IV describes the proposed Gaussian and Hermitian shapers.

II. UWB PULSE GENERATION

A. Pulse Generation Approaches

Well-established techniques for pulse generation are based on up-conversion, pulse shaping, and waveform synthesis [16]. The first technique employs analog lowpass filter to obtain a baseband pulse, which is then converted up to the UWB passband, as shown in Figure 1. In the second technique, the UWB pulse is obtained directly, by using analog bandpass filter called pulse shaper, as illustrated in Figure 2. The third technique employs discrete-time synthesis followed by up-conversion.

It is clear that the lowpass filter is usually less complex than the bandpass filter. However, employing the lowpass filter is paid by additional circuitry implementing mixer and local oscillator. Therefore, we consider here the latter approach.

This paper was supported in part by Ericsson Nikola Tesla d.d. and University of Zagreb, Faculty of Electrical Engineering and Computing under the project ILTERA, and in part by Croatian Science Foundation under the project Beyond Nyquist Limit, grant no. IP-2014-09-2625.

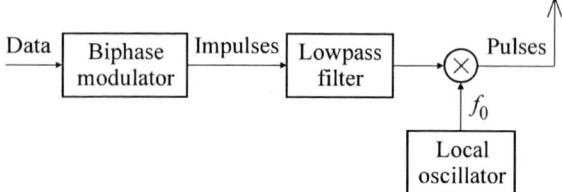

Figure 1. Block diagram of UWB impulse-radio transmitter incorporating lowpass filter and up-conversion.

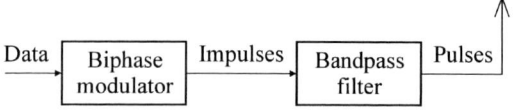

Figure 2. Block diagram of UWB impulse-radio transmitter incorporating bandpass filter as pulse shaper.

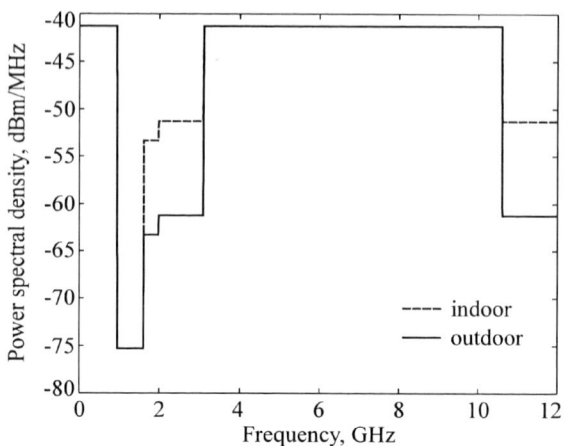

Figure 3. FCC spectral mask for indoor and outdoor UWB communications.

The power spectral density (PSD) of UWB pulses are regularized by FCC masks. Different masks are used for indoor and outdoor communications, as shown in Figure 3 [1]. For both masks, the FCC defines the UWB passband from 3.1 GHz to 10.6 GHz, with the maximum PSD value of −41.3 dBm/MHz.

Common measures of quality for UWB pulse generators are the spectral efficiency and the 10 dB-bandwidth. The spectral efficiency is defined as the average power of pulse normalized to the total allowed power under the FCC mask, that is,

$$\eta = \frac{\int_B |P(f)|^2 \, df}{\int_B S(f) \, df} \times 100 \, [\%] \qquad (1)$$

where $P(f)$ is the Fourier transform of the UWB pulse, $S(f)$ is the FCC spectral mask, and B is the band of interest.

B. Basic Pulse Shapes

Despite many efficient pulse shapes, the derivatives of the Gaussian pulse are still widely used for UWB pulses. In practice, the derivatives up to the seventh order are employed [7].

The Gaussian pulse is defined as

$$g_0(t) = \exp\left[-\left(\frac{t}{\tau}\right)^2\right] \qquad (2)$$

where τ denotes a bandwidth scaling factor. The nth derivative of the Gaussian pulse is called the nth-order Gaussian monocycle. The shape of the nth-order monocycle can be written in an analytical form, as in [23]

$$g_n(t) = \frac{d^n g_0(t)}{dt^n} = (-1)^n \frac{1}{\tau^n} H_n\left(\frac{t}{\tau}\right) g_0(t) \qquad (3)$$

where $H_n(t)$ is the Hermite polynomial of order n. A closed-form expression for the nth order Hermite polynomial is given by [23]

$$H_n(t) = n! \sum_{m=0}^{\lfloor n/2 \rfloor} \frac{(-1)^m}{m!(n-2m)!} (2t)^{n-2m} \qquad (4)$$

where $\lfloor u \rfloor$ denotes the greatest integer equal to or smaller than u.

The central frequency and bandwidth of the Gaussian monocycle are determined by factor τ in (2). For the FCC compliant monocycles of up to the tenth order, the values of τ are given in Table I in [4]. Later, in [7], it is shown that the second- and third-order Gaussian monocycles cannot meet the FCC masks unless a frequency translation is employed. In that sense, we do not consider them in this paper. In the paper referred to, it is also shown that the Gaussian monocycles with order higher than three meet the FCC indoor mask, whereas the monocycles with order higher than five fit the FCC outdoor mask.

Other well-established UWB pulses are modified Hermitian monocycles [2]. Similar to the Gaussian monocycles, these pulses have short duration and good spectral efficiency. The nth-order modified Hermitian monocycle is defined as [2]

$$m_n(t) = \exp\left[\left(\frac{t}{\sqrt{2}\tau}\right)^2\right] g_n(t) \qquad (5)$$

where τ is the same scaling factor as in $g_n(t)$.

In [7], it is shown that the modified Hermitian monocycles with orders greater than one are suitable for pulse shaping. However, to meet the FCC mask, they require additional bandpass filtering, what makes the UWB pulse generation more complex.

III. TIME-DOMAIN SYNTHESIS OF PULSE SHAPERS

A. Problem Formulation

The time-domain synthesis of a pulse shaper approximates the ideal UWB monocycle $p(t)$ with causal impulse response. Since $p(t)$ is noncausal, it should be appropriately delayed and truncated. Furthermore, its amplitude is usually normalized to unity. Such a monocycle is then considered as the desired impulse response. It has the form

$$h_d(t) = \begin{cases} \dfrac{p(t-t_d)}{p_{\max}} & , \quad 0 \le t \le T \\[2mm] 0 & , \quad \text{otherwise} \end{cases} \quad (6)$$

where t_d is the delay, T is a duration to which $p(t)$ is truncated, and $p_{\max}$ is the maximum absolute value of $p(t)$.

Our objective is to design analog pulse shapers whose impulse responses approximate the monocycles in (3) and (5) in the least-squares sense. Therefore, we define the objective function as

$$\varepsilon(\mathbf{x}) = \frac{1}{T} \int\limits_{0}^{T} [h(t,\mathbf{x}) - h_d(t)]^2 \, dt \quad (7)$$

where $h(t, \mathbf{x})$ is the shaper's impulse response and $\mathbf{x}$ is the vector of shaper parameters. It is clear that the design using function in (7) assumes delay t_d in (6) is known in advance.

In the design of shapers, we calculate $\varepsilon(\mathbf{x})$ using $h(t, \mathbf{x})$ and $h_d(t)$ evaluated on uniformly spaced time grid $Q = \{t_q; q=0, ..., Q\}$ defined within $0 \le t \le T$. Hence, the objective function in (7) is approximated with a finite sum, thus obtaining

$$\varepsilon(\mathbf{x}) \approx \frac{1}{Q} \sum\limits_{q=0}^{Q} \left[h(t_q, \mathbf{x}) - h_d(t_q) \right]^2 \quad (8)$$

The optimum shaper parameters that minimize the error in (8) can be found by solving the problem

$$\hat{\mathbf{x}} = \underset{\mathbf{x}}{\arg\min} \left[\varepsilon(\mathbf{x}) \right] \quad (9)$$

To solve the problem in (9), we use the method proposed in [19]. In this method, the optimum parameters are found by iterative procedure in which a second-order cone program is solved in each iteration. Such an approach ensures fast convergence and low sensitivity to optimization starting-point. In addition, the method deals with the zero-pole-gain model of the system, thus enabling simple control of shaper's stability.

B. Model of Pulse Shaper

The transfer function of an Nth order shaper with M zeros is given by

$$H(s) = H_0 \frac{\prod\limits_{i=1}^{M}(s-z_i)}{\prod\limits_{k=1}^{N}(s-p_k)} \quad (10)$$

where H_0 is the gain constant, and p_k and z_i denote transfer function poles and zeros. Since the shaper is uniquely described by p_k, z_i, and H_0, these parameters represent the components of $\mathbf{x}$. To form $\mathbf{x}$ as a real vector, we describe complex pairs of the poles and the zeros by their real and imaginary parts. If the transfer function contains M_1 real zeros, M_2 complex zeros, M_3 imaginary zeros, N_1 real poles, and N_2 complex poles, where $M_1+M_2+M_3=M$ and $N_1+N_2=N$, then $\mathbf{x}$ can be defined as [19]

$$\mathbf{x} = [H_0, \\ z_1,...,z_{M_1}, \alpha_1, \beta_1,...,\alpha_{M_2}/2, \beta_{M_2}/2, \gamma_1,...,\gamma_{M_3}/2, \quad (11) \\ p_1,...,p_{N_1}, \sigma_1, \omega_1,...,\sigma_{N_2}/2, \omega_{N_2}/2]^T$$

where α_i, β_i, γ_k, σ_l, and ω_l are real and imaginary parts of complex zeros and poles, as given in [19]

$$z_{M_1+i} = \alpha_i + j\beta_i, \quad z^*_{M_1+i} = \alpha_i - j\beta_i, \quad i = 1,...,M_2/2 \quad (12)$$

$$z_{M_1+M_2+k} = j\gamma_k, \quad z^*_{M_1+M_2+k} = -j\gamma_k, \quad k = 1,...,M_3/2 \quad (13)$$

$$p_{N_1+l} = \sigma_l + j\omega_l, \quad p^*_{N_1+l} = \sigma_l - j\omega_l, \quad l = 1,...,N_2/2 \quad (14)$$

In the design, the shaper's impulse response, $h(t, \mathbf{x})$, should be known. If the poles are simple and $M < N$, the impulse response can be obtained as

$$h(t) = \sum\limits_{r=1}^{N} K_r \exp(p_r t), \quad t \ge 0 \quad (15)$$

where K_r; $r = 1, ..., N$; are pole residues given by

$$K_r = H_0 \frac{\prod\limits_{i=1}^{M}(p_r - z_i)}{\prod\limits_{k=1, k\ne r}^{N}(p_r - p_k)} \quad (16)$$

According to [19], the choice of the initial point is not critical. However, the convergence rate can be significantly reduced if the initial point is far from the optimum. Furthermore, since the transfer function in (10) vanishes for $H_0 = 0$, it is recommended to start the optimization with H_0 of the appropriate sign, or run the optimizations for both signs.

IV. PROPOSED SPECTRAL-EFFICIENT PULSE SHAPERS

In this section, we describe the transfer functions of the shapers whose impulse responses approximate the Gaussian and modified-Hermitian monocycles. In addition, we provide the spectral efficiencies of the obtained impulse responses measured within the FCC UWB passband.

All transfer functions were obtained by approximating the desired impulse response in (6) which was sampled in 1001 points, that is, with $Q = 1000$. Furthermore, the delay t_d was estimated to ensure the truncated $p(t)$ contains 99.99 % of its total energy. Hence, it was obtained by solving the equation

$$\int\limits_{-t_d}^{t_d} p^2(t)dt = 0.9999 \cdot \int\limits_{-\infty}^{\infty} p^2(t)dt \quad (17)$$

Finally, the time axis is normalized to 1 ns. Consequently, the zeros and poles are normalized to 1 Grad/s.

A. Gaussian Shapers

The Gaussian shapers were obtained by approximating the desired response with $T = 10\tau$. The optimum zeros, poles, and gain constants of the shapers

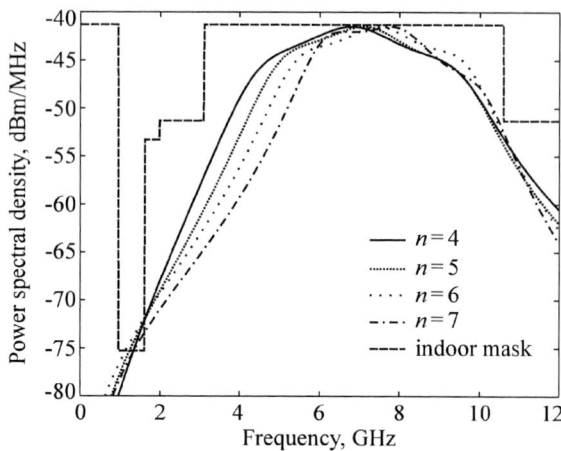

Figure 4. Power spectral densities of impulse responses approximating nth-order Gaussian monocycle obtained by proposed shapers with $N = 6$, together with FCC indoor mask.

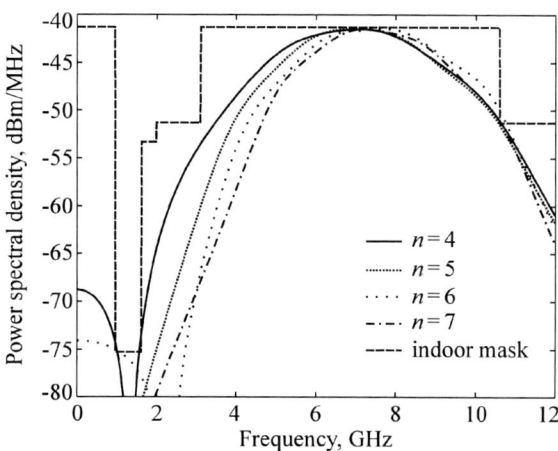

Figure 5. Power spectral densities of impulse responses approximating nth-order Gaussian monocycle obtained by proposed shapers with $N = 8$, together with FCC indoor mask.

approximating the monocycles of orders from $n = 4$ to $n = 7$ are given in Table I. For each n, the sixth- and eighth-order transfer function was obtained. In addition, to ensure the proposed shapers are compliant to the FCC indoor mask, the FCC gain constant C for each shaper is tabulated. This constant should be used in (10) instead of H_0. In the table, the spectral efficiencies of the corresponding impulse responses are also given. It is clear that the obtained spectral efficiency decreases with the order of monocycle. It is expected since the bandwidth of the ideal monocycle decreases with monocycle's order, as it is described in [4].

Figure 4 and 5 show the power spectral densities of the impulse responses of the proposed sixth-and eighth-order Gaussian shapers, together with the FCC indoor mask. It is clear that all shapers are closely compliant to the mask. Namely, outside the UWB passband they almost

TABLE I.
ZEROS, POLES, GAINS, AND FCC CONSTANTS OF PULSE SHAPERS GENERATING GAUSSIAN MONOCYCLE OF ORDER n

$n = 4$ with $\tau = 0.06647$ and $t_d = 0.18688$		
$M = 4$	$24.3772 \pm 63.6343\,j$ $4.41124 \pm 1.52466\,j$	$H_0 = 21.0178$ $C = 7.13571$ $\eta = 81.2\%$
$N = 6$	$-7.05717 \pm 59.4434\,j$ $-8.17639 \pm 42.9447\,j$ $-6.84788 \pm 27.4355\,j$	
$M = 6$	$13.3323 \pm 79.0313\,j$ $\pm 8.36727\,j$ 38.9955 38.9955	$H_0 = -18.2432$ $C = -6.46810$ $\eta = 88.8\%$
$N = 8$	$-9.05537 \pm 66.1232\,j$ $-10.7319 \pm 48.7615\,j$ $-10.4152 \pm 32.5719\,j$ $-9.67026 \pm 13.0463\,j$	
$n = 5$ with $\tau = 0.07212$ and $t_d = 0.20922$		
$M = 4$	$23.2788 \pm 63.4835\,j$ 21.6042 0	$H_0 = 18.2211$ $C = 5.20835$ $\eta = 76.0\%$
$N = 6$	$-6.25793 \pm 59.9713\,j$ $-7.26166 \pm 45.1554\,j$ $-6.09085 \pm 31.1535\,j$	
$M = 5$	$19.1303 \pm 71.3904\,j$ 12.9017 0 0	$H_0 = 998.092$ $C = 296.149$ $\eta = 82.6\%$
$N = 8$	$-7.39071 \pm 65.7296\,j$ $-8.85083 \pm 50.9586\,j$ $-8.58172 \pm 37.4470\,j$ $-6.73907 \pm 24.1066\,j$	
$n = 6$ with $\tau = 0.07495$ and $t_d = 0.22071$		
$M = 3$	$29.9416 \pm 47.8954\,j$ 2.39398	$H_0 = -760.847$ $C = -233.974$ $\eta = 74.3\%$
$N = 6$	$-5.38548 \pm 60.6325\,j$ $-6.42499 \pm 47.1044\,j$ $-5.52264 \pm 33.9556\,j$	
$M = 4$	$40.0656 \pm 57.1491\,j$ $\pm 14.3802\,j$	$H_0 = -28576.0$ $C = -9207.52$ $\eta = 82.4\%$
$N = 8$	$-6.01416 \pm 65.1139\,j$ $-7.55214 \pm 51.7788\,j$ $-7.49202 \pm 39.0740\,j$ $-6.14528 \pm 26.0428\,j$	
$n = 7$ with $\tau = 0.08061$ and $t_d = 0.23093$		
$M = 3$	$20.7203 \pm 64.6888\,j$ 0	$H_0 = -828.225$ $C = -222.469$ $\eta = 70.2\%$
$N = 6$	$-5.15940 \pm 63.5246\,j$ $-5.94215 \pm 50.4177\,j$ $-4.84870 \pm 38.1205\,j$	
$M = 4$	$28.8548 \pm 74.8823\,j$ 1.62664 1.62452	$H_0 = -25834.0$ $C = -7052.51$ $\eta = 74.0\%$
$N = 8$	$-6.17274 \pm 67.1398\,j$ $-7.57758 \pm 54.5552\,j$ $-7.45103 \pm 42.8736\,j$ $-6.00026 \pm 31.3212\,j$	

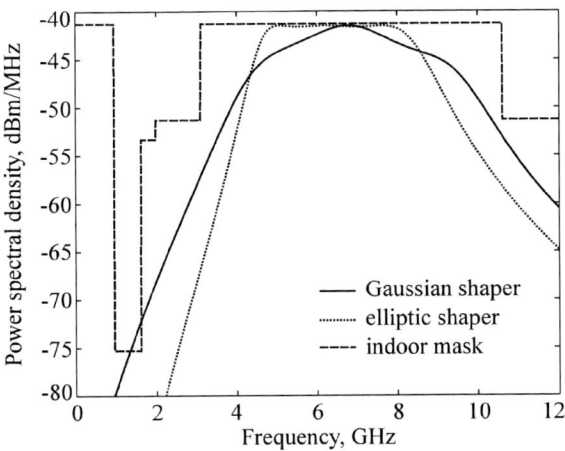

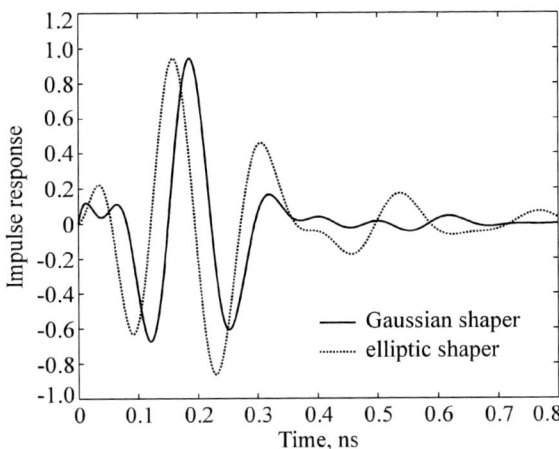

Figure 6. Power spectral density of impulse response approximating fourth-order Gaussian monocycle with $N = 6$ and $M = 4$ together with density obtained by elliptic shaper with $N = 6$ and $M = 5$ [18].

Figure 7. Impulse response of proposed Gaussian shaper with $N = 6$ and $M = 4$ generating fourth-order monocycle and impulse response of elliptic shaper with $N = 6$ and $M = 5$ [18].

fit the mask. However, it is considered tolerable since the UWB antennas have bandpass characteristics and, thus, additionally filter the obtained responses [21], [22].

Let us now compare our shaper with $N = 6$ and $M = 4$ generating fourth-order monocycle with the elliptic shaper having $N = 6$ and $M = 5$, which is proposed in [18]. Figure 6 shows the power spectral densities of the impulses responses of both shapers. Within the UWB passband, the proposed Gaussian shaper ensures $\eta = 81.2\%$, whereas the elliptic shaper results in $\eta = 90.5\%$. However, the proposed shaper ensures better time localization of the impulse response than do the elliptic shaper, as it is illustrated in Figure 7.

B. Hermitian Shapers

Let us now describe the shapers whose impulse responses approximate modified Hermitian monocycles with $n = 2, 3$, and 4. The shapers were obtained by using $T = 20\tau$. In addition, according to [2], the magnitude responses of the modified Hermitian monocycles contain zeros. Therefore, for their approximation we used the transfer functions having the imaginary zeros. The optimum zeros, poles, and gain constants of the Hermitian shapers are given in Table II, together with the corresponding FCC gain constants and spectral efficiencies measured within the UWB passband. It is clear that to achieve good spectral efficiency, higher monocycle's order requires higher transfer function's order.

Figure 8 shows the impulse response of the shaper generating the fourth-order modified-Hermitian monocycle, whereas Figure 9 shows its PSD together with the FCC indoor mask. It is clear that the Hermitian shaper is not FCC compliant. It is expected since the ideal modified-Hermitian monocycles have multiple high sidelobes, and, consequently, do not meet the FCC mask [7]. Therefore, the Hermitian shapers require additional bandpass filters, thus increasing the implementation complexity [7].

V. CONCLUSIONS

The analog pulse shapers generating spectral-efficient high-order Gaussian and modified Hermitian monocycles were proposed. Their transfer functions were obtained by minimizing the mean squared error in the time domain. The functions are represented with the zero-pole-gain model having simple poles only. It is shown that within the FCC UWB passband the proposed Gaussian shapers ensure spectral efficiency up to 89%, whereas the Hermitian shapers provide the efficiency up to 62%.

VI. REFERENCES

[1] "FCC first report and order: Revision of part 15 of the Commission's rules regarding ultra-wideband transmission systems," Washington, DC, FCC 02-48, 2002.

[2] M. Ghavami, L. B. Michael, S. Haruyama, and R. Kohno, "A novel UWB pulse shape modulation system," *Wireless Pers. Commun. J.*, vol. 23, pp. 105–120, Oct. 2002.

[3] B. Parr, B. Cho, K. Wallace, and Z. Ding, "A novel ultra-wideband pulse design algorithm," *IEEE Commun. Lett.*, vol. 7, no. 5, pp. 219–221, May 2003.

[4] H. Sheng, P. Orlik, A. M. Haimovich, L. J. Cimini Jr., and J. Zhang, "On the spectral and power requirements for ultra-wideband transmission," in *Proc. IEEE Int. Conf. Commun.*, Anchorage, Alaska, USA, May 2003, vol. 1, pp. 738–742.

[5] X. Luo, L. Yang, and G. B. Giannakis, "Designing optimal pulse-shapers for ultra-wideband radios," *J. Commun. Networks*, vol. 5, no. 4, pp. 344–353, Dec. 2003.

[6] M. Ghavami, A. Amini, and F. Marvasti, "Unified structure of basic UWB waveforms," *IEEE Trans. Circuits Syst.-II, Exp. Briefs*, vol. 55, no. 12, pp. 1304–1308, Dec. 2008.

[7] B. Hu and N. C. Beaulieu, "Pulse shapes for ultrawideband communication systems," *IEEE Trans. Wirel. Commun.*, vol. 4, no. 4, pp. 1789–1797, Jul. 2005.

[8] L. A. de Avila, R. Kunst, E. Pignaton, J. Rochol, and S. Bampi, "Energy efficiency evaluation of the pulse shapes and modulation techniques for IR-UWB in WBANs," in *Proc. IEEE Int. Conf. Electron. Circuits Syst.*, Monaco, Dec. 2016, pp. 173–176.

TABLE II.
ZEROS, POLES, GAINS, AND FCC CONSTANTS OF PULSE SHAPERS
GENERATING MODIFIED HERMITIAN MONOCYCLE OF ORDER n

$n = 2$ with $\tau = 0.05515$ and $t_d = 0.20074$		
$M = 4$	$23.9001 \pm 44.3637\,j$ $\pm 12.4314\,j$	$H_0 = 28.6981$ $C = 6.20057$ $\eta = 61.8\%$
$N = 6$	$-6.42630 \pm 39.6797\,j$ $-7.06111 \pm 24.3636\,j$ $-6.06555 \pm 7.56470\,j$	
$n = 3$ with $\tau = 0.06223$ and $t_d = 0.24734$		
$M = 6$	$20.6618 \pm 39.4412\,j$ $\pm 19.5214\,j$ $\pm 0.99368\,j$	$H_0 = 33.2854$ $C = 6.77222$ $\eta = 59.5\%$
$N = 8$	$-4.81876 \pm 39.7134\,j$ $-5.07600 \pm 27.0069\,j$ $-4.70509 \pm 13.2715\,j$ $-1.78883 \pm 1.01560\,j$	
$n = 4$ with $\tau = 0.06647$ and $t_d = 0.28410$		
$M = 8$	$\pm 75.9298\,j$ $19.9731 \pm 46.1548\,j$ $\pm 24.8823\,j$ $\pm 7.80364\,j$	$H_0 = 14.5409$ $C = 2.60827$ $\eta = 52.2\%$
$N = 10$	$-5.36191 \pm 49.7136\,j$ $-6.60265 \pm 38.5947\,j$ $-6.10334 \pm 27.7797\,j$ $-6.25259 \pm 16.3914\,j$ $-5.90958 \pm 5.33935\,j$	

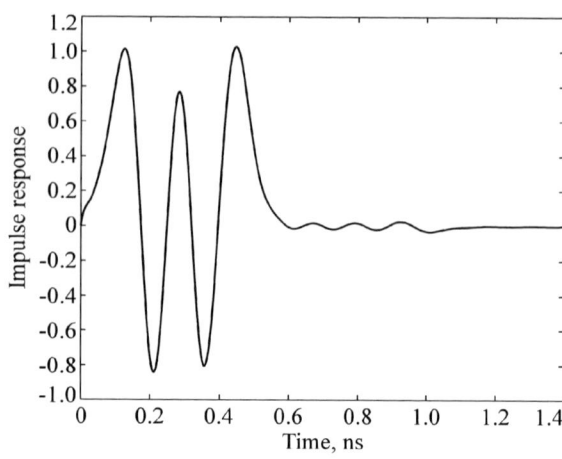

Figure 8. Impulse response of proposed Hermitian shaper generating fourth-order monocycle.

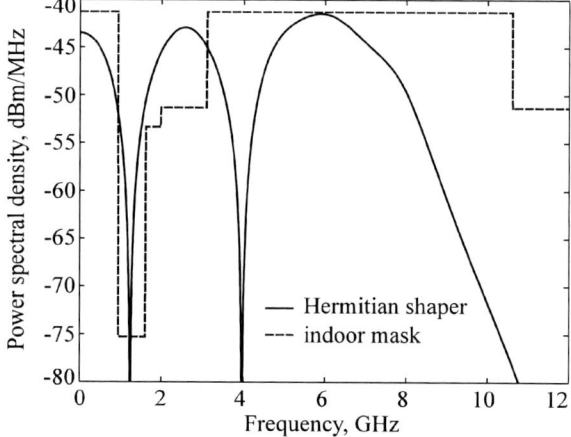

Figure 9. Power spectral density of impulse response approximating fourth-order modified Hermitian monocycle, together with FCC indoor mask.

[9] N. C. Beaulieu and B. Hu, "A pulse design paradigm for ultra-wideband communication systems," *IEEE Trans. Wirel. Commun.*, vol. 5, no. 6, pp. 1274–1278, Jun. 2006.

[10] N. C. Beaulieu and B. Hu, "On determining a best pulse shape for multiple access ultra-wideband communication systems," *IEEE Trans. Wirel. Commun.*, vol. 7, no. 9, pp. 3589–3596, Sept. 2008.

[11] R. Vauche, S. Bourdel, N. Dehaese, O. Fourquin, and J. Gaubert, "Fully tunable UWB pulse generator with zero DC power consumption," in *Proc. IEEE Int. Conf. Ultra-Wideband*, Vancouver, BC, Sept. 2009, pp. 418–422.

[12] A. Popa and N. D. Alexandru, "Waveform and CMOS generator for a pulse designated for UWB European band 6–8.5 GHz," in *Proc. Int. Symp. Signals Circuits Syst.*, Iasi, Romania, Jul. 2015, pp. pp. 1–4.

[13] S. Pohoata, A. Popa, and N. D. Alexandru, "Approximation of the third derivative of the Gaussian pulse", in *Proc. Int. Symp. Signals Circuits Syst.*, Iasi, Romania, Jul. 2011, pp. 1–4.

[14] A. N. Akansu, W. A. Serdijn, and I. W. Selesnick, "Emerging applications of wavelets: A review," *Physical Commun.*, vol. 3, no. 1, pp. 1–18, Mar. 2010.

[15] J. A. Silva and M. L. Campos, "Spectrally efficient UWB pulse shaping with application in orthogonal PSM," *IEEE Trans. Commun.*, vol. 55, no. 2, pp. 313–322, Feb. 2007.

[16] Y. Zhu, J. D. Zuegel, J. R. Marciante, and H. Wu, "Distributed waveform generator: a new circuit technique for UWB pulse generation, shaping and modulation," *IEEE J. Solid-State Circuits*, vol. 44, no. 3, pp. 808–823, Mar. 2009.

[17] S. A. P. Haddad, N. Verwaal, R. Houben, and W. A. Serdijn, "Optimized dynamic translinear implementation of the Gaussian wavelet transform," in *Proc. IEEE Int. Symp. Circuits Syst.*, Vancouver, Canada, May 2004, vol.1, pp. 145–148.

[18] Y. Shamsa and W. A. Serdijn, "A 21pJ/pulse FCC compliant UWB pulse generator," in *Proc. IEEE Int. Symp. Circuits Syst.*, Paris, France, May/Jun. 2010, pp. 497–500.

[19] M. Vucic and G. Molnar, "Time-domain synthesis of continuous-time systems based on second-order cone programming," *IEEE Trans. Circuits Syst. I, Reg. Papers*, vol. 55, no. 10, pp. 3110–3118, Nov. 2008.

[20] L. C. Neves, G. M. de Araujo, J. C. da Costa, and S. A. P. Haddad, "Design of a PSWF impulse response filter for UWB systems," in *Proc. IEEE Int. Symp. Circuits Syst.*, Seoul, Korea, May 2012, pp. 1935–1938.

[21] S. Bagga, A. V. Vorobyov, S. A. P. Haddad, A. G. Yarovoy, W. A. Serdijn, and J. R. Long, "Codesign of an impulse generator and miniaturized antennas for IR-UWB," *IEEE Trans. Microw. Theory Tech.*, vol. 54, no. 4, pp. 1656–1666, Apr. 2006.

[22] M. Mirshafiei, M. Abtahi, and L. A. Rusch, "Ultra-wideband pulse shaping: bypassing the inherent limitations of the Gaussian monocycle," *IET Commun.*, vol. 6, no. 9, pp. 1068–1074, 2012.

[23] I. S. Gradshteyn and I. M. Ryzhik, *Table of Integrals, Series, and Products*, Elsevier, 2007.

MIPRO 2017, May 22- 26, 2017, Opatija, Croatia

Design of Multiplierless CIC Compensators Based on Maximum Passband Deviation

Goran Molnar[1], Aljosa Dudarin[1], and Mladen Vucic[2]

[1]Ericsson Nikola Tesla d. d. Research and Development Centre
Radio Development Unit
Krapinska 45, 10000 Zagreb, Croatia

[2]University of Zagreb Faculty of Electrical Engineering and Computing
Department of Electronic Systems and Information Processing
Unska 3, 10000 Zagreb, Croatia
E-mails: goran.molnar@ericsson.com, aljosa.dudarin@ericsson.com, mladen.vucic@fer.hr

Abstract—Cascaded-integrator-comb (CIC) decimation filters are the simplest multiplierless filters supporting high sample rate conversion factors. However, the magnitude response of high order CIC filters exhibits a high passband droop. Such a droop can be reduced by connecting an FIR compensator in the cascade with the CIC filter. In this paper, we present a method for the design of CIC compensators whose coefficients are expressed as a sum of powers of two (SPT). The method is based on the minimization of the difference between the maximum and the minimum passband amplitude. To obtain the compensator coefficients, we use a global optimization technique, which is based on the interval analysis. In the optimization, the number of SPT terms for each coefficient is specified. The compensators obtained efficiently compensate narrow and wide passbands by using three and five coefficients having small number of SPT terms. For these compensators, multiplierless structures are provided.

I. INTRODUCTION

Cascaded-integrator-comb (CIC) decimation filters are the simplest multiplierless filters supporting high sample rate conversion factors [1]. Consequently, they found their application mainly in digital down converters. However, the CIC filters of high orders introduce a high passband droop, which is not tolerable in multi-standard receivers [2]–[4]. The most popular technique for reducing the droop is compensation. It is based on connecting a low-order finite-impulse-response (FIR) filter called compensator in the cascade with the CIC filter. Since the CIC filters are multiplierless, compensators with a multiplierless structure are preferable.

It is clear that the multiplierless compensation makes a tradeoff between complexity and compensation capability. Therefore, multiplierless narrowband and wideband compensators are separately considered in literature. Various methods for their design have been developed [2]–[13], covering compensators with two [11], three [2]–[10], [12], [13], five [7], and arbitrary number of coefficients [6], [8].

This paper was supported in part by Ericsson Nikola Tesla d.d. and University of Zagreb, Faculty of Electrical Engineering and Computing under the project ILTERA, and in part by Croatian Science Foundation under the project Beyond Nyquist Limit, grant no. IP-2014-09-2625.

In many applications, the CIC compensators with three coefficients are sufficient. However, wideband applications require more complex structures, such as compensators with more than three coefficients, sharpened compensated-CIC filters [4], [10], or multi-stage compensators [14], [15]. Recently, a two-stage multiplierless wideband compensator with high compensation capability has been proposed [16].

In this paper, we consider single-stage multiplierless CIC compensators. We present a method for their design which is based on the minimization of the maximum passband deviation over the sum-of-powers-of-two (SPT) coefficient space. To obtain the coefficients, we use a global optimization technique based on the interval analysis [19]. In the optimization, the number of SPT terms for each coefficient is specified. We show that the compensators obtained efficiently compensate narrow and wide passbands by using three and five coefficients having small number of SPT terms. For these compensators we provide multiplierless structures.

The paper is organized as follows. Section II describes the compensation of CIC filters based on the maximum passband deviation. The method for the design of multiplierless compensators is presented in Section III. Section IV contains examples illustrating the features of the proposed compensator class.

II. COMPENSATION OF CIC FILTERS BASED ON MAXIMUM PASSBAND DEVIATION

A. CIC Compensator

The amplitude response of the Nth order CIC filter is given by [1]

$$H_{CIC}(\omega) = \left[\frac{1}{R} \frac{\sin\left(\frac{\omega R}{2}\right)}{\sin\left(\frac{\omega}{2}\right)} \right]^N \qquad (1)$$

The CIC compensator is connected to the output of the CIC filter. Therefore, it improves the CIC response relative to the low sampling rate. Using the noble identity for decimation, the low-rate CIC response takes the form [1]

$$H_C(\omega) = \left[\frac{1}{R} \frac{\sin\left(\dfrac{\omega}{2}\right)}{\sin\left(\dfrac{\omega}{2R}\right)} \right]^N \tag{2}$$

Our objective is to design a linear-phase FIR compensator with SPT coefficients that improves $H_C(\omega)$ in order to minimize the maximum deviation of the passband response. It is well known that compensators with odd number of coefficients improve the response within a wider band than do the compensators with even number of coefficients [6]. Since the compensators with a low number of coefficients with ability to compensate wide passband responses are preferred, only the compensators with odd number of coefficients are considered in the paper.

B. Compensator Based on Maximum Passband Deviation

The linear-phase FIR compensator with an odd number of coefficients, L, has the amplitude response [6]

$$H(\omega) = c_0 + 2 \sum_{k=1}^{(L-1)/2} c_k \cos(k\omega) \tag{3}$$

To preserve the gain of the CIC filter, we assume $H(0) = 1$. Consequently, the central coefficient c_0 in (3) takes the value

$$c_0 = 1 - 2 \sum_{k=1}^{(L-1)/2} c_k \tag{4}$$

By substituting (4) into (3), we obtain the response

$$H(\omega) = 1 + 2 \sum_{k=1}^{(L-1)/2} c_k \left[\cos(k\omega) - 1\right] \tag{5}$$

Common approach in the design of linear-phase FIR filters is based on the minimization of the maximum absolute passband error. However, in the design of multiplierless compensators, the difference between the maximum and the minimum passband amplitude is more important. Such an approach introduces a gain, which is considered as a desired passband response. Therefore, we define the objective function as

$$\varepsilon(\mathbf{c}) = \max_{\omega \in \Omega} H_C(\omega)H(\omega,\mathbf{c}) - \min_{\omega \in \Omega} H_C(\omega)H(\omega,\mathbf{c}) \tag{6}$$

where $\mathbf{c}$ is the vector of free compensator coefficients given by

$$\mathbf{c} = \begin{bmatrix} c_1 & c_2 & \cdots & c_{(L-1)/2} \end{bmatrix}^T \tag{7}$$

and $\Omega = [-\omega_p, \omega_p]$, $0 < \omega_p < \pi$, is the passband,
In the design of compensators, we calculate $\varepsilon(\mathbf{c})$ using the responses evaluated on uniformly spaced frequency grid $Q = \{ \omega_k; k=0, \ldots, K-1 \}$ defined within Ω. Hence, the objective function is obtained as

$$\varepsilon(\mathbf{c}) = \max_{\omega_k \in Q} H_C(\omega_k)H(\omega_k,\mathbf{c}) - \min_{\omega_k \in Q} H_C(\omega_k)H(\omega_k,\mathbf{c}) \tag{8}$$

III. MULTIPLIERLESS COMPENSATORS

A. Design of Compensator Based on Interval Analysis

To find the optimum SPT coefficients of the CIC compensator, we specify the number of SPT terms per coefficient to a value P. Such a design is described by the optimization problem

$$\hat{\mathbf{c}} = \arg\min_{\mathbf{c}} \left[\varepsilon(\mathbf{c})\right] \tag{9}$$

subject to: $\mathbf{c}$ is SPT representable

CIC compensators described in literature usually have the coefficients whose values exceed the range $[-1, 1)$, as can be seen in [8], [12], and [13]. Therefore, in solving the problem in (9), we assume the elements of $\mathbf{c}$ contain integer and fractional parts. They are given by

$$c_k = \sum_{p=-F}^{S-1} b_{k,p} 2^p \quad ; \quad k = 1, 2, \ldots, \frac{L-1}{2} \tag{10}$$

where S and F are the wordlengths of the integer and the fractional part, and $b_{k,p} \in \{-1,0,1\}$. Each $b_{k,p} \neq 0$ represents one SPT term.

To solve the problem in (8)–(10), we use the global optimization technique based on the interval analysis [17], [18]. Such a technique has already been applied in the design of CIC compensators [8] and FIR filters [19]. In our design, we use the optimization procedure from [19]. This procedure minimizes the interval extension of the objective function in (8).

B. Multiplierless Implementation

Multiplierless compensators with three coefficients ($L = 3$) are often used for the compensation of CIC filters. Considering compensation capability, they significantly improve narrow passbands. However, for wideband compensations with low passband deviation, the compensators with five coefficients ($L = 5$) are usually sufficient.

Figures 1 and 2 show the multiplierless structures of the proposed compensators with three and five coefficients. It is clear that the computational complexity of the compensators depends on the implementations of constant multipliers. Since each SPT coefficient employs $P-1$ adders, the total number of adders required in the structure is

$$A = P + 2 \tag{11}$$

for the compensator with three coefficients, and

$$A = 2(P + 2) \tag{12}$$

for the compensator with five coefficients.

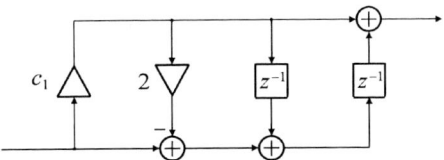

Figure 1. Structure of multiplierless CIC compensator with three coefficients.

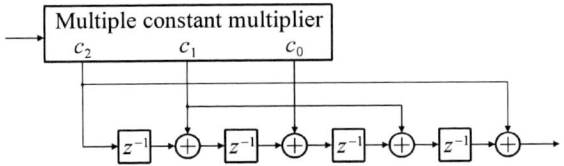

Multiple constant multiplier

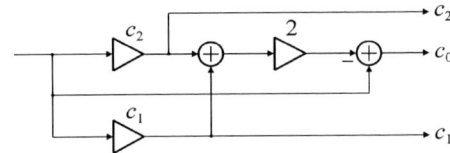

Figure 2. Structure of multiplierless CIC compensator with five coefficients.

IV. DESIGN EXAMPLES

The features of the proposed technique are illustrated with the design of narrowband and wideband compensators. The design is performed for CIC filters with decimation factor $R = 32$ by using $K = 1024$ frequency points. The optimum SPT coefficients are found for $S = 2$ and $F = 16$. The coefficients obtained are given in Table I. They are tabulated for various passband edge frequencies and CIC filter orders.

It is well known that for a given order of the CIC filter, the amplitude response negligibly changes the shape within the passband for $R \geq 10$ [1]. In that sense, the tabulated coefficients can be used for compensation of any CIC filter with $R \geq 10$.

A. Narrowband Compensators

Here, we describe a narrowband compensation of the CIC filter with $N = 5$ and $R = 32$ by using the compensators with three coefficients and $P \leq 3$. For illustration, the passband edge frequency $\omega_p = \pi/5$ is chosen. Figure 3 shows the compensated passband responses. The uncompensated CIC filter introduces the droop of 0.72 dB. However, the compensated CIC filter yields the maximum passband deviation of 0.08 dB for $P = 1$, 0.03 dB for $P = 2$, and 0.02 dB for $P = 3$, resulting in the compensator with three, four, and five adders, respectively.

B. Wideband Compensators

Simple wideband compensators incorporate only three coefficients, as for example in [10] and [13]. Hence, we

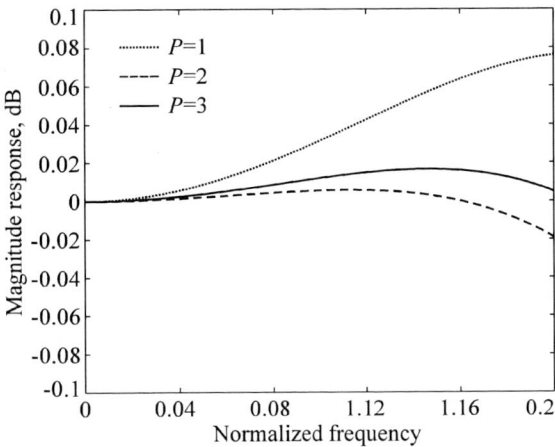

Figure 3. Passband responses at low rate of CIC filter with $N = 5$ and $R = 32$ compensated with proposed compensators having three coefficients and P terms. Passband is $|\omega| \leq \pi/5$.

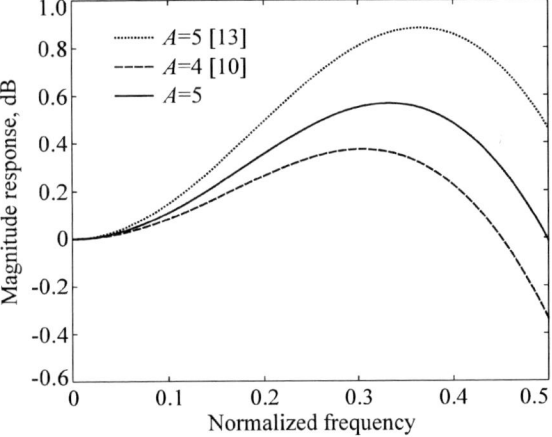

Figure 4. Passband responses at low rate of CIC filter with $N = 5$ and $R = 32$ compensated with proposed compensator having $P = 3$ and with compensators proposed in [10] and [13]. All compensators have three coefficients and A adders. Passband is $|\omega| \leq \pi/2$.

compare our wideband compensator with three coefficients and $P = 3$ with the compensators referred to. Figure 4 shows the obtained passband responses for the CIC filter with $N = 5$ and $R = 32$, assuming $\omega_p = \pi/2$. The uncompensated CIC filter introduces the droop of 4.6 dB. Our response has the maximum deviation of 0.58 dB, whereas the deviation of the compensators in [10] and [13] are 0.71 dB and 0.88 dB. It is clear that the proposed compensation is better. However, in comparison with the compensator in [10], it is paid by increasing the total number of adders by one.

Let us now consider the compensators with five coefficients and $P \leq 3$. For illustration, we describe the compensation of the same CIC filter, but with $\omega_p = 3\pi/5$. Figure 5 shows the obtained passband responses. The CIC filter introduces the droop of 6.6 dB. However, the

121

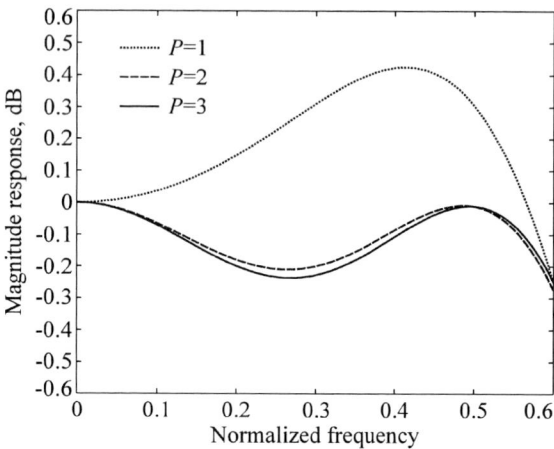

Figure 5. Passband responses at low rate of CIC filter with $N = 5$ and $R = 32$ compensated with proposed compensators having five coefficients and P terms. Passband is $|\omega| \leq 3\pi/5$.

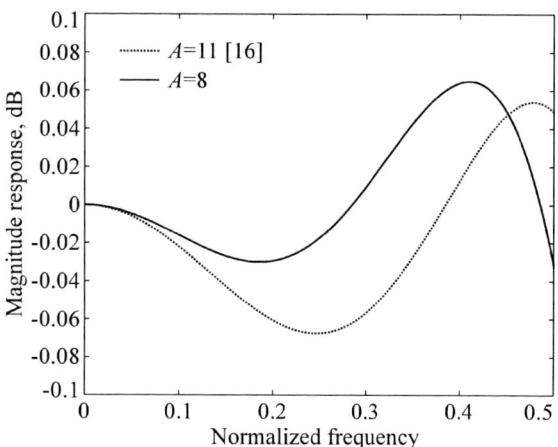

Figure 6. Passband responses at low rate of CIC filter with $N = 5$ and $R = 32$ compensated with proposed single-stage compensator with five coefficients and $P = 2$, and two-stage compensator proposed in [16]. Both compensator structures ensure maximum deviation close to 0.1 dB and have A adders. Passband is $|\omega| \leq \pi/2$.

compensated CIC filter results in the maximum deviation of 0.68 dB for $P = 1$, 0.28 dB for $P = 2$, and 0.25 dB for $P = 3$, which are obtained using six, eight, and ten adders, respectively.

Recently, the multiplierless CIC compensation using the cascade of two linear-phase compensators has been proposed [16]. In this cascade, the first compensator has three coefficients, whereas the second compensator has five coefficients. Here, we compare the proposed single-stage compensator having five coefficients and $P = 2$ with the two-stage multiplierless compensator from [16]. The compensation of the same CIC filter is considered. Figure 6 shows the compensated responses obtained for $\omega_p = \pi/2$. Both responses ensure the maximum passband deviation close to 0.1 dB. However, our compensator employs less number of adders.

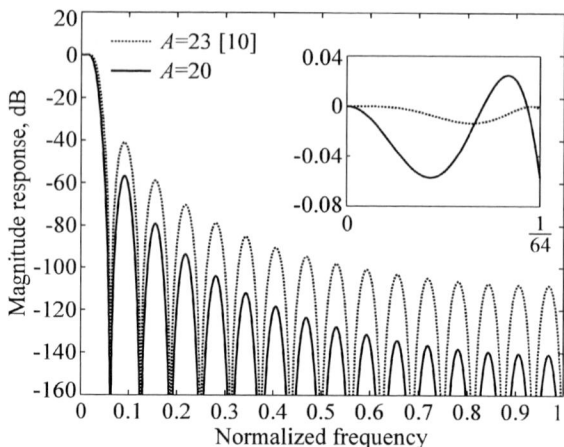

Figure 7. Magnitude response at high rate of proposed compensated CIC filter with $N = 5$ obtained by compensator with five coefficients and $P = 3$, together with response of sharpened compensated-CIC filter with $N = 4$ proposed in [10]. Both filters deal with $R = 32$ and have A adders each. Passband at high rate is $|\omega| \leq \pi/64$.

Let us now compare the proposed compensation with a simple multiplierless sharpening technique proposed in [10]. We use the compensator with five coefficients and $P = 3$, obtained for the CIC filter with $N = 5$. On the other hand, we choose the sharpened filter incorporating the CIC filter with $N = 4$. Such a filter is obtained by substituting $x = [1 + 2^{-1} - 2^{-1}\cos(R\omega)]H_{CIC}(\omega)$ into the sharpening polynomial $p(x) = 2x - x^2$ [10]. In both cases, the filters deal with $R = 32$ and $\omega_p = \pi/2$. Figure 7 shows the compensated and sharpened magnitude responses relative to the high sampling rate. It is clear that both filters have the maximum passband deviation less than 0.1 dB. However, our filter ensures the minimum folding-band attenuation of 52 dB, whereas the sharpened filter provides the attenuation of 36 dB. In addition, our filter has the structure containing less number of adders.

Finally, we describe the application of our compensators for improving the passband of non-recursive CIC-based FIR decimation filters proposed in [20]. These filters provide very high stopband attenuations. However, they introduce rather high passband droop. In the paper referred to, the authors suggest using common CIC compensators to reduce the droop. For illustration, we apply our CIC compensators for improving the passband of the CIC FIR filters. In particular, we design the compensator with five coefficients and $P = 2$ to improve the filter from [20], obtained by $N = 10$ and $R = 10$. In the design, we choose the passband with $\omega_p = \pi/2$. The optimum compensator coefficients are obtained as $c_1 = -2^0 - 2^{-1}$ and $c_2 = 2^{-2} + 2^{-4}$, resulting it the structure containing eight adders. Figure 8 shows the magnitude response of the compensated CIC FIR filter together with the response of the original filter. The CIC FIR filter ensures the minimum folding-band attenuation of 128 dB and introduces the droop of 9.4 dB. However, our compensator significantly improves the passband, resulting in the droop of 0.6 dB.

TABLE I

OPTIMUM SPT COEFFICIENTS WITH P TERMS OF COMPENSATORS FOR CIC FILTER OF ORDER N AND DECIMATION FACTOR $R \geq 10$

	$P=1$	$P=2$	$P=3$	$P=1$	$P=2$	$P=3$	$P=1$	$P=2$	$P=3$
c_k	$N=2$			$N=3$			$N=4$		
$L=3$ and $\omega_p=\pi/5$									
c_1	-2^{-4}	$-2^{-4}-2^{-5}$	$-2^{-4}-2^{-6}-2^{-7}$	-2^{-3}	$-2^{-3}-2^{-7}$	$-2^{-3}-2^{-7}+2^{-14}$	-2^{-3}	$-2^{-3}-2^{-4}$	$-2^{-3}-2^{-4}+2^{-7}$
$L=3$ and $\omega_p=\pi/4$									
c_1	-2^{-4}	$-2^{-4}-2^{-5}$	$-2^{-4}-2^{-5}+2^{-8}$	-2^{-3}	$-2^{-3}-2^{-6}$	$-2^{-3}-2^{-7}-2^{-8}$	-2^{-3}	$-2^{-3}-2^{-4}$	$-2^{-3}-2^{-4}+2^{-9}$
$L=3$ and $\omega_p=\pi/3$									
c_1	-2^{-3}	$-2^{-4}-2^{-5}$	$-2^{-4}-2^{-5}-2^{-9}$	-2^{-3}	$-2^{-3}-2^{-6}$	$-2^{-3}-2^{-6}-2^{-7}$	-2^{-2}	$-2^{-3}-2^{-4}$	$-2^{-3}-2^{-4}-2^{-6}$
$L=3$ and $\omega_p=\pi/2$									
c_1	-2^{-3}	$-2^{-3}+2^{-7}$	$-2^{-3}+2^{-7}+2^{-11}$	-2^{-3}	$-2^{-3}-2^{-4}$	$-2^{-3}-2^{-4}+2^{-9}$	-2^{-2}	$-2^{-2}-2^{-7}$	$-2^{-2}-2^{-7}-2^{-9}$
$L=5$ and $\omega_p=\pi/3$									
c_1	-2^{-3}	$-2^{-3}-2^{-6}$	$-2^{-3}-2^{-6}+2^{-11}$	-2^{-2}	$-2^{-2}+2^{-5}$	$-2^{-2}+2^{-5}-2^{-11}$	-2^{-2}	$-2^{-2}-2^{-5}$	$-2^{-2}-2^{-4}-2^{-7}$
c_2	2^{-7}	$2^{-6}-2^{-10}$	$2^{-6}-2^{-10}-2^{-14}$	2^{-5}	$2^{-5}-2^{-7}$	$2^{-5}-2^{-7}+2^{-12}$	2^{-6}	$2^{-5}-2^{-8}$	$2^{-5}+2^{-7}+2^{-11}$
$L=5$ and $\omega_p=2\pi/5$									
c_1	-2^{-3}	$-2^{-3}-2^{-6}$	$-2^{-3}-2^{-6}-2^{-8}$	-2^{-2}	$-2^{-2}+2^{-7}$	$-2^{-2}+2^{-6}-2^{-10}$	-2^{-2}	$-2^{-2}-2^{-4}$	$-2^{-2}-2^{-4}-2^{-5}$
c_2	2^{-7}	$2^{-6}-2^{-10}$	$2^{-6}+2^{-12}+2^{-14}$	2^{-5}	$2^{-5}+2^{-12}$	$2^{-5}-2^{-9}-2^{-12}$	2^{-6}	$2^{-5}+2^{-8}$	$2^{-4}-2^{-6}+2^{-12}$
$L=5$ and $\omega_p=\pi/2$									
c_1	-2^{-3}	$-2^{-3}-2^{-5}$	$-2^{-3}-2^{-5}+2^{-11}$	-2^{-2}	$-2^{-2}-2^{-8}$	$-2^{-2}-2^{-8}-2^{-10}$	-2^{-2}	$-2^{-2}-2^{-3}$	$-2^{-2}-2^{-3}+2^{-9}$
c_2	2^{-8}	$2^{-6}+2^{-8}$	$2^{-6}+2^{-8}+2^{-11}$	2^{-5}	$2^{-5}+2^{-8}$	$2^{-5}+2^{-8}+2^{-10}$	2^{-8}	$2^{-4}-2^{-8}$	$2^{-4}-2^{-8}-2^{-11}$
$L=5$ and $\omega_p=3\pi/5$									
c_1	-2^{-3}	$-2^{-3}-2^{-5}$	$-2^{-3}-2^{-5}-2^{-6}$	-2^{-2}	$-2^{-2}-2^{-5}$	$-2^{-2}-2^{-5}-2^{-14}$	-2^{-1}	$-2^{-1}+2^{-4}$	$-2^{-1}+2^{-4}+2^{-6}$
c_2	2^{-7}	$2^{-6}+2^{-8}$	$2^{-5}-2^{-8}+2^{-12}$	2^{-5}	$2^{-4}-2^{-6}$	$2^{-4}-2^{-6}+2^{-10}$	2^{-3}	$2^{-4}+2^{-6}$	$2^{-4}+2^{-6}+2^{-9}$
$L=5$ and $\omega_p=2\pi/3$									
c_1	-2^{-3}	$-2^{-3}-2^{-4}$	$-2^{-3}-2^{-4}+2^{-7}$	-2^{-2}	$-2^{-2}-2^{-4}$	$-2^{-2}-2^{-4}+2^{-7}$	-2^{-1}	$-2^{-1}+2^{-4}$	$-2^{-1}+2^{-4}-2^{-6}$
c_2	2^{-7}	$2^{-5}+2^{-8}$	$2^{-5}+2^{-10}-2^{-12}$	2^{-5}	$2^{-4}+2^{-9}$	$2^{-4}-2^{-10}-2^{-11}$	2^{-3}	$2^{-3}-2^{-5}$	$2^{-3}-2^{-5}+2^{-8}$
c_k	$N=5$			$N=6$			$N=7$		
$L=3$ and $\omega_p=\pi/5$									
c_1	-2^{-2}	$-2^{-2}+2^{-5}$	$-2^{-2}+2^{-5}-2^{-7}$	-2^{-2}	$-2^{-2}-2^{-6}$	$-2^{-2}-2^{-6}-2^{-7}$	-2^{-2}	$-2^{-2}-2^{-4}$	$-2^{-2}-2^{-4}-2^{-7}$
$L=3$ and $\omega_p=\pi/4$									
c_1	-2^{-2}	$-2^{-2}+2^{-6}$	$-2^{-2}+2^{-6}-2^{-10}$	-2^{-2}	$-2^{-2}-2^{-5}$	$-2^{-2}-2^{-5}-2^{-8}$	-2^{-2}	$-2^{-2}-2^{-4}$	$-2^{-2}-2^{-4}-2^{-5}$
$L=3$ and $\omega_p=\pi/3$									
c_1	-2^{-2}	$-2^{-2}-2^{-7}$	$-2^{-2}-2^{-7}-2^{-10}$	-2^{-2}	$-2^{-2}-2^{-4}$	$-2^{-2}-2^{-4}-2^{-8}$	-2^{-1}	$-2^{-2}-2^{-3}$	$-2^{-2}-2^{-3}-2^{-8}$
$L=3$ and $\omega_p=\pi/2$									
c_1	-2^{-2}	$-2^{-2}-2^{-4}$	$-2^{-2}-2^{-4}-2^{-5}$	-2^{-1}	$-2^{-1}+2^{-4}$	$-2^{-1}+2^{-4}-2^{-11}$	-2^{-1}	$-2^{-1}-2^{-5}$	$-2^{-1}-2^{-5}-2^{-7}$
$L=5$ and $\omega_p=\pi/3$									
c_1	-2^{-2}	$-2^{-1}+2^{-4}$	$-2^{-1}+2^{-4}+2^{-6}$	-2^{-1}	$-2^{-1}-2^{-5}$	$-2^{-1}-2^{-4}+2^{-7}$	-2^{-1}	$-2^{-1}-2^{-2}$	$-2^{-1}-2^{-3}-2^{-5}$
c_2	-2^{-9}	$2^{-4}-2^{-9}$	$2^{-4}-2^{-7}-2^{-12}$	2^{-4}	$2^{-4}+2^{-7}$	$2^{-4}+2^{-6}+2^{-10}$	2^{-5}	$2^{-3}-2^{-9}$	$2^{-3}-2^{-5}-2^{-9}$
$L=5$ and $\omega_p=2\pi/5$									
c_1	-2^{-2}	$-2^{-1}+2^{-5}$	$-2^{-1}+2^{-4}-2^{-6}$	-2^{-1}	$-2^{-1}-2^{-4}$	$-2^{-1}-2^{-4}-2^{-6}$	-2^{-1}	$-2^{-1}-2^{-2}$	$-2^{-1}-2^{-2}+2^{-6}$
c_2	-2^{-6}	$2^{-4}+2^{-7}$	$2^{-4}+2^{-9}+2^{-13}$	2^{-4}	$2^{-4}+2^{-6}$	$2^{-3}-2^{-5}-2^{-7}$	2^{-5}	$2^{-3}-2^{-9}$	$2^{-3}-2^{-7}+2^{-10}$
$L=5$ and $\omega_p=\pi/2$									
c_1	-2^{-1}	$-2^{-1}+2^{-9}$	$-2^{-1}-2^{-8}+2^{-11}$	-2^{-1}	$-2^{-1}-2^{-3}$	$-2^{-1}-2^{-3}-2^{-5}$	-2^{0}	$-2^{-1}-2^{-2}$	$-2^{0}+2^{-3}+2^{-5}$
c_2	2^{-4}	$2^{-4}+2^{-6}$	$2^{-4}+2^{-6}+2^{-8}$	2^{-5}	$2^{-3}-2^{-5}$	$2^{-3}-2^{-7}-2^{-8}$	2^{-2}	$2^{-3}-2^{-9}$	$2^{-3}+2^{-5}+2^{-9}$
$L=5$ and $\omega_p=3\pi/5$									
c_1	-2^{-1}	$-2^{-1}-2^{-4}$	$-2^{-1}-2^{-4}-2^{-7}$	-2^{0}	$-2^{-1}-2^{-2}$	$-2^{-1}-2^{-2}-2^{-7}$	-2^{0}	$-2^{0}+2^{-6}$	$-2^{0}+2^{-5}-2^{-8}$
c_2	2^{-4}	$2^{-3}-2^{-6}$	$2^{-3}-2^{-7}-2^{-8}$	2^{-2}	$2^{-3}+2^{-5}$	$2^{-3}+2^{-5}+2^{-8}$	2^{-2}	$2^{-2}-2^{-5}$	$2^{-2}-2^{-5}-2^{-8}$
$L=5$ and $\omega_p=2\pi/3$									
c_1	-2^{-1}	$-2^{-1}-2^{-3}$	$-2^{-1}-2^{-3}-2^{-10}$	-2^{0}	$-2^{0}+2^{-3}$	$-2^{0}+2^{-3}+2^{-5}$	-2^{0}	$-2^{0}-2^{-4}$	$-2^{0}-2^{-4}-2^{-6}$
c_2	2^{-4}	$2^{-3}+2^{-6}$	$2^{-3}+2^{-6}+2^{-9}$	2^{-2}	$2^{-2}-2^{-5}$	$2^{-2}-2^{-5}-2^{-6}$	2^{-2}	$2^{-2}+2^{-6}$	$2^{-2}+2^{-6}+2^{-8}$

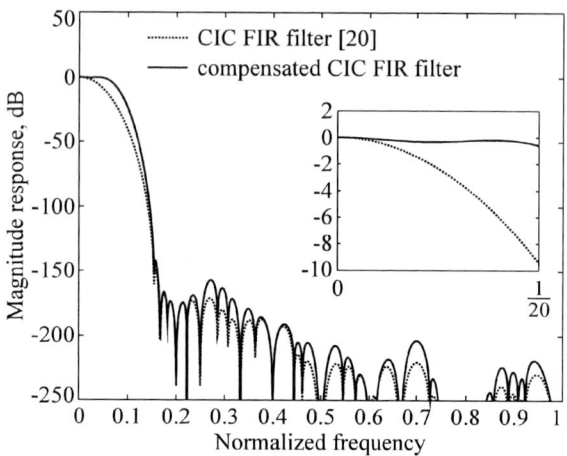

Figure 8. Magnitude response at high rate of non-recursive CIC FIR filter with $N = 10$ and $R = 10$ proposed in [20] compensated with compensator having five coefficients and $P = 2$, together with magnitude response of uncompensated filter. Passband at high rate is $|\omega| \leq \pi/20$.

V. CONCLUSIONS

A method for the design of CIC compensators with SPT coefficients was presented. It is based on the minimization of the difference between the maximum and the minimum passband amplitude, assuming unity DC gain. The optimum coefficients are obtained by using the global optimization technique based on the interval analysis.

The proposed compensators efficiently compensate narrow and wide passbands using three and five coefficients, requiring up to five adders for the compensators with three coefficients and up to ten adders for the compensators with five coefficients.

VI. REFERENCES

[1] E. B. Hogenauer, "An economical class of digital filters for decimation and interpolation," *IEEE Trans. Acoust., Speech, Signal Process.*, vol. 29, no. 2, pp. 155–162, Apr. 1981.

[2] K. S. Yeung and S. C. Chan, "The design and multiplier-less realization of software radio receivers with reduced system delay," *IEEE Trans. Circuits Syst. I, Reg. Papers*, vol. 51, no. 12, pp. 2444–2459, Dec. 2004.

[3] S. Kim, W. C. Lee, S. Ahn, and S. Choi, "Design of CIC roll-off compensation filter in a W-CDMA digital IF receiver," *Digit. Signal Process.*, vol. 16, no. 6, pp. 846–854, Nov. 2006.

[4] G. Jovanovic Dolecek and F. Harris, "Design of wideband CIC compensator filter for a digital IF receiver," *Digit. Signal Process.*, vol. 19, no. 5, pp. 827–837, Sept. 2009.

[5] G. Jovanovic Dolecek and S. K. Mitra, "Simple method for compensation of CIC decimation filter," *Electron. Lett.*, vol. 44, no. 19, pp. 1162–1163, Sept. 2008.

[6] G. Molnar and M. Vucic, "Closed-form design of CIC compensators based on maximally flat error criterion," *IEEE Trans. Circuits Syst. II: Exp. Briefs*, vol. 58, no. 12, pp. 926–930, Dec. 2011.

[7] A. Fernandez-Vazquez and G. Jovanovic Dolecek, "Maximally flat CIC compensation filter: design and multiplierless implementation," *IEEE Trans. Circuits Syst. II: Exp. Briefs*, vol. 59, no. 2, pp. 113–117, Feb. 2012.

[8] M. Glavinic Pecotic, G. Molnar, and M. Vucic, "Design of CIC compensators with SPT coefficients based on interval analysis," in *Proc. IEEE Int. Convention MIPRO*, Opatija, Croatia, May 2012, pp. 123–128.

[9] D. E. T. Romero and G. Jovanovic Dolecek, "Application of amplitude transformation for compensation of comb decimation filters," *Electron. Lett.*, vol. 49, no. 16, pp. 985–987, Aug. 2013.

[10] G. Jovanovic Dolecek and A. Fernandez-Vazquez, "Trigonometrical approach to design a simple wideband comb compensator," *AEU Int. J. Electron. Comm.*, vol. 68, no. 5, pp. 437–441, May 2014.

[11] D. E. T. Romero, G. M. Salgado, and G. Jovanovic Dolecek, "Simple two-adders CIC compensator," *Electron. Lett.*, vol. 51, no. 13, pp. 993–994, Jun. 2015.

[12] N. Saberi, A. Ahmadi, S. Alirezaee, and M. Ahmadi, "A multiplierless implementation of cascade integrator comb filter," in *Proc. IEEE ISSCS*, Iasi, Romania, Jul. 2015.

[13] G. Jovanovic Dolecek and A. Fernandez-Vazquez, "Multiplierless two-stage comb structure with an improved magnitude characteristic," in *Proc. IEEE APCCAS*, Jeju, Korea, Oct. 2016, pp. 607–610.

[14] G. Jovanovic Dolecek, "Simple wideband CIC compensator," *Electron. Lett.*, vol. 45, no. 24, pp. 1270–1272, Nov. 2009.

[15] D. E. T. Romero, "On wideband minimum-phase CIC compensators," in *Proc. IEEE Int. Conf. EIT*, Grand Forks, North Dakota, USA, May 2016, pp. 372–375.

[16] G. Jovanovic Dolecek, R. G. Baez, G. Molina Salgado, and J. de la Rosa, "Novel multiplierless wideband comb compensator with high compensation capability," *Circuits Syst. Signal Process.*, Aug. 2016, pp. 1–19.

[17] R. B. Kearfott, *Rigorous Global Search: Continuous Problems*, Cluver Academic Publishers, 1996.

[18] E. Hansen and G. W. Walster., *Global Optimization Using Interval Analysis*, Second Edition, Marcel Dekker, Inc., 2004.

[19] M. Vucic, G. Molnar, and T. Zgaljic, "Design of FIR filters based on interval analysis," in *Proc. IEEE Int. Convention MIPRO*, Opatija, Croatia, May 2010, pp. 197–202.

[20] B. P. Stosic, D. N. Milic, and V. D. Pavlovic, "Innovative design of CIC FIR filter functions," in *Proc. Int. Conf. TELSIKS*, Nis, Serbia, Oct. 2015, pp. 60–63.

Adaptive State Observer Development Using Recursive Extended Least-Squares Method

Nikola N. Nikolov[*], Mariela I. Alexandrova[*], Vencislav C. Valchev[*] and Orlin P. Stanchev[*]
[*]Technical University of Varna, Varna, Bulgaria

nn_nikolov@tu-varna.bg, m_alexandrova@tu-varna.bg, vencivalchev@hotmail.com, or.stanchev@gmail.com

Abstract – This paper presents a recursive algorithm for adaptive observation of linear single-input single-output (SISO) time-invariant discrete systems. The problem of state observation is of main importance in automatic control, especially for designing modal state controllers when the state variables of the system are unknown. The proposed algorithm is based on the recursive extended least-squares (RELS) method. The adaptive state observer estimates system parameters, initial and current state vector by using the measured input and output system signals. In order to verify the algorithm performance, simulation experiments in the Matlab environment are carried out and provided.

I. INTRODUCTION

The problem of state observation is of main importance in automatic control. Design of state feedback control systems often requires state vector reconstruction. The reconstruction could be done by using measurements of input and output signals values of the controlled system.

The algorithm used for state vector reconstruction is referred to as *state observer*. The observer is a very useful tool for receiving information of the state variables of a system. These variables are otherwise unknown. For this reason, the state observer is widely used in control, estimation and other engineering applications.

Adaptive observation problem includes state observer synthesis and parameter estimators. In the adaptive observers matrices **A** and **b** or **c** (according to the chosen state space canonical representation) are assumed to be unknown. The observation process leads to parameters estimation, determination of the unknown matrices and calculation of the state vector.

A recurrent algorithm for adaptive observation of linear time-invariant (LTI) single-input single-output (SISO) discrete system is presented in this paper. The algorithm is developed by using the recursive extended least-squares method (RELS) [1,2].

In the proposed adaptive observer the parameters estimator is constructed based on the simple mathematical procedure, which includes inversion of the informative matrix given in [3].

The paper is developed in the frames of the project "Model Based Design of Power Electronic Devices with Guaranteed parameters", ДН07/06/15.12.2016, National Scientific Fund.

II. PROBLEM FORMULATION

Let us consider a linear time-invariant SISO discrete system by a state space model of the form:

$$\mathbf{x}(k+1) = \mathbf{A}\mathbf{x}(k) + \mathbf{b}u(k), \quad \mathbf{x}(0) = \mathbf{x}_0, \tag{1}$$
$$y(k) = \mathbf{c}^T\mathbf{x}(k) + f(k), \quad k = 0, 1, 2, \cdots$$

where the state matrices are:

$$\mathbf{A} = \begin{bmatrix} \mathbf{0} & \vdots & \mathbf{I}_{n-1} \\ \cdots & \cdots & \cdots \\ & \mathbf{a}^T & \end{bmatrix}, \tag{2}$$

$$\mathbf{a} = \begin{bmatrix} a_1 \\ a_2 \\ \vdots \\ a_n \end{bmatrix}, \quad \mathbf{b} = \begin{bmatrix} b_1 \\ b_2 \\ \vdots \\ b_n \end{bmatrix}, \quad \mathbf{c} = \begin{bmatrix} 1 \\ 0 \\ \vdots \\ 0 \end{bmatrix}. \tag{3}$$

In the equations given above: n, which denotes the system order, is previously known; $\mathbf{I}_{n-1}$ is a square unit matrix of size $(n-1) \times (n-1)$; $\mathbf{x}(k) \in R^n$ is the unknown current state vector; $\mathbf{x}(0) \in R^n$ is the unknown initial state vector, $u(k) \in R^1$ is a scalar input signal; $y(k) \in R^1$ is a scalar output signal; $f(k)$ is an additive noise signal; $\mathbf{a}$ and $\mathbf{b}$ are unknown vector parameters.

The discrete transfer function corresponding to the state-space model (1) is as follows:

$$W(z) = \frac{h_1 z^{n-1} + h_2 z^{n-2} + \cdots + h_{n-1} z + h_n}{z^n - a_n z^{n-1} - \cdots - a_2 z - a_1}. \tag{4}$$

The relation between the vector elements b_i and the numerator polynomial coefficients h_i of the discrete transfer function (4), in accordance with the chosen canonical form, could be expressed by the following equation [4]:

$$\mathbf{Tb} = \mathbf{h}, \tag{5}$$

where

$$\mathbf{h}^T = \begin{bmatrix} h_1 & h_2 & \cdots & h_n \end{bmatrix},$$

$$T = \begin{bmatrix} 1 & 0 & \cdots & 0 & 0 \\ -a_n & 1 & \cdots & 0 & 0 \\ -a_{n-1} & -a_n & \cdots & 0 & 0 \\ \vdots & \vdots & \ddots & \vdots & \vdots \\ -a_2 & -a_3 & \cdots & -a_n & 1 \end{bmatrix}.$$

The elements a_i of the vector $\mathbf{a}$ are coefficients of the denominator polynomial of discrete transfer function (4) with reversed order and an opposite sign.

The problem to be considered is to estimate the unknown vector parameters $\mathbf{a}$ and $\mathbf{b}$, the initial state vector $\mathbf{x}(0)$ and current state vector $\mathbf{x}(k)$, $k=1, 2, \ldots$

In section III of the present paper is proposed an algorithm for adaptive observation of parameters and state vector of LTI discrete system, which is consisted of 11 steps and is developed based on RELS method. The developed algorithms is call adaptive because it estimates system parameters [5].

III. SOLUTION OF THE CONSIDERED PROBLEM

ALGORITHM FOR ADAPTIVE OBSERVATION BASED ON RELS METHOD

The developed algorithm consists of the following steps:

Step 1: Forming of input-output data arrays:

$$\mathbf{u}_1 = \begin{bmatrix} u(0) & u(1) & \cdots & u(N-2) \end{bmatrix},$$

$$\mathbf{y}_1 = \begin{bmatrix} y(0) & y(1) & \cdots & y(N-1) \end{bmatrix},$$

$$\mathbf{y}_2 = \begin{bmatrix} y(n) & y(n+1) & \cdots & y\left(\frac{N-n}{2}+n-1\right) \end{bmatrix}^\mathsf{T},$$

$$\mathbf{y}_3 = \begin{bmatrix} y\left(\frac{N-n}{2}+n\right) & y\left(\frac{N-n}{2}+n+1\right) & \cdots & y(N-1) \end{bmatrix}^\mathsf{T},$$

$$\mathbf{Y}_{11} = \begin{bmatrix} -y(n-1) & -y(n-2) & \cdots & -y(0) \\ -y(n) & -y(n-1) & \cdots & -y(1) \\ -y(n+1) & -y(n) & \cdots & -y(2) \\ \vdots & \vdots & \ddots & \vdots \\ -y\left(\frac{N-n}{2}+n-2\right) & -y\left(\frac{N-n}{2}+n-3\right) & \cdots & -y\left(\frac{N-n}{2}-1\right) \end{bmatrix},$$

$$\mathbf{Y}_{21} = \begin{bmatrix} -y\left(\frac{N-n}{2}+n-1\right) & -y\left(\frac{N-n}{2}+n-2\right) & \cdots & -y\left(\frac{N-n}{2}\right) \\ -y\left(\frac{N-n}{2}+n\right) & -y\left(\frac{N-n}{2}+n-1\right) & \cdots & -y\left(\frac{N-n}{2}+1\right) \\ -y\left(\frac{N-n}{2}+n+1\right) & -y\left(\frac{N-n}{2}+n\right) & \cdots & -y\left(\frac{N-n}{2}+2\right) \\ \vdots & \vdots & \ddots & \vdots \\ -y(N-2) & -y(N-3) & \cdots & -y(N-n-1) \end{bmatrix},$$

$$\mathbf{U}_{12} = \begin{bmatrix} u(n-1) & u(n-2) & \cdots & u(0) \\ u(n) & u(n-1) & \cdots & u(1) \\ u(n+1) & u(n) & \cdots & u(2) \\ \vdots & \vdots & \ddots & \vdots \\ u\left(\frac{N-n}{2}+n-2\right) & u\left(\frac{N-n}{2}+n-3\right) & \cdots & u\left(\frac{N-n}{2}-1\right) \end{bmatrix},$$

$$\mathbf{U}_{22} = \begin{bmatrix} u\left(\frac{N-n}{2}+n-1\right) & u\left(\frac{N-n}{2}+n-2\right) & \cdots & u\left(\frac{N-n}{2}\right) \\ u\left(\frac{N-n}{2}+n\right) & u\left(\frac{N-n}{2}+n-1\right) & \cdots & u\left(\frac{N-n}{2}+1\right) \\ u\left(\frac{N-n}{2}+n+1\right) & u\left(\frac{N-n}{2}+n\right) & \cdots & u\left(\frac{N-n}{2}+2\right) \\ \vdots & \vdots & \ddots & \vdots \\ u(N-2) & u(N-3) & \cdots & u(N-n-1) \end{bmatrix},$$

where
$\mathbf{Y}_{11}$, $\mathbf{Y}_{21}$, $\mathbf{U}_{12}$ and $\mathbf{U}_{22}$ are Toeplitz matrices and $N=3n+2l$, $l=0, 1, 2, 3, \ldots$.

Step 2: Calculation of the submatrices

$$\mathbf{G}_{11} = \mathbf{Y}_{11}^\mathsf{T}\mathbf{Y}_{11} + \mathbf{Y}_{21}^\mathsf{T}\mathbf{Y}_{21},$$

$$\mathbf{G}_{12} = \mathbf{Y}_{11}^\mathsf{T}\mathbf{U}_{12} + \mathbf{Y}_{21}^\mathsf{T}\mathbf{U}_{22},$$

$$\mathbf{G}_{21} = \mathbf{U}_{12}^\mathsf{T}\mathbf{Y}_{11} + \mathbf{U}_{22}^\mathsf{T}\mathbf{Y}_{21},$$

$$\mathbf{G}_{22} = \mathbf{U}_{12}^\mathsf{T}\mathbf{U}_{12} + \mathbf{U}_{22}^\mathsf{T}\mathbf{U}_{22}.$$

Step 3: Calculation of the covariance matrix $\mathbf{C}(N)$

$$\mathbf{C}(N) = \begin{bmatrix} \mathbf{M}_1 + \mathbf{M}_1\mathbf{G}_{12}\mathbf{M}_2\mathbf{G}_{21}\mathbf{M}_1 & \vdots & -\mathbf{M}_1\mathbf{G}_{12}\mathbf{M}_2 \\ \cdots\cdots\cdots\cdots\cdots\cdots\cdots & \vdots & \cdots\cdots\cdots\cdots\cdots \\ -\mathbf{M}_2\mathbf{G}_{21}\mathbf{M}_1 & \vdots & \mathbf{M}_2 \end{bmatrix}$$

where
$$\mathbf{M}_1 = \mathbf{G}_{11}^{-1},$$

$$\mathbf{M}_2 = \left(\mathbf{G}_{22} - \mathbf{G}_{21}\mathbf{M}_1\mathbf{G}_{12}\right)^{-1}.$$

Step 4: Initial estimation of vectors $\hat{\mathbf{h}}(N)$ and $\hat{\mathbf{a}}(N)$ by using the following vector-matrix system of equations:

$$\hat{\mathbf{p}}(N) = \mathbf{C}(N)\begin{bmatrix} \mathbf{Y}_{11}^\mathsf{T}\mathbf{y}_2 + \mathbf{Y}_{21}^\mathsf{T}\mathbf{y}_3 \\ \mathbf{U}_{12}^\mathsf{T}\mathbf{y}_2 + \mathbf{U}_{22}^\mathsf{T}\mathbf{y}_3 \end{bmatrix}.$$

$$\hat{\mathbf{h}}(N) = \begin{bmatrix} \hat{h}_1 & \hat{h}_2 & \cdots & \hat{h}_n \end{bmatrix}^\mathsf{T} = = \begin{bmatrix} \hat{p}_{n+1}(N) & \hat{p}_{n+2}(N) & \cdots & \hat{p}_{2n}(N) \end{bmatrix}^\mathsf{T},$$

$$\hat{\mathbf{a}}(N) = \begin{bmatrix} \hat{a}_1 & \hat{a}_2 & \cdots & \hat{a}_n \end{bmatrix}^\mathsf{T} = = \begin{bmatrix} -\hat{p}_n(N) & -\hat{p}_{n-1}(N) & \cdots & -\hat{p}_1(N) \end{bmatrix}^\mathsf{T}.$$

Step 5: Calculation of vector $\mathbf{b}(N)$ estimation implementing the following linear algebraic system of equations

$$\hat{\mathbf{b}}(N) = \mathbf{T}^{-1}\hat{\mathbf{h}}(N),$$

where

$$\mathbf{T} = \begin{bmatrix} 1 & 0 & 0 & \cdots & 0 & 0 \\ -\hat{a}_n(N) & 1 & 0 & \cdots & 0 & 0 \\ -\hat{a}_{n-1}(N) & -\hat{a}_n(N) & 1 & \cdots & 0 & 0 \\ \vdots & \vdots & \vdots & \ddots & \vdots & \vdots \\ -\hat{a}_2(N) & -\hat{a}_3(N) & -\hat{a}_4(N) & \cdots & -\hat{a}_n(N) & 1 \end{bmatrix}$$

is lower triangular Toeplitz matrix.

Step 6: Initial state vector estimation

$$\hat{\mathbf{x}}_0 = \left(\mathbf{D}^T\mathbf{D}\right)^{-1}\mathbf{D}^T\left(\mathbf{y}_1 - \mathbf{Q}\mathbf{u}_1\right) = \begin{bmatrix} \hat{x}_{01} & \hat{x}_{02} & \cdots & \hat{x}_{0n} \end{bmatrix}^T,$$

(applicable only in case that $det(\mathbf{D}^T\mathbf{D}) \neq 0$)

$$\mathbf{D} = \begin{bmatrix} \mathbf{c}^T \\ \mathbf{c}^T\hat{\mathbf{A}} \\ \mathbf{c}^T\hat{\mathbf{A}}^2 \\ \vdots \\ \mathbf{c}^T\hat{\mathbf{A}}^{(N-1)} \end{bmatrix}_{(N \times n)} ; \quad \hat{\mathbf{A}} = \begin{bmatrix} \mathbf{0} & \vdots & \mathbf{I}_{n-1} \\ \cdots & \cdots & \cdots \\ & \hat{\mathbf{a}}^T (N) & \end{bmatrix},$$

$$\mathbf{Q} = \begin{bmatrix} 0 & 0 & \cdots & 0 \\ \mathbf{c}^T\hat{\mathbf{b}} & 0 & \cdots & 0 \\ \mathbf{c}^T\hat{\mathbf{A}}\hat{\mathbf{b}} & \mathbf{c}^T\hat{\mathbf{b}} & \cdots & 0 \\ \vdots & \vdots & \ddots & \vdots \\ \mathbf{c}^T\hat{\mathbf{A}}^{(N-2)}\hat{\mathbf{b}} & \mathbf{c}^T\hat{\mathbf{A}}^{(N-3)}\hat{\mathbf{b}} & \cdots & \mathbf{c}^T\hat{\mathbf{b}} \end{bmatrix}_{(N \times (N-1))}.$$

Step 7: Current state vector $\mathbf{x}(k)$ estimation:

$$\hat{\mathbf{x}}(k+1) = \hat{\mathbf{F}}\hat{\mathbf{x}}(k) + \hat{\mathbf{b}}u(k) + \mathbf{g}y(k), \quad \hat{\mathbf{x}}(0) = \hat{\mathbf{x}}_0,$$

$$\hat{\mathbf{F}} = \hat{\mathbf{A}} - \mathbf{g}\mathbf{c}^T.$$

The vector $\mathbf{g}$ could be obtained by solving the pole placement problem (PPP), which is also known as pole assignment problem (PAP). Synthesis of $\mathbf{g}$ should be done according to the following recommendation: eigenvalues of matrix $\hat{\mathbf{F}}$ should be spread into the unit circle closer to the origin than the state matrix $\hat{\mathbf{A}}$ eigenvalues which is basic requirement ensuring good dynamic in terms of observer workability and speed and is also an overall improvement of the quality of the control process [6].

Step 8: Calculation of error vector ε and covariance matrix $\mathbf{C}$ for $k = N-n, N-n+1, ..., N$:

$$\varepsilon(k) = \begin{bmatrix} \varepsilon_1(k) & \varepsilon_2(k) & \cdots & \varepsilon_n(k) \end{bmatrix}^T =$$
$$= \begin{bmatrix} 0 & 0 & \cdots & 0 \end{bmatrix}^T,$$

$$s(k) = \begin{bmatrix} -y(k) & -y(k-1) & \cdots & -y(k-n) & u(k) & u(k-1) & \cdots & u(k-n) & \varepsilon(k)^T \end{bmatrix}^T,$$

$$e(k+1) = y(k+1) - \mathbf{s}(k)^T\hat{\mathbf{p}}(N),$$

$$\varepsilon(k+1) = \begin{bmatrix} e(k+1) & \varepsilon_1(k) & \cdots & \varepsilon_{n-1}(k) \end{bmatrix}^T,$$

$$s(k+1) = \begin{bmatrix} -y(k+1) & \cdots & -y(k+1-n) & u(k+1) & \cdots & u(k+1-n) & \varepsilon(k+1)^T \end{bmatrix}^T.$$

$$C(k+1) = C(k) - \frac{C(k)\mathbf{s}(k+1)\mathbf{s}(k+1)^T C(k)}{1 + \mathbf{s}(k+1)^T C(k)\mathbf{s}(k+1)}.$$

Step 9: Obtain the new $(N+1)$ observation and start the recursive procedure:

- prediction error computation:

$$e(N+1) = y(N+1) - \hat{\mathbf{p}}^T(N)\mathbf{s}(N),$$

- error vector ε update:

$$\varepsilon(N+1) = \begin{bmatrix} e(N+1) & \varepsilon_1(N) & \cdots & \varepsilon_{n-1}(N) \end{bmatrix}^T$$

- parameters vector $\mathbf{p}$ recalculation:

$$s(N+1) = \begin{bmatrix} -y(N+1) & \cdots & -y(N+1-n) & u(N+1) & \cdots & u(N+1-n) & \varepsilon(N+1)^T \end{bmatrix}^T,$$

$$\hat{\mathbf{p}}(N+1) = \hat{\mathbf{p}}(N) + \frac{C(N)\mathbf{s}(N+1)}{1 + \mathbf{s}^T(N+1)C(N)\mathbf{s}(N+1)}e(N+1),$$

$$\hat{\mathbf{h}}(N+1) = \begin{bmatrix} \hat{p}_{n+1}(N+1) & \hat{p}_{n+2}(N+1) & \cdots & \hat{p}_{2n}(N+1) \end{bmatrix}^T,$$

$$\hat{\mathbf{a}}(N+1) = \begin{bmatrix} -\hat{p}_n(N+1) & -\hat{p}_{n-1}(N+1) & \cdots & -\hat{p}_1(N+1) \end{bmatrix}^T,$$

$$\mathbf{T} = \begin{bmatrix} 1 & 0 & \cdots & 0 & 0 \\ -\hat{a}_n(N+1) & 1 & \cdots & 0 & 0 \\ -\hat{a}_{n-1}(N+1) & -\hat{a}_n(N+1) & \cdots & 0 & 0 \\ \vdots & \vdots & \ddots & \vdots & \vdots \\ -\hat{a}_2(N+1) & -\hat{a}_3(N+1) & \cdots & -\hat{a}_n(N+1) & 1 \end{bmatrix},$$

$$\hat{\mathbf{b}}(N+1) = \mathbf{T}^{-1}\hat{\mathbf{h}}(N+1),$$

$$\hat{\mathbf{A}} = \begin{bmatrix} \mathbf{0} & \vdots & \mathbf{I}_{n-1} \\ \cdots & \cdots & \cdots \\ & \hat{\mathbf{a}}^T (N+1) & \end{bmatrix}, \quad \hat{\mathbf{F}} = \hat{\mathbf{A}} - \mathbf{g}\mathbf{c}^T,$$

- current state vector calculation:

$$\hat{\mathbf{x}}(N+1) = \hat{\mathbf{F}}\hat{\mathbf{x}}(N) + \hat{\mathbf{b}}u(N) + \mathbf{g}y(N).$$

Step 10: Covariance matrix $\mathbf{C}(N+1)$ computation:

$$C(N+1) = C(N) - \frac{C(N)\mathbf{s}(N+1)\mathbf{s}(N+1)^T C(N)}{1 + \mathbf{s}(N+1)^T C(N)\mathbf{s}(N+1)}.$$

Step 11: Repeat steps 9 and 10 during the observation process.

IV. SIMULATION RESULTS

Simulations are carried out in MATLAB taking into account the following assumptions:

- for a given observed system description and input signal $u(k)$ (system reference) the output signal $y(k)$ (system response) is to be simulated;

- colored noise $f(k)$ is applied (added) to the system output signal;

- the input signal $u(k)$ and the noise-corrupted response $y(k)$ are used as input data for the observation algorithm;

- by using input-output data, the algorithm calculates the parameters and state vector estimations.

Computer simulation is performed for a given 6th order system described by the following discrete transfer function:

$$W(z) = \frac{0.6z^{-1} + 0.56z^{-2} + 0.2125z^{-3} + 0.308z^{-4} + 0.5488z^{-5} + 0.7221z^{-6}}{1 - 1.4z^{-1} + 0.7875z^{-2} - 0.2275z^{-3} + 0.035525z^{-4} - 0.002835z^{-5} + 0.00009z^{-6}}.$$

The corresponding state vectors are

$$\mathbf{a} = \begin{bmatrix} -0.00009 \\ 0.002835 \\ -0.035525 \\ 0.2275 \\ -0.7875 \\ 1.4 \end{bmatrix}, \quad \mathbf{b} = \begin{bmatrix} 0.6 \\ 0.2 \\ 0.1 \\ 0.3 \\ 0.4 \\ 0.5 \end{bmatrix}, \quad \mathbf{c} = \begin{bmatrix} 1 \\ 0 \\ 0 \\ 0 \\ 0 \\ 0 \end{bmatrix}, \quad \mathbf{x}(0) = \begin{bmatrix} 1 \\ 1 \\ 1 \\ 1 \\ 1 \\ 1 \end{bmatrix}.$$

For the eigenvalues of the system state matrix we obtain:

$$eig(\mathbf{A}) = \begin{bmatrix} 0.4 & 0.3 & 0.25 & 0.2 & 0.15 & 0.1 \end{bmatrix}^{\mathrm{T}}.$$

The input signal $u(k)$ is assumed to be a pseudo-random binary sequence (PRBS), which is generated in Matlab by using the following commands and functions:

$$u = (\text{sign}(\text{randn}(127,1)))*10.$$

The output signal $y(k)$ is noise-corrupted by adding color noise $f(k)$ obtained through white noise filtering. The filter transfer function is:

$$W_f(z) = \frac{1}{1 - 1.4z^{-1} + 0.7875z^{-2} - 0.2275z^{-3} + 0.035525z^{-4} - 0.002835z^{-5} + 0.00009z^{-6}}.$$

The noise level $-\eta$ is calculated when dividing the noise standard deviation σ_f by the output signal standard deviation $-\sigma_y$ according to the following equation:

$$\eta = \frac{\sigma_f}{\sigma_y} 100 = 0 \div 10\%, \tag{6}$$

Vector $\mathbf{a}$ estimation error e_a, vector $\mathbf{b}$ estimation error e_b and state vector $\mathbf{x}(k)$ estimation error e_x are relative mean squared errors (RMSE) determined by the following equations:

$$e_a(k) = -\sqrt{\frac{\sum\limits_{i=1}^{n}\left(a_i(k) - \hat{a}_i(k)\right)^2}{\sum\limits_{i=1}^{n} a_i(k)}}, \tag{7}$$

$$e_b(k) = -\sqrt{\frac{\sum\limits_{i=1}^{n}\left(b_i(k) - \hat{b}_i(k)\right)^2}{\sum\limits_{i=1}^{n} b_i(k)}}, \tag{8}$$

$$e_x(k) = -\sqrt{\frac{\sum\limits_{i=1}^{n}\left(x_i(k) - \hat{x}_i(k)\right)^2}{\sum\limits_{i=1}^{n} x_i(k)}}. \tag{9}$$

In order to verify the algorithm performance, a MATLAB program is developed. It was used to test the performance of the proposed algorithm. The results obtained by the program are posted in figure 1.

In the case of noise-free output signal (i.e. $f(k)=0$) number of the added to matrices $\mathbf{Y_{11}}$, $\mathbf{Y_{21}}$, $\mathbf{U_{12}}$ and $\mathbf{U_{22}}$ rows is $l=0$. Therefore matrices are square and $N=3n=18$. The RMSE $e_a(k)$, $e_b(k)$ and $e_x(k)$ are zero.

In the case of noise-corrupted output signal experiments are carried out for noise level $\eta=10.014\%$ and number of added to the matrices $\mathbf{Y_{11}}$, $\mathbf{Y_{21}}$, $\mathbf{U_{12}}$ and $\mathbf{U_{22}}$ rows $2l=80$ (i.e. $N=3n+2l=98$). Results are shown in Fig.1. The algorithm starts working at the 98th step of calculations and the observation errors are as follows: $e_a(k)<0.051$, $e_b(k)<0.015$, $e_x(k)<0.058$.

V. CONCLUSION

In this work the problem of adaptive state observer for discrete LTI SISO system is discussed. In the proposed algorithm the first estimation iteration is performed by least-squares method (steps 1 to 4). Further, the estimation procedure is continued as a recursive one.

With recursive estimation the regressors vector $\mathbf{s}$ is extended by prediction errors vector $\boldsymbol{\varepsilon}$, which parameters estimation is realized through minimal parameters estimation. The RELS method allows a nonlinear in general estimation task to be adopted as a linear one [1].

The convergence of RELS method is widely discussed in [2]. The necessary and sufficient conditions for convergence are: all the system poles to be spread into the unit circle and to have positive real parts. These conditions ensure method convergence and many authors confirm the fact that it is quite difficult to find cases in which the requirements are not met and method convergence cannot be achieved. The convergence of the proposed adaptive state observer is determined only by the RELS method convergence requirements. Therefore convergence requirements of the observer are the same as the RELS method.

The problem considered in the paper can be solved by using extended Kalman filter (EKF) methods [7] but the proposed algorithm allows for easy software implementation, estimates the initial state vector and is suitable for on-line operation. The developed algorithm is not an alternative of Kalman filter methods but is just another possibility for adaptive observation.

The suggested adaptive observer is suitable for closed-loop observation for the purpose of modal state controllers design. It could also be used to realize vibration control systems design [8], switching power converters design [9], fault diagnosis systems development [10], adaptive battery models development [11], for speed control for PMSM servo systems [12].

ACKNOWLEDGMENT

The paper is developed in the frames of the project "Model Based Design of Power Electronic Devices with Guaranteed parameters", ДН07/06/15.12.2016, National Scientific Fund.

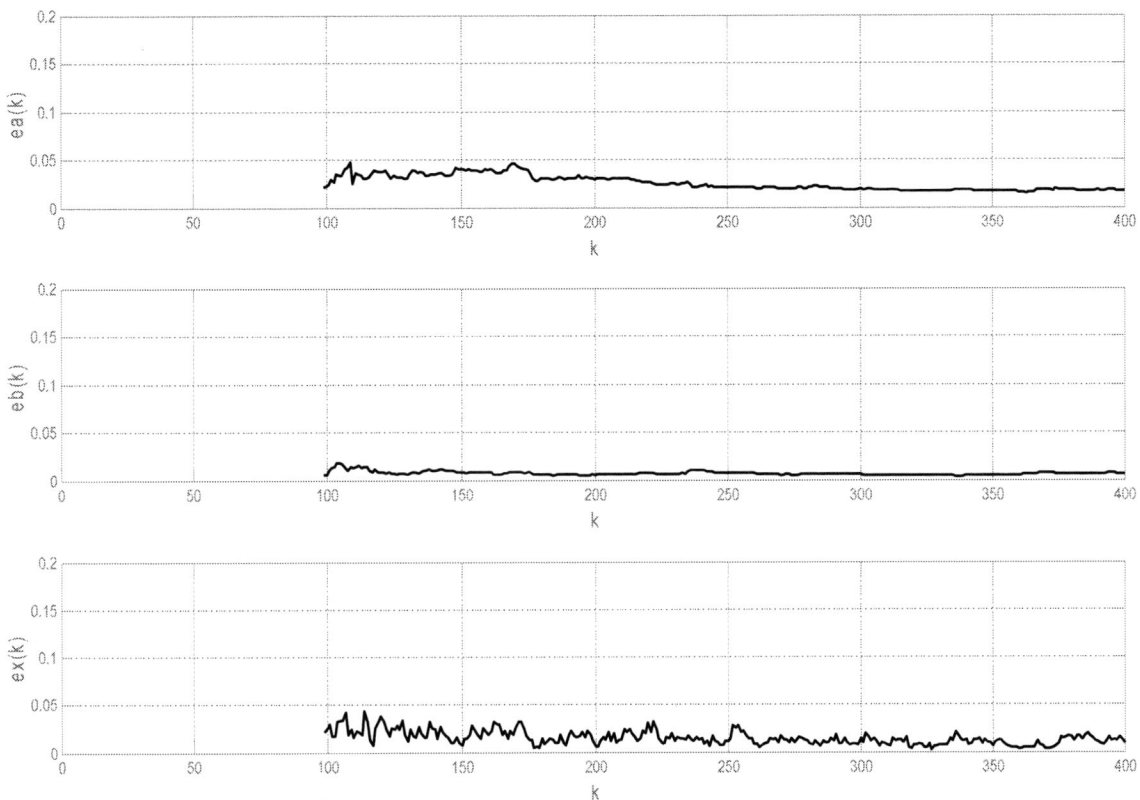

Figure 1 Relative mean squared errors $e_a(k)$, $e_b(k)$ and $e_x(k)$ for the case of noise-corrupted output signal.

REFERENCES

[1] I. Vuchkov, Identification, Sofia ,Yurapel Press, 1996.

[2] L.Ljung, T. Säderström, Theory and practice of recursive identification, MITPress, Cambridge, Mass.,1983.

[3] Sotirov L. N., V. S. Dimitrov, N. N. Nikolov, Discrete Adaptive State Observer for Real-time, International conference "Automatics and Informatics'04", pp. 121-124, Sofia, Oct. 2004.

[4] L.N. Sotirov, Control Theory – part II, Technical University of Varna, 2000.

[5] G. Luders, K.S. Narendra, An adaptive observer, IEEE Transactions on Automatic Control,vol.AC-18, pp.496-499, 1973

[6] R. Isermann, Digital control systems, Springer-Verlag, Berlin Heidelberg, 1989.

[7] Eric A.Wan, Alex T.Nelson, Dual Extended Kalman Filter Methods, Chapter 5. Kalman Filtering and Neural Networks, Edited by Simon Haykin, John Wiley&Sons, 2001.

[8] D. Miljković, Review of Active Vibration Control, Proceedings MIPRO 2009, Vol. III, CTS & CIS pp. 103-108, Opatija, Hrvatska, 25-29 May, 2009.

[9] J. Liu, S. Laghrouche, M. Harmouche, M. Wack, Adaptive-gain second-order sliding mode observer design for switching power converters, Control Engineering Practice, 30, 124-131, 2014.

[10] Ke Zhang, Bin Jiang, V. Cocquempot, Adaptive Observer-based Fast Fault Estimation, International Journal of Control, Automation, and Systems, vol. 6, no. 3, pp. 320-326, June 2008.

[11] C. R. Gould ; C. M. Bingham ; D. A. Stone ; P. Bentley, New Battery Model and State-of-Health Determination Through Subspace Parameter Estimation and State-Observer Techniques, IEEE Transactions on Vehicular Technology, Vol. 58, Issue 8, pp. 3905 – 3916, Oct. 2009.

[12] H. Liu,S. Li, Speed Control for PMSM Servo System Using Predictive Functional Control and Extended State Observer, IEEE Transactions on Industrial Electronics, Vol. 59, Issue 2, pp. 1171 - 1183 Feb. 2012.

Inverter Current Source for Pulse-Arc Welding with Improved Parameters

V. C. Valchev[*], D. D. Mareva[**], D.D. Yudov[**] and R. S. Stoyanov[*]
[*] Technical university of Varna/Department of Electronics, Varna, Bulgaria,
[**] Burgas Free University/Faculty of Computer Science and Engineering, Burgas, Bulgaria
vencivalchev@hotmail.com, yudov@bfu.bg, d_mareva@abv.bg, radkostoianov@gmail.com

Abstract - The feature of power current sources for pulse-arc welding is superimposing of current pulses with a defined shape, size and frequency on the main welding current. The purpose of this paper is to explore the possibilities to control the parameters of superimposed pulses by a particular (additional) welding inverter current source with improved parameters. The current pulse superimposed to the basic current has duration of 1.5 ÷ 3.0 µs and is realized by an additional source. The operation and the characteristics of the scheme are investigated by simulations (PSpice) under various modes and loads. The formation and separation of the drop from the end of the electrode is controlled by the amplitude and duration of the pulse current. Thus, the average welding current can be significantly reduced. Dependences of the parameters of pulses on specific components of scheme are derived. Recommendations are made to improve the performance and utilization of the circuit elements and the technological process. Design recommendations are presented to optimize the parameters of the transformer.

Key words: *welding control, pulse induction arc welding, magnetic components*

I. INTRODUCTION

The advantages of welding inverter sources operating at high frequency are well known [1]. A possible approach of further increasing their performance is the pulse arc-welding. This method is very advantageous with non-ferrous metals and stainless steel.

The pulse arc-welding current sources operate based on a simple principle [2]. Additional current source superimposes a current pulse with defined form, amplitude, power and frequency over the main welding current. The requirements for the current pulse are "solid" or "falling" external V-A characteristic. The polarity of the welding current significantly affects the welding process. The management of the transfer of the metal in this method is been enforced to varies between two levels of current, called base (I_z) and current pulse (II). The current amplitude is selected based on the requirement to ensure the continuity of the arc with small impact on the melting of the electrode. The function of the pulsating current exceeds the critical current for melting the tip of the electrode and form droplets of a certain size and droplet detachment from the end of the electrode under the action of electromagnetic force (pinch effect) [3,4]. An optimal transfer the metal is considered when only a single drop of the metal electrode is formed for each current pulse. The drop formation and separation is controlled by the amplitude and duration of the pulse current, and the average welding current can be greatly reduced (reduction in the frequency of the current pulse or reduction of the base current).

The high frequency (for control of the transfer of the metal) and low frequency (for control of the formation of cavity) pulsating currents are used simultaneously and serve as a basis for the consolidation of this process as "double pulse". The current pulse superimposed over the base current of the arc has duration of 1.5 ÷ 3.0 µs [5]. This current is produced by additional source connected in parallel with the welding source. Simultaneously a high-frequency control (the transfer of the metal) and a low frequency one (for control of the formation of cavity) are performed to obtain so called "double pulse".

This paper is focused at exploring the possibilities to adjust the parameters of the superimposed pulses on a particular hardware implementation.

The studies presented are realized by computer simulation. The models of power semiconductor switches (IGBT, diodes) are based on models provided by the corresponding manufacturers.

II. PULSE ARC-WELDING INVERTER CURRENT SOURCE

A source of current pulses with a large value of the welding current can be obtained if a capacitor charged with a voltage higher than the voltage of the arc is connected in parallel with the load. Converters for charging super-capacitors are presented and analyzed in [6]. Experiments and simulations of the current arc are complex and another solution could be using high precision electronic DC loads [7].

Figure 1 shows a block diagram of the inverter source for pulsed-arc welding.

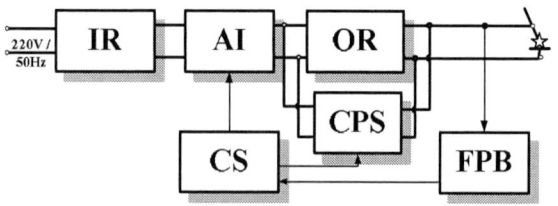

Figure 1. Block diagram of the inverter welding current source,

IR – Input Rectifier, AI – Autonomous Inverter, OR – Output Rectifier, CPS – Current Pulses Source, CS – Control Scheme, FPB – Feedback and Protection Block

The current pulse is presented as:

$$i_i = \frac{V_c - V_z}{R_L} e^{\frac{t_i}{RC_z}} , [A] \tag{1}$$

where: R_L is the resistance of the arc and the connecting wires;
V_c – voltage over the "pulse" capacitor C_z;
V_z – voltage of the arc;
RC – time constant charging the "pulse" capacitor C_z;
t_i – current pulse width.

In a manual arc welding with direct current according to the presented scheme the voltage of arc can be calculated using the following equation:

$$V_z = 19 + 0,04 . I_z , [V] \tag{2}$$

where: I_z – welding current (main current). Its value is chosen as 120 A.

The power electronic part consists of three blocks - rectifier with input and output filter, autonomous inverter and an output block for current pulses.

Figure 2 presents a schematic diagram of a power block of inverter welding source realized according to the block diagram introduced in Figure 1.

The H-bridge autonomous inverter is supplied directly from the grid through a bridge rectifier. In the other diagonal of the inverter a transformer is connected. In the secondary winding of this transformer another rectifier is connected ($D_9 \div D_{11}$), inductive filter (L_f) and a part receiving the current pulses (L_4, S_1, D_{12}, C_z).

The linear part of the arc current I_z corresponds to the transition of the drops from the electrode to the welding cavity. After that the arc current increases to its peak amplitude.

The current pulse can be calculated using the following equation [5]:

$$I_i = (1.5 \div 2.0) I_{cr} , [A] \tag{3}$$

where Icr is the critical current value, while maintaining the current pulse in the interval of time, providing for transmission of the metal flow, according to the following equation:

$$t_i = \frac{\Delta t_a}{\Delta v_{bi}} , [s] \tag{4}$$

where: t_i – duration of the current pulse,
Δt_a – arc length range as a function of pulse duration,

Δv_{bi} – change in the melting rate range of the electrode during the transition from basic current to pulse current.

Then current pulses reduced to the value of the base current, and held it for the following period of time:

$$t_p = \frac{\Delta t_{dp}}{\Delta v_{bi2}} , [s] \tag{5}$$

where: t_p – pause length, securing the length of the arc to a time corresponding to the base current,
Δt_{dp} – arc length duration range during the pause,
$\Delta v_{bi2} = v_{bp} - v_{bi2}$ – electrode melting point speed reduction when the weld with a base current is going through a pause. During this interval, the arc length time is reduced to a value corresponding to an arc with large drops.

The value of speed Δv_{bi1} can be determined by the following expression:

$$\Delta v_{bi1} = (k_{ri}(i_i - i_z) - k_{ru}k_g\Delta t_a) , [m/s] \tag{6}$$

where: k_{ri}, k_{ru} – coefficients of current and voltage self-regulation,
k_g – coefficient linked to the length of the arc.

The base current of the arc I_z is determined by the controllable power source based on inverter. It has a combined V-A characteristic with trapezoidal section in the range of operating currents.

The transition from the base current to the peak current and back is done discretely by switching the trapezoidal section of V-A characteristics of the power source.

Depending on the type and diameter of the electrode, depending on the chemical composition of the welded metal, the current pulses have a frequency of $4 \div 2000Hz$ [5,8,9]. Their amplitude and frequency determine how the drop is separated. Depending on the purpose of the welding unit, its effectiveness (universality) and the voltage across the capacitor C_z the magnitude of the voltage V_3 is determined.

$$Vc_z = V_{2am} + V_{3m}, [V] \tag{7}$$

where: V_{cZ} – voltage over the pulse capacitor C_z;
V_{2am} – the maximum value of the voltage source 2;
V_{3m} – the maximum value of the voltage source 3;

The parameters of the current pulses depend from the capacitance of the C_z capacitor and the active component of the arc impedance R_L.

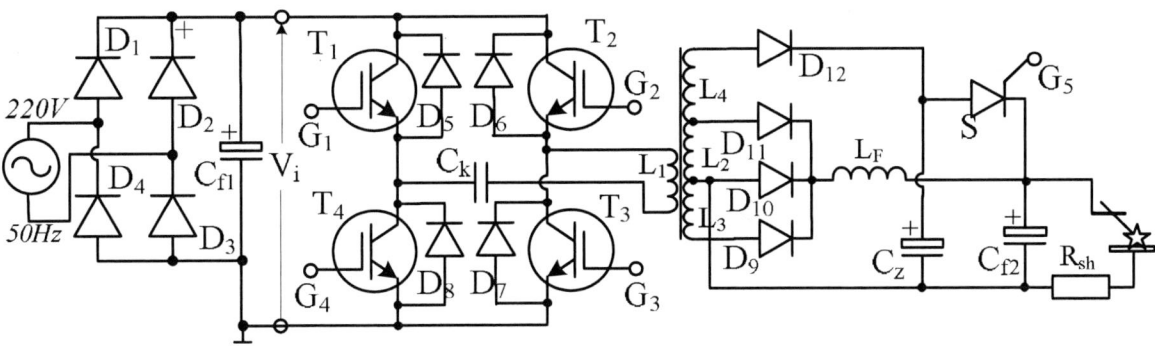

Figure 2.   Schematic diagram of a pulse arc-welding inverter current source, [2]

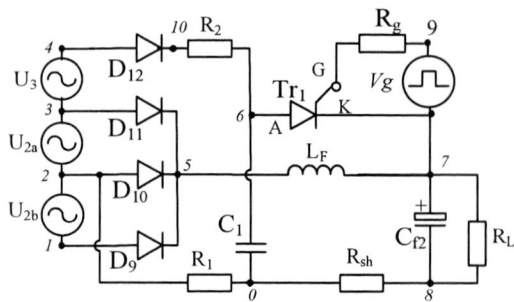

Figure 3. Equivalent scheme of the PSpice model of pulse arc-welding inverter current source

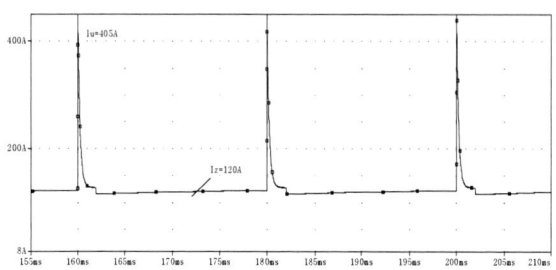

Figure 4. Welding current and current pulse I_i

III. SIMULATION ANALYSIS OF THE PROPOSED PULSE ARC-WELDING INVERTER CURRENT SOURCE

To study the effects of the pulse current a PSpice model of the inverter source for pulse-arc welding with the equivalent scheme presented in Figure 3 is developed. Welding current and current pulse Ii are shown in Figure 4.

The advantage of the proposed scheme is the fact that the amplitude value of the output voltage V_3 of the auxiliary winding of the transformer is considerably greater than the voltage of the arc. This voltage is used to charge the capacitive source for current pulses (C_z). The main voltages on the secondary side of the transformer are V_{2a} and V_{2b}, V_3 is the voltage of the additional voltage source which is generated by the same transformer with an auxiliary winding. The thyristor G5 is controlled by a pulse gererator with frequency of $0,5 \div 200$Hz in the proposed scheme.

In the welding electrical chain the resistance R_l represents the active component of the input impedance of the scheme and the resistance R_L mainly represents the active resistance of arc during welding. Using equation 2, V_z can be calculated ($V_z = 24$ V) and then:

$$R_L = \frac{V_z}{I_z} = 0,2\Omega \qquad (8)$$

The filtering inductance can be approximately calculated using the following equation:

$$\omega L_f = (10 \div 100)R_L, \qquad (9)$$

where the resonant frequency of the inverter is $f_{inv} = 50$kHz.

Based on the carried out simulations, the value of the components used in PSpice simulation model are determined as follows: $L_f = 100\mu$H, $R_s = 0.01\Omega$, $R_L = 0.2\Omega$.

The parameters for the other components are as follows: $R_1 = 0,01\Omega$, $C_z = 300\mu$F, $C_{f2} = 3\mu$F, $R_2 = 0.5 \Omega$ and R_2 is, the resistance connected in series to the diode D_{12} it equalizes the losses in this part of the circuit.

Figure 5 presents the voltage of the arc at the operating point, the current through the filtering inductor and the current through the arc. When suitable values for pulses current supplying capacitor C_z are selected, a current pulse with approximately three times greater amplitude than the welding current is obtained. The waveform of the current pulse (exponent) is determined by C_z and R_z, as a requirement the front edge has a greater slope which depends on the time constant of the RC group C_z and R_2. This leads to a better drop leasing.

The energy of the current pulse depends on the pulse width and frequency and determines the speed of the drop separation. Figure 5 shows the waveform of the current pulse for different values of its length at a constant frequency and pulse generation.

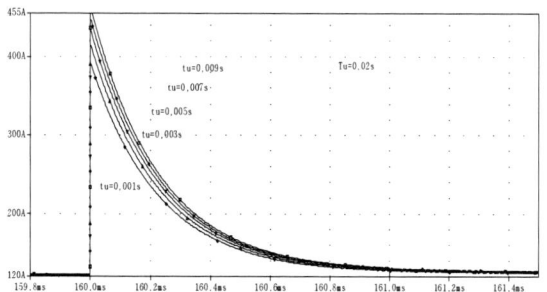

Figure 5. Current pulse dependence on different values of its duration

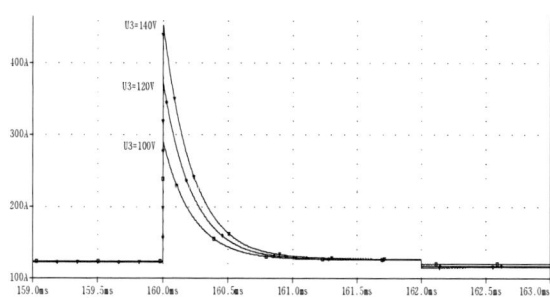

Figure 6. Current pulse as a function of different voltage amplitudes (V_3)

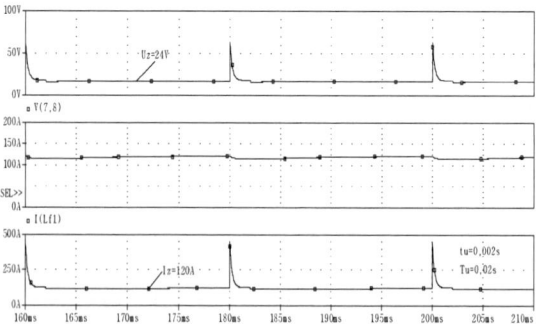

Figure 7. Current flowing through the pulse supplying capacitor, the current through the filtering inductor and arc current

The Figure 5 also shows that when the supply voltage varies, the amplitude and the exponent of the trailing edge of the current pulse changes significantly. Welding with different electrodes is determined by the magnitude of the welding current, and the drop separation is provided by regulating the frequency of the pulses.

The dependencies presented in Figures 6 and 7 are showing the change of the magnitude of the current pulse I_i in function of the pulse duration and its frequency for three different amplitudes of the auxiliary source's voltage V_3.

When changing the pulse width, the maximum current value varies in a linear manner and its amplitude varies in a narrow range if supply voltage drifts. The same phenomenon applies to the frequency response. The maximum current value changes linearly in small limits to a maximum of 70 A in function of the pulse frequency. As a summary, the change in the supply voltage has a bigger impact on the current. The waveform of the pulses is maintained in both control strategies.

Figures 8 and 9 present the relation of the current pulses amplitude I_i when the value of the pulse capacitor C_z is changed for three different values of the pulse duration and frequency.

Figures 10 and 11 present the relation of the current pulses amplitude Im_i when its duration and frequency are changed, while using three different values for C_f.

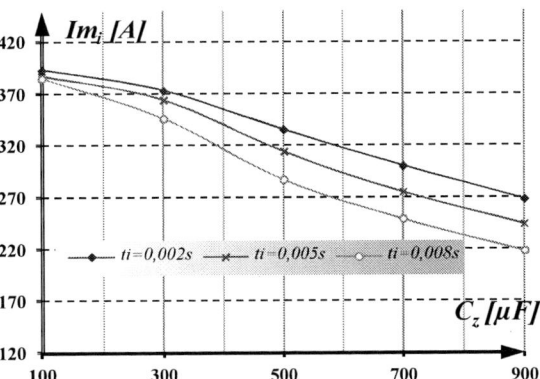

Figure 10. Current pulse magnitude dependence on the values of C_z for three different pulse periods

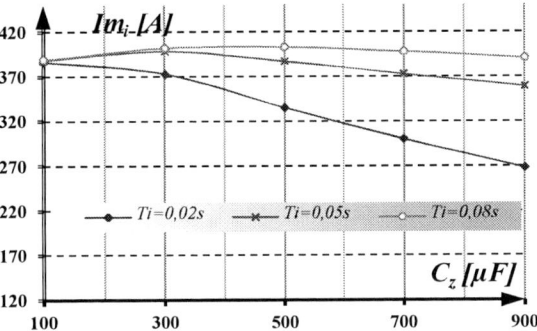

Figure 11. Current pulse magnitude dependence on the values of C_z for three different frequencies

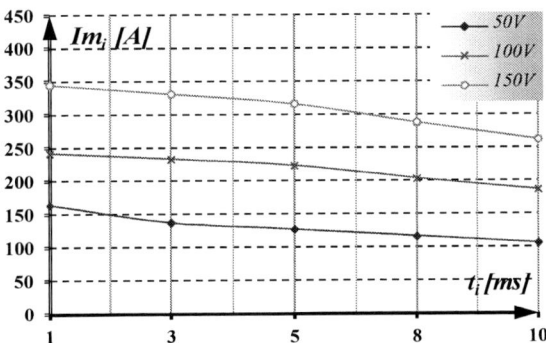

Figure 8. The magnitude of the current pulse I_i as a function of the pulse duration for three different amplitudes of the auxiliary source's voltage V_3

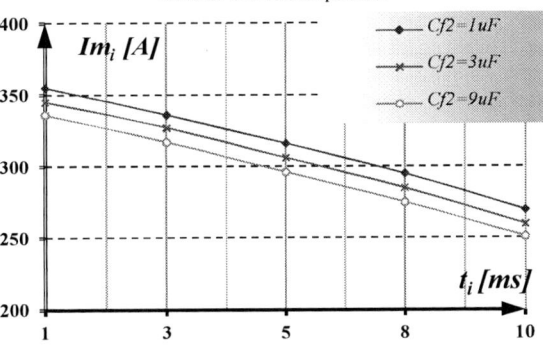

Figure 12. Current pulse magnitude dependence of its duration for three different values of C_z

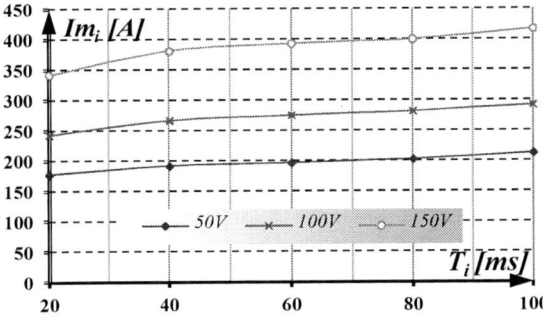

Figure 9. The magnitude of the current pulse I_i as a function of its frequency (1/T) for three different amplitudes of the auxiliary source's voltage V_3

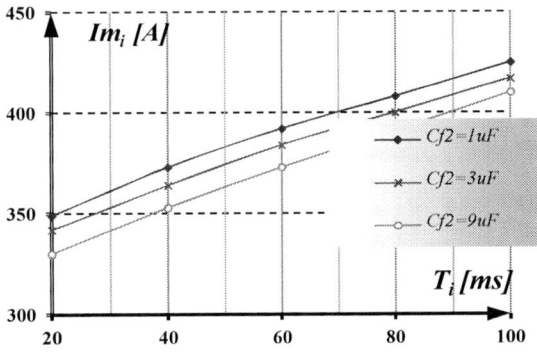

Figure 13. Current pulse magnitude dependence of its frequency for three different values of C_z

TABLE I.	INPUT DATA FOR THE TRANSFORMERS
Primary voltage	300 V
Secondary voltage (no load)	60 V
Secondary voltage (during welding)	25 V
Secondary current (continuous)	140 A
Working frequency	50 kHz

The maximum current value changes linearly in small limits (up to 5μs) corresponding to the pulse width, alongside the change of the capacitor value. Furthermore, the form the pulse is maintained, and the amplitude is regulated smoothly. It can be controlled evenly up to a maximum of 60 A.

The maximum current value changes exponentially in large limits up to $T_i = 60$μs when the frequency of the pulses varies. Then the slope decreases if the value of the capacitor C_z is changed. The form of the pulse is retained and amplitude can be regulated roughly up to a maximum of 130A.

IV. TRANSFORMER DESIGN AND DISCUSSION

In the proposed welding source the transformer is one of the most important components in respect of losses, weight, volume and total performance of the device. The design of the transformer includes the choice of magnetic material for the core. Nowadays the new nanocrystalline alloys prove to be concurrent materials to traditional ferrite cores for power electronics [10, 11]. Their magnetic properties: great permeability (20 000 to 600 000 vs. 10 000 for ferrites) and high saturation induction (1.2 T vs. 0.4 T for ferrites), combined with low core losses, make them a possible choice for presented welding source.

Following the stated considerations three transformers are calculated, suitable for the current application. Two of them are based on ferrite cores (EE 80/38/20, material N87 [12]), (PM 87/70, material N87) and one is based on nanocrystalline cores (2 x F3CC025, material Vitroperm 500, [13]). The purpose of the comparative designs is to derive conclusions regarding using these two main options for the core material of the transformer.

Design algorithm used is based on the design approach proposed in [14] and further improved in [15], [16] consisting of 15 steps.

The calculations are based on the transformer input data, shown in Table I.

All of the transformers are calculated using Litz wire as the secondary winding currents in welding sources are high. This causes respectively significant eddy current losses. In a general case with p parallel wires (Litz wires), the eddy current losses $P_{cu,eddy,litz}$ are found as [12]:

TABLE II.	MAIN SPECIFICATIONS AND POWER LOSSES OF THE COMPARED DESIGNS		
	2 x F3CC025	EE 80/38/20	PM 87/70
Primary winding	1 x (315 x 0.071mm)	1 x (210 x 0.1mm)	1 x (175 x 0.1mm)
Secondary 1 (L4)	3 x (1575 x 0.071mm)	3 x (1575 x 0.071mm)	2 x (1890 x 0.071mm)
Secondary 2 (L3)	3 x (945 x 0.071mm)	2 x (735 x 0.1mm)	2 x (630 x 0.1mm)
Secondary 3 (L2)	1 x (945 x 0.071mm)	1 x (735 x 0.1mm)	1 x (630 x 0.1mm)
Core losses	8.40 W	7.62 W	7.18 W
Copper losses	15.8 W	13.22 W	14.16 W
Total losses	24.20 W	20.84 W	21.34 W
Core set weight	732 g	358 g	770 g
Copper weight (all windings)	235 g	244 g	174 g
Total weight	967 g	602 g	944 g
Copper fill factor,[-]	0.177	0.136	0.147

$$P_{cu,eddy,litz} = P_{cu,eddy,orig}\, p \left(\frac{d_{litz}}{d_{orig}}\right)^4$$
$$= \frac{l_w\, \pi\, \dfrac{d^4}{4}\, N^2}{p\, 48\, \rho_c} \left(\frac{2\pi\, f\, I_{ac}\, \mu_0}{w}\right)^2 \qquad (10)$$

where $P_{cu,eddy,orig}$ are the eddy current losses for a round wire;

l_w is the conductor length of the winding;

N is the turns number;

I_{ac} is the AC component of the current;

w is the winding width;

d_{litz}, d_{orig} are the diameter of the Litz wire and this one of the round wire;

f is the operating frequency.

The main specifications obtained are presented in Table II. All windings are wound in a single layer with Litz wire.

V. CONCLUSION

The purpose of this paper is present and studies the possibilities to control the parameters of welding process with superimposed pulses by a particular (additional) welding inverter current source with improved parameters. The current pulse superimposed to the basic current has duration of $1.5 \div 3.0$μs and is realized by an additional source. The operation and the characteristics of the scheme are investigated by simulations (PSpice) under various modes and loads.

The following conclusions are derived based on the carried out simulations and designs:

• The pulse-arc welding inverter current source studied in the current paper shows significant prospects in functionality. The scheme is capable of regulating the parameters of the current pulse in a wide range.

• The form of the current pulse slightly depends on its frequency.

• The energetic characteristics of the current pulses are controlled stepwise by changing the value of the pulse capacitor. Precise control of the frequency of the drop separation is performed by changing the frequency of the pulses. Pulse width does not affect the amplitude of the current pulse.

• Three different transformer designs are realized and compared. The best parameters are obtained using a EE ferrite core. Design with a PM ferrite core provides improved EMC parameters as the windings are completely inside the core and fringing field is minimized.

The experimental results are envisaged in the future publications. The accuracy of the simulation results is high as the simulation results are obtained utilizing manufacturer's models of the semiconductor devices (IGBTs, diodes).

ACKNOWLEDGMENT

The carried out research is realized in the frames of the project "Model based design of power electronic devices with guaranteed parameters", ДН07/06/15.12.2016, Bulgarian National Scientific Fund.

REFERENCES

[1] Dimitar Yudov, Atanas Dimitrov, Georgi Toshkov, Daniela Mareva, „INVERTER WELDING POWER SOURCE WITH TWO RESONANT FREQUENCIES", 2006, Annuals of BFU, Bourgas, Bulgaria.

[2] Dimitar Yudov, Georgi Todorinov, Daniela Mareva „INVERTER SUPPLY FOR PULSE ARC WELDING", 2006, Annuals of BFU Bourgas, Bulgaria.

[3] Бардин В. М., Борисов Д. А., „ИНВЕРТОРНЫЕ СВАРОЧНЫЕ АППАРАТЫ ПЕРЕМЕННОГО ТОКА ВЫСОКОЙ ЧАСТОТЫ", ГОУВПО «МГУ им. Н. П. Огарева» УДК 621.314.

[4] А. В. Кобзев, В. Д. Семенов, "Формирователь импульсов сварочного тока на основе двухтрансфор-маторного комбинированного преобразователя", ТУСУРа, 2 часть 1, декабрь 2011УДК 621.314, 621.791.

[5] http://www.findpatent.ru/patent/257/2570145.html
© FindPatent.ru - патентный поиск, 2012-2016.

[6] Kraev G., N. Hinov, D. Arnaudov, N. Rangelov, B. Gilev "Serial ZVS DC-DC converter for supercapacitor charging", Proceedings of the XIX-th international symposium on Electrical Apparatus and Technologies "SIELA - 2016", 29 May – 1 June, 2016, Bourgas, Bulgaria

[7] Kanchev, H.C., Hinov, N.L., Arnaudov,. D.D. and Hranov, T.H., Current fed inverter application as a controllable DC load, XXV International Scientific Conference Electronics (ET), 12-14 Sept. 2016, Sozopol, Bulgaria.

[8] А. Ф. Князьков, С. А. Князьков, А. Н. Мусин, „Оптимизация формы импульсов тока для управления переносом электродного металла при сварке в защитной среде аргона", Томский ПУ.

[9] Verdelho P., M Pio Silva, E. Margato, "An Electronic Welder Control Circuit" IEEE, 0-7803-4503-7/98.

[10] Hitachi Metals,Ltd., Nanocrystalline soft magnetic material "FINEMET®", 2005.

[11] Martin Ferch, Application overview of nanocrystalline inductive components in today's power electronic systems, Proc. Soft Magn. Mater. Conf., 2013.

[12] https://en.tdk.eu/download/528882/6a0da25e2745be5c13b587b3d4a8de48/pdf-n87.pdf.

[13] http://www.vacuumschmelze.com/en/research-innovation.html, accessed 2016.

[14] Van den Bossche A. and V. C. Valchev, Inductors and Transformers for Power Electronics, CRC Press, Boca Ration, FL, USA, 2005.

[15] V. C. Valchev, T. P. Todorova and A. Van den Bossche, "Comparison and Design of Power Electronics Transformers in 25 kHz - 400 kHz Range", 19th International Symposium on Electrical Apparatus and Technologies, IEEE SIELA, 29 May - 2 June, 2016, Bourgas.

[16] V. C. Valchev, T. P. Todorova and A. Van den Bossche, "Comparative study of winding arrangements for power electronic transformers", IEEE XXV International Scientific Conference Electronics 2016, 12 - 14 September 2016, Sozopol, Bulgaria.

MIPRO 2017, May 22- 26, 2017, Opatija, Croatia

Power Output Comparison of Three Phase Passive Converter Circuits for Wind Driven Generators

V. C. Valchev*, P. V. Yankov*, A. Van den Bossche**

* Technical University of Varna, Department of Electronic engineering and microelectronics, Varna, Bulgaria
** Ghent University, Department of Electrical Energy, Systems and Automation, Ghent, Belgium
vencivalchev@hotmail.com, plamenvalentinov@gmail.com, alex.vandenbossche@ugent.be

Abstract - The paper presents a performance comparison of two variations of a three phase passive converter circuit for wind driven generators included in patent PCT/EP2010/055637. A brief description of the concept is presented. Then along with the simulation of the basic circuit, already described in previous research, an improved schematic with added magnetizing inductors is proposed and investigated. A methodology for adjustment of the power/frequency curve of both circuits is added. A design approach for dimensioning the components at 100kW power output rating of the wind turbine generator is developed. The OrCAD Capture computer software is used for the simulations of both analog circuits. A graphical comparison of their power/frequency curves is realized in order to assess increase or decrease in power at the range of 15 Hz ÷ 65 Hz. The difference between the two curves is also expressed as percentages in the last column of the table. Considerations are derived to optimize the design of the magnetic components in the circuit. Finally a conclusion is made about the need of a more complex design to achieve higher output results.

I. INTRODUCTION

Wind turbine's electronic equipment and especially power electronic converters (PEC) have the specific task to maximize the performance of the conversion process and to be as reliable as possible. These considerations ensure evolved circuitry within newer, high power, grid connected wind turbines (WT) [1]. At the same time, the electric generators, reach their performance peak with permanent magnet synchronous generator (PMSG). These electrical machines are used in direct driven wind turbines such as Siemens D3 platform. Figure 1 depicts an example of a simplified structure of a modern grid connected WT using PMSG [2, 3, 4].

Both PECs, the rectifier and the inverter, are fully controlled and accomplished with specialized high power electronic switches. This allows to load the electric generator close to ideal performance curve all the time that the WT operates and therefore converts the kinetic energy of the wind most efficient. However this microcontroller operated converters can be damaged

---

The paper is developed in the frames of the project "Model Based Design of Power Electronic Devices with Guaranteed parameters", ДН07/06/15.12.2016, National Scientific Fund.

during storm or just fail due to weakness in some electronic component or internal supply. Another thing to be considered is the cost of the PECs compared to a classic three phase uncontrolled full wave rectifier.

So, in a long term use and/or at a distant installation site the reliability of the system is a significant advantage. The passive converter circuits, proposed by patent PTC/EP2010/055637, improve this parameter [5]. It is an AC/DC converter (its position is marked with dashed line on figure 1), which improves power output characteristics over classic three phase uncontrolled full wave rectifier. Specific to this circuit is bridge rectification of every phase of the generator and externally connected inductors to form external star point gives cost-efficiency and robustness to the design of this PEC. There is also an advantage that paralleling converters with diodes is more reliable than paralleling transistor based converters in a wind park. In a number of previous articles a performance analysis and comparison of basic passive and fully controlled PEC for wind turbines is published, [6, 7].

To assess and compare the theoretical power output of the WT with the following equation (1) is given:

$$P = \frac{1}{2} \rho A C_P V^3 \qquad (1)$$

, where P - output power in [W], ρ – density of the air in [kg/m^3], A – swept area of the blades in [m^2], C_P – power coefficient, V - wind speed in [m/s].

Although the basic passive converter circuit offers comparable performance to the fully controlled PECs at nominal wind speed, the middle section of the power output curve still can be improved.

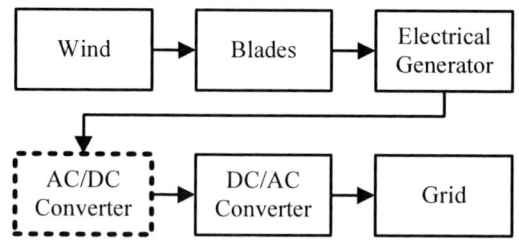

Figure 1. Example topology of a wind turbine with PMSG and PECs operating with full output power of the electric generator

136

II. PASSIVE CONVERTER CIRCUITS AND PROPOSED METHODOLOGY FOR ADJUSTMENT OF THE COMPONENTS

A. Circuit basics

In order to assess and compare the improvement on power output at different wind speed, two passive converter circuits are examined in this article.

The basic schematic includes components marked with blue color on figure 3. Simulation model consists of three main parts – electric generator, rectifiers and external inductors. The electrical machine is represented by three voltage sources (AC 1, AC 2 and AC 3) and three inductors (L1, L2 and L3). Two three-phase uncontrolled rectifiers are connected at both ends of every generator winding. Then all windings are externally connected in star point through external inductors (L4, L5 and L6).

The second investigated circuit includes all the components in blue and red color on figure 3. In addition to the basic structure, inductors (L7, L8 and L9) are added and coupled to form an autotransformer. The turns ratio of the transformer is determined by their values. Other external inductors (L10, L11 and L12) are added in series to the secondary windings of the autotransformer.

All simulation models aim at P_{nom}=110kW power output at nominal wind speed (converted to the specific multipole PMSG it is at f_{nom}=40Hz) and a power/frequency curve that follows as close as possible the ideal cubic dependence from equation (1). The DC bus is set at 1000V. The study uses a parametric analysis in OrCAD Capture to obtain at least twenty points form the power output curve at the range of 0÷100Hz [8]. Frequency is set as global parameter and an electromotive force coefficient of 40V/Hz is chosen, thus the voltage of the three AC sources operate correctly.

B. Reactive Components Adjustment methodology

Adjustment of reactive components values is done with direct optimization method - line search with variable and backtracking step size. The method uses only one limit "a" of the variable "x" (which is the value of the reactive components in mH), at this case "a ≤ x". Our

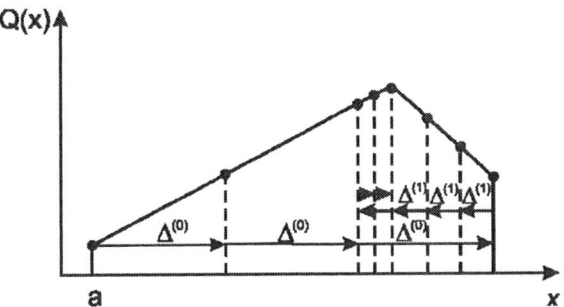

Figure 2. Optimization method with variable step size is used to adjust values of inductors in the simulation

value of the lower limit "a" is set to one mH. The search step size initial value Δ^0 is set at eight mH. When first scan is finished, next step size it is determined by equation 2 until the function $Q_{(x)}$ (which represents optimum power output) reaches the extremum and satisfies the search.

$$\Delta^{(k+1)} = \frac{\Delta^k}{4} \qquad (2)$$

, where k is the iteration number with range of [0÷2] and Δ^k is the step size. Graphical interpretation of the method is depicted on figure 2. One of the advantages is its simplicity of implementation in algorithm capable of solving the problem automatically [9].

The values of inductors L1, L2 and L3 determine the maximum power output of the wind turbine generator. Lower values give higher power output. The external inductors L4, L5 and L6 adjust the power output in the low frequency range. Final values for a basic converter design are 38mH for the generator inductors and 120mH for the externally connected inductors.

The inception point of the curve is set by DC bus voltage Udc, which determines when diodes start to conduct set at 1000V and the electromotive force coefficient Kemf set at 40V/Hz. This is the "cut in" wind speed or frequency of the generator.

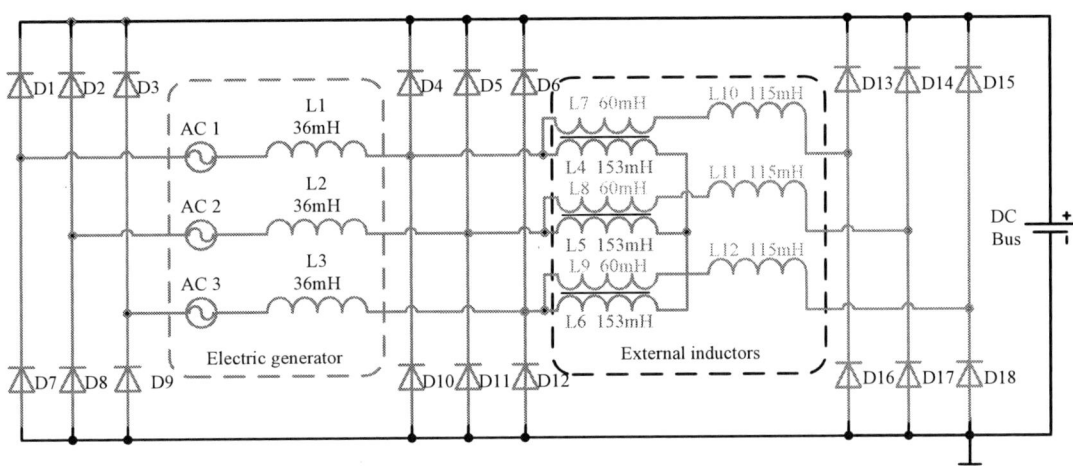

Figure 3. Passive converter circuits: basic – components are marked with blue color, improved – includes blue and red components

It can be expressed with the following equation:

$$f_{cut\_in} = \frac{\dfrac{U_{dc}}{2}}{K_{emf}} = \frac{500}{40} = 12{,}5\,Hz \qquad (3)$$

Figure 4 shows waveforms of instantaneous current and voltage of a phase and the three phase instantaneous power of the passive converter circuit at 30Hz. Colors of different phases correspond to each other. It is an example of extraction of only one point of the power/frequency curve and the data is stored in seventh row of table 1. Those waveforms correspond to the second improved circuit simulations, where P_{autotr} reaches around 51kW.

Table 1 gives all the necessary data from the performed simulations. Last two columns specifically inform about the difference in percent between ideal cubic dependence curve P_{ideal} and both passive convertor circuits power output curves - P_{basic} and P_{autotr}. It is clear that at 30Hz there is around 50% increase with the use of more complex circuit and at 35Hz the power output is 20% greater. These numbers have to be as an improvement to get a smaller deviation closer to the ideal torque speed curve. In fact, the power changes much less than the deviation, certainly if pitch control is used.

The three power output curves are shown on figure 5. First the optimal (ideal) power output P_{ideal} is represented by black dashed curve. Next the power output curve of the basic passive converter circuit P_{basic} is shown with blue color and square markers. The third curve P_{autotr}, which represents the improved design power output is with red color and circle markers.

TABLE I.   TABLE COMPARISON OF POWER OUTPUT OF PASSIVE CONVERTER CIRCUITS

f, [Hz]	P_{ideal}, [kW]	P_{basic}, [kW]	P_{autotr}, [kW]	P_{diff_basic} [%]	P_{diff_autotr} [%]
0	0,0	0,0	0,0	0,0	0,0
5	0,2	0,0	0,0	0,0	0,0
10	1,7	0,0	0,0	0,0	0,0
15	5,8	0,2	0,2	3,4	3,4
20	13,8	18,0	13,8	130,9	100,5
25	26,9	24,0	23,3	89,4	86,6
30	46,4	26,0	51,1	56,0	110,1
35	73,7	58,0	72,2	78,7	98,0
40	110,0	105,0	108,9	95,5	99,0
45	156,6	152,0	151,7	97,0	96,9
50	214,8	190,0	187,8	88,4	87,4
55	286,0	217,0	216,7	75,9	75,8
60	371,3	237,0	237,5	63,8	64,0
65	472,0	252,0	253,3	53,4	53,7
70	589,5	263,0	265,8	44,6	45,1
75	725,1	273,0	275,4	37,7	38,0
80	880,0	281,0	283,3	31,9	32,2
85	1055,5	287,0	289,7	27,2	27,4
90	1253,0	292,0	295,0	23,3	23,5
95	1473,6	296,0	299,5	20,1	20,3
100	1718,8	300,0	303,3	17,5	17,6

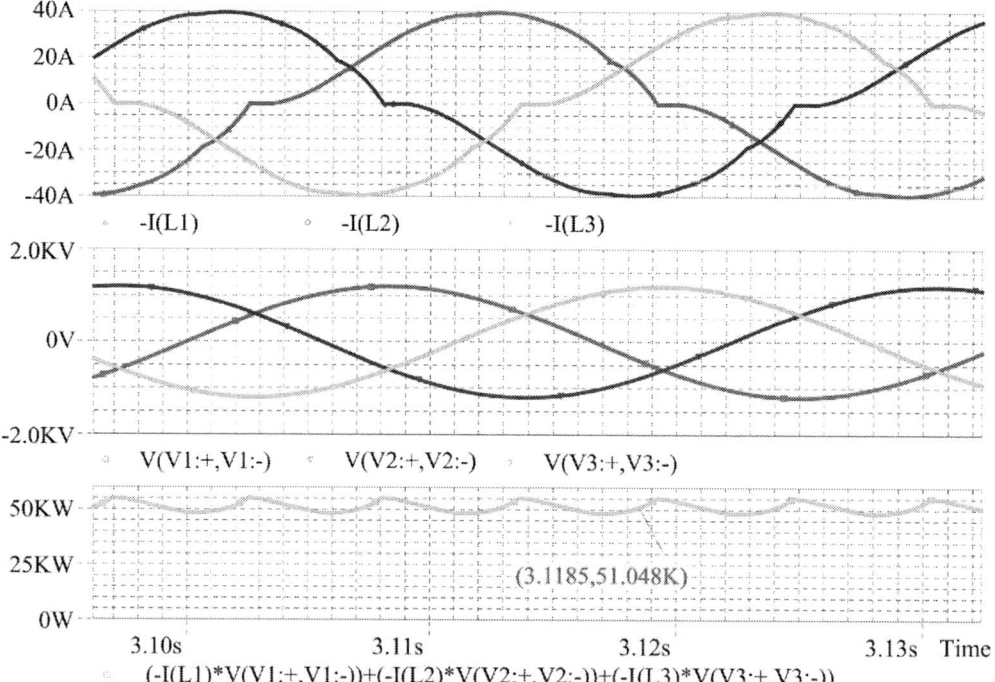

Figure 4. Waveforms of phase instantaneous current and voltage and three phase instantaneous power of the improved passive converter circuit

Data analysis of figure 5 and table 1 allow to conclude that P_{basic} at low and nominal frequency range there is sufficient power output to compete with fully controlled power electronic convertor. However at the mid work range of 25÷35Hz, there is a significant power drop, which on annual basis reduces energy production compared to the ideal power output. The proposed basic circuit is cost effective solution and by the lower number of included components and lack of control unit, improves the reliability.

Since efficiency of the wind turbine depends significantly from variable speed operation, an improvement on the passive converter circuit is considered. The second circuit includes an autotransformer and other inductors L10, L11 and L12 connected to its secondary winding. Values of the coupled inductors which form the autotransformer determine the turns ratio, which is close to 2,5:1.

Those added components affect positively the middle area of the frequency range of the red curve P_{autotr}. Estimated values of all reactive components differ from basic circuit and are as follows:

- Generator inductances L1, L2 and L3 – 36mH,

- Primary winding L4, L5 and L6 – 153mH,

- Secondary winding L7, L8 and L9 – 60mH,

- External inductor to the secondary winding L10, L11 and L12 – 115mH.

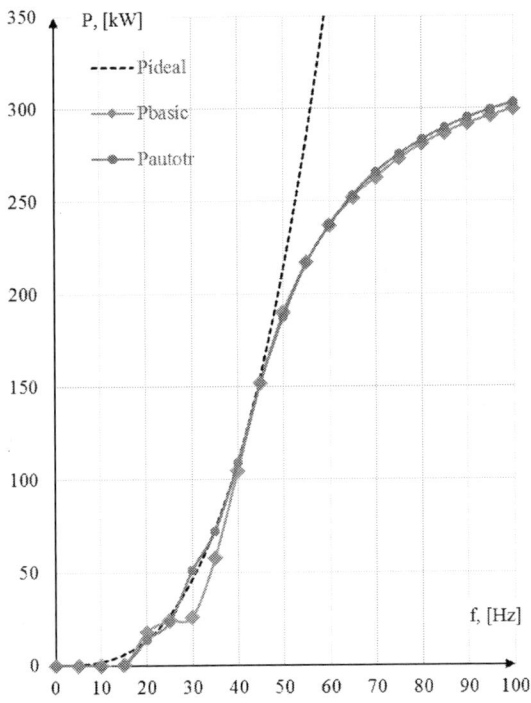

Figure 5. Passive converter circuits: basic – components are marked with blue color, improved – includes blue and red components

The graphical comparison shows clearly that the improved passive converter circuit follows the ideal power curve at all frequency range with insignificant deviations. Even if this passive converter circuit is more complex to design there is still the advantage of improved reliability.

Values of P_{ideal} for the comparison table and graphic are calculated with equation (4):

$$P_{ideal} = P_{nom}\left[\frac{f_a}{f_{nom}}\right]^3 \qquad (4)$$

, where P_{nom} - is the output power at nominal frequency; f_a – is the current frequency; f_{nom} - is the nominal frequency.

III. INDUCTOR DESIGN CONSIDERATIONS

While designing the inductors the following conditions are to be considered:
- the sum of VA rating of the inductors are only a fraction of the total transmitted power;
- the inductor current is a function of the speed and frequency.
- in low frequency operation the whole current goes through the inductors. At high frequency operation the voltage is limited and only a fraction of the current flows through the inductors. Even that current could be omitted by disconnecting the inductors in figure 3.

A comparison of inductor design could start with settling an EI core type as a reference case. Thus, a set of reference parameters can be defined [10].

A few possible core sets are considered:
- CIC type set that uses a cut ring core and I-legs;
- OI type of set core based on punched O rings and an I core with rounded tips;
- OIO set based on two rings and an I core. It has the advantage that the ring cores have not to be cut;
- 4*I core set, the 4*I skewed type set performs better, the 6*I set has a quite good use of the material;
- O cores seem very performing but there drawbacks related to difficulties of providing an air gap.

The main results of the comparison are summarized in Table II. It is better not to correct the power curve (figure 5) to the very low speeds, as then the turbine might stop due to friction torque and start more difficultly due to cogging.

TABLE II. OVERALL COMPARISON OF THE POSSIBLE SHAPES AND SETS BASED ON 4 CONSIDERED PARAMETERS, 5: EXCELLENT; 4: GOOD; 3 FAIR; 2: POOR; 1: BAD

Core set type	Heat transfer copper-ambient	Grain oriented steel	Copper filling factor	Manufacturing simplicity
EI	1	1	3	5
CIC	1	3	4	3
OI	1	1	4	3
OIO	3	3	4	5
4*I skewed	5	5	4	3
6*I skewed	5	5	5	2
O cores	4	4	1	1

The detailed considerations of the comparison between different cores based on a reference EI-inductor, presented in [10], are applicable in the discussed passive control. For instance, for the same filling factor and peak induction, only small differences are found in respect weight. The best solutions are with ring cores and hexagonal types. The CIC type inductor shows good performance parameters and uses grain oriented steel more efficient than in EI-cores. Disadvantage of CIC type inductor compared to EI-shapes is no worse heat transfer. The hexagonal core inductor combines a high peak induction, a high filling factor and a good heat transfer, thus total reduction of weight and losses could be obtained [10]. Further design aspects of power inductors are presented in [11] and [12].

IV. CONLUSION

A comparison of the power output between passive converter circuits is made and expressed in percent in table. An adjustment methodology for the included inductors is proposed. The data from the simulations is presented in graphically.

The following conclusions are derived based on the carried out comparative study:

- An improved design of passive converter circuit competes with active controlled PECs and it can be preferred in cases when the quantity of generated energy is the dominant factor;

- All significant advantages of basic passive control circuit apply to the improved one – cost efficient solution, better reliability and comparable performance;

- In the following research it could be considered to switch the inductors off at full load. It is probably a third knee in the power-speed curve. This could be realized by thyristors used as on/off in the star point of the inductors;

- The proposed system is suitable for paralleling wind turbines in wind parks. Diode bridges are more reliable than transistors. If one turbine fails the other could continue to provide energy flow and to be controlled by the diode bridge;

- Cores based on grain oriented steel are preferably recommended design for large inductors. The hexagonal core inductor is further recommended because of good heat transfer provided and total reduction of weight and losses [10].

ACKNOWLEDGMENT

The paper is developed in the frames of the project "Model Based Design of Power Electronic Devices with Guaranteed parameters", ДН07/06/15.12.2016, Bulgarian National Scientific Fund.

REFERENCES

[1] M. Stiebler, "Wind energy systems for electric power generation", Springer, ISBN: 978-3-540-68762-7, 2008.

[2] D. Vizireanu, "Investigation on Brushless DC Appropriateness to Direct-Drive Generator Wind Turbine", ICREPQ'05, Saragossa, Spain, 2005, paper 293.

[3] M. B. C. Salles; J. R. Cardoso; K. Hameyer, "Dynamic modeling of transverse flux permanent magnet generator for wind turbines", Journal of Microwaves, Optoelectronics and Electromagnetic Applications, Vol. 10, No. 1, ISSN 2179-1074, June 2011.

[4] I. Boldea, "Variable speed generators", Taylor & Francis, ISBN 0-8493-5715-2, 2006.

[5] Patent application number - PCT/EP2010/055637, http://patentscope.wipo.int/.

[6] A. Van den Bossche, P. Yankov, V. Valchev, "Design of Passive Converter for Wind Driven Generators", EPE-PEMC'11, Birmingham, UK, 30 August - 1 September, 2011, pp. 1-10.

[7] P. Yankov, V. Valchev, Performance Comparison of Active and Passive Converters for Wind Driven Generators, ET2016, September 12 - 14, 2016, Sozopol, Bulgaria, pp. 234-237.

[8] M. Rashid: "SPICE for Power Electronics and Electric Power", Second Edition, Taylor & Francis, 2006, pp. 201-237.

[9] A. Van den Bossche, P. Yankov, A. Marinov, 'Automated methodology for adjustment of component values in passive converter circuit for wind turbine generators', EPE-PEMC, 4-6 September, 2012, Novi Sad, Serbia, DS2d.4-1÷DS2d.4-4.

[10] A. Van den Bossche, D. Van de Sype and Y. Gao, "Inductors combining ring cores and I cores", Proceedings of the 11th European Conference on Power Electronics and Applications, Dresden, Germany, 2005, ISBN: 90-75815-08-5.

[11] W.G. Hurley and W.H. Wölfle, "Transformers and Inductors for Power Electronics: Theory, Design and Applications", John Wiley & Sons, 2013.

[12] A. Van den Bossche and V. Valchev, "Inductors and transformers for power electronics", 2005, CRC, Boca Raton, USA.

Dynamic range optimization and noise reduction by low-sensitivity, fourth-order, band-pass filters using coupled general-purpose biquads

Edi Emanović* and Dražen Jurišić*

*University of Zagreb/Faculty of Electrical Engineering and Computing
Unska 3, HR-10 000 Zagreb, Croatia
edi.emanovic@gmail.com; drazen.jurisic@fer.hr

Abstract— In this paper we compare two realisations of fourth-order band-pass filter sections: first is realized as a coupled structure, having negative feedback around two Biquadratic sections in cascade (CO), second is a common design of two Biquadratic sections in cascade (CA). Dynamic range optimization by equating opamp output voltage levels using signal-flow graph is presented for a single second-order GP-Biquad and for a fourth-order coupled Biquads. The reduction of sensitivity and noise is already well-known for the case of CO BP filter, with identical Biquads, when compared to the CA design. In this paper, it is shown that equating opamp output voltage levels reduces output noise even more. In that way, dynamic range is improved substantially. Because we use a general-purpose (GP) Biquad using two integrators, we can use its simple tuning features and simple design; design equations are given. Sensitivity and output thermal noise are simulated using circuit analysis program Pspice.

Keywords: Cascade of Biquadratic sections, Biquartic section, Butterworth approximation, band-pass filters, general-purpose section, sensitivity, output thermal noise, dynamic range.

I. INTRODUCTION

The frequency response of a filter varies from the nominal due to tolerances of passive and active elements, aging, temperature, etc. To maintain the filter's characteristics inside given specifications, the main problem to be solved is to design filters with reduced sensitivity to changes (tolerances) of passive element values.

In this paper, we realize an example of fourth-order Butterworth band-pass (BP) filters as two structures: (i) two Biquads in cascade usually designated as 'Cascaded Biquads' (CA); and (ii) two cascaded Biquads inside negative feedback, designated as 'Coupled Biquads' (CO). In [1] there was used two-integrators general-purpose-1 (GP-1) Biquad, whereas in this paper we use general-purpose-2 (GP-2) Biquad, in accordance with the designation in [2].

In the GP-2 Biquad (also known as direct-form-II realization) we can take out the signal from LP, and BP outputs, and with one additional amplifier, from high-pass (HP), band-rejection (BR), as well as, all-pass (AP) outputs. To build low sensitivities, and low noise filters,

we use such two-integrators Biquads as building blocks for coupled filter because of their considerably lower sensitivities comparing to single-amplifier structures, at the expense of the increased power consumption. Besides, this Biquad is practical because of its common-mode operation, having positive inputs of all opams grounded. In that way, this circuit has a possibility of realization balanced-to-ground symmetrical filter circuits. It also possesses orthogonal tuning features and simple design.

It is long-time well-known that the coupled Biquads (CO) filter in the case of the BP transfer function has significantly reduced sensitivities, particularly within the pass band, in comparison to the commonly used cascade design (CA) [3][4]. They also possess reduced thermal output noise when compared to the cascade, as shown in [1].

In this paper, it is for the first time demonstrated how the *optimization of dynamic range by equating opamp output voltage levels reduces BP filter output noise when coupling is applied.* For the simple case of cascaded filter circuits, reduction of output noise due to equating opamp output voltage levels is not so effective.

Using PSpice program for circuit analysis we simulate frequency characteristic of the realized filter, and check the correctness of the design. Then we compare the sensitivity to the passive components of the two filters by Monte Carlo runs. We also compare thermal noise at theirs outputs.

The filters are presented in the form of voltage-mode signal processing using opamps, and the obtained results can be efficiently applied to the case of current-mode CCII filters, that can be readily realized in the IC form.

II. FOURTH-ORDER BP FILTER DESIGN

In this Section, we briefly summarize the most useful design equations for the realization of BP filters of both CA and CO type in the form of a cookbook. They can be efficiently used by design engineers.

A. Cascaded Biquads (CA)

To design fourth–order geometrically symmetrical BP filter we start from a second-order LP prototype filter with the transfer function given by

$$T_{LP}(S) = \frac{k_{LP} \cdot \omega_{LP}^2}{S^2 + (\omega_{LP}/q_{LP}) \cdot S + \omega_{LP}^2}, \quad (1)$$

where ω_{LP} and q_{LP} are parameters of the complex-conjugate pole pair, and the k_{LP} is pass-band gain. Using common LP–BP transformation

$$S = \frac{s^2 + \omega_0^2}{Bs}, \quad (2)$$

where ω_0 is the desired center frequency, and B is the desired pass-band width of the BP filter, we realize a fourth-order BP transfer function. Substituting (2) into (1) we obtain $T_{BP}(s) = N(s)/D(s)$ in the form

$$T_{BP}(s) = \frac{k_{LP} \cdot B^2 \omega_{LP}^2 \cdot s^2}{s^4 + B\omega_0^2 \frac{\omega_{LP}}{q_{LP}} s^3 + \left(2\omega_0^2 + B^2 \omega_{LP}^2\right)s^2 + B\frac{\omega_{LP}}{q_{LP}} s + \omega_0^4}. \quad (3)$$

Transfer function of the fourth-order BP filter realized as a cascade in Fig. 1(a) has the form

$$T_{CA}(s) = \prod_{i=1}^{2} T_{cai}(s), \quad (4)$$

where $T_{ca1}(s)$ and $T_{ca2}(s)$ are the second-order Biquadratic sections with BP voltage transfer functions given by

$$T_{cai}(s) = \frac{k_{cai} \cdot (\omega_{cai}/q_{cai}) \cdot s}{s^2 + (\omega_{cai}/q_{cai}) \cdot s + \omega_{cai}^2}; \quad i = 1, 2. \quad (5)$$

By equating (4) to (3) we arrive to the parameters of Biquads in (5) realizing a cascaded version of fourth-order BP transfer function given by (4), and the corresponding cascaded realization is shown in Fig. 1(a). Parameters $q_{ca} = q_{ca1} = q_{ca2}$, ω_{ca1} and ω_{ca2} of the cascade realization (Geffe equations) [5] are given by:

$$q_{ca} = \sqrt{\frac{q_{LP}\omega_0}{B\omega_{LP}}} \times \sqrt{q_{LP}X + \sqrt{q_{LP}^2 X^2 - 1}}; \quad (6)$$

$$X = 2\omega_0/(B\omega_{LP}) + B\omega_{LP}/(2\omega_0),$$

$$\frac{\omega_{ca2}}{\omega_0} = \frac{\omega_0}{\omega_{ca1}} = \frac{B\omega_{LP}}{2q_{LP}} q_{ca} + \sqrt{\left(\frac{B\omega_{LP}}{2q_{LP}} q_{ca}\right)^2 - 1}. \quad (7)$$

If we equate numerators in (4) and (3) we obtain

$$k_{ca1} \cdot k_{ca2} = k_{LP} \cdot \frac{q_{ca}^2}{\omega_0^2} \cdot B^2 \omega_{LP}^2. \quad (8)$$

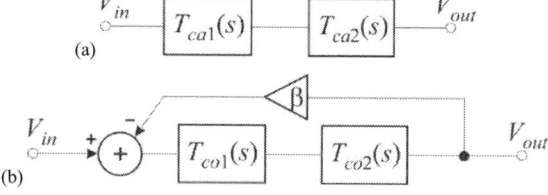

(a)

(b)

Figure 1. Fourth-order filter realiaztion. (a) Cascade. (b) Coupled.

B. Coupled Biquads (CO)

Coupled Biquads or 'Biquart, which realizes fourth-order BP transfer function will be realized by cascading two BP Biquads (of second-order) and applying a negative feedback, as shown in Fig. 1(b) (see [3][4]). Gain β represents the feedback coefficient ($\beta>0$), and $T_{co1}(s)$, $T_{co2}(s)$, are the second-order BP Biquadratic sections with voltage transfer functions given by

$$T_{coi}(s) = \frac{k_{coi} \cdot (\omega_{coi}/q_{coi}) \cdot s}{s^2 + (\omega_{coi}/q_{coi}) \cdot s + \omega_{coi}^2}; \quad i = 1, 2. \quad (9)$$

Voltage transfer function of the Biquartic section in Fig. 1(b) is given by

$$T_{CO}(s) = \frac{V_{out}(s)}{V_{in}(s)} = \frac{T_{co1}(s)T_{co2}(s)}{1 + \beta T_{co1}(s)T_{co2}(s)} = \frac{N_{BQ}(s)}{D_{BQ}(s)}. \quad (10)$$

Since the positions of the poles of $T_{CO}(s)$ is determined by the product $\beta k_{co1}k_{co2}$, i.e. not only by β, we can use the designation β' ($\beta' = \beta k_{co1}k_{co2}$) to simplify the expression. It is evident that for $\beta' = 0$ ($\beta = 0$, k_{co1}, $k_{co2} \neq 0$) the CO becomes the two Biquads in cascade (CA).

Consider the commonly used cascade realization and compare it with the Biquartic section (both of fourth order). If we equate the coefficients multiplying the potentions of complex variable "s" in (10) to (4), we arrive to the parameters of the two Biquads T_{co1} and T_{co2} defined by (9) inside coupling. We readily note that one of the unknowns can be freely chosen (therefore we have one degree of freedom). Thus, we choose both the sections T_{co1} and T_{co2} to be identical [6], and we have:

$$q_{co1} = q_{co2} = q_{co} = \frac{2\sqrt{\omega_{ca1}\omega_{ca2}}}{\omega_{ca1} + \omega_{ca2}} \cdot q_{ca}, \quad (11)$$

$$\omega_{co1} = \omega_{co2} = \omega_0 = \sqrt{\omega_{ca1}\omega_{ca2}}. \quad (12)$$

Feedback factor β takes the value defined by

$$\beta = \frac{q_{ca}^2}{k_{ca1}k_{ca2}} \frac{(\omega_{ca2} - \omega_{ca1})^2}{\omega_0^2}\left(1 - \frac{1}{4q_{ca}^2}\right), \quad (13)$$

and the pass-band gains are defined by

$$k_{co1}k_{co2} = \frac{q_{co}}{q_{ca}} k_{ca1}k_{ca2}. \quad (14)$$

III. EXAMPLE

In this Section, we realize one simple example of the active-RC BP filter: Butterworth with central frequency $f_0 = 1$kHz and bandwidth $B = 500$Hz. (or normalized $\omega_0 = 1$ and $B = 0.5$). Corresponding design parameters that follow from above two sections for the cases of CA and CO filters are given in Table I. They are readily calculated using expressions in Section II. The basic building block is GP-2 Biquad and is presented in Fig. 2.

142

TABLE I. NORMALIZED PARAMETERS OF THE FOUTH-ORDER BP BUTTERWORTH EXAMPLE WITH $B=0.5$.

Param.	CA	CO
ω_{p1}	1.19550	1
ω_{p2}	0.83647	1
q_{p1}	2.87364	2.82843
q_{p2}	2.87364	2.82843
k_1^{*}	1	1
k_2	2.06445	2
β	0	0.5

*We choose $k_1=1$ to obtain maximum dynamic range.

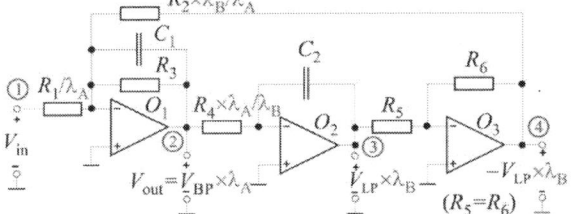

Figure 2. General purpose (GP-2) active-RC Biquadratic section.

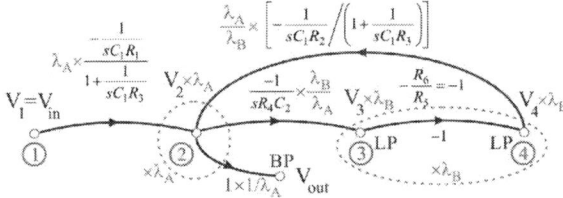

Figure 3. Optimization of GP-2 section in Fig. 2 using sfg.

The transfer function of the Biquad shown in Fig. 2 has the form given by (9) or (5), with parameters given by

$$\omega_p = \sqrt{\frac{R_6}{C_1 C_2 R_2 R_4 R_5}}; \quad q_p = R_3 \sqrt{\frac{C_1 R_6}{C_2 R_2 R_4 R_5}}; \quad (15)$$
$$k_{BP} = R_3 / R_1; \quad k_{LP} = R_2 / R_1.$$

Using (15) and parameters in Table I we calculate normalized elements of our filter example. The step-by-step design procedure for the filter in Fig. 2 is given by:

(i) Choose $C_1 = C_2 = C = 1$ and $R_5 = R_6 = 1$.

(ii) Choose $R_2 = R_4 = R$. Then the design equations (15) become much simpler

$$\omega_p = 1/R; \quad q_p = R_3/R; \quad k_{BP} = R_3/R_1. \quad (16)$$

(iii) From (16) we obtain

$$R = 1/\omega_p; \quad R_3 = R \cdot q_p; \quad R_1 = R_3/k_{BP}. \quad (17)$$

(iv) We denormalize elements to the desired central frequency $\omega_0 = 2\pi f_0$, ($f_0 = 1$kHz) and to the denormalization resistance $R_d = 15.91$kΩ (resistance level) to obtain denormalization capacitance $C_d = 10$nF. Final denormalized elements are obtained by multiplication of the normalized values with R_d and C_d. Note that in this realization we use a separate operational amplifier for negative feedback and that $R_\beta = R_0/\beta$. The active realization of the coupled Biquads is shown in Fig. 5 and the element values are in Table II. Theirs amplitude-frequency characteristics are shown in Figs. 6 and 7.

A. *Equating maximum opamp output voltage levels and its influence on the output noise reduction and dynamic range optimization [7]*

Note that the frequency responses at "point 1" and at "point 2", within the circuit of Fig. 5 (CO), are approx. 1.1185 times higher and 0.637 times lower than that at the output node [see Fig. 7(a)], respectively. A 1V output signal level will therefore cause serious overdrive and distortion within the circuit (e.g. at nodes "point 1" and "point 2"), and consequently at the output. One criterion that is useful to guarantee maximum dynamic range is therefore to specify that *the maximum signal level at any node within the circuit should at no time exceed the signal level at the input or output.* Thus, for a maximum zero-dB gain the signal within the circuit should nowhere exceed 1V (or zero dB). For our circuit in Fig. 5, this means that the signal level at the output of amplifier O_7 and of O_1 must be reduced by a factor $\lambda_1 = 1/1.1185 = 0.8941$ and increased by a factor $\lambda_2 = 1/0.637 = 1.5698$, respectively (see frequency response in Fig. 7(b)). Magnitude levels at the outputs of opamps O_2, O_3 and O_5, O_6 are also optimized to have maximums of 1V, by factors $\lambda_3 = 0.8080$ and $\lambda_4 = 0.90662$, respectively.

A simple way of signal level scaling within a circuit is to consider the equivalent signal-flow graph (sfg). The sfg for the circuit of Fig. 2, is shown in Fig. 3. The so-called critical voltage nodes are at the outputs of each voltage source and are indicated in Figs. 2 and 3 by numbers.

Now, to change (i.e., increase or decrease by an amount λ) the voltage level at any node in an sfg, we must *multiply* every *incoming* branch by the factor λ and *divide* every *outgoing* branch by the same amount. Obviously, λ will be around unity. Thus, if the signal level at node 2 is to be increased by a factor λ_A, all incoming signal paths to node 2 have been multiplied by λ_A, while all outgoing paths from node 2 have been divided by λ_A. Increasing the signal level at node 3 (and 4) by another factor λ_B, we have the additional modification to the sfg as also shown in Fig. 3. To modify the path by some factor λ we must multiply or divide one multiplicative component in the corresponding transmission quantity. We do this by multiplying R_1 and dividing R_4 by λ_A, and multiplying R_2 by λ_B/λ_A, and R_4 by λ_A/λ_B. The factors λ_A, and λ_B are designated in Figs. 2 and 3.

In Fig. 4 it is shown sfg of two coupled Biquads the realization of them is in Fig. 5. To scale the signal levels to be of equal maximums in sfg in Fig. 4, we perform the same optimization technique. The corresponding voltages at the outputs are shown in Fig. 6(b) and 7(b). The optimized element values are given in Table II, too. Optimized elements are marked as bold. Parameters λ_1 and λ_2 are used to optimize Biquads gains inside coupling *and are the most influential.*

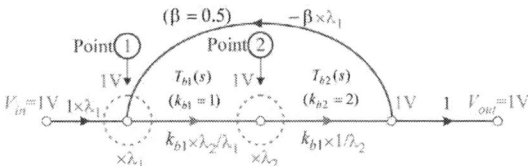

Figure 4. Optimization of CO section in Fig. 5 using sfg.

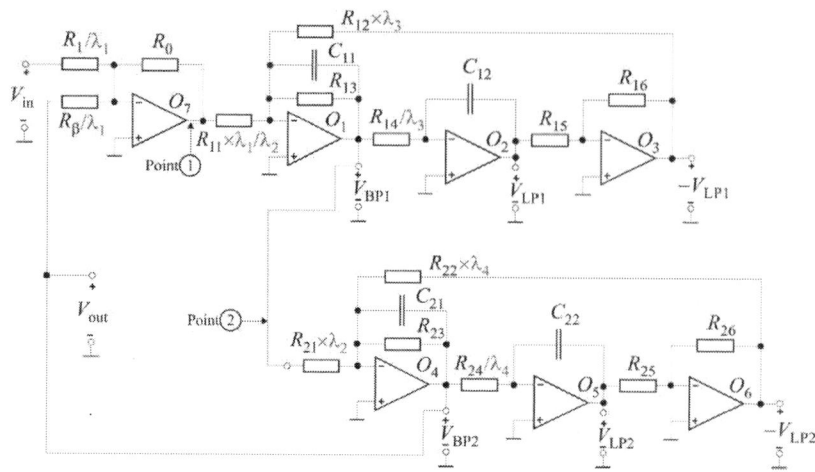

Figure 5. Realization of Biquartic section with two general purpose active-RC Biquadratic sections and feedback.

TABLE II. ELEMENT VALUES OF THE CIRCUIT IN FIG. 5 FOR THE FOUTH-ORDER BP BUTTERWORTH EXAMPLE DENORMALIZED TO 1kHz.

Elements	CA		CO	
	Non-optimized	Optimized	Non-optimized	Optimized
R_{11}	38.256kΩ	38.256kΩ	45.016kΩ	**25.637kΩ**
R_{12}	13.312Ω	**13.057Ω**	15.915kΩ	**12.860kΩ**
R_{13}	38.256kΩ	38.256kΩ	45.016kΩ	45.016kΩ
R_{14}	13.312Ω	**13.573Ω**	15.915kΩ	**19.697kΩ**
R_{15}	10kΩ	10kΩ	10kΩ	10kΩ
R_{16}	10kΩ	10kΩ	10kΩ	10kΩ
C_{11}	10nF	10nF	10nF	10nF
C_{12}	10nF	10nF	10nF	10nF
R_{21}	26.485kΩ	26.485kΩ	22.508kΩ	**35.334kΩ**
R_{22}	19.027kΩ	**20.625kΩ**	15.915kΩ	**14.429kΩ**
R_{23}	54.677kΩ	54.677kΩ	45.016kΩ	45.016kΩ
R_{24}	19.027kΩ	**17.552kΩ**	15.915kΩ	**17.555kΩ**
R_{25}	10kΩ	10kΩ	10kΩ	10kΩ
R_{26}	10kΩ	10kΩ	10kΩ	10kΩ
C_{21}	10nF	10nF	10nF	10nF
C_{22}	10nF	10nF	10nF	10nF
R_0	∞	∞	10kΩ	10kΩ
R_1	∞	∞	10kΩ	**11.185kΩ**
$R_β$	∞	∞	20kΩ	**22.37kΩ**

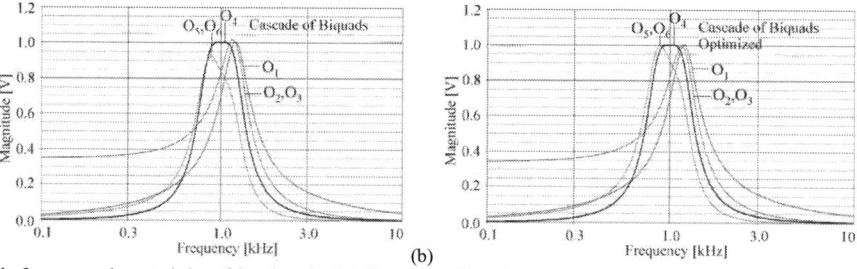

Figure 6. Amplitude-frequency characteristics of fourth-order BP filter CA in Fig 5 (opamp O_7 does not exist). (a) Non-optimized. (b) Optimized.

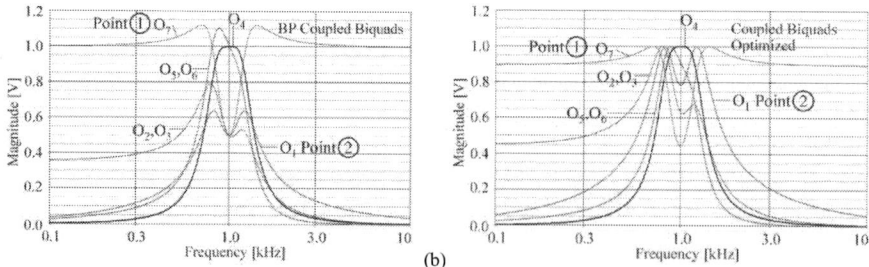

Figure 7. Amplitude-frequency characteristics of fourth-order BP filter with coupling (CO) in Fig 5. (a) Non-optimized. (b) Optimized.

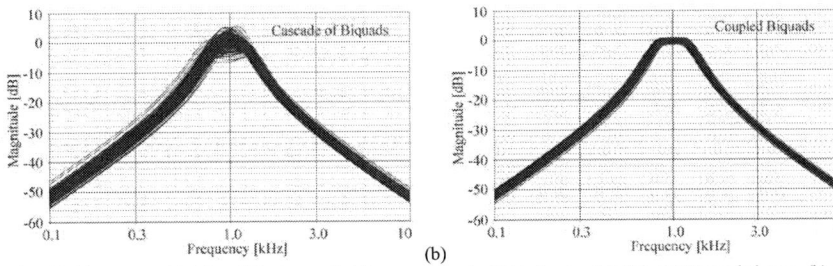

Figure 8. Simulated MC runs sensitivities of Butterworth filter examples in Table I using Matlab. (a) Cascaded case. (b) Coupled case.

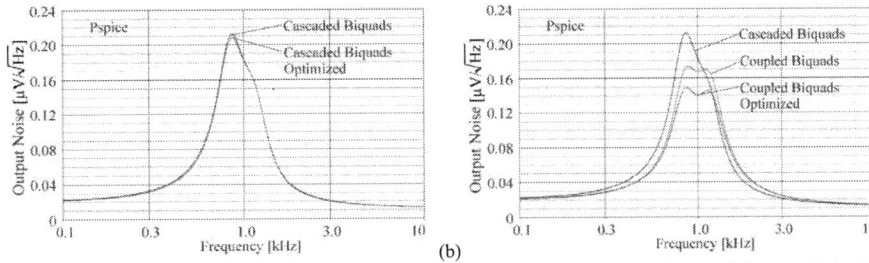

Figure 9. Output thermal noise voltage spectral density. (a) Cascaded non-optimized and optimized. (b) Coupled non-optimized and optimized.

Dynamic range is defined by [8][9]

$$D_R = 20 \log \frac{(V_{out\,rms})_{max}}{(E_{no})_{rms}} [dB], \quad (18)$$

where $(V_{out\,rms})_{max}$ in [V] in the numerator represents the maximum undistorted rms voltage at the output. The denominator $(E_{no})_{rms}$ in [μV] is the noise floor defined by the noise power at the output, obtained by integration of the noise spectral density as in Fig. 9 from zero to infinity. The lower the curve of the output noise spectral density the smaller the rms output noise voltage. On the other hand $(V_{out\,rms})_{max}$ is determined by the opamp slew rate, power supply voltage and the corresponding THD factor of the filter. Keeping the THD factor fixed, for the optimized filter, having opamp outputs as in Fig. 7(b), the value of $(V_{out\,rms})_{max}$ is larger than that for the non-optimized filter with opamp outputs as in Fig. 7(a). For the first approximation, we say that $(V_{out\,rms})_{max}$ is close to the supply voltages of opamps. Using presented optimization, we increase D_R by increasing numerator and decreasing denominator in (18), at the same time.

IV. RESULTS OF SIMULATION

Comparing the sensitivities of two filters examples in Table II with Monte Carlo runs using PSpice, while assuming a zero-mean uniform distribution and 5% standard deviation for all components, we obtain the ensemble of responses shown in Fig. 8. The spread of the ensemble of responses for each filter is an indication of its sensitivity to component tolerances, the lower the spread, the better the filter, because possessing low sensitivity property. We can conclude that the sensitivity in the pass-band is higher for the cascade (Fig. 8a), than for the coupled case with identical sections (Fig. 8b) [6].

Using the PSpice program output thermal noise spectral density of filter examples in Table II was generated and shown in Fig. 9. In the simulation, a model of TL081/TI (Texas Instruments) FET input opamp, is used [8]. From Fig. 9 one can conclude that the lowest noise possesses 'optimized' CO filters. Namely, since the dynamic range is limited from the upper border with the maximum undistorted signal level, and from the lower border by noise floor, by optimization as in Section III.A, the dynamic range has improved *significantly,* when BP filter is designed as coupled Biquads.

V. CONLUSION

We conclude that low-sensitivity and -noise BP filters can be designed using CO filters with equal building blocks. It is shown that the optimization of the dynamic range plays very important role when using three-amplifier Biquads as building blocks. Optimization does not affect sensitivity. Namely, both sensitivity and noise are reduced using coupling, whereas *the output noise can be further reduced by optimization of the opamp signal levels and by that the dynamic range even more improved.*

ACKNOWLEDGEMENT

This work has been fully supported by Croatian Science Foundation under the project (IP-2016-16-1307) Fractional analog and mixed systems for signal processing.

REFERENCES

[1] N. Mijat and D. Jurisic, "Optimized Coupled Band-Pass Filters," in *Proc. of the 37th International Convention* MIPRO 2014, (Opatija, Croatia), May 26-30, 2014, pp. 156–161.

[2] G. S. Moschytz and P. Horn, *Active Filter Design Handbook*, John Wiley&Sons, Chichester, 1981.

[3] N. Mijat and G. S. Moschytz. Multiple-critical-pole coupled active filters. *Int. J. Circ. Theor. Appl.* 12(3), pp. 249–268, July 1984.

[4] S. Fotopoulos and T. Deliyannis. Active RC realization of high-order bandpass filter functions by cascading biquartic sections. *Int. J. Circ. Theor. Appl.* Vol. 12, no. 3, pp. 223–238, July 1984.

[5] P.R.Geffe. Designers guide to active BP filters. *EDN* 1974; 46–52.

[6] N. Mijat and G. S. Moschytz, "Sensitivity of narrowband biquartic BP active filter block," in *Proc. of IEEE Int. Sym. on Network Theory and Design*, Sarajevo, BiH, 1984, pp. 158–163.

[7] G. S. Moschytz. A comparison of continuous-time active RC filters for the analog front end. *Int. J. Circ. Theor. Appl.* Vol. 35, no. 5–6, pp. 575–595, Sept.-Dec. 2007.

[8] D. Jurisic. *Applications of MATLAB in Science and Engineering. Chapter 10: Low-Noise, Low-Sensitivity Active-RC Allpole Filters Using MATLAB Optimization*. Rijeka: InTech, 2011, pp. 197–224.

[9] P. Bowron, K. A. Mezher, and A. A. Muhieddine, "Complete dynamic range for second-order analogue active filters," in *Proc. IEEE Int. Symp. Circ. Syst.*, San Diego, CA, 1992, pp. 847–850.

A Circular Economy for Photovoltaic Waste - the Vision of the European Project CABRISS

W.Brenner*, Nadja Adamovic*

* TU Wien, Institute of Sensor and Actuator Systems, Vienna, Austria
Werner.brenner@tuwien.ac.at

Abstract - Growing Photovoltaic (PV) panel waste causes a new environmental challenge, but on the other hand opportunities to create value and new economic paths. The main vision of the European project CABRISS „Implementation of a circular economy based on recycling, reused and recovered indium, silicon and silver materials for photovoltaic and other applications" is to develop a circular economy mainly for the photovoltaic, but also for electronic and glass industry. CABRISS bundles the efforts of 16 European companies and research institutions. The project consists in the development of: (i) recycling technologies to recover In, Ag and Si for the sustainable PV technology and other applications; (ii) a solar cell processing roadmap, which will use Si waste for the high throughput, cost-effective manufacturing of hybrid Si based solar cells and will demonstrate the possibility for the re-usability and recyclability at the end of life of key PV materials. From the begin the CABRISS consortium decided to have all results in accordance with valid European standards to ease access to market.

I. INTRODUCTION

Photovoltaics, also called solar cells, are electronic devices that convert sunlight directly into electricity. A PV system consists of PV cells that are connected to form a PV module, and the additional components, including the inverter, controls, etc. A wide range of PV cell technologies is on the market today including wafer-based silicon and a variety of thin-film technologies. PV cell technologies are usually classified into three generations, depending on the basic material used and the level of commercial maturity.

TABLE I. GENERATIONS OF PV SYSTEMS

Generation	% of Market, % of Waste
First-generation PV systems	fully commercial – use the wafer-based crystalline silicon (c-Si) technology: 85% of the global PV market and 90% of PV waste
Second-generation PV systems	early market deployment, are based on thin-film PV technologies: 15% of global PV market and 10% of PV waste
Third-generation PV systems	include technologies which are still under demonstration or have not yet been widely commercialized (e.g. concentrating PV (CPV), organic PV cells, novel concepts under development).

II. THE EUROPEAN WASTE DIRECTIVE WEEE

The European legislation proactively pushes forward the development of a green solution that is expected to master the issue of Waste Electrical and Electronic Equipment WEEE including PV waste. WEEE comprises a wide pallet of materials and components that, if not properly treated, can cause major environmental problems. Additionally, the production of modern electronics uses rare and expensive materials (eg worldwide around 10% of total gold is used for electrical and electronic components). To improve the environmental handling of WEEE and to foster a circular economy the complete chain of collection, treatment, reuse and recycling of end-of-life electronics is essential for the world's transition to a sustainable energy future [1]. Since 13 August 2012, the recasted Waste Electrical and Electronic Equipment (WEEE) Directive 2012/19/EU provides a legislative framework for extended producer responsibility of PV modules at European scale. As from 14 February 2014, the collection, transport and treatment of photovoltaic panels are regulated in every single European Union (EU) country. The large majority of collected PV modules that have been collected come from on-site pick-up for large quantities.

Figure1. Container filled with end-of-life solar cells and broken modules [15]

The recasted WEEE directive states that when supplying a new product, distributors are responsible for ensuring that such waste can be returned to the distributor at least free of charge on a one-to-one basis under the condition that the equipment is of equivalent type and of same function as the supplied equipment. Thus PV modules distributors will have an increasing role in the PV collection network. Nowadays, the recycling of crystalline modules is still limited by economic barriers; the economically realistic separation and recovery of silicon remains a challenge.

TABLE II. RESPONSIBILITIES OF PV PRODUCERS ACCORDING TO WEEE [5]

	Responsibilities according to WEEE
1	Yearly report of the PV modules put on the national market: WEEE register.
2	Organizing and financing the take-back and waste management of the photovoltaic modules.
3	Ensuring a financial guarantee of the disposal operations when photovoltaic modules are considered as household WEEE.
4	Achieve mandatory collection and recycling targets.
5	Mark all products with a crossed-out wheelie-bin.
6	Informing treatment facilities of the product's composition, including the potential use of hazardous materials.
7	Informing end customers on how to dispose their end-of-life PV modules.

Being responsible of the PV waste take-back can be costly since solutions to decommission and recycle such waste are not technically mature now. Therefore, the manufacturer pay the decommissioning and the cost to run a waste disposal plant. This cost is one additional factor that reduces their margin and makes their businesses less profitable. On the long run, CABRISS aims at preparing the technologies for recycling even large quantities of PV waste thus allowing a reduction of the operating expenses (OPEX). The liability brought by the legislation cited above gives the basis for a valuable asset of the European PV manufacturers. PV end-of-life management also offers opportunities relating to each of the 'three Rs' of sustainable waste management:

TABLE III. THE 'THREE RS' OF SUSTAINABLE WASTE MANAGEMENT [1]

Category of "R"	Waste Management
Reduce	Research and development (R&D) together with technological advances and a maturing industry allow design and manufacturing of panels which require less raw material
Reuse	Rapid global PV growth is expected to give the basis for a vital secondary market for panel components and materials
Recycle	Current PV installations reach the final decommissioning stage, recycling and material recovery will be preferable to panel disposal

In a circular economy the value of products and materials is maintained for as long as possible. Waste and resource use are minimized, and when a product reaches the end of its life, it is used again to create further value. This can bring major economic benefits, contributing to innovation, growth and job creation [9]. Raw materials are crucial to Europe's economy. Securing reliable and unhindered access to certain raw materials is a growing concern. To identify the most prominent challenges, the European Commission issued a list of Critical Raw Materials (CRMs). CRMs combine a high economic importance to the EU with a high risk associated with their supply [10]. Indium, a prominent resource of thin-film PV is one example of those CRMs. In June 2016 the International Energy Agency Photovoltaic Power Systems Program (IEA-PVPS) [7] in collaboration with the International Renewable Energy Agency (IRENA) [8] published the report: "End-of-Life Management: Solar Photovoltaic Panels" [1]. This report can be seen as the first global projection for future PV panel waste volumes to 2050. It underlines that recycled PV represents an opportunity to create and pursue new economy. This study analyzes national approaches to PV waste management and on the other hand predicts volumes of decommissioned PV panels, opportunities for the „3 Rs" (Reduce, Reuse, Recycle), as well as costs and requirements for an expanded waste management infrastructure.

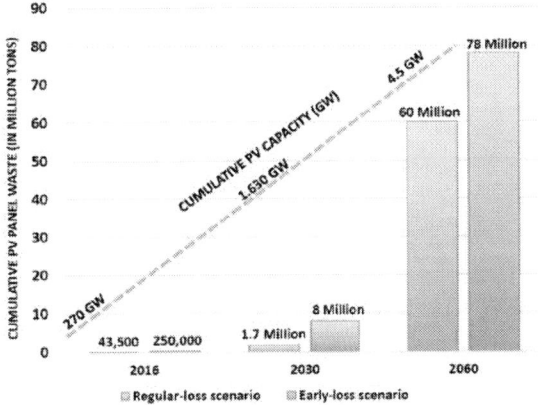

Figure2. Overview of global PV panel waste projections 2016-2050 [1]

Analyzing the documents IEA-PVS and IRENA issues regularly helps to predict and to estimate the future legal and technological challenges. CABRISS aims at pioneering a circular economy dedicated to handle the critical situation of recycling the significant volume of photovoltaic waste. By these means the project aims at benefitting also to electronics, metallurgy and glass industries.

CABRISS' five main objectives:

- Developing industrial symbiosis by providing raw materials such as glass or silver as feedstock for other industries (e.g. glass, electronics or metallurgy).

- Fostering collecting up to 90% of the PV waste throughout Europe compared to the 40% rate in 2013

- Retrieving up to 90% of the high value raw materials from the PV cells and panels: silicon, indium and silver

- Manufacturing PV cells and panels from the recycled raw materials achieving lower cost (25%

less) and at least the same performances (i.e. cells efficiency yield) as the conventional processes thanks to the implementation of a solar cell processing roadmap, which uses Si waste for the high throughput, cost-effective manufacturing of hybrid Si based solar cell.

- Involving the EU citizens and industry into such a sustainable and financially viable new economy. Industrial symbiosis needs proactive coordination between a variety of stakeholders, such as industry, research institutions, civil society organizations, public authorities and policy makers, and an increased awareness of producer responsibility for waste production. The project will rise the awareness of producers on the reduction of the material waste and the potential for recovery of Si, Ag, In. In addition, the project will involve the existing European PV clusters as multiplicators and platforms for exchange of experience made and best-practice.

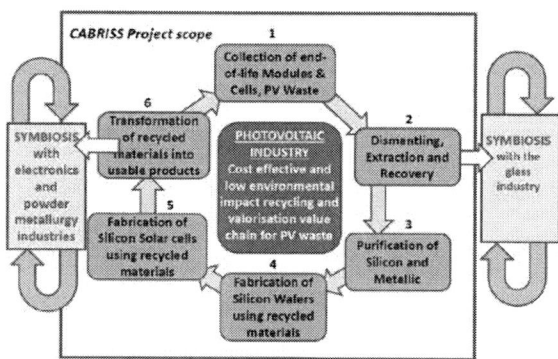

Figure3. CABRISS Project strategic concept [6]

The European Commission's action plan [2] stressed that the transition to a more circular economy requires action throughout a product's life-cycle: from production to the creation of markets for waste-derived raw materials. Waste management is one of the main areas where further improvements are needed and within reach: increasing waste prevention, reuse and recycling are key objectives both of the action plan and of the legislative package on waste.

The developed Si solar cells aim at having a low environmental impact by the implementation of low carbon footprint technologies resulting in low energy pay back (about 1 year). Energy Pay Back Time EPBT is used as an indicator for the environmental impact caused by PV power systems.

$$EPBT = E_{input}/E_{saved} \qquad (1)$$

E_{input} represents the total of energy input during the module lifetime including the energy requirement for manufacturing, installation, energy use during operation, and energy needed for decommissioning whereas E_{saved}

stands for the annual energy savings due to electricity generated by the module.

The originality of CABRISS relates to the cross-sectorial approach associating together different sectors like the powder metallurgy (fabrication of Si powder based low cost substrate), the PV industry (innovative PV cells based on secondary materials) and the industry of recycling (hydrometallurgy and pyrometallurgy) with a common aim: making use of recycled waste materials (Si, In and Ag). CABRISS focuses mainly on a photovoltaic production value chain, thus demonstrating the cross-sectorial industrial symbiosis with closed-loop processes.

III. CABRISS TECHNOLOGICAL APPROACH

The technology development in CABRISS is organized in five workpackages WPs [15].

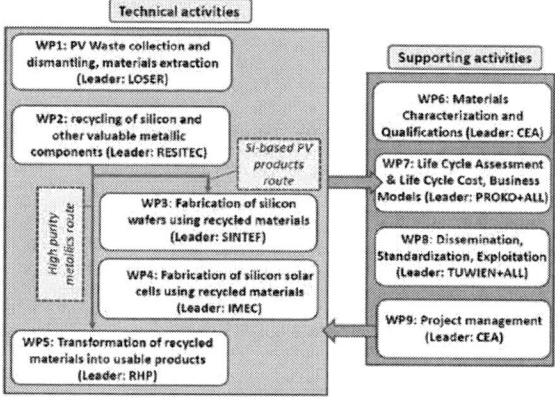

Figure4. CABRISS workpackages

WP1: PV Waste collection and dismantling, materials extraction – The overall objective of WP1 is to develop innovative cost-effective methods for the extraction of silicon, indium, and silver from the different sources of PV waste in order to constitute materials feedstock. From the begin the consortium decided to proactively cooperate with standardization institutions on national and European level.

WP2 Purification of silicon recovered in PV wastes – The main goal of WP2 is to develop methods for the purification of Si feedstock from Si recovered from solar wafers, cells resp. panels. At this stage of the project, the CABRISS partner SINTEF (Norway) is focusing on the growth of multi-Si and single-crystalline Si ingots. In general, Si scrap can consist of: broken wafers, broken solar cell structures or solar cells that are extracted from end-of-life solar modules. In all cases, all non-Si solar cell-base related layers (metallization, antireflection coating, emitter, back surface field) have to be etched away before used for the growth of ingots. An image of a single-crystalline Si ingot of about 20 kg is shown below. Properties of all materials obtained from the extraction and purification processes are ongoingly be analyzed by various analytical methods available within the CABRISS consortium.

148

WP3 Fabrication of silicon wafers using recycled materials – standard and cost-effective processing Si based substrates from the recycled Si feedstock will be developed.

Figure5. Cz Si ingot (~20 kg) grown from Si scraps (source: SINTEF)

WP4 Fabrication of silicon solar cells using recycled materials – Overall objective of this work package is to develop skills, know-how and processes to produce high efficiency cells from Si wafers obtained from a recycling process and by using recycled metals.

WP5 Transformation of recycled materials into usable products – A main objective of this workpackage is to develop processes for using of Ag and In based recycled materials. The demonstration of the recycled materials' quality for making new PV modules will be shown through indoor and outdoor measurements in comparison with standard PV modules.

An additional WP addresses a comprehensive assessment of environmental, sustainability and economic criteria which are essential for the long term economic success of the innovative processes, materials and technologies investigated in CABRISS project. Life Cycle Assessment (LCA), performed according to ISO 14040/14044, the internationally most recognized evaluation method to compare the environmental impacts of different production processes and technologies [11]. This comprises systematic compiling and examining of all inputs and outputs of materials and energy and the associated environmental impacts directly attributable to the function of a product "from cradle to grave" [12], [13]. Results are presented in environmental impact categories, e.g. Global Warming Potential (relating to emission of greenhouse gases), Eutrophication potential (related to the pollution of water by phosphates and nitrates), etc. Life Cycle Costs (LCC) are calculated on the basis of empirical cost micro-data [14]. The LCA and LCC results of the CABRISS project will be compared with results for conventional PV products documented in literature.

IV. WASTES FROM SILICON PV MANUFACTURING AND THEIR ECONOMIC VALUE

Customers more and more favour products that are esteemed to do the least damage of the environment. By this interest manufacturers are encouraged to mark their products to convince consumers that their products have less impact on the environment [4]. Easing market access through environmental labelling makes sense, particularly where companies already have invested in environmental

improvement. CABRISS focuses on a thorough market and competitive analysis, parts of it are presented below [6].

A number of wastes are generated during the PV cell production process.

TABLE IV. CATEGORIES OF SILICON WASTES

Category	Processing
off-cuts	Off-cuts created when cutting either the cast multi-crystalline silicon or the mono-crystalline ingot can be recycled within ingot manufacturing (closed-loop recycling at this point).
Slurry recycling	The wafering step leads to a 45% loss, and this volume is either in the form of dry powder (diamond saw), or in the case of steel wire cutting in the form of silicon powder mixed with silicon-carbide.
Other	„other" comes from cell and module manufacture, including final installation. Three main stages • If the wafer is broken early or before processing: re-melting and reintroduction into the process of ingot manufacture. • In the case that cell processing has begun, but before metallization: the cell can be processed through a series of chemical cleaning and lapping stages and again can be returned for ingot manufacture • after metallization: cell is much more difficult to recycle, but falls into the scope of CABRISS, in the sense that the silicon and the silver can in principal be recovered and used either inside or outside of PV applications.

One of the goals of CABRISS' partner RHP is to develop an industrial process which allows a fast transformation of recycled Silicon powders into ingots by pressure assisted sintering which allows densification of >95%. This requires temperatures which are close to the melting point of silicon. An additional advantage of the used powder metallurgical process is the possibility to tune the electrical conductivity by using doping of the silicon with elements such as boron. First ingots could be processed with a size of 156 mm x 156 mm.

Figure6. Square ingot sized 156 mm x 156 mm (source: RHP)

V. STANDARDS

The European Commission requested the European standardization organisations to develop European standards for the treatment, recovery, recycling and preparing for re-use, of WEEE including separation and removal of key components, such as frames, glass, polymers, metals and cables. Those standards have to reflect the state of the art laying down minimum quality standards [5]. Companies and research institutions can profit from participation in a Technical Committee (TC), which enables them to benefit from the ability to internalize the external information resources and thus to increase innovation skills [20]. Integrating research with standardization is characterized by several advantages for researchers, economy, and for society [19]. Standards can give the foundations for further developments, new research and ultimately new knowledge. The objectives of standardization [16] are to:

- meet the requirements of the global market efficiently;
- ensure world-wide use of PV standards;
- improve the quality of products and services;
- establish the conditions for the interoperability of omplex systems ;
- increase the efficiency of industrial processes;
- contribute to the improvement of human health and safety and the protection of the environment.

Standardization within CABRISS aims at providing a bridge that connects research to industry. This is of importance as investors in new technologies have to overcome the critical phase between demonstration and commercialization. To cross this well-known "valley of death", a good awareness and understanding of all pre-requirements for accessing the market is crucial. Two technology steps are addressed in CABRISS: (i) collection of end-of-life modules, cells and PV waste and (ii) dismantling, extraction and recovery

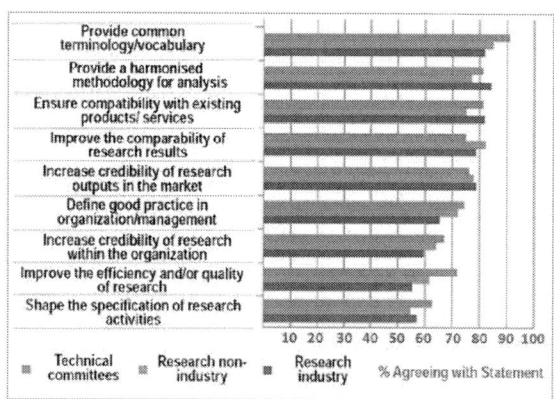

Figure 7.   Benefits of Research of Using Standards [18]

CABRISS has established contacts to SEMI [17] and CEN/CENELEC [18] for a closer interaction and knowledge exchange. SEMI is the global industry association serving the manufacturing supply chain for the micro- and nano-electronics industries, including: Semiconductors, Photovoltaics (PV), High-Brightness LED, Flat Panel Display (FPD), Micro-electromechanical systems (MEMS), Printed and flexible electronics, Related micro- and nano-electronics. Meanwhile,

CABRISS has deepened collaboration with CENELEC's CLC/TC 111X „Environment" which deals with collection, logistics and treatment requirements for WEEE- part 2-4: „Specific requirements for the treatment of photovoltaic panels" and for WEEE 3-5: „Technical specification for de-pollution – photovoltaic panels".

In the context of circular economy the SEMI „International Technology Roadmap for Photovoltaic (ITRPV)" [3] can be seen as one of those forward-looking stratedic documents which inform suppliers and customers about anticipated technology trends in the field of crystalline silicon (c-Si) photovoltaics and which help to stimulate discussion on required improvements and lacks of standardization. The objective of the roadmap is not to recommend detailed technical solutions for identified areas in need of improvement, but instead to emphasize to the PV community the need for improvement and to encourage the development of comprehensive solutions.

TABLE V.   CABRISS CONSORTIUM: 16 INSTITUTIONS FROM 9 COUNTRIES COMPRISING 6 SMEs, 5 INDUSTRIES AND 5 RTOs

1	COMMISSARIAT AL ENERGIE ATOMIQUE ET AUX ENERGIES ALTERNATIVES, France	CEA
2	STIFTELSEN SINTEF, Norway	STIFTELSEN SINTEF
3	INTERUNIVERSITAIR MICRO-ELECTRONICACENTRUM IMEC VZW , Belgium	IMEC
4	LOSER CHEMIE GMBH Germany	LOSER
5	FERROATLANTICA I & D SL , Spain	FAID
6	UAB SOLI TEK R&D, Lithuania	SOLI TEK
7	PYROGENESIS SA, Greece	PYROGENES IS
8	RHP TECHNOLOGY GMBH, Austria	RHP TECHNOLO GY GMBH
9	RESITEC AS, Norway	RESITEC
10	TECHNISCHE UNIVERSITAET WIEN, Austria	TU VIENNA
11	SUNPLUGGED - SOLARE ENERGIESYSTEME GMBH, Austria	SUNPLUGGE D
12	FRAUNHOFER GESELLSCHAFT ZUR FORDERUNG DER ANGEWANDTEN FORSCHUNG EV, Germany	THM FRAUNHOF ER
13	PROJEKTKOMPETENZ.EU - GESELLSCHAFT FUR PROJEKTENTWICKLUNG UND -MANAGEMENT MBH, Austria	PROKO
14	PV CYCLE FRANCE, France	PV CYCLE France
15	INKRON OY, Finland	INKRON
16	ECM GREENTECH, France	ECM GREENTECH

The consortium of CABRISS covers the whole value chain from the collection of end of life PV modules to the transformation into new modules following a dedicated process of extraction and recycling. All the partners of the CABRISS project were selected according to their complementary skills. PVCYCLE will be leading the collection of end-of-life PV modules and cells, and PV waste while LOSER and RESITEC will be in charge of dismantling and extracting materials from these collected modules.

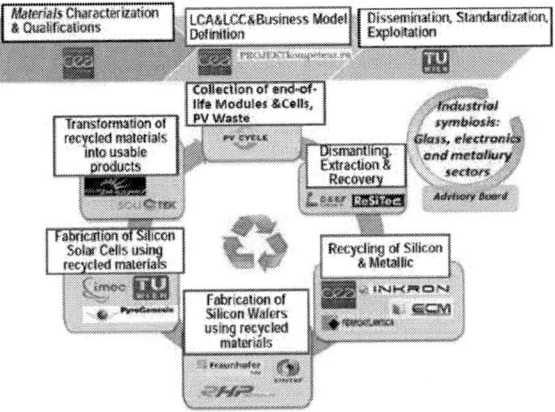

Figure 8 – CABRISS Value chain.

VI. CONCLUSION

Growing Photovoltaic (PV) panel waste causes a new environmental challenge, but on the other hand unprecedented opportunities to create value and to open new economic paths. The European project CABRISS aims at implementing a circular economy based on recycling, reused and recovered Indium, Silicon and Silver materials for photovoltaic and other applications. This approach is inline with the European legislation fostering the development of a green solution to master the issue of Waste Electrical and Electronic Equipment WEEE including PV waste. Collection, de-pollution and treatment technologies for end-of-life PV as prerequisite for recovery and recycling are in the focus of CABRISS. Proactive involvement of European standardization guarantees validity of results for the market.

ACKNOWLEDGMENT

CABRISS has received funding from the European Union's Horizon 2020 research and innovation program, under grant agreement No 641972.

REFERENCES

[1] IRENA and IEA-PVPS, „End-of-Life Management: Solar Photovoltaic Panels", International Renewable Energy Agency and International Energy Agency Photovoltaic Power Systems. IRENA and IEA-PVPS (2016), http://iea-pvps.org/index.php?id=95&eID=dam_frontend_push &docID=3222 .

[2] European Commission, „The role of waste-to-energy in the circular economy", communication from the Commission to the European Parliament, the Council, the European Economic and Social Committee and the Committee oft the Regions, Brussels, 26.1.2017, COM(2017) 34 final .

[3] SEMI, "International Technology Roadmap for Photovoltaic (ITRPV)", 2014 results (Revision 1, July 2015), www.itrpv.net .

[4] ISO Central Secretariat, „Environmental labels and declarations How ISO standards help", Geneve 2012, ISBN 978-92-67-10586-4.

[5] Directive 2012/19/EU of the European Parliament and of the Council of 4 July 2012 on waste electrical and electronic equipment (WEEE) Text with EEA relevance, Official Journal of the European Union L 197/38 , Document 32012L0019 .

[6] H2020 project CABRISS „Implementation of a circular economy based on recycling, reused and recovered indium, silicon and silver materials for photovoltaic and other applications", Website, of the https://www.spire2030.eu/CABRISS .

[7] International Energy Agency Photovoltaic Power Systems Program (IEA-PVPS) - www.iea-pvps.org .

[8] International Renewable Energy Agency (IRENA) - www.irena.org .

[9] European Commission, "Communication from the Commission to the European Parliament, the Council, the European Economic and Social Committee and the Committee of the Regions, Closing the loop", an EU action plan for the Circular Economy COM/2015/0614 final, https://ec.europa.eu/growth/industry/ sustainability/circular-economy_en .

[10] European Commission, "2014 Communication On the review of the list of critical raw materials for the EU and the implementation of the Raw Materials Initiative". https://ec.europa.eu/growth/sectors/raw-materials/specific-interest/critical_en .

[11] International Standards Organization ISO, „Environmental management – life cycle assessment – principles and framework", ISO 14040, together with ISO 14044: Geneva, 2006, Switzerland:, http://www.iso.org/iso/home.html .

[12] O. Andersen, J. Hille, G. Gilpin, Anders S.G. Andrae, "Life Cycle Assessment of Electronics", 2014 IEEE Conference on Technologies for Sustainability (SusTech), pp 22-29 DOI: 10.1109/SusTech.2014.7046212 .

[13] O. Andersen, "Consequential Life Cycle Environmental Impact Assessment," in: Unintended Consequences of Renewable Energy. Problems to be Solved, London: Springer, 2013, pp. 35-45. Lll .

[14] S. Fuller, "Life-Cycle Cost Analysis (LCCA), " National Institute of Standards and Technology (NIST), 2007 .

[15] CABRISS „Implementation of a circular economy based on recycling, reused and recovered indium, silicon and silver materials for photovoltaic and other applications", Grant Agreement number: 641972, unpublished .

[16] D. Blanquet; P. Boulanger; A. G. de Montgareuil; P. Jourde; P. Malbranche; F. Mattera, "Advances needed in standardization of PV components and systems", 3rd World Conference on Photovoltaic Energy Conversion, 2003. Proceedings of, Year: 2003, Volume: 2, pages: 1877 - 1881 Vol.2, IEEE Conference Publications .

[17] SEMI, http://www.semi.org .

[18] CEN (European Committee for Standardization)/CENELEC (European Committee for Electrotechnical Standardization), "Research Study on the Benefits of Linking Innovation and Standardization," Dec 2014, Ref: J2572/CENCEN/CENELEC, http://www.cencenelec.eu/research/news/publications/Publications /BRIDGIT-standinno-study.pdf .

[19] CEN (European Committee for Standardization)/CENELEC (European Committee for Electrotechnical Standardization), "Why Standards," http://www.cencenelec.eu/research/WhyStandards/ Pages/default.aspx .

[20] E. N. Filipovic, "How to support a standard on a multi-level playing field of standardization: propositions, strategies and contributions," Proceedings of the 2014 ITU Kaleidoscope Academic Conference: Living in a converged world - Impossible without standards?, 2014, pp: 207-214, DOI: 10.1109/Kaleidoscope.2014.6858464 referenced in: IEEE Conference Publications .

Smart Farm Computing Systems for Animal Welfare Monitoring

Marcel Caria, Jasmin Schudrowitz, Admela Jukan and Nicole Kemper*
Technische Universität Carolo-Wilhelmina zu Braunschweig
Stiftung Tierärztliche Hochschule Hannover*
Email: {m.caria, j.schudrowitz, a.jukan}@tu-braunschweig.de, nicole.kemper@tiho-hannover.de

Abstract—Smart sensing and computing have become important concepts in the last few years, creating opportunities in the new sector of smart agriculture. A few commercial smart agriculture systems have been introduced to this end, and albeit closed for experimentation, are paving the way for high-tech innovations for crop and livestock agriculture. In this paper, we focus on open and low-cost concepts for smart fog (edge) computing systems to create a smart farm animal welfare monitoring system. We develop an open source system that enables networking and computing of edge devices but also processing of data in a server – all being a connected system that we refer to as smart farm computing system. We propose to use Raspberry Pis as edge devices to monitor the animals and the farm environment, and we let the edge devices communicate with a local farm controller. The proposed farm computing system conceptually creates a fog computing layer and is further connected with cloud computing systems and a mobile application. We demonstrate that a low-cost and open computing and sensing system can effectively monitor multiple parameters related to animal welfare.

I. INTRODUCTION

Smart computing and sensing are becoming common terms for technologies used in our everyday's life to collect, analyse and share data over interconnected communication networks. One of the important areas that could highly benefit from using sensing technologies includes the broad area of human-animal interaction and co-habitation, and within that area of *smart farming*, often also referred to as Precision Livestock Farming (PLF). Currently, smart farming addresses basic needs of farmers, such as helping farmers to automate farming, for example, with feeding control systems [1]. In this sector, animal welfare is currently addressed through animal health care monitoring. At the same time, there is a wide consensus that biomedical and environmental parameters and factors can be easily measured with modern technologies indicating that animals are healthy, pain-free, and live in an environment which is species appropriate and allows them to be active and positively stimulated, broadening the scope from animal health to animal welfare.

As a consequence of the rapid smart livestock farming evolution, multiple specialized Internet- and cloud-enabled technology products are emerging on the market, albeit in form of commercial and proprietary products. The economic factors in agriculture severely affect not only the technology and the farmers, but also the entire society and citizens in form of food prices, – further contributing to the lack of open experimentation, and lack of emphasis on animal welfare factors in today's systems. If technology from other sectors is of any indication, we envision that animal welfare in livestock agriculture can prosper with help of low-cost and open smart systems and technologies on the principles of openness, programmability, interoperability, vendor-neutrality, and data sharing, which are all features that have greatly benefited the evolution of computing and sensing in general. As we are witnessing an ever increasing communication and computing capacity in all domains, we can expect the same evolution in the area of livestock agriculture, for which new thinking and system design is required.

In this paper, we focus on the design of open, programmable and low-cost smart farming systems for animal welfare by leveraging two most recent advances in computing: edge computing (also referred to as fog computing) and low-cost edge devices [4]. In fog computing, all required control, computing and networking capabilities are implemented closer to the edge devices, be it sensors and actuators or smartphones [3]. In our system, we use the popular low-cost Raspberry Pi single-board computers as stationary and mobile edge devices to monitor the animals as well as the stables, and we let the devices communicate with a local work station managed by the farmer. The paper presents an open-source architecture and a prototype of a farm animal welfare framework, ready to be installed in an experimental farm environment for cattle. Among multiple features, we demonstrate in more detail two basic parameters relevant to animal welfare, i.e., the stable temperature and animal movements, and discuss results obtained, as well as the lessons learned in building an open system.

The rest of the paper is organized as follows. Section II discusses the related work. Section III presents the system architecture, including the individual functions. The communication system is explained in Section IV. Section V concludes the paper.

II. RELATED WORK

Our main contribution here is in attempting the build and open-source system, with an open architecture that can enable experimentation and data sharing, further enabling an effective fog and cloud computing integration [4]. In addition, we take an interdisciplinary approach, by considering the real-life needs of one of the species suitable for this system. Even though our system is designed to be applicable to various animal species, our focus is on cattle. This is because current systems may not always comply with regulations and practices for various species, where for instance, due to the exploring behavior of pigs, sensor fixation is extremely difficult and accidental intake of sensors or other material should be avoided. By studying these salient features, our goal is to take a truly interdisciplinary approach in the design. We herewith omit any commercial products as related work, as they are typically closed for experimentation.

Animal farming is becoming an important research focus in engineering and computer science. A general overview about smart farming in livestock agriculture can be found in [7]. The paper divides the issue in three main aspects: robotic milking, automated feeding, and the subject of quality, whereas the latter refers to both, the quality of animal products and quality-of-life for the animals (i.e. animal welfare). In [1], the authors developed a smart farm system that supports the farmer in multiple daily work tasks. The system provides food and water for the animals and can additionally detect fire in the stable. Another issue was camera surveillance of the farm, one of the features we also implement in our system. An environmental monitoring system based on the Raspberry Pi is presented in [5]. The measured data in this approach is uploaded directly to the cloud. However, the work does not focus on smart farming as such, but rather on low-cost environmental monitoring systems in general. Given a myriad of similar projects based on the Raspberry Pi, we do not further review this body of work outside the area of livestock farming. In [6], a health monitoring system for cattle using a Raspberry Pi is presented. The rumination activity, body temperature, and humidity are measured to monitor the health status of the cattle. In contrast to our work, the the authors use the Raspberry Pi as a webserver and the sensors are implemented as (PIC) microcontrollers. The implementation of neither a cloud nor mobile application is foreseen in the project, which we consider as critical features for long-term data analysis.

While many papers refer to one or more parameters related to animal welfare, the papers focusing on animal welfare as such have been few and far between [2]. In [8], an ad-hoc wireless sensor network is used to measure the body temperature of several pigs with the help of implanted sensors. Additionally, the base station measures the temperature of the environment and collects the measured data. The paper discusses the relation between the body and the environmental temperature, and that it's possible to improve the pigs' health with monitoring and controlling the environmental temperature. Another work considering animal welfare aspects is [9]. Here, the authors use sensors to monitor the changes in behavior, like feeding, drinking, and social behavior. The paper furthermore describes the process towards automating monitoring and detection of behavioral changes. In this way, the animal welfare aspects of commercial, housed piggeries should be improved. The authors of [10] discuss the term animal welfare and define a more universal concept of it, without considering specific technologies.

III. SYSTEM ARCHITECTURE

In this section, we present the system architecture of our animal welfare smart farming system, which is depicted in Figure 1. We distinguish between two basic subsystems: the animal-centric subsystem (wearable or non-wearable) and the environmental subsystem. Both are an integral part of the overall system and send their measured data to a common farm controller, which is the central entity of the system (which could be implemented nonetheless in a distributed fashion). This architectural separation is in line with a broader consensus in the animal welfare research community, which identifies two main types of measures to asses animal welfare: the environment and the individual animals, see e.g. in [10]. The farm controller synchronizes the data with a cloud application. The cloud provides the connection with the farmer's mobile device. In our implementation, we implemented the environment and health monitoring subsystems on Raspberry Pis, which are credit-card sized microcomputers with a Linux operating system called Raspbian. The Raspberry Pi offers a less complex and more affordable solution for initial experiments with wireless monitoring. The advantages of the Raspberry Pi are particularly the low-end pricing and the low power consumption. One Raspberry Pi – denoted as *Environment R-Pi* – is used for environmental monitoring and control, which is placed in the stable. In addition, there are multiple other Raspberry Pis – denoted as *Wearable R-Pis* – for animal health monitoring, which are placed directly on the animals, e.g. using collars. The farm controller is implemented as a Java application, which is executed on the *farm work station*. At the current state of our experiments, an of-the-shelf office PC is sufficient for all work station purposes and is assumed to be located

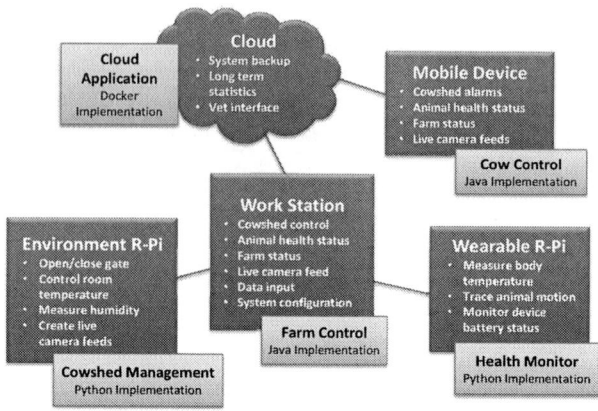

Fig. 1: The system architecture.

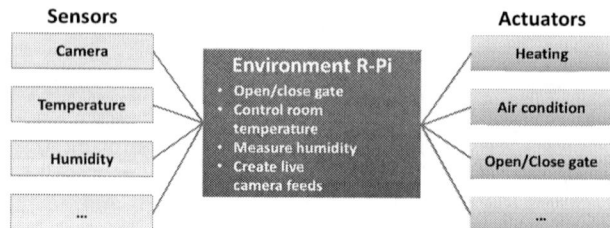

Fig. 2: Environment Raspberry Pi.

on the farm. The cloud implementation is currently run on an external server, but can likewise be operated using e.g. Amazon Web Services. Our mobile application is implemented as a light-weight Android app.

A. Environment R-Pi

The Environment R-Pi is used to monitor and control the environmental parameters. In our implementation, this is realized with a single Raspberry Pi. Please note that our proposed architecture supports multiple Environment R-Pis (which can be connected to the same farm controller), in order to support large farms that may require more than one environmental monitor. Our Environment R-Pi is extended with the so-called *SenseHAT* module, which is an add-on board containing different sensors, including temperature, humidity, accelerometer, gyroscope, magnetometer and barometric pressure. It additionally provides an 8x8 LED matrix and a five-button joystick[1]. We furthermore connected a camera module to the Raspberry Pi, namely the v2 Camera Module[2]. For the environmental monitoring so far, we deployed a temperature and a humidity sensor, and the camera enables live feeds from the stable. However, due to the widespread use of Raspberry Pis, there is a large number of compatible sensors available, which allows the extension of the Environment R-Pi for the measurement of all kinds of additional parameters. The program that sends all measured data to the farm controller is written in Python, and the Raspberry Pi is connected to the network via Ethernet. The measured data is exchanged using a TCP connection. In addition to the sensors and the networking, the Environment R-Pi is also able

to control the environment with actuators: heating, ventilation (or air condition), opening and closing of doors, windows, food dispensers, etc. An overview of the sensors and actuators of the Environment R-Pi is given in Figure 2. The dots indicate that the system can be extended. It should be noted that Environment R-Pi may not be reliable enough for production scenarios, as ours is only a proof of concept.

B. Wearable R-Pi

We measure health-related data with the wearable R-Pis, comparable to other animal health monitoring systems that are typically based on wearables. Our current implementation measures the animal's body temperature and movements using an accelerometer and a temperature sensor. An overview on the sensors and actuators of the Wearable R-Pi is given in Figure 3. Also here, the dots mean that the system is extendible with additional sensors. The LEDs could be used to indicate device- or animal-related information, such as low battery status.

We note that the connection to the work station requires some sort of wireless communication channel (e.g., WiFi or Bluetooth), and that the power supply relies on a battery to keep the Wearable R-Pi mobile. For the wireless communication, either a specific version of the Raspberry Pi is required (i.e. model B, version 3), or a compatible WiFi interface can be attached via USB. The power supply is an open challenge for a real-world deployment, as the battery either lasts only for relatively short time periods (e.g., the farmer would have to recharge the Wearable R-Pi every day) or it is too heavy to be worn on a collar. In terms of energy consumption, we plan to evaluate the suitability of other computing platforms, ranging from alternative low-cost system-on-chips (e.g. the popular Arduino boards) to FPGA boards with extremely low-power requirements (e.g. the Xilinx UltraScale+ SoCs).

Accelerometer: The used R-Pi accelerometer returns 3D acceleration values (including the acceleration of gravity), which are not useful to track permanently. We assume that the *duration* and *type* of body movement are significant indicators to assess

---

[1]https://www.raspberrypi.org/products/sense-hat/, Raspberry Pi Foundation, February 2017

[2]https://www.raspberrypi.org/products/camera-module/, Raspberry Pi Foundation, February 2017

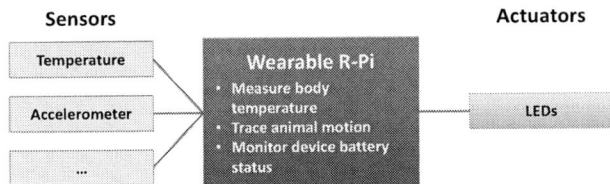

Fig. 3: Wearable Raspberry Pi.

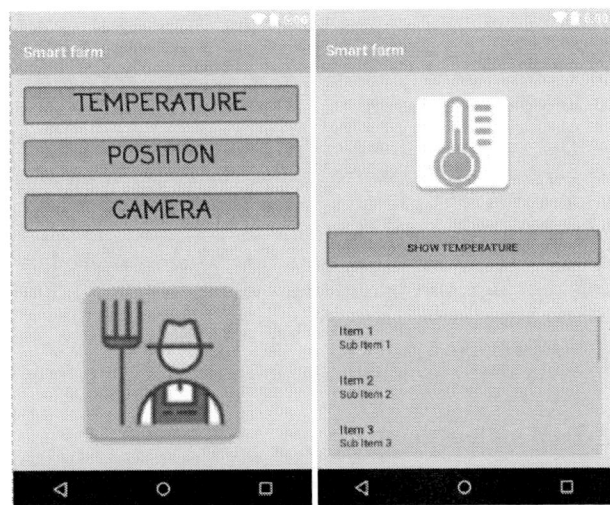

Fig. 4: Design of the mobile application

the animal's wellbeing. As the Wearable R-Pi may twist/shift relatively to the animal (depending on how it is attached), the absolute orientation of the device requires frequent calibration, so that the movement can be computed correctly. Therefore we designed a calibration method for the accelerometer, which allows to delete the acceleration of gravity from the measurements independently from the device's current orientation. The method is based on the length ℓ of the 3D acceleration vector $\vec{a} = (x, y, z)$, which is computed as

$$\ell = |\vec{a}| = \sqrt{x^2 + y^2 + z^2}$$

The value of ℓ is 1, when there is no acceleration except the acceleration of gravity and the sum of all acceleration values caused by the animal's body movements between two standstills is zero. Our method to calibrate the accelerometer therefore works as follows: We start a calibration cycle when the device is not moving ($\ell = 1$), and then measure and add up all acceleration values until the device is not moving again. The resulting mean acceleration from this calibration cycle is exactly the vector of the gravitational force, assuming that the device's orientation didn't change during this cycle. Assuming that the Wearable R-Pi will twist/shift only slowly, we take the estimated vector of the acceleration of gravity to calibrate the device's orientation for the following measurements, and simultaneously compute a new gravity vector. By subtracting the acceleration of gravity from the measured acceleration vector, we get the movement of the animal. Our method has shown to provide a sufficient accuracy for the purpose of measuring an animal's duration and type of body movement during extensively tests.

The Wearable R-Pi samples the animal's body movements with a rate of 30 Hz and transmits it to the farm work station. The further processing and classification of this data is performed in the Farm Controller (and explained explicitly in Section III-E) to avoid a processing overload on the Raspberry Pi's small CPU.

C. Farm Controller

The Farm Controller is the central part of the overall system architecture, which is also the phys-

ical location at which the smart farm application is operated. The system can run on an off-the-shelf office PC, and the smart farm application is implemented in Java. The farm controller is designed as a logical entity that represents the fog computing layer, which allows the farmer to access the system locally, for example for configuration purposes, to add/remove wearable devices from the system, to check the environmental parameters, or to manually control actuators (like open or close windows, doors and gates). For the whole system, the controller plays an important part, as it is in fact the connection layer between the cloud and the edge devices (i.e., the Raspberry Pis). The controller evaluates the measured sensor data locally and/or transmits it to the cloud for further processing, which is the salient feature of our proposed architecture.

D. Cloud and the mobile app

The current version of the cloud and the mobile app is to be regarded as work in progress. So far, we emulate the cloud simply with a web-based database, which saves all measured sensor data and enables the communication between the mobile application and the work station. We are currently implementing further cloud-related functions and features. The processed data includes room temperature, humidity, body temperature values, and body movement of individual animals. It is envisioned that these statistics can be used in a long-term analysis to compare the stable with others rooms, or to compare temperature or other values in different seasons or years. The camera feeds are not saved in the current implementation, as they are only intended to be used as live feeds. We see no significant benefit from saving such a large amount

155

of data additionally, event though we plan to start the live camera transmission in case of some health related or emergency triggers, which may require an off-line data evaluation.

The mobile application is based on the Android Studio framework (version 2.2.2) for Java. The main window allows the farmer to choose the parameters to check from the menu, like room temperature, humidity or a camera live feed. Then, a new window is opened, showing the requested information in more detail. The mobile application's design is illustrated in Figure 4. The left part of the Figure shows the main menu and the right part an individual info screen for the farmer.

E. Data Management

The information is currently stored in a MySQL-Database, which can be accessed via SQL queries by the work station, by the mobile application, and directly by the user, e.g., with a data base management tool. However, due to the modularity of our framework, the data base can be exchanged easily. We defined one table per type of sensor for simplicity. For practical purposes, we currently use the Ethernet MAC address of the Raspberrry Pis as data base identifiers. It is an important design decision which values to store and at which frequency to monitor and store the data. It is for example not useful to store the room temperature or the animal's body temperature once per second, as these values typically do not alter quickly. It is furthermore unnecessary to save the raw data of an animal's body acceleration.

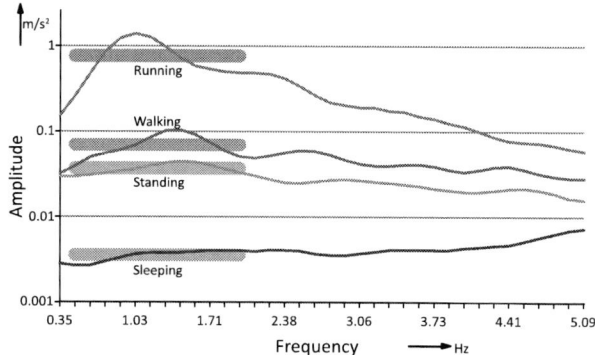

Fig. 5: Classification of motion patterns from the accelerometer.

We use the data measured by the accelerometer of the Wearable R-Pi to classify the *type* of animal movement. This data, if constantly monitored along with other measurable health information like body temperature, can be useful to evaluate the health status of farm animals and could indicate early symptoms

in the course of specific diseases or behavioral changes. We expect that especially the correlation with this type of data with the results of veterinarian examinations can pave the way for new methods in the area of farm animal health care and animal welfare. Figure 5 shows the results of one test run with our Wearable R-Pi, with the acceleration values transformed into the frequency domain. To this end, we used the calibrated acceleration values (transmitted by the Wearable R-Pi) in a Fast Fourier transform and filtered out statistical noise from the sensor and the animal's movements. We repeated the test a few times with several simulated acceleration values and the results show that the amplitude average of the region between 0.5 Hz and 2 Hz (depicted as the transparent bars of each of the four plots) provides a clear differentiation of different animal movement *classes* (e.g., a cow that either sleeps, stands, walks, or runs). Compared to the large amount of raw accelerometer data, the animal movement can be tracked storage-economically per class (e.g. x hours sleeping, y hours walking, etc.). Monitoring the animals' movements for all cows in the same stable can provide new insights to the animals' social behavior, which has been widely recognized as important for animal welfare [9]. Therefore, we decided to save the movement class of an animal in the database every time the class changes.

We expect that this and similar data, monitored on a daily basis along with other body- or environment-related parameters, may become a significant health and welfare status indicator for farm animals. It should be noted that the proposed system is the first step towards early detection of animal welfare issues, and is expected to lead to more sophisticated systems like automated actuation of environmental changes to improve the animals' welfare. For instance, a lack of social interaction, a lack of exploration drive, low illumination in a stable, or all the factors combined, may indicate that visual barriers and walls need to be moved to allow the animals to interact and have more space, while light illumination may need to increase or windows may have to be opened. While smart farms will further rely on human activities to some extend (e.g. for veterinary medicine), the automation of many procedures could lead to a much desired cost reduction in farming, while improving animal and human health at the same time.

IV. COMMUNICATION AND SYSTEM FUNCTIONALITY

The communication architecture is illustrated in Figure 6. The communication between the Environment R-Pi, the Wearable R-Pis, and the work station is currently realized as TCP connections, which has security issues and should be replaced with an encrypted communication layer (e.g., Transport Layer Security)

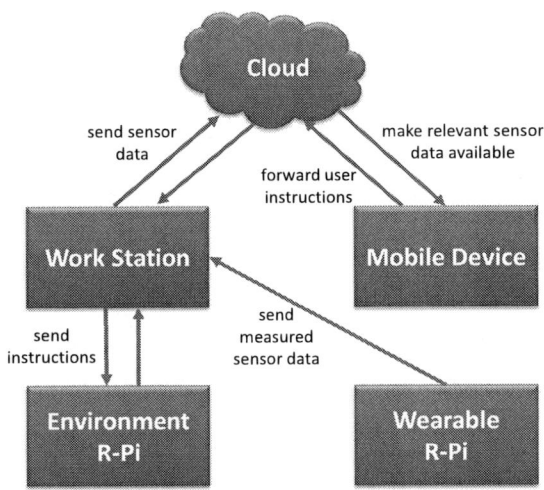

Fig. 6: Logical communication schemes.

before an actual deployment on a real farm. The Raspberry Pis measure several values via different sensors, most of them deliver *float*-values, like temperature and humidity. In order to simplify communication, we currently use a "plain text" communication protocol, i.e. all values are rounded and simply converted into Strings before transmission, and converted back to their respective data types. The work station evaluates the received values and generates response messages for the different clients. A Wearable R-Pi just expects acknowledgments, whereas the Environment R-Pi can receive instructions for its actuators. Another function of our system is the camera live feed for the work station and the mobile app, but this feature is not completely implemented yet.

An important feature of the system is the ability to send alarms from the farm controller to the mobile app. For example, an animal's body temperature can be indicating fever. The farm controller receives all measured values, evaluates and classifies them, and signals alarms in case parameters are outside normal/acceptable ranges. It then sends an alert to the cloud, which pushes it further to the mobile application, which makes an acoustic and optical alarm for the farmer. Such alerts could be used in different ways, as several works indicated. For instance, in [12], it was studied how a wireless system could easily detect, if a cow is pregnant, based on its body temperature. A similar approach is shown in [13]. Here the authors detect an estrus via body temperature measurement with the help of wireless monitoring. Hence, this kind of measurements carry potential to contribute to the multi-dimensional and holistic view of various measured parameters that can be used to assess animal welfare.

V. CONCLUSION

In this paper, we proposed open and low-cost concepts for fog (edge) computing systems to create a smart farm animal welfare monitoring system. The proposed farm computing system conceptually creates a fog computing layer with Raspberry Pis as edge devices, and is further connected with the cloud. We demonstrated that a low-cost and open computing and sensing system can effectively monitor multiple parameters related to animal welfare. While animal welfare remains a broad concept, this paper shows that many parameters relevant to various stakeholders can be measured, collected, evaluated and shared, opening up new possibilities to improve animal welfare and foster high-tech innovations in this sector.

REFERENCES

[1] M. H. Memon, W. Kumar, A. Memon, B. S. Chowdhry, M. Aamir and P. Kumar, *"Internet of Things (IoT) enabled smart animal farm"*, 2016 3rd International Conference on Computing for Sustainable Global Development (INDIACom), New Delhi, 2016, pp. 2067-2072.

[2] A. Jukan, X. Masip-Bruin and N. Amla, *"Smart Computing and Sensing Technologies for Animal Welfare: A Systematic Review"*, CoRR, http://arxiv.org/abs/1609.00627, 2016

[3] F. Bonomi, R. Milito, J. Zhu and S. Addepa, *"Fog Computing and Its Role in the Internet of Things,"* Proceedings of the First Edition of the MCC Workshop on Mobile Cloud Computing, Helsinki, Finland, August 2012

[4] X. Masip-Bruin, E. Marin-Tordera, G. Tashakor, A. Jukan and G. J. Ren, *"Foggy clouds and cloudy fogs: a real need for coordinated management of fog-to-cloud computing systems"*, IEEE Wireless Communications, October 2016

[5] M. Ibrahim, A. Elgamri, S. Babiker and A. Mohamed, *"Internet of things based smart environmental monitoring using the Raspberry-Pi computer"*, 2015 Fifth International Conference on Digital Information Processing and Communications (ICDIPC), October 2015

[6] L. Narayan, Dr. T. Muthumanickam and Dr. A. Nagappan, *"Animal Health Monitoring System using Raspberry Pi and Wireless Sensor"*, International Journal of Scientific Research and Education (IJSRE), Volume 3 Issue 5, May 2015

[7] A. Grogan, *"Smart Farming"*, Engineering Technology, July 2012

[8] I. McCauley, B. Matthews, L. Nugent, A. Mather and J. Simons, *"Wired Pigs: Ad-Hoc Wireless Sensor Networks in Studies of Animal Welfare"*, The Second IEEE Workshop on Embedded Networked Sensors, 2005. EmNetS-II., May 2005

[9] S. Matthews, A. Miller, J. Clapp, T. Ploetz and I. Kyriazakis, *"Early detection of health and welfare compromises through automated detection of behavioural changes in pigs"*, The Veterinary Journal. 21743?51, September 2016

[10] R. Botreau, M. Bonde, A. Butterworth, P. Perny, M. B. M. Bracke, J. Capdeville and I. Veissier, *"Aggregation of measures to produce an overall assessment of animal welfare. Part 2: analysis of constraints"*, Animal 1: 1188-1197, June 2007

[11] A. Patil, C. Pawar, N. Patil and R. Tambe, *"Smart health monitoring system for animals"*, 2015 International Conference on Green Computing and Internet of Things (ICGCIoT), October 2015

[12] A. H. H. Nograles and F. S. Caluyo, *"Wireless system for pregnancy detection in cows by monitoring temperature changes in body"*, 2013 IEEE 9th International Colloquium on Signal Processing and its Applications, March 2013

[13] L. M. Andersson, H. Okada, Y. Zhang, T. Itoh, R. Miura and K. Yoshioka, *"Wearable wireless sensor for estrus detection in cows by conductivity and temperature measurements"*, 2015 IEEE SENSORS, November 2015

MIPRO 2017, May 22- 26, 2017, Opatija, Croatia

Power Management Circuit for Energy Harvesting Applications with Zero-Power Charging Phase

D. Vinko

Faculty of Electrical Engineering, Computer Science and Information Technology Osijek,
Department of Communication, Osijek, Croatia
davor.vinko@etfos.hr

Abstract - Energy harvesting systems are becoming an exciting alternative for powering low-power electronics. For applications where real time response is not mandatory, energy harvesting systems can also be used with electronic devices whose power demand exceeds available power. Hence, it operates in two phases: charging and discharging phase. For switching between these two operating phases a power management circuit is used. This paper proposes power management circuit with zero power consumption during charging phase. The proposed design of a power management circuit is evaluated through measurements on a developed prototype. Compared to other power management circuit designs, switching time of the proposed design is up to 30 times faster, debounce-like behavior during switching is eliminated, it supports wider range of output currents and has a lower level-of-complexity.

I. INTRODUCTION

With limited power available from energy harvester, a well-balanced power management is essential to ensure the best possible performance. In applications where power demand exceeds power available from energy harvester, a charge/discharge scenario is used. Fig. 1. a) shows the block diagram of energy harvesting system which uses power management circuit to switch between charging and discharging operating phase. Energy from energy harvester is stored on a storage capacitor, and when sufficient energy is collected, the energy is transferred to the powered device. Power management circuit (PMC) is depicted as a switch on Fig. 1. a). Powered device operate in short active intervals (during discharge phase) followed by longer inactive intervals (charging phase), Fig. 1. b). The frequency of occurrence of active intervals is determined by the power ratio between energy harvester and powered device, but also by the features of the power management circuit. To achieve higher frequency of active intervals, the power consumption of the power management circuit must be as low as possible and the switching times between charging and discharging phase must be as short as possible. Power consumption of the PMC is only critical for charging phase. With very low power available from energy harvester, any additional power consumption from PMC, decreases power transfer rate from energy harvester to the storage capacitor, Fig. 1. c), which leads to increased duration of the charging phase.

The additional requirement on the power management circuit design is that is must operate without stable DC power supply. The only power supply available for the power management circuit is the voltage across storage

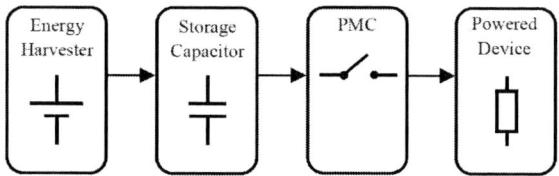

a)   Block diagram of energy harvesting system with power management circuit

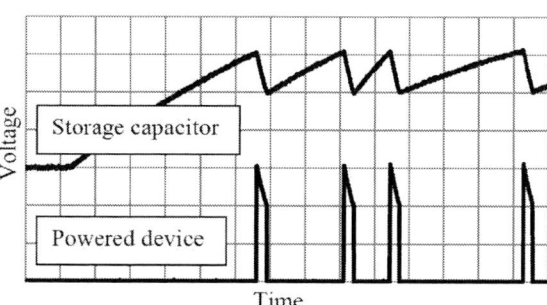

b)   Voltage across storage capacitor and powered device during charging and discharging phases

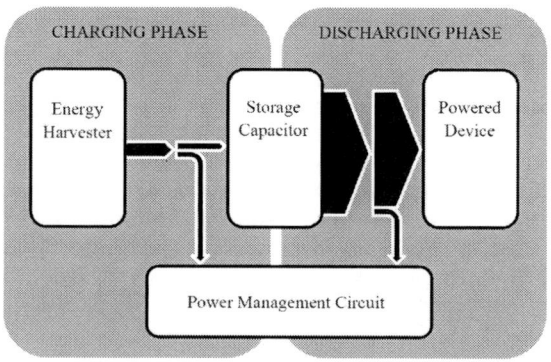

c)   Power disipation flow during charging and discharging phase

Figure 1.   Operation principle of charge/discharge scenario in an energy harvesting system with power management circuit (PMC)

158

capacitor, which changes with the amount of collected energy, Fig. 1. b). With assumption that the energy harvester will have time intervals with no output power, the minimal voltage across storage capacitor is 0 V. Maximal voltage depends on the operating voltage of the powered device, which is commonly between 3.3 V and 5.5 V. Therefore, the power management circuit has to ensure accurate switching between charging and discharging phase, with its power supply voltage changing from 0 V to over 5.5 V.

Fig. 1. b). shows voltage across storage capacitor and powered device for different durations of the charging phase. Energy harvesters are devices that extract small amount of usable energy for the environment (solar, RF, heat [1]-[4]), and as such, the amount of available power varies with the changes in the environment; difference in insolation for solar energy harvester, variations in wind, changes in vibration intensity, increase/decrease of RF radiation, temperature fluctuations, etc. Therefore, the duration of charging phase is greatly influenced by the amount of available power from energy harvester [5]-[7]. The duration of the charging phase decreases with increased energy harvester output power, and vice versa. Such stochastic behavior prevents power management circuit to use predefined time intervals for switching between charging and discharging phase. Fig. 1. b) assumes constant power consumption of the powered device, which can be seen as same duration of all discharging phases. In real life, the power consumption of the powered device can also vary, resulting in different durations of discharging phase as well.

In section II the operating mode of the proposed power management circuit is described. Section III gives the measurement results on the prototype of the proposed design and comparison with previous designs. At the end of the paper, the conclusion and references are given.

II. POWER MANAGEMENT CIRCUIT WITH ZERO-POWER CHARGING PHASE

Proposed power management circuit, Fig. 2, consists of two capacitive dividers (C_1/C_2 and C_3/C_4), latching circuit (P_1, P_2, N_1, N_2) and PMOS switch P_3. Capacitive dividers (C_1/C_2 and C_3/C_4) define threshold voltages for turning the powered device "on" and "off", respectively. Latching circuit uses two resistors (pull-up resistor R_{PU} and pull-down resistor R_{PD}) to hold P_1 and N_1 turned "off" (during charging phase).

Fig. 3 shows the voltage waveforms of the PMC: voltage across the storage capacitor V_C, gate-source voltage of N_2, P_2, N_1, P_1, P_3, and voltage across powered device V_R. The storage capacitor starts its first charging phase at T_0 with all MOSFETs turned off. As the voltage across storage capacitor V_C increases so do the gate-source voltages of N_2 and P_2 which are determined by capacitive dividers C_1/C_2 and C_3/C_4, respectively. At T_1, P_2 turns on (later on, during discharging phase, at this threshold the P_2 will turn off, which will turn off the entire latching circuit). The voltage across storage capacitor increases until channel resistance of N_2 decreases (with respect to pull up resistor R_{PU}) at T_2. This raises gate-source voltage of P_1, which starts conducting at T_3, pulling N_1 in conductions as well. At this point, the P_3 turns on the powered device and discharging phase begins.

The turn-on threshold voltage determined by capacitive divider C_1/C_2 has no affect during discharging phase, because P_1 and N_1 (together with resistors R_{PU} and R_{PD}) are sustaining each other in conduction. During discharging phase the current drain from storage capacitor is higher than the charging current from energy harvester and the voltage across storage capacitor (and both capacitive dividers) decreases until gate-source voltage of P_2 drops below its threshold voltage at T_4. After that, the channel resistance of P_2 increases, (with respect to resistor R_{PD}) lowering gate-source voltage of P_1 and N_1. At T_5, the P_1 turns off, what also turns N_1 off. Switch P_3 turns off the powered device and PMC goes back to the charging phase. During continuous operation, the beginning of each

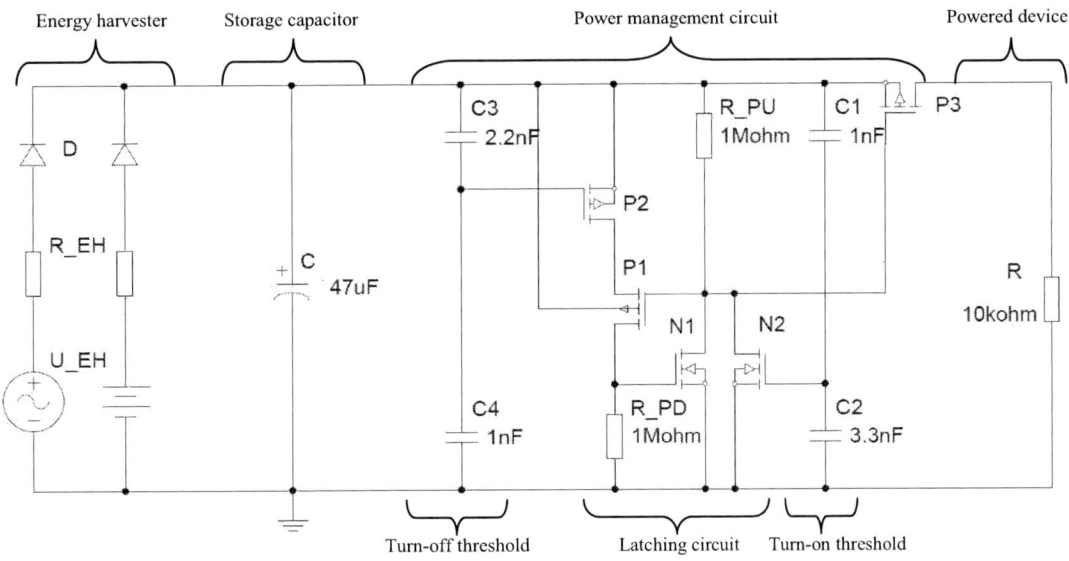

Figure 2.   Power management circuit with zero-power charging phase

159

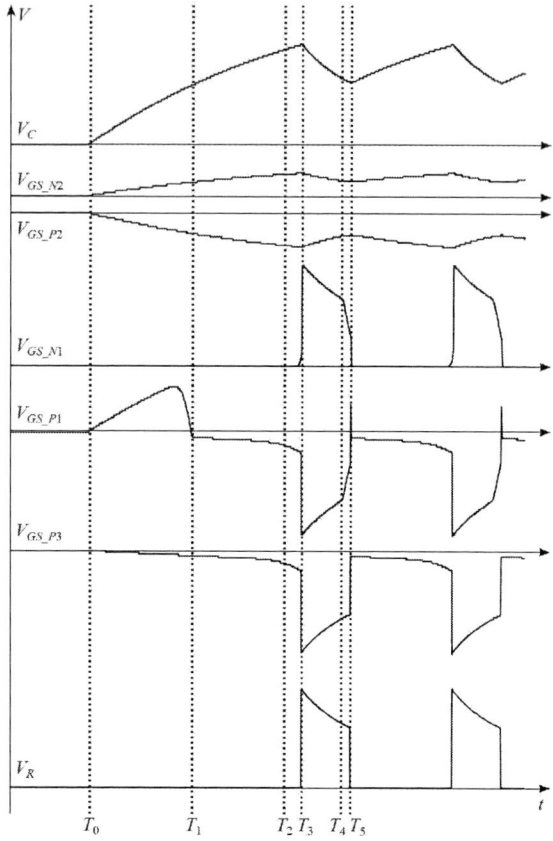

Figure 3. Voltage waveforms of the power management circuit

consecutive charging phase corresponds to T_1 in the first charging phase.

Such working principle of the power management circuit ensures that during charging phase only capacitors are connected to the storage capacitor. If subthreshold conduction of MOSFETs is neglected, this results with zero-power consumption during charging phase, since the charge stored in capacitive dividers contributes to current drawn by powered device during discharging phase.

III. MEASUREMENTS

A measurement setup consists of a discrete realization of PMC given by Fig. 2. For MOSFETs, two CMOS TC4007UBP ICs are used. Laboratory DC voltage source with series resistor is used to mimic the energy harvester. A main problem with two previous designs [8], [9] of the power management circuit is the switch "off" mechanism. The proposed power management circuit is compared to previous designs by comparing following two parameters: switch "off" response time and impact of output current

TABLE I.   PMC SWITCH "OFF" TIME COMPARISON

PMC version	Switch "off" time
2013 [8]	400 ms
2014 [9]	350 µs
2017 [this paper]	12.7 µs

on transient behavior of switch "off" mechanism. Switch "off" response time is measured as the fall time of the voltage across powered device. Additional analysis of the proposed circuit covers the impact of pull-up and pull-down resistors on switch response time, threshold voltage levels and acceptable input current.

Measured values for switch "off" response time are given in Table I. Evaluated power management circuit designs are named by the year of their publication. The proposed design has the fastest switching time, by almost two orders of magnitude. The impact of output current on transient behavior of the switch "off" mechanism is measured for 2014 and 2017 PMC design. Low output currents degrade the transient behavior. Therefore, the transients are measured for two values of resistive load (10 kΩ and 100 kΩ) of powered device (resistor R in Fig. 2.). Results are given in Fig. 4. Proposed design does not have debounce-like behavior during switch "off" which is prominent in previous design, especially at lower output currents.

Switch "off" time of the proposed design is dependent on pull-up and pull-down resistor values, R_{PU} and R_{PD} in Fig. 2, and it can be further reduced. Fig. 5. a) shows the impact of the R_{PU} and R_{PD} resistor values on rise and fall time for the voltage across powered device. Values of pull-up and pull-down resistors have significant impact on

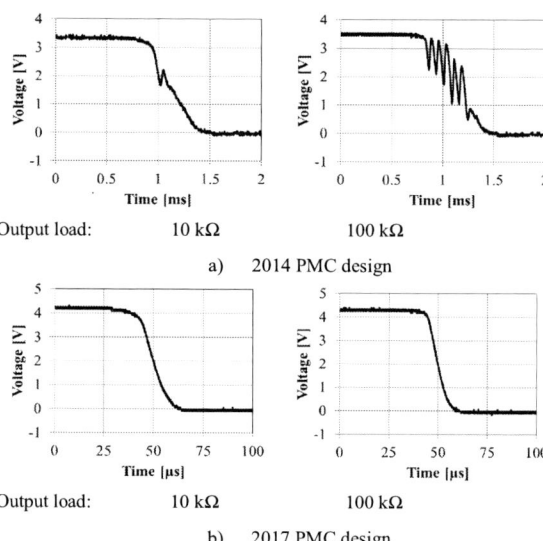

Figure 4. Transient behavior of switch "off" mechanism for a) 2014 and b) 2017 PMC design with 10 kΩ and 100 kΩ output load

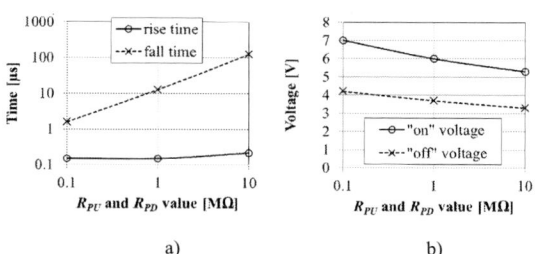

Figure 5. Impact of pull-up R_{PU} and pull-down R_{PD} resistor values on a) switching time and b) threshold voltages

160

the fall time (switch "off" time), ranging from 1 μs to 100 μs for R_{PU} and R_{PD} resistor values from 100 kΩ to 10 MΩ. The change in resistor values impacts the "on" and "off" threshold voltages as well, Fig. 5. b). Changes in threshold voltage levels can be adjusted by changing the ratios in capacitive dividers (C_1/C_2 and C_3/C_4).

There is a negative side effect in the reduction of the pull-up and pull-down resistor values. Fig. 6. shows voltage waveforms across storage capacitor (upper trace) and powered device (lower trace) for power management circuit with 100 kΩ pull-up and pull-down resistors. Waveforms are given for three values of input current (current from energy harvester that charges the storage capacitor during charging phase).

With decrease of the input current the power

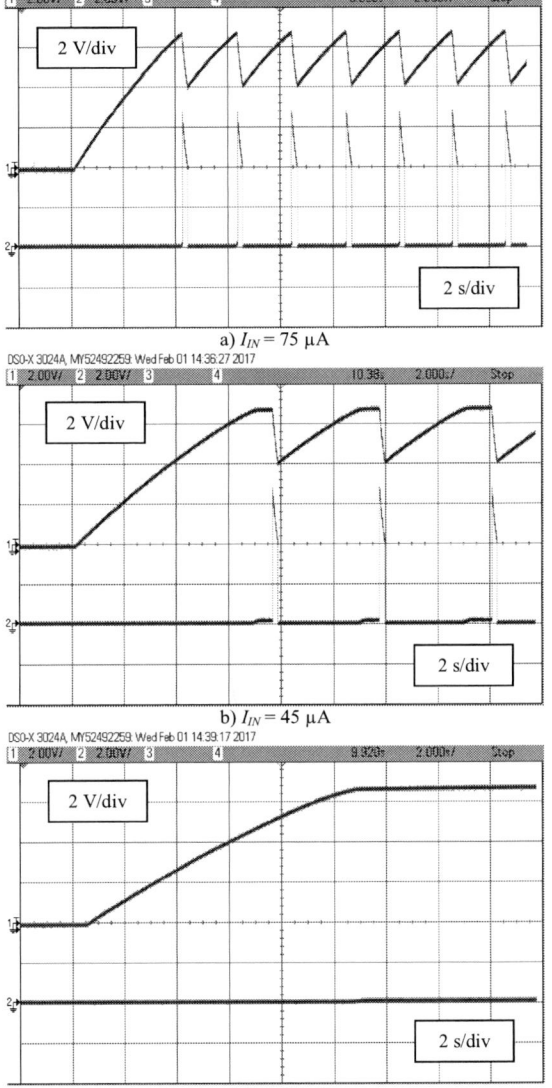

a) $I_{IN} = 75$ μA

b) $I_{IN} = 45$ μA

c) $I_{IN} = 30$ μA

Figure 6. Voltage across storage capacitor and powered device for a) 75 μA, b) 45 μA and c) 30 μA input current. $R_{PU} = R_{PD} = 100$ kΩ

management circuit enters an intermediate state between charging and discharging phase, Fig. 6. b). With further decrease of the input current, the intermediate state prolongs until the power management circuit reaches stalemate point, Fig. 6. c). For the pull-up and pull-down resistor values of 100 kΩ, 1 MΩ and 10 MΩ, the minimal input current values needed for device functionality are 33 μA, 3 μA and 200 nA, respectively.

IV. CONLUSION

Power management circuit with zero-power consumption during charging phase is presented. The performance of the proposed circuit is compared with previously reported power management circuits. Measurements show that proposed circuit it has faster switch "off" time by two orders of magnitude with improved transient behavior. The impact of pull-up and pull-down resistors in latching circuit on switching time, voltage threshold levels and minimal input current is evaluated. The value of pull-up and pull-down resistors represents a trade-off in proposed design. Lower values ensure faster switching but require higher input current for device functionality. Both switching time and minimal input current have fairly linear dependence on the pull-up and pull-down resistor value. To ensure N times faster switching the energy harvester must be able to provide N times larger current to the power management circuit. The future work will focus on removing this trade-off from the power management circuit design.

REFERENCES

[1] T.-H. Tsai, B.-Y. Shiu and B.-H. Song, "A Self-Sustaining Integrated CMOS Regulator for Solar and HF RFID Energy Harvesting Systems," IEEE Journal of Emerging and Selected Topics in Power Electronics, Vol. 2, No. 3, pp. 434-442, 2014.

[2] A. Harbd, "Energy harvesting: State-of-the-art," Journal of Renewable Energy, 36, pp. 2641-2654, 2011.

[3] X. Liu and E. Sanchez-Sinencio, "A Highly Efficient Ultralow Photovoltaic Power Harvesting System with MPPT for Internet of Things Smart Nodes," IEEE Transactions on Very Large Scale Integration (VLSI) Systems, Vol. 23, Issue 12, pp. 3065-3075, 2015.

[4] M. Ashraf and N. Masoumi, "A Thermal Energy Harvesting Power Supply With an Internal Startup Circuit for Pacemakers," IEEE Transactions on Very Large Scale Integration (VLSI) Systems, Vol. 24, Issue 1, pp. 26-37, 2016.

[5] A. S. M. Z. Kausar, A. W. Reza, M. U. Saleh and H. Ramiah, "Energizing wireless sensor networks by energy harvesting systems: Scopes, challenges and approaches," Renewable and Sustainable Energy Reviews, 38, pp. 973-989, 2014.

[6] M. Belleville et al., "Energy autonomous sensor systems: state and perspectives of a ubiquitous sensor technology," Proceedings of the third International Workshop on Advances in sensors and Interfaces IWASI, pp. 134–138, 2009.

[7] M. Belleville et al., "Energy autonomous systems: future trends in devices, technology, and systems," Report. CATRENE Working Group on Energy Autonomous Systems, Table 3, p. 20, 2008.

[8] G. Horvat, D. Vinko and T. Švedek, "LED Powered Identification Tag – Energy Harvesting and Power Consumption Optimization," Proceedings of 36th International Convention MIPRO, Opatija, 2013.

[9] D. Vinko and G. Horvat, "100 nA Power Management Circuit for Energy Harvesting Devices," Proceedings of 37th International Convention MIPRO, Opatija, 2014.

MIPRO 2017, May 22- 26, 2017, Opatija, Croatia

System for early condensation detection and prevention in residential buildings

Mateo Marcelić, Roman Malarić

University of Zagreb, Faculty of Electrical Engineering and Computing, Unska 3, 10000 Zagreb, Croatia
e-mail: mateo.marcelic@fer.hr

Abstract—In this paper a system is described that measures the indoor relative humidity and temperature of a room, outdoors relative humidity and temperature and the temperature of the interior wall in the coldest part of the wall. Moisture and temperature sensors will be used for measurements, and for measurement system and processing the Programmable Wireless Stamp (PWS) employed. The data will be viewed through an LCD display. Out of the measured data, the following information will be calculated: indoor and outdoor dew point temperature and their difference, indoor and outdoor humidity and their ratio and the difference between the wall temperature and indoor dew point. This data will also offer corresponding warnings, advice and instructions. The system will consist out of a central inner unit with a display which will measure the indoor humidity and room and wall temperature, and out of an outdoor weatherproof unit which will measure the outside humidity and temperature. The system will be used for early detection and prevention of condensation on walls and its consequences.

Index Terms—micro-controller, temperature, humidity, data acquisition, dew point

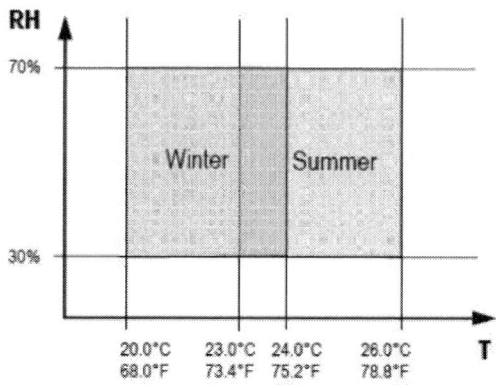

Fig. 1. Comfort diagram

I. INTRODUCTION

In residential buildings, indoor humidity represents a constant problem [1]. Even in buildings with quality isolation, moisture produced by occupants can lead to a condition of excessive indoor humidity. Since humidity represents how much water vapor there is in the air, high humidity means there is a high condensation probability. This condensation problem can appear in any season but winter is the most dangerous. Residents keep their rooms warm but badly ventilated, while outside is very dry and cold and these are perfect conditions for condensation to happen. Old buildings and buildings with bad isolation are highly affected by the problem of condensation and wall deterioration. The wall condensation leads to serious wall damage and moisture leads to development of bacteria hazardous for human health. Since relative humidity is the ratio of how much moisture the air is holding to how much moisture it could hold at a given temperature [2], by controlling humidity conditions in housing we can prevent indoor condensation. For example, in winter the absolute humidity level outdoors is low, since low temperature air can hold small quantities of moisture, so if its exact value is known, we can determine how long an indoor space needs to be ventilated in order to lower the dew point temperature to a non risk value. The same result can also be accomplished by raising the indoor temperature, but raising the temperature means that air can hold more moisture, which in turn means that when

condensation begins to happen, more moisture will condensate on the wall and create more damage. And also, the temperature in the room can only be raised to a certain point not to lower quality of life. Fig. 1 represents RH/T diagram showing the comfort zone according to ISO7730 [3]. Therefore in this paper a system will be described which is used for measuring relative humidity and temperature from which the dew point temperature will be calculated and a solution will be offered to prevent condensation using ventilation of indoor rooms.

II. DEW POINT TEMPERATURE

The dew point temperature is the temperature on which the air becomes saturated with water vapor. Further cooling the air down will condensate the vapor to form liquid dew [4]. On the surface that has a lower temperature than the dew point temperature, water vapor will begin to condensate. One way of determining the dew point temperature is by pyschrometric chart [5] represented with Fig. 2. The pyschtometric chart graphically represents relations between the relative humidity, temperature and dew point temperature and is used to directly acquire dew point temperature. When pyschtometric chart is not available dew point can be acquire from tables or by equations giving the approximate results compared to pyschtometric chart. This system uses equation (1) for determining dew point temperature.

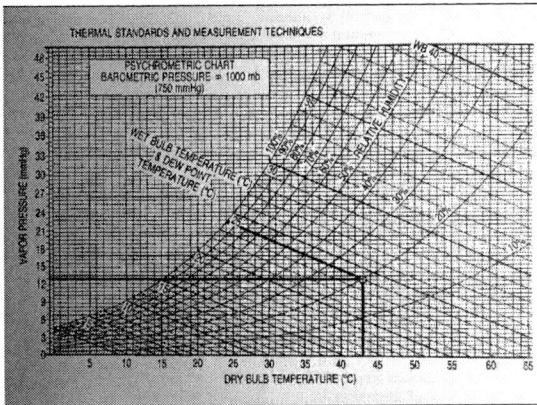

Fig. 2. Pyschrometric chart [4]

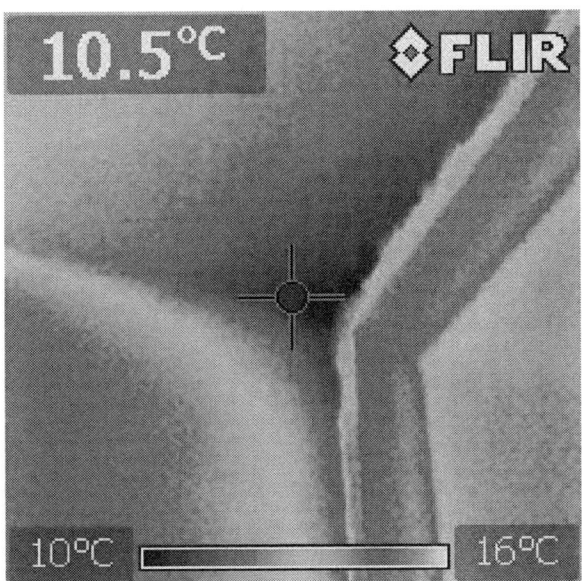

Fig. 3. Thermal picture of the wall

$$T_d = \frac{B_1 \left[\ln \left(\frac{RH}{100} \right) + \frac{A_1 T}{B_1 + T} \right]}{A_1 - \ln \left(\frac{RH}{100} \right) - \frac{A_1 T}{B_1 + T}} \qquad (1)$$

T_d represents calculated dew point temperature given the room temperature T and relative humidity in percentage RH. With constants A1 = 17.625 and B1 = 243.04 °C highly accurate results are obtained [2]. By measuring the relative humidity at a given room temperature we can determine the dew point temperature furthermore by comparing the dew point temperature to wall temperature we can determine if there is a risk of water condensation on indoor wall. If indoor wall temperature is higher than the dew point temperature in the room, condensation is avoided, on the contrary, if indoor wall temperature is lower than dew point temperature in the room, condensation on indoor wall is occurring. In addition, by measuring the outdoors temperature and relative humidity levels we can determine how long to ventilate the room to exchange indoor air to balance temperature and relative humidity levels therefore lowering the dew point temperature of the room. Also, the system can the provide early warning by continuously measuring outside and inner humidity and temperature, for example if the inner wall temperature is approaching the dew point temperature the action an be taken before the actual condensation occurs.

III. OVERVIEW OF THE MEASURING SYSTEM

The measuring system consists of indoor and outdoor sensors and an LCD display. Data communication is carried out through a wireless network established through a central processing unit with an outdoor unit connected to the network used for data transmission. Furthermore, users can connect to the same network so that warnings and data are directly viewed from remote locations, for example a mobile phone. For fulfilling these tasks Programmable Wireless Stamp (PWS) [6] is deployed. PWS is a micro-controller based on ESP-WROOM-02 [7] that integrates an ESP8266EX [8] chip, which is also a low-power micro-controller that can be powered form

a battery, making it suitable for outdoor use. This configuration is well-suited for sensor network, data acquisition, processing and publishing data.

A. Sensor placement and data acquisition

Indoor sensors include temperature and relative humidity sensors and a wall temperature sensor, all of which are connected directly to the central processing unit. The sensors are placed in an isolated area to reduce direct influence of any heat or moisture source. The wall temperature sensor is placed to measure the temperature of the inner side of the exterior wall, since the risk of condensation there is the highest. Fig 3 shows a thermal picture of the wall from which we can determine coldest point on the wall and place a sensor to measure that point on the wall.

Outdoor sensors include temperature and relative humidity sensors which are connected to an outdoor unit in charge of acquiring and publishing data for the central unit. The outdoor sensors are placed in an area close to the exterior wall where they are the least affected by weather conditions. In addition, the sensors are weatherproof so that measurement is not affected by temporary outside conditions and gives accurate values. The best placement is in the shade with rain and wind protection. The data acquisition interval is dynamically set for the indoor central unit with minimum interval being 15 seconds and maximum 3 minutes. Since humidity and temperature cannot change rapidly and stability of conditions is determined, this interval is increased, preserving energy. On the contrary, if conditions begin to rapidly change, especially while the danger of condensation is high, the interval is shortened to be more responsive to changes, giving more accurate and up to date results. On the other hand, the outdoor

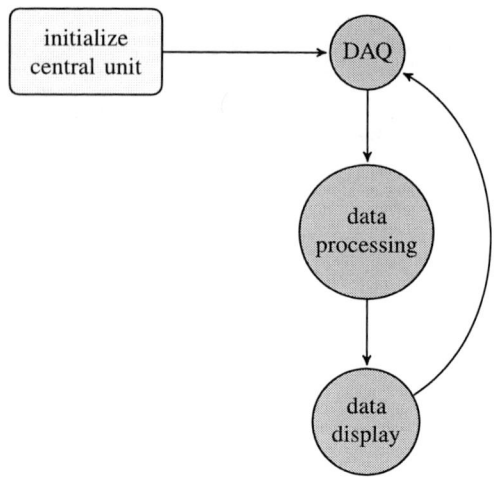

Fig. 4. Block diagram for central unit

unit's interval is constant since weather conditions do not change so often and therefore better energy usage is achieved.

B. Central processing unit

Fig. 4. represents a simple block diagram for the central unit. The central unit reads data directly from the indoor sensors and receives data from the outdoor unit. The central unit calculates the indoors and outdoors dew point temperature in Celsius given the equation (1) and their difference, indoor and outdoor ratio of moisture, difference between wall temperature and the indoor dew point temperature. From all these results it determines the potential threat of condensation and offers corresponding warnings, advice or instructions. In addition to handling all the data, the central unit establishes a wireless network for data transmission and dynamically manages the data acquisition and message publishing interval which directly depends on the current status and indoor conditions.

C. Data display

Fig. 5 shows an example of an LCD display. The display shows the inside and outside temperature in Celsius and relative humidity in percent, wall temperature in Celsius and the inside dew point temperature in Celsius. Additionally it displays status and action messages. Table I represents possible status messages and Table II possible action messages, where first column contains messages and second column a short description of a message. Additionally, Table I contains the conditions when will a certain status message be displayed, where T is difference between the wall temperature and indoor dew point temperature.

IV. PRELIMINARY RESULTS

The system measured the relative humidity in a room for 24 hours where intervals of measurements where set to 1 minute. Fig. 6 represents relative humidity over time in percentage,

	Inside	Outside
Temperature	27 C	33 C
Humidity	30 %	40 %
Wall temperature	12 C	
Dew point temperature	6.6 C	

Status

Action

Fig. 5. LCD display

TABLE I
STATUS MESSAGES

OK	no danger of condensation, T >5
Warning	there is a risk of condensation, 0 <T <5
Danger	condensation is happening, T <0

TABLE II
ACTION MESSAGES

Do not ventilate	ventilating does not help in dehumidification
Ventilate	dehumidification must be carried out

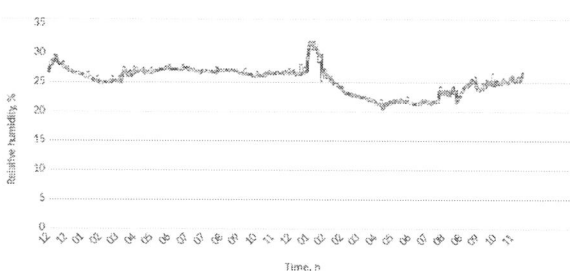

Fig. 6. Humidity

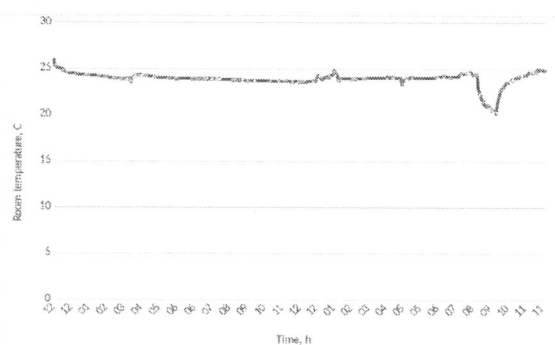

Fig. 7. Temperature

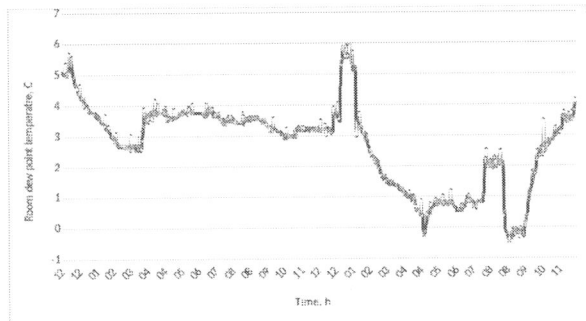

Fig. 8. Dew point temperature

REFERENCES

[1] H. M. Künzel, "Indoor relative humidity in residential buildings–a necessary boundary condition to assess the moisture performance of building envelope systems," *Download: http://www. hoki. ibp. fraunhofer. de/ibp/publikationen/fachzeitschriften/wksb% 20Raumluftfeuchte1_E. pdf*, 2015.

[2] M. G. Lawrence, "The relationship between relative humidity and the dewpoint temperature in moist air: A simple conversion and applications," *Bulletin of the American Meteorological Society*, vol. 86, no. 2, pp. 225–233, 2005.

[3] "Influence of measurement uncertainties on the thermal environment assessment," May 14 2014. [Online]. Available: http://www.azosensors.com/article.aspx?ArticleID=487

[4] InspectApedia.com, "Dew point data for building wall," 2017. [Online]. Available: http://inspectapedia.com/Energy/Dew_Point_Calculation.php

[5] J. S. Roberts, "Dew point temperature," *Encyclopedia of Agricultural, Food, and Biological Engineering (Print)*, p. 186, 2003.

[6] *The Programmable Wireless Stamp (PWS) - User manual*, TSXperts LLC-QWAVE SYSTEMS, 46 Waterbury Ln, Novato CA, 94949, 2105. [Online]. Available: https://tsxperts.com/wp-content/uploads/2014/03/Programmable-Wireless-Stamp-PWS-Specifications.pdf

[7] *ESP-WROOM-02 Datasheet*, Espressif Inc., 2017. [Online]. Available: https://espressif.com/sites/default/files/documentation/0c-esp-wroom-02_datasheet_en.pdf

[8] *ESP8266EX Datasheet*, Espressif Inc., 2016. [Online]. Available: espressif.com/sites/default/files/documentation/0a-esp8266ex_datasheet_en.pdf

[9] A. V. Arundel, E. M. Sterling, J. H. Biggin, and T. D. Sterling, "Indirect health effects of relative humidity in indoor environments." *Environmental Health Perspectives*, vol. 65, p. 351, 1986.

Fig. 7 represents temperature over time in Celsius and Fig. 8 represents the calculated dew point for the room using equation (1) over time. System was left in a room where there was no ventilation and 2 residents were sleeping for the first 12 hours. For the next 6 hours system was left alone in the room, which is seen in relative humidity levels drop. In last 6 hours window was open and room was ventilated to drop the temperature and relative humidity levels. From this charts we can see how the dew point depends on humidity and temperature. When temperature is almost constant, the dew point temperature follows a humidity curve and we can approximate that if humidity changes positively for 1 percent, dew point will positively change for 0.5 Celsius, while if temperature changes approximately 5 Celsius dew point will change for 2 Celsius if humidity stays constant. Therefore humidity has a bigger influence on dew point than temperature. Since humidity is also easier to control than temperature, the system for condensation prevention is focused on ventilation for decreasing the dew point temperature.

V. CONCLUSION

This system is developed to help control humidity level of desired rooms and prevent wall damage as a result of condensation. Residential buildings have indoor relative humidity levels form 20% to 50% since most people agree that in these boundaries they feel most comfortable [9] and temperature levels are mostly kept above 20 degrees Celsius, but living is a dynamic activity and conditions indoor change often, especially in badly ventilated areas, condensation can happen even if we do not know it. The system is used to warn us of sudden changes in our environment and offer quick solutions to avoid serious damage.

ACKNOWLEDGMENT

This paper is fully supported by the Croatian Science Foundation under the project Metrological Infrastructure for Smart Grid IP-2014-09-8826.

MIPRO 2017, May 22- 26, 2017, Opatija, Croatia

FEM analysis and design of a voltage instrument transformer for digital sampling wattmeter

Martin Dadić, Karlo Petrović and Roman Malarić
University of Zagreb/ Faculty of Electrical Engineering and Computing
Unska 3, 10 000 Zagreb, Croatia
e-mail: martin.dadic@fer.hr, karlo.petrovic@fer.hr, roman.malaric@fer.hr

Abstract - A voltage instrument transformer (VIT) or a voltage divider is needed if a National Instruments PXI 4461 Dynamic Signal Analyzer is used in a digital sampling wattmeter application. They assure that the input voltage is lowered to the allowed level of the acquisition card, and VIT also gives the galvanic isolation. This paper discusses possible types of sectionalized windings of a VIT with primary and secondary interleave, which result in lower leakage inductances. On the basis of a detailed finite element method (FEM) analysis, a prototype VIT is designed and produced. The ratio error and the phase angle error are measured on the prototype using a digital calibrator, digital multimeter (DMM) and a phase-angle meter. As well, the performance of the transformer over the wider frequency bandwidth and under the harmonically rich excitation is experimentally investigated.

I. INTRODUCTION

The main goal of the analysis presented in this paper was the design and production of a voltage instrument transformer (VIT) for the digital sampling wattmeter application. Since the digital sampling wattmeter is based on the National Instruments PXI 4461 Dynamic Signal Analyzer cards, where the peak input voltage is constrained to 42.4 V maximum, the mains voltage (i.e. 230 V) has to be lowered using an instrument transformer or a voltage divider. The VIT gives additional benefit of the galvanic isolation, and they generally can be manufactured as more compact and lighter devices, comparing with the inductive voltage dividers (IVD). If a VIT is produced using a standard transformer ferromagnetic core, it further decreases its cost and dimensions, which is paid by a greater ratio and phase-angle errors, especially compared to the IVDs designed using toroidal cores with high permeability.

The target goals in the design of here described VIT can be summarized as follows:

-   Small and compact dimensions, small mass

-   Easy manipulation

-   Use of standard transformer ferromagnetic cut cores

-   Easy assembling of the windings using a standard winding machine

-   Small burden due to a high input impedance of the acquisition card

-   Nominal primary voltage 230 V (RMS) at 50 Hz

-   Class 1 according to IEC 61869-3 [1]

The additional requirement was that the peak value of the secondary voltage at 250 V is approximately equal to 12 V, which is the full scale voltage of the digital multimeter (DMM) Agilent (Keysight) 3458 A at the corresponding DC range. At this range Agilent 3458A has the best accuracy, and true RMS value of an AC voltage can be measured using a GPIB communication and PC processing of the sampled voltages based on Swerlein's algorithm very precisely. This gives the additional possibility to use the VIT in conjunction with the 3458A DMM.

The additional requirement defines the turn ratio, which has to be approximately 29.56. The actual number of turns is defined by the cross section of the core and the magnetic flux density at the nominal primary voltage. The targeted maximum flux density at the primary voltage of 250 V (RMS) was set to 0.6 T, which ensures that the magnetic flux density is far bellow the saturation knee even at the primary voltage equal to 120% of its nominal value. Furthermore it decreases core losses and harmonic distortion of the primary current, at the expense of the higher dimensions and greater mass of the core. Since the chosen transformer core was an older item of standard CM core CM85 (CM85b, older label SM85) core produced by Iskra Sistemi d.d., the targeted number of turns in the primary winding was N_p=1336 and N_s=45 in the secondary winding. The CM cores consist of four core halves, with the coil former on the middle leg. The dimension of the cores can be found in [2]. The sectionalized windings with the different kinds of interleave of the primary and secondary can decrease leakage inductances and distributed capacitances [3-6]. Since the operating bandwidth of the transformer is relatively low (the mains frequency and its harmonics up to 40th), the analysis of the sectionalized winding in the subsequent sections is restricted to the calculation of leakage inductances.

II. SECTIONALIZED WINDINGS

The substantial reduction of leakage inductances in the layer-wound coils can be obtained by interleaving the primary and secondary windings [3]. This technique was introduced in audio-frequency transformers as early as in 1920-ties [4]. Several configurations of sectionalized windings can be found in references such as [3-6]. The configurations of transformer windings that were analyzed regarding their applicability for VIT are depicted in Figs. 1-5. Fig. 1 shows a conventional configuration of the audio transformer windings.

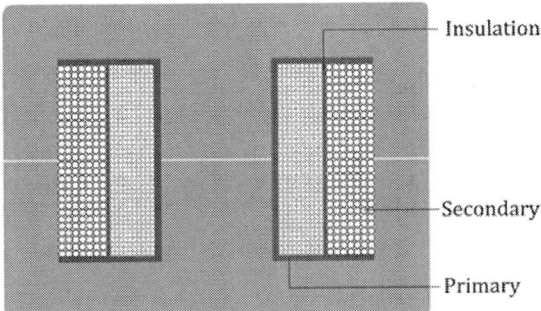

Fig. 1. Conventional design of audio transformer windings

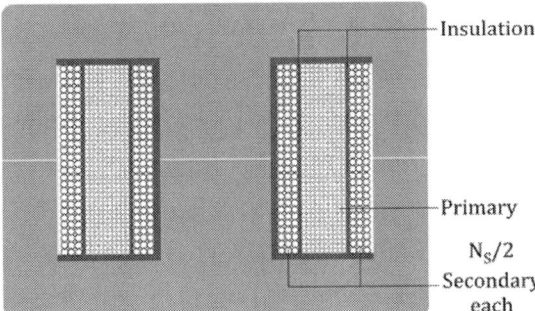

Fig. 2. Configuration with simple interleave – sectionalized secondary

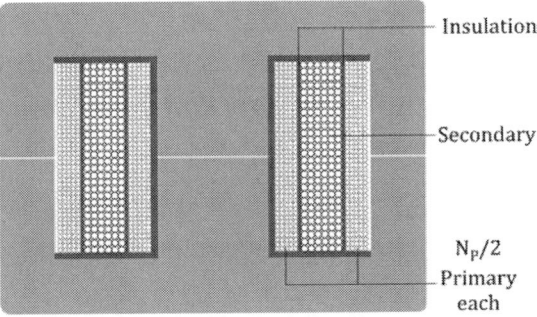

Fig.3. Configuration with simple interleave – sectionalized primary

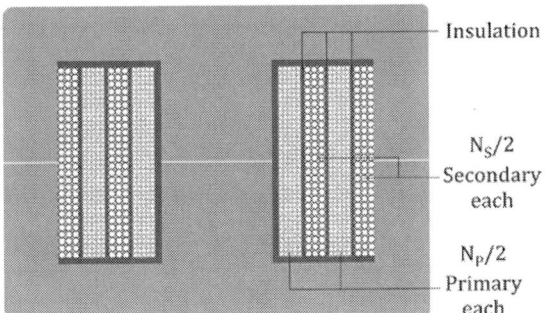

Fig. 4. Primary and secondary interleave

This work was fully supported by Croatian science foundation under the project 1118 Numerical modeling of complex electromagnetic phenomena in transformers

III. CORE CHARACTERIZATION

For the accurate FEM modeling of the transformer core, the B-H curve was experimentally determined. The actual C-core can be efficiently used for the determination of the B-H curve and core losses, because they have reasonably uniform flux density [7]. For this purpose, two windings with the same number of turns $N=28$ and with the wire diameter 1.7 mm were temporarily wound on the core former. The measurement method and measurement set-up was based on the IEEE standard procedures for magnetic cores [8], and it was consisted of a shunt of 0.2 Ω in series with the primary, a regulated transformer, a variable transformer and a RC integrator connected on the secondary (search) coil. The peak values of the magnetizing current and induced flux were measured using a digital storage oscilloscope Agilent DSO 3062A. The operating frequency was 50 Hz. On the basis of the measured values, a 7th order polynomial approximation was determined in the least-squares sense using the computer environment MATLAB. Using this polynomial regression the small variations in the measured curve due to the measuring errors were smoothed. The polynomial representation with odd-order polynomials is a common representation of ferromagnetic cores [9]. The obtained polynomial model was applied in the creation of a custom material in the Infolytica MagNet FEM software. Fig. 5 shows the polynomial model of the BH curve.

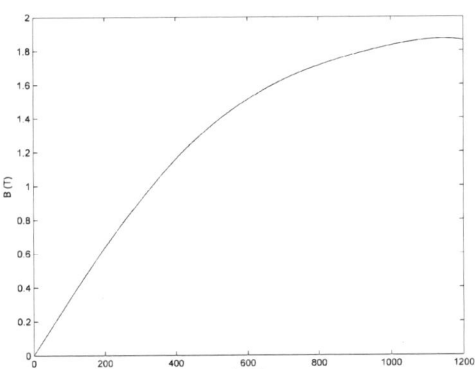

Fig. 5. BH curve

IV. FEM ANALYSIS

Using the actual dimensions of the cores and the windings, four 3D FEM models of the transformer for the winding combinations represented by Figs. 1-4 were created in Infolytica MagNet FEM computer package. The material of the core was custom created using the measured data from Section III. Fig. 6 presents FEM mesh of the transformer. The inductances of the windings were extracted from the FEM analysis, while the leakage inductances were calculated using the following equations [10]:

$$L_{p\sigma} = L_p - \frac{N_p}{N_s} \cdot M \tag{1}$$

$$L_{s\sigma} = L_s - \frac{N_s}{N_p} \cdot M \qquad (2)$$

where $L_{p\sigma}$ denotes leakage inductance of the primary coil, $L_{s\sigma}$ denotes leakage inductance of the secondary coil, L_p and L_s are inductances of the primary and secondary, while M is mutual inductance between primary and secondary.

The results of the FEM simulation for four different sectionalized windings are presented Tables I-IV.

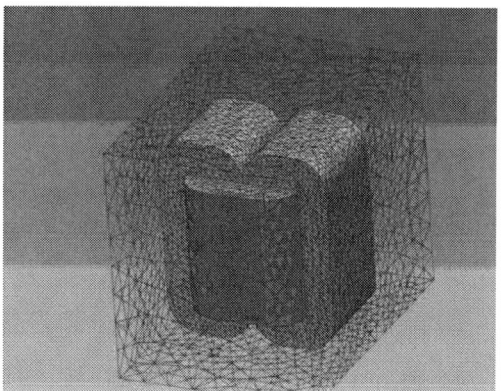

Fig. 6. 3D FEM model of the transformer – mesh of finite elements

The leakage inductances in the case of the sectionalized primary were tending to show very small negative values. This tendency was decreasing with the mesh refinement. Nevertheless, due to the very large number of elements and very long computation times needed for the full 3D model, it was impossible to decrease the element size and increase the computational accuracy to achieve physically realistic positive values. All sectionalized models showed a significant decrease of the leakage inductance of the primary, and the decrease was smallest with the sectionalized primary (Table III). It is reasonable to choose the wire diameter in the secondary big enough to allow the secondary winding to be wound in a complete layer. Due to the high turn ratio, it was possible only with relatively big wire diameters, and the wire diameter of the secondary was chosen to be 1.2 mm. The value of the leakage inductance of the secondary is not so influential as the leakage inductance of the primary, because of the very small burden and secondary currents. Because of all of these reasons, our final choice was a transformer with the simple interleave – sectionalized primary. For the analysis of the configuration with the sectionalized primary, FEM mesh consisted of 7551313 tetrahedra with linear elements and the global maximum element size of 1 mm. The number of the nodes was 1339161. The maximum usage of computer RAM during the solving process was 3.372 GB on a Dell Precision T5500 workstation running at 2.67 GHz with 48 GB RAM installed. The number of elements and the memory requirements were similar in all other presented configurations. A very small burden and a big inductance of the primary coil are constraining the primary currents bellow 20 mA even at 120% of the rated primary voltage. On the other side, very small diameters of the wire

increase the difficulties in the assembling and increase risk of breaking the wire during the assembling. As a compromise, the wire diameter of the primary was chosen as 0.22 mm. In this way, each of two sections of the primary is consisting of 4 complete layers. All layers were mutually isolated using the cellulose paper to allow easier assembling. The final number of turn was slightly different of the targeted, because all layers were wound completely to the edge of the former. Fig. 7 shows the assembled unit with open housing.

Fig. 7. Assembled transformer with the opened housing

TABLE I   CONVENTIONAL DESIGN

Fig. 1.	H
Inductance of the primary coil L_p	44.49
Inductance of the secondary coil L_s	$50.43 \cdot 10^{-3}$
Leakage inductance of the primary coil $L_{p\sigma}$	$13.93 \cdot 10^{-3}$
Leakage inductance of the secondary coil $L_{s\sigma}$	$26.47 \cdot 10^{-6}$
Mutual inductance M	1.497

TABLE II   SECTIONALIZED SECONDARY

Fig. 2.	H
Inductance of the primary coil L_p	44.28
Inductance of the secondary coil L_s	$50.22 \cdot 10^{-3}$
Leakage inductance of the primary coil $L_{p\sigma}$	$8.18 \cdot 10^{-3}$
Leakage inductance of the secondary coil $L_{s\sigma}$	$1.30 \cdot 10^{-6}$
Mutual inductance M	1.491

TABLE III   SECTIONALIZED PRIMARY

Fig. 3.	H
Inductance of the primary coil L_p	44.44
Inductance of the secondary coil L_s	$50.42 \cdot 10^{-3}$
Leakage inductance of the primary coil $L_{p\sigma}$	≈ 0 to the computing accuracy
Leakage inductance of the secondary coil $L_{s\sigma}$	≈ 0 to the computing accuracy
Mutual inductance M	1.497

TABLE IV   PRIMARY AND SECONDARY INTERLEAVE

Fig. 4.	H
Inductance of the primary coil L_p	44.45
Inductance of the secondary coil L_s	$50.42 \cdot 10^{-3}$
Leakage inductance of the primary coil $L_{p\sigma}$	$4.25 \cdot 10^{-3}$
Leakage inductance of the secondary coil $L_{s\sigma}$	$6.90 \cdot 10^{-6}$
Mutual inductance M	1.497

V. MEASUREMENTS

A. Ratio error and phase angle error

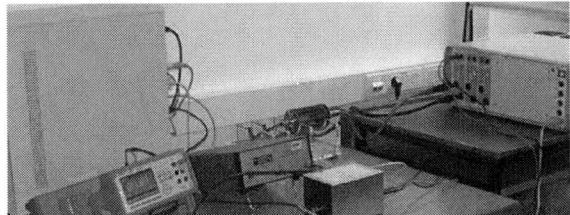

Fig. 8. Measurement set-up

The ratio error (voltage error) as specified in [1] is

$$\varepsilon = \frac{k_R \cdot U_S - U_P}{U_P} \times 100 \left[\%\right] \qquad (3)$$

where U_p is actual primary voltage and U_s is actual secondary voltage when U_p is applied under the conditions of measurement. The measurements of the rated transformation ratio, ratio error and phase angle error were performed using a Calmet C300 three phase power calibrator and tester, a 61/2 digit multimeter Keysight 34465A and the phase-angle meter Krohn-Hite 6200B. Ambient temperature during the ratio and phase angle error measurement was varying between 24.3 and 24.9 °C. All measurements were performed according to [1]. Due to the small burden, the rated output of the VIT was chosen to be 1 VA at power factor 1, which is the lowest standard value for the burden range I. Most of the manufacturers usually set the minimum ratio error at 75% of the rated burden. Due to a very small burden of the acquisition card, the rated transformation ratio was determined by the measurement at 6.25% of the rated burden, e.g. with the secondary terminated by a 1 kΩ resistor. Therefore, the rated transformation ratio was k_R=29.094. The uncertainty of the calibrator C300 is ±0.05% of the settled value (for voltages and ranges which were applied in measurements) and this is uncertainty in the primary voltage. The upper bound of the uncertainty of the secondary voltage can be derived from the last calibration report of the multimeter Keysight 34465A. The multimeter was factory calibrated within one year from the measurements. In the 10 V AC range, the errors were not declared for 50 Hz, and the bounds of the uncertainty were derived from the bigger error between calibration data at 10 Hz and 1 kHz. Taking into account the one year specification of ±0.07% (for the ambient temperature of the laboratory), and recalculating the absolute error to the percentage of the lowest voltage measured at the secondary, the upper bound of the uncertainty in the ratio error is 0.163% of measured values in all cases and frequencies that were applied. The phase angle error was measured using a reference phase from the cage-type AC resistor shunt with the nominal resistance 714 mΩ and with the nominal current 1A that was produced in our laboratory as the part of a set of precise resistor shunts. The upper bound of the phase angle error introduced by the shunt can be estimated from the impedance measurement of the shunt. The impedance of the shunt was measured in slow-mode with the LCR bridge Hameg 8118, applying the averaging and 4-wire connection, a custom-made test fixture and the open-short calibration. The measured phase angle of the shunt was 0.0056° at 1 kHz, while the basic accuracy of the phase angle is ±0.005° at 1 kHz as the secondary value, declared by the LCR bridge manufacturer. The phase angle of the current was set in the calibrator to 0 relatively the voltage, while the accuracy of the phase of the calibrator is 0.05°. The repeatability of the phase angle measurement as declared by the manufacturer of the phase-angle meter is ±1 digit, i.e. 0.1°. The phase angle meter was calibrated using Calmet 300 before and after the measurement campaign for the range ±1° in steps of 0.1°, and the repeatability was always 1 digit declared by Krohn-Hite. Consequently, the uncertainty of the phase angle error measurement was ±0.1606°. The measurement set-up is depicted in Fig. 8. The results of the measurements are summarized in Table V.

TABLE V   ERRORS

Burden	Primary voltage	Ratio error (%)	Phase angle error (°)
Nominal	184 V (80% U_{nom})	-0.58	0.4
Nominal	230 V (100% U_{nom})	-0.57	0.4
Nominal	276 V (120% U_{nom})	-0.59	0.5
0 %	184 V (80% U_{nom})	0.04	0.4
0 %	230 V (100% U_{nom})	0.04	0.4
0 %	276 V (120% U_{nom})	0.03	0.5

It can be concluded that the transformer fully satisfies the limits of a class 1,0 transformer, which are ±1,0% for the ratio error and ±40 minutes for the phase displacement.

B. Wider frequency bandwidth

To investigate the errors for the primary voltage containing several higher harmonics of 50 Hz, the additional measurements were performed. The burden was nominal. The results are summarized in Table VI. It can be seen that the errors are within limits defined at nominal frequency even in wider frequency bandwidth.

TABLE VI   ERRORS

Frequency (Hz)	Ratio error (%)	Phase angle error (^{o})
100	-0.55	0.2
150	-0.54	0.2
200	-0.53	0.1
250	-0.52	0.1
300	-0.52	0.1

C. Harmonically rich excitation

Finally, the primary voltage was enriched with the higher harmonics. The RMS value of the composite voltage was 230 V (nominal) in all cases. The RMS value of the secondary voltage was measured with the true-volt DMM Keysight 34465A, and the ratio error was determined based its value. The percentages of the higher harmonic components were measured as well in the secondary using the power analyzer Chauvin-Arnoux CA8220. The burden was nominal. The results are summarized in Table VII. The transformer is very accurate even for the harmonically rich excitation.

TABLE VII   HARMONICALLY RICH PRIMARY VOLTAGE

2nd harm. (primary)	3rd harm. (primary)	2nd harm. (secondary)	3rd harm. (secondary)	Ratio error (%)
0	40%	-	39.9%	-0.015
40 %	0	40.1%	-	0.038
100 %	0	99.1%	-	0.08
40 %	40%	40.4%	40.1%	0.045

VI. CONCLUSIONS

It is shown that the sectionalized configuration of windings, typically applied in audio transformers for vacuum-tube amplifiers are a good choice for the voltage instrument transformers, because of many similar design demands: relatively high turn ratio, low leakage inductances and wider frequency bandwidth in operation. The FEM analysis has shown that the interleaved configuration of the windings with the sectionalized primary is the best choice for the voltage instrument transformer with the given ferromagnetic core and turn ratio. The analyzed transformer is manufactured, and the measurement has shown that the design demands are fully satisfied and that the transformer shows satisfying results even in wider frequency bandwidth, and with harmonically rich primary voltage.

REFERENCES

[1] IEC 61869-3 International Standard: *Instruments transformers – Part 3: Additional requirements for inductive voltage transformers*, Edition 1.0, 2011-07

[2] V. Bego, *Mjerni transformatori*, Školska knjiga Zagreb, 1977.

[3] C. Wm. T. McLyman, *Transformer and Inductor Design Handbook*, 4th Edition, CRC Press, 2011.

[4] G. Koehler, "The Design of Transformers for Audio-Frequency Amplifiers with Preassigned Characteristics," *Proceedings of the IRE*, vol. 16, no. 12, pp. 1742-1770, 1928.

[5] T. Jelaković, *Transformatori i prigušnice*, Biblioteka časopisa „Elektrotehničar", Zagreb 1952.

[6] F. E. Terman, *Radio Engineer's Handbook*, McGraw, 1943.

[7] P. Rupanagupta, J.S. Hsu, "Determination of Iron Core Losses Under Influence of Third-Harmonic Flux Component," IEEE Transactions on Magnetics, vol. 27, no. 2, pp. 768-777, 1991.

[8] IEEE Std 393 – 1991, *IEEE Standard procedures for magnetic cores*, IEEE 1992.

[9] W. A. Geyger, *Nonlinear-Magnetic Control Devices: Basic Principles, Characteristics and Applications*, McGraw, 1964.

[10] A. Dolenc, *Transformatori*, 1. i 2. dio, Sveučilište u Zagrebu, Elektrotehnički fakultet, 1991.

Random Number Generation with LFSR Based Stream Cipher Algorithms

Taner Tuncer* Erdinç Avaroğlu**
*Fırat Üniversitesi Mühendislik Fakültesi / Bilgisayar Mühendisliği, Elazığ, Türkiye
**Mersin Üniversitesi Mühendislik Fakültesi / Bilgisayar Mühendisliği, Mersin Türkiye
{ttuncer@firat.edu.tr, eavaroglu@gmail.com}

Abstract - Random numbers have a wide range of usage area such as simulation, games of chance, sampling and computer science (cryptography, game programming, data transmission). In order to use random numbers in computer science, they must have three basic requirements. First, the numbers generated must be unpredictable. Second, the numbers generated should have good statistical properties. Finally, the generated number streams must not be reproduced. Random number generators (RNGs) have been developed to obtain random numbers with these properties. These random number generators are classified into true random number generators (TRNG) and pseudo random number generators (PRNG). One of the PRNGs used for generate random numbers is Stream Encryption algorithms. In this paper, random number generation of LFSR based stream encryption algorithms and their hardware implementations are presented. LFSR based stream encryption algorithms have been implemented on Altera's FPGA based 60-nm EP4CE115F29C7 development boards by using VHDL language. The obtained random numbers passed the NIST statistical tests, accepted as standard for cryptographic applications.

I. INTRODUCTION

Randomness means that changes do not follow a certain rule and that the samples occur with the same probability. Random numbers are defined as a series of numbers that are completely independent of each other in a given range with the same probability. Random numbers have a wide range of usage area such as simulation, games of chance, sampling and computer science (cryptography, game programming, data transmission). In particular, random numbers are needed in cryptography. The security of cryptographic systems depends on the actual randomness of the numbers used. Random number generators used for this purpose must have some security requirements. These requirements are listed in Table I.

Various random number generators (RNGs) have been developed in order to obtain random numbers. These random number generators are classified as True Random Number Generators (TRNG) and Pseudo Random Number Generators (PRNG) [1,2]. True random number generators are generated random numbers by means of physical noise sources. PRNGs generate numbers using a deterministic algorithm. The algorithm generates a random number sequence using a random selected seed value. The advantage of PRNG is that it is cheap, easy to implement, fast and does not require equipment. However, numbers generated with PRNGs can be estimated when seed value is detected or when the functions used in the algorithm are not complex enough [1,2]. PRNGs provide R1, R2 and R3 from the requirements specified in Table 1[3,4]. But, since R4 is not provided, the use of PRNGs in cryptographic systems is not appropriate.

TABLE I. Security requirements for random number generators[3,4]

R1	Random numbers should not have statistical weakness
R2	Knowing the subsequences of random numbers should not allow calculation or estimation of the previous and subsequent random numbers.
R3	It shall not be practically feasible to compute preceding random numbers from the internal state or to guess them with nonnegligibly larger probability than without knowledge of the internal state
R4	It shall not be practically feasible to compute future random numbers from the internal state or to guess them with nonnegligibly larger probability than without knowledge of the internal state

Random number generators known in the literature are Blum Blum Shub, Linear Congruential, ANSI X9.17 generator and LFSR. LFSRs are easy to implement and produce random numbers with large periods and good statistical properties[5,6]. However, the numbers generated are easy to estimate. LFSRs consist of XOR and flip flop circuits. The LFSR should have an seed value in number generation with the LFSR. If the seed value is known, it is easy to determine the numbers generate. Therefore, It is possible to generate random numbers with good statistical properties using multiple LFSRs.

One of such random number generation methods is a stream encryption algorithm. There are 5 design strategies for streaming encryption. These are Linear Feedback Shift Register (LFSR), nonlinear feedback Shift Register, block cipher-based, mixing-based and hash function-based, respectively.

In this paper, random number generation is presented by LFSR based stream ciphering. Section 3 presents stream cipher algorithm based on the structure, characteristics and implementation of the LFSR. Section 4 gives NIST test results and evaluates the random numbers generated. Finally, Section 5 presents our conclusions.

II. LFSR AND STREAM CIPHER

LFSRs are often used in key generation. The reason for this is that the hardware implementations are easy, they can generate numbers in large periods, they have good statistical properties and they can be easily analyzed because of their structure. The L-length linear feedback shift register (LFSR) consists of L flip-flops numbered 0, 1, ..., L - 1, each of which can store one bit and have one input and one output. Fig. 1 shows a simple LFSR structure and the characteristic polynomial for L = 4. The LFSR is denoted by <L,C(D)>. Where, C(D) = 1+ $c_1D + c_2D^2 + \cdots + c_LD^L$ is characteristic polynomial and The degree of C (D) is L. The seed value of the LFSR is $[s_{L-1},\ldots ,s_1, s_0]$ with $0 \le i \le L - 1$, $s_i \in \{0, 1\}$. If the seed value of the LFSR is $[s_{L-1},\ldots ,s_1, s_0]$ then the output sequence or generated random number will be s = s_0, s_1, s_2,... according to the euation.1. In this case the period of the LFSR is 2^L-1.

$$s_j = (c_1s_{j-1} + c_2s_{j-2} + \cdots + c_Ls_{j-L}) \bmod 2 , \; j \ge L \quad (1)$$

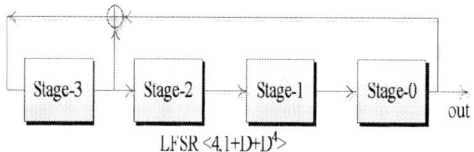

Figure 1. LFSR structure for <4,1+D+D⁴> characteristic polynomial

Stream cipher is an important class of encryption algorithms. Stream ciphers generate as long a periodic and random key sequences as possible with a key (K) and an initial vector (IV - Initialization Vector). The key obtained also inserts a function (usually XOR operation) to obtain the encrypted text. Fig. 2 shows the stream cipher's stages of generating cipher text.

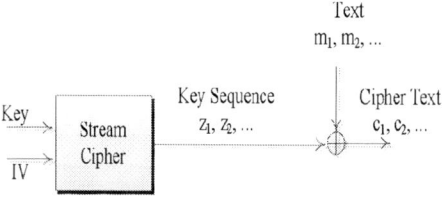

Figure 2. Random key sequence and encryption generated by stream cipher

III. LFSR BASED STREAM CIPHER METHODS AND RANDOM NUMBER GENERATION

A. Nonlinear Combination Generators

One technique for destroying the linearity inherent in LFSRs is to use several LFSRs in parallel. The structure of the nonlinear combining generator with good statistical properties and long periods is as shown in Fig. 3 The LFSRs are combined with a nonlinear logical function so that the nonlinear combining generator can be robust against the cryptographic attack. This function is known as the merge function. The simplest f function is XOR.

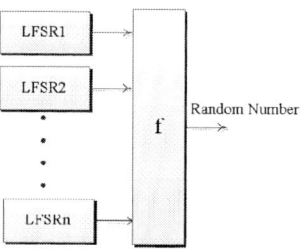

Figure 3. Nonlinear combination generators

The choice of the LFSR structure and function f is important to generate random numbers with the nonlinear combining generator. The f function should be high covariance, nonlinearity, and good correlation. A function f with these properties is given in Eq.2. The lengths of the LFSRs used are 67,71, 73, 79, 83, respectively, and the LFSR polynomials are as in Equation 3.

$$f(x_1,x_2,x_3,x_4,x_5)=1 \oplus x_2 \oplus x_3 \oplus x_4x_5 \oplus x_1x_3x_4x_5 \quad (2)$$
$$P_1(x)=x^{67}+x^5+x^2+x+1$$
$$P_2(x)=x^{71}+x^5+x^3+x+1$$
$$P_3(x)=x^{73}+x^4+x^3+x^2+1 \quad (3)$$
$$P_4(x)=x^{79}+ x^4+x^3+x^2+1$$
$$P_5(x)=x^{83}+x^7+ x^4+x^2+1$$

Figure 4 shows the implementation of a nonlinear combination generator in the FPGA. The numbers generated are stored in a memory unit in order to perform statistical tests of the numbers. The memory unit used has 16 bit address and 1 bit data inputs. In the system is used 50MHz clock. A number is generated for each clock. A counter is used to store the generated numbers to corresponding address in the memory unit. The counter frequency is 50 Mhz and synchronous with the number generation. The LFSR and f functions were implemented using VHDL, and in the design was used schematic representations. Fig. 4 shows that the nonlinear combination generator is implemented on Altera's Cyclone IV EP4CE115F29C7 core.

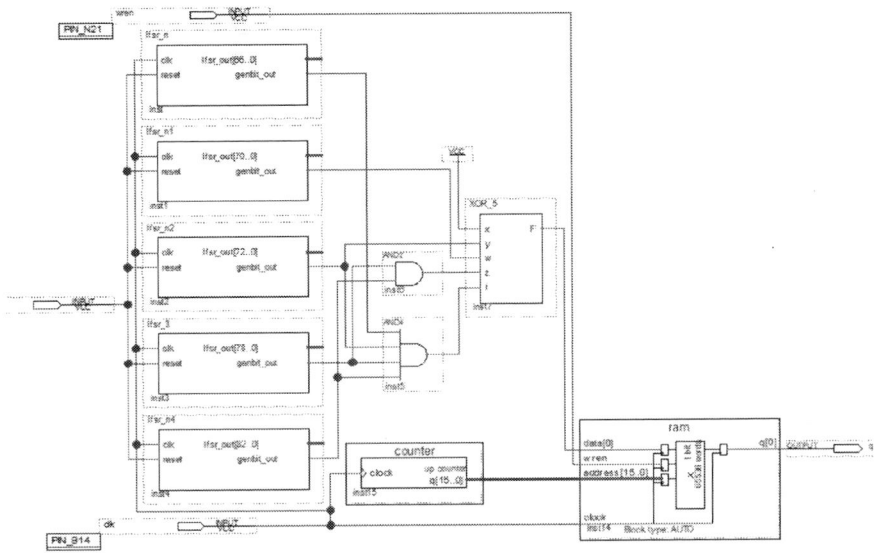

Figure 4. Implementation of nonlinear combination generator in Altera Quartus environment

B. Geffe Generator

Another stream cipher algorithm based on LFSR is the Geffe generator. As shown in Fig. 5, it consists of 3 LFSRs and one logical function. The Geffe generator, which is relatively weak against the cryptographic attack, LFSR2 controls LFSR1 and LFSR3. If the number generated by the LFSR2 is x_2 = 1, the output determines the by LFSR1. Otherwise output determines by the LFSR3. The logical output function for Geffe is as in Equation 4 The degree of LFSR polynomials is L1, L2 and L3. The complexity and period of the Geffe generator are $(2^{L1}-1)(2^{L2}-1)(2^{L3})$, $L_1L_2+L_2L_3+L_3$ respectively.

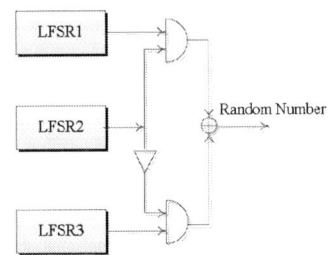

Figure. 5 Generating number with geffe generator

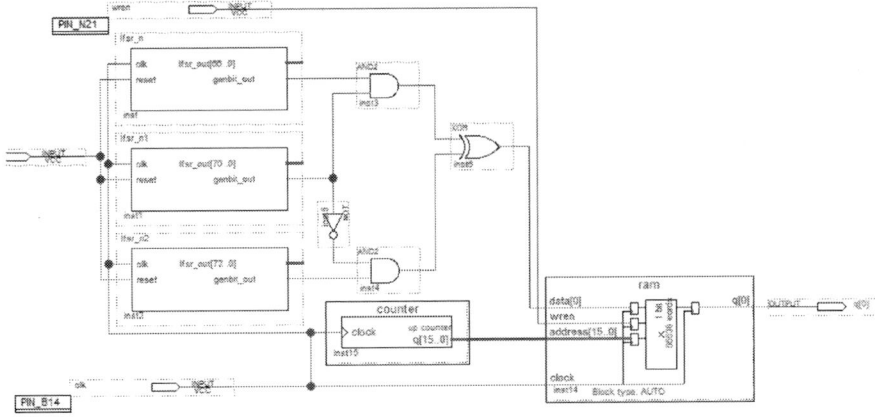

Figure 6. Implementation of Geffe random number generator in Altera Quartus environment

$$f(x_1,x_2,x_3)=x_1x_2\oplus(1+x_2)x_3 \qquad (4)$$

Figure 6 shows the implementation of the Geffe generator in the FPGA. The polynomials used for LFSR1, LFSR2 and LFSR3 are P1 (x), P2 (x) and P3 (x) in Equation 3.. In order to perform the statistical tests of the produced numbers, in the system memory circuit is used and the system frequency is 50 Mhz The LFSR and f functions were implemented using VHDL. Fig. 6 shows that the Geffe generator is implemented on Altera's Cyclone IV EP4CE115F29C7 core.

C. Summation Generator

The combining function in the Summation Generator is based on integer addition. The output in the generator is the least significant bit of the sum. One of the inputs of the f function is carry bit. a and b are two integers where

$$a = a_{n-1}2^{n-1} + ... + a_1 2 + a_0$$
$$b = b_{n-1}2^{n-1} + ... + b_1 2 + b_0.$$

So the total function is a nonlinear function. The total function consists of the sum of the numbers and the carry bit.

$$z_j = a_j + b_j + c_{j-1}$$
$$c_j = a_j b_j + (a_j + b_j)c_{j-1}$$

Period of random numbers obtained by using 2 LFSRs is $(2^{L1} - 1)(2^{L2} - 1)$. Fig. 7 shows the structure of the total generator with n LFSRs. Fig. 8 shows the implementation of the total generator with FPGAs using four LFSRs. The LFSR polynomials used are P1 (x), P2 (x), P3 (x) and P4 (x) in Equation 3.

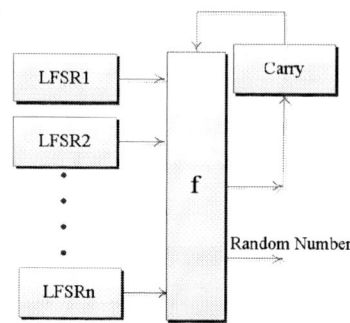

Figure 7. The structure of the Summation Generator

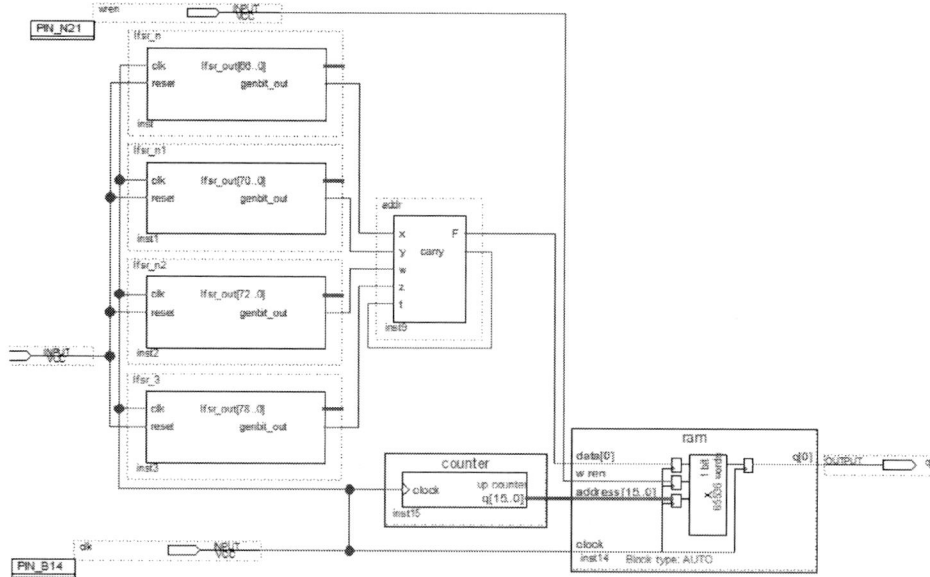

Figure 8. Implementation of Total Generator in Altera Quartus environment

IV. NIST TEST SUITE

Quartus software allows the numbers saved in RAM structures to be uploaded in a text file. 1,048,560 numbers were saved in a text file by generating numbers 16 times totally. Numbers generated by number generators should show good statistical properties. The numbers generated for this should be subject to randomness tests. Diehard, FIPS 140-2, Scale Index, NIST tests are used to determine the randomness of the numbers. A total of 15 tests were performed for the NIST test suite and the tests were given [7]. One of the most important parameters used in tests is the α value. The importance level is used to determine the randomness of random numbers. Another parameter used in the tests is the P-value. P-value is the measure of randomness. If the P-value is equal to 1, the numbers are said to have perfect randomness. If the P-value for each test is greater than or equal to α, the random numbers are said to be successful in the test. Otherwise, the random numbers are unsuccessful from the test and the numbers generated are not random. In the literature, the importance level is selected in the range [0.001, 0.01]. The test results according to the NIST 800.22 test suite are given in Table II.

TABLE II. NIST Test Result

Test Name	Geffe	Nonlinear	Summation
Frequency Test	0.520	0.209	0.807
Frequency Test within a Block Test	0.581	-	0.751
Runs Test	0.531	0.788	0.736
Test for the Longest Run of Ones in a Block Test	0.604	0.917	0.993
Binary Matrix Rank Test	0.260	0.937	0.867
Discrete Fourier Transform	0.614	0.105	0.407
Non-overlapping Template Matching Test	0.404	0.815	0.364
Overlapping Template Matching Test	0.325	0.447	0.543
Maurer's Universal Statistical Test	0.955	0.429	0.501
Linear Complexity Test	0.495	0.292	0.184
Serial Test	0.441	0.240	0.490
	0.046	0.649	0.151
Approximate Entropy Test	0.898	0.086	0.810
Cumulative Sums Test	0.886	0.222	0.713

V. CONCLUSION

Random numbers have a wide range of usage area such as simulation, games of chance, sampling cryptography, game programming and data transmission. In this paper, Stream cipher-based methods are used for random number generation. Each method is implemented in FPGA hardware. NIST 800.22 test suite was applied to the obtained numbers. The statistical properties of the generated numbers should be good so that the generated numbers can be used in the field of cryptography. According to test results, random numbers have good statistical properties. Only, the Nonlinear Combination generator method failed within Frequency Block Test. Geffe and Summation generators are success from all tests.

REFERENCES

[1] M. Jonathan, C. J. Cerda, C. D. Martinez, H. David, K. Hoe, "Random number generators using cellular automata implemented on FPGAs", 44th IEEE Southeastern Symposium Son system Theory, 2012, pp. 67–72.

[2] K. Wold, C. H. Tan, "Analysis and enhancement of random numberGenerator in FPGA based on oscillator ring", *Int. Conf. On Reconfigurable Computing and FPGAs*, 2008, pp. 385–390.

[3] Özkaynak, F.: Cryptographically secure random number generator with chaotic additional input. Nonlinear Dyn. (2014). doi:10.1007/s11071-014-1591-y

[4] T. Tuncer, E.Avaroglu, M. Turk, A. BedriOzer, "Implementation of non-periodic sampling true random number generator on FPGA", *Journal of Microelectronics, Electronic Components and Materials*, vol. 44, no. 4, pp. 296–302, 2014.

[5] E. Erkek, T. Tuncer, "The implementation of ASG and SG random number generators", *ICSSE 2013*, pp. 363–367. http://dx.doi.org/10.1109/icsse.2013.6614692.

[6] Avaroğlu E., "Hardware Based Realization Of Random Number Generator", Phd Thesis, Electrical and Electronics Engineering Fırat, University, 2014

[7] NIST, "NIST Random Number Generation and Testing", 2006.

RAM-based mergers for data sort and frequent item computation

Artjom Rjabov[*], Valery Sklyarov[**], Iouliia Skliarova[**], Alexander Sudnitson[*]

[*] Department of Computer Engineering, Tallinn University of Technology,
Tallinn, Estonia
[**] Department of Electronics, Telecommunications and Informatics/IEETA, University of Aveiro,
Aveiro, Portugal
artjom.rjabov@ttu.ee

Abstract - Data sorting and frequent item computation are important tasks in data processing. The paper suggests an architecture for parallel data sorting with simultaneous counting of every item frequency. The architecture is designed for streaming data and incorporates data sorting in hardware, merging of preliminary sorted blocks with compressing of repeated items with calculating of repetitions in hardware, and merging large subsets received from the hardware in general-purpose software. Hardware merge components of this architecture count and compress repeated items in sorted subsets in order to reduce merging time and prepare the data for frequent item computation. The results of experiments clearly demonstrate advantages of the proposed architectures.

Keywords—High-performance computing systems, Information processing; Sorting networks; Parallel sorting; Partial sorting; reconfigurable computing.

I. INTRODUCTION

Sorting is a procedure that is needed in numerous computing systems. Parallel algorithms for data sorting have been studied in computer science for decades. There are many different parallel sorting algorithms. The most notable of them are Parallel QuickSort, Parallel Radix Sort, Sample Sort, Histogram Sort [1] and a family of algorithmic methods known as sorting networks [2]. To better satisfy performance requirements, fast hardware accelerators have been researched in depth. The sorting networks presents a great interest for hardware acceleration because of their massive parallelism. A sorting network is a set of vertical lines composed of comparators that can swap data to change their positions in the input multi-item vector. The data propagate through the lines from left to right to produce the sorted multi-item vector on the outputs of the rightmost vertical line. Sorting is a very resource expensive and time consuming operation. There are different approaches to overcome the resource limitation. Utilizing iterative networks with reusable comparators permits to process significantly larger data sets, but still to some extent. The combination of iterative network-based sorters with subsequent merging permits to process larger data sets than the sorting network allows. The merge operation can be implemented completely in software or partially in hardware for relatively small sorted data subsets. The merging can be implemented as tree-like structure of merge units. This approach allows us to reduce sorting

time even further by detecting the repeated elements with subsequent recording of how many times the data was repeated and deleting the repeated entries in the sorted list. This approach allows us to reduce sorting time for data sets with repeated elements and prepare the data for frequent item computation.

Data sorting and frequent item computation is required in searching, statistical data manipulation and data mining (e.g. [3, 4]). To describe one of the problems from data mining informally let us consider an example [3] with analogy to a shopping card. A basket is the set of items purchased at one time. A frequent item is an item that often occurs in a database. A frequent set of items often occur together in the same basket. A researcher can request a particular support value and find the items which occur together in a basket either a maximum or a minimum number of times within the database [3]. Similar problems appear to determine frequent inquiries at the Internet, customer transactions, credit card purchases, etc. requiring processing very large volumes of data in the span of a day [3]. Fast extracting the most frequent or the less frequent items from large sets permits data mining algorithms to be simplified and accelerated.

The paper suggests a method and high-performance hardware implementation of data sorting algorithm based on parallel sorting network with subsequent merge. The functionality of the merge units is expanded by adding the operation of compressing the data by counting of the repeated data. The system is designed for working over streaming data. It utilizes Advanced eXtensible Interface (AXI) interfaces and is suggested as a PCI express (PCIe) peripheral.

The remainder of the paper contains 6 sections. Section II analyzes the related work. Section III describes highly parallel networks for sorting and explores hardware co-design. Section IV presents proposed system for data merge and item counting. Section V describes experimental setup and hardware accelerator architecture. Section VI presents the results of experiments and comparisons. The conclusion is given in Section VII.

II. RELATED WORK

Different approaches of hardware sorting units were studied by Marcelino et al. in [5]. They implemented a hardware/software hybrid sorter with a sorting unit based on insertion sorting algorithm and unbalanced merging unit. They also utilized Bathcer's Even-Odd sorting network for software implementation and experimented with different combinations of software (QuickSort, Even-Odd network) and hardware (Insertion sorting, unbalanced merge). They also discussed possibilities of using pipelined sorting networks and balanced merging units. Chen and Prasanna in [6] proposed a hardware/software hybrid solution for accelerating database operations using Field-Programmable Gate Array (FPGA) and Central Processing Unit (CPU). Their sorting algorithm is based on merge-sort algorithm where first few sorting stages are implemented in FPGA as folded bitonic sorting networks. The rest of the algorithm is implemented in CPU.

Hardware acceleration of frequent item computation was explored by Teubner et al. in [7]. They suggested to use FPGAs and proposed three different methods. The first method is a straightfoward implementation of Space-Saving algorithm with min-heap data structure in Block Random-Access Memory (BRAM) for data storage. For the second method instead of BRAMs with min-heap structure they used two search trees implemented in lookup tables in order to get rid of min-heap sorting. The pipelined circuit of their third solution choses the best results in terms of performance and scalability. They achieved throughput four times higher than the best published result.

In our previous works we also explored different approaches of hardware/software systems for high-performance data processing and sorting [8-12]. In [9] we proposed a sorting-network-based hardware sorters with subsequent merge in software as well as different approaches of partial sorting for minimal and maximal subsets extraction which is used in frequent item computation. The latter problem also was explored in [10, 11]. In [12] we proposed a multi-level architecture for minimal/maximal subset extraction which utilizes a general purpose processor of a host PC and the programmable logic and processing system of Zynq device.

III. PIPELINED ITERATIVE PERIODIC SORTING NETWORK

As a basis for our sorting circuit we use a periodic pipelined Odd-Even Transition Sorting Network (also known as Odd-Even Transposition Sorting Network or OETS). A periodic network is a type of network which

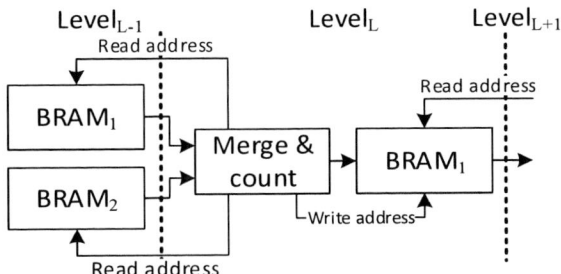

Fig. 2. Interaction between levels of merge operation

consists of identical sequences of comparators. Traditional implementation of OETS is less efficient than Batcher's networks, but it is more reliable and its implementation is simpler. Salloum and Wang proved that OETS has good fault-tolerant properties [13].

Like in our previous works, we use pipelined approach with reusable comparators presented in [14]. K M-bit data items that have to be sorted are loaded (from block RAM) to the feedback register. Sorting is executed in a segment of even-odd transition network composed of two linked lines with even and odd comparators. Sorting is completed in K/2 iterations (clock cycles) at most. Note, that almost always the number of iterations is less than K/2 because of the technique [14] according to which if there are no swaps of data on the right-most line of comparators then sorting is completed. Note that the network [14] possesses significantly smaller combinational delays than networks from [2]. Besides, in the proposed architecture iterations are done at the same time as subsequent data are being received from the inputs. Such parallelism enables delays to be optimally adjusted allowing the total performance to be improved.

The sorter used in the proposed architecture is depicted in Fig. 1. It is based on the iterative sorting network described above and designed as an AXI bus peripheral to receive data from PCI express bus. The AXI slave control unit writes the data to the input register and initiates the sorting operation when the register is full. At the same time it starts writing the next data subset to the input register while the sorting network performs data sorting. After the completion of the sorting the system moves the data to the output register and the merging system copies it into its embedded RAM blocks. All three operations (receiving the data, sorting, copying from the output register) work in parallel in order to achieve maximal possible performance.

IV. PIPELINED MERGING AND ITEM COUNTING

The merging part of the circuit is based on embedded block-RAM. Xilinx Virtex 7 FPGA provide RAM blocks with 36kbit of memory [15], where the data word size can be adjusted for the needs of the system. Every word in RAM blocks of our system contain a pair of data item and its count. The sizes of both the item and the count are also adjustable and should be chosen considering the nature of the input data.

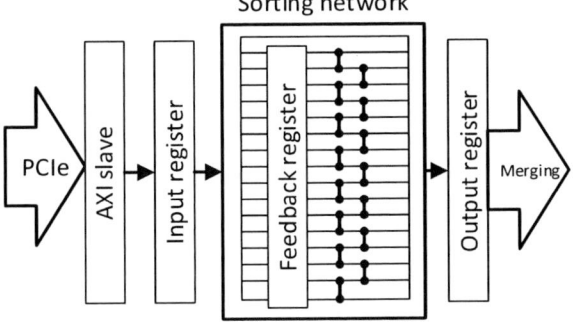

Fig. 1. The circuit for sorting data blocks

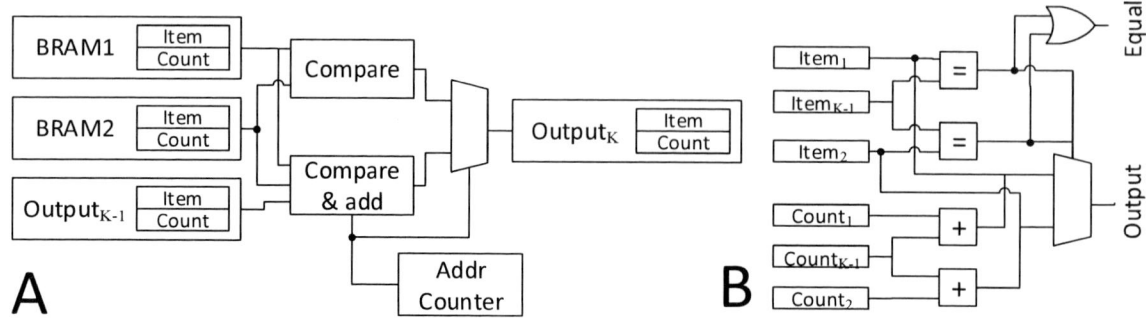

Fig. 3. "Merge and count" architecture: A) General architecture of the merger B) Compare and add operation

Fig. 2. depicts the elements of the merging system and interactions between them. The "merge and count" block of level L receives the input data from two embedded block-RAMs of the previous level(L-1). The circuit compares all the items in two sorted subsets of N data items and merges them into one sorted subset. The maximal size of the final data set is 2×N items. This worst case scenario can occur if no repeated items were found in both input subsets. Every merging element of the system contain two address counters for selecting the data from block RAMs of the previous level and one counter for writing the merged and counted data to the output RAM block.

Although the maximal number of clock cycles for merging N-item blocks is 2×N, our system with compression and item counting require less clock cycles for the data sets with repeated items. Every subsequent level of merging require less clock cycles than the previous one, because the compression and counting was partially done in the previous level.

Merge fragment from Fig. 2. is depicted on Fig. 3(a). The compression and the counting of the items is done in "compare and add" block shown in Fig. 3(b). The system stores the data item which was written after the previous comparison and compares it with both inputs. If the item part of the item/count pair previously written to the RAM

block is not equal to both of them, then the merger writes the item/count pair with larger item value to the output RAM block and increments both write address counter and read address counter for the input with the largest value. Otherwise, the merger does not increment the write address of the block and writes the new count number to the count part of the item/count pair. The new count number is the sum of the count parts of the previously written data item and the count of one of the inputs, which has an item part equal to the previously written one. During the first level of merging every pair have '1' as its count value. All zeros in the count part mean that the total number of repetitions exceeded the capabilities of the RAM block.

The RAM blocks of every item of the merging system are capable of storing all data from the inputs, but if the sorted set supplied to the merger contains repeated items, the system does not fill the RAM blocks completely. The merger reads the value from the write address register of the mergers from the previous level. It informs the merger about how many item pairs were actually written during the previous merge operation.

The proposed architecture is designed to be constructed with any number of merging/counting layers and the size of the initial sorter, but the final implementation always depends on the target device.

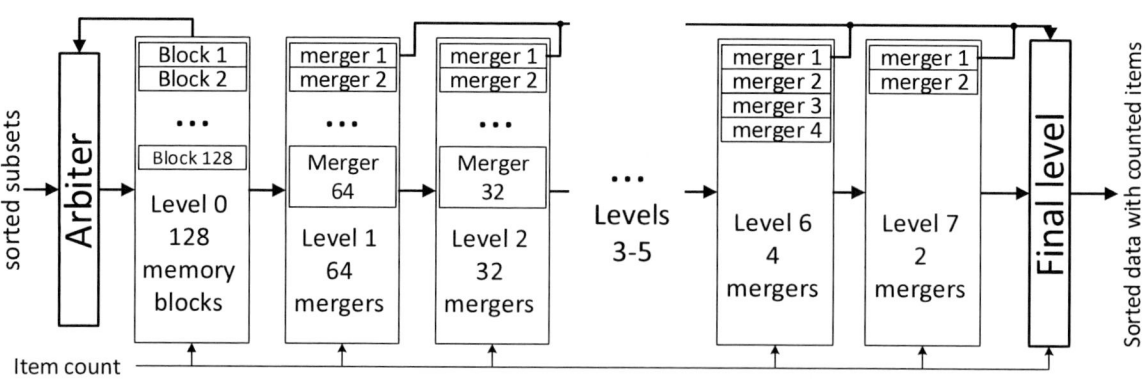

Fig. 4. Merging system

Fig. 4. depicts the merging system implemented for Virtex-7 FPGA device with 8 level of merging/counting. It has one initial level of RAM blocks which contain data subsets sorted with the sorting network described in section III. The sorter in this implementation receives 32-bit items grouped by 4 and writes sorted subsets of 512 elements into the initial level of RAM blocks. The writing to the first RAM blocks of the merging system is also implemented in groups of 4 items in order to keep up with the bus speed. The merging operation is not performed during this step.

The first element of the first level of the merge operation activates when the first two sorted subsets become available on the initial BRAM level. Every merging and counting element of every level starts its operation immediately when two mergers/counters connected to it from the previous level finish merging and counting.

Although the number of merging/counting levels and therefore the size of the final data set is limited by the resource availability of the target device, the architecture was designed to process streaming any volumes of data. The merging levels can be disabled if not all of them are required for the data processing. It can be done by supplying item count value (see Fig. 4). The final level is responsible for the preparation of the data from the last enabled level for the bus transaction and merges the subsets from the previous level if all levels are enabled. The merging/counting system also can merge several data sets in a pipeline for subsequent processing in GPC. When the merge level finishes its operations, the previous level becomes available to process new item pairs since the next level has acquired all necessary data.

V. HARDWARE ARCHITECTURE AND EXPERIMENTAL SETUP

The system was designed as a hardware accelerator for a host PC which communicates through PCI-express interface in Direct Memory Access (DMA) mode. Fig. 5. depicts this architecture.

Software in the host PC runs the 32-bit Linux operating system (kernel 3.16) and executes programs (written in C language) that take results from PCI-express (from the FPGA) for further processing. We assume that the data collected in the FPGA are preprocessed in the programmable logic by applying various highly parallel networks (see Section III), and the results are transferred

to the host PC through the PCI-express bus. To support data exchange through PCI-express, a dedicated driver was developed. The programmable logic uses the Intellectual Property (IP) core of the central direct memory access (Xilinx CDMA) [16] module to copy data through AXI PCI express bridge (Xilinx AXI-PCIE) [17]. Data transfer in the host PC is organized through direct memory access (DMA). To work with different devices, a driver (kernel module) was developed. The driver creates in the directory /dev a character device file that can be accessed through read and write functions, for example write(file, data array, data size). The PC BIOS assigns a number (an address) to the selected base address register (BAR) and a corresponding interrupt number that will be later used to indicate the completion of a data transfer. As soon as the driver is loaded, a special operation (probe) is activated and the availability of the device with the given identification number (ID) is verified (the ID is chosen during the customization of the AXI-PCIE). Then a sequence of additional steps is performed (see [18 pp. 302-326] for necessary details). A number of file operations are executed in addition to the probe function. In our particular case, access to the file is done through read/write operations.

VI. EXPERIMENTAL RESULTS AND COMPARISON

The system for data transfers between a host PC and an FPGA has been designed, implemented, and tested. Experiments were done in the VC707 prototyping board [19] that contains Virtex-7 XC7VX485T FPGA from the Xilinx 7th series with PCI express endpoint connectivity "Gen1 8-lane (x8)". All circuits were synthesized from the specification in VHDL and implemented in the Xilinx Vivado 2016.2 design suite. Software programs in the host PC run under Linux operating system and they were developed in C language. Data were transferred from the host PC to the VC707 and back through PCI express. The host PC is based on Intel core i7 3820 3.60 GHz.

For the experiments two sorting and merging systems were built. The first system is capable of sorting 32-bit data items and does not perform sorting compression and data counting. The second system is based on the first system, but includes compressing and merging functionality. The second system is adjustable for different data item sizes. Both systems are capable of sorting 2^{16} of 32-bit data items (256KB) maximum and require identical number of RAM blocks.

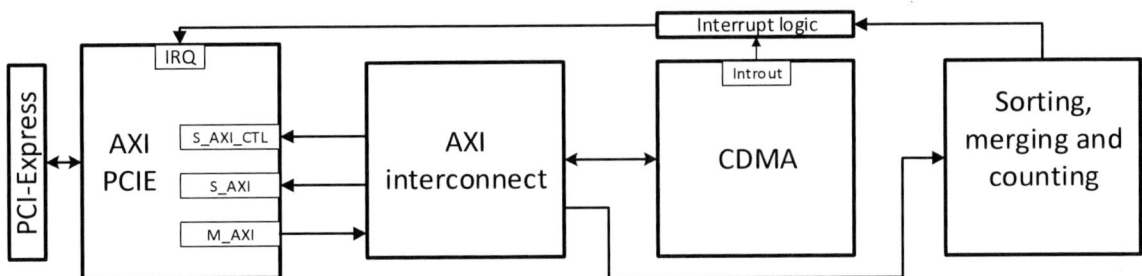

Fig. 5. Basic hardware architecture.

We conducted the experiments for sorting and merging without any item counting only for of 2^{16} 32-bit, since the merge operation is the most time consuming part of the system and it depends only on the number of the inputs and not the size of the item. Merging with item counting was performed for 2^{16} of 32-, 16- and 8-bit items. The 36-bit size of the word for 32-bit items in the BRAM was chosen. It means that the item count part of the word is 4-bit and capable of counting up to 15 repetitions, which is enough for experiments with randomly generated data. The system was configured to work with 32-bit words with 16-bit size of both the item and the count parts for counting and merging of 16- and 8-bit data.

The experiments were conducted with randomly generated numbers. The merging with counting 32-bit items didn't show any noticeable speedup over simple merging, since 2^{16} of randomly generated numbers do not have significant number of repetitions. The merging with counting of the same number of 16-bit data items is 1,45 times faster than the simple merge and merge of 8-bit items is 27,28 times faster.

We experimented with different volumes of 8-, 16- and 32-bit data items and compared them with software sorting. The host PC was used for merging the data sets larger than volume of data that can be processed with the FPGA. In addition to data sorting and merging, PCI express throughput and operating system overhead were also taken into account.

Fig. 6. shows the experimental results of sorting different volumes of randomly generated 32-bit data items in the proposed system compared to the software sorting implemented with C language qsort function. The results clearly demonstrate that the proposed solution is faster. Our experiments did not show any noticeable difference in sorting sets with different size of data items in software with qsort and therefore the results only for 32-bit data items are presented.

Fig. 7. depicts comparison of sorting data sets in the proposed system with 8-, 16-, and 32-bit item sizes.

The experiments show that the sorting throughput for the proposed systems is significantly better than in the host PC. Also it is clearly seen that merge operation with compression of repeated items is faster for data sets with high item repetition. Also the proposed system performs not only data sorting, but provides the number of repetitions of every data item in the set along with completely sorted data. This information can be used for extracting the most frequently encountered items and other subsequent data processing. The subset extraction of the most frequent numbers can be done in hardware by inserting after the last level of merging partial sorters for subset extraction presented in [10].

VII. CONLUSION

The paper suggests hardware-based methods of data sorting with simultaneous counting the repetition of all data items. The merging system performs counting and compression of the sorted subsets, which speeds up

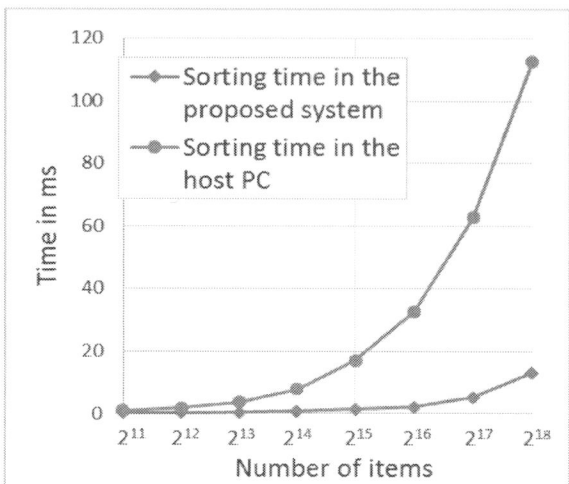

Fig. 6. Experimental results of sorting 32-bit data items in the proposed system and in the host PC.

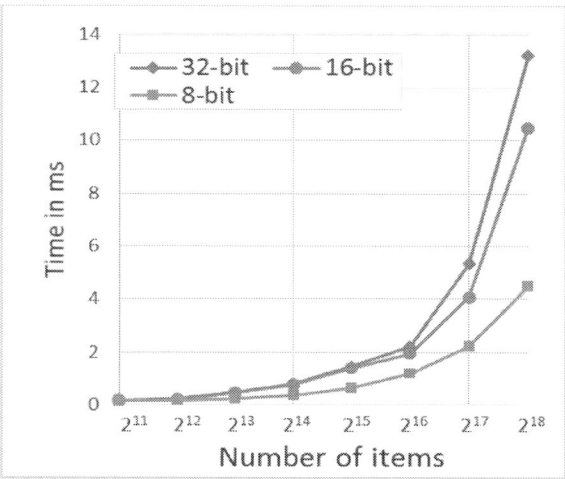

Fig. 7. Experimental results of sorting data sets with different item sizes.

sorting of data sets with large number of repeated items and provides the number of repetitions for all items along with completely sorted data set. The proposed solutions are highly parallel permitting capabilities of programmable logic to be used very efficiently. All the proposed methods were implemented in commercial microchips, tested, evaluated, and compared. The results of experiments have shown significant advantages of the proposed architecture.

ACKNOWLEDGMENT

This research was supported by the institutional research funding IUT 19-1 of the Estonian Ministry of Education and Research, the Study IT in Estonia Programme, and Estonian Association of Information Technology and Telecommunications.

REFERENCES

[1] E.V. Kalé, E. Solomonik, "Sorting," in Encyclopedia of Parallel Computing, Springer Science+Business Media, 2011, pp. 1855-1862.

[2] S.W. Aj-Haj Baddar, K.E. Batcher, Designing Sorting Networks. A New Paradigm., Springer, 2011.

[3] Z.K. Baker, V.K. Prasanna, "An Architecture for Efficient Hardware Data Mining using Reconfigurable Computing Systems," in Annual IEEE Symposium on Field-Programmable Custom Computing Machines, Napa, USA, 2006.

[4] X. Wu, V. Kumar, J.R. Quinlan, et al., "Top 10 algorithms in data mining," Knowledge and Information Systems, vol. 14, no. 1, pp. 1-37, 2014.

[5] Marcelino, R.; Neto, H.C.; Cardoso, J.M.P., "A comparison of three representative hardware sorting units," in *35th Annual Conference of Industrial Electronics*, 2009.

[6] Ren Chen, Viktor Prasanna, "Accelerating Equi-Join on a CPU-FPGA," Ming Hsieh Department of Electrical Engineering – Systems, University of Southern California, Los Angeles, California, 2016.

[7] J. Teubner, R. Muller, and G. Alonso, "Frequent item computation on a chip," IEEE Trans. Knowl. Data Eng, vol. 238, pp. 1169-1181, 2011.

[8] V. Sklyarov, I. Skliarova, A. Rjabov, A. Sudnitson , (2013), Implementation of Parallel Operations over Streams in Extensible Processing Platforms. Circuits and Systems (MWSCAS), 2013 IEEE 56th International Midwest Symposium on (pp. 852-855).

[9] Sklyarov, V.; Rjabov, A.; Skliarova, I.; Sudnitson, A. (2016). High-performance Information Processing in Distributed Computing Systems. International Journal of Innovative Computing, Information and Control, 12 (1), 139−160.

[10] V. Sklyarov, I. Skliarova, A. Rjabov, A. Sudnitson, (2015). Zynq-based System for Extracting Sorted Subsets from Large Data Sets. Informacije MIDEM, 45(2), 142-152.

[11] V. Sklyarov, A. Rjabov, I. Skliarova, A. Sudnitson, (2016). High-performance Information Processing in Distributed Computing Systems. International Journal of Innovative Computing, Information and Control, 12 (1), 139−160.

[12] A. Rjabov, V. Sklyarov, I. Skliarova, A. Sudnitson, (2015). Processing Sorted Subsets in a Multi-level Reconfigurable Computing System. Elektronika ir Elektrotechnika, 21(2), 30-33.

[13] S.N. Salloum, D.H. Wang,, "Fault tolerance analysis of odd-even transposition sorting networks with single pass and multiple passes," in IEEE Pacific Rim Conference on Communications, Computers and signal Processing, 2003.

[14] V. Sklyarov, I. Skliarova,. (2014); High-performance implementation of regular and easily scalable sorting networks on an FPGA, Microprocessors and Microsystems, 38(5): 470-484.

[15] Xilinx, Inc., 7 Series FPGAs Memory Resourcesv1.12, https://www.xilinx.com/support/documentation/user_guides/ug473_7Series_Memory_Resources.pdf, 2016

[16] Xilinx, Inc., AXI Central Direct Memory Access v4.1, http://www.xilinx.com/support/documentation/ipdocumentation/axicdma/v41/pg034-axi-cdma.pdf, 2015.

[17] Xilinx, Inc., LogiCORE IP AXI Bridge for PCI Express v1.06, http://www.xilinx.com/support/documentation/ipdocumentation/axipcie/v25/pg055-axi-bridgepcie.pdf, 2012.

[18] J. Corbet, A. Rubini, G. Kroah-Hartman, Linux Device Drivers, http://lwn.net/Kernel

[19] Xilinx, Inc., VC707 evaluation board for the Virtex-7 FPGA user guide. http://www.xilinx.com/support/documentation/boards_and_kits/vc707/ug885_VC707_Eval_Bd.pdf, 2016 (accessed 08.06.2016).

MIPRO 2017, May 22- 26, 2017, Opatija, Croatia

New Classes of Kochen-Specker Contextual Sets

Norman D. Megill* and Mladen Pavičić†

*Boston Information Group, 19 Locke Lane, Lexington MA 02420, U. S. A.

†Department of Physics, Nanooptics, Math.-Nat. Fakultät, Humboldt-Universität zu Berlin, Germany and
Center of Excellence for Advanced Materials and Sensing Devices (CEMS), Photonics and Quantum Optics Unit,
Ruđer Bošković Institute, Zagreb, Croatia

Email: *nm@alum.mit.edu, †mpavicic@irb.hr

Abstract—**Finding Kochen-Specker contextual sets proves to be essential for quantum information and quantum computation in particular. It is therefore essential to find algorithms and programs which can generate arbitrary Kochen-Specker sets in a nearly-exhaustive manner. In this paper we present such generations for two new classes of Kochen-Specker sets. All sets from one of the classes are completely invisible to standard algorithms and programs from the literature as well as the upper part of sets from the second class. We also describe the methods and programs we used to obtain the sets on supercomputing clusters.**

I. INTRODUCTION

Kochen-Specker (KS) sets are sets of n-tuples of mutually orthogonal vectors from n-dim Hilbert space to which it is impossible to assign 1s and 0s in such a way that
(i) No two orthogonal vectors are both assigned the value 1;
(ii) In any group of n mutually orthogonal vectors, not all of the vectors are assigned the value 0.

KS sets not properly containing smaller KS subsets are called critical KS sets. If any orthogonal basis of a critical KS set is removed, it will no longer be a KS set. Only critical KS sets are relevant for experimental implementations.

KS sets can be represented as hypergraphs in which each vertex represents a vector and each edge an orthogonal basis.

In addition to presenting new results, this paper describes the methods and programs that we used. The collection of programs that helped us find these results are open source and freely available. Many of them are useful for working with hypergraphs generally, not just hypergraphs representing KS sets. The programs are normally run in a Unix or Linux command-line shell. Most of them perform a specific operation or test on a hypergraph. Because they all read and write a common language for hypergraphs, they can be chained (piped) to each other, and standard Unix utilities such as `grep` can be placed in the chain to filter desired results for the next stage. For very large problems such as finding the KS results described here, we can easily create scripts that can distribute the work over many CPUs in a supercomputing cluster.

For our computer representation, we encode hypergraphs using alphanumeric and other printable ASCII characters. We call these character strings *MMP hypergraphs*. Each vertex (vector) is represented by one of the following characters: 1 2 ...9 A B ...Z a b ...z ! " # $ % & ' ( ) * - / : ; < = > ? @ [\ ] ^ _ ` { | } ~ and then

again all these characters prefixed by '+', then prefixed by '++', etc. Skipping of characters in this sequence is allowed.

Hypergraph edges (orthogonal bases) are encoded as a string of concatenated vertices, each edge containing as many vertices as there are dimensions, and edges are separated with commas. The complete hypergraph is terminated with a full stop (period) and is contained on a single line of text regardless of length. The order of the edges is irrelevant as is the order of vertices within each edge. The numbers of vertices and edges are unlimited. We often present MMP hypergraphs starting with edges forming the largest loop to facilitate their possible drawing.

A simple example of a hypergraph expressed with MMP notation is `123,3ab,a#++a.` which has 3 edges with 3 vertices per edge and 7 different vertices named 1, 2, 3, a, b, #, and ++a. (This example is for illustration and does not represent a KS set.)

II. RESULTS

Waegell and Aravind recently constructed a KS set from the Witting polytope [1]. The set has 148 vertices and 265 edges—we denote it as 148-265—and it served us to generate over 300 types of smaller KS critical sets, where a *type* means a KS set with a particular number of vertices and edges. All KS sets that can be generated from this set build a *class* of KS sets which we shall call the 148-265 class. The 148-265 set itself we call the *master set*.

For the generation of KS critical sets from the 148-265, our approach turns out to be indispensable, since a standard approach in the literature, [2]–[8] using what are called parity proofs, turns out to be inapplicable.

A KS set is said to have a *parity proof* if its hypergraph has an odd number of edges and each vertex is common to an even number of edges. Any hypergraph satisfying this condition corresponds to a KS set, because it is impossible to assign an even number of 1s to an odd number of edges if we require exactly one 1 per edge. The converse fails, though; there are KS sets whose hypergraphs don't satisfy this condition.

We obtained over 300 types of KS, and none of them has a parity proof. Thus all the KS sets we found from the 148-265 class are completely invisible to a standard approach using parity proofs. In Fig. 1 we show a graphical representation of the statistics for the obtained KS from the 148-265 class together with two chosen hypergraphs.

182

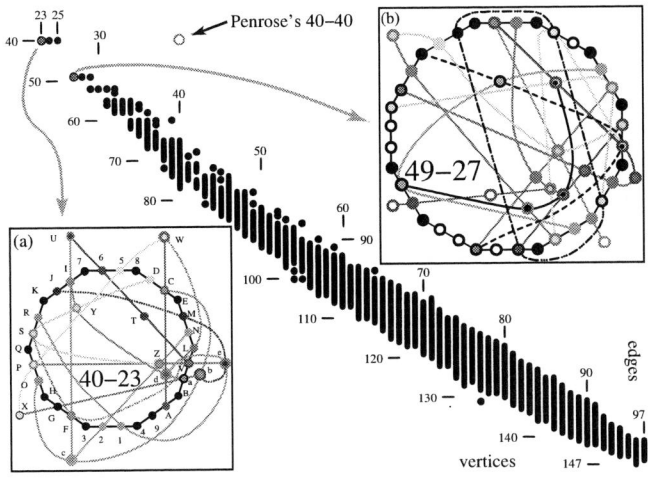

Fig. 1. KS criticals from the 4-dim 148-265 class; 40-40 hollow circle indicates the Penrose 40-40 non-critical KS set; inset (a) shows one of the smallest KS criticals generated from Penrose's 40-40 set; inset (b) shows one of the smallest KS criticals not contained in the 40-40 set.

The other class we considered is also mostly invisible in the standard approach using parity proofs. Waegell and Aravind considered a dual of 600-cell convex regular polytope and obtained a 300-675 master KS set [7]. From it, using parity-proof-based algorithms and programs they obtained 96 types of KS criticals, from 38-19 to 82-41. Then they commented that according to the parity proof method there are no sets smaller than 38-19 and said that they had not found any set larger than 82-41. In Fig. 2 we plotted the statistics of over 260 types we found (top left in red and bottom right in blue) together with their 96 types (top left in cyan).

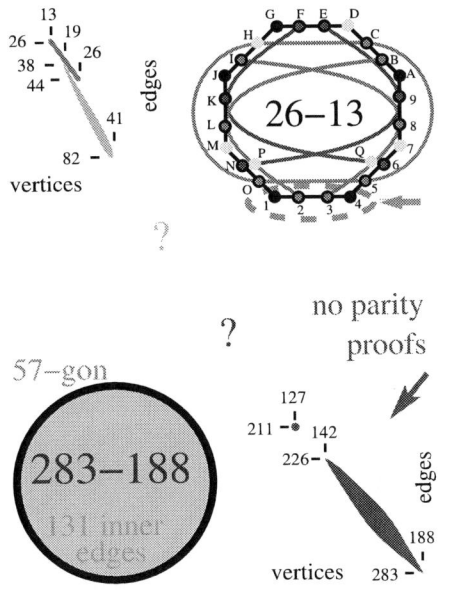

Fig. 2. KS criticals from the 4-dim 300-675 class; there are 250 types of KS critical sets from the upper part of the class (211-127 till 283-188); 26-13 is the smallest KS critical and 283-188 the largest we obtained.

In Fig. 2 we show that, first, there are 250 large KS criticals, from 211-127 to 283-188. Since none of these has a parity proof, Waegell and Aravind did not find them using the parity proof method. We also found the following small types: 26-13, 30-15, 32-17, 33-17, and 34-17, which are all smaller than 38-19. The reason they did not see them is that they are generated directly from large subsets of the master set that do not have parity proofs and are invisible in the parity proof approach. So, our approach is indispensable for generating all possible KS sets, large as well as small.

In particular, we not only revealed the aforementioned huge KS criticals but also showed that the 300-675 class partially overlaps the 60-75 class we generated in [9] (all 26-13,...,44-26 we obtained from the 300-675 master set can also be deduced from the master set 60-75 which defines the 60-75 class). (Their 38-19 is not from the 60-75 class.)

In order to prove that a set has the KS property, we must show that it does not admit an assignment of 0s and 1s such that exactly one vertex on each edge has a 1 assigned to it. Rather than rely on indirect parity proofs, our approach tests this condition directly. Our program states01 (see Sec. IV-A) does this with a backtracking algorithm that either finds such a state (showing the set is not KS) or exhausts all possibilities (showing the set is KS). In the latter case, it means that at least one edge will be left over without it being possible to assign a 1 to any of its vertices, regardless of how the other edges have their 1 assigned. When a KS set is represented graphically as, e.g., 26-13 in Fig. 2, the KS property instances may be visualised (dashed magenta ellipse indicating such an edge). The parity proofs used by the standard programs in the literature are implicitly contained in our general proofs. However, states01 can also reproduce the literature results with an additional option that checks whether a KS set also has a parity proof.

III. ALGORITHMS AND PROGRAMS

A. Overall philosophy

Our programs have a more or less uniform way of specifying the options to control functionality, and learning to use them is straightforward. Most options available for a program perform a single task that is well documented, and the programs can be easily chained via pipes to perform complex functions. All programs written by us have a built-in --help option, e.g. states01 --help, that provides full documentation of the program's options and features as well as providing examples of the program's use. We have been careful to ensure the help documentation accurately describes the program behaviour, so that in principle the programs can be used by other workers without assistance from us.

A great deal of work has been put into making the programs' runtimes as small as practical, often with good results. Not only does this let us obtain results faster, it reduces the expense of the computation resources needed. A common situation with the types of problems that some of our programs work with is that their run times inherently grow exponentially with MMP hypergraph size in the worst case, which as far as we know is theoretically unavoidable. To overcome this limitation in a practical sense, a number of heuristic techniques were developed to suppress the exponential behaviour.

For example, the program `states01` iterates through the edges; at each edge, it either makes a trial 0/1 assignment to the unassigned vertices on the edge or backtracks to try another path when an assignment is not possible. A very successful heuristic, which we call the "clustering algorithm," starts with edges that have the most connections (i.e. shares the most vertices) with other edges. This greatly increases the probability of an early assignment conflict, so that backtracking will exhaust all possible assignment paths more quickly when proving that an MMP hypergraph is a KS set.

Since it seems impossible to avoid exponential behaviour completely, all programs having such behaviour incorporate user-settable timeouts. MMP hypergraphs which timed out are flagged in the output and can be put aside to study later, whether it is running them longer or using them to study and improve the algorithms further.

There are occasional situations where the exponential behaviour manifests itself and becomes problematic, i.e. the program gets "stuck," when the hypergraph traversal coincidentally encounters an unlucky pattern. Typically these arise as a long sequence of edges or vertices of length k that do not have early conflicts, and the program will "count" through 2^k possibilities for each later path it backtracks from when a conflict is eventually found. We don't have a good method for identifying these situations in advance; instead, we put them aside to deal with later when they time out. Empirically, we have found that randomising the edge order with our program `mmpshuffle` will often break the unlucky pattern. This has been quite successful in allowing our programs to run to completion for problematic hypergraphs, although it may require some manual trial and error.

In some cases the number of possible MMP hypergraphs that need to explored are astronomical and even with a supercomputer cluster may be unfeasible to explore. For this reason, some programs such as `mmpstrip` have the ability to take random samples of the search space so that we can statistically characterise the properties of the whole search space. We have put care into the randomisation process in two ways. First, for the seed we combine different sources of entropy such as the time of day and the process ID so that it will be unlikely that parallel runs on a supercomputer cluster will use the same random number sequence. Second, we want our results to be reproducible: all programs using random numbers will print out the seed that resulted from the entropy, and the user can specify that seed to replicate the program exactly run if desired.

B. Standard input and output formats

In order to allow the programs to present their outputs in a standard way and communicate to other programs (as well as to humans interpreting the outputs), additional information is sometimes prefixed and suffixed to the MMP hypergraph. An optional *MMP prefix* is any sequence of printable characters (that may include spaces) that ends with a space. An optional *MMP suffix* is a vector assignment enclosed in

braces $\{\dots\}$; it is described below but in particular does not contain any spaces. An MMP hypergraph with a prefix and/or suffix is called an *MMP line*. An example of an MMP line with both a prefix and suffix is "#2(1/7=22%) fail:: 1234,1567.{{1={0,0,1}}}", where "1234,1567." is the MMP hypergraph proper. If an MMP line has a space, that means it has a prefix which ends at the last space on the line. If an MMP has characters after the full stop ending the MMP hypergraph, those characters are assumed to be an MMP suffix. In all cases, the MMP line ends with a new line character in the computer file containing it (i.e. there must be exactly one MMP hypergraph on each line).

The prefix information depends on what information the program needs to present to the user, and also to allow the program output to be filtered with programs such as `grep` or `sed`, matching for example the string "fail:: ", as part of a Unix pipe. The format of each program's MMP prefix, if it has one, is documented by the program's `--help` command. Some programs preserve the MMP prefix or suffix while others rewrite them with the program's results or strip them off. The unrestricted format of the prefix (other than its trailing space) gives complete flexibility for a program to present its output information in whatever form is convenient as well as to add future information as a program's capabilities are enhanced.

The optional MMP suffix is a list of partial or complete vector assignments to vertices. Each vector assignment consists of a vertex name followed by = followed by a comma-separated list of complex number expressions in braces, with a complex number for each dimension. The assignments are comma-separated, with the entire list surrounded by braces. An example is {a={0,1,0,0},++B={0,0,1,i*sqrt(2)}}. The syntax for the complex number expressions is documented by the `--help` option of the `vecfind` program. Programs that renumber MMP hypergraph vertices or extract subsets of the MMP hypergraph will also modify any vector assignment suffix to keep it consistent.

Overall, the ability to chain programs via pipes and filters, sharing a standard hypergraph "language," allows sophisticated processing to be performed in a relatively simple and robust way. The starting input set can consist of millions or billions of MMP hypergraphs, providing an ideal problem for a supercomputer cluster.

C. Finding KS sets

In order for an MMP hypergraph to correspond to a KS set, there must exist an assignment of vectors to the vertices such that the orthogonality conditions specified by the edges are satisfied. Second, there must not exist an assignment (sometimes called a "colouring") of 0/1 (non-dispersive or classical) probability states to the vertices such that each edge has exactly one vertex assigned to 1 and its others assigned to 0.

For a given MMP hypergraph, we use two programs as filters for these two conditions. The program `vecfind` attempts to find an assignment of vectors to vertices. The program `states01` determines whether or not a 0/1 colouring is possible that meets the above requirement. We will describe their algorithms below.

One basic method in our approach—carried out by means of the program `mmpstrip`—has been to generate successive subsets from a master KS set with many redundancies i.e. that is far from being critical. There are several of these that have been identified in the literature. The goal is to find smaller critical subsets that may be of use in future experiments.

IV. SUMMARY OF PROGRAMS

In this section we summarise the functionality of the programs used in our KS work. In all cases, a program's `--help` option can be consulted for more detail. The input to most programs is a list of MMP hypergraphs (in a file or from the standard input), and the output is zero or more MMP hypergraphs for each input MMP hypergraph. The programs can be chained with Unix pipes to perform more complex filtering and processing of MMP hypergraphs (which we will also call just MMPs in this section for brevity).

Most of our programs are written in the C language for efficiency and designed to operate primarily in Unix environments at the command-line or shell-scripting level. They are written in strict ANSI C so that they will work with different operating systems and compilers. We have been careful to keep the help documentation updated to accurately describe the available options so that they can be used by workers outside of our group. Typically the help includes simple examples that can be reproduced while learning to use the program.

A. Program `states01`

This program determines whether or not there exists an assignment of 0/1 (non-dispersive or classical) probability states to the vertices of an MMP hypergraph such that each edge has exactly one vertex assigned to 1 and its others assigned to 0. If not, then the MMP is a candidate for a KS set.

By default, an exhaustive search, using a backtracking algorithm, is done to determine whether an assignment is possible. Optionally, a faster search can be done using the parity proof method (`-p` option). While MMPs with parity proofs are definitely KS sets, MMPs that don't have parity proofs may or may not be KS sets. One purpose of the `-p` option is to be able to reproduce work done by others that is based on parity proofs.

Since the backtracking algorithm is exponentially slow in the worst case, the `-t` option allows a timeout to be specified. We have found that for typical MMPs of interest, timeouts occur relatively infrequently.

The `-c` option tests whether or not a KS candidate is a critical KS. It does this by first checking that the MMP is a KS set, then checking that it ceases to be a KS set when any edge is removed.

For non-critical KSs, the `-r` option will randomly discard edges until a critical KS is obtained. For highly redundant starting KSs, different critical KSs will be produced each run (unless a fixed random number seed is specified to reproduce a previous run).

B. Program `vecfind`

In order to determine whether a given KS candidate (i.e. one admitting no 0/1 states) is a true KS set, we must show that a non-conflicting assignment of mutually orthogonal vectors to each edge is possible. In general, this is a difficult problem. In some cases, symmetry of a master starting MMP allows assignment from a small collection of possible vectors; for example, the 60-60 MMP of [5] can be assigned with vectors with components from the set $\{0, 1, \tau = (1 + \sqrt{5})/2, 1/\tau\}$ and their negatives.

The `vecfind` program will search for an assignment from a set of vector components, or a set of vectors, provided by the user. It will show an assignment if one is possible, or tell the user that it is impossible. If a set of vector components is provided, internally `vecfind` will compute all possible non-proportional vectors constructed from them and use that set for the search.

The algorithm assigns vectors from the given or computed list of vectors to the vertices. If a further assignment from the list is not possible (meaning it can't achieve a mutually orthogonal assignment of vectors from the list to an edge containing the vertex), it will backtrack and continue with the next vector in the list at the backtrack point.

Unlike the `states01` backtracking algorithm, there may be thousands of vectors to try at each vertex rather than just the two states 0 and 1. Both programs have exponential behaviour worst case, but clearly it is much more severe with `vecfind` since the number of assignments to reject grows as v^n where v is the number of vectors to try and n is the number of vertices. Because of this, we have invested a large effort into heuristics that experimentally have speeded up the search significantly, sometimes orders of magnitude, for typical MMPs and vector sets. For example, we use a "dynamic" version of the clustering algorithm of `states01`, in which the next vertex to try is the one with the largest number of potential conflicts based on the vector assignments so far, the idea being to encourage earlier backtracking. These efforts have proved moderately successful in that we can find vector assignments already known to the literature (such as the 60-60 mentioned above) almost instantly. However, some large MMPs still cause the program to be excessively slow, particularly when no assignment is possible, and therefore `vecfind` also has a `-t` timeout option.

The output of `vecfind` is the input MMP with a prefix indicating whether the assignment was successful and other information, along with a suffix containing a vector assignment to the vertices. If the assignment was successful, all vertices will be in this list, otherwise the best partial assignment it could find is used for the list. An MMP with a partial vector assignment suffix can be used as the input of a second run of `vecfind` with a different set of vectors to try on the unassigned vertices, and the assignments made by the suffix will remain the same.

The vector components are complex numbers that may be specified directly as in "4+5*i" or they may be specified with expressions involving arithmetic operations and common functions and constants such as sqrt, e, and pi, as in "e^(2*pi*i/3)". The syntax and available functions are documented in the --help output. A built-in calculator, invoked with the option -calc, allows the user to check that a given expression evaluates to the expected complex number.

C. Program mmpstrip

Much of our work has involved starting with a very large starting or master KS set, such as the 60-60 (non-critical) KS set mentioned above, and exploring and characterising the smaller critical KS sets that are embedded. This has been a very fruitful effort, allowing us to find millions of non-isomorphic critical KS subsets in this case. This was an unexpected find and means that there is no key critical KS that characterises the 60-60, but instead it provides a rich source of critical KSs that can be used for physical experiments.

To assist studying such master KS sets, the program mmpstrip provides a flexible means of selecting subsets from the master. After subsets are selected with mmpstrip, they are filtered for ones that are KS sets (have no colouring) using states01, then filtered again to eliminate isomorphic ones using the program shortd.

From an input MMP hypergraph with n edges, mmpstrip will strip k edges so as to produce all $\binom{n}{k}$ subsets with a simple combinatorial algorithm. Partial output sets can be generated by means of start and end parameters. An increment parameter specified by option -i can be applied to skip all but every ith output line for partial sampling of the output subsets if the full output is too large. With the option -c1, the program will calculate in advance how many output hypergraphs will result, in order to help the user choose optimal parameter settings.

Alternatively, the option -r will choose random samples from the master set in order to lessen the chance of a biased selection in case there is a subtle pattern that repeats every ith hypergraph. As with our other programs offering random sampling, the random number seed is displayed and may be used to reproduce the run exactly if desired.

The option -u will discard unconnected MMP hypergraphs (i.e. those consisting of two or more disconnected parts). By default, the vertices in output MMPs are relabelled so as to avoid gaps in the vertex naming, although this behaviour may be suppressed with the -n option.

The above options are the ones most commonly used and are described in more detail the mmpstrip --help output. There are several other options also described in the help documentation.

D. Program shortd

Two MMP hypergraphs are *isomorphic* if one can be transformed into the other via reordering edges, reordering vertices within an edge, and renaming the vertices.

The program shortd, written by Brendan McKay, renumbers an input MMP hypergraph into a canonical form, so that two isomorphic MMPs will have the same canonical form and can be easily identified. The program is extremely fast and is based on McKay's extensive theoretical work on graphs and hypergraphs.

We routinely use shortd to eliminate isomorphic MMPs from a list, for example after producing a collection of MMP subsets with mmpstrip.

E. Program subgraph

This program, described in Ref. [10], will check whether an MMP is isomorphic to a subset of a larger MMP. It uses an algorithm suggested by Brendan McKay. If the two MMPs are the same size, it will tell us whether the two MMPs are isomorphic. While shortd will do the same thing much faster, subgraph will show the actual isomorphism in case there is one, rather than providing just a yes/no answer.

F. Program loop

Two edges are *connected* if they share a vertex. A *loop* is a set of connected edges in which each shared vertex is shared with exactly two edges in the set, in other words a chain of connected edges where the last edge connects to the first.

The program loop by default will list all possible loops in an MMP hypergraph. The list of loops can assist drawing the MMP; in particular, a large loop can be selected as the main circle for the drawing, with other edges drawn with lines to complete the MMP.

loop has a number of options that can be seen in its --help listing. In particular, the program can list only the b largest loops that were found (-b option), it can stop searching after m loops are found (-m option), and a timeout can be set to give up and continue from the next edge.

G. Program mmpshuffle

This program is primarily used to randomly scramble order of the edges of an MMP hypergraph and the order of the vertices in each edge, producing an isomorphic MMP with components in a different order. This can be useful to help break exponential behaviour of certain programs caused by unlucky patterns in the MMP. For example, if states01 times out with ./states01 -t100000 < in.mmp > out.mmp, we can try ./mmpshuffle -r < in.mmp | states01 -t100000 > out.mmp to see if a different edge order is more successful. There are also options in mmpshuffle to rename the vertices, eliminating gaps in naming, and to reverse the order of edges and vertices.

When the MMP has a vector assignment suffix, mmpshuffle will also rename the vertices specified in the assignment accordingly.

H. Program mmptag

This is a small utility program with several functions. It can strip off MMP prefixes and suffixes, it can add prefixes specifying the number of edges and vertices, and it can produce a statistical breakdown of the number of edges and vertices in a large collection of MMP hypergraphs.

I. Program mmpxlate

This program can translate from and to several other hypergraph formats found in the literature. We are open to adding more formats to meet the needs of other groups.

V. RUNNING PROGRAMS ON CLUSTERS

We have empirically found that for most tasks it is much more efficient to distribute sequential tasks through, for instance, HTCondor grid scheduling, to nodes in the grid, than to parallelize jobs. An example of our procedures is the following one. HTCondor software allows for two input-output parameters, say i and j. In the first step, we might strip edges from a master set, say 148-265, and filter them with sed, states01, again sed, and shortd, so as to generate non-isomorphic KS sets in, say 148-j, $j = 165, \ldots, 264$, i.e., 100 output files, each cut to a chosen number of lines, say 1 million; each line contains a KS set i-j where i depends on j in a rather involved manner depending of how many vertices, if any, were stripped together with stripped edges. We then grep KS sets into i-j files each of which contains only i-j KS sets. In the next step, we use these files as input files to generate non-isomorphic critical KS sets i-j-c-k, $k = 1, \ldots, 20$, via states01 and shortd. Each line in these output files contains an l-m final critical KS set, where l and m are not related to i and j except for the inequalities $l \leq i$ and $m \leq j$. Some KS sets are so intricate that they ask for parallelizing tasks, and that takes us to the next section.

VI. FUTURE WORK

Most of our work up to now has been with MMP hypergraphs that are small enough so that thousands or millions can be characterised on each CPU in a supercomputer cluster in a few days (so in the end we could end up with billions of non-isomorphic KS sets). However, some recent very large hypergraphs, such as the 300-675 mentioned above, stress the limits of some of our programs in that they may take days or weeks to test just one MMP hypergraph. We are planning to give these programs the ability to partition the run for a single MMP hypergraph into different sections that can be run in parallel on different CPUs in a supercomputer cluster. In particular, we are currently working on states01 and vecfind to add this feature.

We are encouraged towards this goal by our success in earlier work with Hilbert lattice equations, where we were able to verify that a huge equation failed in a very large lattice counterexample (the 7oa equation and the Peres lattice of Ref. [11]) with a run that would have taken years if done on a single CPU. By carefully modifying our program latticeg described there, we able to partition the run into small pieces that were distributed among many CPUs in a supercomputer cluster, leading to the answer in a few days. This established the independence of this new Hilbert lattice equation from earlier members 3oa through 6oa in the OA (orthoarguesian) family. Incidentally, that work used an MMP hypergraph in the completely different role of representing a Greechie diagram (a kind of orthomodular lattice), showing the flexibility of applications for the MMP notation.

VII. DISCUSSION

An important point to take away from this article is the power of a standard, compact, and precisely specified format for representing hypergraphs, which in our case is the MMP hypergraph notation. It provides the common language that allows all of our hypergraph processing programs to communicate with each other. The code needed to parse MMP hypergraphs into internal arrays, and conversely to generate them from internal arrays, is relatively simple and very fast; it can easily be incorporated into other programs whether written by us or someone else.

The ability to pass MMP hypergraphs from one program to another via Unix pipes, along with Unix filters (grep, sed, etc.) in between, has allowed us to automate processing of massive hypergraph collections and divide the work among supercomputer CPUs in an efficient way.

As mentioned in Section VI, MMP hypergraphs can be used for other applications involving hypergraphs. The mmpxlate program can translate from and to several other hypergraph notations, and we are always looking to add more. As the mmpxlate collection grows, we anticipate it will become a generally useful translation tool between hypergraph notations via MMP hypergraphs.

We note that all of the MMP hypergraph processing programs described above, other than states01 and vecfind, can be used for general hypergraph work and are not specialised for the study of KS sets. Most of the programs have been used for many years with a wide variety of MMP hypergraphs, and all are free from known bugs. With the aid of their built-in --help option, they are usually no harder to learn to use than a new Unix command.

ACKNOWLEDGEMENT

M.P. acknowledges a support by the Croatian Science Foundation through project IP-2014-09-7515, and the Ministry of Science, Education, and Sport of Croatia through the CEMS funding. Computational support was provided by the cluster Isabella of the Zagreb University Computing Centre and by the Croatian National Grid Infrastructure.

REFERENCES

[1] M. Waegell and P. K. Aravind, "The Penrose dodecahedron and the Witting polytope are identical in $\mathbb{CP}^3$," *ArXiv:1701.06512*, January 2017.

[2] P. Lisoněk, P. Badziąg, J. R. Portillo, and A. Cabello, "Kochen-Specker set with seven contexts," *Phys. Rev. A*, vol. **89**, pp. 042 101–1–7, 2014.

[3] M. Planat, "On small proofs of the Bell-Kochen-Specker theorem for two, three and four qubits," *Eur. Phys. J. Plus*, vol. **127**, pp. 86–1–11, 2012.

[4] M. Planat and M. Saniga, "Five-qubit contextuality, noise-like distribution of distances between maximal bases and finite geometry," *Phys. Lett. A*, vol. **376**, pp. 3485–3490, 2012.

[5] M. Waegell and P. K. Aravind, "Parity proofs of the KochenSpecker theorem based on 60 complex rays in four dimensions," *J. Phys. A*, vol. **44**, pp. 505 303–1–15, 2011.

[6] ——, "Proofs of Kochen-Specker theorem based on a system of three qubits," *J. Phys. A*, vol. **45**, pp. 405 301–1–13, 2012.

[7] ——, "Parity proofs of the KochenSpecker theorem based on 120-cell," *Found. Phys.*, vol. **44**, pp. 1085–1095, 2014.

[8] ——, "Parity proofs of the KochenSpecker theorem based on the Lie algebra E8," *J. Phys. A*, vol. **48**, pp. 225 301–1–17, 2015.

[9] N. D. Megill, K. Fresl, M. Waegell, P. K. Aravind, and M. Pavičić, "Probabilistic generation of quantum contextual sets," *Phys. Lett. A*, vol. **375**, pp. 3419–3424, 2011.

[10] M. Pavičić, N. D. Megill, and J.-P. Merlet, "New Kochen-Specker sets in four dimensions," *Phys. Lett. A*, vol. **374**, pp. 2122–2128, 2010.

[11] M. Pavičić, B. D. McKay, N. D. Megill, and K. Fresl, "Graph approach to quantum systems," *J. Math. Phys.*, vol. **51**, pp. 102 103–1–31, 2010.

New Method for Determination Complexity Using in AD HOC Cloud Computing

M. Babič[*] and Borka Jerman-Blažič

[*]Jožef Stefan Institute
babicster@gmail.com

Abstract - The paper present new method for determination complexity of using in new hyper-hybrid AD HOC cloud computing. A large body of research has been devoted to identifying the complexity of structures in networks. The study of complex networks is a young and active area of scientific research inspired largely by the empirical study of real-world networks such as computer networks and social networks. Networks are a ubiquitous way to represent complex systems, including those in the social and economic sciences. Understanding complexity and broad systems-level frameworks in the life, physical and social sciences has turned, in recent decades, to issues of network dynamics. It has been shown that many real complex networks share distinctive characteristic properties that differ in many ways from the random and regular networks. We present some challenges problem of AD HOC network. In this paper we introduce a new method for quantifying the complexity of a new hyper hybrid AD HOC cloud network based on presenting the nodes of the network in Cartesian coordinates, converting to polar coordinates, and calculating the fractal dimension. Our results suggest that this approach can be used to determine the complexity for any type of network.

Key words: complexity, cloud computing, *AD HOC* network,

I. INTRODUCTION

In computer networking, an *AD HOC* network [1-4] refers to a network connection established for a single session and does not require a router or a wireless base station. Many ad hoc networks are local area networks where computers or other devices are enabled to send data directly to one another rather than going through a centralized access point. A wireless network that transmits from computer to computer. Instead of using a central base station (access point) to which all computers must communicate, this peer-to-peer mode of operation can greatly extend the distance of the wireless network. To gain access to the Internet, one of the computers can be connected via wire or wireless to an ISP. Basically, an ad hoc network is a temporary network connection created for a specific purpose, such as transferring data from one computer to another. Networks are a ubiquitous way to represent complex systems, including those in the social and economic sciences. Understanding complexity and broad systems-level frameworks in the life, physical and social sciences has turned, in recent decades, to issues of network dynamics. The growth of the World Wide Web, staggering advances in computing power and diminishing computational complexity of network algorithms, and demonstrations of the robustness and ubiquity of networks fitting small-world models helped spark much of the recent cross-disciplinary explosion in network analysis and modelling.

Fig. 1. The internet, the global communication network

A. Open problems and Challenges

In today's world of information science some of the most popular, fascinating and useful structures are *AD HOC* networks. However these have also given rise to many open problems. I would like to present some of these challenging problems:

Problem 1. The characteristics of large *AD HOC* networks vary if we examine them at different magnifications. Questions are: Can we find a relationship between the magnification of big networks and their topological properties? How does the presence of a small network inside big network impact on it? And can we gain insights into the properties of networks by researching small networks inside big networks?

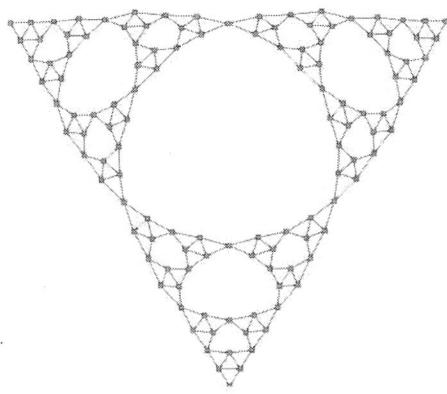

Fig. 2. Fractal graph

Problem 2. An *AD HOC* network of spacecraft around and in transit between the Earth and Mars. Direct communication between Earth and Mars can be strongly disturbed and even blocked by the Sun for weeks at a time, cutting off any future human mission to the Red Planet. Questions is: How to ensure reliable radio communication even when Mars and Earth line up at opposite sides of the Sun, which then blocks any signal between mission controllers on Earth and astronauts on the red surface.

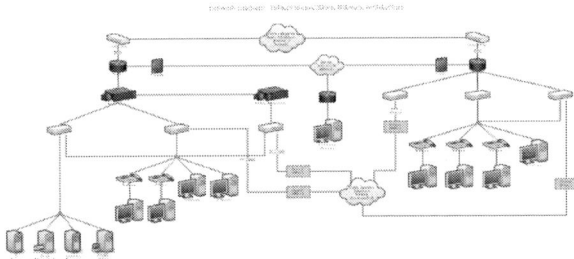

Fig. 3. MER Telecommunication Architecture Diagram

Problem 3. Security is a critical issue of *AD HOC* networks that is still a largely unexplored area. Traditional methods of protecting the data with cryptographic methods face a challenging task of key distribution and refresh. The main objectives of the proposed research are to develop secure authentication and key management protocols specific to the needs of ad hoc networks and to design highly efficient hardware architectures of the proposed protocols.

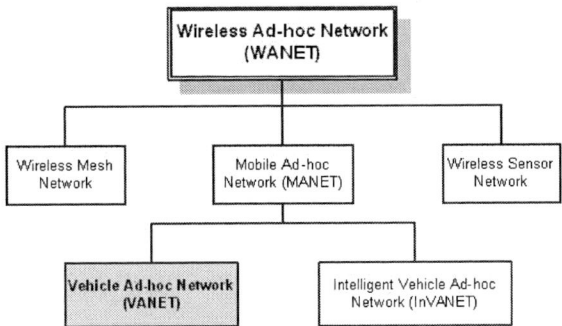

Fig. 4. Wireless Ad Hoc Network

Therefore, the proposed studies for all open problems would involve novel approaches to *AD HOC* network modeling and could contribute to predicting the behavior of *AD HOC* network in the future. All these problem we can solve with mathematical tool; graph theory. Representing a problem as a graph can make a problem much simple. A graph is a way of specifying relationships among a collection of items. In mathematics and computer science, graph theory is the study of graphs, which are mathematical structures used to model pair wise relations between objects. A graph consists of a set of objects, called nodes, with certain pairs of these objects connected by links called edges. Nodes represent computing hosts, and there is an edge joining two nodes in this picture if there is a direct communication link between them. Graphs are useful because they serve as mathematical models of network structures. Graph theory is a suitable playground for the exploration of proof techniques in discrete mathematics, and its results have applications in many areas such as computing, social, and natural sciences. Network analysis has been developing and flourishing for several decades. Network analysis is popular in every type of academic social science, AD DOV network, applied social science (such as marketing), studies of nonhuman social life, branches of mathematics, computer science, and even physics.

II. METHOD

In this paper we introduce a new method for quantifying the complexity of network based on presenting the nodes of the network in Cartesian coordinates, converting to polar coordinates, and calculating the fractal dimension.

First of all, all nodes of the network are presented in the Cartesian coordinate system. Also, determinate coordinates (x,y) of every nodes of network (step 1).

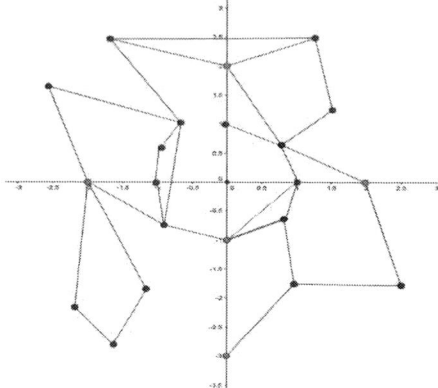

Fig. 5. Step 1

In second step, we transform nodes of the graph into polar coordinate system (step 2). This means that all nodes have two coordinates (r, Φ). So to convert from Cartesian coordinates (x,y) to Polar coordinates (r, Φ):

$$r = \sqrt{x^2 + y^2}$$

$$\tan \Phi = \frac{y}{x}$$

$$\Phi = \arctan(\frac{y}{x})$$

Also, we can describe transformation of the nodes (x,y) into (r, Φ):

$$P(r, \Phi) = P(\sqrt{x^2 + y^2}, \arctan(\frac{y}{x}))$$

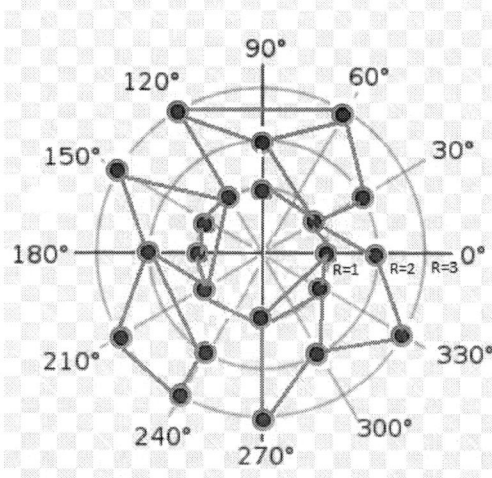

Fig. 6. Step 2

Also, in polar coordinate system points (nodes) are presented as $T_i(r_i, n_i)$. For all r_i, we denote the number of nodes as n_i for all i. All points $T_i(r_i, n_i)$ represent a linear graph $\Gamma(r, n)$.

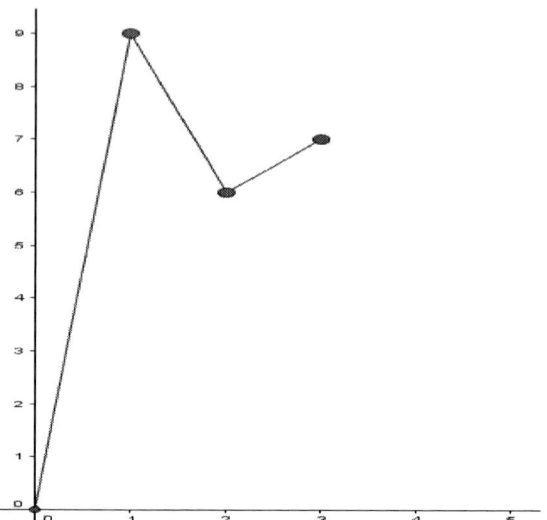

Fig. 7. Linear graph $\Gamma(r, n)$ of network

A random process is statistically evaluated using Hurst parameter H [5] or by determining the distribution function. Hurst parameter H as self-similarity criteria cannot be accurately calculated, but it can be only estimated. There are several different methods [6] to produce estimates of the parameter H, which together more or less deviate. In doing so, we have no criteria to determine which method gives us the best result.

We use R/S method for estimating Hurst exponent H. R/S method Adjusted rescaled range method or adjusted scale is also a graphic method based on the properties Hurst phenomenon. Adjusted scale of the partial summation area time series deviation from the mean. For graph $\Gamma(r, n)$, we estimate the Hurst exponent H (step 3).

The Rescaled Range is calculated for a time series or space component,

$$X = X_1, X_2, ..., X_n$$

as follows:

Calculate the m by equation (1), calculate the cumulative deviate series Z by equation (2), create a range series $R(n)$ by equation (3), create a standard deviation series S by equation (4), and calculate the rescaled range series (R/S) by equation (5).

(1)

$$m = \frac{1}{n}\sum_{i=1}^{n} X_i$$

(2)

$$Z_t = \sum_{i=1}^{t} (X_i - m)$$

(3)

$$R(n) = \max(Z_1, Z_2, ..., Z_n) - \min(Z_1, Z_2, ..., Z_n)$$

(4)

$$S = \sqrt{\frac{1}{n}\sum_{i=1}^{t} (X_i - m)^2}$$

(5)

$$\frac{R}{S} = (\frac{n}{2})^H$$

After this we can calculate the fractal dimension with equation $D = 2-H$ (step 4).The Hurst exponent $H \in (0,1)$, thus the fractal dimension of complex networks are $D \in (1,2)$.

III. RESULT AND DISCUSSION

We present randomly only one network with 111 nodes. This network is very complex and has fractal characteristics.

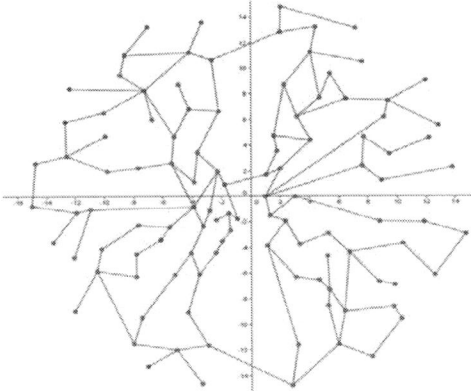

Fig. 8. Randomly network with 111 nodes

Complexity of this randomly network with 111 nodes we calculate with R/S method for estimator of Hurst exponent H. Hurst exponent H for presented network is 0.15, thus fractal dimension of network is 1.85. Also complexity of randomly network with 111 nodes is 1.85. Fig. 16. Present linear graph $\Gamma(r, n)$ of randomly network with 111 nodes. Also, for each r, we determinate number n of nodes.

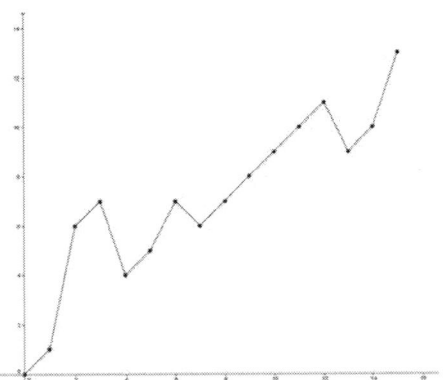

Fig. 9. Linear graph $\Gamma(r, n)$ of randomly network with 111 nodes

In this article, we introduce a new method for quantifying the complexity of a new hyper hybrid *AD HOC* cloud network based on presenting the nodes of the network in Cartesian coordinates, converting to polar coordinates, and calculating the fractal dimension.

IV. CONCLUSSION

In this work we present new method for determination complexity of network apply in cloud computing and *AD HOC* network. Finally, we present ten challenges problem on cloud and *AD HOC* network. One of these problem we can solve with new presented method of complexity of network. The main findings can be summarized as follows:

1) We present ten challenges problem of *AD HOC* network.

2) We present new method for determination complexity of network.

3) Our results suggest that this approach can be used to determine the complexity for any type of network.

REFERENCES

[1] K.U. R. Khan, R. U. Zaman, and A. V. G. Reddy, "Integrating Mobile Ad Hoc Networks and the Internet challenges and a review of strategies," presented at the 3rd International Conference on Communication Systems Software and Middleware and Workshops, COMSWARE, 2008.

[2] M.Suguna and P. Subathra, " Establishment of stable certificate chains for authentication in mobile ad hoc networks," presented at the International Conference on Recent Trends in Information Technology (ICRTIT), 2011.

[3] H.Nishiyama, T. Ngo, N. Ansari, and N. Kato, "On Minimizing the Impact of Mobility on Topology Control in Mobile Ad Hoc Networks," Wireless Communications, IEEE Transactions, 2012.

[4] X.Lv and H. Li, "Secure group communication with both confidentiality and non-repudiation for mobile ad-hoc networks," Information Security, IET, vol. 7, 2013.

[5] Bhattacharya, R. N., Gupta, V. K., and Waymire, E. (1983). The Hurst effect under trend. *Journal of Applied Probability* 20, 649-662.

[6] J. Beran, R. Sherman, M. S. Taqqu, in W. Willinger, Long-range dependence in variable bit rate video traffic, IEEE Transactions on Communications, vol. 43, 1566–1579, 1995.

Neneta: Heterogeneous Computing Complex-Valued Neural Network Framework

Vladimir Lekić* and Zdenka Babić*

* Faculty of Electrical Engineering, University of Banja Luka, 78000 Banja Luka, Bosnia and Herzegovina

Abstract—Due to increased demand for computational efficiency for the training, validation and testing of artificial neural networks, many open source software frameworks have emerged. Almost exclusively GPU programming model of choice in such software frameworks is CUDA. Symptomatic is also lack of the support for complex-valued neural networks. With our research going exactly in that direction, we developed and made publicly available yet another software framework, completely based on C++ and OpenCL standards with which we try to solve problems we identified with already existing solutions.

I. INTRODUCTION

Attention that complex machine learning algorithms are receiving in the recent years is tremendous. Research laboratories are competing in making their sets of training data publicly available [1],[2], along with software frameworks [3],[4],[5], tutorials and courses to use this data. On the other side, scientists, industry professionals and enthusiasts are competing in tuning the available models, and tweaking the algorithm performance. Of course, this is perfectly valid approach, but what somehow stays hidden in this machine learning hype, is that everyone is building machine learning models with the same sets of training data on more or less the same hardware.

The neural network framework "neneta" that will be introduced in this paper is a product of a research we are conducting on complex-valued neural networks [6],[7]. Our goal was never to compete with the state-of-the-art frameworks already available, but to build the tool that will support us efficiently through our research. On the other hand, we believe that presented tool has a potential to attract attention of the broader community. Not only by offering ability to efficiently design and train neural networks on a broader range of GPUs, but also by offering to do that in a more general way by using complex-valued neural networks.

II. DESIGN DECISIONS

Choice of a programming languages and APIs when starting work on a project like this is everything but an easy task. Idea was also to allow software to run on most popular operating systems. Decisions to enable all these goals are as follows.

A. C++11

For all programming tasks related to the network and GPU configuration, input data preprocessing and result presentation we use C++ [8]. There are two main reasons behind

this decision. First, we already had enough knowledge of the language to make significant progress fast. Second, although preprocessing tasks in deep neural networks are relatively simple compared to the network model itself, they are not insignificant, and they can definitely impact the overall performance of the framework. Therefore, we needed programming language with minimum overhead possible, but still with object oriented programming support. C++ was an ideal candidate.

B. OpenCL

We wanted to be able to run our software on wide range of available devices, and of course on the ones yet to come. By this we do not only consider the GPUs having available OpenCL [9],[10] support, but also FPGAs and DSPs utilizing this standard [11].

C. Testing

Due to significant complexity of the software, unit testing and component testing [12] was necessary and done for most of the components.

D. Operating systems support

We use gcc compiler for development with CMake build system. Basically, all operating systems having this compiler and build system support and appropriate OpenCL drivers can run *neneta*. Until now, software has been successfully tested on Microsoft Windows and Linux operating systems.

III. SOFTWARE ARCHITECTURE

Component diagram of *neneta* is shown on Fig. 1. Components are compiled to static libraries and at the end of the linking process linked together to a single executable. As it can be seen, only requirement for the operating system is support for OpenCL. Number of GPUs is not limited and it is completely abstracted away from the neuralnetwork component through interfaces provided by the gpgpu, which is the only component interfacing directly to the GPU.

In the following subsections is given short description of the framework components.

A. confighandler

This component is responsible for parsing the XML configuration files. There are three types of configuration files:

- `configuration.xml` - Holds general configuration information for the logging (log level, log rotation, log

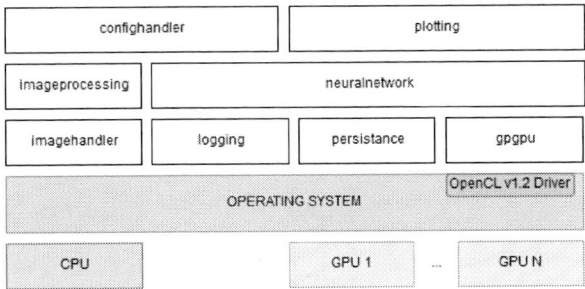

Fig. 1.   neneta Component Diagram

format etc.), plotting, input data sources, persistence, OpenCL kernel sources and GPU.

- `kernels.xml` - Holds profiling configuration for all the OpenCL kernels used by the framework. On startup, all configured OpenCL kernel sources are compiled by the OpenCL driver. This also means that kernels can be added, removed or modified without recompiling any of the framework libraries.
- `network_params_<id>.xml` - Holds configuration for the neuralnetwork component. As an example configuration, one of the complex-valued neural network layers is shown of Fig. 2. Configuration of the layers is parsed automatically. Based on the layer type, appropriate objects are instantiated and enqueued for the execution on the GPU. At the moment following types of the layers are supported:
  - Input layer
  - Convolution layer
  - Fully connected layer
  - FFT layer
  - IFFT layer
  - Projection Layer
  - Softmax Layer
  - Spectral-pooling layer
  - Error calculation layer

```
...
<layer type="conv" id="conv1">
<input>input1</input>
<channels>1</channels>
<kernels>10</kernels>
<kernelsize>5</kernelsize>
<stride>1</stride>
<inputdim>2</inputdim>
<inputsize>28</inputsize>
<actfunc>complextanh</actfunc>
<weightsdev>1</weightsdev>
<weightsmean>0</weightsmean>
<weightstype>complex</weightstype>
<biasre>0.0001</biasre>
<biasim>0</biasim> </layer>
...
```

Fig. 2.   Example of complex-valued neural network layer configuration.

B. plotting

Task of the plotting component is to abstract the data presentation tools for the neuralnetwork component. Normally, for data presentation tools some kind of plotting library is used (for example gnuplot [13], but not limited to it).

C. imageprocessing

Input date comes in various formats. Task of the image-processing components is to convert, adapt, merge or filter input data based on the neuralnetwork component needs. This component runs only on host CPU and it is not desirable that these operations have high complexity. In case input data preprocessing step consumes significant processing time (eg. FFT [14]), additional layer type should be introduced to the neuralnetwork component.

D. neuralnetwork

This is the core component of the *neneta* framework. Basically, entire neural network processing is done within this component. Main features of this component are:

- It is consisted of various types of layers, available within the framework.
- It is straightforward to define and implement a new layer. Framework it self will enqueue it for execution on GPU.
- Performance critical functionality of the layers is transferred to the OpenCL kernels. Changing the functionality within kernel source files doesn't require recompilation, but only application restart.

Fig. 3 shows simplified class diagram of *ConvLayer* layer. In this example, *ConvLayer* implements *IPersistedLayer* interface. Functions *store()* and *restore()* are called for this layer after each training epoch. Other two classes that *ConvLayer* inherits are clearly indicating the relation of the layer to OpenCL execution plan. Being also *IOpenCLChainableExecutionPlan*, allows layer to be linked with other layers. Functions *setInputBuffer(BufferIO)/setBkpInputBuffer(BufferIO)* are called from the left/right layer during forward/back propagation configuration. Forward and back-propagation are configured once during startup, but executed many times during training. Input parameter *BufferIO* is the block in GPU's global memory. Role of this memory block is to pass needed information between layers - what directly means that size of neural network model is directly proportional to the size of the available GPU global memory.

E. imagehandler

Any set of training data can be used to train the modeled neural network. Task of the imagehandler component is to abstract away the training set from the neuralnetwork component. At the moment support for MNIST [2] and IMAGENET [1] is available.

F. logging

During the long training periods some sort of logging system (for example boost logging library [15]) is desirable. This component provides logging capabilities to

193

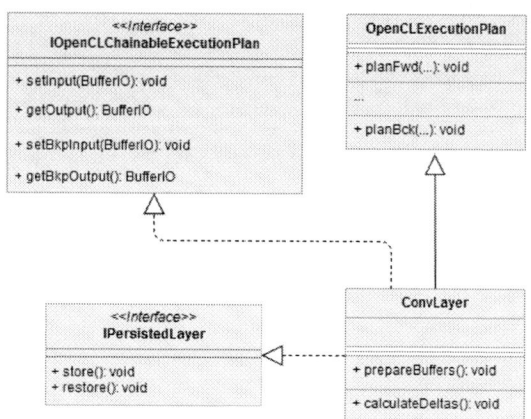

Fig. 3.   Simplified class diagram of ConvLayer

the entire framework. Logging file path, rotation size, logging level and logging format can be configured in `configuration.xml` configuration file.

G. persistence

Role of the persistence component is to ensure that network training execution can be interrupted and continued at will. Persistence interface *store()* can be called at the end of each batch execution or at the end of each training epoch. On the other hand *restore()* is called only once during initialization phase. Persistence data are stored as binary blob on hard disk.

H. gpgpu

Although OpenCL offers C++ interface wrappers [16], we introduced even higher level of abstraction in order to incorporate the GPU execution plan into the model. Component gpgpu offers interfaces to plan kernel execution in predefined order, at the same time giving ability to profile kernel execution if desired.

IV. CONFIGURATION EXAMPLE

As an example we performed training of the network consisted of a single Soft-Max layer on MNIST data-set [2]. Data-set is consisted of 60,000 training images of hand written Arabic numerals and of 10,000 test images. In configuration, we have split training data-set in 50,000 training and 10,000 validation images, as shown on Fig. 4.

An example of such network configuration is show on Fig. 5. Input layer allocates a continuous block of global memory on GPU. Although not relevant for this example, this memory is always split equally to hold real and imaginary data, using parameters *rpipesize* and *ipipesize*. Other parameters in input layer are determined by the input data size (for MNIST these are 28x28 pixel gray-level images).

Layer of type softmax is real-valued layer. Parameters of the layer are descriptive, as show on Fig. 5, and require no further explanation.

For loss calculation we used cross-entropy function, simply defined through errorcalc layer.

```
...
<images source="mnist">
<trainset>
<offset>0</offset>
<size>50000</size>
<minibatchsize>1</minibatchsize>
<path>train-images.idx3-ubyte</path>
<labels>train-labels.idx1-ubyte</labels>
</trainset>
<testset>
<offset>0</offset>
<size>10000</size>
<path>t10k-images.idx3-ubyte</path>
<labels>t10k-labels.idx1-ubyte</labels>
</testset>
<validationset>
<offset>50000</offset>
<size>10000</size>
<path>train-images.idx3-ubyte</path>
<labels>train-labels.idx1-ubyte</labels>
</validationset>
</images>
...
```

Fig. 4.   Configuration of MNIST data-set.

```
<?xml version="1.0"?>
<neneta>
<layer type="input" id="input1">
<rpipesize>101326592</rpipesize>
<ipipesize>101326592</ipipesize>
<inputdim>2</inputdim>
<inputsize>28</inputsize>
<outputsize>10</outputsize>
<inputchannels>1</inputchannels>
</layer>
<layer type="softmax" id="sm1">
<input>input1</input>
<channels>1</channels>
<inputdim>2</inputdim>
<inputsize>28</inputsize>
<outputsize>10</outputsize>
<actfunc>softmax</actfunc>
<weightsdev>1</weightsdev>
<weightsmean>0</weightsmean>
<bias>0.1</bias>
</layer>
<layer type="errorcalc" id="err1">
<input>sm1</input>
<channels>10</channels>
<errorfunc>crossentropy</errorfunc>
</layer>
</neneta>
```

Fig. 5.   Simple Soft-Max layer configuration.

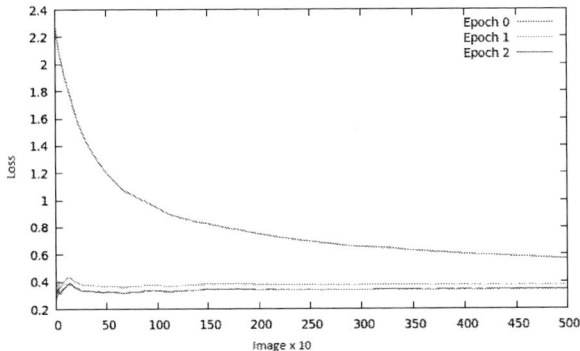

Fig. 6. Three training epochs on MNIST data-set.

Training progress was monitored using the plotting inter-face for gnuplot [13], as shown on Fig. 6. More detailed training results are obtained from the log file and here are presented in Table I.

TABLE I

DETAILS OF THREE TRAINING EPOCHS ON MNIST DATA-SET.

Ep.	Train. Loss	Train. Acc. [%]	Val. Loss	Val. Acc. [%]
1	0.568672	86.36	0.365522	90.32
2	0.374496	89.7	0.325954	91.24
3	0.343442	90.468	0.309137	91.51

V. PERFORMANCE COMPARISION

We compared the performance of the CPU and GPU running the same example configuration described in pre-vious section. Properties of the OpenCL devices used to run simulations are given in Table II.

TABLE II

CPU AND GPU DEVICE PROPERTIES

Property	CPU	GPU
Name	Phenom II X4 965	Radeon HD 5770
Vendor	AMD	AMD
Max. proc. elements	4	800
Max. clock freq. [MHz]	3400	850
Max. gl. mem. size [B]	8371974144	536870912
Max. lo. mem. size [B]	32768	32768
Max. work group size	1024	256
Max. work items sizes	1024,1024,1024	256,256,256

Measured execution time for forward and back-propagation for both devices is shown on Fig. 7. Measure-ment is taken on one entire training epoch (50,000 images). It is obvious that for most of the training epoch, algorithm execution is approximately 10 times faster on the GPU than it is on the CPU. In this simulation, both CPU and GPU were not dedicated computing devices (OS and graphics are running on them). This could explain some of the peaks in the graph.

It is interesting to analyze shortly the parameters of the devices and how they relate to the algorithm performance.

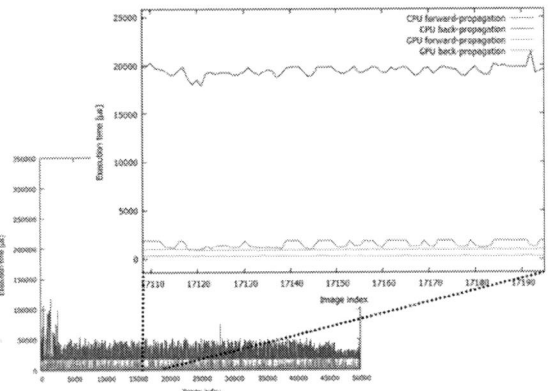

Fig. 7. Performance comparison CPU-GPU.

CPU clock speed is four times of the GPU clock speed, but number of processing elements on the GPU is 200 times higher (AMD Radeon HD 5770 Juniper graphics card has 10 computing elements, each computing element has 16 stream cores and each stream core has 5 processing elements). Based on this, one could expect even higher performance gain when executing the algorithm on GPU. To explain the obtained result, it must be taken into account that GPU has a SIMD (Single Instruction Multiple Data) processor archi-tecture. That means that all processing elements within the given work group are always executing the same instruction. If particular care is not taken during kernel development to avoid problems that can arise due to limitations of such architecture (for example branching divergence) full performance gain of GPU cannot be achieved. Another point to consider (and maybe more relevant for this example) is the well known memory transfer bottleneck that occurs during data transfer between host (CPU) memory and GPU memory. To cope with this problem it is desirable to transfer as much as possible data at a time to the GPU global memory (entire mini batches) and let GPU work on multiple passes through the network on this data.

VI. CONCLUSIONS

Although initially intended to serve as a platform for research of complex-valued neural network, due to its sim-plicity and extensibility *neneta* can be used for training of real-valued neural networks as well. Platform already offers a number of different types of neural network layers, and moreover, with opening the code to the public we hope to attract more researchers contributing to it.

We are aware that some of already implemented OpenCL kernels are far from optimal from execution time point of view. Our goal for the future is to further improve the code base in that sense, and to improve the design and quality aspects of it as well.

REFERENCES

[1] J. Deng, W. Dong, R. Socher, L.-J. Li, K. Li, and L. Fei-Fei. ImageNet: A Large-Scale Hierarchical Image Database. In *CVPR09*, 2009.

[2] Yann LeCun and Corinna Cortes. The mnist database of handwritten digits, 1998.

[3] Alex Krizhevsky, Ilya Sutskever, and Geoffrey E Hinton. Imagenet classification with deep convolutional neural networks. In *Advances in neural information processing systems*, pages 1097–1105, 2012.

[4] Martín Abadi, Ashish Agarwal, Paul Barham, Eugene Brevdo, Zhifeng Chen, Craig Citro, Greg S. Corrado, Andy Davis, Jeffrey Dean, Matthieu Devin, Sanjay Ghemawat, Ian Goodfellow, Andrew Harp, Geoffrey Irving, Michael Isard, Yangqing Jia, Rafal Jozefowicz, Lukasz Kaiser, Manjunath Kudlur, Josh Levenberg, Dan Mané, Rajat Monga, Sherry Moore, Derek Murray, Chris Olah, Mike Schuster, Jonathon Shlens, Benoit Steiner, Ilya Sutskever, Kunal Talwar, Paul Tucker, Vincent Vanhoucke, Vijay Vasudevan, Fernanda Viégas, Oriol Vinyals, Pete Warden, Martin Wattenberg, Martin Wicke, Yuan Yu, and Xiaoqiang Zheng. TensorFlow: Large-scale machine learning on heterogeneous systems, 2015. Software available from tensorflow.org.

[5] Yangqing Jia, Evan Shelhamer, Jeff Donahue, Sergey Karayev, Jonathan Long, Ross Girshick, Sergio Guadarrama, and Trevor Darrell. Caffe: Convolutional architecture for fast feature embedding. *arXiv preprint arXiv:1408.5093*, 2014.

[6] Akira Hirose. *Complex-valued neural networks*. Springer Science & Business Media, 2006.

[7] Danilo P Mandic and Vanessa Su Lee Goh. *Complex valued nonlinear adaptive filters: noncircularity, widely linear and neural models*, volume 59. John Wiley & Sons, 2009.

[8] Bjarne Stroustrup. *The C++ Programming Language*. Addison-Wesley Professional, 4th edition, 2013.

[9] Jonathan Tompson and Kristofer Schlachter. An introduction to the opencl programming model. *Person Education*, 49, 2012.

[10] John E. Stone, David Gohara, and Guochun Shi. Opencl: A parallel programming standard for heterogeneous computing systems. *IEEE Des. Test*, 12(3):66–73, May 2010.

[11] Deshanand Singh. Implementing fpga design with the opencl standard. *Altera whitepaper*, 2011.

[12] Robert C Martin. *Clean code: a handbook of agile software craftsmanship*. Pearson Education, 2009.

[13] T Williams, C Kelley, HB Bröker, J Campbell, R Cunningham, D Den-holm, E Elber, R Fearick, C Grammes, and L Hart. Gnuplot 5.0.5: An interactive plotting program, 2016. *URL http://www. gnuplot. info*.

[14] Keun-Yung Byun, Chun-Su Park, Jee-Young Sun, and Sung-Jea Ko. Vector radix 2×2 sliding fast fourier transform. *Mathematical Problems in Engineering*, 2016, 2016.

[15] Boris Schling. *The Boost C++ Libraries*. XML Press, 2011.

[16] Benedict R Gaster. The opencl c++ wrapper api, 2010.

Cloud-distributed computational experiments for combinatorial optimization

Mario Brcic*and Nikica Hlupic *

* Faculty of Electrical Engineering and Computing, Zagreb, Croatia
mario.brcic@fer.hr, nikica.hlupic@fer.hr

Abstract - The development of optimization algorithms for combinatorial problems is a complicated process, both guided and validated by the computational experiments over the different scenarios. Since the number of experiments can be very large and each experiment can take substantial execution time, distributing the load over the cloud speeds up the whole process significantly. In this paper we present the system used for experimental validation and comparison of stochastic combinatorial optimization algorithms, applied in the specific case of project scheduling problems.

I. INTRODUCTION

Combinatorial optimization (CO) is a subfield of mathematical optimization, that relates to the situations where the best option must be selected from a discrete feasible set. Applications are widespread, such as in scheduling, vehicle routing and auctions, to name a few. Most often, exhaustive search for combinatorial optimization problems is infeasible. Even worse, for some problems it is assumed that computationally efficient algorithms for finding optimal solutions do not exist. Instead, metaheuristics are used to get sufficiently good solutions in a reasonable amount of computational time. However, designing such algorithms as well as proving their ranking in comparison to the other algorithms on the same type of problems is guided empirically, by the experimental results. Experiment consists of experiment units. For deterministic combinatorial problems, each unit is associated to a member in sampled set of problem instances. The experiment conclusions are generalized to a wider set of similar instances. Stochastic and robust problem variants assume for some parameters to be unknown and effectively random. Experiment in that case is composed of units for each combination of selected problem instance and its corresponding sampled scenario for unknown parameters. This combination makes the number of experiment units much greater. The number becomes greater yet if the experiment includes additional factors, for example the used optimization algorithm.

Gathering the empirical data can be expensive operation, depending on the CO problem complexity. However, the whole process is inherently parallelizable across experiment units. There have been many situations where all the experimentation was done on a single computer such as in [1] for stochastic variant of vehicle routing problem and [2] for project scheduling, but the total execution time could have been greatly reduced had there been used greater degree of parallelization. There are

situations where the problem is of such complexity that computer cluster must be employed in order make algorithm practical. Such was the case for fleet operations optimization in [3].

Distributed computation is unavoidable in high-performance computing (HPC). Usually, in-house grids are used, but cloud computing offers flexible and cheap alternative. Studies have been conducted on the matter of using the cloud for high performance computing. It was found in [4] and [5] that coupled applications have potential up to some limit of used computational nodes. Beyond that, the overheads become overwhelming performance detractors. The similar was confirmed in [6] where virtualization, latency and system noise in comparison to the dedicated supercomputers are singled out as the biggest issues. They conclude that research groups with limited access to supercomputer resources and varying demand over time might find cloud computing to offer beneficial choice. It enables flexible renting of practically as many as necessary identical instances. That is is desirable since we would like to have identical experimenting environments for our parallel runs, moving as close as possible to the ideal of running all the units in single environment in a short amount of time.

There are existing tools that help the design and comparison of optimization algorithms. COmparing Continuous Optimisers (COCO) [7] is a platform for continuous optimization. The comparisons are done on standard benchmark problems. Experiment can use Shared Memory Parallelism (SMP), but grid computing is not utilized. ParadisEO [8] is a white-box C++ framework for reusable design of parallel and distributed metaheuristics. It supports many features and components necessary for composing new algorithms. The focus of this framework is enabling the design of topologies that encompass single running system. But, it does not specify how to do so efficiently in the case of distributed running of experiment. Our proposed system could in principle be assembled and implemented in ParadisEO. Similar option to ParadisEO is Java Evolutionary Computation Toolkit [9]. Multi-Objective Evolutionary Algorithm (MOEA) Java-based Framework [10] is focused on multi-objective optimization and it combines the features of aforementioned tools. As stated, COCO is focused on continuous optimizers, while the rest of the tools enable easier and faster design of algorithms through reusability of common mechanisms when the optimization problem has favorable features. The guidelines for extensive experimentation that could, due to significant runtime, utilize distributed computing power in the cloud, are left unspecified.

In this paper we present our architecture of a system for running extensive computational experiments in the cloud. This system could be implemented in some of the mentioned frameworks, if the satisfactory amount of reuse can be achieved. We explain the distributed design of data storage for experimental results that also reduces communication overhead. Storing the results is necessary for further analysis, as well as the mean of efficient scientific scrutiny. Finally, we show application of the proposed architecture on a specific problem of developing and validating algorithms for a complex project scheduling problem.

The paper is organized as follows: in section 2 we present the general architecture of our system for distributed experimental runs. Section 3 details the distributed storage choices. In section 4 we describe the application of the architecture to the real research problem. Section 5 offers concluding remarks and future research ideas.

II. ARCHITECTURE

Our intended use-case has inherently parallelizable task, with low communication requirements between the computational nodes. In such case latency is not an issue, since we only need occasional communication; for sending initial experiment chunk instructions and migrating the final partial results. In both cases, the bandwidth plays the main role and should not present a problem for the distributed experimentation. Aforementioned features make our problem suitable for cloud deployment. We propose using distributed database instances in order to further reduce networking while logging all the necessary details of each experiment unit run. The results, raw or processed, can be pooled periodically or upon the finalization of assigned experiment chunk.

Figure 1 depicts the used pipeline architecture for distributed running of computationally expensive experiments in the domain of combinatorial optimization. In the rest of this section we explain architecture elements sequentially, in order of and according to the numeration in the Figure 1.

1. At this, optional, step we can use standard test set as a source for experimental problem instances. For many established CO problems, such sets are shared within the community of the researchers to boost the consistency of algorithm ranking. In the case of a new CO problem types, a test set for related standard family can be used as template for creation of new types of problem instances.

2. The creation of the actually used test set takes place at this point. If based on step 1, problem selection with required adaptations can be used. "Seed" database is created and populated with all the shared data needed for successful distributed execution and result logging, as detailed in the following section. Part of the shared data may be in the form of the files, if it is

more appropriate, but such files are linked from a DB. Each computational node will have its independent version of the database, initialized from the singular "seed" database. The further database distribution over the nodes is achieved through the horizontal fragmentation of writeable relations.

3. VM image used for computational nodes is set up. It is populated with a copy of the "seed" database as well as all the necessary code for running an experiment chunk. The tuning to the intended VM hardware is done.

4. VM image is migrated to the cloud service, where it is ready to be deployed on created computational instances.

5. The experiments are started with sampled pilot runs on single system to estimate the total experiment CPU run time. Based on that estimate and the input criteria, monetary and/or temporal (desired due date), the number of computational instances N_i is inferred.

6. Experimental chunk is a set of experiment units that are to be executed on a single node. Experiment spawner is a script that, based on the N_i and the total workload, calculates the partition of experiment into disjoint experimental chunks. Each chunk constitutes a load for one of the computational instances which is to be created and fed with it.

7. Experiment spawner creates N_i cloud nodes and sends them over the network the parameters that define their workload chunks, accordingly labeling each node. Depending on the chunk runner, additional instances can be created to form a mini-grid using Message Passing Interface (MPI). Such situation would occur if some of the algorithms require said execution by the design. SMP can be switched on through the runner's parameters. Each node has appointed performance supervisor operating the utilization alarm triggers. The purpose of the triggers is automation of experiment run wrap-up which results in renting costs reduction.

8. Upon finishing the assigned chunk, node migrates the results, processed or raw, to a permanent cloud storage and triggers the supervisor's alarm which terminates the node. Storing the data on cloud leaves options for the analysis in the cloud, should the nature of data require it.

9. Supervisor alarm notifies the user of finalized chunk. Local procedure that listens for such events reacts to the notifications.

10. The local procedure downloads the results from the cloud storage.

11. Analysis is done iteratively over the partial data. After the complete data from all the nodes become available, the concluding results are output.

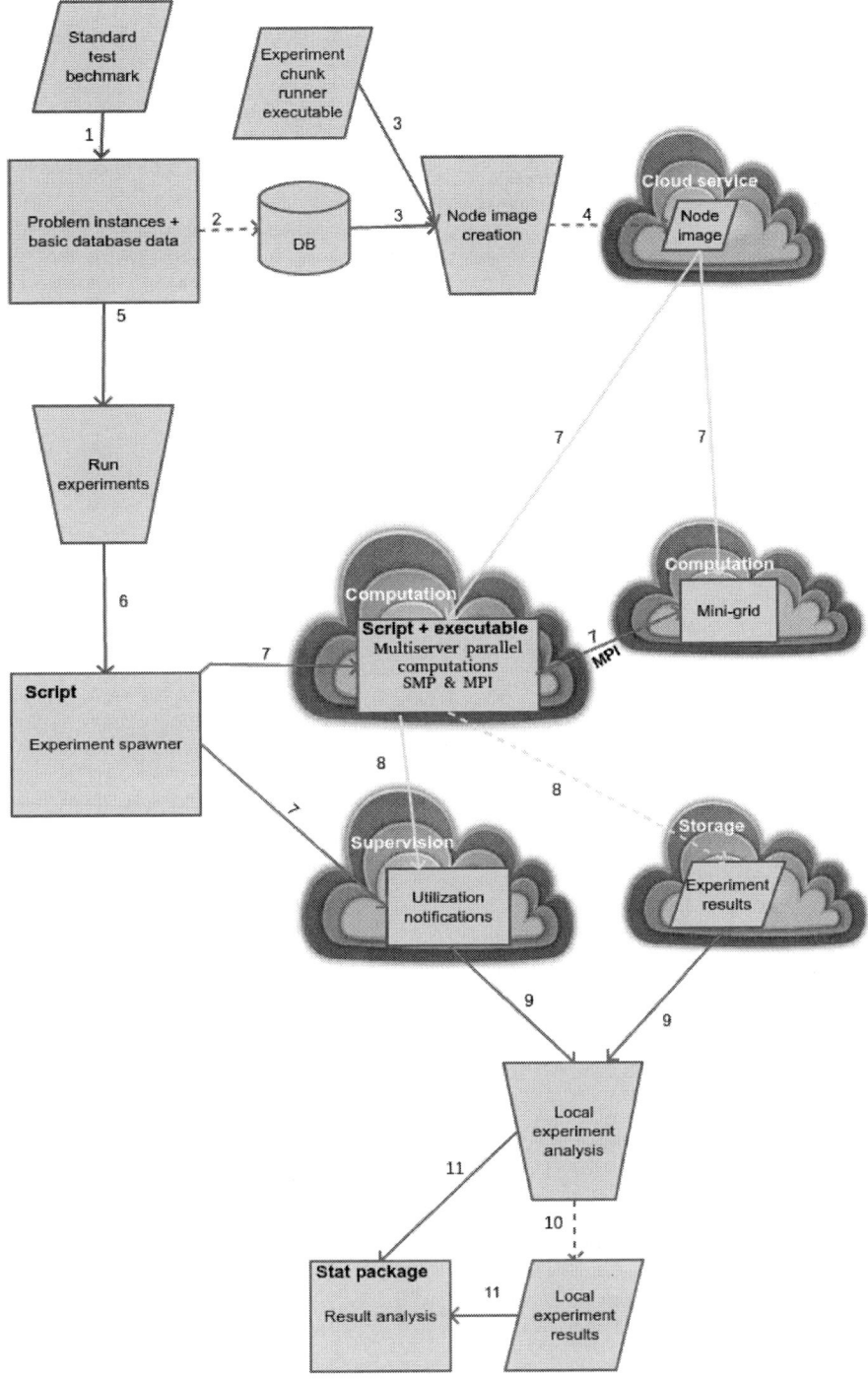

Figure 1. Experimental pipeline

III. DATABASE

The main objective for our system is producing detailed experiment documentation in as short as possible period of time. As the amount of data can be overwhelming, we store everything into the database. The rationale is that it is expensive to run experiments in need of distributed computation, in monetary and/or temporal terms. Because of this, as much data as possible should be stored for future analyses to avoid experiment re-runs.

Practicing scrutiny in computer science can be done by replication studies that simply repeat small sampled set of experiment units, reusing stored settings for pseudo-random number generators (PRNG), algorithms and problem instances. The replication results should match the stored performance values.

In order to reduce the communication overhead between cloud instances, we propose using independent database instances, based on the single "seed" instance. "Seed" database holds the metadata and identification/replication data. The latter are the basic necessary data, such as the sufficient shared information on all the CO problem instances, used optimization algorithms, PRNG types and uncertainty scenarios (in the case of stochastic or robust CO).

Each node is created with a separate database copy, initialized from the "seed" database. The further database distribution over the nodes is achieved through the horizontal fragmentation of writeable relations, each node only writing to its copy the data associated to its disjoint chunk. During chunk execution, each node saves the identification/replication and performance data. The former consists of the settings and parameters for used optimization algorithms, PRNGs and other components for an experiment unit. The performance data track important decisions and performance components, the bread and butter of the experimental runs, upon which final conclusions are made. They quantitatively make up the majority of the data. The distributed data uses the identification scheme based on shared identification/replication information originated from the "seed" database, supplemented with the same type of node-specific information. Such scheme enables unambiguous data aggregation from all the chunks.

IV. APPLICATION

Our primary motivation for developing the aforementioned system was to enable faster experimentation with the algorithms in the new, complex type of project scheduling problem, Cost-based Flexible Stochastic Resource Constrained Project Scheduling Problem (CBF-SRCPSP), defined in [11]. As CBF-SRCPSP is a non-standard CO problem, we had to create our own test set. However, we decided to do so based on the PSPLIB set of instances for deterministic resource constrained project scheduling problems. We used cluster sampling to select a set of J30, J60 and J120 template instances. These templates were expanded to fit CBF-SRCPSP model which resulted in 300, 300 and 200 instances respectively. For each instance, we generated 1000 activity duration scenarios, since activity durations were the only source of uncertainty. All the data was fed into a "seed" database.

We have developed several search algorithms, but some were computationally expensive. Our estimations, based on sampled pilot experiment run were rather grim, predicting three and half months of computational labor on the available hardware. Such timeline was unacceptable due to the research deadline. After some consideration, weighing up the options and taking into account the computational nature of the task, we have decided to use the computation distributed over the cloud.

The selected service provider was Amazon Web Sevices (AWS) [12]. We used Elastic Compute Cloud (EC2) for computations, always employing c3.large instances with 2 dedicated physical cores of Intel Xeon E5-2680 v2 (Ivy Bridge) processors and 3.75 GiB of RAM. Simple Storage Service (S3) was a storage for experimental results and CloudWatch (CW) and Simple Notification Service (SNS) for supervision and utilization notifications.

The used database was sqlite3 as it fit our needs. Its shortcomings in comparison to the more elaborate database management systems traded off well with the simplicity. Three main algorithms were used for the study as well as several auxiliary, keeping the track of all the relevant in the database. Two PRNGs were used: Mersenne twister with careful parameterization [13] and Threefry [14].

The image used on nodes used Linux with gcc and necessary libraries. New algorithms were developed upon the C++ simulation library presented in [15], meaning that the bulk of experimentation and database logging was coded in C++ for speed. We did not use mini-grids as our algorithms utilized only shared memory parallelization using OpenMP to speed up the search. Though, the option of using MPI is opened for future developments.

Experiment spawner was coded in python, using the boto API for accessing AWS and paramiko module for manipulating SSH2 connections. The experiment chunking was done on the basis of problem instances. The sizes of chunks were determined by the necessary number of computational instances estimated from pilot runs' performance. The chunk experiment runner has call parameters that describe the chunk boundaries. At the end of chunk execution, the database with the results was migrated to S3 to await further analysis.

The raw results were downloaded from S3 using python script. We used statistical language R for result analysis. RSQLite was used to query the databases for the relevant data and gather them together from all the sources in order to make conclusions with the complete overview.

We have used up to 26 cloud instances per experiment in order to reduce the total duration to 6 days, meeting the imposed deadlines and having the wealth of experimental data for future analysis. The data was used for definitive statistical conclusions on rankings of new algorithms in comparison to the selected benchmark algorithm. Performance effectiveness was examined in several dimensions, outlining possible future research directions.

V. CONCLUSION

The system for cloud-distributed computational experimentation for combinatorial optimization has been presented. The cloud computing is not a clear-cut choice for high performance computing, due to issues explained in [6]: communication overhead, virtualization and system noise.

The described use case, doing expensive computational experiments in order to validate and guide the design of CO algorithms is a specific problem inasmuch that it can be parallelized with a low coupling between the simultaneous tasks.

The communication between the nodes is kept to minimum by using distributed database with horizontal fragmentation. Each node populates only the data pertinent to its assigned disjoint experimental chunk, hence preserving unambiguity of the data across the whole system without additional effort. Even in the case of using mini-grids for each experimental chunk, mini-grid size is expected to be within the limits established by the studies such as [6] in order to reap the best performance benefits. Additional tuning and advances in cloud computing technology might increase the limits found by the studies.

We described the successful application of our proposed system for the experimentation over the newly developed algorithms for complex stochastic combinatorial optimization problem, CBF-SRCPSP. The utilization of such system cut the experiment duration from estimated several months to several days, helping us meet the deadline. The flexibility of using on-demand cloud instances helped the development of our system with frequent pilot runs during the system implementation, as well as algorithm development.

Possible future research ideas include improvement to the robustness of the total workload estimator. It might take into account the sampling distributions of pilot performance statistics. The creation of component based framework for general experimenting which would increase the reuse of existing components together with code generator for repetitive parts is interesting venue if these types of applications become common. Monetary cloud costs could be reduced by more careful node tracking and possibly using the cheaper spot instances for non-critical computations, such as during algorithm design and prototyping. Further investigation into the performance penalty for the case of using such instances should be done.

REFERENCES

[1] J. C. Goodson, J. W. Ohlmann, and B. W. Thomas, "Rollout Policies for Dynamic Solutions to the Multivehicle Routing Problem with Stochastic Demand and Duration Limits," *Oper Res*, vol. 61, no. 1, pp. 138–154, Jan. 2013.

[2] P. Lamas and E. Demeulemeester, "A purely proactive scheduling procedure for the resource-constrained project scheduling problem with stochastic activity durations," *J. Sched.*, vol. 19, no. 4, pp. 409–428, Aug. 2016.

[3] H. P. Simão, A. George, W. B. Powell, T. Gifford, J. Nienow, and J. Day, "Approximate Dynamic Programming Captures Fleet Operations for Schneider National," *Interfaces*, vol. 40, no. 5, pp. 342–352, Jul. 2010.

[4] C. Evangelinos and C. N. Hill, "Cloud Computing for parallel Scientific HPC Applications: Feasibility of Running Coupled Atmosphere-Ocean Climate Models on Amazon's EC2," in *In The 1st Workshop on Cloud Computing and its Applications (CCA*, 2008.

[5] D. Kondo, B. Javadi, P. Malecot, F. Cappello, and D. P. Anderson, "Cost-benefit analysis of Cloud Computing versus desktop grids,"

in *2009 IEEE International Symposium on Parallel Distributed Processing*, 2009, pp. 1–12.

[6] A. Gupta *et al.*, "The Who, What, Why, and How of High Performance Computing in the Cloud," in *2013 IEEE 5th International Conference on Cloud Computing Technology and Science*, 2013, vol. 1, pp. 306–314.

[7] N. Hansen, A. Auger, O. Mersmann, T. Tusar, and D. Brockhoff, "COCO: A Platform for Comparing Continuous Optimizers in a Black-Box Setting," *ArXiv160308785 Cs Stat*, Mar. 2016.

[8] S. Cahon, N. Melab, and E.-G. Talbi, "ParadisEO: A Framework for the Reusable Design of Parallel and Distributed Metaheuristics," *J. Heuristics*, vol. 10, no. 3, pp. 357–380, May 2004.

[9] D. R. White, "Software review: the ECJ toolkit," *Genet. Program. Evolvable Mach.*, vol. 13, no. 1, pp. 65–67, Mar. 2012.

[10] "MOEA Framework, a Java library for multiobjective evolutionary algorithms." [Online]. Available: http://moeaframework.org/index.html. [Accessed: 05-Feb-2017].

[11] M. Brčić, D. Kalpić, and M. Katić, "Proactive Reactive Scheduling in Resource Constrained Projects with Flexibility and Quality Robustness Requirements," *Comb. Optim.*, pp. 112–124, Aug. 2014.

[12] "Amazon Web Services (AWS) - Cloud Computing Services," *Amazon Web Services, Inc.* [Online]. Available: //aws.amazon.com/. [Accessed: 14-Dec-2014].

[13] M. Matsumoto and T. Nishimura, "Dynamic Creation of Pseudorandom Number Generators," presented at the Proceedings of the Third International Conference on Monte Carlo and Quasi-Monte Carlo Methods in Scientific Computing, 1998, pp. 56–69.

[14] J. K. Salmon, M. A. Moraes, R. O. Dror, and D. E. Shaw, "Parallel random numbers: As easy as 1, 2, 3," in *High Performance Computing, Networking, Storage and Analysis (SC), 2011 International Conference for*, 2011, pp. 1–12.

[15] M. Brčić and N. Hlupić, "Simulation library for Resource Constrained Project Scheduling with uncertain activity durations," in *2014 37th International Convention on Information and Communication Technology, Electronics and Microelectronics (MIPRO)*, 2014, pp. 1041–1046.

Design of Digital IIR Filter using Particle Swarm Optimization

F.Serbet, T.Kaya, M.T. Ozdemir

Firat University /Department of Electrical and Electronics Engineering, Elazig, Turkey

fatmanur.serbet@hotmail.com, tkaya@firat.edu.tr, mto@firat.edu.tr

Abstract - The paper aims to establish a solution methodology for the optimal design of digital Infinite Impulse Response (IIR) filter by integrating the features of Particle Swarm Optimization (PSO). PSO is a method that optimizes a problem by iteratively trying to improve a candidate solution with regard to a given measure of quality to mathematical formula over the particle's position and velocity. The applied method explores the search space locally as well as globally and uses them in the mathematical formulas. The optimal design of IIR filter is realized with the result obtained with PSO.

Keywords: Filter Design, IIR filter, particle swarm optimization (PSO).

I. INTRODUCTION

The electronic filter is a circuit that pass and suppresses some of the signals with different frequencies. Filtering has an important role in eliminating undesired signals such as noise in the signal, or only part of the signal is obtained at the output by filtering.

Filters can be separated into two parts, analog and digital filters,which are quite different from each other in terms of their physical structure and working principle. A digital filter is a method or algorithm that operates on digitized analog signals and converts the input signal to the desired output signal. In an analog filter, analog electronic circuit elements consisting of elements such as resistance, capacitance and operational amplifier are used. Analog filter circuits can be used in many different areas, such as noise filtering and video signal recovery. Digital filters are not affected by the variation of external factors as compared to analog filters, their functions can easily be changed, easily applied and tested, etc. This is why digital filters have been the subject of many researches in recent years. At the same time, digital filters have many applications in the engineering field. Digital filters can be examined in two main categories as Finite Impulse Response (FIR) filters and Infinite Impulse Response (IIR) filters. FIR filters have the following advantages:

- It has a linear phase.
- They are always determined.

- Design methods are usually linear.
- Efficient hardware implementation is possible.

The disadvantage of FIR filters is that they usually achieve the desired level of performance with a very high filter degree compared to IIR filters, and the delay of these filters is very high compared to the equivalent IIR filter. In the IIR filters, the output data is feedback to the input data to provide a continuous dynamic flow. So the final response of the system is determined by the previous outputs as well as the inputs. As a result, IIR filters are much more useful than FIR filters in terms of application area and performance.

In literature, different methods has been proposed for the digital filter design [1-6].The IIR filter design, which is the subject of this study, is provided by an optimization method and it is aimed to realize an alternative filter design to the existing filter. The selected optimization method, Particle Swarm Optimization (PSO), is an optimization technique inspired by swarm intelligence and is designed to deliver the best solution to the system. It has been determined that PSO produces better solutions in terms of convergence speed and performance than Genetic Algorithm (GA) and Differential Evolution Algorithm (DEA).

The content of the paper is as follows: Section 2 explains design of IIR filters and Section 3 describes PSO algorithm. In Section 4, the results of the PSO-based IIR filter design are available and in the last section, some information has been given about future work.

A. IIR FILTER DESIGN

Although digital filters are structures that operate in time form, they are implemented in frequency form. In the frequency form transformations, Laplace and z transform techniques are used and the design is performed in s-plane and z-plane. In this section of the paper, the design of IIR filters, one of the digital filters, is explained[7].

The difference equation for a digital filter, denoted by the input expression x(n), the output expression y(n), is expressed as:

$$a_0 y(n) = b_0 x(n) + b_1 x(n-1) + \cdots + b_M x(n-M) - a_1 y(n-1) \ldots - a_N y(n-N) \quad (1)$$

As this equation becomes more generalized,

$$y(n) = \frac{1}{a_0}\left(\sum_{k=0}^{M} b_k\, x(n-k) - \sum_{k=1}^{N} a_k\, y(n-k)\right) \quad (2)$$

When rearranged;

$$\left(\sum_{k=0}^{N} a_k\, y(n-k) = \sum_{k=0}^{M} b_k\, x(n-k)\right) \quad (3)$$

To find the transfer function of the filter, we first take the Z-transform of each side of the above equation [8].

$$\left(\sum_{k=0}^{N} a_k z^{-k}\, Y(z) = \sum_{k=0}^{M} b_k z^{-k}\, X(z)\right) \quad (4)$$

Transfer function of the filter:

$$H(z) = \frac{Y(z)}{X(z)}$$

$$= \frac{\sum_{k=0}^{M} b_k z^{-k}}{\sum_{k=1}^{N} a_k z^{-k}} \quad (5)$$

Assuming the coefficient $a_0 = 1$, the transfer function of the IIR filter becomes:

$$= \frac{\sum_{k=0}^{M} b_k z^{-k}}{1 + \sum_{k=1}^{N} a_k z^{-k}} \quad (6)$$

or,

$$H(z) = \frac{Y(z)}{X(z)} = \frac{b_0 + b_1 z^{-1} + b_2 z^{-2} + \ldots + b_M z^{-M}}{1 + a_1 z^{-1} + a_2 z^{-2} + \ldots + a_N z^{-N}} \quad (7)$$

The frequency response of a digital filter is found using the conversion between the Laplace transform and the z-transform ($z = e^{sT}$ (T: sampling period)). As is known, a system involving the axis jw of the convergence region can be found by writing the frequency response $s = j\omega$.

$$H(z) = H(e^{sT}) = H(e^{j\omega T}) = H(e^{j\Omega}) \quad (8)$$

According to this expression H (z) is regulated:

$$H(e^{j\Omega}) = \frac{b_0 + b_1 e^{-j\Omega} + b_2 e^{-j2\Omega} + \ldots + b_M e^{-jM\Omega}}{1 + a_1 e^{-j\Omega} + a_2 e^{-j2\Omega} + \ldots + a_N e^{-jN\Omega}} \quad (9)$$

It is sufficient to obtain $H(z)$ for the values of u between 0 and π in (9) because it is periodic and a period between 0 and π. The frequency response of the filter obtained in radians can be converted to Hz by $\frac{\Omega}{2\pi} f_s$ conversion.

B. PARTICLE SWARM OPTIMIZATION

It is a population-based heuristic optimization technique developed by Dr.Eberhart and Dr.Kennedy in 1995[9]. The behaviors of individuals interacting with each other and with their environment have been examined and improved. This concept is also called particle intelligence. PSO, function optimization, scheduling,training of artificial neural networks, fuzzy logic systems, image processing, etc. are widely used in many areas.

C. ALGORITHM OF PARTICLE SWARM OPTIMIZATION

In PSO, each particle represents a bird, and each particle presents a solution. All particles have got fitness values found with the fitness function. Particles have got information similar to speed information that guides birds' flights. The PSO is initiated with a certain number of randomly generated solutions (particles). Potential solutions called particles navigates the problem area by following the best available solutions. The particles are updated to find the optimal solution value [10-11].

Fig. 1 shows the data flow diagram of the PSO

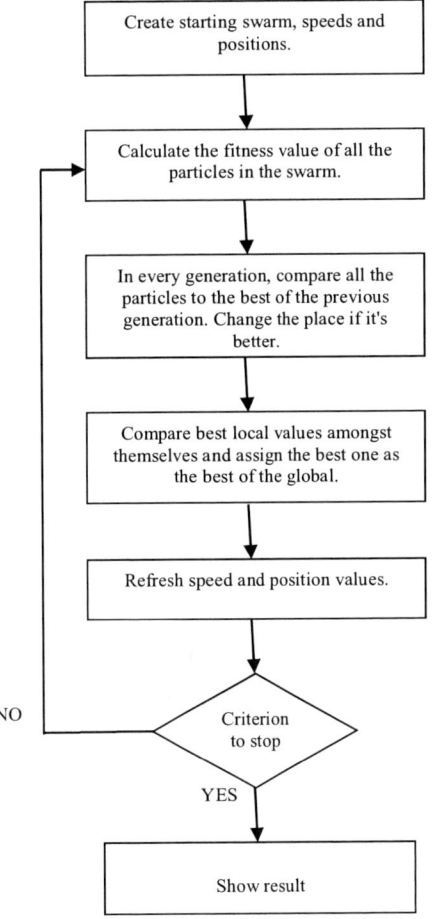

Figure 1.   Data Flow Diagram of the Particle Swarm Optimization

D. RESULTS AND ANALYSIS

In this section are the results of the analysis of the magnitude response of the band-pass digital IIR filter. MATLAB simulation is used to design the IIR filter. The order of the filtration is 4. The PSO algorithm is used to obtain the best filter coefficients. Below are the graphs of the results obtained using the PSO algorithm.

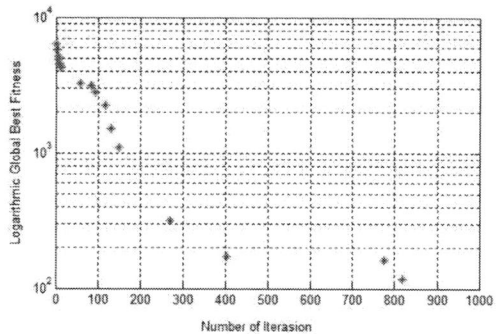

Figure 2. Iteration value of found logarithmic global best fitness values

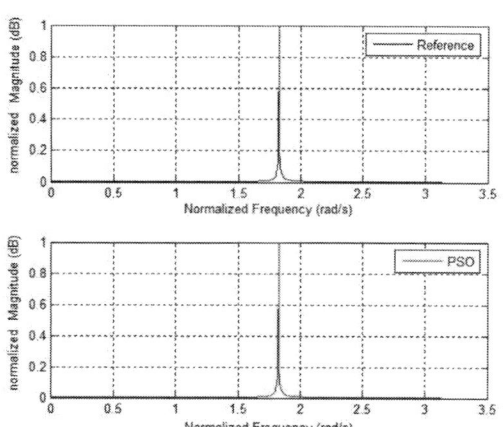

Figure 3.   Magnitude response of band-pass digital IIR filter

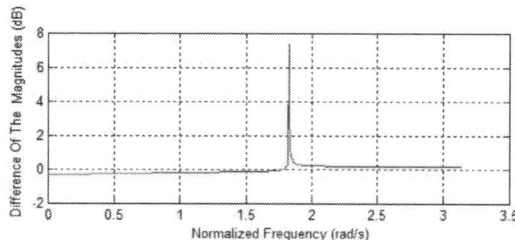

Figure 4.Amplitudes difference of signals

II.   CONCLUSION

This paper presents design of digital IIR filter using particle swarm optimization algorithm. Particle swarm optimization is a successful method for designing digital filters and has enabled many research to achieve effective results. The best result obtained by the PSO method is obtained in the shortest time due to the high convergence speed. Thanks to this feature of the PSO method, it is more advantageous than other optimization methods. With this application, It is possible to create an alternative to the filter taken as reference with the results obtained and to perform the same filtering with different coefficients.

REFERENCES

[1]   T. Kaya andM. C. Ince, The Obtaining of Window Function Having Useful Spectral Parameters by Helping of Genetic Algorithm, 2nd World Conference On Educational Technology Researches Near East University, Nicosia, North Cyprus, 27-30 June 2012.

[2]   T. Kaya and M. C. Ince, The FIR filter design by using window parameters calculated with GA, ICSCCW 2009-Fifth International Conference on Soft Computing, Computing with Words and Perceptions in System Analysis, Decision and Control, 1-4, 2009.

[3]   IIR digital filters with critical monotonic pass-band amplitude characteristic, International Journal of Electronics and Communications (AEÜ), Vol. 69/ 10, pp. 1495–1505, 2015.

[4]   S.K. Saha, R. Kar, D. Mandal, S.P.Ghoshal, Optimal IIR filter design using Gravitational Search Algorithm with Wavelet Mutation, Journal of King Saud University – Computer and Information Sciences, Vol. 27/1,pp. 25–39, 2015.

[5]   A. K. Dwivedi, S. Ghoshn , N.D. Londhe, Low power FIR filter design using modified multi-objective artificial bee colony algorithm, Engineering Applications of Artificial IntelligenceVol. 55, pp 58–69, 2016.

[6]   A. Aggarwal, T. K. Rawat, D.K. Upadhyay, Design of optimal digital FIR filters using evolutionary and swarm optimization techniques, International Journal of Electronics and Communications (AEÜ), Vol. 70/ 4, pp 373–385, 2016.

[7]   Van de Vegte, J. Fundamentals of digital signal processing. New Jersey: Prentice Hall.

[8]   I. Jury, Theory and Application of the Z-Transform Method, New York: Wiley, 1964.

[9]   J., Kenny, and R. Eberhart, Particle swarm optimization. InProceedings of IEEE international conference neural networks pp. 1942–1948, Vol. IV. Perth, Australia,1995.

[10]   M. Y.Ozsaglam, M.Cunkas, Particle Swarm Optimization Algorithm for Solving Optimization Problems. Politelnik dergisi, Vol. 11, Number 4, 2008.

[11]   Performance of swarm based optimization techniques for designing digital FIR filter: A comparative study,Engineering Science and Technology, an International JournalVol. 19/ 3, pp:1564–1572, 2016.

MIPRO 2017, May 22- 26, 2017, Opatija, Croatia

Big data analytics in electricity distribution systems

St. Stoyanov* and N. Kakanakov*

* Technical University of Sofia branch Plovdiv, Dept. Computer Systems and Technologies, Bulgaria
e-mail: stefan@it-advanced.com, kakanak@tu-plovdiv.bg

Abstract - **Many problems in power distribution systems affecting today's technological equipment are often generated locally within a facility from any number of situations, such as local construction, heavy loads, faulty distribution components, and even typical background electrical noise. Penetration of advanced sensor systems such as advanced metering infrastructure (AMI), high-frequency overhead and underground current and voltage sensors have been increasing significantly in power distribution systems over the past few years. To manage the massive amounts of data generated from smart meters and other components of the grid, utility companies need a solution models such as e.g. Apache Hadoop ecosystem that operates in a distributed manner rather than using the centralized computing model. The paper aims to discuss a solution for easily discovering of problems with power quality that have local origin which collects data from AMI and implements distributed computing across clusters of computers using simple programming models.**

I. INTRODUCTION

Our technological world has become deeply dependent upon the continuous availability of electrical power. Sophisticated technology of smart devices has reached deeply into our homes and careers and a minute of downtime due to power interruption can create problems for software recovery operations that may take weeks to resolve. Many power problems originate in the commercial power grid, which, with its thousands of miles of transmission lines, is subject to weather conditions such as hurricanes, lightning storms, snow, ice, and flooding along with equipment failure, traffic accidents and major switching operations [1, 8].

Also, power problems affecting today's technological equipment are often generated locally within a facility from any number of situations, such as local construction, heavy startup loads, faulty distribution components, and even typical background electrical noise. The study of power quality, and ways to control it, is a concern for electric utilities, large industrial companies, businesses, and even home users. The study has intensified as equipment has become increasingly sensitive to even minute changes in the power supply voltage, current, and frequency [1, 8].

National Science Fund of Bulgaria, project number E02/12

A. Smart sensors

Penetration of advanced sensor systems have been increasing significantly in power distribution systems over the past few years. It is estimated that the electricity usage data collected through AMI would double every two years to reach 20 Zettabytes (ZB) by 2020. To unleash full value of these complex data sets, innovative big data algorithms need to be used to show how we can improve in terms of efficiency in the way we discover and avoid power supply problems. One thing is quite certain that even if the hardware technology is getting better and storage is becoming cheaper everyday, there will always be place for reducing the cost of data storage and data processing.

IP Based Networks Internet Protocol based networks are used for data communication involving the smart grid. This allows data transmission over both private (dedicated) and public network (also by local wireless networks). Smart sensors for voltage, current, true power, temperature, humidity & other data transmitted to Internet via Ethernet or Wi-Fi. An advanced sensor system almost always generates output data organized in some semi-structured data formats like CSV, JSON or XML. This data formats are easy to store on HDFS and there are also a big number of tools in the Hadoop's ecosystem for query processing against the data.

Figure 1. Smart sensor nodes

To manage the massive amounts of data generated from smart meters and other components of the grid, utility companies need a solution that operates in a distributed manner. In this paper, we describe such an approach and as possible solution running on the Apache Hadoop framework. Services that are provided by cloud enterprises and infrastructures for smart grid users including providing a storage space and analytic tools may get use of our study and conclusions. However latency time and security of cloud computing might be the reasons for utility companies not to adopt cloud systems and implement their own instead.

205

Collecting of smart meter data is of course very useful for the power distribution companies. Useful analytical information could be fetched from the data to serve planning purposes [2, 8]. Sensor IDs and geographical components of the data log provide relation with any external data, which may be an opportunity for acquiring knowledge, that is difficult to predict in advance. When we have a certain task, algorithms for filtering information from sensors at different levels can be implemented, so as to allow a solution of the problem on a centralized system - operated by Relational Database. Demand for analytical information in large companies is dynamic and different issues appear daily. In many cases, the discovery of various dependencies in seemingly unrelated data may provide answers to some important questions. This gives high value to the data obtained in time and companies are looking for solutions for data storage and opportunities for further processing.

There is a good reason for customers to install such electricity monitoring tools in their homes – it is the ability to analyze and have control over the energy consumption, early discover and take measures (e.g. using UPS devices) to avoid potential losses from power disturbances. A web page that shows a simple consumption data trend line can help them relate power consumption to household activity. Also forecasts which are predictions of future events or values using historical data. For instance, a forecast of power consumption for a new residential subdivision can be created using historical data from similar buildings [1, 2, 8].

B. Apache Hadoop – data storage and processing

In this paper we discuss Apache Hadoop for the management of the massive amounts of data generated from smart meters and other components of the grid.

Apache Hadoop has become a standard for managing Big Data. It provides its own distributed file system and runs MapReduce jobs on servers near the data stored on the file system. In very simple terms, Hadoop is a framework of of programs (open source software) which allows storing huge amount of data, and processing it in a much efficient and faster manner (via distributed processing). So essentially, the core part of Apache Hadoop comprises of two things: a storage part (Hadoop Distributed File System or HDFS) and a processing part (MapReduce algorithm) [9].

C. MapReduce

One of the attractive qualities about the MapReduce programming model is its simplicity. The MapReduce program consists only of two functions, called Map and Reduce, that are written by a user to process key-value data pairs. The input data set is stored in a distributed file system deployed on each node in the cluster. The program is then executed on a distributed processing framework. In general, the Map function reads a set of records from an input file, does any filtering and then outputs a set of intermediate records in the form of new key-value pairs. There are multiple instances of the Map function running on different nodes of a compute cluster. In the second

phase N instances of the Reduce function are executed, where N is typically the number of nodes. Each Reduce combines the records assigned to it and then writes records to an output file in the distributed file system, which forms the computation's final output [11].

D. Apache Hive as a query processing tool

The Apache Hive data warehouse software facilitates querying and managing large data sets residing in distributed storage. Built on top of Apache Hadoop, it provides access to files stored either directly in Apache HDFS and interface for query execution via MapReduce.

Hive is often used as the interface to an Apache Hadoop based data warehouse because of its SQL like query language. Hive is considered friendlier and more familiar to users who are used to using SQL for querying data. However tasks on sensor's data in most cases do not involve complex queries. So most of the MapReduce jobs on such a data could be easily programmed [5, 10].

Some authors [12] also suggest the Hadoop (HDFS and MapReduce) framework as a solution for smart meter data storage and management on Cloud computing resources that could scale easily. Other [13] propose different open source software products that run on Linux like OpenPDC acting as intermediate gateway for data collection from sensors and Cloudera flume agent as a GUI for management of services like Hive. In our study we analyze deeper the volume of typical data extracted from grid by which the usage of centralized computing model is still working better.

II. TEST-BED PERFORMANCE MEASUREMENTS

A. Experimental setup

For the experimental evaluation we collect data from sensors working in the "Virtual Laboratory for Distributed Systems and Networking" at Technical University of Sofia, Plovdiv branch. The sensors and embedded devices are either "cloud-ready" or communicating via cloud adapter software or are connected to an intermediate node called gateway. They provide data from the physical sensors as a service using XML or JSON format. The data is extracted from devices (or gateways) every few minutes as XML using REST web service that makes the data transformation and storage on distributed file system. The new data is actually stored on HDFS as log files in CSV format [6, 7].

In the presented implementation the electrical parameters are measured using energy sensor as a binary device and it's communication with the web service application goes through intermediate gateway (which provides a cloud service). The power meter measures current values of – voltage, current, frequency, active power, reactive power, power factor, active and reactive energy in three-phase electric system. Other sensors for measurement of brightness and temperature are connected either via the gateway or directly by REST web services [6, 7].

The model of the experimental system for remote data collection and storage is shown on Fig. 2.

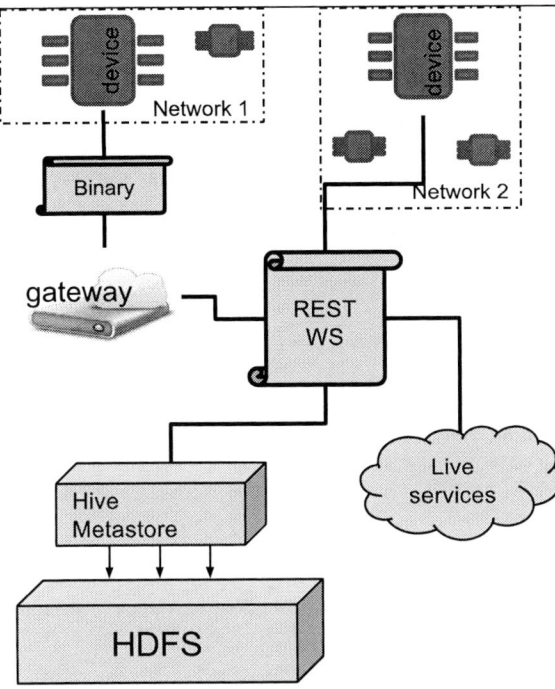

Figure 2. Model for data collection from remote smart sensors

The raw measurements are available through REST interface on the following URLs:

- http://dsnet.tu-plovdiv.bg/website/pst04.jsp

- http://dsnet.tu-plovdiv.bg/bjr.php.

The JSON (and XML) results are stored as CSV log files with lines like the following:

'pst04','2017-02-06 12:26:04','I1','6','A','I2','0','A','I3','0','A','U1','243.95',' V','U2','0','V','U3','0','V','active_power','0.1','W','freq uence','49.98','Hz','geo_latitude','42.139175','static', 'geo_longtitude','24.772743','static','reactive_power ','0','W','status','active'

'xbr','2017-02-06 12:26:12','light','0','brightness','temperature','19.78 ','C'

B. Results and discussion

We have executed measurements on how Hadoop performs and scales with the size of the data set of different number of meters over 4 nodes experimental Hadoop cluster set up: OS – Open Suse 24.2.1-x86_64, file system EXT4, java version 1.7.0, Hadoop-2.6.4, HDFS block size 64MB, Apache Hive 2.0.0, CSV files.

We store uncompressed data on HDFS. Then we execute different queries via Hive calls against the data aiming to determine very high or very low peak values compared to the average value of each single parameter [3]. We compare the results with the performance of relational database on a single machine that is running MySql Server version 5.5.33 on OS Open Suse 24.2.1-x86_64, file system EXT4 and 16GB of RAM.

Under-voltage and over-voltage caused by load changes or utility faults are the most common types of power disturbances [1]. The number of such disturbances in certain time period is a good subject of analysis. Dependencies between number of disturbances and power consumption fluctuations or temperature levels could be searched.

For the test bed experiments we produce report that counts number of records in one day showing values 5% below/over the average value for longer period. It is expected that with the increase in the volume of the data, we can just increase the number of nodes in the Hadoop cluster and keep the time for reports almost constant [2, 4]. However MySQL performs far faster with the same task on our experimental data set. It seems that as long as we do not reach the hard disc capacity of on single server, MySQL will perform better. On the other hand Hadoop offers scalable processing and storage which means that we can keep the performance at same level by adding more nodes into the cluster.

We could easily make the conclusion that centralized model works better as long as our data does not exceed the limits of the modern hard disks. The use of distributed model is not reasonable at one billion records of this type. At just several million records daily, we still may consider a centralized model as a better solution. However sensors in one million homes could produce similar data of several terabytes in just few days. Some authors [11] say that according conservative estimating, the number of smart meter will be about 170 million only in China.

Given this interest in Hadoop (data storage on HDFS, MapReduce processing), it is natural to ask "Why not use parallel SQL database management systems (DBMS) instead?" When it comes to performance, it is stated that at 100 nodes the two parallel DBMSs range from a factor of 3.1 to 6.5 faster than MapReduce on a variety of analytic tasks [11]. However the benefit is still questionable when it comes to data collected by sensors in a single table where we can not benefit much from data relations. Such performance advantage depends much on the proper DBMS configuration for each specific task. All DBMSs require that data conform to a well-defined schema, whereas MapReduce permits data to be in any arbitrary format on HDFS.

Besides the simplicity of implementation and capability for installation on low-end hardware, Hadoop offers flexibility to easily export or transform data or change the schema by adding more information from meters. Internally for intermediate storage, we may consider different data serialization binary file formats like Apache Avro and RCFile (Record Columnar File), ORC (Optimized Row Columnar) formats in order to provide indexing or compression optimizations that would improve the performance [5].

TABLE I.   PERFORMANCE EVALUATION

Time to run required reports and storage space	Hadoop cluster	Relational database
2 million meters count records filtered by one indexed column	20sec to 1 min 112MB	0.48sec 323MB
10 million meters count records filtered by one indexed column	20sec to 1 min 560MB	1.18sec 808MB for data 1.7GB (for indexes)
10 million meters count records filtered by two indexed columns	20sec to 1 min 560MB	1.43sec 808MB for data 1.7GB (for indexes)
100 million meters count records filtered by two columns	1min to 2 min on 4 nodes 2min to 5 min on 2 nodes	28.00sec

All possible requests against big number of similar records that involve aggregates (like *COUNT, MAX, SUM* or *AVG*) may benefit much of parallel computing via MapReduce.

III.   CONCLUSION

There is a variety of smart sensor tools on the market at affordable price that transmit data in standard formats over Internet. Enterprises and home consumers would benefit of information services based on data collected from power distribution smart meter.

The benefit of using the model of distributed storage and processing of data collected from advanced metering infrastructure compared to centralized computing model is the scalability that allows storage of really huge amounts of data records. Since with lowering the costs of smart meter devices, their number will increase dramatically in the near future, this will produce huge amounts of data logs that could not be served by centralized computing systems.

With the variety of tools of Apache Hadoop and with its modular concept, the suggested experimental model has the potential to be a technical solution for distributed

data storage and processing in smart power grid. Hadoop seems to be a competitor of DBMSs because of its simplicity of implementation, capability for changing the data schema and tools developed in the framework for optimization in the computing resource usage.

Some readers may feel that experiments conducted using 4 nodes nodes are not relevant to a real world systems but our point is to show that the distributed system is scalable and we can keep the performance at certain acceptable level by adding more nodes to the installation when our data grows up.

ACKNOWLEDGMENT

The presented work is supported by the National Science Fund of Bulgaria, project "Investigation of methods and tools for application of cloud technologies in the measurement and control in the power system" under contract E02/12.

REFERENCES

[1]   J.Seymour, "The Seven Types of Power Problems", White Paper 18, Revision 1, Schneider Electric white paper library, online: http://www.apc.com/salestools/VAVR-5WKLPK/VAVR-5WKLPK_R1_EN.pdf

[2]   Dr. Darold Wobschall President, Esensors Inc. Designing Sensors for the Smart Grid, Advanced Energy Conference – Buffalo, 2011

[3]   Panda, Mrutyunjaya. "Intelligent Data Analysis for Sustainable Smart Grids using Hybrid Classification by Genetic Algorithm based Discretization." Accepted for Publications in IDT Journal, IOS Press. Netherlands-2017 (In Press)

[4]   Rabi Prasad Padhy, "Big Data Processing with Hadoop-MapReduce in Cloud Systems", Oracle Corp., Bangalore, Karnataka, India, 2013

[5]   Stefan Stoyanov, "Sensor data processing – key factors for choosing file formats in HADOOP", Computer & Communication Engineering, ISSN: 1314-2291, Technical University Sofia, Bulgaria, 2017 [accepted for review]

[6]   M. Shopov, "Practical Implementations of Cloud Computing Technologies for Smart Metering in Electrical Power Systems", Annual Journal of Electronics, vol. 9, pp. 124-127, 2015, ISSN 1314-0078

[7]   N. Kakanakov, M. Shopov, and G. Spasov, "Distributed automation systems based on java and web services," in Proc. International Conference on Computer Systems and Technologies (CompSysTech), 2006, pp.III-A.24-1-6.

[8]   "Managing big data for smart grids and smart meters", IBM Software White Paper, IBM Corporation, 2012.

[9]   Apache Hadoop Project homepage: http://hadoop.apache.org/.

[10] Apache Hive Project homepage: http://hadoop.apache.org/hive/.

[11] A. Pavlo, A. Rasin, E. Paulso, D. J. Abadi, D. J. DeWitt, S. Michael and M. Stonebraker, "A Comparison of Approaches to Large-Scale Data Analysis" - M.I.T. CSAI,

[12] Li Wengting, Zheng Yan, Liu Shaobo, Long Zhaozhi, Li Zhicheng, "Research of Smart Meter Massive Data Storage Based on Cloud Computing Platform", Trans Tech Publications, Switzerland 2013

[13] Mukhtaj Khan, Dr. Maozhen Li, Dr. Gareth Taylor, "Massively Scalable Smart Grid Data Storage and Analysis in Hadoop Virtualized Environments" School of Engineering and Design, Brunel University London

Running HPC applications on many million cores Cloud

D. Tomić*, Z. Car ** and D.Ogrizović***

* University of Rijeka/Department of Informatics and Center for Advanced Modelling and Simulation, Rijeka, Croatia,
e-mail: drasko.tomic@uniri.hr
** University of Rijeka/Faculty of Technology and Center for Advanced Modelling and Simulation, Rijeka, Croatia,
email: zlatan.car@uniri.hr
*** University of Rijeka/Center for Maritime Studies, Rijeka, Croatia, e-mail: dario@uniri.hr

Abstract - Despite the various hardware and software improvements in Cloud architecture, there still exists the huge performance gap between the commodity supercomputers and Cloud when running HPC communication intensive applications. In order to find what is preventing them to better scale on Cloud, we evaluated HPL and NAMD benchmarks on HPE Openstack testbed, and NAMD benchmarks on supercomputer located at Rijeka University Supercomputing Center. Our results revealed two major bottlenecks: the throughput of the interconnect, and Cloud orchestration layer, among other responsible for the management of the communication between Cloud instances. We investigated the influence of jittering, but did not find the significant influence on performance. Our conclusion is that by solely increasing the interconnect throughput, one will not improve the scalability of HPC communication intensive HPC applications in Cloud. This is also backed up with NAMD performed at HP Labs, and with HPL benchmark performed at San Diego Supercomputing Center. We propose two possible scenarios of scalability improvements. One with distributed model of Cloud Orchestration layer; another with bare metal containers. Efficient load balancing remains the must if we want to see HPC applications scaling over many million Cloud cores. For this, we propose novel SLEM based load balancing strategy.

I. INTRODUCTION

Despite the expectations that Cloud will adopt High Performance Computing applications quickly, there is still a huge gap between the performance of applications running on bare metal supercomputers and those on Cloud.

We gratefully acknowledge HPE to grant an access to HPE Openstack testbed, on which part of this work was performed.

Several reasons prevent HPC community to move more quickly towards Cloud, like the investment already done in their datacenters and the non-deterministic nature of the most Cloud environments preventing someone to more accurately estimate the cost of running HPC application in the Cloud. However, the major preventing factor is the limited performance of the virtualized Cloud resources. Although the major players in the Cloud arena offer dedicated resources like the bare metal servers and fast interconnect just for HPC users, the majority of the scientific community sees no advantage to migrate their applications towards Cloud if they cannot exploit the full benefit of the virtualization. Major virtualization benefits are higher availability and possibility of dynamic resource re-allocation, both backed up by VM live migration possibility. Besides scalability issues stated in a survey [1], there are other open questions, like the collaboration between physical and virtual resources and support of the standard HPC tools for virtualized environments, mentioned in [2]. The numerous researchers performed the tremendous work in order to evaluate the performance of HPC applications in Cloud. Due to its popularity, EC2 from Amazon is by far the most evaluated platform. One of the earliest reports on the performance of HPC applications in Amazon EC2 Cloud appeared in [3]. Authors analyzed the performance of the Amazon EC2 platform using micro-benchmarks, kernels, and e-Science workloads. They concluded virtualization might induce significant performance penalties for demanding scientific computing workloads. Following, the conclusion of the group of authors in [4] was that overall results indicate EC2 is six times slower than a typical mid-range Linux cluster, and twenty times slower than a modern HPC system. Another conclusion was that interconnect on EC2 cloud platform severely limits performance and causes significant variability. However, in the same work they stated Amazon Cluster Compute Instance performs significantly better. The similar findings were carried out in [5], and according to the authors, results indicate that the current Clouds need an order of magnitude in performance improvement to be useful to the scientific community. Further examination of I/O overhead in [6] revealed that I/O performance does not keep the pace with the level of CPU efficiency for low latency and high throughput applications, such as MPI or large-scale data processing in para-virtualized and fully virtualized virtual machines. In addition, as noted in [7], because the virtualized environments have the significant influence network overhead, this hampers the adoption of GPGPU communication-intensive applications. Last but not the least, the authors of [8] noticed a big noise in all experiments, the phenomena, which deserves further investigation. That means, they found the big overhead induced by the virtualization layer, however they were not able to find an exact source of it. Besides Amazon EC2, Azure and Nimbus environment were also under investigation, however with the similar conclusion that a huge source of noise in the virtualization layer exist, preventing HPC I/O demanding applications to scale as expected. In [9], researches landed with a similar conclusion that the major impact factor preventing HPC applications to scale is a virtualization layer. To underline the complexity of this topic, we will mention the paper were authors believe the virtualization is not a big problem at all, but the slow interconnect [10]. Another researchers evaluated various hypervisors in a hope to find one with a relative little or no overhead [11]. However, the prevalent statement of most researchers is that virtualization layer and more specifically intensive I/O induces noise and prevents HPC applications to scale even nearly so good as on bare metal platforms [12]. Moreover, the problem of the performance evaluation of HPC applications in the Cloud is much more difficult than on commodity supercomputers, as derived results vary from one experiment to another, even with the equal initial conditions. For example, in [13] and through a case study using the Linpack benchmark, authors show that cache utilization becomes highly unpredictable and similarly affects computation time. Concepts like average, expected performance and execution time, expected cost to completion, and variance, measures traditionally ignored in HPC context, now should complement or even substitute the standard definitions of efficiency. The similar view that virtualization exacerbates the common problems of accurate performance measurement and benchmarking is given in [14].

II. TOWARDS THE CLOUD OPTIMIZATION FOR HPC APPLICATIONS

In addition to the work elaborated in the previous session, which is related to the performance evaluation of HPC applications on the Cloud, various researchers proposed and evaluated the methods for improving the Cloud performance for HPC. There are several approaches how to make Cloud more appealing for HPC. One approach is to modify Cloud job scheduler in a way to make workload balancing more efficient regarding the computation and the communication. That means to distribute load on Cloud compute instances in a way which maximizes computational efficiency, while at the same time minimizes the communication between them. One possibility to reach these optimization goals is to minimize the **S**econd **L**argest **E**igenvalue in **M**agnitude (SLEM) of

the matrix representing the data flow in a certain parallel processing system. This approach was tested in a practice on supercomputer system running HPL benchmark [15], and HPL efficiency increased dramatically. In the same paper, authors expressed their vision to apply this optimization on hybrid Cloud structures. One of the first proposals to modify Cloud in order to support dynamic load balancing of virtual machines across available Cloud resources was [16]. Another group of authors undertook an interesting optimization approach, where they succeed to decrease the cost of the Cloud infrastructure by sixty percent, while at the same lowering performance by only ten to fifteen percent [17]. They also found that the execution overhead in Cloud could be minimized to a negligible level using thin hypervisors or OS-level containers. In [18] the same group of authors stated that network virtualization has multiple negative effects on HPC application performance by increasing the network latency, decreasing the network bandwidth, and by decreasing the application performance by interfering with the application processes. Besides, they found CPU affinity might improve performance. Different static scheduling strategy with topology and homogeneity aware scheduler built on the top of Openstack scheduler is considered in [19], while mixed static and dynamic load balancing strategy for tightly coupled HPC applications is presented in [20]. This strategy infers the static hardware heterogeneity in virtualized environments, and also adapts to the dynamic heterogeneity caused by the interference arising due to multi-tenancy. Through continuous live monitoring, instrumentation, and periodic refinement of task distribution to virtual instances, load balancer adapts to the dynamic variations in Cloud resources. Through experimental evaluation on a private Cloud with 64 virtual instances using benchmarks and a real science application, performance benefits up to 45% were demonstrated. Another interesting scheduling strategy, in contrary to the de-facto overutilization approach in [17], include underutilization of the Cloud resources [21]. The authors found that for the some cluster configurations, the solution is reached almost an order of magnitude earlier on average when the available resources are underutilized. The performance benefits for single node computations are even more impressive: Underutilization improves the expected execution time by two orders of magnitude. Finally, in contrast to unshared clusters, extending underutilized clusters by adding more nodes often improves the execution time due to an increased parallelism even with a slow interconnect. Their findings reinforces the common opinion that the virtualization noise is the main source for the performance degradation of HPC applications in the Cloud. And while container based deployment of HPC applications in the Cloud is the promising technology when it goes around the performance, it still lacks some important security features. An excellent research on containerized HPC applications in the Cloud is provided in [22]. The researchers found that container-based virtualization, aka OS level virtualization, promises more than hypervisor based virtualization regarding the performance. Their benchmarks showed near bare metal performance. However, they also found that poor isolation and the security still exists, thus preventing this technology to become the mainstream in HPC Cloud arena. Last but not the least, and in order to make Cloud more feasible for HPC applications, there are ongoing projects to make necessary modifications both in code and in hardware, for example one described in [23].

III. WHAT AND WHY CLOUD CONTROLLER

By careful examination of the previous work mentioned in the preceding section, we understand that the noise generated from the virtualization layer prevents HPC tightly coupled applications to scale on Cloud even nearly as good as on bare metal infrastructure. This effect prevents them to scale properly even with the most sophisticated schedulers built atop of Cloud orchestration layer. The container technology is not mature enough to fulfill important security requirements; changing hardware is not feasible, as it is not likely that hardware providers will bring Cloud optimized hardware in the near future; improving interconnect bandwidth does not help to bridge the performance gap between Cloud and bare metal infrastructure. This last statement is the most challenging one, however strongly backed up by the work of the researchers from San Diego Supercomputing Center [24], who managed to scale HPL over 32 bare metal compute nodes with 100Mbit/s Ethernet interconnect. And in order to verify the frequent statement that the faster interconnect will speed up HPC applications on Cloud, we decided to run HPL on HPE Openstack testbed with 1GB Ethernet interconnect. Our decision to use HPL is based on the fact that in the last decade HPL became a common way of measuring the ability of HPC clusters to execute applications with heavy demand for computational and communication power. In addition, although HPL puts a certain computing system to its limits like no other application, it is a fair estimation of how this system will perform ordinary HPC applications. Therefore, and in order to estimate the ability of Cloud to run HPC applications, we run first a series of HPL benchmarks on Openstack based Cloud. We performed HPL benchmarking in three steps. First we run benchmarking on 1, 2, 4 and 8 (1 vCPU and 1GB RAM) compute instances respectively. Instances were defined and activated first in the same, than in the different availability zones. We tried the number of HPL input parameters, and found that among others, N=8832 and NB=192 parameters deliver the best Rmax. However, despite our efforts, HPL scaled up to two instances only:

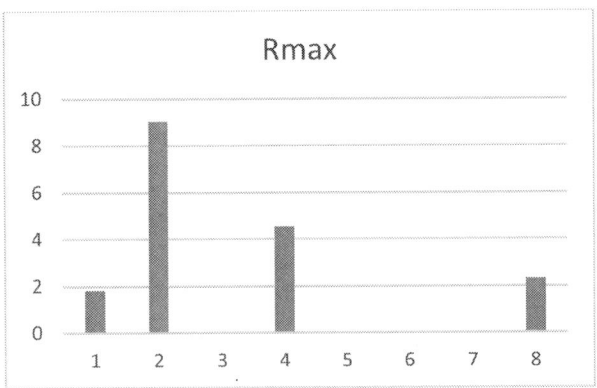

Fig 1: HPL scaling over Cloud virtual instances. We observed the maximal Rmax on two instances, then HPL performance dropped by adding more instances.

As the next step, we run HPL with the same parameters, but this time with each instance defined in another availability zone. This time, HPL scaled again up to two instances, and then dropped when we add more instances. Lastly, we started HPL benchmarking on bigger (2 vCPUs and 2GB RAM) instance. Already on one instance, HPL result was considerably higher than before. However, when adding second instance, result dramatically dropped down, showing that with bigger instances HPL scaling get worse. We also investigated if the difference in the performance, i.e. jittering of virtual instances is the source of these poor HPL performance results, so we measured it by running HPL on each instance:

Instance number	HPL Rmax [Gflops]
1	1.777
2	1.691
3	1.771
4	1.804
5	1.757
6	1.769
7	1.678
8	1.778

Table 1: Cloud virtual instances jittering. The mean value is 1.753 GFLOPS, and the maximal deviation from the mean is 0.126 GFLOPS, i.e. ~ 7.2%.

At the time of HPL benchmarking, 1Gbit links between compute instances were available. However, another group of authors [24] scaled HPL on bare metal cluster successfully over much slower 100 Mbit Ethernet. Because of that, we believe the main bottleneck preventing HPL to scale is not the network, but Cloud orchestration layer; calls from this layer to network Cloud controllers do not keep the pace with HPL inter-process messaging, i.e.

with exchanging of matrix columns and rows. Nevertheless, we explored several other possibilities for improving HPL scalability: By balancing load more efficiently as in [15], and by using Jumbo Ethernet frames as proposed in [26]. Our both approaches failed: We did not even deployed more efficient load balancing, as jittering between the instances was only 7.2% from average, thus scaling improvement would be negligible. In addition, when trying to run HPL over Jumbo packets (9000 MTUs), OpenMPI processes died for unknown reasons. For this reason, we revert back to standard 1500 MTUs packets. Varying HPL input parameters in HPL.dat file also did not improved scalability significantly. For all the reasons mentioned, our conclusion is that for the better scaling of HPL and other communication intensive HPC applications on Cloud, Cloud orchestration layer must be more efficient. Either with the code optimized for the faster handling of inter-process communication, or it should be distributed across more nodes, thus allowing to perform its work more efficiently. However, we propose another approach, which will be discussed in the rest of this paper.

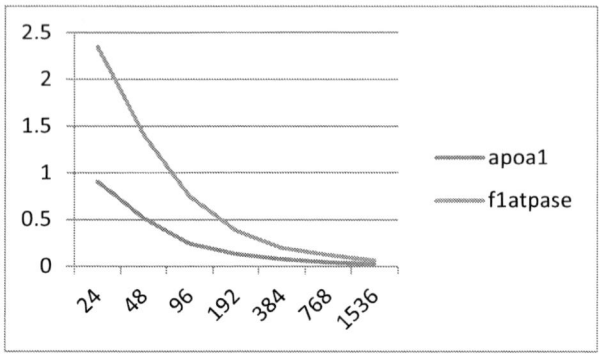

Fig. 3: Scaling NAMD ApoA1 benchmark with 92.224 and ATPase benchmark with 327.506 atoms up to 1536 cores on supercomputer located at the University of Rijeka. Benchmarking results are expressed in days needed per nanosecond simulated. Lower is better, and linear scaling is observed.

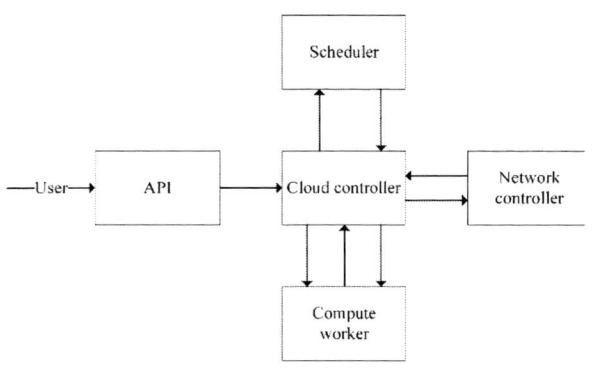

Fig 2: A schematic view of the main Openstack processes, orchestrated by the Cloud orchestration layer, i.e. Cloud controller. The whole communication between other parts of Cloud goes through Cloud controller. If there are many compute instances and HPC application is communication intensive, Cloud controller becomes a bottleneck.

We continued to benchmark Cloud with NAMD, a highly scalable code for complex biomolecular simulations [27]. In order to check NAMD scalability on bare metal, we run NAMD ApoA1, ATPase and STMV benchmarks on HPC cluster Bura with 6912 Intel Xeon E5-2690 v3 processor cores, located at the University of Rijeka. The benchmarking results are expressed as We investigated NAMD scalability up to 1536 cores:

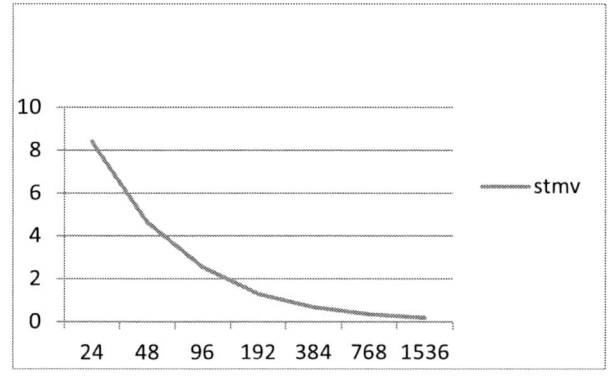

Fig. 4: Scaling NAMD STMV benchmark with 1.066.628 atoms up to 1536 cores on supercomputer located at the University of Rijeka. Benchmarking results are expressed in days needed per nanosecond simulated. Lower is better, and linear scaling is observed.

For some reasons still under investigation, we were not able to scale NAMD benchmarks across multiple Openstack based Cloud virtual instances. However, another group of authors [28] successfully performed ApoA1 NAMD benchmarks on Eucalyptus Cloud with 1Gbit interconnect, Open Cirrus with 10 Gbit interconnect, and Scientific Linux based Taub Cloud with QDR Infiniband. On Taub, NAMD scaled nearly linearly up to 256 cores, on Open Cirrus scaling was also up to 256 cores but far from being linear, while on Eucalyptus scaling stopped already after 64 cores. Another author [29] also investigated NAMD scaling on Openstack with 1Gbit Ethernet. Benchmarks performed were ApoA1, ATPase and STMV. Elapsed times for these benchmarks on 4 VCPUs and 16GB RAM were 246, 719 and 2483 seconds respectively, thus showing that NAMD scales nearly linear within one virtual instance. These two NAMD related research papers, together with our own experience, reinforced our opinion that NAMD is scaling well on Cloud with Infiniband interconnect across multiple instances and within one virtual Cloud instance, but has a performance bottleneck in interconnect if Cloud platform is using 1GB or 10GB Ethernet interconnect.

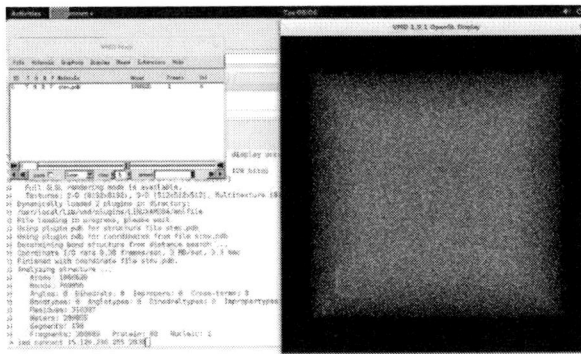

Fig. 5: Output of Cloud console while running STMV standard NAMD benchmark.

IV. CONLUSION AND THE FUTURE WORK

According to the work performed by a numerous researchers, the prevalent position is that Cloud, despite the numerous advances in hardware and software, still does not keep the performance race with bare metal supercomputers when running communication intensive HPC applications. To find out what is preventing these kind of applications for better scaling, we performed several HPL and NAMD benchmarks on HPE Openstack testbed and supercomputer located at the Rijeka University, and compare these results with the research from other authors. We were able to identify several bottlenecks: the Cloud interconnect, the Cloud orchestration layer and the jittering of virtual instances. Our results while running HPL across multiple Cloud virtual instances over 1 Gbit interconnect , as well as the results from [25], confirmed HPL does not scale over more than two virtual Cloud instances. However, in the experiment described in [24], researchers were able to scale HPL on bare metal supercomputer on over 32 nodes and with an order of magnitude slower 100 Mbit interconnect. This is an excellent proof Cloud orchestration layer is the major bottleneck preserving HPL and other highly

communication intensive HPC applications to scale better. For the moderatelly communication intensive HPC applications like NAMD, Cloud orchestration layer is not the most important bottleneck; According to the findings in [28], the major performance limiting factor here is the throughput of Cloud interconnect. Lastly, we investigated the influence of instances jittering on applications scaling. We found jittering is to small to cause the considerable performance degradation, at least in Clouds with uniform compute nodes. The code optimization of the Cloud orchestration layer can provide some performance improvement, but for the highly communication intensive HP applications like HPL the distributed version of the Cloud orchestration layer would be the solution able to deliver more performance. Another way to overcome this bottleneck is to avoid the Cloud orchestration layer at all, by implementing the bare metal containers. Eventually, the load distribution on containers can be performed with the intelligent load balancing strategy like SLEM, first envisioned in [30] and later proved to be very successful in heterogeneous cluster environment [15]. We expect SLEM will improve the performance and scalability by minimizing the communication between containers, especially in hybrid Cloud environments. Our future work will be oriented towards the several goals: to establish the model of the interplay between the number of virtual Cloud instances, amount of the interprocess communication and the scalability of HPC applications in Cloud; to efficiently implement bare metal containers in massive parallel environments; to add SLEM load balancing strategy to the containers. We hope our research will help Cloud to become standard platform for running communication intensive HPC applications.

ACKNOWLEDGMENT

We gratefully thanks Jean-Luc Assor from HPE and Dejan Milojicic from HP Labs who helped in the preparation of this paper.

REFERENCES

[1] S. P. Ahuja, S. Mani, The State of High Performance Computing in the Cloud, Journal of Emerging trends in Computing and Information Sciences, vol. 3, no. 2, (2012), pp. 262-265.

[2] D. Tomic, D. Ogrizovic, Z. Car, Cloud Solutions for HPC: Oxymoron or Realm, Tehnicki vjesnik, vol. 20, No. 1, (2013), pp. 177-182.

[3] S. Ostermann, A. Iosup, N. Yigitbasi, R. Prodan, T. Fahringer, D. Epema, An Early Performance Analysis of Cloud Computing Services for Scientific Computing, In Parallel and Distributed Systems Report Series, Delft University of Technology, 2008, pp. 931-945.

[4] K. R. Jackson, K. R. Ramakrishman, L. Muriki, K.Canon, S. Cholia, S. Shalf, J. Waserman, J. Harvey, N. J. Wright, Performance Analysis of High Performance Computing Applications on The Amazon Web Services Cloud, in Proceedings of the 2nd IEEE Int. Conf. On Cloud Computing Technology, 2010, pp.159-168.

[5] N. Regola, J. Ducom, Recommendations for Virtualization Technologies in High Performance Computing, in the Proceeding of the 2nd IEEE International Conference on Cloud Computing Technology and Science, 2010, pp. 409-416.

[6] A. Iosup, S. Ostermann, N. Yigitbasi, R. Prodan, T. Fahringer, D. Epema, Performance Analysis of Cloud Computing Services for Many-Tasks Scientific Computing, in Parallel and Distributed Systems, IEEE Transactions on , vol.22, no.6, 2011, pp. 931-945.

[7] R. R. Exposito, G.L.Taboada, S.Ramos, J. Tourino, R. Doallo, General-Purpose Computation on GPUs for High Performance Cloud Computing, Concurrency Computat.: Pract. Exper. 2011, pp. 1-18.

[8] O. Litvinski, A. Gherbi, Experimental Evaluation of OpenStack Compute Scheduler, In Proceedings of the 4th International Conference on Ambient Systems, Networks and Technologies, Halifax, Canada, 2013, pp. 25-28.

[9] R. Tudoran, A. Costan, G. Antoniu, L. Bouge, A Performance Evaluation of Azure and Nimbus Clouds for Scientific Applications, In Proceedings of the 2nd International Workshop on Cloud Computing Platforms, Bern, Switzerland, 2012.

[10] Q. He, S. Zhou, B. Kobler, D. Duffy, T. McGlynn, Case Study for Running HPC Applications in Public Clouds, In Proceedings of the 19th ACM International Symposium on High Performance Distributed Computing, Chicago, USA, 2010.

[11] A.J.Younge, R. Henschel, J.T. Brown, G. von Laszewski, J. Qiu, G. C. Fox, Analysis of Virtualization Technologies for High Performance Computing Environments, In Proceedings of the IEEE 4th InternationalConference on Cloud Computing, Washington DC, USA, 2011, pp. 9-16.

[12] G. Devarshi, R. S. Canon, L. Ramakrishnan, I/O Performance of Virtualized Cloud Environments, Lawrence Berkeley National Laboratory, LBNL Paper LBNL-5432E, 2013.

[13] P. Bientinesi, R. Iakymchuk, J. Napper, HPC on Competitive Cloud Resources, Achen Institute for Advanced Study in Computational Engineering Science, technical report, 2010.

[14] P. Luszczek, E. Meek, S. Moore, D. Terpstra, V. M. Weaver, J. Dongarra, Evaluation of the HPC Challenge Benchmarks in Virtualized Environments, In Proceeding of the 2011 International conference on Parallel Processing, Taipei, Taiwan, 2011, pp. 436-445.

[15] D. Tomic, L.Gjenero, E. Imamagic, Semidefinite optimization of High Performance Linpack on heterogeneous cluster, In Proceedings of the 36th Int. Convention MIPRO, Opatija, Croatia 2013, pp. 157-162.

[16] F. Wuhib, R. Stadler, H. Lindgren, Dynamic Resource Allocation with Management Objective: Implementation for an OpenStack Cloud, In Proceedings of the 8th international conference on Network and Service Management, Las Vegas, US, 2012, pp. 309-315.

[17] A. Gupta, L.V. Kale, F. Gioachin, V. March, C.H. Suen, B. Lee, P. Faraboschi, R. Kaufmann, D. Milojicic, Exploring the Performance and Mapping of HPC Applications to Platforms in the Cloud, in the Proceedings of the 21st International symposium on High-Performance Parallel and Distributed Computing, Delft, Nederland, 2012, pp. 121-122.

[18] A. Gupta, L.V. Kale, F. Gioachin, V. March, C.H. Suen, B. Lee, P. Faraboschi, R. Kaufmann, D. Milojicic, The Who, What, Why, and How of High Performance Computing in the Cloud, HP technical report, 2013.

[19] A. Gupta, D. Milojicic, L. V. Kale, Optimizing VM Placement for HPC in the Cloud, in Proceedings of the 9th ACM International Conference on Autonomic Computing, San Jose, US, 2012.

[20] A. Gupta, O. Sarood, L. V. Kale, D. Milojicic, Improving HPC Application Performance in Cloud through Dynamic Load Balancing, In Proceedings of IEEE International Symposium on Cluster Computing and the Grid, Delft, Netherland, 2013, pp. 402-409.

[21] R. Iakymchuck, J. Napper, P. Bientinesi, Improving High-Performance Computations on Clouds Through Resource Underutilization, In proceedings of the 2011 ACM Symposium on Applied Computing, Taichung, Taiwan, 2011, pp. 119-126.

[22] M.G.Xavier, M.V.Neves, F.D.Rossi, T.C.Ferreto, T. Lange, C.A.F. de Rose, Performance Evaluation of Container-based Virtualization for High Performance Environments, in Proceedings of the 21st Euromicro International Conference on Parallel, Distributed and Network-based Processing, Belfast, Ireland, 2013, pp. 233-240.

[23] R. Ledyayev, H. Richter, High Performance Computing in a Cloud Using OpenStack, In Proceedings of the 5th international conference on Cloud Computing, GRIDs, and Virtualization, Venice, Italy, 2014.

[24] P.M.Papadopolous, C.M.Papadopolous, M.J.Katz, Configuring Large High-Performance Clusters at Lightspeed: A Case Study . In International Journal of High Performance Computing Applications, 2004, pp. 317-326.

[25] J. Napper, P. Bientinesi, Can cloud computing reach the top500?, In Proceedings of the combined workshops on Unconventional High performance computing workshop plus memory access workshop, New York, NY, USA, 2009, pp. 17-20.

[26] Dr. Swamy N. Kandadai, Tuning Tips for HPL on IBM xSeries Linux Clusters, An IBM Redbooks paper 2004. , http://www.redbooks.ibm.com/redpapers/pdfs/redp3722.pdf (accessed 08.0.0.2016).

[27] http://www.ks.uiuc.edu/Research/namd/ (accessed 11.02.2016).

[28] A. Gupta, D. Milojicic, Evaluation of HPC applications on Cloud, HP. Laboratories, HPL 2011-132

[29] D. Tomic, Exploring Bacterial Biofilms with NAMD, a Highly Scalable Parallel Code for Complex Molecular Simulations, in Proceedings of the 38th International Convention Mipro, Opatija, Croatia, 2015.

[30] D. Tomic, Spectral Performance Evaluation of Parallel Processing Systems, in J. Chaos, Solitons & Fractals, 13(1), 2002., pp. 25-38.

Gap in pagination due to withheld paper.

Pages 215-220

Modeling heterogeneous computational cluster hardware in context of parallel database processing

K.Y. Besedin[*], P.S. Kostenetskiy[**]

[*] South Ural State University, Chelyabinsk, Russia
besedinki@susu.ru
[**] South Ural State University, Chelyabinsk, Russia
kostenetskiy@susu.ru

Mathematical modeling is an important approach for creating a parallel database management system that could efficiently use capabilities, provided by modern heterogeneous computational clusters, equipped with manycore coprocessors or GPUs. To this day, several models heterogeneous computational systems were proposed, but none of them is suited for modeling database processing.

In this paper, we address this problem by proposing the Heterogeneous Database Multiprocessor Model. Proposed model consists of several submodels, which describe different aspects of database processing on heterogeneous computational systems. This paper describes hardware platform submodel, which describes the hardware of modeled computational cluster and execution submodel that describes the rules of cooperation between hardware submodel components.

I. INTRODUCTION

Computational clusters, equipped with manycore coprocessors and GPUs play key role in modern high-performance computations [1]. Scientific community is interested in using capabilities, provided by such systems not only for computational tasks. Database systems is an important example of possible applications for heterogeneous clusters because information technologies development is creating extremely large amounts of data that needs to be processed. One of the reasons that make the usage of heterogeneous computational systems in Database Management Systems (DBMS) difficult is that heterogeneous systems have a number of characteristics that make them different from traditional CPU-only systems [2]. Those characteristics are need to be taken into account when developing software for those platforms. Very often, differences between traditional and heterogeneous systems require review of existing approaches for database processing and creating the new approaches. Mathematical modeling can ease the process of creating those approaches and their evaluation, but, to the best of our knowledge, there is no existing model that could be used for this purpose.

To address this problem we propose to develop the Database Heterogeneous Multiprocessor Model (DHM). This paper describes an ongoing research on creating this model. We describe to submodels of DHM — the hardware platform submodel and execution submodel. We review the existing mathematical models for heterogeneous systems. Then we describe the DHM model itself and hardware platform and execution submodels.

II. PRIOR WORK

In the last few years, a number of mathematical models of heterogeneous computational systems was created.

The Roofline model [3] is designed for performance and efficiency evaluation of low-level optimizations and is intended to be used for computers with multicore CPUs. Under certain conditions, it also may be used for heterogeneous computers. Its ability to model parallel database processing is limited for a number of reasons: it cannot be applied for distributed computational systems; it does not allow modeling of input/output operations and it uses floating-point performance as one of key metrics.

Paper [4] describes a model for performance and power consumption evaluation of computing clusters, equipped with Intel Xeon Phi coprocessors, working in offload mode or GPUs. In this model, it is assumed that original task is divided into several subtasks. Each subtask is iteratively processed and each iteration consists of two phases: computation and data exchange, which cannot be done at the same time. One limitation of this model is that modeled workload is not typical for database processing. Other limitations is inability to model I/O operations.

The PerDome model [5] is designed for low-level optimizations performance and efficiency evaluation for heterogeneous computational systems. PerDome model is an extension of Roofline model for heterogeneous systems and has the same limitations for database processing modeling.

The Database Multiptocessor Model (DMM) [6] allows to model parallel database processing on computational cluster. The main limitation of DMM model is its inability to model database processing on heterogeneous systems.

This work was supported by Russian Foundation Fund for Basic Research, project No 16-37-00245 and by Act 211 Government of the Russian Federation, contract № 02.A03.21.0011.

III. DHM Model

In this work, we propose to develop the Database Hybrid Multiprocessor model – DHM. As the DMM model, DHM is composed of a number of submodels. Each submodel describes some aspect of database processing on a heterogeneous computational cluster. There are three submodels in DHM model:

1. Hardware platform submodel – describes hardware of modeled computational system

2. Execution submodel – describes the rules of interaction of hardware submodel components and their algorithms

3. Transaction submodel – will describe the rules of parallel transaction execution

This paper describes hardware platform and execution submodels which are described later in this section.

A. Hardware platform submodel

The hardware platform submodule describes the computational system as a DHM-graph. DHM-graph is a connected graph. Vertices of DHM graph represent hardware components of a computational system and called modules. There are three types of modules: computational modules, communicational modules and data storage modules. DHM module must have at least one module of each type. Edges of DHM graph represent communication lines between hardware components.

Computational module $P \in \mathfrak{P}$ is a computational device that is used for database processing. In a real system it can represent a node of a computational cluster, equipped only with CPUs, a node, equipped with CPUs and GPUs, Intel Xeon Phi coprocessor in native mode, etc. In general case, any device that is used to process database queries and is able to initiate data exchange with external devices can be treated as a computational module. In DHM graph, computational module can be connected only with one communicational module.

Data storage module $M \in \mathfrak{M}$ is a device that is used to store database objects. In can be connected only with one communicational module. In a real system, data storage module can represent a hard drive, SSD, network storage device, etc.

Communicational modules $N \in \mathfrak{N}$ are used for data exchange within a computational system. Computational modules and data storage modules can be connected only with communicational modules. In real systems, communicational module can represent network switch or internal bus.

Fig. 1 shows an example of a DHM-graph for a computational cluster that consist of n nodes. Each node is equipped with a central processor, which is modeled by a computational module P_i^1, manycore coprocessors in a native mode, which are modeled by computational modules P_j^2. Computational modules are connected to a PCI-E bus, represented by a communicational module N_k^1.

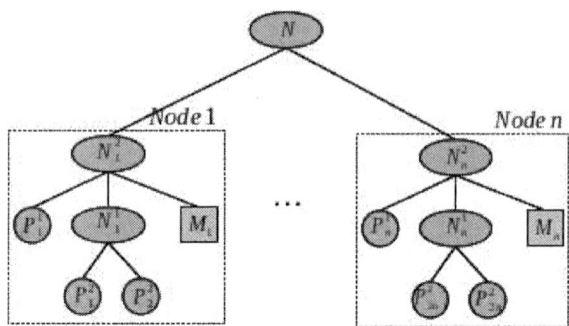

Figure 1. Example of DHM graph

Communicational modules N_m^2 represent nodes system buses and their network adapters. Each node is equipped with a hard drive, which is modeled by a data storage module M_o. Cluster nodes are connected with each other with a network switch, modeled by communicational module N.

B. Execution submodel

In DHM model, packet is the smallest atomic unit of data processing. All packets have same size. Each packet has a header, which includes the address of packet's sender and receiver. In a real DBMS packet can represent one or more tuples, relation column or its fragment, etc. Data exchange operations are initiated by computational modules, which exchange data with data storage modules. Computational module can initiate new data exchange operation without need to wait for a completion of previous data exchange operation. Computational module cannot have more than s_r unfinished read operations and more than s_w unfinished write operations.

In DHM model, database processing is modeled as a loop performing a sequence of steps, called ticks. All ticks have the same duration and does not change during modeling. To each module $m \in \mathfrak{P} \cup \mathfrak{M} \cup \mathfrak{N}$, a performance coefficient $h_m \in \mathbb{N}, 1 \leq +\infty$ is assigned. It determines how many packets module can process within one tick. Every module has packet queue Q, where waiting for processing packets are placed. The size of Q is denoted as $size(Q)$.

1) Computational module

On each DHM tick, each computational module $P \in \mathfrak{P}$ processes packets from its queue Q and initiates operations of packet reading and writing. The number of packet read and write operations is determined by modeled algorithm.

Fig 2. Illustrates the pseudo-code of packet reading operation.

```
if r(P) < sᵣ then
    Put packet E with P as receiver
address to queue of module M
    r(P)++
else
    Wait
```

Figure 2. Pseudo-code of packet reading operation

Suppose that a computational module P needs to read packet E from data storage module $M \in \mathfrak{M}$. If computational module already initiated s_r read operations that were not completed yet, it is placed in a wait state. Otherwise, packet E with P as receiver address and M as sender address is placed to the packet queue of module M. Module P also assigns E a cost coefficient j. This coefficient determines the amount of work, needed by module P to process packet E.

Suppose that a computational module P needs to write packet E to data storage module M. If computational module already initiated s_w write operations that were not completed yet, it is placed in a wait state. Otherwise, packet E with M as receiver address and P as sender address is placed to the packet queue of module N, where N is a communicational module that is connected to computational module P. Pseudo-code of write operation is illustrated on fig. 3, where $r(P)$ is number of unfinished packet read operations of module P.

To process packets, computational module P uses algorithm, illustrated by pseudo-code on fig. 4. Packets are extracted from packet queue Q of module P while this queue is not empty and sum of extracted packet's cost coefficients is less than or equal to computational module's performance coefficient.

2) Communicational module

On each tick, communicational module N extracts packets from its queue and passes them to other modules. For each extracted packet, N chooses the most optimal delivery path and puts it to packet queue of next module in chosen path. The maximum number of packets that can be processed by communicational module within one tick is determined by module's performance coefficient j.

Suppose that X is module-receiver of packet E. If X is adjacent to N then N places packet to packet queue of X. Otherwise, suppose that modules $(N, N_1', N_2', \dots, N_k', X)$ form the shortest path from module N to module X, $N_i' \in \mathfrak{N}$, N_{i+1}' is adjacent to N_i', N_1' is adjacent to N and N_k' is adjacent to X. In this case, packet E is placed to packet queue of module N_1'. Fig. 5 shows the pseudo-code of communication module's algorithm.

3) Data storage module

On each tick, data storage module $M \in \mathfrak{M}$ executes packet read and write operations, initiated by computational modules. Pseudo-code of data storage module is shown in fig. 6, where N_m is a communicational

```
if w(P) < s_w then
    Put packet E with M as receiver
address to queue of module N_p
    r(P)++
else
    Wait
```

Figure 3. Pseudo-code of packet writing operation

```
for i=0; i<h_m and not Empty(Q);i+=j_E do
    E = Front(Q)
    Process(E)
```

Figure 4. Pseudo-code of packet processing

```
for i=0; i<h_m and not Empty(Q); ++i do
    E = Front(Q)
    if X adjacent to N then
        Put E to queue of module X
    else
        Put E to queue of module N'_1
```

Figure 5. Pseudo-code of communication module algorithm

```
for i=0; i<h_m and not Empty(Q); ++i do
    E = Front(Q)
    if α(E) = M then
        --(w(β(E)))
    else
        Put E to queue of module N_p
```

Figure 6. Pseudo-code of data storage module algorithm

module that is adjacent to M, $\beta(E)$ is sender of packet E and $w(\beta(E))$ is number of sender's unfinished packet write operations.

IV.   CONCLUSION

In this paper, we proposed the Database Heterogeneous Multiprocessor Model (DHM). DHM will allow to model database processing on hybrid computational clusters. This model consists of three submodels that describe different aspects of database processing. At this moment, two submodels are described: the hardware platform submodel that describe hardware components of modeled computational system, and execution submodel, which define the rules of interaction between the components of hardware platform submodel. Next steps of our research include:

- development of transaction submodel

- software implementation of model in the form of database emulator

- Conduct experiments using real hardware and developed emulator to validate DHM model.

ACKNOWLEDGMENT

This work was supported by Russian Foundation Fund for Basic Research, project No 16-37-00245 and by Act 211 Government of the Russian Federation, contract № 02.A03.21.0011.

REFERENCES

[1] Kostenetskiy P.S., Safonov A.Y. SUSU supercomputer resources // Proceedings of the 10th Annual International Scientific Conference on Parallel Computing Technologies (PCT2016). Arkhangelsk, Russia, March 29–31, 2016. CEUR Workshop Proceedings, 2016. –Vol. 1576. – P. 561–573.

[2] Besedin K.Y., Kostenetskiy P.S., Prikazchikov S.O. Increasing efficiency of data transfer between main memory and Intel Xeon Phi coprocessor or NVIDIA GPUS with data compression. // Lecture Notes in Computer Science, Springer, 2015. – Vol. 9251.– P. 319–323.

[3] Williams S., Waterman A., Patterson D.A. Roofline: an insightful visual performance model for multicore architectures. // Commun. ACM, 2009. – Vol. 52. – No. 4. – P. 65–76.

[4] Lawson G., Sundriyal V., Sosonkina M., Shen Y. Modeling performance and energy for applications offloaded to Intel Xeon Phi. // Proceedings of the 2nd International Workshop on Hardware-Software Co-Design for High Performance Computing, Co-HPC 2015 November 15 2015, Austin, Texas. – USA: ACM, 2015. – P. 7:1–7:8.

[5] Tang L., Hu X.S., Barrett R.F. PerDome: a performance model for heterogeneous computing systems. // Proceedings of the Symposium on High Performance Computing, part of the 2015 Spring Simulation Multiconference, SpringSim '15 April 12–15 2015, Alexandria, VA. – USA: SCS/ACM, 2015. – P. 225–232.

[6] Kostenetskii P.S., Sokolinsky L.B. Simulation of hierarchical multiprocessor database systems // Programming and Computer Software, 2013. – Vol. 39. – No. 1. – P. 10–24.

Properties of Mathematical Number Model Provided Exact Computing

Valentin Golodov

South Ural State University, School of Electrical Engineering and Computer Science, Chelyabinsk, Russia
golodovva@susu.ru

Abstract—The paper describes the author's experience in identification of computer arithmetic's implementation principles that allow extending mathematical properties of a number on its computer representation. The range of the representable numbers, bitwise identity of the result on the different computer architectures and in parallel computation are very important in the current cloud and parallel computing era. Implementation issues are shown on the example of rational number representation but implementation procedure may be used for any other number representation.

I. INTRODUCTION

The IEEE754 [1] is de facto standard for implementation of real numbers in a computer.

Each format has representations [2] for NaNs (Not-a-Number), $\pm\infty$ (Infinity), and its own set of finite real numbers, all of the simple form

$$2^{k+1-N}n \tag{1}$$

with two integers n (signed *Significand*) and k (unbiased signed *Exponent*) that run throughout two intervals determined from the format thus: (K+1) Exponent bits: $1 - 2^K < k < 2^K$. N Significant bits: $-2^N < n < 2^N$.

Obviously, that most of the real numbers can not be represented in such format, e.g. decimal 0.1 cannot be represented in binary exactly. So it is only approximated.

Rounding rules also lead to different results, e.g. fraction $\frac{1}{3}$ is $\frac{1}{3} \approx 3eaa\ aaab_{16}$ in single precision format, but in double precision $\frac{1}{3} \approx 3fd5\ 5555\ 5555\ 5555_{16}$. We can see, that single precision rounds $\frac{1}{3}$ up, when double precision round it down. More commonly there are no basic arithmetic laws such as commutativity and associativity in discrete set of IEEE754 real numbers.

This critics of IEEE754 are known for a long time. The author's opinion, that IEEE754 critics are not so widespread because of its not-constructiveness. Critics usually provide (if provides at all) weak alternatives to current IEEE754 hegemony. This paper is devoted to alternatives of IEEE754 standard and requirements to these alternatives.

II. PROPERTIES OF CORRECT COMPUTER CUMBER

When someone looking for the theory about numerical method of solving certain tasks there are some amount of numerical method with appropriate mathematical properties, e.g. whether the method is exact or approximate/iterative.

Exact numerical method needs finite number of steps to reach exact solution of the task, e.g., Gauss-Jordan method of solving linear equation system. On other side approximative/iterative method applies a sequence of steps to specify current result, e.g., NewtonRaphson method for finding successively better approximations to the roots (or zeroes) of a real-valued function. But this properties are make sense only up to implementation on the computer.

On the implementation level under IEEE754 [3] restrictions there are no exact or approximate methods, all methods are additionally exposed to round and representation errors. There are no any guaranties of the theoretical precision, no any guaranties of reproducibility of program run results on different computers, in different environments or different number of parallel processes (in general) [4].

So what are the requirements to computer numbers that allow numerical methods being exact, effective and parallel computation world ready? Let's try to describe some common properties that computer number representation must provide to a computer user for the purpose of real world scientific computing.

- Provide range/set of the numbers sufficient for appropriate task and solving method. With overflow and underflow control.
- Provide the portability of the program across architectures in modern heterogeneous computing era.
- Provide the bitwise repeatability of the result on the different computing devices regardless of device architecture and used system software.
- Provide the bitwise repeatability of the result in modern parallel computing era. Different parallel algorithms for same task, different number of nodes in parallel execution must give the same bitwise final results.
- Be as effective as possible in different aspects (memory, energy).

List of requirements above contains no any specification of number representation. Different number representations satisfy to these requirements. One of them but, of course, not only, is rational number field with integer numerator and denominator. The numerator and denominator may be, e.g., represented as an integer number in position number notation. This number representation was implemented by the author see [5]–[8] the further thesis will be illustrated with the examples and the results of this representation.

An absolutely other approach to number representation is given in the John L. Gustafson book named "The end of the error. Enum Computing" [9]. *Enum* representation has properties listed above and even more, it uses the structure consisting of sign, exponent, fraction, u-bit and exponent and fraction sizes fields.

The two mentioned approaches has some common features, in spite of their difference, e.g., both are open-ended, i.e. representation is capable to extend to save bigger or more precise number, all number bits contain meaningful data, etc.

III. IMPLEMENTATION OF THE DECLARED REQUIREMENTS

From the implementation point of view number representation is based on:

1) bits of memory of the number notation,
2) arithmetic algorithms that operate with the bits of the number,
3) input, output and conversion procedures.

Lets consider some implementation aspects. Memory subsystem must provide: 1) access to bits of the number notation, copying of the notation bits, 2) buffer for the result of the arithmetic operation. Implementation of the memory subsystem may vary a lot, it may be bit string, linear array of the bytes, C programming language structure, C++ programming language structure/class, but it is basic prerequisite of any number representation realization. Own implementation of operating with memory leads to successful error control, overflow/underflow control.

After the basic memory subsystem fulfilled prerequisite requirements then any arithmetic algorithms may be implemented. There are usually few alternatives, each preferred for certain device architectures. Arithmetic algorithm layer should be independent from underlayer memory layer.

Input, output operations are exactly conversion between different number notations, so its implementation fully based on conversion algorithms i.e. arithmetic operations.

IV. IMPLEMENTATION EXAMPLE

Consider one example of number representation. Let's check whether implementation is suitable as the "good" computer number described above. Ideas below received approval during workflow on Exact Computation 2.0 library [8].

Number representation. Consider rational number as couple of sign and fraction with unsigned integer numerator and denominator. Rational arithmetic fully based on integer arithmetic so our arithmetic operations will be operations with unsigned integer numbers of arbitrary length. Assume that integer will be presented in radix notation. Note that radix notation is not unique integer number representation and multilayer library implementation provides also other not digit oriented number representation such as system of residual classes and others.

Implementation for Radix Notation Firstly choose the device to carry computation on, e.g., x_86/x_64 CPU. CPU uses

binary notation and 32-bit/64-register words respectively, so optimal radix is power of 2: 2^{16}, 2^{32} or 2^{64}.

Now we need to store our number's bits in the memory. It may be linear array of memory cells with cell size equal to amount of bits in the radix. So an access to given position is simply an access to a corresponding array cell, copying of the number notation is straight copying of array. Memory allocation and reallocation is performed through `malloc` and `free` C-function or `new` and `delete` C++ calls.

Next step is arithmetic operation realization. One of the simplest way is to implement columnar addition, subtraction, multiplication division with remainder in radix system [10]. It may be implemented on low level programming that is important for cryptography purposes.

As mentioned above the pair of memory subsystem and arithmetic operations allows implementing number representation on particular device. Consider other device – nVidia GPU, nowadays GPUs are capable for general purpose computation (GPGPU) [11] and its massive parallel architectures are very compute capable for a certain sort of tasks. CUDA is C-language extension commonly known in science computation world.

For example, memory subsystem may be implemented through `cudamalloc` CUDA-function calls. Most encouraging is using massive parallel GPU chip to perform basic arithmetic operations simultaneously on each position of radix notated number, more about this algorithms the reader can find in [6]. Other encouraging moment is energy saving of GPU chip, at rather distant 2010-th relevant generations of CPU need about 4.2 nJ ("nJ" is nanojoules) to read 8-byte portion of the data from the DRAM [9] and GPU need about 83 nj to read 128-byte data block that is approximately 5.2 nJ per 8-byte. The successive Kepler, Maxwell and Pascal nVidia GPU architectures are much more energy efficiency architectures so not only parallel algorithms but also hardware advantages are possible. Energy saving aspects of computation will be subject of the further particular paper.

After memory operations on the given device and the effective arithmetic algorithms are implemented, the next step is to increase usability. The wrapper such as C++ class allows scientific users to use implementation without any additional coding overhead.

Tables I and II show an example of architecture dependent arithmetic algorithms development. Table I shows time measurement of comparison of two equal numbers of given length on different architectures, CPU Intel Core i7–950 [3.06 Ghz, 6 Gb DRAM]) and nVidia Fermi NVidia GTX460 [700 Mhz, 1 GDDR5]) and Kepler GTX660 Ti [980 Mhz, 2 GDDR5]). Table II shows the addition time on the same devices. Time of the sequential algorithms marked as (Seq), parallel algorithms marked are (Par).

V. RATIONAL AS MATHEMATICALLY CORRECT COMPUTER NUMBER

As mentioned above the computer number named as correct should be a number in mathematical sense and meet the

TABLE I. Comparison of the Integer on Different Architectures

L	Time in milliseconds					Speedup	
	CPU(S)	Fer(S)	Kep(S)	Fer(P)	Kep(P)	Fer(P)/CPU	Kep(P)/CPU
10^1	0.0001	0.130	0.120	0.146	0.148	0.0068	0.0067
10^2	0.0001	0.135	0.130	0.192	0.138	0.0052	0.0073
10^3	0.0001	0.320	0.410	0.155	0.149	0.0065	0.0067
10^4	0.0200	2.260	3.040	0.200	0.210	0.1000	0.0952
10^5	0.1900	21.62	29.91	0.230	0.220	0.8370	0.8636
10^6	1.9000	218.0	301.1	0.521	0.360	3.6538	5.2777
10^7	19.000	—	—	3.170	1.581	6.2776	12.5950
10^8	198.10	—	—	—	—	—	—

TABLE II. Addition of the Integer on Different Architectures

L	Time in milliseconds					Speedup	
	CPU(S)	Fer(S)	Kep(S)	Fer(P)	Kep(P)	Fer(P)/CPU	Kep(P)/CPU
10^1	0.0001	0.300	0.180	0.240	0.240	0.0041	0.0041
10^2	0.0001	0.300	0.200	0.240	0.240	0.0041	0.0041
10^3	0.0001	0.610	0.620	0.240	0.240	0.0041	0.0041
10^4	0.0200	3.910	3.040	0.240	0.240	0.0041	0.0041
10^5	0.5000	37.60	4.010	0.540	0.700	0.9259	0.7142
10^6	6.0100	369.5	42.20	0.880	0.860	6.8181	6.9767
10^7	60.000	—	409.2	4.300	2.76	14.000	21.814
10^8	602.10	—	—	—	—	—	—

following requirements: provide sufficient range, portability of the program codes between architectures, provide the bitwise identity of the results on different architectures, provide the bitwise identity of the results in parallel computations, be effective.

Sufficient range. As allocation operations are device dependent, the range of the rational number is also device dependent, so minimum non zero rational number written in radix notation with radix 2^{32} depends on maximum memory pitch. For example NVIDIA Tesla M40 GPU maximum memory pitch is equal to $2147483647 = 2^{31} - 1$ bytes [12], so minimum non zero number representable as rational on M40 GPU is $1/(2^{(8\cdot(2^{31}-1))+1} - 1)$. Current CPUs allow addressing even more bytes. A range is feasible to numbers stored at current moment. Underflow, overflow are memory allocation errors so they are monitored with hardware indicators and memory subsystem.

This paragraph is impracticable for all fixed size number representation, of course, for IEEE754 too.

Portability between architectures. Implementation of the rational numbers requires only basic memory allocation operations and assembler `add`, `mul` and `div` operation with integer numbers which are basic on all compute capable architectures. After arithmetic is implemented user program may use it capabilities in corpore.

Bitwise identity of results on different architectures. In modern heterogeneous era, cloud computing era, when a user knows nothing about specification of the device where program runs it is very important to be sure that if the program finished correctly its results are bitwise the same as on its own machine. If the computer successfully finished program running then all allocation operations and assembler arithmetic operation were successful, therefore result is always bitwise the same.

Bitwise identity of results in parallel computation. This requirement is close to previous but has its own specifics. Now algorithms assert equivalent accurate within rounding errors only. Scalable parallel program runs in thousand of cores of supercomputers and there is no opportunity to assure repeatability of the run. Each run on the supercomputer costs tangible amount of money and time, requires huge amount of cores in privileged usage, so repeated verificatory runs may be even impossible.

Three paragraphs above ensure that computer representation of the number is the same that mathematical model one.

Effectiveness. Effectiveness (as operations per second) of the IEEE754 floating point numbers based on its fixed size and good hardware support, but effectiveness (as time for solving certain computation task) should be computed precisely. Especially for iterative methods when round errors may be the cause of the method divergence.

C/C++ languages provides extensional effectiveness in right hands, there is no garbage collection during execution so all memory operation are effective as much as coder can do them. Rational numbers have no any wast data during computations, only exact bits of numbers are stored.

Issues of improvement of the rational number representation are partially worked in [6] but question is not closed.

An example of successful usage of rational in scientific computations, is that it allows solving hard computation tasks as linear equation system with Hilbert matrix, Vandermonde matrix etc., rationals are also an instrument for *interval regularization approach* [7], [13] implementation. *Interval regularization* helps to solve ill-conditioned linear equation systems, e.g. in chemistry [13].

VI. FUTURE WORK

The further work will be concentrated on two main directions. The first big part is implementation of the constructor of "good" number representations. It will be an useful for implementing any number representation that can be implemented on the computer without loss of the mathematical features. It leads to more calculation methods using "good" number representations and additional research in number theory area.

Other big area is low level realization of rational arithmetic to provide it on operation system core level for the cryptography purposes with focus on efficiency but with full portability between computer architectures.

VII. CONCLUSION

The requirements to computer numbers that allow numerical methods to be exact, effective and parallel computation world ready have been analyzed, e.g. sufficient range/set, portability across architectures, bitwise repeatability of the result on the different computing devices, bitwise repeatability of the result in parallel computations, efficiency. Common properties that computer number representation implementation must provide to a computer user for the purpose of real world scientific computing have been listed, these are: memory allocation/deallocation layer, arithmetic algorithms layer, input/output and user code layer. Listed properties and implementation have been demonstrated on rational number representation. Efficiency issues have been demonstrated on massive parallel GPU devices. The next step is realization of multilayer library model for non radix mathematical number representation which are capable of errorless computations.

ACKNOWLEDGMENT

The work is supported by RFBR grant 16-31-00108\16 "Solutions to fundamental issues of usability of extended and arbitrary-precision arithmetic".

The work was supported by Act 211 Government of the Russian Federation, contract 02.A03.21.0011.

REFERENCES

[1] IEEE, "754-2008: IEEE standard for floating-point arithmetic," [online], Aug. 2008, http://ieeexplore.ieee.org/servlet/opac?punumber=4610933.

[2] W. Kahan, "Lecture notes on the status of ieee standard 754 for binary floating-point arithmetic," online, University of California, Berkeley CA 94720-1776, October 1997, http://people.eecs.berkeley.edu/ wkahan/ieee754status/IEEE754.PDF. [Online]. Available: http://people.eecs.berkeley.edu/ wkahan/ieee754status/IEEE754.PDF

[3] D. Goldberg, "What every computer scientist should know about floating-point arithmetic," *ACM Comput. Surv.*, vol. 23, no. 1, p. 548, 1991.

[4] V. V. Voevodin, *Mathematical Foundations of Parallel Computing*. World Scientific Publishing Company, 1992.

[5] V. Golodov and A. Panyukov, "Scalable algorithms for the integer arithmetics and rational calulations in heterogeneous computation environment," *Bulletin of South Ural State Univercity. Series "Computational Mathematics and Software Engineering"*, vol. 4, no. 2, pp. 71 – 88, 2015. [Online]. Available: http://vestnik.susu.ru/cmi/article/view/3005

[6] A. Panyukov and V. Golodov, "Parallel algorithms of integer arithmetic in radix notations for heterogeneous computation systems with massive parallelism," *Bulletin of the South Ural State University. Series "Mathematical Modelling, Programming & Computer Software".*, vol. 8, no. 2, pp. 117 – 126, 2015, http://mmp.vestnik.susu.ru/article/en/340.

[7] A. V. Panyukov and V. A. Golodov, "Computing best possible pseudo-solutions to interval linear systems of equations," *Reliable Computing*, vol. 19, pp. 215 – 228, 2013, http://interval.louisiana.edu/reliable-computing-journal/volume-19/reliable-computing-19-pp-215-228.pdf.

[8] V. Golodov and A. Panyukov, "Library of classes exact computation 2.0," programs, Data Bases and Topologies of VLIS. Official bulletin of Russian Ageny of Patients and Trademarks. March, 2013.

[9] J. L. Gustafson, *The End of Error: Unum Computing*, ser. Chapman & Hall/CRC Computational Science. Chapman and Hall/CRC, 2015.

[10] D. E. Knuth, *The Art of Computer Programming*, 2nd ed. Addison-Wesley Longman, 1981, vol. 2.

[11] GPGPU.org, "General-purpose computation on graphics hardware," GPGPU.org, August 2013. [Online]. Available: http://gpgpu.org/

[12] E. Eshelman, "Nvidia tesla m40 24gb gpu accelerator (maxwell gm200) up close," online, April 2016, https://www.microway.com/hpc-tech-tips/nvidia-tesla-m40-24gb-gpu-accelerator-maxwell-gm200-close/.

[13] V. Golodov, "Interval regularization approach to the Firordt method of the spectrophotometric analysis of the non-separated mixtures," *Lecture Notes in Computer Science*, vol. 9553, pp. 201 – 208, 2013.

MIPRO 2017, May 22- 26, 2017, Opatija, Croatia

Simulation of the Parallel Database Column Coprocessor

P.S. Kostenetskiy

South-Ural State University, Chelyabinsk, Russia
kostenetskiy@susu.ru

Abstract - The paper proposes a mathematic model allowing exploration of effectiveness of different hardware cluster computing configurations based on multi-core coprocessors while processing databases using approach of distributed columnar indices.

I. INTRODUCTION

A volume of human-generated data is growing continuously. Meanwhile, solutions to a great number of problems require online processing of cumulative super large databases [3]. Their processing needs in new methods and hybrid hardware architectures containing multi-core coprocessors and graphical accelerators [1], [2].

The paper [3] suggested approach to run super large database queries based on distributed columnar indices and domain-interval fragmentation using multi-core coprocessors. It allows significant raising of the request processing efficiency in parallel DBMS.

Actually, there already exist the models of the parallel computing such as PRAM [6] and BSP [8]. However, they do not consider specificity of the parallel database systems. There also exist the models considering the specificity, for example, DMM [5], but they do not support distributed columnar indices.

Consequently, a current problem is a simulation of the databases that operate on cluster systems equipped with multi-core coprocessors and using distributed columnar indices.

II. MODEL OF COLUMN COPROCESSOR

A model of column coprocessor includes three sub-models such as a model of hardware platform, a model of operating environment, and cost estimation model.

2.1. Model of hardware platform

A set of multiprocessor system modules is divided into three disjoint subsets:

$$\mathbb{M} = \mathbb{C} \cup \mathbb{N} \cup \mathbb{E}, \mathbb{C} \cap \mathbb{N} = \emptyset, \mathbb{N} \cap \mathbb{E} = \emptyset, \mathbb{C} \cap \mathbb{E} = \emptyset, \quad (1)$$

where $\mathbb{C}$ is a set of coprocessor modules, $\mathbb{N}$ is a set of network switch modules, and $\mathbb{E}$ is a set of communications between two devices. Let us denote a module of coprocessor-coordinator as $C_1 \in \mathbb{C}$ while the rest of coprocessor modules $C_n \in \mathbb{C}$, where $n = \{2, 3, \dots\}$, will be specified as coprocessor executive modules.

PCI-Express bus data transfer between coprocessors is a bottleneck stage [4], which significantly slows down data processing. That is why it is reasonable to store and process the distributed columnar indices [7] directly on the coprocessor modules. Consequently, the modules of the drive devices and RAM memory are not simulated.

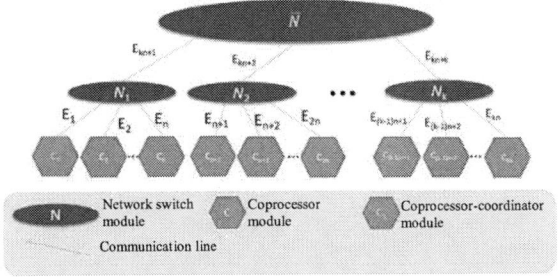

Figure 1. DCH-tree example

We use the term *DCH-tree (Database Coprocessor's Hardware Tree)* to refer to the weighted tree consisting of a set of multiprocessor system modules where there is a selected vertex $\overline{N} \in \mathbb{N}$ called root network hub. The structure of DCH-tree has the following constraints.

- Only a network switch module can be a root of DCH-tree.

- Each ith module of the network switch has a weight w_i.

- Only the coprocessor modules each of which has a coefficient of efficiency p^i can be the leaves of DCH-tree.

- Coprocessor modules can not have child nodes.

- An interconnecting channel being a tree edge with a weight of w_j where j is a number of a module connects any two modules.

An example of DCH-tree is given in Fig. 1. The example corresponds to a computer cluster with k computation nodes which of each is equipped with n coprocessors.

2.2. Model of operating environment

Let us consider a tuple of precomputation table (PCT) to be a minimal unit of data in the proposed model [3].

Time of data processing on the coprocessor module is calculated by a formula

$$t_{\mathrm{pr}} = \frac{f_{\mathrm{perf}}(|I_j^i|, |I_k^i|)}{p^i}, \qquad (2)$$

where f_{perf} is a performance function for a particular operation with the fragments of distributed columnar indices I_j^i, I_k^i disposed on the ith coprocessor module and p^i is a performance of the ith coprocessor module (a number of operations that can be processed by a coprocessor module in a unit time).

Let us assume that the fragments of columnar indices were previously distributed to coprocessors as it was suggested in [3].

A packet is a compressed part of a precomputation table for a single coprocessor module.

Operation time of a model is divided into the cycles. A cycle means a sequence of actions consisting of the following steps:

- processing the ith connection of indices;

- dividing PCT into the parts required for passing to each coprocessor;

- transferring PCT prepared in p.2 to the corresponding coprocessors;

- generating the columnar indices from the received parts of PCT.

The number of cycles is determined by the number of relational *join* operations (p. 2.1) required for a query execution and is equal n − 1 where n is the number of relations involved in relational *join* operations. In the final cycle, PCT is not divided into parts but transferred to the coprocessor module-coordinator.

Time of generating the columnar indices from PCT is calculated as follows

$$t_{ind.gen.} = \frac{f_{sort\ comp.}(|PCT|)}{p^i} N, \qquad (3)$$

where $f_{sort\ comp.}$ is sort complexity function for sorting applied for generating the columnar indices, N is the number of PCT attributes, and p^i is a performance of the ith coprocessor module (a number of operations that can be processed by a coprocessor module in a unit time).

The number of the tuples received on the ith coprocessor executive module as a result of *join* operation execution in the jth cycle is evaluated using a formula

$$n = \frac{\varepsilon_j}{N}, \qquad (4)$$

where ε_j is the number of the tuples received by all coprocessor modules in the jth cycle and N is the number of coprocessor executive modules.

The PCT transfer between the coprocessors is simulated as a packet transfer. A packet is a tuple of PCT. The number of packets is calculated by a formula

$$m = \frac{|PCT_i|}{N}, \qquad (5)$$

where PCT_i is a part of PCT got on the ith coprocessor executive module and N is the number of coprocessor executive modules.

Time of the third step of every cycle is calculated by an algorithm given in Figure 2.

```
function GetTransferTime():
    totalSendTime = 0
    while there are packets on one of modules:
        for each module:
            module SendForward()
            totalSendTime = totalSendTime + 1
    return totalSendTime
```

Figure 2. Algorithm of packet transfer

The SendForward() method in case of packet presence on the module makes the following transfers:

- for an interconnecting channel, the packets in amounts of $\frac{w_j}{q}$ are sent to parent and child modules;

- for a network switch module, the packets in amounts of $\frac{w_j}{q}$ are sent to parent and child modules;

- for a coprocessor executive module, all the packets are sent to a parent communication line;

where w_j is a weight factor of DCH-tree module from which the send takes place and q is the number of attributes of *PCT* received in the previous step.

2.3. Cost estimation model

Operation time of the modelled database system is equal to a sum of time of the model cycles

$$t_{\mathrm{total}} = \sum_1^n t_i, \qquad (6)$$

where t_i is time of cycles and n is the number of cycles.

Time of a cycle consists of a query execution time on the number of fragments of columnar indices, transferring the parts of PCT to every coprocessor module, and generating the distributed columnar indices from the received PCT on every coprocessor module.

2.4. Modelled query

Let the following relations be given:

$\mathbf{R}_1$ consisting of the attributes $A_1^1, A_2^1, \ldots, A_{k_1}^1$. Let us denote $\{A_1^1, A_2^1, \ldots, A_{k_1}^1\}$ as A^1;

$\mathbf{R}_2$, consisting of the attributes $A_1^2, A_2^2, \ldots, A_{k_2}^2$. Let us denote $\{A_1^2, A_2^2, \ldots, A_{k_2}^2\}$ as A^2;

$\ldots$

$\mathbf{R}_n$, consisting of the attributes $A_1^n, A_2^n, \ldots, A_{k_n}^n$. Let us denote $\{A_1^n, A_2^n, \ldots, A_{k_n}^n\}$ as A^n.

Let $\mathbb{A}$ denote a set of all attributes for all relations: $A^1 \cup A^2 \cup \ldots \cup A^n = \mathbb{A}$.

It is necessary to execute a relational query P:

$$P = \pi_{\mathbb{B}}\left(\sigma_{f(\mathbb{C})}(R_1 \bowtie \cdots \bowtie R_n)\right), \qquad (7)$$

where $\mathbb{B} \subset \mathbb{A}$ is a set of attributes composing the resulting relational expression, $f: \mathbb{C} \to \{0,1\}$ is Boolean function specifying the conditions of a sampling, and $\mathbb{C} \subset \mathbb{A}$ is a set of attributes involved in the condition of a sampling.

3. Simulation experiments

Two series of simulation experiments were performed. In the first one, a model setting was done by assigning the weight factors. In the second one, the model verification was executed.

To perform the emulator setting, the experiments were conducted on the «SUSU Tornado» supercomputer [4] which main characteristics are presented in Table 1.

During the second series of the simulation experiments, a *join* of two relations R and S was carried out. A size of the relation R was 630 000 tuples while a size of the relation S was 63 000 000 tuples. The *join* was done using the distributed columnar indices.

TABLE I.   CHARACTERISTICS OF THE SUPERCOMPUTER

Characteristic	Value
Number of computational nodes/processors/coprocessors	480/960/384
Type of processor	Intel Xeon X5680 (Gulftown, 6 cores 3.33 GHz each)
Type of coprocessor	Intel Xeon Phi SE10X (61 cores 1.1 GHz each)
RAM	16.9 TB
Type of system area network	InfiniBand QDR (40 Gbit/s)
Operating system	Linux CentOS

The results of the experiments in performing the natural *join* of relations of R and S on the «SUSU Tornado» supercomputer are given in Fig. 3.

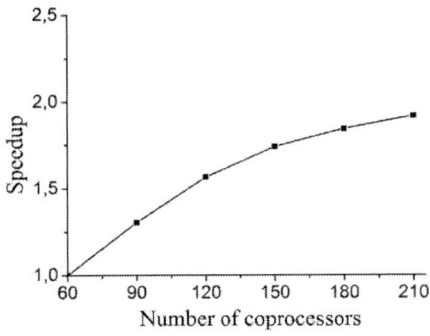

Figure 3.   Query execution on the «SUSU Tornado» supercomputer

The experiments were executed by means of the column coprocessor. The research was carried out with the different numbers of nodes equipped with Intel Xeon Phi coprocessors. The numbers were 60, 90, 120, 150, 180, and 210 nodes.

After that, the developed emulator on DCH-tree describing the architecture of the «SUSU Tornado»

supercomputer simulated the *join*. The results of the experiment are presented in Fig. 4.

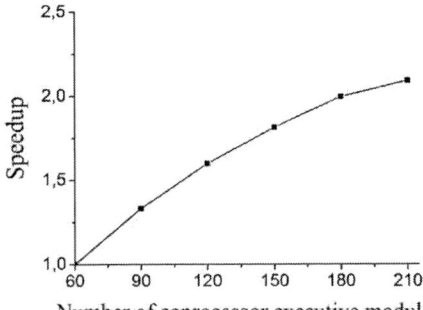

Figure 4.   Query execution on the emulator

Comparison of the graphs on Figures 3 and 4 shows that the emulator simulates the query execution adequately. This proves the DCH model validity.

III.   CONLUSION

The paper considered the problem of simulation of the database column coprocessor architecture. A model of the database column coprocessor was developed. The model was implemented as a software emulator. The developed emulator was applied for simulation experiments to check the model adequacy by comparison with the column coprocessor prototype.

In our future research we are going to deal with the model extension in order to support compression simulation during database processing by using the distributed columnar indices.

ACKNOWLEDGMENT

This work was supported in part by the Russian Foundation for Basic Research within the framework of the research project No. 15-29-07959 OFIM and by Act 211 Government of the Russian Federation, contract № 02.A03.21.0011.

REFERENCES

[1] Besedin K.Y., Kostenetskiy P.S., Prikazchikov S.O. Increasing Efficiency of Data Transfer Between Main Memory and Intel Xeon Phi Coprocessor or NVIDIA GPUS with Data Compression // 13th International Conference on Parallel Computing Technologies, PaCT 2015, Petrozavodsk, Russian Federation, 31 August 2015 - 4 September 2015, Proceedings. Lecture Notes in Computer Science, Springer, 2015. Vol. 9251. P. 319-323.

[2] Besedin K.Y., Kostenetskiy P.S., Prikazchikov S.O. Using Data Compression for Increasing Efficiency of Data Transfer Between Main Memory and Intel Xeon Phi Coprocessor or NVidia GPU in Parallel DBMS // 4th International Young Scientist Conference on Computational Science, Proceedings. Procedia Computer Science, 2015. Vol. 66. P. 635–641.

[3] Ivanova E.V., Prikazchikov S.O., Sokolinsky L.B. Join Execution Using Fragmented Columnar Indices on GPU and MIC // Proceedings of the 1st Ural Workshop on Parallel, Distributed, and Cloud Computing for Young Scientists. CEUR Workshop Proceedings, 2015. P. 1–10.

[4] Kostenetskiy P.S., Safonov A.Y. SUSU Supercomputer Resources // Proceedings of the 10th Annual International Scientific Conference on Parallel Computing Technologies (PCT 2016).

Arkhangelsk, Russia, March 29-31, 2016. CEUR Workshop Proceedings. 2016. V. 1576. P. 561-573.

[5] Kostenetskii P.S., Sokolinsky L.B. Simulation of Hierarchical Multiprocessor Database Systems // Programming and Computer Software. Vol. 39, No. 1. 2013. P. 10–24.

[6] McColl W.F. General purpose parallel computing. Lectures on Parallel Computation. USA: Cambridge University Press, 1993. 496 p.

[7] Prikazchikov S., Kostenetskiy P. Modeling distributed column indexes in the context of parallel database systems // Russian Supercomputing Days International Conference, Moscow, Russian Federation, 28-29 September, 2015. CEUR Workshop Proceedings. 2015. Vol. 1482. P. 586.

[8] Valiant L.G. A bridging model for parallel computation // Communication of the ACM. USA: ACM, 1990. Vol. 33. No. 8. P. 103–111.

Towards Flexible Open Data Management Solutions

Benedikt von St. Vieth*, Jedrzej Rybicki‡ and Maciej Brzeźniak§

*‡Juelich Supercomputing Center (JSC),
Wilhelm-Johnen-Strasse, 52425 Juelich, Germany
§Poznań Supercomputing and Networking Center (PSNC),
ul. Jana Pawła II 10, 61-139 Poznań, Poland
Email: *b.von.st.vieth@fz-juelich.de, ‡j.rybicki@fz-juelich.de, §maciekb@man.poznan.pl

Abstract—Many data sharing initiatives emerged in the recent time. There are various driving factors behind this phenomena subsumed under the terms Open Data and Open Research. The most prominent one is the fact that more and more funding agencies request publishing of all project results including data. Publishing and sharing data enables the verification of obtained results by facilitating repeatability and reproducibility. Also economical aspects are important: Data reuse can reduce the research costs, and redoing of the same experimental work can be avoided (provided sufficient visibility of earlier results). Publishing or Open Data is also contributing to researcher's visibility and reputation. The implementation of data sharing is, however, quite challenging. It demands both storage technologies and services that can deal with the increasing amounts of data in an efficient and cost-effective way while remaining user-friendly and future-proof. We describe the architecture of a data management solution that is able to provide a scalable, extensible, yet cost-effective storage engine with several replication mechanisms. For the sake of user-friendliness and high data visibility, the solution has a mechanism to create flexible namespaces for a efficient organization, searching, and retrieving of the data. Such namespaces can be tailored to the researchers' needs, potential facilitating data reuse across community borders. The presented architecture embodies our experiences gathered in the EU-funded project *EUDAT2020*.

Keywords-Data Management, Object Stores, Data Replication, Namespaces

I. INTRODUCTION

The requirement to share research data is becoming commonplace for many reasons. Those are verification of obtained results by facilitating repeatability, reproducibility, and the reduction of research costs, to name only a few. Data sharing itself is driven by efficient data management solutions. They demand both storage resources and services that can deal with the increasing amounts of data. A very promising technology for storing different kinds of data are object stores. They offer scalability, cost-effectiveness, maintainability, and extensibility. In short, data objects are stored in a distributed system of interdependent storage elements. Objects placement on available resources is randomized by means of hashing. Therefore, workload is spread evenly across the elements of the system and "hot spots" are avoided. The far-reaching autonomy of the storage elements increases fault-tolerance and throughput. Transfers to and from different storages are independent from each other. Although object stores are widely used, there is an aspect of their operation which is not yet properly addressed in many implementations and

commercial services. It is the object replication across different administrative domains, which is quite popular for scientific data. Many times community data centers require additional replicas of their data in generic data centers for data safety reasons. At the same time, they want to keep the local copies to maintain control over their data. We would like to present solutions to this problem which we implemented as prototypes within the EUDAT2020 project [1].

Object stores have many advantages, however, they come at the cost of some limitations. Objects are stored in a "flat" namespace which is not searchable and has only limited metadata capabilities. Namespaces, and more generally metadata, are crucial for efficient data management. They introduce structure into a collection of data and embody knowledge about the domain the data stem from. Such structures are also essential for localizing relevant data. In our solution we address this issue by enabling flexible namespaces for research data. These namespaces are independent from the way the data are stored. We are convinced that it is vital to account for all kinds of storages, especially those which provide only limited namespace functionalities.

One way of creating flexible namespaces is to use graph databases. Graphs are very powerful abstractions often used in computer science. They can represent different domain models i.e., views on the data. Graphs can also be used for efficient search. Currently, the most popular approaches to search through research data are based on indexing of metadata. This is similar to the very beginning of the Internet search engines, where also keyword-based indexes were used. The revolution of the Internet search started by Google was propelled by graph algorithms [2]. Graph-based heuristics are used to identify the most popular web pages containing a phrase searched for. The popularity metrics works very well with web pages but might not be the best option for research data. Researchers could, for instance, be more interested in the least popular data which were not analyzed very often and where they have a larger potential to generate new findings.

This paper is structured as follows. In Section II we will review some exemplary approaches to data management and examine the technologies used for both storing the data and managing the namespaces. The detailed architecture of our system is presented and discussed in Section III, followed by an introduction to a first implementation. An assessment of the performance of the proposed solutions is presented in

Section IV. We conclude our paper with a summary and an outlook on future work.

II. BACKGROUND

Two very influential trends shaping modern science are that it is based on collaboration and that it is increasingly data driven. That makes researchers use commercial services like Dropbox [3] for storing and sharing their data. Such services excel in ad-hoc sharing, they are, however, not sufficient for storing large amounts of data on a long-term. They don't offer either metadata functionality or citeability, making it less suitable for Open Data. To this end, during the recent years several different research infrastructures like EUDAT emerged. Their aim is to provide services to support researchers in management of Open Data, implementing the vision of making (research) data freely and easily accessible to everybody who is interested.

A. EUDAT B2SAFE

EUDAT is a federation of research institutions forming a distributed research infrastructure. EUDAT B2SAFE is a service for data replication and long-term data management. It constitutes a trustworthy, generic storage service for research communities. B2SAFE is based on iRODS (integrated Rule Oriented Data System [4]). iRODS uses rules to imperatively define data management policies. With iRODS it is possible to build and maintain a virtual namespace unifying the view of data objects replicated across several different storage resources within a data center. B2SAFE uses *synchronization* between B2SAFE instances running at different service providers for data replication. To this end, iRODS provides tools to transfer objects between administrative domains. However, it does not offer a virtual namespace spanning across them. The same data object stored in different domains will have different names. To establish a common namespace that will keep track of replicas, EUDAT is using a PID (Persistent Identifier) system [5].

Listing 1. An excerp of a PID using JSON representation

```
GET 11097/88319713-8e7e-4001-a893-4dd81b539b86
[
    {
        "idx": 1,
        "type": "URL",
        "parsed_data": "http://www.mipro.hr/",
        "data": "aHR0cDovL3d3dy5taXByby5oci8=",
        "timestamp": "2017-02-02T12:43:08Z",
        "ttl_type": 0,
        "ttl": 86400,
        "refs": [],
        "privs": "rwr-"
    },
    {
        "idx": 2,
        "type": "CHECKSUM",
        "parsed_data": "d989ccc7c709c7d6d88bfbaac07ccb39",
        "data": "ZDk4OWNjYzdjNzA5YzdkNmQ4OGJmYmFhYzA3Y2NiMzk
            =",
        "timestamp": "2017-02-02T12:43:08Z",
        "ttl_type": 0,
        "ttl": 86400,
        "refs": [],
        "privs": "rwr-"
    }
]
```

A PID system constitutes an indirection layer which maps a persistent, static URL to a (changeable) URL [6]. An example of such a PID structure is shown on Listing 1. One can store additional metadata in a PID record using for example the EPIC HTTP API [7]. The PIDs are essential for the citeablity of data. They can be included in publications and will remain valid (i. e. they will point to the intended data) even if the data will be migrated to other locations. PIDs are also used by communities to verify the implementation of the data management policies. They are agreed between users and resource providers and define e. g., the number of replicas that should be available for a digital object.

B. Object Stores

Suitable research infrastructures like EUDAT have to keep scouting for new technologies, evaluate their applicability, and integrate them in the infrastructure to provide a highest-level of service. Also technologies offering alternatives to the iRODS were evaluated. One promising technology are object stores like OpenStack Swift [8] or Ceph [9]. Those are Open Source implementations, competing with proprietary software like EMC ECS [10] or DDN WOS [11] and a commercial public cloud service from Amazon Inc., called Amazon S3 (Simple Storage Service) [12].

Replication is an inherent part of Swift and Ceph. For Swift, a typical storage clusters consist of a large number of servers that hold at least three copies of the data. In opposite to iRODS, the replication policies are defined declaratively and don't have to be manually implemented as rules. This makes cluster management much easier. It is possible to add new resources, scaling the cluster using additional disks or servers, and the software will reallocate the data to make the most of the available space so that the policy constrains remain fulfilled. The distribution of data across available resources is achieved with the help of consistent hashing, resulting in even load distribution between storage elements. Object stores provide a "flat" namespace, only supporting *accounts* (per user), *containers* (several per account) and *objects* (several per container). Account and container information are stored using *sqlite* databases which are also replicated. Metadata are typically supported in a very simplistic way, it is possible to store key/value pairs at container and object level. Metadata and additional object information are stored using extended file attributes on filesystem level.

In the course of technology evaluation in EUDAT, it was tested how replication of object stores across administrative domains could be implemented. This was, at least for OpenStack Swift, possible using a feature called container synchronization, which we will describe later in this paper. For the sake of data citeablity, support for PIDs was added to Swift [13]. Similarly to the B2SAFE service based on iRODS, PIDs could form an additional namespace that spans across replicated data objects. OpenStack Swift with container synchronization and PIDs is offering the same functionality as B2SAFE, in Section III we will show improvements going beyond the emulation of the existing service.

234

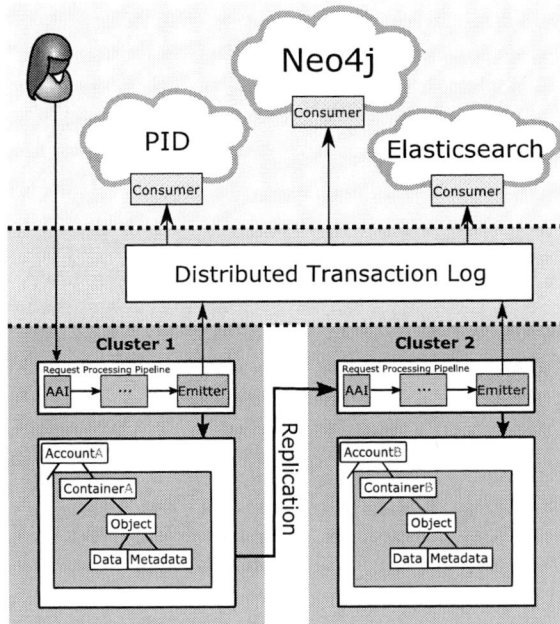

Figure 1. Architecture overview.

C. Graph Databases

In its effort to scout out for new technologies, EUDAT also evaluated approaches to enable flexible, searchable namespaces spanning across all its resources and services [14]. The current approach of building a namespace with the help of a PID system has clear limitations. First, they are not searchable. They only support point queries. For a given PID they can produce a record of fields, including basic metadata and a URL to the data object (see Listing 1). It is also not easily possible to create links between objects or other more complex structures.

Graph databases, on the other hand, can be used for building different views on the same set of data. Instead of a flat namespace, they provide a way to express different aspects of the stored data. It is possible to interlink objects, or make collections of objects based on attributes that are definable in a flexible way. Furthermore, it would be possible for each user to make its own view on the data.

III. PROTOTYPE

The goal of our work was to provide a flexible data management solution. While object stores offer a scalable technology to hold large amounts of data, graph databases can be used to create complex, user-tailored namespaces. To this end, we propose a layered architecture as depicted in Fig. 1. A storage layer, at the bottom (dark gray), is responsible for storing the data objects, while at the top layer (light gray) is a system for creating and managing complex and semantically-rich, user-tailored namespaces. One of the namespaces could be a PID system as currently used in EUDAT, to facilitate citeablity of the data objects.

To enable exchange of information between the layers, we use a distributed transaction log. One could directly move information from the storage layer to a graph database, but this would limit the flexibility of the final solution. The loose coupling of components allows to run more than one namespace and dynamically add or remove namespaces. The transaction log records all the events happening in the system. An example of such an event would be an upload of a data object to the storage layer. Information about the event would then be published to the transaction log, together with some meta information like name and size of the digital object, its location, etc.

A number of entities can subscribe to the transaction log, receive the information, and put it into a namespace. The transaction log stores events ordered, thus it is possible to add namespaces and "replay" all the events to arrive at a final consistent state.

The parts that are exposed to the end users of the systems are: namespaces and storage API. The later is more relevant for the uploading of digital objects. The user has to understand the features of this API to take full advantage of the system. In case of download, we expect the users to interact with one of the namespaces to identify relevant data objects. The namespaces will also store the up-to-date URL of the locations. In our approach we rely on the Swift REST API, which is well-documented [15] and pretty intuitive.

A. Implementation

Our prototype implementation is based on three major technologies: OpenStack Swift used as the storage layer, Kafka [16], for holding the transaction log, and the graph database system neo4j [17], for creating namespaces.

To facilitate the integration of the components we use a particular feature of Swift. Incoming requests have to go through a *Request Processing Pipeline*. The pipeline is based on the Python framework Paste [18] and comprises of a set of *middlewares*, chained together. Several middlewares e. g., for authentication and caching, are already available. It is possible to implement additional middlewares and add them to the request processing pipeline.

To publish events from the storage layer to the Kafka-based transaction log, we implemented a Paste middleware called *Emitter*. It is incorporated into the request processing pipeline as it can be seen in Fig. 1.

We have chosen Kafka as a transaction log because of its key features, namely the possibility of a distributed, horizontally scalable deployment and the guaranteed correct ordering of messages for a given topic. Due to the ordering we are able to make sure that our namespaces reflects the correct order of actions that happened on the storage layer. The availability of this kind of information helps to keep track of the historical changes in data placement and can be used to implement data provenance functionality.

A counterpart of the Emitter is a *Consumer* that subscribes to the transaction log and builds a namespace according to the published events. As mentioned in Section II-C, graph

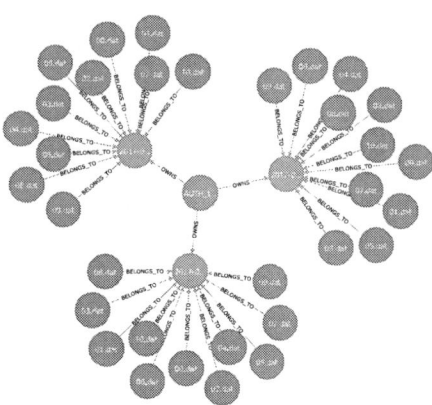

Figure 2. An example of a graph-based namespace.

databases are one technology that provides a flexible, searchable namespace. An example of a simple, graph-based namespace, that was created in our prototype, is depicted on Fig. 2. It shows data objects from a benchmark run available in Swift. Currently, the users can interact with this namespace by issuing queries using the CYPHER language, which is a open standard and has some similarities with SQL, known from relational databases. CYPHER queries are based on pattern matching. Thus, the users can define patterns they are searching for in the namespace and by doing so define their own searching algorithms.

The final part of our prototype was the replication. As already explained, Swift automatically replicates data objects within one cluster. The replication across administrative domains was done using the aforementioned container synchronization feature [19]. When using this way of replication to a remote cluster, a additional server daemon has to be installed on the local cluster. This daemon then regularly checks the metadata entries for all containers and, in case of a enabled container synchronization, asynchronously copies all objects belonging to the container to the remote cluster that was selected. The actual data transmission between Swift clusters is conducted via HTTP and all the metadata of a local cluster are also passing through the request pipeline of the remote cluster, where the copy is created. This allowed us to easily publish the information about the remote replicas into the transaction log, by using exact the same Emitter that was described above. Since the replication events are available in the transaction log, namespace Consumers can take advantage of them and present them to the users.

The source code of our prototype implementation with a deployment specification defined using the *docker-compose* format is available at [20].

At this point it is worth mentioning that we strive to make as little assumptions about the concrete technologies as possible. In fact it would be possible to even use a different technology for the storage layer, it must only be possible to publish the local event information to the transaction log. Also dynamical addition (and removal) of new namespaces should be easily possible, by just subscribing to (or unsubscribing from) the transaction log. A good candidate for a namespace would be Elasticsearch [21], a system that enables fast and efficient keyword-based full-text search.

IV. EVALUATION

In this section we present a preliminary evaluation of our prototype. The tests ran on a system with 16 CPU Cores, 32 GB RAM, and a 1GE network link running CentOS (7.2.1511). We wanted to compare the performance of a plain Swift interface, an interface integrated with the transaction log, and the third comparison was made with a Swift that has a middleware directly communicating with the EPIC PID system. The last system mimics an EUDAT B2SAFE system with iRODS substituted by Swift. We simulated the upload of 50 objects per Swift instance, repeat those uploads ten times, and visualize the average value in Fig. 3a. The uploaded files used here have a size of 512 bytes.

The average request times show that both Swift extensions increase the request processing time. It is caused by a longer run time of the extended Paste pipelines, additional actions take place when a request is processed. Integration with a remote PID system is more costly than integration with the distributed transaction log Kafka. This is caused by the higher latency of the remote system. Furthermore the PID API is only supporting synchronous calls, while the Kafka Emitter can use asynchronous calls. Overall the overhead is acceptable, especially given the fact that in most of the real world scenarios the request times would be dominated by the transfer times of the data objects. In our evaluation with small files we focused on the overhead of the request processing pipeline. A comparison for the upload of different file sizes is given in Fig. 3b, the values shown there are averaging 500 uploads per file size per Swift instance. For a 100 MB large file, the overhead caused by integration with the transaction log amounts to $0.16s$ which is about 9% of the measured request completion time.

The Swift and Kafka deployments for the performance evaluation where done utilizing docker-compose and with this Docker (1.13) to ensure a clean, encapsulated environment and to make our evaluations reproducible. We want to state out that the Swift with PID integration had to connect to an external API while the communication for the transaction log enabled Swift was locally using direct links between Docker containers.

V. CONCLUSION AND FUTURE WORK

In this paper we have described the architecture and implementation of a prototype for a system that is able to provide a safe harbor for research data and at the same time give the ability to build complex, searchable namespaces using different search engines. While the system was built to guarantee access, safety, and searchability of scientific data, it could also be used to enhance operational aspects of data services that data centers offer e. g., by analyzing access patterns.

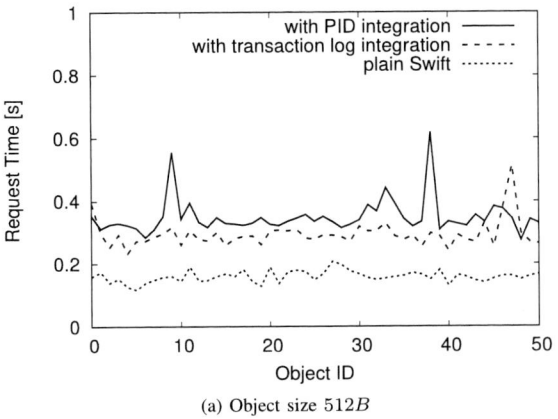

(a) Object size 512B

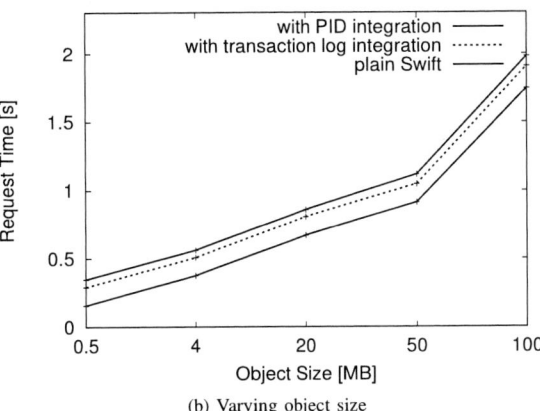

(b) Varying object size

Figure 3. Middleware overhead evaluation.

The evaluation of our first prototype was encouraging. In the future we plan to extend the system with additional namespaces by using different technologies and to provide the users with interfaces for searching through the data even without issuing CYPHER queries. Also operational experiences may influence the design of the system in the future.

ACKNOWLEDGMENT

The work has been supported by EUDAT2020, funded by the European Union under the Horizon 2020 programme - DG CONNECT e-Infrastructures (Contract No. 654065).

REFERENCES

[1] W. Gentzsch, D. Lecarpentier, and P. Wittenburg, "Big data in science and the EUDAT project," in *SRII Global Conference*, Apr. 2014, pp. 191–194.

[2] S. Brin and L. Page, "The anatomy of a large-scale hypertextual Web search engine," *Computer Networks and ISDN Systems*, vol. 30, no. 1-7, pp. 107–117, Jul. 1998.

[3] (2016, Dec.) Dropbox. [Online]. Available: https://www.dropbox.com/

[4] A. Rajasekar, R. Moore, C.-Y. Hou, C. A. Lee, R. Marciano, A. de Torcy, M. Wan, W. Schroeder, S.-Y. Chen, L. Gilbert, P. Tooby, and B. Zhu, *iRODS Primer: Integrated Rule-Oriented Data System*, ser. Synthesis Lectures on Information Concepts, Retrieval, and Services. Morgan & Claypool Publishers, 2010.

[5] R. Kahn and R. Wilensky, "A framework for distributed digital object services," *International Journal on Digital Libraries*, vol. 6, no. 2, pp. 115–123, Apr. 2006.

[6] Benedikt von St. Vieth. (2017, Jan.) Handle representation for a ePIC PID. [Online]. Available: https://hdl.handle.net/11097/88319713-8e7e-4001-a893-4dd81b539b86?noredirect

[7] ePIC Consortium. (2017, Jan.) ePIC – persistent identifiers for eResearch. [Online]. Available: http://www.pidconsortium.eu/

[8] J. Arnold, *OpenStack Swift: Using, Administering, and Developing for Swift Object Storage*. O'Reilly Media, 2014.

[9] S. A. Weil, "Ceph: Reliable, scalable, and high-performance distributed storage," Ph.D. dissertation, University of California,Santa Cruz, 2007.

[10] (2017, Feb.) EMC: Elastic Cloud Storage (ECS). [Online]. Available: https://www.emc.com/collateral/data-sheet/h13079-ecs-ds.pdf

[11] (2017, Feb.) DDN: WOS object storage. [Online]. Available: https://www.ddn.com/download/resource_library/brochures/object_storage/ddn-wos-object-storage-datasheet.pdf

[12] (2016, Dec.) Amazon simple storage service documentation. [Online]. Available: https://aws.amazon.com/documentation/s3/

[13] B. von St. Vieth. (2017, Jan.) Swift persistent identifiers middleware. [Online]. Available: https://github.com/bne86/swift-persistent-identifier/

[14] V. Bunakov, P. D. De Meo, S. Kindermann, A. Queralt, and J. Rybicki, "Graph-based data integration in EUDAT data infrastructure," in *ALLDATA 16: 2nd International Conference on Big Data, Small Data, Linked Data and Open Data*, Feb. 2016, pp. 54–58.

[15] (2017, Jan.) Object Storage API. [Online]. Available: http://developer.openstack.org/api-ref/object-storage/

[16] J. Kreps, N. Narkhede, and J. Rao, "Kafka: A distributed messaging system for log processing," in *NetDB 11: 6th International Workshop on Networking meets Database*, Jun. 2011, pp. 1–7.

[17] J. Webber, "A programmatic introduction to Neo4j," in *SPLASH '12: 3rd ACM Annual Conference on Systems,Programming,and Applications: Software for Humanity*, Oct. 2012, pp. 217–218.

[18] (2017, Jan.) Python Paste. [Online]. Available: http://pythonpaste.org

[19] (2017, Jan.) Swift: Container to container synchronization. [Online]. Available: http://docs.openstack.org/developer/swift/overview_container_sync.html

[20] (2017, Jan.) Swift Messagebus Middleware. [Online]. Available: https://gitlab.version.fz-juelich.de/vonst.vieth1/swift-kafka/

[21] C. Gormley and Z. Tong, *Elasticsearch: The Definitive Guide*. O'Reilly Media, 2015.

Spatial Analysis of the Clustering Process

M. Kranjac*, U. Sikimić, ** J. Salom,***S. Tomić ****

* University of Novi Sad, Faculty of technical sciences/Department for transport, Novi Sad, Serbia
** 3Lateral, Novi Sad, Serbia
***Mathematical Institute of Serbian Academy of Sciences and Arts, Belgrade, Serbia
****Faculty for engineering management, Belgrade, Serbia
mirjana.kranjac@ftn.uns.ac.rs, uros_sikimic@hotmail.com, jakob.salom@yahoo.com, srdjan.tomic@fim.rs

Abstract - Creation of industry clusters is a request of modern economy to reshape itself for permanently changing external and internal conditions. By using industry clusters transition countries, like Serbia, are in a position to act in foreign markets with united offer of products in order to reach critical mass. Governmental support to clustering processes is inevitable and it starts to be important when the mapping of market begins and when clustering territories start to appear. Government of Vojvodina has recognized the importance of clusters and started to give financial aid and political support to first clusters in Vojvodina in 2007. After 9 years of activities, there are still problems and some necessary changes and improvements should be made. There is a need to analyze clustering processes with spatial components. The authors of this paper present visualization of clusters' activities and acquire the results and impacts of governmental support on clustering success. This is performed by using "QGIS" visualization tool. The authors have found very big differences between clusters, big changes during clusters' life cycles, and dependence of these factors on sectors of work. They discuss clustering process, recognize various stages of clusters development, and suggest some visualization tools which should be used to follow a clustering process.

I. INTRODUCTION

GIS technology allows the collecting, organization, manipulation, analysis, and visualization of spatial data. It enables finding and uncovering relationships between entities, their patterns, changes, and time trends. GIS is a useful tool for urban planning [1, 2]. It uncovers many problems in ecology [3], helps in analyzing transportation problems [4] and public health patterns [5]. Demographics are a spatial issue and GIS enables a lot of visualized discoveries [6]. GIS added new value to law enforcement reviews [7] and resource management [8]. It is useful in many other industries. Traditional GIS analysis techniques include spatial queries, map overlay, buffer analysis, interpolation, and proximity calculations [9].

GIS, as a software tool, includes many geostatistical techniques [10] for spatial analysis [11], which have been extended over the years [12]. It, also, offers raster analysis. GIS is enabling analytical methods for business [13]. Its 3D analysis [14] is very often used in network analytics [15], space-time dynamics [16] and techniques specific to a variety of industries [17].

The general goal of the authors in this paper is to research development process of clusters in Serbia. The specific goal is to research effects of the state aid given to clusters, which clusters are the most important, and what errors and problems are preventing intensive clustering and better results of their use.

The authors are using open source software, QGIS, for the visualized presentation of the clustering process in Serbia, with special focus on the autonomous province of Vojvodina.

Clusters are geographic concentrations of interconnected companies and institutions in a particular field. Clusters encompass an array of linked industries and other entities important to competition. They include, for example, suppliers of specialized inputs such as components, machinery, and services, and providers of specialized infrastructure. Clusters also often extend downstream to channels and customers and laterally to manufacturers of complementary products and to companies in industries related by skills, technologies, or common inputs. Finally, many clusters include governmental and other institutions - such as universities, standards-setting agencies, think tanks, vocational training providers, and trade associations - that provide specialized training, education, information, research, and technical support [18].

The concept of clusters has not been defined clearly enough in transition countries, including Serbia. The businessmen in Serbia are keener to join different forms of associations, and there is a degree of skepticism among private entrepreneurs about the role of clusters.

Clusters include companies from one industry or technology area, but also from vertically related areas e.g. producers of complementary products. For example, a cluster in the textile branch might include producers of raw materials, manufacturers of ready-made garments, scientific and educational institutions, and governmental and non-governmental organizations. The goals of clustering are:

- Increasing the competitiveness of domestic producers on domestic and foreign markets,
- Providing conditions for market expansion (especially increase in exports),
- Better and more efficient utilization of domestic resources (nature, production, and personnel),
- Initiate and support cooperation between companies, and between companies and educational and development institutions,

- Better funding of innovative projects,
- Training and education.

Key benefits of clustering are:

- Creation of a wider framework for cooperation,
- Stimulation of economies of scale,
- Development of a higher level of competitiveness,
- Reduction of influence of fears of competition (building trust and cooperation),
- Easier development of innovative products and services,
- Better networking with foreign clusters.

Clusters, by definition, should closely cooperate with the Government, thus giving enterprises possibility to influence development of the legislative and institutional framework for businesses. In this way they can more easily eliminate administrative and other barriers, and subsequently, improve the competitiveness of the entire economy. Cluster, as an idea, which involves making pools of economic, scientific and support organizations together with individual experts, implies defined quality, determines the quantity and continuity of production in order to meet market demand, and define the customers. Therefore, a cluster is strategically oriented and develops within a particular industry sector.

Clusters can become "generators" of new ways of strategic thinking in the national economy. Cluster members should be assured that, when entering into this kind of organization, they retain their independence, individuality, their production, and their market. The concept of clustering relies on the idea of healthy competition and on the assumption that it will be better to develop the competitiveness of the cluster than inside the same number of companies within the whole national economy.

We can hardly expect that the government takes part in forming or organizing a new cluster. However, when the cluster is in the phase of establishment, or has already been created, the state, with its policy and backing, can create an adequate environment (schools, training centers, databases, specialized infrastructure).

How can the state help successful development of clusters? It can make impact on financing conditions by giving focused financial support for clusters development.

As a result of cooperation with the Faculty of technical sciences, The University of Novi Sad, Government of Vojvodina has activated Development program of clusters in Vojvodina. As a start, the Government has established the Center for the Development of Clusters. During 2008 the Government of Vojvodina was a partner in the cross border project APLE.NET between Croatia and Serbia and realized some activities with the goal called "Through networking to economic development". During the following years the Government assisted on setting up and promoting of few new clusters. During 2011 Provincial secretary for economy, for the first time, financially supported 16 clusters with about 60.000 Euros.

In 2014, 15 clusters received financial support of about 60.000 Euros, and in 2015 19 clusters received in total about 130.000 Euros.

The question is how efficiently the state funds have been used and do we have indicators to measure that?

In this paper the authors present the process of starting development of clusters in Vojvodina, what kind of support they were given, which clusters are the most important, what mistakes have been made and what problems have been encountered in trying to have more intensive clustering with better results.

II. ANALYSIS OF THE CLUSTERING PROCESS IN SERBIA

The authors used QGIS to visualize locations of clusters in Serbia which were given governmental support. There are 133 such clusters and some of them have received state aid more than once [19, 20]. Fig. 1 presents towns where the clusters are registered.

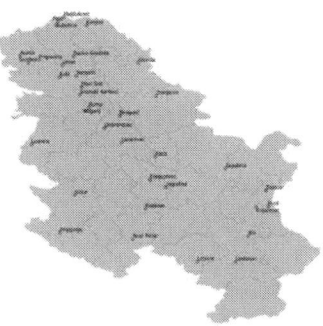

Figure 1 Locations of clusters in Serbia

Fig. 2 presents all clusters that have been set up in Serbia since year 2000. Clusters marked with stars do not exist anymore. Overlapping of layers shows that only clusters that received financial support from government, especially in the first phase of their life cycles, are still functioning. That means that state aid in this phase is of significant importance. It also proves that financial assistance should be harmonized with the phases of clusters' life cycles.

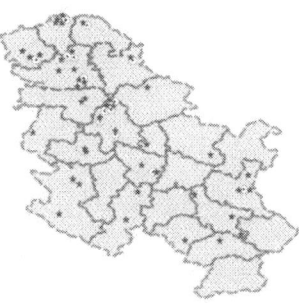

Figure 2 Closed and functioning clusters

Fig. 3 depicts locations of clusters. Sizes of circles and their colors reflect number of clusters on a certain location (bigger diameter and darker color mean more clusters at the location). Belgrade has the most clusters - 30, Novi Sad 19, and Niš 12.

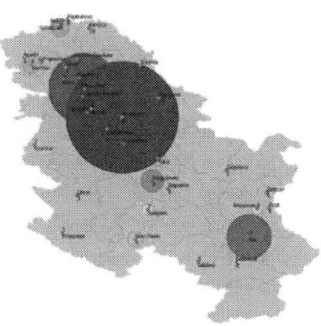

Figure 3 Presentation of number of clusters in towns of Serbia - option 1

The second option to present clusters' numbers is given in Fig. 3 by using NUTS level 2 regions. NUTS - The Nomenclature of Territorial Units for Statistics or Nomenclature of Units for Territorial Statistics (NUTS; French: Nomenclature des unités territoriales statistiques) is a geocode standard for referencing the subdivisions of countries for statistical purposes [21]. The Government of Serbia specified a nomenclature of statistic territorial units in the country and attempt to synchronize the existing statistical division of the country with the Nomenclature of Territorial Units for Statistics of the European Union. In this Act, an additional top level of grouping was introduced, with the territory of Serbia divided into two NUTS 1 regions:

- Serbia-North, comprising
  o Vojvodina
  o Belgrade, and
- Serbia-South, comprising
  o Šumadija and Western Serbia
  o Southern and Eastern Serbia
  o Kosovo and Metohija

The five statistical regions would therefore become NUTS level 2: NUTS 2 regions [22].

NUTS regions are separately created. The sizes of circles around towns present number of clusters in them.

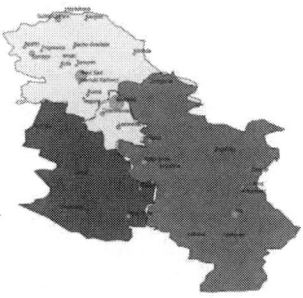

Figure 4 Presentation of number of clusters in towns of Serbia - option 2

Fig. 5 presents the age of clusters by year of their establishment. It is shown by the size and color of a circle around each town. The first clusters were set up in Vojvodina, in Zrenjanin, Subotica and Kanjiža. The Government of Vojvodina was the first that gave financial support to clusters. National level clusters started a few years later.

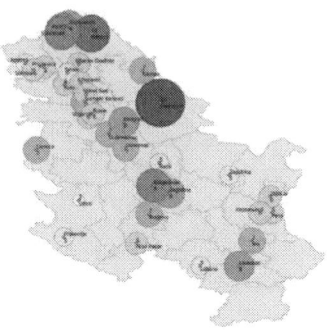

Figure 5 Presentation of clusters age per towns in different NUTS 2 regions - option 1

Fig. 6 is an example of presenting the age of clusters in different NUTS 2 regions with colors that reflect their GDP. The sizes of the circles around towns present average age of clusters in them.

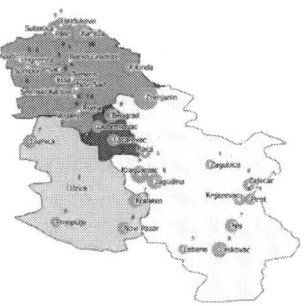

Figure 6 Average age of clusters per towns in different NUTS 2 regions - option 2

Fig. 7 has in the background four NUTS regions of Serbia: Vojvodina, Šumadija and Western Serbia, Southern and Eastern Serbia, and Belgrade. The darkest color shows the region with the most clusters. The brighter colors depict smaller number of clusters. It can be seen that the most clusters are in Vojvodina and the least in Eastern Serbia. The QGIS diagrams are used to give information about number (upper part) and age (bottom part) of clusters at a particular location.

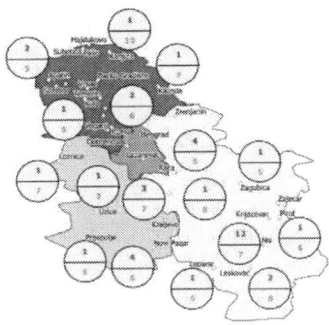

Figure 7   Locations by age and number of clusters

Fig. 8 presents NUTS level 2 in Serbia. Intensity of color shows the value of GDP. The size of circles presents number of clusters in each NUTS level 2I. It can be seen that the number of clusters is synchronized with the value of GDP.

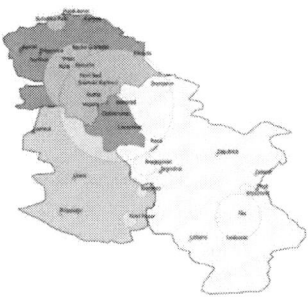

Figure 8 GDP in NUTS 2 and number of clusters
by region - option 1

Fig. 9 has the same content as Fig. 6 presented in a different way. It shows cross cutting of number of clusters per each NUTS region and GDP per each. The small circles present GDP by color intensity. The big circles show value of GDP by their size and color. This figure presents an attractive type to put titles on QGIS map.

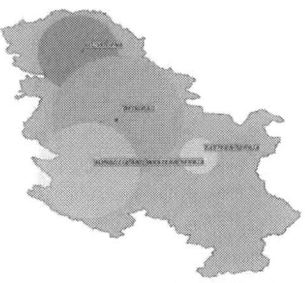

Figure 9 GDP in NUTS 2 and number of clusters
by region – option 2

Fig. 9 is an example of presenting NUTS 2 regions with total sum of clusters in each, together with the region

names. It gives interesting visual opportunities for presentation.

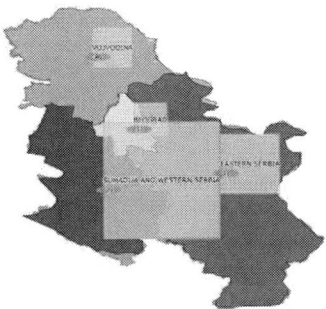

Figure 10 NUTS 2 regions with total sum of clusters
in each, together with the names of the clusters

III. ANALYSIS OF CLUSTERING PROCESS IN VOJVODINA

This part consists of a few examples of possible uses of QGIS for analysis elaborated by authors for the region of Vojvodina. The analysis is made on the OpenStreetMap map what is one of many, practically endless possible ground maps in QGIS.

Fig. 11 is a map of Vojvodina with marked clusters by sectors. Sectors that exist are:

➢ Agriculture
➢ Tourism
➢ ICT
➢ Metal industry
➢ Fashion
➢ Creative industry
➢ Crafts
➢ Construction
➢ Bio
➢ Plastic
➢ Transport
➢ Ecology

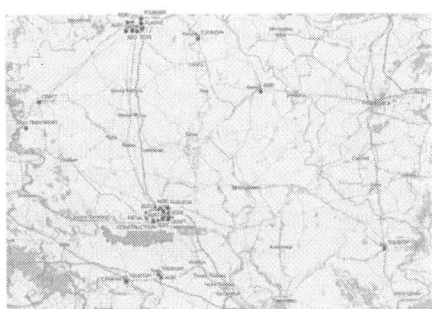

Figure 11 Clusters of Vojvodina by sectors
on OpenStreetMap

Fig. 12 is a result of a "query" function, a useful selection tool in QGIS. It presents only clusters of the agriculture sector in Vojvodina.

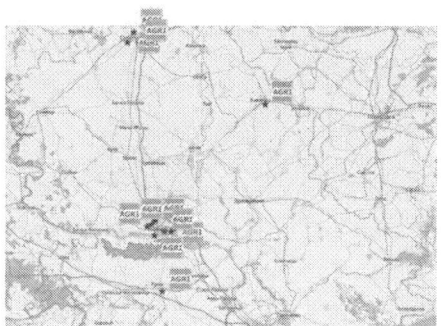

Figure 12 Selected clusters in Vojvodina, sector: Agriculture, on OpenStreetMap

Fig. 13 is a result of a "query" function. It presents only clusters of the tourism sector in Vojvodina.

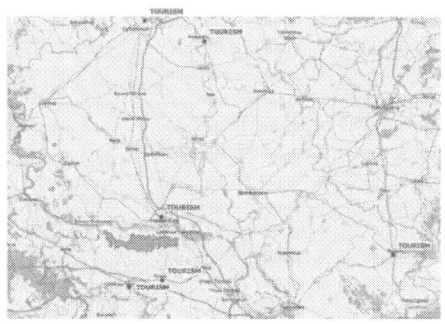

Figure 13 Selected clusters in Vojvodina, sector: Tourism, on OpenStreetMap

Sectors shown in figures 12 and 13 are in focus of Smart specialization in research and innovation strategy in Vojvodina.

Fig. 14 presents spatial locations of members of Vojvodina metal cluster, one of the biggest clusters created in Vojvodina. It has 119 members in all NUTS 2 regions of Serbia, except Eastern Serbia [22].

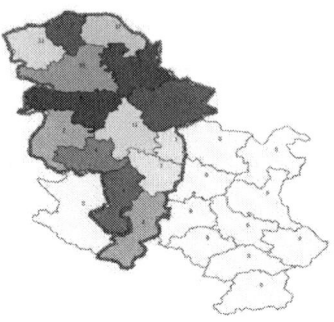

Figure 14 Spatial distributions of members of Vojvodina metal cluster

IV. CONLUSION

Spatial visualization made by QGIS tool brought many conclusions related to state policies toward the development of clusters. This problem is not only local problem of Serbia, it is a general one. Many countries are looking for the tools that will allow them to observe the results of clusters, to motivate them, to organize evaluation of their activities and to support them. European policy is stressing transnational role of clusters, actually networking of clusters. Analysis of results within such structure will be even more challenging.

After 9 years of activities of clusters in Serbia, there are still problems and plenty room for changes and improvements. Spatial analysis show that clusters are mostly located in 3 biggest towns of Serbia, in Belgrade, Novi Sad and Niš. The explanation could be that these are centers of universities whose role is to support clusters activities. The Government of Vojvodina was the first who supported clusters establishment. Thanks to that, Vojvodina has the most of the functioning clusters today. Clusters that were not supported financially by government in the first phase of their existences do not function today. The best working clusters in Vojvodina are from sectors of; agriculture, ICT, and tourism what coincides with smart specialization focus of Vojvodina on these particular sectors.

REFERENCES

[1] Kohsaka H, „Applications of GIS to urban planning and management: Problems facing Japanese local governments", , 2000, vol. 52, issue 3, pp. 271-280.

[2] Maantay J, Ziegler J, „GIS for the urban environment", ESRI, Redlands, 2006.

[3] Johnston C, „Geographic information systems in ecology", Blackwell Science, Malden, 1998.

[4] Thill J. C, „Geographic information systems in transportation research", Elsevier Science, Oxford, 2000.

[5] Cromley E, McLafferty S, „GIS and public health", Guilford, New York, 2002.

[6] Peters A, MacDonald H, „Unlocking the census with GIS", ESRI, Redland, 2004.

[7] Chainey S, Ratcliffe J, „GIS and crime mapping", Wiley, London, 2005.

[8] Pettit C, Cartwright W, Bishop I, Lowell K, Pullar D, Duncan D, „Landscape analysis and visualization: spatial models for natural resource management and planning", Springer, Berlin, Heidelberg and New York, 2008.

[9] Mitchell A, „The ESRI guide to GIS analysis, volume 2: spatial measurements and statistics", ESRI, Redlands, 2005.

[10] P. A. Longley, M. F. Goodchild, D. J. Maguire, and D. WRhind, „Geographic information systems and science", John Wiley, pp. 290-300, Chichester, 2001.

[11] Smith M, Goodchild M, Longley P, „Geospatial analysis", Troubador, Leicester, 2006.

[12] Tomlin D, „Geographic information systems and cartographic modeling", PrenticeHall, New Jersey, 1990.

[13] Pick J, Geo-Business: „GIS in the digital organization", Wiley, New York, 2008.

[14] Abdul-Rahman A, Zlantanaova S, Coors V, „Innovations in 3D geo information systems", Springer, Berlin, Heidelberg and New York, 2006.

[15] Okabe A, Okunuki K, Shiode S, „SANET: a toolbox for spatial analysis on a network", Geogr Anal 38(1):57-66, 2006.

[16] Peuquet D, „Representations of space and time", Guilford, New York, 2002.

[17] Miller H, Shaw S, „Geographic information systems for transportation: principles and applications", Oxford University Press, Oxford and New York, 2001.

[18] Porter, M, "Clusters and the New Economics of Competition", Harward Business review, 1998APV, Report 2016, www.spriv.rs [Accessed on: 10.06.2016].

[19] http://www.lokalnirazvoj.org/en/publications/details/42 [Accessed on: 15.12.2016].

[20] http://www.vmc.rs/ [Accessed on: 07.11.2016].

[21] https://en.wikipedia.org/wiki/Nomenclature_of_Territorial_Units_for_Statistics [Accessed on: 10.01.2017].

[22] https://en.wikipedia.org/wiki/Nomenclature_of_Territorial_Units_for_Statistics [Accessed on: 10.01.2017].

Usage of Android device in interaction with 3D virtual objects

I. Prazina, V. Okanovic, K. Balic and S. Rizvic[*]

Faculty of Electrical Engineering Sarajevo, Sarajevo, Bosnia and Herzegovina
iprazina1@etf.unsa.ba, vokanovic@etf.unsa.ba, kbalic1@etf.unsa.ba, srizvic@etf.unsa.ba

Abstract - Implementation of natural interactive online 3D visualization is difficult process. To keep people interested in virtual heritage the new ways of immersive interaction must be developed. This paper presents a solution for natural interaction with 3D objects in virtual environments using Android device over WiFi. It describes the principles and concepts of functioning of individual components as well as the principles of operation after the integration of all components into a single software solution.

I. INTRODUCTION

Smart devices largely determine and shape the way we live and communicate. In recent years, people are increasingly using smart devices for various purposes. One of the areas where smartphones find their usage is the cultural and historical heritage. Digital technologies are an efficient tool for visualization and presentation of the virtual objects. Accordingly, the combination of smart Android devices and communication (Wi-Fi) technologies, as well as 3D modeling and visualization of virtual objects are a good combination for the presentation of cultural and historical heritage.

Implementing 3D visualization is difficult process. There are different ways of interaction that use new technologies. In this work, interaction with a 3D object using Android device will be presented. Here will be presented the way of interaction with virtual objects where a user can chose in a web page which scene he wants to see. Thus, using their smartphones a user can see the details of the virtual excavation and details of virtual reconstruction of the same excavation.

In section II background and related work is listed. In section III overview of interactions is given. Section IV presents case study and implementation details. Section V gives results of evaluation. And conclusion is given in section VI.

II. BACKGROUND AND RELATED WORK

In papers [1] and [2] two ways of interaction with 3D objects are presented. Both interactions use Leap Motion sensor to control 3D objects' rotation and position. Leap Motion sensor is used to detect hand position and 3D object orientation is set using Leap Motion JavaScript library. This interactions were interesting for user to use but they have problems with precision and can be tiresome for use.

In paper [3] authors show how accelerometer and magnetometer can be used to emulate gyroscope in virtual reality application. They show methods how to calculate angles of inclination using measurements from accelerometer and magnetometer.

In paper [4] authors show methods for calculating head positions and movements using accelerometer and gyroscope to visualise 3D objects on head mounted displays.

III. WAYS OF INTERACTION

The main goal of the application is to provide immersive and interesting ways of interaction. The application is web page which consists of two 3D scenes, and user has control which scene she/he wants to interact with. One scene has historical artifact in it, and another has its reconstruction. When a user chooses a scene she/he can rotate and view an object in the scene from different angles. This application gives user opportunity to see and move object which if she/he was in a museum she/he could not touch or move.

In the first version of the application a user controlled a scene with mouse using click and move actions. When mouse was dragged over the scene object was rotating in the direction of the mouse movement. This type of interaction is sufficient but not immersive enough. Idea was to use Android device to break mental barrier between reality and virtual scene. It is easy to use Android device as proxy for object orientation in real world because most Android devices already have gyroscope sensors to determine device's position in the space. A user rotating and moving the device rotates virtual object on a computer screen. This way of interaction replicates user movements in virtual scene making users to feel like they are directly controlling virtual scene.

IV. CASE STUDY

A. The White Bastion project

Interactions implemented in this work are trying to save cultural heritage from oblivion. The White Bastion project is collection of digitized material from medieval fortress found on the hills of Sarajevo [5].The project consists of multiple 3D scenes of the fortress from different periods, 3D objects of artifacts, linked video stories and web presentation. In the application created in this work only digitized 3D models of artifacts, their reconstructions and short textual stories are used. Each artifact has its reconstruction and both reconstruction and artifact are 3D objects shown in web browser. User can view objects from different angles. The 3D object orientation can be changed using computer mouse or Android device.

B. *Interaction implementation*

Implementation of the idea consists of web back-end in Node.js, web front-end in HTML, CSS and Three.js [6] and Android mobile application. Web back-end listens for requests from Android application which sends device's orientation data and sends them to front-end. Front-end updates orientation of the selected 3D object using data from back-end.

This implementation requires: a desktop PC, a wireless router and an Android device. Back-end and front-end are installed on the PC. Android application is installed on the Android device. PC and the Android device are connected to the same network. Android device is connected through Wi-Fi and PC is connected through Ethernet cable.

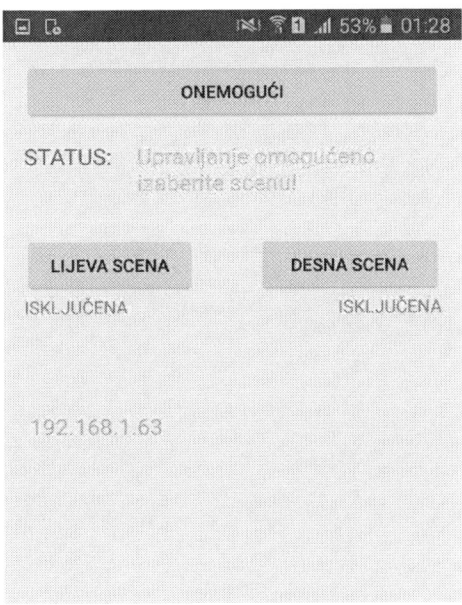

Figure 1. Android application user interface

Communication between Android app - back-end - front-end is made using Socket.io framework. Socket.io enables real-time bidirectional event based communication [7]. Data from Android app is collected through socket in JSON[1] format and using socket emitted to front-end in the same format. On the front-end using JavaScript socket is instantiated and event listener is set to get the data emitted from back-end. When data is emitted event listener gets the event, collects data through event's callback function and sets 3D object's rotation using collected data. In the event listener callback two events must be considered, one if user chooses to control left scene, and another if user chose right scene.

---

[1] JSON - JavaScript Object Notation

Android application has two functions:

- it gives user interface which user uses to connect to the back-end and to choose object to control

- it collects position sensors' data and emits it to the back-end

In the Android application magnetic field and accelerometer sensor are used. When sensor detects a change in position it emits event, event listener in Android application gets the event and in the callback function orientation of the device is calculated using rotation matrix. Rotation matrix uses data from both accelerometer and magnetic field sensor. Orientation data is then encoded in JSON format. When data is ready to be send connection to the web server is established through socket using ip address which user entered in application's interface.

Android application interface is shown in Figure 1. In the interface user can enable/disable control, can choose left or right scene with two buttons and has text field where he can enter PC's ip address. User has feedback from the application in the status text where he can see if control is enabled and which scene he is controlling.

Web front-end is shown in Figure 2. User can see two 3D models side by side, and short text about them. Objects are displayed in web browser using Three.js 3D library. Each object has its own Three.js scene and can be controlled separately.

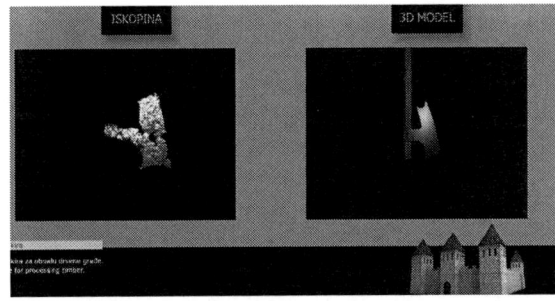

Figure 2. Web application front-end

V. EVALUATION

Interaction is evaluated using qualitative user evaluation. In the evaluation group of five people was given Android device and computer mouse. They had to view all sides of 3D object and answer questions about their experience.

In the evaluation two hypotheses were evaluated:

1. Android device has enough precision for users to control and view 3D object from all sides
2. Android device has comparable ease of use as the computer mouse

Most of the participants of the evaluation had no or had little experience with virtual 3D scenes. All of them recognized importance of preserving cultural heritage and they see this type of projects as good way of doing that.

Some of the participants noted that they have more control over object rotation when they are using Android device than mouse. They didn't notice significant problem with precision and they felt control was more enjoyable and easier to use than computer mouse.

After user evaluation hypotheses weren't denied and it showed promising results for this type of interaction. For more relevant results larger user group is needed.

VI. CONCLUSION

In this paper interaction with 3D virtual objects using Android device was presented. The interaction doesn't need expensive equipment and can be used in any museum or institution which preserves cultural heritage. Required equipment is standard desktop PC and wireless router and museum visitors' Android mobile phone. Users don't need prior training or preparation which is shown in user evaluation. The Android application presented in this work could be improved with better looking user interface and IP address configuration could be automatized.

REFERENCES

[1] I. Prazina, and S. Rizvić, "Natural interaction with small 3D objects in virtual environments", Central European Seminar on Computer Graphics, Smolenice, Slovakia, 2016.

[2] I. Prazina, K. Balić, K. Pršeš, S. Rizivić, and V. Okanović, "Interaction with virtual objects in a natural way", MIPRO 2016 39th International Convention on Information and Communication Technology, Electronics and Microelectronics, Opatija Croatia, 2016.

[3] Delporte, Baptiste, et al. "Accelerometer and magnetometer based gyroscope emulation on smart sensor for a virtual reality application." *Sensor and Transducers Journal* 14. Special Issue ISSN 1726-5479 (2012): p32-p47.

[4] Tolle, Pinandito, et al. "Virtual reality game controlled with user's head and body movements detection using smartphone sensors", Journal of Engineering and Applied Sciences 10(20):9776-9782, November 2015.

[5] S. Rizvic, V. Okanovic, I. Prazina, and A. Sadzak,"4D Virtual Reconstruction of White Bastion Fortress", 14th Eurographics Workshop on Graphics and Cultural Heritage, Genova, Italy, 2016.

[6] Three.js (https://threejs.org/, visited online: February 2017.)

[7] Socket.io (https://github.com/socketio/socket.io, visited online: February 2017.)

Generating Virtual Guitar Strings Using Scripts

Luka Kunić and Željka Mihajlović
University of Zagreb, Faculty of Electrical Engineering and Computing,
Zagreb, Croatia
luka.kunic@fer.hr, zeljka.mihajlovic@fer.hr

Abstract—**Artists who create 3D models usually rely on the traditional method of direct mesh manipulation using basic operations such as translation, rotation, scaling, and extrusion. In some cases, creating a model in this manner requires performing many repetitive and precise actions, which makes a fully manual approach suboptimal. This paper explores an alternative concept of 3D modeling using scripts, which aims to automate parts of the modeling process.**

Existing procedural algorithms use scripts to generate objects based on mathematical models (e.g. generating realistic terrains using fractals), or to build complex structures using simple template models (*model synthesis*). Unlike those methods, which rely on stochastic behavior to generate pseudo-randomized shapes, the goal of this method is to write scripts that generate parametric objects which would be either difficult or time-consuming to model by hand. The example described in this paper is a script that generates animated strings for musical instruments. Although the basic shape of a string is quite simple, animating string vibrations and bending can be quite a tedious task, especially because of the various shapes and sizes of string instruments.

I. INTRODUCTION

With the advancements in computer graphics, 3D modeling has become a major part of many industries. Artists who create 3D models most commonly use dedicated software solutions that include a 3D viewport, which allows them to directly modify the mesh data using basic operations such as translation, rotation, scaling and extrusion. While this traditional method can be used to produce very good results, those results are often difficult to achieve due to large amounts of manual work required. This is especially true for large-scale environments such as terrains or entire cities used in movie and video game industries, which are practically impossible to generate manually in a reasonable time frame. However, many tasks involved in this process can be entirely or at least partially automated, allowing for a faster and less repetitive workflow.

Automating a task can be as simple as creating a custom tool that performs some commonly used or repetitive actions. This allows users to focus on the creative aspect of modeling instead of being hindered by the lack of software features. On the other hand, scripts can be used to generate complex objects based on a set of parameters and constraints, which can be given by the user or determined by the choice of the algorithm being used. Obviously, these scripts must be tailored to each specific problem, which brings into question whether the gains of using a script justify the time spent on development. However, if a task can be generalized or parametrized, writing a script to perform that task can greatly reduce time spent working on different areas of the model. Furthermore, variations of algorithms developed for such generic tasks can be used for solving various modeling problems in different domains.

This paper describes a method for automatically generating and animating 3D models of musical strings, which will be used for creating interactive virtual string instruments. Existing procedural modeling techniques, briefly covered in Section II, mostly rely on some form of stochastic behavior in order to achieve the final shape of the generated model. We present a more deterministic approach, in a sense that running the algorithm repeatedly with the same set of parameters will always result in the same model. The primary gain of this method is not in the model creation, since the actual mesh is quite simple, but rather in automatically adding complex behavior to the model through morphing and animations. Another significant advantage over a manual process is the ability to easily generate arrays of objects using various parameters. This is quite important for rapid prototyping, where the user iteratively creates variations of the model until they are satisfied with the result. Further discussion about the motivation for this method can be found in Section III.

The string generation process, described in detail in Section IV, includes: creating the string mesh based on a set of input parameters, adding vertex animations to the mesh in order to simulate string bending, adding and attaching an armature to the mesh, and animating the string vibrations. The results are presented in Section V.

II. RELATED WORK

The following subsections present several valuable methods for automatically generating 3D content: procedural modeling, template-based methods and interactive simulations.

A. Procedural modeling

Procedural techniques are algorithms that are used to determine the characteristics of an object or effect [3]. An algorithm is implemented as a procedure which contains an abstract representation of the desired features for the output model. Executing the procedure calculates the features based on input parameters and constraints set by the user and generates a finished model which satisfies those constraints. This allows a high level of control and flexibility. The procedures often incorporate a form of stochastic behavior in order to introduce a level of randomization into the output models. This is useful when modeling objects which are represented in nature, so that the finished product appears more organic and realistic.

Procedural modeling techniques have been used in computer graphics since the early beginnings of the field. At first, procedures were used to generate images and textures resembling materials found in nature, as it was found that many natural shapes follow distinct patterns which could be described by mathematical models. A prominent example of such textures are fractals, introduced by Mandelbrot in 1982 [7]. Fractals can be used to create very natural looking objects and structures because of their self-similarity property, which is why they are widely used in computer graphics.

A different approach to procedural modeling uses formal languages and grammars to describe the output model. One such model was introduced by Lindenmayer who used his L-systems [6] to formally describe the growth of plants. The system consists of an alphabet of symbols and set of production rules, which are used to generate strings of symbols that define a plant. This technique is often used in computer graphics because of its high versatility, and because it can be easily extended with additional functionality [8], [13], [14].

Procedural techniques are also commonly used to create and animate natural volumetric phenomena that cannot realistically be modeled with mesh geometry, such as gasses, clouds or fire [3], [5]. Attributes like color and transparency can be controlled using procedures that simulate air turbulence and noise in the space occupied by the model volume. Similar techniques can also be applied to particle systems which are also guided algorithmically and can be affected by some external influence.

B. Model synthesis and template-based modeling

Certain objects cannot entirely be described using mathematical models. This applies to most man-made structures such as buildings or machines that are often found in 3D environments. For example, when creating a large virtual city, it is important to include enough diversity so that the city does not seem artificial, but at the same time most buildings should have similar features so that they visually fit together. One of the solutions would be template-based modeling, a set of techniques that take simple objects as input and use the objects' distinct features to generate various other objects that incorporate those features.

One such algorithm named *model synthesis* is presented by Merell and Manocha [10]. Their algorithm takes an object and uses it as a template to create complex structures. For a given point in space for the model being generated, their algorithm looks at the surrounding area of the point and calculates the possible geometry samples which can be generated at that point. Their algorithm also takes into account various constraints: predetermined dimensions of an object, ratios between dimensions, connectivity constraints, and the general macroscopic shape and scale of the output model. Furthermore, this algorithm can take multiple objects as input, which results in output models with a high degree of variety.

A template-based approach has also been used by Zhou et al. for generating terrain using their *terrain synthesis* algorithm [18]. As input for their algorithm, they used height maps of real-world terrains and mountain ranges, combined with user-made sketches. The algorithm would identify important features from the terrain height map and the sketch. It would use the sketch to determine the general shape for the output model and would apply the extracted terrain features from the height map to that general shape. The result, as shown in Fig. 1, would be a terrain that resembles the real terrain used for the height map, but has the shape of the user-made sketch.

C. Interactive simulations

Simulating how virtual objects would behave in the physical world is a common task. Typical examples include cloth simulations [1], [17], simulating collisions between objects, simulating fluids [12] and force fields, determining how objects

Fig. 1. The result of the terrain synthesis algorithm. On the right is the finished terrain which was generated based on the height map of Mount Jackson in Colorado, USA (center left) and a user defined sketch (upper left). The combined height map generated by the algorithm is shown in the lower left corner.

interact with natural forces like wind, etc. These types of simulations are often used for animating objects in video games and other virtual environments. Simulations used for animating the objects typically have many parameters and obey certain laws of physics which require performing complex calculations. These calculations can either be done in real-time, which often sacrifices accuracy for computation performance, or they can be precomputed, in which case the calculated simulation will not be able to adapt to the surrounding objects in the virtual environment during execution.

III. MOTIVATION

Throughout history, music has been a prevalent source of entertainment in our society. Professional musicians have always been praised for their high technical abilities which they had developed over years of dedication and daily practice. Some of those musicians were also engineers, so they started using their expertise in both areas to create robotic instruments like self-playing pianos, guitars, and other instruments [2], [15], [16]. In the modern era of technology, with continuous advancements in the field of computer graphics, it is natural to explore the possibilities of placing musical instruments into virtual environments.

In order to create a self-playing virtual instrument, the 3D model of the instrument must first be created and animated. In the case of a string instrument like a guitar, this includes modeling the body of the instrument, creating and animating the strings, and adding fretting and picking mechanisms which will be used to press the strings onto the fretboard and to pick the strings as notes are played.

The most challenging part of this process is modeling and animating the strings. This task can be approached in several ways. The animations could be done manually, but that would require an immense amount of work so it is better to find an automated alternative. Because the strings obey certain laws of physics while vibrating, their behavior can be simulated at runtime. This approach offers most flexibility and most realistic behavior, but it can be computationally demanding, which might decrease runtime performance. A good compromise would be to precompute a set of animations and store them, so that they can be played back as required. This preserves the realism gained by the simulation without sacrificing performance. The main idea is to write a script that will generate the mesh of the string, attach an armature to the mesh and automatically compute the animation keyframes and morphing data that will be used in the final animations.

Reusability and the drastic reduction of time required for modeling are the primary advantages of using scripts as part of the modeling process. The main aspect of reusability is the use of input parameters in order to define the characteristics of the generated model. In the case of a guitar string, parameters are used to define string length, diameter, mesh complexity (number of vertices), and the number of frets.

IV. GENERATING MUSICAL STRINGS

This section describes the *StringGenerator* script. The script was developed as a Blender add-on using Blender's Python scripting API.

A. Blender scripting API

Blender is a popular open-source 3D software suite that provides tools for the entire 3D model creation pipeline – modeling, texturing, animation, rendering, compositing and sequence editing. Due to its open-source nature and powerful Python scripting API [19], it is continually being improved by members of the community, both by revising features of the core system and by developing add-ons which add new functionality.

Blender has an integrated Python console that allows users to write and execute scripts directly in the viewport. This is useful for testing and debugging, but also for writing short commands that can easily accelerate the workflow. The API provides references to all objects in the scene, as well as functions for performing operations on those objects. This can be used to perform the same operation on multiple objects in the scene at once, instead of repeating the operation manually for each object. Python scripting in Blender can also be used for generating and modifying meshes, adding textures and animations, implementing new tools and operations, as well as automating tasks involved in other steps of the model creation pipeline, such as rendering or compositing.

Add-ons in Blender are packaged pieces of Python code that can be loaded into Blender in order to seamlessly include additional functionality. Of course, while users can achieve the same functionality by executing the script directly from the integrated console, add-ons provide the convenience of having the functions easily accessible from the UI menus, which greatly increases usability. Also, they can easily be distributed and shared among users, which facilitates collaboration.

B. Planning and design decisions

The model guitar strings should be animated, which refers to animating string vibrations when a string is plucked, and bending the string towards a fret depending on where it is pressed. It is immediately clear how difficult it would be to create these animations by hand because of the complexity of realistic string vibration and the number of frets on the guitar, so the only viable alternative would be to automate these tasks. This will be done by implementing a script which (1) generates a string based on given parameters such as length and diameter, (2) creates shape keys[1] which are used for bending the string,

---

[1]Shape keys are used to deform the mesh into a new shape without the need for an armature. A shape key saves vertex positions for a deformed mesh and allows interpolation between the initial vertex positions and the deformation.

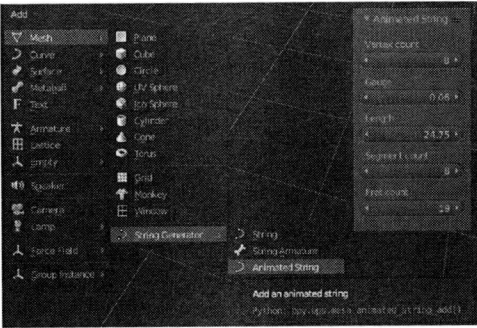

Fig. 2. The custom operation for generating animated strings incorporated into the Blender's *Add mesh* menu. Clicking the menu item calls the custom *animated_string_add* function which generates and animates a string with the given parameters.

and (3) animates the string vibration by calculating vertex positions for each keyframe. The script will be packaged as a Blender add-on.

The fretboard for the guitar can also be generated using a script because exact fret positions depend on string length and are determined using a mathematical formula. The same formula is already used for calculating the string bending shape keys, so a part of the code can be reused.

C. Generating the string mesh

The *StringGenerator* add-on provides a new tool for generating models of strings for musical instruments. Each generated string consists of two components: the string mesh that defines the visible geometry of the string, and the armature which is used for animating string vibration.

As shown in Fig. 2, the tool allows users to modify several properties of the generated string: length, gauge, fret count, vertex count and segment count. String length and gauge are both given in inches, and fret count is used for generating the shape keys required for bending the string. If the fret count is set to zero, only string vibration is animated and no shape keys are created. Vertex count determines the number of vertices in the cross-section of the string and segment count determines the number of segments along the length of the string. These two properties are used to control the complexity of the model. Higher values give better quality and smoother animation which is suited for high-detail offline rendering, but cause slower rendering times due to the increased polygon count. On the other hand, lower values give fast rendering times required for real-time execution, at the expense of reduced quality. However, the reduction in mesh quality is not very noticable, even when the model is being viewed from close distance, because the string shape itself is very simple and proper shading can provide the necessary detail required for a more realistic appearance.

The mesh for the string is generated in three phases. The first phase creates string segments based on the given parameters. Each segment is a closed loop of vertices connected to the neighboring segments by 4-sided faces, forming a cylinder. The segment count and string length properties are used to determine the distance between neighboring segments. The second phase is used to create vertex groups for each of the segments. Vertex groups are simple collections of vertex indices that will be used to easily map vertices to their corresponding bones in the armature.

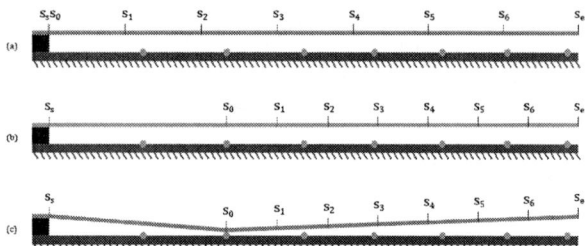

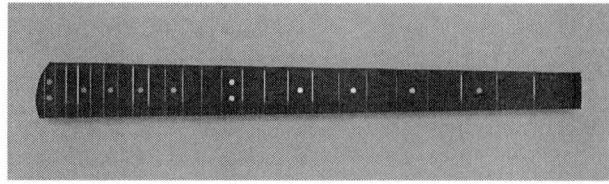

Fig. 4. The finished fretboard for the model. A script was used to calculate the fret positions and to scale the frets to match the width of the neck.

Fig. 3. Repositioning the segments s_i using shape keys. Figure (a) shows the initial setup where the segments are equally distributed along the length of the string. When a fret needs to be pressed, the segments are translated along the string so that s_0 lines up with the fret (b), and when the string is pressed the segments are translated towards the fretboard (c).

The third phase of mesh creation is creating the shape keys. The goal is to have a robust system for animating string bending on each fret. In order to achieve this goal, shape keys are used to reposition the segments based on the fret at which the string is bent. The idea is illustrated in Fig. 3. Segments s_s and s_e indicate the fixed endpoints of the string and will not be affected by the shape keys. The other segments will be used for animating the string bending and vibration and therefore need to be repositioned for each fret. A shape key for each fret translates the segments along the length of the string so that the initial segment s_0 lines up with the corresponding fret. All other segments are equally distributed between s_0 and s_e. Bending the string towards a fret is controlled by one additional shape key which can be used in combination with any of the other shape keys. It bends the string by placing s_0 directly on top of the fret and distributing the other segments to form a straight line from s_0 to s_e.

D. Determining fret positions

A guitar string produces a note of a certain frequency depending on the length of the string and the string tension. Tension is used to set the root note of a string, which is the note produced by playing an open string without pressing down on any fret. This means that the only way to play a different note on the same string is to shorten the string. The chromatic scale consists of 12 notes, so the string pressed down on the 12th fret should produce the note that is an octave higher than the root, which means that the string should be shortened to half of its original length. Similar logic applies to the positions of other frets. The exact position for the fret i is given by

$$d_i = l - \frac{l}{2^{i/12}} \tag{1}$$

where d_i is the distance from the nut to the fret and l is the length of the string.

This calculation can also be used to generate the fret models. The script uses a manually modeled fret as template for generating the entire fretboard. It calculates the position for each fret in the fretboard, duplicates the template and places it in the calculated position. An additional consideration when generating frets is the neck width. In order to compensate for the wider frets towards the head of the guitar, the neck becomes narrower as it goes from the guitar body towards the head in order to improve playability. This means that the frets also need to be scaled to the width of the neck at the calculated position. The fretboard with the generated frets is shown in Fig. 4.

E. Animating the string

The second component created by the add-on is the armature used for animating the string. The armature consists of a set of bones, each of which is used for controlling a group of vertices. A bone is created for each segment of the mesh, and the vertex group containing the vertices of each segment is attached to the corresponding bone.

String vibration is caused by a stationary wave traversing the string, which produces a sound based on the frequency of the wave. In order to simulate a realistic vibration, the wave needs to be broken down into its base components, which are called harmonics. Each harmonic of the wave can be represented by

$$y = A \sin(\omega x) \tag{2}$$

where A is the amplitude of the wave, ω is the radial frequency and x is a longitudinal position on the string. String vibration for a single harmonic can be simulated by varying the amplitude over time, using a sine function to control the oscillation

$$y(t) = A \sin(k\pi t) \sin(\omega x) \tag{3}$$

where k determines the oscillation frequency. The harmonics can then be summed up to get a more realistic result. Lastly, dampening needs to be added so that the vibration fades over time. This is done using the δ factor. The resulting equation is given by

$$Y(t) = \sum_{i=1}^{N} y_i(t) = A \sum_{i=1}^{N} \sin(k_i \pi t) \sin(\omega_i x) \tag{4}$$

$$Y_d(t) = Y(t) e^{-t/\delta} \tag{5}$$

Frame rate is set to 60 frames per second to allow a more detailed simulation. Two harmonics are used to calculate the positions for each bone over 5 seconds (300 frames). Higher degree harmonics are not used for this simulation because the frame rate would need to be much higher for them to make a significant impact on the result. Also, for simplicity, string vibration is only animated along a single axis, which produces very good results despite not being entirely realistic[2]. The final position of each bone at the time t is given by

$$y(x,t) = A(\sin\frac{t\pi}{4}\sin\frac{x}{2} + \frac{1}{4}\sin\frac{t\pi}{2}\sin x)e^{-t/\delta} \tag{6}$$

[2]A musitian who plays the guitar would never pick the string perfectly vertically, so the string would also receive a horizontal vibration component.

The second harmonic was multiplied by the factor 1/4 in order to reduce its maximum amplitude, resulting in a more believable animation.

Two animations are created, one for the downstroke and one for the upstroke when picking the string, and both are added as animation sequences to the armature. One animation is given directly by the equation (6) and the other contains the opposite movement, so the equation result is merely negated. These animations can be used in combination with the shape keys do simulate vibration when the string is pressed to a fret. This is enabled by the fact that bones only control the vertical movement of segments and shape keys translate the segments along the length of the string.

F. Texturing the strings

Each of the strings was UV-unwrapped and assigned a simple striped texture in order to mimic the look of wound[3] strings. While the texture contributes to the realistic appearance of the string, an issue arises when using the shape keys to simulate bending of the textured string. Because the base mesh of the string was UV-unwrapped, each vertex was assigned a fixed (u, v) coordinate of the texture. When the mesh is deformed using the shape keys, the vertices move in 3D space, but not in texture space. This causes stretching in the texture, which is of course unrealistic.

This issue can be solved by splitting the mesh material into three slots and assigning the same texture to each of these slots. One slot would contain all static vertices which form the string endings that were added manually. The second slot would contain the the cylinder made by the first two segments of the string (s_s and s_0 in Fig. 3), and the final slot would contain all faces of the vibrating section of the string (s_0 to s_e in Fig. 3). Splitting the material in this manner allows the texture to be scaled individually for each of those sections, in turn enabling the use of a texture even when the string mesh is deformed by shape keys. When a shape key is used to press a fret, the texture of the second material would be scaled down to increase the level of detal, while the third material would have its texture scaled up. This would preserve the overall textured appearance of the string mesh and eliminate undesirable stretching.

V. RESULTS

Fig. 5 shows the finished model of a 5-string bass guitar. The strings were generated using the *StringGenerator* add-on, and the frets were placed using the script described in the previous section. The remaining components of the model were mostly created manually in Blender.

The parameter values used for generating the strings were based on actual values for common bass strings. For this model, the string diameters were set to values ranging from .040 to .130 inches, and the string length was set to 34 inches. Since the script only generates the vibrating portion of the string, the missing geometry on both ends – the start of the string at the bridge of the guitar, and the string ending which wraps around the tuning post – were modeled manually for each string and appended to the generated models. This had no

---

[3]Wound strings consist of a round wire wrapped in a tight spiral around either a round or hexagonal core. This type of construction allows the string to produce a much lower pitch than regular plain strings, which allows them to be much thinner and easier to play.

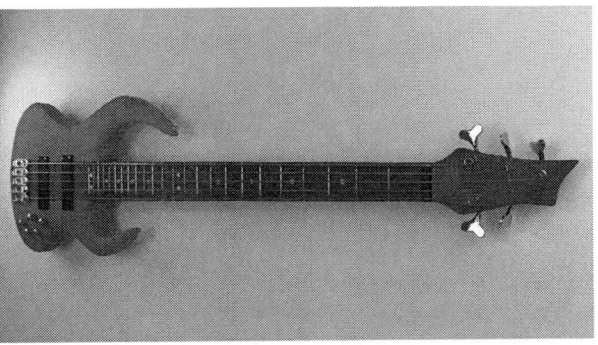

Fig. 5. The finished bass guitar model.

effect on the animation data because the generated geometry had not been modified.

The fretboard was generated using a manually modeled fret template, which was duplicated and placed using the previously described script. The string length was used to determine the exact fret positions, and each fret was scaled vertically to match the neck width at the calculated position.

VI. FUTURE WORK

As stated in Section III, the main motivation behind making this model was building self-playing virtual musical instruments. The idea is to create a 3D model of an instrument, add specific mechanisms that would be used to visualize playing the instrument, such as picking and fretting fingers for string instruments, and finally implement the required functionality for playing a given set of notes.

The process described in this paper can easily be used to create various models of string instruments, but in order for those instruments to become 'self-playing', they need to be programmed. This task can be accomplished using one of the popular game development engines such as *Unreal Engine* [21] or *Unity* [20]. These powerful tools are primarily designed for creating 3D and 2D games, but they can also be used for developing quite complex, realistic, and interactive visualizations.

As a way to feed note data into the self-playing instrument visualization, one could define a custom data stream that would be interpreted and used to play the required animations corresponding to each given note. This method would allow fast playback and high flexibility for defining the exact way each note should be played. However, building an entire song in that manner would be quite time-consuming, mainly because one would need to manually define how to play each note. For a string instrument, this would mean defining which left-hand finger presses which string on which fret, and which right-hand finger picks the string. An alternative method would be to use a common data source like a MIDI file to provide the note data, and then perform an optimization task over the entire song in order to determine the optimal fingering and picking sequences to be used during playback.

Fig. 6 shows the bass model imported into Unreal Engine. The right-hand picking mechanism consists of five mechanical fingers whose behavior has been programmed using a state machine. The current state of each finger is used to determine which finger will be used to pick the next note in the sequence, with the aim to reduce the cumulative movement of all fingers. The left-hand fretting mechanism could have been modeled as a human hand [4], but the note positioning optimization

Fig. 6. The bass model with picking and fretting mechanisms imported into Unreal Engine. The strings are colored red and blue in order to help visualize the material change when a string is bent to a fret using a shape key.

algorithm would have to be more complex and require many more contraints related to physical limitations of a human hand. That is why the left hand was implemented as four sliding fingers, each of which has five moving pins – one for each string. The fingers slide across the fretboard and their pins are used to press the strings, bending them at the fret where the finger is currently positioned.

VII. CONCLUSION

The introduction of this paper mentioned several interesting techniques that make use of scripts to generate virtual objects. It was shown how procedural techniques can be good for generating complex, organic and natural-looking geometry, which is not easy to achieve by manual modeling. These techiques served as inspiration for using scripts to generate parametrized 3D models of guitar strings. A generated string includes shape keys used for deforming the mesh when bending the string, as well as an armature with vibration animations. Using a script for this task provides a fast and repeatable method which can be used for modeling various musical string instruments.

It was shown how this method can be used to easily create strings for a 3D model of a bass guitar, and an idea has been presented on how that model can be used to implement a self-playing instrument visualization. These types of visualizations can be quite useful for educational purposes, especially if the fretting and picking mechanisms can be modeled as actual human hands performing the movements. This would open doors for further investigation regarding visualization of optimized fretting sequences, and automatic creation of human-playable tablature from music sheets or structured audio formats such as MIDI files.

REFERENCES

[1] D. Baraff, A. Witkin, *Large steps in cloth simulation*, ACM SIGGRAPH, 1998.

[2] R. B. Dannenberg, B. Brown, G. Zeglin, R. Lupish, *McBlare: A Robotic Bagpipe Player*, Proceedings of the International Conference on New Interfaces for Musical Expression, 2005.

[3] D.-S. Elbert, F.-K. Musgrave, D. Peachey, K. Perlin, S. Worley, *Texturing & Modeling: A Procedural Approach*, 3rd ed., Morgan Kaufmann Publishers, 2002.

[4] G. ElKoura, K. Singh, *Handrix: Animating the Human Hand*, Eurographics/SIGGRAPH Symposium on Computer Animation, 2003.

[5] J. Kniss, S. Premoze, C. Hansen, D. Ebert, *Interactive Translucent Volume Rendering and Procedural Modeling*, IEEE Visualization, 2002.

[6] A. Lindenmayer, *Mathematical Models for Cellular Interaction in Development*, Journal of Theoretical Biology, 1968.

[7] B. B. Mandelbrot, *The Fractal Geometry of Nature*, W. H. Freeman, 1982.

[8] R. Mech, P. Prusinkiewich, *Visual Models of Plants Interacting with Their Environment*, ACM SIGGRAPH, 1996.

[9] P. Merrell, *Example-Based Model Synthesis*, Symposium on Interactive 3D Graphics (i3D), 2007.

[10] P. Merrell, D. Manocha, *Model Synthesis: A General Procedural Modeling Algorithm*, IEEE Transactions on Visualization and Computer Graphics, 2010.

[11] F. K. Musgrave, C. E. Kolb, R. S. Mace, *The Synthesis and Rendering of Eroded Fractal Terrains*, ACM SIGGRAPH, 1989.

[12] M. Müller, D. Charypar, M. Gross, *Particle-Based Fluid Simulation For Interactive Applications*, ACM SIGGRAPH 2003.

[13] P. Müller, P. Wonka, S. Haegler, A. Ulmer, L. Van Gool, *Procedural Modeling of Buildings*, ACM SIGGRAPH, 2006.

[14] Y. Parish, P. Müller, *Procedural Modeling of Cities*, ACM SIGGRAPH, 2001.

[15] F. A. Saunders, *The Mechanical Action of Violins*, Journal of Acoustic Society of America, 1937.

[16] E. Singer, K. Larke, D. Bianciardi, *LEMUR GuitarBot: MIDI Robotic String Instrument*, Proceedings of the International Conference on New Interfaces for Musical Expression, 2003.

[17] P. Volino, N. Magnenat Thalmann, *Implementing Fast Cloth Simulation With Collision Response*, Computer Graphics International, 2000.

[18] H. Zhou, J. Sun, G. Turk, J. Rehg, *Terrain Synthesis from Digital Elevation Models*, IEEE Transactions on Visualization and Computer Graphics, 2007.

[19] Blender Foundation, *API documentation*, built October 2nd, 2016, https://www.blender.org/apiblender_python_api_2_78_release, February 12th, 2017.

[20] Unity, *About the Unity Editor*, https://unity3d.com/unity/editor, February 12th, 2017.

[21] Epic Games, *What is Unreal Engine 4*, https://www.unrealengine.com/what-is-unreal-engine-4, February 12th, 2017.

Remote Interactive Visualization for Particle-based Simulations on Graphics Clusters

Adrian Sabou, Dorian Gorgan
Computer Science Department
Technical University of Cluj-Napoca
Str. G. Baritiu 28, 400027, Cluj-Napoca, Romania
Email: {adrian.sabou, dorian.gorgan}@cs.utcluj.ro

Abstract—Particle-based models are widely spread in the field of Computer Graphics, and mainly used for real-time simulations of soft deformable bodies. However, simulations including high-resolution models have a great computational cost and, when adding the need for real-time rendering and interaction, they fall way outside the range of applications that traditional computing architectures can accommodate. Graphics clusters can offer the raw computing power needed for such simulations but, due to their physical design and operating mode, introduce a series of challenges that must be overcome, such as efficient distributed rendering and remote visualization and interaction with the simulated scenes. This paper presents a solution to interactive visual particle-based simulations on graphics clusters using an optimized in-situ distributed rendering approach which, coupled with state-of-the-art remote visualization and interaction techniques and tools, provide efficient means for highly scalable interactive simulations.

I. INTRODUCTION

Modeling and simulating three-dimensional dynamic surfaces represents one of the main research areas of Computer Graphics. The most prominent techniques for such simulations are physically-based methods like particle-based modeling, where surfaces are approximated through sets of discrete points having various physical properties such as mass, volume, speed, acceleration, and the behavior of such surfaces along with their interaction with the environment is governed by the laws of physics, more specifically by the forces that act upon particles. Particle-based modeling is a natural choice since, in the real world, interactions concerning deformable surfaces occur at a molecular level and, in theory, given a sufficient number of particles, any such surface can be accurately modeled.

High performance computing (HPC) architectures such as graphics clusters can offer the raw power required for this task but, in the context of interactive visual simulations, new and innovative techniques are required to allow such applications to run on multicore distributed systems. Due to physical design and operating mode, both shared memory and distributed memory paradigms apply when designing simulations that will run on GPU clusters, thus introducing challenges mainly at application development level. Adding the real-time attribute that such simulations usually require, along with the need to transform raw processed data into a form visually meaningful to the users, we end up with probably one of the most complex class of applications attempted to be implemented on such architectures.

This paper addresses two main challenges when dealing with parallel, distributed simulations, namely distributed rendering and remote visualization and interaction. To obtain performant centralized visualization, we propose an optimized in-situ parallel rendering technique. Based on a sort-last approach to parallel rendering and coupled with a Region-of-Interest algorithm, this technique greatly improves simulation performances when visualization of the entire scene and model is required on a single display device. Remote visualization and interaction is ensured by using a solution based on the VNC protocol that offers great performances for running interactive remote graphics applications even when lacking physical display devices as is the case of most GPU clusters. The rest of the paper is organized as follows: Section II highlights important related work. Section III illustrates the proposed distributed rendering and remote interaction solutions in the context of previously developed parallelization techniques for particle-based simulations on graphics clusters. We report on performance measurements in section IV.

II. RELATED WORKS

Molnar et al. [1] identified three broad classes of parallel rendering methods, based on where the sort from object-space to screen space occurs. *Sort-first* methods aim to distribute primitives early in the rendering pipeline, during geometry processing, to individual graphics processors which will do the remaining rendering calculations. In *sort-middle*, primitives are redistributed in the middle of the rendering pipeline, between geometry processing and rasterization. *Sort-last* defers sorting until the end of the rendering pipeline, after primitives have been rasterized into pixels, samples, or pixel fragments.

A great number of general purpose parallel rendering concepts and optimizations have been introduced in existing research literature, such as parallel rendering architectures, parallel compositing, load balancing, data distribution, or scalability. However, only a few generic APIs and parallel rendering systems exist [2].

VR Juggler [3] is a virtual platform for the creation and execution of immersive applications that provides a virtual reality system-independent operating environment. It allows a user to run an application on almost any VR system. VR

Juggler is scalable from simple desktop systems like PCs to complex multi-screen systems running on high-end work stations and super computers. Chromium [4] is a system for interactive rendering on clusters of workstations. It is a completely extensible architecture, so that parallel rendering algorithms can be implemented on clusters with ease. It intercepts the OpenGL calls and processes them, typically to send them to multiple rendering units driving a display wall. Equalizer [5] is an open source rendering framework and resource management system for multipipe applications. Equalizer provides an API to write parallel, scalable visualization applications which are configured at run-time by a resource server. OpenSG [6] is an open source scene graph system that provides parallel rendering capabilities, especially on clusters. It hides the complexity of parallel multi-threaded and clustered applications and supports sort-first as well as sort-last rendering.

All the generic APIs enumerated above offer robust and efficient parallel rendering capabilities. However, no parallel rendering APIs offer integration with GPGPU programming through high-level languages such as CUDA or OpenCL, though the Equalizer project has this as one of its main research directions. The same API is overall preferred by Eilemann et al. [2] following their analysis of the asynchronous parallelization of the rendering stages due to its scalability and configuration flexibility.

III. PARTICLE-BASED SIMULATIONS ON GRAPHICS CLUSTERS

A. Parallel, Distributed Simulation of Particle-based Models

Some of our previous papers discuss in detail the parallel techniques developed for accelerating particle-based simulations on graphics clusters. They cover the most important issues for parallel particle-based simulations such as model decomposition and distribution at GPU level and at graphics cluster level, hybrid CPU/GPU parallelism and parallel numerical integration. For model partitioning we decided on a static domain decomposition method that ensures minimal length frontiers and thus minimizes network traffic required for synchronization [7]. We also proposed a technique that ensures minimal modification to the kernels used for the single machine approach by keeping an extended model on each processing node. CPU/GPU parallelism was employed in order to minimize idle times for the CPU while waiting for GPU tasks to complete [8]. By employing a technique which decouples several steps of the simulation process, we were able to run the simulation in an out-of-phase manner and carry out CPU-based synchronization tasks while the GPU executed the complex computation tasks. Parallel explicit numerical integration is achieved using a two-pass data parallel approach, while, for implicit integration, we rely on a parallel version of the Conjugate Gradient algorithm. To further accelerate computation and reduce memory requirements, we developed an efficient technique to update the large sparse matrices involved in implicit integration directly into the Compressed Sparse Row storage format, by exploiting their regular structure [9].

B. Distributed Rendering

1) *Parallel Particle-based Simulation as a Distributed Rendering Problem:* Even though parallel particle-based simulation on graphics clusters is not entirely a parallel rendering problem, it greatly resembles one if we consider the analogy between the great volumes of data involved in typical parallel rendering and the high demanding real-time computation involved in simulations. Moreover, the rendering process strictly falls into the category of problems aforementioned, thus the concepts of sort-first and sort-last also apply in our case, although the type of primitive sorting that can be employed strictly depends on data locality for the computation process. The challenge is to apply these distributed rendering concepts to our problem, which combines traditional rendering with GPGPU computation.

In terms of parallel rendering strategies applied to the problem of parallel particle-based simulations, a pure sort-first approach would imply the decomposition of the visualization task into several sub-tasks, each responsible for a subset of pixels of the final image. Particles would have to be sorted based on their position on screen and redistributed, such that each rendering processor receives all particles that fall in their respective portion of the final image. Each processor would then perform the remaining transformation and rasterization steps for all its particles and send the finished pixels to be displayed at a central visualization station.

However, due to the GPGPU/GPU rendering duality and the interactivity and dynamics of such simulations, pure sort-first approaches are highly inefficient due to the necessity of resorting and redistributing the particles between processing nodes each time particles cross the borders of different pixel areas. Moreover, efficient load balancing would be very difficult to achieve for both the computing step and the rendering step, since balancing one would imply a loss of balance in the other. For rendering, the screen has to be split differently every frame as the model evolves in order to keep the number of primitives (particles) in each screen portion equal. This would totally unbalance the computation step, for which particles would have to be redistributed as to ensure equally weighted tasks between nodes, most probably leading to a deviation from minimal-border model original model partitioning. Moreover, this kind of balancing would generate a huge amount of supplementary network traffic, further slowing the simulation. Thus we turn our attention towards the sort-last technique.

A sort-last approach would imply the decomposition of the input data into a collection of smaller components, each being processed by a different rendering node. Since our domain decomposition data parallel approach for particle-based simulation splits the computation into equally weighted parts between processing nodes, such an approach seems to offer the best agreement between simulation and rendering components with regard to load balancing. Moreover, a sort-last approach largely benefits from the rendering capabilities of processing nodes since each one implements a full rendering pipeline for their subset of primitives, the central node being responsible for final image composition, based on z coordinate information. The network traffic generated by this pure sort-last approach does not directly depend on the resolution used for the cloth model, but it does depend on the resolution of the final raster image, since pixel information must be communicated through the network.

2) An Optimized Sort-last Approach to Parallel Rendering for Particle-based Simulation: The intrinsic properties of parallel particle-based simulation allow us a certain degree of optimization when considering the parallel rendering process. Distributed rendering only becomes an issue when the number of primitives to be rendered is overwhelming for a single GPU. Usually, when simulating particle-based models, the complexity of the rendered scene is concentrated in the model itself, other objects composing the final photorealistic representation of reality posing less difficulty to a single graphics processor. This allows us to propose optimizations focused on the particularities of the particle model and not on the entire scene.

The domain partitioning scheme for particle models that was described in [7] exhibits several interesting particularities that influence the rendering process. For once, in order to minimize the number of *ghost points*, model partitions are chosen as to represent continuous disjoint parts of the whole model. This continuity ensures that, after being rendered, each partition can be enclosed in a single rectangular two-dimensional bounding box. Thus, without loss of consistency, we can apply a *region-of-interest* (ROI) approach to the sort-last rendering algorithm, since, each new locally rendered frame will generate relevant pixel data only for the small area enclosed by the bounding box. The ROI algorithm tailored for parallel particle-based simulation splits each local frame buffer into parts with potentially active pixels and excludes blank areas, with maximum benefit obtained, since we have already stated that each node renders to a compact region in the frame buffer. It is called after local rendering has been finalized, right before pixel read-back by the central visualization server.

ROI Identification: While sort-last does not communicate primitive data between processing nodes and visualization server after the initial primitive distribution and does not depend on model size, ROI identification does. Therefore, in order for such an approach to be efficient, the computational cost for determining the ROI must be smaller than the communication cost to send full pixel data through the interconnecting network.

To determine the coordinates, we must pass the points representing the position of each particle through all geometric transformations along the graphics pipeline, namely the model, view, projection and viewport transformations. Since this approach actually depends on model resolution, it combines concepts from both sort-first and sort-last rendering, thus it is expected that experimental performance evaluation will depend on both the number of particles and the number of nodes used. A method for determining the optimal number of nodes for a certain model size may also be derived from empirical data, providing that there are a sufficient number of

---

Algorithm III-B.1 Region-of-interest selection

---

initialize ROI {Start from an initial estimation that the ROI is *null*}
for each particle in the model partition (including nearest neighboring *ghost points*) **do**
 project particle on screen {multiply the particle's position with matrices in all matrix stacks (model-view, projection) and apply viewport transformations to transform its position from object space to screen space}
 if current particle's projection is outside the ROI **then**
  expand the ROI {minimally expand the rectangular region that will be the ROI as to contain the current particle's projection}
 end if
end for

---

nodes available in the graphics cluster used for testing.

Algorithm III-B.1 shows the steps required to identify the region-of-interest for a processing node rendering to its frame buffer. Since a particle model is fully enclosed by particles at the extremities, than it is safe to assume that no part of the model will be rendered outside the area delimited by these particles. Unfortunately, due to the fact that particles usually approximate soft, deformable surfaces and rendering depends on viewing parameters, it nearly impossible to estimate what particles will be placed at extremities without iterating through the entire set. Therefore, for regions-of-interest close in size to the whole window, this process will introduce significant overhead when comparing with just sending the entire frame buffer and thus should be avoided. The decision to use or to turn off ROI can be taken dynamically following successive ROI selections comparable to the size of the rendering window. ROI can be reattempted following a predefined delay. The maximum size for which the ROI algorithm offers performance improvement depends on several factors such as network transmission speed and can be determined either through trial and error or empirically. An example of two locally rendered frames and their composition, along with each ROI can be seen in figure 1.

Depth sorting: Sort-last sorting first renders frames locally on each individual rendering nodes and collects local frames on a central visualization server for final frame composition. Frame composition is achieved on the visualization server following a depth sorting process. As can be seen in figure 1, this algorithm ensures that polygons that should be hidden will not be rendered in front of visible polygons, exactly like the hidden faces removal algorithm would work if rendering would be done on a single node.

Further increase in performance can be obtained when considering the particularities of numerical integration techniques applied to particle-based models. Explicit integration such as the popular Verlet [10] technique is conditionally stable, which implies that the differential system of equation diverges for large time steps. Similarly, even with the unconditional stability attribute, very large time steps used in implicit nu-

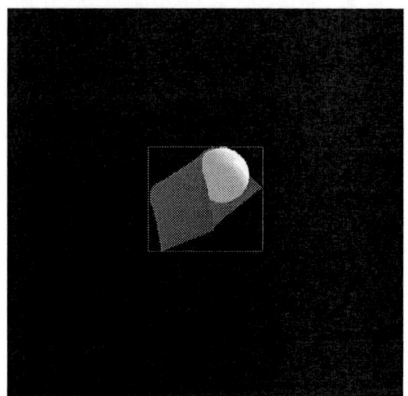 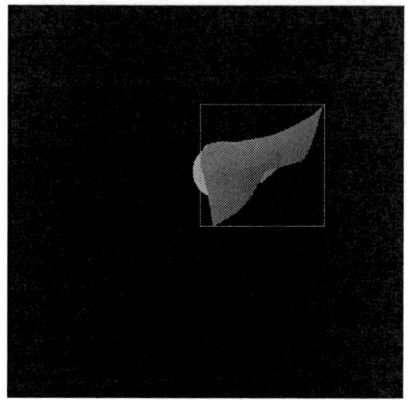

(a) ROI from processing node 1   (b) ROI from processing node 2   (c) ROI overlapping on composite frame

Fig. 1. Sort-last region-of-interest identification

meric integration can result in unwanted visual artifacts when the modeled system exhibits a very high level of dynamism (e.g. very frequent collisions with other objects). This in turn allows us to assume that a dynamic particle-based model which is simulated in real-time (universally accepted as 30 FPS or more) will change its state relatively little between two consecutive frames, thus making it possible to skip the communication of pixel data every two frames when the central visualization server is also a processing node. This assertion is based on the assumption that, at 30+ FPS, the small change in positions on the central node can visually account for the change on other processing nodes without altering the perception of animation fluidity for the entire model. One may take this optimization technique even further and skip pixel synchronization more often when the frame rate permits it.

C. Remote Visualization and Interaction

Remote visualization is one key aspect of designing powerful graphic cluster based applications. Efficient transportation of rendered data or of visual images has been long researched in order to allow users to visualize and interact with a remote scene. The basic idea behind remote visualization is the same one regardless of its implementation: data must be sent over the network from the GPU on the server to the client GPU and finally to the display device of the user.

Different task require different types of remote visualization. Lietsch et al. [11] propose a classification based on three classes for grouping the most common existing systems. Based on the type of remote visualization required for a specific task, this classification identifies the *client-side rendering* class, containing systems where applications run on the server side but graphic objects are decomposed into primitives (polygons meshes, textures, volumes) and sent to the client to be rendered, the *server-side rendering* for 2D and administration class containing systems mainly used for server administration and remote control and the *server side rendering* for 3D applications class containing systems dedicated to displaying

visual results and ensuring interactivity with remotely executed 3D applications (e.g. OpenGL applications).

Particle-based simulations are highly-interactive three-dimensional applications, thus in in order to ensure visualization remotely we must aim towards server side rendering. One of the most popular open source programs that redirects 3D rendering commands from Unix and Linux OpenGL applications to 3D accelerator hardware in a dedicated server and displays the rendered output interactively to a thin client located elsewhere on the network is VirtualGL [12]. It was designed to overcome the main problems that graphical desktop sharing system such as the VNC poses, namely that they either do not support running OpenGL applications at all or force the OpenGL applications to be rendered without the benefit of OpenGL hardware acceleration. VirtualGL uses TurboVNC [12] as an $XProxy$ for image transport that transmits the keyboard and mouse events from one computer to another, relaying the graphical screen updates back in the other direction, over a network.

VirtualGL and TurboVNC are most commonly used both running on a single machine. While it is possible to run them on different machines, this would move us away from our goal to minimize network transfer. However, applying this visualization solution to a GPU cluster can prove a little more difficult than applying it to a single machine, due to the distributed nature of the rendering process and the lack of physical displays in general purpose GPU clusters. While it may seem natural to have a single VirtualGL server, running on the central visualization node, the sort-last approach to distributed rendering, that has been identified as the best solution for our method for parallel particle-based simulation, assumes that each processing node renders locally its partition of the model. Thus the local rendering process requires a display in order to be able to negotiate a window with the underlying operating system. This is where VirtualGL/TurboVNC solution works to our advantage. By providing an $XProxy$ that renders in a virtual framebuffer, each node can request and obtain a virtual window from VirtualGL, thus bypassing the presence

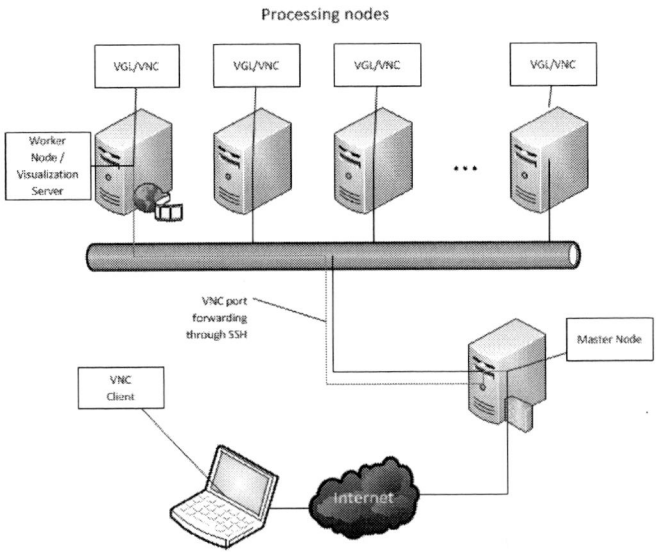

Fig. 2. Using VirtualGL/TurboVNC for remote visualization and interaction

of a physical display dependency that a true X11 server must satisfy.

In order to be able to render locally to virtual windows, a VirtualGL server must be running on each processing node, not just on the visualization server. The mapping of such a remote visualization and interaction solution on a GPU cluster can be seen in figure 2. Each processing node runs a VirtualGL/TurboVNC server and renders locally to a virtual window. The visualization server (dedicated server or one of the processing nodes) composes the final scene using pixel information received from the other nodes and relays the visual information directly to the VNC client through SSH port forwarding of VNC ports. Forwarding the VNC port also allows direct communication of user input to the visualization

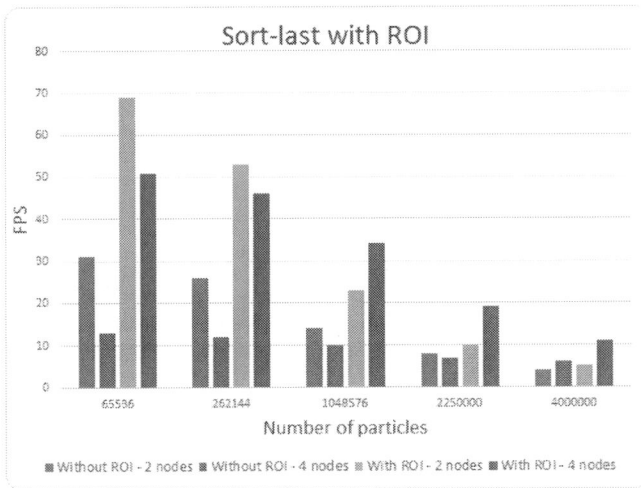

Fig. 3. Computed frame rates for sort-last with ROI compared to full frame merging

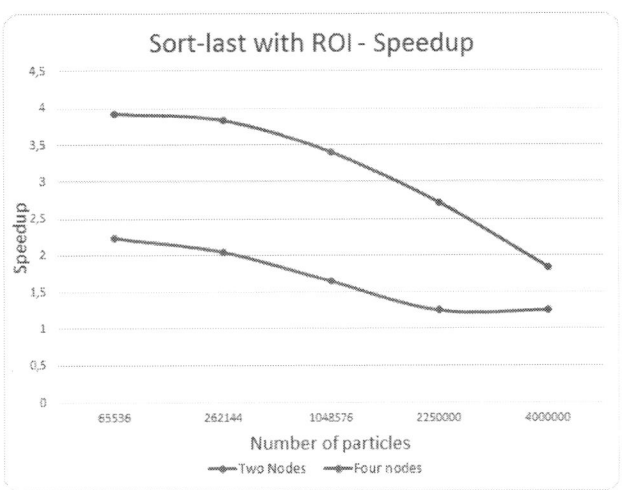

Fig. 4. Computed speedup for sort-last with ROI compared to full frame merging

server, which encodes it and transmits it to all other nodes.

IV. EXPERIMENTS

The experiments were carried out on a graphics cluster consisting of four nodes with either NVIDIA GeForce GTS 250 or NVIDIA GeForce 8800 GTS GPUs and Intel Core 2 Duo E8400 CPUs running on Debian 6.0.5 Squeeze, interconnected by a Gigabit network. The scenarios devised were aimed at evaluating the performance of the proposed solution taking into account factors such as the resolution of the model (i.e. number of particles) and the number of nodes used for the simulation. They try to evaluate the increase in performance obtained by using the sort-last rendering approach, with partial instead of full frame merging (i.e. the performance improvement obtained with applying a Region-of-interest algorithm).

Figure 3 shows the measured frame rates for both the original and the optimized method, with regard to model resolution (i.e. number of particles). The results prove that it is more efficient to obtain centralized visualization by using only those areas of each processing node's frame buffer which hold pixel data belonging to its own cloth patch than to have full frame buffer merging, despite the considerable computational overhead introduced by the need to continually recompute the coordinates of all those rectangular regions (i.e. ROI identification). Experiments have also shown that, for explicit numeric integration, which allows for small simulation steps, it is possible to skip this recomputing process for up to four frames without noticeable visual consequences.

Figure 4 shows the computed speedup for the optimized ROI method when comparing to full-frame composition. As can be seen, the performance improvement is at its peak with smaller models and continually decreases with the model size. This is, however, to be expected, since, the ROI algorithm depends on model size. Providing that sufficient nodes are available for testing, a method for specifying model sizes

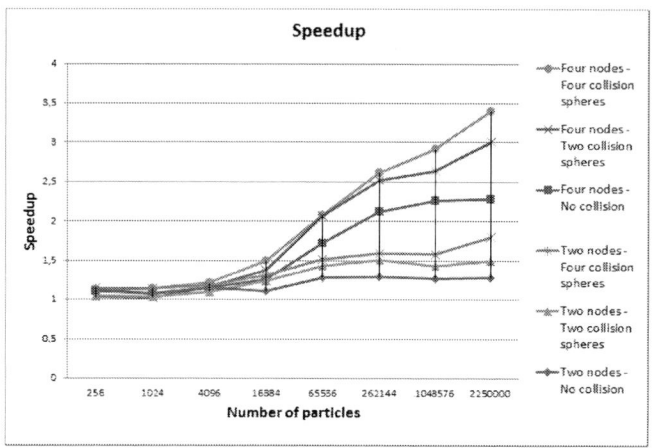

Fig. 5. Computed speedup for test cases including no centralized rendering [7]

beyond which the ROI algorithm performs poorer than the full-frame composition method may be derived empirically.

In order to further evaluate the performance of the proposed solution, we compared results with previously conducted experiments [7], which use no centralized rendering of the entire model, as to observe the impact that this operation has on the overall simulation.

Figure 5 illustrates the speedups obtained for a cloth draping scenario with different model resolutions when using no centralized rendering, while figure 6 illustrates the speedups obtained for the same draping scenario (cloth colliding with four spheres) with centralized rendering. When each node renders its own piece of the model on its display device, either physical or virtual, without visual composing it on a single, central display device, the speedups obtained top 1.7 and 3.4 on two and four nodes respectively. After centralized rendering is implemented, the speedups top 1.5 and 2.7 on two and four nodes respectively. Since the computational effort to centralize the visual data on a single display device is a considerable one,

the relatively small difference in values for speedups and the consistency of the speedup curve (for four nodes) proves that the proposed solution is scalable and robust.

V. CONCLUSION

In this paper we presented a solution to interactive visual particle-based simulations on graphics clusters using an optimized in-situ distributed rendering approach which, coupled with a Region-of-interest algorithm for final frame composition provides a scalable rendering technique for distributed visual simulations. We also integrated state-of-the-art remote visualization and interaction techniques and tools in order to be able to visualize and interact remotely with the simulated scenarios, thus fully utilizing both the computing and the rendering capabilities of the remote HPC architecture.

REFERENCES

[1] S. Molnar, M. Cox, D. Ellsworth, and H. Fuchs, "A sorting classification of parallel rendering," *IEEE Comput. Graph. Appl.*, vol. 14, no. 4, pp. 23–32, Jul. 1994.

[2] S. Eilemann, A. Bilgili, M. Abdellah, J. Hernando, M. Makhinya, R. Pajarola, and F. Schrmann, "Parallel rendering on hybrid multi-gpu clusters." in *EGPGV*, H. Childs, T. Kuhlen, and F. Marton, Eds. Eurographics Association, 2012, pp. 109–117.

[3] A. Bierbaum, C. Just, P. Hartling, K. Meinert, A. Baker, and C. Cruz-Neira, "Vr juggler: A virtual platform for virtual reality application development," in *Proceedings of the Virtual Reality 2001 Conference (VR'01)*, ser. VR '01. Washington, DC, USA: IEEE Computer Society, 2001, pp. 89–.

[4] G. Humphreys, M. Houston, R. Ng, R. Frank, S. Ahern, P. D. Kirchner, and J. T. Klosowski, "Chromium: a stream-processing framework for interactive rendering on clusters," *ACM Trans. Graph.*, vol. 21, no. 3, pp. 693–702, Jul. 2002.

[5] S. Eilemann, M. Makhinya, and R. Pajarola, "Equalizer: A scalable parallel rendering framework," *IEEE Transactions on Visualization and Computer Graphics*, vol. 15, no. 3, pp. 436–452, May 2009.

[6] G. Voß, J. Behr, D. Reiners, and M. Roth, "A multi-thread safe foundation for scene graphs and its extension to clusters," in *Proceedings of the Fourth Eurographics Workshop on Parallel Graphics and Visualization*, ser. EGPGV '02. Aire-la-Ville, Switzerland, Switzerland: Eurographics Association, 2002, pp. 33–37. [Online]. Available: http://dl.acm.org/citation.cfm?id=569673.569679

[7] A. Sabou, C. Mocan, and D. Gorgan, "Particle based modelling and processing of high resolution and large textile surfaces," in *Intelligent Computer Communication and Processing (ICCP), 2012 IEEE International Conference on*, 30 2012-sept. 1 2012, pp. 355 –360.

[8] A. Sabou and D. Gorgan, "Physical simulation of 3d dynamical surfaces on graphics clusters," in *Information Communication Technology Electronics Microelectronics (MIPRO), 2013 36th International Convention on*, 2013, pp. 292–297.

[9] A. Sabou, D. Gorgan, and I. R. Peter, "Parallel implicit time integration for particle-based models on graphics clusters," in *Information and Communication Technology, Electronics and Microelectronics (MIPRO), 2014 37th International Convention on*, May 2014, pp. 336–341.

[10] L. Verlet, "Computer "experiments" on classical fluids. I. Thermodynamical properties of lennard-jones molecules," *Phys. Rev.*, vol. 159, pp. 98–103, Jul 1967. [Online]. Available: http://link.aps.org/doi/10.1103/PhysRev.159.98

[11] S. Lietsch and O. Marquardt, "A cuda-supported approach to remote rendering," in *Proceedings of the 3rd international conference on Advances in visual computing - Volume Part I*, ser. ISVC'07. Berlin, Heidelberg: Springer-Verlag, 2007, pp. 724–733. [Online]. Available: http://dl.acm.org/citation.cfm?id=1779178.1779261

[12] (2014) VirtualGL 3D without boundaries. [Online]. Available: http://www.virtualgl.org/

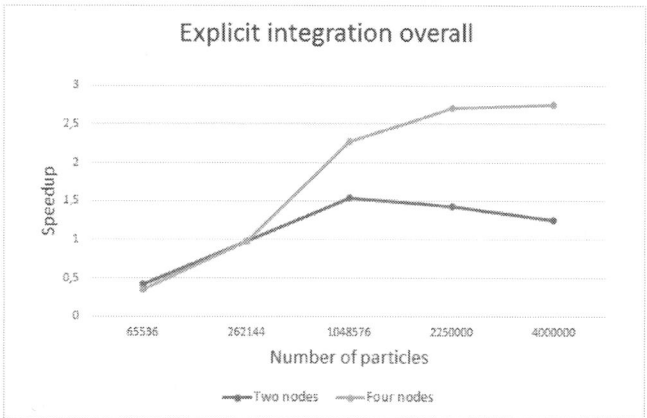

Fig. 6. Computed speedup for parallel explicit integration with centralized rendering

Collaborative view-aligned annotations in web-based 3D medical data visualization

Primož Lavrič, Ciril Bohak, Matija Marolt

University of Ljubljana, Faculty of Computer and Information Science, Ljubljana, Slovenia
pl9506@student.uni-lj.si, ciril.bohak@fri.uni-lj.si, matija.marolt@fri.uni-lj.si

Abstract - the paper presents our web-based 3D medical data visualization framework with emphasis on user collaboration. The framework supports visualization of volumetric data and 3D meshes in web browsers. The paper focuses on integration of user-shareable 3D view-aligned hand drawn or written annotations into the visualization framework. Annotations are created on separate transparent canvases which are aligned with selected views. View parameters are part of annotations and can be shared with other users over the network. Our implementation allows for real-time sharing of annotations during creation. Annotations from the same or different users can be overlaid within the same view. Annotations were implemented through adaptation of the framework's rendering pipeline, which allows for combining multiple visualization layers into a unified final render. View aligned annotations were added in addition to text annotations pinned to 3D locations on the displayed model. In the framework, users can list through all annotations, whereby upon selection of a 3D view-aligned annotation the camera is positioned according to the stored parameters and the annotation is displayed.

I. INTRODUCTION

Visualization of 3D data is an already well established way of supporting work in many different fields, including medicine. In the paper we are focusing on visualization of volumetric data, which can be obtained with techniques such as: Computed Tomography - CT [1, 2], Magnetic Resonance Imaging - MRI [3], Ultrasound [4] and Positron Emission Tomography - PET [5]. Different techniques are suitable for capturing details of different tissues. The common property of all volumetric data is that the data is presented as three dimensional scalar or vector field containing property values for individual blocks of the scanned volume.

Such data can be visualized with indirect or direct rendering techniques. In first case the data is first converted to 3D mesh models [6, 7, 8] and then rendered [9], while with direct rendering, different volumetric rendering techniques can be used [10, 11, 12].

Most of 3D medical visualization systems were developed as standalone applications and require high-performance hardware for real-time display of data. In the past, we have developed a web-based volumetric medical visualization framework - Med3D [13], which allows users to visualize volumetric data in a web browser. The framework exploits the use of local and remote processing power for processing as well as rendering purposes. It's user interface is presented in Figure 1.

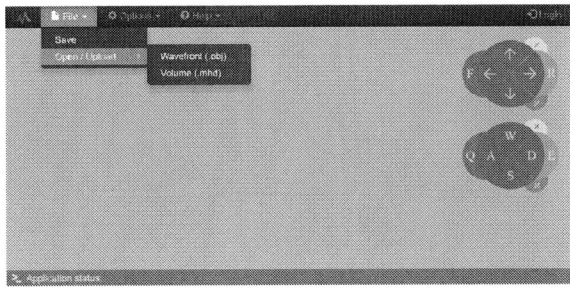

Figure 1: *User interface of Med3D - a web based volumetric data visualization framework.*

It is commonly accepted that user collaboration aids problem solving, and this is also the case in medical diagnosis, where second or even third opinions from experienced colleagues may be needed. In general, doctors with similar expertise rarely work in the same institution or even country, making collaboration slow and ineffective.

We have previously presented the benefits of remote collaboration integrated in our web-based 3D medical visualization framework [14] that allows for sharing: (1) visualization data, (2) camera view, (3) 3D localized user annotations and (4) text-based chat.

In this paper we present an extension of the Med3D framework. We first describe the implementation of render passes and render queue in Section II., in Section III. view-aligned hand-drawn annotations, in Section IV. the implementation of annotation sharing, in Section V. we describe how multiple users can concurrently produce annotations aligned with same view and in Section VI. we present the conclusions and future work.

II. RENDER PASSES AND RENDER QUEUE

To be able to easily extend the visualisation framework and achieve good performance, it is of crucial importance that the underlying rendering pipeline is well designed and implemented efficiently.

In Med3D framework we extended the basic rendering with multiple render pass design. This design emphasises the deferred rendering approach and allows us to easily define the input data (uniforms, buffers and textures), the output (textures, screen) and the shader for each render pass. Data binding and shader selection is performed in the prepossessing function that executes prior to the rendering. This step also allows us to reflect the input data on the user input, the output data of previous render passes and the global state of the framework. Render passes can be grouped together in the render queue as shown in Figure 2.

Formed render queues allow us to execute the render passes in the desired order. Each render pass also has access to the queue's global intermediate render data which allows the data such as textures and other variables to be forwarded to the subsequent render passes. The last render pass can either output the texture to the screen or return it as a queue execution result along the global render data.

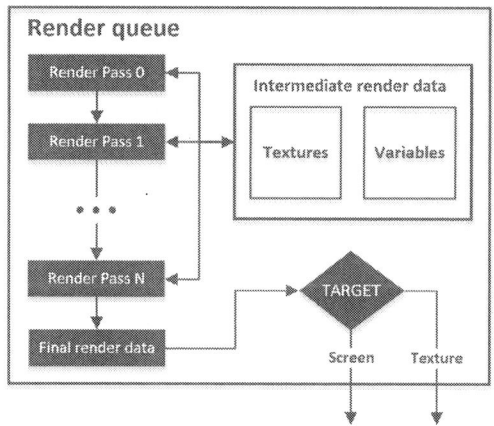

Figure 2: *Figure presents the structure of the rendering queue implemented in the Med3D framework.*

Our rendering pipeline design allows us to easily add new and expand the existing visualisation functionalities of the Med3D framework. Example of such functionality is the overlay multi-layer drawing which we integrated in the framework.

III. VIEW-ALIGNED HAND-DRAWN USER ANNOTATIONS

The Med3D framework allows users to add annotations on the displayed 3D data. Originally we implemented the textual annotations, which can be pinned on to desired spot on 3D data, making annotation connected with specific part of the visualized data. We have already presented such annotations implementation in [14] and an example of such annotations is presented in Figure 3.

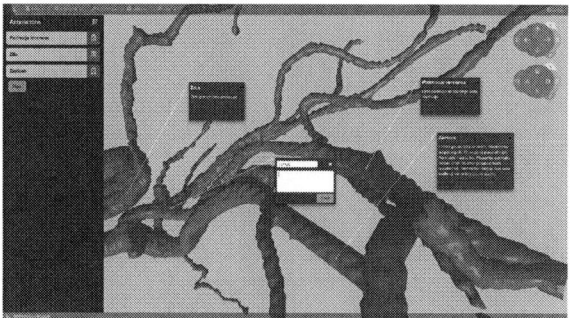

Figure 3: *Figure shows annotations pinned to the selected locations on the model of data.*

When we interviewed end users (doctors) about the initial implementation of annotations in the Med3D framework, they suggested that the implemented annotations are good, but that we should add the possibility of hand-drawn sketches on top of the visualized data.

We therefore expanded the Med3D framework with a drawing functionality where the user can either use a mouse, drawing tablet or touch screen to sketch the annotations. To start the sketching, the user first needs to create a new drawn annotation in the annotation sidebar shown left in the Figure 4. This sidebar contains a list of all the annotations that can be shown, as well as the brush tools such as color, thickness and hardness selector. To create a new drawn annotation user first needs to align the view to capture the point of interest and then create a new annotation. Upon creating the annotation the view is fixated on the current position and camera parameters (position and rotation) are stored so that the view can later be realigned. If the user wishes to view any of the previously created annotations, they can select them from the sidebar. Upon selecting the annotation the view is animated to the right orientation by interpolating the camera parameters (position, rotation) from the current to target values. After the camera is correctly positioned and oriented, the drawn annotation starts rendering on top of the data. This gives a smooth user experience when reviewing previously drawn annotations.

Each annotation can hold up to 20 drawing layers. Each layer holds a texture on which the data is rendered. When rendering the final render to the screen these textures are overlaid based on the order (bottom to top) in which layers are listed in the sidebar. The user can reorder this list using the arrows that appear next to the listed layer while hovering over it with the cursor, consequently changing the overlaying order. Layers can be added, deleted, renamed and hidden/shown using the provided user interface. To begin sketching, the user needs to select the target layer by pressing the "pen button" present on all the layers that are not hidden.

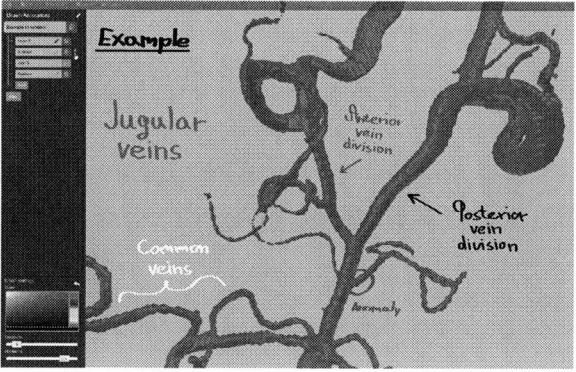

Figure 4: *Figure shows several hand-drawn annotations sketched on different layers and on the left the drawn annotations side bar. On the top of the sidebar is a list of annotations and a list of layers for the selected annotation. On the bottom are brush tools that are used to configure brush color, thickness and hardness.*

The user can draw by dragging the mouse or pen across the canvas. While dragging, a line segment is drawn between each pair of points representing the current and previous cursor position. To draw a line segment, we need to transform the points to the texture coordinate system and pass them to a shader which renders the line segment to a texture with the selected color, thickness (determines line width) and hardness (determines the distance from the line after which it starts to fade off). Color is selected from a color picker located in the bottom part of the annotation sidebar (brush settings) which also holds two sliders that are used to configure thickness (ranging from 1 to 32 pixels) and hardness (ranging from 0 to 1 where 0 represents the maximal fade off).

Because we might need to redraw the lines (in case of window resizing, undoing and annotation sharing) we store the points in a line structure. This line structure represents the combined line segments from cursor press to cursor release and their color, thickness and hardness. Each layer may contain multiple lines. We also need to normalise the stored points with the current aspect ratio as the aspect ratio might not be the same after resizing or on a device of a different user with whom we shared the annotation. We only need to normalise the x coordinate as the view projection is set so that it scales the height to fit the whole canvas whereas the width must always be equal to the height to represent all of the coordinates in a normalised space. Normalised x position can be obtained as $x_n = (x - 0.5) * w/h$ where x represents the position in a texture coordinate system and w and h represent the current width and height of the canvas. When we need to redraw the point, the normalised position x_n is transformed back to the texture position as $x = x_n * w'/h' + 0.5$ where h' and w' represent the new canvas dimensions. Using this process we can store the line segments that are invariant to the screen aspect ratio.

Any layer can be redrawn at any time using the stored information. We do the redrawing using multiple render passes where in each pass up to 251 line segments are drawn. This limitation comes due to the fact that we can only pass up to 1024 float uniforms into the shader to support all of the devices that are compatible with WebGL 2.0[1]. But even with this limitation the redrawing process is still very fast and the redrawing itself does not need to occur very often so it does not affect users on slower devices.

Our implementation of the drawn annotations allows users to easily manipulate their sketches. It is designed so that it is intuitive and easy for the user to add or delete the annotations or the layers. This results in good user experience and can be easily extended to allow for sharing of annotations between users.

IV. ANNOTATION SHARING

The Med3D framework already has many user collaboration functionalities such as the sharing of visualization data, views and text annotations. Sharing of hand drawn annotations is implemented as an extension of the latter functionality. It allows users to present their opinion in an intuitive way and share it with others.

Similar to other Med3D collaboration functionalities, sharing of hand-drawn annotations is done over a remote server on which all of the shared data is stored for easy and fast access. To start sharing the annotations, users must first create or join an existing session. This is already supported by the Med3D framework. Creating a session allows multiple users to view and interact with the same data in real time.

When the user creates a new session, all of the already existing annotations are uploaded to the server. The data consists of a list of annotations where each annotation contains a title, camera parameters (position and rotation) and the list of all layers. Each layer also contains a title, a list of lines where each line consists of color, thickness, hardness and points. Because the drawn annotation data is composed of only Javascript objects and primitives, we can easily send the data to the server using Websockets[2]. Because of the simplicity we're using the Socket.io[3] framework, which handles the transmission of binary data in an efficient manner. After all the data are uploaded, other users may join the session. When a new user joins, the session data (visualization data and annotations) are downloaded from the server. The downloaded annotations and layers are stored and equipped with an additional field that holds the username of the owner. By default the shared annotations are not shown right after they are downloaded. The user first needs to select them for display. At this point the layers are rendered to a new texture that binds to each shared layer and is used for all of the subsequent drawing. Because the line points are aspect ratio normalised, we can easily transform the positions to match the window aspect ratio.

After the user is synchronised with a server, he may add new annotations or layers to his own or to the shared annotations. All the changes that are made are being recorded and sent to the server in a dynamic time interval based on connection quality. After the server receives the changes it applies them to its copy of the data and broadcasts them to all of other session users. This reduces the work of the clients as they only need to send the data to the server and allows for new users to get the data directly from the server without requesting them from the session host. The whole communication process is presented in the figure 5.

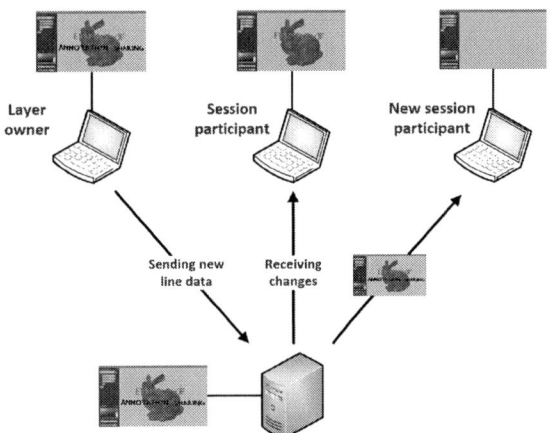

Figure 5: *Figure shows the communication process of sharing the hand-drawn annotations. On the top left there is the session host, which already synchronised the data with the server (bottom) and is now sending recent annotation changes. In the middle (top) there is a participant of the session that already downloaded all of the data and is now listening for changes as well as sending his own. On the top right is a new session participant that is downloading all of the latest data from the server without straining the session host.*

[1]https://www.khronos.org/registry/webgl/specs/latest/2.0/
[2]https://www.websocket.org/
[3]http://socket.io/

Our implementation of communication used for remote collaboration allows users to share their data and annotation changes in real time. This provides good user experience and efficient collaboration between multiple users with very little delay mostly dependent on the quality of the network connection.

V. COLLABORATIVE ANNOTATION

The most important aspect of hand-drawn annotations is that the users can make short notes and markings on the visualization, as well as interact with other users in real-time. This is also emphasised by allowing multiple users that are participating in the same collaboration session to overlay their layers and combine their sketches. Layers of current and all other users are listed together below the corresponding annotation. Each user can discern layers of other users by the user button that replaces the delete button. Hovering over this button displays the username of the layer owner in a tooltip as shown in image 6. Each user can, independently of other users, select the layers that he wants to see by clicking on the sidebar layer item. Users can also order the layers of each annotation using the arrow keys that appear while hovering over the layer to produce the desired overlay order. This allows for comparison of the data present on multiple annotation layers. Users can also use different colors, thickness and hardness allowing them to more easily emphasise the importance of different parts of the annotation as well as to distinguish between different parts of annotation.

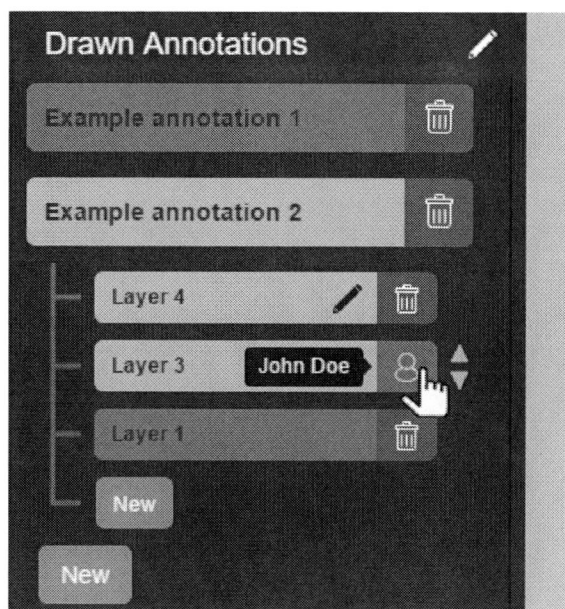

Figure 6: *Figure shows a list of drawn annotations in the sidebar. The second annotation in the list is selected and its three layers are displayed. Two layers are local and belong to the current user, while and one is shared by the user "John Doe".*

Because the layers are updated with short delay, users can see the changes being made in real-time. They can thus interact with each other while sketching with either concurrent drawing on different parts of the visualization or chatting via the text chat provided by the framework. In the future we also intend to add a voice chat that will in combination with hand-drawn annotations further improve the usefulness of the framework.

VI. CONCLUSION

In this work we presented an implementation of view-aligned hand drawn annotations into our Med3D visualization framework. The sharing of such annotations greatly improves the usability of remote collaboration and is an intuitive way for the end users (doctors) to express their views.

The next step in our work will be to evaluate the annotation interface with the doctors and implement improvements based on their feedback. We also aim to add a voice chat to further enhance the collaboration options.

REFERENCES

[1] C. R. Crawford and K. F. King. Computed tomography scanning with simultaneous patient translation. *Medical Physics*, (17):967 – 982, 1990.

[2] W. A. Kalender, W. Seissler, E. Klotz, and P. Vock. Spiral volumetric CT with single-breath-hold technique, continuous transport and continuous scanner rotation. *Radiology*, (176):181 – 183, 1990.

[3] P. A. Rinck. *Magnetic Resonance in Medicine. The Basic Textbook of the European Magnetic Resonance Forum. 9th edition*, volume 9.1. TRTF, 2016. E-Version.

[4] D. Krakow, J. Williams, M. Poehl, D. L. Rimoin, and L. D. Platt. Use of three-dimensional ultrasound imaging in the diagnosis of prenatal-onset skeletal dysplasias. *Ultrasound in Obstetrics and Gynecology*, 21(5):467–472, 2003.

[5] J. M. Ollinger and J. A. Fessler. Positron-emission tomography. *IEEE Signal Processing Magazine*, 14(1):43–55, Jan 1997.

[6] W. E. Lorensen and H. E. Cline. Marching cubes: A high resolution 3d surface construction algorithm. *SIGGRAPH Comput. Graph.*, 21(4):163–169, August 1987.

[7] D. Lesage, E. D. Angelini, I. Bloch, and G. Funka-Lea. A review of 3d vessel lumen segmentation techniques: Models, features and extraction schemes. *Medical Image Analysis*, 13(6):819 – 845, 2009.

[8] M. T. Dehkordi, S. Sadri, and A. Doosthoseini. A review of coronary vessel segmentation algorithms. *J Med Signals Sens*, 1(1):49–54, 2011.

[9] W. J. Bouknight. A procedure for generation of three-dimensional half-toned computer graphics presentations. *Commun. ACM*, 13(9):527–536, September 1970.

[10] R. A. Drebin, L. Carpenter, and P. Hanrahan. Volume rendering. *SIGGRAPH Comput. Graph.*, 22(4):65–74, June 1988.

[11] M. Levoy. Display of surfaces from volume data. *Computer Graphics and Applications, IEEE*, 8(3):29–37, 1988.

[12] E. P. Lafortune and Y. D. Willems. Bi-directional path tracing. In *Proceedings if Third International Conference on Computational Graphics and Visualization Techiques (COMPUGRAPHICS '93*, pages 145–153, 1993.

[13] C. Bohak, P. Lavrič, and M. Marolt. Web based visualisation framework with remote collaboration support. In *Proceedings of 25th ERK 2016*, pages 43–46, Portorož, Slovenia, September 2016.

[14] C. Bohak, P. Lavrič, and M. Marolt. Remote interaction in web-based medical visual application. In *Human-computer interaction in information society : proceedings of the 19th* *International Multiconference Information Society - IS 2016*, pages 5–8, Ljubljana, Slovenia, October 2016.

Feasibility of biometric authentication using wearable ECG body sensor based on higher-order statistics

Sebastijan Šprager [*], Roman Trobec [**] and Matjaž B. Jurič [*]

[*] University of Ljubljana, Faculty of Computer and Information Science, Ljubljana, Slovenia
[**] Jožef Stefan Institute, Ljubljana, Slovenia
sebastijan.sprager@fri.uni-lj.si

Abstract - Besides its principal purpose in the field of biomedical applications, ECG can also serve as a biometric trait due to its unique identity properties, including user-specific deviations in ECG morphology and heart rate variability. In this paper, we exploit the possibility to use long-term ECG data acquired by unobtrusive chest-worn ECG body sensor during daily living for accurate user authentication and identification. Therefore, we propose a novel framework for wearable ECG-based user recognition. The core of the framework is based on the approach that employs higher-order statistics on cyclostationary data, already efficiently applied for inertial-sensor-based gait recognition. Experimental data was collected by four subjects during their regular daily activities with more than 6 hours of ECG data per subject and then applied to the proposed framework. Preliminary results (equal error rate from 6% to 13%, depending on the experimental parameters) indicate that such authentication is feasible and reveal clear guidelines towards future work.

I. INTRODUCTION

We are living in an emerging era of Internet of Things (IoT) paradigm. Therefore, smart devices and wearable sensors and their potential applications are not only widely investigated within corresponding research communities, but are also becoming more and more indispensable in daily living. Low power consumption, connectivity due to low-cost and omnipresence of mobile data enable continuous data collection, transfer and processing (i.e. in a cloud) from sensor nodes of different modalities. Thus, there are many challenges open in this field of research, namely achieving efficient and loose-coupled integration into existing information systems and platforms, intercommunication between wearable nodes and systems, processing of continuous data streams and knowledge extraction. Consequentially, security and privacy are two aspects that cannot be overlooked in this context. For example, in the domain of medicine, the wearables can produce sensitive data that could be abused either from the perspective of privacy or even worse – if data directly influences on the user's health (i.e. information on the detected heart anomalies to medical staff). In this case, a malicious man-in-the-middle attacker could take advantage on this information which could result in hazardous impact on end-user's health. Thus, much effort is given into the exploitation of the possibilities to circumvent such scenarios and, consequently, novel biometric approaches

have emerged in the context of wearable sensing. Such approaches, applied on different underlying biometric traits (i.e. speech, gait, etc.), have many advantages over classical biometric trait-based approaches (i.e. fingerprint and iris recognition) since they are in the context of continuous wearable sensing ever-present and do not require any special actions or attention from the end-users. Besides that, with continuous data processing and collection, such biometric approaches could also adapt to long-term variations in particular biometric trait, thus ensuring its permanence. One of such biometric traits is also electrocardiogram (ECG), which represents a golden standard for assessing myocardial activity and anomalies. Since ECG measurement devices are possible to be implemented as wearables, the exploitation of its potential to be used for the biometric purposes presents a logical subsequent step. We originate from several assumptions and advantages that ECG has among over biometric traits and could represent a basis for efficient user recognition (authentication and identification) if properly exploited. First one is ECG morphology that is specific for an individual user. Besides that, heart rate variability (HRV) is another property that characterizes an individual user. Finally, heart anomalies (i.e. ECG responses to arrhythmias) can also reveal an individual.

The significance of wearable biometrics and latest crucial findings in this field of research have been presented in [1], exposing ECG-based recognition besides other biometric traits that are being investigated in the last period (i.e. gait recognition using wearable sensors [2]) as one of the most prominent for the future applications. First investigations on that problem that had significant impact on the community have been performed few years ago [3]. The authors have proposed several methodological approaches, first evaluated on datasets acquired in regular way [4]. By raising popularity of wearable sensors and body area networks, first attempts on ECG-based authentication and cryptography have been investigated [5]. It is important to point out that many significant works on ECG-based recognition have been published just recently. In particular, there has been a lot of research effort focused on the problem of ECG authentication on mobile devices [6], body sensor networks [7], as well as in smart environments [8]. Some investigations have been made on addressing authentication problem on noisy ECG data [9], [10] or using special methodology that does not rely on the extraction of fiducial points from ECG signals [11]. The work, that addresses inter-subject variability and intra-subject reproducibility of ECG recognition metrics on 12-lead ECG is also significant since indicates that ECG authentication on a long-term basis is meaningful [12]. The problem of continuous authentication using ECG was also addressed in [13].

After considering the most recent state-of-the-art, there are still many research challenges open that still need to be properly addressed to make ECG-based user recognition by employing wearable sensors applicable in daily life and widely accessible. In this paper, we present a feasibility study by employing our framework for wearable ECG-based user recognition that tackles with some of crucial challenges in this field of research. As we employ chest-worn ECG device sensor that enables continuous and completely unobtrusive measurements of ECG data, we focus on longer, continuous measurements that are collected during daily living and are thus exposed to many aggravating factors, including motion artefacts and noise, as well as varying heart rate directly related to user's physical activity (work, rest, etc.). Furthermore, we also implicitly consider the influence of heart anomalies on the recognition accuracy. This is important since such events (i.e. arrhythmic beats) are quite common and do not have constant distribution in the uniform time intervals. Finally, we also need to mention the influence of the on-body sensor placement as the shape of measured ECG signal is directly related to the exact body position where the sensor is attached and reattaching sensor to the very same position is very unlikely. However, the issue of sensor placement is out of the scope of this paper and is left as one of the principal activities during our future investigations.

The rest of the paper is structured as follows. In Section II, we present the proposed framework for wearable ECG-based user recognition. In Section III, the details on experimental work and results are given while in Section IV we discuss our findings and conclude the paper.

II. FRAMEWORK FOR WEARABLE ECG-BASED USER RECOGNITION

The entire pipeline of the proposed framework is depicted in Figure 1. It is based on our developed framework for analysis of stochastic time series based on higher-order statistics which is already proven and sufficiently applied to the problem of gait recognition using inertial sensors [14]. The framework proposed in this paper consists of the following phases: a) ECG data acquisition by chest-worn ECG sensor; b) data processing module which processes incoming ECG data and extracts discriminative patterns from ECG sequences suitable for the recognition procedure; c) data storage used for storing the enrolled patterns; d) recognition module which performs either user authentication or identification on incoming patterns extracted from ECG sequences and e) evaluation module, used to assess the performance of the recognition procedure. All components of the proposed framework are in detail described in the following.

A. ECG Body Sensor

The ECG measurements have been obtained with medical graded ECG body sensor Savvy [15], marketed by

Saving d. o. o., Ljubljana, Slovenia. The sensor provides long-term ECG measurements, e.g., a week or more, during normal daily activities, because of its small dimensions, flexible design, extremely low power consumption (more than seven-day autonomy) and simplicity of use. ECG measurements run in the background and do not interfere with the usual mobile phone functions. However, lost radio connection, removed sensor during user's activities, like showering, or skin irritation under electrodes, can occasionally interrupt measurements. They can also be corrupted by known ECG artefacts like baseline wandering or signal from muscular activities.

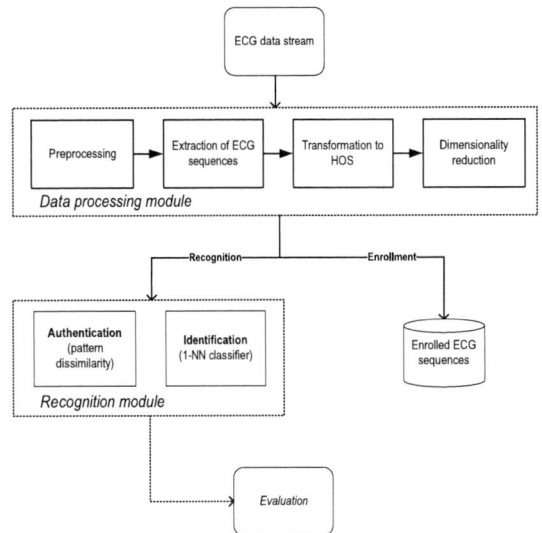

Figure 1. Pipeline of the proposed framework for wearable ECG-based user recognition.

A moderate sampling rate of 125 Hz with 10 bit analogue/digital converter is used, that is a compromise between medical value and amount of generated data. Such measurements suffice for accurate rhythm monitoring and analysis. A mobile application on a smart phone coordinates ECG data transfer from the sensor to the phone storage using low power wireless connection (BT4). The application provides on-line visualization of the measured ECG with a robust adaptive real-time beat detector for the calculation of minute beat rate (BPM).

B. Data processing module

Data processing phase represents a core of the proposed framework and is further divided into four successive steps: preprocessing, extraction of ECG sequences, transformation to discriminative patterns based on higher-order statistics and dimensionality reduction of resulting patterns.

1) Preprocessing

Incoming ECG signals are exported in their raw form with all eventual artefacts. The ECG contains low-frequency drift mostly as a consequence of user's motion as well as high-frequency noise. In order to suppress the influence of drift and noise on the recognition performance, we employed fourth-ordered Butterworth bandpass filter with cut-off frequencies set to preserve frequency content 1 Hz and 4 Hz as suggested in [13].

265

2) Extraction of ECG sequences

Continuous incoming ECG data stream is split into smaller units – ECG sequences. These are signal segments with either specific length or content that undergo further processing steps. We have designed an extraction procedure that is able to deliver different types of sequences. These can be divided into two main categories: sequences with fixed length and sequences containing specific number of successive heartbeats. While the extraction procedure of the first type is straightforward, the extraction of successive heartbeats relies on special algorithms that are able to accurately determine fiducial points in ECG signal (PQRS complexes), i.e. Pan-Tompkins algorithm as one of the most popular [16]. For this purpose, we have employed a custom QRS detector that is based on the analysis of ECG first derivative that automatically removes the base line wandering and reflects fast changes of amplitudes in QRS complexes. A dynamic amplitude threshold is then applied on the obtained derivative for optimal distinction between muscular activity signal and QRS signals. Finally, a quadratic interpolation is applied around the detected peak, to improve the temporal resolution of the detector. At this point it should be mentioned that single extracted heartbeat can also be considered as a single ECG sequence. Furthermore, our implementation of ECG sequence extraction procedure can also perform alignment of ECG sequences to equal length either by linear interpolation or dynamic time warping (DTW), which allows for experimenting with different types of extracted ECG sequences, i.e. suppressing the influence of HRV and thus focusing on the signal shape only.

3) Transformation to feature space

The main part of the recognition procedure transformation of the input ECG sequences into discriminative patterns. We relied on the transformation procedure based on the higher-order statistics (HOS) proposed in [14]. Such approach is very powerful when dealing with random cyclostationary signals with high degree of non-linearity and non-Gaussianity. Since ECG signals convey carry these properties, the application of the proposed procedure for ECG-based authentication is reasonable. One of the principal advantages is that the proposed approach extracts the features in implicit form and does not require any information about fiducial points in ECG signals. Furthermore, it can operate directly on signal sequences with arbitrary length. The only thing that this procedure requires is a-priori knowledge of frequency content of the observed phenomena and statistical measures to consider (i.e. chosen cumulants). The proposed transformation procedure produces output feature vectors having high level of discriminability and providing alternative insight into recognition problem. In order to obtain more details on the proposed transformation procedure, the interested reader is referred to [14].

4) Dimensionality reduction

Since the significant amount of redundancy can appear in transformed feature vectors, applying one of the dimensionality reduction techniques is reasonable. We employ the transformation of feature vectors (patterns) into their approximated versions by employing singular

value decomposition (SVD). These are obtained by projecting feature vectors onto the top of e obtained eigenfeatures. We implicitly define reduced dimension e by proportion of variance covered by top e eigenfeatures. The details on that are also given in [14].

C. Recognition module

Depending on the applicability, user recognition can be performed either as authentication or identification. By authentication, user presents his identity and biometric pattern and recognition module determines whether he is genuine user or imposters. However, by user identification, the recognition module receives user's biometric pattern without any information on his identity. Based on that, recognition module reveals the identity of the user. In both cases, recognition module relies on stored feature vectors enrolled by users accompanied with user's label. Thus, enrolled feature vectors serve as a basis for recognition (i.e. they serve as the comparing patterns in case of user authentication or as the training set for classification algorithms in case of user identification).

1) Authentication

User authentication is performed in terms of comparing incoming probe patterns with all previously enrolled patterns. In its simplest form, authentication is performed by estimating dissimilarity score between incoming probe pattern and enrolled user. In this case, dissimilarity score d between any probe pattern p and enrolled owner i is determined by applying the following minimum rule:

$$d(\mathbf{E}_i, p) = \min_{j_i} \varphi(g_{i,j_i}, p)$$

where $\mathbf{E}_i$ stands for all enrolled patterns of i-th owner. If dissimilarity score d does not exceed predefined global acceptance threshold, p is with correspondence to g_i accepted as owner i and rejected otherwise. Dissimilarity score is computed using suitable distance function φ.

2) Identification

By identification procedure, recognition module employs on classification or machine learning approaches. In this case, enrolled patterns serve as a training set, each of them being labelled by the corresponding class (pattern owner). Followed by the learning procedure, classification process assigns one of the labels defined within the training set to each of the input probe pattern. For the preliminary investigations presented in this paper, we relied on 1-nearest neighbor (1-NN) classifier in order to assess the performance of the proposed framework when applied on one of the simplest classification methods.

III. EXPERIMENTS AND RESULTS

A. Experimental set-up and parameters

Four volunteering subjects (2 healthy subjects with sinus rhythm and 2 subjects with occasional irregular heartbeat in terms of premature supraventricular beats), with average age of 62 years, average height of 179 cm and average weight of 73 kg participated in the experiment. We have collected 6 hours of continuous ECG data by a wireless chest-worn sensor presented in Section II.A. The measurements cover regular daily activities, e.g., sitting, walking and resting, and sleep.

Collected ECG data was the subject to investigation by proposed framework for ECG-based user recognition. In order to get a proper insight into the problem of the proposed and to properly address our research baselines described in section I, we experimented with the three main types of parameters: a) portion of ECG sequences extracted from the incoming 6-hours-long data stream that has been employed for user enrollment; b) ECG sequence type and c) length of chosen ECG sequence type. All information on these experimental parameters are gathered up in TABLE I. .

TABLE I. EXPERIMENTAL PARAMETERS USED FOR THE PRESENTED FEASIBILITY STUDY.

Portion of data enrolled as history ECG sequences (from 6-hour-long measurements)	ECG sequence types	Considered lengths of ECG sequence types [seconds / no. of heartbeats]
❖ 5 min (1.4 %) ❖ 20 min (5.6 %) ❖ 1 h (16.7 %) ❖ 3 h (50 %)	❖ Sequences with fixed lengths ❖ Consecutive heartbeats ❖ Consecutive beats with unified heartbeat lengths	❖ 3 ❖ 6 ❖ 12 ❖ 24 ❖ 48 ❖ 96

In order to investigate the influence of the amount of enrolled data on the recognition performance, we have split set of extracted ECG sequences for each subject into two disjoint, chronologically ordered sets: first one applied to enrollment and second one for the probe patterns. We considered four different splitting points: 5 min-355 min (1.4% - 98.6%), 20 min-340 min (5.6%-94.4%), 1 h-5 h (16.7%-83.3%) and 3 h-3 h (50%-50%). By varying this parameter, we aimed to find out the minimum amount of enrolled patterns that would sufficiently cover pattern discriminability in terms of morphology and heart rate variability. Secondly, concerning ECG sequence type, we experimented with three different types. Sequences with fixed lengths were considered in order to determine how the proposed approach reacts to the unaltered input data, implicitly containing the information on HRV. In the same manner, we considered ECG sequences with equal number of consecutive heartbeats. Unlike previous type, here the sequences are of different lengths depending on HRV. Following that, we also considered ECG sequences with equal number of consecutive heartbeats where the lengths of each heartbeat was unified by linear interpolation to 125 samples (1 second). In such way, we focused on heartbeat morphology only thus excluding the influence of HRV. Finally, for each chosen ECG sequence type, we considered different lengths (units in seconds or numbers of heartbeats, regarding to chosen type). Short sequences are more appropriate for the recognition trials on demand since the data collection step does not take much time. However, short sequences can result in reduced discriminability and decreased recognition performance. In the contrast, by continuous user verification, we can afford longer ECG sequences for the recognition. Thus, it is reasonable to evaluate the performance of the proposed approach by different sequence length. We experimented with the values of 3, 6, 12, 24, 48 and 96 (seconds or the number of consecutive heartbeats, depending on particular chosen ECG sequence type).

For the evaluation purposes, we performed both authentication and identification in different evaluation scenarios – each combination of these tree parameters represented one evaluation scenarios. Thus, we considered 72 $(4 \cdot 3 \cdot 6)$ different evaluation scenarios. Concerning the parameters for HOS-based core algorithm we considered first four orders of cumulants while a-priori knowledge on the lower bound of the frequency content was set to 2/3 Hz, corresponding to the heart rate of 40 bps. Dimensionality of patterns was reduced by selecting parameter e such that represents the number of top eigenfeatures that cover 96% of variance in data. Finally, correlation distance was used as a distance function φ for dissimilarity score d since it yielded in best performance, as revealed in [14].

B. Evaluation metrics and results

The performance of user authentication by proposed framework was determined by ROC curve. It represents the trade-off between false acceptance rate (FAR) and false rejection rate (FFR) by varying global threshold. Based on that, equal error rate (EER) represents a quantitative performance measure determined as a rate at which FAR and FRR in the corresponding ROC curve are equal. Concerning the performance of user identification by employing the proposed approach, we use identification accuracy as a quantitative metric that is determined by ratio of correctly classified probe patterns and all classified probe patterns.

The results of authentication performance on the collected experimental by the proposed framework are shown in Figure 2. , where ROC curves for particular experimental parameter combination are depicted. Plots by rows represent three different ECG sequences while columns determine four different splits of enrolled and probed patterns. For each plot that combines these two parameters, ROC curves for all six lengths of ECG sequences are shown. Furthermore, authentication performance is also revealed by overall EER's for each combination of experimental parameters. These are shown in Figure 3.

The results of identification, performed by the proposed framework supported by 1-NN classifier on our experimental dataset, are shown in Figure 4. Overall identification accuracies are visualized for each combination of experimental parameters in the same way as by overall EER's.

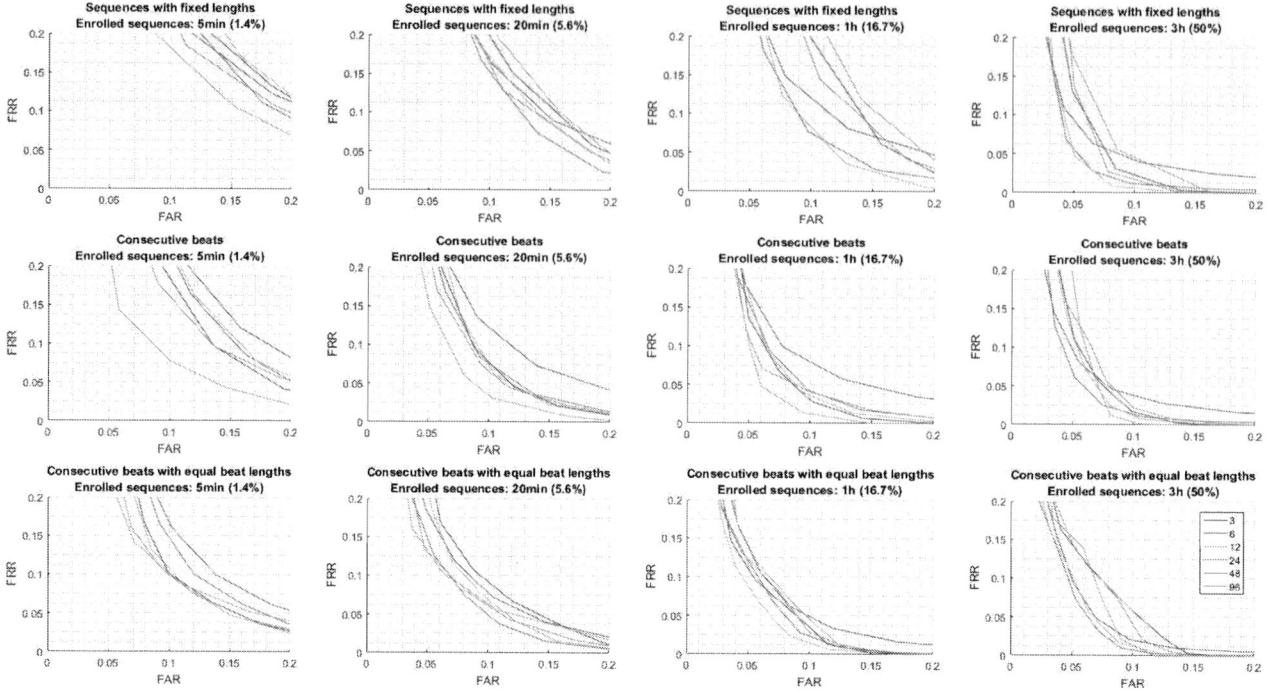

Figure 2. Resulting ROC curves after authentication procedure on ECG sequences for different evaluation scenarios.

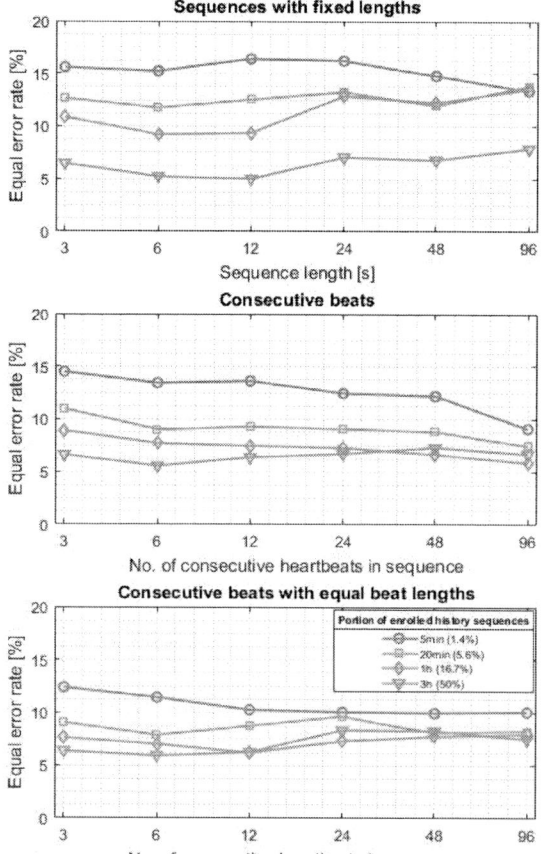

Figure 3. Equal error rates obtained as the result of authentication procedure on ECG sequences for different evaluation scenarios.

IV. DISCUSSION AND CONLUSION

Overall results of the have revealed high recognition performance regardless to the aggravated factors that were induced within the processed data and were not addressed implicitly: motion artefacts and mishits of algorithm for detection of R-waves which led to improper segmentation. Nevertheless, if we consider overall performance of user authentication, we can see that the values of EER are decreasing hand in hand with the amount of enrolled historical sequences. This is expected since in such way we can cover more distinguishable patterns and subject-specific HRV and heart anomalies. It also seems that the ECG sequences with fixed lengths result in decreased performance as the sequences with consecutive heart beats. However, it should be mentioned that the same values of length parameter should not be compared directly since the number of heartbeats included in sequences with fixed lengths is directly related to instantaneous heart rate. Furthermore, it is interesting that the best performance resulted when considering ECG sequences with consecutive heartbeats that were aligned to equal lengths. In such case, the influence of HRV and heart anomalies is covered solely by the shape of heartbeats (i.e. arrhythmias are the most obvious example). However, this matter will be investigated during our future work. To sum up, by increasing ECG sequence lengths and the amount of enrolled data we can observe that overall EER's converge towards 5% while in demanding circumstances, EER's revolve around 13%. It is interesting that similar conclusions can be made for the identification performance, just in terms of identification accuracies instead of EER's.

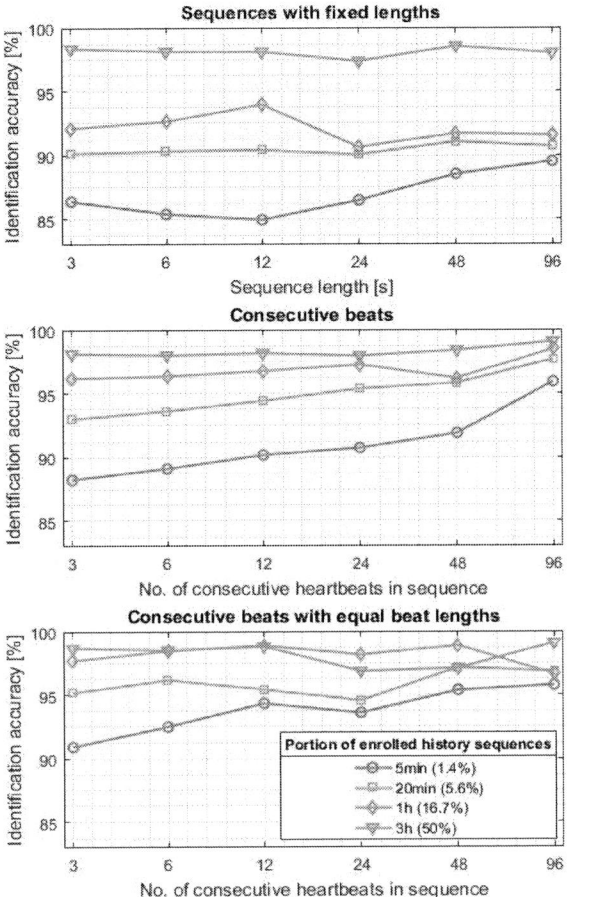

Figure 4. Accuracies obtained as the result of identification procedure on ECG sequences for different evaluation scenarios.

The obtained results are clearly indicating the potential of user recognition during long-term monitoring by using chest-worn ECG sensor and the proposed framework. Based on findings presented in this paper, several further steps towards the development of accurate and robust methodology will be made. That includes the examination of influences of HRV and heart anomalies on recognition performance in explicit way. Furthermore, the influence of sensor position variability by reattaching will be also investigated. Finally, to properly address all open questions, extensive experimental measurements will be performed. Obtained experimental dataset will contain experimental data collected by large number of subjects containing both regular and irregular heartbeats during long-term monitoring over wide time frame (in terms of months).

REFERENCES

[1] J. Blasco, T. M. Chen, J. Tapiador, and P. Peris-Lopez, "A survey of wearable biometric recognition systems," *ACM Comput. Surv.*, vol. 49, no. September, 2016.

[2] S. Sprager and M. B. Juric, "Inertial sensor-based gait recognition: A review," *Sensors (Switzerland)*, vol. 15, no. 9. MDPI AG, pp. 22089–22127, 02-Sep-2015.

[3] I. Odinaka *et al.*, "ECG Biometric Recognition: A Comparative Analysis," *IEEE Trans. Inf. Forensics Secur.*, vol. 7, no. 6, pp. 1812–1824, 2012.

[4] S. I. Safie, J. J. Soraghan, and L. Petropoulakis, "Electrocardiogram (ECG) biometric authentication using pulse active ratio (PAR)," *IEEE Trans. Inf. Forensics Secur.*, vol. 6, no. 4, pp. 1315–1322, 2011.

[5] Z. Zhang, H. Wang, A. V. Vasilakos, and H. Fang, "ECG-cryptography and authentication in body area networks," *IEEE Trans. Inf. Technol. Biomed.*, vol. 16, no. 6, pp. 1070–1078, 2012.

[6] J. S. Arteaga-Falconi, H. Al Osman, and A. El Saddik, "ECG Authentication for Mobile Devices," *IEEE Trans. Instrum. Meas.*, vol. 65, no. 3, pp. 591–600, 2016.

[7] S. Peter, B. P. Reddy, F. Momtaz, and T. Givargis, "Design of secure ECG-based biometric authentication in body area sensor networks," *Sensors (Switzerland)*, vol. 16, no. 4, 2016.

[8] M. S. Islam and N. Alajlan, "Biometric template extraction from a heartbeat signal captured from fingers," *Multimed. Tools Appl.*, 2016.

[9] H. S. Choi, B. Lee, and S. Yoon, "Biometric Authentication Using Noisy Electrocardiograms Acquired by Mobile Sensors," *IEEE Access*, vol. 4, pp. 1266–1273, 2016.

[10] S. J. Kang, S. Y. Lee, H. Il Cho, and H. Park, "ECG Authentication System Design Based on Signal Analysis in Mobile and Wearable Devices," *IEEE Signal Process. Lett.*, vol. 23, no. 6, pp. 805–808, 2016.

[11] M. Hejazi, S. a. R. Al-Haddad, Y. P. Singh, S. J. Hashim, and A. F. Abdul Aziz, "ECG biometric authentication based on non-fiducial approach using kernel methods," *Digit. Signal Process.*, vol. 52, pp. 72–86, 2016.

[12] I. Jekova *et al.*, "Intersubject variability and intrasubject reproducibility of 12-lead ECG metrics: Implications for human verification," *J. Electrocardiol.*, vol. 49, no. 6, pp. 784–789, 2016.

[13] W. Louis, M. Komeili, and D. Hatzinakos, "Continuous Authentication Using One-Dimensional Multi-Resolution Local Binary Patterns (1DMRLBP) in ECG Biometrics," *IEEE Trans. Inf. Forensics Secur.*, vol. 11, no. 12, pp. 2818–2832, 2016.

[14] S. Sprager and M. B. Juric, "An Efficient HOS-Based Gait Authentication of Accelerometer Data," *IEEE Trans. Inf. Forensics Secur.*, vol. 10, no. 7, pp. 1486–1498, Jul. 2015.

[15] "Savvy d. o. o." [Online]. Available: www.savvy.si. [Accessed: 28-Feb-2017].

[16] J. Pan and W. J. Tompkins, "A Real-Yime QRS Detection Algorithm," *IEEE Trans. bio-medical Eng. Eng.*, vol. BME-32, no. 3, pp. 230–236, 1985.

Bio-medical analysis framework

Miha Mohorčič and Matjaž Depolli

Jožef Stefan Insitute, Department of Communication Systems, Ljubljana, Slovenia

mmohorcic@ijs.si, matjaz.depolli@ijs.si

Abstract – Monitoring of ECG signal is used in medicine for multiple purposes. Measurements can be taken at any stage of patient's medical care, either as a preventative, diagnostical or recovery monitoring. Current wearable technology enables users and doctors to produce so far unprecedented amount of information. Processing of such measurements is usually a laborious and time consuming manual task. Automatic processing of such measurements is neither well defined nor thoroughly tested. In this work we focused on the needs of both, health care professionals, and IT engineers developing software for processing of long term multi-sensor measurements. Taking into account future expandability of multi-sensor gadgets, we propose a new framework, which is able to show the data measured by wearable ECG monitor, process it, and compare algorithms for automatic processing. We determined possible signal sources along with their values, units and time continuity. We propose suitable file formats for storage of such measurements, keeping in mind future expandability, size demands and usage of the formats. Suggested framework can therefore be used to display, automatically process and store discrete and continuous biomedical signals beside ECG, producing additional value to gathered measurements.

I. INTRODUCTION

Recent advancements in the field of personal and mobile health monitoring brought numerous improvements in collecting and processing of body measurements. This advancements could in the future lower medical costs and shorten diagnostic and recovery time. People with chronic diseases or hard to detect repeating anomalies in body function will benefit most from personal constant monitoring [1].

Amount of gathered data opens multiple new data sources for researchers and health workers, since long term measurements of patient's state can now include data spanning multiple months, and can be performed on a large population. Personal gadgets that enable long term medical and non-medical measurement collection are being introduced to the market [2]. During the development of such novel wearable ECG sensor at Jožef Stefan Institute (JSI), a need for a better software support to measurement processing has been detected. The burden of processing long measurements should be lifted from the health care workers by designing an automatic process, that can quickly and reliably process large amount of data. Research personnel who assist the medical staff also noted the lack of proper tools to develop, test and compare automatized algorithms for signal processing. Both user groups raised some common issues, triggering the design of a framework that would offer desired functionality.

This paper will outline current state of technology, the identified requirements for developing framework and possible solutions. As a continuation of previous work, current experiences and know-how will be included. Focusing primarily on the ECG signal, a comparison of existing software solutions for displaying, processing and storing the measurements will be performed. In the future new kinds of continuous or discrete signals might be included in the framework. Such signals will be discussed and their specifics determined.

First section of the paper describes the state of technology, divided into subsections containing information about wearables, user groups and state of the art available solutions. Next section will present our proposal and reasoning behind its design. Last section explains expected improvements, added value of framework and future development goals.

II. CURRENT STATE

To determine functionality and design of the new framework, we looked over the type of input measurements that will be processed, user expectations and currently available solutions.

A. Wearables

Development and miniaturization of hardware in the last couple of years caused an increase in the amount of data to be processed. New devices enable capture of medically certified measurements, that can give attending physician more information about patient's state over a longer period of time. While Holter monitors [3,4] are a de facto standard for long term ECG monitoring, there are gadgets available that offer similar functionality. They are, in comparison to Holter monitors, usually cheaper, less obtrusive to the user and offer longer continuous measurements [5]. On the other hand, these devices usually do not provide multi-channel data, offer lower resolution and sampling frequency and inferior software support. While mobile monitoring devices represent significant progress, most are not medically certified, delaying their integration into the medical and other health care institutions.

An example of such device is developed at JSI, now known under the market name *Savvy*. This device already offers long term measurement of ECG signal, but is planned to be upgraded with additional sensors, providing additional information about the patient state at any given moment during measurement. This additional information could help better understand long term measurements and better understand various events seen on the ECG. Planned sensoric upgrades include detection of patient's and ambient temperatures, and patient's activity. Currently the smartphone application that connects to Savvy uses a proprietary file format, that is able to hold streaming ECG measurement and corresponding meta data. Processing and storage of Savvy additional sensoric data are defined in proprietary protocol, but not yet implemented.

B. File storage

Standardized storage and processing of ECG with additional sensoric data is not yet available. Manufacturers of new gadgets therefore invent new proprietary file formats and protocols, fragmenting and limiting measurement usage scope. New hardware and usage patterns also bring new problems to measurement format specification. Different devices contain different hardware sensors, meaning that the general file format should be completely adaptable to accept any kind of data, e.g., different sampling rates and resolutions. Devices can store data locally, stream them to the Internet, or store them on a connected external device, introducing a possibility of data lost in transmission, transmission errors, or sampling errors during the measurement capture. Most currently used file formats do not account for such issues.

One of the problems of existing file formats is also their non-streaming nature. They are either in an XML format or have strictly coded headers and footers. This is not desirable, given the nature of wearable measurements, where data can be corrupted at any time (e.g: device runs out of battery, user forgets that the measurement is active...). Embedded system are also limited in available memory capacity and computational capabilities. They must optimize battery usage and bridge performance deficit in comparison to modern computer. File format should therefore provide option to compress, or at least minimize, overhead of saving and transferring files, while still remain simple enough to be produced and saved on embedded systems.

C. Health care practitioners

Health care practitioners do not use standardized software for viewing and processing of captured files. Most physicians review short segments of in-hospital measured ECGs printed on paper, or use proprietary software solutions supplied by the ECG equipment manufacturer. Such software is usually limited in functionality, locked for use in combination with the given measuring equipment, expensive to procure and does not evolve through time. Transfer of data is also limited to sending measurement copies or other documents via CDs or other media, provided that receiving party has compatible software. Health care professionals wish for an automatized system, that offers quick and condensed overview of patient's health during long periods of time. Some health care professionals require only a global state of patients health, which can be enlightened through a long term ECG measurement, while others might dwell into the details of the same ECG measurement for irregularities of heart beats. They also desire for results to be easily and safely transferable to their colleagues or patient.

Future goal of proposed design is to become a basis for a useful tool in medical application. To achieve it, we need to develop stable, verifiable software that is capable of quick and unsupervised signal processing. Ease of use and powerful automation are desirable to minimize manual labour and filter out useful information from the long term mea-

surements efficiently. Automated tasks must be properly verified and tested to ensure the same result as manual processing. Since software is to be used in medical settings, not only automated tasks, but all implementations of the software must pass necessary tests, ensure privacy and safety of data and be verifiable. Furthermore, software should also have tracking of source code changes and pass security audit, if data transfer will be included.

D. Algorithm developers

Framework should facilitate additional testing and development demands for ECG research teams and algorithm developers. Algorithm developers are able to design efficient and novel ways to tackle issue of beat detection and classification. After the verification of successfulness of algorithm, great percentage of the algorithms stay as proof of concept, and are not practically implemented. Reasons for this vary, but mostly algorithms are developed in laboratory setting, not meant for production environment. Main reason is their incompleteness and lack of robustness when used on measurements captured in real life. Medical certification is also a hurdle that is sometimes hard to pass. Many algorithms, however, could be adapted or used together with currently used processing technologies to aid processing of captured data. Bringing together production and developer environment could improve amount and speed of knowledge exchange between developers and also medical practitioners.

Currently most widely tool used by developers is PhysioNet toolkit [6]. It provides a variety of tools used for communication with public ECG databases, beat detection algorithms and comparison tools. These tools enable standardized testing environment and are themselves thoroughly tested. Tools can be used in conjunction with multiple programming languages and platforms. Using PhysioNet toolkit, the developers test and confirm their implementations on existing publicly accessible databases. Widespread usage of this toolkit ensures that developers can compare themselves to current state of the art. For developers, integration of PhysioNet toolkit's functionality is a must.

E. State-of-the-art algorithms

Despite being in use for decades [7], QRS detection algorithms are still actively being developed. Most known and well researched algorithms produce very good detection rates (more than 99%) on test databases. Signals taken with wearable devices are presenting new challenges. Sampling frequency is usually lower, as is the signal resolution, while the noise level is higher. Existing algorithms still work on such signals, however, their detection rate is lower and more false beats are detected. Newly developed algorithms reduce this effects on quality of detection, via filtering and new approaches to detection [8–10]

Approach to development of detection algorithm is dependent on a type of ECG signal and the goal of detection. Development of an algorithm that has very high detection rate can differ from an algorithm that aims to produce very exact RRI (R-R interval), where detection rate is less important than the signal quality. In such cases developers are interested in different calculated statistic results. Developers have limited selection of tools to analyse such differences. There are existing programs (e.g. PhysioNet's tool bXb [6]), that produce statistical differences between algorithms when comparing with predefined annotation files. They give user ability to create detailed reports, however, they are limited to usage of PhysioNet databases, with their detailed annotation files.

A useful feature to have during algorithm development is a visual representation of the algorithm faults. PhysioNet [6] toolkit does provide such features, however, the developers are limited to comparing only two algorithms or algorithm and manual annotation at a time. To compare multiple algorithms, or different parameters, users have to manually distribute comparison into multiple comparisons between two algorithms at a time, unless they create their own extensions of tools.

F. Available solutions

Standardized testing databases are used by the developers to ensure that their test data is annotated correctly so that different algorithms can be tested on it repeatedly and produce consistent and statistically relevant results. Most known and used are the PhysioBank databases [6, 11]. These databases contain measurements in WFDB (Waveform Database interface library) toolkit compatible format, as well as manual annotations and additional meta information about the measurement. Each database provides measurements of different lengths, noise levels and frequency of irregular events in the signal. This gives developers good comparison between use cases and an algorithm evaluation tool.

Using PhysioNet's toolkits and databases is now widespread practice in research and development of automatic analysis. Documentation of most tools is adequate and provides ample usage options (via Matlab or Octave plugins, standalone command line tools, programming libraries, code snippets...). However, some use cases are not documented well, and usage of tools is not always novice friendly. Tools, while supporting multiple file formats, are not expandable with new additional signal types, limiting compatibility with user-made measurements.

There are multiple solutions for viewing of ECG files. Medical institutions however, are mostly limited to their existing equipment and protocols. Software is therefore dictated by proprietary file formats that are produced by the medical recording devices. As such, software is usually expensive and thoroughly tested; medical institutions are thus not likely to switch immediately. If a common tool would enable users to view wide array of file types and process them, such software could bridge the gap between users using different file formats and equipment.

1) VisECG

VisEcg [12] is an open source program, that was build at JSI primarily for research purposes. It was originally designed to view and process short measurements, up to 30 minutes long, and consisting of multiple ECG channels. Through its lifespan, additional functionalities were added to the program. In its current form, it is capable of capturing measurements from specialized devices, and can load data from a variety of open source, and some proprietary ECG file formats. The program has been upgraded and developed further in recent time, and is now being used as the main

program for processing measurements made with the Savvy wireless ECG monitor. It is able to automatically process gathered measurements and produce short reports of them. Example of such report is shown in Figure 1. Reports are exported to PDF format in full signal resolution. As shown in Figure 2, printed out or viewed on screen, report gives physician all relevant information in a layout similar to a standard Holter output.

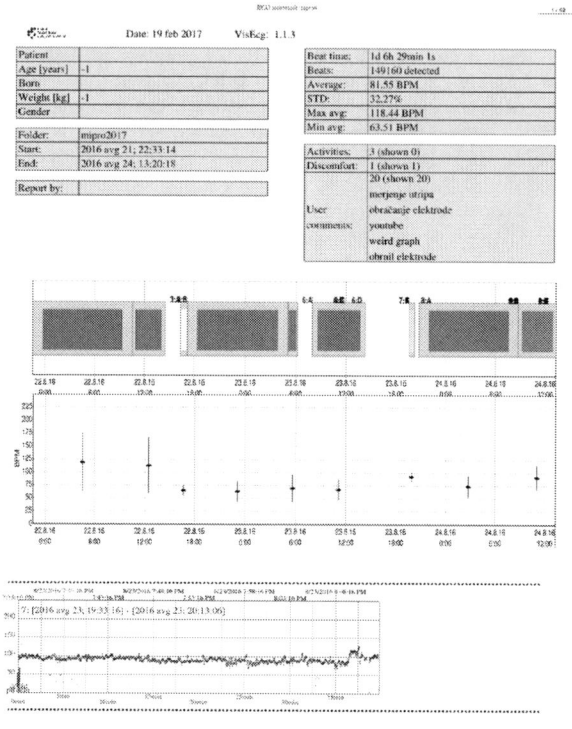

Figure 1: Example of condensed report overview output of VisEcg program.

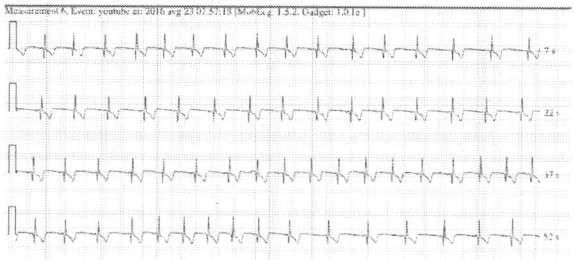

Figure 2: A sample of ECG as seen on the report.

It contains simple processing algorithms for continuous and discrete signals, and offers access to more advanced algorithms. Example of the overview of processed measurement files can be seen in Figure 3. Different colours and annotation markers give quick overview of statistic created by automatic analysis. Sizes of squares mark the length and state of measurement. Users and medical practitioners can quickly judge, by the length of measurements and statistic shown, whether a measurement should be looked through thoroughly.

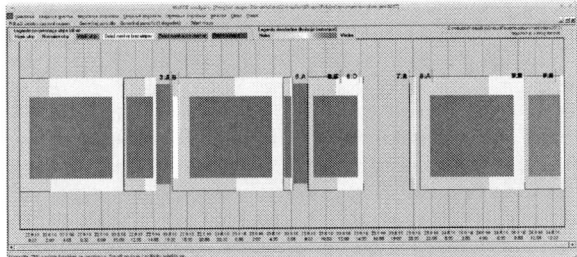

Figure 3: Overview of multiple processed files in the VisEcg program.

While reading multiple open source file formats is supported, VisEcg is able to store measurements only in its own file format. This is enforced because the program is able to generate additional signals and data through analysis, which can not be saved to existing file formats.

User interface and functionality were determined by feedback from researchers and medical experts. Newest additions to the program is an option to show an overview of long term measurements, where medical personnel can quickly assess broader state of patient In addition automatic processing of multiple measurements can be done. It can process raw data from Savvy device, determine heart beats and produce statistic of each file. This reduces the amount of work physician must do, to obtain an overview of recorded measurements.

VisEcg is being ported to newer version of programming standard, however, the application is not scalable and can be adapted to new demands only with difficulties. Using the knowledge gathered from the development of VisEcg and problems that were experienced during that time, new and improved program should be developed.

III. PROPOSED SOLUTION

We propose an open source extendible framework that is able to adapt to multiple use cases. Simplified proposed structure can is shown in Figure 4. The core functionality is composed of well defined and tested program core. It provides basic structures used throughout all the applications and ensures data verification.

API layer is exposed to the application developers for communication with lower layers. Developers can access the provided basic functionality and parameters and use them to enhance functionality (e.g. chain API commands to simplify repetitive tasks), or create standalone functionality (e.g. add additional file format support).

The last layer of the framework represents the user interface. This layer is able to use all, or just a subset of available API functionality, to provide user with access to the information from the measurements. It also uses API to send the commands issued by user to lower layers.

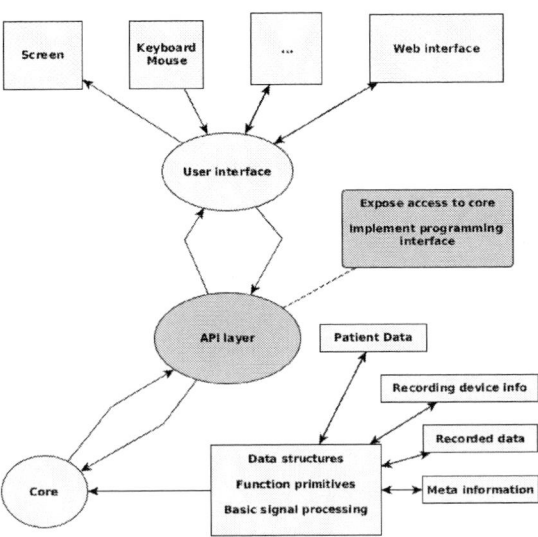

Figure 4: Proposed framework

We can draw parallels of the proposed solution with the Model-View-Controller (MVC) software design pattern. The core functionality of framework corresponds with the MVC model, user interface uses API layer as MVC controller, and results are seen as output of API layer and the user interface changes (MVC View). This enables developers to separate the back end from the front end of the application, making it easier to verify, add or remove functionality. Such separation also greatly eases potential migration to additional platforms. Added benefit is also easier code review and testing since the components are independent. This can greatly ease verification and safety audit of software.

A. Core functionality

Core of the framework must be verifiable, thoroughly tested and quick. Its main function is to provide the developers with standardized functionality and data structures for the application. It can serve also as a library, providing expected basic behaviour for third party applications. Packed as such, it should be available to use in multiple programming environments and hardware architectures. It provides main structure definitions for:

- Patient
  - relevant medical history
  - birth date
  - current prescriptions
- Recording device information
- Recorded data
  - multiple signals (discrete and continous types)
  - signal processing functionality
  - human added annotations
  - extendable signal descriptions

- Meta information
  - date of measurement
  - measurement description
  - comments

The implementation of core layer should also include already existing open source solutions where possible. In addition to the implementation of functionality for working with measurements produced by Savvy, the core layer should implement also complete functionality of PhysioNet toolkit. With ability to read, write and process WFDB file formats, the usefulness of core layer will be increased greatly for the developers.

B. API layer

Implementation of API calls enables the usage of core application functionality and the options to extend it. It exposes the core of application to different programming languages and environments. Multiple API layers can therefore be implemented to fully or only partially support core functionality, e.g. create only interface for reading files, or build upon it to expand it (e.g. integration into other software solutions). It can expand core functionality with additional processing, support for file formats or platform specific demands.

C. User interface

User interface can be implemented using multiple technologies. Communication with the core of framework is done through the API layer. This separation enables that application functionality remains the same, while the user can interact with data on different devices. With advances in web browser performance and supported technologies, web based applications are becoming mainstream. Implementation of user interface on a web platform also solves problems of communication with colleagues and processing speed. ECG processing can be offloaded to the server side, so user hardware is not being tasked.

Application running on a local machine can communicate with the core of framework directly. This provides good performance and very little overhead. However, when bypassing the API layer, the application developers have greater difficulty in ensuring the correctness of implementation. Future enhancement could also bring the functionality of running application locally, while another person is simultaneously connected and interacts with same instance over remote protocol. This layer could also bring functionality of new technology into ECG viewers.

Popularity of virtual reality platforms is growing. Virtual reality is already being used in medicine as a training and interaction tool. In the future, a novel interaction with ECG measurements using such technology could be implemented.

D. File format

For storage of ECG signal and complementary data, there are numerous file formats available. In case of measurements from Savvy, these are stored on the mobile device in binary form appropriate for streaming. When converting measurement files on the personal computer, a new file is created, which supports modified measurements and inclusion of metadata. This file contains a mixture of text and binary stored data. Ideally, this conversion would not be necessary, and one file format would be sufficient.

For usage in our framework open sourced specification file format are preferred. This ensures that future changes to the format can be transparently implemented. When comparing multiple ECG storage formats [13], the biggest difference is observed between text and binary based files. Text based (including XML) file formats usually take significantly more storage space than corresponding binary based file formats [13, 14]. Framework is primarily focused on very long measurements, and file formats that are not optimized for such files are not ideal. Some of the existing standards limit the number of allowed values in the supported fields. As such, some text based file formats limit the measurement length (e.g., HL7 format limits the length to 35656 ECG samples), or impose restrictions on type of value that can be saved (e.g. saving annotations or discrete events).

Binary files, however, are inferior to text files in some other properties. Main issue is inability of a person to read plain text of such files. This is not only important for humans, but also negatively affects computer indexing of files. Readability of text files can also be used to overview, or manually fix corrupted file, when errors occur. Second, to ensure the correct parsing of binary file, its structure must be strict. This limits future expandability and ease of software maintenance. Binary file formats usually do not impose the size limitations, and are generally faster to write directly to memory and require less processing. Binary files can also more easily support streaming, in comparison to text based XML, adding an important fail-safe in case something happens during capture. They can be unfinished, meaning that the data streams can end abruptly without compromising the file readability before the event. Small gadgets can often run out of battery or experience force majeure that abruptly ends the capture. Using stream files enables the user to recover captured measurement with minimal data loss.

To ensure the best user experience, the proposed framework will need to support multiple file types. Framework will implement readers and writers for some open source formats. Where applicable existing test implementations will be used (WFDB toolkit). For the environment built around the Savvy device, both proprietary standards will remain and will be developed further. The binary streaming file will be used where space is premium and some information might be harder to interpret. The file format implemented in VisECG will be further used where the additional post processing is done and size optimization is not as important. Such solution enables best combination of performance, future expansion and safety of information.

IV. RESULTS AND FUTURE WORK

When implementing a large framework, multiple goals must be considered. Our design is aimed at two main user groups, further expanding demands the framework has to fulfil. We believe that the proposed framework design is able to support all of the functionality.

So far we have implemented reading, writing and conversion of files from the device created binary format to the final format and basic signal processing functions. Next step is to include complete PhysioNet toolkit functionality and provide compatibility between PhysioNet formats, designed framework and output files. Following that, the work will be verified and tested before continuing with the API implementation. At this stage, an evaluation of the usability in comparison to existing solutions should be concluded, to verify that the developers find the framework easy to use. When implemented in full, the framework will be opened to public.

Our decision to continue using two file formats from the workflow of Savvy devices in the new version of program is based on the advantages of each format. On the mobile devices, we will continue using the streamable binary file, reducing changes to existing implementations. This enables an efficient use of space and fast transmissions of files, as well as backward compatibility. When captured data is transferred to the personal computer, file sizes are not as important. Using file format that is partially text based and partially binary produces best results in our use case. It enables manual indexing and fixing of corrupted files, while ensuring that long measurements are still smaller by a large factor compared to using text only files.

REFERENCES

[1] Denis Jabaudon, Juan Sztajzel, Katia Sievert, Theodor Landis, and Roman Sztajzel. Usefulness of ambulatory 7-day ecg monitoring for the detection of atrial fibrillation and flutter after acute stroke and transient ischemic attack. *Stroke*, 35(7):1647–1651, 2004.

[2] S Suave Lobodzinski and Michael M Laks. New devices for very long-term ecg monitoring. *Cardiol J*, 19(2):210–214, 2012.

[3] Norman J Holter. New method for heart studies. *Science*, 134(3486):1214–1220, 1961.

[4] Emil Jovanov, Pedro Gelabert, Reza Adhami, Bryan Wheelock, and Robert Adams. Real time holter monitoring of biomedical signals. In *DSP Technology and Education conference DSPS*, volume 99, pages 4–6, 1999.

[5] Paddy M Barrett, Ravi Komatireddy, Sharon Haaser, Sarah Topol, Judith Sheard, Jackie Encinas, Angela J Fought, and Eric J Topol. Comparison of 24-hour holter monitoring with 14-day novel adhesive patch electrocardiographic monitoring. *The American journal of medicine*, 127(1):95.e11, 2014.

[6] Ary L Goldberger, Luis AN Amaral, Leon Glass, Jeffrey M Hausdorff, Plamen Ch Ivanov, Roger G Mark, Joseph E Mietus, George B Moody, Chung-Kang Peng, and H Eugene Stanley. Physiobank, physiotoolkit, and physionet. *Circulation*, 101(23):e215–e220, 2000.

[7] Jiapu Pan and Willis J Tompkins. A real-time qrs detection algorithm. *IEEE transactions on biomedical engineering*, (3):230–236, 1985.

[8] Primož Lavrič and Matjaž Depolli. Robust beat detection on noisy differential ecg. In *Information and Communication Technology, Electronics and Microelectronics (MIPRO), 2016 39th International Convention on*, pages 381–386. IEEE, 2016.

[9] Jure Slak and Gregor Kosec. Detection of heart rate variability from a wearable differential ecg device. In *Information and Communication Technology, Electronics and Microelectronics (MIPRO), 2016 39th International Convention on*, pages 430–435. IEEE, 2016.

[10] Miha Mohorčič and Matjaž Depolli. Heart rate analysis with nevroekg. In *Information and Communication Technology, Electronics and Microelectronics (MIPRO), 2016 39th International Convention on*, pages 467–472. IEEE, 2016.

[11] George B Moody, Roger G Mark, and Ary L Goldberger. Physionet: a web-based resource for the study of physiologic signals. *IEEE Engineering in Medicine and Biology Magazine*, 20(3):70–75, 2001.

[12] M. Šterk B. Meglič V. Švigelj. R. Trobec, V. Avbelj. Neurological data measuring and analysis software based on object oriented design. *In proceedings of The 7th European Federation of Autonomic Societies (EFAS) Meeting*, 2005.

[13] Raymond R Bond, Dewar D Finlay, Chris D Nugent, and George Moore. A review of ecg storage formats. *International journal of medical informatics*, 80(10):681–697, 2011.

[14] Wilfred Ng, Wai-Yeung Lam, and James Cheng. Comparative analysis of xml compression technologies. *World Wide Web*, 9(1):5–33, 2006.

Finding a Signature in Dermoscopy: A Color Normalization Proposal

Marlene Machado, Jorge Pereira, Miguel Silva & Rui Fonseca-Pinto

* Instituto de Telecomunicações, *Multimedia Signal Processing- Lr*, Leiria, Portugal
** Polytechnic Institute of Leiria, Leiria, Portugal
{mmachado; jpereira; msilva}@co.it.pt, rui.pinto@ipleiria.pt

Abstract - Digital image methodologies related with Melanoma has become in the past years a major support for differential diagnosis in skin cancer. Computer Aided Diagnosis (CAD) systems, encompassing image acquisition, artifact removal, detection and selection of features, highlight Machine Learning algorithms as a novel strategy towards a digital assisted diagnosis in Dermatology.

Although the central role played by color in dermoscopic image assessment, Machine Learning algorithms mainly use texture and shape features, derived from gray level images, obtained from the true color images of the skin. Since the acquisition conditions are key for the color characterization and thus, central for the quantification of different colors in a dermoscopic image, this work presents a strategy for color normalization, joint with its use in the calculation of the number of colors of a dermoscopic image.

This methodology will contribute to the uniformity in the use of color features extracted from different datasets in CAD systems (acquired by distinct dermoscopes) possibly presenting distinct illumination characteristics. This normalization proposal can also be applied as an image preprocessing step, aimed to achieve higher scores in the standard metrics in ML algorithms.

I. INTRODUCTION

Dermoscopy is a non-invasive imaging technique used to obtain digital images of the skin surface to assess pigmented structures or vessels, (i.e. clues in the epidermis and superficial dermis), constituting a valued source of signs in skin cancer differential diagnosis [1-5]. The use of dermoscopy has proven to be successful in terms of increasing diagnostic sensitivity and specificity when compared with human eye assessment [6]. Automatic classification systems mostly use texture and shape features to characterize images and to train the classifiers. Still, color information is a source of valuable evidence regarding image classification (with special relevance in the classification of melanocytic lesions whose differential diagnosis encompasses melanoma in the diagnostic list), in particular, by the identification of the number of distinct colors in a given image.

The use of color features in Machine Learning (ML) algorithms, although being a source of relevant information, can be tailored by a color normalization strategy, and be used in non-uniform illumination conditions, enabling by this way the use of datasets obtained by different dermoscopes.

This work presents a color normalization approach used to define a color signature in dermoscopy, followed by its application using in real examples to identify the number of colors in a given dermoscopic image. Besides the proposed normalization procedure (allowing uniformity in the input data to be used in ML) the proposed methodology establishes a quantification strategy to define the number of colors (whose information is determinant to the classification algorithms).

II. COLOR SIGNATURE PROPOSAL

When working in dermatology diagnosis, often one needs to compare images obtained at different times (in the context of the follow-up in suspicious melanocytic skin lesions). These images can be acquired using different illumination conditions, but also using different dermoscopes. Thus, on assessing to color information of the lesions, some values on the RGB space may present variations due to the illumination and possibly to other technical issues. Moreover, the RGB color space is characterized by its lack of linearity in the perceptually definition of colors (sRGB), making the color identification as a challenging task. In order to avoid these issues, it is key implement a normalization strategy.

A. Methodology

The proposed method is based on the principle that linear illumination, reflecting a specific color, can change individual values of RGB space, however the algebraic difference (RG=R-G, RB=R-B, GB=G-B) between channels stays roughly constant. Hence, the histogram of the newly defined channels can be used to create a color signature based on unimodal histograms.

The values of the channels difference may vary among -255 and 255 (in this work), achieving a total range of 511 different values. In order to improve the computational time, while preserving the discriminative feature between several colors, this range is grouped into 30 different sets of color classes. By applying this procedure, every image is now represented by a color signature curve defined by a 30 class array that contains the frequency of each class, as can be observed in observed in Figure 1.

When using real images it is common a multimodal signature (typically, one lesion has at least two colors) due to some overlapping between colors in one or more channels.

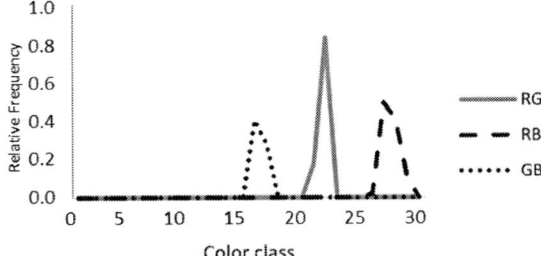

Figure 1: Example of a unimodal color signature

In order to avoid this overlapping issue, a practical solution is proposed and described in the following. Every image is divided into 300 sub-images, decreasing by this way the probability of analyzing areas with high variation of colors and increasing the probability of computing unimodal color areas. It is worth emphasizing that a given color is fully defined by the concordant combination of the three channels.

To implement the proposed methodology, for each different image data set, one image is selected and manually elected one region having unimodal color signature for each of the common six colors in dermoscopy (B-black, Bl-blue, Br-brown, Db-dark brown, R-red and W-white). Next, each image to be analyzed is subdivided (as explained above) and compared with the predefined signatures for the six colors.

B. Results and Conclusion

Using the approach presented in this work, it is possible to apply the methodology in each sub-image and then produce a histogram to establish the number of distinct colors in the whole image (in the example presented in Figure 2, the light-brown color was identified). In order to illustrate the performance of this approach, in the following some examples will be shown.

In Figure 3 is possible to find a typical melanocytic image, in which is possible to found two brown tones and a blue like area. In fact, as is possible to observe, this color identification is achieved by the method.

In Figure 4 a heterogeneous image presenting tones for vascularization but also deep melanin deposition is submitted to analysis. In this case, as is shown in the bar plot, 5 different colors were detected.

As is possible to observe by the two examples in Figure 3 and Figure 4, the definition of the initial color signature is determinant to the succeeded output of the number of colors presented in the image (especially in images with blurred color zones). To address this issue, besides the initial color signature, a ±5% and also a ±10% range was added to the code, to give some flexibility to the algorithm.

Marlene Machado, Jorge Pereira and Miguel Silva were with Instituto de Telecomunicações, in DERMCALSS project, cofounded by MaisCentro under the EU Program CENTRO-07-ST24-FEDER-002022.

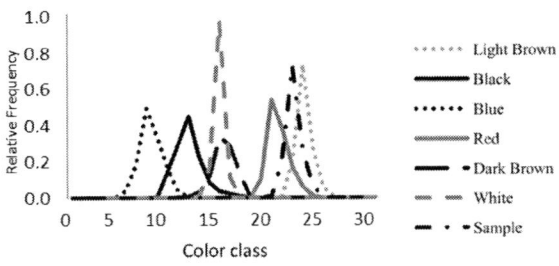

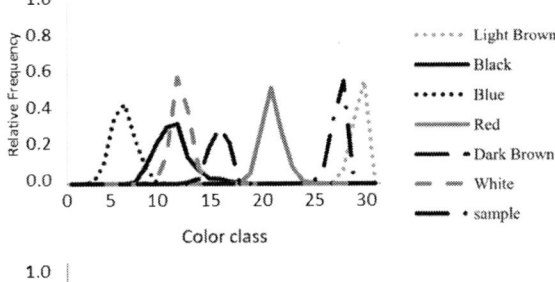

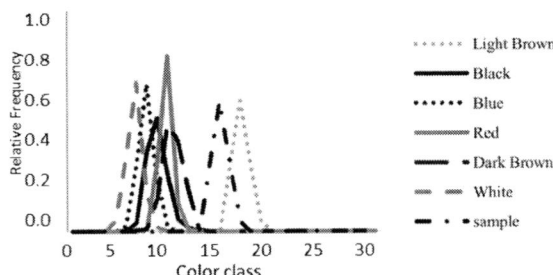

Figure 2: Example of color signature representation. Top: RG, Meddle: RB and Bottom: GB.

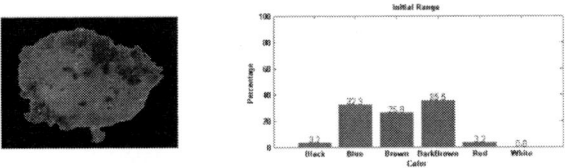

Figure 3: Dermoscopic image at left and color percentage at right.

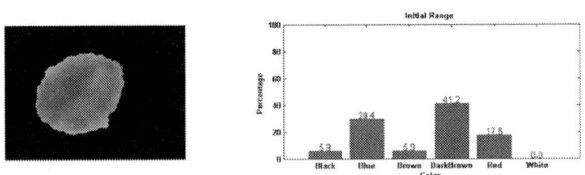

Figure 4: Dermoscopic image at left and color percentage at right.

In Figure 5 it can be found the study with the initial range and the other 4 identified alternatives for another dermoscopic image.

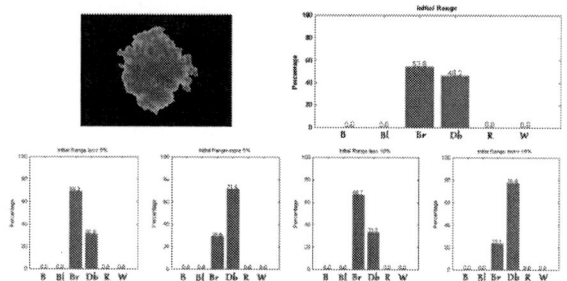

Figure 5: Lesion with 2 colors and graphics with percentage of color detected in the five ranges of color. *Legend: B-black, Bl-blue, Br-brown, Db-dark brown, R-red and W-white.*

In this case reported at Figure 5, all five tested options for initial cluster color variations conduct to the same number of colors, resulting in two (B-brown and Db-dark brown).

The number of colors in a dermoscopic image is an important clue to establish a set of discriminative features to the early detection of melanoma. This methodology establishes a color normalization strategy and can be used to increase the number of features in an automatic classification algorithm.

As a future work, this normalization will be conducted as a pre-processing step and tested against other methods to assess its contribution in the performance of Machine Learning algorithms used in dermoscopy.

REFERENCES

[1] Q. Abbas, M. E. Celebi, C. Serrano, I. Fondón García, and G. Ma, "Pattern classification of dermoscopy images: A perceptually uniform model," *Pattern Recognit.*, vol. 46, no. 1, pp. 86–97, Jan. 2013.

[2] G. Argenziano, C. Catricalà, M. Ardigo, P. Buccini, P. De Simone, L. Eibenschutz, A. Ferrari, G. Mariani, V. Silipo, I. Sperduti, and I. Zalaudek, "Seven-point checklist of dermoscopy revisited," *Br. J. Dermatol.*, vol. 164, no. 4, pp. 785–790, Apr. 2011.

[3] M. Machado, J. Pereira, and R. Fonseca-Pinto, "Classification of reticular pattern and streaks in dermoscopic images based on texture analysis.," *J. Med. imaging (Bellingham, Wash.)*, vol. 2, no. 4, p. 44503, Oct. 2015.

[4] J. Pereira and R. Fonseca-Pinto, "Segmentation Strategies in Dermoscopy to Follow-up Melanoma: Combined Segmentation Scheme," *Online J. Sci. Technol.*, vol. 5, no. 3, 2015.

[5] A. I. Mendes, C. Nogueira, J. Pereira, and R. Fonseca-Pinto, "On the geometric modulation of skin lesion growth: A mathematical model for melanoma," *Rev. Bras. Eng. Biomed.*, vol. 32, no. 1, pp. 44–54, 2016.

[6] D. Piccolo, A. Ferrari, K. Peris, R. Diadone, B. Ruggeri, and S. Chimenti, "Dermoscopic diagnosis by a trained clinician vs. a clinician with minimal dermoscopy training vs. computer-aided diagnosis of 341 pigmented skin lesions: a comparative study.," *Br. J. Dermatol.*, vol. 147, no. 3, pp. 481–6, Sep. 2002.

A Textured Scale-based Approach to Melanocytic Skin Lesions in Dermoscopy

Rui Fonseca-Pinto & Marlene Machado

* Instituto de Telecomunicações, *Multimedia Signal Processing - Lr*, Leiria, Portugal
** Polytechnic Institute of Leiria/School of Technology and Management, Leiria, Portugal
rui.pinto@ipleiria.pt; mmachado@co.it.pt

Abstract - Melanoma is the most dangerous and lethal form of human skin cancer and the early detection is a fundamental key for its successful management. In recent years the use of automatic classification algorithms in the context of Computer Aided Diagnosis (CAD) systems have been an important tool, by improving quantification metrics and also assisting in the decision regarding lesion management. This paper presents a novel and robust textured-based approach to detect melanomas among melanocytic images obtained by dermoscopy, using Local Binary Pattern Variance (LBPV) histograms after the Bidimensional Empirical Mode Decomposition (BEMD) scale-based decomposition methodology.

The results show that it is possible to develop a robust CAD system for the classification of dermoscopy images obtained from different databases and acquired in diverse conditions. After the initial texture-scale based classification a post-processing refinement is proposed using reticular pattern and color achieving to 97.83, 94.44 and 96.00 for Sensitivity, Specificity and Accuracy.

I. INTRODUCTION

A particular challenge for dermatologists is the detection of melanomas at an early stage, before the beginning of the tumor cells dissemination through the body, whose manifestation is the metastatic process. Despite being a very aggressive form of cancer, the melanoma is also associated with high survival rates when early detected. The prognosis is directly related to the early detection of lesions, which should be excised as soon as possible when the signs of malignancy are present [1].

The imaging technique used to screen for melanocytic lesions is a non-invasive imaging procedure applying an external light source and a magnifying lens, known as dermoscopy. Dermoscopy enables the detection of surface skin structures, which are not possible to identify with the naked eye, by identifying clues to evaluate skin lesions assessed by diagnostic algorithms and scores (e.g. ABCD rule [2], Menzies method [3], and the seven-point checklist [4]).

Due to the limitations of the human eye perception, and to the variability of the intra- and inter-observer diagnostics, the use of computational algorithms constitutes a major help in the identification of differential structures with clinical correlation. In fact, the textured pattern holds characteristics that have been successfully used in recent research studies.

These Computer Aided Diagnosis (CAD) systems are an important tool to assist dermatologists in clinical diagnosis, especially in the stratified analysis using step algorithms. Its use allows for a criterious lesion segmentation, to artifact removal, to lesion borders detection and its associated metrics (e.g. regularity, symmetry, axis, eccentricity,...) whose quantification is vital to the lesion scoring.

Recently several CAD approaches to recognize melanoma in dermoscopy images were proposed [5-8]. These approaches use single features (e.g. color, shape or texture) or combined features (e.g. color and texture or shape and texture) to characterize the lesion, reporting suitable results in terms of accuracy. However, the comparison between different approaches is not trivial, as these works either uses datasets with distinct technical characteristics, or the used assessment metrics are not uniform.

This paper presents an approach for the automatic classification of images from melanocytic skin lesions, obtained from three different databases, whose preprocessing avoids lesion segmentation. The proposed methodology starts by decomposing the image in a set of empirically derived components using a scale-dependent method, the Bidimensional Empirical Mode Decomposition (BEMD). Next, the Local Binary Pattern Variance (LBPV) technique is applied to extract, from BEMD components, a histogram of texture features used to train a classifier, and to perform a binary classification: melanoma or benign lesion. After this initial classification, a post-processing step using color and differential structures was applied to refine the obtained results in the automatic classification. The remainder of this paper is organized as follows: in Section 2 BEMD and LBPV algorithms are presented; in Section 3 the method used to evaluate the performance of the melanoma classification based on texture is described. The experimental results are shown in Section 4 and finally in Section 5 this work ends with a conclusion section.

II. TEXTURE ANALYSIS METHODOLOGIES

A textured image possesses visual patterns formed from features such as brightness, color, shape and scale. The most commonly used multi-scale methods to extract texture features in dermoscopic images are Gabor filters [9], Wavelets [10] and Curvelets [11] transforms. In the past decades, some authors have explored the multi-scale BEMD approach and LBP statistics for image texture analysis, as reported in [10-13] . Here an adaptation to the methodology is applied in dermoscopy.

A. Bidimensional Empirical Model Decomposition

The Empirical Mode Decomposition (EMD) is a multi-scale analysis methodology proposed by Huang et al. [14] for non-linear and non-stationary time series. The main idea is to obtain a decomposition of the signal into a set of signal derived basis functions, known as Intrinsic Mode Functions (IMFs) or empirical modes. The IMFs are obtained at each scale from fine (first IMF) to coarse (last IMF), by an iterative procedure called sifting process. The use of EMD has been applied successfully to solve the assumption of stationarity and linearity properties in classical signal processing. This property has encouraged the use of EMD in many practical applications in several research areas, and leads to the extension of the idea to image processing. The first use in image processing was presented in [15] in the context of dermoscopic artifact removal, as a simple extension of 1D process. The methodology was extended to the 2D case and was formally established by Bidimensional Empirical Mode Decomposition (BEMD) [12].

For a given image $I(x,y)$, the BEMD algorithm could be computed as an extension of de 1D signal, once the original image can be reconstructed as presented in (1), where d_k is the kth 2D IMF(mode) and r_k is the residual of d_k.

$$I(x, y) = \sum_{k=1}^{K} d_k(x, y) + r_k(x, y) \quad (1).$$

Details regarding the decomposition and the sifting process can be found in [12].

B. Feature descriptor: Local Binary Pattern Variance

Local Binary Patterns (LBP) were originally presented to describe local structures in grey level texture images. The LBPs descriptors are computed by comparing each image pixel with its circular neighborhood, being defined by (2), with s as unit step function.

$$LBP_{P,R} = \sum_{p=0}^{P-1} s(g_p - g_c) 2^p \quad (2)$$

In (2), P and R are the number of involved neighbors and the radius of neighborhood respectively; g_c and g_p are the value of the central pixel and the value of its neighbors. The LBP pattern of each pixel (i,j) of an image with M×N resolution is used to identify the image texture represented by the histogram of LBP defined in (3) with and here K is the maximal LBP pattern value.

$$H(K) = \sum_{i=1}^{M} \sum_{j=1}^{N} f(LBP_{P,R}(i, j), K) \quad (3)$$

Contrast and spatial structure can also be used to represent texture. The characteristics to define the spatial structure and the contrast can be, respectively, LBP and the variance (VAR) of local image texture. Local Binary Pattern Variance (LBPV), a combined LBP and contrast distribution method, does not intervene in the LBP histogram calculation (3). Indeed, the variance $VAR_{P,R}$ is used as an adaptive weight to adjust the contribution of the LBP code at the histogram calculation, as defined in (4), (5).

$$LBPV_{P,R}(k) = \sum_{i=1}^{M} \sum_{j=1}^{N} w(LBP_{P,R}(i, j), k) \quad (4)$$

$$w(LBP_{P,R}(i,j),k) = \begin{cases} VAR_{P,R}(i,j), & LBP_{P,R}(i,j) = k \\ 0 & otherwise \end{cases} \quad (5)$$

III. METHODOLOGY: LBPV HISTOGRAMS OF 2D IMFS

The BEMD approach, due to its data driven property, has the potential to produce further representative results than classical Fourier based methods, or even, when using Wavelet and other decomposition algorithms to extract intrinsic components of texture [12]. In this paper an efficient usage of BEMD combined with LBPV for texture classification of melanomas in dermoscopy images is presented.

A. Image Preprocessing

Different from other methodologies used in dermoscopic machine learning algorithms, the process of detecting and removing artifacts, and the previous segmentation, it was not followed in this approach. As a preprocessing step the image conversion from RGB into a gray-level image was performed by selecting the highest entropy channel in accordance to the suggestion presented in [7]. Next, a cropping operation to the image is applied to reduce the background interference. Accordingly, all images in this work present a fixed resolution of 512×512.

B. Feature extraction and image classification

The set of features extracted from an image, settled in a vector form, constitutes a representation of the image itself. Among the class of features described in the literature (color, texture and shape), the texture form was primary elected in this work. Local features using LBPV descriptor was used to identify structures from the output of the image decomposition in modes in the form of 2D IMFs. The process is described in the following steps: i) firstly, the image is decomposed into three 2D IMFs. (ii) Secondly, a new 2D IMF image is created through the sum of the second and third 2D IMFs. The first 2D IMF contains the finest details of image, such as noise and artifacts, so it is discarded in this study. In the second 2D IMF remains the detail of the image, and the third 2D IMF represent information in large scale. (iii) Thirdly, the LBPV histogram of the new 2D IMF image is obtained and the feature vector is extracted. (iv) Finally, an automatic classifier is applied to the feature vector to obtain the training model.

The AdaBoost algorithm was chosen for this purpose as it simultaneously selects the most appropriate features and trains the classifier. From the initial set of features, the AdaBoost algorithm selects a small subset, the weak classifier, to building a final strong and well-performing classifier.

IV. RESULTS AND DISCUSSION

A. Dataset and Evaluation Metrics

For the construction of the training model, images from the Derm101 [16] and PCDS databases [17] were used. Three training models, with different number of images were used to ascertain the influence of the training set size and the effectiveness of using only texture-based features in the classification of melanoma lesions.

To test and evaluate the discrimination power of the proposed approach, extensive experiments were implemented using images from three databases: Derm101 [16], Hosei Dataset [18] and PCDS Database [17]. From these online databases, two quite heterogeneous sets were used: one with 200 RGB images for test and other with 240 RGB images for validation. The images were converted into a gray-level image, cropped and resized for 512×512.

In order to demonstrate the robustness of the approach, a stratified 10-fold cross validation procedure was applied, dividing the dataset (240 images) into ten subsets, each one with approximately the same number of melanomas and benign lesions. Nine folds were used for training the classifier and one is used for testing.

For training and validation purposes each image was labeled as melanoma or benign (ground truth label). These images were obtained by dermatologists during clinical exams and stored in JPEG format. To evaluate the performance of the present approach, three metrics were computed: Sensitivity (SE), Specificity (SP) and Accuracy (ACC).

B. Experimental Results

According to the definition in (2), there are several possible values of P and R to extract the LBPV histogram from the IMF image. Among all possibilities, the combination of P and R values used in the work were (8,1), (16,3), and (24,5) being the last one the combination achieving better results.
The influence of the training set size was estimated by means of three experiments. In the Experiment 1, the initial set was formed by 70 images (32 melanomas and 38 benign lesions) for training the classifier. In the Experiment 2 and the Experiment 3, 70% (10 melanomas and 38 benign lesions) and 50% (10 melanomas and 25 benign lesions) images of the initial set were respectively used. From these experiments (see Table 1) it was possible to verify that: (i) increasing the number of melanomas in the training set decreases SP; (ii) only 2.5% of benign images are necessary to obtain a performance of 88% in ACC; and (iii) the best result was obtained in the Experiment 2 with SE = 92,39% and SP = 87,96%. The results of this experimental procedure show a dependence on the choice of the images that are selected for training, highlighting the importance of the training set composition.

The robustness of this approach was reinforced by performing a cross validation process (10-fold cross validation repeated 100 times), conducting to promising results. The lower obtained value of ACC was 75%, the higher was 84% and the greatest classification results in terms of SE-SP relationship was SE = 87,25%, SP = 80,43% and ACC = 83,33 (see Table 2).

TABLE 1: CLASSIFICATION RESULTS FOR DIFFERENT TRAINING SETS. M-MELANOMA, NM – NON MELANOMA

	# Training images		#Tested images	SE (%)	SP (%)	ACC (%)
	M	NM				
Exp 1	32	38	200	90.22	78.70	84.00
Exp 2	10	38	200	92.39	87.96	90.00
Exp 3	10	25	200	90.22	86.11	88.00

To achieve the real performance of this methodology in dermoscopic images, a comparison with other texture-based methods described in the literature for melanoma classification is central. A qualitative assessment of such results is difficult as the databases and the number of images used for validation are different, as is possible to observe in [5], [7], [9-11]. An empirical comparison can be performed by observing Table 2. Regarding the evaluation using cross validation, in the presented work, improved results were obtained when comparing this study with Situ et al. [9] in terms of accuracy, and those of Barata et al. [7] in terms of specificity. The results from Garnavi et al. [10] report stronger accuracy. Regarding the evaluation in a single test set, this approach shows better results than Mahmound et al. [11] and similar results when compared to Sheha et al. [5]. Furthermore, the results reported here were obtained using images from quite heterogeneous databases, no preprocessing artifact removal were implemented, and using a large (when compared with others) number of images in the test set.

B1. Performance improvement

Although the classification results above presented were promising, these findings can be improved with a post-processing step, following a clinical guided hallmark for the lesion classification: number of colors and the presence of differential structures.

TABLE 2: RESULTS OF THE TEXTURE-BASED PUBLICATIONS IN THE DETECTION OF MELANOMAS.

Method	#Tested images	SE (%)	SP (%)	ACC (%)	Coss Validation
Situ et al.	~ 23	-	-	82	8-fold
Sheha et al.	~ 26	-	-	92	-
Barata et al.	1	96	59	-	Leave-one-out
Garvani et al.	~10	-	-	88	10-fold
Mahmoud et al.	90	-	-	70	-
Proposed	200	92.39	87.96	90	-
Proposed	24	87.25	80.43	83.33	10-fold

In order to increase the performance of the approach presented here, an analysis of the wrongly classified lesions in terms of the number of colors and presence of the differential structures (reticular pattern) was performed.

The number of used colors was based on six colors of ABCD rule algorithm (light brown, black, blue, red, dark brown and white).

Regarding the lesions incorrectly classified as melanomas (false positives) it was applied the texture approach based on Curvelet transform to the detection of reticular pattern, as we presented in [19]. To the malign lesions classified as benign (false negative) an approach to counting the number of colors of the lesion was applied. The final results after this post-processing classification are presented in Table 3.

TABLE 3: POST-PROCESSING RESULTS. M-MELANOMA, NM – NON MELANOMA

	# Training images		#Tested images	SE (%)	SP (%)	ACC (%)
	M	NM				
Exp 1	32	38	200	97.83	86.11	91.50
Exp 2	10	38	200	97.83	94.44	96.00
Exp 3	10	25	200	97.83	88.89	93.00

V. CONLUSION

The use of automatic classification in dermoscopy is a strong tool to access dermatologists in clinical diagnosis and in the early detection of skin cancer. In this work, a novel approach is introduced for melanoma classification using LBPV histograms of and a 2dIMF derived image. This approach achieves encouraging results with SE = 87.25%, SP=80.93% and ACC = 83.33% when a single test set and the cross validation process is used for evaluation. After a post-processing procedure using the number of colors and the presence of reticular pattern the classifier performance achieved 97.83, 94.44 and 96.00 respectively for SE, SP and ACC. These results were obtained using images from three quite heterogeneous databases using the original image (without previous preprocessing and segmentation) achieving robustness against the presence of distortive artifacts.

REFERENCES

[1] D. C. Whiteman, W. J. Pavan, and B. C. Bastian, "The melanomas: a synthesis of epidemiological, clinical, histopathological, genetic, and biological aspects, supporting distinct subtypes, causal pathways, and cells of origin," *Pigment Cell Melanoma Res.*, vol. 24, no. 5, pp. 879–897, Oct. 2011.

[2] W. Stolz, A. Riemann, and A. Cognetta, "ABCD-rule of dermatoscopy: a new practical method for early recognition of malignant melanoma," *Eur J Dermatol*, vol. 7, pp. 521–28, 1994.

[3] S. W. Menzies, C. Ingvar, K. A. Crotty, and W. H. McCarthy, "Frequency and morphologic characteristics of invasive melanomas lacking specific surface microscopic features.," *Arch. Dermatol.*, vol. 132, no. 10, pp. 1178–82, Oct. 1996.

[4] G. Argenziano, C. Catricalà, M. Ardigo, P. Buccini, P. De Simone, L. Eibenschutz, A. Ferrari, G. Mariani, V. Silipo, I. Sperduti, and I. Zalaudek, "Seven-point checklist of dermoscopy revisited," *Br. J. Dermatol.*, vol. 164, no. 4, pp. 785–790, Apr. 2011.

[5] M. S.Mabrouk, M. A.Sheha, and A. Sharawy, "Automatic Detection of Melanoma Skin Cancer using Texture Analysis," *Int. J. Comput. Appl.*, vol. 42, no. 20, pp. 22–26, Mar. 2012.

[6] Q. Abbas, M. E. Celebi, C. Serrano, I. Fondón García, and G. Ma, "Pattern classification of dermoscopy images: A perceptually uniform model," *Pattern Recognit.*, vol. 46, no. 1, pp. 86–97, Jan. 2013.

[7] C. Barata, M. Ruela, M. Francisco, T. Mendonca, and J. S. Marques, "Two Systems for the Detection of Melanomas in Dermoscopy Images Using Texture and Color Features," *IEEE Syst. J.*, vol. 8, no. 3, pp. 965–979, Sep. 2014.

[8] M. Machado, J. Pereira, R. Fonseca-Pinto, M. Machado, J. Pereira, and R. Fonseca-Pinto, "Reticular pattern detection in dermoscopy: an approach using Curvelet Transform," *Res. Biomed. Eng.*, vol. 32, no. 2, pp. 129–136, Jun. 2016.

[9] N. Situ, X. Yuan, J. Chen, and G. Zouridakis, "Malignant melanoma detection by Bag-of-Features classification," in *2008 30th Annual International Conference of the IEEE Engineering in Medicine and Biology Society*, 2008, vol. 2008, pp. 3110–3113.

[10] R. Garnavi, M. Aldeen, and J. Bailey, "Computer-Aided Diagnosis of Melanoma Using Border- and Wavelet-Based Texture Analysis," *IEEE Trans. Inf. Technol. Biomed.*, vol. 16, no. 6, pp. 1239–1252, Nov. 2012.

[11] M. Mahmoud and A. Al-Jumaily, "The automatic identification of melanoma by wavelet and curvelet analysis: study based on neural network classification," *Intell. Syst. (HIS), ...*, 2011.

[12] J. C. Nunes, S. Guyot, and E. Delechelle, "Texture analysis based on local analysis of the Bidimensional Empirical Mode Decomposition," *Mach. Vis. Appl.*, vol. 16, no. 3, pp. 177–188, May 2005.

[13] J. Pan and Y. Tang, "Texture Classification based on Bidimensional Empirical Mode Decomposition and Local Binary Pattern," *Int. J. Adv. Comput. Sci. Appl.*, vol. 4, no. 9, 2013.

[14] N. E. Huang, Z. Shen, S. R. Long, M. C. Wu, H. H. Shih, Q. Zheng, N.-C. Yen, C. C. Tung, and H. H. Liu, "The empirical mode decomposition and the Hilbert spectrum for nonlinear and non-stationary time series analysis," *Proc. R. Soc. London A Math. Phys. Eng. Sci.*, vol. 454, no. 1971, 1998.

[15] R. Fonseca-Pinto, P. Caseiro, and A. Andrade, "Image Empirical Mode Decomposition (IEMD) in Dermoscopic Images: Artefact Removal and Lesion Border Detection," in *Proceedings of Signal processing pattern recognition and applications (SPRRA)*, 2009, pp. 6978–84.

[16] "DERM101," [Online]. Available: http://www.derm101.com. [Accessed 2014].

[17] "PCDS - Primary Care Dermatology Society," 1994. [Online] http://www.pcds.org.uk.

[18] H. Zhou, "Medical Imaging," [Online] http://howardzzh.com

[19] M. Machado, J. Pereira, and R. Fonseca-Pinto, "Classification of reticular pattern and streaks in dermoscopic images based on texture analysis.," *J. Med. imaging (Bellingham, Wash.)*, vol. 2, no. 4, p. 44503, Oct. 2015.

Remarks on Visualization of Fuzziness of Cardiac Data

J. Opiła* and T. Pełech-Pilichowski*
* AGH University of Science and Technology, Krakow, Poland
jmo@agh.edu.pl, tomek@agh.edu.pl

Abstract - In contemporary health science sophisticated apparatus delivers a lot of data on vital processes in patients. All of them are processed as a bulk of numbers not suitable directly for diagnosing or research purposes. Moreover, which is common in biomedical sciences, measured data are intrinsically inaccurate, i.e., fuzzy. In order to overcome these deficiencies a set of visualization methods has been developed as well as dedicated file formats. In the paper authors discuss selected formats and imaging techniques useful for cardiologists. Problems of medical data processing is outlined. Strengthens and weaknesses of raw STL file format are analyzed. Visualization styles of data fuzziness using experimental package ScPovPlot3D based on POVRay are proposed and discussed.

I. INTRODUCTION

A wide application of ICT solutions in medicine is related to advanced data retrieving, processing and data visualization. In particular, in cardiology reliable data visualization and related calculations results in a possibility of making an accurate diagnosis thus successful treatment. Despite a number of advanced hardware and software solutions, one may indicate an impediment with data synchronization issues got from heterogeneous sources. Additionally, in many cases, apart from skills of technical stuff, applied algorithms and an expert experience, the real problem is ability of processing not only sharp shapes (as an effect of calculations typically encumbered with errors, but also an image obscured with a noise. Thus, the goal is to seek new ways of advanced visualization of fuzzy data. Note, that algorithms designed for data visualization are, in fact, able to provide data useful for further purpose, such as an automated classification of stenoses,

Today one can observe a persistent quest for novel methods of visualization in order to get insight into complex phenomena in different scientific domains including cardiology. While numerous research teams achieved excellent results, still there are some issues connected with imaging of fuzzy data thus visualization of uncertainty. In following the experimental visualization toolkit ScPovPlot3D is proposed as both simple data (pre)processing tool and sophisticated generator of input file required by POVRay [23].

Main research aim of presented work was to search

The work was supported by AGH University of Science and Technology.under grant: 11/11.200.327, job 6.

for most effective manner of displaying fuzziness of data gathered by means of variety cardiac vessels imaging tools.

II. VISUALIZATION OF SURFACES

A. ScPovPlot3D as a tool for visualization of surfaces

In recent years a set of templates for POVRay, named ScPovPlot3D (Scientific POVRay Plots in 3D) has been developed [30]. Beginning with simple histogram chart some years ago it can now help to prepare a range of 3D charts including raw or smoothed 3D surfaces, wireframe "surfaces" and surfaces based on irregularly dispersed data [11],[12],[13],[14] (employing simple kriging method [6]) available in the stable branch. Experimental modules include procedures for i.a. fuzzy numbers (IntervalPoints.inc), hybrid charts (Histogram.inc, Potential.inc), vector field and advanced potential visualization (modules Vector.inc and Potential.inc [10],[32]). Moreover helper modules deliver easy to use items like smart cameras (Cameras.inc library), coordinate systems (CoordsSys.inc) as well as text with enhanced formatting (TextExt.inc). In the following chapters three prototyped visualization styles of fuzzy surfaces using the ScPovPlot3D templates are discussed.

A problem of the fuzzy surfaces visualization has been originally addressed in the paper [14]. In the range of problems uncertainty is distributed over surface z = f(x, y), or more generally f(x,y,z)=0. usually defined over rectangular domain. The geology is one of sciences widely employing surface visualization to presentation of distribution of mineral resources. Such a distribution may be sampled using wide range of exploratory geophysical techniques. Unfortunately, usually such mapping could not be done over rectangular grid required by common visualization software thus measurement points are irregularly distributed over survey field, not necessarily rectangular. By applying algorithm widely known as kriging ([6], [14] and fig. 1) a regular grid may be evaluated but values computed in every node are subject to uncertainty resulting from uncertainty of measured values, round-off errors and fuzziness imprinted into kriging algorithm. Such uncertainty may be expressed differently, including standard deviation measure as well as fuzzy number paradigm [16]. Possible solution is to draw "thick surface" in contrast to "infinitely thin surface" commonly used. Package ScPovPlot3D cannot explicitly implements "thick" or "fat" surfaces but can simulate this quality utilizing two features of POVRay: prism object and media statement (note, media is a specific to POVRay implementation of particles). Additionally in search for the balance between efficiency and quality variant of "media" statement is proposed, ie. "pseudoparticles design".

Three examples of application of fuzzy surfaces are presented in the work. The first one shows surface defined by fuzzy relationship $z = z(x, y)$ where z denotes fuzzy number. The second one shows lesion deteriorating blood flow deposited in a segment of cardiac vessel and third illustrates pseudoparticles concept.

The common practice in 3D computer graphics is splitting objects into meshes composed of tiny triangles, named facets. Specially crafted functions (shaders) are responsible for computation of smooth tonal transitions as well as simulation of "smooth" look of the surface. By replacing every facet by prism with base accordingly aligned one can obtain "thick" surface elements comprising whole thick surface object. Obviously, an information on a thickness of every element ought to be delivered for every facet separately. Some variants of thick facets are shown in the figure 2. Unfortunately due to surface curvature adjacent prisms may not fit to each other. Thus, in order to avoid visible cracks in generated surface heights of prisms have to be kept within reasonable limit, see Fig. 3.

Media statement introduces smoke like interior of an object with controlled density. Thick facets may be rendered with or without media statement, and accordingly with opaque or transparent surface, so there is a lot of different combinations. Some of them helps overcome cracked surface problem described above.

B. Visualization of fuzziness of cardiac vessels

In treatment of arteriosclerotic vascular disease, assessment of cardiac arteries plays a crucial role. Typically, after implementation of imaging procedure data gathered is presented to an expert, and a decision concerning further treatment is worked out. Many methods has been developed in order to make

Figure 1.   A surface computed by kriging based on measurement points (larger spheres). Rectangular grid (smaller spheres) has been computed using simple kriging algorithm.Source: [14], own by ScPovPlot3D.

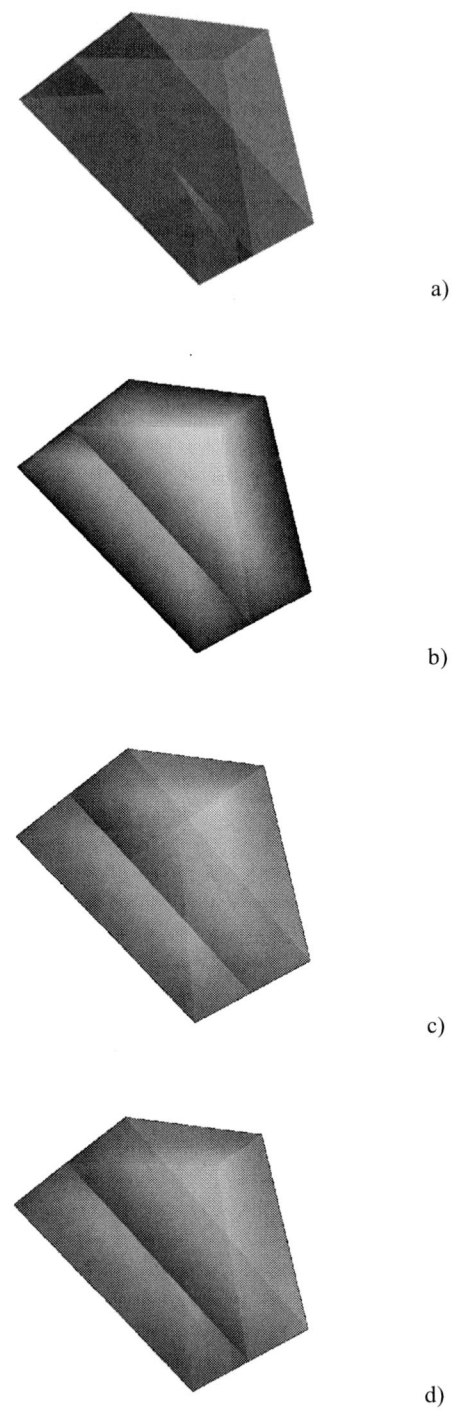

a)

b)

c)

d)

Figure 2.   Examples of fuzzy surface facets. Different settings has been applied to the same geometry. a) texture only and central triangle, b) media only and central triangle, c) media and texture without central triangle, d) mix of all components. Uncertainty range is represented by distance between triangular bases of the prism. Source [16], produced using ScPovPlot3D package

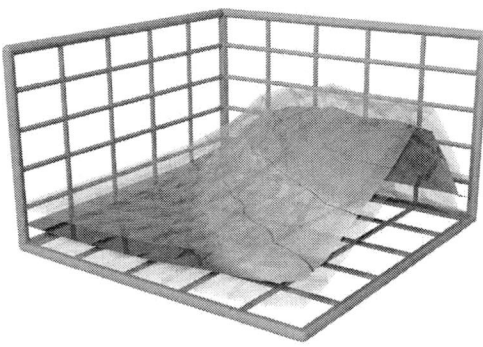

Figure 3. Example of visualization of a fuzzy surface. Min-Max range is represented by two separate surfaces, lower and upper displayed as grayscale thick surface while surface representing supposed exact value is drawn between them with height color mapping. Source: [14], own.

visualization of coronary vessels as precise as possible. Various diagnostic methods commonly used may be more or less suitable for automated medical data processing (for example, segmentation of stenoses) and visualization of such data (inside of lumen of blood vessel), thus for generating reliable information for an expert. This problem may be viewed as decision support process where digital processing of images are to generate useful information for decision-maker. Such an information - when combined with knowledge acquired during work experience - gives an ability to improve the accuracy of diagnosis.

One of the oldest methods widely used by cardiologists is an angiography. Contemporary angiography is a routine diagnostic method oriented toward a visualization of geometry of coronary blood vessels, including assessment of state of walls of arteries. Unfortunately, this method requires intravenous administration of radiocontrast agents followed by registration of X-Ray images in classical or Computed Tomography (CT) setup [1], [7], [8], [18].

A coronary angiography (CAG) is a variant of classical angiography. In this case, the imaging is done using catheter introduced into suspected coronary vessel. It enables localization and assessment of extent of lesions and severity of coronary stenoses [3]. CAG is very invasive and dangerous for the patient, therefore it is performed only if really necessary.

Raw angiography does not allow for collecting detailed information on extent and severity of development of given lesion, for example irregular or hidden in the wall of a blood vessel [7], [18]. However, it is commonly used for preliminary assessment of stenosis of coronary vessels, changes in walls of coronary arteries and anomalies in their placement as well as orifices [1]. Angiography, as opposed to coronary angiography (CAG) [5], is assumed to be relatively safe even for high risk patients [15]. However, a risk connected with prolonged exposition to X-Rays and intravascular administration of radiocontrast agent still remains particularly in patients with a kidney disorder or failure.

Coronary angiography may also be enhanced by dedicated computer algorithm (QCA, Quantitative Coronary Angiography) which allows increasing the reliability of assessment of state of vessel wall and evaluation of profile of vessel's cross section [2]. Another improvement method is use of Computed Tomographic Angiography (CTA) [18]. Among disadvantages of above methods usually uncertainty in interpretation of imaging is mentioned, in particular in case of presence of extensive stenoses where possible contours of cross section may exhibit severe eccentricity as well as heavy reduction of the lumen of the vessel examined.

By using methods of visualization of coronary arteries briefly described above only approximated path and cross section of the vessel, involving some level of fuzziness of resulting diagnosis can be shown. Thus dedicated visualization techniques and tools should be developed in order to enable cardiologists to evaluate and visualize possible deviations from computed vessel surface, thickness of artery wall and atheromatous plaques. Similarly, they may be also used for uncertainty estimation (upper and lower boundary) of FFR (Fractional Flow Reserve [18]) measurement which is one of essential indicators for stent placement.

III. VISUALIZATION BASED ON MEDICAL IMAGING STANDARDS

Ability of processing data received from medical devices depend on digital representation of images. To store information on a recorded object, dedicated file formats are used. They allow an interoperability between applications and systems as well as enable or facilitate data interpretation and visualization also for telemedicine, i.a. cardiac, stomatologic, orthopedic consultations.

The DICOM standard (Digital Imaging and Communications in Medicine) [19] has a wide applicability in many branches of health. It ensures high quality visualization of medical data both static and dynamic [17]. Moreover, supplementary data are recorded in relevant attributes what enables extensive postprocessing and unique identification of the patient. These remarks are also valid for integration of information produced by dedicated applications, delivering data for processing in heterogeneous systems for storage and processing of medical data (EHR systems, Electronic Health Record). DICOM is helpful, in particular, during transferring of patient's data between various medical centers equipped with imaging systems delivered by diverse manufacturers, and, what is connected, diverse applications for processing and data archiving but it can supply redundant or useless data. At the same time definitions of graphical objects may remain incomplete due to filling improper fields or supplying non relevant or erroneous data or even omitting crucial fields. Problems also arise from data processing by apparatus calibrated with different sets of parameters what may deteriorate quality of images, for example by affecting contrast or gamma profiles [9].

Note, many popular applications able to deal with DICOM files convert input data into specific, temporary file format. It may results in difficulties during analysis including loss of information or quality. For example, GIMP graphics editor [24] is able to import images from DICOM files but at the same time drops crucial data on their geometry and topology.

DICOM files store spatial information mainly in a raster format. They comprise several, and often numerous, 2-dimensional grayscale raster images (layers, slices) produced by a variety of imaging systems, like MRI, CT, CAT, CTA, CAG and supplementary information on their mutual layout. These layers are assumed parallel and laid one over the other separated by constant distance resulting from scanning resolution (fig. 4). This is why raw DICOM files may be treated as useless for automated, computer analysis of spatial relations, especially for assessment of state of coronary artery.

The STL format (STereoLithography) [20], [21] is widely used i.a. for cardiology diagnostics purposes. It can be classified as 3D vector graphics format. The surface of the object is defined in 3D space as raw set of triangles. STL definition comprises both a textual (ASCII) variant as well as a binary one [4]. The surface is composed from flat, usually triangular, facets encoded separately (thus coding is redundant). Note that in computer graphics all surfaces, even very complex, are built from myriads of triangle facets in order to minimize rendering times.

Definition of every single facet includes vector normal to the surface of the triangle, directed "outside" of the object, and three additional vectors defining vertices of the triangle in counterclockwise order when watching in the direction opposite to the normal vector. Thus, every triangle has right-hand orientation.

Applications enabling conversion from DICOM into STL format are available including both free and commercial licenses [25], [26]. There is also Python module - STLTools [31] designed for handling STL files.

Specification of ASCII-STL file is presented in Listing 1. A binary counterpart is defined accordingly, but using machine level float point number representation. Keywords like „facet" have to be written with small letters (see: Listing 1). The very first item in STL file is keyword „solid" followed by name of choice of elaborated object. This name may but does not have to be used later. The last item in the file is keyword „endsolid". All characters following symbol "#" are treated as comment and silently ignored.

Next keyword - "facet" - begins declaration of the first triangle which is finished by the keyword "endfacet". Number of facets in the STL file may easily exceed hundreds of thousands. The "facet" keyword is followed by components of the normal vector. Then there is a definition of three vertices in the section beginning with words "outer loop" and terminated by words "endloop". Declaration of components of every given vertex is opened by a keyword "vertex" and then three floats and End Of Line character are expected (EOL is represented on Windows™ systems by two ASCII characters: 'CR', *carriage return* (ASCII '13') and 'LF', *line feed* (ASCII '10'); on Unix systems there is one character - 'CR').

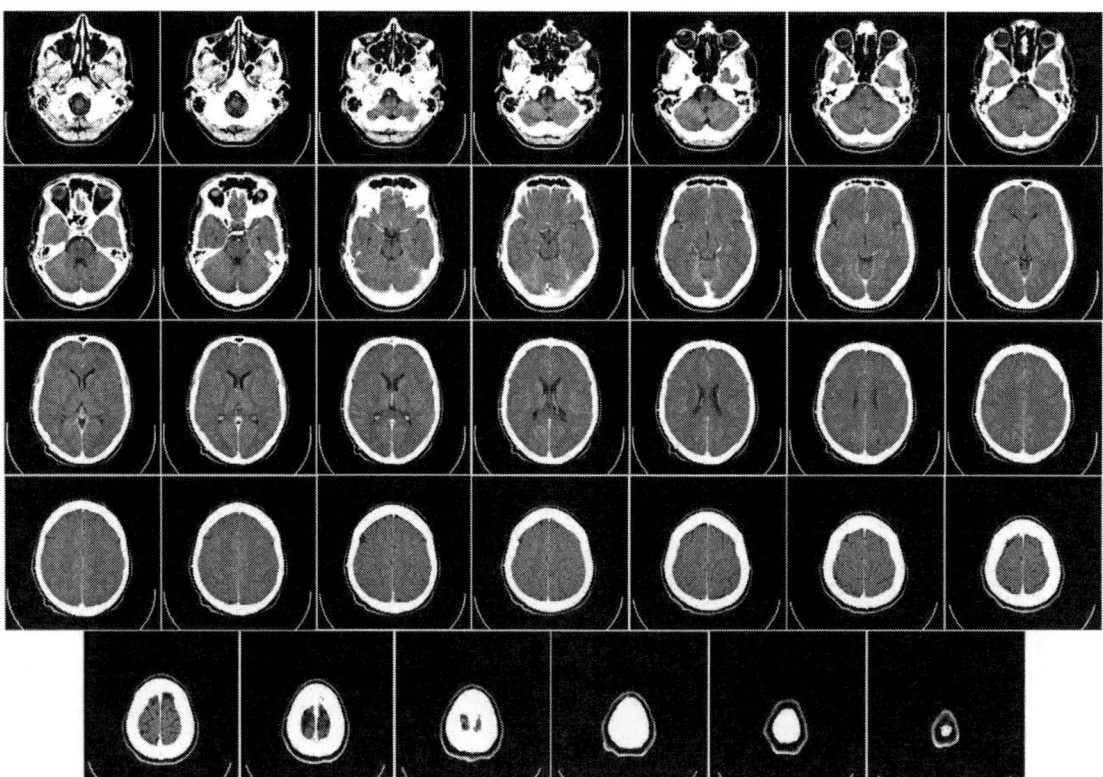

Figure 4. Subsequent layers of scan of a human head (Source: Wikipedia Commons [22]).

While original definition of the standard forbids the use of negative values of vertices component, it is not always complied. It is worth to say, that this restriction is artificial and is not derived from programming principles.

```
solid ascii
  facet normal 0.940 -0.039 0.337
    outer loop
        vertex -38.000 -18.549 -8.320
        vertex -37.970 -18.496 -8.399
        vertex -37.990 -18.367 -8.326
    endloop
  endfacet
  facet normal 0.941 -0.040 0.333
    outer loop
        vertex -38.000 -18.549 -8.320
        vertex -37.990 -18.367 -8.326
        vertex -38.020 -18.420 -8.247
    endloop
  endfacet
  facet normal 0.921 0.0503 0.385
    outer loop
        vertex -38.02 -18.42 -8.24
        vertex -37.99 -18.36 -8.32
        vertex -38.02 -18.23 -8.25
    endloop
  endfacet
  ...
  ...
endsolid
```

Listing 1. Example of ASCII-STL file.
Developed by authors based on [20].

In the "outer loop" section at least three vertices have to be present. Though the STL format specification allows for more, in practice only three (ie. triangles) are used due to computing efficiency. Subsequent numbers (floats) are separated by spaces (at least one space, ASCII code '32') - keywords are put without quotations around. There is no scale definition in the file and all units are relative. A computer program performing tessellation (ie. generating facets) has to obey the rule that vertices of one facet cannot be on the edge of another. So, if two triangles are adjacent they should have exactly one common vertex or one whole edge. However, due to round up errors a couple of vertices belonging to the neighboring triangles usually do not coincide exactly. Thus, a tiny cracks may be spotted in the tessellated surface. Furthermore, STL format does not allow for sophisticated processing of geometry, for example evaluation of cross sections profiles or Fluid Dynamics computation [18] because it is unstructured. This may be overcome only using more robust, indexed data structure [12],[13].

STL file structure resembles the XML file structure, however tags in STL file are not enclosed in acute parentheses. It should be noted, that the STL standard had been elaborated before the XML draft.

Relatively ascetic STL format defines object geometry only. It contains three paired tags (opening and closing) and one odd tag (opening only). Note, that simplicity of the definition is an advantage of the STL format as this format is suitable for fast processing on contemporary, graphics cards, designed for rendering hundreds of thousands of triangles per second. After an adjustment it can be supplied as input data file for thick surface procedures. Example of resulting image of cut-out of coronary vessel is given in the fig. 5.

A. Thick surface and media design

Despite mentioned above STL format deficiencies, it is satisfactory as a base for visualization of 3D fuzzy surface. However, coronary vessel can be visualized only

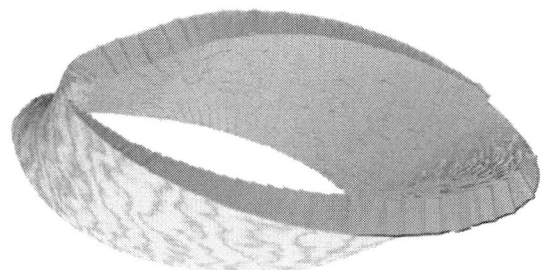

Figure 5.   A piece of cardiac vessel rendered using "thick facets". On the internal surface visible is a pattern created by individual prisms. External surface is patterned using marble like texture for better perception and by no means reflects real look of blood arteries. Source: own, produced with STL.inc module from ScPovPlot3D package.

in tiny slices in order to avoid screening of one part by another. Whole vessel may be then presented using animation created in a standalone or interactive manner. The latter requires additional presentation software which is beyond of the scope of this paper.

B. Pseudoparticles design

Rendering of a media enhanced object usually takes a long time. To make computations faster (reduce computation time down to reasonable limit), density of media "particles" may be reduced or, instead media statement may be simulated using *pseudoparticles* defined as tiny spheres or ovals distributed randomly due to carefully selected probability density function. Even large number of such objects may be rendered hundreds times faster than "true" media. In the figure 6, an example of a surface visualized that way is presented. This design may be easily applied to hybrid designs, as color, eccentricity and size of pseudoparticles may represent space dependent properties, for example gradient or value of the scalar field under consideration [10], ex. blood speed, pressure or temperature.

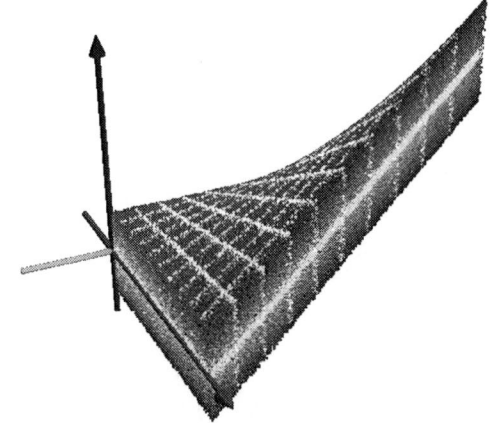

Figure 6.   Thick surface rendered using pseudoparticles design. Uncertainty range is simulated using tiny spheres stochastically distributed within min-max limit. This method is significantly faster than media or particles. White stripes were added to enhance 3D look of the surface. Source: own, produced with STL.inc module from ScPovPlot3D package.

Although problem with screening of parts of the surface persists it can be overcome using static slices or some sort of animation. Using more efficient renderers, for example GPU based FurryBall [27],[28] or OpenGL enabled VTK+ [29] may also allow for interactive operation.

IV. CONCLUSION

Efficient interpretation of cardiac data needs an application of novel algorithms aimed at valuable 3D-visualization, thus increasing the accuracy of a diagnosis made by an expert. Moreover, it's important to gain a possibility of automated digital conversion of information aimed at advanced data processing, in particular - reliable classification of stenoses. One approach is to utilize an alternative graphical tools and lossless data transformation.

ScPovPlot3D package may be used not only for an exploratory analysis of the STL file structure, but also accurate rendering of the surfaces contained within, including "thick surface" design.

The STL file format is relatively simple and well designed for fast rendering and 3D printing but not well suited for computer analysis of coronary vessels including FFR due to the lack of indexation of vertices and facets.

A method of a conversion from the DICOM to the STL format as well as the STL file structure should be revised and improved.

REFERENCES

[1] S. Achenbach, J. Walecki, M. Zawadzki, M. Witulski, "Aplikacje kliniczne wielowarstwowej tomografii komputerowej w kardiologii". Adv Interv Cardiol, 2006; vol. 2(2) pp. 160–168

[2] A. Barańska-Kosakowska, M. Hawranek, M. Gąsior, A. Spatuszko, R. Przybylski et al., "Serial quantitative coronary angiography (QCA) in the assessment of transplant coronary artery disease (TxCAD)". Polish Journal of Thoracic and Cardiovascular Surgery 2010; vol. 7 (3), pp. 319–324.

[3] B. Bartoszek, M. Mościński, T. Niklewski, B. Szyguła-Jurkiewicz, A. Lekston, "Wirtualna histologia — nowoczesna metoda oceny tętnic wieńcowych". Folia Cardiologica Excerpta 2011 vol 6(3), Via Medica, pp. 203–209.

[4] M. Burns, "Automated Fabrication". Prentice Hall 1993.

[5] J. Kubica, R. Gil, P. Pieniążek, "Wytyczne dotyczące koronarografii". Polish Heart Journal, vol. 63.supl III, 2005, pp. 491-500.

[6] D.G. Krige, "A statistical approach to some mine valuations and allied problems at the Witwatersrand", Master's thesis of the University of Witwatersrand, 1951.

[7] J. Legutko, J. Jąkała, B. Mrevlje, S. Bartuś, D. Dudek, "Rewaskularyzacja serca pod kontrolą cząstkowej rezerwy wieńcowej" Advances in Interventional Cardiology, vol. 7(3), 2011, pp. 228–24.

[8] N.R. Mollet, F. Cademartiri, K. Nieman et al., "Multislice spiral computed tomography coronary angiography in patients with stable angina pectoris." J Am Coll Cardiol, vol. 43, 2004, pp. 2265-2270.

[9] M. Mustra, K. Delac, M. Grgic, "Overview of the DICOM Standard", 50th International Symposium ELMAR-2008, 10-12 September 2008, Zadar, Croatia

[10] J. Opiła, "Prototyping of visualization styles of 3D scalar fields using POV-Ray rendering engine", Proc. of the MIPRO 2015: 38th International Convention on Information and Communication Technology, Electronics and Microelectronics: May 25-29, 2015, Opatija, Croatia, pp. 328–333.

[11] J. Opiła, "Some remarks on visualisation of irregular or dispersed data" (pol. Kilka uwag o wizualizacji danych nieregularnych lub rozproszonych) in: Diagnozowanie stanu środowiska, red. J. K. Garbacz. Bydgoszcz: BTN, 2012, ISBN: 978-83-60775-34-9

[12] J. Opiła, T. Pełech-Pilichowski, "Selected topics in visualization of data in cardiology". Proc. of the XV International Conference on Business Management, 21–22 November 2013, Kraków, Poland, AGH-UST

[13] J. Opiła, T. Pełech-Pilichowski, "Visualisation of dispersed or irregular data", Zeszyty Naukowe WSEI, Kraków, nr.8., 2012, pp. 151-178, ISSN 1734-5391

[14] J. Opiła, I. Skalna, T. Pełech-Pilichowski, "Interval arithmetic for irregularly distributed data visualisation by kriging". Proc. of the XIV International Conference on Business Management, 22–23 November 2012, Kraków, Poland, AGH-UST

[15] P. Paluszek, P. Pieniążek, P. Musiałek et al. "Angioplastyka objawowych zwężeń tętnic kręgowych z zastosowaniem stentów konwencjonalnych i pokrytych lekiem antymitotycznym"., Post Kardiol Interw 2009; 5, 1 (15): 1-6

[16] I. Skalna, B. Gawel, B. Rebiasz, B. Basiura, J. Duda, T. Pelech-Pilichowski, J. Opila, "Advances in Fuzzy Decision Making", Springer 2016,

[17] "Strategic Document, Digital Imaging and Communications in Medicine" (last revised 23.10.2013), URL: http://goo.gl/X7f7C4 (accessed: 2017-02-06),

[18] C.A. Taylor, T.A. Fonte, J.M. Min, "Computational Fluid Dynamics Applied to Cardiac Computed Tomography for Noninvasive Quantification of Fractional Flow Reserve. Scientific Basis", J.Am. Coll. of Card., Vol. 61, No. 22, 2013, ISSN 0735-1097/$36.00, Elsevier Inc. http://dx.doi.org/10.1016/j.jacc.2012.11.083

[19] Dicom Library, http://goo.gl/dGTMFv (accessed 2017-02-06),

[20] "The STL Format. Standard Data Format for Fabbers. Fabbers.com Your Digital Fabrication Portal", http://goo.gl/fyzBX6, (accessed 2017-02-06).

[21] "What is Stereolithography?", URL: http://goo.gl/IKPWO6 (accessed 2017-02-06).

[22] Wikipedia Commons, URL: http://goo.gl/VDe1sg (accessed 2017-02-06).

[23] The Persistence of Vision Raytracer, Manual, URL: http://www.povray.org (accessed 2017-02-06).

[24] GIMP - Gnu Image Manipulation Program, URL: http://www.gimp.org, (accessed 2017-02-06).

[25] 3D Slicer, URL: https://www.slicer.org, (accessed 2017-02-06).

[26] DeVIDE, URL: https://github.com/cpbotha/devide, (accessed 2017-02-06).

[27] "FurryBall" GPU renderer, URL: http://furryball.aaa-studio.eu, (accessed 2017-02-06).

[28] "FurryBall, speed comparison", URL: http://goo.gl/mC1Lzj, (accessed 2017-02-06).

[29] Visualization Toolkit, VTK+, URL: http://www.vtk.org, (accessed 2017-02-06).

[30] J. Opiła, "ScPovPlot3D Templates", URL: http://scpovplot3d.sf.net, (accessed: 2017-02-06)

[31] STLTools, development page on github, URL: https://goo.gl/uWrVph, (accessed: 2017-02-06).

[32] J. Opiła, "Prototyping of visualization designs of 3D vector fields using POVRay rendering engine", 2016, 39th International Convention on Information and Communication Technology, Electronics and Microelectronics (MIPRO), Opatija, 2016, pp. 343-348. doi: 10.1109/MIPRO.2016.7522164.

Abdominal Fetal ECG Measured with Differential ECG Sensor

Aleksandra Rashkovska and Viktor Avbelj

Department of Communication Systems, Jožef Stefan Institute, Ljubljana, Slovenia
aleksandra.rashkovska@ijs.si, viktor.avbelj@ijs.si

Abstract - Abdominal ECG is a non-invasive method for monitoring the cardiac activity of a fetus. A complementary method is the detection of the fetal heart rate with an ultrasound. In this paper, we present and analyze abdominal ECG measurements obtained with a differential ECG body sensor. Abdominal ECG was measured in different months of pregnancy within two subjects: one caring a single fetus and another caring twins. The fetal ECG signal measured on the abdomen during pregnancy is superimposed to the mother's AECG and has a very small amplitude, which is smaller than the amplitude of the mother's ECG in that part of her body. The interference from the power grid is not present in the signal, which is crucial for further analysis. The recordings demonstrate the remarkable potential of the sensor for abdominal ECG measurements.

I. INTRODUCTION

Nowadays fetal heart is observed predominantly by Doppler ultrasound. However, this technique exposes the fetus to ultrasound irradiation and it has not been proven to be completely safe. The risk may increase with unnecessary prolonged exposure to ultrasound energy, or when untrained personnel operate the device [1]. There are two other noninvasive monitoring techniques of fetal cardiogram. One technique is fetal magnetocardiography, where faint magnetic fields, resulting from the current sources in the fetus heart, are measured outside the maternal abdomen by super sensitive magnetometers (SQUID – Superconducting Quantum Interference Devices) in magnetically shielded rooms [2, 3, 4]. The other noninvasive technique is the measurement of abdominal ECG (AECG) with electrodes placed on the maternal abdomen.

The last mentioned technique was the first one used more than a century ago. However, despite the significant advances in adult electrocardiography, its practical application is still pending because of several technical difficulties [5, 6, 7]. First of all, the heart of the fetus is very small and it produces low voltage on the skin of the maternal abdomen, so the signal-to-noise ratio is low. The second difficulty is the maternal ECG measured at the abdomen, which has nearly the same frequency range, but larger amplitude as the fetal ECG. The third difficulty is the vernix caseosa – a thin layer on the skin of the fetus

that forms around the 28th-32nd week of pregnancy and that electrically isolates the fetus from the surrounding. This further reduces the amplitude of the fetal ECG on the maternal abdomen, although the fetal heart during those weeks is becoming larger. Another issue is the placement of the electrodes. If there are several of them, they can be placed on fixed locations on the abdomen. However, if there is only one channel, the position should be carefully selected and the best placement should be searched. A study with multiple electrodes is presented in Verdurmen et al. [8].

Recently, many small ECG body sensors have emerged on the market. In our study, an advanced ECG body sensor was used by pregnant subjects in their home environments with the goal to obtain measurements with visible fetal ECG in the second half of the pregnancy. In this paper, we present the measurement equipment, the recommendations for effective measurements given to the expecting parents, the results of their efforts and the conclusions for further work in the field of abdominal ECG.

II. MATERIAL AND METHODS

A. Differential ECG Sensor

The AECG was measured with a Savvy sensor [9] with additional signal amplification. The sensor is the core of a system for personal cardiac monitoring. It is a small (dimensions: 130 x 35 x 14 mm) and light (weight: 21 g) body gadget fixed to the skin of the user by two standard self-adhesive electrodes. The sensor has a long autonomy (7 days) and a low power wireless connection (Bluetooth 4) to a Smartphone or other personal device. Part of the system are also a mobile application (MobECG) for visualization and interpretation of measurements, and a standalone software for visualization and basic analysis of the measured ECG (VisECG) on a

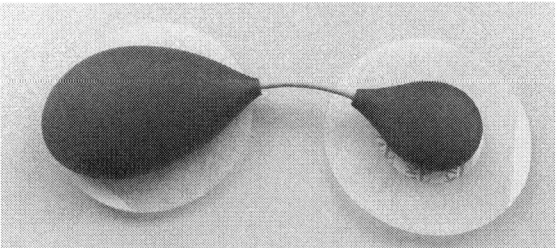

Figure 1. Savvy sensor.

The authors acknowledge the financial support from the Slovenian Research Agency under the grant P2-0095 and the EkoSMART project, grant No. C3330-16-529007, financed by the European Regional Development fund.

safe storage server. The Savvy sensor measures a single lead ECG, differentially between the two electrodes at the distance of 8.5 cm. The sensor is covered with a waterproof and biocompatible plastic housing. The ECG is recorded with a moderate resolution of 125 samples per second. The Savvy sensor is presented in Fig. 1.

The Savvy sensor is the commercial version of a previous differential ECG body sensor prototype intended for personal cardiac monitoring [10]. The device can support solutions to every-day problems of the medical personal in hospitals, health clinics, homes for the elderly and health resorts. Its exceptionally lightweight design allows for unobtrusive use also during sports activities or during exhaustive physical work. The ECG measurements obtained with the sensor are already proven to be suitable for medical use, e.g., monitoring after cardiac surgery [11] and reconstruction of the standard 12-chanell ECG [12-14]. Even though the ECG is obtained only form one differential lead, it has been demonstrated that the quality of the signal is sufficient for basic ECG analysis, like heart beat detection [15] and detection of heart rate variability [16], or more advanced ECG analysis, like clustering of heart beats [17]. Moreover, the sensor has been used also in veterinary medicine for successful monitoring of the cardiac activity in dogs [18, 19].

B. AECG measurement procedure

For performing AECG measurements, the expecting parents were equipped with a Savvy sensor with additional appropriate signal amplification. Namely, the gain of the input amplifier was further increased for a factor of 4.7 or 12.1. The sensor was placed on the abdomen of the mother, in a position appropriate for acquiring AECG with sufficient amplitude of the fetal ECG. For effective measurement, it was recommended that the sensor is shield to avoid electrical interference from the surrounding. The advised method was covering the sensor with both hands while performing measurements.

C. Subjects

AECG was measured in the second half of the pregnancy within two subjects: one caring a single fetus and another caring twins. The mother with a single fetus started with the measurement trials earlier in the second half of the pregnancy, after the fifth month, while the mother with twins started with the measurement trials in later pregnancy, after the seventh month.

III. RESULTS

The mother of a single fetus started to measure earlier in the second half of the pregnancy (from the fifth month on). Fig. 2 shows the AECG recorded in the fifth month of pregnancy with a single fetus. The sensor was positioned in the center of the abdomen, 5 cm below the umbilicus. The gain of the input amplifier was further increased for a factor of 4.7. The raw signal is sampled at a frequency of 125 Hz and a resolution of 1.264 µV. The fetal ECG with the heart rate of 150 beats per minute (BPM) is superimposed to the mother's AECG with the heart rate of 62.5 BPM. Since the fetus is very small, its ECG signal

measured on the abdomen during pregnancy has a very small amplitude, which is smaller than the amplitude of the mother's ECG in that part of her body. The AECG peak-to-peak QRS amplitude of the mother is approximately 40 µV, while the QRS amplitude of the fetal ECG is about 15 µV. According to the literature [20], the size of the heart of a fetus in the fifth month of pregnancy is around 13 mm (length of the left ventricle).

The AECG measured from the same subject in the sixth month of pregnancy is shown in Fig. 3. The fetal ECG was not clearly visible as in the previous month, although the gain of the input amplifier was further increased for a larger factor then the previous measurement, i.e. 12.1. The result is most probably due to the vernix caseosa starting to form in that period. The AECG peak-to-peak QRS amplitude of the mother is approximately 60 µV, while the QRS amplitude of the fetal ECG is significantly smaller, about 6-7 µV. Nevertheless, it was still possible to approximate the fetal hearth rate (app. 153 BPM) and the mother's heart rate (app. 66 BPM).

The subject carrying twins did not manage to acquire proper fetal ECG in the beginning of the measurement trails (after the seventh month of pregnancy). The fetal ECG was not clearly visible, most probably due to the vernix caseosa forming in that period. Nevertheless, a quality AECG was successfully acquired in the eight month of pregnancy (Fig. 4). The gain of the input amplifier was further increased by 12.1. The raw signal is sampled again at a frequency of 125 Hz, but with a resolution of 0.491 µV. The fetal ECG with the heart rate

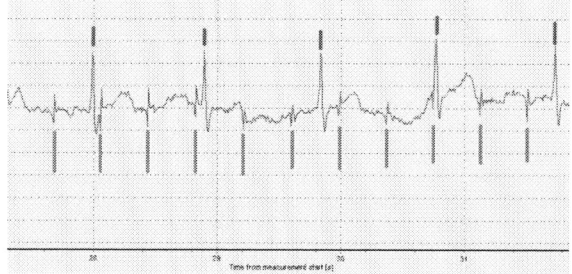

Figure 2. Abdominal ECG (red signal) recorded in the fifth month of pregnancy with a single fetus. The heart rate of the fetus (blue markers) is approximately 2.5 times greater than the heart rate of the mother (red markers).

of 120 BPM is superimposed to the mother's AECG with the heart rate of 74 BPM. The AECG peak-to-peak QRS amplitude of the mother is approximately 40 µV, while the QRS amplitude of the fetal ECG is about 9 µV.

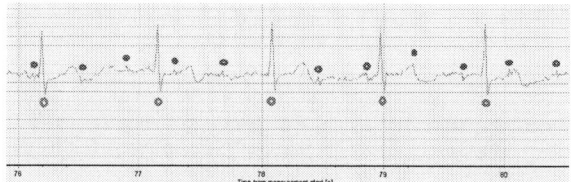

Figure 3. Abdominal ECG (red signal) recorded in the sixth month of pregnancy with a single fetus. The heart rate of the fetus (blue markers) is approximately 153 BPM, while the heart rate of the mother (red markers) is approximately 66 bpm.

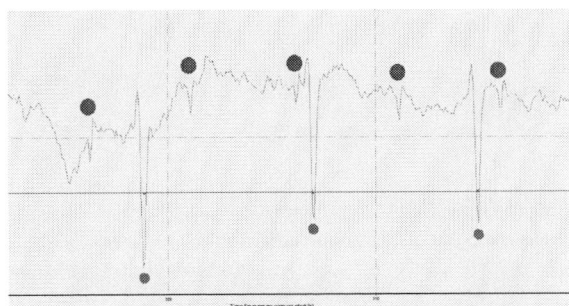

Figure 4. Abdominal ECG (red signal) recorded in the eight month of pregnancy with twins. The heart rate of the fetus (blue markers) is approximately 120 BPM, while the heart rate of the mother (red markers) is approximately 74 BPM.

IV. CONCLUSION

We have demonstrated that a small ECG device could be efficiently used in the intimacy of a home environment to see the fetal heart beats by expecting parents without the presence of a medical staff. They obtained valuable recordings of fetal ECG without power line interference [21] and with low interference from other sources, even when the fetal QRS signal was well below 10 μV peak-to-peak, which is crucial for further analysis. The recordings demonstrate the remarkable potential of the ECG sensor for abdominal ECG measurements. In future, the procedure for finding the best position of the sensor should be prepared in advance.

REFERENCES

[1] fda.gov, "Ultrasound Imaging," 2016. [Online]. Available: https://www.fda.gov/Radiation-EmittingProducts/RadiationEmittingProductsandProcedures/Medicallmaging/ucm115357.htm#patients. [Accessed: February 28, 2017].

[2] C. Kähler, E. Schleussner, B. Grimm, A. Schneider, U. Schneider, H. Nowak, and H. J. Seewald, "Fetal magnetocardiography: development of the fetal cardiac time intervals," Prenat Diagn, vol. 22, pp. 408–414, 2002.

[3] M. J. Lewis, "Review of electromagnetic source investigations of the fetal heart," Med Eng Phys, vol. 25, pp. 801–810, 2003.

[4] J. Stinstra, E. Golbach, P. van Leeuwen, S. Lange, T. Menendez, W. Moshage, E. Schleussner, C. Kähler, H. Horigome, S. Shigemitsu, and M. J. Peters, "Multicentre study of fetal cardiac time intervals using magnetocardiography," BJOG: An International Journal of Obstetrics & Gynaecology, vol. 109, pp. 1235–1243, 2002.

[5] I. Peterfi, L. Kellenyi, and A. Szilagyi, "Noninvasive Recording of True-to-Form Fetal ECG during the Third Trimester of Pregnancy," Obstetrics and Gynecology International, vol. 2014, Article ID 285636, 5 pages. doi: http://dx.doi.org/10.1155/2014/285636

[6] R. Sameni and G. D. Clifford, "A Review of Fetal ECG Signal Processing; Issues and Promising Directions," Open Pacing Electrophysiol Ther J., vol. 3, pp. 4–20, 2010.

[7] M. A. Hasan, M. B. I. Reaz, M. I. Ibrahimy, M. S. Hussain, and J. Uddin, "Detection and Processing Techniques of FECG Signal for Fetal Monitoring," Biol Proced Online, vol. 11, pp. 263–295, 2009. doi: 10.1007/s12575-009-9006-z.

[8] K. M. Verdurmen, C. Lempersz, R. Vullings, C. Schroer, T. Delhaas, J. O. van Laar, and S. G. Oei, "Normal ranges for fetal electrocardiogram values for the healthy fetus of 18-24 weeks of gestation: a prospective cohort study," BMC Pregnancy Childbirth, vol. 16, pp. 227, 2016. doi: 10.1186/s12884-016-1021-x.

[9] Savvy sensor [Online]. Available: http://savvy.si/. [Accessed: March 10, 2017].

[10] M. Depolli et al., "PCARD Platform for mHealth Monitoring," Informatica, vol. 40, pp. 117–123, 2016.

[11] J. M. Kališnik et al., "Mobile health monitoring pilot systems," Proceedings of IS 2015, 18th International Multiconference Information Society, October 9-12, 2015, Ljubljana, Slovenia, pp. 62–65.

[12] R. Trobec and I. Tomašić, "Synthesis of the 12-lead electrocardiogram from differential leads," IEEE Transactions on Information Technology in Biomedicine, vol. 15, pp. 615–621, 2011.

[13] I. Tomašić and R. Trobec, "Electrocardiographic systems with reduced numbers of leads-synthesis of the 12-lead ECG," IEEE Reviews in Biomedical Engineering, vol. 7, pp. 126–142, 2013.

[14] I. Tomašić, S. Frljak, and R. Trobec, "Estimating the Universal Positions of Wireless Body Electrodes for Measuring Cardiac Electrical Activity," IEEE Transactions on Biomedical Engineering, vol. 60, pp. 3368–3374, 2013.

[15] P. Lavrič and M. Depolli, "Robust beat detection on noisy differential ECG," Proceedings of MIPRO 2016, 39th International Convention, May 30-June 3, 2016, Opatija, Croatia, pp. 381–386.

[16] J. Slak and G. Kosec, "Detection of heart rate variability from a wearable differential ECG device," Proceedings of MIPRO 2016, 39th International Convention, May 30-June 3, 2016, Opatija, Croatia, pp. 430–435.

[17] A. Rashkovska, D. Kocev, and R. Trobec, "Clustering of heartbeats from ECG recordings obtained with wireless body sensors," Proceedings of MIPRO 2016, 39th International Convention, May 30-June 3, 2016, Opatija, Croatia, pp. 481–486.

[18] A. Krvavica, Š. Likar, M. Brložnik, A. Domanjko-Petrič, and V. Avbelj, "Comparison of wireless electrocardiographic monitoring and standard ECG in dogs," Proceedings of MIPRO 2016, 39th International Convention, May 30-June 3, 2016, Opatija, Croatia, pp. 416–419.

[19] M. Brložnik and V. Avbelj, "Wireless electrocardiographic monitoring in veterinary medicine," Proceedings of MIPRO 2015, 38th International Convention, May 25-29, 2015, Opatija, Croatia, pp. 375–378.

[20] G. Sharland and L. Allan, "Normal fetal cardiac measurements derived by cross-sectional echocardiography," Ultrasound in Obstetrics and Gynecology, vol. 2, pp. 175–181, 1992.

[21] D. D. Țarălungă, G. M. Ungureanu, I. Gussi, R. Strungaru, and W. Wolf, "Fetal ECG extraction from abdominal signals: a review on suppression of fundamental power line interference component and its harmonics," Comput Math Methods Med., vol. 2014, 2014. doi: 10.1155/2014/239060.

Synchronization of time in wireles ECG measurement

Andrej Vilhar and Matjaž Depolli
Jožef Stefan Institute, Department of Communication Systems, Ljubljana, Slovenia
andrej.vilhar@ijs.si, matjaz.depolli@ijs.si

Abstract – Wireless devices for ambulatory ECG monitoring are becoming increasingly popular. One of the challenges that wireless monitoring has to deal with is clock synchronization between the wireless sensor device and the controller device. The monitoring device is usually kept as simple as possible to maximize its autonomy, and is thus limited in its clock accuracy. In this article, we describe a method of off-line synchronization of data from a completely asynchronous sensor device. The sensor device sampling frequency is allowed to deviate from its declared value by more than 100 ppm, while the controlling device time is assumed to be absolutely accurate. Data synchronization is based on using two sources of timestamps for the data – the sensor device oscillator and the controller device local time. In our case the controlling device is an Android mobile phone, and the wireless technology used is Bluetooth LE. The challenges are: limited precision of time reported by the sensor device, clustering of messages on Android, and lost messages due to lossy transmission mode. The presented synchronization technique is able to achieve several orders of magnitude improvement in estimation of sample time.

I. INTRODUCTION

Electrocardiogram (ECG) is a recording of electrical activity of the heart by means of measuring the potential differences with electrodes placed on the skin. From the beginning of the 20th century, the field of ECG measurement and analysis has matured into todays standard 12-lead ECG (which uses 10 electrodes placed on the skin), multichannel ECG body surface mapping systems [1], Holter monitors [2] (most common ambulatory recorder), and the wireless implantable loop recorder (device is inserted under the skin) [3].

Long-term ECG monitoring with a Holter monitor has been used in medicine since the 60s of the last century. The advancement of electronics has enabled the ambulatory monitoring devices to become ever smaller and to record high-quality ECG signals gathered by a small number of electrodes. Currently, the measurements are made available for detailed analysis after one or more days of recording, when the ambulatory monitor is detached from the subject and attached to the computer. The advances in telecommunications, however, already enable the wireless data transmission [4] from miniature sensor devices to nearby personal terminals (smartphone, tablet) with access to the Internet. This in turn enables the provision of a wide range of mobile health services, from patient monitoring in hospitals [5], remote patient monitoring [6, 7], and remote medical support, to sports, recreation, and entertainment. Current research efforts are focused on the development of de-

vices and instruments which are small, simple to use, reliable, and offer medical-grade measurements.

The mobile health (mHealth), which enables telemonitoring, telecare, and other distant services, is a new way of extending the established health care to cover a larger part of patients. Mobile communications and the Internet are accessible almost everywhere, and are further promoted by the rapid expansion of the smart personal devices, which are becoming necessities in everyday life.

The backbone of any mHealth platform is the ICT technology, which is already mature enough and readily available on the market. Wearable devices with sensors, which are the other key element, on the other hand, are still under rigorous development. Among these is a multifunctional wearable sensor device, which is being developed at Jožef Stefan Institute as a part of the *PCARD* mHealth platform [8].

A. Wearable sensor device for ECG monitoring

The sensor device used in PCARD mHealth platform is a small and lightweight device, comprising an ECG analog sensing circuitry, micro-controller, and Bluetooth LE radio transmitter. Its intended use is to be put on the body, usually on the torso in the vicinity of the heart, where it measures ECG and transmits it wirelessly to a smartphone. The sensor device has no own human interface, no storage capacity and very low processing capabilities, and therefore cannot operate in isolation, that is, without a wireless connection to the smartphone. Its design uses a trade-off between the long-term usage – it is possible to perform measurements for more than a week on a single charge, and data integrity – measurements are allowed to contain missing data. Such a trade-off was selected because the focus of the platform is on long-term measurement and in this context it is perfectly acceptable to experience several seconds, minutes, or sometimes even hours of missing data within a weekly measurement.

A methodology has been developed to synthesize 12-lead ECG measurement from 3 concurrent differential ECG measurements [9]. Such measurements can theoretically be produced by three sensor devices connected to the same smartphone. This ability of the wearable ECG sensor has a great applicability in extending the procedure for measuring 12-channel ECG from health care institutions that have the required equipment available, to anywhere, provided a smartphone and three sensor devices. The goal of this paper is to analyze options for supporting the 12-lead ECG synthesis via accurate synchronization of multiple ECG measurements.

II. THE PROBLEM

Keeping accurate time, which is required for synchronizing events measured on different devices, is not trivial. While the smartphone is a fairly complex platform with plenty of processing power and an Internet connection, it can maintain quite an accurate time representation. The wearable on the other hand is much simpler and its internal representation of time depends purely on the quartz oscillator, which is estimated to be accurate to under 100 ppm [10], and the communication with the smartphone. The unaltered quartz accuracy allows for several seconds of time drift to accumulate during a single week of measurement. While this time drift is perfectly acceptable from the long-term measurement point of view, it is not acceptable for making multiple concurrent measurements that have to be synchronized in time. The synchronous measurements form the basis for 12-channel ECG synthesis, which further imposes a very strict synchronization demand, for the timing error to be below a single sampling period length. Therefore, time accuracy must be improved with the help of the smartphone to satisfy the requirements. If the concurrent measurements are not all performed on the same smartphone, then the synchronization should be extended among the multiple smartphones as well, but we consider this option out of scope for now.

Since the ECG body sensor and the smartphone could be considered as a simple distributed system, one possible solution to time synchronization could be to translate it to clock synchronization and then use one of the many clock synchronization techniques for distributed systems [11]. To keep the software on the sensor device simple and power efficient, though, time keeping is simplified to the point that such techniques cannot be implemented. Time is not kept at all, the signal from the on-board oscillator is used as a clock and to trigger signal sampling.

Therefore the problem is re-defined as follows: The smartphone time is considered as a reference that is absolutely correct, and synchronization is applied on the recorded measurement only, keeping the internal representation of time on the sensor device intact. The recorded measurement is time synchronized by deducing the reference time of each sample on the smartphone, using the data gathered so far. The data consists of the 10-bit counters taken from each received packet, the times recorded when packets were received, the sampling period, and the measurement start time.

A. More problems

A custom protocol on top of the Bluetooth LE is used for for wireless data transmission. Data transmission is performed by enumerated packets of 19 Bytes each, with no retransmission on error. This means that some packets will get lost in the transmission, but the order of the received packets is not affected and identification of missing data is possible. The contents of a single packet are 14 10-bit samples, a 10-bit time stamp that also serves as the packet identifier, and a 2-bit operation mode identifier.

The procedure for data transmission starts with the digital data sampling on the sensor device. Current time is kept as the number of sampling periods that passed since the measurement was started. Therefore, current time on

the sensor device will hereafter be termed counter, to avoid mixup with the time kept on the smartphone. The sampling procedure is triggered in a periodic manner by a quartz oscillator on the sensor device. Each 10-bit sample is then stored in one of three buffers. Once that a buffer is full, i.e. it contains 14 samples, it is stamped with current time (mod 2^{10}), marked for transmission and another buffer is selected for caching further samples. Data packet marked for transmission is transmitted to the smartphone in the next communication interval defined by the Bluetooth radio, and the delay between marking a packet and the completion of transmission can be considered a random variable.

The problem is further complicated by the following three factors. First, application on the smartphone has to receive the packets and record when they were received (assign them a reference timestamp). Android, which is the most proliferated smatphone operating system, does not give the application direct access to Bluetooth hardware but rather only access to the communication layer via callbacks. Since the time of receiving each packet is not recorded by the system, it has to be done in those callbacks. This, however, can be problematic, since the communication layer does not seem to trigger callback for received message straight away when the message is received. Callbacks seem to be called in a periodic fashion which indicates that the operating system is most likely periodically polling Bluetooth hardware and does not communicate with it in real time. An example is shown in Figure 1, where the points represent delays between two subsequent calls to a callback, which are clearly quantized. Such behavior makes it hard to estimate the time a packet was received.

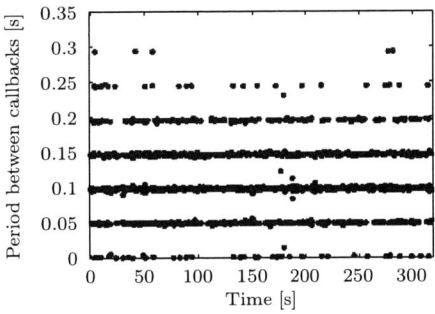

Figure 1: Example of how the Bluetooth callback calls are quantized on a typical Android device.

Second, handling unreliable connection has its own set of problems attached. These can be classified in lost packages and disconnection events. In both cases one or more packets of data is lost in transmission, which has to be handled on the Android side. If the Bluetooth connection drops, which can happen if the Android device and sensor device get too far from each other for too long, e.g. 30 seconds, then the sensor device goes to low power mode and does not continue time keeping. Thus, after a reconnect, the counter values received will not indicate all the missing data correctly.

Third, smartphones were in several occasions found to severely delay the forwarding of received packets to the application, thus making the receive timestamps severely unreliable. These are the occasions when the smartphone application receives packages from the Bluetooth communication layer of the Android operating system in a large cluster, instead of in regular intervals as they are received by the hardware. An example of data where the receive timestamps are distorted in such a way is given in Figure 2. Although the relation between the sensor device and reference times should be linear, several *steps* are visible on the graph, where the reference time changes very little while the time on the wearable changes as expected.

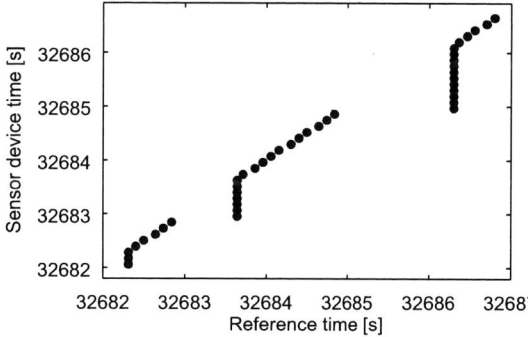

Figure 2: Example of how the Bluetooth packets get clustered on some Android devices. Reference time is the time a packet is received, while sensor device time is the local time on the wearable sensor device.

All the mentioned factors contribute to a fairly complex and innovative method for synchronization of the measurement time, that is, the time individual samples were taken, and Android time. The problem also contains some randomness, and is even not well specified, since the implementation of Bluetooth LE communication on Android devices varies wildly.

III. Method

A new method is proposed, which estimates sample times from the provided approximate sampling frequency, the reference timestamps of when the packets were received, and the counters in the packets. This method was developed to be used to extend the NevroEKG program [12] for use with wearable sensors.

A. Decomposition into continuous blocks

This part of the method represents an effort to eliminate the effects of disconnection events and long sections of missing data, which could introduce large errors into the estimation of true sampling times. Both mentioned artifacts are represented as pauses, i.e., long intervals of missing data, in the measurement. This step splits the measurement into blocks with no pauses by performing detection of pauses in a iterative manner.

The procedure is as follows. The intervals between consecutive reference timesteps are checked and when a pause is detected, the measurement is split. A so called continuous block of measurement is created as the part of the measurement from measurement start to the start of the pause. Each continuous block is guaranteed to contain measurement with no disconnection events and with manageable

amount of missing data. The second part of the measurement that starts as the pause ends and lasts to the end of the measurement is then input into the next iteration of the detection of pauses. This way the pause is eliminated from further processing with no ill effects. When the detection of pauses is done, another continuous block of measurement is created as the last part of the measurement.

Pauses are detected using the following criterion. The resolution of the counters is 10 bit, which results in 2^{10} possible counter values. A *safe difference* between consecutive packets Δr_s is then defined as one half of the time interval in which the counter values start repeating itself, which is caused by the overflow in counter value. Counter values start repeating in 2^{10} packets, therefore the safe difference can be expressed in counter units as:

$$\Delta c_s = \frac{1}{2} 2^{10} = 2^9,$$

and safe difference in reference time can then be derived from the expected sampling frequency f_s as:

$$\Delta r_s = \frac{\Delta c_s}{f_s}$$

Therefore, the pause is declared when a disconnect occurs, or the two received packets reference timestamps differ by more than Δr_s, or the two received packet counters differ by more than Δc_s.

B. Processing of individual block

This part of the method is applied on each continuous block of measurement separately. The relation between the counters and the time stamps is estimated from the data of the block. The relation is composed of frequency and offset (or the slope and the intercept). Frequency estimate from each block will later be used to estimate the mean frequency across the measurement, while the offset will be used for deriving the sample times of this block only.

In the first step, the unreliable data are filtered out. Individual differences between consecutive packets are examined for the ratio between the counter c and reference timestamp r: $f_i = \frac{c_{i+1} - c_i}{r_{i+1} - r_i}$, where c_i and r_i are the counter and reference timestamp of the i-th packet. This ratio should be close to the sampling frequency for all samples but will not be accurate because of the timestamp quantization errors. On the other hand, the error should not be very large either. Therefore a threshold filter is applied that marks packages i and $i + 1$ to be ignored in further processing if f_i appears either too small or too large. The threshold for declaring f_i either too small or large was in our case set to 2, which means that packages i and $i + 1$ are removed, if the condition $\frac{f_s}{2} \geq f_i \geq 2f_s$ is not satisfied. The method is not very sensitive to the selection of threshold since it is designed mostly to weed out packets that were received in clusters. The timestamps of clustered packets are very close to each other and usually several orders of magnitude closer together that the timestamps of correctly received packets.

Linear regression is then used to determine the relation between the counters and timestamps of the remaining packets. Regression coefficient is taken as the estimated sampling frequency f_b of this block.

Counters are also extended into 32-bit form at this time. Since the blocks contain data that is guaranteed not to have pauses longer than r_s and counter differences larger than c_s, the procedure is quite simple:

```
offs=0
for each i:
    if (c_i < c_{i-1}) then offs=offs+2^{10}
    c_i = c_i+offs
```

The result of this part is that the counter overflows are accounted for, and the sampling frequency is estimated on this block.

C. Averaging the sampling frequency

Individual sampling frequencies of blocks f_b are gathered, weighted by the sample size on which they were made, and averaged to produce the final sampling frequency estimate for the whole measurement $\hat{f}_s$. The effects of disconnection events are completely eliminated from $\hat{f}_s$, and the weighted averaging should also produce best possible estimate based on the given data. In future, more aggressive weighting or even thresholding might be necessary, since benefit of including very short blocks into the averaging is questionable.

D. Estimating sampling times

In the final step, the estimated sampling frequency is used to derive sampling time for each received packet. A two-step algorithm is used, with initial estimate of sampling time $\hat{t}_i$ and the final sampling time estimate t_i.

In first step, an estimate of the sampling times is made, which uses the estimated sampling frequency:

$$\hat{t}_i = r_0 + i * (c_i - c_0) * \hat{f}_s.$$

The second step builds upon the property of timestamp errors e, that they can only be positive, which can be written formally as:

$$r_i - t_i > 0.$$

Therefore, estimated timestamps are uniformly shifted back in time so that $t_i \leq r_i$ for $\forall i$ within each measurement block:

$$t_i = \hat{t}_i - \max(t_i - r_i).$$

Blocks are again kept separate at the final estimate, since the sampling frequency is the only parameter that is constant throughout the whole measurement.

IV. EXPERIMENTS AND RESULTS

An example of an extremely difficult measurement to handle is given and the proposed algorithm is applied on it. The measurement is 34 hours and 52 minutes long but contains only 3 hours of data. First part of the algorithm distributes this data into 2045 short blocks, the longest of which is only 7.5 minutes long. Such a measurement is unfortunately unusable for monitoring or diagnostic purposes. It does, however, represent an excellent testing facility for the proposed algorithm.

The results of the second algorithm part, where sampling frequency is estimated for individual blocks, are plotted in Figure 3. There is a marker on the figure for each block, with its x position representing the number of data points used to estimate frequency estimate and y position representing the error in frequency estimate. It is clear that the frequency estimates that are made from small number of data points contain large errors. These are the result of timestamp quantization that was mentioned before. The errors are reduced quite quickly, though, with larger number of data points.

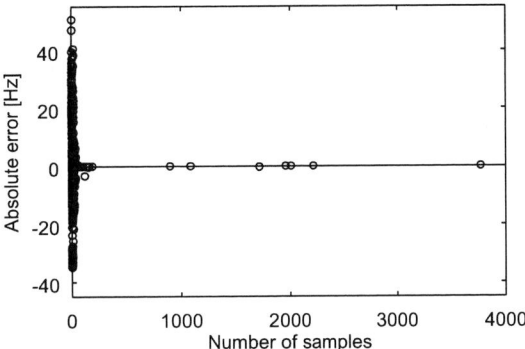

Figure 3: Relation between the number of samples taken for the calculation of frequency (linear regression) and the absolute error in frequency.

Another observation based on Figure 3 can be made regarding the measurement itself. It can be seen that the majority of frequency estimates is made on very low number of data points. Analysis shows than more than 76 % frequency estimates are made on 10 data points or less, and 67 % of frequency estimates are made on 5 data points or less. Over the whole measurement, 69 % of data points are discarded by the method and not used at all. A closer look into the measurement shows that the majority of data is really not useful at all, since packets were received in large clusters, as shown before in Figure 2.

The third step of the method averages the frequency to 125.0645, which is close to the real sampling frequency of 125.0687. On the largest block of the measurement, which is 7.5 minutes long, the error in frequency leads to 0.015 s of error in sampling time. Considering the quality of input data, for which the quantization error alone is 0.05 s, and error from packet clustering is several seconds, this is a large improvement. It is insufficient though for measurement synthesis, since it is about twice the length of a single sampling period, which equals $\frac{1}{125} = 0.008$ s.

The last step of the algorithm, which estimates the sampling times is not tested here, since this particular measurement was made in isolation and the data for comparison does not exist.

V. CONCLUSION

An example of how the modern wearable technologies push the available hardware and software to its limits has been presented. When combining multiple concurrent measurements for a measurement synthesis, an important first problem to solve is assuring the participating devices have their clock synchronized. This problem was transformed into a simpler one – to the off-line mapping of a single measurement to a reference time. Thus the synchronization of multiple device clocks was avoided, which enables the sensor devices to have simpler hardware and greater autonomy.

A solution for measurement synchronization was proposed, which manages to handle some difficult measurements. Nevertheless, the remaining error is still too great in the most demanding measurement and improvements will have to be made.

In future work, the last step of the algorithm will be tested by performing special experiments in which the measured signal will be controlled. Also more experiments will be made on all kind of measurements to gather some statistically significant results.

REFERENCES

[1] Roman Trobec. Computer analysis of multichannel ECG. *Comput Biol Med*, 33(3):215–226, May 2003.

[2] SCHILLER AG, Switzerland. *ECG Measurements and Interpretation Programs, Physician's Guide*, 2009.

[3] C. Zellerhoff, E. Himmrich, D. Nebeling, O. Przibille, B. Nowak, and A. Liebrich. How can we identify the best implantation site for an ECG event recorder? *Pacing Clin Electrophysiol*, 23:1545–1549, 2000.

[4] C. De Capua, A. Meduri, and R. Morello. A Smart ECG Measurement System Based on Web-Service-Oriented Architecture for Telemedicine Applications. *IEEE Trans on Instr & Measur*, 59(10):2530–2538, 2010.

[5] P. Bifulco, M. Cesarelli, A. Fratini, M. Ruffo, G. Pasquariello, and G Gargiulo. A wearable device for recording of biopotentials and body movements.

In *Proceedings of the IEEE International Workshop on Medical Measurements and Applications Proceedings (MeMeA)*, pages 469–472, 2011.

[6] Alexandros Pantelopoulos and Nikolaos G. Bourbakis. A Survey on Wearable Sensor-Based Systems for Health Monitoring and Prognosis. *IEEE Transaction on Systems, Man, and cybernetics–part C: Applications and reviews*, 40(1), 2010.

[7] M. Lindén and M. Björkman. Embedded sensor systems for health - providing the tools in future healthcare. *Stud Health Technol Inform*, 200:161–163, 2014.

[8] Matjaz Depolli, Viktor Avbelj, Roman Trobec, Jurij Matija Kalisnik, Korosec Tadej, Antonija Poplas Susic, Uros Stanic, and Ales Semeja. Pcard platform for mhealth monitoring. *Informatica*, 40(1):117, 2016.

[9] I. Tomašić and R. Trobec. Electrocardiographic systems with reduced numbers of leadssynthesis of the 12-lead ECG. *IEEE Reviews in Biomedical Engineering*, 7:126–142, 2014.

[10] C. S. Lam. A review of the recent development of MEMS and crystal oscillators and their impacts on the frequency control products industry. In *Proccedings of the IEEE International Ultrasonics Symposium*, pages 694–704, 2008.

[11] Martin Horauer. *Clock synchronization in distributed systems*. Südwestdeutscher Verlag für Hochschulschriften SVH, 2009.

[12] R. Trobec, V. Avbelj, M. Šterk, B. Meglič, and V. Švigelj. Neurological data measuring and analysis software based on object oriented design. *In proceedings of The 7th European Federation of Autonomic Societies (EFAS) Meeting*, 2005.

Long-term Follow-up Case Study of Atrial Fibrillation after Treatment

M. Jan[*] and R. Trobec[**]

[*] Department of Cardiovascular Surgery, University Medical Centre, Zaloška 2, Ljubljana, Slovenia
[**] Department of Communication Systems, Jožef Stefan Institute, Jamova 38, Ljubljana, Slovenia
matevz.jan@kclj.si; roman.trobec@ijs.si

Abstract - A case study is presented, based on long-term ECG measurements of a patient with diagnosed persistent atrial fibrillation (PeAF) that undergone the classical diagnostic procedures. The long-term measurements have been performed with an ECG body sensor. Based on the European Heart Rhythm Association (EHRA) guidelines for treatment of atrial fibrillation the left atrial catheter cryo-ablation with an endpoint of pulmonary vein isolation was performed. After the cryo-ablation PeAF still persists, therefore an additional catheter radiofrequency ablation was performed. After the second procedure and in combination with antiarrhythmic drugs the atrial fibrillation (AF) was controlled on the level of relatively rare and short documented AF episodes. A detailed analysis of a long-term measurement has enabled detection of a large spectrum of arrhythmias, which have been documented over a ten-week period of measurements. Those include atrial extrabeats and nonsustained atrial tachycardias that might be the initial triggers for AF. The initial study motivates new hypotheses about the long-term impact of ablation procedures and antiarrhythmic drugs on the outcome of medical therapies, which deserves to be further elucidated with a larger and more systematic study.

I. INTRODUCTION

The atrial fibrillation (AF) is a cardiac arrhythmia with various symptoms, e.g., fatigue, palpitations, syncope, heart failure, etc. but can also exists with no symptoms [1]. It affects more than 4 millions of people in the European Union and about 100 millions worldwide. The AF is associated with more frequent hospitalizations because of stroke, transient ischemias and heart failure [2, 3] therefore it is maintained with some success with antiarrhythmic drugs or with atrial catheter ablation [4]. The AF can be classified by frequency and duration of its episodes as: paroxysmal, persistent, or long-standing persistent. It has been demonstrated that the consequences of AF have been correlated with episodes' duration and with the total amount of AF time, which is termed often as AF burden [5]. The AF burden is therefore one of the most important clinical indicator for the selection and outcome of therapeutic approaches.

Diagnosis of AF is usually based on patients' clinical history, physical examination, and is confirmed by ECG. However, it has been shown in several studies that the diagnosis, based on symptoms, is not always effective for confirmation and medical management of AF [6, 7], because the yield of ambulatory 12-lead ECG monitoring

is limited. The correctly detected amount of AF burden using the prolonged rhythm monitoring is proportional to the duration of the continuous rhythm monitoring. As a consequence, in cases with asymptomatic, undetected events of AF, patients are exposed to increased risk of ischemic stroke or thromboembolic complications [8].

Long-term electrocardiographic (ECG) recordings are recommended from the European Society of Cardiology (ESC) and European Heart Rhythm Association (EHRA) [9] for detection and maintenance of AF and other threatening arrhythmias that could influence the heart rhythm. External cardiac monitors, such as (i) non-invasive but obtrusive Holter monitors, (ii) implantable cardiac monitors (ICM) or loop recorders [10], and currently, (iii) wireless ECG body sensors [11, 12], incorporated into a system of mobile cardiac patients monitoring [13], are among the most common approaches. While the first two options are already matured with known advantages and limitations, the body sensors could be a complementary future approach for efficient long-term assessment of the AF burden.

By the first option, a Holter monitor with a reduced number of electrodes that are connected with wires to a small portable recorder, acquires continuous ECG measurement that can last from 24 hours to maximum a week. Usually one to three wired leads are utilized for automatic ECG reporting, including heart rhythm analysis and detection of myocardial ischemia, which is an advantage in comparison with the remaining options.

The second option, implantable loop recorders, are invasive and heart rhythm can only be analyzed on-line, because of their limited storage capacity. Additionally, a limited storage of energy prevents to record a high resolution ECG that is paramount in the case of AF detection. Next, a complicated equipment for data transfer from an ICM to external device for analysis of results, requires frequent visits of specialistic ambulance and expensive manipulation of expert personnel.

Finally, the long-term unobtrusive ECG recordings are possible with an ECG body sensor that is wirelessly connected with a smart phone or other personal device that is in constant use by the patient. Such a sensor measures a potential difference between two proximal electrodes on the skin, which enables monitoring of several vital functions, e.g., heart activity and respiration [14].

In this paper, a case study is presented, based on long-term ECG measurements in a patient with diagnosed

---

The author Roman Trobec acknowledges the financial support from the Slovenian Research Agency under the grant P2-0095 and and the EkoSMART project, grant No. C3330-16-529007, financed by the European Regional Development fund.

persistent atrial fibrillation (PeAF) that has undergone classical diagnostic and therapeutic procedures. Long-term measurements have been performed with a single ECG body sensor, placed in the vicinity of heart atria. The remaining of this paper is organized as follows. In Section II the methodology of long-term measurements is described with a short description of the novel wireless body ECG sensor. In Section III results are presented and discussed, and finally, in Section IV the paper is concluded.

II. METHOD

A. Monitoring Period

The patient, monitored in this case, had a persistent atrial fibrillation (PeAF) diagnosed on December 2014. Based on the European Heart Rhythm Association (EHRA) guidelines for treatment of atrial fibrillation (AF) the left atrial cryo-baloon catheter ablation with an endpoint of pulmonary vein isolation was performed at the end of March 2015. After the cryo-ablation the PeAF was interrupted and the heart rhythm improved, but the paroxysmal atrial fibrillation (PAF) still persisted. Therefore, an additional catheter radiofrequency ablation was performed at the end of October 2015. After the second procedure the PAF episodes were shorter and less frequent. However, the PeAF returned at the end of 2015. In January 2016 the electric conversion of the irregular rhythm was performed followed by application of antiarrhythmic drugs. Since then the AF was controlled at the level of relatively rare and short documented AF episodes. The heart rhythm was monitored by the ECG body sensor during the whole period of rehabilitation, from December 2014 to March 2017, in order to obtain evidence-based insight into heart rate patterns.

A detailed analysis of rhythm monitoring was performed during a ten-week period from the end of December 2016 to the beginning of March 2017. Five weeks before the end of analyzed period, on 2017-02-03, the antiarrhythmic drug has been reduced by 100% in order to assess eventual heart rhythm changes. We have expected that based on the obtained results from these measurements, an optimal dose of antiarrhythmic drugs will be determined to minimize their negative effects, and improved diagnostic procedures and medical treatment strategies will be planned.

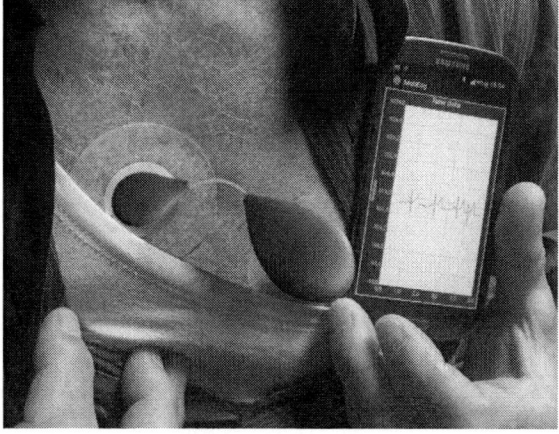

Figure 1. Wireless ECG sensor fixed on the chest above heart atria.

B. ECG Data Acquisition

The measured ECG data has been obtained from the medical graded ECG body sensor Savvy [15] (see Figure 1), marketed by Saving d.o.o., Ljubljana, Slovenia. The sensor is unobtrusive for users because of its small dimensions, flexible design, more than seven-day autonomy and simple use, and therefore appropriate for continuous measurements during normal daily activities. Users can mark specific events that describe their activity, comfort, sensor position and similar. The body ECG sensor can also support solutions in every-day monitoring in hospitals, in health clinics, in homes for elderlies and in health resorts. Its exceptionally lightweight design allows for unobtrusive use also during sports activities or during exhaustive physical work.

The placement of sensor electrodes on the chest can be easily fine-tuned to maximize the quality of the ECG recording [16, 17]. A moderate sampling rate of 125 Hz with 10 bit analogue/digital converter, that is a compromise between medical value and amount of generated data, suffices for accurate rhythm monitoring. User can easily master the mobile application on a smart phone that coordinates the data transfer from the sensor to the phone storage using sensor's low power wireless connection (BT4). The application provides on-line visualization of the measured ECG with a robust real-time beat detector for the calculation of minute beat rate (BPM) [18].

ECG measurements run continuously in the background and do not interfere with the usual mobile phone functions. However, short interruptions in measurement can always be present, for example, due to lost radio connection. Besides, user can either interrupt the current measurement, in the case of other activities, e.g., showering, or simply stop the measurement, in the case that one does not wish to be measured. Long-term measurements with skin electrodes are often a reason for skin irritation, which can be overcome by applying new electrodes for appropriate shift of the sensor position.

C. ECG Analysis

The measured ECG data, that are continuously stored in the mobile phone memory, can at any time be transferred to a personal computer, also while the current measurement is still running. The further analysis is currently possible with an open source program VisECG [19] (available on http://www.savvy.si) that was devised from its research prototype version. VisECG comprises the following basic functions:

(i) conversion of binary ECG files to VisECG native format,

(ii) time-overview of measured data files, marked with basic statistical properties, e.g., amount of lost data, mean values, etc.,

(iii) signal preprocessing and adaptive beat detection,

(iv) visualization of ECG measurements in their full resolution, and with corresponding events, e.g., QRS time with R-R intervals, and

(v) generation of different ECG reports in the form of pdf files.

Date: 28 feb 2017   VisEcg: 1.1.3

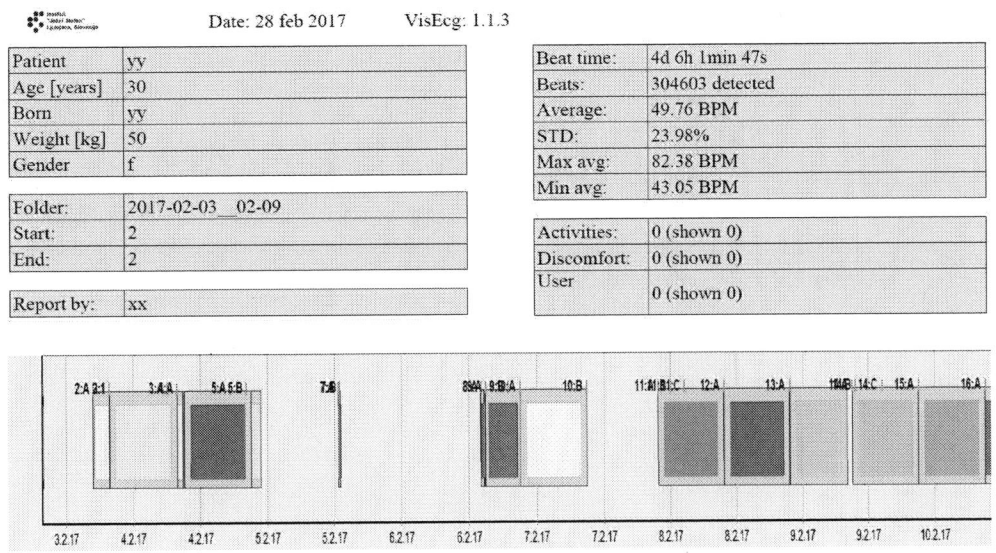

Patient	yy
Age [years]	30
Born	yy
Weight [kg]	50
Gender	f

Folder:	2017-02-03__02-09
Start:	2
End:	2

Report by:	xx

Beat time:	4d 6h 1min 47s
Beats:	304603 detected
Average:	49.76 BPM
STD:	23.98%
Max avg:	82.38 BPM
Min avg:	43.05 BPM

Activities:	0 (shown 0)
Discomfort:	0 (shown 0)
User	0 (shown 0)

Figure 2. A typical one-week overview of ECG of measurements from week 6 (2017-02-03 to 2017-02-09).

(vi) There are some additional advanced functions that enable, for example, study of heart rate variability, ECG derived respiration, frequency domain analysis, customized filtering and similar.

The VisECG program is not a replacement for standard Holter analysis programs that are able to perform a complete ECG analysis and present the results in the means of a few characteristic numbers that can be used for fast diagnosis. On the contrary, the VisECG approach provides a simple graphical presentation of BPM curve and assume an active engagement of medical expert that can devise a fast diagnosis from the long-term measurements by visual inspection. For example, a sinus rhythm can be identified by a smooth BPM curve, which indicates that no sudden change in consecutive R-R intervals is present. Alternatively, a visible change in BPM on the beat-to-beat basis could indicate a premature beat, or a bolded BPM curve could indicate an AF sequence because of unstructured significant changes in consecutive R-R times. Other ECG phenomena, e.g., start of multifocal atrial depolarizations that is reflected either in momentary change in BPM and in variations in P-wave morphology, cannot be accurately detected by computer program. On the other hand, such jumps in BPM will be easily noticed and reported by an expert supervisor.

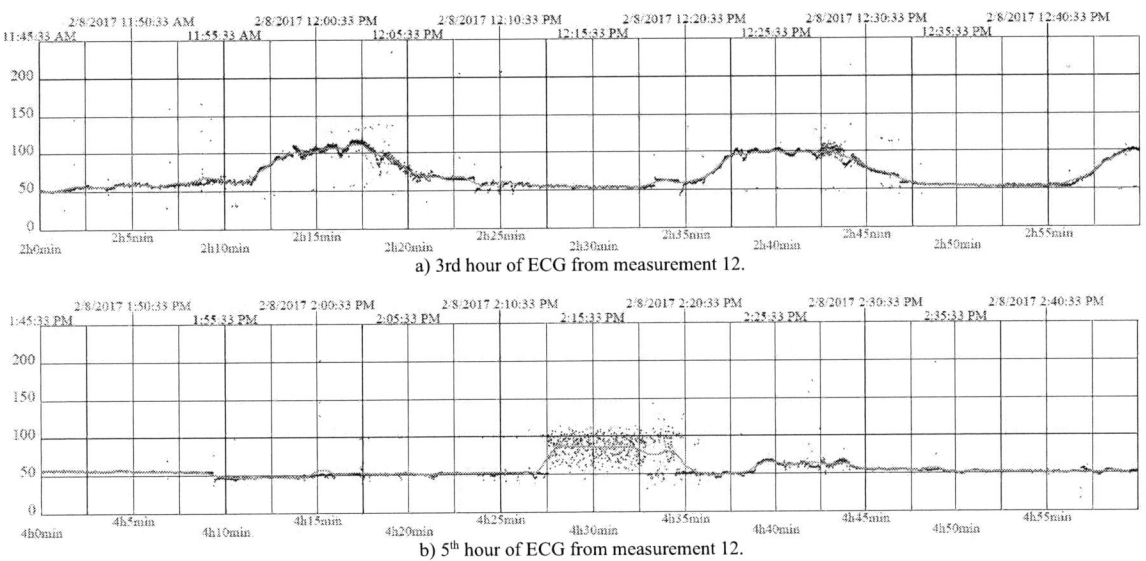

a) 3rd hour of ECG from measurement 12.

b) 5th hour of ECG from measurement 12.

Figure 3. Two one-hour BPM segments from ECG report of measurement 12 (2017-02-07_21.45.29_part_2) from week 6.

III. RESULTS

A. ECG Measurements Overview

The analyzed ECG measurements have been grouped in folders by weeks and analyzed correspondingly. A typical week-overview of 16 ECG measurements in week 6, from 2017-02-03 to 2017-02-09, is shown in Figure 2. Beside the demographic data, colored rectangles graphically represent particular measurements. If a measurement is longer than 12 hours, it is divided in 12-hour segments, denoted in its filename by a suffix _part_i, where i is the corresponding 12-hour segment. This enables faster visualization and easier manipulation. The color of a rectangle and its edge color indicate mean BPM and the standard deviation (SD) of measurement, respectively. Marked events that have been inserted by user or by examiner are marked with letters. Note that there are also periods with no measurements, which could have a minor impact on the overall results. However, usually the amount of time of missing measurements is significantly shorter than the total

occasional single beats reflect significantly increased BPMs (possibly atrial or ventricular extra beats).

By examining line-by-line of BPM a long-term tabular summary report was generated in the form shown in Table I. Gray lines represents five weeks after the reduction of antiarrhythmic drugs. We see that the total amount of analyzed time was 67.8% and the number of all detected beats was about 3.5 million. The AF burden was 132 minutes in 10 episodes, 118 sequences of severe arrhythmias in total duration of 1735 minutes have been identified. Severe arrhythmias are defined as continuous sequences, longer than 30 second, with the number of premature atrial beats (PAB) in the same range as the number of sinus beats. Such arrhythmias have often preceded AF sequences and could therefore be a trigger for the AF. A small increase in arrhythmias, before and after the reduction of the antiarrhythmic drugs can be noticed, however, a longer monitoring period would be needed for a reliable determination of statistical significance of this result.

TABLE I.   ECG SUMMARY REPORT, FROM WEEK 1: 2016-12-30_01-05 TO WEEK 10: 2017-03-03_03-09

Week number	Week dates	Analyzed time [min]	Analyzed time [%]	No. of detected beats	Mean BPM	Standard deviation of BPM [%]	Maximum smoothed BPM	Minimum smoothed BPM	Mean No. PAB /hour	No. of arrhythmic episodes	Duration of arr. episodes [min]	No. of PAF episodes	Duration of PAF episodes [min]
1	2016-12-30__01-05	5390	53,5	279100	51,8	30,2	61,8	43,5	6,9	0	0	0	0
2	2017-01-06__01-12	4409	43,7	229823	52,1	24,1	63,4	42,6	6,9	3	121	0	0
3	2017-01-13__01-19	6784	67,3	383883	56,6	29,0	73,9	44,6	9,0	21	180	2	45
4	2017-01-20__01-26	6686	66,3	352958	52,8	26,6	62,1	45,9	8,0	6	130	0	0
5	2017-01-27__02-02	6555	65,0	333970	50,9	25,9	74,0	40,8	9,8	0	0	0	0
6	2017-02-03__02-09	6121	60,7	304603	49,8	24,0	82,4	43,1	10,6	13	187	3	20
7	2017-02-10__02-16	9077	90,0	437219	48,2	24,7	59,0	41,8	7,1	9	147	0	0
8	2017-02-17__02-23	7263	72,1	379952	52,3	23,1	67,6	41,4	6,5	29	705	2	7
9	2017-02-24__03-02	9363	92,9	466108	49,8	21,2	56,8	42,2	7,4	15	170	3	60
10	2017-03-03__03-09	6689	66,4	344757	51,5	23,2	70,6	42,4	9,2	22	95	0	0
	TOTAL	68337	67,8	3512373	51,6	25,2	82,4	40,8	8,1	118	1735	10	132

measuring time.

B. Arrhythmias Identification

The complete ECG measurements with segmented heart beats have been examined by a trained person on the basis of automatically generated BMP visualization arranged in one-hour intervals in each line. An example of 3rd and 5th hours in part_2 of ECG measurement 12 from Figure 2, which starts at 2017-02-07_21:45:29, is shown in Figure 3. Dots represent a beat-to-beat BPM and the green graph represents a 10 seconds window average of BPM.

In Figure 3. a) two activity cycles are present, with a period of 25 minutes and maximum BPM greater than 100, while in Figure 3. b) an ECG at rest is shown. During most of the measurement time a sinus rhythm is present with an average of 50 BPM. On the 3rd hour segment, at 33th minute, a sudden small change of rhythm appears (possibly an alternate ectopic rhythm). On the 5th hour segment b) at 27th minute a seven-minute period of unstructured beats is present (possibly paroxysmal AF), at 48th minute an unexpectedly high BPM of 200 can be noticed (possibly an artefact), while throughout the whole ECG measurement

C. Types of Arrhythmic Events

During the analysis of ECG measurements for ECG summary report some typical cardiac events are identified and often also several unexpected behaviors in the BPM graph. The examiner can always check the original ECG signal in its full resolution to resolve the noticed phenomena for correct interpretation. For example, four before mentioned phenomena from the analyzed measurement shown in Figure 3 will be presented in more detail. After a click on the rectangle that corresponds to measurement 12, the ECG signal is zoomed around the noticed time in hour 3 and 5. The results are shown in Figure 4 by strips of ECG from VisECG report. Time scale is in seconds after the measurement start.

In Figure 4. a) a ten-second strip of ECG from 3rd hour segment is shown around a PAB that results in a sudden change of P-wave morphology and BPM, from 54 to 65, at 9197 s after the start of measurement. We confirmed by visual examination that the mean frequency of such events is 20 times/day. With further detailed inspection a short paroxysmal AF sequence was confirmed on panel b) of Figure 4. On panel c) the sudden change of BPM is due to

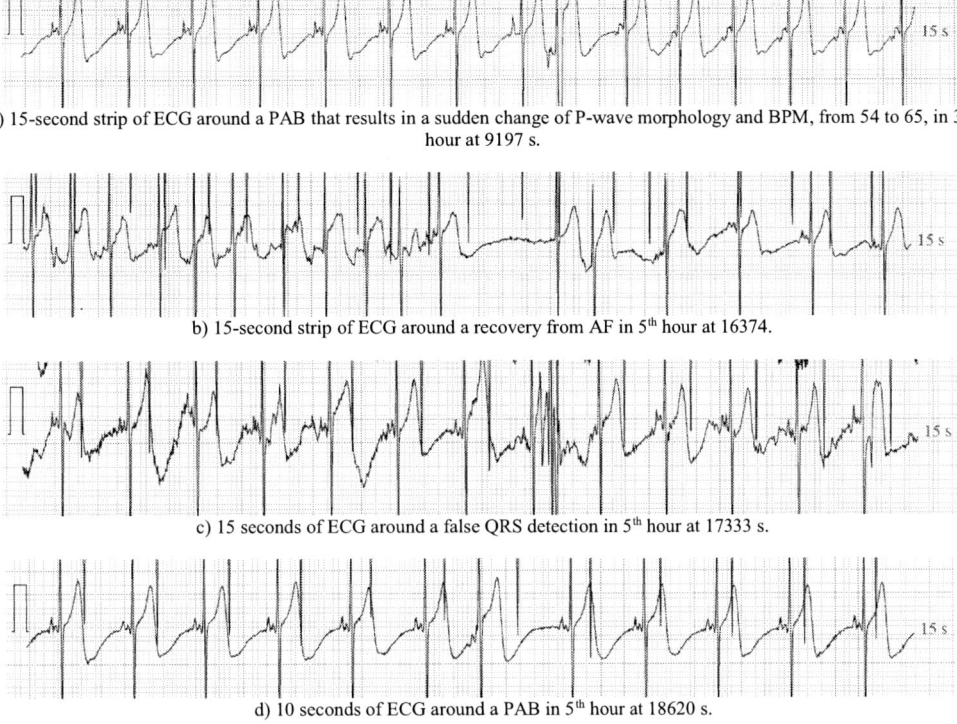

a) 15-second strip of ECG around a PAB that results in a sudden change of P-wave morphology and BPM, from 54 to 65, in 3rd hour at 9197 s.

b) 15-second strip of ECG around a recovery from AF in 5th hour at 16374.

c) 15 seconds of ECG around a false QRS detection in 5th hour at 17333 s.

d) 10 seconds of ECG around a PAB in 5th hour at 18620 s.

Figure 4. Detailed interpretation of four phenomena noticed in the 3rd and 5th hour of ECG measurement 12. Note that some parts of S-waves from previous 15 s are visible in the upper part of all ECG stripes.

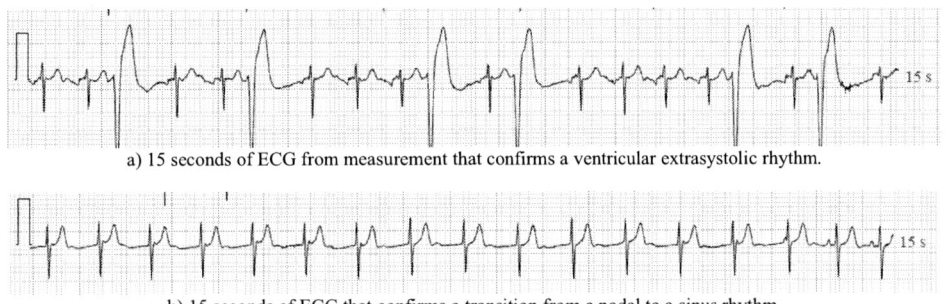

a) 15 seconds of ECG from measurement that confirms a ventricular extrasystolic rhythm.

b) 15 seconds of ECG that confirms a transition from a nodal to a sinus rhythm.

Figure 5. Two additional phenomena detected in other measurements of the measuring period.

an artefact from movement or muscular activity. Finally, on panel d), a PAB is confirmed with no change in P-wave morphology and with no change in BPM.

Two other rarely detected phenomena, from other measurements of the analyzed period, are shown in Figure 5. On panel a) a 15-second of ECG is shown, from measurement 2017-02-20_23:14:03 at 21041 s, that confirms a ventricular extrasystolic rhythm. On panel b) an ECG strip from measurement 2016-12-21_01:00:00 at 35265 s, documents a smooth transition from a nodal to a sinus rhythm.

IV. CONCLUSION

In this paper, we present a case of PeAF that was constantly monitored for two and half year, which is to our knowledge first long-term report with complete data set of ECG measurements. A detailed long-term analysis was done for a ten-week period in order to determine the AF burden and the rhythm status after two catheter ablations. Further, an optimal dose of anti-arrhythmic drugs can be determined and plans for eventual further actions can be established based on evidences from the ECG measurements. The results show that we were able to precisely identify all AF periods and consequently the AF burden. However, these methods should be accounted only as assistance for the medical professional and should not be trusted blindly. The acquired ECG measurements enable also more complex methods [20] for the assessment of hearth rhythm.

In further work, we plan to improve the interpretation and analysis software that must be tailored to big-data sets because the current version of the ECG body sensor produces about 15 MB of data per day. Measures for the improving accuracy of the interpretation, based on the concurrent measurement of acceleration, i.e., user's activity, should be implemented. Segmentation of ECG measurements can be supported by the accelerometer data [21]. Various clustering methods will be tested, which will form the base for automatic annotation of the ECG towards the final task of automatic ECG beats classification.

REFERENCES

[1] Stewart S, Hart CL, Hole DJ, McMurray JJ. A population-based study of the long-term risks associated with atrial fibrillation: 20-year follow-up of the Renfrew/Paisley study. Am J Med 2002;113:359–364.

[2] Wattigney WA, Mensah GA, Croft JB. Increased atrial fibrillation mortality: United States, 1980–1998. Am J Epidemiol 2002;155:819–826.

[3] Wong CX, Brooks AG, Leong DP, Roberts-Thomson KC, Sanders P. The increasing burden of atrial fibrillation compared with heart failure and myocardial infarction: a 15-year study of all hospitalizations in Australia. Arch Intern Med 2012;172:739–741.

[4] Piccini JP, Fauchier L. Rhythm control in atrial fibrillation. Lancet 2016;388:829-40.

[5] Botto GL, Padeletti L, Santini M, et al. Presence and duration of atrial fibrillation detected by continuous monitoring: crucial implications for the risk of thromboembolic events. J Cardiovasc Electrophysiol 2009;20:241–248.

[6] Strickberger SA, Ip J, Saksena S, Curry K, Bahnson TD, Ziegler PD. Relationship between atrial tachyarrhythmias and symptoms. Heart Rhythm 2005;2:125–131.

[7] Verma A, Champagne J, Sapp J, Essebag V, Novak P, Skanes A, Morillo CA, Khaykin Y, Birnie D. Discerning the incidence of symptomatic and asymptomatic episodes of atrial fibrillation before and after catheter ablation (DISCERN AF): a prospective, multicenter study. JAMA Intern Med 2013;173:149–156.

[8] Flaker GC, Belew K, Beckman K, Vidaillet H, Kron J, Safford R, Mickel M, Barrell P, AFFIRM Investigators. Asymptomatic atrial fibrillation: demographic features and prognostic information from the Atrial Fibrillation Follow-up Investigation of Rhythm Management (AFFIRM) study. Am Heart J 2005;149:657–663.

[9] Kirchhof P, at al. 2016 ESC Guidelines for the management of atrial fibrillation developed in collaboration with EACTS. Eur Heart J 2016; 37(38):2893-2962.

[10] Sanders P, at al. Performance of a new atrial fibrillation detection algorithm in a miniaturized insertable cardiac monitor: Results from the Reveal LINQ Usability Study. Heart Rhythm 2016;13:1425-1430.

[11] A. Rashkovska, I. Tomašić, K. Bregar and R. Trobec, "Remote Monitoring of Vital Functions - Proof-of concept System," Proceedings of 35th International Convention, May 21-25, Opatija, Croatia, 2012, pp. 446–450.

[12] M. Etemadi et al., "A Wearable Patch to Enable Long-Term Monitoring of Environmental, Activity and Hemodynamics Variables," IEEE Transactions on Biomedical Circuits and Systems, 2016;10:280-288.

[13] J. M. Kališnik et al., "Mobile health monitoring pilot systems," Proceedings of IS 2015, 18th International Multiconference Information Society, October 9-12, 2015, Ljubljana, Slovenia, pp. 62–65.

[14] R. Trobec, V. Avbelj, and A. Rashkovska, "Multi-functionality of wireless body sensors," The IPSI BgD transactions on internet research, vol. 10, pp. 23–27, January 2014.

[15] www.savvy.si, last time visited on 28. February, 2017

[16] R. Trobec and I. Tomašić, "Synthesis of the 12-lead electrocardiogram from differential leads," IEEE Transactions on Information Technology in Biomedicine, 2011;15:615–621.

[17] I. H. Hansen, K. Hoppe, A. Gjerde, J. K. Kanters, and H. B. Sorensen, "Comparing Twelve-lead Electrocardiography with Close-To-Heart Patch Based Electrocardiography," Proceedings of IEEE EMBC 2015, 37th Annual International Conference, August 25-29, 2015, Milan, Italy, pp. 330–333.

[18] P. Lavrič, M. Depolli, "Robust beat detection on noisy differential ECG", Proceedings of 39th International Convention May30-June 3, Opatija, Croatia, 2016, pp. 401–406.

[19] M. Mohorčič, M. Depolli, " Heart rate analysis with NevroEkg", Proceedings of 39th International Convention May30-June 3, Opatija, Croatia, 2016, pp. 487–492.

[20] J. Kšela, V. Avbelj, J.M. Klišnik, "Multifractality in heartbeat dynamics in patients undergoing beating-heart myocardial revascularization", Computers in Biology and Medicine, 2015;60:66-73.

[21] H. Gjoreski, A. Rashkovska, S. Kozina, M. Luštrek, M. Gams, "Telehealth using ECG sensor and accelerometer," Proceedings of 37th International Convention, May 26-30, Opatija, Croatia, 2016, pp. 270–274.

MIPRO 2017, May 22- 26, 2017, Opatija, Croatia

A Case Report of Long-Term Wireless Electrocardiographic Monitoring in a Dog with Dilated Cardiomyopathy

M. Brložnik*, V. Avbelj**

* Small animal clinic, Veterinary faculty, University of Ljubljana, Slovenia
** Department of Communication Systems, Jožef Stefan Institute, Ljubljana, Slovenia

Correspondence: viktor.avbelj@ijs.si

Abstract — **Wireless electrocardiographic (ECG) sensor attached to the skin and connected to a smart device via low power Bluetooth technology has been used to record more than 500 hours of ECG data in a German shepherd dog with dilated cardiomyopathy (DCM). Wireless ECG monitoring has been used for a period of 6 months. With the wireless body electrodes, the ECG data were obtained while the dog was resting, walking, playing and eating. Atrial fibrillation, ventricular premature complexes, occasional ventricular tachycardia and multiform ventricular beats were observed. Numerous standard 6-lead ECG recordings have been compared to the recordings obtained with wireless body electrodes. Instantaneous and average heart rates and standard duration measurements evaluated with the two devices were identical in all cases. The extended ECG monitoring time with the wireless device increased the diagnostic yield of arrhythmias.**

The dog was treated with diuretics, positive inotropes, ACE inhibitor and antiarrhythmics for 2 years. Influence of various drugs, dog's activities, and environmental factors on ECG data was investigated. During the 6 months period dog's condition was changing substantially and long term ECG monitoring excluded arrhythmias as the cause for dog's weakness.

The wireless device, which proved to be reliable and simple to use, enables an excellent option of long-term monitoring of canine cardiac rhythm in real-world environment.

I. INTRODUCTION

Canine dilated cardiomyopathy (DCM) is a primary myocardial disease characterized by chamber dilation and a decrease in myocardial contractility. Beside the predominant systolic dysfunction of one or both ventricles, diastolic dysfunction is also present. DCM is the most common form of canine cardiomyopathy. It is an inherent disease of large and medium sized dogs [1 - 6]. In dogs with DCM arrhythmias occur commonly. Due to their intermittent nature a long-term electrocardiographic (ECG) monitoring in real-world environment is most suitable for the diagnostics [7 - 9].

Arrhythmias frequently require treatment. Treatment success and possible adjustment of dosages is best evaluated with follow-up long-term electrocardiographic monitoring.

In this case report we present a German shepherd dog with DCM and various arrhythmias, which were recorded with a wireless ECG sensor. This wireless device was described and used previously [7, 8, 10, 11].

II. PRESENTATION

A. Materials and Methods

A 10-year old male German shepherd dog with 40 kg (Fig. 1) presented with exercise intolerance. The dog has been previously diagnosed with degenerative joint disease. Tachyarrhythmia, increased respiratory rate, weak pulse and distended abdomen were noticed with clinical examination. Abdominocentesis revealed modified transudate. With thoracic radiography enlarged round cardiac silhouette and pulmonary edema were documented. Abdominal ultrasound revealed free intra-abdominal fluid, increased size of liver and spleen, and cystic hyperplasia of prostate. Urinalysis was unremarkable. Hematology and biochemistry of serum were normal except for mildly increased values of liver enzyme alanine aminotransferase. Atrial fibrillation with frequency 230 beats per minute and severely prolonged QRS (88 ms) were diagnosed with standard ECG. Echocardiography revealed severely enlarged left atrium, moderate mitral valve regurgitation, severely enlarged end-diastolic and end-systolic diameter of left ventricle (severely decreased myocardial contractility). Taurine and thyroxine serum concentrations were normal.

Diagnosis of dilated cardiomyopathy (DCM) was made and treatment with furosemide, spironolactone, ramipiril, pimobendan and digoxin was initiated.

Fig. 1: A 10-year old German Shepherd Srečko

The frequency of atrial fibrillation lowered to 180 beats per minute. To further lower heart rate diltiazem (30

303

mg/8h) was added to therapy. Heart rate reduced to 130-150 beats per minute.

Dog's condition appeared stable, but ECG measurements revealed occasional slow solitary ventricular premature contractions (VPCs). Later, also VPC couplets and triplets were observed with standard ECG. Atenolol and sotalol were administered, but were not tolerated. Half a year into DCM diagnosis, surgery had to be performed due to gastric torsion.

In following months frequent weakness, depression and anorexia have been observed and long-term electrocardiographic monitoring of cardiac rhythm was advised to exclude arrhythmia like ventricular tachycardia as a possible cause. Long term ECG monitoring was unavailable for another year, when a possibility of wireless ECG monitoring was introduced.

Wireless ECG sensor described previously [7, 8, 10, 11] consists of an electronic module with battery and two self-adhesive electrodes with a distance of 9 cm (Fig. 2).

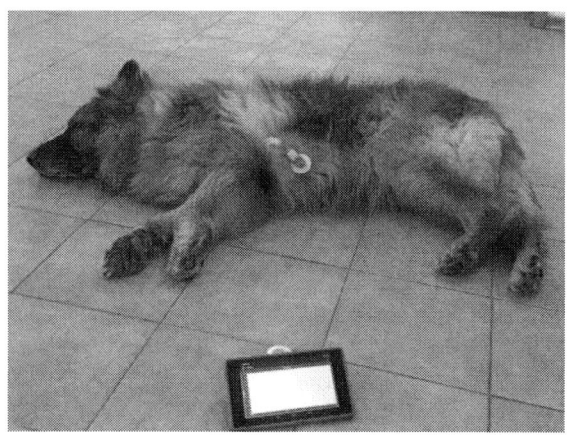

Fig. 2: Wireless body ECG sensor placed on the left side of the thorax, negative electrode near atria and positive electrode at the apex of the left ventricle.

The ECG sensor records one bipolar lead. Wireless communication with the sensor via low power Bluetooth technology allows the display of the ECG signal on a smart device (tablet or smart phone), which records real-time data from the electrodes. Evaluation of the ECG recordings was performed with VisECG software (Jožef Stefan Institute, Ljubljana). The software can also extract the respiration rate based on the amplitude changes of QRS complexes [12].

The device was used to record more than 500 hours of ECG data over a period of 6 months. To prevent detachment, the ECG sensor was occasionally bandaged (Fig. 3). With the wireless body electrodes, the ECG data were obtained while the dog was resting, walking, playing and eating. Numerous standard ECG recordings were compared to the recordings obtained with wireless body electrodes.

Influence of various drugs, dog's activities, and environmental factors on ECG data was investigated.

During the 6 months period dog's condition was changing substantially and the purpose of long term ECG monitoring was to exclude arrhythmias as the cause of

debilitating condition.

Fig. 3: Wireless body ECG sensor during activities in real time environment

B. Results

Heart rate and rhythm from the wireless ECG data were compared with the data from the standard ECG. Instantaneous and average heart rates measured with both devices were identical in all cases. Whenever compared standard duration measurements (P wave width, QRS width, PR interval and QT interval) were congruent. All arrhythmias documented with standard ECG were observed with wireless sensor. The extended ECG monitoring time of the wireless device increased the diagnostic yield of arrhythmias. Atrial fibrillation (Figs. 4 – 9), fusion beats, VPCs (Figs. 5 – 8), and ventricular tachycardia (Fig. 9) were observed. VPCs were solitary (Fig. 5), couplets (Fig. 8), bigemini (Fig. 7), and multiform (Fig. 6).

Atrial QRS complexes were usually notched, which represents an indication of an asynchronous depolarization of ventricles. The notches are clearly visible in Fig. 4.

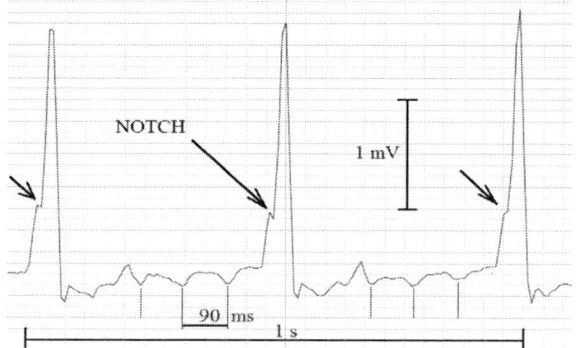

Fig. 4: Atrial fibrillation with a dominant frequency of approx. 650 fibrillations/min (see 90 ms cycle length) and ventricular frequency of approx. 130 beats/min. Presumably, every 5th fibrillation in the vicinity of atrio-ventricular node is conducted to the ventricles. Note the notched QRS complexes (arrows).

It was estimated that approximately 10% of the recordings were composed of artifacts (motion artifacts or loss of signal). During strenuous physical activity (running, jumping, etc.) interpretation of ECG signal was not possible.

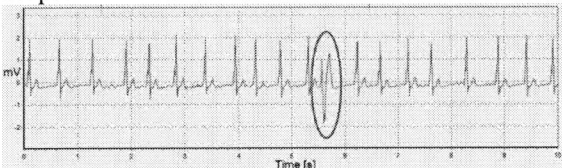

Fig. 5: Atrial fibrillation with ventricular frequency of 120 beats/min with a solitary ventricular premature complex (encircled).

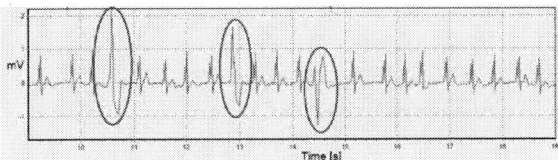

Fig. 6: Atrial fibrillation with multiform VPCs (encircled).

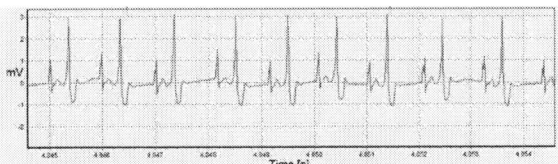

Fig. 7: Atrial fibrillation with bigemini (every 2nd beat is VPC).

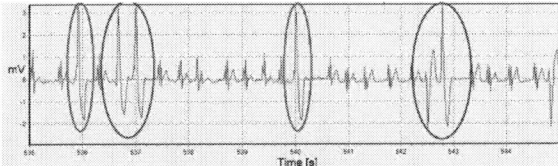

Fig. 8: Atrial fibrillation with multiform VPCs. Couplets are encircled in blue and solitary VPCs are incircled in red.

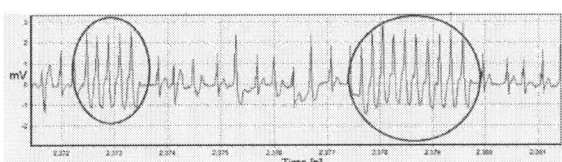

Fig. 9: Two bouts of non-sustained ventricular tachycardia (encircled).

Due to frequent bouts of non-sustained ventricular tachycardia, other antiarrhythmics for ventricular arrhythmia have been recommended (carvedilol, amiodarone or mexilletine), but were declined by the owner due to possible side effects.

Immediate influence of administering various drugs was not documented with wireless ECG. It was, however, noted that increasing the dose of diltiazem to 60 mg/8h during treatment decreased the heart rate to 120-130 beats per minute.

Influence of dog's activities on frequency of arrhythmias was not observed with wireless ECG.

Meteoropathy and other changes in dog's condition were observed frequently, but association of dog's status with arrhythmias was not. Even in bouts of non-sustained

ventricular tachycardia, dog's condition did not seem to worsen.

However, an interesting observation was made with wireless ECG sensor: in first two months of measurements all ventricular arrhythmias were very frequent, while in the last months of dog's life heart rhythm appeared more steady and slower, and only solitary VPCs were documented.

Due to cardiac cachexia and kidney failure the owner opted for dog's euthanasia 2 years after diagnosis of DCM (6 months after initiation of wireless ECG monitoring).

C. Discussion

Diagnosis of DCM is based on the identification of myocardial dysfunction and myocardial eccentric hypertrophy with the exclusion of other congenital or acquired cardiac disease [2 - 4]. Patients with DCM frequently present with decreased exercise tolerance, as was the case in this dog. Other frequent clinical manifestations are poor appetite, lethargy, generalized weakness, cough and syncope [1, 2, 5, 6]. The primary morphologic change in DCM is ventricular eccentric hypertrophy. This occurs in response to a functional systolic contractile failure. Histopathological findings in canine DCM include fatty infiltration with degeneration and/or attenuated wavy fibers [2]. Atrial fibrillation is a common arrhythmia in patients with DCM and was noted at the presentation of this case. The dog presented with congestive heart failure (CHF). CHF most commonly develops at a certain stage of the disease and is mostly present at the time of diagnosis [1, 2]. The pathophysiology of CHF is no longer considered a mere haemodynamic consequence of a pump dysfunction, but a complex clinical syndrome with release of many neurohormones, which are believed to have impact on the progression of the disease [3]. The diagnosis of overt DCM is usually straightforward [4]. Therapy of the dogs with DCM and heart failure consists of inotropic support (pimobendan, digoxin), ACE inhibitors, diuretics, and antiarrhythmics, if indicated [4]. In this case the frequency of atrial fibrillation was initially decreased with digoxin, but diltiazem had to be added. Therapy for ventricular arrhythmias was either not tolerated (sotalol, atenolol) or was declined by the owner (carvedilol, amiodarone or mexilletine) due to possible side effects.

Prognosis of DCM is likely to be dependent on the underlying cause. In one study, the median survival time in the dogs was 19 weeks, with the survival rate at one year 28% and at two years 14% [1]. In another study survival time ranged from 2 to 1108 days with median survival time 671 days [6]. Certain negative predictors of survival time were proposed including age, breed, pleural effusion, pulmonary edema, ascites, arrhythmias, severely increased end-systolic volume and ejection fraction, a restrictive pattern of transmitral flow, increased duration of QRS, etc. [5, 6, 13]. In our case, pulmonary edema, ascites, severely enlarged heart with poor contractility, various arrhythmias, and prolonged QRS (88 ms, reference < 60 ms) were identified. However, the dog was living a quality life for two years after diagnosing DCM, which is a favorable outcome considering the negative prognostic factors.

Long-term ECG monitoring can prolong survival of the patients with cardiac diseases, because it can exclude or diagnose arrhythmias as the cause for debilitating condition. Although the condition of this dog was worsening over the last few months of his life, wireless sensor showed that heart rhythm was more steady and slower with less arrhythmias compared to the first months of wireless ECG measurements. This enabled more confident approach to treatment and support.

Wireless ECG devices have been used in animals previously [7, 8, 14 - 18].

In this case the results of wireless ECG monitoring were compared with numerous standard ECG recordings and a conformation was made that the device is accurate and highly reliable. All arrhythmias documented with standard ECG were also documented with wireless sensor. Instantaneous and average heart rates, and standard duration measurements, determined with the two devices, were identical in all cases. This has been reported for the device previously [7, 8, 10]. Furthermore, the accuracy of this device has been confirmed with the synthesis of a 12-lead ECG [19 - 22]. Other devices to monitor ECG wirelessly have also proved to be accurate and reliable [23 – 32].

A higher diagnostic yield of arrhythmias was documented with wireless sensor due to prolonged monitoring. This has been reported previously in veterinary medicine for standard Holter and Event monitors [9, 33, 34], and also for the used wireless device [7, 8]. In human medicine the wireless devices were compared to 24-hour Holter monitoring and these reports are emphasizing the advantage of wireless devices to diagnose arrhythmias due to possibility of extended monitoring [23 - 25]. Longer monitoring is enabled with wireless devices due to their wearing comfort (they are waterproof, suitably small, and there are no wires attached to the body). Low power consumption of these devices contributes not only to prolonged lifetime, but also to system miniaturization, because the size of the battery occupies most of the system volume [31, 32, 35]. It is reasonable to expect that further development of electronics will enable even more monitoring of vital functions in clinical and home settings [36, 37] and that the wireless devices will become an important tool in human and veterinary medicine.

Beside the heart rate, duration measurements and arrhythmia recognition, the wireless device was also able to identify details like notched QRS complexes and predominant frequency of atrial fibrillation. This is another indicator of device's precision and it suggests that wireless sensor might have additional diagnostic value and might not be used as a mere heart rate and rhythm monitor.

Atrial fibrillation (AF), which was the prevailing heart rhythm of the dog in our case, most commonly occurs secondary to serious underlying cardiac disease. The onset of AF usually coincides with deterioration in clinical status and AF is associated with a high mortality rate [38, 39]. Different mechanisms of AF have been proposed, including a single focus firing rapidly and causing fibrillatory conduction, and multiple re-entrant wavelets with random propagation over the atria [40]. In persistent AF, the prevailing theory regarding its mechanism involves coexistence of multiple random wavelets of activation, which create a chaotic cardiac rhythm [41]. Dominant frequency analysis of atrial electrograms has been used to understand the pathophysiology of AF and it has been shown that the dominant frequency of AF is an effective tool in estimating activation rate during AF condition. In our case the dominant frequency of approximately 650 fibrillations/min was observed with the wireless sensor. It is noteworthy that dominant frequency of AF is different at various localized atrial sites, which has been shown with spectral analysis and dominant frequency mapping of AF [41– 43].

III. CONCLUSIONS

Wireless long term ECG monitoring was beneficial in diagnostics and treatment of the dog in our case, because arrhythmias were excluded as the cause of dog's debilitating condition. This enabled more confident approach to treatment.

The wireless device used in this case report is now readily available [44]. With additional validation, software development, and experts offering readings and interpretation, the wireless devices might replace conventional Holter and Event monitors in human and veterinary medicine in due course.

ACKNOWLEDGMENT

This work was partially supported by the Slovenian Research Agency under grant P2-0095 (V. Avbelj).

REFERENCES

[1] M. W. Martin, M. J. Stafford Johnson, B. Celona, "Canine DCM: a retrospective study of signalment, presentation and clinical findings in 369 cases", J Small Anim Pract 2009; 50: 23–29.

[2] A. Tidholm, J. Häggström, M. Borgarelli, A. Tarducci, "Canine idiopathic DCM. Part I: aetiology, clinical characteristics, epidemiology and pathology", Vet J 2001; 162: 92–107.

[3] M. Borgarelli, A. Tarducci, A. Tidholm, J. Häggström, "Canine idiopathic DCM. Part II: pathophysiology and therapy", Vet J 2001; 162: 182–195.

[4] J. Dukes-McEwan, M. Borgarelli, A. Tidholm, A. C. Vollmar, J. Häggström, "Proposed guidelines for the diagnosis of canine idiopathic DCM", J Vet Cardiol 2003; 5: 7–19.

[5] M. W. Martin, M. J. Stafford Johnson, G. Strehlau, J. N. King, "Canine DCM: A retrospective study of prognostic findings in 367 clinical cases", J Small Anim Pract 2010; 51: 428–436.

[6] M. Borgarelli, R. A. Santilli, D. Chiavegato, G. D'Agnolo, R. Zanatta, A. Manelli, A. Tarducci, "Prognostic indicators for dogs with dilated cardiomyopathy", J Vet Intern Med 2006; 20: 104-110.

[7] M. Brložnik, V. Avbelj, "Wireless electrocardiographic monitoring in veterinary medicine", Proceedings of 38th International Convention MIPRO 2015, Opatija, Croatia, 2015: 356–359.

[8] A. Krvavica, Š. Likar, M. Brložnik, A. Domanjko-Petrič, V. Avbelj, "Comparison of Wireless electrocardiographic monitoring and Standard ECG in Dogs", Proceedings of 39th International Convention MIPRO 2016, Opatija, Croatia, 2016: 416–419.

[9] G. Wess, A. Schulze, N. Geraghty, K Hartmann, "Ability of a 5-minute electrocardiography (ECG) for predicting arrhythmias in Doberman Pinschers with cardiomyopathy in comparison with a 24-hour ambulatory ECG", J Vet Intern Med 2010; 24: 367–371.

[10] A. Rashkovska, I. Tomašić, R. Trobec, "A telemedicine application: ECG data from wireless body sensors on a smartphone", Proceedings of MIPRO 2011, Opatija, Croatia, 2011: 293–296.

[11] R. Trobec, M. Depolli, V. Avbelj, "Wireless network of bipolar body electrodes", Proceedings of the 7th International Conference on Wireless On-demand Network Systems and Services, WONS 2010: 145–149.

[12] R. Trobec, A. Rashkovska, V. Avbelj, "Two proximal skin electrodes – a respiration rate body sensor", Sensors (Basel) 2012; 12: 13813–13828.

[13] B. M. Pedro, J. V. Alves, P. J. Cripps, M. J. Stafford Johnson, M. W. S. Martin, "Association of QRS duration and survival in dogs with dilated cardiomyopathy: A retrospective study of 266 clinical cases", J Vet Cardiol 2011; 13: 243–249.

[14] R. Brugaloras, J. Dieffenderfer, K. Walker, A. Wagner, B. Sherman, D. Roberts, A. Bozkurt, "Wearable Wireless Biophotonic and Biopotential Sensors for Canine Health Monitoring", IEEE 2014.

[15] E. Aguirre, P. Lopez-Iturri, L. Azpilicueta, J. J. Astrain, J. Villadangos, D. Santesteban, F. Falcone, "Implementation and Analysis of a Wireless Sensor Network-Based Pet Location Monitoring System for Domestic Scenarios", Sensors 2016; 16: 1384–1404.

[16] A. A. Uddin, P. P. Morita, K. Tallevi, K. Armour, J. Li, R. P. Nolan, J. A. Cafazzo, "Development of a Wearable Cardiac Monitoring System for Behavioral Neurocardiac Training: A Usability Study", JMIR Mhealth Uhealth 2016; 4: e45.

[17] M. S. Kraus, F. C. Brewer, M. Rishniw, A. R. Gelzer, "Comparison of the AliveCor® ECG device for the iPhone with a reference standard electrocardiogram", ACVIM Seattle 2013: abstract.

[18] H. Mongue-Din, A. Salmon, M. Y. Fiszman, Y. Fromes, " Non-invasive restrained ECG recording in conscious small rodents: a new tool for cardiac electrical activity investigation", Eur J Physiol 2007; 454: 165–171.

[19] I. Tomašić, R. Trobec, V. Avbelj, "Multivariate linear regression based synthesis of 12-lead ECG from three bipolar leads", Proceeding of the 3rd International Conference on Health Informatics, HealthInf 2010: 216–221.

[20] R. Trobec, I. Tomašić, "Synthesis of the 12-lead electrocardiogram from differential leads", IEEE Trans Inf Technol Biomed 2011; 15: 615–621.

[21] I. Tomašić, R. Trobec, "Electrocardiographic systems with reduced numbers of leads-synthesis of the 12-lead ECG", IEEE Rev Biomed Eng 2014; 7: 126–142.

[22] I. Tomašić, S. Frljak, R. Trobec, "Estimating the Universal Positions of Wireless Body Electrodes for Measuring Cardiac Electrical Activity", IEEE Trans Biomed Eng 2013; 60: 3368–3374.

[23] P. M. Barret, R. Komatireddy, S. Haaser, S. Topol, J. Sheard, J. Encinas, A. J. Fought, E. J. Topol, "Comparison of 24-hour Holter monitoring with 14-day novel adhesive patch electrocardiographic monitoring", Am J Med 2014; 127: 95 e11–7.

[24] D. Schreiber, A. Sattar, D. Drigalla, S. Higgins, "Ambulatory cardiac monitoring for discharged emergency department patients with possible cardiac arrhythmias", West J Emerg Med 2014; 15: 194–198.

[25] M. Rosenberg, M. Samuel, A. Thosani, P. Zimetbaum, "Use of a noninvasive continuous monitoring device in the management of atrial fibrillation: a pilot study", Pacing Clin Electrophysiol 2013; 36: 328–333.

[26] S. S. Lobodzinski, "ECG patch monitors for assessment of cardiac rhythm abnormalities", Prog Cardiovasc Dis 2013; 56: 224–229.

[27] M. P. Turakhia, D. D. Hoang, P. Zimetbaum, J. D. Miller, V. F. Froelicher, U. N. Kumar, X. Xu, F. Yang, P. A. Heidenreich. "Diagnostic utility of a novel leadless arrhythmia monitoring device", Am J Cardiol 2013; 112: 520–524.

[28] S. L. Higgins, "A novel patch for heart rhythm monitoring: is the Holter monitor obsolete?", Future Cardiol 2013; 9: 325–333.

[29] G. Hindricks, E. Pokushalov, L. Urban, M. Taborsky, K. H. Kuck, D. Lebedev, G. Rieger, H. Pürerfellner, XPECT trial investigators, "Performance of a new leadless implantable cardiac monitor in detecting and quantifying atrial fibrillation: Results of the XPECT trial", Circ Arrhythm Electrophysiol 2010; 3: 141–147.

[30] T. Vezzosi, C. Buralli, F. Marchesotti, F. Porporato, R. Tognetti, E. Zini, O. Domenech, "Diagnostic accuracy of a smartphone electrocardiograph in dogs: Comparison with standard 6-lead electrocardiography", Vet J 2016; 216: 33–37.

[31] M. S. Kheertana, A. E. Manjunath, "A survey on wearable ECG monitoring using wireless transmission of data", International journal of advanced research in computer and communication engineering 2015; 4: 277–279.

[32] F. Miao, Y. Cheng, Y. He, H. Qingyung, Y. Li, "A wearable context-aware ECG monitoring system integrated with built-in kinematic sensors of the smartphone", Sensors 2015; 15: 11465–11484.

[33] R. H. Miller, L. B. Lehmkuh, J. D. Bonagura, M. J. Beall, "Retrospective analysis of the clinical utility of ambulatory electrocardiographic (Holter) recordings in syncopal dogs: 44 cases (1991–1995)", J Vet Intern Med 1999; 13: 111–122.

[34] J. M. Bright, J. V. Cali, "Clinical usefulness of cardiac event recording in dogs and cats examined because of syncope, episodic collapse, or intermittent weakness: 60 cases (1997–1999)", J Am Vet Med Assoc 1999; 216: 1111–1114.

[35] M. Schrivastav, S. Padte, V. Arora, M. Biffi, "Pilot evaluation of an integrated monitor-adhesive patch for long-term cardiac arrhythmia detection in India", Expert Rev Cardiovasc Ther 2014; 12: 25–35.

[36] A. Rashkovska, I. Tomašić, K. Bregar, R. Trobec, "Remote monitoring of vital functions - Proof-of-concept system", Proceedings of MIPRO 2012, Opatija, Croatia, 2012: 463–467.

[37] P. H. Chan, C. K. Wong, Y. C. Poh, L. Pun, W. W. Leung, Y. F. Wong, M. M. Wong, M. Z. Poh, D. W. Chu, C. W. Siu, " Diagnostic Performance of a Smartphone-Based Photoplethysmographic Application for Atrial Fibrillation Screening in a Primary Care Setting", J Am Heart Assoc 2016; 5: 1–7.

[38] J. M. Bright, J. M. Martin, K. Mama, "A retrospective evaluation of transthoracic biphasic electrical cardioversion for atrial fibrillation in dogs", J Vet Cardiol 2005; 7: 85–96.

[39] B. J. J. M. Brundel, P. Melnyk, L. Rivard, S. Nattel, " The pathology of atrial fibrillation in dogs", J Vet Cardiol 2005; 7: 121–129.

[40] A. Arenal, T. Datino, L. Atea, F. Atienza, E. Gonzáles-Torecilla, J. Almendral, L. Castilla, P. L. Sanchez, F. Fernández-Aviles, "Dominant frequency differences in atrial fibrillation patients with and without left ventricular systolic dysfunction", Europace 2009; 11: 450–457.

[41] J. Jalife, O. Berenfeld, "Atrial fibrillation: A question of dominance", Hellenic J Cardiol 2004; 45: 345–358.

[42] A. Elvan, A. C. Linnenbank, M. W. van Bemmel, A. R. R. Misier, P. P. H. M. Delnoy, W. P. Beukema, J. M. T. de Bakker, "Dominant Frequency of Atrial Fibrillation Correlates Poorly With Atrial Fibrillation Cycle Length", Circ Arrhythm Electrophysiol 2009; 2: 634–644.

[43] S. Gojraty, N. Lavi, E. Valles, S. J. Kim, J. Michele, E. P. Gerstenfeld, "Dominant frequency mapping of atrial fibrillation: comparison of contact and noncontact approaches", J Cardiovasc Electrophysiol 2009; 20: 997–1004.

[44] Savvy - Monitor the activity of your heart anytime and anywhere, http://www.savvy.si/en/ (accessed February 19, 2017).

SaaS Solution for ECG Monitoring Expert System

Aleksandar Ristovski
Innovation Dooel, Skopje, Macedonia
aleksandar.ristovski@innovation.com.mk

Marjan Gusev
Ss. Cyril and Methodius University, Skopje, Macedonia
marjan.gushev@finki.ukim.mk

Abstract—The advances in embedded systems and sensor technology have only recently unlocked the potential of wearable ECG monitoring expert systems. So far, the ECG monitoring software solutions that deal with the processing of harnessed ECG data essentially consist of straight-forward desktop applications. Although there are a number of proposed telecardiology concepts, very few of them are capable of replacing the extensive capabilities of Holter monitoring services in the form of a SaaS application. The SaaS functional infrastructure presented is a fully operational commercial service intended to replace the standard desktop setup in a scenario where the ECG sensor is capable of transmitting the data to the cloud.

Index Terms—*Telecardiology*; *Cloud Computing*; *Software as a Service*; *ECG Monitoring*; *Expert System*; *Medical Services*;

I. INTRODUCTION

There are very good reasons behind the tendency of substituting the standard desktop applications for the cloud hosted SaaS systems: paying exclusively for the services required, information and functionalities that are accessible from anywhere and anytime, as well as platform independent client side. This comes at the cost of the overhead for provisioning security, reliability, compliance and integration; requisitions which the industry has managed to tackle with success[1]. The challenge lies within the following: will the SaaS application surpass the consumer standards so that the consumer is swayed to leave the commodity of the desktop application?

In addition to the standard benefits the SaaS solution offers, there are unique assets that come from telemedicine, which is the the conjunction of cloud computing and bioinformatics[2]. By combining these two scientific fields, the standard of living can be significantly increased, since what telemedicine offers is a real-time tracking of the individual's health condition and immediate intervention. Notably, ischaemic heart disease and stroke are the world's biggest killers, accounting for a combined 15 million deaths in 2015, which makes up 26.6% of the world's mortality quota[3]. Technological progress in cloud computing capabilities and wearable sensors is the reason why telemedicine has gained in popularity in recent years, i.e. strives to serve the living standards' improvement by bringing a new lease of life where possible.

The paper is organized as follows. The background of the proposed SaaS solution for ECG monitoring is provided in Section II. Section III gives an overview of related work in the field of SaaS architectures that deal with telecardiology. Description of the SaaS infrastructure is given in Section IV. Section V discusses the challenges faced during the application development, the pitfalls and how they were overcome. The conclusion and future work is presented in Section VI.

II. BACKGROUND

ECGalert is an expert system being developed by the R&D team of Innovation DOOEL[4], in Microsoft ASP.NET Framework. The system aims to provide continuous computer assisted diagnosis for ECG recordings harnessed by wearable ECG sensors, connected to the cloud through an intermediary Bluetooth connection with a smartphone. The data is processed in the cloud, with the extracted findings available to suitable extent to both the patient and medical experts. Although the range of the analysis, i.e. the number of anomalies the expert system will be able to detect, is by a great deal still being explored, the goal is to provide pervasive monitoring service available in the form of a SaaS application. The scope of the service offered is intended to fully replace the standard desktop software used by the non-invasive cardiologists in the process of analysis of long-term ECG recordings.

The patient is also given access to the findings of the monitoring services, and is offered the capability to make comments on a particular and flexible time frame regarding his personal perceptions that concern the cardiac condition for the corresponding time frame, which has proven to be of great importance to the medical professionals in the process of interpreting the findings.

Two important advantages over the standard monitoring solutions, besides the already mentioned SaaS convenience in comparison to desktop applications are:

- not making a compromise between the extensiveness of the analysis and the real-time delivery of the findings
- capabilities of large-scale data mining

The proficiency in computer assisted ECG analysis is a challenge on it own. Several research papers have been published that deal with the subject regarding ECGalert[5][6][7]. However, the subject of interest of the presented research lies within the functional SaaS infrastructure that incorporates the computer assisted diagnosis and the monitoring interfaces.

III. RELATED WORK

The benefits of telecardiology have been present for some time in the medical community. However, few telecardiology concepts have been put into practice. Such is the example of the usability discussion of S. Spinsante et al.[8]. A general architecture of SaaS application is presented in the work of G. Fortino et al.[9]; this work deals with general concepts of

Body Sensor Networks (BSNs) whose data stream is processed in the cloud and does not cover in detail the specific scenario incorporated by the ECGalert monitoring service.

A 12-lead ECG telemedicine service has been devised by J. Hsieh et al.[10]. Although this service has been carried out into action, its nature is very different from the ECGalert monitoring service, since the services of ECGalert are intended to provide continuous monitoring in contrast to the emergency ad-hoc monitoring, as well as computer assisted diagnosis that is afterwards revised by a medical expert, in contrast to the alert monitoring present at the referent system.

A related SaaS software solution, regarding the nature of the problem, has been published by B. E. Reddy et al.[11], R. G. Lee et al.[12] and M.A. Al-Zoube et al.[13]. Although there is much similarity in the approach, especially in the workflow of computational steps, the nature of the problems and goals differs significantly. A similar SaaS products have been proposed for mobile monitoring [14][15], which are closely related to *ECG Cloud Digital Cardiology Service*[16]. The functionalities of the latter, along with SPYDER[17], are very close to ECGalert. However, very little information can be found, solely on the Internet, on what these products are offering, and even less on their SaaS functional design.

The diagnostic capabilities of the SaaS solution proposed offer more advanced capabilities than the related work, and offer state-of-the-art visual interfaces for data presentation, designed for professional use. The goal is to compete with affirmed expert software such as CardioScan[18] and ZYMED[19].

IV. SaaS Infrastructure

A. The ECG Analysis Routine

The ECGalert expert system is designed to offer interoperability regarding the type of sensor used for the ECG recording. Therefore, each sensor type has its own data stream format. The data stream is sent via Bluetooth to a smart phone, and from there it is rerouted towards a designated Streaming Unit (SU) via the Internet. In order to avoid continuous streaming between the smart phone and the SU, the ECG data is packed into a ECG record of fixed duration. The SU reroutes the ECG record to a Data Processing Unit (DPU), where the data is reconsolidated into a DB record.

Before writing the ECG record into the DPU DB, the signal is processed and is given a diagnosis, hence, the capability of real-time assessment of the existence of potential anomalies in the hearts physiology. In case such anomalies are detected, the suitable alert/notification mechanisms will be triggered. For the purpose of security and information preservation, the user personal data is sored in a different server, the Main Unit (MU). This concludes the general SaaS architecture scheme, as shown in Figure 1, and has been designed so that it meets with big data requirements.

However, this processing workflow provides no means of rendering the necessary standard output the professional medic would need in order to have a comprehensive understanding of the patient's overall condition, nor the analysis is detailed

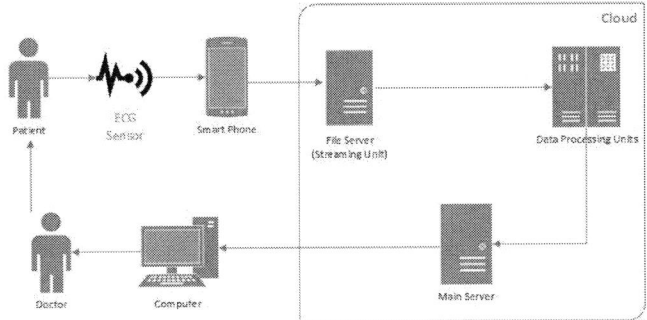

Fig. 1. General SaaS architecture scheme

enough for detection of anomalies that require thorough reasoning.

Non-invasive cardiologists are accustomed with a typical form of ECG data representation that comprises of ECG history of at least several days. That is so because to accurately set the diagnosis for a number of cardiac conditions it is necessary to look retrospectively into the patients' ECG history. Not only a parametrized set of findings are imperative for the cause, but it is also required to have the ECG history visualized in a pragmatic and efficient format, organized into segments, each with a duration of 24h. These interfaces are further referred to as monitoring interfaces. Therefore, the functional SaaS architecture has been designed to meet these requirements, thus, enabling the full capacity of a professional desktop ECG monitoring application. When visualizing the ECG signal on a medium sized display, it is somewhat optimal to display a signal with a duration of 30s.

However, it is practically impossible to expect that the sensor would transmit all the 2880 signal segments of 30s within a day. Some of these segments are missing due to lost Bluetooth connections between the ECG sensor and the smart phone, and lost Internet connection between the smart phone and the SU. In addition, the sensor user might willingly choose to terminate the recording for an indefinite amount of time, or the sensor might run out of battery. Whatever the reason, for the most of the time the SU ends up admitting less than 2880 ECG signal segments within 24h. Besides these random periods of missing ECG signal stream, it is rarely the case that the signal will start at 0s or 30s relative to a minute start.

Despite these circumstances, it is necessary to format the ECG records on the DPU in a structured manner. That is why the unstandardized collection of 30s ECG signal segments is re-factored into a continuous 24h ECG signal, by paying attention to the timestamps the 30s ECG signal segments have. To indicate which section is valid and which one is not, the signal values are accompanied by a validation mask.

Then, this 24h ECG signal is partitioned into segments that are conclusive, meaning that these segment have continuous valid ECG sensor readings. They consequently undergo analysis, i.e. the diagnostic routine is performed. The advantage of analysis of continuous signal segments over the analysis of 30s segments lies within the benefit of the dynamic algorithm

309

used for the feature extraction routine. Namely, the results of the analysis when run on a record of greater duration are more accurate because of the nature of the dynamic feature extraction algorithms, such as the Pan-Tompkins algorithm used for the detection of the R waves of the ECG signal[20].

Another advantage are the computational resources saved for metadata initialization and maintenance. Since the findings of the thorough analysis are not that time-sensitive, the diagnostic routine can go through the patient history and look for potential anomalies whose nature does not require immediate intervention to the patient, but if found might raise suspicion of a potential troubling deviations. To speed up the process of processing the patient history of detected anomalies, there is a metadata structure for the weekly, monthly and overall findings.

Although there are already records in the DPU database of the findings of the initial real-time analysis, which occurred when the sensor data was delivered from the SU to the DPU, the findings are now re-entered. In advance to data re-entry, the conclusive ECG signal fragments are again split into segments of 30s. Before the thorough analysis, there were only DB records exclusively for the periods of time when the sensor was streaming data. After the thorough analysis, the DPU DB holds record for each 30s segment starting 0s or 30s relative to the beginning of each minute of the day.

This kind of thorough analysis that re-structures the metadata into suitable format for further the specific visualization of the daily recorded data will trigger automatically at the end of the day. Before it is triggered, the information derived from the real-time analysis will not be by default available for the particular kind of visualization needed. However, the real-time findings are available to the doctor and the user, but in a different presentation format. Yet, this kind of thorough analysis can be invoked on user request before it is automatically triggered, and it takes several moments before the procedure is finished and the current day findings are available for visualization in the specific monitoring interfaces.

The monitoring interfaces have incorporated several other embedded visualization functionalities, besides drawing the ECG signal representation for a period of 24h. These embedded functionalities include the visualization of the Beats Per Minute - BPM graph, which itself comprises average, minimum and maximum graph series for the BPM findings. The metadata for the BPM graph is referred to as the Heart Rate Record. Also, the monitoring interfaces have distinctive color representation for some of the detected anomalies. The metadata that stores the information for the detected anomalies is referred to as the Afflicted Areas Record. The Heart Rate Record and the Afflicted Areas Record are written into the DPU DB and have expire policy that matches the expire policy of the corresponding ECG signal itself.

Besides the sensor data which consists of the ECG signal stream, the mobile application running on the smart phone allows the user to enter anamnesis information, i.e. the way the user felt at a particular time while wearing the sensor, which might be relevant to the medical expert when setting

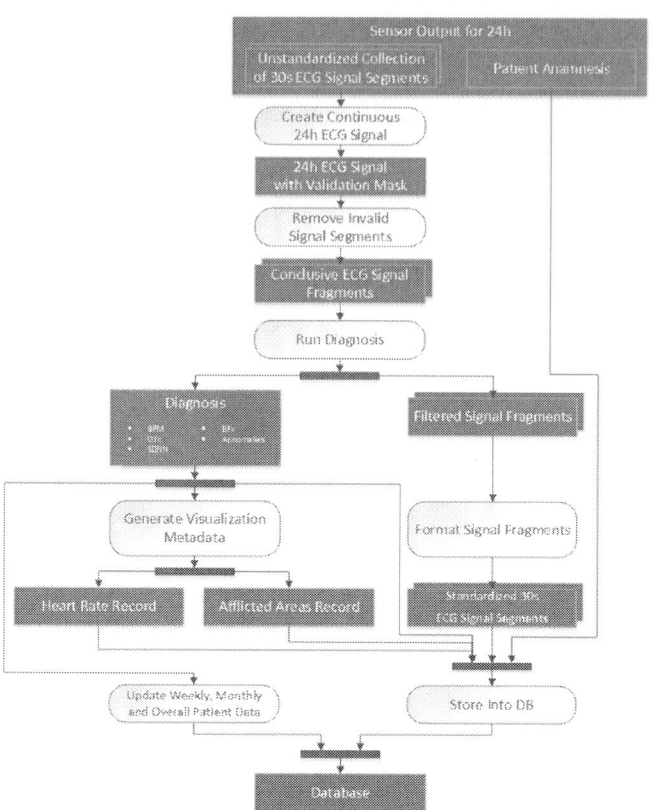

Fig. 2. Workflow diagram of the sensor output data until the point of DB store

a diagnosis in addition to the diagnosis findings from the ECGalert expert system. The anamnesis information, just as the Heart Rate Record and the Afflicted Areas Record, is written into the DPU DB with the same expire policy. This concludes the ECG analysis routine. Its steps are shown in Figure 2.

B. Rendering the Monitoring Interfaces

The monitoring interfaces of the ECGalert expert system are designed to intuitively respond to the inquiry of the medic professionals in the process of determining the final diagnosis. Although the results from the ECG analysis routine are very likely legitimate, it is still necessary for a cardiologist to confirm the findings for the more complex cardiovascular diseases and pathological conditions. Hence, the cardiologist, by going through the findings of the computer assisted diagnosis and by assessing their accuracy, is by a great deal assisted in the diagnosis assessment. In order to be as useful as possible, the monitoring interface design must not differ by a great deal from the existing standards in design, be fast and responsive, have clean and coherent layout and make great use of the interfaces' space.

In spite of the principles of the web application design, the ECG monitoring interfaces follow the design principles of desktop applications and make use of the whole browser

310

tab. This way, maximum workspace has been obtained. The look of the monitoring interfaces provided to the doctor role is shown in Figure 3.

The first section gives information on the name of the patient and the date of the 24h record that is currently visualized. A drop-down list offers fast switching between the daily records. In case the ECG sensor of the patient is currently active and is streaming data, on user request that data can be analyzed and processed so that can be displayed within the monitoring interfaces within moments, as explained in Section IV-A.

The next section of the ECG Monitoring interfaces is the hour selector section, which enables selection of the hour relevant to the current daily data. The hour selector in its default and initial form provides information on the BPM of the patient. The x axis represents the time with a tick on each 30s, while the y axis represents the BPM count. With two distinctive colors three series are drown: the average BPM, the maximum BPM and the minimum BPM values. This chart representation gives the user information of the heart pace, and can easily be determined in which time of the day the patient has had drastic changes in the pace or has been tachycardic or bradycardic.

Alternative visual representations of the hour selector section include histogram of one of three of the most prominent physiological anomalies: Premature Atrial Contractions (PACs), Premature Ventricular Contractions (PVC) and Sinus Arrests (SAs).

The next section is the half minute selector section, which enables selection of the half minute relevant to the begging of the currently selected hour. In case particular half minute has anomalies present in it, it is colored in accordance with the anomaly present.

The next section is the list of half minute charts, each with duration of 30s. The visualized charts belong to the currently selected hour and have timestamps for better temporal orientation. The currently selected half minute from the list of half minute charts is marked with red indicator rectangle on both the half minute selector and the list of half minute charts.

The focus chart section displays the currently selected half minute in greater detail, as it offers 3 levels of magnification. In case of maximum magnification behind the chart, the standard ECG chart grid is drawn; in case the level of magnification is other than X1, the focus chart is draggable and its position is indicated by a semi-transparent rectangle on the corresponding chart from the list of half minute charts. Because the focus chart is most likely to be used for examination, its beginning is 2s prior to the half minute segment start, and its end is 2s after the half minute end, so that it provides maximum transparency of the signal morphology.

The final section is the diagnosis section and it contains information on the diagnosis and the patient anamnesis. In this section the doctor can alter the diagnosis, enter notes, write down the treatment and read the anamnesis.

The design of the monitoring interfaces is fully responsive in therms of web responsiveness[21]. All the design components will scale up with the size of the browser window. The design components themselves contain very little standard html elements. Most of the design elements are in fact Scalable Vector Graphics (SVG)[22]. The SVG graphic has been chosen for the purpose since it is highly reusable, it is highly cross-browser compatible and its user interaction can be easily obtained by using JavaScript, since the SVG is XML structured. The responsive design is backed up with collapsible 'div' content for the focus chart section and the diagnosis section, to maximize the transparency in the interface design.

What makes this functional SaaS solution unique is the way the monitoring interfaces are generated. The SaaS application is Model-View-Controller (MVC) based and every time the Controller for the monitoring interfaces is accessed, not only it generates the Model, but it also somewhat generates the View as well. The SVG contents are generated within the Controller and are passed to the View though the ViewBag entity. The View itself consists of a structural html frame of 'div' elements whose content is then filled with the suitable SVG content collected from the ViewBag.

In order to generate the monitoring interfaces, several sets of data need to be read from the DPU DB. These include the standardized 30s ECG signal fragments, the Afflicted Areas Record, the Heart Rate Record, the diagnosis findings, the patient's anamnesis and the doctor's diagnosis and treatment. Since the focus chart section displays ECG signals whose duration is 34s instead of the standard 30s, the set of 2880 30s records needs to be reformatted into set of 2880 34s fragments.

Then, the Controller can begin generating the SVG graphics for the monitoring interfaces. The half minute selector only needs the Afflicted Areas Record for its content, while the hour selector only needs the Heart Rate Record for its content. The list of half minute charts needs both the list of 34s ECG signal fragments and the Afflicted Areas Record and the focus chart needs the list of 34s ECG signal fragments, the Afflicted Areas Record and the diagnosis findings for its content.

The content of the hour selector, the half minute selector, the list of half minute charts and the focus chart is then added to the ViewBag entity. One other thing that is also inserted into the ViewBag entity is the Afflicted Areas Record, because its content will need to be accessed in case the user requests change of hour or changes the magnifying level of the hour selector. The patient's anamnesis and the doctor's diagnosis and treatment need be passed through the Model, since their content might need be altered and sent back to the Controller.

A technique that has been used to improve the Controller performance is controller caching. The content of everything that was inserted into the ViewBag can be cached on the side of the controller, so that the SVG content rendering in near future access can be avoided. Surprisingly, the time needed to generate the SVG content is almost insignificant and does not affect the overall SaaS response time. However, what matters to the performance factor is the formating of the 34s ECG signal fragments. Hence, caching the 34s ECG signal fragments is the main reason the controller caching technique has been used, while all the other cached content is almost

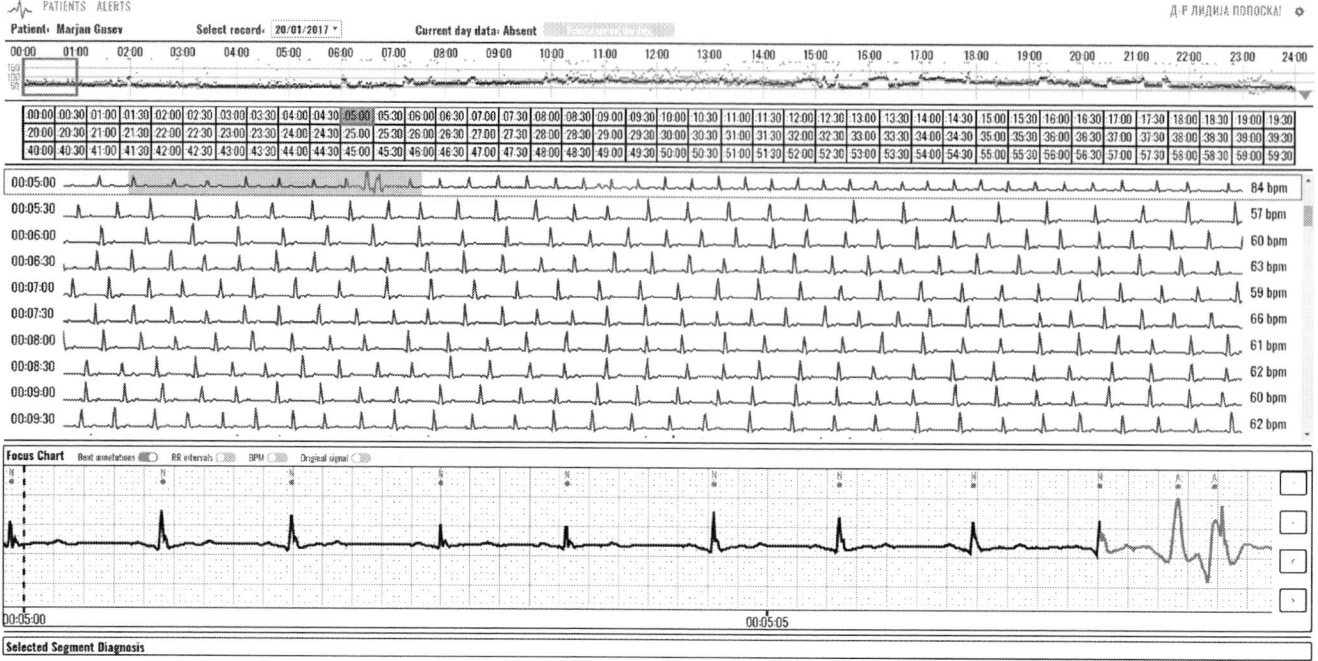

Fig. 3. Snapshot of the monitoring interface

unnecessary considering the small computing resources and cache storage space it requires. The user cache data is managed according to user cache plan policies. The workflow diagram of the generation of the monitoring interfaces content is shown in Figure 4.

Note that for the rendering of the contents of the focus chart only one 34s ECG signal segment is needed and for the rendering of the contents of the list of half minute charts only the first 120 34s ECG signal segments are needed. That is why the contents of the monitoring interface weight approximately 700KB. If all the necessary content for the interfaces was generated in advance, the monitoring interfaces would have weighted approximately 120MB, which is unacceptable network traffic load.

Instead of overloading the network, the monitoring interfaces are using two services for obtaining the contents of the list of half minute charts on change of the hour, and for obtaining the content of the focus chart on change of magnifying level. These services access the cache of 34s ECG signal fragments and generate the graphics on the server side before sending it to the client. In addition, the client side has incorporated caching system on its own so no duplicate service requests are ever made, hence minimizing the network traffic.

V. Discussion and Lessons learned

A significant challenge when designing such a SaaS solution for ECG monitoring expert system is finding the approach that will keep the network traffic at minimum. Several other approaches have been implemented before the current solution, one of which was transferring all the ECG data on the client side and rendering the needed SVG graphics via JavaScript.

While the SVG rendering did not affect the performance on the client side, the network traffic of 70MB for the 24h recording was unacceptable.

Another option that was also considered was to render all possible SVG graphic on the server side and then send it to the client. While the rendering of the SVG graphics was not a computational overhead, the network traffic would almost double: 120MB. Thus, the current solution with implemented on user request service calls to the server gives an optimal performance regarding the network traffic and the server cache size. The current functional SaaS approach is a product of the optimal computational strategy.

However, there are a number of other SaaS design challenges, but principles as well, that need be taken into consideration. A number of them have been discussed by J. Hsieh et al.[23], in a more general sense. Also, it is inevitable to neglect the fundamental principles of quality of service a telemedicine application should incorporate[24].

VI. Conclusion and Future Work

The bottlenecks of the current solution are made up from delays in the current DB and File System (FS). Some possible issues regarding this bottleneck have been localized and optimization is planned for near feature. A flowable algorithm design for the initial ECG processing would greatly reduce the workload by excluding a thorough analysis afterwards: a change that notably wold affect the functional infrastructure. Another approach which would benefit the performance factor is to decrease the file size by resampling the ECG signal to a lower sampling frequency by factor that does not impact the diagnosis.

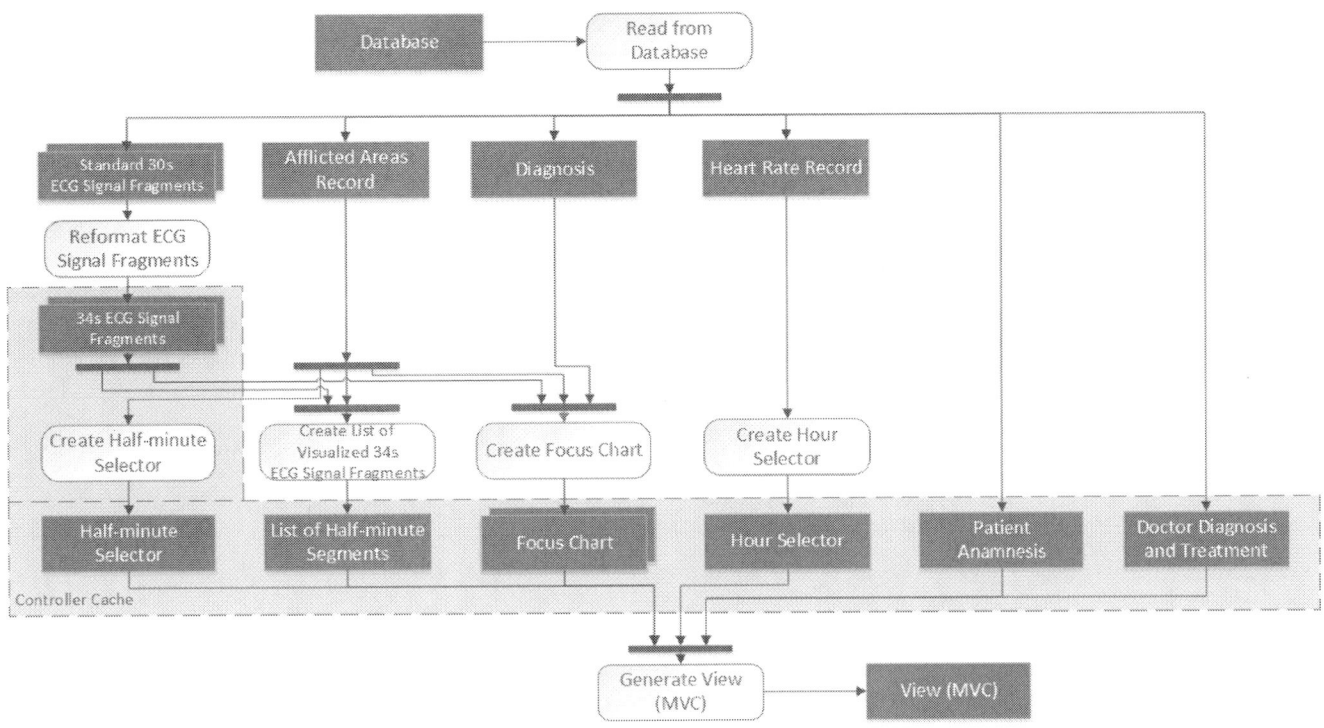

Fig. 4. Workflow diagram of the data retrieved from the DB towards creating the ECG monitoring interface

An important thing to have in mind is that although the diagnostic capabilities of the expert system itself have advanced diagnostic features, they should not be the absolute authority. The computer assisted diagnostic is indeed state-of-the-art technology, but when it comes to serious matters such as medicine, the artificial intelligence is still not ready to make the final verdict.

REFERENCES

[1] M. Cusumano, "Cloud computing and saas as new computing platforms," *Communications of the ACM*, vol. 53, no. 4, pp. 27–29, 2010.

[2] D. A. Perednia and A. Allen, "Telemedicine technology and clinical applications," *Jama*, vol. 273, no. 6, pp. 483–488, 1995.

[3] W. H. Organization *et al.*, "The top 10 causes of death. 2015," *Reference Source*, 2015.

[4] I. DOOEL. (2016) ECGalert project description. [Online]. Available: http://ecgalert.com/

[5] M. Gusev, A. Stojmenski, and I. Chorbev, "Challenges for development of an ecg m-health solution," *Journal of Emerging Research and Solutions in ICT*, vol. 1, no. 2, pp. 25–38, 2016.

[6] J. Tasic, M. Gusev, and S. Ristov, "A medical cloud."

[7] M. Gusev, A. Ristovski, and A. Guseva, "Distribution of premature heartbeats," in *Telecommunications Forum (TELFOR), 2016 24th*. IEEE, 2016, pp. 1–4.

[8] S. Spinsante, R. Antonicelli, I. Mazzanti, and E. Gambi, "Technological approaches to remote monitoring of elderly people in cardiology: a usability perspective," *International journal of telemedicine and applications*, vol. 2012, p. 3, 2012.

[9] G. Fortino, D. Parisi, V. Pirrone, and G. Di Fatta, "Bodycloud: A saas approach for community body sensor networks," *Future Generation Computer Systems*, vol. 35, pp. 62–79, 2014.

[10] J.-c. Hsieh and M.-W. Hsu, "A cloud computing based 12-lead ecg telemedicine service," *BMC medical informatics and decision making*, vol. 12, no. 1, p. 1, 2012.

[11] B. E. Reddy, T. S. Kumar, and G. Ramu, "An efficient cloud framework for health care monitoring system," in *Cloud and Services Computing (ISCOS), 2012 International Symposium on*. IEEE, 2012, pp. 113–117.

[12] R.-G. Lee, C.-C. Lai, S.-S. Chiang, H.-S. Liu, C.-C. Chen, and G.-Y. Hsieh, "Design and implementation of a mobile-care system over wireless sensor network for home healthcare applications," in *Engineering in Medicine and Biology Society, 2006. EMBS'06. 28th Annual International Conference of the IEEE*. IEEE, 2006, pp. 6004–6007.

[13] M. A. Al-Zoube and Y. A. Alqudah, "Mobile cloud computing framework for patients'health data analysis," *Biomedical Engineering: Applications, Basis and Communications*, vol. 26, no. 02, p. 1450020, 2014.

[14] P. Guzik and M. Malik, "Ecg by mobile technologies," *Journal of Electrocardiology*, vol. 49, no. 6, pp. 894–901, 2016.

[15] P.-C. Hii and W.-Y. Chung, "A comprehensive ubiquitous healthcare solution on an android mobile device," *Sensors*, vol. 11, no. 7, pp. 6799–6815, 2011.

[16] Technomed. (2017) ECG Cloud Digital Cardiology Service description. [Online]. Available: https://technomed.co.uk/products/holter-analysis

[17] W. B. P. Ltd. (2017) Spyder product. [Online]. Available: http://www.web-biotech.com/spyder-bt

[18] D. Software. (2017) CardioScan Holter ECG Systems software solution. [Online]. Available: http://www.holterdms.com/CardioScan_HolterECG_Systems.php

[19] Philips. (2017) PHILIPS ZYMED holter software. [Online]. Available: http://www.usa.philips.com/healthcare/product/HC860292/holter-monitoring-software-holter-analysis-software

[20] J. Pan and W. J. Tompkins, "A real-time qrs detection algorithm," *IEEE transactions on biomedical engineering*, no. 3, pp. 230–236, 1985.

[21] J. W. Palmer, "Web site usability, design, and performance metrics," *Information systems research*, vol. 13, no. 2, pp. 151–167, 2002.

[22] J. Ferraiolo, F. Jun, and D. Jackson, *Scalable vector graphics (SVG) 1.0 specification*. iuniverse, 2000.

[23] J.-C. Hsieh, A.-H. Li, and C.-C. Yang, "Mobile, cloud, and big data computing: contributions, challenges, and new directions in telecardiology," *International journal of environmental research and public health*, vol. 10, no. 11, pp. 6131–6153, 2013.

[24] A. Zvikhachevskaya, G. Markarian, and L. Mihaylova, "Quality of service consideration for the wireless telemedicine and e-health services." in *WCNC*, 2009, pp. 3064–3069.

Wavelet-Based Analysis Method for Heart Rate Detection of ECG Signal Using LabVIEW

Duygu Kaya, Mustafa Türk, Turgay Kaya

Firat University /Department of Electrical and Electronics Engineering, Elazig, Turkey

{dgur, mturk, tkaya}@firat.edu.tr

Abstract - LabVIEW is a graphical programming language that uses a dataflow model instead of sequential lines of text code. LabVIEW allows multiple operations to work in parallel. So, designers spend less time than a text based programming language. Application areas such as signal processing, image processing and data analysis are available. In this paper, wavelet analysis is used for the elimination of undesired frequency noise. In the obtained noise-free signal, the accurate heart rate was determined with helping of the program developed in the LabVIEW environment. The performance of the system was tested with different wavelet types and satisfactory results were obtained.

Keywords – LabVIEW, wavelet analysis, bioelectrical signal, ECG, heart rate detection.

I. INTRODUCTION

Electrocardiography (ECG) is the process of recording the electrical activity of the heart using electrodes placed on the skin surface [1]. The ECG provides a significant opportunity for the diagnosis of some diseases such as cardiac arrhythmias, myocardial disease, myocarditis and heart attack. Early diagnosis is important in the diagnosis of heart diseases as it is in many diseases. ECG is noninvasive, also an effective method that keeps vital signs.

A normal ECG has electrical activity, rhythmic regulation, and a ventricular rate of 60-100 bpm. Each cycle in the ECG consists of a P wave, a regular narrow QRS wave and followed by T waves. In addition to these waves, PR, ST, QT and RR segments also provide information about diseases [2].

As with other bioelectrical signals, various signal processing methods can be used to analyze and interpret the ECG signal [3]. As a matter of fact, studies on this subject are available in the literature. In [4], for different arrhythmia discrete wavelet transform was used to extract feature. In [5], a robust single-lead electrocardiogram (ECG) delineation system based on the wavelet transform was developed. In [6], QRS detection and P, T using a complex wavelet function was done. In [7], two wavelet was used for both QRS detection and P, T detection. In [8], wavelet based new system developed for online processing ECG. In [9] with multiresolution, peak detection was done. In [10] ECG signal analyze was done both MATLAB and LabVIEW. In present work, wavelet-based ECG signal analysis was performed with

LabVIEW, which is the graphical programming language. The heart rate of the signals is obtained by the signal analysis. By looking at the numerical values obtained, it is possible to distinguish between normal and tachycardia signals.

II. MATERIAL AND METHODS

A. Wavelet Transform

Wavelet is a mathematical function that divides data into different frequency components. So, each component is evaluated a frequency range.

The wavelet analysis is a time-frequency analysis method of signal, with the characteristics of multi-resolution analysis. Its principles are similar to Fourier analysis, but have advantages in case of sharp rise and discontinuity [11,12]. This method is used to extract ECG signal features and to detect heart rate in this work. Heart rate gives whether a signal is normal or not.

It is extremely important to choose the wave that will represent the signal appropriately. For this, choosing the most similar wave to the original signal is effective for analysis. In this paper, Daubechies (db06) was chosen because of its similarity to the ECG signal. Also, different wavelet types, such as biorthogonal, morlet, coiflets, were tried, but satisfactory results was obtained by Daubechies (db06). Another important issue in wavelet transform is to determine the level of decomposition. The sampling rate of the data we use is 512 Hz, the decomposition level is 9 selected.

$$\text{Decomposition Level} = \log_2 N = \log_2 512 \qquad (1)$$

So, decomposition level was chosen 9.

B. LabVIEW

LabVIEW is a graphical programming language that uses a dataflow model instead of sequential lines of text code. LabVIEW allows multiple operations to work in parallel [13,14]. It provides that acquires of bioelectrical signal, preprocesses and analyzes them.

LabVIEW is a software based graphical programming language that consists of front panel and block diagram. The place where the user interface is prepared and the code is written correspond to front panel and block diagram, respectively. LabVIEW is also referred to as Virtual Instrument (VI) because of its similarity to the physical instruments used in laboratories. Whereas traditional instruments such as oscilloscopes and waveform generators are expensive and performs some specific tasks and cannot be customized, virtual instruments is PC-based, portability and can access to the internet. Also the greatest advantage of VI is that it can be made to the desired design [15].

The biomedical workbench toolkit in LabVIEW allows applications such as recording, by sensor with DAQ hardware, and viewing bioelectrical signal, heart rate variability (HRV) analysis, image processing. By file format converter, the files are converted into different types of file formats for the various applications. For instance, it supports .hea, .tdms, .mat, .rec extensions file [16].

III. ECG SIGNAL ANALYZE

For ECG signal analysis, LabVIEW's own database is used. One of the signals used is a signal of a normal ECG signal, and the other is a patient of a tachycardia, which means an increase in heart rate. The data used is sampled at 512 Hz. The heart rate of a person with tachycardia is above 100 bpm while the heart rate is normally considered to be between 60-100 bpm. The fact that heart rate is normally lower or higher is important as many other heart conditions. Much of it can lead to serious problems like heart attacks in the future

At a normal ECG, each cycle consists of a P wave, a regular narrow QRS wave followed by a T wave (Fig. 1). For this purpose, we first use wavelet transform to remove undesired signals (such as respiration, some artifacts) from the ECG signal in order to detect heart rate more accurately in this study.

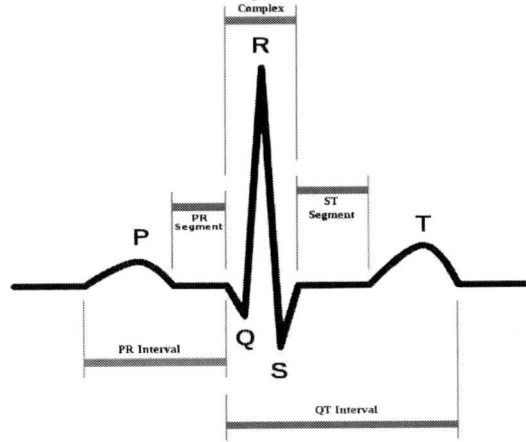

Figure 1. ECG Wave [17]

Since the similarity of signal shape of the selected filter is important, the db06 wave from the Daubechies family was selected for wavelet transform in this study. The next step after the coefficient decomposition is to apply the appropriate threshold values. Energy is spread over a limited number of coefficients in the wavelet domain,

while noise energy is distributed over all coefficients [18]. Therefore, after wavelet decomposition, the wavelet transform coefficients become larger than the noise wavelet coefficients and the signal and noise are successfully separated from each other [19]. In the study, soft thresholding was used that is one of the most popular denoising method [20]. Also, there are lots of methods to determine the threshold. In this paper, SUER was used as a threshold selection rule.

These methods are performed with LabVIEW. The block diagram shown in Figure 2. Firstly, the .tdms extended ECG signal is read, the wavelet transform is applied to the read signal, and finally the heart rate is obtained from the extracted signal. Table 1, which is first column defines general situations, second column defines this study results, shows the normal, tachycardia and bradycardia heart rate values. Figure 3 shows normal raw ECG, figure 4 shows extracted feature raw ECG, figure 5 shows tachycardia raw ECG, figure 6 shows extracted feature tachycardia ECG.

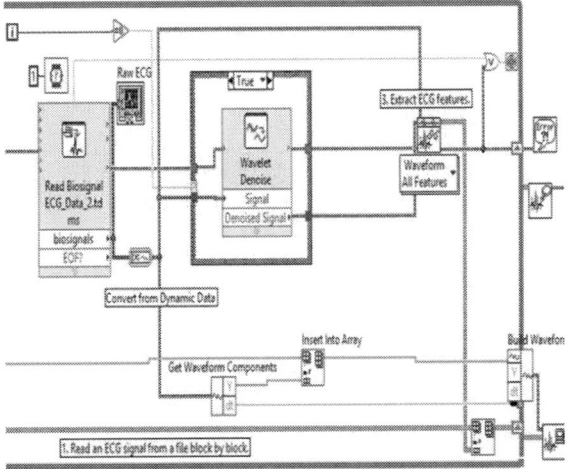

Figure 2. LabVIEW Block

TABLE I.   ECG HEART RATE

State	ECG Heart Rate (bpm)	
	General	*In this work*
Normal	60-100	77,8
Tachycardia	> 100	114,2
Bradygardia	< 60	-

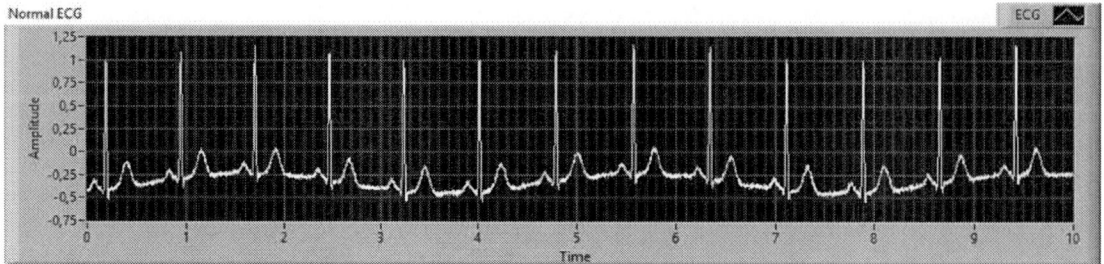

Figure 3. Normal ECG

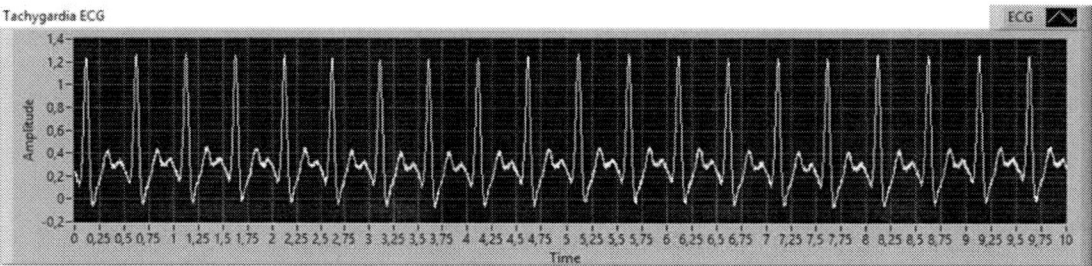

Figure 4. Extracted Feature Normal ECG

Figure 5. Tachycardia ECG

Figure 6. Extracted Feature Tachycardia ECG

IV. CONLUSION

In the current study, heart rate rates of different ECG data were determined using a LabVIEW based program without using text-based programs. The raw signal used in this study was first filtered out of undesired components by wavelet transform, then the heart rate was investigated to determine the impairment of rhythm. These obtained numerical values are compared with general values and it is determined which data is included in which class. Thus, it is considered that when unknown data is entered, it helps to determine whether there is a negative condition for health or not. For a complete diagnosis, it may be useful to examine PR, QT and ST segments along with heart rate together with QRS time and heart rate.

REFERENCES

[1] M. H. Sazlı, "EKG Sinyallerinin Korelasyon Analizi ile Bazı Kalp Aritmilerinin Belirlenmesi," KSU Journal of Science and Engineering 10(1), pp. 69-76, 2007.

[2] R. Çekik and S. Telçeken, "Classification of ECG Signals Using Rough Sets Theory" Applied Sciences and Engineering 15(2), pp: 125 – 135, 2014.

[3] X.,Jiang, L. Zhan., ECG Arrhythmias Recognition System Based on Independent Component Analysis Feature Extraction, Department of Computer Science and Engineering, Shanghai Jiaotong University,China, 2002.

[4] Z. Yılmaz, M.R. Bozkurt, "Ayrık Dalgacık Dönüşümü Kullanarak Aritmilere Ait Özniteliklerin Çıkarılması ", Akademik Bilişim 2013, 23-25 January 2013 – Akdeniz University, Antalya

[5] Juan Pablo Martinez, Rute Almeida, Salvador Olmos,,"A Wavelet Based ECG Delinator: Evaluation on standard data bases", IEEE Transactions on Biomedical Engineering. 2004, Vol 51, No (4),570-581.

[6] A.Schuck Jr, J.O.Wisback, "QRS Detector pre-processing using the complex wavelet transform", Proceedings of 25th Annual International Conference of IEEE, 2003, 2590-2593.

[7] S.C.Saxena, V.kumar and S.T.Hamde , "Feature Extraction From ECG signals Using Wavelet Transform for disease diagnosis", International journal of system science, 2002 , volume 33, number13, 1073 -1085.

[8] J. S. Sahambi, S. N. Tandon and R. K. P. Bhatt, A new approach for on-line ECG characteristics, pp. 409-411.

[9] C.Bhyri, V. Kalpana., S.T.Hamde, and L.M.Waghmare, "Estimation of ECG features using LabVIEW", TECHNIA – International Journal of Computing Science and Communication Technologies, VOL. 2, NO. 1, July 2009. (ISSN 0974-3375)

[10] M. K. Islam, A. N. M. M. Haque, G. Tangim, T. Ahammad, and M. R. H. Khondokar, "Study and Analysis of ECG Signal Using MATLAB & LABVIEW as Effective Tools", International Journal of Computer and Electrical Engineering, Vol. 4, No. 3, June 2012.

[11] S.K., Mitra, 2004. Digital Signal Processing A Computer - Based Approach, 2nd Ed.pp.1, 50-52.

[12] S.G. Aydın,T. Kaya, H. Güler, "Heart Rate Variability (HRV) Based Feature Extraction for Congestive Heart Failure", International Journal of Computer and Electrical Engineering (IJCEE). Vol.8(4): 275-285 ISSN: 1793-8163, DOI: 10.17706/IJCEE.2016.8.4.275-285.

[13] N. Kehtarnavaz and S. Mahotra, "Digital Signal Processing Laboratory: LabVIEW-Based FPGA Implementation", pp.7

[14] S.G. Aydın,T. Kaya, H. Güler, "Wavelet-based study of valence–arousal model of emotions on EEG signals with LabVIEW", Brain Informatics, vol.3, pp.1-9.

[15] http://www.ni.com/white-paper/4752/en/

[16] A. Deshmukh and Y.Gandole "ECG Feature Extraction Using NI Lab-View Biomedical Workbench ", International Journal of Recent Scientific Research 6(8), pp.5603-5607, August, 2015.

[17] http://www.todayifoundout.com/index.php/2011/10/how-to-read-an-ekg-electrocardiograph/

[18] M.,Y Han,. J. Xi Liu, , and W.Guo, "Noise smoothing for nonlinear time series using wavelet soft threshold", IEEE Signal Process. Lett., 14(1),pp. 62–65, 2007.

[19] W.Bingsheng, ChaozhiC.,"Wavelet Denoising And Its Implementation In LabVIEW", Image and Signal Processing, 2009. CISP '09. 2nd International Congress October 2009.

[20] D.L. Donoho and J.M. Johnstone, "Ideal spatial adaptation via wavelet shrinkage", Biometrika, 81(3), 425-455.

Parallelization of Digital Wavelet Transformation of ECG Signals

Ervin Domazet and Marjan Gusev

Ss. Cyril and Methodius University, Faculty of Computer Science and Engineering,

1000 Skopje, Macedonia

e-mail: ervin_domazet@hotmail.com, marjan.gushev@finki.ukim.mk

Abstract—The advances in electronics and ICT industry for biomedical use have initiated a lot of new possibilities. However, these IoT solutions face the big data challenge where data comes with a certain velocity and huge quantities. In this paper, we analyze a situation where wearable ECG sensors stream continuous data to the servers. A server needs to receive these streams from a lot of sensors and needs to star various digital signal processing techniques initiating huge processing demands. Our focus in this paper is on optimizing the sequential Wavelet Transform filter. Due to the highly dependent structure of the transformation procedure we propose several optimization techniques for efficient parallelization. We set a hypothesis that optimizing the DWT initialization and processing part can yield a faster code. In this paper, we have provided several experiments to test the validity of this hypothesis by using OpenMP for parallelization. Our analysis shows that proposed techniques can optimize the sequential version of the code.

Index Terms— *Wavelet Transform; ECG; Heart Signal; Parallelization; OpenMP.*

I. INTRODUCTION

Advances in the Electronics and ICT industry have initiated lots of possibilities for IoT industry. One such innovation is the real time processing of the Electric Health Records (EHRs). Nowadays, it is scientifically proven that several arrhythmia can be efficiently detected [1], [2]. Any ECG processing algorithm requires three phases, i.e data preprocessing, feature space reduction and feature extraction. Optimization is inevitable due to the big data challenge.

In our previous research [3], we focussed on parallelizing Digital Signal Processing (DSP) filters by considering the tremendous power of dataflow cores. Our analysis showed speedup values linearly proportional to the kernel size of the filter. We have also considered using GPU cores for optimizing the DSP filters [4], where the GPU code achieved linear speedup values, much more efficient than the classical single processor sequential processing. Our next research [5], has contributed to the CUDA GPU optimization strategies for the noise elimination on ECG heart signals.

Wavelet Transformation is being used in many signal processing applications. It has been successful in the area of signal compression, data compression and detection of ECG characteristics. It mainly generates time-scale representation of an ECG signal, thus making it possible to accurately extract features from a non-stationary ECG signal [6], [7].

Wavelet Transformation is basically a linear operation, which decomposes a signal into various scales according to their frequency components. Each of these scales is further analyzed with a predefined resolution [8].

Wavelet Transform of a continuous signal is by definition a sum of the signal multiplied by scaled and shifted versions of the wavelet function, used to divide a continuous-time function into wavelets with the ability to construct a time-frequency representation of the signal. In many practical applications, though, Discrete Wavelet Transformation (DWT) is sufficient, as it provides only the vital information of the signal in a significantly faster manner.

This paper aims at optimizing the sequential Discrete Wavelet Transform (DWT) used for DSP filtering and feature extraction. DWT is a highly dependent structure with numerous dependencies between data. We set a hypothesis that optimizing the DWT initialization and processing parts can yield a faster code. Our analysis shows that proposed optimization techniques provide faster code.

In this paper, we use OpenMP as a parallel computing model for shared memory multiprocessors, which uses a set of compiler directives to support shared memory parallelism. The main aim behind this choice is primarily to minimize the complexity by adding parallel structures. Additionally, it supports incremental parallelism with the ability to parallelize bottlenecks of the application part by part [9]. Moreover, since all threads share a common address space, we expect a decrease in the overhead required by the introduced parallelism.

The paper is organized as follows: DWT Algorithm is analyzed in Section II. Section III describes the dependency analysis of the DWT and the parallel algorithm implementation. The experimental methodology is described in Section IV. The evaluation of results and discussion are presented in Section V. Related work and comparison is given in Section VI. Finally, the paper is concluded in Section VII with directions for future work.

II. DWT ALGORITHM ANALYSIS

The DWT algorithm contains two phases determined as initialization and processing phases, as presented in Fig. 1. Table I presents the variables used in the algorithm.

The *initialization* phase creates the base context for the processing part of the DWT. It is formed of 4 stages, which can be listed as Variable Initializations, Reading Wavelet Coefficients, Output Delay Calculation and Step Array Calculation.

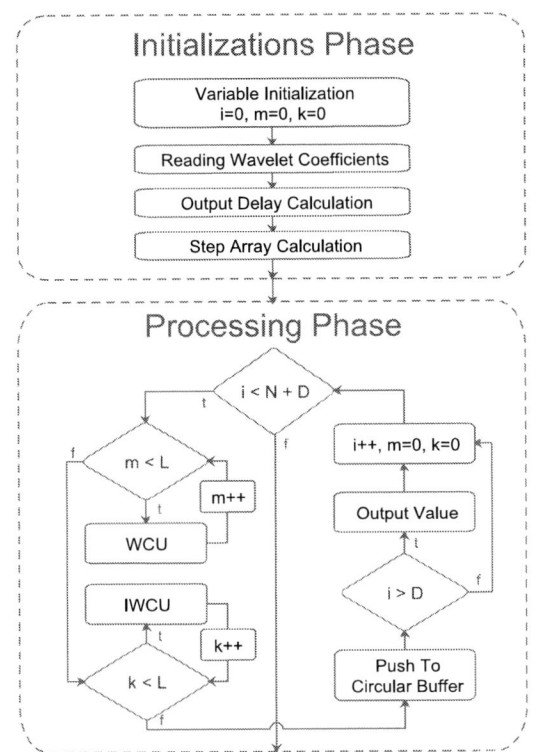

Fig. 1. A high-level abstraction of the DWT algorithm.

TABLE I. Variables used in the DWT algorithm

variables	meaning
i, m, k, K, and P	loop index variables
t and f	boolean evaluation , i.e true and false
D	delay for the first output
L	number of Wavelet levels
N	data array input size (ECG data samples)
F	filter length
WCU	wavelet compute and update
$IWCU$	inverse wavelet compute and update

Variables are declared and initialized in the first stage, whereas wavelet coefficients are read on the second. In wavelet transformation algorithms, there is a delay for the first output. This is calculated in stage 3. Finally, a vital *step* matrix for the wavelet computation is calculated.

On the other hand, the *processing* phase is responsible for the transformation itself. For each input element, the Wavelet Compute and Update (WCU) module realizes the decomposition of the ECG signal into signal approximation and detailed information.

The Inverse Wavelet Compute and Update (IWCU) removes the low-frequency components and regenerates the signal. The output of IWCU is pushed to a circular buffer of length D. The first baseline drift eliminated signal is actually produced after D steps [10].

Let L be the number of wavelet levels to compute. Assuming an input ECG signal using a 500 Hz sampling frequency,

the highest frequency component that exists in the signal is 250 Hz (each DWT level processing divides the band in two parts). Since the ECG baseline drift removal needs a high pass filter of 0.5 Hz then the DWT algorithm requires $L = 9$ wavelet levels.

Our research is concentrated on the parallelization of DWT algorithm. Analysis shows that DWT has highly dependent structure. This in turn makes direct parallelization inconvenient. In this section, we provide the dependency analysis and propose an efficient solution for parallelizing DWT algorithm.

Profiling the code showed that execution time is mostly spent in two segments of the code. These are the *Step Array Calculation* stage and the *Processing* phase.

Algorithm 1 Step Initialization and Calculation Algoritm

1: $P, K \leftarrow 0$
2: **while** $P < L$ **do**   ▷ Initialization
3:      **while** $K < 2^L$ **do**
4:          $Step[P][K] \leftarrow 0$
5: $P \leftarrow 0$
6: **while** $P < L$ **do**   ▷ Calculation
7:      $M, K \leftarrow 0$
8:      **while** $K < P$ **do**   ▷ Calculate M
9:          $M \leftarrow M * 2$
10:      **while** $K < 2^L$ **do**
11:          $Step[P][K] \leftarrow 0$
12:          **if** $K\%M == 0$ **then**
13:             $Step[P][K] \leftarrow Step[P][K] + 1$
14:          **if** $K\%(M * 2) == 0$ **then**
15:             $Step[P][K] \leftarrow Step[P][K] + 1$

Algorithm 1 shows the operations executed for the *Step Array Calculation* stage. At a glance, it is seen that the complexity of this stage is $O(L * 2^L)$. Especially on high Wavelet levels, this part becomes a serious bottleneck.

The operations start with initializing the two-dimensional *step* array, of length L and 2^L. A 2-level loop is used, though the second level executes 2^L iterations. Once initialized, the calculations are performed on the second loop, requiring 2^L iterations for each input.

In the *Processing* phase, a loop iterates $N + D$ times. Calculation of the delay D is presented in Algorithm 2. We can conclude that D is proportional to $C_L * 2^L$, where C_L is a constant number depending on level L. As the number of levels increases significantly, the constant C_L and N can be neglected resulting in an algorithmic complexity of $O(L*2^L)$.

Algorithm 2 Delay Calculation

1: $D, P \leftarrow 0$
2: **while** $P < L$ **do**
3:      $D \leftarrow 2 * D + (F - 1)$

From the high-level algorithm presented in Fig. 1, we observe that in each iteration WCU and IWCU iterate L

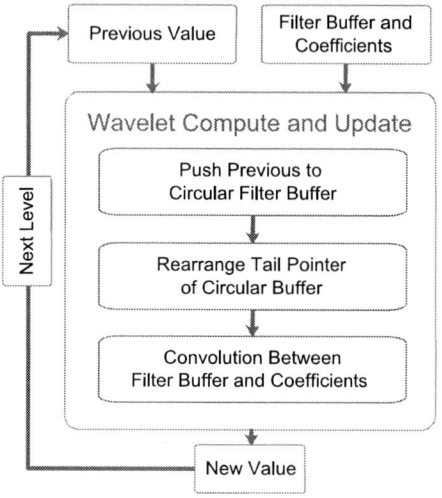

Fig. 2. A high-level view of Wavelet Compute and Update.

times. Algorithm 3 presents the inner structure of the WCU. Operations start from the first level and repeatedly execute until the last level. The input to this module is the data value computed as result of the previous WCU. WCU actively performs operations with the values of the dynamic wavelet filter stored in the filter buffer.

Algorithm 3 Wavelet Compute and Update operations

1: $Circular[Tail] \leftarrow Previous$
2: $Tail \leftarrow Tail - 1$
3: $Next, P \leftarrow 0$
4: **while** $P < L$ **do**
5:      $Next \leftarrow Circular[P] * Coefficients[P]$

Each WCU operation starts with pushing the previously calculated value by a preceding WCU to the circular buffer. The next step is to rearrange the tail pointer of the circular buffer. Finally, the coefficients are convolved with the buffer. The output of this operation is used as input to the next WCU. WCU is performed in a sequence for all wavelet levels L.

IWCU practically contains same operations except that it uses $Step$ array as an indicator whether the operations will be performed or bypassed in the current iteration. If $Step$ array is 1, then the previously computed value by a preceding IWCU is updated to 0.

From the profiled code we observed that WCU part is the most important bottleneck having a highly dependent nature.

III. DEPENDENCY ANALYSIS AND PARALLELIZATION

The previous section presented algorithmic details, especially for the main bottlenecks which are *Step Array Calculation* stage and the *Processing* phase. This section discusses the data dependence between loop iterations.

Starting from the *Step Array Calculation* stage, it is observed that iterations are independent, which is important for

efficient parallelism. On the other hand, when we analyze the *Processing* phase, we see that it has a highly dependent structure, especially due to the fact that current input to the WCU or IWCU, depends on the output of preceding WCU or ICWU computation.

To realize a visual presentation of the data dependence we will use that $A \rightarrow B$ means B depends on A. Fig. 3 shows the data dependency of the *Processing* phase implementation and gives an initial idea about the sequence of computations that need to be processed. It also gives an idea how to arrange computations in a parallel environment exploiting concurrent computations.

Each of the nodes presented in Fig. 3 stands for both WCU and IWCU operations. The result of WCU computing the input N and level L is basically the *detailed approximation* of the signal at L'th decomposition level. This information is then transferred to the next $L + 1$'th level of the N'th input signal for further processing.

The computations in each node realize a WCU operation, that is composed of several steps. The first step inserts previously computed data sample. The second step is to increment the tail of the circular buffer and lastly convolve coefficients with the buffer, where the output of the convolution is passed as input to the next WCU.

For a single input element, the detailed approximation for decomposition level L starts by computing the approximation at first level. Detailed approximation at any level is computed by performing a convolution with an orthonormal wavelet basis. This information then is transferred to the next node, which is the right node in Fig. 3. In this manner, by repeatedly computing and passing the detailed approximation to the right node, next level detailed approximation can be computed.

Once the level L is reached, the same procedure is applied to the next input element which is at the bottom of the starting node. In this manner, the algorithm flows from right to left when going at higher decomposition levels, and top to bottom when computing the next input elements. One can observe that the algorithm has highly dependent nature.

On the first sight, it is observable that dependency prevents direct parallelization. However, several methods can rearrange the presented structure and parallelize the execution. One possible way rearranges the nodes, such that the calculation of values for a certain node assumes that previous (left and upper) nodes are already calculated. We use the Pipeline-Parallel-Processing methodology for parallelizing DWT.

Fig. 4 shows the organization and flow of computations in the existing data flow arrangement of the nodes. One can observe that computation waves can flow with 45 degrees to the axes. Each wave contains independent computations and can be executed simultaneously at a given time stamp. This ensures that previous nodes (found on the left) are already calculated. Due to this pipelined structure, the first output will be ready after L iterations, which is 9 in this case.

When iteration size is relatively bigger, the overhead due to the opening and closing phases of the pipeline can be neglected. Next section will outline the implementation strategies

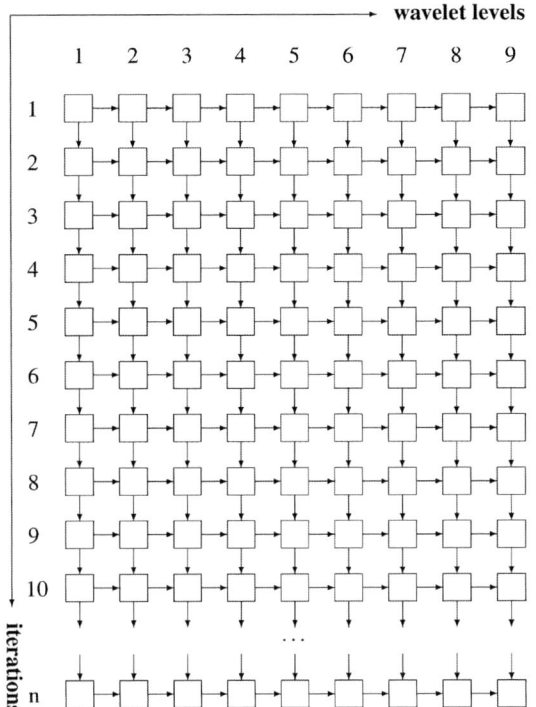

Fig. 3. Dependency analysis of the Sequential DWT code. Each node presents a WCU or IWCU operation.

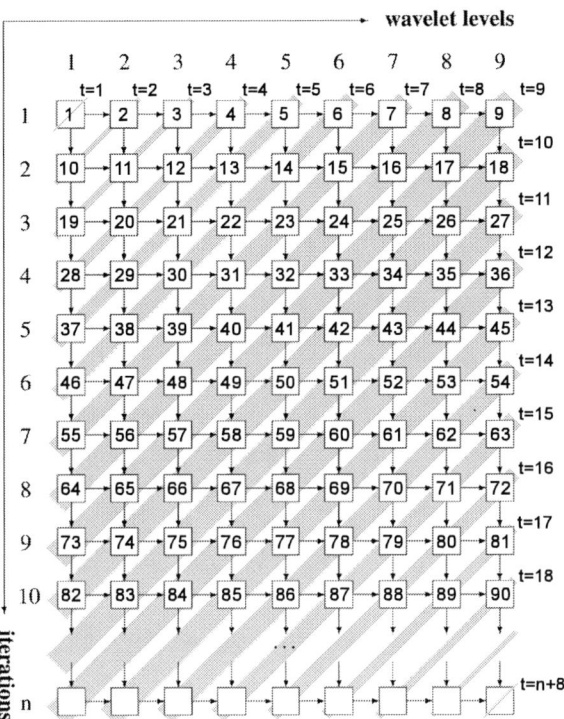

Fig. 4. Simultaneous Execution of nodes on the DWT code for 9 decomposition levels.

of the following algorithm to different platforms.

In this paper, we use OpenMP library to implement the parallel algorithm. The *Step Array Calculation* stage does not have dependencies between loops, thus iterations are directly parallelized by OpenMP loop directives.

On the other hand, for the *Processing* phase, each node in Fig. 3 computes the WCU, by performing a set of complex operations. The idea behind is to allocate a separate thread to each of the nodes. The main consideration would be to order the execution of threads in the right manner, with the aim to produce a correct output.

The nodes (representing computations) found on the wavefront can be processed independently when the previous nodes (on the left) are calculated and results are transmitted to the neighbor. These threads are synchronized once they finish their execution, and continued to the next iteration.

Our analysis is based on using cores more than the maximum number of algorithm nodes that can be simultaneously executed. This is critical in order to eliminate delays. Let's consider a reverse case of having less cores than needed, such as 4 cores and decomposition level 9. Fig. 4 shows the numbered WCU nodes. In the first time step $t = 1$, only one thread will execute the node number 1 and other threads will be idle. The next time step $t = 2$ addresses execution of 2 threads (nodes 2 and 10). This is followed by $t = 3$, where three cores will execute the code (nodes 3, 11, 19) and only one core will be idle. Starting with the fourth time step

($t = 4$) the cores will fully execute the algorithm without idle moments. However, in the next timestamp ($t = 5$) 5 nodes should be executed simultaneously and there are however there are only 4 available cores. Thus, this will require 2 cycles in order to complete, such that nodes 5, 13, 21, 29 will be executed in one cycle simultaneously, and, only the node 37 will execute in the latter cycle, while the other cores will be idle. This will increase the delay, and decrease the performance of the application seriously. The best performance for decomposition level of L is to use at least L cores.

By parallelizing the both bottlenecks, theoretically, the algorithm can achieve a speedup of L on L cores. The next sections present experimental research of the proposed optimizations.

IV. TESTING METHODOLOGY

Denote the response time required to process the sequential algorithm be denoted by T_s, and the response time required to process the parallel algorithm with p cores, be denoted by T_p. Then, the speedup is defined as the ratio of the execution times by (1).

$$S_P = \frac{T_s}{T_p} \qquad (1)$$

The sequential and parallel code are tested on an Amazon C3 c3.8xlarge instance. It consists of a high-frequency Intel Xeon E5-2680 v2 (Ivy Bridge) Processor with 32 cores, 60GB of memory. The performance of the code is tested for various wavelet levels.

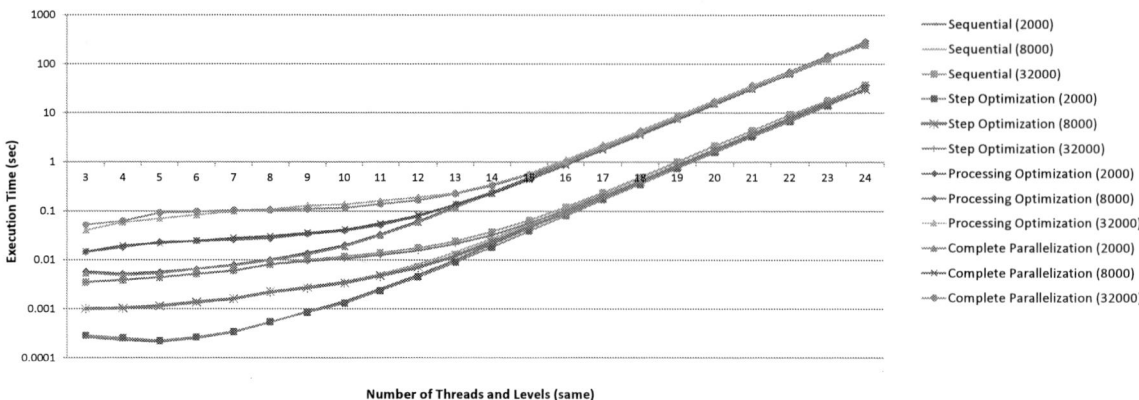

Fig. 5. Execution times of running the three proposed optimizations, with input size of 2000, 8000 and 32000. Number of cores are identical to number of Wavelet Levels. Values are presented on a logarithmic scale of base 10.

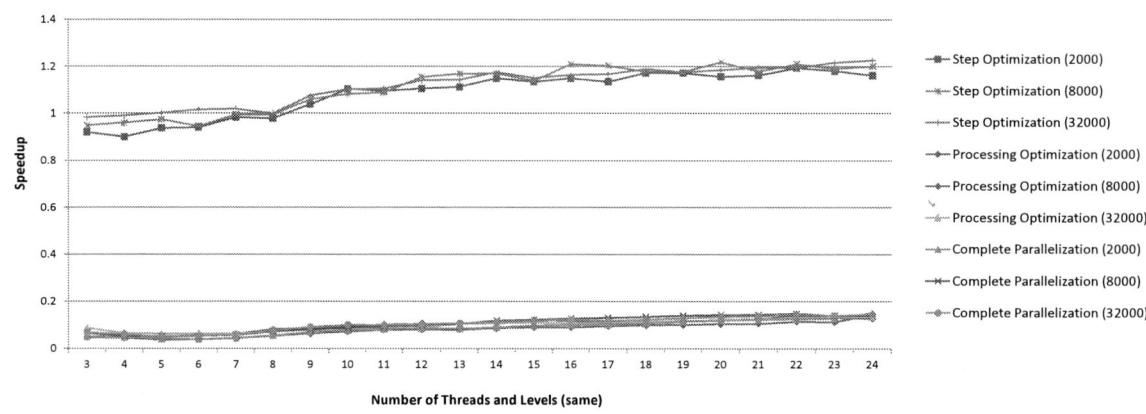

Fig. 6. Speedup of three proposed optimizations, with input size of 2000, 8000 and 32000. Number of cores are identical to number of Wavelet Levels.

OpenMP library is used for shared memory parallelism [9] without prebuilt optimizations from the OpenMP library. The static scheduling method is used to evaluate the sequential results.

Three optimization approaches are tested:

O1: Parallelization of *Step Array Calculation*

O2: Parallelization of *Processing*

O3: Complete Parallelization

The experiments were defined by testing the each previously described optimization approaches in a combination with the other approaches. Each experiment had several test runs for various sizes of the input data stream that consist of 2.000, 8.000 and 32.000 samples of an ECG signal. The tested wavelet levels consists of 3 up to 24, with incremental steps of 1 level. Daubechies filters with length are used in the experiment.

Although being theoretically possible, it is difficult to calculate behind 24 wavelet levels practically, due to memory and architectural constraints.

Each test run for the experiments was tested at least five

times and an average value of measured times was calculated and used for further processing. Moreover, functional verification was conducted to verify the functional characteristics of the sequential and optimization parallel algorithm executions obtain identical results.

V. EVALUATION AND DISCUSSION

Fig. 5 presents the execution times on a logarithmic scale with base 10 since the algorithm complexity is $O(L * 2^L)$. Additionally, Fig. 6 presents the speedup values for the proposed optimization approaches. Input sizes are selected as 2000, 8000 and 32000 samples of ECG activity. Wavelet levels vary from 3 to 24, with incremental steps of 1. The increase of wavelet levels demands more cores.

It can be observed that **O1** optimization approach tends to give positive results. The performance of the code is increased by roughly 20% for increased number of wavelet levels and cores.

The results with the **O2** and **O3** optimization approaches are not efficient. This is due to the barrier synchronization, used to synchronize the nodes.

This can be neglected only when the number of nodes (that can simultaneously execute wavelet levels) is relatively high.

Speedup values presented in Fig. 6 show that the optimization approach **O1** gives up to 20% faster code. This was the expected case since the *Step Array Calculation* phase does not contain any data dependencies between loop iterations. Considering that an input ECG signal is using a 500Hz sampling frequency, then eliminating the baseline drift will require 9 or 10 wavelet levels. In this case, the proposed optimization approach will yield $10 - 15\%$ faster code.

Moving forward to the optimization approach **O2**, it is clearly seen that this approach is not attractive for significantly low number of wavelet levels. As wavelet levels increase, the performance of the algorithm increases. The main reason not to obtain a higher speedup is the overhead of using barrier synchronization for synchronizing nodes.

Since using higher wavelet levels is practically not possible we conclude that this strategy is not attractive. Same arguments can be made for the optimization approach **O3**, except that it is yielding a bit faster code compared to the optimization approach **O2**.

We tried using explicit synchronization strategies, without any success. Such strategies are only efficient on a high level of iterations.

Based on the results presented, we can conclude that the hypothesis set in this paper is partially confirmed. Even though the proposed parallel algorithm for the processing part was not efficient on small wavelet levels, parallelization of the initialization part resulted with a faster code.

VI. RELATED WORK

We use the circular buffer solution proposed by Milchevski and Gusev [11], where the authors have reported an average speedup of 15 over the conventional buffer, which is using data shifting in each DWT level.

Sava et al. [12] have developed two parallel solutions for generic Wavelet Transform for signal processing, without addressing specifically the concept of baseline drift of the signal. Their algorithm is based on pipeline processing farming. They conclude that the performance of algorithm increases as filter length and data length increase. However, authors have not provided any information about the impact of core numbers, parallelization platform, and the scalability.

Rajmic and Vlach [13] have proposed a real-time algorithm via segmented wavelet transform analysis, presenting only the principle without practical implementations.

Stojanović et al. [14] have proposed optimized algorithms for biomedical signal processing. Their results are 2-4 times faster than the sequential implementation, though being incomparable with our work.

VII. CONCLUSIONS

This work contributes OpenMP parallelization for the baseline drift elimination of ECG heart signals. Three optimization strategies were provided.

Results obtained showed that parallelizing the initialization part gives a speedup of nearly 1.2. On the other hand, the parallelization approach for processing part was based on transforming loop iterations, in a manner that they become independent. Pipelined parallel algorithms were developed, and tested.

Our observation is that, on low wavelet levels, the parallel algorithm for the processing part is not efficient. This is primarily due to barrier synchronization between iterations.

We also tested the effect of input sizes. Unless the delay is bigger, input size plays a huge role in the speedup. The higher the input size is, the higher the speedup.

In both cases, it can be noted that the provided algorithm is scalable. Theoretically, if we run the algorithms on higher orders of magnitude, the achieved speedup will be higher.

We conclude that hypothesis we set is confirmed only for the initialization part, and this does not hold for the processing part when the number of wavelet levels is small. As a future work, we plan to find a more appropriate solution, taking advantage of GPU and Dataflow cores.

REFERENCES

[1] P. Laguna, N. V. Thakor, P. Caminal, R. Jane, H.-R. Yoon, A. Bayés de Luna, V. Marti, and J. Guindo, "New algorithm for qt interval analysis in 24-hour holter ecg: performance and applications," *Medical and Biological Engineering and Computing*, vol. 28, no. 1, pp. 67–73, 1990.

[2] T. S. Lugovaya, "Biometric human identification based on ECG," 2005.

[3] E. Domazet, M. Gushev, and S. Ristov, "Dataflow DSP filter for ECG signals," in *13th International Conference on Informatics and Information Technologies*, in press, Bitola, Macedonia, 2016.

[4] ——, "CUDA DSP filter for ECG signals," in *6th International Conference on Applied Internet and Information Technologies*, in press, Bitola, Macedonia, 2016.

[5] ——, "Optimizing high-performance CUDA DSP filter for ECG signals," in *27th DAAAM International Symposium*. in press, Mostar, Bosnia and Herzegovina: DAAAM International Vienna, 2016.

[6] C. Saritha, V. Sukanya, and Y. N. Murty, "ECG signal analysis using wavelet transformation," *BulgJ Physics, pp-68-77,(35)*, 2008.

[7] M. Alfaouri and K. Daqrouq, "ECG signal denoising by wavelet transform thresholding," *American Journal of applied sciences*, vol. 5, no. 3, pp. 276–281, 2008.

[8] P. S. Addison, "Wavelet transforms and the ECG: a review," *Physiological measurement*, vol. 26, no. 5, p. R155, 2005.

[9] OpenMP, "Openmp application program interface version 3.0," 2008, Jan. 2014], http://www.openmp.org/mp-documents/spec30.pdf.

[10] I. Ara, M. N. Hossain, and S. Y. Mahbub, "Baseline drift removal and de-noising of the ECG signal using wavelet transform," *International Journal of Computer Applications*, vol. 95, no. 16, 2014.

[11] A. Milchevski and M. Gusev, "Improved pipelined wavelet implementation for filtering ECG signals," University Sts Cyril and Methodius, Faculty of Computer Sciences and Engineering, Tech. Rep. 27/2016, 2016.

[12] H. Sava, M. Fleury, A. Downton, and A. Clark, "Parallel pipeline implementation of wavelet transforms," *IEE Proceedings-Vision, Image and Signal Processing*, vol. 144, no. 6, pp. 355–359, 1997.

[13] P. Rajmic and J. Vlach, "Real-time audio processing via segmented wavelet transform," in *Proc. of the 10th Int. Conference on Digital Audio Effects (DAFx-07), Bordeaux, France*. Citeseer, 2007.

[14] R. Stojanović, S. Knežević, D. Karadaglić, and G. Devedžić, "Optimization and implementation of the wavelet based algorithms for embedded biomedical signal processing," *Computer Science and Information Systems*, vol. 10, no. 1, pp. 503–523, 2013.

MIPRO 2017, May 22- 26, 2017, Opatija, Croatia

Hilbert Transform Based Paroxysmal Tachycardia Detection Algorithm

Ivana Čuljak*, Mario Cifrek*
*University of Zagreb, Faculty of Electrical Engineering and Computing, Zagreb, Croatia
ivana.culjak@fer.hr

Abstract—Paroxysmal tachycardia (supraventricular and ventricular) is an episodic condition with an abrupt onset and termination followed by a rapid heart rate, usually between 140 and 250 beats per minute. Paroxysmal tachycardia can be discovered by detecting a QRS complex in ECG signals. Supraventricular tachycardia is characterized by a narrow QRS complex, and on the other side, ventricular tachycardia is characterized by a broad QRS complex. Detecting paroxysmal tachycardia can prevent pathogenesis of a heart disease. Implementation of an algorithm for QRS detection based on properties of the Hilbert transform is proposed in this paper. The results of the algorithm were compared with the Pan-Tompkins algorithm and the detection efficiency of the implemented algorithm on the used signals was 97.5 %. Both algorithms were tested using the recordings from the MIT-BIH Arrhythmia database.

Keywords: Paroxysmal tachycardia, ECG, QRS complex, Hilbert transform, MATLAB, Pan-Tompkins algorithm

I. INTRODUCTION

Paroxysmal tachycardia is an episodic condition with an abrupt onset and termination [7]. The heartbeat frequency is usually between 140-250 beats per minute, but often it can be higher or lower than that. Paroxysmal tachycardia can be divided into two main groups. The first group, supraventricular tachycardia (PSVT), are characterized by a narrow QRS complex. They are generally benign. PSVT can occur during vigorous physical exercises or activity. The most common type of PSVT is nodal reentrant supraventricular tachycardia, and it makes about 90% of all PSVT. It occurs often in otherwise healthy patients. On the other side, the ventricular tachycardia (VT) is characterized by the broad QRS complex. Most commonly, they are the result of significant heart disorder. Episodes of VT (or PVT) that last more than a few seconds can be very dangerous and escalate into (fatal) ventricular fibrillation [6].

In this research, an algorithm for paroxysmal tachycardia (PT) detection was implemented. The first step for detecting PT is the detection of the QRS complex in ECG signals.

Detection of the QRS complex is one way of monitoring the heart functionality. Detection of the QRS complex has been the subject of many studies in recent decades, but there is still no algorithm that can detect 100 % of the QRS complex. Also, many algorithms are unsuitable for hardware implementation.

As QRS shows the electrical activity of the heart, the time of its appearance as well as its shape provides us with a lot of information about a heart condition. Software QRS detection provides all the important information about the patient's condition and provides diagnoses for many diseases.

The Hilbert transform based QRS detection algorithm is proposed in this paper. Most of the algorithm steps are taken from the article [2]. The results of the proposed algorithm are compared with the Pan-Tompkins algorithm [5].

II. ALGORITHM IMPLEMENTATION

The first step of the algorithm is filtering using the bandpass filter [5]. The reason for selecting the filtering as the first step is the noise in the input signal. The noise is caused by the poor contact between electrodes and the skin, muscle and skin resistance, supply voltage and many other disorders. To emphasize the slope of the QRS complex and remove the baseline due to the artifacts, the first derivation of the filtered signal is used [2]. After derivation, the signal is divided into segments of 1000 samples [2]. The Hilbert transform and the detection of the R peak are implemented above each fragment [2]. After derivation and the Hilbert transform, a detector with adaptive thresholds is used to find the R peak [2,4]. The rule that two R peaks cannot be detected within 200 ms (72 samples) is implemented. If it is detected, it means that only one of them is the right one. This condition prevents false detection of the QRS complex. Figure 1 shows the algorithm block diagram. The presence of PT is detected based on the heart rate and frequency of used ECG signals and annotations in the database. The recordings with supraventricular/ventricular tachycardia were extracted from three databases that were used in the algorithm.

A. QRS Detection

The input signal is filtered by a band-pass filter in the frequency range from 5 to 15 Hz for the recordings from MIT-BIH Arrhythmia database, and range 5-50 Hz for the recordings from MIT-BIH Malignant Ventricular Arrhythmia and Creighton University Ventricular Tachyarrhythmia database. The filter is realized with the cascade of the low-pass and high-pass digital biquad filters.

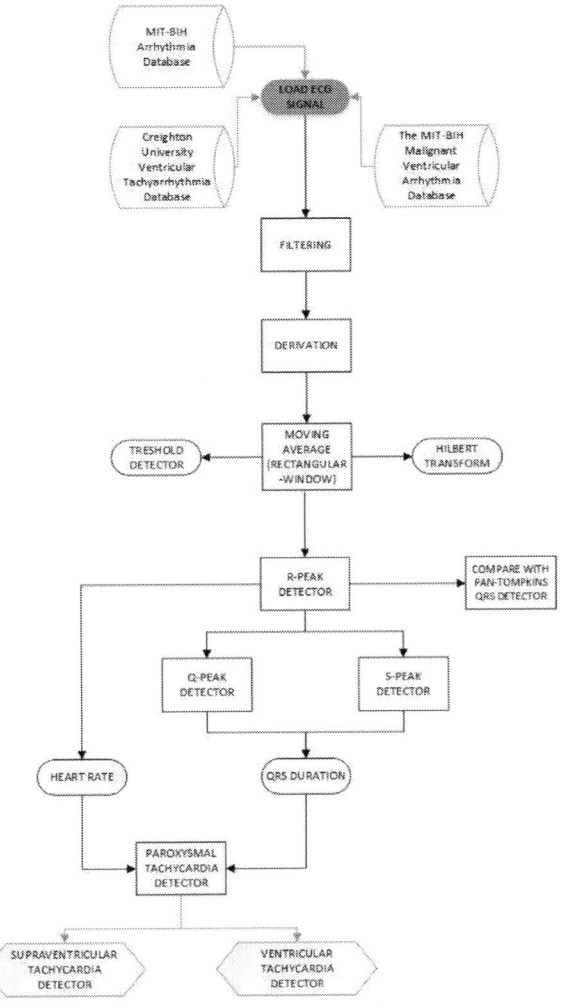

Figure 1. The algorithm block diagram

Its transfer function is defined as follows in equation (4):

$$H(z) = \frac{b0 + b1z^{-1} + b2z^{-2}}{a0 + a1z^{-1} + a2z^{-2}} \qquad (4)$$

The coefficients of low-pas and high-pass biquad filter were calculated using equations given in [1].

Filtering is followed by derivative. The first derivative of the filtered ECG signal equation is given in article [2].

Derivative signals are divided into 1000 samples. This part of the algorithm is a function *hilbertDetectQRS ()*, that is implemented in MATLAB. If this is not the end of the signal, then the starting position moves for 1000 samples. If there is an end of the signal, then it is moved to the end [2,4]. Next, the

hilbertECG () function is called, and it returns the output signal after the Hilbert transform [2,4].

Hilbert transform of the real function $x(t)$ is defined as [2,3]:

$$H\{x(t)\} = \frac{1}{\pi} \int_{-\infty}^{+\infty} x(\tau) \frac{1}{t - \tau} d\tau \qquad (5)$$

The main properties of the Hilbert transform [3]:

- A signal $x(t)$ and $H\{x(t)\}$ are mutually orthogonal

$$\int_{-\infty}^{+\infty} x(t)H\{x(t)\}dt = 0 \qquad (6)$$

- Linearity:

$$[k_1 f_1(t) + k_2 f_2(t)]^\wedge = k_1 \hat{f}_1(t) + k_2 \hat{f}_2(t) \qquad (7)$$

Where: k_1, k_2 – arbitrary scalars and $f_1(t), f_2(t)$ – signals

- Time-shifting and time-dilation:

If HT of $f(t)$ is $\hat{f}(t)$, then HT of $f(t - t_0)$ is $\hat{f}(t - t_0)$.

HT of $f(at)$ is $sgn(a)\hat{f}(at)$ (where $a \neq 0$)

- Time derivation:

$$H\left[\frac{d}{dt}f(t)\right] = \frac{d}{dt}H[f(t)] \qquad (8)$$

- Hilbert transform is an odd function

The Hilbert transform can distinguish between dominant peaks among other peaks in the signal, thereby improving R peak detection. The Hilbert transform is implemented over each signal's section which was previously differentiated [2].

The implementation of this step is conducted in MATLAB in function *hilbertECG ()*. How it appears after the Hilbert transform is shown in Figure 2.

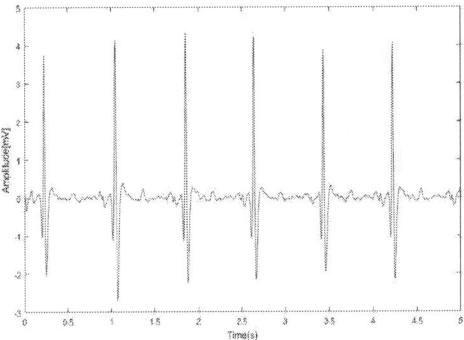

Figure 2. ECG signal after the Hilbert transform

After the Hilbert transform, a detector with adaptive thresholds is used for finding peaks. The adaptive thresholds detectors are proposed in [2] and are provided in equation (9).

This step of the algorithm is implemented in MATLAB in function *hilbertDetectQRS ()*.

$$threshold = \begin{cases} 0.39max(i) & , RMS(i) \geq 0.18\max(i) \\ 0.39\max(i-1) & , \max(i) < 2\max(i-1) \qquad (9) \\ 1.6RMS(i) & , RMS < 0.18\max(i) \end{cases}$$

Next, the *thresholdCalculation ()* function is used. This function returns all the peaks that were greater than the previously calculated thresholds. Once we find the highest value of each segment, we need to remove the false QRS complex, due to the conditions that we mentioned above (two QRS complexes cannot be detected within 200 ms – refractory period). In the last step, the indexes of the QRS complexes that are marked in the database MIT-BIH were compared with the QRS complexes obtained in the algorithm.

B. Paroxysmal Tachycardia Detection

When the QRS complex is narrow (QRS < 120 ms) there are no differential diagnostic difficulties in distinguishing supraventricular from ventricular tachycardia because VPT with the narrow QRS complex is a rarity. The problem occurs with the broad QRS complex (QRS > 120 ms). Broad complexes may be ventricular (the most often) or supraventricular due to aberrant conduction of supraventricular complexes [6].

Supraventricular tachycardia typical characteristics [6]:
- Abnormal P wave (or negative/or either no P wave)
- 150-220 beats per minute
- QRS <0.12 s

Ventricular tachycardia typical characteristics [6]:
- Normal P wave
- 140-200 beats per minute
- QRS >0.12 s
- If the amplitude of the R peak in tachycardia is lower than in normal sinus rhythm it is considered to be ventricular tachycardia.
- Three or more episodes of a rapid heart rate (tachycardia) that come from the ventricles it is considered to be ventricular tachycardia.

Referring to the characteristic above after the QRS detection, PT is detected in ECG recordings from the MIT-BIH databases specified in Chapter III.
For monitoring beats rhythm, the heart rate variability is calculated on every ECG signal segment.

$$HRV = \frac{60 * f_s}{diff(R\ index)} \left[\frac{beats}{minute}\right] \qquad (10)$$

where fs is sampling frequency.
Figure 3 shows the heart rate variability in record 215 of the MIT-BIH Arrhythmia database.
Figure 4 and Figure 5 show the VT episodes and sinus rhythm in ECG signal, respectively.

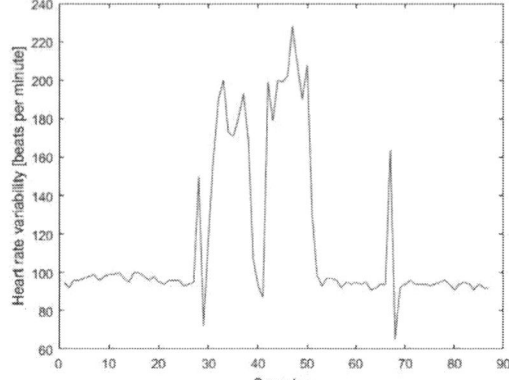

Figure 3.Heart rate variability in record 205 (MIT-BIH Arrhythmia database) – time duration: 280 s-330 s

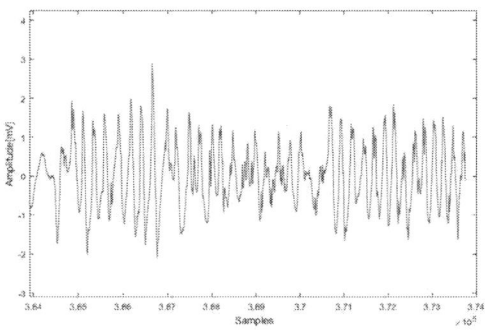

Figure 4. VT episodes in the ECG signal from the MIT-BIH Malignant Ventricular Arrhythmia database

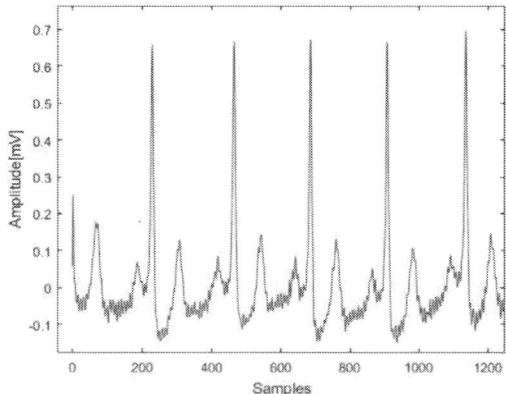

Figure 5. Sinus rhythm in the ECG signals from the MIT-BIH Malignant Ventricular Arrhythmia database

C. Application

A MATLAB GUI application (Fig. 6) is implemented. The application shows the results of certain phases of the algorithm and demonstrates the success of the algorithm.

It consists of a panel with buttons, which represent certain stages of the algorithm, and of the panel with input settings and a time domain signal plot.

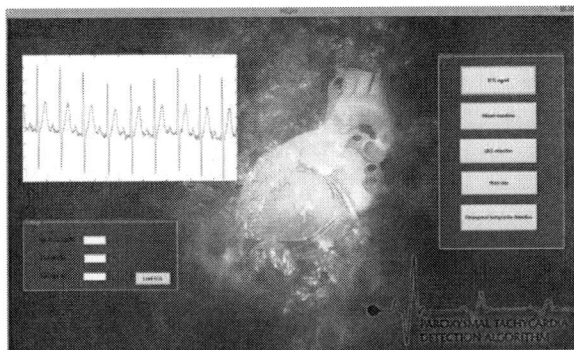

Figure 6. MATLAB GUI application; **right**: panel with options buttons; **left**: time domain signal plot (depends on the selected stage of the algorithm) **down left**: Input settings.

III. DATABASES

A. The MIT-BIH Arrhythmia Database

The MIT-BIH Arrhythmia database (Massachusetts Institute of Technology-Beth Israel Hospital Arrhythmia Laboratory) contains forty-eight 30 minutes ECG recordings. The recordings were sampled at 360 Hz with 11-bit rate resolution and ± 5 mV range [8].
Table 1 presents the signals that were used in this paper.
ECG recordings with tachycardia where the heart rate was higher than 140 beats per minute were used in this paper. The database is available on the Physionet webpage [8].

Table 1. MIT-BIH Arrhythmia recordings description

MIT-BIH Arrhythmia Database	
Record	**Description**
203	Ventricular tachycardia - Rate: 124-189 - Episodes: 21 - Duration: 0:33 Points of interest: - **5:00** Ventricular tachycardia, 4 beats and 9 beats - **26:39** Ventricular tachycardia, 7 beats - **27:15** Ventricular tachycardia, 3 beats
205	Ventricular tachycardia - Rate: 79-216 - Episodes: 6 - Duration: 0:23 Points of interest: - **4:57** Ventricular tachycardia, 8 beats and 10 beats - **15:22** Ventricular tachycardia, 3 beats - **24:18** Ventricular tachycardia 12 beats, start of VR, 13 beats - **24:30** End of VR, 3-beat VT (continued from the previous strip)

210	Ventricular tachycardia - Rate: 103-161 - Episodes: 2 - Duration: 0:06 Points of interest: - **6:56** Ventricular tachycardia, 6 beats - **17:57** Ventricular tachycardia, 6 beats
215	Ventricular tachycardia - Rate: 174-177 - Episodes: 2 - Duration: 0:2 Points of interest: - **2:55** Ventricular tachycardia, 3 beats - **20:25** Ventricular tachycardia, 3 beats

B. The MIT-BIH Malignant Ventricular Arrhythmia Database

This database includes 22 half-hour ECG recordings. The recordings were sampled at 250 Hz. The database is available on the Physionet webpage [8].
Recordings provided in Table 2 were used in this paper.

Table 2. The MIT-BIH Malignant Ventricular Arrhythmia recordings

Record	Points of interest		
	Time	Sample	Aux.
420	23:50.152	337538	(VT
	26:34.692	39867	(VT
	26:58.612	404653	(VT
	Time	Sample	Aux.
421	14:54.152	223538	(VT
	..	...	...
	22:15.844	333961	(VT
	Total (VT appearance = 50		
	Time	Sample	Aux.
422	22:12.844	333211	(VT
	22:15.768	333942	(VT
	Time	Sample	Aux.
426	27:47.844	416961	(VT
	29:00.844	435211	(VT
	29:30.152	442538	(SVTA
	29:37.076	444269	(VT
	29:52.460	448115	(SVTA
	30:34.088	458522	(VT
	30:39.164	459791	(SVTA
	Time	Sample	Aux.
607	2:13.844	33461	(VT
	26:21.844	395461	(VT
	30:48.536	462134	(VT
	31:00.536	465134	(SVTA
	31:26.076	471519	(VT
	31:34.692	473673	(SVTA

611	Time	Sample	Aux.
	0:00.152	38	(SVTA
	19:56.228	299057	(VT

615	Time	Sample	Aux.
	5:26.844	81711	(VT
		...	
	33:33.536	503384	(VT

C. Creighton University Ventricular Tachyarrhythmia Database (CUDB)

This database includes 35 eight-minute ECG recordings. The recordings were sampled at 250 Hz with 12-bit rate resolution and ± 5 mV range. The database is available on the Physionet webpage [8].

The record provided in Table 3 was used in this paper.

Table 3. Creighton University Ventricular Tachyarrhythmia Database

Record	Points of interest		
	Time	Sample	Aux.
cu02	3:12.408	48102	(VT
	3:16.908	49227	(VT
	8:08.708	122177	(VT
	8:12.436	123109	(VT
	8:16.308	124077	(VT

IV. RESULTS

Results obtained by detecting the QRS complex based on the Hilbert transform were compared with the Pan-Tompkins algorithm in this chapter.

The characteristics that evaluate the performance of an algorithm are sensitivity (Se), positive prediction (+P) and detection error rate (DER) [5]. They are calculated according to the equations below:

$$Se = \frac{TP}{TP + FN} \qquad (11)$$

$$+P = \frac{TP}{TP + FP} \qquad (12)$$

$$DER(\%) = \frac{FP + FN}{Total\ number\ of\ QRS\ complexes} \qquad (13)$$
$$\cdot 100$$

Where:
- TP (true positive) = total number of QRS complexes that are correctly located by the algorithm
- Number of FN (false negative) and FP (false positive) QRS complexes.

The sensitivity, positive prediction, and detection error rate were calculated using the recordings (that were used in this paper) from the MIT-BIH Arrhythmia database and Creighton University Ventricular Tachyarrhythmia Database. They were

not calculated for the MIT-BIH Malignant Ventricular Arrhythmia database because the reference annotation (.atr) files contain only rhythm labels (no beat labels) [8].

Table 4 shows the results of QRS complex detection based on the Hilbert transform implemented in MATLAB.

Table 4. Results of implemented algorithm on the MIT-BIH Arrhythmia database

Record	Se[%]	+P[%]	DER[%]
203	96.453	98.713	4.8
205	99.625	100	0.4
210	97.833	99.923	2.2
215	98.495	99.970	1.5
Total:	97.7818	99.703	2.5

Table 5 shows the results of the Pan-Tompkins algorithm for the same recordings.

Table 5. Results of the Pan-Tompkins algorithm on the MIT-BIH Arrhythmia database

Record	Se[%]	+P[%]	DER[%]
203	95.807	99.899	4.3
205	99.549	100	0.5
210	97.721	99.962	2.3
215	98.495	100	0.4
Total:	98.061	99.972	1.76

The Pan-Tompkins algorithm works better than implemented algorithm by 0.74 %.

Table 6. Results of algorithms on the Creighton University Ventricular Tachyarrhythmia Database for 120 seconds

Record	Se[%]	+P[%]	DER[%]
The algorithm based on Hilbert transform			
cu02	94.742	99.675	5.567
The Pan-Tompkins algorithm			
cu02	94.95	99.35	5.67

After QRS detection, the PT detector was applied on signals. In the vector variable were saved the indexes where is annotated appearance of VT and SVT in databases. Next, the indexes from the annotation variable were compared with the obtained results. The output of the resulting PT detector are indexes of the beginning of VT and SVT. The detector is applied to over 12 signals from the three databases (MIT-BIH Arrhythmia Database, MIT-BIH Malignant Ventricular Arrhythmia Database, Creighton University Ventricular Tachyarrhythmia Database).

The Q and S peaks are easily located by finding slope reversals on either side of R-peak [4]. Every 10 seconds were calculated heart rate and QRS duration. It is sufficient time to capture most of the necessary information of the currently cardiac condition.

Table 7 shows the results of the implemented algorithm. The record 203 from MIT-BIH Arrhythmia database and record 421 from the MIT-BIH Malignant Ventricular Arrhythmia database have the worst result because there is the presence of considerable noise, including muscle artifact and baseline shifts. In the recordings from the MIT-BIH Malignant Ventricular Arrhythmia database, there is a constantly signal droop. The MIT-BIH Malignant Ventricular Arrhythmia does not have a beat annotation and because of that, we cannot calculate episodes of VT.

Table 7. The results of implemented PT algorithm

Record	Annotation	Detect	FP	FN
cu02	5	VT = 2	4	3
420	3	2	0	1
421	50	VT = 12	27	38
426	7	VT = 4 SVT = 3	32	0
607	6	VT =4 SVT = 2	19	0
611	2	VT = 1	18	1
615	5	VT= 4	15	1
203	21	VT = 18	34	3
205	6	VT = 1	2	3
210	2	VT = 1	1	1
215	2	VT = 2	0	0

V. CONCLUSION

The paroxysmal tachycardia detection algorithm based on the Hilbert transform was implemented in this paper. The proposed algorithm was applied to ECG signals from the three MIT-BIH databases. The QRS detection efficiency is 97.5 %. The PT detect algorithm shows a worse result because of a large presence of noise and constantly signal droop in most of the recordings.

REFERENCES

[1] Bristow-Johnson, Robert. "Cookbook formulae for audio EQ biquad filter coefficients", http://www.musicdsp.org/files/Audio-EQ-Cookbook.txt

[2] D. Benitez, P.A. Gaydecki, A. Zaidi, A.P. Fitzpatrick, "A new QRS Detection Algorithm Based on the Hilbert Transform", Computers in Cardiology, 27: 379-382, 2000

[3] Frank R. Kschischaug, The Hilbert Transform, The Edward S. Rogers Sr. Department of Electrical and Computer Engineering, University of Toronto, October 22, 2006

[4] Hossein Rabbani, M. Parsa Mahjoob, E. Farahabadi, A. Farahabadi, R Peak Detection in Electrocardiogram Signal Based on an Optimal Combination of Wavelet Transform, Hilbert Transform, and Adaptive Thresholding, Journal of Medical Signals&Sensors, Vol 1., Issue 2, May - Aug 2011

[5] J. Pan and W.J. Tompkins, "A real-time QRS detection algorithm" IEEE Trans. Biomed. Eng., vol. 32, pp. 230-236, 1985

[6] Lj. Barić, LIBELLI MEDICI, Elektrokardiografija u praksi, 2. dopuljeno i prerađeno izdanje, Volumen VII, Zagreb 1992

[7] Monika Gugneja, MD Consulting Staff, Department of Emergency Medicine, William Beaumont Hospital, "Paroxysmal Supraventricular Tachycardia", Updated: Dec 30, 2015, Available at: http://emedicine.medscape.com/article/156670-overview

[8] PhysioBank Databases: https://physionet.nlm.nih.gov/physiobank/databas

Biomedical time series preprocessing and expert-system based feature extraction in MULTISAB platform

Alan Jovic*, Davor Kukolja*, Kresimir Friganovic*, Kresimir Jozic**, and Sinisa Car***

* University of Zagreb Faculty of Electrical Engineering and Computing, Zagreb, Croatia
** INA - industrija nafte, d.d., Zagreb, Croatia
*** Dom zdravlja Dubrovnik, Ambulanta Babino polje, Babino polje, Croatia
Corresponding author: alan.jovic@fer.hr

Abstract - In this paper, we review the current state of implementation of the MULTISAB platform, a web platform whose main goal is to provide a user with detailed analysis capabilities for heterogeneous biomedical time series. These time series are often encumbered by noise that prohibits accurate calculation of clinically significant features. The goal of preprocessing is either to completely remove the noise or at least to ameliorate the quality of the recorded series. The focus of this paper is on the description of an expert feature recommendation system for electrocardiogram analysis. We demonstrate the process through which one arrives at a point where significant expert features are proposed to a platform user, based on time series at hand, analysis goal, and available length of the time series. We also provide the description of implemented preprocessing techniques and feature extraction procedure within the platform.

I. INTRODUCTION

The need for web-based biomedical software is continuously growing in the healthcare community [1,2]. Although, according to [1], web based solutions could, among many other things, reduce the cost of healthcare expenditure and facilitate access to a better healthcare, specialized software developed for medical professionals is vastly limited to continuous monitoring of biomedical time series (BTS) [3]. Hence, in the currently running Croatian Science Foundation research project HRZZ-MULTISAB[1], we pursue the development of an efficient and upgradeable BTS analysis system for automatic classification of human body disorders in a form of an integrative web platform. The aims of such a system are to help medical specialists in diagnostics and early detection of various diseases and to improve scientific investigation of BTS disorders.

In our previous work, we described the early efforts in designing such a system, which included an architectural overview, an overview of used technologies, detailed description of the MULTISAB project's frameworks [4] and use case based scenarios specification [3].

The architectural overview of the platform is shown in Fig. 1. MULTISAB project is divided into three

---

[1] This work has been fully supported by the Croatian Science Foundation under the project number UIP-2014-09-6889.

subprojects: frontend on the client side; backend and processing subproject on the server side. The frontend and backend subprojects support communication between end users and servers, enabling a user (client) to send requests to the server computer through his web browser. Besides processing requests from users, the backend subproject serves for communication with a database and with data analysis hosts, on which processing subproject is running. The processing subproject is designed for heterogeneous BTS data analysis. It contains various frameworks for data handling, signal visualization, signal preprocessing, feature extraction, and data mining. Such web-based architecture provides users with the ability of distant access and enables biomedical data upload to the server, where efficient analysis is performed. Also, a web platform enables easier maintenance and a larger user base.

BTS data analysis process, described in detail in [3], is divided into 8 steps, some of which may be skipped, depending on the user: 1) analysis type selection, 2) scenario selection, 3) input data selection, 4) records inspection, 5) records preprocessing, 6) feature extraction, 7) model construction, and 8) reporting.

The aim of this paper is to review the current state of the MULTISAB platform implementation with the focus on the description of an expert feature recommendation system for ECG analysis and its initial implementation through predefined analysis scenarios. We also present the implemented preprocessing techniques and feature extraction procedure within the platform.

With the reported progress on the MULTISAB platform implementation, we have achieved the functionality of the first six analysis process steps (finishing with feature extraction). However, it should be noted that the current implementation is still in the test phase and that the user interface (UI) is in Croatian language. Nevertheless, we expect that in a few months a fully operational version will be available in Croatian and in English. The functionality will be limited to the first six steps, ending with the possibility of download of calculated features to a local file, thus enabling further independent data mining and reporting.

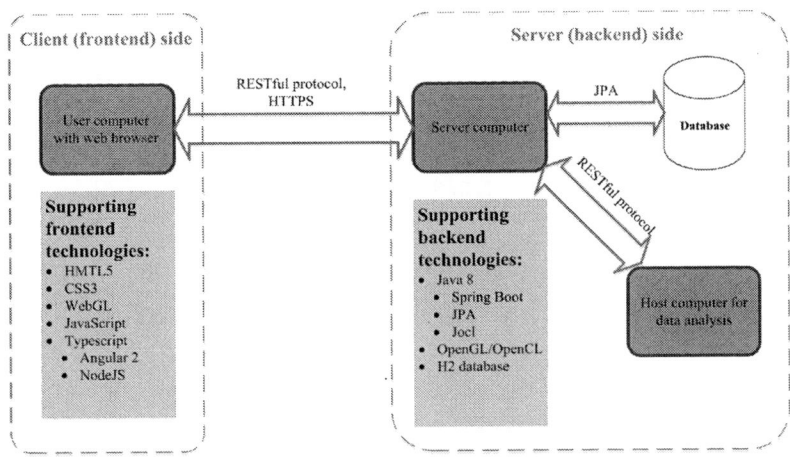

Figure 1. Architectural overview of the MULTISAB project web platform (modified from [4])

II. PREDEFINED ANALYSIS SCENARIOS WITH EXPERT FEATURES

Scenario selection UI, shown in Fig. 2, enables the user to select a predefined scenario ("Odabir spremljenih scenarija") and the construction of a completely new scenario through full customization ("Specifikacija novog scenarija"). A current list of predefined analysis scenarios includes: detection of acute myocardial ischemia ("Detekcija akutne miokardijalne ishemije"), detection of congestive heart failure ("Detekcija kongestivnog zatajenja srca"), and detection of atrial fibrillation ("Detekcija fibrilacije atrija") from ECG signals. The user can confirm a scenario with a button ("Dalje"). In later steps of the analysis, it is possible to change some of scenarios' aspects, such as preprocessing methods, feature extraction methods, or model construction methods. The predefined scenarios contain necessary preprocessing steps and expert features that are to be extracted from signals, which are selected by the medical expert system specifically designed for this purpose.

Implementation of the medical expert system is a continuously evolving process that gradually introduces new predefined scenarios. New scenarios are based on currently valid medical guidelines, standards, consultations with medical specialists, and relevant medical and biomedical engineering literature.

The current initial version of the MULTISAB expert system, implemented in cooperation with cardiologists, includes the recommended features for detection of acute myocardial ischemia (AMI), atrial fibrillation (AF) and congestive heart failure (CHF) from ECG signals.

In the following subsections, we demonstrate the process of designing the predefined analysis scenarios with expert features for detection of specified disorders. To facilitate readers' understanding of the underlying terminology, we provide an image of characteristic ECG points and waves, with a particular focus on ST elevation, in Fig. 3 (courtesy of Rob Kruger and ECGpedia.org).

A. Acute Myocardial Ischemia

According to the medical guidelines [5], criteria for diagnosis of AMI (in absence of left ventricular hypertrophy and left bundle branch block) are:

1. ST elevation: New ST elevation at the J point in two contiguous leads with the cut-points: ≥ 0.1 mV in all leads other than leads V_2–V_3,

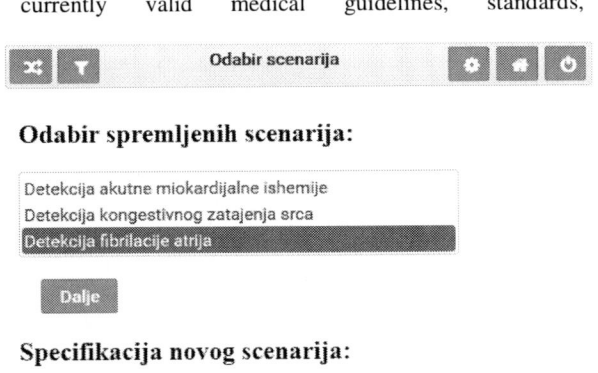

Figure 2. Scenario selection user interface; currently in Croatian language

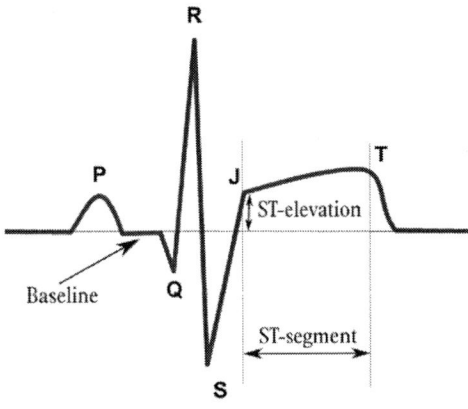

How to measure ST elevation?

Figure 3. Characteristic ECG points and waves

where the following cut points apply: ≥0.2 mV in men ≥40 years; ≥0.25 mV in men <40 years, or ≥0.15 mV in women.

2. ST depression and T wave changes: New horizontal or down-sloping ST depression ≥0.05 mV in two contiguous leads and/or T inversion ≥0.1 mV in two contiguous leads with prominent R wave or R/S ratio >1.

From criteria for diagnosis of AMI, we can easily define expert morphological ECG features:

Morphological ECG features: ECG signal value (in regard to the baseline) at the J point, R/S ratio, T wave amplitude, and ST segment slope (for all leads).

According to guidelines [5], dynamic changes in the ECG waveforms during AMI episodes often require acquisition of multiple ECGs, particularly when the ECG at the initial presentation is non-diagnostic. Serial recordings in symptomatic patients with an initial non-diagnostic ECG should be performed at 15 - 30 min intervals or, if available, continuous computer-assisted 12-lead ECG recording.

MULTISAB platform currently supports only continuous signals, so dynamic changes in the ECG can only be monitored if a segment of ECG signal of at least 15 min exists. According to [5,6], ECG manifestation of AMI includes dynamical ST segment. For quantification of the ST segment dynamics, we can use some measure of dispersion, such as standard deviation, which was recommended by cardiologists.

Dynamical ECG features: standard deviation of ECG signal value at the J points (for all leads).

In the case of a continuous recording, we can also analyze heart rate variability (HRV) and QT Interval Variability (QTV). These time series could be of great importance, because, according to the guidelines [5], further ECG signs associated with AMI include: cardiac arrhythmias, intraventricular and atrioventricular (AV) conduction delays, and loss of precordial R wave amplitude. According to [7], comparison of a normality zone against the ischaemic episode in the same record has showed increases in power spectral density for low frequency (LF) and high frequency (HF) bands.

HRV features: LF, HF.

There are also indications that QTV can have clinical meaning [8]. In [9], normalized QTV (QTVnorm) and QT variability index (QTVI) were greater during ischemic episodes than during nonischemic episodes.

QTV features: QTVnorm, QTVI.

For AMI detection, metadata which would include age and sex and additional information, like, e.g. body mass index (BMI) would also be useful, but MULTISAB currently does not support import of anamnesis and metadata. We plan to add this feature in the future.

Additional expert features, which demand expanding the current frameworks with new operations, could include calculation of energy and entropy measures in wavelet domain [10] or additional information in the ST

segments, such as shape, width, height, curvature, etc. In [11], a five-order polynomial fitting is applied to extract additional features from ST segments.

B. Atrial Fibrillation

According to the medical guidelines [12], and with up to date medical research [13], AF is associated with the following changes in ECG:

1. Lack of discrete P waves.

2. Fibrillatory or f waves are present at a rate that is generally between 350 and 600 beats/minute.

3. Ventricular response follows no repetitive pattern; the variability in the intervals between QRS complexes is often termed "irregularly irregular."

4. The ventricular rate, especially in the absence of AV nodal blocking agents or intrinsic conduction disease, usually ranges between 90 and 170 beats/min.

5. The QRS complexes are narrow, unless AV conduction through the His Purkinje system is abnormal due to functional (rate-related) aberration, pre-existing bundle branch or fascicular block, or ventricular preexcitation with conduction down the accessory pathway.

From ECG manifestation, we can define the following expert morphological ECG features:

- P wave absence - to detect a lack of discrete P waves

- Fibrillatory rate - must be between 350 and 600 beats/minute

- QRS complexes duration - to detect narrow QRS complexes

Morphological ECG features: P wave absence, fibrillatory rate, QRS complexes duration.

In the case of continuous recording, we can also analyze HRV, because one of AF manifestations is that ventricular response follows no repetitive pattern. Many researchers [14-16] have studied HRV features for AF detection and also gave their pathophysiological meaning. For expert features, we have chosen those with best discriminatory capability and those with the most important objective according to [14].

It should also be noted that the choice of HRV features depends on the length of the analyzed segment. The features are usually categorized into short-term HRV features, which can be calculated from segments duration of several minutes (normally 5 minutes), and long-term HRV features, which are calculated from 24 hours of recordings [17].

Short-term HRV features: Heart rate, standard deviation of all NN (normal beat to normal beat R-R) intervals (SDNN), the square root of the mean of the sum of the squares of differences between adjacent NN intervals (RMSSD), percentage of the number of pairs of adjacent NN intervals differing by more than 50 ms (pNN50), LF, HF, LF/HF ratio, approximate entropy (ApEn), sample entropy (SampEn).

Long-term HRV features: total power spectral density (TP), very low frequency band power spectral density (VLF), 1/f power-law exponent α.

Additional expert features, which demand expanding the current frameworks with new operations, could include calculation of wavelet entropy (WE) [18] or relative wavelet energy (RWE) [19], which are based on the wavelet transformation of the TQ interval. Both features can perform successfully even under heart rates with no variability. For example, in [19], AF was detected in less than 7 beats, with an accuracy higher than 90%.

C. Congestive Heart Failure

ECG manifestations of CHF are unspecific. In the guidelines for the diagnosis and treatment of acute and chronic heart failure [20], there is only a statement that an abnormal ECG increases the likelihood of the diagnosis of CHF.

The medical diagnosis of CHF is based on presenting signs and symptoms, patient's prior clinical history, physical examination and resting ECG. If all the elements are normal, CHF is highly unlikely. If at least one element is abnormal, plasma natriuretic peptides (NPs) should be measured, to identify those who need echocardiography, because echocardiography is the most useful and widely available test in patients with suspected CHF used to establish the diagnosis.

For CHF detection from ECG signals, expert morphological ECG features were selected based on ECG manifestations of various heart diseases described in [21] and [22].

Morphological ECG features: Q wave amplitude, PR interval duration, P peak amplitude, R peak amplitude, QRS complex duration, ratio of QT segment and heart rate corrected QT segment (QT/QTc), S wave duration in leads I, V_5, V_6, V_1 and V_2, R wave duration in leads V_5 and V_6.

For expert HRV features, we have chosen those with the best predictive values according to [23-26] and from recommendations by cardiologists.

Short-term HRV features: Heart rate (or 1/RRmean), short term detrended fluctuation analysis coefficient α1 (DFA α1), symbolic dynamics one variation pattern (1VP), LF, SDNN.

Long-term HRV features: VLF, 1/f power-law exponent α.

Although some features of heart rate turbulence (HRT), such as turbulence slope (TS), have significant predictive value [26], they are not used in the expert system because they cannot be calculated from all ECG recordings.

III. PREPROCESSING

Records preprocessing assumes various procedures for signal filtering, characteristic waveform detection, and

data transformations as we can see in the Preprocessing ("Predobrada signala") UI, Fig. 4, where the user can select various methods for signal filtering, characteristic ECG waveform detection and data transformation.

Biomedical time series are often encumbered by noise that prohibits accurate calculation of clinically significant features. The goal of preprocessing is either to completely remove the noise or at least to ameliorate the quality of the recorded series. Signal filtering includes noise filtering (e.g. PLI notch filters for 50 / 60 Hz), baseline wandering corrections (e.g. for ECG because of breathing, body movements or electrodes impedance changes), and records segmentation into windows of certain width prior to applying preprocessing method [18]. Window width may be chosen explicitly by the user or it may be implicit, determined by the selected method preprocessing method.

The user can also select which characteristic waveforms are needed to be detected in an ECG and the corresponding detection methods, because locating waveforms is sometimes necessary for domain specific feature calculation. For example, if R peak annotations are not uploaded together with the ECG data signals, it is necessary to detect R peaks in ECG signals for HRV domain specific feature calculation (e.g. with Pan-Thompkins or Elgendi methods [27]).

Data transformations are performed after the detection of characteristic waveforms. The user can select various time (e.g. principal component analysis - PCA), frequency (e.g. fast Fourier transform, AR Burg method, Lomb-Scargle periodogram), time-frequency (e.g. wavelet transform, Hilbert Huang transform) and other types of data transformations.

The goal of data transformations is to obtain the data in a form from which it is easier to calculate precise domain or general features for describing specific subjects' states. For example, it is possible to transform ECG signal into wavelet domain and then apply some of the common signal extractors (see the list in [4]) in order to calculate features from the transformed signal.

IV. FEATURE EXTRACTION

Feature extraction is the central step in the analysis of biomedical time-series. In the Feature ("Značajke") tab of Feature extraction ("Izlučivanje značajki") UI, Fig. 5, the user will already be provided with a list of (expert) features selected (during construction of a completely new scenario) or predefined (for predefined analysis scenarios) in the scenario selection phase.

Fig. 5 shows how Feature extraction UI looks like for AF detection analysis scenario. Regardless of the scenario, the user can select additional features ("Dodatne značajke") in the feature extraction phase. User can also specify parameters for calculation of a particular feature. For parameters specification, the user must click on the feature to open a dialog box where it is possible to specify parameters for those features that are parametric. Before starting feature extraction, the user must specify signal segmentation parameters ("Prozor primjene").

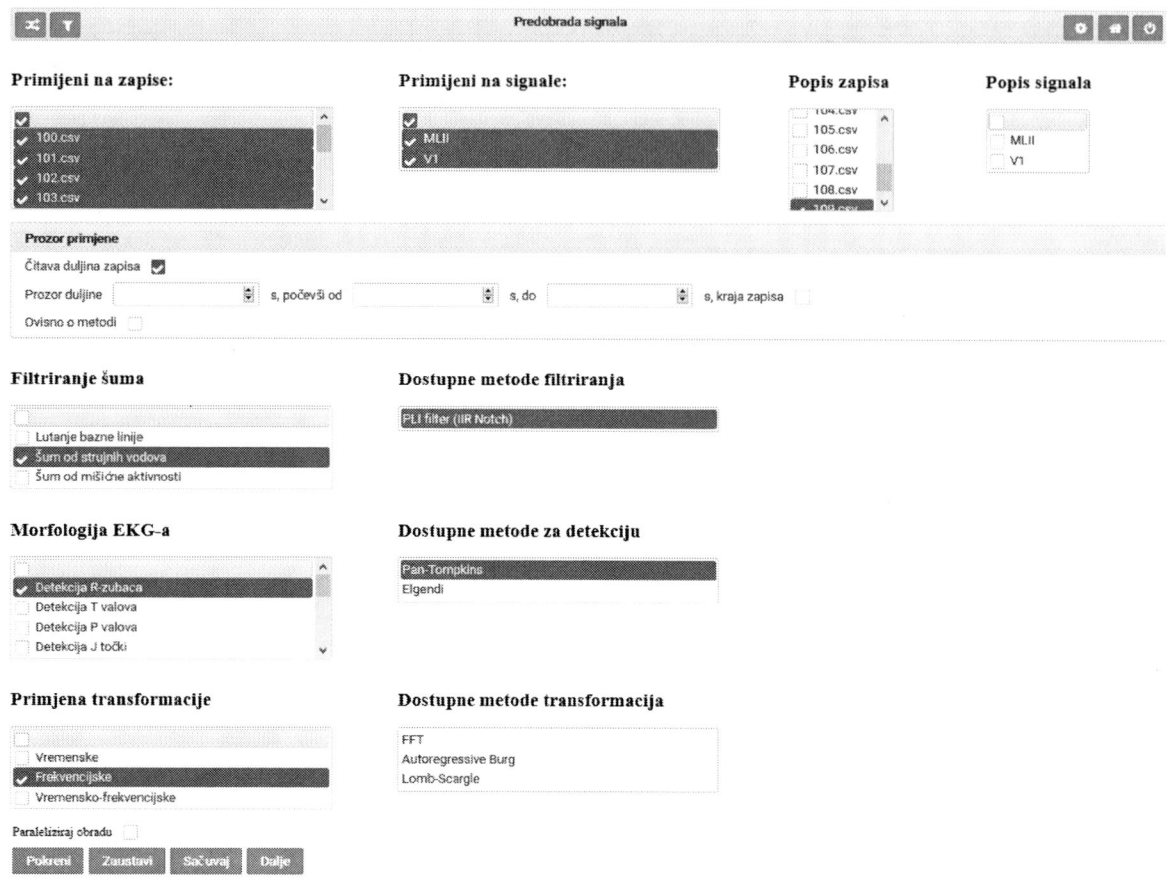

Figure 4.   Preprocessing user interface; the user interface is currently implemented in Croatian language

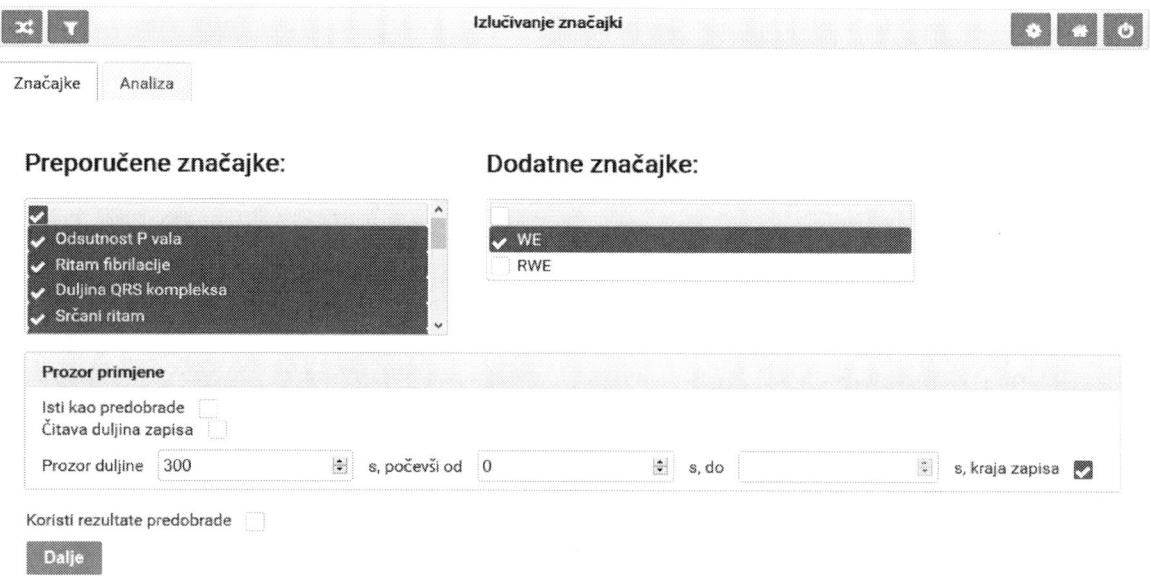

Figure 5.   Feature extraction user interface; the user interface is currently implemented in Croatian language

In the Analysis ("Analiza") tab, the user can specify the type of calculation parallelization that will be performed. After feature extraction, one can save the extracted feature vectors in a file for future analysis, if one wants to have the intermediate results recorded.

V. CONCLUSION

In the paper, we have described an expert feature recommendation system for ECG analysis through predefined analysis scenarios. Initial version of the MULTISAB expert system is implemented in cooperation with cardiologists and includes the detection of AMI, AF, and CHF from ECG signals.

We have also reported the progress on the MULTISAB platform implementation. Currently, we have achieved the functionality of the first six analysis process steps, ending with the possibility of calculated features download. However, it should be noted that the current implementation is still in the test phase, which we will strive to complete in a few months.

ACKNOWLEDGMENTS

We would like to thank Goran Krstačić, MD, PhD, FESC (from the Institute for Cardiovascular Diseases and Rehabilitation) for his valuable assistance with the expert system design.

REFERENCES

[1] J. Oster, J. Behar, R. Colloca, Q. Li, Q. Li, and G. D. Clifford, "Open source Java-based ECG analysis software and Android app for atrial fibrillation screening," in: Computing in Cardiology Conference (CinC) 2013, IEEE, pp. 731–734, 2013.

[2] A. Szczepanski and K. Saeed, "A mobile device system for early warning of ECG anomalies," Sensors, vol. 14, no. 6, pp. 11031–11044, 2014.

[3] A. Jovic, D. Kukolja, K. Jozic, and M. Cifrek, "Use case diagram based scenarios design for a biomedical time-series analysis web platform," in: Proceedings of the MIPRO 2016 Conference, P. Biljanovic, Ed., Rijeka: MIPRO Croatian Society, pp. 326–331, 2016.

[4] K. Friganovic, A. Jovic, K. Jozic, D. Kukolja and M. Cifrek, "MULTISAB project: a web platform based on specialized frameworks for heterogeneous biomedical time series analysis - an architectural overview," in: Proceedings of the CMBEBiH 2017 Conference, A. Badnjevic (Ed.), IFMBE Proceedings vol. 62, Springer Nature Singapore, pp. 9–15, 2017.

[5] K. Thygesen, J. S. Alpert, A. S. Jaffe, et al., "Third universal definition of myocardial infarction," European Heart Journal, vol. 33, pp. 2551–2567, 2012.

[6] E. Burns, "Myocardial ischaemia," Life in the fastlane, http://lifeinthefastlane.com/ecg-library/myocardial-ischaemia/, accessed on: 2017-02-20.

[7] L. G. Gamero, J. Vila, and F. Palacios, "Wavelet transform analysis of heart rate variability during myocardial ischaemia," Medical & Biological Engineering & Computing, vol. 40, no. 1, pp. 72–78, 2002.

[8] Y. Peng and Z. Sun, "Characterization of QT and RR interval series during acute myocardial ischemia by means of recurrence quantification analysis," Medical & Biological Engineering & Computing, vol. 49, no. 1, pp. 25–31, 2011.

[9] T. Murabayashi, B. Fetics, D. Kass, et al., "Beat-to-Beat QT Interval Variability Associated With Acute Myocardial Ischemia," Journal of Electrocardiology, vol. 35, no. 1, pp. 19–25, 2002.

[10] E. S. Jayachandran, K. P. Joseph, and R. U. Acharya, "Analysis of myocardial infarction using discrete wavelet transform," Journal of Medical Systems, vol. 34, no. 6, pp. 985–992, 2010.

[11] L. Sun, Y. Lu, K. Yang, and S. Li, "ECG analysis using multiple instance learning for myocardial infarction detection," IEEE Transactions on Biomedical Engineering, vol. 59, no. 12, pp. 3348–3356, 2012.

[12] P. Kirchhof, S. Benussi, D. Kotecha, et al., "2016 ESC Guidelines for the management of atrial fibrillation developed in collaboration with EACTS," European Heart Journal, vol. 37, pp. 2893–2962, 2016.

[13] B. Olshansky, "The electrocardiogram in atrial fibrillation," UpToDate, http://ultra-medica.net/Uptodate21.6/contents/mobipreview.htm?25/42/26278, accessed on: 2017-02-20.

[14] A. Bollmann, D. Husser, L. Mainardi, et al., "Analysis of surface electrocardiograms in atrial fibrillation: techniques, research, and clinical applications," Europace, vol. 8, no. 11, pp. 911–926, 2006.

[15] D. G. Shin, C. S. Yoo, S. H. Yi, et al., "Prediction of paroxysmal atrial fibrillation using nonlinear analysis of the R-R interval dynamics before the spontaneous onset of atrial fibrillation," Circulation Journal, vol. 70, no. 1, pp. 94–99, 2006.

[16] R. Alcaraz and J. J. Rieta, "A review on sample entropy applications for the non-invasive analysis of atrial fibrillation electrocardiograms," Biomedical Signal Processing and Control, vol. 5, no. 1, pp 1–14, 2010.

[17] R. Sassi, S. Cerutti, F. Lombardi, et al., "Advances in heart rate variability signal analysis: joint position statement by the e-Cardiology ESC Working Group and the European Heart Rhythm Association co-endorsed by the Asia Pacific Heart Rhythm Society," Europace, vol. 17, pp. 1341–1353, 2015.

[18] J. Ródenas, M. García, R. Alcaraz, and J. J. Rieta, "Wavelet entropy automatically detects episodes of atrial fibrillation from single-lead electrocardiograms," Entropy, vol. 17, no. 9, pp. 6179–6199, 2015.

[19] M. García, J. Ródenas, R. Alcaraz, and J. J. Rieta, "Application of the relative wavelet energy to heart rate independent detection of atrial fibrillation," Computer Methods and Programs in Biomedicine, vol. 131, pp. 157–168, 2016.

[20] P. Ponikowski, A. A. Voors, S. D. Anker, et al., "2016 ESC Guidelines for the diagnosis and treatment of acute and chronic heart failure," European Heart Journal, vol. 37, pp. 2129–2200, 2016.

[21] J. Malmivuo and R. Plonsey, "Bioelectromagnetism: principles and applications of bioelectric and biomagnetic fields," chapter 19: "The basis of ECG diagnosis," Oxford University Press, New York, 1995.

[22] DailyCare BioMedical Inc., "Introductory guide to identifying ECG irregularities," http://www.dcbiomed.com/proimages/materials/EKG_Introduction/portableEKG_weitere_Informationen.pdf, accessed on: 2017-02-20.

[23] S. Guzzetti, M. T. La Rovere, G. D. Pinna, et al., "Different spectral components of 24 h heart rate variability are related to different modes of death in chronic heart failure," European Heart Journal, vol. 26, pp. 357–362, 2005.

[24] T. H. Makikallio, H. V. Huikuri, U. Hintze, et al., "Fractal analysis and time- and frequency-domain measures of heart rate variability as predictors of mortality in patients with heart failure," American Journal of Cardiology, vol. 87, issue 2, pp. 178–182, 2001.

[25] R. Maestri, G. D. Pinna, A. Accardo, et al., "Nonlinear indices of heart rate variability in chronic heart failure patients: redundancy and comparative clinical value," Journal of Cardiovascular Electrophysiology, vol. 18, no. 4, pp. 1–9, 2007.

[26] M. T. La Rovere, G. D. Pinna, R. Maestri, et al., "Autonomic markers and cardiovascular and arrhythmic events in heart failure patients: still a place in prognostication? Data from the GISSI-HF trial," European Journal of Heart Failure, vol. 14, issue 12, pp. 1410–1419, 2012.

[27] M. Elgendi, "Fast QRS detection with an optimized knowledge-based method: evaluation on 11 standard ECG databases," PLOS ONE, vol. 8, no. 9: e73557, 2013.

Evaluation of chronic venous insufficiency with PPG prototype instrument

Matej Makovec*, Uroš Aljančič** and Danilo Vrtačnik**

* General Hospital Novo mesto, Šmihelska cesta 1, 8000 Novo mesto, Slovenia,
** University of Ljubljana, Faculty of Electrical Engineering, Laboratory of Microsensor Structures and Electronics - LMSE, Tržaška 25, 1000 Ljubljana, Slovenia

E-mail: matej.makovec@sb-nm.si

Abstract - Photoplethysmography (PPG) is a non-invasive optical technique for measurement of blood volume changes inside an organ or body part. In presented study, a chronic venous insufficiency (CVI) of 25 individuals (38 legs) was measured using PPG prototype instruments. The results of PPG evaluation were compared to results of Doppler ultrasound, which is the gold standard for diagnosis of CVI. The sensitivity and specificity of PPG prototype instrument in amount of 82% and 81%, respectively was calculated.

INTRODUCTION

Plethysmography is the term given to the recording of changes in limb size due to tissue fluid or pooled blood within the veins. This measurement can be undertaken in a variety of ways, fluid displacement, electrical impedance, electronic strain gauge, gravimetric methods and PPG [1]. Historically, the diagnosis of venous insufficiency was performed by invasive ambulatory venous pressure measurements (AVP), which has been described as the ideal diagnostic standard [2,3] or as the 'gold standard'[4]. Early investigations using infra-red radiation to identify fluctuations in dermal blood flow resulted in the introduction of PPG. Hertzman [5] described the method of measuring circulation through the skin using photoelectric plethysmography by relating the blood content of the skin to the amount of light reflected. AVP is a comparatively invasive technique, as it involves

cannulation of the dorsal foot vein. It has been described as painful and cumbersome [6], associated with complications such as bleeding or haematoma formation [7] and unsuitable for repeated use on the same patient or for screening purposes [8]. In contrast, PPG is described as easy to undertake, without risk and user-friendly [9]. PPG has increased in popularity due to the ease and speed of the investigation; this method depends on the absorption of light by haemoglobin in the red cells. Increasingly this was developed to investigate the venous haemodynamics of the lower limb and was renamed light reflection rheography (LRR). A light emitting diode is placed 5-10 cm above the medial malleolus to measure the speed at, which the capillary bed becomes filled with blood following calf muscle exercise. In the normal subject refill time may take between 20 and 45 seconds and a reduction in this refill time identifies degrees of venous insufficiency.

Ineffective venous return from the lower legs leads to a condition of venous hypertension in the superficial venous system [10,11,12], which frequently results in ulceration11. It has been reported that 1% of the adult population suffer from leg ulcers [10], which has placed a large financial burden on the health service [13].

METHOD

Thirtyeight legs in 25 patients with evidence of CVI were studied prospectively by physical examination, PPG and duplex ultrasound scanning in Vascular laboratory of General Hospital Novo mesto during a 12-month period from October 2015 to September 2016. Data were recorded after a single examination. The patients were currently seeking care by a vascular surgeon for worsening of CVI symptoms. They were examined by one senior vascular surgeon.

PPG was performed with our prototype instrument. After an area of skin 5 cm superior to the medial malleolus was cleaned, a small piece of tape was used to secure the transducer to this site. Patients wore loose-fitting clothing to prevent impedance of outflow from the leg, and all patients were barefoot. The leg to be examined was placed in a dependent position over the side of the examination table for a few minutes to allow lower extremity blood volume to stabilize. The patient was then instructed to forcefully dorsiflex the foot ten times in rapid succession and then to relax the foot (Fig.1). The time taken from the beginning of relaxation until reestablishment of the baseline was documented as the venous refilling time (VRT). A refill time of greater than 20 seconds was used as the normal value.

A PPG prototype instrument for CVI was designed, fabricated and characterized together with customized readout electronics. Readout electronics provides photovoltaic operating mode of silicon photodiode, transforms photocurrent to measured voltage and serves as constant current source for LED. After laboratory characterization of each component, fabricated PPG prototype was clinically

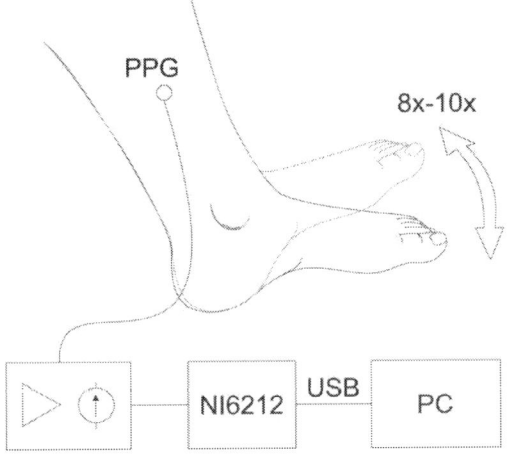

Figure 1: Measurement setup during clinical test

tested on 53 patients. Details of PPG prototype fabrication and characterisation are published elsewhere [14].

Duplex scanning with Doppler colour flow imaging of the leg was performed with a GE scanner equipped. Duplex scanning is considered as "gold standard" for detecting CVI. Patients were placed in a 30-degree reverse Trendelenberg position with the limb externally rotated. Scanning was performed with a linear array transducer equipped with a 5 MHz pulsed-wave Doppler scanning. The saphenofemoral junction and the saphenous vein were insonated and evaluated for the presence of superficial venous reflux (SVR) with the Valsalva maneuver. SVR was defined as pathologic finding for CVI.

The results of PPG refill times were evaluated and compared with the duplex ultrasonography results and were evaluated for sensitivity, specificity and efficiency.

RESULTS

In period between October 2015 and September 2016 38 limbs of 25 patients were prospectively evaluated with PPG and duplex ultrasonography. The group consisted of 16 women and 9 men with age range from 34 to 85 years (a mean of 61,4 years). Clinical examination was performed at initial presentation with the standard criteria.

The analysis results of PPG show the CVI in 18 limbs and missed the disease in 4 limbs. In addition, PPG results show 13 normal limbs, but missed 3 limbs.

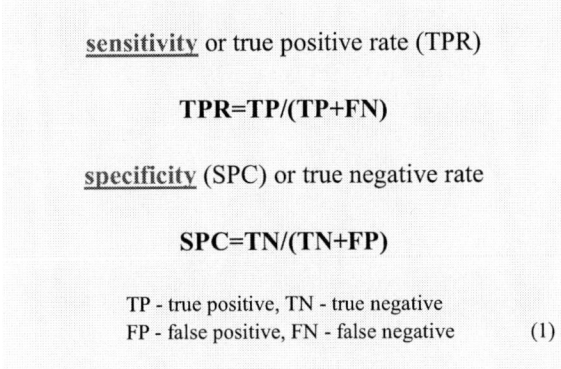

sensitivity or true positive rate (TPR)

$$TPR = TP/(TP+FN)$$

specificity (SPC) or true negative rate

$$SPC = TN/(TN+FP)$$

TP - true positive, TN - true negative
FP - false positive, FN - false negative   (1)

The calculated sensitivity and specificity of the PPG was 82 % and 81 % respectively (1). The efficiency of the PPG method was calculated in amount of 82 %.

DISCUSSION

Venous function of the lower limb has proved to be a difficult concept to quantify. Many tests were developed in an attempt to separate normal from abnormal function, including ambulatory venous pressures, foot volumetry, photoplethysmography, and air plethysmography. The presence of complications of venous disease provides a means for a crude classification of venous disease and is the "best" "gold standard," on the assumption that the presence and severity of clinical disease reflects venous function. In particular, the accepted gold standard of ambulatory venous pressure demonstrates a large overlap in the values take. The VRT as measured by photoplethysmography not only reflects reflux but also efficiency of calf pump function because inadequately emptied veins will need a shorter period of time to refill [4]. In clinical practice, however, we found a large degree of overlap of the results with subsequent poor separation of normal from clinically abnormal limbs thus reducing the test's usefulness. We therefore undertook to improve the PPG trace in an attempt to enable a more accurate analysis. The goal of our study was to evaluate the new PPG prototype instrument for detecting chronic venous insufficiency.

Bays [15] showed low specificity (60%) of PPG refill times and low κ statistic for correlation between SVR and VRT. It was only 0.47. Darvall [16] showed the correlation between abnormal VRT correlated and presence of SVR on duplex (sensitivity 75%). Iafrati [17] showed that PPG provides a sensitivity of approximately 75%, with specificity near 90%. In our study the abnormal VRT correlated well with the presence of SVR on duplex (sensitivity 72%, specificity 81%).

CONCLUSION

Although PPG can provide an assessment of the overall physiologic function of the venous system, it is most useful as a relatively simple and non-invasive measurement to detect the presence of venous reflux. Duplex ultrasonography provides detailed information on segmental reflux, and PPG provides an estimate of the global effect of reflux on the limb. Because of its inability to reliably grade the severity of CVI, PPG has limited utility for assessing the results of corrective venous surgical procedures. Therefore, PPG is a reasonable measure of the presence or absence of CVI that is best used when no further information concerning the venous hemodynamic situation is desired. If information on the severity of CVI or evaluation of improvement after venous surgery is required, a quantitative test will be more useful [18].

REFERENCES

[1] King BM. (2004) A review of the literature on the use of photoplethysmography (PPG) as an assessment tool to identify the presence of venous insufficiency and in screening for DVT. In: King BM, ed. Using photoplethysmography to assess for venous insufficiency and screen for deep vein thrombosis (DVT). Cardiff: Huntleigh Healthcare Limited, 1-7.

[2] Norris, S.C., Beyrau, A., Barnes, R.W.(1983) Quantitative photoplethysmography in chronic venous insufficiency: A new method of non-invasive estimation of ambulatory venous pressures. Surgery, 94: (5) 758-764.

[3] Martino Neumann, H.A., Boersma, I. (1992) Light reflection rheography a non-invasive diagnostic tool for screening for venous disease. Journal Dermatol Surgical Oncology,18: 425-430.

[4] Sarin, S., Shields, D.A., Scurr, J.H., Coleridge Smith, P.D. (1992) Photoplethysmography: A valuable noninvasive tool in the assessment of venous dysfunction? Journal of Vascular Surgery, 16:154-62.

[5] Hertzman, A.B. (1938) The blood supply of various skin areas as estimated by the photoelectric plethysmograph. The American Journal of Physiology, 124: 328-340 cited in: Martino Neumann, H.A. Boersma, I. (1992) Light reflection rheograpy a non-invasive diagnostic tool for screening of venous disease. Journal Dermatol Surgical Oncology, 18: 425-430.

[6] Fdafaf Abramowitz, H.B., Queral, L.A., Flinn, W.R. et al (1979) The use of photoplethysmography in the assessment of venous insufficiency: A comparison to venous pressure measurements. Surgery, 86: (3) 434-440.

[7] Shepard, A.D., Mackey, W.C., O'Donnell, T.F. Correlation of Venous Pressure Measurements with Light Reflection Rheography. Boston, M.A: Tufts New England Medical Centre, Division of Vascular Surgery.

[8] Struckmann, J. (1994) Venous investigations: The current position. Angiology, 45: (6 part 2) 505-509.

[9] Fronek, A. (1995) Photoplethysmography in the diagnosis of venous disease. Dermatol Surg. 21: 64-66.

[10] Callum, M.J., Ruckley, C.V., Dale, J.J., Harper, D.R. (1985) Chronic ulceration of the leg: extent of the problem and provision of care. British Medical Journal, 290: 1855-1856.

[11] Cornwall, J.V., Dore, C.J., Lewis, J.D. (1987) Graduated compression and its relation to venous filling time. British Medical Journal, 293: 1087-1090.

[12] Cullum, N., Roe, B. (1995) Nursing assessment of patients with leg ulcers. Primary Health Care, 5: (5) 22-24.

[13] Bosanquet, N. (1992) Costs of venous ulcers: from maintenance therapy to investment programs. Phlebology, 7: (suppl), 44-46.

[14] Aljančič U, Makovec M, Resnik D, Možek M, Pečar B, Vrtačnik D. (2015) Photoplethysmograph prototype for chronic venous insufficiency. In: Trontelj J (ed.), Topič M (ed.), Sešek A (ed.). Conference proceedings, 51th International Conference on Microelectronics, Devices and Materials and the Workshop on Terahertz and Microwave Systems, September 23 - 25 2015, Bled, Slovenia. Ljubljana: MIDEM - Society for Microelectronics, Electronic Components and Materials, 2015, p. 234-238.

[15] Bays RA, Healy DA, Atnip RG, Neumyer M, Thiele BL. (1994) Validation of air plethysmography, photoplethysmography, and duplex ultrasonography in the evaluation of severe venous stasis. J Vasc Surg, 20(5):721-7.

[16] Darvall KA, Sam RC, Bate GR, Adam DJ, Silverman SH, Bradbury AW. (2010) Photoplethysmographic venous refilling times following ultrasound guided foam sclerotherapy for symptomatic superficial venous reflux: relationship with clinical outcomes. Eur J Vasc Endovasc Surg, 40(2):267-72.

[17] Iafrati MD, Welch H, O'Donnell TF, Belkin M, Umphrey S, McLaughlin R. (1994) Correlation of venous noninvasive tests with the Society for Vascular Surgery/International Society for Cardiovascular Surgery clinical classification of chronic venous insufficiency. J Vasc Surg, 19(6):1001-7.

[18] Lal BK. (2010) Chapter 16, Vascular Laboratory: Venous Physiologic Assessment. In: Cronenwett JL, Johnston KW. Rutherford's Vascular Surgery, 7th edition. Philadelphia: Saunders Elsevier, 257-64.

Highly parallel online bioelectrical signal processing on GPU architecture

Z. Juhasz

University of Pannonia,
Department of Electrical Engineering and Information Systems,
Veszprem, Hungary
juhasz@virt.uni-pannon.hu

Abstract – Signal processing is of central importance in biomedical systems, in which pre-processing steps are unavoidable in order to reduce noise, remove unwanted artefacts, segment time series into smaller epochs, or extract statistical and other descriptive features that can be used in consecutive classification stages. The high sampling rates and electrode counts used e.g. in advanced EEG or body-surface potential mapping ECG systems, result in very large data sets, which require considerable time to process. While long computation times can be tolerated in research settings, in many application areas, e.g. brain computer interfaces, online feature detection (e.g. in epilepsy or heart monitoring) or clinical diagnostics, fast – often real-time – execution speed is required. This paper examines whether and how GPUs can be used efficiently to implement parallel signal processing algorithms. The performance critical factors of GPU programs used for real-time processing are identified then we show that high performance is achievable only if great attention is given to the hardware execution characteristics of the GPU. Overlapped data transfer and computation, optimized use of the GPU memory hierarchy, using enough threads to hide memory transactions are used to obtain better results than sequential CPU implementations.

I.   Introduction

Bioelectrical signal processing focusing on signals heavily contaminated by noise and other artefacts is a computationally intensive task. This is particularly true for modern multi-channel EEG and ECG systems that can use hundreds of electrodes and sample rates above 1 kHz. The processing of these signals requires several steps; after a pre-processing stage (noise filtering, artefact removal), various signal features are extracted and fed into the classification or recognition module. Many software packages and libraries assist users and developers to carry out these tasks with minimum effort. While these packages save precious developer time, the computational performance is typically sub-optimal. This is due to the fact that most of these packages still rely on sequential algorithms.

While long computation times can be tolerated in research settings, in many application areas, e.g. brain computer interfaces, online feature detection (e.g. in epilepsy or heart monitoring) or clinical diagnostics, fast – often real-time – execution speed is required. Traditional, multi-core CPUs provide only a partial solution to this problem as the small number of cores severely limits the achievable speedup and scalability of our algorithms. GPUs, on the other hand, have tremendous 'horse power' with several teraflops of raw computational power and

thousands of processing cores per card, even in embedded variants. It seems natural to use these devices for the demanding EEG and ECG signal processing tasks.

In this paper, a set of representative algorithms (time and frequency domain FIR filters, correlation, power spectrum, wavelet transformation, statistical features) is implemented in parallel and executed in a pipeline fashion first on a multi-core CPU to provide baseline performance figures. Then the performance of GPU version of the algorithms is examined in terms of the above optimization techniques and compared with the CPU results. It is shown that with in-depth profiling and performance analysis, using an incremental development strategy, very efficient single processing algorithm implementations can be created.

The contribution of the paper is the analysis of performance critical factors and a demonstration of their effect on the performance of the algorithms. While the emphasis is on EEG and ECG algorithms, the techniques and results shown can be of benefit for other signal processing application areas as well. The structure of our paper is as follows. Section II describes the traditional implementation of sequential signal processing systems. Section III provides a brief summary of the GPU hardware and computing model. Section IV explores the parallel implementation strategies for signal processing algorithms, investigates the performance limiting factors and shows how these can be overcome to create efficient GPU programs. Section V discusses our results to date. The paper ends with the conclusions.

II.   Bioelectrical signal processing

A typical biomedical signal processing process may use several pre-processing stages, such as various filters (low, high or band pass), artefact removal, baseline correction, sub-sampling, Fourier or Wavelet transformations [1]–[6]. Often, we use high sampling frequencies (up to 2-4 kHz) and many input channels (128-256). There are two distinct classes of the processing strategies; (*i*) the offline method processes signals after the measurement is finished, the input is the data stored in files, (*ii*) the online method, however, processes data during the measurement; continuously arriving data packets are processed until the measurement is finished. Fast processing speed is equally important in both cases but it becomes crucial in online processing if the system cannot keep up with the incoming data arrival rate. Due to this reason, we are focusing on online, stream data processing in this article.

A. Traditional CPU implementation

A typical signal processing pipeline is shown in Fig. 1. The CPU executes pre-processing, feature extraction and classification steps in a succession. The structure of a program implementing this pipeline is show in the code section below.

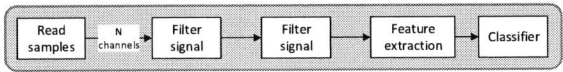

Figure 1. Typical sequential signal processing pipeline including pre-processing, feature extraction and classification stages.

The stages of the pipeline are executed by successive function calls. The body of the loop fetches the next sample and passes it to a series of operations.

```
while (has samples){
    double s = readNextSample();
    double data = filter_A(s);
    data = filter_B(data);
    data = filter_C(data);
    data = filter_C(data);
    ...
    extract signal feature;
    classify signal;
}
```

The sequential implementation running on modern CPUs delivers acceptable performance up to a limited number of pre-processing steps at medium sampling rate and channel count. As shown in Table I, computationally more demanding filters and their combinations can easily lead to execution times that exceed online/real-time processing deadlines.

TABLE I. CPU FILTERING TIME OF 128 CHANNEL × 1 SECOND (SAMPLED AT 2048 HZ) INPUT DATA (Intel Core i7-3820QM 2.7GHz)

	Execution time (msec)	
Pipeline configuration	Sequential C	Sequential Java
1 FIR filter (tap=101)	18.343	22.008
4 FIR filters (tap=101)	71.433	88.182
1 FIR filter (tap=1001)	143.680	177.073
4 FIR filters (tap=1001)	577.304	705.133
8 FIR filters (tap=1001)	1151.971	1413.402

There are several parallelization strategies for the above signal processing algorithm [7]. One option is to execute each task (processing stage) in a separate step (Fig. 2.A). This is the well-known parallel pipeline structure in which the pipeline stages execute the original sequential steps in an overlapped fashion. The next level is to extend the pipeline with parallel stage implementations (Fig. 2.B). This strategy uses data-parallelism to execute filters for individual channels or channel groups in parallel. The advantage of both methods is that the internal processing algorithms (filters, feature extraction, classifier) remain sequential, consequently, existing code and libraries can be used with limited programming effort to create the parallel implementation. Since the number of cores is typically less than 8 in current desktop or mobile systems, the favoured implementation strategy on CPUs for online processing is the parallel pipeline.

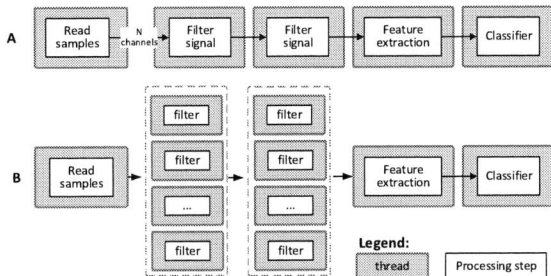

Figure 2. Exploiting different levels of parallelism in the processing pipeline [7]: **(A)** each stage of the pipeline is executed by a separate thread, **(B)** preprocessing is performed in a stream-parallel fashion; each thread filters 1 or N/P channels, where P is the number of CPU cores.

Regardless the chosen parallelisation strategy, the achievable speedup is limited by the number of cores, hence a four- or eight-fold performance is the maximum in CPU-based systems. We do not consider supercomputers having a large number of CPU cores among the potential platforms, since due to their access and usage restrictions, they cannot be used for real-time processing applications.

As a more promising alternative, potentially providing orders of magnitude larger performance, in the remainder of this paper we explore how graphical processing units (GPUs) can be used for signal processing.

III. GPU COMPUTING OVERVIEW

Graphical processors were once dedicated solely to performing graphics tasks. Today, they are high-performance general-purpose processors used in diverse application fields, such as scientific computing, artificial intelligence, automotive industry, health care, to name a few. GPUs contain a large number of simple processing cores and operate in a *SIMT* (Single Instruction Multiple Threads) fashion.

NVIDIA GPUs comprise of multiple Streaming Processors (*SP*s) each having 192/128/64 cores (Kepler/Maxwell/Pascal architecture, respectively). At the time of this writing, the top NVIDIA Titan X card has 3584 cores giving a staggering 10 teraflops compute performance. The *SP*s execute instructions in groups of 32 cores (*warp*s). *Thread scheduler*s select the warps that are ready for execution. Context switching is hardware assisted and can be performed at no cost.

The CUDA programming model executes threads in parallel. Due to the warp execution granularity, 32 threads are executed synchronously in one step. The data-parallel threads are launched by special CUDA functions, called *kernel*s. The number of threads is controlled by the developer. Due to the potentially large number of threads, threads are organised into a hierarchy. At the lowest level, threads are grouped into one-, two- or three-dimensional *block*s. At the next level, blocks can be organised into one-, two- or three-dimensional *grid*s. The dimensionality and the hierarchy should reflect the underlying data structures and/or the level of parallelisation.

The CUDA program launching the kernels runs on the CPU. Before the kernel can execute, the main program should also copy the input data to the GPU device memory over PCI-e bus, and upon completion, copy the result back from the device to the host memory. On the device side, the GPU has a hierarchical memory system comprising the off-chip global memory, the on-chip shared and texture memory as well as the internal registers. Since global memory access is an expensive operation (taking several hundreds of cycles), the number of threads should well exceed the number of cores in order to hide the memory latency by executing instructions from other threads.

IV. GPU IMPLEMENTATION STRATEGIES

This section investigates various strategies for implementing signal processing algorithms on GPUs. The focus of the treatment is on the degree of parallelism in the implementation, on identifying and analysing performance critical factors and iteratively fine-tuning performance with the goal of comparing CPU and GPU execution times and efficiency. Interestingly, the number of published papers on GPU signal processing algorithms is quite small. Most of the papers are covering individual algorithms in the audio and acoustics field [8], [9], [10], [11]. Bioelectrical applications appear in [12].

For simplicity, we will use a 101-tap time-domain FIR filter implementation as a case study for the online filtering of a 128-channel EEG data stream sampled at 2048 Hz. Using a simple back-of-the-envelope calculation, the number of multiply-add instructions required for filtering 1 sec of data is $C = 128 \times 2048 \times 101 = 26{,}476{,}544$. The theoretical performance of the GPU is given as $\mathrm{perf}_{GPU}^{GFlops} = cores \times f_{clock}^{MHz} \times 2$. From these, we can estimate the theoretical peak execution time for various GPUs. Table II lists the peak performance of three NVIDIA GPU cards used in our tests, as well as the predicted 1 second filtering time, $t = C \times 10^{-6} / \mathrm{perf}_{GPU}^{GFlops}$ (msec), and the per sample (128 channel) filtering time (μsec). When compared to the CPU results in Table I, the performance advantage of the GPU is evident. The rest of the paper examines whether this can be achieved in reality.

TABLE II.   GPU PERFORMANCE AND PREDICTED EXECUTION TIME

	Quadro K2000M	GTX 980	Tesla K40
performance (GFlops)	572	4612	4291
predicted 1 sec filter time (msec)	0.0463	0.0057	0.0062
per sample time (μsec)	0.023	0.003	0.003

The sampling interval at 2048 Hz is 488.28 μsec, this is the hard real-time deadline for performing all the signal processing steps for each sample.

A. CUDA program structure

The signal processing program follows the pattern of a generic CUDA kernel loop. The host data, `h_input`, is copied to the device memory, `d_input`, then the kernel is executed using *grid × blocks* threads, and finally the result, `d_output`, is copied back to the host variable, `output`. The loop continues while there are samples to process.

```
while (1) {
    cudaMemcpy(d_input, h_input, input_size);
    kernel<<<grid, block>>>(d_input, d_output);
    cudaMemcpy(output, d_output, output_size);
}
```

In our tests, 128 float values (128 channels, 1 sample per channel) are copied to and from the device in each sample cycle.

B. Naïve parallel implementation

The simplest approach to run the processing steps in parallel is to use a channel-based data-parallel approach. Assuming N input channels, we execute N threads on the GPU. Each thread works on one channel and performs a complete time-domain FIR filter computation step (which can easily be replaced by a moving-average, correlation, feature extraction, FFT, etc. routine). The kernel code for the kernel implementing a sequential convolution is as follows (samples are stored in `buffer`, output is stored in `result`):

```
__global__ void FIRfilter(float* result, const
float* coeffs, float* buffer, int position)
{
    int idx = blockIdx.x;
    float output = 0.0f;
    int index = position;
    for (int j = 0; j < coeff_size; j++) {
        output += coeffs[j] *
                buffer[i*128 + index];
        index--;
        if (index < 0) {
            index = 100;
        }
    }
    result[i] = output;
}
```

The results, as shown in Table III, are disappointing. Kernel execution is several thousand times slower than the predicted time. The execution time also varies with the thread grid configuration. Having multiple blocks result in better performance, since multiple blocks gives more freedom to the thread scheduler.

Executing this kernel 2048 times for the entire 1 second sample window results in **342.643 msec** which is much slower than the original sequential algorithm. The main source of the kernel inefficiency is the limited parallelism. Only 128 threads are launched while the GPUs have 384, 2048 and 2880 cores (K2000M, GTX 980, K40). As Fig. 4 illustrates, low performance is due to the inefficient kernel but also to the data transfer time between the host and the device.

TABLE III.   NAÏVE FIR FILTER EXECUTION TIME

Kernel configuration blocks × threads	Kernel execution time (μsec)	
	K2000M	GTX 1080
128 × 1	70.726	24.277
1 × 128	105.42	29.561

C. Host-device data transfer

Before exploring ways to increase parallelism at the kernel level, let us examine how to reduce data transfer time. Using asynchronous host-device transfer and multiple streams (concurrent kernel execution 'threads'), one can overlap data transfer and kernel execution as shown in the following code skeleton.

```
int stream = 0;
while (1) {
    cudaMemcpyAsync(d_input, h_input,
            input_size, stream % 2);
    kernel<<<grid, block, 0, stream % 2>>>(
            d_input, d_output);
    cudaMemcpyAsync(output, d_output,
            output_size, stream % 2);
    stream++;
}
```

Asynchronous transfer reduces the overall elapsed time to **171.466 msec**. Fig. 5 illustrates the timeline of the overlapped data transfer and kernel execution.

D. Parallel filter kernel

As the next step in the performance optimisation process, the level of parallelism is increased in the FIR filter. The convolution is computed by 101 (filter length) threads in parallel. To avoid shifting data in device memory, a circular buffer is used in order to execute the convolution in-place. In addition, the filter coefficients, the circular buffer content and the temporary output values are stored in fast shared memory. The final summation is performed by the CUDA atomicAdd() function in the shared memory, and finally the first thread in each block writes the output value into the result area.

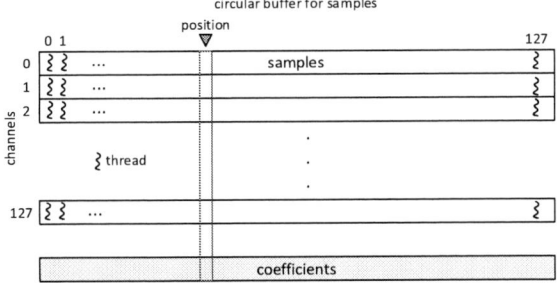

Figure 3 2D data structure and thread configuration for the parallel filter. 128 x 128 threads are launched.

```
__global__ void filter_2(float* result, const
float* coeffs, float* buffer, int position)
{
    int i = blockIdx.x ; // channel index
    int j = threadIdx.x; // sample index
    if (j >= 101)
        return;

    __shared__ float coefficients[128];
    __shared__ float output[128];
    __shared__ float buf[128];
    coefficients[j] = coeffs[j];
    output[i] = 0.0f;
```

```
    buf[j] = buffer[i*128 + j];
    __syncthreads();

    int index = position - j;
    if (index < 0)
        index += 101;
    float value = coefficients[index] *
            buf[j];
    atomicAdd(&output[i], value);
    __syncthreads();
    if (j==0)
        result[i] = output[i];
}
```

Another possibility for performing the final summation is to use a reduction operation on the partial results. The kernel is modified at the atomicAdd() line to first call a warp-level reduce function, than add the per-warp subtotals using an atomicAdd() operation. The performance of the two kernel versions are shown in Table IV. The more efficient reduction-based kernel results in further time reduction; the elapsed 1 second sample processing time is **56.245 msec** on the K2000M GPU.

Warp reduce version

```
__inline__ __device__
float warpReduceSum(float val) {
    for (int offset = warpSize/2; offset > 0;
        offset /= 2)
        val += __shfl_down(val, offset);
    return val;
}

__global__ void filter_2_reduction(float* result,
const float* coeffs, float* buffer, int position)
{
    ... // same as before
    float value = coefficients[index] * buf[j];
    float sum = warpReduceSum(value);
    if (j % 32 == 0)
        atomicAdd(&result[i], sum);
}
```

TABLE IV. EXECUTION TIME OF THE PARALLEL FIR FILTER KERNELS

Kernel configuration blocks × threads	Kernel execution time (μsec)	
	K2000M	GTX 1080
128 × 128 (atomic add)	75.601	24.3
128 × 128 (warp reduction)	9.921	2.86

Execution on the K2000M and GTX 980 GPUs show large variance in execution time. This is the result of the GPU multi-tasking between the display and compute tasks. Consequently, these cards can introduce unwanted delays in the online processing task (see Fig. 6). If hard real-time deadlines must be met, compute-only GPU cards have to be used. Execution on the Tesla K40 GPU verifies this behaviour; filter execution is time deterministic (Fig. 7).

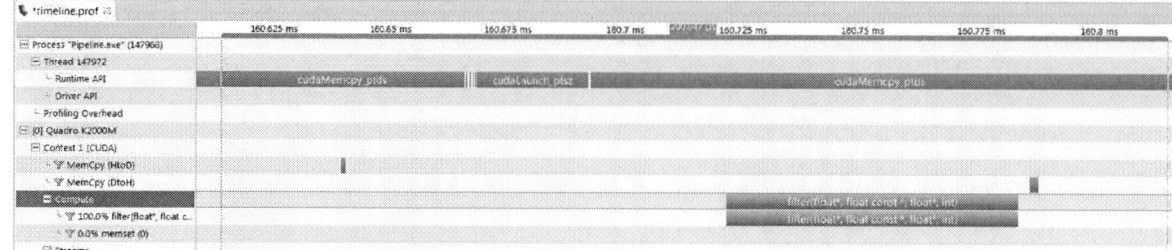

Figure 4. Timeline view of the FIR filter execution in the NVIDIA Visual Profiler. The high cost of the data transfer operations is evident.

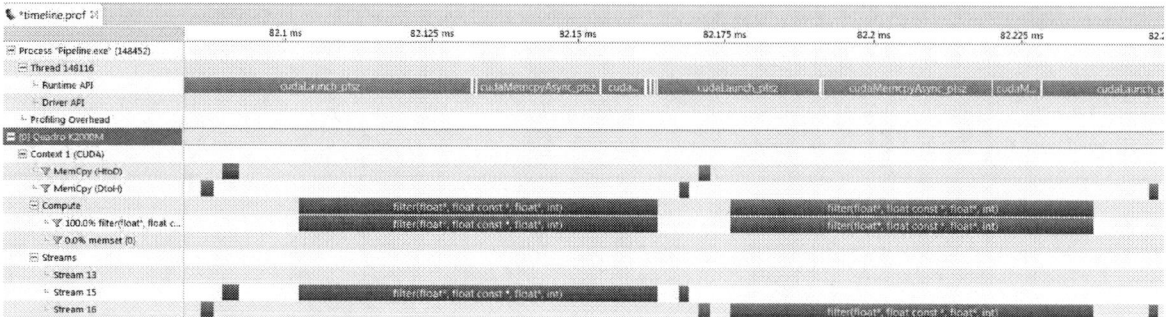

Figure 5. Timeline view of the FIR filter execution using asynchronous copy operations. Note the reduced delay (overlap) between the host-device memory copy and the kernel execution.

The overall performance of the filter operation is still worse than that of the CPU, which is the consequence of a number of factors: under-utilised GPU, synchronisation delays, low compute/data transfer ratio. The performance of small filters cannot be increased further but there are ways to increase the entire system performance by increasing the work to be performed. There are several options for doing this:

- Executing multiple filters within the same kernel
- Executing more intensive kernels (e.g. longer filters)
- Executing several kernels in a sequential order
- Executing several kernels in a pipelined fashion using overlapped streams.

The first option is based on the fact that up to 1024 threads can be used in a block. So far only 1D blocks have been used; 1 FIR filter per block. By using 2D blocks, we can perform multiple filter operations at the same time. The kernel will execute the same convolution, but we can pass different filter coefficient arrays to each row of the 2D thread block. Using 128 threads per filter, up to 8 filters can be specified in the y block direction.

For space considerations, the last three options are described and analysed in combination. Executing several complex kernels (1001 tap FIR filter) in sequence using multiple streams are presented. Fig. 7 illustrates the execution details. Results of the two examples are in Table V and VI.

TABLE V. EXECUTION TIME OF 8- FIR FILTER KERNELS

Kernel execution time (μsec)		Elapsed time (msec)	
K2000M	Tesla K40	K2000M	Tesla K40
86.082	15.686	184.179	39.62

TABLE VI. EXECUTION TIME OF 4×1001-TAP FIR FILTERS IN SEQUENCE USING 8 CONCURRENT MULTIPLE STREAMS

Kernel execution time (μsec)		Elapsed time (msec)	
K2000M	Tesla K40	K2000M	Tesla K40
54.842	17.389	461.766	59.658

V. RESULTS

The results of the various parallelisation options are summarised in Table VII. The sequential C filter execution time is compared to the K40 GPU parallel implementations. As shown, the execution of a single short filter (101 taps) is slower than the sequential version. As discussed above, this is the result of the host-device data transfer overhead and the sub-optimal utilisation of the GPU. As we increase the complexity or size of the operations (longer or more filters) the GPU implementations perform better than the sequential ones. In addition, even in the worst case (8×1001-tap filters), the per-sample GPU processing time is 103 μsec, which is well below the 488.28 μsec hard real-time deadline. Consequently, a set of complex pre-processing operations can be executed on the GPU while leaving time for feature extraction and potentially classification steps too.

TABLE VII. EXECUTION TIME OF 4×1001-TAP FIR FILTERS IN SEQUENCE USING 8 CONCURRENT MULTIPLE STREAMS

Pipeline configuration	Execution time (msec)		Speedup
	Sequential C	Parallel Tesla K40	
1 FIR (tap=101)	18.343	40.672	0.45
8 FIR (tap=101)	71.433	39.62	1.80
4 FIR (tap=1001)	577.304	59.658	9.67
8 FIR (tap=1001)	1151.971	208.933	5.51

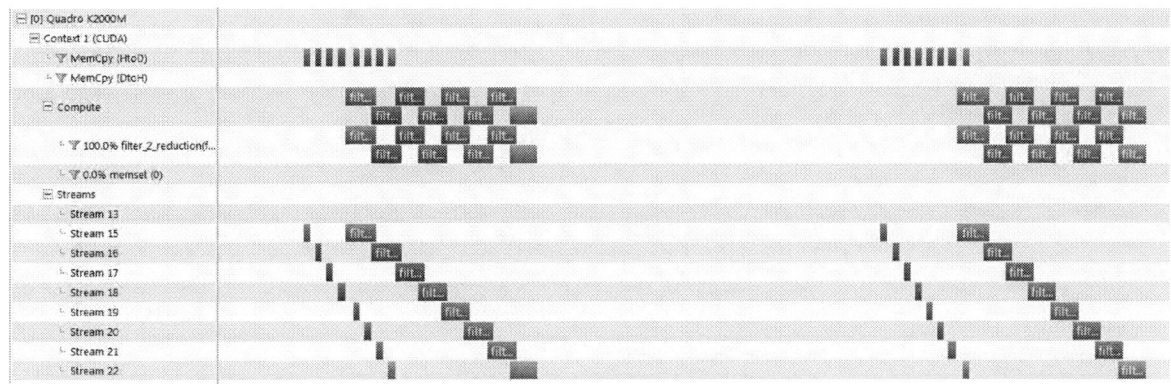

Figure 6 Timeline view of the 1001-tap FIR filters in 8 concurrent streams on the Quadro K2000M card. Note the large gap between the two groups of kernels which is the result of the card performing display task concurrently with computations.

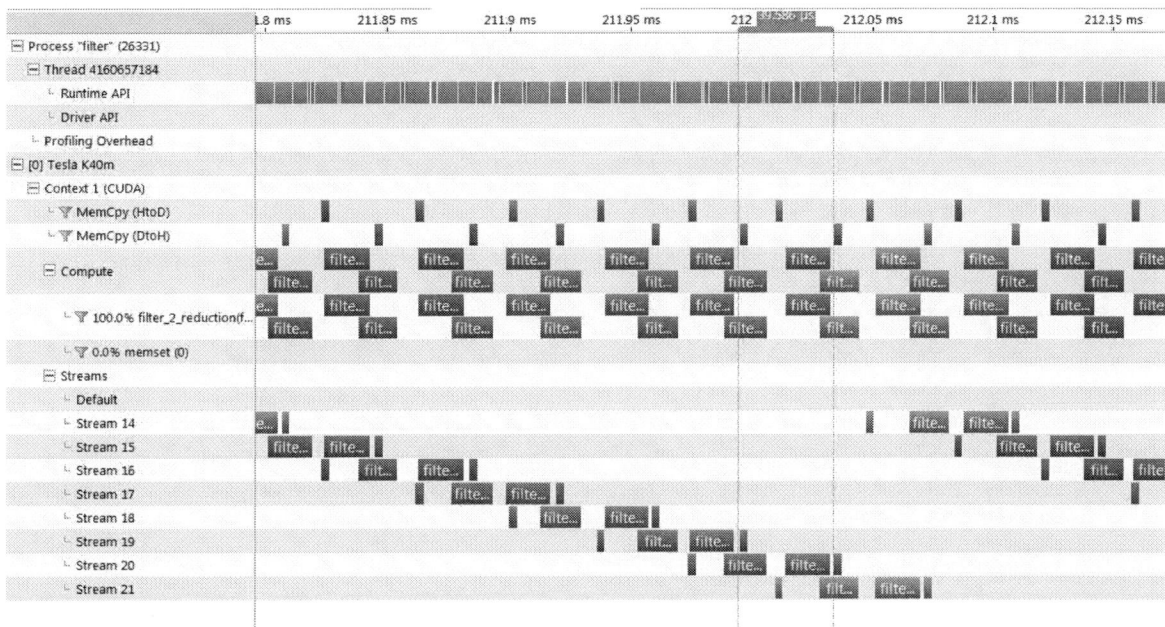

Figure 7 Timeline view of two 1001-tap FIR filters executed in sequence in 8 concurrent streams on the Tesla K40 dedicated accelerator card. Note the deterministic execution of the kernels without context-switch delays.

VI. CONLUSIONS

This paper examined the use of GPUs for bioelectrical signal processing tasks in the context of online, ideally real-time, stream processing of EEG or ECG signals. High-resolution EEG and ECG coupled with high sampling rates present processing difficulties for traditional sequential or multi-core CPU implementations. Using a FIR filter as example, we have examined various algorithm parallelisation strategies and investigated what hardware and software factors might limit the achievable performance. While the raw performance of today's state-of-the-art GPU chips are several orders of magnitude higher than that of CPUs, we have shown that achieving peak performance is difficult in streaming applications.

Various overheads, such as host-device data transfer, insufficient parallelism and/or low utilisation of the GPU resources, frequent memory operations can all degrade device performance. This paper demonstrated, however, that groups of operations can be executed efficiently and performed within hard real-time deadlines.

The results are promising and suggest that a complete signal processing pipeline (from pre-processing to classification) can be execute real-time on GPUs which opens up possibilities for real-time monitoring and diagnostics applications, such as advanced BCI systems, focal epilepsy diagnosis, high-resolution brain mapping and body surface potential mapping cardiac diagnostics. The next development steps in this project is a complete implementation of one end-to-end processing pipeline from the above areas and its evaluation in clinical settings.

ACKNOWLEDGMENT

The support of the Hungarian NIIF Supercomputing Centre is gratefully acknowledged for granting access to their NVIDIA Tesla K40 GPU cards.

REFERENCES

[1] S. Motamedi-Fakhr, M. Moshrefi-Torbati, M. Hill, C. M. Hill, and P. R. White, "Signal processing techniques applied to human sleep EEG signals—A review," *Biomed. Signal Process. Control*, vol. 10, pp. 21–33, Mar. 2014.

[2] A. Ahmadi, R. Jafari, and J. Hart Jr., "Light-weight single trial EEG signal processing algorithms: computational profiling for low power design," *Conf. Proc. ... Annu. Int. Conf. IEEE Eng. Med. Biol. Soc. IEEE Eng. Med. Biol. Soc. Annu. Conf.*, vol. 2011, pp. 4426–4430, 2011.

[3] D. J. McFarland, D. J. Krusienski, and J. R. Wolpaw, "Brain–computer interface signal processing at the Wadsworth Center: mu and sensorimotor beta rhythms," *Prog. Brain Res.*, vol. 159, pp. 411–419, 2006.

[4] A. Roman-Gonzalez, "EEG signal processing for BCI applications," *Adv. Intell. Soft Comput.*, vol. 98, pp. 571–591, 2012.

[5] K.-J. Huang, J.-C. Liao, W.-Y. Shih, C.-W. Feng, J.-C. Chang, C.-C. Chou, and W.-C. Fang, "A real-time processing flow for ICA based EEG acquisition system with eye blink artifact elimination," in *SiPS 2013 Proceedings*, 2013, pp. 237–240.

[6] A. Bashashati, M. Fatourechi, R. K. Ward, and G. E. Birch, "A survey of signal processing algorithms in brain-computer interfaces based on electrical brain signals," *J Neural Eng*, vol. 4,

no. 2, pp. R32-57, 2007.

[7] J. Fischer, T. Milekovic, G. Schneider, and C. Mehring, "Low-latency multi-threaded processing of neuronal signals for brain-computer interfaces.," *Front. Neuroeng.*, vol. 7, no. January, p. 1, 2014.

[8] J. Owens and S. Sengupta, "Assessment of Graphic Processing Units (GPUs) for Department of Defense (DoD) Digital Signal Processing (DSP) Applications," *Dep. Electr. Comput.*, 2005.

[9] F. Trebien, "An Efficient GPU-based Implementation of Recursive Linear Filters and Its Application to Realistic Real-Time Re-Synthesis for Interactive Virtual Worlds," no. August, 2009.

[10] J. A. Belloch, B. Bank, L. Savioja, A. Gonzalez, and V. Välimäki, "Multi-channel IIR filtering of audio signals using a GPU," *ICASSP, IEEE Int. Conf. Acoust. Speech Signal Process. - Proc.*, pp. 6692–6696, 2014.

[11] F. Wefers and J. Berg, "High-performance real-time FIR-filtering using fast convolution on graphics hardware," *Proc. 13th Int. Conf. Digit. Audio Eff.*, pp. 1–8, 2010.

[12] J. A. Wilson and J. C. Williams, "Massively Parallel Signal Processing using the Graphics Processing Unit for Real-Time Brain-Computer Interface Feature Extraction.," *Front. Neuroeng.*, vol. 2, no. July, p. 11, 2009.

[13] F. Ries, "Heterogeneous multicore systems for signal processing," University of Bologna, 2011.

[14] S. Anwar and W. Sung, "Digital Signal Processing Filtering with GPU," vol. 2.

The Dawn of Dew

Dew Computing for Advanced Living Environment

Zorislav Sojat, Karolj Skala

Ruđer Bošković Institute, Zagreb, Croatia

Zorislav.sojat@irb.hr, skala@irb.hr

Abstract – Our present technological civilisation always aims towards new and new goals, and has a strong urge to integrate all what is available in a huge interrelated and as much as possible autonomous system. Thousands and thousands of people try to imagine a problem and get a solution to it, hoping that in our common effort some of their work will show to be beneficial. Entering the Age of Information after a, historically, extremely long period of materialism (not only the last few centuries), and the many ages of not understanding the meaning, or even the existence, of information itself, we are, naturally, inspired, intrigued, confused, elated and stunned by the wast area of possibilities, the enormous applicability and a vision of a future world which would be perfect for a human. Or would it? In this work we will primarily explore the application of Dew Computing paradigm in integrating the lowest level, physical-edge human environment-controlling devices into higher level behavioural systems. An optimistic vision of the future is given in a form of a short "science-future" story.

INTRODUCTION

In this everyday effort to think out something we would like to have, and integrate the new and new application ideas into the explosively expanding amount of tiny specks of dotted silicon, which just a few decennia ago we realised can be organised in such a way that our wildest dreams of what a machine can do never before could even faintly imagine, we need to give Names to what we do. Our need to create has found a very new and a very unexplored way of realisation in its own world somewhere between the imagination and the material reality. And we try, somehow similar to what Adam did, to give each of the new ideas, things, paradigms, principles, algorithms individuality, by whose name we could recognise about what we are talking. So it is natural that, as we see more and more opportunities, by gaining more knowledge, and sometimes even insight, new and new approaches, and therefore new names emerge. Presently, in the area of Computer Science, we envision a new world of little machines all over the planet, in every home, car, street, road, pocket, on hands and as pendulums. As monitors, assistants, caretakers... (don't we trust ourselves a little too much that we would really succeed to make all that for the betterment of human life?). A lot of different terms are mixed up, and often yet we did not have enough time to settle down what we actually mean when we say some of the information-processing related words.

We explore the application of Dew-Computing paradigm in integrating the lowest level, physical-edge human environment-controlling devices into higher level behavioural systems. However, attention of the reader is driven to the scientific responsibility we all have, as not only the How, but very much also the Why is here also an essential question, although often we answer this question in an almost derogatory way. Are we certain Where we will End up as free human beings?

Unfortunately we are all aware of the undeniable fact that even the most beneficiary inventions often get used for "squeezing" the individuals into unwanted/unwaranted lower or even low quality living conditions of any kind. Technologies often rush, and the society lags behind, this discrepancy frequently leading to the misuse of technology. As scientists aiming towards promoting the quality of the human life-experience and the personal and social freedoms essential for the health (mental and physical) of each and every Human being, we are aware of the fact that sometimes it is important to imagine a future world as a living environment, trying to "empathise" with the vision, and than, in a kind of "retrograde" analysis, try to find at least some important elements which shall or shall not be part of an emerging new Technological Paradigm, as the Dew-Computing is [1], [7]. Specifically the Dew-Computing Paradigm grabs deep into the immediate physical environment of humans, and could be consequently dangerous if misused.

Therefore already at the level of Architectural design it is essential to analyse what precautions have to be made, and that way many stray paths may be escaped. Many pessimistic artistic visions of the same type of technologies (global hierarchical integrated information and environment supervision/control systems) can be found in much written and visual artwork and in widespread popular fears. To counterbalance the negative, one possible optimistic artistic vision of the future using the same type of technologies is presented here. It is said that often a "Picture is worth more than thousands of Words", here we offer the interested reader exactly the inverse: "Thousands of Words are worth more than a Picture" - a fictional Science-Future Story[1] to present a Vision of the Future towards which the present developments aim, and also to generally picture a Vision of what the Dew-Computing Paradigm may mean in everyday human life. Therefore here we will first use Artistic and not Scientific means of expression. So, we hope, an 'arty' "story from the future" can picture the Larger Vision of the technological Aims much better than heaps of 'sciency' diagrams explaining present and future technologies. Or: let the 'story written by somebody in the

---

[1] Though the introductory text "A Blog Post From the (Near) Future" is completely unscientific Fiction (at least until we invent a Time-Machine).

future' talk about What-Could-Be, and the 'technological explanations' talk about How-To-Attain it. On the other hand, the following story can also be viewed as a statement of a humanist/artist on the 'hows and whys' of his possible feeling well in the future living environment of our emerging fully globalised civilisation, and poetically also as a statement of warning not to forget the basic human needs - freedom, knowledge, art, decision making, science, reading, enjoying company, starring at stars...

A VISION OF FUTURE

A Blog Post From the (Near) Future

Recently found on Internet while browsing through the not so far future, the here presented short story is an unedited copy of the post which will be published by the Known Web-Site, specially popular for regularly publishing best excerpts from the Known Artist's blog "My Life"; if that will ever happen.

"On a Cold Winter Day", from the blog "My Life" by a Known Artist
(please note: the views expressed are author's own, not necessarily also the publisher's)

The winter is really cold. The clouds are low and the fog penetrates everywhere. In the morning mist even the dew comes in the form of frost. The wind gusts from time to time rattling around the solar panels on the frozen roof. Keeping the rhythm of the icy blasts nearby windmills are at moment frantically trembling and flapping their propeller-shaped hands, and at moment standing as still as high metal beams of their construction generally should. On the Home-monitor I saw that not much of the power generation is done by solar and wind sources nearby, and our own solar panels hardly see any light. Though, my Home did add a short statement that the wave-power generators are working full capacity on the other side of Europe. So, with no interesting news, I left the Home-Environment system to care about energy on its own.

I've just finished a synesthetic conversation with a friend of mine living on another continent. He is a renown cooking artist and allowed me to smell some of the ingredients he prepared for the "Taste of Art" cake printing competition tomorrow at the Baking Art Exhibition in my Town. He promised later to share some of his own renowned BOP ("baking-oven printer", for those unfamiliar with cooking terminology) recipes, designs and knowledge with my wife and our oven (I, however, personally had never interest in cooking, and we do not own a food printer, so the form artistry will be only my wife's). Anyway, I did not see him for some time, and promised to pick him up at arrival to the airport, and to have a meal with him before his exhibition. I am quite certain there will be no traffic problems tomorrow for us, as my Home already informed the Traffic system of our intended destinations and time schedules. And, naturally, informed our Car to collect the necessary electric energy, if possible.

In the morning we were woken up by tender music and waking light, and my Coffee-machine had the coffee ready just as we came down to the living room, as it was informed by our Home that we will wake up at the time just fine for me and my wife to be ready in time. For years my wife used to cook the coffee for me, but that is a real sacrifice. So now my coffee-machine does that for her.

Our Home informed me that during the night the outside temperature was extremely low, and that our Car asked for additional heating in the garage, as to best preserve the battery power and enable easy morning start-up. Though, when I came into the garage, it was exactly the same temperature as any time I plan to use the car. I long ago asked the Home to preheat or precool the garage when I plan to use it.

As many know, I am quite keen on the science-of-the-whole, the ecology, and I like to follow all kinds of short and long term processes around me and in the Global Environment. Knowing this, our Home presented me with the Area Environment News statement on the power-distribution situation. It said that although most of the local power production is down to meagre outputs, the general system has enough power, and the sudden lowering of night temperatures, leading consequently to a much higher demand on energents, did not cause any disturbing consumption peak, due to the self-organisation and consumption-time-spreading of the individual consumption elements in our Town and our Homes. I attentively (or perhaps meditatively?) looked at the different presentation diagrams, and played a little with the information by processing and cross-referencing it, while my wife took a bath.

As I supposed, most of the traffic lights on the route to the airport and back were open to our Car. Though, I must admit, the traffic was quite congested, as we had to wait two times on a semaphore coming back from the airport to the town centre. But, then, it was not over-congested; otherwise the Route we took would suggest another instead of itself. Approaching the town our Car informed me about several free parking slots in the mesh of little streets around the restaurant we wanted to go to. I love that restaurant (I did not say I do not like food, I do not like cooking myself!), because it did not change in atmosphere from when I was a child. So we went there, turned off all of our mobile communicators, sat at a nice and cosy table and ordered food prepared by hand and cooked on real fire. In friendly communication the hours quickly passed. I paid with cash, as I always do. We, in our Homeland, call it "Solar" payment or "take the Cash out into the Sunlight!" payment. The best way for everyday financial transactions, you always know how much you have :). And surprisingly, the food in that restaurant is quite cheap, not much more expensive than good ingredients for an oven printed meal would be (well, at least if the food design is not by some known cook, they charge sometimes exorbitantly for their food printer forms and even more if you order their ingredients! - but, as I mentioned, we actually do not own/need/like? the food printer). Anyway, after the marvelous meal we parted ways with my friend...

Though I was looking, as so often, to see the long-necked guy with the strange and funny hat somewhere in the town, as many of you, lovers of Queneau's Exercises in Style do, I have to inform "all interested parties" that I still never saw him, and have no idea what happened to the button. I will continue waiting on him, although he may be Godot. (If you have no clue what I am talking about, just forget what I just wrote about the overcoat button!)

Well, in the evening my wife and I participated in a multi-locational Parallel Art Performance (the ever more popular PAPs!) - and there was even a projection of Beckett walking through Berlin. As the performance started I felt a jolt of remembrance, as I forgot to turn on my Communicator. But I am obviously old-fashioned (I will not admit old), as it, naturally, turned itself on as soon as the time of the performance was nearing! :).

The PAP was very successful. There was quite a huge amount of public from all around the world, and many of them actively enhanced the artistic value of the common experience.

Late at night, returning home, the Town was empty. The street lights and semaphores were turned off, except on our Route and in the few rare side Streets were somebody was walking. I adore this time of night in the town, when nobody is around, and I will be granted the wish to turn the street lights off. Suddenly you see all the beauty of the stars above. When I was young a town-dweller would never see more than sometimes a glimpse of the moon or the few close planets. Believe it or not, in that time the whole town would be lighted as if somebody was walking on every street, and driving on every avenue!

Well, this was such a starry night, it was freezing but the skies were clear, and the stars must have shined intensely. So we decided to stop in a park-forest, and informed our favourite park-Bench to preheat itself as we were approaching. In the park we asked the Lights go off, and as nobody else was there, they did. Oh, the view of the endlessness of the Universe above...

Lost in the depths of Space, suddenly we were "woken" by the path lights going on. We saw a young couple hurrying on the path, trembling and cuddling in the breezing cold. The sky-magic broken, we went Home.

From the clear sky a few of the first snowflakes came down. And what a beautiful surprise when we came Home! On the Replicator (or, if you prefer, our multi-material 3D-plotter/printer) we found a beautiful sculpture hand made and hand-scanned by our dear friend from afar (he still prefers hand scanning!). He sent it to our Home as his gift for my wife's and my round Meeting-Aniversary (No, I will not admit the count. Period.). For that occasion our Home, with the expert help of our far away friend, made a special lighting and sound arrangement of his sculpture which emerged from the Replicator... No smell though. He is still an old style sculptor, although he was among the very first to use bright and abstract colours on sculptures times ago.

And while the late night fire in the fireplace made delicate wood-burning sounds, the fire shadows played a lively dance and the smoky smell evoked the atmosphere of mountain dwellings, the windows slowly, slowly got embroidered with tiny exquisite crystals making intricate weaving, never same, never repeating, never ending...

Life is Beautiful, however strongly we try not to see it! And just finally, I sincerely wish that all the dreams and wishes of each and everyone of you, my dear Readers, come true, but do, please, always be very careful and mindful of what you really wish and want, because not all dreams and wishes result in greater good to some or even all of us.

THE CLOUDS AND THE FOG

Naming new scientific ideas often evokes associative thinking, and recently we got several new, quite romantic, terms. Firstly, how did we get Clouds in computing? Well, it has been shown that both from the aspect of computer science and economy the idea that for an individual or collective user it is actually completely irrelevant where their processing is done and where their data is actually kept, as long as they can access their "virtual machine" or their data as if they where locally – is very viable, enabling individual and collective work, collaboration, communication and entertainment independent of the location of the users and the machines. Many companies, scientific and other institutions and individuals very quickly accepted this approach. So we got a lot of Clouds.

It was not hard to imagine that the Cloud will soon have to lower itself into the Fog. The Clouds are, obviously, very far away from the Edge. Through the massive increase in processing and storage capacities, as well as human interface developments of the edge data gatherers, processors and communicators, diverse "intelligent" devices like telephones, tablets, palmtops, laptops, and, naturally, desktops, integrated homes, individual health-monitoring equipment, "infotainment" cars, including those cars which slowly start to learn how to self-drive, i.e. those devices which allow "Edge computing", the user needs and wishes for the usage of Clouds increased drastically. However, here new problems emerged. We got an "exponentially" growing amount of data in all forms and of all derivations and provenances which we would like to be communicated, processed, collected, shown, safely stored, indexed, modeled, the myriad of information we would like to learn, understand and not forget, and then also a plentiful of social, scientific, health-related, security-concerned, artistic, entertainment-oriented, gambling-fascinated, financial and other intentions, applications and wishes which have to be satisfied.

In this "mess" of all kinds of interconnections and data-flows it becomes obvious that the Cloud is too far away from the user-edge, so we have problems of latency and time-lag, of connectivity persistence and availability, Quality of Service (QoS), data amounts, compatibility and enormous heterogeneity of individual devices. Heterogeneity from the level of the processor type, memory and storage size, processing and communication abilities, complexity of the input/output system, human interfaces (and "Augmented reality" is strongly going into mainstream already), up to the different operating systems and programming languages. It was really not hard to imagine that the Cloud soon lowered itself and touched the Edge by Fog. The "lowering" of Cloud into Fog is a serious theoretical and practical problem, and a lot of computer scientists are tackling it from different aspects, as to find out the principles on which such a hierarchical connection can be seamlessly attained. The mentioned vast spread and heterogeneity of different Edge devices, including the big data quantities which they may generate could enormously change the human environment in a large area of possible applications. However, the transfer of big data quantities into scattered centralised processing centers, which are the foundation of present-day Cloud-computing, will not be sustainable in the (near) future, due to bandwidth, latency etc. The programming models which presuppose that the "intelligence" and "processing/storage capacity" necessary for data processing lays in the processing center of some Cloud will not be adapted to the new necessities of emerging applications. The Fog-Computing paradigm has to solve these problems by introduction of scalability, elasticity, quality of service control, services migration, understanding the Value of Information, extended security and privacy etc., as well as by a different paradigmatic approach to programming, programming models and application development. By decomposing higher level processing functions (up to the level of wide applications), into "micro-services", and by their allocation/reallocation and selective engagement based on elasticity and

349

scalability, and by differentiating the control of network and security related aspects from the control of user data and application related aspects, it is possible to perform a kind of "osmotic" flow of micro-services (which themselves together make one or more macro-services, by that being the foundation for user-level applications) by their penetration horizontally and vertically inside the Fog and Cloud. This recently suggested "Osmotic computing" approach takes as the basic idea the chemical osmosis, by which molecules from the solution of higher concentration flow into the solution of lower concentration, finally balancing the "load". Therefore the "molecules" of the Fog (i.e. the micro-services) pass from the more congested area to the less congested area, maintaining an active balance, where "congestion" is defined in the terms of Quality of Service, and those are primarily security, throughput, availability, reputation, privacy, price etc. The introduction of the principles of self-organisation, multi-criteria decision-making, filtering, multi-criteria optimisations, Value of Information etc. will enable the redistribution of those "processing-molecules" according to the abovementioned criteria.

The Dawn of Dew

The Computer-science notion of Dew emerged as it became obvious that there is an extreme amount of very important apparata and devices which functionally necessitate self-sufficiency, but which could globally be even more effective if they would cooperate. This is the reason why in Dew-Computing there are two basic notions, which do not exist in the rest of the hierarchy, in the Fog and in the Cloud: *self-sufficiency* and *co-operation*.

The biggest amount of tiny processing elements, micro-controllers, is employed on this lowest level, in direct contact with the human physical environment, controlling a multitude of elementary or just slightly complicated environmental processes. These are the microcontrollers in gas boilers, air-conditioning and heating equipment, in traffic lights, home entertainment systems, car motors, light and lighting controllers, solar power controllers etc. These are devices for which it is completely obvious that they must posses self-sufficiency and that they may not receive any kind of "order" from anybody/anything, except from the rightful human owner (as with private "things") or responsible person (as with public services). Due to their boundedness by human perception of their proper functioning, these devices may accept only "suggestions" from their neighbors or higher hierarchical levels, but even those "suggestions", as we have seen in the "science-future" story told, can enable, by cooperation with devices of the same or higher level, fascinating possibilities of energy saving, transport and energy load balancing, traffic congestion and pollution reduction, faster and more effective emergency service reaction, health-monitoring, artistic expression and romantic outings. Further, by higher-level processing many scientific investigations and predictive model developments will be possible, specifically in cooperation with the modern-day developments in the areas of "Artificial intelligence" and "Cognitive computing".

This will be established through horizontal self-organising coordination of individual Dew-droplets (i.e. the dew

devices with communication / self-organisation possibilities) and vertical Fog/Cloud integration based on the newest developments of the Fog-Computing paradigm. The collaboration of Dew-droplets shall primarily be established on the level of their information exchange and self-organising coordination by (intra-) networking them, using proper ontologies/protocols and security/privacy and QoS based channels. Such self-coordinating and collaborating systems of individual Dew devices will then, as microservices, be seamlessly integrated higher up into the computing hierarchy through Internet gateways into the Fog and Cloud, enabling the coordination of the physical edge control through common strategies and multi-level intelligent behaviour, including adaptive-learning, predicting and other aspects of cognitive computing, with multi-modal user interfaces to facilitate appropriate Human/Computing interaction.

The basic Dew-Computing element ("molecule") is the Dew-droplet, which consists of a self-organising co-operational communication layer, an ontological "interpreter" and individual physical Dew-devices, which contain all necessary sensors, effectors and needed algorithms to be able to independently perform their job, as well as a physical communication layer. This does not meen that the Dew-devices must be able to execute the "software" of a Dew-droplet. In other words, individual physical microcontrollers may or may not produce information (i.e. ontologically contextfull messages), but minimally have to have the ability to receive "suggestions" and transmit data. It is naturally obvious that during next several years, while the networking of "things" is slowly approaching these devices, there will be no industrial production of standardised, or at least easily obtainable, microcontrollers which will be easy to directly incorporate into such collaborative structures, and that specific application solutions will have to be developed to physically and informationally (through the use of specific programmable interfaces, additional "piggy-back" microcontrollers) connect Dew-devices through Dew-droplets in the Dew-Computing Ecosystem.

By these approaches the Dew-Computing paradigm will, by including the physical-edge self-sufficient and independent devices into a collaborative structure, lead towards the full development of a Distributed Information-Services Ecosystem.

The Rainbow

As already mentioned, names and term in sciences can sometimes be quite romantic. And so it is quite logical that when integrating the Dew, the Fog and the Cloud we have to get a Rainbow. Though in this article the existing, well established and very much used Cluster and Grid technologies were not specifically discussed, we shall not forget that the whole of the present and (at least near) future processing environment will in many areas of human endeavor be heavily dependent on those.

As we could see from the above text, in each of the hierarchical areas from Clouds down to Dew we have specific problems still to solve. Though the Cloud is presently well the most developed in technological sense and most penetrating in the social sense, by introduction of the Fog Computing elasticity, scalability and heterogeneity and Dew Computing independence and cooperation, it will be unavoidable to introduce an "all-penetrating" paradigm/architecture to seamlessly integrate the vast spectrum of prerequisites, functional principles, programming models... of the full computer-driven and human-controlled processing "haystack".

Such an all-penetrating "force" through Dew, Fog and Clouds will give a Rainbow. The basic problems Rainbow-Computing will have to solve are the unification of all those approaches into a consistent whole, based on new principles of both human and computer understandable communication (which can be attained by a standardised common ontology, which must be adapted both to human and computer understandability), based on new multimedial and intermedial approaches towards user interfaces, and also on significant ergonomical changes. It is obvious that many present-day and future Cloud, Fog and/or Dew applications will perform their duties without the inclusion into the Rainbow. However, the needs and wishes of the development of a Global Information-Processing Environment will significantly enhance the usability of those applications and slowly introduce them as selfstanding micro- or macro-services by introducing an interface layer. One of the most important conceptual (or better to say paradigmatic) changes which we will have to do to enable further development is (re-)introducing the notion of information and devaluating the notion of data. Though the words like Informatics, Systemics, Cybernetics in all kinds of contexts and derivatives get associated to Computer Science, the word Information is much more used in news-language than in Computeristics. Naturally the media uses this word, as their main objective is actually the transfer of Information to the population, humans, and the term "information" we can, folk-like, reasonably well define as "something new", as otherwise it would be either redundant (when it is "nothing new"), or it would be not recognised/understood, therefore not being inside the word-meaning of "information". What does it mean? It means that information has to be in full sense a sign, i.e. it has to have both the "expression" and "content", or to say it in deSaussure's words, the "significator" and the "significant". For an information to be received, be it redundant or not, it is necessary that it can be "incorporated" into its appropriate subsystem of the understanding system, that is that it "comes" from the same context which is recognisable and known to the receiver. The internal structure of the sequence of individual expression-fractions (e.g. impulsemes, codemes, phonemes, morphemes, words...) must follow rules recognisable by the receiver, and individual "contents" (i.e. ontological links in the structure of the context entirety) must in a great part be identical to the ontological links in the appropriate sub-structure which the receiver recognises as the reception-context. "Information" is actually the difference of its full structural position inside a context of transmission in relation to the context inside which it is received / perceived. In other words, information is the non-isomorphism of two sub-structures which contextually concord with the same supra-structure.

It is obvious that without Context there is no Understanding. But the notion of context itself is in computer sciences so much neglected that even almost all "programming languages" strictly adhere to the rules of context-less grammar, and then sometimes some contextness is "glued on", according to needs and the capabilities of some language, in a myriad of ways, without a general system or common rules. Programming languages which remember the context of individual strings of expressions are almost non-existing. Those few which allow contextfull programming/processing, so important for the future relationship of the human and the machine and the productivity of this relationship, can be put into the category of "High Productivity Computing", and will be essential in further development of Rainbow-Computing.

Therefore it is not strange in present day that there is talk only about data, big data, data-flows, data-throughput, data-storage, data-processing... But we shall be deeply conscious of the fact that a "datum" by itself really has no meaning at all, as it is only the "expression", only the "significator", without any content, anything which would be actually "expressed". We become painfully aware of that when the data on the data on our disk get intermixed. Not only there are no file names any more, but, worse, due to disk fragmentation almost all operating systems do, we do not know even which recorded segment comes after which one, or is it a segment of music, video, picture or text. No wonder restoration of data is so very expensive – there is no context whatsoever about individual data segments.

Almost as an excuse the so-called meta-data, data on data, are used. But the meta-data are coming from a completely separated and from the data which they "describe" independent context – the context in which we thought them out to help us or the machines in their classification, categorisation and searching. Meta-data are not the "content", the "significant"!

By the introduction of an enormous quantity of all kinds of equipment, what we like to call "things", and by introducing those devices which only Dew-Computing can "absorb", the notion of Information will necessarily have to take its merited place in the notion-hierarchy of the science which would like to make the "Internet of Things" and the "Internet of Everything". The only way to convert data and meta-data into information is by progressive introduction of a consistent and scalable, expandable ontology of the Global Information-Processing Environment. Ontology, the science of the nature of being, becoming, the existence, in the context of Information sciences develops a model, a structure which represents / makes notions inside some area through interrelationships of those notions, defining by that a contextual supra-system and for the Signs all necessary sub-systems of "contents", or "significants". This way it enables the use of information and not data, in the sense explained earlier. Thus such ontology enables comprehension inside a given area, in fact defining what is existing in that information system.

Another important area of future development is the algorithmics. The prevailing frame-of-mind in programming, a heritage of the very first days of calculating and computing, is the serial processing approach to algorithmic expression, and therefore also to our programming methods and languages. As much as we try for a long time, this frame-of-mind – to regard the world we model as a series of individual "dots" which we then process one after the other – is still the prevailing one. Only recently the penetration of graphical processors into mainstream computing did shift this mental frame somewhat towards the mindset of parallel algorithmics.

The Rainbow will have to, both as an architecture and as an integrated system-approach, systematically solve the problem of the necessity of a new approach towards parallel, multi-computer, multi-device, multi-micro-/macro-service programming through the development of a novel paradigmatic approach towards "programming" languages, by transforming them towards active computer linguistic systems, which are contextfull and process information and not data (therefore knowing what can be done with them and where and in which form they exist). Only this will allow seamless and sustainable inclusion of multiprocessor and multicomputer Clusters and Grids, huge Clouds, the Fog and the Dew to form an unified "programmable" and "controllable" Global Information-Processing System, hopefully for the betterment and not detriment of our life.

Do We Know Where We Go?

Our age is full of buzzwords. And unfortunately even many of those words which in their naming intention have very important scientific insights and relevance often fall, by overuse and frequent not-know-really-use, into the realm of buzzwords. Is it not true that the "Cloud" became recently one of them? "Internet" certainly has (with much popular misunderstanding of the term). So has "Web". So, as many other fields, the field we are writing about also contains a vast amount of Names which do describe it from different perspectives, but too often, unfortunately, from the marketing aspect, or just because they sound trendy and fashionable. Phrases like "Internet of Things", "Web of Things", "Cyber physical Systems", "Virtual Machines", "Containers", than "Platform as a Service", "Infrastructure as a Service", "Software as a Service", then different "storages", "servers", "applications", "interfaces", not to mention the acronyms (IoT, WoT, VM, PaaS, IaaS, SaaS, API, SOE, QoS, AES...), even up to the now popular "Internet of Everything" (IoE), actually do, by their over-usage and buzzwordiness, often obscure their real meaning and relevance even to the scientists themselves. And they certainly easily obscure their actual denotational or associative meaning. However, as Confucius would say: "If words do not mean what they really mean, nothing can be established." We have to be aware of what words really mean, or what the connotations or associations they awaken want to say, as opposed to what we think they mean or they ought to mean. Much confusion in terming comes out of terms (names) which in their name have an over-wide object of naming. As an example, just think for a moment of the basic denotation of some of those Names. What are really Cyberphysical systems? I am. You are. The World is! What Things make the "Internet of Things" and what is the Everything we want to include into the "Internet of Everything"? The glass? A coffee-bean? The Moon? Thoughts? An oak? Intentions? Each and every strawberry? Our emotions? Like in a (hopefully never future) remark: "How come you still did not connect your feelings with the Internet of Everything? You must do that, you know it's the Law!".

So than, please, let us not forget the "Three laws of responsible robotics" by Isaac Asimov, and a vast area of ethic, moral, philosophical, anthropological, sociological and other questions which the coming introduction of "Cognitive computing" through the recent high rate of development in the field of so called "Artificial intelligence" and the "Ubiquitous computing tendency" have to raise. It is on us, the scientists and artists, to always deeply reflect and carefully step on this dangerous path, so our road would be paved with right decisions and empathetic visions, and not with good intentions and nice wishes! It is on us, the artists and scientists, to always and again perceive the world around us with soberness and emotion, and to give to the society and the science a multi-faceted, but therefore also a truthful, mirror. As there are many faces to the truth, it is essential to look at them as individual facets, as different standpoints, as carefully as possible. To learn from errors others did, even if those others were literary or cinematographic characters, is much cheaper and safer than learning from our own errors. We can not proud ourselves that in this our consumeristic and hyperpoductivistic civilisation we did not make too many unnecessary errors, and very often got stuck due to them, and we constantly grind against quite serious problems. If we entangle ourselves too much, in vain will we wait for some Godot to come and, like the "deus ex machina" of Roman theatres, pull us out of our own world which we are creating.

Acknowledgment

The work was partially supported by Croatian Centre of Research Excellence for Data Science and Advanced Cooperative Systems and EGI Engage and INDIGO DC Horizon 2020 projects.

References:

[1] Karolj Skala, Davor Davidovic, Enis Afgan, Ivan Sovic, Zorislav Sojat: "Scalable Distributed Computing Hierarchy: Cloud, Fog and Dew Computing", Open Journal of Cloud Computing (OJCC), 2(1), Pages 16-24, 2015, https://www.ronpub.com/publications/OJCC_2015v2i1n03_Skala.pdf

[2] Sinisa Marin, Mihajlo Ristic, Zorislav Sojat: "An Implementation of a Novel Method for Concurrent Process Control in Robot Programming", Third International Symposium on Robotics and Manufacturing: Research, Education and Application, ISRAM '90, Burnaby, BC/CA; 1990

[3] Janko Mršić-Floegel, Derek Reynolds, Zorislav Šojat, Marco Biancessi, Stefano Sala: Data Communications, Patent WO2002025897 A1, priority 13/9/2000, http://www.google.com/patents/WO2002025897A1, retreived 29/2/2016.

[4] Zorislav Šojat: "Operating System Based on Device Distributed Intelligence", 1st Orwellian Symposium, Baden Baden, Germany; 1984

[5] Zorislav Sojat, Karolj Skala: "Multiple Programme Single Data Stream Approach to Grid Programming", Hypermedia and Grid Systems, Opatija, Croatia; 2004

[6] Zorislav Šojat, Tomislav Ćosić, Karolj Skala: "Virtue — A different approach to human/computer interaction" Information and Communication Technology, Electronics and Microelectronics (MIPRO), 2014, 37th International Convention on. IEEE, 2014, http://grgur.irb.hr/Library/Shoyat.Cyosicy.Skala.Virtue_A_different_approach_to_human_computer_interaction.pdf

[7] Y. Wang, "Cloud-dew architecture," International Journal of Cloud Computing, vol. 4, no. 3, pp. 199–210, 2015.

Service-oriented application for parallel solving the parametric synthesis feedback problem of controlled dynamic systems

G.A. Oparin, V.G.Bogdanova, S.A. Gorsky, and A.A. Pashinin
Matrosov Institute for System Dynamics and Control Theory of SB RAS, Irkutsk, Russia
gorsky@icc.ru

Abstract. Currently, there is an intensive development and complication of controlled dynamic systems. Despite the large amount of existing methods, the problem of analysis and synthesis of various classes of such systems remains relevant. In particular, binary dynamic systems (Boolean network) represent a large interest to study for the theory and practice. The necessary requirements for controlled systems are the demands of their stability. We suggest a service-oriented application PSF (Parametric Synthesis Feedback) for the parametric synthesis of stabilizing controller for controlled continuous and binary dynamic objects. This application was implemented with High Performance Computing Service Oriented Multiagent System (HPCSOMAS) Framework developed by authors. This tool provides the functionality for automated creation and multiagent control of execution of service-oriented applications. PSF integrates the functionality represented by services that implement new methods for the synthesis of linear static (or dynamic) state (or output) feedback. These proposed methods allow a natural data parallelism and scalability with increasing the dimension of the problem. Based on integration of problem-oriented services PSF interacts with distributed computational environment through the system services of HPCSOMAS Framework. We demonstrate some examples of the parametric synthesis of stabilizing controller for different classes of linear dynamic control objects.

I. INTRODUCTION

Currently, scientific computational experiments require the use of high-performance calculation in heterogeneous distributed computing systems (HDCS). The complexity of modern HDCS prevents the wide use of their capabilities by field experts and facilitates the development of technologies for integrating heterogeneous resources into a single HDCS, and specialized approaches to the development of distributed applications that allow an end user to abstract from the specifics of their technical implementation and execution on computational resources. This tendency actualizes the development of technologies of the service-oriented access to the resources environment and scientific applications. The researches connected with designing of platforms for creating and performing the service-oriented scientific application [1-3] are actively developing. During recent decades, such scalable distributed computing paradigms as Cloud Computing, Fog Computing and Dew Computing have been developed [4-7].

The great resources potential of such computing infrastructure as Grids and Clouds provided the development more complicated scientific applications. Many of these applications are designed as workflow (WF). By this reason, the tendency of Workflow Management System (WMS) creation is observed [8]. Despite the variety of tools of this kind (for example, known WMSs [8-12]), a number of problems is still a challenging issue. This problems concerned with the integration of decentralized management which are more preferable in distributed environments for providing scalability, of the access methods to the application components, of the tools of subject area specification, and of the tools of description logic computing schemes not by program but declarative way, of the treatment of static as well as dynamic WF structure.

We describe the use of High Performance Computing Service Oriented Multi-agent System (HPCSOMAS) Framework [13] for creating and using of service-oriented applications in HDCS, which integrate availabilities enumerated above. We presented an example of application designing for solving the task of parametric synthesis stabilizing feedback for controlled dynamic objects.

Currently controlled dynamic systems are actively developing and improving. Despite the large number of existing methods, the task of analyzing and synthesizing different classes of such systems is still a challenging issue. Particularly, binary dynamic systems (Boolean networks) are of great interest in theory and practice. The class of linear binary dynamic system (BDS) as the object of scientific research having great practical significance was developed owing to works [14, 15] in the last quarter of the past century. The great number of that time publications as well as modern ones are devoted to problems of analysis of linear BDS. The issues of linear static (or dynamic) state- (or output-) feedback synthesis were given less attention. In earlier works, the linear BDS models in the form of Control flow graph [16] and of transfer functions [17] were used to solve the problems of synthesis. In past two decades, the approach connected with using of algebraic space state models is prevalent [18]. This work offers new logical synthesis method of linear feedback [19], developed by authors, when in the general case the state equation is used to describe linear object dynamics and linear regulator, and specification of required dynamic characteristic of closed-loop system (in these case it is stability feature) sets by the language of formal logic. In this circumstance, the problem of synthesis is reduced to examination of quantified Boolean formulas validity (task TQBF [20]) with the following searching of feedback matrix (task SAT [20]). The offered method provides natural data parallelism and high scalability with increasing the problem dimension in the process of problem running in HDCS.

Program tools for solving tasks of controlling complicated dynamic systems have been developing for a long time [21]. The system MATLAB [22] is very well known, but, as indicated in [23] this system doesn't provide an expert with a necessary level of automation of computation work and is designed mostly for conducting large-scale tests of various algorithms and methods rather than effective solution of complex engineering problems of controller design. Therefore program tools automated solution of tasks at hand, oriented on different categories of users, and providing ready-for-use applications for typical project solutions as well as tools for creating new applications, are being developed along with this system, as an example [24]. However, the issues, connected with using high-capacity resources for solving resource-intensive problems in the considered subject area and organization of control of their parallel solution in HDCE, still have not been solved.

In the review [25] some classes of tasks of linear control theory, the non-convexity and NP-complexity of which make getting a simple solution impossible were described. Thus, in the work [26] NP-complexity of the task of State Output Feedback (SOF) was showed.

A large number of topical tasks in contemporary control theory, which require high-capacity resources, are connected with the problem of parametric synthesis of linear regulator for controlled dynamic objects. In the work [27] the method of controlled search for considered problems solving in high-performance environment, designed by authors, is described. This method provides natural parallelism during solution searching into space of regulator parameters, varied at defined intervals. The application Parametric Synthesis Feedback (PSF), created on HPCSOMAS Framework (hpcsomas application) ensure automated solving of considered problems based on the logical method and method of controlled search designed by authors, afford the tools of task formulation on the content level to an end user.

II. THE DESIGN OF HPCSOMAS APPLICATION

The application created on HPCSOMAS Framework consists of two main parts – system part, independent from the solving problem class, and subject-oriented. The system application part consists of set of components, which are the agents of hierarchical role-based multiagent system (MAS). The user agent is on the top level, hpcsomas agent – on the middle level, system reactive agents – on the lower level: an authentication agent, file management agent, executive agent. The basic system agent – is hpcsomas agent, acts as: the agent-manager, which coordinates resource distribution and as the local agent, which provides the resource. The hpcsomas agent

gets the function of the agent-manager to which the user inquires. The self-organization of agents is realized on the request analysis into virtual community (VC). It includes the agent-manager and local agents, which offer necessary hardware and software resources for task solving. The VC of agents performs decentralized management of distributive performing of user's request in HDPE, decomposition of request to subtasks, distribution of resources and monitoring of subtasks execution, informing the user about the process of request execution. If the calculating resource of the agent-manager fails, one of the local agents takes its duties.

Subject-oriented part of the application is implemented on the set of application user programs and transforms into calculating agents. Computational model of subject area is a bipartite graph $<P, O, E>$, part P contains set of parameters, part O – set of operations of application program package, and part E contains directed connections between parameters and operations.

Application agents are implemented in the form of services. The services can be atomic and composite. The operations are transformed into atomic services, which subsequently are constructed into composite services computational schemes for task solving (WF). The performance of composite services is interpreted under control of VC agents. The atomic services are able to function autonomously, similar to micro-services, the using of which is represented as an advantage of Dew Computing in the work [7]. The application development scheme with using of automation subsystem ABCSW is shown on Fig. 1.

III. THE APPLICATION DEPLOYMENT TO COMPUTATION RESOURCES

Scalable application can be deployed on server side providing an access to components with the help of web-interface, on client side providing local access to components (fig. 2).

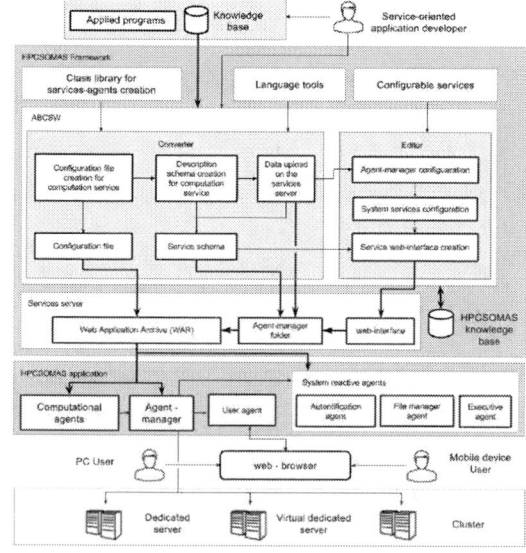

Figure 1. The scheme of development of hpcsomas-application

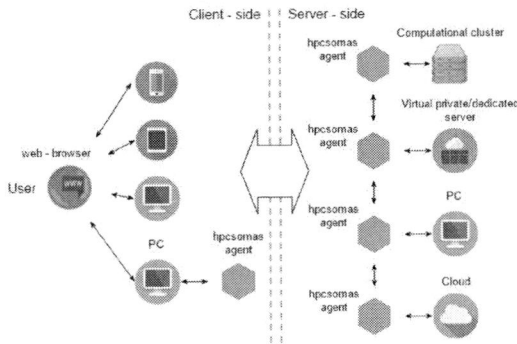

Figure 2. Methods of deployment hpcsomas application

Additionally the connection between components installed on calculating nodes with frontend access and nodes of local network call only. To provide safety, the components' interaction is realized by using access keys assigned in the configuration settings. Planning of complex components' use is realized on the base of calculating subject area model, rather execution of designed workflow is performed by decentralized multi-agent menedgement. The distribution of this workflow tasks on HDCS resources on user application level is performed cooperatively by hpcsomas agents installed on calculating resources.

IV. AN EXAMPLE OF PSF APPLICATION DEVELOPING

Developed hpcsomas application consolidates objects of the subject domain, which are correspondent to the family of research methods, similar in terms of parameters and operations. In the PSF application, the following services are grouped:

- Parametric synthesis of the static/dynamic output controller for a linear continuous dynamic system.

- Parametric synthesis of a static controller for a linear discrete dynamical system.

- Parametric synthesis of a static controller for a BDS.

- Constructing a region of stability in the controller parameters space (if the number of parameters exceeds three, then a number of parameters are fixed).

A. The service of parametric synthesis dynamic output feedback for linear continues dynamic system

The example of developing of the PSF application composite service for parametric synthesis dynamic output feedback for linear continues dynamic system (PS DOF LCDS) is described below. This service intended for the following problem: for system

$$\dot{x} = Ax + Bu \,, \; y = Cx, \qquad (1)$$

where $x \in R^{n_x}$ – the state vector of dynamic object, $u \in R^{n_u}$ – the control vector, $y \in R^{n_y}$ – measure output, n_x, n_u and n_y – dimensions of x, u, and y vectors, find out whether it is possible to stabilize it with the help of the following feedback type:

$$\dot{x}_r = A_r x_r + B_r y \,,$$

$$u = C_r x_r + D_r y \,,$$

where $x_r \in R^{x_r}$ - the state of dynamic regulator, interval limitations are imposed on A_r, B_r, C_r, D_r matrixes. With $k \neq 0$ the equation of the closed-loop system is the following: $\dot{x}_c = A_c x_c$, where $x = \mathrm{col}(x, x_r)$.

In the process of the PS DOF LCDS service design on the base of applied programs and knowledgebase, at first, atomic subject-oriented services of the closed-loop system A_c matrix, and the specter calculation of the closed-loop system A_c matrix, were designed. The web-interface of the PS DOF LCDS service was generated. Afterwards the agent-manager, user agent and the agents for system functions performing were developed, the agents' configuration under specific computing resources was made as well. The user enters input task data (dimension of vectors, matrices of coefficients, interval limits) and requests to execution. The agent-manager initiates the generation of VC agents. The VS of agents forms taking into account the available agents, the availability of the required resources and the size of the task.

For this task solving the following steps based on the method of directed search need to be performed:

1. Decomposition of grid space to 2^l segments, where l is a number of regulator parameters;

2. Calculation into center of every segment of A_c matrix and segment priority;

3. Determination of stability of closed-loop system with A_c matrix;

4. If the closed-loop system with A_c matrix is stable, extinction of the calculation and issuing the response;

5. Ordering of segments in the ascending order of their priorities;

6. Circular parallel processing of segments (step 2-4 repetition), from the first segment to the last segment within the interval limits with decreasing of the step for every regulator parameter.

The workflow of this task has a complicated structure. Multivariate calculations (depending on the result of calculations) are followed by a cycle in which multivariate calculations are performed again (for other dynamic parameter values). Subtasks of multivariate calculations require the processing of a condition, the execution of which leads to the removal of the entire computational layer of subtasks.

Consider an illustrative example; the matrices of system (1) have the form [29]:

$$A = \begin{pmatrix} 0 & 1 \\ 1 & 0 \end{pmatrix}, B = \begin{pmatrix} 0 \\ 1 \end{pmatrix}, C^T = \begin{pmatrix} 1 \\ 0 \end{pmatrix}.$$

Interval limitations are following:

$$-2 \le A_r \le 2; \quad -2 \le B_r \le 2; \quad -2 \le C_r \le 2; \quad -2 \le D_r \le 2.$$

The parameter number of the regulator of first order is $n_p = 4$. The set of parameters regulator is:

$$p = \{ p_1 = A_r, p_2 = B_r, p_3 = C_r, p_4 = D_r \}.$$

On the Figure.3 the page of HPCSOMAS Framework site is observed, where the structure of experimental HDCS is represented used at the moment for solving the task of parametric synthesis with the help of directed search controlled by this system. On this page, you can also see the list of denominations of calculating resources and installed HPCSOMAS-agents on these resources. In HDCS the clusters (matrosov, blackford), the dedicated cloud Server (1stvds), virtual machines (VM) (teslav and 4 VM for imitative modeling), workstations (angarsk01, angarsk02) are combined together. The agent matrosov has no external access. The connection with it is performed by tesla-agent. The agent Imitation Modeling is not needed for considered task solving, what is why there is no access point (by color highlighting) and agents' status information for it. The agents angarsk01 and angarsk02 are engaged for other purposes, that is why they are inaccessible for considered task solving as well. In status line active agents are marked by green circle and inactive – by red one. The hpcsomas-agent installing on workstations is performed by thick client technology (Client-side), all the rest resources - by thin client technology (Server-side).

The access to hpcsomas-agents is also available through web-browser. Only registered at the appropriate computing resource and in HPCSOMAS Framework the user is able to connect to the agents. After the registration process, the user needs to get rights on using a group of applications from system administrator. The users of different application groups observe information according to their rights stated in profile. There is a guest input for non-authorized users for familiarization with HPCSOMAS Framework and observation of demonstrative examples.

As a result of the service, the desired regulator is described by equations:

$$\dot{x}_r = -1,5x_r - 1,45y,$$

$$u = -0,4x_r - 1,5y.$$

The matrix of a closed system with such feedback has the following eigenvalues:

$$\lambda_{1,2} = -0,1492 \pm j0,3453, \text{ and } \lambda_3 = -1,2016.$$

B. The service of stubility region construction

These services intended for constructing two-dimensional and three-dimensional stability region in the space of the controller of a closed-loop control system within the given ranges. By varying the selected parameters (two or three correspondently), a numerical grid is constructed. Multivariate calculations are used to determine the stability of the matrix of a closed-loop system in each point of this grid.

The result of these services represents tabular data and the graphic image of the stability region. These tabular data allow selecting the stable point, corresponding to the given parameters of the control quality. Setting the task by the user for all services of the PSF application only requires the input of initial data.

Fig. 4 shows the result of the service of constructing the stability region in the neighborhood of the solution found above. First regulator parameter (A_r) is fixed. Other three parameters vary in the interval [-1.5; 0.5] with step 0.05. Values of parameters B_r, C_r, D_r are on the x, y, z axes correspondently.

C. The service of parametric synthesis of a static controller for a BDS

Service PS SC BDS for parametric synthesis of static controller for a binary dynamic system (PS SC BDS) is designed to solve the following problem. We consider a linear BDS, where the vector-matrix equation is as follows:

$$x^t = Ax^{t-1} \oplus Bu^{t-1}, \tag{2}$$

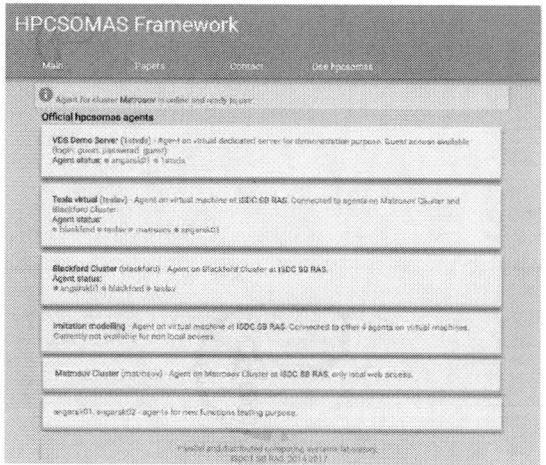

Figure 3.   Agents deploying on the experimental GRBC

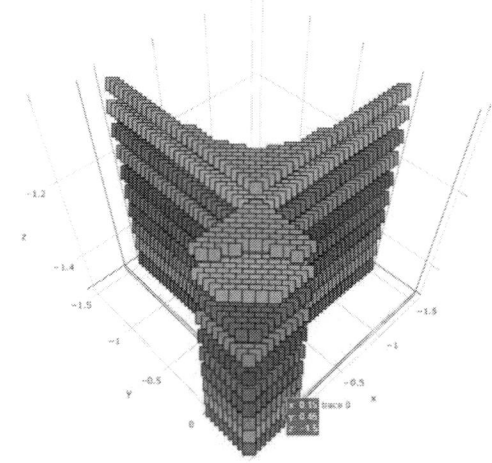

Figure 4.   The region of stability in the neighborhood of the point (-1.5, -2; -0.1; -1.4). The step of incrising value is 0.05.

where x, u – vector of state and control vector, respectively, ($n_x \in B^n$ $n_u \in B^m$, $B \in \{0, 1\}$), $t \in T = \{1, 2, ..., k\}$ – discrete time (measure number), A – ($n \times n$) binary matrix of state, B – ($n \times m$) binary input matrix, operations of addition and multiplication are performed by mod2 . The problem of static controller synthesis for (2) is in the choice of the control law from the class of inverse linear connections by state as follows

$$u^{t-1} = Px^{t-1},$$

where P – binary matrix of controller parameters of the corresponding order, which provides consistent balance x = 0 of the closed-loop system

$$x^t = A_c x^{t-1} = (A \oplus BP)x^{t-1}. \quad (3)$$

Balance position $x = 0$ of autonomous system (3) is considered stable, if for each $x^0 \in B^n$ there is such a moment of time $t \in T$, that trajectory $x(t, x^0)$ for t time steps reaches zero state: $x(t, x^0) = 0$. It is evident, that $x(t, x^0) = 0$ for all following moments of time $t > k$.

This service provides automated recording of the stability parameter in the language of formal logic in the form of a quantified Boolean formula (QBF). The synthesis problem concludes in checking the trueness of QBF (task TQBF) with the following search of feedback matrix (task SAT). A parallel solver, which includes tools of automated creation of the original task model in TQBF and SAT formats as well as methods of establishing solvability of the controller synthesis problem and the search of feedback matrix using these models, has been developed. The mathematical derivation of reducing to TQBF, and SAT problems is given in [19]. In this work the following illustrative example is considered. Let the non-linear binary dynamic object have the following form:

$$x^t = F(x^{t-1}) \oplus Bu^{t-1}, u^{t-1} = Px^{t-1}, n = 3, m = 1, T = \{1,2,3\},$$

$$B = col(1,0,0), P = (p_1, p_2, p_3),$$

$$F(x_1, x_2, x_3) = col(x_1 \overline{x_3}, \overline{x_1} x_2 x_3, x_1 x_2).$$

As a result of the service, we obtain the feedback matrix $P = (1, 0, 0)$. Thus, the dynamics of a closed system will be determined by the following system of equations

$$x_1^t = x_1^{t-1} \cdot x_3^{t-1}; x_2^t = \overline{x_1^{t-1}} \cdot x_2^{t-1} \cdot x_3^{t-1}; x_3^{t-1} = x_1^{t-1} \cdot x_2^{t-1}.$$

The state diagrams for this example are shown in Fig. 3.

The created logical method is applicable for solving the problem of static output regulator and also of dynamic output or state regulator. Moreover, the logical method is applicable for nonlinear BDS with additive input of linear controlling actions. The following development of the logical method is connected with its usage for synthesis problem solving of stabilizing feedback for monomial BDSs, which are of great importance in the search of gene regulatory networks [15], and correspondingly for qualitative analysis of the other dynamic features, which are typical for a closed-loop control system.

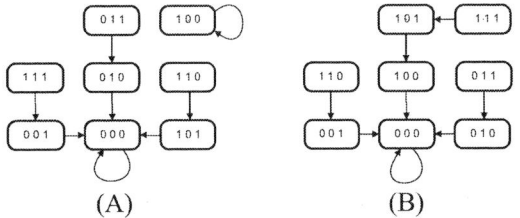

Figure 5. The state diagram of an open-loop (A) and closed-loop (B) system

The task formulation for all PSF application services needs to enter input data only. The means of logic description of service performance of HPCSOMAS environment are embedded on creation phase and provide an ability to use cycles, logical conditions, multi-variant calculations. In the process of stability area creation the experimental data is stored. It allows, at a later time, to choose a stable point according to predefined criteria of qualitative characteristics. If extensible set of offered by the system services and experimental data are stored on the cloud agent, the problem of offline Internet access appears. Nowadays a new trend is developed, dew computing [4]. Such advantages of this paradigm as scalability and memory caching seem perspective for the further developing of the considered HPCSOMAS Framework in the direction of ensuring an effective and stable user researches.

V. CONLUSION

We examined the use of tools of automation service creation and organization of high-capacity problem-oriented computations with the help of decentralized multiagent management in HDCS. The experimental results demonstrate high level of scalability and effectiveness of calculations using scientific services, performed on the base of new methods of parametric synthesis of stabilizing feedback for dynamic controlled objects.

ACKNOWLEDGMENT

The study was supported by Russian Foundation of Basic Research, project no. 15-29-07955-ofi_m.

REFERENCES

[1] P. Östberg, A. Hellander, B. Drawert, E. Elmroth, S. Holmgren, and L. Petzold, "Reducing Complexity in Management of eScience Computations," Proceedings of CCGrid 2012 - The 12th IEEE/ACM International Symposium on Cluster, Cloud and Grid Computing, 2012, pp. 845-852.

[2] C. Vecchiola, S. Pandey and R. Buyya, "High-Performance Cloud Computing: A View of Scientific Applications," 10th International Symposium on Pervasive Systems, Algorithms, and Networks, Kaohsiung, 2009, pp. 4-16. doi: 10.1109/I-SPAN.2009.150.

[3] O. Sukhoroslov, S.Volkov, and A. Afanasiev, "A Web-Based Platform for Publication and Distributed Execution of Computing Applications," 14th International Symposium on Parallel and Distributed Computing (ISPDC). IEEE, 2015, pp. 175-184.

[4] I. Foster, Y. Zhao, I. Raicu, and S. Lu, "Cloud computing and grid computing 360-degree compared," Grid Computing Environments Workshop, 2008. GCE'08, 1-10.

[5]   F. Bonomi, R. Milito, J. Zhu, and S. Addepalli, "Fog Computing and its Role in the Internet of Things," In Proc of MCC (2012), pp. 13-16.

[6]   Y. Wang, "Cloud-dew architecture," International Journal of Cloud Computing, vol. 4, no. 3, pp. 199–210, 2015.

[7]   K. Skala, D. Davidovic, E. Afgan, I. Sovic, and Z. Sojat, "Scalable Distributed Computing Hierarchy: Cloud, Fog and Dew Computing," Open Journal of Cloud Computing (OJCC), RonPub, vol. 2, no. 1, pp. 16–24, 2015. DOI: 10.19210/1002.2.1.16.

[8]   D. Talia, "Workflow systems for science: concepts and tools," ISRN Software Engineering. Vol. 2013, Article ID 404525, 15 pages. http://dx.doi.org/10.1155/2013/404525.

[9]   K. Wolstencroft, R. Haines, D. Fellows, A. Williams, D. Withers, S. Owen, and C. Goble, "The Taverna workflow suite: designing and executing workflows of Web Services on the desktop, web or in the cloud," Nucleic Acids Research, 41(Web Server issue), 2013, pp. 557–561. http://doi.org/10.1093/nar/gkt328

[10] E. Deelman, K. Vahi, M. Rynge, G. Juve, R. Mayani, and R. Ferreira da Silva, "Pegasus in the Cloud: Science Automation through Workflow Technologies," IEEE Internet Computing, vol. 20, iss. 1, pp. 70-76, 2016.

[11] R. Ferreira da Silva, E. Deelman, R. Filgueira, K. Vahi, M. Rynge, R. Mayani, and B. Mayer, "Automating Environmental Computing Applications with Scientific Workflows," in Environmental Computing Workshop (ECW'16), 2016.

[12] D. Nasonov, N. Butakov, M. Balakhontseva, K. Knyazkov, A. Boukhanovsky, "Hybrid Evolutionary Workflow Scheduling Algorithm for Dynamic Heterogeneous Distributed Computational Environment," International Joint Conference SOCO'14-CISIS'14-ICEUTE'14, pp. 83-92.

[13] Bychkov, G. Oparin, A. Feoktistov, V. Bogdanova, and A. Pashinin, "Service-Oriented Multiagent Control of Distributed Computations," Automation and Remote Control, vol. 76, no. 11, 2015, pp. 2000–2010.

[14] A. Gill, "Linear Sequential Circuits –Analysis, Synthesis, and Applications," McGraw-Hill, New York 1967.

[15] R. Faradzhev, "Linear sequential machines," ["Линейные последовательностные машины"], Moscow: Soviet Radio, 1975. 248 p.

[16] A. Gill "Analysis and Synthesis of Stable Linear Sequential Circuits," J. ACM. 1965. Vol. 12, No. 1. pp. 141-149.

[17] R. Faradzhev, "On the equations of synthesis of linear sequential machines," ["Об уравнениях синтеза линейных последовательностных машин" // Automation and telemechanics. 1980. № 9. pp. 81-90.

[18] J.Reger, and K.Schmidt "A Finite Field Framework for Modeling, Analysis and Control of Finite State Automata," Mathematical and Computer Modeling of Dynamic Systems. 2004. Vol. 10(3-4). pp. 253-285.

[19] G. Oparin, A. Feoktistov, V. Bogdanova, S. Gorsky, A. Pashinin, and I.A. Sidorov, "Parallel solution of static regulator problems for binary dynamic systems," ["Параллельное решение задачи о статическом регуляторе для двоичных динамических систем"], Parallel Computing Technologies - XI International Conference, PCT2017, Moscow. Kazan, April 3-7, 2017 Short articles and descriptions of posters. Chelyabinsk: Publishing Center of SUSU, 2017. pp.416-426.

[20] M. Garey, and D. Johnson "Computers and intractability; A Guide to the Theory of NP-Completeness," W. H. Freeman & Co., New York, NY, USA, 1979.

[21] Y. Somov, S. Butyrin, G.A. Oparin, V.G. Bogdanova, "Methods and Software for Computer-Aided design of the Spacecraft Guidance, Navigation and Control Systems", MESA, vol. 7, no. 4, CSP - Cambridge, UK; I&S - Florida, USA, 2016, pp. 613-624.

[22] "MathWorks," http://www.mathworks.com/, [online, accessed: 31-Jan-2017].

[23] A. Aleksandrov, R. Isakov, and L. Mikhailova, "Structure of the software for computer-aided logical design of automatic control," Automation and Remote Control, vol. 66, iss. 4, 2005, pp. 664–671.

[24] A. Aleksandrov, L. Mikhailova, and M. Stepanov, "GAMMA-3 system and its application," Automation and Remote Control, vol. 72, iss. 10, 2011, pp. 2023–2030. DOI: 10.1134/S0005117911100031

[25] B. Polyak, and P. Shcherbakov, "Hard Problems in Linear Control Theory: Possible Approaches to Solution," Autom. Remote Control, vol. 66, iss. 5, 2005, p.p. 681-718. doi:10.1007/s10513-005-0115-0

[26] A. Nemirovskii, "Several NP-hard problems arising in robust stability analysis," Mathematics of Control, Signals, and Systems, vol. 6, 1993. pp. 99 – 105.

[27] G. Oparin, A. Feoktistov, V. Bogdanova, and I. Sidorov, "Automation of multi-agent control for complex dynamic systems in heterogeneous computational network," AIP Conference Proceedings 1798, 020117, 2017, http://doi.org/10.1063/1.4972709

[28] "Irkutsk Supercomputer Center of SB RAS," http://hpc.icc.ru/ [online, accessed: 31-Jan-2016].

[29] D. Balandin, M. Kogan, "Synthesis of nonfragile controllers on the basis of linear matrix inequalities," Automation and Remote Control, vol. 67, iss. 12, 2006, pp. 2002–2009. doi:10.1134/S0005117906120125.

Augmented Coaching Ecosystem for Non-obtrusive Adaptive Personalized Elderly Care on the Basis of Cloud-Fog-Dew Computing Paradigm

Yu.Gordienko[1]*, S.Stirenko[1], O.Alienin[1], K.Skala[2], Z.Sojat[2], A.Rojbi[3], J.R.López Benito[4], E.Artetxe González[4], U.Lushchyk[5], L.Sajn[6], A.Llorente Coto[7], G.Jervan[8]

[1] National Technical University of Ukraine "Igor Sikorsky Kyiv Polytechnic Institute" (NTUU KPI), Kyiv, Ukraine
[2] Ruder Boskovic Institute, Zagreb, Croatia
[3] University of Paris 8, Paris, France
[4] CreativiTIC Innova SL, Logroño, Spain
[5] Medical Research Center "Veritas", Kyiv, Ukraine
[6] University of Ljubljana, Ljubljana, Slovenia
[7] Private Planet, London, United Kingdom
[8] Tallinn University of Technology, Tallinn, Estonia
* yuri.gordienko@gmail.com

Abstract - The concept of the augmented coaching ecosystem for non-obtrusive adaptive personalized elderly care is proposed on the basis of the integration of new and available ICT approaches. They include multimodal user interface (MMUI), augmented reality (AR), machine learning (ML), Internet of Things (IoT), and machine-to-machine (M2M) interactions. The ecosystem is based on the Cloud-Fog-Dew computing paradigm services, providing a full symbiosis by integrating the whole range from low level sensors up to high level services using integration efficiency inherent in synergistic use of applied technologies. Inside of this ecosystem, all of them are encapsulated in the following network layers: Dew, Fog, and Cloud computing layer. Instead of the "spaghetti connections", "mosaic of buttons", "puzzles of output data", etc., the proposed ecosystem provides the strict division in the following dataflow channels: consumer interaction channel, machine interaction channel, and caregiver interaction channel. This concept allows to decrease the physical, cognitive, and mental load on elderly care stakeholders by decreasing the secondary human-to-human (H2H), human-to-machine (H2M), and machine-to-human (M2H) interactions in favor of M2M interactions and distributed Dew Computing services environment. It allows to apply this non-obtrusive augmented reality ecosystem for effective personalized elderly care to preserve their physical, cognitive, mental and social well-being.

I. INTRODUCTION

A. Background

The advances in medicine and living standards in the last century have resulted in a significant increase in the number of elderly people in Europe and most other developed countries in the world. Over the next decades, the worldwide number of older people will further increase dramatically. In Europe, this development is even more pronounced: for example, in Portugal, Spain, Croatia and other European countries, the old age dependency ratio, which gives the quotient of people 65+ will reach ~30-36% with pan-European average value up to 29.6% in 2050 [1]. These demographic changes have drastic structural, societal and economic implications, and challenge elderly care stakeholders like policymakers, families, businesses and healthcare providers alike. The ever increasing percentage of old people in the most advanced Western and Eastern countries is posing a great challenge in social healthcare systems. The effort required by formal caregivers for supporting older people can be enormous, and this requires an increase in the efficiency and effectiveness of today care. One way for achieving such a goal is the use of information and communication technologies (ICTs) for supporting and assisting people in their own homes.

Older generations need to be included as active and integral pillars of our society instead of being isolated in the special elderly care facilities. They should remain active members of the work force as long as possible, since the traditional assumption that retirement equals the worker's final exit from the labor force does not hold true any longer. The required transition of society can only be successful if huge efforts are made on various levels to foster independence of this age group, from more flexible employment arrangements, remote services in care giving (telecare), support of independent living (ambient assisted living - AAL), access to information, access to transportation (accessibility), to specific communication services and devices as well as entrepreneur approaches in educational offers like life-long learning (LLL).

ICT is believed to play a key role in all these fields. However, ICT can successfully contribute to their individual well being, and help to meet the challenges of an aging society in general, if ICT could be non-obtrusively adapted to the older adults' knowledge, needs, and abilities. Furthermore, the whole of our society can gain enormous benefits by integrating the knowledge and skills and high degree of experience the elderly can provide to the coming generations, in all aspects of living, from technological expertise in any field, to everyday living experiences. Current ICTs range from systems for reminding appointments and activities [2], for medical assistance and tele-healthcare [3], to human-computer interfaces for older persons or people with special needs [2]. Usually, these ICTs incorporate application dependent sensors, such as sensors, cameras or microphones. Many studies [4] have demonstrated that people prefer non-invasive sensors, such as microphones, over cameras and wearable sensors, and this drove the scientific community to develop systems and technologies based on non-invasive approaches only.

ICTs have become an integral component of everyone's life, including older adults, to continue education, obtain health information, communicate and exchange experiences, as well as online banking/shopping etc. Though recent research has shown that older adults are receptive to using ICTs, a commonly held belief is still prevalent that supports the idea that older adults are unwilling to use ICTs due to bodily and cognitive decline in working memory, attention, and spatial abilities [5,6].

The main problem is that despite the current progress of elderly care facilities the vast majority of EU older people wish to live independently at home as long as possible; meeting their needs can be a major challenge [7]. The different providers often work under conditions of poor coordination among ICT experts, elderly caregivers, patients, and their families [8-9].

B. State of the Art (Similar Works)

ICTs are promising for the long-term care of elderly people. As all European member states are facing an increasing complexity of health and social care, good practices in ICTs should be identified and evaluated. Recently, several projects funded by DG CNECT were related to Active and Healthy Ageing (AHA). They provided: independent living and integrated services — BeyondSilos (http://beyondsilos.eu), integrated care coordination, patient empowerment and home support — CareWell (www.carewell-project.eu), set of standard functional specifications for an ICT platform enabling the delivery of integrated care to older patients — SmartCare (http://pilotsmartcare.eu/). Some successful initiatives were initiated in Europe and supported by EU, for example, European Rosetta project [11], research network for design of environments for ageing (GAL) [10], assisted living environment for independent care and health monitoring of the elderly (ENRICHME), responsive engagement of the elderly promoting activity and customized healthcare (REACH), digital environment for cognitive inclusion (DECI), integrated intelligent home environment for the provision of health, nutrition and mobility services to the elderly (MOBISERV), unobtrusive smart environments for independent living (USEFIL), open architecture for accessible services integration and standardization (OASIS) and others.

C. Unresolved Problems

These innovations can improve health outcomes, quality of life and efficiency of care processes, while supporting independent living. However, in the face of new challenges some disruptive innovations should be proposed and implemented, and the new challenges/problems should be addressed. The potential radically new solution should take into account the following additional set of aspects/problems related to quite different (1) targeted communities; (2) level of functional (technical/computer/digital) literacy of the targeted communities; (3) realistic time of massive implementation of the proposed technologies for these communities with people of various functional literacy; (4) differences in national and geographical mentality as to elderly care in Europe.

Targeted communities in the context of elderly care consist of:

- individuals — self-directed elderly care, where elders control both the objectives and means of elderly care;

- families, i.e. individuals inside family and/or supported by family — informal elderly care, where elders control the means/tools, but not the objectives of elderly care;

- assisted elderly care — non-formal elderly care, where elders control the objectives but not the means/tools of elderly care;

- specialized elderly care facilities — formal elderly care, where elders have no or little control over the objectives or means/tools of elderly care.

Level of functional/computer/digital literacy of the targeted communities (in the order from the lowest to highest): absolute computer illiteracy, digital phobic, basic computer literacy, digital immigrants [12], intermediate computer literacy, digital visitors [13], proficient computer literacy, digital residents [13], digitally native [12].

The proposed time of massive implementation of the proposed technologies/environments depends on the maturity of the available solutions and the functional/computer/digital literacy level of the targeted community: now (the current mature technologies can be applied immediately), in the nearest future (the perspective technologies can be mature in the nearest 2-3 years), in the much later future (the perspective technologies can be mature at unknown time).

Differences in national and geographical mentality as to elderly care in Europe were observed and reported elsewhere [14,15]:

- informal care is more common in South than in North Europe;

- informal care is more common in the "new" member states in the "East" than in the "old" member states in the "West";

- informal care provision to someone outside the household is comparatively rare in the Mediterranean countries, elderly care to someone in the home is more common in these countries than in the EU-states on average;

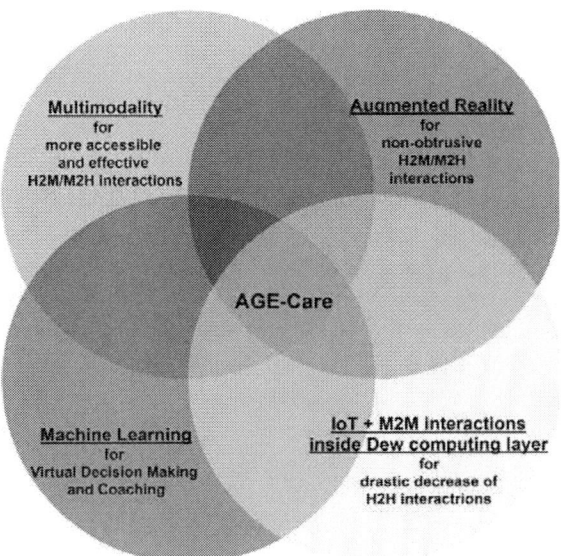

Figure 1. The integration concept of non-obtrusive augmented reality learning and coaching ecosystem for effective personalized elderly care.

- the low proportion of people providing care within households is explained by the rarity of multigenerational households in Nordic Europe.

The proposed Augmented Coaching Ecosystem for Non-obtrusive Adaptive Personalized Elderly Care (AGE-Care) is focused on the provision of the virtual care, support, and coaching to elderly people in the various targeted communities and with different functional/computer/digital literacy of the targeted communities. It will be achieved by enhancement of available ICT-enabled elderly care services, development of new ones, and their application with the tight coordination, monitoring, self-management and caregivers involvement inside the proposed AGE-Care ecosystem.

II. CONCEPT, MAIN AIMS, AND BASIC PRINCIPLES

A. General concept

The proposed AGE-Care ecosystem is assumed to be based on the integration of the several new ICT approaches and available ones, which should be enhanced by the radically new ICT based technologies concepts (shown in Figure 1) in favor of the elderly care stakeholders. They include multimodal user interface (MMUI), augmented reality (AR), machine learning (ML), Internet of Things (IoT), Internet of Everything (IoT), machine-to-machine (M2M) interactions, based on the Cloud-Fog-Dew computing paradigm services, providing a full symbiosis by integrating the whole range from low level sensors up to high level services using integration efficiency inherent in synergistic use of applied technologies.

The AGE-Care ecosystem is assumed to penetrate any organizational, national, mental, gender, and cultural division lines, boundaries, and limits. It will use the most appropriate available resources and elderly care, healthcare, and social care services. The AGE-Care ecosystem will be based on open standards, multi-vendor

interoperability, collaboration with ICT suppliers and ICT-related service providers.

B. Main aims

The main aims of AGE-Care ecosystem are as follows:

- to develop, test, and validate radically new ICT based concept of non-obtrusive augmented reality learning and coaching ecosystem for effective personalized elderly care to improve and maintain their independence, functional capacity, health status as well as preserving their physical, cognitive, mental and social well-being,

- to develop and implement the synergetic user-centered design of intuitive human-to-machine (H2M) and machine-to-human (M2H) interactions on the basis of information and communication technologies (ICTs) including internet of things (IoT), multimodal augmented reality (AR), and predictive machine learning (ML) approaches,

- to decrease the physical, cognitive, and mental load on elderly care stakeholders by decreasing the secondary human-to-human (H2H), human-to-machine (H2M), and machine-to-human (M2H) interactions in favor of machine-to-machine (M2M) interactions and distributed Dew Computing services environment,

- to overcome cognitive, mental, institutional, regional, and national barriers enabling delivery of integrated elderly care on the European scale by joining efforts across governmental, non-governmental, and volunteer elderly care organizations and individuals.

The following radically new ICT based main concepts and approaches are planned to be used to reach these aims (Figure 1):

- multimodal user interface (MMUI) — for the more accessible and effective intuitive H2M/M2H interaction on the basis combination of creative "artistic" approaches;

- augmented reality (AR) — for non-obtrusive H2M/M2H interactions,

- machine learning (ML) — for virtual decision making and virtual guidance of users,

- Internet of Things (IoT) + Internet of Everything (IoT) + machine-to-machine (M2M) interactions encapsulated inside Dew computing layer — to hide "behind the curtains" the mental and cognitive overloads, and shift them from H2H to M2M interaction zone.

C. Basic Principles

The proposed open AGE-Care ecosystem is based on the several basic principles:

- dominance of machine-to-machine (M2M) interaction over human-to-human (H2H);

- multimodal instead of single-modal interactions;

361

- non-obtrusive augmented reality feedback instead of obtrusive direct communication with numerous high-tech sensors, actuators, devices, and gadgets;

- virtual decision making and coaching by machine learning instead of real human-related services,

- short adaptive learning curve by selection of specific and context-related virtual coaching methods based on LLL principles instead of the obsolete and awkward "user guide" and "context help" approaches;

- highly distributed service oriented local and distance communication and service facilities.

III. STRUCTURE, WORKFLOWS, AND SOME EXAMPLES

A. Hierarhical Structure

This basic hierarchical structure of the AGE-Care ecosystem is virtualized at different levels and visually presented in Figure 2. In contrast to the current concept of elderly care (Fig. 2a), the proposed concept (Fig. 2b) will allow stakeholders:

- to decrease significantly (and avoid in the most situations) the level of H2H interactions —by emphasis on the M2M interactions for the basic technological scenarios;

- to avoid technological H2H interactions, but emphasize emotional H2H interactions in favor of emotional positive feedback from elderly people due to involvement of augmented multimedia channels like observed and even performed art, music, dance, etc.;

- to increase efficiency of H2M/M2H interactions — by introduction of multimodal communication channels like audio, visual, tactile, odor, etc., so-called Augmented Reality Human-to-IoT (ARH2IoT) interactions;

- to increase the acceptance level of the available ICT technologies for elderly care — by providing their functional abilities through non-obtrusive augmented reality pathways;

- to eliminate the gap between the newest available ICT technologies for elderly care and computer literacy of the targeted communities — by context-related, problem-based, and personalized virtual AR-related coaching;

- to decrease the market entry threshold for the future ICT technologies for elderly care — by providing the related open platform specifications based on the best practices and lessons learned during the project;

- to provide more security and privacy — by the localization of the personal consumer data at the lower scales of the AGE-Care ecosystem.

B. Workflows and Network Layers

Inside of AGE-Care ecosystem all workflows are encapsulated in the following network layers:

- Dew computing layer: the raw sensor data and basic multimodal actuator actions are concentrated, pre-processed, and resumed in the smallest scale local network (Dew) at the level of the IoT-controllers (individuals) and shared with the upper Fog computing layer;

- Fog computing layer: the resumed IoT-controller data and advanced actuator actions are located in the medium scale regional network unit (Fog) at the level of the IoT-gateway (family/room/office) and shared with the lower Dew computing layer and upper Cloud computing layer;

- Cloud computing layer: the accumulated IoT-

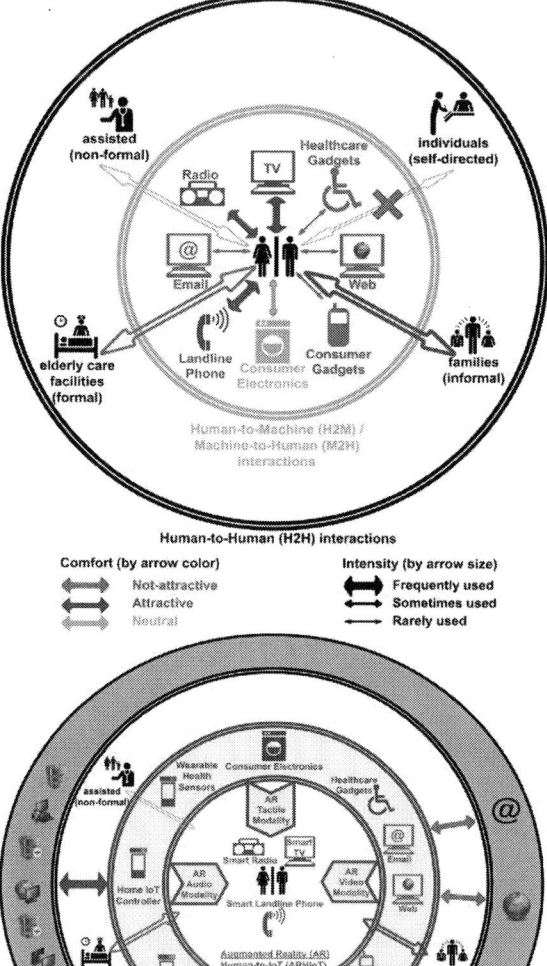

Figure 2. The current concept of elderly care (a, top), and the proposed concept of Augmented Coaching Ecosystem for Non-obtrusive Adaptive Personalized Elderly Care (AGE-Care) (b, bottom).

gateway data are thoroughly analyzed by ML methods to provide virtual decisions and coaching advices in the highest scale global network (Cloud) at the level of the global computing centers (hospitals, healthcare authorities, associations, corporations, etc.) and delivered to the lower Fog and Dew Computing layer.

C. Communication Flows and Interactions

The typical communication flows inside the AGE-Care ecosystem are schematically shown in Fig. 2b by arrows, where the higher emphases (in contrast to the current concept of elderly care) are placed on:

- ARH2IoT interactions — under Dew computing layer: green arrows depict the main dataflows from/to consumers by the familiar communication channels and devices, but with context-sensitive information provided by the multimodal augmented reality;

- M2M interactions — mainly inside Dew computing layer: light blue circle depicts the undercover dataflows among sensors and actuators, which are laid in the base of the multimodal augmented reality in ARH2IoT interactions;

- Cloud-Fog interactions — between Cloud and Fog computing layers: red arrows denote the familiar dataflows between the global computing centers and the medium scale network unit (Fog) at the level of the IoT-gateway (family/room/office);

- Cloud-Dew interactions — between Cloud and Dew computing layers: blue arrow denotes the dataflows between the global computing centers and the IoT-controllers.

It will allow to decrease cognitive overload on the stakeholders, because in the current concept of elderly care (Fig. 2a) the stakeholders are overwhelmed by the everyday increasing variety of the newest ICT technologies, the related devices and unusual practices. In the current paradigm of eHealth and elderly care, the stakeholders have to go by the long, complicated, and non-familiar learning curve to leverage the new ICT technologies. In contrast to it, the AGE-Care ecosystem proposes them to use the familiar information pathways (devices like television and radio broadcasting, landline phone communication), that seem to be the same old things, but actually enhanced by newest AR and AI technologies under the hood.

Instead of the "spaghetti connections" to the numerous sensors, actuators, devices, and gadgets with sporadic dataflows, "mosaic of buttons", and "puzzles of output data" for each device/technology, etc. (Fig. 2a), the AGE-Care ecosystem will provide the strict division in the following dataflow channels (Fig. 2b):

- consumer interaction channel — by allowing feedback data from all applied eHealth and elderly care ICT technologies through augmented reality pathway only at ARH2IoT layer;

- machine interaction channel — by integration of all sensor/actuator technologies and isolation of their raw data at Dew computing layer,

- caregiver interaction channel — by integration of Dew, Fog, and Cloud computing layers.

In general, the AGE-Care ecosystem will decrease the high cognitive load on customers, increase the efficiency of caregivers, and provide a unified way for incorporation of any future ICTs by division of dataflows into the above mentioned consumer, machine, and caregiver channels. This work will include the necessary formalization procedures: standardization, definitions of customer and stakeholder interfaces, identification of data models and data processing tools, and privacy and security policies and recommendations.

The necessary conditions for incorporation of the available and future ICTs to the AGE-Care ecosystem are mostly related with adaptation to the paradigms of:

- multimodal augmented reality (AR) data output for consumers;

- Dew computing (and available M2M standards inside it) for basic and automatic decision making;

- multilayer interaction between Cloud, Fog, and Dew computing for advanced (mostly automatic and limited manual) decision making.

D. Some Implemented Combinations of Components

Several combinations of the new ICTs (which are actually the components of the AGE-Care ecosystem) are already implemented by authors and their detailed explanation and related background can be found elsewhere in the related publications, for example:

- *Frameworks for Integration of Workflows and Distributed Computing Resources*: gateway approaches in science and education [16-18];

- *Dew (+ Fog + Cloud) computing + IoT + IoE*: the conceptual approach for organization of the vertical hierarchical links between the scalable distributed computing paradigms: Cloud Computing, Fog Computing and Dew Computing, which decrease the cost and improve the performance, particularly for IoT and IoE [19];

- *AR + visual + tactile interaction modes*: to provide tactile metaphors in education to help students in memorizing the learning terms by the sense of touch in addition to the AR tools [20,21];

- *ML + visual + tactile interaction mode*: to produce the tactile map for people with visual impairment and recognize text within the image by advanced image processing and ML [22];

- *IoT for eHealth (wearable electronics) + ML + AR + brain-computing interface + visual interaction mode*: to monitor, analyze, and estimate the accumulated fatigue by various gadgets and visualize the output data by AR means [23-25].

IV. CONCLUSIONS

The proposed integrated ecosystem provides the basis for effective personalized elderly care by introduction of multimodal personalized communication channels. It allows end users to get cumulative effect from mixture of ICTs like IoT/IoE, multimodal AR, and predictive ML approaches. As a result, it could exclude obtrusive H2M/M2H technological interactions by delivering them to M2M interactions encapsulated in Dew Computing layer, and enhancing the pleasant multimedia H2M/M2H intuitive interactions. It hides "behind the curtains" the mental and cognitive overloads by: shifting the most portion of ICT-related interactions from H2H to M2M zone; using AR pathways for delivering status information and advices for elderly end users; increasing AR-readiness of the available ICTs for AR-output of data for non-obtrusive H2M/M2H interactions, and improving every-day communication and service needs. It could be the integral platform and paradigm for overcoming cognitive, cultural, mental, gender/ethical, institutional, regional, and national barriers and enabling the targeted delivery of integrated elderly care on European and worldwide scale by joining efforts across governmental, non-governmental, and volunteer elderly care organizations and individuals. In this way elimination of any kinds of "borders" between people at European (and worldwide) scale by targeted efforts can strengthen the relationships between the different age categories of people and various elderly communities despite their intrinsic or imposed differences.

ACKNOWLEDGMENT

The work was partially supported by Ukraine-France Collaboration Project (Programme PHC DNIPRO) (http://www.campusfrance.org/fr/dnipro), EU TEMPUS LeAGUe project (http://tempusleague.eu), and Croatian Centre of Research Excellence for Data Science and Advanced Cooperative Systems.

REFERENCES

[1] Vienna Institute of Demography (2016) (http://www.populationeurope.org)

[2] Boll, S., Heuten, W., Meyer, E. M., & Meis, M. (2010). Development of a multimodal reminder system for older persons in their residential home. Informatics for health and Social Care, 35(3-4), 104-124.

[3] Lisetti, C., Nasoz, F., LeRouge, C., Ozyer, O., & Alvarez, K. (2003). Developing multimodal intelligent affective interfaces for tele-home health care. International Journal of Human-Computer Studies, 59(1), 245-255.

[4] Ziefle, M., Rocker, C., & Holzinger, A. (2011). Medical technology in smart homes: exploring the user's perspective on privacy, intimacy and trust. In IEEE Proc. 35 Annual Computer Software and Applications Conference Workshops (pp. 410-415).

[5] Czaja, S. J., & Lee, C. C. (2007). The impact of aging on access to technology. Universal Access in the Information Society, 5(4), 341-349.

[6] Smith, A. (2014). Older adults and technology use: Adoption is increasing but many seniors remain isolated from digital life. Pew Research Center (http://www.pewinternet.org/2014/04/03/older-adults-and-technology-use).

[7] Dimitrova R. (2013). Growth in the intersection of eHealth and active and healthy ageing. Technol Health Care; 21(2):169-72.

[8] Hernandez C, Alonso A, et al. Integrated care services: lessons learned from the deployment of the NEXES project. Int J Integr Care. 2015; 15.

[9] Frenk J. (2009). Reinventing primary health care: the need for systems integration. Lancet. 374 (9684): 170-3.

[10] Haux R, Hein A, Kolb G, Kunemund H, Eichelberg M, Appell JE, et al. (2014). Information and communication technologies for promoting and sustaining quality of life, health and self-sufficiency in ageing societies--outcomes of the Lower Saxony Research Network Design of Environments for Ageing (GAL). Inform Health Soc Care, 39(3-4):166-87.

[11] Meiland FJ, Hattink BJ, Overmars-Marx T, de Boer ME, Jedlitschka A, Ebben PW, et al. (2014). Participation of end users in the design of assistive technology for people with mild to severe cognitive problems; the European Rosetta project. IntPsychogeriatr., 26(5):769-79.

[12] Prensky, M. (2001). Digital natives, digital immigrants part 1. On the horizon, 9(5), 1-6.

[13] White, D. S., & Le Cornu, A. (2011). Visitors and Residents: A new typology for online engagement. First Monday, 16(9).

[14] Haberkern, K., & Szydlik, M. (2010). State care provision, societal opinion and children's care of older parents in 11 European countries. Ageing and Society, 30(02), 299-323.

[15] Jacobs, M. T., van Groenou, M. I. B., Aartsen, M. J., & Deeg, D. J. (2016). Diversity in older adults' care networks: the added value of individual beliefs and social network proximity. J.Gerontol. B, Psychol. Sci. Soc. Sci, DOI: 10.1093/geronb/gbw012.

[16] Davidović, D., Lipić, T., & Skala, K. (2013), AdriaScience gateway: Application specific gateway for advanced meteorological predictions on croatian distributed computing infrastructures. In Proc. IEEE 36th International Convention on Information & Communication Technology Electronics & Microelectronics (MIPRO), (pp. 217-221).

[17] Gordienko, Y., Bekenov, L., Baskova, O., Gatsenko, O., Zasimchuk, E., & Stirenko, S. (2015). IMP Science Gateway: from the Portal to the Hub of Virtual Experimental Labs in e-Science and Multiscale Courses in e-Learning. Concurrency and Computation: Practice and Experience, 27(16), 4451-4464.

[18] Gordienko, Y., Stirenko, S., Gatsenko, O., & Bekenov, L. (2015). Science gateway for distributed multiscale course management in e-Science and e-Learning—Use case for study and investigation of functionalized nanomaterials. In Proc. IEEE 38th International Convention on Information and Communication Technology, Electronics and Microelectronics (MIPRO) pp. 178-183.

[19] Skala, K., Davidovic, D., Afgan, E., Sovic, I., & Sojat, Z. (2015). Scalable distributed computing hierarchy: Cloud, fog and dew computing. Open Journal of Cloud Comp. (OJCC), 2(1), 16-24.

[20] E. Artetxe González, F. Souvestre, J.R. López Benito (2016), Augmented Reality Interface for E2LP: Assistance in Electronic Laboratories through Augmented Reality, in Embedded Engineering Education, Advances in Intelligent Systems and Computing (Volume 421, Chapter 6), Springer.

[21] I. Kastelan, J.R. Lopez Benito, E. Artetxe Gonzalez, J. Piwinski, M. Barak, M. Temerinac (2014), E2LP: A Unified Embedded Engineering Learning Platform, Microprocessors and Microsystems, Elsevier, Vol. 38, Issue 8, Part B, pp. 933-946.

[22] Nizar Bouhlel and Anis Rojbi (2014), New Tools for Automating Tactile Geographic Map Translation, Proc. 16th Int. ACM SIGACCESS Conf. on Computers & Accessibility, pp.313-314.

[23] Gordienko, N., Lodygensky, O., Fedak, G., & Gordienko, Y. (2015). Synergy of volunteer measurements and volunteer computing for effective data collecting, processing, simulating and analyzing on a worldwide scale. In Proc. IEEE 38th International Convention on Information and Communication Technology, Electronics and Microelectronics (MIPRO) (pp. 193-198).

[24] Gordienko, N. (2016). Multi-Parametric Statistical Method for Estimation of Accumulated Fatigue by Sensors in Ordinary Gadgets, arXiv preprint arXiv:1605.04984.

[25] Stirenko, S., Gordienko, Yu., Shemsedinov, T., Alienin, O., Kochura, Yu., Gordienko, N., Rojbi, A., López Benito, J.R., Artetxe González, E. (2017) User-driven Intelligent Interface on the Basis of Multimodal Augmented Reality and Brain-Computer Interaction for People with Functional Disabilities, IEEE Ukraine Conf. on Electrical and Computer Engineering (UKRCON-2017).

Cloud-Dew Computing Support for Automatic Data Analysis in Life Sciences

Peter Brezany*, Thomas Ludescher[†] and Thomas Feilhauer[‡]

* Research Group Scientific Computing, Faculty of Computer Science, University of Vienna, Vienna, Austria
and SIX Research Centre, Brno University of Technology, Brno, Czech Republic
Email: peter.brezany@univie.ac.at
[†] Department of Computer Science, University of Applied Sciences, Dornbirn, Austria
Email: thomas@ludescher.at
[‡] Department of Computer Science, University of Applied Sciences, Dornbirn, Austria
Email: thomas.feilhauer@fhv.at

Abstract—In this paper we show how the technologies associated with the evolution of Cloud computing to Dew computing can contribute to the advancing scientific computational productivity through automation. In the current big data paradigm developments, there is growing trend towards automation of data mining and other analytical processes involved in data science to increase productivity of associated applications. There are already several efforts to create automated data science platforms. However, these platforms are prevalently oriented towards business and engineering application domains. This paper addresses the automatic data analysis enabled by Cloud-Dew computing in the context of the life-science sector, in particular, in two application domains: breath gas analysis and brain damage restoration.

Keywords-automatic scientific data analysis, Cloud-Dew computing, breath gas analysis, brain damage restoration

I. INTRODUCTION

Cloud-Dew architecture is a fundamental computing architecture that concerns the distribution of workloads between Cloud services and on-premise computers [1]. Cloud-Dew Architecture and its applications also lead to the creation of a completely new area: **Dew Computing**. We believe that the technologies associated with the evolution of Cloud computing to Dew computing could significantly contribute, among others, to the advancing scientific computational productivity through automation.

In the current big data paradigm developments, there is growing trend towards automation of data mining and other analytical processes involved in data science to increase productivity of associated applications. There are already several efforts to create automated data science platforms. A dedicated web site http://www.kdnuggets.com/software/automated-data-science.html brings up-to-date information about this development that, however, takes mostly in business and engineering application domains.

This paper addresses the automated data mining issues in the context of the life-science sector, in particular two application domains:

- *Breath gas analysis.* It tries to detect lung and oesophageal cancers in a non-invasive manner. Other related research areas, e.g., testing breath gas reactions to different swallowed chemical substances, are addressed as well [2].

- *Brain damage restoration.* Brain disorders occur when our brain is damaged or negatively influenced by injury, surgery, disease, or health conditions. Close cooperation of diverse medical, therapy, IT, and other specialists is needed to find methods that improve and support healing processes as well as to discover underlying principles [3].

Before starting the design of our automatic analysis framework (**AAF**), we conducted an exhaustive analysis of existing automatic analysis systems, categorized them, identified potential gaps in them, and proposed an original taxonomy that can be also helpful for other potential developers of such frameworks. This effort is presented in Section II. Based on this taxonomy/systematic start, we developed a hybrid Cloud-based AAF and applied it to the data repositories captured in the breath gas analysis domain; Section III addresses this effort. Section IV visionary proposes reengineering of the Cloud-based AAF to a platform based on Dew Computing; some design decisions profit from the taxonomy elaborated in Secttion II. Finally, we briefly conclude in Section V.

II. AUTOMATIC DATA ANALYSIS TAXONOMY

Our taxonomy characterizes and classifies approaches of all well-known automatic analysis systems (Figure 1). It is based on six elements that describe an automatic analysis system: (a) automation mode, (b) start process, (c) input data, (d) workflow construction, (e) algorithms used, and (f) algorithms selection. The following subsections describe each element and its taxonomy.

A. Start Process

The *Start Process* determines how an automatic analysis system can be triggered.

Manually: A user starts the process by hand. For example, the research group leader can start the process using all available input data.

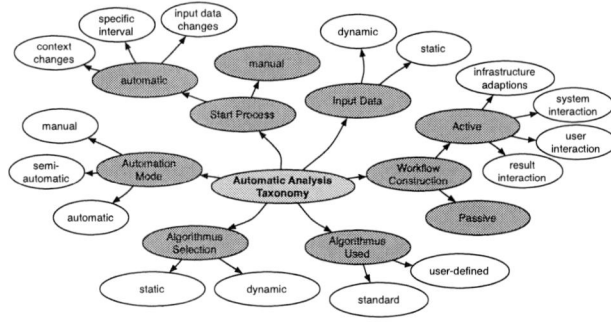

Figure 1. Automatic Data Analysis Taxonomy

Automatically: There exist several different possibilities to start an automatic analysis system automatically. Triggering events can be divided into (a) input data changes (e.g., multiple database updates/inserts), (b) a specific time interval (e.g., once a week), (c) a context change (e.g., accident with chemicals, stock marked crash), or a concept drift or shifts are detected; this phenomenon is typically associated with modeling data streams [4].

B. Input Data

The element *Input Data* characterizes the behavior of the input data in the scientific analysis with regard to changes.
Static: If an automatic analysis system uses the same specific input dataset (reload) in each workflow execution, the input data is called static.
Dynamic: If the input data (deep pull) is changing (e.g., new data, update data) after an execution, the automatic analysis system generates new results, which can be differing from the previous execution. The system prepares the input data automatically. This procedure is called dynamic.

C. Workflow Construction

In [5] the *workflow construction* is divided into *passive* and *active*.
Passive: This approach is based on the assumption that the workflow composer is able to compose a reasoning-based complete workflow description involving all possible scenarios of the workflow engine behavior and reflecting the status of the involved Grid/Cloud resources and task parameters provided by the user at the workflow composition time. The composition of the running workflow will not be adapted and therefore a passive system does not reflect the 'state of the world'.
Active: This approach assumes a kind of intelligent behavior by the workflow engine supported by an inference engine and the supporting knowledge base. Workflow composition is done in the same way as in the passive approach, but its usability is more efficient because it reflects a state

of the world (e.g., changes of the involved infrastructure resources).

D. Algorithms Used

Algorithms Used determines the application area of the mathematical method.
Standard: Common data mining algorithms can be used in almost all applications, such as linear methods, neural network, principal component analysis, etc. Automatic analysis systems, which are using common data mining algorithms, are generally applicable (e.g., the Weka system[1]).
User-defined: In some research areas, standard algorithms don't achieve the desired quality or don't work at all. In such a case, the researchers must develop their own algorithms (user-defined) and the automatic analysis system must be able to deal with these algorithms.

E. Algorithms Selection

An automatic analysis system provides different algorithms, which can be used in the execution process. The selection of the algorithms can be *static* or *dynamic*.
Static: If the user of the system must choose the provided algorithms by hand (manual), it is called *static*.
Dynamic: If the system chooses the suitable algorithms automatically, it is called *dynamic*.

F. Automation Mode

The *automation mode* is the cross selection of all involved automatic components in such a system. These components are located at (a) the *start process* (p_S), (b) the *input data* (p_I), (c) the *workflow construction* (p_W), and (d) the *algorithms selection* (p_A). Each component can be either *manual* or *automatic*. The different combinations of these processes $P = (p_S, p_I, p_W, p_A)$ can result in (a) automatic, (b) manual, or (c) semi-automatic mode.
Manual: The *automation mode* is manual if all decisions must be done manually. The researcher must (a) start the process by hand, (b) select the input data, (c) choose the workflow and the corresponding machine, and (d) select the provided algorithms.

$$\forall p \in P \mid p = \text{manual} \implies \text{mode} = \text{manual}$$

Automatic: In comparison to *manual*, the automation mode is *automatic*, if all processes work completely stand alone, which means, without any further user interaction.

$$\forall p \in P \mid p = \text{automatic} \implies \text{mode} = \text{automatic}$$

Semi-automatic: A combination between at least one *manual* and one *automatic* process is called *semi-automatic*.

$$\forall (p_1, p_2) \mid p_1 \in P, p_2 \in P,$$
$$p_1 = \text{manual} \ \land \ p_2 = \text{automatic}$$
$$\implies \text{mode} = \text{semi-automatic}$$

---

[1]http://www.cs.waikato.ac.nz/~ml/weka/

III. AUTOMATIC ANALYSIS FRAMEWORK

In this section, we explain the design and application of the AAF having different levels of automatization. On level one, a researcher manually starts the analysis process, which automatically uses different data mining algorithms and multiple parameter sets (AAF-1). This analysis can be used to (a) evaluate the quality of the data or (b) to analyze this data with well-known common data mining algorithms (e.g., standard classification algorithms, such as linear regression). On the second level, the Automatic Analysis Framework (AAF-2) is executed automatically, using the newest available input data. This trigger automatically fires, for example, (a) once a week, (b) through database changes (e.g., ten new datasets), or (c) through context changes (e.g., accident with chemicals, stock marked crash, etc.). The AAF-2 can be useful for several different approaches. For example a research institute collects new patient data with corresponding sensor values (e.g., breath samples) on a daily basis. It is possible that new research areas will be detected. For example, if the newly added data consists of e.g., lung cancer proband data, it is possible that common data mining algorithms are able to separate (classify) patients with lung cancer from healthy test persons automatically. On the third level of the Automatic Analysis Framework (AAF-3) existing domain knowledge is reused during the whole analysis process, to achieve better results than with common data mining algorithms (AAF-2, which uses no domain knowledge). Design principles of all AAF-levels are described in details in [6].

A. General Workflow

The AAF uses a specific process to generate new knowledge automatically. We designed and developed the workflow of the AAF, which contains (a) data preparation, (b) data analysis, and (c) result presentation (Figure 2). For example, in the data preparation step the researcher selects the dependent and independent variables, if the classification technique should be applied. The data analysis step contains all supported analytical methods (e.g., common data mining algorithms such as neural network, linear methods, principal component analysis or already stored domain dependent algorithms, etc.), which can be deactivated by the researcher if a method is not promising. All algorithms used will be evaluated (e.g., percentage of correctly classified samples, over-fitting) and its output ranks the corresponding model in the result presentation step. In many cases the output of the result presentation step is a simple HTML-report, a decision tree, a neural network, etc. The researcher uses this report for further investigations.

B. System Overview and Implementation

The whole AAF executes a lot of independent data mining algorithms (e.g., generate a model to classify smoker and none-smoker with multiple different independent variables).

All algorithms are executed in the Cloud using our **Code Execution Framework**, which can execute code of different problem solving environments, such as MATLAB, R and Octave, in parallel [7], [8]. Most problem solving environments are implemented as single threaded programs; because of this constraint, the execution cannot use the power of current computers with multiple cores. The framework supports different Cloud infrastructures, such as Amazon EC2 and Eucalyptus. Therefore it is possible to use hybrid Cloud infrastructures, e.g. a private Cloud based on Eucalyptus for general base-level computations using the available local resources and additionally a public Amazon EC2 for peak-load and time-critical calculations. The approach is to provide a secure platform that supports multiple problem solving environments, execute code in parallel with different parameter sets using multiple cores or machines in a Cloud environment, and support researchers in executing code, even if the required problem solving environment is not installed locally.

Figure 2 displays all sub systems involved. The Code Execution Controller (CEC) is the entry point of the Code Execution Framework (CEF). The controller provides several different Web services to add new calculations or monitor running calculations. The CEC is further responsible for (a) generating all sub-calculations (e.g., one calculation for each parameter-set), (b) start/stop/monitor virtual machines, (c) start queued sub-calculations on idle virtual machines, and (d) monitor running sub-calculations. The advantage of using our Cloud-based CEF is that the resources (Virtual Machines) can be added on the fly, depending on the required CPUs, required memory, or number of waiting calculations.

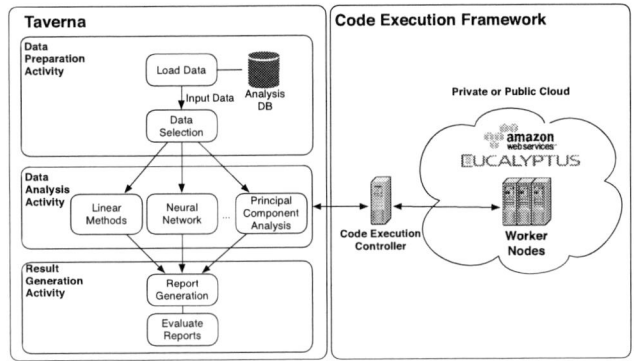

Figure 2.   System diagram of AAF and workflow

We use **Taverna**, an open source and domain-independent Workflow Management System [9]. Taverna provides several different activities for data analysis (e.g. classification, prediction, clustering). The common activities use the local or remote machines for calculations. We provide several new activities. Our statistical algorithm activities use the CEF for the calculations and therefore a Cloud based infrastructure is used. With Taverna the AAF-workflow can be adopted and used for many different purposes.

C. Continuous Automatic Analysis (AAF-2) in action

This section describes how the Automatic Analysis Framework (second level) can be used (a) to gain new insights into current research areas or (b) to detect new topics of interest. The AAF-2 is similar to the AAF-1. The workflow and the Taverna activities can be used one-by-one. The difference to the AAF-1 is (a) that the execution will be started automatically (e.g., once a week, after a certain number of database updates/inserts, etc.) and (b) that all possible useful dependent variables and independent variables are selected via the activity *Data Preparation*.

The AAF-2 workflow (Figure 2) is executed continuously to preprocess and analyze all newly collected data. To execute the workflow continuously we use the Taverna Command Line Tool in combination with a standard Unix cronjob. The *Load Data* part is responsible to request all data from the database and converts this data into a valid format for further automatic analysis. The activity *Data Selection* contains different configuration possibilities compared to the activity used in AAF-1. In general, the AAF-2 tries to describe each data column with all other columns (independent variables), all different combinations and available analytical methods. In many cases, the huge amount of all different combinations must be reduced to save computation time. This can be done by user input or automatically by learning algorithms. The following list outlines five possibilities to address this problem: 1) The researcher is able to define which data column should not be classified (exclude list). For example, the input data of the breath community contains (a) personal related data and (b) breath data (substances with their concentration). The community tries to classify personal related data (e.g., smoker, lung cancer) with the breath data. Here, it makes no sense to try to classify substances/compounds; 2) The researcher is able to define a maximum number of independent variables to shrink the total number of calculations and reduce the likelihood of overfitting; 3) The researcher can remove non-promising algorithms (e.g., linear regression) manually; 4) Depending on the type of the dependent variable, not all statistical algorithms can be used and therefore can be skipped; 5) The activity *Data Selection* automatically evaluates each column if it is a potential dependent variable. This will be done by (a) looking at the data type or (b) statistical analysis. For example, if the datasets contain only one smoker and all others are none-smokers, it is useless to try to separate smokers from non-smokers.

The activity *Evaluate Reports* is responsible for (a) comparing the new results with the previous results from, e.g., the last week, (b) detecting new topics of interest, and (c) informing the responsible person about the result by email.

For example, if the AAF detects a significant difference between the actual result and the result from last week, an automatic email will be sent to the researcher. The researcher can then further investigate the new input data and possibly improve the existing model. A subsequent automatic email will be sent, if the AAF detects a new topic of interest that promises a good result.

To detect new topics of interest, the AAF-2 generates a ranked output of all calculated models. If a high number of correctly classified samples of a new research topic (e.g., distinguishing healthy persons from persons with cancer) can be achieved, the research leader will be notified by email. The administrator can configure the threshold of the minimum percentage of correctly classified samples. Furthermore, the ranked output can easily be displayed to the researchers to support further investigations.

The AAF is able to calculate the independent calculations (e.g., classifications) in parallel in the Cloud. If the researcher would like to try all different combinations of independent variables and algorithms, the number of different calculations is rapidly increasing. For example, if the input data consists of five possible dependent variables and ten possible independent variables and the researcher would like to try all combinations with ten different algorithms, the AAF must execute 51150 different independent calculations. If the number of independent variables is 20, the AAF system must execute over 52 million calculations. The AAF evaluates each individual calculation separately and generates a ranking. The researcher can use these ranked results for further investigations. This automatic analysis reduces the required time for the involved researchers dramatically.

IV. AUTOMATIC ANALYSIS FRAMEWORK AND DEW COMPUTING

Dew computing is a computing paradigm that sets up on top of Cloud computing and overcomes some of its restrictions like the dependency on the Internet.

A. Principles of Dew Computing

There are different definitions of Dew computing [1] [10]. For the context of this elaboration, we follow Yingwei Wang's definition of Dew computing: *"Dew computing is an on-premises computer software-hardware organization paradigm in the Cloud computing environment where the on-premises computer provides functionality that is independent of Cloud services and is also collaborative with Cloud services. The goal of Dew computing is to fully realize the potentials of on-premises computers and Cloud services"* [11]. This guarantees that the offered services are independent of the availability of a functioning Internet connection. The underlying principle for this is a tight collaboration between on-premise and off-premise services based on an automatic data exchange among the involved compute resources.

368

The Dew computing paradigm can also be used to improve the AAF. From the very beginning, its design has been based on services and the services of the CEC are implemented in form of micro-services that can be invoked from all its clients (Taverna, worker nodes). It has always been a central principle of the AAF to use internal and external compute resources. While Taverna and the CEC is typically installed on on-premise machines, the data analysis activities can be started on worker nodes deployed on on-premise or off-premise machines by using a common Cloud infrastructure. Still there is the problem of failures in the external part of our system over which we do not have immediate control. It would be an extreme benefit for the AAF if the analysis activities could continue even if the Internet connection to the Cloud breaks or if some of the worker nodes in the Cloud fail. In the case of a broken Internet connection, the current architecture of the AAF would require to restart all analysis activities running on external worker nodes, which for long running jobs could imply an unacceptable waste of time and money for the costly compute resources.

B. Intensified usage of on-premise resources

Our AAF can be extended corresponding to the principles of Dew computing. An idea for a stronger involvement of on-premise computers is to directly use these resources when running idle. While the computers in the private Cloud are reserved exclusively for the analysis activities of the AAF, many institutions have a certain number of (workstation) computers that are typically used only part-time. When the on-premise computers are not used for other tasks (if they run idle), the installed CEF-client notifies the CEC and signals that it is available for analysis activities. The CEC then integrates this on-premise computer into its set of worker nodes and provides it with analysis activities. If the on-premise computer is later on required for other tasks, CEC tasks are terminated and the results stored into the local data source.

C. Adapting AAF towards DBiD

In [11] Yingwei Wang suggests to provide redundancy among on-premise and off-premise computers along with synchronization of the data sets to cope with the problems of hardware and network failures. In our case it means that data sources (analyzed DB) have to be provided redundantly, one on-premise DB and an off-premise DB for each public Cloud region in which the AAF is currently active. The off-premise DBs will automatically synchronize their data with the local DB. We can use standard mechanisms for one-way synchronization of databases [12]. If the connection to the Cloud breaks or the Cloud data source fails, the CEC still can access the intermediate results of the analysis activities completed before the failure occurred and can restart the uncompleted analysis activities on on-premise computers or

in another Cloud. Yingwei Wang [11] terms this automatic synchronization between databases on on-premise computers and databases on off-premise computers "Database in Dew" (DBiD).

D. Adapting AAF towards PiD

If the AAF should be extended so that the worker nodes in the Cloud can be enabled to complete their analysis activities with a broken Internet connection, then the CEC must also be provided redundantly with an additional CEC in the Cloud region of the corresponding worker nodes. The on-premise CEC then forwards the activities to the off-premise CEC which distributes them among its worker nodes. If the Internet connection between on-premise CEC and off-premise CEC breaks then the off-premise CEC can still continue to coordinate the execution of the activities. If there is also an off-premise copy of the CEC installed in the AAF-system, then all components are redundantly implemented on- and off-premise with all settings and application data being dynamically synchronized. According to Yingwei Wang [11] we could characterize this version of the AAF as "Platform in Dew" (PiD).

To illustrate the complete suggested Dew computing extension of the AAF, Figure 3 shows an updated version of the original AAF system diagram from Figure 2.

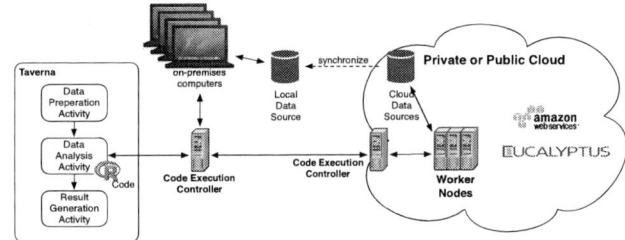

Figure 3.   AAF architecture supported by Dew computing

By making the AAF compliant to Dew computing, we can improve the reliability of the system and by adding on-premise computers, the performance can be enhanced while the costs will be reduced.

E. Application in the Balance Disorder Rehabilitation

Balance disorders create a significant category of brain damage impairments. In the context of the innovative medicine research project, we develop a biofeedback-based balance disorder rehabilitation training [13]. The patient is standing on a measuring instrument called a force platform [15] that is wireless connected to a computing platform and following balance stimulation tasks presented on the display. Besides video/audio feedback he/she and a therapist are receiving, a set of body sensors is streaming physiological signals that are processed, integrated with other data and analysed. Based on the knowledge extracted, the therapy

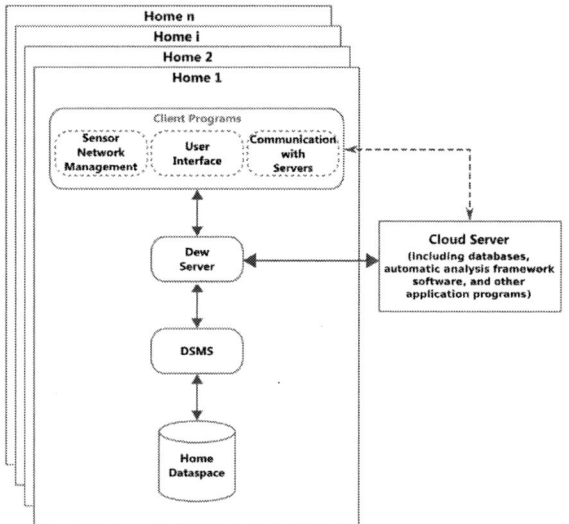

Figure 4. Cloud-Dew architecture for balance disorder training, where Dew is represented by the home infrastructure. Based on [13], [14]

scenario is continuously adapted by either the therapist or automatically by intelligent software modules. Continuous analysis of the latest status of the available data sets (dataspace) provided by the AAF can be a significant contribution to the therapy efficiency. Another issue is the sustainability need of the improvement achieved. The traditional model assuming rehabilitation conducted in highly-specified rehabilitation centers (clinics) that involves therapy pauses due to the high costs doesn't support this aim. Therefore, there are already efforts to provide therapy at homes equipped with appropriate technical resources, too. Our long-term goal is to create a network of cooperating nodes, where a node denotes a Cloud server associated with a rehabilitation center steering a set of home infrastructures. Each home system works autonomously, even if the Internet connection is not available at the moment, and in specific intervals exchanges collaborative information (analysis results, data mining models, etc. produced by the AAF) with the center involved in the Cloud. This Cloud-Dew concept is reflected in Figure 4. The Sensor Network Management steers the dataflow from sensors to the Home Dataspace managed by the Data Space Management System (DSMS) [7].

V. CONCLUSIONS

We provided the definition of the term automatic analysis and proposed the novel automatic analysis taxonomy based on six categorization elements we specified. The taxonomy steered the design of the three-level Automatic Analysis Framework (AAF-1,2,3). In the context of the AAF-2 and the Breath Gas Analysis domain, we introduced the features and application power of this framework that was realized on the basis of several (incl. Cloud) technologies. Next,

migration to the framework realization within the ongoing Dew Computing development was discussed; the provided rationales were underpinned by a brief vision outline of the novel Dew-enabled balance disorder rehabilitation approach.

REFERENCES

[1] Y. Wang, "The initial definition of dew computing," *Dew Computing Research*, 2015.

[2] A. Amann and D. Smith, "Breath analysis for clinical diagnosis and therapeutic monitoring," World Scientific, 2005.

[3] P. Brezany *et al.*, "Optimized management of big data produced in brain disorder rehabilitation," in *Big Data Optimization: Recent Developments and Challenges*, ser. Studies in Big Data, A. Emrouznejad, Ed. Springer, 2016, pp. 281–318.

[4] M. Garofalakis and J. Gehrk, "Data stream management: Processing high-speed data streams (data-centric systems and applications)," 2016.

[5] P. Brezany, I. Janciak, and A. M. Tjoa, "Ontology-Based Construction of Grid Data Mining Workflows," In: H. O. Nigro at al. (eds), Data Mining with Ontologies, pp. 182-210, Information Science Reference, 2008.

[6] T. Ludescher, "Towards High-Productivity Infrastructures for Time-Intensive Scientific Analysis," Doctoral Dissertation, Faculty of Computer Science, University o Vienna, 2013.

[7] I. Elsayed *et al.*, "Data Life Cycle Management and Analytics Code Execution Strategies for the Breath Gas Analysis Domain," *Procedia Computer Science*, vol. 9, pp. 156 – 165, 2012.

[8] T. Ludescher, T. Feilhauer, and P. Brezany, "Cloud-Based Code Execution Framework for scientific problem solving environments," *Journal of Cloud Computing: Advances, Systems and Applications*, vol. 2, no. 1, 2013.

[9] "Taverna - open source and domain independent Workflow Management System," http://www.taverna.org.uk, 2012.

[10] K. Skala, D. Davidovic, E. Afgan, I. Sovic, and Z. Sojat, "Scalable distributed computing hierarchy: Cloud, fog and dew computing," *Open Journal of Cloud Computing (OJCC)*, vol. 2, no. 1, pp. 16–24, 2015.

[11] Y. Wang, "Definition and categorization of dew computing," *Open Journal of Cloud Computing (OJCC)*, vol. 3, no. 1, pp. 1–7, 2016.

[12] P. A. Alsberg and J. D. Day, "A principle for resilient sharing of distributed resources," in *Proceedings of the 2nd international conference on Software engineering*. IEEE Computer Society Press, 1976, pp. 562–570.

[13] P. Brezany *et al.*, "A novel big data–enabled approach individualizing and optimizing brain disorder rehabilitation," in *Big Data for the Greater Good*, ser. Studies in Big Data, C. Vincent and A. Emrouznejad, Eds. Springer, 2017.

[14] Y. Wang, "Cloud-Dew architecture," *Open Journal of Cloud Computing (OJCC)*, vol. 3, 2015.

[15] "Force platform," https://en.wikipedia.org/wiki/Force_platform, Accessed February 2017.

MIPRO 2017, May 22- 26, 2017, Opatija, Croatia

Distributed Database System as a Base for Multilanguage Support for Legacy Software

Nenad Crnko

University of Zagreb, University Computing Centre – SRCE

Adding user interface with multi language support into legacy software not planed during the development phase for such type of using, can be very challenged process. In this paper is suggested innovative procedure for solving mentioned problem using a Distributed Database System as a part of the Dew Computing Paradigm. One of the key advantages of distributed database system in the building multi language extension of legacy software is related to performance.

I.   INTRODUCTION

Today many companies are using some kind of legacy software. Legacy software remains useful for several practical purposes even though it is now far beyond what its original purpose was intended to be.

Higgins [7] described the main problems with legacy software:

a.   "Information relating to implementation and features isn't complete, accurate, current, or in one place. Often it is missing altogether. Worse still, the documentation that does exist is often filled with information from previous versions of the application that is no longer relevant and therefore misleading.

b.   A lot of intellectual property is "embedded" in the application and no place else.

c.   There's incomplete or sometimes no documentation, and the accuracy and currency of the documentation that does exist is suspect.

d.   The original designers are no longer around.

e.   There have been many "surgeons" who've performed a variety of operations over the years, but none bothered to take notes on what these operations were.

f.   There's incomplete or sometimes no documentation, and the accuracy and currency of the documentation that does exist is suspect.

g.   The application is based on older technologies (languages, middleware, frameworks, interfaces, etc.)

h.   Skill sets needed to work with the old technologies are no longer available."

Additional problem is when the legacy software should be used with user interface translated to another language.

This paper will suggest the solution for the last of the problems listed above - preparing the user interface of the legacy software in another language. The solution is suitable for Windows COM (Component Object Model) based legacy software introduced by Microsoft in 1993 with available source code for rebuilding solution. Very important parts of the solution are the distributed databases organized on the principles adopted from the Dew Computing.

II.   HOW TO TRANSLATE LEGACY SOFTWARE

Very common approach for solving the problems with legacy software (especially with a user interface) is developing "wrapper" around the legacy software.

An identical approach could be used in translating legacy user interface to another (one or more) languages.

If the legacy software is Microsoft Windows COM compatible software, and the source code of the legacy software is available, then all objects from the user interface can be exposed to the wrapper COM based component [6]. The wrapper component can change all displayed text values from the original language to the destination language.

In such software environment (COM based main software and COM based wrapper component for translation) with available source code for main software, it is relatively easy to call wrapper component for every part of the legacy user interface that requires translation. Also explained in [6].

In addition to the COM based wrapper component, second equally important part for the translation process of the legacy software is the translation depository in one or more languages. One of the supported database systems within the development tool for developing COM wrapper can serve as a depository.

The existence of the database with translations is necessary, but not enough prerequisite when using translation legacy software for translation of one language to another. Using a database in combination with COM wrapper component in an inadequate way could dramatically decrease the performance of the whole system. The rcsult can be translated legacy system with very poor performance. The right place for using Dew Computing principles is within the database organization.

371

III. PROBLEMS WITHOUT DEW COMPUTING

It is very common that the COM based legacy software is developed using the client / server model. The client application via its own user interface (the goal

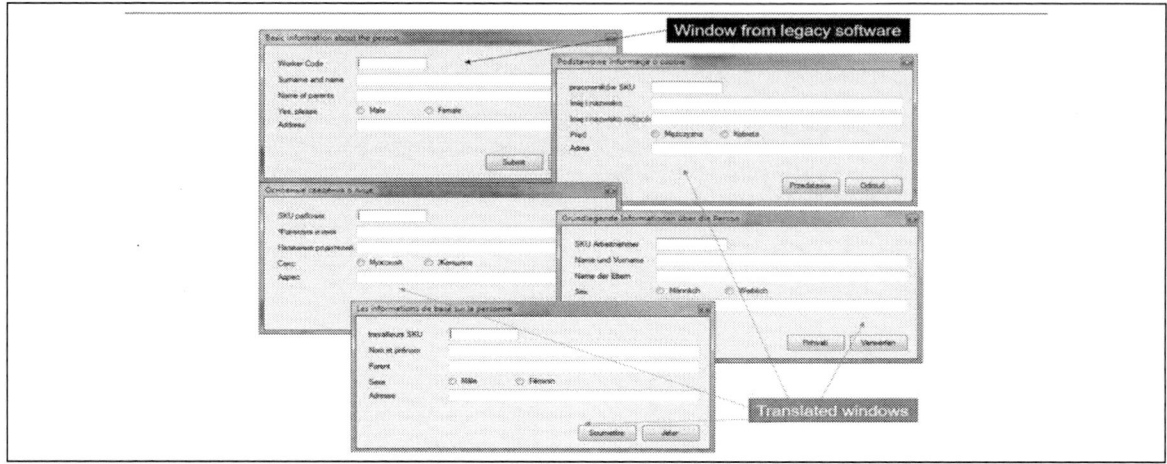

Example of English legacy software translation to several languages

of translation) sends request to server, and then shows the result in the same user interface.

If the database for translation is installed on the same server as the main database, every request for user data from the main database will at the same time trigger several requests for translation from the same server. However, it will slow down the whole system compared to the before the added translation, especially in the situations where many clients use the same server, and some clients are not part of the local area network (use internet for data access).

The second type of solution could be a local database with all translations on the same computer with the client as a part of the legacy application. In such configuration, wrapper component for translation will send and receive requests for translation within the same computer thus avoiding network traffic. Regardless of the potentially huge number of requests for the translation, the speed of the whole system stays at an acceptable level. But, this configuration can result in another problem - maintenance of huge number of local databases with updated version of translation, e.g. during the implementation of the new language to the user interface.

IV. DEW COMPUTING SOLUTION

Based on [9] - "Normally, the dew database is a partial replica of the central database. The central database contains data of all users, but the dew database can only contain data of the current user. Therefore, the dew database is a subset of the central database. In special cases, should the application require, it is possible for the dew database to contain data beyond the central database."

The organization of the whole translation system can be transformed to Dew Computing solution. There are two database types in regard to this solution: local database within the same computer as the legacy client and a

central database on the server. That can be the same server used for the main database with user data, or some other server used just for the purpose of translation. Using the local database solves the problem of speed, and using the central database which allows data synchronization with

several copies of the local databases, solves the problem of the maintenance.

Various database system can be used in both cases, but in this example SQLite is used as the local database and MySQL as the central database.

SQLite is chosen for the client due to its small size, speed and transaction support. "SQLite is a compact library. With all features enabled, the library size can be less than 500KiB, depending on the target platform and compiler optimization settings." [13] "SQLite generally runs faster the more memory you give it. Nevertheless, performance is usually quite good even in low-memory environments." [13] On the other side, as central database for translation, MySQL is chosen as it is " the most popular Open Source SQL database management system" [14], proven in huge number of different projects regardless of their size.

The structure of the whole system is shown on the second picture. The main components of the systems are a couple of the old components and four new components.

Old components:

a) Old legacy software - old software that should be translated in user interface part to one or more new languages. The source code of the legacy software should be available and the software should be Windows COM based.

b) Old database - database with the user data used in old legacy software. This part has no influence on the translation process.

New components:

a) Local database with translation for every client with the legacy software

b) Central database with translations on the server

c) Local component (wrapper) for translating user interface of the legacy software and integrated communication functions within central database.

translation is located within the local database (of course much slower). At the same time, the system saves this translation in the local database for future quick access.

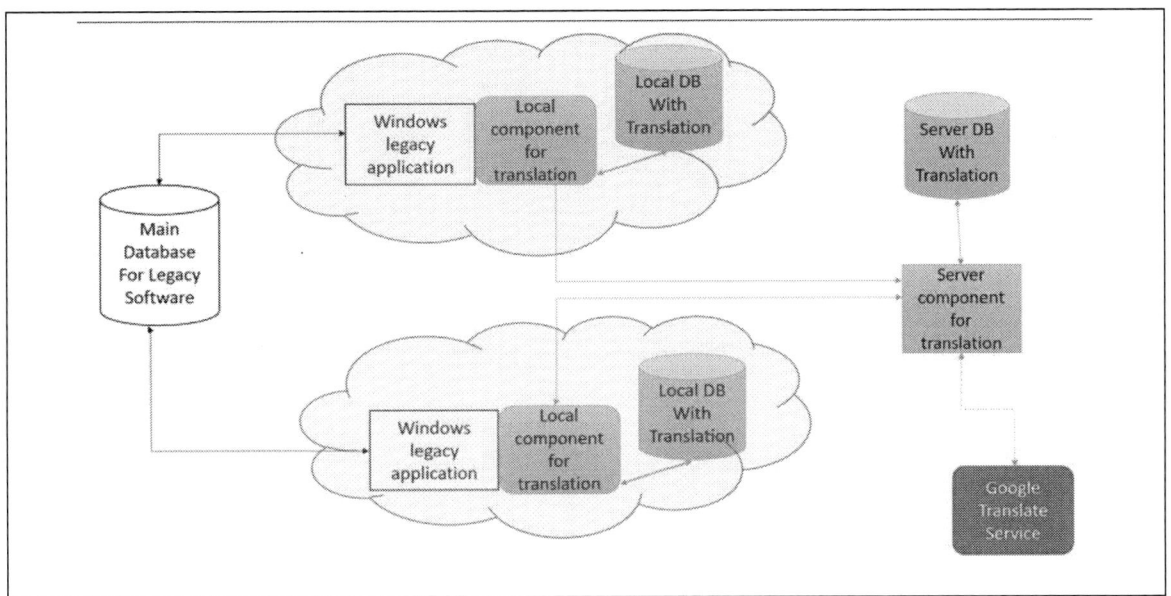

The structure of the system for translation legacy

d) Server component for processing requests from the local component and read/save data to the central database.

V. FUNCTIONING OF THE SYSTEM

There are many operations that should be performed correct and fast simultaneously while using the system that includes translation. That can be achieved by basing the whole system on the Dew Computing structure.

Typical operations displayed while showing one window from the user interface in legacy software are as follows:

a) Legacy software prepares a standard window in original language.

b) Before showing the window to the user, legacy software calls the new COM based wrapper for translation.

c) The COM wrapper can access all exposed objects and textual values in the original language. First, it tries to find a translation for values in local translation database.

d) If the translation exists there, it replaces the original value in user interface with the value from the local database. In that case, the process of translation is finished.

e) If the translation does not exist in the local database, local component sends the request for translation to the central database via software component on the server side.

f) Server component returns the translation via the local component of the legacy software as if the

At the first request for translations, there is a possibility that all translations can be copied from the central to local databases (slow process), if the local databases were empty at the beginning. After that, every client with the legacy software can work fast and independently of the central database.

VI. ADDITIONAL POSSIBILITIES IN THE SYSTEM

There are several possible extensions of the described translation system:

a) Procedure for forced copy of the whole central database to local databases and overwrite existing translations can be implemented within the translation components. This is a very useful option for initial setup of the local databases, or synchronization of the translations after correction in translations is being made within the central database.

b) The special user interface for entering initial values for translation can be implemented via local or server component for translation. In the case of the local implementation, local component should contain the code for reverse copying of the translation - from local to central database.

c) In cases where translation cannot be found in local or central database, the component for translation on the server can implement additional procedure for searching translation via Google Cloud Translation service [12]. Every translation received with such request may not be ideal, but it will almost certainly be better than the value in the original language without translation. Value received from the Google Cloud Service can be saved into the central base for translation, and later, via the standard procedure for translating, copied to the local database.

VII. CONCLUSION

The described system based on Dew Computing implementation of the databases can be very effective when used in the translation of the user interface in the legacy software. The necessary premises for that are the Windows COM based legacy software and the existence of the source code of the system itself. The source code is needed for the integration with the wrapper COM component for translation without the need for deeper analysis of the legacy software functioning.

The same framework described in this paper can be used for creating other kind of Dew Computing software based on the distributed databases.

REFERENCES

[1] A. Rindos, Y. Wang, "Dew Computing: the Complementary Piece of Cloud Computing", 2016 IEEE International Conferences on Big Data and Cloud Computing (BDCloud), Social Computing and Networking (SocialCom), Sustainable Computing and Communications (SustainCom), 15-20, 2016.

[2] K. Skala, D. Davidović, E. Afgan, I. Sović, Z. Šojat, "Scalable Distributed Computing Hierarchy: Cloud, Fog and Dew Computing", Open Journal of Cloud Computing (OJCC) Volume 2, Issue 1, 16-24, 2015.

[3] M. Abdelshkour, "IoT, from Cloud to Fog Computing," http://blogs.cisco.com/perspectives/iot-fromcloud-to-fog-computing, 2015.

[4] M. C. Mazilu, "Database Replication"; Database Systems Journal, Vol. I, No. 2, 33—38, 2010.

[5] M. T. Özsu, P. Valduriez, "Principles of Distributed Database Systems", Spinger Science+Business Media, LLC 2011.

[6] N. Crnko, "Information technology in optimization of life-long education for various language areas", 2nd International Conference "Vallis Aurea", 257-261, 2010.

[7] T. Higgins, "How to maintain, enhance legacy applications", http://searchsoftwarequality.techtarget.com/tip/How-to-maintain-enhance-legacy-applications, 2017.

[8] Y. Wang, "Cloud-Dew Architecture", Int. J. Cloud Computing, Vol. 4, No. 3, 199-210, 2015.

[9] Y. Wang, " Cloud-dew architecture: realizing the potential of distributed database systems in unreliable networks", Int. J. Cloud Computing, Vol. 4, No. 3, 199-210, 2015.

[10] Y. Wang, Y. Pan, "Definition and Categorization of Dew Computing", Int'l Conf. Par. and Dist. Proc. Tech. and Appl. | PDPTA'15, 85-89, 2015.

[11] Z. Šojat, K. Skala, "Views on the Role and Importance of Dew Computing in the Service and Control Technology", MIPRO 2016, May 30 - June 3, 2016.

[12] Google, "Google Cloud Translation API Documentation", https://cloud.google.com/translate/docs/, 2017.

[13] SQLite, "Documentation", https://www.sqlite.org/docs.html, 2017.

[14] Oracle, "MySQL Documentation", https://dev.mysql.com/doc/, 2017.

[15] Microsoft, "What is COM?", https://www.microsoft.com/com/default.mspx, 2017.

Gap in pagination due to withheld paper.

Pages 375-380

Toward a Framework for Embedded & Collaborative Data Analysis with Heterogeneous Devices

Mathieu Goeminne and Mohamed Boukhebouze
CETIC Research Center, B-6041 Charleroi, Belgium
Email: {mathieu.goeminne, mohamed.boukhebouze}@cetic.be

Abstract—The Internet of things (IoT) has emerged in numerous domains for collecting and exchanging large datasets in order to ensure a continuous monitoring and realtime decision-making. IoT incorporates sensors for carrying out raw data acquisition, while data processing and analysis tasks are addressed by high performance computational facilities, such as cloud-based infrastructures (remote processing approach). However, in several scenarios, the data export incurred by a remote processing approach is not desired, due to privacy issues, bandwidth limitations, or the lack of a reliable communication channel, among others. This paper presents MODALITi, a framework we develop for facing these recurring technical and social issues. In this framework, sensors collaboratively carry out a prediction algorithm by processing locally collected data. Their resources limitations are taken into account by relying on a context-aware adaptation of their behavior, as well as an optimised data exchange and processing.

I. INTRODUCTION

The Internet of Things (IoT) designates the interconnection of a variety of everyday things or objects (e.g. smart home, smart car, wearable devices, ...), that are able to interact for collecting and exchanging data [2]. Today, the IoT plays a major role in the ubiquitous collection of large datasets in many domains including e-health, smart buildings, intelligent transportation, and smart cities. To fulfill their role, these objects incorporate a huge variety of sensors – such as physiological sensors (e.g. EEG, ECG), environmental sensors (e.g. temperature monitor), and location sensors (e.g. GPS) – that catch and transmit useful events or measurement information.

Over the past few years, the growing ubiquity of the IoT and the data deluge it produces caused increasing needs in terms of analysis capacity. These needs are typically met by cloud-based analytical tools designed to massively scale up when required. However, these so-called *remote processing* solutions are not appropriate for all scenarios involving IoT data processing. Internet access restrictions, privacy-sensitive data mining applications, limited bandwidth, and more generally technical, social, or managerial limitations may make massive remote processing an inappropriate choice [3], [19], [10]. To deal with these scenarios, IoT devices must be used not only for collecting data, but also for executing *on-board* data analysis algorithms (local processing). As a result, IoT devices must be adapted for supporting ubiquitous and offline data stream mining. Moreover, data analysis tasks can advantageously be distributed in a device network, in such a way the involved devices can carry out real-time data processing in a collaborative way. The approach of an on-board and collaborative

data analysis over a network of devices can be used in many scenarios, including the continuous and automatic monitoring of epileptic patients. Despite its partially unpredictable nature, some risk factors can be considered as accurate warning signs of an epilepsy seizure [1]. Interconnected wearable devices (such as EEG, sleeping and stress trackers) can be used to continuously monitor the patients and to notify them of the risk of imminent seizure based on the detection of specific patterns in values collected by these devices. Therefore, the on-board and collaborative data analysis approach could improve the mobility and the safety of epileptics and favor a better life quality.

Despite the recent progress in IoT technology and ubiquitous computing, many issues must be addressed in order to implement efficient on-board data mining solutions, including resource limitations. Because they are intended to be deployed in various environments, including as wearable material, IoT devices have to adopt a small form factor, and are therefore limited in terms of computational power, energy consumption, storage and memory capacity and bandwidth. Another strong constraint IoT objects must deal with is the volatility of the environment in which they operate. Ubiquitous data stream mining solutions must adapt to changes in their operating context (for instance, an unexpected change in a patient's behavior) and must be resilient to unpredictable device disconnections.

To deal with these issues, we propose a new framework called MODALITi (Fra**Mew**Ork for Embe**D**ded & Coll**A**borative data ana**L**ys**I**s with He**T**erogeneous Dev**I**ces), which is currently under development. MODALITi carries out a decision tree based prediction algorithm over a network of sensors in order to facilitate the development of pervasive and privacy-sensitive applications.

MODALITi deals with the resource limitations incurred by the IoT sensors by combining both remote and local processing approaches. The framework proposes a training phase for creating machine learning models based on remote processing facilities (such as cloud amenities), because this phase typically involves resource-consuming tasks. MODALITi offers a tool set for automatically training and deploying machine learning models into heterogeneous embedded devices. This tool set transforms models into configuration documents, which can be deployed into various devices. Additionally, MODALITi provides an inter-device communication protocol for efficiently and reliably running deployed machine learning models.

MODALITi copes with context changes in the sensor environment by taking into account contextual indications provided by specific sensors. These indications are combined with the rest of the collected data for predicting the properties of interest. The integration of contextual indications increases the prediction accuracy, and therefore reduce the number of irrelevant remote alerts. This minimizes the energy consumption associated to remote communications and enhances the protection of personal privacy.

MODALITi also takes into account the heterogeneity of the devices it manages. It deploys predictive models into devices and proposes a communication protocol in such a way that parts of models are only sent to devices that can process them, and devices only communicate with devices with capacity to run the model further.

The reminder of this paper is structured as follows. Section II introduces the concepts involved in MODALITi. Section III details how the framework trains predictive models and deploys them into a network of sensor devices. Section IV explains how devices communicate in order to run deployed models. An extension of the basic model approach is also detailed. Section V discusses a prospective use case for validating our framework. Finally, Section VI compares MODALITi with the existing related approaches, and Section VII concludes.

II. FRAMEWORK OVERVIEW

In order to be able to process incoming data in a generic way, we propose a framework consisting of a processing methodology accompanied by tools implementing the proposed approach. This framework, visually presented in Figure 1, supports all the aspects of value prediction based on a network of devices equipped with specialized sensors.

The framework is based on a set of **sensors**. Each sensor is able to produce, at any time, the value of a particular **metric**. While sensors can intuitively correspond to physical sensors, the framework operates at a higher level of abstraction, in such a way that a software system using a remote web service, or the aggregation of values produced by physical sensors, are also considered as a valid sensor. In order to harmonize data representation, and therefore to simplify data processing, we consider that all metric values are scalar numbers. Sensors are considered as resource-limited components in the framework. In particular, they are supposed to lack the necessary resources for storing and processing an amount of data large enough to carry out machine learning processes.

Devices manage one or many sensors and can ask them the current value of their metric. Consequently, each module can provide on demand a list of metrics they relate to.

Devices also have a component that enables inter-device communications. As discussed below, these communications require a moderated bandwidth since most information to exchange consists in values among a pre-agreed list of symbols and scalar numbers. Implementations of the communication components could therefore be based on lower-energy technologies, such as Z-Wave and ZigBee [11].

A device also has limited resources, and is not supposed to have the necessary resources for carrying out machine learning processes. However, its resources are considered as sufficient enough for supporting inter-device communications, for maintaining and querying associative tables, and for performing simple number comparisons.

Devices typically have a lightweight form factor, and can adopt various designs including wearable modules, embeddable microchip board, etc. Depending on the particular situation in which they are deployed, some of them are therefore likely to spatially move over time.

By using their communication component, devices that are closed to one other can share information in a peer-to-peer fashion.

III. MODEL PREPARATION

In order to carry out predictions based on the observed values extracted by the sensors, the framework prepares a machine learning model with a two-phase process. This process requires the presence of a **collector device** on the device network. In comparison to ordinary devices, a collector device needs some extra resources for storing and processing the collected information.

A. Model Training

Collection. First, the framework collects feature values from an installed device network. During this phase, the devices regularly communicate the current values of their associated sensors to a central collector device. These observed values are completed with samples of the variable to predict, in such a way the resulting dataset can be used by any machine learning algorithm for training a model that predicts the value of the target variable. This phase is visually represented in Figure 1 by the dotted arrows marked as ①.

Model Training. At any time, the populated data frame can be exploited in order to create a predicting model. Because the model must be efficiently distributed in the device network, not all machine learning models can appropriately be trained with this framework. At this time, only models based on decision trees are taken into account by the framework.

The result of training is a decision tree in which each interior node corresponds to a predicate on the value of a particular metric. Depending on the value of the evaluated predicate, either the left or the right child of the node is considered as the next decision node to follow for predicting the target variable. The leaf nodes represent the predicted value of the target variable.

In Figure 1, model training is represented by ②. This approach is flexible enough to train classification and regression models, for instance by using the CART [6] (Classification And Regression Tree) methodology.

Since the framework relies on state-of-the-art techniques for carrying out the collection and the model training phases, details that relate to the algorithms used for training machine learning models are beyond the scope of this paper.

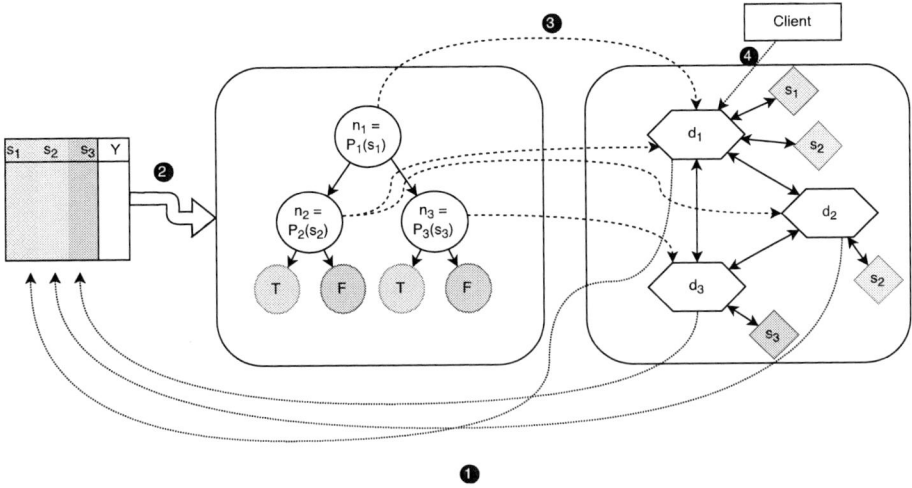

Fig. 1. Model representation and processing phases of the MODALITi framework.

B. Model Deployment

The result of the machine learning is typically a file representing the trained model, *i.e.* a decision tree. Our framework proposes a tool for automatically creating a representation of this model that can be understood and executed by resource-limited devices, and for deploying this representation on the device network. Dashed arrows marked as ③ in Figure 1 represent this deployment.

The framework deploys a machine learning model one node at a time. If a node n belonging to decision tree t is based on predicate p which operates on metric m, the following operations are carried out.

First, the devices managing sensors that are able to produce values for s are listed. For instance, in Figure 1, devices d_1 and d_2 are listed for the node n_2.

Secondly, the tuple $(t, n, p, m, answer_{true}, answer_{false})$ is communicated to each of the listed devices. Each device stores the received tuple in its *node table* that associates n to the entire tuple.

Finally, if n is an internal node, tuples containing the node and the metric identities of both left and right children of n are also communicated to each of the listed devices. In that case, $answer_{true}$ and $answer_{false}$ represent the best result that can be retrieved at this point of the tree traversal, for each possible value of the current predicate. If, otherwise, n is an leaf node, $answer_{true}$ and $answer_{false}$ represent the possible final result of the tree traversal for the possible values of the current predicate.

In order to avoid name collisions, any node, metric, device, and tree, must have a unique identifier. An approach based on UUID [15] might be used to ensure the universal unicity of any generated id.

IV. MODEL PROCESSING

Devices have to communicate in order to efficiently traverse the decision trees that are deployed on their network. We present a message-based approach for creating a network of collaborating nodes, and for traversing any deployed model and retrieving the values predicted by this model.

A. Device Announcement

When a device enters in a network, and periodically after that, it broadcasts a list of metrics for which it owns an appropriate sensor. Based on these announce messages, the other devices continuously keep track of devices that are able to evaluate any declared metric in a so-called *metric table*. Each device inserts its own identifier in this its metric table for the metrics it can evaluate.

B. Initiating a Model Evaluation

Decision trees can be exploited by submitting a **tree traversal query** to any device belonging to the network in which the considered tree has been deployed beforehand. Such a traversal query is parameterized by

- a unique query identifier;
- the identifier of the root node of the tree to be traversed;
- the identifier of the metric associated to this node;
- the identifier of the **collecting device** to which the final result must be transmitted. Depending on the considered use case, the devices that initiate the query and the collecting devices could be specialized in decision-making or alerting, for instance.
- a *hop counter* set to a positive value or zero.

C. Device Selection

When a device receives a traversal query, it determines which of its neighboring devices should be responsible for proceeding the evaluation of the predicate associated to the tree node mentioned in the query, as well as the continuation of the tree traversal. In a perpetually changing network topology, determining the best candidate, *i.e.* the device that minimizes the risk that the tree traversal does not reach one of the tree leaves while maximizing the traversal performances, is not a realistic goal. As an approximation, we propose the following selection heuristic:

1) All devices associated to the considered metric in the metric table of the queried device are considered as candidates.

2) If the queried device belongs to the candidate list, it chooses itself, in such a way the predicate evaluation is certainly carried out without the need of communicating with any other device.

3) Otherwise, the candidate is selected on the basis of a *quality value* associated to each candidate device. Various strategies may apply for determining the quality value of a candidate device. For instance, the value can result from a communication quality metric, the time since the last receipt of an announce message from the candidate, etc.

4) If no candidate devices are available, and the hop counter of the query is strictly positive, a device belonging to any entry of the metric table (*i.e.* any neighboring device) is randomly selected. In that case, the query is modified by decreasing its hop counter by one. The hop counter helps to limit the message flooding over the device network, while allowing neighboring devices to act as relays when no direct neighbor of the processing device can process the submitted query.

In order to increase the likelihood of a complete tree traversal, multiple devices can be selected instead of a single one. This alternative could, however, lead to duplicate processing, and the receiving device(s) could receive multiple final results for the same traversal query. The multiplicity of the results can be easily managed, though, by keeping track of the identifiers of the terminated traversal queries.

Once a device is selected, the processing device forwards it the traversal query.

D. Node Processing

A device that decides to process a traversal query browses its node table in order to fetch the predicate associated to the node the query refers to. By exploiting the appropriate sensor, it produces the current value of the metric associated to the node specified in the query, and evaluates the predicate on that value.

If the considered node is a leaf of the traversed decision tree, the evaluation of the predicate determines the predicted value of the target variable, and the processing device communicates this result to the collecting device(s) specified in the query.

Otherwise, the node is internal and the next decision node to be traversed is determined by the value of the predicate. The processing device replaces the node parameter of the processed query by the identifier of the next decision node and selects a device for the modified query. This allows the device network to traverse the decision tree further.

E. Tree Removal

The ability to remove a previously deployed tree is not required for allowing the evolution of the exploited predictive models. Each deployed tree being associated to a unique identifier, updating a tree could simply consist in deploying a new tree and using it instead of the previous one. However, this approach has the disadvantage that information that relates to unused tree nodes unnecessarily remains in memory of devices involved in the traversal of obsolete trees.

The proposed framework supports tree removal by sending a **remove order** to all reachable devices. The unique parameter of this order is the identifier of the tree to be removed. When receiving such an order, a device reacts by removing all the tuples that refer to the considered tree from its node table.

In order to increase the probability that a tree for which a removal order has been submitted is actually removed from device memory, a message relay strategy similar to the one proposed for selecting the processing device of a traversal query could be implemented.

F. Incomplete Tree Traversal

The proposed device selection heuristic does not guaranty a complete decision tree traversal. In other words, a tree traversal could be in a situation for which no reachable devices are available for traversing further, and therefore no leaf nodes of the decision tree are reachable at that point.

Besides the already discussed possibility to multiply the number of candidates to which traversal queries are forwarded, the *answer* parameters associated to decision nodes and transmitted to the processing device during the model deployment could be used to provide a prediction value despite an incomplete traversal. Depending on the value of the last evaluated predicate, either $answer_{true}$ or $answer_{false}$ is selected as the final result and communicated to the collecting device(s). An advantage of this approach is the certainty that the submission of a traversal query will always result in (at least) one attempt to communicate a final result to a collecting device. Moreover, processing devices that take this approach when they are unable to traverse the tree further can provide results with minimal resource consumption.

Generally speaking, predicted values resulting from partial tree traversals are likely to be less accurate than values resulting from complete traversals. In order to help the collecting devices assess result accuracy, the fact that a traversal has not been completed could be transmitted to collecting devices with the final result. Furthermore, accuracy-related metrics, such as the node *impurity*, could be measured during model training and transmitted during deployment, in such a way a device can transmit these metrics along with the predicted value.

G. Extending Models to Ensembles

Decision trees are standard and popular predictive models that gave rise to derived machine learning approaches. In particular, ensemble methods, such as random forests [12], have been proposed for improving the quality of predictions provided by decision trees.

Because random forest models are essentially sets of decision trees, the results of which are *aggregated* to obtain final results, the framework discussed in this paper naturally extends to such predictive models.

In order to support random forest models, the proposed framework uses state-of-the-art machine learning tools for training multiple decision trees that are deployed on a device network as defined in Section III-B.

Minor amendments are required to the proposed communication protocol for offering support for random forest models. A traversal query must be submitted to the device network for each of the tree belonging to the model. Because any set of trees can be traversed simultaneously, our approach offers a highly parallelized and distributed approach for exploiting decision trees and random forest models. Similarly, the removal of a random forest model from a device network consists in submitting a removal message for each of the decision tree belonging to the model. Collecting devices must take into account the fact that results of multiple tree traversal have to be collected and aggregated in order to get the predicted value.

Ensemble techniques are abundantly described in the literature [5], [20], [24]. Compared to the evaluation of a single decision tree, an incomplete tree traversal has lower impact on the evaluation of a random forest model, because the final aggregation of evaluation results can be used to mitigate the accuracy loss due to missing results. Therefore, approaches that could be adopted in order to prevent incomplete tree traversal are less relevant in that case, and their benefits should be evaluated in light of the extra resource consumption they incur.

V. USE CASE

To validate our proposed framework, we are considering epilepsy as a case study in particular, refractory epilepsy patients who do not respond to conventional treatment. Thirty percent of people with epilepsy have refractory epilepsy a very debilitating form of the disease since the seizures are random and not controllable with medication [14].

Our proposed framework can be used to continuously monitor refractory epilepsy patients and notify them of seizure risks. These notifications contribute to improve the safety and quality of life of individuals affected by refractory epilepsy. Furthermore, the framework can provide context-aware recommendations to reduce seizure risk by analysing the triggering factors (e.g., stress level), consequently helping to reduce seizure frequency. Finally, the collected data could be used by epilepsy specialists to better identify appropriate treatment.

To achieve this objective, the different steps of MODALITi approach will be implemented:

1) **Machine learning model preparation.** In this step, the early detection seizure patterns should be developed by tracking different data of epileptic patients using wearable devices (e.g. EEG, sleeping and stress trackers). We also plan to rely on the several works have been already done to develop such as patterns like [14], [19], [1] and European projects like FP7- EPILEPSIAE. The developed patterns will be deployed on a programmable wearable devices in order to enable to a continuously monitor.

2) **Machine learning model processing.** In this step, network of collaborating variables will realtime analyse an epileptic events or measurement in order to early diseases detection based on prediction models. The patient should receive a notification in a useful time if a seizure is predicted.

VI. RELATED WORK

Collecting, processing and analyzing of the data generated by IoT and sensors is not a new challenge. Many approaches have been proposed to manage the huge volume, velocity and variety of data produced by these sensors.

Zomaya and Sakr [26] propose to centralize the storage, processing and analysis of sensor data in a high performance processing infrastructure like a cloud infrastructure (remote processing). This approach ensures a good scalability to store and process data. However, in this approach, many threats must be considered, including fault tolerance, efficient task distribution, usability and processing speed. For this reason, many works have been conducted to propose new data models (e.g. NoSQL models [7]), new storage file systems (such as HDFS [21]), new programming models (e.g. MapReduce [9]) and new data processing frameworks (including Hadoop [25] and Spark [18]). Even if a centralized approach seems to be the de facto solution to processing and analysing of the data generated by IoT, it suffers from some threats in terms of data privacy and connectivity as discussed above.

A second approach of the IoT data management suggests that a part or full of data processing operations is done locally on the device (local processing). Aggregations, including calculating maximum, minimum, and average values, can be carried out at device level [16], [13]. Embedding predefined and primitive aggregation operations on devices allows to optimize the volume of IoT data collected and therefore optimizes the bandwidth and the storage space consumption. However, only simple aggregation formulas are supported by such operations, that are not appropriate for carrying out complex machine learning models such as decision trees.

Alternative approaches have been proposed for implementing machine learning algorithms over sensor networks. A peer to peer topology is usually considered for this implementation due to its ability to deal with different distributed sources of voluminous data, and multiple computing nodes [23]. For example, Bhaduri [4] proposes a framework for local data distribution in large scale peer to peer systems. Gaber et al. [22] propose the *Pocket Data Mining Framework* that carries out data mining algorithms in mobile environment. Datta et al. [8] propose a decentralized implementation of the K-Means Clustering on large and dynamic networks, while Mehyar et al. [17] focus on the implementation of data mining algorithms over a asynchron, distributed and dynamic network topology. However, contrary to the MODALITi framework, these works do not address how to prepare and deploy machine learning models over a device network in an optimized way. Indeed, as discussed above, the framework we propose deploys a machine learning model by taking into account the capacity and the specificities of the involved devices, in such a way these devices will only receive part of the models they can process.

VII. CONCLUSION

IoT applications may incur technical and social threats that prohibit the remote processing of the collected datasets or that make it inefficient. On the other hand, on-board solutions are energy efficient and facilitate the preservation of privacy, but are not adapted for resource-consuming tasks such as machine learning model training. This paper introduced MODALITi, an hybrid framework that supports predictive model training on a remote environment and that deploys and runs these models on a resilient network of low-energy devices.

So far MODALITi has no concrete implementation. In order to determine to which extent it can help to define and deploy a real-time on-board data mining process, we intend to carry out a simulation of the elements belonging to the framework. Among others, we will simulate model deployments and query submissions on various device networks in order to assess the impact of device selection strategies on the success rate of tree traversals, the number of simultaneous queries a given device network can execute simultaneously, and the number of queries a network can execute before energy shortages put it in a state that prevents it from carrying out subsequent queries.

We would also propose a research project, the purpose of which would be to create a concrete instance of MODALITi for predicting imminent seizures among epileptic patients.

REFERENCES

[1] Kris Cuppens Anouk Van de Vel a, Bert Bonroy, Milica Milosevic, Katrien Jansen, Sabine Van Huffel, Bart Vanrumste, Lieven Lagae, and Berten Ceulemans. Non-EEG seizure-detection systems and potential SUDEP prevention: State of the art .

[2] Kevin Ashton. That 'internet of things' thing. *RFiD Journal*, 22(7):97–114, 2009.

[3] Kanishka Bhaduri. *Efficient Local Algorithms For Distributed Data Mining In Large Scale Peer To Peer Environments: A Deterministic Approach*. PhD thesis, April 2008.

[4] Kanishka Bhaduri. *Efficient Local Algorithms for Distributed Data Mining in Large Scale Peer to Peer Environments: A Deterministic Approach*. PhD thesis, Catonsville, MD, USA, 2008. AAI3316035.

[5] Leo Breiman. Bagging predictors. *Machine Learning*, 24(2):123–140, 1996.

[6] Leo Breiman, J. H. Friedman, R. A. Olshen, and C. J. Stone. *Classification and Regression Trees*. Wadsworth, 1984.

[7] Rick Cattell. Scalable sql and nosql data stores. *SIGMOD Rec.*, 39(4):12–27, May 2011.

[8] Souptik Datta, Chris Giannella, and Hillol Kargupta. K-means clustering over a large, dynamic network. In Joydeep Ghosh, Diane Lambert, David B. Skillicorn, and Jaideep Srivastava, editors, *Proceedings of the Sixth SIAM International Conference on Data Mining, April 20-22, 2006, Bethesda, MD, USA*, pages 153–164. SIAM, 2006.

[9] Jeffrey Dean and Sanjay Ghemawat. Mapreduce: Simplified data processing on large clusters. *Commun. ACM*, 51(1):107–113, January 2008.

[10] Mohamed Medhat Gaber, João Bártolo Gomes, and Frederic Stahl. Pocket data mining. *Big Data on Small Devices. Series: Studies in Big Data*, 2014.

[11] Drew Gislason. *Zigbee Wireless Networking*. Newnes, Newton, MA, USA, pap/onl edition, 2008.

[12] Tin Kam Ho. Random decision forests. In *Proceedings of the Third International Conference on Document Analysis and Recognition (Volume 1) - Volume 1*, ICDAR '95, pages 278–, Washington, DC, USA, 1995. IEEE Computer Society.

[13] Márk Jelasity, Alberto Montresor, and Ozalp Babaoglu. Gossip-based aggregation in large dynamic networks. *ACM Trans. Comput. Syst.*, 23(3):219–252, August 2005.

[14] Patrick Kwan and Martin J Brodie. Early identification of refractory epilepsy. *New England Journal of Medicine*, 342(5):314–319, 2000.

[15] P Leach, M Mealling, and R Salz. A Universally Unique IDentifier (UUID) URN Namespace. RFC 4122 (Proposed Standard), July 2005.

[16] Kun Liu, H. Kargupta, and J. Ryan. Random projection-based multiplicative data perturbation for privacy preserving distributed data mining. *IEEE Transactions on Knowledge and Data Engineering*, 18(1):92–106, Jan 2006.

[17] M. Mehyar, D. Spanos, J. Pongsajapan, S. H. Low, and R. M. Murray. Asynchronous distributed averaging on communication networks. *IEEE/ACM Transactions on Networking*, 15(3):512–520, June 2007.

[18] Xiangrui Meng, Joseph Bradley, Burak Yavuz, Evan Sparks, Shivaram Venkataraman, Davies Liu, Jeremy Freeman, DB Tsai, Manish Amde, Sean Owen, Doris Xin, Reynold Xin, Michael J. Franklin, Reza Zadeh, Matei Zaharia, and Ameet Talwalkar. Mllib: Machine learning in apache spark. *Journal of Machine Learning Research*, 17(34):1–7, 2016.

[19] Shyamal Patel, Hyung Park, Paolo Bonato, Leighton Chan, and Mary Rodgers. A review of wearable sensors and systems with application in rehabilitation. *Journal of Neuro Engineering and Rehabilitation*, 9(1):21, 2012.

[20] Robert E. Schapire. *The Boosting Approach to Machine Learning: An Overview*, pages 149–171. Springer New York, New York, NY, 2003.

[21] Konstantin Shvachko, Hairong Kuang, Sanjay Radia, and Robert Chansler. The hadoop distributed file system. In *Proceedings of the 2010 IEEE 26th Symposium on Mass Storage Systems and Technologies (MSST)*, MSST '10, pages 1–10, Washington, DC, USA, 2010. IEEE Computer Society.

[22] Frederic Stahl, Mohamed Medhat Gaber, Max Bramer, and Philip S. Yu. Pocket data mining: Towards collaborative data mining in mobile computing environments. In *Proceedings of the 2010 22Nd IEEE International Conference on Tools with Artificial Intelligence - Volume 02*, ICTAI '10, pages 323–330, Washington, DC, USA, 2010. IEEE Computer Society.

[23] Rekha Sunny T and Sabu M. Thampi. Survey on distributed data mining in P2P networks. *CoRR*, abs/1205.3231, 2012.

[24] Joaquín Torres-Sospedra, Carlos Hernández-Espinosa, and Mercedes Fernández-Redondo. *Using Bagging and Cross-Validation to Improve Ensembles Based on Penalty Terms*, pages 588–595. Springer Berlin Heidelberg, Berlin, Heidelberg, 2011.

[25] Tom White. *Hadoop: The Definitive Guide*. O'Reilly Media, Inc., 1st edition, 2009.

[26] Albert Y. Zomaya and Sherif Sakr, editors. *Handbook of Big Data Technologies*. Springer, 2017.

A Dew Computing Solution for IoT Streaming Devices

Marjan Gusev
Ss. Cyril and Methodius University
Faculty of Computer Sciences and Engineering
Skopje, Macedonia
Email: marjan.gushev@finki.ukim.mk

Abstract—Most people refer to modern mobile and wireless ubiquitous solutions as an IoT application. The advances of the technology and establishment of cloud-based systems emerge the idea of the connected world over Internet based on distributed processing sites. In this paper, we discuss the dew computing architectural approach for IoT solutions and give an organizational overview of the dew server and its connections with IoT devices in the overall cloud-based solutions. Dew servers act as another computing layer in the cloud-based architecture for IoT solutions, and we present its specific goals and requirements. This is compared to the fog computing and cloudlet solutions with an overview of the overall computing trends. The dew servers are analyzed from architectural and organizational aspect as devices that collect, process and offload streaming data from the IoT sensors and devices, besides the communication with higher-level servers in the cloud.

Index Terms—Mobile Cloud Computing, Cloudlet, Edge computing, Fog computing, computation offload

I. INTRODUCTION

Streaming sensors are those that supply continuous data by sampling the signal on regular time periods. In order to clarify the difference between streaming data and static sensors we define that if the sampling frequency is higher than 100 samples per second then the sensor is streaming data [1]. Any IoT sensor that samples on lower frequencies than 1 Hz is defined to be static.

Streaming sensors can be implemented in industrial applications, such as power distribution; environmental and smart home applications, such as cameras; or mHealth applications, such as ECG, blood pressure, glucose level or other health related parameter. Advances in technology increase their production and applications.

Static sensors are most common conventional systems. They can sense some environmental measures that gracefully change their values in relatively longer time frames, such as temperature or pressure, or detection of OFF/ON or open/close status.

Streaming sensors supply high volumes of data with certain velocity that needs to be processed. Data values can change frequently defining also properties of variety, variability and veracity. This classifies the corresponding solution as a Big Data processing solution.

World of IoT is based on information exchange between different cloud servers. IoT devices usually transmit data to servers and they exchange to get a more comprehensive picture of the whole ecosystem.

This paper presents an architecture and organization of a dew computing solution for IoT devices, especially for streaming sensor. The solution consist of adding a new architecture level on the path between the IoT device and higher-level servers hosted in the cloud. The dew server can be standalone computing and communication box, such as smart home processing center or a mobile phone. It needs to communicate to the IoT device, to the high-level servers and enable human computer interface to the user.

The idea of ubiquitous computing [2] to enable computing everywhere and anytime is practically realized by the dew computing concept. Pervasive computing [3], [4] is another similar term that explains the growing trend of embedded processors in everyday objects and enable an environment to exchange the information. Actually, the IoT has largely evolved out of pervasive computing.

In this sense, the dew computing concept makes the things in the IoT definition become "smart" by bringing processing closer or embedding it. Instead of offloading the processing to the clouds, this concept brings the computing back to the sources of data.

Skala et al. [5] concludes that significant improvement can be achieved by applying the dew computing approach as bringing the processing and communication requirements to physical-edge devices.

The rest of the paper follows the next structure. Section II presents the related work about cloud, cloudlet, fog and dew computing architectural principles. The proposed architecture and organization of a dew computing solution for IoT streaming device is presented in Section III and the corresponding description of micro services in Section IV. Relevant discussion is given in Section V with elaboration of the need for the dew server. Section VI presents the conclusions and future work directions.

II. RELATED WORK

The merits of seamlessly integrating "dew" devices into the Cloud - Fog - Dew Computing hierarchy are enormous, for individuals, the public and industrial sectors, the scientific community and the commercial sector, by bettering the physical and communication, as well as the intellectual, immediate human environment [5].

Dew computing is referred as a complementary piece of cloud computing [6] and is related to definitions of cloudlets and fog/edge computing.

A. Cloudlet and fog computing

Cloudlet computing uses the same philosophy as dew computing by bringing the processing closer to the users in a typical cloud client-server organization. However, unlike dew computing, cloudlets are based on small and medium-sized servers, while the dew servers can be micro and very small servers, including tablets, smart mobile phones, various controllers, embedded processors and similar processing boxes.

The idea of cloudlets originates by Satyanarayanan et al. [7]. They define cloudlets as trusted, resource-rich computers, located near mobile users. In this sense, our definition of dew servers, actually present end users in the cloudlet architecture.

The motivation behind setting cloudlets, as smaller servers closer to the users is to distribute the processing and eliminate WAN latency in the provision of services. It is well known [7] that local area networks can be superior in latencies when compared to long latencies to the cloud, and up to certain degree, one can still benefit even if the response of the smaller cloudlet server is higher than the cloud.

Some researchers define the concept of cloudlets as a special type of mobile cloud computing [8]. Cloudlet is an intermediate level, where the server is less powerful, smaller and less expensive, and can be situated in common areas, such as coffee shops, or shopping centers. Initially, they solve high latency and bandwidth problems.

Several papers [9], [10] present several cloudlet challenges and overview cloudlet vs cloud architectures, including a summary of essential characteristics to offload computations, using fat and thin clients, etc.

We have discussed cloudlets as an architecture that sets a new layer in front of the cloud and closer to the users. This idea was also used by the mobile operators to build fog/edge computing solutions. The main approach is the same, but instead of hosting small or medium servers by independent Internet providers or users, the mobile network operators position their servers at the edge of a network (their base stations) [11].

Bonomi et. al [12] discuss the benefits of setting servers at the base stations to reduce the latencies and distribute the processing in a number of IoT applications and scenarios that include connected vehicles, smart grids and wireless sensor and actuator networks.

B. Dew computing

Several papers [13], [14], [6], [15], [16] define the dew computing as a new type of computer architecture and organization that extends the convenient cloud and classical client-server architecture. In this new architecture, a new dew device is positioned between the cloud and the end-user, with a goal to provide micro services. In this paper, we define the dew

computing layer as an intermediate between the IoT device and high-level servers.

Wang [14] defines the cloud-dew architecture as an extension of the client-server architecture, and defines the dew servers as web servers that reside on users' local computers and have a pluggable structure so that scripts and databases of websites can be installed easily.

Here, we extend this concept and apply it in the IoT, by defining a new dew computing layer, where the dew server is not necessarily a web server, but an intermediate server, with function to collect weak signals from nearby IoT devices, store, process and transmit data and results to high-level servers. Depending on the sampling frequency and data precision, the dew server may be a simple IO controller, independent processing unit, modern smart mobile phone/tablet, stand-alone personal computer, or small size server.

Rindos and Wang [6] propose dew computers, as another abstraction of the same dew computing level.

Wang [13] defines independence and collaboration features of a dew application. Wang and LeBlanc [17] discuss Software as a Service (SaaS) and Software as a Product (SaaP) concepts in relation to the dew computing. They argue that the independent work of dew server is satisfied by SaaP and collaboration by SaaS, and offering both features, as done in dew computing will have a greater benefit. Our dew computing solution of a streaming IoT actually fits to this argument, enabling the dew server to act as a service, delivering data and results to the high-level servers, and as an independent software to act as an IoT device controller with extended services.

III. ARCHITECTURE AND ORGANIZATION

Fig. 1 presents an architecture of a dew computing solution for cloud-based IoT streaming data processing. It describes a high-level hardware organization of connecting the IoT sensor, dew server, intermediary devices (servers) and cloud-based server.

The IoT sensor is the device that converts the environmental signals to data, so one can process them and obtain relevant info about a specific physical, chemical or biological measure. The IoT sensor not necessarily measures natural signals, but can also be used to sense an occurrence of a specific event, status or other digital information. The basic functions of the IoT sensor are:

- sensing,
- transforming the signal into digital data, and
- transmission of digital data.

The dew server is a device positioned on an intermediary architecture level between the IoT sensor and high-level servers. It communicates with the IoT sensor and high-level servers, and performs basic processing and controlling functions. The basic functions are categorized to be:

- IoT device communication management,
- Internet communication management,
- data collection,
- data storage,

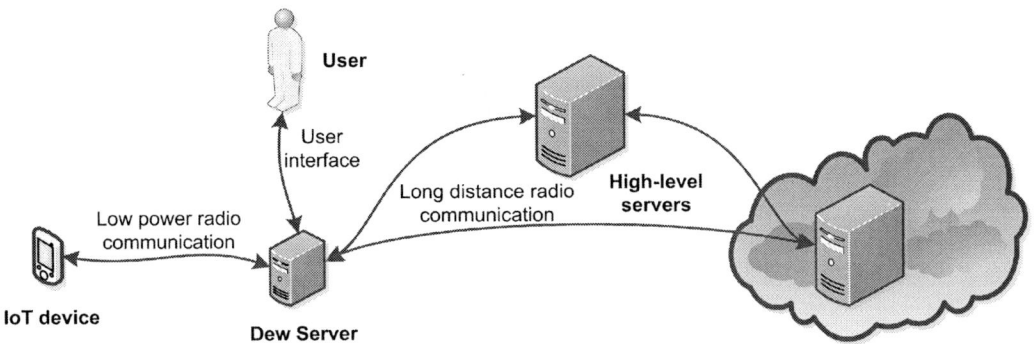

Fig. 1. Architecture of a dew computing solution for cloud-based IoT streaming data processing.

- data transmission,
- data processing,
- data visualization,
- IoT device control, and
- user interface.

The high-level servers are usual cloud servers that realize final data processing, by deploying web services and web applications to the intended users.

Introducing an intermediate dew computing level should not be mixed with the intermediate cloudlet and fog servers. In our model, cloudlets and fog servers are representing high-level servers situated in front of the clouds, while dew servers are found very close to the IoT sensors. A lot of scenarios in fog computing think of a smart mobile phone as a user, and in our model, this device is dew server to other IoT sensors found nearby the smart mobile phone (as in usual ubiquitous computing scenario).

Therefore, a more detailed cloud-based IoT architecture builds on both the dew computing and cloudlet/fog computing layers on the path between the IoT sensor and the cloud servers.

A. The need of an intermediary level

The intermediary layer is needed for several reasons. The IoT sensor can not use technologies for long-distance transmission if it is not connected by a cable. In reality, a lot of IoT sensors are moveable and, therefore, the Internet can be accessed by a WiFi connection. There are also IoT sensors, which need to be light and small, and their battery life needs to be preserved as long as possible, due to their size and position. Examples of these sensors are light wearable mHealth sensors that need to be attached to the human's body. in this case, WiFi is not a preferred solution for these sensors, but Bluetooth or other low power energy radio communication. In all of these cases, an intermediary device needs to accept low power radio communication and then transmit WiFi signal or be attached to Internet by other media.

A lot of IoT sensors perform a controlling function, such as, activating some signals or triggering an event. These actuators connected to the IoT sensor also need a processing device and approves the existence of the intermediary layer device. The

data processing may include small processing needs or act as a pipelined processing device capable of performing calculations on streaming data.

IoT devices are connected to the Internet and provide sufficient data to cloud servers. However, the previously discussed functions that require an intermediary level device, can be solved by smaller servers on the path between the IoT sensor and cloud-based servers. These solutions include

- *servlets*, as smaller servers positioned in the local area network of the IoT sensor with the goal to collect data, perform initial processing and transmit data to the server,
- *fog computing servers*, as servers positioned on the edge of the mobile operator,
- *dew computing servers*, as smaller servers or devices, positioned even closer to the IoT sensor (in a very close distance).

These architectural concepts to implement a server between the sensors and the cloud-based server, actually follow an interesting pattern. More close to the IoT sensor they are, smaller they are. This means, that the mobile operator will host medium sized servers at the base stations (edge of the network), private companies even smaller sized servers at the premises of their LAN and dew computing servers will be even smaller devices, positioned close to the IoT servers.

Note that the small dew server can be present in all designs, even though another intermediate server is used, such as cloudlet or fog computing server. In some cases, the dew server may act as an IoT controller or actuator next to the IoT sensor, or even an embedded processor in the same box.

B. Interconnections and power supply

Energy supply and Internet connection are the most important resources to enable proper functioning of the IoT device. We can categorize the IoT device according to the power supply to:

- device with permanent power supply, or
- battery operated device

The connection to the dew server ensures establishing an Internet connection, usually by a local area network. It can be realized by a direct cable (including fiber) connection, or by a WiFi or other radio communication.

When analyzed with its geographic position, the IoT sensor can be moveable and, therefore, the connection needs to be mobile. A wireless connection is needed in most of the cases when wires can not be used.

A stationary IoT sensor usually has access to uninterruptible energy and Internet supply, while the IoT sensors that move dynamically in the environment are mostly battery operated and use WiFi or other radio communication technology. This makes a lot of difference since battery operated IoT sensors need very efficient pipelining algorithms to save the energy and usually use low power Bluetooth interconnection. On contrary, those IoT sensors that have proper power supply usually use WiFi or direct cable connection to establish an Internet connection.

The dew server is the device, which is found close to the IoT sensor and can perform initial data processing. Similar to the previous categorization, the dew server can also be moveable and use wireless networking, similar to the IoT sensor.

Although this paper analyzes a general dew computing solution of an IoT device, we give more details on streaming sensors, since they require a specific architecture capable of dealing with the Big Data related higher data volumes and streams.

IV. DESCRIPTION OF MICRO SERVICES

The dew server performs as a server that delivers micro services. It communicates to three instances:

- *IoT device* (sensor, actuator or combined device) via low power personal area network,
- *high-level servers* hosted on Internet via local area network,
- *human users*, that monitor and control the IoT device.

Fig. 2 describes a high-level abstraction of software modules and organization of micro services identified for a dew server connected to a streaming IoT sensor or any IoT device. Next, we give a detailed elaboration of these software modules:

- *Communication manager* that manages the communication of the IoT device with the dew server and high-level servers, including at least the following functions:
  - pairing,
  - connecting,
  - disconnecting, and
  - reconnecting.

Pairing is realized in the initialization process, and whenever the IoT device is relatively close to the dew server it starts the connection. When the IoT device is out of the reach of the radio coverage of the dew server, it disconnects and tries to reconnect and establish communication whenever it gets the signal from the paired device. The same principles apply for the connection to the higher level servers.

Practically, there are two instances of the communication manager, one for the low power radio communication with the IoT device, such as Bluetooth or other personal area network radio communication, and the other for the connection to the higher-level servers on Internet, such as WiFi, 3G/4G or other similar radio connection.

- *Device control* is needed by the dew server to monitor the IoT device status and control its functioning. The basic functions supported by this module is to:
  - get sensor status,
  - set a specific variable
  - trigger specific actions, and
  - reset the function.

The IoT device sends its status to the nearby dew server, so it gives a relevant information and the dew server can react upon received status. Since the dew server controls the IoT sensor, it can set up a specific variable to initiate a more accurately sensing and measurement of the appropriate signal. Sometimes, the dew server can reset the IoT device, and for example, start a new measurement of the IoT sensor or, in the case of an IoT actuator, it can trigger specific actions that control its performance.

- *Streaming data* sent by the IoT sensor needs further pipelined stream processing. The essential functions that target the received streaming data for the dew server are:
  - data collection,
  - data storing, and
  - data transmission to higher level servers.

Data collection is the most important action of the dew server. Usually, the IoT sensor streams data without checking if it is collected or not. Additionally, in the case of higher sampling frequencies, the IoT sensor can not store data, so these data may be lost forever. Actually, the dew server realizes the data storing function of the IoT sensor, especially in the case when it is battery operated and is a light moveable device, without any wires that can enable more secure connection, such as the wearable sensors are.

In some sense, the dew server, also performs the function of a digital repeater, since it buffers the received data stream on a low power radio connection, and generates a signal with a higher power to transmit it to higher level servers hosted on Internet.

- *Data processing*, which is complementary to the data collection and storage. Depending on the size of the dew server, data processing may include at least:
  - preprocessing,
  - analyzing,
  - classification (decision making), and
  - visualization.

The incoming data stream may contain a lot of noise, so usually, data preprocessing uses digital filters or similar techniques to preprocess data for further processing. Data analysis is the essential step for each received data sample to make a good estimation of the sensed signal and further conclusions.

Usually, conclusions are made based on a classification algorithm, with a goal to determine the class of the incoming data stream. Some dew servers enable user

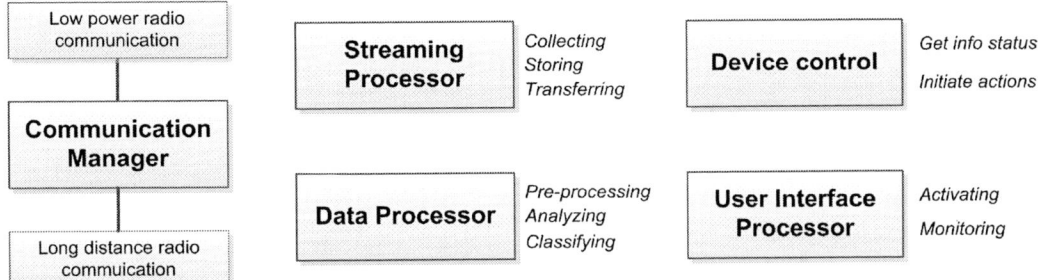

Fig. 2. Micro services of a dew server connected to an IoT streaming sensor.

interfaces, so a user can monitor the sensed data stream. For this purpose, data is processed to enable visualization.

- *User interface* of the dew server is found on most of devices, so one can monitor and initiate certain actions for the IoT device. This human computer interface realizes at least the following functions:
  - activating commands (via menu options and control buttons),
  - setting various parameters,
  - real-time visualization, and
  - navigation through recorded data.

Although the dew server may run specific algorithms and work autonomously, users can control the dew server. In addition, they have the advantage to monitor the visualized data and activate various parameters or actions on the IoT device.

V. DISCUSSION

Application of direct cloud connection, fog computing, and cloudlets is not possible without the dew computing layer inserted in the overall architecture of the solution.

Direct connections of IoT devices to the cloud need a stable power supply and Internet connection. This is only possible for non-moveable IoT sensors and devices, to cope with higher energy demands that direct connection requires.

Solutions that involve cloudlets or fog computing servers can solve partially the increased data throughput and processing demands, but still, the sensors need a good power supply to enable WiFi or 3G/4G connection.

In this paper, we analyze IoT devices that are battery operated and can be moveable and wearable. In order to communicate to the remote server, they will need more energy resources to be capable of transmitting data to the local area network. The proposed solution adds a dew server to realize the essential data collecting, storing, processing and further transmitting.

A typical streaming IoT device requires storage of at least 1KB per second, which sums up to more than 80 MB per day, 2.5 GB per month or 1TB per year. Memory is not the only problem when an IoT device is analyzed. Probably a higher problem is the energy consumption. In order to make the IoT device smaller, moveable and wearable, it requires a small battery and the designers have to compromise whether to use

a larger battery or embed a smaller number of functionalities. Most of the producers conclude that due to these constraints the IoT device can not store or process data, it needs just to transmit to the nearby device, which can perform these functions instead.

Communication speed is at least 8Kbps for the analyzed data stream, which is satisfied by most of the today's communication links, but it generates a connection bottleneck on the server side if hundreds or thousands similar sensors are connected to a particular server. This is also a problem when analyzing the processing requirements. If the server processes a huge number of data streams, then it needs more processing power. One can conclude that offloading data streams to remote high-level servers will require very powerful servers with high communication demands.

More details on performance evaluation of various architectural approaches are described in [18]. An example of a comprehensive analysis of the storage, communication and processing requirements along with the energy consumption for a wearable ECG sensor is given in [1].

The previous analysis actually concludes that in the case of streaming IoT devices, distributing the processing power closer to the data sources is a better solution.

Analyzing the "ubiquitous" idea of Mark Weiser [2] we realize that the tabs, pads, and boards (and now the smartphones) are actually a realization of a small dew server in the proposed architecture.

Analyzing the Wang's definition of a dew server [14], we agree on all specifications (1.light-weight server, 2.data storing, 3.easy disappearance, 4.recreation from cloud data, and 5.local accessibility and independent work). In addition, instead of cloud-dew applications provided by a web server, we extend this to be micro services provided by the dew server. Our definition of a dew server, practically means the server can be a miniature controller, smart mobile phone, tablet or any stand alone processing device.

Wang [13] categorizes at least two essential characteristics for the dew device, that is to work independently and collaborate with the server. We have specified a set of basic functionalities for a dew server in the context of IoT.

Rindos and Wang [6] define the dew applications analyzing their independence and collaboration. Their model of IoT system with dew applications, belongs to the Web in Dew

(WiD) category [13], based on establishing a dew web server, which works independently and collaborates with the server providing information in a typical client-server scenario.

Our architectural model is rather different. It is, rather, based on a push principle and, although the dew server works independently, it tries to synchronize with the high-level servers. In this sense, we propose a new category **IoTD** with a simple dew server described in this paper with the key function to collect, store, process and transmit streaming data from an IoT device.

The main characteristics of dew computers, as defined by Rindos and Wang [6] are: 1.low energy cost 24/7 work, 2.OS supports dew computing, 3.collaboration support, 4.security databases, and 5.WiD applications support. Our IoT model do not use all these features, our definition of dew servers does not need to rely on databases, if the corresponding OS supports existence of a file system and supports the TCP/IP protocol.

VI. CONCLUSION

We have modeled a dew computing solution for a streaming IoT device and higher-level cloud-based servers. The proposed solution enables efficient streaming data collection, storing, processing and transmitting, even in the cases where the sensor needs to be a battery operated wearable device without any wired connection.

The dew computing layer is based on a realization of a corresponding dew server, that can be a mobile smartphone or standalone smart home processing and communication box. We have presented functions and micro services the dew server provides with elaboration why the corresponding solution is preferred, comparing it to the conventional direct cloud solutions, fog computing or cloudlet solutions.

The new defined specific dew server realizes a new IoTD category of the already existing dew computing categories.

The overall concept actually pushes back the processing closer to the devices to enable a more independent processing and organization.

REFERENCES

[1] M. Gusev, "Going back to the roots: The evolution of Dew computing," University Sts Cyril and Methodius, Faculty of Computer Sciences and Engineering, Tech. Rep. 01/2017, 2017.

[2] M. Weiser, "The computer for the 21st century," *Scientific american*, vol. 265, no. 3, pp. 94–104, 1991.

[3] M. Satyanarayanan, "Pervasive computing: Vision and challenges," *IEEE Personal communications*, vol. 8, no. 4, pp. 10–17, 2001.

[4] F. Mattern, "Wireless future: Ubiquitous computing," in *Proceedings of Wireless Congress*, 2004, pp. 1–10.

[5] Z. Sojat and K. Skala, "Views on the role and importance of dew computing in the service and control technology," in *Information and Communication Technology, Electronics and Microelectronics (MIPRO), 2016 39th International Convention on*. Croatian Society MIPRO, May 2016, pp. 164–168.

[6] A. Rindos and Y. Wang, "Dew computing: The complementary piece of cloud computing," in *Big Data and Cloud Computing (BDCloud), Social Computing and Networking (SocialCom), Sustainable Computing and Communications (SustainCom)(BDCloud-SocialCom-SustainCom), 2016 IEEE International Conferences on*. IEEE, 2016, pp. 15–20.

[7] M. Satyanarayanan, P. Bahl, R. Caceres, and N. Davies, "The case for vm-based cloudlets in mobile computing," *Pervasive Computing, IEEE*, vol. 8, no. 4, pp. 14–23, 2009.

[8] N. Fernando, S. W. Loke, and W. Rahayu, "Mobile cloud computing: A survey," *Future Generation Computer Systems*, vol. 29, no. 1, pp. 84–106, 2013.

[9] A. Bahtovski and M. Gusev, "Cloudlet challenges," *Procedia Engineering*, vol. 69, pp. 704–711, 2014.

[10] M. Satyanarayanan, "Mobile computing: the next decade," in *Proceedings of the 1st ACM workshop on mobile cloud computing & services: social networks and beyond*. ACM, 2010, p. 5.

[11] I. Stojmenovic, "Fog computing: A cloud to the ground support for smart things and machine-to-machine networks," in *Telecommunication Networks and Applications Conference (ATNAC), 2014 Australasian*. IEEE, 2014, pp. 117–122.

[12] F. Bonomi, R. Milito, J. Zhu, and S. Addepalli, "Fog computing and its role in the Internet of things," in *Proceedings of the first edition of the MCC workshop on Mobile cloud computing*. ACM, 2012, pp. 13–16.

[13] Y. Wang, "Definition and categorization of dew computing," *Open Journal of Cloud Computing (OJCC)*, vol. 3, no. 1, pp. 1–7, 2016.

[14] ——, "Cloud-dew architecture," *International Journal of Cloud Computing*, vol. 4, no. 3, pp. 199–210, 2015.

[15] K. Skala, D. Davidovic, E. Afgan, I. Sovic, and Z. Sojat, "Scalable distributed computing hierarchy: Cloud, fog and dew computing," *Open Journal of Cloud Computing (OJCC)*, vol. 2, no. 1, pp. 16–24, 2015.

[16] S. Ristov, K. Cvetkov, and M. Gusev, "Implementation of a horizontal scalable balancer for dew computing services," *Scalable Computing: Practice and Experience*, vol. 17, no. 2, pp. 79–90, 2016.

[17] Y. Wang and D. LeBlanc, "Integrating SaaS and SaaP with dew computing," in *Big Data and Cloud Computing (BDCloud), Social Computing and Networking (SocialCom), Sustainable Computing and Communications (SustainCom)(BDCloud-SocialCom-SustainCom), 2016 IEEE International Conferences on*. IEEE, 2016, pp. 590–594.

[18] M. Gusev, "Modeling the performance behavior of Dew computing solutions for streaming IoT devices," University Sts Cyril and Methodius, Faculty of Computer Sciences and Engineering, Tech. Rep. 02/2017, 2017.

MIPRO 2017, May 22- 26, 2017, Opatija, Croatia

3D-based Location Positioning Using the Dew Computing Approach For Indoor Navigation

D. Podbojec, B. Herynek, D. Jazbec, M. Cvetko, M. Debevc and I. Kožuh

University of Maribor, Faculty of Electrical Engineering and Computer Science, Maribor, Slovenia
(damir.podbojec, bronja.herynek, denis.jazbec1, mitja.cvetko)@student.um.si
(matjaz.debevc, ines.kozuh)@um.si

Abstract - In the field of indoor navigation, there is still a lack of a unified system which could be applied to different buildings. As far as existing indoor navigation solutions are concerned, there is a recognised need for developing an application which would allow users to navigate within large buildings, especially when educational institutions are in question. Thus, in our study, we designed, developed and evaluated a web-based application "Virtual FERI" which allows indoor navigation of the G3 building of the Faculty of Electrical Engineering and Computer Science (FERI) in Maribor, Slovenia. Following the 3D model of the building, it simplifies the process of seeking the specific classroom in the building which could help newcomer students and guest lecturers to navigate within the building. To evaluate the application, we conducted an experiment with 25 students from the University of Maribor who had not been in building G3 before. Usability and user experience of the application were assessed with a User Experience Questionnaire and System Usability Scale. Time was measured as well. The findings revealed that navigating within the buildings can be faster by using the "Virtual FERI" where usability was sufficient and user experience was satisfactory.

I. INTRODUCTION

In everyday life, people constantly seek for new places and, due to the fast pace of life, there is a need for time-efficient navigation, especially when large buildings are in question. Usually, signs and other graphic symbols are used for navigation, although the efficiency of these signs could be lower when people search for a specific place within large buildings.

In existing research, there have been many attempts to design, develop and evaluate the indoor navigation systems of particular buildings. Firstly, Ozdenizci developed the indoor navigation system based on the Near Field Communication (NFC) stickers positioned all around in the building, so that the users are able to navigate within the shopping centre. Secondly, another study presented the system with Wi-Fi indoor navigation by Dongsoo et al., while, thirdly, Chin Gee and Yunli developed a system using augmented reality, where Wi-Fi, GPS and camera are combined, so that the users can utilise the camera to recognise a building or even an indoor place [1][2][3].

However, in previous studies, despite many advantages, there are disadvantages as well. NFC technology is simple and convenient to use, but is complicated for the end user [1]. The user has to search and scan NFC codes, which are often hard to find. There are mobile phones which do not support NFC as well. As far as the user is concerned, Wi-Fi can be more efficient [2]. The problem with actually getting the positon of a user is in the size of the building itself. There's good practice in using

augmented reality, which uses the phone's camera and other sensors for actually recognising the user's surroundings and to get the position of the user [3]. In our application, we have primarily concerned ourselves with user friendliness and simplicity.

Thus, we developed a user-friendly web-based application "Virtual FERI" which allows users to navigate through the particular 3D virtual model of the G3 building of the Faculty of Electrical Engineering and Computer Science. The aim of the web-based application is to help newcomer students, guest lecturers and other university staff to navigate within the building. Another aim was to take advantage of Dew Computing, where a large number of different interactive devices are joined in peer-to-peer virtual processing environments [4].

The 3D model was modelled with real measurements of the actual Faculty, which makes the whole experience more realistic. The added value of the application is its user friendliness, which adds to the whole user experience.

In this paper, we describe the "Virtual FERI" application, its features and the findings of the evaluation where user-experience and usability were tested. Afterwards, the findings are discussed in relation to the findings of the previous studies, while the paper concludes with a thorough description of the application's limits and key advantages of the application.

II. APPLICATION "VIRTUAL FERI"

The "Virtual FERI" is a web-based application which allows users to navigate virtually through the Faculty's premises. The application was developed for students, especially newcomers, and other Faculty members who search for specific locations within the new building G3. A working internet connection and a modern web-browser which supports WebGL are required in order to be able to use the application. The model of the G3 building was modelled entirely in the Sketchup Pro, while the application was developed with the game engine Unity. Currently the web application is still under development, so the users can only navigate in one out of three floors.

The "Virtual FERI" application enables full movement in the 3D modelled space of the Faculty by using a mouse and a keyboard. The user can move across the 3D model easily from the first-person viewpoint. What helps the user to navigate from a certain point towards a specific classroom within the floor are the coloured lines on the ground. For each classroom, there is a different colour used, so that the lines do not confuse the user.

The aim of the application is to help students, guests and other Faculty members find a certain classroom before they actually visit the Faculty in person. We assume that our application will help users avoid the confusion of finding a certain classroom and, therefore, enable a more efficient way of knowing where a certain classroom is located. Since the 3D model of the Faculty was developed in such a way that it resembles the actual physical space as closely as possible, we expect that it will be easier to navigate through the Faculty's premises if you've seen the model beforehand.

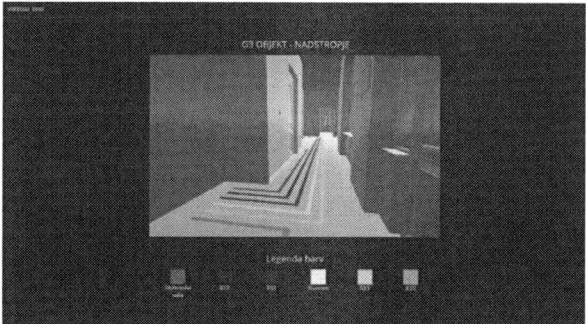

Figure 1. User interface of the application "Virtual FERI"

III. RELATED WORKS

In existing research, there have been a few attempts where the authors discussed the newly developed solutions for the indoor navigation systems. In what follows, we present the approaches used, as well as the strengths and weaknesses of these studies.

Ozdenizci, Coskun & Kerem presented the indoor navigation system based on NFC stickers which are placed around an entire shopping centre [1]. By tapping these NFC stickers, the visitor allows the mobile application to track their location. The advantages of this system are low budget, simplicity, safety and privacy. On the other hand, the problems occur because not every single person has an NFC feature on his/her phone. In contrast to our solution, our application does not support real-time navigation, although it supports 3D building exploring and step-by-step navigation, so that the user can check their path before visiting the building.

The second solution is based on a Wi-Fi connection. Dongsoo et al. presented a system which is intended to be used in the COEX complex in Korea [2]. It is advantageous that the Wi-Fi connection is used since almost every person with a mobile device has Wi-Fi support. Wi-Fi network coverage was used and almost all possible sensors on mobile devices were utilised to achieve high accuracy of the user location. Some issues were found with recognising the user's floor, which was solved by using the barometer on the mobile device. When compared with our solution, the user interface of our application seems to be more simplified, while also supporting 3D building exploring.

The third solution represented by Chin Gee and Yunli allows users to use navigation within the Sunway University campus [3]. Due to the size of the campus, the authors decided to use all possible sensors on a user's mobile device. The application

allows navigating inside and outside the building. The technology of augmented reality is used, where the users can be informed about the buildings by using the camera on their mobile devices. Accordingly, the system can recognise the object and can locate the user with other sensors. The advantage of this application is the simple user interface; however, it can be inconvenient for users a to jump around in public with their mobile devices always in their hands. In contrast, our application allows users to see the location of rooms before getting to the place, so that they can spend less time at the location by searching for the right location of a particular room. Moreover, our application is based on the web platform and it can be reached easily by users through any web browser without pre-installing any plug-in or app.

IV. METHODS

A. Experimental Design

In the experiment, two different groups were included – the experimental and the control group. The experimental group was planned to use the app Virtual FERI first, complete predefined tasks and then to find the particular room in the G3 building physically. On the contrary, the control group had to find the particular room in the G3 building without using the application.

We defined the following tasks for the experimental group: "You are a student who is not familiar with the G3 building of the Faculty of Electrical Engineering and Computer Science at the University of Maribor. You want to attend a lecture in the classrooms Shannon and Diplomska soba.

- Using your web browser, please open the web application "Virtual FERI".
- When you have opened the web application, please read the instructions on how to use the application.
- Please find the classroom called Shannon, which is located on the 1st floor of the G3 building.
- Please find the classroom called Diplomska soba, which is located on the 1st floor of the G3 building.
- After finding both classrooms in the application, turn it off and go to your lecture which is being held in the classroom Shannon."

Likewise, for the control group, the tasks was defined in a way that the participant is a student who is not familiar with the G3 building. Due to the lectures held in the classrooms Shannon and Diplomska soba, (s)he has to find both classrooms which are located on the first floor of the G3 building.

B. Participants

25 first-year students of Media Communications participated in the experiment. Out of these, 12 were female and 13 were male. On average, they were 20 years old. Participants were asked to self-assess their knowledge of Information Communication Technology (ICT) on a scale from 1 (very poor) to 5 (excellent). The average knowledge of Information Communication Technology was 3.9.

Participants were divided randomly into two different groups – the experimental and the control group. The experimental group included 15 students, while the others were classified in the control group.

C. Procedure

The experiment was held on 18th January, 2016, at the Faculty of Electrical Engineering and Computer Science at the University of Maribor. Before participating in the experiment, participants signed informed consent. Both the experimental and the control groups had to find the classroom in the G3 building by walking – the experimental group after using the application "Virtual FERI" and the control group without prior use of the application. We measured the performance of each participant by measuring the time when the task of finding the classroom Shannon was completed successfully. Both groups started their search on the ground floor of the G3 building, while the classrooms that they had to find were located on the first floor of the same building. Once the participant touched the door of the classroom Shannon, measuring time ended.

After completing all tasks, participants in both groups completed the demographic questionnaire and the questionnaire about their experience with finding the classrooms. The experimental group completed also the System Usability Scale (SUS) questionnaire and the User Experience Questionnaire (UEQ) to assess the application "Virtual FERI".

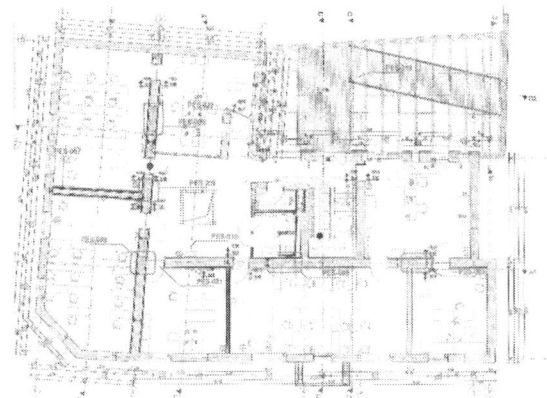

Figure 2. Floor plan where the Shannon and Diplomska Soba classrooms are. The blue circle indicates the starting point of the search in the application "Virtual FERI". The red circle indicates where the Shannon classroom is, and the yellow circle indicates where Diplomska Soba is.

D. Instruments

In the experiment, we used four different questionnaires. Firstly, we used a demographic questionnaire, where we asked participants about their gender and birth year. They were also asked to self-assess their knowledge of ICT (evaluated on the scale, ranging from 1 to 5).

Secondly, an SUS questionnaire was used to assess the usability of the application "Virtual FERI" [5]. It is a commonly used, freely distributed, and reliable questionnaire consisting of 10 items consisting of a Likert's scale from 1 to 5 where 1 means "strongly disagree" and 5 "strongly agree". The final SUS score could be between 0 and 100. The closer the score is to 100, the better the usability of the evaluated application is. Fig. 3 shows acceptability ranges for the application based on the SUS score.

The final SUS score is shown with three indicators. The first indicator is to establish user friendliness, the second indicator shows the user's grade of the application with letters (F, D, C, B, A, where F is the lowest possible value and A is the best) and the third indicator shows on a scale from 0 to 100 the categorization of usability where 0 - 25 is the worst imaginable, 25-39 is poor, 40-51 is ok, 52-72 is good, 53-85 is excellent and 86-100 is the best imaginable.

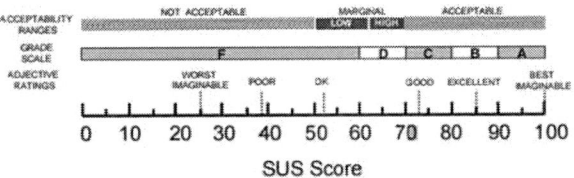

SUS Score

Figure 3. The ratio of indicators when using the SUS method. The final SUS score ranges from 1 to 100 and is divided in three different testing grades (acceptability ranges, grade scale and adjective ratings). The acceptability range indicates if the tested application is not acceptable, marginal or acceptable from an user experience's viewpoint. The grade scale is as follows: F for 0 to 60, D for 60 to 70, C for 70 to 80, B for 80 to 90 and A from 90 to 100. The adjective ratings are from worst imaginable (25) to best imaginable (100). [1]

Thirdly, we used a UEQ to evaluate the user experience of the application "Virtual FERI" [10]. The questionnaire consists of 26 items showing six different user experience categories:

1. Attractiveness - the general feel of the product, because it is important to establish if the user likes the product or not.
2. Perspicuity - is easy to understand and to learn how to use.
3. Efficiency - the users can use the product with ease.
4. Dependability - the users feel that they have control while using the application.
5. Stimulation - the product is exciting to use.
6. Novelty - the product is innovative and attracts the user's interest.

Compared to other questionnaires, the advantage of the UEQ questionnaire is the focus on aesthetics and the more subjective viewpoint of the user interface. It also consists of objective viewpoints, such as efficiency and reliability. UEQ offers us a final grade of the application by each category. The grading is based on existing grades of other solutions. These solutions include data from 9,905 users, who have participated in 246 studies.

Fourthly, we used the questionnaire about participants' experience with finding the classrooms. Two different versions were used – one for the experimental group and one for the control group. The version of the questionnaire which was intended to measure the experience with finding the classrooms in the experimental group comprised six Likert type question items where participants were asked to answer on the scale ranging from 1 (strongly disagree) to 5 (strongly agree) to what extent they agree with the question items. Three question items referred to the ease of using the application, while the other three items referred to searching for the classroom by walking in the G3 building. Item examples: "It was easy for me to find the classroom Shannon in the Virtual FERI." and "It was easy for me to find the classroom Shannon while walking in the G3 building."

Besides questionnaires, we also used an instrument for measuring time performance while participants were searching the classroom in the G3 building by walking.

V. RESULTS

A. Time performance

Results of measuring time revealed that participants in the control group needed 24,22 seconds to reach classroom Shannon and 6,22 seconds on average more to walk from the Shannon classroom to the Diplomska soba classroom. Altogether, it took them 30,44 seconds to reach the classroom Diplomska soba.

The experimental group needed 20,08 seconds to find the Shannon classroom and 6,44 seconds on average more to walk from the Shannon classroom to the Diplomska soba classroom. Altogether, it took them 26,52 seconds to find the Diplomska soba classroom.

These results demonstrate that the experimental group, where students used the application for navigating within the building, was 17,09% faster than the group which did not utilise the application. Moreover, when the measured time for finding both classrooms is considered, the experimental group was 14,8% faster than the control group.

Fig. 4 shows comparisons of time across both groups. On average, the participants who were using the application "Virtual FERI" were faster in finding the Shannon classroom by 4,14 seconds and by 3,52 seconds finding the Diplomska soba classroom. Accordingly, participants who used the application "Virtual FERI" spent less time than users without prior use of the application.

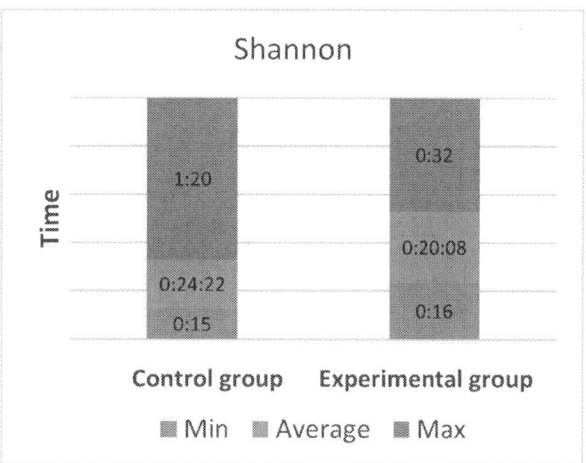

Figure 1. Time comparison between two control groups for Shannon. Time comparison between the control group and experimental group for Shannon. The minimum time required to find Shannon is shown with the orange color. The maximum time needed to find Shannon is shown with the blue color and the average is shown in gray.

Figure 4.

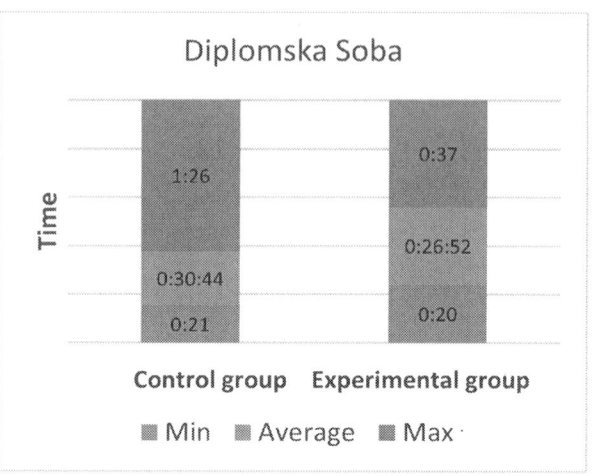

Figure 5. Time comparison between two control groups for Diplomska Soba. Time comparison between the control group and experimental group for Diplomska Soba. The minimum time required to find Diplomska Soba is shown with the orange color. The maximum time needed to find Diplomska soba is shown with the blue color and the average is shown in gray.

B. SUS score

The results of the SUS method for assessing user friendliness have showed that the application "Virtual FERI" was successful in testing and is graded as usable. The average result of the third indicator was 80 points, which is right in the middle of very good and excellent. The grade of the second indicator was in between B and C, which shows positive motivation from participants to use the application "Virtual FERI".

C. Results of UEQ Questionnaire

The results have shown that judging by attractiveness, efficiency, novelty and perspicuity, there are 10% of results which are better, and 75% which are worse. Only for the category of dependability and stimulation were there 25% of results which were better, and 50% which were worse. As one can see from Fig. 6, participants who used the "Virtual FERI" application assessed the application with the grade good in four out of six categories. In the other two categories, the application was graded with above average.

Participants perceived the "Virtual FERI" application as attractive, they understood it and could learn what it does easily. Moreover, they used it with relative ease and graded it with an above average for overall control and interaction. Stimulation was also graded with the grade of above average which shows that the application is exciting to use and that there is also room for improvements in order to achieve better results.

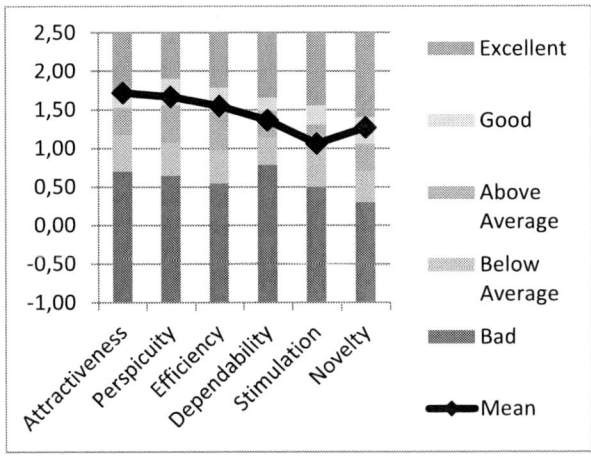

Figure 6. The final grades of the application.

VI. DISCUSSION

The aim of this study was to design, develop and evaluate a functional application for 3D-based navigation within the building G3 of the Faculty of Electrical Engineering and Computer Science at the University of Maribor and using the Dew Computing approach.

The application used in the evaluation was a prototype which showed the first floor of the G3 section of the Faculty only. Since all floors of the G3 object are quite small, our study has shown that there have been only small, but still visible, time differences between both groups (group using the application and group without using the tools). The group which used the application has, overall, been faster by 17,09% and 14,8% at locating the right classroom. We have observed that the application would be impacted significantly if we will include all floors and applied the solution to larger objects.

Our study had some additional limitations. Firstly, the application in its current state can only be run by using the latest version of Firefox, because the application is based on Web GL, which is, unfortunately, not supported by all browsers.

Secondly, the size of the building was not large. As the building is only about 30 meters broad and the way from the starting measured point to the final measured point was quite short, there may be different results when the experiment takes place in a larger building. For the experimental groups, for example, the minimum measured time was 16 seconds and the maximum measured time was 26 seconds. On the contrary, in the control group, the minimum measured time was 15 seconds, while the maximum measured time was 1 minute and 20 seconds. These results demonstrate that the differences in measured time when the application 'Virtual FERI' was used were smaller in the experimental group compared to the measured time in the control group where the span of differences in measured time were greater. It is possible that the effects we have observed may change significantly in favour of application use if navigation for the whole building was developed. This is clearly an area for further work.

According to the results, the SUS score of the application "Virtual FERI" was 80 out of 100, which demonstrates good conditions for usability of the system. The score may be higher when the indoor navigation for all other floors would be implemented as well and when the building would be larger.

Concerning the UEQ score for user experience rating, lower results were achieved in the stimulation category only, but even this category is still above average. Altogether, we have achieved satisfactory results, but there is still room for improvement. In particular, we could increase the number of displayed floors and the application could be optimised for other browsers and not only Firefox.

The findings of our study complement the findings of previous studies [1][2][3] where the authors found in their study that the indoor navigation with a smartphone application could improve user experience in a building significantly and could lower stress when searching for appropriate rooms.

The findings encouraged the authors of this study to start developing the indoor navigation further for the remaining floors of the building as well.

VII. CONCLUSION

The findings of our study revealed that indoor navigation could be faster when the application "Virtual FERI" is used, since the usability and user experience were satisfactory.

Specifically, the experimental group is more likely to use indoor navigation "Virtual FERI" compared to the group without the application. When the application was used, users demonstrated high interest in using the indoor navigation system, although it is still under development.

It is expected that new indoor navigation technologies will continue to emerge, especially in the field of Dew Computing, that will continue to make distributed computing more powerful and efficient.

Perhaps the most open question is still how to improve the user experience of the proposed application. Here we recommend, as an important step in this process, the need of real-time indoor navigation in conjunction with the various types of markers and sensors used in Dew Computing.

ACKNOWLEDGMENT

We would like to thank the students of Media Communications from the Faculty of Electrical Engineering and Computer Science at the University of Maribor for their participation in the study.

REFERENCES

[1] B. Ozdenizci, V. Coskun and K. Ok, "NFC Internal: An Indoor Navigation System". Sensors, 2015. Available on: http://www.mdpi.com/1424-8220/15/4/7571/

[2] H. Dongsoo, L. Minkyu, Y. Giwan and J. Sukhoon, "Building a Practical Wi-Fi-Based Indoor Navigation System". IEEE Pervasive Computing, 2014. Available on: https://www.researchgate.net/publication/262528341_Building_a_Practical_Wi-Fi-Based_Indoor_Navigation_System

[3] L. Chin Gee and L. Yunli, "Interactive Virtual Indoor Navigation System using Visual Recognition and Pedestrian Dead Reckoning Techniques". International Journal of Software Engineering and its

Applications, Vol. 9, No. 8, 2015. Available online: http://www.sersc.org/journals/IJSEIA/vol9_no8_2015/2.pdf

[4] K. Skala, D. Davidović, E. Afgan, I. Sović, Z. Šojat. (2015). Scalable Distributed Computing Hierarchy: Cloud, Fog and Dew Computing, Open Journal of Cloud Computing (OJCC), Vol. 2, Issue 1, pp. 16-24.

[5] J. Brooke, (1996) "SUS: a "quick and dirty" usability scale", in: P. W. Jordan, B. Thomas, B. A. Weerdmeester in A. L. McClelland. Usability Evaluation in Industry. London: Taylor and Francis.

[6] A. Bangor, P. Kortum, J.A. Miller, The System Usability Scale (SUS): An Empirical Evaluation, International Journal of Human-Computer Interaction, 24, (2008), 6, pp. 574–594.

[7] B. Laugwitz, T. Held, M. Schrepp, Construction and Evaluation of a User Experience Questionnaire, Chapter HCI and Usability for Education and Work, Volume 5298 of the series Lecture Notes in Computer Science, pp. 63-76, 2008.

[8] Finstad, K. (2006). The System Usability Scale and non-native English speakers. Journal of Usability Studies, 1, 185--188.

[9] M. Debevc, J. Saša, J. Lapuh Bele. (2016). User Experience Questionnaire (UEQ). Available on: http://www.ueq-online.org/

Architecting a Hybrid Cross Layer Dew-Fog-Cloud Stack for Future Data-Driven Cyber-Physical Systems

Marc Frincu

e-Austria Research Institute and West University of Timisoara
Department of Computer Science
Timisoara, Romania
Email: marc.frincu@e-uvt.ro

Abstract—The Internet of Things is gaining traction due to the emergence of smart devices surrounding our daily lives. These cyber-physical systems (CPS) are highly distributed, communicate over wi-fi or wireless and generate massive amounts of data. In addition, many of these systems require near real-time control (RTC). In this context, future IT platforms will have to adapt to the Big Data challenge by bringing intelligence to the edge of the network (dew computing) for low latency fast local decisions while keeping at the same time a centralized control based on well-established scalable and fault tolerant technologies brought to life by cloud computing.

In this paper we address this challenge by proposing a hybrid cross layer dew-fog-cloud architecture tailored for large scale data-driven CPSs. Our solution will help catalyze the next generation of computational platforms where mobile and dynamic IoT platforms with energy and computational constraints will be used on demand for storing and computing Big Data located nearby in near real-time for local decisions and extend to cloud systems for fast orchestrated centralized decisions. The proposed architecture aims to leverage the advantages of both cloud and dew systems to overcome the challenges and limitations of modern communication networks.

We also discuss two real-life life solutions from the field of smart grids and smart transportation systems.

1. Introduction

The Internet of Things (IoT) is growing at a fast pace especially with the increasing presence of networked interacting components present in everyday activities such as transportation, distributed robotics, medical monitoring, automatic pilot avionics, and the emerging smart city. In 2016, 6.4 billion connected devices were, according to Gartner [1], present around the world with forecasts predicting up to 30 billion by 2020. Many of these devices are part of autonomous cyber-physical systems (CPS), e.g., smart grids, self-driving cars, intelligent traffic management systems, and require near real-time control (RTC) for coordination and critical decision making. Thus, IoT is already impacting our society in ways that will revolutionize our lives and how we interact with each other and with the CPSs themselves from a cultural and socio-economic perspective. However,

migrating towards a fully interconnected world poses both scientific and technological challenges which need to be met in the short term by novel solutions if the vision of an IoT world is to become reality. These systems will no longer be controllable through complex linear and nonlinear systems of equations but through **data-driven machine learning algorithms**. In particular, the Big Data problem together with the low energy requirements and computational resources of IoT devices need to be mitigated through novel platforms which enable the *extension* of these systems to large scale processing environments such as the clouds which already posses the necessary software stack for data analytics of heterogeneous data streams.

The large amount of high speed sensor data arriving in variable data rates poses serious design and cost challenges to existing on-edge cyber-infrastructures. The stress of having tens of thousands of devices communicating by using wi-fi or wireless over vast distances with centralized data centers placed in the cloud is already limiting the ability of existing smart grids to send real-time data (e.g., most smart grids can only send aggregated data every 8 hours or even daily which means that decisions at finer granularity are hard if not impossible to achieve). This greatly limits the ability to efficiently control these large distributed CPS in near RTC. The problem is that the cost of upgrading the data transmission medium (e.g., wi-fi, wireless, PLC, bluetooth) to handle city wide CPS infrastructures of millions of devices makes the current approaches which centralize data and control inefficient. Thus, cloud computing alone will not be able to address the issue of CPS generated Big Data.

One solution would be to move the processing towards the edge of the network closer to the data source. We identify here three layers of computation [2]: edge or dew (on device), fog (local area with partial view of the CPS), cloud (global view of the CPS). *Combining these three layers in a single coherent scalable data-driven solution is the main challenge that IoT will have to overcome to truly become ubiquitous.* While clouds offer the premise for on-demand scalable computing they need the data delivered to them before proceeding. Battery powered edge devices already have some processing power and storage capacity to preprocess local data before sending them up the hierarchy for centralized control and optimizations. In this setup the network itself becomes the bottleneck. *A hierarchical approach where intelligence is brought near the edge and combined with the advantages of using clouds is therefore required.* In this context, the limitations of the network become a problem especially for near RTC based CPSs.

Fundamental research is warranted in several core areas ranging from hardware infrastructure, networking and communication to programming models, scalable and resilient distributed execution frameworks as well as domain specific software optimizations. This paper addresses the latter end of the spectrum. The objectives are twofold: 1) propose an architecture for generic near RTC based CPS to leverage fog and cloud computing, and 2) map the proposed architecture on two use cases with different requirements, demonstrating the wider applicability of our solution.

The rest of the paper is structured as follows: Section 2 details existing solutions, Sect. 3 describes the proposed cross layer architecture, Sect. 4 describes the software stack envisioned to be deployed on top of the architecture, Sect. 5 describes two possible use cases, and finally Sect. 6 outlines the key points of the paper.

2. Related Work

While much research has been conducted on designing secure and scalable CPSs [3], [4], [5], [6], [7] few focus on utilizing cloud infrastructure to enable scalability and RTC [8], [9], [10]. Most of the existing work is geared towards challenges in specific CPSs such as vehicular networks [8] or smart grids [11] and employ several domain specific strategies to achieve RTC.

A framework to integrate IoT and CPS is proposed in [12]. However, it does not consider the resources of clouds to enable fast coordinated control over large scale distributed CPSs.

We propose to abstract the features common across an array of CPSs and develop a generic hierarchical architecture and middleware with scalability, reliability, and adaptability as its fundamental principles.

Finally, studies [9], [13], [14] have been conducted on evaluating the feasibility of using clouds in the context of CPS and on identifying challenges, performance issues and bottlenecks. By building on these results we develop an architecture that will is beneficial not only in the context of CPSs but also in other domains where near RTC is crucial.

3. Proposed Architecture

We assume CPSs to comprise of low energy and memory distributed devices which need to share their information in order to globally optimize the entire CPS. Examples of such systems include smart grids where each smart meter and sensor needs to send its information to the utility which then decides on the critical areas of the power grid. However, clients can still optimize their consumption by taking decisions based on local information.

Orchestrating large scale CPSs is challenging due to the data deluge coming from their sensors which needs to be timely processed and analyzed for decision making. to avoid flooding the network with data several choices exist. First, the sampling interval can be decreased but this can lead to crucial loss of information which cannot be accurately reconstructed through interpolation techniques. Second, data could be preprocessed in situ before sending *meta-information* to the *central controller* for the heavy processing part of global decision making. The *local processing* could act as a local optimization where decisions requiring a local or partial view of the entire CPS are taken without involving the central controller.

Focusing on the second approach we argue that in order o make the process efficient the central controller could be employed only when localized control fails to keep the system within predefined parameters as given by the *control model*. The control model is a data-driven model used by machine learning algorithms to take local or global optimizations of the CPS. In a smart grid context this means that while certain users could optimize their consumption reducing the stress on the power grid in demanding days (i.e., hot days when A/C is responsible for a large portion of the total energy consumption) they do not sum up to form the critical mass required to reduce the demand below the utility generation capacity.

Figure 1 depicts the envisioned architecture for such a hybrid system. The entire system is event based where the execution of each component is triggered by the occurrence of one or more events. There are three key components we discuss next in relation with the events used for triggering them.

Big Data preprocessing and local control component which enables dew computing. Data is generated locally on the edge devices. Based on the recorded data data pruning takes place and historical data is stored within the limited device memory of each device. Then, data energy efficient and low memory footprint machine learning algorithms can be used to predict the future state of each device assuming no outside world interaction. This simple model can be extended by allowing nearby devices – in the fog – to share their information and feed their data-driven algorithms with fog data. Section 5 will further discuss the implications of this approach on two different use cases. This preprocessed data (either by each device or by a group of nearby devices) is sent to the RTC control component. By sending only partial data we can relieve some of the stress on the network and enable better horizontal scalability of the CPS. Depending on the CPS the preprocessed data could be sent only when a global deterioration of the overall system is noticed. This situation could be triggered by an event from a module monitoring the control accuracy in the RTC control component.

RTC component for fast cloud processing. To efficiently control large scale CPSs global information is required. This information can be exchanged through a P2P approach among edge devices or can be handled centrally. It is obvious that for near RTC this information can easily saturate,

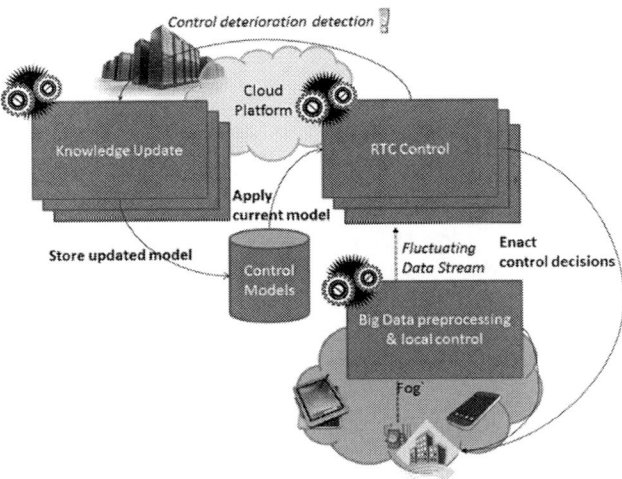

Figure 1. Envisioned hybrid cross layer architecture for dynamic data-driven CPSs.

in both cases, the network pushing the need for alternative methods which do not require constant near RTC data delivery. In our case we argue for a cloud based approach due to the elasticity, reliability, and support (i.e., software like Apache Storm, Spark, MapReduce) of such systems to process Big Data. This component is responsible for the management of the CPS and acts the central controller. Based on the control model and recent data from the edge devices it updates the global view of the CPS and takes decisions to globally optimize the system. Recent works [15] have also shown that it is possible to not rely on information from all edge devices but rather use causality to reduce the search space and still be able to take accurate decisions.

Knowledge update component. This component is responsible for updating the control model used by the RTC control component. The control model can refer to a pair <machine learning algorithm, data> where it has been shown [16] that the efficiency of a machine learning algorithm depends on the input data. Hence, based on the characteristics of the data arriving from the edge devices different algorithms could be used and changed online. The update in the control model could be triggered by the RTC component when significant changes in the data properties are noticed or when a control deterioration is observed despite using up to date data from all edge devices. the entire collection of control models is computed offline based on benchmarks on real-life CPS data.

4. Envisioned Software Stack

Based on the proposed architecture, a hybrid dew-fog-cloud control system for near RTC of CPSs would have to implement the following software stack comprising of on-edge algorithms, middleware, and elastic machine learning on clouds.

4.1. Low Energy and Memory On-Edge Algorithms

Most edge devices have limited memory and energy consumption limitations due to battery lifetime. As example, the current state of the art smart meters have a memory of only 92KB on which they store temporary data and perform minimal data preprocessing. The remaining memory, while extremely low could be used for data pruning and simple data-driven predictions. The predictions could be done either on the smart meter or at the data concentrator (gateway to a set of customers) in the fog. Alternatively, some of the data could be sent to customers mobile devices – if in the local area – for processing.

Performing simple machine learning on these edge devices will require analysis and redesign of existing algorithms to tailor them for memory and energy constrained environments without significantly impacting their accuracy. Furthermore, the algorithms will have to consider data scarcity and veracity.

4.2. Dew-Fog-Cloud Middleware

Bridging fog and clouds requires a middleware capable of leveraging the benefits of clouds to improve the control of edge devices. The three-tier middleware will enable:

- Dew level:
  - Local data preprocessing and control decisions by using on on-edge algorithms for low energy and low memory environments;

- Fog level:
  - Preprocessed data aggregation and processing to reduce network overhead when communicating updates to the cloud;

- Cloud level:
  - Periodic control model updates to train the machine learning analysis on incoming data;
  - Near RTC centralized decisions based on the current control model.

The middleware consists of three key pluginnable components as depicted in Fig. 1, each acting as a plug-in for easy customization. Inter-component communication needs to be reliable and light to reduce chances for network congestion and to enable horizontal device scalability to support large scale distributed CPSs. Hence, some tasks will have to be migrated towards the edge either directly on the device (dew computing) or in a local area aggregator (fog computing).

Since two key components, the RTC and the Knowledge update, are deployed on clouds, several key aspects will have to be overcome. These include reliability, variability in cloud resource performance, data privacy, and timely delivery of the control decisions to meet the near RTC requirements. In [13] several cloud performance issues of a community seismic application running on virtual machines were identified. These include variable load and deadline misses for

401

processing requests. While the analysis was done on a non elastic cloud environment running on Google AppEngine it does outline some key problems encountered by applications requiring near RTC which require the redesign of applications and algorithms to support elastic environments for fast processing. Building such an infrastructure requires taking into account the cost impact. The cost model will have to link virtual machine performance fluctuations, infrastructure reliability, network latency, and cloud elasticity with the time needed to perform knowledge training, behavior forecasting, and control decision and enactment. Cost will therefore measure the penalty for performance deterioration due to late decisions in soft and hard RTC CPSs. The performance model will also have to be tightly linked to the economics of clouds which allows significant cost reductions through their pay-per-use approach since it may be more affordable to deploy the middleware on public clouds rather than on private infrastructures.

4.3. Elastic Machine Learning on Clouds

An important part of the middleware is represented by the machine learning algorithms used by the central controller and selected based on the current control model. These algorithms will have to be adapted for elastic cloud infrastructures and suitable for deployment in streaming software such as Storm and Spark. While public providers such as Amazon and Azure already offer machine learning tools, some of the algorithms in use by CPSs may be newly designed for the specific use cases, improved versions, or ensemble models of several existing algorithms. In these cases, these algorithms which are not normally designed for elastic and environments will have to be re-engineered.

5. Use Cases

In this section we focus on describing two use cases in relation to our proposed architecture by detailing functional aspects and particularities in each of them. The objective is to demonstrate the generality of our proposed solution.

The first use case deals with smart grids which rely on homogeneous data transmitted periodically using limited network transmission environments for global consumption optimization. The second use case is a smart transportation system where the route of a car must be optimized by considering heterogeneous sources of information as well as the behavior of other drivers. Heterogeneity is not the only difference between the two. Smart grids are inherently static while vehicles are constantly moving. An generic cross layer architecture must therefore cope with both cases efficiently. In the smart transportation case processing data in the nearby fog could lead to data inconsistency and wrong decisions as a car which shared information might leave the fog after a decision is taken but before it is transmitted to the drivers.

5.1. Smart Grids

An increasing number of utilities migrate their traditional power grids to automated smart grids. These systems rely on smart meters to receive consumption data. As seen most of these meters have inbuilt limited memory allowing them to store historical data and to run simple algorithms. Information from smart grids can be used both locally by clients to reduce their consumption and globally by utilities to balance the supply-demand of energy during peak hours. In both cases there is a need for constant monitoring and adaptation based on latest data. The problem is challenging especially for the utility which needs to constantly monitor clients and select those that are most likely to respond to energy curtailment actions without impacting too much their comfort. Customer selection algorithms and machine learning algorithms for predicting the energy curtailment of each selected customer are needed and during peak hours they need to execute within the sampling period of the smart meters. Obviously to maximize efficiency the sampling rate must be as high as possible.

When mapping this use case on our architecture we have the smart meters acting as edge devices and the utility which relies on cloud computing to process the incoming data. As seen in Sect. 1 data networks currently limit the transmission rate of smart meters. Therefore, bringing some of the centralized decision toward the edge could greatly impact the efficiency of the smart grid for near RTC control. Such localized control (by the Big Data preprocessing and local control component running on the smart meter) could include optimization of individual customers' consumption based on their own historical usage and behavior. Furthermore, these predictions for the consumption of each customer could be sent to the cloud for taking global control actions and to target customers most likely to take part in the energy curtailment event (i.e., demand response). Sending predictions to the RTC component instead of raw data based on which the RTC component would make its own predictions enables us to remove some of the stress in both the network and the component itself. In addition, each smart meter could be remotely programmed with a different prediction method tailored to each customer's behavior. The RTC component would then use these predictions to select the most suited customers and to monitor the impact on the supply-demand balance. When an increasing negative impact (in terms of predicting the aggregated consumption in the next time frame for instance) would be noticed the component would adapt by changing both the customer selection and the consumption prediction algorithms by calling the knowledge update component.

5.2. Smart Transportation

Recent years have witnessed the installation of intelligent traffic lights and speed control systems in major cities. Combined with GPS information and smartphone applications which constantly monitor traffic and cars that exhibit increasingly intelligence when it comes to driving efficiency it becomes clear that integrating the three systems (traffic, car, and smartphone) can reduce CO_2 emissions and traffic congestion. Such an integrated system would consider traffic speed, fuel consumption, traffic lights, and driver preferences to pick the optimal route to a given point by leveraging information from other cars, drivers, and the traffic management system.

When mapping this use to our architecture we have the sensors on cars, the smartphones, and the sensors on the street as edge devices. This heterogeneous CPS will have to share data and coordinate to reduce CO_2 emissions, improve traffic speed and reduce congestion. In this scenario, each car would take local decisions on how to adapt speed and path based on the preferences of the driver, nearby vehicles, and traffic speed and situation taken from Google Maps for instance. This decision could be taken in a car controller or by the smartphone application and then presented visually to the driver both playing the role of the by the Big Data preprocessing and local control component. Then, by relying on information from other cars it would further optimize its path by considering the direction (and possibly the destination) and speed of other drivers. In this way traffic jams and average speed can be improved by predicting where congestions are likely to happen. Given the limited processing and storage power of car sensors and controllers, the energy constraints of smart phones, and mobility of nearby vehicles this global optimization based on swarm intelligence can be taken centrally on a cloud system through the RTC component. Information from the CPS is then processed and analyzed and path optimization algorithms based on traffic and driver behavior can be applied to suggest the best routes. In this case the knowledge update component could pick from the control model between various algorithms for stream based route selection and prediction algorithms based on incoming data properties.

6. Conclusion

In this paper we have argued that due to the complexity of IoT systems cross layer dew-fog-cloud systems should be developed. However, since the data transmission network limits both the scalability and data rate of edge devices some of the processing usually done on clouds needs to be migrated towards the fog by leveraging the storage and processing capabilities of edge devices. Based on these we have proposed a hybrid cross layer architecture and a possible software stack consisting of low-energy low-memory on-edge machine learning algorithms, middleware, and elastic machine learning algorithms for clouds. Two use cases have been investigated with references to the proposed architecture.

Future work will include building a prototype and implementing some machine learning algorithms for edge and cloud environments.

Acknowledgments

This work has been partially funded by a grant of the Romanian National Authority for Scientific Research and Innovation, CNCS/CCCDI - UEFISCDI, project number PN-III-P3-3.6-H2020-2016-0005, within PNCDI III, and by the EU H2020 CloudLightning project under grant no. 643946.

References

[1] A. Nordrum. (2017) Popular internet of things forecast of 50 billion devices by 2020 is outdated. (accessed January 3, 2017). [Online]. Available: http://spectrum.ieee.org/tech-talk/telecom/internet/popular-internet-of-things-forecast-of-50-billion-devices-by-2020-is-outdated

[2] K. Skala, D. Davidovic, E. Afgan, I. Sovic, and Z. Sojat, "Scalable distributed computing hierarchy: Cloud, fog and dew computing," *Open Journal of Cloud Computing (OJCC)*, vol. 2, no. 1, pp. 16–24, 2015. [Online]. Available: https://www.ronpub.com/OJCC_2015v2i1n03_Skala.pdf

[3] Y. Xin, I. Baldine, J. Chase, T. Beyene, B. Parkhurst, and A. Chakrabortty, "Virtual smart grid architecture and control framework," in *2011 IEEE International Conference on Smart Grid Communications (SmartGridComm)*, 2011, pp. 1–6.

[4] A. Banerjee, K. K. Venkatasubramanian, T. Mukherjee, and S. K. S. Gupta, "Ensuring safety, security, and sustainability of mission-critical cyber-physical systems," *Proceedings of the IEEE*, vol. 100, no. 1, pp. 283–299, 2012.

[5] P. Bogdan and R. Marculescu, "Towards a science of cyber-physical systems design," in *2011 IEEE/ACM Second International Conference on Cyber-Physical Systems*, 2011, pp. 99–108.

[6] S. Karnouskos, "Cyber-physical systems in the smartgrid," pp. 20–23, 2011.

[7] Z. Wang, Y. Zhang, and K. Du, *Cyber-Physical Traffic Systems: Architecture and Implementation Techniques.* Berlin, Heidelberg: Springer Berlin Heidelberg, 2013, pp. 490–500.

[8] H. Abid, L. T. T. Phuong, J. Wang, S. Lee, and S. Qaisar, "V-cloud: Vehicular cyber-physical systems and cloud computing," in *Proceedings of the 4th International Symposium on Applied Sciences in Biomedical and Communication Technologies*, ser. ISABEL '11, 2011, pp. 165:1–165:5.

[9] M. Kim, M. O. Stehr, J. Kim, and S. Ha, "An application framework for loosely coupled networked cyber-physical systems," in *2010 IEEE/IFIP International Conference on Embedded and Ubiquitous Computing*, 2010, pp. 144–153.

[10] A. Thiagarajan, L. Ravindranath, K. LaCurts, S. Madden, H. Balakrishnan, S. Toledo, and J. Eriksson, "Vtrack: Accurate, energy-aware road traffic delay estimation using mobile phones," in *Proceedings of the 7th ACM Conference on Embedded Networked Sensor Systems*, ser. SenSys '09, 2009, pp. 85–98.

[11] A. J. Conejo, J. M. Morales, and L. Baringo, "Real-time demand response model," *IEEE Transactions on Smart Grid*, vol. 1, no. 3, pp. 236–242, Dec 2010.

[12] T. S. Dillon, H. Zhuge, C. Wu, J. Singh, and E. Chang, "Web-of-things framework for cyber–physical systems," *Concurr. Comput. : Pract. Exper.*, vol. 23, no. 9, pp. 905–923, Jun. 2011.

[13] M. Olson and K. M. Chandy, "Performance issues in cloud computing for cyber-physical applications," in *2011 IEEE 4th International Conference on Cloud Computing*, July 2011, pp. 742–743.

[14] Y. X. Junhua Zhao, FushuanWen and Z. Lin, "Implementing an essential computing platform for future power systems," *Automation of Electric Power Systems*, vol. 34, no. 15, pp. 1–8, 2010.

[15] S. Aman, C. Chelmis, and V. K. Prasanna, "Influence-driven model for time series prediction from partial observations," in *Proceedings of the Twenty-Ninth AAAI Conference on Artificial Intelligence*, ser. AAAI'15. AAAI Press, 2015, pp. 601–607.

[16] S. Aman, M. Frincu, C. Chelmis, M. Noor, Y. Simmhan, and V. K. Prasanna, "Prediction models for dynamic demand response: Requirements, challenges, and insights," in *2015 IEEE International Conference on Smart Grid Communications (SmartGridComm)*, 2015, pp. 338–343.

Future Applications of Optical Wireless and Combination Scenarios with RF Technology

Erich Leitgeb

Institute of Microwave and Photonic Engineering, Graz University of Technology Graz, Austria, erich.leitgeb@tugraz.at

Abstract— **Optical Wireless Communications, (OWC), also known as Free Space Optics (FSO) is presented in this contribution for future applications. OWC in combination (as hybrid transmission method) with other telecommunication technologies (including WLAN and satellite communications) is shown as innovative approach for alternative network scenarios. Modular communication systems are considered, which allows worldwide access to the Internet or other networks by combining satellite communications, FSO, Wireless LAN, Local Multipoint Distribution System (LMDS) and DVB-T (terrestrial digital video broadcast). Current and future applications of OWC, including deep space missions and autonomous driving systems (cars, ships, planes) combined with 5G networks are shown.**

Keywords— *Hybrid networks, Free Space Optics, Optical Wireless Communications, Satellite Communications, Wireless LAN, Microwave link, weather conditions, reliability and availability, LMDS, DVB-T, Civil-Military-Cooperation, 5G*

I. INTRODUCTION

OWC and RF links in combination (as hybrid solutions) will offer high data rates for future and innovative applications. Wireless LAN offers connectivity to mobile users in a network cell and Free Space Optics allows quick installation of broadband fixed wireless links instead of cables. Additional traditional satellite communications provides a backbone between distant locations in the world. Also DVB-T as the current video broadcast standard (instead of former analogue TV) can be used for Internet-access (SEE TV-WEB). Different scenarios (and results) using modular wireless technologies are shown. Hybrid solutions are also used more and more for data communication on deep space missions and in applications for autonomous driving (in combination with 5G networks).

Because of the increasing need of high data rates in different applications, the Optical Wireless technology has become more and more important within the last years. High capacity transmission technologies all over the world are necessary to connect the user to high speed internet for various multimedia applications. Not only bad weather conditions will be counteracted by a combination of FSO as well as other wireless and RF-technologies, also natural disasters or wars can profit on combined networks, because such national and global incidents generate a lack of infrastructure (including the telecommunications facilities). In our current century the usual type of wars and also the catastrophes have changed their faces. Wars have become asymmetrical fights and battles (visible by terror attacks), causing different of reactions in the warfare. Connected to this change in our world very often the typical natural disaster has given way to human catastrophes after war and terror attacks. So nowadays the world population needs a combination of different network technologies to survive within the various threats.

II. OPTICAL WIRELESS COMMUNICATION AND RF

A brief introduction shows the advantages and disadvantages of optical wireless (also called Free Space Optics, FSO) compared to fibre and RF technologies. The first part of the invited talk is focused on increasing the interest on Free Space Optics. A rough overview to the physical/electrical description of the various components, notably emitters (light sources), receivers (light detectors) and the transmission medium and -techniques are given. A look into the basics, describes the main influences on the reliability and availability of Free Space Optics units using this technology in the atmosphere. The effects like molecular absorption, scattering on small particles and atmospheric turbulences are discussed and the main limiting factors of FSO are demonstrated in this talk. As mentioned high data rate applications (for tele-medicine, disaster recovery, tele-teaching, sports, music, events etc.) require broadband access in our current ages.

To increase the reliability and availability of optical wireless links is achievable on the one hand with special coding techniques, auto-tracking methods and automatic gain control and on the other hand with combined hybrid networks (FSO and microwave systems). Microwave and FSO links have similar properties regarding offered data rates and flexibility of setup, but operate under different conditions, with their benefits and challenges [1, 6]. The benefit of a combination of communication systems operating at millimetre waves and optical waves is the complementary behaviour of each technology during different weather conditions. Rain is the dominant cause for temporary variable attenuation in the microwave link, whereas fog is the most important cause for attenuation in the optical wave link. For a stand-alone FSO system, fog can cause attenuations of 100 dB/km in the climate around Graz [3], while rain at a thunderstorm at a rain rate of 150 mm/h only can generate attenuations of 25 dB/km [5]. The same rain rate can cause up to 50 dB/km attenuation for a microwave link [4] (up to 35 dB/km for 40 GHz), while fog does not particularly matter, and increased humidity causes less than 5 dB/km. To achieve high availability, a high link margin is necessary for each single technology. Bad weather conditions are limiting Free Space Optics applications mainly within the last mile access area and are requiring high output power for a longer range microwave system. In the hybrid system, FSO only needs to overcome rain attenuation and the microwave system needs to overcome attenuation caused by increased humidity, fog and clouds. This allows operating the hybrid link at lesser margins for both communications systems, or at extended distance.

In general in FSO systems any optical wavelength can be used, but because of the atmospheric conditions and the laser safety regulations the longer wavelengths (e.g. 1,550 nm) should be preferred. FSO links through the troposphere are mainly influenced by weather conditions [2]. As mentioned rain does not influence optical transmissions drastically (attenuation of 3 dB/km), because raindrops size is a few millimetres, much larger than the operating optical wavelengths (1,550 nm), thus causing minimal scattering of the laser energy. However, FSO links are affected dramatically [2, 3] by heavy fog (more than 30 dB/km) and clouds; because the fog and cloud droplets have comparable size, as the used wavelengths, causing much scattering of the laser energy as the fog and clouds become thicker. Another major influence on the FSO transmission is the scintillation, which is caused by small-scale fluctuations in the refraction index of the atmosphere, mainly influencing coherent transmission techniques in FSO. Within an ESA contract and former international EU COST action (IC0802 and IC1101) investigations on different FSO applications and on various weather effects were carried out.

A study under the ESA project (AO/1-5718/08/NL/US "Feasibility Assessment of Optical Technologies & Techniques for Reliable High Capacity Feeder Links") [9] has evaluated different usable wavelengths for the FSO between earth and satellite. 850 nm is the oldest system (the 1st "optical window") for optical fibres, so the cheapest and best evaluated components are available. Also the well-known 1,300 nm technology (the 2nd "optical window" for optical fibres) is cheaper than 1,550 nm technology (the 3rd "optical window"), but more expensive than 850 nm. For all the 3 mentioned wavelengths standard components for fibre based systems are available for use which is important for interfaces and connecting to other networks. However, first experiments show that the 10 μm technology is much better suited to overcome the bad weather conditions. But components for this wavelength are still in the research phase. The use of 10 μm in FSO would also lead to reduction of redundancy and network nodes for the same reliability, since the system itself is much better in counteracting the bad weather conditions. In literature [9] FSO data links at 10 μm are documented in details.

III. COMBINATION SCENARIOS AND APPLICATIONS

As mentioned nowadays the various applications and the increasing data rates for high speed internet need a combination of different network technologies. A few examples are shown, which have been evaluated in former experiments and activities or will be established successful in future applications.

A. FSO combined with LMDS (Microwave-System)

One of the oldest but very important experiments at TU Graz the combination of LMDS (Local Multipoint Distribution System) with FSO shows excellent results on the availability measurements of the combined system [1]. The very interesting analysis of the availability of a one year period shows the expected general behaviour of each technology and quite a good overall availability for the hybrid

combination. In [1] it is documented that the overall hybrid availability for a 1 year period is 99.926%, whereas the availabilities of both single links are lower than around 96%.

Obviously the optical link is highly influenced by fog and snow fall, which can be seen in winter months, like in December 2002. In contrast the microwave link does not have excessive losses during this period. The hybrid system promises very high availability [1].

B. FSO combined with Satellite Communications

Applications in tele-medicine, disaster recovery, tele-teaching, sports, music, social events or exploration at unusual locations require broadband access to the rest of the world. In the absence of terrestrial infrastructure, connections based on geostationary satellite technology (GEO), cover the whole world with 3 satellites at minimum, except for the polar zones.

A few scenarios, as example, FSO and satellite applications at the civil-military exercise in Styria have been successfully demonstrated. In this CIMIC exercise a mobile satellite earth station (equipped with FSO and WLAN) was used for video-conferencing between military and civil organisations, see [1, 2]. The mobile satellite ground station was able to transmit the data from the station via FSO and / or WLAN into the conference room. Similar setups were used for tele-medicine for UN support and recovery missions in 2001 and 2002 [2, 7].

The emerging WLAN systems according to the IEEE 802.11 standards family offer wireless shared connectivity of up to 11 or 54 Mb/s overall data rate in the 2.4 and 5.6 GHz radio frequency bands. Other combination possibilities of FSO and WLAN are shown in subchapters D and E (with DVB-T).

C. Long range FSO combined with Satellite

In [9] it is documented that long distance FSO could be used as uplink to feed satellites with high data rates and big broadcast-data to distribute TV- and radio programmes via satellite. The proposed idea behind this combination is for double the usable capacity for the broadcasted programmes. If the uplink is done by the high capacity optical feeder link, the regular downlink plus the additional uplink RF-spectrum can be used for satellite TV and radio. By using the high capacity optical feeder link as uplink, the RF uplink can be spared and used for additional downlink capacity. Optical carrier frequencies in the order of 200 THz (1,550 nm) or 350 THz (850 nm) are free of any license requirements worldwide and cannot interfere with satellite or other RF equipment. To counteract the bad weather conditions (fog and clouds) for FSO systems, site diversity and longer wavelengths have to be used.

In a current ESA contract (ESTEC 4000115256/15/NL/FE) a System Study of Optical Communications with a Hybridised Optical/RF Payload Data Transmitter is carried out by a consortium, to investigate future combination possibilities of RF and FSO for Communication in Deep Space missions.

D. FSO combined with WLAN and GSM

Additional application scenarios of FSO can be realised and demonstrated with WLAN and Global System for Mobile Communication (GSM). On big events and also for natural disasters this usage is an excellent possibility to have quick available networks with high transmission capacity. As we know the WLAN systems offer license-free wireless shared connectivity. This is a difference to FSO systems, where the full data rate is dedicated to one directed connection, and moreover there is also a change in the protocol and coding standard from Ethernet 802.3 to WLAN 802.11, see also [11].

As example in Austria one of our telecom providers offers mobile (nomadic) base stations, which can be used for big events or disasters to ensure the mobile telephony. Such a mobile base station was installed, when in June and July 2012 in the alpine region in Styria a very big flood occurred. FSO can be used for linking a base station to a neighboured one or to some access point to the relevant network (MAN, WAN).

E. FSO and WLAN combined with DVB-T system

DVB-T (Digital Video Broadcasting-Terrestrial) is the current technology for broadcasting of digital data services. It transmits a compressed video/audio stream using the OFDM modulation. The compression of the data is realised with MPEG-2 or MPEG-4 (H264/AVC) algorithms. DVB-T is typically used to broadcast digital TV channels in a frequency range between 470 and 862 MHz (UHF Bands IV and V). The technology for one channel is comparable to the technology used in satellite TV broadcasting (DVB-S) for one transponder.

In case the user has a direct line of sight connection to the corresponding WLAN or FSO unit, a WLAN access point/antenna or FSO unit and a DVB-T antenna have to be mounted onto the user's residence to connect to the forward channel (downlink). The user's PC or notebook has to be equipped with a device that is capable to receive DVB-T data (downlink) and a network interface card which is connected either to the WLAN access point or to the FSO unit (uplink). The WLAN access points simply can be exchanged with FSO units, if used for the uplink/backchannel.

An appropriate technology could be meshed networks to connect the uplink gateway with directed radio antennas up to 20 km or with an FSO system up to 2 km of distance [8]. Interesting work on using DVB-T for Internet access was done within the finalized special strategic EU project called "SEE TV-WEB". The main objective of SEE TV-WEB was to tackle the Digital Divide by developing a joint, coordinated and viable initiative which will increase the accessibility and availability of information services offering at least minimal internet access to elderly or disabled persons and for regions with weak telecommunications infrastructure [11].

F. OWC, 5G and RF for Communication with Vehicles

Current and future applications of OWC and hybrid solutions with 5G and RF links are also under investigation for Vehicle to Vehicle (V2V) and Vehicle to Infrastructure (V2I) Communications (for cars, ships and planes). The combination of 5G and FSO is well suited for autonomous driving systems. In such application also other types of combinations of optics and RF are an important fact in sensor and measurements systems. As example radar (working on RF) can be combined with Lidar for safety reason. Both systems can "see" and detect different types of targets and obstacles.

OWC in combination with 5G will be the next innovative step in future Communications technology. Also Visible Light Communication (VLC) (well known for Data Broadcast within buildings) can be used for V2V (and cars). New possibilities for the back-channel for VLC will give new aspects in future network mobility.

IV. CONCLUSIONS

Within this contribution different applications were proposed and demonstrated, which allows worldwide access to the Internet by combining FSO with other wireless technologies (like satellite communications, Wireless LAN, LMDS and DVB-T). Wireless LAN offers good connectivity to mobile users in a network cell and for backchannel solutions. Broadband mobile access with IP networking technologies was demonstrated without the use of terrestrial infrastructure. The combination of different wireless technologies, satellite communications, FSO and WLAN shows good performance and interoperability in the complete installation. The system concepts presented in this paper were realized and investigated by TU Graz and successfully demonstrated.

The contribution illustrated that different propagation behaviours and contrary weather effects lead to complementary advantages of microwave and FSO. The demonstration at various scenarios has shown that combined use of different wireless technologies can be used for reliable and quick installable communication links [11].

REFERENCES

[1] Leitgeb E., Gebhart M., et al., High Availability of Hybrid Wireless Networks, SPIE's Intern. Symposium Photonics Europe, 2004, Straßburg

[2] Leitgeb, E.; Birnbacher, U.; et al., Hybrid Wireless Networks combining WLAN, FSO and Satellite Technology for Disaster Recovery, IST Mobile and Wireless Communications Summit 2005, p. 13–18

[3] Gebhart M., Leitgeb E., et al., Ethernet access network based on free-space optic deployment technology, Free Space Laser Communication Technologies, Proc. SPIE Vol. 5338 (2004)

[4] Birnbacher, U.; Schrotter, P.; et al., Broadband Wireless Communication for Telemedicine, EMBEC'02 (2002), p. 1390 - 1391, European Medical and Biological Engineering Conf.; 2002

[5] Leitgeb E., Gebhart M., et al., Impact of atmospheric effects in Free Space Optics transmission systems, SPIE Photonics West, LASE 2003

[6] Nadeem, F.; et al., Comparing the Cloud Attenuation for Different Optical Wavelengths, International Conf. on Emerging Techn.; 2009

[7] Gebhart M., Schrotter P., et al., Satellite Communications, Free Space Optics and Wireless LAN combined: Worldwide broadband wireless access independent of terrestrial infrastructure, 12th IEEE Med. Electrotechnical Conference (MELECON 2004) Dubrovnik, Croatia

[8] Mandl, P.; Schrotter, P.; Leitgeb, E.: Hybrid Systems Using DVB-T, WLAN and FSO to Connect Peripheral Regions with Broadband Internet Services, Intern. Conf. on Telecom. (ConTEL) 2009, p. 67 - 71

[9] Leitgeb, E., Plank, T.; et al., Analysis and Evaluation of Optimum Wavelengths for Free-Space Optical Transceivers, Intern. Conf. on Transparent Optical Networks; ICTON 2010, Munich

[10] Leitgeb, E., Plank, et al., Integration of FSO in Local Area Networks – Combination of Optical Wireless with WLAN and DVB-T for Last Mile Internet Connections; Int. Conf. on Networks and Optical Communications; NOC 2014, Milano

[11] Leitgeb, E.; Plank, T.: Combination of Free Space Optics (FSO) and RF for Different Wireless Application Scenarios; 9th European Conference on Antennas and Propagation (EuCAP) (2015)

The Golden Ratio in the Age of Communication and Information Technology

Y. I. Doychinov[*], I. S. Stoyanov[**] and T. B. Iliev[***]

[*]University of Ruse, Department of Industrial Design, Ruse, Bulgaria
[**]University of Ruse, Department of Electrical and Electronic Engineering, Ruse, Bulgaria
[***]University of Ruse, Department of Telecommunication, Ruse, Bulgaria
e-mail addresses: doychinov@uni-ruse.bg, stoyanov@uni-ruse.bg, tiliev@uni-ruse.bg

Abstract - This paper is based on two main theses. The first one, according to a research of Gustav Fechner, people have sustainable preferences to forms with proportions, close to Golden ratio. The second these is that human perception and aesthetical criteria can be developed and trained. Having mind the fact, that in this moment the lifestyle and especially dynamic presentation and perception of the information (television, monitors, mobile devices) by people are absolutely different from the moment when the Fechner's experiment is made (before more than 100 years). It is possible a change to be expected in the human's preferences. In the context of the subject matter may be noted the fact, that in March 2011 the 16:9 resolution 1920×1080 became the most common used resolution among Steam's users. In the present paper are shown the results from repeating of Fechner's experiment, made between 65 persons at the age of 18-20. According to the data obtained, respondents indicated the most harmonious and preferred proportions are 1:1, 2:1 and 1,78:1 (which in practice is well known aspect ratio 16: 9). Golden ratio takes the fourth position.

I. INTRODUCTION

It is assumed that the "Great theory of beauty" was founded by the ancient Greeks and it is based on the ratios of the whole and its parts. Based on this, they accept the order and proportions as wonderful. The golden ratio is one of the most significant and stable appearances of the harmony of nature. This conception keep during the centuries and finds its place in different fields of contemporary art and science.

In contemporary design conceptions, the proportions are one of the main parts of composition. It is assumed that "the man often has some stable preferences to proportions of its environment" [1]. The knowledge of these preferences and their adjustment in the process of projecting of different types of items or product would help the creation of more admired and preferred forms.

This stability of the preferences is proved by Gustav Fechner by means of a test with rectangles that have ratio of the sides from 1:1 to 1:2. The results he achieved are from the all ten shapes one is most preferred. It is with ratio of the sides 21:34 (0,617674), that in fact matches

the Golden ratio (0,618). This is known as "Golden rectangle".

Gustav Theodor Fechner lived in the beginning of XIX century (1801–1887). That means that from his experience has passed more than hundred years. During all this period the results he achieved are taken as absolute truth.

It is normal to evoke the question what would be the results at this moment in the beginning of XXI century?

The reason for asking this is the higher level of dynamic with which the environment, fashion, the main tendencies and aesthetical criteria in the perception of contemporary people are changing. All these processes are strongly affected by the fast development of the technologies and means of communication which full a significant part of the environment of the contemporary society and even replace partly the „face to face" communication.

The present paper has interdisciplinary character. It affects problems, connected with design, marketing, technologies, human perceptions and the means of information transfer (TV-s, monitors, mobile phones) that become "standards" of aesthetics in visual communication.

The main aim of the present paper is to check (prove or reject) actuality of Fechner's experiment and to establish the stability of people's preferences to certain proportions in the conditions of continually impact of modern communication means on the perceptual characteristics of humans.

The rest of the paper is organized as follows. Section 2 presents repetition of the Fehner's experiment. Section 3 gives possible reasons for a change of perception of the golden ratio in the age of communications. Section 4 concludes the paper.

II. REPETITION OF FEHNER'S EXPERIMENT

To explore the contemporary preferences for certain proportions has developed a test that repeat the research done by Fechner.

To carry out the experiment are set up two groups of ten rectangles with proportions from 1:1 to 1:2. The difference between them is that for the first group the rectangles increase vertically, as the horizontal side is constant (Fig. 1), while for the second group the height of the rectangles is constant, but increase the horizontals (Fig. 2).

The separation of horizontally and vertically oriented forms is done to achieve a higher level of objectivity due to an important ability of human perception, according to which arises the difference in assessing the dimensions in both directions.

For example, the same line looks longer if it is positioned vertically, rather than if placed horizontally. Therefore, it is possible to expect divergence in answers to the most preferred form of group with horizontal direction and the group with vertical direction.

Objects with the same numbering in the two groups have the same aspect ratio, the difference between them is the direction

Rectangle 1 has a ratio of 1: 1

Rectangle 2 has a ratio of 1: 1.1

Rectangle 3 has a ratio of 1: 1.22

Rectangle 4 has a ratio of 1: 1.33

Rectangle 5 has a ratio of 1: 1.44

Rectangle 6 has a ratio of 1: 1.56

Rectangle 7 has a ratio of 1: 1.67

Rectangle 8 has a ratio of 1: 1.78

Rectangle 9 has a ratio of 1: 1.89

Rectangle 10 has a ratio of 1: 2

Respondents are placed with the following task: "From each group rectangles numbered 1 to 10, choose and highlight the number of the form with the most harmonious proportions (the most beautiful form)."

An experiment covers 65 (sixty five) people, most of them (55 people) are students aged between 18 and 23 years.

The results, initially are presented in two separate groups (Fig. 3 for vertically oriented rectangles and Fig. 4 for horizontal oriented rectangles). And then in Fig. 5 are presented the consolidated results.

In a vertically oriented group of rectangles (Fig. 1) most preferred form appears to be the number 1 with an aspect ratio of 1: 1 - the square. It is listed as the most harmonious of 18 respondents. Second place with 8 votes share rectangle 3 aspect ratio 1: 1.22, rectangle 8 with a ratio of 1: 1.78 and rectangle 10 with an aspect ratio of 1:2. In third place with 6 votes is rectangle 7 with ratio 1:1.67 (which is close to the golden ratio 1: 1,618).

Most disliked rectangles (with 2 votes) are number 2 with ratio of 1: 1.1, number 4 with ratio of 1: 1.33 and number 6 with the proportions 1: 1, 56 (which is also close to the golden ratio 1: 1,618).

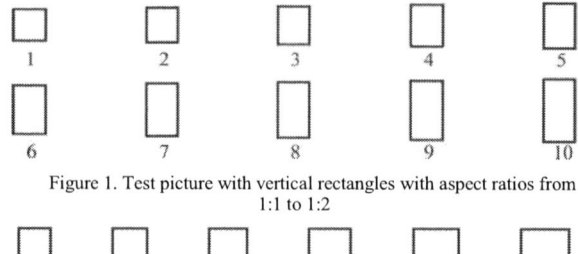

Figure 1. Test picture with vertical rectangles with aspect ratios from 1:1 to 1:2

Figure 2. Test picture with horizontal rectangles with aspect ratios from 1:1 to 2:1

In horizontally oriented set of rectangles (Fig. 4) results observed at first glance differ from those in vertically organized group, but at the same time are very close.

The most preferred rectangle turns out to be rectangle 2 aspect ratio 1:1.11 (in the first group was rectangle 1). It is listed as the most harmonious of 10 respondents. Second place with 8 votes share the rectangle 8 (as in the previous group) with an aspect ratio of 1: 1.78, rectangle 9 aspect ratio 1:1.89 and rectangle 10 with an aspect ratio of 1:2. In third place with 6 votes were rectangle 1 aspect ratio 1:1, rectangle 6 with ratio of 1:1.56 and rectangle 7 with an aspect ratio of 1:1.67 (which is closer to the golden section 1:1.618).

The most disliked rectangles (3 votes) again (as in the previous group) is rectangle number 4 with an aspect ratio of 1:1.33.

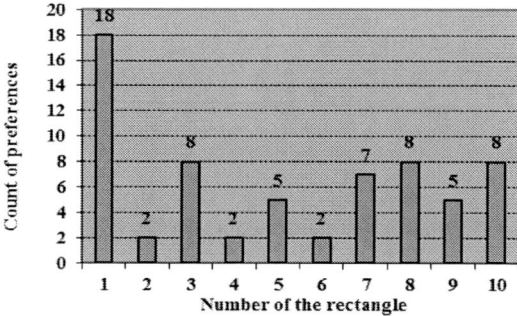

Figure 3. Number of preferred vertical rectangles

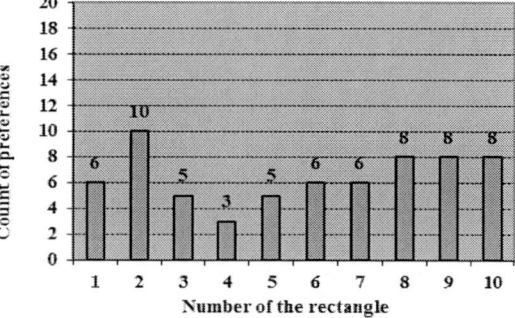

Figure 4. Number of preferred horizontal rectangles

The earlier hypothesis that the difference arises in assessing the dimensions horizontally and vertically in practice received its confirmation by the made test. There is clearly pronounced divergence in preferences for the most harmonious rectangle, depending on whether it is vertically or horizontally oriented. While in the first case the amplitude among the most liked and most disliked rectangle is quite large - 16 votes, in the second case the difference is only 7 votes.

The sum of the results of the most preferred rectangles of the two groups is shown in Fig. 5.

What impresses is the following - the sum of preferences for the two rectangle closest to the golden section (Fig. 6 and Fig. 7) is less than the number of respondents that indicated rectangle 1 for most harmonious.

Special attention deserves the ranked second rectangles with number 8 and 10. They occupy second place in both groups.

At the third place are rectangles with numbers 3, 7 and 9 with the same number of votes.

The most disliked rectangle in all three rankings is number 4 with ratios of 1: 1.33.What could be the factors leading to these results?

It can be assumed that influenced from modern minimalist trends in art and fashion, the majority of respondents indicated square as the most harmonious shape. The same reasoning can be attributed to the rectangle number 10, who has actual proportions of the two squares adjacent to each other.

It is also probable hypothesis that the human brain seeks and prefers simple and elementary forms to be able to cope with our overwhelming avalanche of information.

Of interest is rectangle 8 which has a ratio of 1: 1.78 and is at the second place of preferences in the two groups, and the total score.

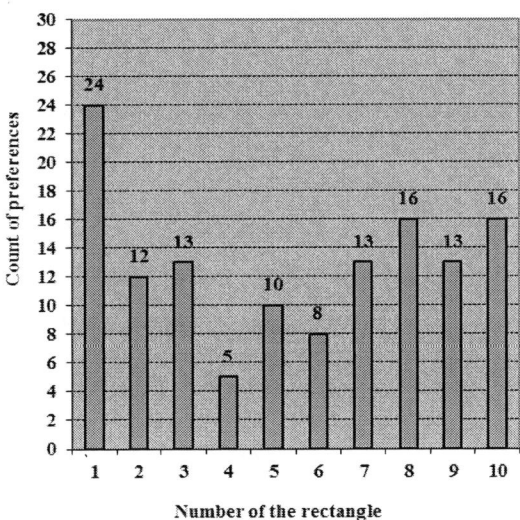

Figure 5. Number of preferred rectangles

III. THE GOLDEN RATIO IN THE AGE OF COMMUNICATIONS

An interesting fact is that currently the most common means of information and communication - TVs and computer monitors and mobile phones are with displays with resolutions, with a ratio of 16: 9, which is in fact the same proportion 1:1.78. Reasonable question arises how their style and proportions influence preference for harmonic proportions.

In January 2017 resolution 1366 × 768 became the most common worldwide used desktop, tablet & console screen resolution according Statcounter (Fig. 6).[6]

So far these are the results (facts) of the experiment. To reach or at least approach the reasons for these results we will expose several fact relating to human perception or related to the topic:

- Over 90% of the information (according to some authors around 95%), reaching our brain is through the sight.

- According to information from the 2015, survey of TNS shows that consumers aged between 16 and 30 spend an average of 3.2 hours a day, using their smartphones. The same material indicates that we take an average day look at our smartphones - 150 times. [7]

- Online survey of Samsung Techonomic Index from 2015 of 18 000 respondents aged over 16 in 18 European countries shows that Europeans watch TV average of 3 hours and 7 min. per day [8]

- According data from the National Statistics Institute in Bulgaria 87.2% of those aged 18-24 and 82.5 percent of those aged 25-34 regularly use the Internet. [9]

- According to its own investigation of the report's authors conducted among 450 people in 2014, 84% responded that they use the Internet every day.

Given the facts of the time that person spends using modern communication devices cannot help but make the connection between them and the new aesthetic preferences and criteria that are formed.

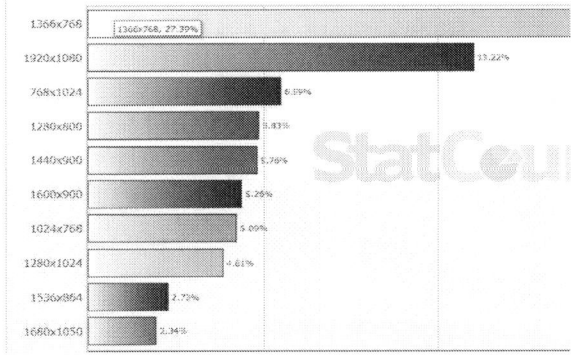

Figure 6. Most popular screen resolutions in January 2017 according Statcounter

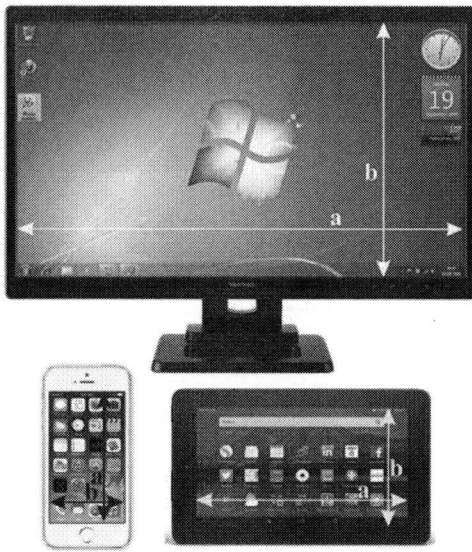

Figure 7. Devices with display aspect ratio a/b =1,78

Although various types of used modern devices have different screen sizes, they present us important information for us in proportions that appear common constant 1: 1.78 (Fig. 7). Whether to perform work activity on a computer, watching television or communication over phone and tablet, always our information is presented in the same frame within the screen. Thus we begin to connect the important things for us with the format in which they are presented to us.

Surely the process of replacing or development of aesthetic criteria is a fact and at the same time depends on many factors. The use of modern means of communication is one of many factors, probably even the most important.

The aim of this study is not to reveal all of them rather to focus people's attention to subtle and slow processes that occur to us as a result of communication with the environment, which surround us.

In the end of 19 century Gottfried Semper says that "... science and technological advances provide artistic practice with such materials and methods for their treatment, which still are not utilized aesthetically. Time limits required for aesthetic assimilation of innovations in the process of historical development constantly cut ... "[3].

Obviously this problem, formulated by Semper is valid today.

IV.   CONCLUSION

It is quite possible the sample of 65 people, among which is the study is not fully representative and provides insufficient objective information on the state of the problem, but the fact is that there is no proven by Fechner and cited by later sources uniformity of preferences in favor of the golden section.

However, based on the obtained results can be drawn the following conclusions and assumptions:

The fact is that human perception and aesthetic criteria are subject to change and training. From this perspective, it is very normal, under the influence of artificial habitat, the man to lose to some degree the preferences for proportions close to the golden section.

Based on this survey can be announced that under the constant influence of electronic means of information and communication is quite possible that a new aesthetic criteria that form the preferences and attitudes of modern consumers had formed.

The most logical direction of development of the study would be to continue to cover a wider group of people to be able to make considerably more accurate conclusions. Its results would be of great benefit to people involved in design, advertising, software developers and other professionals involved in the structuring and visualization of presenting electronic information.

Even now the Internet is full of lessons and recommendations on how to use the Golden Ratio in the creation of graphic compositions or entire websites in order for them to be harmonious and attractive to consumers.

It is reasonable to ask whether this really is.

REFERENCES

[1]   S. Delchev, Fundamentals of the design in architecture. Sofia, 1993. [in Bulgarian]

[2]   S. Vodchic, Aesthetics proportion design. Moscow, 2005. [in Russian]

[3]   S. Mihaylov, History of design. Moscow, 2002. [in Russian]

[4]   I. Shevelev, A. Marutaev, I. Shmelev. Golden ratio. Moscow, 2009. [in Russian]

[5]   http://store.steampowered.com/hwsurvey

[6]   gs.statcounter.com/screen-resolution-stats/desktop-tablet-console/worldwide/#monthly-201701-201701-bar

[7]   http://www.news.com.au/technology/gadgets/mobile-phones/new-research-shows-how-many-days-per-year-users-spend-staring-at-mobile-phone-screens/news-story/a105e9a6d9ca27628cd72b3d24a4ca27

[8]   http://www.samsung.com/bg/news/local/home-the-new-culture-capital/

[9]   http://www.nsi.bg/bg/content/2814/%D0%BB%D0%B8%D1%86%D0%B0-%D1%80%D0%B5%D0%B3%D1%83%D0%BB%D1%8F%D1%80%D0%BD%D0%BE-%D0%B8%D0%B7%D0%BF%D0%BE%D0%BB%D0%B7%D0%B2%D0%B0%D0%B0%D1%89%D0%B8-%D0%B8%D0%BD%D1%82%D0%B5%D1%80%D0%BD%D0%B5%D1%82

MIPRO 2017, May 22- 26, 2017, Opatija, Croatia

Wireless machine-to-machine communication for intelligent transportation systems: Internet of Vehicles and Vehicle to Grid

Ntefeng Ruth Moloisane*, Reza Malekian*,‡‡, Dijana Capeska Bogatinoska**

*Department of Electrical, Electronic and Computer Engineering
University of Pretoria, Pretoria, South Africa
** Faculty of Computer Science and Engineering, University of Information Science and Technology "St. Paul the Apostle", Ohrid, Republic of Macedonia
‡‡Corresponding author: Prof. Reza Malekian, Reza.malekian@ieee.org

Abstract - Machine to machine communication in intelligent transportation is a technology that aims to interconnect various components such as sensors, vehicles, road infrastructures and wireless networks. The significance thereof is to solve problems such as road congestion, road accidents and high vehicle fuel consumption. This paper gives an overview of how Machine-to-machine (M2M) communication can be used in intelligent transportation systems (ITS) to improve road safety and efficiency, where Vehicular ad-hoc networks (VANETs) play a major role. These applications include traffic light control, fleet management and smart grid systems. Some of M2M architectures that have been devised are also discussed.

Index Terms -- Intelligent Transportation Systems (ITS), Machine to machine (M2M), Traffic light control, Vehicular ad-hoc networks (VANETs, Smart Grid, Vehicle to Grid.

1. INTRODUCTION

Transportation related problems and road accidents have been increasing worldwide, due to an increase in the number of automobiles on roads [1]. Traffic congestion is one of the major problems in transportation that has been experienced in urban areas. This problem has caused a number of undesirable events such as increased road accidents, high vehicle fuel consumption and an increase in gas emissions from vehicles. Road intersections are more prone to accidents due the requirement for multiple vehicle drivers, pedestrians and infrastructure [2] to interact, when right of way is required. Between 30% and 60% of fatal road accidents have been reported to occur at road intersections [1].

Fixed time traffic light control is currently a common method used to regulate traffic at road intersections. This method is not efficient as it does not take dynamic and unpredictable road changes such as accidents and peak time traffic congestion into account. Intelligent transportation systems (ITS) have been studied and developed to improve transportation safety, reliability and efficiency [3]. Intelligent transportation integrates different technologies to establish an interconnected communication between vehicles, road infrastructures and pedestrians [4]. The integration of these technologies can thus change and improve the traditional way of interaction between vehicles, infrastructure and people significantly [5].

ITS can be realized, modeled and developed by incorporating Machine-to-machine communication (M2M). M2M is an automated communication between a central management system and multiple remote machines for the automation and monitoring of real-time processes [6]. M2M is based on the idea that a network of different entities, referred to as machines can be more useful than an isolated system and that the efficiency of automated systems can be improved if more machines are interconnected [7]. This phenomenon thus results in various applications of M2M which are not only limited to traffic control, but also include health, remote sensing, home security and other applications as shown in Fig1.

M2M can be viewed as a component of the Internet of Things (IoT) [8], which like the Internet of Vehicles (IoV) aims at the development of applications and systems that can allow for large scale interconnected networks between digital devices and physical things [8]. The IoV is an

Figure 1.   Applications of M2M

emerging technology and integrates the internet, mobile and wireless communication technologies to realize vehicular ad-hoc networks (VANETS) through Vehicle-to-Vehicle (V2V), Vehicle-to-Road (V2R), vehicle-to-human (V2H) and vehicle-to-sensor (V2S) interactions [9]. V2R is sometimes also referred to as Vehicle-to-Infrastructure (V2I).

2. MACHINE-TO-MACHINE ARCHITECTURE

One of the challenges in the development of M2M communications is the need for an adoption of a standard architecture that can be employed. Some architectures of M2M such as oneM2M and standards based on the European Telecommunications Standards (ETS) have thus been developed. A high level or general M2M architecture with three domains, namely the M2M domain, Network domain and an Application domain was devised [10] and is shown in Fig2. The M2M domain is a group of nodes or machines interconnected in a network and send data to a gateway. In intelligent transportation these M2M nodes would typically be vehicles, sensors and road infrastructure such as traffic lights. The gateway may be a wireless communication interface; it may form part of a wireless or wired network which sends data received from M2M nodes to a remote server for processing and analysis. The network domain can utilize short range communication protocols such as Wifi, ZigBee and Bluetooth. A table showing a comparison of the bitrates and communication range of wireless communication protocols that may be used in M2M communication for intelligent transportation is presented below.

Different projects and organizations such as 3GPP (Third Generation Project), European Telecommunications Standards Institute (ETSI) and Internet Engineering Task Force (IETF), have worked on devising M2M standards.

A. 3GPP

3GPP is an organization that develops standards for mobile communication protocols [11] such as GSM, HSPA and LTE. The 3GPP architecture uses mobile networks for transmitting M2M data. These networks are ideal for use when relatively small data transmission rates are required due to the cost involved. GSM for instance uses Short Messaging (SMS), which can be expensive when incorporated in M2M transportation systems [11]. Mobile communication networks have an advantage of connection reliability and good network coverage which can allow for M2M devices to communicate over a wide distance range [12].

B. OneM2M

The architecture proposed by the oneM2M Global Initiative is based on the general M2M architecture shown in Fig2. The oneM2M architecture is divided into an infrastructure domain and a field domain. Different components or devices which form part of the M2M system are then classified into four nodes , infrastructure node, middle node, application service node and application dedicated nodes. These nodes are then organized in a layered network with an application layer, common service layer and an underlying network layer [13].

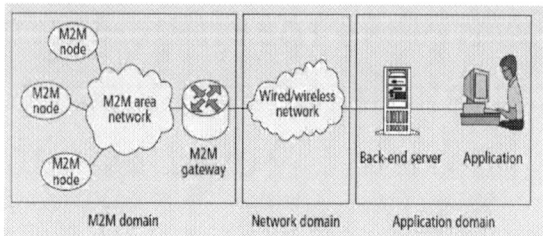

Figure 2.   M2M architecture (From [11]).

Comparison of short range wireless communication protocols.
(From [4], [10])

Protocol	Range	Bitrate
IEEE 802.11 a/b/g/n (WIFI)	120 m – 250 m	5.5 Mbps - 75 Mbps
IEEE 802.11 p (Mesh)	< 1 Km	6 Mbps – 27 Mbps
IEEE 802.15.1 (Bluetooth)	10 m – 100 m	1 Mbps
IEEE 802.15.4 (ZigBee)	10m – 75 m	20 – 250 Kbps

3. INTELLIGENT TRANSPORTATION

Research and development of intelligent transportation systems has been done by incorporating machine-to-machine in the form of vehicular ad-hoc networks with V2V and V2I interactions. The problem of road intersections being prone to accidents can be mitigated by implementing intelligent traffic light control.

C. Traffic Light Control

An intelligent traffic light control algorithm that used scheduling was developed [14]. VANET was used to get real-time data of traffic flow at a road intersection. The algorithm was tested in a simulated environment and it was found that the overall throughput of vehicles at a road intersection increased and the queuing time decreased in comparison to a fixed time traffic light control method. The system was also designed to send a warning signal via a wireless network, to caution approaching vehicles to reduce speed when the traffic lights turned red.

GPS and speed information from on-board vehicular sensors was collected in a VANET based system to optimize waiting period of vehicles at road intersects [15]. Data was sent from a vehicle to an intelligent traffic light controller through wireless communication, where V2V and V2I communications were established. Traffic congestion was modeled as Oldest Job First scheduling problem. The scheduling algorithm resulted in reduction of the waiting time of vehicles by more than two times compared to a fixed time traffic light scheduling algorithm.

A traffic light control system was designed using V2V communication [16]. The system calculated the estimated vehicle density and the total number of vehicles queuing at a traffic light. This information was used in an algorithm developed to reduce the waiting time and queue length of vehicles. A simulation tool was used and it was found that the average queuing time of vehicles was reduced from the fixed time traffic light approach.

A control algorithm to dynamically change and optimize the minimum and maximum green light time for actuated traffic light control was developed [17]. The queue length of vehicles was to be obtained from detector sensors which acted as actuators. The queue length was used in a statistical approach to compute optimum minimum green time. The maximum green time was optimized by using the queue length and a stochastic model that used a Poison distribution to model the arrival time behavior of vehicles at a road intersection. An improvement and efficiency in traffic flow was achieved.

Machine-to-machine communication was employed in the form of V2V and V2I to solve the problem of traffic congestion and improve traffic flow [18]. Traffic congestion was reduced by developing a system that reduces the average stopping time of vehicle at traffic lights thereby reducing the total fuel consumption. A simulation environment demonstrating that the fuel consumption of a vehicle may be reduced by avoiding unnecessary stopping was developed. This required an optimal speed advisory algorithm which was developed to advice a driver of a suitable speed when approaching an intersection. The real life implementation of this system could be done by obtaining the current vehicle speed from on-board sensors and then send this information wirelessly to a controller interfaced with a traffic light using a technology such as 802.11. The traffic light can then send back a recommended travelling speed to the vehicle which will mitigate unnecessary stops. A disadvantage of this approach is that a driver may be distracted if a lot of information is constantly communicated.

RFID was used for an Intelligent Traffic Light Control system. This system was designed to solve the problem of traffic congestion, emergency vehicle clearance and detection of stolen vehicles [19]. Vehicles were equipped with RFID tags which were detected by an RFID reader placed at a certain distance from the traffic light. The congestion volume of vehicles approaching the intersection could then be calculated and used to determine the total green light time. Emergency vehicles were equipped with a Zigbee transmitter which communicated with a Zigbee receiver located at the intersection. When communication was established, the traffic light was changed from red to green to give the emergency vehicles right of way. Stolen vehicles were tracked by checking the read RFID against a recorded list of stolen RFIDs. If a stolen vehicle was detected a signal was sent to turn the traffic light red and an alert message was sent to the police. This system automates the process of traffic control by a human and is thus more efficient.

D. Intersection safety

INTERSAFE-2 is a project aimed at improving the safety at road intersections, by using a camera and image processing to obtain information about the state of the road at an intersection [20]. An algorithm which ensured that warning alerts were sent to approaching vehicles in the event of accidents or dangerous situations was developed. Data from on-board sensors was sent wirelessly to a control unit interfaced with a traffic light. An effective mechanism to resolve the issue of road safety during unpredictable circumstances was achieved through interconnectivity.

E. Virtual traffic light

A self-organizing traffic control system was proposed in [21], where an ad-hoc communication between vehicles approaching an intersection was established, without interaction or communication with any infrastructure. This method was based on the concept of a virtual traffic light (VTL). Short range communication technology incorporated in vehicles and VANET were considered. The system was designed to improve the manner in which traffic flow was directed in the event of accidents, so as to allow for emergency vehicles to reach the accident scene as fast and as safe as possible.

The proposed algorithm was based on the assumption that vehicles sent hello packets with information about their current position and speed. A local map of the current state of the road would then be calculated by each vehicle to detect possible conflicts between vehicles approaching an intersection from different directions. If a conflict was detected, leading vehicles from all sides of the intersection were identified. The vehicle closest to the intersection was then elected as a VTL. The VTL sent a red message to an approaching vehicle in its direction and a green message to a vehicle orthogonal to it. A handover process then finally happened after a certain time period, where the orthogonal vehicle was selected as the VTL. Simulation results of the proposed system showed a 60 percent increase in traffic flow from the use of traditional traffic lights [21]. Fig3. shows a graph of the results of the average travel time obtained. A comparison between the performance of physical traffic lights and VTL was made.

Artificial intelligence has also been used in research relating to traffic light control. Fuzzy logic was used in [22] to control a traffic light at a road intersection which reduced the average waiting time of vehicles.

F. Fleet Management

M2M communication also has some advantages in intelligent transportation other than traffic light control. Different M2M use cases were reported in [23] which included fleet management and logistics. M2M has been

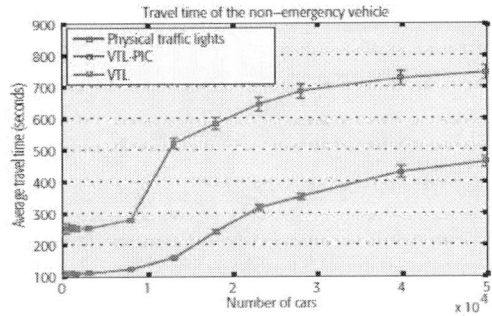

Figure 3. Travel time of non-emergency vehicles using VTL and pyhsical traffic lights. (From [21]).

used to send various vehicle parameters such as speed, distance travel and fuel consumption to road infrastructures and remote data management systems. Vehicle tracking systems are also an application of M2M where a GPS device for instance can communicate to various satellites and the device in turn sends data to a remote module wirelessly.

G. Smart Grid and Electric Vehicles

Energy efficiency, energy conservation and reduction of the environmental carbon foot print are some of the challenges faced in the Utility industry due to the electricity grid that is currently employed [24].Smart grid power technology is thus evolving and mainly employs renewable energy sources such as water, wind and solar. These energy sources are very intermittent so mechanisms to conserve energy are thus necessary [25]. Electric vehicles (EV) have been discovered to be capable of storing energy through charging and can thus be used for energy storage, where after this stored energy can be supplied back to the smart grid by discharging when it is needed [26].

M2M communication was used in an architecture demonstrating the interaction of smart grid and EVs [27]. The architecture had three parties involved, which were the power grid integrated with smart grid, Smart Communities (SCs) and the EVs. The function of the power grid was to generate energy consumed by SCs as and EVs through charging. The EVs not only consumed energy but also supplied energy to the grid. Different forms of M2M communication were used to realise the interaction of the different parties of the architecture and these were, vehicle-to-grid (V2G/G2V), vehicle-to-community (V2C/C2V), vehicle-to-vehicle (V2V), and

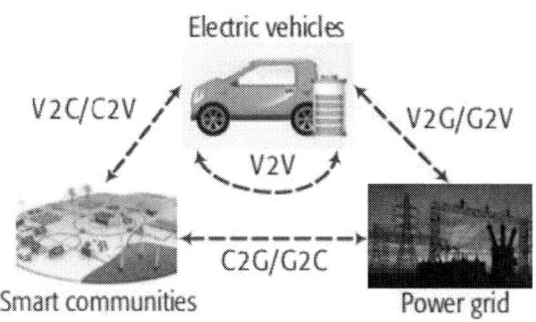

Figure 4.   Smart grid and M2M communication. (From [27]).

community-to-grid (C2G/G2C) [27] as shown in Fig4.

A study showing the use of different artificial intelligence (AI) approaches for V2G systems was done [26]. It was shown that artificial intelligence could be used for charging or discharging EVs and to solve the problem of network congestion. A solution to the problem of charging and discharging Plug-in Hybrid vehicles (PHEVs) was proposed in [28]. It was based on allowing PHEVs and PHEV parking bays to interact via V2V [28-

33] and V2G communications [34,35]. A hierarchical architecture of this interaction was to be devised so as to maintain scalability of the size of the network or interaction.

4.   CONCLUSION

Safety, efficiency and traffic control are currently a major concern in transportation. Research and projects have been done on Intelligent Transportation systems with the aim of improving the current state of road safety and transportation. This paper presented how machine-to-machine communication can be used in ITS. This mainly entailed integrating different technologies such as sensors and wireless communication with intelligent algorithms in order to establish an interconnected network between vehicles, humans and road infrastructure. A challenge with developing M2M, IoV and IoT systems is that a standard architecture has not yet been devised, so the development of such systems requires a broad and an open ended evaluation of already existing knowledge. Testing and qualification of ITS is also a challenge due to a possible danger that might be imposed on motorists and pedestrians. Despite these challenges, ongoing research and development of intelligent transportation systems, machine type or M2M communications, is proof that transportation safety and efficiency can indeed be improved.

ACKNOWLEDGMENT

The subject is supported by and National Research Foundation, South Africa (grant numbers: IFR160118156967 and RDYR160404161474).

REFERENCES

[1]   L. Qi, M. Zhou, and W. Luan, "Emergency traffic-light control system design for intersections subject to accidents," *IEEE Transactions on Intelligent Transportation* Systems, vol. 17, no. 1, pp. 170-183, Jan. 2016.

[2]   L.-W. Chen, P. Sharma, and Y.-C. Tseng, "Dynamic traffic control with fairness and throughput optimization using vehicular communications", *IEEE Journal on Selected Areas in Communications*, vol. 31, no. 9, pp. 504-512, Sep. 2013.

[3]   L. Qi, "Research on intelligent transportation system technologies and applications," in *Power Electronics and Intelligent Transportation System*, 2008.

[4]   F. Qu, F.-Y. Wang, and L. Yang, "Intelligent transportation spaces: vehicles, traffic, communications, and beyond," IEEE Communications Magazine, vol. 48,no. 11, pp. 136-142, Nov. 2010.

[5]   I. Turcanu, P. Salvo, A. Baiocchi, and F. Cuomo, "An integrated vanet-based data dissemination and collection protocol for complex urban scenarios," *Ad Hoc Networks*, vol. 52, pp. 28-38, July. 2016.

[6]   S. Whitehead, "Adopting wireles machine-to-machine technology," *Computing and Control Engineering*, vol. 15, no. 5, pp. 40-46, Oct. 2004

[7]   V. B. Misic and J. Misic, *Machine-to-machine communications: architectures, tech nology, standards, and applications.* CRC Press, Feb. 2014.

[8] M. A. Razzaque, M. Milojevic-Jevric, A. Palade, and S. Clarke, "Middleware for internet of things: a survey," *IEEE Internet of Things Journal*, vol. 3, no. 1, pp. 70-95, Feb. 2016.

[9] N. Liu, "Internet of Vehicles: Your next connection," *Huawei WinWin*, vol. 11, pp. 23–28, 2011.

[10] R. Lu, X. Li, X. Liang, X. Shen, and X. Lin, "Grs: The green, reliability, and security of emerging machine to machine communications," *IEEE communications magazine*, vol. 49, no. 4, pp. 28-35, Apr. 2011.

[11] T. Korhonen, T. Vaaramaki, V. Riihimaki, R. Salminen, and A. Karila, "Selecting telecommunications technologies for intelligent transport system services in helsinki municipality," *IET intelligent transport systems*, vol. 6, no. 1, pp. 18-28, Nov. 2012.

[12] T. Taleb and A. Kunz, "Machine type communications in 3gpp networks: potential, challenges, and solutions," *IEEE Communications Magazine*, vol. 50, no. 3, pp. 178-184, Mar. 2012.

[13] H. Lee, S. Vahid, and K. Moessner, "A survey of radio resource management for spectrum aggregation in lte-advanced," IEEE Communications Surveys & Tutorials, vol. 16, no. 2, pp. 745-760, 2014.

[14] K. Pandit, D. Ghosal, H. M. Zhang, and C.-N. Chuah, "Adaptive traffic signal control with vehicular ad hoc networks," *IEEE Transactions on Vehicular Technology*,vol. 62, no. 4, pp. 1459-1471, May. 2013.

[15]] M. B. Younes and A. Boukerche, "Intelligent traffic light controlling algorithms using vehicular networks," IEEE Transactions on Vehicular Technology, vol. 65, no. 8, pp. 5887-5889, Aug. 2016.

[16] K. Katsaros, R. Kernchen, M. Dianati, D. Rieck, and C. Zinoviou, "Application of vehicular communications for improving the efficiency of traffic in urban areas," Wireless Communications and Mobile Computing, vol. 11, no. 12, pp. 1657-1667, 2011.

[17] G. Zhang and Y. Wang, "Optimizing minimum and maximum green time settings for traffic actuated control at isolated intersections," *IEEE Transactions on Intelligent Transportation Systems*, vol. 12, no. 1, pp. 164-173, Mar. 2011.

[18] N. Maslekar, J. Mouzna, M. Boussedjra, and H. Labiod, "Cats: An adaptive traffic signal system based on car-to-car communication," *Journal of network and computer applications*, vol. 36, no. 5, pp. 1308 -1315, June. 2012.

[19] R. Sundar, S. Hebbar, and V. Golla, "Implementing intelligent traffic control system for congestion control, ambulance clearance, and stolen vehicle detection," *IEEE Sensors Journal*, vol. 15, no. 2, pp. 1109-1113, Feb. 2015.

[20] P. Pyykonen, M. Molinier, and G. Klunder, "Traffic monitoring and modelling for intersection safety," in *Intelligent Computer Communication and Processing (ICCP), 2010 IEEE International Conference on*, pp. 401-408, IEEE, 2010.

[21] O. K. Tonguz and W. Viriyasitavat, "A self-organizing network approach to priority management at intersections," *IEEE Communications Magazine*, vol. 54, no. 6, pp. 119-127, Jun. 2016.

[22] E. Azimirad, N. Pariz, and M. B. N. Sistani, "A novel fuzzy model and control of single intersection at urban traffic network," IEEE Systems Journal, vol. 4, no. 1,pp. 107-111, Mar. 2010.

[23] Y. Mehmood, S. N. K. Marwat, K. Kuladinithi, A. Forster, Y. Zaki, C. Gorg, and A. Timm-Giel, "M2m potentials in logistics and transportation industry," *Logistics Research*, vol. 9, no. 1, p. 15, Jul. 2016.

[24] H. Farhangi, "The path of the smart grid," *IEEE power and energy magazine*, vol. 8, no. 1, Mar. 2010.

[25] W. Kempton and J. Tomic, "Vehicle-to-grid power fundamentals: Calculating capacity and net revenue," Journal of power sources, vol. 144, no. 1, pp. 268-279, 2005.

[26] E. S. Rigas, S. D. Ramchurn, and N. Bassiliades, "Managing electric vehicles in the smart grid using artificial intelligence: A survey," *IEEE Transactions on Intelligent Transportation Systems*, vol. 16, no. 4, pp. 1619-1635, Aug. 2015.

[27] R. Zhang, X. Cheng, and L. Yang, "Energy management framework for electric vehicles in the smart grid: A three-party game," *IEEE Communications Magazine*, vol. 54, no. 12, pp. 93-101, Dec. 2016.

[28] R. Yu, J. Ding, W. Zhong, Y. Liu amd S. Xie, "PHEV and discharging cooperation in V2G networks: A coalition game approach," *IEEE Internet of Things Journal*, vol.1, no 6, pp. 578 – 589, Dec. 2014.

[29] R. Malekian, N. Moloisane, Lakshmi Nair, BT Maharaj, Uche A.K. Chude-Okonkwo, "Design and Implementation of a Wireless OBD II Fleet Management System", IEEE Sensors Journal, Vol 17, no.4, pp. 1154-1164, 2017.

[30] Jaco Prinsloo, "Accurate Vehicle Location System Using RFID, an Internet of Things Approach", Sensors MDPI, Vol.16, No.6, pp.1-24,Article no.825, 2016.

[31] R. Malekian, Alain Kavishe, B.T. Maharaj, P. Gupta, G. Singh, H. Waschefort, "Smart vehicle navigation system using Hidden Markov Model and RFID Technology", Wireless Personal Communications, Springer, Vol. 90, issue, 4, pp. 1717-1742, 2016.

[32] Ning Ye, Yingya Zhang, Ruchuan Wang,"Vehicle trajectory prediction based on Hidden Markov Model", KSII Transactions on Internet and Information Systems, Vol. 10, No.7, pp. 3150-3170, 2016. (Impact factor: 0.365),

[33] Zhong-qin Wang, Ning Ye, Ru-Chuan Wang, "A Hidden Markov Model Combined with RFID-Based Sensors for Accurate Vehicle Prediction, International Journal of Ad Hoc and Ubiquitous Computing, Inderscience, Vol.23, No.1/2, pp.124-133

[34] Zhongqin Wang, Ning Ye, Ruchuan Wang, Peng Li,"TMicroscope: Behavior Perception Based on the Slightest RFID Tag Motion", Elektronika ir Elektrotechnika (Impact factor: 0.561), Vol.22, No.2, pp.114-122, 2016.

[35] Yingya Zhang, Ning Ye, Ruchuan Wang, "A Method for Traffic Congestion Clustering Judgment Based on Grey Relational Analysis", ISPRS International Journal of Geo-Information, MDPI, Vol.5, No.5, pp.1-15, article ID: 71, 2016.

Power Control Schemes for Device-to-Device Communications in 5G Mobile network

T. B. Iliev*, Gr. Y. Mihaylov*, E. P. Ivanova*, I. S. Stoyanov**

*University of Ruse, Department of Telecommunication, Ruse, Bulgaria
**University of Ruse, Department of Electrical and Electronic Engineering, Ruse, Bulgaria
e-mail addresses: tiliev@uni-ruse.bg, gmihaylov@uni-ruse.bg, epivanova@uni-ruse.bg, stoyanov@uni-ruse.bg

Abstract - Device-to-device (D2D) communications integrated into cellular networks is a means to take advantage of the proximity of devices and thereby to increase the user bitrates and system capacity. D2D communications has recently been proposed for the 3GPP Long Term Evolution (LTE) system as a method to increase the spectrum- and energy-efficiency. Device-to-device communication has the potential of increasing the system capacity, energy efficiency and achievable peak rates while reducing the end-to-end latency. To realize these gains, are proposed resource allocation (RA) and power control (PC) approaches that show near optimal performance in terms of spectral or energy efficiency.

I. INTRODUCTION

Increased number of mobile devices and systems that need wireless communications are leading to congestion of radio spectrum in cellular networks. For that reason, effective use of spectrum has become more significant and new technologies are required for this purpose. Device-to-device communication (D2D) has been announced as a key technology to LTE-Advance networks [1]. In D2D, users with short distance and high signal-to-interference-plus-noise ratio (SINR) ratio may directly communicate with each other without sending the information through base station (BS), BS only sends the control signals to these users. These users can either use license excused bands or licensed frequency bands. In [2] authors have investigated different resource allocation approaches for D2D, where D2D connection can use dedicated resource or using resources of one or more than one users. Interference management from D2D communication to mobile users and from cellular communication network to D2D link is the key challenge.

One of the main technology component of D2D is mode selection (MS), which selects the cellular or direct communication mode for a D2D pair based on issues such as the current resource condition, traffic load, and level of the interference signals [2].

Device-to-device (D2D) communication in cellular spectrum supported by a cellular network allows direct communication between pieces of user equipment (UE). The advantages of D2D communication over traditional cellular transmission include not only the proximity gain in terms of improved link budget, but also the so called reuse and hop gains [3]. In order to use efficiently D2D, are proposed efficient algorithms for scheduling, resource allocation and power control, that help realize the gains of near communications. These algorithms protect at the same time the cellular layer from interference caused by local traffic. The 3rd Generation Partnership Project (3GPP) has recently specified the technology components of incorporating D2D communications in long term evolution advanced (LTE-A) networks [4, 5].

With use of D2D communications can be met the growing requirements of fifth generation mobile network (5G). Due to the limited spectrum resources and the requirement on providing broadband services must be ensured both spectral and energy efficiency requirements.

The main goal for communication technologies have always been to enable people to talk to each other more flexible and convenient. Now that this goal seems to be achieved and the foundation is set (product acceptance and convenient service coverage for mobile devices) new goals have been set for the next generation 5G.

To this end, we structure the paper as follows. The next section discusses the defined 5G network requirements and D2D resource usage. Next, in Section III we present the power control options based on LTE mechanisms and the proposed mode selection and random resource allocation algorithm. Section IV concludes the paper.

II. FIFTH GENERATION NETWORK (5G)

The Next Generation Mobile Networks (*NGMN*) Alliance is a mobile telecommunications association of mobile operators, vendors, manufacturers and research institutes.

NGMN Alliance defined 5G network requirements as:

- Data rates of several tens of Mb/s should be supported for tens of thousands of users.

- 1 Gb/s to be offered, simultaneously to tens of workers on the same office floor.

- Up to Several 100,000's simultaneous connections to be supported for massive sensor deployments.

- Spectral efficiency should be significantly enhanced compared to 4G.

- Coverage should be improved.

- Signaling efficiency enhanced.

Although these requirements seem quite vague they are targeting three foreseen problems for year 2020, which is the aimed time frame for succeeding the development:

- Avalanche of traffic:

- Explosion of the number of connected devices. From 5 billion (2010) to 50 billion (2020).

- Large diversity of use cases and requirements:

- Guaranteed service level, reliability, latency, device capabilities, Device-to-Device

- Communications Car-to-Car Comm.

A. Device-to-Device Communication

Device-to-device communication is introduced as a key technology to LTE-A. In D2D, UEs with short distance and high SINR ratio can directly communicate with each other. Base station (BS) sends only the control signals to these users.

Device-to-device communications in cellular spectrum supported by a cellular infrastructure has the potential of increasing spectrum and energy efficiency as well as allowing new peer-to-peer services by taking advantage of the so called proximity and reuse gains. [6] In fact, D2D communications in cellular spectrum is currently studied by the *3GPP* to facilitate proximity aware internetworking services [7], national security and public safety applications [3] and machine type communications [8].

Obviously, D2D communications utilizing cellular spectrum poses new challenges, because relative to cellular communication scenarios, the system needs to cope with new interference situations. For example, in an orthogonal frequency division multiplexing (*OFDM*) system in which user equipments (*UE*) are allowed to use D2D (*LTE direct mode*) communication, D2D communication links may reuse some of the OFDM time-frequency physical resource blocks (RB). Due to the reuse, intracell orthogonality is lost and intracell interference can become severe due to the random positions of the D2D transmitters and receivers as well as of the cellular UEs communicating with their respective serving base stations [6, 9].

B. Smart Mobility Management for D2D Communications

We assume that the D2D resource usage and coordination are under the network's control. This is due to the fact that in-band D2D operation, as an underlay for cellular communications, requires the network's control on D2D radio resources in order to provide optimized resource utilization, minimized interference among D2D links and from D2D links to cellular link, as well as more robust mobility.

Enabling very low latency data communications between end-users is one of the distinctive advantages expected from Device-to-Device communications. However, when several base stations (*BSs*), which are connected to each other via a non-ideal backhaul, are involved in the D2D radio resource control, the quality of service requirement in terms of latency may not be

satisfied due to large backhaul delay. Furthermore, the additional control overhead is expected due to the exchange of necessary information between controlling nodes as depicted in Fig. 1. Therefore, we propose two smart mobility management solutions that can be used to minimize the negative impacts (e.g., larger latency and additional signaling overhead) of multi-site radio resource control on D2D communications by controlling the D2D control handover and cell selection during the mobility of D2D UEs (DUEs):

- D2D-aware handover solution,

- D2D-triggered handover solution.

Here, it should be noted that D2D control handover and regular cellular handover could be executed separately, such as in dual connectivity [9].

D2D-aware handover solution

D2D-aware handover solution is introduced to minimize the End-to-End (E2E) latency in D2D communications and reduce the network signaling overhead in case of DUE mobility.

D2D-triggered handover solution

With D2D-triggered handover solution, we propose to cluster the members of a D2D group within a minimum number of cells or BSs in order to reduce the network signaling overhead caused by the inter-BS information exchange, such as related to D2D radio resource usage. The solution targets the scenarios where D2D groups are dynamically formed by more than two DUEs (Fig. 2). The solution can be applied when DUEs taking part in a D2D group are varying in time, for instance, due to the mobility. [10]

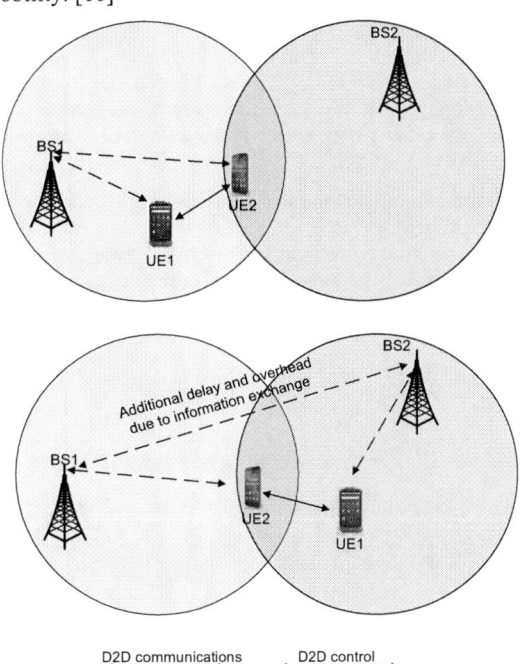

Figure 1. D2D control and communications before and after regular cellular handover execution

417

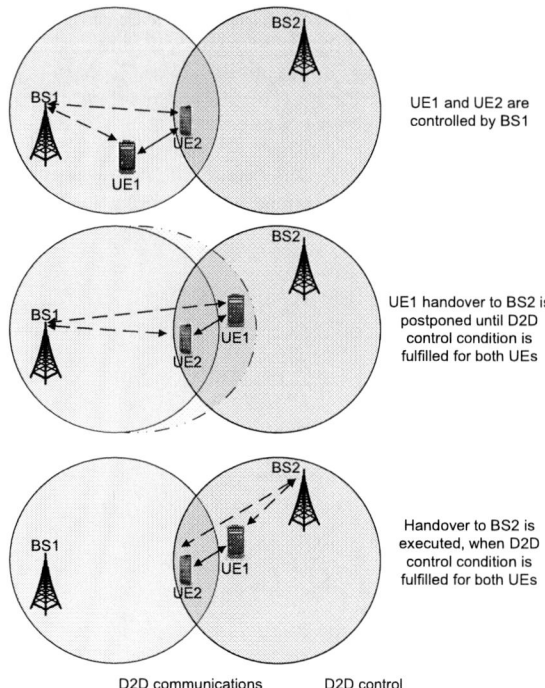

Figure 2. D2D control and communications during the DUE mobility between different sites

III. POWER CONTROL OPTIONS BASED ON LTE MECHANISMS

It is natural to base a power control (*PC*) strategy for D2D communications underlying [11] an LTE network on the LTE standard uplink *PC* mechanisms. Building on the already standardized and widely deployed schemes facilitate not only a smooth introduction of D2D enabled user equipment, but would also help to develop inter-operable solutions between different devices and network equipment.

The LTE *PC* scheme can be seen as a 'toolkit' from which different *PC* strategies can be selected depending on the deployment scenario and operator preference [12]. It employs a combination of open-loop (*OL*) and closed-loop (*CL*) control to set the UE transmit power (up to a maximum level of PMAX = 24 dBm) as follows:

$$P^{UE} = P_0 - \alpha G + \Delta_{TF} + f(\Delta_{TPC}) + 10\log_{10} M, \quad (1)$$

where the P_0-αG is *OL* operating point, which allows for path loss (*PL*) compensation and the dynamic offset (Δ_{TF}+$f(\Delta_{TPC})$) can further adjust the transmit power taking into account the current modulation and coding scheme (MCS) and explicit transmit power control (*TPC*) commands from the network. The bandwidth factor (10log$_{10}$M) takes into account the number of scheduled RBs (*M*). For the OL operating point, P_0 is a base power level used to control the SNR target and it is calculated as [13]:

$$P_0 = \alpha(\gamma^{tgt} + P_{IN}) + (1-\alpha).(P_{MAX} - 10\log_{10} M) \quad (2)$$

where α is the *PL* compensation factor, *G* is the path gain between the *UE* and the base station and P_{IN} is the estimated noise and interference power. For the dynamic offset, Δ_{TF} is the transport format (*MCS*) dependent component, $f(\Delta_{TPC})$ represents the explicit TPC commands. Alternatively, an LTE *PC* scheme can be carried out with a closed-loop operation. In this case, a variable Δ is introduced such that:

$$P = \min\{P_{\max}, P_0 + 10\log_{10} M + \alpha L + \Delta\} \quad (3)$$

where Δ represents a tuning step, which can be either fixed or dynamic.

A. Utility-Maximization Power Control

Utility-maximization power control is able to maximize the total utility of the system and minimize the total transmit power at the same time. The utility itself is logarithmically proportional to the transmission rate of link-l defined as $u_l(x)=ln(x)$, $\forall l$. The objective of the utility maximizing power control scheme is expressed by:

$$\begin{aligned} &\sum_l u_l(s_l) - \omega \sum_l P_l \\ &s_l \le c_l(p), \\ &p, s \succ 0 \end{aligned} \quad (4)$$

where *p* and s represent *UE* transmit powers and transmission rates respectively and ω is a design parameter that can tune the spectral versus power efficiency tradeoff-lower ω promotes higher spectral efficiency in exchange of higher transmit power. This problem is solved by an iterative approach using nested loops [12], such that the so called inner loops calculate the optimum transmit power for a given *SINR* target while the outer loops update the *SINR* targets based on local measurements by the receivers.

B. Mode Selection and Resource Allocation

Mode selection and resource allocation are intertwined, because the availability of all resources has an impact on what communication modes (cellular or D2D mode) can be chosen for D2D candidates. This is because of the fundamental requirement of maintaining orthogonality among cellular *UE*s while allowing for resource reuse between the cellular and D2D layers [14].

Resource mode selection and allocation scheme is proposed, that relaxes the need for full gain matrix knowledge at the *BS* at the expense of potential performance degradation (Fig. 3). This algorithm requires the availability of Channel state information (CSI) between the D2D candidates and between the BS and all (cellular and D2D) transmitters.

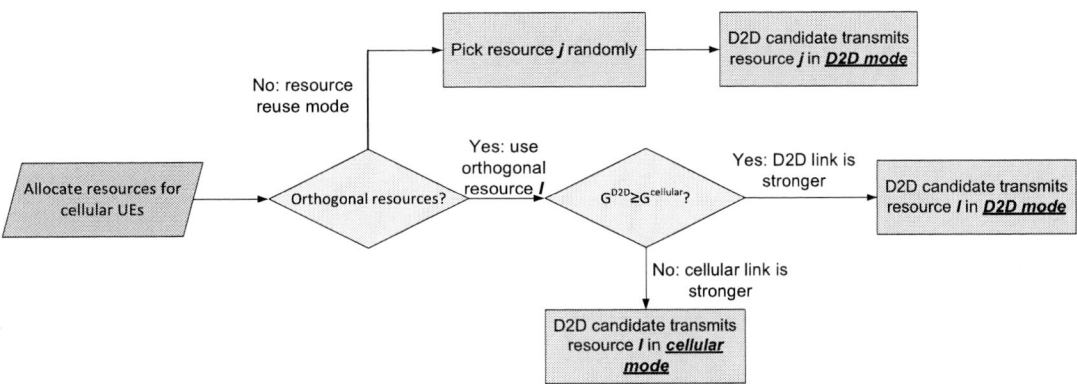

Figure 3. Mode selection and random resource allocation

This set of CSI reports is already maintained by cellular BSs to support mobility (handover) decision. This information is required for the MS decision, whereas resources are allocated by selecting randomly out of the available resource blocks. The rationale for this RA for the D2D layer is that *RA* to D2D pairs by the *BS* in practice is expected to operate on a much coarser time scale (e.g. 500 ms) than what is used for physical resource block scheduling in LTE systems [9].

IV. CONLUSION

Technology components for interference management, resource allocation, mobility management and other system level techniques enable the efficient D2D operation. A power control scheme is proposed, in which the cellular UEs continue to use the legacy LTE *PC* scheme, while the D2D UEs are allowed to use either LTE *PC* scheme or a utility-optimal *PC* scheme [14, 15] with additional practical constraints: a power-limiting threshold and/or reduced iterations.

The vast amount research, that had to be done to comply with the desired results are not finished yet, but some detail about the Power Control Schemes for Device-to-Device Communications can already be presented. Future work may consider different mobility scenarios in cellular radio networks, for example, vehicular communications.

REFERENCES

[1] K. Doppler, M. Rinne, C. Wijting, C. Ribeiro, and K. Hugl, "Deviceto-device communication as an underlay to lte-advanced networks," Communications Magazine, IEEE, vol. 47, no. 12, pp. 42–49, Dec 2009.

[2] B. Wang, L. Chen, X. Chen, X. Zhang, and D. Yang, "Resource allocation optimization for device-to-device communication underlaying cellular networks," in Vehicular Technology Conference (VTC Spring), 2011 IEEE 73rd, 2011, pp. 1 6.

[3] G. Fodor, E. Dahlman, G. Mildh, S. Parkvall, N. Reider, G. Miklos, and Z. Turanyi, "Design aspects of network assisted

device-to-device communications," Communications Magazine, IEEE, vol. 50, no. 3, pp. 170 –177, march 2012.

[4] 3GPP, "Scenarios and requirements for general use cases and national security and public safety," May 2013, Tech. Report 22.80

[5] T. Iliev T., Gr. Mihaylov, 3GPP LTE system model analysis and simulation of video transmission, Proceedings in Advanced Research in Scientific Areas, Slovak Republic, 2012, pp. 2016-2021

[6] T. Jämsä, P. Kyösti, H. Taoka, V. Nurmela, V. Hovinen, J. Medbo, METIS Propagation Scenarios, European Cooperation in Science and Technology Cooperative Radio Communications for Green Smart Environments (COST IC1004), 2013

[7] J. F. Monserrat, H. Droste, Ö. Bulakci, J. Eichinger, O. Queseth, M. Stamatelatos, H.Tullberg,V. Venkatkumar, G. Zimmermann, U. Dötsch, A. Osseiran, Rethinking the Mobile and Wireless Network Architecture: The METIS Research into 5G, European Conference on Networks and Communications (EuCNC), 2014

[8] G. Fodor, Sorrentino S., S. Sultana, Smart Device to Smart Device Communication, Springer, 2014

[9] M. Botsov, M. Klugel, W. Kellerer, P. Fertl, Location Dependent Resource Allocation for Mobile Device-to-Device Communications, IEEE Wireless Communications and Networking Conference (WCNC), 2014, pp. 1679 – 1684

[10] V. Venkatasubramanian, F. S. Moya, K. Pawlak, Centralized and Decentralized Multi-cell D2D Resource Allocation using Flexible UL/DL TDD, IEEE WCNC, 2015

[11] W. Sun, E. G. Strom, F. Brannstrom, Y. Sui, K. C. Sou, D2D-based V2V Communications with Latency and Reliability Constraints, IEEE Global Communications Conference, Exhibition and Industry Forum (GLOBECOM), 2014

[12] A. Pradini, G. Fodor, G. Miao, M. Belleschi, Near-Optimal Practical Power Control Schemes for D2D Communications in Cellular Networks, EuCNC, 2014

[13] E. Ternon, P. Agyapong, L. Hu, A. Dekorsy, Energy Savings in Heterogeneous Networks with Clustered Small Cell Deployments, ISWCS, 2014

[14] N. Pratas, P. Popovski, Underlay of Low-Rate Machine-Type D2D Links on Downlink Cellular Links, IEEE International Conference on Communications, ICC 2014 (ICC'2014) Workshop, 2014

[15] R. Ratasuk, A. Prasad, Z. Li, A. Ghosh, M. A. Uusitalo, Recent Advancements in M2M Communications in 4G Networks and Evolution Towards 5G, 18th International ICIN Conference 2015

MIPRO 2017, May 22- 26, 2017, Opatija, Croatia

Brain Computer Interface Communicator : A Response to Auditory Stimuli Experiment

Guruprasad Madhale Jadav, Luka Batistić, Saša Vlahinić and Miroslav Vrankić

Tehnički Fakultet Sveučilište u Rijeci, Rijeka, Croatia
jguruprasad@riteh.hr

Abstract – This paper describes a communication system, designed to help people with disabilities to communicate with others. In this case, it is built on Event Related Potentials (ERP) which occur as a response to stimuli. The stimuli are auditory answers to the asked questions, where one of the stimulus is the correct answer. The stimulus which is the possible correct answer is the target stimulus and other stimuli are considered as non-targets. The only reliable information available is the onset of the stimuli, based on this information the spatio-temporal pattern for the target and non-targets is expected to be uncorrelated to the ongoing brain activity and with each other. This allows us to train the spatial filter coefficients for the know targets, so that it can use this trained dataset to correctly classify targets. Results are obtained by optimizing the spatial filter parameters prior to classification, and the classification results are analyzed for several healthy participants.

I. INTRODUCTION

Brain computer interface (BCI) communicator is a system which uses signals produced in the brain to interact with computer or devices. Such communication device could allow people with disabilities to communicate with their caregivers for their everyday needs. The described BCI communicator is built on Event Related Potentials (ERP) which occur as a response to stimuli.

Robustness of the algorithm to detect response to stimuli is one of the primary concerns, when it comes to classifying ERPs of the data recorded from people in locked-in state (LIS). In such a state person although is aware of his surroundings has no reliable mode of communication to respond to his caregiver. Although these cases are rare, the people in LIS state with partial or complete blindness have poor quality of life. It is obvious that the caregiver cannot attend to the correct needs of the LIS patient due to the communication barrier.

The BCI communicator will therefore serve to bridge the communication barrier between people in LIS state and their caregivers. To build such a communicator, experiments to test the devised algorithm is crucial to deduce its reliability. At this stage of experiment, only healthy people participated in the experiment. The auditory stimuli however do not expect the participant or subject to have ability to see. The audio played by the system are a combination of target and non-target stimuli which are possible responses to the question that is been asked by the caregiver. The target stimulus is among one of these stimuli to which the subject responds during the

TABLE I.   XDAWN ALGORITHM

Steps	Algorithm
1	Compute $\hat{\Sigma}_X = \dfrac{X^T X}{N_t}$ and $\hat{\Sigma}_i = \dfrac{\hat{P}_i^{(c)^T} \hat{D}_i^{(c)^T} \hat{D}_i^{(c)} \hat{P}_i^{(c)}}{N_t}$
2	Compute GEVD of $(\hat{\Sigma}_i, \hat{\Sigma}_X) \Rightarrow (\Lambda_i, \Theta_i)$ where, $\Lambda_i^{(s)}$ is eigen valuse and $\Theta_i^{(s)}$ is eigen vectors
3	Select the N_{f_i} components assiciated with the N_{f_i} largest generalized eigenvalues $\Lambda_i^{(s)}$
4	Finally $\hat{U}_i = \Theta_i^{(s)}$ where, $\hat{U}_i$ are the eigen vectors for the N_{f_i} largest eigen values
5	Estimate enhanced signals $\hat{S}_i = X\hat{U}_i$

EEG recording. The change in EEG activity around 300 ms post target stimuli is called P300. The effects of response to stimuli is generally visible on the electrodes at the parieto-occipital region [1]. The signal immediately after the onset of the stimulus or event until 600 ms is considered for distinguishing the target from non-targets.

The BCI communicator described in this paper uses xDAWN algorithm which aims at maximizing signal to signal plus noise ratio by estimating the temporal signature and spatial distribution of the ERPs. This unsupervised method enhances the P300 evoked potential by estimating P300 subspace from the raw EEG data. Asymptotically optimal results are obtained for this method as the prior knowledge of the onset of the stimuli is the only reliable information for training the spatial filter [2].

II. EXPERIMENTAL FRAMEWORK

The auditory BCI communicator experiment was built in such a way that all the stimuli are played in the form of segments, while making sure that none of the stimulus is repeated immediately.

The EEG signals are obtained with Emotiv's EEG headset. Using this device does not just give the advantage of being cost effective and mobile but also eases the electrodes placement, meanwhile maintaining the 10-20 international standards of electrode placement. The data recorded from this headset is arguably of good EEG quality according to the results of previous experiments [3].

The OpenVIBE's Acquisition Server is used to obtain raw EEG data, where the dataset from first trial was used for training and the following datasets were considered for testing. The algorithm for EEG data processing was built using the OpenVIBE Designer [4]. The explanation on use of signal processing methods and parameters in the algorithm is been detailed as follows.

The EEG recorded was composed of ocular and muscular artifacts, and other ongoing brain activity in addition to spatio-temporal pattern due to each stimulus. The signal recorded was pre-filtered with a lower cutoff of 1 Hz and a higher cutoff of 20 Hz, using a 4th order Butterworth band-pass filter with a pass band ripple of 0.5 dB. The interval of the signal considered for analysis or epochs were therefore limited in frequency range between 1 to 20 Hz. The spatio-temporal pattern for target and non-targets are expected to be uncorrelated to the ongoing brain activity and with each other.

xDAWN algorithm from Table I [5], aims at estimating the spatial filter using the epochs created for training in such a way, that the signal-to-signal-plus-noise ratio (SSNR) of the target epochs from the test dataset is maximized. This algorithm is based on the assumption that the temporal differences exist among the same class of ERP's and also among N_e different classes of ERP's. The model of the recorded signal X consists of common temporal pattern (index (c)) and random temporal pattern (index (r)) and can be written as in (1), with spatial distribution over sensors W_i and noise N taken into consideration

$$X = \sum_{i=1}^{N_e} (D_i^{(c)} A_i^{(c)} + D_i^{(r)} A_i^{(r)}) W_i^T + N \qquad (1)$$

where $A_i^{(c)}$ is common temporal pattern and $A_i^{(r)}, D_i^{(c)}, D_i^{(r)}$ are matrices of white centered Gaussian random variables. After stochastic estimation of the spatial covariance matrix of i-th ERPs $\hat{\Sigma}_i$ and spatial

covariance matrix of measured signal $\hat{\Sigma}_X$, the generalized eigenvalue decomposition (GEVD) gives eigenvalues $\Lambda_i^{(s)}$ and corresponding eigenvectors $\Theta_i^{(s)}$ which are used as spatial filter coefficients $\hat{U}_i$ [6]. These spatial filter coefficients are used with the training dataset to train the Linear Discriminant Analysis (LDA) classifier [7] which computes a discriminant vector that separates the target and non-target classes. The data obtained for classification in the scenario consists of 20% of targets and 80% of non-targets. Since using balanced classes for training classifier might give over optimistic results, the classifier thus trained does not use balanced classes.

The output of the classifier was analyzed by changing the spatial filter dimension for the same datasets.

III. MATERIALS AND EXPERIMENT

A. Participants

10 participants 8 males and 2 females aged between 22 and 29 where chosen, none of them suffer any past neurological or psychiatric disorder. The cognitive session comprises of subject actively responding to the stimulus of his choice by counting the number of times the stimulus was played.

B. Device specifications

The EEG headset consists of 14 gold plated contact sensors which make direct contact with the detachable electrode tips. These electrode tips have foam which is soaked in saline solution and sits on the scalp adapting to the scalp topology. The saline solution acts as a good conductive medium for small voltage fluctuations, thus giving an efficient contact area. The electrodes are pre-attached to the headset with fixed arms to point at the locations: AF3, F3, F7, FC5, T7, P7, O1, O2, P8, T8, FC6, F8, F4, AF4, Common Mode Sense (CMS) at the left mastoid M1 and Driven Right Leg (DRL) at right mastoid M2. The electrodes when marked with green circles, which in this case represent good signal quality as in Fig. 1. The headset has onboard amplifier, analog to digital converter and filters due to which the signals from electrodes were high pass filtered at 0.16 Hz, pre-amplified and low pass filtered at 43 Hz cut-offs giving the effective bandwidth of 0.16-43 Hz. The analogue signals were digitized at 2048 Hz which is the fundamental sampling frequency of the device. The digitized signal is filtered using a 5th order sinc notch filter (50-60 Hz) and low pass filter and finally down sampled.

C. Auditory stimuli

The auditory stimuli used were Croatian words "Da", "Ne", "Ja", "Ti" and "Mi", and together they form one segment. "Da" and "Ne" (Yes or No) are possible targets, while other words are "dummy" words, inserted in order to reduce the number of occurrences of targeted response. These stimuli were chosen with an intention to keep the duration of the stimuli short and similar. During the experiment the stimuli were played at a moderate sound

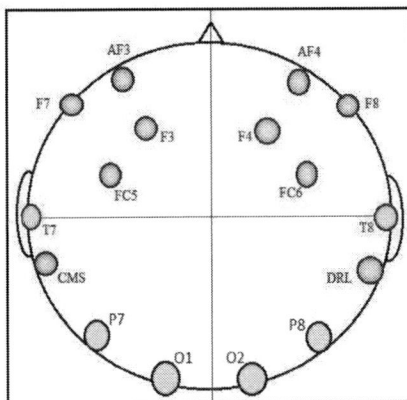

Figure 1. Scalp locations of the electrodes placed using Emotiv's research EEG headset[3]

TABLE II.   SPATIAL FILTER DIMENSION 3

Session	Trials	Result (dimension 3)
S	S-T1	Training
	S-T2	1
	S-T3	1
	S-T4	1
	S-T5	N
	S-T6	1
	S-T7	0
	S-T8	1

TABLE III.   SPATIAL FILTER DIMENSION 4

Session	Trials	Result (dimension 4)
S	S-T1	Training
	S-T2	N
	S-T3	N
	S-T4	1
	S-T5	1
	S-T6	N
	S-T7	N
	S-T8	1

level, with an inter stimulus interval of 1 second, and the stimuli were randomly shuffled in a segment. Each time the target stimuli appeared the subject had to respond to it by mentally counting it and thus incrementing the count whenever he heard the target stimulus.

D. Experimental procedure

The subject was briefed about the experiment and was asked to keep the movements minimum to avoid artifacts. They were seated in a faraday cage and were asked to focus at a fixed point, to make sure that there are minimum eye movement artifacts. The experiment has a training phase and an online phase

- During the training phase, the subject is asked a question to which answer was known a priori, and that dataset is used to train spatial filter and classifier. This training phase was conducted only once in the beginning of the session.

- During the online phase where the EEG dataset was classified in real time, the subject was just asked question and the classifier would classify the data based on the subject's selection. At the end of each trial subject is asked to verify the answer.

- A series of 5 sessions were conducted with the same subject on different days, and each session comprised of 5 questions. Once a question was asked the answers were played by the system in the form of 30 random segments, in such a way that the last stimuli of the previous segment was never the same as the first stimuli of the next segment. This and shuffling of stimuli in the segment ensured that the subject was not able to predict the played stimuli.

- As there were 30 audio events for each kind of stimuli, there were a total of 150 stimuli per question which means 150 epochs were considers for classification.

E. Data acquisition and analysis

The data was obtained and saved in the global data format (GDF) for training phase. Data from all the channels was selected for training, which were pre-filtered

to reduce signal bandwidth between 1 and 20 Hz. The signal decimation is done at a decimation factor of 4 to reduce the sampling frequency of the data from 128 samples to 32 samples per second. The epoch of 600 ms is considered. The spatial filter thus trained has been tested with different filter dimensions among which the filter dimension 3 and 4 had better results. The original number of electrodes being 14, the filter dimension 3 and 4 roughly corresponds to reduction by factor 4 and 3. Thereby comparing the results of these two spatial filter dimensions provided interesting results.

During the online phase, real time results of the ongoing trial were obtained. The Table II and Table III, show the effect of change in dimension of the spatial filter on the data. Where "S" is the session and "T" is the trials in the session. This clarifies that the best results obtained were when dimension of the spatial filter was 4 times less than the total number of electrodes.

The value "1" in tabular result indicates the intended target and the selected target were same, where as "0" indicates that the intended target and selected target were different. The "N" indicates that although the intended target was selected correctly the non-target also had a similar count as the target giving a neutral result.

IV. CONCLUSION

The developed BCI communicator proves to be robust and reliable in classifying between targets and non-targets. Tuning the spatial filter to find the suitable dimension, which is 4 times less than the total number of EEG channels gave better classification results. It was also deduced that training of the classifier at the beginning of every session gave more promising results. It would be interesting to evaluate classification results using methods like independent component analysis (ICA), in order find P300 ERP components in classified target epochs. Efforts will be towards further improving algorithm and testing it to make it more robust. The research and development goals also involve measurements with people in LIS state, to compare the data and thus results.

REFERENCES

[1] S.J.Luck, An introduction to the event-related potential technique, The MIT Press, 2005.

[2] Bertrand Rivet, Antoine Souloumiac, Virginie Attina, and Guillaume Gibert. "xDAWN algorithm to enhance evoked potentials : application to brain computer interface", IEEE Transactions on Biomedical Engineering, Institute of Elecrtical and Electronics Engineers, 2009.

[3] Guruprasad Madhale Jadav, Miroslav Vrankić, and Saša Vlahinić. "Monitoring cerebral processing of gustatory stimulation and perception using emotiv epoc", 38th International Convention on Information and Communication Technology, Electronics and Microelectronics, MIPRO 2015 – Proceedings, pages 643-645, 2015

[4] Yann Renard, Fabien Lotte, Guillaume Gibert, Marco Congedo, Emmanuel Maby, Vincent Delannoy, Olivier Bertrand, and Anatole Lé. „OpenViBE: An open-source software platform to design, test, and use brain-computer interfeaces in real and virtual environments", Presence: Teleoperators and Virtual

Enviornments/Presence Teleoperators and Virtual enviornments, MIT press, 2010.

[5] Bertrand Rivet, Antoine Souloumiac. "Optimal linear spatial filters for event-related potentials based on a spatio-temporal model: Asymptotical performance analysis", Signal Processing, Elsevier, 2013.

[6] Bertrand Rivet, Antoine Souloumiac, Virginie Attina, Guillaume Gibert. "xDAWN algorithm to enhance evoked potentials:application to brain-computer interface", IEEE Transactions on Biomedical Engineering, Institute of Electrical and Electronics Engineers, 2009, 56 (8),pp.2035-43.

[7] U.Hoffmann, J.-M. Vesin, T. Ebrahimi, and Diserens, "An efficient p300-based brain-computer interface for disabled subjects". Journal of Neuroscience Methods, vol. 167, no. 1, pp. 115-125, January 2008.

Determination of origins and destinations for an O-D matrix based on telecommunication activity records

Saša Dešić[1], Mia Filić[2], Renato Filjar[1]

[1]Ericsson Nikola Tesla, Krapinska 45, 10000 Zagreb, Croatia
[2]Faculty of Science, University of Zagreb, Croatia

Abstract - *The origin-destination (O-D) matrix is a known indicator of individual and group mobility in transport science, an invaluable input for socio-economic activity assessment of the community, and a tool for strategic planning and policy developments. The abundance of mobility-related data allows for expansion of the ODM towards the description of the over-all socio-economic activity, of which the transport is just a component, thus rendering policy development and strategic planning more efficient. Determination of the sources (origins) and targets (destinations) of the socio-economic activities remains an unresolved point in the ODM estimation process. Here we present a novel method for activity spot (origins and destinations) identification from the anonymised records of telecommunications activity, and publicly available data on telecommunication network access points. Using the spatial tessellation algorithm, the origins and destinations can be effectively identified as the focal points of socio-economic activity, thus establishing the foundation for more efficient urban policy development and strategic planning.*

I. INTRODUCTION

The position-location duality in separated (physical and contextual) domains, introduced by one of us in [6], has gained the attraction among the researchers aiming to explore the integration of spatial and Information and Communication Technology (ICT) data for generation of completely new groups of systems and services.

Here we present a novel approach in advanced utilisation of telecommunications activity data (context) matched with spatial data (physical world) in advanced determination of areas of attractions

(origins and destinations) for Origin – Destination (O-D) matrices estimation.

The manuscript is organised as follows.

Section II outlines the problem of determination of origins and destinations of an O-D matrix, and depicts the traditional approach to solve it. Section III describes the process of the O-D matrix estimation using telecommunications activity records. Section IV presents a novel approach in determination of origins and destinations using telecommunications activity records and a mathematical method called Voronoi tessellation. Section V presents results of the concept validation with the actual telecommunications activity record, and discusses advantages and shortcomings of the proposed approach. The manuscript concludes with the summary of research results and proposal for future research in Section VI.

II. PROBLEM DESCRIPTION

The origin-destination (O-D) matrix is a known indicator of individual and group mobility in transport science [10, 13], an invaluable input for socio-economic activity assessment of the community [11, 7, 5], and a tool for strategic planning and policy developments [7, 9]. An O-D matrix represents the socio-economic-driven intensity of local migration between the previously estimated areas of socio-economic attraction (residential areas, office areas, manufacturing areas, shopping malls, sport centres, concert halls etc.) [12, 5].

An O-D matrix is a scale-dependent tool. Its design begins with a proper determination of areas of attraction (origins and destinations) of local migrations [10]. The scale-dependency nature of an O-D matrix introduces constraints in definition of areas of attractions (origins and destinations). In traditional O-D matrix determination process, the origins and destinations are defined based on naïve estimations, interviews and examinations of cadastral data. Such an approach inevitably introduces uncertainty and noise, thus rendering the estimated O-D matrix less accurate and less fitted to the purpose.

Furthermore, a proper scaling of areas of interest requires skills and experience. Otherwise, it can produce a misleading evidence by either rendering areas of attraction too large (thus merging several actual areas of attraction into one), or too small (thus converting an O-D matrix into matrix of individual- or group-migration and loosing the original substance of an O-D matrix).

Therefore, a systematic and methodical approach in determination of origins and destinations of an O-D matrix, that will efficiently integrate the contextual knowledge with a proper scaling, is required.

III. O-D MATRIX DETERMINATION FROM TELECOMMUNICATIONS ACTIVITY DATA

The methodology of O-D matrix estimation comprises the following tasks: data preparation, determination of areas of activity (origins and destinations), migration identification (paring origins with destinations), O-D matrix estimation and graphical presentation [10, 5].

Let assume that we have n pre-defined disjoint areas, where n is a natural number. Let them count $1, 2, ..., n$. An O-D matrix is a square real matrix with non-negative elements. An (i, j)-element of an O-D matrix represents the number of migrations from i-th to j-th areas of attraction (origin-to-destination) in the observed time-span. Matrix rows represents origins, while columns represent destinations. It should be noted that, (i, j) element represents the number of migrations from the origin i to destination j, and (j, i) from origin j to destination i, with the two elements not necessarily equal (i.e. the O-D matrix is generally not symmetrical). An origin and a destination (comprising one migration O-D pair) are defined as the first and the last stop of the one migration event. In order to estimate the O-D matrix, the areas of observation and the number of migrations (migration connection) in predefined time-window between two areas from earlier defined set of areas of migration observations should be determined.

In a recently introduced approach [9], an O-D matrix can be estimated using the data sets describing telecommunications activity: voice calls, short-messages exchange, and data traffic over mobile internet. Summed up in internal Charging Data Records (CDRs) utilised by telecommunication network operators to charge the users for telecom services usage, those data sets can be used for distillation of the information necessary for the O-D matrices estimation procedure. Charging Data Record (CDR) is a collection of information about a chargeable telecommunication events (performed phone call, internet usage from mobile devices/units and Short Messaging Services (SMS), and ID of the base station involved in telecommunication activity) which are used by telecommunication network operators for user charging purposes. In a crucial development, a base station ID has been matched with the base station position in physical world, thus allowing for definition of transformation between physical and contextual (information) domain [9, 5, 4]. An O-D matrix based on telecommunications activity record provides more detailed and accurate insight into socio-economic-driven urban migrations that the classical O-D matrices based on road traffic data. Based on the CDR data set as the only source of information on migration of local population, we created the rules for determination of the areas of observation and mutual migration connection number [5].

The traditional approach in origins and destinations determination for telecom-based O-D matrix calls for the spatial partition founded on the estimation of the base station coverage. The base station radio coverage areas are not strictly separated in practice, allowing for a frequent overlapping of adjacent areas, thus creating a considerable ambiguity in origins and destination determination.

IV. DETERMINATION OF ORIGINS AND DESTINATIONS FROM CDR USING VORONOI DIAGRAM

A CDR is a massive data set generated during usage of a telecommunication network by a user. The set comprises numerous data that does not necessarily contribute to the ODM development, Additionally, the original CDR contains data from which the user's ID can be derived easily, pointing to his or her record of movement, among the other features. In due course, for the purpose of the ODM estimation the original CDR must be firstly anonymised in order to preserve the telecom users' privacy, followed by the data set reduction in the sense of extraction the ODM-relevant data only. The research presented here were conducted using the anonymised and ODM-related extract of the original CDR data set, a sub-set referred to as the CDR data in the remaining part of the manuscript. The CDR data sub-set comprised the following ODM-related data: (anonymised) user ID, time-stamp of telecom service initiation, user position approximated by position of the base station. A publicly available CDR data sub-set was used in this research [14].

A Voronoi diagram of a set S of n uniquely described points, called *seeds* or *sites*, in $\mathbf{R}^d$ is a partition of space into Voronoi cells, in a manner that each cell consists of all the points closer to a particular seed than to any other. [1]. A Voronoi cell corresponding the seed with co-ordinates *(x, y)* (if the 2-dimensional Euclidean space is considered, or more generally: $(x_1, x_2, ..., x_n)$ for n-dimensional Euclidean space) bounds the area around *(x, y)* that is closest to the *(x, y)* relative to the other seeds of the same Voronoi diagram. The Euclidian metrics for distance determination is the most common one used, while the other metrics (such as: Manhattan and Mahalanobis distance metrics) may be used for particular case scenarios) [2, 8].

The Euclidian metrics, utilised in this research, determines the distance between the points A (x_A, y_A) and B (x_B, y_B) using the model depicted in (1).

$$d(A,B) = \sqrt{((xB - xA)^2 + (xB - xA)^2)} \quad (1)$$

Each Voronoi cell is uniquely represented by its seed (the base station coordinates, in the case of telecommunication activity record utilisation) and the distance determination metrics (it determines the shape of a Voronoi cell). A procedure for the space partition using Voronoi diagram is often referred to as Voronoi tessellation.

Data preparation process comprises the extraction (identification) of all the base stations involved in telecommunications activity. Indirectly, the partition of the observed area (defined by CDR data set itself by observing the area around the base stations that CDR set comprises) onto Voronoi cells determines the areas of socio-economic activities exploiting the direct relationship between the position of base stations and the concentration of the mobile communication network users.

Each Voronoi cell corresponds to only one base station, or its seed. Following the change of base stations ID in the ordered set of CDRs which belongs to one user, a user migration can be determined. By determination of all migrations of all users referred to CDR remembering the first and the last stop of each migration, the number of migrations between specific areas can be determined. The spatial separation according to Voronoi cells represents the data arrangement in spatial domain.

Furthermore, the data preparation includes the extraction of information about the neighbouring Voronoi cells for each Voronoi cell in observation (set as an origin, for instance). A neighbouring cell of cell C is a cell which seed is from the C no more than D = 1km away. This way we define all neighbouring cells of cell C as all cells that can, theoretically, take over the voice call/Short Messaging Service (SMS)/data transfer from base station seed C due to its overload. The D value was set to 1 km in order to accept the common practice in telecom networks to disperse the base stations (in third-generation (3G) mobile communication networks) at 1 km – 3 km separations. [5]. The

CDR set utilised in this project originates from urban area.

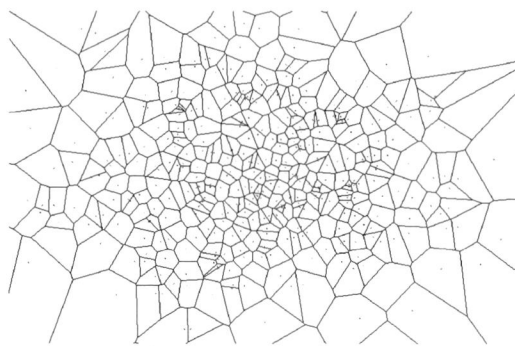

Fig 1 Voronoi tessellation

V. A CONCEPT VALIDATION AND DISCUSSION

The concept of determination of origins and destinations using telecommunication activity records and Voronoi tessellation was validated using openly available data set comprising anonymised CDR data sub-set collected in a 3G mobile network in Shenzhen, China [15]. The Voronoi tessellation algorithm based on Euclidian metrics was developed and implemented in the R programming environment for statistical computing [12, 3]. Spatial data provided through Google Maps API was used for identification of spatial objects and graphical presentation. The results of the Voronoi tessellation for the case presented are depicted in Fig 2.

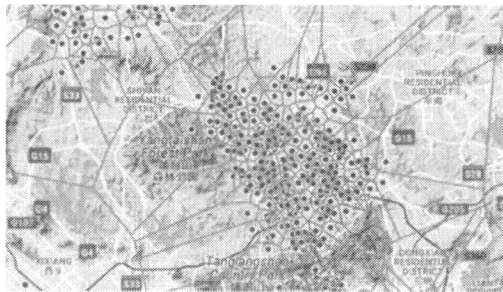

Fig 2 A Voronoi tessellation of Shenzhen, based on anonymised CDR

A provisional visual inspection of a digital map presentation revealed a good spatial separation of areas of socio-economic attraction by exploitation of the nature of radio network planning, that favours the situation of telecommunication network base station in centres of clusters of objects that attract the socio-economic activity. Still a more formal methodology for measurement of such a spatial fit remains to be developed in order to provide an objective indicator for the quality of spatial partition based on socio-economic activity.

In comparison with the previous research accomplishments [7-10,12,13], we have identified the contextual value in matching the base station-centred Voronoi tessellation with the spatial data on socio-economic activities. This finding is to be explored further through the assessment of the other concepts in Voronoi tessellation and the advanced algorithms for identification of regions of attraction through spatial analytics of contextual data.

Additionally, the utilisation of the 3G-network base stations as the Voronoi cell seeds assures the proper scaling, while rendering a set of Voronoi-determined origins and destinations a proper foundation for an accurate determination of the O-D matrix. The proposed concept may be improved further with consideration of positions of the objects of interest (concert halls, shops, restaurants etc.) and tailoring the origins and destinations algorithm determination in order to evaluate the influence of small-sized attraction spatial objects.

VI. CONCLUSION AND FUTURE RESEARCH

A novel approach in determination of origins and destinations for an O-D matrix, based on the utilisation of telecommunications activity records and the Voronoi spatial tessellation is presented in this manuscript. Validated with the real 3G telecommunications activity record data using the R-based software developed by our team, the proposed concept delivered improved quality of origins and destinations determination (in a sense of ability to identify the areas of socio-economic activity) over the traditional approach.

Future research will focus on the proposed concept improvement through exploration of the additional contextual and spatial relations between the Voronoi cell seeds (locations of base stations) and points of socio-economic activity (contextual data on shops, concert halls, residential areas etc.).

REFERENCE

[1] Arya, S, Malamatos, T, and Mount, D M. (2002). Space-Efficient Approximate Voronoi Diagrams. Proc of the 34th annual ACM symposium on Theory of computing, 721 – 730. Montreal, Canada.

[2] Aurenhammer, F, and Klein, R. (2000). Voronoi Diagrams. In Handbook of Computational Geometry, Chapter V, (J. Sack and G. Urrutia, editors), 201-290. Elsevier Science Publishing. Amsterdam, The Netherlands.

[3] Crawley, M J. (2013). The R Book (2nd ed). John Wiley & Sons. Chichester, UK.

[4] Filić, M, Filjar, R, and Vidovic, K. (2016). Graphical presentation of Origin-Destination matrix in R statistical environment. *Proc of KoREMA Automation in Transport 2016 Conference*, 26-30. Krapina, Croatia.

[5] Filjar, R, Filic, M, Lucic, A, Vidovic, K, and Saric, D. (2016). Anatomy of Origin-Destination Matrix Derived from GNSS Alternatives. *Coordinates*, **12**(10), 8-10.

[6] Filjar, R, Jezic, G, and Matijasevic, M. (2008). Location-Based Services: A Road Towards Situation Awareness. *J of Navigation,* **61**, 573 – 589.

[7] Florez, M et al. (2016). Measuring the impact of economic well being in commuting networks – A case study of Bogota, Colombia. Available at: http://bit.ly/2cVpihm, accessed on 20 March, 2017.

[8] Groff, E R, and McEwen, T. (2006). Visualization of Spatial Relationship in Mobility Research: A Primer (Final Report). Institute for Law and Justice. Alexandria, VA. Available at: http://bit.ly/2cdC9Li, accessed on 19 March, 2017.

[9] Gonzalez, M C, Hidalgo, C A, and Barabasi, A-L. (2008). Understanding individual human mobility patterns. Nature, **453**, 779-782. doi:10.1038/nature06958

[10] Peterson, A. (2007). The Origin-Destination Matrix Estimation Problem – Analysis and Computations (MSc thesis). Linkoeping University. Norrkoeping, Sweden.

[11] R Development Core Team (2016). R: A language and environment for statistical computing. R Foundation for Statistical Computing, Vienna, Austria. ISBN 3--900051--07--0. Available at http://www.r-project.org, accessed on 5 March, 2017.

[12] Toole JL, Herrera-Yaque C, Schneider C M, Gonzalez, M C. (2015). Coupling human mobility and social ties. *J. R. Soc. Interface* **12**: 20141128. http://dx.doi.org/10.1098/rsif.2014.1128. Available at: http://bit.ly/2d0GWBp, accessed on 10 March, 2017.

[13] Wang, P et al. (2012). Understanding Road Usage Patterns in Urban Areas. *Scientific Reports*, **2**, e01001. doi: 10.1038/srep01001.

[14] Zhang, D et al. (2015). UrbanCPS: A Cyber-Physical System bascd on Multi-source Big Infrastructure Data for Heterogeneous Model Integration. 6th ACM/IEEE International Conference on Cyber-Physical Systems (ICCPS'15), 2015. Available at: http://bit.ly/2bTEkWk, accessed on 1 March, 2017.

What Factors Influence the Quality of Experience for WebRTC Video Calls?

J. Baraković Husić*, S. Baraković**, *** and A. Veispahić****

* Faculty of Electrical Engineering, University of Sarajevo, Sarajevo, Bosnia and Herzegovina
** Faculty of Transport and Communication, University of Sarajevo, Sarajevo, Bosnia and Herzegovina
*** American University in Bosnia and Herzegovina, Sarajevo, Bosnia and Herzegovina
**** Anovis IT, Services and Trading GmbH, Vienna, Austria
jasmina.barakovic@etf.unsa.ba, barakovic.sabina@gmail.com, aida.veispahic@anovis.com

Abstract - Web real-time communication (WebRTC) based services introduce the new way of communication and collaboration. These types of services allow real-time communication with voice, video, messaging, and data sharing via browser. Their success and adoption depend on a number of different factors that influence user's experience and its quality. As WebRTC based services are growing, it is important to identify these influence factors (IFs) and investigate their impact on quality of experience (QoE). In this paper, the focus is on users' experience with WebRTC video calling. The objective is to determine the most and least influential factors in general according to examined users' opinions in the context of WebRTC video calling, and all for the successful management of QoE. With this in focus, this paper may serve as a basis for further research in the field in order to increase the utilization and adoption of WebRTC technology.

I. INTRODUCTION

Services allowing the real-time communications have acquired the great popularity over the last years. Some of the well-known services today supporting voice/video calls, messaging, and data sharing are Viber, Skype, Facebook Messenger's video calling feature, and FaceTime [1]. In order to access these services, the plug-ins, log-in, or particular device is required. In contrast, the Web real-time communication (WebRTC) based services require no installation and are in-browser applications easily accessed from a broad range of devices (e.g. smartphone, tablet, and laptop). These devices have to utilize a web browser supporting WebRTC (e.g. Google Chrome, Mozilla Firefox, or Microsoft Edge), so that users can conduct voice/video calls, messaging, and data sharing without having to purchase or download extra software [2]. This may attract both experienced and non-experienced individuals to access these services much easier.

The greatest difference between WebRTC and other web technologies is the user experience [3]. The success and adoption of WebRTC based services is strongly affected by a number of different factors that influence user's experience and its quality. There exist a number of attempts to define the quality of experience (QoE) both in standards and literature [4]. Recent definition given by [5] describes QoE as "the degree of delight or annoyance of the user of an application or service. It results from the fulfilment of his or her expectations with respect to the utility and/or enjoyment of the application or service in the light of the user's personality and current state."

In order to provide the best possible QoE, there is a need for deep understanding of various influence factors (IFs) which are according to [5] defined as "any characteristic of a user, system, service, application, or context of use". In the context of communication services, QoE may be affected by numerous IFs [6] which are categorized into three major groups, i.e., human, context, and system IFs, where all three together need to be considered for the QoE evaluation. In this paper, the focus is on the most relevant factors that influence QoE in the context of WebRTC video calling. The objective is to identify the most and least influential factors according to examined users' opinions and investigate their impact on QoE in this context. This gives the contribution in the process of creating prerequisites for successful adoption of WebRTC solutions for video calling.

The behavior of WebRTC is not enough documented in terms of factors that influence the user's experience and overall quality, especially for video calls. This is important research topic because it offers useful insight for service providers of any type how users experience the quality in order to provide the highest possible success and acceptance of WebRTC based services. The implementation of these services brings also the benefits for service providers which can be grouped into five categories [7]: (1) ease of use, (2) cost reduction, (3) security, (4) fast time to market, and (5) simplified device integration. Moreover, it leads to a new phase in the competition between over the top (OTT) players and operators in the core communication business, i.e. voice/video calls, data sharing, and messaging. With WebRTC, operators gain the opportunities to effectively compete with OTT players and redesign the user's experience of real-time communication services which are outside the traditional fixed and mobile device context [7].

The rest of the paper is organized as follows. Section II introduces the influence factors affecting the QoE and provides the review of related studies which consider their impact on user's experience and its quality during WebRTC video calling. For the sake of quantitatively identifying the most and least influential factors that affect QoE according to users' opinion in the context of WebRTC video calling, an end user survey has been conducted in form of an online questionnaire. Section III describes the research methodology and results, which are focused on exploring the users' satisfaction, opinions, habits, and expectations related to WebRTC video calling via wide range of smart devices. Section IV gives the concluding remarks.

II. QUALITY OF EXPERIENCE INFLUENCE FACTORS

As stated in previous section, factors influencing the QoE can be classified into three categories, i.e. human, context, and system IFs. Therefore, in this section, we provide a brief description of IFs categories and related research studies considering their impact on QoE. Note that this overview should not be considered exhaustive.

A. Human Influence Factors

The human influence factors (HIFs) are defined like any predictable or unpredictable property or human characteristic and could be divided in two subgroups, i.e. the low level influence factors (e.g., age, gender, personality, and mood) and high level influence factors (e.g., socio-economic conditions, educational background, needs, previous experience) [8].

HIFs are very unpredictable factors and until now they are taken into the account in a very limited fashion for QoE estimation. They are poorly understood due to their complexity and interrelation. Some research studies showed that HIF should not be avoided in a process of QoE estimation because 24.5% of total QoE could be predicted by the human factors [9].

Until now, the QoE requirements are considered equally for every user, no matter on their age, gender, facial attractiveness, emotions or cultural behavior. But nowadays, due to high market competition, it is important to provide personalized and adjusted multimedia services considering aforementioned differences. The results of research studies show slightly different trends in perception and behavior between male and female participants [10].

When it comes to the occurrences in synchronization's errors, it is shown that younger females are the most sensitive for such skews while on the other hand older male are the least sensitive [11]. Other research study shows that older adults are more critical than younger users and the elderly people may have higher requirements for QoE [12]. It has also been found that facial appearance has more important influence when communication is realized based on web-video services than on a traditional way of communicating. Moreover, it is shown that nonverbal cues, the attractiveness or opposite gender, for conducted research in sales, resulted in higher evaluations of customer satisfactions in daily business tasks [13].

Therefore, HIFs should be more investigated in order to provide better QoE. Besides the aforementioned factors, human's personality, interests and even a cultural background could have the huge impact. Hence western cultures are usually more inclined and focused on the salient objects independently from the context, whereas Asian cultures more likely search for the relationship between the objects and the context [14]. The personality influence can be shown as an IF at least on the user performance part on QoE. Neurotic people are less able to switch the viewed

program or change the volume on their first attempt compared to agreeable people and/or people with technical competence or enthusiasm [15].

QoE is triggered by the user's personal interest in video content. Users tend to value a video with the higher QoE when they are more interested in the content of the video [16]. Since HIFs are complex for the investigation, their evaluation is usually not possible just by conducting objective predictions. Therefore, the impact of these factors is usually perceived and verified through subjective tests [17], such as emotion-perception-emotion user model that could enable higher user satisfaction [18].

B. Context Influence Factors

The context influence factors (CIFs) describe user's environment picturing physical, temporal, social, economic, and technical dimension. Their classification includes six sub-groups: (1) physical factors, (2) time factors, (3) social factors, (4) economic factors, (5) factors associated with assignments, and (6) technical factors [8].

CIFs do not represent an easier task when it comes to predicting their impact on QoE. Research studies showed how these factors could gain even higher or more significant influence on QoE regardless of other influential determinants. It is shown that the ease of interaction with other people, ease of using new systems by allowing participants to speak interactively no matter of their current environment, time of the day, and their social environmental factors (normative beliefs, subjective norms, intention) could drastically increase or decrease QoE and users expectation [19].

The user's physical environment can drastically affect the QoE through the number of different factors (e.g., the seating position of the user, lighting conditions, viewing distance or disturbance that could occur in a user's environment – incoming calls, social-network notifications, etc.). The shorter viewing distance makes user more involved in the context which he/she is watching and provide the better visibility. The economic context has as well significant influence on QoE. If the user's expectations are not fulfilled, they usually lose their interest in repeating their experience. Users' social context (e.g., family, friends, strangers, etc.) also affects the QoE. It is demonstrated that user's satisfaction with the watched context can be increased if he/she is surrounded with friends or family, or if he/she is able to communicate, record, and share his/hers favorite programs with friends even when they are not physically co-located [20].

The physical location is also important consideration factor. Even nowadays when humans have the chance to be constantly reachable by smart devices and view the content of interest anytime and everywhere, there are certain locations which are more or less suitable for such purposes (in the waiting halls, home, park, school, office, in a train, during breaks for lunch, etc.). Time factors should be investigated more as well when it comes to QoE evaluation. Typical viewing time is couple of minutes (10-15) up to maximum 40 minutes. Also, the prime time should be taken into the consideration, i.e., early morning (while driving to school, university, job), midday (during the lunch break, or while standing in a line for bills payment), and evening (before dinner time, after or before sleeping) [21].

Moreover, the user's usage must include as well the transitions between the context and within the context (e.g., personal or shared use, temporally, between waiting or hurrying, from walking to standing, sitting, or between multi- and uni-tasking) [22]. All of this is pointing us on hard dynamic nature of user's context and its significant impact on QoE which should be more investigated and reported during the system's QoE evaluation.

C. System Influence Factors

The system influence factors (SIFs) define the properties and characteristics which determine the technical quality of applications or services. These factors could be divided into four basic sub-categories: (1) content factors, (2) medium factors, (3) network factors, and (4) device factors [8].

In order to ensure the optimal usage of the network resources and to increase the video flow quality, many different factors should be taken into account. The video flow quality depends on the network indicators, such as delay, jitter, loss or bandwidth. It is shown that it is better to delay the audio flows in order to ensure the synchronization with video flows because the tolerance towards the delay is higher [23]. This is one of the possible ways of dealing with the delay issues but only if the overall delay is below 600 ms. In a case that delay is greater than 800 ms, quality is considered as unacceptable [1].

However, to provide the high level of QoE, it is necessary to take care about other system factors that are not network oriented. Here, it is possible to speak about the features of devices as well as content of the media which is transferred via the networks. The traditional approaches usually allocate the bandwidth as a measure of application's quality. But this does not apply and longer, since different multimedia applications require not only capacity but also other guarantees (e.g., delay, jitter, loss) which are considered as QoE killers since result in "video freezing" and the quality insurance in a stable mode [24]. Therefore, they are worthy of further and in depth investigation.

Video calls are generally observed as delay sensitive service. The traditional transmission control protocol (TCP) is not suitable to deliver the traffic which is generated by the real-time applications. On the other hand, user datagram protocol (UDP) leaves the congestion control to the application layer. The results of different research studies showed up that developing and using appropriate algorithms, like the one developed by Google, i.e., Google Congestion Control (GCC), can provide the congestion detection and adapt the sending rate in order to track the link capacity [25].

A high level of QoE depends not only on the issues which need to be addressed by the software/network, but hardware as well. The QoE for video calls depends on device factors, such as central processor unit (CPU) speed, display resolution/size, memory, type of operating system (OS)/browser, etc. The research study investigating the impact of the terminals' technical specification on QoE based on their collected subjective scores declares that minimum hardware requirements for multimedia services/videoconferencing are 2GB random access memory (RAM) and quad-core 2.5 GHz processor [26].

Better video resolution can contribute to the better video quality based on the results obtained in the subjective studies [27-28]. It is shown that higher video resolution demands higher processing requirements on the system. All these demands lead to the higher congestion under certain bandwidth limitations. Obtained results showed a small difference in the perceived QoE between 480x320 and 640x480 resolutions, but that insight into bandwidth requirements for such services can provide a better input for the network resource allocation mechanism [27]. The final results show that terminals need to meet high CPU requirements for videoconferencing services. Additionally, the overall QoE could be improved by implementing other users' functionalities like possibility of recording conversation, sharing data, or texting [26].

III. END-USER SENSITIVITY TO QoE INFLUENCE FACTORS

All influence factors discussed in previous section play the role in the process of QoE creation and evaluation. Therefore, it is interesting to determine which of them is less or more important when it comes to QoE evaluation. However, these complex IFs are interrelated. That is why they should never be investigated like isolated factors. Accordingly, this paper provides thoroughly identification and joint investigation of factors influencing user's experience in the context of WebRTC video calling. In general, there are two common way of describing user's experience, i.e. objective (rational) which is based on property generalization, and perceptual (individual) which is based on human's evaluation of some service/application [29]. This paper is focused on the perceptual way of describing user's experience for the WebRTC video call service.

A. Research Methodology

There are very few systematic studies related to QoE of WebRTC video calling [30]. Therefore, the end user survey has been conducted with the aim of identifying of the most and least influential factors that affect QoE for WebRTC based video calling service according to the users' opinion. A wide range of factors belonging to all three aforementioned groups (i.e., HIF, CIF, and SIF) were selected based on the overview of existing literature [9-28] and considered through the case study questionnaire. Many IFs, such as audio quality, image quality or quality of service were considered as composite factors given that they depend on multiple sub-factors, which will be in focus of our future studies. The intention was to test and verify relevant qualitative findings and quantitatively determine the degree to which the subject factors affect the QoE in a specified context. The survey contains sets of questions that intend to identify users' habits and opinions regarding the WebRTC based video services via different smart devices.

The online questionnaire has been prepared in Bosnian and English by Google platform for online surveys. It contains 22 questions including Yes-No questions, questions with one or multiple possible answers, and short answer questions. The investigation form was distributed by the social networks (e.g., Facebook) and via e-mail. It was completed by 140 users in a 10-day period. The users have been selected randomly in order to achieve representatives of the sample.

Collected demographic data related to the age, gender, educational level, and origin country of the examinees, as well as smart device usage describes the group that approached the questioning:

- 68% of examinees fit into the category of age 21 to 30, 14% fit into the category age 31 to 40, 10% fit into the category of age below 20, and 8% fit into the category of age above 41;
- 53% of examinees were male and 47% of them were female;
- 20% of examinees reported as having a high-school diploma, 29% reported as being students, 30% as having faculty degree, 20% of them as having a Master of Science degree, and 1% of examinees as holding a Doctor of Philosophy degree;
- 87.57% of examinees are from Bosnia and Herzegovina (50%) and Austria (33.57%);
- 35% of examinees use smart phones, 30% of them use laptops, 15% of them use personal computers (PCs), 9% of them use tablets, whereas 11% of them use all of these devices on a daily basis for communication purposes.

B. Results and Discussion

The influence of each factor has been rated on a 5-point scale where 1 means "does not affect", 2 "weakly affects", 3 "moderately affects", 4 "affects", and 5 "strongly affects". In order to identify the most and the least important IFs according to users, statistical analysis of obtained results has been conducted by using percentages and descriptive statistics, i.e., measures of central tendency and variability (Table I).

In order to determine which factors have the strongest effect on users' satisfaction in the context of WebRTC based video call service (based on participant opinions), the ones with means higher than 3.6 and coefficient of variance under 36% have been considered. Namely, a mean of 3.6 means that 4s and 5s are the most frequent rates of participants in terms of the impact of the factor, while the second condition means that the average distance from the mean of all rates for each factor is less than 36%, i.e., the values do not vary more than 36%. According to this, the first seven influence factors from Table I have been found to be the most influential: **audio quality, image quality, quality of service, price of accessing the service, loss of video frames, ease of use**, and **procedure of accessing web environment**, going from more to less influential.

Further on, in order to supplement the results, the frequencies of each rate, i.e., the mode has been considered as well. Namely, the factors that 67% of questioned users have rated with 4s and 5s have been considered as most influential. According to participants' ratings, those factors are: **audio quality, image quality** and **quality of service**, as well going from more to less effective.

However, what differentiates this paper in comparison to the literature is that we have provided a list of IFs that were ordered by user perception of the strength of their impact on user's QoE. This may serve as a useful input, in terms of what to focus on and what to enhance for stakeholders dealing with WebRTC based video calls (network providers, service providers, content providers, device manufacturers, etc.) when trying to improve their performance. Most of these IFs fall into category of system IFs, thereby confirming existing literature's qualitative conclusions that the most influential factors in terms of QoE in this context comes from this class. However, having service price on a fourth place in the list, together with location, noise or movement in the first half of the IFs list, launches context IFs, as a group, shoulder to shoulder with system IFs, and should additionally motivate research community and stakeholders to investigate these factors and enhance their positive and suppress the negative effects.

In addition, in the subject questionnaire, we gave the opportunity to participants to give their opinion on the most influential factor and to suggest additional factors that may affect their satisfaction with the service. Gathered answers indicate that the list of examined factors is comprehensive and that factors identified as the most influential by the statistical analysis have been identified as most influential by the participants as well.

In the same way, we have determined that the least influential IFs are the ones that can be classified in human IF group: **difficulties in using modern technologies, emotional state, time of the day, hand injuries,** and **speech difficulties**, going from less to more influential. This justifies the lack of research which addresses this category of IFs (in comparison to system and context IFs) in context of user experience and QoE.

IV. CONCLUSION AND FUTURE WORK

The advent of WebRTC enables the service providers of all types to rapidly and efficiently differentiate their business and service offerings. As this promising technology continues to proliferate, it becomes important to gain an in-depth understanding of the numerous influence factors which affect the user's experience and its quality in the context of WebRTC video communication. To prevent users from getting unsatisfied with WebRTC video calling, we have identified the most and least influential factors according to examined users' opinions. The list of influence factors that affect the QoE in the context of WebRTC video calling has been created and sorted by the impact strength. The list shows the dominance of SIFs, but launches CIFs on the same level, while confirming the relapsing importance of HIFs. This may be useful for stakeholders dealing with WebRTC based services to acquire knowledge that is necessary for improving the service performance. All this up-to-date information should motivate research community and stakeholders to consider these factors in more detail.

Therefore, the further investigation of key influence factors for different WebRTC based services, as well as understanding how they mutually correlate is planned for the future work due its importance for satisfying end user needs and expectations. Moreover, since the most influential factors have been considered as composite factors (e.g., audio quality, image quality, quality of service), it is desirable to decompose them into components, and examine their impact on QoE (e.g., image quality may be decomposed into encoding, resolution, sampling rate, etc.). The knowledge of the effects of each component (sub-factor) can provide a deeper understanding which can be further utilized by interested stakeholders in order to improve user's QoE.

Since today's users are still not familiar with WebRTC based services, there is a reason to engage more effort in research activities to discover and remove existing disadvantages of this technology according to end user needs and expectations. Such research activities give the contribution in the process of adopting the WebRTC based services and successfully managing QoE in this context including all its components, i.e. modelling, monitoring, and optimization.

REFERENCES

[1] M. R. Melhoos, "Dashboard for Quality of Experience Studies of WebRTC based Video Communication," M.S. thesis, Department of Telematics, Faculty of Information Technology, Mathematics and electrical Engineering, Norwegian University of Science and Technology, Trondheim, Norway, 2016.

[2] E. Fosser and L. Nedberg, "Quality of Experinece of WebRTC based Video Communication," M.S. thesis, Department of Telematics, Faculty of Information Technology, Mathematics and electrical Engineering, Norwegian University of Science and Technology, Trondheim, Norway, 2016.

[3] S. Baraković and L. Skorin-Kapov, "Multidimensional Modelling of Quality of Experience for Mobile Web Browsing," Computers in Human Behaviour, vol. 50, pp. 314-332, 2015.

[4] S. Baraković and J. Baraković Husić, "Web categorization and end user survey addressing mobile web," J. BH Elektrotehnika, vol. 10, pp. 36-45, 2016.

[5] P. Le Callet, S. Möller, and A. Perkis, Qualinet White Paper on Definitions of Quality of Experience, European Network on Quality of Experience in Multimedia Systems and Services (COST Action IC 1003) Version 1.2, 2013.

[6] S. Ickin, K. Wac, M. Fiedler, L. Jankowski, J. H. Hong, and A. K. Dey, "Factors Influencing Quality of Experience of Commonly Used Mobile Applications," IEEE Communication Magazine, vol. 50, no. 4, pp. 48-56, 2012.

[7] GSM Association, White Paper on WebRTC to complement IP communication Services, Version 1.0, 2016.

[8] U. Reiter et al., "Factors Influencing Quality of Experience," in Quality of Experience: Advanced Concepts, Applications and Methods, S. Moller, A. Raake, Eds., Switzerland: Springer International Publishing, 2014, ch. 4, pp. 55-72.

[9] M. J. Scott, S. C. Guntuku, Y. Huan, W. Lin, G. Ghinea, "Modelling Human Factors in Perceptual Multimedia Quality: On The Role of Personality and Culture," in Proceedings of the 23rd ACM international conference on Multimedia, Brisbane, Australia, 2015, pp. 481-490.

[10] M. Hyder, K. R. Laghari, N. Crespi, and C. Hoen, "Are QoE requirements for Multimedia Services different for men and women? Analysis of Gender Differences in Forming QoE in Virtual Acoustic Environments," in Emerging Trends and Applications in Information Communication Technologies, B.S. Chowdhry, F.K. Shaikh, D.M.A. Hussain, M.A. Uqaili ,Eds., Germany, Springer Berlin Heidelberg 2012, pp. 200-209.

[11] N. Murray, Y. Qiao, B. Lee, G. M. Muntean, A. K. Karunakar, "Age and gender influence on perceived olfactory & visual media synchronization," in Proceedings of the 2013 IEEE International Conference on Multimedia and Expo(ICME), San Jose, California, USA, 2013.

[12] K.M. Wolters, K.P. Engelbrecht, F. Gödde, S. Möller, A. Naumann, and R. Schleicher, "Making it easier for older people to talk to smart homes: The effect of early help prompts," Universal Access in the Information Society, vol. 9, pp. 311-325, 2010.

[13] R. Mccoll and Y. Truong, "The Effects of Facial Attractiveness and Gender on Customer Evaluations During A Web-Video Sales

Encounter," Journal of Personal Selling & Sales Management, vol. 33, no. 1, pp. 117–128, 2013.

[14] R. E. Nisbett and Y. Miyamoto, "The influence of culture: holistic versus analytic perception," Trends in Cognitive Sciences, vol. 9, pp. 467-473, 2006.

[15] I. Wechsung, M. Schulz, K. P. Engelbrecht, J. Niemann, and S. Moller, "All Users Are (Not) Equal – The Influence of User Characteristics on Perceived Quality, Modality Choice and Performance," in Proceedings of the Paralinguistic Information and its Integration in Spoken Dialogue Systems Workshop, R. L. C. Delgado, T. Kobayashi, Eds., Springer New York, 2011, pp. 175–186.

[16] J. Palhais, R. S. Cruz, and M. S. Nunes, "Quality of Experience Assessment in Internet TV," in Mobile Networks and Management, K. Pentikousis, R. Aguiar, S. Sargento, R. Aguero, Eds., Germany, Springer Berlin Heidelberg, 2012, pp. 261-274.

[17] H. G. Msakni and H. Youssef, "Is QoE estimation based on QoS parameters sufficient for video quality assessment?," in Proceedings of the 2013 IEEE 9th International Wireless Communications and Mobile Computing Conference (IWCMC), Cagliari, Sardinia - Italy, 2013.

[18] F. Pereira, "Sensations, Perceptions and Emotions: Towards Quality of Experience Evaluation for Consumer Electronics Video Adaptations," in Proceedings of the 1st International Workshop on Video Processing and Quality Metrics for Consumer Electronics, Scottsdale, AZ, USA, 2005.

[19] K. L. Hsiao, "Exploring the Factors that Influence Continuance Intention to Attend one-to-some Online Courses via Videoconferencing Software," The Turkish Online Journal of Educational Technology, vol. 11, no. 4, pp. 155-163 , 2012.

[20] Y. Zhu, I. Heynderickx, and J. A. Redi, "Understanding the role of social context and user factors in video Quality of Experience," Computers in Human Behaviour, vol. 49, pp. 412-426, 2015.

[21] S. J. Pyykko and T. Utriainen, "A Hybrid Method for Quality Evaluation in the Context of Use for Mobile (3D) Television," Journal Multimedia Tools and Applications, vol. 55, no. 2, pp. 185-225, 2011.

[22] S. J. Pyykko and T. Vainio, "Framing the Context of Use for Mobile HCI," International Journal of Mobile Human Computer Interaction, vol. 2, no. 4, pp. 1-28, 2010.

[23] D. Ammar, K. De Moor, M. Xie, M. Fiedler, and P. Heegaard, "Video QoE killer and performance statistics in WebRTC-based video communication," in Proceedings of the 2016 IEEE Sixth International Conference on Communications and Electronics (ICCE), Novotel, Ha Long, Vietnam, 2016, pp. 429-436.

[24] J. Seppaenen, M. Varela and A. Sgora, "An autonomous QoE-driven network management framework," Journal of Visual Communication and Image Representation, vol. 25, no. 3, pp. 565-577, 2014.

[25] G. Carlucci, L. De Cicco, S. Holmer, and S. Mascolo, "Analysis and Design of the Google Congestion Control for Web Real-Time Communication (WebRTC)," in Proceedings of the 7th International Conference on Multimedia Systems (MMSys'16), Klagenfurt, Austria, 2016.

[26] D. Vučić and L. Skorin-Kapov, "The impact of mobile device factors on QoE for multi-party video conferencing via WebRTC," in Proceedings of the 13th International Conference on Telecommunications (ConTEL), Graz, Austria, 2015.

[27] D. Vučić, L. Skorin-Kapov, and M. Sužnjević, "The impact of bandwidth limitations and video resolution size on QoE for WebRTC-based mobile multi-party video conferencing," in Proceedings of the 5th ISCA/DEGA Workshop on Perceptual Quality of Systems (PQS 2016), Berlin, Germany, 2016.

[28] G. Berndtsson, M. Folkesson, and V. Kulyk, "Subjective quality assessment of video conferences and telemeetings," in Proceedings of the 19th International Packet Video Workshop (PV), Munich, Germany, 2012.

[29] A. Raake and S. Egger, "Quality and Quality of Experience," in Quality of Experience: Advanced Concepts, Applications and Methods, S. Moller, A. Raake, Eds., Switzerland: Springer International Publishing, 2014, ch. 2, pp. 11-33.

[30] Y. Suying, Y. Guo, Y. Chen, F. Xie, C. Yu, and Y. Liu. (2016). Enabling QoE Learning and Prediction of WebRTC Video Communication in WiFi Networks. [Online]. Available: http://eeweb.poly.edu/faculty/yongliu/docs/yishuai_icc17.pdf.

TABLE I. DESCRIPTIVE STATISTICS FOR COLLECTED RATINGS (USER OPINIONS ON HOW A GIVEN FACTOR AFFECTS THEIR SATISFACTION ON A SCALE OF 1-5 (1 – DOES NOT AFFECT; 2- WEAKLY AFFECTS, 3 – MODERATELY AFFECTS, 4- AFFECTS, 5 – STRONGLY AFFECTS)) AND FREQUENCY OF SCORES 1-5 FOR EACH INFLUENCE FACTOR

No.	Influence factor	Mean	Median	Variance	Standard deviation	Coefficient of variation (%)	Freq. of 1	Freq. of 2	Freq. of 3	Freq. of 4	Freq. of 5	%
1.	Audio quality	4.05	5	1.57	1.25	30.99	10	10	20	28	77	72.41
2.	Image quality	4.00	5	1.60	1.26	31.62	12	5	25	28	71	70.21
3.	Quality of service	3.95	5	1.65	1.29	32.57	11	9	28	22	74	67.66
4.	Service price	3.76	4	1.89	1.37	36.59	16	14	22	32	63	64.62
5.	Loss of video frames	3.73	4	1.54	1.24	33.27	8	17	33	30	53	58.86
6.	Ease of use	3.65	4	1.73	1.32	36.08	13	16	29	33	50	58.86
7.	Procedure of accessing web environment	3.60	4	1.61	1.26	35.18	14	13	34	41	44	58.21
8.	Video frame delay	3.57	4	1.83	1.35	37.93	15	18	30	30	51	56.25
9.	Battery consumption	3.55	4	1.76	1.33	37.38	15	17	32	34	46	55.55
10.	Web environment object load time	3.50	4	1.31	1.14	32.63	9	18	40	47	31	53.79
11.	Screen resolution	3.49	4	1.57	1.25	35.84	15	11	45	35	39	51.03
12.	Processor speed	3.48	4	1.36	1.16	33.53	8	21	42	38	34	50.35
13.	Video frame shift speed	3.48	4	1.63	1.28	36.68	13	20	34	37	39	53.14
14.	Video call purpose	3.47	4	1.99	1.41	37.77	16	14	38	33	42	52.44
15.	Location	3.42	4	1.48	1.21	35.63	15	15	36	49	28	53.84
16.	Noise	3.39	4	1.54	1.24	35.29	13	18	39	45	28	51.04
17.	Screen size	3.32	3	1.83	1.35	40.82	20	20	37	32	37	47.26
18.	Screen lightness	3.22	3	1.44	1.20	37.29	18	16	50	40	22	42.46
19.	Web environment object loading order	3.16	3	1.63	1.27	40.32	19	25	35	40	23	44.36
20.	Web environment object visual organization	3.16	3	1.47	1.21	38.41	17	21	50	32	23	38.46
21.	Movement	3.12	3	1.42	1.19	38.69	17	26	46	36	21	39.04
22.	Prior experience	3.08	3	1.63	1.28	41.49	25	15	45	37	20	40.14
23.	Service accessibility	3.06	4	1.94	1.39	38.63	19	12	31	30	54	57.53
24.	Web environment object size	3.05	3	1.84	1.35	44.51	26	25	40	28	28	38.09
25.	Used browser	3.03	3	1.63	1.27	42.12	24	23	44	34	21	37.67
26.	Visual impression	2.99	3	1.34	1.16	38.68	17	29	49	32	15	33.09
27.	Visual difficulties	2.94	3	1.84	1.35	46.11	28	28	36	28	24	36.11
28.	Personal relations with the interlocutor	2.87	3	2.07	1.43	50.08	38	16	38	24	25	34.75
29.	Weather	2.81	3	1.43	1.19	44.12	25	34	47	21	18	26.89
30.	Keyboard type	2.74	3	1.61	1.27	46.37	32	26	48	21	16	25.87
31.	Lightning	2.74	3	1.46	1.21	43.55	29	30	49	28	11	26.53
32.	Background and text colour	2.64	3	1.58	1.25	47.6	36	26	47	21	13	23.77
33.	Speech difficulties	2.64	3	1.78	1.33	50.57	37	31	31	25	15	28.77
34.	Hand difficulties	2.63	3	1.58	1.26	47.91	37	27	42	26	11	25.87
35.	Time of the day	2.60	3	1.72	1.31	54.24	47	22	41	14	22	24.65
36.	Emotional state	2.54	3	1.71	1.31	51.42	45	23	40	24	12	25.00
37.	Difficulties with using modern technologies	2.52	2	1.85	1.36	53.88	45	28	33	19	16	24.82

Is there any impact of human influence factors on quality of experience?

J. Baraković Husić*, S. Baraković **, *** and S. Muminović****

* Faculty of Electrical Engineering, University of Sarajevo, Sarajevo, Bosnia and Herzegovina
** Faculty of Transport and Communication, University of Sarajevo, Sarajevo, Bosnia and Herzegovina
*** American University in Bosnia and Herzegovina, Sarajevo, Bosnia and Herzegovina
**** East Codes and Tours d.o.o., Sarajevo, Bosnia and Herzegovina
jasmina.barakovic@etf.unsa.ba, barakovic.sabina@gmail.com, sabina@east.ba

Abstract - The increasing popularity of real-time communication services has driven a rising interest in understanding the factors influencing the quality of experience (QoE). Many research activities aiming to accomplish this goal have addressed different influence factors (IFs), which may be classified into three categories, i.e., human IFs, context IFs, and system IFs. Although the importance of human IFs (HIFs) has been already emphasized, these factors have been considered to a limited extent, and due to lack of empirical evidence, they still remain not well understood. Therefore, this research has been motivated by the challenge to provide a deeper understanding of HIF's impact on QoE in the context of the real-time communication services. In order to accomplish this and verify the conclusions derived from related work, we have conducted two experimental studies which consider voice over internet protocol (VoIP) and unified communication (UC) services. Experimental results show no existence of impacts of selected HIFs on the user's overall QoE while using either VoIP or UC services. However, this highlights the need for further research on this topic due its importance for satisfying end user needs and expectations in the context of real-time communication services.

I. INTRODUCTION

The telecom industry is under a significant transformation. The real-time communication services are under a pressure as telecom market demands richer, higher value services available on any device, across any network, and with satisfactory performance. Hitherto, performance in the telecom industry has been typically addressed in terms of quality of service (QoS). However, the recent research in the field of quality of experience (QoE) [1]-[3] has shown that the traditional QoS mechanisms are not sufficient to satisfy end users, prevent their churn, and attract new customers or incline them to adopt new complex services and thereby support further technology development. Therefore, in order to improve users' satisfaction and offer maximized quality while at the same time minimizing the costs, a deep understanding of the influence factors (IFs) that affect QoE is needed.

A QoE IF has been defined as "any characteristic of a user, system, service, application, or context whose actual state or setting may have influence on the quality of experience for the user" [4]. Although there are several approaches to the categorization of QoE IFs [1], [5], [6]

we consider the classification proposed by EU Qualinet community which groups them into following three categories [4]: (1) human IFs; (2) system IFs; and (3) context IFs. In this paper, the focus is on the human IFs (HIFs) which present "any variant or invariant property or characteristic of a human user. The characteristic can describe the demographic and socio-economic background, the physical and mental constitution, or the user's emotional state." [4]. These factors are divided into two sub-categories [7]: (1) low-level processing influence factors related to the physical, emotional, and mental constitution (e.g. gender, age, lower-order emotions, user's mood, personality traits, motivation, attention level, etc.), and (2) higher-level cognitive processing influence factors (e.g. socio-economic situation, educational background, attitudes and values, expectations, needs, knowledge, previous experiences, etc.).

The low-level processing influence factors, especially those that are related to human perception of exterior stimulus, have the highest impact on QoE [7]. However, the previous research activities on HIFs were limited to the relatively easy to understand factors, such as gender, age, visual acuity, etc. These factors were not considered either as the main research focus or as independent variables whose influence on QoE is in detail investigated [8]. Therefore, this research has been motivated by the challenge to provide a deeper understanding of the selected HIFs, i.e. gender, age, prior experience, and emotional state, and their impact on QoE in context of real-time communication services. In order to accomplish this goal and verify the conclusions made by the authors addressed in the related work [9]-[35], we have conducted two experimental studies which consider voice over internet protocol (VoIP) and unified communication (UC) services.

The rest of the paper is organized as follows. Section II provides a state-of-the-art literature review in the field of HIFs and their impact on QoE. Section III describes the design of the experimental studies performed in order to examine the impact of selected HIFs on QoE in the context of the VoIP and UC services. It contains two sub-sections which give hypotheses, describe the experimental environment and provide result analysis. Finally, Section IV summarizes the conclusions and presents open issues for further research.

II. A SURVEY OF HUMAN INFLUENCE FACTORS

Human influence factors may be defined as "the overall assessment of human needs, feelings, performance, and intentions" and thereby may be classified as subjective and objective factors on the basis of psychological and physiological factors [2]. The subjective factors are associated to both quantitative and qualitative aspects of human needs and requirements (e.g. ease of use, joy of use, usefulness, etc.). By their nature, they are psychological factors that take into account human perceptions, intentions and needs. In order to understand these factors, different psychological models (e.g., technology acceptance model (TAM), theory of planned behavior (TPB), etc.) may be used according to the nature of the service and environment [37]. On the other side, the objective factors are based on physiological (e.g., brain waves, heart rate, blood pressure, etc.) and cognitive aspects (e.g., memory, attention, human activity, human task performance, language, etc.). Since these factors are quantitative in nature, different physiological tools (e.g., body sensors, galvanic skin response (GSN), etc.) and human performance models (e.g., goals, operators, methods, selection rules (GOMS) model) may be used for their capturing.

On the basis of the foregoing, it is clear that human influence factors are highly complex because of their subjectivity and relation to internal states and processes. In addition, they are strongly interrelated and may also deeply interplay with other groups of IFs. Therefore, the more recent research activities on HIFs have gain the popularity resulting in the growth of literature and number of studies dealing with the influence of human factors on QoE. Nevertheless, the understanding of more complex HIFs and their impact on QoE is still relatively limited due to the fact that the wide range of these factors was margined by the research community for a long period of time [8].

The importance of HIFs' influence on user satisfaction, experience, and its quality has been emphasized by several research studies [9]-[35]. Table I surveys a number of references concerning HIFs in order to obtain a "broader picture" on this topic and provide the brief analysis of several commonly considered HIFs. Although this survey should not be considered exhaustive, it aims to identify and select HIFs which are relevant for our further consideration. Based on the state-of-the-art literature review, the following HIFs are identified as commonly analyzed in research studies: cultural background [10]-[13], [26], [29] age [13], [15], [19], [26], [28], [33], [35], gender [13]-[22], [28], [32], [35], sensation [9], [23], perception [9], [11], [23], emotional state [20], [21], [23], [27], [31], [34], personality [10], [24], [25], [27] and prior experience [30]. However, these factors have been considered to the limited extent, and still remain not well comprehended.

Therefore, we have selected four HIFs, i.e., age, gender, prior experience, and emotional state, to be our main research focus and the independent variables whose influence on QoE will be investigated in the context of VoIP and UC services. The HIFs are selected on the basis that, according to Table I, age, gender, and emotional state are the most commonly analysed HIFs, whereas the prior experience is the rarest considered HIFs.

TABLE I.   SURVEY OF REFERENCES CONCERNING THE IMPACT OF HIFs ON USER EXPERENCE AND ITS QUALITY

Authors of the reference [Number of the reference]	Culture Background	Age	Gender	Sensation	Perception	Emotional State	Personality	Prior Experience
E. B. Goldstein [9]				×	×			
M. A. Scott, S. C. Guntuku, H. Yang, G. Ghinea [10]	×						×	
R. Nisbett and Y. Miyamoto [11]	×				×			
L. D. Setlock, P. A. Quinones, S. R. Fussel [12]	×							
Z. Zhu, I. Heynderickx, J. A. Redi [13]	×	×	×					
M. Hyder, K. R. Laghari, N. Crespi, C. Hoen [14]			×					
K. R. Laghari [15]		×	×					
I.C. Zuendorf, H.O. Karnath, J. Lewald [16]			×					
L. Sax [17]			×					
D. Kimura [18]			×					
N. Murray, Y. Qiao, B. Lee, G. M. Muntean, A. K. Karunakar [19]		×	×					
C. C. Tossell, et al. [20]			×			×		
A. Wolf [21]			×			×		
C. Lee [22]			×					
F. Periera [23]				×	×	×		
J. A. Redi, Y. Zhu, H. de Rdder, I. Heynderickx [24]							×	
I. Wechsung, M. Schulz, K. P. Engelbrecht, J. Niemann, S. Möller [25]							×	
A. B. Naumann, I. Wechsung, J. Hurtienne [26]		×						
S. Chandra, M. Scott, W. Lin, G. Ghinea [27]	×					×	×	
Y. C. Yen, C. Y. Chu, S. L. Yeh, H. H. Chu, P. Huang [28]		×	×					
T. Daengsi, N. Khitmoh, P. Wuttidittachotti [29]	×							
S. Ickin, K. Wac, M. Fiedler, L. Janowski, J. Hong, A. Dey [30]								×
R. Gupta, K. Laghari, H. Banville, T. H. Falk [31]						×		
R. McColl and Y. Truong [32]			×					
K. M. Wolters, et al. [33]		×						
I. Tjostheim, et al. [34]						×		
B. Weiss, S. Moller, M. Schulz [35]		×	×					

Giving the fact that a distinction can be made between factors that are stable and factors that have more dynamic nature [8], our research considers two stable factors, i.e., age and gender, and two more dynamic factors, i.e., prior experience and emotional state. As will be described in Section III, these factors are acquired at the beginning of each experimental study, since our research examines their impact on QoE in the context of VoIP and UC service (e.g., one research question is how emotional state affects the QoE, and not how VoIP or UC service usage affects the emotional state).

A comparative research studies on various real-time communication services show that there exist both age and gender differences in forming QoE. For example, as far as age is concerned, older individuals are more critical than younger ones, which may suggest that elderly people have higher requirements to QoE [15], [33]. However, the opposite results illustrate that older individuals tend to rate the quality more positively than younger ones [26], [35]. On the other hand, the research activities considering gender show that better performances of females are indicated in hearing sensitivity [17], memory localization facilities [18], or synchronization's errors sensitivity [19], whereas the research results are in favour of males for localizing target sounds in a multi-source sound environment [16]. Nevertheless, it is shown that age and gender differences are insignificant in terms of QoE [28].

As indicated above, examples of HIFs that have more dynamic nature are emotional state and prior experience. The human emotional state may have a great impact on QoE and thereby it has recently gained importance which is demonstrated, for example, by incorporating it into speech QoE models [31] or investigating it through play quiz games [34]. Despite its strong relevance for QoE [30], the impact of prior experience and the way in which it interlace with or depend on other IFs is still insufficiently understood. Giving the fact that prior experience, as well as other selected HIFs may interplay with other groups of IFs, their influence may be increased or decreased in accordance with the appearance of specific other IFs. Therefore, the motivation for this research lies in measuring the selected HIFs, in disentangling them, and understanding their influence on QoE.

III. EXPERIMENTAL STUDIES

To examine the impact of HIFs on QoE and verify the conclusions made by the authors addressed in the related work we have conducted two studies which included the testing of VoIP and UC services. The following two sub-sections provide more detailed discussion on the experimental studies and their results.

A. The impact of HIFs on VoIP

a. Introduction and Hypothesis

Based on the related work and reviewed literature, we derived the following four hypotheses. The hypotheses try to illuminate the existence of impacts of selected HIFs, i.e., age, gender, prior experience, and emotional state on user's overall QoE while he or she is using the VoIP service. We expect to find no impact of these factors on QoE, meaning that QoE will not change if these factors change.

H1.1: There is no impact of age on user's QoE when using VoIP service.

H1.2: There is no impact of gender on user's QoE when using VoIP service.

H1.3: There is no impact of prior experience on user's QoE when using VoIP service.

H1.4: There is no impact of emotional state on user's QoE when using VoIP service.

b. Description of the testing environment

The total of 30 participants has participated in the study (Fig. 1). Mostly those were the family members, friends, and colleagues and they have participated in the experiment on a good will basis, in their free time and for free. Experiments were conducted at authors' homes (in Sarajevo, Tuzla, and Zenica).

All participants were given a task of registering, establishing, and terminating one VoIP call by using JITSI (https://jitsi.org/) after which they were asked to express their opinion regarding their QoE while using the VoIP service. The subjective evaluation of the test VoIP calls was performed by using the electronic evaluation questionnaire, which contained the part that was completed at the beginning of the experiment and it included the questions that covered the information related to the participant's personal data, previous experience with VoIP service usage, and emotional state of the participant (considered HIFs), and the part that deals with the participant's rating of the statement related to overall QoE when using VoIP service. The latter statement was a simple mean opinion score (MOS) scale used as the de facto standard in QoE Studies and specified in ITU-T Recommendation P.800.1. It must be noted that subjective QoE evaluation are to-date most commonly reported in terms of a single MOS value [36].

The experiment procedure lasted about 15-20 minutes and has included the three steps [36]: (i) introduction and clarification of the experiment tasks that need to be performed by the participants (8 minutes), (ii) participant training (5 minutes), and (iii) testing and rating of the experimental VoIP calls (5 minutes). All participants were asked not to think about their feelings during evaluation, but to be intuitive. As it could be concluded, the considered HIFs were differed in several groups. Namely, the factors of age and prior experience were considered in five abovementioned groups, while gender and emotional state in two.

c. Results analysis

In order to address the hypotheses (H1.1-H1.4) stating that differences in the age, gender, prior experience, and emotional state do not impact user's QoE while using the VoIP service, we refer to the results of four one-way ANOVAs given in Table II. Independent variables (IVs) used for conducting this statistical examination are our selected HIFs, while the dependent variable (DV) is user's QoE. As previously mentioned, IVs are manipulated as follows: age and prior experience in five levels (groups), while gender and emotional state in two (Fig. 1). Based on the obtained results, one may conclude that there does not exist a strong and statistically significant impact ($p > 0.001$) of user's age, gender, prior experience, and emotional state on QoE when he or she is using the VoIP service. This implies that the hypotheses H1.1–H1.4 are supported.

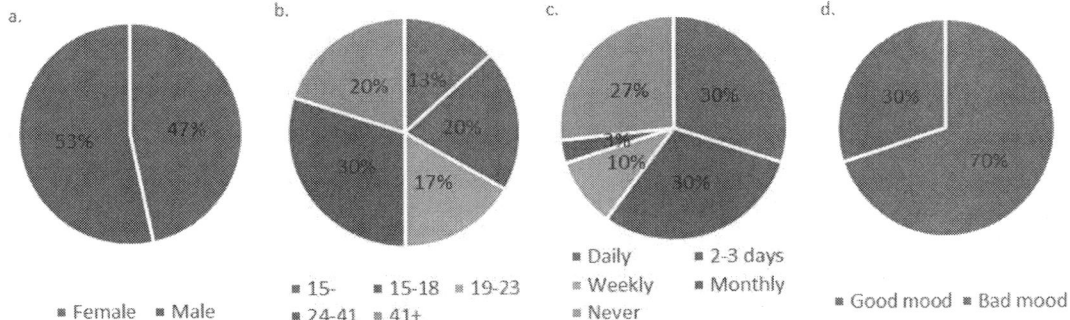

Figure 1. Demographic data of the VoIP study sample: a. Gender groups, b. Age groups, c. Prior experience groups, d. Emotional state groups.

TABLE II.   ANOVA RESULTS INDICATING NO IMPACT OF CONSIDERED HIFS ON QoE FO VoIP SERVICE.

Factor	Mean (SD) 1st group/ female/ good mood	Mean (SD) 2nd group/ male/ bad mood	Mean (SD) 3rd group	Mean (SD) 4th group	Mean (SD) 5th group	F	p-value
Age	3.17 (1.765)	3.09 (1.719)	3.38 (1.642)	2.99 (1.714)	2.94 (1.753)	0.502	0.734
Gender	3.10 (1.707)	3.08 (1.723)				0.007	0.932
Prior experience	3.11 (1.764)	3.11 (1.707)	3.13 (1.738)	3.25 (1.685)	2.86 (1.730)	0.498	0.737
Emotional state	3.13 (1.656)	3.10 (1.701)				0.565	0.652

B. The impact of HIFs on UC

a. Introduction and Hypothesis

Based on the related work and reviewed literature, we derived four hypotheses as well. The set tries to illuminate the existence of impacts of selected HIFs, i.e., age, gender, prior experience, and emotional state on user's overall QoE while he or she is using the UC service. We expect to find no impact of these factors on QoE, meaning that QoE will not change if these factors change.

H2.1: There is no impact of age on user's QoE when using UC service.

H2.2: There is no impact of gender on user's QoE when using UC service.

H2.3: There is no impact of prior experience on user's QoE when using UC service.

H2.4: There is no impact of emotional state on user's QoE when using UC service.

b. Description of the testing environment

The total of 96 participants has participated in the study (Fig. 2). As in the study described in previous subsection, mostly participants were the family members, friends, and colleagues and they have participated in the experiment on a good will basis, in their free time and for free. Experiments were conducted at authors' homes (in Sarajevo, Tuzla, and Zenica).

All participants were given a task to register for using the UC service – Linphone application (https://www. linphone.org/), after which they had to: (i) change their presence status, (ii) send the message "How are you?" and receive the answer, (iii) establish an audio call lasting 30 seconds, and (iv) establish a video call lasting 30 seconds. Then, they were asked to express their opinion regarding their QoE while using the UC service. The subjective evaluation of the test UC service was performed by using the electronic evaluation questionnaire, which contained the part that was completed at the beginning of the experiment and it included the questions that covered the information related to the participant's personal data, previous experience with UC service usage, and emotional state of the participant (considered HIFs), and the part that deals with the participant's rating of the statement related to overall QoE when using UC service. As in previous study, the latter statement was the MOS scale.

The experiment procedure lasted about 15-20 minutes and has included the three following steps [36]: (i) introduction and clarification of the experiment tasks that need to be performed by the participants (10 minutes), (ii) participant training (5 minutes), and (iii) testing and rating of the experimental UC calls (2 minutes). All participants were asked not to think about their feelings during evaluation, but to be intuitive.

As it could be concluded, the considered HIFs were differed in several groups. Namely, the factors of age, emotional state, and prior experience were considered in three and five abovementioned groups, while gender in two.

c. Results analysis

In order to address this set of hypotheses (H2.1-H2.4) stating that differences in the age, gender, prior experience, and emotional state do not impact user's QoE while using the UC service, as well as in previous subsection, we refer to the results of four one-way ANOVAs given in Table II.

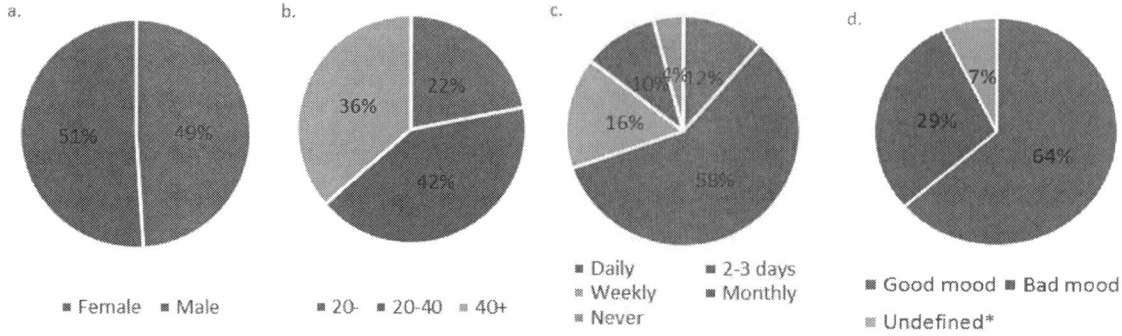

Legend: *the mood that users defined neither as good nor as bad.

Figure 2. Demographic data of the UC study sample: a. Gender groups, b. Age groups, c. Prior experience groups, d. Emotional state groups.

TABLE III.   ANOVA RESULTS INDICATING NO IMPACT OF CONSIDERED HIFS ON QOE FOR UC SERVICE.

Factor	Mean (SD) 1st group/ female/ good mood	Mean (SD) 2nd group/ male/ bad mood	Mean (SD) 3rd group/ undefined*	Mean (SD) 4th group	Mean (SD) 5th group	F	p-value
Age	4.62 (0.498)	4.63 (0.490)	4.69 (0.471)			0.187	0.830
Gender	4.63 (0.487)	4.66 (0.479)				0.074	0.786
Prior experience	4.75 (0.500)	4.50 (0.527)	4.67 (0.488)	4.59 (0.496)	5.00 (0.000)	2.058	0.093
Emotional state	4.70 (0.460)	4.50 (0.509)	4.71 (0.488)			1.852	0.163

Legend: *the mood that users defined neither as good nor as bad.

Independent variables (IVs) used for conducting this statistical examination are our selected HIFs, while the dependent variable (DV) is user's QoE. As previously mentioned, IVs are manipulated as follows: gender in two, age and emotional experience in three, and prior experience in five levels (groups) (Fig. 2). Based on the obtained results, one may conclude that there does not exist a strong and statistically significant impact ($p>0.001$) of user's age, gender, prior experience, and emotional state on QoE when he or she is using the UC service. This implies that the hypotheses H2.1–H2.4 are supported.

IV. CONLUSION AND FUTURE WORK

An enhanced understanding of human influence factors is a clear prerequisite for the successful management and optimization of end user experience. This means that there is a need to understand user's QoE in this attractive and contemporary context, in order to maintain a satisfied customer base, prevent customer churn, and respond to upcoming challenges relevant to real-time communication services. Therefore, here lies the motivation for this paper which tries to provide a deeper understanding of the influence of human factors on QoE. The aim has been to experimentally investigate the existence of impact of HIFs on user's overall QoE by using the cases of VoIP and UC service.

In order to accomplish this, a brief survey of the state-of-the-art literature in the field of the HIFs has been prepared. Further on, two experimental studies have been conducted to obtain data for analysing the influence of selected HIFs, i.e. age, gender, prior experience, and emotional state, on user's QoE when using VoIP and UC services. The statistical analysis of data collected within these experiments imply that there does not exist strong and statistically significant impact of user's age, gender, prior experience, and emotional state on QoE either he or she is using VoIP or UC service.

However, the conducted experimental studies have certain limitations that may be overcome in the future work. Giving the fact that the impact of HIFs selected for our research study was found to be statistically insignificant, it is necessary to additionally investigate their impact when larger number of participants is included in the study in order to draw non-misleading conclusions. In addition, broader range of different HIFs may be taken into the consideration, as well as the relations between them in order to draw conclusions which may be helpful in a practical sense.

REFERENCES

[1] S. Baraković, J. Baraković, and H. Bajrić, "QoE Dimensions and QoE Measurement of NGN Services," in 18th Telecommunication Forum 2010 (TELFOR 2010), Belgrade, Serbia, 2010.

[2] K. R. Laghari and K. Connelly, "Toward total quality of experience: A QoE model in a communication ecosystem," in IEEE Communications Magazine, vol. 50, no. 4, pp. 58-65, 2012.

[3] P. Brooks and B. Hestnes, "User Measures of Quality of Experience: Why Being Objective and Quantitative is Important," in IEEE Network Magazine, vol. 24, no. 2, pp. 36-41.

[4] P. Le Callet, S. Möller, and A. Perkis, "Qualinet White Paper on Definitions of Quality of Experience," European Network on Quality of Experience in Multimedia Systems and Services (COST Action IC 1003) Version 1.2, Lausanne, Switzerland, 2013.

[5] R. Stankiewicz and A. Jajszczyk, "A Survey of QoE Assurance in Converged Networks," Computer Networks: The International Journal of Computer and Telecommunications Networking, vol. 55, no. 7, pp. 1459-1473, 2011.

[6] L. Skorin-Kapov and M. Varela, "A Multi-Dimensional View of QoE: the ARCU Model," in 35th Jubilee International Convention

of Information and Communication Technology, Electronics and Microelectronics (MIPRO 2012), Opatija, Croatia, 2012.

[7] U. Reiter, et al., "Factors Influencing Quality of Experience," in Quality of Experience: Advanced Concepts, Applications and Methods, S. Moller, A. Raake, Eds., Switzerland: Springer International Publishing, 2014, ch. 4, pp. 55-72.

[8] M. Varela, L. Skorin-Kapov, K. De Moor, and M. Reichl, "QoE – Defining a User-Centric Concept for Service Quality," in Multimedia Quality of Experience (QoE): Currnet Status and Future Requirements, C. W. Chen, P. Chatzimisios, T. Dagiuklas, L. Atzori, Eds.,United Kingdom: John Wiley & Sons, Ltd., 2016, ch. 2, pp. 5-28.

[9] E. B. Goldstein, Sensation and Perception, Wadsworth, Cengage Learning, 2009.

[10] M. A. Scott, S. C. Guntuku, H. Yang, and G. Ghinea, "Modelling Human Factors in Perceptual Multimedia Quality: On The Role of Personality and Culture," in 23rd ACM International Conference on Multimedia, Brisbane, Australia, 2015.

[11] R. Nisbett and Y. Miyamoto, "The influence of culture: holistic versus analytic perception," Trends in Cognitive Sciences, vol. 9, no. 10, pp. 467-473, 2005.

[12] L. D. Setlock, P. A. Quinones, and S. R. Fussel, "Does Culture Interact with Media Richness? The Effects of Audio vs. Video Conferencing on Chinese and American Dyads," in 40th Annual Hawaii International Conference on System Science (HICSS), Hawaii, 2007.

[13] Z. Zhu, I. Heynderickx, and J. A. Redi, "Understanding the role of social context and user factors in   quality of experience," Computers in Human Behaviour, vol. 49, pp. 412-426, 2015.

[14] M. Hyder, K. R. Laghari, N. Crespi, and C. Hoene, "Are QoE requirements for Multimedia Services different for men and women? Analysis of Gender Differences in Forming QoE in Virtual Acoustic Environments," in Emerging Trends and Applications in Information Communication Technologies, B.S. Chowdhry, F.K. Shaikh, D.M.A. Hussain, M.A. Uqaili ,Eds., Germany, Springer Berlin Heidelberg, 2012, pp. 200-209.

[15] K. R. Laghari, "On quality of experience (QoE) for multimedia services in communication ecosystem," Ph.D. dissertation, Universite Pierre et Marie Curie, Informatique et Telecommunications, Paris, France, 2013.

[16] I.C. Zuendorf, H.O. Karnath, and J. Lewald, "Male advantage in sound localization at cocktail parties," in Cortex, vol. 47, no. 6, pp. 741-749, 2010.

[17] L. Sax, "Sex Differences in Hearing: Implications for Best Practice in the Classroom," in Advances in Gender and Education, vol. 2, pp. 13-21, 2010.

[18] D. Kimura, "Sex, sexual orientation, and sex hormones influence human cognitive function," in Current Opinion in Neurobiology, vol. 6, pp. 259-263, 1996.

[19] N. Murray, Y. Qiao, B. Lee, G. M. Muntean, A. K. Karunakar, "Age and gender influence on perceived olfactory & visual media synchronization," in Proceedings of the 2013 IEEE International Conference on Multimedia and Expo(ICME), San Jose, California, USA, 2013.

[20] C. C. Tossell, P. Kortum, C. Shepard, L. H. Barg-Walkow, A. Rahmati, L. Zhong, "A Longitudinal Study of Emoticon Use in Text Messaging from SmartphonesA Longitudinal Study of Emoticon Use in Text Messaging from Smartphones," in Computers in Human Behavior , vol. 28, pp. 659-663, 2012.

[21] A. Wolf, "Emotional Expression Onine: Gender Differences in Emoticon Use," CyberPsychology & Behaviour, vol. 3, no. 5, pp. 827-833, 2000.

[22] C. Lee. (2003). How Does Instant Messaging Affect Interaction Between the Genders? [Online]. Available: http://citeseerx.ist.psu.edu/viewdoc/download?doi=10.1.1.557.218&rep=rep1&type=pdf

[23] F. Periera, "Sensations, Perceptions and Emotions Towards Quality of Experience Evaluation for Consumer Electronics Video Adaptations," in Proceeding of the Second International Workshop on Video Processing and Quality Metrics for Consumer Electronics (VPQM), Scottsdale, Arizona, 2005.

[24] J. A. Redi, Y. Zhu, H. de Rdder, I. Heynderickx, "How Passive Image Viewers Became Active Multimedia Users," in Visual Signal Assessment: Quality of Experience (QoE), L. Ma, W. Lin, K. N. Ngan, C. Deng, Eds., Switzerland: Springer International Publishing, 2014, ch. 2, pp. 31-72.

[25] I. Wechsung, M. Schulz, K. P. Engelbrecht, J. Niemann, S. Möller, "All Users are (not) Equal – The Influence of User Characteristics on Perceived Quality, Modality Choice and Performance," in Proceedings of the Paralinguistic Information and its Integration in Spoken Dialogue Systems Workshop, R. L. C. Delgado, T. Kobayashi, Eds., New York: Springer, 2011, pp. 175-186.

[26] A. B. Naumann, I. Wechsung, and J. Hurtienne, "Multimodal Interaction: A suitable Strategy for Including Older Users?," in Interacting with Computers, vol. 22, no. 6, pp. 465-747, 2010.

[27] S. Chandra, M. Scott, W. Lin, and G. Ghinea, "Modelling the Influence of Personality and Culture on Affect and Enjoyment in Multimedia," in the Proceedings of the 6th International Conference on Affective Computing and Intelligent Interaction (ACII), Xi'an, China, 2016.

[28] Y. C. Yen, C. Y. Chu, S. L. Yeh, H. H. Chu, P. Huang, "Lab Experiment vs. Crowdsourcing: A Comparative User Study on Skype Call Quality," in the Proceedings of the 9th Asian Internet Engineering Conference (AINTEC), Chiang Mai, Thailand, 2013.

[29] T. Daengsi, N. Khitmoh, and P. Wuttidittachotti, "VoIP quality measurement: subjective VoIP quality estimation model for G.711 and G.729 based on native Thai users," in Multimedia Systems, vol. 22, no. 5, pp. 575-586, 2016.

[30] S. Ickin, K. Wac, M. Fiedler, L. Janowski, J. Hong, and A. Dey, "Factors influencing Quality of Experinece of commnly used mobile applications," IEEE Communication Magazine, pp. 48-56, 2012.

[31] R. Gupta, K. Laghari, H. Banville, T. H. Falk, "Using affective brain-computer interfaces to characterize human influential factors for speech quality-of-experience perception modelling," in Human-centric Computing and Information Sciences, vol. 6, no. 5, pp. 1-19, 2016.

[32] R. McColl and Y. Truong, "The Effects of Facial Attractiveness and Gender on Customer Evaluations During Web-Video Sales Ecounter," in Journal of Personal Selling and Sales Management, vol. 33, no. 1, pp. 117-128, 2013.

[33] K. M. Wolters, K. P. Engelbrecht, F. Gödde, S. Möller, A. Naumann, and R. Schleicher, "Making it easier for older people to talk to smart homes: The effect of early help prompts," Universal Access in the Information Society, vol. 9, pp. 311-325, 2010.

[34] I. Tjostheim, W. Leister, T. Schulz, N. Regnesentral, and A. Larssen, "The Role of Emotion and Enjoyment for QoE – a Case Study of a Science Centre Installation," in Proceedings of the 7th International Workshop on Quality of Multimedia Experience (QoMEX), Costa Navarino, Messinia, Greece, 2015.

[35] B. Weiss, S. Moller, and M. Schulz, "Modality of Preferences of Different User Groups," in Proceedings of the 5th International Conference on Advances in Computer-Human Interactions (ACHI), Valencia, Spain, 2012.

[36] S. Barakovic and L. Skorin-Kapov, "Multidimensional Modelling of Quality of Experience for Mobile Web Browsing," Computers in Human Behaviour, vol. 50, pp. 314-332, 2015.

[37] K. ur Rehman Laghari et al., "QoE Aware Service Delivery in Distributed Environment," in IEEE Workshopps of International Conference on Advanced Information Networking and Applications (WAINA), Biopolis, Singapore, 2011.

Gap in pagination due to withheld paper.

Pages 440-444

MIPRO 2017, May 22- 26, 2017, Opatija, Croatia

Development Trends of Telecommunications Metrics

Nataša Banović-Ćurguz and Dijana Ilišević
Telekomunikacije RS A.D., Banja Luka, Bosnia and Herzegovina
natasa.banovic-curguz@mtel.ba, dijana.ilisevic@mtel.ba

Abstract — **The transition from voice to data services has led to new trends in telecommunication business, and therefore the need to adjust legacy metrics and KPIs (Key Performance Indicators) that aims to point out the effects of these changes. In this paper, we give the review of development trends of new KPIs that provide greater insight into the use of data based services, as well as estimating the share of revenue that is obtained from these services. The importance of customer segmentation and correlation with KBOs (Key Business Objectives) is discussed. Special attention was paid to the situation of the market in Bosnia and Herzegovina and the need for trends monitoring in this area.**

I. INTRODUCTION

Telecommunications companies are faced with the key changes in the market such as the explosive growth of data, a constant increase in the number of smart mobile devices and new applications, as well as changes in customers' behavior and expectations. Although changes in the telecommunications industry brought increased demands for the transfer of data, there has been a decreasing in income and stagnation in the number of mobile users. Operators' revenues are under the pressure because of the increasing rise in sophisticated demands of users and regulators and due to increasing competition of OTT (Over-The-Top) providers. As the market becomes saturated, a new investment business model cannot be based on the penetration and the number of new users. The legacy services such as voice and SMS are only applications in the world of data transfer. Thus the legacy metrics (per-user), such as: average revenue per user ARPU (Average Revenue per User), Churn rate, and Net Additions, do not give a real picture of the network. These metrics are relevant when voice was the main service and when each customer has owned a single mobile device. However, as customers have multiple types of mobile devices and use a variety of services, the company do not provide that existing metrics to obtain a complete picture of how subscribers use their devices and services. Operators will need to track success with the right KPIs that can shape their strategy for the long period.

The necessity of adapting existing KPIs to dynamic changes in the market has prompted many authors to deal with this problem. This paper presents the development trends of new KPIs in telecommunications companies. The second section summarizes the characteristics of the modern telecommunications market. In the third chapter is given an overview of KPIs evolution from the period when voice was dominant service to the period of maturity of data services. After that the importance of customer segmentation is discussed. At the end is given a use case with the example of new metrics selection in the specific market, as well as the benefits and challenges of multi-dimensional metrics selection.

II. MARKET TRENDS

Modern telecommunications market is characterized by the following features:

- The use of data based services is becoming the dominant mobile service. Monthly global mobile data traffic will be 30.6 EB (Exabyte EB=1 billion GB) by 2020, [1].
- The increase of the amount of data transferred does not mean an increase in revenues from data based services, Fig. 1.
- The total number of smartphones will be nearly 50% of global devices and connections by 2020. Because of increased usage on smartphones, smart traffic will cross four-fifths of mobile data traffic by 2020, Fig. 2.

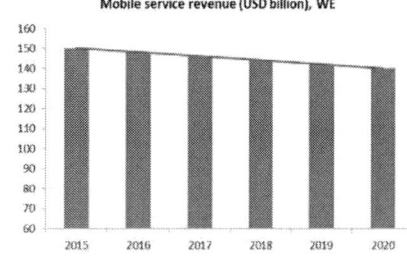

Figure 1. Mobile service revenue, [1].

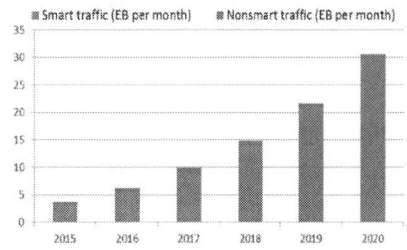

Figure 2. Smart/nonsmart traffic in EB, [1].

- 4G technology is becoming the dominant technology on global market with 40.5% of connections and 72% of total traffic, Fig.3.

445

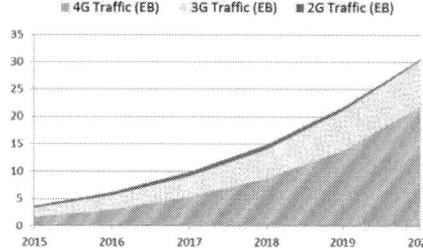

Figure 3. Global mobile traffic per connection type in EB, [1].

- The number of M2M (Machine-to-Machine) connections increases rapidly as the number of multi-SIM service users.
- The impact of OTT applications and providers is becoming increasingly serious.
- The improving of the user experience is becoming more and more important.

The evolution of the telecommunications market has led to an environment with the saturation of voice and the rapid growth of data, which has imposed the need to define new business models and development strategies. The growing use of services such as the exchange of multimedia content, social networking, games, shopping over the Internet, and meteoric growth of OTT applications and rapid penetration of smartphones drove out services such as voice, text and data for less demanding applications. Telecommunications companies have large investments in infrastructure and technology, but they do not have the expected profit. Most of the operators are based their business strategy on reducing churn, and they offer flat-rate tariff models, fixed broadband, various service packages etc., which led to great use services based on high-speed data for a very small fee. This has led to decreasing ARPU in all world markets. With the loss of revenue from voice service as the growing use of VoIP, operators found that urgently need to develop strategies that help to monetize the increasing use of data based services. One way is better understanding of customer behavior and more efficiently CEM (Customer Experience Management).

As it is shown in [2] analysis of customer behavior and manage customer experience have become priorities for telecommunications companies, and investments are focused on improving the user experience higher than ever. The operators are changing their business models from a network-oriented to customer-oriented models [3]. This situation resulted in the redefinition of metrics that were indicators of quality of service, and performance indicators of the company.

III. THE EVOLUTION OF METRICS AND KPIS

A. Limitations of legacy KPIs

The evolution of the market is followed by evolution of existing KPIs and metrics in the new metrics that provide greater insight into the use of data based services, as well as estimating the share of revenue that is obtained from these services. The increasing use of data based services does not mean higher revenue from these services.

According to [4], the limitations of existing KPIs are as follows:
- Network Coverage, the legacy metric, which is losing its relevance due to the high penetration of smartphones and high-tech innovation solutions in developed markets, which are already saturated.
- Minute of use services (MoU), the legacy metric, which losing its relevance due to the growing popularity of VoIP traffic, flat-rate tariff models and various service packages.
- ARPU (Average Revenue per User) is a KPI that represents the average revenue per user, which is not the appropriate metric for multi-SIM users.
- Net Mobile Connection Additions is the number of new users reduced by the number of users who have left the network. It is a legacy metric, which losing its relevance due to the decreasing number of new subscribers. New business users in VPN groups, multi-SIM users and M2M connections do not mean an increase in this KPI by its definition.
- Market Penetration was initially defined as a KPI that measures and presents growth of core services (voice and SMS) on the market, in order to access to the market potential for the introduction of new services. But, this legacy metric does not successfully represent the full potential of the market for new services.
- Because of market development and the transition to data services, users can not be viewed in the same way as earlier because of new trends in the growth of M2M connections and the use of multiple devices per person.

B. Development of new KPIs

The legacy metrics and KPIs in the telecommunications industry are subject to significant changes due to external factors, but also because of the internal strategy of the company. The external factors include: new technologies, competition of OTT providers, market regulation, price reduction, increasing number of service providers and MVNO (Mobile Virtual Network Operator), changing in customers' requirements. Internal strategies include new data-centric business models and new data services.

To overcome the limitations of existing KPIs that does not give a true picture of the emerging markets, in [4] and [5] are proposed the following KPIs:
- Data share of revenue indicates revenue share of data traffic in relation to the total revenue, which is very important since the service data represent a major part of the mobile operators offer.
- RGU (Revenue Generation Unit) metric scales revenue based on the number of services used by users (Multi-Service per User). This is very important from the point of analysis increase cross-selling and up-selling service offer.
- Since the MoU is losing its relevance, data usage per user becomes a metric that provides an insight on the amount of transferred traffic per user. This metric is particularly interesting from the point of billing.

- Machine to Machine Average Revenue per SIM - M2M ARP-SIM is an important metric to monitor the arrival of a growing share of M2M connections for that is expected a rapid growth in the future.
- As business metrics, instead of the ARPU, which has become irrelevant because of multiple devices per customer account, some operators as a measure of profitability of its offer measure the revenue with a new metric ARPA (Average Revenue per Account).

As an illustration, in Table 1 is presented the evolution of KPIs for wireless services in three phases: from the period in which voice was the dominant service to the period in which voice and data are equally presented, to the third phase of the evolution when data services are dominant with voice as an application. Acquisitions of customers, service delivery and business results are considered. For the first phase with voice as a primary service, as well as for the second stage of evolution when data services are getting more important, measuring the quality of service is still remained at the network level than at the application level. Only at the third stage of development services, new metrics are gaining in importance in the assessment of all aspects of the business, from the acquisition of customers to daily operations, [6].

TABLE 1: THE EVOLUTION OF METRICS AND KPIs.

	Customer acquisition	Service delivery	Business results
Voice only	Market share% User penetration % Net Additions	Network coverage% Voice quality Call drop rate% Call setup SR% Handover SR %	ARPU(voice)% MoU
Voice & Data	Churn rate% SAC/SRC (Subscriber acquisition/retention cost) % Smartphone penetration %	Data coverage% Data roaming coverage% Latency% Packet loss% Jitter % Data traffic per user	ARPU (data)% ARPA (per account) %
Data only (Voice as an application)	UES (User Experience Score) Smartphone/tablet penetration % Number of M2M connections	QoS (Quality of service) per application QoE (Quality of Experience) Quality of coverage Content per device	Data share of revenue% RGU (Revenue Generating Unit) Revenue per application Service usage per user% M2M ARP-SIM

Operators should consider a holistic approach to measuring service performance, because the end-to-end service quality not only relies on the performance of the network, but also other factors such as device and application performance. In a multi-dimensional approach, metrics and KPIs about network performance as well as segmentation of customer data (IMSI, time of the day, geo-location, age group, type of device, type of application, usage, habits, price sensitivity) etc. are necessary to enable operators to effectively assure services.

C. Customer segmentation

In [6] is proposed that operators have to access the segmentation of their customer base, and to control and analyze user behavior and expectations. When one user has multiple mobile devices and access to more services, operators must provide consinstent user experience. Information about the type of content can be obtained from measurements such as the contents of the device (Content per Device) and the number of M2M connections per user (M2M Connections per User). Traditionally, customers were segmented on postpaid and prepaid customers. Another type of segmentation is per tariff models to business customers and individuals what allowed the operators a better understanding of the expected quality. The third type of segmentation is based on factors that are specific to the use of the service (per usage). To define a user-experience score, factors to consider include the user's location, age group, price sensitivity, and frequently used applications. Network operators can construct a well-planned strategy to optimize their networks based on the regional user-experience score, while giving customers the best experience. These factors are as follows:

- Location: The types of applications and the level of service quality expected vary vastly between locations. The location can be seen as an urban, suburban or rural, or as indoor or outdoor. Data transmission has a different meaning in relation to the situation when the voice was the dominant service. That is because data has a visual aspect and the probability of accessing data services is greater in an indoor than an outdoor setting. For example, the probability of a request for transfer of data and the expected quality is the highest in indoor urban locations.
- Age group: Subscribers can be segmented by generations into seniors (birth year before 1946), baby boomers (birth year 1946-mid 1960s), Gen X (birth year mid 1960s-early1980s), Gen Y or millennials (birth year 1980s-2000s) and Gen Z (birth year after 2000s). This segmentation can give to operator good insights into the quality of service the network must deliver to the most important age group in that region, Figure 4.

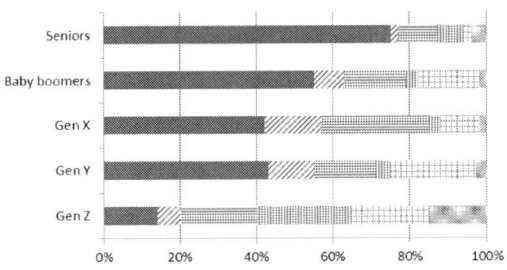

Figure 4. Segmentation per age group, [6].

- Price sensitivity: According to the consumption subscribers are divided into VIP customers and others who are viewed through consumption and habits, which correlates with the expected quality. For example, for a group of business users in a specific region the operator must provide the highest quality services and the most reliable secure connection.

- Usage pattern: According to the use of the service users are divided into: users who predominantly use voice as a service and minimally data. Light users are checking e-mails and browsing the Internet. Medium users actively use social networks and often access the network. Heavy users extensively use video streaming and OTT applications.

Operators also can benefit from network analytics using customer micro-segmentation. With rapidly changing customer behavior and expectations, and increasing demand for personalization, operators need a better way to segment the market. One way is micro-segmentation. Micro-segments gather considerably smaller numbers of customers into groups based on application usage, device usage and geo-location. Operators can be better served their customers in various areas of their interest such as handling complaints, troubleshooting QoE issues, or providing micro-services tailored to their exact needs, [7]. Customer micro-segmentation technology empowers marketers to achieve deeper customer understanding and more effective customer marketing. Based on above given factors and using appropriate metrics and KPIs including user-experience score, the operator can get the insights for optimizing and tuning the network in accordance with the requirements of the users and identify services to target user groups, [8].

IV. CASE STUDY - SELECTION OF NEW METRICS

The selection of new metrics is affected by numerous factors which are determined by the specifics of the local and regional markets, technological developments and the competitive environment. In accordance with that appropriate KPIs are selected. The market in Bosnia and Herzegovina is very specific, [9]. It is characterized by the following factors: the three dominant national operators, the only telecommunications market in the region that is not implemented 4G/LTE technology, weak economic development and low ARPU in relation to the average ARPU in EU. At the end of 2015 the level of penetration of fixed telephony amounted to 20.45% and in the mobile telephony 89.63% with the dominant share of pre-paid users. There has been growth in the number of Internet users and at the end of 2015, penetration was 72.41%. Broadband services are increasing, so the number of broadband subscribers reached 99.72% of the total number of Internet subscribers. The most subscribers have access speeds of 2-10 Mbit/s. Data traffic is growing continuously, and the number of smartphones and OTT providers and applications are increasingly common in the market. Taking into account the above factors, it can be concluded that there is potential for further growth of the market. In that context, as the indicators for this specific emerging market can be taken the KPIs that are given in Table 2. They were chosen with a view to control customers and networks from the perspective of efficiency, the use of service and business results.

TABLE 2: THE EXAMPLE OF NEW KPI SELECTION FOR B&H OPERATORS.

	Customers	Network
Customer acquisition & Network Efficiency	Smartphone/tablet penetration % UES (CEM metrics) Churn rate% Call center/Customer care department availability%	Availability % Call/session setup SR % Call/session drop rate % BTS down time % Paging channel & RRC congestion %
Service delivery	QoE QoS per application Data coverage% Data traffic per user Service usage per user	Network utilization % On-net traffic % OTT share % The number of connections with good voice quality
Business results	RGU per user Data share of revenue% M2M ARP-SIM	Revenue per cell/location FTTH penetration % VDSL penetration % IPTV penetration %

Each of KPIs from Table 2 has to be correlated to the strategic objectives of the company described with Key Business Objectives (KBO): improving business processes, improving service quality and customer experience, increasing sales and operating income, strengthening the brand etc. Previously listed key performance indicators are the detailed indicators measured in real time that are measurable and support directly the KBO via Key Performance Objectives KPO, [10]. The example of that KPI-KBO mapping is given in Table 3.

TABLE 3: KBO-KPO-KPI MAPPING.

Key Business Objective - KBO	Key Performance Objective - KPO	Key Performance Indicator - KPI
Business processes improvement	Efficiency Productivity Availability Innovation Downtime	Availability% Utilization% On net traffic% Service usage per user% BTS down time
Quality of service improvement	QoS Accessibility Coverage Retainability Mobility Reliability	QoS per application Quality of coverage% Data coverage% Paging SR% RRC Congestion% Call/session setup SR% Call/session drop rate % Number of connections with good voice quality
Sales increase (growth)	Penetration Retention (Up selling, Cross selling)	IPTV penetration% FTTH penetration% VDSL penetration% Smartphone penetration% RGU per user%
Revenue growth	Integrity Reliability Retention	Churn rate% Revenue per cell M2M ARP SIM Data traffic per user Data share of revenue% Share of OTT data in total data% ASR (Answer to Seizure Ratio)%
Improvement of customer experience	QoE Consistency Notification Security	QoE Call center/customer care availability% UES metrics: Net Promoter Score NPS, Customer Satisfaction Score CSA, Customer Effort Score CES

448

Some of these metrics are already measured using tools for monitoring signaling and network control. The examples of service based metrics are given in the Figure 5. They are related to QoS per application and were obtained in real network environment using MasterClaw monitoring tool.

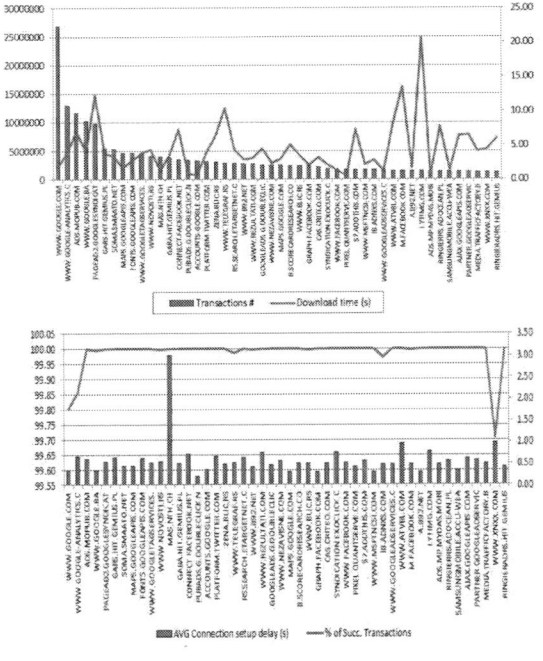

Figure 5. The examples of service based metrics, QoS per application.

V. BENEFITS AND CHALLENGES OF MULTI-DIMENSIONAL METRICS SELECTION

A multi-dimensional approach to monitoring, metrics selection and assuring the end user's service quality can be the differentiator on competitive telecommunication market. Advanced analytics capabilities can support better capacity planning and traffic management as well as more effective service assurance to deliver a customer experience that retains subscribers and increases revenue. This approach should consider the following aspects: end-to-end network performance, including the end device, access, core network, the application servers, geo-location, age profile, the applications or services accessed by the customer, the customer's identity, the device being used and the time period of usage. Since operators have large amounts of data about their customers, they can model the behavior of their users based of these data. In this area there are numerous challenges, because the data are collected and aggregated from various systems and processes. To take advantage of the enormous potential of this information, the operator must have a significant investment in the areas of data mining, warehousing and Big Data solutions.

Specificities of the market that affect the selection of appropriate KPIs are also barriers to compare the performance of telecommunications companies across the different markets. To avoid this disadvantage it is necessary to adopt a consensus on the choice of KPIs that are consistently able to follow on a global level. In this sense, it is not expected only from regulatory authorities to impose new metrics and methods of their measurements, but also the initiative of the operators, investors and market analysts to support the move toward greater consistency.

VI. CONCLUSION

In accordance with changes in the market, operators ·must select the appropriate metrics and KPIs that will follow new trends in the telecommunications industry to achieve long-term strategic goals and targets of the company. The selection of suitable KPIs depends on the operator's ability to actually measure the indicators. KPIs that take into account the specificities of the market and customer requirements for quality of service are crucial in differentiating a company in a competitive environment. In very specific B&H telecommunication market it is very important continually monitoring the trends in this area and investing in appropriate tools that can support the implementation of new metrics and KPIs. Only a multi-dimensional approach in network analytics across multiple capabilities ensures a true end-to-end view of the service, enabling operators to improve service quality and overall customer experience. Modelling of new business processes and improving of KPI-KBO correlation are the challenges in our future work with the intention to realize greater business value.

REFERENCES

[1] "Cisco Visual Networking Index: Global Mobile Data Traffic Forecast Update, 2015–2020.", February 2016.
[2] Mika Uustalo, "Customer Experience Management in Telecom Operator Business", Master's thesis, Helsinki Metropolia University of Applied Sciences, May 2012.
[3] N. Banovic-Curguz and D. Ilisevic, "Moving from Network-Centric toward Customer-Centric CSPs in Bosnia and Herzegovina", Mipro, Opatija, May 2016.
[4] J. Sujata and other, "Transforming Telecom Business: Scaling the Shift using Predictive Analytics", Indian Journal of Science and Technology, Vol. 8, February 2015, pp. 34-43.
[5] http://www.ey.com/Publication/vwLUAssets/Metrics_transformatio n_in_telecommunications/SFILE/Metrics_transformation_in_teleco mmunications_EF0117.pdf, 2013.
[6] S. Ranganna and other, "New metrics for a changing industry", http://www.pwc.com/gx/en/industries/communications/publication s/communications-review/new-metrics.html, 2014.
[7] A.Rao "Improving mobile customer retention through multi-dimensional analysis of QoE", January 2016. www.analysismason.com
[8] https://www.scribd.com/document/257602372/Analysys-Mason-CEM-Framework-Retain-Apr2013-RMA15
[9] "Annual report of Communication Regulator Agency", http://www.rak.ba, 2016.
[10] ITU-T E.419 "Business oriented key performance indicators for management of networks and services".

Implementation and Testing of Cisco IP SLA in Smart Grid Environments

J. Horalek, F. Holik, V. Hurtova

University of Pardubice, Faculty of electrical engineering and informatics Pardubice, Czech Republic
josef.horalek@upce.cz, filip.holik@student.upce.cz

Abstract - Smart grid networks are becoming more and more commonly deployed due to their undisputed benefits. On the other hand, there is a high demand for reliability and functionality of these networks. This paper is analysing usage of the IP SLA for monitoring network state and collecting important information for potential problem detection and solving. The practical part of the paper presents implementation of the IP SLA into the smart grid network environment and its testing. The results from several simulated scenarios with different QoS classes, used within the smart grid networks, are discussed.

I. INTRODUCTION

Securing performance parameters and monitoring of network functionalities are two of the most important tasks of intelligent networks, in order to ensure their effective usage. Most of the backbone and local smart grid networks are using MPLS L3 VPN [1-5] due to its reliability and security. Implementation and testing of this technology in smart grid networks is the main point of this paper. Presented measurement can be used for confirming the SLA and for proactive problem solving of potential issues. The goal of the paper is to prove, if the Cisco proprietary solution for measuring performance parameters is a suitable technology in the industrial environment of smart grid networks built on Cisco devices [6,7]. Two types of measurements were conducted. In the first case, the priority class with g.729a codec, simulating demanding communication of Intelligent Electronic Devices (IEDs); and in the second case, default class simulating standard communication of the same devices.

The paper is further organized as follows: the second section describes the implementation details of the Cisco IP SLA. This implementation is then thoroughly tested in the third section. The paper is concluded in the last section.

II. IMPLEMENTATION OF THE CISCO IP SLA

Cisco Internet Protocol Service Level Agreement (IP SLA) is a proprietary technology introduced by Cisco for effective monitoring of network traffic. It can be used for measuring network performance and performance critical parameters like packet loss, delay, and jitter. IP SLA can therefore detect and prevent problems, which can influence network functionality and performance [8,9]. This is one of the most important tasks in the environment of intelligent energetical networks. Effective monitoring and measurement of the complete network can be done using Cisco RTTMON with management information

base (MIB) together with SNMP and IP SLA statistics. IP SLA can also be used for policy-based routing. This type of routing can adjust the direction of packet flows based on actual statistics, and therefore better utilize each link and ensure availability of critical parts of the network.

In order to conduct a measurement, the topology has to contain one Cisco router for packet generation (monitor) and one host, acting as a responder. The responder can be any IED with IP address [10], able to reply to requests (ICMP echo, or HTTP GET). These devices are common in smart grid networks. In the case that the responder is also a Cisco router (in the energetical networks typically a gateway between different areas), IP SLA can be better utilized because a larger number of critical parameters can be measured. The following data collection and presentation is realized with Network Management System (NMS). After successful configuration, the router is collecting results of each operation and save the results in a form of IOS RTTMON statistics. The router is then using SNMP NMS to collect proper information from MIB. From the technology perspective, IP SLA is using a concept displayed in Figure 1. Every operation is defining a type of packet generated by the router, source and destination address, and other values. The configuration also contains time, when each operation should be executed.

A. IP SLA Monitor (Generator)

Tests are defined on the IP SLA monitor. Based on the configured parameters of each test, the IP SLA is generating specific traffic, analysing the results and saving them for a future analysis over CLI or SNMP. The IP SLA monitor can be every Cisco router having IOS with a proper set of functions, depending on the chosen type of test. Processor load on the IP SLA monitor is a critical part for measuring different metrics, especially for recording timestamps. For this reason, a proper methodology has to be used in order not to exceed 30% of the router's CPU utilization. It is therefore recommended to use a dedicated router just for the measurement, so the data traffic would not be influenced and the measurement will get more precise results.

B. IP SLA Responder

IP SLA responder is reacting on tests generated by the IP SLA monitor. The responder creates timestamps with packet received and packet send time and then includes them in the payload. These timestamps will allow the elimination of processing time on the responder from the

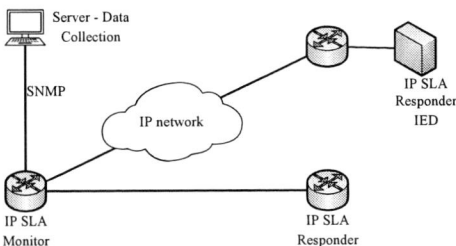

Figure 1. Principles of IP SLA monitoring

final measurement time as shown in Figure 2. As in the case of the monitor, the CPU utilization of the responder should not exceed 30%, so it is important to carefully choose the testing methodology.

$$RTT = T4 - T1 - \Delta \qquad (1)$$

where: RTT = Round trip time
         T1 = Timestamp 1
         T4 = Timestamp 4

C. Multioperations Scheduler of IP SLA

Cisco IP SLA allows the use of a multioperations scheduler, which can monitor complex networks containing large number of probes, and is ideal for smart grid networks (containing tens to hundreds of IEDs). This scheduler can be turned on with the "ip sla group" IOS command. The scheduler allows the planning of a sets of IP SLA operations, which allows the monitoring of traffic in a uniformly distributed timeframe. The realization requires the specification of a range (ID) of each probe and the function can then being run at once. This feature helps minimize the CPU utilization and therefore to increase the network scalability. The function is using the following configuration parameters:

- Operation ID numbers – the list of all IP SLA probes and their IDs within a particular group.

- Group operation number – configuration parameter, containing the number of a particular group.

- Schedule period – the amount of time, for which the group of IP SLA operations is planned.

- Ageout – specify for how long the operations actively collecting information are held in a memory.

- Life – the amount of time for operation to actively collect information.

- Frequency – time after which every IP SLA is repeated.

- Start time – a time when the operation will start to collect information.

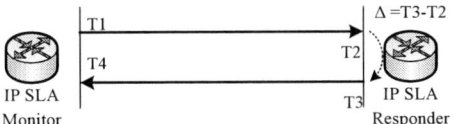

Figure 2. The system timestamps

TABLE I.   PARAMETERS OF UDP JITTER TEST

Parameter	SLA-Voice settings	SLA-Normal settings
codec	g.729a	none
packet size	32B	20B+12B
the number of packets	550	100
interval	100 ms	500 ms
frequency	60 s default	60 s default
timeout	5000 ms default	5000 ms default
threshold	5000 ms default	5000 ms default
type of service	184(EF)	0

III.   THE TESTING

A. The Methodology of Testing

The tests were based on ICMP and UDP implemented in IOS IP SLA. Because most of the smart grid networks are based on MPLS, the UDP jitter operation was selected for testing. This operation is primarily used for diagnosis of real-time application availability, which is essential for smart grid networks. This type of test is also the only one, able to measure with micro-seconds precision, which is important for critical infrastructure. The UDP jitter test is generating sequential information and timestamps for both the sending and the receiving sides. We have chosen two variants of UDP jitter for measuring performance metrics in MPLS L3 VPN infrastructure. These two tests can relevantly simulate proper data flows in smart grid networks. This includes link congestion when collecting data from IED devices and high priority control commands. These tests are:

- UDP jitter with g.729a codec, which is used for measuring in a priority class in a priority traffic (SLA-Voice).

- UDP jitter without a codec, used for measuring in non-prioritized classes (SLA-Normal).

In the case of SLA-Voice variant, data traffic can be separated from control traffic, making the measurement more relevant. In the case of SLA-Normal variant, direct behaviour in the class using a class-default queue can be observed. This corresponds with the process of IED data collection.

Table 1 shows that in the SLA-Voice variant, packets with 32B size will be generated for a 55 seconds with 100 ms intervals between packets and 5 seconds space between each test. This means, that the measurement is taking more than 90% from the complete measurement time length of 60 seconds. This test is not influenced by data flows between end devices like IEDs, but only by the link state. The test will be used only for measurement between substations due to its complexity.

In the SLA-Normal variant, packets with 32B length will be generated for 50 s with 500 ms intervals between

451

each packet and 10 seconds break between each test. This will ensure, that the measurement will run in more than 80% of the total timeframe of 60 seconds. This test is used for simulation of consumer traffic and parameters in the class-default (with DSCP CS0). This test will be applied in all experiments.

B. Design and Parameters

The measurement was realized on the testing topology depicted in the Figure 3.

Measurement type Provider Edge (PE) - Customer Edge (CE) measured SLA metrics between end stations, CE routers and core PE routers. For this type of measurement, hub and spokes design was chosen. The hub was represented by the IP SLA router (PE router) and spokes were represented by the substation end routers (CE routers). The goal of this measurement was to detect problems with communication technology, which can be rented from an external provider of network connectivity. This measurement can be therefore used for solving connectivity problems (like QoS transparency or packet loss) with the external connectivity provider. In the PE-CE type of measurement, only the SLA-Normal variant was chosen due to the possibility of high CPU utilization on the SLA monitor. This would result in a large increase of identical measurements. Classification of the IPv4 traffic will be conducted based on a DSCP parameter located in the IP header. QoS configuration was done using the same approach as in the case of a service provider – with the minimal setting. As a use case, the consumer would order services and the required speed. In this case, two policy-maps were chosen, with the direction to each substation, and one policy-map of the consumers input (Table 2). One of the policy-maps contains class-default shaping and it is parent for the second policy-map. The second policy map contains classes for traffic marking.

Traffic coming from a customer will not be remarked at the PE. Traffic in the SLA-Voice class will go through input policing, discarding all the traffic exceeding 50% of the total bandwidth of the interface. This is a standard measure in MPLS networks.

Measurement type Provider Edge (PE) - Provider Edge (PE) measured SLA metrics amongst PE routers in the MPLS network. This type of measurement used full-mesh, where every PE router was connected to all the other PE routers. From the MPLS point of view, the measurement was not realized on the global routing process level. Instead, the dedicated MPLS VPN was created, so tests in each core QoS classes could be defined. This test was aimed at detecting and solving problems within the core infrastructure. QoS in the core

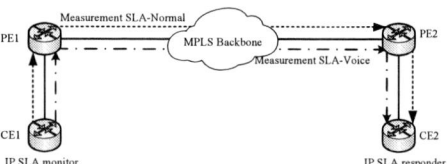

Figure 3. Topology design

infrastructure was similar to PE-CE. The OUT-MPLS policy-map was created and applied on the output interfaces between PE routers. This policy-map contained three classes displayed in the Table 3.

Measurement type Customer Edge (CE) - Customer Edge (CE) was used for measurement of SLA metrics between end points and IEDs. The whole link was therefore monitored – from a single substation, via the whole infrastructure of the provider, to the next substation. This test can be used for the specification of maximum latency, jitter, or packet loss between the central system and a substation. In our case, the router CE1 was used as the monitor and CE2 as the responder. CE1 generated both types of measurements (SLA-Voice and SLA-Normal).

C. Results

Every scenario for diagnosis of Cisco IP SLA behaviour in the environment of smart grid networks on a simulated topology of energetical company, was tested after the configuration. A reference values were collected during the standard traffic. Tables 4 and 5 show parameters gathered from IP SLA probes. Measured data also shows times when the operation was conducted, the number of successful and unsuccessful operations, and the lifetime of the operation. Lastly, the one-way statistics are also available, allowing to analyse information for solving conn connectivity problems of the transport network.

TABLE II.   CLASS SERVICE FOR CUSTOMERS

Class	DSCP	CoS/EXP	Note
SLA-Voice	CS3, EF	5	Priority class
Critical	CS6	3	Critical traffic, packet loss sensitive
class-default	0	0	Other traffic

TABLE III.   CLASSES OF OPERATIONS FOR BACKBONE TRAFFIC

Class	DSCP	Co S/E XP	Guaranteed bandwidth / exceed action
SLA-Voice	CS3, EF	5	50% (of the total BW) / Packet drop
Critical	CS6	3	Remaining 60% / Can exceed if the capacity is available
class-default	0	0	Remaining 40% / Can exceed if the capacity is available

TABLE IV.   SUMMARY TABLE FOR MEASURING SLA-VOICE VARIANT OF NORMAL DATA TRAFFIC

SLA Voice	Measurement PE1-PE2	Measurement CE1-CE2
RTT (avg)	25 ms	35 ms
Latency S->D	7 ms	19 ms
Latency D->S	28 ms	16 ms
Jitter S->D	14 ms	8 ms
Jitter D->S	7 ms	6 ms
Packet loss	0	0
Mean Opinion Score	4,06	4,06
IPCIF	11	11

IPCIF = Calculated Planning Impairment Factor

Results for the scenario: Utilization data lines

The next scenario tested data link usage by an IED. In the high utilization, latency will increase, jitter will fluctuate more, and there can be some packet loss, influencing functionality and effectivity of data centrals. QoS policy with the 5Mbit bandwidth between PE1-CE1 and PE2-CE2 was used for sufficient trustworthiness of the measurement. Data traffic was simulated using pings with the size of 1500 bytes and 0 time limit for the reply. This ensured a link congestion resulting in packet drops in a direction between CE1 and PE1.

Table 6 shows, that in a default class, there is packet loss due to the high traffic load. Unlike in original values, latency and jitter also increased. A direction in which packets are lost can be also detected – in our case it is from CE1 to CE2. Priority class SLA-Voice shows almost no change, proving good conditions of the link without any packet drops in a core or transit infrastructure.

Results for the scenario: Utilization of voice lines

The next tested scenario is focusing on lowering quality of the priority line – the voice in our case. This situation can happen if the communication between IEDs is using more bandwidth than what is assigned to the prioritized traffic. Bandwidth in the test was set to 100Kbit. A typical data flow with G729a codec is using approximately 32Kbit/s. That means, that three parallel transmissions can be realized at once and be fully functional. Simulation was again conducted with a ping tool and packets marked with DSCP EF. Collected data shows, that the priority class became saturated and packets from this class were dropped. The default class transferred practically no traffic, so there were no packets dropped there. The Voice class had a priority over the class-default, resulting in a possibility of delayed packets in the class-default. Despite the possible delay, no packets were dropped in this class.

The results of PE1-PE2 measurement in the SLA-Voice class are present in Table 7 and show increased latency. This however presents only a simulated situation, in the real environment, such traffic should not influence core infrastructure.

The measurement of SLA-Voice between CE1 and CE2 (Table 8.) shows decreased performance parameters for voice technologies and consequent packet loss. Latency rapidly increased to an average of 146 ms, 14 packets were lost, and MOS decreased while IPCIF increased.

On the other hand, SLA-Normal measurement between CE1 and CE2 clearly shows no packet loss. But as already mentioned, the situation where latency in the SLA-Voice priority class will increase, can happen as it happened in our case – to the average of 147 ms.

In both cases it is clear, that the voice traffic was generated from the consumer CE1, because the data drops and latency increased in the direction from the source (CE1) to the destination. This measurement evaluated each state, which can happen on a link. We can then detect in which traffic class is a potential problem and therefore to proactively react.

TABLE V.   SLA-NORMAL VARIANT OF NORMAL DATA TRAFFIC

SLA Normal	Measurement PE1-CE1	Measurement CE1-CE2
RTT (avg)	17 ms	41 ms
Latency S->D	10 ms	29 ms
Latency D->S	10 ms	12 ms
Jitter S->D	12 ms	7 ms
Jitter D->S	7 ms	7 ms
Packet loss	0	0

TABLE VI.   A COMPARISON OF NORMAL AND LOADED STATE

CE1-CE2	Normal state		Loaded state	
Parameter	SLA Normal	SLA Voice	SLA Normal	SLA Voice
RTT (avg)	41 ms	35 ms	211 ms	34 ms
Latency S->D	29 ms	19 ms	165 ms	19 ms
Latency D->S	12 ms	16 ms	45 ms	15 ms
Jitter S->D	7 ms	8 ms	7 ms	13 ms
Jitter D->S	7 ms	6 ms	8 ms	11 ms
Packet loss	0	0	23	0
Packet loss S->D	0	0	23	0
Packet loss D->S	0	0	0	0
MOS	X	4,06	X	4,06
IPCIF	X	11	X	11

TABLE VII.   PE1-PE2 COMMUNICATION

PE1-PE2	Normal state	Loaded state
Parameter	SLA Voice	SLA Voice
RTT (avg)	25 ms	60 ms
Latency S->D	7 ms	28 ms
Latency D->S	28 ms	32 ms
Jitter S->D	14 ms	5 ms
Jitter D->S	7 ms	5 ms
Packet loss	0	0
Packet loss S->D	0	0
Packet loss D->S	0	0
MOS	4,06	4,03
IPCIF	11	12

TABLE VIII.   CE1-CE2 COMMUNICATION

CE1-CE2	Normal state			Loaded state
Parameter	SLA Normal	SLA Voice	SLA Normal	SLA Voice
RTT (avg)	41ms	35ms	147ms	146ms
Latency S->D	29ms	19ms	101ms	103ms
Latency D->S	12ms	16ms	47ms	44ms
Jitter S->D	7ms	8ms	9	6ms
Jitter D->S	7ms	6ms	9ms	5ms
Packet loss	0	0	0	14
Packet loss S->D	0	0	0	14
Packet loss D->S	0	0	0	0
MOS	X	4,06	X	3,12
IPCIF	X	11	X	20

IV. CONCLUSION

The goal of the paper was to show suitability of Cisco IP SLA implementation in the intelligent environment of smart grid networks. These networks, built on MPLS technology are realizing access into each sub-areas of the smart grid and also providing core data traffic forwarding. Measuring performance characteristics with the IP SLA is important for solution of problems, which can happen in these networks. The conducted measurement scenarios and their results clearly shows, that the IP SLA is a very effective technology for problem solving, while at the same time is providing detailed information about different communication parameters for various data types. This effect was tested during parameter measurement in specific traffic types. It was proven, that the IP SLA is a very complex tool for network monitoring. It allows us to supervise large number of services and traffic commonly used within intelligent networks like smart grid.

ACKNOWLEDGMENT

This work and contribution is supported by the project of the student grant competition of the University of Pardubice, Faculty of Electrical Engineering and Informatics.

REFERENCES

[1] R. Froom, B. Sivasubramanian, and E. Frahim, "Implementing Cisco IP switched networks (SWITCH): foundation learning guide," Indianapolis: Cisco Press, 2010. ISBN 978-1-58705-884-4.

[2] A. Indurkar, A. Patil, and B. Pathak, "Performance failure detection and path computation," in: Proceedings - IEEE International Conference on Information Processing, ICIP, 2015, art. no. 7489357, pp. 90-95.

[3] A. Rayes, and K. Sage, "Integrated management architecture for IP-based networks," in: IEEE Communications Magazine, 2000, pp. 48-53. DOI: 10.1109/35.833556. ISSN 0163-6804

[4] H. Nemati, A. Singhvi, N. Kara, and M. El Barachi, "Adaptive SLA-based elasticity management algorithms for a virtualized IP multimedia subsystem," in: 2014 IEEE Globecom Workshops, GC Wkshps 2014, art. no. 7063377, pp. 7-11.

[5] M. Jiang, J. Byrne, K. Molka, D. Armstrong, K. Djemame, and T. Kirkham, "Cost and risk aware support for Cloud SLAs," in: CLOSER 2013 - Proceedings of the 3rd International Conference on Cloud Computing and Services Science, 2013, pp. 207-212.

[6] V. Gungor, D. Sahin, T. Kocak, S. Ergut, C. Buccella, C. Cecati, and G. Hancke, "Smart Grid Technologies: Communication Technologies and Standards," in: IEEE Transactions on Industrial Informatics. 2011, 7(4), pp. 529-539.

[7] L. Wenpeng D. Sharp, and S. Lancashire, "Smart grid communication network capacity planning for power utilities," in: IEEE PES T&D, 2010 pp. 1-4. DOI: 10.1109/TDC.2010.5484223. ISBN 978-1-4244-6546-0.

[8] F. Holik, J. Horalek, S. Neradova, S. Zitta, and O. Marik, "The deployment of Security Information and Event Management in cloud infrastructure," in: 2015 25th International Conference Radioelektronika (RADIOELEKTRONIKA), 2015, pp. 399-404. DOI: 10.1109/RADIOELEK.2015.7128982. ISBN 978-1-4799-8117-5.

[9] F. Holik, J. Horalek, S. Neradova, S. Zitta, and M. Novak, "Methods of deploying security standards in a business environment," in: 2015 25th International Conference Radioelektronika (RADIOELEKTRONIKA) 2015, pp. 411-414. DOI: 10.1109/RADIOELEK.2015.7128984. ISBN 978-1-4799-8117-5

[10] J. Horalek, J. Matyska, V. Sobeslav, and P. Suba, "Energy efficiency measurements of data center systems," in: 2014 ELEKTRO. pp. 41-45, DOI: 10.1109/ELEKTRO.2014.6847868. ISBN 978-1-4799-3721-9.

MIPRO 2017, May 22- 26, 2017, Opatija, Croatia

Fault management and Management Information Base (MIB)

O. Jukić, I. Heđi, I. Špeh
Virovitica College, Virovitica, Republic of Croatia
{oliver.jukic | ivan.hedi | ivan.speh}@vsmti.hr

Abstract - Fault management function mostly rely on SNMP protocol and SNMP agent capabilities, where SNMP agents are implemented over different network elements. Structure of management information is described in Management Information Base (MIB) for every single SNMP agent class. Hence, integrated management systems needs to implement different SNMP managers capable to communicate with number of different SNMP agents, obtaining management information from real telecommunication network. Those managers are often called access modules.

In order to decrease delivery time for SNMP management solutions, number of generic fault management MIBs are designed (e.g. Alarm Management Information Base). Theoretically, if different SNMP agents rely on the same management information base, only one SNMP manager (access module) is enough to manage different network elements. It is very logical way of thinking, if we consider that in real life different network elements share the same management information structure. For instance, alarm message from many different network elements consists of start time, probable cause, additional information, affected network element, etc. However, state of the art is not so idyllic, from the integrator's point of view. Number of MIBs exist in real telecommunications network. In this paper we will make short research on MIBs implemented on different network elements, in order to detect real integrator position in network management solution's development. At the end, solution is proposed for SNMP agent software development.

I. INTRODUCTION

"Fault management is one of the most relevant functional areas when we are talking about customer experience of service quality. When problem appears, network operator's reaction time depends on many factors. One of the most important is to recognize problem root-cause, based on unsolicited events coming from the network" [1].

Two crucial terms within network management functional area are *Manager* and *Agent*. Their interaction is rather simple: manager controls number of managed objects via agent located close to those managed objects. That is way how management of network resources is achieved [2]. Manager-agent interaction is shown on figure 1.

Manager is software application serving as interface between the real world (e.g. network personnel dedicated for network management) and management information that agent will understood. Network management applications include some kind of graphical user interface usually. Management information is generally standardized and available to any network manager,

respecting security issues (e.g. passwords). Agent acts as an interface between management information and real telecommunications network.

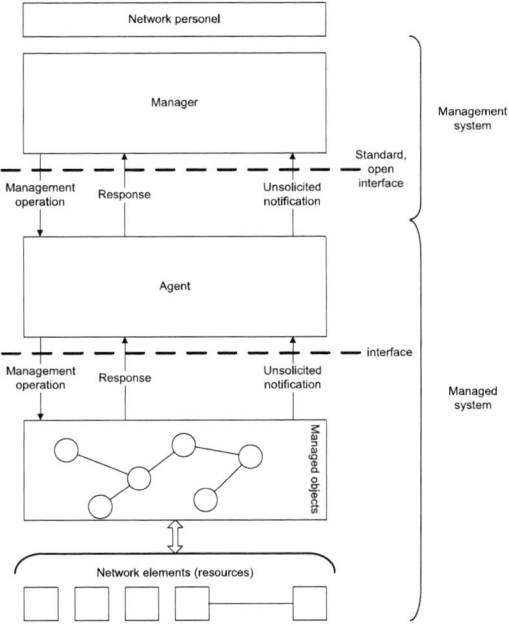

Figure 1.   Manager-agent concept

Manager and agent communicate to each other interchanging management information. Manager sends management commands, while agent receives that commands. Agent interprets commands, mapping them into suitable format and forwards commands to managed network objects (resources).

When response from network resources is received by agent, information received is sent back to the manager. Agent serves as a proxy, converting management information to managed objects.

This way of communication is synchronous. Manager sends management messages (e.g. GET) and receives responses. There is asynchronous way of communication also. In that case, agent will send message on their own, spontaneously, under some specific circumstances. Typical situations are managed object failures (alarms) or change of managed object configuration. Those messages are very often called "unsolicited messages". Sometimes,

455

they are called event reports. Unsolicited messages sending implicates agent's built-in intelligence.

Manager is sometimes called managing system, while agent is called managed system [3].

For consistent manager-agent communication, following conditions must be fulfilled:

- there is communication infrastructure between manager and agent,

- communication protocol is defined,

- management information format is precisely defined and it is known to both manager and agent.

A. Simple Network Management Protocol

SNMP (Simple Network Management Protocol) is *de facto* standard in network management domain. It allows very simple set of network management operations [4]. First version, SNMPv1, was developed in 1988th by IETF (*Internet Engineering Task Force*). Next version, SNMPv2, is improved, but with a security issue unsolved completely. Finally, third version, SNMPv3, coped with that security issue.

```
-- The Interfaces table contains information on the entity's
-- interfaces. Each sub-layer below the internetwork-layer
-- of a network interface is considered to be an interface.
ifTable OBJECT-TYPE
    SYNTAX       SEQUENCE OF IfEntry
    MAX-ACCESS   not-accessible
    STATUS       current
    DESCRIPTION
           "A list of interface entries.  The number of entries
           is given by the value of ifNumber."
    ::= { interfaces 2 }

ifEntry OBJECT-TYPE
    SYNTAX       IfEntry
    MAX-ACCESS   not-accessible
    STATUS       current
    DESCRIPTION
           "An entry containing management information applicable
           to a particular interface."
    INDEX    { ifIndex }
    ::= { ifTable 1 }

IfEntry ::=
    SEQUENCE {
        ifIndex            InterfaceIndex,
        ifDescr            DisplayString,
        ifType             IANAifType,
        ifMtu              Integer32,
        ifSpeed            Gauge32,
        ifPhysAddress      PhysAddress,
        ifAdminStatus      INTEGER,
        ifOperStatus       INTEGER,
        ifLastChange       TimeTicks,
        ifInOctets         Counter32,
        ifInUcastPkts      Counter32,
        ifInNUcastPkts     Counter32,  -- deprecated
        ifInDiscards       Counter32,
        ifInErrors         Counter32,
        ifInUnknownProtos  Counter32,
        ifOutOctets        Counter32,
        ifOutUcastPkts     Counter32,
        ifOutNUcastPkts    Counter32,  -- deprecated
        ifOutDiscards      Counter32,
        ifOutErrors        Counter32,
        ifOutQLen          Gauge32,    -- deprecated
        ifSpecific         OBJECT IDENTIFIER -- deprecated
    }
```

Figure 2.   MIB - sample

SNMP network management is based on manager-agent model. Terms SNMP manager and SNMP agent are introduced. In order to be capable to communicate and cooperate during network management process, manager and agent use the same information model of network management information. In SNMP protocol, those model is defined in so called MIB (Management Information Base).

MIB is management information database having tree structure, very similar to folder structure on hard disk. That structure describes managed object. MIB structure is described in textual file ("MIB file" or just: "MIB"), which is written in ASN.1 format (*Abstract Syntax Notation #1*) [5].

One example is given on figure 2 [6]. Knowing MIB structure, manager knows structure of management information that can be used during communication with agent. Visualization of that structure is shown on figure 3.

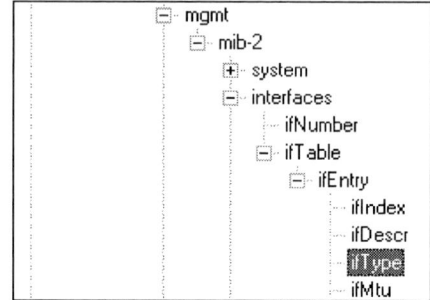

Figure 3.   MIB visualization

Every node within MIB structure has its own number; specific node's position can be unique defined as a "list" of all node's numbers leading from the root to the specific node. Numbers within list are separated by sign ".". That list of node numbers separated by full stop sign is called OID (*Object Identifier*). One example is: 1.3.6.1.4.1.14103.2.4.3.

Every company or organization is able to "reserve" its own branch within MIB root. Within own branch organization has full autonomy regarding sub-branches creation, deletion, naming and organization. MIB branch assignment is managed by international organization IANA (*Internet Assigned Numbers Authority*). Centralization of MIB branch assignment ensures unification of MIB branches on world level.

Message sequence in SMP protocol is very simple and intuitive. SNMP manager sends management commands to SNMP agent (SET-REQUEST) or retrieve some management information from SNMP agent (GET-REQUEST). Those requests are followed by RESPONSE message from SNMP agent. On the other hand, when SNMP agent wants to send unsolicited message, it uses SNMP TRAP message to do it.

II.   SNMP AND FAULT MANAGEMENT

A. Fault management

"Fault management primarily covers the detection, isolation and correction of unusual operational behaviors of telecommunication network and its environment" [7]. On network problem's appearance, network generates large number of unsolicited notifications carrying information about malfunction; these notifications are also called events; in the fault management functional area these notifications are called alarms. For instance, in the case of transmission link failure, nodes from both sides of transmission link will generate alarm (e.g. "Loss of signal") [8]. All notifications are potential entries in a network management system.

B. Fault management notifications

ITU-T recommendation X.733 [9] provide the detail for the general parameters of the event reporting service, which is used to report events ("alarms" actually). The most important parameters are:

- event type;
- event information;

Event type categories the alarm. Five basic categories of alarm are specified. These are: communications alarm, quality of service alarm, processing error alarm, equipment alarm, and environmental alarm.

Event information carry notification specific information, processed later by network management system. The most important are:

- Probable cause
- Specific problems
- Perceived severity
- Additional text/information
- Notification identifier

Event type and event information, in addition to general event reporting parameters, such as managed object class, managed object instance and event time, are used to notify network management system about network alarm.

Consider the generic content of alarm message mentioned above, it is reasonable to expect that management information structure on different network element types may be the same. Implementation of any specific management agent may vary, but general alarm information structure is common.

C. Related work

There are number of papers that are focused on integrated fault management [1], [3], [8]. However, main idea in integrated management is to ensure appropriate "proxy" modules that will handle with different management information formats performing mediation function. Further, in [3] integrated management is considered as hierarchical system where system intelligence is spread among different management planes. However, we have not found relevant work handling with different MIB type comparison.

Hence, we've decided to convey small survey on different SNMP Management Information Base, for different network element types, and to compare it.

III. MIB COMPARISON

A. INC-MIB-AL

INC-MIB-AL is implemented on Nokia Siemens Network @vantage Commander v11.0 [10], which is part

of core network. System alarms are forwarded in form of trap messages, while alarms are stored in alarm table. Alarm table entry has following format:

```
AlarmTableEntry ::= SEQUENCE {

    tiAlarmDateTime DISPLAY STRING,

    tiAlarmReportingObject OBJECT IDENTIFIER,

    tiAlarmFaultyObject OBJECT IDENTIFIER,

    tiAlarmEventTypeId INTEGER,

    tiAlarmSeverity INTEGER,

    tiAlarmErrorID INTEGER,

    tiAlarmEndKey INTEGER,

    tiAlarmDescription OCTET STRING,

    tiAlarmSequenceNumber INTEGER,

    tiAlarmSourceName DISPLAY STRING,

    tiAlarmSymbNEname DISPLAY STRING,

    tiAlarmNEtype DISPLAY STRING,

    tiAlarmNotificationID INTEGER,

    tiAlarmTransferID INTEGER,

    tiAlarmRepairText OCTET STRING,

    tiAlarmLongText OCTET STRING

}
```

Current alarm resynchronization on SNMP manager startup or reconnection is implemented. In that case, re-transmission of the active alarms should be requested from SNMP agent (using SNMP SET-REQUEST).

B. X733GROUP-MIB

Siemens fixed telephony exchange EWSD is managed by Net Manager system [11]. SNMP agent implements X733GROUP-MIB [12]. As for @vantage commander, system alarms are forwarded in form of trap messages, while alarms are stored in alarm table. Alarm table entry has following format:

```
snmpAlarm NOTIFICATION-TYPE OBJECTS {

    neName,

    notificationId,

    severity,

    eventType,

    eventTime,

    probableCause,

    specificProblems,

    managedObjectClass,

    managedObjectInstance,

    ipAddress,

    trapName,

    originalAlarm

}
```

Current alarm resynchronization on SNMP manager startup or reconnection is implemented. In that case, re-transmission of the active alarms should be requested from SNMP agent (using SNMP SET-REQUEST).

Further, it is allowed to request alarm with specific notification Id. It ensures unbroken alarm sequence on SNMP manager side. Finally, this SNMP agent sends alarm summary periodically to SNMP manager, or upon specific request.

C. OPENMIND-MOS-MIB

This MIB is implemented at SNMP agent on one implementation of SMS center module.

There is no active alarm table [13]. Hence, alarm resynchronization on SNMP manager startup or reconnection is not implemented. In the case of trap loss, there is no mechanism to recover missing alarm information.

System alarms are forwarded in form of trap messages. There is no unique format of alarm; every specific alarm type has its own parameters encapsulated within SNMP trap message.

D. TNMS-MIB

TNMS-MIB is used on Telecommunication Network Management System (TNMS), product by Siemens covering, among other, management of SDH multiplexers in transmission network.

System alarms are forwarded in form of trap messages, while alarms are stored in alarm table. Alarm table entry has following format (specific variable types are described in MIB document [14]):

```
EnmsAlarmEntry ::= SEQUENCE{

    enmsAlAlarmNumber Integer32,

    enmsAlSeverity PerceivedSeverity,

    enmsAlProbableCause ProbableCause,

    enmsAlClass AlarmClass,

    enmsAlServiceAffect Boolean,

    enmsAlState AlarmState,

    enmsAlTimeStampFromNE Boolean,

    enmsAlTimeStamp EnmsTimeStamp,

    enmsAlEntityString DisplayString,

    enmsAlEntityType EntityType,

    enmsAlNEId NEId,

    enmsAlPortId PortId,

    enmsAlTPIdH TPId,

    enmsAlTPIdL TPId,

    enmsAlTPName DisplayString,

    enmsAlModuleId ModuleId,

    enmsAlProbableCauseString DisplayString,

    enmsAlNELocation DisplayString

}
```

Current alarm resynchronization on SNMP manager startup or reconnection is implemented. In that case, SNMP manager should make "walk" through MIB alarm table using SNMP-GET-NEXT command.

E. MIB comparison

Every of these four MIBs follows general alarm message structure, as described in section "Fault management notifications". All alarms described above contain probable cause, perceived severity, affected managed object info, event time, additional information about alarm etc.

However, agents that implement management function are different applications, developed by different vendors. In parallel with SNMP agent development, MIB structure is defined also. Hence, for different network element types, there are four different MIB structures, regardless of fact that information content is almost the same.

Some SNMP agents ensure reliable message transmission, with resynchronization function implemented, some of they don't. Short summary is shown in table 1:

TABLE I
SNMP MIB STRUCTURE COMPARISON

	INC-MIB-AL	X733GROUP-MIB	OPENMIND-MIS-MIB	TNMS-MIB
Alarm table	Y	Y	N	Y
Resync function	Y	Y	N	Y
Probable cause	Y	Y	Y	Y
Specific problems	Y	Y	Y	Y
Perceived severity	Y	Y	Y	Y
Additional info	Y	Y	Y	Y
Managed object	Y	Y	Y	Y

IV. PROBLEM SOLUTION PROPOSAL

SNMP agents are pieces of software handling several functions. We will try to detect these functions in domain of fault management. First, it is necessary to handle all information from network element's hardware and software. For instance, SNMP agent should recognize high CPU temperature, link synchronization problem or crash of software component. It must communicate with hardware. Further, based on information from network element, any unusual operational behavior should be recognized. SNMP agent should know when to trigger network element alarm and when to cancel alarming condition. Third, alarms should be stored in alarm database. It can be implemented as relational database (e.g. SQLite), or simple as a file. Finally, communication between SNMP manager and SNMP agent in both directions must be supported. SNMP manager request management information from SNMP agent (SNMP-GET, SNMP-GET-NEXT), sets some management information (SNMP-SET) or receives unsolicited messages from SNMP agent (SNMP-TRAP).

Our proposal is to implement API (Application Programming Interface) that will be able to cope with SNMP support and alarm handling functions (figure 4). API will take care of alarm forwarding to SNMP manager, alarm database handling and processing of SNMP sets and requests from SNMP manager. For fault management function, API will use predefined and unique MIB. However, if SNMP agent handles any other data such as configuration, all of these data can remain in additional, specific MIB.

API should be integrated into existing SNMP agent software. All SNMP agent's parts that are specific for monitored network element (e.g. communication with hardware and software as well as alarm condition detection) can remain unchanged (interrupted line on figure 4). However, after alarm start or end is recognized, appropriate API functions are called (programming interface I_1 on figure 4).

Central part of API should be unified alarm structure. For instance, it can be realized as C++ class or data structure containing all alarm information. These information are mostly common for all MIBs analyzed in this paper, as we mentioned.

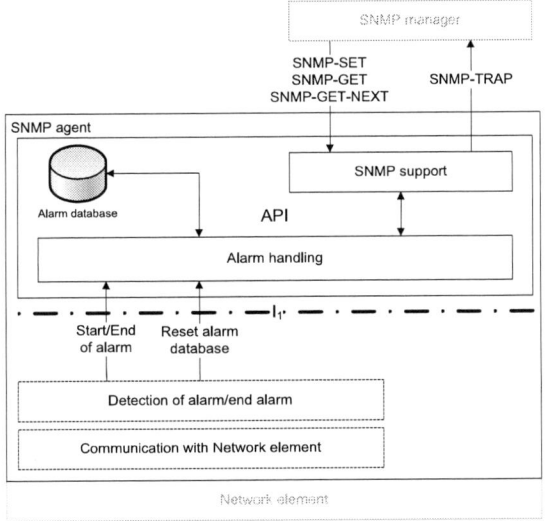

Figure 4. SNMP agent and proposed API

Since all SNMP agent's parts that are specific for agent implementation may remain the same, minimum of changes in any SNMP agent software is needed to use proposed API structure. It means that cost of API implementation within existing agents is minimal.

V. CONCLUSION

In this paper we have made comparison between four different SNMP MIBs as well as SNMP agent functionalities implementation. Although all SNMP agents handle almost the same alarm information structure, all of MIBs analyzed have different format.

It requires different SNMP manager application for every single network element type. Consequence is increased delivery time for network management systems. It has influence on service delivery time plan which is related to telecom operator business results.

As recommendation for network elements' and SNMP agents' vendors, conclusion of this paper can be that unique Management Information Base could and should be used for fault management functional area. Thinking in that way, we have proposed API structure that can ensure unique SNMP support as well as management information format in fault management functional area, with minimum implementation cost.

Further work should be focused on API component development in order to prove that concept.

REFERENCES

[1] O. Jukic and M. Kunstic, "Logical inventory database integration into network problems frequency detection process", *2009 10th International Conference on Telecommunications*, Zagreb, 2009, pp. 361-365.

[2] ITU-T: Recommendation M.3010: Principles for a Telecommunications Management Network, May, 1996

[3] O. Jukić: "Formal specification of telecommunications problems and solving agents", Doctoral thesis, Faculty of electrical engineering and Computing, University of Zagreb, 2012.

[4] Uyless, B.: Network Management Standards – SNMP, CMIP, TMN, MIBs and object libraries, McGraw-Hill 1994.

[5] Stallings, W.: "SNMP, SNMPv2 and CMIP – The practical guide to network-management standards", Addison-Wesley 1993.

[6] Management Information Base for Network Management of TCP/IP-based internets: MIB-II, International Engineering Task Force, Network Working Group, RFC 1213, 1991.

[7] D. K. Udupa: "TMN – Telecommunications Management Network", McGraw-Hill Telecommunications, New York, 1999.

[8] O. Jukić and I. Heđi, "Service monitoring and alarm correlations", 4th International congress on ultra-modern telecommunications and control systems, pp. 330-334, Saint Petersburg, 2012.

[9] ITU-T: Recommendation X.733: Information technology – open systems interconnection – systems management: alarm reporting function, Geneve, 1992.

[10] Nokia Siemens Networks: "@vantage Commander v11.0, development manual: Interface description; NBI SNMP FM Interface", 2010.

[11] Siemens: "Information – Net Manager – System Description", Siemens AG, 2001.

[12] RFC 3877: Alarm Management Information Base (MIB), IETF, 2004.

[13] OPENMIND MOS MIB, OpenMIND Networks Ltd., 2004.

[14] TNMS MIB, Nokia Siemens Networks, 2013.

Distributed threat removal in software-defined networks

D. Samociuk*, A. Chydzinski*

* Institute of Informatics, Silesian University of Technology, Gliwice, Poland
e-mail address: {dominik.samociuk, andrzej.chydzinski}@polsl.pl

Abstract - We propose an architecture for distributed threat removal in software-defined networks. This is a novel design of a large network, in which security analysis must be performed. In the classic paradigm, the security analyzer is an entry device, connected serially with the rest of the topology. Obviously, this device may suffer from a high processing load. Therefore, it may create a bottleneck, when the arriving traffic is waiting for the security verification, before being forwarded to next devices in the network. In the proposed architecture, traffic is immediately forwarded towards all destinations, while the security analysis is carried out in parallel, resulting in offloading the entry security device. We show that the proposed solution reduces the bottleneck in the topology and increases the rate of the carried traffic, while ensuring the same, as in the classic approach, security level.

I. INTRODUCTION

Modern computer networks require several security solutions. Analyzers, qualifiers, classic firewalls and web-application firewalls, intrusion prevention systems and others, are used to ensure high level of security. They must be located on the route of the traffic (in-line deployment), what introduces delays in the delivery of packets, if they need to be in-depth analyzed for security purposes.

In general, the security of end users can be achieved using various mechanisms. For instance, the monitoring tools can be used to detect events before they cause safety hazards, the control mechanisms can help to delete hazardous components before they reach their destinations, the limiting mechanisms can facilitate cleaning up the topology and locating the threat. Companies must decide, which of the above mechanisms should be implemented in order to achieve the needed level of security.

At the same time, safety has to be carefully balanced in the network core. In particular, the impact of security mechanisms implemented in the network devices on the network performance has to be taken into account. As it was shown in [1], these mechanisms may induce non-negligible delays in data transmissions, mainly due to their serial location in relation to transfer path.

In [2], the dogma of the firewall was challenged, especially as a security mechanism for corporations. Obviously, the complete withdrawal of firewalls was not recommended, but some dubious situations were pointed,

in which the safety mechanism create a bottleneck, or worse, a single point of failure of the entire network.

Web services, exposed to the outside of our network, are most vulnerable to security attacks (e.g., the denial of service). At the same time, these applications have a unique requirement for access to anyone, even unauthorized users. When the network service does not have the access control, it raises the question of how to implement the firewall. The firewall has not ceased to fulfill its role in the network architecture, but has become a heavy burden on the infrastructure performance. In modern networks, it is important to ensure security while preserving the performance and avoiding drastic delays.

Yet another issue is the fact, that modern networks are constantly moving the boundaries of their capabilities. It has not been that long since 1Gbps Ethernet was introduced. In the recent years however, technologies such as SaaS, IaaS, and PaaS have become popular, causing the extension of Gigabit Ethernet to the speed of 10Gigabit and 100Gigabit per second. There are also many factors driving the spike in performance requirements of network devices - from the Internet of Things, to the increasing demands for services of the *aaS type. However, as it turns out quite often, organizations undertake the infrastructure improvements and increase the throughput of the intermediate devices, only to find out that their security systems are not able to maintain transmission of data with the required speed.

We address this problem, by presenting a novel, parallel solution, in which the traffic is simultaneously being forwarded towards destinations, while the security analysis of packets is being carried out. Such parallel forwarding and analyzing of the data results in offloading the single device at the entry point, while ensuring the same, as in the classic approach, security. In our solution the entry device, instead of waiting for analysis to be finished, performs the forwarding immediately. If the traffic is malicious, it will be dropped before the final delivery in one of the intermediate or edge switches, after they have received the appropriate control data with analysis results.

The paper is organized as follows. In Section 2, we discuss the related work. In Section 3, the novel architecture is presented and compared with traditional architectures. In Section 4, an OMNeT++ based simulator of the proposed threat removal mechanism is described in detail and the motivation behind the topology used in the tests is presented. In Section 5, we present and discuss the simulation results. Finally, Section 6 concludes the paper and presents our plans of future work related to the subject.

II. RELATED WORK

The combined topics of security and software-defined networking aspects have been studied in the following papers.

In [3,4], possible solutions to the problem of the lack of authentication, access control and creation of secure channel network architecture in programmable networks, were discussed. Namely, the applicability of three different protocols: Transport Layer Security, Secure Shell Tunnel and Host-to-Host IPSec, as the secure channel medium in software-defined networks with the OpenFlow protocol, was studied. The articles show that the implementation of one of the proposed solution would increase safety and reduce risks of attacks on an SDN topology. This is especially important when using the software-define network paradigm in wide-area networks.

Safe architectures of programmable network topologies, based on the OpenFlow protocol, were studied in [5,6].

FlowVisor [5] works as a transparent proxy between controllers and switches, in order to limit the rules created by the applications running on the controllers. The FlowVisor mechanism creates virtual topology divisions (slices), as combinations of switch ports, MAC addresses, IP addresses, Layer 4 ports or message types (ICMP), and then delegates control of these fragments to the various controllers. The role of FlowVisor in SDN topology is insulation the impact sent rules to specific part of the network. This means, that a rule created for one part cannot affect the traffic in the next division.

A similar concept is FortNOX [6]. This is a software extension, developed on the NOX controller, for checking (in real time) the flow of contradictory rules. It uses authentication, based on the roles of applications, using the OpenFlow protocol. (It concerns applications that want to modify the traffic in the network using the OpenFlow protocol, e.g. firewalls, intrusion prevention systems, etc.).

The difference between FortNOX and FlowVisor is that FortNOX is a single application controller, which operates in parallel, while FlowVisor operates independently of the controller (usually as a separate device in the network). Both of these solutions limit the possibility of introducing a security risk by untrusted controllers and applications. However, they are both based on the usage of the OpenFlow protocol and the assumption, that its communication channels are secure, which is not provided by default. This could be accomplished, however, by implementing the solutions proposed in [3,4].

The subject how to ensure the security of networks using the methodology of software-defined networks and the OpenFlow protocol was researched in [7,8].

NICE [7] is a system of protection against distributed attacks on the limitation of service. The tool for the detection of attacks is based on the graphic analytical models, capable of reconfiguring topologies in an emergency situation.

Moving Target Defense [8] is based on the fact that in the OpenFlow environment we can change frequently IP addresses of the internal devices, to prevent attacks and to reduce the reconnaissance carried out from external networks.

It must be stressed that to the best of the authors' knowledge, there are no published results on analyzers in software-defined networks. Especially, we are not aware of works on the utilization of the SDN paradigm for enhancing the performance and functionality of classic analyzers and on the reduction of the bottleneck effects they generate now.

Moreover, there are no articles on the security-related behavior of software-defined networks in the wide area.

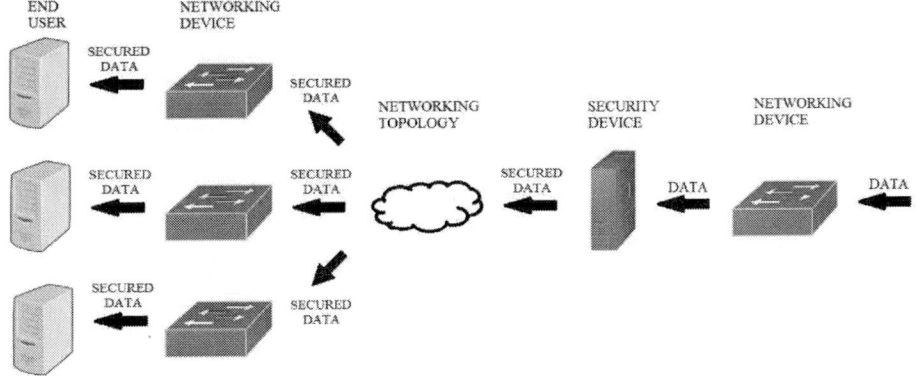

Figure 1. Classic network architecture and data flow through security device.

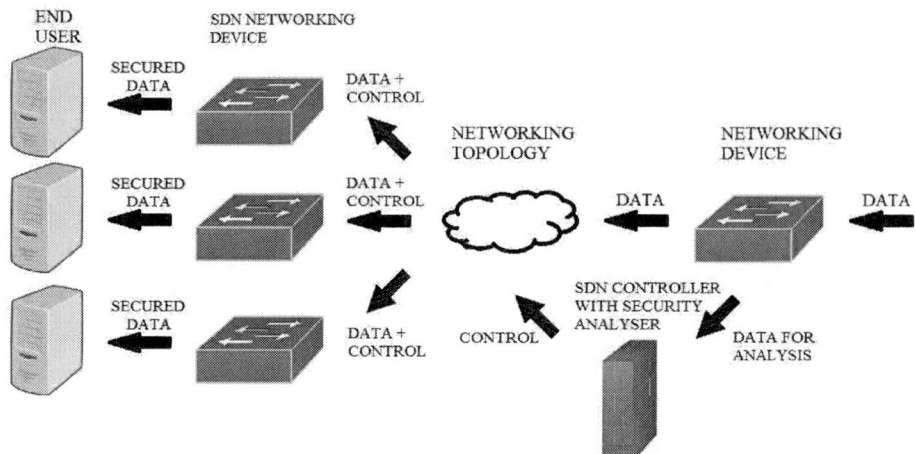

Figure 2. Proposed distributed treats removal architecture for software-defined networks.

III. ARCHITECTURE

Traditional network architectures are not very well suited to meet the requirements of modern corporations, service providers and end users.

The classic data flow (Fig. 1) is as follows. The traffic entering the network (from the right) goes to the first switch and then to the analyzer, in which the first packet from every flow is analyzed (security check), while the rest of the flow is waiting in a queue. If the positive decision is made, the whole flow is forwarded and delivered to the destination. Otherwise, the whole flow is dropped.

In the proposed architecture (Fig. 2), we use the SDN paradigm, that decouples the network control and forwarding functions. Namely, the traffic entering the network (on the right) passes through the first switch, but a copy of the first packet from every new flow is forwarded to the analyzer. While the analysis is being performed, the whole flow is being forwarded towards the destination, without waiting for the analyzer's decision. As soon as the analyzer reaches the decision about the new flow (drop or deliver), this decision is spread over the control plane, to all the programmable switches in the network. Therefore, a malicious flow can be dropped in every intermediate switch in the network. In the worst case, it is dropped in the last switch, before delivery to the end user - it is not allowed to deliver a flow to the end user without analyzer's decision.

In this way, different flows may be queued in several different locations in the network, instead of one (entry point), while waiting for the analyzer decisions. These decisions are forwarded in control packets, carrying information whether a flow can be delivered, or must be dropped.

Comparing the new solution with the classic architecture, we may observe a few important differences, which are:

- volume of data that passes through the analyzer (only a copy of the first packet of a flow goes to the analyzer, instead of all data);

- type of packets forwarded (data + control data with the security decision sent once per flow, instead of pure data in the classic approach);

- physical connections (additional link between the first network switch and the rest of the network in the proposed solution, as presented in Fig. 2);

- queueing behavior of switches (every queue has three possible actions for each data flow: "deliver", "drop" and "block". The default action for a new flow is "forward" in all intermediate switches and "block" in the last-before-destination switches. After the analysis of the new flow is finished, every switch receives the control packet with the final decision on the flow, which is "deliver" or "drop");

- switches, where the packets are queued waiting for the analyzer decision (multiple network switches instead of a single, entry device).

The main advantage of the proposed architecture is that the network is not idle (it is forwarding data) while waiting for the analyzer to finish its work. This reduces significantly the possibility of the bottleneck at the network entry point.

The cost we have to pay for this is the need for the programmable switches and the SDN controller.

IV. SIMULATION DESIGN

To simulate the proposed solution, the OMNeT ++ discrete event simulation framework was used (see *https://omnetpp.org*).

In fact, two separate simulators were implemented – one for the classic solution (Fig. 3) and one for the SDN-based solution (Fig. 4). For fair comparison, both simulators have common topology and network parameters (propagation times, link throughputs, buffer sizes etc.) .

The topology used in the simulators was created to mimic the new laboratory of wide area network, named

PL-LAB2020 [9-11]. PL-LAB2020 is an experimental network connecting six geographically dispersed, Polish research centers, namely:

- National Institute of Telecommunication;
- Warsaw University of Technology;
- Poznan Supercomputing and Networking Center;
- Silesian University of Technology;
- Gdansk University of Technology;
- Wroclaw University of Technology.

In the simulators, these research centers are denoted using their Polish acronyms, namely *il, pw, pcss, psl, pg* and *pwr*, respectively (see Figs. 3 and 4),

The network core consists of six Juniper ACX switches (*acx_il, acx_pw, acx_pcss, acx_psl, acx_pg* and *acx_pwr* in Figs. 3 and 4). These switches are connected via dedicated 10Gbit/s links, according to the schemes in Figs. 3 and 4. All the remaining, local links (e.g. *acx_pw-sdn_pw*) are of 1Gb/s throughput.

To simulate the geographical distribution of the PL-LAB2020 centers, the following core link propagation times were assumed: *psl-pw:* 25ms, *pw-il:* 19ms, *pw-pcss:* 21ms, *il-pg:* 29ms, *pg-pcss:* 29ms, *pcss-pwr:* 24ms, *pwr-psl:* 25ms. All the local links (e.g. *acx_pw-sdn_pw*) were assumed to have zero propagation time.

The entry point, as well as the analyzer and the SDN controller, are located in *psl* location, while four other locations are used as flow destinations: *il, pw, pcss,* and *pwr*.

In every destination, there is a programmable switch (in the classic architecture, it can be a standard switch) and a sink, for simulating an end user. The *pg* node is used only for forwarding the traffic.

Therefore, the final simulators, as shown in Figs. 3 and 4, consist of the following elements:

- *generator*, which generates multiple flows of packets, with predefined packet interarrival time distribution, flow size distribution, packet sizes and percentage of malicious flows;

- *sdn2_** devices – programmable network switches, needed in the new architecture, with modified queuing capabilities and predefined buffer sizes;

- *sdn1_** devices, *acx_**– standard network switches with predefined buffer sizes;

- *analyzer,* which performs the security verification of packets. Namely, the first packet in every new flow is analyzed. This takes a random time, according to some predefined distribution. In the classic architecture, all packets are queued in the analyzer, despite the fact that only the first packet in the flow has to be analyzed. When the decision based on the first packet is made, the whole flow is forwarded or dropped. In the proposed architecture, only a copy of the first packet of each flow is transmitted to the analyzer, while the original, complete flow is forwarded immediately. When the decision is made, the control packet is created and sent to all the switches, changing the default action for this flow to either "deliver" or "drop";

- *sink_** - simulates an end user.

Before the simulation results are presented, it is worth mentioning that we have considered also evaluating the performance of the proposed architecture using analytical methods. In particular, the potential method, [13-16], for finding characteristics of queues of packets in the system, was considered. Unfortunately, due to multiple queueing and control mechanisms involved, the mathematical analysis of the system is extremely hard – it seems to be beyond the current capabilities of the queueing theory. Therefore, we performed the analysis of the performance of the system using the simulations only.

In the future however, we are planning to implement and test the proposed solution in the PL-LAB2020 laboratory.

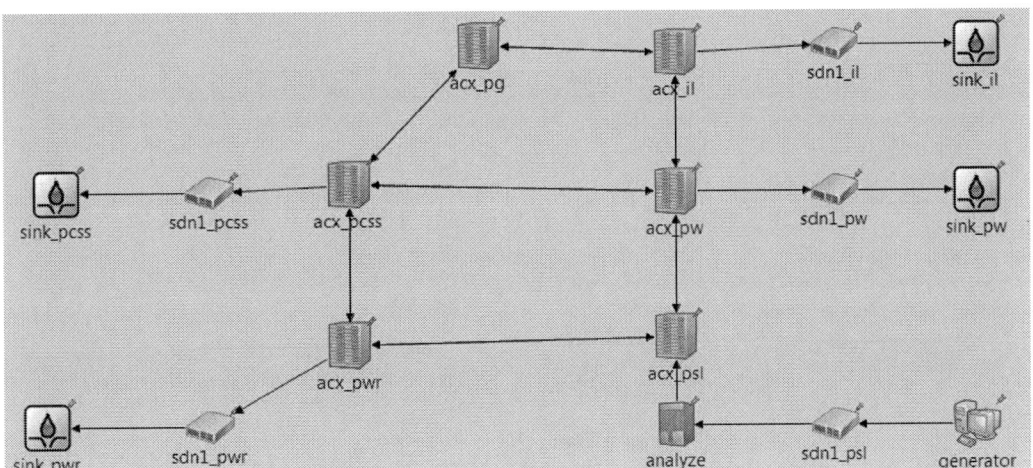

Figure 3. Classic architecture simulated in OMNeT ++.

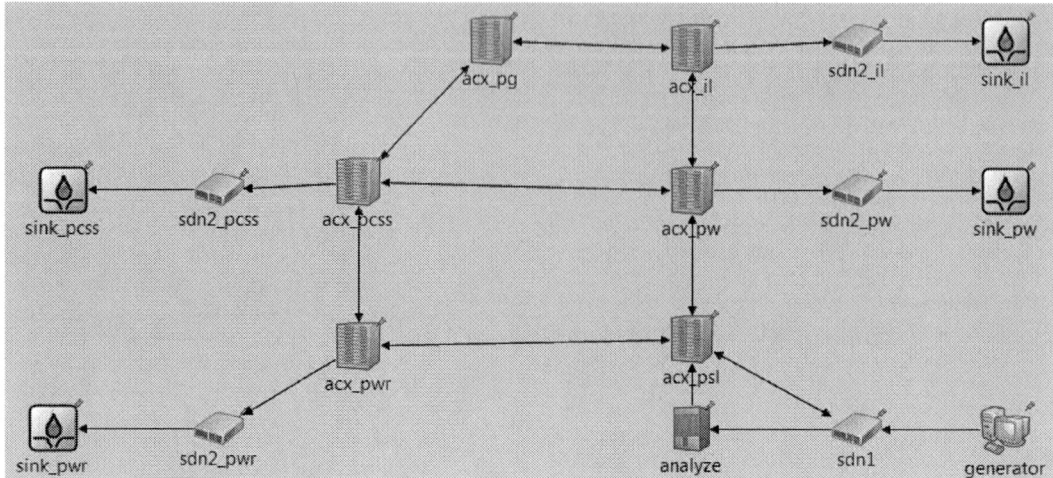

Figure 4. Proposed SDN architecture simulated in OMNeT ++

V. SIMULATION RESULTS

The main purpose of all simulations was to find the maximum rate (in Mb/s) of the traffic arriving to the network, that does not cause losses of safe packets, due to buffer overflows. Naturally, both the classic and the new architecture were tested.

It is quite obvious, that the maximum allowed arrival rate depends strongly on the average time of analysis. Moreover, it depends on the average length of the flow, as only the first packet in each flow is analyzed. Therefore, three values of the average flow length were used (100, 200 and 1000), and two different average analysis times (0,003 and 0,0001s).

On the other hand, the maximum allowed arrival rate depends very little on the percentage of malicious flows. This is due to the fact, that the analysis time statistically does not depend on whether the packet is safe, or not.

In detail, the following parameters were used in simulations:

- packet interarrival time distribution: exponential;
- total arrival rate: varied from 0 to 1Gb/s.
- flow size distribution: exponential;
- average flow length (pkts): 100, 200 or 1000;
- packet size (bytes): 1500 or 9000 (Jumbo);
- control packet size (bytes): 64;
- buffer size at every switch (pkts): 10000;
- distribution of the security analysis time: exponential;
- average analysis time (s): 0,003 or 0,0001;

As the typical average value of the analysis time 0,003s was used. This value was chosen according to the speed of security devices common in contemporary enterprise networks.

The buffer size of 10000 packets was chosen following the bandwidth-delay product rule, which is well-known in buffer sizing. Assuming the packet size of 1500bytes and a 10Gb/s link, this size allows for 120ms of traffic buffering.

The total arrival intensity was manipulated by changing the average interarrival time between consecutive packets. The destinations of the arriving flows were assigned randomly, with equal probabilities for all available destinations.

Table I shows the maximum possible rate, for which we do not lose safe packets, in the case of 1500-bytes-long packets, the average analysis time of 0,003 second and three average flow lengths: 100, 200 and 1000 packets.

TABLE I. Comparison of the maximum allowed generator rate between the classic and the proposed solution. Packet size 1500bytes, average analysis time 0,003s.

mean flow size	classic solution	proposed solution
100 packets	282 Mb/s	359 Mb/s
200 packets	440 Mb/s	664 Mb/s
1000 packets	793 Mb/s	Line speed

The table demonstrates clearly, that the proposed solution provides significantly higher maximum rate for both short and long flows, assuming typical packet sizes and typical analysis times. Moreover, as can be can be noticed, where the classic architecture meets its limits for long flows (1000 packets), the proposed solution can still run with full line speed.

Having checked the typical case, we tested also the architectures in the case of very large packets.

Table II presents the maximum possible arrival rate, for which we do not lose safe packets, in the case of large frames, the average analysis time of 0,003 second and three average flow lengths: 100, 200 and 1000 packets.

TABLE II. Comparison of the maximum generator rate between the classic and the proposed solution for Jumbo frames. Packet size 9000bytes, analysis time 0,003s.

mean flow size	classic solution	proposed solution
100 packets	707 Mb/s	Line speed
200 packets	806 Mb/s	Line speed
1000 packets	953 Mb/s	Line speed

As we can see, the new solution outperforms the classic one in this case as well.

We also conducted some tests to find out, if our solution can be useful in use-cases other than security, e.g. in situations, where the average time of the analysis is very short. As a real-life example, the classification and/or marking of flows may serve - such operation can take about 0,0001s on average. Results of comparison of the maximum rate in such case are presented in Table III.

TABLE III. Comparison of the maximum generator rate between the classic and the proposed solution for fast operations. Packet size 1500bytes, analysis/operation time 0,0001s.

mean flow size	classic solution	proposed solution
100 packets	927 Mb/s	Line speed
200 packets	958 Mb/s	Line speed
1000 packets	990 Mb/s	Line speed

Similarly, the new architecture outperforms the classic architecture, though not that much as in the previous setups.

VI. CONCLUSIONS

Nowadays, network security mechanisms are becoming more important than ever. Moreover, the security devices must meet the rapidly increasing demands for serving large volumes of traffic. Inappropriate design of security mechanisms, devices or architecture may cause a huge decline in overall network performance, no matter how fast the network switching and transmitting mechanisms are.

We presented a possible solution for the bottleneck problem occurring in classic network topologies, caused by the security device.

Simulations proved that in distributed, SDN-based topologies, the proposed solution performs better than the classic solution. In all the tests, the level of security was the same in the classic and proposed architecture, but the maximum carried traffic was higher in the new design.

The benefit depends on the flow size distribution – the shorter the flows are, the more we gain. It also depends on the packet sizes in the same manner. Finally, it can be deduced that the benefit varies also with the number of destinations. Generally, the higher number of destinations is, the bigger the difference between the classic network and our solution is, in favor of the SDN.

As for the future work, the authors are planning to implement and test the proposed solution in a wide area networking laboratory.

ACKNOWLEDGEMENT: The infrastructure was supported by "PL-LAB2020" project, contract POIG.02.03.01-00-104/13-00. The work was carried out within the statutory research project of the Institute of Informatics (RAU2).

REFERENCES

[1] D. Samociuk, B. Adamczyk, A. Chydzinski. "Impact of Router Security and Address Translation Mechanisms on the Transmission Delay". INTERNET 2015, The Seventh International Conference on Evolving Internet p. 1-5, 2015.

[2] Challenging the firewall data center dogma. URL: https://devcentral.f5.com/articles/challenging-the-firewall-data-center-dogma} [29-01-2017].

[3] D. Samociuk. "Secure Communication Between OpenFlow Switches and Controllers". The Seventh International Conference on Advances in Future Internet (AFIN 2015), p.39-46, 2015.

[4] D. Samociuk. "Metody zapewniania bezpieczenstwa komunikacji pomiedzy przlacznikami i kontrolerami OpenFlow" ("Methods for securing OpenFlow controller-switch communications"), Przeglad Telekomunikacyjny+ Wiadomosci Telekomunikacyjne, vol. LXXXVIII, no. 8-9/2015, p.1529-1536, 2015.

[5] R. Sherwood et al., "Flowvisor: A network virtualization layer", OpenFlow Switch Consortium, Tech. Rep, pp. 1-13, 2009.

[6] P. Porras et al., "A security enforcement kernel for openflow networks," in Proceedings of the First Workshop on Hot Topics in Software Defined Networks. New York: ACM, pp. 121-126, 2012.

[7] C.-J. Chung et al., "Nice: Network intrusion detection and countermeasure selection in virtual network systems," IEEE Transactions on Dependable and Secure Computing, no. 4, pp. 198-211, 2013.

[8] J. H. Jafarian, E. Al-Shaer, and Q. Duan, "Openflow random host mutation: transparent moving target defense using software defined networking," in Proceedings of the First Workshop on Hot Topics in Software Defined Networks. New York: ACM, pp. 127-132, 2012.

[9] A. Binczewski et al. "Infrastruktura PL-LAB2020" ("PL-LAB2020 infrastructure"). Przeglad Telekomunikacyjny+Wiadomosci Telekomunikacyjne, no. 12/2015, pp. 1399-1403, 2015.

[10] A. Binczewski, D. Samociuk, et al. "Laboratorium SDN" ("SDN Laboratory"). Przeglad Telekomunikacyjny+ Wiadomosci Telekomunikacyjne, no. 12/2015, pp. 1413-1418, 2015.

[11] B. Belter et al. "Rozlegle sieci badawcze dla testowania rozwiazan nowych generacji Internetu" ("Wide area research network for New Generation Internet solutions"). Przeglad Telekomunikacyjny+ Wiadomosci Telekomunikacyjne, vol. LXXXVIII, no. 8-9/2015, pp. 681-690, 2015.

[12] Horizon 2020 - The EU Framework Programme for Research and Innovation, URL:http://ec.europa.eu/programmes/horizon2020/en/area/ict-research-innovation [29-01-2017]

[13] A. Chydzinski. "Transient Analysis of the MMPP/G/1/K Queue". Telecommunication Systems, vol. 32, n. 4, pp. 247-262, 2006.

[14] A. Chydzinski. "Duration of the buffer overflow period in a batch arrival queue". Performance Evaluation, vol. 63, issue: 4-5, pp. 493-508, 2006.

[15] A. Chydzinski. "Queue Size in a BMAP Queue with Finite Buffer". Lecture Notes in Computer Science, vol. 4003, pp. 200-210, 2006.

[16] A. Chydzinski, R. Wojcicki and G. Hryn. "On the Number of Losses in an MMPP Queue". Lecture Notes in Computer Science, vol. 4712, pp. 38-48, 20

Performance Analysis of Virtualized VPN Endpoints

D.Lacković*, M.Tomić**

*University of Rijeka Faculty of Engineering/Department of Computer Engineering, Rijeka, Croatia
e-mail: dario3lackovic@gmail.com
** University of Rijeka Faculty of Engineering/Department of Computer Engineering, Rijeka, Croatia
e-mail: mtomic@riteh.hr

Abstract - Virtual Private Networks (VPN) are an established technology that provides users a way to achieve secure communication over an insecure communication channel, such as the public Internet. It has been widely accepted due to its flexibility and availability on many platforms. It is often used as an alternative to expensive leased lines. In traditional setups, VPN endpoints are set up in hardware appliances, such as firewalls or routers. In modern networks, which utilize Network Functions Virtualization (NFV), VPN endpoints can be virtualized on common servers. Because data encryption and decryption are CPU intensive operations, it is important to investigate limits of such setups so that feasibility of endpoint virtualization can be evaluated. In this paper, we analyze performance of two industry standard VPN implementations - IPSec and OpenVPN. We examine TCP throughput in relation to encryption algorithm used and packet size. Our experiments suggest that moving VPN endpoints from a specialized hardware appliance to a virtualized environment can be a viable and simple solution if traffic throughput requirements are not too demanding. However, it is still difficult to replace high-end appliances with large throughput capabilities.

I. INTRODUCTION

VPN has been an important technology since its creation in the mid-1990s. Point-to-Point Tunneling Protocol (PPTP) marked a starting point in the ability to connect remote computers to a common network. Since then, there have been many breakthroughs and innovations in the field, which resulted in competing technologies. As the technologies mature, it is important to focus on their comparison and evaluation of their viabilities in new networking use cases and paradigms. One such paradigm is Network Functions Virtualization (NFV) [1]. The NFV assumes moving network functions, such as a firewall, VPN endpoint, load balancer, etc. from specialized appliances to commercial off-the-shelf (COTS) servers. Often times the raw processing capabilities present a barrier to successful network function virtualization [2], [3].

Currently, there are two prevailing technologies used in the enterprise environments: IPSec and SSL/TLS based OpenVPN. Although IPSec is a de facto standard in VPN access, OpenVPN has been steadily gaining traction and is the only mature and widely used alternative to IPSec. The web proxy SSL/TLS based VPN access is not a network layer VPN tunnel and will not be discussed here.

Both, OpenVPN and especially IPSec, have been researched extensively, however, not much work has been done in comparing performance of the two in virtualized environments and on multi-core processors. For instance, Jaha and Libya in [4] evaluated relative performance of several VPN protocols on two operating systems, over wireless networks. Along with raw performance, they also investigated some QoS metrics. The authors findings were inconclusive as the results varied widely. Depending on the operating system in use, IPSec was from 8% slower than OpenVPN to up to 25% faster. This could suggest problems in their testbed or software and operating system implementation or configuration. Kotuliak et al. in [5] investigated differences between IPSec and OpenVPN, for different encryption algorithms. They showed that when using the 3DES encryption, IPSec was approximately 25% slower than OpenVPN. In case of Blowfish encryption, they were basically on par, but in case of AES, IPSec was 45% faster. The 45% difference is of greatest significance since AES is the new encryption standard and is replacing the aging 3DES and Blowfish. Instead of measuring raw network throughput, Voznak in [6] compared difference in bandwidth requirements for voice calls over OpenVPN and IPSec tunnels using AES encryption. He showed that packet payload size greatly influenced tunnel efficiency. Depending on voice codec used, payload size varied significantly and for very small payloads, IPSec required considerably less bandwidth than OpenVPN. In [7], Raumer et al. researched similar topics as this paper, but they limited their experiments to IPSec. They showed that, expectedly, network throughput is heavily dependent on packet size and that Intel's AES-NI instruction set [8] brings huge throughput gains – in case of maximum size Ethernet frames, almost threefold.

In this paper, we will analyze and compare raw network throughput of IPSec and OpenVPN on COTS hardware to assess the viability of VPN endpoint virtualization. We will study several encryption algorithms and analyze the influence of packet size and hardware acceleration on the throughput.

This work has been supported in part by the University of Rijeka under the project number 16.09.2.2.05.

II. TEHNOLOGY OVERVIEW

There are two types of network layer VPN connections: remote access and site-to-site. Remote access VPNs are generally used to connect teleworkers or set up temporary tunnels, while site-to-site VPNs are used to set up permanent secure network layer tunnels between distant networks. A special case of VPN access are web based SSL/TLS VPNs [9]. They are based on establishing a secure SSL/TLS session between any web browser a user might have and a VPN gateway. Upon session establishment, the gateway offers client a list of available web applications from internal network for which a user has access privileges. The gateway establishes session to the selected web application and relays traffic between application server and the client. Although widely used and useful, the approach does not set up a secure network layer tunnel between two networks and is therefore not examined further. Instead, two most commonly technologies for setting up network layer tunnels will be discussed and compared, namely, IPSec and OpenVPN.

A. IPSec

Internet Protocol Security (IPSec) is a collection of security protocols [10] designed by the Internet Engineering Task Force (IETF), with the aim of providing secure communication channel that spans, possibly, public or otherwise insecure Internet Protocol (IP) based networks. IPSec works by authenticating and encrypting each IP packet of a session. It provides data integrity and basic authentication and encryption services, which protect data from modification or unauthorized viewing. It operates on layer three (network layer) of the Open Systems Interconnection model (OSI) model and uses TCP protocol on the transport layer.

IPSec has become a de facto standard for setting up VPNs. Vendor acceptance has been wide and fast and it is currently available on mostly all hardware appliances with VPN capabilities. It offers high performance and its modular architecture allows for seamless integration of new algorithms. Since it is an open standard there are no vendor lock-in problems.

There are three primary components of IPSec:
- Authentication Header (AH)
- Encapsulating Security Payload (ESP)
- Internet Key Exchange (IKE)

AH is responsible for authentication and data integrity check, while ESP provides data confidentiality and encryption services. IKE serves as a protocol for negotiating algorithms, keys, and protocols and establishing security associations. Selection of algorithms for each of the components is up to a user to configure (Fig. 1).

Over the years, there were many software implementations of IPSec, most notable of which are FreeS/WAN and its forks Openswan and strongSwan, as well as Libreswan which is a fork of Openswan. In this paper, we will use strongSwan [11] as it is currently the most mature and actively developed IPSec implementation. It is an open source software, originally designed for Linux operating system, but also ported to

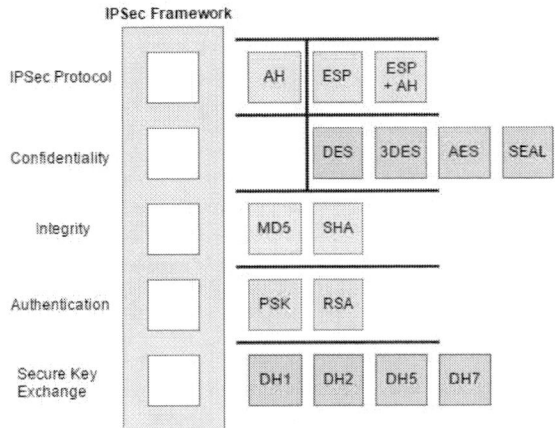

FIGURE 1: IPSec FRAMEWORK

Windows, Android, Mac OS X and other platforms. StrongSwan has full support for both IKEv1 and IKEv2 protocols.

B. OpenVPN

OpenVPN [12], [13] is an open source implementation of VPNs, which is, like IPSec, capable of setting up network layer tunnels in both site-to-site and remote access configurations. It appeared later than IPSec and, without the backing of large vendors, its development and acceptance has been slow, but in time it evolved and matured. Lack of support on network appliances has for long been a large disadvantage and limited its acceptance, nevertheless, it has still been steadily gaining traction. In light of the NFV paradigm, hardware support becomes less relevant and OpenVPN can now more easily compete with IPSec.

Unlike IPSec, OpenVPN can setup tunnels using both UDP and TCP protocols. This is an important consideration, as sending traffic over UDP tunnel can result in better latency when compared to a TCP tunnel [14]. Using UDP tunnels can also alleviate problems induced by IPSec on high delay and error prone links, such as satellite links [15]. Another advantage of OpenVPN is the ease of setup. Where IPSec is known for its complexity and steep learning curve, OpenVPN can be setup very quickly. It is easier to handle dynamic addresses and traverse firewalls and network address translation (NAT). The use of common networking protocols (TCP and UDP) for traffic tunneling makes it a desirable alternative to IPSec in situations where an ISP might intentionally block IPSec protocol.

Another difference to IPSec is that OpenVPN SSL/TLS tunnels operate at the transport and session layers of the OSI model. OpenVPN uses a custom protocol based on SSL/TLS for key exchange. It delegates services of both encryption and authentication to the OpenSSL library. This allows OpenVPN to use all the ciphers available in the OpenSSL package.

OpenVPN supports any operating system with an OpenVPN compatible VPN client, which nowadays is almost every OS, including Android.

C. AES-NI

Encryption is generally computationally very demanding. Current preferred standard for symmetrical key encryption is the Advanced Encryption Standard (AES). Traditional high throughput network appliances use special processors to offload AES encryption. The same approach was needed in COTS hardware if high performance were to be reached. For that purpose, Intel introduced AES-NI instruction set. Using AES-NI instructions can result in multiple fold increase in encryption performance. The instruction set is supported on all AMD x86 and x64 architecture processors since AMD Bulldozer Family 15h, and many Intel x86 and x64 processors. The full list of supported Intel processors can be found in [16]. Since both strongSwan and OpenVPN utilize AES encryption, we will analyze performance benefits of AES-NI on both technologies.

III. TEST SETUP

Testing was conducted on two HP ProLiant servers. Network connection was established by back-to-back connection using 10Gbps ethernet network interface cards (NIC). One machine was configured as a VPN server, while the other was used as a VPN client. Hardware and software specifications of the machines are shown in Table 1, while Fig. 2 shows the test setup.

To test performances in a virtualized environment, virtual machines were created to host tunnel endpoints. Since both physical servers ran on Linux operating system, KVM/QEMU [17] has been chosen as a virtualization solution. It is a full virtualization solution for Linux on x86 hardware. For operation, it requires a processor supporting Intel VT or AMD-V virtualization extensions.

Virtual machines (VM) hosting VPN endpoints were identical (Table 1). Virtual NICs were connected to a 10Gbps physical port using Linux bridge. Each VM has been allowed to use two CPU cores. One core was assigned to a traffic generator, while the other was assigned to a VPN tunnel. As a traffic generator, the *iperf3* tool was used.

Testing was performed on commonly used ciphers: AES, Blowfish, Camellia, and 3-DES. All test ciphers were in Cypher Block Chaining mode of operation (CBC). Testing procedure consisted of establishing a VPN tunnel between the server and client virtual machines, starting *iperf3* server on the VPN server machine and sending TCP traffic over the encrypted tunnel from the client machine. Traffic generation was set to last for 60 seconds. Network throughput was measured for packet sizes in the range of 100 to 1500 bytes (the largest frame that can be sent over ethernet channel without getting fragmented).

The first set of tests was performed with AES-NI hardware acceleration disabled. In order to evaluate performance benefits of AES-NI, another set of tests was performed, but with AES-NI enabled.

Lastly, we tested how well do IPSec and OpenVPN scale on multi-core processors. To eliminate any influence

TABLE I.   HARDWARE SPECIFICATION OF VPN SERVER CLIENT

Specification	VPN Server	VPN Client	Virtual machines
Model:	ProLiant ML350e Gen8 v2	ProLiant DL320e Gen8	-
OS:	Ubuntu Server 16.04.2 LTS	Ubuntu Server 16.04.2 LTS	Ubuntu Server 16.04.2 LTS
CPU:	Intel(R) Xeon(R) CPU E5-2407 v2 @ 2.40GHz 4 Core	Intel(R) Xeon(R) CPU E3-1220 V2 @ 3.10GHz 4 Core	Depends on host 2 Core
Memory:	4 GB	8 GB	1 GB
NIC:	Ethernet controller: Intel Corporation Ethernet Controller 10-Gigabit X540-AT2 (rev 01)	Broadcom Limited NetXtreme II BCM57810 10 Gigabit Ethernet (rev 10)	Virtio network device

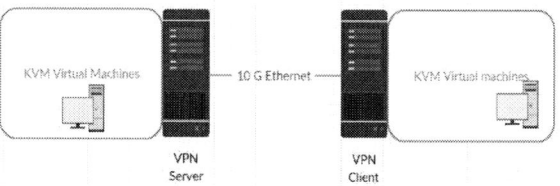

FIGURE 2: TEST SETUP

of virtualization on test results and evaluate scaling efficiency of the software, the tests were performed by establishing VPN tunnels between physical machines, instead of virtual machines.

In all the tests, throughputs have been low enough such that a network card virtualization did not affect performance. As memory usage for VPN tunnels is low, maximum throughput is mostly constrained by a virtual CPU performance.

IV. RESULTS AND DISCUSSION

A. IPSec with AES-NI disabled

Figure 3 shows results for IPSec with AES-NI disabled. Depending on packet size, best performing ciphers were AES and Blowfish. At packet size of 1500 bytes, AES 128, AES 192, AES 256 and Blowfish achieved 415 Mbps, 387 Mbps, 370 Mbps and 338 Mbps, respectively. Camellia 256 was slower, with 303 Mbps, while 3-DES could only manage 116 Mbps. In 3-DES, to achieve reasonable security, DES algorithm is applied three times for each packet (encrypt-decrypt-encrypt). This procedure introduces processing overhead that results in a significant drop of network throughput.

As in all later tests, it can be seen that network throughput for small packets is significantly smaller than for large packets. For such types of traffic patterns, much larger number of packets needs to be sent to achieve the same throughput and, consequently, processing overhead grows exponentially.

B. OpenVPN with AES-NI disabled

In Fig 4., network throughputs achieved by OpenVPN with AES-NI disabled are shown. Again, AES is the fastest, while 3-DES is the slowest cipher. AES 128 and AES 192 are similar in performance and peak at 266Mbps, but AES 256 is about 5% to 10% slower, depending on packet size. Blowfish and Camellia ciphers performed similarly and at 186Mbps are about 30% slower than AES ciphers, while 3-DES could only manage 85Mbps. It can be seen that OpenVPN is about 30% to 50% slower than IPSec, depending on the cipher used.

C. IPSec with AES-NI enabled

IPSec throughput with AES-NI hardware acceleration enabled is shown in Figure 5. AES ciphers most benefited from the hardware acceleration. For packet size of 1500 bytes, AES 128, AES 192 and AES 256 ciphers achieved 607 Mbps, 605 Mbps and 600 Mbps, respectively. An increase of between 40% and 64% in performance. What is also important to see is that previously noticeable difference in performance between AES 128 and AES 256 is significantly reduced. Camelia improved by 24% to 376Mbps, while Blowfish showed little improvement and 3-DES showed no improvement at all from AES-NI.

D. OpenVPN with AES-NI enabled

Test results for OpenVPN with AES-NI enabled are shown in Figure 6. Camellia, Blowfish and 3-DES did not improve as a result of AES-NI being enabled. Throughput with AES ciphers is improved but not as much as in the case of IPSec. Overall, there is little difference in performance for different key lengths. For small packets, the difference is up to 10%, between AES-128 and AES-256. For packet size of 1500 bytes, AES 128, AES 192 and AES 256 ciphers resulted in 303 Mbps, 308 Mbps and 309 Mbps, respectively. The improvements over test case with AES-NI disabled are about 16%. Such a small improvement suggests that in the case of OpenVPN encryption is not the most limiting factor and that the bottleneck is somewhere else in a data path. StrongSwan's IPSec implementation achieved almost double the throughput of OpenVPN.

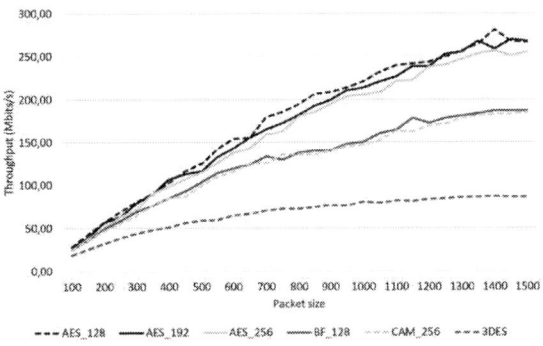

FIGURE 4: OPENVPN WITH AES-NI DISABLED

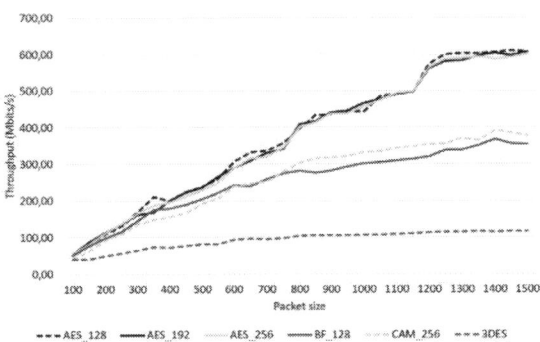

FIGURE 5: IPSEC WITH AES-NI ENABLED

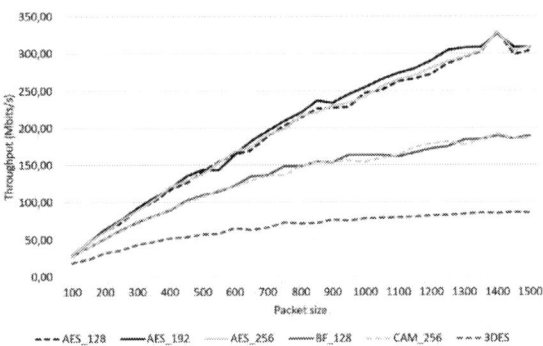

FIGURE 6: OPENVPN WITH AES-NI ENABLED

E. Software Scaling

To test how well strongSwan's IPSec implementation and OpenVPN scale, we modified the test setup such that traffic generation was taken off the servers. *Iperf3* was run on separate machines, which were connected over 10Gbps network switch. For strongSwan, three tests were run. In the first run, a single CPU core was made available to the VPN tunnel. In the second run, all 4 cores were made available to the tunnel, and only a single tunnel (a single client) was used for all traffic. In the last run, 4 IPSec clients connected to a single IPSec server, which had all 4 cores available. Figure 7 shows achieved network

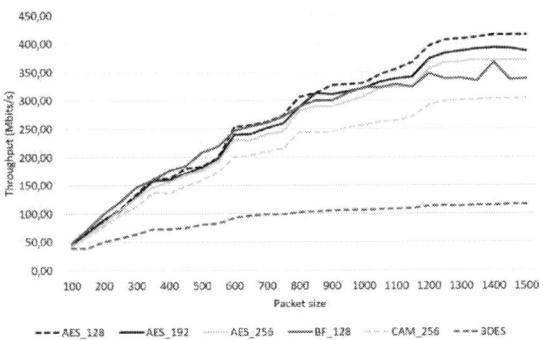

FIGURE 3: IPSEC WITH AES-NI DISABLED

throughputs. It can be seen that there is little difference in performance if a single tunnel or multiple tunnels/clients are used. Scaling with multiple clients is easier to solve, as each client can get a separate thread, but the results show that the strongSwan software efficiently scales even in a more difficult case of a single tunnel communication. Although all 4 CPU cores get utilized equally, the throughput is not proportional to the number of cores used and peaks at 1.4Gbps.

OpenVPN is not as good at scaling. It uses a single process and a single thread for all VPN tunnels of the same OpenVPN server instance. It means that it can only utilize a single CPU core. Although discussed by both developers and community, no multithreaded implementation has yet been released. Scalability can only be achieved by setting up one OpenVPN server instance per each available CPU core. The setup introduces configuration complexity and also requires careful distribution of VPN clients to server instances, such that traffic load on the instances is well balanced, which often makes the solution impractical. If only a single high throughput VPN connection is necessary, OpenVPN is not a viable option. Figure 8 shows achieved throughput in case of 4 OpenVPN server instances to which 4 clients are connected. Maximum throughput is about 27% less than is the case for strongSwan IPSec.

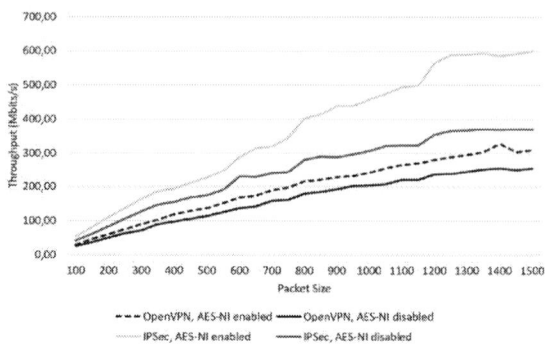

FIGURE 9: IPSEC AND OPENVPN AES 256 COMPARISON

F. Results Summary

The test results showed that, in terms of raw network throughput, strongSwan IPSec implementation is clearly more capable than OpenVPN. As summarily showed in Figure 9, in the analyzed case of virtual VPN endpoints, it achieved higher throughput for each tested cipher, both when AES-NI enabled and AES-NI disabled. Of the tested ciphers, AES yielded the best throughput in all test scenarios. We showed that using processors with AES-NI instruction set can improve network throughput by 62% in case of strongSwan's IPSec implementation and 16% when using OpenVPN.

On multi-core systems, strongSwan scales much better than OpenVPN. It can utilize all available CPU cores, regardless if only a single tunnel is used or traffic load is distributed between separate VPN tunnels. However, network throughput is not proportional to the number of CPU cores available. OpenVPN is single threaded so it does not scale, however, if applicable, it is possible to set up independent OpenVPN server instances, which run in separate processes and can, therefore, execute on different CPU cores.

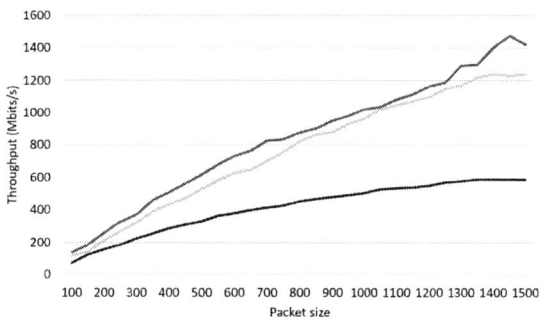

FIGURE 7: IPSEC ACHIEVED NETWORK THROUGHPUTS

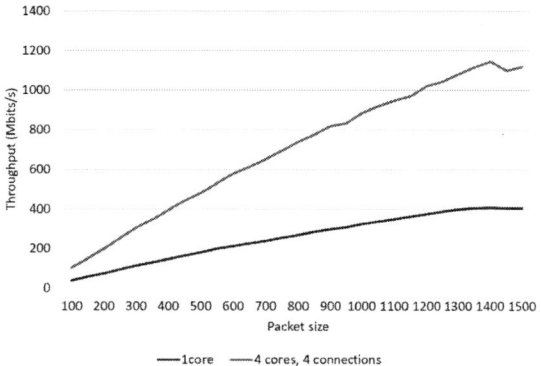

FIGURE 8: OPENVPN ACHIEVED NETWORK THROUGHPUTS

V. CONCLUSION AND FUTURE WORK

VPN endpoints are traditionally deployed on specialized network appliances, such as routers or firewalls/security devices. In this paper, we explored the viability of the endpoints virtualization on COTS hardware, with benefits in line with the NFV paradigm.

We find that VPN functionality of entry level appliances (up to about 1Gbps) can be easily virtualized even on low end servers. In most cases, both IPSec and OpenVPN will suffice. Although IPSec generally provides better throughput, OpenVPN has an advantage of setting up tunnels over UDP, which can lower latency and in special cases, such as satellite links, improve overall throughput.

On mid-range appliances, we expect throughputs of up to about 25Gbps. OpenVPN does not scale on multi-core systems so it is not suitable for such high requirements. StrongSwan's IPSec implementation does scale and it can be used in the lower range of throughput requirements. Depending on a specific use case, and with proper configuration and optimization, it, possibly, might also be used in the upper range.

When it comes to high-end security appliances, we currently cannot see any benefits of virtualization of their functionality on COTS hardware and we find it not to be a viable option.

In a future work, this study should be expanded to cover jitter and latency tests, as well as investigate network throughputs when UDP traffic is tunneled.

REFERENCES

[1] Published E2E Arch, REQ, Use Case, Terminology documents in ETSI NFV Open Area

[2] J. DiGiglio, D. Ricci. "High Performance, Open Standard Virtualization with NFV and SDN." *White paper, Intel Corporation and Wind River* (2013).

[3] M. Falkner, A. Leivadeas, I. Lambadaris, G. Kesidis "Performance analysis of virtualized network functions on virtualized systems architectures." Computer Aided Modelling and Design of Communication Links and Networks (CAMAD), 2016 IEEE 21st International Workshop on. IEEE, 2016.

[4] A. A. Jaha, M. Libya, Performance Evaluation of Remote Access VPN Protocols on Wireless Networks, International Journal of Computer and Information Technology, Volume 04, Issue 02, 2015.

[5] I. Kotuliak, P. Rybár, and P. Truchly. "Performance comparison of IPsec and TLS based VPN technologies." Emerging eLearning Technologies and Applications (ICETA), 2011 9th International Conference on. IEEE, 2011.

[6] M. Voznak "Speech bandwith requirements in IPsec and TLS environment." *WSEAS International Conference. Proceedings. Recent Advances in Computer Engineering*. Eds. N. E. Mastorakis, et al. No. 13. WSEAS, 2009.

[7] D. Raumer, S. n Gallenmül, P. Emmerich, L. Märdian, F. Wohlfart, G. Carle, "Efficient serving of VPN endpoints on COTS server hardware." Cloud Networking (Cloudnet), 2016 5th IEEE International Conference on. IEEE, 2016.

[8] Intel, https://software.intel.com/sites/default/files/article/165683/aes-wp-2012-09-22-v01.pdf

[9] R. Stanton, Securing VPNs: comparing SSL and IPsec, Computer Fraud and Security September 2005.

[10] S. Kent, S. Karen, "RFC 4301: Security Architecture for the Internet Protocol. 2005." URL: https://tools.ietf.org/html/rfc4301 (2006).

[11] strongSwan, https://www.strongswan.org/documentation.html

[12] OpenVPN Official Documentation, OpenVPN Technologies, https://openvpn.net/index.php/open-source/documentation.html

[13] C. Hosner, "OpenVPN and the SSL VPN revolution." 2009203205]. http://www. openvpn. net/index. php/open2 source/documentation. html.

[14] I. Coonjah, P. C. Catherine, and K. M. S. Soyjaudah. "Experimental performance comparison between TCP vs UDP tunnel using OpenVPN." *Computing, Communication and Security (ICCCS), 2015 International Conference on*. IEEE, 2015.

[15] P. Chitre, K. Manish, and M Hadjitheodosiou. "TCP in the IPSEC environment." *22nd AIAA International Communications Satellite Systems Conference & Exhibit 2004 (ICSSC)*. 2004.

[16] Intel, http://ark.intel.com/search/advanced/?s=t&AESTech=true

[17] KVM, http://www.linux-kvm.org/page/Documents

A Big Data Solution for Troubleshooting Mobile Network Performance Problems

K. Skračić, I. Bodrušić
Ericsson Nikola Tesla, Zagreb, Croatia
E-mail: kristian.skracic@ericsson.com

Abstract - Big Data has become a major competitive advantage for many organizations. The analytical capabilities made possible by Big Data analytics platforms are a key stepping stone for advancing the business of every organization. This paper illustrates the development of a big data analytics system for mobile telecommunication systems. The authors developed a solution for analyzing data produced by mobile network nodes which contain data relevant for predictive maintenance and troubleshooting purposes. The solution is built around the problem of working with small files in the Hadoop environment. The logs collected from mobile network nodes are small binary files between 5 and 15MB in size. These binary log files need to be decoded to a readable format, and then analyzed to extract useful information. In this paper, the authors provided a benchmark of various scenarios for collecting and decoding the binary log files in a Hadoop cluster. As a result, the scenario with the highest performance has been used in the implementation of our solution. The developed solution has been built and tested on a live Hadoop cluster using real-world data obtained from several telecom operators around the world.

I. INTRODUCTION

The telecommunications industry is one of the largest and fastest growing industries in the world. Over the past few decades we have witnessed the change from 2G, 3G, 4G and now 5G in the near future [1]. With the rising demand for constant connectivity, the availability of telecommunication systems and their various components has never been more important. This trend is particularly apparent in the case of mobile networks. As shown in [2], during 2016 there were 7.5 billion mobile subscriptions in the world. The report in [2] estimates that in 2022 there will be 8.9 billion mobile subscriptions, 8 billion mobile broadband subscriptions and 6.1 billion unique mobile subscribers in the world. Mobile networks allow a subscriber to consume various telecommunication services via mobile phone from any place of coverage. Thus, they have become one of the most important resources today. This paper proposes a solution for analyzing data produced by mobile network nodes which contain information relevant for predictive maintenance and troubleshooting purposes.

As telecommunication systems are becoming larger and more advanced, it is necessary to constantly monitor and measure the performance of their various subsystems. However, larger networks and higher Internet access speeds carry with them the need to analyze larger amounts

of data in a short period of time, in order to prevent network outages as soon as possible. Such requirements can be addressed by leveraging the analytical capabilities made possible by Big Data analytics platforms [3]. Apart from improved performance, Big Data can enable deeper analytics by providing access to historical data. For example, by storing network performance data it becomes possible to compare the current results with those obtained in past measurements. By storing the insight obtained through troubleshooting, it becomes possible to predict and prevent the same failures from happening again, or at least to shorten the response time when the same or similar failures present themselves again [4].

In this paper, the authors developed a Big Data solution for analyzing data produced by mobile network nodes, which contain important information used for troubleshooting purposes. The solution ingests large amounts of logs which contain network event information. The logs are gathered through an event-based monitoring (EBM) system, which is an embedded recording tool in the Ericsson EPG, SGSN and MME nodes [5].

This paper is organized as follows. Section II provides an overview of existing research on small files in the Hadoop environment, as well as the research on the applicability of Big Data solutions in the telecommunications industry. Section III shows the architecture of the developed solution and how the small files problem in Hadoop impacts it. Section IV shows the benchmark results for the scenarios laid out in Section III.

II. RELATED WORK

This section provides an overview of the results of existing research in the field of applying Big Data in the telecommunications industry, as well as previous work done on the analysis of small files in Hadoop.

A. Big Data Analytics Platforms in the Telecommunications Industry

Big Data solutions have become an important part of today's industry for all types of businesses, such as finance [6], law enforcement [7], education [8] and others. To show the applicability of Big Data solutions in the telecommunications industry, we briefly summarized some existing use cases and their impact on the industry.

Telecom operators have access to large amounts of valuable data that can be used for various analytical use cases. The research in [9] shows a way to leverage Big

Data analytics for classifying subscribers based on their movement. Reusing existing data for new use cases such as this is the first step in data monetization, which is expected to become a major source of income for all types of businesses in the near future [10]. The work in [11] is able to predict customer churn for telecom operators, which has a direct impact on the operator's profitability. Similarly, the work in [12] predicts customer experience when using over-the-top (OTT) applications such as WhatsApp or Viber.

As the Internet of Things (IoT) is evolving, it is expected to have an impact in the way telecommunications providers analyze the large amounts of sensor data such systems bring with them [13]. We would also like to note that the move to Big Data has a big impact on network infrastructure evolution, as such systems require higher link speeds to transfer the data form one node to another [14].

The solution developed in this paper is focused on mobile network performance troubleshooting. Thus, it is used to calculate various key performance indicators (KPIs) relevant for this domain. KPI measurement is frequently used by mobile operators and mobile network infrastructure vendors as a means to systematically search and identify network bottlenecks and anomalies [15].

B. Analyzing Small Files in the Hadoop Environment

Hadoop is an open-source software framework used for distributed storage and processing of very large data sets [16]. The Hadoop distributed file system (HDFS) has been widely adopted as the standard for storing data in Hadoop based clusters [17]. In the Hadoop ecosystem, access to stored data is handled by a system called Namenode, which manages the file system namespace and regulates client access. First the client asks the Namenode for instructions on where to find the files it needs to read, as well as the location of a free block it can write to [18]. Figure 1 illustrates this process. DataNodes provide block storage and serve I/O requests from clients.

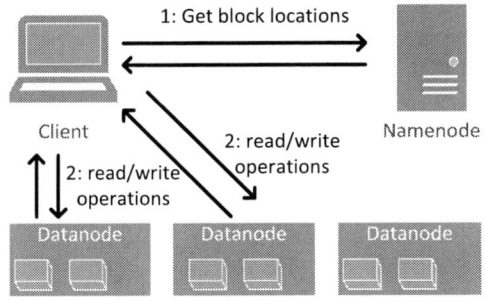

Figure 1. HDFS system overview

A major drawback of HDFS is its poor performance with large numbers of small files, which has attracted significant attention [19]. According to the research in [19], the main reasons for such lower performance are:

- large numbers of small files impose a heavy burden on NameNode memory;

- correlations between small files are not considered for data placement;

- no optimization mechanism, such as prefetching, is provided to improve I/O performance

We would like to note that when small files are stored on HDFS, disk utilization is not a bottleneck. The research in [20] shows that a small file stored on HDFS does not take up any more disk space than is required to store its contents. More precisely, a 6 MB file stored with an HDFS block size of 128 MB uses 6 MB of disk space, not 128 MB.

HDFS is designed to read/write large files, and provides no optimization for handling small files. In cases where large amounts of small files are accessed directly in HDFS, a mismatch of accessing patterns will emerge [21]. HDFS will ignore the optimization offered by the native storage resource, which will lead to local disk access becoming a bottleneck [22]. Additionally, in such a scenario data prefetching is not employed to improve access performance for HDFS [22].

The research in [21] considers all files smaller than 16MB as small files, although no justification or proof were provided as to why this size was chosen as the cut-off point between large and small files in the context of HDFS. The research in [19] has quantified this cut-off point through experimentation. The study indicates that access efficiency starts to drop significantly with files smaller than 4.35 MB.

The small file processing problem in Hadoop has seen many different solutions with various levels of success, depending on the nature of the data. One of these is the merging of multiple small files into a single bigger file, which has shown some significant performance improvements [21], [23]. This paper explores different scenarios in which this solution can be applied to mobile network data. The scenarios are explained in more detail in the following sections.

III. SYSTEM OVERVIEW

This section provides an overview of the developed Big Data solution for mobile network performance troubleshooting.

A. Data Collection

Event data has been used for various troubleshooting purposes [5]. The event data is collected from EPG, SGSN and MME nodes within the core network. The environment is based on the Evolved Packet Core (EPC) [24], as sown on Figure 2. This study uses only event data generated by these nodes. The authors used an event-based monitoring (EBM) system, which is an embedded recording tool in the Ericsson EPG, SGSN and MME. We collected events on 2G, 3G and 4G networks.

The event data is collected in small log files which are between 5 and 15 MB in size, depending on the configuration. The overall size and and velocity of the logs depends on the size of the network (e.g. number of base stations/eNodeB, number of EPG, SGSN and MME nodes, overall network throughput). An average operator will generate around 200 GB of logs per day.

The log files are stored in a binary format which needs to be decoded to text (usually CSV) in order to be processed. After the decoding process, the files are up to 10 times larger than in their binary format. EBM logs contain information that documents successful and unsuccessful events for completed mobility and session management procedures.

As shown in the following section, the developed solution was tested on several small and large networks around the world.

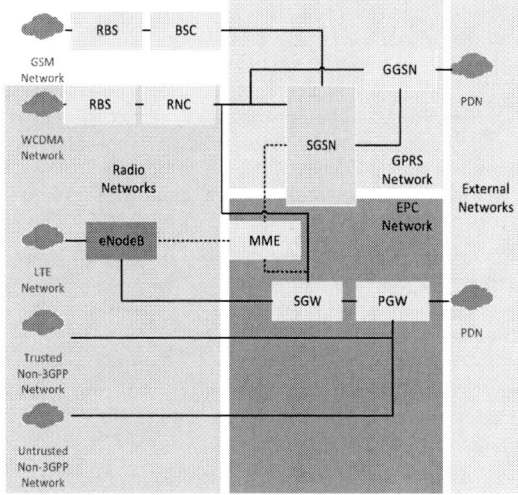

Figure 2 - Evolved Packet Core (EPC) schema

B. Architecture

Figure 3 shows the architecture of the developed solution. Apache Flume is used to transfer the data from a network location to HDFS. Flume was chosen because it is a widely used distributed, reliable and available service for efficiently collecting, aggregating and moving large amounts of streaming event data [25]. The binary logs are usually dumped to a server, which is usually somewhere in the operator's network. Using Flume, the binary logs are transferred to the cluster that hosts the proposed solution. Although the proposed solution can be deployed within the operator's network, we argue that a centralized off-site deployment is more appropriate. A centralized approach enables data aggregation from various networks into a single cluster, and thus enriches the data with variety that comes from different network configurations and environments.

HDFS is used to store the raw binary log files until they are decoded. A MapReduce job is used to decode the binary files into CSV. The decoding process is explained in more detail in the following section.

The decoded CSV files are imported into an Apache Hive database [26]. Hive offers an SQL-like query language which enables data access. The developed solution has a number of Hive queries that calculate various KPI's, which provide insight about mobile

network bottlenecks. This is a quick method of finding out which parts of the network are worth looking into during troubleshooting. The results, or KPI's calculated from such queries, are stored in a separate relation database, which is based on PostgreSQL. External applications can also connect to the Hive database and query the data to calculate KPI's relevant for mobile network troubleshooting. One of the goals for this solution is to enable data mining. Mobile network experts can connect to the Hive database either though the Hive shell or by using a visualization tool like Tableau. Using the original measurement data stored in Hive, mobile network experts can extract new insight and get to the root cause of a problem. This is often not possible with aggregated KPI data because it hides much of the information it is derived from.

Hive provides several mechanisms for optimizing the storage of the data and query performance. The developed solution makes use of partitioning and bucketing functionalities offered by Hive. Partitioning in Hive is the process of horizontally dividing the data into a number of smaller and more manageable slices. Every partition is stored as a directory within a data warehouse table in Hive. The developed solution partitions the decoded CSV data based on the event identifier (or event name) attribute.

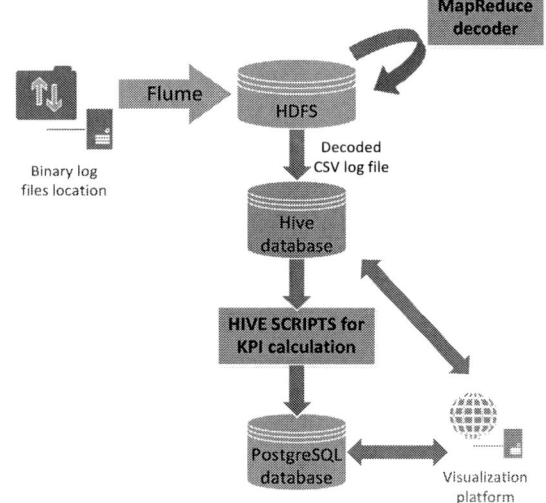

Figure 3 – Architecture of the developed solution

Bucketing is another technique of decomposing data into more manageable parts. This optimization method distributes the data evenly across multiple files. It is used to distribute and organize the table or partition data into multiple files so that similar records are present in the same file. The value of this column will be hashed into buckets by a user-defined number. Bucketing has many performance benefits, most notably faster Map side joins, more efficient grouping using "Group By" statements and more efficient sampling. As the developed solution is used by mobile network experts, such statements are used very often when accessing the Hive database directly.

C. Log Decoder and the Impact of Small Files in Hadoop

The developed solution implements a MapReduce job to decode the binary files into the CSV format. The MapReduce job first reads the binary files in memory and then decodes them in parallel. The number of parallel decoding jobs is defined by the number of input splits in Hadoop. This is where the small files problem influences the developed solution. If each raw log file is decoded separately, it will have a negative performance impact. In the developed solution, we tried to influence the number of input splits by combining multiple raw log files into one larger file. More precisely, we tested the performance impact of using small files with the following scenarios:

- Scenario 1: the raw logs are stored as small files in HDFS and they are directly used as input splits in the MapReduce job. A custom Hadoop input reader is used to read the binary log files, which is based on the native RecordReader class in Hadoop
  (*org.apache.hadoop.mapreduce.RecordReader*)

- Scenario 2: the raw logs are combined into larger files, stored in HDFS and then decoded using MapReduce jobs. Each line in the combined file is a hexadecimal representation of the binary log file. In this scenario, Flume is used to combine the raw log files. Thus, a native Hadoop input reader is used (located in *org.apache.hadoop.mapreduce.lib.input.TextInput Format*)

- Scenario 3: the raw logs are combined into larger files, stored in HDFS and decoded using MapReduce jobs with only mappers and no reducers. Like in Scenario 2, each line in the input file contains a single file, and Flume is used to combine the raw log files

IV. RESULTS

For the purposes of this study, a Hadoop-based cluster was used to evaluate the developed solution. The cluster is based on the Hortonworks Data Platform (HDP) [27] and is composed of 10 servers (2 masters and 8 slaves). The master nodes have 2 model E5-2630 CPU-s, 128GB of RAM and 6 hard disk drives (HDD), each with 3TB of space. The slave nodes have 1 model E5-2623v3 CPU, 64GB of RAM, and 8 HDDs, each with 2TB of space. Each node runs on top of CentOS v7, which is installed on a separate SSD disk which is not part of the HDFS.

The input data was collected from several small and large networks around the world, including operators from Europe, North and South America, and Southeast Asia.

A. Small Files Decoding Benchmark

The combined files in Scenario 2 and 3 were grouped into larger files. Several performance tests were carried out on batches of 2, 9 and 33 GB of raw log files.

Table 1 shows how the decoder performs in Scenario 1, when the small log files are used directly. It can be seen

that the MapReduce decoder is having difficulties processing even the smaller batches of 2 and 9 GB of raw logs. As stated in previous sections, the reason for this is the large number of small files that is imposing a heavy burden on NameNode memory. In contrast, Scenario 2 (Table 2) shows a significant performance improvement of the MapReduce decoder. Also, we used 18 reducers in Scenario 2, one for each event type available through the logging system.

Undoubtedly, the best performance was achieved in Scenario 3 by combining the input files into larger files and using map-only jobs (Table 3). The reason for such an improvement is that both the shuffle-sort and the reduce phases of the MapReduce job are skipped, thus drastically reducing the amount of processing power and memory needed to decode a large input file. For the largest batch of 33 GB, we can see that there is a 37% improvement compared to Scenario 2. We would like to note that this improvement increases with batch size (Figure 4). Figure 5 shows the performance gain for each scenario. The performance improvements rise with the batch size, which is traditionally not the case for solutions that are not based on Big Data.

The drawback of using the approach in Scenario 3 is that the output of the decoder is split into several smaller files which need to be imported into Hive. This is due to the way MapReduce jobs work. The intermediate results of the mappers are shuffled and sorted before being delivered to the reducers. By having only map jobs in our decoder, the unsorted intermediate results become the output. In contrast, when the reducers are used we can influence the number of files that will be generated. For example, each event type could be stored into a separate file. However, the files can easily be merged within Hive, as the output is textual (CSV). Also, the partitioning and bucketing in Hive restructures the physical layout of the data, so that the output of the map-only decoder does not influence the performance of the Hive queries.

TABLE 1. DECODER BENCHMARK FOR SCENARIO 1 – USING SMALL LOG FILES AS INPUT

Raw log size (GB)	Seconds	Minutes
2.00	843	14.05
9.00	3206.00	53.43
33.00	13740	229.00

TABLE 2. DECODER BENCHMARK FOR SCENARIO 2 – USING 18 REDUCERS AND SEVERAL LARGER COMBINED INPUT FILES

Raw log size (GB)	Seconds	Minutes
2.00	428.00	7.13
9.00	1405.00	23.41
33.00	7282.00	121.36

TABLE 3. DECODER BENCHMARK FOR SCENARIO 3 – USING ONLY MAPPERS AND SEVERAL LARGER COMBINED INPUT FILES

Raw log size (GB)	Seconds	Minutes
2.00	348.00	5.80
9.00	883.00	14.71
33.00	3194.00	53.23

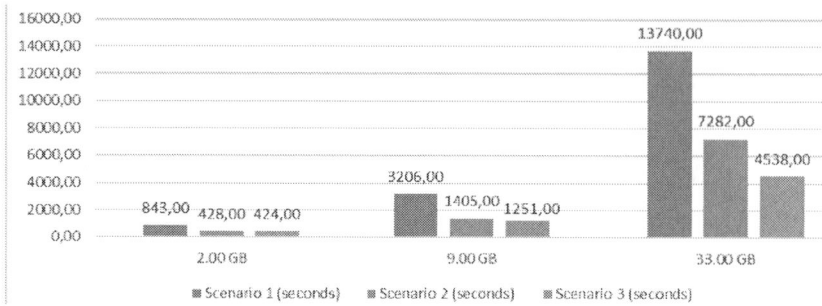

Figure 4 – Batch processing benchmark for 3 considered Scenarios

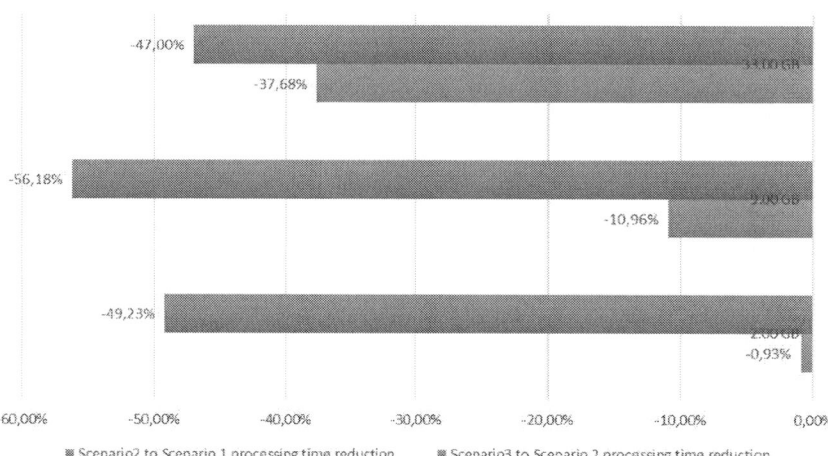

Figure 5 – Scenario to scenario processing time reduction comparison

B. Comparison with existing solutions

In order to clearly show the performance benefits when using big data solutions and technologies in this scenario, the study also shows the benchmark for decoding EBM logs without the proposed solution. Table 4 shows the performance benchmark on a single server, with the same hardware as the master node in the HDP cluster.

As shown in table Table 4, the proposed solution is up to 6 times faster than existing solutions. The largest set of logs was unable to be measured since it was not possible with the existing solutions.

We note that the performance gains represent only one benefit of the proposed solution. The main advantage of the proposed solution is the ability to process a much larger set of logs than was possible with legacy solutions. Also, the proposed solution provides a way to continuously gather and store logs for deeper analytics, which was no the case with legacy solutions.

TABLE 4 EXISTING SOLUTIONS BENCHMARK

Raw log size (GB)	Seconds	Minutes
2.00	1264.00	21.07
9.00	5538.00	92.30
33.00	N/A	N/A

V. CONCLUSION

Big Data analytics can provide insight into the data available within the telecommunications industry. This paper demonstrates one use case in which analytics can be leveraged to improve the efficiency and value of troubleshooting in mobile networks. The main advantage of the developed solution is its ability to adapt to any new analytical requests, as well as the ability to adapt to changing input file sizes. More precisely, the results of this study show that the developed solution is capable of processing small files in an efficient manner within the Hadoop environment, which was not built for processing large amounts of small files. Additionally, the study shows that by skipping the reduce phase we can decrease the execution time of the MapReduce job used for decoding. This was shown to be the most time-consuming process. The developed solution was tested using log data collected in various small and large networks from around the world, which demonstrates its applicability for mobile network troubleshooting. The developed solution has proven that Big Data platforms are suitable for processing large batches of mobile network data, and that they bring significant performance and scalability improvements compared to traditional solutions. Future research may include the use of other Big Data tool that run on the Hadoop platform. Most notably, the use of Apache Cassandra instead of the Hive, and the use of Apache Spark as a replacement to the MapReduce decoder job.

ACKNOWLEDGMENT

This study was fully funded by Ericsson Nikola Tesla. The authors thank their leadership team for providing an environment in which it was possible to combine research and industry into one.

REFERENCES

[1] G. Fettweis and S. Alamouti, "5G: Personal mobile internet beyond what cellular did to telephony," *IEEE Communications Magazine*, vol. 52, no. 2, pp. 140–145, Feb. 2014.

[2] C. Patrik, L. Anette, and J. Peter, "Ericsson Mobility Report," EAB-16:018498 Uen, Revision A, Nov. 2016.

[3] D. Šipuš, "Big data analytics for communication service providers," in *2016 39th International Convention on Information and Communication Technology, Electronics and Microelectronics (MIPRO)*, 2016, pp. 513–517.

[4] L. Yang, G. Kang, W. Cai, and Q. Zhou, "An Effective Process Mining Approach against Diverse Logs Based on Case Classification," in *2015 IEEE International Congress on Big Data*, 2015, pp. 351–358.

[5] I. da Silva, Y. Wang, F. Mismar, and W. Su, "Event-based performance monitoring for inter-system cell reselection: A SON enabler," in *2012 International Symposium on Wireless Communication Systems (ISWCS)*, 2012, pp. 6–10.

[6] A. Munar, E. Chiner, and I. Sales, "A Big Data Financial Information Management Architecture for Global Banking," in *2014 International Conference on Future Internet of Things and Cloud*, 2014, pp. 385–388.

[7] A. Jain and V. Bhatnagar, "Crime Data Analysis Using Pig with Hadoop," *Procedia Computer Science*, vol. 78, pp. 571–578, Jan. 2016.

[8] F. Xhafa, D. Garcia, D. Ramirez, and S. Caballé, "Performance Evaluation of a MapReduce Hadoop-Based Implementation for Processing Large Virtual Campus Log Files," in *2015 10th International Conference on P2P, Parallel, Grid, Cloud and Internet Computing (3PGCIC)*, 2015, pp. 200–206.

[9] B. Furletti, L. Gabrielli, C. Renso, and S. Rinzivillo, "Analysis of GSM calls data for understanding user mobility behavior," in *2013 IEEE International Conference on Big Data*, 2013, pp. 550–555.

[10] H. Cao *et al.*, "SoLoMo analytics for telco Big Data monetization," *IBM Journal of Research and Development*, vol. 58, no. 5/6, p. 9:1-9:13, Sep. 2014.

[11] H. Li, D. Yang, L. Yang, YaoLu, and X. Lin, "Supervised Massive Data Analysis for Telecommunication Customer Churn Prediction," in *2016 IEEE International Conferences on Big Data and Cloud Computing (BDCloud), Social Computing and Networking (SocialCom), Sustainable Computing and Communications (SustainCom) (BDCloud-SocialCom-SustainCom)*, 2016, pp. 163–169.

[12] E. Diaz-Aviles *et al.*, "Towards real-time customer experience prediction for telecommunication operators," in *2015 IEEE International Conference on Big Data (Big Data)*, 2015, pp. 1063–1072.

[13] S. Din, H. Ghayvat, A. Paul, A. Ahmad, M. M. Rathore, and I. Shafi, "An architecture to analyze big data in the Internet of Things," in *2015 9th International Conference on Sensing Technology (ICST)*, 2015, pp. 677–682.

[14] I. Tomkos, C. Kachris, P. S. Khodashenas, and J. K. Soldatos, "Optical networking solutions and technologies in the big data era," in *2015 17th International Conference on Transparent Optical Networks (ICTON)*, 2015, pp. 1–1.

[15] S. Singh, Y. Liu, W. Ding, and Z. Li, "Evaluation of Data Mining Tools for Telecommunication Monitoring Data Using Design of Experiment," in *2016 IEEE International Congress on Big Data (BigData Congress)*, 2016, pp. 283–290.

[16] K. Shvachko, H. Kuang, S. Radia, and R. Chansler, "The Hadoop Distributed File System," *2010 IEEE 26th Symposium on Mass Storage Systems and Technologies (MSST)*, pp. 1–10, May 2010.

[17] W. Tantisiriroj, S. Patil, and G. Gibson, "Data-intensive File Systems for Internet Services: A Rose by Any Other Name... (CMU-PDL-08-114)," *Parallel Data Laboratory*, Oct. 2008.

[18] S. Bende and R. Shedge, "Dealing with Small Files Problem in Hadoop Distributed File System," *Procedia Computer Science*, vol. 79, pp. 1001–1012, Jan. 2016.

[19] B. Dong, Q. Zheng, F. Tian, K.-M. Chao, R. Ma, and R. Anane, "An optimized approach for storing and accessing small files on cloud storage," *Journal of Network and Computer Applications*, vol. 35, no. 6, pp. 1847–1862, Nov. 2012.

[20] T. White, *Hadoop: The Definitive Guide*. O'Reilly Media, Inc., 2009.

[21] X. Liu, J. Han, Y. Zhong, C. Han, and X. He, "Implementing WebGIS on Hadoop: A case study of improving small file I/O performance on HDFS," in *2009 IEEE International Conference on Cluster Computing and Workshops*, 2009, pp. 1–8.

[22] J. Shafer, S. Rixner, and A. L. Cox, "The Hadoop distributed filesystem: Balancing portability and performance," in *2010 IEEE International Symposium on Performance Analysis of Systems Software (ISPASS)*, 2010, pp. 122–133.

[23] P. Gohil, B. Panchal, and J. S. Dhobi, "A novel approach to improve the performance of Hadoop in handling of small files," in *2015 IEEE International Conference on Electrical, Computer and Communication Technologies (ICECCT)*, 2015, pp. 1–5.

[24] G. Kuhn, J. Eisl, and H. Becker, "Co-operative handover in 3G System Architecture Evolution," in *32nd IEEE Conference on Local Computer Networks (LCN 2007)*, 2007, pp. 643–650.

[25] P. B. Makeshwar, A. Kalra, N. S. Rajput, and K. P. Singh, "Computational scalability with Apache Flume and Mahout for large scale round the clock analysis of sensor network data," in *2015 National Conference on Recent Advances in Electronics Computer Engineering (RAECE)*, 2015, pp. 306–311.

[26] G. P. Haryono and Y. Zhou, "Profiling apache HIVE query from run time logs," in *2016 International Conference on Big Data and Smart Computing (BigComp)*, 2016, pp. 61–68.

[27] K. K. Gadiraju, M. Verma, K. C. Davis, and P. G. Talaga, "Benchmarking performance for migrating a relational application to a parallel implementation," *Future Generation Computer Systems*, vol. 63, pp. 148–156, Oct. 2016.

Overlapping Blocks in Reconstruction of Sparse Images

Isidora Stanković, Miloš Daković, Irena Orović

Faculty of Electrical Engineering, University of Montenegro
Dzordza Vasingtona bb, 20 000 Podgorica, Montenegro
[isidoras,milos,irenao]@ac.me

Abstract—**Images are commonly analysed by the discrete cosine transform (DCT) on a number of blocks of smaller size. The blocks are then combined back to the original size image. Since the DCT of blocks have a few nonzero coefficients, the images can be considered as sparse in this transformation domain. The theory of compressive sensing states that some corrupted pixels within blocks can be reconstructed by minimising the blocks sparsity in the DCT domain. Block edges can affect the quality of the reconstruction. In some blocks, a few pixels from an object which mostly belongs to the neighbouring blocks may appear at the edges. Compressive sensing reconstruction algorithm can recognise these pixels as disturbance and perform their false reconstruction in order to minimise the sparsity of the considered block. To overcome this problem, a method with overlapping blocks is proposed. Images are analysed with partially overlapping blocks and then reconstructed using their non-overlapped parts. We have demonstrated the improvements of overlapping blocks on images corrupted with combined noise. A comparison between the reconstructions with non-overlapping and overlapping blocks is presented using the structural similarity index.**

Keywords—**compressive sensing, image reconstruction, overlapping blocks, gradient algorithm, noisy image**

I. INTRODUCTION

An image is said to be sparse if it consists of only few nonzero coefficients in a transformation domain. A sparse image can be reconstructed with a reduced set of pixels. The processing and reconstruction of such images are examined within the theory of compressive sensing (CS) [1]–[9]. The theory of CS is widely used in various applications in the area of digital signal processing, since many real signals are sparse in a certain transformation domain. Numerous reconstruction algorithms for different kinds of signals have been developed within this field. They can be divided in several groups. The algorithm considered in this paper is from the group of algorithms based on the minimisation of the sparsity measure by using the gradient of the L_1-norm [10]. In this algorithm, the image is reconstructed in the spatial (pixels) domain. The corrupted pixels are detected, declared as unavailable (missing) and considered as the minimisation variables. This property

This work is supported by the Montenegrin Ministry of Science, project grant funded by the World Bank loan: CS-ICT "New ICT Compressive sensing based trends applied to: multimedia, biomedicine and communications".

Authors are with the University of Montenegro, Faculty of Electrical Engineering, Dzordza Vasingtona bb, 20 000 Podgorica, Montenegro.

Corresponding author is Isidora Stanković (e mail: isidoras@ac.me).

makes the algorithm suitable for denoising of corrupted pixels in a noisy environment.

Common images have a small number of nonzero coefficients in the two-dimensional discrete cosine transform (2D-DCT) space. A reduced set of pixels can be used to reconstruct sparse images. Different reasons can cause that only a reduced set of pixels is available. One reason can be in heavy pixels corruption. Corrupted pixels may be declared as unavailable. Then the image is reconstructed using the CS methods. The impulsive noise is example of such a disturbance. It can appear due to analog to digital conversion errors, communication errors, dead pixels in image acquisition equipment, etc. Here we will consider a form of impulsive noise known as the salt and pepper noise with an addition of noise whose values are within the range of the original image pixels.

The paper is organised as follows. In Section II, theoretical background of sparse signal processing is presented. In Section III, the reconstruction algorithm is introduced. Section IV explains the idea of adding the overlapping block step. In Section V, the results and comparison using the structural similarity index is shown. In Section VI, conclusions are presented.

II. THEORETICAL BACKGROUND

Let us consider a grayscale image $I(m,n)$ of size $N \times N$. Based on the JPEG standard, we will split the image into blocks of size $B \times B$. We will assume that a image block starting at pixel (m_0, n_0) is defined as

$$x(m,n) = I(m_0+m, n_0+n), \quad m,n = 0,1,\ldots,B-1. \quad (1)$$

Its 2D-DCT representation is denoted by $X(k,l)$. The vector notations of the image blocks and their 2D-DCT are

$$\mathbf{x} = \mathbf{\Psi X} \text{ and } \mathbf{X} = \mathbf{\Phi x} \quad (2)$$

where $\mathbf{\Psi}$ and $\mathbf{\Phi}$ are the transform and inverse transform matrices with rearranged elements of the 2D-DCT. Vectors $\mathbf{x}$ and $\mathbf{X}$ are obtained by stacking columns of the corresponding block pixels and 2D DCT transform.

The sparsity of one image block is $K < (B \times B)$. From the compressive sensing theory we know that, if an image is of a sparsity K, it can be reconstructed from less than $B \times B$ pixels/measurements. The measurements are denoted by $\mathbf{y}$

$$\mathbf{y} = [x(m_1,n_1), x(m_2,n_2), \ldots, x(m_M,n_M)] \quad (3)$$

where M is the number of pixels/measurements used in a block. The relation between the values is $K < M < B \times B$. Our goal, in the sparse signal processing sense, is to reconstruct an image using the available pixels/measurements by minimising the sparsity, i.e.

$$\min \|\mathbf{X}\|_1 \qquad \text{subject to} \quad \mathbf{y} = \mathbf{A}\mathbf{X} \qquad (4)$$

where $\mathbf{A}$ is an $M \times N$ measurement matrix obtained from matrix $\mathbf{\Psi}$ by selecting rows that correspond to the available pixels. The measurements are the available (uncorrupted) pixels in the image

$$y(i) = x(m_i, n_i). \qquad (5)$$

Positions of the available pixels/measurements are $(m_i, n_i) \in \mathbb{M} = \{(m_1, n_1), (m_2, n_2), ..., (m_M, n_M)\}$. If we have an 8-bit $B \times B$ image block, corrupted with salt and pepper noise, the block can be then written as

$$x_a^{(0)}(m,n) = \begin{cases} x(m,n), & \text{for } (m,n) \in \mathbb{M} \\ 0 \text{ (or } 255), & \text{elsewhere} \end{cases} \qquad (6)$$

where 0 and 255 are salt and pepper noise. If an uniform noise is used then the values are between 0 and 255, In the next section we will present an algorithm used for the recovery of the noisy pixels.

III. RECONSTRUCTION ALGORITHM

The algorithm was introduced in [10], [11]. It is based on the minimisation of the gradient of corrupted pixels. Consider a corrupted image block as presented in equation (6). In the initial stage, we add an arbitrary value $\pm\Delta$ to the corrupted pixels

$$\begin{aligned} x_a^+(m,n) &= x^{(p)}(m,n) + \Delta\delta(m - m_i, n - n_i) \\ x_a^-(m,n) &= x^{(p)}(m,n) - \Delta\delta(m - m_i, n - n_i) \end{aligned} \qquad (7)$$

where p is the iteration index. In the initial stage $p = 0$.

The arbitrary value Δ is usually the maximal uncorrupted pixel value, i.e. $\Delta = \max_{m,n}(\mathbf{y})$. For the corrupted pixels at positions $(m_i, n_i) \notin \mathbb{M}$ the gradient value is estimated as

$$g(m_i, n_i) = \frac{1}{2\Delta}\left(\left\|\mathbf{X}_a^+\right\|_1 - \left\|\mathbf{X}_a^-\right\|_1\right) \qquad (8)$$

where $\mathbf{X}_a^{\pm}$ are the 2D-DCT domain values of the signals in (7). Note that the gradient value for the uncorrupted pixels will be zero. Based on the gradient value, the corrupted pixel $x(m_i, n_i)$ is updated. Each corrupted pixel value is changed in the direction opposite of the gradient for a step μ

$$x_a^{(p)}(m_i, n_i) = x_a^{(p-1)}(m_i, n_i) - \mu g(m_i, n_i). \qquad (9)$$

Because of the shape of the gradient and norm-one sparsity measure, when the values are close to the true signal values, they will oscillate around the solution. The oscillations are proportional to the step size. When the oscillation is detected, the step sizes Δ and μ are reduced. These new parameters continue approach to the true signal values until a new precision is reached. The procedure is repeated until

the desired reconstruction accuracy is achieved. One of the stopping criterion for the algorithm can be if the change in two successive iteration is smaller than some desired accuracy.

Note that only corrupted pixels (which are changed during reconstruction steps) contribute to this change. This is the basic reconstruction algorithm when the positions of the corrupted pixels are known (if, for example, the pixels are distinguishably corrupted).

If other noise types are used, then an additional step is proposed in [12], [13]. Since the positions of the corrupted pixels are unknown, we repeat Eq. (7) and (8) for all pixels (corrupted and uncorrupted). Each time we take the pixel with the largest gradient, reconstruct it and eliminate it from the array of possible values. This will be repeated until the error of two successful iterations is below an acceptable level.

IV. OVERLAPPING BLOCKS

Let us consider an image of size $N \times N$ and that we split the image in number of blocks of size $B \times B$. Each block has M available/uncorrupted pixels. Within blocks we have corrupted pixels with salt and pepper noise and a uniform noise, whose values are similar to the uncorrupted pixels.

Assume that there are a few uncorrupted pixels at the edges which are not of the same or similar value as the other pixels in the block. These pixels mainly belong to the object of a neighbouring block. The compressive sensing theory looks for the sparsest possible solution as the reconstruction result. The sparsest solution of the considered block would be obtained by taking the uncorrupted pixels (which are part of the neighbouring object) as the corrupted ones, since these pixels significantly differ from the majority of the other uncorrupted pixels in that block. In this sense, a method will be falsely reconstruct these pixels as the pixels of similar values to the other pixels within the block.

To overcome this problem, we introduce the overlapping blocks. The idea is to take a bigger block to analyse the objects in surrounding blocks. Then we reconstruct the block and use only a smaller central part of the analysed block in the final image reconstruction. The method of overlapping blocks is suitable as an addition to the algorithms which are based on the detection and reconstruction of the noisy pixels itself, not the transformation domain coefficients.

As an illustration, let say that the block for analysis is of size $B \times B = 32 \times 32$. The size $B_o \times B_o$ will denote the size of the part which will be used in the final reconstruction. Obviously it must hold that $B_o < B$. We assume that we will use the central block of size $B_o \times B_o = 16 \times 16$. Illustration of this kind of blocks is presented in Fig. 1. Bigger blocks represent the blocks for analysis, which are of size $B \times B$, and smaller blocks are the blocks for final reconstruction. They are of size $B_o \times B_o$. Note that for the blocks which are at the edges of the whole image, we use the reconstruction from the edge analysis blocks.

V. RESULTS

In this section we will present reconstruction results using the method presented in the previous sections. Note that the

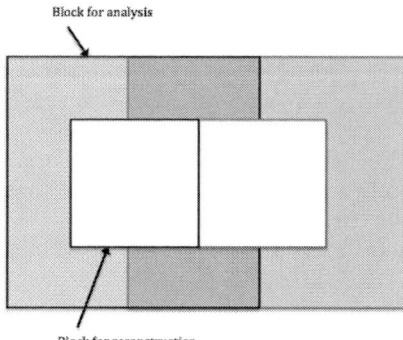

Fig. 1. Illustration of two overlapping blocks. Bigger block is used for the analysis and the smaller one is used for the final reconstruction

idea of overlapping blocks can be considered as an improvement to the reconstruction algorithms for the edges of objects in images. The algorithm was tested using overlapping and non-overlapping blocks. It was tested on a grayscale 512×512 image "Lena" with 50% of salt and pepper noise and 12.5% of random noise. It is considered that 10% of components in each block are nonzero in the 2D-DCT domain. The original image and the noisy image are presented in Fig. 2. The reconstruction without using overlapping blocks is shown in Fig. 3 (top). In the overlapping case the blocks for analysis are of size 32×32 and the part used for the final reconstruction is the central part of size 16×16. The result is shown Fig. 3 (bottom). The images zoomed in to the upper right corner are shown in Fig. 4.

A. Comparison

The comparison between the reconstructions using non-overlapping and overlapping blocks is presented using the localized structural similarity (SSIM) index. The SSIM index is a comparison parameter between two images. It is defined in [14] as

$$\text{SSIM}(\mathbf{x}_o, \mathbf{x}_r) = \frac{(2\mu_{x_o}\mu_{x_r} + c_1)(2\sigma_{x_o x_r} + c_2)}{(\mu_{x_o}^2 + \mu_{x_r}^2 + c_1)(\sigma_{x_o}^2 + \sigma_{x_r}^2 + c_2)} \quad (10)$$

where $\mathbf{x}_o, \mathbf{x}_r$ are the original and the reconstructed image, respectively. The values μ_{x_o}, μ_{x_r} are the mean values of the images, $\sigma_{x_o x_r}$ is the covariance between the two considered images, and $\sigma_{x_o}^2$, $\sigma_{x_r}^2$ are the variances of the two images. The values c_1, c_2 are used as stabilisation variables. If SSIM index is close to 1 the images are similar, if it is close to 0 they are not similar. The SSIM index of the zoomed images from Fig. 4 are shown in Fig. 5.

VI. Conclusions

A method for improving the reconstruction of noisy images using overlapping blocks is proposed. It is an improvement of the methods for reconstruction algorithms which are based on the detection of the corrupted pixels in spatial domain. The reconstruction of the images using non-overlapping and overlapping blocks is shown. The use of overlapping blocks improved results in the denoising of images.

Original image

Noisy image

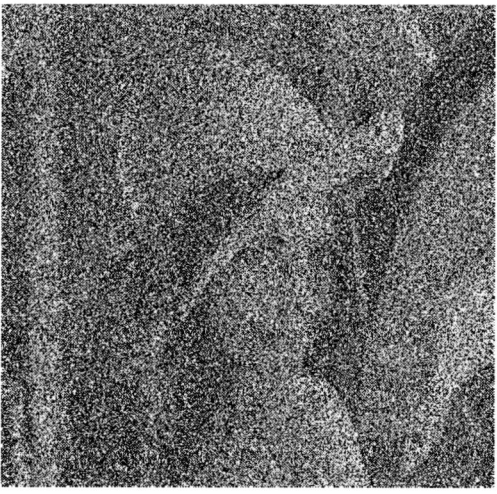

Fig. 2. Original image (top); Noisy image (bottom) used for the reconstruction

References

[1] R. Baraniuk, "Compressive Sensing," *IEEE Signal Processing Magazine*, vol. 24, no. 4, pp. 118-121, July 2007.

[2] E.J. Candes, and M. Wakin, "An Introduction to Compressive Sampling," *IEEE Signal Processing Magazine*, vol. 25, no. 2, pp. 21-30, March 2008.

[3] E.J. Candes, J. Romberg, and T. Tao, "Robust uncertainty principles: Exact signal reconstruction from highly incomplete frequency information," *IEEE Transactions on Information Theory*, vol. 52, no. 2, pp. 489-509, February 2006.

[4] D.L. Donoho, "Compressive Sensing," *IEEE Transactions on Information Theory*, vol. 52, no. 4, pp. 1289-1306., April 2006.

[5] S. Stanković, I. Orović, LJ. Stanković, "An Automated Signal Reconstruction Method based on Analysis of Compressive Sensed Signals in Noisy Environment," *Signal Processing*, vol. 104, Nov 2014, pp. 43 - 50, 2014.

[6] LJ. Stanković, S. Stanković, M. Amin, "Missing Samples Analysis in Signals for Applications to L-estimation and Compressive Sensing," *IEEE Signal Processing Letters*, vol. 94, Jan 2014, pp. 401-408, 2014.

[7] M. Elad, *Sparse and Redundant Representations: From Theory to Applications in Signal and Image Processing*, Springer, 2010.

Reconstructed with non–overlapping blocks

Reconstructed with overlapping blocks

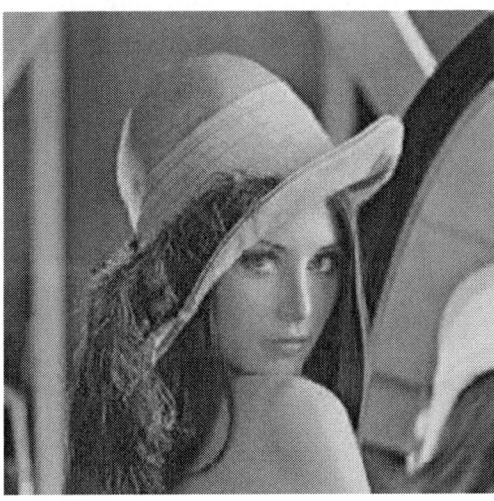

Fig. 3. Reconstructed images: with 32×32 non-overlapping blocks (top); with overlapping blocks (bottom)

[8] LJ. Stanković, *Digital Signal Processing with Selected Topics*. CreateSpace Independent Publishing Platform, An Amazon.com Company, November 4, 2015.

[9] S. Stanković, I. Orović, and E. Sejdić, *Multimedia signals and systems*. Springer - Verlag, 2012.

[10] LJ. Stanković, M. Daković, and S. Vujović, "Adaptive Variable Step Algorithm for Missing Samples Recovery in Sparse Signals," *IET Signal Processing*, vol. 8, no. 3, pp. 246 -256, 2014. (arXiv:1309.5749v1).

[11] I. Stanković, I. Orović, and S. Stanković, "Image Reconstruction from a Reduced Set of Pixels using a Simplified Gradient Algorithm," *22nd Telecommunications Forum TELFOR 2014*, Belgrade, Serbia

[12] I. Stanković, I. Orović, S. Stanković, M. Daković, "Iterative Denoising of Sparse Images," 39th International Convention on Information and Communication Technology, Electronics and Microelectronics MIPRO, May 2016.

[13] I. Stanković, I. Orović, M. Daković, S. Stanković, "Denoising of Sparse Images in Impulsive Disturbance Environment," Multimedia Tools and Applications, in print, 2017.

[14] Z. Wang, A. C. Bovik, H. R. Sheikh, and E. P. Simoncelli, "Image quality assessment: From error visibility to structural similarity," *IEEE Transactions on Image Processing*, vol. 13, no. 4, pp. 600-612, April 2004.

Reconstruction with non–overlapping blocks

Reconstruction with overlapping blocks

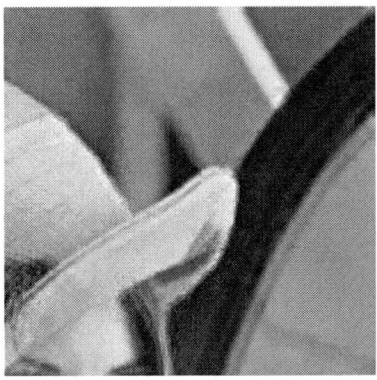

Fig. 4. Zoomed reconstructed images: with 32×32 non-overlapping blocks (top); with overlapping blocks (bottom)

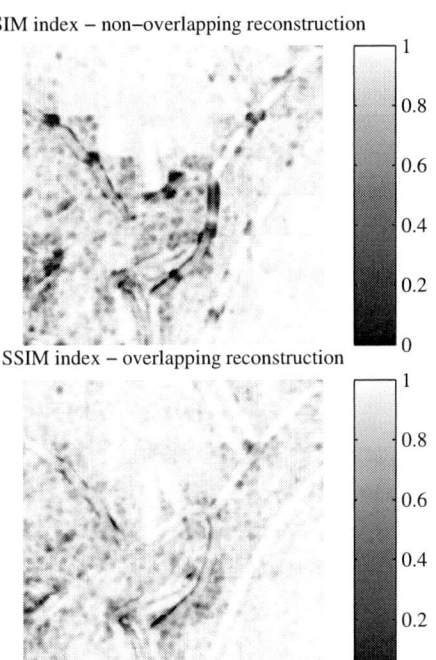

Fig. 5. SSIM index of the zoomed reconstructed images: with 32×32 non-overlapping blocks (top); with overlapping blocks (bottom)

MIPRO 2017, May 22- 26, 2017, Opatija, Croatia

Sparse Signal Reconstruction Based on Random Search Procedure

Miloš Daković, Isidora Stanković, Miloš Brajović, Ljubiša Stanković
Faculty of Electrical Engineering, University of Montenegro
Dzordza Vasingtona bb, 20 000 Podgorica, Montenegro
[milos,isidoras,milosb,ljubisa]@ac.me

Abstract—A method for reconstruction of sparse signals is presented in this paper. It is an improved version of the direct-search method for finding the set of non-zero coefficients representing the solution in sparsity domain. The proposed random search procedure is performed assuming the largest possible number of non-zero coefficients still satisfying the available measurements system. In the sparse signal processing and compressive sensing theory, this number should be smaller than or equal to the number of measurements. For each possible arrangement of the examined non-zero coefficients, the reconstruction is done by solving the system of equations in the least square sense, until the solution is found. Benefits of the proposed method are discussed. The calculation complexity improvement of the proposed method, compared to the direct-search, is analytically expressed. It depends on the total number of signal samples, number of measurements and the signal sparsity. The presented theory is confirmed with numerical examples.

Keywords — compressive sensing, signal reconstruction, direct search, sparsity

I. INTRODUCTION

A signal with a small number of nonzero coefficients (sparse signal) can be reconstructed from a reduced set of available samples/measurements [1]–[11]. The reduced number of measurements can be a consequence of various circumstances. It can occur as a result of a sampling strategy developed with the aim to reduce the storage requirements for the data. Measurements can be unavailable due to physical constraints or their intentional omitting due to a high noise corruption [11]. The analysis and reconstruction of sparse signals was the topic of many research papers [1]– [15]. Numerous reconstruction theorems and algorithms were developed [1], [2], [4]–[8].

Many signals in real applications can be considered as sparse in a certain transformation domain, meaning that the idea of sparse signal reconstruction can be exploited in different areas of signal processing. For example, digital images can be considered as sparse in the domain of discrete cosine transform (DCT), whereas radar ISAR data are sparse in the domain of two-dimensional Fourier transform [16].

One of the challenging topics in the compressive sensing is the optimal sampling strategy that will allow to reconstruct the signal with smallest possible number of available samples

This work is supported by the Montenegrin Ministry of Science, project grant funded by the World Bank loan: CS-ICT "New ICT Compressive sensing based trends applied to: multimedia, biomedicine and communications".

Authors are with the University of Montenegro, Faculty of Electrical Engineering, Dzordza Vasingtona bb, 81 000 Podgorica, Montenegro. Corresponding author is Isidora Stanković (e mail: isidoras@ac.me).

[12]. Various approaches are used to this aim, like those that minimize the coherence index of the isometry constant for a given signal transform. The direct search method provides exact reconstruction results if the unique reconstruction is theoretically possible. However, this method is computationally complex. For a large dimension of the reconstruction problem, number of available samples and sparsities, it is NP-hard and therefore not computationally feasible.

In this paper, we introduce an improvement in the computation of signal reconstruction by modifying the direct search strategy. It is known that for an observed sparsity level, compressive sensing algorithms, for instance ℓ_1-norm based or iterative, greedy, and other proposed algorithms [5], require a larger number of available measurements than the one required in ℓ_0-norm minimization. The aim is to overcome this issue by exploiting the possibility to reduce the number of trials in the direct search reconstruction procedure.

The paper is organized as follows. Basic compressive sensing definitions are presented in Section II. In Section III the direct search reconstruction algorithm is presented. The proposed, random search method, is introduced in section Section IV. Results and comparison are shown in Section V whereas the paper ends with concluding remarks.

II. BASIC DEFINITIONS

Let us consider a complex-valued discrete signal $x(n)$ of length N and its corresponding transformation domain $X(k)$

$$x(n) = \sum_{k=0}^{N-1} X(k)\psi_k(n), \qquad X(k) = \sum_{n=0}^{N-1} x(n)\varphi_n(k), \quad (1)$$

or in vector form $\mathbf{x} = \mathbf{\Psi X}$ and $\mathbf{X} = \mathbf{\Phi x}$. The inverse and the direct transform matrices are denoted as $\mathbf{\Psi}$ and $\mathbf{\Phi}$, respectively. We say that a signal is K-sparse in the transformation domain if the number of nonzero coefficients K is much smaller than the total length of signal N, i.e., $K \ll N$. Then a sparse signal can be reconstructed with $M < N$ measurements. The signal with M samples/measurements available is denoted as $y(m)$

$$y(m) = \sum_{k=0}^{N-1} X(k)\psi_k(m). \quad (2)$$

Previous definition can be written in a vector form as

$$\mathbf{y} = \mathbf{AX} \quad (3)$$

where $\mathbf{A}$ is the measurement matrix of size $M \times N$. It is formed based on the matrix $\mathbf{\Psi}$, containing rows that correspond to the positions of the available measurements/observations, whereas the rows corresponding to the missing samples are omitted.

The sparse signal reconstruction can be defined as the solution of the optimization problem

$$\min \ \|\mathbf{X}\|_0 \ \text{subject to} \ \mathbf{y} = \mathbf{AX}. \tag{4}$$

Having an undetermined system of linear equations defining the available measurements, the solution of the signal reconstruction problem is the one satisfying this system of equations, and being the sparsest possible. That is, the aim is to minimize the sparsity of $\mathbf{X}$ using the available measurements $\mathbf{y}$. This is achieved by exploiting a sparsity measure. A natural choice for this measure is the so-called ℓ_0-norm which counts the number of nonzero coefficients in $\mathbf{X}$, although not satisfying norm properties in a strict mathematical sense. However, this function is not convex and its minimization could be done only through a combinatorial search. Moreover, it can be easily shown that a direct combinatorial search is not computationally feasible for a reasonable length of the considered signal, its sparsity and number of available samples. This pseudo-norm is also very sensitive to the noise influence and quantization errors. This is the reason why, in practice and theory, more robust norms are exploited as sparsity measures.

The ℓ_1-norm is the most frequent used norm since it is closest convex function to the ℓ_0-norm. It is equal to the sum of absolute values of $\mathbf{X}$. However, all norm-one reconstruction methods, as well as other standard algorithms developed within the fields of sparse signal recovery and compressed sensing, require more samples/measurements than the minimal possible number that can provide a unique signal reconstruction in theory. Motivated by this fact, we try to reduce the computational cost of the combinatorial approach, with the aim to obtain the results similar to the direct search in sense of the minimal required number of measurements needed for a successful unique reconstruction.

III. DIRECT SEARCH RECONSTRUCTION

Any problem described with (4) can be solved by a direct search over the whole set of possible values of nonzero coefficient positions. This procedure is defined as the direct search minimisation of the ℓ_0-norm. Assume a vector $\mathbf{X}$ with sparsity K. We try to detect indices of the nonzero values $k \in \{k_1, k_2, ..., k_K\}$ out of the set of all possible indices between 1 and N

$$k \in \mathbf{K} \subset \mathbf{N} \tag{5}$$

where $\mathbf{K} = \{k_1, k_2, ..., k_K\}$, $\mathbf{N} = \{1, 2, ..., N\}$. The vector $\mathbf{X}_K$ contains assumed K nonzero elements of $\mathbf{X}$ at the positions from set $\mathbf{K}$. The system

$$\mathbf{y} = \mathbf{A}_K \mathbf{X}_K \tag{6}$$

with $M > K$ equations is solved by minimising the least square error

$$e^2 = (\mathbf{y} - \mathbf{A}_K \mathbf{X}_K)^H (\mathbf{y} - \mathbf{A}_K \mathbf{X}_K) =$$
$$\|\mathbf{y}\|_2^2 - 2\mathbf{X}_K^H \mathbf{A}_K^H \mathbf{y} + \mathbf{X}_K^H \mathbf{A}_K^H \mathbf{A}_K \mathbf{X}_K. \tag{7}$$

The minimum of the error is found from

$$\frac{\partial e^2}{\partial \mathbf{X}_K^H} = -2\mathbf{A}_K^H \mathbf{y} + 2\mathbf{A}_K^H \mathbf{A}_K \mathbf{X}_K = 0. \tag{8}$$

The solution is calculated as

$$\mathbf{A}_K^H \mathbf{A}_K \mathbf{X}_K = \mathbf{A}_K^H \mathbf{y}$$
$$\mathbf{X}_K = \left(\mathbf{A}_K^H \mathbf{A}_K \right)^{-1} \mathbf{A}_K^H \mathbf{y}. \tag{9}$$

For all solutions we check the error $\mathbf{y} - \mathbf{A}_K \mathbf{X}_K$. The reconstruction of the signal $\mathbf{X}$ is exact when the mean square error is equal to zero. The reconstruction is not unique if there is more than one solution.

IV. RANDOM SEARCH RECONSTRUCTION

In the direct search procedure we should check all combinations of K nonzero out of N coefficients in total. To find all possible combinations of $\{k_1, k_2, ..., k_K\} \subset \mathbf{N}$, the total number of combinations is equal to

$$\binom{N}{K} \tag{10}$$

and it could be very large. The expected number of checked combinations is

$$T_d = \frac{1}{2} \binom{N}{K}. \tag{11}$$

Even though the direct search is an accurate method, it is computationally not feasible to get a solution for a large signal. For the random search procedure, we will consider a system with M unknowns. Taking $M - 1 > K$ equations in a combination, less trials will be needed to find the solution. Let us consider a random combination of $M-1$ nonzero positions. The new system is then

$$\mathbf{y} = \mathbf{A}_{sel} \mathbf{X}_{sel} \tag{12}$$

with M equations. If the considered combination includes all K nonzero positions from $\mathbf{X}$, then the system (12) of M equations with $M-1$ unknowns have the unique solution. Note that only K coefficients in the solution are nonzero and remaining $M - 1 - K$ coefficients are zero valued.

Considering the new system (12), the probability that we find the solution is

$$P_s = \frac{\binom{N-K}{M-1-K}}{\binom{N}{M-1}}. \tag{13}$$

The expected number of trials can be estimated as

$$T_r = \frac{1}{P_e} = \frac{\binom{N}{M-1}}{\binom{N-K}{M-1-K}}. \tag{14}$$

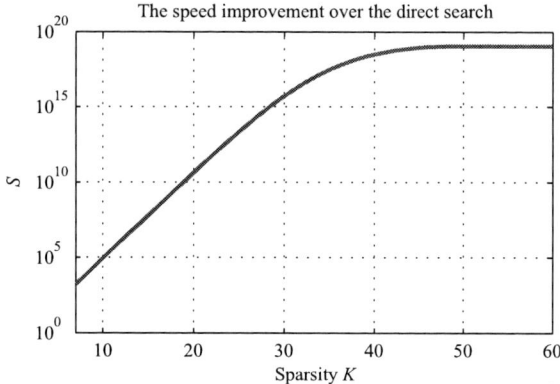

Fig. 1. The improvement in the calculation complexity over the direct search, when the random search is applied on signals with various sparsity K, length $N = 128$ and number of available samples $M = 2K + 1$.

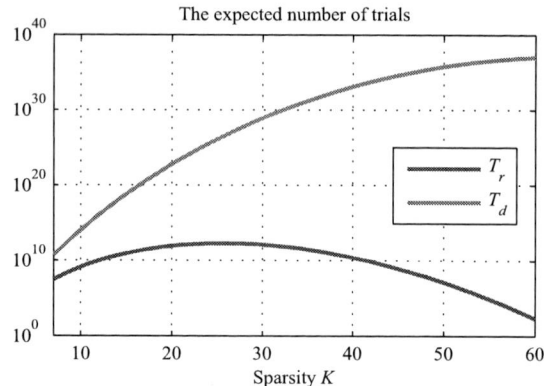

Fig. 2. The expected number of trials in direct search and random search for signal of length $N = 128$ with various sparsity K and number of available samples $M = 2K + 1$.

The improvement in speed of the proposed random search procedure compared to the direct-search is

$$S = \frac{T_d}{T_r} = \frac{1}{2} \frac{\binom{N}{K}\binom{N-K}{M-1-K}}{\binom{N}{M-1}}. \qquad (15)$$

The amount of the speed improvement will be illustrated by several examples in the next section.

V. EXAMPLES

Example 1: Let us consider a signal of length $N = 128$, having $M = 15$ available samples and sparsity $K = 7$. The total number of direct-search combinations for this case is

$$\binom{N}{K} \sim 10^{10}.$$

Probability that we guess solution by proposed method is

$$P_s \approx 3.6 \times 10^{-8}$$

whereas the expected number of trials equals

$$T_r \approx 2.75 \times 10^7.$$

The proposed method is $S \approx 1700$ times faster than direct-search (in average).

Let us now observe a signal having the same length $N = 128$, and $M = 31$ available samples with sparsity $K = 15$. Following the previous analysis, the calculation speed-up is

$$S \approx 7.7 \times 10^7.$$

The expected number of trials in the random search procedure applied in this case is

$$T_r \approx 8.5 \times 10^{10}.$$

Example 2: For signal of length $N = 128$ we vary sparsity K from 7 to 60 and assume that the number of available samples is $M = 2K + 1$ for each observed sparsity. The speed improvement of the proposed method over the direct search,

calculated according to (15) is shown in Fig. 1. The expected trials number is also calculated for both procedures, according to (11) and (14). The results are shown in Fig. 2.

Example 3: The main motivation to introduce an improved version of the direct search lies in the fact that standard sparse reconstruction algorithms, including those based on ℓ_1-norm require a larger number of available samples for the recovery of missing samples than the reconstruction based on the corresponding direct search aiming to minimize the ℓ_0-norm. In order to illustrate this issue, we observe a $N = 20$ length signal K-sparse in the discrete Fourier transform domain. The signal has the following form

$$x(n) = \sum_{i=1}^{K} A_i e^{j2\pi k_i \frac{n}{N}}, \qquad (16)$$

with amplitudes and frequencies having random values with uniform distribution, satisfying $0 \leq A_i \leq 2$ and $0 \leq k_i \leq N - 1$. Number of available samples was varied from $M = 1$ to $M = 19$ whereas for each number of available samples sparsity was varied from $K = 1$ to $K = 19$. Note that in cases when $K > M$ the reconstruction is not possible.

The proposed random search is compared with Orthogonal Matching Pursuit (OMP), a representative algorithm from the compressive sensing framework introduced in [5]. The random search procedure terminates when the solution is found, or the number of trials exceeds $\binom{N}{M-1}$. The experiment was conducted based on 100 independent realizations of signals with random missing samples positions and the probability of successful reconstruction is calculated.

The results are shown in Figs. 3 and 4. Comparing these results, it can be seen that the OMP-based reconstruction requires a larger number of available samples M (for a given sparsity K) than the corresponding random search procedure. For example, in the OMP case with $K = 2$ accurate reconstruction in 100% of trials requires exactly $M = 9$ available

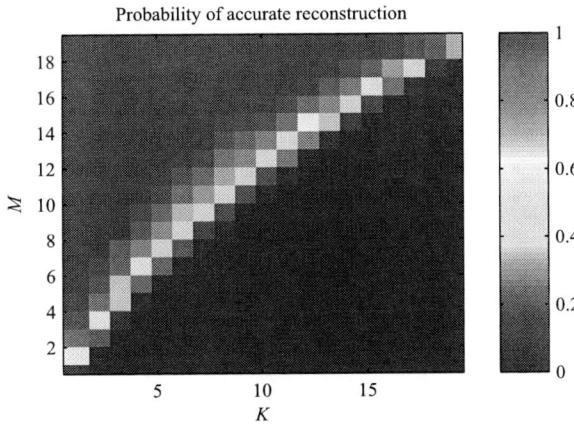

Fig. 3. The probability of successful reconstruction using OMP as an example of standard sparse signal recovery algorithms. The results, shown for various K and M are obtained based on 100 independent realizations of signals with random missing samples positions, amplitudes and frequency positions. The signal length is $N = 20$.

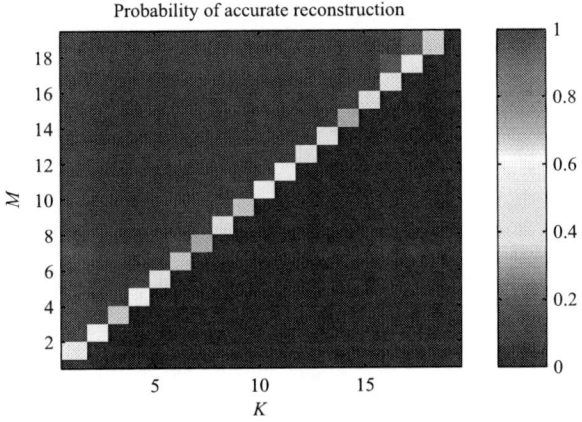

Fig. 4. The probability of successful reconstruction using the proposed random search procedure. The results, shown for various K and M are obtained based on 100 independent realizations of signals with random missing samples positions, amplitudes and frequency positions. The signal length is $N = 20$.

samples, whereas the proposed procedure requires only $M = 4$ available samples.

VI. CONCLUSIONS

In this paper we propose a method for sparse signal reconstruction based on the random search for K nonzero coefficient positions in the sparsity domain. Since we have M available measurements, in each trial $M - 1$ randomly selected positions are considered. If the considered combination includes all the nonzero coefficients, then the reconstruction is done successfully. Transform coefficients with wrongly assumed nonzero values are automatically set to zero using the partial sensing matrix pseudo-inversion involved in the signal reconstruction.

Taking more positions in one trial will converge to the solution faster than the direct-search procedure. The random search method is compared with the direct-search, showing noticeable improvement in the calculation cost. The basic motivation for this research is the fact that standard algorithms from the compressed sensing framework require a larger number of available samples for the successful reconstruction than it is required by the direct search based reconstruction. This issue is illustrated in comparison with OMP algorithm. The obtained results confirm that the calculation improvements in the direct search based signal recovery represent important and open topics for further research.

REFERENCES

[1] M.A.T. Figueiredo, R.D. Nowak and S.J. Wright, "Gradient Projection for Sparse Reconstruction: Application to Compressed Sensing and Other Inverse Problems," *IEEE Journal on Selected Topics in Signal Processing*, 2007

[2] E. Candes, J. Romberg and T. Tao. "Robust uncertainty principles: Exact signal reconstruction from highly incomplete frequency information," *IEEE Trans. on Information Theory*, vol. 52, pp. 489–509, 2006.

[3] E. J. Candès, "The restricted isometry property and its implications for compressed sensing," *Comptes Rendus Mathematique*, vol. 346, no. 9, pp. 589-592, 2008.

[4] L. Stanković, M. Daković, S. Vujović, "Adaptive variable step algorithm for missing samples recovery in sparse signals," *IET Signal Processing*, vol.8, no.3, 2014, pp.246–256, doi: 10.1049/iet-spr.2013.0385

[5] D. Needell, J. A. Tropp, "CoSaMP: Iterative signal recovery from incomplete and inaccurate samples," *Applied and Computational Harmonic Analysis* vol.26, no.3, 2009, pp.301–321.

[6] T. Blumensath, "Sampling and reconstructing signals from a union of linear subspaces," *IEEE Transactions on Information Theory* vol.57, no.7, 2011, pp.4660–4671.

[7] L. Stanković, M. Daković, and S. Vujović, "Reconstruction of Sparse Signals in Impulsive Noise," *Circuits, Systems and Signal Processing*, vol. 2016. pp. 1–28

[8] L. Stanković, "A measure of some time–frequency distributions concentration," *Signal Processing*, vol. 81, pp. 621–631, 2001

[9] E. Sejdić, A. Cam, L.F. Chaparro, C.M. Steele and T. Chau, "Compressive sampling of swallowing accelerometry signals using TF dictionaries based on modulated discrete prolate spheroidal sequences," *EURASIP Journal on Advances in Signal Processing*, 2012:101 doi:10.1186/1687–6180–2012–101

[10] L. Stanković, S. Stanković, and M. Amin, "Missing Samples Analysis in Signals for Applications to L-estimation and Compressive Sensing," *Signal Processing*, vol. 94, Jan 2014, pp. 401-408

[11] L. Stanković, I. Orović, S. Stanković and M.G. Amin, "Robust Time-Frequency Analysis based on the L-estimation and Compressive Sensing," *IEEE Signal Processing Letters*, May 2013, pp.499–502.

[12] J. Ender, "Compressive Sensing and the Real World of Sensors," Keynote Speech, *4th International Workshop CoSeRa 2016*, 19-22 September 2016, Aachen, Germany

[13] M. Elad, *Sparse and Redudant Representations: From Theory to Applications in Signal and Image Processing*, Springer, 2010

[14] H. Rauhut, "Stability Results for Random Sampling of Sparse Trigonometric Polynomials," *IEEE Trans. on Information theory*, 54(12), pp. 5661–5670, 2008

[15] L. Stanković, and M. Daković, "On the Uniqueness of the Sparse Signals Reconstruction Based on the Missing Samples Variation Analysis," *Mathematical Problems in Engineering*, vol. 2015, Article ID 629759, 14 pages, 2015. doi:10.1155/2015/629759

[16] M. Daković, L. Stanković, and S. Stanković, "A Procedure for Optimal Pulse Selection Strategy in Radar Imaging Systems," *4th International Workshop CoSeRa 2016*, 19-22 September 2016, Aachen, Germany

A fast noise level estimation algorithm based on adaptive image segmentation and Laplacian convolution

Emir Turajlić

Faculty of Electrical Engineering, University of Sarajevo, Sarajevo, Bosnia and Hercegovina
emir.turajlic@etf.unsa.ba

Abstract - This paper proposes a fast algorithm for additive white Gaussian noise level estimation from still digital images. The proposed algorithm uses a Laplacian operator to suppress the underlying image signal. In addition, the algorithm performs a non-overlapping block segmentation of images in conjunction with the local averaging to obtain the local noise level estimates. These local noise level estimates facilitate a variable block size image tessellation and adaptive estimation of homogenous image patches. Thus, the proposed algorithm can be described as a hybrid method as it adopts some principal characteristics of both filter-based and block-based methods. The performance of the proposed noise estimation algorithm is evaluated on a dataset of natural images. The results show that the proposed algorithm is able to provide a consistent performance across different image types and noise levels. In addition, it has been demonstrated that the adaptive nature of homogenous block estimation improves the computational efficiency of the algorithm.

I. INTRODUCTION

Numerous image processing and computer vision algorithms require some means of accurate and computationally efficient estimation of noise variance from still images. Some of the applications of noise level estimation algorithms include image denoising [1-3], feature extraction [4] and edge detection [5]. In this paper, it is assumed that digital images are corrupted by additive white Gaussian noise (AWGN). Although, digital images can be corrupted by a range of different types of noise, such as the shot noise, the salt-and-pepper noise and the Poison noise, the estimation of AWGN variance from images remains the problem where much of the research effort is focused on.

Typically, the additive white Gaussian noise occurs as a result of image transmission, but it can also arise from image acquisition, storage and during some image processing operations. This type of noise is particularly challenging to estimate as it affects the entire image. In practice it is very difficult to discern to which degree the observed local variations can be attributed to the underlying image signal and to which degree they are associated with the corrupting noise signal. Due to the importance and the challenging nature of this research problem, over the years, various approaches for noise variance estimation have been developed. The most prominent approaches in spatial domain include block-based [6] and filtering-based noise estimation methods [7]. Block-based methods tessellate an image into a number of non-overlapping blocks. Commonly, they attempt to establish the homogenous image patches with little or no visual activity from which the local noise level estimates are obtained. Subsequent statistical analysis of these local noise level estimates leads to the final noise level estimation value. Conversely, the filter-based methods use some type of image filtering to suppress the underlying image signal and in doing so facilitate a more accurate noise level estimation. On the other hand, transform domain methods use a decorrelating transform to attain spectral decomposition of images and enable noise level estimation [8]. Here, the energy compactness of the decorrelating transform constitutes an important issue.

This paper presents a novel block-based algorithm for a fast and accurate noise estimation across a range of noise levels. The proposed algorithm is essentially a hybrid algorithm that relies on filtering as well as on block segmentation of images. Specifically, the proposed algorithm uses a zero mean Laplacian operator to suppress the underlying image signal from the noise corrupted image. Subsequently, a non-overlapping block segmentation is employed to estimate homogenous image regions. Here, the estimation process is adaptive in nature and is implemented in two distinct stages whereby the size of tessellation blocks is related to the initial noise level estimate. In this paper, the optimal relation between the size of the tessellation blocks and the observed noise signal is ascertained. The performance of the proposed method is evaluated on a dataset of natural images [9] and it is compared to the fast noise estimation of method, presented in [10], and the block-based noise estimation in singular value decomposition (SVD) domain [11].

The remainder of this paper is organized as follows. In section II, a brief overview of the fast noise estimation method [10] is presented. An overview of the block-based noise estimation in SVD domain [11] is presented in section III. Section IV presents the proposed noise estimation algorithm. Section V presents and discusses simulation results. Section VI concludes the paper.

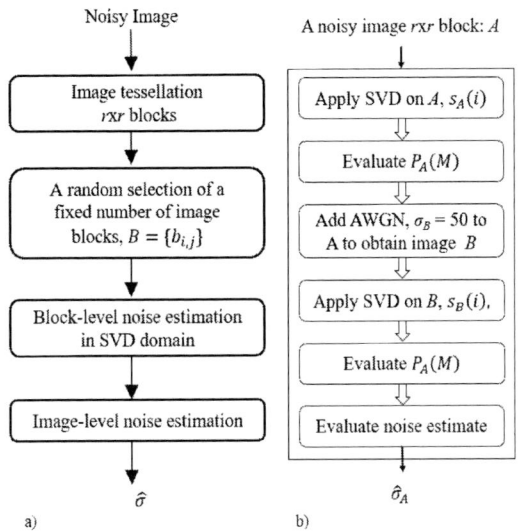

Figure 1. a) Block diagram of the block-based noise estimation in SVD domain; b) Local noise level estimation from a single image patch.

II. FAST NOISE ESTIMATION ALGORITHM

The fast noise estimation algorithm [10] is a simple algorithm that entails only two steps. In the first step, a zero mean Laplacian operator is used to suppress the underlying image signal and to estimate the noise signal. The noise estimation operator is described by a following 3x3 mask:

$$h = \begin{matrix} 1 & -2 & 1 \\ -2 & 4 & -2 \\ 1 & -2 & 1 \end{matrix} \qquad (1)$$

In the second step, the estimate of the standard deviation of Gaussian noise is obtained using the following relation [10]:

$$\hat{\sigma} = \sqrt{\frac{\pi}{2}} \frac{1}{6(W-2)(H-2)} \sum_{image} |I(x,y) * h| \qquad (2)$$

Clearly, this algorithm is designed with fast execution in mind. The simplicity of estimation comes at the cost of lower accuracy and estimation reliability.

III. NOISE ESTIMATION IN SVD DOMAIN

Singular value decomposition (SVD) is particularly applicable to the problem of noise estimation from images as the underlying image signal and additive noise can be well dissociated in the SVD domain.

A block diagram of the block-based noise estimation in SVD domain [11] is presented in Fig. 1 a). This method entails image segmentation via rectangular non-overlapping $r \times r$ blocks. Subsequently, a fixed number of randomly selected blocks is used to obtain a set of local noise level estimates. The procedure for obtaining local

noise level estimates is essentially the same as the one described in [12], except that small rectangular image patches are used instead of the entire image.

The procedure is also illustrated in Fig. 1 b). Here, singular value decomposition is performed on an image block A. An average of M trailing singular values $s_A(i), 1 \leq i \leq r$ is evaluated to obtain a parameter $P_A(M)$, where M constitutes a user defined parameter in a range $1 \leq M \leq r$. Subsequently, the noisy image block is further corrupted by AWGN with $\sigma_B = 50$ to produce a block B. The previously described procedure is repeated on the block B to yield $P_B(M)$. Finally, the block-level estimate is evaluated according to:

$$\hat{\sigma}_A = \frac{\alpha \sigma_B^2}{2(P_B(M) - P_A(M))} - \frac{P_B(M) - P_A(M)}{2\alpha} \qquad (3)$$

Here, the parameter α describes the slope of linear relationship between $P_A(M)$ and a noise level σ. Parameter α is required to be defined experimentally.

The implementation of this block-based noise estimation method in SVD domain requires a number of options and parameter values to be defined by a user. In this paper, these values are defined according to the recommendations made in [11] and [12]. For each image a total of 25 blocks are used to obtain local noise level estimates [11]. Here, each tessellation block is of fixed size 64x64 [11]. In addition, the number of considered singular values that are used in the process of producing a local noise estimates is set to $M=3r/4=48$ [11], [12]. Finally, the parameter α is defined as 4.86 [11].

IV. PROPOSED NOISE ESTIMATION ALGORITHM

A diagram of the proposed noise level estimation algorithm is presented in Fig. 2. The proposed method adopts features that are usually associated with both block-based and filtering-based approaches. Specifically, the proposed method implements the Laplacian convolution based on the operator h that is defined in (1), to suppress the underlying image signal. This feature is clearly associated with the filtering-based methods. On the other hand, the proposed method also employs rectangular block-based image tessellation to evaluate the homogenous image patches. However, the proposed algorithm evaluates the homogenous image patches in two distinct steps. In that sense it is different than the traditional block-based methods. It is inherently adaptive to the perceived noise level.

In the first step, the result of Laplacian convolution is tessellated into 64x64 non-overlapping image patches. From each image patch, the averaging procedure described in (2) is used to produce a local noise level estimate. The only difference is that the averaging is performed over a 64x64 image patch, instead of the entire image. The main results of the first stage in the noise estimation process are the initial noise level estimate $\hat{\sigma}_n^{(1)}$ and the initial estimate of homogenous image regions $B^{(1)}$. The initial set of homogenous image patches $B^{(1)}$ corresponds to image blocks that exhibit the smallest local noise level estimates.

487

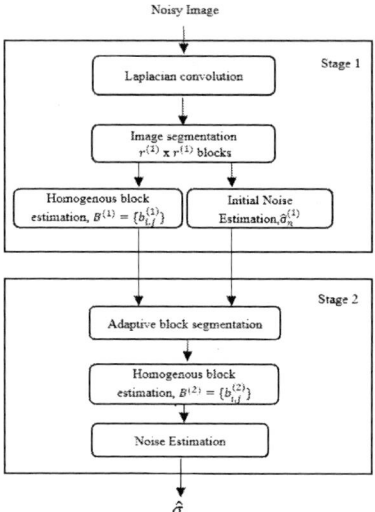

Figure 2. Block diagram of the proposed noise estimation algorithm with two distinct stages: 1) Laplacian convolution and initial noise estimation; 2) Adaptive homogenous area estimation and noise estimation.

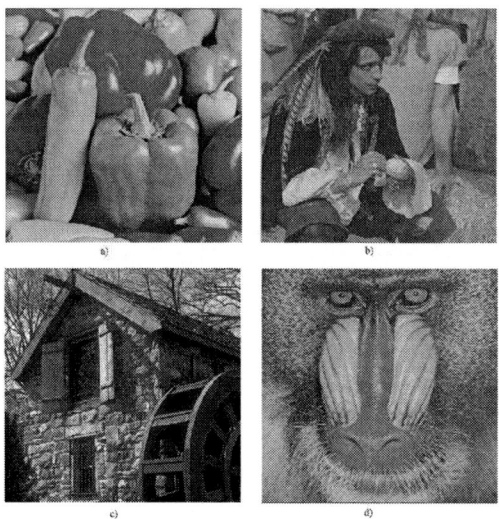

Figure 3. Test images a) Peppers; b) Pirate; c) House; d) Baboon.

Each local noise level estimate is associated with a particular homogenous image patch. Here, the number of homogenous image blocks is selected so that the total homogenous area corresponds to roughly (rounding) 15% of the total image area. On the other hand, the initial noise level estimate $\hat{\sigma}_n^{(1)}$ is ascertained as the minimum value among the entire set of local noise level estimates.

In the second stage of the noise estimation process, the proposed method uses the initial noise level estimate $\hat{\sigma}_n^{(1)}$ as the basis to perform further segmentation of homogenous image patches $B^{(1)}$. Each rectangular block is segmented into even smaller image patches. The size of the new image segments, $r^{(2)} \times r^{(2)}$, is directly related to the initial noise level estimate $\hat{\sigma}_n^{(1)}$ obtained in the first

stage of the noise estimation process. Specifically, the relationship is of the form $r^{(2)} = k \times \hat{\sigma}_n^{(1)}$, where the optimal value of proportionality constant k is experimentally evaluated. In order to have meaningful image segmentation the $r^{(2)}$ values are also constrained to $3 \leq r^{(2)} < r^{(1)}$. The adaptive image segmentation ensures that when smaller amount of noise is observed, the image is tessellated into smaller blocks. Conversely, the higher the value of the initial noise level estimate, the bigger the size of segmentation blocks. Following the adaptive block segmentation, a set of local noise level estimates is obtained. In the similar manner as in the first stage of the estimation process, a new set homogenous of image blocks $B^{(2)}$ is identified. The only difference is that the final number of homogenous blocks is set to denote 85% of all blocks that are derived over the area corresponding to the set of blocks, $B^{(1)}$. The averaging over the set of local noise level estimates associated with image patches $B^{(2)}$ produces the final noise level estimate.

The proposed algorithm enables a very efficient two-stage estimation of homogenous image patches. Initially, a course segmentation based on the large 64x64 blocks is performed. A fraction of these blocks is identified as initial homogenous image patches. In the second stage, these homogenous image patches are further segmented based on the initial noise level estimates. This second stage image segmentation is performed over a fraction of the total image area as only the initial estimates of homogenous image patches are involved. The presented hierarchical two-stage image segmentation is expected to reduce the overall computational complexity of the proposed algorithm. In the following section, we will demonstrate that variable block size also improves the accuracy of noise estimation.

V. SIMULATION RESULTS AND DISCUSSION

In this paper, experiments are conducted on a subset of 40 images from the database of natural images [9]. Prior to any experiments, these RGB images are converted to grayscale. In addition, some experiments are performed on four standard test-images, presented in Fig. 3.

A. Proportionality Constant

The objective of the first experiment is to establish an optimal relation between the initial noise level estimate, as evaluated during the course image segmentation, and the size of the image blocks that are to be used in the second stage of image segmentation. It is assumed that this relation is linear, where $r^{(2)} = k \times \hat{\sigma}_n^{(1)}$, and it is only the value of the proportionality constant that need be ascertained.

The optimal value of the constant k is evaluate in the following manner. The value of k is varied between $k=1$ and $k=4$ in increments of 0.1 and the performance of the proposed algorithm is measured objectively, using the mean square error across the dataset of 40 images and the considered range of noised levels as the principal quality measure.

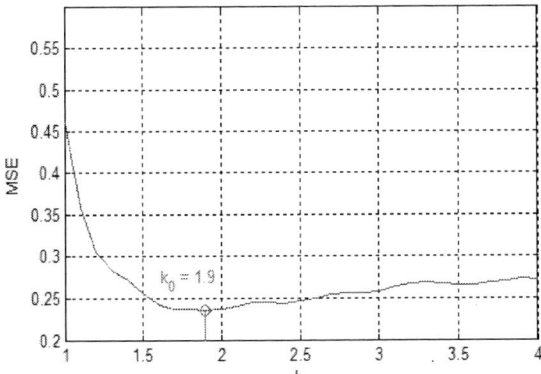

Figure 4. The mean square estimation error as a function of proportionality constant k.

Here, MSE is defined as:

$$MSE_k = \frac{1}{L \times N} \sum_n \sum_{i=1}^{L} (\hat{\sigma}_{i,n} - \sigma_n)^2 \qquad (4)$$

, where $L=40$ denotes the size of the image data set and N describes the number of considered noise levels. Subscripts i and n denote the image index and noise index, respectively. Here, noise between $\sigma_1 = 1$ and $\sigma_{12} = 23$ in increments of 1. The results of the experiment showing MSE vs k curve is shown in Fig. 4. Clearly, the optimal value of the proportionality constant is $k=1.9$ for which the MSE curve attains a global minimum value. Since, $r^{(2)} = 1.9 \times \hat{\sigma}_n^{(1)}$, we can ascertain the optimal linear relationship between area of the segmentation blocks and the initial noise variance value: $r^{(2)} \times r^{(2)} = 3.61 \times \left(\hat{\sigma}_n^{(1)}\right)^2$. The proportionality constant value $k=1.9$ is employed in all subsequent experiments and on all images to define the relationship between the initial noise level estimates and the size of the segmentation blocks used during the second stage of the noise estimation algorithm.

B. Estimation of Homogenous Image Regions

The homogenous image region estimation and the adaptive nature of image segmentation process is illustrated in Fig. 5. The presented results are based on the optimal proportionality constant value $k=1.9$ and AWGN at $\sigma=3$.

Here, the framed blocks denote the results of the initial homogenous region estimation via a course image segmentation, whereas the filled image patches denote the homogenous regions as estimated during the second stage of the noise estimation procedure. Note that, in each case, the initial homogenous region corresponds to the 15 % of total image region, whereas the final homogenous patches denote 85 % of the initial estimates. Visual inspection of the images, presented Fig. 5 and the corresponding results of Laplacian convolution indicate that the relative ranking among the local noise level estimates is to large extent correlated with the perceived level of visual activity. Hence, the estimated homogenous image patches are perceived as flat image regions, where most of the intensity variations can be attributed to the noise signal.

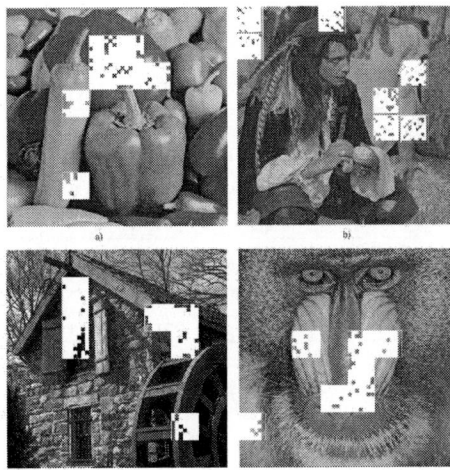

Figure 5. Estimated homogenous patches under the course and fine image segmentation. All test images are degraded by AWGN at $\sigma=3$.

C. Performance Comparison

In the next set of experiments, the performance of the proposed noise level estimation algorithm is compared to the fast noise level estimation algorithm, presented in section II and the SVD domain based noise estimation algorithm, presented in section III. From here on, these two algorithms will be denoted as FAST and SVD, respectively. On the other hand, the proposed algorithm will be denoted as PR. The comparative analysis between the different noise level estimation algorithms involves a broad range of noise levels. Specifically, in each of the following experiments, the standard deviation of the additive white Gaussian noise is varied from $\sigma = 1$ to $\sigma = 23$ in increments of two.

Table I and Table II report the average estimation error results attained for each of the four test images. Here, for each image and considered noise level, the reported results denote the average estimation error obtained over 30 independent simulation runs. The bolded values indicate the best performing method for a particular noise level.

TABLE I. COMPARISON OF AVERAGE ESTIMATION ERROR FOR PEPPERS AND PIRATE IMAGES

	Peppers			Pirate		
	SVD	FAST	PR	SVD	FAST	PR
1	2.01	4.32	**1.98**	2.62	3.41	**0.56**
3	**1.08**	3.02	1.52	1.78	2.24	**0.29**
5	**0.55**	2.23	0.93	1.27	1.59	**0.10**
7	**0.33**	1.73	0.67	1.01	1.21	**-0.04**
9	**0.18**	1.40	0.44	0.67	0.97	**-0.12**
11	**0.08**	1.18	0.32	0.63	0.81	**-0.21**
13	**-0.02**	1.01	0.16	0.55	0.69	**-0.31**
15	-0.06	0.89	**0.00**	0.52	0.61	**-0.37**
17	0.04	0.79	**-0.02**	0.42	0.53	**-0.41**
19	**-0.06**	0.70	-0.17	**0.41**	0.48	-0.58
21	**-0.14**	0.64	-0.29	**0.39**	0.43	-0.60
23	**-0.11**	0.58	-0.39	**0.37**	0.39	-0.62

TABLE II. COMPARISON OF AVERAGE ESTIMATION ERROR FOR HOUSE AND BABOON IMAGES

	House			Baboon		
	SVD	FAST	PR	SVD	FAST	PR
1	5.52	5.92	**2.45**	9.81	9.59	**2.40**
3	4.19	4.47	**1.54**	8.16	7.96	**1.45**
5	3.27	3.48	**1.09**	7.06	6.66	**0.91**
7	2.76	2.79	**0.81**	6.05	5.65	**0.63**
9	2.24	2.31	**0.63**	5.43	4.87	**0.41**
11	1.92	1.96	**0.50**	4.94	4.23	**0.29**
13	1.68	1.70	**0.37**	4.44	3.75	**0.15**
15	1.52	1.48	**0.30**	4.24	3.33	**0.06**
17	1.36	1.32	**0.19**	4.24	3.00	**0.05**
19	1.25	1.20	**0.09**	3.95	2.73	**-0.09**
21	1.15	1.07	**0.00**	3.78	2.49	**-0.14**
23	1.23	·1.00	**-0.10**	3.54	2.31	**-0.19**

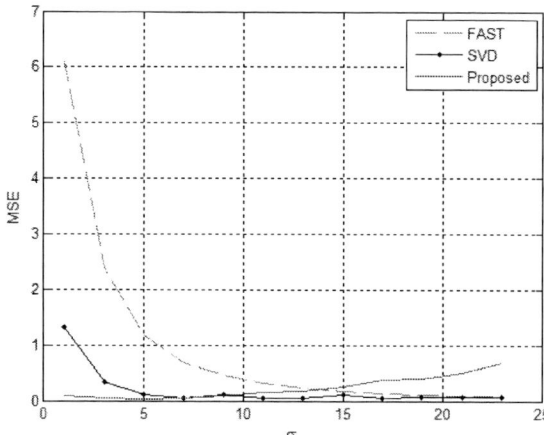

Figure 6. Mean square error evaluated across images for different noise levels.

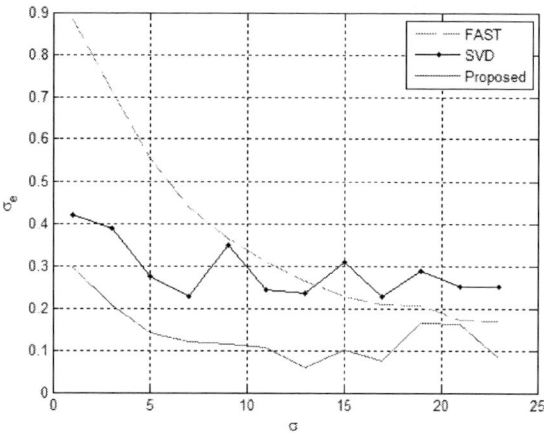

Figure 7. Standard deviation of estimation error evaluated across images for different noise levels.

The results reported Tables I and II demonstrate that the proposed method outperforms the other two methods. In each case, these improvements are mostly evident when the amount of corrupting image noise is relatively low. The SVD algorithm offers the best performance for 'Peppers' image. On the other hand, the proposed algorithm

constitutes the best option for all other images. In the case of test images with the highest degree of texture and underling visual activity ('House' and 'Baboon'), the proposed algorithm offers the best performance over the entire range of the considered noise levels.

Since these types of images are traditionally challenging for the types of algorithms that employ block-based noise estimation, the accuracy of noise level estimation can be attributed to the adaptive nature of image segmentation as well as to image filtering and suppression of the underlying image signal.

A more systematic evaluation of the considered noise level estimation algorithms is presented in the following experiment. Here, the noise level estimation algorithms are evaluated on a subset of 40 images from the dataset of natural images [9]. Prior to experiment, these RGB images are converted to grayscale. Again, the standard deviation of additive white Gaussian noise is varied from 1 to 23 in increments of two.

The performance of each method is evaluated in terms of the mean square estimation error (MSE) and the standard deviation of estimation error. The MSE is used to quantify the accuracy of the considered algorithms, while the standard deviation of estimation error is used to objectively evaluate the reliability of the considered algorithms. The results are reported in Fig. 6 and Fig. 7. The results clearly demonstrate that the proposed method, in comparison to the other considered algorithms' exhibits higher levels of both accuracy and reliability. When considering the attained MSE values, it is clear that the proposed algorithm offers significant improvements in the accuracy of noise level estimation at nose levels. This property has been is also observed as part of our analysis of four test images. In addition, for most of the considered noise levels, the proposed algorithm attains the lowest standard deviation in estimation error values. This results indicates high levels of reliability of the proposed algorithm and its ability to provide a consistent performance across different image types and noise levels.

In the final experiment, the execution time of the proposed method is evaluated. The proposed method and the SVD method both employ image tessellation and use local noise estimates to ascertain the overall noise image estimate. In general, these methods are associated with a higher degree complexity compared to the FAST method. Here, FAST method entails only simple filtering and averaging operations, only. Consequently, they offer a much more accurate and consistent noise level estimation across images.

The execution times of the noise level estimation algorithms is evaluated on a dataset of ten 512x512 images Here, images are corrupted by AWGN with $\sigma = 10$ and the total time required for each algorithm to perform noise estimation over the entire set is recorded. The result of the FAST method is used as the performance benchmark as it is expected to require the least amount of time to complete the task. The ratio of execution time of the SVD domain method compared to the FAST method is 3.31:1. On the other hand, the attained execution time ration of the proposed method in relation to the FAST method is 1.84:1. Clearly, the proposed algorithm is nearly twice as fast as the block-based noise level estimation in SVD domain. The computational efficiency can be attributed to the efficient estimation of homogenous image patches. In comparison to the FAST estimation, the proposed algorithm, offers significantly higher levels of accuracy and reliability at the cost somewhat slower performance.

Thus, the proposed algorithm offers a high quality performance that is better or in certain cases at least comparable to the SVD domain method, while offering a faster execution time.

VI. CONLUSION

This paper proposes a novel block-based algorithm for a fast and accurate noise level estimation. The proposed algorithm is in fact a hybrid algorithm that exhibits features that can be associated with both filter-based methods and block-based methods. The proposed method employs a zero mean Laplacian operator to suppress the underlying image signal from the noise corrupted image. In addition, it uses a non-overlapping block tessellation and adaptive image segmentation to ascertain homogenous image patches. These homogenous patches are used to obtain the local noise level estimates from which the final noise level estimate is produced.

The experimental results demonstrate that proposed algorithm offers high levels of accuracy and reliability across different images and noise levels. The ability to adaptively very the size of the image segmentation blocks is of particular benefit when the images are corrupted by a small amount of noise, where the traditional block-based methods tend to overestimate the noise levels in such instances.

REFERENCES

[1] M. Lebrun, A. Buades, J. M. Morel, "A nonlocal Bayesian image denoising algorithm," SIAM Journal on Imaging Sciences, 6(3):1665–1688, 2013.

[2] E. Turajlić, V. Karahodzic. "An Adaptive Scheme for X-ray Medical Image Denoising using Artificial Neural Networks and Additive White Gaussian Noise Level Estimation in SVD Domain." CMBEBIH 2017: Proceedings of the International Conference on Medical and Biological Engineering 2017. Vol. 62. Springer,pp. 36-41, 2017.

[3] E. Turajlić, "Application of neural networks to denoising of CT images of lungs," 2016 XI International Symposium on Telecommunications (BIHTEL), pp. 1-6, 2016

[4] D. Lowe, "Object recognition from local scale-invariant features," In Proc. IEEE Int'l Conf. Computer Vision, pages 1150–1157, 1999.

[5] N. Senthilkumaran, R. Rajesh, "Edge Detection Techniques for Image Segmentation and A Survey of Soft Computing Approaches", International Journal of Recent Trends in Engineering, Vol. 1, No. 2, PP.250-254, May 2009.

[6] J. S. Lee, K. Hoppel, "Noise modeling and estimation of remotely sensed images," in Proc. 1989 Int. Geoscience and Remote Sensing, Vancouver, Canada, June 1989, vol. 2, pp. 1005-1008.

[7] K. Rank, M. Lendl, R. Unbehauen, "Estimation of image noise variance," IEE Proceedings - Vision, Image and Signal Processing, vol. 146, pp. 80-84, Apr. 1999.

[8] C. Tang, X. Yang, G. Zhai, "Dual-transform based noise estimation", Proc. IEEE Int. Conf. Multimedia Expo (ICME), pp. 991-996, Jul. 2012.

[9] Olmos, A., Kingdom, F. A. A. (2004), A biologically inspired algorithm for the recovery of shading and reflectance images, *Perception*, 33, 1463 - 1473.

[10] J. Immerkær, "Fast Noise Variance Estimation", Computer Vision and Image Understanding, Vol. 64, No. 2, pp. 300-302, Sep. 1996.

[11] Liu, Wei. "Additive white Gaussian noise level estimation based on block SVD." Electronics, Computer and Applications, IEEE Workshop, 2014.

[12] Wei Liu, Weisi Li, "Additive White Gaussian Noise Level Estimation in SVD Domain for Images," IEEE Transactions on, 22(3):872–883,201

Advanced regulation approach: Dynamic rules for capturing the full potential of future ICT networks

Dijana Ilisevic, Natasa Banovic-Curguz[*]

[*] Mtel/Division for Technics, Banja Luka, Bosnia and Herzegovina
dijana.ilisevic@mtel.ba, natasa.banovic-curguz@mtel.ba

Abstract - **Software Defined Networks (SDN) and Network Function Virtualization (NFV) could be taken as a different expression of the overall transformation trend toward network softwarization which is deeply impacting and bringing Telecom and ICT Industry. These changes will have a broader impact on society, including aspects of regulation, policy, social impacts and new business model. A number of new stakeholders are involved in the realization of SDN/NFV driven architecture. Telecom companies are often regulated more than Internet players and this asymmetry represents a problem in terms of competitiveness and capability to deliver innovative services. In this article we have provided short overview of the most relevant stakeholder roles in future ICT industry. Afterwards, we have presented importance of regulation in future ICT environment which place huge impact on foster innovation and service delivery.**

I. INTRODUCTION

The Fourth Industrial Revolution – digitization and internetization are already underway and is far cleverer than the previous because it is connect emerging technologies. This is blurring boundaries between physical, digital as well as transforming production, management and governance system in nearly every industry in every country. Digital transformation (DT) is a major driver for economic growth and has the potential to catalyze a country's innovation, growth and development. Governments can accelerate or break this transformation and guide new digital mindset in this fast-changing area using ongoing digital transformation experiences.

All companies and governments will have to think about how to deal with this digital transformation and what their role in the newly growing digital ecosystems might be. Digital transformation takes beyond managing internal Information Technology (IT) applications. They need forecasting and business insight; they need to fully understand what transformation really means; and they need to understand the technology and how it will change their business value. Digital transformation is all about data, but its full value can only be delivered when adequate IT and technology form the infrastructure that seamlessly connects that data and enables its exchange – anytime and anywhere. This growing demand for connectivity will require companies to have broadband access and a high-speed infrastructure, which would result in a rise in spending for ubiquitous broadband (BB)

access. A key foundation in digital society will be powerful Information and Communications Technology (ICT) infrastructure that support service delivery and foster innovation.

BB infrastructure puts telco sector in the heart of transformation process, but telcos are having a hard time because growth and profitability across the industry are in a downward spiral. First, consumers think that data services are too expensive and the experience is inadequate. Their needs for connectivity, bandwidth, reliability, and security are still not being met. Second, Internet service providers use telecom networks to serve their customers, but they think they are paying too much for bandwidth and data traffic throughput. Third, the governments are trying to cut the prices of telecom services. The EU is in the process of cancellation mobile roaming fees within Europe. Other governments around the world are trying to find ways to lower prices. The telecom industry can only become healthy when telcos themselves are healthy. In the same way, the industry can only complete its digital transformation once telcos have successfully gone digital. Telco companies and systems are connection centric. Barriers have built up between their marketing, network management, and IT functions, and their front-end and back-end systems aren't integrated. Contrarily, Internet companies keep their focus on their users, and have updated their organizations to quickly respond to user demands. As a result, a service that takes Internet companies a few months to develop and launch might take some telco companies one or two years. With such a long time-to-market, telcos can hardly stay ahead of the competition in the Internet era, where everyone is fighting for rapid iteration and innovation. DT will not only help telcos sharpen their competitive edge in new markets such as Internet of Things (IoT), video, and cloud, it will also support the cloudification of networks and operations systems to make telcos more agile. Telcos need the support of external enablers, and policy makers need to define issues of transitional and long-term period. Dynamic rules for capturing the full potential of future ICT networks can require redisinging or building new institutions, transformation of regulatory practice and governance, new skills and competencies, and significant changes to mindset and culture (legacy systems, institutions and processes).

From a technology point of view, we are going to see virtualized autonomous networks that are obviously moving into 4G, 5G, and it will be data centric and security-centric. The telcos of course are in danger of being pushed down into a dumb pipe because it is the Internet players that are developing increasingly strong customer relationships. Telecom companies are often regulated more than Internet players and this asymmetry represent a problem in terms of competitiveness and capability to deliver innovative services. Software Defined Networks (SDN) and Network Function Virtualization (NFV) could be taken as a different expression of the overall transformation trend toward network softwarization which is deeply impacting and bringing Telecom and ICT Industry. These changes will have a broader impact on society, including aspects of regulation, policy, social impacts and new business model. Detailed following of technological changes and trends, with understanding consequences it has on society as a whole, is the first step towards creation of advanced regulatory frameworks that can answer to modern trends in ICT sector.

II. FORECAST TOWARD ICT NETWORKS

A. Tehnology vison of future ICT networks

With the increase in traffic, content and applications, connectivity has become part of our work and life. From technology perspective competing standards requires much more complex networks scenario. The future network infrastructure have to support complete redesign of services and service capabilities, architectures, functions, access as well as connection security. The important challenges that must be addressed by 5G networks is higher capacity, lower E2E latency, massive device connectivity, reduced capital and operation cost and consistent Quality of experience (QoE) provisioning, [1]. New 5G architecture consisting of three generic services: extreme mobile broadband, massive of machine type communication and a spectrum toolbox, [2].

Moreover, the technological advances proposed for beyond 4G and 5G mobile networks still mostly focus on capacity increase, which is fundamentally constrained by the limited radio spectrum resources as well as the investment efficiency, and therefore will always lag behind the growth rate of mobile traffic. Unlike the communication resource, which is fundamentally limited by the bandwidth and power, the computing and caching (i.e., memory) resources are abundant, economical, and sustainable. Various attempts have been made to accommodate expanding mobile services through the use of sustainable non-communication resources. For example, to deliver multimedia contents (which constitute most of mobile traffic), proactive pushing through data caching has been proposed at both base stations and mobile terminals. In addition to use caching to provide "individualized" services to mobile users, savings in communication resources can also achieved through computing, in which contents intended for different mobile users are logically processed through coded multicasting or similar techniques, [3]. The three types of resources must be treated as equally essential and intrinsic in order for mobile systems to be scalable and sustainable. Instead of only the communication dimension, the capabilities of future mobile systems should be visualized by a native mobile 3C cube in which different services are supported by the 3C core functionalities. In order to analytically characterize the mobile 3C cube, first their metrics and mode of operations so that the impact of 3C can be quantified must be defined. In the native 3C form, the capability of the mobile network is determined by three vectors: Communication, Cache, and Computing.

The "communication vector" of the mobile system pertains to its ability to deliver information streams over an imperfect channel with given bandwidth and power. Its capability is measured by the data rate R (unit: bit per second per Hertz), with a relation with the system bandwidth and the signal power to noise ratio is well understood by Shannon's capacity formula. Note that communication operations (e.g., modulations and channel coding) only protect but do not alter the information stream upon delivery. The "caching vector" of the mobile system pertains to its ability to buffer or store a certain amount of information (i.e., bitstreams) at nodes within mobile networks. Its capability is typically measured by the memory size (unit: byte). While neither protecting nor altering the information streams, caching operations introduce non-causality or time-reversal into the system, in turn increasing the system's ability to deliver information over longer periods of time. The "computing vector" of the mobile system pertains to its ability to perform logic or algebraic operations across information streams. Unlike communication or caching operations, computing operations alter the information bits such that additional logic operations must be performed at the destination in order to recover the original information streams. By formulating "caching" as a form of non-causal operations and "computing" as a form of logic operations across information streams, the three native and complementary operations for mobile 3C systems has been established.

B. Stakeholders analyses

Softwarization will be a radical change of paradigm. Current telecommunications infrastructures have been exploited with purpose-built equipment designed for specific functions. In the future, network functions and services will be virtualized software processes executed on distributed horizontal platforms mainly made of standard hardware resources. Cost savings alone will not be enough to assure the future sustainability of the telecommunications industry (it is key to enable innovative service paradigms). Standards development organizations (SDOs), exist to assure the development of consensus-based, quality standards. These formal standards are needed in the telecommunications market to achieve functional interoperability. The standardization process takes years, and then a vendor still needs to implement the resulting standard in a product. This prevents service providers (SPs) who are willing to venture into new domains from doing so at a fast pace. With the development SDN and NFV, open source

493

technology is emerging as a new option in the telecommunications market. In contrast to SDOs, open-source software (OSS) communities create a product that may implicitly define a de-facto standard based on market consensus. Therefore, SPs are drawn to OSS, but they face technical, procedural, legal, and cultural challenges due to their lack of experience with open software development. The question therefore arises, how the interaction between OSS communities, SDOs and Industry Fora (IF) can be organized to tackle these challenges. A number of stakeholders are involved in the realization of this SDN/NFV-driven architecture, [4].

Most actors will perform more than one role at the same time. For example, traditional ISPs fulfill the role of infrastructure provider, virtual service infrastructure provider, and service provider. Users: Users, i.e. end/enterprise users, retail, or over-the-top (OTT) providers, request, and consume a diverse range of services. In general, users have no strong opinion about how the service is delivered as long as their quality of experience expectations is satisfied.

Service Providers (SPs) accommodate the service demand from users by offering one or multiple services, including over-the-top service and X-play services (e.g. triple play). The service provider realizes the offered services on a (virtualized) infrastructure via the deployment of virtualized network functions (VNFs). Virtual Service Infrastructure Providers (VSIPs) deliver virtual service infrastructure to SPs, meeting particular service level requirements by combining physical network and cloud resources into service infrastructure meeting particular *SLA* requirements implemented through NFV-enabled network applications. These network applications might involve resources (or network functions) that are either implemented in traditional network hardware, or as virtualized network functions, (NF). These are the result of an orchestration system that interacts with the network control system as well as the cloud control system. Infrastructure Providers (InPs) own and maintain the physical infrastructure and run the virtualization environments. By virtualizing the infrastructure, they open up their resources to remote parties for deploying virtual NFs. The reusable physical resources comprise all possible resource options (computing, storage, and networking), and they span the entire service delivery chain from the end-user gateway and set-top-box over the access, aggregation, and core network up to the cloud. Hardware Vendors provide the physical devices that are deployed by the infrastructure providers. The shift away from specialized equipment toward reusable, industry- standard high-volume servers, switches, and storage devices can reduce the total costs of InPs (furthermore, they cost less than manufacturer-designed hardware and increase flexibility). The hardware must provide an interface toward the controller systems. Software Vendors, including OSS developers, deliver the implementation of the logic that is used to optimally deploy the services on the physical

infrastructure. Today a patchwork of specialized software products exists to realize that functionality. The most relevant software for the SDN/NFV architecture is those that focus on the following, as given in Figure 1, [4].

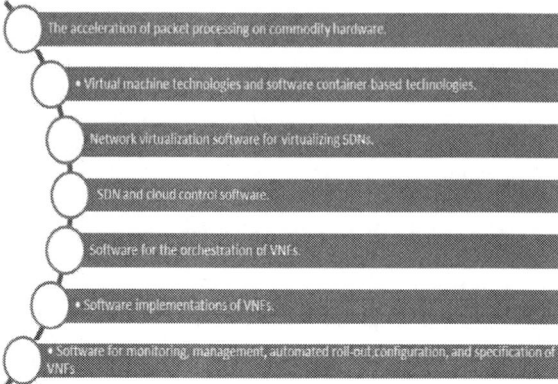

Figure 1. Future software oriented SDN/NFV architecture

For each of these, OSS communities have developed or are developing viable alternatives to proprietary software. SDOs and IF: The networking industry today is very much standards-driven to make a product or service safe (safety standards) and interoperable (interface standards), while making the industry as a whole more efficient.

C. Content delivery networks

How to effectively utilize these existing resources for massive content distribution is a fundamental question current and future networks must address. It is proposed that the leveraging of ubiquitous caching and computing at the wireless network edge will radically change future mobile network architecture, and can potentially solve the current bottleneck for massive content delivery. The recent trend toward network softwarization is transforming the networking industry into an open ecosystem, with three main stakeholders: users, operators and content providers, Figure 2, [5].

The business of wireless caching involves three key stakeholders that together form a complex ecosystem. The users of telecommunication services are primarily the customers and consumers of the content, but in the case of wireless caching they are also active stakeholders. Users might be requested to help in the form of contributing with their own resource (e.g., in the case of coded caching it will be memory and processing, or in device-to-device (D2D), caching it will also be relaying transmissions, and they will end up spending energy for the benefit of better performance. On the other hand, one could envision users employing D2D technology to enable caching without the participation of other stakeholders. Due to the complexities mentioned above, however, efficient wireless caching will require heavy coordination and extensive monitoring/processing. Hence, D2D approaches will be limited to restricted environments. The operators of telecommunication networks are well placed for wireless caching. Due to the particularities of coded caching and multi-access caching, operators are in a unique position to implement new protocols in base stations, affect the standards for new mobile devices, and develop big data processing infrastructure that can realize wireless caching. Nevertheless, for reasons related to encryption, privacy, and global popularity estimation, operators might not be able to install these technologies without the cooperation of the other two stakeholders.

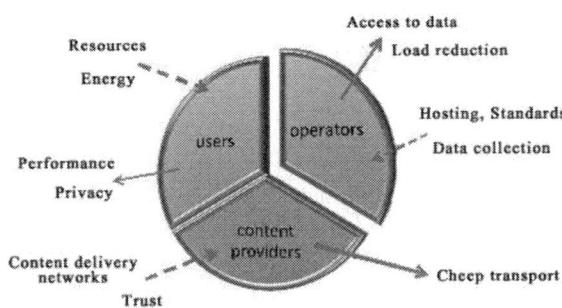

Figure 2. Stakeholder's analysis

The providers of Internet content are champions of trust from the user community. Apart from the security keys, they also hold extensive expertise in implementing caching techniques in core networks. From this advantageous position, they can positively affect the progressive evolution of caching in wireless networks. It promises to qualify network owners and providers, to increase the pace of innovation, diversify the supply chain for networking hardware and software, and drive the transformation of mobile networks into a highly capable platform in supporting emerging IoT and data science applications, among others. Mobile transport networks will play a vital role in future 5G and beyond networks. In particular, access transport networks connecting radio access with core networks are of critical importance. They will be required to support massive connectivity, super high data rates, and real-time services in a ubiquitous environment. To attain these targets, transport networks should be constructed on the basis of a variety of technologies and methods, depending on application scenarios, geographic areas, and deployment models.

D. Net Neutrality

5G is not just the next evolution of 4G technology; it is a paradigm shift. 5G is not only evolutionary (providing higher bandwidth and lower latency than current-generation technology);more importantly, 5G is revolutionary, in that it is expected to enable fundamentally new applications with much more stringent requirements in latency (e.g. real time) and bandwidth (e.g. streaming). 5G should help solve the last-mile/ last-kilometer problem and provide broadband access at much lower cost because of its use of new

spectrum and its improvements in spectral efficiency. Flexibility, ease of use, the dynamic nature of the network, Quality-of-Service, and anytime/anywhere availability, are some of the benefits for end users in this move to 5G. 5G is an enabler of exciting use cases that will transform the way people live, work, and engage with their environment. In the short term, 5G can support exciting use cases such as the IoT, smart transportation, eHealth, smart cities, entertainment services, etc. [6]. Net neutrality is "the principle that Internet service providers should enable access to all content and applications equally, regardless of the source, without favoring or blocking particular online services or websites" (*Oxford Dictionary*), [7]. Net neutrality is a highly politicized topic. Net neutrality debate should be extended to all the actors involved in the Internet delivery chain. Content delivery networks play a key role by storing content closer to users, thereby reducing transit costs and improving performance for that content; for device builders may introduce biases through the features of their products (possibly colluding with some other actors); and search engines (seen as service providers) directly affect the accessibility (visibility) of content. Regulatory Agencies therefore face the delicate task of defending fairness and universality principles in this complex ICT ecosystem, where ever changing technical and business conditions prevent (or considerably complicate) analysis and comparison.

III. ADVANCED REGULATION APPROACH

Digital transformation is accompanied by profound social and institutional changes. As a result, realizing the full potential of future ICT networks can therefore require redesigning of regulatory framework and even radical changes in adoption of innovation. The role of policymakers is to lead regulatory framework change, empower innovations and support effective adoption of new technologies. Ensuring affordable and competitive BB access (fixed and mobile) is essential. Managing of risks in DT is part of regulation framework: impact on job market, education, Public Safety, Green Energy, Privacy and Security, as given in Figure 3.

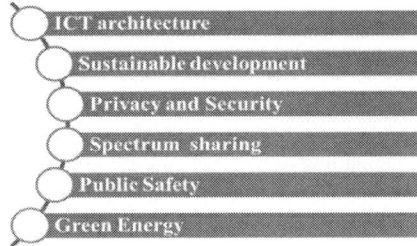

Figure 3. Regulation framework

Understanding future technology, policymakers must identify the key stakeholders/actors in creating ICT society. Policymakers must be aware of the critical role in recognizing limitations in current regulatory model and to support fast growing ICT industry. Reforms are essential to capture the ongoing technology changes and ensure their integration in society as a whole to achieve max transformation impact. With the explosive growth of mobile devices, collecting user mobility information will generate huge amounts of data. Thus, big data analytics to extract the required mobility information is another challenge in mobility-aware caching. In order to take advantage of the user mobility pattern, some personal information (e.g., home locations and work place locations) may be divulged in the collected mobility information. This will certainly cause some concerns on privacy issues. Thus, how to extract useful user mobility information without touching individual privacy is important, [8]. Spectrum sharing is area where max efficiency is needed. 5G services will require novel and more complex ways of interaction and collaboration among operators. Spectrum sharing is one example of increased cooperation among operators, [2]. The use of exclusively licensed spectrum builds the basis for mobile network operators to deliver good quality of service to their users. This type of spectrum usage should be prioritized, especially for applications requiring high reliability of data transmissions. However, as we are facing an exponential increase in the volume of wireless data traffic, we need to be prepared for supplementary solutions to serve the mobile users demand in space and time. The spectrum sharing enablers include, for example, spectrum opportunities detection and dynamic frequency selection/dynamic channel selection, and use of a geo-location database. The energy efficiency can be considered at a device side or a network side, though the distinction will be blurred by the dual role of the mobile wireless devices in 5G. As networks densification continues to meet the capacity demands, it becomes increasingly important to implement new lean signaling procedures, to be able to activate and deactivate network nodes depending on the traffic load, or to switch off some of the node functionality in low load modes. The move to IP-based telecommunications expands the functions and features those telecommunications applications and services can provide. Nowadays, citizens are getting used to new ways of agile communications supporting media-enriched and context-aware information. However, the adoption of these evolved technologies in emergency communications between citizens and public authorities faces a series of barriers, including the lack of harmonized and interoperable solutions. Different initiatives worldwide are addressing the need for specifying a stable IP-based next generation emergency communications framework, [9].

A. Case study – regulation in Bosnia and Herzegovina

Telecommunications companies in Bosnia and Herzegovina are in quite unique position, since delivery of content and application services is possible only in 3G networks. Increasing user demands and competitive market creates a need to improve offered services as well as the degree of utilization of existing resources. In Bosnia and Herzegovina OTT providers are threats to three leading telecom operators. Decoupling of the service from the network have arisen and raised questions over the regulation framework that could answer to all future challenges. Innovative OTT services are cheaper than traditional but with low quality. The first regulatory issue

refers to defining activities of OTT providers and to define online service such as voice telephony, SMS and television that can be regarded as potentially substituting for traditional telecommunications services such as voice, SMS and television. Specially designed regulatory legal act must protect users, considering that current acts in Bosnia and Herzegovina do not apply to the activities of OTT providers. Furthermore, detailed informing can contribute to user security.

IV. CONCLUSION

The role of a regulating agency is to recommend policies that guarantee widely accepted principles. The regulatory framework needs to evolve along with industry it regulates. It is up to the national policymakers to provide successful creation of ICT future that has a potential to transform society as a whole. This paper highlights the need for dynamic rules for capturing the full potential of future ICT networks, and can require redisinging or building of a better regulatory framework of fairness and the guarantee of fair competition among actors and the preservation of the motivation for investment toward satisfying end users better. Regulatory Agencies therefore face the delicate task of defending fairness and universality principles in this complex ICT ecosystem, where ever changing technical and business conditions prevent (or considerably complicate) analysis and comparison.

REFERENCES

[1] D. Ilisevic and N. Banovic- Curguz, ``Effective Strategies for Transport Network Deployments to Support Future Internet Services``, Infoteh Jahorina, pp. 280-283, March 2016.

[2] H. Tuulberg, P. Popovski, Z. Li, M. A. Uuisiatlo, A. Hoglund, O. Bulacki, M. Fallgren and J. Moneserrat, ``The METIS 5G System Concept: Meeting the 5G requirements``, IEEE Communication Magazine Vol. 54, pp. 16-22, September 2016

[3] H. Liu, Z. Chen, and L. Qian, ``The Three Primary Colors of Mobile Systems``, IEEE Communication Magazine Vol. 54, pp 16-22, September 2016

[4] B. Naudts, W. Tavernier, S. Verbrugge, D. Colle, and M. Pickavet, ``Deploying SDN and NFV at the Speed of Innovation: Toward a New Bond Between Standards Development Organizations, Industry Fora, and Open-Source Software Projects``, IEEE Communication Magazine-Communication Standards Suplement, pp. 46-53, March 2016

[5] G. Paschos, E. Bastug, I. Land, G. Caire, and M.Debbah, ``Wireless Caching: Technical Misconceptions and Business Barriers``, IEEE Communication Magazine Vol. 54, pp. 16-22, August 2016

[6] P. Marsch, I. Da Silva, O. Bulakci, M. Tesanovic, S. E. Ayoubi, T. Rosowski, A. Kaloxylos and M. Boldi, ``5G Radio Access Network Architecture Design Guidelines and Key Considerations``, IEEE Communication Magazine Vol. 54, pp. 24-32, May 2016

[7] P. Maillé, G. Simon, and B. Tuffin, ``Toward a Net Neutrality Debate that Conforms to the 2010s `` IEEE Communication Magazine Vol. 54, pp. 94-99, March 2016

[8] R. Wang, X. Peng, J. Zhang, and K. B. Letaief, ``Mobility- Aware Caching for Content-Centric Wireless Networks: Modeling and Methodology``, IEEE Communication Magazine Vol. 54, pp .77-83, August 2016

[9] F. Liberal, J. O. Fajardo, C. Lumbreras and W. Kampichler, ``European NG112 Crossroads: Toward a New Emergency Communications Framework``, IEEE Communication Magazine Vol. 55, pp .132-138, January 2017

MIPRO 2017, May 22- 26, 2017, Opatija, Croatia

LTE eNB Traffic Analysis and Key Techniques Towards 5G Mobile Networks

T. B. Iliev[*], Gr. Y. Mihaylov[*], T. D. Bikov[*], E. P. Ivanova[*], I. S. Stoyanov[**] and D. I. Radev[***]

[*]University of Ruse, Department of Telecommunication, Ruse, Bulgaria
[**]University of Ruse, Department of Electrical and Electronic Engineering, Ruse, Bulgaria
[***]University of Telecommunications and Posts, Department of Telecommunication, Sofia, Bulgaria
e-mail addresses: tiliev@uni-ruse.bg, gmihaylov@uni-ruse.bg, cbikov@gmail.com, epivanova@uni-ruse.bg,
stoyanov@uni-ruse.bg, dradev@abv.bg

Abstract - With the fast growth of mobile data services, rich services deliver a brand new experience to end users, and also bring about new opportunities for operators. Some applications, such as vehicular and industrial applications, demand a level of reliability that wireless communication systems typically are not able to guarantee. This paper provides a framework that enables these applications to make use of wireless connectivity only if the transmission conditions are favorable enough. In this paper, we will partially compare the fourth and future fifth generations of mobile networks and their applicability and were carried out measurements of data traffic amount and quality characteristics.

Keywords: Ultra-Reliable Communication, Fifth-Generation (5G), Mobile communication systems

I. INTRODUCTION

Few past years, there was been a tremendous surge in the demand for mobile and wireless connectivity, which is forecasted to grow exponentially within the near future. However, the Fifth-Generation (5G) mobile communication system [1] that is currently under discussion will not only have to cope with an increasing demand of traffic volume, but also provide a wider range of applications with new requirements in terms of reliability, availability and efficiency. Instead of integrating new radio access concepts such as Device-to-Device (D2D), Vehicle-to-Vehicle (V2V) communications, Massive Machine Type Communications (MMC), or Moving Networks (MN), the support of Ultra-Reliable Communication (URC) is seen as a key enabler for 5G communication systems [2].

Link reliability is the ability of a radio link to transmit and receive a certain amount of information successfully within a predefined deadline. System reliability is the ability of a system to accurately indicate the absence of link reliability to the application, and at the same time, to ensure the presence of link reliability as often as possible when required by the application [3, 4]. URC services such as road safety applications require very high rates within low deadlines circa 100 ms. Due to the sensitive nature of these applications, it is from very high importance to warn the application about the lack of link reliability according to the specific requirements of each application. Wireless communication systems are not

designed to provide reliability at all times and in every reception scenario, as this would result in an overdesigned system with a very ineffective air interface in terms of data rate and power consumption. With this approach may harm the acceptance of URC services and restrict their usage. In this paper we define the URC concept in a general manner, i.e., without relying on details of any air interface requirement or radio transmission scheme. This method is motivated by the fact, that definition from the applications point of view is required in order to allow URC services to be deployed in a wide variety of scenarios. From this point of view, the implementation details related to the wireless communication system are not included in the suggested URC concept.

The rest of the paper is organized as follows. Section 2 gives a brief overview and theoretical representation of the system concept for Ultra-Reliable Communication. Section 3 presents the conducted experimental measurements of data traffic amount and quality characteristics of two Bulgarian mobile operators. Section 4 concludes the paper.

II. KEY ENABLERS FOR 5G COMMUNICATION SYSTEMS

Despite the fact that it is practically impossible to ensure error-free wireless communication - it is feasible to derive boundary conditions for the transmission success. As illustrated in Fig. 1, the system concept for URC is based on a "Reliable Transmission Link" (RTL) that is set to transmit packets successfully and within a predefined deadline, and an "Availability Estimation and Indication" (AEI) mechanism that is able to reliably calculate the availability of the RTL under the corresponding conditions. Availability Indicator (AI) signals the outcome of the AEI to the application. An application requests an RTL by sending an Availability Request (AR) to the AEI. Depending on the implementation characteristics, the AR contains information such as the packet size, the maximum acceptable delay until successful reception or the maximum allowable error probability.

For the availability estimation, the AEI needs to monitor the channel conditions, e.g., by evaluating the Signal-to-Noise and Interference Ratio (SINR) and/or the ACK/NACK statistics of the retransmission protocols used at link level.

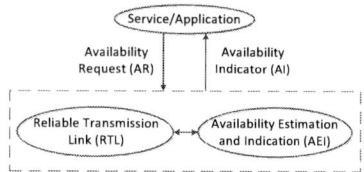

Figure 1. System concept of URC

Typically, the AI is a binary value, i.e., either RTL available ($AI=1$) or unavailable ($AI=0$). After indicating the RTL availability, the application will be able to use it by transmitting data packets over the RTL (it is not shown in Fig 1).

A. Mathematical representation

In this section we formulate mathematically the URC concept by adopting a simple time-slotted model, in which each time slot (τ) corresponds to the time interval [t, $t + D_{AR}$), where D_{AR} is the maximum delay tolerated by the application. According to this definition is defined:

$$RTL_{(\tau)} = \begin{cases} 1, & \text{transmission is successful} \\ \\ 0, & \text{transmission is not successful} \end{cases} \quad (1)$$

For the availability indication is used a simple binary signaling format per time slot:

$$AI_{(\tau)} = \begin{cases} 1 \\ 0 \end{cases}, \quad (2)$$

where AI refers to availability of RTL for time slot τ ($AI_{(\tau)}=1$) and the non – availability of RTL for time slot τ ($AI_{(\tau)}=0$).

On Fig. 2 is shown the URC state transition probabilities divided into two stages. The principle of the URC concept is that an application should rely on the wireless communication only in those situation in which the link reliability is guaranteed with a certain probability. The AI indicates the availability of a RTL for time slot τ to the application, the probability of successful transmission for the time slot τ must be over a certain value, P_{UR}, according to the application requirements. This criteria is refer as the ultra-reliable requirement, which can be explained by (3):

$$P_{1|1} = \Pr\left(RTL_{(\tau)} = 1 \middle| AI_{(\tau)} = 1\right), \quad (3)$$

where $P_{1|1} \geq P_{UR}$. The idea of the URC system design is to maximize the availability of the RTL under the predefined requirement:

$$\max P_1 = \Pr\left(AI_{(\tau)} = 1\right) \\ \text{s.t } P_{1|1} \geq P_{UR} \quad , \quad (4)$$

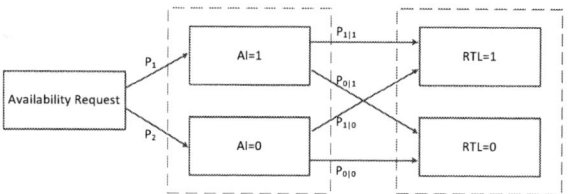

Figure 2. URC State Transition Probabilities

According to Bayes' rule and from Fig.2, we have:

$$P_1 = \Pr\left(AI_{(\tau)} = 1\right) = \\ = \frac{\Pr\left(RTL_{(\tau)} = 1\right)\Pr\left(AI_{(\tau)} = 1 \middle| RTL_{(\tau)} = 1\right)}{P_{1|1}}, \quad (5)$$

It can be concluded that there are two possibilities in order to improve the URC concept:

✓ improving the transmission model ($\Pr(RTL_{(\tau)} = 1)$);

✓ improving the estimation of the AI ($\Pr(AI_{(\tau)} = 1 | RTL_{(\tau)} = 1)$).

The first possibility can be achieved by more effective modulation and forward error correction (FEC) schemes, whereas the second possibility can be accomplished through more precise channel estimation and prediction methods.

It is interesting to note that the two probabilities $P_{0|1}$ and $P_{1|0}$ in Fig. 2 correspond to the Type I and Type II errors [5] in statistical probability analysis:

$$P_{0|1} = \Pr\left(RTL_{(\tau)} = 0 \middle| AI_{(\tau)} = 1\right) \quad (6)$$

$$P_{1|0} = \Pr\left(RTL_{(\tau)} = 1 \middle| AI_{(\tau)} = 0\right) \quad (7)$$

B. URC system concept

In this section we will explain the URC concept by using a simple example based on the predicted SINR. We assume that the SINR for time slot τ is $\Gamma(\tau)$ and define that the transmission is successful for time slot τ if the corresponding SINR is larger than or equal to a given threshold Γ_ι ($\Gamma(\tau) \geq \Gamma_\iota$), which is given by the used modulation and FEC scheme including the use of retransmissions. The AEI signals to the application AI=1 if the predicted SINR for time slot τ, $\Gamma_p(\tau)$, is larger than or equal to the threshold Γ_ι ($\Gamma_p(\tau) \geq \Gamma_\iota$). We can formulate the optimization problem of the URC system model as:

$$\max P_1 = \Pr\left(\Gamma(\tau) \geq \Gamma_l\right) \\ \text{s.t } P_{1|1} = \Pr\left(\Gamma(\tau) \geq \Gamma_l \middle| \Gamma_p(\tau) \geq \Gamma_l\right), \quad (8)$$

where the Type I and Type II error can be expressed as:

$$P_{0|1} = \Pr\left(\Gamma(\tau) < \Gamma_l \middle| \Gamma_p(\tau) \geq \Gamma_l\right) \qquad (9)$$

$$P_{1|0} = \Pr\left(\Gamma(\tau) \geq \Gamma_l \middle| \Gamma_p(\tau) < \Gamma_l\right). \qquad (10)$$

In this model, it would be possible to improve the URC concept by decreasing Γ_l by means of more robust modulation and FEC schemes, or by optimize the computation of $\Gamma_p(\tau)$ in order to predict more accurately the availability of the RTL.

III. DATA USAGE AND TRAFFIC ANALISYS

In this section we will compare two Bulgarian Mobile operators and their data traffic amount and quality characteristics, will describe the benefits of developing and implementing next generation mobile network standards and what are the key point which eventually needs improvement. To highlight the differences, the research was done in two different locations as follows:

- ✓ **Service Provider 1** – Mid-size City with approximately 76 000 citizens

- ✓ **Service Provider 2** – Capital City

Used for the study Base Stations (*BS*), from both Service Providers (*SP*) are from one and the same manufacturer. This approach, to compare two base stations from two different cities has selected to highlight the traffic usage, throughput and TBF Drop Rates in 2G/3G and LTE standards used in the selected BSs.

So far for Service Providers, the ultimate goal of network optimization is the improvement of user experience and enhance user satisfaction. Nonetheless, network optimization does not perfectly matches Quality of Service (*QoS*) improvement. Traditional network Key Performances Indicator (*KPI*) optimization largely deals with signaling, such as the Temporary Block Flow (*TBF*) establishment success rate and attach success rate. Although operators have been paying attention to them, these indexes cannot accurately reflect QoS. QoS indicators like "slow page loading", "slow download rate" and "failure to open page" are indexes on the data service layer. As mentioned above, the TBF drop is KPI which is related to 2G data called GPRS/EDGE in GSM Network. TBF drop indicate how often the services is disconnected when GPRS or EDGE data Services are used [6].

A. 3G/UMTS

On September 1995, officially first GSM network in Bulgaria took place. It has been started working with six base stations covering a few central parts of the Capital City. The total number of active 3G/UMTS base stations, twenty years later in this country is nearly 5000. The 3G coverage reaches 96% of the areas in the country and 99% of its population.

The diagrams of current TBF drop rates in downlink and respectively uplink direction are shown on Fig.3 and Fig.4.

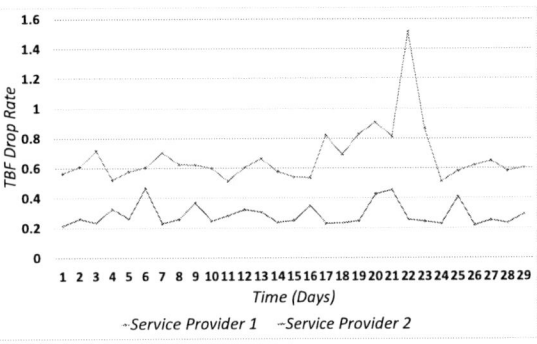

Figure 3. Measured TBF Drop Rates in Downlink direction for a month

As mentioned above, the statistics are from BSs, physically located in two different geographical locations.

As shown in Fig.3, the second Service Provider's BS is either not optimally utilized or very precisely planned and implemented in the selected region, because of the low volume of TBF Drop Rate in downlink direction. Another parameter, which is from major importance for consumers and in which Service Providers are working to improve, is throughput. Large number of different throughput values for 3G/UMTS standard are known.

However, on Fig.4 and Fig.5 are depicted diagrams of currently measured throughput in Mbps.

The metrics are measured in uplink and downlink direction for a period of one month. In both figures, Service Provider 2 notably have wider throughput. In downlink, SP2 has peak value of 164 Mbps, where SP1 has just 20,5 Mbps and in uplink, SP2 has maximum value of 73,7 Mbps where SP1 has almost constant 12 Mbps with peak from only 14 Mbps.

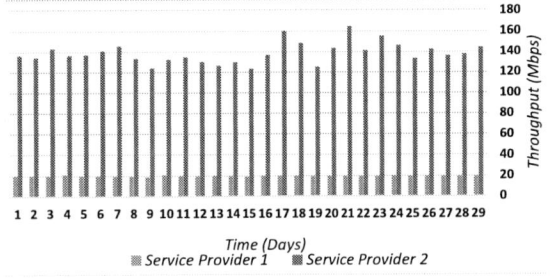

Figure 4. Throughput, measured in Megabits/s in Downlink direction for a month

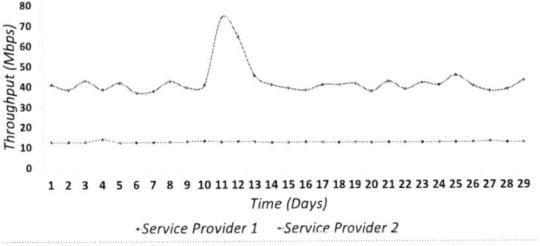

Figure 5. Throughput, measured in Megabits/s in Uplink direction on monthly basis

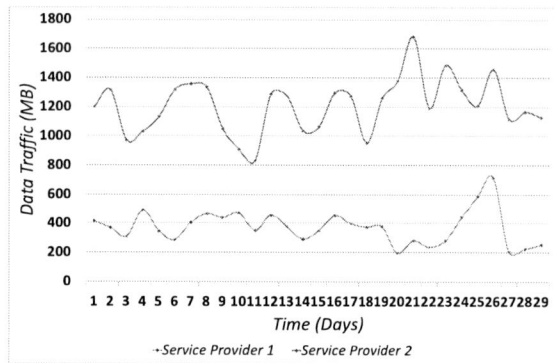

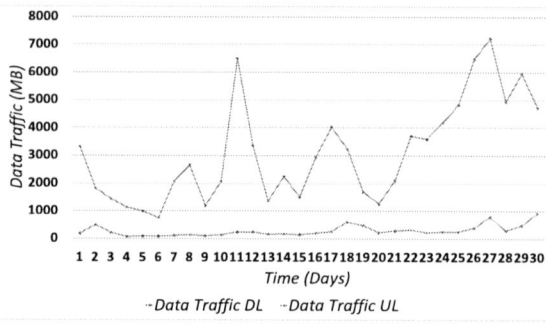

Figure 8. Data Traffic amount measured in Uplink and Downlink direction for one month

Figure 6. Data Traffic amount measured in Megabytes for one month

Last, but not least, Fig.6 represent graphical results of studied and analyzed Data Traffic from the two Service Provider`s Base Stations. Measured data traffic volume is for period of one month in Megabytes.

B. Long Term Evolution

Since the summer of 2014 the fourth generation mobile internet (*4G*) started to be offered in Bulgaria. As of September 2015, Bulgarian customers using Long Term Evolution (*LTE*), which is Fourth Generation (*4G*) wireless broadband technology, are 0.3% of all Internet using mobile users. According to offered plans, the high-speed internet access can provide download speeds of up to 75 Mbps and upload speeds of up to 25 Mbps. The bandwidths used by the fourth generation mobile networks are 900, 1800, and 2100 MHz. In 2015, the 900 MHz band is divided among the largest three operators in the country and in early September 2015 the Commission for Regulation of Communications (*CRC*) announced that it will hold a tender for permits for using the 2,50-2,69 GHz band.

With high data speed, the LTE can provides downlink peak rates of 300 Mbps. Some of the advantages of LTE over 3G/UMTS are High Speed, High Capacity, better Spectral Efficiency and others [7, 8]. To compare the two standards and to present the advantages of using the LTE, we analyze the data from one Base Station running LTE and represent it graphically (Fig.7 and Fig.8).

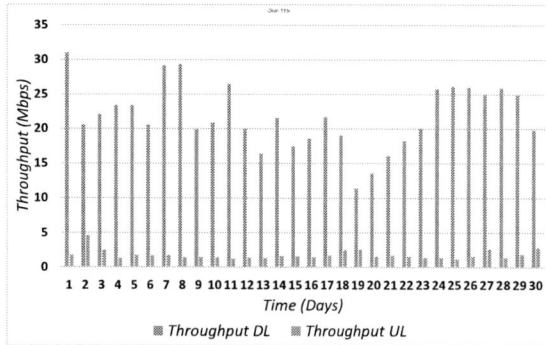

Figure 7. Throughput, measured in Megabits/s in Uplink and Downlink direction on monthly basis

On both charts above is shows that the throughput and the Data Traffic amount, passed through the specified BS running LTE, for a period of one month is significantly higher than those passed through the same BS, using the 3G/UMTS standard.

IV. CONLUSION

Comparing the future fifth-generation mobile networks with current and the previous generations of mobile services were not so extremely sensitive in manners like delay. Expected characteristics for 5G networks are data rates of tens of megabits per second for tens of thousands of users; data rates of 100 megabits per second for metropolitan areas and significantly enhanced spectral efficiency, compared to 4G. To meet these specific requirements contemporary mobile networks should be improved in signaling efficiency, better coverage and to offer lower latency.

The expectations of fifth generation mobile networks are the mobile subscribers to grow even more, including D2D, V2V communication and Internet of Things (*IoT*), which demands a lot of improvements.

REFERENCES

[1] "METIS - Mobile and wireless communications Enablers for the Twenty-twenty Information Society." [Online]. Available:https://www.metis2020.com/

[2] D. Gozalvez Serrano, H. D. Schotten, P. Fertl, R. Sattiraju and Z. Ren, "Availability Indication as Key Enabler for Ultra-Reliable Communication in 5G"

[3] F. B. F. Bai and H. Krishnan, "Reliability Analysis of DSRC Wireless Communication for Vehicle Safety Applications," 2006 IEEE Intelligent Transportation Systems Conference, 2006.

[4] A. Birolini, Reliability Engineering: Theory and Practice, 2010.

[5] S. Kay, Fundamentals of Statistical Signal Processing: Detection theory, ser. Prentice Hall Signal Processing Series. Prentice-Hall PTR, 1998

[6] Sadinov S., D. Koleva, Software Research the Quality of Signals in Cellular Network with the Software ASSET, International Scientific Conference Unitech'15, Gabrovo, Bulgaria, Vol.2, ISSN 1313-230X, pp.142 ÷ 146

[7] J. F. Monserrat, , H. Droste, Ö. Bulakci, J. Eichinger, O. Queseth, M. Stamatelatos, H.Tullberg,V. Venkatkumar, G. Zimmermann, U. Dötsch, A. Osseiran, "Rethinking the Mobile and Wireless Network Architecture: The METIS Research into 5G", EuCNC 2014, June 23-26, Bologna, Italy, 2014

[8] I. Silva, G. Mildh, M. Säily, S. Hailu, "A novel state model for 5G radio access networks", 5G RAN design Workshop, IEEE ICC 2016, 27 May 2016

IoT network protocols comparison for the purpose of IoT constrained networks

I. Heđi, I. Špeh, A. Šarabok

Virovitica College, Virovitica, Republic of Croatia

{ ivan.hedi | ivan.speh | antonio.sarabok }@vsmti.hr

Abstract - In this paper, a comparison of Internet of Things protocols used for data transfer in Internet of Things constrained networks is presented. Setting up such a network with a large number of physical interconnected IoT devices can be a challenge. In the IoT world, one of the key challenges is to efficiently support M2M communication in constrained networks. This can be achieved using MQTT (Message Queuing Telemetry Transport) and CoAP (Constrained Application Protocol) protocols. Choosing the appropriate protocol can be difficult while developing IoT application. There are several conditions that need to be considered while determining which protocol should be used. In this paper, we will evaluate performance and compare these protocols through different scenarios.

Keywords: Internet of Things, data protocols, MQTT, CoAP, M2M communication

I. INTRODUCTION

Everyday growing number of objects connected to the Internet Worldwide has promoted the Internet of Things technologies and protocols as one of the most commonly used in the modern systems. IoT refers to the networked interconnection of everyday objects which are often equipped with electronic circuits and sensors [1]. In the IoT sense, these objects can refer to a wide variety of small devices embedded with electronics, software, sensors and network connectivity often integrated into larger systems, for example, transportation systems, smart cities, telecommunication networks, healthcare industry and many others. According to [2] Gartner says that 8.4 billion connected things will be used worldwide in 2017., up 31 percent from 2016. A total number of connected objects will reach 20.4 billion by 2020. Implementation of the IoT technologies extends system's ability to gather, analyze and distribute data and thus improve efficiency. Furthermore, IoT has made Internet sensory (temperature, pressure, vibration, light, moisture, stress) [3] thereby achieving improved monitoring, analyzing and tracking systems. During the implementation of the IoT technology, several things must be considered of which the most important are a hardware platform, network protocol and data protocol. Hardware requirements can be broken up into five main components: a power source and power management, processor and memory storage, ability to gather data from sensors and implemented modules for wireless communication. On the side of the connectivity between IoT devices, there are a variety of communication standards and protocols used, of which most widespread are IEEE 802.15.4, Internet Protocol

version 6 (IPv6) and IPv6 over Low-Power Wireless Personal Area Network (6LoWPAN) [4]. Regardless the specific communication protocol used to deploy IoT network, all the IoT objects should make their data available to the other side, which can be another IoT object, application or the Internet. This can be achieved by sending the data to a web server or by employing the cloud through the use of Application Programming Interfaces with built-in functions for end-users [5]. This paper focuses on the data protocols that handle the communication and data transfer between the objects or applications. These protocols are application layer protocols that are used to send latest data or commands to servers which are responsible for forwarding or processing information. Choosing the appropriate protocol can be difficult while developing IoT application. There are several conditions that needs to be considered while determining which protocol should be used of which most important are the amount of data to send, the frequency of data transfer and hardware platform. Also, most of the IoT devices are installed in an isolated area where connection to the Internet is limited because it is realized over slow DSL connection or cellular network. In such environment, unreliability which reflects high packet loss rate is one of the main disadvantages [6].

II. IOT PROTOCOL STACK

There are several protocols proposed for M2M/IoT communication with a focus on mentioned constrained environments. Most frequently adopted protocols are MQTT (Message Queue Telemetry Transport) and CoAP (Constrained Application Protocol), XMPP (Extensible Messaging and Presence Protocol), RESTFUL Services (Representational State Transfer) and AMQP (Advanced Message Queuing Protocol). IoT applications can be simplified in the way of the error handling which can be done using these protocols [6].

Our main task is to test the performance of data protocols and compare them in different scenarios. Results can be helpful in determining which protocol should be used. After a briefly description of MQTT and CoAP protocol, testing results are presented.

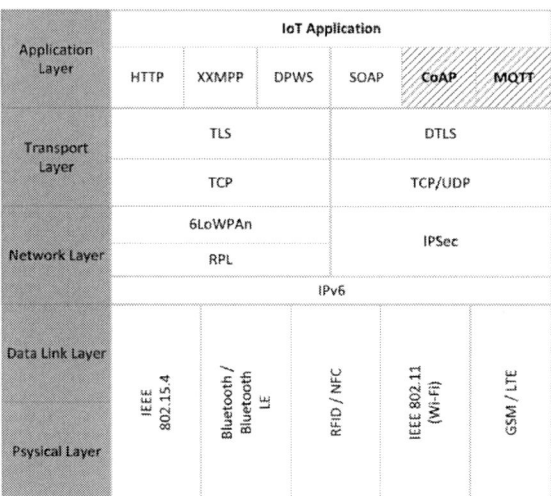

Figure. 1. *The IoT Stack*

A. Message Queue Telemetry Transport (MQTT)

MQTT is a machine-to-machine (M2M)/Internet of Things connectivity protocol for use on top of the TCP/IP protocol stack which was designed as an extremely lightweight broker-based publish/subscribe messaging protocol for small code footprints (e.g., 8-bit, 256KB ram controllers), low bandwidth and power, high-cost connections and latency, variable availability, and negotiated delivery guarantees [7]. In the hub and spoke model of Message-Oriented Middleware messaging server forwards messages from sensor devices to monitor devices [8]. In such architecture, a device whose main task is to continuously produce and send data to the server is defined as publisher. The Central server, an MQTT broker, collects messages from publishers and examines to whom the message needs to be sent. On the other side, every device which had previously registered its interests with a server will keep receiving messages until the subscription is canceled (Fig. 2).

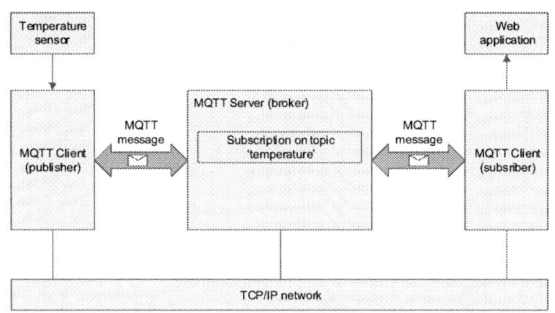

Figure. 2. *Message queuing telemetry transport protocol model*

Using this architecture, publishers and subscribers do not need to know for each other which is one of the major advantages of this protocol. Devices that send data need not to know who are the clients that are subscribed for receiving data and conversely [9]. Further to this, the publishers and subscribers do not need to participate in

the communication at the same time and do not need to be familiar with each other [10]. It is intended for devices with limited power and memory capabilities, where the network is expensive, has low bandwidth or is unreliable. One of the key requirements of an Internet of Things concept is low network bandwidth used to send data and minimal device resource requirements. While attempting to ensure reliability and delivery, MQTT has met these requirements [11]. This protocol has been applied in a variety of embedded systems. For example, hospitals use this protocol to communicate with pacemakers, oil and gas companies use it to monitor oil pipeline thousands of miles away. Facebook uses this protocol for messaging applications [12].

B. Constrained Application Protocol (CoAP)

Usage of RESTful (Representational state transfer) Web services on the Internet has become an essential part of building applications. The work on Constrained RESTful Environments (CoRE) targets the realization of the REST architecture in a convenient form for the most constrained nodes (e.g., 8-bit microcontrollers with limited RAM and/or ROM) and networks (e.g., 6LoWPAN), Constrained networks (e.g., 6LoWPAN) supports the IPv6 packets fragmentation into small link layer frames, but this causes major reduction in probability of packet delivery. This effect solves Constrained Application Protocol (CoAP). CoAP aims to keep message overhead small, thus limiting the need of fragmentation.

Main goals of CoAP is to design a generic web protocol for the special requirements of this constrained environment, focusing on energy, building automation, and other machine-to-machine (M2M) applications. The goal of CoAP is to but realize a subset of REST common with HTTP, optimized for M2M applications. Although CoAP could be used for remodeling simple HTTP interfaces into a compact protocol, and more importantly it offers features for M2M such as multicast support, built-in discovery and asynchronous message exchanges.

CoAP has the following main features:

• Web protocol fulfilling M2M requirements in constrained environments

• UDP binding with optional reliability supporting unicast and multicast requests.

• Asynchronous message exchanges

• Low header overhead and parsing complexity.

• URI and Content-type support.

• Simple proxy and caching capabilities.

• A stateless HTTP mapping, allowing proxies to be built providing access to CoAP resources via HTTP in a uniform way or for HTTP simple interfaces to be realized alternatively over CoAP

• Security binding to Datagram Transport Layer Security (DTLS)

III. NETWORK ARCHITECTURES

A. MQTT client

Any IoT object can be MQTT client that sends or receives telemetry data. This could be any device from a microcontroller up to a server. The type of the MQTT client (subscriber or publisher) depends on its role in the system. It can produce or collect telemetry data. In both cases, MQTT client should first connect to a messaging server using a particular type of message explained in the following chapter. After the connection is successfully established, a client must declare himself whether he is a subscriber or publisher. To distinguish data sent by the publisher, a topic string is used. For example, a client can publish temperature and humidity values using a different string for each value (e.g. temp/25 or hum/40). On the other side, if an MQTT client wants to receive the data, he must subscribe to the specific topic. To make an MQTT client from a device, an MQTT library must be installed and connection to an MQTT broker must be established over any kind of network. For example, MQTT client can be a small computer (Arduino or Raspberry Pi) connected to a wireless network with a library strapped to the minimum or a typical computer running a graphical program. The client implementation of the MQTT protocol is simplified and very straightforward [13]. Client libraries can simplify the process of writing MQTT client applications. MQTT libraries are available for different types of programming languages and platforms, for example, JavaScript, PHP, C, C++, Android, iOS etc. For the initial implementation, Python library will be used.

B. MQTT broker

An MQTT broker is a central device in mentioned hub and spoke model. The main responsibilities of an MQTT broker are handling communication between MQTT clients and distributing messages between them [6]. A broker can handle up to thousands connected MQTT client at the same time. When the message is received, the broker must find all the clients that have a subscription to the received topic [13]. It is responsible for receiving messages from the sensors connected to the device with implemented MQTT client library and forwarding it to the designated remote device. There are plenty of other tasks and responsibilities that are handled by the broker. First of all, there is authentication and authorization of the clients for the security purposes. A client can send username and password within connect message that will be checked by the broker. On the broker side topic permission is implemented to restrict the client to publish or subscribe. Furthermore, for encrypted communication between the broker and the clients, TLS (Transport Layer Security) and SSL (Secure Sockets Layer) encryption are used, the same security protocol that can be used by the HTTP protocol [5]. It is also possible to implement custom authentication or authorization logic into the system [13]. There are several message brokers that implement MQTT protocol like

Mosquito - open source MQTT v3.1/v3.1.1. Broker [14] and HiveMQ – an enterprise MQTT broker [13].

C. MQTT over WebSockets

Using MQTT over WebSockets every browser can be MQTT device. Due the publish/subscribe pattern of MQTT a real-time communication between end device (e.g. temperature sensor) and monitor device (e.g. a web application) is achieved. Using QoS 1/2 message will arrive on the client or broker at least once/exactly once. The broker will queue all messages which client misses when it is not connected. Messages which are retained on the server are delivered when a client subscribes to one of the topics instantly. Paho.js [15] implements this behavior and will be used in initial testing.

D. CoAP cloud-based system

Transport layer protocol which is used by CoAP is UDP (User Datagram Protocol) so each CoAP message occupies the data section of one UDP datagram. CoAP messages are exchanged asynchronously between CoAP endpoints. Packets order and retransmission depends on software stack. As already mentioned CoAP was designed according to the REST architecture and because of that it is similar to HTTP protocol so CoAP makes use of GET, PUT, POST, and DELETE methods in a similar manner like HTTP. Reliable message transmission is provided by marking a message as confirmable (CON). Recipient sends acknowledgment message (ACK), shown in figure 2.

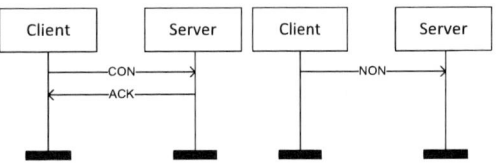

Figure. 2. *Reliable and non-reliable transmission with CoAP*

A message that does not require reliable transmission (for example single measurement of sensor data) can be sent as a non-confirmable message (NON). CoAP also supports the use of multicast IP destination addresses, enabling multicast requests [16].

CoAP was usually designed to connect low power electronic devices (e.g. wireless sensors) with Internet based systems. In many cases data is collected for subsequent processing. A CoAP device is connected to a cloud-based system via an HTTP proxy using a standard CoAP-HTTP mapping, figure 3. Using a proxy/bridge adds an additional communication overhead and increases message latency.

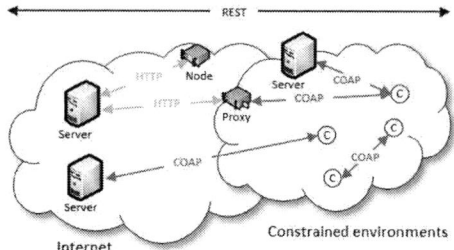

Figure. 3. *CoAP cloud-based system*

In CoAP, a sensor node is typically a server, not a client or it can be both. The sensor provides resources which can be accessed by clients to read or alter the state of the sensor. As CoAP sensors are servers, they must be able to receive inbound packets. To function properly behind NAT, a device may first send a request out to the server, as is done in LWM2M, allowing the router to associate the two. Although CoAP does not require IPv6, it is easiest used in IP environments where devices are directly routable.

IV. COMPARISON

Fig. 4 shows the frames sent in MQTT for one exchange. To establish a connection, TCP uses a three-way handshake (SYN, SYN-ACK, ACK). After TCP connection is established, MQTT protocol uses different messages for communication. There are 14 message types specified in 3.1 version of MQTT protocol. CONNECT (client request to connect to server), PUBLISH (publish message), PUBACK (publish acknowledgment), SUBSCRIBE (client subscribe request) are some of the message types.

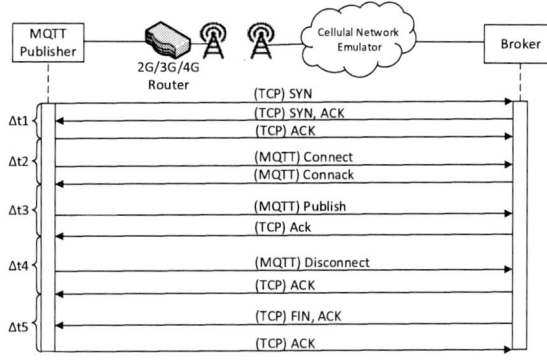

Figure. 4. *MQTT data exchange (Payload 2500 bytes)*

TABLE 1. ANALYSIS OF MQTT DATA EXCHANGE

Section	Bytes	Time [ms]	Remark
Δt_1	186	305	TCP Connection Establishment
Δt_2	150	278	MQTT Connection Establishment
Δt_3	2729	517	MQTT Payload
Δt_4	116	105	MQTT Connection Termination
Δt_5	114	107	TCP Connection Termination
Sum	**3195**	**1312**	

The bytes sent and the time needed for each section of the data transfer are shown in Table I.

TABLE 2 ANALYSIS OF COAP DATA EXCHANGE

Section	Bytes	Time [ms]	Remark
Δt_1	1137	623	CoAP Payload 1/3
Δt_2	1137	618	CoAP Payload 2/3
Δt_3	565	534	CoAP Payload 3/3
Sum	**2839**	**1775**	

Fig.5. shows the frames sent in CoAP for one exchange. The bytes sent and the time needed for each section of the data transfer are shown in Table II.

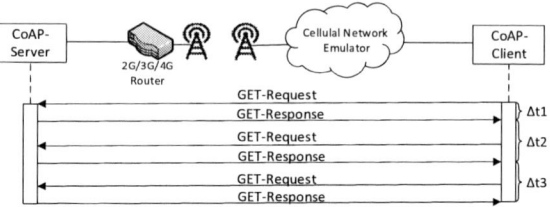

Figure. 5. *CoAP data exchange (Payload 2500 bytes)*

One of the main differences between MQTT and CoAP can be found in the transport layer of the OSI model. MQTT runs on top of the Transmission Control Protocol while CoAP runs on top of the User Datagram Protocol. Although UDP is not reliable, CoAP provides its own reliability mechanism using „confirmable messages" and „nonconfirmable messages". Former require an acknowledgment while latter do not need an acknowledgment [17].

MQTT is a many-to-many communication protocol for exchanging messages between multiple clients through a central broker. MQTT decouples clients by letting publishers to send messages and having the broker to decide to which subscriber the message will be sent. On the other side, CoAP network is primarily a one-to-one protocol for transferring state information between client and server. However, it supports one-to-many or many-to-many multi-cast requirements. This is inherent in CoAP because it's built on top of IPv6, which enables multicast addressing for devices in addition to their normal IPv6 addresses [18].

For the purpose of the security, a username and a password can be sent inside of an MQTT packet. This can help simplify the authentication of individual clients in a network by reducing the number of keys that need to be distributed and managed in comparison to an exclusively key based system. Exchanged data can be encrypted using SSL or TLS, independently from the MQTT protocol [19]. Since CoAP is built on top of UDP, it cannot rely on SSL/TLS to provide security capabilities. To achieve security, CoAP uses Datagram Transport Layer Security (DTLS) which provides the same assurances as TCP.

In a simple point-to-point configuration between objects, MQTT and CoAP have similar performance characteristics, although broker-based routing requires an additional overhead when compared to a broker-less infrastructure such as HTTP or CoAP request. Additional overhead can be found in a request to a CoAP resource from HTTP client because it has to be forwarded via a

CoAP/HTTP proxy. It is recommended to use CoAP in systems where real-time performance and low latency are not a requirement. The single CoAP client can send the same request to multiple CoAP servers concurrently, which gives multicast characteristic to CoAP protocol.

At the side of the hardware, RaspberryPi model 3 is used. Main differences between RPi model 3 and prior versions are integrated wireless module (802.11n), integrated Bluetooth module (Bluetooth 4.1) and 1.2GHz 64-bit quad-core ARMv8 CPU. Using this type of RPi there is no need for additional modules for connecting end device to the network. For communication between clients and broker, Python language with Paho Python MQTT library is used. The library implements a client class that can be used to add MQTT support to RPi. For the purpose of CoAP testing, WebIOPi framework is used.

V. CONCLUSION

In this paper, a comparison of Internet of Things data protocols is given. IoT architecture is presented by describing the parts where application layer protocols are needed to handle communication. We have presented the most widespread application layer protocols in the IoT through the comparison among each other. Choosing the best appropriate protocol depends on several facts of which most important are: environmental conditions, network characteristics, the amount of data to be transferred, security levels, quality of service requests. CoAP network is primarily a one-to-one protocol for transferring state information between client and server while MQTT is a many-to-many communication protocol for exchanging messages between multiple clients. CoAP runs over UDP which means that communication overhead is significantly reduced. If constrained communication and battery consumption is not an issue, RESTful services can be easily implemented and interact with the Internet using the worldwide HTTP [5]. On the battery-run devices MQTT is more suitable. Additionally, if the targeted final applications require massive updates of the same value, MQTT protocol is more suitable. In a simple point-to-point configuration between objects, MQTT and CoAP have similar performance characteristics. Future work will be aimed at implementing MQTT and CoAP protocols in a lab environment and obtain an experimental comparison among them.

REFERENCES

[1] F. Xia, L. T. Yang, L. Wang and A. Vinel, "Internet of Things", International Journal of Communication Systems, Vol. 25, pp. 1101-1102, 2012.

[2] http://www.gartner.com/newsroom/id/3598917

[3] Evans, "The Internet of Things – how the next evolution of the Internet is changing Everything", April 2011.

[4] S. Katsikeas, „A lightweight and secure MQTT implementation for Wireless Sensor Nodes", Technical University of Crete, June 2016

[5] V. Karagiannis, P. Chatzimisios, F. Vazquez-Gallego, J. Alonso-Zarate „A Survey on Application Layer Protocols for the Internet of Things", Transaction on IoT and Cloud Computing 2015

[6] L. Dürkop, B. Czybik and J. Jasperneite, "Performance Evaluation of M2M Protocols Over Cellular Networks in a Lab Environment", Intelligence in Next Generation Networks (ICIN), pp. 70-75, February 2015.

[7] V. Gazis, M, Görtz, M. Huber, A. Leonardi, K. Mathioudakis, A. Wiesmaier, F. Zeiger and E. Vasilomanolakis. "A Survey of Technologies for the Internet of Things", International Wireless Communications and Mobile Computing Conference, pp. 1090-1095, August 2015.

[8] Benchmark of MQTT servers, version 1.1, January 2015.

[9] D. Thangavel, X. Ma, A. Valera, H. Tan, C. K. Tan, "Performance Evaluation of MQTT and CoAP via a Common Middleware", Intelligent Sensors, Sensor Networks and Information Processing, April 2014.

[10] S. K. Shriramoju, J. Madiraju and A. R. Babu, "An approach towards publish/subscribe system for wireless networks", International Journal of Computer and Electronics Research, Vol. 2., pp. 505-508, August 2013.

[11] V.Lampkin, W. T. Leong, L. Olivera, S. Rawat, N. Subrahamanyam, R. Xiang, "Building Smarter Planet Solutions with MQTT and IBM WebSphere MQ Telemetry", First Edition, September 2012.

[12] E. G. Davis, A. Calaveras and I. Demirkol, "Improving Packet Delivery Performance of Publish/Subscribe Protocols in Wireless Sensor Networks", Sensors, 2013.

[13] http://www.hivemq.com

[14] http://mosquitto.org/

[15] https://eclipse.org/paho/clients/js/

[16] RFC 7252: The Constrained Application Protocol (CoAP), IETF, 2014.

[17] D. Thangavel, X. Ma, A. Valera and C. K. TAN, "Performance Evaluation of MQTT and CoAP via a Common Middleware", IEEE Ninth International Conference on Intelligent Sensors, Sensor Networks and Information Processing (ISSNIP), April 2014

[18] http://electronicdesign.com/iot/mqtt-and-coap-underlying-protocols-iot

[19] A. Foster, „Messaging Technologies for the Industrial Internet and the Internet of Things", white paper, PrismTech, November 2013.

Digital Forensic Analysis Through Firewall For Detection of Information Crimes in Hospital Networks

Ayhan AKBAL[*], Erhan AKBAL [**]
[*] Firat Univesity/ Electrical-Electronics Engineering, Elazig, Turkiye
[**] Fırat University/ Digital Forensics Engineering, Elazig, Turkiye
ayhanakbal@gmail.com

Abstract - Digital forensics analysis was done by taking a view of Firewall on the Firewall used in the hospitals, and the data that could create a criminal element were determined. As is known, all network traffic on the networks is over the firewall. For this reason, the traffic on the entire network is recorded on the firewall. When these records need to be analyzed in terms of forensic information and criminal elements should be detected, the records on the firewall should be analyzed without deterioration. For this purpose, the image of the firewall needs to be taken. However, in order to obtain images, it is necessary to calculate MD5 and SHA-1 HASH values with international validity, which confirm the integrity of the image. For this purpose, the Juniper SSG 550 firewall device used in Firat University Hospital will be analyzed. For analysis, FTK Imager program which is developed by AccessData firm and offered for free use will be used. This image will be analyzed with forensic tools such as forensics explorer.

I. INTRODUCTION

In recent years, rapid developments in information technologies have been used rapidly in the healthcare sector. Depending on the developments in technology, the health sector has also started to carry all patients' data electronically[1]. Many hospitals that carry data to digital systems have to protect their personal health data to have fast and timely medical intervention. [2,3]. The ability to control devices that are used on the network, especially in intensive care and which are vital for patients, makes data security and forensic information problems important for the safety of personal data [2, 5].

In this study, digital forensics analysis was done by taking a view of Firewall on the Firewall used in the hospitals, and the data that could create a criminal element were determined. As is known, all network traffic on the networks is over the firewall. For this reason, the traffic on the entire network is recorded on the firewall. When these records need to be analyzed in terms of forensic information and criminal elements should be detected, the records on the firewall should be analyzed without deterioration. For this purpose, the image of the firewall needs to be taken. However, in order to obtain images, it is necessary to calculate MD5 and SHA-1 HASH values with international validity, which confirm the integrity of the image. For this purpose, the Juniper SSG 550 firewall device used in Fırat University Hospital will be analyzed.

For analysis, FTK Imager program which is developed by AccessData firm and offered for free use will be used. This image will be analyzed with forensic tools such as forensics explorer.

II. DIGITAL FORENSIC AND ANALYSIS METHODOLOGY

A. Cyber-crime in Healtcare

With the digitization of the health sector, the health sector is increasing with cyber-crime [4]. In the last five years there have been a lot of cyber-attacks in the healthcare sector, and these cyber-attacks cost only US $ 6 billion [4]. People's personal information, address, credit card, health information about the patient is very important information. This information is different from other health information. The credit card information of the patient can be changed by canceling the credit card of the stolen patient. However, medical information about the patient can not be changed at all. This makes medical data more important.

In the healthcare industry, hospitals or healthcare organizations have begun to use information technology more quickly to perform transactions. The systems followed by patients in intensive care have become accessible via the internet on many medical devices such as patient monitors, respiratory devices, temperature monitors.

However, these institutions are a clear target for cyber-attacks because their areas of expertise are health. The wired and wireless use of devices that store patient information, provide remote access, and the lack of specialist expertise or the management of non-expert systems leave health systems vulnerable. For this reason, it is also a very important issue to determine the criminal elements that control the data of health systems.

In this study, an example of how digital forensics analysis of firewalls, which are connection points of the hospitals to the outside world, is done.

B. Firewall

Firewall is a hardware based network security system that controls outgoing packet traffic on a network based on a rule set. It is a device that keeps Internet traffic under control, with many different filtering features, including incoming and outgoing packets of computer and network. All incoming and outgoing packets pass through the firewall. For this reason, packets must be passed through the firewall during all attacks from outside the network. In this study, Juniper SSG 550 (Fig.2.1) firewall used in Firat University Hospital was used.

Figure 2.1.a. Juniper SSG 550M Firewall

Figure 2.1.b. Juniper SSG 550M Firewall(Upside)

Juniper SSG550 line of secure services gateways consists of high-performance security platforms for regional branch office and medium-sized, standalone businesses that want to stop internal and external attacks, prevent unauthorized access and achieve regulatory compliance. l firewall performance and 500 Mbps of IPsec VPN performance, while the Juniper Networks SSG550M Secure Services Gateway provides 650 Mbps of state full firewall performance and 300 Mbps of IPsec VPN performance[6].

C. FTK Imager

It is a continuously developed digital data analysis and analysis program written by FTK ACCESSDATA. This program, which is very successful in transforming

evidence analysis and evidence into concrete reports, is being used extensively all over the world, especially by security forces. The program is considered to be highly productive and functional, as well as evidence of the reports and data submitted to the court by all courts through this program.

FTK Imager Hash value found in terms of information, image date etc. Information. With the FTK Imager software, copies can be made in raw (dd), E01 (Expert Witness, Encase) and AFF formats. These images can then be analyzed in terms of forensic science and support filing formats such as FAT, NTFS, ext2, ext3 [7].

In this study, the image of Firewall was taken with FTK Imager. For this purpose, a write-locked UltraBlock USB3.0 Forensic Card Reader, shown in Figure 1, used in Forensic Computing Engineering Laboratories of Fırat University [8].

Figure 2.2. UltraBlock USB3.0 Forensic Card Reader

D. Forensic Explorer

Forensic Explorer is a tool for the analysis of electronic evidence. Primary users of this software are law enforcement, corporate investigations agencies and law firms. Forensic Explorer has the features you expect from the very latest in forensic software. Inclusive with Mount Image Pro, Forensic Explorer will quickly become an important part of your forensic software toolkit [9]. Forensic Explorer combines a flexible graphic user interface (GUI) with advanced sorting, filtering, keyword searching, previewing and scripting technology. It enables investigators to: Manage the analysis of large volumes of information from multiple sources in a case file structure, Access and examine all available data, including hidden and system files, deleted files, file and disk slack and unallocated clusters, Automate complex investigation tasks, Produce detailed reports and Provide non forensic investigators a platform to easily review evidence[9].

E. Analysis Methodology

In this study, the file structure of the operating system used by the Juniper firewall device and the data analyzed by the firewall were analyzed. For this purpose, the storage unit of the active Juniper SSG 550 firewall device (Fig.2.3) Firewall uses 512 MB Compact Flash as the disk unit.

Figure 2.3 Juniper Firewall System Disk

Images were taken with the FTK Imager program developed by ACCESSDATA. To do this, the Compact Flash disk on the Firewall is first removed and the UltraBlock USB3.0 Forensic Card Reader with write lock is inserted so that the evidence can not be changed. FTK Imager program The disk image operation has been started with the Create Disk Image menu. Later, source selection was made for the image to be taken in the program. At this stage, the physical source on which the software belonging to Juniper on the computer connected CF Reader is selected. After this process, CF Flash memory is selected from the list of physical drives in the computer. The CF image is taken in RAW (dd) format recognized by Forensic Explorer (Fig 2.4).

```
Created By AccessData® FTK® Imager 3.4.2.6

Case Information:
Acquired using: ADI3.4.2.6
Case Number: 1
Evidence Number: 1
Unique description: 1
Examiner: 1
Notes: 1

--------------------------------------------------

Information for C:\Users\aakbal\Desktop\image\Juniper_FW_SMG550_write_lock:

Physical Evidentiary Item (Source) Information:
[Device Info]
Source Type: Physical
[Drive Geometry]
Cylinders: 63
Tracks per Cylinder: 255
Sectors per Track: 63
Bytes per Sector: 512
Sector Count: 1.015.056
[Physical Drive Information]
Drive Model: Generic- USB3.0 CRW-CF/MD USB Device
Drive Serial Number: 2012062914345300
Drive Interface Type: USB
Removable drive: True
Source data size: 495 MB
Sector count: 1015056
[Computed Hashes]
MD5 checksum:  4629ffa00cd5732b5e2c09441bbc78c9
SHA1 checksum: 77753113155f00c524499a297c497115e3b11e9c

Image Information:
Acquisition started:  Thu Feb 02 19:04:19 2017
Acquisition finished: Thu Feb 02 19:05:05 2017
Segment list:
C:\Users\aakbal\Desktop\image\Juniper_FW_SMG550_write_lock.001
```

Figure 2.4. FTK Imager alınan imaja ait bilgiler.

The image of the disc structure on the disc was examined by using the Forensic Explorer software and the file was analyzed(Fig. 2.5).

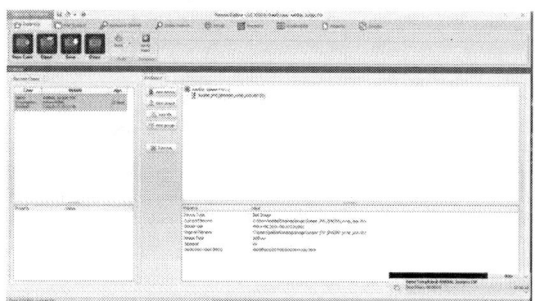

Figure 2.5. Forensic Explorer GUI

III. STUDY

In this study, when the disk of Juniper Firewall which is taken as an image is connected to any computer under normal conditions, the first partition in the diskette can not access the 488 MB part that the operating system belongs to while the first partition is 6 MB FAT. This partition is hidden as shown in Figure 3.1.

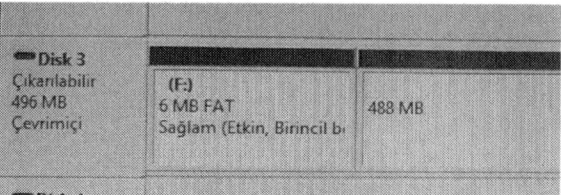

Figure 3.1. Juniper Disk

For this reason, the CF disc image in RAW format has been analyzed with Forensic Explorer. The disk structure as a result of the analysis is shown in figure 3.2. The resulting files are shown in figure 3.3.

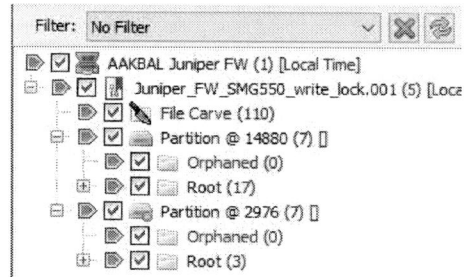

Figure 3.2. Juniper FW Disk Structure

Figure 3.3. Juniper FW File Structure

As shown in Figure 3.3, the file structure of the Firewall has been obtained in detail. The contents of these files can be viewed as HEX. When these files are examined, *"golerd.rec"* file contains log information to the firewall [11]. The firewall user information is stored in the *"node_secret.ace"* file [13]. The file *"crushdump.dmp"* is a file that records the data during execution for examining events occurring in the firewall. The *"syscert.cfg"* file contains system certificates. The *"prngseed.bin"* file and *"license.key"* file contains the license information for the Firewall [14]. The *"envar.rec"* file contains the licenses received with the system when the system is being installed [11]. The *"ns_sys_config"* file contains all the settings that the users have defined on the firewall [12]. In the *"dnstb.rec"* file, all DNS information registered by the firewall can be accessed (Figure 3.4).

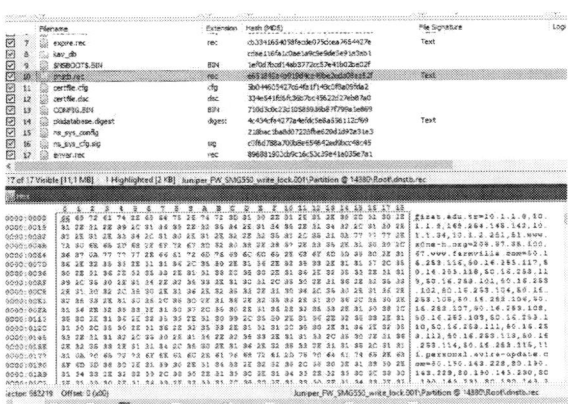

Figure 3.4. dnstb.rec file detail

IV. CONLUSION

Forensic Computing has become an important element in recent years. For this reason, this has increased in studies. However, the high number of devices in the IT sector makes it difficult to analyze the products, which are a wide variety of manufacturers. For this reason, a forensic analysis of the firewall has been done in the literature.

As a result, the safety of the data is very important in hospitals where patient records are kept. Every year attacks are carried out in these data. When working on these data, it is necessary to perform forensic examinations in order that the data or configurations in the devices are correct and that the necessary work on the existing system can be done more easily and more accurately when a criminal element occurs. In this frame, the forensic examination of the Juniper Firewall SSG 550M device within the Fırat University Hospital was made. As a result of this review, the Juniper Firewall file structure was accessed the device and analyzed without any changes to the data. All data of the Firewall has been

accessed to the configuration that is active in the device, and to the data of users entirety. The file structure of the Firewall has been detected.

REFERENCES

[1] Carrier, B. "Digital Investigation Foundation", File System Forensic Analysis, Editor: Carrier, B. , Addison Wesley Professional, NJ, 12-21, 2005.

[2] Jones, K. J. & Bejtlicj, R. & Rose, C.W. "Forensic Analysis Techniques", Real Digital Forensics: Computer Security And Incident Response, Editors: Jones, K.J., Bejtlicj R., Rose C.W., Addison Wesley Professional, NJ, 205-246, 2006.

[3] Altheide, C. & Carvey, H. & Davidson, R. "Disk and File System Analysis", Digital Forensics with Open Source Tools, Editor: Davıdson, R., Elsevier Inc, MA, 39-67, 2011.

[4] http://www.healthcareitnews.com/blog/growing-pains-cybercrime-plagues-healthcare-industry

[5] Altheide, C. & Carvey, H. & Davidson, R. "Disk and File System Analysis", Digital Forensics with Open Source Tools, Editor: Davıdson, R., Elsevier Inc, MA, 39-67, 2011.

[6] Juniper, "SSG500 Line of Secure Services Gateways - Juniper Networks",https://www.juniper.net/us/en/local/pdf/datasheets/100 0143-en.pdf , Date: 08.01.2017

[7] AccesData, "FTK Imager version 3.4.3", https://ad-df.s3.amazonaws.com/Imager/3_4_3/FTKImager_UG.pdf, Date: 08.01.2017

[8] Digitalintelligence, "UltraBlock USB 3.0 Forensic Card Reader", https://www.digitalintelligence.com/products/forensic_card_reader ", Date: 08.01.2017

[9] GetData Forensics, "Forensic Explorer", "http://download.getdata.com/support/fex/documents/forensic-explorer-user-guide.tr.pdf ", Date: 08.01.2017

[10] Juniper Networks, Juniper Firewall ScreenOS Console, https://kb.juniper.net/infocenter/index?page=content&id=KB5477, Date: 09.01.2017

[11] Juniper Networks, Juniper Firewall ScreenOS Console, https://kb.juniper.net/infocenter/index?page=content&id=KB5943, Date: 09.01.2017

[12] Juniper Networks, Juniper Firewall ScreenOS Console, https://kb.juniper.net/infocenter/index?page=content&id=KB1278 Date: 4, 09.01.2017

[13] Juniper Networks, Juniper Firewall ScreenOS Console, https://kb.juniper.net/infocenter/index?page=content&id=KB9556, Date: 09.01.2017

[14] Juniper Networks, Juniper Firewall ScreenOS Console, https://kb.juniper.net/infocenter/index?page=content&id=KB2559, Date: 9.01.2017

Gap in pagination due to withheld paper.

Pages 510-514

Mine Safety System Using Wireless Sensor Networks

Valdo Henriques[*], Reza Malekian[*╫], Dijana Capeska Bogatinoska[**]

[*] Department of Electrical, Electronic and Computer Engineering, University of Pretoria, Pretoria 0002, South Africa
[**] Faculty of Computer Science and Engineering, University of Information Science and Technology "St. Paul the Apostle", Ohrid, Republic of Macedonia
╫Corresponding author: Prof. Reza Malekian, Reza.malekian@ieee.org

Abstract: The activity of mining always had an element of risk associated with it. Miners often had to encounter challenging underground environment which sometimes result in injuries and fatalities. Some portion of them can be attributed to error caused by human carelessness. More often the cause of these accidents can be attributed to the ambient conditions. In this paper, we design a mine safety system using wireless sensor networks with measurement of parameters such as temperature, air-flow, humidity, noise, dust, and gas concentration. These six ambient characteristics have been identified as hazardous to the health and safety of the mine worker. From the experimental results obtained, an accurate mine safety system is achieved by design of various sensors. The temperature sensors, humidity sensor, airflow sensor, and noise sensor achieved an accuracy of 94.45%, 98.55%, 85.4%, 99.14% alternatively. Besides, the dust sensor has a resolution of $0.003mg/m^3$ and the gas sensor show a resolution of 0.9ppm which is close to the required resolution of 1 ppm.

Keywords: mining industry, sensor systems, temperature measurement, humidity measurement, airflow measurement, noise sensor measurement, wireless sensor networks

I. INTRODUCTION

The concept of mining has been around since the beginning of civilization; our ancestors used stones, ceramics and later metals found close to earth's surface to make tools and weapons. The activity of mining always had an element of risk associated with it. Miners often had to encounter challenging underground environment which sometimes result in injuries and fatalities. Some portion of them can be attributed to error caused by human carelessness. More often the cause of these accidents can be attributed to the ambient conditions underground.

The ambient conditions underground are extremely difficult to monitor without placing someone's life at risk [1]. This type of first hand monitoring endangers the life of the observer which is not ideal. The International Organization of Standard proposes the ISO 45001 standard which benefits both miners and the economy [2].

Wireless sensor technology if implemented in mines can improve the safety of mines considerably by eliminating the need for human testers inside the mines. A wireless sensor network consist of a number of sensor nodes [3] [4] distributed around various locations in the mine communicating with each other [5] [6]; they form a non-invasive mine safety system. Moreover, wireless sensors provide a considerable reduction in wiring [7] which can be damaged during mine blasting; sensors also provide real-time measure of the mining environment [8].

Due to the dangerous and potentially unsafe mining practices there exists a relationship between hazard, latent danger and accident [9].

The wireless sensor network discussed in this research work considers star and mesh network topologies. The ambient characteristics measured in our research work include temperature, air-flow, humidity, noise, dust, and gas concentration. These six ambient characteristics have been identified as hazardous to the health and safety of the mine worker. In other words, each of these characteristics can be solely or collectively responsible for incurring a risk to a mine worker. The idea of our system is not to mitigate an accident but rather to detect a hazard early and provide a control output to alleviate latent danger. This output could be a decision by the active mine manager to call all miners out of the mine for a temporary period, or a visual alarm system that can provide a timely warning when conditions become unstable.

Our research work takes advantage of available wireless technologies, which will not only monitor the conditions inside the mine but also be used to provide a noise mapping feature which will output a noise protection scheme for the workers, and provide ventilation switching to regulate air in the mine. Current related works in place do not consider noise or humidity as a parameter for measurement as in [10] [11]. It also provides an enhanced snapshot of the condition inside the mine at any given time.

II. PROPOSED SYSTEM

Our proposed system consists of Measurement nodes and a Data collections station. The measurement node consists of different sensors which detects the values of temperature, humidity, gas concentration, air-flow and noise levels. These measured values are transferred to the data collection center using a wireless network. Each of these sensors has their own respective signal processing circuits. The input from the sensor is sampled by the micro controller, which forms the central processing Unit (CPU) of the measurement module. The microcontroller also has a timer which determines the delay between two sensor sample readings. A wireless communication module is also integrated into the microcontroller for transmitting data to the data collection station. The proposed system uses ZigBee 802.15.4 wireless protocol for wireless communication.

The data collection station has also a wireless communication module and a microcontroller. The data collection center has the necessary hardware and software capabilities to process, display and store these data values. Data collection stations display the measured sensor values in the GUI of the laptop. The data collection station's tasks also include developing safety schemes for the mine workers using the sensor data collected.

The three control outputs that will be created by this proposed system are:

- Visual Indicator on Central Computer

- Noise Protection Scheme

- Ventilation Switching

Controlling a hazardous situation involves using the results of monitoring and performing some control output driven by these results to avoid latent danger.

The first two outputs will assist the mining supervisor or equivalent to employ existing safety policies by showing the output of the collected data at the data collection center at the central computer.

The noise protection scheme will be implemented purely in software. This scheme will take a predefined visual map of the mine and combine it with the noise values measured in each different segment of the mine. This scheme will then subsequently provide the miners with a guide of which areas in the mine are noisy. The switching module will use the data from the system namely the outputs from the gas and air-flow sensors to provide a switching mechanism to switch on a ventilation unit (a fan). This module will take a control signal from the microcontroller and a domestic AC input and provide an AC power point where the ventilation unit can be powered from. The temperature, air flow and humidity sensors are created using the first principles. The dust and gas sensors used in our work are off the shelf components.

III. SYSTEM DESIGN

A. Temperature Sensor

This sensor is used to measure ambient air temperature. There are a variety of different methods which can be used to design temperature sensor from first principles, the two alternatives which were considered are the use of a thermistor and the use of a thermocouple. After evaluating these two alternatives in the context of this project it was decided that a thermistor based temperature sensor will be designed.

Thermistor we used has a resistance of 2000 ohms at $25°$ C. We measured the resistance of the thermistor at three different temperatures. In order to find the correct load to provide the best range of measurement a range of different values were simulated for the load. It was found that a 1000 ohm resistor was best. This is shown in Figure 1.

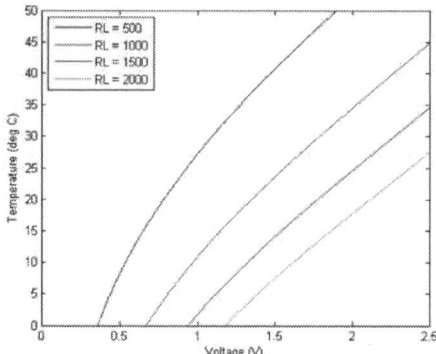

Figure 1. Temperature to Voltage Relationship of Temperature Sensor for Multiple Loads

B. Humidity Sensor

Humidity sensor is used to measure the ambient humidity in its surroundings. There are different methods which can be used to design humidity sensor from first principles, the two different methods we considered are resistive and capacitive type humidity sensors. Since resistive type sensors have a higher dependence upon exposed temperature and are more vulnerable to damage from dust and pollution a capacitive type approach has been chosen for the humidity sensor.

The sensor used in our work is the Honeywell HIH-4000 series humidity component. This sensor has a typical current draw of 200 µA [12]. The voltage across the sensor will vary depending on the water vapour it is exposed to. The microcontroller used in our work can has an operating voltage of 5V, so we improved on the off the shelf humidity sensor with voltage range (0.8V – 3.8V) by designing an operational amplifier. This increased the resolution of the measured humidity. The figure given below, Figure 2, shows the amplifier circuit current drain simulation circuit.

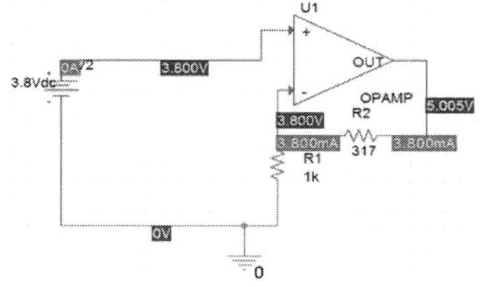

Figure 2. Humidity Sensor Amplifier Circuit Current Drain Simulation

C. Air Flow Sensor

Air flow sensor is used to measure the ambient air flow in its surroundings. Different techniques can be used to design air-flow sensor from first principles. The rotation based approach has been chosen for our work.

The air flow sensor is built from the first principle by using a three cup anemometer and slotted optical switch.

The air flow sensor can be divided into two parts the electronic component and the mechanical component consisting of the cup anemometer. The slotted optical switch used in our work is OPB804.

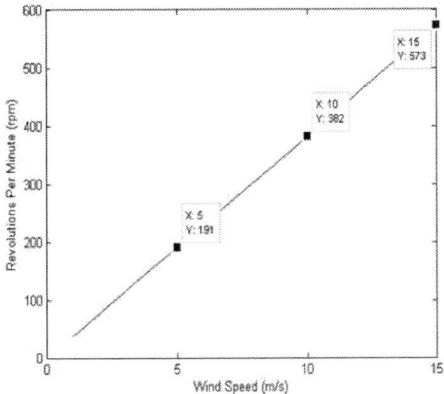

Figure 3. Theoretical relationship between linear wind speed and RPM

The optical switch consists of a phototransistor and an LED. The near IR light emitted by the LED is detected by the photo transistor which causes a voltage change from low to high. The mechanical part has a small metal piece mounted on the bearing of the anemometer. When the arm rotates, the metal crosses the slot of the optical switch, each time this happens one revolution is observed. The timer in the microcontroller is used to calculate the corresponding rotations per minute (RPM). Theoretical relationship between linear wind speed and RPM is given in Figure 3. The airflow sensor calibration setup is shown in Figure 4.

Figure 4. Airflow Sensor Calibration Setup

D. Noise Sensor

A noise sensor is used to determine the ambient sound levels in its surroundings. It is designed and implemented using the first principle.

In our system in order to convert noise (sound) into a usable metric for the system, a microphone is used to capture the analog variances with sound changes. The output of the microphone is fed into the filter and amplification circuit which gives a constant voltage which can be measured by the controller. The measured sound

levels are considered noise because they negatively affect the employees of the mine.

E. Gas Sensor

Gas sensor is used to measure the concentration of ambient gas in its surroundings. In a mine we measure the concentration of methane gas. We implement gas sensor using an off the shelf component called Figaro TGS2611-C00.

The heater (RH) voltage is applied to the heater element in order to maintain the sensing element at a specific temperature which is optimal for sensing. The circuit voltage is applied to allow for a voltage to be measured across the load resistor. Figure 5, shows Gas Sensor Resistance to Gas Concentration. The measured voltage is converted into gas concentrations using mathematical modeling.

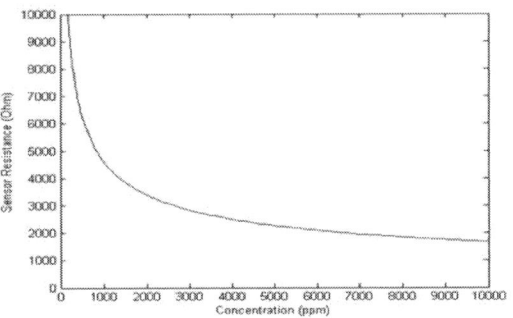

Figure 5. Gas Sensor Resistance

F. Dust Sensor

Dust sensor is used to measure the concentration of dust in its surroundings. We use an off the shelf optical type dust sensor Sharp GP2Y1010AU0F. This sensor measures the dust concentration in the air by assessing the intensity of light.

The sensor requires a square wave input, to achieve

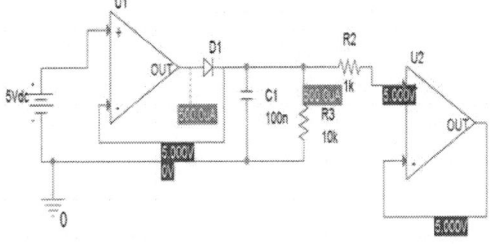

Figure 6. Dust Sensor Peak Detector Current Drain Simulation

this we use the PWM output of the microcontroller. Figure 6, shows the simulation on OrCAD to find the current draw of the peak detector. Figure 7 shows, Dust Sensor Driving Signal and Output.

G. Design of the System

The final design [15] is implemented using a set of PCBs. The sensor nodes needs to be mobile so they are battery operated. The collection node is static so they are powered by a suitable power supply. Each node utilises a dsPIC30F4011 microcontroller [16,17] which has the required hardware and software functionalities. XBee Series 1 Pro acts as the ZigBee wireless interface [18,19, 20] for data transmission. The graphical user interface provides the visual output of the parameters measured by

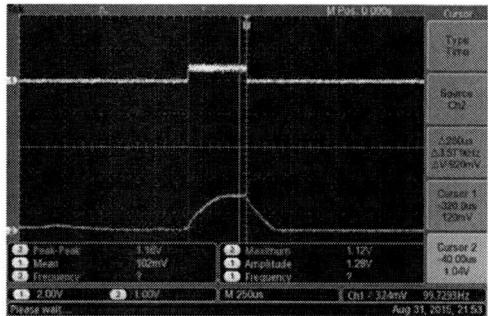

Figure 7. Dust Sensor Driving Signal and Output

TABLE I.   SENSOR NODE CURRENT CONSUMPTION

Component	Current Consumption	Component	Current Consumption
Temperature Sensor	2.857mA	Noise Sensor	21mA
Humidity Sensor	4mA	Microcontroller	20mA
Airflow Sensor	5mA	XBee idle/receive	55mA
Dust Sensor	10.5mA	XBee transmit	150mA
Gas Sensor	56mA		

our system.

IV.   RESULTS, OBSERVATION AND DISCUSSION

We subjected our system to various tests which focuses on a specific subsystem. Different tests on various sections are discussed in the subsections below.

A. Testing the Temperature Sensor

The parameters that are considered for this test include the reference temperature measured by the digital thermometer and the temperature measured by the prototype system [21, 22]. The reference value measured is compared against the prototype's measured value. We exposed our test setup to different environments. These environments provided different ambient conditions. The ambient values are sampled from the prototype system and compared against the reference value measured on the digital thermometer. The results obtained from the experiment for testing the accuracy of the measured temperature are shown in Table 2 and Table 3 below.

The first observation made is regarding the error obtained on each temperature measurement made, when compared to the reference temperature. The error percentage of the measurement made will be calculated using the equation for relative error given below.

$$Relative\ Error = \frac{measured\ value - reference\ value}{reference\ value} \times 100 \quad (1)$$

TABLE II.   TEMPERATURE SENSOR ACCURACY TEST RESULTS WITH REF. TEMP 22.3^{0}C

Reference Temperature = 22.3			
Environment 1	Measured Temp. (°C)	Error (°C)	Relative Error Percentage (%)
Sample 1	24.42	2.12	9.51
Sample 2	23.98	1.68	7.53
Sample 3	24.42	2.12	9.51
Sample 4	24.75	2.45	10.99

Using Sample 1 in the equation we get a relative error of 9.51%, if the same calculation is done for sample number 6 the result is 5.55%. This shows that the temperature measurement from the prototype system has a worst case accuracy of 89.01% and a best case accuracy of 94.45%.

This resolution obtained from the result is 0.105°C. The obtained resolution of the temperature exceeds the specification of 0.5°C. The range of the temperature obtained (2-45°C) also exceeds the specification of the range of 20-35°C. This range is determined while the accuracy is being tested.

TABLE III.   TEMPERATURE SENSOR ACCURACY TEST RESULTS WITH REF. TEMP 26.5^{0}C

Reference Temperature = 26.5			
Environment 2	Measured Temp. (°C)	Error (°C)	Relative Error Percentage (%)
Sample 5	28.18	1.68	6.34
Sample 6	27.97	1.47	5.55
Sample 7	28.49	1.99	7.51
Sample 8	28.39	1.89	7.13

B. Testing the Humidity Sensor

The parameters that are considered for this test include the reference humidity measured by the digital hygrometer and the humidity measured by the prototype system. For resolution test the ambient humidity is varied in small increments. The ambient values are sampled from the prototype system and compared against the reference value measured on the digital hygrometer. The result of the humidity sensor accuracy test is shown in Table 4 and Table 5. The results show that, the humidity sensor measures humidity in the required range with a worse case accuracy of 98.55% accuracy which is in line with the required specifications.

TABLE IV.   HUMIDITY SENSOR ACCURACY TEST RESULTS WITH REFERENCE HUMIDITY 41.3

Reference Humidity = 41.3			
Environment 1	Measured Humidity (%RH)	Error (%RH)	Relative Error Percentage (%)
Sample 1	41.18	0.12	0.29
Sample 2	40.82	0.48	1.16
Sample 3	40.7	0.6	1.45
Sample 4	40.7	0.6	1.45

TABLE V. HUMIDITY SENSOR ACCURACY TEST RESULTS WITH REFERENCE HUMIDITY 42.2

Reference Humidity = 42.2			
Environment 2	Measured Humidity (%RH)	Error (%RH)	Relative Error Percentage (%)
Sample 5	41.65	0.55	1.30
Sample 6	42.61	0.41	0.97
Sample 7	42.13	0.07	0.17
Sample 8	42.73	0.53	1.25

C. Testing the Air Flow Sensor

The parameters that are considered for this test include the reference airflow measured by the digital anemometer and the airflow measured by the prototype system. The airflow is varied in small increments. The results show that the airflow sensor measures the magnitude of airflow in the range of (0 - 15m/s). The magnitude of the error increases with the increase in the wind speed. For 5m/s the worst case error percentage is 2.4% (average = 1.45%). For 10m/s the worst case error percentage is 11.5% (average = 10%). For 15m/s the worst case error percentage is 14.6% (average = 12.5%). This resulted in a worst case accuracy of 85.4% (average = 90.5%). This worst case accuracy will only be applicable for airflow magnitudes near the 15m/s magnitude with lower magnitudes showing better accuracy. Therefore this sensor does meet the specification for range of measurement. The air flow resolution test is completed using 3 trials of 2 measurements. The results across the trials gave the similar airflow resolution output. So the test-retest reliability of the airflow resolution is high for this system. The result shows that an average resolution of 0.045m/s which exceeds the expectation of 0.5m/s.

TABLE VI. AIRFLOW SENSOR RESOLUTION TEST RESULTS

	Measured Value 1 (m/s)	Measured Value 2 (m/s)	Airflow Step (Resolution, m/s)
Trial 1	4.93	4.88	0.05
Trial 2	9.12	9.08	0.04
Trial 3	13.00	12.95	0.05

D. Testing the Noise Sensor

The parameters that are considered for this test include the reference noise measured by the sound level meter and the noise measured by the prototype system.

During testing different sounds are made to elevate the sound level. This level is different for a set of measurements to measures different values. The reference sound is noted and compared to the sound level measured from the prototype system. The results show that sound level measurement from the prototype system has a worst case accuracy of 89.53% and a best case accuracy of 99.14%.

The fluctuation in the observed sensitivity of the noise measurement system shows that only 80% of the tests conducted produced a relative error percentage lower than 5%. This shows a promising trend for the sensor. There are however the remaining 20% of the results which are greater than 5%. The lowest value that can be measured is 50 dB SPL, below this value, the sound level is almost silent. This noise sensor is able to measure to 95 dB SPL, this meets the required specification. The noise sensor resolution test is completed using 3 trials of 2 measurements. The result of the resolution test is given in Table 7. The result shows a resolution of 0.23 dB SPL which exceeds the specification of 5 dB SPL.

TABLE VII. NOISE SENSOR RESOLUTION TEST RESULTS

	Measured Value 1 (dB SPL)	Measured Value 2 (dB SPL)	Sound Level Step (Resolution, dB SPL)	Voltage Step (V)
Trial 1	57.02	56.86	0.16	0.0050
Trial 2	62.00	61.8	0.20	0.0069
Trial 3	81.36	81.13	0.23	0.0170

E. Testing the Gas Sensor

The parameter that is considered for this test is the measured gas concentration. The measured value is obtained from the electronic circuitry and software implemented. The mine safety system is setup inside a fume cabinet, with the gas meter in close proximity to the sensor array to provide the highest level of comparability between the two measured values. The equipment used for this test included the prototype mine safety system, a fume cabinet, a gas cylinder (cocktail of gases, methane concentration = 50%LEL (lower explosive limit), 2.5% volume) and a gas meter. During experiment the gas concentration is varied in small measurements. The graph of the range tests are shown in Figure 8.

Figure 8 shows the result when the tube is not directly aimed at the sensor. The results show that our proposed gas sensor has a range of (0 – 2554.75 ppm) which

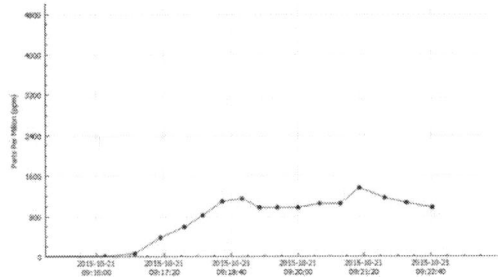

Figure 8. Gas Sensor Range Test Results

TABLE VIII. GAS SENSOR RESOLUTION TEST RESULTS

	Measured Value 1 (ppm)	Measured Value 2 (ppm)	Gas Concentration (Resolution, ppm)
Trial 1	13.16	12.22	0.94
Trial 2	12.22	11.32	0.9
Trial 3	11.32	10.46	0.86

exceeds the original specification of 0 – 70ppm. The

result of the gas sensor's resolution test is given in Table 8. The result shows a resolution of 0.9ppm which is close to the required resolution of 1 ppm.

F. Testing the Dust Sensor

The parameter that is considered for this test is the measured dust concentration. The measured value is obtained from the electronic circuitry and software implemented. The results obtained from the experiment for testing the accuracy of a measured dust concentration is shown in Figure 9. This figure shows the functioning of the dust sensor as well as the maximum value that could be measured using this experiment. The result of the resolution test for the dust sensor is given in Table 9. The results show a resolution of 0.003mg/m^3 which is higher than the required expectation of 0.5 mg/m^3. The result is still good as the threshold value for the dust concentration/crystalline silica is limited to 0.1 mg/m^3 [13][14].

V. CONCLUSION

In this paper, we designed wireless sensor networks consisting of temperature, air-flow, humidity, noise, dust, and gas sensors to improve the safety of mines considerably

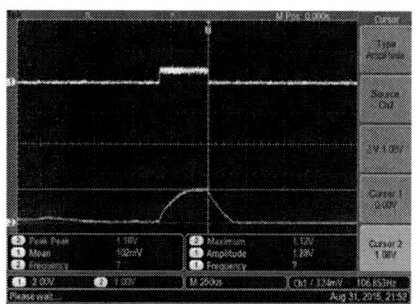

Figure 9. Dust Sensor Range Test Results Trial 1

TABLE VII.   DUST SENSOR RESOLUTION TEST RESULTS

	Measured Value 1 (mg/m^3)	Measured Value 2 (mg/m^3)	Dust Concentration (Resolution, mg/m^3)
Trial 1	0.141	0.138	0.003
Trial 2	0.138	0.136	0.002
Trial 3	0.133	0.130	0.003

by eliminating the need for human testers inside the mines. Each of these characteristics can be solely or collectively responsible for incurring a risk to a mine worker. The wireless sensor network discussed in this research work considers star and mesh network topologies. We also takes advantage of available wireless technologies, which will not only monitor the conditions inside the mine but also be used to provide a noise mapping feature which will output a noise protection scheme for the workers, and provide ventilation switching to regulate air in the mine. From the experimental results obtained, an accurate mine safety system is achieved by design of various sensors. The temperature sensors achieved an accuracy of 94.45%, humidity sensor achieved an accuracy of 98.55%. The airflow sensor measures accuracy of 85.4%, the noise sensor has a worst case accuracy of 89.53% and a best case accuracy of 99.14%. The dust sensor has a resolution of 0.003mg/m3 which is higher than the required expectation of 0.5 mg/m. The result

of the gas sensor shows a resolution of 0.9ppm which is close to the required resolution of 1 ppm.

REFERENCES

[1] R. Nutter, "Hazard evaluation methodology for computer-controller mine monitoring/control systems," *IEEE Transactions on Industry Applications,* vol. 3, pp. 445-449, 1983.

[2] International Organization for Standardization, Occupational health and safety ISO 45001.

[3] X. Chen and P. Yu, "Research on hierarchical mobile wireless sensor network architecture with mobile sensor nodes," in *International Conference on Biomedical Engineering and Informatics, IEEE,* 2010.

[4] P. Deshpand, et. al. "Techniques improving throughput of wireless sensor network," in *International Conference on Circuit, Power adn Computing Technologies, IEEE,* 2015.

[5] "Network Topology and Extent," in *Communication Networks: Principles and Practice,* McGraw Hill Education, 2005 .

[6] Y. Choi, et. AL. "A study on sensor nodes attestation protocol in a Wireless Sensor Network," in *International Conference on Advanced Communication Technology, IEEE,* 2010.

[7] Y. Tan and K. Tseng, "Low-Voltage, DC Grid–Powered LED Lighting System with Smart Ambient Sensor Control for Energy Conservation in Green Building," in *Smart Grid Infrastructure and Networking,* McGraw-Hill, 2013.

[8] J. Dickens, R. Teleka, "Mine safety sensors: Test results in a simulated test stope," in *6th Robotics and Mechatronics Conference,* 2013.

[9] W. Bing, X. Zhengdong, Z. Yao and Y. Zhenjiang, "Study on coal mine safety management system based on "Hazard,Latent Danger and Emergency Responses"," *Procedia Engineering,* vol. 84, pp. 172-177, 2014.

[10] C. Zhao, F. Liu and X. Hai, "An Application of Wireless Sensor Networks in Underground Coal Mine," *International Journal of Future Generation Communication and Networking,* vol. VI, no. 5, pp. 117-126, 2013.

[11] S. Pindado , J. Cubas and F. Sorribes-Palmer, "The Cup Anemometer, a Fundamental Meteorological Instrument for the Wind Energy Industry," *Sensors,* no. 14, pp. 21418-21452, 2014.

[12] Honeywell, HIH-4000 Series Humidity Sensor Datasheet, 2010.

[13] W. H. Organization, Concise International Chemical Assessment Document 24 - Crystalline Silica,Quartz, Geneva, 2000.

[14] Association Advancing Occupational and Environmental Health, *Material Safety Data Sheet-Natural Gas,* 2012.

[15] Valdo Henriques, R Malekian,"Mine Safety System Using Wireless Sensor Network",IEEE Access, Vol.4, pp. 3511-3521, 2016.

[16] Bo Liu, et. al., "Groundwater Mixing Process Identification in Deep Mines Based on Hydrogeochemical Property Analysis", Applied Sicence, Vol. 7, Issue.1, Paper ID: 42, 2017.

[17] Benyu Su, R. Malekian, et al. "Electrical Anisotropic Response of Water Conducted Fractured Zone in the Mining Goaf", IEEE Access, Vol.4, pp.6216-6224, 2016.

[18] Zhongqin Wang, Ning Ye, R. Malekian,et. Al. Li,"TMicroscope: Behavior Perception Based on the Slightest RFID Tag Motion", Elektronika ir Elektrotechnika, Vol.22, No.2, pp.114-122, 2016.

[19] Zhang Xin,et al."An Improved Time-Frequency Representation based on Nonlinear Mode Decomposition and Adaptive Optimal Kernel",Elektronika ir Elektrotechnika,Vol. 22,No.4,pp.52-57, 2016.

[20] Xiangjun J, et al. "Modeling of nonlinear system based on deep learning framework", Nonlinear Dynamics, Springer, Vol.84, No. 3, pp.1327-1340, 2016.

[21] Yaping Huang, Mingdi Wei, Xiaopeng Zhen, "CBM Reservoir Rock Physics Model and Its Response Characteristic Study, IEEE Access, (ISI/SCIE, Impact factor: 1.270), In press.

[22] Bo Liu, Jinpeng Xu, "Groundwater Mixing Process Identification in Deep Mines Based on Hydrogeochemical Property Analysis", Applied Science (ISI, Impact factor: 1.726),, Vol. 7, Issue.1, Paper ID: 42, 2017.

Replication of Virtual Network Functions: Optimizing Link Utilization and Resource Costs

Francisco Carpio, Wolgang Bziuk and Admela Jukan

Technische Universität Braunschweig, Germany

Email:{f.carpio, w.bziuk, a.jukan}@tu-bs.de

Abstract—Network Function Virtualization (NFV) is enabling the softwarization of traditional network services, commonly deployed in dedicated hardware, into generic hardware in form of Virtual Network Functions (VNFs), which can be located flexibly in the network. However, network load balancing can be critical for an ordered sequence of VNFs, also known as Service Function Chains (SFCs), a common cloud and network service approach today. The placement of these chained functions increases the ping-pong traffic between VNFs, directly affecting to the efficiency of bandwidth utilization. The optimization of the placement of these VNFs is a challenge as also other factors need to be considered, such as the resource utilization. To address this issue, we study the problem of VNF placement with replications, and especially the potential of VNFs replications to help load balance the network, while the server utilization is minimized. In this paper we present a Linear Programming (LP) model for the optimum placement of functions finding a trade-off between the minimization of two objectives: the link utilization and CPU resource usage. The results show how the model load balance the utilization of all links in the network using minimum resources.

I. INTRODUCTION

Network Function Virtualization (NFV) is a new paradigm that virtualizes the traditional network functions and places them into generic hardware and clouds, as opposed to the designated hardware. The placement of the virtual network functions (VNFs) can happen either in remote data centers (DC) or by deploying single servers or clusters of servers. Placing VNFs in remote data center can lower the cost of deployment, but is known to typically increasing the delay and create churns of network load, due to the fix and often remote location. Installing new mini data centers inside the network can mitigate the distance-to-data center problem. At the same time, the deployment of new servers forming small data centers in regular nodes requires new investment costs, which requires a gradual upgrade of the network. Therefore, the optimal placement of these servers in the network is a must for network operators to reduce the operational costs.

While most of the current work concentrates on the optimal placement of VNFs under some specific objective minimizing the network costs under some specific resources constraints, e.g. costs of power consumption or the number of physical servers, less effort has been on addressing the network load balancing problem with VNF placement. In this paper, we address, jointly, the network load balancing and resource cost problem with VNF placement, where the concept of VNF replications is used to find a trade-off between network load balancing and network costs. The Fig. 1 illustrates the idea,

whereby we assume a service chain, which is composed of non-replicable VNFs and a variable number of replicas, may be split to one or more parallel sets of an ordered sequence of VNFs towards the service end-point. The major advantage is to split the traffic flows in a controlled way such that network traffic load balancing can be optimized when the service is running. As commonly in network services, the service chain starts with a non-replicable VNF, e.g. a load balancer or gateway, and is allocated in a dedicated data center which generates service requests, while the rest of the functions are allocated on small servers, maintaining the sequence order. To find the optimum placement of servers required for the deployment of VNFs, we formulate the problem as an Integer Linear Programming (ILP) model.

The rest of the paper is organized as follows. Section II presents related work. Section III describes the reference architecture. In Section IV, the related optimization models are described. Section V analyzes the performance and Section VI concludes the paper and discusses future research.

II. RELATED WORK

Early work in [1] studies the optimal VNFs placement in hybrid scenarios, where some network functions are provided by dedicated physical hardware and some are virtualized, depending on demand. They propose an ILP model model with the objective to minimize the number of physical nodes used, which limits the network size that can be studied due to complexity of the ILP model. In [2], a context-free language is proposed for the specification of VNFs and a Mixed Integer Quadratically Constrained Program (MIQCP) for the chaining and placement of VNFs in the network. The paper finds that the VNF placement depends on the objective, such as latency, number of allocated nodes, and link utilizations. In mobile core networks, [3] discusses the virtualization of mobile gateways, i.e., Serving Gateways (S-GWs) and Packet Data Network Gateways (P-GWs) hosted in data centers. They analyze the optimum placements by taking into consideration the delay and network load. In [4], the authors also propose the instantiation and placement of PDN-GWs in form of VNFs.

Unlike previous work, we solve the VNF placement problem by considering VNF replications which is the novel idea that we already proposed in [5]. In this work, we extend [5] to solve the optimum placement of VNFs by minimizing the network cost while maximizing load balancing with constraints on the available resources in mobile core networks. Therefore,

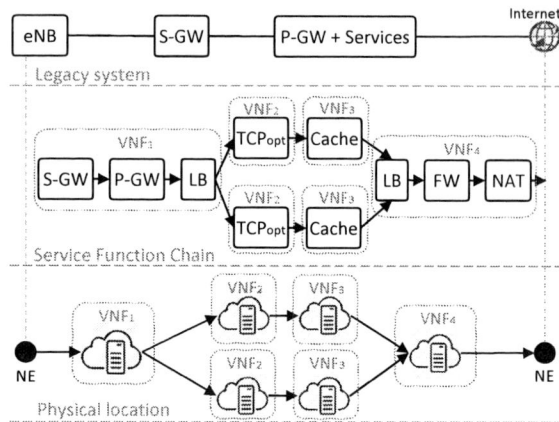

Fig. 1: Use case on Mobile Core Networks

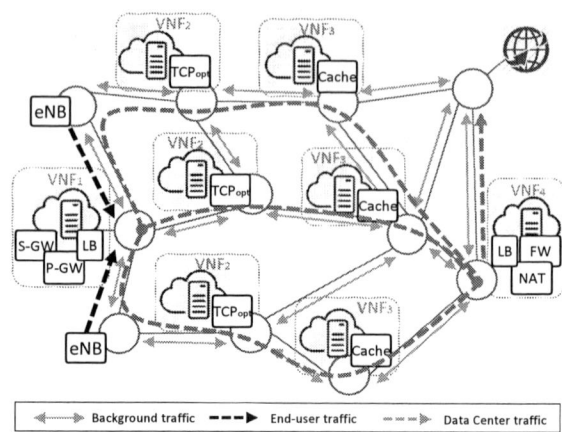

Fig. 2: Network traffic considerations

we propose a more realistic approach to enable an scalable growth of the mobile data traffic over years. This paper also includes a new traffic model of end-user traffic generated in the Radio Access Network, which previous paper did not consider. We also consider multiple VNFs placement per node with replications, which is a novel idea.

III. REFERENCE ARCHITECTURE

The NFV architecture is basically described by three components: Services, NFV Infrastructure (NFVI) and NFV Management and Orchestration (NFV-MANO). A service is the composition of VNFs that can be implemented in virtual machines running on operating systems or on the hardware directly. The hardware and software resources are provided by the NFVI that includes connectivity, computing, storage, etc. Finally, NFV-MANO is composed of the orchestrator, VNF managers and Virtualized Infrastructure Managers responsible for the management tasks applied to VNFs.

In NFV-MANO, the orchestrator performs the resource allocation based on the conditions to perform the assignment of VNFs chains on the physical resources. The sub-task running in the orchestrator, known as VNF Forwarding Graph Embedding (VNF-FGE) or VNF placement problem, tries to find the optimum place to allocate VNFs with regard to some specific objective, such as minimization of computation resources or power consumption, network load balancing, etc.

A. Gateway Virtualization: The EPC Use-Case

In the focus area of VNF placement problem, we propose to study a concrete scenario from legacy mobile systems, namely the Evolved Packet Core (EPC) networks, where we believe that the VNF placement in Service Function Chaining can bring most benefits. The proposed case study is shown in Fig. 1. The Serving Gateway (S-GW) and PDN Gateway (P-GW) are connected to e-NodeB and send the end-user traffic towards Internet. This traffic usually requires various additional services, currently deployed using traditionally embedded network functions, such as load balancers (LB), TCP optimizers (TCP_opt), firewalls (FW) and NATs. Considering the virtualization of the S-GW and P-GW on small data

centers, as proposed in [3], both the data-plane and control-plane functions of the current legacy gateways are moved to an operator's datacenter. At the places of the legacy gateways, an off-the-shelf network element (NE) is used to redirect the traffic from the origin access point of the Radio Access Network (formely done by the S-GW) to the DC and from the DC to the external backbone interface, which is, in the most cases, the former place of the P-GW. Typical VNFs related to control-plane functionalities of the S-GW and P-GW can not be further distributed on the network. In contrast, due to the large traffic volumes handled by the data-plane, parallel transmission paths can be used in the transport network, and thus, VNFs related to data-plane functions with high intensives tasks, e.g. TCP optimizers or Performance Enhancement Proxies (PEP), may also be replicated and be used in parallel at different network locations, as proposed in [6]. Thus, we study the chaining and virtualization of the additional functions related to the data-plane on different physical locations in the mobile core network. The number of required replicas will be in relation with the network traffic demands. Therefore, by knowing how many replicas are necessary we can place them to maintain an optimum network load balancing. On the other hand, the usage of additional network locations increases the number of required DCs, increasing the network costs. To this end, we define the problem as finding the optimum placement for these functions subject to the network costs while at the same time load balancing the network.

B. Network Traffic Model

In a real network, not all services will be virtualized at once. Thus in this paper, we assume two kinds of network traffic, defined as the *background traffic* and *data center traffic*. This is illustrated in Fig. 2. The background traffic is related to the legacy not virtualized services and the traffic is generated from each core node to the rest of nodes and routed by a traditional network core protocol, e.g. IP routing. The virtualized services are responsible for the data center traffic. This traffic is generated at the S-NEs (former S-GW places) and has to be transported towards the Internet gateways (P-NE's at former P-GW locations), as shown in Fig. 1. Since the latter category of traffic, usually TCP connections, is generated by end users, it has to traverse a set of network functions to match the required service before accessing to Internet.

In our approach, we assume that the background traffic can be generated randomly and forwarded following the rules of a specific Traffic Engineering (TE) model defined by the traditional single path destination oriented IP routing. The TE model written in form of ILP formulation (detailed in the next section) minimizes the link utilization of all links in the network using a linear cost functions approach. Once the background traffic is load balanced, it will not be affected by the control of the data center traffic, but it has to be considered as a fixed input parameter for the next model called Resource Allocation (RA). This model is used to allocate optimally VNFs in the network trying to minimize the cost associated to the used resources, maximizing the network load balancing. The optimum placement of VNFs and replicas can provide the optimum locations for the data centers, which will be responsible for the instantiation of VNFs in the network.

IV. OPTIMIZATION MODELS

This section formulates the Link Capacity Dimensioning, TE and RA models as optimization problems subject to a set of constraints, as described next. The notation of all parameters and variables is summarized in Table I.

A. Link Capacity Dimensioning Model

This model allows the initial link dimensioning with the aim to minimize the required capacities for a given topology and a given traffic matrix. The set of available capacities are defined as an input parameter, such as, 10, 40 and 100 Gbps:

$$\text{Minimize: } \sum_{l \in \vec{L}} \sum_{t \in \vec{T}} C_t^l \qquad (1)$$

The constraint (2) assures that for the given link traffic (left side) the capacity of each link is dimensioned according to an over-provisioning factor ϑ, commonly used for fault tolerant operation or demand forecast. So, $\forall l \in \vec{L}$:

$$\sum_{\lambda \in \vec{\Lambda}_b} \sum_{p \in \vec{P}_\lambda} \lambda \cdot R_p \cdot t_p^l \leq \vartheta \cdot \sum_{t \in \vec{T}} C_t^l \cdot t \qquad (2)$$

Each link can only take exactly one of the available capacities:

$$\forall \ell \in \vec{L} : \sum_t C_t^l = 1 \qquad (3)$$

Finally, the next constraint assures that every traffic demand exactly uses one admissible path:

$$\forall \lambda \in \vec{\Lambda}_b : \sum_{p \in \vec{P}_\lambda} R_p^\lambda = 1 \qquad (4)$$

B. Traffic Engineering Model

The TE model minimizes the utilization cost of all links in the network given as

TABLE I: Notation

Parameter	Meaning
$\vec{N} = \{n_0, n_1, ..., n_{(N-1)}\}$	set of all nodes
$\vec{L} = \{l_0, l_1, ..., l_{(L-1)}\}$	set of all links
$\vec{P} = \{p_0, p_1, ..., p_{(P-1)}\}$	set of all paths
$\vec{T} = \{t_0, t_1, ..., t_{(T-1)}\}$	set of available capacities
$\vec{Y} = \{y_0, y_1, ..., y_{(Y-1)}\}$	set of linear cost functions
$\vec{S} = \{s_0, s_1, ..., s_{(S-1)}\}$	set of service chains
$\vec{V_s} = \{v_0, v_1, ..., v_{(V_s-1)}\}$	set of VNFs in service chain s
$\vec{\Lambda} = \{\lambda_0, \lambda_1, ..., \lambda_{(\Lambda-1)}\}$	set of all traffic demands
$\vec{\Lambda}_b \subseteq \vec{\Lambda}$	subset of background traffic demands
$\vec{\Lambda}_s \subseteq \vec{\Lambda}$	subset of demands of service chain s
$\vec{P}_\lambda \subseteq \vec{P}$	subset of paths for a specific λ
$\vec{P}_s \subseteq \vec{P}$	subset of paths for service chain s
$t_p^l \in \{0,1\}$	1 if path p traverses link l
$r_v \in \{0,1\}$	1 if function v can be replicated
r_{max}	max. number of replicas per chain
w_{max}	maximum number of VNFs per DC
c_l	maximum capacity of link l
ϑ	over-provisioning capacity ratio

Variable	Meaning
K_ℓ	utilization cost of link ℓ
$R_p \in \{0,1\}$	1 if path p is being used
$R_p^\lambda \in \{0,1\}$	1 if traffic demand λ is using path p
$C_t^l \in \{0,1\}$	1 if link l takes capacity type t
$R_p^s \in \{0,1\}$	1 if service chain s is using path p
$R_p^{\lambda,s} \in \{0,1\}$	1 if traffic demand λ from service chain s is using path p
$F_n^{v,s} \in \{0,1\}$	1 if VNF v from service chain s is allocated to node n
$F_n \in \{0,1\}$	1 if node n is running some VNF

$$\text{Minimize} : \sum_{l \in \vec{L}} K_l \qquad (5)$$

where the cost of every link is related to its link utilization U_l^{TE} and defined by the resulting value from all linear cost functions $y_i(U_l^{TE}) = a_i \cdot U_l^{TE} - b_i$ as follows:

$$\forall l \in \vec{L}, \forall y \in \vec{Y} : K_l \geq y\left(U_l^{TE}\right) \qquad (6)$$

where constants a_i and b_i are chosen in a way that the slope of the incremental cost values $y_i \in \vec{Y}$ approximately follows an exponential function [7]. The link utilization is given by

$$U_l^{TE} = \sum_{\lambda \in \vec{\Lambda}_b} \sum_{p \in \vec{P}_\lambda} \frac{\lambda \cdot R_p^\lambda \cdot t_p^l}{c_l} \qquad (7)$$

The summation takes into account each traffic demand λ out of the set of all background demands $\vec{\Lambda}_b$ whose specific path $p \in \vec{P}_\lambda$ is traversing link l, divided by the link capacity. The only routing constraint for this model assures that every traffic demand exactly uses one admissible path:

$$\forall \lambda \in \vec{\Lambda}_b : \sum_{p \in \vec{P}_\lambda} R_p^\lambda = 1 \qquad (8)$$

523

C. Resource Allocation Model

The second optimization model, called RA model, uses a similar objective function as the previous model, but adding a new term using the binary variable F_n (1 means the node is used by at min 1 VNF, 0 no VNF is assigned) which minimizes the number of nodes that are allocating VNFs subject to the constraints explained below:

$$Minimize : \alpha\left(\sum_{l \in \vec{L}} K_l\right) + \beta\left(\sum_{n \in N} F_n\right) \qquad (9)$$

Both terms are weighted by α and β parameters to enable the desired tradeoff between load balancing and network costs. In this paper we restrict to use them as selection parameter for one of the objective functions. Although, we only consider these two variables to reduce computing complexity, additional parameters such as network delay or power consumption, already considered for future extensions of this work, could be added to the model.

Differently to the TE model, in this model, for each service chain $s \in \vec{S}$ a total number of $\|\vec{\Lambda}_s\|$ traffic demands $\lambda \in \vec{\Lambda}_s$ are defined, where each of them could not be split and has to be forwarded over one specific path $p \in \vec{P}_s$, which is taken into account by binary variable $R_p^{\lambda,s}$ (is only 1 if path p is used by demand λ). Furthermore, the number of paths $\|\vec{P}_s\|$ may be different than the number of demands. Thus, without the usage of VNF replicas all demands of a service chain must use the same path. As a result, the link utilization due to the RA model is given by:

$$U_l^{RA} = \sum_{s \in \vec{S}} \sum_{\lambda \in \vec{\Lambda}_s} \sum_{p \in \vec{P}_s} \frac{\lambda \cdot R_p^{\lambda,s} \cdot t_p^l}{c_l} \qquad (10)$$

In the RA model, the linear utilization cost functions take into account the superposition of the fix given background TE traffic with the RA traffic specified as:

$$\forall l \in \vec{L}, \forall y \in \vec{Y} : K_l \geq y\left(U_l^{TE} + U_l^{RA}\right) \qquad (11)$$

Because each demand has to be assigned to a path, the routing constraint is given by

$$\forall s \in \vec{S} : \sum_{\lambda \in \vec{\Lambda}_s} \sum_{p \in \vec{P}_s} R_p^{\lambda,s} = \|\Lambda_s\| \qquad (12)$$

Then, to know if a specific node is being used or not, the binary variable F_n will only be 1 if ≥ 1 VNFs are assigned to a node n, which is assured by the constraint,

$$\forall n \in \vec{N} : \frac{\sum_{s \in \vec{S}} \sum_{v \in \vec{V}} F_n^{v,s}}{W} \leq F_n \leq \sum_{s \in \vec{S}} \sum_{v \in \vec{V}} F_n^{v,s} \qquad (13)$$

,where the binary variable $F_n^{v,s}$ indicates, if VNF v from service chain s is allocated to node n. W is a large constant to assure, that the left side of the equation is always less than 1. The next constraint (14) assures that a certain traffic demand λ can only use a path p if the requested service chain

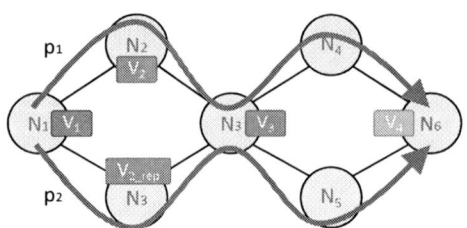

Fig. 3: VNF replication strategy

is also using the same path. Thus (14) takes into account, that more than one demand can use the same path. On the other hand, constraint (15) guarantees, that a service schain s can not establish an active path without traffic assigned to it.

$$\forall s \in \vec{S}, \forall p \in \vec{P}_s, \forall \lambda \in \vec{\Lambda}_s : R_p^{\lambda,s} \leq R_p^s \qquad (14)$$

$$\forall s \in \vec{S}, \forall p \in \vec{P}_s : R_p^s \leq \sum_{\lambda \in \vec{\Lambda}_s} R_p^{\lambda,s} \qquad (15)$$

Furthermore, the number of admissible paths for each service chain s is constrained by the number of replicas. Then, for $[0, 1, 2, ..., r_{max}]$ replicas, each service chain can use at maximum $[1, 2, 3..., (r_{max} + 1)]$ paths to forward traffic, i.e.,

$$\forall s \in \vec{S} : 1 \leq \sum_{p \in \vec{P}_s} R_p^s \leq r_{max} + 1 \qquad (16)$$

Therefore, with no replicas, a certain service chain can only use one path, while increasing number of replicas, the number of admissible paths proportionally increases. Then, for each activated path in service chain s defined by R_p^s, the next constraint allocates all VNFs of the service chain:

$$\forall s \in \vec{S}, \forall p \in \vec{P}_s, \forall v \in \vec{V}_s : R_p^s \leq \sum_{n \in p} F_n^{v,s} \qquad (17)$$

On the other hand, the sequence order of VNFs in the service chain has to be maintained. Then, for a certain path p, the function v can not be allocated in the node n, if the previous function $v - 1$ is not already allocated in any of the previous nodes of the same path. So, $\forall v \in \vec{V}_s, \forall p \in \vec{P}_s, \forall n \in p$:

$$\left(\sum_{m=0}^{n} F_m^{(v-1),s}\right) - F_n^{v,s} \geq R_p^s - 1 \quad when \quad v > 0 \qquad (18)$$

The next two constraints limit the maximum number of VNFs that can be allocated in the network. First, the maximum number of VNFs allocated in some specific node n is constrained by the parameter w_{max}:

$$\forall n \in \vec{N} : \sum_{s \in \vec{S}} \sum_{v \in \vec{V}_s} F_n^{v,s} \leq w_{max} \qquad (19)$$

Second, if a certain function v can be replicated r_v, then, the maximum number of replicas is constrained by the maximum number of active paths R_p^s. If the function can not be replicated, then can only be placed once. So, $\forall s \in \vec{S}, \forall v \in \vec{V}_s$:

$$\sum_{n \in N} F_n^{v,s} \leq r_v \sum_{p \in \vec{P}_s} R_p^s + 1 - r_v \qquad (20)$$

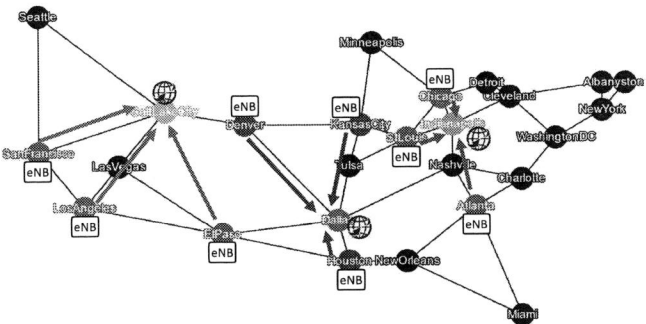

Fig. 4: Janos-us network

To improve the load balancing feature using replicated VNFs, we introduce an additional constraint. In the proposed replication model, we exclude the case where two selected paths p_1 and p_2 from the service chain s are choosing the same shared node n to place the same function (original and replica). So, following the example shown in Fig. 3, v_2 and v_{2_rep} can never be allocated in the same node n. Then, for $\forall v \in \overrightarrow{V_s}, \forall p_1 \in \overrightarrow{P_s}, \forall p_2 \in \overrightarrow{P_s}, \forall n_1 \in p_1, \forall n_2 \in p_2$:

$$R_{p_1}^s + R_{p_2}^s + 2F_n^{v,s} \cdot r_v \leq 3 \quad \left\{ \begin{array}{l} n_1 = n_2 = n \\ v \neq 0 \neq V_s - 1 \end{array} \right. \quad (21)$$

This constraint is added for all pairs of admissible paths of each service chain. To reduce the computing complexity, only link dis-joint paths are taken into account.

V. PERFORMANCE EVALUATION

In this section, we show the results of four different scenarios: 1) network load balancing (minLB), 2) minimization of network costs (minNC) given by the number of required data centers (DCs), 3) minimization of network costs with limited number of VNFs per DC (minNC_constr) and 4) network load balancing with limited number of DCs to place VNFs and limited number of VNFs per DC (minLB_constr). The LP models are implemented using the Gurobi Optimizer [8] and the topology (26 nodes and 84 links) is chosen from SNDLib website [9] and shown in Fig. 4.

The background traffic is randomly generated for each source-destination node pair within the interval [1, 4] Gbps. Since the allocation of VNFs is known to be NP-hard, we restrict to 9 eNBs and 3 internet gateways placed in the network, which are responsible for the data-center traffic (see Fig.4). Unlike other proposals, where the traffic can not be split, each eNB generates 10 traffic demands of 4.4 Gbps to its associated gateway, which can be routed along different paths. For the link dimensioning, the background and data-center traffic is optimally load balanced by solving the TE model, where the GWs are not virtualized and placed at the original locations. Based on the traffic flows, we assume an overprovisioning factor of 1.2, which are chosen from different granularities (2.5, 10, 40, 100 and 200 Gbps).

For the RA model the background traffic is routed over the single path derived above, which yields the partial link

utilization U_l^{TE}. Only the DC traffic is now optimally routed, possibly using parallel paths. All traffic demands between NEs are handled by one service chain. The service chain analyzed is shown in Fig. 2, which is composed of VNF_1 (i.e. S-GW, P-GW, LB), which can not be replicated, VNF_2 (i.e. TCP-optimizer) and VNF_3 (i.e. Cache PEP), which can be replicated, and finally VNF_4 (i.e. LB, FW, NAT), which again can not be replicated. Depending of the optimization objective, all VNFs may be placed in the same or in different DCs. We assume the location of the DCs and the assignment of the VNFs to DCs are the variables to optimize.

A. Network Load Balancing vs Minimum Network Node Costs

We study optimization results obtained for the exclusive minimization of the load balancing (minLB, $\alpha = 1, \beta = 0$) and exclusive minimization of the network costs (minNC, $\alpha = 0, \beta = 1$). In the Fig. 5, we show the comparison of the average link utilization, maximum link utilization, number of used DCs and maximum number of VNFs per DC with no replication, with one and two replicas, respectively. Due to Eq.(16) the number of replications are restricted to the maximum value r_{max}, later shown in the figures as replicas.

For the minNC scenario we observe expected results. Due to the cost optimization, all VNFs are placed in 3 DCs traversed by the same single path between NEs. Thus the average and maximum link utilization is maintained constant for any case. The maximum number of VNFs per DC for the NC scenario changes if one VNF replication can be used (see Fig. 5c), although the optimized value of the objective function remains the same. This comes from the fact that the solution is only unique with respect to the number of used DCs and not to the number of assigned VNFs per DC, which is an important behavior and for further studies.

On the other hand, in the minLB scenario, the average link utilization increases with replication (Fig. 5a) because of the optimization model tries to decrease number of overloaded links using longer paths with underutilized links, as can be seen from the reduction of the maximum link utilization (Fig. 5b). In more detail Fig. 6 shows a histogram of the link utilization, which verifies, that an increment of the available parallel paths to forward traffic decreases the number of overloaded links due to replication. The improvement between one and two replicas is not significant, but it is totally dependent on the chosen topology. The effect of the objective function (9) is quite important, as can be seen from the chart embedded in Fig. 6. On the other hand, as shown in Fig. 5c, the number of used DCs increases, due to for longer paths more DCs are required to allocate replicas.

B. Network Optimization under DC and VNF Constraints

In the minNC case, the objective is to minimize the number of used DCs without restricting the number of VNF per DC. However, in a realistic case the number of VNFs allocated to a virtual maschine (VM) will be restricted by the available computation resources, e.g. the service processing time. Furthermore, the number of VMs per DC will be constrained as well. However, due to our small scale example we will

525

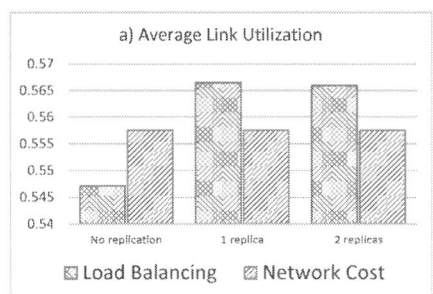 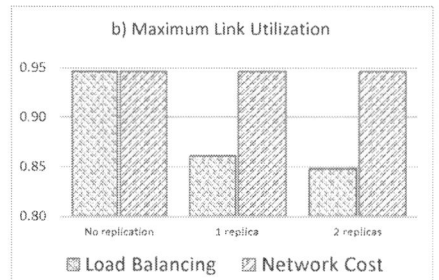 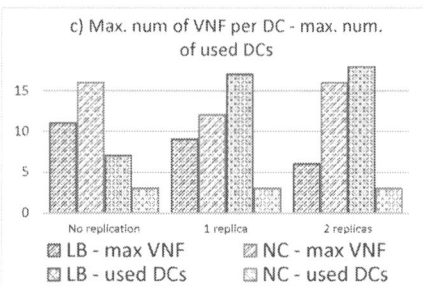

Fig. 5: Comparison between Load Balancing and Network Cost models

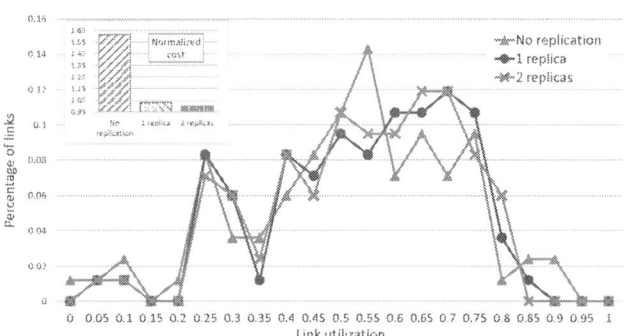

Fig. 6: Link Utilization of minLB Model

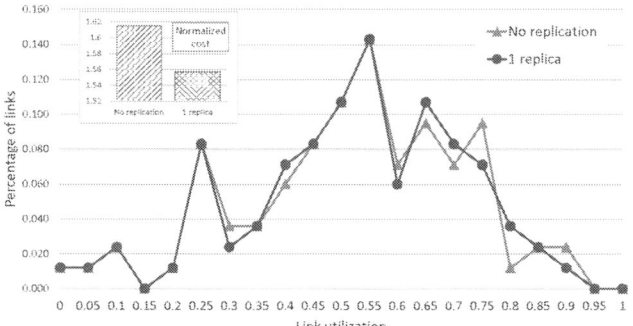

Fig. 7: Link Utilization of minLB_constr model

TABLE II: Node Cost model with at most 8 VNFs per node

Case	avg utilization	path length	avg_vnf
No replication	0.568	2.22	1.38
With replication	0.55	2	1.38

deal with this problem by constraining the maximum number of VNFs per DC to 8 VNFs. The solution of the constraint optimization problem minNC_constr results into 6 DCs, which doubles the network costs compared to the previous case (Fig.5c). Due to the increased number of DCs, the use of VNF replications offers some advantages, as shown in Table II. The constraint reduces the average number of VNFs (avg-vnf) per node and further, the usage of replicas allows to use some alternative paths in parallel, reducing the average path length (in hops). Consequently, the total network traffic decreases as well as the average link utilization.

Finally we want to optimize the load balancing under constraints given by the maximum number of usable DCs as well as the maximum number of VNFs per DC, where the constraints are taken from the optimal results given by the minNC_constr problem. In Fig. 7, we show the results for the histogram of the link utilization. Here, we can see that one replica is still able to improve the load balancing, but due to the small example with less effect.

VI. CONCLUSIONS

Since VNFs can only be placed onto servers located in data centers, the traffic directed to these DCs has not only significant impact on the network costs but also on the load balancing. To enable a cost efficient DC placement jointly with network load balancing, we introduce the VNF placement with replications in this paper, and especially how the replications of VNFs can help to load balance the network. For future work, we will extend this model to allow more variables to optimize and work with more scenarios.

ACKNOWLEDGMENT

This work has been performed in the framework of SENDATE-PLANETS (Project ID C2015/3-1), and it is partly funded by the German BMBF (Project ID 16KIS0470).

REFERENCES

[1] H. Moens and F. De Turck, "VNF-P: A model for efficient placement of virtualized network functions," *Proceedings of the 10th International Conference on Network and Service Management, CNSM 2014*, 2015.

[2] S. Mehraghdam, M. Keller, and H. Karl, "Specifying and placing chains of virtual network functions," *2014 IEEE 3rd International Conference on Cloud Networking, CloudNet 2014*, pp. 7–13, 2014.

[3] A. Basta, W. Kellerer, M. Hoffmann, H. J. Morper, and K. Hoffmann, "Applying NFV and SDN to LTE mobile core gateways, the functions placement problem," *AllThingsCellular '14*, pp. 33–38, 2014.

[4] M. Bagaa, T. Taleb, and A. Ksentini, "Service-aware network function placement for efficient traffic handling in carrier cloud," *IEEE Wireless Communications and Networking Conference, WCNC*, 2014.

[5] F. Carpio, S. Dhahri, and A. Jukan, "VNF Placement with Replication for Load Balancing in NFV Networks," pp. 1–6, 2016. [Online]. Available: http://arxiv.org/abs/1610.08266

[6] W. Haeffner, J. Napper, M. Stiemerling, D. Lopez, and J. Uttaro, "Service Function Chaining Use Cases in Mobile Networks," *Internet-Draft*, 2016. [Online]. Available: https://datatracker.ietf.org/doc/draft-ietf-sfc-use-case-mobility/

[7] M. Caria, T. Das, A. Jukan, and M. Hoffmann, "Divide and conquer: Partitioning OSPF networks with SDN," in *IFIP/IEEE International Symposium on Integrated Network Management*, 2015.

[8] Gurobi, "Gurobi Optimizer." [Online]. Available: http://www.gurobi.com

[9] "SNDlib." [Online]. Available: http://sndlib.zib.de

Impact of human resources changes on performance and productivity of Scrum teams

Dino Alić, Almir Djedović, Samir Omanović, Anel Tanović

University of Sarajevo/Faculty of Electrical Engineering, Sarajevo, Bosnia and Herzegovina

alic.dino@gmail.com, almir.djedovic@infostudio.ba, somanovic@etf.unsa.ba, atanovic@etf.unsa.ba

Abstract – This paper presents results of an analysis of the impact of the human resources changes in Scrum teams. Four Scrum teams were tracked (two developments and two quality assurance) along with their productivity and performance. Analysis showed that human resources changes have a significant impact on the entire team and its behavior. Their effort increased by adding overtime hours. In the same time, their performance and effective work decreased, which is reflected on the quantity of work that can be billed to the client. The analysis shows that it takes, in average, three sprints (each lasting fourteen days) for new team members to fully adjust to the team development process and acquire a business knowledge needed for maximum productivity. Teams whose members have been working together longer period and who have more senior members can adjust to team shifts more quickly. The analysis also showed a correlation between quality assurance and development team – when development team had extra utilization due to overtime, quality assurance team had an increase in overtime hours almost proportionately.

I. INTRODUCTION

In 2001, 17 software development leaders created the "Manifesto for Agile Software Development" [1]. One of the principles behind the Agile manifesto is that the best architectures, requirements, and designs emerge from self-organizing teams[2]. Manifesto addresses many points in software development but the main point is that teams require autonomy to achieve excellence. However, the Agile Manifesto does not prescribe steps on achieving the goals [1][3]. Instead, it is necessary for an organization to use a more specific framework like extreme programming, Scum, Kanban, etc. Scrum is the most popular one[4][5].

Scrum is designed in an attempt to not only increase productivity of the team and the quality of the end product but to increase the job satisfaction as well by creating a comfortable workplace. The goal of Scrum is to shift focus to the team itself in an attempt to address team members concerns which, in turn, should bring higher quality product [2].

The six core Scrum principles are [3]: empirical process control, self-organization, collaboration, value-based prioritization, time-boxing and iterative development. Scrum principles are non-negotiable and must be applied [3]. It is obvious that all Scrum principles are easier to follow with a strong and coherent team, which has been working together for longer period. Job related skills are crucial but soft skills and ability of the team to create a friendly work environment is just as important for Scrum team.

A key strength of Scrum lies in its use of cross-functional, self-organized, and empowered teams [4]. To fully utilize those strengths, it is crucial to enhance and later preserve the productivity and efficiency of the team. During the relatively short history of Scrum, many collaboration tools have been developed in an attempt to track and evaluate the team performance. They help the team and management in organization, planning, track the project progress and detecting bottlenecks. However, there are variables in Scrum that are not that easy to track, measure and react to. In general, control of some aspects of agile software development is based on experience from practice, like stable velocity [6] or predicting change requests [7], etc. Human resource (HR) issues is an example of those be a challenging factor for every Scrum project[8][9].

Preserving a stable and well organized Scrum team for longer period on dynamic job market (especially in IT sector) is becoming increasingly difficult. Scrum teams, at some point, are facing changes in human resources. This study is focused on analyzing their ability to react on these changes. New team member can disrupt agile principles. It is difficult for a team member to prioritize work on the bases of business value without proper business knowledge. Team can have problems with self – organization, as they are unaware of the knowledge and capacity of new members. Collaboration can suffer as new members may lack soft skills etc. [10][11]. Coherent and stable team will ultimately deliver higher quality products.

The Scrum aims for a quick delivery of viable product – team must be able to quickly respond to any changes in business process and implement the support for the business flow inside the application they are building. This concept can result in an absence of detailed documentation [12]. Newcomers most overcome this obstacle before becoming a useful team member. The absence of documentation means that the team members are the sole owners of business logic [13]. Shifts in human resources inside the team could result in a cascade effect of productivity loss on team level as team members might be forced to dedicate additional time to compensate during the knowledge transfer phase [13]. On top of that, the quality of the product can drop since an increase in number of bugs can happen due to the lack of business logic knowledge.

Considering all the above, an argument can be made that human resources changes have a negative impact on the team. This case study attempts to determine how exactly this affects the team by using credible metrics to measure the team performance during this period.

II. CASE STUDY

To determine how the HR changes are affecting Scrum teams we have conducted an analysis where we tracked four teams during the course of 19 sprints. Two of the teams were development (DEV) teams while two were Quality Assurance (QA) teams. All teams worked on the same project for the same client. Team changes were tracked along with their productivity and their billable utilization (percentage of work which can be actually billed to client).

The focus of the analysis was on the development teams. QA teams were tracked to show the correlation between the two. Reasoning for HR changes was not considered - analysis focuses only on parameters that could be precisely measured.

A. Team organization and Scrum in case study

Sprint lasts for 14 days for all teams analyzed. They are synchronized so all teams begin and end their sprint at the same time. First day of the sprint includes sprint retrospective where team evaluates previous sprint and establishes points of improvement. New members are most often presented on the first day of the sprint. Sprint planning occurs day before the end of the sprint or on the last day. Sprint demo, where new features developed are presented to client, occurs at the end of each sprint. Team members from all teams (both DEV and QA) are present. After the product demo, development team implements feedback from client, if any. After that the product will be deployed to QA environment for testing. Once the quality assurance is completed, all significant notes from the QA teams implemented the process continues on UAT (staging) environment. In this environment, all items are tested again and regression testing takes place. This is the phase where client can conduct their own tests. Any item can be returned to development team for additional work from any of these stages.

Production release occurs every month and includes development work from two sprints. The code is thoroughly tested and ready to be shipped to the end users. Figure 1 shows the release cycle and responsibilities of each team during the process. Some of the involvement is amended for clarity (for example, client is involved in development and QA phase if required by the team).

B. Metrics used

In this analysis, we measured three key parameters: resource utilization (referred simply as utilization), billable utilization and quality for the development teams.

- **Resource utilization** – a percentage of planned work vs. actual work done on team level. Planned utilization takes into account all planned activities and absence of team members (holidays and unpaid and annual leaves). It does not, however, include any unplanned activity such as sick leaves or unplanned personal leaves. Ideal utilization would be 100% but it is often difficult to achieve, especially in teams with more team members. For teams analyzed in this study, fluctuations in utilization percentage are not a cause for concern for management as long as it does not affect the billable utilization. However, if the team utilization raises above 100% this means that overtime work hours were needed and should be addressed to determine the cause.

- **Billable utilization** – utilization streamed toward fulfilling client requirements for the software. All activities from which client does not gain short term benefit are considered as non – billable. These may include company administration activities, in – company meetings, internal and external trainings (unless explicitly requested by client to comply with the immediate needs) etc. It is important to note that Scrum activities count as billable utilization (daily Scrum meetings, sprint retrospective, sprint demoes etc.). Ideal billable utilization is 100% which is the maximum. Non – billable utilization does not represent the time "wasted". It only means that this time cannot be invoiced to client.

- **Quality** – is calculated for the development team as a measure of backlog items (BI) completed in a sprint without QA or end user finding any issues in implementation. The formula for calculating quality is: **100** – (BI done / BI Failed). For the QA team, it is calculated with the same formula but using the number of bugs found by QA team and those found by client (usually on UAT environment and in production). QA team quality was not considered in this analysis.

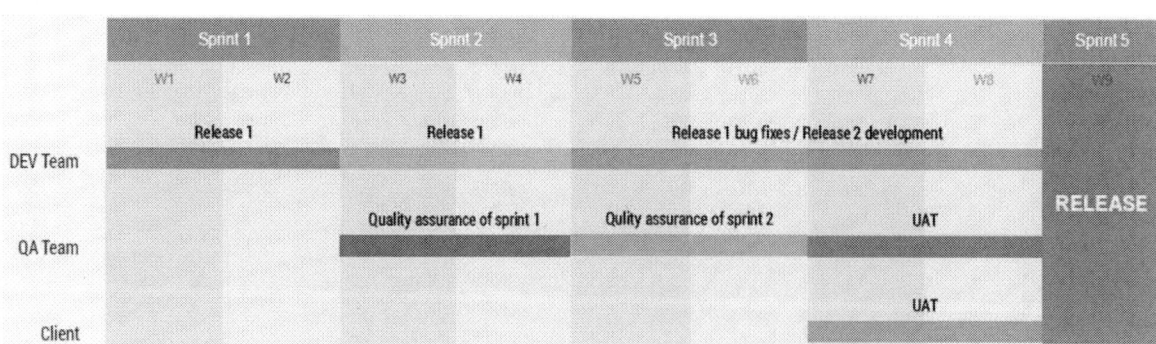

Figure 1. Release cycle for the Scrum teams in case study

TABLE I. HUMAN RESOURCES CHANGES DURING THE ANALYSIS, PER TEAM AND SPRINT. NUMBER OF INCOMING (NEW) TEAM MEMBERS ARE MARKED IN GREEN. TEAM MEMBESR LEAVING THE TEAM ARE MARKED IN RED.

Sprint / Team	1	2	3	4	5	6	7	8	9	10	11	12	13	14	15	16	17	18	19	Totals	Start HR	End HR
DEV 1		1			3		1		1				1	1						8	3	6
(DEV 1)					2		1						1			1				5		
DEV 2										1								1		2	5	5
(DEV 2)		1															1			2		
QA 1	1											1								2	2	4
(QA 1)																				0		
QA 2				1									2							3	4	4
(QA 2)				1								2								3		

(DEV 1 total HR changes: 13)

There are several important points to note on Table I.

All teams had HR changes during the time analyzed. All teams, except QA 1 had team members leaving the team. DEV 1 team not only had most changes in team but the team size doubled during the time analyzed (from three to six). There was a total of 13 HR changes in this team, one more than all other teams combined. DEV 2 and QA2 had smaller team changes and kept the original number of team members. QA1 team had an increase of team members by 50% - two new team members joined and none left.

Also, we had two occurrences of multiple team member changes in a single sprint. In sprint 5 DEV 1 experienced major changes with two team members leaving and three joining in. QA2 team had similar situation in sprints 12 and 13 with two members leaving and two joining the team.

Table II shows all measured parameters per team and sprint, along with the HR changes count for comparison. It can be seen that with every team members change there is a notable change in parameters measured.

TABLE II. CUMULATIVE DATA COLECTED FOR EACH SPRINT. RU – RESOUCE UTILIZATON, BU – BILLABLE UTILIZATION, Q – QUALITY, HRC – HUMAN RESOURCE CHANGES (IN AND OUT)

TEAM / SPRINT	DEV TEAM 1					DEV TEAM 2					QA TEAM 1				QA TEAM 2			
	RU	BU	Q	HRC		RU	BU	Q	HRC		RU	BU	HRC		RU	BU	HRC	
1	100	86	89			100	93	100			100	74	1		100	98		
2	100	76	89	1		88	98	95		1	95	80			93	95		
3	100	77	94			99	99	93			101	70			95	96		
4	82	75	96			100	100	87			97	100			93	79	1	1
5	107	73	87	3	2	103	100	89			102	100			110	82		
6	98	92	100			96	96	98			102	98			101	87		
7	103	84	99	1	1	100	98	100			99	100			101	82		
8	104	99	94			98	100	93			101	100			98	72		
9	88	81	100	1		88	97	98			100	100			92	78		
10	100	83	100			100	80	100	1		102	97			100	73		
11	102	87	97			101	98	86			100	99			99	75		
12	103	91	97			95	83	100			101	75	1		95	54		2
13	104	92	93	1	1	102	78	90			102	87			103	74	2	
14	102	66	91	1		100	95	100			101	94			94	69		
15	100	83	95			103	95	100			102	98			100	64		
16	107	94	96		1	90	98	100			91	99			103	74		
17	106	87	85			110	86	89		1	103	100			98	77		
18	103	93	94			101	92	89	1		105	90			98	66		
19	109	85	83			127	93	89			116	100			128	80		
AVERAGE	101	84	94			100	94	95			101	93			100	78		

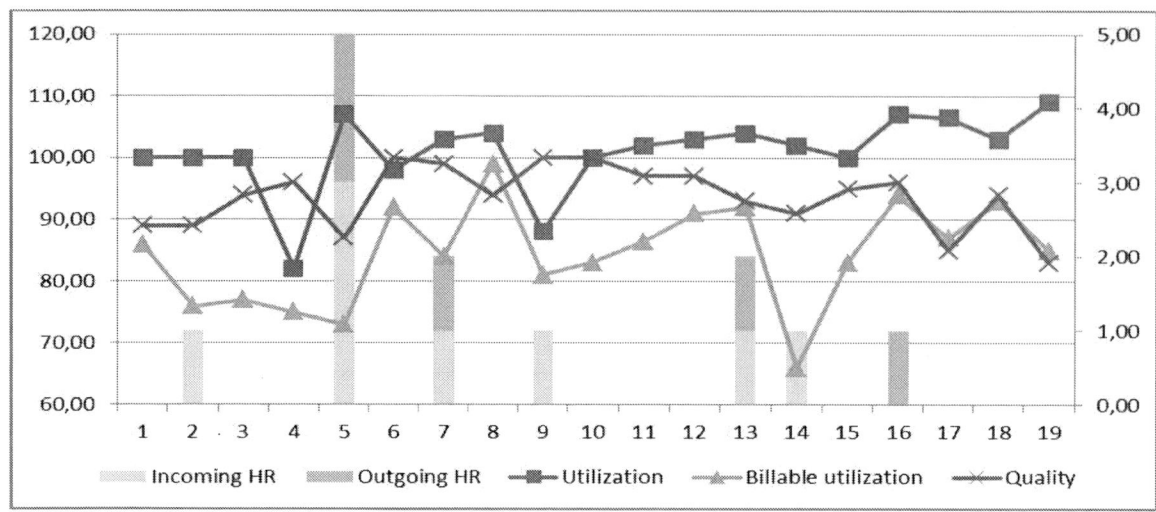

Figure 2. Trend of measured parameters in comparison with HR changes per sprint for DEV 1 team

Figure 2 shows utilization, billable utilization and quality for DEV 1 during the 19 sprints which were analyzed. DEV 1 is especially interesting as it had the most HR changes, as stated previously. It can be seen that each change was followed by decline of billable utilization. Also, for each outgoing member (four occurrences) we have an increase of utilization when compared to previous sprint, which indicates overtime work. Billable utilization is decreased with each new incoming HR in average by 16%, while utilization increased by 4% in average.

When it comes to team members leaving the team, there were no significant changes in billable utilization (negative trend of -0,4%). It also had an effect on utilization which was increased by 12%. Quality dropped by 6% in average when new team members joined, while it dropped by 1,5% when team members left the team.

The most significant is sprint 5 when team experienced a major HR changes, followed by another two in following two sprints. Major fall of billable utilization, increase of utilization and a notable drop in quality can be

seen. Parameters were slowly stabilizing until the sprint 13 when new changes were made.

DEV team 2 had less HR changes which can be observed on Figure 3. As a result, the team had significantly less oscillations in parameters measured. However, even though this team consists of more experienced members when compared to DEV 1, it also had troubles adjusting to HR changes.

In average, billable utilization dropped by 5,5% when new team members joined while utilization increased by 1,5%. When team members left the team, billable utilization dropped by 2%, in average, while utilization was increased by 1,5%. Meanwhile, quality actually increased by 1% in both cases, meaning that team leveraged the seniority of both incoming and existing team members.

However, although the HR changes affected the DEV2 team, it can be observed that the team adapted faster when compared to less experienced DEV1, even managing to increase quality, while DEV1 had significant drop.

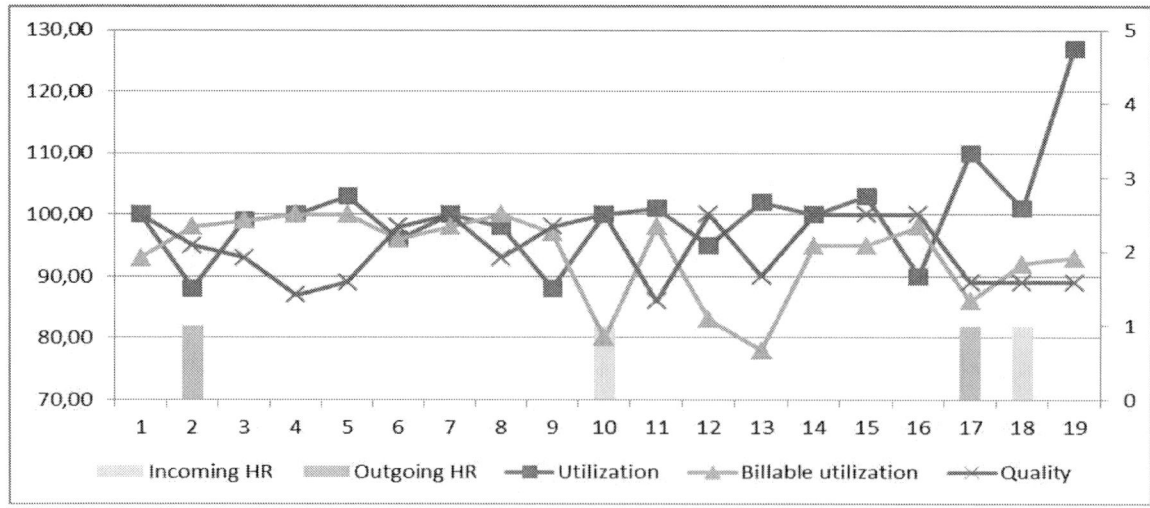

Figure 3. Trend of measured parameters in comparison with HR changes per sprint for DEV 2 team

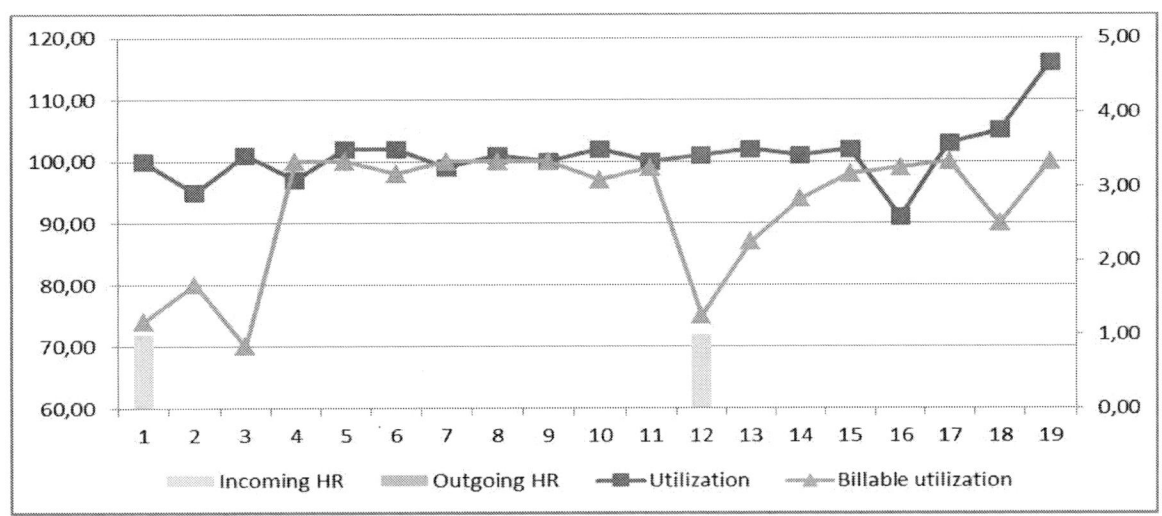

Figure 4. Trend of measured parameters in comparison with HR changes per sprint for QA 1 team

QA teams showed similar trends to DEV teams. QA1 team had an average drop in billable utilization by 24% when new members joined, which can be observed on Figure 4. Utilization increased by 1% at the same time. In average, three sprints were needed for new members to reach billable utilization of the team members already in place. No team members left the team during the period analyzed.

QA2 team had the most HR changes after DEV1 team, with three members joining and three leaving the team. As expected, billable utilization dropped by 11,1% in average with new team members but it increased by 1,5% when they had team members leaving. Utilization was reduced by 5,5% and increased by 3% for team for incoming and outgoing team members respectively.

III. DISCUSSION

Team with more experienced members handles the HR changes better and manages to "recover" faster. Their utilization does not increase significantly while less experienced team has more issues and suffer from increased utilization during the sprints following the HR changes.

Utilization, measured as a percentage of planned work and work done, is an important parameter to observe. Fluctuation in any direction is not desirable, especially when monitoring the HR changes. Rise in utilization means that team members spent more time introducing the new members to business processes and as a consequence, they worked overtime to compensate so that the sprint goals can be achieved. Lower utilization, in most cases, means that sprint goals were not achieved.

Although utilization above 100% may not look like a serious indicator, a simple calculation proves otherwise. For example, in a team consisting of four members, assuming that the sprint has 10 work days with 8 work hours each, it brings us to a total of 320 work hours per sprint which represents the ideal utilization of 100%. If the utilization is increased by 10% that means that each team member has additional work day on top of 10 existing.

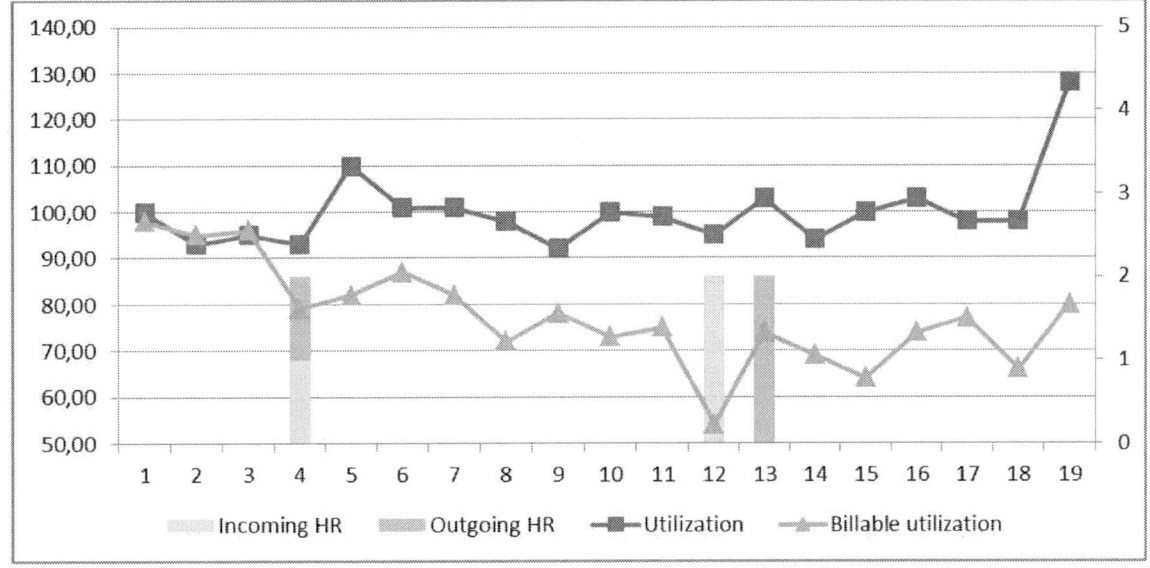

Figure 5. Trend of measured parameters in comparison with HR changes per sprint for QA 2 team

Impacts that this might have is a motivation for future research.

It is also important to observe a correlation between the development team's utilization and the one of the quality assurance, which is to be expected. This can be clearly seen during the last three sprints – whenever the development teams were over – utilized, quality assurance were over – utilized almost proportionately.

Apart from the impacts which are measurable and presented above, HR changes, especially unexpected and unplanned, happening during the sprint had a negative effect in fulfilling the sprint goals and commitment made to the client. In all cases analyzed this lead to the sprint re-planning and, in some cases, adjusting the scheduled release dates. This can reduce the client satisfaction since their own plans for the end product might be affected as a result.

IV. CONLUSION

Conclusions of the case study are presented in bullets below:

- Changes in human resources have significant negative effect on the team as a whole.

- Team utilization change after HR changes goes both ways, depending on whether the team choses to work overtime to compensate. In three out of four team's utilization was increased when compared to previous sprints.

- Billable utilization (work actually billed to client) drops by 14% in average in the sprint when new team members are introduced but tends not to significantly change when team members leave.

- Quality for the development teams drops by 5,5% in average in sprint when new team members are introduced, while it drops by 1,5% when team members leave.

- A correlation between QA and DEV team's utilization has been established - when DEV team was significantly over utilized, QA was over - utilized almost proportionately.

- It takes, in average, three to four sprints for the team to consolidate and return to previous values measured, provided that no new HR changes take place.

- To maintain product quality, high billable utilization and reduce the risk of over-utilization it is crucial to maintain the stable and coherent Scrum team for as long as possible.

V. FUTURE RESEARCH

This analysis did not account for in-sprint re-planning nor does it include any mid – sprint requirements which might affect all the parameters analyzed above. Motivation for future research is to track all related variables to provide more insights into performance fluctuations of scrum teams. As previously noted, effects of over – utilization on product quality and ultimately, job satisfaction should be analyzed.

Also, the reasoning between the human resources changes was not considered. The study can be improved by comparing the impacts of team members leaving the team voluntary (team/project change, pursuing job opportunities in other companies etc.) in comparison to leaving because they failed to meet company quality standards.

REFERENCES

[1] Beck, K.; Beedle, M.; Bennekum, A. van; Cockburn, A.; Cunningham, W.; Fowler, M.; Grenning, J.; Highsmith, J.; Hunt, A.; Jeffries, R.; Kern, J.; Marick, B.; Martin, R. C.; Mellor, S.; Schwaber, K.; Sutherland, J.; Thomas, D., "Manifesto for Agile Software Development" [Online]. Available: http://agilemanifesto.org. [Accessed 01-Feb-2016]

[2] Scrum Alliance, "The Agile Manifesto Principles" [Online], Available: https://www.scrumalliance.org/community/articles/2013/novembe r/the-agile-manifesto-principles-what-do-they-mean [Accessed: 21-Jan-2016]

[3] ScrumStudy, "Scrum Principles" [Online], Available http://www.scrumstudy.com/WhyScrum/Scrum-Principles [Accessed: 01-Feb-2017]

[4] Jeff Sutherland, "The Scrum Papers" [Online], Available: http://jeffsutherland.org/scrum/ScrumPapers.pdf [Accessed: 04-Feb-2017]

[5] Kenneth S. Rubin, Essential Scrum: "A Practical Guide to the Most Popular Agile Process", Addison-Wesley, 2012

[6] S. Omanovic and E. Buza: "Importance of Stable Velocity in Agile Maintenance", XXIV International Conference on Information, Communication and Automation Technologies (ICAT) 2013 Sarajevo.

[7] S. Omanovic and E. Buza: Predicting Future Change Requests in Agile Software Engineering, NAUN International Journal of Computers, Volume 7, Issue 3, pp.91-98, 2013.

[8] E. Hossain et al.: "Using Scrum in global software development: A systematic literature review". Fourth IEEE International Conference on Global Software Engineering , 2009, 175-184.

[9] Janeth López-Martínez et al.: "Problems in the Adoption of Agile-Scrum Methodologies: A Systematic Literature Review", 4th International Conference in Software Engineering Research and Innovation (CONISOFT), pp. 141-148, 2016

[10] Maarit Laanti, "Agile and Wellbeing – "Stress, Empowerment, and Performance in Scrum and Kanban Teams", 6th Hawaii International Conference on System Sciences, 2013

[11] D. Turk, R. France, B. Rumpe. "Assumptions Underlying Agile Software Development Processes". In: Journal of Database Management, Volume 16, No. 4, pp. 62-87, Idea Group Inc., 2005

[12] Cho, Juyun Joey, "An Exploratory Study on Issues and Challenges of Agile Software Development with Scrum" (2010). All Graduate Theses and Dissertations. Paper 599

[13] Juyun Cho: "Issues and Challenges of Agile Software Development with Scrum", Issues in Information Systems, volume IX No. 2, pp. 188-195, 2008

Synergy of ITIL methodology and help users systems

Ivan Ivosić

Croatian Electricity Company d.d.

Zagreb, Croatia

ivan.ivosic@hep.hr

Abstract - Information and communication technology, services and corporate security shape modern working places creating new added values. The development of technology markets and new innovations impose phrase how to lead IT and thereby be successful. The article describes and develops pragmatism and interdisciplinary of information systems of companies with the help of ITIL methodology. The construction of the help users system can improve business, but it is not easy to provide safety and legal economic regulator. Social and political system defines a new user and its function and role and the challenge is to balance customer needs and IT engineering support in a globalized world where is the thin dividing line between work and private.

I. INTRODUCTION

In recent years office management is made easier with the help of information and communication technology. The synergy of different professions is wanted to act efficiently and cost – effectively in a working environment that is created and developed. The challenge for computer scientists and managers is how to design, develop and maintain working place in a manner that allows safe and proper relationship with colleagues, tasks and used tools. The computer with paper and stationery has become irreplaceable in many corporations today and it may be said that the workers are aware of the significance and importance of provided information communication services. Computer scientists are thinking about the methods that create new added value for the company so they invented ITIL[1] methodology that is not standard, but the framework by which it can create and maintain a knowledge base of different IT solutions across the different processes that are implemented in the company. (IT – Information Technology – generally refers to management of information systems with the help of technological solutions, ITIL – Information Technology Infrastructure Library - a set of knowledge and business experience for IT Service Management (ITSM) that performs IT services in accordance with business needs)

II. SERVICES AND ITIL

II. 1. Service

Services are from immemorial time a term that describes the effort. It is not measurable, but it is inevitable in the sense that the time is used for well-being and sometimes to negative. In the last twenty years information and communication technology was respected bond and guiding community in the work, that the computer, cell phone and Internet services change the paradigm of behaviour of staff members and customers. ITIL methodology is not the standard, but quality demystifies services of client and computer information system.

II. 2. Service strategy [2]

Strategic analysis of the industry, public and state-owned enterprises, banks and other companies can't ignore the fact that the information and communication technology is the future business and therefore the investment in the development of local computer networks contributes to a successful corporate social responsibility. The system of the company can be created, developed and built with the help of ITIL methodology. The production strategy is basis where services are designed to provide the support to financial management and demand management process where the experts analyse the market and form a flexible portfolio of services.

II. 3. Service design [3]

Architects and moderators devote most attention to the design of specific service that works on the foundations of operational work through service level management process to the relations with suppliers when it comes to interaction with the external environment and availability and capacity management processes where the internal control is established criteria for the same pragmatism. A quality of the methodology is established for the business continuity and information security management as high as possible. The specific phase is spreading knowledge about a specific service and its possible range in the business information system of the company through the service catalogue management process. Synergy of existing design processes develops services and benefits for the system by reducing the negative impact on the enterprise work.

II. 4. Service Transition [4]

It is necessary to provide the logistics and security of the communication channel between the working premises of computer experts and end users where they aren't be deprived of pleasant and comfortable work after returning to their workplace. An important link between design and service operation is transfer of technical and business knowledge for solving the specific situations in the company. Change management process in coherence with release and deployment management manages time and allows to IT experts space for the smooth and save processing tasks for each individual case. Database of client equipment staff builds information systems in the company and creates intellectual capital of organizational units, while an IT worker with help of the service validation and testing process has the opportunity to test life cycle of IT service.

II. 5. Service Operation [5]

IT operations personnel can control their services a number of ways. The first and basic process of control is event management which creates services of certain value in the portfolio companies. Related processes are request fulfilment and access management where user's aspirations are being modelled in accordance with the system policies in the establishment of a functional work environment and in synergy with key enterprise business processes. But we should not neglect the incident and problem management process that reveals the unwanted situations of hardware and software topics in the company.

II. 6. Continual service improvement [6]

Phase continual service improvement is working on cooperation and development of all phases of the ITIL methodology. Management on the one hand through various reports can measure the usefulness of each process and can estimate value of investment returns whereas computer scientists improve process by use of different methodologies (e.g. PDCA – Plan Do Check Act management method for the control and quality improvement of processes and products[7], CMMI – Capability Maturity Model Integration – a model of maturity and quality of application support[8]).

III. THE CONSTRUCTION AND DEVELOPMENT OF A HELP USER SYSTEM

Happy user is a satisfied customer and each working position for which is provided the use of information and communication technology requires careful consideration and analysis. Service Desk can be a function of the organizational unit that relieves the work of certain teams and integrates catalogue of IT services. Based on the specific services defined in the IT strategy and in accordance with business requirements it is desirable to define the specific working teams who are willing to help in telecommunications systems, system support system, in technical assistance to end users and in the development program activities.

The responsibilities of individual levels of support can be analysed in organized and structured manner in the establishment of a modern and information conscious companies. Service Desk team can be a logical organization who establishes new services or improves existing ones. One level of responsibility provides insight into the events that are introduces into the system of enterprises (for example – procurement and configuration of computer, phone or cell phone). The purpose of that level is management of authentication process, respectively the preparation and training of workers for the next level of IT support. On the different level technically trained staff allows authorized end-user work in a safe and proper way acting normally on the user's location. The next level support is ultimate invisible for the end user, but not unimportant as it allows system and application support and the inclusion of equipment in the safe and correct way in local area network. The synergy of the levels support enables full control of the working process with the help of information and communication technology that establishes the definition and evaluation of standard procedures for a more comfortable work as computer scientists and so end users.

Each help user system in their own way systematize mutual relations, powers and the principle of action and it is not possible to standardize the methodology. The following shows the possible organization of IT business support in the company:

a. Computer technical support works on processes and activities:

- Training of workers for the first work with computer equipment
- Replacement of worn-out computer equipment with new where workers create new added value for the company
- Hardware and software upgrade in the client system of the company
- The disposal of used equipment and preparing the same for the process of repeated allocation, expenses or donations

b. Telecommunications technical support works on processes and activities:

- Training of workers for the first work with telecommunication equipment
- Replacement of worn-out telecommunications equipment with new where workers create new added value for the company
- Hardware and software upgrade in the client system of the company
- The disposal of used equipment and preparing the same for the process of repeated allocation, expenses or donations

c. System support works on processes and activities:

- Inclusion of user authorization in the computer system of the company
- Control mechanisms for safe and proper end-user work on the client equipment
- Hardware and software upgrade in the computer system of the company
- Exclusion of user authorization in the computer system of the company

d. Application support works on processes and activities:

- Inclusion of user authorization in the specialized applications of the company
- Control mechanisms for safe and proper end-user work on the specialized applications
- Software upgrades of applications
- Exclusion of user authorization in the specialized applications of the company

The challenge of any IT-conscious company is work on the definition of boundary parameters of authority and responsibility for the different IT teams. Continuous work on yourself and on your team allows better interpersonal relationships and creates the framework of normal communication with members of other teams. Only then it can be formed the basis of quality help that is a precondition for the correct IT support.

IV.   THE USER IN THE CLIENT INFORMATION SYSTEM

The technology of the present is changing the way of thinking and business approach. Innovative models of new and existing enterprises are created in relation to the finances of the owner and the mission and vision of the company. We distinguish large business information systems that are geographically dispersed and that build infrastructure of computer system in an organized and disciplined way and smaller companies where it is all centralized in one place. But, one thing never changes and that is the system user or employee.

Client information system is simply for planning and maintenance, but it requires the continuous development of communication skills of informatics experts for comfortable direct relationship with the customer during the establishment of certain service. The next time flow chart demystifies specific development phases of user desires and possible scenarios of the life cycle of equipment (Picture 1).

a) End user will get specific equipment for use with the help of standardized communication channels – entrance, exit, requirements management
b) During the handover of equipment employee in theory by his signature on the assignments document guarantees and warrants that he will use the equipment for work in a safe and correct manner – booking of equipment in the books of companies, amortization
c) In the initial phase of establishing a new service it is possible that computer scientists (application and system support) are working on concrete settings for authorized work on the system and with specialized applications – the establishment of a functional work environment and user workplace
d) In the next phase the user (lawyer, economist, sociologist, psychologist..) on a stand-alone mode uses the default privileges for continued work in the computer network – the maintenance of equipment
e) The user during work may be faced with unpleasant situations where it is desirable that the user solves the incident and at the very least looks for help in the

case of endangering the business continuity process – the setting up of business continuity

f) While working user can specifically improve their business – hardware or software upgrade

g) In the final stage of using the particular equipment the user will be discharged with same – the disposal of the equipment

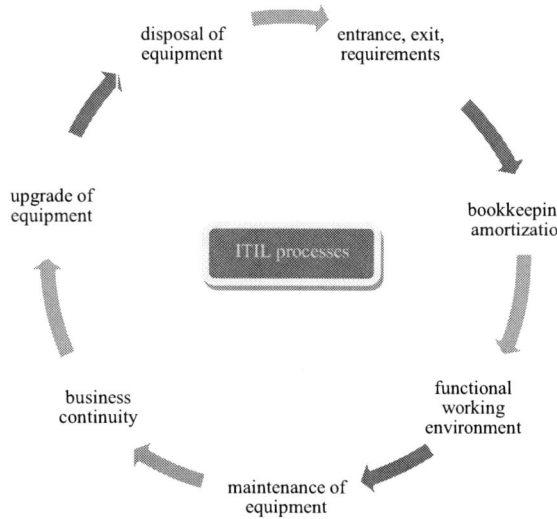

Picture 1. Possible scenarios for the life cycle of IT equipment

The described practice in ITIL development within the companies generally shows a various range of imaginative and innovative requirements of end users and the challenge is arranging regulations with the legal aspects of using the same. Presented subjective SWOT analysis is showing a possible pattern of employee's behaviour in a multicultural environment that comes during the work with the information and communication technology. The nature of the analysis is subjective and this model doesn't present the usual paradigms of the working environment. (SWOT – Strengths Weaknesses Opportunities Threats – a method for the analysis of internal and external factors of the company with the conditions in the environment)

Strengths

- Work in a prestigious company with the most modern equipment
- Continuous IT improvement and undergoing additional training
- Facilitated communication with various parties

Weaknesses

- Unprofessional and unknown work with information and communication technology
- (not)ethical working staff
- Lack of experience in time management and crisis situations

Opportunities

- Security and control in the management of business processes
- Better organization
- Pragmatism at work

Threats

- The physical alienation of equipment
- Unauthorized work in the Wide Area Network
- Trading with confidential information

The possible range of analyzes of user rights and personal desires leads to the rational conclusion that user don't need to be afraid of stories about good and bad sides of using technology, already he should be continuously directed on the good. With own professional approach and the control of the working process end user will contribute to the quality in business at mutual pleasure of his superiors and computer experts.

V.   THE SYNERGY OF ITIL, IT AND BUSINESS

The principles of efficiency, economy and security of information systems[10] raise the quality level of complete business system. The legal view of the development of technical and functional processing activities of information and communication technology enhances the reputation and honour of the system and the same is better prepared for the work and activities in the corporate environment. At some point the end user has a need for computer equipment, but he doesn't have the sense and knowledge of the effort that is invested in the strategy, design, preparation and transition of the same. The legal aspect of informatics experts with information and communication technology is reflected in the handover of equal or higher quality than the existing one. ITIL provides guidance to information officer for work, but it takes work on a continuous adjustment of the ITIL processes with key business processes of the company. Work on IT principles contributes to the creation of the better and more efficient decision support systems and business will be easier to control with the help of a sophisticated information and communication technology (picture 2).

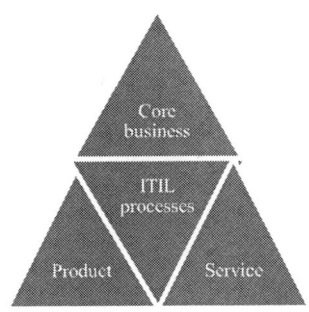

Picture 2. Integration of ITIL processes into core business

VI. CONCLUSION

The business information systems (human resources) are built in a way which each company has its own story and system activity. ITIL processes can help in the integration of products. The article describes the ITIL methodology that describes the processes of creating, maintaining and abolition of IT services. The service architecture presented in the paper is designed to build highly important trust and loyalty between computer experts and end users. The possibility for the development of client responsibilities (in cooperation with interdisciplinary professional development of informatics experts) is envisaged.

The future is in the innovative services and the question is how they can be incorporated into daily operations without diminishing the effort and the will of information and engineering staff for pragmatic management of the same.

The challenge is to develop the business potential of the marginal utility of information and communication technologies with legal – economic regulation use of the same.

LITERATURE

[1] Panian Ž, Spremić M., et al. (2007) Korporativno upravljanje i revizija informacijskih sustava, Zgombić&partneri, Zagreb.

[2] – [6] ITIL Training Zone. Available at http://itsm.zone/

[7] Bosilj Vukšić Vesna, Kovačić Andrej (2004) Upravljanje poslovnim procesima, Sinergija, Zagreb.

[8] Bosilj Vukšić Vesna, Hernaus Tomislav, Kovačić Andrej (2008) Upravljanje poslovnim procesima – organizacijski i informacijski pristup, Školska knjiga, Zagreb.

[9] Buble, M., et. al. (2005) Strateški menadžment, Sinergija, Zagreb.

[10] Dražen Dragičević (2015) Pravna informatika i pravo informacijskih tehnologija, Narodne novine, Zagreb.

Gap in pagination due to withheld papers.

Pages 538-546

Simulation Study of M-ARY QAM Modulation Techniques using Matlab/Simulink

S. M. Sadinov

Technical University of Gabrovo/Department of Communications Equipment and Technologies, Gabrovo, Bulgaria
e-mail address: sadinov.tc@abv.bg

Abstract - Recent theoretical studies of communication systems show much interest on high-level modulation, such as M-ary quadrature amplitude modulation (M-QAM), and most related works are based on the simulations. In this paper, a simulation model to study various M-ary QAM modulation techniques is proposed. The simulation model is implemented in Matlab/Simulink environment and BERTool in conjunction with the model is also used. Simulation for 64-QAM and 256-QAM modulation techniques is done. The effect of the phase noise on the constellation diagram for both M-QAMs is examined. The impact of changing the power of the input signal, phase noise and frequency offset on Bit Error Rate (BER) performance of 64-QAM and 256-QAM is also studied. The simulation results in terms of the constellation diagram and the BER curve under various conditions are presented and analyzed.

Index Terms—Additive White Gaussian Noise (AWGN), Bit Error Rate (BER), BERTool, M-ARY Quadrature Amplitude Modulation (QAM), Matlab, phase noise, Simulink.

I. INTRODUCTION

As the digital communication industry continues to grow and evolve, the applications of modulation techniques continue to grow as well. This growth, in turn, spawns an increasing need to seek automated methods of analyzing the performance of digital modulation types using the latest mathematical softwares. The Amplitude Shift Keying (ASK), Frequency Shift Keying (FSK) Phase Shift Keying (PSK), and Quadrature Amplitude Modulation (QAM) are the basic forms of the digital modulation techniques [1]-[3]. With the fast development of modern communication techniques, the demand for reliable high data rate transmission is increased significantly, which stimulate much interest in modulation techniques. The requirements for spectral and power efficiency in many real world applications of the modulation lead to the necessity of using the modulation techniques other than binary modulation. Modern modulation techniques exploit the fact that the digital baseband data may be sent by varying both envelope and phase/frequency of a carrier wave. Because the envelope and phase offer two degrees of freedom, such modulation techniques map baseband data into four or more possible carrier signals. Such modulation techniques are known as M-ary modulation [4], [5], since they can represent more signals than if just the amplitude or phase were varied

alone. In an M-ary signaling scheme, two or more bits are grouped together to form symbols and one possible signal is transmitted during each symbol period. Usually, the number of possible signals is $M = 2^m$ where m is an integer. Depending on whether the amplitude, phase or frequency is varied, the modulation technique is M-ary ASK, M-ary PSK, or M-ary FSK [4]-[6]. The modulation that alters both, amplitude and phase is M-ary QAM [4], [7], [8], [11]. Different bandwidth efficiency at the expense of power efficiency can be achieved using M-ary modulation techniques. M-ary modulation schemes are one of most efficient digital data transmission systems as it achieves better bandwidth efficiency than other modulation techniques and give higher data rate. In order to improve the performance of M-ary modulation techniques, always a need of studying and analyzing the unwanted effects caused by different factors on their characteristics exists.

In this paper, two variants of M-ary QAM, namely 64-QAM and 256-QAM, are studied in Matlab/Simulink environment. The studies are done in terms of effects caused by such factors as input signal power, Additive White Gaussian Noise (AWGN), phase noise and frequency offset on behavior of these modulations. Constellation diagrams and Bit Error Rate (BER) curves for mentioned variants of M-QAM are obtained and analyzed.

II. M-ARY QAM

QAM is the encoding of the information into a carrier wave by variation of the amplitude of both the carrier wave and a „quadrature" carrier that is 90° out of phase with the main carrier in accordance with two input signals. That is, the amplitude and the phase of the carrier wave are simultaneously changed according to the information needed to transmit. It is such a class of non-constant envelope schemes that can achieve higher bandwidth efficiency than M-PSK with the same average signal power [4].

Mathematically, M-ary QAM can be written as shown in [9]

$$s_{kl}(t) = A_k \cos(2\pi f_c t + \theta_l), \qquad (1)$$
$$k = 1, 2, ... M_1, \ l = 1, 2, ... M_2,$$

where A_k is the signal amplitude, θ_l is the signal phase, M_1 is the number of possible amplitudes of the carrier, M_2 is the number of possible phases of the carrier, and f_c is the carrier frequency.

The combined amplitude and phase modulation, results in the simultaneous transmission of $\log_2(M_1 M_2)$ bits per symbol.

M-ary QAM is one of the widely used modulation techniques because of its efficiency in power and bandwidth [4], [6]. A variety of forms of QAM are available and some of the more common forms include 16-QAM, 32-QAM, 64-QAM, 128-QAM, and 256-QAM. QAM is in many radio communications and data delivery applications [6], [7]. Some variants of QAM are used in some specific applications and standards [4]. For domestic broadcast applications [10] for example, 64-QAM and 256-QAM are often used in digital cable television and cable modem applications. Variants of QAM are also used for many wireless [5], [12] and cellular [5], [13] technology applications.

Because of the widely use in modern communication systems, M-QAM modulation technique needs to be continuously studied, especially in terms of factors that affect its performance.

III. M-QAM Performance Measures

A. Constellation Diagram

The constellation diagram provides a graphical representation of the complex envelope of each possible symbol state, and also is an important visual tool in evaluating a performance of M-QAM modulations. The x-axis of the constellation diagram represents the In-phase component of the complex envelope and the y-axis represents the Quadrature component of the complex envelope. For

$$I\,x\,J\ \left(M = I\,x\,J,\ I = 2^{(m-1)/2},\ and\ J = 2^{(m+1)/2}\right)$$

rectangular QAM constellation, each message point can be represented as

$$x = x_I + jx_Q \tag{2}$$

where x_I is the In-phase component and x_Q is the Quadrature component of the point x.

The components, x_I and x_Q are given by

$$x_I \in \left\{\pm d,\ \pm 3d,...(I-1)d\right\} \tag{3}$$
$$x_Q \in \left\{\pm d,\ \pm 3d,...(J-1)d\right\} \tag{4}$$

If $2d$ is the Euclidean distance between two adjacent signal points and E_b is the bit energy, d can be expressed in terms of E_b, I and J as [9]

$$d = \sqrt{\frac{3E_b \log_2(I\,x\,J)}{I^2 + J^2 - 2}}. \tag{5}$$

For the case of M-ary square QAM (5) becomes

$$d = \sqrt{\frac{3E_b \log_2(M)}{2(M-1)}}. \tag{6}$$

The distance between the signals on the constellation diagram relates to how different the modulation waveform are, and how well a receiver can differentiate between all possible symbols when random noise is present.

B. Bit Error Rate Probability

BER is a performance measurement that specifies the number of bit corrupted or destroyed as they are transmitted from its source to its destination. Several factors that affect BER include bandwidth, signal-to-noise ratio (SNR), transmission speed and transmission medium. The definition of bit error rate can be translated into a simple formula:

$$BER = \frac{Number\ of\ bit\ errors}{Total\ number\ of\ bits\ sent}. \tag{7}$$

BER can also be defined in terms of the probability of error (POE) [13] and expressed as

$$POE = \frac{1}{2}(1 - erf)\sqrt{\frac{E_b}{No}} \tag{8}$$

where erf is the error function, E_b is the energy in one bit and N_0 is the noise power spectral density. At that, the error function erf is different for the each of the various modulation methods. This is because each type of modulation performs differently in the presence of noise [14]. It is important to note that POE is proportional to E_b / N_o, which is a normalized form of SNR.

The performance of each modulation is measured by calculating its BER or POE with assumption that systems are operating with additive white Gaussian noise (AWGN). Knowledge of the BER enables other parameters of the communication systems, such as power, bandwidth, etc to be tailored to enable the required performance to be obtained.

C. Noise

The term noise refers to unwanted electrical signals that are always present in electrical systems. Particularly, in terms of communication systems, noise can be defined as any unwanted energy tending to interfere with the proper reception and reproduction of transmitted signals.

In digital communication systems, noise may produce unwanted pulses or perhaps cancel out the desired ones.

Noise may introduce serious mathematical errors in signal analysis and limit the range of systems for a given transmitted power. It can also affect the sensitivity of receivers by placing restrictions on the weakest signals to be amplified. All these are some of the effects of noise on signals and communication systems at large [15].

1. Additive White Gaussian Noise (AWGN)

In communication systems, the most common type of noise added over the channel is the Additive White Gaussian Noise (AWGN). AWGN is the effect of thermal noise generated by thermal motion of electron in all dissipative electrical components i.e. resistors, wires and so on [16]. Mathematically, AWGN is modeled as a zero-mean Gaussian random process where the random signal "z" is the summation of the random noise variable "n" and a direct current signal "a" that is

$$z = a + n . \tag{9}$$

The probability distribution function for this Gaussian noise can be represented as follows

$$p(z) = \frac{1}{\sigma\sqrt{2\pi}} \exp\left[\left(-\frac{1}{2}\right)\left(\frac{z-a}{\sigma}\right)^2 \right] \tag{10}$$

where σ^2 is the variance of n.

The higher the variance of the noise, the more is the deviation of the received symbols with respect to the constellation set and, thus, the higher is the probability to demodulate a wrong symbol and make errors.

The model of this noise assumes that its power spectral density $G_n(f)$ is flat for all the frequencies and is denoted as

$$G_n(f) = N_0 / 2 . \tag{11}$$

The factor 1/2 in (11) is included to indicate that the power spectral density is a two-sided power spectral density and indicates that half the power is associated with positive frequencies and half with negative frequencies.

This type of noise is present in all communication systems and is the major noise source for most systems with characteristics of additive, white and Gaussian. It is mostly used to model noise in communication systems which are simulated to determine their performance. This noise is normally used to model digital communication systems which can be replaced with other interference schemes.

2. Phase Noise

Phase noise is a type of noise that affects the phase of carrier signals. Here, the noise (whose phase is random) causes the phase of the carrier signal to be random. When phase noise is introduced into a carrier signal, it causes angular displacement of the carrier signal. Phase noise is associated with oscillators. In M-QAM modulations, phase noise is usually generated when the constellation points oscillate.

Phase noise like other noises can only be described statistically, because of its random nature and is typically expressed in units of dBc/Hz at various offsets from the carrier frequency. This is usually referred to as Phase Noise Level Density (PNLD) [17] and it can be expressed mathematically as

$$PNLD = -10\log(I / I_0) \tag{12}$$

where I = noise intensity level in dB and $I_0 = 10^{-12}$ is a reference noise intensity level.

IV. M-ARY QAM SIMULATION MODEL

Simulink [18], developed by Math Works, is an environment for multi-domain simulation and model-based design for dynamic and embedded systems. It provides an interactive graphical environment and a customizable set of block libraries that enable to design, simulate, implement, and test a variety of time-varying systems, including communications, controls, signal processing, image processing. Simulink is integrated with Matlab [19], providing immediate access to an extensive range of tools for algorithm development, data visualization, data analysis, and numerical computation.

In this paper, a baseband simulation model of M-ary QAM is implemented in Matlab/Simulink environment. The model is shown in Fig. 1. The Simulink model is a graphical representation of a mathematical model of a communication system that generates a random signal, converts it into a binary representation, modulates it using QAM, adds noise (AWGN and phase noise) to simulate a channel, demodulates the signal and finally converts it into an integer representation. The model also contains blocks for calculating and displaying the BER and SER and the scatter plot of the modulated signal. In addition, the BERTool in conjunction with the Simulink model is used to evaluate the performance of each QAM technique through plotting the BER versus the Eb/No. BERTool is an interactive Graphical User Interface (GUI) that invokes the simulation for Eb/No specified range, collects the BER data from the simulation, and creates a plot. In BERTool, three ways of simulations are possible, that are: theoretical, semi-analytical and Monte Carlo. To invoke the BERTool, the command "BERTool" needs to be entered in main command window of Matlab.

The blocks and the lines in the model describe mathematical relationships among signals and states.

The simulation model allows studying the effects of input signal power, phase noise and frequency offset on performance of various M-QAM modulation techniques in terms of their constellation diagram and BER curve.

V. SIMULATION RESULTS AND DISCUSSIONS

The model, shown in Fig. 1 is extracted into BERTool and tested in Monte Carlo for 64-QAM, as well for 256-QAM.

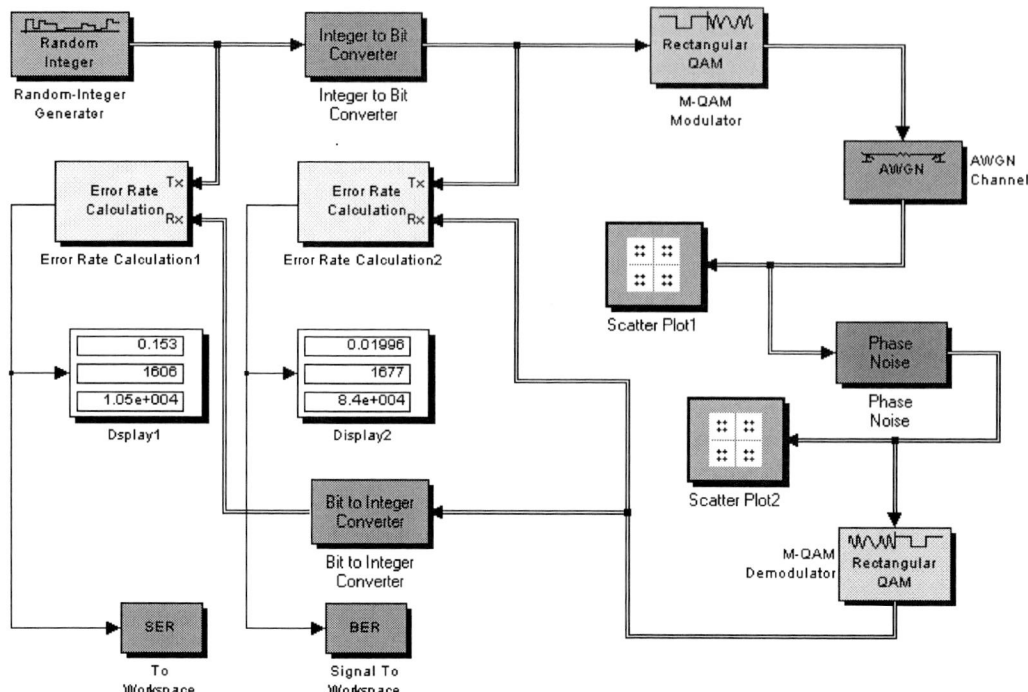

Figure 1. Simulink model for M-QAM

To take into account the presence of phase noise, in addition of AWGN, phase noise is also added as a source noise, assuming that PNLD = -60 dBc/Hz. Furthermore, the input signal power is set to 1 W and the frequency offset is set to 200 Hz, respectively. The only exceptions are cases, where the studies are done with more than one value of these parameters.

Before running the simulation to study the effect of phase noise on constellation diagram for each of two modulations, the value of signal-to-noise ratio (Eb/No) in the AWGN channel is set to 100 dB, such that the channel itself will not have any effect on the modulated signal.

A. Results for 64-QAM

The first Scatter Plot Scope1 displays the constellation diagram shown in Fig. 2. Because of high value of Eb/No (Eb/No=100 dB) in the channel, this constellation diagram is almost the same as the original constellation diagram which is presented in [3].

The effect of phase noise on the modulated signal is illustrated by the result of the second Scatter Plot Scope2. The constellation diagram for PNLD = [-60 -70] dBc/Hz is presented in Fig. 3 and Fig. 4, respectively.

The constellation diagram in Fig. 3 shows a significant angular displacement/distortion of the modulated signal for PNLD of -60 dBc/Hz. The reduction of the phase noise level density with 10 dBc/Hz (from -60 dBc/Hz to -70 dBc/Hz) results in the reduction in the angular displacement, i.e., to a higher quality of the constellation diagram, which is obvious in Fig. 4.

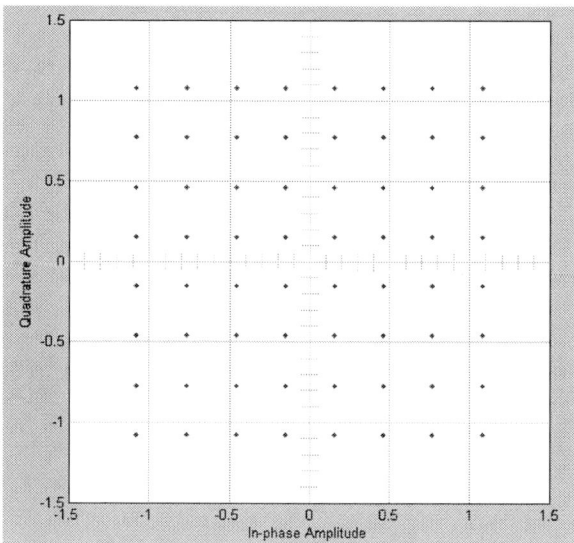

Figure 2. Constellation diagram of 64-QAM for Eb/No=100dB and without the presence of phase noise.

To study the impact of changing the power of the input signal on the noise variance and therefore on the bit error rate, BER as a function of Eb/No is simulated. The simulation is performed for values of the input signal power [0.5 1.0 2.0 4.0] W in the presence of phase noise (PNLD=-60 dBc/Hz). For each value of the input signal power a result in the form of BER graph, as shown in Fig. 5, is obtained. The simulation result illustrates that as the power of the input signal increases, the error rate also

550

increases. This is due to the proportional relation between the signal power and the noise variance [14] which relation is implemented in the AWGN Channel block (see Fig. 1).

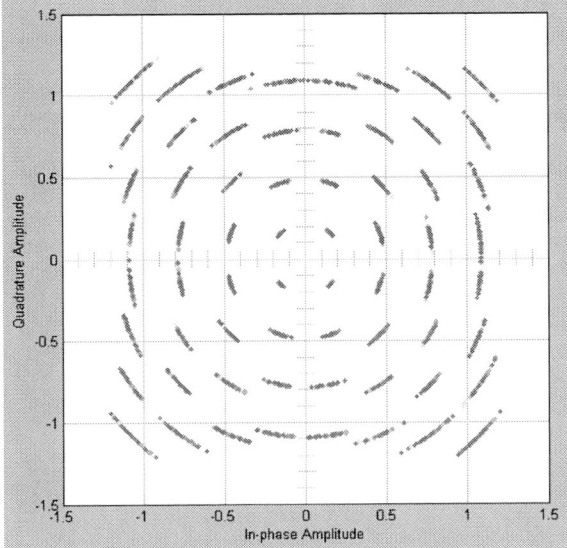

Figure 3.   Constellation diagram of 64-QAM for Eb/No=100 dB and PNLD= = -60 dBc/Hz.

The effect of phase noise on BER performance of 64-QAM is studied, as the values of the PNDL are varied from -50 dBc/Hz to -70 dBc/Hz, in step of -10 dBc/Hz. For each value of PNLD, as well, for case without phase noise (in the presence of AWGN only), a BER curve versus Eb/No is obtained. The result is represented in Fig. 6.

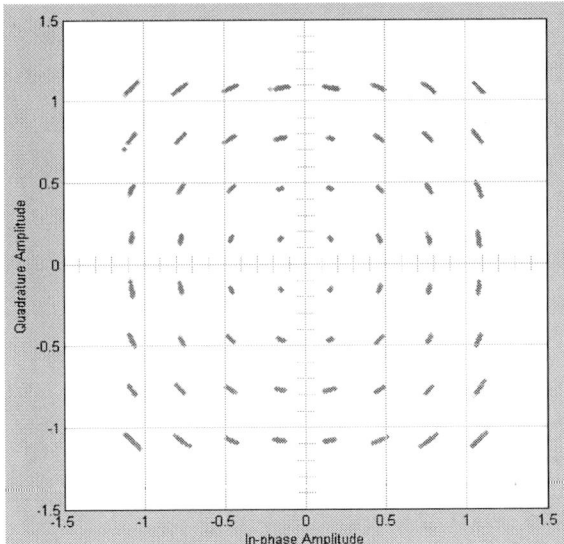

Figure 4.   Constellation diagram of 64-QAM for Eb/No=100 dB and PNLD = = -70 dBc/Hz.

As can see in Fig. 6, the higher the phase noise level density, the higher the BER simulated value at a fixed

Eb/No value. For example, at Eb/No=18 dB, the value of BER obtained for PNLD of -50dBc/Hz is 6.3476E-2 while if the noise level density is reduced to -70 dBc/Hz, there is a significant reduction of the BER value (the value obtained equals 1.5873E-5). Moreover, it is observed from the simulated curves, that the BER curve for PNLD of -70 dBc/Hz gets close to the curve without phase noise (in the presence of AWGN only).

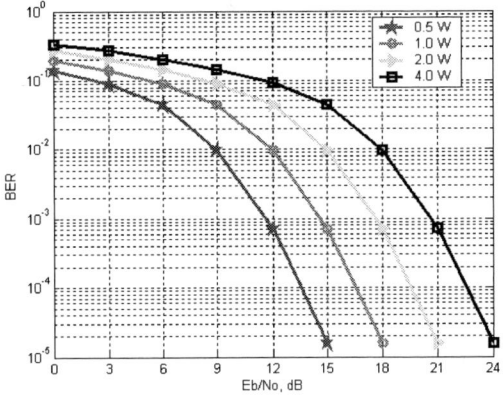

Figure 5.   BER vs. Eb/No for 64-QAM for different values of the input signal power.

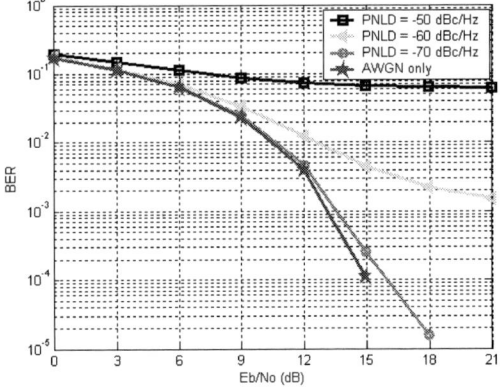

Figure 6.   BER vs. Eb/No for 64-QAM at different phase noise level densities and without the presence of phase noise.

BER as a function of the Eb/No ratio, at various offsets from the carrier frequency is simulated, too. The values of the frequency offset are varied from 100 Hz to 300 Hz, in step of 100 Hz. The simulation is performed with PNLD equal to -60 dBc/Hz. For each value of the frequency offset a BER as a function of Eb/No is obtained. The result, in the form of graphs, is displayed in Fig. 7. How is seen in Fig. 7, at low SNRs (Eb/No under of 8 dB), the three BER curves almost coincide. For values of a Eb/No ratio over 10 dB, the obtained BER curves more and more get away from each other with increasing Eb/No. As is expected, it is observed in Fig. 7, that an increase of the frequency offset increases the BER value at a fixed Eb/No ratio. In other words, the BER curve for a frequency offset of 100 Hz outperforms the BER curve for a frequency offset of 200 Hz and 300 Hz, respectively. For example, the change of the frequency offset from 100 Hz to 300 Hz at Eb/No=18 dB and PNLD=-60 dBc/Hz, causes the BER value increase from 5.5531E-4 to 7.1251E-3.

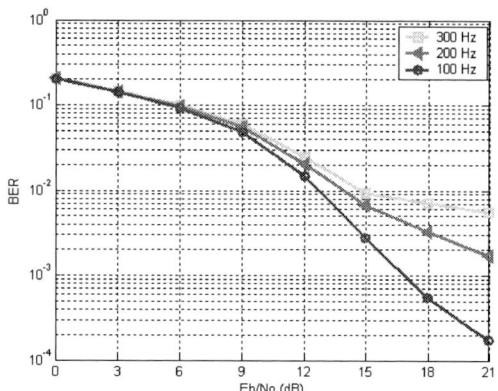

Figure 7. BER vs. Eb/No for 64-QAM for different values of the frequency offset.

B. Results for 256-QAM

Constellation diagrams and BER curves for 256-QAM, are simulated under the same conditions as for 64-QAM using Simulink model in Fig. 1.

The constellation diagram displayed by the first Scatter Plot Scope1 in Fig. 8 is almost the same as the original constellation diagram that is shown in [1]. The similarity is due to the high value of the Eb/No ratio in the channel (Eb/No=100 dB).

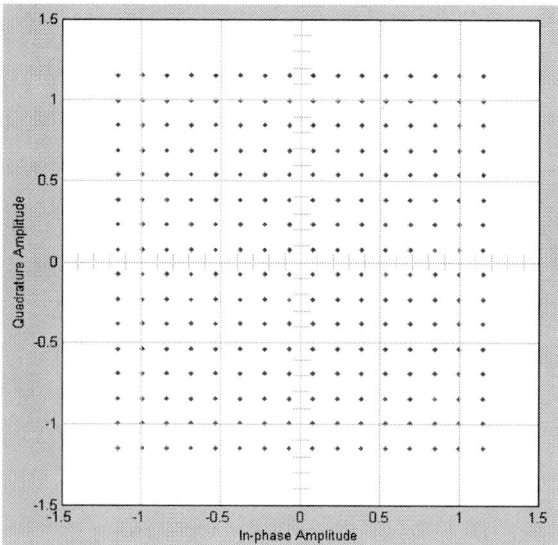

Figure 8. Constellation diagram of 256-QAM for Eb/No=100 dB and without the presence of phase noise.

The Scatter Plot Scope2 placed after the Phase Noise block gives different result for each PNLD value. Resulting constellation diagrams for PNLD=-60 dBc/Hz and PNLD=-70 dBc/Hz are displayed in Fig. 9 and Fig. 10, respectively. It is obvious from Fig. 9 and Fig.10 that the reduction of the phase noise level density from -60

dBc/Hz to -70 dBc/Hz, results in a significant reduction in the angular displacement.

The dependence of the BER on Eb/No ratio for different values of the input signal power is also examined. The value of the input signal power is varied from 0.5 W to 4 W. With each variation, the model is simulated. The set of the simulated BER curves for values [0.5 1 2 4] W of the input signal power is presented in Fig. 11. It can be seen in Fig. 11 that with increasing the input signal power, the BER also increases. This is due to the proportional relation between the signal power and the noise variance [14] which is equivalent to proportionality between the signal power and BER ratio. The numerical results, obtained with Eb/No=18 dB and PNLD = -60 dBc/Hz, show that the increase of the input signal power from 0.5 W to 4 W, results in an increase of the BER value from 1.3095E-4 to 5.8251E-2.

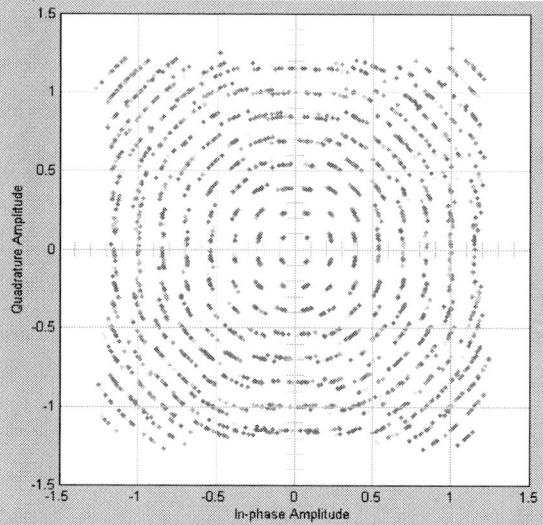

Figure 9. Constellation diagram of 256-QAM for Eb/No=100 dB and PNLD=-60 dBc/Hz.

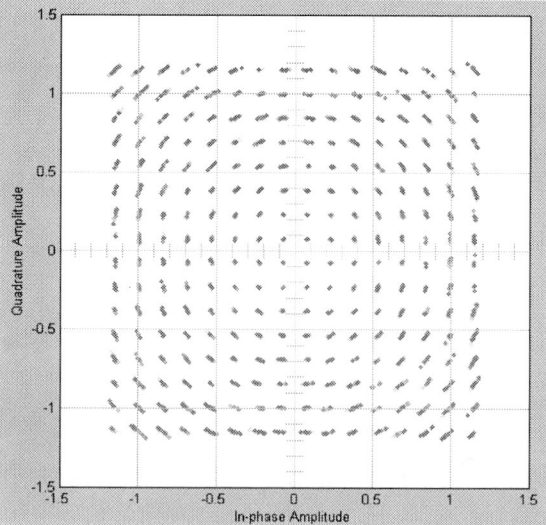

Figure 10. Constellation diagram of 256-QAM for Eb/No=100 dB and PNLD=-70 dBc/Hz.

To study the effect of phase noise on BER performance of 256-QAM modulation technique, the phase noise level density is varied in the same way as for 64-QAM modulation technique. Simulation of BER curve for each of values [-50 -60 -70] dBc/Hz of a PNLD is performed and the result is shown in Fig. 12.

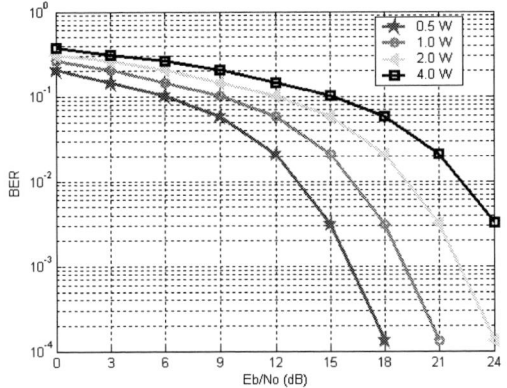

Figure 11. BER vs. Eb/No for 256-QAM for different values of the input signal power.

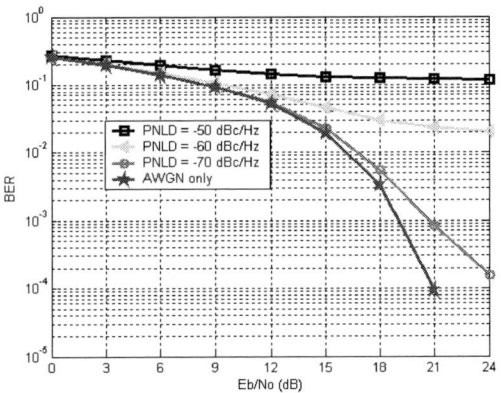

Figure 12. BER vs. Eb/No for 256-QAM at different phase noise level densities and with no presence of phase noise.

The first three BER curves (from up to down) correspond to the three values of PNLD that are given above. The last curve corresponds to so-called AWGN case (in the presence of AWGN only). It is obvious from Fig. 12, that with the phase noise reduction, the BER curve comes closer to the curve with AWGN only. In other words, with the phase noise level density reduction, the value of BER also reduces. The numerical results for BER obtained at 18 dB Eb/No for different values of PNLD are as follows: BER=[1.2482E-1 3.0048E-2 5.5238E-3] for PNLD=[-50 -60 -70] dBc/Hz, respectively.

BER performance of 256-QAM for different values of the frequency offset is also simulated. The simulation is done for PNLD equal to -60 dBc/Hz by varying the value of the frequency offset from 100 Hz to 300 Hz in step of 100 Hz. The result is obtained in the form of BER curve for each of the three values of the frequency offset, as shown in Fig. 13.

As can be seen, the BER curve is moved upward with increasing of the frequency offset. That means the increase of the frequency offset affects negatively the BER curve. It is also evident, that the three BER curves are nearly identical up to Eb/No=10 dB. For values of Eb/No, which are over 12 dB the curves are gradually moving away one from another.

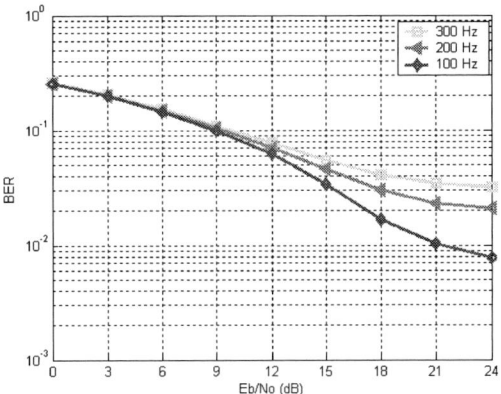

Figure 13. BER vs. Eb/No for 256-QAM for different values of the frequency offset.

VI. CONCLUSIONS

A simulation model to study 64-QAM and 256-QAM modulation techniques is built using the Simulink tool that is integrated in the Matlab environment. Also, the BERTool under Matlab is used to evaluate the performance of both modulation techniques through plotting the BER versus Eb/No. The model can not only be used for 64-QAM and 256-QAM modulation techniques, it can also be used for the other QAM order (for example, for M = [4 8 16 32 128 512, etc]), as well for a platform to simulate other modulation techniques, such as M-PSK, DPSK, etc.

The simulation results show that the phase noise affects constellation diagram. The effect of phase noise on constellation diagram, results in an angular displacement of the modulated signal. It can be seen from Figs. 3, 4, 9 and 10 that the points do not lie exactly on the constellations because of the added noise. The radial pattern of points is due to the addition of phase noise, which alters the angle of the modulated signal. Moreover, the effect of phase noise on constellation diagram increases with an increase of phase noise level density, as well with an increase of the modulation order.

The simulations show that the BER for both type (64-QAM and 256-QAM) modulations decreases with increasing value of Eb/No. The higher the Eb/No value, the better the BER based on data generated by the computer simulation model.

According to the results of the simulations it is possible to conclude that as the QAM order increases, the simulated BER values also increase. In case of 256-QAM the BER is greater than in case of 64-QAM as constellation points come closer.

It follows from the simulation results, that all the

specified factors (AWGN noise, input signal power, phase noise, and frequency offset) affect the BER performance of M-QAM (M=64 and 256) modulation techniques. At that, increasing each one of them results in a BER increase. In fact, how it can be seen in Fig. 6 and Fig. 12, for a PNLD value greater than -60 dBc/Hz there is no a significant change in the BER value with increasing the Eb/No.

Finally, it can be concluded that 64-QAM has better BER and phase noise performance than 256-QAM.

REFERENCES

[1] S. Haykin and M. Moher, Introduction to Analog & Digital Communications. John Wiley & Sons, Inc., 2007.

[2] J. G. Proakis and M. Salehi, Communication Systems Engineering, 2nd ed. Pearson Education, 2002.

[3] B. Sklar, Digital Communications: Fundamentals and Applications. Prentice-Hall, 2nd ed., 2001.

[4] T. S. Rappaport, Wireless communications: Principles and Practices. 2nd ed., Prentice-Hall, New Jersey, 2003.

[5] S. Haykin and M. Moher, Modern Wireless Communications. Pearson Prentice Hall, 2005.

[6] F. Molisch, Wireless Communications. John Wiley & Sons, 2005.

[7] W. Ho. Sam, Adaptive Modulation, (QPSK, QAM): Intel Communications Group, 2004.

[8] "Quadrature Amplitude Modulation", digital Modulation Techniques" www.digitalmodulation.net/qam.html.

[9] Tamer Youssef and Eman Addelfattah. Performance Evaluation of Different QAM Technologies Using Matlab/Simulink. Systems, Applications and Technology Conference (LISAT), 2013 IEEE, Long Island, 2013, pp. 1-5.

[10] K. Konov. Digital Radio and Television Broadcasting. Dios, Sofia, 2014. (Bulgarian)

[11] Sadinov S., K. Koitchev, P. Penchev, K. Angelov, Simulation Evaluation of BER Characteristics for M-PSK and M-QAM Modulations used in the Reverse Channel of Cable TV Nets, Journal "Electronics and Electrical Engineering" Vol. 7 (95), pp.71-76, ISSN 1392-1215, TECHNOLOGIJA Kaunas, Lithuania, 2009.;

[12] Mihaylov, Gr.Y., Iliev, T.B., Ivanova, E.P., Stoyanov, I.S., Iliev, L., Performance analysis of low density parity check codes implemented in second generations of digital video broadcasting standards, Proceedings of 39th International Convention on Information and Communication Technology, Electronics and Microelectronics, MIPRO 2016, Opatija, Croatia, 2016, pp. 499 – 502.

[13] Andrea Goldsmith. Wireless Communications. Artech House, London, 2005.

[14] Gary Breed. High Frequency Electronics, 2003 Summit, Technical Media LLC "Bit Error Rate: Fundamental Concepts and measurement issues".

[15] J. G. Proakis and M. Salehi. Digital Communications. 5th ed., McGraw-Hill International, 2008.

[16] M. Samsuzzannan, M. A Rahman and M. A Masud. "Bit Error Rate Performance Analysis on Modulation Techniques of Wideband Code Division Multiple Access." Journal of Telecommunications, vol. 1, no. 2, pp. 22-29, March 2010.

[17] E. L. Efurumibe and A. D. Asiegbu. Computer-Based Study of the Effect of Phase Noise in256-Quadrature Amplitude Modulation Using Error-Rate Calculation Block - A Comparative Study. Journal of Information Engineering and Applications, vol. 2, no. 11, 2012, pp. 35-41.

[18] M. Samsuzzannan. Modeling and Simulation in Simulink for Engineers and Scientists. Bloomington, Indiana: AuthorHouse, 2004.

[19] Matlab Help Documents. 2014.

Multiscale and Multiobjective Modelling: a Perspective for Mastering the Design and Operation Complexity of IoT Systems

Khalil DRIRA

LAAS-CNRS, Université de Toulouse
CNRS, INSA, UPS,
Toulouse, France
khalil.drira@laas.fr

Abstract— Modeling IoT systems behaviors and architectures requires new approaches to be elaborated for handling the challenging issues such as: large scale interaction and real-time reconfiguration. This can help both during the design and the operation steps for ensuring a correct design and for delivering a performant service. Different studies have been recently conducted and new initiatives, at the national, European and international levels, have been taken for the domain of IoT systems and for the more general domain of System of Systems. Different modelling approaches have been elaborated. Bridging the models can lead to powerful solutions for modelling the next generation systems of systems of the family of IoT complex systems. Cross-disciplines initiatives have to be taken to enable the emergence of new Multiscale Multi-objective modelling approaches and frameworks.

Keywords—IoT, architecture; modelling; multiscale; systems-of-systems

I. INTRODUCTION

We tackle the complexity of architectural design for the Internet of Thing (IoT) complex systems and the future smart systems of systems built on top of IoT platforms. Internet of Thing platforms will connect billions of devices deployed on different geographic locations and manage different kinds of traffic generated by smart applications such as smart cities and connected vehicles. Design and modelling of such systems is a complex task that requires different levels of abstraction to be distinguished according to the properties to be validated. Recently, new architectural modelling techniques have been defined [4, 6]. These techniques handle the structural properties of communication architectures. The most used in IoT and distributed systems is the mediated communication pattern such as publish/subscribe brokers that decouple information producers and consumers, and load-balancers that distribute the load on different servers. Dynamic architecture modelling (e.g. graph-based dynamic structures), and analytic or quantitative approaches of behavioral performance modelling (e.g. Stochastic modelling, queueing theory) if integrated, can constitute a powerful modelling technique for a wide category of IoT systems. Such approaches address respectively, the

problems of architectural reconfiguration strategies (service composition and deployment, network topology, etc.), and service provisioning policies (intelligent load balancing, resource allocation, etc.). The design of new integrated methods can be a promising objective that enables the elaboration of efficient multi-model solutions for self-configuring, self-healing, self-optimizing and self-protecting the IoT systems and the corresponding smart applications. Our objective, here, is to review the different studies and approaches in order to motivate the emergence of new modelling initiatives. We present the modeling challenges for Next Generation Systems in section II. We present the principles of multiscale modelling in section III. We consider the case of architectural models in section IV. We give a summary of our statements in section V.

II. MODELING CHALLENGES FOR NG SYSTEMS

Next Generation systems of systems such as IoT complex systems will be composed of a large number of interacting entities, possessing decision-making autonomy, and whose behavioral evolution is difficult to predict deterministically. A first cause may arise from the ignorance of certain parameters governing the evolution of the system or the random nature of their variation. These parameters may be intrinsic (internal variables), or context-dependent (external factors). Another cause may be the impossibility of performing, accurately or completely, within a reasonable time, the execution of the computation required to make it possible to determine the evolution of the state of the system. Communication networks (satellites, routers, gateways, protocol stacks), connected objects (sensors, actuators, devices, servers, appliances, machines, etc.) and intelligent services (discovery, security and protection, autonomy, etc.) as well as Smart Homes and Buildings, Smart Cities, Smart Power Grids, constitute an important part of the next generation complex systems that introduce important challenges for the design and operation steps.

The modeling of the next generation complex systems is faced with scalability problems, heterogeneity of models, in particular structural and behavioral, and importance of both qualitative (correctness) and quantitative (performance or Time constraints). Elaborating the appropriate approaches can rely on modeling the evolution of a complex system by characterizing rather than enumerating or exploring all the configurations of the system architecture or its states space. To do this, models should be concerned with both the structural dimension and the behavioral (or functional) dimension of complex systems for analysis of correctness and performance properties.

Structural models have to be elaborated to describe a software architecture reconfigurable by dynamic integration of components or real-time composition of services. These models can also describe the topology of a network for which the graph represents the information propagation paths at various levels ranging from a representation of the neighborhood links in a sensor network to the relations of acquaintance in social networking or collaborative network platforms. The modeling can be based on conceptual graphs and graph grammars [2]. We can consider an initial graph and a set of transformation rules that characterize the set of the possible reachable configurations without their explicit enumeration. The transformation can operate by modifying the attributes associated with the nodes and the edges of the graph or by transforming its structure by adding or deleting nodes and arcs.

Behavioral models have to be elaborated to describe the functioning of the system. These models can describe the dissemination or the propagation of information according to an epidemic model, taking into account the influence of external factors of types interactions with users, and phenomenon of abandonment, etc. We can also proceed by a stochastic modeling of the system, the interactions between its components and its environment and we handle the scalability challenge through approximation models such as: fluid limits (e.g. for load balancing management) or the average field (e.g. for routing data in sensor networks). The challenge is to design systems where performance is associated with intelligence for the routing of information and its autonomous interpretation.

III. PRINCIPLES OF MULTISCALE MODELING

Multiscale modeling enables to look at a problem simultaneously from different scales and different levels of detail. It takes advantage of data available at distinct scales by modeling interaction between those scales, accordingly managing the complexity of behavior involved [3,7]. Practically this can be achieved by decomposing a problem into a set of single scale models that exchange information across the scales. In this context Borgdorff et al [1], gives a definition of a sub-model multiscale as a component model which describes only one scale of the system. He also considers a multiscale model as a composite model formed from two or more sub-models that describe different behaviors at different scales.

A. Multiscale modeling strategies

The key issue in multiscale modeling is the order in which the multiscale model is constructed. There are four strategies, discussed by [5], to establish such multiscale models:

– Bottom-up: Complex Systems can be understood on the higher scale by analyzing lower-scale mechanisms. A model is developed to describe the finest scale of interest, then models at increasing scales are constructed in turn; time or length scales may be used.

– Top-Down: A large scale model is constructed. It is refined by successively adding smaller scale models until detail and accuracy goals are reached.

– Middle-out: In some multiscale biological applications 'middle-out' modelling is favored. This refers to constructing a multiscale model by starting with the scales that are richest in data and best understood, and then working 'outwards' from there, to smaller and larger scales.

– Concurrent: All levels in the process hierarchy should be attacked simultaneously, from the microscopic level to the macroscopic level.

B. Multiscale modelling steps:

With the four strategies presented above, three steps are involved [5]:

– Step 1: identifying and selecting scales to include in the multiscale model

– Step 2: adopting or developing appropriate sub-models at each scale of interest

– Step 3: linking, or integrating, the sub-models into a coherent multiscale model. There are several broad ways of linking sub-models into a multiscale model.

An important part of multiscale modeling is how the scales of different behaviors relate to each other. Multiscale simulation enables coupling of behaviors at various scales from the quantum scale to the molecular, mesoscopic, device, and plant scale [5].

Two approaches for linking sub-models at different scales are considered: sequential multiscale modeling and concurrent multiscale modeling.

With sequential multiscale modeling, the smallest (finest) scale model is solved first, and its results are passed to the larger (coarser) scale. For instance, some details of the macroscale model are precomputed using microscale models.

Concurrent multiscale modeling is preferred when the macroscale model depends on many variables, and becomes difficult to extract by precomputing from microscale models. In addition, in some cases, the one-way coupling is inadequate, and fully coupled models across scales are needed, i.e., two-way information traffic exists. There are two types of multi-scale models:

IV. MULTISCALE MODELING OF IOT SYSTEMS

Most problems in IoT are multiscale in nature. Things are made of sensors, actuators, gateways and servers at the atomic scale, and at the same time are characterized by their own architectural composition, geographic distribution as well as networking and processing capacities that have a larger order of magnitude.

A. Scale concepts

A scale is characterized by two major concepts: the grain and the extent. The grain is the finest spatial resolution; the resolution refers to the granularity used in the sub-system modelling. The extent refers to the structural or functional scope covered by the global system modelling. In the context of IoT systems modelling, the extent scale can refer to the abstract description considering a sub-system of the system architecture. Variation in extent can be used, for example, to describe a given description level or a given communication layer in the IoT infrastructure. It allows the architect to describe the necessary details to understand the system behavior and validate the associated functional and structural properties. Besides, the grain scale refers to the level of details and precision pertaining to the abstract description, providing more details of a given current description, such as composition and interactions in a given system description.

B. Top-down scale transformations

The top-down scale transformation process, much like regular refinement, begins with a high-level model of a system, which we describe as a whole. Then, scale changes are applied to obtain a more detailed description, by describing components that compose the subsystems and their connections. An iterative modeling allows to refine IoT systems descriptions: A vertical refinement can be applied to add the architecture composition details iteratively and to obtain a more detailed description by zooming on previously defined components. A horizontal refinement is needed to add details on the interconnections between components and their interfaces, and to establish the compatibility of interfaces by determining interfaces that can satisfy all possible sequences of required and provided relationships.

V. CONCLUSION

Multi-scale modelling of IoT systems architectures can be used to validate their behavioral properties both from the functional and non-functional points of views. We can consider static and dynamic structures with spatio-temporal properties. The abstraction or refinement related to grain and extent can address both components composition, routing or processing functions and exchanged data structure. The modelling strategies can be conducted in a bottom-up way allowing to characterize emerging properties in IoT systems viewed as systems of systems. We can also proceed by a top down modelling until we reach the necessary detail level for proving a given property such as information propagation in networked services.

REFERENCES

[1] J. Borgdorff, J.-L. Falcone, E. Lorenz, C. Bona-Casas, B. Chopard, and A. G. Hoekstra. Foundations of distributed multiscale computing: Formalization, specification, and analysis. Journal of Parallel and Distributed Computing, 73(4):465 – 483, Jan 2013.

[2] C. Eichler, T. Monteil, P. Stolf, L. Grieco, K. Drira. Enhanced graph rewriting systems for complex software domains. Software and Systems Modeling, Vol.15, N°3, pp.685-705, July 2016, doi 10.1007/s10270-014-0433-1 ; 2015 Best papers award.

[3] W. Einer and B. Engquist. Multiscale modeling and computation. Notices Amer. Math. Soc, 50(50):1062–1070, 2003.

[4] A. Gassara, I. Bouassida, M. Jmaiel, K. Drira. A Bigraphical multi-scale modeling methodology for system of systems. Computers and Electrical Engineering, Vol.58, pp.113-125, Feb. 2017.

[5] G. Ingram, I. Cameron, and K. Hangos. Classification and analysis of integrating frameworks in multiscale modelling. Chemical Engineering Science, 59(11):2171 – 2187, 2004.

[6] I. Khlif, A. Hadj Kacem, A. Hadj Kacem, K. Drira. A multi-scale modelling perspective for SoS architectures. European Conference on Software Architecture (ECSA) , Vienna (Austria) August 2014, LNCS 8627.

[7] E. Weinan and L. Jianfeng. Seamless multiscale modeling via dynamics on fiber bundles Communications in Mathematical Sciences. Volume 5, Issue 3 (2007), 649-663.

Topological data analysis and applications

Joao Pita Costa
Institute Jozef Stefan
joao.pitacosta(at)ijs.si

Abstract—**Topological data analysis is interested in problems relating to nonlinear systems, large scale data and development of more accurate models, that contribute to a high level research. In the study of time-series data we can identify problems that focus aspects of that nature, with applications to digital disease detection, gene expression, viral evolution, etc. In particular, it describes the qualitative aspects of the data, focusing the topological features with longest lifetimes. To encode the latter, it uses novel mathematical methods that can provide diagrams which are a clear and practical tool that allow us the detection of outliers and to capture the dynamics of the system. In this paper we review this methodology and discuss several applications that can lead to new perspectives to modelling and simulation in system behaviour.**

I. INTRODUCTION

A. Encoding the shape of data

Today, there is an increasingly urgent need to efficiently manipulate Big Data. To predict an outbreak of the next epidemics taking profit of the information present in the social media we need to have a fundamental mathematical knowledge that permits us to do so. Today's interest in Big Data analytics is a powerful source of motivation of mathematical flavor [18]. Computational topology can successfully retrieve information from large and complex data sets. In particular, topological data analysis (TDA) has applications from social media analysis to cancer research, providing a multi-scale framework to study the shape of data. Its essential tool – persistent homology (also known as persistence) - combines the mathematical methods of algebraic topology with techniques of computational geometry to identify a global structure of a given point cloud.

In the past years TDA has been a vibrant area of research due mainly to developments in applied and computational algebraic topology. Essentially, it applies the qualitative methods of topology to problems of machine learning, data mining and computer vision. In particular, persistent homology is an area of mathematics interested in (a) identifying the global structure of data by deducing high-dimensional structure from low-dimensional representations; and (b) studying properties of a continuous space by the analysis of a discrete sample of that space. When considering a certain metric (i.e., a notion of distance on the space), one gets a perspective of the space under different scales. In that, small features will eventually disappear and be considered noise. Persistence allows us to compute the homology at all scales, thereby giving us the ability to find ranges of scales where the structure of the space is stable [12].

Techniques of persistence can be used to infer topological structure in data sets. Moreover, certain variations of the method can be applied to study aspects of the shape of point clouds. By considering all possible scales, one can infer the correct scale at which to look at the point cloud simply by looking for scales where the persistent homology is stable. Multi-scale methods enable us to study the topology of point clouds as a route for approximating topological features of an unobservable geometric object. For these reasons, multi-scale signatures are of great importance to the application of this research towards the development of new techniques in data science. Such theory leads to topological data analysis particularly for large, high-dimensional data where topological methods allow us to transfer information from local to global structures. A good introduction to persistent homology can be found in [4] and [7], and to general algebraic topology with a computational flavor in [24].

Evolving complex software systems contribute to a rapidly growing range of applications, products and services supporting daily human activities in all economic sectors. Due to this central role, their reliable operation became one of the key quality attribute. Hence, these software systems are developed in a sequence of releases aiming to adapt to continuously changing environmental needs. These systems are often safety critical and their reliable operation becames one of the key quality attribute. The existing mathematical models to model software reliability are numerous [17], [16], [19] and extensively used in industrial practice to predict effort and time needed to bring the system to required reliability state [13]. Though, the main problem still remains and is related to early determination of the best mathematical reliability growth model early enough during the development process [23], [1]. In this paper we present the potential of TDA as a new approach to this problem, thorough the related study of the behaviour of biological systems such as epidemics and genetic variety. These serve us as motivation to later explore the potential to extract the topological features of system failure time series, in the context of their influence on failure distributions and, consequently, on reliability.

B. Related Work

The analysis of data in a time line is present in many applications beyond mathematics. Time series are present in applications of a wide range from medicine to aquaculture, being a framework to develop novel capabilities. The persistence approach to time series data analysis was motivated by the study of new ways to find

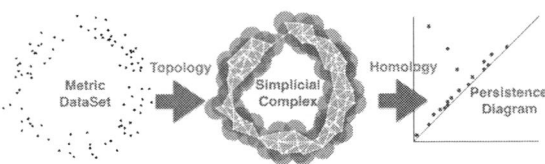

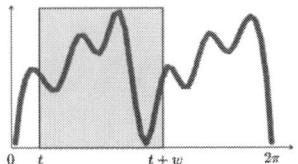

Fig. 1. Encoding persistent topological features of a pointcloud (on the left) into a persistence diagram (on the right) by approximating the space through simplicial complexes (in the middle) [4].

Fig. 2. The sliding window persistence method applied to time series data, where the window size is w [20].

periodicity and quasi-periodicity in signals in [20]. The studied method is based on sliding windows providing time-delay reconstructions useful in dynamical systems and engineering applications. It consists in the topological study of a point cloud in dimension N extracted from a given discrete sample of a time series by considering windows (i.e. closed intervals) with N elements each that cover all the time series by moving forward one time step at the time. In that, each window provides the coordinates of a point in the point cloud (the images of each time step in the given time series). The persistence pipeline is then applied to that pointcloud in order to extract topological features that are further encoded in a persistence barcode, or equivalently, in a persistence diagram (see Figure 1). Each feature of the pointcloud is thus tracked by this encoding where the size of the corresponding bar in the barcode (i.e. the lifetime of the feature) is recorded. In that, noise is typically noticed by topological features with short lifetime, and relevant features are characterized by its persistence consequent of a long lifetime. This study shows in [20] that maximum persistence occurs when the window size corresponds to the natural frequency of the signal. Moreover, the discovery of periodicity in gene expression time series data across different processes is a central computational problem that can be tackled using this methodology. The quantification of time series periodicity in [21] using direct measurements evaluates the circularity of a high-dimensional representation of the signal, typical for gene expression data. Furthermore, this method was used in preliminary studies of epidemiological data in [5] to analyze the incidence of the Influenza virus across Europe and compare flu seasons in different countries. That studies show the complementary information that topological data analysis can provide towards classical quantitative methods such as Fourier analysis or dynamic time warping.

Moreover, the scientific advances in the analysis of time series contributes to progress both inside and outside of mathematics. Time series data is present in diverse areas of science from biology and medicine to engineering. A deeper insight in the features that can be retrieved from such data will permit a better understanding of the data itself in those various fields. The pioneering nature of this approach is the introduction of state-of-the-art methods of persistence in the context of the sliding window method in order to be able to manipulate more general data structures in the considered applications. An example of this is to consider the simultaneous study of several time series and

extract topological features permitting us to access global information provided by that data. This approach to the study of time series permits innovation in the analysis of time series data and will contribute to the advances on global dynamics, enabling diverse applications to social media data (detecting temporal patterns from historical time-dependent data and project the detected patterns) or epidemiological data (comparing the real-time data of the flu seasons in different countries). The contributions with open source R and Python libraries to a wider research community are available in [11] and [14].

II. TOPOLOGICAL DATA ANALYSIS OF TIME-SERIES

The analysis and forecasting of time series data has many valuable applications such as monitoring the evolution of an epidemics, monitoring an industrial process, or tracking corporate business metrics. By a time series we mean a sequence of data points indexed by a linear ordered set (usually representing a time stamp). Many successful time series analysis methods exist, aiming to extract quantitative characteristics of the data (e.g. Fourier analysis). Though, when we want to compare the behaviour of two time series under the same linear referential, the qualitative analysis of such data can provide a valuable source of complementary information.

In order to use TDA to proceed with time-series analysis, the data needs to be projected to higher dimensions where an appropriate study of it's shape is made possible. The method of sliding window persistence (also known as SW1PerS) has been successful in the study of time series data (see [20] and [6]). Given a time series $g_0, g_1, \ldots, g_S$ measured at times $t_0, t_1, \ldots, t_S$, we consider the graph of g restricted to the interval (i.e. the window) $[t_i, t_{i+w}]$, where $i = 0, \ldots, S - w$ and w is the length of the window, and we consider the pointcloud

$$\{(g_{t_i}, \ldots, g_{t_{i+w}}) \mid i = 0, \ldots, S - w\}.$$

This method is illustrated in Figure 2.

The sliding window method can then provide a point-cloud of a corresponding input time series, by projecting that data to w dimensions, being w the size of the considered window. Let us now take balls centered in each point of the pointcloud (for an appropriate metric) together with a real function - named height function. Such function permits us to consider growing radii, independent of a fixed threshold. In that we are assuming a topological filtration for that pointcloud. The topological information of the filtration can be encoded in a set of closed intervals - named a barcode - where the birth (first time stamp)

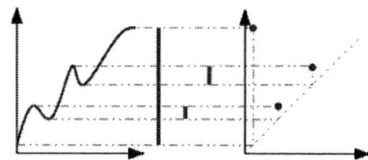

Fig. 3. Encoding persistent topological features of a space (on the left) into a persitence barcode (in the center) and its equivalent representation as a persistence diagram (on the right) [7].

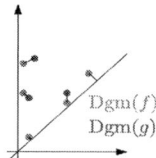

Fig. 4. The bottleneck distance comparing two persistence diagrams [8].

and death time (last time stamp) of every connected component of the sample of a given topological space is recorded. Each barcode can also be represented by a multiset of vertices in a plane, eventually some of them with multiplicity bigger than 1, constituting an equivalent representation to the latter barcode – named a persistence diagram – where the first coordinate corresponds to the appearance – birth time – and the second coordinate corresponds to the disappearance – death time – of a topological feature (see Figure 3).

The standard method to compare two persistence diagrams is based on a bijection between the points - named bottleneck distance - and is therefore always at least the Hausdorff distance between the two diagrams. For points $p = (p_1, p_2)$ and $q = (q_1, q_2)$ in $\mathbb{R}^2$, let $\|p - q\|_\infty$ be the maximum of $|p_1 - q_1|$ and $|p_2 - q_2|$. Let X and Y be multisets of points. Every persistence diagram is such a multiset. The bottleneck distance between X and Y is defined as

$$d_B(X, Y) = \inf_{\eta} \sup_{x \in X} \|x - \eta(x)\|_\infty,$$

where the infimum is taken over all bijections η from X to Y. Each point with multiplicity k in a multiset is interpreted as k individual points, and the bijection is interpreted between the resulting sets.

These persistence diagrams can be compared at several simultaneous levels using functional summaries for data known as persistence landscapes [2]. These techniques are

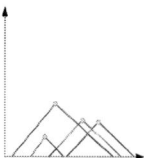

Fig. 5. The comparison between persistence diagrams, each one of them representing the topological features of the reliability behaviour during evolution of the considered software products [2].

easily combined with tools from statistics and machine learning due to their quantitative nature.

Persistence landscapes are techniques of TDA that permit us to measure the pairwise distance between persistence diagrams at several different levels [2]. It extends bottleneck distance to a wider concept, providing us with more granularity to enhance the comparison. This method is illustrated in Figure 5. Formally, the persistence landscape is a function $\lambda : \mathbb{N} \times \mathbb{R} \to \overline{\mathbb{R}}$, where $\mathbb{N}$ is the set of positive integers and $\overline{\mathbb{R}} = [-\infty, +\infty]$ the extended real numbers. Alternatively, it may be thought of as a sequence of functions $\lambda_k : \mathbb{R} \to \overline{\mathbb{R}}$, where $\lambda_k(t) = \lambda(k, t)$. Define:

$$\lambda_k(t) = \sup\{m \geq 0 \,|\, \beta^{t-m, t+m} \geq k\},$$

with $\lambda : R^2 \to R$ given by

$$\lambda(m, h) = \begin{cases} \beta m - h, m + h & \text{if } h \geq 0 \\ 0 & \text{otherwise} \end{cases}$$

where $\beta^{a,b} = \dim(\text{Im}(M(a \leq b)))$, for any persistence module M. The persistence module M attached to persistent homology of a time series in degree i, as in our study, consists of degree i homology spaces $M_r = H_i(X_r)$ of the Vietoris–Rips simplicial complex at level $r \in \mathbb{R}$ and the linear maps $M(r \leq s) : H(X_r) \to H(X_s)$ induced by the inclusion of simplicial complexes $X_r \hookrightarrow X_s$. Hence, the image $\text{Im}(M(r \leq s))$ is the persistence homology space, and $\beta^{r,s}$ is its Betti number.

Efficient algorithms for calculating persistence landscapes and distances between their averages are provided in [3].

III. APPLICATIONS

In the following we shall present a few applications of topological data analysis to areas of research that can contribute to the study of system behaviour due to their nature.

A. Gene expression

Due to the recent advances on genetic sequencing and its huge socio-economic impact, the bioinformatics field of gene expression is of growing importance. Since the availability of Next-Generation Sequencing, the genetic decoding is much faster and cheaper than it ever was. It also provides new Big Data problems to handle due to the size and complexity of the data extracted. In that, topological data analysis can provide important insights to that data (such as signal periodicity or finer clustering).

Often the study of time-series can lead us to multiple unknown signal shapes. These can be systemic indicators of imperfections like noise, trending, or limited sampling density. A central problem of such nature in computational biology is to identify periodically expressed genes across different processes. The design biases of existing methods for detecting periodicity, can limit their applicability. The authors in [21] used the method of sliding window persistence tagged as SW1PerS - mentioned in Section II - to quantify periodicity in a biological time series. The measurement was performed directly, in a shape-agnostic

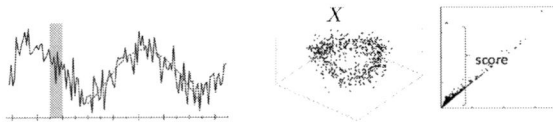

Fig. 6. Periodic and Non-Periodic signals in the synthetic gene expression data (in the upper left) ; and the sliding window persistence method applied to time series data, where the window size is w (in the upper right) [20].

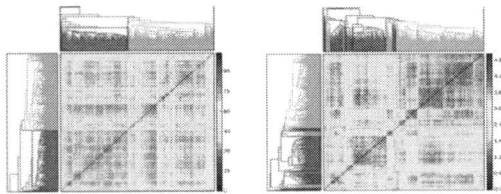

Fig. 7. Single linkage clustering of raw data relating real human sequence and phenotype data for family data set and unrelated individuals (on the left), and the clustering of that same data based on the distance between persistence diagrams that represent each of the clinical study participants [15].

manner (i.e., without presupposing a particular pattern) by evaluating the circularity of a high-dimensional representation of the genetic signal (see Figure 6). The authors concluded that in biological systems with low noise (i.e., with more frequent interestingly shaped periodic signals), SW1PerS can be used as a powerful tool in exploratory analyses. Moreover, the lists of top 10% genes ranked with SW1PerS recover up to 67% of those generated with other popular algorithms in computational biology.

The impact of the progress in the study of gene expression data in clinical applications is growing due to the also growing interest in personalised medicine. The author in [15] shows that topological data analysis can help us providing a finer single-linkage clustering of the gene expression data (represented in Figure 7) after measuring the distance between persistence diagrams rather than clustering the raw data [15]. In that, the author compares persistence diagrams representing the encoded input data for two patients. Then, all persistence diagrams corresponding to all the patients are clustered using hierarchical clustering considering the earlier mentioned bottleneck distance on persistence diagrams. In the raw data, each patient is represented by a vector of numbers extracted from gene expressions. When comparing the clustering of this raw data with Euclidean distance, and the clustering of the correspondent persistence diagrams with bottleneck distance, the latter provides better visible clusters. The results on the biological level are still to be explored.

B. Digital epidemiology

The system Influenzanet monitors online the activity of influenza-like-illness (ILI) with the aid of volunteers via the internet. It has been operational for more than 10 years, and at the EU level since 2008. Influenzanet obtains its data directly from the population, contrasting with the traditional system of sentinel networks of mainly

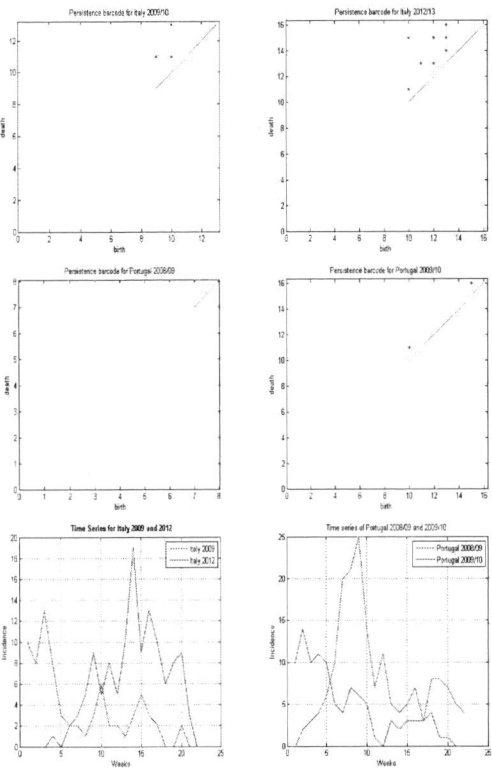

Fig. 8. Comparing the flu seasons using persistence diagrams for dimension 1 for: Italy 2009/10 (on the upper left), Italy 2012/13 (on the upper right), Portugal 2008/09 (on the center left), and Portugal 2009/10 (on the center right). These are identified as particular cases, described by the time-series comparing the seasons 2009/10 with 2012/13 in Italy (on the lower left), and the seasons 2008/09 with 2009/10 in Portugal (on the lower left) [5].

primary care physicians. Influenzanet is a fast and flexible monitoring system whose uniformity allows for direct comparison of ILI rates between countries [9].

Our goal with this study in [5] was to analyze the Influenzanet data using persistence, identifying topological features relevant to the epidemiological study. To do so, we identified data noise, distinguished higher dimension features and looked at the overall structure of the disease as well as its evolution during the flu season in Portugal and Italy. In particular, this provides a way to test agreement at a global scale arising from standard local models.

We used multidimensional scaling to identify outliers within the flu seasons analyzed in this study. TDA provides a qualitative analysis of the time series of the incidence of influenza, looking in particular at the peaks and dramatic changes. In that perspective, the time series of Italy 2009/10 and 2012/13 plotted in Figure 8 describe very different flu seasons with very different peaks. On the other hand, the flu seasons of Portugal 2008/09 and 2009/10 are identified being very close with very similar peaks, although the behavior of the curve being different. The knowledge on secondary attack rates in the influenza season is of importance to access the severity of the seasonal epidemics of the virus, estimated recently with information extracted from social media in [25]. Here lies a strong point of TDA where it can provide relevant contribution complementing other methods.

The persistence diagrams in Figure 8, correspondent to the identified flu seasons of Italy 2009/10 and Italy 2012/13, and Portugal 2008/09 and 2009/10. They encode the lifetimes of the topological features of the curves of the time series of those seasons. Persistence diagrams are a clear and practical tool that allows us the detection of outliers and to capture the qualitative features of the dynamics of the system. These ideas provide a new approach to the analysis of the seasons in the epidemiology of Influenza.

IV. CONCLUSION AND FURTHER WORK

The recent research on sliding window persistence provides a wide field of open problems that bring many challenges. This is mainly due to the novel approach of TDA, to the advances in the sliding window persistence methodology, and the further understanding of multidimensional persistence. The mathematical study of time series itself and its generalizations can profit from the contributions proposed in this project for the sliding window persistence method. Moreover, the modeling and prediction for topological time series analysis is today an area of study that was not yet tackled to date and where much can be developed, in particular with the recent progress of topological data analysis by itself and in relation to other areas of science. Being an emerging area in mathematics, computational topology can also take profit of the results of this project opening up novel areas of study with problems motivated by time series analysis.

In that, we shall focus further work on the application of this scientific progress to the study of the evolution of complex software systems. In such context, we consider as input the time series data of software failures in the several releases of software products. In Figure 9 we show as proof of concept the time series and topological data analysis of two open source software products from Eclipse community (JDT and PDE) and several releases of an industrial system from telecommunication core network product provider (MSC). This approach allows us to extend the application of the underlying mathematical models to study the behaviour of complex software systems, comparing reliability properties and existing reliability models. That will permits us to explore the application of persistent homology on reliability behaviour of software system, and to identify similarities between topological features within complex systems. Moreover, topological data analysis can prove itself useful for characterising software system behaviour early enough, and for early determination of several properties of system reliability as in [22].

Furthermore, the proposed application to more general time-based sets and their persistence structures will provide a wider framework where other data structures can be considered, sharing that same lifetime analysis flavor. This is the case of system behaviour, where time-series structures can be considered in a wider framework. In

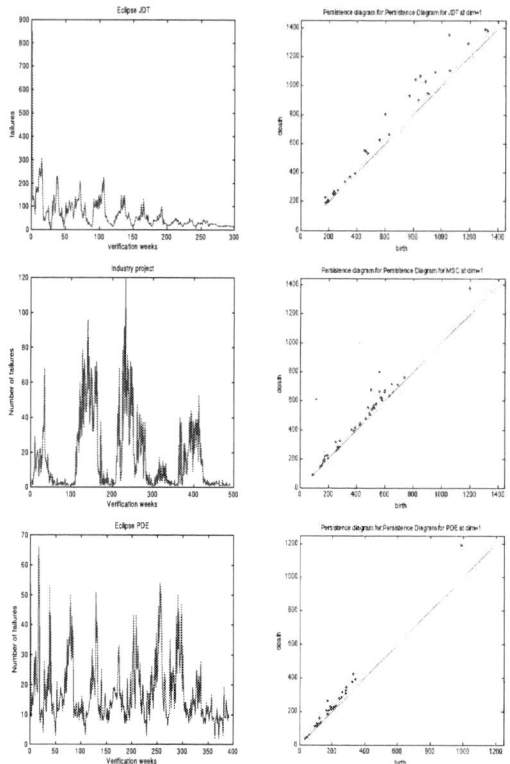

Fig. 9. Time series and persistence diagrams of dimension 1 for the topological data analysis of the software faults data in the several versions and updates of the software products JDT (above), MSC (in the center) and PDE (below) [22].

further research we will contribute to the software library RedHom, built for efficient computation of the homology of sets. This library implements algorithms implemented in C++ with bindings for python, based on geometric and algebraic reduction methods (see [14]). We will also contribute to the R package TDA which includes functions providing topological information about a given space (see [11] and [10]).

ACKNOWLEDGMENTS

The author would like to thank to: Janez Kokosar, Rafael Rosengarten and Luka Avsec for the discussions on gene expression; to Primož Škraba for the discussions on the applications of topological data analysis to time series data; to Mateusz Juda for his support with the computation of the bottleneck distance between diagrams, and the related discussions on his work about clustering gene expression data using TDA; and to Tihana Galinac Grbac for the discussions on the applications of topological data analysis to the evolution of software systems in the context of EVOSOFT. The author would also like to thank to the funding of the EU project TOPOSYS (FP7-ICT-318493-STREP) that made possible part of the research developed in this work.

REFERENCES

[1] Carina Andersson. A replicated empirical study of a selection method for software reliability growth models. *Empirical Softw. Engg.*, 12(2):161–182, April 2007.

[2] Peter Bubenik. Statistical topological data analysis using persistence landscapes. *Journal of Machine Learning Research*, 16(1):77–102, 2015.

[3] Peter Bubenik and Paweł Dlotko. A persistence landscapes toolbox for topological statistics. *Journal of Symbolic Computation*, 78:91–114, 2016.

[4] G. Carlsson. Topology and data. *Bulletin of the American Mathematical Society*, 46(2):255–308, 2009.

[5] J. Pita Costa and P. Škraba. A topological data analysis approach to the epidemiology of influenza. In *SIKDD15 Conference Proceedings*, 2015.

[6] Vin de Silva, Primož Škraba, and Mikael Vejdemo-Johansson. Topological analysis of recurrent systems. *Workshop on Algebraic Topology and Machine Learning, NIPS*, 2012.

[7] H. Edelsbrunner and J. Harer, editors. *Computational topology: an introduction*. Am. Math. Soc., 2010.

[8] Herbert Edelsbrunner and John Harer. Persistent homology-a survey. *Contemporary mathematics*, 453:257–282, 2008.

[9] D. Paolotti et al. Web based participatory surveillance of infectious diseases: the influenzanet participatory surveillance experience. *Clinical Microbiology and Infection*, 20(1):17–21, 2014.

[10] B. T. Fasy, J. Kim, F. Lecci, and C. Maria. Introduction to the R package TDA. *arXiv:1411.1830*, 2014.

[11] B. T. Fasy, J. Kim, F. Lecci, C. Maria, and V. Rouvreau. TDA package for R. *cran.r-project.org/web/packages/TDA, Accessed 18/12/2016*, 2016.

[12] R. Ghrist. Barcodes: The persistent topology of data. *Bulletin of the American Mathematical Society*, 45(1):61–75, 2008.

[13] Daniel R. Jeske and Xuemei Zhang. Some successful approaches to software reliability modeling in industry. *J. Syst. Softw.*, 74(1):85–99, January 2005.

[14] M. Juda and M. Mrozek. Capd: Redhom v2-homology software based on reduction algorithms. *Math. Software–ICMS 2014*, page 160–166, 2014.

[15] Mateusz Juda. Topological structures in gene expression data. *unpuplished work presented at the Genetic Analysis Workshop 19*, 2014.

[16] D. Kececioglu, editor. *Reliability engineering handbook, vol. 2*. Prentice-Hall, Englewood Cliffs, NJ, USA, 1991.

[17] Michael R. Lyu, editor. *Handbook of Software Reliability Engineering*. McGraw-Hill, Inc., Hightstown, NJ, USA, 1996.

[18] E. Morley-Fletcher. Big data: What is it and why is it important? *Digital Agenda for Europe. European Comission*, 2013.

[19] John D. Musa. *Software Reliability Engineering: More Reliable Software Faster and Cheaper*. Authorhouse, 2004.

[20] Jose A. Perea and John Harer. Sliding windows and persistence: An application of topological methods to signal analysis. *Foundations of Computational Mathematics*, 15(3):799–838, 2015.

[21] J. A. Perea et al. Sw1pers: Sliding windows and 1-persistence scoring; discovering periodicity in gene expression time series data. *BMC bioinformatics*, 16(1):257, 2013.

[22] Joao Pita Costa and Tihana Galinac Grbac. The topological data analysis of time series failure data in software evolution. *Proceedings of the 8th ACM/SPEC on International Conference on Performance Engineering Companion, ACM*, pages 25–30, 2017.

[23] C. Stringfellow and A. Amschler Andrews. An empirical method for selecting software reliability growth models. *Empirical Softw. Engg.*, 7(4):319–343, December 2002.

[24] M. Mrozek T. Kaczynski. Conley index for discrete multivalued dynamical systems. *Top. & Appl.*, 65:83–96, 1995.

[25] et al. Yom-Tov, Elad. Estimating the secondary attack rate and serial interval of influenza-like illnesses using social media. *Influenza and other respiratory viruses*, 9(4):191–199, 2015.

Modelling of Pedestrian Groups and Application to Group Recognition

D. Brščić[*], F. Zanlungo[**] and T. Kanda[**]

[*] University of Rijeka, Faculty of Engineering, Rijeka, Croatia
[**] Advanced Telecommunications Research Institute International (ATR), Kyoto, Japan
dbrscic@riteh.hr

Abstract - In analysing the movement and interaction of pedestrians it is very important to take into account the social groups they form. In this paper we describe the work we have done on the modelling of the spatial formations of socially interacting groups of pedestrians. We than explain the application of the obtained statistical models to the automatic detection of pedestrians who are in a group. The results show that a high detection accuracy can be achieved.

I. INTRODUCTION

Knowing the behaviour and state of people in an environment is important for applications such as security, surveillance [1,2], providing services [3] or the use of socially interacting robots and other autonomously navigating vehicles, like wheelchairs or delivery carts [4,5]. A large part of pedestrians move in groups [6] and present a specific behavioural and dynamical pattern [7]. For most of these applications it would be very useful to recognise pedestrians that move as part of groups, both for robot navigation purposes (since groups move in a different way and thus robots have to take into account such characteristic behaviour) or service and surveillance tasks (since groups exhibit specific behaviour and interests that are different from those of individuals).

Automatic detection and tracking of pedestrians using cameras [8] or range sensors [9] has recently become possible, but the automatic recognition of groups is still an open problem. In recent years a few works performing group recognition based on analysis of similarity and proximity in motion [10-12] have been proposed. The shortcoming of this approach is that it does not allow for identifying pedestrians who are actually *socially interacting* (and not just moving together), which are in general the target of the surveillance, service, and, since interacting groups have a strong tendency to keep on moving together, navigation applications. Furthermore, human observers generally do not need to examine a full trajectory to recognise interacting groups, since they can hint their nature based on visual clues such as conversation and gazing [13-15].

Although such clues are usually not available to tracking systems, the work in [7] showed that interacting groups move with a specific spatial and velocity structure, and thus their identification is possible also through a common tracking system. The works by Yücel et al. [16-19] already investigated the possibility of using pedestrian

---

This work was supported by CREST, JST.

---

pair relative positions and velocity to recognize interacting groups and distinguish them from unrelated pedestrians or human-object (e.g. carts) pairs. Nevertheless, their works 1) combined a set of observations to obtain a binary evaluation of the nature of the pair, 2) did not take advantage of the theoretical and empirical insights like the ones described in [7], and 3) operated recognition in a fixed square area.

In this work we will build on the results of [7] to obtain a system that, in an arbitrary pedestrian environment and possibly based on a single observation, may provide us with the *probability* that a pedestrian pair is part of a group or not. Having the probability instead of just a binary yes-no classification is useful because it allows for better dealing with ambiguous situations. The recognition will be based on pedestrian pairs because walking groups of size larger than 4 do not present long lasting interaction and usually split in groups of 2 or 3 [20]. Furthermore we will assume that the structure of 3-people groups may be inferred by pairwise interaction, as suggested in [7]. This assumption will be examined again in the results section.

II. PEDESTRIAN GROUPS DATASET

The pedestrian data that was used in this work was collected by tracking pedestrians during 2 days in two large straight corridors which connect the *Diamor* shopping centre in Osaka, Japan, with the railway station. The tracking was done using an automatic tracking system described in [21], which used multiple laser range finders distributed in the environment. The collected pedestrian data was then analysed by one coder, who manually labelled pedestrians that were part of a group.

The whole dataset with pedestrian trajectories and labelled groups is freely available and can be downloaded from http://www.irc.atr.jp/sets/groups/. (It also contains pedestrian data taken in a different environment, but that one was not used in this work.)

III. GROUP MODELLING

Analysis in [20] has shown that two pedestrians who are socially interacting in a walking group tend to have a stable abreast formation, i.e. they walk in such a way that the direction of movement is perpendicular to the direction towards the partner. In [7] a theoretical model for this formation was proposed, which considered the

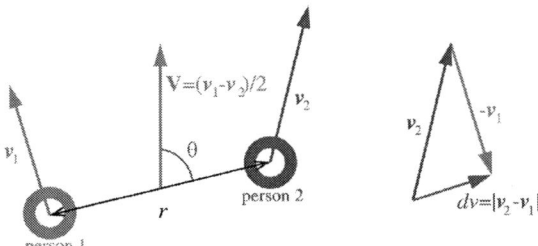

Figure 1. Definition of used observables.

necessity of members of interacting groups to keep both their walking goal and interaction partner in their field of view, and this was compared and validated against empirical observations of pedestrian groups. In the resulting model if the relative position of a pedestrian pair is studied using polar coordinates in the "centre of mass frame" (or better just the average position frame, since to each pedestrian is assigned a fictitious mass $m=1$), the spatial structure of a 2-people interacting group assumes a simple form. Following Fig. 1, we define r to be the distance between pedestrians and $\theta \in [-\pi \ \pi]$ the angle between the line connecting the pedestrians and the group velocity vector $\mathbf{V}$, equal to the mean of the velocities of the two pedestrians, i.e. $(v_1+v_2)/2$. The probability distribution for the interacting pedestrian's relative position can then be written as proportional to the following expression [7]:

$$p_{\mathrm{I}}(r,\theta) \propto r \exp(-R(r)) \exp(-\Theta(\theta)). \quad (1)$$

(Note that this also shows that r and θ are statistically independent variables.) The functions R and Θ in (1) are given by:

$$\begin{aligned} R(r) &= A(r_0/r + r/r_0), \\ \Theta(\theta) &= B(\theta - \mathrm{sgn}(\theta)\pi/2)^2, \end{aligned} \quad (2)$$

The parameters A, B and r_0 depend on the environment and pedestrian density [22-24]. For example, when the density of pedestrians is higher, people in a group will tend to stand closer to each other, which would result in a different function R. We therefore do not try to model the functions in (2) explicitly but rely on the empirical distributions computed as frequency histograms. In other words, for each environment we gather pedestrian data and for the interacting pedestrian pairs we calculate the observed probability distributions $p_{\mathrm{I}}(r)$ and $p_{\mathrm{I}}(\theta)$. The probability in (1) then becomes:

$$p_{\mathrm{I}}(r,\theta) = p_{\mathrm{I}}(r)\,p_{\mathrm{I}}(\theta). \quad (3)$$

For non-interacting groups, assuming isotropy of the environment and taking in account the finite size of the human body, we may model the probability distribution as a uniform one for $r > \delta$ (= 0.5 m), i.e. using the polar Jacobian:

$$p_{\mathrm{NI}}(r,\theta) = \begin{cases} C\,r & \text{if } r > \delta \\ 0 & \text{otherwise} \end{cases} \quad (4)$$

so that r and θ are again independent. Although the isotropy condition is quite strong, it appears to work reasonably well also for a bidirectional corridor like the *Diamor* environment, and we decided to use it due to its simplicity (there is no need of calibration or learning from

collected pedestrian data). Since $p_{\mathrm{I}}(r) \approx 0$ when pedestrians are sufficiently far away from each other, we can assume pedestrians to be non-interacting if their distance is larger than a fixed value R_{MAX} (= 3 m), which can be used to speed up the calculations.

Paper [7] studied also absolute velocity distributions, which were indeed different between groups and individuals, but their difference did not result to be strong enough for a stable recognition. We thus decided to compute, for both interacting and non-interacting pedestrians, the probability distribution of the magnitude of velocity difference dv (Fig. 1), since this value turned out to be clearly different for the interacting and non-interacting pairs:

$$dv = |v_1 - v_2|. \quad (5)$$

Using this value can be compared to the approach of [16], which used the scalar product of velocities ($v_1 \cdot v_2$). However the velocity difference allows us to recognise pedestrians moving in the same direction but at different velocities (such as one pedestrian overtaking the other one) even with a single observation. One may still wish to use the scalar product as a preliminary binary condition (i.e. if the scalar product of the pedestrian velocities is negative they are clearly non-interacting) for computational economy.

The variable dv is also assumed independent from both r and θ, so with the addition of the probability distribution $p_{\mathrm{I}}(dv)$ (which also needs to be calculated from observations of pedestrians), the total probability for interacting groups becomes:

$$p_{\mathrm{I}}(r,\theta,dv) = p_{\mathrm{I}}(r)\,p_{\mathrm{I}}(\theta)\,p_{\mathrm{I}}(dv). \quad (6)$$

Apart from non-interacting and interacting pairs (denoted below as 'non-group' and 'group' relationships), we considered also another type of pair – that of a person pushing a wheelchair, a baby buggy or pram, or a cart. Compared to the interacting pairs, this type is different in that the two people are moving one after the other. We name this a 'cart' relationship. (Although carts are not people we included them here because our tracking system did not distinguish between humans and non-humans.) The probability of belonging to a 'cart' type of group $p_{\mathrm{c}}(r, \theta, dv)$ is calculated in a similar way as for interacting groups in (6), where the corresponding probabilities are again estimated based on observed data.

Fig. 2 shows an example of the probability distributions of the observables θ, r, dv, for each of the three types of relationships we considered. They were obtained based on the collected data described in section II, except for the angle θ and radius r in the 'non-group' which were based on the theoretical model (5). There is obviously a large difference in the distributions between the 'non-group' and the other two cases, which suggest that it should be easy to distinguish between them. On the other hand the 'group' and 'cart' relationships mainly differ in the angle θ, which reflects the fact that interacting group pairs walk abreast while the 'cart' pairs move one after the other.

565

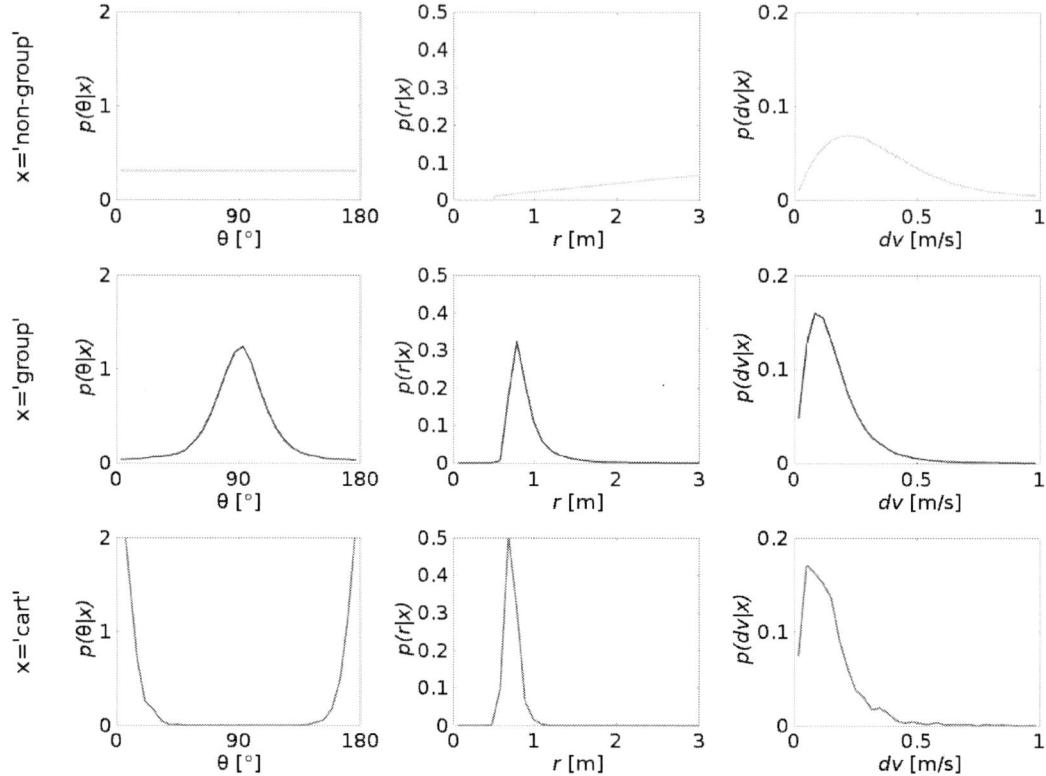

Figure 2. Learned conditional probabilities.

IV. ESTIMATION OF GROUP MEMBERSHIP

We wish to estimate the group membership probabilities, or to be more precise, the conditional probabilities $p(x|z)$, where x is one of 'non-group', 'group' or 'cart', and z is used here to denote the "measurement", i.e. $z = \{\theta, r, dv\}$. Making use of the Bayes' formula the group membership probability can be estimated as:

$$p(x|z) \propto p(x)p(z|x). \qquad (7)$$

Here $p(x)$ is the prior probability for the group membership and $p(z|x)$ is the conditional probability (likelihood) of obtaining specific values for the observables depending on the case x. (The proportionality in (7) means that the final probabilities need to be normalized so that their sum is 1.) Using the fact that θ, r, and dv are independent variables, as discussed in the previous section, we have that:

$$p(z|x) = p(\theta|x)\, p(r|x)\, p(dv|x). \qquad (8)$$

The conditional probabilities on the right-hand side correspond to the ones plotted in Fig. 2. As described in the previous section they can be obtained by collecting pedestrian data and calculating the resulting probability distributions for θ, r, and dv, separately for pedestrians not in a group, pedestrians in a group and cart-pedestrian groups. Concerning the prior probability $p(x)$, since it represents the probability of group membership without any external knowledge, a good value for it can be the observed relative frequency of non-groups, groups and

carts, which can also be obtained from the collected and labelled pedestrian data.

At each instant, based on the observed values of θ, r and dv for a specific person and people surrounding her, it is possible to use (7) and (8) to calculate the probabilities for group membership. This will be called the instantaneous estimate in the following text.

Another approach is to have a recursive Bayesian estimation, where instead of using the same fixed prior $p(x)$ at every instant one can make use of the estimated probability from the previous time step as a better representative of our prior knowledge about the probability at the current instant. In that case (7) becomes:

$$p(x_k|z_k) \propto p(x_k|z_{k-1})\, p(z_k|x_k), \qquad (9)$$

where the index k stands for the time instant. The factor $p(x_k|z_{k-1})$ in this expression is the predicted probability at step k based on the measurements up to the previous step k-1. For that we use the estimated group membership probability from the previous step. However, we reset it partially toward the default prior value $p(x)$, to reflect the fact that the we are less certain when predicting the probabilities for one step ahead:

$$p(x_k|z_{k-1}) = (1-\alpha)\, p(x_{k-1}|z_{k-1}) + \alpha\, p(x). \qquad (10)$$

Here $\alpha \in [0\ 1]$ is a factor which defines how quickly the estimator "forgets" the previously estimated value. Larger values of α correspond to quicker forgetting, and for $\alpha = 1$ the recursive estimate is equivalent to the instantaneous

Relationship		Correct classification	
		instantaneous	recursive
Non-group		96.6%	96.6%
Groups	2-person	90.6%	91.9%
	3-person	69.6%	73.0%
Cart		92%	96%

TABLE I.   RESULT OF DETECTION OF GROUPS

one. Combining (8), (9) and (10) the estimated probability can be recursively updated from step to step (recursive estimate).

Finally, in order to make a classification based on the group membership probabilities it is enough just to chose as classification result the one type of relationship x for which the estimated $p(x|z)$ is higher then the other two cases.

V.   RESULTS

To evaluate the estimation of the probability of belonging to a group we first trained the model on 20% of the dataset. The conditional probabilities plotted in Fig. 2 were obtained like that. Based on the observed data the prior probability $p(x)$ was set to the relative frequencies of non-group, group and cart pairs, which were 0.61, 0.33 and 0.05, respectively.

Table I shows a summary of the classification results. The percentage of correct results was high for pedestrians not in a group, in 2-people groups, and for people with carts. Recursive estimation (with forgetting factor in (10) $\alpha = 0.05$) gave a better classification rate than single instant estimates, even though the difference was not large.

The accuracy of classifying 3-people groups was lower. This was not completely unexpected, since they were not explicitly modelled but instead we used the model for 2-people groups also for them. As shown in [7], 3-people groups tend to walk in a V-shaped formation and the spatial relations between two adjacent pedestrians in a 3-people group are typically somewhat different than for 2-people groups. In particular, the angle θ is different and can often fall out of the range typical for 2-people groups, which is why the classification frequently failed (in almost all error cases the pedestrians were classified as not being in a group).

An example of a successful estimation of group membership is shown in Fig. 3. The trajectories of the two pedestrians that were walking along the corridor (from right to left) are shown in Fig. 3a, and the corresponding estimated probabilities of non-group and group membership using both the instantaneous and recursive method are shown in Figs. 3b and 3c, respectively (the probability for being a cart-group stayed almost 0 all the time so it was omitted to keep the figure simple). The probability for being in a group was high at almost all instances resulting in correct classification.

At the end of the analysed trajectory the pedestrians slightly changed the direction of movement which affected the spatial formation and temporarily lowered the probability of being a group. Since it went below 50% and became smaller than the probability for non-group, the pair was misclassified as non-group for a few steps. This did not happen when using the recursive Bayesian estimator, which kept the group membership probability above 50%. This shows how the memory of the previous probability makes the recursive estimate more resilient to this kind of transitional behaviours, which is the reason why it gave better classification results in total.

An example of a situation where the classification gave a result different from how the coder labelled a group is shown in Fig. 4. Here again we had a group of two pedestrians walking along the corridor (from left to right). However after a while they suddenly stopped and started moving in the perpendicular direction (up) towards a store. The results in Figs. 4b and 4c show that although they were classified as group at the beginning of the trajectory, after changing the direction of movement they were classified as no longer being a group. The reason is that, as can be seen in the figure, they did not walk in an abreast formation any more but one of them was following the other, resulting in a larger angle θ. Arguably, because of that they should no longer be considered a socially interacting group, so the estimated probabilities were actually correctly updated. Since the coder only labelled whole trajectories, cases like this where pedestrians were part of the time in a group and part of the time not necessarily lead to discrepancies between the estimate and the label and is one reason why 100% classification rates cannot be obtained.

VI.   CONCLUSION AND FUTURE WORK

In this work we presented the modelling of spatial formation of groups of pedestrians and a method for using this model for detecting pedestrians who are part of a group. The method allows us to estimate the probability that a pair of pedestrians form a group (or part of group) based on the observed spatial and dynamical relationship between them and to continuously update this estimate. The results show that the proposed method allows to accurately distinguish between group and non-group pedestrian pairs, as well as a person with a wheelchair or cart.

The results presented here are still preliminary and there are several areas of possible improvement. First, we did not attempt to deal with groups of 3 or more people in a straightforward manner, but only looked at the pairs of pedestrians inside the group. While this allowed us to obtain a fairly simple method, it ignored the fact that in 3-people groups the spatial relationships are somewhat different. As a result the classification did not work so well. This could be improved by either including the pairs inside larger groups in the learning of the model or by making a more complex model which directly uses the statistics for 3 pedestrians.

The method was tested on only one environment and it still needs to be tried in other environments as well. It would also be interesting to test if a single model can be found that works well across many environments. Moreover, one could apply the knowledge on the

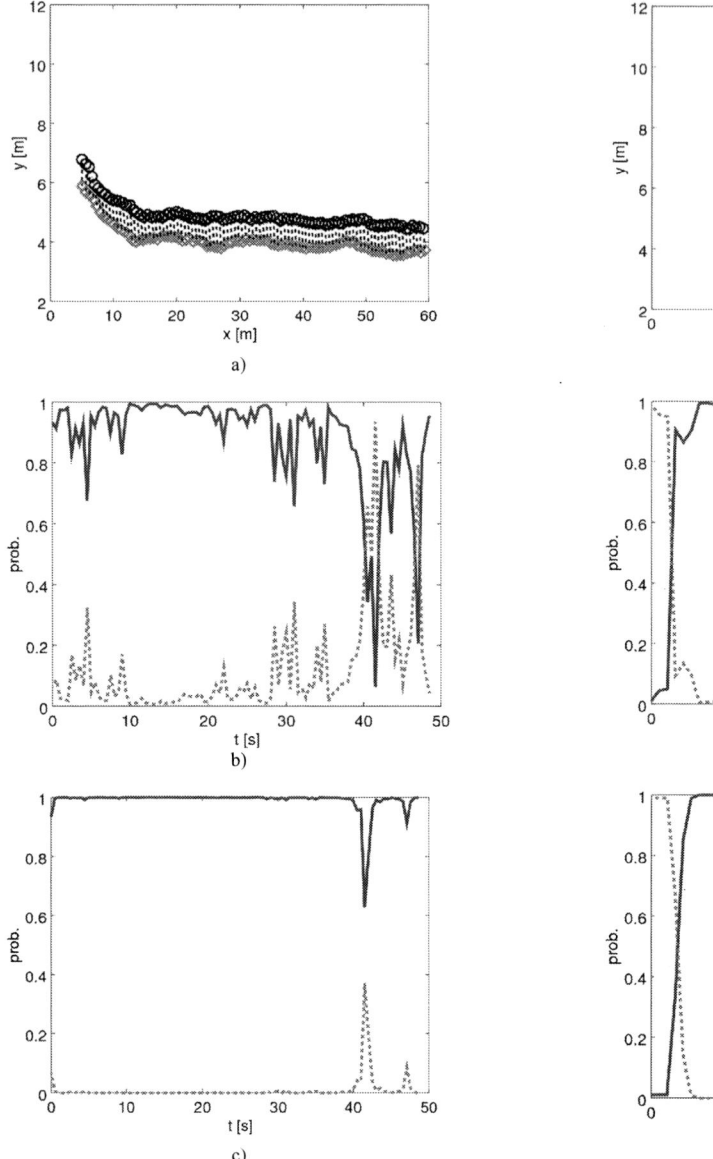

Figure 3. Example of group estimation: a) trajectories of the people in the group; b) estimation result using instantaneous estimation; c) estimation result using recursive estimation. (probabilities: 'group' - blue solid line; 'non-group' - red dashed line)

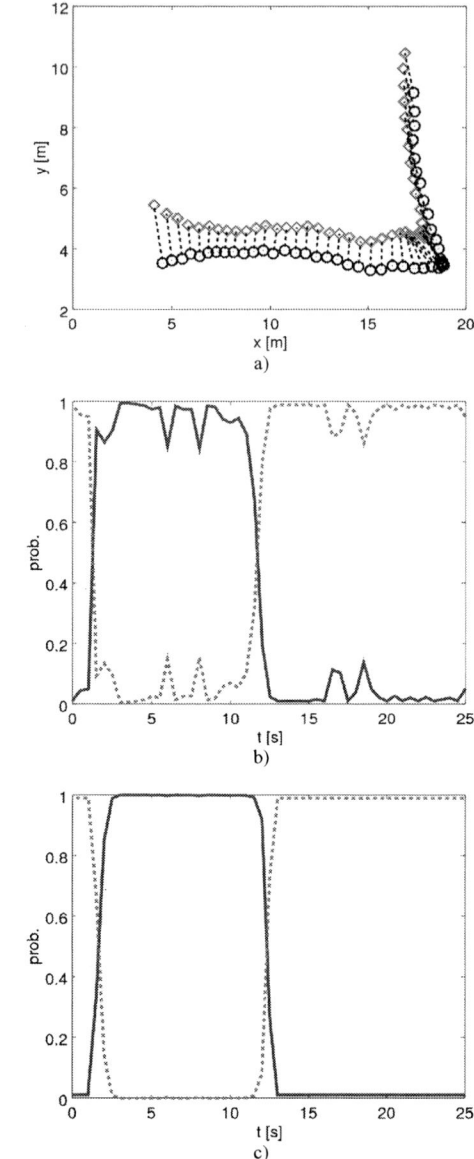

Figure 4. Example of dynamical re-estimation of group membership: a) trajectories of the people in the group; b) estimation result using instantaneous estimation; c) estimation result using recursive estimation. (probabilities: 'group' - blue solid line; 'non-group' - red dashed line)

dependence of conditions such as people density, which were studied e.g. in [22-24], to automatically adapt the model (i.e. the conditional probability densities like the ones in Fig. 2) to a specific environment and current conditions.

REFERENCES

[1] M.C. Chang, N. Krahnstover, W. Ge, Probabilistic group-level motion analysis and scenario recognition, IEEE International Conference on Computer Vision, 2011, pp. 747-754

[2] T. Yu, S.N. Lim, Monitoring, recognizing and discovering social networks, IEEE Conference on Computer Vision and Pattern Recognition, 2009, pp. 1462--1469

[3] I. Haritaoglu, M. Flickner, Detection and tracking of shopping groups in stores, IEEE Computer Society Conference on Computer Vision and Pattern Recognition, 2001, pp. I-431-I-438

[4] T. Kanda, M. Shiomi, Z. Miyashita, H. Ishiguro, N. Hagita, An affective guide robot in a shopping mall, ACM/IEEE International Conference on Human-Robot Interaction, 2009, pp. 173-180

[5] M. Shiomi, F. Zanlungo, K. Hayashi, T. Kanda, Towards a socially acceptable collision avoidance for a mobile robot navigating among pedestrians using a pedestrian model, International Journal of Social Robotics, 6, 3, pp. 443-455, 2014

[6] M. Moussaïd, N. Perozo, S. Garnier, D. Helbing, G. Theraulaz, The walking behaviour of pedestrian social groups and its impact on crowd dynamics, PLoS One, 5, 4, e10047, 2010

[7] F. Zanlungo, T. Ikeda, T. Kanda, Potential for the dynamics of pedestrians in a socially interacting group, Phys. Rev. E, 89, 1 , 021811, 2014

[8] G. Shu, A. Dehghan, O. Oreifej, E. Hand, M. Shah, Part-based multiple-person tracking with partial occlusion handling, Proc. IEEE Conference on Computer Vision and Pattern Recognition, 2012, pp. 1815-1821

[9] D. Brščić, T. Kanda, T. Ikeda, T. Miyashita, Person tracking in large public spaces using 3D range sensors, IEEE Transactions on Human-Machine Systems, Vol. 43, No. 6, pp. 522-534, 2013

[10] W. Ge, R.T. Collins, B. Rubak, Vision-based analysis of small groups in pedestrian crowds, IEEE Transactions on Pattern Analysis and Machine Intelligence, 34, 5, pp. 1003-1016, 2012

[11] V. Bastiani, D. Campo, L. Marcenaro, C. Regazzoni, Online pedestrian group walking event detection using spectral analysis of motion similarity graph, 12th IEEE International Conference on Advanced Video and Signal Based Surveillance, 2015, pp. 1-5

[12] C. Stephen, Dynamic phase and group detection in pedestrian crowd data using multiplex visibility graphs, Procedia Computer Science, 53, pp. 410-419, 2015

[13] M. Knapp, Nonverbal communication in human interaction, Cengage Learning, 2012

[14] C. Kleinke, Gaze and eye contact: a research review, Psychological bulletin, 100, 1, 78, 1986

[15] M. Argyle, J. Dean, Eye-contact, distance and affiliation, Sociometry, pp. 289-304, 1965

[16] Z. Yücel, T. Ikeda, T. Miyashita, N. Hagita, Identification of Mobile Entities Based on Trajectory and Shape Information,

IEEE/RSJ International Conference on Intelligent Robots and Systems, 2011, pp. 3589-3594.

[17] Z. Yücel, F. Zanlungo, T. Ikeda, T. Miyashita, N. Hagita, Modeling Indicators of Coherent Motion, IEEE/RSJ International Conference on Intelligent Robots and Systems, 2012, pp. 1-8.

[18] Z. Yücel, T. Miyashita, N. Hagita, Modeling and Identification of Group Motion via Compound Evaluation of Positional and Directional Cues. 21st International Conference on Pattern Recognition, 2012, pp. 1-8.

[19] Z. Yücel, F. Zanlungo, T. Ikeda, T. Miyashita, N. Hagita Deciphering the Crowd: Modeling and Identification of Pedestrian Group Motion, Sensors, Vol. 13, pp. 875-897, 2013

[20] F. Zanlungo, T. Kanda, Do walking pedestrians stably interact inside a large group? Analysis of group and sub-group spatial structure, Cognitive Science Society Conference, 2013

[21] D. Glas, T. Miyashita, H. Ishiguro, N. Hagita, Laser-based tracking of human position and orientation using parametric shape modeling, Advanced robotics, Vol. 23, No. 4, pp. 405-428, 2009

[22] F. Zanlungo, D. Brščić, T. Kanda, Pedestrian group behaviour analysis under different density conditions, Transportation Research Procedia, 2, pp. 149--158, 2014

[23] F. Zanlungo, D. Brščić, T. Kanda, Spatial-size scaling of pedestrian groups under growing density conditions, Physical Review E 91 (6), 062810, 2015

[24] F. Zanlungo, T. Kanda, A mesoscopic model for the effect of density on pedestrian group dynamics, Europhysics Letters, 111, 38007, 2015

eCST to Source Code Generation - an Idea and Perspectives

N. Rakić*, G. Rakić **, N. Sukur** and Z. Budimac**
* Lucid Code Industries, Novi Sad, Serbia
** University of Novi Sad, Faculty of Sciences, Novi Sad, Serbia
nrakic90@live.com, {goca, nts, zjb}@dmi.uns.ac.rs

Abstract - eCST (enriched Concrete Syntax Tree) is introduced as a fundament of SSQSA (Set of Software Quality Static Analyzers) platform for consistent static quality analysis across the input languages. It is based on the concept of enrichment of the complete syntax tree representing the input program by universal nodes. Universal nodes are based on the idea of imaginary nodes in an abstract syntax tree, but unified, so that one single node is used for all languages where it is applicable. In this paper, we describe a translation of eCST back to source code. At this moment, this is only translation to the original language in which code is written. Moreover, the translation of eCST to a code written in the original language can have a wide spectre of applications such as in (semi-)automated code refactoring and transformations.

I. INTRODUCTION

Nowadays, when software is rapidly developing and changing and the time for its delivery is quite short, the quality of that product is seldom high. Therefore, it is necessary to have some automated approach that will focus on the flaws in the design, efficiency, usage of resources and other important software characteristics.

When software metrics are applied, it shows us characteristics of the analysed software in numeric values. It is a very important practice in the process of software development - if these values are not within certain boundaries, it is a sign that something needs to be changed. In many cases, source code needs to be refactored [1][2] due to lacks in its structure, such as bad design decisions, copy-paste actions (code clones), inappropriate level of coupling and cohesion, and similar.

Furthermore, software evolves continuously and in order to follow this trend properly, it is often necessary to make changes to it. Reengineering is a complex process which consists of reverse engineering, performing changes, and forward engineering [1]. It can be demanding, but in order to keep a project "alive", it is an activity that usually needs to be performed from time to time. It enables us to fix all upcoming and possible errors in their early stages, while it is still not so costly.

Software projects are mostly not simple programming solutions, but rather very complex and they consist of a large number of programming languages and technologies. Maintenance of these projects requires an attention to their quality where abiding by the guidelines that were mentioned is crucial [3]. In this process, it is advisable to have a tool which will provide necessary analyses that are consistently applicable to all programming languages that were used [2][4][5].

In most cases, reengineering process is performed on internal representations of the source code, which means placing source code elements in some special structures such are trees, graphs, meta-models and similar [6][7][8]. The most commonly used structure is the tree, which can also vary on the level of abstraction, i.e. the amount of information extracted from the original source code. The idea behind using these internal representations is enabling automation or at least semi-automation in the reengineering process. That is why it is necessary to have a tool (or a set of tools) which will provide this kind of functionality for us [1][2].

SSQSA [9] is a platform which is created to integrate tools for software quality analysis. Its main goals are focused on consistency in source code analysis, reached by an innovative intermediate representation, enriched Concrete Syntax Tree (eCST) [9]. SSQSA is extending its functionalities to perform and monitor code changes, such as reengineering and code refactoring, not only to point out errors, but also to help the analysed software evolve. It is language independent and the main benefit of such approach is the ability to uniformly analyse and improve software systems upon flaw detection, regardless of how many languages their components are implemented in.

Prior to manipulating eCST or any other intermediate representation, it is necessary to generate it. A general recommendation in SSQSA is to use generated parsers with inbuilt mechanism to build uniform trees for all languages. Our first choice is ANTLR parser generator (ANother Tool for Language Recognition) which provides very intuitive notation for modifications of the trees to be generated [10].

This paper describes the possibility of transforming eCST back to source code. It describes general guidelines to follow in order to have eCST with all characteristics required to achieve successful translation. This is important as an illustration and the first step to a more general reengineering and refactoring process: read the code, refactor it in eCST (which will be implemented only once for all supported languages), and put it back into source code.

Paper outline: In the next section we will introduce basics of ANTLR parser generator which importantly facilitated our task. Afterwards, we introduce SSQSA platform and its internal representation, eCST and illustrate how translation will be used in the reengineering and refactoring process. Section 4 describes the basic principles to follow in order to achieve all characteristics of the eCST required for the successful translation and the translation itself. Section 5 briefly describes the related work. Finally, in the last section we conclude this research and describe plans for the future work. The eCST was not complete and the only thing that was missing were comments. They could be used, for example, to indicate existence of bad smells and it was necessary to include them into eCST.

II. ANTLR

ANTLR[1][2] is a tool that automatizes generation of language processing tools, i.e. lexers and parsers. These tools recognize, analyse and transform input code, based on a grammar of the language that we have previously defined. ANTLR provides automation and flexibility in work with a variety of languages, their grammars and translation rules. All the necessary corrections regarding the structure and the content of the resulting tree are specified in the grammar [10].

Based on a language grammar, ANTLR generates corresponding language lexer (scanner) and parser. Lexers take a byte stream as an input and provide a stream of tokens, based on rules defined in the grammar. Aside from generating specified tokens, they are able to place some of them, such as whitespaces, in a separate channel that can be marked as hidden. Parsers later mostly use the main token stream, usually provided by lexers, to match phrases using the rules that are specified and performs some semantic actions for those phrases. At the end of these phases, it is common to generate syntax trees, which are useful for performing additional processing.

Notation used in rule expressions is very similar to EBNF (Extended Backus–Naur form) notation [11]. Basically, a grammar rule consists of three parts: rule name, parsing rule expression and tree rewriting rule expression. A rule notation is as follows.

```
ruleName:
        parsing rule expression
                -> tree rewriting expression
```

The third part of the rule is optional. If nothing is specified there, the resulting tree will be generated in the default structure, without adding new nodes or omitting any of the existing ones. However, some basic manipulations on the tree can be done even in the second part of the rule. For example, "!" is used to exclude some nodes from the tree, while "^" is used to define a parent-children order in the subtree. Still, some modifications of the tree, e.g. insertion of additional (imaginary) nodes, are possible only in the tree rewriting expression.

ANTLR parser generator is not the only available option for generating parsers that can create and walk

---

[1] http://www.antlr.org/

[2] The version of ANTLR which is used is ANTLR 3, since it provides simplicity in enriching the resulting tree

---

parse trees. They could also be generated manually. This approach is recommended for integration of a new language in SSQSA framework to reduce the time and effort required to complete this process.

III. SSQSA CONCEPTS AND PRINCIPLES

Set of Software Quality Static Analyzers (SSQSA) is a platform for building a quality analysis framework. It is focused on consistency of static software analysis based on independency of the input language. The idea behind it is the following: the source code of one of the supported languages is translated into an enriched Concrete Syntax Tree (eCST). eCST is a universal intermediate representation created by using a predefined set of nodes, called universal nodes, to enrich a complete concrete syntax tree. Enriched Concrete Syntax Tree (eCST) is an intermediate structure to which all the source code is translated to. The whole concept of SSQSA universality and language independence is based on it.

A. Enriched Concrete Syntax Tree (eCST)

Syntax trees are frequently used for source code analysis. SSQSA defines its own special kind of trees, a combination of abstract syntax trees and concrete syntax trees, which differ on the abstraction level of the available data.

Enriched Concrete Syntax Tree (eCST) [9] is a syntax tree that contains all the concrete tokens. It is enriched by universal nodes. The universal nodes are imaginary nodes, which are added in the phase of generating abstract syntax trees. The idea behind using them is to denote the semantic of a concrete syntax in its subtree. These nodes are used to represent all the necessary source code elements. Set of these universal nodes is as minimal as possible and it is the same for all supported languages. For example, LOOP_STATEMENT node is used to represent all kinds of statements that imply repetition of execution, such as for, loop, while, do-while... The way this set of nodes was selected was by observing some characteristics that are in common for most of the programming languages.

eCST universal nodes are parents of the subtrees with the specific meaning. Information about this meaning is in the universal node. The subtree contains other subtrees and the concrete source code elements, which are usually the leaf nodes. That gives us the exact information that we need for the source code reconstruction, since universal nodes tend to generalize parts of code (i.e. otherwise we would not know if LOGICAL_OPERATOR node represents || or && or any other kind of logical operators).

B. SSQSA Components

SSQSA components have different roles. The first kind of components is taking care of providing necessary structures for future analysis. The eCST Generator has a crucial role in this task, since it generates eCST from source code. Furthermore, there is a set of generators of derived internal representation, such as eGDN (enriched General Dependency Network) and eCFG (enriched Control Flow Graphs). These generators are only implemented in order to provide structures for easier further analysis. They are also language independent, since they create internal representations by extracting necessary information already contained in eCST. The other type of components are analysers, the ones operating on these structures. They are calculating software metrics, detecting code clones, and performing many other calculations and measurements concerning software quality. Both kinds of these components are implemented only once, independently of input language. The way components communicate is by exchanging XML files.

C. Extendability and adaptability of SSQSA

The meaning of extendability in terms of SSQSA was easy support of new analyses by implementing a certain algorithm on one of the existing intermediate representations or adding another language [9].

The advantage of the approach that is used in creating and developing SSQSA is the possibility to develop components exactly once. Newly introduced analysers are applicable to all supported languages because they operate either directly on eCST or indirectly, on eCST-based representations, and not on the source code itself.

Apart from that, adaptability of SSQSA reflects the possibility to add a support for a new input language. If a new language is introduced to SSQSA, source code written in this language can easily be translated to eCST and analysed by the existing analysers.

The extension of SSQSA in the sense of adding a new analyser or introducing a new language always meant adding exactly one implementation for all analysers to support all languages or for all languages to be analysable by all analysers. That is because eCST serves as a mediator between them. Therefore, it is clear that eCST plays an important role in the concept of extendability and adaptability in SSQSA.

For the purpose of new implementations, one might find some pieces of information missing. In the case of writing eCST back to its original source code without omitting any information, it meant including elements missing in the eCST, e.g. comments.

D. Transformations

Additional transformations could be performed on eCST before returning it to the original source code. These transformations would be unique and the source code could be generated back from the transformed tree. Some of these transformations could be implemented for the purpose of refactoring. Generally, there are many formalisms used for refactoring, such as assertion, graph transformation [1][12][13] and software metrics [1]. In our approach, the following should be done to refactor the code:

1. generate eCST from source code

2. transform eCST using implemented, language independent transformations

3. generate source code from the transformed eCST

We can illustrate this process by a very simple example. Let us assume that we want to transform the following fragment of source code:

```
a = b + x*y; //to be factored
```

We want to split the complex expression at the right-hand side of the assignment operator to simple expression as follows:

```
/*factored*/
z = x*y;
a = b+z;
```

To meet this goal automatically, we first generate the eCST (Fig. 1).

Afterwards we transform the tree according to our goal – simplifying the expression. This means creating an additional assignment statement, moving multiplication part to this subtree and replacing this part of the original subtree with the variable that gets the value of this multiplication. In some languages, it will be required to declare the new variable first, and this task will involve some learning algorithm. It is important to notice that this transformation does not require only moving the tree nodes, but also intervention on numbers of lines and columns in the concrete token nodes. Universal nodes have generic values for these attributes as they do not figure in the source code. Result of these transformations is given in Fig. 2.

Finally, we generate the code from the eCST. We can notice that, apart from transformed expressions, we also changed the comment. Now it is block comment, moved in front of the code fragment which is also done as a part of the eCST transformation.

The external tool eCSTAntiGenerator serves to generate source code from eCST. By inspection of the implementation it was established that the code generation works independently of the source code. It is based only on the information about position of the tokens in the source code. This means that the same tool can be used for all combinations of programming languages that are supported by the SSQSA environment. However, in order to enable eCSTAntiGenerator to successfully generate code from eCST, it was necessary to place all concrete code elements, including comments, in the tree.

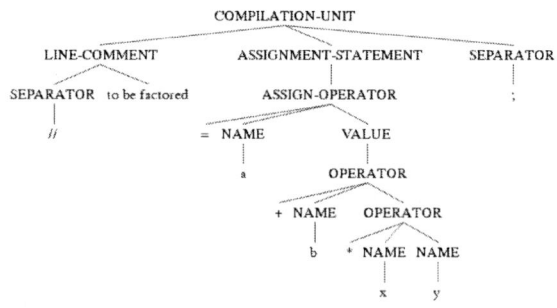

Figure 1. eCST which is generated from the original source code before the transformation

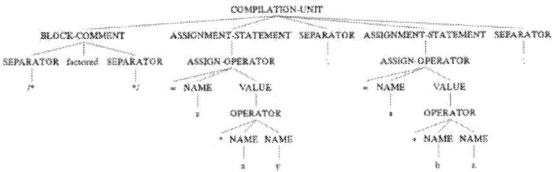

Figure 2. Illustration of the resulting eCST after having performed the transformation

IV. COMPLETENESS OF THE ECST

Most of intermediate language representations (including earlier version of eCST) contain only those data from the source code that are important for the purpose of the representation (e.g., compilation, measurements, transformations, etc...). Thus, most of the intermediate representations do not contain comments but also various kinds of concrete code elements such as separators, etc., depending on the purpose. For example, one of previous versions of eCST for the previously introduced example (the original version before transformation) would look as illustrated by the Fig. 3. In the previously implemented analyses we completely omitted all whitespaces including comments, but also separators. For purposes of Halstead metrics calculation, we included separators, while now, for code generation, we need all information contained in the source code to be represented by eCST. This is the main rule to be followed when generating the eCST. Additionally, we have a set of rules for enriching the tree by universal nodes, but these are irrelevant in this context.

In order to retain the completeness of the eCST we have to follow certain rules already while writing a language grammar (or parser if developed manually). There are two main rules to be followed:

(1) Keep all concrete tokens in the tree.
(2) Stay aware of whitespaces.

Practically, there are different approaches to follow them, and different aspects of the rules when a certain approach is chosen. We will focus on writing ANTLR grammar as it is the recommended choice.

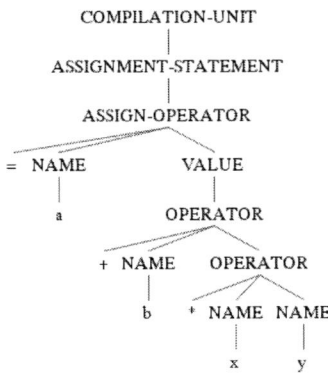

Figure 3. The structure of earlier version of eCST before introducing comments and separators

A. Program syntax completness

We have to consider two situations to keep all concrete tokens in the tree.

a) *The rule does not contain the tree rewriting rule expression (the right-most part).*

These rules are to be left as they are in a sense that nothing should be excluded from the tree (by using the "!") However, restructuring the tree without omitting anything is allowed. Example for expression list rule and the accompanying expression (Modula-2 grammar) follows:

```
expList :
    expression (comma expression)*;
expression :
    simpleExpression
    (relation^ simpleExpression)? ;
```

In some trees, the separator "comma" in "expList" rule, would be omitted, but it is not in eCST. Furthermore, we did not do any restructuring if this subtree as all expressions should appear listed at the same level in the tree. In the "expression" rule we, again, do not omit anything but we build the subtree whose parent is a relation and children are simple expressions participating in this relation.

This is an ANTLR specific step and it means that the resulting tree will be automatically built with changes or interventions related to the content of the tree. In some other compiler generators, rules should be set in such a way to explicitly build the resulting tree (or any other intermediate representation).

b) *The rule does contain the tree rewriting rule expression.*

The rewriting rules should only restructure (i.e., "pack") the input language structure into an eCST and universal nodes should be built in it (following the certain rules). The following example illustrates one case, where the Modula-2 language construct for assignment statement is transformed into an appropriate universal eCST node.

```
Assignment :
    qualidentWithTail ':=' expression
    ->
    ^(ASSIGNMENT_STATEMENT
        ^(ASSIGN_OPERATOR ':=')
        qualidentWithTail
        ^(VALUE expression)
    ) ;
```

B. Whitespaces awareness

Apart from the abovementioned rules to preserve all language elements, whitespaces are usually a separate problem. There is usually no need to represent whitespaces in the intermediate program representations, because they are typically of no value for any of possible analyses. On the other hand, the definition of whitespaces in the language grammar can be cumbersome and is usually done on the level of the scanner that will typically ignore them. To define whitespaces in ANTLR (and similar compiler generators) we are using 'channels'.

Channel is the filter that defines which tokens will be directed to which specific parser. By default, all 'important' tokens are sent to the main parser, while tokens that constitute whitespaces will be sent to other channels to be handled separately, usually ignored.

In our context, the only whitespaces of importance are comments. Other whitespaces (spaces, new lines, etc.) can be closely reconstructed from the positions of tokens around them, while comments should be processed separately. All comments can be divided in three main categories: block comment, line comment and documentation comment. Accordingly, we introduce three universal nodes to denote comments.
BLOCK_COMMENT, representing comments that can take more than one line.
LINE_COMMENT, representing one line comments
DOC_COMMENT, representing comments that are used for documenting program.
Comments belonging to these three different types will be directed to three different channels: 11, 12, and 13, respectively. The following example shows definitions for block and line comments in Java, while documentation ones are very similar to block comments:

```
blockComment :
    blockCommentStart
        ( options {greedy=false;} : . )*
        blockCommentEnd
        { $channel = 11; };
lineComment :
    lineCommentStart
        ~('\n'|'\r')* '\r'? '\n'
        { $channel = 12; };
```

Directing comments to separate channels will enable their collecting, but they are still not included in the generated eCST. We still have to process them and to postprocess the tree in order to integrate corresponding subtrees to the eCST. This process is illustrated by Fig. 4.

1) Processing the comments
Once the comments are defined according to their types and sent to appropriate channels/parsers, they should be merged in eCST tree that contains the program code. To do so, one could intercept the communication between lexer (scanner) and parser. The usual relationship between lexer and parser (in ANTLR) is the following:
The lexer is initialized for the provided input. Then the common token stream is created based on it, and passed to initialize the parser based on it. The parser iterates the common token stream, using the iteration method implemented in the lexer, and processes tokens. To influence this usual flow of tokens between lexer and parser and to intercept tokens in between, one needs to implement an alternative token collector which will be able to process also the alternative streams. The alternative token collector should be able to recognize, save, and manipulate the tokens collected from the predefined channels. This means to be able to access to the initialized default token collector and to override the token iteration method. Additionally, it should contain the map of channels and token types that are expected at those channels and implement all operations needed to integrate

streams. Effectively, when token given by the token iterator is the one whose channel belongs to the set of predefined set of channels (in our case 11, 12, 13), the token is saved in internal collection of the alternative token collector. Afterwards, collected tokens are packed in subtrees of corresponding universal nodes of eCST, depending on the channels from which they were collected. Now we have collected all the tokens, including comments but in a separate collection.

2) Postprocessing eCST to integrate the comments
The remaining problem is how to put collected comments to a proper place in the eCST tree. Architecturally it is the best to do so in a separate component in the form of postprocessor[3].

The proper place for the comment will be defined as a child of lowest common ancestor of two tokens, one of which is immediately before and the other one immediately after the comment. In other words, the position of the comment in a subtree of lowest common ancestor should be between ancestors of tokens that are around the comment in the source code. The outline of the algorithm is the following:
- Find the preceding token;
- Find the succeeding token;
- Find the lowest common ancestor;
- Find the proper position in a subtree with respect to above-mentioned principles;
- Insert the comment to the found proper position.
After this process, we have a complete eCST and we can perform translation.

V. RELATED WORK

The motivation for reengineering has not been present only in SSQSA. Research on a language-independent meta-model for modelling object-oriented software for the purposes of reengineering was described in [14], later used in **MOOSE** platform[4]. The advantage of SSQSA, in this case, is that it is not limited to a specific family of languages – it supports object-oriented, procedural and functional languages. **Rigi** [15] is a reverse engineering environment that analyses source code and can visualize software networks. However, it is also not language independent. **Bauhaus** [16] is a tool suite which is also oriented towards reverse engineering on different levels of abstraction. Although it also does not operate directly on the source code and uses special representations, it is not fully language independent and falls behind SSQSA in that way. According to [9], SSQSA is the only platform which stands out with its consistency, because of a consistent representation and consistent way of translation to that representation.

VI. CONCLUSION

Reading the source code, transforming it to intermediate representation and then writing it back in the same or analogous form, proved to be important as a pre-requisite to source code transformations, refactoring, etc.,

---

[3]SSQSA architecture has components involved in manipulating eCST. One of them is eCST Adaptor, a specific eCST post-processor.
[4] The MOOSE book, 2014 http://www.themoosebook.org/book

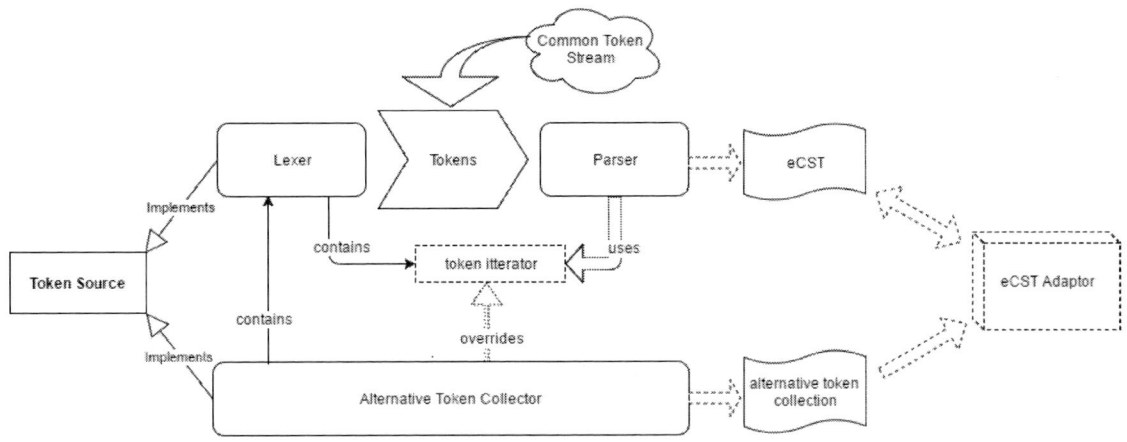

Figure 4. The flow and logic behind intercepting comments and including them into eCST

and not so easy. For this task it is also important to preserve the comments. In this paper we showed how it can be done using eCST, ANTLR, Modula-2 and Java as case studies. We understand that results described in this paper are the first step that will lead to a more general tool which will perform already mentioned tasks. By using eCST as an intermediate representation, we will have the possibility to implement code transformations only once and then it will be applicable to all supported languages.

Also, it is possible to use this converter for the purpose of developing new features in SSQSA. They could involve some transformations, such as refactoring. This feature would be useful in everyday software development, but also in improving legacy code and optimizations. Since principles of refactoring are similar in most of the languages, a logical step is to try to integrate them in SSQSA by creating single implementations performed on eCST, rather than on each language separately.

Without tree transformations, it could also be possible to perform translation, i.e. to generate eCST from source code written in one language and later transform eCST to source code of another one, but it would have to involve much more research.

ACKNOWLEDGMENT

The work is partially supported by Ministry of Education and Science of the Republic of Serbia, through project no. ON 174023: Intelligent techniques and their integration into wide-spectrum decision support.

The authors also gratefully acknowledge Igor Manojlović, who took part in implementation of eCSTAntiGenerator.

REFERENCES

[1] Tripathy, P. and Naik, K., 2014. Software evolution and maintenance. John Wiley & Sons.

[2] Mens, T., Wermelinger, M., Ducasse, S., Demeyer, S., Hirschfeld, R. and Jazayeri, M., 2005, September. Challenges in software evolution. In Principles of Software Evolution, Eighth International Workshop on (pp. 13-22). IEEE.

[3] Godfrey, M.W. and German, D.M., 2008, September. The past, present, and future of software evolution. In Frontiers of Software Maintenance, 2008. FoSM 2008. (pp. 129-138). IEEE.

[4] Tonella, P., Torchiano, M., Du Bois, B. and Systä, T., 2007. Empirical studies in reverse engineering: state of the art and future trends. Empirical Software Engineering, 12(5), pp.551-571.

[5] Stevens, P. and Pooley, R., 1998, November. Systems reengineering patterns. In ACM SIGSOFT Software Engineering Notes (Vol. 23, No. 6, pp. 17-23). ACM.

[6] Bhattacharya, P., Iliofotou, M., Neamtiu, I. and Faloutsos, M., 2012, June. Graph-based analysis and prediction for software evolution. In Software Engineering (ICSE), 2012 34th International Conference on (pp. 419-429). IEEE.

[7] Le Gear, A., 2004. Thematic Review of Software Reengineering and Maintenance. Technical Report UL-CSIS-04-3, University of Limerick, Plassy, Castletroy, Co. Limerick, Ireland.

[8] Kagdi, H., Collard, M.L. and Maletic, J.I., 2007. A survey and taxonomy of approaches for mining software repositories in the context of software evolution. Journal of software maintenance and evolution: Research and practice, 19(2), pp.77-131.

[9] Rakić, G., 2015. Extendable and Adaptable Framework for Input Language Independent Static Analysis.

[10] Parr, T., 2007. The Definitive ANTLR Reference. Pragmatic Bookshelf, p.384.

[11] 'ISO/IEC 14977:1996 Information Technology -Syntactic Metalanguage - Extended BNF', 1996. International Organization for Standardization, ed.,.

[12] Mens, T., Van Eetvelde, N., Janssens, D. and Demeyer, S., 2003. Formalising refactorings with graph transformations. Fundamenta Informaticae, p.69.

[13] Mens, T. and Tourwé, T., 2004. A survey of software refactoring. IEEE Transactions on software engineering, 30(2), pp.126-139.

[14] Tichelaar, S., 2001. Modeling object-oriented software for reverse engineering and refactoring (Doctoral dissertation, PhD thesis, University of Berne).

[15] Kienle, H.M. and Müller, H.A., 2010. Rigi — An environment for software reverse engineering, exploration, visualization, and redocumentation. Science of Computer Programming, 75(4), pp.247-263.

[16] Raza, A., Vogel, G. and Plödereder, E., 2006, June. Bauhaus–a tool suite for program analysis and reverse engineering. In International Conference on Reliable Software Technologies (pp. 71-82). Springer Berlin Heidelberg.

Design and Development of Contactless Interaction with Computers Based on the Emotiv EPOC+ Device

B. Šumak*, Matic Špindler* and M. Pušnik*

* Faculty of Electrical Engineering and Computer Science, University of Maribor, Slovenia
bostjan.sumak@um.si, matic.spindler@gmail.com, maja.pusnik@um.si

Abstract - Devices and methods for interaction with a computer do not change very often. However, with technology advancements, solutions for new ways of human-computer interaction are possible. Especially for physically disabled users, solutions for contactless interaction with a device promise new possibilities for independent computer use and management. In this study a solution for a contactless interaction with a computer was proposed based on a brain-computer interface (BCI) device Emotiv EPOC+. The EPOC+ device is able to detect brain signals, facial expressions or levels of engagement (e.g. frustration, mediation, etc.), eye movement and head position through the built-in gyroscope sensors. The contactless interaction solution proposed in this study was based on (1) mapping facial expressions into user interface commands, (2) head-movement detection for controlling the mouse cursor, and (3) a virtual keyboard, which was developed to enable text input. The proposed solution was tested in a quasi-experimental setting, where non-disabled users had to complete several computer tasks. After completing the tasks, perceived usefulness and perceived ease of use of the proposed solution were assessed. At the end, limitations and future work are discussed.

I. INTRODUCTION

It is estimated that there is over a billion of people that are living with disability [1]. Disability is a complex and multidimensional human condition that almost everyone is going to experience at some point in life. Disability is a general term for different types of impairments. It is expected that almost everyone will get temporary or permanently impaired and older people are going to experience increasing difficulties in functioning [1].

In this paper we are addressing people with motor disabilities especially. The Slovenian Paraplegic Association [2], founded in 1969, is a major national organization that represents and protects paraplegics and quadriplegics and enforces their rights and interests. The most common causes of disability traffic are accidents, falls from heights, jumping into water, spinal cord surgery, etc. In 2015 there were 1049 people registered in the organization, which is approximately 0.051% of population living in Slovenia. Of these, more than half were people diagnosed with paraplegia and a third with tetraplegia [3].

Person's quality of life, especially in case of persons with motor impairments, can be affected by different factors such as physical health, psychological state, level of independence, social relationship, etc. [4]. Motor impaired users are mostly not able to use the standard computer interfaces (e.g. mouse, keyboard) due to their health condition preventing them to use one or both hands. By being able to use a computer, motor impaired users can gain in their quality of life through the access to different online services, such as communication, education, socialization, e-commerce, e-banking, etc. The ability to control computers and other devices needed for accessing such services has become a necessary part of our lives in order to be able to live modern life [5].

Latest advancements in different fields such as electronics, computers, wireless technologies, signal processing and classifications have enabled development of new non-invasive or user-friendly wireless devices, which can be used for implementing innovative solutions for human-computer interaction (HCI), providing capabilities for computer controlling through speech recognition, brain activity, eye-tracking, etc. Such accessories and solutions for assistive technology (AT) have enabled to design and implement new ways for controlling computers with hands-free solutions [5].

People with motor impairments are confined within a home environment, unable to use devices like TV, lights, computers, kitchen accessories, etc. This study is focused on designing an AT solution, which will enable motor impaired users to independently control and use computers based on a brain-computer interface (BCI) device. The BCI device provides users ability to control a computer by using brain activity only [6]. The BCI devices can be categorized based on their degree of recording invasiveness of the electroencephalography (EEG) method into one of the three categories [7]: invasive (scalp EEG recording), partially or medium invasive (electrocorticography), and highly invasive (intracortical recording).

The BCI is one of the most growing fields of interdisciplinary research, involving disciplines such as neuroscience, computer science, engineering and clinical rehabilitation [8]. In the past few years several commercial BCI headsets were launched, which accelerated the research and development of this technology for various home environment settings. Recent advancements in the field of EEG technology and BCI have offered acceptable quality-to-cost ratio and easy-to-use, out-of-the-box equipment with commercial BCI headsets, which can be used for variety new applications [9].

The rest of the paper is organized as follows. In next section, existing research based on the Emotiv EPOC+ device is summarized. Section that follows, describes the Emotiv EPOC+ device and corresponding software packages that were used in this study for developing the solution for the contactless interaction with the computer. In section four, the proposed solution is presented. Results of a pilot study, which aimed to test the proposed solution, are discussed in section five. The last section concludes the paper together with directions and ideas for future research.

II. RELATED WORK

Results of several years of research are promising with solutions with BCI applications such as BCI systems for environmental control [10], speller systems [11,12], interfaces for robotic wheelchair steering [13–15], gaming [16–18], etc. In existing literature, several studies were published reporting results of evaluation of the Emotiv EPOC+ device and its application. For example, the EPOC+ device was used for designing and developing a BCI-based system for generation of synthesized speech, which works on eye-blinks detected from EEG signals [19].

The EPOC+ device was also used for improving a speech interaction system by integrating the BCI device for the purpose of associating the brain signals with commands for controlling wheel chairs for parallelized people [20]. A study conducted by Vourvopoulos and Liarokapis demonstrated, that that commercial BCIs like EPOC+ can be used for effective and natural robot navigation both in the real and the virtual environments [8].

Motor imagery is also an interesting topic in BCI research, which examines brain activity when imagining motoric activities such as moving the left hand. EPOC+ device could be used for acquiring brainwaves, however a recently published study showed that the device is not recommended for implementing motor imagery application [21].

In the field video games, the EPOC+ device can be used for assessing user experience and cognitive processes caused by various stimulus modalities and gaming events while playing video games producing [22]. As demonstrated in [23], the EPOC+ device can also be used for analyzing how the frontal lobe of the brain (executive function) works in terms of prominent cognitive skills during games playing.

To the best of our knowledge, this study is a first attempt to develop and evaluate a solution for contactless interaction with a computer based on the EPOC+ device and it capabilities for providing information about facial expressions, head-movement and others.

III. EMOTIV EPOC+ DEVICE

Emotiv EPOC+ is a 14 channel wireless non-invasive EEG and electromyography (EMG) device, which was designed for contextualized research and advanced BCI applications [24]. Electrodes of the EPOC+ headset are located at AF3, AF4, F3, F4, F7, F8, FC5, FC6, P7, P8, T7, T8, O1, O2 with two additional sensors that serve as CMS/DRL reference channels (one for the left and the other for the right hemisphere of the head) [22].

The EPOC+ headset is able to detect brain signals and facial expressions or levels of engagement (e.g. frustration, mediation, etc.), eye movement and head position through the built-in gyroscope sensors. The 9-axis inertial motion sensor allows several motion/positional tracking and monitoring. The sampling rate of motion data is adjustable by the user (128Hz, 64Hz, 32Hz, or off). The device can be trained in order to recognize cognitive neuro-activities. Emotiv provides several software solutions for receiving and analyzing EEG data from the headset together with APIs and detection libraries:

- *Performance Metrics & Emotional States* suite provides capabilities for monitoring user's emotional states in real time and enables an extra dimension in interaction by allowing the computer to respond to user's emotions. Using this suite user's state of mind can be monitored and analyzed and information used for different innovative solutions. For example, music and lighting can be adjusted according to user's experience in real time. Next, while in game, the difficulty level of the game can be tailored and adjusted according to the user's gaming experience. This suite can be used in combination with eye tracking devices and other solutions which together provide rich information for real time evaluation of the user's experience while using the computer or mobile device. Information provided by this suite can be used also for implementing adaptive user interfaces that can adjust according to user's engagement, excitement, and other states.

- *Facial expressions* enables interpretation of the signals measured by the EPOC+ headset into facial expressions in real time. This suite can be used for example for implementing avatars that react to user's facial mimics in real time.

- *Mental Commands* is a suite that can read and interpret user's conscious thoughts and intent. With this suite solutions for manipulation of virtual (e.g. an object in a computer game) or real objects (e.g. robotic arm) based on user's thoughts detection can be implemented.

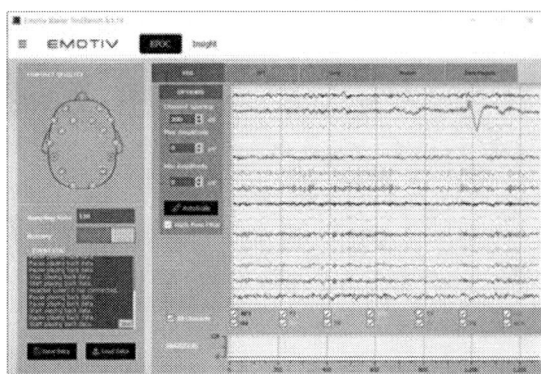

Figure 1. The Emotiv Xavier TestBench

Emotiv TestBench (see Fig. 1) is a software package that is used for displaying data from the Emotiv EPOC+ headset, including EEG signals, contact quality, FFT, motion detection signals, wireless packet acquisition or loss display, marker events and headset battery level. The software is able to record and replay files in a binary EEGLAB format. The binary format can also be converted into a .csv format for analyzing using open source EEG data analysis software such as OpenViBE, BCI2000, Matlab libraries, etc.

In EEG systems it is difficult to separate the muscle signals from the brain signals, therefore most medical EEG systems require that patients don't move and stay completely still while recording. Due to the sensitivity of these EEG recording systems the recordings can be biased through eye blinks and swallowing. However, Emotiv solutions are able to filter these noises and actually use these noise information as their advantage when compared with other EEG solutions. Eight of the fourteen EEG sensors of the EPOC+ device are placed in the frontal head lobe, which can read signals from the facial muscles and eyes. Most EEG systems treat these signals as noise and these are being filtered and eliminated from further analysis. Emotiv has adjusted the detection of brain signals and based on their classification algorithms they are able to detect which muscles have caused the noise. Facial expressions such as left eye wink, right eye wink, raising eyebrows, smile, etc. can be detected through the analysis of noises caused by different muscles. Detection of the eye movement is also possible because eye is electrically polarized and its movement causes electrical signals, which in turn can be detected.

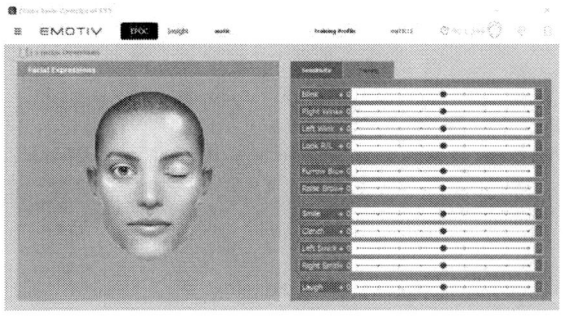

Figure 3. Facial expression detection in Emotiv Xavier Controlpanel

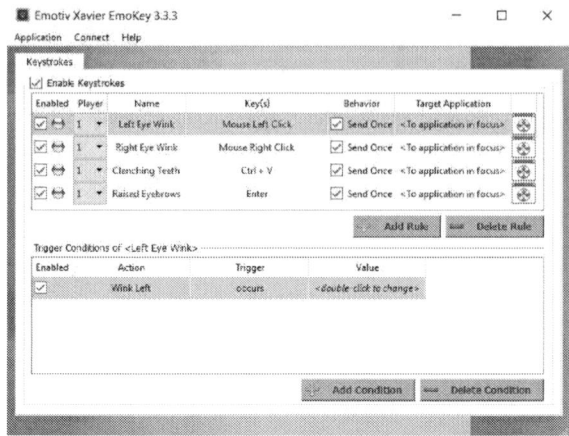

Figure 2. Specifying the keystrokes based on facial epxressions using the Emotiv Xavier EmoKey tool

All the above described features provided by the Emotiv EPOC+ devices can be used for implementing innovative human-computer interaction solutions that can replace standard computer devices such as the mouse and the keyboard completely.

IV. DESIGN OF THE SOLUTION FOR CONTACTLESS COMPUTER INTERACTION BASED ON EPOC+

For designing the solution for hands-free HCI, together with the Emotiv EPOC+ headset following software was required:

- Emotiv Xavier EmoKey (see Fig. 2) enables to specify the translation map between specific facial expressions and computer commands. It can be used for simulating the use of mouse buttons, keyboard buttons or combinations of keys.

- Emotiv Xavier ControlPanel (see Fig. 3), which is the basic software package for managing the EPOC+ device. This software is necessary when first placing the headset on the head and establishing the communication link between the device and the computer. It enables to check the quality and strength of the signals, the ability for facial expression detection, the cursor movement ability through the gyroscope sensors, etc. The ControlPanel software can also be used for training purposes in operating the computer through mental commands.

- Emotiv EPOC Brain Activity Map – is a solution that shows brain activity on individual sensors and provides additional graphical information about it.

Solutions for hands-free interaction with the computer usually also need virtual keyboard similar to the keyboards on mobile devices with touchscreens. There are several solutions for virtual keyboard available such as OptiKey. There are also solutions that enable writing based on the speech recognition. However, due to the compatibility issues with Emotiv software and because speech recognition didn't work for the Slovenian language, we developed a custom virtual keyboard. The developed keyboard allowed inserting special characters

TABLE II.   THE TRANSLATION MAP

Facial Expression	Computer command
Left eye wink	Click of the left mouse button
Right eye wink	Click of the right mouse button
Raised eyebrows	Pressing the button "Enter"
Clenching teeth	Pressing the combination of keys "Ctrl+V"
Raised righ corner of the mouth	Pressing the button "Backspace"
Head movement	Moving the mouse cursor

such as 'š', 'č', etc. Some extra buttons/functionalities were added as well such as "copy all", "delete all", "clear clipboard", etc.

To be able to use the computer for tasks such as writing an e-mail, browsing on the web, watching videos, etc., the solution for hands-free computer use has to enable triggering actions equal to traditional input interfaces. The EmoKey software enables to emulate computer keys and mouse buttons, which are triggered based on the specified facial expression. In this study, several facial expressions were translated into computer commands using the EmoKey software (see Table I for definitions). With combination of the virtual keyboard, the proposed solution enables to opening applications, positioning the mouse cursor on the specific button or application area, etc.

V.   EVALUATION OF THE PROPOSED SOLUTION: A PILOT STUDY

The proposed solution was evaluated at the faculty in a controlled lab environment. For testing purposes, six computer tasks were specified, which users had to complete using the proposed solution only. The specified tasks were: (1) writing a short "Hello"-type e-mail using a web browser and an Gmail account, (2) finding a photo in the photo gallery on the computer using the Windows explorer, (3) downloading an e-book (PDF) from a specific web address, opening the e-book and finding a specific chapter in the book, (4) making a short Skype call, (5) finding a YouTube video and viewing the video at

Figure 4.   The evaluation of the solution

a specific time frame, and (6) completing an online purchase in an online shop. Prior executing the tasks, users were logged in to online services and they didn't have to register and/or login. During the execution of the tasks users were not allowed to use the mouse and the computer keyboard.

In the pilot study, which objective was to evaluate the proposed solution, ten non-disabled users participated. At the beginning of the testing session we explained them the purpose of the experiment and explained how the Emotiv EPOC+ device functions. Then, all sensors of the device were moistened and the headset was put on the head of the participants. After the quality of the signals was checked, the participants entered the training phase of the experiment (see Fig. 4). The training phase is important for users as well as for the Emotiv software and its machine learning algorithms. For training purposes the Emotiv Xavier Control Panel was used. After users felt comfortable with using the headset and the sensitivity of

TABLE I.   DECRIPTIVE STATISTICS OF PU AND PEOU

Item	median	1	2	3	4	5
Perceived Usefulness (PU) (adapted from [25]*)*						
PU1 - Using the EPOC+ based solution in my job would enable me to accomplish tasks more quickly.	1.5	5	3	1	0	1
PU2 - Using the EPOC+ based solution would improve my job performance.	1	6	3	0	1	0
PU3 - Using the EPOC+ based solution in my job would increase my productivity	1	6	2	1	1	0
PU4 - Using the EPOC+ based solution would enhance my effectiveness on the job.	1	6	1	2	1	0
PU5 - Using the EPOC+ based solution would make it easier to do my job.	1	6	3	0	1	0
PU6 - I would find the EPOC+ based solution useful in my job.	2	4	4	1	0	1
Perceived Ease of Use (PEOU) (adapted from [25]*)*						
PEOU1 - Learning to operate the EPOC+ based solution would be easy for me.	3.5	1	0	4	5	0
PEOU2 - I would find it easy to get the EPOC+ based solution to do what I want it to do.	2.5	1	4	4	1	0
PEOU3 - My interaction with the EPOC+ based solution would be clear and understandable.	4	0	0	1	6	3
PEOU4 - I would find the EPOC+ based solution to be flexible to interact with.	3.5	2	1	2	4	1
PEOU5 - It would be easy for me to become skillful at using the EPOC+ based solution.	4	0	3	1	5	1
PEOU6 - I would find the EPOC+ based solution easy to use.	3	1	0	5	4	0

the facial expressions detections were adjusted according to individuals' facial expression capabilities, participants got instructions for completing the six tasks.

The main purpose of this study was to evaluate users' perceptions about usefulness (PU) and ease of use (PEOU) of the proposed solution. Times needed for completing tasks were not measured. After users completed the tasks, they answered a questionnaire consisting of items that were adapted from the Technology Acceptance Model [25]. PU and PEOU items were measured using a 5-point Likert type scale with values from 1 ("Strongly Disagree") to 5 ("Strongly Agree"). Descriptive statistics about the users' perceptions about the usefulness and ease of use of the proposed solution are presented in Table II. It was expected that users won't perceive the proposed solution very much useful and the scores for all PU items were very low. However, assessments of users' perceptions in easiness of use of the proposed solution showed that with exception of PEOU2 users didn't experience difficulties in learning and using the proposed solution. This outcome is promising, because the solution is intended for disabled users that are not able to use one or both hands. For such users, the proposed solution can be an alternative to their existing AT solution which they already use.

Next step in evaluating the proposed solution will be testing the solution using real disabled users and to measure their efficiency and perceptions in using the proposed solution. We expect that disabled users will have higher perceptions about the usefulness of the proposed solution.

VI. CONLUSION

Motor impaired users based on their condition usually have difficulties in using standard computer interfaces and input devices such as the keyboard, mouse, etc. because of the lack of controlling hands. Therefore, different AT solutions were designed and developed that enable alternative interactions with a computer. Such AT require design and development of innovative hardware and software solutions. This paper reports results of a pilot study, in which a solution for contactless interaction with computer was designed and developed using the Emotiv EPOC+ device. The EPOC+ device is an affordable non-invasive EEG device that can be used for innovative solutions based on brain activity analysis and facial expressions detection. It also provides head motion sensors, which together can fully replace standard input devices such as the computer mouse and keyboard. Emotiv software for the EPOC+ device and custom software packages (e.g. virtual keyboard, speech recognition, etc.) represent a basis for developing new and innovative AT for users, who are not able to use standard computer devices and interfaces.

In addition to the development of different AT solutions it is very important to fully understand how such solutions impact real users and their experience in using computers. The main limitation of this study is that involved only non-disabled users. Next, only users' perceptions about usefulness and ease of use of the solution were assessed. There are also other factors such

as self-efficacy, compatibility, user experience quality, etc., which need to be considered and assessed. In order to fully understand the potential of the proposed solution, future research is needed that will include real disabled users. This way, we will be able to understand what are the positive and/or negative effects of the proposed solution and how can we improve it. Future research is also needed in order to understand whether the proposed solution is in some way better if compared to existing AT solutions that disabled users are using in their daily computer use activities. We also plan to compare the outcomes between the non-disabled and disabled users. We assume that disabled-users will perceive the solution more useful than non-disabled users, since they need such AT solutions more.

ACKNOWLEDGMENT

The authors acknowledge the financial support from the Slovenian Research Agency (research core funding No. P2-0057).

REFERENCES

[1] The Lancet. World Report on Disability. Lancet 2011;377:1977. doi:10.1016/S0140-6736(11)60844-1.

[2] Slovenian Paraplegic Association n.d. http://zveza-paraplegikov.si/eng/.

[3] Kastelic D, Ermenc H. Letno poročilo za leto 2015 2016:1–34.

[4] Siegel C, Dorner TE. Information technologies for active and assisted living—Influences to the quality of life of an ageing society. Int J Med Inform 2017;100:32–45. doi:10.1016/j.ijmedinf.2017.01.012.

[5] Kumar DK, Arjunan SP. Human–Computer Interface Technologies for the Motor Impaired. 2016. doi:10.1201/b19274.

[6] Wolpaw JR, Birbaumer N, McFarland DJ, Pfurtscheller G, Vaughan TM. Brain-computer interfaces for communication and control. Clin Neurophysiol 2002;113:767–91.

[7] Pires G, Nunes U, Castelo-Branco M. Evaluation of Brain-computer Interfaces in Accessing Computer and other Devices by People with Severe Motor Impairments. Procedia Comput Sci 2012;14:283–92. doi:10.1016/j.procs.2012.10.032.

[8] Vourvopoulos A, Liarokapis F. Evaluation of commercial brain–computer interfaces in real and virtual world environment: A pilot study. Comput Electr Eng 2014;40:714–29. doi:10.1016/j.compeleceng.2013.10.009.

[9] Göhring D, Latotzky D, Wang M, Rojas R. Semi-autonomous Car Control Using Brain Computer Interfaces, 2013, p. 393–408. doi:10.1007/978-3-642-33932-5_37.

[10] Cincotti F, Mattia D, Aloise F, Bufalari S, Schalk G, Oriolo G, et al. Non-invasive brain–computer interface system: Towards its application as assistive technology. Brain Res Bull 2008;75:796–803. doi:10.1016/j.brainresbull.2008.01.007.

[11] Pires G, Nunes U, Castelo-Branco M. Comparison of a row-column speller vs. a novel lateral single-character speller:

Assessment of BCI for severe motor disabled patients. Clin Neurophysiol 2012;123:1168–81. doi:10.1016/j.clinph.2011.10.040.

[12] Pires G, Nunes U, Castelo-Branco M. Statistical spatial filtering for a P300-based BCI: Tests in able-bodied, and patients with cerebral palsy and amyotrophic lateral sclerosis. J Neurosci Methods 2011;195:270–81. doi:10.1016/j.jneumeth.2010.11.016.

[13] Millan J d. R, Galan F, Vanhooydonck D, Lew E, Philips J, Nuttin M. Asynchronous non-invasive brain-actuated control of an intelligent wheelchair. 2009 Annu. Int. Conf. IEEE Eng. Med. Biol. Soc., IEEE; 2009, p. 3361–4. doi:10.1109/IEMBS.2009.5332828.

[14] Lopes AC, Pires G, Vaz L, Nunes U. Wheelchair navigation assisted by human-machine shared-control and a P300-based Brain Computer Interface. 2011 IEEE/RSJ Int. Conf. Intell. Robot. Syst., IEEE; 2011, p. 2438–44. doi:10.1109/IROS.2011.6048355.

[15] Pires G, Castelo-Branco M, Nunes U. Visual P300-based BCI to steer a wheelchair: A Bayesian approach. 2008 30th Annu. Int. Conf. IEEE Eng. Med. Biol. Soc., IEEE; 2008, p. 658–61. doi:10.1109/IEMBS.2008.4649238.

[16] Pires G, Torres M, Casaleiro N, Nunes U, Castelo-Branco M. Playing Tetris with non-invasive BCI. 2011 IEEE 1st Int. Conf. Serious Games Appl. Heal., IEEE; 2011, p. 1–6. doi:10.1109/SeGAH.2011.6165454.

[17] Wang Q, Sourina O, Nguyen MK. EEG-Based "Serious" Games Design for Medical Applications. 2010 Int. Conf. Cyberworlds, IEEE; 2010, p. 270–6. doi:10.1109/CW.2010.56.

[18] Holz EM, Höhne J, Staiger-Sälzer P, Tangermann M, Kübler A. Brain–computer interface controlled gaming: Evaluation of usability by severely motor restricted end-users. Artif Intell Med 2013;59:111–20. doi:10.1016/j.artmed.2013.08.001.

[19] Soman S, Murthy BK. Using Brain Computer Interface for Synthesized Speech Communication for the Physically Disabled. Procedia Comput Sci 2015;46:292–8. doi:10.1016/j.procs.2015.02.023.

[20] Al-Hudhud G. Affective command-based control system integrating brain signals in commands control systems. Comput Human Behav 2014;30:535–41. doi:10.1016/j.chb.2013.06.038.

[21] Fakhruzzaman MN, Riksakomara E, Suryotrisongko H. EEG Wave Identification in Human Brain with Emotiv EPOC for Motor Imagery. Procedia Comput Sci 2015;72:269–76. doi:10.1016/j.procs.2015.12.140.

[22] McMahan T, Parberry I, Parsons TD. Modality specific assessment of video game player's experience using the Emotiv. Entertain Comput 2015;7:1–6. doi:10.1016/j.entcom.2015.03.001.

[23] Mondéjar T, Hervás R, Johnson E, Gutierrez C, Latorre JM. Correlation between videogame mechanics and executive functions through EEG analysis. J Biomed Inform 2016;63:131–40. doi:10.1016/j.jbi.2016.08.006.

[24] Inc E. Emotiv EPOC+ 2017. https://www.emotiv.com/epoc/.

[25] Davis FD. Perceived Usefulness, Perceived Ease of Use, and User Acceptance of Information Technology. MIS Q 1989;13:319. doi:10.2307/249008.

Patterns for Improving Mobile User Experience

M. Pušnik*, D. Ivanovski* and B. Šumak*

* University of Maribor, Faculty of electrical engineering and computer science, Institute of informatics, Slovenia
maja.pusnik@um.si, dimitar.ivanovski@student.um.si, bostjan.sumak@um.si

Abstract - Mobile telephone sales are achieving a constant growth on the global level as well as locally in Slovenia, where stores have a record of approximately fourfold increase of mobile device sales in the last 6 years. Accordingly, the development of mobile applications and services has also escalated, however, not every mobile application received the anticipated acceptance rate. The reason for poor success of mobile applications can be low user experience. The paper focuses on the concept of a mobile user experience definition and enhancement, addressing the questions (1) what is mobile user experience, (2) how to measure the mobile user experience and (3) how to improve the mobile user experience based on mobile design patterns. Results of a case study are included, recording the behavioural characteristics of 27 users using a mobile application for checking receipts, evaluating its ease of use, usefulness and other user's needs and requirements.

I. INTRODUCTION

Today, more than 780 million people use exclusively mobile devices [1] and less time on desktop computers. Indicators in recent years consistently show global growth in sales of mobile phones and similar trends are being recorded also in Slovenia. The Slovenian online store "*Mimovrste.com*" for example recorded a 381 percent [2] of increase in sales of mobile devices between 2010 and 2014. With mobile devices, the use of mobile applications increases as well [15].

In order to increase the use and promote mobile applications and services, developers of mobile software need to understand the needs of mobile users, focusing on providing a rich and satisfactory mobile communication experience to end-users [3]. Factors that contribute to good mobile user experience (mUX), include interaction with mobile devices and applications, which are natural, intuitive, easy to use, comfortable, easy to remember and flexible to the wishes and needs of individuals.

The ultimate goal from the perspective of the user is an effective and efficient interaction with the user interface. For this reason, it is important that the user interface is intuitive, requiring minimal learning time [16]. To enable a joyful application handling, it is necessary to implement its functionality in a manner, that users do not get tired due to the constant re-keying or repeating steps.

Because of listed suggestions it is important to examine the mUX within the development of mobile applications and user interfaces. Research focus is to identify *mobile design patterns for improving user experience.*

A mobile design pattern is a solution to a problem that may supports the developers when developing mobile applications [14]. It is a formalized best practice guide or proposal for designers, developers and managers of products to use at solving common problems while implementing mobile applications [21]. Different mobile design patterns are focused on the user interface development as well as on actions of users and hardware. Regardless of the pattern focus, the mobile design patterns allow mobile developers the opportunity to create the mUX through a variety of methods and techniques to meet the objectives of the user [4][5][6].

Quality assurance of mUX has many challenges, since it is a subjective, dynamic and complex phenomenon [15]. Various features of the product, its context, relations with other people, and others, have an impact on how the product is perceived [7]. mUX is regarded as an integrated approach that involves the use of the product [8], feelings [9], meaningfulness of the product [10], consumer attachment to the product [11] and how the product is practically applicable [12]. Research to understand the concepts of UX has been conducted, but in literature, the general theory is still missing, especially in the field of mobile UX.

The purpose of this paper is to review literature and identify common mobile design patterns and best practices for efficient development of mobile applications with quality mUX. In addition, one of the goals of this research is to answer the question "*How to improve the planning and changing the mobile application, its entry page, navigation, architecture, design, menu items and other elements to make it more useful and logical to users, improving the mobile user experience?*".

II. USER EXPERIENCE

In existing literature, several definitions for UX can be found, however they can significantly differ based on the industry domain. In the era of traditional desktop applications, companies generally did not invest a lot of effort and money in the analysis of user habits, and tended to reject the cost, which was not absolutely necessary for a good final product. Lee (1996) was the first to mention the UX term in the context of interactive products [13]: "*UX covers all aspects of how people use interactive products: how they feel when they take the product in their hands, how well they understand how it works, how they feel when they use it, how well it serves its purpose, and how well it fits into the overall context in which they use it*". UX awareness increased in 2001 when iPod came out and in 2007 with the beginning of iPhone. Apple dominated the market and paved the way for other companies and the future of UX. Don Norman [14], while working in Apple in the 90', stated: "*UX covers all aspects of interaction with the end-user companies and their services and products*". International Standardization Organization ISO (9241-210:2010) provides the following definition: "*UX are perceptions and responses of individuals, arising from the use or expected use of the product, system or service*".

Recent advances in mobile, universal, social and computer technology have spread human-computer interaction (HCI) theory and concepts in practically all fields of human activities, creating new fields of user experience including the mobile user experience.

A. Mobile user experience

Go-Globe [15], an organization that provides statistics and trends on the use of mobile applications, claims that mobile applications are growing exponentially. Statistics show that people use 52% of total time on digital media using mobile applications and it is estimated that by 2017, global revenues from mobile applications will be doubled. Currently, mobile solution providers are interested in better understanding of consumer behaviour, their wishes and which mobile devices are used, in order to be able to develop a strategy for developing successful mobile applications providing quality user experience.

The average time spent on mobile applications grew by 21% compared to last year (2016); used on the following mobile applications [15]: games (43%), social networks (26%), entertainment (10%), aids (10%), news (10%), productivity (2%), health and fitness (1%), lifestyles (1%), and others (5%). Among them, music (79%), health and fitness (59%) and social networks (41%) are the three most growing categories. In addition, the percentage of mobile applications that are used only once, decreased by 2%, while the number of all applications used increased 11-times in the last year.

Mobile applications create a special domain to monitor, measure and improve user experience. There are significant differences between users (their skills), devices and the limitations of the technical infrastructure (screen size), as well as the specifics of the platform as well as guidelines and contexts of use. According to Kuusinen and Mikkonen [16], interaction for mobile devices should be designed in a way, so that the time span of the actions of users is shorter than with desktop computers. Moreover, the interaction must be done with ease, using a minimum number of pressing buttons or keys, because in real life, people often walk, drive or engage in other activities, while using mobile applications.

B. Elements of mobile user experience

In the paper »The Elements Of The Mobile User Experience« [21] the authors present key elements of mUX (Figure 1). They argue that the creation of mobile UX that satisfies all users, forces to rethink the design of mobile application because it is quite different from a desktop user experience. When designing mUX, several restrictions related to mobile device must be considered, such as display sizes, significant differences in the features of the device, restrictions of the use and connectivity, as well as constant changing of mobile context.

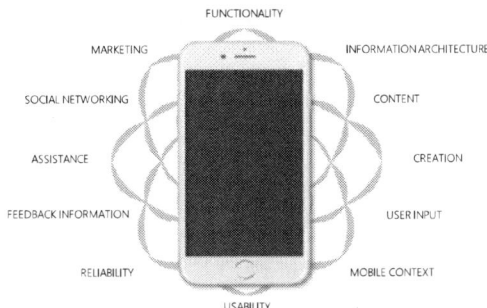

Figure 1.   Elements of mobile UX [21]

Since the authors in [21] presented the key elements of mobile UX and argued that the creation of mobile UX forces developers to rethink the design of mobile applications, which is quite different from the desktop UX. Different elements are: *functionality, information architecture, content, creation, user input, mobile context, usability, reliability, feedback information, assistance, social networking and marketing*. In order to have a positive mobile UX for all users, developers of mobile applications should focus on each individual element. The meaning of each element will change, depending on the type of device as well as on the user interface.

The literature review also included papers focused on mobile UX improvement ([22][23][24][25][26]). Based on the findings, some guidelines for measuring mUX were provided:

- "*Let's meet our goals*" is the first guideline, trying to understand the purpose of its application in terms of customer's needs.

- "*Identify users' goals*" includes the use of user characters, scenarios, currents, mockups and others.

- "*Measure feedback*" where prototyping is used with the ultimate aim to get initial feedback from real users.

- "D*esign and development*" includes iOS and Android guidelines, which have to be considered to accompany the existing mobile design patterns and anti-patterns, and occasionally verify the consistency of the application of formative research.

- "*Measurement*" of UX applications through summative research.

The guidelines are presented as a continuous iterative process, taking into account the "real data" obtained from real users, At the same time, we must realize that UX of applications can improve with time, when the user becomes more and more proficient.

UX measuring consists of five steps [5]:

- The first step depends on the life cycle of application development and includes the activity "*realize the objectives of the research*". More precisely, we must first decide whether to measure UX of new applications or do we want to improve UX existing applications. According to the decision, there are two options: formative or summative research.

- Second step includes "*identify the goals of the users*" and can measure two aspects of mobile UX: performance and satisfaction. It is important to emphasize that efficiency and satisfaction are not always a match, however the implications from [22] indicate we need to measure both.

- Depending on whether we are measuring performance or satisfaction, appropriate "*metric selection*" must be completed.

- After selecting the right metric, the next step "*choice of research methods*" follows. There is no best way to carry out the research, and the choice of research method depends largely on the nature of the research question.

- When research method (or combination of methods) is selected, we continue with the process' *execution of research methods*.

Similarly, as in the previous section, a measurement of mobile UX is presented as a continuous iterative process.

III. MOBILE DESIGN PATTERNS

Mobile UX design pattern is a formalized best practice guide, proposed by designers, developers and managers of products, that can be used to solve common problems while developing mobile applications [4][5][6]. In existing literature [26][27] several mobile design patterns were identified for different mobile platforms, only few of them are presented in this paper due to space limitations. The differences in mobile platforms (Android, iOS, Windows) and their mobile design patterns were examined. All listed mobile platforms adopted similar, minimalistic UX design patterns (bright colours, moving from left to the right side of the home screen). However, each mobile platform is substantially different and with different guidelines:

- Currently, iOS is the most profitable platform for developers [19]. Apple Store charges more for the transfer of applications compared to Google Play, and it is easier to develop iOS applications. For this reason, even the most innovative applications generally first appear on Apple devices. On the other hand, the biggest drawback of the Apple Store are their strict guidelines and their complex and time consuming process for approving new applications [17].

- Unlike iOS or Windows, Android mobile platform is completely open source, which means that any developer can have access to the code [18]. As a result, we get a number of different applications from the Google Play Store. However, the quality may not be so high. Furthermore, due to the lack of uniform standards in the creation of applications, it is possible when buying a new phone, some apps are not compatible.

- Windows is the youngest of the three systems and has the smallest customer base with less than 5% market share [19]. as a result, they cannot attract such a large developer community, as there is for iOS and Android. Windows platform is still located in its early stages, compared with Apple and Android, which are highly advanced. They also have no guidelines for native mobile applications, only guidelines for applications to be carried out on several different platforms [20].

Since most developed applications are for the two platforms, the research was focused on Android and iOS mobile design patterns. Apple and Google have developed solutions for users to interact with applications in a particular way for many years, by encouraging application developers to follow their guidelines.

By conducting a literature review, we identified mobile design patterns with different impacts on the perception of mobile UX. Based on several authors ([4][5][6]) a categorization in five categories of "best practice patterns" was made: (1) *basic patterns*, (2) *social networking*, (3) *image and media support*, (4) *support and feedback* and (5) *anti-patterns* for counterproductive practices. A set of mobile design patterns was defined, including mobile design pattern, which affect minimally one or more of mobile UX elements (Figure 1.). Within the five categories, 13 mobile design patterns and 2 anti-patterns were included in this research. Simplified examples of each pattern are presented below.

A. Basic mobile design patterns

- **Interacting** - mobile user wants to know how to use the mobile app to perform tasks as simple as possible.

- **Data entry** - mobile user wants to quickly enter data that will be most successful in their work.

- **Navigation** - mobile user wishes to use various features of the application and find where they need to go in less time.

- **Flat design** - mobile user wishes a visually pleasing user interface.

- **Personalization** - Mobile user wishes customized content regarding their interests, needs and location in several interaction aspects.

- **Gamification** - Mobile user wants to feel happy and to satisfy his competitive nature.

- **Content marketing** - Mobile user wants to buy something at a discount in order to save money.

B. Social networking

- **Social** - Mobile user wants to monitor and be up to date with changes, as well as keep in touch with friends.

- **Widgets** - Mobile user wants quick access to information, needed to open an application.

C. Image and media support

- **Camera** - Mobile user wants to capture content anytime and anywhere and share it with their friends.

- **Streaming** - Mobile user wants to get an instant overview of recent activities of any content in the application as quickly as possible.

D. Support and feedback

- **Help** - Mobile user wants to learn how to use the app with minimum errors.

- **Constructive feedback** - Mobile user wants to know what is happening with the application, to decide what his next step will be.

E. Anti-patterns

- **Complexity** - Mobile user wants to do more in less time.

- **Horizontal segmentation** - Mobile user wants more different things to be satisfied.

Based on selected patterns, a preliminary case study was conducted, examining the use of listed patterns and user experience of the final product.

IV. THE CASE STUDY

The main objective of the case study was to examine the proposed patterns in case of an existing mobile application and investigate their impact on elements of UX. The following research questions were defined, focusing on the case study, presented in this research:

- How the use of mobile design patterns affects the *attitude towards the use* (ATU) of mobile applications?

- How the use of mobile design patterns affects the *ease of use* (PEOU) of mobile applications?

- How the use of mobile design patterns affects the *usefulness* (PU) of mobile applications?

We conducted a summative research on existing mobile application through a case study and statistics. The case study section is divided in 5 subsections: (1) application selection, (2) planning, (3) data collection, (4) analysis and (5) limitations.

A. Application selection

In order to encourage consumers to require and take the receipt in shops and stores, the Financial Administration of Slovenia launched a prize game, called "Check your receipt". It is open to all who collect at least ten receipts of different issuers and send them by using mobile application "*Vklopi razum, zahtevaj račun*". (Figure 2.). The application verifies data and sends it to the Financial Administration which further verifies its parameters. When a consumer collects ten receipts of various issuers, a package is automatically created and the consumer can participate in the prize game lottery.

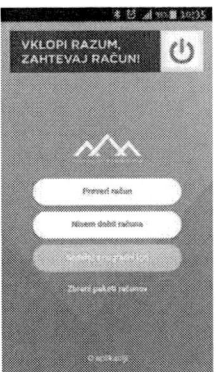 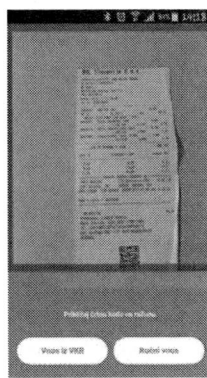

Figure 2.   Screenshot of "Check your receipt" application

The idea of the case study was not to participate in the competition, but to explore the ease of participation in it. Since "*ease of use*" largely depends on the input of receipts in the application and use of QR code scanning technology, the following mobile design patterns were addressed: Interacting, Data Entry, Navigation, Flat design, Camera, Assistance, Constructive feedback. All 7 mobile design patterns were measured with following metrics: *task success, task execution time, errors, effectiveness, acceptance, awareness, subjective metrics* in form of a survey. Table 1 presents the connection between metrics and mobile design patterns, presenting how each mobile design pattern was measured (metric details are not included in this paper).

TABLE I.   IDENTIFIED PATTERNS IN "CHECK YOUR RECEIPT" APPLICATION AND MEASUREMENT APPROACHES

	Task success	Task execution time	Errors	Effectiveness	Acceptance	Awareness	Subjective metrics
Interacting	X	X	X	X	X		X
Data Entry PU_{DE}	X	X	X	X	X		X
Navigation PU_{NA}	X		X	X	X		X
Flat design PU_{FD}			X			X	X
Camera PU_{CA}	X	X	X	X	X	X	X
Assistance	X		X	X			X
Constructive feedback PU_{CF}	X		X	X			X

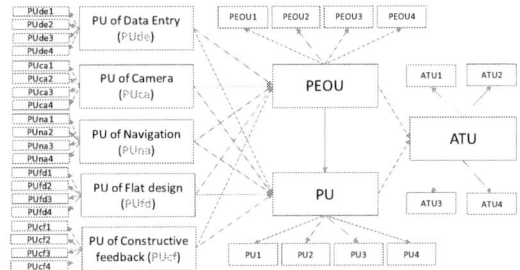

Figure 3.   Theory model

Based on identified mobile design patterns, a theory model was defined (Figure 3.). As basis, a modified TAM (technology acceptance model [28] model was used, including variables PEOU, PU and ATU and with mobile design patterns connected variables (without patterns assistance and interacting): PU of data entry, PU of camera, PU of navigation, PU of flat design and PU of constructive feedback.

A. The planning

The case study investigated the use of mobile application of 27 users. It included quantitative and qualitative data gathering and took place in a controlled environment at the Faculty of Electrical Engineering and Computer Science at the University of Maribor. Each participant had to perform two tasks (to scan and check receipt number 1 and receipt number 2). After finishing the tasks, each tester evaluated the application through a web questionnaire, not completely included in this paper due to space constraints, however presented in section D. Analysis.

B. Collecting data

After completing the tasks of mobile application "Check your receipt", the users responded to an online questionnaire (simplified version is presented in TABLE II.). The basic results of dependent variables are presented in TABLE III.

TABLE II.   ADAPTED QUESTIONNAIRE FOR DEPENDENT (PEOU, PU, ATU) AND INDEPENDENT VARIABLES ($PU_{DE}, PU_{CA}, PU_{NA}, PU_{FD}, PU_{CF}$)

PEOU1-4	I believe that the application is easy to use. I believe that the application is simple to use. I believe that the application of the includes insufficient number of steps required to achieve my final goal. I believe that I can easily become adept at using the application.
PU1-4	Application allows me a faster entry of receipts. Application allows me to simplify the entry of invoices. Using the increases the effectiveness of the verification of invoices. I believe that the use of the application is useful in checking receipts.
ATU1-4	Using the application is a good idea. Application makes checking receipts more interesting. Work with application is fun. I like to work with the app.
$PU_{DE\,1-4}$	Data entry is simple and useful. Data entry with one hand is simple. The application enables me fast receipt checking. The application optimizes the receipt checking.
$PU_{CA\,1-4}$	It is easy to use the camera. Camera simplifies receipt checking. Camera enables a faster receipt checking. Camera optimizes the receipt checking.
PU_{NA1-4}	It is easy to navigate in the application. I don't always find it easy to navigate the application. A lot of steps are necessary to navigate. It is simple to complete the secon task.
PU_{FD1-4}	User interface resambles other applications. The design of the graphic symbols is sufficient. The interface of this system is pleasant. Visual display does not distract me from completing the tasks.
$PU_{CF\,1-4}$	The information provided within the application is clear. The information is effective. System feedback is good. The software documentation is very informative.

C. Analysis

Analysis of data was carried out in programs MS Excel and XLSTAT. In general, the typical user who participated in the case study was male, aged between 23 and 26 years, self-proclaimed as very experienced with the use of smart phones. In addition, the average tester had no experience with the use of applications "Check your receipt".

TABLE III. presents the descriptive values of dependent variables ATU, PU and PEOU. Due to lack of space, variables, measuring specific mobile design patterns, influencing PU and PEOU cannot be presented in detail in the paper and only partial analysis results are included.

TABLE III.   FREQUENCY OF DEPENDENT VARIABLES PU, PEOU AND ATU

Construct	Identifiers	Strongly disagree	Disagree	Neither	Agree	Strongly agree
PEOU	PEOU1	0%	0%	7%	52%	41%
	PEOU2	0%	4%	11%	41%	44%
	PEOU3	4%	7%	19%	26%	44%
	PEOU4	0%	4%	15%	15%	66%
PU	PU1	0%	0%	11%	30%	59%
	PU2	0%	0%	11%	33%	56%
	PU3	0%	11%	7%	26%	56%
	PU4	0%	0%	4%	33%	63%
ATU	ATU1	0%	0%	15%	15%	70%
	ATU2	4%	7%	11%	22%	56%
	ATU3	7%	26%	22%	19%	26%
	ATU4	0%	7%	44%	30%	19%

To investigate correlations between independent and dependent variables, Spearman correlation test was made. Values of the Spearman correlation coefficient r_s indicate, which identified mobile patterns in "Check your receipt" effected PU and PEOU variables (see TABLE IV.). Results are significant only for the PEOU, PU and ATU.

TABLE IV.   ANALYSIS OF VARIABLE CORRELATIONS BASED ON SPEARMAN CORELATION COEFICIENT R_s (MODIFIED TABLE)

		Correlations		Comment
		PU	ATU	
PEOU	r_s	1.000	0.400	PEOU has positive impact on PU
	p	0.0001	0.750	PEOU has positive impact on ATU
		ATU		
PU	r_s	0.400		PU has positive impact on ATU
	p	0.750		
		PEOU	PU	
PU_{DE}	r_s	0.400	0.400	PU_{DE} has positive impact on PEOU
	p	0.750	0.750	PU_{DE} has positive impact on PU
		PEOU	PU	
PU_{CA}	r_s	-0.400	-0.400	PU_{CA} has negative impact on PEOU
	p	0.750	0.750	PU_{CA} has negative impact on PU
		PEOU	PU	
PU_{NA}	r_s	-0.400	-0.400	PU_{NA} has negative impact on PEOU
	p	0.750	0.750	PU_{NA} has negative impact on PU
		PEOU	PU	
PU_{FD}	r_s	0.435	0.435	PU_{FD} has positive impact on PEOU
	p	0.419	0.419	PU_{FD} has positive impact on PU
		PEOU	PU	
PU_{CF}	r_s	-0.132	-0.132	PU_{CF} has negative impact on PEOU
	p	0.803	0.803	PU_{CF} has negative impact on PU

D. Limitations

The research had some limitations. One of the main limitations was the use of only one mobile application, therefore the results cannot be generalized. The second main restriction is the sample of test users, which was not random, as well as the small number of users, included in the case study (27). In addition, we were limited to the use of 13 patterns and 2 anti-patterns (several other mobile design patterns are not included in this research). Lastly, only two mobile platforms were examined.

V. CONCLUSION

In this study, the primary objectives were to determine what mobile design patterns exist and how they affect mUX. Previous research has shown that the mobile UX is different from the desktop computer UX, and several mobile design patterns can improve mobile UX. Based on research results and statistic analysis we concluded that 5 identified mobile design patterns in "Check your receipt" had a positive effect on perceived ease of use and perceived usefulness of the applications, however a larger sample is needed for a higher significance of the results.

This research can serve as a basis for further studies in the field of mobile UX as well as mobile design patterns. Since the results are based on a single mobile application evaluation, future work will contain research of other mobile applications and include the following tasks: examination of more design patterns (such as welcome experience, search, map and location, the tray and others), determining how to measure mobile UX more complexly, how each pattern affects the mobile UX, and what are the challenges of measurement for each pattern.

ACKNOWLEDGMENT

The authors acknowledge the financial support from the Slovenian Research Agency (research core funding No. P2-0057).

REFERENCES

[1] B. Kerschberg, "5 Elements of A Killer Mobile App," Forbes, 2015. [Online]. Available: http://www.forbes.com/sites/benkerschberg/2015/02/10/5-elements-of-a-killer-mobile-app/. [Accessed: 15-Mar-2015].

[2] Mimovrste, "Prodaja mobilnih telefonov še vedno konstantno raste, prodaja tablic se umirja," Mimovrste, 2014. [Online]. Available: https://www.mimovrste.com/sj-2014-08-20. [Accessed: 15-Mar-2015].

[3] S. R. Subramanya and B. K. Yi, "Enhancing the User Experience in Mobile Phones," Computer (Long. Beach. Calif)., vol. 40, no. 12, pp. 114–117, Dec. 2007.

[4] T. Neil, Mobile Design Pattern Gallery: UI Patterns for Smartphone Apps, 2nd Edition. 2014.

[5] UXPin, Mobile UI Design Patterns 2014 - A Deeper Look At the Hottest Apps Today. 2014.

[6] A. Mendoza, Mobile User Experience: Patterns to Make Sense of it All. Newnes, 2013.

[7] M. Buchenau and J. F. Suri, "Experience prototyping," in Proceedings of the conference on Designing interactive systems processes, practices, methods, and techniques - DIS '00, 2000, pp. 424–433.

[8] P. Dourish, "What we talk about when we talk about context," Pers. Ubiquitous Comput., vol. 8, no. 1, pp. 19–30, Feb. 2004.

[9] Fjord, "The nature of emotions and their meaning for design," Fjord, 2015. [Online]. Available: https://www.fjordnet.com/conversations/the-nature-of-emotions-and-their-meaning-for-design/. [Accessed: 15-Mar-2015].

[10] P. Desmet and P. Hekkert, "Framework of Product Experience," IJDesign, vol. 1, no. 1, pp. 57–66, 2007.

[11] H. N. J. Schifferstein and E. P. H. Zwartkruis-Pelgrim, "Consumer-product attachment: measurement and design implications," IJDesign, vol. 2, no. 3, pp. 1–13, Jan. 2008.

[12] A. Faris, "Quality of Experience," AlbenFaris Inc., 1996. [Online]. Available: http://www.albenfaris.com/publications/pub_qofe.shtml. [Accessed: 15-Mar-2015].

[13] L. Alben, "Quality of experience: defining the criteria for effective interaction design," interactions, vol. 3, no. 3, pp. 11–15, May 1996.

[14] D. Norman and J. Nielsen, "The Definition of User Experience (UX)," Nielsen Norman Group Publication, p. 1, 2016.

[15] I. Go-Globe, "Mobile Apps Usage – Statistics and Trends [Infographic]," 2015.

[16] K. Kuusinen and T. Mikkonen, "On Designing UX for Mobile Enterprise Apps," in 2014 40th EUROMICRO Conference on Software Engineering and Advanced Applications, 2014, pp. 221–228.

[17] Apple, "App Store Review Guidelines - Apple Developer." p. 12, 2015.

[18] Android, "Introduction to Android," 2016. [Online]. Available: https://developer.android.com/about/android.html

[19] F. Richter, "Android and iOS Are the Last Two Standing," Statista, 2016. [Online]. Available: https://www.statista.com/chart/4431/smartphone-operating-system-market-share/.

[20] Microsoft, "User experience guidelines for Universal Windows Platform (UWP) apps," 2015.

[21] L. Cerejo, "The Elements Of The Mobile User Experience," Smashing Magazine, 2012. [Online]. Available: http://www.smashingmagazine.com/2012/07/12/elements-mobile-user-experience/. [Accessed: 15-Mar-2015].

[22] S. Griffiths, "Mobile App UX Principles," 2015.

[23] B. Ohlund and C. Yu, "Threats to validity of Research Design."

[24] C. Wohlin, P. Runeson, M. Höst, M. C. Ohlsson, B. Regnell, and A. Wesslén, "Are the Perspectives Really Different? Further Experimentation on Scenario-Based Reading of Requirements," in Experimentation in Software Engineering, Berlin, Heidelberg: Springer Berlin Heidelberg, 2012, pp. 175–200.

[25] B. Šumak, "Domenski model ocenjevanja sprejetosti in uporabe e-storitev." UM FERI, Maribor, p. 431, 2011.

[26] R. Allen, "Flat Design 2.0," 2014. [Online]. Available: http://articles.dappergentlemen.com/2014/12/03/flat-design-2/.

[27] G. Bohak, "Analiza uporabniških vmesnikov na mobilnih napravah," Univerza v Mariboru, 2012.

[28] F. D. Davis, R. P. Bagozzi, P. R. Warshaw, "User acceptance of computer technology: A comparison of two theoretical models", Management Science 35, p. 982–1003,1989.

Drawing Process Recording Tool
for Eye-hand Coordination Modelling

V. Giedrimas, L. Vaitkevicius and A. Vaitkeviciene

Siauliai University, Siauliai, Lithuania

vaigie@mi.su.lt, lukas.vaitkevicius@su.lt, menas885@gmail.com

Abstract - The process of drawing is one of the main activities during childhood. The drawing dynamics, drawing techniques are on the focus of education science. It is known that the result of the drawing process depends on eye-hand coordination. Testing of the psycho-motoric reactions allows classifying motion and positioning problems. Psychomotor reaction data, compared with the characteristics of the drawing such as line thickness and plasticity, allows to clarify the stages of drawing process as well as to establish correlations between the arm motion, plasticity of the drawing and the eye characteristics. In order to perform such research, to make models of the Eye-hand Coordination special tool for drawing process date acquisition, recording and analysis is needed. In this paper one possible implementation of such tool is presented as well as the formal foundations and premises on which the tool is operating.

I. INTRODUCTION

Starting from middle of the XIX century child drawing is in the focus of the international researchers. First attempts have tried to answer such questions as: what the child is drawing? Why he/she is drawing? To whom is dedicated this drawing? The researchers still do not have unambiguous answers for these questions. In XX century the researchers [1, 3, 4, 6] have pointed out that the drawing appears and is changing because of the reason for motion, because of the evolution of eye-hand coordination. After analysis of the changes in children drawings the researchers made taxonomy of the children's drawings evolution [3]; made typology of scribbles (which includes 20 different types) [4]; revealed and explained the process of perception of visual forms (shapes) and the projection of 3D real-life objects to 2D paper sheet [1]; defined future trends for drawing analysis [6]. However the claims of researches about the evolution of child drawings in relation to the beginning of early drawings and the perception of first lines and shapes are very different. For example [3, 4] states that first scribbles can be made by the 2 years (approximately) old child. In contrast to it [6] claims that even one year old child can begin to make some scribbles. Author of the taxonomy of scribbles [4] presents that first scribbles began from random points, then a child discovers horizontal line, a little bit later – vertical one, and finally he\she find a diagonals. Opposite opinion comes from [6], where is claimed that the drawing process began from free, natural pendulum-like motions from right to left and have diagonal direction.

So when and how child drawing begins? What is the sequence of the scribbles inside the drawing? When do the schematic drawings appear and how they evolve during different ages? The scientific community still has a lack for these answers.

It's possible that scientists disagree not only on early drawings. It's possible that the conclusions of children drawings research seems not reliable and strong enough [5] because of the mismatches between the conclusions and claims of different researchers. In other hand the authors [1,3,4,6] do not disclose all the details on the methodologies used in their approaches. As a result of this the doubts about the correctness of the sample size or about the reliability of the used methods can arise. On the basis of few details [6] it is possible to judge that e.g. D. Widlocher made the conclusions and insights one the basis of summary of the carefully selected and critically evolved works of other authors. The authors of [3] do not provide any information about their methodology. We do not know what the survey sample was or how the taxonomy of the children's drawings evolution was made. Only R. Kellogg discloses [4] that the typology of scribbles was made after analysis of 100 000 drawings. Impressive sample size could lead to the conclusion, that the findings are very reliable, however R. Kellogg does not comment on the discrepancies between the conclusions of different researchers, in the aspects of age and line direction precedence.

E. Do, M. D. Gross [2] has defined the following set of features of the drawing (as a child activity): overtracing, speed, pressure, erase, shape specification, shape generalization, symbols, hatching. These features of drawing, as an activity, are related to drawing perception and interpretation as well as with the original intended image, idea or plan for the drawing that the author usually has at first (except earliest child drawings).

The relation between drawing process, drawing perception and the drawing intention reveals the complexity of child drawing phenomena, when attempts to look to drawing from adult perspective or the attempts to help to evolve child' abilities (on the aspect of intention of the drawing' view) are made. This relation also exposes that the drawing is more important for adult than for child itself; the drawing perception, interpretation and pedagogic help for the child during drawing depends on the knowledge about drawing as a process.

In order to understand drawing process in other approaches scientists usually collected and analyzed the drawings, watched drawing children, tried to record this process using available means (e.g. notebooks, photo cameras, video camcorders). The video camcorder was invented and used by the end of the XIX century, however scientists had only little desire to use it, because of possible side effects (distraction, extra anxiety, and possibility of non-natural behavioral during recording). Additional disadvantage or this method is that video recording or photography can be performed only in some angle, so important details can be not recorded because they will be hidden under other objects (e.g. arms, hairs, and clothes). Moreover, important indicators, such as the pressure, erasure or thickness of the line cannot be acquired at all even in analog form.

In order to get more objective and more reliable data about the drawing process, it is necessary to have drawing process recording tools and methods for the gathering of maximal number of drawing process details. The need for drawing process digital recording software is essential in this context.

The rest of the paper is organized as follows: Section 2 exposes the idea ant the implementation details of drawing process recording tool, Section 3 presents the testing methodology and the test results. Section 4 compares our approach to the state-of-art in the area. Finally conclusions are made and future steps of the eye-hand coordination modelling project are discussed.

II. THE TOOL

A. Prerequisities

As is mentioned before, different hardware can be used for the drawing process recording and acquisition. Our drawing recording tool is based on digitizer tablets. This type of device was chosen because it is more sensitive for pen operations [7] compared to other tablets, smartphones or personal digital assistants (PDA). Moreover not all models of smartphones are equipped with a pen (drawing can be performed only using fingers), so they cannot be used at all. Smart devices (smartphones and tablets) probably would be easy to use for drawing at home, the data could be sent to backend, where it can be shared and used by scientists. However it would be very complicated to do playback and analysis of such drawing. Moreover some models of digitizer tablet are integrated in monitors and gives to users the feel of "real drawing".

Initially the term "tablet" had the same meaning as "digitizer tablet" – i.e. a computer peripheral pointing device enabling to acquire the movements of user's hand of lower granularity, e.g. pen operations, handwriting. Later on when Apple introduce revolutionary product – iPad the term "tablet" started to use with such type of smart devices (autonomous and wireless).

In this paper we are using terms "digitizer tablet" and "tablet" interchangeably, because we reject using smart devices and always have in mind digitizer tablet.

B. The implementation

Technically there is no problem to get coordinates of the pen/mouse/finger and track its changes. However much harder task is to get pressure, pen angle etc. The tablet is necessary for the acquisition of such drawing aspects and conveniently, on the Windows operating systems tablets share a common API, simplifying data acquisition

During the years each manufacturer of tablets (Wacom and others) has provided special device driver. In addition

computer assisted design (CAD) software was not device-independent. Each CAD tool was prepared to support closed list of devices. Each device was either supported (via included bridging tool) or not supported at all. This became not acceptable for the users of operating systems with standard GUI, so the industrial standard for tablets WinTab [9] was created.

The *contexts* are used for the data transfer from tablet to software WinTab API. There are two types of the contexts [9]: *digitizing context* and *system context*. While digitizing context provides acquired data directly, system context performs changes on pointer position (in fact tablet is performing the role of mouse). The contexts allow using simultaneously the tablet by different computer programs, to manage context-specific active zone.

When using table, WinTab collects packets of information about pen position in the queue. The sharing of the packets between different contexts is forbidden. The programs are working with the packets with strictly specified hierarchy as well. In the case of violation of software hierarchy, all programs within it can start to act unpredictably.

The configuration of tables can be changed using WinTab API also. This eliminates the need for the operations in the lower device-driver layer. Any WinTab-based program can change configuration of the tabled easy, just by using API.

WinTab API is the library written using C++ programming language. In order to use it in C# program special wrappers are required. There are two options: either to use .NET *InteropServices* or WintabDN [10]. The latter is uses in our tool. Because of using WinTab API [9], WintabDN [10] library wrapper, C# and.NET Framework our tool is device-independent.

Current implementation of drawing process recording tool is able to perform the following functions:

- Simple drawing functions. The drawing user interface is not very rich in terms of the special effect because our target group is children. Richer user interface would distract them and the data about the drawing process would be not natural, as usually, most child drawings will be done on paper, without digital image editing tools;

- The drawing process recording and saving functions. The drawing (including the pressure, speed etc.) and color changing actions are recorded. The data is saved in binary format and can be used for different purposes, starting with statistical analysis ending with playback of the drawing process.

- Playback functions. The drawing process can be played/repeated many times, because main aspects of drawing dynamics are recorded. It is also possible to pick playback speed: line-by-line, frame-by-frame, original, increased speed, fast forward. Line-by-line mode redraws the drawing gradually. After each line the "Drawing player" waits for the user' actions (button press, click or similar. Not new drawing actions.) and only then continues.

589

Although our tool records the drawing process by the actions, not by the frames, it is possible to perform frame-by-frame playback. The frame is creating dynamically on the basis on drawing dynamics data. Increased speed mode allows watching drawing process in accelerated way. All the drawing actions are presented. In contrast to this mode, using fast forward mode, the playback is "silent", the user only seas the result of drawing process at some intermediate (or final one) stage. None of the actions, which were required to produce this drawing, are shown. This mode can be used when the scientists would like to skip large parts of long drawing process and focus only on some particular part of the process.

- Export to still image (JPEG, BMP or PNG). It is possible not only to acquire dynamics of the drawing but to save drawing as still image for further comparison (with the drawings of other children, with earlier/further drawings of the same child etc.) or just as handout/prize for the encouragement reasons.

- The statistical data about the drawing process. After the drawing, in main window of the tool it is shown the following statistical data: the usage of the colors (how intensively each color is used comparing to other colors, this factor depends on the number of lines drawn using particular color), average length of the lines (in the period from pushing pen down till release), the average time for one line.

The drawing data is recorder in the special binary PHR file. The following data is saved to the header of the file: file format and version, size of the file in bytes, the dimensions (height and width) of the final image, the duration of drawing process, the data about the drawer (anonymized) and the comments of the researcher. The rest part of the file is dedicated to binary data of drawing process. Although the drawing data can be serialized using various formats (e.g. JSON, XML), the binary format is chosen for this tool, because of high ration of compression. The structure of the file is exposed in Figure 1 (each small rectangle means one byte), where

1. File format I ASCII coding;
2. Version of the format;
3. The size of metadata in bytes;
4. The size of data segment in bytes;
5. Metadata;

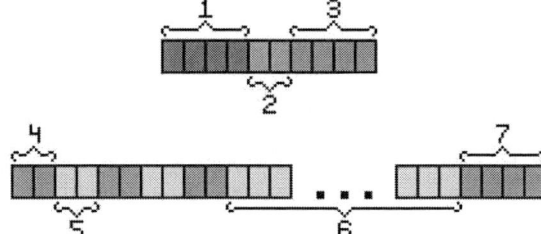

Figure 1. Binary compression of drawing process record

6. Scientists comments (in UTF-8 format);
7. Header end marker with the information about the size of remaining part (body);
8. (not shown in Figure 1) Body, including all persisted data about drawing process object.

The file format allows assuring backwards compatibility with previous versions of the tool.

The tool is tested on Huion and Wacom tablets and showed similar results and functionality. This once again proved our hypothesis that using WinTab interface can make the tools platform-independent.

C. Main issues

During the development and testing several issues demanded of attention. Most important of they are the following:

- **Render flickering**. *WinForms* framework is oriented to static applications, so it is not very suitable for frequent dynamic redrawing actions. *Panel* control element is redrawing few times per second, however human eye is sharp enough to see the changes, so we see the undesirable effect of flicker. The problem is solved using double buffer for *Panel* control element. Double buffer for entire application does not worked well in some operating systems (e.g. Windows 10).

- **Drawing process recording and playback**. The drawing process can be recorded either as the sequence of drawing actions or frame-by-frame view changes. The former way is changed in order to spare memory and have possibility to playback drawing process in high precision.

- **Drawing, pressure and transparency**. It is not enough just to having a set of coordinates of points, and the features of connecting lines. Computer and its devices are using discrete values, so the line drawn by hand (Figure 2, part 1) can be drawn in the screen with some spaces (Figure 2, part 3). We are using the filling points with the coordinates of the line connections (Figure 2, part 2) in order to solve the problem of undesired spaces between opaque lines. However this solution cannot be

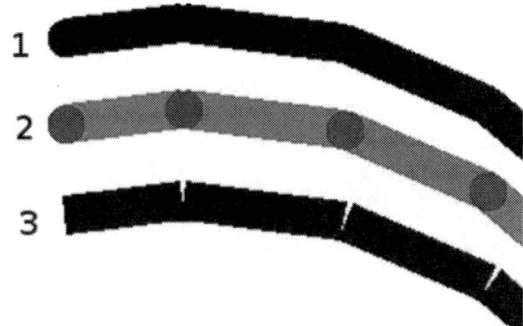

Figure 2. Line and its elements

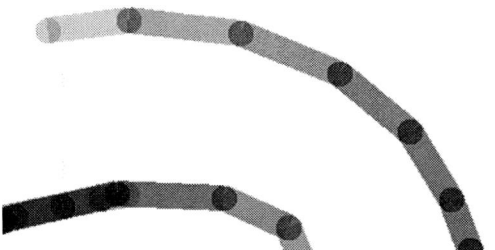

Figure 3. The problem of transparent/gradient lines

applied on the tablets supporting different pen pressure levels and transparent lines. The attempts of use this method for transparent lines is giving not natural look and feel (Figure 3). One of possible solutions is to draw overall trace in separate layer as gradient line using the values of pressure in start and end points. However this solution can be applied in all the cases. For example child can press pen few times while drawing one trace.

All the issues, except the transparent lines are solved in current version of the tool.

III. TESTING

The precision of drawing process playback is measured comparing the "recreated" drawings to original ones. For each pair (recreated image and original) in the sets the following analysis actions using GIMP image processing tool have been done:

1. The images is imported as separate layers;

2. The *blend* feature is set to *difference*. As a result of point-to-point blending equal (by the color) dots are changed to black dots, opposite dots – to white dots;

3. The layers are merged;

4. When the *Threshold* feature is set to 0, and *Antialiasing* is turned on, the *Select by Color* tool of the GIMP enable user to select area with similar color.

If the images are exactly the same, all the image (actually it looks like black rectangle) is select. If some differences are observed and few white dots or lines exists, the tool show them explicitly. For most pairs of testing drawings we observed exactly much or partial mismatch (in terms of only few dots).

With the help of the program it is possible to record the drawing process from the first stroke to the last. Having objective data of the drawing process will allow a look into the drawing - not as the final result of a thought but as a dynamic and changing process. The recording can be viewed at different points in its timeline, showing how elements of the drawing (lines, overtracing, shapes, symbols) are created and change one another. This permits the accurate detection of the moment when a certain element of the drawing was made and its place in the order of the element creation. Recording these drawing features will allow deeper insight into the drawing author's idea or concept for their drawing.

IV. RELATED WORK

H. S. M. Beigi [11] presents the algorithms of data preprocessing before handwriting text recognition. The set of possible forces for the pen is analyzed and used for handwriting result normalization. In our approach such precision isn't necessary because our focus is drawing not text handwriting.

The paper [8] is probably closes to our approach. It presents major techniques, main functions and data structures of WinTab, for developing an e-pen-whiteboard application. Similar to our case they are using Wintab API, and the result is acquired (from whiteboard) and displayed (in the computer screen) drawing, handwriting elements etc. However the purpose of the tool presented in [8] is different. It is designed in order to facilitate education. In contrast to it we are focusing on the acquisition of primary data for future scientific analysis and eye-hand coordination modelling. Because of different purpose in the approach [8], the precision of its tool is tested in different way. The attempt to enhance the lectures using tablets are described in [12, 13] also. However in contrast to our tool, the approach of [12] is based on screen recording of lectures (in mathematics), without any possibility to perform analysis or redraw the dynamics of e.g. formulae writing.

Quite interesting document is Google standard for handwriting recording devices [14]. It describes the details how electronic device for touch recognition should act in order to record touch event most precisely. Similar like in [11] it is implicitly oriented to handwriting recognition. In our case we do not make any new device, instead of this we are using standard means – the digitizer tablet and the software.

V. CONLUSION

Drawing Process Recording Tool presented in this article is device-independent because of the use of WinTab API.

The tool is able not only to record but to playback drawing process. The experiments have shown high precision of playback process. This is because of decision to record drawing process as the sequence of drawing actions instead of recording of view changes as a video movie. Video movie approach would be far more complex in analysis aspect.

Moreover our tool enables to collect data about drawing styles. Because the recorder is able to acquire such aspects of drawing as line thinness, pressure and coloring, this data can be used to identify drawing styles, perform other judgements. The results of this approach is first step in eye-hand coordination modelling project, so the future improvements of the tool are possible including different user interfaces for youngest drawers, for a little bit older children and for the researchers.

REFERENCES

[1] R. Arnheim. Art and Visual Perception: A Psychology of the Creative Eye. Berkeley and Los Angeles: University of California Press, 1974.

[2] E. Do, M. D. Gross, "Drawing as a means to design reasoning". In Artificial Intelligence in Design '96 Workshop on Visual Representation, Reasoning and Interaction in Design. [online] https://depts.washington.edu/dmgftp/publications/pdfs/aid96-mdg.pdf., 1996.

[3] V. Lowenfeld, W. Lambert Brittain. Creative and Mental Growth. The Macmillan Company, New York. Collier'Macmillan Limited, London, 1964.

[4] R. Kellogg. Understanding children's art. In P. Cramer (Ed.), Readings in Developmental Psychology Today. Delmar, CA: CRM, 1970.

[5] D. Nasvytiene. "The analysis of psychometric properties of human figure drawings' test" (In Lithuanian). Psichologija, 2007, Vol. 36, pp. 61–73, 2007,

[6] D. Widlocher "L'interprétation des dessins d'enfants" Liege: Mardaga, 1998.

[7] C. Peiper Mobile Computing: A Primer for Selecting a Mobile Device. June 2011 [online] URL http://74.91.182.85/local/uploads/files/TabletPCWP_final20110609.pdf

[8] R. Zhu and H. Duan, "Techniques for Developing Pen Tablet Aided Instruction Applications," 2007 First IEEE International Symposium on Information Technologies and Applications in Education, Kunming, 2007, pp. 407-410.

[9] R. Poyner. WinTab Interface Specification 1.1: 16- and 32-bit API Reference[S]. Revised May 9, 1996

[10] R. Cohn. WintabDN specification. [online] URL https://sourceforge.net/projects/wintabdn/

[11] H. S. M. Beigi. "Pre-Processing The Dynamics Of On-Line Handwriting Data, Feature Extraction And Recognition". In Proceedings of the International Workshop on Frontiers of Handwriting Recognition. 1996, pp. 255-258.

[12] P. Bonnington et al. "A report on the use of tablet technology and screen recording software in tertiary mathematics courses". In Vision and change for a new century, proceedings of Calafate Delta 7 (2007): pp 19-32.

[13] D.J. Radosevich, P. Kahn. "Using tablet technology and recording software to enhance pedagogy." In Journal of Online Education 2.6 (2006): 3.

[14] A.Green, X.Y.Huang, T.L Schneider. Writing tablet information recording device. Google Patents. 2010/10/21/. [online] URL https://www.google.ch/patents/US20100265214

Gap in pagination due to withheld paper.

Pages 593-597

Modelling of variable shunt reactor in transmission power system for simulation of switching transients

Alan Župan *, Božidar Filipović-Grčić ** and Ivo Uglešić **

* A. Župan is with the Croatian Transmission System Operator Ltd., Kupska 4, 10000 Zagreb, Croatia
e-mail: alan.zupan@hops.hr
** B. Filipović-Grčić and I. Uglešić are with the University of Zagreb, Faculty of Electrical Engineering and Computing, Unska 3, 10000 Zagreb, Croatia
e-mail: bozidar.filipovic-grcic@fer.hr, ivo.uglesic@fer.hr

Abstract - This paper describes a model of three-phase variable shunt reactor (VSR) for simulation of switching transients in EMTP-RV software. Inrush currents caused by VSR energization and overvoltages caused by de-energization are analysed. For this purpose, a model of VSR, substation equipment and electric arc in SF₆ circuit breaker was developed in EMTP-RV software.

Key words: variable shunt reactor, inrush currents, switching overvoltages, electric arc, EMTP-RV

I. INTRODUCTION

Shunt reactors are used in power transmission system for consuming an excessive reactive power generated by overhead lines under low-load conditions. These conditions can increase system voltages above the maximum operating voltage due to the Ferranti effect. By connecting shunt reactors to transmission system, voltages can be maintained within the prescribed limits, which is important for normal operation of high voltage equipment. Fixed shunt reactors are quite often switched on and off, following the load situation in the system. Instead of having two or more shunt reactors with fixed power ratings, a single variable shunt reactor (VSR) could be used for compensation of reactive power.

Energization and de-energization of VSR on frequent basis cases high mechanical and electrical strains for VSR and substation equipment [1]. Occasionally at maximum consumption in electric power system, it is necessary to do the VSR de-energization which causes overvoltage due the small inductive current chopping. Small inductive current chopping is a complex appearance which requires detail modelling of circuit breaker and electric arc.

Overvoltage caused by de-energization may cause an insulation breakdown of VSR. To be protected from this risk, surge arresters are used in VSR bay [2]. Unlike de-energization, VSR energization may cause inrush current with high magnitudes and long-time constants. If VSR has a solidly grounded neutral, this switching operation causes zero-sequence current flow which can activate zero-sequence current relays [2].

To avoid the appearance of high inrush current and overvoltage during VSR de-energization, it is required to

perform controlled switching, a method which eliminates harmful transients via time controlled switching operation. Controlled switching reduces mechanical and dielectric stress of circuit breaker and VSR, and reduces the probability of restrike phenomena in circuit breaker [2], [3].

This paper describes model of a three-phase 400 kV VSR in 400/110 kV substation for simulation of switching transients in EMTP-RV software. Except VSR model, this paper contains model of circuit breaker with electric arc and other high voltage equipment in VSR bay. Also, in this paper inrush currents and overvoltages are calculated caused by VSR switching.

II. MODEL FOR SIMULATION OF VSR SWITCHING TRANSIENTS

VSR bay inside 400 kV substation was modelled in detail including model of VSR, circuit breaker and other high voltage equipment. Five limb core VSR was considered in this paper, with delta connected windings and solidly grounded neutral point. Technical data of VSR are given in Table 1 [4].

TABLE I. TECHNICAL DATA OF VSR

Rated voltage	400 kV	
Rated frequency	50 Hz	
Core type	Five limb	
Reactive power	150 MVAr	75 MVAr
Rated current	216.95 A	108.25 A
Total losses (at 400 kV)	240 kW	150 kW
Zero sequence impedance	1200 Ω per phase	2400 Ω per phase
Capacitance of winding to ground	3.8 nF per phase	

Each phase of VSR was represented by winding inductance L connected in series with resistance representing copper losses R_{Cu}. Resistance R_{Fe} representing iron losses was added in parallel with the winding branch as shown in Figure 1.

Copper and iron losses are calculated from total losses P_{Tot} using the following expressions [5]:

$$P_{Cu} \approx 0.75 \cdot P_{Tot}, \tag{1}$$

$$P_{Fe} \approx 0.25 \cdot P_{Tot}. \tag{2}$$

Resistances representing copper and iron losses in the model are determined from the following expressions:

$$R_{Cu} = \frac{P_{Cu}}{3 \cdot I_n^2}, \tag{3}$$

$$R_{Fe} = \frac{U_n^2}{P_{Fe}}. \tag{4}$$

Zero sequence inductance L_0 of VSR is determined by using the following equation:

$$L_0 = \frac{Z_0}{\omega} \tag{5}$$

The magnetic coupling between the star connected phases was represented by a zero-sequence inductance which provides a path for the zero-sequence current [6].

The calculation of inrush currents requires an adequate modelling of the nonlinear flux–current curve which describes the magnetizing characteristics of the VSR iron core. Recorded RMS voltage–current curves obtained from manufacturer were converted into instantaneous flux–current saturation curves (Figure 2 and Figure 3) which were used in the nonlinear inductance model in EMTP-RV [6] and approximated with two segments.

400 kV SF$_6$ circuit breaker in VSR bay with two breaking chambers was modelled in EMTP-RV considering nonlinear behaviour of electric arc (Figure 4).

Grading capacitors (500 pF) are connected in parallel with breaking chambers.

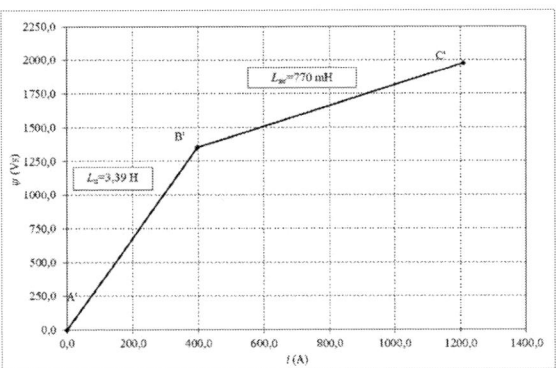

Figure 2. Instantaneous flux-current saturation curve (150 MVAr)

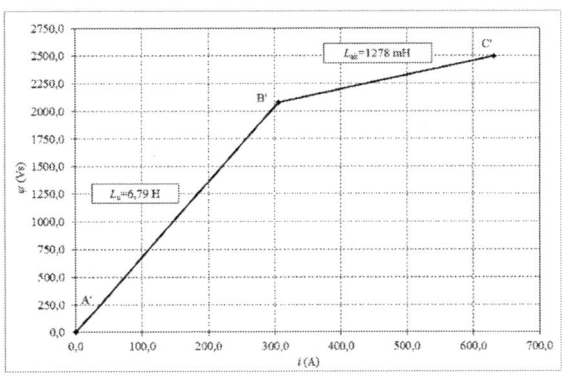

Figure 3. Instantaneous flux-current saturation curve (75 MVAr)

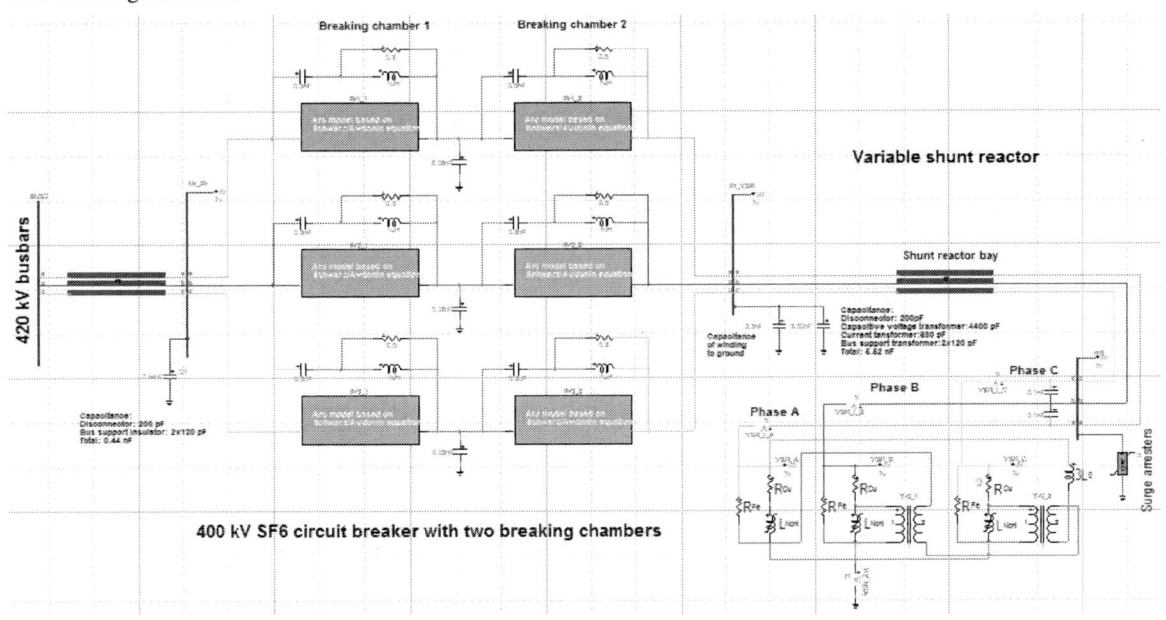

Figure 1. Model of VSR in EMTP-RV

Electric arc was mathematically described with Schwarz/Avdonin differential equation [8], [9], which was solved by using numerical integration in EMTP-RV [7].

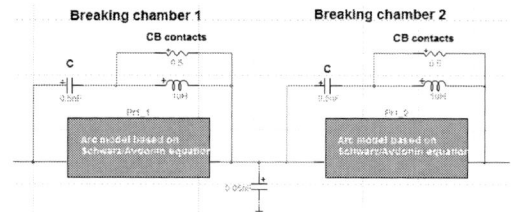

Figure 4. Model of electric arc in EMTP-RV: 400 kV SF$_6$ circuit breaker with two breaking chambers

Since switching of VSR produces high frequency transients, other substation equipment in VSR bay were represented by capacitance to ground (Table II) [10], [11]. Surge arresters were modelled by nonlinear U-I characteristic obtained from manufacturer data.

TABLE II. CAPACITANCE TO GROUND OF HIGH VOLTAGE EQUIPMENT

High voltage equipment	Capacitance (pF)
Disconnector	200
Circuit breaker	60
Current transformer	680
Capacitive voltage transformer	4400
Bus support insulator	120

III. SIMULATION OF SWITCHING TRANSIENTS

Switching transients were simulated in case of 150 MVAr and 75 MVAr reactive power.

A. Uncontrolled energization of VSR

Current waveforms in case of uncontrolled energization of 150 MVAr reactive power at time instants t_A=15 ms, t_B=14 ms, t_C=16 ms are shown in Figure 5 and Figure 6 [4].

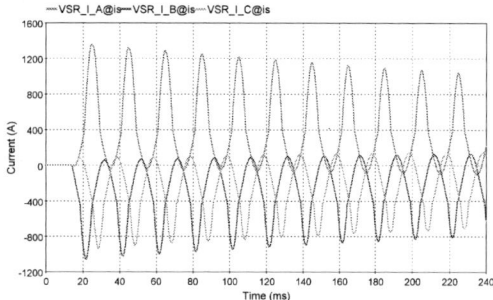

Figure 5. VSR currents: I_{Amax}= 1362.0 A, I_{Bmax}= -1059.0 A, I_{Cmax}= -936.0 A

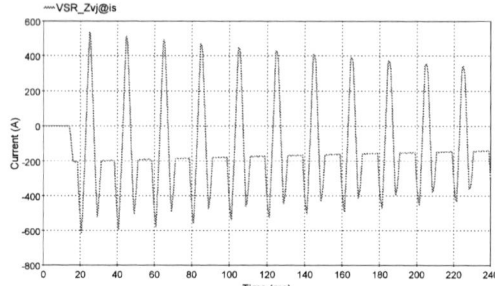

Figure 6. VSR zero-sequence current, I_{Zmax}= -614.0 A

Uncontrolled energization produces inrush currents and zero sequence currents of high amplitudes with relatively long duration. This event may trigger unwanted operation of overcurrent protection relays. Figure 7 and Figure 8 show current waveforms in case of uncontrolled energization of 75 MVAr reactive power.

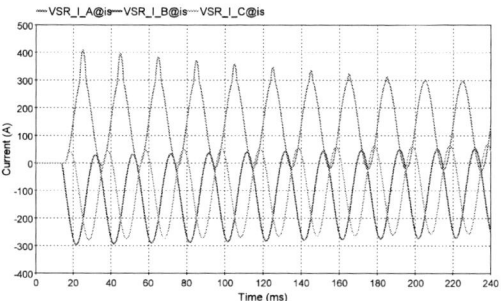

Figure 7. VSR currents: I_{Amax}=408.4 A, I_{Bmax}=-295.1 A, I_{Cmax}=-279.1 A

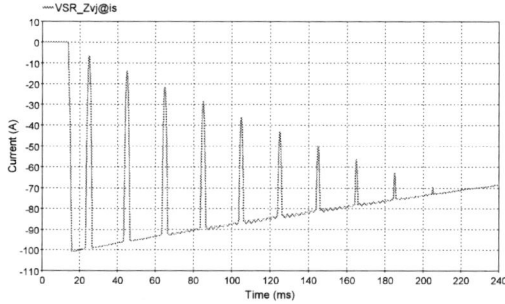

Figure 8. VSR zero-sequence current, I_{Zmax} =-100.9 A

Inrush current amplitudes are lower in case with 75 MVAr reactive power.

B. Controlled energization of VSR

Controlled switching at optimum time instant corresponding to voltage peak value in each phase reduces inrush currents significantly. Current waveforms during controlled switching for 150 MVAr reactive power are shown in Figure 9 and Figure 10, while for 75 MVAr reactive power are shown in Figure 11 and Figure 12.

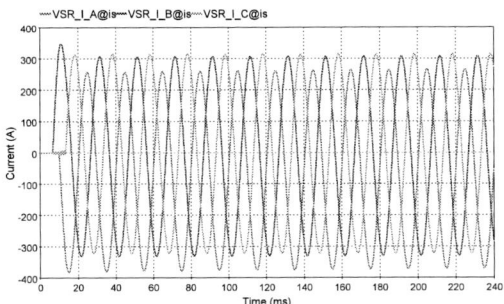

Figure 9. VSR currents: I_{Amax}= -381.0 A, I_{Bmax}=344.4 A, I_{Cmax}= -322.6 A

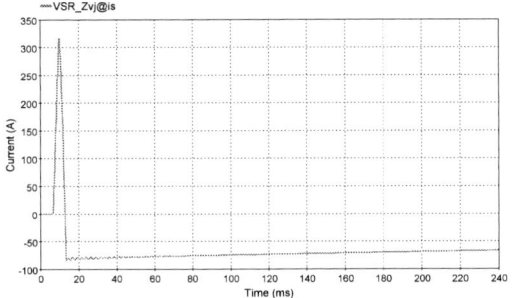

Figure 10. VSR zero-sequence current, I_{Zmax}= 315.5 A

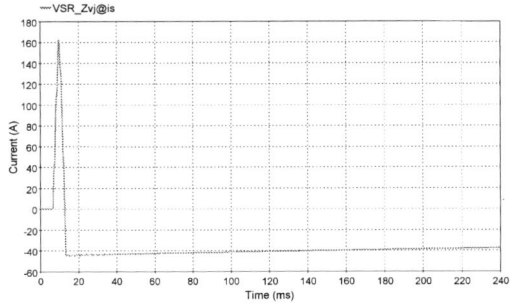

Figure 11. VSR currents: I_{Amax}=-193.2 A, I_{Bmax}=176.9 A, I_{Cmax}=-165.3 A

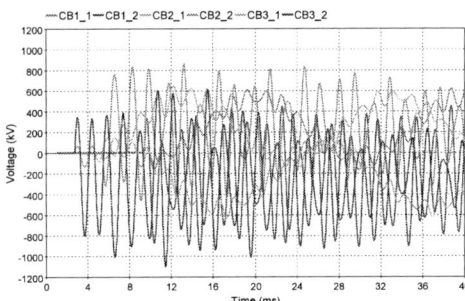

Figure 12. VSR zero-sequence current, I_{Zmax}=162.3 A

C. Deenergization of VSR

Overvoltages across VSR were determined in case of maximum/minimum reactive power. Previously described circuit breaker model was used in simulations. Effect of surge arresters on overvoltage reduction is shown in Table

III. Higher overvoltages appear in case when reactive power of VSR is at minimum value (75 MVAr).

TABLE III OVERVOLTAGES ON VSR

Reactive power (MVAr)	Surge arresters in VSR bay	VSR overvoltages (kV)		
		Phase A	Phase B	Phase C
150	No	550.6	667.0	642.1
150	Yes	-501.1	471.3	516.8
75	No	834.8	880.2	978.9
75	Yes	-540.3	535.2	-542.4

Overvoltage amplitudes across the breaking chambers of circuit breaker during VSR deenergization are shown in Table IV. Surge arrester are not considered in this case.

TABLE IV OVERVOLTAGES ACROSS THE BREAKING CHAMBERS OF CIRCUIT BREAKER

REACTIVE POWER (MVAR)	Voltages across breaking chambers (kV)					
	Phase A		Phase B		Phase C	
	U_{1max}	U_{2max}	U_{1max}	U_{2max}	U_{1max}	U_{2max}
75*	568.9	583.8	561.7	551.0	580.3	576.9
75**	620.1	927.1	632.7	866.1	654.5	1103.3

*with grading capacitors, ** without grading capacitors

Figure 13 and Figure 14 show voltage waveforms across breaking chambers in case with and without grading capacitors.

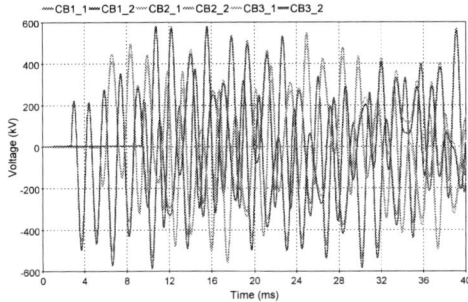

Figure 13. Voltages across breaking chambers (without grading capacitors)

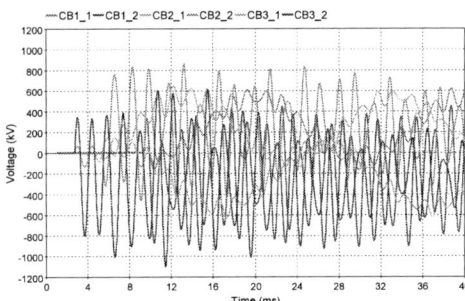

Figure 14. Voltages across breaking chambers (with grading capacitors)

Simulation results show that grading capacitors equalize potential distribution across breaking chambers of circuit breaker. Overvoltage amplitudes across the breaking chambers are higher in case of 75 MVAr reactive power. In case without grading capacitors [12], nonlinear voltage distribution across breaking chambers increases the probability of restrike occurrence inside the circuit breaker which produces steep overvoltages.

Surge arresters reduce both overvoltages on VSR and overvoltages across breaking chambers. Overvoltages on VSR in case of 75 MVAr without surge arresters are shown in Figure 15 [4].

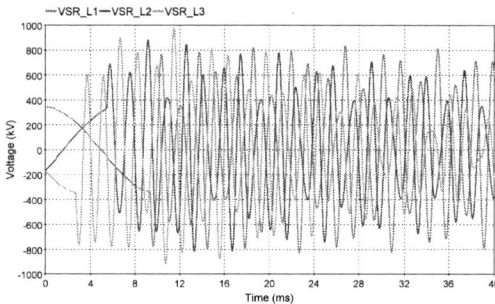

Figure 15. VSR overvoltages: U_{Amax}=834.8 kV, U_{Bmax}=880.2 kV, U_{Cmax}=978.9 kV

Harmonic analysis of VSR overvoltages show that 12th and 17th harmonic have highest amplitudes corresponding to dominant frequencies of voltage oscillations (Figure 16).

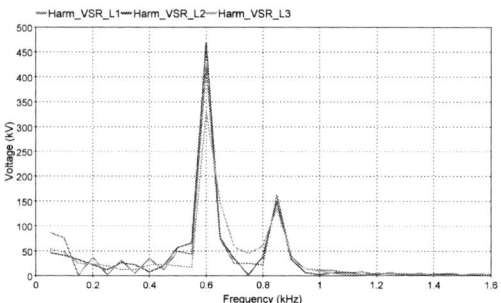

Figure 16. Harmonic analysis of VSR overvoltages

IV. CONLUSION

This paper describes the model of 400 kV variable shunt reactor with reactive power 75-150 MVAr for analysis of switching transients. Apart from variable shunt reactor, SF_6 circuit breaker with an electric arc was modelled in detail in EMTP-RV software.

Transients calculation was performed for controlled and uncontrolled VSR energization at its lowest and highest reactive power. Controlled energization significantly reduces both inrush currents and zero-sequence currents. Simulation shows that inrush currents in case of 75 MVAr are significantly lower compared to 150 MVAr.

Overvoltages on VSR and on circuit breaker were calculated during VSR de-energization. Overvoltages on VSR are significantly higher during de-energization in

case of 75 MVAr. In all cases surge arresters reduce overvoltages on VSR and on breaking chambers.

Calculation shows that grading capacitors equalize potential distribution across the breaking chambers, which reduces the probability of restrike occurrence inside the circuit breaker.

ACKNOWLEDGMENT

This work has been supported in part by the Croatian Science Foundation under the project "Development of advanced high volt-age systems by application of new information and communication technologies" (DAHVAT).

REFERENCES

[1] H. A. Hamid, "Transients in reactors for power systems compensation", School of Engineering, Cardiff University, 2012

[2] A. Župan, B., Filipović-Grčić, D. Filipović-Grčić, "Transients Caused by Switching of 420 kV Three-Phase Variable Shunt Reactor", Electric power systems research. Available online 6 January 2016 (2016)

[3] I. Uglešić, B. Filipović-Grčić, S. Bojić, "Transients Caused by Uncontrolled and Controlled Switching of Circuit Breakers", The International Symposium on High-Voltage Technique "Höfler's Days", 7–8 November 2013, Portoroz, Slovenia.

[4] A. Župan, "Izbor parametara I redosljeda sklapanja regulacijskih prigušnica u visokonaponskoj prijenosnoj mreži", Doctoral Thesis, Faculty of Electrical Engineering and Computing, Zagreb, Croatia, 2016.

[5] M. Khorami, , "Application Fields and Control Principles of Variable Shunt Reactors with Tap-Changer", Investigation of Possible Control Strategies of Variable Shunt Reactors through Simulation of a Power System under Different Operating Conditions, Master of Science Thesis, Department of Energy and Environment, Division of Electric Power Engineering, Chalmers university of technology, Göteborg, Sweden 2011

[6] J. Vernieri, B. Barbieri, P. Arnera, "Influence of the representation of the distribu-tion transformer core configuration on voltages during unbalanced operations", International Conference on Power System Transients (IPST), Rio de Janeiro,2001.

[7] EMTP-RV, documentation, WEB: http://emtp-software.com/

[8] CIGRE WG 13.01 Of Study Committee 13, "State of the art of circuit-breaker modelling", December, 1998.

[9] B. Filipović-Grčić, D. Filipović-Grčić, I. Uglešić, "Analysis of transient recovery voltage in 400 kV SF6 circuit breaker due to transmission line faults", Int. Rev.Electr. Eng. 6 (5 Part B) (2011) 2652–2658.

[10] IEC 60071-4:2004, "Insulation co-ordination – Part 4: Computation guide to insulation co-ordination and modeling of electrical networks"

[11] Ali F. Imece, Daniel W. Durbak, Hamid Elahi, Sharma Kolluri, Andre Lux, Doug Mader, Thomas E. McDemott, Atef Morched, Abdul M. Mousa, Ramasamy Natarajan, Luis Rugeles, and Eva Tarasiewicz, "Modeling guidelines for fast front transients", Report Prepared by the Fast Front Transients Task Force of the IEEE Modeling and Analysis of System Transients Working Group, IEEE Transactions on Power Delivery, Vol. 11, No. 1, January 1996

[12] B. Filipović-Grčić, I. Uglešić, S. Bojić, "Analysis of transient recovery voltage on SF6 circuit breakers when switching unloaded 400 kV transmission lines", The 12th International Symposium on High-Voltage Technique "Höfler's Days", 12–13 November 2015, Portorož, Slovenia.

An Extended Model of a Level and Flow Control System

H. Šiljak, J. Hivziefendić and J. Kevrić

* Electrical and Electronics Engineering Department, International Burch University Sarajevo, Bosnia and Herzegovina
{harun.siljak, jasna.hivziefendic, jasmin.kevric}@ibu.edu.ba

Abstract—**FESTO Compact Workstation is a well known didactic tool in process control. This paper aims at providing an improved transfer function model of this system's level and flow control loops. This higher order model is compared to existing first order system approximations of the level control loop in various input-output scenarios to verify its applicability and superiority. Results are obtained using MATLAB System Identification Toolbox after data acquisition in LabVIEW. MATLAB Simulink is used for cascade PI and single loop PI experiments to show the improvement cascade control on the new model brings. Together with the practical value the results have, the procedure conducted here can serve as a primer and a tutorial for system identification class using this or similar apparatus.**

Keywords—process control; modeling; simulation; level control; flow control

I. INTRODUCTION

The challenge of system identification is often introduced in engineering curricula hand in hand with the challenges of control. In this context, modeling of didactic tools for control systems education is important so the students can use the models for preparation of their control strategies, understand the process of identification, linearity and nonlinearity of models.

The FESTO Compact Workstation [1] as a practical model for process control using four basic variables (level, flow, pressure and temperature) has been extensively used in teaching [2,3] and research [4-7] alike. Efforts have been made to provide models of the workstation, leading to first order systems [4, 5] and second order systems [6] for the level control loop, as well as models for the flow control loop [6,7]. These models served as a basis for control algorithm design for the Workstation.

This paper aims at providing a new model for the level control loop, based on combining the results from level and flow measurements, hence improving the model relevance. Using MATLAB's System Identification Toolbox, several proposed models are compared and the results are discussed in order to explain the differences between the models fitting the real data collected from the Compact Workstation. It is demonstrated that this model can be used for cascade control as well, hence improving control features of the system as well.

Another important contribution of this work is its suitability for educational use, as control engineering students can use it as a tutorial in system identification and readily apply the process described in the paper to the same or similar plant and examine similarities and differences.

The paper is organized as follows. The next two sections give an overview of important modeling aspects in process control and introduce the FESTO compact workstation with the input and output signals collected from it to be used in system identification. Fourth section presents the process of system identification and its results through a comparison of fitting quality. Fifth section discusses the applicability of the model for control (cascade in particular, using PI controllers), while the last section discusses limitations of the study and possible future work.

II. OVERVIEW: MODELING IN PROCESS CONTROL

While modeling and control of processes were considered to be separate disciplines in process engineering, the two have merged long ago [8], with the advent of control algorithms heavily relying on good process models which have taken the leading role [9].

Proportional-integral-derivative control (PID) has been used for almost a century and represents a classical reference algorithm both in theory and practice. The use of plant models for PID tuning has been recognized long time ago [10] as it allows both use of mathematical methods and simulation for determination of optimal controller settings.

Model predictive control (MPC) has been introduced 40 years ago [11] and aims at utilizing reliable models to achieve superior control performance. Ever since its inception, it has been widely applied in the industry [12]. Another method using plant models is the internal model control [8]: both these methods have led to convergence of modeling and process control.

It is interesting to note that a recent survey among control professionals ranked PID, MPC and system identification as the top three control technologies in terms of industry impact, leaving modern control theoretical concepts such as nonlinear, adaptive, intelligent or robust control far behind [13]. This implies the necessity of good models for the industry and control theory research alike.

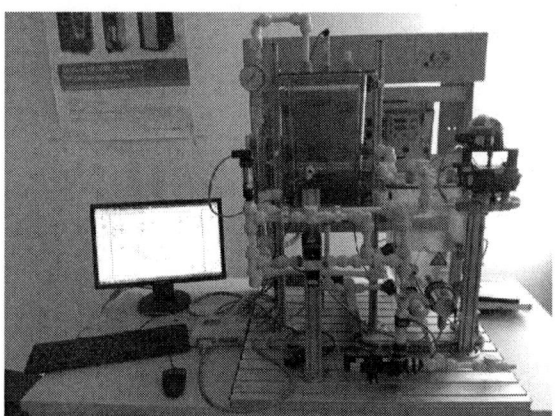

Fig. 1. The laboratory setup

The difficulties in nonlinear [14] and linear [15] system identification are well-known: system order, type of nonlinearity and the everlasting problem of measurement noise.

Even though the water tank system is an inherently nonlinear model, previous research cited in the previous section suggests the possibility for the use of linear models (transfer functions).

The results presented in this paper aim at improving the models used for fluid tank plants by accounting for previously ignored flow dynamics. While it can be useful for experiments done on this particular setup [16], it can also be readily applied to other water tank processes, both with single [17] and multiple tanks [18].

An extension of this work would account for the nonlinearity and produce a model based on a nonlinear differential equation, but again taking into account the flow dynamics.

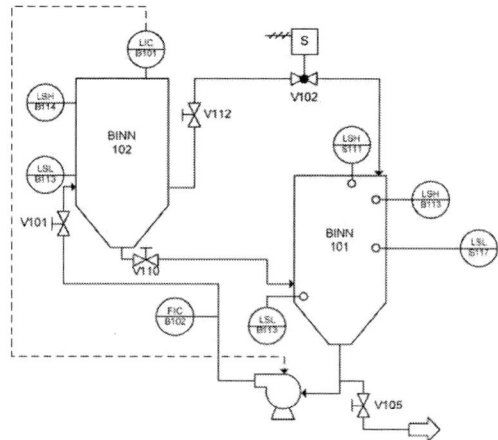

Fig. 2. The level control loop [1]

III. FESTO COMPACT WORKSTATION

A. The Workstation

FESTO Compact Workstation is manufactured by FESTO's division FESTO Didactic as a helping tool in process control education, providing a small-scale, but complete system with possibility of measurement and control of four basic control variables: level, flow, pressure and temperature [1]. The Workstation is easily connected to various controllers and devices, including PC, as shown in Fig. 1. It is consisted of two water tanks, a pump and a set of valves. The pump is used to fill the upper tank with the water from the lower, buffer tank. Different valves in the system can be used to switch between different control loops and/or introduce disturbance in the system, as well as to control the system if the pump voltage is kept constant.

For the purpose of this study, we have focused on the level control loop, the piping and instrumentation diagram of which is shown in Fig. 2 with the addition of FIC (flow instrumentation, i.e. flow sensor) B102. In the experiments conducted, valves V102, V105 and V112 are fully closed, V101 fully open and the return valve V110 is roughly 50% open (i.e. at 45° angle). Data is collected from sensors B101 (level, LIC) and B102 (flow, FIC). At this point it is worth noting that models developed here depend on the state of V110 in a nonlinear fashion, so the exact numerical values shown here are valid just for the position used in the experiment. The overall conclusions about usability of 3rd order system, however, hold for all positions of the valve.

The system is operated through a LabVIEW application developed using a library for Compact Workstation, developed by ADIRO. The same application is easily used for various types of control. It is worth mentioning that FESTO provides standalone applications for control solutions based on LabVIEW, but development of customized control applications using ADIRO libraries is a straightforward task.

B. Signal Collection

The only system input is 0-10 V voltage which is amplified to the 0-24 V (in reality, 0-22 V [5]) input of the pump. The outputs are 0-10 V voltages corresponding to flow and level measurements. Throughout this paper, the sensor readings will be kept in volts, i.e. there will be no conversion to level or flow values.

In order to provide relevant data for system identification, various open loop scenarios were created and data was collected. Namely, the following data collection setups were used:

1. 50 sample hold without offset – for 50 samples at varying sampling rate (between 5 and 10 Hz) a constant random value of input voltage between 0 and 10 V was applied and level and flow sensor voltage was collected. This experiment ended once the buffer tank was left empty.

2. 50 sample hold with offset – once the first experiment was conducted, it was noticed that for low voltages, flow is non-existent. In order to provide realistic data,

another experiment was made under same conditions, but with the random input voltage ranging from 2 to 10 V.

3. 100 sample hold without offset – same as the first experiment, just with the hold time doubled.

4. 100 sample hold with offset – same as the second experiment, just with the hold time doubled.

5. 2V step, non-zero initial condition – the upper tank was more than half full when 2V input voltage was applied.

6. 4V step, zero initial condition – the upper tank was empty when 4V input voltage was applied.

Fig. 3 shows an example of the input/output signals collected (the first experiment).

IV. SYSTEM IDENTIFICATION

A. Physical considerations

From the work done by researchers before [4-7], it is possible to extract useful physical information about the system.

Namely, the system has an inherent nonlinearity caused by the output flow (valve V110) which is proportional to the square root of the level in the upper tank, where the coefficient of proportionality depends nonlinearly on the state of valve V110. This fact can be derived from the Bernoulli's equation [6]. As suggested in [5], the system could be modeled in nonlinear fashion as a feedback with an integrator in the direct branch and the square root nonlinearity in the return branch.

In our study, we are observing the effect of linearization over the whole possible range of output values, not focusing on pointwise linearization. Rationale behind this approach is the aim to provide a model usable in the wide interval of level and flow conditions.

Let us observe a two stage model of the level control system in open loop shown in Fig. 4. which corresponds to what is seen in the P&ID model in Fig 2.

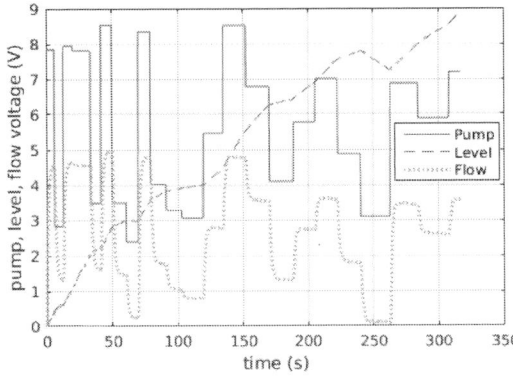

Fig. 3. Input and output voltages in first scenario

In [5-7], the part of the plant model whose input is flow and the output is level was determined to be a first order model following from its derivation from mass balance equation and approximating the square root linearly. However, in [5,6] the dynamics between voltage and flow is neglected (i.e. considered to be a constant gain), while in [7] this part of the model is considered to be a first order system (with a time constant of one second), leading to the entire system being a second-order block.

Our experience in flow control in the same loop suggests that the voltage-flow relationship is (at least) a second order system, due to its oscillatory behaviour and hence we are targeting a third order model for the entire system.

At this point it is worth mentioning that in [5], a nonlinearity in the amplifying stage before the pump is reported. Since the useful range of pump voltages is measured to be 0-22 V, it would be expected that the relationship between the input voltage and voltage applied to the pump is $u_p = 2.2 \cdot u_i$. However, it is shown that the characteristic becomes nonlinear at approximately 8.1 V:

$$u_p = \begin{cases} 2.2\,u_i \text{ for } u_i < 8.1\,V \\ 2.2\,u_i + 5\left(u_i - 8.1\,V\right) \text{ for } 8.1\,V < u_i < 8.68\,V \\ 22\,V \text{ for } 8.68\,V < u_i < 10\,V \end{cases}$$

To make our model as linear as possible, the transfer functions in the identification stage were not level vs. input voltage, but level vs. voltage applied to the pump.

B. Identification in MATLAB

The first scenario (50 sample hold without offset) was used as the basis for model forming, because of its fast changes. After removing few initial samples because of transient disturbances, two models were found using MATLAB System Identification Toolbox [19], a first order and a third order system with a zero:

$$G_1(s) = \frac{0.9262}{222.36\,s + 1}$$

$$G_3(s) = \frac{0.84689\left(0.21248\,s + 1\right)}{\left(221.79\,s + 1\right)\left(0.017111\,s + 1\right)\left(10^{-6}\,s + 1\right)}$$

As one may note, the third order system is actually the first order system with very small time constants corresponding to additional poles and zeros. As such, it does not reflect the model in Fig. 4. Hence, a new model was created by identifying one block from Fig. 4 at a time. The transfer functions obtained for the level output and the flow output, respectively were

$$G_F = \frac{6.1541}{383.4671\,s + 1}, \quad G_L = \frac{0.1414\left(779\,s + 1\right)}{\left(323.0581\,s + 1\right)\left(1.4582\,s + 1\right)}$$

605

so the transfer function of the whole system is given by the expression

$$G_{3F} = \frac{0.8702(779s+1)}{(383.4671s+1)(323.0581s+1)(1.4582s+1)}$$

Note that the level transfer function somewhat corresponds to findings in [7], if we observe only the fast-changing component and cancel the slow zero and pole. Analysis of flow signal shape suggests that an even more precise model of the dynamics would use a saturation block between the two blocks in Fig. 4, limiting flow to positive values.

To compare these models with those in [4-7], the first approximation can be simply multiplying the transfer functions with 2.2 (assuming the amplifier linearity).

C. Model comparison and discussion

After the three models were obtained, their performance was measured on the data sets collected from the Compact Workstation. The results are shown in Table 1 in terms of quality of fitting.

The experiment number five (falling step) results are anomalous because the fitting algorithm could not find proper initial conditions for the 1st and 3rd order models generated by MATLAB. On the other hand, generated 3rd order system's good performance on the data it was generated on may easily be a consequence of overfitting.

As the step function test is a traditional way of assessing models in system identification [19], the last scenario is shown graphically as well in Fig. 5, demonstrating that the auto-generated models either have an undershoot or an overshoot in the open loop step response.

The results in Table 1 suggest that the proposed model (3rd order F, where F stands for flow as the intermediate variable) gives a good approximation of system behaviour in a wide range of situations and therefore approximates the nonlinearities of the system well without strict limitations to small changes around a working point. The results also show that the rich dynamics of the system cannot be entirely covered by a first order system.

V. CASCADE CONTROL APPLICATION

Once a model is found, a control algorithm can be made based on it. While a transfer function representation allows nonlinear and/or adaptive controllers to be built upon it [20], in this particular implementation we will keep the controller linear and use the PI (proportional integral) controller.

Since the model obtained through our identification is composed of two transfer function, this enables us to make a cascade control model, such as the one shown in Fig. 6.

This Simulink model contains additional saturation blocks to ensure the range of inputs and outputs is limited to 0-10 V, but otherwise it is just the two transfer functions identified in the previous section.

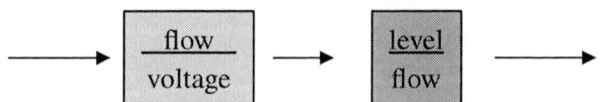

Fig. 4. The two-stage model

TABLE I. MATLAB FITTING RESULTS

Scenario	Fitting		
	1st order	3rd order	3rd order F
50 sample hold without offset	76.42%	**89.89%**	87.72%
50 sample hold with offset	95.72%	84.67%	**97.35%**
100 sample hold without offset	82.43%	89.82%	**93.09%**
100 sample hold with offset	82.57%	**93.30%**	92.72%
2V step, non-zero initial condition	-49.72	-33.89	**83.38%**
4V step, zero initial condition	89.85%	84.09%	**96.30%**

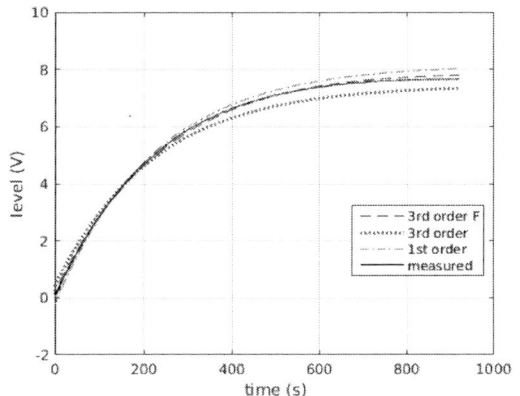

Fig. 5. 4V step, zero initial condition

PI controllers in the outer and inner loop can be tuned using genetic algorithm. This was done by a slight modification of procedures demonstrated in [21], with the cost function aiming at reducing static error, rise time and the overshoot (cost function is equal to the mean square error if the overshoot is less than a predefined threshold, infinite otherwise). In case of first order system approximation, a single PI controller was used as there is no way to extract the flow information. This controller was also tuned and results are shown in Fig. 7.

The cascade control algorithm has obviously produced a fast transient without an overshoot, accelerating the system dynamics significantly, with respect to time constants which can be read from the transfer function. In case of the single loop control of first order system, the acceleration of the transient is not as significant, and an overshoot couldn't be avoided, pursuing a relatively short rise time.

It is worth noting that it is very hard to tune controllers involved in cascade control on the actual plant. Hence, having a model of the system enables us to tune the controllers on the model and then simply enter the controller parameters on the software or hardware controllers used on the actual plant. A model also enables model reference adaptive control or model predictive control to be applied, but these applications are out of scope of our present study.

VI. CONCLUSION AND FUTURE WORK

We have presented a novel higher order linear model of a well-known process and demonstrated its applicability. This model can be improved further: other identification methods can be used to fine tune the time constants and gains of the system, and potential additional sources of high order dynamics can be investigated. The model developed has been used for cascade control as well, once again emphasizing its applicability and usefulness.

Furthermore, the model can be used for design of different control algorithms for this particular plant, as already suggested. This is also an excellent exercise for both undergraduate and graduate students in their process control courses, as they have a chance to prepare control algorithms offline and test them on the real plant. The model can also be used for a graphical demonstration of the plant operation.

The model can also be extended with nonlinear components to closely match the real plant, and further work will be done to identify the effect of the return valve and approximate it linearly. In this particular study, that valve has been kept in the same state throughout the identification process. Additionally, we note that the model used is linear, and efforts will be made to make a new higher order nonlinear model as well.

The approach shown here can be extended and applied to other water tank (flow and level) processes, extending it to MIMO systems in multiple tank environment.

ACKNOWLEDGMENT

The first author wishes to thank ADIRO Automatisierungs-technik GmbH for their excellent software support in the data acquisition segment.

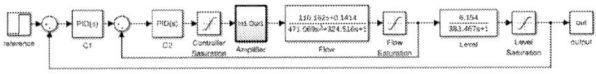

Fig. 6. Cascade Control Simulink model

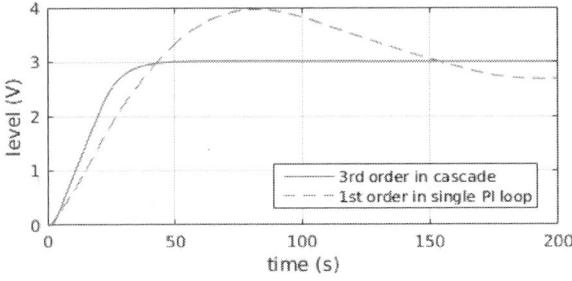

Fig. 7. Cascade control vs. single loop PI control

REFERENCES

[1] J. Helmich, Festo MPS PA Compact Workstation Manual, 2008

[2] E. Brindfeldt, A. Grinko and M. Müür, "Course of automation in industrial processes based on the blended learning approach," 10th International Symposium „Topical Problems in the Field of Electrical and Po wer Engineering" Pärnu, Estonia, pp. 187-192, 2011

[3] D. Sendrescu, M. Roman and D. Selisteanu. "Interactive teaching system for simulation and control of electropneumatic and electrohydraulic systems," Proceedings of the 24th EAEEIE Annual Conference (EAEEIE), pp. 151-156. IEEE, 2013.

[4] V. Skopis, I. Uteshevs and A. Pumpurs, "Advanced Control System Development on the Basis of Festo Training Laboratory Compact Workstation," Journal of Energy and Power Engineering 8(5), 2014

[5] I. Naşcu, R. De Keyser, I. Naşcu and T. Buzdugan. "Modeling and simulation of a level control system," IEEE International Conference on Automation Quality and Testing Robotics (AQTR), vol. 1, pp. 1-6. IEEE, 2010.

[6] P.U.M. Liduário, D. Colón, Á. M. Bueno, E. P. Godoy and J. M. Balthazar. "Identification and practical control of a fluid plant," ABCM Symposium Series in Mechatronics - Vol. 6 , pp. 1095-1104, 2014

[7] E. M. Arruda, E. B. Cavalca and M. S. M. Cavalca. "Evaluation of the characteristics of a robust model-based predictive control approach: a case study of a flow process control with dead zone, noise and power loss.", 22nd International Congress of Mechanical Engineering (COBEM 2013), pp. 3543-3554, 2013.

[8] D.E. Rivera, M. Morari and S. Skogestad. "Internal model control: PID controller design." Industrial & engineering chemistry process design and development, 25(1), pp.252-265. 1986

[9] M. Annergren, C.A. Larsson, H. Hjalmarsson, X. Bombois and B. Wahlberg. "Application-Oriented Input Design in System Identification: Optimal Input Design for Control." IEEE Control Systems Magazine, 37(2) 31-56. 2017

[10] T.S. Schei. "Automatic tuning of PID controllers based on transfer function estimation." Automatica, 30(12), pp.1983-1989. 1994.

[11] J. Richalet, A. Rault, J.L. Testud and J. Papon, "Model predictive heuristic control: Applications to industrial processes." Automatica, 14(5), pp.413-428. 1978.

[12] S.J. Qin and T.A. Badgwell. "A survey of industrial model predictive control technology." Control engineering practice, 11(7), pp.733-764. 2003.

[13] T. Samad. "A Survey on Industry Impact and Challenges Thereof". IEEE Control Systems, 37(1), pp.17-18. 2017.

[14] G.J. Gray, D.J. Murray-Smith, Y. Li, K.C. Sharman and T. Weinbrenner. Nonlinear model structure identification using genetic programming. Control Engineering Practice, 6(11), pp.1341-1352. 1998.

[15] K. Steiglitz and L. McBride. "A technique for the identification of linear systems." IEEE Transactions on Automatic Control, 10(4), pp.461-464. 1965.

[16] M. Patrascu and A. Ion. "Evolutionary Modeling of Industrial Plants and Design of PID Controllers." In Nature-Inspired Computing for Control Systems (pp. 73-119). Springer International Publishing. 2016

[17] S. Krivić, M. Hujdur, A. Mrzić and S. Konjicija. "Design and implementation of fuzzy controller on embedded computer for water level control." In MIPRO, 2012 Proceedings of the 35th International Convention (pp. 1747-1751). IEEE. 2012.

[18] K.H. Johansson. "The quadruple-tank process: A multivariable laboratory process with an adjustable zero." IEEE Transactions on control systems technology, 8(3), pp.456-465. 2000

[19] L. Ljung, System identification: theory for the user. PTR Prentice Hall, Upper Saddle River, NJ, 1999.

[20] Ž. Jurić, and H. Šiljak. "One variant of self-tuning DC motor adaptive controller." 36th International Convention on Information & Communication Technology Electronics & Microelectronics (MIPRO), pp. 936-941. IEEE, 2013.

[21] T-F. Chan and K. Shi. *Applied intelligent control of induction motor drives.* John Wiley & Sons, 2011.

Procedure for Modelling of Soft Tissues Behavior

M. Franulovic*, K. Markovic ** and S. Pilicic***
Faculty of Engineering, University of Rijeka, Croatia
*marina.franulovic@riteh.hr, **kristina.markovic@riteh.hr, ***stjepan.pilicic@riteh.hr

Abstract - The procedural steps for the efficient modelling and simulation of material behavior of human cervical spine ligaments have been presented in the paper. They are based on the mechanical principles incorporated in the material model, suitable for the description of soft tissues behavior. The material parameters set which have to be identified for the presented model has been additionally expanded here to make possible better calibration of the model by the application of genetic algorithm. The genetic algorithm has been recognized in this investigation as an efficient tool to overlap the bridge among the non-linearity of material behavior and material parameters of the chosen model and thus make possible material behavior modelling which follows materials response as accurately as possible. The basic process steps have been developed here for the efficient optimization of soft tissues behavior modelling.

I. INTRODUCTION

Modelling the behavior of mechanical systems and simulation of their behavior in operating conditions has become an unavoidable process in optimization of their design. This refers to both mechanical components which are used in industrial processes to make possible foreknowledge of their mutual interactions and the modelling and simulation of materials behavior out of which mechanical components are produced. Modelling of material behavior consists of description of phenomena that occur within material when subjected to the loading, usually by taking into account possible crystal defects and dislocations on the atomic level [1]. In order to take into account these phenomena, the appropriate constitutive material models should be used, which have incorporated appropriate physical principles and material parameters. Constitutive models are basically mathematical simplifications which take into account the specific dependencies among variables, which characterize the observed system [2]. These models are usually complex non-linear systems and their calibration should be based on the experimentally obtained results in order to make possible their validation [3], which means to approve the accuracy of the systems simulated by the calibrated model.

Today more and more attention is given to the usage of some innovative materials, such as biomaterials [4]. Besides their usually advantageous mechanical properties, which can be used in modelling of technical structures in different operating conditions, the possibility to model biomechanical systems and simulate their behavior in specific conditions on the basis of calibrated material

models, became the point of interest of various investigations [5]–[7]. The specific direction in biomechanical systems investigations became the modelling of behavior of soft tissues [8], which have unique structure, and thus complex physical principles are expected to be applied for their modelling.

The procedure for calibration of material model for soft tissues behavior, in this case human cervical spine ligaments is the focus of this investigation. Although there are several material models developed for modeling material behavior of this biomaterial, the commonly used one [9] is based on the theory of hyperelasticity. The presented model can be furtherly expanded to take into account damage growth and failure. One of the prerequisites to accurately calibrate this material model is determination of procedures for the material parameter identification.

II. MATERIAL MODEL

The aim is to define material model suitable for modelling and simulation of behavior of the human cervical spine ligaments when human body is subjected to some specific movements. Ligaments are connective tissues that are bone to bone linkages with function of restriction of relative motion between connected bones. Human cervical spine ligaments are complex fiber-reinforced composite structures with a predominant natural fiber direction.

The mechanical behavior of soft tissues strongly depends on structural arrangement and concentration of major phases of mentioned material which are collagen fibers, elastin fibers, proteoglycans, cells and interstitial fluid phase [6]. Arrangement of these phases determines material properties of considered materials such as anisotropy, viscoelasticity and ability to undergo large deformations [10]. As mentioned, human cervical spine ligaments are fiber-reinforced and the fibers that give tensile strength to the tissue are collagen fibers.

As known, human cervical spine ligaments have non-linear stress-displacement response until failure when load increases uniformly, which is recorded during tensile tests. The characteristic stress-displacement curve for investigated material is shown in Figure 1, which is adopted from [6].

It consists of three characteristic regions called toe region until the point A, linear region between points A and B and damage (failure) region behind the point B, which corresponds to the traumatic phase.

This work has been supported by Croatian Science Foundation under the project number IP-2014-09-4982 and also by the University of Rijeka under the projects number (13.09.1.2.09) and (13.09.2.2.18).

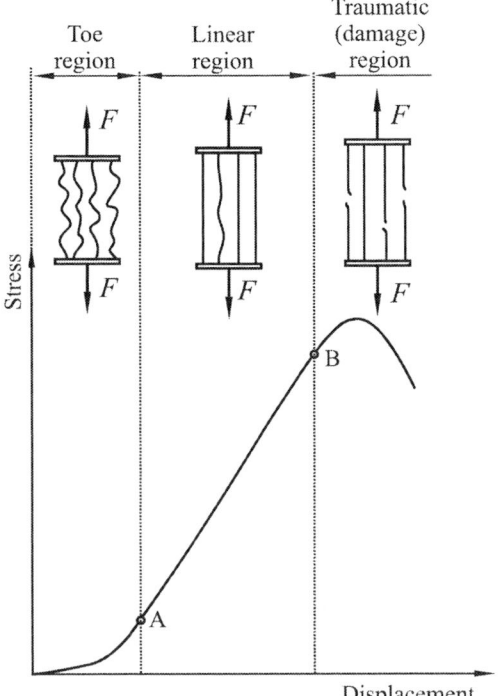

Figure 1. Characteristic force - displacement material response, adopted from [6]

In a relaxed position (no load applied) the collagen fibers are crimped. During nonlinear toe region collagen fibers are uncramping and tend to line up with load direction. While uncramping collagen fibers stiffness is low. In linear region collagen fibers are straighter, mutually aligned and stiffer. When ultimate tensile stress is reached, the fibers start to break which refers to the existence of the damage region.

In this paper only toe region and linear region have been considered for the development of the procedure for the modelling of soft tissues behavior. In order to simulate material behavior as accurate as possible, the hyperelastic material model has been used. Hyperelastic model is suitable for numerical modeling of soft tissues because it allows anisotropy and large deformations.

Mathematical formulation of mentioned model [11] is given by the expressions (1) for toe region and (2) for linear region:

$$\sigma_1 = C_1(\lambda^2 - \frac{1}{\lambda}) + 2\lambda C_3(\exp(C_4(\lambda^2 - 1)) - 1), \quad (1)$$

$$\sigma_2 = C_1(\lambda^2 - \frac{1}{\lambda}) + 2\lambda(\frac{C_5}{\lambda} + \frac{C_6}{\lambda^2}), \quad (2)$$

where stresses σ_i are derived from the force values, deformation is denoted by λ, while C_i are material

parameters, which calibration influence the numerical model accuracy.

Although the presented material model is suitable for modelling and simulation of investigated material behavior in two characteristic regions of materials life, the position of the point A is neglected in the mathematical formulation. Nevertheless, the importance of its position is significant, because it defines the applicability of given expressions on the chosen range of data points. Therefore, the proposed material model is upgraded by the additional characteristic value Λ, which is introduced as the deformation point which mark off toe and linear region. The value Λ has to be determined on the basis of condition $\sigma_1 = \sigma_2$.

III. NUMERICAL PROCEDURES

In order to assess mechanical properties of the investigated material, tensile testing has to be implemented, together with the use of specifically developed numerical procedures. Data acquired by testing and the results obtained using a numerical methods are the basis for the conduction of the analysis, which is in turn, together with the previously defined material model, the basis for material parameter identification. Parameters identification can't be effectively completed (or at least feasible values can't be obtained) without the use of corresponding controls. The controls include both controls for the accuracy of testing and controls related to the use of numerical methods. The process of evaluation is shown schematically in Figure 2.

For the conduction of the numerical procedure, it is decided that an evolutionary algorithm shall be used for presented case it would be genetic algorithm. In order to define effective genetic algorithm for the parameter identification process, inverse analysis should be used. Inverse analysis itself represents a method of searching an unknown characteristic(s) of a sample by observing its response on a given stimulating signal.

Inverse analysis consists of three steps or phases (not counting the initial definition of a problem, which has been done before), all of which are shown in form of the flow diagram in Figure 3. The first step deals with the characterization of the system. That means the minimal set of model parameters must be defined, considering that the set must wholly characterize the system. After the characterization, the so-called „forward modelling" (the second step) is used to foresee the test results, as well as the behaviour of the system. Forward modelling is founded upon mechanical principles of the behaviour of the chosen material. The third step consists of what is called „inverse (backward) modelling". Inverse modelling is based on the materials response recorded through tests, in this case obtained by tensile testing, which are used to influence the values of the model parameters in order to define the system as closely as attainable [12], [13]. In order to do this, an objective function must be set, as evaluation of the parameters would be impossible without it.

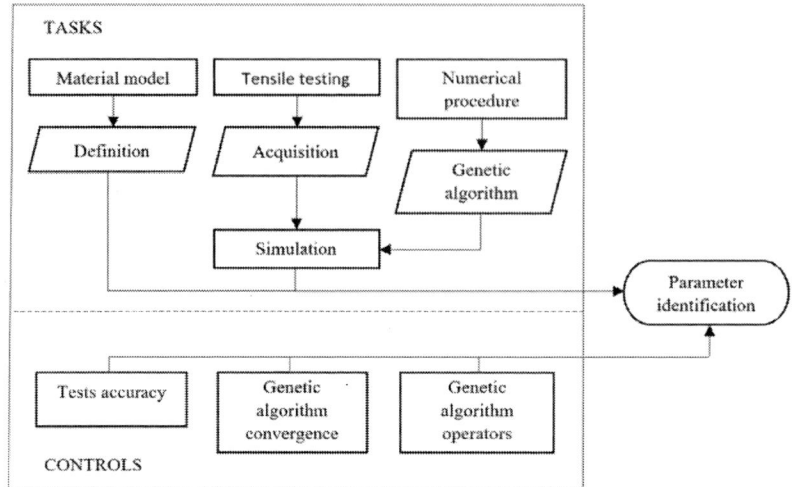

Figure 2. Process of material parameter identification

In this case, the objective function is defined on the basis of:

$$\sigma = \hat{\sigma}(\lambda; a_i), \qquad (3)$$

$$a_i = [C_1, C_3, C_4, C_5, C_6, \Lambda], \qquad (4)$$

where $\hat{\sigma}$ reflects to the mapping function which couples stress-displacement relationships of both the experimental and computed data points with the material parameters.

The objective function here may be chosen among different variants [14]–[16], which are all expected to be compatible with the genetic algorithm procedure for the material parameter identification. Therefore, the application of simple objective function is expected to be valid in the form:

$$f = \sum_{i=1}^{n} \left[\frac{\sigma_i^* - \hat{\sigma}(\lambda_i^*; a)}{\sigma_i^*} \right]^2 , \qquad (4)$$

where n refers to the number of data points, while asterisk denotes the experimental values.

To achieve the solution, the genetic algorithm (flow diagram shown in Figure 4) is used, as was stated previously. The first step of the genetic algorithm is the generation of the initial population of solutions, which means generation of different sets of parameters. This must be done having in mind the point of interest (the problem for which the genetic algorithm is applied) and the capacity of the equipment (computer hardware and software) on which the algorithm is run. After the initial population is generated, the fitness evaluation must be done [13]. This is done using the previously mentioned objective function (or fitness function).

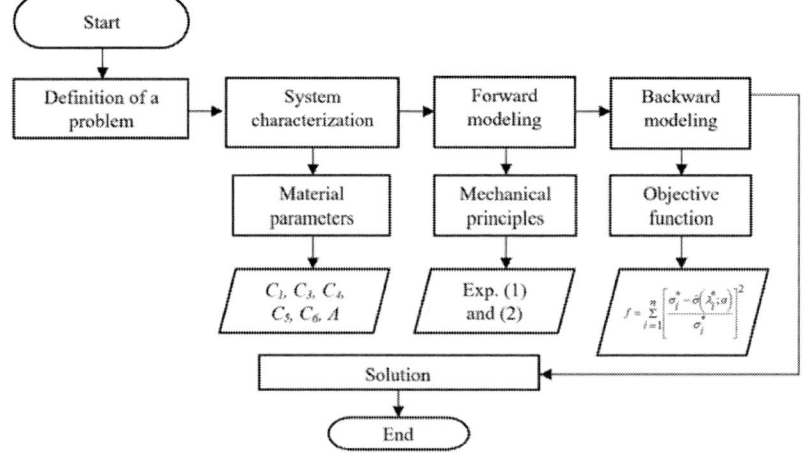

Figure 3. Inverse analysis

610

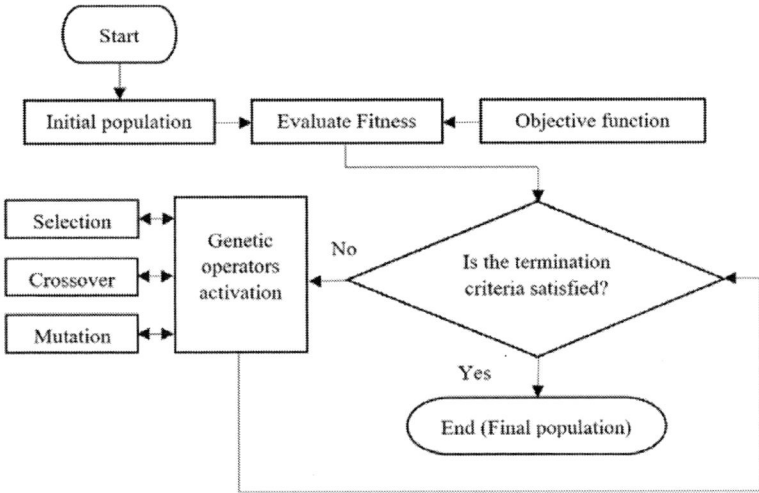

Figure 4. Genetic algorithm flow diagram

Once the fitness is evaluated, there rises a question of the termination criteria. In the best case, termination happens when the optimal solution is reached (in other words, that the solution satisfies the fitness function). However, several other criteria shall be set, to prevent endless run of the algorithm in case the solution is not found. Those criteria include termination after a certain number of evaluations, after a diversity level of the population is too low (below the chosen threshold that depend on the problem). As with the population size, the termination criteria are set with having in mind the equipment capacity.

If the initial population's fitness doesn't satisfy the criterion based on fitness, what is almost always the case, the genetic operators must be activated in order to create a new population (or new generation). The term „genetic operators" encompasses three different operators: selection, crossover and mutation. Selection is a process of choosing which of the members of the population will be chosen to participate in the creation of the next generation, either as „parents" of different „offsprings", created by crossover or mutation, or by themselves (as clones – identical offsprings). Selection process needs to be connected with the fitness and several different approaches exist. Usually, scaling or ranking is required to conduct selection as planned. After the selection, if crossover and/or mutation criteria are satisfied and the respective operators activated, new members of the population are created. Same applies if there is a criteria for cloning (usually reserved for the best individuals) [17].

Crossover is a process of recombination of the previously selected parents. The easiest to understand its definition would be that it is a replacement of some of the first parent's alleles with the alleles of the second parent. For crossover to be valid, the alleles must be the alleles of the same gene. The third (and final) genetic operator is the mutation. The process of mutation is less complicated than the process of crossover and consists of the selection

of a gene or a subset of them and the change of the value of the chosen gene's allele. It should be noted that mutation always produces a different string, but a crossover can create clones of their parents if the corresponding alleles are same in both parents, although chance for this reduces if more alleles are replaced [12].

Once all the genetic operators are applied, a new population is created and it is again evaluated in the same way as the initial population was evaluated before any of the operators were applied. If the new population satisfies the fitness condition(s), the process ends and the population at the end is named the final population. If it does not satisfy it, the process of applying the genetic operators is repeated until one of the previously mentioned termination criteria is satisfied.

IV. CONCLUSION

In order to make possible efficient calibration of material model suitable for the modelling and simulation of soft tissues behavior, the numerical procedure has been developed. Soft tissues, in this case human cervical spine ligaments, are expected to demonstrate highly non-linear behavior and undergo large deformations in the loading conditions. Therefore, material model which is suitable for material behavior description consists of several mathematical relations, which cover different characteristic behavioral regions of material life. Consequently, in addition to the material parameters of the chosen well-known material model, another material parameter has been introduced, which will make possible fast and reliable system characterization and its calibration.

The mechanical principles for the material behavior of the investigated material model have been chosen for the successful inverse analysis. This allowed the definition of the objective function for the given problem and thus possibility to apply an evolutionary method, such as genetic algorithm for the optimization of identification of material parameters.

The procedural steps for the efficient genetic algorithm for the parameter identification have been detailed in this investigation. They are the basis for the development of the genetic operators which should be suitable for a given problem. They have to be developed in a way that experimental data from the tensile tests can be compared with the materials response calculated by the possible result sets gained through the genetic algorithm procedure. The deviation between simulated and experimental materials response should be minimized, which can be assured by the proper choice of objective function and evaluated through the developed procedure.

The validation of the developed procedural steps should follow by using datasets obtained through experimental procedures, which will allow to further develop suitable genetic operators.

ACKNOWLEDGEMENTS

This work has been supported by Croatian Science Foundation under the project number IP-2014-09-4982 and also by the University of Rijeka under the projects number (13.09.1.2.09) and (13.09.2.2.18).

REFERENCES

[1] J. (Jean) Lemai tre and J.-L. Chaboche, Mechanics of solid materials. Cambridge University Press, 1990.

[2] N. S. Ottosen and M. Ristinmaa, The mechanics of constitutive modeling[Recurso electrónico]. 2005.

[3] L. E. Schwer, "An Overview of the PTC 60 / V&V 10 Guide for Verification and Validation in Computational Solid Mechanics," 2007.

[4] Z. Qin, L. Dimas, D. Adler, G. Bratzel, and M. J. Buehler, "Biological materials by design.," J. Phys. Condens. Matter, vol. 26, p. 73101, 2014.

[5] A. Trajkovski, S. Omerović, M. Hribernik, and I. Prebil, "Failure properties and damage of cervical spine ligaments, experiments and modeling.," J. Biomech. Eng., vol. 136, no. 3, pp. 31002-1-31002–9, 2014.

[6] N. Yoganandan, S. Kumaresan, and F. A. Pintar, "Biomechanics of the cervical spine. Part 2. Cervical spine soft tissue responses and biomechanical modeling," Clinical Biomechanics, vol. 16, no. 1. pp. 1–27, 2001.

[7] M. Nordin and V. H. Frankel, "Biomechanics of Articular Cartilage," in Basic Biomechanics of the Musculoskeletal System, Lippincott., 2001, pp. 60–97.

[8] M. Freutel, H. Schmidt, L. Dürselen, A. Ignatius, and F. Galbusera, "Finite element modeling of soft tissues: Material models, tissue interaction and challenges," Clin. Biomech., vol. 29, no. 4, pp. 363–372, 2014.

[9] J. a Weiss and J. C. Gardiner, "Computational modeling of ligament mechanics.," Crit. Rev. Biomed. Eng., vol. 29, no. 3, pp. 303–371, 2001.

[10] Y.-C. Fung, Biomechanics Mechanical Properties of Living Tissues. Springer, 1993.

[11] J. a. Weiss, B. N. Maker, and S. Govindjee, "Finite element implementation of incompressible, transversely isotropic hyperelasticity," Comput. Methods Appl. Mech. Eng., vol. 135, no. 1–2, pp. 107–128, 1996.

[12] M. Franulovic, R. Basan, and K. Markovic, "Material behavior simulation of 42CrMo4 steel," in SIMULTECH 2016 - Proceedings of the 6th International Conference on Simulation and Modeling Methodologies, Technologies and Applications, 2016, pp. 291–296.

[13] M. Franulović, R. Basan, and I. Prebil, "Genetic algorithm in material model parameters' identification for low-cycle fatigue," Comput. Mater. Sci., vol. 45, no. 2, pp. 505–510, 2009.

[14] T. Furukawa, T. Sugata, S. Yoshimura, and M. Hoffman, "An automated system for simulation and parameter identification of inelastic constitutive models," Comput. Methods Appl. Mech. Eng., vol. 191, no. 21–22, pp. 2235–2260, 2002.

[15] R. Fedele, M. Filippini, and G. Maier, "Constitutive model calibration for railway wheel steel through tension–torsion tests," Comput. Struct., vol. 83, no. 12–13, pp. 1005–1020, May 2005.

[16] D. Szeliga, J. Gawąd, and M. Pietrzyk, "Parameters identification of material models based on the inverse analysis," Int. J. Appl. Math. Comput. Sci., vol. 14, no. 4, pp. 549–556, 2004.

[17] C. R. Reeves and J. E. Rowe, Genetic Algorithms—Principles and Perspectives. Kluwer Academic Publishers, 2002.

Analysis of ERASMUS Staff and Student Mobility Network within a Big European Project

Miloš Savić*, Mirjana Ivanović*, Zoran Putnik*, Kemal Tütüncü**,
Zoran Budimac*, Stoyanka Smrikarova***, Angel Smrikarov***

* University of Novi Sad, Faculty of Science, Department of Mathematics and Informatics, Novi Sad, Serbia
** Selcuk University, Technology Faculty, Konya, Turkey
*** University of Ruse, Department of Computing, Ruse, Bulgaria

milos.savic@dmi.uns.ac.rs, mirjana.ivanovic@dmi.uns.ac.rs, zoran.putnik@dmi.uns.ac.rs, ktutuncu@selcuk.edu.tr,
zoran.budimac@dmi.uns.ac.rs, SSmrikarova@ecs.uni-ruse.bg, asmrikarov@ecs.uni-ruse.bg

Abstract - The academic mobility is one of key factors that enable the globalization of research and education. In this paper we study the network of ERASMUS staff and student exchange agreements between academic institutions involved in FETCH – a big European project oriented towards future education and training in computer science. The structure of the network was investigated relying on standard metrics and techniques of social network analysis. Obtained results indicate that the network is in a mature phase of the development in which none of the institutions has a critical role to the overall connectedness of the network. Additionally, the network has a clear core-periphery structure with an active core and mostly inactive periphery.

I. INTRODUCTION

Over the past years, we have witnessed a continuous trend of enlargement of number of exchanges for students and lecturers between European countries, under various exchange programs, and especially under the Erasmus programme. The main goal of the Erasmus programme is to promote and facilitate mobility in higher education. The programme was inaugurated in 1987, and today it is the largest programme for the academic mobility in Europe: it has enabled more than 2 million students to temporary study in another European county [5], [6]. Moreover, there are a large number of signed bilateral ERASMUS agreements between European Universities for exchange of both students and lecturers. These agreements, and therefore exchanges, help not only as a tool for personal development and introduction to other cultures, but also improve future chances for easier and better opportunities for employability, especially for students from central and eastern European countries [15]. In turn, this leads to improved chances for future social cohesion and economic development not only at local and regional, but also at the European level, promoting the further development of labor market.

In this paper we study a ERASMUS staff and student mobility network formed by academic institutions participating in the FETCH project (Future Education and Training in Computing: How to support learning at anytime anywhere) [4]. The project encompasses 67 universities and companies from 35 European countries. The principal objective of the project is to raise the quality of computing education with novel methodologies,

didactical theories, learning models and innovative technologies. More specifically, the main goals of the project are the development of strategic and evaluation frameworks for modern computing education and training, preparation of recommendations for future digital curricula, and the development of theories and models for social media education. The project is divided into 9 working packages where one of them, Work Package 8 (WP8), is partially devoted to the creation of a European Erasmus network for student and lecturer exchange in the field of computing education. The main goal of this paper is to evaluate the current state of the ERASMUS mobility network among institutions participating in WP8.

The rest of the paper is structured as follows. The second section gives the insight into related works performed on similar topics. The third section explains the data set used to extract the network. The methods used in the analysis of the network are presented in the subsequent, fourth section of the paper. The obtained results are presented and discussed in the fifth section. The last section concludes this article and outlines directions for possible future research work.

II. RELATED WORK

Analysis of topologies of real-world networks, and in particular Erasmus networks of student exchange has been performed in the past. For example, one of the interesting papers dealing with this subject is [3]. The authors constructed a directed and weighted graph of Erasmus collaboration among universities determined by Erasmus student mobility realized in 2003. Then they analyzed the structure of its non-directed and non-weighted (NDNW) projection. The results of the analysis showed that the NDNW projection of the network contains a giant connected component exhibiting the small-world property. The degree distribution analysis revealed that the NDNW projection of the network does not belong to the class of scale-free networks, but to the class of random graphs with exponential degree distributions. Finally, the authors showed that the configuration model of random networks can reproduce and explain empirically observed structural characteristics of the network.

Another important article giving valuable insight into student mobilities is [16]. This article tries to analyze interaction patterns of Erasmus students. Among other things, authors conclude that "...our study shows that students' networks abroad are already formed before actual departure." Also, authors claim that they "...provide empirical evidence that institutional as well as group practices encourage or impede interaction between exchange and local students."

The fact and consequences of "...studying abroad can directly or indirectly influence students' career paths and potentially offers additional job opportunities..." is investigated in [1]. Authors were using the similar method as the one used in this article, and claim that "...with the help of an advanced network analytic method – the island approach – we were actually able to determine the groups of institutions in the network which collaborate most extensively inside of Erasmus student exchange programme."

A very interesting conclusion concerning the similar researched subject is given in [14]. Namely, the author discovered that "...although studying abroad led to increased socializing with other Europeans, contact with host country students remained limited." Considering the so-called "European identity" and the influence of studying abroad on it, there are two conclusions given in a mentioned article:

- ERASMUS does not strengthen students' European identity, on the contrary, it can have an adverse effect on it, and

- increased socializing with Europeans has a positive, though modest, impact on European identity.

As a continuation of this practical kind of research of mobilities induced by ERASMUS programmes, it is also important to mention article [11] that investigates "...the factors influencing these student flows." As might be expected, "...country size, cost of living, distance, educational background, university quality, the host country language and climate are all found to be significant determinants." Yet, the main finding that authors mention is that "...despite the financial support granted by the EU and other institutions, the cost of living differences and distance are still relevant when explaining European student mobility (ESM) flows." Based on this finding, authors conclude that "...it is evident that more economic support could enhance ESM."

Finally, research on the similar subject is often conducted within one country. An example of research of this kind can be found in [10]. The paper "...describes two projects based on the issue of virtual mobility...." where "...special attention is devoted to the efficiency of virtual mobility." One of the interesting conclusions authors gave is that "...beside gained knowledge they enrich their students' lives with worthy experience collected by their study in different cultural environment."

Contrary to previously mentioned studies, in this paper we analyze an Erasmus mobility network encompassing institutions involved in a large, but specific project related to computer science education. The primary goal of the study is to examine whether the project itself imposed a stimulating environment for establishing and realizing Erasmus exchanges among participating institutions.

Therefore, we investigate the structure of the network using social network analysis methods in order to assess its cohesiveness, compactness and robustness. Additionally, we analyze the structure of active links in the network in order to identify key institutions in the actual realization of established Erasmus agreements.

III.   DATA SET

This study is based on the data about ERASMUS staff and student exchange agreements reported from 37 institutions involved in the Work Package 8 (WP8) of the FETCH project. The members of the project involved in activities of WP8 were asked to deliver the following information for their institutions:

1.   the list of institutions with whom they have settled ERASMUS exchange agreements in computer science domain (including both FETCH partners and institutions that do not participate in the FETCH project),

2.   the number of student exchange agreements and the number of staff exchange agreements per each ERASMUS partner, and

3.   the numbers of realized student and staff exchange agreements for the last three school years per each ERASMUS partner.

The FETCH institutions who responded to the survey are from the following European countries: Albania (1 institution), Austria (1), Bulgaria (8), Croatia (1), Cyprus (1), Czech Republic (1), Denmark (1), Estonia (1), Finland (1), Germany (1), Greece (1), Island (1), Italy (3), Liechtenstein (1), Lithuania (1), Luxembourg (1), FYR Macedonia (1), Portugal (1), Romania (1), Serbia (1), Slovakia (1), Spain (1), Sweden (1), Turkey (3) and UK (2).

IV.   METHODS

From the collected data we reconstructed the FETCH network of ERASMUS bilateral agreements. The nodes of the network correspond to the institutions that responded to the data collection survey. Two institutions are connected in the network if they signed one or more student or staff ERASMUS exchange agreements. The nodes in the network are connected by undirected links since signed ERASMUS exchange agreements between the institutions participating in the FETCH project are bilateral. The link between A and B is considered active if there were student or staff exchanges between *A* and *B* in the last three school years. Consequently, an institution in the network is considered active if the corresponding node in the network is incident to at least one active link.

Standard techniques and metrics used in social network analysis are employed to analyze the FETCH network of ERASMUS bilateral agreements. The structure of the network is investigated using connected component analysis and *k*-core decomposition [8], [9], [13]. A node is reachable from some other node if there is a path connecting these two nodes. A connected component of a network is a maximal set of mutually reachable nodes. If a component encompasses a vast majority of nodes then we say that the network has a giant connected component [9]. Connected components can be identified using classical graph traversal algorithms such as BFS (breadth first search) and DFS (depth first search). The absence of a giant connected component in the FETCH network of ERASMUS agreements indicates that the mobility network of institutions involved in the FETCH project is still in an early phase of its development and that it consists of several, small independent clusters among which there are no mobility flows. A node is called articulation point if its removal from the network increases the number of connected components. A giant connected component can be considered robust if there is no articulation point whose removal disintegrates the component into several non-giant connected components. A robust giant connected component implies that none of the institutions in the network has the critical role to the overall connectedness of the network.

The degree of a node in the network is the number of links incident to the node. In our case the degree of a node is also equal to the number of other nodes to which the node is directly connected since the network does not contain parallel links (different links connecting the same pair of nodes). A node is called isolated if its degree is equal to zero. Isolated nodes in the analyzed network actually represent FETCH institutions that have not established ERASMUS exchange agreements with other FETCH partners.

To quantify compactness of connected components and their connected sub-networks we rely on two network statistics: characteristic path length and diameter [8]. The distance between two nodes is defined as the length of the shortest path connecting them. The characteristic path length is the average distance of all pairs of nodes in the component, i.e.

$$l = \frac{N(N-1)}{2} \sum_{i,j \in V, i \neq j} d(i,j),$$

where V denotes the set of nodes in component, N their number and $d(i, j)$ is the distance between nodes i and j. On the other hand, the diameter of the component is the maximal distance between nodes, i.e.

$$D = \max_{i,j \in V} d(i,j).$$

The transitivity of links in the network is measured by the clustering coefficient [17]. The clustering coefficient is a probability that two neighbors of a randomly selected node are neighbors among themselves, or, equivalently the density of links among neighbors of a randomly selected node. More formally, the clustering coefficient of node i can be expressed as

$$C(i) = \frac{2}{k_i(k_i - 1)} \left| \{ e_{jk} : j, k \in N_i, \ e_{jk} \in E \} \right|,$$

where E denotes the set of links in the network, k_i is the degree of node i and N_i is the set of nodes connected to i.

Node centrality metrics can be employed to identify the most important nodes in a network considering its structure. In our analysis we employ betweenness, closeness and eigenvector centrality measures [9] to rank and identify the most important institutions within the network. The betweenness centrality of a node z, denoted by $BET(z)$, is the extent to which z is located on the shortest paths connecting two arbitrary nodes different than z, i.e.

$$BET(z) = \sum_{\substack{x,y \in V \\ x \neq y \neq z}} \frac{\sigma(x,y,z)}{\sigma(x,y)},$$

where $\sigma(x, y)$ denotes the number of shortest paths connecting x and y and $\sigma(x, y, z)$ is the number of shortest paths between x and y that pass through z. The closeness centrality of z, denoted by $CLO(z)$ is inversely proportional to the cumulative distance between z and other nodes in the network, i.e.

$$CLO(z) = \left(\sum_{x \in V, x \neq z} d(x,z) \right)^{-1}.$$

The intuition behind the eigenvector centrality measure is that a node can be considered important if it is surrounded by important nodes. This means that the eigenvector centrality of z, denoted by $EV(z)$, is proportional to the sum of eigenvector centralities of its neighbors, i.e.

$$EV(z) = c \sum_{x \in N(z)} EV(x),$$

where c is a constant and $N(z)$ is the set of neighbors of z.

A k-core of a network is a maximal sub-network of the network such that each node in the sub-network has degree higher or equal to k [13]. A k-core therefore can be obtained by recursively deleting all nodes whose degree is less than k until all nodes in the remaining network have degree at least k. A node has shell index k if it belong to some k-core of the network, but not to a $(k+1)$-core. It is important to emphasize that the shell index is not the same as the degree of a node. For example, a node with a high degree connected to nodes of degree 1 has shell index equal to 1. The maximal core of a network encompasses nodes with the highest value of the shell index. If the maximal core is a relatively large, densely connected sub-network then we can conclude that the network possesses a core-periphery structure [11]. The existence of a core-periphery structure in the FETCH network of ERASMUS agreements implies that the network is organized around a subset of institutions that have the crucial role for the development and cohesiveness of the whole network.

To identify cohesive clusters in the network we use the Louvain method for community detection [1]. This method is based on a greedy multi-resolution approach to maximize the Girvan-Newman modularity measure [1], [9] starting from the partition in which all nodes are put in different communities. When the modularity is optimized locally the algorithm builds the network of communities and repeats the local optimization step until no increase of modularity is possible.

The network encompassing only active links can be studied as a directed graph according to the actual realization of Erasmus exchange agreements. In our analysis we use the HITS link analysis algorithm [7] to identify the most important hub and authority institutions in realized Erasmus agreements. In our case, important hubs correspond to institutions having a high out-going mobility to the most important authorities, while important authorities are institutions with a high in-coming mobility from the most important hubs. The HITS algorithm assigns two mutually recursive scores, hub and authority scores, to each node in the network. The authority score of a node z is equal to the sum of hubs scores of nodes pointing to z, while the hub score of z is the sum of authority scores of nodes referenced by z.

V. NETWORK ANALYSIS - RESULTS AND DISCUSSION

Institutions involved in the FETCH project which reported data about ERASMUS exchange agreements established 2138 student and 930 staff exchange agreements in total (with both FETCH and NON-FETCH institutions). Approximately 18% of student exchange agreements and 28% of staff exchange agreements are agreements between FETCH partners. An average institution involved in the FETCH project established student/staff agreements with approximately 5 FETCH and 16 NON-FETCH partners. The average number of ERASMUS student exchange agreements among FETCH partners is equal to 10.65, while the average number of ERASMUS staff exchange agreements is slightly lower and it is equal to 7.08.

The FETCH network of ERASMUS bilateral agreements consists of 37 nodes (FETCH institutions) and 89 links (agreements between FETCH institutions). The connected component analysis revealed that the network contains 4 connected components: 3 isolated nodes and 1 giant connected component encompassing the rest of the nodes (see Figure 1 (a)). Isolated nodes represent FETCH institutions that have not established ERASMUS bilateral agreements with other FETCH partners. However, all 3 isolated nodes in the network represent institutions that have ERASMUS student and staff exchange agreements with NON-FETCH institutions. On the other hand, the existence of the giant connected component implies that the vast majority of institutions are either directly or indirectly connected in the network, further suggesting that the mobility network among FETCH institutions is in a mature stage of the development and that it transcends regional boundaries.

The diameter of the giant connected component is equal to 4, while the characteristic path length is equal to 2,12. This means that the network possesses the small-world property – the average distance between randomly selected nodes is significantly smaller than the number of nodes in the network. The network has a high clustering coefficient (probability that two neighbors of a randomly selected node are neighbors among themselves) equal to 0,439. The clustering coefficient of a comparable random graph is equal to $C_{rand} = 2L/N(N-1) = 0.13$, where N and L denote the number of nodes and links in the network, respectively. Since empirically observed clustering coefficient is three times higher it can be concluded that the network exhibits the small-world phenomenon in the Watts-Strogatz sense [17].

The giant connected component of the network contains 5 articulation points – nodes whose removal from the network disintegrates the giant connected component into two or more connected components. The removal of each articulation point breaks the giant connected component into exactly two components: one isolated node and one non-trivial component containing the rest of the nodes. This means that the network still has a giant connected component after an articulation point is removed. Therefore, we can conclude that the giant

connected component is highly robust: there are no FETCH institutions whose existence is critical to the overall connectedness of the network. In other words, the network is not built around a small number of institutions that keep a large number of other institutions connected.

The k-core decomposition of the giant connected component revealed that the maximal shell index of the nodes is equal to 4. The maximal core of the network (4-core) is shown in Figure 1 (b). It consists of 17 nodes (46% of the total number) and 53 links (60% of the total number). The diameter of the maximal core is equal to 2 which implies that each two nodes from the maximal core are either directly connected or indirectly connected via common neighbors that also belong to the maximal core. It can be observed that the maximal core is a densely connected sub-network. Therefore, we can conclude that the network has a core-periphery structure where highly connected nodes in the network form a compact, densely connected core, while loosely connected nodes from periphery are attached to nodes from the core.

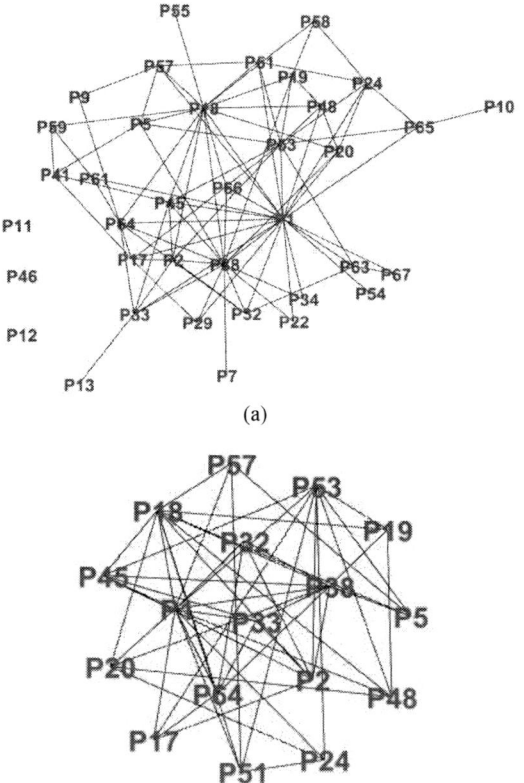

Figure 1. The giant connected component of the network (a) and the maximal core (4-core) of the giant connected component (b).

We used three different node centrality metrics to rank and identify the most important institutions in the network. The top five institutions having the highest rank by betweenness, closeness and eigenvector centrality measures are shown in Table 1. It can be noticed that four institutions can be considered as the most important by all three considered node centrality metrics. Those institutions are P1 (University of Rousse, Bulgaria), P18

616

(Czech Technical University in Prague), P38 (Vilnius University, Lithuania, and P53 (Linnaeus University, Sweden). Additionally, P33 (University of Pavia, Italy) and P45 (University of Coimbra, Portugal) are among the top 5 highest ranked institutions according to betweenness and eigenvector centrality, respectively.

TABLE I.   THE TOP FIVE INSTITUTIONS IN THE NETWORK ACCORDING TO DIFFERENT CENTRALITY MEASURES (BET – BETWEENESS CENTRALITY, CLO – CLOSENESS CENTRALITY, EV – EIGENVECTOR CENTRALITY).

BET		CLO		EV	
P1	220.26	P1	0.75	P1	1.00
P18	96.55	P38	0.63	P38	0.76
P38	84.31	P18	0.62	P18	0.75
P53	34.14	P64	0.56	P53	0.57
P33	33.33	P53	0.55	P45	0.51

We identified communities in the giant connected component using the Louvain method. The Girvan-Newman modularity of the obtained partition is equal to 0.256 which means that the network exhibits a moderately strong community structure. The partition of the network into communities is shown in Figure 2 and it can be seen that it consists of 2 large clusters each of them encompassing 12 institutions and 2 relatively small clusters (one of them consists of 6 and another of 4 nodes). The first large community is organized around P1 (University of Rousse, Bulgaria) and P38 (Vilnius University, Lithuania). The best connected nodes in the second large community are P18 (Czech Technical University in Prague) and P53 (Linnaeus University, Sweden). In other words, the two largest communities are organized around the most central nodes in the network (see Table 1). A common characteristic of two small

institutions from Bulgaria, Cyprus, Luxembourg, Lichenstein and Turkey, while institutions from Austria, Bulgaria, Italy and Portugal belong to the second small cluster.

Tables 2 and 3 show the summary statistics of realized student and staff exchange agreements, respectively, for the last three school years. A link in the network is considered active if there were students or staff exchanges between connected FETCH institutions during the aforementioned time period. Not all links in the network were active implying that there are unrealized Erasmus agreements among FETCH institutions. The number of active links is equal to 43 (48.31% of all links in the network) and they are shown in Figure 3.

TABLE II.   THE SUMMARY STATISTICS OF REALIZED STUDENT EXCHANGE AGREEMENTS.

Year	Total	Between FETCH institutions	With non-FETCH institutions
2013/12	377	103	274
2014/15	875	63	812
2015/16	447	48	399
Sum	1699	214 (12.6%)	1485 (87.4%)

TABLE III.   THE SUMMARY STATISTICS OF REALIZED STAFF EXCHANGE AGREEMENTS.

Year	Total	Between FETCH institutions	With non-FETCH institutions
2013/12	166	29	137
2014/15	181	36	145
2015/16	125	22	103
Sum	472	87 (18.4%)	385 (81.6%)

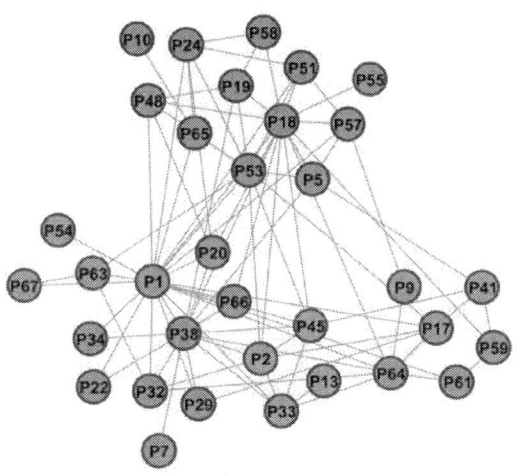

Figure 2.   Communities in the giant component

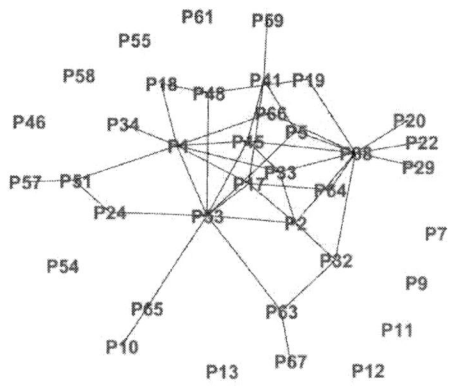

Figure 3.   The network without inactive links.

clusters is that they do not contain two institutions from the same country: the first small cluster encompasses

The network of active links reflecting realized student and staff mobility among institutions participating in the FETCH project can be analyzed as a directed graph. We applied HITS link analysis algorithm to this network in order to determine the most prominent hubs and authorities in mobility flows among FETCH partners. The results are summarized in Table 4. The institution with the largest hub score is actually the leading institution of the FETCH project (P1 – University of Rousse, Bulgaria). The institution exhibiting the largest authority score is P53 (Linnaeus University, Sweden). Also it is interesting to notice that the most important hubs are institutions from different countries (Bulgaria, Lithuania, Cyprus, Italy, and Luxembourg). The same holds for the most important authorities which are institution from Sweeden, Portugal, Turkey, Cyprus and Spain. Therefore, we can conclude that the most important institutions in the Erasmus student and staff mobility flows among FETCH institutions are not regionally localized.

TABLE IV.   THE TOP FIVE HUBS AND AUTHORITIES IN THE NETWORK OF ACTIVE LINKS.

Hubs		Authorities	
P1	0.70	P53	0.47
P38	0.40	P45	0.46
P17	0.26	P66	0.32
P33	0.26	P17	0.28
P41	0.24	P51	0.27

VI. CONCLUSIONS AND FUTURE WORK

In this paper we studied the ERASMUS mobility network formed by the institutions involved in the FETCH project. The network was reconstructed from the data about student and staff ERASMUS exchange agreements reported by 37 institutions from 25 European countries. Techniques commonly used in social network analysis were employed to investigate the structure of the network. The analysis of connected components in the network showed that the network is not in an early stage of the development and that it transcends regional boundaries. Moreover, the network is a compact, small-world in which none of the institutions has a critical role to the overall connectedness of the network. The k-core decomposition of the network revealed that the network has a core-periphery structure with compact, densely connected core composed of active institutions that realize established ERASMUS agreements with other FETCH partners. On the other hand, the periphery of the network is mostly inactive indicating that there is a huge space to improve current mobility flows among institutions involved in the FETCH project.

In this work we have shown that techniques used in social network analysis can be beneficial to the understanding of the structure of academic mobility flows. Therefore, in our future work we plan to investigate academic mobility at a larger scale that is not restricted to institutions involved in a particular scientific project. Also, we will model and analyze mobility networks as temporal, spatial, directed and weighted graphs in order to

(1) investigate spatial and temporal patterns of academic mobility, (2) analyze the strength of academic mobility considering spatial and temporal aspect of the network, and (3) address issues related to the incoming and outgoing academic mobility.

ACKNOWLEDGMENT

This work is partially supported by two projects: the Serbian Ministry of Education, Science and Technological Development through project no. OI174023 and Tematic network project: "FETCH - Future Education and Training in Computing: How to support learning at anytime anywhere".

REFERENCES

[1] Blondel, V., Guillaume, J.-P., Lambiotte, R., Lefebvre, E., Fast unfolding of communities in large networks, 2008 (10), P10008, 2008.

[2] Breznik, K., Gologranc, G., Erasmus Mobility On The Institutional Level, International Conference "Management, Knowledge and Learning", Portorož, Slovenia, 2014.

[3] Derzsi, A., Derzsy, N., Káptalan, E., Néda, Z. Topology of the Erasmus student mobility network, Physica A, Vol. 390, pp. 2601–2610, 2011.

[4] ETN FETCH – Future Education and Training in Computing: How to support learning at anytime anywhere. Official website of the project: http://fetch.ecs.uni-ruse.bg/?cmd=gsIndex (accessed July 17th 2016)

[5] Gonzales, C. R., Mesanza, R. C, Mariel P. The determinants of international student mobility flows: an empirical study on the Erasmus programme. Higher Education, 62(4), pp. 413-430, 2011.

[6] Janson, K., Schomburg, H., Teichler, U. The Professional Value of ERASMUS Mobility – The Impact of International Experience on Former Students' and on Teachers' Careers. Bonn: Lemmens Medien GmbH, 2009.

[7] Kleinberg, J., Authoritative sources in a hyperlinked environment. Journal of the ACM. 46 (5): 604–632, 1999.

[8] Newman, M. E. J. The Structure and Function of Complex Networks. SIAM Review, 45(2), pp. 167-256, 2003

[9] Newman, M. E. J. Networks: An Introduction. Oxford University Press, 2010.

[10] Poulová, P., Černá, M., Svobodová, L. University Network – Efficiency of Virtual Mobility, 5th WSEAS/IASME International Conference on Educational Technologies (EDUTE' 09),

[11] Rodríguez González, C., Bustillo Mesanza, R. & Mariel, P. The determinants of international student mobility flows: an empirical study on the Erasmus programme, High Educ, 62: 413. doi:10.1007/s10734-010-9396-5, 2011.

[12] Rombach, M. P., Porter, M. A., Fowler, J. H., and Mucha, P. J. Core-periphery structures in networks. SIAM Journal of Applied Mathematics, 74(1), pp. 167-190, 2014.

[13] Seidman, S. B. Network structure and minimum degree. Social Networks, 5, pp. 269-287, 1984.

[14] Sigalas, E. Cross-border mobility and European identity: The effectiveness of intergroup contact during the ERASMUS year abroad, European Union Politics, 11(2) 241–265, DOI: 10.1177/1465116510363656, 2010.

[15] Teichler, U., Jason, K. The Professional Value of Temporary Study in Another European Country: Employment and Work of Former ERASMUS Students. Journal of Studies in International Education, 11(3), pp. 486-495, 2007.

[16] Van Mol, C., Michielsen, J. The Reconstruction of a Social Network Abroad. An Analysis of the Interaction Patterns of Erasmus Students, Mobilities Vol. 10, Iss. 3, pp. 423-444, 2015

[17] Watts, D. J., Strogatz, S. H. Collective dynamics of "small-world" networks. Nature, 393, pp. 440-442, 1998.

Modern education and its background in cognitive psychology: Automated question creation and eye movements

Margit Höfler[1], Gudrun Wesiak[1], Paul Pürcher[1], Christian Gütl[2,3]

[1]University of Graz/Department of Psychology, Graz, Austria
[2]Graz, University of Technology/Institute for Information Systems and Computer Media, Graz, Austria
[3]Curtin University of Technology, Perth, Western Australia

In modern education, the automatic generation of test items out of texts becomes increasingly important. We have recently presented an enhanced automatic question creator (EAQC) that is able to extract the most relevant concepts out of a text and to create test items based on these concepts. Furthermore, from eye-movement research it is known that learners fixate potentially relevant information longer and more often. In the present study, we investigated whether the measurement of (individual) relevance of EAQC-concepts is related to the frequency and duration of fixations from learners when they read a text, learn it or manually extract concepts out of it. Overall, results showed a non-significant tendency that participants needed more fixations when learning the text as compared to when reading it or extracting concepts. Saccade lengths were reliably longer when participants extracted concepts than when they read them; no other differences were found. How long and how often a concept was fixated did not vary with the subjective relevance of the concept according to participant's ratings. Future research is therefore necessary to investigate a possible relationship between eye movement behavior and automated concept extraction.

I. INTRODUCTION

According to [1], personal computers as well as the internet have led to substantial changes in our community, whereas educational methods have changed little. Moreover, in [1] it is also argued that e-learning will change all sorts of learning and education in the 21st century and those who are seriously seeking to improve learning and teaching cannot ignore it.

One increasingly important part of modern education is the usage of automated question creation: There is a bunch of research on how question items can be created automatically from a given learning content, bringing together experts from technology, (cognitive) psychology, education and linguistics. For example, [2] used automatically generated questions to assess vocabulary. They reported that the vocabulary skills assessed with automatically created questions correlated significantly with vocabulary skills tested with human-written questions and standardized vo-

cabulary tests. Furthermore, in [3] it was pointed out that automatic question generation can also support people with other mother tongues than English because non-native speakers often lack grammatical knowledge and foreign language vocabulary in order to prepare efficient questions.

We have recently developed an enhanced automatic question creator (EAQC) [4, 5, 6] that aims to generate individualized test items easily with small effort. It extracts concepts (i.e., the most relevant information such as key words and key phrases) out of a given text, by combining part-of-speech tagging, word statistics and semantic analysis. With these concepts, four types of questions can be created, namely open-ended items, completion exercises, true or false items, and multiple-choice items. The validity of the created items was already assessed in previous evaluation studies, demonstrating that the quality of the automatically extracted concepts and created questions is comparable to those generated by humans [4, 7]. Interestingly, and important for the current study, this was not only true for test items that were generated fully automatically (i.e., concepts and test items were created by the EAQC) but also for test items that were created semi-automatically (i.e., the EAQC generated questions based on concepts extracted manually from a text, e.g. by learners, [8]).

Creating test items semi-automatically ensures that the resulting test items address the needs of the learners, because the concepts extracted by the EAQC might not always be of high relevance for the learners. For instance, although a comparison of the concepts extracted manually by participants with the concepts automatically extracted by the EAQC showed a large overlap of concepts, there were also manually created concepts that had not been considered by the EAQC and vice versa [8]. This suggests that the individual relevance of a concept for a learner might not always correspond to the "generalized" relevance of a concept as computed by a tool such as the EAQC (not to mention that there are also individual differences between learners with regard to the perceived relevance of a concept and contextual background).

To sum up, the (individual) relevance of the underlying concepts for a learner seems to be one crucial factor for successful automated question generation and should be emphasized in future research. As a starting point, we investigated in the current study whether the measured individual relevance of a concept is reflected in the eye movement behavior of a learner. More precisely, one might assume that the more relevant a concept is, the more often and longer it will be fixated while a learner studies a text. If there is such a relationship, analyzing the eye movements during reading and/or learning a text would allow us to determine which concepts are individually relevant for a learner (and which not). In turn, these concepts might then be used to feed the EAQC for the generation of questions that are consequently even more adjusted to the learner's need.

The aim of the present study was therefore to investigate, in a first step, whether the relevance of a concept might be reflected in how long and often a concept is fixated by a learner. In other words, we were interested in whether eye tracking measurements provide a valid indicator for the relevance of key concepts.

One important variable when measuring eye movement behavior using eye tracking is the fixation duration. How long a word is fixated depends on several factors: For instance, it is well known that we fixate a word for about 225-250 ms [9, 10] during the reading process, i.e. our eyes keep on fixating a word for this period of time in order to process the information it provides. Such fixation durations can increase to about 260-330 ms when we view a scene and decrease to about 180-275 ms when we search for something in the environment. Within reading, previous research has also shown that infrequent words are fixated longer than frequent words (word-frequency effect; e.g., [11]) and that fixation durations also vary with the difficulty of the text (e.g., longer fixation durations were reported for Biology texts or Physics texts than for newspapers; [12]).

Another variable that is typically analyzed when eye movements are measured is the saccade length (i.e., the angular distance the eye gaze travels during the movement from one to another fixation). Saccade lengths do not only vary with the task (e.g. about 4° visual angle for scene viewing, 3° for visual search and 2° for reading [13]), but again also with text difficulty (e.g. shorter saccade lengths were observed when reading Biology texts or Physic texts and longer when reading newspapers; [12]). Crucial for our current research question is that there are some experiments showing that also syntactic language categories can be predicted from reading behavior. For example, we know that not all words are fixated when reading a text: While content words are fixated 85 % of the time, function words are only fixated about 35 % of the time [9, see also 14].

So far, we have shown that eye movement measurements (i.e. fixation duration and saccade lengths) vary with the task and correlate with the meanings of a text and the types of words in the text. Based on this, we were interested in whether some of these measurements might also be useful when defining the relevance of a concept. To our best knowledge, this has not been investigated yet. In the present study we therefore had participants read a text first and then learn it in preparation for a knowledge test. Furthermore, they were required to extract the main concepts (key words) out of the text. During all of these activities we recorded their eye movements. This allowed us not only to investigate whether the fixation durations and saccade lengths differ with the task but also how long and often they fixated each word/concept (as provided by the EAQC) during the various tasks. Using this information, we were able to investigate whether there is a relationship between these eye movement variables and the relevance of EAQC-extracted concepts (rated by the participants). In particular, we expected longer fixation durations and a higher number of fixations for those concepts extracted by the EAQC that are evaluated as more relevant by the learner.

II. METHODS

A. Design

In the current study participants were required to perform several tasks on a given text. These tasks required reading the text (Phase 1), learning it in preparation for a knowledge test (Phase 2) and extracting the main concepts out of it (Phase 3). Furthermore, participants were required to rank the self-extracted concepts with regard to their importance (Phase 4) as well as to assess the relevance of concepts which were extracted by the EAQC out of the same text (Phase 5). Additionally, in Phase 5 participants answered a short knowledge test about the learned text and some general questions about the experiment. Note that for the purpose of the present paper, only Phases 1-3 and the relevance assessment of Phase 5 are taken into consideration although we will present some overall findings also for Phase 4. In the first three phases the eye movements of the participants were recorded in order to analyze how long and how often each EAQC-based concept was fixated. These measures were then correlated with the ratings of the concepts extracted by the EAQC.

B. Participants

Nine female participants (students of Psychology) took part in the study (mean age: 22.6 years). They had normal or corrected-to-normal (contact lenses) vision and German was their native language. However all of them had English as a second language.

C. Stimuli and Procedure

The text we used for the tasks was a selection of an English research article about eye tracking, written by [15]. The selected parts contained four different aspects of eye tracking: general introduction about eye tracking (1), coarse overview on eye tracking's physiological background (2), most important methods of analysis (3) and the usage of eye tracking for usability research (4). We slightly simplified some of the sentences to make them better understandable for the (German-speaking) readers. During the first three phases (i.e., reading, learning, concept extraction) participants were seated in a dimmed and soundproof booth while their eye movements were monitored with an EyeLink 1000 eye tracker (SR Research, Canada). Data were collected from the dominant eye with a sampling rate of 1000 Hz.

The text was presented on four successive slides using Experiment Builder (SR Research, Canada). Each slide consisted of 149.5 words on average. The text was written in black Arial font (22 px) on white background, the line spacing was 1.5.

At the beginning of a trial participants were asked to read the text (i.e. all four slides) carefully (Phase 1). Each of the four slides were then presented for maximally two minutes, but participants were able to proceed to the next slide once they had finished reading the slide by pressing a button.

After this reading phase, participants were told that they should learn the text for an upcoming knowledge test (Phase 2). During this learning phase each slide was presented again consecutively for maximally two minutes. When participants completed the learning of all slides they had the opportunity to go through all slides for one more time, in order to allow better learning. During the learning phase the participants were informed regularly by the experimenter about how much time was left until the next slide was presented.

After the learning phase, participants were given a break of several minutes. Then, they were asked to extract the, in their opinion, 1 to 5 most important concepts separately for each slide while their eye movements were tracked (Phase 3). To this end, the four slides were presented again consecutively (2 minutes per slide). Participants were required to speak out the concepts loudly, without moving their head, while the experimenter took them down separately on a lab sheet.

After the extraction phase, the participants were seated into another quiet room (as eye tracking was no longer required for the following tasks) and asked to rank the previously extracted concepts regarding their relevance (Phase 4). Finally, in Phase 5 participants were asked to fill in a questionnaire created with Limesurvey (www.limesurvey.org). This questionnaire consisted of two parts: Part 1 included knowledge questions on the text (16 questions; 8 questions created by the EAQC, 8 questions created by the experimenter), as well as an assessment of participant's general attitude and experiences during the task. Part 2 consisted of a rating of relevance (1 [not relevant] – 5 [very relevant]) for the 11 concepts which had been extracted by the EAQC before the experiment started. These 11 concepts were *web pages, thinking aloud, the natural situation, gaze plot, eye tracking, computer monitor, fixations and saccades, area of interest analysis, hot spot, information*, and *mental model of system* (for an overview of the procedure of this experiment, see Figure 1).

III. RESULTS

For the first three phases (reading, learning, concept extraction), eye movement data was analyzed separately for each of the four slides from the start of each slide until the button press (or time-out). As described above, during the learning phase, participants had the possibility to go through the slides once again; seven out of nine participants took this opportunity. However, for a better comparability across the participants we only analyzed the first run of the learning phase.

A. General eye-movement measures (Phases 1 - 3)

We first analyzed the eye movement data of the first three phases with regard to "traditional" measurements in order to be able to integrate the findings in the general literature. Pooled across all four slides, participants made on average 728.3 fixations (SD = 77.4) during the reading phase, 1,316.6 fixations (SD = 467.2) during the first learning phase and 973.7 fixations (SD = 214.6) during the extraction phase.

Per each slide, participants made 182.1 fixations (SD = 19.9) during the reading phase, 329.1 fixations (SD = 158.2) during the first learning phase and 243.4 fixations (SD = 65.5) during the extraction phase. A one-way repeated ANOVA calculated over the number of fixations made per slide did not reveal a significant difference between the three phases, F (1.13, 9.01) = 3.63, p = .09. There was, however, a tendency (p = .09) showing that more fixations were made during the learning phase than during the reading or the extraction phase, whereas no difference was observed between the latter two (p > .05).

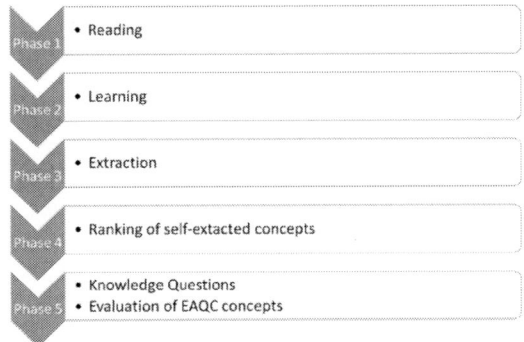

Figure 1. Procedure of the experiment (Phases 1-5).

Mean saccade lengths were 4.03 (SD = .39) for the reading phase, 4.46 (SD = .61) for the learning phase and 5.05 (SD = .55) for the extraction phase. A one-way repeated-measures ANOVA showed a significant effect of task, F (2, 16) = 12.81, p < .001, η_p^2 = .62). Posttests (with Bonferroni-corrected alphas) revealed that saccadic lengths were reliably longer when participants extracted concepts than when reading or learning the concepts. No difference was found between the other tasks (p > .05).

B. Participant's concept extraction

Participants extracted 65 concepts in total and 1.8 concepts (SD = 0.5) per slide on average. The most popular concepts regarding the number of extractions were "thinking aloud" (8 times), "hot spot (maps/visualizations)" (7 times), "fixations and saccades" (4 times) and "scan path" (4 times). The concepts extracted by the participants were consistent with some of the concepts extracted by the EAQC (see above), but there were also other manually extracted concepts that were not extracted by the EAQC, which is consistent with prior literature [8].

C. Concept evaluation of EAQC concepts (Phase 5)

Participants had to rate the relevance of concepts extracted by the EAQC using a 5-point Likert scale (1 = not relevant, 5 = very relevant). One participant's data had to be excluded due to a computer error, so finally the data of eight participants were taken into account.

The results showed that the concepts "eye tracking" (M = 4.8, SD = 0.4), "fixations and saccades" (M = 4.1, SD = 0.8) and "hot spot" (M = 3.9, SD = 0.8) received the highest ratings. The lowest ratings were addressed to the concepts "web pages" (M = 2.4, SD = 0.9) and "information" (M = 2.6, SD = 1.1). All other concepts were rated from 3.0 to 3.6, not differing substantially from each other.

D. Correlations between EAQC concepts and eye tracking measures

In order to investigate a possible relationship between the relevance of the EAQC concepts and participants eye movements we correlated the participants' rating for each of the 11 EAQC-extracted concepts in Phase 5 with two cognitive measures from the eye tracking data. These two measures were (1) how long the participants fixated a concept on average when they fixated it for the first time (average duration of first fixation, ADOFF) and (2) how often they fixated the concept during each phase (average fixation count, AFC).

Figure 2 provides the main findings of this analysis. In the reading phase, both the correlation between the concept ratings and ADOFF and between the concept ratings and the AFC were not significant (all ps > .05). This suggests that there is no relationship between the relevance of a concept and how long it was fixated. However, not depicted in Figure 2, the corre-

lation between ADOFF and AFC was at the threshold to statistical significance (r = .601, p = .05). This would suggest that the longer a concept is fixated when looking at it for the first time during the reading phase, the more often it is fixated during the same phase.

Furthermore, for the learning phase and the concept extraction phase, the correlations between ADOFF and concept rating, as well as between AFC and concept rating did not reach statistical significance (all ps > .05; see Figure 2). However, in both the learning phase and the extracting phase, we found again a significant correlation between ADOFF and the AFC (learning phase: r = .630, p = .038; concept extraction phase: r = .783, p = .004), revealing that the longer a concept is fixated when looking at it for the first time, the more often it is fixated.

IV. CONCLUSION

In the current study we wanted to investigate whether eye movement measures might be useful in determining the perceived relevance of concepts out of a given text. To this end, participants read a text, learned it in preparation for a knowledge test and finally extracted concepts out of it while we were recording their eye movements.

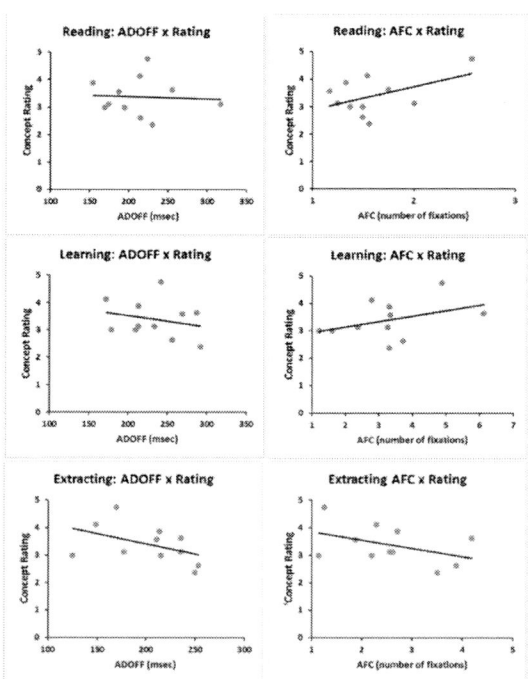

Figure 2. Correlations of ADOFF x Concept Rating and AFC x Concept Rating for all three phases: Reading (1st line), Learning (2nd line) and Extracting (3rd line). All six correlations did not reach statistical significance (all ps > .05).

Overall, the experiment showed that, as expected, different tasks lead to different eye movement's behavior on the same text [12]: A tendency of a higher number of fixations arose during the learning phase as compared to the reading and extraction phase, while saccade lengths were the longest during the extraction phase as compared to the other two phases. However, with regard to our main research question, the correlation of eye tracking measurements with the participants' ratings did not provide a clear picture, urging for future research. The correlations between the average fixation duration of a concept and the individual ratings for the EAQC-extracted concepts were not significant, neither were the correlations between the average number of fixations on a concept and its rating. These findings suggest that there is no reliable connection between the participants' eye movements towards certain concepts and the subjective importance of these concepts. Only the correlation between the average fixation duration and the average number of fixations on a concept showed the expected significance.

There are a number of reasons that might have led to the current results and might be in this way also a starting point for future research. First, our sample size was rather small. Although such small sample sizes are very common within standard eye tracking research (given the large number of eye tracking data from each participant), future research on the validation of automated question creators with physiological eye measurements should include a larger sample size in order to receive distinct results.

Second, the text used in this experiment was a scientific text about eye tracking. Probably, a more general topic would result in a broader and more concise image of eye movements during the different exercises. Moreover, the text we used was written in English, whereas the mother tongue of all participants in our experiment was German (although all participants reported a sound knowledge of English). It would be nevertheless interesting in future research to investigate whether results differ for mother-tongue texts.

Finally, prior experiments as well as our experiment showed that participants' perception towards what is relevant and what is not, is very widespread. This fact has to be further investigated as well in future research, perhaps by trying to find a relationship between the eye-movement measures and those concepts that were extracted by the participants themselves. This analysis was however out of the scope of the current study.

In sum, our experiment opened a door to link the modern and upcoming field of automated question creation with the field of classical research on cognitive processes using eye tracking measurements, aiming to illuminate and allow a validation and enhancement of automated question creation. A broad body of future research is needed to further investigate this connection, whereas this experiment provides first steps and important lessons learned for future research.

V. REFERENCES

[1]. D. R. Garrison, E-learning in the 21st century: A framework for research and practice. London: Taylor & Francis, 2011.

[2]. J. C. Brown, G. A. Frishkoff, and M. Eskenazi, Automatic question generation for vocabulary assessment. In Proceedings of the conference on Human Language Technology and Empirical Methods in Natural Language Processing, pp. 819-826, Association for Computational Linguistics, October 2005.

[3]. T. Goto, T. Kojiri, T. Watanabe, T. Iwata, and T. Yamada, Automatic generation system of multiple-choice cloze questions and its evaluation. Knowledge Management & E-Learning: An International Journal (KM&EL), 2(3), 210-224, 2010.

[4]. C. Gütl, K. Lankmayr, J. Weinhofer, and M. Höfler, Enhanced Automatic Question Creator-EAQC: Concept, Development and Evaluation of an Automatic Test Item Creation Tool to Foster Modern e-Education. Electronic Journal of e-Learning, 9(1), 23-38, 2011.

[5]. M. Höfler, M. AL-Smadi, and C. Gütl, Investigating content quality of automatically and manually generated questions to support self-directed learning. In CAA 2011 International Conference, University of Southampton, 2011.

[6]. M. Höfler, M. Al-Smadi, and C. Guetl, Investigating the suitability of automatically generated test items for real tests. International Journal of e-Assessment, 2(1), 2012.

[7]. G. Wesiak, R. H. Rizzardini, H. Amado-Salvatierra, C. Gütl, and M. AL-Smadi, Automatic Test Item Creation in Self-Regulated Learning: Evaluating quality of questions in a Latin American experience. In Proceedings of CSEDU 2013 – 5th International Conference on Computer Supported Education, 351-360, 2013. DOI: 10.5220/0004387803510360

[8]. M. Al-Smadi, M., Hoefler, and C. Gütl, An Enhanced Automated Test Item Creation Based on Learners Preferred Concept Space. INTERNATIONAL JOURNAL OF ADVANCED COMPUTER SCIENCE AND APPLICATIONS, 7(3), 397-405, 2016.

[9]. K. Rayner, Eye movements in reading and information processing: 20 years of research. Psychological Bulletin, 124(3), 372, 1998.

[10]. K. Rayner, Eye movements and attention in reading, scene perception, and visual search. The quarterly journal of experimental psychology, 62(8), 1457-1506, 2009.

[11]. M. A. Just, and P. A. Carpenter, A theory of reading: from eye fixations to comprehension. Psychological review, 87(4), 329, 1980.

[12]. K. Rayner, and A. Pollatsek, The psychology of reading. Englewood Cliffs, NJ: Prentice Hall, 1989.

[13]. R. van der Lans, M. Wedel, and R. Pieters, Defining eye-fixation sequences across individuals and tasks: the Binocular-Individual Threshold (BIT) algorithm. Behavior Research Methods, 43(1), 239-257, 2011.

[14]. M. Barrett, and A. Søgaard, Reading behavior predicts syntactic categories. CoNLL 2015, 345, 2015.

MIPRO 2017, May 22- 26, 2017, Opatija, Croatia

Learning to program – does it matter where you sit in the lecture theatre?

Aidan MC GOWAN*, Philip HANNA*, Des GREER*, John BUSCH*, Neil ANDERSON*.

*School of Electronics, Electrical Engineering and Computer Science, Queens' University, Belfast, Northern Ireland. BT9 6AY Tel: +44 (0)28 9097 1185, Email: aidan.mcgowan@qub.ac.uk

ABSTRACT - Seating position in university lectures is commonly linked with student grade performance. Sitting at the front of lecture theatre is generally reported to have a positive effect on final grade. This study investigates and analyses the seating positions, course engagement, prior programming experiences and academic abilities of students throughout a 12 week Java programming university programme and relates these themes to the students' final grade performances. Unlike other studies in this area it did not control the students' seating arrangements. This required the development of a mobile and web based software tracking system which enabled a unique unrestricted study of the effects of lecture theatre seating on assessment performance. It finds that the best assessment results were achieved by the students in the front row and that assessment score degraded the further students sat from the front. While the most engaged were found to regularly sit at the front the same was not true for the most academically able or those with the greatest prior programming experience.

Keywords

Computing; Attitudes; Assessment performance; Seating Lecture theatre;

I.   INTRODUCTION

The performance effect of where students sit during university lectures has received some limited research attention [1]. Conventional teaching experience and most of the research in the area has suggested those students that regularly sit at the front of the lecture theatre tend to achieve higher grades than those that sit elsewhere ([2]; [3]; [4]). These studies have tended to conclude that higher grades are achieved by the regular front row students because the academically best able students tend to voluntarily position themselves there. Giles [5] found a direct relationship between test scores and seating distance from the front of class: students in the front, middle, and back rows of class scored 80.0%, 71.6%, and 68.1% respectively on course exams.

Decision to sit at front – attitudes and abilities

Other studies have sought to further identify why there appears to be a relationship between grades and seating distance from the lecturer. Some studies have sought to establish the causation factors in the decision making of students with regard to the decision of where to sit. These studies have included an analysis of the attitudes of the students including attendance [6],

attention [7] and motivation [8]. Generally they suggest that those that sit at the front are indeed the best attenders, most attentive and motivated. Becker [9] even suggests that those students that opt to sit at the front have a more positive regard for the lecturer. These studies conclude that the best and most engaged students are to be found at the front.

However, the convention that the best students are found at the front and this group will achieve the highest grades was challenged by Perkins [10]. In their seating experiment they randomly allocated seats to the students thereby preventing any potential natural clustering of the academically better able and engaged students at the front of the lecture theatre. Nevertheless, the study concurs with the majority of other studies. The same pattern of degradation of final grade occurs the further back the students sat. While the authors failed to identify why sitting in the front led to better grades, their results would suggest that there are other factors affecting performance and seating than simply student ability or positive attitudes.

Better learning experience

Regardless of ability or engagement, perhaps the students that sit at the front may simply have a better learning experience and consequently end up with better grades. Part of this better experience may involve better note taking. Traditional lecturing involves the presenter speaking almost continuously for an extended period of time while the students' job is to listen and take notes [11]. The importance of accurate notes is not lost on students with [12] and [13] reporting that the majority of test questions on college exams come from the lectures and that students who take better class notes get better course grades. This in turn places demands on students' ability to listen carefully and take notes that are accurate and complete. Conventional wisdom would also suggest the note taking is better done at the front rather than the back of the lecture theatre due to better sight lines and being easier to hear the lecturer.

Participation

Marx [14] links active participation of students in a lecture with a more positive learning experience and resulting in a strong positive influence on attention and long-term memory storage. Students that sit up front are generally less inhibited in asking questions and are able to make better eye contact with the lecturer and are regularly the most participative [15]. The ability to interact [16] and participate [17] in the lecture are important factors influencing the learning experience during the lecture. Traditional lecturing would tend to offer better participation opportunities at the front. It would follow then that if active participation opportunities are extended beyond the traditional front rows of the lecture hall by using active learning approaches then this would negate the perceived advantage to sitting up front. An analysis of these active teaching styles and seating effect was conducted by Perkins [18]. Nonetheless, the results displayed the same degradation of grade the further the students sat from the front. This would suggest that even with active learning styles there are still performance gains to be made by sitting at the front.

There are some other obvious logistical reasons a student may opt to sit at the front. These include where their friends are sitting or if the student were late to class and the front happens to be all that is available [19] or better visibility and improved ability to hear at the front [20].

Issues and gaps in the research

While most studies report a correlation between seating position and performance, a small number of studies including [21] and [22] suggest that seating position has no effect on student attainment. There are some areas that have received little or no research attention. The methodology employed in most of the seating studies involved students either self-selecting their seats or being allocated seats at the beginning of the course. The students were then restricted to the same seat for the remaining number of lectures. This is not the natural or the usual occurrence on university courses. It is arguable that this control restriction improperly influences the research outcomes. It also follows that any temporal factors that may influence a student to change seats could not be accessed or reported on in these studies. For instance if a student that normally sits towards the back receives a low in-term grade does this influence them to start sitting at the front? Other unanswerable questions from these restricted movement studies include analysis of natural migrations over time, for example do those that sit at the front at the start of the course remain there throughout? What would cause a student to move? Do students tend to move towards the front as exam or submission dates approach in the hope to glean exam hints from the lecturer? Additionally while almost all studies have concentrated on the effects of front versus back seating few have considered the possible effects of seating to the left or right of the lecturer.

II. RESEARCH OBJECTIVES

This research study was designed to use an unrestricted and accurate lecture theatre seat tracking system to enable an investigation and identification of possible patterns relating to academic performance of students in a university Java programming course.

The key research questions are:

1. Do the most academically able students regularly sit at the front of lecture theatres?
2. Do the most engaged students regularly sit at the front of lecture theatres?
3. Do students with prior programming experience regularly sit at the front of lecture theatres?
4. Is there a relationship between student performance and seating position?
5. Do the best assessment results come from the front row students?

The answers to these questions will gain insights in the effects of seating on student academic performance. This has possible consequences for teaching delivery styles, class sizes, lecture venues and the architectural design of lecture theatres.

III. METHODOLOGY

The study was conducted with a cohort of 91 postgraduate students taking a compulsory module in Java programming in semester one of a one year Masters course in Software Development. The demographics of the cohort are shown in Table 1. The mode of allocating seats for the vast majority of previous studies in this area was in the initial lecture to either allow the students to sit where they wish or allocate seating at random. The students were then compelled to sit in the same seat for the remaining weeks' lectures. Other variants on this included reversing seating positions of the students at some module during the module or simply asking students, in a post course questionnaire, where they sat. In this research the students were allowed to sit wherever they wished and were free to move seats throughout the twelve week module. In order to record each session seating arrangements a mobile application (*PinPoint*) was developed by the researchers.

Category	Subcategories and frequency
Gender	Male (78), Female (13)
Previous degree classification	1st (17), 2.1 (48), 2.2 (26)
Previous programming experience	None (50), Some (41)

Table 1 – Demographics of students in the study

Student seating tracking – PinPoint

The students were encouraged to use PinPoint (Figure 1), a mobile and web app to register their selected seating zone at the start of each lecture. To ensure authenticity each student's profile was initially registered and authenticated with the PinPoint system. Subsequently each student simply selected the zone they were sitting in during each lecture. It was not possible for the student to select a zone outside of the lecture time, nor was it possible to submit more than once per lecture. The zones were front, middle and back further subdivided into left, centre and right. The lecture rooms all had excellent Wi-Fi facilities and the mobile data strength in most reached 4G levels.

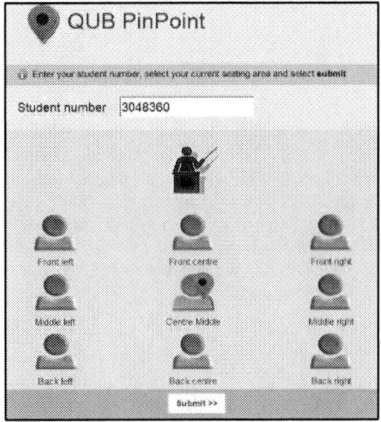

Figure 1 – PinPoint mobile app user interface

Other data collection areas

The assessment of other factors that may affect seating position was grouped into several thematic areas, namely; student performance, course engagement, academic ability and relevant programming experience. There were two summative assessment periods at the end of weeks six and twelve. A survey was completed by each student at the beginning of the course to capture individual data such as previous programming experience. In addition a weekly non-assessed quiz relating to the previous week's taught content was made available for each student. The score from the quiz for each student and the date submitted was recorded. All course materials were placed online via the course VLE and the individual student access statistics were collected. A questionnaire with various questions relating to seating position choice was given to the students at the end of the course.

Theme	Data type	Source
Student performance	Assessment results	Summative assessments at weeks 6 and 12
Course Engagement	Attendance at lectures	PinPoint App
	Engagement with weekly tests	VLE
	Engagement with online module resources	VLE
Academic ability	Previous degree classification	University records
Relevant programming experience	Subject knowledge	Student demographic information survey(week 1)

Table 2 - Thematic areas of investigation with data types and data source.

IV.   RESULTS AND CONCLUSIONS

The use of PinPoint was on a voluntary basis and all the 91 students actively participated in its use throughout the course. While PinPoint exactly recorded the seating positions in nine zones the results presented here are accurately aggregated into front, middle and back rows. In addition to enable comparisons with other studies, in instances where previous research attention has been focused on front row versus other seating locations the results were further aggregated in front row and other rows (middle and back).

1. Do the most academically able students regularly sit at the front of lecture theatres?

The students on the course with previous First Class Honours (n=17) degrees were for the purposes of the research considered to be the most academically able. While the researchers acknowledged that the previous degree may not have been directly related to programming themes or even STEM (Science, technology, engineering, and mathematics) based, the previous degree standard was considered a mark achieved through a rigorous independent academic process and likely a better measure of academic performance than a pre-course assessment.

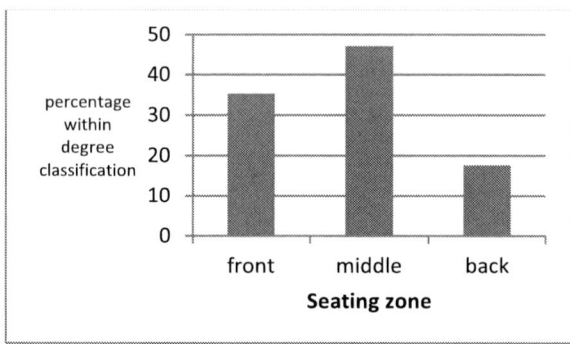

Figure 2 – Average seating location in weeks 1 to 12 for students with 1st class honours degrees. (n=17)

The tracking of seating positions, as illustrated in Figure 2, shows that of the First Class degree students (35.3%) regularly sat in the front rows. The remaining 47.1% sat in the middle rows with 17.6% regularly sitting on the back row seats. While significantly more of these students sat in the front than the back, the majority actually sat in the middle.

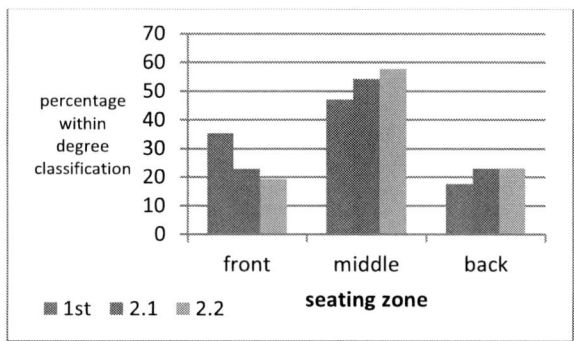

Figure 3 – Percentage within each degree classification and average seat location in weeks 1 to 12.

Contrasting the First Class honours degree students with the other degree classifications (Figure 3) shows that proportionally more students with a First Class degree regularly sat at the front. The majority of students within each classification sat in the middle rows. The students with Upper Second Class degrees (2.1) were proportionally evenly distributed with the majority in the middle and comparable numbers in the front and back. The Lower Second Class degree (2.2) students also mostly sat in the middle but tended to aggregated more at the back than the front.

Recalling that there are disproportionate numbers of previous degree classifications within the cohort (1st (**17**), 2.1 (**48**), 2.2 (**26**)) it is of interest to review the actual numeric breakdown by seating areas and degree classifications (Table 3).

	Front row	Middle row	Back row
1st (n=17)	35.3% (6)	47.1% (8)	17.6% (3)
2.1 (n=48)	22.9% (11)	54.2% (26)	22.9% (11)
2.2 (n=26)	19.2% (5)	57.7% (15)	23.1% (6)
Total in rows	22	49	20

Table 3 -Percentage and actual numbers within each degree classification and aggregated seating position during the 12 week course.

The percentage breakdown by classification would therefore suggest that if the intake of the course had even numbers in each classification then the makeup of the front row would generally consist of the higher performing students. In reality for most similar courses this is unlikely to be the case. Indeed it is worth noting that in this occasion from total students in the front row (n=22) the majority (n=16) held the lower degree classification. This is in contrast with much of the previous research that has suggested that the best final results will come from the front rows based on the assertion that the best students are there. In this case the majority of the top end academic performers (64.7%) were not located in the front rows.

2. Do the most engaged students regularly sit at the front of lecture theatres?

The measure of engagement (*engagement factor*) was based on a mapping scale of: 1) access hits for online course materials, 2) attendance and 3) voluntary engagement with the weekly formative tests. Scaling was from 5 (high engagement) to 0 (no engagement). Historically the course attracts students that are considered to be highly motivated with high attendance levels, high attainment and observed high engagement with course materials. As such it was not surprising that the engagement mappings placed the majority of students within the range 3 to 5. The breakdown of engagement factor showed that the front row on average contained the most engaged students (3.50) followed by the middle (3.35) row and then the back rows (2.98).

The trend line in Figure 4 suggests that engagement declines from front to back rows. It should be noted that there were a large number of highly engaged students (an engagement factor greater than 4.0) clustered in the middle rows but also a large number of low engaged (an engagement factor less than 2.5) which affects the overall average engagement for this zone. The back row contains none of the most engaged students and proportionally the highest number of the least engaged. Based on the averages achieved per zone it can therefore be concluded that the most engaged students voluntarily clustered at the front throughout the course.

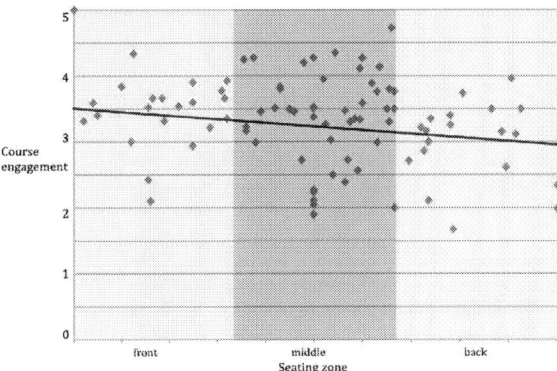

Figure 4 – Course engagement ranging from 5 (high engagement) to 0 (no engagement) plotted against seating zone. Each point represents a student in respect to their calculated engagement in the course and their average seating position throughout the course.

3. Do students with prior programming experience regularly sit at the front of lecture theatres?

Historically the course attracts a number of students that have some programming experience but a larger proportion that have none. This is also a current and comparable theme experienced in first year undergraduate Computer Science courses worldwide. Whereas in the past the majority of students entering first year computing degree courses would have had little or no programming knowledge, more recently and largely due to the recent uptake of computing qualifications in Secondary Level education, universities have seen an increase in the prior programming experience of students.

In this study and perhaps not surprisingly, the average assessment scores of the students that had some prior programming experience (72%) was higher than achieved by those without any programming experience (68%).

Considering the mixture of previous programming experience (n=41) and none (n=50), it is of relevance to report on the seating positions of these groups and also the breakdown per seating zone to judge if the most experienced and best academic performers sat in the front rows.

	Front	Middle	Back
Some programming experience (n=41)	22.0% (9)	48.8% (20)	29.2% (12)
No programming experience (=50)	26.0% (13)	58.0% (29)	16.0% (8)

Table 4 - previous programming and no programming experience groups as shown by actual numbers per seating zone and percentage breakdown per group within each zone.

The figures presented in Table 5 point to a comparably equal proportional representation of the two groups in the front rows, with the majority of both groups sitting in the middle rows. It can therefore be concluded that the more experienced programming students did not opt to sit predominantly in the front rows.

4. Is there a relationship between student performance and seating position? & 5. Do the best assessment results come from the front row students?

There were two equally weighted assessment points in the course which made up the final assessment score. Both were closed book assessments taken under exam conditions consisting of theory and application of the taught course content. The distribution of the overall assessment percentage scores shows normal distribution (p=0.193, Kolmogorov-Smirnov Z=1.081). Table 6 illustrates that the front row students returned the higher assessment scores in comparison to the students sitting elsewhere in the lecture theatre.

	Overall assessment score	Standard Deviation
Front rows	75.11%	19.24
Other rows (middle and back)	68.18%	21.63

Table 5 – Assessment scores per seating zones.

To test the statistical significance of the assessment scores from the students in the two seating zones groups (front and other) an analysis using an Independent-Samples T Test was conducted. It revealed that despite the small sample sizes involved that the difference between scores achieved by the front row students and the other students was approaching statistical significance (p=0.059, t=1.900, df=180).

A more detailed view is presented in Figure 5 illustrating each student's overall assessment score and their average seating position during the course. It demonstrates that there was a general trend where the assessment score degraded the further a student sat from the front. The majority of top grades come from those at the front but not exclusively, there were two students that sat at the back and scored over 90%. The numbers of students that sat at the front and back zones are comparable. Yet, examining the assessment scores highlights just one failing student at the front but four failing students at the back.

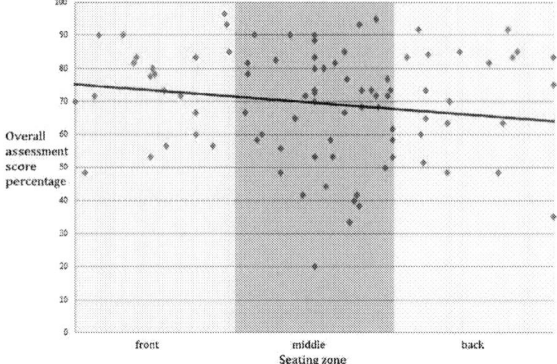

Figure 5 – Overall student assessment scores and average seating position and trend line illustrating degradation in assessment score from front to back.

It can therefore be concluded that in this case there is a relationship between the seating position and assessment performance, with a general degradation in scores from the front to the back rows. Furthermore the best assessment results were achieved by the front row students.

V. DISCUSSION AND FUTURE WORK

General conclusions

The results of this research confirm the findings of much of the previous study in the area that indeed the best assessment results did come from the students that regularly sat in the front rows. Additionally the assessment results tended to degrade from the front to the back of the lecture theatre. Conversely, unlike other studies it finds that the assessment score patterns in relation to seating are not necessarily due to the best academics or those with prior experience in programming sitting at the front. The study did however identify that the most engaged students repeatedly sat in the front rows.

This suggests that there is a benefit to sitting at the front regardless of academic ability or prior subject knowledge. It points to engagement being a potentially significant factor determining assessment outcome. It may also indicate that other untested factors may be at work in positively influencing the front row performances.

One such factor is the influence of the physical environment of the lecture theatre. All the lecture theatres used in the study utilised addition screens to aid vision of the projected screen for those at the back. All the theatres had excellent audio capabilities; the quality of the audio at the back was as good as the front. However responses from the seating choice questionnaire highlighted that a number of students stated the ability to see and hear better at the front as a determinant of sitting position. The influence of the entrance / exit points in a lecture theatre may also be a factor affecting seating decisions. There may well be a considerable difference in seating effect when contrasting a lecture theatre that has the entrances at the front compared to one with the entrances at the back. In the

former a student turning up late may well be happy to avoid attention and sit at the lower assessment performing back rows whereas in the latter the student may be forced to sit up front with the top performers.

The influence of the lecturer is also a relatively untested factor in previous research in relation to seating position. The style of teaching came up as an influence in seating choice in the student survey, with some responding that they actively avoided the front rows to actively avoid having to answer questions. The majority of the lectures in this study involved individual and student programming activities. Over 95% of the students brought and used their own laptops for these activities during the lecture. In that sense the learning experience should have been reasonably homogenised irrespective of seating position. Yet, the results point to a better performance at the front.

It is perhaps that sitting at the front promotes a better learning experience. The cognitive activity of those at the front may be higher than those sitting elsewhere. As acknowledged by many of the seating survey responses, being at the front appears to promote more potential cognitive activity, be that with peers or the lecturer. There is certainly more chance of eye contact with the lecturer, or opportunity to ask or be asked questions by sitting at the front.

Get more students to the front?

It would appear that having more students regularly sitting at the front may be of benefit to their assessment score. There are a number of lecture theatre architectures aimed at facilitating greater numbers of students at the front or closer to the position of lectern. While these structures are perhaps of reasonable consideration for new lecture theatres the adaption of existing shaped theatres to these designs is likely cost and logistically prohibitive.

A reduction in class size would enable more students to sit at the front. Decreasing class sizes would effectively increase the ability of the front rows to potentially seat more students. However a reduction in class sizes for university programming classes is unlikely to happen any time soon. The demand for graduate computing professionals has steadily increased over the past numbers of years with the IT sector regularly requiring 140,000 entrants each year in the UK [23]. However, according to the Higher Education Statistics Agency [24] there are only 16,000 computing graduates per annum, leaving a shortfall of 120,000. Consequently UK universities have responded by increasing their intake in computing degrees. While the demand remains it is reasonable to expect that class sizes will remain high.

Perhaps it may be as simple as the lecturer encouraging the students to sit at the front or if the benefits of sitting at the front were explained then they may choose to sit there. Although in this research it should be noted that even with the lecture theatre at half capacity the front row regularly had some empty seats. It would appear that given the option some students will still likely chose to sit elsewhere. In that regard the decision of where to sit is personal and generally consciously made. The student questionnaire on seating positions found that most of the front row students sat there because of better vision, being less distracted at the front and the perception of being forced to pay more attention if closer to the lecturer. Those that sat at the back did so mostly because they wished to listen and not have to be involved in the class activities; importantly there was evidence to indicate that this was a particularly strong influence for students with diagnosed anxiety issues. Some others at the back explained that arriving late meant they could get to a seat at the back with least disruption to others. Many students responded that they tended to sit wherever their friends sat.

Limitations and future work

As with many predominantly quantitative based studies the understanding of the responses would be further enhanced with a more comprehensive qualitative study. The ability to interview students, individually or in focus groups, or getting them to record their seating choice decision for each lecture using the PinPoint app would provide a greater understanding of the findings. Replication of the study using the same methodology would of course test the repeatability of the results. It would be of interest to re-examine this issue with a more traditional less interactive style lecture.

It would also be of interest to analyse the intra group assessment performance. Does the performance of students with similar standard academic ability, engagement and prior subject knowledge remain consistent regardless of seating position?

As is common in many programming modules in universities there were a limited number of females in the cohort. This inhibits the study of any potential gender influences other than to report in this instance that there was normal distribution of the females throughout the lecture theatre.

While the study concentrated on highlighting differences between front, middle and back rows there would be merit in a further analysis of the zonal areas to include the left, middle and right areas of the lecture theatre.

The cohort in this study were generally regarded to be highly motivated, it would be if interest to extend the study to a less motivated group.

Although there were some occurrences of movement between seating zones over the twelve weeks it would be of interest analyse if there were any significant changes in seating due to an occasion, such as an exam revision lecture.

While this study was with a cohort of programming students the authors feel that the accuracy of the data collection methods used could easily be transferrable to other subjects areas.

REFERENCES

[1] Weinstein, C.S. (1979). The physical environment of the school: A review of the research. Review of Educational Research 49 (4): 577–610.

[2] Benedict, M.E., and J. Hoag. 2004. Seating location in large lectures: Are seating preferences or location related to course performance? Journal of Economic Education 35 (3): 215–31.

[3] Holliman, W.B., and H.N. Anderson. 1986. Proximity and student density as ecological variables in a college classroom. Teaching of Psychology 13 (4): 200–03.

[4] Pedersen, D.M. 1994. Personality and classroom seating. Perceptual and Motor Skills 78 (33): 1355–60.

[5] Giles, R. M. et al. (1982). Recall of lecture information: A question of what, when, and where. Medical Education, 16(5), 264-268.

[6] Stires, L. (1980) Classroom seating location, student grades, and attitudes: Environment or self-selection. Environ. Behav, 12, 241-254.

[7] Schwebel, A.L. and Cherlin, D.L. (1972) Physical and social distancing in teacher pupil relationships. J. Educ. Psychol. 63, 543-550.

[8] Burda, J. M., and Brooks, C. I. (1996). College classroom seating position and changes in achievement motivation over a semester. Psychological Reports, 78, 331-336.

[9] Becker

[10] Perkins, K.K., and C. Wieman. 2005.The surprising impact of seat location on student performance. ThePhysics Teacher 43 (1): 30–33

[11] Bligh, D. (2000), What's the Use of Lectures?, San Francisco: Jossey-Bass.

[12] Brown, R. D. (1988). Self-quiz on testing and grading issues. Teaching at UNL, 10(2), pp. 1-3. The Teaching and Learning Center, University of Nebraska-Lincoln

[13] Kierwa, K. A. (2000). Fish giver or fishing teacher? The lure of strategy instruction. Teaching at UNL, 22(3), pp. 1-3. Lincoln, NE: University of Nebraska-Lincoln.

[14] Marx, R. W. (1983). Student perception in classrooms. Educational Psychologist, 18, 145-164.

[15] Cuseo, J., Fecas, V. S., & Thompson, A. (2007). Thriving in College & Beyond: Research-Based Strategies for Academic Success and Personal Development. Dubuque, IA: Kendall/Hint.

[16] Adams, R. & Biddle B. (1970). Realities of teaching: Explorations with video tape. New

[17] Cinar,A. Classroom geography: who sit where in the traditional classroom? J Int Sco Res. 2010;3:200–12.

[18] Perkins, K.K., and C. Wieman. 2005.The surprising impact of seat location on student performance. ThePhysics Teacher 43 (1): 30–33

[19] Mastrine, J. (2012) Does where you sit in the classroom say a lot about you? Available at http://college.usatoday.com/2012/01/05/does-where-you-sit-in-class-say-a-lot-about-you/

[20] Martin, J (2012) It doesn't matter where I sit does it? The Effect of Changing Classroom Seating Position on Student Performance and Motivation Factors. Online Educational Research Journal. Available at : http://www.oerj.org/View?action=viewPDF&paper=51

[21] Kalinowski, Steven; Taper, Mark L 2007 The Effect of Seat Location on Exam Grades and Student Perceptions in an Introductory Biology Class. Journal of College Science Teaching, v36 n4 p54-57 Jan-Feb 2007

[22] York: Holt, Rinehart, and Winston Armstrong, N. and Chang, S. (2007) Location, Location, Location: Does seat location affect performance in large classes? Journal of College Science Teaching 37 (2), 54-58.

[23] eskills, 2013. e-Leadership:e-Skills for Competitiveness and Innovation Vision, Roadmap and Foresight Scenarios Vision Report. Available at: http://eskillsvision.eu/fileadmin/eSkillsVision/documents/VISION%20Final%20Report.pdf

[24] Higher Education Statistics Agency: Home - HESA (2016), Available at: https://www.hesa.ac.uk/ (Accessed: 6/12/2016).

Knowledge and Skills: A Critical View

Mario Radovan
University of Rijeka, Department of informatics,
Rijeka, Croatia
mradovan@inf.uniri.hr

Abstract - The paper puts forward a critical view on knowledge and education in the present techno-economic culture. This culture stresses the importance of knowledge and education, but it promotes only the instrumental knowledge that the present techno-economy directly needs. Such practice displaces other dimensions of the human knowledge and cognitive abilities, which are needed for the progress of people and humanity. Business forces are raising the means to the level of the end, and shape education in the way that serves business aim, rather than the promotion of knowledge. Information technology has brought many excellent means and services to education , but it should not be used in the way that reduces people to automatons that perform prescribed procedures, without understanding or asking anything beyond that.

I. INTRODUCTION

Let us describe some of the concepts that we use in the paper. Contemporary economic activities strongly depend on technology, and the development of new technological systems is powered by economic aims. The concept of *techno-economy* points to the connection of technology and economy and to their mutual dependence. The way people work and live is determined by the socio-economic system, more than by technology itself. This system shapes the technology we use and the environment in which we live: it shapes the physical and mental space of our existence, and our understanding of existence [1].

In computer terminology, a precisely defined finite process is called *algorithm*. A computer program implements an algorithm; when active, the program performs the process defined by that algorithm. Knowledge and activities of contemporary people are getting more and more algorithmic or procedural. Knowledge is reduced to the ability of performing precise procedures, and human cognitive activity is reduced to a constant learning and performing of such procedures. We are surrounded with screens and buttons, and to do anything, from buying bananas to flying an airliner, we must perform appropriate procedures with those screens and buttons. People have always wished to possess the power to control and steer the world; technology brought them plenty of such power. However, the means that give people power, shape their cognitive abilities and their mental space. Technology creates immense opportunities,

but it compels people to think and behave in ways that it requires and imposes.

The concept of *instrumental reason* refers to that dimension (capacity) of human cognitive system, which allows people to perform specific procedures with the aim to produce specific effects. In other words, instrumental reason is the human reason used as a means (instrument) for carrying out a specific task. People have always used reason in this way in their endeavours to satisfy their needs and achieve their ends. The intense use of technology in all spaces of human activity has led to the complete domination of instrumental thinking. Technological means compel people to constantly think and act in instrumental way: in this way, those means gradually reduce human cognitive abilities to the instrumental dimension. For the procedural (instrumental) people, to know means to know *how:* they learn and think with the aim to *do* something, not with the aim to enjoy the understanding. They acts in the framework of the paradigm with which they are steered, without questioning its values, aims and effects.

We can speak about four kinds of human reasoning or about four dimensions of human cognitive abilities: creative, instrumental, critical and integrative. Creative reason explores the space of values, aims and visions of the human life and activities. Instrumental reason deals with the means and methods of the realization of the chosen aims and visions. Critical reason examines and evaluates the wider effects of the activities of the former two dimensions. Finally, integrative reason aims to keep the previous three components in balance, because only their harmonious coexistence can allow people to achieve their best possibilities. Education as well as public discourse should promote all four dimensions of human cognitive abilities. The present-day education promotes only instrumental reasoning and abilities, and public discourse, reduced to stupefying infotainment, promotes none. This reduces people to servants of the techno-economic system, who do not think about its aims and effects in a critical way.

We assume that emotions are the moving force behind all human reasoning and activities, creative and destructive ones. Reason by itself has no ends: reason chooses and evaluates ends on the basis of emotions. We must not blame instrumental reason when our emotions and passions make it serve destructive aims. Instrumental reason serves all aims: the problem is that human feelings and passions can lead to the adoption of destructive aims.

II. THE QUESTION OF KNOWLEDGE

Techno-economic society calls itself the *knowledge society*, because knowledge is essential for production and consumption. However, this society promotes only instrumental knowledge and procedural (algorithmic) thinking, which are needed for the performance of technological processes and business activities, so that this society does not deserve to be called the knowledge society. A more modest name, such as the *skill society*, would be more appropriate for this society. A descriptive name, such as the society of *specific skills and general ignorance* would describe the present society even better [2]. For the development of complex technological systems, as well as for their control and use, an huge amount of instrumental knowledge is needed, but all this knowledge belongs to only one dimension of human mental abilities.

It is important to make difference between knowledge and skills. A culture that mistakes skills for wisdom, and measures its progress in terms of production and consumption instead of in terms of cooperation and compassion, is doomed to failure: such culture "condemns itself to death", says Chris Hedges [3] (p. 103). Such claims may seem too strong, but life in such a culture is one-dimensional and it seems empty. Knowledge is something more comprehensive and less operative than algorithmic thinking and instrumental skills are. To make the world better, it is necessary to promote understanding, not only instrumental thinking, and solidarity, not only competition and struggle. General ignorance is dangerous, because it lowers the level of public discourse and makes manipulation easier. This usually leads toward radical views and aggressive behaviour.

Ignorance is present in techno-economically leading countries more than we are aware of. Let me mention an example. Craig Venter, a famous geneticist, delivered a lecture on the role of science and the challenges of the present world, which was broadcast by BBC television. In that lecture, Venter put forward many interesting data, some of which regarded the education and ignorance in the present-day world. Among other things, he said that 25 percent of American citizens do not know that the earth revolves around the sun. This means one out of four citizens. Hedges [3] says that "nearly a third" of the American population is "illiterate or barely literate" (p. 44). Let me add to this that "in the secular United States", 66 percent of people "believe in the Devil" [4] (p. 77). I am not sure what this believing actually means, but it does not seem encouraging in terms of knowledge and education.

Mark Bauerlein [5] puts forward data which show that general knowledge and literacy have been in decline among young people. Bauerlein mentions various reasons for such a bad situation, but he blames primarily the young because they do not read books enough, and teachers because they do not do their job well enough. He criticizes "the custodians of culture, the people who serve as stewards of civilization and mentors to the next generation"; those custodians and mentors are primarily "teachers, professors, writers, journalists, intellectuals, editors, librarians, and curators", as well as the institutions for which they work (p. 161). Bauerlein concludes that "they have let down the society that entrusts them to sustain intelligence and wisdom and beauty" (p. 161). We argue that Bauerlein's criticism misses the source of the problem. In ruthless capitalism, teachers and journalists are pawns that are compelled to behave according to the rules set by the system. Bauerlein blames the young and teachers, but he says nothing about the socio-economic system that shapes their reality. Our education and our ignorance are shaped by the socio-economic system; teachers and similar "stewards of civilization" do not have the power to change this system.

Bauerlein says that the young read less and less, especially when it comes to more demanding books. The consequence of this is a decline in the knowledge of history, culture, art, and science. The knowledge about current events, at the local and global level, is also in decline. This is bad, but it is wrong to blame the young and the "stewards of civilization" for such a situation. The young read less and less books because the industry produced new means and methods of efficient recording and distribution of all kinds of information contents. The reason for the decline in general knowledge is not in the fact that the young do not read books, but in the fact that the means and contents that have displaced and replaced books are used in the ways that are not good from the educational point of view. The information industry makes everything fast, exciting and shallow. Anyway, the behaviour of the young is the product of the society in which they have been raised; hence, the causes of the weaknesses in their behaviour must be sought in the weaknesses of the society, which Bauerlein does not do.

Socio-economic system sets the values and determines the structure of power in society. With this, the socio-economic system shapes the behaviour of people, especially of the young. It makes no sense to blame the young for the fact that they behave in the way they do, because they behave in the way that the world created by the older encourages and requires them to behave. It makes no sense to blame teachers and journalists, because business values and principles shape the rules according to which we all must behave. Teachers, journalists, librarians and similar creatures do not have the power to change the rules and values that have been set by the ruling paradigm (ruthless capitalism and consumerism) and by power-holders who shape the structure of power and the relationships in society and in the world. Bauerlein blames pawns, but he neglects the rules of the game, as well as the masters who play the game.

Images have displaced written contents, and *Homo online* has displaced the old, conceptually oriented, *Homo sapiens*. These are very relevant changes, but the socio-economic system has a much larger impact on the life of people and society than the means by themselves have. The decline in general knowledge is a product of the socio-economic system that shapes education and public discourse. Every economic system shapes people and society in its own image, and so does the present techno-economic system. I studied and graduated computer science because I wished to learn how these machines work under the surface. My students are not curious about such things; they are interested in what they can *do* and *achieve* with those machines. Procedural people know virtually nothing about the internal structure of the systems they use, and even less about the theoretical principles on the basis of which these systems have been constructed and work. And they do not show much curiosity about that. Procedural people are masters of procedures, and ignoramuses about nearly everything else.

Finally, people spend more and more time on the learning of how to use the means; frequent upgrading makes the situation even more difficult. Devices and systems are getting more and more complex; new versions are overloaded with "capacities" and "intelligence" that most people do not need, but they make the process of learning the system and its functions more demanding. Frequent upgrading of tools compels us to spend more and more time and energy on the learning of how to use new tools for doing the same things we were doing with the previous versions of those tools. New and better things are developed with time, so that it is normal that new and upgraded versions of tools are produced and released. However, upgrading and changes are often made primarily for business reasons: to compel people to buy new things and to make them more and more dependent on the industry and its knowledge of how to use those things. Upgrading is unavoidable and welcome to a certain extent, but it has often produces a huge waste of time and energy, and do not bring much good.

III. THE ROLE OF HIGHER EDUCATION

The intense development of industry in the nineteenth and twentieth century created a need for specialized knowledge and research. Higher education was required to satisfy those needs, so that the main task of higher education became to supply society with the labour force that possesses the instrumental knowledge that techno-economy needed. The demands about the autonomy of higher education and about critical analysis and discourse were displaced by economic reasons, and they have fallen into oblivion. The present higher education produces instrumental knowledge that techno-economy and corporate business ask it to produce. It shapes its teaching programs and research activities in accordance with business principles and the demands of the market. The industry of higher education is large and growing; however, the education is not moved by the authentic human *desire to know*, but by business principles and market value of the instrumental knowledge it produces [6].

Corporate business and the ruling socio-economic system have compelled universities to behave like business companies. Instrumental knowledge has been in demand, and universities have been compelled to produce this kind of knowledge. Corporate business shapes education in the way that serves its needs and promotes its aims. Knowledge gets value on the market, and this value is determined by the market. Higher education has become an industry that produces instrumental knowledge that the techno-economy needs and for which there is a demand on

the labour market. Such education is reduced to the training for a job, and it does not promote a critical thinking about the ruling paradigm, its values and practice.

Leaders of the best known universities warn that the material dependence of universities on corporate business may have negative effects on their research and teaching activities. A former president of Harvard university said that the "commercialization of universities" is the most severe threat that the contemporary higher education faces. The growing material dependence of the university activities on corporate money can seriously compromise those activities. Teaching contents and research results can become partial and biased in favour of those who pay for them [7] (p, 263-64).

Higher education has always served the socio-economic system, rather than questioned and criticized it; it have taught students how to find a comfortable place in the system, rather than how to change it. Universities have seldom been sites of intense social criticism, because they existentially depend on the ruling system and power-holders. If they criticize the system and power-holders (economic, political, cultural) they may be punished for that. Universities have sometimes been places of intellectual resistance to the ruling narrative, but they have usually been socially conservative and politically passive. They did nor participate much in large social movements and changes that took place in modern history. The struggle for the abolition of slavery, the workers' struggle for better working conditions, and the struggle for the liberation of colonies were not influenced much by academic institutions and activities [6] (p. 2). The academic world has followed historical changes, more than it created them.

At the present time, the social impact of the university seems even lower than in the past. Instead of being a space of free and critical enquiry about all socially relevant issues, universities have become corporations and servants of corporate business that they financially depend on. Universities have adopted values and practice of the corporations they serve [3] (p, 110). They have become part of the world shaped by the corporate mind, values and aims. Their aim has become the efficient production of marketable instrumental knowledge, not the promotion of a wide and deep understanding [8]. Students are given a narrow specialized knowledge, without the insight into social values, ethical principles, and common good. Such students are only capable of serving the existing practice and structure of power. Universities produce formally educated people who should be considered illiterate by traditional standards, says Hedges [3] (p. 96). Young people have not been taught about the structure of social power; they know very little about their own civilization, and they do not seem to be able to maintain and advance it.

In the present techno-economic society, knowledge about history, culture, art, and social issues is in decline. The complete domination of instrumental education diminishes the abilities and readiness of new generations to assess their socio-economic system and its practice in a critical way and to make them better. Society needs professional training, but education should promote

632

analytical and critical thinking about all narratives and systems, values and aims, because such thinking is necessary for the progress of people and humanity. Education must produce the instrumental knowledge that society needs, but it should not be reduced to this task: it should teach more than instrumental knowledge and operative skills. It should teach the rational and objective way of thinking; it should present major traditional and contemporary narratives and historical events that shape our reality, in an impartial way. Higher education and scientific institutions must promote the understanding and critical discourse about every narrative and power that shape our lives, society and the world, or has the ambition to do so. Science and education cannot assume political leadership in a society, but they must enquire and evaluate every power, politics and behaviour. In this way, science and education can contribute essentially to the development of a more reasonable and more just world, and of new visions of the peaceful progress of humanity. On the other hand, limited to the production of instrumental knowledge, science and education only serve the preservation of the existing paradigm and structure of power, with all their drawbacks and weaknesses.

Education has always shaped new generations according to the values and needs of the dominant narrative and practice. Enthusiasts say that education must free young people from the tyranny of the truths and values of their time. But education has always aimed to subdue new generations to the tyranny of the truths and values of their times, and it has taught them how to best serve that tyranny. The basic aim of education must be to reduce ignorance, superstition, and suffering, says Neil Postman [9]; the aim of the present-day education is to mould new generations in accordance with the needs of the techno-economic system (p. 70). Education must teach young people about the human creative aspirations and destructive inclinations, and promote understanding, goodness, and beauty. Education must promote universal values and principles, openness, social justice, and human dignity. In this way, education promotes constructive and cooperative behaviour of people and communities, and creates a sense of coherence of human endeavours. Such education would make a great contribution to the solution of many problems and to the creation of better people and a better world. All ages needed such an education, but they did not have it; the present age is not an exception in this regard.

IV. THE MEANS AND THE CONTENTS

Information technology gives students the access to an immense amount of information contents from which they can learn. But the quantity of textbooks and learning materials has not been the problem in recent decades, even before the appearance of the internet and its services. On the other hand, the *quality* has been the problem, and the proliferation of information technology has made this problem worse. A huge increase of the quantity of information contents of all kinds has been accompanied with the fall in the average quality of those contents. Books used to be written by those who really knew the

subject they were writing about, and who had a passion for explaining things in a clear and inspiring way. The present-day learning materials aim to be "fun", but I seldom find a real understanding and good explanations in those materials. As a former passionate computer programmer and the author of a book on programming, I can say that the level of teaching and of student's knowledge of computer programming is notably lower today than it was forty years ago. Programming is a demanding activity that requires knowledge and cognitive effort. Various video materials (films and cartoons) do not help in a serious teaching of computer programming. The same probably holds for many other subjects of education.

The learning and understanding of demanding things, such as formal theories and complex systems, requires a high level of concentration and a serious mental effort. Screens do not help concentration nor do they encourage students to make a serious mental effort. I have been working with screens for decades, but the reading from screen, especially of larger and demanding texts, has remained different for me than the reading of printed matter. On screen, everything looks superficial and ephemeral. Printed matter makes contents more physically present and lasting than screens do. The medium is *not* the message, but the medium matters. Learning from the web seems even more problematic. Web stimulates superficiality rather than concentration and cognitive effort. Links lure students to press them and seek knowledge somewhere else.

The world is inundated with information contents and messages of all kinds, but the sense of quality has been lost. The internet offers students a huge amount of contents of dubious quality, and stimulates surfing instead of concentrated and engaged reading and thinking. The internet contains countless useful data and information, but the learning from the internet is superficial and it shapes students in its own image. Information technology facilitated the production and distribution of an immense amount of information contents; to survive in this noisy jungle, we must constantly select. But with the increase of the quantity of information contents an efficient selecting is getting more and more difficult.

Computers must be used in teaching those things that are done by means of computers. All sorts of computations and designs are done by means of computers and software tools: it is normal to teach students how to do such things by means of computers and software tools. However, we hold that technology does not help in learning the *principles and laws* on which computations and designs are based. A bridge can be designed by means of computer and a software tool. But for such a tool to exist, somebody must *produce* it. To produce such a tool, it is necessary to know the laws and principles on the basis of which bridges are constructed. To be able to produce tools, we must know *more* than how to use them: we must know laws and principles that are implemented by those tools. Finally, we the ones who must constantly evaluate the *aims* of our individual and collective endeavours. Information technology is the means (tool) that increases the efficiency of *doing*, but it may obstruct the process of learning and understanding the basic principles and aims. If education is reduced to the art of handling the means, people will become ignorant about laws and principles, and unable to chose and evaluate their aims.

V. THE ART OF TEACHING

Techno-economy aims to eliminate direct (live) lectures held by professors, and to replace them by learning systems based on information technology. I consider this idea bad for several reasons. I admit that at least half of the professors I have known as a student or later, have not seemed good to me. But some professors *are* good, and that is what matters here. Direct encounters with such people who know, and who know how to present knowledge, leave a lasting impression on the young people who want to learn and understand, not only to get a degree. A good professor presents a relevant content in a clear way, and shows a way of thinking about the topic he or she speaks about. If such people are eliminated because of business reasons, new generations will lose the possibility to encounter such people and to learn essential things from them. A physical encounter and a direct discourse with a person are quite different from the encounter over the screen. It has been said that "the excellence in teaching remains an art that cannot be simulated by machines" [10] (p, 221). This is probably true, but even if this art can be replicated by machines to a notable extent, why should this be done? If young people will be educated only by machines, they will become like machines. They will not be able to explore the possibilities of *human* progress; they will think only about the advance of technology. To hand over education to machines may be good for business, but it is bad for people.

Automated education will transform people into machines that learn from machines how to handle machines, and who understand and ask nothing beyond that. This will bring about a complete domination of instrumental reasoning over creative and critical reasoning, which is bad, because this will narrow the space of human abilities and experience. The automation of education is based on the bad assumption that money is the supreme value and that every activity must be evaluated exclusively according to the economic criteria. Virtual universities, without professors and campuses, may cost less, but such systems of higher education will lack the best that traditional universities offer: direct encounters and interactions between people.

Universities are peculiar *physical places* because of their specific atmosphere, which cannot be replicated by means of the internet. In spite of all communication technology, physical encounters are important and they will remain important, as long as we are physical beings. If professors and lecturing are eliminated, the understanding of life and the attitudes of new generations will be shaped only by celebrities and politicians, which is not good. Finally, being a professor is not so bad, although it is not as good as it may seem. Being a classical student is the best job I have ever known. Why should these jobs be eliminated? With the aim to lower the costs and increase profitability, ruthless capitalism has been replacing good jobs with machines, so that only bad jobs remain for people. This is not something I consider a progress of people and humanity.

VI. CONCLUSION

It is necessary to be "contemporary" to a certain extent to be able to live, but those who uncritically accept the ruling narrative of their time often act in suboptimal way. We must constantly examine the world in which we live, its values, means and aims. Information technology facilitates many excellent things and offers many good services in the space of education. But we must not *raise the means to the level of the end*, and shape the education in the way that serves corporate business, but does not promote a wider knowledge and understanding.

By thinking and living only at the instrumental level, people diminish their experience of existence, and with this they diminish themselves. We do not advocate a spirituality here, but a reflective and *poetical* attitude towards the world in which we live and toward existence. Instrumental reason has achieved great things at the operative level, but those whose mind has been reduced to the instrumental dimension do usually not see other things, without which the efficient functioning at the operative level do not bring much good.

Education cannot solve all problems of this world, especially not if people and communities do not behave in the way that promotes cooperation, mutual respect and solidarity. However, if education cannot solve all the problems, ignorance makes them much more dangerous and harmful. It is difficult to change dominant paradigm and practice, but we must preserve the awareness that things can be different; otherwise, we will *create and accept* a totalitarian world, in spite of many nice slogans that we constantly produce.

REFERENCES

[1] M. Radovan, Communication and Control: The shaping of reality and people, Amazon / Kindle, 2015

[2] M. Radovan, "ICT and Human Progress", *The Information Society*, Vol. 29(5), pp. 297-306, 2013.

[3] C. Hedges, *Empire of Illusion*, New York: Nation Books, 2009.

[4] S. A. Diamond, *Anger, Madness, and the Daimonic*. New York: State University of New York Press, 1996.

[5] M. Bauerlein, *The Dumbest Generation*, New York: Tarcher/Penguin, 2009.

[6] G. Delanty, *Challenging Knowledge: The University in the Knowledge Society*, Buckingham: SRHE and Open University Press, 2001.

[7] F. Webster (ed), *The Information Society Reader*, London: Routledge, 2004.

[8] R. Hassan, *The Information Society*, Cambridge: Polity Press, 2008.

[9] N. Postman, Technopoly: The Surrender of Culture to Technology, New York: Vintage Books, 1993.

[10] W. H. Dutton (ed), *Society on the Line: Information Politics in the Digital Age*, New York: Oxford University Press, 1999.

Today is the future of yesterday; what is the future of today?

H.Jaakkola[*], J.Henno [**], J.Mäkelä [***] and B.Thalheim [****]

[*] Tampere University of Technology, Pori, Finland
[**] Tallinn University of Technology, Tallinn, Estonia
[***] University of Lapland, Rovaniemi, Finland
[****] Christian Albrechts University, Kiel, Germany
Corresponding author email: hannu.jaakkola@tut.fi

Abstract - In the educational context, understanding the future is important for two reasons. First, we are educating people for *future tasks*, which need skills that are useful in the future. Secondly, educators have to be able to select the most promising *tools and technologies to apply in their work*. The problem is that there is no clear way to weigh the importance of the alternatives – what the real importance of a certain technology will be in the near future and especially in the long term. In our paper, we focus on analyzing selected technologies. Our approach applies the framework developed by the authors. The promising technologies are reviewed by a systematic literature study, focusing on and restricted to the information and communication technology (ICT) sector. The findings are classified according to their importance and the time span of their effectiveness. The question we answer is "*What should every educator know about changes in technology?*"

I. INTRODUCTION

Technological changes and progress in technology are enablers and accelerators for wider changes in our society and economy. Ultimately, the consequences are seen not only in products, but also in processes, business models, and common behavioral patterns. It is justified to say that now one of the biggest change factors is information and communication technology (ICT). We do not wish to underrate other fields of technology, but in everyday life, the changes made possible by ICT in particular can be perceived the clearest. Forecasting is the highest level of being prepared for the future – especially if the prognoses are accurate. Having forecasts provides us with the means to be *proactive* – prepared in advance for situations that happen, or also *preactive* – having the means to affect in advance the alternatives that may happen.

For educators, understanding the future is important because they are educating people for future tasks (the "product" we are producing), and because they have to be able to select the most promising tools and technologies to be applied in their work (the optimal process and environment to "produce our product"). Both of these aspects are changing just as fast as the surrounding society.

The aim of the paper is to build a synthesis – not complete (which is actually not possible at all) but a list of a variety of technological changes that are worth recognizing now. The *research question* handled by our paper is *"What should every educator know about changes in technology?"* Our approach applies the principles of the *Technology Change Analysis and Categorization (TCAC) framework* introduced by Jaakkola et al. in [1]. The material used in the analysis covers the publicly available technological forecasting sources that are the best-known and most referred to. In spite of the fact that the material covers a limited scope of sources, we consider the picture it gives to be relatively reliable.

We have approached this study topic with two papers (Figure 1). The principles of the analysis method are introduced in Paper I [1]. The aim of the present paper (Paper II in Fig. 1) is to apply the method in ICT-related technologies and to give a synthesis – a list of the technological changes that are worth recognizing now (*ICT Change Analysis and Categorization – ICAC*).

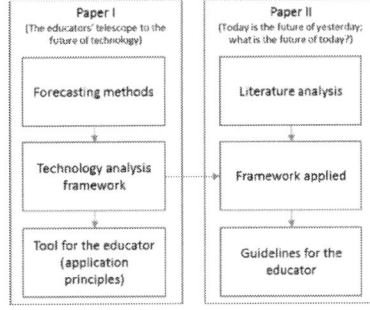

Figure 1.   Structure of the problem solving – two interconnected papers.

The *research question* handled by our paper is *"What should every educator know about changes in technology?"* The time span of our analysis covers mainly the next five years, but related aspects with a longer time span are also discussed. We selected this split approach to separate method development and its application from each other and provide the means for a deeper discussion than only one paper allows. We urge the reader to study both of these papers.

Figure 2 introduces the principles of the TCAC framework. It also reflects the structure of our paper. The method is based on Paper I [1]; the details can be found there.

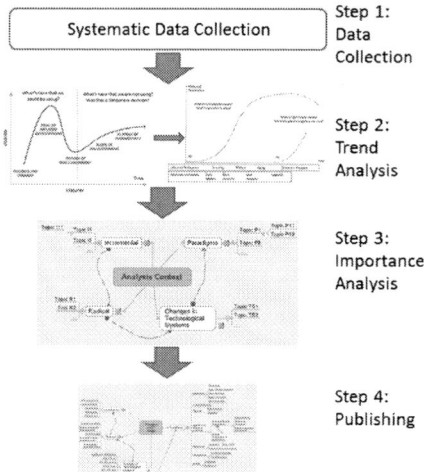

Figure 2.   TCAC framework – summary of the components of the analysis.

The starting point is data collected on the study context (in our case ICT-related changes) – Step 1. Two methods are used to analyze the data: hype cycle based analysis of pre-embryonic technologies to understand the delay of their commercial appearance, and life cycle model based analysis to understand the life cycle and renewal power of the technology analyzed – Step 2. There is also an arrow between the hype cycle curve and life cycle curve. Since the hype cycle represents the pre-embryonic phases of the phenomena analyzed, it ultimately feeds the embryonic phase of the life cycle analysis; pre-embryonic becomes embryonic provided that the exit (not valid for mass and commercial use) has not been realized before it. Finally, the changes are classified according to their importance in four categories – Step 3. This phase also includes "fine-tuning" the interpretations of the importance and effectiveness classes, i.e., to give them exact semantics. The result is the classification of the selected technologies (Publishing – Step 4).

The paper is structured in the following way. Section 2 covers the first step of the analysis method – collecting the material for further analysis and classification according to the principles of the systematic literature review. The findings are organized in table format, having topic areas and publishers as the dimensions. Every topic area is handled in detail in Section 3, which is divided into sub-sections according to the topics. Section 4 covers a summary of the classification of the findings. The results are published and visualized in mind map format. This section also validates our analysis method in a practical situation. Section 5 reflects the findings in terms of the education sector and gives answer(s) to our original research question. Section 6 concludes the paper.

II.   WHAT WILL OUR NEAR FUTURE BE LIKE?

The main purpose of writing this paper is to make a review of the ICT-related technological changes that are relevant today. We specified the research problem to be discussed as "*What should every educator know about changes in technology?*" Especially in higher education, we

must be prepared for the assignments of the future. Every educator should be an innovator in the adopter categories, or at least an early adopter (see [1]; Paper I).

We have approached the answer to the research question following the steps of the framework introduced in Fig. 2.

1. *Data Collection*: We collected the "raw data" for our analysis from fourteen analyst sources (Business Insider, Frogdesign, CSC, Fjord, PWC, Quantumrun, Howstuffworks, WEF, Wired, Forbes, Gartner, Deloitte, Forrester, IDC) and sixteen reports. It is augmented by two additional analyses (Cisco Trends and Analysis [5; 6]; Gartner Emerging Technologies 2016 [15]). The reports were selected according to the principles of systematic literature search. The search was restricted to publicly available material, also accepting indirect (from reliable sources) analysis material (due to the difficult availability and high prices of the original reports). The relevance and the reliability of the sources were analyzed by the researchers (authors) based on their long-term experience in the study field (expert opinion). Additional material was also reviewed without any remarkable new findings (penetration rule of the systematic literature study was achieved).

2. Trend Analysis: The findings were organized in seven main groups covering 22 primary topics to merge the overlapping findings of the original sources for trend analysis. Table 1 summarizes our findings from the reports.

3. *Importance analysis*: Classification of the importance and effectiveness of the reported technology is based on the discussion in Section 3.

4. *Publishing*: The final result - ICT Change Analysis and Categorization (ICAC) - is published in Fig. 4.

The reports used in the analysis are listed in the references of this paper. To avoid complexity, we have not used positioned source references in the text; these can be found indirectly from the Table 1 column headings.

Most of the studies analyzed were focused on the current time: what are the leading technologies for 2017; long-term studies covered the period ten years ahead from now. When examining these, it is, of course, good to keep in mind that changes are usually slow. Today's technology provides the basis for the following ten years. In principle, everything that will be new in ten years' time is already known today; ultimately, even long-term predictions do not provide anything dramatically new. The only thing worth recognizing is the uncertainty built into the trends, especially the ability to separate the hype effect from the reality. In the following review, we satisfy ourselves with a superficial examination; details are available in the original material.

TABLE I. TECHNOLOGY TRENDS SUMMARIZED

Group	Category	Business Insider [2]	Frog design [11]	CSC [5]	fjord [8]	PwC [16]	Quantum run [17; 18]	Howstuff works [14]	WEF [20]	Wired [19]	Forbes [9]	Gartner [12]	Deloitte [16]	Forrester [7]	IDC (62000) [15]
AI	Artificial Intelligence	Logistics					Human work replaced				Repetitive jobs replaced	System intelligence	Cognitive computing	Personalization Predictive analytics	
	Machine Learning	Support for humans	Automated tasks, assistant									Intelligence of the apps			
Robots	SW robotics: intelligent chatbot, problem solving applications	Automated tasks Business bots			Digital services	Service robots Modular structure	Powerful speech-controlled virtual assistant								
	HW Robotics:		Wearable intelligence (materials)	Automated tasks		Service robots Modular structure	Wearable robots (for disabled) Microrobot (medical)	Avatars, surrogates, robots Zero size intelligence		"Techy" textiles Health wearables Nano particles	Physical-Digital integration	Digital twins			
Data Analysis	Video analytics						3D/360 degree cameras Online analytics								
	Big Data	Medical Data	Precision medicine				Variety of sources Daily new 10**18 B	Mass data	Data driven healthcare	Cognitive behavioral therapy	Humanized big data		Big data analytics		
UI	Advanced UI technologies	Audio	Audio				Open air gesture, speech, AR/VR; haptic UI Digital streaming	Neuro hacking Universal translators				Conversational systems Natural language			
	VR, AR, MR	Entertainment Therapy	Therapy VR-on-demand	Next wave digital interface	Mass market, integrated to daily life	Consumer level Enterprise level	Consumer applications Teaching	Avatars, surrogates			Mature to apply in real life	High value in a variety of apps		AR & VR	
Processes & Structures	User led innovations Importance of user experience				Fast time to market, digital innovations						Everything on-demand	Evolution of the user experience New elements	Flexible consumption business	Virtual, physical and digital experience harmonized	
	Hyperconnected lifestyle				Social networks, video, image presence					Massive messaging					
	Renewed business			Technology enabled changes		Distributed trust Smart contracts		IoT embedded in business		Casual programming		Intelligent digital mesh	Digital economy Stategy driven technology	Business restructuring	Digitally enhanced products, services and experiences
	Open protocols			SW defined networking		Growth of non-structured data						Development platforms	Enterprise architecture Open platforms		
Key Technologies	Internet						In ten years: 1000 times faster Global traffic 452 EB Mobile web 126 EB		Humanized Internet interfaces everywhere	Mobilizing the next 4 Billion					
	Cloud			Simplifying cloud platforms									Cloud computing	Cloud	Cloud-based IT infrastructure
	Block chain									Digital currency		Distributed ledgers			
	NoSQL data														
	Security and privacy				Digital ethics, Digital Cannibalism			Dark networks Cyber attacks			Control of super intelligence Digital feudalism	Security attacks Adaptive security architectures	Effective solutions needed		
Key Applications	Drones	Human work assistance	Cost reduction Data gathering			Commercial drones	Police (car) drones Distribution tasks								
	3D and 4D printing					Rapid prototyping Complete systems	Continuing exponential growth		Bioprinting	4D printing, transforming objects					
	Autonomous Vehicles	Interacting cars	Intelligent operations		Close to becoming daily life		Smart car 2026: driverless bus 2026: 3D fast bus								
	IoT			Maturing of IoT				Convergence of digital and physical world			Mature for early stage applications			IoT	
	Intelligent home / space / things	Sensors Analytics	Adapting space		Wi-fi enabled controllers		Smart home			Ambient intelligence	Lack of seamless applications	Intelligent things	Emerging phase		

III. HELPFUL HINTS

A. General findings

A common finding in all the reports analyzed is the rising importance of artificial intelligence (AI) – either as it is but mainly embedded in a wide variety of products and processes (robotics, intelligent analysis, intelligent sensors, networks, etc.). Embedded intelligence is seen no longer as a support, but as an alternative for human work. AI beats human intelligence because of its ability to manage masses of information and repeat the same routines carefully time after time. AI also has an unlimited memory capacity that can be retrieved quickly in decision making, as well as the capability to store large masses of information either directly or by remote access.

B. Artificial Intelligence and its Applications

Cognitive computing provides the means for "human kind of thinking" and machine learning (deep learning) for "human kind of learning" and adaption in new situations. *Virtual assistants* (software based chatbots; conversational systems) are used in the role of service robots and advisors (Apple (Siri), Google (Now), Amazon (Alexa) and Microsoft (Cortana) are examples of such tools). The real *electromechanical robots* help or replace people in a variety of (simple) human tasks. Voice activated speakers (also called Smart Speaker Hubs, Voice Activated Speakers) are Internet connected virtual assistants that use conversational systems as a user interface to activate demanded services. Such speakers are available e.g. from Amazon (Echo), Google (Now), LG, Harman Kardon, Lenovo and Sonos. When the opportunity to use more versatile communication than before provided by the new user interfaces (speech, haptic control, augmented reality (AR)) is connected here, a robot (soft or hard) becomes a part of ordinary processes, both in business and in private life. In robotics, such sub-areas as wearable intelligence (intelligent materials), wearable robots (e.g., to support the disabled in their daily life), the use of avatars and surrogates to replace humans (digital twins; physical-digital integration) in certain situations, are recognized in the reviewed analysis. One of the special areas in robotics, which has a promising future, is microrobotics. This is based on nano particles (zero size intelligence) that are small in size and possible to use, e.g., in medical care, to provide (non-destructive) access to such parts of the human body that are unreachable in traditional human hand touch-based medicine. . Appearance of artificial heart is also reported related to the field of medicine.

C. Data and Analysis

One key technology area of today is big data. Quantumrun reports on a study by IBM: Every single day human beings create 2.5 quintillion (10**18) bytes of data. Individual human-related data is produced in a variety of ways: cellphone signals, social media, on-line shopping, credit card usage, web usage, service usage, health wearables are all examples. The data may be open (providing open access) or closed (access restricted and controlled). Individual-related data has in principle a closed characteristic, but in increasing amounts is kept open by the _agreements defined by the "data owners" (e.g., most of the

social media services). Big data technologies are used to manage and handle large amounts of data, centralized and distributed, closed and open. An increasing amount of data is signal data (audio, video, mixed format) that provide an important source for data analytics – to an increasing extent in real time.

Several analysts have pointed out the importance of big data technologies to support medical care – a good example of this is the Watson medical diagnostics system. Their reports are also worried about *data usage ethics and ownership of data*. The terms digital ethics, digital cannibalism, digital feudalism, security attacks are found in the reports analyzed. All of these focus on questions of privacy and data security and the use of data (access permitted or not permitted). *Adaptive security architectures* (based on intelligent, scalable architectures or platforms) are seen as a solution. . One specific area of data analysis was handled in the report of IEEE [18] – face recognition. This technology provides enormous opportunities for recognizing peoples face from raw data, both in live stream and especially in stored data. In turn it further increases privacy related problems; technology itself starts to be reasonable mature for mass use.

D. Work Landscape

"Robots and AI will take our jobs." It is estimated that between 35 and 50 percent of jobs that exist today are at risk of being lost to automation. Repetitive, blue-collar type jobs might be first, but even professionals — including paralegals, diagnosticians, and customer service representatives — will be at risk. The problem is that the jobs that will remain will require high levels of education and creativity, and there will be fewer of them to go around. We need to start thinking about what kinds of jobs the rest of the population will be doing – those that are no longer employable in their traditional jobs. Life-long learning and transfer to new jobs is not an answer to this societal problem.

E. User Interfaces

One sector widely referred to in the reports relates to new user interfaces. Virtual Reality (VR), Augmented Reality (AR), and Mixed Reality (MR) are seen to be at the breakthrough point for the mass market and applications. Promising use of these is seen in entertainment, therapy, usage as a digital interface, integrated in a variety of daily life and business applications, teaching and training. Glasless AR – 3D imagery, based on the Magic Leap AR technology (https://www.magicleap.com/) is also reported to be close to the embryonic phase. The use of audio-based interaction and natural language are a rising trend. This covers audio control, conversational systems, and real-time speech-to-speech language translation. Fold-up flexible screens start to be available for display purposes. Gesture-based and open-air haptic interfaces are mentioned as new interfacing technologies, as well as the rising importance of digital streaming as a replacement for traditional media distribution. The term "neurohacking" describes the technologies that are used to pay attention to the technologies that are used to identify the "inner signals" of the object – e.g., the signals from a human brain; ; the signals are retrieved from the brain connected sensors or remotely by retrieving the electromagnetic signals produced by human body. It can be seen as a first step towards the computer – brain interface that can be found in the "innovation trigger" phase of the Gartner hype cycle (see details below in this paper).

F. Device Innovations

In the category of device innovations, the reports mention that autonomous interacting vehicles are close to being ready for mass use; these cover autonomous cars, robotaxis, and also fly robotics. Most of the reports also recognize the fast growth of electricity as the power source of cars. It is also expected that in ten years autonomous buses managing 3D driving will be reality – at least in restricted usage. 3D printing is seen as an ordinary, but fast improving application. Transfer towards application areas that are more versatile than at present is going on: printing of complete systems, bioprinting (printable bio materials), and printing of transforming objects (4D printing based on materials adapting in time and situation). The IoT (Internet of Things) is in its emerging phase and mature for early stage applications. Intelligent home and adapting space are based on improved (Wifi) networked sensors (as an application area of IoT) and the analytics of the data produced by them. However, it is seen to suffer from the *lack of seamless solutions* (standardized interfaces and commonly accepted development platforms). Drones (UAVs – Unmanned Aerial Vehicles) show a lot of promise. A wide variety of application areas cover assistance of human work, surveillance tasks, supervision tasks, distribution tasks, and a variety of commercial tasks. A new sector of flight traffic control is also needed to control and guide the drone traffic.

G. Key Technologies

The reports list several key technologies. The fast growth of the *Internet* – both its use and its transmission capacity – is maybe the most important of these. The trend towards *mobilization* (mobilization of the 4th billion of the world population) and wide coverage (Internet everywhere) relate to this issue. Fast transfer towards *cloud* architectures will continue; cloud-based ICT infrastructure is becoming mainstream in companies. Simplifying (standardized) cloud platforms are awaited. In the data management area, the transfer toward NoSQL (non-structured) data handling provides opportunities for new data-oriented applications, as well as *block chain* technology. In addition to digital currency, application areas related to the latter cover wide opportunities in distributed trust creation (smart contracts, distributed ledgers).

H. Process and Business Landscape

The benefit of technology innovation is ultimately achieved only if it is adopted in *daily actions* and in *business processes*. Many of the reports point out the important role of the *user and consumer*: Customer satisfaction, customer experience, fast time to market, on-demand based services, customization, digital services, and combining virtual and physical experience are the factors mentioned in the reports. People live in a *hyperconnected world*, in which social presence and *experience sharing* play an important role. The transition from face-to-face communication to the use of virtual communication channels is accelerating. New phenomena appear (AR-based games like Pokemon Go), new jobs are created (bloggers for example), and massive messaging - an increasing amount based on video - are an essential part of the current lifestyle. Everything can be shared almost in real time; nothing remains a secret (if so desired, sometimes even when that is not the case). *Characteristics of business* are changed by the opportunities provided by technology. New businesses are born and some exiting ones disappear. The IoT is seen as changing business processes. Future companies are a part of an *intelligent digital mesh,* which implements the idea of Connected Intelligence Everywhere. It applies data science technologies and allows the creation of intelligent physical and software-based systems that collaborate as a member of the digital mesh (a term introduced by Gartner). The use of AI in operations and big data in management increases productivity. The concept of product is changing: digitally enhanced products and services provide new experiences for customers, increasing their satisfaction and diversifying their experience.

I. Managing System Complexity

As business is to an increasing extent based on networking, even the systems used to support business implement the same idea. Instead of large monolithic information systems, the future is for solutions that provide *open interfaces*, *modular* structure, and versatile *interoperability* support; these can be called *complex systems of systems*. The role of widely accepted and applied frameworks – *development platforms* – is growing. These are used to structure and *layer* the complexity following generally accepted principles. They also guide development experts to use the right architectural components in system implementation. The practical implementation of system components partially remains the responsibility of end-users (casual programmers). Expert work focuses more on conceptualization and structuring (new skill profile).

J. Mobile Data and Internet Traffic

The future of the telecommunication landscape can be envisaged based on two reports by Cisco [5; 6]. These reports provide a wide detailed view of the progress during the next five years. In this paper we provide a very general synthesis and encourage the reader to study the reference reports in detail. There will be a rapid transfer from traditional (wired) Internet traffic to *mobile*.

Global *mobile data traffic* in 2015 reached 3.7 exabytes ($10^{**}15$) per month; in 2020 it is expected to be 30.6 exabytes. The *traffic* has grown 4,000-fold over the past 10 years and almost 400-million-fold over the past 15 years. More than half a billion mobile *devices* were added in 2015. *Smartphones* accounted for most of that growth. The total *number of smartphones* will be nearly 50 percent of global mobile devices and connections by 2020 and will transmit four-fifths of mobile data traffic. Global mobile *devices and connections* in 2015 grew to 7.9 billion and are projected to grow to 11.6 billion by 2020 (exceeding the world's projected population for that time of 7.8 billion). By 2020, aggregate *smartphone traffic* will be 8.8 times greater than it is today, with a CAGR (Compound Annual Growth Rate) of 54 percent. Annual global IP traffic (wired and wireless) will surpass the 1 ZB (zetta, 10**21) threshold in 2016, and reach 2.3 ZB by 2020. Its CAGR is calculated to be 22 percent from 2015 to 2020. *Smartphone traffic* will exceed *PC traffic* by 2020. Smartphones will account for 30 percent of total IP traffic in 2020, up from 8 percent in 2015. In general, traffic from wireless and mobile devices will account for two-thirds of total IP traffic by 2020, wired devices 34 percent. There will be 3.4 *networked devices per capita* by 2020. Globally, IP video traffic will be 82 percent of all consumer Internet traffic by 2020. Virtual reality traffic in 2015 was 17.9 PB (10**15) per month; it will increase 61-fold between 2015 and 2020, a CAGR of 127 percent. The recent issue of IEEE Spectrum [18] reports a study of a daily usage profile of selected applications in one day period (July 11th, 2016): 33.4% Pokemon Go (AR), 22,1% Facebook, 18,1% Snapchat, 17,9% Twitter an average Iphone user spent; this confirms the fast growth of AR traffic also in real life.

Summarizing the numbers above indicates *exponential growth* both in total IP traffic and especially in *mobile traffic*. The transfer toward *more intelligent and mobile devices* is clear (*mobilization*), and the exponentially growing amount of users and increasing complexity of the data format (video, AR) is also clear. Network operators are responding to this by providing faster transmission technologies. The figures also confirm the findings discussed above in the forecast analysis part of this paper.

K. The Appearance and Speed of Changes

Gartner publishes an annual technology forecast in the form of "The Hype Cycle of Emerging Technologies." It provides a good overview of the expected changes in the time span from two to over 10 years; innovations are classified in five categories according to the delay in reaching the "Plateau of Productivity." The Hype Cycle 2016 is given in Fig. 3 [15].

Figure 3. Gartner Hype Cycle of Emerging Technologies 2016 [15].

The Hype Cycle distills insights from more than 2,000 technologies into a succinct set of must-know emerging technologies and trends that will have the single greatest impact on strategic planning. These technologies show promise in delivering a high degree of competitive advantage to organizations over the next five to 10 years. The technologies included in the hype cycle also summarize (partially) the earlier analysis given in this paper, and put some of the findings in the right position in the time span. Since the same analysis is published annually, comparisons between years also provide valuable information for the reader about the progress and changes identified. One good example of radical changes in the analysis relates to "big data," which was located on the 2014 cycle in the "Trough of Disillusionment" with a time span of 5-10 years. In the 2015 Cycle, it had disappeared with the comment "Big Data is out, Machine Learning is in." Instead of total disappearance, the phrase indicates the importance of AI as part of analytics, rather than the data itself, which is seen as an ordinary part of a variety of technologies.

The 2016 report indicates that technology will continue to become more human-centric. It will introduce transparency between people, businesses, and things. The evolution of technology is becoming more adaptive, contextual, and fluid within the workplace, at home, and interacting with businesses and other people. Critical technologies include 4D Printing, Brain-Computer Interface, Human Augmentation, Volumetric Displays, Affective Computing, Connected Home, Nanotube Electronics, Augmented Reality, Virtual Reality, and Gesture Control Devices. *Smart machine technologies* will be the most disruptive class of technologies over the next 10 years. Enablers for this progress are the radical growth of computational power, near-endless amounts of data, and unprecedented advances in deep neural networks. Smart machine technologies harness data in order to adapt to new situations and solve problems that no one has encountered previously. The following technologies are seen to play key roles: Smart Dust, Machine Learning, Virtual Personal Assistants, Cognitive Expert Advisors, Smart Data Discovery, Smart Workspace, Conversational User Interfaces, Smart Robots, Commercial UAVs (Drones), Autonomous Vehicles, Natural-Language Question Answering, Personal Analytics, Enterprise Taxonomy and Ontology Management, Data Broker PaaS (dbrPaaS), and Context Brokering. The shift from technical infrastructure to ecosystem-enabling platforms is laying the foundations for entirely new *business models,* which are forming the bridge between humans and technology. Organizations must proactively understand and redefine their strategy to create platform-based business models. The key platform-enabling technologies include Neuromorphic Hardware, Quantum Computing, Blockchain, IoT Platform, Software-Defined Security, and Software-Defined Anything (SDx).

The paragraph above explaining some details related to Fig. 3 is quoted from the original source with minor changes and modifications.

IV. SUMMARY OF CHANGES IN ICT

The purpose of TCAC framework Step 3 is to classify the changes analyzed in four categories: *Incremental changes, Radical changes, Changes in technological systems,* and *Changes in paradigms* (explained in Paper I; Jaakkola et al. 2017) to enable the publishing of the results (Step 4). Figure 4 summarizes the technology analysis discussion of this paper in the form of the ICT Change Analysis and Categorization (ICAC) framework.

640

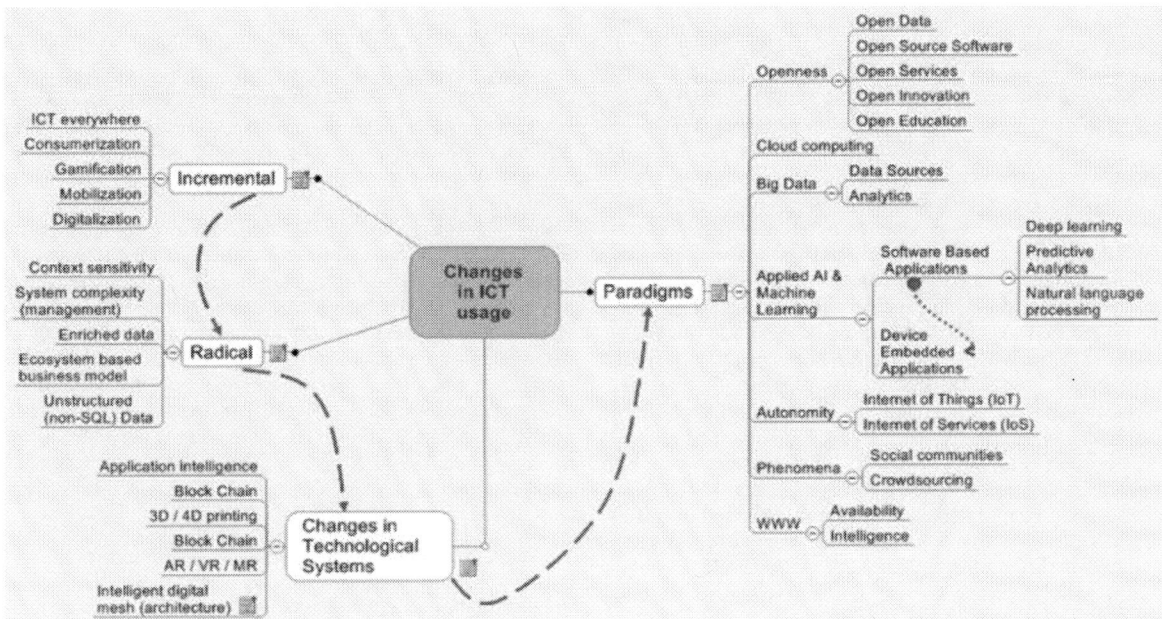

Figure 4. ICT Change Analysis and Categorization (ICAC) framework

The classification may cause criticism – we accept that. The approach is partially subjective. However, it reflects our experience-based findings. Our aim is first to provide an expert opinion for those that are interested in the topic. Secondly, we want to provide a tool for applied use for those who are ready to do this kind of.

How to benefit from the analysis? This is the subject discussed in the following section.

V. HOW DOES IT AFFECT THE EDUCATION SECTOR?

At the beginning of our paper we promised to answer the question "What should every educator know about changes in technology?" The answer can be divided into two parts – effects on the education process itself, and effects on the goals of education from the content point of view.

The *education process* part is the easier sub-question to answer. Every educator should recognize the current situation and be prepared to use tools and technologies that support the study goals in the best way and suit the selected pedagogical approach. An additional aspect that must be taken into account is the audience, i.e., students. Education must motivate them and take into consideration their *built-in behavioral patterns*. At the moment,

- students in higher education start to be members of *generation Z* (birth years that range from the mid-1990s to the early 2010s; coming of age today); for the details of the generation classification see e.g. [3; 2]. Note: in different sources the year limits are varying a bit.

- The educators are representatives of earlier generations. The youngest belong to *generation Y* (birth from the mid-1970s to mid-1990s; coming of age late 1990s to early 2000s),

- the majority to *generation X* (birth from the middle 1960s to the late 1970s; coming of age late 1980s to late 1990s) and

- the oldest to the generation known as *baby boomers* (birth from the late 1940s to the late 1960s).

Recognizing the age profiles is important. Every generation has its own attitudes and values, technical skills, attitude to work, goals in life, interests and basic (professional) skills inherited from their education and adopted through their life experience. The educational methods should follow the needs of generation Z, which can be characterized by the term "*digital native.*" They are used to the Internet, computers, mobile devices, network-based communication, etc. They are nomads (used to working independently of time and place) and the concept of time differs from that of the older generations. The ICT skills of non-generation Z are not native but learned and variance seems to be wide. Consequently, every educator – either digitally native or learned – should be aware of the opportunities that technology provides for education both now and in the future. Therefore, understanding technological changes is necessary. New media supported learning methods based on distance participation, the blended learning approach (mixture of videos, face-to-face classes, self learning), time independence, and the use of modern technology is worthy of consideration. Answers to the question ""What and how to use?" can be supported by the findings of our analysis.

The content of education is a slightly more complicated aspect. The perspective in education is partially on today and partially on the future. The time span from studies to work varies between education levels. In *vocational education* the time span is close to today; the objective is to provide skills and readiness for immediate utilization. In contrast, in higher education, the utilization span is longer but forgetting the ability for short-term applicability is not

641

wise. We (university lecturers) are teaching the professionals of the future based on the knowledge and understanding of today. At all educational levels it is important to remember that even the skills for today must be adaptable to serve *future needs*. It means that education must provide *permanent skills and knowledge* that are independent of the changes in tools and technologies – skills that are *renewable*. In addition to practical skills, we have to be concerned especially about their theoretical foundations. As noticed in our technology analysis, human work will be replaced by robots. Intelligent chatbots, service robots and physical robots will be more productive and skilled in a variety of routine work – even in the work that needs "imitated" human intelligence. Current industrial automation has been the first step in replacing humans in assembly kind of work. Human-type robots, like *Sony's Pepper robot*, are already in use, e.g., in some restaurants and hospitals in assistance work. *Telepresence* can be implemented by a surrogate robot (i.e., a digital twin) in situations where an interactive presence is needed to fulfill the goals of interactive work. A wide variety of robots are available on the commercial market (VGO, Giraff, Double, Kubi, AMY, TeleMe2; see e.g. https://telepresencerobots.com/; https://telepresencerobots.com/comparison).

In the education sector, the key question is "What professions will we really need in the future?" New professions are born, old ones are disappearing, and the rest are changing as they adapt to changes in the environment. Now we have a high demand for *data analysts*. This demand is reflected in education: many universities have included the study topic in their curricula. Fortunately, many of these are focusing on developing analysis algorithms and tools, and providing expertise in development work of intelligent systems; this work will remain. Pure analyst work is disappearing: according to Gartner, 40% of the work of analysts is disappearing and moving to substance-area experts (citizen data scientists). The same goes for routine software development. Casual programming – programming work done by the end-users – is increasing. Routine software development work can be conducted by software robots, as a lot of other routine work in the software engineering area (testing automation is a good example, service support transferred to virtual assistants, etc.). In banking, customer advisory work is conducive to robots, as is investment advisory work. Fact and rulebook management is much more suitable for (software) robots than for humans. The whole banking branch is in the transitional stage. Slow, non-adaptable processes are being replaced by flexible intelligent automated processes that are available on a 24/7 basis, providing better (routine) services and freeing people up for more demanding work.

VI. CONCLUSIONS

The purpose of our paper is to analyze the changes caused by ICT-related technologies. We have followed the principles of our TCAC framework and reviewed a wide variety of technological forecast reports, analyzed their contents, classified the findings, and finally published the results in the form of a mindmap (ICAC). The aim is to act as guidance for educators in developing the content and structures of their courses and curricula.

The problem with technological forecasting is that finally, when we have found a good solution for everything, something unexpected happens: everything is sensitive to radical changes in our environment – the climate, political situation, etc., belong to factors outside our sphere of influence. Forecasting the future is no easy task; there is also a saying that neither is the forecasting of the past (what would have happened if we had selected an alternative path from the past to today; what would be our today then). One thing at least is sure; computerization in all its forms is continuing rapidly. A lot of tasks are done better and faster by computers: mathematical tasks, data management, combining information chunks to form knowledge, management of detailed data, understanding foreign languages, etc. It has been predicted that the computer-brain interface will become reality in around 10 years (still in the hype phase, however). We already have trials to transfer brain signal data for analysis by computers and interpretation as human feelings. This would be the first step.

What will the kernel of human skills be in the future? Detail management and fast calculations are no longer so important. The world and systems are becoming more and more complex. We have to be able to manage the whole - modularization and interfacing the pieces to collaborate. It is good to finish the paper with the saying of Larry Page (CEO of Google): "The main reason why companies fail is that they missed the future". So – be very worried!

REFERENCES

[1] H. Jaakkola, J. Henno, B. Thalheim and J. Mäkelä, "The educators' telescope to the future of technology", Paper submitted to the MIPRO 2017 conference, 2017.

[2] H. Jaakkola, P. Linna and J. Henno, "(Social) networking is coming - are we ready?" In the Proceedings of the MIPRO 2011 Conference, Opatia, M. Cicin-Sain, I. Uroda, IU. T. Prstacic and I. Sluganovic, Eds. Mipro and IEEE, pp. 170-176, 2011.

[3] W. J. Schroew, "Generations X,Y, Z and the Others", Available at http://socialmarketing.org/archives/generations-xy-z-and-the-others/, Retrieved on January 26th, 2017.

[4] Business insider, "11 tech trends that will define 2017", Available at http://www.businessinsider.com/tech-trends-that-will-define-2017-2016-12?r=US&IR=T&IR=T/#buildings-will-harness-the-powers-of-nature-1, Retrieved on January 26th, 2017.

[5] Cisco, "Cisco VNI Mobile Forecast (2015 – 2020) – Cisco", Available at http://www.cisco.com/c/en/us/solutions/collateral/service-provider/visual-networking-index-vni/mobile-white-paper-c11-520862.html, Retrieved on January 26th, 2017.

[6] Cisco, "The Zettabyte Era — Trends and Analysis – Cisco", Available at http://www.cisco.com/c/en/us/solutions/collateral/service-provider/visual-networking-index-vni/vni-hyperconnectivity-wp.html, Retrieved on January 26th, 2017.

[7] CSC, "6 Technology Trends to Watch in 2017", Available at http://www.csc.com/innovation/insights/139169-6_technology_trends_to_watch_in_2017, Retrieved on January 26th, 2017.

[8] Deloitte, "2016 Technology Industry Outlook", Available at https://www2.deloitte.com/us/en/pages/technology-media-and-telecommunications/articles/2016-technology-industry-outlook.html, Retrieved on January 26th, 2017.

[9] Forrester, "Predictions 2017", Available at https://go.forrester.com/research/predictions/, Retrieved on January 26th, 2017.

[10] Fjord, "Trends 2017", Available at https://trends.fjordnet.com/trends/, Retrieved on January 26th, 2017.

[11] Forbes, "The 5 Most Worrying Technology Trends For 2017 And Beyond", Available at http://www.forbes.com/sites/bernardmarr/2016/12/23/the-5-most-worrying-technology-trends-for-2017-and-beyond/#13f2307a6a5e, Retrieved on January 26th, 2017.

[12] Forbes, "7 Technology Trends That Will Dominate 2017", Available at http://www.forbes.com/sites/jaysondemers/2016/11/16/7-technology-trends-that-will-dominate-2017/#6058c4371b2a, Retrieved on January 26th, 2017.

[13] Frogdesign, "Tech Trends 2017" Available at http://www.frogdesign.com/techtrends2017, Retrieved on January 26th, 2017.

[14] Gartner, "Top 10 Strategic Technology Trends for 2017". Available at https://www.gartner.com/doc/3471559?refval=&pcp=mpe#-363727574, Retrieved on January 26th, 2017.

[15] Gartner, "Gartner's 2016 Hype Cycle for Emerging Technologies Identifies Three Key Trends That Organizations Must Track to Gain Competitive Advantage", Available at http://www.gartner.com/newsroom/id/3412017, Retrieved on January 26th, 2017.

[16] Howstuffworks, "10 Futurist Predictions in the World of Technology", Available at http://electronics.howstuffworks.com/future-tech/10-futurist-predictions-in-the-world-of-technology.htm, Retrieved on January 26th, 2017.

[17] IDC, "Top 10 Tech Predictions For 2017 From IDC", Reported by Forbes at http://www.forbes.com/sites/gilpress/2016/11/01/top-10-tech-predictions-for-2017-from-idc/#1149ef5c2790, Retrieved on January 26th, 2017.

[18] PWC, "Technology forecast", Available at http://www.pwc.com/us/en/technology-forecast/landing.html, Retrieved on January 26th, 2017.

[19] Quantumrun, "State of technology in 2017. Future Forecast", Available at http://www.quantumrun.com/future-timeline/2017/future-timeline-subpost-technology, Retrieved on January 26th, 2017.

[20] Quantumrun, "State of technology in 2026. Future Forecast", Available at http://www.quantumrun.com/future-timeline/2026/future-timeline-subpost-technology, Retrieved on January 26th, 2017.

[21] Wired, "15 Predictions for Tech and Design in 2015", Available at https://www.wired.com/2015/02/frog-design-predictions/, Retrieved on January 26th, 2017.

[22] World Economic Forum, "14 tech predictions for our world in 2020", Available at https://www.weforum.org/agenda/2014/08/14-technology-predictions-2020/, Retrieved on January 26th, 2017.

Interdisciplinary utilization of IT

S. Neradová, S. Zitta

University of Pardubice, Faculty of Engineering and Informatics, Pardubice, Czech Republic
sona.neradova@upce.cz

Abstract - This article describes a project realized in laboratory seminars. The aim was to practically show students utilization of information technologies in a non-technical discipline. The project involved observing and monitoring breathing of chosen plants. Recorded data were subsequently analysed into several conclusions. The goal was to design and practically build a solution which would enable to record the collected data from the sensors. The designed system contained wireless module and sensors for measuring temperature, humidity and CO_2 concentration. The collected data were transferred to the web server ThingSpeak. Data values could be studied through an internet browser and also downloaded for further analysis. The major founding is that interdisciplinary skills led to successful creation of financially undemanding functional unit, which enabled automation of the data collection. The students were given the opportunity to design an IT platform, to implement and to analyse collected data. Individual components of the project required combination of the skills from informatics, physics, biology and mathematics.

I. INTRODUCTION

Generally, education programs and concepts of all schools should preparing students for the "real life", for different tasks in different companies and organisations. Logically, the primary target of every school is to provide students the best education and knowledge of the studied disciplines. This is not an easy task, especially in such dynamic discipline like information technologies. Students have to learn many things, the education programs are focused on programming languages, networking, data security, hardware... The IT world is changing very fast, technologies, which were new when students started are often obsolete when they graduate. This is also very demanding for teachers, who should follow this rapid development and create and continuously update content of their lessons. Although teachers have to focus on the content of the study and how to overhand the knowledge to students, they shouldn't forget that it is also very important to guide students, how they can practically use their knowledge and interconnect it with other, totally different disciplines, what tasks they will face after leaving the school. And, last but not least, an interesting project and its' successful finishing can create some sort of "positive feedback loop"- motivate students for self-study of the given topic and thus deepen their knowledge.

As mentioned in the previous text, a part of the education program of any school with IT as a subject of the study should be a project, where students can prove their ability to practically use what they have learned. Students should be able:

- Understand and analyse the given problem.

- Design a solution (with possible variants) on a theoretical level.

- Choose the best variant with regard to the feasibility, delivery time and given budget.

- Practically realize the chosen variant.

- Evaluate the result, conclude what they have learned.

II. THE PROJECT

General description

The described project has been realized with students of the High school of electrical engineering (author is also teaching in this school). The aim was to show students an example, how the IT can be used in a real application for measuring and collecting non-electrical values. To demonstrate their ability to fulfill such task, students had to design a measuring system to ascertain and document development of plants in closed system of breathing gasses (oxygen and carbon dioxide) in natural light conditions with different leaf surfaces and in different stages of their development. An important part of the tasks' definition was to work with very limited financial budget to motivate students to carefully consider the price / performance ratio as it is usually required in the commercial environment.

Further processing and evaluation of the collected data was not a part of this project. However, the collected data were used in another students' project (students from another school).

- Monitoring CO_2 concentration, temperature and humidity in a closed environment with experimental plants (there were 4 plants to be observed).

- Storing the collected data for later analysis and evaluation (the evaluation of the collected data was not a part of the project for students of the electro technical school).

- Keeping the costs of the designed system as low as possible.

Analysis of the project task

At the beginning of the project, it was necessary to analyse the task and determine the particular steps. Initially, the following questions had to be answered:

- How will be the desired quantities measured?

- How long will be the interval between particular measurements?

- How will be the measured data transferred to the data storage?

- Where will be the collected data stored?

- How will be the technical layout designed?

Solution for the particular steps

Once the particular steps were defined, students could start with their realisation.

Measuring the desired quantities

Selecting the most suitable sensors was a crucial part success of the project. Students had to consider several parameters – accuracy of the measurement, digital or analog output, separate or integrated sensors and obviously the price. As there wasn't an absolute accuracy required (more important was to monitor temperature changes during day), there were sensors DHT11 and DHT22 chosen for their excellent performance vs price ratio. These sensors provide acceptable accuracy; integrate temperature and humidity measurement in one unit at low price. There were both types selected to get the opportunity to compare them in a real environment. The situation with CO_2 sensors was different: available sensors are mainly dedicated for industrial applications; their price was therefore too high for a student's project. The only sensor acceptable for the project budget was the MQ135, although this sensor has some drawbacks that students had to overcome.

Determining the intervals of measurement

Students have to find the balance between the amount of collected data and reliability of the measurement. As compromise there was an interval of 10 minutes chosen: it doesn't generate too large data but still provides enough data to compensate failures (caused by interferences, voltage drops etc.).

Data transfer to the data storage

There were two options considered, how to connect the sensors: wired and wireless. Students had to consider pros and cons of both solutions. Wired connection offers higher reliability, but there would be a dedicated computer (with additional interface) needed for connecting the sensors. For wireless connection, there were just Wi-Fi modules required. Potentially lower reliability of the connection could be easily compensated by certain redundancy of measured data and calculating the average. As the most suitable Wi-Fi module and interface for the sensors there was the ESP8266 chosen. Students considered also using Arduino or Raspberry, but they wouldn't offer any advantage in this application compare to the ESP8266.

Storing the collected data

Also for this particular task, students had two options: storing the collected data on a local server or using one of the publicly available cloud services. Using a local server would require certain skills for programming web accessible database and require an additional HW.

Generally, it wouldn't bring any advantage comparing to a public cloud, especially because the Thingspeak cloud provides services, which were specifically designed for collecting data using IoT. Using Thingspeak automatically solved tasks like programming a database and backup the collected data. The limits of the free of charge version of this service (which is available for non-commercial projects) weren't affecting this project.

III. TECHNICAL LAYOUT

On the basis of the previously described analysis, technical documentation was created. The following table shows the required numbers and types of particular components.

TABLE I.   TECHNICAL DOCUMENTATION

Nr.	Technical documentation		
	Name of the component	*Description*	*Qty*
1	Experimental plants	Experimental plants – cress, pea	4
2	ESP8266	Wi-Fi module, interface for sensors	2
3	PSU	Power supply for ESP8266 (phone charger)	2
4	MQ135	Gas concentration sensor	2
5	DHT11	Temperature and humidity sensor	1
6	DHT22	Temperature and humidity sensor	1

Wiring diagram is shown on the next picture.

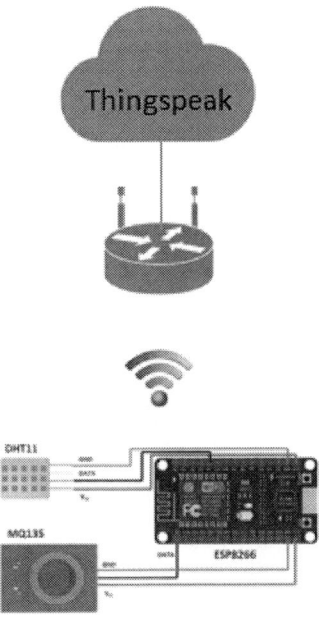

Figure 1.   Schematic wiring diagram

The experimental plants were planted into pots and covered with 0,7 litre clear glasses. Along with the plants, sensors MQ135 [2] for measurement of CO_2 concentration and sensors DHT11 [3], DHT22 [4] for measurement of temperature and humidity were situated under the glasses. The sensors were connected to the module ESP8266 with short cables. The module provided reading of the entries from the sensors in 10 minute intervals. Relatively short interval of 10 minutes was chosen to enable elimination of drop-outs caused by random interference. The sensor MQ135 does not provide data of measured temperature but voltage equal to the temperature, therefore the module ESP8266 also performed the conversion. Analog input of the module ESP8266 was used to connect the sensor MQ135. Dependence of the module MQ135 on the temperature is not neutral, therefore correction constant for each of the sensors was established by measuring the voltage at 0°C and 20°C. The constant was afterwards used during the conversion. The module ESP8266 also dispatched the collected data about CO_2 concentration, temperature and humidity via local Wi-Fi network to server Thingspeak.com where the data were saved (including dates and times). After the experiment, the collected, saved data were downloaded from the server in csv format and subsequently evaluated in program MS Excel.

I. THE ESP8266 MODULE DESCRIPTION

Figure 2.   The ESP8266 module

The module ESP8266 is an affordable module with control microprocessor, programmable GPIO, HSPI/UART/PWM/I2C/I2S interfaces and integrated Wi-Fi unit. The module exists in multiple variations which differ with number and type of inputs / outputs and with RAM. The variation ESP8266 ESP-12E Development Board with integrated USB interface and 4MB RAM was used in this experiment. The integrated USB interface facilitates programming of the module and it can be also used for power supply of the module. The firm Espressif Systems is the producer of the module. Detailed data and technical parameters about the module are accessible at the producer's webpage https://espressif.com.

Many programme libraries and even actual firmware for the module ESP8266 are accessible. Its programmes can be written in multiple programming languages. In this experiment, firmware NodeMCU 1.5.4.1 was uploaded to the module, and scripting language Lua was used for programming. Programmes in the language Lua can be written in any text editor. In this project, a specialized editor LuaEdit was used (free-downloadable e.g. at http://luaedit.sourceforge.net) which provides expanded options for editing and tuning of programmes written in the Lua language [5]. The firmware and the actual service programme for communication with the sensors, the

temperature calculation and the data dispatch were uploaded to the module via interface ESPlorer (see http://esp8266.ru) which is also freely available, and is specially designed for the ESP8266 modules [6].

II. THE MQ135 MODULE DESCRIPTION

The module MQ135 is sensor-based module designed for measurement of gas concentration. The gas sensor, operative repeater LN393 for output amplification, analogue voltage output and digital output with TTL level are attached to the module. The voltage on the analogue output is directly proportional to the gas concentration. The conversional characteristics between the gas concentration / voltage and additional parameters are stated in the producer's datasheet.

Figure 3.   The module MQ135

The DHT11 and DHT22 modules description

The DHT11 and DHT22 modules are designed for temperature and humidity measurement. There is one data output accessible for the connection to the control microprocessor. Also, two power supply pins are present. The difference between the DHT11 and DHT22 types occurs in the range of measurement and the sensitivity. The DHT11 type measures temperature in the range 0 – 50 degrees Celsius with accuracy ±1 °C and humidity in the range 20 – 90 % with accuracy ±4 %. The advanced module DHT22 enables to measure the temperature in the range -40 to +80 degrees Celsius with accuracy ±0,5 °C and the air humidity in the range 0-100 % with accuracy ±2 %. Parameters of the modules are stated in the producer's datasheet.

Figure 4.   The DHT11 and DHT22 modules for temperature and humidity measurement

III. THE THINGSPEAK SERVER

In this project, the Thingspeak.com server was used as the data storage. This server does not offer only the storage and basic data visualisation but also an analysis via MATLAB programme, warning dispatch, Twitter messages dispatch and other functions. Some of the functions are available only in the paid version,

646

nevertheless the server is for non-commercial use (with limited number of entries 3 million / year) free of charge.

To use the server, it is necessary to create a user account. After logging in to the account, it is possible to create so called channels where the data are saved. The channels or more precisely the saved data can be tagged and their development can be observed via charts. Data from 8 devices is the maximum which is possible to store in one channel. The saved data are downloadable in csv format at any time. It is also possible to upload data to the server (again in the csv format). The channels can be established as private, thereafter it is necessary to log in for the access to them. The second option is to establish the channel as public, thereafter the view of the channel is freely accessible. When creating a channel, a unique 16 digit alphanumeric chain, so called API-Key, is generated. This chain serves for identification of the channel and it is necessary to enter it when programming the log in procedure of the device connected to the Thingspeak.com server.

Extensive documentation, tutorials, examples of the programming code (including links to extern servers like GitHub or Instructables) and community forum, where it is possible to ask other users and search for solutions, are at disposal at the server.

Following pictures show the data, which were stored on the Thingspeak server.

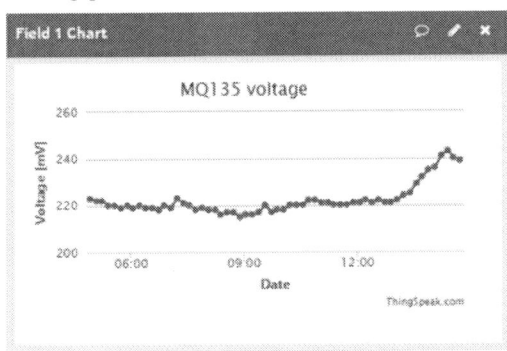

Figure 5.   Output voltage

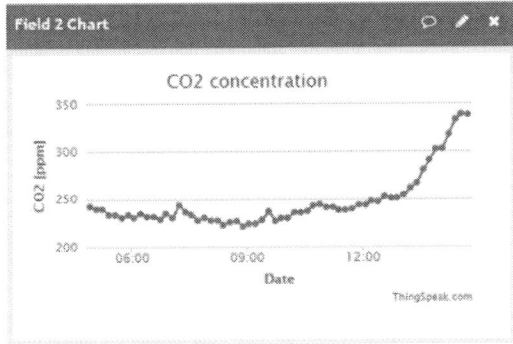

Figure 6.   Calculated CO_2 concentration

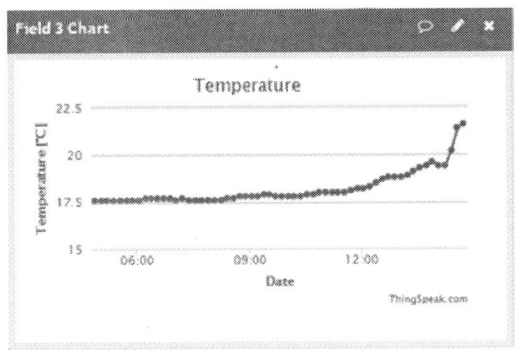

Figure 7.   Measured temperature

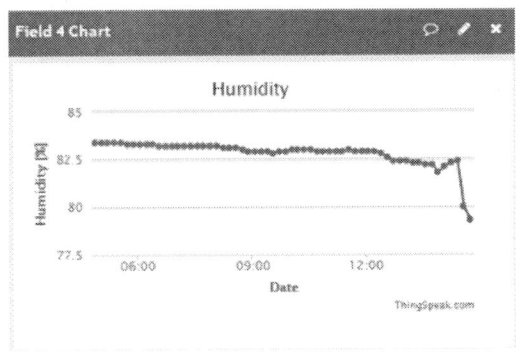

Figure 8.   Measured humidity

IV.   CONCLUSION

The goal of the project was fulfilled. The students designed and tested solution, which enabled collecting data from the sensors. Students have learned how to analyse the given task, split it into particular steps and consider possible variants of the solution for each particular step. They have learned that the solution should be the best compromise between the technically ideal solution and the practically available resources.

The students were acquainted with settings of hardware, software installation and data storage at the Thingspeak server. During the project, the students learned how to connect the real word with the digital one. They tried the measuring of physical quantities via sensors in practise. The mutual cooperation between the students during solving the project leads to the cognition that interconnection between different disciplines is very important.

ACKNOWLEDGMENT

This work and contribution is supported by the project of the student grant competition of the University of Pardubice, Faculty of Electrical Engineering and Informatics "Usage of ambient intelligence in the area of information systems and their communication".

REFERENCES

[1]   ThingSpeak, "IoT Analytics – ThingSpeak," (jan 2017) [online]. Available: https://thingspeak.com/

[2] Olimex, "SNS-MQ135," (jan 2017) [online]. Available:goo.gl/prBtGw

[3] DHT11 Humidity & Temperature Sensor, "Micropik.com," (jan 2017) [online]. http://www.micropik.com/PDF/dht11.pdf

[4] Aosong(Guangzhou) Electronics, "Digital-output relative humidity & temperature sensor/module. Adafruit," (jan 2017) [online]. Available: https://cdn-shop.adafruit.com/datasheets/DHT22.pdf

[5] LuaEdit, "LuaEdit - Free IDE/Debugger/Editor for Lua," (jan 2017) [online]. Available: http://luaedit.sourceforge.net/

[6] ESP8266, " esp8266 – Сообщество разработчиков," (jan 2017) [online]. Available: https://esp8266.ru/

[7] Auxins, Cytokinins, and Gibberellins, " Boundless learning," (jan 2017) [online]. Available: goo.gl/1ZttYo

Maximizing Quality Class Time Using Computers for a Flipped Classroom Approach

C. P. Fulford, fulford@hawaii.edu and S. Paek, spaek@hawaii.edu
Learning Design and Technology, University of Hawaii at Manoa, Honolulu, Hawaii, USA

Abstract - **This study presents a two-course sequence of graduate instructional design courses that were developed and implemented using a flipped classroom model for both online and face-to-face contexts. This model is an instructional strategy that reverses what is traditionally done at home and what is done at school, such as watching video lectures at home and working authentic problems in class. The presentation will 1) demonstrate how the two courses were designed and developed, 2) share findings from student responses on surveys regarding perceptions and satisfaction with the new model, and 3) discuss the successes and challenges of implementing these courses.**

I. INTRODUCTION & PROBLEM

Computers in education have reached a new level of capability. With connectivity so widespread, teachers have new opportunities to utilize them in more creative ways. One way to use them strategically is through "flipping the classroom". The idea behind flipped classrooms is that work traditionally completed as homework, such as problem solving, writing, and group work is better undertaken in class with the support and guidance of classmates and the instructor [1] In contrast, activities such as listening to a lecture or watching videos are better accomplished at home [2]. Thus, a truly flipped classroom uses in-class time for active learning and problem solving on the part of students.

While the benefits and drawbacks of flipped classrooms have been explored elsewhere [1, 3, 4,]. Bishop and Verleger [5] explain that the flipped classroom model "represents a unique combination of learning theories once thought to be incompatible—active, problem-based learning activities founded upon a constructivist ideology and instructional lectures derived from direct instruction methods founded upon behaviorist principles" (para.1). Drawing on this powerful idea, the authors designed and developed a two-course sequence of instructional design courses in a graduate program in Hawaii.

Instructional design (ID) is a "construct that refers to the principles and procedures by which instructional materials, lessons, and whole systems can be developed in a consistent and reliable fashion" [6, p. 574]. It is the practice of creating "instructional experiences which make the acquisition of knowledge and skill more efficient, effective, and appealing." [7, p. 2] Teaching students about the principles and procedures of ID requires an emphasis on both theory and practice.

Accordingly, many ID courses rely on project-based learning models, ones that organize learning around design and technology projects [8]. Although effective in many contexts [9,10], it can be difficult for students, particularly those starting a Master's program in ID, to start working on a project in the beginning of their program, as they often have little to no background. This reality can be a challenge for instructors teaching introductory ID courses at the graduate level.

For example, how can an ID course be designed to balance theory, research, and practice? How can an ID course cover enough breath of content versus providing enough depth? How can an ID course provide authentic experiences while being able to use well thought out academic cases? Essentially, how can more content be included in already full courses?

With these challenges in mind, the authors, faculty members in a graduate program that emphasizes instructional design, developed two courses using a flipped classroom instructional model. What the instructors hoped to do was to create a studio course where students could work on ID projects while the instructor was available to assist. This purpose of this paper is to discuss how the courses were developed, and investigate how they changed over a two-year period.

II. LITERATURE REVIEW

A. Benefits of Flipped Classrooms

Flipping the classroom involves providing instructional resources for students to use outside of class so that class time is freed up for other instructional activities [11]. Literature shows that there are many benefits to flipped classrooms. For example, it encourages active and independent learning thereby increasing student engagement, strengthening team-based skills, personalizing student guidance, and focusing classroom discussion [1,12].

Flipped classrooms may also provide for better interaction and communication since it may allow teachers greater insight into students' grasp of information and learning as a result of increased student to teacher interaction. When a flipped classroom is designed well, it can increase flexibility and provide nearly unlimited access to learning materials [4].

Flipped classroom pedagogy can also address situations in which students miss lectures due to illness or who are engaged in university-supported activities such as athletics. It allows students to move at their own pace, access instruction at any time, and obtain expertise from multiple people [4,13].

B. Concerns About Flipped Classrooms

Though, there are some concerns about flipped classrooms. First, teachers must ensure that students have connectivity and access to the materials being provided online [14]. And with a growing movement towards no homework, can teachers justify the increased time requirements outside of class without improved pedagogy in class?

Use of lectures to provide instruction may disregard individual student learning styles unless ample materials are used to overcome this deficit. Students experienced with the out-of-class instruction may not be as interactive as a lecturer who is able to directly answer students' questions and engage them in discussions. The online prerecorded lectures don't have the same ability to monitor comprehension and provide just-in-time information when needed [15].

Flipped classrooms do not work just by moving everything to home tasks. The lack of adapting the classroom environment to reflect the flipped classroom's ability to support student-centered learning can hinder the methods effectiveness. This paired with the lack of accountability for students to complete the out-of-class instruction, can limit the desire of some teachers to try this new strategy. [16].

III. METHODOLOGY

A. Development

Winter [17] proposed that "Flipped learning is composed of two integral but inherently different learning spaces (p. 68)". The "individual space" is now the place where students receive direct instruction, such as recorded lectures, web-based materials, and readings. The students themselves generally manage how they approach these materials.

The "group space" is in the classroom, which can provide a rich source of interaction and communications. Here they may work on assignments, engage in activities, and dialog with the teacher and other students. The instructors used this idea, and both behaviorist and constructivist theories and models in developing both group learning spaces and individual learning spaces for this project [17].

The first year was a heavy development period that progressed week by week, as the individual learning space was developed. The platform for the course was a university-based version of Sakai, a learning management system (LMS). Within this system, weekly modules were created that contained all of the materials.

Over the first year, online lectures of narrated slideshows were produced that were carefully recorded and edited to increase the production value and decrease the length of the videos. A YouTube channel was created for open access to these for other instructors of ID at

https://www.youtube.com/channel/UCPqBgPU1IGL-xdN9QhpuC0g.

Research literature, web resources, professional examples, and past exemplary student projects were found and posted. To ensure accountability for students were accessing and using the required materials, a discussion group was set up for students to post their weekly reflections on the material they had watched and read. Each student was required to post one reflection with a relevant question at the end and respond to two other students' reflections and questions.

The group learning space was during the scheduled class time. The instructors made short presentations to follow-up on the week's topic, answer questions, note interesting comments or questions from students' reflections, and clarify the assignments for the next week. Then, activities were used to engage the students in active learning of the topic. Students were given time to work in their teams on their course projects, and then present progress on their projects and have it critiqued. Peer review sessions in both courses encouraged students to learn from each other and learn designer roles used in the workplace. They also made end of project presentations.

B. Setting

The courses chosen for redevelopment in this project were the first two ID courses that students take in the masters program for Learning Design and Technology. Both are core courses that prepare ID professionals. One is taken in the first semester, and the second in the next semester.

LTEC 600 - Theory and Practice in Educational Technology covers foundational processes, methods, theories, and strategies, and discusses how these are put into practice in the field. It uses academic case studies and teamwork. The products and outcomes of the course are: a front-end analysis developed using student's new knowledge of systems theory, need assessment, change theory, task analysis; and a technology project using the new models and methods they have learned. They also develop their professional skills including teamwork, grant writing, and presentation skills.

LTEC 613 - Instructional Design and Development covers the complete process of ID and uses authentic projects and teamwork. The products and outcomes of the course are a technology-based individualized instructional module developed using the students' practical application of design methods, processes, and strategies. They also develop test and survey instruments, apply visual design principles, and engage in a practical experience with formative evaluation. They continue to develop their team and presentation skills.

C. Participants

Participants were in LTEC 600 and 613, the two sequential courses for first year Master's students. There was one section taught on campus and the other taught online. A full professor taught campus section and an assistant professor taught the online section. There was one teaching assistant shared with both sections each semester who made up the team. This team met once a week for two years to develop the course and conduct the research.

Because of slightly different schedules in the first year, the online section could be mostly flipped one semester earlier than the campus section. Even now it is almost impossible to make the two sections exactly identical. Class activities due to the class time difference and the limited face-to-face interaction for online students are often modified.

These courses were for graduate students, and relied solely on a flipped classroom model for both online and face-to-face sections. Two sections of each course were offered in two different platforms over a two-year period. There were 25 students in the first year and 22 in the second year.

IV. PROCEDURES

This paper evaluates student perceptions regarding the design and use of the flipped classroom model particularly investigating differences between students in the first year of development and that of the second year.

A. Research Instrument

A questionnaire was distributed at the end of the 2nd year. There were 10 sections including instructions and demographics. In six of the sections, there were 48 Likert type questions with the "not" side equal to 1 and the "very" side equal to 5. In all cases except the section on "challenges" "not" signified negative and "very" signified positive.

There were 14 open-ended questions, typically one at the end of each section for additional comments on the topic and 4 in the final section for opinions, comments, and suggestions.

The response rate was very high at 77%. Of the 47 students in the courses, 36 responded. The sample consisted of 65% females and 35% males. Because this was a master's degree and professional program, there were a relatively high percentage of older students.

B. Independent variables

There were two independent variables: year of the trial and delivery method. (Please note, data regarding comparisons of online versus campus will be reported in another paper. This paper compares only within group data.)

There were a total of 36 students in the sample. They were fairly evenly distributed. (See Table 1.)

TABLE 1. STUDENTS IN EACH YEAR AND DELIVERY

Delivery	Year	Number	Percentage
On Campus	1	9	25.0
	2	11	30.6
Online	1	7	19.4
	2	9	25.0

n=36

C. Dependent variables

Dependent variables were drawn from the various sections of the questionnaire. Each had Likert style questions and an open-ended question. These were:

- Engagement – 8 questions,
- Effectiveness (compared to a traditional class) – 7 questions,
- Benefits – 8 questions,
- Individual space – 11 questions,
- Group space – 8 questions.

The individual space and group space questions targeted specific strategies that were used in the design of these particular courses.

V. RESULTS AND ANALYSIS

To examine if there were any differences between Year 1 and Year 2 in terms of how students perceived the flipped classroom model, the survey data was analyzed using independent samples t-tests. Results of this study are reported using a 2-tailed significance at the ($p<.05$) level.

A. Engagement & Effectiveness

There were no significant differences with engagement between the first and second year students with regard to engagement, though Year 2 was marginally higher. There were however, three significant findings with regard to engagement within campus students. Second year campus students reported significantly higher enjoyment, $t(18)=-2.22$, $p=.040$, efficiency, $t(18)=-2.12$, $p=.045$, and engagement, $t(18)=-2.821$, $p=.011$,) than the first year campus students. (See Table 2.)

TABLE 2. ENGAGEMENT - YEAR COMPARISON CAMPUS DELIVERY

Question	Year	N	Mean	SD
Enjoyment	1	9	3.44	.73
	2	11	4.18*	.75
Efficiency	1	9	3.22	.97
	2	11	4.09*	.83
Engagement	1	9	3.00	1.12
	2	11	4.18*	.75
Average Engagement	1	9	3.35	.80
	2	11	4.04	.80

* Significant at $p<.05$

When comparing the effectiveness of flipped classrooms to traditional ones, all Year 2 students perceived engagement compared to traditional classes was only marginally higher than all Year 1 students. Though within campus students, those in Year 2 reported the flipped classroom significantly higher, $t(18)=-2.30$, $p=.033$ overall, especially with regard to it being more engaging, $t(9.88)=-2.712$, $p=.022$ and efficient, $t(18)=-2.72$, $p=.014$ and marginally more effective than a traditional class, $t(11.18)=-2.11$, $p=.058$. (See Table 3.)

651

This quote shows the typical sentiment of some students.

At first I was quite resistant to the idea and it forced me to change my way of thinking and learning, but now that I can read backwards so to say it makes a lot of sense. Highly efficient and effective. For someone who has a full time job and going to school at night with limited amounts of energy, this method was a LOT more effective and conducive. Year 2, Campus Student

TABLE 3. EFFECTIVENESS OF FLIPPED CLASSROOM COMPARED TO TRADITIONAL CLASSROOM - YEAR COMPARISON CAMPUS DELIVERY

Question	Year	N	Mean	SD
Engagement Compared to Traditional Classrooms	1	9	3.11	1.45
	2	11	4.50*	.53
Efficiency Compared to Traditional Classrooms	1	9	2.89	1.364
	2	11	4.27*	.91
Effectiveness Compared to Traditional Classrooms	1	9	3.00	1.41
	2	11	4.09*	.70
Average Engagement Compared to Traditional Classrooms	1	9	2.96	1.26
	2	11	3.96	.77

* Significant at $p<.05$

B. Benefits

There were no significant differences in benefits overall between Year 1 and Year 2, however there were differences within delivery methods for years in both the campus delivery and the online delivery. There were significant differences between the first and second year campus group with two questions regarding: having the ability to rewind and repeat the videos, $t(9.72)=2.35$, $p=.041$, and not having to listen and follow the lecture in class $t(18)=-2.76$, $p=.013$. (See Table 4.) There were also significant differences $t(18)=-2.26$, $p=.036$ with higher average benefits for Year 2. Year 2 also showed marginally higher differences regarding: having materials in various formats and having more class time to work in groups.

TABLE 4. BENEFITS - YEAR COMPARISON CAMPUS DELIVERY

Question	Year	N	Mean	SD
Materials in various formats	1	9	4.11	.782
	2	11	4.64	.505
Ability to rewind and repeat	1	9	4.22	.833
	2	11	4.91*	.302
Not have to listen and follow	1	9	4.11	.782
	2	11	4.73	.647
Average Benefit	1	9	3.33	.866
	2	11	4.45*	.934

* Significant at $p<.05$

Interestingly the opposite occurred in online sections. Students in Year 1 ($M=5.00$, $SD=.00$) strongly agreed that having more class time to work in groups was beneficial, which was significantly higher, $t(14)=2.53$, $p=.035$ than the responses from Year 2 students ($M=4.56$, $SD=.53$). Overall, students in Year 1 ($M=4.80$, $SD=.307$) valued benefits of flipped classroom higher than Year 2 students ($M=4.41$, $SD=.315$), and the difference was statistically significant, $t(14)=2.44$, $p=.029$.

This student quote is an example of challenges compared to benefits.

The workload was a bit much especially in a flipped classroom setting, as it feels like there's a lot more to stay on top of compared to a traditional classroom. Flipped classroom discussions are great in a sense that students will be able to provide better ideas after taking some time to come up with their own thoughts. Year 2, Campus Student

C. Challenges

Overall between the two years, students in different years found different things challenging. Sixteen student sin Year 1 ($M=3.38$, $SD=.957$) reported significantly higher levels of challenge with recorded lectures that are less detailed, $t(34)=2.170$, $p=.037$) than 20 students in Year 2 ($M=2.60$, $SD=1.42$). Though the students in Year 2 ($M=3.40$, $SD=883$) found the schedule significantly more challenging, $t(34)=-2.14$, $p=.039$) than the Year 1 group. ($M=2.75$, $SD=9.31$).

Within the campus group, Year 1 students were significantly more challenged than the Year 2 group with regard to two survey items and the overall average challenge. The survey items were: Recorded lectures are less detailed than in class lectures, $t(15.06)=2.84$, $p=.013$, and working in class in groups with the instructor watching, $t(18)=2.88$, $p=.010$). The average of all survey items on challenge was also significant, $t(18)=2.25$, $p=.037$). (See Table 5.)

TABLE 5. CHALLENGES YEAR COMPARISON CAMPUS DELIVERY

Question	Year	N	Mean	SD
Recorded lectures less detailed	1	9	3.78*	.667
	2	11	2.45	1.368
Working with instructor watching	1	9	2.78*	1.202
	2	11	1.55	.688
Average Challenge	1	9	3.29*	.426
	2	11	2.69	.694

* Significant at $p<.05$

Within the online group, Year 2 also reported marginally higher levels of challenge with the schedule. To highlight the results regarding challenges, here are three student quotes.

Recorded lectures allow you to work at your own pace, but allow you to ask questions in the next face-to-face meeting, this might work for some but might not for others. Year 1, Campus Student

652

Sometimes the schedule was confusing since not all classes are flipped and knowing what was due for what class was sometimes hard. Year 2, Campus Student

D. Individual Learning Space

With regard to strategies used in the individual learning space, there was one significant result at the p<.05 level. First-year trial students reported higher, $t(34)=2.07$, $p=.046$ usefulness for reflections discussions ($M=3.81$, $SD=1.109$) than second year students ($M=3.00$, $SD=1.21$). In the online group there were marginally higher scores for students in the first year with regard to reflections. Overall the students seemed happy with the individual learning space. Here are some typical quotes.

Recorded lectures are great since you can pause and rewatch as many times as needed, and do some research at the same time. I've learned a lot more watching recorded lectures than traditional lectures since it is hard to keep up with traditional lectures and is easier to lose focus. We can always ask questions by emailing or talking to the instructor in person in class, so that was never an issue. Year 2, Campus Student

I feel as though I might not have had access to so many resources and information had the class not been flipped. Year 1, Online Student

E. Group Learning Space

With regard to the group learning space there were no significant differences between Year 1 and 2, only marginal differences. First year students felt more strongly about the ability to share their work and have it critiqued, and second year students about the short instructor presentations to review content and preview the next week.

Within delivery methods there were three topics that were significantly different at the p<.05 level: in class group time, the ability to share and critique projects, and peer reviews.

First-year online students reported significantly higher usefulness for in-class group time to work on projects, $t(8.00)=2.530$, $p=.035$, group presentations to share and critique work, $t(8.00)=3.162$, $p=.013$ and peer reviews of projects, $t(8.00)=2.530$, $p=.035$, than second year students. Second year campus students reported significantly higher usefulness for in-class group time to work on projects than first year campus students, $t(8.00)=-3.50$, $p=.008$. These students also marginally preferred the short instructor presentations, $t(18)=-1.743$, $p=.098$. (See Table 6.)

Here are some thoughts from students.

Excellent vehicle to accommodate varying paces and skill level students. Makes more efficient use of class time assuming that all students have exposed themselves to the course content before it is talked about in class. Online Year 1

I think spending time in class actually working on group work and applying the information learned through videos and readings is a more efficient

use of time than being lectured to. Year 2, Campus Student

TABLE 6. GROUP SPACE YEAR & DELIVERY COMPARISON

Method - Question	Year	N	Mean	SD
Online - In class group time	1	9	5.00*	.000
	2	11	4.56	.527
Online – Group to share & critique	1	9	5.00*	.000
	2	11	4.44	.527
Online – Peer reviews	1	9	5.00*	.000
	2	11	4.56	.527
Campus – Short instructor presentation	1	9	4.00	.866
	2	11	4.55	.522
Campus – In class group time	1	9	4.22	.667
	2	11	5.00*	.000

* Significant at p<.05

VI. DISCUSSION

Some of the not so surprising differences occurred between the first and second year groups. There are particularly more significant differences between the campus students' Year 1 and Year 2 particularly with regard to engagement, benefits, and challenges. Obviously there were some things that needed to be worked on, most especially expectations and understanding of how flipped classrooms work. Online students are generally expecting a lot of work online outside the classroom. Campus students don't seem to have this expectation. Since online courses are already closer to the flipped model, it is not surprising that the campus students had a greater appreciation of its effectiveness compared to a traditional class once it was fully implemented. They appeared to see better engagement, more benefits and have fewer challenges the second year.

In an effort to improve issues of engagement and effectiveness, an orientation to flipped classrooms was added in the second year to create a better understanding of the process and to help set more positive expectations. In the future we will also use the results of this study to discuss the benefits and some of the challenges they will face in the flipped classroom.

In the study there appeared to be a clear lack of understanding of the course schedule using the flipped classroom model. In addition to being one of the challenges, this came up in the comments and in class in Year 2. As a result, the schedule has been reworked and more effort put into clarifying it in the short presentation each week at the beginning of each class. This seems to be working. The third year students will also be surveyed for continual improvement.

Within the campus group, the second year students perceived greater benefit from the ability to watch and rewind the videos than the first year students, most likely because they given the full lecture and then the video during much of the first semester. In the department's online courses, video recordings of the class proceeding are generally posted after class, so these students are more accustomed to having videos of lectures available.

Though it was not as clear from the data, the video lectures had a number of appreciative comments. This was gratifying as a lot of effort were put into making them as professional as possible. Producing the lecture videos is very time consuming, but is well worth the trouble. Student reflections often commented on their usefulness, as do others who view them through the YouTube.

Although the group space had higher significance in some areas in the first year online group, overall the group space means are high, particularly time to work on projects with teammates engaged in discussions and critiques and peer reviews has a positive response. Since online students are not physically together, and sometimes not even on the same island, it is understandable that having time for group interaction is essential. To provide even more group time, the presentation at the beginning of class has been shortened even more.

VII. CONCLUSION

The project has met its goals in being able to add more content to the courses while providing better support for student projects. Another important result was reducing the amount of time students needed to schedule and meet with their groups outside of class. It really helps to have a partner to share in the development. Recruiting advanced students to serve as teaching assistants helps with the development and seeing the course through the eyes of student. For compensation, course credits were given to teaching assistants for their work on the project. The synergistic energy of creative ideas and reduced development time is worth working together. A team approach is definitely worthwhile.

The biggest change that has occurred over time has been increasing the amount of student work time in class and providing instructor facilitation. It is really hard to give up the mindset and control one has in a traditional classroom. It is a little like trying to drive on the other side of the road. You have always prepared for class in a particular way and now your role is completely different. It took time evolve these courses into truly flipped classrooms. Although it does take substantial time and effort to develop everything needed, the model does appear to have considerable benefits as well as a few challenges.

ACKNOWLEDGMENT

We would like to thank our Learning Design and Technology students who were patient in the process of developing these courses, especially those that subsequently participated in the research. We would especially like to thank our teaching assistants Dainan M. Skeem, Kimberly Suwa, Elon Ng, and Bhonna Gaspar who served on our team and contributed their ideas during the development process.

REFERENCES

[1] J. Enfield, "Looking at the impact of the flipped classroom model of instruction on undergraduate multimedia students at CSUN," TechTrends, 2013, 57(6), 14-27.

[2] C. F. Herreid, & N. A. Schiller, "Case studies and the flipped classroom," Journal of College Science Teaching, 2013, 42(5), 62-66.

[3] M. Horn, "The transformational potential of flipped classrooms," Education Next, 2013, 13(3), 78-79.

[4] A. Roehl, S. L. Reddy, & G. J. Shannon, "The flipped classroom: An opportunity to engage millennial students through active learning strategies," Journal of Family & Consumer Sciences, 2013, 105(2), 44-49.

[5] J. L. Bishop, & M. A. Verleger, "The flipped classroom: A survey of the research," In ASEE National Conference Proceedings, Atlanta, GA., 2013, June.

[6] M. Molenda, C. M. Reigeluth, & L. M. Nelson, L.M. "Instructional Design," In L. Nadel (Ed.), Encyclopedia of Cognitive Science. Vol. 2, 2003 pp. 574 - 578. London: Nature Publishing Group.

[7] Merrill, M. D., Drake, L. D., Lacy, M. J., & Pratt, J. A. (1996). "Reclaiming instructional design," Educational Technology, 36(5), 5-7.

[8] J. W. Thomas, "A review of research on project based learning," 2000, (http://www.autodesk.com/foundation/news/pblpaper.htm)

[9] G. Clinton & L. P Rieber, "The studio experience at the University of Georgia: An example of constructionist learning for adults," Educational Technology Research and Development, 2010, 58(6), 755-780.

[10] N. Dabbagh, & C. W. Blijd, "Students' perceptions of their learning experiences in an authentic instructional design context," Interdisciplinary Journal of Problem-based Learning, 2010, 4(1), 6-29.

[11] B. Tucker "The flipped classroom: Online instruction at home frees class time for learning," Education Next, Winter 2012.

[12] E. Millard, "5 Reasons Flipped Classrooms Work," University Business, 2012 p.26-29.

[13] K. Fulton, "Upside Down and Inside Out: Flip Your Classroom to Improve Student Learning," Learning & Leading with Technology, June/July 2012

[14] Nielsen, L. (2012) "Five Reasons I'm Not Flipping Over The Flipped Classroom," Technology and Learning, p. 46.

[15] N. Milman, "The flipped classroom strategy: What is it and how can it be used?" Distance Learning, 2012, 9(3),85-87.

[16] S. M. P. Schmidt &, D. L. Ralph, "The Flipped Classroom: A Twist On Teaching," Contemporary Issues in Education Research, 2016, 9(1), 1-6.

[17] J. W. Winter, "Flipped Learning in a Middle School Classroom: Analysis of the individual and group learning spaces," University of Hawaii at Manoa, Honolulu, Hawaii, 2016.

MIPRO 2017, May 22- 26, 2017, Opatija, Croatia

Quantitative structured literature review of research on e-Learning

L. Abazi-Bexheti*, A.Kadriu* and M.Apostolova*
* South East European University/Faculty of Contemporary Sciences and Technologies, Tetovo, Macedonia
l.abazi@seeu.edu.mk, a.kadriu@seeu.edu.mk, m.apostolova@seeu.edu.mk

Abstract – In this study is explored the changing path of e-Learning discipline since its appearance to nowadays. E-Learning has evolved in different ways in different sectors therefore it might have different implications in certain periods of time in certain sector. A small number of things have changed in the last half century as much as the way we learn. Consequently, a significant research is going on throughout these periods in identifying cases of success in the application of e-learning. In this context, the study presented in this paper is about the analysis of the evolution on research papers related to e-learning and any form of computer assisted education published in the last five decades. We show the methodology for gathering these papers from the DBLP database, including their authors and conferences where they are published. We present the results from this study showing how published work regarding e-Learning is distributed through years, which are the conferences where these papers show up mostly and what is the collaboration community in this framework.

I. INTRODUCTION

Defining e-Learning is a challenge having in view the fact that the common definitions are so extremely wide-ranging that almost any kind of learning that involves electronics seems to fit the definition. Although initially used in the business sectors for computer-based or online training, the term 'e-Learning' has increasingly been taken up within education. E-Learning generally refers to learning that is prepared, delivered, or managed using a variety of technologies which can be set out either locally or globally.

However, it is essential to note that e-Learning has evolved in different ways in different sectors therefore it might have different meanings in certain periods of time in certain sector. According to Paul Nicholson in his article about the History of e-Learning "The History of e-Learning across all sectors is best summed up as: 'Opportunities multiply as they are seized.' as for the past 40 years, educators and trainers at all levels of Education, Business, Training and the Military made use of computers in different ways to support and enhance teaching and learning" [12]. As a result, the contemporary use of the term 'e-Learning' has different meanings in different contexts and different periods of time. In other words, although the prefix 'e' in e-Learning gives us information that the course is digitized and can be stored in electronic form, it was the rise of Internet that brought the prefix 'e' into popular usage. To be more specific, almost in the same time with the increased usage of Internet we noticed the dawn of new disciplines such as e-commerce, e-business and e-government, what makes

even more clear that it is the Internet that is really the defining technology in e-Learning. Actually, it is the online aspect that makes e-Learning different and if an institution is dealing with the essential concerns that e-Learning raises for education, it needs primarily to focus on the particular characteristics of the Internet. Learning from CD-ROMs has been in use for quite a long period but because the offline mode of delivery they carry out the same functionality as books and they do not tackle the issues that the Internet raises.

Hence, a more appropriate characterization presenting the current use of e-Learning would be as learning experience that utilizes Internet-related technologies to some degree. This definition highlights the Internet as the primary medium with regards to e-Learning but does not leave out combination with other media and approaches.

Teachers are using more and more technology and technological tools to improve their teaching practice aiming to enhance the learning process for their students. There are a lot of research papers which illustrate various aspects that describe tools and methodologies for improving the learning process, together with the obtained results after applying these approaches. On the other hand, although new century frameworks are thought to advocate new types of knowledge, little has changed and significant changes related to how technologies change all three types of knowledge - foundational, meta, and humanistic need to be conveyed [8]. For example, it has been investigated how digital games would benefit from a systematic program of experimental work, examining in detail which game features are most effective in promoting engagement and supporting learning [9]. It has been shown also that playground (hole-in-the-wall) computer kiosk, made freely available to children, without supervision, indicate uniform improvement in the computing skills of the children who used these kiosks [11]. Many projects explore a wide choice of innovative instructional strategies, such as integrating technology into teaching and differentiating learning for students with special needs [5]. Researchers in [6] discuss the characteristics of Web 2.0 that differentiate it from the Web of the 1990s, describe the contextual conditions in which students use the Web today, and examine how Web 2.0's unique capabilities and youth's proclivities in using it influence learning and teaching. e-Learning can also support the learning principles of problem-based learning, in groups working either face-to-face or online [10].

In this article, we will describe the methodology used to extract papers that have content related to the e-Learning, using a specific list of keywords. This database of gained articles, has information about paper titles, authors, years of publication and conference/journal where they appear. A quantitative analysis of this database is made based on the different used keywords. Despite this it has been examined how the e-learning research is distributed through years, conferences, and authors, showing when the first e-learning articles started to show up, when their number started to increase rapidly, which are the main e-learning conferences and who are the authors with the largest numbers of papers and collaborators in the e-Learning community.

II. EXTRACTING DBLP DATA

For the purposes of this research we have used DBLP Computer Science Bibliography, downloading the entire DBLP dataset and then converting this dataset, which is in XML format, to a relational schema, which has the following three main relations:

- authors, with information about name and paper(id), for example:

  Chris Rizos (name), conf/3dgis/ArsanaRS06 (paper id)

- papers, with information about paper id, title, year of publication, conference, for example:

  conf/aaai/MichalowskiBN11((paper id), Bayesian Learning of Generalized Board Positions for Improved Move Prediction in Computer Go (ttile), 2011 (year), AAAI (conference name)

- conferences, with information about name and description, for example:

  AAAI (conference name), Proceedings of the Twenty-Fifth AAAI Conference on Artificial Intelligence, AAAI 2011, San Francisco, California, USA (description)

III. EXTRACTING DATA OF INTEREST

The power of e-Learning is too vast and enthusiastic to be used only locally, thus currently we are facing a strong and successful intersection with a variety of other areas as well. One could draw a Venn diagram with e-Learning overlying with other topics in education, such as lifelong learning, education in developing countries, assessment approaches, learning institutions' role, intellectual property, flexibility in education, etc.

However, since we were interested only on the papers that concern computer assisted learning, we extracted these data using keywords, which are the most intuitive to our perception about computer assisted learning.

All our keywords contain two or just one word. Keywords that we used are shown in Table 1, together with the number of paper titles that contain these keywords.

As it can be seen, the most used keyword is e-learning (2895 titles), followed by the keyword computer learning (1068 titles).

For the keyword google classroom we have only 2 papers (one in 2014, and the other one in 2015).

Similar is the situation with flipping classroom and e-books learning, which is not surprising since all of them are concepts and methodologies that started to show up recently.

In this way in total we extracted 8786 papers[1]. These new dataset is used for further analysis on the papers regarding tendencies in the era of e-learning.

From these relations, we have created views on researchers that are authors of our extracted papers, associating to each of them an 'id' that will be used later for easier processing. Also, we have created a list of authors and papers, which is a subset of the authors relation. This list is much easier for further processing compare to the initial authors relation, which has even 5303478 entities.

TABLE I. LIST OF USED KEYWORDS TOGETHER WITH THE NUMBER OF GAINED TITLES FOR EACH OF THEM

USED KEYWORD	NUMBER OF GAINED PAPERS
e-learning	2895
computer learning	1068
distance learning	868
computer education	766
technology learning	685
software education	628
computer teaching	473
technology education	398
learning management system	392
elearning	322
technologies learning	317
technologies education	184
m-learning	135
virtual classroom	112
mobile learning education	62
technological learning	47
technological education	40
mlearning	34
flipping classroom	6
e-books learning	6
google classroom	2

---

[1] This number is not equal to the sum of papers gained for different keywords, since more keywords can appear in a same paper.

656

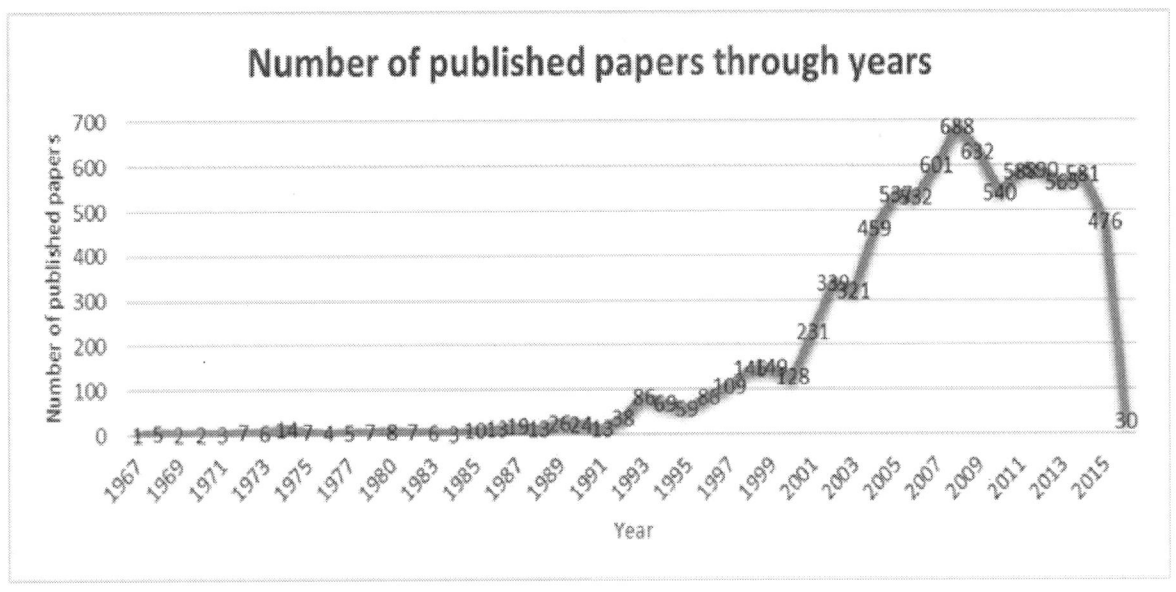

Figure 1. e-Learning research evolution through years

increase considerably, having the peak in 2008, when 688 papers were published.

IV. ANALYSING RESULTS

As part of the research we have done different investigations on different aspects that concern computer assisted learning. The first matter of interest understandably is about the timing – when did computers started to assist in learning and how the research regarding this issue distributed through years. Fig. 1 shows this distribution for almost a half century:

A paper entitled "Analog & hybrid computers in education: a panel session" was published in early 1967, marking so the beginning of a new era where machines will be used to assist in teaching math concepts: "Since the analog computer is ideally suited to solving differential equations, its use in teaching that subject is well established. With the advent of inexpensive, reliable, and accurate function generators and multipliers the analog computer can also be effectively used to teach other topics in mathematics" [1].

Following five papers in the next year, none of which is gained as false positive (containing provided keywords, but not linked to the concepts of e-Learning). Until the early 90's, there are continual publications in each year, but there is no substantial increase in terms of total published papers.

Exception is the year 1974, when even 14 papers regarding computer-based learning were published. In the late 90's the number of published papers started to

Another aspect that is of interest is related to the conferences that hosted this type of papers. The top conference is IACLT conference (IEEE International Conference on Advanced Learning Technologies) which hosted 599 papers, followed by SIGCSE (283 papers), ITiCSE (191) and WebNet (171 papers).

Seven conferences hosted over 100 papers, other fifteen hosted over than 50 papers, twenty conferences hosted more than 30 papers, 113 conference hosted more than 10 papers and the other 1430 conferences hosted less than 10 papers in total.

V. COAUTHORSHIP NETWORK IN E-LEARNING CONTEXT

The main idea behind specialized networks created for specific category of experts and researchers is to group them around a topic which could be of help to each other. In such a network, professionals can communicate with each other and share their achievements [7]. They can follow different experts in their field of interest and subfields in that area, can upload papers or make/answer questions. This is all about knowledge sharing.

A lot of work has been done at analyzing scientific collaboration networks. From the time of the pioneering article of Price [13] and Beaver & Rosen [2][3][4], a considerable amount of research articles has put emphasis on different ways and functions of scientific collaboration in particular scientific areas.

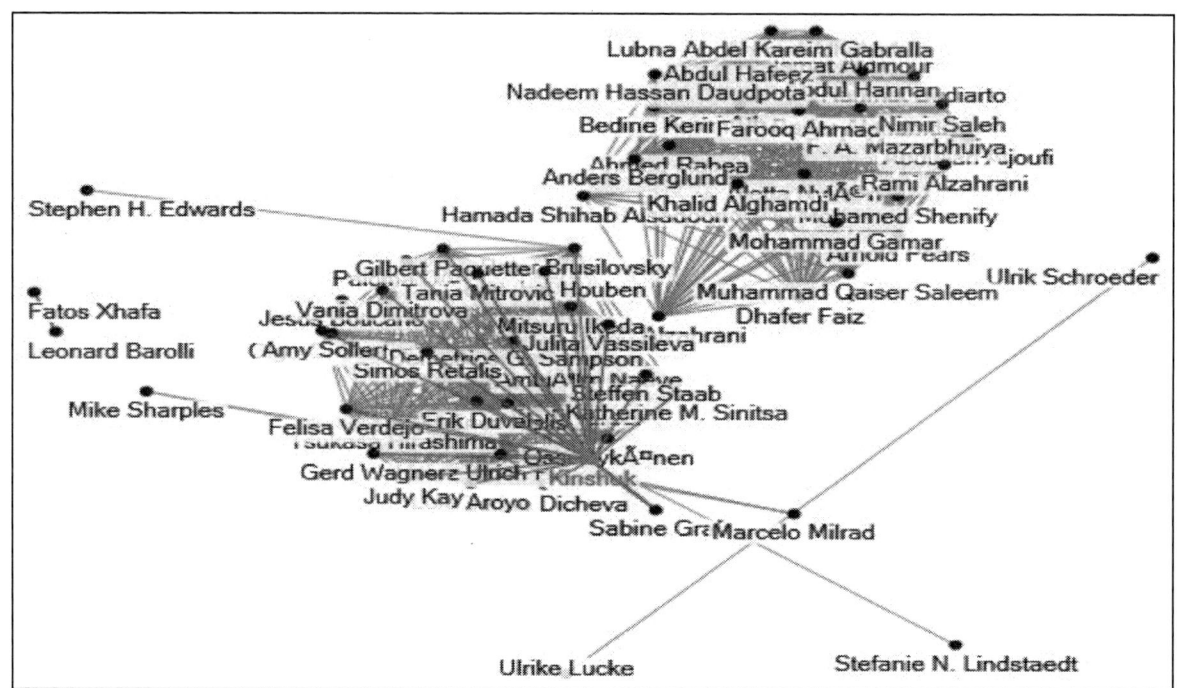

Figure 2. Coauthorship network for authors with more than 25 coauthors

In the context of our research, we analyzed the coauthorship network of authors engaged in e-learning research. The extracted data from the DBLP database, regarding has no direct information for the coauthorship network, but we have the list of papers (identified by their paper key) and a list of authors, each of them having information about published papers. The second list is the total list of all DBLP authors. In this way, from these two lists we extracted only authors that published papers from the list gained with the keywords explained in Table 1.

In this way, we have identified 18196 authors, which form in total 34154 one-to-one coauthorship relationships. From these relationships, 27445 are unique, which means that these authors have only one common paper. The "strongest" relationship have Ivan Ganchev and Mairtin O'Droma, which have 51 common papers.

The highest number of coauthors has Kinshuk, who in total has 63 coauthors in context of this research. Fig.2 shows a subgraph of the overall graph of the coauthorship network, where this researcher is denoted with red color. The subgraph is only for authors with more than 30 coauthors, that is way not all the authors are shown in this figure. The same researcher has the largest number of paper which is 25. He is followed by Toshio Okamoto, who has 24 papers and 29 different coauthors in our (extracted) paper titles.

VI. CONLUSION AND FUTHER WORK

Although e-Learning is present for pretty long period and is making significant progress, even if we take these

results into account, more study, research and application work is compulsory in order to produce more cases of success and to generalize e-Learning in training and education.

In this paper is presented the investigation that was made through the evolution of e-Learning research community, bringing here the main actors of this community such as conferences and authors.

We plan to further extend this research with exploration of paper content. Using different natural language processing algorithms, we can observe how concepts and topics have evolved through past decades in this community.

REFERENCES

[1] Boylea A. E, Haineyb, T., Connollyb, M. T. , Graya, G., Earpc, J., Ottc, M., Limd, T., Ninause, M., Ribeirof, C., Pereiraf, J. An update to the systematic literature review of empirical evidence of the impacts and outcomes of computer games and serious games. Computers & Education, Volume 94, Pages 178–192, 2016.

[2] Beaver, D. & Rosen, R. Studies in scientific collaboration.1.Scientometrics, 1: 6584, Sage, 1978.

[3] Beaver, D. & Rosen, R. Studies in scientific collaboration. 2. Scientometrics, 2: 231245, Sage, 1979.

[4] Beaver, D. & Rosen, R. Studies in scientific collaboration. 3. Scientometrics, 3: 133149, Sage, 1979.

[5] Darling-Hammond, L., & Adamson, F. (Eds). Teacher and leader effectiveness in high-performing education systems. Washington, D.C.: Alliance for Excellent Education and Stanford/Stanford Center for Opportunity Policy in Education, 2014.

[6] Greenhow, C., Robelia, B., & Hughes, J. (2009). Learning, teaching, and scholarship in a digital age: Web 2.0 and classroom

research: What path should we take now? Educational Researcher, 5(38), 246–259

[7] Kadriu, A. Discovering Value in Academic Social Networks: A Case Study in ResearchGate. Proceedings of ITI 2013, Cavtat, Croatia.

[8] Kereluik, K., Mishra, P., Fahnoe, C., & Terry, L. (2013). What knowledge is of most worth: Teacher knowledge for 21st century learning. Journal of Digital Learning in Teacher Education, 29(4), 127–140.

[9] Kovach, L. Analog & hybrid computers in education: a panel session. Proceedings of AFIPS '67 (Spring). Proceedings of the April 18-20, 1967, ACM New York, NY, USA (AFIPS 1967).

[10] M. L. Verstegen M. L. D, Jong, N., Berlo, J., Camp, A, Könings, D. K., Merriënboer, J. G. J., Donkers, J. How e-Learning Can Support PBL Groups: A Literature Review. How e-Learning Can Support PBL Groups: A Literature Review. Educational Technologies in Medical and Health Sciences Education. Volume 5 of the series Advances in Medical Education, pp. 9-33, 2016.

[11] Mitra, S., Dangwal, R., Chatterjee, S., & Jha, S. (2005). A model of how children acquire computing skills from hole-in-the-wall computers in public places. Information Technologies and International Development Journal, 2(4), 41–60.

[12] Nicholson, P. (2007). A history of E-learning. In J. M.-P.-P.-R.-R. Baltasar Fernández-Manjón, *Computers and education* (pp. 1-11). Dordrecht: Springler Netherlands.

[13] Solla Price, D. J. Little science, big science, Columbia University Press, New York, 1963.

The educators' telescope to the future of technology

H.Jaakkola[*], J.Henno [**], B.Thalheim [***] and J.Mäkelä[****]

[*] Tampere University of Technology, Pori, Finland
[**] Tallinn University of Technology, Tallinn, Estonia
[***] Christian Albrechts University, Kiel, Germany
[****] University of Lapland, Rovaniemi, Finland
Corresponding author email: hannu.jaakkola@tut.fi

Abstract - **We live in a world of accelerating changes, where technology plays an important role as an enabler. Looking ahead means being prepared for these changes. Preparedness may be** *reactive* **– reacting to the situation at the moment something happens;** *proactive* **– being prepared in advance for a situation that may happen; or** *preactive* **– being able in advance to affect something that may happen in the future and how it happens.** *Forecasting* **the future helps us to be prepared for new situations. It is based on making predictions that are derived from understanding past and present data. Known data is organized in the form of trends and further extrapolated to cover the future. From the technical point of view, there are a variety of approaches for forecasting: algorithmic, simulation, statistical analysis etc. The methods used may be** *quantitative* **(future data is seen as a function of past data) or** *qualitative* **(subjective, based on the opinion or judgment of the target group used in the analysis). Technology is an essential part of education, both in supporting effective learning and as a content of teaching itself. As a result, every educator needs skills to analyze the future of relevant technologies. In this paper, we introduce a framework that can be used in analysis of the importance of technological changes in education and as a part of curricula. The approach is based on trend analysis and classification of the relevant technologies to take into account the time span of their effects in society. The question we answer in this paper is "How can an educator analyze the consequences of technological changes in their work?".**

I. INTRODUCTION

Technological forecasting deals with the characteristics of technology. It is also used to survive in the future, in the long term and from the strategic point of view (being *preactive*, or at least *proactive*; to recognize the opportunities and risks; to recognize the weaknesses and to be able to benefit from the strengths). The forecasts provide insight into future opportunities and the processes applying them. The methods of technological forecasting are commonly known and available for all that are interested in them. The approaches may be *quantitative* (exact) or *qualitative* (heuristic); the methods cover a wide variety of alternatives representing algorithmic, simulation based, statistical analysis based techniques, or systematic ways to collect data to represent the opinion or judgment of the target group. Making one's own (usually focused) forecasts is done in a closed manner (by companies, organizations, individuals) but a lot of such knowledge is publicly

available from open sources. The latter are in most cases freely available and applicable for different purposes.

Although technological changes seem to be problematic, there is a lot of *regularity* built into the changes. In the short term, changes base on the innovations known today. Even in the middle to long term, the changes base on the expected evolution of the currently known innovations. Forecasts rarely reach further than 15-20 years ahead.

The study methods used in technological forecasting are usually based on known trends of the past and their continuum to the future. The aim of this paper is to provide a framework that helps educators to analyze the technological changes surrounding them and to help them to be prepared in the future changes, in their course contents, curricula structure and relevance, learning methods and the tools used. The framework is based on commonly known models that are integrated to provide several viewpoints to the phenomena related to changes in the technology landscape. The key elements of our framework cover the principles of the *innovation process* and *technology adoption*, the concept of *technology life cycles*, and the importance of the technologies in the society, including the expected *adoption delay* of them.

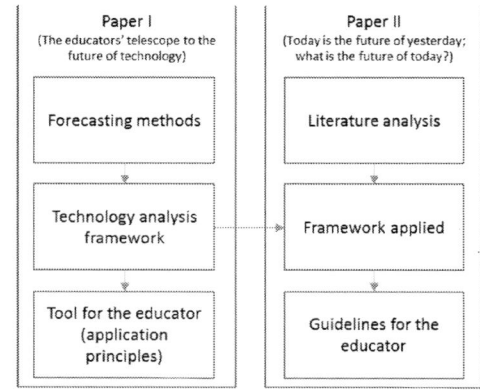

Figure 1.   Structure of the problem solving – two interconnected papers

We approach this study topic with two papers (Figure 1). We selected this split approach to separate method development and its application from each other. This approach provides a means for a slightly deeper discussion on the study method than only one paper allows, and more space to the discussion on the technological changes and their importance. In spite of this "paired characteristic", both are also self-standing research papers.

The *research question* handled in this paper (Paper 1 in Fig. 1) is "*How can an educator analyze the consequences of technological changes in their work?*" To answer this question we will introduce our analysis framework. In Paper II (Fig. 1; [10], the framework is applied for analyzing the changes caused by Information and Communication Technology (ICT) related innovations.

This paper is structured in the following way. Section 2 links the paper to the theoretical foundations of it and in our earlier studies. In Section 3 we discuss the analysis tools on a general level. Sections 4-6 cover the components of our model: importance and effectiveness classification (4), life cycle model (5), and hype cycle (6). Section 7 integrates the model components in the form of the *Technology Change Analysis and Categorization (TCAC)* Framework and its application process. Section 8 summarizes the paper.

II. BACKGROUND OF THE STUDY AND RELATED STUDIES

In this paper we will focus on the *life cycle* of an innovation. In technology research, there is an analogy between life cycle models and the *diffusion models* that describe the adoption of a new technology. [11; 12; 4] Whereas a *life cycle model* focuses on the maturity of a technology, the *diffusion approach* is focused on its adoption among *potential users*. In both cases the distribution curve follows the same *S-shaped figure*.

The life cycle model analysis presented in this paper has its roots in the early 1990s in [4]. The concept of the "*Heuristic Diffusion Model*" as a technology analysis method approaches the problems of technological changes from the aspect of *phenomena understanding* rather than exact mathematical modeling. Since that, the authors have published several research papers having different focus in this topic. The recent years, studies have focused on the *analysis of technology driven changes in society and the business environment*, the papers related to this work are the basis of the current pair of papers. The first step in this "foundation building" is the paper [5] which introduced the main principles of *trend based analysis* and *interpretation of the trends*. The paper ends with a discussion covering a wide variety of trend analyses (based on public data sources), without any synthesis.

The focus of the paper [6] was on the use of *Open Data* and the role of *open ecosystems* in the current society. It also handled *open ecosystem related trends*, culminating in the first (very preliminary) version of the Technological TCAC Framework; we call it the "*ICT Change Analysis and Categorization (ICAC)*" framework; Paper II [10]. It applies the theory of *Freeman & Perez* [1] to the findings of ICT related change factors. The same theme was continued in papers [6; 7] focusing on *open data related changes* in society, and further in the paper [8] focusing on *data driven ecosystems*, the role of social networking, security related problems, and scalable business models

(*hyperscalability*). Both of these papers cover new improved versions of the ICAC.

One more version, still based on minor iterative and incremental changes to the earlier version, was published in a conference presentation [9]; the topic was related to the *value of data and its privacy / ownership*. An interesting observation arose when preparing the talk – the trend discontinuity of some major trends, i.e. the fact that the importance level of some phenomena may change fast; something that seems to be an important driver of future changes suddenly takes on the role of embedded and indirect technology driver. The crystallized analysis method used in the analysis, TCAC, is introduced in this paper, and its application in ICT related changes, the final ICAC version, is presented in Paper II [10].

Technological forecasting is a topic of long term and wide research interest. It covers both innovation related topic areas and modelling of diffusion. One of the "bibles" related to forecasting methods is the book of J.P. Martino [13]. His work in advances in technological forecasting is available in his paper [14]. In the area of diffusion models and innovation adaption remarkable work is done by Rogers (adoption) and Majahan (adoption of innovations, innovation life cycles) [11; 12]. Because our aim is to introduce our analysis framework, we have excluded wider systematic review of earlier studies. However, most relevant references are listed in the following sections of this paper embedded in each topic context. The topics cover classification of the importance of innovations [1], trends, diffusion and adoption [11; 12; 4; 5] and Gartners Hype Cycle of embryonic technologies [2].

III. COMPONENTS OF THE ANALYSIS FRAMEWORK

Forecasting helps us to be prepared in the future. In most cases, predictions are based on analyzing past and present data and extrapolating the existing trend to the future. However, there is a problem and a paradox: in spite of understanding the past, the future does not follow the trend. The time span is also important. It is not so difficult to see into the near future, but seeing far ahead is no easy task. As a result, published analyses usually cover "important trends of the next year" as near future analysis. These are usually based on the reviews of expert opinions more than on the use of systematic fact-based analysis approaches. Consequently, the results are not uniform, and look at the same phenomenon from different viewpoints. The fact is, however, that when one makes technology prognoses, nothing radical can happen in the next ten years if the analysis environment remains stable, especially in the technology context. All we apply in ten years from now has more or less already been invented.

Why then make these predictions? The answer is: most of us can neither see the near future, nor recognize the importance of the technologies already in our use. Something that we feel important (from our subjective point of view) may be of very little importance on the global scale, and vice versa. The technology analysis makes us focus our thinking on the right things – it answers the question "*What kind of progress should we take into account?*"

An additional question is *"What is the importance of each technology?"* There is no clear way to weigh the real *importance* and *effectiveness* of a certain technology neither in the near future and especially not in the long term. Our *TCAC framework* gives at least a partial answer to this question. The following sections contain some "tools" – ways to analyze future changes, to help the reader in the analysis. It covers the following viewpoints: (1) classification of the innovation importance (Section 4), (2) trend and life cycle model principles – the S-shape (Section 5), and (3) the hype cycle of pre-embryonic technologies (Section 6). The final framework (Section 7) is built of these components.

IV. CLASSIFICATION OF INNOVATIONS

The classification of innovations answers the question *"What is the importance of each technology?"* Our work applies the idea presented by Freeman & Perez [1]. According to them,

- *Incremental changes* appear continuously in existing products and services (continuing the existing trend);

- *Radical changes* appear when new research findings are applied in products to transfer their properties or performance to a new step or cycle (movement to a new trend);

- *Changes in technological systems* are caused by combinations of several incremental and radical innovations in societal and organizational systems;

- *Changes in paradigms* are revolutionary and lead to pervasive effects throughout the target system under discussion (in our case the information society).

The external importance and effectiveness of tehcnologies grows from top to bottom.

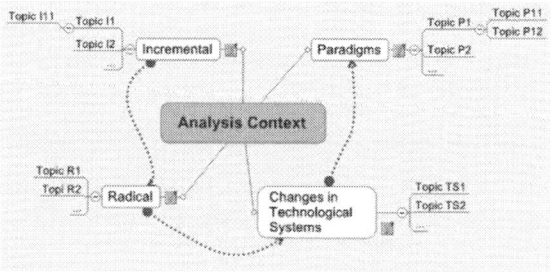

Figure 2. Technology category analysis illustrated

Classifying the importance of technologies according to this classification is not easy. We follow the principles introduced by Freeman and Perez in [1] and in more detailed by Jaakkola in [4]:

- *Incremental changes* accelerate the existing change on the existing trend;

- In the case of *radical changes* the trend continues, but at a certain moment there is an *upward shift* (break in the trend) caused by the innovation.

- *Changes in technological systems* are combinations of several technological innovations providing means for

changes in societal systems, *daily ways of operating and business models*; the key aspect is increasing *competitiveness* and fast growth of *productivity* for early adopters.

- *Changes in paradigms* indicate *permanent changes* in societal systems, daily ways of operating and business models replacing the old ones.

There are several ways to illustrate the categorization. In our framework we have selected a mind map (Fig. 2) to illustrate the categorization and to group the topic areas belonging to each category. The arrows between the categories indicate the transfer of technologies from categories of lower importance to ones of higher importance.

V. TRENDS, LIFE CYCLE ANALYSIS AND DIFFUSION

A trend shows the changes in the target phenomenon as a function of time. In *technology analysis* the trend represents the *life cycle of an innovation* from embryonic phase towards maturation. It follows the S-shape curve, illustrating simultaneously both the maturity and the adoption of the technology. Fig. 3 represents the generalized life cycle / diffusion model.

The analogy between life cycles and the adoption process can be explained by the human decision-making process. According to Rogers [12], the *adoption (diffusion) of a product (technology)* among potential adopters can be divided into four classes: *innovators* (2.5%), *early adopters* (13.5%), *early majority* (34%), *late majority* (34%), and *laggards* (16%); see lower text box in Fig. 3. The cumulative adoption curve follows logistic distribution, most commonly analyzed by the mathematical model developed by Bass (Bass's diffusion model, see e.g. the paper [11]). The adoption decision is based on the uncertainty related to the technology: the more mature it is, the more experience-based information is available of it; this accelerates the (imitation based) adoption of it towards the full penetration (full user potential reached). Because of that the use of technology indicates its maturity (depth of the use, importance to the users; only a mature technology is widely spread among the potential users).

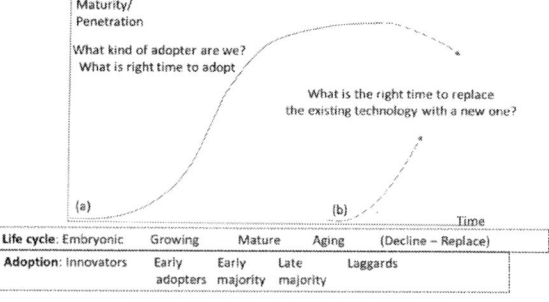

Figure 3. Technology life cycle / diffusion model

When we look at the same phenomenon from the technology maturity (life cycle) point of view, its stability increases along the life cycle. The life cycle can be divided into four phases (upper text box in Fig. 3): *Embryonic* (not stabilized), (accelerated) *growing*, mature (slowing growth), and *aging* (old technology). Finally, it is followed by the *decline* phase, in which the technology is replaced by a new one or by a new generation of the existing one. Analysts can use the life cycle model to support their decision making with two questions: *What is the right time to adopt it? What is the right time to replace it?* The embryonic phase is usually a period of uncertainty – we do not yet exactly know its future and the technology is not yet fully stabilized; several competing alternatives exist. This part is described usually by a *hype cycle* (discussed below), providing the means for more detailed analysis covering the transfer from the *pre-embryonic* (hype) to embryonic phase.

In life cycle analysis, the curve can also be interpreted as a consumption curve of the *innovation power* of a technology. The innovation potential is consumed along the life cycle and fully used at the end of the aging phase. It also describes the existence of *competitive edge* and productivity benefits to its users: the innovation power (ability to provide competitive edge to its users) will be big at the beginning of the lifespan, but also the uncertainty level and the risk of the failure are high.

VI. HYPE CYCLE

The hype cycle (also called hype slope) is an illustration of technology promises. It was introduced by Gartner Group, which is one of the best-known long-term actors in the area of technology studies and analysis. It illustrates the very first part of the life cycle of a technology (*pre-embryonic phase* in Figure 3) and answers the questions: *When do new technologies make bold promises? How do you discern the hype from what's commercially viable? When will such claims pay off, if at all?* Gartner annually publishes a variety of hype cycles in different technology sectors. These provide an insight into managing the decision to deploy selected technologies to meet specific business goals.

Gartner introduces the hype cycle as a research methodology for *pre-embryonic technologies* [2]. The structure is illustrated in Figure 4.

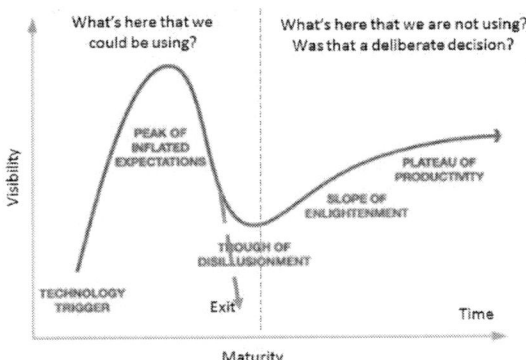

Figure 4.   The hype cycle – interpreting technology hype

Each Hype Cycle drills down into the five key phases of a technology's life cycle (following the definitions of Gartner [2]):

- *Technology Trigger*: A potential technology breakthrough kicks things off. Early proof-of-concept stories and media interest trigger significant publicity. Often no usable products exist and commercial viability is unproven.

- *Peak of Inflated Expectations*: Early publicity produces a number of success stories — often accompanied by scores of failures. Some companies take action; many do not.

- *Trough of Disillusionment*: Interest wanes as experiments and implementations fail to deliver. Producers of the technology shake out or fail. Investments continue only if the surviving providers improve their products to the satisfaction of early adopters.

- *Slope of Enlightenment*: More instances of how the technology can benefit the enterprise start to crystallize and become more widely understood. Second- and third-generation products appear from technology providers. More enterprises fund pilots; conservative companies remain cautious.

- *Plateau of Productivity*: Mainstream adoption starts to take off. Criteria for assessing provider viability are more clearly defined. The technology's broad market applicability and relevance are clearly paying off.

- *Exit* (added to the original source by the authors) occurs if ultimately the technology has not been able to prove its validity for commercial and mass use – to be ready to transfer towards the embryonic phase of the life cycle.

The Gartner Hype Cycle illustrates the maturity and adoption of technologies and applications in graphical form. It also shows how they are potentially relevant to solving real business problems and exploiting new opportunities. For decision makers, it is an analysis tool about the promise of an emerging technology; for risk takers, it opens the door to the frontier of new technologies, and for the decision-maker who appreciates safety, it will lead to a decision not to use the possibilities of the technology at this stage. It also supports preparedness for the appearance of new technology – being proactive and preparing the organization for future changes (being preactive).

VII. FROM COMPONENTS TO THE MODEL

Figure 5 summarizes the components of our approach and provides a process model to produce the TCAC. The final result of the analysis process introduced in Fig. 5 is the classification structure of the relevant factors of the analysis context. In our pair of papers, the context is ICT-related technological changes (ICAC). The analysis steps are the following:

1. The starting point is *data collected* on the study context (in our case ICT-related technological changes; Paper II). Everything is based on the data collected from valid and relevant sources; these depend on the study topic. In validating the data sources, our recommendation is to apply the principles of *systematic literature review* to guarantee that all

important sources are found and that new ones no longer provide the kind of new information that should be taken into account.

2. Two methods are used to analyze the data. *Hype cycle* based analysis (Section 6) is used to find pre-embryonic technologies to understand the delay of their commercial appearance. *Life cycle model* based analysis (Section 5) is used to understand the life cycle and renewal power of the technology analyzed. There is also an arrow between the hype cycle curve and life cycle curve. Since the hype cycle represents the pre-embryonic phases of the phenomena analyzed, it ultimately feeds the embryonic phase of the life cycle analysis; pre-embryonic becomes embryonic, unless the exit (not valid for mass and commercial use) is realized before it.

3. The *changes are classified* according to their importance by using the four categories of Perez and Freeman (Section 4).

4. The last part of the analysis is publishing. Based on the analyst's best knowledge as an expert in the topic area, the result phenomena (in our case ICT-related technologies) are published in visually informative form. We have selected a mind map.

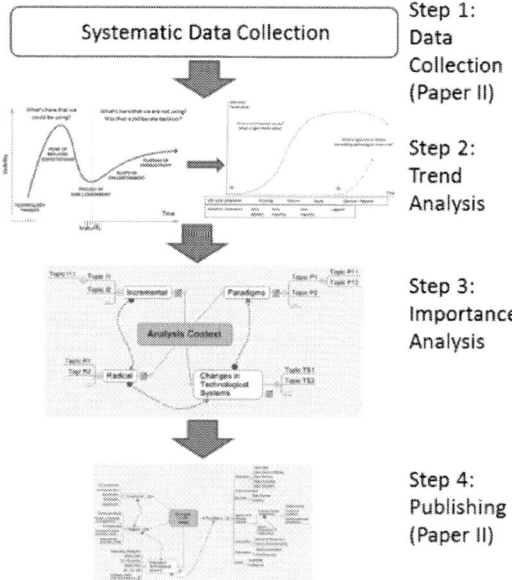

Step 1:
Data Collection
(Paper II)

Step 2:
Trend Analysis

Step 3:
Importance Analysis

Step 4:
Publishing
(Paper II)

Figure 5.   Technological Change Analysis and Categorization (TCAC) Framework – summary of the components of the analysis

As part of the publishing – because in any case the analysis remains the subjective opinion of the analyst (group) – it is good to implicitly define the interpretations of the four importance classes.

VIII.   CONCLUSION

The paper introduces a framework that provides a systematic approach for analyzing real world phenomena from their life cycle point of view. It is the pair of another

paper published in the same forum as this one (Paper II; [10]), in which we have applied the TCAC framework to analyze ICT-related technological changes. However – the model is context-free, and it can be used for a variety of analysis contexts and topic areas. The paper introduces the background of the work, its components, and its use – without real application in this paper.

As the model represents an artifact, the method used to develop the model follows the principles of *design science*. Hevner et al. [3] state that in design science research the novel knowledge and understanding come from the design process of an artifact. Hevner et al. give very clear guidelines for the design science research process, defining seven steps. The steps both provide a design path for the researcher and provide the *means to validate* the work. We have included both aspects in the following list of design science guidelines:

1. *Design as an artifact*. The artifact is a model.

2. *Problem relevance*: The model handles a relevant problem in a society, e.g. in the education sector.

3. *Design evaluation*. The design is evaluated by several earlier pilot models. The final model is an iteration of these, taking into account the weaknesses and strengths of the earlier versions, and finally synthesized into the process introduced in this paper.

4. *Research contributions*: The model provides a practical tool for analysts in different (life cycle) analysis tasks.

5. Research rigor: The model is rigorously validated by the developers.

6. Design as a search process. During the research process the authors have, in addition to a lot of fun in collaboration, increased their understanding both of the innovation diffusion process and the phenomenon analyzed.

7. *Communication of research*: The earlier versions of the work are communicated in several earlier publications in order to get feedback. The feedback and the developers' own critical findings are taken into account in the result model.

Finally, our model will be validated in a practical case in Paper II in a real analysis environment. This proves its usefulness and also its use in creating new knowledge related to the ICT-related technological changes in the education environment.

REFERENCES

[1] C. Freeman and C. Perez," Structural Crises of Adjustment, Business Cycles and Investment Behavior", In Technical Change and Economic Theory, G. Dodi, C. Freeman, R. Nelson, G. , Silverberg and L. Soete, Eds. London: Pinter Publishers, 1988.

[2] Gartner, "Research Methodologies", Available at http://www. gartner.com/technology/research/methodologies/hype-cycle.jsp, Retrieved on January 26th, 2017.

[3] A. R. Hevner, S. T. March, J. Park, J. and S. Ram, "Design science in information systems research", MIS Quarterly Vol. 28, No 1, 2004, pp. 75-105.

[4] H. Jaakkola, An Analysis of Diffusion of Information Technology in Finnish Industry, PhD Thesis, Tampere: Tampere University of Technology Publications 70, 1990.

[5] H. Jaakkola, Brumen, J. Henno and J. Mäkelä, "Are We Trendy?", In the Proceedings of the MIPRO 2013, Opatija, P. Biljanović, Ed. Mipro and IEEE, pp. 649-657, 2013.

[6] Jaakkola, H., Mäkinen, T., Henno, J. and Mäkelä, J., "Open^n", In the Proceedings of the MIPRO 2014 Conference, Opatia, P. Biljanović, Ed., Mipro and IEEE, 2014, pp. 726-733.

[7] H. Jaakkola, T. Mäkinen and A. Eteläaho, "Open Data – Opportunities and Challenges", In the Proceedings of the 15th International Conference on Computer Systems and Technologies - CompSysTech 2014, Ruse, Bulgaria, B. Rachev and A. Smrikarov, Eds. ACM, 2014, pp. 25-39.

[8] H. Jaakkola, J. Henno and J. Soini, "Data Driven Ecosystem - Perspectives and Problems", In the Proceedings of the 4th Workshop on Software Quality Analysis, Monitoring, Improvement, and Applications (SQAMIA 2015), Maribor, Slovenia, Z. Budimac and M. Hericko, Eds., CEUR Workshop Proceedings 1375, CEUR-WS.org and University of Maribor, 2015, pp. 17-26.

[9] H. Jaakkola, "Data, and only Data, is important, or is it?" Presentation in the 3rd Keio Creativity Initiative Symposium for the Top Global University Project, TGU Symposium on Smart and Connected Cities, July 25, Yokohama, Japan, 2016, restricted availability.

[10] H. Jaakkola, J. Henno, J. Mäkelä and B. Thalheim, "Today is the future of yesterday; what is the future of today?", **Paper submitted** to the MIPRO 2017 Conference, 2017.

[11] V. Majahan, E. Muller and F.M. Bass, "Dynamics of Innovation Diffusion: New Product Growth Models in Marketing", Journal of Marketing, Vol. 54,No. 1 (January), pp. 1-26, 1990.

[12] E. Rogers, Diffusion of Innovations. New York: The Free Press, 1983.

[13] J. P. Martino, Technological Forecasting for Decisionmaking, McGraw Hill, 1993.

[14] J. P. Martino, "A review of selected recent advances in technological forecasting", Technological Forecasting and Social Change, Vol 70, No. 8, pp. 719-733, 2003.

ICT Support for Promotion of Nature Park

Andreja Žiško*, Andrej Šorgo**, Marjan Krašna***

* University of Maribor, Faculty of Education, Maribor, Slovenia
** University of Maribor, Faculty of Natural Sciences and Mathematics, Maribor, Slovenia
*** University of Maribor, Faculty of Arts, Maribor, Slovenia
andreja.zisko@um.si; andrej.sorgo@um.si; marjan.krasna@um.si

Abstract – Around the World and particularly in the Europe the conservation of the habitats becomes more important. For that reason, the project of connecting nature and people in Europe promotes the European Green Belt that stretch across 12500 km. A part of that European Green Belt is also the three state nature park (Goričko-Raab-Örség) spanning Slovenia, Austria and Hungary.

Inhabitants the Goričko nature park are important factor for the park's the long time survival and preservation. But living in the park also presents multiple limitations and young people see them as hampering constraints of their life. Therefore, the activities for constant promotion and value of the park needs to become the daily routine. The old fashion promotional activities are not effective for the youngsters, which are accustomed to the fast life style and daily use of ICT. The study was made how to incorporate ICT into the promotion and daily life of the nature park and empower the activities to create visitors own footprint in the park and park's cyber space. The promotional activities, educational activities and competition between youngsters are covered in the article. The ICT is the backbone for the teamwork; individual's curiosity; and professionals.

I. INTRODUCTION

ICT gadgets have become a part of our daily life. If the older generations see them as necessary evil, the young see them as extension/upgrade of their bodies. The use of cell phones, tablets and computers have changed the perspective on time, motivation, and learning. Traditional methods of studying; though effective; are seen boring to the young generations of students and regretfully they make them less inquisitive about their surroundings. "Going out" and playing in the backyards, forests, meadows, climbing trees, seems very odd to them. The term digital natives denotes generations born into the ICT reach environment, and should be distinguished from digital immigrants – generations born before ICT rich environments [1]. It is generally accepted that digital natives possess knowledge and skills to safely navigate in digital worlds, but there are signs that this assumption is only wishful thinking [2]. We want to teach digital natives to respect and protect the nature and all its biodiversity.

Sadly, people often seems to care less for things that seems granted to them. Therefore, the inhabitants of the nature parks sees them more as the limitations than the wonders of the nature that needs to be protected for the posterity. Nature parks "health" depends on their inhabitants and these were sufficient reasons to do the research and uncover the coupling between the nature park

Goričko (the Park), schools, and professionals employed for nature park conservation. In the year 2016, the research about the Educational potential of landscape park Goričko (see Figure 1) was prepared and presented to the public [1].

The research used online survey among teachers in the area of the park and the results are presented in this article. The objectives of our study were to show multiple aspects of cooperation and complementarity between educational institutions and park administration [1].

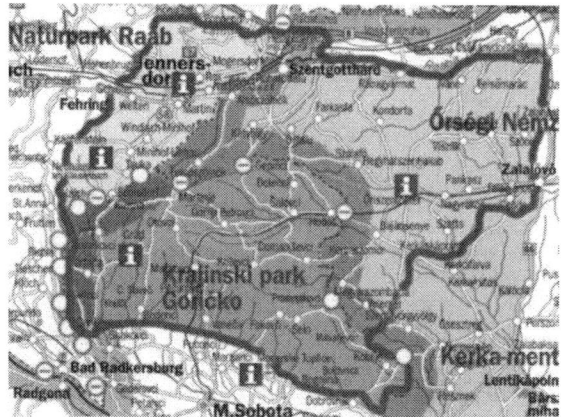

Figure 1: Tree-stare part Goričko-Őrség-Raab

II. EDUCATIONAL BACKGROUND

The Slovenian educational system have defined quality of education; educational processes; and educational goals in the White paper on education in Republic of Slovenia [2]. Slovenian view is in-line with descriptors of the quality of knowledge defined by different authors [3] [4] [5]. However, in practice the quality of knowledge could be the mash between educational strategies with non-cognitive, emotionally motivational and value components [6].

Outdoor education in natural and landscape parks often combines teaching of the schoolteachers and professionals from the park. While teachers are trained for teaching contents according to the curricular goals, this is not always true for employees of the park, who are not always trained educators. For them teaching in the real environment is always only a part of their regular work. Because of their limited pedagogical content knowledge [9] their teaching is usually confined to lower cognitive levels, which they consider as high quality. But for the retention of the knowledge, real world education should consider also higher cognitive processes, as they are described in the revised Bloom taxonomy (Figure 2). Learning by doing [7] is still considered as one of the most effective principle for learning of skills in the combination with the factual knowledge. The highest levels achievements occurs in the context of goals and objectives that are relevant, meaningful, and attract the students. The motivation is the essential ingredient that allow full utilization of the real world experiences. The motivation of the students have changed in the course of the years and what was once motivational today is considered boring for the "touch screen" generations. Traditional ICT skills are in decline [8] and we need to find another paradigm to activate the students.

The educational activities in the Park are summoned in the educational worksheets where different questions and educational assignments guide students in the park. Analysis of the worksheets, using revised Bloom taxonomy (Figure 2) [9], shows that in most cases the questions and the assignments are on the lower order thinking skills. Rarely the assignment require higher order thinking (cognitive processes) and metacognitive knowledge dimensions. It is somehow understandable that this is the case because park's administration do not have a clear understanding of students' previous knowledge and higher order of thinking cannot be achieved without the solid ground understanding. Therefore, for better quality work assignment the cooperation between schools and the Park's professionals is almost mandatory. But it is not that bad as it may have seen because it is possible to upgrade the current worksheets into new form with fairly simple procedures known from the past [10].

Objective of the research were:

- Acquire the current state of the cooperation between the Park's professionals and the teachers of local schools.
- Discover the ways to enhance the cooperation between the Park's professionals and the teachers of local schools.
- Provide blueprint principles to activate the digital natives to participate in the nature protection.

The Knowledge Dimension	Cognitive processes					
	1. Remember	2. Understand	3. Apply	4. Analyse	5. Evaluate	6. Create
Factual						
Conceptual						
Procedural						
Metacognitive						

Figure 2: Revised Bloom taxonomy [9]

III. SURVEY

In the course of 2014 the teachers of natural sciences from primary schools in the area of the Nature Park Goričko were surveyed [1]. The invitation to participate in online survey was send to the addresses of twelve (12) primary schools situated in the area of the Park. The survey had 24 questions. Most of them were closed type questions; some were also five (5) level Likert type questions; and some were open type questions.

Some questions were optional and connected with answers from previous questions [1]. The survey questions covers four main topic of the research and the gathered data provide us valuable feedback [1]:

- about the knowledge of teachers about educational activities of the park officials;
- motivation of the teachers to cooperate with the park officials in the design and in the production of learning topics;
- obstacles to visit the Nature park Goričko; and
- teachers' position to the educational activities in the park.

In the course of the research, the semi-structured interview was used on three members of the Park professionals who were assigned for the cooperation with the schools. The details of the interviews are not part of this article but can be find the other article [14].

We have received the responses from 11 schools but six (6) responses about the teachers' affiliation were missing, therefore it is possible that we have answers from all schools in the area of the Park. Total number of responses was 59. Most of the teachers were familiar with the activities (73.7 %) that were prepared for the education by the park administration. Half of the remaining teachers (~13 % of all responses) were unaware that the park has its own web page with all relevant data available. Acquired responses also shows that the most effective way to disseminate the educational activities are web site and recommendations of colleagues (55.6 %).

Table 1: Information acquisition

Where have I found the information	f	f (%)
Park's web page	31	29.2 %
recommendations of colleagues	28	26.4 %
professional associations	17	16.0 %
e-mail	13	12.3 %
leaflet	10	9.4 %
radio or tv	3	2.8 %
poster	1	0.9 %
other	3	2.8 %

If we add to the score the scheduled professional associations and e-mail we cover the 83.9 % of the interested teachers population. Traditional dissemination channels (radio, TV, press) are waste of time and resources. Because the Ministry of Education use the electronic communication only for a decade now, we expected such results. The only not entirely expected result was the proportion of teachers that would like to have regular monthly e-mail update (81.4 %). We thought that this proportion would be even higher because school usually cannot react quickly to unscheduled activities.

Most of the teachers (96.4 %) would like to cooperate with the park administration in their educational activities. Only around a third (34.6 %) would like to cooperate in the preparation of learning materials. The rest, 61.8 % have not thought about this possibility.

Despite good intentions of the teachers, the reality is somehow different. Professionals from the Park expressed, in the interviews, that their effort in the preparation of educational activities is not always recognized by schools and much more educational activities could be scheduled in the park [1]. However, teachers and park's professionals agree that the biggest obstacle is the low budget of the educational institutions. Optimal educational activities are between 1 and 2 hours long for elementary education (6 to 10 year) and 2 and 4 hours for primary schools education (11 to 14 year) [1].

The factors of the highest importance for the educational activities in the Park would be [1]:

- Real examples from domestic environment.
- Direct contact with the real world.
- The development of a responsible attitude towards the environment.
- Increased motivation for learning.

In general, teachers have good experiences with the Park's professionals. Amazingly, 91.9 % of those who have already been in the Park, with their students, grade the learning activities as good or very good and no negative grade was issued by any teacher [1].

IV. MOTIVATING ACTIVITIES

For the design of the contemporary education, the key competences [11] and the generic competences [10] need to be considered. Learning activities should cover all competence spectrum. The generic competences in our case are [1] [10]:

- the ability to collect information,
- the ability of analysis and organization of information,
- the ability to interpret,
- the ability to synthesize conclusions,
- the ability to learn and solve problems,
- transfer theory into practice,
- application of mathematical ideas and techniques,
- adaptation to the new realities,
- quality awareness,
- the ability to work independently and participate in the teamwork,
- organizing and planning work,
- verbal and written communication,
- interpersonal interaction, and
- safety at work.

Therefore, the revised Bloom taxonomy and the competences provide the suitable framework for tailoring the content of learning. However, we need to find the motivation for the students inside the same framework. The ability to motivate is like an art of applied psychology because it differentiates substantially between individuals and between cultures [12] [13]. What motivates the students in one country generally does not apply to the students in the others [14].

The students in questions are primary school kids living in the Park. They having access to or possess all contemporary home ICT equipment. Like most modern world youngster, they often consider old educational principles obsolete because it provides information too slow and often in strictly predefined schedule [15] [16]. They do not consider learning by heart as the mean to train

their brain and expand their brains capacities and capabilities; but as unnecessary, old school, obsolete, nonsense, stupid, time-consuming activity because everything is available on Google in just a few seconds.

A. Application of ICT in the work assignment

There are numerous occasions where ICT can be used in the education. Lately many educational specialists strive to use it just where it is most appropriate. The appropriateness of the use of ICT does not depend only on teachers' skills and feeling but also the students. Contemporary students have different demands for the learning materials and the educational principles. Even the best topics that are appropriate for their age and attracts them become boring if they are not presented properly. The previous knowledge of the students is also important in the educational processes. The right assignment for the right students is much more motivating and challenging [6].

Traditionally the assignments are delivered in the paper formats (worksheets). Such worksheets have many benefits and also some limitations. The paper never run out of electricity and can be used for extended duration of the learning in the park. Paper can be copied and distributed to all participants and even if some papers fall to the ground, into the mud or water, no harm is done. Obvious benefits of the paper is overwhelming and it is wise to use it. Despite papers' benefits there are also some drawbacks of the paper. Paper is passive and cannot provide the feedback. However, the questions (quiz type) on the paper can be easily transformed to the digital form and their use on the portable electronic devices (phone, tablet or laptop) could provide instant feedback. The quiz's results are immediately stored and no additional effort is needed to achieve them. The paper can present text and images but not the sound and video. For the identification of the natural wildlife in the park sometimes even the optimal printed image is not enough for unexperienced users. However, the video with the sound (e.g. bird's song) or even interactive 3D model of the specimens would enables young students to achieve much better results.

The important part of learning is also preparing the sketches of the wildlife. Today's children and even older students perceive such assignments are total waste of time because they could use their phone to make a photo almost in an instance. Even the sensors on the portable devices could be used for measurements (temperature, atmospheric pressure, humidity, altitude, light intensity ...). Why don't we use the phones and tablets if they are so great? The problems arose with their diversity. For the educational purposes, each student should have adequate (standard) user experience. The screen resolution, processing power, storage capacity, sensory array, operating system ... makes the effective use of students' devices a logistic nightmare. The effective use of ICT depends not only on technology but also on communication infrastructure. The sufficient cheap and reliable bandwidth is mandatory.

BYOD (Bring your own device) [17] works only occasionally therefore a combination of both, paper and portable devices, are still a wise decision.

B. The digital library

The Park is important biodiversity environment. The cataloguing processes of all wild life and plants could take years. It would be wise to encourage visitors to participate in this enterprise. Especially this could be motivating for the students who lives in the Park and could provide images and videos all year round. A geotagging of the data could be used for different analysis and distribution of the acquired data to the individual visitors who want to explore the Park. For the ease of use of portable devices with cameras, the QR codes could be used since they were already tested in the education [18] [19] [20] [21]. Throughout the park, a weather resisting QR code could be posted (see Figure 4). Visitors and students could read them with their devices with the camera and get the online data. Since the Park is located in the three states (see Figure 1) the information could be translated into different languages further improving the experience of the students and enhance their key competence – communication in the foreign language.

Figure 3: QR code sample

Figure 4: A place where QR codes could be used.

Most mobile devices have web browsers and are able to display text and images. The interactive elements processed on the devices (using script languages) are not so reliable and other multimedia elements can be even more problematic. If design require multimedia element to be displayed in the context of the web page but the device play multimedia stream in full screen the intended user experience is lost [22]. Despite obvious observed shortcoming this is just temporary fault that will fade out with the new devices.

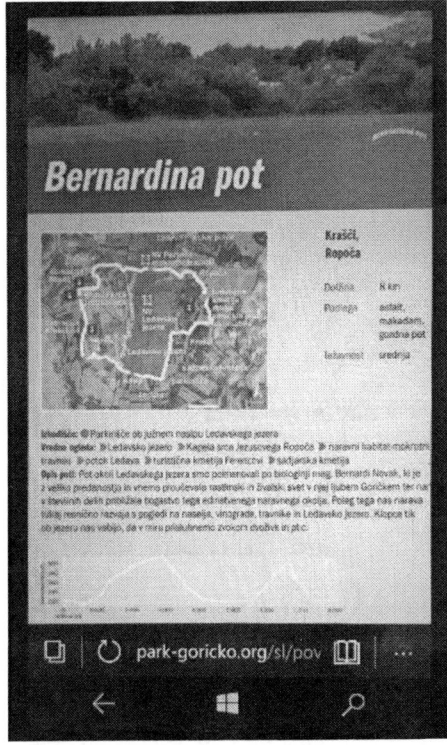

Figure 5: The digital content available to the visitors

C. The multimedia library

A subsection of a digital library is a multimedia library. In general, it all seems the same but the technical characteristics are different. Images takes much less disk space and even moderate web server could serve hundreds of users. However, the multimedia library with audio and video materials is much more storage demanding and requires huge bandwidth for pleasant user experience. Video can be stored in various bitrates or on the fly transcoded to the user's specific bandwidth or devices' capabilities. Both principles are possible and both require a reliable funding for the system operation.

The multimedia library could be used everywhere and particularly in the classroom education. Multimedia can act as a learning material or promotion. Logistics for students that wants to visit the Park becomes more complex with the distance therefore better motivation is needed for the decision "to go". Multimedia can provide the overview of the Park's educational activities and influence teachers and headmasters to schedule the educational activities there. However, the use of multimedia in the park on the portable devices a necessary ICT infrastructure must be constructed and maintained. Internet access and powering stations should be available upon major routes and points of interest.

D. Competitions

One of the motivation factor for the students to visit the Park and learn there could be the competitions between their peers. The value system to encourage the learning should be established and the best should be promoted on different levels.

There are different possibilities for the competitions:

- individuals knowledge,
- schools competition (teamwork),
- scoring the products (picture, video, presentation, ...).

Good cooperation between countries could culminate even in the financial support for the international competition. Governments of three countries could apply their professionals to prepare the EU funded project even lowering their costs.

E. Proposed improvements

The research gives us the better understanding of the present situation of the educational activities in the Park. Based on the analysis in this topic we can summarize the proposed improvements and we suggest:

- Cooperation of the Parks professionals and teachers should become operational.
- Use of revised Bloom's taxonomy and competences framework for the design of the work assignments.
- Design of the learning materials for dual purpose (paper and electronic).
- Prepare activities where students would actively participate and their results would be available online using Web 2.0 paradigm.
- Build the infrastructure and user interface for the digital/multimedia library.

The assignments for students could be modified to encourage their active cooperation and stimulate their creativity and innovation. The findings can be applied in different situations of outdoor educational activities and seemingly embed the ICT into the existing and future information infrastructure of the (nature) parks, educational trails or other similar installations.

V. CONCLUSION

The nature protected environments and especially Nature Park Gorčko live and flourish with its inhabitants. The awareness of the importance of biodiversity should start with the kids and continue through their life. Therefore, this never-ending project requires dedicated professionals, motivated teachers, and their cooperation. The legislative representatives should provide sufficient funding and benefits to the park's inhabitants who should not consider living in the park as constraints.

The Park's professionals spent considerable effort to enhance educational activities in the Park and cooperate with the public educational institutions. Our analysis have shown that ICT could be used to promote the Park even more.

The modern educational paradigm demands students' active participation in the learning processes. This could be achieved with establishment of digital library that works as public domain. Additional financial investment could be recovered in a short time. More visitors and students could boost the local economy and further provide the benefits to the conservation of the nature. There are also additional benefits of use of social media for promotion and education, but this requires additional human resources because social media are much more vibrant and requires quick responses and regular updates not to be considered "dead".

REFERENCES

[1] A. Žiško, „Educational potential of nature park Goričko," University of Ljubljana, Biotechnical Faculty, Ljubljana, 2016.

[2] Ministry of Education and Sport, Bela knjiga o vzgoji in izobraževanju v Republiki Sloveniji (eng. White paper on education in the Republic of Slovenia), Ljubljana: National Education Institute Slovenia, 2011.

[3] UNICEF, „Defining Quality in Education," 2000. [Elektronski]. Available: http://www.unicef.org/education/files/QualityEducation.PDF. [Poskus dostopa 23 11 2015].

[4] W. Van de Grift, „Quality of teaching in four European countries: a review of the literature and application of an assessment instrument," Educational Research, Izv. 49, št. 2, pp. 127-152, 2007.

[5] D. Raynolds, P. Sammons, B. De Faine, J. Van Damme, T. Townsend, C. Teddlie in S. Stringfield, „Educational effectiveness research (EER): a state-of-the-art review," School Effectiveness and School Improvement, Izv. 25, št. 2, pp. 197-230, 2014.

[6] B. Marentič Požarnik, „Kaj je kakovostno znanje in kako do njega?," Sodobna pedagogika, pp. 28-50, 2011b.

[7] R. C. Schank, T. R. Berman in K. A. Macpherson, „Learning by Doing," v Instructional-Design Theories and Models: A New Paradigm of Instructional Theory, New York, Routledge, 1999, pp. 161-182.

[8] ACARA - Australian Curriculum Assessment and Reporting Authority, „2014 NAP – ICT literacy report shows a decline in ICT literacy," 17 11 2015. [Elektronski]. Available: http://www.acara.edu.au/news-and-media/news-details?section=201511170907#201511170907. [Poskus dostopa 2 2 2017].

[9] D. R. Krathwohl, „A Revision of Bloom's Taxonomy: An Overview," Theory Into Practice, Izv. 41, št. 4, pp. 212-218, 2002.

[10] University of Maribor, Faculty of Natural Science and Mathematics, „Development of Science Competences," University of Maribor, Faculty of Natural Science and Mathematics, Maribor, 2008-2011.

[11] European Commission, „Lifelong learning — key competences," EU law and publications, 10 10 2016. [Elektronski]. Available: http://eur-lex.europa.eu/legal-content/EN/TXT/?uri=URISERV:c11090. [Poskus dostopa 3 2 2017].

[12] C. Ames, „Classrooms: Goals, structures, and student motivation.," Journal of Educational Psychology, Izv. 84, št. 3, pp. 261-271, 1992.

[13] M. L. Maehr in C. Midgley, „Enhancing Student Motivation: A Schoolwide Approach," Educational Psychologiest, Izv. 26, št. 3-4, pp. 399-427, 1991.

[14] D. H. Lim, „Cross Cultural Differences in Online Learning Motivation," Educational Media International, Izv. 41, št. 2, pp. 163-175, 2004.

[15] N. Hoic-Bozic, V. Mornar in I. Boticki, „A Blended Learning Approach to Course Design and Implementation," *IEEE Transaction on Education,* Izv. 52, št. 1, pp. 19-30, 2008.

[16] H. Beetham in R. Sharpe, An Introduction to Rethinking Pedagogy. Rethinking Pedagogy for a Digital Age: Designing for 21st Century Learning, New York: Routledge, Taylor & Francis Group, 2013.

[17] J. Keyes, Bring Your Own Devices (BYOD) Survival Guide, Boca Raton: Taylor & Francis Group, 2013.

[18] F. M. Tretinjak, „The Implementation of QR Codes in the Educational Process," v *MIPRO 2015*, Opatija, 2015.

[19] H.-C. Lai, C.-Y. Chang, W.-S. Li, Y.-L. Fan in W. Ying-Tien, „The implementation of mobile learning in outdoor education: Application of QR codes," *BRITISH JOURNAL OF EDUCATIONAL TECHNOLOGY,* Izv. 44, št. 2, pp. E57-E62, 2013.

[20] K. Dourda, T. Bratitsis, E. Griva in P. Papadopoulou, „Combining Game Based Learning With Content and Language Integrated Learning Approaches: A Case Study Utilizing QR Codes and Google Earth in a Geography-Based Game," v *PROCEEDINGS OF THE 7TH EUROPEAN CONFERENCE ON GAMES BASED LEARNING*, Porto, 2013.

[21] J. Hvorecky, „An Integral Approach to Online Education: An Example," v *PROCEEDINGS OF THE 12TH EUROPEAN CONFERENCE ON E-LEARNING (ECEL 2013)*, Sophia Antipolis; France, 2013.

[22] M. Krašna, Izobraževanje v digitalnem svetu, Izv. 108, M. Jesenšek, Ured., Maribor, Bielsko-Biała, Budapest, Kansas, Praha: ZORA - Mednarodna založba Oddelka za slovanske jezike in književnosti, Filozofska fakulteta, Univerza v Mariboru, 2015.

Experiences in Using Educational Recommender System ELARS to Support E-Learning

Martina Holenko Dlab

Department of Informatics, University of Rijeka, Rijeka, Croatia
mholenko@inf.uniri.hr

Abstract – **Educational recommender systems have the potential to support e-learning by personalizing learning process to the individual characteristics of students. To achieve this, such systems can recommend e-courses, learning materials, learning paths, collaborators and similar. This paper describes experiences in using educational recommender system ELARS for the e-course "Hypermedia Supported Education". The system provides personalization in the context of e-learning activities (e-tivities) by recommending optional e-tivities, possible collaborators, Web 2.0 tools and by offering advice. Analysis of students' academic achievements and results of conducted questionnaire are presented together with teacher's observations and guidelines for further development of the ELARS system.**

I. INTRODUCTION

One of the trends in web personalization is to provide a tailored user experience using recommender systems [1], [2], [3]. Web personalization presumes the presentation of a content to match a specific user's preferences. The use of recommender systems expands the possibilities for web personalization by providing navigation support in order to help users to find appropriate items (e.g. web pages, books, restaurants, music videos, etc.) [4]. Personalization, as key element of recommender systems, can be applied to support education process as well [5].

The recommendation process is based on data regarding target users (*to whom* is recommended), data regarding items (*what* is recommended), and recommendation techniques (*how* is recommended). The following techniques are in use: content-based recommendations, collaborative filtering, and knowledge-based recommendations. In most cases, these techniques are combined and hybrid recommenders are developed [6]. Personalization techniques and items that are recommended vary depending on whether e-learning is moderated or non-moderated. Moderated e-learning refers to e-learning through a set of structured course activities defined by a teacher, usually performed with the help of learning management systems – LMSs. On the other hand, non-moderated e-learning is self-regulated so the students decide where, how and when they want to learn [7].

The development of learning environments that support e-learning is influenced by so-called Web 2.0 [8] and e-learning 2.0 [9]. E-learning 2.0 promotes interaction between students that can be accomplished during collaborative e-learning activities, called e-tivities [10]. E-tivities are usually performed with the help of Web 2.0 tools (e.g. collaborative writing in Wikispaces, blogging in Blogger, mental mapping in MindMeister). The

changes driven by the development of Web 2.0 bring new possibilities for personalization that will ensure the achievement of learning outcomes in so-called "2.0" environments [11]. All the above mentioned motivated the development of the educational recommender system ELARS (E-Learning Activities Recommender System) [12]. ELARS was developed during the research project "E-Learning Recommender System" in order to enable personalization of e-tivities in "2.0" learning environment. During the project, didactical models for different types of moderated e-courses were developed [13]. All developed models promote acquiring knowledge in line with the constructivist theory of learning and advantages of Web 2.0 technologies for e-learning.

This paper presents experiences from several years of using educational recommender system ELARS for the e-course "Hypermedia Supported Education". Results of analysis of students' academic achievements and conducted questionnaire are presented as well as teacher's observations and guidelines for further development of the system. The paper is organized as follows: Section II presents review of the related work. Section III presents means of personalization in the ELARS recommender system. Section IV describes experiences from the e-course "Hypermedia Supported Education" and the results of course evaluation while section V brings conclusions.

II. STATE OF THE ART

Among existing educational recommender systems, systems that recommend learning objects (e.g. additional teaching materials) prevail. Recommended learning objects can be published on the Web or available in a learning management system [14], [15]. Similarly, the recommendations can support access to appropriate learning materials within learning objects repositories, as is the case with the system DELPHOS [16] and the MERLOT repository [17]. Recommending learning objects may also include determining personalized learning paths [18], [19] and recommendations of university courses. Examples are CourseRank system, used at the Stanford University [20], and smart e-course recommender [21]. Instead of recommending learning objects, fewer systems recommend activities that accompany the process of e-learning [5], for example, during learning of programming. For example, the system WeHelp warns on the most common mistakes Part of available types of recommendations covered by the domain independent approach, TORMES methodology, are intended to support e-tivities (e.g. it is recommended which blog posts target student should comment) [23].

Development of systems that are not restricted to recommending learning objects is especially important in the context of e-learning 2.0 where the emphasis is not on the reproduction of the learning content. Students should create their own versions of subject matter through collaboration with colleagues, which can be achieved by organizing collaborative e-tivities. Additionally, for supporting collaborative learning, it is necessary to develop approaches to recommend collaborators of peers for learning activities. Example is the approach presented in [24] which proposes a model for recommendation of peers to communicate and/or collaborate with based on students' preferences about some important features of the learning process. Authors in [25] describe a model for recommending collaborators to work in pairs. Collaborators are recommended to target student in respect with information that include knowledge, social and technical aspect of potential collaborator. When used in the context of non-moderated or self-directed e-learning, recommendation techniques have the potential to support creation of personal learning environments (PLE) or mash-up applications that combine several data sources in one Web service. For example, in the system ReMashed [26] users can combine content from different Web 2.0 services based on the received recommendations.

Existing educational recommender systems take into account that recommendations in the educational domain cannot be determined solely based on interests and preferences. Therefore, together with these characteristics [27], data regarding student's knowledge [1], communication skills [23], and learning styles [1], [15] are used in the recommendation process. Regarding the used techniques, the most of the systems rely on hybrid techniques so, for example, combine content-based recommendations with collaborative filtering. An exception are systems for self-directed learning within social networks where usually only collaborative filtering is used [5].

III. PERSONALIZATION IN THE ELARS SYSTEM

To foster personalization within moderated e-learning scenarios in ELARS system, educational recommendation strategy was implemented [28]. It takes into account that e-learning is an active process which includes interaction not only between a student and prepared learning content, but also between students themselves. The strategy relies on learning design specification which describes the sequence of e-course activities and on data representing characteristics of target users and items.

A. Learning design

Learning design represent a workflow of course activities that are classified into six categories (Table I) and grouped into learning modules. The most important categories for the recommendation process are e-tivity using Web 2.0 tool and decision activity. E-tivities can be optional (i.e. teachers can plan several e-tivities among which student will choose one). Recommendations are presented to students within decision activities. For that reason, this type includes recommendation criterion that should be set in the process of defining learning design (i.e. select user's characteristics that will be used in the recommendation process) [13]. For example, optional e-tivities can be recommended depending on the activity level reached for preceding e-tivity and/or preferences of tools offered for its realization.

B. Target users

In order to achieve personalization and enable recommendations of various items, four groups of characteristics are used to represent students and groups: preferences of learning styles according to VARK model (visual, read/write, aural, kinesthetic) [29], preferences of Web 2.0 tools, knowledge levels, and activity levels. All characteristics (Fig. 1) are represented as continuous variables with values ranging from -1 to 1 and used in the recommendation system. Data regarding students' preferences towards Web 2.0 tools are identified using questionnaire available in the ELARS system. To collect data regarding VARK learning styles preferences, students are asked to solve online questionnaire [30] and enter the results in the ELARS system. Knowledge and activity level are automatically determined based on student's results for testing activities and e-tivities. Activity level represents quantitative aspects of students' engagement in e-tivities. Calculation of activity level is based on automatically collected activity data from Web 2.0 tools using APIs [31]. Collected data is analyzed in order to determine quantity and continuity of contributions. Activity level is calculated for each formed group, as well. Other characteristics are not determined for the group as a whole. If needed for the recommendation process, data representing group members are aggregated.

C. Items

ELARS enables four types of recommendations to personalize collaborative learning: optional e-tivities, Web 2.0 tools, collaborators, and advice [31].

TABLE I.   DESCRIPTION AND EXAMPLES OF ACTIVITY CATEGORIES

Abbr.	Category	Description	Example
FA	F2f activity	Involves face-to-face (f2f) instruction.	Attending the lecture "Web 2.0".
CA	Learning content activity	Foster students' interaction with learning materials (are associated with learning resources).	Studying the lesson "Moderating collaborative e-tivities".
TA	Testing activity	Used for assessment of student's knowledge of subject matter.	Solving online test for self-assessment.
eLA	E-tivity using Web 2.0 tool	Interaction of a student with the learning environment and with other participants as a response to a specific task, oriented towards intended learning outcomes.	Collecting and tagging additional resources regarding topic "Distance learning" using Diigo or Google+.
DA	Decision activity	Used for presenting recommended items.	Selecting optional e-tivity for revision.
SA	Support activities	Support the realization of e-tivities (not oriented towards learning outcomes).	Reading the instructions for writing a seminar paper.

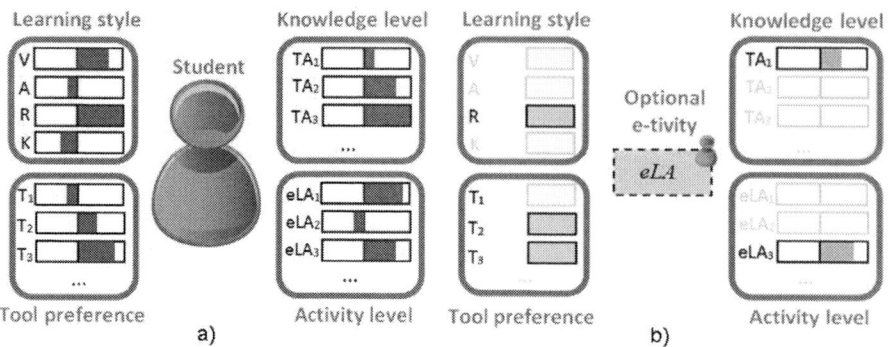

Figure 1. Examples of representation of: a) student, b) optional e-tivity

E-tivities are recommended if they are optional. List of characteristics that represent an e-tivity includes Web 2.0 tool(s) that is/are offered for its realization, learning styles for which its task is appropriate for, teacher's expectations regarding quantitative aspects of participation in terms of expected quantity of contributions. *Web 2.0 tools* are recommended if more than one tool is offered for carrying out an e-tivity. Characteristics included in the representation include possible categories of contributions (updates, comments, tagging, sharing), appropriateness for VARK learning styles and similar. *Collaborators* are recommended in case an e-tivity is defined as group-based. Potential collaborators are those students who participate in the e-tivity and not have been grouped yet. Potential collaborators are at the same time the users (students) and are represented with a set of characteristics specified in the previous subsection. *Advice* is used to encourage students and groups regarding four different aspects of active participation. Each piece of the advice is accompanied with the appropriate symbol indicating whether the recommendation is based on (un)satisfactory activity level or it recommends encouraging participation of collaborators. The text includes explanation that contains variable parts (e.g. due dates, names of active collaborators) and recommended action (Table II). Besides, students can see their activity level and activity levels for their collaborators and other groups.

D. Recommendation techniques

Recommendation techniques used are adapted to the educational domain and to the chosen set of items. Based on the data about items and/or target users, offered e-tivities, Web 2.0 tools and collaborators are ranked and appropriate advice is displayed [28].

Optional e-tivities and *collaborators recommendations* are determined using content-based (attribute-based) recommendations during which student's characteristics and matching characteristics of e-tivity or potential collaborator are compared (Fig. 2) and similarity measure is calculated. So for example, an optional e-tivity that has characteristics the most similar the target student's characteristics will be at the top of the list of recommended e-tivities. *Web 2.0 tools recommendations* are determined using hybrid approach that combines content-based (attribute-based) recommendations and collaborative filtering. For *providing advice*, knowledge-based (constraint-based) recommendation technique with "if...then..." rules is used.

IV. EXPERIENCES IN USING ELARS FOR E-COURSE "HYPERMEDIA IN EDUCATION"

ELARS was developed to foster achievement of learning outcomes and increase the level of user's motivation and participation during moderated e-courses. On completion of each course, evaluation is carried out to determine the extent to which students are satisfied with the course and with the ELARS system. Students' feedback and teachers' experiences are used to improve the course and the system. The course "Hypermedia in Education" was one of the first e-courses enhanced by the ELARS system. It is a blended learning course designed for students in the graduate program in Computer Science major at the Department of Informatics, University of Rijeka, Croatia.

TABLE II.   ADVICE EXAMPLES FOR DIFFERENT ASPECTS OF ACTIVE PARTICIPATION

Aspect	Symbol	Text
Participation with contributions		You did not contribute to the e-tivity [*e-tivity name*]. Participation in e-tivities is necessary for the achievement of learning outcomes and is therefore recommended that you try to be more active in the following e-tivities.
Continuous participation		The activity level of your group is greater than for the previous interval. Keep it up with the continuous participation in order to achieve a successful result in this e-tivity.
Participation with various categories of contributions		By now you've contributed with all expected categories: [*categories*]. If you continue in this way your results will be even better!
Encouraging collaborators to participate		There are members of yours group whose final activity level is not satisfactory. These are: [*not_active_collaborators*]. In order to achieve is better result as a group, it is recommended that in the following e-tivities your try to encourage your collaborators to engage to a greater extent.

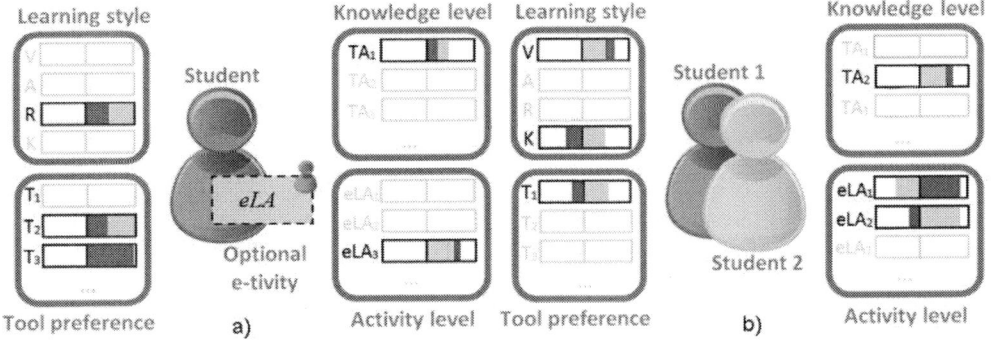

Figure 2.   Comparison of characteristics for: a) student and optional e-tivity, b) two students

A. E-learning environment

The course "Hypermedia in Education" is managed by two teachers. Besides ELARS, teachers and students use MudRi LMS and a set of Web 2.0 tools that are not integrated in the LMS, but available online [31]: Blogger, Diigo, Google Drive, MindMeister, SlideShare, Wikispaces, and YouTube. Using the LMS, teachers prepare learning materials, conduct online assessments and communicate with students. They also perform preparatory tasks in Web 2.0 tools. In the ELARS, teachers define course activities and their workflow, customize recommendation rules, and monitor students' activity levels after a certain e-tivity starts. Students, on the other hand, study learning content, read instructions, solve online tests, and communicate with teachers in the LMS. Students carry out planned e-tivities using Web 2.0 tools. They use ELARS before e-tivities to select among recommended items and during e-tivities to read advice.

B. Course content and learning design

The overall course objective is that students acquire fundamental theoretical knowledge about e-learning and the usage of hypermedia learning materials. The students are trained to implement information and communications technologies (ICT) in traditional (f2f) education as well as to use contemporary e-learning approaches. The course is organized in the following modules: *ICT and hypermedia in education, E-learning and distance learning, Web 2.0 and collaborative e-tivities,* and *Final test*. To enable students to achieve learning outcomes of the course, the first three aforementioned learning modules contain learning materials, self-assessment tests, and e-tivities. During the module *Final test*, one testing activity in the classroom is planned. Students are supposed to solve an online test with multiple-choice questions in the LMS. Initially, the course learning design included the following personalized (collaborative) e-tivities [31]:

1. *Seminar 1* – students are expected to write analysis of potential use of ICT in Education (individually)

2. *Mind mapping* – students are expected to draw mind map which shows the key concepts of the assigned topic (groups with 4 or 5 members)

3. *Wiki* – students are expected to create wiki document with analysis of selected distance learning courses (groups with 4 or 5 members)

4. *Optional e-tivity 1* – students are expected to summarize subject matter in the middle of the course (groups with 4 or 5 members)

5. *Seminar 2* – students are expected to write a seminar with description of assigned Web 2.0 tool and identify its potential use in education and create and publish presentation (in pairs)

6. *Optional e-tivity 2* – students are expected to summarize subject matter at the end of the course (individually)

After the first year of the execution of the course, learning design was slightly changed. Based on teachers' observations and feedback received from students, e-tivities *Mind mapping* and *Optional e-tivity 1* were left out in order to reduce the total number of e-tivities. In addition, students suggested that user interface of the ELARS system should be changed since they were not able to browse through recommendations on different hand-held devices like tablets and smartphones, but just on laptops and desktop computers. Therefore, user interface was redesigned using Bootstrap framework for developing responsive Web applications.

C. Evaluation method and participants

The purpose of presented research was to determine is there any significant difference in students' course results and attitudes after the changes in the course learning design. In addition, the aim was to gather additional experiences and comments from the students that will encourage further improvements of the course or the ELARS system. The first group of participants consisted of students who participated in the course during the academic year 2013/2014 (N=21). Those students participated in six e-tivities. The second group consisted of students who participated in the course after the changes in the learning design, during the academic years 2014/2015 and 2016/2017 (N=26). Those students participated in four e-tivities and were expected to engage to a greater extent and during longer period of time than students from the first group.

A comparison of the course results for students from the first and the second group was carried out. Points from final test were compared using the t-test for parametric independent samples. Final course points were compared using the Mann-Whitney U test for nonparametric independent samples. In both cases, D'Agostino-Pearson test was used prior comparison to accept or reject normality of data. Furthermore, students' attitudes towards applied learning model and the use of Web 2.0 tools and the ELARS system for e-tivities were tested using a questionnaire. Students expressed their attitudes towards statements shown in Table III using the Likert scale (1 – Strongly Disagree, 5 – Strongly Agree). A couple of open-ended questions for students' comments and suggestions regarding the course and the ELARS system were also included in the questionnaire.

D. Results and discussion

Comparison of students' academic achievements (points) showed that there is no statistically significant differences in the results on final test (p=0,12) nor in the final course results (p=0,91). In the first group, average result on final test was 84% and final course result was 82% while in the second group, average result on final test was 85% and final course result was 86%. These numbers indicate that students from both groups achieved course learning outcomes (regardless the changes in the learning design) and that they were motivated to acquire knowledge through planned e-tivities.

Table III shows average results with standard deviations for the most important statements from the questionnaire. Values are approximately the same, which suggests that performed changes in the course learning design did not affect students' attitudes. Students from both groups have a positive attitude towards the implemented learning model. They consider it effective for their learning, especially regarding freedom to choose optional e-tivities, collaborators and tools for e-tivities. These results were expected since different students have different needs and preferences, which imply the necessity for flexibility in the learning design.

Students' attitudes about the usefulness of Web 2.0 tools were positive, which is important, considering that all e-tivities were performed using Web 2.0 tools. In general, students did not have problems with using tools, but it should be noted that students who are not studying computer science could have difficulties and require additional instructions. Students also expressed affirmative attitude regarding usefulness of the ELARS system for e-tivities. The most of them believe that the system positively influenced on their motivation and engagement level. It was observed that some students are

well aware of the benefits of personalization while others are not. In their comments, some students stated that they would prefer to receive more recommendations of optional e-tivities, which indicates that they perceive the advantages of personalization. Recommendations of optional e-tivities include a high level of personalization because the task of each offered e-tivity is tailored to a particular learning style. For example, student can revise subject matter at the end of the course by drawing a mind map (visual LS) or writing a summary (read/write LS).

When asked about the criterion for generating recommendations, the most of the students were neutral regarding the fact that items are recommended based on criterion defined by the teacher. It is not surprising that students would like to receive recommendations based only on their preferences. This is especially the case with collaborators recommendation when most of the students know each other well and have no doubt about who they want to collaborate with. In the future, implementation of new recommendation rules that will include student's preferences towards their colleagues should be considered. Additionally, is necessary to improve the presentation of recommendations, to make the used recommendation criterion more clear. Some students stated that it would be more convenient if the system would be integrated within the LMS. This suggestion will be also considered during the further development of the system.

V. CONCLUSIONS

Implementation of educational recommender system in the e-learning should presume ongoing evaluation regarding system's effectiveness, in term of determining the influence of the system on students' academic achievements, as well as students' satisfaction with the system and received recommendations. Evaluation results for the course "Hypermedia in Education" showed that e-tivities included in the learning design enable students to gain knowledge through collaboration with their colleagues. Despite minor changes, the students from both groups achieved high results and expressed positive attitudes regarding the e-course and the ELARS system.

Several years of experience on this e-course showed that, although the ELARS support moderated e-learning, to some extent non-moderated or self-directed e-learning is supported as well. By reading recommendations, students received potentially useful information, applicable outside the context of e-tivities included in this

TABLE III.   QUESTIONNAIRE RESULTS

No.	Questionnaire statements	1st group (N=18)		2nd group (N=21)	
		AVG	SD	AVG	SD
1	You are satisfied with realized e-tivities and communication within the e-course.	4,30	0,49	4,38	0,67
2	You find positive the freedom to choose collaborators and tools for e-tivities within the course.	4,70	0,47	4,76	0,44
3	You consider Web 2.0 tools useful for realization of e-tivities.	4,60	0,49	4,47	0,66
4	You consider ELARS recommender system useful for e-tivities.	3,76	0,93	3,64	0,90
5	You are satisfied with recommendations received in ELARS system.	3,61	0,95	3,67	0,86
6	You believe that the use of ELARS recommender system positively affected the level of your motivation and engagement during e-tivities.	3,44	0,86	3,52	0,98

e-course. This refers to colleagues they would like to collaborate with in the future and on Web 2.0 tools or e-tivities and they found useful (for example, for revision of subject matter).

Students' comments indicate that the future work should focus on improvements of collaborators recommendations. These recommendations are perceived as least useful, so implementation of new algorithms that will be used for dividing students to groups based on their characteristics and the criteria defined by the teacher is planned. Thus, the possibilities for personalization of e-tivities in the ELARS systems will be extended by offering recommendations to teachers. After receiving the recommendation, the teacher will have the opportunity to accept formed groups or to revise the grouping criterion and restart the process. Further ELARS development the will also include changes in the user interface in order to enable new functionalities and support teachers while forming groups with the help of the system.

ACKNOWLEDGMENT

The research has been conducted under the project "E-learning Recommender System" (reference number 13.13.1.3.05) supported by University of Rijeka (Croatia).

REFERENCES

[1] A. Klašnja-Milićević, B. Vesin, M. Ivanović, and Z. Budimac, "E-Learning personalization based on hybrid recommendation strategy and learning style identification," *Comput. Educ.*, vol. 56, no. 3, pp. 885–899, Apr. 2011.

[2] G. Adomavicius and A. Tuzhilin, "Context-aware recommender systems," in *Recommender Systems Handbook: A Complete Guide for Research Scientists and Practitioners*, Springer, 2010.

[3] S. Berkovsky and J. Freyne, "Web Personalization and Recommender Systems," in *Proceedings of the 21th ACM SIGKDD International Conference on Knowledge Discovery and Data Mining - KDD '15*, 2015, pp. 2307–2308.

[4] M. K. Khribi, M. Jemni, and O. Nasraoui, "Recommendation Systems for Personalized Technology-Enhanced Learning," in *Ubiquitous Learning Environments and Technologies*, R. H. Kinshuk, Ed. Springer Berlin Heidelberg, 2015, pp. 159–180.

[5] H. Drachsler, K. Verbert, O. C. Santos, and N. Manouselis, "Panorama of recommender systems to support learning," in *Recommender Systems Handbook*, 2015, pp. 421–451.

[6] G. Adomavicius and A. Tuzhilin, "Toward the next generation of recommender systems: a survey of the state-of-the-art and possible extensions," *IEEE Trans. Knowl. Data Eng.*, vol. 17, no. 6, pp. 734–749, 2005.

[7] N. Manouselis, H. Drachsler, R. Vuorikari, H. Hummel, and R. Koper, "Recommender Systems in Technology Enhanced Learning," in *Recommender Systems Handbook*, Boston, MA: Springer US, 2011, pp. 387–415.

[8] P. Anderson, "What is Web 2.0? Ideas, technologies and implications for education," *Technology*, vol. 60, no. 1, pp. 1–64, 2007.

[9] S. Downes, "E-learning 2.0," *Int. Rev. Res. Open Distrib. Learn.*, vol. 6, no. 2, 2005.

[10] G. Salmon, *E-tivities: the key to active online learning*. Psychology Press, 2002.

[11] M. Bieliková, M. Šimko, and M. Barla, "Personalized web-based learning 2.0," in *Proc. of 8th Int. Conf. on Emerging eLearning Technologies and Applications*, 2010, pp. 5–10.

[12] "ELARS Home page," *(in Croatian)*, 2015. [Online]. Available: http://elars.uniri.hr/elars. [Accessed: 01-Feb-2017].

[13] J. Mezak, N. Hoic-Bozic, and M. Holenko Dlab, "Personalization of e-tivities using Web 2.0 tools and ELARS (E-learning Activities Recommender System)," in *2015 38th International Convention on Information and Communication Technology, Electronics and Microelectronics (MIPRO)*, 2015, pp. 669–673.

[14] M.-I. Dascalu, C.-N. Bodea, M. N. Mihailescu, E. A. Tanase, and P. Ordoñez de Pablos, "Educational recommender systems and their application in lifelong learning," *Behav. Inf. Technol.*, vol. 35, no. 4, pp. 290–297, Apr. 2016.

[15] O. Bourkoukou, E. El Bachari, and M. El Adnani, "A Personalized E-Learning Based on Recommender System," *Int. J. Learn. Teach.*, vol. 2, no. 2, pp. 99–103, 2016.

[16] A. Zapata, V. H. V. Menéndez, M. E. M. Prieto, and C. Romero, "A framework for recommendation in learning object repositories: An example of application in civil engineering," *Adv. Eng. Softw.*, vol. 56, pp. 1–14, 2013.

[17] M. Sicilia and E. García-Barriocanal, "Exploring user-based recommender results in large learning object repositories: the case of MERLOT," *Procedia Comput. Sci.*, vol. 1, no. 2, pp. 2859–2864, 2010.

[18] G. Durand, N. Belacel, and F. Laplante, "Graph theory based model for learning path recommendation," *Inf. Sci. (Ny).*, vol. 251, pp. 10–21, 2013.

[19] T.-C. Hsieh, M.-C. Lee, and C.-Y. Su, "Designing and implementing a personalized remedial learning system for enhancing the programming learning," *Educ. Technol. Soc.*, vol. 16, no. 4, pp. 32–46, 2013.

[20] G. Koutrika, B. Bercovitz, F. Kaliszan, H. Liou, and H. Garcia-Molina, "CourseRank : A Closed-Community Social System Through the Magnifying Glass," in *3rd Int'l AAAI Conference on Weblogs and Social Media (ICWSM 2009)*, 2009, pp. 98–105.

[21] M. M. El-Bishouty, T.-W. Chang, S. Graf, Kinshuk, and N.-S. Chen, "Smart e-course recommender based on learning styles," *J. Comput. Educ.*, vol. 1, no. 1, pp. 99–111, Mar. 2014.

[22] S. Sheth, N. Arora, C. Murphy, and G. Kaiser, "weHelp : A Reference Architecture for Social Recommender Systems," *Architecture*, pp. 46–47, 2010.

[23] O. C. Santos and J. G. Boticario, "Modeling recommendations for the educational domain," *Procedia Comput. Sci.*, vol. 1, no. 2, pp. 2793–2800, 2010.

[24] S. Nowakowski, I. Ognjanović, and M. Grandbastien, "Two recommending strategies to enhance online presence in personal learning environments," *Syst. Technol. Enhanc. Learn.*, pp. 227–249, 2014.

[25] Y. Zheng and Y. Yano, "A framework of context-awareness support for peer recommendation in the e-learning context," *Br. J. Educ. Technol.*, vol. 38, no. 2, pp. 197–210, 2007.

[26] H. Drachsler *et al.*, "ReMashed – Recommendations for Mash-Up Personal Learning Environments," *Work*, vol. 5794, pp. 788–793, 2009.

[27] M. K. Khribi, M. Jemni, and O. Nasraoui, "Automatic Recommendations for E-Learning Personalization Based on Web Usage Mining Techniques and Information Retrieval," *Educ. Technol. Soc.*, vol. 12, no. 4, pp. 30–42, 2009.

[28] M. Holenko Dlab and N. Hoic-Bozic, "Increasing students' academic results in e-course using educational recommendation strategy," in *Proceedings of the 17th International Conference on Computer Systems and Technologies 2016 - CompSysTech '16*, 2016, pp. 391–398.

[29] N. D. Fleming, "I'm different; not dumb. Modes of presentation (VARK) in the tertiary classroom," in *Research and Development in Higher Education, Proceedings of the Annual Conference of the Higher Education and Research Development Society of Australasi*, 1995, pp. 308–313.

[30] N. D. Fleming, "VARK - The Younger Version," 2011. [Online]. Available: http://www.vark-learn.com/english/page.asp?p=younger. [Accessed: 01-Feb-2017].

[31] N. Hoic-Bozic, M. Holenko Dlab, and V. Mornar, "Recommender System and Web 2.0 Tools to Enhance a Blended Learning Model," *IEEE Trans. Educ.*, vol. 59, no. 1, pp. 39–44, Feb. 2016.

Influence of Accuracy of Simulations to the Physics Education

R. Repnik[*],[**], G. Nemec[*] and M. Krašna[*],[***]

[*] Faculty on Natural Sciences and Mathematics, University of Maribor, Maribor, Slovenia
[**] Association for Technical Culture of Slovenia, Ljubljana, Slovenia
[***] Faculty on Arts, University of Maribor, Maribor, Slovenia

robert.repnik@um.si

Abstract - Regular discussion in the LLL (Life Long Learning) courses for the teachers in primary and secondary education gives us insight into their oppinion. We discovered that teachers of primary and secondary education in our country believe that the highest retention of knowledge and understanding of mechanics can be achieved in the real world environment (in lab and outdoor) observing the natural physics phenomena and repeating the experiments that led to the physics laws. We know that this is just partially true. It is our belief that simulations using as supplement to real laboratory experiments increase students' understanding of mechanics. Physical processes are often beyond the capability of humans to observe them in schools. Teaching students we have seen that adequate understanding using lab experiments is difficult even in the simplest mechanic phenomena (ball roling on the slope). Students were unable to comprehand the full concept of the physics phenomena from the experiment unless we teach them theory first. In this article, we explain how we used simulations that allow students to adjust the experiments to their comprehantion abilities. We compared the accuracy of three different freeware simulations to the highly accurate proprietary one. We have also analyzed the the observed simulation software to be customized to the accurate transformation of 3D world experiment to the 2D simulated presentation (e.g. the sphere and the cylinder cross-section look equal in 2D screen but the physics background is different).

I. INTRODUCTION

The use of computers in physics has brought many new opportunities for demonstration of physical phenomena. Computers are able to present static objects and dynamic phenomena in different ways depending on the required complexity. Presentation can be from simple 2D images on the transparencies (PPT), to the videos or simulations of 3D interactive models. Computers can control the experiments and/or acquire measurements. When the experiment of specific physical phenomena can not be done in the lab or we wish to enhance its clarity we can use computer's simulations. Simulations allow us to change parameters and observe their interactive physics response. On the internet, there are numerous simulations, which are ready for use in the classroom [1]. Each of them is ready to treat specific physical phenomena, or their part. However, the described simulations can not be used to treat all physical phenomena [2] [3]. In the teaching of physics the observation of the real physical phenomena ist of great importance, but in some

cases this is not possible (it´s time consuming, not repeatable etc.). In such situations the physics teacher prepare school experiments, where he or she demonstrate the physics phenomena to the students in order to analyze the influences of specific factors on the experiment dynamic. The experiment should not be substituted but supplemented with ICT tools, e.g. simulations [4]. In frame of Didactics of Physics we teach the university students that by following the strategy of introducing the simulation into the theaching of physics they need first to execute the real experiment and secondly they have to show the same physics experiment in the simulation environment and compare the results. The students can perhaps observe the differences which arising ither from measurement errors or inaccuracy of simulations. If the difference is too great, the trust of students in simulation can not be achieved. If the results match, the trust in simulation is present and we can use the simulation in new physics situations that we can not access with real experiment. The accuracy of the simulation is obviously crucial.

In this paper we have limit ourselves to the free accessible simulation software: Algodoo [5], Step [6] and Physon [7]. They all allow a virtual space where we are free to move and insert arbitrary objects. Movement and interaction(s) between objects obey the physical laws. Such software is called **Physics sandbox**. **Physical engine** of the simulations assure accurate objects' interactions based on mathematical modeled of physical laws. The virtual space where the objects are inserted is named the **interactive simulation environment**. A common feature of all simulation environments is a transformation between 3D space into the 2D projection (on screen). Simulation environments are build by inserting objects (blocks, circuits, any polygonal wire, springs, and other building blocks), which can be combined to the complex structures, such as a trolley with the engine.

The software simulations can be used in teaching physics on different occasions and educational topics, both in elementary school (primary school), [8] and high school (secondary school). [9] The potential user can use our experience of the three public available software simulation and use them in the classroom.

A. Application of the computer in the physics education

The experimental observation of the phenomenon can not be replaced by any other method, because it enables the control of conditions for the observation of the physical phenomenon [4] [10] [11]. Various reasons lead to the fact that the experimental observations in practice are not always possible. Some of the reasons are listed below [2] [12]:

- Research subject is too big or too small,
- The observed process is too fast for human,
- Financial expenses are too high,
- Experiments are dangerous,
- The school experimental environment is typically limited to specific place or area properties, (we cannot test the experiment on other planets/moons with different gravitational constant).

B. Strategy of the simulation

Three main factors occurred in the treatment of the concept of strategy of the simulation: **real system**, **model**, and **simulation**. The **real system** or part of the real world is what we want to present in a graphic way. The **model** is a theoretical description of the real system, which must be precise enough to explain the crucial behavior of the real system. The model can be used to test the theory too. **Modeling** is the process of adequate describing the relationship between the model and the real system. **Simulation** can be defined as an abstract model, which is implemented as a computer program. Simulation is a demonstrational computer model in which the phenomena dynamics is calculated and presented in time steps (the certain time increments) [10].

One of the main advantages of using simulations in teaching of physics is that students can freely change the parameters of the simulation. Setting the value of the simulation's parameters the entire course of the simulation changes and that enables verification of the various hypotheses. The simulation strategy in education is particularly suitable for analysing the functional interaction(s) between the parts of the complex system. In the physics teaching the experiment is indispensable, but is does not allow us to control all the environmental or inter-object influences. Therefore, it is wise to add the simulations to the experiment for better students' understanding of underlaying concepts of physics phenomena. Simulations in the education allows the teacher can overcome some real world constraints and explain the behaviour of physics phenomena in the articifialy changed environments. The simulation is not suitable for everything therefore should be logically included to the educational processes. The teacher should be trained in the use of simulation strategy and should encourage the discussion during the use of simulation. The students could use the simulations for understanding the background of the physics phenomena and understand why some differences between simulation and real world phenomena can occur [2] [10].

C. Possibilities of the application of simultation in the education

We estimated that the teaching methods where simulation environments can be used are as following [4] [10]:

- Demonstration,
- Method of written and graphic work,
- Practical work – laboratory work.

Simulation can be used in all phases of education, from introductory demonstrations, to introducing the new concepts at new content teaching, and to verify acquired knowledge.

D. Execution of the experiment is followed by the preparation of the simultation

We chose to show a simple experiment: the calibration of the spring [13]. Accessories that are needed in the experiment: a spring, a ruler, and a variety of weights. Students hung a roler next spring (for measuring the distance). With the addition of the weights (10 g) to the spring, they record the extension of the spring. Acquired data are later used to draw a graph and the graph can be used to see the relationship between weight (or even better, the force) and elongation for any given weight (Figure 1). Expected is linear dependency, if we not exceed the limit elastic values. The result of the calibration is the determination of the spring constant in N/m.

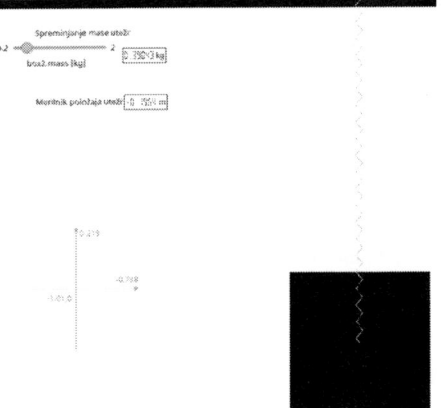

Figure 1: Simple test for calibration of the spring in Step.

II. OVERWIEV OF INTERACTIVE SIMULATION ENVIRONMENTS

In this section, we present the details of the specific simulation software and Simulation environment. We also introduce the basic characteristics of the physical engine.

A. Algodoo

Simulation environment allows adding the basic objects to the simulated world: blocks, rings, ropes, and springs. From individual basic objects, we built complex (composite) structures, e.g. a trolley with the engine. Various physical properties can be modifyed for any individual building part; we also can change the properties of the environment. When the simulation is ready to run, we can observe a static or dynamic behavior of each object, influenced by other objects in its vicinity and surrounding.

Algodoo simulate mechanical systems defined by Newton's equations. This does not begin with a form of Newton's laws, from which to get associated differential equations that could be used for discreet method of calculation. The equations are written in the Lagrange formulation of mechanics. Lagrange's formulation uses a discrete time and position format. Lagrange's formulae enables us to use the differences in intervals to walk through the system of equations. This approach has been used to describe particles' system, a system of rigid bodies with the extremities and motors, collisions, contacts, dry friction, viscous incompressible fluids, and elastic-plastic materials. To stabilize the extreme edge, control, and physical engine, the calculations used framework SPOOK [14]. At each timeframe, simulation environment calculates a large system of equations, which is comparable in size to the number of extreme edge and contacts of bodies.

The constraints of the software simultation are the following:

- The simulation runs at the time interval of 1/60 sec (60 Hz). It is not possible to model the events that are happening in a shorter time interval. It is therefore not advisable to generate simulations for the bodies that are less than a centimeter wide.

- When the simulation is running, it calculates a large system of equations using a numerical method. In order to ensure smooth interactive simulation events the fast approximations are used. Unfortunately, this can lead to the differences between simulated results and the physics theory.

- We must be aware that the simulation software is complex and may contain bugs or faults in the program.

- Algodoo authors' claim that they are trying to make the simulation software to conduct physical simulations at a high level, but legally there is no assurance for this.

B. Step

Simulation software is available under GNU GPLv2 open access license [15]. The design of the simulations was done in the following steps: we add the body and the force of gravity to the simulated model. By pressing the button "Simulate" the simulation started and bodies started to move according to the calculated physical laws. Bodies and forces attributes can be changed even during the simulation and they are immediately used in the calculations.

Simulation software Step uses software library StepCore for physical engine. This engine is used inside the simulation environment for all simulation calculations (even the compex one) but can also be used in our own software for which necessary programming skills are required. The design of the library, which can be disseminated and adjusted, has enabled us to implement precise simulations. Capabilities of the physical engine and the simulation environment are:

- Particles, spring damping, various forces, gravitational force, gravitational and Coulomb force can be added to the simulated model.
- Rigid bodies can be added.
- Collision of the bodies can be detected.
- Soft bodies, which can deform and consist of particles - connected with springs, can be created.
- Molecular dynamics is based on the Lennard-Jones potential. This allows simulations of gas and liquid, condensation and evaporation, and the calculation of macroscopic liquids.
- Measurement errors (i.e. 1.2 ± 0.3) for any property can be specified. Calculations of all values are done using statistical formulas.
- There are many differential solvers which user can choose. Solver component in the simulation environment allows stepwise calculation. Most of them are based on GSL-library [16].
- Errors of calculation component solver are added to those made by the user.

C. Physion

An online community maintains the project. Therefore, a number of examples, tutorials and videos can be found in the forum or in the YouTube. Physionet offers the possibility of inserting program code in JavaScript.

Physionet uses many open-source technologies and is based on a physical engine Box2D library [17]. Physical engines that are designed primarily for use in computer games are focus on fast and fluid calculations. That means heavy use of fast approximations that can generate significant differences between physics theory and real world. Our hypotesis is that these approximations will not significantly influence the results for the school's educational use. Based on the capabilities of the library physical engine we discovered that we can simulate in Physion similar simulation as in the previous two simulation environments.

III. COMPARISON OF THE SIMULATIONS

We know that the tests we have performed (more our tests are available in [18]) are not suitable for a general review but we are trying to make at least elementary impartial review, presented in the following table (Table 1).

Table 1 Conclusion and comparison of experiencs in the moddeling and simulations' use.

	Algodoo	Step	Physion
Accurate modeling of the objects	grid	attributes of the objects	grid and attributes of the objests
Editing and modifying of the objects (size, position, rotation,...)	with the use of software tools	yes	yes
Composition of the complex objects	yes	no	yes
Size of history of undo.	flawless	some actions are not recorded in the history	flawless

		and cannot be repealed.	
Software stability	yes	often halts	yes
Addition of the objects during the simulations' calculations	yes	no	yes

A. Testing the quantitative credibility of the simulation environments

In high school physics, a significant increase of the need for quantitative use of simulations emerges. It is therefore important that the measured results from the experiments and calculated results from the simulations are comparable. For the quantitative credibility test of simulation environments, we prepared the same simulation in all three simulation environments and compared the results with the model that we have created with the reference software Berkeley Madonna.

For each simulation environment, we prepare the equal model and start the simulation with the same parameters. We have used a damped oscillation model for the test. In the simulations, we allow the plotting of graphs (position / time). Calculated data from each simulation were compared with the data we have obtained by use of the reference software Berkeley Madonna.

B. Physics background of damped oscillation

The oscillation is damped, when another body (e.g. surrounding media), which inhibits dampening (presence of the frictional force, proportional to the velocity), influences the body that is swinging [19]. The block with mass m were fixed with a spring coefficient k and the own damping coefficient b (Figure 2). We observe the damped oscillations due to springs own damping coefficient. Potential energy (consequently the kinetic energy) of the spring oscillation transmites to the internal energy of the spring.

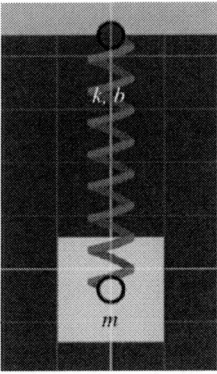

Figure 2: The model used for the dumped oscillation.

Due to damping properties of the spring the oscillating system was damped with a force $F_d = -bv$, where the b is the damping coefficient and the v is the velocity of movement of the block. The block were operated by a force of the spring $F_{vz} = -kx$ (the x denotes the extension of the spring) and the force of gravity F_g. We ommit the influence od the F_g on the movement because it had no effect on the oscillations (only define the stable-equilibrium position). For all forces acting on the block, we wrote:

$$ma = -bv - kx. \tag{1}$$

We rewrite the (1) by replacing the velocity v with dx/dt and acceleration a with d^2x/dt^2 and get

$$m\frac{d^2x}{dt^2} = -b\frac{dx}{dt} - kx. \tag{2}$$

The solution of the equation is the following

$$x(t) = x_m e^{-bt/2m} \cos(\omega't + \varphi), \tag{3}$$

where x_m is the starting amplitude and ω' angular frequency of damped oscillation. The angular frequency is defined with the following equation:

$$\omega' = \sqrt{\frac{k}{m} - \frac{b^2}{4m^2}}. \tag{4}$$

Equation (3) implies that because of the dampening coefficient b the amplitude x_m exponentialy decrease with time t. The angular frequency ω' also decrease because of (4). The physical unit for the b is [Ns/m] or [kg/s].

C. Preparation of the simulations

We can compare the data if we create the similar conditions for all three models and simulations. All three simulations work on the numeric step-by-step calculations. For comparison, we set the time step 0.02 (or 50 Hz or 50 calculations per second) to calculate the values. With this setup, we obtained the same number of data from each simulations. In the design of the model we use the block with the mass $m = 1$ kg, the spring coefficient of $k = 35$ N / m and springs own damping coefficient $b = 0.5$ kg / s. We tried to standardize the maximum amplitude x_m as much as possible. The duration of the simulation was 10 seconds.

D. Preparation of the simulations – Algodoo

We prepared the simulation in Algodoo as seen in the Figure 3. We delete the lower surface and added a new one to forms a ceiling. We secure the spring to the block and on the ceiling. The block size was set on 0.4 m on all sides, and its weight to 0.5 kg. Block was placed (hanged) at a distance of 1 m from the ceiling. The spring constant was set on 35 N / m. In Algodoo we had some problems by defining the b directly, so we defined the dumping ratio ς, which we calculated for given data $b=0.5$ Ns/m, $m=0.5$ kg and $k=35$ N/m with [19]:

$$\varsigma = \frac{b}{2\sqrt{mk}}. \tag{5}$$

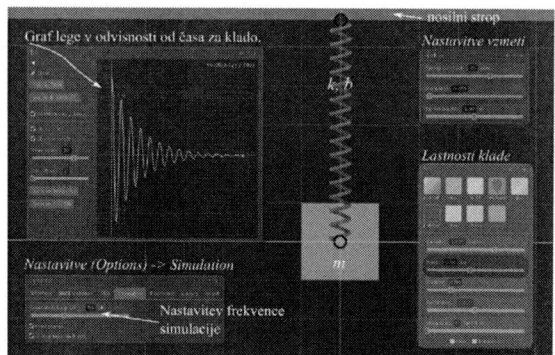

Figure 3: Simulation of dumpening oscillation in Algodoo.

We set simulation to run for 10 seconds. Data were displayed on the graph and there are built-in function to export the data in CSV (comma separated value) text file.

E. Preparation of the simulations – Step

The modeling in the Step started with the position of the block 1 m from the center of the coordinate system – origin (which represent the ceiling) (Figure 4). We set the coordinates for the block (0, 1) m (first number present x and second y coordinate). We secure the block on both ends to the simulation background with the object *Pin*. In the coordinate center (0, 0) m we set the block with size (0.4, 0.4) m and set its weight to the 0.5 kg. We attached the spring to both blocks (the upper one was fixed, the lower one can oscilate). In the attributes of the spring we set *localPosition1/2* on both ends to enshure that spring is connected to the center axes of the blocks. With the object *WeightForce* we add the force of gravity.

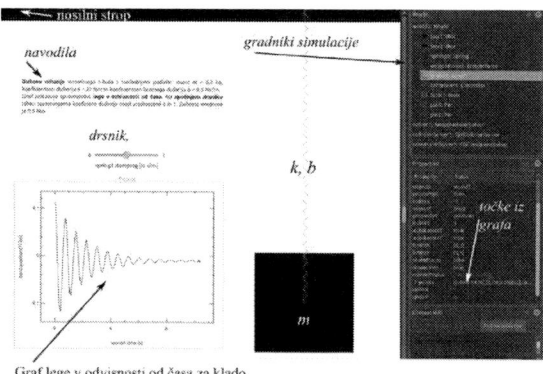

Figure 4: Simulation of dumpening oscillation in Step.

Again, we set the simulation to run for 10 seconds, before we save the calculated data.

F. Preparation of the simulations –Physion

When we opened the software Physion there were a initial new environment with three blocks that have been deleted. First, we create a ceiling (Figure 5), which was made up of the block size (10.0, 0.2) m at a height of 1 m from the coordinate center (0, 0) m. We added the block size (0.5, 0.5) m (the size 0.4 was changed to 0.5 because we could define the mass of the block only by changing the blocks density). We add the mass of the block by setting its

density to 2 (which equals to 0.5 kg). We connect both blocks with the spring for which we set **k** and **b** (equal to models in other simulations). We need to set gravitational force to 9.81 m/s² and frequency of the calculation to 50 Hz.

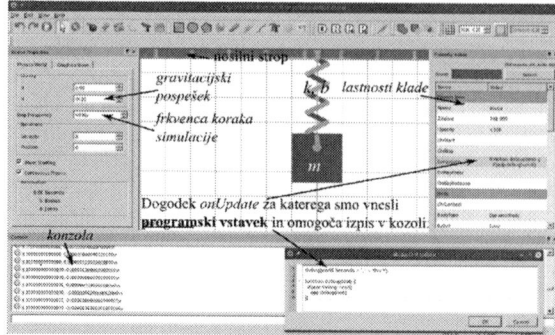

Figure 5: Simulation of dumpening oscillation in Physion.

G. Preparation of the simulations –Berkeley Madonna

Berkeley Madonna is a software package for mathematical modeling [20] developed by the University of Califorina at Berkeley. With this software, it is possible to numerically solve the differential equation [20]. The software is used in many fields of science for research and learning. Calculations ware performed with Runge-Kutta 4th order (RK4). In frame of measurement errors, the Berkeley Madonna simulation results perfectly match with experimental results. Consequently, we used this software as the reference.

To create a mathematical model, we need an equation in differential form. We transformed (1) for dampening oscillation by seting instead of *a* the dv/dt, and thus we get

$$\frac{dv}{dt} = -\frac{bv}{m} - \frac{kx}{m}. \qquad (6)$$

Equation (6) was therefore in a form ready for use in the software. In (6) are two variables *v* and *x*, therefore, we have yet to define $v = dx/dt$. Into the program we inserted the settings; the equations; and the initial value (see Table 2). For the initial values, we used the same values as were used in the other three simulations. In the simulation we used the calculation method RK4.

Table 2. Data used for modeling in Berkeley Madonna.

Initial values	
$m = 0{,}5$ kg; $k = 35$ N/m	
Description of the lines	Model – damped oscillation
	$b = 0{,}05$
1 Method for calculations	1 METHOD RK4
2 start time	2 STARTTIME = 0
3 end time	3 STOPTIME = 10,7
4 time step	4 DT = 0.01
5 equation for calculation	5 d/dt (x) = v
6 equation for calculation	6 d/dt (v) = - (b*v)/m - (k/m)*x
7 initial x value	7 init x =2.07
8 initial v value	8 init v = 0

9 constant b	9 b = 0,05
10 constant k	10 k = 35
11 constant m	11 m = 0.5

After running the simulation for 10 s we get the graph (Figure 6) of displacement (position) and velocity of the oscilating block as function of time. We have saved the data to the CSV file. We used the data for the analysis of the position of the block in time.

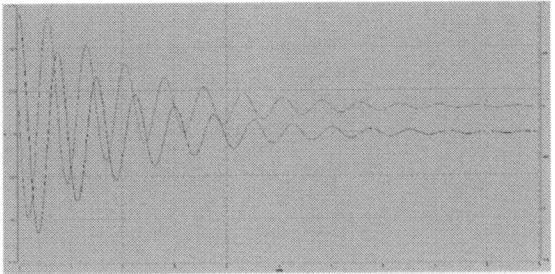

Figure 6: Graphs from Berkeley Madonna: position (black), velocit (red).

IV. COMPARISON OF THE RESULTS OF THE SIMULATIONS

We tested the accuracy of three free accessible simulation environments with the reference simulation. The data of the damped oscillation collected from each simulation environment and in adition the Berkeley Madonna data as reference, were combined in a single graph (Figure 7). From the graph, we conclude that all the simulation environments successfully passed our accuracy test. Simulated results form the damped oscillations models in Algodoo, Step, and Physion and mathematical model in Berkeley Madonna were comparable, since the peaks of the individual cycles were well matched. The matching of results is satisfactory for displacement axis (small differences only for results of Step), but even more for the time axis (excellent matching). All three simulation environments (Algodoo, Step and Physion) prove to be quantitatively credible and consequently adequate for use in classroom. We are aware that this is just one example, but it was planned carefully as possible. This took into account some specific physics and computer knowledge and skills. The results of our credibility test were positive and confirms the appropriateness of the simulation environments.

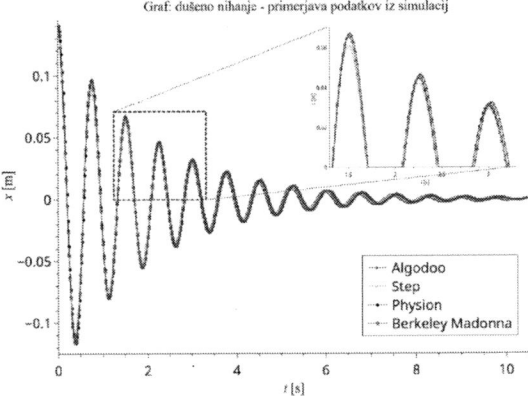

Figure 7: Cumulative graph from all simulations: Algodoo (blue), Step (green), Physion (black), and Berkeley Madonna (red). The overlapping of curves is excellent.

V. CONCLUSION

In support of operating systems, we found that most of desktop operating systems supports Algodoo, because it could be installed on Windows operating systems, Mac OS X and GNU / Linux using Wine program. Step can be installed only by users of GNU / Linux operating system. For Physion, it was possible to install it on Windows and GNU / Linux using Wine software. As we can see the users of GNU / Linux could install all three simulation environments which is very favourable for the free software.

Physical content coverage analyzis showed that the three reviewed simulation environments mostly overlap. It is possible to create models to simulate numerous physical phenomena in mechanics. In addition, the Algodoo enables simulations in content of light, buoyancy and swimming. The Step enables simulations in contents relating to electrical charge. Physion and Algodoo are the most comparable, but in modeling of simulations using Algodoo we can add also the effect of air resistance.

In our analyzis, we find out that the three reviewed public available simulation environments (Algodoo, Step, Physion) provide the necessary software supplement to the experimental work at physics teaching. The quantitative testing of the credibility of the simulation environments, based on the analyzing the almost perfectly overlapping of the time dependency of the displacement of the dumped oscillators, proves the adequacy of them. They cover most of the typical content in primary and secondary schools; they may be also adequate for university level education of physics. In the future, we can expect the development of even more advanced simulation software for teaching physics and for other disciplines in school. The accuracy of the analyzed simulation environments is appropriate, consequently all three simulation environments are adequate for using them in the teaching of physics in elementary and high schools.

ACKNOWLEDGMENT

Acknowledgement: The work was financially supported by the donation of the company nT-BROG d.o.o.

[21] to the Faculty of Natural Sciences and Mathematics to support the interdisciplinary development of didactics of physics.

References

[1] List of computer simulation software, https://en.wikipedia.org/wiki/List_of_computer_simulation_soft ware .

[2] I. Gerlič, Sodobna informacijska tehnologija v izobraževanju, Ljubljana: DZS, 2000.

[3] „Phet interactiv simulations," [Electronic]. Available at: https://phet.colorado.edu/ (18.08.2016).

[4] „Razvoj naravoslovnih kompetenc," [Electronic]. Available at: http://kompetence.uni-mb.si/ (04.08.2016).

[5] Algodoo, „What is it?," [Electronic]. Available at: http://www.algodoo.com/what-is-it/ (11.08.2016).

[6] V. Kuznetsov, „Kde Step," [Electronic]. Available at: https://edu.kde.org/step/ (21.07.2016).

[7] D. Xanthopoulos, „Physion," [Electronic]. Available at: http://physion.net/ (19.07.2016).

[8] Člani predmetne komisije, UČNI načrt. Program osnovna šola. Fizika, Ljubljana: Ministrstvo za šolstvo in šport: Zavod RS za šolstvo, 2011. [Electronic]. Available at: http://www.mizs.gov.si/fileadmin/mizs.gov.si/pageuploads/podr ocje/os/prenovljeni_UN/UN_fizika.pdf (25.08.2016).

[9] G. Planinšič, B. Ruben, I. Kukman, and M. Cvahte, UČNI načrt. Fizika, Gimnazija, Splošna Gimnazija, Ljubaljana: Ministrstvo za šolstvo in šport : Zavod RS za šolstvo, 2008. [Electronic]. Available at: http://eportal.mss.edus.si/msswww/programi2010/programi/med ia/pdf/un_gimnazija/un_fizika_gimn.pdf (25.08.2016)

[10] I. Gerlič, Metodika pouka fizike v osnovni šoli, Maribor: Pedagoška fakulteta, 1991.

[11] B. Vybral, „Physics Experiment in Teaching, Education and Competition",XXIX International Colloquium on the Management of Educational Processes, PT 2, Brno, Czech Republic, pp. 551-560, May 2011

[12] I. Gerlič and V. Udir, Problemski pouk fizike v osnovni šoli, Ljubljana: Zavod RS za šolstvo, 2006.

[13] M. Ambrožič , Planinšič Gorazd, E. Karič, S. Kralj, M. Slavinec and A. Zidanšek, Fizika, narava, življenje, 1. del, Ljubljana: DZS, 2000.

[14] C. Lacoursière, Ghosts and Machines: Regularized Variational Methods for Interactive Simulations of Multibodies with Dry Frictional Contacts, [Electronic]. Available at: http://umu.diva-portal.org/smash/record.jsf?pid=diva2%3A140361&dswid=-7447 (25.08.2016)

[15] F. S. Foundation, „GNU General Public License," [Electronic]. Available at: http://www.gnu.org/licenses/gpl.html (22.07.2017).

[16] „GSL - GNU Scientific Library," [Electronic]. Available at: http://www.gnu.org/software/gsl/ (10.08.2016).

[17] „Box2D - A 2D Physics Engine for Games," [Electronic]. Available at: http://box2d.org/ (10.08.2016).

[18] G. Nemec, "Interaktivna simulacijska okolja Algodoo, Step in Physion pri pouku fizike", diploma, Faculty of Natural Sciences and Mathematics, University of Maribor, 2016.

[19] D. Haliday, R. Resnick and J. Walker, Fundamentals of Physics, 7th edition, Cleveland: John Wiley & Sons, 2005.

[20] B. Madonna, „Berkeley Madonna," [Electronic]. Available at:http://www.berkeleymadonna.com/ (19.07.2016).

[21] Company nT-BROG d.o.o., [Electronic]. Available at: http://www.ntbrog.com/

Competence-oriented model of representation of educational content

O.N. Ivanova, N.S. Silkina
South Ural State University, Chelyabinsk, Russia
onivanova@susu.ru, silkinans@susu.ru

Abstract - The paper is devoted to a competence-oriented model of representation of educational content. The model is based on the requirements to results of educational programs in higher education and illustrates the approach to developing e-learning courses as structured sets of didactical units containing evaluation funds. Didactical units of training material are the content elements of e-learning encyclopedias. Evaluation funds of didactical units allow determining the level of graduates' competence. The multi-variant presentation of the training material helps to build individual learning trajectories for persons with disabilities, for students with different backgrounds and with different individual characteristics of personal and cognitive spheres.

I. INTRODUCTION

There is currently a widespread use of network forms of education when the implementation of educational programs. In some cases, we see even a full transition to network forms. The commonly used representation of e-learning materials is a variety of the e-learning courses including MOOC. Not only academics are engaged in creating e-learning courses; also special departments of industrial companies interest in the growth of their staff; teachers in schools and just individuals share their knowledge with the widest possible range of the interested respondents. Nevertheless, this task still remains the most urgent for the universities whose primary purpose is the provision of educational services.

At the same time, a university is an accredited institution, which has the right to issue a state diploma. It must provide a high-quality level of the e-learning courses. When the full transfer of education to a network form, universities need to solve the problem of matching the learning results to the requirements of regulatory documents.

A modern approach to the representation of learning materials is a competence approach [1-4]. However, at present, there is no single approach to assessing the quality and the compliance of the learning materials with the regulatory requirements, as there are no proper factors and metrics of such assessment.

The aim of this work is to describe a competence-oriented model of representation of educational content based on the compliance of the learning materials with the regulatory requirements of the Russian federal state educational standards.

The work was supported by Act 211 Government of the Russian Federation, contract № 02.A03.21.0011

II. REQUIREMENTS OF FEDERAL STATE EDUCATIONAL STANDARD OF HIGHER EDUCATION TO THE LEVEL OF LEARNING OUTCOMES OF GRADUATES

The founding document of the higher education in the Russian Federation is a federal state educational standard of higher education (FSES HE). The Ministry of Education and Science of Russian Federation defines the enlarged groups of specialties, each of them contains the definite sub-fields. The Ministry calls such sub-fields "directions". The examples of the enlarged groups of specialties are "Mathematics and Mechanics", "Computer and Information Sciences", "Physics and Astronomy", etc. The examples of directions within the enlarged group "Computer and Information Sciences" are "Fundamental Informatics and Computer Sciences", "Mathematics and Computer Sciences", "Mathematical Software and Administration of Information Systems", etc. For each direction, the Ministry approves a separate FSES HE. This standard contains requirements for the learning outcomes of the graduates.

These requirements are a list of competencies that each graduate must achieve. The standard defines three types of competences: the general cultural, the general professional and the professional competencies. There are several groups of the professional competences according to the activities, towards which the educational program is oriented: research activities, design and production activities, management activities, methodical activities, and others. The university choose all or several activities for each educational program and provide only those competences, which belong to the chosen activities. It leads to a significant variety of bachelor and master's programs depending on the profile of the university. For example, the educational program in the pedagogical university differs significantly from the program in the same direction in a classical or a technical university.

Thus, the bachelor educational direction "Computer Science" in South Ural State University (National Research University) [5] presents two bachelor programs: "Calculative Machines, Complexes, Systems and Networks" and "Information and Analytical Support of Management in Social and Economic Systems". The master educational direction "Fundamental Informatics and Information Technologies" includes two programs: "Technology of high load systems" and "Database Technologies".

For each educational program, the university develops the model of a graduate meeting the requirements of FSES, the needs of the regional labor market and its own scientific and educational interests. The model specifies definite competences for each subject in a curriculum.

The teacher decides which learning materials he will use to achieve the given set of the competences for his subject. Therefore, the learning materials of the same subject can vary greatly among teachers. These learning materials form the educational content of the whole educational program.

Thus, the university has the actual problem of verification of compliance of the learning materials developed by a teacher with the model of a graduate.

III. VERIFICATION OF COMPLIANCE OF EDUCATIONAL CONTENT WITH THE REQUIREMENTS OF FSES HE

Traditionally teachers divide the learning materials into several sections and sub-sections (topics). A didactical unit is a minimum indivisible unit of the learning material, e.g., a definition, an example, an exercise. We understand a topic as a nonempty relatively independent set of didactical units combined by semantic logical relations.

For each didactical unit the teacher can prepare the evaluation fund of its achievement. An evaluation fund is a set of questions, assignments, exercises and other tasks for student, which allow checking his knowledge. It can include different types of control tools: quiz questions, practical tasks for independent or group implementation, case studies, essay topics, reports, etc. One control tool can check the achievement of several didactical units; several control tools can check one didactical unit.

According to the interpretations of the concept of a competence [6], any competence can be implemented at various levels: knowledge, understanding, and skills. To determine the minimum required level of achievement of the competence the university needs first to establish the compliance of the evaluation fund of all subjects with the competences. The teacher must specify such compliance for all didactical units within his course.

If you have a general bank of didactical units, which can overlap or reenter in different courses, the incomplete comprehension by the student of all didactical units will not constitute the incomplete or insufficient mastery of the competencies defined in the model of the graduate. A graduate can master all competencies, even missing some material. Indeed, not all students master all the learning material at the highest level.

The aim of the university is to find the decision of two problems:

1) Define a minimum required level of mastering the competencies by a graduate;

2) Prepare the curriculum that would achieve and exceed this minimum level.

It is impossible to formalize the process of solution of both problems. However, there are the opportunities for the partial automation, at least at the stage of compliance of the existing curriculum with the requirements of FSES HE. We can compare FSES requirements with the entire set of evaluation funds of all courses of the curriculum.

IV. ELEMENTS OF EDUCATION CONTENT REPRESENTATION

In terms of the compliance of the curriculum with the requirements of FSES, we operate the specific concepts: the module of the education content and the component of this module.

The module of the education content is a set of logically interrelated concepts that form a holistic unit of information. The closest concept to the module in terms of a course is a topic.

Each module contains a set of related components. A didactical unit is the closest term to a component. A component represents an item of the module too, but, in contrast to the didactical unit, it may contain a set of elements, equivalent by content but different by the form of representation. For example, a definition may be given in the form of a text, a diagram, an audio or a video file. Different forms of representation of information will allow each student choosing the most convenient method of obtaining the learning material according to his leading psychological type of perception (auditory, visual, kinesthetic), which will undoubtedly have a beneficial effect on the learning outcomes [7].

The frequently used component types are:

- *theory* – a detailed theoretical description of the concept;

- *summary* – a brief theoretical description of the concept, the definition or the formula;

- *example* – an example to illustrate the concept;

- *solution* – an example of the typical task solution;

- *exercise* – an practical assignment for the theoretical material;

- *bibliography* – a bibliography on a given topic.

For each component, the teacher prepares the evaluation fund containing the tasks for checking the competencies.

The most frequently used types of tasks in the evaluation fund are:

- *questions* – a list of open-type questions;

- *test* – a list of closed-type questions;

- *assignment* – a task to check the practical skills and abilities.

It seems appropriate to combine modules into a thematic set according to their content. We call it an e-learning encyclopedia (ELE). A teacher can use the e-learning encyclopedias to develop a new course.

V. MODEL OF REPRESENTATION OF EDUCATIONAL CONTENT

To develop a new course, a teacher needs to specify both a list of modules, and their chronological sequence. An e-learning course (ELC) is a directed graph, whose nodes are modules, and arcs specify the sequence of the study of two connected modules. A teacher chooses

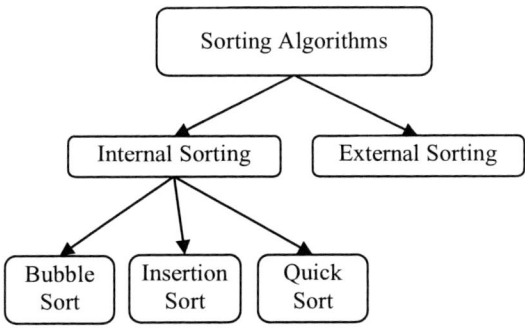

Figure 1. Fragment of ELC "Basics of programming"

modules which he wants to include into his course. It provides a variability of the course content created by different teachers.

Fig. 1 shows a fragment of the ELC "Basics of Programming". Each node in the graph corresponds to a module chosen by a teacher from the e-learning encyclopedia.

The traditional representation of the sequence of learning materials in the form of a tree graph has many disadvantages [9], the main of which, in our opinion, is the inability to build an individual learning path for a student. If a teacher wants his students to study several algorithms of the internal sorting, it is no matter which algorithm they will study first, as well as the order of all other algorithms. Therefore, the e-learning course should provide various trajectories for the module.

Fig. 2 presents a fragment of the graph illustrating the different trajectories of achievement the module C: A – B – C; A – D – C; A – B – E – C. Here modules in a double border mean the mandatory modules required from a student. Modules A and C are mandatory in this figure. A background shows modules, which the student did not pass yet. In fig. 2 they are modules E, C and D.

The comparison of the paths B – C and B – E – C is interesting. Indeed, after the study of module B, a student may transfer directly to the module C, but may first meet the modules B, which will not be mandatory to achieve the module C.

A basic trajectory for his course is for those students who are ready to pass only the minimum amount of topics. Those who want to study the course deep can pass the additional modules. The students will have the ability to choose that trajectory which seems to him the most

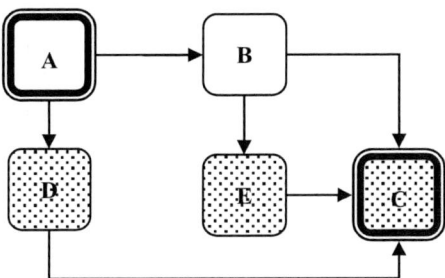

Figure 2. A fragment of the graph
with different trajectories of module C achievement

appropriate. The reasons to miss some modules may be different: the student already knows some of the modules, he is not interested in all of them, the modules are too difficult for him, or he does not have time for the full course.

When forming the graph a teacher can specify the property of necessity of each module. An attribute "Necessity" of a module means that this module is required for the successful completion of the course. Since each module has evaluation tasks, we get the whole evaluation fund for the course.

Additionally, a system can control the property "Available" for each module. It will change the status for the modules after the student passed the income module.

The general scheme of the competence-oriented model of educational content is presented in fig. 3. "M" means "module", "CMN" means "component" and "CMT" means "competence" here. According to the model, the curriculum consists of e-learning courses. An e-learning course is associated with a number of modules. One module can be included in many ELC. The modules can

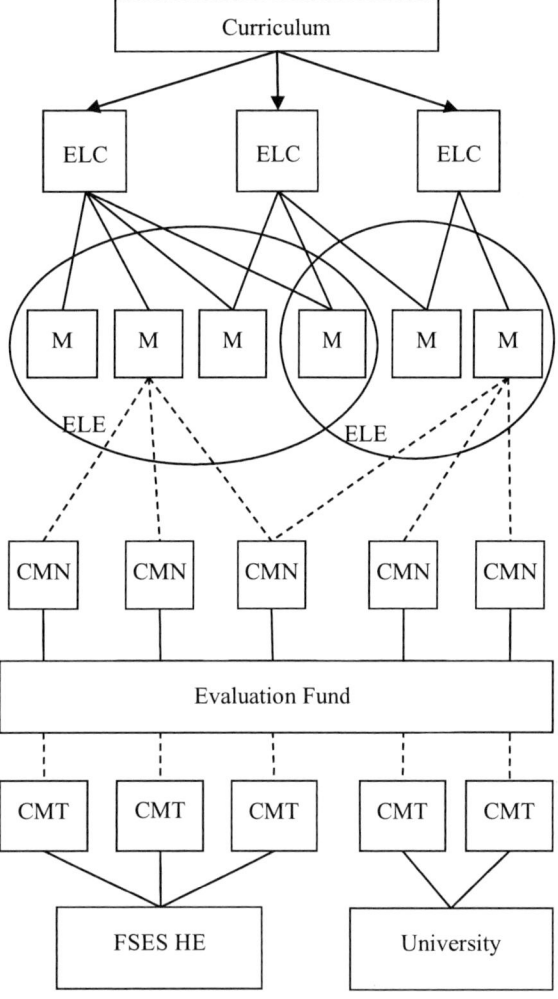

Figure 3. Competence-oriented model of representation
of educational content

be grouped into an e-learning encyclopedia (ELE). One module can belong to several ELEs. Each module is a nonempty set of components associated with the set of evaluation tasks. All evaluation tasks together create a total evaluation fund of the curriculum. FSES HE and the university provide a list of competencies, which graduates must achieve. Now the verification of the competencies achievement is possible through the evaluation fund of all courses.

The presented model is a basis for specifying requirements to the software for verification of the compliance of the curriculum with the requirements of FSES HE.

VI. DEFINITION OF REQUIREMENTS FOR A SOFTWARE IMPLEMENTING A COMPETENCE-ORIENTED MODEL OF REPRESENTATION OF EDUCATIONAL CONTENT

Let us describe the requirements to the software implementing a competence-oriented model of representation of educational content.

The basic software functionality is the following:

1) To create and modify of an e-learning course;

2) To create an e-learning course based on an existing one;

3) To create and modify an e-learning encyclopedia;

4) To create a new type of a module component;

5) To create a new module component;

6) To create a new module;

7) To build the trajectories of the e-learning course;

8) To search components by keywords.

Interesting features of the designed software are the following.

1) Automatic generation of documentation by selection of types of components included in the ELC. For example, the system will form a guide for performing the practical assignments (by selecting the components of the type *solution*), or a glossary (by selecting the components of the type *summary*), etc.

2) Verification of compliance of the curriculum to the requirements of FSES HE.

3) Verification of learning outcomes of graduates to the competencies.

4) Construction of the individual study trajectory.

We see the possible ways for the further development of the model by introducing the following functions into the system:

1) Development of the algorithm of analysis of the competencies in FSES HE.

2) Development of a fuzzy algorithm of the complexity evaluation of the course by its information content, hierarchal structure, and heterogeneity [10, 11].

VII. CONCLUSIONS

The proposed model of representation of educational content shows the possibility of verifying the students' learning outcomes to the requirements for graduates of a definite educational direction. This verification is possible due to a logical connection of the competencies, which a graduate must achieve, and the content of learning materials.

All e-learning courses in our model consist of the modules. Modules, in turn, consist of the components of the learning material each of which has a set of evaluation tasks. The connection between the content of learning materials and the competencies through the evaluation fund provides a variation of possible courses in a curriculum. Thus, the presented model could become a means to check whether the content of the e-learning courses meets the requirements to a graduate of the university.

REFERENCES

[1] L. O. Filatova, "Competence-based approach to the construction of the learning content as a factor in the development of continuity of school and university education," Supplementary education, vol. 7, pp. 9–11, 2005.

[2] V. A. Bolotov and V. V. Serikov, "Competence model: from idea to educational program," Pedagogics, vol. 10, pp. 8–14, 2003.

[3] A. G. Bermus, "Problems and prospects of implementation of competence approach in education," Int. J. "Eidos", 10.09.2005. [Electronic resourse]. Access mode: http://www.eidos.ru/journal/2005/0910-12.htm.

[4] N. V. Chigirinskaya and R. E. Gorelik, "Competence-based model training engineer: continuity and prospects of introduction," Modern problems of science and education, vol. 2, pp. 166, 2014.

[5] South Ural State University (National Research University): official site. [Electronic resource]. Access mode: http://www.susu.ru.

[6] A. E. Fedorov, S. E. Metelev, A. A. Solov'ev, and E. V. Shlyakova, "Competence approach in educational process," Omsk: Omskblankizdat, 2012.

[7] I. S. Zenkin, R. A. Timokhin, P. G. Kozlov, and R. S. Fediuk, "The development of the leading representative system of information perception in the learning process," J. Psychology, sociology and pedagogics, vol. 11 (62), pp. 21–24, 2016.

[8] N. S. Zhigal'skaya (Silkina), and L. B. Sokolinskiy, "Standardization of the content of e-learning courses and encyclopedias based on the structural-hierarchical approach," J. New information technologies in education: materials of international scientific and practical conference, vol. 1, pp. 84–89, 2008. Ekaterinburg: Publishing house of Russian state professional pedagogical university.

[9] A. M. Dvoryankin and G. G. Gerkushenko, "Model for automated synthesis of the learning objects using AND-OR trees," J. Quality. Innovation. Education, vol. 1(9), pp. 49–52, 2004.

[10] "Informatization of the general secondary education: scientific-methodological tutorial," D. Sh. Matros Ed. Moscow: Pedagogical society of Russia, 2004.

[11] O. N. Ivanova, "Optimizing the selection of educational methods using the fuzzy set theory," Theses of speeches of participants of the international conference "Informatization of general and pedagogical education – the main condition of their modernization", pp. 28, 2004. Chelyabinsk: Chelyabinsk state pedagogical university.

Gap in pagination due to withheld paper.

Pages 689-693

Developing curiosity and multimedia skills with programming experiments

J. Henno*, H. Jaakkola** and J. Mäkelä***
* Tallinn University of Technology, Estonia
** Tampere University of Technology, Pori
*** University of Lapland, Rovaniemi, Finland
jaak@cc.ttu.ee

Abstract Browsers have become the most common communication channel. We spend hours using them to get news and communicate with friends, far more time than communicating face-to face. WWW-based communication and content-creation for www will be the most common job in future work life for students specializing in software engineering.

We expect our screens to be colorful and animated, thus students should understand technologies, which are used for e.g. for painting jumping Mario to screen. But massive flow of new software engineering ideas, technologies and frameworks which appear in all-increasing temp tend to make students passive receivers of descriptions of new menus and commands without giving them any possibility to investigate and understand, what is behind these menus and commands, killing their natural curiosity. There should be time to experiment, compare formats, technologies and investigate their relations. In the presentation are described experiments used for investigating, how different formats for describing animation in HTML5 document influence animation rendering speed.

I. INTRODUCTION

Curiosity, active interest, desire to know more about something is a precondition for brain for learning [1]. Curiosity makes us consider, ponder different aspects of new and this is needed to understand it. But constant, all the time accelerating flow of new information, new tools and technologies what we are supposed to introduce to our students is effectively making the whole teaching process a mechanical dumping of new facts, procedures, technologies without leaving students time for investigation, discovering the real nature of new technologies or tools.

Especially rapid is the flow of updates, new technologies, new tools in the field of Software Engineering (SE). This creates constant pressure for SE students, who are in danger to drown in constant flow of new menus and commands without time to investigate, what is behind these new menus. And half of this what they learn now becomes in two-three years already obsolete [2].

More important than just teaching menus and commands of new frameworks is developing analytical, inquiring attitudes, ability to seek to understand how new technologies work, what is in them really new and what – just a marketing noise, how different technologies are intervened, support each other (or not) and why they so quickly replace old ones.

Especially intense is this problem in SE new fields: mobile, web and cloud programming. Here a programmer is working with minimal resources for understanding inner workings, debugging, but at the same the code should not only work, but should be memory-efficient and work quickly. Even small delays in page loading does seriously reduce website's user experience and can mean turning customers away [3].

Current SE students should understand the technical side of this media, this will essential part of their future work.

II. DIGITAL COMMUNICATION

Communication, the base of the advancement of mankind, is rapidly moving online. Online communication is visual, colorful and animated, e.g. Microsoft replaced in Windows 8 and Windows 10 the old start menu with start screen with animated live tiles. Games, game news and game discussion are a permanent section in all social media channels, but also in many old 'solid' news channels.

The major mechanism for presenting common digital content on screen is browser. Only very specialized programs (Photoshop, Excel) or big games still use their own windowing systems but even those have already many browser-based analogues. Browser is the first program what we usually open when we start computer and browsers are already considered as the basis of a whole Operating System (OS) [4].

A browser-based interactive media is essentially an frame-based video. Processor calculates next frame content and accompanying sounds, from these bits is then rendered text/images on screen and played sounds/music. For best results developers want maximal color depth (bpp – Bits Per Pixel, determines color quality) and framerate (fps – Frames Per Second), which determines animation smoothness and game responsiveness. But both these features increase processor load, memory use and decrease speed, thus may make animation jumpy or sound disrupted.

Development of browser content is based on use of the HTML language. With HTML are strongly tied several other languages – JavaScript, CSS (current version – CSS3), SVG (Scalable Vector Graphics, a graphics vector language), Flash, WebGL (for 3D graphics). Due to history and current situation with of development of these technologies their interplay is not transparent, it is often difficult to select one of several possible development strategies.

III. BROWSERS, HTML5, CSS3

Human society and culture is based on communication. For long time all communication was based on natural languages. But in the second half of the last century appeared programming languages as means to communicate with 'tools for intelligence' – computers and other devices. With development of techno- and info-sphere communication is more and more moving into Internet and the widely used method is browser-based communication using the HTML language, which allows to use all traditional communication modes – text, images, sound, video, but has added a new feature – it is interactive, messages receiver can actively participate in forming received content.

HTML was not designed as a programming language – the original 18 tags permitted only most-basic text layout options for distributing research reports of scientists working in the Nuclear Research Center in Cern. But it contained one important new tag - the hyperlink; this was the revolutionary concept that created the current-day Internet. And when Netscape invented new programming language JavaScript, which transformed until then passive www-pages into dynamic, interactive media, the web development and use exploded. JavaScript is used by 94.4% of all the websites [5] and is the programming language with highest ranking [6].

Modern web browsers have become a complicated multichannel translators. They accept input from several channels – HTML/XML/SVG text with links to other sources, images, video, sound tags, keyboard and mouse or/and touch input; input may be provided from several files – the HTML-document itself and data from linked external sources, e.g. external CSS (Cascading Style Sheets) and JavaScript files, images, sound/video files (see Figure 1). With these different types of inputs browsers compile complex output for computer/mobile screen, create voice and play sound and video.

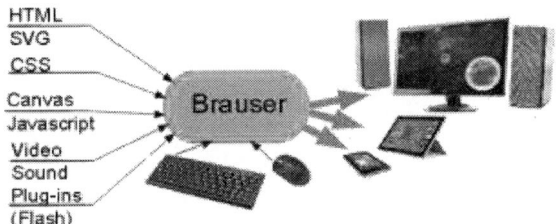

Figure 1. Inputs-outputs of common web browser

Thus in principle browser works like a programming language's translator – it transforms information presented in format/syntax of input channels into format/syntax required by output channels/devices. But its task is essentially more complex – the information presented in inputs is interlinked, depends on each other, e.g. 3D-placement on screen of a paragraph of text depends on size/placement of text/pictures before this paragraph (the CSS box model) and all CSS-formatting rules, which

could be introduced in several different places - the 'cascading' feature of CSS.

New features are added to browsers and its input languages in frantic temp. All major browsers – Google Chrome, Microsoft Internet Explorer (IE), Mozilla's Firefox (FF), Opera - publish several major updates per year, minor updates appear in every 6-8 weeks [7] and most of them are already 'evergreen' - they automatically update themselves. In recent years have appeared also several new browsers: Microsoft Edge (for Windows 10 only) [8], Vivaldi [9], Brave [10], Yandex browser [11], Maxton cloud browser [12].

With such a frantic temp of development it is difficult to maintain overall balance and consonance of all parts of HTML-documents - DOM (Domain Object Model [13],all elements of a HTML document) tree, JavaScript, CSS, add-on's) and new features, which are developed by different parties.

Quite illustrative is development of Cascading Style Sheets (CSS) language. This was simple and natural format to describe styling of web pages, when it was introduced 20 years ago. But innovators found, that "CSS is primitive and incomplete" [14] if we could not have borders with rounded corners, thus should be improved and so appeared frameworks - layers of code re-organizing use of CSS features. First was introduced an extension of CSS - Sass (Syntactically Awesome Style Sheets [15]), where are introduced variables, nested rules, inline imports and other features and which has two syntax options. However, to use it you should first know CSS quite well and also have the programming language Ruby compiler installed in your computer, since the extension is a Ruby program. Then Sass was extended with simple scripting language SassScript. But for some people Sass still is not good enough so we get open-source CSS authoring framework Compass [16]. Now writing style rules - something what could be done with any text editor, Notepad or (for highly technical persons) Notepad++ and which few would call programming requires two interdependent frameworks, compiler for the Ruby language and several Ruby Gems to output CSS. And to see the result this should be uploaded to server.

CSS frameworks may be useful for professional developers of big web sites, but even there their use is questionable [17]. For students they are an overkill.

CSS3 introduced several methods for animating elements of HTML document. For instance, CSS property `transform` manipulates the size, shape, and position of a CSS box and its contents through rotating, skewing, scaling and translating. CSS3 animations allows animation of most HTML elements without using JavaScript or Flash [18], but this is a rather restricted type of animation – CSS3 does not allow variables, thus everything should be fixed before, animation cannot be changed on run-time. CSS3 is introducing also several other very advanced features, e.g. media queries, which allow to apply CSS rules depending on properties of device (e.g. screen width) which is used to present the content. Most of these features already work in many browsers; however, currently (Jan 2017) CSS3 is not yet implemented in

Microsoft browsers IE 11 and Edge, these browsers 'know' only CSS2.

IV. TOO MUCH IS CONFUSING

Animation, movement is the major way to make WWW documents more attractive, thus most of developers want to use this feature. For a long time the only technologies allowing to show movement in WWW were video and Flash. Video is a non-interactive media, thus the main technology for developing interactive, dynamic web content, games and portals was Flash. When Steve Jobs declared in 2010 Flash a 'persona-non-grata' on Apple devices, its popularity decreased also on other platforms. But because of its (very) good performance it is still widely used, so rumors about its death seem to be strongly over-accelerated [19]. Microsoft even included Flash player in its new browser Microsoft Edge - and dropped its own inventions, the ActiveX components and Silverlight. Thus ActiveX and Silverlight become another dying-out stars in quickly changing landscape of web technologies.

The major technology platform for developing interactive web content - games, communication and business portals is currently latest version of HTML – the HTML5. HTML5 introduced a new element – canvas – an area in HTML-document where JavaScript commands can draw and thus create animations.

HTML5 allows to implement animations using several formats and technologies:

- showing animated .gif images either as a part of HTML-document (i.e. with HTML-code) or drawing with JavaScript on canvas; this old image format contains series of frames for storing short animations, is restricted to 8 bpp (bits per pixel) color resolution (i.e. image can have max 256 colors) and animation speed cannot be controlled by browser, it has to be pre-set when creating the .gif animation file, but animation (image) is easy to scale (make smaller, making it bigger destroys quality);

- animation on HTML5 canvas with JavaScript: showing/moving images, possibly clipping them to some figure;

- procedural texture – merging texture images changing their opacity [32]; here this was used to produce Sun's lava texture from only one image;

- CSS3 allows to create frame-based spritesheet animation without using JavaScript [20];

- SVG; SVG elements can be manipulated like HTML elements using transform functions, but many commands and attributes do not work the same way on SVG elements as they do on HTML elements, JavaScript feature detection fails, the local coordinate system of an element works differently for HTML elements and SVG elements, the CSS properties of SVG elements have different names, e.g. instead of *background-color* should be used *fill* etc.

When HTML5, CSS and JavaScript arrived, several browsers, especially Microsoft Internet Explorer (IE) browser versions 6..10 did not follow standards, thus web content developers had insert into their code special checks for browser version, e.g. in HTML:

```
<!--[if IE 7 ]>
```

In order to unify development and eliminate browsers incompatibilities was in 2006 introduced a cross-browser JavaScript library jQuery, which provided a consistent interface that works across different browsers, i.e. also in Microsoft's browsers. With time jQuery has included many properties, e.g. fade ins and fade outs (a 'visual sugar') and animations by manipulating CSS properties. Since JavaScript libraries are a non-transparent layer of code (it is very difficult to check what went wrong is an error occurs in a used library – they act like a compiled module), libraries should be introduced only after students have acquired solid JavaScript skills. But since jQuery is currently very popular, we considered also animation created with jQuery.

V. WHAT TO USE ?

Browser input consists of several parts – the HTML-document (HTML code), CSS stylesheet, JavaScript file(s) (there may be several), image, sound, video files. These are handled by different browser's sub-programs – the browser's layout engine with CSS interpreter and the JavaScript engine. All major browsers use their own, independently developed engines:

TABLE I.   BROWSERS AND THEIR LAYOUT AND JAVASCRIPT ENGINES

Browser	Layout Engine	JavaScript interpreter
Firefox	Gecko	SpiderMonkey
Chrome	Blink (developed from WebKit)	V8
Internet Explorer	Trident	Chakra
Microsoft Edge	EdgeHTML	Chakra
Opera	WebKit	V8
Safari	WebKit	JavaScriptCore

They all interpret HTML+CSS+JavaJcript code a bit differently and the 'inner working' even of major browser engines are mystery, explanations cover only one particular browser [21], [22], [23] and are presented in rather general terms. There is not yet 'browser theory', which were comparable to IT subject 'Translator theory' [24], developed for classical programming languages. This creates many questions. Performance of some components - JavaScript engines, CSS layout calculation, screen renders etc. may be rather different [25]. How this influences the overall performance, in which order are applied CSS rules, if determining DOM tree element attributes is better from HTML-text or from JavaScript, levelling browser's build-in defaults (e.g. different default margins) – these practical issues are unexplained.

For instance, when developing a Christmas game "Santa in Wild Forest" we wanted to introduce a light effect – torch/moonlight moving together with Santa, see Figure 1. It turned out, that the effect can be implemented using different technologies/formats, but they all had some problems (image on mobile screen look a bit dirty) and some solutions worked in different browsers differently.

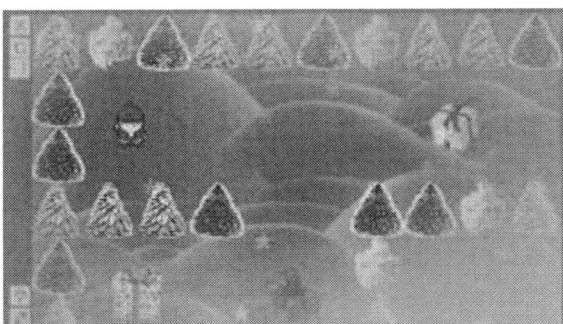

Figure 2. A sceenshot from mobile game "Santa in Wild Forest", developed as a part of the course, captured from mobile phone screen in original mobile screen resolution – 768x1220px; - Santa Clause has to collect all parcels, which Krampus throw into dark forest; we investigated possibilities to add torch/moonlight effect, i.e. create a half-transparent circle covering Santa and moving together with Santa

Thus instead of following intense flow of marketing shouts: "Use JQuery!" [26], "Use Angular!" [27], "Use TypeScript!" [28], "Use Facebook's React Native!" [29], "Use Intel SDK!" [30] we decided first to test with a practical application the real value, first of all – speed – of different animation technologies. Implementing test applications and performing tests give students much better understanding of value of different technologies then just implementing something in one (usually rather randomly selected) format.

VI. THE TEST APPLICATIONS

For comparing different animation formats we implemented a scheme of Lunar Eclipse (actually happened during the course on 20.03.2015), which contained several animated and/or half-opaque elements, see Figure 3.

Figure 3. The test application (captured from mobile browser) and its animated objects:Sun (upper left corner with animated lava texture), Earth (lower right corner, rotating), Moon (small grey circle close to earth), Moon shadow (larger half-transparent grey circle on Earth surface), Earth atmosfere (light half-transparent halo surrounding Earth).

To investigate influence of different animation formats and opacity change technologies on animation speed and memory requirements we prepared tests T1..T9 [31], which all are versions of this animation.

TABLE II. TEST ANIMATIONS

Test files	Animation description
T1: eclipse1.htm	The whole screen is covered with canvas with cosmos as the CSS-defined background image; Earth is animated with JavaScript (texture is constantly moved behind a clipping circle, drawn on canvas by JavaScript); all other objects are images in HTML document placed using CSS attribute z-order over the canvas: Sun is animated .gif image (32 frames) with transparent background, Moon, Moon shadow, Earth atmosphere – images, transparency is 'built-in' to images with Photoshop (is not adjusted in the HTML-document); placement of images is defined with CSS using position attribute value 'fixed'.
T2: eclipse2.htm	The whole screen is covered with canvas with cosmos as the CSS-defined background image; Earth is animated with JavaScript (texture is constantly moved left-to-right behind a clipping circle, drawn by JavaScript using the 2D-graphics context of canvas); all other objects are images in HTML document placed using CSS attribute z-order over the canvas: Sun is animated .gif image (32 frames) with transparent background, transparency of Moon shadow and Earth atmosphere images is defined with CSS rules
T3: eclipse3.htm	The whole screen is a DIV with cosmos as the CSS-defined background image; Sun is animated .gif image, minimal canvas is used only behind Earth, which is animated with JavaScript (texture is constantly moved behind a clipping circle drawn by JavaScript using the 2D-graphics context of canvas); all other objects are images in HTML document placed using CSS attribute z-order over the canvas, 50% transparency of Moon shadow and Earth atmosphere images is defined in Photoshop
T4: eclipse4.htm	As previous, but 50% transparency of Moon shadow and Earth atmosphere images is defined with CSS rules
T5: eclipse5.htm	The whole screen is a series of DIV-s (no canvas); the main DIV with cosmos as the CSS-defined background image covers the whole screen, smaller DIV-s for Earth, Moon, Moon shadow and Earth atmosphere images are placed over it using the CSS position, width/height and z-order attributes; Sun is animated with CSS3 rules (the 32 frames of the sprite sheet were obtained from the animated .gif image), Earth is also animated with CSS (texture is constantly moved behind a CSS clipping circle); transparency of Moon shadow and Earth atmosphere images is defined in Photoshop
T6: eclipse6.htm	The whole screen is a series of DIV-s (no canvas); the main DIV with cosmos as the CSS-defined background image covers the whole screen, smaller DIV-s for Earth, Moon, Moon shadow and Earth atmosphere images are placed over it using the CSS position, width/height and z-order attributes; Sun is video in WebM format, the video is clipped by CSS clipping circle (works correctly only in Firefox; Chrome has its own propietary format for alpha transparency in WebM-video); Earth is animated with CSS (texture is constantly moved behind a CSS clipping circle); transparency of Moon shadow and Earth atmosphere images is created in Photoshop
T7: eclipse7.htm	Sun texture is procedurally generated by JavaScript and jQuery on minimal canvas using two additional canvases (the idea from [32]); all other elements are as in previous example
T8: eclipse8.htm	As in previous but jQuery library (253 kb) was removed and replaced with JavaScript
T9: eclipse9.htm	Texture of Sun is procedurally generated (without jQuery), Earth is JavaScript animation on a separate canvas, thus together there are 4 canvases

VII. RESULTS

All test animations *eclipse1.htm..eclipse9.htm* were HTML5-documents looking similar on screen, but since implemented using different technologies/formats they also produced different results: time needed for constant number of animations (10 rotations of Earth) and RAM memory used by browser. They all had a small JavaScript script, which measured the time was used to run 10 rotations of the Earth; the results were shown on screen in a separate small DIV and stored using the HTML5 local storage feature. For memory performance there is not yet general standards; in Chrome is available a proprietary method `performance.memory` [33], but this can be used only if Chrome is started with specific switch and in our tests Chrome reported always the same values. IE allows to see some memory statistics with the UI Responsiveness tool [34]:

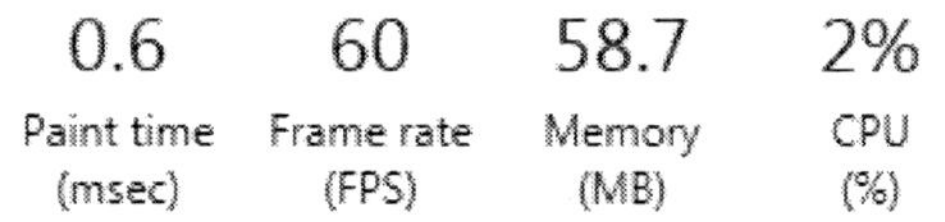

$$0.6 \qquad 60 \qquad 58.7 \qquad 2\%$$

| Paint time (msec) | Frame rate (FPS) | Memory (MB) | CPU (%) |

Figure 4.   The window of the Microsoft UI-responsiveness tool

The values produced by this tool varied 4-10% and therefore not presented here; the only more or less constant change was 3-4% increase in used memory when the `jQuery` library was used (the test *eclipse8.htm*).

Thus the main measured parameters were time-based – the framerate FPS (Frames Per Second) and total time needed to execute a constant-size animation (10 rotations of Earth). A JavaScript script measured several parameters of test (FPS, time for one rotation, number of frames rendered) and at the end of animation showed the main result – total time for the animation (10 rotations of Earth) in milliseconds on screen (it was also stored using the HTML5 feature `localstorage`). In order to eliminate computer speed and network latency, all tests were done locally under a local Wamp server [35]. These tests produced lot of numbers; in order to gain better understanding of these were results normalized, using the *eclipse1.htm* from Firefox as the control case, i.e. in the following table the first number is *time*(Ti) - total time for this test Ti with this browser and the second – percentage of this time of the 'etalon' time, i.e. calculated with formula *time*(Ti)*100/*time*(T1), where *time*(T1) is the time reported for test T1 by Firefox.

TABLE III.   RESULTS OF TESTS

Test	Browser				
	FF 51	*Chrome 55*	*IE 11*	*Edge*	*Opera*
T1	66773ms 100%	68281ms 102%	60008ms 98%	65097ms 98%	65046ms .97%
T2	66744ms 99%	68230ms 102%	59996ms 98%	66007ms 98%	65070ms 97%
T3	66755ms 99%	67874ms 101%	60036ms 99%	66764ms 99%	65065ms 97%
T4	66720ms 99%	68016ms 102%	60032ms 99%	67065ms 100%	65065ms 98%
T5	20011ms 29%	20004ms 29%	CSS clipping with circle does not work	CSS clipping with circle does not work	19974ms 29%
T6	20023ms 29%	19837ms 29%	IE 11 does not play WebM video, CSS clipping does not work	-	20010ms 29% no video clipping
T7	20010ms 29%	19999ms 29%	CSS clipping does not work	-	20006ms 29%
T8	20010ms 29%	19992ms 29%	CSS clipping does not work	-	20004ms 29%
T9	66721ms 99%	67070ms 99%	59626ms 98%	60101ms 99%	64057ms 97%

We performed similar tests also with a mobile phone browsers in an android mobile (LG E975a, Android 4.4.2 'KitKat'); here were used the phone OS built-in browser, Chrome and UC cloud browser (made in China), which currently is a 'rising star' in the landscape of mobile browsers [36]. In the following table are presented the raw results (test times in milliseconds) and results after normalizing using Chrome as the etalon.

TABLE IV.   RESULTS OF TESTS IN MOBILE BROWSERS

Test	Browser		
	Chrome	*LG OS-browser*	*UC browser*
T1	66972ms 100%	80060ms 119%	78215ms 117%
T2	66882ms 100%	83453ms 125%	98413ms 145%
T3	66973ms 100%	85029ms 124%	69619ms 127%
T4	67308ms 100%	82938ms 124%	85169ms 127%
T5	19894ms 30%	20060ms 30% – no clipping	19109ms 29% no clipping
T6	19835ms 30%	No WebM	19930ms 30% no clipping
T7	19838ms 30%	-	19129ms 29% no clipping
T8	19993ms 30%	-	19835ms 30% no clipping
T9	68086ms 100%	-	71027ms 106%

Although mobile browsers show more differences both between browsers and between tests, the general tendencies were rather similar.

VIII. CONCLUSIONS FROM TESTS

These results allowed students to draw several conclusions:

- there are no essential differences in speed between major browsers, but Microsoft browsers do not (yet) implement CSS3 (only CSS2)

– HTML5 canvas+JavaScript allows to create complicated animations, but reduces animation speed ca three times (tests T5,T6,T7 did not use canvas). This result was for students surprising, since canvas is commonly considered the main element in all graphics-intense web applications (games, portals etc.) but it becomes understandable if one thinks what actually is loaded as the canvas 2D context – this is an interpreter 2D graphics commands, which builds its own name table etc.

- using several canvases does not make application slower (test T9);

- changing opacity of bitmaps with JavaScript does not make application slower and allows better to control result (tests T2, T4)

- CSS3 animations and CSS3 clipping (with circle) are quick, but quite difficult to scale (changing size of frame-based CSS animation is very error prone) and did not work in Microsoft browsers

- video in WebM format with transparent background (test T6) can be achieved (using CSS3 clipping) only in Firefox (Chrome has for this a proprietary extension [37]);

- results of tests T7, T8, T9 indicate, that jQuery was officious – big, especially for mobile applications (current version 3.1.1 – 261 kB) and did not have any advantages. The jQuery library was introduced to make Microsoft browsers IE6..IE10 to understand standards, but currently Microsoft has also started to follow them, so jQuery is (mostly) not needed. But jQuery introduces rather difficult to understand cryptic syntax (what has to be learned) and is and changing custom semantics, e.g. cryptic *jQuery* command to get canvas object:-

```
var $canvas = $('#canvas');
```

returns array (#canvas suggests, that there are several canvas objects having the same id?!); equivalent to this, but more understandable plain JavaScript command

```
var $canvas =
document.getElementById('canvas');
```

returns 'flat' variable, thus all uses of these variables also require different syntax. Use of jQuery also increased memory requirements (as measured in IE 11). It was relatively easy to remove all dependencies of jQuery - we actually had to change only 8 lines to convert the script into 'clean' JavaScript, where jQuery was not used.

Students discovered even more, e.g. the speed of canvas animation depends essentially on size of animated objects – scaling page down decreased rendering time.

IX. CONCLUSIONS FROM THE PROJECT

The main benefit of the project was not discovery of dubious properties of use of canvas or other technical results – this may change with the next updates of browsers. The main benefit of the project for students was in making them think and explore, showing exploratory attitude for use of software technologies instead of following blindly the next marketing hype and learning dumbly commands of some commercial framework (e.g. Intel XDK - 643 MB, 35800 files, development for Android phones requires also Android SDK – 27.2 GB, 154999 files). Browsers are the most important communication channel of future and students should understand well the browser technology and for this they should experiment and investigate.

REFERENCES

[1] Scientific American. Curiosity Prepares the Brain for Better Learning. https://www.scientificamerican.com/article/curiosity-prepares-the-brain-for-better-learning/

[2] How fast is our world becoming obsolete? https://www.ericsson.com/thinkingahead/the-networked-society-blog/2014/01/30/how-fast-is-our-world-becoming-obsolete/

[3] A.B. King. Website Optimization: Speed, Search Engine & Conversion Rate Secrets. O'Reilly Media; 2008, 398 pp

[4] Browser Based Operating Systems Reviewed. http://www.tech-tweak.com/browser-based-operating-systems/

[5] Usage of JavaScript for websites. https://w3techs.com/technologies/details/cp-javascript/all/all

[6] The RedMonk Programming Language Rankings: June 2015. http://redmonk.com/sogrady/2015/07/01/language-rankings-6-15/

[7] https://en.wikipedia.org/wiki/Timeline_of_web_browsers

[8] Microsoft Edge. https://www.microsoft.com/en-us/windows/microsoft-edge

[9] Vivaldi. https://vivaldi.com/

[10] Brave. https://brave.com/

[11] Yandex browser. https://browser.yandex.com/desktop/main/

[12] Maxton. http://www.maxthon.com/

[13] JavaScript HTML DOM. https://www.w3schools.com/js/js_htmldom.asp

[14] An Introduction to CSS Pre-Processors: SASS, LESS and Stylus. https://htmlmag.com/article/an-introduction-to-css-preprocessors-sass-less-stylus

[15] Sass – CSS with superpowers. http://sass-lang.com/

[16] Compass. http://compass-style.org/

[17] You might not need a CSS framework. https://hacks.mozilla.org/2016/04/you-might-not-need-a-css-framework/

[18] CSS3 animations. http://www.w3schools.com/css/css3_animations.asp

[19] Usage of Flash for websites. https://w3techs.com/technologies/details/cp-flash/all/all

[20] Using CSS animations. https://developer.mozilla.org/en-US/docs/Web/CSS/CSS_Animations/Using_CSS_animations

[21] WebCore Rendering I – The Basics. https://webkit.org/blog/114/webcore-rendering-i-the-basics/

[22] How Browsers Work: Behind the scenes of modern web browsers. https://www.html5rocks.com/en/tutorials/internals/howbrowserswork/

[23] High Performance Animations. https://www.html5rocks.com/en/tutorials/speed/high-performance-animations/

[24] Alfred V. Aho, Jeffrey D. Ullman Theory of Parsing, Translation and Compiling. Prentice-Hall 1972, 542 pp, ISBN-13: 978-0139145568

[25] A Guide to JavaScript Engines for Idiots. http://developer.telerik.com/featured/a-guide-to-javascript-engines-for-idiots/

[26] jQuery. https://jquery.com/

[27] Angular. https://angularjs.org/

[28] TypeScript. https://www.typescriptlang.org/

[29] ReactNative. https://facebook.github.io/react-native/

[30] Intel XDK. https://facebook.github.io/react-native/

[31] Eclipse. http://deepthought.ttu.ee/users/jaak/slideshow/

[32] Experiment - HTML5 Canvas Nebula. http://www.professorcloud.com/mainsite/canvas-nebula.htm

[33] Static Memory Javascript with Object Pools. https://www.html5rocks.com/en/tutorials/speed/static-mem-pools/

[34] Improving UI responsiveness. https://msdn.microsoft.com/en-us/library/dn255009(v=vs.85).aspx

[35] Wampserver. http://www.wampserver.com/en/

[36] Browser Market Share Worldwide. January 2016 to Jan 2017. http://gs.statcounter.com/

[37] Alpha transparency in Chrome video. https://developers.google.com/web/updates/2013/07/Alpha-transparency-in-Chrome-video

MIPRO 2017, May 22- 26, 2017, Opatija, Croatia

Informetrics:
the Development, Conditions and Perspectives

A. Papić*

* Faculty of social sciences and humanities/Department of information sciences, Osijek, Croatia
apapic@ffos.hr

Abstract - The paper gives an insight into the development of informetrics from its beginnings. Informetrics is a part of information sciences dedicated to measuring of information phenomenon. Informetrics can be defined as the discipline which studies quantitative aspects of information in any form not only within scientific community but also within any other social community. The term informetrics is an umbrella term for several similar but different disciplines such as bibliometrics, scientometrics, webometrics, altmetrics etc. Therefor the paper firstly explains the terminological issues related to different metric disciplines. Also, the paper aims to explore global and local conditions of informetrics and its position across other disciplines. According to the obtained research results the possible directions for the development of informetrics in Croatia are suggested.

I. INTRODUCTION

Nowadays a great attention is given to different kinds of metrics such as journal level metrics, article level metrics, author level metrics, university level metrics etc. in purpose to rank and evaluate those entities. Today many players exist at the market regarding measuring different metrics indicators and there is a quite terminological chaos dealing similar but different terms of metrics such as informetrics, bibliometrics, scientometrics, webometrics, altmetrics etc. The oldest terms were bibliometrics and scientometrics. The historical development of those two disciplines was almost synchronous. The first equivalent of term bibliometrics so called bibliometrie was firstly mentioned by Otlet in his work Traite de documentacion in 1934 year. In 1969 year Pritchard has introduced the English term bibliometrics defined as application of mathematical and statistical methods to books and other communication media [5]. In the same year two Russians Nalimov and Mulchenko have firstly coined the term naukometrya which was equivalent for scientometrics [4]. The term scientometrics became very popular from 1978 year when Tibor Braun in Hungary established the journal called Scientometrics. According to Wilson scientometrics studies quantitative aspects of science about science, scientific communication and science policy [2]. The newest term introduced by Björneborn and Ingwersen is webometrics which main purpose is to analyze WWW with bibliometrics and informetrics approaches [8]. Altmetrics is the latest metrics term and encompasses different kinds of alternative ways of article level metrics such as views, downloads, tweets, likes etc. In 1979 year almost at the same time Nacke [3] and Blackert & Siegel [1] introduced

the German term informetrie as equivalent of the term informetrics. The term informetrics is the widest term and tends to encompass all other metrics. According to Tague-Sutcliffe the informetrics studies quantitative aspects of information in any kind and within any social community not only scientific community [7]. The literature review conducted within this paper also starts from this very point of view. This paper aims to explore global and local conditions of informetrics and its position across other disciplines thus the research questions raised in this paper are the following:

(1) How much informetric disciplines correlate to other traditional disciplines?

(2) In which geographical regions and universities are informetric disciplines mostly employed?

(3) Which are the most authoritative authors and journals within informetric disciplines?

After the second section which deals with explanation of the method used in this paper, the paper is structured in a way that each of the following sections are in fact answers to the raised research questions. Namely, the third section explains correlation of informetrics to other traditional disciplines, the fourth section describes informetrics across geographical regions and the fifth section highlights the most authoritative authors and journals within informetric disciplines. Concluding remarks are summarized at the end of the paper.

II. METHOD

Scopus is bibliographic and citation database which indexes journals, book series and conference proceedings from all scientific areas. Scopus includes sources from all scientific areas and within Scopus database more than 130 Croatian journals are indexed. Citation data within Scopus database are available from 1996 year till today. Scopus contains more than 40 million records and even 70 percent of them have an abstract. Scopus is a product of Elsevier corporation. The method used in this paper is bibliometric analysis of literature within Scopus database according to following search strategy:

TITLE-ABS-KEY (informetrics) OR TITLE-ABS-KEY (bibliometrics) OR TITLE-ABS-KEY (webometrics) OR TITLE-ABS-KEY (altmetrics) OR TITLE-ABS-KEY (scientometrics) AND (EXCLUDE (PUBYEAR,

2017)). Total of 11 051 documents according to this search strategy were found in time period from 1974 year till 2016 year (without 2017 year) and analyzed for the purpose of answering to the research questions.

III. INFORMETRICS ACROSS DISCIPLINES

According to data available through the Scopus database informetrics correlates mostly with Medicine (49.2%), Social sciences (26.5%) and Computer science (22.1%).

Some of the other disciplines which correlate with informetrics in smaller extent are Biochemistry, Genetics and Molecular biology (5.3%), Decision sciences (4.9%), Mathematics (4.1%), Engineering (3.7%) etc. Fig. 1. shows intensive raise of informetrics in scientific communication from the year 2000 till today. It can be seen at the Fig. 1. that the peak in number of published units dealing with informetric disciplines mostly in methodological sense was in the year 2014.

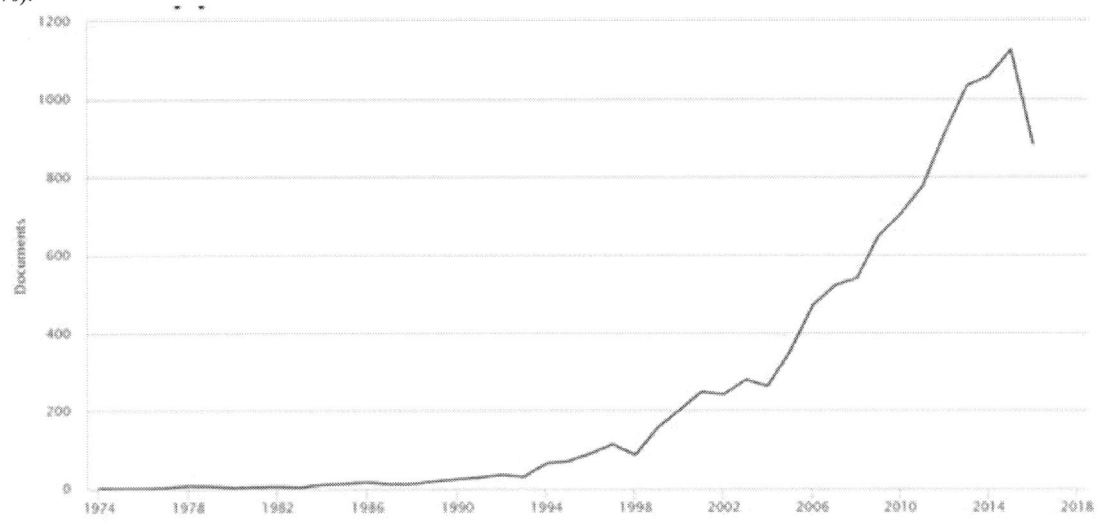

Figure 1. Distribution of documents by year regarding different informetric disciplines [6]

IV. INFORMETRICS ACROSS REGIONS

Distribution of documents regarding informetric disciplines across different countries at Fig. 2. shows that the most documents were published in United States (2342),

United Kingdom (972), Spain (846), China (775), Brazil (526), Germany (526), Canada (507), Australia (376), Netherlands (374) and India (357).

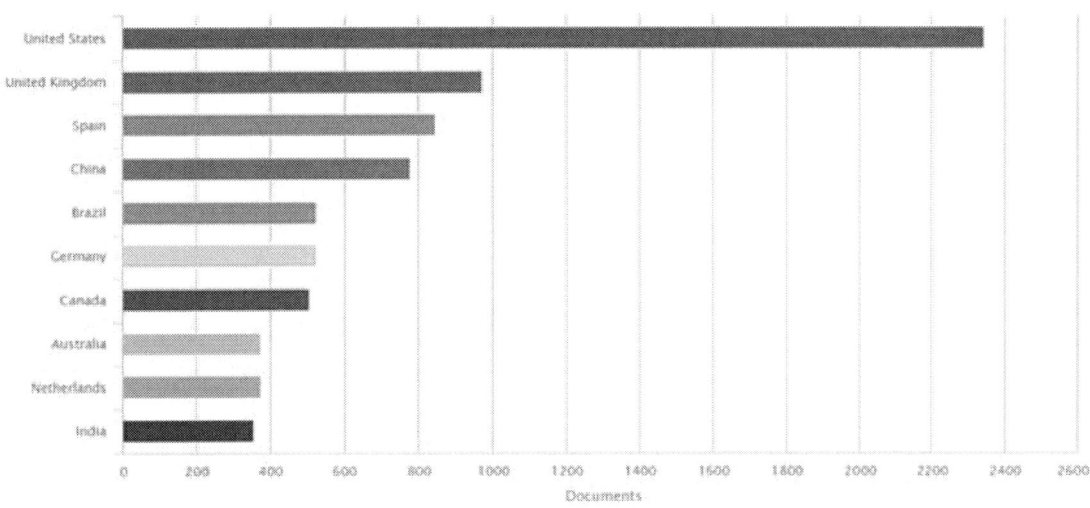

Figure 2. Distribution of documents regarding different informetric disciplines across countries [6]

Fig. 3. shows the most productive universities in the world regarding informetric disciplines. At the first place is Universidade de Sao Paulo (114), Indiana University (98), KU Leuven (98), Universidad de Granada (93), Universitat de Valencia (89), University of Wolverhampton (84), Administrative Headquarters of the Max Planck Society (83), Wuhan University (83), Universiteit Antwerpen (80) and University of Toronto (79).

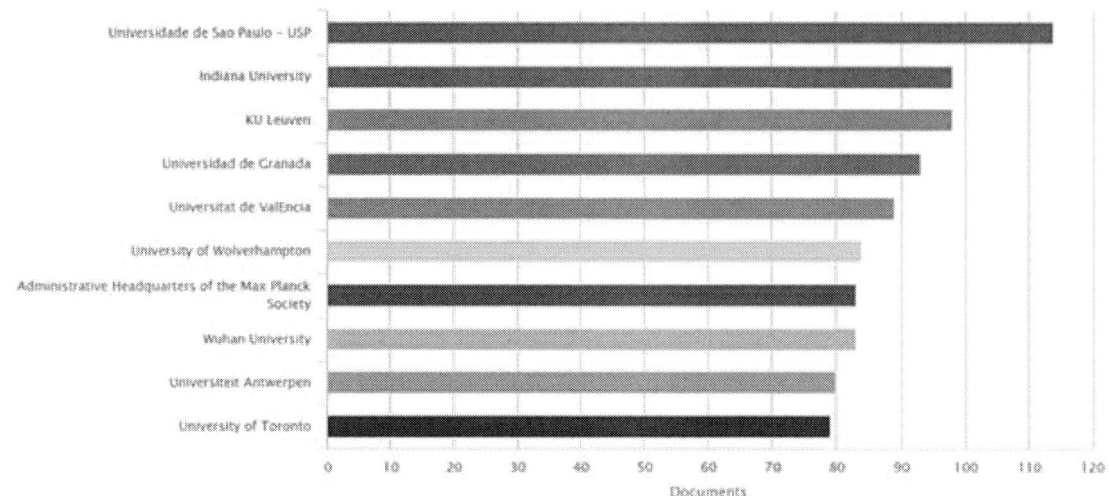

Figure 3. Distribution of documents regarding different informetric disciplines by author affiliation [6]

V.   AUTHORITIES IN INFORMETRICS

Fig. 4. shows the most prestigious journals in the world which publish content regarding different informetric disciplines. The most prestigious journals are Scientometrics (808), Nature (216), Journal of Informetrics (182), of Journal of the Association for Information Science and Technology (117) and Plos One (113).

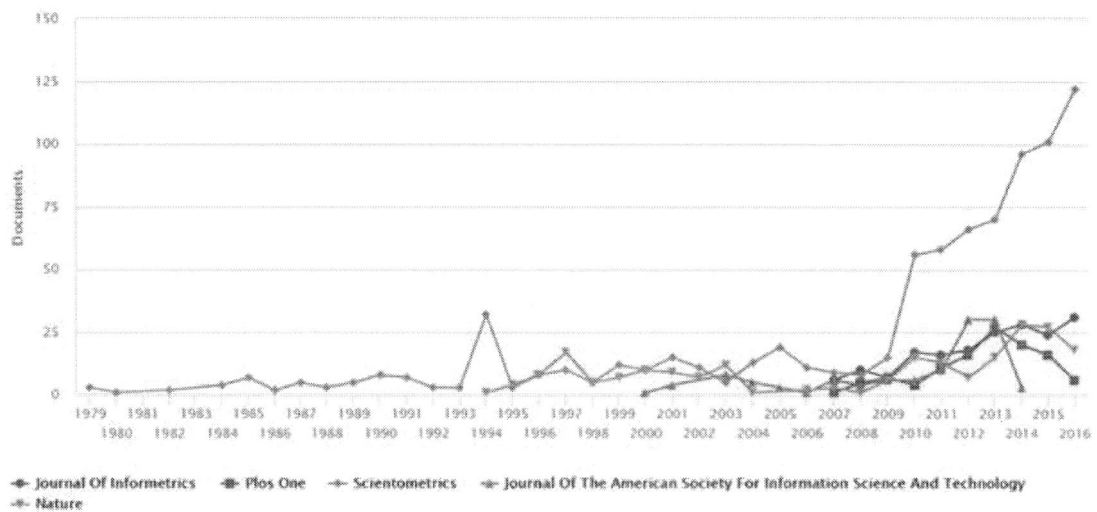

Figure 4. The most prestigious journals regarding different informetric disciplines [6]

Scientometrics is an international journal for all quantitative aspects of the science of science, communication in science and science policy produced by Springer. The journal Scientometrics is concerned with the quantitative features and characteristics of science and scientific research. The scope of the journal Scientometrics encompasses investigations about the development and mechanism of science studied by statistical mathematical methods. Nature is the weekly, international and interdisciplinary journal of science. Nature is a weekly international journal publishing the peer-reviewed research in all fields of science and technology on the basis of its originality, importance, interdisciplinary interest, timeliness, accessibility, elegance and surprising conclusions. The journal Nature provides rapid and authoritative news and interpretation of trends affecting science and scientists. Journal of Informetrics is produced by Elsevier. Journal of Informetrics publishes high-quality research on quantitative aspects of information science. The scope of the journal encompasses topics in bibliometrics, scientometrics, webometrics, and altmetrics as well as contributions studying informetric problems using methods from other quantitative fields, such as mathematics, statistics, computer science, economics and econometrics, operations research, and network science. Journal of the Association for Information Science and Technology (JASIST) is a fully refereed scholarly and technical periodical and has been published continuously since 1950. JASIST publishes reports of research and development in a wide range of subjects and applications in information science and technology. Plos One is the world's first multidisciplinary Open Access journal. *Plos One* provides a platform to publish primary research, including interdisciplinary and replication studies. *Plos One* facilitates the discovery of connections between research whether within or between disciplines.

Fig. 5. shows the most influential authors which publish content regarding different informetric disciplines. The most influential authors are Bornmann, L. (92), Thelwall, M. (78), Abramo, G. (76), D'Angelo, C. A. (75), Aleixandre-Benavent, R. (61), Glänzel, W. (57), Egghe, L. (52), Leydesdorff, L. (52), Ho, Y. S. (46) and Rousseau, R. (45). The great authorities who should be accentuated within this field are certainly authors such as Glänzel, Thelwall, Leidesdorff and Bornmann. Wolfgang Glänzel is full professor based at KU Leuven, Belgium where he is Director of the Centre for R&D Monitoring, one of the leading research centres in bibliometrics. Professor Glänzel is amongst the most prolific and highly cited researchers in his area. He was awarded the Derek de Solla Price Medal in recognition of his contributions. Among many other roles, Professor Glänzel serves as editor-in-chief of Scientometrics. He is also affiliated with the Institute for Research Organisation of the Hungarian Academy of Sciences and holds doctorates in both mathematics and social science. Michael Thelwall is professor of Information Science and leader of the Statistical Cybermetrics Research Group at the University of Wolverhampton, which he joined in 1989. He is also Docent at the Department of Information Studies at Åbo Akademi University, and a research associate at the Oxford Internet Institute. His PhD was in Pure Mathematics from the University of Lancaster. His current research field includes identifying and analysing web phenomena using quantitative-led research methods, including altmetrics and sentiment analysis, and has pioneered an information science approach to link analysis. He is an associate editor of the Journal of the Association for Information Science and Technology. Loet Leidesdorff holds Ph.D. in Sociology, M.A. in Philosophy and M.Sc. in Biochemistry. He is professor in the Dynamics of Scientific Communication and Technological Innovation at the Amsterdam School of Communications Research of the University of Amsterdam.

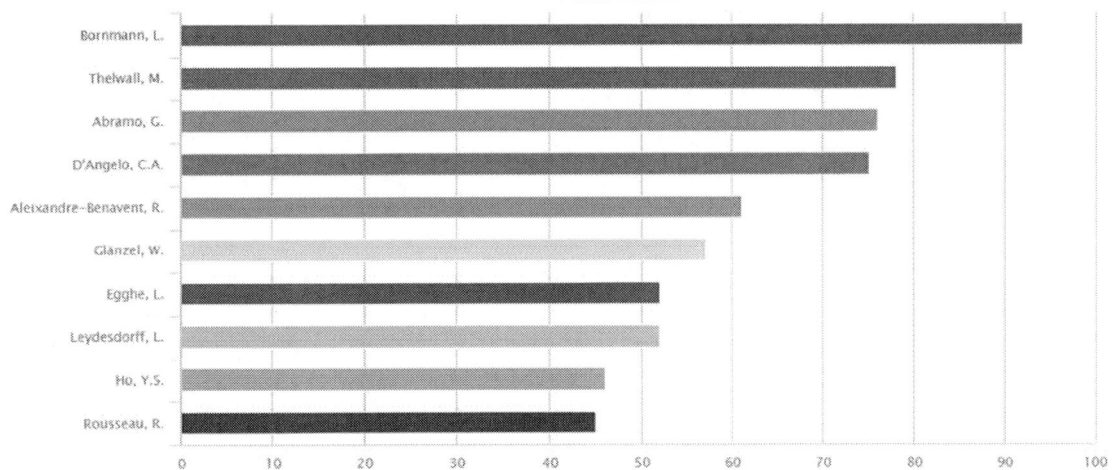

Figure 5. The most influential authors regarding different informetric disciplines [6]

Leydesdorff has published extensively in systems theory, social network analysis and scientometrics. He received the *Derek de Solla Price Award* for Scientometrics and Informetrics in 2003. Bornmann Lutz works as a sociologist of science at the Division for Science and Innovation Studies in the Administrative Headquarters of the Max Planck Society in Munich, Germany. His current research interests include research evaluation, bibliometrics and altmetrics. He is a member of the editorial board of *Journal of Informetrics* (Elsevier), *PLOS ONE*, *Scientometrics* (Springer), and *Journal of the American Society for Information Science and Technology* (Wiley).

VI. CONCLUSION

Informetrics is very young discipline but lately became very popular and employed in scientific communication. The main purpose of informetric disciplines is to inform other traditional and well established disciplines such as Medicine, disciplines within Social sciences such as library and information sciences, sociology of science and history of science mostly with scientometric and bibliometric approaches, disciplines within Computer science mostly with webometric and altmetric approaches. Although informetric disciplines are popular at the global scene the position of Croatia at that scene is almost invisible. This paper highlights the importance of positioning Croatian authors and Croatian journals at the local and global scene regarding informetrics trends.

For example first step toward better visibility of Croatian authors in terms of informetrics indicators could be integration of bibliometric and altmetric indicators such as h-index of authors, altmetric score and other article level metric indicators within CROSBI-Croatian Scientific Bibliography. Researches supported by different informetric disciplines could have great benefits for decision makers and creators of science policy and national educational policy in general.

REFERENCES

[1] L. Blackert and S. Siegel, "Ist in der wissenschaftlich-technischen Information Platz für die Informetrie?" Wissenschaftliches Zeitschrift TH Ilmenau, vol. 25(6), pp. 187-199, 1979.

[2] W. W. Hood and C. S. Wilson, "The literature of bibliometrics, scientometrics and informetrics". Scientometrics, vol. 52(2), pp. 291-314., 2001.

[3] O. Nacke, "Informetrie: eine neuer Name für eine neue Disziplin". Nachrichten für Documentation, vol. 30(6), pp. 219-226, 1979.

[4] V. V. Nalimov and Z. M. Mul'čenko, Naukometrija. Nauka, Moskva, USSR, 1969.

[5] A. Pritchard, "Statistical bibliography or bibliometrics ?" Journal of Documentation, vol. 25, pp. 348-349, 1969.

[6] Scopus database, http://www.scopus.com, 18-2-2017

[7] J. Tague-Sutcliffe, "Quantitative methods in documentation". In: Fifty Years of Information Progress: a Journal of Documentation Review, pp. 147-188, Aslib, London, UK, 1994.

[8] M. Thelwall, L. Vaughan and L. Björneborn, "Webometrics, Annual Review of Information Science and Technology", vol. 39, pp. 81-135, Medford, NJ: Information Today, 2005.

Structuring E-Learning Multi-Criteria Decision Making Problems

Nikola Kadoić, Nina Begičević Ređep, Blaženka Divjak
Faculty of Organization and Informatics, Varaždin, Croatia
{nkadoic, nbegicev, bdivjak}@foi.hr

Abstract - Problem structuring is one of the most critical phases of decision making process. A well-posed problem has direct impact on effective decision making, especially when we use the multi-criteria decision making methods. There are different decision making methods that have been used for decision making on e-learning issues in higher education, but the most suitable method for this kind of problems is the Analytic Network Process (ANP). ANP meets all the theoretical requirements of decision making in higher education, but policy makers use it very rarely in practice because of its implementation weaknesses. One of the weaknesses is a lack of support in structuring problem in the form of a network. This paper brings an overview of several problem structuring methods and approaches, such as simple top-down and bottom-up approaches, the PrOACT approach, ISM (Interpretative Structural Modelling), DEMATEL (Decision Making Trial and Evaluation Laboratory) and the PAPRIKA structuring method. It also brings analysis of how those structuring methods and approaches help overcome some of the ANP weaknesses. Finally, we provide some recommendations of how to design a new problem structuring method that fits the ANP needs.

I. INTRODUCTION

The following quote *"A good solution to a well-posed problem is almost always a smarter choice than an excellent solution to a poorly posed one."* [1] is a very popular quote in decision making field. It highlights the importance of good decision making problem analysis before making any strategic decision. When we make decisions on different strategic e-learning issues and challenges, a real problem analysis before decision making is also requested in the e-learning field.

There are different approaches that we can use for decision making problem structuring. Depending on characteristics of the field in which we make decisions, different problem structuring method(s) are appropriate. E-learning belongs to the field of education in general, but here it will be related to the higher education (HE) field. To identify the most suitable method for decision making on e-learning issues, we follow the next steps:

(1) firstly, we analyse characteristics of decision making in e-learning and HE,

(2) secondly, we analyse the characteristics of decision making methods to be applicable in e-learning and HE field, and

(3) finally, we analyse decision making methods that shall apply in HE and e-learning field to identify

their demands regarding structuring of decision making problem.

Some of those steps are already partly investigated. In [2] author identified characteristics of decision making in e-learning and HE. In [3], [4] authors identified characteristics of decision making methods which would be applicable in the area of HE and e-learning. These features are: problem structuring when multiple perspectives and levels of decision making have to be involved, modelling influences between decision making elements, supporting both – qualitative and quantitative scales (criteria), supporting group decision making, enabling sensitivity analysis including risk, opportunities, benefits and costs. The only method that fits all the demands and characteristics is the Analytic Network Process (ANP) [3], [5].

ANP is a multi-criteria decision making method introduced by Saaty [6] as a generalisation of the Analytic Hierarchical Process (AHP). AHP method is one of the most widely exploited multi criteria decision-making methods in cases when the decision (the selection of given alternatives and their prioritising) is based on several tangible and intangible criteria (sub-criteria). The process of complex decision problem solving is based on the problem decomposition into a hierarchy structure which consists of the goal, the criteria, the sub-criteria and the alternatives. In ANP, decision making problems are structured in the form of a network. The basic structure of ANP is an influence network of clusters and nodes (criteria) contained within the clusters [7]. A network has clusters of elements, with the elements in one cluster being connected to elements in the other cluster (outer dependence) or the same cluster (inner dependence). In outer influence one compares the influence of elements in a cluster on elements in the other cluster with respect to a control criterion; and in inner influence one compares the influence of elements in a group on each other.

Priorities in a network are established in the same way as in AHP using pairwise comparisons and judgments based on the Fundamental Scale (1 to 9 scale of absolute numbers) [6] and deriving priorities as the eigenvector of the judgment matrices. The outline of the ANP steps can be found in [7].

II. KEY FEATURES OF PROBLEM STRUCTURING METHODS TO FIT THE ANP NEEDS

Even though in the ANP general directions are given of how to structure decision making problem, in practice policy makers often experience some issues related to problem structuring as well as applying other ANP steps. According to the literature and authors' experience in the ANP implementation, some of those issues are:

- Structuring decision making problem into optimal number of clusters and elements in clusters is a challenge,

- Problem structuring procedures in the ANP do not include identifying the weights of influences between criteria which makes criteria pairwise comparisons difficult, sometimes not even understandable for decision makers,

- Similarly, on cluster level, the cluster pairwise comparisons are also difficult, especially in the situations when policy makers have to pairwise compare same clusters with respect to several different clusters [8], [9],

- Problem structuring can result with clusters which are strongly inter-connected (many influences between criteria of various clusters), but weakly intra-connected (small number of influences between criteria in the same cluster). That makes cluster comparison more confusable (see Figure 1: pairwise comparison of clusters 1-2-3-4-5 and A-B-C is more confusable in the second example than in the first),

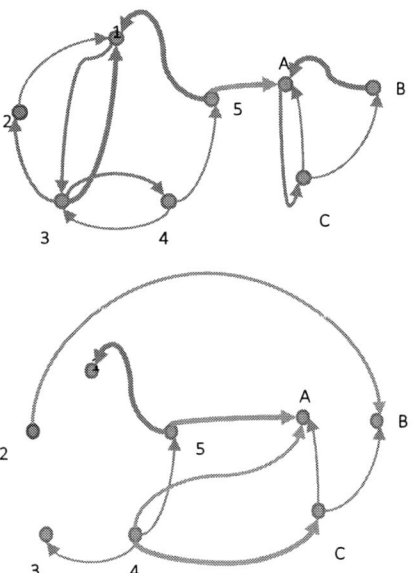

Figure 1. Some possible inter- and intra- connections between clusters 1-2-3-4-5 and A-B-C

- Decision making problem structuring can result with clusters which contain criteria that belong to different fields of expertise. Policy makers cannot make accurate pairwise comparisons and judgements if they are not experts in all needed fields of expertise, e.g. it is difficult to compare ICT and accounting criteria for expert in the ICT field. So, we would like for clusters to contain same-profession criteria, like in the paper [10],

- Finally, decision making problem structure has a direct influence on the number of pairwise comparisons [11].

To conclude, we identified several key features of problem structuring features to fit the ANP needs. These features are:

(1) identifying criteria (nodes in clusters);

(2) modelling influences between criteria (causality);

(3) identifying weights of influences between criteria;

(4) forming network structure;

(5) forming clusters;

(6) forming clusters with strong inter-connection and weak intra-connection; and

(7) forming same-profession clusters.

Some of the examples of the ANP problem structures can be found in [12] (case: strategic planning and decision making on e-learning implementation on the institutional level), [13] (case: evaluating e-Learning platform) and [14] (case: ODL system selection). The above mentioned decision making problem structures are created based on the literature analysis and brainstorming. Details about that approaches will be given in III.-A.

In next section of the paper, we will present several structuring methods and approaches, describe how to use them and analyse them regarding the ANP needs. In section 4 we will summarise results of analyses from section 3 in table form. Finally, in section 5 we will give some recommendations of possible structuring method which will combine advantages of the existing structuring methods.

III. ANALYSIS OF DIFFERENT PROBLEM STRUCTURING APPROACHES AND METHODS

Problem structuring methods that we will analyse in this paper are (1) general top-down and bottom-up approach, (2) the PrOACT approach, (3) the Interpretive structural modelling (ISM), (4) the Decision making trial and evaluation laboratory (DEMATEL) and (5) the PAPRIKA structuring method.

A. General top-down and bottom-up approaches

Top-down and bottom-up are basic and the simplest approaches of decision making problem structuring:

- By using a top-down approach, first, we identify the decision making goal. Then we choose networks that will be analysed. Decision making problems can be analysed from a position of four merits: benefits (B), opportunities (O), costs (C) and risks (R); and depending on the problem we can analyse one or more networks. Then we identify control criteria for each merit (and sub-criteria, if applicable), clusters for each control (sub-) criterion and finally identify criteria for each cluster [7].

- Opposite approach, bottom-up approach means doing all steps like in top-down approach, but in reverse order, starting from criteria, through clusters, control (sub-)criteria and merits to goal.

Decision makers often combine those two approaches and structure problem in both directions simultaneously. Structuring procedure finishes when results from applying both approaches "meet each other". Depending on e-learning problem complexity, some levels of problem structure will or will not be present. More complex problems might have all mentioned levels, but less complex problems might not contain e.g. control criteria (or sub-criteria).

When we analyse BOCR aspects of decision making problem, some of those aspects can be modelled as networks and some as hierarchies. For example, benefits (B) can be modelled as a network, and costs can be modelled as a hierarchy, like in [15]. Of course, the ANP method will be applied to network models and the AHP to hierarchy models.

Regarding fitting the ANP needs, we can conclude that top-down, bottom-up or combined approaches fit some of the ANP needs. Firstly, we can identify decision making criteria, but we cannot be sure that we covered all the important criteria. Secondly, we can model influences between criteria, but not determine the weights of those influences. Criteria are formed into clusters by the decision maker. There are no procedures that will ensure an appropriate number of criteria per cluster, consider influences between criteria when forming clusters or group same-profession criteria in the same cluster. All of this depends on the decision maker and his/her e-learning problem knowledge, analytics capabilities and how much knowledge (s)he has on the ANP method.

B. The PrOACT approach

The PrOACT approach (also known as a proactive approach) represents decomposition of decision making problem on several main elements [1]:

1. Pr (Problem). Problem is an entity which poses a barrier for a particular group of people in certain time and place;

2. O (Objectives). Objectives are goals that we want to achieve by solving the problem. Objectives can be created by using top-down or bottom-up approaches. Also, a method called problem tree can be used in a way that for the defined problem (Pr) we identify problem-sources and problem-consequences. When we get the whole list of problems, then we can define objectives which we will respond to problems. Finally, after objectives are defined, the criteria such as measures of objectives and their scales are determined.

3. A (Alternatives). Alternatives are possible decisions, choices and between them, we want to choose the best one. Some decision making problems have clear alternatives, and in some cases, we must analyse the problem and goal deeply to make the right definition. Methods that

can be helpful in the phase of creating alternatives are: brainstorming/brainwriting, case studies, focus groups, nominal group technique, DELPHI, morphological analysis, Theory of Solving Inventive Problems (Russian: *Theoria Resheneyva Isobretatelskehuh Zadach*, TRIZ).

4. C (Consequences). Consequences are values that alternatives achieve per each criterion. Usually, multi-criteria decision making problem structures are described in a table form (decision making matrix or table of values).

5. T (Trade-offs). Trade-offs mean expressing the values of a certain criterion in terms of another criterion. Trade-offs are mainly used in Even Swaps method [1].

The PrOACT approach has been introduced by authors of Even Swaps method, and it has been designed for purposes of Even Swaps method [1]. However, this approach can be used with other methods which require decision making table as their input. For example, Electre, Topsis, Promethee and some other methods can benefit from applying PrOACT in the problem structuring phase. Still, most of them would also need some additional data that is not already provided in the table of values that was formed as a result of applying the PrOACT approach.

The PrOACT approach can be used for identifying criteria, and similarly to top-down, bottom-up and combined approaches. However, we cannot be sure that we covered all the criteria. Regarding fitting of the ANP needs, the PrOACT approach is not very useful, especially not for complex decision making problems such as strategic e-learning decision making problems. Namely, the PrOACT approach is a one-level approach: there are no procedures that will guide us to define merits, control criteria, clusters, and criteria. Also, the PrOACT approach does not model influences between criteria which means that the resulted model is not the network. All generated criteria form one cluster.

C. Interpretative Structural Modelling (ISM)

The ISM method is almost always combined with Delphi method. Therefore, Delphi can be considered as the first step in the ISM process. The role of Delphi in the ISM process is to ensure a complete list of criteria that describe strategic e-learning decision making problem [16]. Conducting the ISM includes active involvement of decision making problem experts and literature review.

The ISM is useful for analyzing the complex socioeconomic systems. It also helps to impose order and direction on the complexity of relationships among elements of a system. The ISM has two components [16]:

1. Building the hierarchical relationship – modelling influences between elements by using basics of graph theory.

2. Analysis using the MICMAC matrix (fra. *Matriced' Impacts Croise's Multiplication Appliquée a UN Classement*) which consists of ties between elements (criteria). It is used to analyse the driving power and dependence power

of elements (criteria). It further helps to find the key criteria that are driving the whole process. MICMAC provides valuable insights about the relative importance and interdependencies among the criteria.

Steps of ISM (shortened according to [16], [17], [18]):

1. Identification of decision making elements – conducting Delphi method with experts and literature review to get a full list of criteria.

2. The creation of Reachability Matrix - this is a quadrate matrix of criteria with influences between them. In the matrix, on the address (x, y) can be 0 or 1. If there is 1, it means that criterion x has an influence on criterion y, and if there is 0, it means that x has no influence on x. Also, the concept of transitivity is applied.

6. Partitioning the Reachability Matrix- Reachability Matrix is partitioning into levels (clusters) to get a hierarchy of relationships between criteria.

7. MICMAC analysis: for each criterion, we should calculate *driving power* (summing rows) and *dependence power* (summing columns in Reachability Matrix).

8. Building ISM model – a hierarchy of relationships in decision making problem is built.

The main advantages of the ISM are [19]: systematic procedure; efficiency (when the ISM is software-supported); the ISM results and the model are understandable to users; it focuses users to think about only one aspect of the problem at the time. Disadvantages are: in the case of a vast number of elements, the process of conducting the ISM can be very tiring; experts from decision making problem domain are mandatory.

ISM becomes very desirable regarding fitting the ANP needs, but does it fit all the needs? Surely, it identifies the criteria, and here users are advised to conduct Delphi and literature review to be sure that they covered all decision making problem aspects (criteria). Also, systematically we can identify influences between criteria, but the weights of those influences are still missing. On the other hand, the ISM model is not a network structure with a cluster that is required in the ANP. However, the partitioning procedure can be interesting regarding creating clusters with strong inter-connection and weak intra-connection. However, to achieve that, a further adaption of the method is needed. Finally, the process of partitioning of Reachability Matrix does not consider the field of expertise (profession) of criteria when levels are created; so, it is not ensured that levels contain same-profession criteria.

D. The Decision Making Trial and Evaluation Laboratory (DEMATEL)

There are some similarities between ISM and DEMATEL [20]. The main goals of DEMATEL method are similar to the purpose of ISM. There are two main results of DEMATEL:

1. Casual diagram (also known as impact-relation map, [21]) – diagram of only significant influences between elements (criteria in decision making problem), which shows only influences that are over threshold value [22].

2. Relation Matrix – describes influences between criteria. Weights of influences are included now. In most often cases, five-scale 0-4 is used: 0, 1, 2, and 3 represent 'No influence', 'Low influence', 'High influence', and 'Very high influence' respectively [23].

Like ISM, DEMATEL has also been already applied in combination with ANP. DEMATEL and ISM provide a systematic, logical reasoning process to determine causality. They clearly delineate the relationships of the complex elements at the system level, direction and impact [20]. DEMATEL can propose the most important criteria which affect other criteria. DEMATEL can reduce the number of criteria [24].

Regarding fitting the ANP needs, by applying DEMATEL, we can model influences between criteria, as well as calculate weights of influences between criteria. Like in top-down, bottom-up and combined approaches, as well as in the PrOACT, we cannot be sure if we covered all the relevant criteria. The result of DEMATEL, impact-relational map (IRM), has a network structure, but this structure is not usable for ANP. However, network structure that would come as a result of drawing relation matrix would be very interesting regarding ANP. On the other hand, the procedure of how input data for IRM are calculated might be helpful in creating a cluster with strong inter-connection and weak intra-connection relationships because in IRM we draw only the strongest relationships. Other procedures for forming cluster are not available, and neither are procedures for creating same-profession clusters.

E. The PAPRIKA method

1000Minds applies patented PAPRIKA method – an acronym for *Potentially All Pairwise RanKings of all possible Alternatives* [25]. This method requires special decision making problem structure, so we will explain it here and later analyse in related to ANP. 1000Minds is an online suite of tools and processes to help individuals and groups make decisions, and also to understand other people's choices. 1000Minds has tools for decision-making, prioritisation and discovering stakeholders' preferences via conjoint analysis [25]. Depending on the application, 1000Minds can also help the user to think about the value for money of alternatives and allocate budgets or other scarce resources.

To be able to apply PAPRIKA method, structuring of a decision making problem consists of several steps:

1. identifying all criteria for decision making,

2. identifying all values that some alternative can achieve per certain criterion,

3. making pairwise comparisons of all possible pair combinations of possible alternatives' values (e.g. what do you prefer: a hypothetical alternative with values *x* on criterion 1 and *y* on criterion 2 OR a hypothetical alternative with

values z on criterion 1 and v on criterion 2) to get criteria weights and alternative priorities.

This decision making problem structuring reminds on modelling in Dex method. DEX is a qualitative multi-criteria method, in which all criteria are represented by qualitative (symbolic, verbal) attributes. The attributes are structured into a hierarchy, and the evaluation of alternatives is governed by decision rules [26].

This problem structuring method has many weaknesses regarding fitting the ANP needs. First of all, just like the PrOACT, criteria are identified in one-level approach. Then, there are no hierarchical or network levels. Finally, this method does not fit most of the ANP demands, and the features of the method do not contribute to the ANP needs.

IV. RESULTS: HOW DIFFERENT PROBLEM STRUCTURING METHODS FIT THE ANP NEEDS

In Table II we have summarised how different problem structuring methods and approaches fit the ANP needs.

TABLE II. HOW DIFFERENT DECISION MAKING METHODS FIT HE AND E-LEARNING DEMANDS

ANP demands	Top-down, bottom-up	PrO ACT	IS M	DEMA TEL	1000 Minds
Identify criteria	+/-	+/-	+	+	+/-
Modelling influences between criteria	+/-	-	+	+	-
Weights of influences	-	-	-	+	-
Network structure	+	-	+	+	-
Hierarchy of criteria sets	+/-	-	-	-	-
Forming clusters	+/-	-	+/-	+/-	-
Strong inter- and weak intra-connection	-	-	+/-	+/-	-
Same-profession clusters	+/-	-	-	-	-
Cluster size	+/-	-	-	-	-

If we simply count number of fits, we can say that the method that the best fits the ANP needs is DEMATEL. Also, if the top-down or bottom-up approach is implemented by experts in both, decision making problem field and ANP method, the results might be even better than in DEMATEL - if experts pay attention to (1) profession of criteria (create same-profession clusters) and (2) connections between criteria when forming clusters (strong inter- and weak intra-connection). On the other hand, there is room for improvement of the structuring method.

V. RECOMMENDATIONS FOR UPGRADED PROBLEM STRUCTURING METHOD

In this section, we give some recommendations for upgraded problem structuring method which will combine good sides of the presented model structuring approaches and upgrade them with some additional features:

1. The starting point of the upgraded method is Delphi and literature review. Those methods

ensure a list of all criteria that are relevant for certain e-learning problem.

2. All identified criteria will be grouped by experts or Q-sorting into several groups that are related according to four merits: B, O, C and R, and further analysis goes separately for each merit.

3. To all identified criteria, we join a profession (field of expertize). A profession will be considered in step 5 as a factor for creating same-profession clusters.

4. Now, in each merit, we identify influences between criteria and calculate their weights (as in DEMATEL).

5. This step requires the development of clustering procedure that will separate weighted network of criteria into clusters. Those clusters will consist of same-profession criteria (as a feature with the highest priority) and will have weak inter- and strong intra- connection between criteria. Cluster size should be between 5 to 9 (number of criteria in the cluster) in most of the clusters. To develop that procedure, algorithms that resulted in ISM model and IRM model can be analysed and possibly reused, as well as different cluster algorithms, such as algorithms in the Pajek [27], an algorithm for affinity analysis [28] and others.

After developing the method, it should be evaluated by using qualitative and quantitative analysis, including software implementation which will simplify new structuring method applications. The proposed method will fit the best into ANP needs, but it requires considerable resources in expertize (use of experts), time and professional guidance.

VI. CONCLUSION

Dealing with different e-learning challenges systematically requires making the proper strategic decisions related to those challenges. Prerequisite for making a right decision requires a good structure of decision making problem. The method that meets the most characteristics of decision making in HE is the ANP method.

In this paper, we presented several problems structuring approaches and described how they fit the ANP needs. We conclude that the most suitable structuring method regarding the ANP needs is the DEMATEL. However, the DEMATEL still has weaknesses, so we gave some recommendations for upgraded problem structuring method that will be based on the DEMATEL.

ACKNOWLEDGMENT

Croatian Science Foundation has supported this work under the project Higher Decision IP-2014-09-7854.

REFERENCES

[1] J. Hammond, R. Keeney, and H. Raiffa, *Smart Choices: A Practical Guide to Making Better Decisions*. 1999.

[2] B. Divjak, 'Challenges of Strategic Decision-Making within Higher Education and Evaluation of the Strategic Decisions', in *Central European Conference on Information and Intelligent Systems*, 2016, pp. 41–46.

[3] B. Divjak and N. Begicevic, 'Strategic Decision Making Cycle in Higher Education: Case Study of E-learning'. International Conference on E-learning 2015, p. 8, 2015.

[4] B. Divjak, 'Development of a methodological framework for strategic decision-making in higher education – a case of open and distance learning (ODL) implementation (project application)', Varaždin, 2014.

[5] R. Wudhikarn, 'An efficient resource allocation in strategic management using a novel hybrid method', *Management Decision*, vol. 54, no. 7, pp. 1702–1731, Aug. 2016.

[6] T. L. Saaty, *Decision Making with Dependence and Feedback: The Analytic Network Process : the Organization and Prioritization of Complexity*, Second and. New York: RWS Publications, 2001.

[7] T. L. Saaty and B. Cillo, *A Dictionary of Complex Decision Using the Analytic Network Process, The Encyclicon, Volume 2*, 2nd ed. Pittsburgh: RWS Publications, 2008.

[8] M. A. Ortíz, H. A. Felizzola, and S. N. Isaza, 'A contrast between DEMATEL-ANP and ANP methods for six sigma project selection: a case study in healthcare industry', *BMC Medical Informatics and Decision Making*, vol. 15, no. S3, p. S3, Dec. 2015.

[9] M. M. Tavakoli, H. Shirouyehzad, and R. Dabestani, 'Proposing a hybrid method based on DEA and ANP for ranking organizational units and prioritizing human capital management drivers', *Journal of Modelling in Management*, vol. 11, no. 1, pp. 213–239, Feb. 2016.

[10] E. Mu and H. A. Stern, 'The City of Pittsburgh goes to the cloud: a case study of cloud solution strategic selection and deployment', *Journal of Information Technology Teaching Cases*, vol. 4, no. 2, pp. 70–85, Jan. 2015.

[11] M. Castillo and R. Zarama, 'APPLICATION OF THE ANALYTIC NETWORK PROCESS (ANP) TO ESTABLISH WEIGHTS IN ORDER TO RE-ACCREDIT A PROGRAM OF A UNIVERSITY', in *Proceedings of the International Symposium on the Analytic Hierarchy Process*, 2009, pp. 1–14.

[12] N. Begicevic, B. Divjak, and T. Hunjak, 'Comparison between AHP and ANP: Case Study of Strategic Planning of E-Learning Implementation', *Development*, vol. 1, no. 1, pp. 1–10, 2007.

[13] S. Sadi-Nezhad, L. Etaati, and A. Makui, 'A fuzzy ANP model for evaluating e-learning platform', *Lecture Notes in Computer Science (including subseries Lecture Notes in Artificial Intelligence and Lecture Notes in Bioinformatics)*, vol. 6096 LNAI, no. PART 1, pp. 254–263, 2010.

[14] Z. KAMIŞLI ÖZTÜRK, 'USING A MULTI CRITERIA DECISION MAKING APPROACH FOR OPEN AND DISTANCE LEARNING SYSTEM SELECTION', *Anadolu University Journal of Science and Technology-A Applied Sciences and Engineering*, vol. 15, no. 1, p. 1, May 2015.

[15] B. P. M. S. (BPMSG), 'Analytic Network Process ANP - Introduction', *Lecture on Youtube*, 2011. [Online]. Available: https://www.youtube.com/watch?v=ow-BUs7ojaQ.

[16] A. K. Bhadani, R. Shankar, and D. V. Rao, 'Modeling the barriers of service adoption in rural Indian telecom using integrated ISM-ANP', *Journal of Modelling in Management*, vol. 11, no. 1, pp. 2–25, Feb. 2016.

[17] R. Attri, N. Dev, and V. Sharma, 'Interpretive Structural Modelling (ISM) approach : An Overview', *Research Journal of Management Sciences*, vol. 2, no. 2, pp. 3–8, 2013.

[18] P. Sharma, G. Thakar, and R. C. Gupta, 'Interpretive Structural Modeling of Functional Objectives (Criteria ' s) of Assembly Line Balancing Problem', *International Journal of Computer Application*, vol. 83, no. 13, pp. 14–22, 2013.

[19] U. Khan and A. Haleem, 'Improving to Smart Organization', *Journal of Manufacturing Technology Management*, vol. 26, no. 6, pp. 807–829, Jul. 2015.

[20] Y. Shih-Hsi, C. C. Wang, L.-Y. Teng, and Y. M. Hsing, 'Application of DEMATEL, ISM, and ANP for key success factor (KSF) complexity analysis in R&D alliance', *Scientific Research and Essays*, vol. 7, no. 19, pp. 1872–1890, 2012.

[21] E. Falatoonitoosi, S. Ahmed, and S. Sorooshian, 'Expanded DEMATEL for Determining Cause and Effect Group in Bidirectional Relations', *The Scientific World Journal*, vol. 2014, pp. 1–7, 2014.

[22] J. Shao, M. Taisch, M. Ortega, and D. Elisa, 'Application of the DEMATEL Method to Identify Relations among Barriers between Green Products and Consumers', *17th European Roundtable on Sustainable Consumption and Production - ERSCP 2014*, pp. 1029–1040, 2014.

[23] Y. Yang, H. Shieh, J. Leu, and G.-H. Tzeng, 'A novel hybrid MCDM model combined with DEMATEL and ANP with applications', *International Journal of Operations Research*, vol. 5, no. 3, pp. 160–168, 2008.

[24] B. Chang, C. W. Chang, and C. H. Wu, 'Fuzzy DEMATEL method for developing supplier selection criteria', *Expert Systems with Applications*, vol. 38, no. 3, pp. 1850–1858, 2011.

[25] P. Hansen and F. Ombler, '1000Minds.com', *https://www.1000minds.com*, 2017. [Online]. Available: https://www.1000minds.com. [Accessed: 01-Feb-2017].

[26] M. Bohanec, 'Qualitative MultiCriteria Modelling Method DEX: Approach, Recent Advances and Applications', in *Book of Abstracts, 16th International Conference on Operational Research, KOI 2016*, 2016.

[27] A. Ferligoj and V. Batagelj, 'Some types of clustering with relational constraints', *Psychometrika*, vol. 48, no. 4, pp. 541–552, Dec. 1983.

[28] J. Brumec, 'Optimizacija strukture složenih informacijskih sustava', *Journal of Information and Organizational Sciences*, vol. 17, pp. 1–23, 1993.

Introducing Gamification into e-Learning University Courses

A. Bernik*, D. Radošević*, and G. Bubaš*

* University of Zagreb, Faculty of Organization and Informatics, Varaždin, Croatia

andrija.bernik@foi.hr; danijel.radosevic@foi.hr; goran.bubas@foi.hr

Abstract - Research on educational e-courses that contain only a series of motivating elements of computer games but do not include playing computer games has intensified since 2010 [1] [6]. This field of research is called *gamification* and represents the *use of game elements* (mechanics, dynamics and aesthetics) in a field (education, marketing etc.) that is not a computer game. A review of literature related to the field of teaching with online courses in information technology (e.g. programming, software engineering) shows that the topic of *gamification* has so far been inadequately explored, with the lack of theoretical and empirical research that would involve gamification methodology. Previous studies have shown that gamification can have a positive impact on the pedagogical and psychological aspects of e-learning. In this paper empirical research is presented regarding the use of gamification in online teaching of programming. A gamified e-course was designed for the lectures in programming, and a possible positive effect was examined on the usage of learning materials in an experimental group of students who will use a gamified e-course (online system).

I. INTRODUCTION

The idea of using an e-learning system with the attributes of a computer game, but without game playing, has been a rapidly growing trend since 2010 [2][5][6]. Recent research on a similar topic of *serious games* confirms a positive attitude of subjects (learners) who used interactive dynamic systems (with an emphasis on learning) like computer simulations (economic, political, military etc.) where a student plays according to a predefined scenario and monitors the outcomes of his/her decisions on goal realization and learning outcomes. For this purpose commercial games can be used, as well as specialized educational games designed for specific fields of study.

The term "gamification" represents the use of elements (mechanics, dynamics and aesthetics) of computer games in the field that is not a computer game [2][5][6]. Gamification as a rising trend has been recognized by many researchers and institutions like *Gartner Research* [8][14][19][24][25].

In 2013 gamification was positioned as a technological innovation whose development should be followed with the exceptional importance (see: [4][9][15]). Previous studies have focused on proving that this approach gives positive results in various fields like business, marketing and education [12][17][20]. A simple search using the scientific literature search engine *Google Scholar* (http://scholar.google.com; February 2017) reveals that in the documents related to the year 2010 the term "gamification" appears only 173 times, while in those for the year 2013 and 2016 it appears 3,780 times and 8,410 times, respectively.

About a decade ago it was demonstrated that the use of elements of computer games can have a positive impact on the psychological characteristics of students and learning behaviour (for instance, see: [22]). A review of literature related to the field of teaching information technology (e.g. programming, software engineering) revealed that, until recently, there have been very few empirical studies related to gamification of respective online courses. Also, it has not been clearly defined what all the elements of computer games are that should be specifically taken into account when designing online courses, as well as how they can be effectively implemented in the (re)design of existing online learning systems.

In our study experimental research methodology will be used. A *traditional* (for the *control* group) and *gamified* (for the *experimental* group) online course will be designed to investigate the influence of the use of elements of computer games on learning outcomes in an online environment.

When creating a gamified online course, according to Iosup and Epema [11], teachers, designers and/or administrators of gamified online courses are expected to ensure at least *one week* for the consideration of the computer game elements which are included in the e-course, and at least *one day* for creating educational content. Furthermore, Iosup and Epema [11] state that it is necessary to set aside at least *two hours* to analyze questions for each teaching unit as well as *two days* for entering the results of each knowledge assessment. Finally, it is important to ensure one week for analysis of the transition to gamification.

II. RESEARCH BACKGROUND

With today's pervasive use of technology in everyday life, there is a growing need for technological advancement in the use of online education that would be based on pedagogical and psychological processes which positively influence the student's perception of the teaching content and their motivation for active participation, research and cooperation [1][6][7]. An earlier example is the use of 3D virtual worlds in online learning where *Second Life* was the most commonly used product.

Second Life enables online connectivity and teaching/training in 3D virtual space with the use of educational methods/metaphors that can be used in many other technological forms in the future.

A turning point for the use of gamification was the "GSummit 2014" conference held in San Francisco with numerous speakers, investors and members of the world's leading companies presenting and seeking gamified solutions for their products and systems [7]. Another conference entitled "Gamification World Conference 2016" was held in Madrid, Spain, with the presence of the most famous researchers of gamification methodology and technology in the world, such as [3], [13], [16], [26], [29].

According to Souza-Concilio and Pacheco [23], the implementation of elements of computer games is getting more visible in various fields including education, health and fitness, task management, environmental sustainability, science, user generated content and others. About 40 years ago Malone et al. [15] emphasized the need to make learning more interesting. As a confirmation of their claim, it must be noted the value of gamification market had risen to 513 million dollars by 2013, with its value increasing to 980 million dollars in 2014, amounting to as much as 2.8 billion dollars in 2016 [19][27][28].

The importance of gamification can also be illustrated by the recent EU project call in the Horizon 2020 funding scheme (ICT-21-2014) for research of gamification technologies with funding opportunity of eight million euros in total [9]. A more recent call was launched in 2016 for proposals that could be funded with up to one million euros (ICT-24-2016) [10].

The previously presented market value and research funding opportunities indicate the relevance of gamification research. In this paper the authors will focus on the context of online education, especially in the field of informatics and software engineering. Therefore related theoretical and empirical analyses are presented as well as a proposal for standardization of elements of computer games that might come into consideration for implementation with the learning management system (LMS) Moodle.

According to Nielson [18] and Schonfeld [21], the following 24 types of **gamification mechanisms** that are currently recognized and accepted in practice have been most frequently cited in the literature (see *Table 1*).

TABLE 1. MECHANICS AND AESTHETICS IN
A GAMIFICATION SYSTEM

Achievements	Bonuses	Countdown	Endless duration of the game
Duties / Challenges	Introduction with the information	Uncertainty / Detection	Levels
Behavioural momentum	"Combo" effect x3	Epic meaning	Loss of aversion
Productivity	Joint collaboration	Surprise	Conscious risk
Ownership	Regular rewarding	Advancement	Optimism
Points	Status	Tasks and challenges	"Addiction" / Commitment to the game

Each element in Table 1 can be categorized according to three attributes: (1) the *mechanics* of a game, (2) *benefits* and (3) *personality*. Mechanics of the game can be divided into *behaviour*, *feedback* and *promotion*. According to the categorization by Bartle [3], personality can be divided into *winner*, *wordsmith*, *collector* and *researcher* subtypes, all of which represent basic types of users or players in a game. The components of computer games (i.e. *mechanics*) shown in Table 1 are not new. It is the use of information and communication technology to support a more effective and visually attractive creation/application of a game that represents today's novel modality for gamification. In that respect, it should be noted that not all of the *elements* of computer games are appropriate for all *types of players*. However, most of those elements can find some application in business or education systems.

III. RESEARCH HYPOTHESIS

For this research paper only one directional hypothesis is defined:
• **H**: *An online course which is pedagogically designed with the application of the elements of computer games (i.e. gamified) will have a greater effect on the amount of use of online teaching materials in comparison with a course with the same educational content, but without the presence of elements of computer games.*

To confirm the hypothesis H1, first a research of literature was conducted focusing on the topic of *motivation of the participants in gamified online courses* and other related positive effects. The focus was also placed on online courses related to information and communication technologies in higher education institutions. In the *empirical phase* of our research an investigation was performed of the influence of gamification on the use of e-learning materials. In the analysis of empirical data the log report of participants' activities in the Moodle LMS was used for both the *experimental* and the *control* group of subjects.

IV. CURRENT STATE OF THE USE OF ELEMENTS OF COMPUTER GAMES IN E-LEARNING

The research that is presented in this paper began by collecting the views of teachers in two higher education institutions (HEI) from two Central European universities. A total of 43 correctly completed survey forms were collected from the HEI teachers. All of the subjects/respondents used the Moodle LMS in their academic teaching. Their courses were mostly delivered in the second or third year of an undergraduate study.

It must be noted that our survey respondents were not using the customization functionality of the Moodle LMS and their answer to the question "Do you use a special custom graphics template?" was "No" or "We have no choice". Also, most of the respondents (53%) were not familiar with the learning systems of *Khan Academy*, *Duolingo* or similar learning systems. This was not a positive indicator of how broad their knowledge of e-learning was since the aforementioned systems had received prestigious awards as well as introduced some innovative approaches to knowledge transfer via e-learning. Furthermore, in their response to the survey question regarding the use of external links (plugins/functionalities) that can be added into the Moodle LMS system *(Facebook, Twitter, Yahoo, YouTube, Gmail, G-search, Wiki)* as many as 58% of respondents chose the response "none of the above".

Other questions in our survey among academic teachers were related to the mode of presentation of learning materials in an online course. Our survey revealed that the majority of respondents preferred traditional *(offline)* teaching materials, but in their online course material they also frequently used static text that is accompanied with a PowerPoint presentation or a PDF document/article. In the Moodle LMS system the most widely used functionalities by our respondents were the following:

1. Forum (86%)
2. Achieved current points (67%)
3. Questionnaires (65%)
4. Multimedia (58%)
5. Bonus teaching materials (49%)
6. Editing of profile & avatar (28%)

According to the surveyed teachers, the most frequent activities that their students performed in the LMS environment were:

1. Use of quizzes and assignments (53%)
2. Feedback from students to the professor (53%)
3. Student cooperation on problem solving tasks (26%)
4. Individual voluntary casual tasks (26%)
5. Personalization of user interface (14%)

Most of the teachers agreed that story and motivational elements visible in computer games can have a positive impact on the interest of students regarding the teaching subject. Also, a considerable percent of respondents (49%) stated that they were not familiar with the *flow theory*, which is important in interpretation of *gamification* of e-learning, while only 37% of them replied that they were partially informed about flow theory.

Our survey of teachers at two academic institutions in Croatia revealed that the variety of their implementation of pedagogical elements related to computer games in their e-learning courses was, on average, rather low. As it is previously listed, the most widely used pedagogical features were discussion forums, questionnaires, quizzes, assignments, multimedia, bonus teaching materials, as well as feedback from the instructor to students, and *vice versa*. The surveyed teachers were not familiar with e-learning systems and products for e-learning like *Khan Academy* or *DuoLingo*. It would be of great value to the teachers to be included into an online course covering the possibilities of e-learning and digital networking tools which are more game-oriented.

In the continuation of this paper the authors will try to demonstrate that the use of elements of computer games (i.e. *gamification*) can positively affect the use of teaching and non-teaching material that is available to students within an e-learning system like Moodle.

V. PROCEDURE AND SUBJECTS

For the purpose of our study a new installation of Moodle 2.7. LMS was used with additional gamification components. The gamification elements were added in form of plugins because they were not available in the initial

version of Moodle 2.7 system. A conceptual model with potentially useful gamification elements was created on the basis of a survey of academic teachers who used e-learning, available plugins for Moodle 2.7 system, and *Octalysis Gamification Framework* [http://yukaichou.com/gamification-examples/octalysis-complete-gamification-framework/]. For the *experimental group* of students a *gamified* e-learning course for teaching computer programming (on the topic "Batch and Stack") was designed in Moodle with the use of selected elements of computer games. Conversely, for the *control group* of students, a traditional *non-gamified* online course was developed in parallel and with equal content in the Moodle system that was located on a separate physical server. The traditional non-gamified course had only three elements that were set besides the educational materials: profile and avatar area, use of forums, and nonlinear access to educational materials.

To complete the development of both e-learning courses (*gamified* and *traditional*), the teaching content and teaching materials of all the basic teaching material that was adapted for this research was standardized (made equal) for both e-learning courses and reduced to HTML text accompanied by pictures or videos. In other words, identical basic educational content and materials were placed in both e-learning systems. Also, the teaching materials and topics that were used *online* in both e-learning courses were *not used* for lecturing in *traditional classroom* face-to-face environments, including exercises in computer laboratories. Students were instructed to use online teaching materials in the gamified and non-gamified course completely alone, without any intervention from the teacher in the physical environment. In this manner, the subjective influence of teachers on students in both the *experimental* and the *control* group was considerably reduced, increasing the credibility and reliability of the experimental procedure. It must especially be noted that the teaching materials used in the course did not differ in its content or in the order in which they were listed the Moodle system. After the two weeks during which the students were involved with the online course materials their log entries were analysed to draw conclusion regarding hypothesis H1.

The subjects in this study were students of an informatics college in the Republic of Croatia. All of the students attended the course "Programming 2" at the undergraduate level of study of informatics in the winter semester of the academic year 2015/2016. The total number of subjects who were voluntary participants in the study was 201. The subjects were divided into two groups: the *experimental* group and the *control* group. Of the total number of subjects, 44 (or 21.9%) were female and 157 (or 87.1%) were of male gender. Their average age was 20 years. Information on our convenience sample of subjects is presented in more detail in *Figure 1*. The permission for students' participation in the study was requested and obtained by the relevant higher education institution authorities.

VI. RESULTS AND DISCUSSION

The analysis of the initial written test of prior knowledge (pre-test) in *Table 2* provides insight into the distribution of the subjects/respondents in the *control* and *experimental* group. The data presented in *Table 2* indicate that there was no statistically significant difference between the *experimental* (G_E) and the *control* (G_K) group regarding the results of *prior knowledge* testing (pretest). The value of the *t-test* was 0.57, with p>0.56. Also, there was only a slight difference between the mean results of the two groups (about 5% of the standard deviation). It can be concluded that the two groups were suitable for performing the subsequent experimental research procedure.

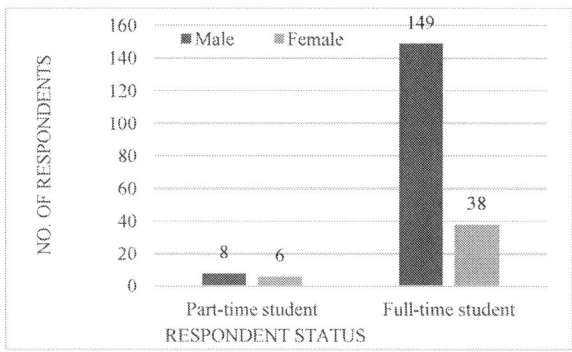

FIGURE 1. STATUS OF THE COURSE RESPONDENTS

TABLE 2. TESTING OF THE STATISTICAL SIGNIFICANCE OF DIFFERENCE IN *PRIOR KNOWLEDGE* (PRETEST RESULTS) BETWEEN EXPERIMENTAL (G_E; N=99) AND CONTROL (G_K; N=102) GROUP

Group	N	Mean	SD	t	p
G_E	99	15.57	4.17	0.57	0.5658
G_K	102	15.25	3.72		

In the continuation of this paper a report is presented on student activities that was generated based on the log entries from the Moodle system. This represents the final part of our data analysis which will enable the confirmation or rejection of our research hypothesis H1.

TABLE 3. REPORT ON THE NUMBER OF REGISTERED ACTIVITY INSTANCES (LOG ENTRIES) IN THE MOODLE SYSTEM FOR *EXPERIMENTAL* (N=88) AND *CONTROL* (N=102) GROUP OF STUDENTS

Activity name	Sum	Mean	
Learning outcomes	342	3.89	Activity of the *experimental* group
Forum	1214	13.80	
List of terms	838	9.52	
Learning outcomes	41	0.40	Activity of the *control* group
Forum	252	2.47	
List of terms	22	0.22	

* Mean is calculated as Sum/N

Table 3 provides a list of activities that were accessed by students within the *gamified* and *non-gamified* online course. For the *experimental* group of students, the average number of 9.07 access attempts was registered in log entries for all of the activities listed in *Table 3*, while for the students in the *control* group the average number of registered instances of access was only 1.03. The results in *Table 3* indicate a significant difference in favor of the *experimental* group. It can be concluded that the *experimental* group showed a much greater average interest in the available activities in the online course such as "Forum" (5.6 times greater), "Learning outcomes" (9.7 times greater) and "List of terms" (43.3 times greater).

TABLE 4. COMPARISON OF THE FREQUENCY OF ONLINE ACCESS TO BASIC EDUCATIONAL AND BONUS MATERIALS REGARDING THE LEARNING MATERIAL OF THE GENERAL E-COURSE TOPIC "BATCH AND STACK"

Activity name (learning content and bonus materials)	Experimental group		Control group	
	Sum	Mean	Sum	Mean
Teaching section 1a: Batch	430	4.89	184	1.8
Teaching section 2a: Charging and implementation of piles using fields	391	4.44	135	1.32
Teaching section 3a: Deleting a root and implementation of piles using integer fields	358	4.07	94	0.92
Teaching section 4a: Charging stacks	359	4.08	84	0.82
Teaching section 5a: Sequential deletion root piles (Heap Sort)	354	4.02	71	0.70
Bonus teaching content: Old and new names of standard C++ library	301	3.42	64	0.63
Teaching section 1b: Stack	251	2.85	108	1.06
Teaching section 2b: Adding (PUSH), reading and deleting the top of the stack (POP)	236	2.68	77	0.75
Teaching section 3b: Print whole and therefore deallocation	234	2.66	76	0.75
Bonus teaching content: The relationship between C and C ++	159	1.81	72	0.71
Total sum for all activities	**3073**	**34.92**	**965**	**9.46**

In case of this research, bonus learning materials were present in both e-courses. *Table 4* shows the access frequency / number of instances of access to course materials by respondents in the *experimental* and the *control* group with regard to *core learning content* and *bonus materials*. The *experimental* group of respondents used both the basic *teaching materials* and *bonus materials* which were added to the course to motivate students for additional

learning and to access content more frequently. It can be seen in *Table 4* that the *experimental* group of respondents accessed the teaching and bonus materials on 34.92 occasions on average, while the same kind of materials were accessed by the *control* group only on 9.46 occasions on average. Therefore, it can be calculated that regarding the data in *Table 4* the summative indicators of frequency of access to all of the listed learning activities and bonus materials were 3.69 times higher in favor of the *experimental* group of respondents. In other words, the students of the *experimental* group, who were involved with the *gamified* e-course, used learning and bonus materials *3.69 times more often* than the *control* group, which was involved in the use of the conventional *non-gamified* e-course.

The data presented in *Figure 2* demonstrate that the *experimental* group of respondents had continuous access to learning materials with decreased motivation in subsequent topics and learning activities (also see *Table 4*) as the e-course was approaching its end. A similar trend in accessing learning materials is visible for the *control* group of respondents, but to a somewhat lesser extent. It must be emphasized that, at one moment, the *experimental* group of respondents had nearly 5 times higher access frequency to teaching materials in comparison to the *control* group, which is a substantial indicator of the effect of gamification on student activity and motivation in e-learning.

However, it must be noted that some of the results of comparison presented in *Figure 2* could be attributed to the greater visual quality of the *gamified* e-course, which had a more appealing appearance to the *experimental* group of respondents. Still, in other aspects the core learning materials for both the *experimental* and the *control* group were almost identical.

From the indicators that are presented in *Table 3* and *Table 4* as well as in *Figure 2*, it can be concluded that the *experimental* group of respondents, which used the gamified version of the online course in computer programming (on the topic "Batch and Stack"), had a significantly greater motivation to access and use the online learning material, which is visible by the analysis of the registered log entries.

In fact, the experimental group had a greater frequency of access within the Moodle e-course to all available teaching, non-teaching and bonus materials. The conclusion is that hypothesis H1 is confirmed and that *an online course which is pedagogically designed with the application of elements of computer games (i.e. gamified) will have a greater effect on the amount of use of online teaching materials in comparison with a course with the same educational content, but without the presence of the elements of computer games.*

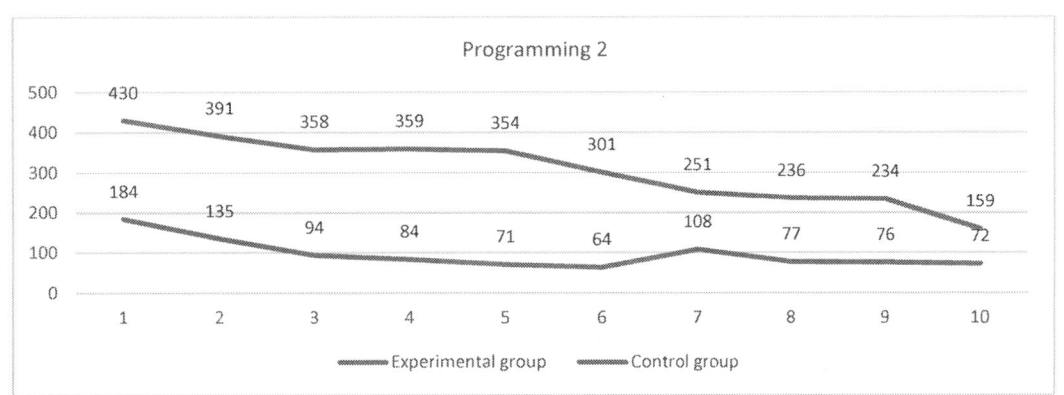

FIGURE 2. FREQUENCY OF ACCESSING EDUCATIONAL MATERIALS IN 10 SUBSEQUENT COURSE TOPICS
(LISTED IN TABLE 4) FOR EXPERIMENTAL AND CONTROL GROUP

VII. CONCLUSION

Use of the elements of computer games (*gamification*) can be a meaningful way to increase student motivation and improve the educational effectiveness of online courses. The authors of this study conclude that the pedagogical design based on *gamification* places greater emphasis on student motivation in the learning process and makes an online course more interesting, as well as increases students' willingness to learn and their engagement with course materials.

In our study an *experimental research procedure* was used since two separate and equivalent groups of students were engaged in learning equal core learning material in two pedagogically different online learning environments.

At the beginning of our empirical procedure, the *experimental* and the *control* group of students were subjected to a written test of prior knowledge. Based on these results (average number of points) the students were assigned to either the experimental or the control group. Since the 'Programming 2' course has a total of 14 groups in computer labs, seven of them were treated as *control* and the other seven as *experimental* study groups. Online learning materials related to the topic "Batch and Stack" were created within the 'Programming 2' course with a total of 10 subtopics with teaching and bonus materials which were available to students for a period of 14 days.

The hypotheses H1 for our study stated the following: *An online course which is pedagogically designed with the application of elements of computer games (i.e. gamified) will have a greater effect on the amount of use of online teaching materials in comparison with a course with the same educational content, but without the presence of elements of computer games.*

In the central empirical part of our study it was confirmed that the e-learning course which is designed by applying elements of computer games (i.e. *gamified*) can have a positive effect resulting in greater frequency of use of teaching materials compared to the course with the same educational content, but without the presence of elements of computer games. Therefore, the hypotheses H1 was confirmed. This is clearly evident in the graphical analysis of objective indicators of online activity of *experimental* and *control* group of subjects regarding their frequency of access to learning materials placed in the Moodle system (see *Figure 2*).

The main contribution of this paper is related to the verification of the hypothesis that gamification can lead to greater usage of the gamified e-course educational materials measured by objective indicators that are present within the Moodle system in terms of participants' activity logs.

LIMITATIONS

The main limitations of our study are related to the following:

• **Students' obligations.** Our research was conducted during regular classes at the higher education institution. In addition to having to use the experimental e-learning system, students had obligations in other courses. It is important to emphasize that participation was voluntary. Conducting the experiment was planned at a time when students did not have a mid-term or final exam. Individual tasks and obligations at the level of the study program could not be included in the analysis of each student workload, but it is highly possible that students had other obligations related to other courses during the use of the experimental system.

• **Time period of the use of online courses.** Research activities for each course were planned with particular regard to the academic syllabus and other educational assignments. It was decided that the measurement should be carried out in a short time-period during the two to three weeks after the students had gained access to learning materials in the online courses that were designed for the experiment.

REFERENCES

[1] J. M. Adeel, "The Benefits of Game Based Learning", Microsoft Partners in Learning, 2014.

[2] F. Aparicio, F. L. G. Vela, J. L. G. Sánchez, J. L. Isla, "Analysis and application of gamification", ACM Interaccion'12, pp. 113-126, Elche, Alicante, Spain, 2012.

[3] A. R. Bartle, "Players Who Suit MUDs", Journal of MUD Research, vol. 1, no. 1, 1996.

[4] B. Bruke, "Gartner says get gamification working for you", Gartner Report, Report ID:G00245563, URL: http://thisiswhatgoodlookslike.com/2013/03/24/gartner-says-get-gamification-working-for-you/, Accessed 1 February 2017.

[5] S. Deterding, "Gamification: Designing for motivation", interactions, vol. 19, no. 4, 14–17, 2012.

[6] S. Deterding, D. Dixon, R. Khaled, L. Nacke, "From game design elements to gamefulness: Defining gamification", MindTrek '11, Proceedings of the 15th International Academic MindTrek Conference: Envisioning Future Media Environments, pp. 9-15, Tampere, Finland, 28-30 September 2011.

[7] "Gamification Summit", San Francisco, CA 2014, URL: http://sf14.gsummit.com/, Accessed 1 February 2017.

[8] "Gartner Says By 2015, More Than 50 Percent of Organizations That Manage Innovation Processes Will Gamify Those Processes", Gartner Press Release, Egham, UK, 2011, URL: www.gartner.com/it/page.jsp?id=1629214, Accessed 1 February 2017.

[9] ICT 2014 - Information and Communications Technologies, Sub call of: H2020-ICT-2014, URL: http://ec.europa.eu/research/participants/portal/desktop/en/opportunities/h2020/topics/90-ict-21-2014.html#tab2, Accessed 1 February 2017.

[10] ICT 2016 - Information and Communications Technologies, Sub call of: H2020-ICT-2016, URL: https://ec.europa.eu/research/participants/portal/desktop/en/opportunities/h2020/topics/5088-ict-24-2016.html, Accessed 1 February 2017.

[11] A. Iosup, D. Epema, „An Experience Report on Using Gamification in Technical Higher Education", CM Technical Symposium on Computer Science Education, pp. 27-32, USA, 2014.

[12] J. T. Kim, W. H. Lee, "Dynamical model for gamification of learning (DMGL)", Multimedia Tools and Applications, vol. 74, no. 19, 8483-8493, 2015.

[13] J. M. Kumar, M. Herger, "Gamification at Work: Designing Engaging Business Software" [Online book], Interaction Design Foundation, Denmark, 2013, URL: https://www.interaction-design.org/literature/book/gamification-at-work-designing-engaging-business-software, Accessed 1 February 2017.

[14] H. LeHong, J. Fenn, "Key Trends to Watch in Gartner 2012 Emerging Technologies Hype Cycle", Gartner Inc. 2012, URL: https://www.forbes.com/sites/gartnergroup/2012/09/18/key-trends-to-watch-in-gartner-2012-emerging-technologies-hype-cycle-2/, Accessed 1 February 2017.

[15] T. W. Malone, M. R. Lepper, "Making learning fun: A taxonomy of intrinsic motivations for learning", In Snow, R. & Farr, M. J. (Eds.), Aptitude, Learning, and Instruction, vol. 3: Conative and Affective Process Analyses, pp. 223-253, Hillsdale, NJ, 1987.

[16] A. Marczewski, "A New Perspective on the Bartle Player Types for Gamification", 2013, URL: http://gamification.co/2013/08/12/a-new-perspective-on-the-bartle-player-types-for-gamification, Accessed 1 February 2017.

[17] J. McGonigal, "Reality is Broken: Why Games Make Us Better and How They Can Change the World", New York, NY: Penguin Press, 2011.

[18] B. Nielson, "Gamification Mechanics vs. Gamification Dynamics", Your Training Edge, 2013.

[19] P. Petridis, K. Hadjicosta, I. Dunwell, P. Lameras, T. Baines, V. Guang Shi, K. Ridgway, J. Baldin, and H. Lightfoot, "Gamification: Using gaming mechanics to promote a business". Proceedings of the Spring Servitization Conference, Aston University, Birmingham, UK, 2014.

[20] M. Rauch, "Gamification is Here: Build a Winning Plan", STC Intercom, Society for Technical Communication, pp. 7–12, 2012.

[21] E. Schonfeld, "SCVNGR's Secret Game Mechanics PlaydeckQ", 2010. URL: http://techcrunch.com/2010/08/25/scvngr-game-mechanics, Accessed 1 February 2017.

[22] M. Sharples, P. McAndrew, M. Weller, R. Ferguson R., E. FitzGerald, T. Hirst, M. Gaved, "Innovating Pedagogy 2013: Exploring New Forms of Teaching, Learning and Assessment to Guide Educators and Policy Makers", Open University Innovation Report 2, 2013.

[23] I. A. Souza-Concilio, B. A. Pacheco, "Games and Learning Management Systems: A Discussion About Motivational Design and Emotional Engagement", SBGames, 2013.

[24] J. Van der Meulen, R. Rivera, "Gartner's 2013 Hype Cycle for Emerging Technologies Maps Out Evolving Relationship Between Humans and Machines", Stamford, 2013.

[25] E. N. Webb, "Gemification in enterprise: Gartner Hype Cycles 2013", Gartner Inc., 2013., URL: http://gartner.com/newsroom/id/2575515, Accessed 1 February 2017.

[26] K. Werbach, D. Hunter, "For the Win: How Game Thinking Can Revolutionize Your Business", Wharton Digital Press, University of Pennsylvania, Philadelphia, 2012.

[27] Xu Y., "Literature Review on Web Application Gamifcation and Analytics", Collaborative Software Development Laboratory Technical Report, 11-05, University of Hawaii, 2011.

[28] M. Yana, "Applicability of the Concept "Gamification" Within Business Organizations", Master Thesis at University St. Kliment Ohridski, Faculty of Economics and Business Administration, Sofia, 2013.

[29] G. Zichermann, C. Cunningham, "Gamification by Design: Implementing Game Mechanics in Web and Mobile Apps", O'Reilly Media, Inc., 2011.

Perceived Security and Privacy of Cloud Computing Applications Used in Educational Ecosystem

Tihomir Orehovački *, Darko Etinger * and Snježana Babić **

*Juraj Dobrila University of Pula, Department of Information and Communication Technologies, Pula, Croatia
{tihomir.orehovacki, darko.etinger}@unipu.hr
** Polytechnic of Rijeka, Department of Business, Rijeka, Croatia
snjezana.babic@veleri.hr

Abstract – When employed in educational settings, cloud computing applications enable users to create, store, organize, and share divergent artefacts with their peers. As an outcome, they have a large number of users worldwide which makes them vulnerable to a variety of security and privacy related threats. With an aim to examine the extent of the perceived security and privacy in the context of cloud computing applications that are most commonly used for educational purposes, an empirical study was carried out. Participants in the study were students from two Croatian higher education institutions. Data was gathered by means of the post-use questionnaire. Study findings uncovered pros and cons of examined cloud computing applications with respect to the manner they are addressing security and privacy concerns of their users.

I. INTRODUCTION

The introduction of Cloud Computing technologies in higher education institutions improves the efficiency of existing resources usage, as well as the reliability and scalability of software tools and applications, enabling users to create, store, organize, and share divergent artefacts with their peers.

According to the National Institute of Standards and Technology (NIST) definition, "the cloud computing is a model for enabling convenient, resource pooling, ubiquitous, on-demand access which can be easily delivered with different types of service provider interaction" [1]. The cloud computing follows simple "pay as you go" (PAYG) model, where you pay for the services you've used [2].

Cloud computing is an emerging new computing environment for delivering computing services which can be categorized regarding its basic components, deployment models and service delivery models. An overview of Cloud computing components is presented in Figure 1. [3].

The demand for cloud computing applications in educational ecosystem builds on the promises of free to low-cost alternatives to expensive tools. Cloud computing applications are especially suitable for educational institutions lacking technical expertise [4] to support their own IT infrastructure. According to [5] and [6], students and higher education institutions benefited from the advantages and effectiveness that cloud computing applications provided them.

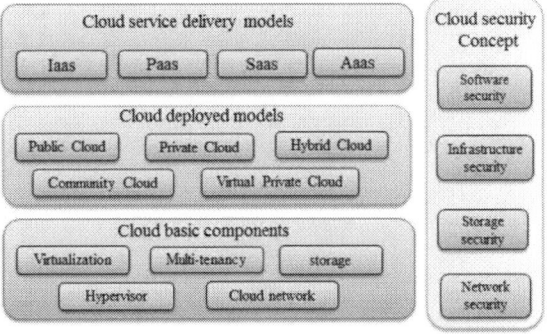

Figure 1. Cloud Computing Framework [3]

While cloud computing applications present a great opportunity for educational institutions, its usage raises concern about a variety of security and privacy threats, by placing a very large amount of student, teacher and institution data into the hands of a third-party service providers [7].

The objective of this paper is to examine the degree of perceived security and perceived privacy of cloud computing applications commonly used in educational settings.

The remainder of the paper is structured as follows. Brief theoretical foundation of our study is provided in next section. Findings of an empirical study are presented and discussed in third section. Concluding remarks, study limitations, and future work plans are offered in last section.

II. BACKGROUND TO THE RESEARCH

Most cloud computing applications used in the educational ecosystem today are SaaS cloud services that operate in the "public" cloud. These applications include productivity suites like Microsoft Office 365 and Google Apps along with data storage services such as Microsoft OneDrive and Google Drive.

One of the most significant barriers to cloud computing adoption are security and privacy [8], that relate to risk areas such as external data storage, dependency on the "public" internet, lack of control, multi-tenancy and integration with internal security.

Among the main privacy challenges for cloud computing are complexity of risk assessment in a cloud environment, emergence of new business models and their implications for consumer privacy and achieving regulatory compliance.

The key elements and attributes of security issues are categorized by authors such as [9], [10], [11], [3] and they include availability (certify information is available when needed), integrity (sanctuary information integrity) and confidentiality (prevent unauthorized disclosure). They further elaborate and identify security apprehensions that a cloud computing user should discourse with cloud computing providers before approving: regulatory compliance, user access, data segregation and location, disaster recovery and long-term viability.

The concept of trust is elaborated by [11], with numerous trust objects and measures that can operationalize the impact of trust on the adoption of technological innovations. They identified the following items as appropriate for the operationalization of the security & trust factor in the context of cloud computing: data security, trustfulness of the cloud service provider, contractual agreements and geographical location where data is stored and processed. According to [11], security is viewed as a composite notion, namely "the combination of confidentiality, the prevention of the unauthorized disclosure of information, integrity, the prevention of the unauthorized amendment or deletion of information, and availability, the prevention of the unauthorized withholding of information".

A mapping of cloud service and security requirements was carried out by [12], while [13] composed a list and description of cloud computing threats, compromised attributes and related studies.

Topic areas in information privacy research include among others, information privacy concerns, information privacy attitudes, trust and information privacy and information privacy practices [13].

Examining the individuals' security and privacy concerns with their intention to use mobile applications, [14] developed a research model from the principal tenets of the theory of planned behaviour (TPB) and protection motivation theory (PMT). The study by [15] has provided early empirical support for a model that explains the formation of privacy concerns from the CPM theory perspective. The authors state that the globalization of economies and information technology and the ubiquitous distributed storage and sharing of data puts the issue of privacy on the forefront of social policies and practices. Drawing on the CPM theory, [15] developed a model suggesting that privacy concerns form because of an individual's disposition to value privacy, or situational cues that enable one person to assess the consequences of information disclosure.

III. EMPIRICAL STUDY

A. Procedure

The study was conducted during the winter semester of the academic year 2016. /17. in controlled lab conditions and was composed of two main parts: (1) scenario-based interaction with two cloud computing application for managing artefacts and (2) evaluation of their perceived security and privacy by means of the post-use questionnaire. Upon arriving to the lab, the participants were welcomed and briefly acquainted with the study. At the beginning of the scenario performance session, each participant received the form containing a list of 12 representative steps of interaction. Participants were asked to carry out all scenario steps twice – first with Google Drive and thereafter by means of Microsoft OneDrive (both shown in Figure 2). Upon finishing all the scenario steps with both cloud based applications, the participants were asked to complete the post-use questionnaire. At the end of the study, respondents were debriefed, and thanked for their participation. The duration of the study was 40 minutes.

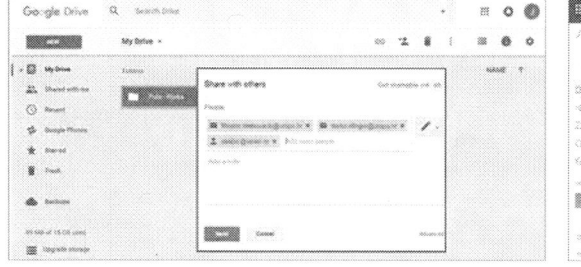 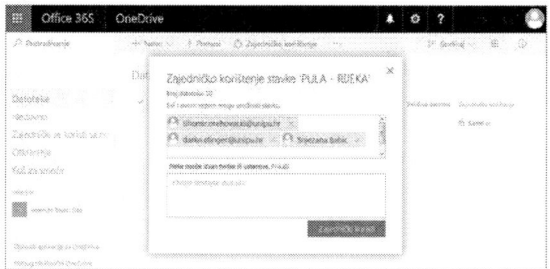

Figure 2.   Examples of screenshots that indicate which level particular student reached within predefined time interval (left: Google Drive, right: Microsoft OneDrive)

B. Apparatus

The post-use questionnaire was administrated online by means of the KwikSurveys questionnaire builder. The questionnaire comprised 16 items related to participants' demography and 35 items meant for measuring facets of perceived security and privacy. Items on perceived security

and perceived privacy were adopted from Cheung and Lee [16], Flavián and Guinalíu [17], Janda et al. [18], O'Cass and Fenech [19], and Ranganathan and Ganapathy [20]. Responses to the post-use questionnaire items were modulated on a five point Likert scale (1- strongly agree, 5 – strongly disagree). The psychometric features of the measuring instrument were examined with respect to the construct validity and reliability [21]. Convergent and discriminant validity, as indicators of construct validity, were explored by means of a principal component analysis (PCA) with equamax rotation and Keiser normalization. With an aim to verify that the requirements for factor extraction were met, the Kaiser-Meyer-Olkin test of sampling adequacy and Bartlett's test of sphericity were evaluated. As a criterion for identifying the number of factors, an eigenvalue greater than one was employed. Only items with loadings above .40 and cross-loadings below .40 were retained [22]. Reliability in terms of the internal consistency of extracted factors was measured with Cronbach's Alpha coefficient.

C. Participants

A total of 318 subjects took part in the study. They ranged in age from 18 to 48 years (M = 21.03, SD = 4.197). The sample was composed of 67.30% male and 32.70% female students. At the time study took place, majority of them (50.31%) were students at Juraj Dobrila University of Pula, Department of Information and Communication Technologies while remaining 49.69% studied at Polytechnic of Rijeka. Most of the study participants (80.50%) were full-time students. When the computer literacy is considered, respondents are proficient users of both computers and the Internet. More specifically, they have between 2 and 29 years (M = 11.82, SD = 3.559) of experience in employing computers and between 2 and 20 years (M = 9.76, SD = 3.092) of experience in using the Internet. In addition, 74.21% and 82.08% of participants believe that their computer skills and Internet skills, respectively, are at least very good. When the frequency of using the Internet for different purposes is taken into account, 69.50% of respondents is employing it for communication at least 11 hours per week, 60.06% of students is using the Internet for educational purposes between 4 and 20 hours per week, 71.07% of participants is using the Internet for fun more than 11 hours per week, and 41.82% of students is using the Internet for business purposes at least one hour per week. Study participants had also been loyal users of popular social Web applications. Namely, 65.55% respondents have been socializing on Facebook for more than 6 years, 52.86% of them have been podcasting on YouTube for more than 7 years, whereas 67.96% of students have been sharing their moments with a community for less than 2 years. Regarding the length of using Google Drive and Microsoft OneDrive, 49.16% of participants have been using them for more than one year, while 12.04% have not used aforementioned cloud computing applications prior to this study.

D. Findings

The Kaiser-Meyer-Olkin measure of sampling adequacy (KMO = .936, KMO = .944) and Bartlett's test of sphericity ($\chi2$ = 7638.851, p = .000; $\chi2$ = 8026.838, p = .000) confirmed that the data in the case of both Google Drive and Microsoft OneDrive, respectively, have met the requirements for conducting the principal component analysis (PCA). During the purification procedure, eight items (SCR6, SCR7, SCR8, SCR11, SCR14, SCR15, PRV10, and PRV11) were dropped. As presented in Table 1 and Table 2 (see Appendix), the final iteration of PCA uncovered two dimensions of perceived security and four

dimensions of perceived privacy, respectively. They accounted for 69.137% and 66.021% of the sample variance in the case of Google Drive and Microsoft OneDrive, respectively. Values of the Cronbach's Alpha coefficient were in range from .787 (in the case of measuring the Confidentiality of Google Drive) to .934 (in the context of evaluating Integrity of Microsoft OneDrive) thus indicating that reliability of scales was deemed adequate. Items marked with asterisk are reverse coded.

Results of data analysis indicate that 66.67% and 57.55% of study subjects believe that Google Drive and Microsoft OneDrive, respectively, have built-in high-quality mechanisms that protect users' artefacts from unauthorized use (SCR1). It was also found that 63.84% and 58.49% of students reported that Google Drive and Microsoft OneDrive, respectively, have integrated good security measures that protect their personal information (SCR2). In addition, 61.64% and 53.77% of study participants stated that Google Drive and Microsoft OneDrive, respectively, protect the security of all activities carried out by their employment (SCR3). The collected data also imply that 61.01% and 53.14% of students perceive that Google Drive and Microsoft OneDrive, respectively, have good protection mechanisms that prevent the theft of their identity by a third party (SCR4). Moreover, it was discovered that only 38.99% and 34.91% of respondents believe that Google Drive and Microsoft OneDrive, respectively, are protected to the extent that no third party cannot falsely introduce oneself to their users (SCR5).

Results of data analysis also indicate that 69.50% and 62.58% of subjects believe that Google Drive and Microsoft OneDrive, respectively, have built-in mechanisms that prevent unauthorized changes to information about the user (SCR9). Furthermore, 68.24% and 61.95% of students agree that Google Drive and Microsoft OneDrive, respectively, have built-in mechanisms that prevent unauthorized modification of stored documents (SCR10). Similarly, 75.47% and 71.70% of subjects is convinced that Google Drive and Microsoft OneDrive, respectively, verify user's identity before granting access to personal data and documents (SCR12).

The data gathered from study participants revealed that 76.73% and 71.07% of them believe that Google Drive and Microsoft OneDrive, respectively, is taking care of the protection of personal data and documents that are stored on it (SCR13). In addition, results of data analysis imply that 69.81% and 60.38% of students think that Google Drive and Microsoft OneDrive, respectively, are secure cloud computing applications (SCR16). Finally, according to data presented in Table 1, 63.52% and 57.23% of study subjects believe that Google Drive and Microsoft OneDrive, respectively, have implemented all the required security mechanisms (SCR17).

Data displayed in Table 2 imply that 32.30% and 33.65% of study participants is concerned that Google Drive and Microsoft OneDrive, respectively, will use their personal data for other purposes without their permission (PRV1). Furthermore, the findings of the pilot study imply that 39.94% and 36.48% of respondents think that Google Drive and Microsoft OneDrive, respectively, collect too

719

much information about their users (PRV2). It was also discovered that 31.45% and 32.70% of study subjects is concerned about the privacy of their personal information. when using Google Drive and Microsoft OneDrive, respectively (PRV3). In addition, 32.08% and 31.45% of students is concerned that Google Drive and Microsoft OneDrive, respectively, will give their personal information to third parties without their permission (PRV4). The collected data also indicate that 65.09% and 58.18% of study participants think that Google Drive and Microsoft OneDrive, respectively, take care of the privacy of their users (PRV5). Moreover, it was found that 52.52% and 46.86% of respondents feel that their privacy is protected when storing their personal information and documents on the Google Drive and Microsoft OneDrive, respectively (PRV6).

According to the results of data analysis, 70.13% and 65.72% of study subjects think that Google Drive and Microsoft OneDrive, respectively, comply with the laws and regulations on the protection of users' personal data (PRV7). In addition, 48.43% and 47.80% of students agree that Google Drive and Microsoft OneDrive, respectively, collect only the information about users that is necessary for their use (PRV8). Data gathered from students revealed that 63.84% and 61.01% of them think that while collecting data about users, Google Drive and Microsoft OneDrive, respectively, respect their rights (PRV9). Students' responses are also implying that 38.68% and 37.74% of them is afraid to provide Google Drive and Microsoft OneDrive, respectively, with their personal information because they do not know what these cloud computing applications could do with them (PRV12).

Considering the results of data analysis, 43.40% and 41.51% of students think is risky to provide Google Drive and Microsoft OneDrive, respectively, with their personal information (PRV13). It was also discovered that equal number of students (25.47%) believe that if they provide Google Drive or Microsoft OneDrive with their personal information, they could be faced with unexpected problems (PRV14). Moreover, 25.47% and 23.59% of respondents think that Google Drive and Microsoft OneDrive, respectively, could use their personal information in an inappropriate fashion (PRV15). Results of the data analysis also indicate that 34.59% and 35.54% of study participants think there is a very strong correlation between the potential loss of privacy and disclosure of personal information to Google Drive and Microsoft OneDrive, respectively (PRV16). Moreover, 43.71% and 39.31% of students believe that they have control over who has the access to their personal data collected by Google Drive and Microsoft OneDrive, respectively (PRV17). Finally, as can be observed from students' responses, 40.25% and 36.48% of them think they have control over how Google Drive and Microsoft OneDrive, respectively, are using their personal information (PRV18).

IV. CONCLUSION

The aim of this paper was to examine the perceived security and perceived privacy of cloud computing applications. For that purpose, an empirical study was carried out. Based on the data collected from study participants, psychometric features of measuring instrument were evaluated. Construct validity was tested with the use of a principal component analysis whereas Cronbach's Alpha coefficient was employed for assessing the construct reliability. As an outcome, data analysis uncovered two dimensions of perceived security (Integrity and Confidentiality) and four facets of perceived privacy (Privacy Concerns, Privacy Protection, Privacy Risks, and Privacy Control).

As with all empirical studies, some limitations which require further examination have to be acknowledged. The first one deals with the homogeneity of participants. Although students in our study are a representative sample of cloud-based applications users, perceived security and perceived privacy might vary if it would be evaluated by more heterogeneous group of users. The second limitation is that the findings cannot be generalized to all types of cloud computing applications except to the ones involved in the study. Keeping the set forth limitations in mind, study outcomes should be interpreted with caution.

Takin into account that this study is a part of an ongoing work, our future work efforts will be focused on exploring the extent to which identified aspects of perceived security and perceived privacy contribute to users' behavioural intentions regarding the employment of cloud computing applications.

REFERENCES

[1] S. Subashini and V. Kavitha, "A survey on security issues in service delivery models of cloud computing," J. Netw. Comput. Appl., vol. 34, no. 1, pp. 1–11, 2011.

[2] D. Zissis and D. Lekkas, "Addressing cloud computing security issues," Futur. Gener. Comput. Syst., vol. 28, no. 3, pp. 583–592, Mar. 2012.

[3] S. Singh, Y. S. Jeong, and J. H. Park, "A survey on cloud computing security: Issues, threats, and solutions," J. Netw. Comput. Appl., vol. 75, pp. 200–222, 2016.

[4] M. Al-zoube, "E-Learning on the Cloud," Int. Arab J. e-Technology, vol. 1, no. 2, pp. 58–64, 2009.

[5] H. Bicen, "Effects of Training on Cloud Computing Services on M-learning Perceptions and Adequacies," Procedia - Soc. Behav. Sci., vol. 116, pp. 5115–5119, 2014.

[6] T. Lis and B. Paula, "The Use of Cloud Computing by Students from Technical University – The Current State and Perspectives," Procedia Comput. Sci., vol. 65, pp. 1075–1084, 2015.

[7] S. Mutkoski, "Cloud Computing, Regulatory Compliance, and Student Privacy: A Guide For School Administrators and Legal Counsel," John Marshall J. Inf. Technol. Priv. Law, vol. 30, no. 3, p. 3, 2014.

[8] U. J. Bora and M. Ahmed, "E-Learning using Cloud Computing," Int. J. Sci. Mod. Eng., vol. 1, no. 2, pp. 9–13, 2013.

[9] B. Hari Krishna, S. Kiran, G. Murali, and R. Pradeep Kumar Reddy, "Security Issues in Service Model of Cloud Computing Environment," Procedia Comput. Sci., vol. 87, pp. 246–251, 2016.

[10] R. V. Rao and K. Selvamani, "Data Security Challenges and Its Solutions in Cloud Computing," Procedia Comput. Sci., vol. 48, no. Iccc, pp. 204–209, 2015.

[11] M. Stieninger, D. Nedbal, W. Wetzlinger, G. Wagner, and M. A. Erskine, "Impacts on the organizational adoption of cloud computing: A reconceptualization of influencing factors," Procedia Technol., vol. 16, pp. 85–93, 2014.

[12] S. A. Hussain, M. Fatima, A. Saeed, I. Raza, and R. K. Shahzad, "Multilevel classification of security concerns in cloud computing," Appl. Comput. Informatics, 2016.

[13] F. Bélanger and R. Crossler, "Privacy in the digital age: a review of information privacy research in information systems," MIS Q., vol. 35, no. 4, pp. 1–36, 2011.

[14] G. Garrison, S. Kim, and X. Xu, "Consumer adoption and use of mobile applications: Do privacy and security concerns matter?", Issues Inf. Syst., vol. 17, no. II, pp. 56–64, 2016.

[15] H. Xu, T. Dinev, J. Smith, and P. Hart, "Information Privacy Concerns: Linking Individual Perceptions with Institutional Privacy Assurances," J. Assoc. Inf. Syst., vol. 12, no. 12, pp. 798–824, 2011.

[16] C.M.K. Cheung, and M.K.O. Lee, "Trust in internet shopping: instrument development and validation through classical and modern approaches", Journal of Global Information Management, vol. 9, no. 3, pp. 23-35, 2001.

[17] C. Flavián, and M. Guinalíu, "Consumer trust, perceived security and privacy policy: three basic elements of loyalty to a web site", Industrial Management & Data Systems, vol. 106, no. 5, pp. 601-620, 2006.

[18] S. Janda, P. Trocchia, and K. Gwinner, "Consumer perceptions of internet retail service quality", International Journal of Service Industry Management, vol. 13, no. 5, pp. 412-431, 2002.

[19] A. O'Cass, and T. Fenech, "Web retailing adoption: exploring the nature of internet users web retailing behaviour", Journal of Retailing and Consumer Services, vol. 10, no. 2, pp. 81-94, 2003.

[20] C. Ranganathan, and S. Ganapathy, "Key dimensions of business-to-consumer web sites", Information & Management, vol. 39, no. 6, pp. 457-465, 2002.

[21] D. Straub, M. Boudreau, and D. Gefen, "Validation Guidelines for IS Positivist Research. Communications of the Association for Information Systems", vol. 13, no. 1, pp. 380–427, 2004.

[22] J.F. Hair Jr., W.C. Black, B.J. Babin, and R.E. Anderson, "Multivariate Data Analysis", 7th edn. Prentice Hall, Englewood Cliffs, 2009.

APPENDIX

TABLE I.   RESPONSES OF STUDY PARTICIPANTS TO QUESTIONNAIRE ITEMS RELATED TO THE PERCEIVED SECURITY

Perceived Security Items	Google Drive		Microsoft OneDrive	
	Mean	SD	Mean	SD
Integrity (Cronbach's α = .919 and .934 in the case of Google Drive and Microsoft OneDrive, respectively)				
SCR1. I believe this application has integrated high-quality mechanisms that protect my documents from unauthorized use.	2.26	.951	2.40	.980
SCR2. The application has implemented good security measures that protect my personal information.	2.33	1.015	2.47	1.025
SCR3. I think this application protects the security of all activities carried out by its use.	2.36	.982	2.48	.985
SCR4. I think this application has good protection mechanisms that prevent the theft of its identity by a third party (other organizations or individuals).	2.35	.964	2.47	.991
SCR5. I think this application is protected to the extent that no third party (individual or organization) cannot falsely introduce oneself to its users.	2.86	1.173	2.93	1.175
SCR16. I think this application is secure.	2.16	.939	2.32	.981
SCR17. I believe that the application has implemented all the required security mechanisms.	2.30	.954	2.40	.951
Confidentiality (Cronbach's α = .787 and .792 in the context of Google Drive and Microsoft OneDrive, respectively)				
SCR9. The application has built-in mechanisms that prevent unauthorized changes to information about the user.	2.19	.853	2.30	.881
SCR10. The application has built-in mechanisms that prevent unauthorized modification of stored documents.	2.20	.846	2.30	.846
SCR12. Before granting access to personal data and documents, the application verifies user's identity.	1.99	1.054	2.02	1.073
SCR13. The application is taking care of the protection of personal data and documents that are stored on it.	1.99	.870	2.08	.913

TABLE II.   RESPONSES OF STUDY PARTICIPANTS TO QUESTIONNAIRE ITEMS RELATED TO THE PERCEIVED PRIVACY

Perceived Privacy Items	Google Drive		Microsoft OneDrive	
	Mean	SD	Mean	SD
Privacy Concerns (Cronbach's α = .858 and .864 in the case of Google Drive and Microsoft OneDrive, respectively)				
PRV1. I am concerned that the application will use my personal data for other purposes without my permission.*	3.10	1.167	3.04	1.142
PRV2. I think the application collects too much information about its users.*	2.78	1.109	2.85	1.038
PRV3. When using the application, I am concerned about the privacy of my personal information.*	3.14	1.126	3.06	1.107
PRV4. I am concerned that the application will give my personal information to third parties (organizations or individuals) without my permission.*	3.20	1.216	3.18	1.179
Privacy Protection (Cronbach's α = .903 and .901 in the context of Google Drive and Microsoft OneDrive, respectively)				
PRV5. I think that the application takes care of the privacy of its users.	2.29	.926	2.37	.951
PRV6. When storing personal information and documents on the application, I feel that my privacy is protected.	2.54	.977	2.64	.955
PRV7. I think that the application complies with the laws and regulations on the protection of users' personal data.	2.14	.897	2.21	.883
PRV8. I think the application collects only the information about users that is necessary for its use.	2.67	1.106	2.67	1.055
PRV9. I think that while collecting data about users, the application respects their rights.	2.31	.961	2.36	948

* reverse coded items

TABLE II. CONTINUED

Perceived Privacy Items	Google Drive		Microsoft OneDrive	
	Mean	SD	Mean	SD
Privacy Risks (Cronbach's α = .863 and .862 in the case of Google Drive and Microsoft OneDrive, respectively)				
PRV12. I am afraid to provide the application with all my personal information because I do not know what it could do with them.*	2.95	1.168	2.92	1.143
PRV13. I think is risky to provide the application with my personal information.*	2.84	1.154	2.84	1.153
PRV14. If I provide the application with my personal information, I could be faced with unexpected problems.*	3.14	1.051	3.11	1.029
PRV15. I think the application could use my personal information in an inappropriate fashion.*	3.22	1.067	3.25	1.068
PRV16. I think there is a very strong correlation between the potential loss of privacy and disclosure of personal information to the application.*	2.98	1.076	2.94	1.065
Privacy Control (Cronbach's α = .880 and .861 in the context of Google Drive and Microsoft OneDrive, respectively)				
PRV17. I believe that I have control over who has the access to my personal data collected by the application.	2.86	1.067	2.91	1.078
PRV18. I think that I have control over how the application is using my personal information.	2.91	1.108	2.96	1.080

* reverse coded items

Estimating profile of successful IT student: data mining approach

Dijana Oreški, Mario Konecki and Luka Milić
University of Zagreb, Faculty of Organization and Informatics, Varaždin, Croatia
dijana.oreski@foi.hr; mario.konecki@foi.hr; luka.milic@foi.hr

Abstract - The study presented in this paper aims to explore students' characteristics and to determine student groups based on their previous education and socio-demographic characteristics. Descriptive data mining method, cluster analysis, is applied in the analysis process. Data used in the research is collected among first, second and third year IT students. Research results indicate profile of successful IT student. As such, research results provide useful insight into both micro and macro level aspects of educational process, which can benefit both students and academic institutions. Data mining has shown promising results in educational domain and a substantial potential to serve as a tool for improvement of quality in education.

Keywords - **educational data mining, academic performance, cluster analysis, educational strategy**

I. INTRODUCTION

The goal of higher education institutions is to provide a quality education process to all enrolled students. One way to achieve this goal is by discovering knowledge about students' performance and students' characteristics. This knowledge can in many cases be extracted from different data sets by using various data mining techniques. In this paper a research that includes evaluation of cluster analysis, a data mining approach, for the purpose of IT student's performance description is presented and elaborated. Thus, in this paper specific educational setting is used – 217 university undergraduates in computer science. Presented research is focused on the students' socio-demographic characteristics and high school grades. Although some authors argue that high school grades are unreliable variable for prediction because there are no common grading standards across schools or across courses in the same school [1], others have proven that previous grades are the most important factor of academic performance [2]. Furthermore, Petersen, Louw and Dumont indicated transition period from high school and adjustment to the university environment as an important factor in predicting university outcomes [3]. Hillman found differences in academic performance among students from different types of high schools [4]. The aim of this paper is to understand the profile of IT students in order to build students learning strategies and to enhance institutional policy making.

II. RESEARCH METHODOLOGY

Based on the results of the previous studies, the research presented in this paper is constructed around main hypothesis which states: "Type of high school and higher grades in high school are associated with higher university grades".

This chapter describes in detail methodology applied in order to test the hypothesis and gives an overview of the data collection.

A. Cluster analysis

Data mining is one of the fastest growing fields in ICT sector. Educational data mining, as a subfield of data mining, is an emerging interdisciplinary research area that develops methods for data exploration in educational domain [5]. Its main objective is to analyze educational data sets in order to resolve educational research issues [6]. There are two main tasks of educational data mining: descriptive and predictive modelling. Whereas predictive modelling methods seek to predict or classify data according to the target variable, descriptive modelling methods are not dependent on any driving objective function and their aim is to explain and describe data. Cluster analysis is the most common technique used for descriptive modelling. [7]. It is applicable in the wide range of domains, including social science, like: education [8; 9; 10; 11; 12; 13], strategic management and economics [14; 15; 16], natural sciences like medicine [17], and genetics [18], or technical science in the domain of environment protection [19].

Clustering performs grouping of cases into classes of similar objects. As a result, clusters are obtained. Cluster refers to a collection of records that are similar to one another and dissimilar to records in other clusters [20]. Clustering algorithms are classified into one of two groups: hierarchical or nonhierarchical clustering. Hereinafter, nonhierarchical algorithm, k-means clustering, is applied, since it is more frequently used in analytics industry [21].

B. Data description

The data were collected via a questionnaire among first, second and third year students of the undergraduate study programme Information and Business Systems at the Faculty of Organization and Informatics, University of Zagreb. Questionnaire comprised of 11 attributes (see Table 2). Questionnaire was paper-based and 217 questionnaires were collected from a stratified proportional data sample. Data sample was representative considering the total number of students enrolled in each year of the study programme as shown in Table I. The attribute upon which the population is stratified (Year of the study) is statistically significantly correlated with the dependent attribute, grade point average (GPA). This relationship justifies selection of the attribute for stratification.

TABLE I.   REPRESENTATIVE DATA SAMPLE DETAILS

	1st year	2nd year	3rd year	Σ
Sample	89	78	50	217
Population	320	264	224	808
Sample	41,01%	35,94%	23,04%	100%
Population	39,61%	32,67%	27,72%	100%

Most of the respondents were male, more than 80%. This distribution is in line with Faculty of Organization and Informatics distribution. Almost half of respondents came from gymnasiums (either general or mathematical programme). One third of respondents came from technical or electrotechnical high schools (see Table II).

TABLE II.   DISTRIBUTIONS OF ANSWERS

Variable	Distribution of answers
Gender	Female =18,89% Male =81,11%
Year of study	First year = 41,01% Second year = 35,94% Third year = 23,04%
High school	General gymnasium = 40,09% Mathematical gymnasium = 17,51% Technical high school = 29,03% Economic high school =7,84% Electrotechnical high school = 5, 53%
Math high school grade	2 = 12, 90 % 3 = 29,03 % 4 = 31,34 % 5 = 26,73 %
Informatics high school grade	3 = 10,10% 4 = 23,50 % 5 = 66,40 %
English high school grade	2 = 3,23 % 3 = 11,06 % 4 = 32,26 % 5 = 53,45 %
Math 1 faculty grade	No grade = 6,91 % 2 = 45,62 % 3 = 33.64 % 4 = 11,52 % 5 = 2,31 %
Informatics 1 faculty grade	No grade = 11,52% 2 = 39,63 % 3 = 41,01% 4 = 5,99 % 5 =1,85 %
Programming 1 faculty grade	No grade = 18,43 % 2 = 42,86 % 3 = 32,72 % 4 = 5,07 % 5 = 0,92 %
English 1 faculty grade	2 = 21,66 % 3 = 35,48 % 4 = 32,72 % 5 = 10,14 %
GPA	Mean = 3,1; St. Dev. = 0,6

The lowest high school grades were from the Math course (less than a third of respondents had 5), whereas the highest high school grades were from the Informatics course (85% of the students had grades 4 or 5). Some of the first year students still did not pass Informatics 1 course (11,52%), Mathematics 1 (6,91%) or Programming 1 (18,43%). The highest faculty grades that students have achieved were the grades from English 1 course.

III.   RESEARCH RESULTS

In order to test the stated hypothesis, first clustering was performed based on the following attributes: Gender, Year of study, High school, Mathematics high school grade, Informatics high school grade and English high school grade. Different number of clusters was varied and finally number 8 was selected as the optimal number of clusters. Number of students in each cluster is represented in Table III.

TABLE III.   NUMBER OF INSTANCES IN CLUSTERS

	Number of instances
Cluster 1	18
Cluster 2	42
Cluster 3	32
Cluster 4	30
Cluster 5	42
Cluster 6	15
Cluster 7	15
Cluster 8	23

Clusters 5 and 2 are the largest, whereas clusters 6 and 7 are the smallest. In addition, the characteristics of students in each cluster have been examined and relationships of clusters with performance of the students have been investigated.

Since "grades are often seen as the gold standard measure of success in education, and determining the factors that affect academic achievement has been a common endeavor by educational researchers" [21], attributes regarding high school grades are in the center of this research.

TABLE IV.   CLUSTER ONE CHARACTERISTICS

	Cluster 1
Gender	Male
Year of study	Second
High school	Technical
Mathematics HSG	5
Informatics HSG	5
English HSG	5

Cluster 1 consists of second year male students which achieved excellent grades from all courses at the high school, as shown in Table IV.

TABLE V.   CLUSTER TWO CHARACTERISTICS

	Cluster 2
Gender	Male
Year of study	Second
High school	Technical
Mathematics HSG	3
Informatics HSG	4
English HSG	4

Cluster 2 consists of the students whose faculties GPA is the lowest. Their characteristics are shown in Table V.

TABLE VI.   CLUSTER THREE CHARACTERISTICS

	Cluster 3
Gender	Male
Year of study	First
High school	General Gymnasium
Mathematics HSG	3
Informatics HSG	5
English HSG	4

Unlike the first two clusters, in the third cluster there are students from general gymnasiums, with very diverse high school grades: varying from 3 (Mathematics) to 5 (Informatics), as shown in Table VI.

TABLE VII.   CLUSTER FOUR CHARACTERISTICS

	Cluster 4
Gender	Female
Year of study	Second
High school	General Gymnasium
Mathematics HSG	5
Informatics HSG	5
English HSG	5

Cluster 4 represents the prototype of the successful IT students. Those are students with the highest GPA on the faculty, mostly second year female students from general gymnasiums with highest grades from all high school courses, as shown in Table VII.

TABLE VIII.   CLUSTER FIVE CHARACTERISTICS

	Cluster 5
Gender	Male
Year of study	Second
High school	Technical
Mathematics HSG	4
Informatics HSG	5
English HSG	5

Cluster 5 shows typical IT student: male from technical high school with excellent Informatics grade, but lower Mathematics grade, as shown in Table VIII.

TABLE IX.   CLUSTER SIX CHARACTERISTICS

	Cluster 6
Gender	Male
Year of study	First
High school	Technical
Mathematics HSG	4
Informatics HSG	4
English HSG	5

Cluster 6 includes first year male students from technical high schools (see Table IX).

TABLE X.   CLUSTER SEVEN CHARACTERISTICS

	Cluster 7
Gender	Male
Year of study	Third
High school	Economic
Mathematics HSG	4
Informatics HSG	5
English HSG	4

Students from the cluster 7 come from the economic high schools and their characteristic is lower English and Mathematics grade and high Informatics grade (Table X).

TABLE XI.   CLUSTER EIGHT CHARACTERISTICS

	Cluster 8
Gender	Male
Year of study	First
High school	General Gymnasium
Mathematics HSG	3
Informatics HSG	5
English HSG	5

Cluster 8 (see Table XI) consists of male students from general gymnasiums with low Mathematics background.

The clustering procedure serves as a good starting point for students' performance observation. The presented findings identify subgroups of students with different academic success. GPA has been used as a measure of success of IT students (students with GPA 4 and above were considered as successful). Finally, to test hypothesis regression analysis was performed. Results indicated statistically significant relationship between type of high school and high school grades with faculties GPA. The magnitude of the prediction was found to be higher than the one with socio-demographic characteristics. When the relationship between academic success and characteristic of specific subgroups was examined, high school programme was found to be a significant predictor of academic success. The magnitude of the prediction was found to be higher than the one with socio-demographic characteristics. The greater predictive ability of type of the programme for university achievement may be the result of more diversity in curriculum in the high school sample. Such results are compatible with the results of the McKenzie and Schweitzer who have identified that previous academic performance is the most significant predictor of the university performance [2]. In this paper, significant relationship between high school programme and academic performance has been identified: general gymnasium provides a good starting point for IT students.

This research tackles another significant issue: gender based performance differences in information and communication technologies. Research results refute hypothesis that "Hi-tech = guy-tech" [23]. Exactly the opposite, cluster with the highest GPA consists of female students. These results are in line with previous research results that have confirmed woman success in the STEM field [24; 25; 26]. Results of this study did not confirm the pattern observed in the literature [27] which states that women underperform on difficult (Mathematics and Informatics) courses.

IV. CONCLUSION

The research presented in this paper deals with developed student clusters, based on students' previous education, which provide a basis for IT students' success prediction. Interpretation and employment of the obtained results can be very useful for instructors to predict the performance of new students and to make decisions about helping students. Findings of this research consist of the following:

- Research results have practical implications for advising students and support in institutional strategy definition,
- "One size fits all" model cannot be effectively applied to all students and the solution is to segment students and perform different approach for each group.

Presented study has several limitations, mostly concerning the nature of the sample and the attributes included in the questionnaire. The research tackles only one, very specific type of students, IT students. Guidelines for further research are to perform comprehensive researches on bigger data set, several years in duration, with different students of different study programmes.

REFERENCES

[1] W. Camara, and M. Michaelides. "AP use in admissions: A response to Geiser and Santelices". College Board Research Note, 2005.

[2] K. McKenzie, and R. Schweitzer. Who succeeds at university? Factors predicting academic performance in first year Australian university students. Higher education research & development, 20(1), pp. 21-33, 2001.

[3] I. H., Petersen, J. Louw, and K. Dumont. Adjustment to university and academic performance among disadvantaged students in South Africa. Educational Psychology, 29(1), pp. 99-115, 2009.

[4] K. Hillman. "The first year experience: the transition from secondary school to university and TAFE in Australia". LSAY

Research Reports. Longitudinal surveys of Australian youth research report; n.40, http://research.acer.edu.au/lsay_research/44, 2005.

[5] C. Romero, and S. Ventura. "Educational data mining: a review of the state of the art". IEEE Transactions on Systems, Man, and Cybernetics, Part C (Applications and Reviews) 40(6), pp. 601-618, 2010.

[6] T. Barnes, M. Desmarais, C. Romero, and S. Ventura, S. Educational Data Mining 2009: Proceedings of the 2nd International Conference on Educational Data Mining. Cordoba, Spain, 2009.

[7] W. Hämäläinen. Descriptive and predictive modelling techniques for educational technology. Licentiate thesis, Department of Computer Science, University of Joensuu, 2006.

[8] R. E. Myers and J. T. Fouts. A cluster analysis of high school science classroom environments and attitude toward science. Journal of Research in Science Teaching, 29(9), pp. 929-937, 1992.

[9] G. J. Conti. Using cluster analysis in adult education, 1996.

[10] C. J. Wangand S. J. Biddle. Young people's motivational profiles in physical activity: A cluster analysis. Journal of Sport and Exercise Psychology, 23(1), pp. 1-22, 2001.

[11] S. Ullrich-French and A. Cox. Using cluster analysis to examine the combinations of motivation regulations of physical education students. Journal of Sport and Exercise Psychology, 31(3), pp. 358-379, 2009.

[12] N. Erdogmus and M. Esen. Classifying Universities in Turkey by Hierarchical Cluster Analysis. Egitim ve Bilim, 41(184), 2016.

[13] T. Dietrich, S. Rundle-Thiele, L. Schuster, J. Drennan, R. Russell-Bennett, C. Leo, and J. Connor. Segmenting Australian High School Students Utilising a Two-Step Cluster Analysis: Differential Effects Following the Game on Know Alcohol Program. In Rediscovering the Essentiality of Marketing. Springer International Publishing, pp. 413-414, 2016.

[14] D. J. Ketchen and C. L. Shook. The application of cluster analysis in strategic management research: an analysis and critique. Strategic management journal, pp. 441-458, 1996.

[15] K. Maršić and D. Oreški. Estimation and Comparison of Underground Economy in Croatia and European Union Countries: Fuzzy Logic Approach. Journal of Information and Organizational Sciences, 40(1), pp. 83-104, 2016.

[16] M. S. Ugur. A cluster analysis of multidimensional poverty in Turkey. Economic and Social Development: Book of Proceedings, 12, 2016.

[17] E. J. Woytowicz J. C. Rietschel, R. N. Goodman, S. S Conroy, J. D. Sorkin, J. Whitall, and S. McConmbe Waller. Determining levels of upper extremity movement impairment by applying cluster analysis to upper extremity Fugl-Meyer assessment in chronic stroke. Archives of Physical Medicine and Rehabilitation, pp. 1-18, 2016.

[18] A. Brazma and J. Vilo. Gene expression data analysis. FEBS letters, 480(1), pp. 17-24, 2000.

[19] G. Shyamala, and J. Jeyanthi. Application of Integrated Hydrochemical Model and Cluster Analysis in Assessing Groundwater Quality. International Journal of Ecology & Development™, 31(4), pp. 34-45, 2016.

[20] D. T. Larose, and D. C. Larose, Data mining and predictive analytics, Wiley, 2015.

[21] A. Bansal, M. Sharma, and S. Goel, Improved K-mean Clustering Algorithm for Prediction Analysis using Classification Technique in Data Mining, International Journal of Computer Applications, 157(6), 2017.

[22] R. H. Stupnisky, R. P. Perry, R. D. Renaud, and S. Hladkyj, S. Looking beyond grades: Comparing self-esteem and perceived academic control as predictors of first-year college students' well-being, Learning and Individual Differences, 23, pp. 151-157, 2013.

[23] N. Selwyn. Hi-tech= guy-tech? An exploration of undergraduate students' gendered perceptions of information and communication technologies. Sex Roles, 56(7-8), pp. 525-536, 2007.

[24] L. Jones. Confidence and mathematics: A gender issue?. Gender and education, 7(2), pp. 157-166, 1995.

[25] C. Gunn, M. McSporran, H. Macleod, and S. French. Dominant or different: Gender issues in computer supported learning. Journal of Asynchronous Learning Networks, 7(1), pp. 14-30, 2003.

[26] S. J. Spencer, C. M. Steele, and D. M. Quinn. Stereotype threat and women's math performance. Journal of experimental social psychology, 35(1), pp. 4-28, 1999.

[27] C. M. Steele, and J. Aronson. Contending with a stereotype: African-American intellectual test performance and stereotype threat. Journal of Personality and Social Psychology, 69, pp. 797-811, 1995.

Comparison of Game Engines for Serious Games

Sanja Pavkov, Ivona Franković, Nataša Hoić-Božić
Department of Informatics, University of Rijeka, Rijeka, Croatia
spavkov@student.uniri.hr
ifrankovic@inf.uniri.hr
natasah@inf.uniri.hr

Abstract – Serious educational games are specially designed computer games which are used in an educational setting, in other words, they are interactive competitive lessons with defined learning outcomes which allow students to have fun during learning. The importance of serious games in contemporary educational practice is increasing. Applying serious games in teaching, students facilitate the learning process, adopt new skills and abilities, show more interest in learning, are more focused and more active in a class, and better understand and apply lessons learned. The complexity of serious games requires large efforts for their development. For the development of serious games teachers mostly use commercial game engines. One of the important parts of game development is a selection of appropriate development tool. Due to the range of available tools the choice of platforms for serious games is a challenge, whose selection often has considerably different goals and technical requirements depending on context and usage. The aim of this paper is to propose criteria which should be considered before selecting a game engine for serious games and shows results of a comparison of evaluated most popular game engines. Some recommendations for teachers about choosing the most suitable game engines for serious games development are also presented.

Keywords: serious games, game engine, developing serious games

I. INTRODUCTION

The 21st century is marked by the use of information technology like Internet, computers, laptops, tablets, and other smart devices to almost all areas of human activity. We are surrounded by technology from all sides, from the jobs, government and public administration, schools, universities, hospitals, banks, through to our homes. Since the computer has become an indispensable part of our everyday life, its use in education is increasing and more important. Evaluation and monitoring of the impact of computer-assisted learning have shown that the combination of learning the traditional way and learning using computers is good and useful and provides a better understanding and memorized material. As a consequence, the need arose for developing new skills to prepare students for new jobs and technologies [1]. According to Prensky technology is important to live, survive and thrive in the 21st century. He points out that youth today communicate, search information and socialize differently than their predecessors [2]. Kids of the digital age are generations of students who are familiar with digital technology and using digital devices from an early age. Information and communication technology (ICT) has almost become a native language to them or a language with which they communicate, express themselves, understand, and perceive the world around them.

Today is necessary to have a new set of skills to succeed in learning, working and living. Those competencies are beyond ICT literacy and include communication, collaboration, social and cultural skills, creativity, critical thinking, problem-solving, productivity, flexibility, risk-taking, conflict management, and a sense of initiative. They should be developed by everyone, from primary school to lifelong students. For the acquisition of those skills, the effectiveness of adopting active learning methodologies is widely acknowledged; they involve activities that accentuate the development of learner skills. Active learning enhances students engagement in learning tasks and enables them to deal with new challenges, solve problems, and adapt to changes in technology and knowledge [3].

Serious games are becoming more and more popular form of active learning. Serious games are specially designed computer games which are used in an educational setting. They enable the learner to have some control of the game activity and engage in interaction [4]. The complexity of serious games requires large efforts for their development and needs a lot of planning. One of the important parts of game development is a selection of appropriate development tool or game engine. Teachers mostly use a commercial engine for serious games design and it should be chosen to provide developing the game with all necessary education goals set for students.

This paper proposes criteria which should be considered before selecting a game engine for serious games. The second part introduces the serious games, while the third part is about game engine. The fourth part describes criteria and provides comparison results of evaluated game engines. In the last, fifth part we draw the conclusion.

II. SERIOUS GAMES

Serious games in teaching and education encourage students to participate in the learning process. It is important that these games are stimulating and motivating and that they develop students' creativity [4]. The game in education can be used in all stages of the teaching process, from the introductory part as a motivation for students to the evaluation and formative or summative assessment of students. Serious games allow students to complete tasks, acquire knowledge and reasoning of certain principles while having fun. They represent a simulation that includes the challenge of solving a particular task important for students [5]. Important elements that contribute to the educational values of the game are stimuli, fantasy, challenge, and curiosity.

Games in general can be classified into seven

categories: adventure, role-playing games, shooting games, simulation, puzzles, strategy and sports games [3]. Games in education are: quiz games, word games, puzzle games with branching scenarios, various simulations with the tasks, simulation environments, simulations that require personal answers toy and game playing role [5].

Serious Games are primarily used for learning and adopting the learning content in a fun way aiming to increase student interest and motivation. During the development of the game, it is necessary to pay attention that all the learning outcomes are fulfilled. Every serious game has to have well-defined learning objectives and promote the development of important skills in order to increase cognitive and intellectual abilities of students.

The process of serious games creation is based on the methodology SADDIE which include **S**pecification of the game, **A**nalysis, **D**esign, **D**evelopment, **I**mplementation, and **E**valuation [3]. Creating a didactic computer game requires choosing an appropriate tool or game engine during the analysis phase. There are a number of different types of programming tools and several different alternative systems for creating computer games, such as programming languages, multimedia development environments, environments designed specifically for game development, development tools for e-learning software and the modding environment (some gaming environments come with an additional creation engine that allows the development of extensions to the game such as new storylines known as "modding") [6].

III. GAME ENGINES

Today there are many types of software tools for developing computer games, so it is a challenging task to choose the appropriate game engine for educational games. There are many dilemmas which tool to choose because they all have strengths as well as problems. Some of them are fast for development, some may have performance issues, some require a programming skills, and some come with a developer-friendly interface [7].

For the purposes of this study, we evaluated five different game engines, these are respectively: Adventure Game Studio, Construct 2, e-Adventure, GameMaker: Studio, Phaser Editor. Each of these tools is downloaded, installed on the computer and used for game development in order to choose those engines in which 2D adventure game can be realized. Below is an overview of the examined tools for creating computer games with general information and a brief description of each of them.

A. Adventure Game studio

Adventure Game Studio (AGS) engine is a Windows-based integrated development environment (IDE 2) which allows users to create 2D "point and click" adventure games. It primary target group is not educators, although can potentially be used to create educational games. Among the biggest disadvantages to the development of educational games using this engine are: his games cannot easily be integrated with Learning Management Systems (LMS) or inserted in a SCORM package and does not have educational features to allow assessment of student performance [8] [9]. On the other hand, Adventure Game Studio supports high-resolution graphics, has a fully integrated audio and video. Games created with it can be run on multiple operating systems, supports the use of thousands of sprite's (sprite is a two-dimensional bitmap graphic that is integrated into a larger scene, it can either be a static image or an animated graphic that plays a specific role), creating hundreds of rooms, an unlimited number of characters, dialogue, GUI and other interface elements [10].

B. Construct 2

Programming tool Construct 2 is HTML5 tool designed specifically for 2D games and provides game development without coding to all users regardless of previous knowledge of programming. Currently is available installation on a Windows platform only but games created with it can run on a very large number of different platforms which is a big advantage of this tool. It offers a simple Drag-and-Drop (DnD) interface, is easy and intuitive to use, and it is intended primarily for people who have no prior experience with programming [11].

C. e-Adventure

The e-Adventure engine has emerged within the research project at the University of Madrid (Universidad Complutense de Madrid) in Spain with an aim to facilitating the integration of educational and simulation games in the educational [12]. The tool is specially designed to create a Point & Click adventure type of educational games [13]. The three main objectives of development e-Adventure tool are: reducing the development cost for serious games, the inclusion of educational specific features in a game engine and games integration with existing educational materials in virtual learning environments

The platform is composed of two applications - a game authoring editor (for creating the educational games) and a game engine (for execution of the games). The editor is completely instructor oriented, it does not require any technical background or programming skills to be used [14]. Through a simple and intuitive graphical interface tool, it enables the creation of games by creating scenes, objects, characters, dialogue and conditions that define the logic of the action in the game. The e-Adventure supports the creation of two different types of adventure games: third-person and first-person games [13].

Using this engine, teachers can create their own educational video games or adapt some existing popular games to add educational value. According to that, e-Adventure provides educational features that other popular game engines usually don't. The two most important educational features of this tool are the assessment reports and the integration with popular LMS (e.g. Moodle) [15].

As an example of an application in education, the e-Adventure platform is using for making educational computer games at the Faculty of Education, University of Ljubljana [3] and at the Department of Informatics, University of Rijeka.

D. GameMaker: Studio

GameMaker: Studio is a development tool for creating a 2D video game. It is also possible to develop 3D video games, but the documentation notifies about the possibility of having several problems with depths, views and other attributes [16]. This engine is used by teachers in several schools to create digital games that suit their curriculum or to improve pupils' programming skills. This engine enables development of games without programming skills. It offers an intuitive and easy Drag-and-Drop (DnD) interface action icon that provides a very quick start development of its own games. Actions are used to define how, when and where we want to something happened during the game. It is possible to load and create images, sounds and other graphical elements, and to add them to the objects in the scene. While creating actions, GameMaker: Studio in the background creates code of those actions which this tools make appropriate for users with and without programming experience [17]. GameMaker: Studio allows to create games for various platforms, including Windows, MAC OS, Web, Android and iOS [18]. The feature of this engine includes a sprite editor, timelines, paths, room editor, a custom programming language GameMaker Language (GML) and connection with external databases [19].

E. *Phaser editor*

Phaser Editor is an integrated development environment for creating HTML5 2D games. For a development of games in this tool, it is necessary to know programming. It offers a lot of examples, different Physics libraries and some plugins [20]. Programming is not required only for adding objects to the scene since it is done using a simple drag-and-drop option. However, everything else must be programmed by a user therefore knowledge of JavaScript programming language is required. [21].

IV. CRITERIA, TABLES AND COMPARISON RESULTS OF SELECTED ENGINES

Selected five tools that are primarily designed for making 2D games were compared in five criteria which are important for the development of the game. For the purposes of this research, we developed our own set of criteria and divided them into five groups:

- basic features of the engine including price
- support, flexibility, interoperability and usability
- engine system requirements and installation
- functionality and the ability to export
- multimedia support and working environment.

Among them, we considered as the most important criteria: functionality, basic features of the engine including price, and flexibility. For each of these groups a table with detailed engine comparison is presented. We also introduce short descriptions of the comparison results for other criteria.

A. *Basic features of the engine including price*

The first criterion is split into two groups - the basic features and tools price. Basic features of the engine give basic information about every of them like the first version, the current version, is the tool open source or not etc., while price describes whether the engine is free or a commercial tool.

The comparison of selected engines based on the first criteria is given in Table 3.1.

After the comparison according to the criteria of the basic engine features and the price, considering the basic features, we can see that the oldest among them is GameMaker: Studio and the newest is Phaser Editor released in 2015. Adventure Game Studio, e-Adventure and Phaser Editor are open source engines, and Construct 2 and GameMaker: Studio is not. The programming language of the compared engines is relatively similar - Java, JavaScript, C #, C ++, except for GameMaker: Studio, which is based on GML, his own programming language. Considering the criterion of price, Adventure Game Studio and e-Adventure are totally free tools and Construct 2, GameMaker: Studio and Phaser Editor are commercial tools, although they have free demo versions.

B. *Support, flexibility, interoperability and usability*

Support criterion includes categories that describe customer support, a community of users and documentation. Flexibility criterion is about how much the engine is adapted to some special situations. Interoperability includes the ability to use ASCII and UTF-8 characters in the compared engines. Usability implies how easily and quickly the engine can be learned and how simply is to develop a game.

The comparison of engines based on the second criteria is given in Table 3.2.

After the comparison according to the criteria of support, mostly all engines offer support for customers, complete documentation tools, access to finished examples of games and video tutorials for its community of users, access to forums, blogs and social networks. With regard to the criterion of flexibility, all compared engines are accessible to the wider community of users.

Creating multiplayer games allow only Construct 2, GameMaker: Studio and Phaser Editor. Of all benchmarked engines, only e-Adventure programming tool has multi-language support and, support for SCROM 1.2 (**S**harable **C**ontent **O**bject **R**eference **M**odel – a collection of standards and specifications for web-based e-learning) reference model and the possibility of integration with the LMS systems. Also, is the only tool among compared that is tailored to work with people with special needs.

According to the criterion of interoperability, all compared engines support the ASCII character set, while UTF-8 character set only Adventure Game Studio does not support. Regarding the criterion of usability, e-Adventure, GameMaker: Studio and Construct 2 enable creating games without programming. GameMaker: Studio has an advanced option that enabled programming in a way that allows users can create more game options. Adventure Game Studio requires programming scripts that define certain actions in the game.

C. *Engine system requirements and installation*

Engine system requirements include which operating system is required for installation and some properties related to hardware. Examination shows that most of the comparison tool requires minimum Windows XP, Windows Vista or newer operating system. Considering the compared data of selected engines, the least system requirements for installation has Adventure Game Studio. Installation criterion describes how easy engine installation is, and testing has shown that all engines are simple to install and have appropriate installation instructions.

TABLE 3.1. COMPARISON BASED ON THE BASIC FEATURES OF THE ENGINE AND PRICE CRITERIA

CRITERIA	CATEGORY	Adventure Game Studio	Construct 2	e-Adventure	GameMaker: Studio	Phaser Editor
BASIC FEATURES OF THE ENGINE	*First version and release year*	Adventure Creator version 1, 1997.	Construct 2 r45, 2011.	e-Adventure 0.1 b, 2009.	Animo, 1991.	Phaser Editor 15.11 RC, 2015.
	Current version and release year	Adventure Game Studio 3.4.0. 2016.	Construct 2 r239,2016.	e-Advenure 1.5., 2012.	GameMaker: Studio 1.4.1763, 2016.	Phaser Editor 1.2.1., 2016.
	Open source	Yes	No	Yes	No	Yes
	Programming language	Scripting language similar to Java and C#	C++, JavaScript	Java	GML programming language	JavaScript
PRICE	*Free engine*	Yes	Free version, but without all the features	Yes	Free version, but without all the features	Available a free version for 15 days without all the features
	Commercial version	No	Personal and Business version of the engine	No	Professional and Master Collection	1 year, 2 year or Lifetime License

D. Functionality and the ability to export

Functionality analysis the features that are specific for game development, while ability to export lists possible ways of exporting games for different platforms. The comparison of selected engines based on the functionality criteria are given in Table 3.3 After the comparison according to the criteria of functionality, we found out that all tools support the development of 2D computer games quite good. eAdventure allows making first-person games where the main character is not present, and third-person where the game is playing with the main character (avatar), Adventure Game Studio allows creating games in the third person, while other engines do not have limited modes for games creation. In the Adventure Game Studio and eAdventure navigation through the game is with the mouse pointer, and in other tools keyboard controls can also be used. Construct 2 and GameMaker: Studio have built-in Box2D Simulator Editor by which elements like friction and gravity are define. Adventure Game Studio and eAdventure engines have built-in dialogue editors, while GameMaker: Studio has a separate external editor that saves dialogues which are implemented in the project in .txt files. In the eAdventure dialogue can be displayed normally, as a whisper or as a thought, and dialogue text can be converted to speech. It provides the ability to change the main character, NPC (a non-player character/non-person character/non-playable character is any character that is not controlled by a player) or items appearance in scenes. These changes are flags conditioned. It has a built-in book subject that facilitates the information entry or giving instructions to a player. Built-in item inventory has Adventure Game Studio and eAdventure, while in others are necessary to program the inventory. In Adventure Game Studio and eAdventure are necessary to set up the main character to every scene, regardless of whether it is needed there or not. Out of all tools, only eAdventure offers game statistics view. Assessment and evaluation through a various question can be made in Adventure Game Studio, eAdventure and GameMaker: Studio, while in other tools this feature is not optimized in the best way. From all compared tool, the best option to create multiple-choice questions and assessment provides eAdventure. Regard to the criterion ability to export the game the most options for exporting offers Construct 2, and at least Phaser Editor engine.

Regarding the ability to export the game the most options for exporting offers Construct 2, and at least Phaser Editor engine, but we should consider that it's created in 2015. and is still in development phase. All engines can export games for Windows and Linux platforms. eAdventure and Phaser Editor can't export for Android, iOS nor consoles' platforms and Phaser can't export for Mac OS X. Only eAdventure offers the ability to export g games as learning object where the project is saved as educational content.

E. Multimedia support and working environment

Criterion multimedia support is related to which multimedia elements engine supports, while engines working environment analyzes if an engine is user-friendly. After the comparison according to multimedia criterion, it seems that all compared engines allow the inclusion of multimedia elements like images, sound, video and animation. Compared engines mainly support common formats for image, sound, video and animation, while GameMaker: Studio supports vector graphics. The e-Adventure can create animations of the main characters and NPC only for standing, talking, walking and the use of objects, while other tools can create animations on request. Considering the category of engine working environment all compared engines have an intuitive graphical interface with Drag-and-Drop options that can be adjusted if necessary. Also, all of them allow testing of the project as well as the detection of errors that occur when creating games.

F. Review of work with selected software tools for comparison

The research involved five engines for creating computer games which are primarily designed for making 2D games. For this study, we have developed our criteria and according to them compared all selected engines. In every engine, we made a basic version of the game, while in GameMaker: Studio a full version of the game was developed. For each compared tool we've created a new project in order to examine the functionality and features of the engine. We adapted designed game scenario to the possibilities of particular engine while some elements could not be realized in each of them. Further, in GameMaker: Studio, Construct 2 and Phaser Editor free engine versions was not possible to use

TABLE 3.2 COMPARISON BASED ON THE FLEXIBILITY AND SUPPORT CRITERIA

CRITERIA	CATEGORY	Adventure Game Studio	Construct 2	e-Adventure	GameMaker: Studio	Phaser Editor
SUPPORT	*Customer support*	Yes	Yes	Yes	Yes	Yes
	Complete documentation	Yes	Yes	Yes	Yes	Yes
	Community	Yes	Yes	Yes	Yes	Yes
	Forum/Blog	Yes	Yes	Yes/No	Yes	No
	Facebook/Twitter	Yes/No	Yes	No	Yes	Yes
FLEXIBILITY	*Adapted to work with people with special needs*	No	No	Yes	No	No
	Accessible to the wider community	Yes	Yes	Yes	Yes	Yes
	Integration with LMS systems	No	No	Yes	No	No
	SCORM 1.2	No	No	Yes	No	No
	Multilingual engine	No; English only	No; English only	Yes;	No; English only	No; English only
	Multiplayer	No	Yes	No	Yes	Yes

all the features and functionalities. After the comparison, GameMaker: Studio was chosen for the development of the full game "Explore the solar system". since it met all the required criteria. We decided to use it because it's free engine for developing 2D adventures games for PCs. It has an intuitive graphical interface and complete engine, and it's relatively easy to create game with this engine. It has very good user community, offers various examples, tutorials, manuals, forums, developers personal experience.

The serious game "Explore the solar system" is intended for 5th grade students and is associated with teaching unit Earth in space in Geography course. The game can be used at the beginning of the lesson as an introduction in order to provide students' motivation for learning. Also, it can provide the repetition about the Solar system in a funnier and more interesting way for students. The game type that we developed is a 2D point and click adventure.

V. CONCLUSION

One aspect of using ICT in the education is game-based learning and using serious computer games. Implementing games encourage students to participate and develop their creativity. They are primarily used for successful material completion in a more interesting and fun way. Serious games represent an opportunity for enhancing education from primary school through to lifelong learning.

The creation of a serious game is a complex project, and one of the most challenging task is to choose the appropriate game engine. Within this study, we examined five tools for game development. The comparison showed that each tool has certain advantages and disadvantages. Customers choose development platform based on engine characteristics, the experience they have in game development and their needs. This paper proposed criteria that will allow the developer to select the engine based on these characteristics.

For game development, we chose GameMaker: Studio since it met all the criteria set. It's very affordable engine with intuitive graphical interface and simple mode which facilitate game development.

GameMaker: Studio is suitable for beginners in game development, or for those who do not have much programming experience because in a very short period they can make a simple game.

TABLE 3.3 COMPARISON BASED ON THE FUNCTIONALITY CRITERIA

CATEGORY	Adventure Game Studio	Construct 2	eAdventure	GameMaker: Studio	Phaser Editor
Game types	*Point & Click* adventures	Different HTML5 2D games	*Point & Click* adventures	Different 2D games	Different HTML5 2D games
Creating a game in a certain mode	Yes, in the third person	Not limited	Yes, in the first or third person	Not limited	Not limited
Including finished scripts in tools	Yes	Yes	No	Yes	Yes
Navigation through the game	Mouse pointer	Mouse pointer or keyboard	Mouse pointer	Mouse pointer or keyboard	Mouse pointer or keyboard
Programming code editor	Yes	Yes	No	Yes	Yes
Box 2D Simulator Editor	No	Yes	No	Yes	No
Game logic	Scripts defined	Defined with certain events and actions	Defined with flags and variables	Programming	Programming
Behavior of the characters and games objects	Yes	Yes	Yes	Yes	Yes
Dialogue editor	Yes	No	Yes	External editor	No
Embedded options for defining the movement of the main character and the NPC	Yes	Yes	Yes	Yes	No
Turning another elements look (the main character, NPCs, items)	Yes	No	Yes	No	No
Input the instructions	No	No	Yes	No	No
Items inventory	Yes	No	Yes	No	No
Creating icon object for games inventory	No	No	Yes	No	No
Embedded options to save, load and exit the game	Yes	No	Yes	No	No
Timer option	Yes	No	Yes	Yes	No
The random appearance of objects	No	Yes	No	Yes	Yes
The game statistics view option	No	No	Yes	No	No
Creating multiple-choice questions	Yes	No	Yes	Yes	No
Examination and assessment through various questions	Possible to make	Possible, but not the best optimization	Well optimized	Possible to make	Possible, but not the best optimization

ACKNOWLEDGMENT

The research has been conducted under the project "E-learning Recommender System" (reference number 13.13.1.3.05) supported by University of Rijeka (Croatia).

REFERENCES

[1] G. J. Hwang, L. H. Yang, and S. Y. Wang, "A concept map-embedded educational computer game for improving students' learning performance in natural science courses," *Comput. Educ.*, vol. 69, no. August, pp. 121–130, 2013.

[2] M. Prensky, "Digital Natives, Digital Immigrants," *Horiz.*, vol. 9, no. 5, pp. 1–6, 2001.

[3] J. Rugelj, "Serious computer games in computer science education," *EAI Endorsed Trans. Game-Based Learn.*, vol. 2, no. 6, p. 150613, 2015.

[4] S. Erhel and E. Jamet, "Digital game-based learning: Impact of instructions and feedback on motivation and learning effectiveness," *Comput. Educ.*, vol. 67, no. March 2016, pp. 156–167, 2013.

[5] W. Horton, "Games and simulations," in *E-Learning by Design*, vol. 45, no. 5, 2012, pp. 323–398.

[6] N. Whitton, *Learning with digital games*. 2009.

[7] C. Peng, "Introductory game development course: A mix of programming and art," *Proc. - 2015 Int. Conf. Comput. Sci. Comput. Intell. CSCI 2015*, pp. 271–276, 2016.

[8] E. J. Marchiori, J. Torrente, Á. Del Blanco, P. Moreno-Ger, P. Sancho, and B. Fernández-Manjón, "A narrative metaphor to facilitate educational game authoring," *Comput. Educ.*, vol. 58, no. 1, pp. 590–599, 2012.

[9] C. D'Apice, C. Grieco, R. Piscopo, and L. Liscio, "Design of an Educational Adventure Game to teach computer Security in the working environment," *21st Int. Conf. Distrib. Multimed. Syst. DMS 2015 A4 - Knowl. Syst. Inst. Grad. Sch. KSI Res. Inc.*, pp. 179–185, 2015.

[10] S. Machines, »Adventure Game Studio - Members List,«

Simple Machines, 2013. [Online]. April 21, 2017. Available: http://www.adventuregamestudio.co.uk/forums/index.php?action=mlist.

[11] SCIRRA, "Construct 2." 2016.

[12] J. Torrente, P. Moreno-Ger, B. Fernández-Manjón, and Á. del Blanco, "Game-Like Simulations for Online Adaptive Learning: A Case Study," *Int. Conf. Technol. E-Learning Digit. Entertain.*, pp. 162–173, 2009.

[13] J. Torrente, Á. Del Blanco, and ... E. M., "<E-Aventure> Introducing Educational Games in the Learning Process," *Educ. ..., 2010 -. ieeexplore ieee org.*, 2009.

[14] J. Torrente, Á. del Blanco, P. Moreno-Ger, I. Martínez-Ortiz, and B. Fernández-Manjón, "Implementing accessibility in educational videogames with <e-Adventure>," *Proc. first ACM Int. Work. Multimed. Technol. distance Learn.*, pp. 57–66, 2009.

[15] Á. Blanc, J. Torrente, E. J. Marchiori, I. Martínez-ortiz, P. Moreno-ger, and B. Fernández-Manjón, "Easing Assessment of Game-based Learning with <e-Adventure> and LAMS," *Contract*, pp. 25–30, 2010.

[16] D. Bigas Ortega, "Machine Learning Applied to Pac-Man Final Report," 2015.

[17] YoYo Games Ltd., "GameMaker_ Studio _ Documentation." 2015.

[18] B. Froz, A. Busson, R. Martins, and A. Oliveira, "Comparison between two Development Platforms of Games to E- learning Courses," *Int. Conf. Futur. Educ. 3rd*, 2013.

[19] E. Johansson and T. Andersson, "A closer look and comparison of cross-platform development environment for smartphones," 2014.

[20] A. Fornaris, "Phaser Editor _ HTML5 2D Games IDE." .

[21] A. M. de Miguel, "HTML5 2D videogame programming," 2016.

Case Study of Online Resources and Searching for Information on Students' Academic Needs

Antonela Čižmešija*
University of Zagreb, Faculty of Organization and Informatics, Varaždin, Croatia
acizmesi@foi.hr
Violeta Vidaček-Hainš **
University of Zagreb, Faculty of Organization and Informatics, Varaždin, Croatia
vvidacek@foi.hr

Abstract: The rapid development of digital technology and the constantly increasing number of online information sources have meant that students' skills related to information seeking are crucial for academic success. This Paper presents the selected findings of a case study with an emphasis on identifying differences and similarities between international and Croatian students' information seeking skills in an online environment. Internet search engines, E-books, online databases and research networks are the online resources which were analysed in this study in order to discover information on the literacy of two groups of students. Quantitative research was conducted in March of 2015 via an online survey and the sample consisted of 70 international and 69 Croatian students. The following research fields were identified in line with the literature review and the expected research outcomes: (1) discover similarities and differences in usage of online information for academic purposes between international and Croatian students, (2) compare the frequency of usage of different online sources between international and Croatian students, and (3) describe the most common online sources of information used by the two groups of students. The results of the study could be used to raise awareness of the need to develop students' online information skills and to improve courses within the Croatian higher education system by harmonizing them with those from European Union.

Key words: Online Resources, Information Seeking, Students

I. INTRODUCTION

Rapid technology development, the Internet era and ever-increasing amounts of available information in the on-line environment within and academic context require that students develop skills connected to information literacy. Generally, information literacy is defined by the American Library Association as the abilities that enable people to 'recognize when information is needed and have the ability to locate, evaluate and effectively use information'[1]. Information literacy skills are closely connected to technological innovations and the Internet so student competencies should be oriented on the successful

navigation through on-line resources, knowledge production and collaboration with colleges [2].

Nowadays students often tend to accept information without checking its accuracy[3], and therefore it is important to develop their research skills. In an on-line environment, on-line search engines are a primary step in seeking information, and when resources such as E-books, on-line library databases and research networks for scientists are used, time and skills are demanded in an academic context. Even students who feel confident searching for information on-line have an overdependence on trivial information sources from search engines, and this can be problematic in information diffusion in an on-line context. The results of research that focused on identifying E-resources such as E-books, journals, databases, articles, search engines and websites show that students often cannot identify resources correctly or at least not on a level that is academically acceptable [4]. In order that student on-line information search processes for academic purposes could be better understood, the SCOOP research tool was developed to identify crucial information related to students on-line search behaviour and the problems they came across while searching for information on-line [5]. Students not only just consume on-line information, they also generate information through their academic work, and the use of on-line sources that are suitable within an academic context is crucial for developing a learning society on an international level.

Research in Croatia was conducted which connected the use and perceptions of on-line academic databases among university teachers and researchers. These results show that use of this on-line source of information is not satisfactory, especially in cases of younger academic staff, such as teaching assistants at university, who need instructions and trainings how to use on-line databases to retrieve information adequately [6]. These results also show that even those working in academic society who teach students have insufficient information literacy skills, so it can be assumed that students' skills for on-line information sources are not sufficiently developed. The manner in which students search for and use information is a part their academic skillset and research competences. Different student strategies for using information resources have different benefits for student academic achievement and they depend on the actual needs of the user's understanding. Improving access methods can produce better information query efficacy and help students to gather information and attain academically in general [7]. The nonlinear model of information-seeking behaviour among academics includes three core processes (which are linked to opening, orientation and consolidation) and three levels of contextual interaction. Contextual interactions within the nonlinear model are linked to external and internal context and depends on a cognitive approach. Information seeking processes within the nonlinear model are similar to the processes on an artist's palette, which means that their preparation and starting and finishing points are not fixed. All information-seeking processes are nonlinear, dynamic, holistic, and flowing [8]. The elements of this model are connected to the development of information literacy curricula.

Research results [9] show that seeking scientific information is a complex process for which only few students have sufficient knowledge. The search process includes library competences. Researchers [9] divided respondents into three groups, according to their different levels of competences for scientific information searches: novice, intermediate and expert users. According to the results for all three groups, collaboration between faculty and librarians is necessary in order to help students to gather information for their learning outcomes. Research conducted among postgraduate engineering students shows that there is a linear relationship between a user's cognitive style and information seeking. Different user profiles depend on the differences in a user's information seeking skills [10]. Database experts and students that attend the college information management course, in which they gather trend transition skills, have better learning performances in comparison with users of other databases. A comparison of students who use conventional search engines with database experts shows that the latter can easier target the issue [11]. International students and students from different cultures, such as Chinese students, mostly use two search engines: Baidu and Google. Baidu is more popular because students are familiar with their interface and the Chinese resources. Google is used more often for global information seeking and information provided in English language, whereas BAIDU is used for everyday communication and networking [12].

II. METHODOLOGY

Research was conducted amongst international and Croatian students in March of 2015 via an on-line

survey. Students were asked to answer anonymously and on a voluntary basis.

A. Respondents

The Research's core sample (N=139) consisted of two independent samples: 69 international students and 70 Croatian students. Details on the distribution of respondents according to gender, age and type of study is shown in Table 1.

TABLE 1. DISTRIBUTION OF RESPONDENTS ACCORDING TO GENDER, AGE AND TYPE OF STUDY

Chara-cteristic	International (N=69)		Croatian (N=70)	
	Frequency	Percentage	Frequency	Percentage
Gender				
Female	34	24.46%	32	23.02%
Male	35	25.18%	38	27.34%
Age				
18-20	9	6.47%	7	5.04%
21-23	21	15.11%	42	30.22%
24-26	16	11.51%	18	12.95%
27 or more	23	16.55%	3	2.16%
Type of Study				
Under graduate	39	28.06%	23	16.55%
Graduate	20	14.39%	43	30.94%
Post-graduate	10	7.19%	4	2.88%

Some meaningful differences between the two student groups will be mentioned in following paragraph. From data shown in Table 1, it can be stated that most of the international respondents were from the following countries: the United Kingdom, Netherlands, Germany, Norway, Switzerland, Portugal, Greece, France, Spain, Romania, Latvia, Italy, Ukraine, Austria, Macedonia and Poland. 16.55% were 27 years or older and 28.06% were undergraduate students. In case of Croatian students, 30.22% of respondents belonged to age group from 21 to 23 and 30.94% of them were graduates.

B. Questionnaire

The questionnaire was created in English and Croatian and contained 31 items. Results of both questionnaires were summarised and one data set was used for answer analysis.

The first part of the questionnaire consisted of questions about the respondent's socio-demographic characteristics: gender, age, type of study and home country (for European students) or university (for Croatian respondents). The second part included 27 items about on-line information resources and student attitudes about their use for academic purposes. A 5 point Likert scale was used to measure student attitudes or opinions by level of agreement or disagreement with a particular item. The Cronbach alpha coefficient was used to check the internal consistency of the measuring instrument. For this questionnaire, Cronbach alpha was α=.0717 for 31 items, which indicates a high level of internal consistency, and it can be concluded that this scale is reliable and acceptable in social researches[13],[14].

The Main limitations of this study relate to its sample size (N=139), and a larger and differently structured sample size would ensure a more representative distribution of the population. For example, respondents in the Croatian student sample were mostly from the University of Zagreb's Faculty of Organization and Informatics, and students from other universities were not included. One limitation of this study's result is also the effect of pseudorandom sample (N=139) that it is not representative for whole faculty population. Still, the selected sample includes students from all levels of study (undergraduate, graduate and postgraduate levels) and amounts to approximately 5% of the core sample which includes 3,000 students from the Faculty. In the case of international students, respondents were from different countries and universities (19 various home countries were listed in the responses). A better approach would have been to compare student attitudes to on-line information searches from just from one foreign country. Furthermore, the data was collected in 2015, so student opinions could have changed in the interim. To find further differences in usage of different on-line scientific resources for academic purposes, more detailed questions should have been included in the questionnaire.

III. RESULTS AND INTERPRETATION

Three research questions were determined to identify similarities and differences between international and Croatian students' information seeking skills within an on-line environment.

A. Similarities and Differences between International and Croatian Students Regarding On-line Information Usage

In order to compare whether there are differences in on-line information usage habits between international and Croatian students for academic purposes, two types of analyses were performed for the following types of on-line information sources: a) Internet search engines (Google, Yahoo and Bing), b) On-line scientific databases (EBSCO, Science Direct, ProQuest and Wiley On-line Library), c) E-books (Google books, Scribd, Intechopen) and d) social networking sites for scientists and researchers (ResearchGate, Academia.edu).

Levene's Test for Equality of Variances was first used to determine if the two groups had about the same or different variability between scores. The results are shown in Figure 1. Since the Sig. value in

the four analysed resources in Levene's Test is > 0.05, it can be concluded that variability in two observed groups of students is not significantly different.

TABLE 2: HABITS OF USAGE THE ON-LINE INFORMATION AMONG INTERNATIONAL AND CROATIAN STUDENTS
Legend: CRO = Croatian Students, INT = International Students

	International / Croatian students	N	Mean	Std. Deviation	Std. Error Mean
On-line database	CRO	70	2.89	1.17	0.14
	INT	69	3.36	1.18	0.14
Internet search engines	RH	70	4.76	.71	0.09
	EU	69	4.74	.50	0.06
E-books	RH	70	3.10	1.14	0.14
	EU	69	2.84	1.08	0.13
Research networks	RH	70	2.23	1.29	0.15
	EU	69	2.51	1.12	0.14

An independent samples t-test was used to compare the means determine the differences in habits of on-line information source usage between two unrelated groups of students (Croatian and international students) for academic purposes. The results of t-test show that there is a statistically significant difference between international and Croatian students regarding usage habits of on-line scientific databases (t=2.39, df=137, p<0.05). It can be concluded that international students use databases more than Croatian students for academic purposes (M_{int}=3.36, sd_{INT} = 1.18, M_{cro}=2.89, sd_{CRO}= 1.17).

The t-test for results of another three analysed sources of on-line information shows that there is not a statistically significant difference between international and Croatian students regarding Internet search engine usage habits (t=0.17, df=137, p>0.05), E-books (t=0.547, df=137, p>0.05) and social networking sites for scientists and researchers (t=0.11, df=137, p>0.05).When comparing the means for each on-line source, it can be concluded that Croatian students are slightly more likely to use Internet search engines (M_{CRO}=4.76, sd_{CRO}= 0.711, M_{INTt}=4.74, SD_{INT} = 0.50) and E-books (M_{CRO}=3.10, sd_{CRO}= 1.14, M_{INT}=2.84, sd_{INT} = 1.08) than international students, but the differences are not statistically significant. On the other side, international students have a habit to use research social networks for researchers and scientists slightly more that Croatian students, but the differences are not statistically different (M_{INT}=2,51, sd_{INT} = 1.12, M_{cro}=2,23, sd_{CRO}= 1.29)

The results show there is need to improving the skills required to use scientific databases among Croatian students because currently they have a habit of using simple, non-scientific sources of on-line information rather than those which are more complex and better suited for academic purposes. The reasons for such results could be following: Croatian students are not familiar enough with this kind of information and their importance in producing higher quality student papers and researchers as they consider them more skill and time intensive, and another issue could be that there is no access to relevant scientific databases for academics at a national level or that the prices for specific articles is relatively high.

Other results in general show no differences in habits of Internet search engine, E-books and social networking sites usage for scientists and researchers. The main reasons for such on-line seeking behaviour could be that these are sources students use in everyday life, they consider them appropriate enough for academic purposes, they are intuitive to use and do not demand special information literacy skills and they are also free and easy to access.

B. Comparative Frequency of Usage of Different On-line Sources by International and Croatian Students

In order to determine comparative frequency of usage of different on-line sources by international and Croatian students, two types of analyses were performed for the following types of on-line information sources: a) Internet search engines (Google, Yahoo and Bing), b) On-line scientific databases (EBSCO, Science Direct, ProQuest and Wiley On-line Library), c) E-books (Google books, Scribd, Intechopen) and d) social networking sites for scientists and researchers (ResearchGate, Academia.edu). A 5 scale Likert scale was used to determine the frequency of usage of these four on-line resources and the possible answers were: 1 – Never used, 2 – Used once 2, 3 – Using 2 or 3 times a month , 4 - Using every week, 5 – Using every day.

TABLE 3. FREQUENCY OF USAGE DIFFERENT ON-LINE SOURCES FOR INTERNATIONAL AND CROATIAN STUDENTS
Legend: CRO = Croatian Students, INT = International Students

	International / Croatian Students	N	Mean	Std. Deviation	Std. Error Mean
On-line Database	CRO	70	3.17	1.26	0.15
	INT	69	3.88	1.13	0.13
Internet Search Engines	RH	70	4.10	.75	0.09
	EU	69	4.03	.84	0.10
E-books	RH	70	3.33	1.00	0.12
	EU	69	3.26	.98	0.12
Research Networks	RH	70	2.97	1.15	0.14
	EU	69	3.09	.10	0.12

Likewise for the first research question, Levene's Test for Equality of Variances was used to determine if the two groups had about the same or different amounts of variability between scores. Since the Sig. value in the four analysed resources in Levene's test is > 0.05, it can be concluded that variability in two observed groups of students is not significantly different.

The t-test results of another three analysed sources of on-line information show that there is not a statistically significant difference between international and Croatian students regarding to frequency of use of Internet search engines ($t=0.53$, $df=137$, $p>0.05$), E-books ($t=0.73$, $df=137$, $p>0.05$) and social networking sites for scientists and researchers ($t=0.53$, $df=137$, $p>0.05$). When comparing the means for each on-line source, it can be concluded that Croatian students use Internet search engines ($M_{CRO}=4.10$, $sd_{CRO}= 0.09$, $M_{INT}=4.03$, $sd_{INT} = 0.10$) and E-books ($M_{CRO}=3.33$ $sd_{CRO}= 1.00$, $M_{INT}=3.26$, $sd_{INT} = 0.98$) more often than international students. International students more often find information for their academic purposes on research networks than Croatian students ($M_{INT}=3.09$, $sd_{INT} = 0.10$ $M_{CRO}=2.97$, $sd_{CRO}= 1.15$). An independent sample t-test was used to compare the means and find differences in frequency of usage of on-line information sources between international and Croatian students. The results of this t-test show that there is a statistically significant difference between international and Croatian students regarding to frequency of usage of on-line scientific databases ($t=3.50$, $df=137$, $p<0.05$). It can be concluded that international students use databases more frequently than Croatian students for academic purposes ($M_{INT}=3.88$, $sd_{INTT} = 1.13$, $M_{CRO}=3.17$, $sd_{CRO}= 1.26$). The t-test results of another three analysed sources of on-line information show that there is no statistically significant difference between international and Croatian students regarding frequency of Internet search engine usage ($t=0.528$, $df=137$, $p>0.05$), E-books ($t=0.729$, $df=137$, $p>0.05$) and social networking sites for scientists and researchers ($t=0.52$, $df=137$, $p>0.05$).

When comparing the means for each on-line source, it can be concluded that Croatian students use Internet search engines ($M_{CRO}=4.10$, $sd_{CRO}= 0.09$, $M_{INT}=4.03$, $sd_{INT} = 0.10$) and E-books ($M_{CRO}=3.33$ $sd_{CRO}= 1.00$, $M_{INT}=3.26$, $sd_{INT} = 0.98$) more often than international students. International students find information for their academic purposes slightly more often than Croatian students on research networks, but there is not a statistical difference ($M_{INT}=3.09$, $sd_{INT} = 0.10$ $M_{CRO}=2.97$, $sd_{CRO}= 1.15$).

The results of second research question are similar to the first and show that international students have a preference for using scientific resources (on-line scientific databases and on-line networks for researchers) for information more often than Croatian students do. The key reason for such a difference could be that European counties have more developed practices at a national or institutional level than Croatia, and these measures improve awareness of the importance of using scientific on-line sources and developing need skills in an academic context.

C. On-line Sources of Information Used by International and Croatian Students – Knowledge and Skills

A third research question was used to determine how students evaluate their knowledge and skills in usage of the most common on-line sources of information. Possible answers ranged from 1, which indicated a low level of knowledge and skills, and 5, which indicated a very high level of usage of some information source for academic purposes. Results are shown in Figure 2.

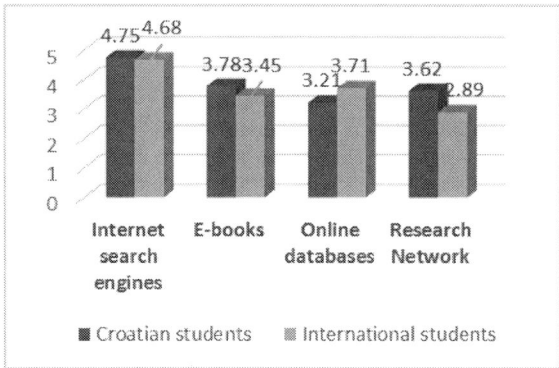

FIGURE 1: KNOWLEDGE AND SKILLS LINKED TO THE ON-LINE SOURCES OF INFORMATION BETWEEN INTERNATIONAL AND CROATIA STUDENTS
Legend: Access x: Type of On-line Sources; Access y: Answers on the Likert Scale (1-5)

From the data represented on this graph, Croatian students from this sample evaluated their skills and knowledge more positively in cases of Internet search engines, and E-books and international students consider themselves more skilful in using on-line scientific databases and on-line networks for researchers and scientists. It can be concluded that both group of students have highly developed skills for Internet search engine usage, but there is room for improvement regarding other on-line sources, especially in case of the Croatian group of students who have lower scores than international students.

IV. CONCLUSION

The results of this study show that students from Croatia and international students use similar sources to gather information for academic purposes. Those sources are mostly linked to Internet search engines, E-books (such as Google books, Scribd, Intechopen...), on-line scientific databases (such as EBSCO, Science Direct, ProQuest, Wiley On-line Library...) and on-line networks (such as Research Gate, Academia.edu...).

There is the slightly more frequent usage of databases by international students in comparison with Croatian students. Reasons for this could include possible access to data bases and different research skills. Further education in the field of attaining academic skills and the development of information seeking behaviour could be improved with specific courses and training. Closer collaboration between librarians, staff and students is a necessary prerequisite for using academic sources of information for research purposes. In general, higher education should include information literacy in curricula and focus more on developing requisite research skills so that student can use appropriate on-line information sources for academic tasks. Further research could include more students from other universities so that this core sample could be more relevant.

V. REFERENCES

[1] H.B. Rader, „Information literacy and the undergraduate curriculum", Library trends, vol. 44, No. 2, p.p. 270-278, 1995.

[2] J.Sin, I.A. Mokhtar, „Information literacy skills of humanities, arts, and social science tertiary students in Singapore". Reference & User Services Quarterly, Vol.53, No.1, p.p. 40-50, 2013.

[3] D. Jackson, „The trouble with digital natives", New Zealand Management, vol 62, No. 2, p.p. 56-58, 2015.

[4] A. Buhler, T.Cataldo, „Identifying e-resources: an exploratory study of university students", Library Resources & Technical Services, Vol. 60, No.1, 2016.

[5] M. Zhou, „SCOOP: A measurement and database of student online search behavior and performance", British Journal of Educational Technology, vol 46, No. 5, p.p. 928-931, 2015.

[6] D. Dukić, „Use and perceptions of online academic databases among Croatian University teachers and researchers". Libri, vol. 64, No. 2, p.p. 173-184, 2014.

[7] A.Biaz, A. Bennamara, A. Khyati, and M. Talbi, „Informational strategies and the use of information systems by doctoral students: A case study at the University of Hassan ii Mohammedia, Casablanca", Procedia - Social and Behavioral Sciences, vol. 116, p.p. 3598-3604, Feb. 2014.

[8] A.Foster, „A nonlinear model of information-seeking behavior", Journal of the American Society for Information Science and Technology, vol. 55, No. 3, pp. 187–279, Feb. 2004.

[9] L. Karlsson, L. Koivula, I. Ruokonen, P. Kajaani, L. Antikainen, and H. Ruismäki, „From novice to expert: Information seeking processes of University students and researchers", Procedia - Social and Behavioral Sciences, vol. 45, pp. 577-587, 2012.

[10] M. Salarian, R. Ibrahim, and K. Nemati, „The Relationship between Users Cognitive Style and Information Seeking Behavior among Postgraduate Engineering Students", Procedia - Social and Behavioral Sciences, vol. 56, pp. 461-465, Oct. 2012.

[11] C. Yin, , H.-Y. Sung, G.-J. Hwang , S. Hirokawa, H.-C. Chu, B. Flanagan, and Y. Tabata, „Learning by Searching: A learning environment that provides searching and analysis facilities for supporting trend analysis activities". Educational Technology & Society, vol. 16 no. 3, pp. 286–300, 2013.

[12] T. Fu, and K. Karan, „How Big is the World you can Explore? A Study of Chinese College Students' Search Behavior via Search Engines", Procedia - Social and Behavioral Sciences, vol. 174, pp. 2743–2752, Feb. 2015

[13] J.C. Nunnally, "Psyhometric theory", New York: McGraw-Hill, 1978.

[14] R.A.Bernardi, "Validating research results when Cronbach's alpha is below. 70: A methodological procedure", *Educational and Psychological Measurement*, vol. *54*, No. 3, p.p. 766-775, 1994.

Recursions and how to teach them

Predrag Brođanac

V. gimnazija, Zagreb

predrag.brodanac@skole.hr

Summary – Learning programming is difficult. This refers to primary and secondary school students, but to university student beginners at programming too. These particular reasons stand as a challenge to teachers, as well. Most of the authors agree that recursion presents one of the most difficult concepts at programming, whereas some authors think it is the most difficult concept in programming. On the other hand, it is no longer only an important mathematical concept, but a programming technique, a way of algorithm thinking, and a tool to problem solving. There have been many attempts how to teach recursions. Hereinafter, one can find certain approaches for which authors claim to bring a better understanding of this important concept.

Key words: *recursion, learning models, learning styles.*

I. INTRODUCTION

Recursion is a technique within which a program function, a procedure or a method calls itself; and a recursive algorithm is the one self-referencing [1] If a function continuously calls itself, the process would never stop, therefore a recursive function should meet three of the following conditions [2]:

- recursive calling termination condition,
- recursive relation,
- consecutive recursive calling should be converged according to the recursive calling termination condition.

The basic example of a recursive function is the function for product of first *n* positive integers, known as the factorials. In the *Python* programming language function is as follows:

```
def fact(n):
    if n == 1:                        #1
        return 1
    else:
        return n * fact(n - 1)        #2
```

Command (#1) has a defined *recursion termination condition.* The issue here is a basic case with a trivial problem solution. In the upper example, a basic case is the one in which one should multiply the first 1 positive integer and that product is 1. Should the basic case not be the situation here, the function performance is done recursively (#2). Namely, the product of the first *n* positive integers can be attained in the manner that the product of the first *n-1* positive integers is multiplied by *n*. This statement is evident, for example, at multiplication of the first 5 positive integers:

$$fact(5) = 1 \cdot 2 \cdot 3 \cdot 4 \cdot 5 = \underbrace{1 \cdot 2 \cdot 3 \cdot 4}_{fact(4)} \cdot 5 = fact(4) \cdot 5$$

Command (#2) has, therefore, defined the *recursive relation.* Since the value, used to call the function, constantly decreases, it stands clearly that recursive calling converges to the basic case, i.e. *n = 1.*

Recursion is a special case of the *Divide-Conquer-Glue* technique [3], where the problem addressed is divided into smaller problems, which later are solved independently, and, finally, by „gluing" the solutions we come to the initial problem solution. Recursion here is specific in the way that smaller problems solved here are the same as the initial problem, only they differ in „size", which we also call the *instance*, i.e. *dimension.* For example, *fact(5)* and *fact(4)* are two instances of the problem of multiplication of the first *n* positive integers, wherein *fact(5)* instance represents the product of the first 5 positive integers, whereas *fact(4)* instance represents the product of the first 4 positive integers.

At the same time, we can observe recursion as a loop generalization [4]. A loop consists of a set of commands that repeat. The most common loops used in programming languages are the *for* and *while* loops. Commonly, the loop has one of the following two shapes:

```
command-1
command-2
...
command-n        (decision    on    repeated
performance)
```

or

```
command (decision on one-time execution)
command-1
command-2
...
command-n (decision on repeated execution)
```

One can easily notice that decision on executing of a block shall be made at the end or at the beginning of the block of commands. Recursion can be observed as a case where a decision on repeated execution of a block of commands can be made at any point within the block:

```
block-1

command
...
Jump to block-1 (condition satisfied)
...
command
end of the block
```

During executing of the *block-1*, the execution comes to a stop and the commands of the same block are called again. This repeated call may cause another call, etc. After the commands of one block have been executed fully, we return to the following command of the block that has called the block execution, wherein the executed block's results may be used in the future for executing of the block that called the block executed. It is important to mention that while calling the new block, the status of the block that called it should be filed, and then later on loaded, and continue to execute. It is realized by *runtime stack* [5]. Each time a recursive subprogram has been called, variables, parameters, return address and other required data are stored on the stack. At the moment when a subprogram is returning the control over to the subprogram which called it, the related data are popped from the stack.

A large number of programs, that are usually solved recursively, are possible to be solved by using loops, as well. For example, such is the program for multiplication of the first *n* positive integers. Its non-recursive solution in the Python has the following form:

```
def fact(n):
    t=1
    for i in range(2, n + 1):
        t *= i
    return t
```

There are recursive problems which cannot be solved by loops or the loop-based solution is much more complex, and this may serve as motivation for recursions [6]. Sometimes, instead of the recursive, one can implement a problem solution which, using the stack, simulates recursion, e.g. for labyrinth. Similar to the loops, usage of the stack disturbs the elegance of the program. Recursive solutions will, at least, be as elegant as any other solution. Recursions enable elegant explanation of complex algorithms and data structures [7]. There are situations in which recursive solution will not be efficient. Such example are the *Fibonacci numbers*[1].

The appropriate recursive solution in the *Python* would be:

```
def b_z(n):
    if n < 3:
        return 1
    else:
        return  b_z(n-1) + b_z(n-2)
```

Picture 1: *Recursive computing of the 5th Fibonacci number*

As one can see, certain recursive calls, e.g. *b_z(2)*, occur on several occasions, which shall result in slower function performance than the following non-recursive solution:

```
def b_z(n):
    a = b = 1
    for i in range(3, n + 1):
        a, b = b, a + b
    return b
```

One should mention that recursive solution of the nth Fibonacci number computation may be upgraded in the way that the already calculated problem solutions be stored in the data structure, therefore avoiding the multiple computing

[8]. Numerous researches show that the concept of recursions is highly complex for programming beginners. Problems arise at defining the basic case, already [9]. A far greater problem is pinpointing the recursive relation in the problem set [10] and [11]. Furthermore, it is no simpler to follow the recursive relation performance and establishing the recursive function value for the defined values of parameters [12], [13] and [14]. One of the possible reasons of having a hard time understanding the recursions is non-existence of adequate analogies in the real world, [1] and [14]. In order to make this concept understandable to beginners, a set of tools has been created for recursion visualization, motivational games have been created in which a beginner subconsciously develops their recursive thinking, furthermore real-life examples have been chosen to help understanding recursions, and methodologies have been invented how to teach recursions, and other things as well.

II. CLASSIFICATION OF RECURSIONS

Recursive algorithms can be classified on multiple grounds. One of such categorizations has been stated by [15]:

- *linear* – has only one recursive call. A typical example of such a recursion is computing factorials. A special type of linear recursion is *tail recursion*. This recursion has recursive call as the last command within the function.
- *Multiple* or *exponential* – doesn't have only one recursive call, but it can have more of them. An example of such a recursion is generating all permutations of the set from *n* various elements.
- *nested* – recursive call is one of its arguments. *Ackermann's function*[2] is an example of such a recursion,
- *mutual* – this one doesn't call only itself, but another function, which then calls the first function again. We can have more than two functions. An example of such a recursion is a slightly different point of view on computing the nth element of the Fibonacci sequence [15]

$$F_i = A_i + B_i, B_i = \begin{cases} 1 & i = 1 \\ A_{i-1} & i \neq 1 \end{cases},$$

$$A_i = \begin{cases} 0 & i = 1 \\ A_{i-1} + B_{i-1} & i \neq 1 \end{cases}$$

There is another division of the recursive programs defined by [16]. He addresses three types of recursions:

- *structural* – at recursive call, the arguments, by which the recursion is called, are either unchanged or closer to the basic case. If the parameter of the recursive function be, for example, data structure, the recursive call will happen over the real subgroup of that structure. That way, when binary tree traverse, one recursive call traverses only one tree branch,

- *generative* – the arguments used to call the recursive function are computed each time over again and there is no guarantee that a change of parameters leads to the basic case. Examples to such recursions is the *Collatz Problem*[3] and Euclide algorithm for

---

[1] The first and second Fibonacci number are 1, and each number is the sum of the two preceeding.

[2] $A(m,n) = \begin{cases} n + 1 & m = 0 \\ A(m - 1, 1) & m > 0 \ i \ n = 0 \\ A(m - 1, A(m, n - 1)) & m > 0, n > 0 \end{cases}$

calculating the largest common divisor of two positive integers,

- *accumulative* – in these recursions, among the parameters of recursions, there is one more or several additional parameters containing partial solutions of problems.

III. ARE RECURSIONS TRULY THAT DIFFICULT?

Almost every author writing about recursions, already in their introductions claims that it is one of the most difficult concepts for programming beginners. Even [17] and [18] agree with the fact that it is the most difficult concept. According to the research described in [18], including 559 students and 34 teachers at 6 European universities, the most difficult concepts for the students were: *recursions, pointers, abstract data types,* etc.

Having finished learning the recursions, students will be using them seldom for any other further problem solution, except if it has been explicitly demanded from them to do so [19], and even in situations when they have an offered solution to the same problem by the loop, i.d. by recursion, they will get a better understanding of the solution implemented by the loop [20].

Recursion problem is linked to constructivism [1], is argumented by the fact that making references to the already existing knowledge is really important for gaining new knowledge, whereas recursions have nothing to refer to on account of them being so specific.

IV. CONCEPTUAL MODELS FOR TEACHING RECURSIONS

Many an author has made an attempt to make the recursions more appealing to beginner programmers. Some authors think that suitable preparatory actions, such as playing specially designed games, can help in later understanding of recursions. Whereas, others are of the opinion that analogies to problems that students are familiar with are of much importance; and there are some who think that visualization of the recursion flow is essential, while there are those whose opinion is that following the flow is not necessarily connected to development of recursive thinking, but they regulate a line of steps how to develop a recursive way of thinking.

Mental model defines a certain way of understanding abstract concepts or certain real systems [21]. It determines our way of thinking and acting, as well. *Conceptual model* stands for a manner of presenting new contents by an individual. Understanding of the content can be described as possessing a mental model on the content. Conceptual model represents an important tool for understanding of and teaching new contents. One of the teacher's tasks is to build a conceptual model enabling students develop a suitable mental model. According to the *Structural Mapping Theory* an important factor in adopting new content are *analogies*. Analogies represent a connection between a known (*base*) and unknown (*target*) domain. The conceptual model, within that context, would represent the base we take the analogies from for the unknown domain (*target*). Considering the area which we take the analogies from, we can talk about

- *abstract conceptual model* – its base is abstract, for example the mathematical model,

- *concrete conceptual model* – its base is concrete, for example a concrete object.

There is a set of models for teaching recursions, [21] bring 5 models:

- *Russian Dolls* – the Russian doll can contain a smaller doll within itself, which contains a smaller one, etc. (recursion),

- *Process Tracing* – tracing recursive calls. It describes the functioning of recursion,

- *Stack Simulation* – recursion mechanism is simulated from the computer perspective, where each recursive call creates a new record on the runtime stack,

- *Mathematical Induction* – recursions from the formal, mathematical aspect,

- *Structure Template* – explains recursions based on the example of recursive programs, describing therefore the basic case and recursive relation. The template for solving recursive problem comes down to: *searching for recursive relation* and *defining the basic case.*

In their work, the authors refer to *learning styles*, making reference to the Kolb's definition of *learning styles*, according to which a manner of information processing is one of the basic factors having influence on mental model design. Combination of manners, in which we perceive information and in which we process them, defines our *learning style*. Speaking of Kolb's learning styles we can observe them through two dimensions:

(1) *way of adopting knowledge: concrete experience and abstract conceptualization*

(2) *way of experience transformation: reflexive observation and active experimenting*

School students and university students with different learning styles will have a different reaction to various teaching methods, therefore teaching methods should be suitable to students' learning style. The authors have performed a research showing that students with abstract learning style are more successful at learning recursions than students with concrete learning style. No research has shown that the abstract type has a better understanding of recursions if they have been explained by abstract model, nor that the concrete type has a better understanding of recursions if they have been explained by a concrete model.

Two mental models for teaching recursion describe [7]:

- *The Little People model* – there is a large number of little people in the computer, each of them being an expert for a specific task (function) performance. When there is a certain problem, it is to be divided into smaller pieces and the *little people* hired to solve these smaller pieces. An example of such a model from perspective of a manager and a worker has been described even in [3].

- *„Top-Down Frames" model* – enables tracing recursive calls on paper. A new window is created for each recursive call that is within the window which has called it. Within the window there is recursive function program code with concrete parameters values, whereas at the bottom of the window there is a value that the appropriate recursive call returns.

[3] For the positive integer *n* we define transformations in the following manner: if the number is even, it becomes *n / 2*, otherwise it becomes *3 * n + 1*. The question is

how many of transformations, such as this one, over the *n* is to be performed in order to reach number 1.

The author [5] addresses the following recursion teaching models in his work:

- *Induction* – recursion is explained as a function that calls itself. The Fibonacci numbers are set as an example. It is an important approach, however it lacks a deeper understanding of how recursions function.

- *Runtime stack* – a simulation runtime stack is created in computer, which stack is an occurrence in recursive calls. The shortfall of this manner of tracing occurrences in the runtime stack is it being done on paper.

- *Function performance tracing* – each time there is a recursive call, a line is written with the name of the subprogram being called, and with suitable parameters. The same thing happens at returning from the recursive call. At output, one has to take care of the indents in order to visualize recursive calling. Tracing can be realized on paper, as well, and it makes a runtime stack simplification. Presenting recursive functions that call themselves iteratively may not be the perfect solution, e.g. functions for computing the Fibonacci numbers, i.e. Ackermann's function.

- *Recursion tree* – tree whose nodes represent the momentary surrounding, encompassing the parameters and local variables. At that point, the nodes are identified with recursive calls, and a node's children are recursive calls occurring in a suitable surrounding. Some of the shortfalls of the recursive tree are: (1) it does not contain the value returned by the recursive call; (2) it is not easy to draw a tree in case when we have more than two recursive calls within a function.

- *Activation tree* – a combination of the runtime stack and recursive tree. Each node contains, besides the name of the function and parameters, the function's return value. The author is of the opinion that this manner has no shortfalls.

Name of the function parametre 1 parametre 2 ...	return value

Picture 2*: Activation Tree Node*

A. Preliminary work that improves recursion learning

One of the ways that contributes to a better understanding of recursions later on, is playing a well thought computer game that will develop the recursive way of thinking. Authors [22] and [2] describe one of these games, called *Cargo-Bot*, showing that it helps in recursion learning.

B. Concrete Teaching Model Examples

The largest number of authors has been studying some of the learning models.

[23] provides 3 examples that can serve as introductory examples, of which each may have its drama version:

- *Opening of presents* (closed brackets problem) – a friend has given you a watch, however, in order to keep the suspense going, the watch had been placed inside a box, inside another box, etc. The objective is to describe the gift opening algorithm in one's own words.

- *Chain design* (factorials) – there are N links with springs defined, which are to make a chain together. The idea of recursive solutions is to give one's friend

to make a chain of $N-1$ length, and then add one more link at the end of the chain.

- *Eating chocolate* (searching for elements on the list) – at our disposal we have a bar of hazelnut chocolate. We are to eat only those squares containing the hazelnuts.

The authors of the [24] have an interesting way to describe how to adopt the recursion concept. It is their opinion that in times of object-oriented programming one should change their approach to teaching recursion, as well. They, also, think that recursion learning should be implemented over data structures. This is illustrated by an abstract data structure list *OurList* in the *Python*, whose every instance had two properties: *_head* – preserves the element value, and *_rest* – all the remaining list elements. The authors have been able to explain this kind of a structure to the students, and, through role playing, the students were to come up with certain basic operations over the lists: counting how many time does a certain element appear on the list, *verifying whether or not the defined element exists on the list, adding an element at the end of the list,* and so on. Each student represents an instance of the list and the teacher has established which student represents the whole list. Only the students sitting next to each other are allowed to speak to each other.

The authors [25] provide an example to recursion teaching of children at the age of 11 to 14. Recursions are described on examples not directly connected to programming. The following problems are stated as examples:

- *Genome sorting* – the students were handed envelopes containing integers (from 1 to 14) and were to line up the numbers without opening the envelopes. They were using only one student, *comparator*. He was the only one able to open the envelopes and say which envelope contains the larger number. The comparator's service costs one monetary unit for each comparison. The objective is to sort the envelopes the way to spend as little money as possible.

- *Visual recursion* – a picture containing the same picture in one of its parts, etc. Students were given several pictures and they were to determine which of these pictures contained recursions, and to provide their opinions through discussion.

- *Sierpinski Carpet* – one of the students is drawing, and the other one is describing. The one describing was given the picture of the *Sierpinski Carpet* and is to describe, to the one who is drawing, how to draw the picture.

- *Line Up* – students are lined up, as in a queue, and their eyes have been blind-folded. They are to determine how many students are there in front of them in the queue, without ever leaving the line.

Parallel car parking principle, along the road can also be a good introductory example to illustrate recursions [26]. For example, there is a street 10 units long and we would like to know how many cars can fit into these 10 units. After a car has been parked in a random space, it occupies a single unit and it „divides" the street into two parts. The total number of cars parked is going to be: the number of cars that can be parked in front of the parked car, the number of cars that can be parked behind the car parked, increased by one parked car.

743

The author [27] describes an interesting example of introducing the recursion through examples from the computer world. Recursions have been described on the example of:

- Computer system of files and folders - going through the tree like structure of files and folders by using the *Python os* module.

- Web searching – using *urllib.parse* module, by which we can reach each all the URLs on a single page and then recursively visit all those URLs.

Recursions are not natural in imperative languages [28], [29] whereas functional languages are stated to be the alternative.

Numerous works ([13], [30], [31]) deal with models of stack simulation, i.e. with visualisation of recursion tracing. To be able to understand recursions it is key to understand the principal of a subprogram execution termination, and transfer of the execution flow onto another subprogram [13]. In this context it is the moments that are important when a subprogram calls another subprogram, i.e. itself in a case of recursion. The author calls this moment *active transfer of the execution flow*. The moment the called subprogram has been executed till the end, the execution control returns to the subprogram that called it – *execution flow passive transfer*. These, but many other procedures have been visualised in the *EROSI system* (*Explicit Representer Of Subprogram Invocations*). These authors' work demonstrates that usage of this program improves understanding, and writing of recursive programs, as well.

Besides o computer, recursion mechanism visualisation is possible on paper, too. [31] speaks of the recursion graph notion – *RGraph*. The idea for this kind of an approach has come to life as a reaction to the *recursive tree*, which, according to its author, doesn't have an accentuated passive transfer and value return at the end of the recursive call; which is solved in *RGraph*. *Square nodes* in the graph represent recursion call, while *oval nodes* represent preliminary actions for recursion call, i.e. actions after recursion call. Moreover, there are two types of edges: *dashed* line and *full* line. The dashed ones are those that go from the higher level onto the lower level, whereas the full lines represent the edges among the nods on the same level.

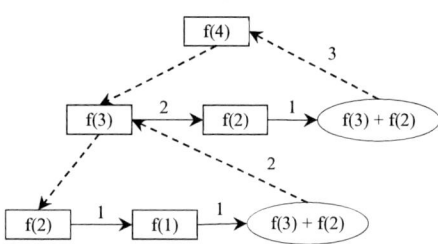

Picture 3: *Visualisation of Execution of the 4th Element of the Fibonacci Sequence Recursive Computing by the* RGraph

Hybrid visualisation model is the one trying to unify more conceptual teaching models. One of such has been described by [12]. Their idea was to unify the conceptual model of *stack simulation* and *structure templates*. The result of it is creation of a system for helping students learning recursions, called *ChiQat-Tutor*. Stack simulation model is based on the graphic variant of the *RGraph*. The system offers a few standard problems that enable studying of recursions (*factorials, palindrome,* etc.), and provides recursions tracing, complementing of recursive algorithms, creating the RGraph for a set recursive code, animating of recursion execution, etc. After using this tool, a significant leverage in solving recursive problems has been observed.

C. Abstract Model Examples

Concrete model of recursion teaching is not enough for a deeper understanding of recursions [10], [32], [33]. Recursion execution tracing is the basic mechanism that is going to help the recursions-inexperienced students to be successful at computing the recursive function value, however that does not mean that they will be able to understand the recursion fully, and especially the passive transfer of execution control that is happening at the moment of our exiting the function's recursive call.

Programming represents a specific way of thinking and addresses problem solving [10], where the focus is placed on figuring out algorithms, and not on the programming language syntax.

Algorithm to approach the recursive problem solving has been described by [10] and [11]. According to [10], the basic steps are:

(1) *Finding the basic case* – the smallest instance of the problem with a basic solution,

(2) *Rules simplification* – manner to reduce the size of a problem,

(3) *Natural recursion* – link between a simplified and an actual solution to the problem,

(4) *Finish* – merging of the previous steps into a single recursive function.

V. LOOPS BEFORE RECURSIONS OR VICE VERSA?

There is also a debate on when to teach recursions. Several authors think that they should be taught after loops, while another group thinks that it is better to show the recursions before the loops.

The research according to which it is better to learn loops first, and then recursions, has been presented in [34]. On the other hand, [35] and [36] propose learning recursions before iterations. The authors [35] state several arguments to their hypothesis: *recursion makes a special case of Divide-Conquer-Glue method; to understand recursions, it is enough to know functions and branching, whereas, to understand loops, one needs to understand a brand new context; the basic reason for not understanding recursions arises due to the created mental model of the loop.*

VI. CONCLUSION

Most researchers will agree with the statement that, on one hand, recursions make an extremely important concept in computing, and an exceptionally complex one for beginners, on the other hand. There are various methodologies for teaching recursions, which can generally be divided into concrete and abstract. One group of researchers point out that concrete methods are better, whereas another group states that one should use abstract methodology in order to understand recursions fully. The truth is, surely, somewhere in the middle. One should use various concrete methods, their visualisations, etc., to be able to understand the recursion mechanisms. However, this is not the way the students will develop a recursive manner of thinking. They will develop it only if they have a deep knowledge of recursions, if they will be using them in various situations during their education, if the teacher has the habit of leading them into thinking could the problem on the table be solved recursively, and so on.

BIBLIOGRAPHY

[1] D. Levy and T. Lapidot, »Recursively speaking: analyzing students' discourse of recursive phenomena,« u *SIGCSE '00*, Austin, Texas, USA, 2000.

[2] J. Tessler, B. Beth and C. Lin, »Using cargo-bot to provide contextualized learning of recursion,« u *ICER '13*, San Diego, San California, USA, 2013.

[3] J. Edgington, »Teaching and viewing recursion as delegation,« *Journal of Computing Sciences in Colleges,* Vol. 23, no. 1, pg. 241-246, 2007.

[4] G. Ford, »A framework for teaching recursion,« u *ACM SIGCSE*, New York, 1982.

[5] S. M. Haynes, »Explaining recursion to the unsophisticated,« u *ACM SIGCSE*, New York, 1995.

[6] B. Stephenson, »Using graphical examples to motivate the study of recursion,« *Journal of Computing Sciences in Colleges,* vol. 25, no. 1, pg. 42-50, 2009.

[7] O. Hazzan, T. Lapidot i N. Ragonis, Guide to Teaching Computer Science, An Activity-Based Approach, London: Sprenger-Verlang, 2011.

[8] L. Budin, P. Brođanac, Z. Markučič i S. Perić, Napredno rješavanje problema programiranjem u Pythonu, Zagreb: Element, 2014.

[9] B. Haberman and H. Averbuch, »The case of base cases: why are they so difficult to recognize? student difficulties with recursion,« u *ITiCSE '02*, Aarhus, 2002.

[10] R. Sooriamurthi, »Problems in comprehending recursion and suggested solutions,« u *ITiCSE '01*, Canterbury, United Kingdom, 2001.

[11] M. Rubio-Sánchez, »Tail recursive programming by applying generalization,« u *ITiCSE '10*, Bilkent, Ankara, Turkey, 2010.

[12] O. AlZoubi, D. Fossati, B. D. Eugenio, N. Green, M. Alizadeh and R. Harsley, »A Hybrid Model for Teaching Recursion,« in *SIGITE '15*, Chicago, 2015.

[13] C. E. George, »EROSI—visualising recursion and discovering new errors,« u *SIGCSE '00*, Austin, Texas, USA, 2000.

[14] S.-H. Tung, C.-T. Chang, W.-K. Wong i J.-C. Jehng, »Visual representations for recursion,« *International Journal of Human-Computer Studies,* vol. 54, no. 3, pg. 285-300, 2001.

[15] M. Rubio-Sanchez, J. Urguiza-Fuentes i C. Pareja-Flores, »A gentle introduction to mutual recursion,« u *ACM SIGCSE*, Madrid, 2008.

[16] M. T. Morazán, »Functional video games in CS1 II: from structural recursion to generative and accumulative recursion,« in *TFP'11*, Madrid, Spain, 2011.

[17] J. Gal-Ezer i D. Harel, »What (else) should CS educators know?,« *Communications of the ACM,* pg. 77-84, 1998.

[18] E. Lathinen, K. Ala-Mutka and H.-M. Jarvinen, »A study of the difficulties of novice programmers,« u *ITiCSE '05*, Caparica, Portugal, 2005.

[19] D. Ginat, »Do senior CS students capitalize on recursion?,« in *ITiCSE '04*, Leeds, United Kingdom, 2004.

[20] R. McCauley, B. H. S. Fitzgerald and L. Murphy, »Recursion vs. Iteration: An Empirical Study of Comprehension Revisited,« in *SIGCSE '15*, Kansas City, Missouri, USA, 2015.

[21] C.-C. Wu, N. B. Dale and L. J. Bethel, »Conceptual models and cognitive learning styles in teaching recursion,« in *SIGCSE '98*, Atlanta, 1998.

[22] E. Lee, V. Shan, B. Beth and C. Lin, »A structured approach to teaching recursion using cargo-bot,« u *ICER '14*, Glasgow, Scotland, 2014.

[23] M. Ben-Ari, »Recursion: from Drama to Program,« 1996.

[24] M. Goldwasser and D. Letcher, »Teaching strategies for reinforcing structural recursion with lists,« in *OOPSLA'07*, Montreal, Quebec, Canada, 2007.

[25] K. Gunion, T. Mildford and U. Stege, »Curing recursion aversion,« in *ITiCSE '09*, Paris, France, 2009.

[26] M. Wirth, »Introducing recursion by parking cars,« *ACM SIGCSE Bulletin,* vol. 40, no. 4, pg. 52-55, 2008.

[27] A. Settle, »Reaching the 'aha!' moment: web development as a motivator for recursion,« in *SIGITE '13*, Orlando, Florida, USA, 2013.

[28] J. A. Velazquez-Itubide, »Recursion in gradual steps (is recursion really that difficult?),« in *SIGCSE '00*, Austin, Texas, USA, 2000.

[29] G. D. Weber, »Drawing and understanding recursive functions,« *Journal of Computing Sciences in Colleges,* pg. 50-59, 2013.

[30] L. Stern and L. Naish, »Visual representations for recursive algorithms,« in *SIGCSE '02*, Cincinnati, Kentucky, 2002.

[31] W.-J. Hsin, »Teaching recursion using recursion graphs,« *Journal of Computing Sciences in Colleges,* vol. 23, no. 4, pg. 217-222, 2008.

[32] T. Scholtz and I. Sanders, »Mental models of recursion: investigating students' understanding of recursion,« in *ITiCSE '10*, Bilkent, Ankara, Turkey, 2010.

[33] D. Ginat and E. Shifroni, »Teaching recursion in a procedural environment — how much should we emphasize the computing model?,« in *SIGCSE '99*, New Orleans, Louisiana, USA, 1999.

[34] S. Wiedenbeck, »Learning recursion as a concept and as a programming technique,« in *SIGCSE '88*, Atlanta, Georgia, USA, 1988.

[35] F. Turbak, C. Royden, J. Stephan and J. Herbst, »Teaching recursion before loops in CS1,« *Computing in Small Colleges,* vol. 14, no. 4, pg. 86-101, 1999.

[36] C. Mirolo, »Is iteration really easier to learn than recursion for CS1 students?,« in *ICER '12*, Auckland, New Zealand, 2012.

Gap in pagination due to withheld paper.

Pages 746-749

Machine Learning Techniques in the Education Process of Students of Economics

J. Bucko*, L. Kakalejčík*
*Department of Applied Mathematics and Business Informatics
Faculty of Economics, Technical University of Košice, Slovakia
jozef.bucko@tuke.sk, lukas.kakalejcik@tuke.sk

Abstract - owadays, in the period of the digitali ation and nowledge economy development, ma ority of activities result in the increase of data that should be captured. n each area of business, there is an increasing urge to e tract the nowledge from data in a timely manner in order to be able to ma e shifts that ensure a competitive advantage. Thus, the nowledge of methods and techniques of big data processing is currently a tool of a modern economist. Courses oriented to how to use machine learning tools in the economic practice should become a part of a study plan at universities of economics and business. n our paper, we propose the concept of teaching of machine learning techniques using software e a as a possible solution to integrate this issue into the study plan based on the e ample of problem-based learning.

I. INTRODUCTION

Digitalization of the surrounding reality has become a matter of course of the current information society. Modeling of real situations and events in the modern economy allows faster and more effective dealing with rapidly-increasing requirements regarding their fast processing. Properly designed model of real events allows predicting the future values. This feature is among fundamental assumptions of decision support systems which have taken their place in almost every aspect of life. Nowadays, information represents the most valuable commodity and this is why knowledge and the ability of its extraction from the amount of collected data is not only useful but in the today's competitive environment it could be marked as must-have, or existential.

Processing of huge amount of data and information extraction from given data set is a scope of research of Business Intelligence. Business Intelligence includes processing of Big Data, methods and techniques of Data mining, design of prediction models and decision support systems. In order to master the techniques of data processing, it is necessary to link mathematical skills, statistical knowledge and computing skills. This knowledge and ability to use abovementioned methods should be in the skillset of the modern economist and thus teaching of these fields as a part of educational process of universities of economics seems to be the natural trend. 81% of companies with appropriate analytical talent claimed that business analytics creates a competitive advantage [1]. However, report of Universities UK [2] adverts that there is a shortage of graduates with the right combination of analytical skills in terms of transforming data into meaningful insights that has a business value for the employer. Moreover, study by Manyika et al. [3] projects the shortage of talent, particularly of people with deep expertise in statistics and machine learning, and the managers and analysts who know how to operate companies by using insights from big data. In 2018, the demand of deep analytical skills in the US is projected to be 50-60% higher than supply. This equals to 140,000-190,000 professionals. In Canada, there is a talent gap estimate between 10,500 and 19,000 with these skills for the roles like Chief Data Officer, Data Scientist, and Data Solutions Architect. Moreover, there is a gap for professionals with solid data literacy at 150,000 (for the roles Business Analyst or Business Manager) [4]. Based on these predictions, it is obvious that students with particular skills in machine learning can gain a significant competitive advantage on the labour market.

II. THE CHOICE OF MACHINE LEARNING TECHNIQUES AND DATA SETS

Machine learning is a multidisciplinary area involving artificial intelligence (AI), probability, statistics, information theory, philosophy, psychology and neurobiology. The goal of machine learning is to solve real-world problems with the use of model that provides a good data approximation [5]. Currently, there are plenty of machine learning methods that could be integrated into the study plan of universities of economics. In their paper [5], Dhage – Raina presented an overview of basic machine learning techniques. In terms of our university course, we decided to include the methods that (1) we think are understandable and manageable by students; (2) are supported by proper software available free of charge; (3) are useful for common economic practice; (4) are described in the case studies which ease both the teaching and learning process.

When it came to selection of the right techniques, mathematical and statistical background were another criteria we took into account. As our students previously took part of courses Mathematics I., Mathematics II., Probability and Statistics and Statistical Methods in Economic Science, we assumed they have enough knowledge to start with methods proposed by us without necessity to learn fundamentals. From the wide range of methods, for the purpose of course, we decided to choose the following:

a. Linear regression

b. Logistic regression

c. Factor analysis

d. Cluster analysis (hierarchical and non-hierarchical)

e. Naive Bayes algorithm

f. k-nearest neighbours

g. Market basket analysis using apriori association rules

As the abovementioned methods are commonly known and widely used, the details regarding these methods could be found in fundamental machine learning books and studies.

In order to successful acquirement of proposed techniques by students, we assume that it is important to motivate them by stressing the information/value/insight they can gain from conducted analysis. We also assume the option to compare the results with the results of the example problem is important in order to have a reference about steps taken during the analysis. The selection of problem sets was important in order to arouse the interest and curiosity of students. Although the course is intended for students of faculty of economics, the problem sets and selected data sets were not selected with the emphasis on economic problems. On the contrary, their character is more or less sociological while describing general events which might seem more attractive for the students of the 2nd grade.

Based on the abovementioned arguments, we decided to embrace well-known and verified data sets that have been analysed broadly and there are many information and additional comments regarding their usage. To name a few, we decided to choose data sets Titanic, Affairs and data set from website Kaggle (www.kaggle.com), such as European Soccer Database, IMDB 5000Movie Dataset, 2016 US Election, World University Ranking, as much as well-known data set Iris Species.

III. THE CHOICE OF SOFTWARE TOOLS

There are many suitable software tools to teach machine learning techniques, such as SaS, The R Project, Rapidminer, Weka etc. In order to select the right software tool, we evaluated several criteria that are critical in case of software implementation in the academic environment. As students meet with the machine learning techniques for the first time, simplicity and user friendly user interface are basic criteria that have to met. Availability and ease of installation were among evaluated factors, too. In order to assemble the course that is perceived to be useful, it is necessary that user base is large enough and technical support of software developers is available reasonably. The existence of large amount of study materials, tutorials, demo problem sets and examples is among the most significant factors we evaluated. This kind of accessibility is important in terms of further development of downstream courses or personal development of students themselves. In case they will consider the course to be valuable for their further career, they can extend their

knowledge in the field. Regarding the software and platform independence, we selected software that can be installed and executed on the operation system Windows. The last but not least, the price was the last criterion we took into the account. In case the software license would have to be paid, we assume that students would be less motivated to install the software at home (out of the scope of the course). Moreover, the free solution could help them in their career life, as to prove their knowledge, it will be easier to get management buy-in.

After careful considerations, we decided to choose the solution The Waikato Environment for Knowledge Analysis (WEKA). It is well-known and broadly used software developed by University of Waikato in New Zeland. Dhage – Raina [5] consider Weka to be a platform that embodies best practices for process, configuration and implementation of machine learning algorithms and in addition, it meets all abovementioned criteria.

Weka is a comprehensive software solution containing many state-of-the-art machine learning and data mining algorithms included in the Java class libraries. Weka could be downloaded from the Internet free of charge. Moreover, accompanying book explaining the included algorithms has been published [6]. Weka can be installed and run on any computer that supports Web browsing regardless of the computer platform. As the software is written in Java, the choice of computer platform is up to the user's preferences.

Classification algorithms, decision trees, association rules and data clustering algorithms are among primary methods included in the software. The core package contains classes that are accessed from almost every other class in Weka. The most important classes are Attribute, Instance, and Instances. An object of class Attribute represents an attribute - it contains the attribute's name, type, and in case of a nominal attribute, all possible values.

Ability to select relevant functions to be included in the model induction is an essential component of applied machine learning system. The Weka software provides three feature selection systems: a locally produced correlation based technique [7], the wrapper method and Relief [8].

Weka contains an implementation of the Apriori learner for generating association rules, a commonly used technique in market basket analysis [9].

The objective of clustering methods available in The Weka is not to predict the particular classes. Instead they try to divide the data into homogenous groups called clusters. The Weka contains implementations that could be used for variety of tasks, e. g. EM algorithm (commonly used for unsupervised learning) or Naïve Bayes (for more information refer to [10]).

IV. EDUCATIONAL PROCESS

There are several previous studies describing the machine learning teaching process:

• students of Artificial Intelligence, Computer Science and Information Science [11];

- students of design [12];
- diverse audience [13].

However, we weren't able to find any paper describing the teaching process in terms of the students of economics or business administration.

Machine learning techniques are part of the process of the knowledge extraction from databases (data sets). These techniques are fundamental for the Data mining process. Datamining is considered to be one of the six steps regarding the knowledge discovery in databases [14]. The process of knowledge discovery in databases could be considered as a part of Business Intelligence which is crucial for leveraging the intellectual assets of companies by creating, storing and sharing obtained knowledge for effective decision making. It is a process of obtaining the knowledge from the databases that are valid in the statistical meaning, currently unknown, and potentially useful for specific purposes. [15], [16], [17].

Process of KDD may consist of different number of steps and their specification vary based on the author [18], [19], [20], [21] and [22]. Paralič [14] defines KDD as a process of the following 6 steps:

1. Definition and analysis of the task
2. Gathering of relevant data and its understanding
3. Data pre-processing
4. Data mining
5. Evaluation of examined patterns and identification of knowledge
6. Application of knowledge obtained and evaluation of its use

In the educational process of machine learning techniques we focus on the 3rd and 4th step of KDD. However, for the complex understanding of the problem and for motivation, it is necessary to address each step of KDD.

The first phase of the educational process is connected to the determination of the problem or research task by lecturer. Afterwards, the relevant dataset is made available. Students will use the dataset to complete their research tasks.

The second phase of the educational process is analysis of the research task in order to fully understand it. As a part of this phase, decomposition of the research task and analysis of the dataset should be done. The main focus of this phase is to choose a proper machine learning technique/techniques that will be used for the data analysis and the consultation about the method selection with the lecturer.

The third phase of the educational process is pre-processing of the available dataset. In this phase, students are going to use software program MS Excel. This phase will result in cleansed dataset in the format suitable for processing in Weka. We chose MS Excel for the variety of reasons for this phase. MS Excel is broadly used tool and students are familiar with its basic functions. Moreover, by pre-processing the dataset in MS Excel, we can improve the knowledge in terms of its use by students.

The fourth phase is the application of machine learning techniques on pre-processed dataset with the use of Weka software. The outcome of this phase should be results of analysis.

The last phase of the educational process is the evaluation of obtained results and their comparison based on the use of particular machine learning methods.

The course is intended to be a part of the educational process of the students of the 2nd grade of the study major Finance, Banking and Investment at Technical university of Košice, Faculty of Economics. Currently, there are several courses of statistics and econometrics available at Faculty of Economics. However, we have chosen to propose the new course based on the following:

- statistical courses in the 1st grade cover only basic methods and lack touch with business reality;
- statistical courses in the 2nd and 3rd grade miss machine learning methods and are taught using software that is not freely available. The same applies for courses in the higher grades;
- econometrics course in 4th grade is taught (in our opinion) too late. Moreover, in order to perform analysis, R is used. It involves programming and causes troubles for students in this very first step.

The course is an elective subject. In the pilot run, there will be 12 students involved in this course. Based on the number of students in the course, we selected educational methods that allow individual approach towards students. We consider this approach to be the most suitable regarding the character of the course. This method has been marked as expository-problem method and research method [23]. Expository-problem method will be used in the first half of the course in which we will outline the problem situation and research task. The expository part will be focused on the introduction of the fundamentals of knowledge extraction from databases with the theoretical framework oriented toward machine learning techniques. Students will learn how to work with the dataset during the pre-processing phase (with the use of MS Excel and Weka). In the second half, we decided to apply the method of research. Students will select the research task that will be specified by the particular dataset. The emphasis will be put in knowledge extraction with the use of selected machine learning techniques. When research task is handled properly, students will present the method they chosen, the procedure and results obtained by analysis.

V. EVALUATION OF EDUCATION PROCESS

The success of this pilot course and its sequel strongly depends on the course and methods rating of students themselves. Questionnaires and feedback polls are inevitable in order to improve the course or make changes that will lead to the better overall perception of the course by students. We decided to put the "startup approach" in place. The primary aim of the pilot project is to set some sort of the MVP (most viable product) - the university course that will be iteratively improved based on the feedback of the customers – the students. Due to the small sample of course participants, we decided to use semi-structured interview in order to determine student's expectations regarding the course. At the end of the course, we will distribute the questionnaires among students that will try to map the following areas connected to the educational process:

- the suitability of selected problem sets (datasets and techniques);

- the suitability of selected teaching methods;

- the difficulty of the course content;

- the practical applicability of the course;

- the suitability of chosen software tools and their difficulties;

- the evaluation of the student's results by lecturer;

- what was the degree of satisfaction with the course in terms of meeting the students' expectations.

Results obtained from questionnaire survey will be compared to the results of the interviews conducted in the beginning of the course. The survey will not be anonymous as we would like to follow up with the students that can provide us with additional insights regarding the course quality improvement. It is a vital step not to expect students to be the only ones working with the data. It is also our duty to practice what we teach and responsibility for the proper intellectual development of the future generations of economists.

VI. CONCLUSION

As was presented in the beginning of our paper, there is a shortage of people with data mining skills. The use of machine learning methods is currently necessary in all fields, including economics and business. Based on the abovementioned state on the labour market, we assume the machine learning skills to be a competitive advantage of students when applying for the job. In our paper, we propose the design of the course with the focus on machine learning techniques. The course is designed for the students of Faculty of Economics. The course consists of several steps necessary in order to execute proper data analysis. We also suggest a particular range of methods that would be taught and executed in software Weka. Weka is a significant step in the transfer of machine learning technology into the workplace. We selected the software based on the criteria that was set prior the course. In order to improve the satisfaction of students and utility of the course, we would like to apply selected methods for obtaining valuable feedback that will help us adapt the course.

REFERENCES

[1] RANSBOTHAM, S. et al.: The talent dividend: Analytics talent is driving competitive advantage at data-oriented companies. In:

MIT Sloan Management Review, April 2015. ISSN 1532-9194, (2015).

[2] UNIVERSITIES UK.: Making the most of the data: Data skills training in English universities. London: Universities UK. ISBN 978-1-84036-339-5, (2015).

[3] MANYIKA, J. et al.: Big data: The next frontier for innovation, competition, and productivity [online]. [cited 27-12-2016]. Available online: http://crono.dei.unipd.it/~dm/MATERIALE/MGI_big_data_full_r eport.pdf

[4] STEELE, P.: Closing Canada's Big Data Talent Gap [online]. [cited 27-12-2016]. Available online: http://www.ryerson.ca/content/dam/provost/pdfs/ryerson_ccbdtg.p df

[5] DHAGE, S. N., RAINA, C. K.: A review on Machine Learning Techniques. In: International Journal on Recent and Innovation Trends in Computing and Communication, vol. 4, no. 3, pp. 395-399. ISSN 2321-8169, (2016)

[6] WITTEN, I. H., FRANK E.: Data Mining: Practical Machine Learning Tools and Techniques with Java Implementations, Morgan Kaufmann, San Francisco, (1999).

[7] HALL, M.A., SMITH, L.A.: "Practical feature subset selection for machine learnings." roc. Australian Computer cience Conference 181-191, Pert, Australia, (1998).

[8] KIRA, K., RENDELL, L.A.: "A practical approach to feature selection." Proc 9th Int Conf on Machine Learning, 249-256, (1992).

[9] AGRAWAL, R., IMIELINSKI, T. and SWAMI, A.N.: "Database mining: a performance perspective." IEEE Trans Knowledge and Data Engineering, Vol. 5, 914–925, (1993).

[10] PATIL, T. R. et al.: Performance analysis of Naive Bayes and J48 classification algorithm for data classification. In: nternational ournal of Computer cience and Applications, vol. 6, no. 2, s. 256-261. ISSN 0972-9038, (2013).

[11] VAN SOMEREN, M.: Teaching Machine Learning at University of Amsterdam, (2016). [online]. [cited 27-12-2016]. Available online: https://dtai.cs.kuleuven.be/events/Benelearn2010/submissions/ben elearn2010_submission_22.pdf.

[12] VAN DER VLIST, B. et al.: Teaching machine learning to design students. In: International Conference on Technologies for E-Learning and Digital Entertainment. China: Springer Berlin Heidelberg, vol. 5093, pp. 206-217. ISBN 978-3-540-69736-7, (2008).

[13] LIN, H. T. et al.: Teaching machine learning to a diverse audience: the foundation-based approach. In: Teaching Machine Learning Workshop at the 25th International Conference on Machine Learning (ICML), (2012).

[14] PARALIČ, J.: Objavovanie znalosti v databázach. Elfa, Košice. ISBN 80-89066-60-7, (2003).

[15] BRACHMAN, R.J.; ANAND, T.: The Process of Knowledge Discovery in Databases. In Advances in Knowledge Discovery & Data Mining, Fayyad, U.M. - Piatetsky-Shapiro, G. - Smyth, P. - Uthurusamy, R., Eds. AAAI/MIT Press, Cambridge, Massachusetts, (1996).

[16] FAYYAD, U.M. and IRANI, K.B.: "Multiinterval discretization of continuous-valued attributes for classification learning." Proc IJCAI, 1022–1027. Chambery, France, (1993).

[17] HAN, J., KAMBER, M.: Data Mining – Concepts and Techniques. Morgan Kaufmann Publishers, (2000)

[18] ESTER, M., SANDER, J.: Knowledge Discovery in Databases – Techniken und Anwendungen. Springer Verlag, (2000)

[19] KLÖSGEN, W., ZYTKOW, J.M.: The Knowledge Discovery Process. In Handbook of Data Mining and Knowledge Discovery. Oxford University Press, pp. 10 – 21, (2002)

[20] MANNILA, H.: Methods and Problems in Data Mining. In the proceedings of International Conference on Database Theory, Afrati, F. - Kolaitis, P., Delphi, Springer-Verlag, (1997)

[21] SIMOUDIS, E.: Reality Check for Data Mining. IEEE EXPERT, Vol.11, No.5, (1996)

[22] KHAN, R. A.: KDD for Business Intelligence. In: *ournal of nowledge anagement ractice*, vol. 13, no. 2. ISSN 1705-9232, (2012).

[23] BLAŠKO, M.: Kvalita v systéme modernej výučby, Katedra inžinierskej pedagogiky Technickej univerzity Košice 2013, ISBN: 978-80-553-1281-1, (2013).

Moodle-based data mining potentials of MOOC systems at the University of Szeged

G. Kőrösi*, F. Havasi**
* University of Szeged, Institute of Informatics, Hungary
** University of Szeged, Department of Software Engineering, Hungary
*korosig@inf.u-szeged.hu, **havasi@inf.u-szeged.hu

Abstract - In today's world virtual online educational platforms emerge literally on daily bases and many offer MOOC-based courses. With the appearance of MOOC, educational platforms have gained an additional boost, a new aspect in their evolutionary process, which has opened a new field of research thanking to the extraction of logging information within the frames of data mining. It has become clear that educators will be able to tailor their courses by merging the two previously mentioned fields and by carrying out MOOC-based data mining, targeting pedagogical aspects. This field of research seems promising and important, thus a faculty at the University of Szeged has created its own MOOC educational platform which has been set to facilitate data mining by implementing a wide range of logging algorithms. The data would be processed through a complex Artificial Intelligence program, which, in the short term, could reveal new and exciting pedagogical findings, while in the long run, the supervisors could put together a platform that would help and notify educators about relevant information. It would become possible to create adaptive educational materials, as well. This work aims at clarifying how such platforms function and what the steps of data collection and evaluation are.

I. INTRODUCTION

In today's world virtual online educational platforms emerge literally on daily bases and many offer MOOC-based courses. Mushrooming as a scalable lifelong learning paradigm, Massive Open Online Courses (MOOCs) have enjoyed significant limelight in recent years, both in industry and academia [10]. With the appearance of MOOC, educational platforms have gained an additional boost, a new aspect in their evolutionary process, which has opened a new field of research thanks to the extraction of logging information within the frames of data mining. Plenty of research at various institutions in the last 8 years seem more than promising with a significant breakthrough. Its origin dates back to the Educational Data Mining conference in 2008, where the idea of educational data mining of MOOC courses first emerged. All the findings since then have been published in the form of research papers and scientific journals underpinning that it is worth digging even deeper.

It might be asked what the reasons are that brought this interdisciplinary field of sciences to life. The main reason is that MOOCs are open education platforms, in which the participants are self-motivated to complete courses [24]. However, learning outside the framework of an educational institution and the supervision of a teacher may bring about certain obstacles. Learning in a MOOC requires that students apply self-regulation [23]. Among

the most debatable issues are the high drop-out rates, this has been proven by dozens of researches. [3][7][9][11][13][23][25]. From this standpoint, one could easily question the existence of MOOC courses; nevertheless, there is pressure in higher educational institutions to provide up-to-date information to achieve institutional effectiveness [20]. Online platforms are among the most important tools to gain insight into how online education functions.

Ironically, the problem can be solved through the online platform itself because its structure allows all around logging of student activities, which may lead us to some so far unknown tools of pedagogical research. This idea has been grounded in the investigation of Romero & Ventura who think that learning management systems accumulate a great deal of log data about students' activities [20]. The system can automatically record whatever student activities are involved, such as reading, writing, taking tests, performing various tasks, and even communicating with peers [26]. In short, MOOC big data is a gold mine for analytics [18]. On the other hand, data mining technology has been proved effective in CMS pedagogical research as well [13].

Science that deals with verifying such data is called EDM (Educational Data Mining), with many prominent names of the field and outstanding research achievements. Berland et. al suggest that EDM may have the potential to support research that is meaningful and useful both to researchers working actively in the constructionist tradition but also to wider communities [16]. Data collected from learning systems can be aggregated over large numbers of students and can contain many variables that data mining algorithms and techniques can explore for model building. [22] Working from student data can help educators both track academic progress and understand which instructional practices are effective [5].

Educational data mining (EDM) is a research area which utilizes data mining techniques and research approaches for understanding how students learn [22]. In recent years, there has been an increasing interest in the use of data mining to investigate scientific questions within educational research, an area of inquiry termed educational data mining [2]. The scope of educational data mining includes areas that directly impact students [19]. The emerging field of educational data mining (EDM) examines the unique ways of applying data mining methods to solve educationally related problems [19].

It must be stated that the primary goal of research is not just to obtain information but to keep as many students as possible signed up to our courses. Through his investigation into the relevant papers, Huebner reveals works that suggest how learners can be kept in the learning environment, efficient educational techniques, and better course books in the future which may help reducing the drop-out rate in a predictive way [19]. Along this line of thought, researchers at Bowie State University have assigned risk factor points to each learner which demonstrated who would have difficulties [6].

The authors' research takes this route to test e-learning platforms by putting together two MOOC courses (Conscious and Safe Internet Usage - TÉBIA, Database Management) a logging platform to register online activities. This paper has been written to investigate EDM opportunities and to develop the authors' own logging system dealing with how the steps of data recording, cleaning, and pre-processing are done.

II. WHAT IS EDM AND WHY IS IT IMPORTANT?

While thousands of students have been attracted to large online classes, keeping them motivated has become a challenging endeavor [13]. Thus, it is of paramount importance to understand student motivation or why it is lost. A tool to gain access to such answers in Data Mining. In order to find out what this notion is one may quote Baker who states that Data mining, also called Knowledge Discovery in Databases (KDD), is the field of discovering novel and potentially useful information from large amounts of data [2]. Knowing that EDM has existed for only a decade, it is advisable to take a close look at it to reveal what it really means. Its meaning depends on how it is defined, however a common meeting point has been established, which seems to be digital education. Educational data mining is a research area which utilizes data mining techniques and research approaches for understanding how students learn [22]. EDM is an emerging tool and technique used to comprehend and represent educationally related data [19]. Furthermore, data mining is a series of tools and techniques for uncovering hidden patterns and relationships among data [27]. Data mining is a multidisciplinary area in which several computing paradigms converge: decision tree construction, rule induction, artificial neural networks, instance-based learning, Bayesian learning, logic programming, statistical algorithms, etc. And some of the most useful data mining tasks and methods are: statistics, visualization, clustering, classification, association rule mining, sequential pattern mining, text mining, etc. [21]. Educational data mining is an emerging discipline that focuses on applying data mining tools and techniques to educationally related data [1]. A large number of researchers within EDM focus directly on course management systems and how they can be improved to support student learning outcomes and student success [19]. Data mining is the process of efficient discovery of non-obvious valuable patterns from a large collection of data [12].

Interactive e-learning methods and tools have opened up opportunities to collect and scrutinize student data, to ascertain patterns and trends in those data, and to formulate new discoveries and test assumptions about how students learn [22]. Researchers have found that they can

apply data mining to rich educational data sets that come from course management systems such as Angel, Blackboard, WebCT, and Moodle. Numerous studies have shown that data mining can be used to discover at-risk students and help institutions become much more proactive in identifying and responding to those students [14]. Educational data mining is defined as the area of scientific inquiry centered around the development of methods for making discoveries within the unique kinds of data that come from educational settings, and using those methods to better understand students and the settings which they learn in [2]. Online learning systems log student data that can be mined to detect student behaviors that correlate with learning [22].

Four main axes can be identified along which EDM methods may be helpful for constructionist research:

- EDM methods do not require constructionists to abandon deep qualitative analysis for simplistic summative or confirmatory quantitative analysis;

- EDM methods can generate different and complementary new analyses to support qualitative research;

- By enabling precise formative assessments of complex constructs, EDM methods can support an increase in methodological rigor and replicability;

- EDM can be used to present comprehensible and actionable data to learners and teachers in situ.

- In order to investigate those axes, the first step is to describe one's perspective on compatibilities and incompatibilities between constructionism and EDM [16].

The strengths of EDM systems can be traced back to their tools, primarily logging methods that provide information to researchers, who would in turn reveal so far unknown pedagogical conclusions. Baker sums up (Table I.) what is known about those tools and result [2].

TABLE I.   THE PRIMARY CATEGORIES OF EDUCATIONAL DATA MINING

Category of Method	Goal of Method	Key applications
Prediction	Develop a model which can infer a single aspect of the data (predicted variable) from some combination of other aspects of the data (predictor variables)	Detecting student behaviors (e.g. gaming the system, off-task behavior, slipping); Developing domain models; Predicting and understanding student educational outcomes
Clustering	Find data points that naturally group together, splitting the full data set into a set of categories	Discovery of new student behavior patterns; Investigating similarities and differences between schools
Relationship Mining	Discover relationships between variables	Discovery of curricular associations in course sequences; Discovering which pedagogical strategies lead to more effective/robust learning
Discovery	A model of a	Discovery of relationships

with Models	phenomenon developed with prediction, clustering, or knowledge engineering, is used as a component in further prediction or relationship mining.	between student behaviors, and student characteristics or contextual variables; Analysis of research question across wide variety of contexts
Distillation of Data for Human Judgment	Data is distilled to enable a human to quickly identify or classify features of the data.	Human identification of patterns in student learning, behavior, or collaboration; Labeling data for use in later development of prediction model

III. E-LEARNING LOGGER MODULE

The system presented in this paper was built on the basis of Moodle which is an open-source, free, well supported, popular e-learning platform. It has a long history, given that the first version came out in 2002. The platform is known for its robustness, though its user interface is little less modern than it is expected in these days. This is the underlying reason for completely replacing its font-end and develop a new one, which calls the Moodle's back-end. One module of its front-end is responsible for logging, which is the focus of this article.

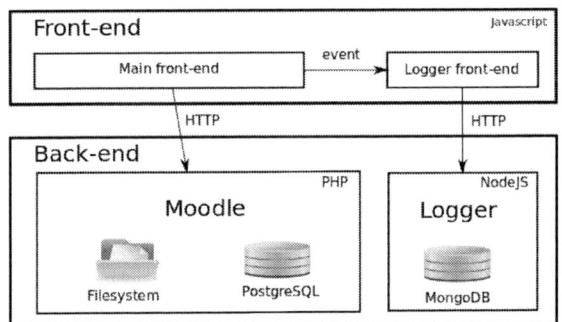

Figure 1. Front-end logging system

This logger front-end collects and process events, and calls its back-end part via HTTP to store them. This back-end part – which is completely independent from the back-end of the Moodle - is developed in NodeJS and uses MongoDB to store events (Figure 1.).

Every log entry is an event. Each of the events classified the system into different types, and depending on the type, they store different parameters for it. For example, a "textinput" event has a parameter, which stores the typed text, called text:

```
{
    type: "textinput",
    data: {
        target: "search-target",
        text: "database"
    },
    time: "2017.01.23. 16:01:28.242",
    page: "https://...",
    userid: 1876
```

```
}
```

There are some parameters, which are stored for all events:

- userid: the ID of the user, who has executed the operation, or 0, if there was an anonymous user

- time: the date of the event

- page: the URL of the page, where the event was happened

- type: the type of the event (see below)

The type of the events, and their parameters can be the following ones:

- load, unload, focus, blur: they are generated in the case of loading or unloading of the page, and getting and losing the focus.

- resize: it means resizing the browser window. It has two parameters: x, y (the new size of the windows).

- click: it represents a mouse click. Its parameters are x and y.

- testClick: it signifies a mouse click to an answer of a quiz. It is a preprocessed event: this javascript event handler automatically recognizes if the mouse click happened over an answer, and generates a testClick event, not a simple click event. Its parameters are: question, answer, correct, choiceCleared.

- download: it is generated in the case of downloading a file. Parameter: filename.

- textinput: this event represents a change of a text input field. Parameters: target (id of the text element), text (actual value of the text element).

- textinput_focus, textinput_blur: they are generated when a text input gains or loses the focus. Parameters: target (id of the text element), text (actual value of the text element).

- passwordinput, passwordinput_focus, passwordinput_blur: similar to the previous ones, but because of security consideration, the value of the password text input is not stored. Parameters: target (id of the text element), and length (of the text element).

- mousemove: mouse moving event. Parameters: x, y, xDistance, yDistance, realDistance. The system stores only two mouse events in a second.

- scroll: means scrolling the page. Parameter: top. The system stores only two mouse events in a second.

There are video events, as well. The system supports two kinds of video: html5 video element and embedded YouTube video. Events:

- videoSeek: means seeking in the html5 video element. Parameters: seekTime, videoId, totalTime, src.

- videoPlay, videoPause, fullscreenOn, fullscreenOff: html5 video playing events. Parameters: actualTime, videoId, totalTime, src

- volumeChange: html5 video element volume change. Parameters: actualTime, videoId, totalTime, src, newVolume.

- youtubePlay, youtubeEnd, youtubePause, youtubeBuffering: youtube video playing events. Parameters: actualTime, videoId, totalTime, src.

- youtubeQuality. changing youtube video quality settings. Parameters actualTime, videoId, totalTime, src, quality

- youtubeRate. parameters: actualTime, videoId, totalTime, src, rate

IV. DATA

Two courses have been created in order to test the logging platform.. In the first part of the research, a pilot study was conducted between the dates of March 1 and May 30, 2016, while the second study was recorded in the interval of October 1 to December 10, 2016. Altogether 163 students took part in the pilot study and 347 student signed up for the two courses in the Autumn semester. The details about the course are presented in Table II below. The learning material for both courses comprised a three week study period. One of the courses, which ran under the name 'TÉBIA,' included 4 + 1 (embedded) videos, while the other course, with the name 'Databases' had 7 (embedded/Youtube links) videos with attached embedded texts, or external links.

The primary point of interest for the researchers lay not in the drop-out rate, instead the aim was to discern how the platform functioned and how the learners would behave. It can thus be concluded that 99.8% of the learners who had signed up for the course, had also completed it.

TABLE II.   COURSE CONTENTS

Course name	TÉBIA	Databases
Content	Basics of Conscious and Safe Internet Usage	Basics of Databases
Time frame	3 weeks	3 weeks
Parts of the Learning Material	Introduction: Video (3.37 min., Embed);	The concepts of databases: Video (2.12 min, Embed); Video (2.12 min, Youtube link); HTML embedded text;
	Digital footprint: Video (14.04 min, Embed); HTML embedded text;	Database handling systems: Video (3.52 min, Embed); Video (3.52 min, Youtube link); HTML embedded text; HTML embedded text;
	Conscious and Safe Internet Usage: Video (13.07 min, Embed); HTML embedded text; External link;	Transaction, closing methods: Video (3.28 min, Embed); Video (3.28 min, Youtube link); HTML embedded text;
	Online bullying: Video(13.31 min, Embed); HTML embedded text;; Extra video (11.55 min, Embed);	Basics of handling databases: Video (3.24 min, Embed); Video (3.24 min, Youtube link); HTML embedded text;
		Database models: Video (2.51 min, Embed); Video (2.51 min, Youtube link); Video (4.11 min, Embed); Video (4.11 min, Youtube link); HTML embedded text;
		Relational database models: Video (3.17 min, Embed); Video (3.17 min, Youtube link); HTML embedded text;;

The logging system during the two courses registered 4.663.120 logs, out of which 26 variables were generated and assigned to the users. These were the following:

Data, Page, Pid, Time, Type, User, Data.realDistance, Data.x, Data.xDistance, Data.y, Data.yDistance,

Data.Text, Data.Top, Data.Target, Data.Filename, Data.Length, Data.ActualTime, Data.Scr, Data.TotalTime, Data.VideoId, Data.SeekTime, Data.NewVolume, Data.Ip Adress, Data.Quality, IP

V. PRE-PROCESSING

The e-learning platform of the University of Szeged is a website which provides a wide variety of services, including video lessons to every courses, however it does not have a complex platform such as Coursera or edX. This is the reason why is was not possible to close the e-learning portal after the pilot study and the two courses. Thus, as a consequence, the platform used for this research could not only be accessed by those students who had signed up, but it was accessible to but external users, as well. This led to the recording of 1,443,817 (404 Mb) logs, while during the second time the system had 3.219.303 (936 MB) recorded logs. The portal recorded

1229 students and lecturers, out of which only 513 were relevant. The contaminated raw data had to be put to serious data cleaning procedures. The file was cleaned and sorted out by examining students' IDs, user behavior, and sign-in tendencies. Table III demonstrates a simple but effective algorithm that filters relevant and usable data:

TABLE III.   USER BEHAVIOUR, AND SIGN-IN TENDENCIES

Dataset	Lectures	Lecture Videos	Video Length (min)	Quizzes	Users	Clickstream Events
Pilot	11	11	65.74	3	163	1,443,817
Autumn	11	11	65.74	1	347	3.219.303

A deeper study of this JSON file revealed further important information. This includes the user activities showing that 54% of the logs were generated by 10% of the users. The comprehensive diagram of user activities shows the distribution of active and passive participants (Figure 2).

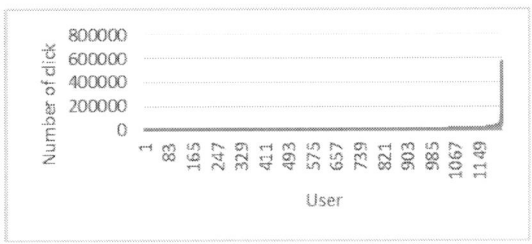

Figure 2. Number of clicks by user

This fact is not surprising since one of the side effects of a MOOC course is the uneven distribution of active and passive students. Anderson et al. created 5 categories [28]:

- Viewers, in the left mode of the plot, primarily watch lectures, handing in few if any assignments.

- Solvers, in the right mode, primarily hand in assignments for a grade, viewing few if any lectures.

- All-rounders, in the middle mode, balance the watching of lectures with the handing in of assignments.

- Collectors, also in the left mode of the plot, primarily download lectures, handing in few assignments, if any. Unlike the Viewers, they may or may not be actually watching the lectures.

- Bystanders registered for the course but their total activity is below a very low threshold.

If the student activities and other data are converted to percentages, one would gain deeper knowledge in this topic. Figure 3 shows these findings. In order to complete a test, an average user generates approximately 400-2000 log files, which demonstrates that 58% of the learners do not aim at having a thorough understanding of the material

but want to complete to course as soon as possible, while only 10% can be categorized as superusers according to Zhu et al.[24].

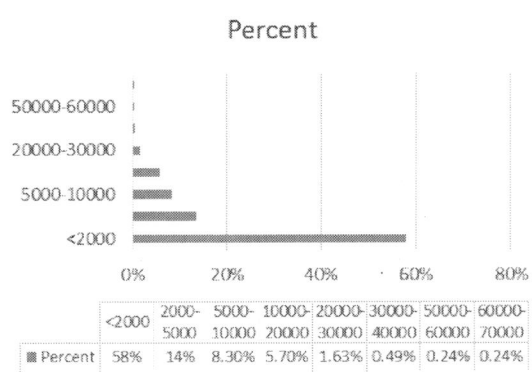

Figure 3. Student activities/ logging data

VI.   CONLUSION

At this stage of the studies, the authors were able to add a functioning logging system to a Moodle platform with weak tools of analyses, which would serve well for similar portals to live up to current measuring requirements. The data obtained and analyzed from students' logging could reveal some unexpected pedagogical aspects thus helping educators and learners in the process of course planning and learning. The analysis of the pile of data amounting to millions of logs recorded during the pilot studies and the autumn courses could help the authors design an artificial intelligence (machine learning) that would automatically process input data without human intervention and which could intervene if extreme values emerge.

The aim of the redesigned and modified website is to enhance student motivation, learning achievements, and output results. After examining the relevant literature, the authors were able to sort out errors and potential opportunities that were unknown to them, like predictive analysis through clickstream [4], [7], [8], [15], [17] or feedback buttons by Chang et al. [11] which improved student concentration in the long run. A total of 1.2 GB of data was collected, enabling the authors to make the next step of designing a suitable mathematical model to their MOOC system. This would help to provide a full-scale predictive background support to educators who would upload their learning materials, and would help learners who sign up to a course. It is expected that the spring semester of 2017 will bring more users to this platform, which would double the amount of collected data. After quantitative and qualitative data analysis the new findings will also be published in a research paper.

REFERENCES

[1]   R. Baker, K. Yacef, The State of Educational Data mining in 2009: A Review and Future Visions, Journal of Educational Data Mining vol. 1, no. 1, pp. 1–14, fall 2009.

[2]   R. Baker, Data Mining for Education. In McGaw, B., Peterson, P., Baker, E. (Eds.) International Encyclopedia of Education (3rd edition), vol. 7, pp. 112-118. Oxford, UK: Elsevier, 2010.

[3] Y. Bergner, S. Droschler, G. Kortemeyer, S. Rayyan, D Seaton,D. Pritchard, E Model-Based Collaborative Filtering Analysis of Student Response Data: Machine-Learning Item Response Theory. International Educational Data Mining Society, Paper presented at the International Conference on Educational Data Mining (EDM), pp. 95–102 (5th, Chania, Greece, Jun 19-21, 2012)

[4] C. G. Brinton, R. Rill, S. Ha, M. Chiang, R. Smith, and W. Ju, "Individualization for Education at Scale: MIIC Design and Preliminary Evaluation," Transactions on Learning Technologies, vol. 8, no. 1, pp. 136–148, 2015.

[5] L. Cao, B. RAMESH, M. ROSSI, Are Domain Specific Models Easier to Maintain Than UML Models? IEEE Software, vol. 26, pp. 19–21, 2009.

[6] F. Chacon, D. Spicer, A. Valbuena, Analytics in Support of Student Retention and Success (Research Bulletin 3, 2012 ed.). Louisville, CO: Educause Center for Applied Research, pp. 1–9, 2012.

[7] C. G. Brinton ,Mung Chiang, Mooc performance prediction via clickstream data and social learning networks, In: Computer Communications (INFOCOM), 2015 IEEE Conference on Computer Communications, IEEE, pp. 2299–2307, 2015.

[8] C. G. Brinton, S. Buccapatnam, M. Chiang, H. V. Poor, "Mining MOOC Clickstreams: Video-Watching Behavior vs. In-Video Quiz Performance", Signal Processing IEEE Transactions on, vol. 64, pp. 3677–3692, 2016.

[9] J. Guan, W. Nunez, J. Welsh, Institutional strategy and information support: the role of data warehousing in higher education, Campus-Wide Information Systems, vol. 19, no. 5, pp.168–174, 2002.

[10] S. Haggard, S. Brown, R. Mills, A. Tait, S. Warburton, W. Lawton, T. Angulo, Haggard S, The maturing of the MOOC: Literature review of massive open online courses and other forms of online distance learning. Department for Business, Innovation and Skills, UK Government. 2013 Sep 20.

[11] J. Cheng, C. Kulkarni, S. Klemmer, Tools for predicting drop-off in large online classes. In: Proceedings of the 2013 conference on Computer supported cooperative work companion. ACM, pp. 121–124, 2013.

[12] W. Klosgen, J. Zytkow, M. Jan, Knowledge discovery in databases: the purpose, necessity, and challenges. In: Handbook of data mining and knowledge discovery. Oxford University Press, Inc., pp. 1–9, 2002.

[13] D. Liang, J. Jia, X. Wu, J. Miao, J. Wang, Analysis of learners' behaviors and learning outcomes in a massive open online course. Knowledge Management & E-Learning, vol. 6, no. 3, pp. 281–298, 2014.

[14] J. Luan, Data Mining and Knowledge Management in Higher Education – Potential Applications. Paper presented at the Annual Forum for the Association for Institutional Research, Toronto, Ontario, Canada, pp. 1–13, 2002.

[15] M. Speiser, G. Antonini, A. Labbi, J. Sutanto, On nested palindromes in clickstream data, KDD '12 Proceedings of the 18th ACM SIGKDD international conference on Knowledge discovery and data mining p 1460-1468, 2012.

[16] M. Berland, R. S. Baker, and P. Blikstein, Educational Data Mining and Learning Analytics: Applications to Constructionist Research, Technology, Knowledge and Learning, July 2014, vol. 19, no. 1, pp. 205–220, 2014.

[17] R. Mazza and C. Milani, Exploring usage analysis in learning systems: Gaining insights from visualisations. In Workshop on Usage analysis in learning systems at 12th International Conference on Artificial Intelligence in Education, New York, USA, pp. 1–6, 2005.

[18] U. O'Reilly and K.Veeramachaneni, Technology for Mining the Big Data of MOOCs, Research & Practice in Assessment, vol. 9, pp. 29–37, 2014.

[19] R. A. Huebner, A Survey of Educational Data-Mining Research, Research in Higher Education Journal, vol. 19, pp. 1–19, April 2013.

[20] C. Romero and S.Ventura, Educational Data Mining: A Review of the State of the Art. Systems, Man, and Cybernetics Part C: Applications and Reviews, IEEE Transactionson, vol. 40, no. 6, pp. 601–618, 2010, doi: 10.1109/tsmcc.2010.2053532.

[21] C. Romero, S. Ventura, and E. García, Data mining in course management systems: Moodle case study and tutorial. Computers & Education, vol. 51, no. 1, pp. 368–384, 2007, doi: 10.1016/j.compedu.2007.05.016.

[22] S. Lakshmi Prabha and A.R.Mohamed Shanavas, Educational data mining applications ,Operations Research and Applications: An International Journal (ORAJ), vol. 1, no. 1, pp. 1–6, August 2014.

[23] T. Sinha, P. Jermann, N. Li and P. Dillenbourg, Your click decides your fate: Inferring Information Processing and Attrition Behavior from MOOC Video Clickstream interactions, EMNLP Workshop on Modelling Large Scale Social Interaction in Massive Open Online Courses, pp. 3–14, 2014.

[24] T. Zhu, W. Wang, W. Zhao and Riming Zhang, Participation Prediction and Opinion Formation in MOOC Discussion Forum, International Journal of Information and Education Technology, vol. 7, no. 6, pp. 417–423, June 2017.

[25] P. Esztelecki, G. Korosi, N. Namesztovszki, L. Major, The comparison of impact offline and online presentation on student achievements: A case study, 39th International Convention on Information and Communication Technology, Electronics and Microelectronics (MIPRO), 2016, pp. 802–806, DOI: 10.1109/MIPRO.2016.7522249 IEEE Conference Publications

[26] J. Mostow, J. Beck, H. Cen, A. Cuneo, E. Gouvea and C. Heiner, An educational data mining tool to browse tutor–student interactions: Time will tell! In Proceedings of the workshop on educational data mining, pp. 15–22, 2005.

[27] M.H. Dunham, Data mining introductory and advanced topics. Upper SaddleRiver, NJ: Pearson Education, Inc., 2003.

[28] A. Anderson, D. Huttenlocher, J. Kleinberg and J. Leskovec (Engaging with Massive Online Courses, WWW '14 Proceedings of the 23rd international conference on World wide web, pp. 687–698, 2014.

MIPRO 2017, May 22- 26, 2017, Opatija, Croatia

View on Development of Information Competencies and Computer Literacy of Slovak Secondary School Graduates

L. Révészová

Faculty of Economics Technical University of Košice
Department of Applied Mathematics and Business Informatics, Košice, Slovakia
libusa.reveszova@tuke.sk

Abstract - The degree of including information, knowledge and technologies into all human activities is very high. New skills for new society are not only classical literacy (reading, writing, counting), but also information competencies and digital/computer literacy. According to ACM Europe Working Group in our "digital world", information is available anywhere at any time, computer power is ubiquitous, communication of vast amounts of information is almost instantaneous, and storage capacities seem infinite. But these powerful possibilities only benefit those who have learned to use them effectively. That is why all students need to be educated in both: fluency with computer tools and the Internet and the science behind information technology, to use ICT and its devices intelligently. Educational system should be able to react appropriately to the development and trends, however, the required state and the reality do not meet. This article summarizes the results of the research of information and digital competence/literacy of secondary school graduates who entered the first grade at the Faculty of Economics, Technical University in Košice. Our research problems are as follows: What are the information competencies and students' knowledge gained during their school informatics education? How can they use this knowledge and their competencies?

I. INTRODUCTION

The Information and Communication Technology (ICT) sector plays an important role in a country's economy and welfare. The ICT sector relies on highly performing technical infrastructures but also needs skilled people who are able to understand its complexities and are fully capable of making the best use of its potential [1].

The role of ICT on productivity and standards of living seems to be critical, because e-skills shortages, gaps and mismatches as well as a persistent digital divide will affect negatively productivity growth, competitiveness, innovation, employment and social cohesion in Europe[5].

As we can read in [6] and [7], Europe is facing a growing lack of e-skills and it is predicted that there will be an overall shortage of supply of about 800 000 ICT professionals in the EU by the year 2020. The European Commission confirms a growing demand for ICT professional skills.

New skills for new knowledge society are not only classical literacy - reading, writing, counting, but also

computer and digital literacy and good creative, logical and critical thinking.

Computing enables and empowers new methods of data and information processing that have led to monumental changes across disciplines, from art to business to science [19].

The educational system should be able to react appropriately to the development and trends. The most recent National reform program in the Slovak Republic, adopted in 2014, presents actions to fulfill targets contained in the Europe 2020 strategy [16]. Another government document released in 2013 is Digipedia - a concept of informatization of the educational resort with an outlook until 2020. The concept presents digitization as one of the most effective tools for Slovak teachers and scientists how to achieve better results in education and research [13].

The most important targets and priorities of society development should be imported into the basic curricular documents as well as into the Educational framework which adjusts lessons allocation so that the declared efforts could be transformed into educational process. However, if we take a look at the, for example, Educational framework for secondary schools [14], the real situation does not testify to the proclaimed interests and efforts.

II. PROBLEMS OF RESEARCH AND DATA COLLECTION METHODS

In our research we focus on the real ICT skills and informatics knowledge – computer and digital literacy of secondary school graduates at the age of 18/19 years. Our research problems are as the follows: What are the information competencies/digital literacy and students knowledge gained during their school informatics education? How can they use these knowledge and competencies? What is the level of computer/digital literacy of secondary school graduates?

As a research tool we use a questionnaire administrated on the first seminars of Informatics I Course. We search for both the extent and content of compulsory education in the field of informatics/computer science and ICT at secondary schools. We have created the questionnaire entries based on legislation, pedagogical documentation, and standards [14], [15], [3]. Our questionnaire is divided into several parts. Within the first part of the questionnaire our students are asked the se questions: What type of secondary school did you attend? How many obligatory informatics lessons did you take per week at your secondary school in particular school years? In the second part of the questionnaire there are the following questions: Did you meet these terms in the informatics lessons at the secondary school? (Possible answer yes/no.) We focus on students´ knowledge in the field of basic terms of informatics, hardware architecture, operation systems, text editors etc. In the third part we want students to describe their experience with information systems.

Our aim is to compare results, developments and improvements in a longer time period. For this reason we did not modify the questionnaire entries during the research and we provided some formal modifications in a minimum possible way.

The sample of our research consists of all secondary school graduates who entered their first year at Faculty of Economics Technical University of Košice in 2003 - 2015. Students in the studied sample are from the Eastern Slovakian Region in more than 95 % of cases. Numbers of respondents in the research period, and type of graduates´ secondary schools are presented in Tab. 1 in detail. Letter "G" stands for Grammar School, "B" for Business School, and "O" for Other/Vocational School (low number of these graduates in our sample is due to the practical focus of these schools).

TABLE I.   NUMBER OF RESPONDENTS AND TYPES OF SECONDARY SCHOOLS

Year	Total number	Type of school		
		G	B	O
2003	147	113	32	2
2004	171	144	25	2
2005	158	140	17	1
2006	154	136	18	0
2007	163	141	20	2
2008	166	149	16	1
2009	177	166	11	0
2010	165	140	22	3
2011	158	149	9	0
2012	144	125	17	2
2013	152	129	22	1
2014	166	137	25	4
2015	136	118	18	0
	2057 100 %	1787 87 %	252 12 %	18 1 %

Within the first part of the questionnaire our students were asked the question: How many obligatory

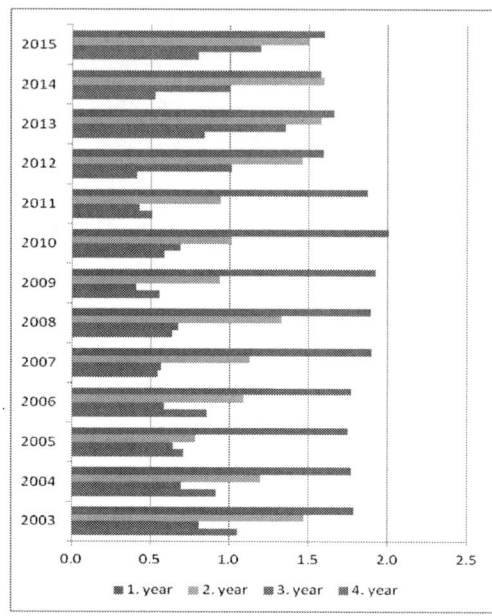

Figure 1.   The average number of obligatory informatics lessons at secondary schools

informatics lessons (lesson = 45 minutes) per week did you get over at secondary school in particular school years? As we can see in Fig.1, despite the increasing importance of Informatics and ICT competencies, the number of lessons devoted to this subject at secondary schools is still very low.

The conceptual framework which has been used to analyze further data follows [1], [2], [12]:

- Computer literacy means skills that are fundamentally necessary for the effective use and application of common ICT systems, devices and software tools in support of our own work and personal interests. Broadly speaking, these cover the term "ICT user skills", which refers to the confident and reflected use of ICT for work, leisure, learning and communication.

- Information competencies/digital literacy represent skills that are necessary for researching, designing, developing, planning strategically, managing, producing, consulting, marketing, selling, integrating, installing, administering, maintaining, supporting and servicing ICT systems. These also refer to the abilities needed to exploit strategic opportunities by using ICT technology (especially the Internet), to assure stronger performance of organizations and to research capabilities for new ways of improving or implementing business, administrative and organizational processes.

III.   EVALUATION OF STUDENTS´ COMPUTER LITERACY

Standard of European Computer Driving License (ECDL, [4]) version 5.0 and basic pedagogical documents mentioned above were the base for designing the next part of the questionnaire concerned on finding out the level of

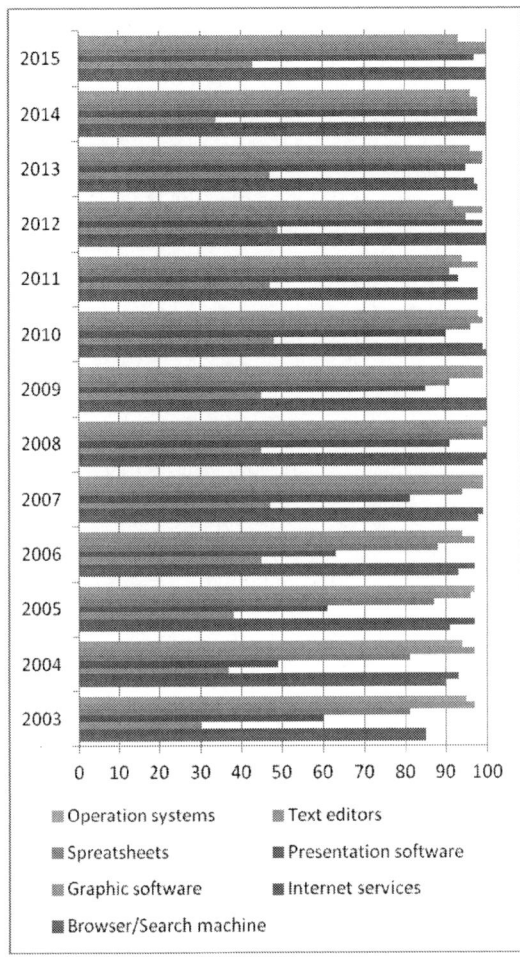

Figure 2. Percentage of students who worked with the operation systems, text editors, spread sheet programs, presentation and graphic software, used browsers and internet services

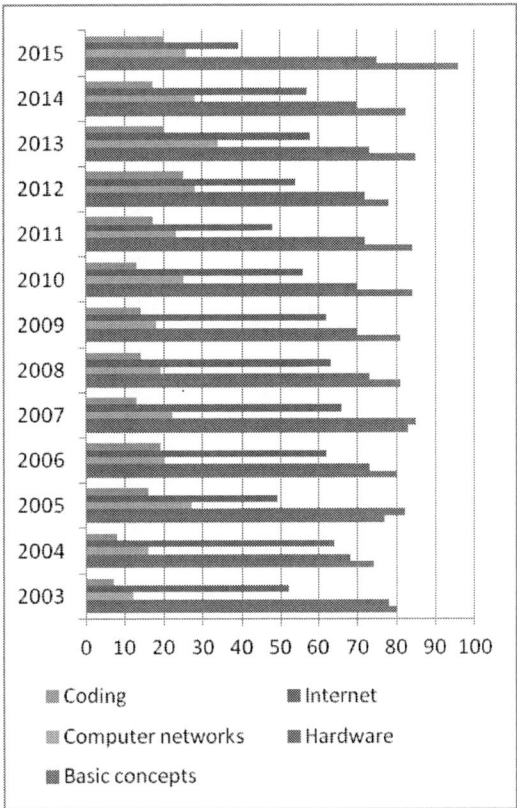

Figure 3. Percentage of students who are familiar/worked with coding topology and typology of computer networks, basic concepts of Informatics, internet protocols and computer architecture/hardware

students´ computer literacy. In this part we focused on working with basic applications and programs: operation systems, text processing, spread sheet programs, presentation software, graphic programs, web browsing and network communication via internet services. Questions in this part of the questionnaire were formulated as follows: Did you deal or work with these programs and applications?

Within Fig. 2 we present achieved percentage of students who had some experience in the mentioned programs and applications. We show an evaluation of the students' answers expressed as a percentage number of response, "yes".

As we can see in Fig. 2, in the last years of our research gradually almost all secondary school students worked with text editors and spreadsheets, used the internet services (email, www), and the application for web browsing. Fig. 2 also shows an increase in the ratio of students who worked with software designed to make presentation. The proportion of students who work with graphic programs is basically the same in the research period – only about 40 %. Surprisingly, working with the

operation systems in recent years, has not been indicated by 100 % of the students.

The graph shown above illustratively presents shifts which occurred or did not occur during estimated time periods in particular fields. Based on our findings and consistent with other surveys (e.g. [23]), we can say that in our sample the proportion of students working with basic applications and programs falling under the computer literacy is increasing. On this basis, we can assume that the basic level of computer literacy is growing.

IV. ASSESSMENT OF STUDENTS´ INFORMATION COMPETENCE/DIGITAL LITERACY

In this section we focus on working with applications requiring higher level of the cognitive demands and complexity of the tasks. At the same time, the range and nature of the tools and applications that the students are required to use to come to a solution are more complex. We examined whether the students had dealt with basic concepts of informatics (data, information, knowledge), with Von Neumann computer architecture, topology and typology of computer networks, the functioning of the Internet (internet protocols), coding and encryption, as we can see in Fig. 3. Our research shows that a higher level of ICT competency/digital literacy achieved significantly

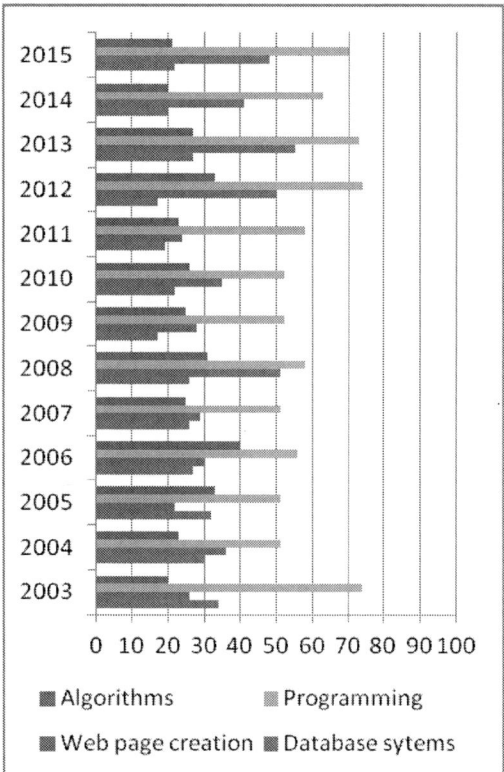

Figure 4. Percentage of students who are familiar/worked with algorithms, programming, web page creation and data base systems

lower proportion of high school graduates, as in the case of computer literacy.

In the area of coding, knowledge of the computer/hardware architecture, topology and typology of computer networks as well as the functioning of the Internet, the situation has not been improved. Slight improvement occurs in the ratio of students familiar with the basic concepts of Informatics.

Working with program languages - programming, algorithm creation, web sites creation and work within the database system represented the highest level in our investigations. The difficult tasks in these parts tended to involve transferring information from one application to another, knowledge application and requiring the student to follow a relatively complex sequence of actions. The results of our findings are presented in Fig. 4. As we can see, the situation observed for relatively long period is not improved dramatically in any of the monitored areas.

In the next part of the questionnaire the students had to indicate whether they had worked with any information system (IS), and if they had met with the definition or typology of IS. On average only about 20 % of the students had already encountered with the term information system and about 10 % with information systems typology. Despite the fact that nowadays most secondary schools use electronic classroom books, electronic pupils' books, electronic library systems, and IS is widely used in everyday life (e-shops, transport, search for information, entertainment etc.), 80 % of the students,

on average, had failed to give an example of a particular IS.

In general, development and output of our educational system in the field of Informatics and using ICT at national level can be seen for example in [21]. Slovakia can be compared with other countries via results of international surveys, for example OECD PIAAC Survey of Adult Skills - Problem solving in technology rich environments [17].

Our research shows that a higher level of information competencies/digital literacy achieves significantly lower proportion of secondary school graduates as in the case of computer literacy.

V. DISCUSSION

Information society starts to transform into a different, more organized form, so called knowledge society. From the development so far we can expect, that this form will be characterized by an urge in making the information more and more available for public, by using not only information but also knowledge stored and processed by information technology according to users' requirements. Computer technology will obviously dominate even in discoveries, formulation and gaining of new knowledge [11].

Nowadays informatics education is essential for all of us - citizens of information and upcoming knowledge society. That is why Informatics, not only computer literacy is essential for all students of all types and all levels of schools.

Although computer science has been a regular subject at school for a relatively long time, there is still more and more discussion on how to teach computer science and what to teach: Should computer science education be oriented more towards its applications or more towards its fundamentals or more towards its social effects [20]?

Many authors make a distinction between information technology - deals with the use of the computer and its applications and Computer Science/Informatics - deals with the design of informatics products or more theoretical objects. This is often related to difference between users and designers with underlying idea that designers are „those who know" and users are „those who don't need to know". We agree with the author of [9] that it is necessary to go beyond this dichotomy, which is undesirable from the education point of view. According to [9] „Informatics education for all" means informatics education also for students who will not necessarily become computer science experts.

"Computational thinking could be added to the traditional three Rs: reading, writing and arithmetic as an additional basic skill needed especially by university students – they will be better prepared to choose a future career not only as a computer specialist but also as a specialist in other disciplines, who professionally, according the needs of her/his profession, can use computing methods and tools professionally." [21]

If we consider changes in the ICT use and the Internet communication - network security issues, and the issue e.g. concerning of promotion ideas of extremism etc. - the development of critical thinking via increasing the level of digital literacy is considered crucial. Critical thinking is a complex process of deliberation which involves a wide range of skills and attitudes. It includes, according to [3], identifying other people's positions, arguments and conclusions; evaluating the evidence for alternative points of view; weighing up opposing arguments and evidence fairly; being able to read between the lines, seeing behind surfaces, and identifying false or unfair assumptions; recognizing techniques used to make certain positions more appealing than others, such as false logic and persuasive devices; reflecting on issues in a structured way, bringing logic and insight to bear; drawing conclusions about whether arguments are valid and justifiable, based on good evidence and sensible assumptions; synthesizing information: drawing together your judgments of the evidence, synthesizing these to form your own new position; presenting a point of view in a structured, clear, well-reasoned way that convinces others.

The expansion of the digital literacy can stimulate the development of creativity, which can be understood as a multidimensional phenomenon that manifests itself in many fields and contexts, from arts and crafts to design, science, research and entrepreneurship. It is regarded as a cognitive ability, but it is not the same as intelligence. It involves the ability to synthesize and combine data and information, but also requires confidence to take risks. In general, we can say, that creativity is the process of having original ideas that have value [8].

The benefits of better ICT skilled users and employees for companies and organizations can be noticed for example in the following areas:

- Higher productivity and performance of the company/organization.

- Possibility of implementation new organizational forms, e.g. development of business nets, participation in supply chains.

- Increased added value of the product or services.

- Entry to new markets. Utilization of new business channels.

- New products or services, changing business processes.

- Responding to new business activities of competitors [18], [22].

In recent years, several advanced countries have declared the intention to transform their ICT school subject into a new Computer Science, Informatics, or Computing education, with the emphasis on developing computational thinking, programming, designing computational systems and other basic concepts of Informatics. In Slovakia the theoretical conception of the school Informatics at the primary and secondary stages can be approximately characterized by the following principles:

- It is not the ICT in education, but Informatics. Both components are necessary in modern

education: integration of ICT into learning processes, and also a separate subject focused on the concepts of Informatics/Computing.

- Informatics is a general (opposed to vocational) subject, for every student, without regard to his or her future professional profile and the degree of achieved education.

- Programming is considered an inseparable and core constituent of the school Informatics [10].

It is necessary to transform these ideas into the real informatics education.

All of these suggestions imply that the learners, teachers and educators, employers and policy makers feel jointly responsible for improving the quality of existing training and developing new initiatives in the field of increasing computer and digital literacy level [1].

VI. CONCLUSION

From the questionnaire evaluation we derived that the education of Informatics at secondary schools is focused on handling different packages of applications' programs. There is a lack of understanding the basic terms, working with data on higher levels and understanding the informatics principles through the creation of programs and algorithms. Algorithmic approach, using of knowledge for creating "something new" and interdisciplinary relations are often missing.

Students are not able to identify that they are working with IS regardless the fact, that they work with it in some form "every day" in school and in the common life.

Especially in the field of higher degree - digital literacy, the situation does not change, improvements are negligible, according to our findings. We can see that the real situation does not testify to the proclaimed interests and efforts. The subject Informatics has the second lowest lessons allocation which is only 3 lessons in the 1st up to the 4th year in the higher secondary educational stage. Insufficient is the ICT integration into the tuition of other subjects, acute is also lack of competent qualified teachers in the field of informatics and ICT use.

Our students will probably face several paradigm changes in their future careers and life. Their acquired school knowledge and skills might become obsolete within a short time. That is why they must be robust enough to meet the challenges of the latest fashion, and also enable the students to cope with changes [20].

We hope that our work has helped to raise public awareness and to put the higher levels of digital literacy topic on the agenda of education policy makers, academics and teachers.

REFERENCES

[1] S.J. Behrens, "A conceptual analysis and historici overview of information literacy." College and Research Libraries. 1994. vol. 35, no. 4, s. 309-322

[2] N. Binsfeld, J. Whalley, and L. Pugalis, "Competing through e-skills: Luxembourg and its second level digital divide," 27th European Regional Conference of the International

Telecommunications Society (ITS), Cambridge, United Kingdom, 7th - 9th September 2016

[3] S. Cottrell, "Critical thinking skills, Developing Effective Analysis and Argument," Second edition, Palgrave MacMillan, 2011

[4] ECDL Foundation. "ECDL Standard 5.0" Retrieved from www.ecdl.org

[5] European Commission. "e-skills for jobs in Europe: Measuring progress and moving ahead" final report, february 2014

[6] European Commission., "e-SKILLS IN EUROPE" Luxembourg Country report, 2014

[7] European Schoolnet. "THE e-SKILLS Manifesto" (2015th ed.). Brussels: European Schoolnet, 2015

[8] eu2008.sk, "Lifelong learning for creativity and innovation - A Background Paper". 2008, 19 p., Retrieved from http://www.sac.smm.lt/wp-content/uploads/2016/01/12en-Vertimas-SAC-Creativity-and-innovation-SI-Presidency-paper-anglu-k-2007.pdf

[9] P. Hubwieser,"The Darmstadt Model: A first step towards a research framework for computer science education in schools," in Informatics in Schools, sustainable Informatics education for Pupils of all ages, 6th International Conference on Informatics in schools: Situation, Evolution, and Perspectives, ISSEP 2013, Oldenburg, Germany, Februeary/March 2013, Proceedings, Springer

[10] I. Kalaš, and M. Gujberová, "Designing productive gradations of tasks in primary programming education" 2013 Retrieved from http://dl.acm.org/citation.cfm?id=2532750

[11] J. Kelemen et al., "Invitation to the knowledge society" (Pozvanie do znalostnej spoločnosti). Iura Edition, Bratislava, 2007, 266 p.,

[12] S. Mclaughlin, S., M. Sherry, E. Doherty,M. Carcary,C. Thornley, and Y. Wang, "e- Skills: The International dimension and the Impact of Globalisation." Brussels. 2014 Retrieved from http://eprints.maynoothuniversity.ie/5559/1/CT_e_Skills_report.pdf

[13] Ministry of education, science, research and sport of the Slovak republic, "Digipedia" – A concept of informatization of the educational resort with an outlook till 2020", Bratislava 2013

Retrieved from https://www.iedu.sk/digipedia/Documents/4796.pdf

[14] Ministry of education, science, research and sport of the Slovak republic, "Educational framework for secondary schools" 2011-7915/18752:1-922, 2011. Retrieved from www.statpedu.sk/files/documents/svp/gymnazia/rup3_sjog.pdf

[15] National Institute for Education. "The national education program for high school ISCED-3a", 2012 Retrieved from www.statpedu.sk/sites/default/files/dokumenty/statny-vzdelavaci-program/informatika_isced3a.pdf

[16] National reform programme in the Slovak Republic 2014 Retrieved from http://ec.europa.eu/europe2020/pdf/csr2014/nrp2014_slovakia_sk.pdf

[17] OECD. "Skills Outlook 2013: First Results from the Survey of Adult Skills", OECD Publishing. 2013, Retrieved from http://dx.doi.org/10.1787/9789264204256-en

[18] D. Paľová, "Experience with Usage of LMS Moodle not Only for the Educational Purposes at the Educational Institution" In: MIPRO 2016. - Rijeka p. 1006-1011.

[19] Seven big ideas of Computer Science Retrieved from https://csprinciples.cs.washington.edu/sevenbigideas.html

[20] A. Schwill, "Computer Science education based on fundamental ideas", 1998, Retrieved from http://ddi.uni-muenster.de/didaktik/Forschung/Israel97.pdf

[21] M. M. Syslo, and A. B. Kwiatkowska, "Informatics for all high school students – a computational thinking approach" in Informatics in Schools, sustainable Informatics education for Pupils of all ages, 6th International Conference on Informatics in schools: Situation, Evolution, and Perspectives, ISSEP 2013, Oldenburg, Germany, February/March 2013, Proceedings, Springer

[22] M. Vejačka, and D. Paľová, "FASTER Platform - an online tool for EU accountants education", In: MIPRO 2015. - Rijeka : Croatian Society for Information and Communication Technology, Electronics and Microelectronics , 2015 p. 838-843.

[23] M. Velšic, "Digital literacy Slovakia 2013 - Digitálna gramotnosť na Slovensku 2013", Inštitút pre verejné otázky, Bratislava 2013

Digital learning as a tool to overcome school failure in minority groups

D. Paľová*, N. M. Novak** and V. Weidinger***

* Department of Applied Mathematics and Business Informatics, Faculty of Economics, Technical University of Košice, Slovakia
** Institute of Software Technology and Interactive Systems, Vienna University of Technology, Austria
***Verein Offenes Lernen, Vienna, Austria
dana.palova@tuke.sk, niina.novak@tuwien.ac.at, valerie@talkademy.org

Abstract - In the European Union development strategy formulated in the Europe 2020 document (European Commission, 2015) it was indicated, that the smart growth of the EU as a whole should be reached through the realization of three priorities: the increase in employment, the increase of productiveness and the social cohesion and specialized agendas: Digital Agenda, Education and Learning, E-skills and Employment. The main documents identify main weaknesses and risk areas. One of the most significant was described as the early school leaving of Roma minority members. Roma constitute Europe's largest transnational ethnic minority with an estimate of ten million people. Learning outcomes of this minority are significantly lower than outcomes of the majority. As one of the reasons for early school leaving of Roma, insufficient understanding of learning materials is identified. The result is that most of the Roma community members drop out of education before attending a secondary school and continue their lives as unemployed or enter the labor market as unskilled workers. Within the paper will be presented the CloudLearning project that represents an alternative and innovative educational method: the way of the SOLE method implemented in their education. This paper will include partial results from the pilot tests realized these days.

I. INTRODUCTION

The European Commission (EC) has identified three key drivers for EU growth: *smart growth*, *sustainable growth* and *inclusive growth* within the Europe 2020 document [14]. One of the headline targets that is hoped to be achieved is pushing the rate of the early school leavers below 10%, because educational skills acquired during the higher stages of education (vocational and upper secondary), improve employability and reduce poverty [12]. Also for this reason, educational systems should be able to react appropriately to the developments and trends in the todays knowledge society in close connection to the targets contained in the Europe 2020 strategy [16].

The ability to work with information and its transformation into knowledge can massively affect the further development of everyone [17]. In this regard, ICT plays the role of a mediator in education – it is helpful for cognitive development, enhancing the acquisition of generic cognitive competencies as essentials for life in our knowledge society. Students using ICTs for learning purposes are more involved in the process of learning and more and more students use computers as information sources and cognitive tools [6].

Roma people represent Europe's largest ethnic minority (with an estimate of 10 – 12 million people). Despite of different strategies and action plans signed in and after 2010 and concerns with improving their fundamental rights and advancement their social integration (for example "European Commission communication on social and economic inclusion of Roma" [11], "EU Framework for National Roma Integration Strategies up to 2020" [9], [8]) and further progress evaluation documented in [15], [11], [13], many Roma still face severe poverty, profound social exclusion, barriers to exercising their fundamental rights and discrimination. These problems affect their access to quality education, which, in turn, undermines their employment and income prospects, housing conditions and health status, curbing their overall ability to fully exploit their potential. [15] In particular, areas with high proportions of Roma in the population show high rates of early school leaving, and it is estimated that 89 % of Roma leave school early [15]. Based on results of PISA tests [18] it can be concluded, that about 80 – 95% of Romani-speaking students have not acquired basic cognitive skills and competencies and have thus limited possibilities to find qualified employment and cope with the complex demands of today's societies [20], [10].

Early school leaving of Roma is therefore a critical issue with direct social and economic impact. There are a number of reasons for early school leaving like e.g. cultural or cognitive differences. These lead to insufficient understanding of learning materials because of unacquaintedness with linear text, inappropriate ways of presenting knowledge by teachers to this group, and lack of interest, energy and motivation of teachers to use innovative teaching methods. Based on [15], the socio-economic reasons (e.g. poor infrastructure and shortages of equipment, geographical distance to schools and the lack of available public transport), individual reasons (language and communication problems, low confidence in schools, early marriage and childbirth or the necessity of contributing to household income) of early drop-out of education are often aggravated by teaching styles or curricula that do not resonate with the real-life experiences of Roma children.

Improving the educational situation of Roma is a critical test of the EU's ability to achieve progress in the inclusion of all extremely marginalized and socially excluded groups.

II. PROJECT CLOUDLEARNING

The mentioned challenges of the target group in education were the motivation of the consortium of the project "Head in the Clouds: Digital Learning to overcome school failure" (short "CloudLearning"; https://brainsintheclouds.eu) to design a program, which would use an alternative educational approach to research its potential of improving the above numbers in school failure. If successful, the approach implemented should prevent students from leaving school and increase their basic and transversal skills. It should enhance digital education in the work with the target group, mainly members of the Roma minority group. In the long run this project therefore wishes to support a breaking of the vicious circle created when educational disadvantages and exclusion are later manifested in an exclusion from society in general and specifically from the labor market.

The project (an Erasmus+ strategic partnership in the field of school education) was initiated and is coordinated by the Vienna University of Technology, which implements it together with the Verein Offenes Lernen (Austria), Technical University of Kosice (Slovakia), GAIA (Kosovo), Fundatia Crestina Diakonia Filiala Sfantu Gheorghe (Romania), Sukromna zakladna skola (Slovakia) and www.scio.cz SRO (Czech Republic). It is being implemented in three locations in Kosovo, Romania and Slovakia by three of the partners who were already before the project directly working with the target group and therefore were familiar with their educational challenges and needs [1].

A. Educational Approach

To reach the above described aims the program is based in the education method of Self-Organised Learning Environments (SOLE) [1], which was developed by Sugata Mitra initially with a "hole in the wall" experiment starting in 1999. In walls in villages and urban Indian slums Mitra incorporated computers with big screens at a height that made it easy for 8 to 13 years old children to use them. Next to these computers he placed signs telling children that they could use them freely. Over a time period of 5 years Mitra observed how the children learned by themselves how to use the computers, downloaded media, played games and researched information online. Collaborating in groups the children showed educational achievements while working in this unsupervised environment. What can be described as interactions in a chaotic manner therefore

turned out to be what Mitra named "self-organizing systems" [7].

The reason for the CloudLearning team to choose this approach is the desire to engage students in their own learning process [1], because a learning process driven by students themselves is curious, collaborative, engaged, self-organized and facilitated by adult encouragement. In the case of SOLE this means that students, encouraged by the educators, use the internet to find answers to what Mitra calls "big questions". He refers to questions without an easy way to answer them. They have to be tricky and open, and many times they connect two subjects. The objective is not finding a clear answer, but to capture the children's attention and encourage them to discuss a topic deeply, from which they learn working collaboratively and thinking critically [21]. In this context the learning objectives of an activity are not predefined [19], but SOLE among other things aims at enhancing computer literacy. It also aims at improving skills in presenting and interpersonal skills [21], which belong among the transversal and basic skills the CloudLearning project wishes to improve [5].

In a SOLE per four students there should be one computer, of which all have to have a connection to the internet. Since there is not one computer per person the students will automatically have to work together [7]. To create an environment for free exploration of a topic it is furthermore important that the process is characterized by self-discovery, spontaneity, sharing [21], openness and flexibility [19]. There are only 5 rules: 1. the children receive a big question, or are encouraged to come up with their own big question; 2. the groups in which students work are chosen by them and throughout the activity they are free to change groups at any time; 3. they are free to move, interact and exchange ideas with each other; 4. the students can explore the topic in whichever direction they choose, as there is not only one correct answer; 5. and in the end of the activity the groups have to present their learning points [21]. They also have to hand in a report of one-page length summarizing their findings. In case the educator wishes to add to these learning points he or she can do so at another moment [7].

The SOLE experiments have come to the conclusion that learning in a group helps memory recall [21]. Crawley and Mitra, e.g., in a study show the tendency of students to retain the knowledge that they have gained even three months after the SOLE activity. They mostly test higher on the subjects three months after, which Crawley and Mitra referred to as an "anomalous expansion of understanding". The reason for this is that students, after the lesson, kept discussing the subject with others in their class as well as their mother and father, and did further research in their free time [7]. By implementing a SOLE activity, students should get motivated to study a variety of ideas and subjects [21], and the mentioned data seems to confirm that. Research has also shown that learning in unsupervised groups can make up for inadequate teaching in school [22] and a study in England showed that many students react to

SOLE positively and they view it as different from what they perceive as a normal lesson [19].

The authors of [19] observe that SOLE can be perceived as an educational innovation in two aspects:

1. Technological: The educator is not on the central stage anymore and needs to incorporate technology into his/ her teaching practices; and
2. Autonomy: The students are expected to be more autonomous because the approach is based on enquiry.

The teacher therefore has more the role of a mediator than an instructor in the learning process of his/ her students. SOLEs are considered to make it more possible to have a curriculum closer related to the experiences, interests and questions of the students [19].

The SOLE experiments indicate that children often have the capability to understand more than the adults would have thought. Working with the internet and the right encouragement children are able to answer almost any question [21]. It has been experimented with in for example in Argentina, England, Italy, Australia, Chile, Brazil, India, Uruguay, China and the USA. In all these places the children seemed to be able to work on big questions that were on a level usually considered a couple of years above their age, as long as they had internet and could work in unsupervised groups [7]. After many years of research Sugata Mitra was the first to receive the TED Prize [21] 2013, because he has inspired educators around the world with his ideas [19], just like he inspired the CloudLearning team.

B. NeedsAssessment

The first phase of the CloudLearning project was fully devoted to the needs assessment of the participating partners and students. The implementing partners gathered information about the students, their families and living circumstances, knowledge of computers, IT in general as well as language skills. Another aim of the needs assessment phase was to learn about the partners' expectations from the outcome of the CloudLearning project. This information played a vital role in developing the content of the SOLE-boxes in order to assess the local context and provide material that addresses the shortcomings of the students' current education standard.

1) Romania

In the case of the Romanian partners, students and their families were asked to fill out a questionnaire in order to gain insights into the students' living circumstances and educational background. None of the students from Romania have access to computers or IT at their schools, while one student out of five has access to a computer at home. The students speak in the Hungarian and Romanian languages. Their knowledge in the Roma and English languages are very minimal. Among the expectations from the Romanian Project partners they hope to achieve at the end of the program one can find capabilities such as developing computer skills and familiarity with technology for both students and teachers, increasing students' level of creativity, advancement in English, and the implementation of new learning and teaching methods [2].

2) Slovakia

The Slovak Partners' needs assessment showed that the students' native language is Roma and the school curriculum is also taught in the Roma language. The students struggle with the Slovak language and do not speak or understand English. The participating school is very eager to start the program in order to: increase students' computer literacy, incorporate new technologies in the learning process, and receive new PCs—current PCs are depreciated and often not compatible with software. The Slovak Partners expect the project to focus on computer science, geography, English, music, and the Roma language. The school would like the children to learn how to search for information online and develop long-term thinking habits with an emphasis on thinking about the future, not just living in the moment. Many families migrate to England or Belgium for employment and better living circumstances. Therefore, linguistic skills in English are very beneficial for the students [2].

3) Kosovo

The needs assessment from the project partners from Kosovo showed a strong desire to incorporate music into the SOLE-boxes. With regards to expectations they would like their students to develop music skills and connect with other children in the project in addition to society at large. Moreover, they would like their students to enhance their creativity, self-expression, and video making skills. This will strengthen the voice of the children and youth. The project partners from Kosovo want their students to explore the world through the CloudLearning project as the students suffer from isolation and they have no resources for travelling. Moreover, it is important for the participating organization that the students learn about renewable energy. They hope the outcome will reduce the gap between ethnic groups and promote the inclusion of children from majority ethnic groups into minority groups; promoting a different approach [2].

4) Overall Outcomes

The needs assessment process clearly indicated that there are five common expectations that need to be implemented in the CloudLearning project. First, the development of computer skills and learning how to obtain information online. Second, learning English, the country's official language, and their native language. Third, teaching students more about real life topics such as basic hygiene, steps necessary to get a passport, environmental protection etc. Fourth, enhancing their creativity and stimulate their thought process. Fifth, encouraging the students to work together and develop social skills.

C. Project Progress

Based on the needs assessment, the SOLE approach and additional research the team discussed how best to structure and to implement the educational program for the target groups [5]. It was decided to divide the program in eight topics (educational "boxes"), each of them lasting two months [4]: two boxes deal with video making and editing, one with the English language, one with the environment and one with "real life challenges" like getting a passport, health etc. While all these boxes automatically increase digital literacy because of the use of technology and internet, three of the boxes are specifically aimed at digital literacy and approach for example the topic of programming [5].

To ensure a smooth implementation of the project and establish the direct contact and exchange between the educators directly working with the target group, the developers of the educational boxes and the team evaluating and documenting the outcomes, a training for teachers and youth workers took place in Romania a month before the start of the work with children and youth. During the training the team also had the chance to meet with the local group of children that would later participate in the project and visit the location [3].

In October 2016 the implementation was kicked off with the first box being brought to the students. Since then in all location between 3 and 6 hours a week are dedicated to this project in the after school hours or on weekends. In Slovakia 12 children attend the sessions regularly, in Kosovo 37 and in Romania 30. Both in Romania and Kosovo the children are of different ages between 8 and 13, in Kosovo also some of them above, the eldest being 17 years. In Slovakia the groups consist of children attending the 6th or the 7th grade [5].

1) Tracking the Progress

In his trials of SOLE, Mitra usually performs tests with the students before they participate in a SOLE activity and then does tests with them after [19]. The same is being done with our target group. The project partner SCIO has developed a detailed and specifically for this project designed evaluation plan. It consists of

- a demographics questionnaire,
- two creative activities to learn the hobbies of participants (beginning and end),
- a tool to support teachers and youth workers in observing developments in the learner autonomy of each student,
- one evaluation of the free time of the students,
- an accomplishment questionnaire and a personality questionnaire to be filled after each educational box
- as well as structured interviews with the educators every two months.

Additionally, to observing the achievements of the students and therefore evaluating the success of the approach tested, the team uses a hand-in application designed and implemented by the partner Verein Offenes Lernen in which the students regularly upload their answers to questions, whether these are in picture form, written or documented as a video. The uploads make it possible for the team to document the learning activities, showing not only what are the results of the students' research, but also giving the information of who did what activity when and with whom. Each task is also connected to a set of competences that the student develops in the set task. The competences are based on standards such as the ATC 21st century skills and the CEFR English competences. The data is furthermore connected to a system of learning analytics which makes it easy to observe all this information and draw conclusions from them [5]. SOLEs should empower students in taking ownership of their own learning experience [21], which is a central goal of the project. This ownership is supported through the ability of the students to access the documentation of their work and see the progress they have made [5].

2) Case: Videobox in Košice

As mentioned above, developed SOLE boxes need to be tested in real educational settings. One of the pilot testers is the project partner Súkromna Základná Škola, placed in Košice, Slovakia [23]. The school is focused on the education of Roma pupils. The school has 206 students, aged from 6 to 17 years, who study in 11 classes from grade zero until the 9th grade.

The pilot testing of SOLE boxes started by testing the first box, dealing with video making. (The first experiences of Roma students with ICT and SOLE boxes was recorded and is available at [4].)

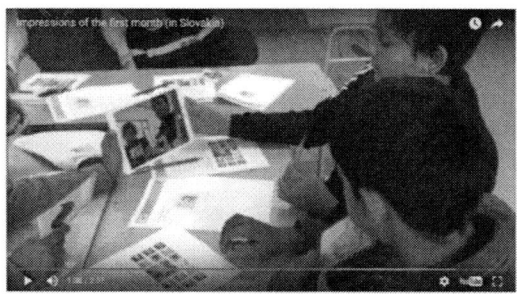

Figure 1.   First experieces with ICT-SOLE Box; Source: [4]

In the pilot testing 12 students and 2 teachers were involved. During 8 lessons the box covered a series of topics, e.g.:

- *The "Getting Started"* tasks to encounter and understand QR codes, mobile devices, taking and editing pictures or movies using mobile devices.
- *Do you want to be a YouTuber?* – tasks for registering a Google account and YouTube channel, downloading the YouTube application, recording and publishing video.
- *Why are some videos good, and others are not?* - tasks about making scene plans, creation of scenarios, planning and recording the video, discussion on how to improve it.
- *Be a professional movie editor!* - movie editing with other videos, sounds, subtitles and pictures tasks.
- Finally, they will give a feedback to other classmates like *a professional movie reviewer* and at same time they also receive the feedback from others.

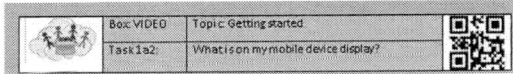

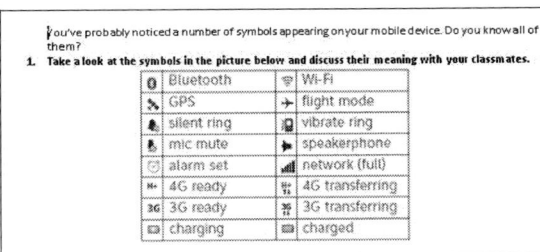

Figure 2. Example of one of the videobox tasks

The tasks were, as previously mentioned, created in a way that children could be able to work on them independently without the intervention of a teacher. The teacher in this case acts only as a mentor, especially in cases where children can't cope with the task alone, i.e. do not understand the text of the task and respectively they can't find appropriate information to solve the task on the Internet.

After finishing the pilot testing an interview with the teachers was done in order to learn about their experiences and impressions from the first SOLE box implementation. The most essential findings from the Slovakian partner are the following:

- *"The pilot testing started slowly, because there was a problem with internet access. This problem is in process of resolving. But that was the reason why we needed more time for the particular tasks. The teachers had to stay after the lessons at school and upload the files instead of children to the portal."*

- Using the SOLE box was smooth, tasks were written clearly and were very interesting for the children. They tried to solve them individually (within their skills and abilities).

- Some children have problems with reading comprehension and therefore help was needed.

- It was best to create small groups of children, preferably pairs.

- The most interesting task was picture editing – creating of picture collage.

3) Observations made

As the project is currently only in the fourth month no quantitative conclusions can be drawn yet, but some qualitative observations in all three implementing groups have been done. Also it should be considered that, as Mitra himself observed, it can take students some time to adapt to the new way of learning [21]. The CloudLearning team has been working with the children and youth for almost four months now, and while it is continuously learning to adapt the tools to the needs of the participants, most of them very much enjoy the approach already [5].

Four main challenges have been identified so far:

1. In his SOLE toolkit Mitra mentions the improvement of students in the fields behavior, abilities to solve problems, language, creativity and reading comprehension [21]. Our target groups are specifically challenged in reading comprehension and language [5]. Mitra also foresaw this problem happening in some places and suggested to encourage students to consult with groups that have a higher reading level and to look for alternative solutions of finding answers on the internet. By doing this he hopes that children will not see a lack of reading capabilities as such a big barrier in the future which can reduce their presentation anxiety [21]. In fact, e.g. in the Kosovarian group the phenomenon of supporting each other is very present. The children with higher reading levels help the others and the others also look for their support.

2. Initial internet problems because of remote locations and because of weak routers are being solved or have already been solved.

3. On a similar level challenging in the beginning was the wrong expectations as to how many smartphones the students would have. E.g. in the Slovakian group the educators had expected that half of the group would have smartphones and as it turned out almost none of them actually did. Steps to solve this problem have been taken.

4. The fourth problem is an intercultural challenge, as some of the tasks given were not understood because of concepts that were included in the tasks like "famous person", "person on TV" or "role model". Also this is being discussed in the team designing the boxes together with the implementing partners. Each newly written task is evaluated based not just whether the language is simple enough but also whether the concepts behind the questions will be understood by the students.

Generally, it can be said that the students are excited about the new, different and surprising way of learning. Some of the tasks posed to them so far they loved, some they did with less enthusiasm and some they were not interested in. The team is evaluating the common points of the favorite tasks to learn from these best practices for the next boxes, and for the moment it seems students like those tasks the most where there is a concrete outcome. This refers to tasks where in addition to understanding something new the students also produce something, like a video, a drawing etc. This current assumption will have to be confirmed or proven wrong in the future.

Very positively we can also mention the observations of educators of how much children have already learned, e.g. in the field of digital literacy just by needing to learn what a QR code is and how it works to be able to use the hand in application. The teachers and youth workers also report that after an initial encouragement and some explanations or hints where needed the students are mainly able to do tasks working independently without further guidance from the educators [5].

III. CONLUSION

Based on the experience of other educators of SOLE and its effects the team considers the approach to be the right fit in creating a learning environment and support for students who like Roma are statistically more likely to leave school. In creating an environment that encourages a different, more active and self-organized learning and encouraging learning through new methods the project hopes to spark an interest in learning and to fulfill learning needs that the children and youth might not feel answered in the school lessons. The project develops the transversal and basic skills of participants and aims at sparking interest for learning. By doing this it hopes to lower school drop-outs and improve employability of a disadvantaged group of European society. The results will show the possibilities and impact of this approach.

The CloudLearning team believes that through methods of computational thinking, programing skills, group activities, board games, and the utilization of stand of the art technology such as robots, the educational boxes can represent a curriculum that would address the concerns mentioned above in addition to introducing long-term innovative tools and methods to their learning process. The team believes in a playful and exploratory approach to education throughout the project.

ACKNOWLEDGMENT

This work has been kindly funded by the European Commission under the ERASMUS+ Programme Project CloudLearning: Head in the Clouds: Digital Learning to Overcome School Failure (No. 2015-1-AT01-KA201-005024). This publication reflects the views only of the authors, and the National Agency and the European Commission cannot be held responsible for any use which may be made of the information contained therein.

REFERENCES

[1] CloudLearning Consortium, "Head in the Clouds: Digital Learning to Overcome School Failure", [online], Available at <https://brainsintheclouds.eu>, [Accessed 23.January 2017].

[2] CloudLearning Consortium, "Needs Assessment Report", June 2016, unpublished.

[3] CloudLearning Consortium, "Trainings", [online], Available at <https://brainsintheclouds.eu/?page_id=79>, [Accessed 23.January 2017].

[4] CloudLearning Consortium, "Working with children and youth", [online], Available at <https://brainsintheclouds.eu/?page_id=83>, [Accessed 23.January 2017].

[5] CloudLearning Consortium, Collection of internal communication and meeting minutes between 15th of March 2016 and 25th of January 2017, unpublished]

[6] D. H. Jonassen, T.C. Reeves, "Learning with technology: Using computers as cognitive tools". Handbook of research for educational communications and technology (1st ed.), Macmillan Library Reference USA, 1996.

[7] E. Crawley, S. Mitra, "Effectiveness of Self-Organised Learning by Children: Gateshead Experiments", [online], Available at <http://jehdnet.com/journals/jehd/Vol_3_No_3_September_2014/6.pdf>, [Accessed 23.January 2017].

[8] e-RR, "EU Framework for National Roma Integration Strategies up to 2020", [online], Available at <http://www.eroma-resource.com/e-rr-europe/policies-and-legislation/eu-framework-for-national-roma-integration-strategies-up-to-2020.html>, [Accessed 23.January 2017].

[9] European Commission, "An EU Framework for National Roma Integration Strategies up to 2020", [online], Available at <http://ophrd.government.bg/view_file.php/21131>, [Accessed 19.January 2017].

[10] European Commission, "Education and Training, Literacy", [online], Available at <http://ec.europa.eu/education/literacy/about/what-is-it/index_en.htm>, [Accessed 15.January 2017].

[11] European Commission, "EU and Roma", [online], Available at <http://ec.europa.eu/justice/discrimination/roma/index_en.htm>, [Accessed 19.January 2017].

[12] European Commission, "Progress in reducing early school leaving and increasing graduates in Europe, but more efforts needed", Press release, [online], Available at <http://europa.eu/rapid/press-release_IP-12-577_en.htm>, [Accessed 15.January 2017].

[13] European Commission, "Report on the implementation of the EU framework for National Roma Integration Strategies", [online], Available at <http://ec.europa.eu/justice/discrimination-/files/roma_implement_strategies2014_en.pdf>, [Accessed 19.January 2017].

[14] European Commission: "EUROPE 2020 - A strategy for smart, sustainable and inclusive growth", [online], Available at <http://eur-lex.europa.eu/LexUriServ/LexUriServ.do?uri=COM:2010:2020:FIN:EN:PDF>, [Accessed 23.January 2017].

[15] European Union Agency for Fundamental Rights (FRA): "Roma survey – Data in focus Education: the situation of Roma in 11 EU Member States", 2014 [online], Available at <http://fra.europa.eu/sites/default/files/fra-2014_roma-survey_education_tk0113748enc.pdf>, [Accessed 15.January 2017].

[16] L. Révészová, "Designing modern informatics education for future managers and advanced users according to their knowledge base.", In: E + M: Ekonomie a Management. Vol. 19, no. 4 (2016), p. 186-201, ISSN 1212-3609

[17] L. Révészová, "How to integrate ICT into the education?" In: Information and Communication Technology in Education: proceedings: Rožnov pod Radhoštěm, Czech Republic, 11th-13th September 2012. - Ostrava: University of Ostrava, 2012 P. 219-228. - ISBN 978-80-7464-135-0

[18] OECD (2013), PISA 2012 Results: Excellence Through Equity: Giving Every Student the Chance to Succeed (Volume II), PISA, OECD Publishing. [online], Available at <http://dx.doi.org/10.1787/9789264201132-en>, [Accessed 15.January 2017].

[19] P. Dolan, D. Leat, L. Mazzoli Smith, S. Mitra, L. Todd, K.Wall, "Self-Organised Learning Environments (SOLEs) in an English School: an example of transformative pedagogy?", [online], Available at <http://www.ocrj.org/View?action=view Paper&paper=109>, [Accessed 23.January 2017].

[20] S. Bloem, C. Brüggemann, "Student Performance and Inequality in Central and South Eastern Europe: Cross-country Comparison and a Case Study on Romani-speaking Students in Slovakia." Working Paper No. 5., Roma Education Fund: Budapest., 2016, [online], Available at <https://www.romaeducationfund.hu/sites/default/files/publications/bloem_bruggemann_ref_working_paper_2016_en_web.pdf, [Accessed 28.January 2017].

[21] S. Mitra, "SOLE TOOLKIT. How to bring self-organized learning environments to your community", [online], Available at <https://s3-eu-west-1.amazonaws.com/school-in-the-cloud-production-assets/toolkit/SOLE_Toolkit_Web_2.6.pdf>, [Accessed 24.January 2017].

[22] S. Mitra, M. Quiroga, "Children and the Internet – A Preliminary Study in Uruguay", [online], Available at <https://www.researchgate.net/publication/273447437_Children_and_the_Internet__A_Preliminary_Study_in_Uruguay>, [Accessed 23.January 2017].

[23] Súkromná Základná Škola, "Official web page" [online], Available at <http://szsgalakticka.edupage.org>, [Accessed 20.January 2017]

C based laboratory for teaching Digital Signal Processing to Computer Engineering Undergraduates

D. Bokan*, M. Temerinac*, Z. Lukac**, and S. Ocovaj**
* University of Novi Sad, Faculty of Technical Sciences, Novi Sad, Serbia
** RT-RK Institute for Computer Based Systems, Novi Sad, Serbia
dejan.bokan@rt-rk.uns.ac.rs

Abstract – This paper describes experience and lessons learned from teaching introductory Digital Signal Processing course as a part of Computer Engineering curricula using C based laboratory exercises. Matlab based laboratory exercises are substituted with a series of hands-on experiments which include implementation of signal processing algorithms using programing language C and DSP development boards, and the analyses of implemented systems. The theoretical content of the course and the topics covered by the laboratory exercises have remained unchanged. Main goal of a new set of exercises was to enhance student performance, motivation, as well as their understanding of theory. Results are measured by comparing student final grades on this course with grades of students who enrolled in Matlab based course. After finishing the course both student groups were given a survey on experience and satisfaction with the course. Survey results indicated that students who enrolled in C based laboratory course rated the course better and were more interested in Digital Signal Processing topics.

I. INTRODUCTION

Computer engineering (CE) is a field which covers design, construction, and implementation of both hardware and software components of computer-based systems. It is traditionally regarded as a combination of Computer science and Electrical engineering, which managed to evolve as a separate discipline. To solve problems covered by computer engineering area, one needs a wide knowledge on hardware, software and the way those two interact. That is why designing a CE curriculum represents a challenging task. A joint task group formed by ACM and IEEE created curriculum guidelines for undergraduate degree programs in CE. First version [1] was released in 2004 and the second, updated version [2] in 2016. Both [1] and [2] introduce DSP (Digital Signal Processing) as one of the core disciplines incorporated in CE curriculum.

One of the main tasks of modern engineering education is providing students with essential knowledge and skills, and enabling them to meet industry demands. Rapid technology changes are particularly present in the field of computer engineering and computer science.

Alongside with the constant development of new learning processes and improvements of the existing ones, it creates a necessity for frequent revision, changes and upgrades of academic courses. Documenting experience and results of different teaching approaches in different universities is essential for further research in the field of CE education.

In this paper experience and lessons learned from teaching introductory Digital Signal Processing course as a part of computer engineering curricula at the Faculty of Technical Sciences, University of Novi Sad are described. Results are obtained with the students which participated in the course in years 2014/2015 and 2015/2016. In between these two school years a major change in teaching process was made. The change includes using programming language C instead of Matlab for implementation of digital signal processing algorithms as a part of laboratory exercises. Matlab based simulations were replaced by execution of implemented algorithms using real hardware. The results of the above mentioned changes were obtained by comparing students' grades in two successive years, and by using a survey on experience and satisfaction with the course.

The rest of the paper is organized as follows: section 2 gives the motivation for introducing C based laboratory exercises into existing course and choosing DSP hardware. Section 3 shows a short overview of an introductory DSP course. Section 4 gives detailed description of laboratory exercises. Section 5 contains description of evaluation method and obtained results. Finally, section 6 gives some concluding remarks.

II. MOTIVATION

Digital Signal Processing is a mature field. There is a range of textbooks which cover DSP fundamentals. Next to the theories and algorithms, a significant knowledge for any computer engineer is DSP implementation and its application in practice. In order to meet industry demands students need to learn how to implement, optimize and apply different signal processing algorithms. That is why a number of authors expanded their books with a set of projects and simulations for demonstrational purposes, as in [3-5].

This work was partially support by Ministry of Education and Science of Republic of Serbia, under the project No III44009.

In [6] different approaches to teaching signal processing are analyzed. It reports on lessons learned from observations made by Texas Instruments over the past 25 years. It highlights providing practical experience with both algorithmic and real-time signal processing concepts as one of the key components in learning process. One method of providing practical experience is using mathematical software packages like Matlab or Mathematica. Such is demonstrated in [7]. Similar approach was used in Digital Signal Processing course in Faculty of Technical Sciences in the years 2011-2016.

The main motive for reorganizing the course laboratory experience was to increase students' motivation for participating in the course. A study by Huettel [8] shows how introducing hands-on experience with DSP hardware in laboratory increased student's level of interest in the field of signal processing for 80% of students. Although there is a short learning curve required to become familiar with hardware, it allows students to become aware of many real-time concepts at an early stage. Using hardware also shortens the time of simulation in comparison to software simulation.

An approach Huetell used was using development environment which enables compiling Matlab code to binary files which are downloadable to hardware. Students who represent a test subject, take introductory Digital Signal Processing course as part of the fifth semester. At this point CE students have a basic knowledge of Matlab and a very good knowledge of programming language C. Also, students which passed the Matlab based laboratory course tended to struggle with implementation of the same algorithms in C with real-time constraints as a part of the following courses. In order to resolve this problem and to eliminate the learning curve needed to get familiar with Matlab, a set of new laboratory exercises using programming language C for algorithm implementation was designed, which uses programming language C for algorithm implementation.

In order to choose a hardware platform to be used, several already existing approaches were considered. Features taken in consideration, apart from hardware capabilities, were price and development tools availability. There is a recent work [9] showing the usage of ARM-based development boards in combination with CMSIS-DSP library to create a low-cost hardware laboratory setup for teaching DSP. A number of universities, as shown in [7], are using TI- based discovery kits as a central hardware component. Experience with using different TI signal processors for teaching Digital Signal Processing is shown in [10-12].

In order to deliver hands-on experience described in this paper, the authors have used TMS320C5535 [13] fixed-point DSP processor hosted on a Spectrum Digital eZ-DSP evaluation board (shown in Fig. 1). This DSP architecture achieves high performance while being suitable for low power applications. Development board is equipped with TI TLV320AIC3204 stereo codec which supports two channels AD/DA with sampling rate up to 48 kHz. It also contains buttons, switches and LCD display which enable interaction with student during code run-time. Connection between processor and a host computer

for the purpose of downloading or debugging application is achieved using XDS100 USB connector which is mounted on the board. The board is supported with an integrated development environment and a set of runtime libraries for controlling processor peripherals.

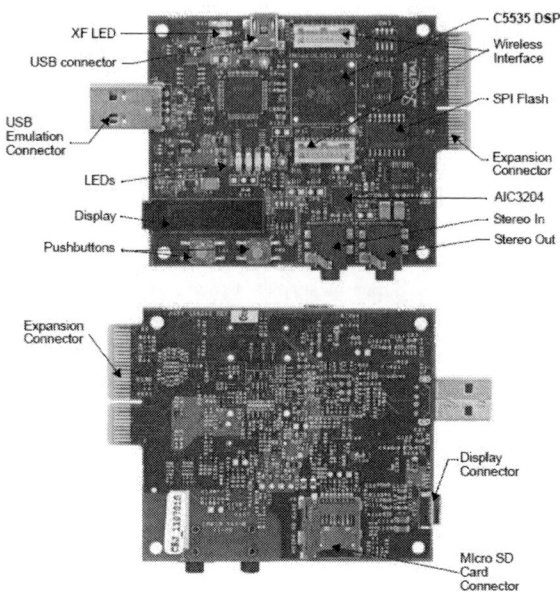

Figure 1. TMS320C5535 eZ-DSP evaluation board

III. COURSE OVERVIEW

As already mentioned, introductory Digital Signal Processing course is positioned within fifth semester. The course covers the fundamental topics in Digital Signal Processing. The course outline is given below:

- Introduction, Continuous signals and systems, spectral analysis of continuous time systems and signals

- Sampling, aliasing and reconstruction.

- Discrete-time signal representation, quantization

- Discrete Fourier Transform, spectral analysis of discrete-time signals

- Fast Fourier Transform

- Z – transform

- Discrete-time linear systems, discrete convolution, frequency response of discrete-time linear systems

- Filter theory, Finite Impulse Response (FIR) filters

- Infinite Impulse Response (IIR) filters

- Adaptive filters and application

The course consists of traditional lecture and laboratory experience. Laboratory exercises are designed in such way that they require students to be actively involved. Students have the possibility to collect 40% of total exam points in laboratory. The rest of the points are collected through tests and final exam. The grading system is presented in Table I.

TABLE I. POINT DISTRIBUTION IN THE COURSE

No	Exam	Points	Minimum to pass
1	Laboratory exercise	40	-
2	Test 1	15	7.5
3	Test 2	15	7.5
4	Final exam	30	-

IV. LABORATORY EXERCISES

This section describes the new set of laboratory exercises developed for the course. A set of exercises consists of a tutorial on using environment followed by 8 different exercises each containing demonstrational part (which is given) and a set of tasks which are implemented by students on their own. After completing these 8 exercises student is given a final project which needs to be realized during 4 weeks.

Single laboratory station is equipped with a PC, TMS320C5535 eZ-DSP evaluation board, headphones, microphone and a 2 x 3.5 mm audio cable. In addition, audio spectrum analyzer is used for debugging purposes. Main software package used in this course is Code Composer Studio (CCS) integrated development environment (available at [14]). It is a proprietary software package developed by Texas Instruments. It includes support for C code development, an integrated toolchain for TMS320C5535 processor and a set of tools for code debugging and signal visualization (frequency and time domain). Additional software package which is used is Audacity (available at [15]), an open source application which enables audio playback, recording, signal generation and analysis.

As mentioned, first exercise represents an introduction to laboratory, where students are being introduced to development environment and evaluation board. The rest of the exercises are described as follows.

A. Basic discrete-time signals

The goal of this exercise is for students to see how elementary signals are presented in discrete-time domain and how they can be generated using programming language C. Signals which students generate and analyze are: sine wave signal, square signal, Dirac impulse, multitone signal and a frequency sweep signal (linear and logarithmic). Generated signals are sent to DA convertor so students can hear those using headphones, or record and analyze using PC. This exercise helps students get familiar with circular buffering concept and signal generation using look-up table.

B. Sampling and aliasing

In this exercise a sampling theory is introduced to students. Signals generated in the previous exercise are given as an input, and are sampled using different sampling rates. Students analyze the signal in time and spectral domain. Aliasing effect and the impact of sampling rate to spectral resolution are demonstrated.

C. Quantization and signal format

This exercise demonstrates the impact of signal quantization on signal quality. Students are introduced to different signal presentation formats. Different signals are being quantized using different number of bits and using linear and non-linear quantization. Students calculate quantization noise and signal to noise ratio (SNR), and compare the values with memory needed to store uncompressed signal. Finally SNR values are compared to hearing experience.

D. Discrete Fourier Transform

In this exercise students are introduced to a DFT implementation in C. They learn how to use DFT for spectral analysis in real-time. The impact of using different window functions is shown. An example of FFT implementation (Radix-2 with decimation in time) is also given. Students compare computational requirements of the two algorithms. The last part of exercise is reconstruction of signal from spectral coefficients using the overlap-add method.

E. Discrete-time systems

Goal of this exercise is getting to know basic discrete-time linear time-invariant systems, and how they can be implemented using programming language C. Given examples are audio effects based on delay: echo and reverberation. Discrete convolution is introduced with convolutional reverberation effect. Students are taught how to analyze given systems by calculating impulse response, magnitude and phase response.

F. FIR filters

During this exercise students are introduced to practical use of signal filtering. Different order low-pass, high-pass and band-pass filters are applied to input signals, and students can analyze the output. Students are given a task to create their own Nth order FIR filter implementation using programming language C. Finally, students are shown how an array of filters can be used to split signal into frequency sub-bands (simple filter bank).

G. IIR filters

In this exercise IIR filters are introduced. Students get to know first and second direct form implementation of second order IIR filter. After that a cascade realization of higher order filter using second order sections is shown. Magnitude and phase response, so as computational requirements are compared to FIR filters. All-pass filters are introduced. A way of using all-pass filters to create simple half-band structure is demonstrated.

TABLE II – SURVEY FOR EVALUATION OF LABORATORY EXPERIENCE

	2014-2015	2015-2016
Q1	On scale 1-5 rate the quality of laboratory exercises: 1 2 3 4 5	On scale 1-5 rate the quality of laboratory exercises: 1 2 3 4 5
Q2	On scale 1-5 rate how interesting you consider laboratory exercises: 1 2 3 4 5	On scale 1-5 rate how interesting you consider laboratory exercises: 1 2 3 4 5
Q3	On scale 1-5 rate the quality of project assignment: 1 2 3 4 5	On scale 1-5 rate the quality of project assignment: 1 2 3 4 5
Q4	On scale 1-5 rate how interesting you consider project assignment: 1 2 3 4 5	On scale 1-5 rate how interesting you consider project assignment: 1 2 3 4 5
Q5	Laboratory experience with Matlab simulations provided me with a better understanding of DSP concepts: SA A N D SD	Laboratory experience with real hardware provided me with a better understanding of DSP concepts : SA A N D SD
Q6	Using Matlab introduced additional difficulties in learning process: SA A N D SD	Using programming language C introduced additional difficulties in learning process: SA A N D SD

SA = Strongly agree; A = Agree; N = Neutral; D = Disagree; ND = Strongly Disagree

H. Adaptive filters

In this exercise a concept of adaptive filtering is demonstrated. An implementation of LMS and NLMS algorithms is shown. One task for students is to use adaptive filter for simple echo cancellation. The second one is to use adaptive filter to approximate response of unknown system watched as a black box.

I. Project assignmnet

After finishing previously described laboratory exercises students are given a project assignment. Each student needs to complete his assignment in 4 weeks. Idea behind the project is to make student use the knowledge obtained during previous weeks to create a software implementation for a given signal processing algorithm. Projects are designed in such way that they include signal generation, signal transport between two DSPs (using DA/AD), filtering and signal analysis.

One example of given project assignment is DTMF (dual tone multiple frequency) generator and receiver. DTMF represents a method used for communication between telephone equipment and other communication devices and switching centers. It uses voice-frequency band. Student assignment consists of the following tasks:

- Create DTMF signal generator. It consists of two sine generators and one addition unit.

- Create DTMF signal detection unit based using FFT.

- Add noise generators to transmitter.

- Design filter to remove generated noise. Apply filtering at the receiver side, before executing DTMF signal detection algorithm.

- Examine the success rate of DTMF tone detection depending on the value of SNR.

After successfully finishing a project assignment, a student is awarded maximum of 30 points, which is 75% of all the points achievable through laboratory exercises.

V. EVALUATION AND ASSESMENT

A set of described laboratory exercises were integrated into the introductory DSP course in the CE curricula at Faculty of Technical Sciences in Novi Sad in school year 2015/2016. In order to evaluate the impact of introducing new laboratory exercises students were asked to fill a survey developed for this purpose. The similar survey was given to the students who were enrolled in the course a year before (2014-2015) and attended Matlab based laboratory exercises.

TABLE III. SURVEY RESULTS

	2014-2015					2015-2016				
	A1	A2	A3	A4	A5	A1	A2	A3	A4	A5
Q1	0	7	36	21	36	0	4	26	37	33
Q2	0	7	36	36	21	0	15	11	22	52
Q3	0	0	21	43	36	0	11	22	22	44
Q4	0	7	36	29	29	0	7	15	26	52
Q5	0	7	21	43	29	4	7	22	30	37
Q6	36	14	21	21	7	63	19	11	7	0

The survey is shown in Table II. It was explained to students that the term "quality" refers to the quality of accompanying materials, environments to perform exercises and the presentation by the lecturer. Group of students who attended the course in 2014-2015 and took the survey counted 14 students, while the group of students who attended the course in 2015-2016 and took the survey counted 27 students. Table III contains survey results. Q1-Q6 stands for the question (as in Table II). A1-A5 stands for each of the 5 offered options for the given question. A value in each cell $[Q, A]$ shows percentage of students who have selected answer A for a question Q.

Students who were part of 2015-2016 group rated the laboratory exercises quality with average grade of 4.0, while other group of students gave average grade of 3.86. When it comes to the second question, C based laboratory achieved average grade of 4.11, while Matlab based laboratory was given average grade of 3.71. A conclusion that can be drawn based on these results is that although both sets of exercises were prepared with similar quality from students' point of view, C based laboratory exercises which included real hardware were more interesting to CE students. When it comes to project assignment students from the 2015-2016 group gave an average grade of 4.0 for quality and 4.22 on how interesting it was. Group which took the Matlab based laboratory gave an average grade of 4.14 on quality and 3.79 on how interesting it was. So although the quality of Matlab simulation project was considered to be higher than the C based project, it proved to be less interesting to students.

When asked if laboratory experience provided better understanding of DSP concepts, both groups gave similar answers. In Matlab group 72% agreed (29% SA and 43% A), 21% was neutral while the rest 7% disagreed (0% SD). 67% of those students who participated in the C based laboratory agreed to previous statement (37% SA and 30% A) while 22% were neutral. 11% of students disagreed (7% D and 4% SD). But when the overhead introduced by chosen technology was in question C based laboratory achieved more positive result with 63% who strongly disagree, 19% who disagree and 11% who were neutral. Only 7% agreed. Students from the other group answered 36% for SD, 14% disagreed, 21% were neutral, while 30% agreed (21% SA and 7% A).

Finally, grades of students who passed the course were compared. Final grades of students who have passed the course, but haven't participated in the survey were also taken in consideration. Total number of students whose grades were taken in calculation is 18 in 2014-2015 and 33 in 2015-2016. Students who participated in the course during 2014-2015 achieved an average grade of 7.94, while the students who participated in course during 2015-2016 achieved an average grade of 7.74. It is important to notice that an additional factor that influenced final grades of the students is that in 2015-2016 there weren't lower boundary for number of points in the final exam, while in 2014-2015 it was 50% percent (or 15 points). The result was that in 2015-2016 39% of students had a minimal final grade (6), while in the 2014-2015 that percent was 17%. As far as the number of students who achieved the highest grade (10) is concerned, in 2015-2016 that number reached 30% of the total number of students, while in 2014-2015 it was 17%.

VI. CONCLUSION

This paper described a new set of laboratory exercises developed with the purpose to provide hands on experience in teaching DSP concepts to computer engineering students. Having C based laboratory exercises which included real hardware was positively accepted by students. Evaluation showed that students considered this approach more interesting than the laboratory exercises based on Matlab simulation. More students were motivated to achieve the highest grade. Finally, evaluation results showed that using programming language C instead of Matlab shortens the learning curve, and reduces overhead in learning process for CE students. In order to get more reliable conclusions, continuous evaluation of described laboratory exercises should be performed over the following years.

REFERENCES.

[1] Joint Task Force on Computing Curricula IEEE Computer Society, Association for Computing Machinery,"Curriculum Guidelines for Undergradute Degree Programs in Computer Engineering", 2004.

[2] Joint Task Force on Computing Curricula IEEE Computer Society, Association for Computing Machinery,"Curriculum Guidelines for Undergradute Degree Programs in Computer Engineering", 2016.

[3] S. Mitra, *Digital signal processing*, 1st ed. New York, NY: McGraw-Hill, 2011.

[4] Embree, Paul M. *C Algorithms For Real-Time DSP*. 1st ed. Upper Saddle River, N.J.: Prentice Hall PTR, 1995.

[5] S. Kuo and B. Lee, *Real-time digital signal processing*, 1st ed. Chichester: Wiley, 2004.

[6] C. Wicks, "Lessons learned: teaching real-time signal processing [DSP Education", *IEEE Signal Processing Magazine*, vol. 26, no. 6, pp. 181-185, 2009.

[7] McClellan, R. Schafer and M. Yoder, "Experiences in teaching DSP first in the ECE curriculum", *1997 IEEE International Conference on Acoustics, Speech, and Signal Processing*.

[8] L. Huettel, "A DSP Hardware-Based Laboratory for Signals and Systems", *2006 IEEE 12th Digital Signal Processing Workshop & 4th IEEE Signal Processing Education Workshop*, 2006.

[9] M. Wickert, "Using the ARM Cortex-M4 and the CMSIS-DSP library for teaching real-time DSP", *2015 IEEE Signal Processing and Signal Processing Education Workshop (SP/SPE)*, 2015.

[10] J. Cadena and A. Beex, "DSP education by fixed-point implementation & measurement", *2015 IEEE Signal Processing and Signal Processing Education Workshop (SP/SPE)*, 2015.

[11] I. Abdel-Qader, B. Bazuin, H. Mousavinezhad and J. Patrick, "Real-time digital signal processing in the undergraduate curriculum", *IEEE Transactions on Education*, vol. 46, no. 1, pp. 95-101, 2003.

[12] A. Kwasinski, "In-class demonstrations with a portable laboratory for teaching DSP to Computer Engineering majors", *2011 IEEE International Conference on Acoustics, Speech and Signal Processing (ICASSP)*, 2011.

[13] "TMS320C5535 Fixed-Point Digital Signal Processor | TI.com", *Ti.com*, 2017. [Online]. Available: http://www.ti.com/product/TMS320C5535. [Accessed: 20- Feb-2017].

[14] "CCSTUDIO Code Composer Studio (CCS) Integrated Development Environment (IDE) | TI.com", *Ti.com*, 2017. [Online]. Available: http://www.ti.com/tool/CCSTUDIO. [Accessed: 20- Feb- 2017].

[15] "Audacity®", *Audacityteam.org*, 2017. [Online]. Available: http://www.audacityteam.org/. [Accessed: 20- Feb- 2017].

Analysis of video views in online courses

Esztelecki Péter

University of Szeged, Faculty of Science and Informatics

epeter@inf.u-szeged.hu

Havasi Ferenc

University of Szeged, Faculty of Science and Informatics

havasi@inf.u-szeged.hu

Abstract - Nowadays, in our fast-paced world there are countless MOOC courses in the Internet with various topics that have been designed to broaden our knowledge. One of the most powerful tools for effective learning are online videos. Many case studies have been carried out in order to specify the qualities of a good educational video. Philip J. Guo, Juho Kim and Rob Rubin published an article (How Video Production Affects Student Engagement - An Empirical Study of MOOC Videos) on their experience during edX courses. They analyzed the optimal length of a video and determined which videos were preferred by the students.
Our research group from the University of Szeged has developed a logging system which records student activities in online courses to analyze learning styles and behaviour when watching a video. In our research, we did a literature review about on-line videos and introduced the system to log student activities, focusing on video views.

Key words: on-line videos, video views, MOOC courses, electronic curriculum

I. INTRODUCTION

E-learning (Electronic Learning), in our times, is perceived as a new technology in the field of education, which makes possible sharing and reaching learning content, practicing and measuring acquired knowledge, and finally communicating between members of courses. Education in case of E-learning is computer supported which allows individual learning pace cancelling traditional school timetables for the benefit of the most convenient time to acquire knowledge. In order to develop E-learning, not only the promotion of LLL (lifelong learning) was crucial, but it was necessary to promote quality education, to advocate international education; though the most important factor was the accelerated progress of information technologies [1].

Another notion has also gained ground that advocates learning through mobile devices, the so called M-learning. It emphasizes that knowledge is not only dependent on time but can be acquired regardless of location [2]. Such technological advancements could simplify the achievement of lifelong learning, since advanced studies will become accessible side by side with full- or part-time jobs. An increasing number of universities launch online courses, which have penetrated even to secondary and primary schools where teachers incorporate experimental online courses in their curriculum.

We may differentiate between virtual and mixed methods. Sharing learning materials and communication with this virtual method comes off in the virtual learning environment (VLE) with the help of sharing messages and participating in forum discussions. Thus, learners do not meet with their tutors or with each other, which is not a problem in adult education, but is not convenient for the K-12 age-group that cannot neglect supervision and meeting peers for mutual learning. In this latter case a mixed method, Blended learning, would be more beneficial, since it would imply classical school lessons within an accepted curriculum, however some elements still can be allowed to be acquired through online methods and individual time-tables regardless of location. Blended learning makes it possible for learners to deepen their knowledge that can be measured by testing and seems to be more durable [3]

II. AUDIO-VISUAL CONTENTS IN ONLINE COURSES

Audio-visual contents target listening and visual capacities in order to trigger information acquisition. Such tools and learning materials appeared after the I. World War as object lessons, though in those times slide-projectors were already in use accompanied by a recorded audio tape. The mass popularization of the television in the 60s and 70s peaked just in those times on the Hype cycle representing increased expectations. Most of the studies conducted in the United States tried to prove that education through watching television is more efficient than traditional methods, but it was quickly revealed that a single most effective system does not exist [4]. Later, as personal computers appeared in households and Internet coverage spread out covering the entire globe, new opportunities came to life in the 90s, when viewing and sharing video materials helped establishing the first online courses.

First generation digital learning materials were nothing else than scanned traditional school books and digitalized educational videos, while second generation tools were put together exclusively using digital media and content optimized for computer use. Criteria for the third generation digital educational materials were the following: interactivity, inclusion of a wide range of motion pictures and audio recordings, the use of links for enhanced navigation, methodological guidance, all of which created an independent learning material. The fourth generation puts emphasis on interactivity and collaboration whose products did not originate from a single author, but could be derived from various multicultural communities [5]. Real breakthrough in this field was noticeable after the launch of online video sharing sites, like YouTube, since thereafter educational videos were easily accessible. For instance, video shares by Khan Academy were visited 663 million times on YouTube since 2006. The expression MOOC (Massive Open Online Course) became wide known around this period which comprises web-based courses with unlimited participation and accessibility for free.

We can differentiate between four types of educational videos:

- classroom lessons, which are basically records of school lessons or university lectures

- "think talk" type recorded videos during which tutors give lectures from their office teaching a pre-established learning material

- digital table file format, popular at Khan Academy with animated digital tables for teaching

- traditional PowerPoint presentations [6]

The studies of Guo, Kim, and Rubin [6] provide an insight into the following important findings:

- Shorter videos are more appealing to students. The optimal length of videos have been established at 6 minutes (3 and 9 minute are still acceptable). Longer videos lower the amount of time spent on them proportionally with their length.

- Learners, surprisingly, found those videos more interesting that visualize not only presentations but also tutors. It is purposeful to record the tutor as well as the presentation by turns to break monotony.

- The target audience prefers homemade videos that seem more personal than professional studio recordings

- An environment that resembles to digital tables of the Khan Academy is more popular than simplistic presentations.

- Recordings from school lessons that were not intended for a MOOC course are not popular at all.

- Also more support is given to recordings in which lecturers are skilled orators and can catch the attention of the audience.

- There is also a qualitative difference between lecture like videos and presentations, so called tutorial videos, since lectures are usually watched only once, while presentations are consulted several times.

It has been established that visual and audio information strengthen each other, still videos have a quality of showing operations in a chronological order simulating how a user would see them on their own device. This approach creates a so called easy-to-follow model [7]. Van der Meij and Van der Meij [7] carried out a study measuring fifth and sixth grade learners (average age: 11.8 years), altogether 111 children (56 boys and 56 girls). Four groups were established of whom two groups worked following a mixed method (videos and paper-based), while the other two groups respectively used videos or paper-based activities exclusively for learning, after which they were given a test. Those students who were learning with the help of paper-based materials achieved a test score of 65.4%, while those who were learning by consulting videos, on average, scored 89.7%. The latter group seems to have acquired deeper and long lasting knowledge also shown by a post-test (76.8% to 55.6%).

In general, the appearance of the communication media did not alter how people use learning platforms, since they are similar to already existing ones, for instance, digital materials in the beginning of television broadcasting were nothing more than filmed radio programs, while digital books were scanned books. This traditional approach was maintained when the first tutoring videos appeared being recorded lectures. They were later refined adding more creativity in order to get the most out of this wide-spread media [6]. According to Don McIntosh, videos may enhance the effectiveness of the learning momentum, however the price/benefit ratio must be thoroughly examined [8].

So far, we have dealt with so called on-demand videos stored locally and available from the Internet, however it is necessary to mention video conferences as a category of real-time audio-visual contents, which are also suitable in an educational environment. Though, a webcam can only forward visual information, devices at video conferences are equipped with broader interactive possibilities connecting participants more successfully.

Studies carried out by Athabanca University in 2002 revealed some weak points of this system, namely software support and bandwidth; furthermore they noticed a considerable amount of functional errors in forwarding video and audio content due to a slow internet connection when they tested free services and products. We shouldn't neglect the fact that similar errors occurred with some payable pieces of software, too [9]. Nowadays, most of the users have access to a broadband Internet connection, and software in general has undergone a huge development penetrating into the field of education.

III. TECHNICAL BACKGROUND OF THE VIDEO LOGGER MODULE

Our research group and the team of the Software Engineering Department at the University of Szeged prepared a Moodle based system which is capable of creating a log file of learner activities to be later saved for further study and evaluation. The system uses a JavaScript solution that records individual user activities (click, download, mouse move, video views, etc.) with the help of a userId, the type of action initiated, the exact web location and time of the activity. A study carried out by Esztelecki Peter (2016) may provide some further information related to this topic. A paper on Big Data in Education are published in the same research [10].

We used Moodle as a e-learning framework extended with own modules. One of these modules is responsible for logging and is composed of two parts: a client-side part and a server-side part. Both of them are implemented in JavaScript: in the first case we used jQuery, in the second NodeJS with Express.

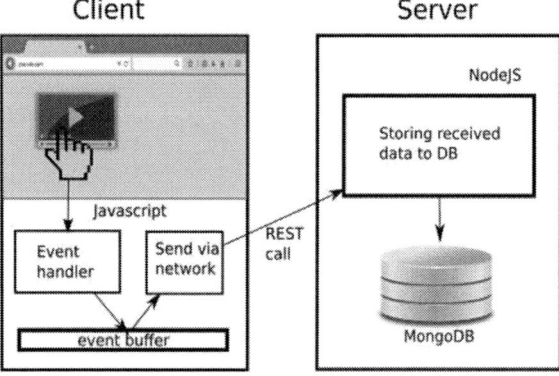

Figure 1. Logging the video events

The client part of the code handles all JavaScript events, processing and storing them in a buffer (event buffer). If the buffer is full, or the user closes the page, or it is passive for 5 minutes, it sends the content of the buffer to the server-side part via an HTTP request. The server-side part is very simple: it just stores the given events into a Mongo database for further processing.

Our system supports two kinds of videos: html5 video elements and embedded Youtube videos:

```
<video id="video1" controls>
<source src="movie.mp4" type="video/mp4">
</video>
<iframe    id="video2"    data-videoid="62rWxTfjcKc"
src="https://www.youtube.com/embed/62rWxTfjcKc">
</iframe>
```

The system is able to log both of them by catching JavaScript events, processing and storing them in a JSON format. For example here is a videoPlay event:

```
{
type: "videoPlay",
data: {
actualTime: 95.458972,
videoId: "video1",
totalTime: 956.48,
src: "/elearningdata/oktatok/sd/13.mp4"
},
time: "2016.07.18. 09:04:57.884",
page: "https://...",
userid: 942
}
```

The event has a type parameter, and the following parameters:

- userId: the ID of the user who has executed the operation, or 0, if there was an anonymous user
- time: the date of the event
- page: the URL of the page where the event occured
- type: the type of the event (see below)
- data: optional parameters, depending on the type of the event.
- In the case of an html5 video element, the system catches the following JavaScript events for processing:
- play: in this case a videoPlay event is generated, with the following parameters: actualTime, totalTime, videoSrc and videoId, as in the above example.
- seeked: the system generates a videoSeek event, with the following parameters: seekTime, totalTime, videoSrc, videoId.
- pause: a videoPause event is generated. Parameters: actualTime, totalTime, videoSrc, videoId.
- webkitfullscreenchange, mozfullscreenchange, fullscreenchange: it would mean a fullscreenOn or a fullscreenOff event. Parameters: actualTime, totalTime, videoId.
- volumechange: it generates a volumeChange event, with the same parameters as above having a newVolume parameter, as well.

In the case of embedded Youtube videos using the JavaScript API of the Youtube, the system catches the following events and generates events with the following parameters: actualTime, videoId, totalTime and src:

780

- onStateChange: depending on the given event object:
- YT.PlayerState.PLAYING: youtubePlay
- YT.PlayerState.ENDED: youtubeEnd
- YT.PlayerState.PAUSED: youtubePause
- YT.PlayerState.BUFFERING: youtubeBuffering
- onPlaybackQualityChange: youtubeQuality
- onPlaybackRateChange: youtubeRate

As an example, we are providing the structure of the YT.PlayerState.PLAYING code:

```
function onPlayerStateChange(event) {
if (event.data == YT.PlayerState.PLAYING) {
pushEvent({
type: 'youtubePlay',
data: {
actualTime: event['target'].getCurrentTime(),
videoId: event['target']['a']['id'],
totalTime: event['target'].getDuration(),
src: event['target'].getVideoUrl()
} }); } ...}
```

These events are stored in the "event buffer" (using the pushEvent method), and later submitted to the server to be written into the database.

IV. PILOT COURSES AND LOGGED DATA

Our team comprised pilot courses for students at the University of Szeged touching upon two topics: Conscious and Safe Internet Use and Databases. We did the best to fix any errors that might have occurred during the pilot research in order to build a stable and reliable logging system. Even at the test phase, we were able to accumulate a log file with a volume of 1.2 GB recording more than 4.5 million lines of raw data. It has to be noted that beside collecting data from the two pilot studies other logs were also saved that were coming from Moodle courses announced and executed by the University.

During the evaluation of the JSON file, our team revealed a system error which was related to the process of filling in the tests and playing the videos, namely these events were not recorded. Later, this error was fixed allowing us to save such data to a log file during a live course session in the future.

Soon we found out that raw data had to be cleaned and changed appropriately despite the fact that they were not analyzed using Data Mining algorithms. Esztelecki Péter, for this reason, developed a PHP based program that converts a json file into a csv (comma separated values) file, which is a format that database management and data mining tools can both easily handle. The program allows users to choose which courses or course data should be converted to a csv file; furthermore it is also possible to decide which events are to be put into the output file. The conversion process generated a stats file that indicates how much data was processed from an event and how much time it would consume to convert data, etc. To convert a file of 1.2 GB that contains the information of all the courses and their events on an Intel (R) Core (TM) i7-2630QM with 2.00GHz processor and 6.00 GB RAM took us 6 minutes. This time volume could be furhter reduced if an SSD system replaced the current 5400 RPM SATA II harddisk composition since most of the consumed time was spent on harddrive processes (writing and retreiving data).

The method of data mining will be used to process raw data and to acquire new pieces of information in order to map learners' behavior and achievement more precisely. We would furthermore classify and put participants into clusters to create groups where students would be taught in a more personalized environment and later to establish a platform for adaptive learning.

Using regressive calculations, it is possible to envisage the outcome results of a participant with high precision based on learner routines. The same process can be done using statistical matching which assigns a learner or learners from the database to the appropriate subject.

Our future plan is to move our research courses to secondary schools in Southern Hungary and Vojvodina in Serbia to enlarge the pool of participants to a couple of thousand making it possible to analyze a larger volume of data.

V. CONCLUSION

Broadcasting media offer a wide range of learning opportunities in public education; however it must be stated that the appearance of the audio-visual content meant a giant leap in the field of distance learning. Another great achievement was the spread of the Internet that brought about tools provided by Web 2.0. Finally, the founding and popularization of video sharing websites have to be also mentioned. Guo, Kim and Rubin [6] used different methods to record their educational videos, and later came to the conclusion that shorter videos are more appealing to learners; furthermore they prefer lectures where tutors and presentations follow each other. Nevertheless, much depends on the communicative skills and dynamics of the tutors how they deliver their lectures. The authors also realised that studio recordings were less popular than homemade recordings; moreover learners did not cherish classroom recordings that were clearly not intend for a MOOC platform.

The logging system created by our team makes it possible to entirely reconstruct learner activities in the Virtual Learning Environment opening the field to data mining tools to analyze learners' behaviour and their learning styles.

We share the expectations of Mark Van Rijmenam according to which Big Data analysis will revolutionize the way students learn and teachers teach [10][11].

REFERENCES

[1] Pfaff A. (2013), Az E-learning az Európai Unióban és Magyarországon. URL:

http://www.oktopusz.hu/domain9/files/modules/module15/26831 CB34D1124D.pdf (2017.01.10.)

[2] Mohamed A. (2009), Mobile Learning Transforming the Delivery of Education and Training. AU Press, Athabasca University, Athabasca, Canada. URL: http://www.zakelijk.net/media/boeken/Mobile%20Learning.pdf (2017.01.10.)

[3] Kőrösi G., Esztelecki P., Muhi B. (2014). A blended learning-en alapuló saját, inverz oktatási modell alkalmazásának lehetőségei a számítástechnika oktatásában. Eruditio Educatio, Selye János Egyetem Tanárképző Kar, Szlovákia, Komárom

[4] Anderson C. M. (1972). In search of a visual rhetoric for instructional television. A V Communication Review, 1972, 20, 43–63.

[5] Szepesi J. (2012). .Elektronikus tananyag készítése. URL: katalogus.nlvk.hu/html/szte/tananyag_szerkesztes.pdf (2017.01.10.)

[6] Guo J. P., Kim J. Rubin R. (2014). How Video Production Affects Student Engagement: An Empirical Study of MOOC Videos. Retrieved on 20th September 2015 URL: https://groups.csail.mit.edu/uid/other-pubs/las2014-pguo-engagement.pdf (2017.01.10.)

[7] Van der M. H.., Van der M. J. (2014). A comparison of paper-based and video tutorials for software learning. Computers & Education 78 (2014) 150-159.

[8] Mcintosh D. (2013). Video in Training and Education. URL: http://www.trimeritus.com/Video (2017.01.10.)

[9] Craven P., Keppy B., Baggaley J. (2002). Online Videoconferencing Products.International Review of Research in Open and Distance Learning.Vol 3, Number 2. URL: http://www.irrodl.org/index.php/irrodl/article/download/88/582 (2017.01.10.)

[10] Esztelecki P. (2016). Big Data in Education, Information Technology and Education Development (ITRO). URL: http://www.tfzr.rs/itro/Zbornik%20ITRO%202016.pdf (2017.01.10)

[11] Rijmenam van M. (2013). Big Data Will Revolutionize Learning. Smart Data Collective.

URL:http://www.smartdatacollective.com/bigdatastartups/121261/big-data-will-revolutionize-learning

Author index

Abazi-Bexheti, L.	655	Berkovic, I.	1132, 1178
Adamovic, N.	146	Bernik, A.	711
Adzinets, D.	1126	Besedin, K.Y.	221
Aglic Cuvic, M.	794	Bhattacharya, J.	1084
Akagic, A.	1104, 1195	Bikov, T.D.	497
Akbal, A.	506	Bjazic, T.	1551
Akbal, E.	506, 1241	Blazic, G.	1491
Aksentijevic, S.	1454	Blech, A.	57
Alajbeg, T.	910, 944	Boban, M.	1486
Aleksi, I.	1011	Bodrusic, I.	472
Alexandrova, M.I.	125	Bogdanova, V.G.	353
Alic, D.	527	Bohak, C.	259
Alienin, O.	359	Bokan, D.	773
Aljancic, U.	336	Bosilj Vuksic, V.	1355, 1391
Alkan, A.	1094	Boukhebouze, M.	381
Allred, P.	43	Brajovic, M.	482
Amelio, A.	1110	Brcic, M.	197
Anderson, N.	624	Brenner, W.	146
Andjelkovic, M.	887	Brezany, P.	365
Androcec, D.	1285	Brezovec, I.	88, 93
Antolic, G.	1379	Brkic Bakaric, M.	1546
Antulov-Fantulin, B.	920	Brkic, K.	7
Apostolova, M.	655	Brkic, L.	1379, 1528
Arslan Tuncer, S.	1094	Brloznik, M.	303
Artemkina, S.B.	48	Brodic, D.	1110
Artetxe González, E.	359	Brodjanac, P.	740
Asenbrener Katic, M.	1221	Brscic, D.	564
Astrova, I.	215	Brtka, E.	1178
Avaroğlu, E.	171	Brtka, V.	1178
Avbelj, V.	289, 303	Brumen, B.	1275
Babic, M.	188	Brunetti, G.	967
Babic, S.	717	Brzeźniak, M.	233
Babic, Z.	192	Bubas, G.	711
Bach, L.M.	1522	Bucko, J.	750
Baggini, M.B.	1038	Budimac, Z.	570, 613
Bakalar, G.	1038	Bundalo, D.	103
Bakalar, S.G.	1038	Bundalo, Z.	103
Balaz, Z.	1211	Bunja, D.	1437
Baldini, G.	1269, 1292	Burdukovskii, V.F.	27
Balic, K.	244	Buric, M.	1098
Balota, A.	593	Busch, J.	624
Balug, M.A.	1562	Busch, K.	57
Banovic-Curguz, N.	445, 492	Buza, E.	1104, 1195
Barakovic Husic, J.	428, 434	Bychkov, I.V.	1116
Barakovic, S.	428, 434	Bziuk, W.	521
Baric, A.	13, 88, 93	Candrlic, S.	836, 1221
Batistic, L.	420	Cano Pons, E.	1269
Begicevic Redjep, N.	705	Capeska Bogatinoska, D.	411, 515
Berdinsky, A.S. 27,	48	Car, S.	330

Car, Z.	209	Drira, K.	555
Carapina, M.	853	Drmic, A.	995
Caria, M.	152	Dudarin, A.	119
Carpio, F.	521	Duic, I.	1309
Carrato, S.	1084	Duic, M.	818, 824
Cavrak, I.	979	Dundjer, I.	1207, 1217
Cegar, S.	1431	Dzapo, P.	818
Chiasera, A.	21	Dzeng, R.J.	1027
Chiussi, S.	37	Dzuzdanovic, S.	1044
Chydzinski, A.	460	Eickemeyer, C.	215
Cifrek, M.	324	Emanovic, E.	141
Cizmesija, A.	734	Esztelecki, P.	778
Cobic, A.	853	Etinger, D.	717
Colston, G.	43	Fedorov, V.E.	27, 48
Coric, R.	1182	Feilhauer, T.	365
Crnko, N.	371	Feoktistov, A.	1138
Culjak, I.	324	Ferrari, M.	21
Cupec, R.	1120	Fertalj, K.	1501
Cupic, M.	7	Filic, M.	424
Curkovic, J.	944	Filipovic Tretinjak, M.	746
Cuzzocrea, A.	1337	Filipovic-Grcic, B.	598
Cvejic, R.	887	Filjar, R.	424
Cvetanovic, M.	895	Filko, D.	1120
Cvetko, M.	393	Fischer, I.A.	37, 57
Cvetkovic, D.	865, 891	Fonseca-Pinto, R.	276, 279
Cvetkovic, L.	1501	Fosic, I.	1298
Cvijic, B.	103	Franjkovic, D.	1566
Cvrtila, V.	1309	Frankovic, I.	728
Dadic, M.	166	Franulovic, M.	608
Dajak, L.	1546	Frey, H.	1084
Dakovic, M.	478, 482	Frieiro, J.L.	37
Damasevicius, R.	1373	Friganovic, K.	330
Davydov, A.	1161	Frincu, M.	399
Debevc, M.	393	Fruk, M.	1557
Dedic, V.	887	Fulford, C.P.	649
Delac, G.	995	Funk, H.S.	31
Depolli, M.	270, 292	Galzerano, G.	21
Desic, S.	424	Gascic, D.	30
Devcic, K.	905	Gavran, M.	1557
Dinjar, D.	1397	Gecevic, D.	1551
Divjak, B.	705	Geneiatakis, D.	1269, 1292
Djedovic, A.	527	Giedrimas, V.	588
Djumic, M.	1182	Goeminne, M.	381
Djurovic, P.	1120	Golodov, V.	225
Dogan, S.	1241	Golub, B.	1302
Dogru, N.	1314	Gordienko, Y.	359
Domazet, E.	318	Gorgan, D.	253
Dorosz, D.	21	Gorjanac, V.	1280
Dovedan Han, Z.	1221	Gorsky, S.	353, 1138
Doychinov, Y.I.	407	Gracak, Z.	1528
Draganic, A.	1227	Granic, A.	1540
Drazic, A.	979	Grba, B.	876

Grbic, R.	1120	Jahandar, P.	43
Greer, D.	624	Jaklic, J.	1355
Gregorio, A.	1073	Jakobovic, D.	1182
Gros, S.	1262	Jakopec, T.	1516
Grubjesic, I.	915	Jakovic, B.	1367, 1476
Grubljesic, T.	1355	Jaksic, D.	1401, 1491, 1496
Gruicic, S.	1233	Jakstas, A.	1373
Grzunov, L.	824	Jakupovic, A.	1221
Gumzej, N.	1424	Jamic, M.	812
Gusev, M.	308, 318, 387	Jan, M.	297
Gütl, C.	619	Jazbec, D.	393
Hajdarevic, K.	1314	Jerman-Blazic, B.	188
Halili, A.	1189	Jervan, G.	359
Hanna, P.	624	Jovanovic, V.	1401
Hashad, Y.	57	Jovic, A.	330
Hasic, H.	1195	Jozic, K.	330
Hausknecht, K.	1233	Jugovic, A.	1454
Havasi, F.	755, 778	Juhasz, Z.	340
Hebrang Grgic, I.	842	Jukan, A.	152, 521
Hedji, I.	455, 501	Jukic, O.	455
Henno, J.	635, 660, 694	Jumic, J.	1251
Henriques, V.	515	Jung, C.H.	1090
Herceg, M.	109	Juric, M.B.	264
Herynek, B.	393	Juricic, B.	920
Hivziefendic, J.	603	Jurisic, D.	141
Hlupic, N.	197	Kadoic, N.	705
Ho, C.W.	1027	Kadriu, A.	655
Hocenski, Z.	1011	Kakalejcík, L.	750
Höfler, M.	619	Kakanakov, N.	205, 1001
Hoic-Bozic, N.	728	Kalpic, D.	440
Holenko Dlab, M.	672, 836	Kamimura, N.	31
Holik, F.	450, 1256	Kanda, T.	564
Holjevac, N.	1465	Kaplar, A.	1144
Horalek, J.	450	Kavcic, A.	848
Horvat, I.	1298	Kaya, D.	314
Horvat, M.	1207	Kaya, T.	202, 314
Hsueh, H.H.	1027	Kazi, L.	1132
Hurtova, V.	450	Kazi, Z.	1132
Hyrynsalmi, S.	991, 1442	Kemper, N.	152
Iliev, T.B.	407, 416, 497	Kenzin, M.Y.	1116
Ilisevic, D.	445, 492	Kersten, J.	215
Indihar Stemberger, M.	1391	Kevric, J.	603
Ivanda, M.	21	Kholkhoev, B.C.	27
Ivanjko, T.	915, 1309	Kisasondi, T.	1285
Ivanova, E.P.	416, 497	Klasinc, J.	1407
Ivanova, O.N.	685	Klemo, V.	995
Ivanovic, M.	613, 901, 1144	Knezevic, B.	1476
Ivanovski, D.	582	Knezevic, K.	1324
Ivasic Kos, M.	1098	Knezevic, T.	72
Ivkovic, N.	1149	Kochemazov, S.	1166, 1172
Ivosic, I.	533	Kocijan, K.	806
Jaakkola, H.	635, 660, 694	Koerner, R.	57

Komen, V.	375	Llorente Coto, A.	359
Konecki, M.	723	López Benito, J.R.	359
Konjevod, B.	824	Luburic, N.	1144
Koren, A.	510	Ludescher, T.	365
Koricic, M.	77, 83	Lugovic, S.	1207, 1217
Kőrösi, G.	755	Lukac, A.	1534
Koschel, A.	215	Lukac, D.	689
Kosec, G.	1049	Lukac, Z.	773
Kostecki, K.	57	Lukovac, B.	1079
Kostenetskiy, P.S.	221, 229	Lukowiak, A.	21
Kostromin, R.	1138	Lushchyk, U.	359
Kounelis, I.	1292	Machado, M.	276, 279
Kovacevic, I.	1385	Madhale Jadav, G.	420
Kovacevic, Z.	1471	Magerl, M.	88, 93
Kovacic, B.	876, 905, 1418	Makari, T.	1275
Kovács, V.	1506	Mäkelä, J.	635, 660, 694
Kozlova, M.N.	48	Mäkinen, T.	1448
Kozuh, I.	393	Makotchenko, V.G.	27
Krajcek Nikolic, K.	1566	Makovec, M.	336
Kramaric, I.	1331	Maksimkin, N.	1032, 1116
Kranjac, M.	238	Malaric, R.	162, 166
Krasna, M.	666, 678	Malcic, G.	1055
Kresoja, M.	901	Malekian, R.	411, 515
Krivec, S.	66	Mandic, T.	13
Krpan, D.	800	Manojlovic, M.	1546
Krstic, V.	937	Maras, J.	794
Kruglov, A.	57	Marasovic, K.	1005
Kruzic, S.	1015	Marcelic, M.	162
Kuhar, U.	1049	Mareva, D.D.	130
Kukolja, D.	330	Maric, M.	1227
Kuman, S.	1262	Maris, M.	1073
Kunic, L.	247	Markic, Z.	1459
Kurdija, A.S.	995	Markovic, K.	608
Kurent, P.	858	Marolt, M.	259, 848
Kurtaj, L.	1201	Marsi, S.	1073, 1084
Kuzle, I.	1465	Marsic, D.	1055
Kuznetsov, V.A.	27, 48	Márton, M.	538
Lackovic, D.	466	Martucci, A.	21
Larionov, A.	1161	Masetic, Z.	1314
Lasic-Lazic, J.	915	Matic, T.	109
Lavric, P.	259	Mauher, M.	1471
Lazic, N.	882	McGowan, A.	624
Ledneva, A.Y.	48	Medic, B.	865, 891
Lee, H.-C.	1090	Medved, D.	1431
Lee, S.G.	1090	Megill, N.D.	182
Lee, S.H.	1090	Mekterovic, I.	1349, 1385, 1510
Leitgeb, E.	404	Merluzzi, A.	967
Lekic, V.	192	Mestrovic, A.	870
Lerga, R.	836	Meznaric, D.	1566
Limani, I.	1201	Mihajlovic, Z.	7, 247
Linna, P.	1442, 1448	Mihaljevic, B.	1005, 1522
Lisek, J.	932	Mihaylov, G.Y.	416, 497

Mijatovic, M.	865	Ovseník, Ľ.	538
Mijatovic, M.	865	Ozdemir, M.T.	202
Mikuc, M.	1262	Paek, S.	649
Mikulic, J.	88, 93	Pale, P.	17
Milasinovic, B.	1501	Paľová, D.	767
Milenkovic, M.	1319, 1412	Papic, A.	700
Milic, L.	723	Pasalic, D.	103
Miljkovic, D.	1061, 1067	Pashinin, A.A.	353
Milos, A.	113	Pavelin, G.	1437
Miskovic, T.	1459	Pavic, I.	1465
Mladenovic, S.	794, 800	Pavicic, M.	182
Mlinac, F.	1437	Pavkov, S.	728
Mohorcic, M.	270	Pavlic Sipek, Z.	1476
Molnar, G.	113, 119	Pavlic, M.	1221
Moloisane, N.R.	411	Pavlinic, A.	375
Mouetsi, S.	53	Pavlinovic, M.	920
Mrvica, A.	1459	Pejcinovic, B.	1
Muminovic, S.	434	Pejic Bach, M.	1355, 1367
Music, J.	1015	Pełech-Pilichowski, T.	283
Muzaffar bin Baharudin, A.	991	Pereira, J.	276
Myronov, M.	43	Pesek, M.	848
Nagul, N.	1155, 1161	Pesic, D.	895
Nai-Fovino, I.	1292	Pesut, D.	941
Nanver, L.K.	72	Petrovic, J.	17
Neisse, R.	1292	Petrovic, K.	166
Nemec, G.	678	Pezer, M.	882
Nenadic, K.	1280	Pilicic, S.	608
Neradová, S.	644, 1256	Pita Costa, J.	558
Neskovic, A.	1465	Pivac, M.	1540
Ng, J.	31	Pobar, M.	1098
Nikolov, G.T.	98	Podbojec, D.	393
Nikolov, N.N.	125	Podkonjak, M.	830
Novak, N.M.	767	Polancec, D.	1510
Obarcanin, K.	1044	Poljak, M.	66
Ocevcic, H.	1298	Poljak, R.	1496
Ocovaj, S.	773	Poscic, P.	1401, 1491, 1496
Odak, M.	882	Póser, V.	1506
Offel, N.	215	Prazina, I.	244
Ognjenovic, V.	1178	Prcic, V.L.	440
Ogrizovic, D.	209	Predavec, D.	1211
Okanovic, V.	244	Prkic, S.	961
Okresa Djuric, B.	1149	Prohaska, Z.	949
Omanovic, S.	527, 1104	Prohaska, Z.	949, 1481
Oparin, G.A.	353	Pticek, M.	1361
Opiła, J.	283	Puligheddu, M.	1073
Orehovacki, T.	717	Pürcher, P.	619
Orescanin, D.	1397	Pusnik, M.	576, 582
Oreski, D.	723	Putnik, Z.	613, 901
Orovic, I.	478, 1227	Radev, D.I.	497
Osmakcic, K.	806	Radivojevic, Z.	895
Ostojic, R.	1044	Radman Pesa, A.	1481
Ostreika, A.	1373	Radonic, M.	1349

Radosevic, D.	711	Shopov, M. 974,	1001
Radovan, A.	1522	Sikimic, U.	238
Radovan, M.	1302	Silic, M.	995
Radovan, M.	630	Siljak, H.	603
Radulovic, B.	1132	Silkina, N.S.	685
Rajh, A.	812	Sillberg, P.	985
Rakic, G.	570	Silva, M.	276
Rakic, N.	570	Simovic, V.	910, 944
Ramljak, D.	542	Simunic, D.	440, 510, 1079
Ramljak, M.	1245	Skala, K.	347, 359
Ramljak, T.	1298	Skendzic, A.	905, 1418
Ramponi, G.	1084	Skliarova, I.	176
Ramponi, R.	21	Sklyarov, V.	176
Rantanen, P.	985	Skracic, K.	472
Rashkovska, A.	289	Slamic, M.	1471
Rasic, M.	1412	Smrikarov, A.	613
Rechem, D.	53	Smrikarova, S.	613
Repnik, R.	678	Sneler, L.	109
Révészová, L.	761	Soini, J.	985
Rexhepi, A.	1189	Sojat, Z.	347, 359
Righini, G.C.	21	Sokele, M.	910
Ristic, D.	21	Solic, K.	1280, 1298
Ristov, P.	1459	Sorgo, A.	666
Ristovski, A.	308	Soric, I.	1397
Rizvic, S.	244	Speh, I.	455, 501
Rjabov, A.	176	Speranza, G.	21
Rodin, R.	1021	Spes, M.	538
Rojbi, A.	359	Spindler, M.	576
Rolseth, E.G.	57	Spoljaric, T.	1562
Romanenko, A.I.	27, 48	Sprager, S.	264
Rybicki, J.	233	Sretenovic, M.	1418
Saari, M.	991	St. Vieth, B.	233
Sabou, A.	253	Stajcer, M.	1397
Sadinov, S.M.	547	Stajduhar, I.	1021
Sajn, L.	359	Stanchev, O.P.	98, 125
Salom, J.	238	Stancic, H.	812
Samociuk, D.	460	Stancic, I.	1015
Sarabok, A.	501	Staneviciené, E.	1373
Saric, A.	1005	Stanic, J.	1055
Sarlija, N.	1367	Stankovic, I.	478, 482
Savic, M.	613	Stankovic, L.	482
Schatten, M.	1149	Stankovic, S.	1227
Schatzberger, G.	88, 93	Stefanyuk, A.Y.	27
Schlipf, J.	37	Steri, G.	1292
Schudrowitz, J.	152	Sterle, U.	954
Schulze, J.	31, 37, 43, 57	Stirenko, S.	359
Scotognella, F.	21	Stjepic, A.M.	1355
Semenov, A.	1166, 1172	Stoyanov, I.S.	407, 416, 497
Senthil Srinivasan, V.S.	57	Stoyanov, R.S.	130
Serbet, F.	202	Stoyanov, S.	205
Serra, C.	37	Strnad, B.	862, 958
Shatri, V.	1201	Subasic, M.	1534

Sudnitson, A.	176
Sukur, N.	570
Suligoj, T.	66, 72, 77, 83
Sumak, B.	576, 582
Susa Vugec, D.	1391
Susac, F.	1011
Susak, T.	1486
Svigelj, A.	1049
Sylejmani, K.	1189
Taccheo, S.	21
Tanovic, A.	527
Tapiska, S.	901
Tchernykh, A.	1138
Temerinac, M.	773
Thalheim, B.	635, 660
Tijan, E.	1454
Tomas, B.	1285
Tomic, D.	209
Tomic, M.	466
Tomic, S.	238
Tomicic, I.	1149
Tretinjak, M.	746
Trobec, R.	264, 297
Tucakovic, M.	932
Tuncer, T.	171
Turajlic, E.	486
Turán, J.	538
Türk, M.	314
Tütüncü, K.	613
Uglesic, I.	598
Ul'yanov, S.	1032
Ungermanns, C.	789
Uroda, I.	949, 1481
Vaccari, A.	21
Vaitkeviciene, A.	588
Vaitkevicius, L.	588
Valchev, V.C.	98, 125, 130, 136
Valligatla, S.	21
Van den Bossche, A.	136
Varas, S.	21
Vasilchenko, I.	21
Veispahic, A.	428
Vejacka, M.	783
Velki, T.	1280
Vidacek-Hains, V.	734
Vidakovic, M.	1144
Vilcek, T.	1516
Vilhar, A.	292
Vinko, D.	158
Vlahinic, S.	420
Vojkovic, G.	1319, 1412
Vojvodic, S.	1431
Vrana, R.	830, 926
Vrankic, M.	420
Vrbanec, T.	870
Vrdoljak, B.	1361
Vrtacnik, D.	336
Vucic, M.	113, 119
Vuckovic, N.	1079
Vujisic, G.	1557, 1562
Vukovic, M.	1251, 1298
Weidinger, V.	767
Weiser, M.	57
Weisshaupt, D.	43
Wendav, T.	57
Werth, W.	789
Wesiak, G.	619
Xie, Y.-H.	31
Yakovleva, G.E.	48
Yankov, P.V.	136
Yrjönkoski, K.	1448
Yudov, D.D.	130
Zagar, I.	979
Zaharija, G.	800
Zaikin, O.	1166, 1172
Zailskaitė-Jakstė, L.	1373
Zajgar, T.	961
Zanlungo, F.	564
Zhabinski, A.	1126
Zhabinskii, S.	1126
Zidar, M.	1465
Zilak, J.	77, 83
Zisko, A.	666
Zitta, S.	644
Zoroja, J.	1367
Zouach, F.	53
Zufic, J.	961
Zupan, A.	598
Zur, L.	21
Zymbler, M.	1343

IEEE
445 Hoes Lane
Piscataway, NJ 08854-4141

ISBN 978-1-5090-4969-1

2017 40th International Convention on Information and Communication Technology, Electronics and Microelectronics (MIPRO 2017)

Opatija, Croatia
22-26 May 2017

Pages 783-1569

IEEE Catalog Number: CFP1739K-POD
ISBN: 978-1-5090-4969-1

2017 40th International Convention on Information and Communication Technology, Electronics and Microelectronics (MIPRO 2017)

Opatija, Croatia
22-26 May 2017

Pages 783-1569

IEEE Catalog Number: CFP1739K-POD
ISBN: 978-1-5090-4969-1

Copyright © 2017, Croatian Society for Information and Communication Technology, Electronics and Microelectronics (MIPRO)
All Rights Reserved

*** This is a print representation of what appears in the IEEE Digital Library. Some format issues inherent in the e-media version may also appear in this print version.*

IEEE Catalog Number: CFP1739K-POD
ISBN (Print-On-Demand):   978-1-5090-4969-1
ISBN (Online):   978-953-233-090-8

Additional Copies of This Publication Are Available From:

Curran Associates, Inc
57 Morehouse Lane
Red Hook, NY 12571 USA
Phone: (845) 758-0400
Fax: (845) 758-2633
E-mail: curran@proceedings.com
Web: www.proceedings.com

2017 40th International Convention on Information and Communication Technology, Electronics and Microelectronics (MIPRO)

May 22 – 26, 2017
Opatija, Croatia

Proceedings

Edited by:
**Petar Biljanovic
Marko Koricic
Karolj Skala
Tihana Galinac Grbac
Marina Cicin-Sain
Vlado Sruk
Slobodan Ribaric
Stjepan Gros
Boris Vrdoljak
Mladen Mauher
Edvard Tijan
Filip Hormot**

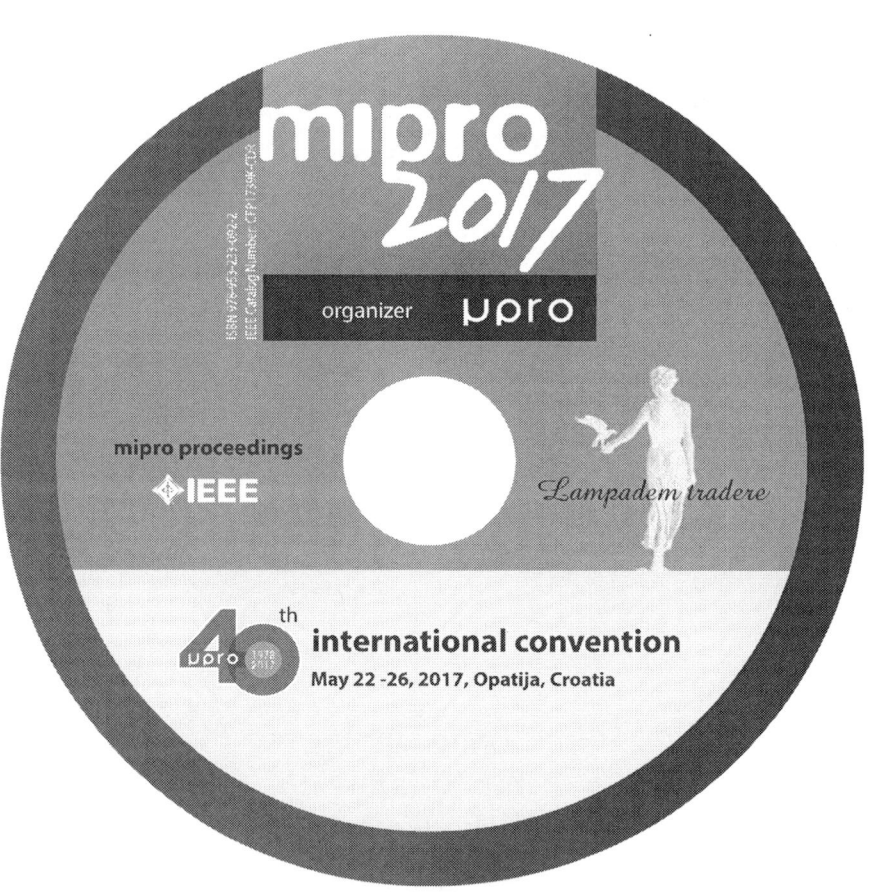

Introduction

The Conference Proceedings contain papers accepted for the 2017 40th International Convention on Information and Communication Technology, Electronics and Microelectronics (MIPRO) held from 22 to 26 May 2017 at the Grand Hotel Adriatic Congress Centre and Hotel Admiral in Opatija, organized by MIPRO Croatian Society and technically co-sponsored by IEEE Region 8.

Papers are from the following fields:

Microelectronics, Electronics and Electronic Technology /MEET
Distributed Computing, Visualization and Biomedical Engineering /DC VIS
Dew Computing /DEWCOM
Telecommunications & Information /CTI
Modeling System Behaviour/MSB
Computers in Education /CE
Computers in Technical Systems /CTS
Intelligent Systems /CIS
Information Systems Security /ISS
Business Intelligence Systems /miproBIS
Digital Economy and Government, Local Government, Public Services /DE-GLGPS
MIPRO Junior - Student Papers /SP

The authors are from industry, education, academia and public administration.

MIPRO 2017 Convention was held under the patronage of the Government of Croatia.
It was supported by many sponsors and patrons among which we single out HEP-Croatian Electricity Company Zagreb, Ericsson Nikola Tesla Zagreb, Koncar-Electrical Industries Zagreb, T-Croatian Telecom Zagreb, City of Opatija, InfoDom Zagreb, IN2 Zagreb, Transmitters and Communications Company Zagreb, King-ICT Zagreb, Hewlett Packard Croatia Zagreb, Storm Computers Zagreb, Danieli Automation Buttrio, VIPNet Zagreb, Mjerne tehnologije Zagreb, Selmet Zagreb, Institute SDT Ljubljana, Nomen Rijeka, EuroCloud Croatia, University of Zagreb, University of Rijeka, IEEE Croatia Section, IEEE Croatia Section Computer Chapter, IEEE Croatia Section Electron Devices/Solid-State Circuits Joint Chapter, IEEE Croatia Section Education Chapter, IEEE Croatia Section Communications Chapter, University of Zagreb Faculty of Electrical Engineering and Computing Zagreb, Rudjer Boskovic Institute Zagreb, University of Rijeka Faculty of Maritime Studies, Faculty of Engineering and Faculty of Economics, Faculty of Tourism and Hospitality Management, Faculty of Organization and Informatics Varazdin, University of Applied Sciences Zagreb, Croatian Regulatory Authority for Network Industries, CISEx, Kermas energija Zagreb, Business center Silos Rijeka and River Publishers.

The purpose of the Proceedings is to present the state of development and work within the ICT, electronics and microelectronics field in the world, particularly in countries of Southeast Europe – we hope we have been successful in doing that.

In Opatija/Rijeka/Zagreb, June 20th 2017

Professor Petar Biljanovic
MIPRO 2017 General Chair

The Government of the Republic of Croatia is a Patron of the convention

organized by
MIPRO Croatian Society

technical cosponsorship
IEEE Region 8

under the auspices of
Ministry of Science and Education of the Republic of Croatia
Ministry of the Sea, Transport and Infrastructure of the Republic of Croatia
Ministry of Economy, Entrepreneurship and Crafts of the Republic of Croatia
Ministry of Public Administration of the Republic of Croatia
Central State Office for the Development of Digital Society
Croatian Chamber of Economy
Primorje-Gorski Kotar County
City of Rijeka
City of Opatija
Croatian Regulatory Authority for Network Industries
Croatian Power Exchange - CROPEX

patrons
University of Zagreb, Croatia
University of Rijeka, Croatia
IEEE Croatia Section
IEEE Croatia Section Computer Chapter
IEEE Croatia Section Electron Devices/Solid-State Circuits Joint Chapter
IEEE Croatia Section Education Chapter
IEEE Croatia Section Communications Chapter
T-Croatian Telecom, Zagreb, Croatia
Ericsson Nikola Tesla, Zagreb, Croatia
Koncar - Electrical Industries, Zagreb, Croatia
HEP - Croatian Electricity Company, Zagreb, Croatia
VIPnet, Zagreb, Croatia
University of Zagreb, Faculty of Electrical Engineering and Computing, Croatia
Rudjer Boskovic Institute, Zagreb, Croatia
University of Rijeka, Faculty of Maritime Studies, Croatia
University of Rijeka, Faculty of Engineering, Croatia
University of Rijeka, Faculty of Economics, Croatia
University of Zagreb, Faculty of Organization and Informatics, Varazdin, Croatia
University of Rijeka, Faculty of Tourism and Hospitality Management, Opatija, Croatia
University of Applied Sciences, Croatia
EuroCloud Croatia
Croatian Regulatory Authority for Network Industries, Zagreb, Croatia
Selmet, Zagreb, Croatia
CISEx, Zagreb, Croatia
Kermas energija, Zagreb, Croatia
Business Center Silos, Rijeka, Croatia
River Publishers, Aalborg, Denmark

general sponsor
HEP - Croatian Electricity Company, Zagreb, Croatia

sponsors
Ericsson Nikola Tesla, Zagreb, Croatia
Koncar-Electrical Industries, Zagreb, Croatia
T-Croatian Telecom, Zagreb, Croatia
City of Opatija
InfoDom, Zagreb, Croatia
Hewlett Packard Croatia, Zagreb, Croatia
IN2, Zagreb, Croatia
King-ICT, Zagreb, Croatia
Storm Computers, Zagreb,
Croatia Transmitters and Communications Company, Zagreb, Croatia
VIPnet, Zagreb, Croatia
Danieli Automation, Buttrio, Italy
Mjerne tehnologije, Zagreb, Croatia
Selmet, Zagreb, Croatia
Institute SDT, Ljubljana, Slovenia
Nomen, Rijeka, Croatia
EuroCloud, Croatia

donor
Erste&Steiermärkische bank, Rijeka, Croatia

International Program Committee

Petar Biljanovic, General Chair, Croatia
S. Amon, Slovenia
V. Andjelic, Croatia
M.E. Auer, Austria
S. Babic, Croatia
A. Badnjevic, Bosnia and Herzegovina
M. Baranovic, Croatia
B. Bebel, Poland
L. Bellatreche, France
E. Brenner, Austria
G. Brunetti, Italy
A. Budin, Croatia
Z. Butkovic, Croatia
Z. Car, Croatia
M. Colnaric, Slovenia
A. Cuzzocrea, Italy
M. Cicin-Sain, Croatia
M. Cupic, Croatia
M. Delimar, Croatia
T. Eavis, Canada
M. Ferrari, Italy
B. Fetaji, Macedonia
R. Filjar, Croatia
T. Galinac Grbac, Croatia
P. Garza, Italy
L. Gavrilovska, Macedonia
M. Golfarelli, Italy
S. Golubic, Croatia
F. Gregoretti, Italy
S. Gros, Croatia
N. Guid, Slovenia
J. Henno, Estonia
L. Hluchy, Slovakia
V. Hudek, Croatia
Z. Hutinski, Croatia
M. Ivanda, Croatia
H. Jaakkola, Finland
L. Jelenkovic, Croatia
D. Jevtic, Croatia
R. Jones, Switzerland
P. Kacsuk, Hungary

A. Karaivanova, Bulgaria
M. Koricic, Croatia
T. Kosanovic, Croatia
M. Mauher, Croatia
I. Mekjavic, Slovenia
B. Mikac, Croatia
V. Milutinovic, Serbia
N. Miskovic, Croatia
V. Mrvos, Croatia
J.F. Novak, Croatia
J. Pardillo, Spain
N. Pavesic, Slovenia
V. Persic, Croatia
S. Ribaric, Croatia
J. Rozman, Slovenia
K. Skala, Croatia
I. Sluganovic, Croatia
M. Spremic, Croatia
V. Sruk, Croatia
S. Stafisso, Italy
U. Stanic, Slovenia
N. Stojadinovic, Serbia
M. Stupicic, Croatia
J. Sunde, Australia
A. Szabo, IEEE Croatia Section
L. Szirmay-Kalos, Hungary
D. Simunic, Croatia
Z. Simunic, Croatia
D. Skvorc, Croatia
A. Teixeira, Portugal
E. Tijan, Croatia
A.M. Tjoa, Austria
R. Trobec, Slovenia
S. Uran, Croatia
T. Vámos, Hungary
M. Varga, Croatia
M. Vidas-Bubanja, Serbia
M. Vranic, Croatia
B. Vrdoljak, Croatia
D. Zazula, Slovenia

List of paper reviewers

Aksentijevic, S.	(Croatia)	Gamberger, D.	(Croatia)
Antolic, Z.	(Croatia)	Gamulin, O.	(Croatia)
Antonic, A.	(Croatia)	Garza, P.	(Italy)
Asenbrener Katic, M.	(Croatia)	Geric, S.	(Croatia)
Avbelj, V.	(Slovenia)	Giedrimas, V.	(Lithuania)
Babic, D.	(Croatia)	Gilhespy, M.	(Netherlands)
Babic, S.	(Croatia)	Glavas, J.	(Croatia)
Bacmaga, J.	(Croatia)	Golfarelli, M.	(Italy)
Bakalar, G.	(Croatia)	Golub, M.	(Croatia)
Bako, N.	(Croatia)	Golubic, S.	(Croatia)
Bala, P.	(Poland)	Gracin, D.	(Croatia)
Banek, M.	(Croatia)	Gradisnik, V.	(Croatia)
Baotic, M.	(Croatia)	Grbac, N.	(Croatia)
Bebel, B.	(Poland)	Grd, P.	(Croatia)
Begusic, D.	(Croatia)	Grguric, A.	(Croatia)
Bellatreche, L.	(France)	Gros, S.	(Croatia)
Biondic, I.	(Croatia)	Gulic, M.	(Croatia)
Bogdan, S.	(Croatia)	Gumzej, N.	(Croatia)
Brcic, M.	(Croatia)	Henno, J.	(Estonia)
Brezany, P.	(Austria)	Holenko Dlab, M.	(Croatia)
Brkic Bakaric, M.	(Croatia)	Horvat, M.	(Croatia)
Brkic, K.	(Croatia)	Hrkac, T.	(Croatia)
Brkic, L.	(Croatia)	Hudek, V.	(Croatia)
Brodnik, A.	(Slovenia)	Huljenic, D.	(Croatia)
Brscic, D.	(Croatia)	Humski, L.	(Croatia)
Budin, A.	(Croatia)	Hyrynsalmi, S.	(Finland)
Budin, L.	(Croatia)	Inkret, R.	(Croatia)
Butkovic, Z.	(Croatia)	Ipsic, I.	(Croatia)
Candrlic, S.	(Croatia)	Ivasic-Kos, M.	(Croatia)
Capko, Z.	(Croatia)	Ivosevic, D.	(Croatia)
Cavrak, I.	(Croatia)	Jaakkola, H.	(Finland)
Cicin-Sain, M.	(Croatia)	Jakobovic, D.	(Croatia)
Costa, J.	(Portugal)	Jakopovic, Z.	(Croatia)
Cubrilo, M.	(Croatia)	Jaksic, D.	(Croatia)
Cuzzocrea, A.	(Italy)	Jakupovic, A.	(Croatia)
Delac, G.	(Croatia)	Jardas, M.	(Croatia)
Djerek, A.	(Croatia)	Jelenkovic, L.	(Croatia)
Djerek, V.	(Croatia)	Jevtic, D.	(Croatia)
Dobrijevic, O.	(Croatia)	Jezic, G.	(Croatia)
Domitrovic, A.	(Croatia)	Joler, M.	(Croatia)
Dundjer, I.	(Croatia)	Josanov, B.	(Serbia)
Dzanko, M.	(Croatia)	Jovanovic, V.	(United States)
Dzapo, H.	(Croatia)	Jovic, A.	(Croatia)
Eavis, T.	(Canada)	Jurcevic Lulic, T.	(Croatia)
Fertalj, K.	(Croatia)	Jurdana, M.	(Croatia)
Filjar, R.	(Croatia)	Juricic, V.	(Croatia)
Frankovic, D.	(Croatia)	Jurisic, D.	(Croatia)
Frid, N.	(Croatia)	Kalafatic, Z.	(Croatia)
Galinac Grbac, T.	(Croatia)	Kalpic, D.	(Croatia)

Kastelan, I.	(Serbia)	Miletic, V.	(Croatia)
Katanic, N.	(Croatia)	Milic, L.	(Croatia)
Kaucic, B.	(Slovenia)	Milicevic, M.	(Croatia)
Kazi, Z.	(Serbia)	Milicic, S.	(Croatia)
Keto, H.	(Finland)	Min Tjoa, A.	(Austria)
Kljajic Borstnar, M.	(Slovenia)	Miskovic, N.	(Croatia)
Knezevic, K.	(Croatia)	Mohorcic, M.	(Slovenia)
Konecki, M.	(Croatia)	Molnar, G.	(Croatia)
Koricic, M.	(Croatia)	Nacinovic Prskalo, L.	(Croatia)
Kovacevic, T.	(Croatia)	Nikitovic, M.	(Croatia)
Kovacic, B.	(Croatia)	Nikolovski, S.	(Croatia)
Kovacic, Z.	(Croatia)	Ocko, M.	(Croatia)
Kozak, D.	(Croatia)	Oletic, D.	(Croatia)
Kragic, D.	(Sweden)	Oreski, D.	(Croatia)
Krapac, J.	(Croatia)	Orsulic, J.	(Croatia)
Krasna, M.	(Slovenia)	Pale, P.	(Croatia)
Krivec, S.	(Croatia)	Palestri, P.	(Italy)
Krois, I.	(Croatia)	Palomaki, J.	(Finland)
Krpic, Z.	(Croatia)	Pandzic, I.	(Croatia)
Kruzic, S.	(Croatia)	Pavcevic, M.	(Croatia)
Kuhar, U.	(Slovenia)	Pecar-Ilic, J.	(Croatia)
Kurdia, A.S.	(Croatia)	Pejcinovic, B.	(United States)
Kusek, M.	(Croatia)	Pelin, D.	(Croatia)
Kuusisto, M.	(Finland)	Peric Hadzic, A.	(Croatia)
Leppaniemi, J.	(Finland)	Petkovic, T.	(Croatia)
Lerga, J.	(Croatia)	Petrovic, T.	(Croatia)
Lesic, V.	(Croatia)	Pintar, D.	(Croatia)
Lucic, D.	(Croatia)	Poljak, M.	(Croatia)
Lukac, D.	(Germany)	Poscic, P.	(Croatia)
Ljubic, S.	(Croatia)	Prokopec, G.	(Croatia)
Magdalenic, I.	(Croatia)	Pusnik, M.	(Slovenia)
Makinen, T.	(Finland)	Rantanen, P.	(Finland)
Malaric, R.	(Croatia)	Rashkovska, A.	(Slovenia)
Malekovic, M.	(Croatia)	Repnik, R.	(Slovenia)
Males, L.	(Croatia)	Ribaric, S.	(Croatia)
Mandic, T.	(Croatia)	Rolich, T.	(Croatia)
Marcetic, D.	(Croatia)	Rupnik, R.	(Slovenia)
Markovic, I.	(Croatia)	Salamon, K.	(Croatia)
Marovic, M.	(Croatia)	Sarolic, A.	(Croatia)
Marsic, D.	(Croatia)	Schatten, M.	(Croatia)
Matetic, M.	(Croatia)	Seder, M.	(Croatia)
Matic, T.	(Croatia)	Sevrovic, M.	(Croatia)
Matijasevic, M.	(Croatia)	Silic, M.	(Croatia)
Mausa, G.	(Croatia)	Sillberg, P.	(Finland)
Mekovec, R.	(Croatia)	Simunic, D.	(Croatia)
Mekterovic, I.	(Croatia)	Skala, K.	(Croatia)
Mestrovic, A.	(Croatia)	Skocir, P.	(Slovenia)
Mihajlovic, Z.	(Croatia)	Skvorc, D.	(Croatia)
Mikac, B.	(Croatia)	Slivar, I.	(Croatia)
Mikuc, M.	(Croatia)	Sluganovic, I.	(Croatia)
Milanovic, I.	(Serbia)	Sluganovic, I.	(Croatia)
Milasinovic, B.	(Croatia)	Smuc, T.	(Croatia)

Soler, J.	(Denmark)	Trobec, R.	(Slovenia)
Solic, K.	(Croatia)	Trost, A.	(Slovenia)
Sprager, S.	(Slovenia)	Trzec, K.	(Croatia)
Spremic, M.	(Croatia)	Uroda, I.	(Croatia)
Stajduhar, I.	(Croatia)	Vasic, D.	(Croatia)
Stanic, U.	(Slovenia)	Vidacek-Hains, V.	(Croatia)
Staresinic, D.	(Croatia)	Vinko, D.	(Croatia)
Stipcevic, M.	(Croatia)	Vladimir, K.	(Croatia)
Stojkovic, N.	(Croatia)	Vlahovic, N.	(Croatia)
Struc, V.	(Slovenia)	Vojkovic, G.	(Croatia)
Subasic, M.	(Croatia)	Vranic, M.	(Croatia)
Sumak, B.	(Slovenia)	Vrdoljak, B.	(Croatia)
Sunde, V.	(Croatia)	Vrhovec, S.	(Slovenia)
Supic, H.	(Bosnia and Herzegovina)	Vukadinovic, D.	(Croatia)
Susac, F.	(Croatia)	Vukovic, M.	(Croatia)
Suznjevic, M.	(Croatia)	Vukovic, M.	(Croatia)
Svelec, D.	(Croatia)	Wrembel, R.	(Poland)
Tijan, E.	(Croatia)	Yrjonkoski, K.	(Finland)
Tomic, M.	(Croatia)	Zagar, D.	(Croatia)
Topic, M.	(Slovenia)	Zilak, J.	(Croatia)
Toth, Z.	(Hungary)	Zonja, S.	(Croatia)
Trancoso, I.	(Portugal)	Zulim, I.	(Croatia)

2017 40th International Convention on Information and Communication Technology, Electronics and Microelectronics (MIPRO)

Microelectronics, Electronics and Electronic Technology

Active Learning, Labs and Maker-spaces in Microwave Circuit Design Courses — 1
B. Pejcinovic

A Learning Tool for Synthesis, Visualization, and Editing of Programming for Simple Programmable Logic Devices — 7
M. Cupic, K. Brkic, Z. Mihajlovic

Active-Learning Implementation Proposal for Course Electronics at Undergraduate Level — 13
T. Mandic, A. Baric

Decision Trees in Formative Procedural Knowledge Assessment — 17
J. Petrovic, P. Pale

Glass Based Structures Fabricated by Rf-Sputtering — 21
A. Chiasera, F. Scotognella, D. Dorosz, G. Galzerano, A. Lukowiak, D. Ristic,
G. Speranza, I. Vasilchenko, A. Vaccari, S. Valligatla, S. Varas, L. Zur, M. Ivanda,
A. Martucci, G.C. Righini, S. Taccheo, R. Ramponi, M. Ferrari

Piezoresistive Effect in Composite Films Based on Polybenzimidazole and Few-Layered Graphene — 27
V.A. Kuznetsov, B.C. Kholkhoev, A.Y. Stefanyuk, V.G. Makotchenko,
A.S. Berdinsky, A.I. Romanenko, V.F. Burdukovskii, V.E. Fedorov

Local Growth of Graphene on Cu and $Cu_{0.88}Ni_{0.12}$ Foil Substrates — 31
H.S. Funk, J. Ng, N. Kamimura, Y.-H. Xie, J. Schulze

Growth of Patterned GeSn and GePb Alloys by Pulsed Laser Induced Epitaxy — 37
J. Schlipf, J.L. Frieiro, I.A. Fischer, C. Serra, J. Schulze, S. Chiussi

Impact of Sn Segregation on $Ge_{1-x}Sn_x$ epi-Layers Growth by RP-CVD — 43
D. Weisshaupt, P. Jahandar, G. Colston, P. Allred, J. Schulze, M. Myronov

Tungsten Dichalcogenides as Possible Gas-Sensing Elements — 48
V.A. Kuznetsov, A.Y. Ledneva, S.B. Artemkina, M.N. Kozlova, G.E. Yakovleva,
A.S. Berdinsky, A.I. Romanenko, V.E. Fedorov

Flicker Noise in AlGaAs/GaAs High Electron Mobility Heterostructure Field-Effect Transistor at Cryogenic Temperature   53
S. Mouetsi, F. Zouach, D. Rechem

Device Performance Tuning of Ge Gate-All-Around Tunneling Field Effect Transistors by Means of Gesn: Potential and Challenges   57
E.G. Rolseth, A. Blech, I.A. Fischer, Y. Hashad, R. Koerner, K. Kostecki, A. Kruglov, V.S. Senthil Srinivasan, M. Weiser, T. Wendav, K. Busch, J. Schulze

Band-Structure of Ultra-Thin InGaAs Channels: Impact of Biaxial Strain and Thickness Scaling   66
S. Krivec, M. Poljak, T. Suligoj

Perimeter Effects from Interfaces in Ultra-Thin Layers Deposited on Nanometer-Deep p^+n Silicon Junctions   72
T. Knezevic, L.K. Nanver, T. Suligoj

Analysis of Hot Carrier-Induced Degradation of Horizontal Current Bipolar Transistor   77
J. Zilak, M. Koricic, T. Suligoj

Impact of the Local p-well Substrate Parameters on the Electrical Performance of the Double-Emitter Reduced-Surface-Field Horizontal Current Bipolar Transistor   83
M. Koricic, J. Zilak, T. Suligoj

Characterization of Measurement System for High-Precision Oscillator Measurements   88
I. Brezovec, M. Magerl, J. Mikulic, G. Schatzberger, A. Baric

Temperature Calibration of an On-Chip Relaxation Oscillator   93
J. Mikulic, I. Brezovec, M. Magerl, G. Schatzberger, A. Baric

Model of High-Efficiency High-Current Coupled Inductor Two-Phase Buck Converter   98
V.C. Valchev, O.P. Stanchev, G.T. Nikolov

Analog to Digital Signal Converters for BiCMOS Quaternary Digital Systems   103
D. Bundalo, Z. Bundalo, D. Pasalic, B. Cvijic

Ultra-Wideband Pulse Generator for Time-Encoding Wireless Transmission   109
L. Sneler, M. Herceg, T. Matic

Spectral-Efficient UWB Pulse Shapers Generating Gaussian and Modified Hermitian Monocycles   113
A. Milos, G. Molnar, M. Vucic

Design of Multiplierless CIC Compensators Based on Maximum Passband Deviation — 119
G. Molnar, A. Dudarin, M. Vucic

Adaptive State Observer Development Using Recursive Extended Least-Squares Method — 125
N.N. Nikolov, M.I. Alexandrova, V.C. Valchev, O.P. Stanchev

Inverter Current Source for Pulse-Arc Welding with Improved Parameters — 130
V.C. Valchev, D.D. Mareva, D.D. Yudov, R.S. Stoyanov

Power Output Comparison of Three Phase Passive Converter Circuits for Wind Driven Generators — 136
V.C. Valchev, P.V. Yankov, A. Van den Bossche

Dynamic Range Optimization and Noise Reduction by Low-Sensitivity, Fourth-Order, Band-Pass Filters Using Coupled General-Purpose Biquads — 141
E. Emanovic, D. Jurisic

A Circular Economy for Photovoltaic Waste - the Vision of the European Project CABRISS — 146
W. Brenner, N. Adamovic

Smart Farm Computing Systems for Animal Welfare Monitoring — 152
M. Caria, J. Schudrowitz, A. Jukan, N. Kemper

Power Management Circuit for Energy Harvesting Applications with Zero-Power Charging Phase — 158
D. Vinko

System for Early Condensation Detection and Prevention in Residential Buildings — 162
M. Marcelic, R. Malaric

FEM Analysis and Design of a Voltage Instrument Transformer for Digital Sampling Wattmeter — 166
M. Dadic, K. Petrovic, R. Malaric

Random Number Generation with LFSR Based Stream Cipher Algorithms — 171
T. Tuncer, E. Avaroğlu

RAM-Based Mergers for Data Sort and Frequent Item Computation — 176
A. Rjabov, V. Sklyarov, I. Skliarova, A. Sudnitson

Distributed Computing, Visualization and Biomedical Engineering

New Classes of Kochen-Specker Contextual Sets – *Invited Paper*   182
N.D. Megill, M. Pavicic

New Method for Determination Complexity Using in AD HOC Cloud Computing   188
M. Babic, B. Jerman-Blazic

Neneta: Heterogeneous Computing Complex-Valued Neural Network Framework   192
V. Lekic, Z. Babic

Cloud-Distributed Computational Experiments for Combinatorial Optimization   197
M. Brcic, N. Hlupic

Design of Digital IIR Filter Using Particle Swarm Optimization   202
F. Serbet, T. Kaya, M.T. Ozdemir

Big Data Analytics in Electricity Distribution Systems   205
S. Stoyanov, N. Kakanakov

Running HPC Applications on Many Million Cores Cloud   209
D. Tomic, Z. Car, D. Ogrizovic

DBaaS Comparison   N/A
I. Astrova, A. Koschel, C. Eickemeyer, J. Kersten, N. Offel

Modeling Heterogeneous Computational Cluster Hardware in Context of Parallel Database Processing   221
K.Y. Besedin, P.S. Kostenetskiy

Properties of Mathematical Number Model Provided Exact Computing   225
V. Golodov

Simulation of the Parallel Database Column Coprocessor   229
P.S. Kostenetskiy

Towards Flexible Open Data Management Solutions   233
B. von St. Vieth, J. Rybicki, M. Brzeźniak

Spatial Analysis of the Clustering Process   238
M. Kranjac, U. Sikimic, J. Salom, S. Tomic

Usage of Android Device in Interaction with 3D Virtual Objects   244
I. Prazina, V. Okanovic, K. Balic, S. Rizvic

Generating Virtual Guitar Strings Using Scripts   247
L. Kunic, Z. Mihajlovic

Remote Interactive Visualization for Particle-Based Simulations on Graphics Clusters   253
A. Sabou, D. Gorgan

Collaborative View-Aligned Annotations in Web-Based 3D Medical Data Visualization   259
P. Lavric, C. Bohak, M. Marolt

Feasibility of Biometric Authentication Using Wearable ECG Body Sensor Based On Higher-Order Statistics   264
S. Sprager, R. Trobec, M.B. Juric

Bio-Medical Analysis Framework   270
M. Mohorcic, M. Depolli

Finding a Signature in Dermoscopy: A Color Normalization Proposal   276
M. Machado, J. Pereira, M. Silva, R. Fonseca-Pinto

A Textured Scale-Based Approach to Melanocytic Skin Lesions in Dermoscopy   279
R. Fonseca-Pinto, M. Machado

Remarks on Visualization of Fuzziness of Cardiac Data   283
J. Opiła, T. Pełech-Pilichowski

Abdominal Fetal ECG Measured With Differential ECG Sensor   289
A. Rashkovska, V. Avbelj

Synchronization of Time in Wireles ECG Measurement   292
A. Vilhar, M. Depolli

Long-Term Follow-Up Case Study of Atrial Fibrillation after Treatment   297
M. Jan, R. Trobec

A Case Report of Long-Term Wireless Electrocardiographic Monitoring in a Dog with Dilated Cardiomyopathy   303
M. Brloznik, V. Avbelj

SaaS Solution for ECG Monitoring Expert System   308
A. Ristovski, M. Gusev

Wavelet-Based Analysis Method for Heart Rate Detection of ECG Signal Using LabVIEW   314
D. Kaya, M. Türk, T. Kaya

Parallelization of Digital Wavelet Transformation of ECG Signals   318
E. Domazet, M. Gusev

**Hilbert Transform Based Paroxysmal Tachycardia Detection
Algorithm**   324
I. Culjak, M. Cifrek

**Biomedical Time Series Preprocessing and Expert-System Based Feature
Extraction in MULTISAB Platform**   330
A. Jovic, D. Kukolja, K. Friganovic, K. Jozic, S. Car

**Evaluation of Chronic Venous Insufficiency with PPG Prototype
Instrument**   336
M. Makovec, U. Aljancic, D. Vrtacnik

**Highly Parallel Online Bioelectrical Signal Processing on GPU
Architecture**   340
Z. Juhasz

Dew Computing

The Dawn of Dew: Dew Computing for Advanced Living Environment   347
Z. Sojat, K. Skala

**Service-Oriented Application for Parallel Solving the Parametric Synthesis
Feedback Problem of Controlled Dynamic Systems**   353
G.A. Oparin, V.G. Bogdanova, S.A. Gorsky, A.A. Pashinin

**Augmented Coaching Ecosystem for Non-obtrusive Adaptive Personalized
Elderly Care on the Basis of Cloud-Fog-Dew Computing Paradigm**   359
Y. Gordienko, S. Stirenko, O. Alienin, K. Skala, Z. Sojat, A. Rojbi,
J.R. López Benito, E. Artetxe González, U. Lushchyk, L. Sajn,
A. Llorente Coto, G. Jervan

**Cloud-Dew Computing Support for Automatic Data Analysis in Life
Sciences**   365
P. Brezany, T. Ludescher, T. Feilhauer

**Distributed Database System as a Base for Multilanguage Support for
Legacy Software**   371
N. Crnko

Sag/Tension Dynamical Line Rating System Architecture   N/A
A. Pavlinic, V. Komen

**Toward a Framework for Embedded & Collaborative Data Analysis with
Heterogeneous Devices**   381
M. Goeminne, M. Boukhebouze

A Dew Computing Solution for IoT Streaming Devices
M. Gusev
387

3D-based Location Positioning Using the Dew Computing Approach for Indoor Navigation
D. Podbojec, B. Herynek, D. Jazbec, M. Cvetko, M. Debevc, I. Kozuh
393

Architecting a Hybrid Cross Layer Dew-Fog-Cloud Stack for Future Data-Driven Cyber-Physical Systems
M. Frincu
399

Telecommunications & Information

Future Applications of Optical Wireless and Combination Scenarios with RF Technology – *Invited Paper*
E. Leitgeb
404

The Golden Ratio in the Age of Communication and Information Technology
Y.I. Doychinov, I.S. Stoyanov, T.B. Iliev
407

Wireless Machine-to-Machine Communication for Intelligent Transportation Systems: Internet of Vehicles and Vehicle to Grid
N.R. Moloisane, R. Malekian, D. Capeska Bogatinoska
411

Power Control Schemes for Device-to-Device Communications in 5G Mobile Network
T.B. Iliev, G.Y. Mihaylov, E.P. Ivanova, I.S. Stoyanov
416

Brain Computer Interface Communicator : A Response to Auditory Stimuli Experiment
G. Madhale Jadav, L. Batistic, S. Vlahinic, M. Vrankic
420

Determination of Origins and Destinations for an O-D Matrix Based on Telecommunication Activity Records
S. Desic, M. Filic, R. Filjar
424

What Factors Influence the Quality of Experience for WebRTC Video Calls?
J. Barakovic Husic, S. Barakovic, A. Veispahic
428

Is There Any Impact of Human Influence Factors on Quality of Experience?
J. Barakovic Husic, S. Barakovic, S. Muminovic
434

Extended AODV Routing Protocol Based on Route Quality Approximation via Bijective Link-Quality Aggregation
V.L. Prcic, D. Kalpic, D. Simunic
N/A

Development Trends of Telecommunications Metrics — 445
N. Banovic-Curguz, D. Ilisevic

Implementation and Testing of Cisco IP SLA in Smart Grid Environments — 450
J. Horalek, F. Holik, V. Hurtova

Fault Management and Management Information Base (MIB) — 455
O. Jukic, I. Hedji, I. Speh

Distributed Threat Removal in Software-Defined Networks — 460
D. Samociuk, A. Chydzinski

Performance Analysis of Virtualized VPN Endpoints — 466
D. Lackovic, M. Tomic

A Big Data Solution for Troubleshooting Mobile Network Performance Problems — 472
K. Skracic, I. Bodrusic

Overlapping Blocks in Reconstruction of Sparse Images — 478
I. Stankovic, M. Dakovic, I. Orovic

Sparse Signal Reconstruction Based on Random Search Procedure — 482
M. Dakovic, I. Stankovic, M. Brajovic, L. Stankovic

A Fast Noise Level Estimation Algorithm Based on Adaptive Image Segmentation and Laplacian Convolution — 486
E. Turajlic

Advanced Regulation Approach: Dynamic Rules for Capturing the Full Potential of Future ICT Networks — 492
D. Ilisevic, N. Banovic-Curguz

LTE eNB Traffic Analysis and Key Techniques towards 5G Mobile Networks — 497
T.B. Iliev, G.Y. Mihaylov, T.D. Bikov, E.P. Ivanova, I.S. Stoyanov, D.I. Radev

IoT Network Protocols Comparison for the Purpose of IoT Constrained Networks — 501
I. Hedji, I. Speh, A. Sarabok

Digital Forensic Analysis through Firewall for Detection of Information Crimes in Hospital Networks — 506
A. Akbal, E. Akbal

Topology Analysis for Energy-Aware Node Placement in Wireless Sensor Networks of Home Sensing Environments — N/A
A. Koren, D. Simunic

Mine Safety System Using Wireless Sensor Networks — 515
V. Henriques, R. Malekian, D. Capeska Bogatinoska

Replication of Virtual Network Functions: Optimizing Link Utilization and Resource Costs
F. Carpio, W. Bziuk, A. Jukan
521

Impact of Human Resources Changes on Performance and Productivity of Scrum Teams
D. Alic, A. Djedovic, S. Omanovic, A. Tanovic
527

Synergy of ITIL Methodology and Help Users Systems
I. Ivosic
533

Design of Optical Fiber Gyroscope System in Program Environment OptSim
M. Márton, Ľ. Ovseník, J. Turán, M. Spes
N/A

Value Based Service Design Elements in Business Ecosystem Architecture
D. Ramljak
N/A

Simulation Study of M-ARY QAM Modulation Techniques Using Matlab/Simulink
S.M. Sadinov
547

Modeling System Behaviour

Multiscale and Multiobjective Modelling: a Perspective for Mastering the Design and Operation Complexity of IoT Systems – *Invited Paper*
K. Drira
555

Topological Data Analysis and Applications
J. Pita Costa
558

Modelling of Pedestrian Groups and Application to Group Recognition
D. Brscic, F. Zanlungo, T. Kanda
564

eCST to Source Code Generation - an Idea and Perspectives
N. Rakic, G. Rakic, N. Sukur, Z. Budimac
570

Design and Development of Contactless Interaction with Computers Based on the Emotiv EPOC+ Device
B. Sumak, M. Spindler, M. Pusnik
576

Patterns for Improving Mobile User Experience
M. Pusnik, D. Ivanovski, B. Sumak
582

Drawing Process Recording Tool for Eye-Hand Coordination Modelling
V. Giedrimas, L. Vaitkevicius, A. Vaitkeviciene
588

Model of Calculating Indicators of Power System Reliability
A. Balota .... N/A

Modelling of Variable Shunt Reactor in Transmission Power System for Simulation of Switching Transients
A. Zupan, B. Filipovic-Grcic, I. Uglesic 598

An Extended Model of a Level and Flow Control System
H. Siljak, J. Hivziefendic, J. Kevric 603

Procedure for Modelling of Soft Tissues Behavior
M. Franulovic, K. Markovic, S. Pilicic 608

Analysis of ERASMUS Staff and Student Mobility Network within a Big European Project
M. Savic, M. Ivanovic, Z. Putnik, K. Tütüncü, Z. Budimac, S. Smrikarova, A. Smrikarov 613

Computers in Education

Modern Education and Its Background in Cognitive Psychology: Automated Question Creation and Eye Movements
M. Höfler, G. Wesiak, P. Pürcher, C. Gütl 619

Learning to Program – Does it Matter Where you Sit in the Lecture Theatre?
A. McGowan, P. Hanna, D. Greer, J. Busch, N. Anderson 624

Knowledge and Skills: A Critical View
M. Radovan 630

Today is the Future of Yesterday; What is the Future of Today?
H. Jaakkola, J. Henno, J. Mäkelä, B. Thalheim 635

Interdisciplinary Utilization of IT
S. Neradová, S. Zitta 644

Maximizing Quality Class Time Using Computers for a Flipped Classroom Approach
C.P. Fulford, S. Paek 649

Quantitative Structured Literature Review of Research on e-Learning
L. Abazi-Bexheti, A. Kadriu, M. Apostolova 655

The Educators' Telescope to the Future of Technology
H. Jaakkola, J. Henno, B. Thalheim, J. Mäkelä 660

ICT Support for Promotion of Nature Park 666
A. Zisko, A. Sorgo, M. Krasna

Experiences in Using Educational Recommender System ELARS to Support e-Learning 672
M. Holenko Dlab

Influence of Accuracy of Simulations to the Physics Education 678
R. Repnik, G. Nemec, M. Krasna

Competence-Oriented Model of Representation of Educational Content 685
O.N. Ivanova, N.S. Silkina

Using CODESYS as a Tool for Programming and Simulation in Applied Education at University .... N/A
D. Lukac

Developing Curiosity and Multimedia Skills with Programming Experiments 694
J. Henno, H. Jaakkola, J. Mäkelä

Informetrics: the Development, Conditions and Perspectives 700
A. Papic

Structuring e-Learning Multi-Criteria Decision Making Problems 705
N. Kadoic, N. Begicevic Redjep, B. Divjak

Introducing Gamification into e-Learning University Courses 711
A. Bernik, D. Radosevic, G. Bubas

Perceived Security and Privacy of Cloud Computing Applications Used in Educational Ecosystem 717
T. Orehovacki, D. Etinger, S. Babic

Estimating Profile of Successful IT Student: Data Mining Approach 723
D. Oreski, M. Konecki, L. Milic

Comparison of Game Engines for Serious Games 728
S. Pavkov, I. Frankovic, N. Hoic-Bozic

Case Study of Online Resources and Searching for Information on Students' Academic Needs 734
A. Cizmesija, V. Vidacek-Hains

Recursions and How to Teach Them 740
P. Brodjanac

Learning Management System (LMS) Software Comparison: Edmodo vs Schoology .... N/A
M. Filipovic Tretinjak, M. Tretinjak

Machine Learning Techniques in the Education Process of Students of Economics 750
J. Bucko, L. Kakalejcík

Moodle-Based Data Mining Potentials of MOOC Systems at the University of Szeged 755
G. Kőrösi, F. Havasi

View on Development of Information Competencies and Computer Literacy of Slovak Secondary School Graduates 761
L. Révészová

Digital Learning as a Tool to Overcome School Failure in Minority Groups 767
D. Paľová, N.M. Novak, V. Weidinger

C Based Laboratory for Teaching Digital Signal Processing to Computer Engineering Undergraduates 773
D. Bokan, M. Temerinac, Z. Lukac, S. Ocovaj

Analysis of Video Views in Online Courses 778
P. Esztelecki, F. Havasi

Education in the Field of Electronic Financial Services of Its Future Users 783
M. Vejacka

Didactic Concepts of Modern Data Analysis 789
C. Ungermanns, W. Werth

Extending the Object-Oriented Notional Machine Notation with Inheritance, Polymorphism, and GUI Events 794
M. Aglic Cuvic, J. Maras, S. Mladenovic

Mediated Transfer from Visual to High-Level Programming Language 800
D. Krpan, S. Mladenovic, G. Zaharija

Story of a 'Storyline Visualization' in High School Readings 806
K. Osmakcic, K. Kocijan

Impact of ICT on Archival Practice from the 2000s Onwards and the Necessary Changes of Archival Science Curricula 812
H. Stancic, A. Rajh, M. Jamic

Publishing of Personal Information on Facebook with Regard to Gender: Comparison of Pupils and University Students 818
P. Dzapo, M. Duic

Web Sources of Literature for Teachers and Researchers: Practices and Attitudes of Croatian Faculty toward Legal Digital Libraries and Shadow Libraries Such as Sci-Hub 824
M. Duic, B. Konjevod, L. Grzunov

Supporting Mobile Learning: Usability of Digital Collections in Croatia for Use on Mobile Devices — 830
R. Vrana, D. Gascic, M. Podkonjak

The Use of ICT in the English Language Classroom — 836
R. Lerga, S. Candrlic, M. Holenko Dlab

LIS Students and Plagiarism in the Networked Environment — 842
I. Hebrang Grgic

A Platform for Supporting Learning Process of Visually Impaired Children — 848
A. Kavcic, M. Pesek, M. Marolt

Web Application for Time Telling — N/A
A. Cobic, M. Carapina

Integration of the Future Technologies to High Schools and Colleges — 858
P. Kurent

Preschool Children and Computers: Who Lives in a Meadow? — 862
B. Strnad

Web Service Model for Distance Learning Using Cloud Computing Technologies — 865
D. Cvetkovic, M. Mijatovic, M. Mijatovic, B. Medic

The Struggle with Academic Plagiarism: Approaches Based on Semantic Similarity — 870
T. Vrbanec, A. Mestrovic

Personal Learning Environment as Support to Education — 876
B. Grba, B. Kovacic

Free and Open Source Software in the Secondary Education in Bosnia and Herzegovina — 882
M. Pezer, N. Lazic, M. Odak

E-Learning of Mathematics, Problem and a Possible Solution — N/A
V. Dedic, R. Cvejic, M. Andjelkovic

Research Methodology in the 21st Century — 891
D. Cvetkovic, B. Medic

A Survey and Evaluation of Free and Open Source Simulators Suitable for Teaching Courses in Wireless Sensor Networks — 895
D. Pesic, Z. Radivojevic, M. Cvetanovic

Excessive Internet Use among Elementary School Students in Vojvodina — 901
S. Tapiska, M. Kresoja, Z. Putnik, M. Ivanovic

E-Learning on Polytechnic Nikola Tesla – Analysis and Comparison
B. Kovacic, A. Skendzic, K. Devcic

905

Determination of Time Criteria for Assessment in Learning Management Systems
T. Alajbeg, M. Sokele, V. Simovic

910

Using Moodle in English for Professional Purposes (EPP) Teaching at the University North
J. Lasic-Lazic, T. Ivanjko, I. Grubjesic

915

Air Traffic Controllers' Practical Part of Basic Training on Computer Based Simulation Device
M. Pavlinovic, B. Juricic, B. Antulov-Fantulin

920

The Perspective of Use of Digital Libraries in Era of e-Learning
R. Vrana

926

ArTeFact – Digitization of Archives of Technical Faculty from the Period 1919 – 1956
M. Tucakovic, J. Lisek

N/A

Introductory Physics Course for ICT Students: Computer-Programming Oriented Approach
V. Krstic

937

Advocacy of Born-Digital Materials: an ESP Course Example
D. Pesut

N/A

Theoretical and Practical Challenges of Using Three Ammeter or Tree Voltmeter Methods in Teaching
V. Simovic, T. Alajbeg, J. Curkovic

944

National Competition of Photography as Visual Art in Croatian Primary Schools, High Schools and Schools of Applied Arts
Z. Prohaska, Z. Prohaska, I. Uroda

949

Generating Large Random Test Data Table for SQL Training
U. Sterle

954

Programming Lego Mindstorms for First Lego League Robot Game and Technical Interview
B. Strnad

958

Children Online Safety
J. Zufic, T. Zajgar, S. Prkic

961

Computers in Technical Systems

Metals Industry: Road to Digitalization – *Invited Paper*
A. Merluzzi, G. Brunetti
967

IoT Gateway for Smart Metering in Electrical Power Systems - Software Architecture
M. Shopov
974

Application Models for Ubiquitous Systems with Sporadic Communication Availability
I. Cavrak, I. Zagar, A. Drazic
979

Towards the Utilization of Crowdsourcing in Traffic Condition Reporting
P. Rantanen, P. Sillberg, J. Soini
985

Survey of Prototyping Solutions Utilizing Raspberry Pi
M. Saari, A. Muzaffar bin Baharudin, S. Hyrynsalmi
991

Evaluating Robustness of Perceptual Image Hashing Algorithms
A. Drmic, M. Silic, G. Delac, V. Klemo, A.S. Kurdija
995

Adaptive Models for Security and Data Protection in IoT with Cloud Technologies
N. Kakanakov, M. Shopov
1001

Making a Smart City Even More Intelligent Using Emergent Property Methodology
A. Saric, B. Mihaljevic, K. Marasovic
1005

Digital Chess Board Based on Array of Hall-Effect Sensors
F. Susac, I. Aleksi, Z. Hocenski
1011

Influence of Human-Computer Interface Elements on Performance of Teleoperated Mobile Robot
S. Kruzic, J. Music, I. Stancic
1015

The Challenge of Measuring Distance to Obstacles for the Purpose of Generating a 2-D Indoor Map Using an Autonomous Robot Equipped with an Ultrasonic Sensor
R. Rodin, I. Stajduhar
1021

Automated Posture Assessment for Construction Workers
R.J. Dzeng, H.H. Hsueh, C.W. Ho
1027

Software Toolbox for Analysis and Design of Nonlinear Control Systems and Its Application to Multi-AUV Path-Following Control
S. Ul'yanov, N. Maksimkin
1032

Remote Alarm Reporting System Responsive to Stoppage of Ballast Water Management Operation on Ships
G. Bakalar, M.B. Baggini, S.G. Bakalar
1038

Parameters for Condition Assessment of the High Voltage Circuit Breakers Arcing Contacts Using Dynamic Resistance Measurement
K. Obarcanin, R. Ostojic, S. Dzuzdanovic
1044

Measurement Noise Propagation in Distribution-System State Estimation
U. Kuhar, G. Kosec, A. Svigelj
1049

Application of Integrated Fail-Safe Technology for Safe and Reliable Natural Gas Distribution
J. Stanic, D. Marsic, G. Malcic
N/A

Brief Review of Self-Organizing Maps
D. Miljkovic
1061

Fault Detection for Aircraft Piston Engine Using Self-Organizing Map
D. Miljkovic
1067

Intelligent Systems

PicoAgri. Realization of a Low-Cost, Remote Sensing Environment for Monitoring Agricultural Fields through Small Satellites and Drones
S. Marsi, A. Gregorio, M. Maris, M. Puligheddu
1073

Architecture of an Information System for Biological Sampling of Reservoir Microhabitats
B. Lukovac, D. Simunic, N. Vuckovic
N/A

Feeding a DNN for Face Verification in Video Data Acquired by a Visually Impaired User
J. Bhattacharya, S. Marsi, S. Carrato, H. Frey, G. Ramponi
1084

Acquiring ISAR Images Using Measurement Instruments
H.-C. Lee, S.G. Lee, S.H. Lee, C.H. Jung
1090

Segmentation of Kidneys and Abdominal Images in Mobile Devices with the Android Operating System by Using the Connected Component Labeling Method
S. Arslan Tuncer, A. Alkan
1094

An Overview of Action Recognition in Videos
M. Buric, M. Pobar, M. Ivasic Kos
1098

Pothole Detection: An Efficient Vision Based Method Using RGB Color Space Image Segmentation
A. Akagic, E. Buza, S. Omanovic
1104

Classification of the Hand-Printed and Printed Medieval Glagolitic Documents Using Differentiation in Orthography — 1110
D. Brodic, A. Amelio

An Evolutionary Approach to Route the Heterogeneous Groups of Underwater Robots — 1116
M.Y. Kenzin, I.V. Bychkov, N.N. Maksimkin

Low Cost Robot Arm with Visual Guided Positioning — 1120
P. Djurovic, R. Grbic, R. Cupec, D. Filko

Symbolic Tensor Differentiation for Applications in Machine Learning — N/A
A. Zhabinski, S. Zhabinskii, D. Adzinets

Reasoning with Air Pollution Data in SWI-Prolog — N/A
Z. Kazi, L. Kazi, I. Berkovic, B. Radulovic

Knowledge Elicitation in Multi-Agent System for Distributed Computing Management — 1138
A. Feoktistov, A. Tchernykh, S. Gorsky, R. Kostromin

Improving a Distributed Agent-Based Ant Colony Optimization for Solving Traveling Salesman Problem — 1144
A. Kaplar, M. Vidakovic, N. Luburic, M. Ivanovic

Towards an Agent-Based Automated Testing Environment for Massively Multi-Player Role Playing Games — 1149
M. Schatten, I. Tomicic, B. Okresa Djuric, N. Ivkovic

On the Properties of Discrete-Event Systems with Observable States — 1155
N. Nagul

The Formal Description of Discrete-Event Systems Using Positively Constructed Formulas — 1161
A. Davydov, A. Larionov, N. Nagul

Runtime Estimation for Enumerating All Mutually Orthogonal Diagonal Latin Squares of Order 10 — 1166
S. Kochemazov, O. Zaikin, A. Semenov

Improving the Effectiveness of SAT Approach in Application to Analysis of Several Discrete Models of Collective Behavior — 1172
S. Kochemazov, O. Zaikin, A. Semenov

Effects of the Distribution of the Values of Condition Attribute on the Quality of Decision Rules — 1178
V. Ognjenovic, E. Brtka, V. Brtka, I. Berkovic

Complexity Comparison of Integer Programming and Genetic Algorithms for Resource Constrained Scheduling Problems — 1182
R. Coric, M. Djumic, D. Jakobovic

Balancing Academic Curricula by Using a Mutation-Only Genetic Algorithm 1189
K. Sylejmani, A. Halili, A. Rexhepi

A Hybrid Method for Prediction of Protein Secondary Structure Based On Multiple Artificial Neural Networks 1195
H. Hasic, E. Buza, A. Akagic

Hardware-in-the-Loop Architecture with MATLAB/Simulink and QuaRC for Rapid Prototyping of CMAC Neural Network Controller for Ball-and-Beam Plant 1201
V. Shatri, L. Kurtaj, I. Limani

Primary and Secondary Experience in Developing Adaptive Information Systems Supporting Knowledge Transfer 1207
S. Lugovic, I. Dundjer, M. Horvat

The Captology of Intelligent Systems 1211
Z. Balaz, D. Predavec

Automatic Information Behaviour Recognition 1217
S. Lugovic, I. Dundjer

Adjective Representation with the Method Nodes of Knowledge 1221
M. Pavlic, Z. Dovedan Han, A. Jakupovic, M. Asenbrener Katic, S. Candrlic

Information Systems Security

Identification of Image Source Using Serial-Number-Based Watermarking under Compressive Sensing Conditions 1227
A. Draganic, M. Maric, I. Orovic, S. Stankovic

Anti-Computer Forensics 1233
K. Hausknecht, S. Gruicic

Analysis of Mobile Phones in Digital Forensics 1241
S. Dogan, E. Akbal

Security Analysis of Open Home Automation Bus System 1245
M. Ramljak

Analysis of Credit Card Attacks Using the NFC Technology 1251
J. Jumic, M. Vukovic

Vulnerabilities of Modern Web Applications 1256
F. Holik, S. Neradová

An Experiment in Using IMUNES and Conpot to Emulate Honeypot Control Networks 1262
S. Kuman, S. Gros, M. Mikuc

A Wireless Propagation Analysis for the Frequency of the Pseudonym Changes to Support Privacy in VANETs 1269
E. Cano Pons, G. Baldini, D. Geneiatakis

Resilience of Students' Passwords against Attacks 1275
B. Brumen, T. Makari

Empirical Study on the Risky Behavior and Security Awareness among Secondary School Pupils - Validation and Preliminary Results 1280
T. Velki, K. Solic, V. Gorjanac, K. Nenadic

Interoperability and Lightweight Security for Simple IoT Devices 1285
D. Androcec, B. Tomas, T. Kisasondi

Security and Privacy Issues for an IoT Based Smart Home 1292
D. Geneiatakis, I. Kounelis, R. Neisse, I. Nai-Fovino, G. Steri, G. Baldini

Towards Overall Information Security and Privacy (IS&P) Taxonomy 1298
K. Solic, H. Ocevcic, I. Fosic, I. Horvat, M. Vukovic, T. Ramljak

Trends in IoT Security 1302
M. Radovan, B. Golub

International Cyber Security Challenges 1309
I. Duic, V. Cvrtila, T. Ivanjko

Cloud Computing Threats Classification Model Based on the Detection Feasibility of Machine Learning Algorithms 1314
Z. Masetic, K. Hajdarevic, N. Dogru

Legal Framework Issues Managing Confidential Business Information in the Republic of Croatia 1319
G. Vojkovic, M. Milenkovic

Combinatorial Optimization in Cryptography 1324
K. Knezevic

Taxonomy of DDos Attacks N/A
I. Kramaric

Business Intelligence Systems

Multidimensional Mining of Big Social Data for Supporting Advanced Big Data Analytics – *Invited Paper* 1337
A. Cuzzocrea

Accelerating Dynamic Itemset Counting on Intel Many-Core Systems 1343
M. Zymbler

ETLator – a Scripting ETL Framework 1349
M. Radonic, I. Mekterovic

The Role of Alignment for the Impact of Business Intelligence Maturity on Business Process Performance in Croatian and Slovenian Companies 1355
V. Bosilj Vuksic, M. Pejic Bach, T. Grubljesic, J. Jaklic, A.M. Stjepic

MapReduce Research on Warehousing of Big Data 1361
M. Pticek, B. Vrdoljak

Selection of Variables for Credit Risk Data Mining Models: Preliminary research 1367
M. Pejic Bach, J. Zoroja, B. Jakovic, N. Sarlija

Brand Communication in Social Media: the Use of Image Colours in Popular Posts 1373
L. Zailskaitė-Jakstė, A. Ostreika, A. Jakstas, E. Stanevicienė, R. Damasevicius

Recommender System Based on the Analysis of Publicly Available Data 1379
G. Antolic, L. Brkic

Alternative Business Intelligence Engines 1385
I. Kovacevic, I. Mekterovic

Insights into BPM Maturity in Croatian and Slovenian Companies 1391
V. Bosilj Vuksic, M. Indihar Stemberger, D. Susa Vugec

Efficient Social Network Analysis in Big Data Architectures 1397
I. Soric, D. Dinjar, M. Stajcer, D. Orescanin

Integrating Evolving MDM and EDW Systems by Data Vault Based System Catalog 1401
D. Jaksic, V. Jovanovic, P. Poscic

Digital Economy and Government, Local Government, Public Services

Distributed Governance of Life Care Agreements via Public Databases N/A
J. Klasinc

Using Public Private Partnership Models in Smart Cities – Proposal for Croatia 1412
M. Milenkovic, M. Rasic, G. Vojkovic

The Platform for the Content Exchange between Internet Music Streaming Services and Discographers 1418
M. Sretenovic, B. Kovacic, A. Skendzic

Law and Technology in Data Processing: Risk-Based Approach in EU Data Protection Law and Implementation Challenges in Croatia 1424
N. Gumzej

Social and Economic Effects of Investments in Primorsko-goranska County Broadband Network 1431
S. Vojvodic, S. Cegar, D. Medved

Mobile Applications in Communication of Local Government with Citizens in Croatia N/A
D. Bunja, G. Pavelin, F. Mlinac

The Role of Applications and their Vendors in Evolution of Software Ecosystems 1442
S. Hyrynsalmi, P. Linna

Open Data Based Value Networks: Finnish Examples of Public Events and Agriculture 1448
P. Linna, T. Mäkinen, K. Yrjönkoski

Financial Impact of Forensic Proceedings in ICT 1454
S. Aksentijevic, E. Tijan, A. Jugovic

Reliability, Availability and Security of Computer Systems Supported by RFID Technology 1459
P. Ristov, T. Miskovic, A. Mrvica, Z. Markic

Transportation and Power System Interdependency for Urban Fast Charging and Battery Swapping Stations in Croatia 1465
I. Pavic, N. Holjevac, M. Zidar, I. Kuzle, A. Neskovic

Croatian Qualification Framework – Data Model and Software Implementation in Higher Education 1471
Z. Kovacevic, M. Mauher, M. Slamic

Internet as a Purchasing Information Source in Children's Products Retailing in Croatia 1476
B. Knezevic, Z. Pavlic Sipek, B. Jakovic

Valuation of Common Stocks Using the Dividend Valuation Approach and Excel 1481
Z. Prohaska, I. Uroda, A. Radman Pesa

The Interconnection between Investment in Software and Financial Performance – The Case of Republic of Croatia 1486
M. Boban, T. Susak

Mipro Junior – Student Papers

Data Warehouse Architecture Classification — 1491
G. Blazic, P. Poscic, D. Jaksic

Comparative Analysis of the Selected Relational Database Management Systems — 1496
R. Poljak, P. Poscic, D. Jaksic

A Tool for Simplifying Automatic Categorization of Scientific Paper Using Watson API — 1501
L. Cvetkovic, B. Milasinovic, K. Fertalj

The Role of Redundancy and Sexual Reproduction in the Conservation of the Genetic Information Tested on a Cellular Automaton — 1506
V. Kovács, V. Póser

Developing MOBA Games Using the Unity Game Engine — 1510
D. Polancec, I. Mekterovic

Comparative Analysis of Tools for Development of Native and Hybrid Mobile Applications — 1516
T. Vilcek, T. Jakopec

Exploring HTTP/2 Advantages and Performance Analysis Using Java 9 — 1522
L.M. Bach, B. Mihaljevic, A. Radovan

Software Supporting International Student Exchange Program in Higher Education — 1528
Z. Gracak, L. Brkic

Blood Vessel Segmentation Using Multiscale Hessian and Tensor Voting — 1534
A. Lukac, M. Subasic

Storytelling in Web Design: A Case Study — 1540
M. Pivac, A. Granic

Idioms in State-of-the-Art Croatian-English and English-Croatian SMT Systems — 1546
M. Manojlovic, L. Dajak, M. Brkic Bakaric

Contactless Control of Sanitary Water Flow and Temperature — 1551
D. Gecevic, T. Bjazic

PI Controller for DC Motor Speed Realized with Arduino and Simulink — 1557
M. Gavran, M. Fruk, G. Vujisic

Laboratory Model of the Elevator Controlled by ARDUINO Platform — 1562
M.A. Balug, T. Spoljaric, G. Vujisic

System for Acquisition and Processing of Pressure Data Around Body in Airflow

D. Meznaric, K. Krajcek Nikolic, D. Franjkovic

Education in the Field of Electronic Financial Services of its Future Users

M.Vejačka*

* Department of Applied Mathematics and Business Informatics
Faculty of Economics, Technical University of Košice, Slovakia
martin.vejacka@tuke.sk

Abstract - Information and communication technologies revolutionized the ways of providing financial services. Modern technologies instigated the rise of electronic banking and electronic finance and their development continues to bring new forms financial services provision. Students of finance, banking and investment fields of study should keep in touch with newest forms of electronic finance to stimulate their interests in the area of the newest technologies in banking and finance. In the process of education in electronic finance, the use of information and communication technologies is undoubtedly required. The aim of this paper is to present how modern electronic financial services are introduced to students in educational process at our faculty in a novel way. A course named Electronic Services in Banking is provided to students, during which students gain theoretical knowledge and practical skills in the area of electronic finance and banking. The course uses multiple eLearning methods such as using learning management system, virtual bank, training videos, online demos etc. The course is provided for several years already and is continuously updated and adjusted as new technologies emerge. Students have the opportunity to come into contact with various forms of electronic services in finance. During the course is emphasis laid on the security of electronic finance usage, including the principles of digital signature and biometrics. Before and after the completion of the course, a survey is conducted to detect students' awareness of electronic finance services and their advances in knowledge of the subject. The evaluation and a comparison of the survey results in recent years is the main output of this paper. Furthermore, the results serve as a feedback, which provides important suggestions for changes in the educational process.

I. INTRODUCTION

Information and communication technologies (ICT) revolutionized the ways of providing financial and banking services. Modern technologies instigated the rise of electronic banking and electronic finance and their development continues to bring new forms financial services provision.

The main characteristic feature and the biggest advantage of modern electronic financial and banking services is a continuous all-time access for all clients. It gives the opportunity to carry out fast and safe payments, get loans regardless of whether the client is home, in work or on holiday, virtually from anywhere in the world. These services are independent of opening hours of institutions providing them and allow convenient and rapid communication between provider and client or user [1]. Electronic finance represents providing and using financial products or services by electronic means including electronic payments, electronic money, deposits, lending, financial advisory and many other. Modern financial services sometimes even erase difference between users and providers of financial services, for example when using peer-to-peer lending services [2].

When speaking of electronic financial services, the communication between financial service provider and its user is conducted by electronic communication media like a mobile network, a public data network or the Internet. Electronic financial services are generally more cost- and time-effective for both involved sides. Financial services' providers might charge lower fees or interests for all the operations carried out electronically than the fees for traditional branch operations [1].

Important preconditions of electronic financial services are the technologies allowing their provision. Modern electronic financial services require hardware (computer, smart device etc.) and software equipment (e.g. application compatible with operating system of users' device). Use of electronic financial services in Slovakia is lagging behind EU average, for example in the case of using internet banking, the EU average is 49% while usage Slovakia is at level of 46% [3]. Many other electronic financial services are often even less penetrated in Slovak population [4].

Fast changes in the field of electronic financial services must be reflected in education of students of finance, banking and investment. At our faculty, the course of Electronic services in banking are taught for several years already [1]. Emergence of new electronic financial services in recent years invoked changes in educational process of this course also. The use of ICT brings many opportunities to innovate educational process. New technologies such as contactless payments, p2p lending and payments or cryptocurrencies are now included in curricula of the course.

In general, ICT is intensively used in educational process at our faculty. Information and communication technologies support process of education not only in informatics, but also in multiple other subjects of study provided to the students at our faculty [5]. Information technologies facilitated creation of e-learning courses that are used in our distant learning form courses [6]. Similarly ICT certainly supports daily based face-to-face education at our institution to help future professionals in the field of finance, banking and investment to develop their skills in all necessary areas of their expectant professions.

In this paper, the new form of the Electronic Services in Banking (ESIB) course is presented to provide education in the field of electronic financial services to future professionals in areas of finance, banking and investment at our faculty. It aims to inspire innovative ways of e-learning in area of banking.

II. THE ELECTRONIC SERVICES IN BANKING COURSE

The course of Electronic Services in Banking (ESIB) is provided at our faculty to students in the specialization of Finance, Banking and Investment for several years already. Turbulent changes in the field of electronic financial services must be expressed in education process of students of the area, too. New ways of providing banking and financial services developed during recent years invoked continuous changes in content of ESIB course. Initially, various forms of electronic banking were in the scope of the educational process in this course. It included getting practical experience with e-banking forms like SMS banking, Homebanking, WAP and GSM banking that were current at the time. These topics were novel at the time and the course provided first touch with these technologies for the students. However during following years a lot of content became outdated and was amended. However, the name of the course was not changed yet.

Now, the course of Electronic Services in Banking provides combination of skills' acquisition in e-learning environment with theoretical seminars on topics related with electronic financial services and partially in public administration. Any additional topics of students' interests are discussed in forms of essays as integral parts of this course. These essays reflect students' real-life experience with chosen topic related to electronic financial services. Students still get to know various forms of electronic finance in both theoretically and practically. Great importance is attributed to the security of electronic services' usage, as it is significant factor of trust-building process [7]. Safe setting-up and usage of electronic financial services is an integral part of their security [8] and therefore is emphasized in the course. Various forms of authorization elements are introduced to students regarding all forms of electronic financial services. An accent is further laid on the use of a digital signature as a very secure form of authentication and authorization.

Multiple studies (e.g. [7], [9]) indicate that perceived usefulness and awareness of given technology positively influence acceptation of electronic banking or similar technologies. Furthermore amount of information available on e-services significantly affects their adoption [10]. Furthermore security and safety of various electronic services supports their adoption by potential users [11], [12]. All these aspects are therefore discussed and analyzed during our course to support the adoption of e-services in finance and banking by the course attendees.

Moodle learning management system is used to allow e-learning support of this course (see [13]). Moodle is used as a standard e-learning platform at our faculty and other courses use it in the same or lesser extent with long-term positive experience not only for the educational purposes [14]. For a provision of certificates for digital signing, the faculty's certification authority based on EJBCA open source authority is used. It provides the possibility of digital signature certificates issuance and their management and strongly supports part of the course dedicated to digital signatures and electronic communication with financial institutions [15]. Furthermore, virtual bank is used for e-learning purposes of ESIB course, which supports several electronic banking forms. Similarly, virtual bank supports digital signature authorization. However, the banking software of the virtual bank is outdated and therefore virtual bank serves only to show some of its features comparable with current technologies (internet banking, digital signature authorization). Its old features like homebanking, WAP banking or SMS banking are only briefly demonstrated. The ESIB course further involves usage of open source software GPG4win for digital signature certificate management. Several other software applications are used to support or demonstrate various forms of electronic financial services.

Then, the main themes of the Electronic Services in Banking course are following:

A. Security of electronic financial services,

B. Internet banking,

C. Smart banking (Mobile banking),

D. Modern electronic financial services.

A. Security of electronic financial services

Main emphasis within the course is given to the security of electronic financial services. Both theoretical knowledge and practical skills in the field of security are possible to gain by students during our course. Secure communication during provision of a financial service between service provider and its user is the very important precondition of e-finance services provision and usage [16]. Students acquaint in this part of the course with topics of authentication and authorization processes, symmetric and asymmetric encryption (RSA), digital and electronic signature, Public Key Infrastructure (PKI), its legislation and software applications.

Following forms of authentication and authorization elements used in e-finance and e-banking are theoretically introduced practically used during the course:

- Static passwords and codes

- One-time passwords and dynamically generated codes (both software or hardware generated)

- Biometric elements (fingerprints, voice biometrics, biometric signature etc.)

- Digital signature

Their combination with provider's predefined limits of e-services usage creates the level of security of given forms of electronic financial services [8]. Usage of these authorization elements is discussed from the security and amenity points of view.

Multiple options of digital signature usage are theoretically introduced and further discussed. Signing, signature verification as long as encryption and decryption are practically demonstrated and checked out by students. Email encryption/decryption and signing/verification is demonstrated with use of Mozilla Thunderbird. Same processes with electronic files are exhibited and then tested by students using GPG4win freeware. The process of acquisition of certificate for digital signature is showed using faculty's certification authority. The course's students apply for a certificate and get it from the certification authority to test its usage practically. All study materials regarding this topic in course are available to students in electronic form in Moodle LMS and at certification authority webpage.

Further students get information on detection of valid e-banking websites and applications and prevention of credential thefts when using electronic financial and payment services. Several applications for secure communications on mobile devices (e.g. OpenKeychain, Decrypto etc.) are checked out by students during the course.

B. Internet banking

Internet banking makes banking services accessible from anywhere in the world when using with an internet connection. In this topic, the safe usage of internet banking is discussed. Principles of user's safe behavior when using it are explained in this part of the course. Students experience several authentication forms used in internet banking, such as GRID card, SMS code, OTP token or digital signature stored on a safe device. Authorization is traditionally secured by user name (login) and password assigned after registration in the virtual bank. Students then practically examine all aspects of usage of internet banking available in faculty's virtual bank (including delayed payments, permanent payments), collections, electronic statements etc.). Besides the payment services, also financial services available in real-life internet banking (such as e-investment, e-deposits, e-loans etc.) are presented. Obligatory practical task in this part of the course is to accomplish a pair of opposite transactions (payment and collection) over student's account authorized by a digital signature via internet banking within our virtual bank to prove gained practical skills. All theoretical information on internet banking is provided via LMS Moodle in form of text or instructional videos.

Some of ESIB course students come to touch with internet banking for the first time, because of their young age and their lack of experience with any bank or financial services before. Many of the students already have some real life experience with internet banking, when they take in the ESIB course, but still they indicate to gain new and valuable knowledge and skills from this part in feedback gathered at the end of the course.

C. Smart banking (Mobile banking)

Smartphones and smart devices became common object of a daily use in recent years. Besides the changes in communication they found broad application in e-commerce and e-banking [17]. Group of the electronic services commonly indicated as mobile banking undergone rapid development since its first introduction. This fact is related with fast changes in technologies available in area of mobile devices. In this part of Electronic Services in Banking course students get in touch with the most modern form of mobile banking called in Slovakia the Smart banking. This represents the form of electronic banking services provided on smart devices (smartphones, tablets etc.) via dedicated application using online data transfers over the internet. This form of e-banking became the most popular in recent years in Slovakia overcoming the usage of internet banking in a number of log-ins over period of time [8]. Smart banking is not supported by faculty's virtual bank therefore it is practically tested on demo application. Students are encouraged to use a real life smart banking application to fulfill voluntary task for extra points. Safety and security precautions for using e-banking are accentuated in this part of the course. The usage of a biometric authentication in smart banking is demonstrated to the students.

Besides these two major forms of e-banking other forms are mentioned and briefly demonstrated to show fast progress in this area of banking and finance. Specifically Homebanking, WAP banking, SMS banking and GSM banking are briefly introduced to students allowing them to remark rapid development in the field. Various demos, emulators and simulators of these forms are further used in these parts of the course.

D. Modern electronic financial services

Furthermore, modern electronic financial services are outlined during the ESIB course. Particularly peer-to-peer (p2p) lending or p2p payment services are mentioned. Definition and usage of cryptocurrencies is shortly noted. Another important area mentioned are contactless payments both card or smartphone based. Pros and cons of these modern electronic financial and banking services' usage are discussed. Further students prepare a semestral essay on the topic of modern electronic financial service usage which is then presented by author to other students in the study group. This essay should not be only of a theoretical character, but it should contain the real experience of a student with given topic expressing his own opinion regarding the theme. Furthermore, all semestral essays are available through LMS Moodle to all other students to allow experience sharing between them. Students' essays are a source of many reflections of electronic services from basic user's point of view and provide feedback on the course contents in this area.

It is understandable that it is not possible to provide practical real-life experience with all modern e-services in finance and banking to the course attendees with limited sources of a public educational institution. However, the course creators try to provide any free demos, video tutorials, free software applications or any suitable materials on given topics to students.

III. THE ESIB COURSE FEEDBACK

During each year of the ESIB course provision, feedback on the course is gathered. This feedback contains a survey committed before the beginning of ESIB course and after its completion investigating students' experience and knowledge in the field of electronic financial services. Questions in the survey are about attendees' knowledge about electronic services, their experience and opinion on their usage. The willingness to use and actual usage of electronic financial services is investigated before and after the course completion. Component part of the survey is getting reactions from students on the ESIB course. Relevant suggestions from respondents' feedback are implemented into educational process in following years of the course provision.

In 2016 following results of survey were gathered. The survey was conducted in electronic form using Google Docs questionnaire. Respondents of the survey were students attending the ESIB course in both distance learning and face-to-face learning forms of the ESIB course. They were addressed with the questionnaire through LMS Moodle. The usable answered questionnaires were gathered from 217 respondents what represents over 98 percent of students in the given year of the ESIB course. Therefore, the results of the survey very high informative value.

Firstly the demographic data of respondents were gathered and are presented in following table:

TABLE I.   DEMOGRAPHIC DATA OF SURVEY RESPONDENTS

Gender	Count	Percentage
Men	96	44.2
Women	121	55.8
Age		
16 - 25	187	86.2
26 - 35	26	11.98
36 and over	4	1.82
Residence		
Large-sized city (over 100 000 citizens)	135	62.2
Medium-sized city (40 000 to 100 000 citizens)	17	7.8
Small town (5 000 to 40 000 citizens)	29	13.4
Village (under 5 000 citizens)	34	15.7

Source: own survey

Almost 56 percent of survey respondents were women. The gender composition of the course attendees reflects the trend of slightly prevailing female students at the faculty. Both internal and external study form students were included in the survey, what caused a bigger variety of respondents' ages, since students of distant or external form of study are often of higher age in comparison with internal students. However, over 86 percent of respondents were students below 25 years of age, thus high school graduates, who immediately continue to study at the university. The residence of students in Slovakia might affect availability of broadband internet services to them. Almost 30 percent of respondents are from smaller towns under 40 000 citizens, where the availability of broadband internet is not perfect in Slovak conditions. This fact might affect their attendance in the course in distant form as long as their usage of electronic financial services.

Since the survey was conducted before the course completion and after it, the pre-course results will be presented firstly and then compared with the results after the completion of the ESIB course. General usage of electronic services in finance and banking were investigated at first group of questions. Almost 80 percent of the course attendees adduced using of some form of electronic financial service. The most of them, 74 percent of all respondents, used internet banking at least once before the course, mostly for basic services as payments or checking the account balance. Over 17 percent indicated usage of smart banking application in some of their smart devices.

The respondents that did not use any e-services, where asked for the causes. Above 5 percent of all respondents admitted that they do not use financial and banking services at all yet. All of them were in the age group from 16 to 25 years of age. It suggest that these students did not have any non-cash income yet before. Almost 9 percent adduced mistrust to e-services expressing their concerns about security and safety. Over six percent of respondents admitted preference of usage of financial and banking services in person and not via electronic communication channels. The frequency of e-services in finance usage was further investigated. Around 49 percent of respondents use electronic services at least once per a week and almost 22 percent at least once per month. The rest of respondents use it less frequently. Over 72 percent of respondents prefer electronic (card or mobile) payment forms before cash payment. Over 45 percent of respondents are willing to use e-deposit services, over 31 percent credit e-services and 27 percent investment services via electronic media. Approximately 57 percent of the ESIB course attendees adduced that e-services are important for their decision about financial services provider.

Another part of the questionnaire was aimed at the area of electronic services security. The preference of quick and user-friendly authorization solutions was detected. Over 61 percent of students admitted usage of static codes preference, followed by dynamic passwords (in any form of delivery) with 24 percent, 12 percent for biometric authorization and only 3 percent by digital signature technology.

Survey respondents mostly failed in identifying the level of security of each element used for authorization and authentication. Over 29 percent of respondents stated as the most secure and at the same time simple way of securing of e-services the static codes or passwords, 38 percent would prefer dynamically generated codes, 21 percent biometric elements and 12 percent digital signature.

Respondents' concerns with using electronic financial services were investigated in addition. The most of them (72 percent of respondents) fears of abuse by a third party, 31 percent has anxiety of technical imperfections of IT solutions of e-services and 18 percent fears of abuse by employees of financial institution. Almost 95 percent considers protection against computer viruses and other harmful software as important for e-services safety. On the other hand, over 55 percent of respondents admits their own failure to adhere or even to know the safety rules when using e-services.

The usage of digital signature for electronic communication by respondents was also investigated. Over 60 percent of respondents considers digital signature as suitable for e-government applications, but only 47 percent expect its usage in various electronic financial services provision. Further stated results were acquired from survey after the ESIB course completion. These results are further compared with above stated results of the survey gathered before the course. Demographic data remained the same, while for comparison were used only the surveys from students who answered it both before and after the course (above 98 percent of attendees).

After the attending the ESIB course the percentage of users of electronic financial services between attendees of our course rose to 86 percent. Several forms of electronic financial services were used by more respondents after the course completion than before its start. Over 80 percent of respondents adduced internet banking usage. Now more than 30 percent of respondents uses smart banking increased from 17 percent before the course. Almost 10 percent of users indicated shifting to more secure forms of authorization in any financial e-service used like dynamically generated codes or biometrics. Mistrust of electronic provision of financial services was indicated only by less than 3 percent of respondents, what is significantly lower than above 9 percent before the course completion. There was detected increase in frequency of financial e-services usage to 58 percent of users on weekly basis.

Almost 9 percent of respondents indicated usage of a digital signature in electronic communication or for e-services authorization. Respondents' concerns with using electronic financial services decreased in all observed cases. Fear of abuse by a third party decreased among the respondents from 72 percent to approximately 58 percent, 25 percent has still anxiety of technical imperfections of IT solutions of e-services in comparison to pre-course 31 percent and only 8 percent fears of abuse by employees of financial institution, what represents decrease of 10 percent from pre-course level of 18 percent. All respondents indicated protection against harmful software as important for e-services safety (increase from 95 percent). After the course completion, less than 42 percent of respondents admitted their failure in adherence of the safety rules for financial e-services usage presented and discussed during the ESIB course. These results suggest positive shifts in students' knowledge, skills and behavior thanks to the course attendance. However the results

might be biased by students' remembrance of the questions in the questionnaire from the first pre-course responding in the survey.

Fig. 1 contains illustration of main positive shifts in results of the survey in comparison of answers before the EIS course completion and after it.

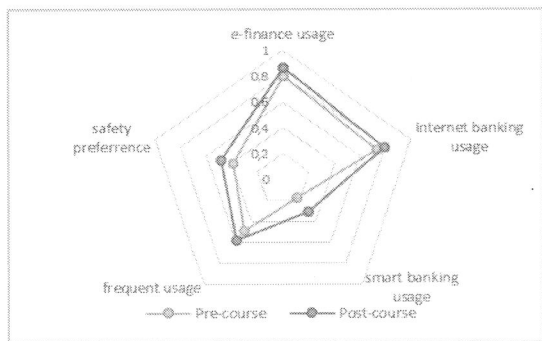

Figure 1.   Positive effects of the ESIB course completion

Further feedback gained from students on the ESIB course attendees was acquired within the survey after the course completion. Students mainly suggested to abandon any remarks of outdated forms of e-services. However even in this area were contradictory opinions gathered. Many respondents would welcome practical demonstrations of mobile contactless payments using NFC supporting devices, this however is beyond technical possibilities of hardware equipment available at our faculty. They are however encouraged to adapt this technology in praxis during the course. Similar situation happened in the areas of p2p lending e-services and e-payments. Any applicable suggestions will the source of improvement of the ESIB course in following years. The respondents indicated being satisfied with the ESIB course in 82 percent. Such feedback is very valuable for improvements in the course.

Besides this survey, there was also the theoretical and practical knowledge tested before and after course completion to measure students' improvement. The test consisted of 20 questions with one or multiple correct answers. This test is used since the year of 2009 with continuously updated composition of questions. Fig. 2 illustrates success rate of students in this test before and after the course during the years of the ESIB course provision.

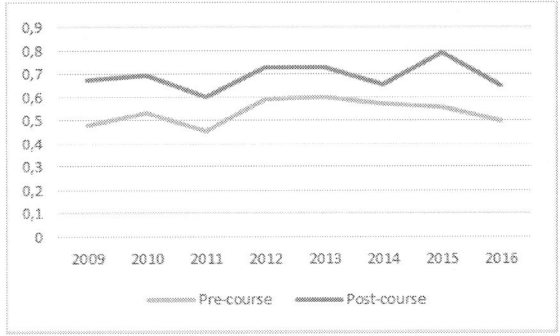

Figure 2.   Test success rates of the ESIB course students

787

The average success rate in the year of 2016 rose from less than 50 percent before course to level slightly above 65 percent after it. In all years of testing the increase of knowledge was detected in the area of electronic banking and electronic finance among the ESIB course attendees. These results suggest that our course improves the knowledge and skills of its attendees in the field of electronic financial services and further increases their adoption of these services.

IV. CONLUSION

Students of our faculty find their job placements mainly in the area of banking and finance after the graduation. They often lack practical experience in particular area, in this case the experience with electronic financial and banking services [6]. Innovated course of Electronic Services in Banking provided at our faculty aims at education of future professionals in areas of finance and banking and provides them knowledge and practical skills development in the field of electronic financial services with a support of e-learning methods. Changes in the course during the years of its provision were invoked by rapid changes in ICT usage in area of e-services in finance. The course attendees get the opportunity to test multiple electronic services in finance and banking using virtual bank, certification authority, several demos, simulation software and video materials. Furthermore, the course is built in the way that students in both forms of study (face-to-face and distant learning) get virtually the same learning experience. This paper aimed to share the experience from ESIB course and to present results achieved by this novel form of the course provided by our faculty.

In 2016 was the questionnaire survey conducted within the course to reveal students' knowledge and skills development in two time moments: before the course and after its completion. Its results show several positive improvements in students' knowledge and suggest practical skills acquisition. The course further increases the awareness of electronic services in finance and banking among the students and their consequential adoption. The survey was a source of important feedback including suggestions for changes in the educational process of the course in following years. The testing of students and their average results over the recent years again show progress in knowledge of the area in each year.

The course combines multiple information and communication technologies to provide education in the area of banking and e-finance in a novel way. Contribution of the course consists in a combination of usage of virtual bank, private certification authority and real-life e-finance applications. Emphasis laid on security of electronic services appreciate course attendees even in their private usage of given services.

REFERENCES

[1] M. Vejačka, "Introduction to electronic banking by elearning methods," MIPRO 2012 - 35th International convention : proceedings: Opatija, Croatia, 2012, pp. 1406-1411.

[2] M. Vejačka, "Consumer acceptance of contactless payments in slovakia," in: Journal of Applied Economic Sciences. Vol. 10, no. 5, 2015, pp. 760-767.

[3] Eurostat, "Individuals using the internet for internet banking", Eurostat 2016, [online] available at: http://ec.europa.eu/eurostat/tgm/table.do?tab=table&init=1&language=en&pcode=tin00099&plugin=1

[4] European Central Bank, "Payment statistics," 2016, [online] available at: http://sdw.ecb.europa.eu/reports.do?node=1000001958

[5] L. Révészová, "Informatics education and requirements of current practice" in: MIPRO 2015. - 38th International convention proceedings: Opatija, Croatia, pp. 868-873.

[6] L. Révészová and D. Paľová, "ICT challenges, opportunities and choices in the knowledge society development," in: ICT as a Determinant Factor in the Development of the Public Services in the Knowledge Society - Targu Mures: Petru Maior University, 2013, pp. 13-37.

[7] M. Vejačka, "Customer acceptance of electronic banking: Evidence from Slovakia," in: Journal of Applied Economic Sciences. Vol. 9, no. 3, 2014, p. 514-522.

[8] J. Bucko, "Security of smart banking applications in Slovakia," in Journal of Theoretical and Applied Electronic Commerce Research, Universidad de Talca - Chile, vol. 12, iss. 1, 2017, pp. 42-52.

[9] T. M. Qureshi, M. K. Zafar and M.B. Khan, "Customer Acceptance of Online Banking in Developing Economies", in: Journal of Internet Banking and Commerce, Vol. 13, No.1, 2008, pp.1-9.

[10] P. G. Schierz, O. Schilke and B.W. Wirtz, "Understanding consumer acceptance of mobile payment services: An empirical analysis," in Electronic Commerce Research and Applications, Vol.9, 2010, pp. 209–216

[11] K. T. Geetha, and V. Malarvizhi, "Acceptance of E-Banking among Customers (An Empirical Investigation in India)", in: Journal of Management and Science, Non Olympic Times, Vol. 2, No.1, 2011, pp 1-9.

[12] Alsajjan, B. and Dennis, C., (2010), Internet banking acceptance model: Cross-market examination, in: Journal of Business Research, Advances in Internet Consumer Behavior& Marketing Strategy, Vol. 63, Iss. 9–10, pp. 957–963.

[13] Moodle.org, "Moodle – Community driven, globally supported" [online] available at: http://www.moodle.org.

[14] D. Paľová, "Experience with Usage of LMS Moodle not Only for the Educational Purposes at the Educational Institution," MIPRO 2016 - 39th International convention proceedings: Opatija, Croatia, 2016, pp. 1006-1011.

[15] EJBCA, "EJBCA PKI Certificate Authority," [online] available at: https://www.ejbca.org/.

[16] T. Bálint and J. Bucko, "Comparative analysis of handwritten, biometric and digital signature," in International Review of Social Sciences and Humanities, vol. 4, no. 2, 2013, pp. 43 - 53.

[17] J. Bucko, L. Kakalejčík, and Ľ. Nastišin, "Use of smartphones during purchasing process," in: CEFE 2015 Central European Conference in Finance and Economics proceedings, Košice Slovakia, 2015, pp. 91-97.

Didactic Concepts of Modern Data Analysis

C. Ungermanns, W. Werth
Carinthia University of Applied Sciences (CUAS)
Europastraße 4, 9524 Villach, Austria
Phone: +43 4242 90500
c.ungermanns@fh-kaernten.at
w.werth@fh-kaernten.at

Abstract - Nowadays, huge amounts of data are produced within shortest time. Especially modern production and measurement systems daily generate a big volume of data, which is growing exponentially. In many cases the challenge is not just obtaining the data, but an appropriate analysis, which is necessary to extract the significant information for enterprises and institutions. The added value of information is zero, if the employees of a company are not able to effectively evaluate the collected data.

In this paper a didactic concept is presented, how master students become familiar with methods of modern data analysis. A complete and systematic approach starting with data acquisition, database management, extraction, up to evaluation and interpretation is shown. For better understanding the used complex and theoretical multivariate algorithms are applied to illustrative examples. This gives the students the opportunity to establish a relationship with these methods and the ability to transfer them to practical industrial problems.

I. INTRODUCTION

Many publications and teaching methods at higher schools and universities are merely theoretical introductions to the subject of statistics and often require a high degree of abstraction from the student. However, the application of statistical methods is becoming more and more important due to the increasing amount of data and performance of today's computers. To a large extent, especially multivariate evaluation procedures can only be executed effectively by the enormous calculating power of current technologies.

The present article proposes a different, practice-oriented approach to selected industry relevant topics of data analysis, which should help to overcome the reservations and prejudices towards statistics being a "secret science".

The basic idea of the educational concept is to prepare the groups for statistical thinking by describing a real problem and motivate them to solve it initially without external help. This creates a close relationship to the respective subject and improves the attention for the solution significantly.

Based on numerous examples, the groups interactively learn different mathematical techniques and

the corresponding field of application. Therefore an education in mathematics and computer science with bachelor level is a necessary prerequisite for the course. Occasionally, group members are surprised by the information hidden in special datasets. The goal is to enable the master students to uncover the maximum of information concealed in the sample data.

II. CONCEPT IMPLEMENTATION

An essential pillar of the didactic concept is to explain the entire data management chain occurring in industry. Fig. 1 gives an overview of all important parts.

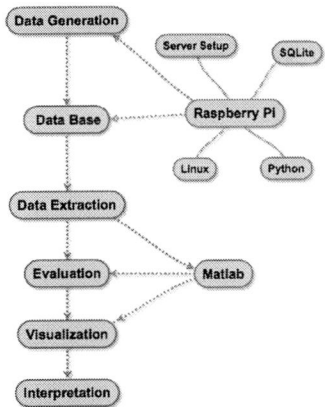

Figure 1. Overview of data management chain occurring in industry.

The process starts by generating a dataset. For this purpose, a Raspberry Pi is used, which provides several advantages:

- It is a low-cost hardware solution, which opens up a considerable potential for hardware expansion.

- It is a Linux-based system. Many professional industrial systems are based on Unix systems.

- There are a lot of open source software and hardware solutions that can be used like a modular construction system, allowing individual solutions to be realized in a short period of time.

- Especially a full working data management system including a server operation can be set up straightforward.

In the first session, the students have to prepare the Raspberry Pi for the tasks to be solved. In detail, the following topics were addressed [1]-[5]:

- Installation of Raspbian OS
- Basics of Linux and Python
- Apache2 Webserver
- PHP plugin
- SQLite
- PhpLite Admin DBM

The next step was the implementation of an ultrasonic

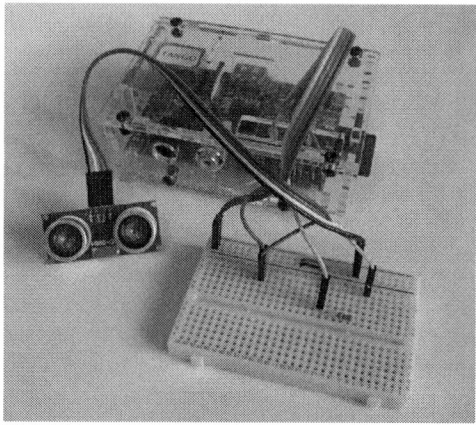

Figure 2. Hardware setup of ultrasonic distance measurement [1].

distance sensor including the corresponding Python code. This experimental setup was taken from [6] and adapted to our purposes.

The objective of this experiment was to check, if the measured ultrasonic sensor data could be immediately and automatically stored in a previously defined SQLite-data base, which was also realized with the Raspberry Pi. In addition to this data, real (anonymized) production datasets were considered as well. After setting up a client–server architecture, the students were able to extract any dataset out of the database by applying a specific SQL (Structured Query Language) command, sent by a remote computer. The necessary SQL skills were also trained in the class. SQL is a widespread industry standard.

At this point, a challenging exercise must be tackled: the connection from the database to MATLAB, our evaluation software. Although MATLAB offers a SQL modul in the database toolbox, this tricky problem could not be solved satisfactorily until now and was outsourced to a different project. Instead, a workaround was found.

Now, a new section of the course was started: the statistical evaluation of different types of datasets with MATLAB. The primary job was the preparation of data. This is a typical situation in industry. The extracted dataset is often not directly suitable for the evaluation due

to missing values, incorrect entries or outliers. Sometimes datasets from other sources must be merged with entries of the database. Furthermore, in certain cases computed columns must be appended. All these preparatory activities could be summarized under the heading "basic data manipulation". This also includes z-standardizing of matrix, ranking, grouping, eliminating missings and winsorizing, just to name a few examples. MATLAB provides a lot of options for this task.

Now, the actual evaluation was performed. Starting with univariate statistics, basic calculations like central tendency and dispersion as well as the corresponding graphical representations, such as histogram, stem and leaf plot, boxplot, cumulative frequency and probability plot, were carried out [7].

The investigation of bivariate datasets led the students to the fundamental concepts of variance and covariance. The terms "(linear) statistical dependence" and "independence" were defined and could be demonstrated by the basic properties of variance. Especially the discussion on the variance of the sum of correlated variables results in interesting opinions of the participants. An appropriate simulation with MATLAB clarified the question.

The presentation of an industry related example at the end of the section deepened the understanding of covariance and correlation matrix. The student's task was to automatically find a predetermined pattern within a group of samples. Such an issue occurs in the quality control of surfaces, where scratches must be detected and the product concerned must be scrapped.

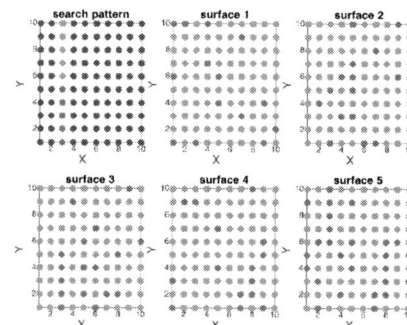

Figure 3. Automatic scratch pattern detection as an example of correlation analyis.

Fig. 3 shows the scratch pattern scenario. The first subplot defines the search pattern. Blue areas are marked as "not relevant" and must be suppressed by the automatic analysis. The green part of the search pattern labels the non-affected area, whereas the red points represent the scratch. Subplots 2-6 show five different measured (fictitious) surfaces. The algorithm, presented in the lecture and worked out by the students, is able to assign surface 3 (subplot 4) to the predetermined pattern with a correlation coefficient of 1.

790

Now, the next typical work in daily business of companies was treated: several machines of a production line, which perform a particular process step, should lead to the same process result. A suitable method to attack this problem is the ANOVA (analysis of variance). A fundamental requirement for understanding this statistical technique is the knowledge of distribution functions. The theory of distribution functions is mathematically demanding and not really helpful for solving the specific issue. Instead, a MATLAB simulation based on simple considerations was started and compared with the exact mathematical solutions. The advantage of this procedure is an intuitive access to statistical distribution functions. Fig. 4 gives an impression of the comparison between the simulation and the exact solution, using the example of the F-function.

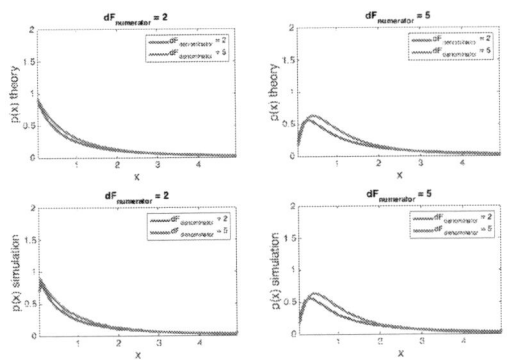

Figure 4.   Comparison of simulated and theoretical curve of F-distribution, displayed for two degrees of freedom, dF=2, dF=5.

Moreover, other distribution functions like the Student-t- and the Chi²-distribution were explained in a descriptive manner and subsequently simulated.

After this preparatory work, the initial ANOVA-task could be solved. Fig. 5 displays a boxplot of a characteristic measurement value grouped by the

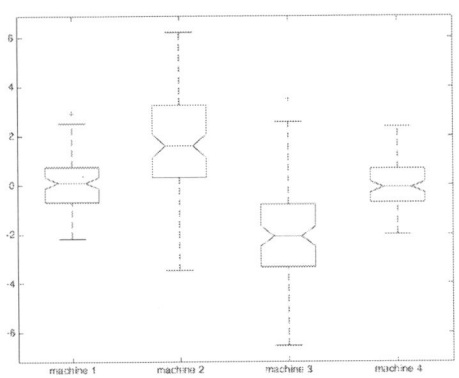

Figure 5.   Graphical MATLAB-output of One-Way-ANOVA: different machine behaviour is clearly visible. The significance is estimated by the F-value.

equipment used.

The question arises, whether there are significant differences between the machines or not. The appropriate statistical method is the F-test, whereby the test value is calculated as the ratio between variance of the treatments and variance within the treatments. The participants of the class immediately grasped the mathematical approach, because they already completely understood the F-distribution. For didactic reasons, a manually performed analysis was done and compared with the results of the "ANOVA"-function of MATLAB. This assures that the students could interpret the results of the software package.

In the last section of the course, the outstanding importance of multivariate statistics was discussed based on complex datasets. A typical task in this context is the question, which and how strong variables are correlated with other variables. Often engineers are interested in the (very complex) influence of many predictors on the yield of a product. The method of choice is the so-called "multiple regression". However, a major challenge for the application of multiple regression is to overcome the problem of multicollinearity, which means, that two or more predictors are highly correlated [8].

In the lecture, this was shown by an easy example: it was assumed that two variables x_1 and x_2 linearly depend on each other with a correlation coefficient of r=1, e.g. $x_1=x_2$ for simplicity. Now the response variable y, e.g. yield, was also assumed to be directly correlated with x_1, e.g. $y=x_1$. Consequently, x_2 is directly correlated with y as well. Obviously, any linear combination $y=ax_1+bx_2$ with a+b=1 represents the response variable y. It is immediately clear that the problem cannot be solved uniquely.

A solution approach is based on the idea of finding new predictor variables w_1, w_2, ..., w_p as linear combinations of the original predictors variables x_1, x_2, ..., x_p with the property that w_1, w_2, ..., w_p are pairwise uncorrelated. This task can be managed by the mathematical formalism of a suitable orthogonal transformation.

A two-dimensional example was presented in the course. The original dataset is shown in Fig. 6.

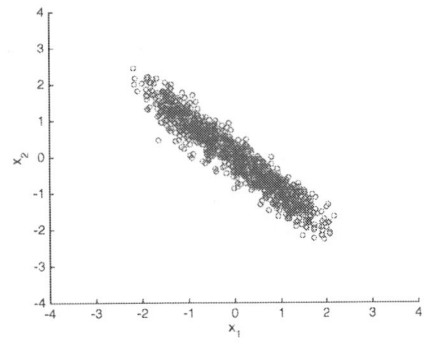

Figure 6.   Original two-dimensional dataset, x_1 and x_2 are highly correlated

After the application of an orthogonal transformation, which is a 45° rotation in this case, the variables w_1 and w_2 are uncorrelated, see Fig. 7.

In high dimensional spaces, it implies the solution of an eigenvalue problem. The underlying mathematical formalism is called "Principal Component Analysis", abbreviated PCA, and the transformed coordinates w_1, w_2, ..., w_p are called principal components PC_1, PC_2, ..., PC_p [9].

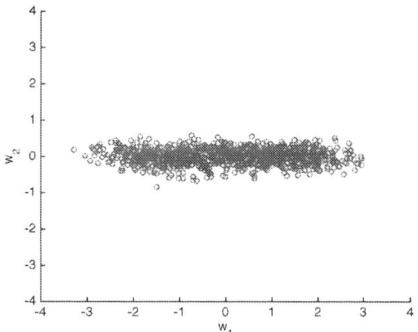

Figure 7. Rotated dataset, w_1 and w_2 are uncorrelated after rotation.

This algorithm can be used to reduce the dimension of the data space. Directions with a low level of information, in Fig. 7 the w_2 coordinate or PC_2, can be neglected for the benefit of a lower memory consumption without losing essential information. Especially in image processing, the principal component analysis is also called "Karhunen-Loève transformation". The higher the involved principle component, the smaller the loss of information.

In the class, the PCA-algorithm was programmed in MATLAB by the participants with support of the lecturer and applied to an image of TRUDI, our rescue robot, see Fig. 8.

Figure 8. Original image of TRUDI (Third Robot for Urban Disaster Intervention), developed at CUAS by the group of Wolfgang Werth.

The result of the PCA is shown in Fig. 9 for a different number of principal components. The effect of data reduction and the improvement of significant information with increasing PC can be clearly seen from the images.

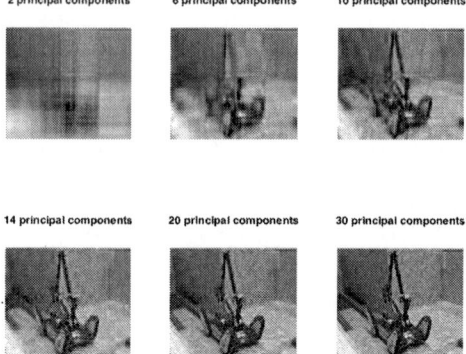

Figure 9.   Application of PCA to image of TRUDI

At the end of the course the principal component algorithm was applied to an audio file as well. The acoustic result was impressing for the students and demonstrated the wide field of application of principal component analysis.

III.   SUMMARY

In this article, the basic ideas of our new data analysis course including the practical implementation were presented. The concept of the course was and will also be in future strictly based on industry-related issues. At the beginning, the installation and configuration of a database were performed. A Raspberry Pi served as the hardware platform with the advantage of being a cost-effective solution. In addition to its task as a database server, the Raspberry Pi was used to generate datasets in combination with an ultrasonic distance sensor. The complete system including data storage and extraction was successfully tested. A data connection to MATLAB could be established.

The second part of the course dealt with the evaluation and interpretation of the data. Knowledge of univariate, bivariate and multivariate techniques was imparted to the participants. A lot of practical exercises performed with MATLAB enabled the students to independently select the correct method adapted to the problem and apply the mathematical concepts.

IV.   FEEDBACK

Finally a few statements of the students on the course are cited:

"This is an interesting course from the practical aspect..."

"Good entry in the world of big data and data science..."

"The data science course is done in a practical manner close to real applications..."

"Also the use of state-of-the-art software is important to stay in touch with current technologies that are used in industry..."

"... I think the MATLAB as well as the Raspberry Pi part were really good prepared..."

V. FUTURE ASPECTS

A concept for measuring the learning outcomes will be developed in cooperation with industry partners. Also pedagogical improvements are planned for future courses, especially a time slot for more interactive, workshop-like collaboration with the participants is intended.

VI. CONLUSION

We are convinced that the practical mastery of statistical evaluation procedures is an increasingly necessary skill in engineering education due to the information content of mass data. Therefore, it is of tremendous importance for companies, to extract significant information at an early stage in order to obtain advantages over competitors.

ACKNOWLEDGMENT

The authors thank Christian Madritsch for the implementation of all Raspberry-Pi-related topics and Marcus Ungermanns for reviewing this article.

REFERENCES

[1] S. P. Wallace and M. Richardson, *Getting started with raspberry pi.* Cambridge, USA: O'Reilly Media, 2013.

[2] B. Ward, *How Linux works: What every Superuser should know.* New York, NY, United States: No Starch Press,US, 2014.

[3] J. A. Kreibich, *Using SQLite.* United States: O'Reilly Media, Inc, USA, 2010.

[4] A. Ford, *Apache 2 pocket Refernce: For Apache programmers and administrators (pocket reference (O'Reilly) series)*, 2nd ed. United States: O'Reilly Media, Inc, USA, 2008.

[5] J. Lockhart, *Modern PHP: New features and good practices.* United States: O'Reilly Media, Inc, USA, 2015.

[6] Matt, "Ultrasonic distance measurement using python – part 1," 2012. [Online]. Available: http://www.raspberrypi-spy.co.uk/2012/12/ultrasonic-distance-measurement-using-python-part-1/. Accessed: Feb. 8, 2017.

[7] B. L. Agarwal and Agarwal, *Basic statistics revised 4/E*, 4th ed. New Delhi: New Age International Pvt Ltd Publishers, 2006.

[8] D. A. Belsley, E. Kuh, and R. E. Welsch, *Regression diagnostics: Identifying influential data and sources of Collinearity*, 20th ed. New York, NY, United States: John Wiley & Sons, 1980.

[9] I. T. Jolliffe, *Principal component analysis*, 2nd ed. New York: Springer-Verlag New York, 2002.

MIPRO 2017, May 22- 26, 2017, Opatija, Croatia

Extending the Object-Oriented Notional Machine Notation with Inheritance, Polymorphism, and GUI Events

Marin Aglić Čuvić*, Josip Maras ** and Saša Mladenović*
* Faculty of Science, University of Split, Croatia
maragl@pmfst.hr, sasa.mlad enovic@pmfst.hr
** Faculty of electrical engineering, mechanical engineering,
and naval architecture, University of Split, Croatia
josip.maras@fesb.hr

Abstract - Learning to program is a challenging task. Novices need to have an accurate understanding of the program execution at the conceptual level provided by the programming language. This level of execution is often referred to as the *notional machine,* which is often easier to understand through program visualizations.

Currently one of the most popular programming paradigms is object-oriented programming, which introduces a number of advanced concepts. In addition, in order to increase student engagement, teachers have started to introduce graphical user interface (GUI) applications into programming courses. This brings its own set of challenges, mostly related to a significantly larger number of application states, which are more difficult to keep track of. However, most existing programming visualizations do not cover all necessary concepts for teaching object-oriented programming, nor have they considered visualizing complex GUI applications. For this reason, we present our own concept of a visualization system that addresses these shortcomings.

In this paper, we have: i) extended an existing notation for a notional machine to support learning of advanced object-oriented concepts inheritance and polymorphism; ii) presented our own concept of a visualization system that introduces the source code into the notional machine, thereby making the relationship between source code and the visual representation more concrete; and iii) proposed solutions for reducing the cognitive overload introduced by GUI applications.

I. INTRODUCTION

Learning to program is difficult [1, 2]. It requires novices to learn logical and computational techniques, the syntax of a new programming language, as well as developing the understanding of how a concrete program is actually executed by the machine [3]. During the acquisition of these techniques, novices often develop incomplete or incorrect understanding of programming concepts, called misconceptions.

Currently, one of the most popular programming paradigms in the software industry is object-oriented (OO) programming, with Java as one of the most popular programming languages taught at universities. At some

universities OO programming is taught as a part of introductory programming [4]. However, for novices this increases the cognitive load, as OO brings about its own concepts such as *classes, objects, encapsulation, inheritance,* and *polymorphism.* A study on misconceptions related to inheritance and polymorphism reveals that even teachers with years of experience in procedural programming have difficulties in understanding those concepts [5].

A common technique to increase the motivation and student engagement is the introduction of Game-based learning [2, 6, 7]. Game-based learning facilitates the acquisition of programming concepts through developing and playing games rather than solving often tedious and repetitive tasks. Although text-based games exist, they are not as popular and have a lower motivation potential than those based on a graphical user interface (GUI). Therefore, students often learn to program GUI applications before moving on to games. This alone might increase student motivation as they might have a greater sense of achievement by building programs similar to those they interact with every day. However, GUI applications, and particularly games, often have a significantly larger number of states which may cause cognitive overload for novices.

To achieve programming proficiency, novices need to understand how the program is executed on the level of abstraction provided by the programming language [3]. This abstraction provides a sufficiently detailed understanding of program execution and is often referred to as the *notional machine* [8]. It is a construct formed from concepts provided by the programming language [3], which allows programmers to understand and explain the execution of a program in terms of those concepts. When discussing imperative and procedural programming, Du Boulay recognized the notional machine as a source of most programming misconceptions (e.g. related to assignment, variables, and arrays) [9]. Therefore, it is important that novices construct a valid mental model. This becomes increasingly important when learning OO programming.

When addressing learning OO programming, Sorva [8] discusses the need for two notional machines. The first notional machine should be an extended version of the notional machine used in imperative or procedural programming languages, and should feature variable assignment, expression evaluation, call stack, object reference, etc. The second notional machine should abstract out these details and focus on object interactions during program execution.

In general, the majority of difficulties faced by novices can be attributed to the lack of understanding of how programs are executed [8, 10]. One way of facilitating the understanding of the dynamic process of program execution is through program visualizations. Program visualizations are often software tools that graphically represent the execution of a program. Most existing program visualizations for programming education have focused on helping students in introductory programming courses, which teach only a limited set of OO concepts, such as classes and objects.

In this paper, we present a concept of a visualization system with the goals of: i) helping students comprehend important OO concepts, such as classes, objects, inheritance, and polymorphism; ii) making the relationship between the source code and the visual representation more concrete; iii) facilitating the understanding of GUI assignments, both classic as well as game development. In order to fulfil these goals, we have: i) expanded the notation for the notional machine introduced in [11] with notations for inheritance and polymorphism, ii) designed a visualization system that overlays the source code with the notional machine, and iii) discussed and proposed solutions for visualizing applications that have a large number of objects and executions, such as GUI and especially game applications.

This paper is structured as follows: in section 2 we introduce existing program visualizations, and section 3 discusses inheritance and polymorphism along with related misconceptions. In section 4 we present our prototype visualization of a notional machine for OO concepts. Finally, section 5 concludes the paper and presents possible future work.

II. RELATED WORK

As expressed in [3], software visualization refers to the use of graphical elements in order to provide assistance with the construction of mental representations of software. According to their purpose, software visualizations can be classified as either algorithm, code, or program visualizations. Algorithm visualizations are used to visualize programs at a high level of abstraction with the goal of explaining how the algorithm works. Code visualizations are used to visualize source code and its features. Finally, program visualizations are graphical representations of the notional machine. They are our primary focus in this paper.

The literature review conducted by Sorva [12] identified forty-six different program visualizations since 1979 to 2013, out of which only fifteen were reported as active at the time and only nine supported OO programming concepts to a certain degree. This was expanded with a recent literature review [3] that identified seven new program visualizations that appeared in the period between 2013-2016. We focus on a few that we have found particularly interesting: UUhistle [10], Online

Python Tutor [13], Jeliot 3 [14], JaguarCode (formerly JavelinaCode) [15], JIVE [16], and Novis [17].

UUhistle [10] is a visualization system that provides two operating modes. In the first one, the system visualizes the execution of a Python program. The user can step back and forth through the visualization and observe program changes at a low level. In the second mode, the system allows the user to take on the role of the machine and manipulate the graphics in order to execute a given program. Sorva refers to this exercise as *visual program simulation*.

Another very popular visualization tool is the Online Python Tutor [13]. Currently, it supports the visualization of both versions of Python, and six other programming languages. The system itself is web-based and can be embedded into other web pages. Furthermore, Online Python Tutor allows users to visualize their code line by line or an entire program automatically via a live programming mode.

Jeliot 3 [14] is a visualization tool for Java programs aimed to help students learn procedural and OO programming. Program execution is visualized line by line. One of the newer visualization systems is JaguarCode [15] which is a web-based visualization system for Java programs. The system visualizes the class diagram and the execution of source code by line, similar to Online Python Tutor and Jeliot 3. JIVE [16] is a popular Eclipse plugin used to generate Unified Modeling Language (UML) diagrams from code. However, JIVE is targeted primarily at professional programmers and is not adapted for novices.

Novis [17] is a visualization tool that has been integrated into BlueJ, an integrated development environment (IDE) designed for educational use. The tool visualizes a class diagram and notional machine of a Java program. The visualized notional machine consists of objects and their fields and values assigned to them. In case a field's value is a reference to another object, the field and referenced object are connected by an arrow. Novis also supports viewing the notional machine at different levels of detail, where at the lower level only objects of complex types are displayed. As opposed to the previously mentioned visualizations, the minimum step granularity in Novis is that of a method call. The system visualizes only the methods that are part of the currently active call stack.

Although the aforementioned visualization systems support the use of inheritance and polymorphism in the program code, they do not explicitly visualize them. This makes them inappropriate for providing conceptual models for these concepts.

Furthermore, none of the visualization systems mentioned above have taken into consideration the possibility of visualizing GUIs. Because GUIs tend to consist of a large number of objects, the visualizations provided by these systems would become increasingly cluttered, which would make them difficult to comprehend.

We present our ideas on how to tackle these shortcomings in section 4. However, before we describe our proposed extensions to the notional machine, in the next section we discuss advanced OO concepts: inheritance and polymorphism, as well as misconceptions related to them.

III. INHERITANCE AND POLYMORPHISM

Inheritance and polymorphism are considered advanced OO programming concepts [5]. Inheritance is a mechanism which allows the definition of a certain relationship between classes, through which one class, called the *derived class*, reuses the behaviors of another, called the *base class* [18]. Through these relationships a class hierarchy is constructed [19]. In statically typed programming languages, this allows a variable to refer to an instance of its subtype, which is called *subtype polymorphism*.

Although inheritance allows the reuse of behavior, there are situations when a reused method needs to be re-implemented. This is done through *overriding*. Hence, it is up to the runtime to decide which method implementation in the class hierarchy should be called. This is another form of polymorphism called *dynamic dispatch*.

Liberman et al. [5] studied the difficulties related to inheritance and polymorphism. They categorized the identified misconceptions into four clusters. We briefly summarize their findings.

The first cluster contained misconceptions related to "simplistic alternative programming models", which were probably caused by the participant's prior knowledge of procedural programming [5]. Within this cluster, four alternative models related to overriding and variable types were identified. In one alternative models the variable type determines which method will be dispatched. However, this model ignores the effect of overriding on dynamic dispatch. Another model caused participants to view variable types as irrelevant, and believed that method invocation depends only on the class from which the object was instantiated. Other models included beliefs that overriding is redundant, and that overriding eliminates the method that was overridden.

In the second cluster, the authors placed misconceptions related to using analogies. These included analogies to everyday life and to procedural programming [5]. Related to the latter, some participants believed that an array can store references of any type, unrestricted by the class hierarchy. This is probably due to an assumption that all references are of the same type.

The third cluster contains misconceptions related to inheritance [5]. Some participants thought that inheritance is used to change the values of fields, rather than for creating relationships between classes and reusing behaviors. A misunderstanding of this relationship caused participants to mistake class composition for inheritance.

In the final cluster, the authors placed misconceptions caused by misunderstandings of previous OO concepts [5]. These may be related to overloading, object creation, etc. Some misconceptions may be caused by certain teaching methods, e.g. using a single example to explain a concept and its uses.

We believe that some of these misconceptions could be resolved by visualizations that facilitate the acquisitions of valid mental models related to inheritance and polymorphism. In the remainder of this paper, we extend the notional machine notation proposed in [11]. Our extensions have been designed with the aforementioned misconceptions in mind.

IV. EXTENDING THE NOTIONAL MACHINE

In this section, we outline our extensions to the notional machine notation in order to support novices learning of polymorphism and inheritance. We further present a visualization system which introduces source code into the visual representation of the notional machine as needed. This should provide a tighter integration of source code and the notional machine visualization. We end this section with a discussion on how GUI applications may be visualized.

A. Extending the notional machine notation

An existing notation for a notional machine [11] provides a convenient and compact view of various instantiated objects, their states and call chain. However, it lacks some details required to visualize advanced OO concepts. For example, the existing notation isn't sufficient to visualize the execution that includes inheritance and polymorphism. We extend the notation by including details such as property types, inheritance and dynamic dispatch, in order to tackle the misconceptions discussed in the previous section.

Our first extension to the proposed notation of the notional machine is concerned with the visualization of inheritance. Although the basic notation visualizes the state of an instantiated object, it doesn't differentiate the ownerships of the inherited fields and methods. We tackle this by visually separating the fields and methods that originate from different classes. An example, along with the corresponding code, is depicted in Fig. 1. The example is taken from an assignment that students needed to complete in their OO programming course. The code in Fig 1. is abbreviated for clarity. It can be seen that the object was instantiated from the class *Cat* which inherits the property *Age* from its base class called *Animal*. Methods are, as in Novis, shown only when they are a part of the currently active call stack [11, 17].

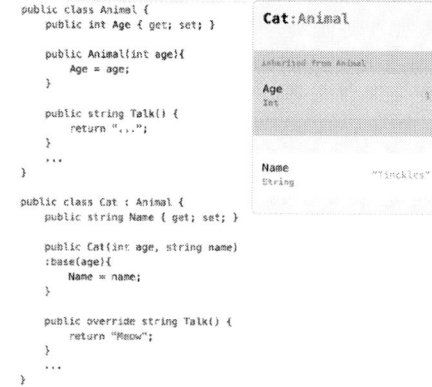

Figure 1.   An example implementation of a class (left) along with its visualization (right)

Unlike inheritance, object composition is depicted as an instance of one class referencing an instance of another via some field. An example is given in Fig 2. The difference in notation should help novices distinguish between inheritance and class composition. It can be seen that an instance of *Form1* holds a reference to an instance of a *Cat*, and *Cat* inherits from *Animal*. Note that object composition by itself is visualized the same way as in Novis [17].

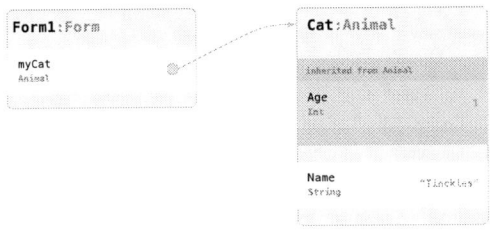

Figure 2.   Notation of object composition and subtype polymorphism

Before moving on to inheritance and polymorphism, novices tend to learn procedural programming and the basics of OO programming, such as classes and objects. In these early stages, the variable's or field's type is uniquely determined by the type of value or object assigned to it. However, with the introduction of inheritance, a variable may reference an object of a different type (subtype polymorphism). Therefore, in our extension each field is accompanied by its type, as shown in Fig. 2.

B. Integrating the source code into the visualization

The visualization that we have mentioned in the previous section are developed in a way that separates the source code that is being visualized and the visualization of the notional machine itself. However, this approach may be unsuitable to clearly visualize the second type of polymorphism, dynamic dispatch. Dynamic dispatch requires novices to understand which method will be invoked on a class instance, which depends on both the variable type and the class instance it refers to. Hence, both information need to be visualized clearly. Because of this, when visualizing the active call stack, we integrate the source code into the currently active method call, as shown in Fig 3.

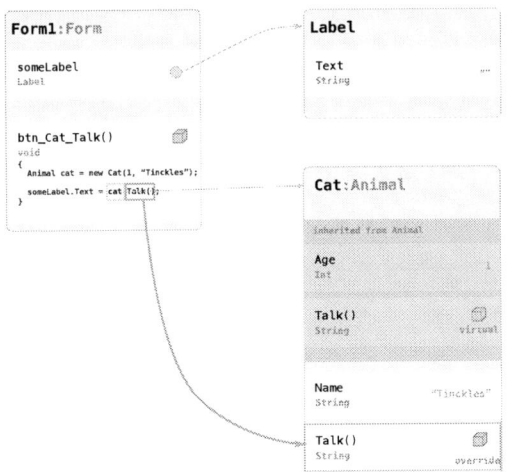

Figure 3.   Notation for visualizing which method will be invoked by dynamic dispatch once the highlighted expression is executed

The figure shows the call of a method with multiple implementations. The currently executing expression (*cat.Talk()*) is emphasized; the details about the callee object (*cat*) are shown (*Cat* in Fig 3.), as well as the dispatched implementation of the method *Talk()*. Both methods are visible in order to emphasize the fact that overriding doesn't eliminate the overridden method.

Integrating the source code into the visualization of the currently active method should reduce the gap between the code itself and the program state. Furthermore, this should reduce cognitive load as it allows the programmer to focus only on the part of the code that is currently being executed.

C. Visualizing applications with a graphical interface

Due to their simplicity and low overhead, console applications are often used by novices when learning the basics of programming and program execution. In turn, based on our experience most existing program visualizations for education are designed to visualize relatively simple console applications. Recently, educators have started to include GUI applications into their curricula, with the goal of increasing student engagement and motivation. However, GUI applications often have a much larger number of objects, program states, and executions. This means that existing program visualizations are often not particularly adapted to visualizing GUI applications, and they can overburden the novice with too many, often irrelevant information.

When tackling the complexity of visualizing GUI applications, we need to decide i) which subset of program executions we wish to visualize; and ii) which objects and fields are relevant for a particular execution.

Our visualization consists of two views. The first view shows the whole program execution as a series of events, where each event transforms the internal state of the program. Showing each change of the program state during event execution would cause cognitive overload in students due to a large number of changes. Because of this, the view shows only the transformations made by the event handler, that is for each object modified by the event handler we show the initial and final states of that object. This shows how the event-handler has transformed the program state.

In certain cases, we may wish to know how a series of events transforms the program state without looking at each individual event. Hence, a sequence of events may be grouped to show how the whole sequence transforms the program state. This view is depicted in Fig. 4.

The second view provides the students with the opportunity to view each step of the execution of a single event. This corresponds to standard visualizations of the notional machine. However, objects that are influenced by a single event may have a large number of fields, which is why our visualization displays only those that participate in the event handler execution.

The use of two views corresponds to Sorva's discussion [8] related to the use of two notional machines for OO programming. Because GUI applications are driven by the user's actions, the first view can be considered a higher-level notional machine for event-driven applications,

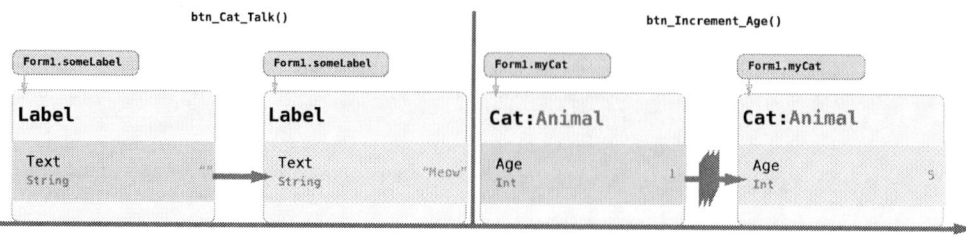

Figure 4. Notation for visualizing how a single event (left) and a sequence of events (right) transform program state

whereas the second is a more detailed notional machine of the execution of a particular event.

V. CONLUSION

In this paper, we have discussed the importance of visualizations in facilitating the understanding of the notional machine as a conceptual representation of how programs are executed. We have reviewed state-of-the-art visualization tools and have highlighted some of their shortcomings for advanced OO concepts, as well as applications with a large number of states, such as event-driven GUI applications. In order to tackle these shortcomings we have presented our own concept of a visualization system that: i) introduces notation for inheritance and polymorphism into the notional machine, where each object visualizes the ownership of inherited fields and methods, and each variable and field show their current type to emphasize the difference between reference type and instance type; ii) integrates the source code into the notional machine by displaying the source code of the currently active method; and iii) includes two views on program execution to reduce cognitive load. The first presents a high-level view of program execution by visualizing the transformations carried out by a group of events on a set of objects. The second gives a detailed view of the execution of a particular event by visualizing the method call stack along with the objects and fields that are relevant to the execution.

For future work, we plan to make a pilot study with students in order to determine the influence of our proposed notional machine extensions on their understanding of event-driven, object-oriented programs. We expect to further re-evaluate and develop the visualization in accordance with the prototyping approach in software development [20].

We will follow that with the development of a tool that would automatically visualize the notional machine in the context of arbitrary event-driven object-oriented programs. Finally, we will evaluate the effects of the visualizations provided by the tool in an educational setting.

We also plan to investigate the use of the high-level view as a notional machine for visualizing other programming paradigms.

REFERENCES

[1] A. Gomes and A. J. N. Mendes, "Learning to program-difficulties and solutions," in *International Conference on Engineering Education*, 2007, pp. 1–5.

[2] S. Mladenović, D. Krpan, and M. Mladenović, "Using Games to Help Novices Embrace Programming: From Elementary to Higher Education," *Int. J. Eng. Educ.*, vol. 32, no. 1, pp. 521–531, 2016.

[3] J. Hidalgo-Céspedes, G. Marín-Raventós, and V. Lara-Villagrán, "Learning principles in program visualizations: a systematic literature review," in *Frontiers in Education Conference (FIE), 2016*, 2016.

[4] A. Ferrari, A. Poggi, and M. Tomaiuolo, "Object Oriented Puzzle Programing," in *Didattica Informatica - Didamatica 2016*, 2016, pp. 1–10.

[5] N. Liberman, C. Beeri, and Y. Ben-David Kolikant, "Difficulties in Learning Inheritance and Polymorphism," *ACM Trans. Comput. Educ.*, vol. 11, no. 1, pp. 1–23, 2011.

[6] A. Fowler and B. Cusack, "Kodu Game Lab: Improving the motivation for learning programming concepts," in *Foundations of Digital Games*, 2011, pp. 238–240.

[7] A. Wilson, T. Hainey, and T. Connolly, "Evaluation of Computer Games Developed by Primary School Children to Gauge Understanding of Programming Concepts," in *Proceedings of the European Conference on Games Based Learning*, 2012, pp. 549–558.

[8] J. Sorva, "Notional machines and introductory programming education," *ACM Trans. Comput. Educ.*, vol. 13, no. 2, pp. 1–31, 2013.

[9] B. Du Boulay, "Some Difficulties of Learning to Program," *J. Educ. Comput. Res.*, vol. 2, no. 1, p. 57, 1986.

[10] J. Sorva and T. Sirkia, "UUhistle: a software tool for visual program simulation," in *Proceedings of the 10th Koli Calling International Conference on Computing Education Research Koli Calling 10*, 2010, pp. 49–54.

[11] M. Berry and M. Kölling, "The state of play," in *Proceedings of the 2014 conference on Innovation & technology in computer science education*, 2014, no. December, pp. 21–26.

[12] J. Sorva, V. Karavirta, and L. Malmi, "A Review of Generic Program Visualization Systems for Introductory Programming Education," *ACM Trans. Comput. Educ.*, vol. 13, no. 4, 2013.

[13] P. J. Guo, "Online python tutor: Embeddable web-based program visualization for cs education," in *Proceedings of the 44th ACM Technical Symposium on Computer Science Education*, 2013, pp. 579–584.

[14] A. Moreno, N. Myller, and E. Sutinen, "Visualizing Programs with Jeliot 3," in *Proceedings of the working conference on Advanced visual interfaces*, 2004, pp. 373–376.

[15] B. Earwood, J. Yang, and Y. Lee, "Impact of Static and Dynamic Visualization in Improving Object-Oriented Programming Concepts," in *Frontiers in Education Conference (FIE)*, 2016.

[16] P. Gestwicki, B. Jayaraman, and B. Hall, "Methodology and Architecture of JIVE," in *Proceedings of the 2005 ACM symposium on Software visualization*, 2005, vol. 1, no. 212, pp. 95–104.

[17] M. Berry and M. Kölling, "Novis: A notional machine implementation for teaching introductory programming," in *Learning and Teaching in Computing and Engineering (LaTICE), 2016 International Conference on*, 2016.

[18] M. Vujošević-Janicic, D. Tošic, M. Vujosevic-Janicic, and D. Tosic, "The role of programming paradigms in the first programming courses," *Teach. Math.*, no. 21, pp. 63–83, 2008.

[19] A. L. Santos, "Novel interaction metaphors for object-oriented programming concepts," in *Proceedings of the 14th Koli Calling International Conference on Computing Education Research - Koli Calling '14*, 2014, pp. 117–126.

[20] R. S. Pressman, *Software Engineering A Practitioner's Approach 7th Ed - Roger S. Pressman*. New York: McGraw-Hill, 2009.

Mediated Transfer from Visual to High-level Programming Language

D. Krpan*, S. Mladenović* & G. Zaharija*
* Faculty of Science, University of Split, Split, Croatia
divna.krpan@pmfst.hr, sasa.mladenovic@pmfst.hr, goran.zaharija@pmfst.hr

Abstract - Visual programming languages (VPLs) are becoming more popular and making the transition from the informal to conventional educational settings. One of the important features of VPL is that novices are not required to remember a list of commands or complex syntax since everything they need is just there in the environment. The objective of introductory computer programming courses at the university is to teach students how to develop solutions in high-level computer programming languages such as C#. However, they also need to acquire problem-solving skills. Since computer programming and problem-solving are both challenging, schools and universities often make use of VPLs combined with game-based programming. Students will eventually need to transfer programming concepts learnt from VPL into a high-level programming language. A transition from VPL to the text-based high-level programming language is not seamless and additional tools and efforts are required. This paper presents prototypes we have developed for undergraduate university students to enable mediated transfer from VPL to high-level programming language by using the idea of mini-languages.

I. INTRODUCTION

Visual programming languages are gaining popularity as the possible solution for novice programmer's difficulties. The purpose of this paper is to provide a better understanding of the advantages and limitations of using visual programming languages in formal educational context and to suggest a potential model for teaching programming at the University undergraduate programming courses. The model is using a visual programming language, and it is based on the established didactical strategies such as didactic reduction and transfer of learning [1, 2].

Teaching programming at the university, specifically teaching introductory programming courses does not promote novices to expert programmers, and while researchers argue that such process may even take ten years [3, 4], the more concerning problem is students' poor performance after completing the first course [5]. Teachers may not expect such poor performance, but it is not uncommon. Sometimes the problem lies in the belief that all students should be able to program, or that weaker students cannot learn to program, so teachers eventually let them pass the course. Although some authors claim that some students will never learn programming, others, in the context of computer literacy, argue that most of them should be able to acquire fundamental programming skills and knowledge without expecting them to become experts [6].

The next section contains related work and didactical strategies that motivated us in designing the prototypes described in this paper. The pilot testing described here is the follow-up research to the three years study on the influence of visual programming languages and game-based learning on the students' achievement in programming courses. The study was conducted at the Faculty of Science, the University of Split from 2010 to 2013 [7]. Researchers expected that transfer of learning is going to happen spontaneously, but the influence of game development in visual programming languages on programming courses taught in professional text-based programming languages was weak. As suggested by the previously mentioned research, we have developed prototypes and conducted pilot testing with the students majoring in computer science (CS) who will become teachers of CS upon graduation. Because teachers learn from other teachers, this is also the opportunity to transfer the experience [8].

II. RELATED WORK

Case studies in computer science education are often small-scale studies within one course, and many studies are reported in "Marco Polo" style while arguing for one approach over the other [9]. Long-term studies dealing with the implications across different courses and long-term effects are not that common [7]. To better understand the whole process of learning and teaching programming it is important to put things into perspective, or a model. The conventional model for teaching programming is represented by "didactical triangle". The didactical triangle consists of (i) *student* (who to teach), (ii) *teacher* (how to teach) and (iii) *content* (what to teach) [9, 10]. The whole triangle also belongs to the context [10].

For example, if a teacher chooses the strategy of problem-based learning for teaching programming, selected problems should be based on real world situations whenever possible. Otherwise, besides abstract nature of programming concepts and language constructs, students also have to deal with the task of analyzing and understanding some unknown abstract problem. Game-based learning (GBL) or gamification is often used as motivation f [7]. Playing educational games and game development are two different types of GBL. Most of the students consider games as fun.

However, while thinking of using game design in the course of object-oriented programming, we have realized that even simple game development in general-purpose languages like C# or Python is very complex for novices. It usually requires using extensive additional libraries, so besides basic language constructs and programming concepts they struggle with, novices are burdened with other subject matter. The focus of the next section is the first part of the triangle related with our educational settings, that is undergraduate university students.

A. Teaching undergraduate students programming

The student in the triangle represents learning process, student's previous knowledge, skills and experience and other characteristics that influence learning. Teaching children and adults is different. Learning programming at the university is concerned about adult students who require a different approach, more student-centered, active learning and problem-based learning which shifts teacher's role into facilitator or guide [11].

Besides learning some programming language, the important goal of introductory programming courses is to develop problem-solving skills [12]. Computer programming used for problem-solving skills is the part of computer literacy [13].

Most of the first year undergraduate students at the Faculty of Science (FOS) at the University of Split are novices without any previous experience. Novices often acquire superficial knowledge, they approach programming "line by line", have trouble planning and understanding programming concepts focusing on the specific programming language syntax. That is somewhat understandable since introductory programming textbooks are often structured according to the programming language constructs [14]. However, students born after the 1980s belong to the new millennial or net generation that requires a different approach and teaching strategies with visual incentive [15]. Even if teachers were successful in learning programming using console applications, that doesn't mean that students will share the same enthusiasm.

B. Teaching strategies for programming

Teacher in the didactic triangle represents overall teaching process, choosing strategies, decisions about content selection and presentation while considering both the content and the student. Novice programmers must learn various new and abstract concepts in programming which is difficult [16, 17]. Teaching object-oriented programming is considered difficult, and some of the identified problems are teachers' lack of experience, unfamiliarity with the paradigm, lack of textbooks, unsuited software tools or environments, focus shift from algorithms to software engineering point of view [18].

One of the instructional strategies we have considered applicable for our settings is known as *didactic reduction*, and it represents the teaching approach where some aspects of learning are simplified [1, 19]. The strategy is applied when learning a complex subject and implies that additional details are discovered subsequently step by step. Computer programming is complex, and we believe this strategy might help students to avoid cognitive overload [20]. The idea of using didactic reduction comes in handy with the transfer of learning strategies or teaching for *transfer*.

1) Transfer of learning

Transfer of learning is the ability to apply knowledge and skills acquired in one context to new contexts. It occurs when learner recognizes common features, links information in memory and sees the value of using familiar concepts from one learning situation to another [2]. Positive transfer occurs when one context improves performance in the other. Negative transfer is the opposite, but it happens in early stages of learning and learners usually correct it through the experience. Two strategies help with the transfer of learning: *hugging* and *bridging*. Hugging exploits low road transfer when conditions in both contexts are similar, and bridging exploits high road transfer when learner must create abstractions and consciously search for connections between contexts. Bridging is the more complicated process, and instead of waiting for students to achieve it spontaneously, teacher interferes and mediates the transfer. In this paper, we believe that *mediated transfer* should include hugging as well since the combination of the both strategies is expected to produce a better effect.

One of the approaches in the attempt to explain teaching programming is called *semiotic ladder* which emphasizes tools (programming or modelling language) with learning stages [21]: (i) syntax, (ii) semantics and (iii) pragmatics. Semiotics studies the signs and symbols in linguistics and communication. The syntax is about combining symbols into programming language instructions, semantics about instructions' meaning or functionalities, and pragmatics is about the application. Each step of the ladder is based on the previous. Although very simple, this approach is mostly about a programming language. It is not the critical issue since all parts of the triangle are important, but studies have demonstrated that students struggle with syntax far more than we expect which leads them to focus on writing code without understanding its functionality [22]. Tools which students use for programming (language and the environment) may have a significant influence on their thinking [13]. The first encounter with the programming is sometimes the crucial moment when teachers win or lose students' interest in the programming, so there is significant responsibility for teachers in choosing the appropriate language for introductory programming courses.

C. First programming language

The content in the triangle represents the learning objectives, evaluation of the achievement and actual content which in programming consists of basic concepts such as variables, conditions, loops, decisions, etc. including programming language specifics. Sometimes the content is identified with a programming language. For example, if we ask students or teachers what they learn or teach in the programming course, the answer is often just the name of programming language. The choice of the first programming language is important for motivation [23]. Tightly coupled with the language choice is also the programming paradigm which affects the way student analyzes the problem and creates the solution [24]. Students should learn programming concepts independent of programming language or the environment [24].

The standard approach for teaching programming using professional programming environment and language results in low effectiveness because of the following difficulties [25]:

- language complexity – large amount of language concepts and its specifics,

- inappropriate feedback for novices – the process of program execution is hidden which makes understanding semantics difficult for novices (some sort of visual feedback would be preferable),

- inappropriate feedback for novices – the process of program execution is hidden which makes understanding semantics difficult for novices (some visual feedback would be preferable),

- problem types – language orientation on number or symbol processing makes developing interesting "real-world" applications difficult because it requires learning a substantial amount of language concepts, mastering programming environment and writing complex programs.

Many general-purpose programming language constructs are too abstract for novices. Papert developed Logo programming language with the purpose of simplifying learning programming for beginners [26]. Such smaller languages are called *mini-languages* [25]. Designers of mini-languages are not limited by the syntax or semantics of any complex language. Mini-languages resolve all three difficulties: they are small with simple syntax and semantics, and preferably with visual feedback. Visual feedback is usually represented by a visual *actor* (e. g. Logo turtle) that exposes semantics while performing visible actions. Logo and turtle graphics influenced many other mini-programming languages that are being used for teaching programming to different age groups ranging from elementary to university students. Successors of Logo are *visual programming languages* (VPLs) such as Scratch, Snap! [27], Blockly [28], Alice, Greenfoot etc. VPLs can be entirely visual where commands are represented by blocks and are referred to as *block-based languages* or hybrid with text-based commands. Greenfoot has a visual run-time environment (with visual actors), but coding is performed in Java programming language.

Scratch [29] is developed at the MIT Media Lab and launched in 2007 [30]. Scratch 2.0 can be used directly from a web browser, and no installation is required although the offline version is available. Scratch environment consists of sprite list, block palette, script area and stage. Blocks in the palette are grouped by their purpose into different colored categories like motion, looks, sound, etc. Each sprite has its script area which may contain multiple scripts. Users design scripts by dragging blocks from palettes and assembling them like puzzles in the script area. Shape enforces the syntax.

Potential weaknesses are simple data structures and functions, but it is not an oversight, language designers are following the idea of Logo to keep it simple and more intuitive for children. Scratch supports advanced CS concepts in the simple two-dimensional environment, while Alice three-dimensional world is more complex [31].

Since universities are interested into drawing more students into programming, one form of motivation is

lowering the first barrier by avoiding complex syntax. Courses using Scratch have been taught at Harvard University [32] and the College of New Jersey. The extension for Scratch was developed at the University of California, Berkely called BYOB (Build Your Own Blocks) with the advanced handling of functions. BYOB was intended for university students with the support for first-class data and more explicit support for object-oriented programming. The successor of BYOB is "Snap!" which was redesigned in JavaScript. In Scratch 2.0 users can create custom blocks that represent functions in conventional programming languages. However, custom blocks in "Snap!" are far more customizable. "Snap!" lists and functions are both first-class data elements and both can be assigned to variables, passed as arguments to other functions, and also returned from functions as a result.

Since VPLs are getting more popular, it is important to understand how to effectively include them into conventional classrooms [33, 34]. The goal of introductory programming courses at the universities is to introduce students to programming quickly. University teachers aim to teach students some of the professional text-based programming languages, and they are interested in the transfer of learning from block-based to text-based languages. Although some VPLs had success motivating these transitions, it is not clear that any of these languages are the best means to support it [22] [12]. An important advantage for VPLs is low barrier. Students without any prior experience in programming can quickly start programming in VPL [35].

Lack of syntax is a key feature that makes VPLs more appropriate for young novice programmers by focusing on the problem solving, but some research shows that it merely delays syntax problems or even makes students believe that syntax errors are not that important causing them to submit assignments that could not be compiled [36].

Google Blockly can are translated into different programming languages from block representation such as JavaScript and Python. Some researchers suggested that transfer in both directions is better [37] because students with lower self-evaluation of their programming skills choose block-based editor more often if given the opportunity, but most students gradually migrate to text-based programming language [12]. However, introducing text-based interface and free typed text in these environments would re-introduce syntax errors. Scratch supports over 60 languages, so the transition to text would complicate translation [38].

The research in [13] was focused on high school students where researchers have conducted the experiment with "Snap!" using programming in JavaScript. Students later transferred to Java programming language. A majority of students perceived block-based programming easier than text-based. Students emphasized some features as advantages, e. g. there is no need to memorize commands and the text on the blocks was in natural language. Students had difficulties with designing more complex programs.

There is a difference in motivation between younger students in elementary school and older students in high school and university. Older students are concerned that block-based programming languages are not "real" [37], although they use a variety of programming structures and concepts [22]. On the other side, a group of students perceived this environment as friendly and playful causing them to feel more relaxed and thus become more productive. Design space which blends text and block-based features should be more explored since students' transition to text-based programming environment is not straightforward [13]. "Drag and drop" was found to be time-consuming since users had to look for the appropriate block in the particular palette [39]. Some of the users avoided using global variables and message broadcasting to trigger some code parts and duplicated code snippets. Such solutions created problems in editing multiple copies of the code.

Graphical blocks used in VPLs take more space than text. That is preferable if blocks are intended to be used by young children using touch screens, but is not so practice for adults [38]. It is harder to navigate and often impossible to search for a particular statement in the code. The palettes with blocks help novices to discover all available commands, but more experienced programmers complain that after learning that rather small set of commands, dragging blocks out of palette takes more time than typing. Searching for the desired block in visual languages involves visual scanning of the palette, and the process may interfere with student's thinking about a problem.

Although Scratch is used as the motivational tool for engaging novice students in programming, it is clearly not suitable for professional development, and some researchers report difficulties even with the transfer from Scratch to pseudo code and flowcharts [40]. Students were confused and their test results were much lower than in Scratch.

Snap! is more complicated for young students in elementary school but offers more advanced features. A simple improvement is searching for blocks to avoid visual scanning of the palette and hiding of the useless blocks thus supporting the idea of didactical reduction. Researchers in [35] exploited some of the advanced features in the effort to introduce concepts of parallel programming, and some of those features are also available in Netsblox [41]. Advanced handling of custom blocks (equivalent for functions) with the lists and functions as first-class data provided the opportunity to design a custom environment for students. Using "Snap!" experimental codification support that enables block mapping into text-based programming language [27] we have developed custom blocks that generate text snippets of code that can be copied into targeted environment which is described in the next section.

III. PROTOTYPE FOR MEDIATED TRANSFER

Students at the Faculty of Science (FOS), University of Split, as novice programmers have similar problems as any other adult novices. They differ based on their study majors: mathematics (M), Physics (P), Technical science (T), Informatics (I), and also combinations such as: MI or IT. All first-year student with majors mentioned above learn Python and procedural programming during course *Programming I* (P1), and in the second semester during course *Programming II* (P2) they are introduced to object-oriented programming paradigm and C# programming language. Since both the language and switch to the object-oriented paradigm are complex, the course P2 is also considered introductory. Students with major in informatics have additional course *IT Project I* (IP1) in the first semester and learn VPLs and simple game development using languages from Scratch 1.4, BYOB, Scratch 2.0 to "Snap!". First version of Scratch was simpler and lacked some of the functionalities. BYOB introduced custom blocks, recursions and clones (run-time sprite instances) and was more appropriate for university students.

In the third semester, students enroll the course *Object-oriented programming* (OOP) with the project-based teaching based on GBL. Using additional game development frameworks such as Unity proved to be too complex, so we have developed our own two-dimensional (2D) game development framework in C#.

Although some students learn VPL and design 2D games during the first semester, OOP course is in the third semester, so the time frame of the whole semester is too long to expect any significant transfer. Hence, in this paper, we present prototypes of the environment designed to support the transfer of learning from visual language to a text-based professional language, namely Python and C#. "Snap!" programming environment has a built-in feature for code mapping, but it is necessary to design each command while respecting the syntax rules. The Fig. 1. a) depicts for-loop blocks developed in "Snap!" environment and an example of mapping those blocks into Python and Fig. 1. b) represents a snapshot of custom designed blocks and the result of the block "C#". Green blocks named "C#" and "method BearFollow" are parts of the implementation. However, mapping feature was not enough to obtain desired features for C# translation, so some additional effort was required. JavaScript language was used with the block "JavaScript function" for running JavaScript code. In Fig. 2 there is a screenshot of developed framework running in the Microsoft Visual Studio (MVS) professional environment using Python tools and Pygame module. Fig. 3 represents C# game development framework.

The goal is to keep students in familiar environments, both "Snap!" and professional environment of the MVS without creating additional applications and students have access to the source code. The result windows represented in the Fig. 2 and Fig. 3 are windows that look a lot like Scratch or Snap! stage. Both frameworks run in MVS environment, and classes and methods mirror "Snap!" programming constructs. For example, students can use methods like *bear.Move()* or *bear.GotoMousePoint()* in

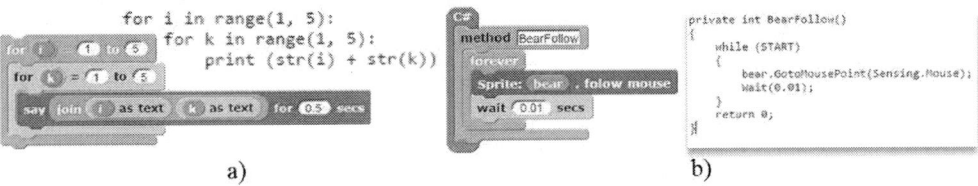

Figure 1.   Translation of Snap! blocks to a) Python and b) C#

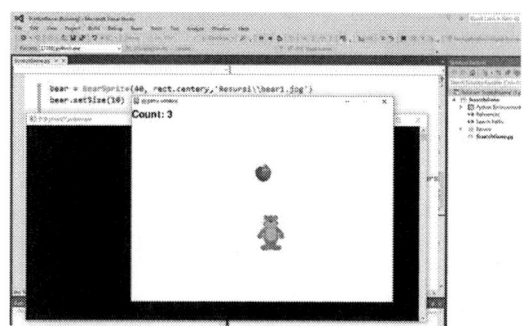

Figure 2. Python game development framework in MVS

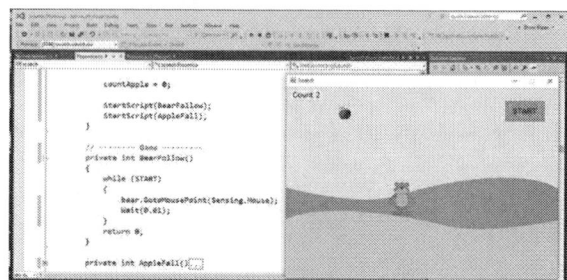

Figure 3. C# Game development framework

both frameworks and blocks with same functionalities are available in Snap! too. However, MVS does not support a visual designer for forms and controls in Python and students are not able to "drag and drop" controls on the stage like they are used to in VPLs and sprites are visible only during runtime (like in Fig. 2). C# sprites are windows controls and students can position those controls and resize them in MVS designer.

The C# framework was presented to the 29 first year students in the last three weeks of the course IP1 after 12 weeks of experience in Scratch and no experience in C#. Students were first introduced to "Snap!" and the transition from Scratch was almost seamless. Students only asked where to download their solution because the menu was different. They were introduced to advanced features of custom blocks in "Snap!" and had a task of designing for-next loop which does not exist in Scratch or "Snap!". This was a very complex task since students were required to pass scripts as arguments, save the script into variable and use run block to run. Naming the variable as "action" was helpful because they understood that the block would "run actions", and they had to put those actions into *C-shaped* block.

During the second week, C# game framework was presented to the students including "Snap!" assignment with the limited set of custom blocks that were translatable into C#. The Snap! project contained only custom blocks and a subset of the standard block set necessary for the solution. The task was to create a simple game consisting of two sprites, one bear and red apples. The red apple was called "good" because bear gets points for catching it, and if he misses three apples, the game is over. Students were required to solve the known problem in this limited environment and use special blocks to generate C# code and copy into MVS environment. They were familiar with this type of game because they solved few similar assignments in Scratch. However, students had different ideas about a solution than teacher predicted and they were missing some standard blocks such as the block labelled "stop-all" to stop all activity when the game finishes, and some of them were frustrated while searching through changed block palette and not finding what they expected.

In the final week, students were provided with the same environment, similar assignment, and prepared blocks for "Snap!" solution in the script area, so students did not have to browse the palette, just to reassemble the script and make it work. The game had one more sprite, brown apple which

was named "bad" because bear loses points for catching it, but he was allowed to drop it. The other part of the assignment was again to transfer the solution into the C# environment. The students submitted 15 correct solutions in "Snap!" and 10 in C#. One of the students added some code to make all sprites in the C# solution stop when the game was over event it was not the part of the assignment. Students did not know C# syntax, and the reason for this testing was to examine the performance of the prototype in real circumstances with students designing possibly unpredicted solutions. The Python prototype was not tested at the time since it required an installation of new libraries and different Python version 2.7 that would possibly interfere with their current knowledge especially since students used Python 3.5 in a procedural way without learning object-oriented programming concepts.

Although the sample was small, a short anonymous online questionnaire was conducted to examine students' attitudes towards their VPLs experience. One of the interesting questions was: *Which programming languages you would choose for the IP1 course?* A total of 25 answers was collected, and 56% of students chose Scratch, while the next choice was *"Snap!" with C# transfer* with 16%. The next question was to select the reason for their choice and most of the students selected: *Simple for learning and using.* Teacher expected student would prefer "Snap! with C#" for the benefits in next programming course, but they picked an easier path for the current course.

IV. CONLUSION

Game development frameworks were developed as a follow-up research to exploit the effect of the mediated transfer and the theory of didactic reduction. The developed prototypes translate "Snap!" blocks into C# and Python. Python was selected since it is widely used in Croatian schools (elementary and high school) as the introductory programming language, and C# is also popular in the industry. Python is an interpreter, and C# compiler language, but both can run in the MVS environment. The suggested model for transfer is expandable for other languages. The important advantage of this model as opposed to the example of Google Blockly is that the generated code can run in the professional environment with the same visual actors (sprites).

However, since block-based and text-based environments are not isomorphic and targeted languages are more complex, after discovering more complex details in the background students will need to include abstract thinking and develop connections.

REFERENCES

[1] G. Futschek, "Extreme Didactic Reduction in Computational Thinking Education," in *X World Conference on Computers in Education*, 2013, pp. 1–6.

[2] D. N. Perkins and G. Salomon, "Transfer of learning," *Int. Encycl. Educ. Second Ed.*, vol. 2, pp. 1–11, 1992.

[3] A. Eckerdal, "Novice Programming Students â€™ Learning of Concepts and Practise," Uppsala University, 2009.

[4] N. Truong, "A Web-Based Programming Environment for Novice Programmers," Queensland University of Technology, Australia, 2007.

[5] R. Lister and J. Leaney, "Introductory programming, criterion-referencing, and bloom," *ACM SIGCSE Bull.*, vol. 35, p. 143, 2003.

[6] M. E. Caspersen, "Educating novices in the skills of programming," Department of Computer Science, University of Aarhus, 2007.

[7] S. Mladenović, D. Krpan, M. Mladenovic, and M. Mladenović, "Using Games to Help Novices Embrace Programming: From Elementary to Higher Education," *Int. J. Eng. Educ.*, vol. 32, no. 1, pp. 521–531, 2016.

[8] J. D. Bransford, A. L. Brown, and R. R. Cocking, *How people learn: Brain, mind, experience, and school: Expanded edition*. Washington DC, USA: National Academies Press, 2000.

[9] J. Bennedsen, "Teaching and Learning Introductory Programming - A model-Based Approach," University of Oslo, Norway, 2008.

[10] M. A. Tchoshanov, *Engineering of Learning: Conceptualizing e-Didactics*. Moscow: UNESCO Institute for Information Technologies in Education, 2013.

[11] M. W. Shreeve, "Beyond the didactic classroom: educational models to encourage active student involvement in learning," *J. Chiropr. Educ.*, vol. 22, no. 1, pp. 23–28, 2008.

[12] Y. Matsuzawa, T. Ohata, M. Sugiura, and S. Sakai, "Language Migration in non-CS Introductory Programming Through Mutual Language Translation Environment," in *Proceedings of the 46th ACM Technical Symposium on Computer Science Education*, 2015, pp. 185–190.

[13] D. Weintrop, "Minding the gap between blocks-based and text-based programming: Evaluating introductory programming tools," in *Proceedings of the 46th ACM Technical Symposium on Computer Science Education*, 2015, p. 720.

[14] A. Robins, J. Rountree, and N. Rountree, "Learning and teaching programming: A review and discussion," *Comput. Sci. Educ.*, vol. 13, no. 2, pp. 137–172, 2003.

[15] L. B. Nilson, *Teaching at Its Best: A Research-Based Resource for College Instructors*, vol. 2nd. Jossey-Bass Wiley Imprint, 2003.

[16] E. Nuutila, S. Törmä, and L. Malmi, "PBL and computer programming—the seven steps method with adaptations," *Comput. Sci. Educ.*, vol. 15, no. 2, pp. 123–142, 2005.

[17] E. Lahtinen, K. Ala-Mutka, and H.-M. Järvinen, "A study of the difficulties of novice programmers," in *ACM SIGCSE Bulletin*, 2005, vol. 37, no. 3, pp. 14–18.

[18] M. Kölling and J. Rosenberg, "BlueJ - The Hitch-Hikers Guide to Object Orientation," Odense, Denmark, 2002.

[19] P. Menck, *Looking into classrooms: papers on didactics*. Greenwood Publishing Group, 2000.

[20] J. Sweller, P. Ayres, and S. Kalyuga, *Cognitive Load Theory*, 1st ed., vol. 10. New York: Springer-Verlag New York, 2011.

[21] J. J. Kaasbøll, "Exploring didactic models for programming," in *NIK 98--Norwegian Computer Science Conference*, 1998, pp. 195–203.

[22] E. Pasternak, "Visual Programming Pedagogies and Integrating Current Visual Programming Language Features," vol. 29, no. 3, pp. 1–30, 2009.

[23] L. Grandell, M. Peltomäki, R.-J. Back, and T. Salakoski, "Why complicate things?: introducing programming in high school using Python," in *Proceedings of the 8th Australasian Conference on Computing Education-Volume 52*, 2006, pp. 71–80.

[24] N. Ragonis and M. Ben-Ari, "A long-term investigation of the comprehension of OOP concepts by novices," *Comput. Sci. Educ.*, vol. 15, no. 3, pp. 203–221, 2005.

[25] P. Brusilovsky, E. Calabrese, J. Hvorecky, A. Kouchnirenko, and P. Miller, "Mini - languages: A Way to Learn Programming Principles," *Educ. Inf. Technol.*, vol. 2, no. 1, pp. 65–83, 1997.

[26] S. Papert, *Mindstorms: Children, Computers, and Powerful Ideas*. Basic Books, Inc., 1980.

[27] B. Harvey and J. Mönig, "SNAP! Reference Manual," *online*, 2017. [Online]. Available: https://snap.berkeley.edu/SnapManual.pdf. [Accessed: 01-Jan-2017].

[28] N. Fraser, "Google Blockly - a Visual Programming Editor," *Google*, 2012. [Online]. Available: https://developers.google.com/blockly/. [Accessed: 01-Feb-2017].

[29] M. Resnick *et al.*, "Scratch: Programming for All," *Commun. ACM*, vol. 52, no. 11, pp. 60–67, Nov. 2009.

[30] B. Harvey and J. Mönig, "Bringing 'No Ceiling' to Scratch: Can One Language Serve Kids and Computer Scientists?," in *Constructionism*, 2010, pp. 1–10.

[31] M. Armoni, O. Meerbaum-Salant, and M. Ben-Ari, "From Scratch to 'Real' Programming," *ACM Trans. Comput. Educ.*, vol. 14, no. 4, pp. 1–15, Feb. 2015.

[32] D. J. Malan and H. Leitner, "Scratch for Budding Computer Scientists," in *Sigcse 2007: Proceedings of the Thirty-Eighth Sigcse Technical Symposium on Computer Science Education*, 2007, pp. 223–227.

[33] O. Meerbaum-Salant, M. Armoni, M. Ben-Ari, and M. (Moti) Ben-Ari, "Learning Computer Science Concepts with Scratch," *Comput. Sci. Educ.*, vol. 23, no. 3, pp. 239–264, 2013.

[34] O. Meerbaum-Salant, M. Armoni, and M. Ben-Ari, "Habits of programming in scratch," in *Proceedings of the 16th annual joint conference on Innovation and technology in computer science education*, 2011, pp. 168–172.

[35] A. Feng and W. C. Feng, "Parallel programming with pictures in a snap!," in *Proceedings - IEEE 28th International Parallel and Distributed Processing Symposium Workshops, IPDPSW 2014*, 2016.

[36] K. Powers, S. Ecott, and L. M. Hirshfield, "Through the looking glass: Teaching CS0 with Alice," *SIGCSE 2007 38th SIGCSE Tech. Symp. Comput. Sci. Educ.*, vol. 1, pp. 213–217, 2007.

[37] D. Weintrop and U. Wilensky, "Bringing Blocks-based Programming into High School Computer Science Classrooms," in *Annual Meeting of the American Educational Research Association (AERA 2016)*, 2016.

[38] J. Mönig, Y. Ohshima, and J. Maloney, "Blocks at your fingertips: Blurring the line between blocks and text in GP," *Proc. - 2015 IEEE Blocks Beyond Work. Blocks Beyond 2015*, pp. 51–53, Oct. 2015.

[39] E. Tanrikulu and B. C. Schaefer, "The Users Who Touched the Ceiling of Scratch," *Procedia - Soc. Behav. Sci.*, vol. 28, pp. 764–769, 2011.

[40] J. Chetty and G. Barlow-Jones, "Bridging the Gap: the Role of Mediated Transfer for Computer Programming," in *International Proceedings of Computer Science & Information Technology*, 2012, vol. 43.

[41] B. Broll, P. Völgyesi, J. Sallai, and A. Lédeczi, "NetsBlox: a Visual Language and Web-based Environment for Teaching Distributed Programming," 2016.

Story of a 'Storyline Visualization' in High School Readings

K. Osmakčić* and K. Kocijan*
* University of Zagreb, Faculty of Humanities and Social Sciences / Department of Information and Communication Sciences, Zagreb, Croatia
kosmakci@ffzg.hr, krkocijan@ffzg.hr

Abstract - Storyline visualization, as a process of illustrating data that has a course of events via a visual medium, has been used in the area of film making for a very long time. Not so long ago, it has moved from the paper version to the digital word allowing for a wider usage. In this paper we propose its usage as a teaching tool in the area of literature reading for the Croatian class (primary language). We have conducted a preliminary research in five Croatian high schools of a different profile to see how storyline visualization, and visualization of school materials in general, affects students understanding of the material being studied. Each school participated with two groups of students where one group was exposed to the storyline visualization of a novel *Prokleta avlija* by Ivo Andrić [N=103 in total] during the reading period, and the other one was reading without the visualization [N=93 in total]. We will present our results taking into account students' gender and type of a school.

I. INTRODUCTION

Many researchers agree that an eye is an important information receiver and that our brain does a good job in finding patterns in visual presentations of data [10], [2], [5], [3]. Thus, it is by no surprise that, in this information era that we live in, we are trying to find alternative ways to present information in a visually more understandable (comprehensive) way in order to absorb more information in a much faster manner. In the recent age, as we are surrounded with more and more data, and presence of Big Data is becoming overwhelming, this obsession of visual presentation of data has become more common, and surely even expected in business domain.

But, what about the arena of education? How can we help our students to better understand and better incorporate all the information that we are feeding them with in schools? Can we use visualization as a helping tool? Obviously, the answer is '*yes*'. Visualization is not a novel thing to education and different types of visualizations have been used for different lessons trying to convey the relevant information [5]. Some of the most widely used ones are geographical maps, periodical table of elements, Venn diagram, cyclic graphs, mind maps, UML diagrams, pie charts, bar charts, tree maps etc. and their usage depends mostly on the topic being taught.

And what about storyline visualization? Is it helpful in mastering the school curriculum? In what subjects can it help? Is it for everyone? These are just few of the

questions that motivated us in this project. The main characteristic of this type of visualization is that it helps to demonstrate when and where did certain characters meet during the course of the story. So far, it has mainly been used to show movie character interactions. The professional illustrators would draw a board with characters moving through time and space to get an insight into the story flow. These lines where usually drawn by hand, but recently several software tools have been introduced that are able to draw the story line automatically after parameters like *characters, places* and *time* are provided as an input [17], [18].

In order to test our hypothesis that storyline visualization can be used in classrooms and can help in better understanding of a story, we have conducted a small scale research on 196 high school students. In the following sections, we will first provide some information on visualization giving a special attention to storyline visualization and its usage throughout history. Second, we will describe in more details how we designed a storyline visualization for a novel *Prokleta avlija* by Ivo Andrić and how we conducted our research. Before we conclude, we will present and discuss our results.

II. STORYLINE VISUALIZATION

A. Short History

Storytelling through visualization is not a novel approach to telling a story. It can be traced all the way back at the very beginnings of humanity. We can talk about visualizations when we talk about drawings in the old caves like Altamira or Lascaux, and later in the old Egyptian pyramids, and up to the more modern times in children story-books. Some of the most known visualizations are Galileo's sun spot map, Descarte's coordinates system, *New Chart of History* by J. Priestly, John Snow's map of cholera appearances in London, or Florence Nightingale's diagram of cause of death of soldiers fighting in Crimean war.

In spite the fact that most visualizations are understandable to different language groups, and that visualization can be regarded as a universal language, it is surprising that its usage was not always as popular or as widely spread as today. Although since 15th century it has been mostly on a raise, since the end of the 19th and up to the middle of the 20th century it had its modern dark age period when it showed a sudden decline, probably caused by two Wars [8]. Speedy raise in science and technology, and production of immense quantities of data since the 2nd half of the 20th century made a fertile ground for the raise of visualizations as well.

This brings us to the modern age of computers which are almost unthinkable without different graphical presentations, from disc free/occupied space and disc defragmentation process, or schema of folders and documents to visualization of different social network happenings and phenomena. All of these examples use very large numbers of information and our brain is much better in finding patterns and making sense of those information if presented visually than via text or pure numbers [10], [13]. The main intention behind these visualizations lies thus in accentuating data's features, finding "patterns, and simultaneously showing features that exist across multiple dimensions" [2]. It is about informing in a fun manner in order to motivate an action.

B. From Visualization to Storyline Visualization

A new approach to storytelling through visualization is a method called storyline visualization. Although it shares a common 'DNA' with other visualizations [5], it differs in the number of details it can show, but also in the type of information we can learn from it. Its main task is to give us an overview of a number of events and/or people over a period of time. Tanahashi and Kwan-Liu [18] define it as "a technique that portrays the temporal dynamics of social interactions by projecting the timeline of the interaction onto an axis". Almost like a specialized timeline in which lines presenting characters converge where characters interact.

The storyline visualization strives to show the main structure of a story flow with the accent on interactions between characters without giving any sentiment or interpretation of a story. During the design process, it is important to refine the data and select only events and characters that are important for the story. Then follow the selection of a software for visualization, choice of colors, positions, sizes and intuitive symbols that will be used, and finally a legend where appropriate.

The most obvious application of a storyline visualization can be in the movie-making industry, where film directors, actors and script adapters can benefit from it [14]. However, that is not the only area that may find it informative. Visualization of linguistic shift [15], genealogical data [11], presentation of dynamic relationships of politicians over time or visual summary of events in informative context in data-driven journalism [14] have also been proposed. One of the most famous storyline visualizations is associated with the war related data, particularly Napoleon's troops marching toward Russia by Charles J. Minard and it dates back to 1862. A very thorough description of this visualization is provided by Sandberg [9] and with six different types of data it provides, it is well suited for both history and geography classes. But, as we are proposing in this paper, these two subjects are not the only ones where storyline visualization can find its place. Display of a historical or a fictitious story through storyline visualization should be the same since we have the same type of parameters

needed for this type of visualization. Thus, it is our strong belief that storyline visualization can find its place in portraying a story flow of a novel as well as in showing how different genres appeared throughout the time of literature, or how different theories in biology, chemistry and other sciences appeared with their main influencers interactions.

III. BUILDING A STORYLINE VISUALIZATION FOR A NOVEL

The use of technology in class does not necessarily mean that students and teachers have to use computers or tablets during lectures. It can also include preparation for class such as making materials which will help students to understand the school subject in an easier way and later to help students review the subject matter before exams.

Until recently, it was only possible to draw storyline visualization either by hand or by using a simple drawing software like *Paint* or some other similar program. In recent years, several authors have either proposed software built specifically for the purposes of storyline visualizations [7] or have proposed some modifications towards improving the layout of a visualization [4], [16], [18]. Since the readability of a *storyline visualization* is of an extreme importance, it is not surprising that much thought needs to be put into ordering the characters among themselves and on the visual presentation as a whole.

We have tried to incorporate all the best practices in our visualization of a novel *Prokleta avlija* (Figure 1) in order to best show the flow of the main story and two additional stories that happen in the middle. In this, we have tried to follow the three main principles for constructing storyline visualization proposed in [18].

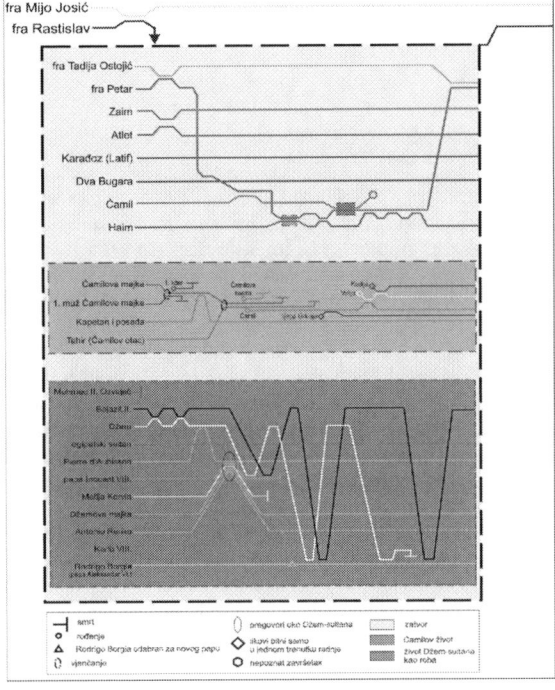

Figure 1. - Storyline visualization of a novel Prokleta avlija

A large amount of information can be obtained from just looking at the visualization and analyzing it, without even reading the book. So, for example, one can see how many plots exist in the book since every plot is positioned in its own rectangle in a different color, who are the characters and which of them meet, and finally where and when they meet in the book. The legend provided at the bottom of the visualization contains explanations of the symbols used in the visualization such as deaths, marriages and births of characters where this information is given through the novel.

IV. PREPARATIONS AND SURVEY IMPLEMENTATION

Our research was conducted through five distinct phases. The first phase included a search for high schools that would be willing to participate. We asked each school that was willing to participate in the research to send us a list of readings they planned to do in the 1st half of the 2nd term in order to find the one book they had in common. Since it was important for us to have different profile of high schools, and that there are at least two classes in each school, we included those parameters as well while selecting our final five schools (one Natural Sciences High school, two Technical High schools and two Gymnasiums). One class from each school would receive the visualization with instructions on how to use it - test group, while the second class was to read the book without the visualization and would receive the visualization after they take the test – control group (in order to answer some questions from the questionnaire).

During the second phase we had to reread the novel and draw the visualization by hand trying to define all the character relations that we believed to be important for the story. The book that turned out to be in common for our schools was *Prokleta avlija*, a historical novel written by Ivo Andrić. This novel is also one of the mandatory books on the *matura* exam that Croatian students need to take in order to get their high school diploma. This was an additional bonus for choosing this particular book since we believed that students would be more willing to read it and participate in the research.

The third phase included the process of transferring the visualization from the paper version to its digital format using graphic design software *CorelDraw X7*. After the visualization was finalized we prepared two sets of test questions and two sets of questionnaires together with the visualizations for each student. Both tests consisted of 16 questions, 8 of which were linked to the visualization (could be answered with its help) and 8 were not. Although two variations of the test were made, the difference between them was in the ordering of questions and possible answers. However, we did add one extra question that was different for each group and required them to recall which two characters from the list we provided met during the story. Half of the students in each class received the test A and the other half received the test B. Two different sets of questionnaires were prepared in order to accommodate the two different groups of students: those who used the visualization during the reading time and those who were shown the visualization after they took the test. We used results of both groups in our research. However, since the questionnaires were anonymous, we were not able to connect the test results with the types of answers provided in the questionnaires. All the materials, together with the instructions for the teacher on how to conduct the testing of each group were delivered to teachers of Croatian language in each of the schools.

In the fourth phase, 103 students who were reading with the visualization have been given the printed copies of the storyline visualization together with the instructions on how to use it. The other group of students (93) did not receive any information on the visualization or the research being conducted. Originally, 217 students were to participate in the research, but 21 student was absent from the school at the test day, so their tests were returned empty. On the test day, both groups of students were first given the tests. The test group received the questionnaires immediately after finishing the test. The control group was first given the visualization after finishing the test and then the questionnaires.

In the final phase, after tests and questionnaires were returned to us, we transferred the answers to our database and prepared them for further data analysis. Our findings with a short discussion are presented in the following section.

V. RESULTS AND DISCUSSION

The main purpose behind this pilot research was to find out if storyline visualization can help students better understand and memorize the timeline of the story and movement of its characters throughout the novel than the conventional way of just reading a book and then analyzing its content in class. We also wanted to see if students of different gender and type of the high school would perform differently on the test. Additionally we were interested to learn how young people today react to an unconventional visual data representation, or in this case, to a storyline visualization of a novel and if they see some other subjects where its usage would be rewarding.

Our main hypothesis was that that the students who had the visualization while reading the novel would have better understanding of the story and its characters and would thus have better test results than the students who were reading without visualization. Also, we expected male students to perform better since research shows that they are better in visual-spatial thinking as well as the students of technical profile high schools who are more accustomed to visual presentations due to the nature of their studies.

A. Test results

There were 16 test questions, eight of which have been shown through the visualization. We will discuss only them. Five of these questions were multiple choice (MC) questions, two were true/false (TF), and one required a short answer (SA).

The results for the first MC question (*Which Pope was in the historic part of the novel?*) were better for the students from the test group (pink sections = left sections). Also, male (M) students (54.14%) performed better on this question than female (F) colleagues (45.86%) (Figure

2). In the visualization, we have used a special symbol showing a change of the pope to mark this occurance.

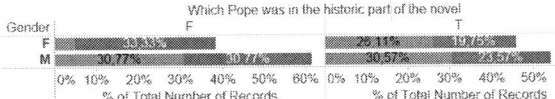

Figure 2. - Distribution of results for the 1st MC question where gender (F/M) and group type (test/control) is taken into account

The second MC question (*How many times did Camil's mother got married and how many children did she have*) gave similar results: test group performed better and male students performed better (Figure 3). A special symbol was also used to show a wedding of a character.

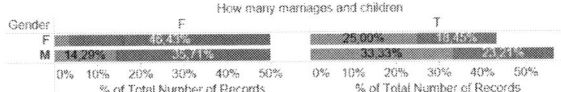

Figure 3. - Distribution of results for the 2nd MC question where gender (F/M) and group type (test/control) is taken into account

The third MC question was about Cem Sultan, the son of Sultan Mehmed II. His fate was a tragic one and at one point in his life he had been sold to the enemies of the Ottoman Empire. The final three MC questions are about his first owner (*Who is the first owner of Cem Sultan*), the number of people who wanted to buy him (*How many people want to buy Cem Sultan*) and what happened to him in the story (*What happens to Cem Sultan*). The graphs (Figure 4, Figure 5 and Figure 6) show the results for these three questions in the same order as afore-mentioned.

Figure 4. - Distribution of results for the 3rd MC question where gender (F/M) and group type (test/control) is taken into account

How many people want to buy Cem Sultan?

Figure 5. - Distribution of results for the 4th MC question where gender (F/M) and group type (test/control) is taken into account

What happens to Cem Sultan?

Figure 6. - Distribution of results for the 5th MC question where gender (F/M) and group type (test/control) is taken into account

There are two TF type of questions. The first one was used to find out if there are characters in the story whose fate was not revealed to us (a special symbol was used to show this in the storyline visualization) (Figure 7).

Figure 7. - Distribution of results for the TF question where gender (F/M) and group type (test/control) is taken into account

In the 2nd TF question, a set of five statements was given that needed to be marked as true or false. These five statements, marked 11.a through 11.e, tested if two characters met in the story or not, and if they have met after a specified event. Results are shown in Figure 8.

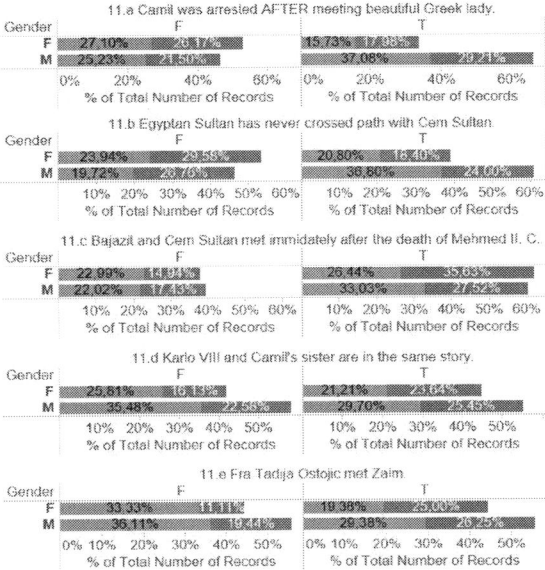

Figure 8. - Distribution of results for the set of TF questions where gender (F/M) and group type (test/control) is taken into account

In the SA category of questions there was only one question where students were asked to write how many stories there were inside the main story. In this historical novel there is one main story inside which three other stories appear. Each story in the visualization was marked with another color. We used white background for the main story, yellow for the jail story inside which two more stories occurred (orange was used for story about Camil and dark pink for Cem Sultan's story). Although a high percentage of students without visualization performed well on this question, still there were more students with the visualization with the correct answer and among them, more male students (Figure 9).

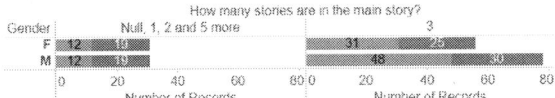

Figure 9. - Distribution of results for the SA question where gender (F/M) and group type (test/control) is taken into account

Our results show that we have been right to assume that students with visualization would perform better on the test and also that male students would perform better than female ones. However, overall test scores show that technical school students did not perform better on the test, as we have expected them to.

B. Questionnaire results

We prepared two sets of questionnaires: one for the test group and the other for the control group. The main goal was the same – to find out how students feel about presenting data from a novel through storyline visualization and what other subject areas they feel it could be useful for. The test group questionnaire also had questions about the visualization (if it was presented in a clear manner, if they find it helpful for preparing for the test, during the reading of a novel and for better understanding of relations between characters). Since the control group was given storyline visualization only after they took the test, their distinctive questions were whether they think it would have helped them during the reading and in better understanding of relations between characters. First, we will present our findings from the selection of questions common to both groups, and then the questions present only in the test group followed by the questions distinctive of the control group.

It is important to say that 82.49% of all tested students believe to be visual types, or to be more precise 94.52% of gymnasium students, 80.39% natural science students and 75.82% of technical students. If we take a look at their overall test grades per school type (Figure 10), than gymnasium and natural sciences students performed accordingly. However, there is a discrepancy between the declaration of technical students and their test scores since the students from the control group performed better than their colleagues from the test group. This came as a surprise to us since 75.82% of these students declared to be visual type and 64.83% believe that visualizations could help better understand other subjects as well.

Figure 10. – Distribution of average scores per school type and their declaration about being visual types

If we take gender into account, than more male (45.16%) then female students (37.33%) declare themselves as visual types. This is in contradiction with their test results since overall grade for male students (15.45) is lower than for female students (15.69). However, both male and female students from the test groups had outperformed their colleagues from the control groups.

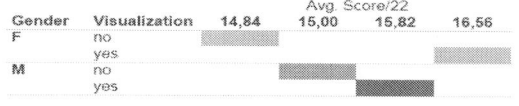

Figure 11. - Distribution of average scores per gender and their declaration about being visual types

When asked if a storyline visualization would help them when reviewing for a test, 78.03% students

answered affirmative, and 7.94% of them are students who originally declared themselves not to be visual types. There are also 13.08% of visual type students who do not find such a visualization as a good tool for test reviews.

There are 86.92% of students (10.28% of nonvisual type) who believe that storyline visualization would be helpful in better understanding even the more complex novels and 73.96% of students (6.98% of nonvisual type) who believe that it would help them in some other subjects as well. When asked in which subjects they could find such a visualization as a helping tool to better understanding of the material being studied 50 listed history, 23 listed biology, 18 listed chemistry, 15 listed physics, 15 listed mathematics and 36 of them stated that it could be useful in all subjects. There were up to 9 students that also listed one or at least one of the following subjects: sociology, geography, technical subjects, psychology, logics, art of music, visual arts and philosophy.

A larger portion (81%) of test group students found the storyline visualization to be presented in a clear manner. Among them, there are more male (60%) than female students and more students from technical schools (41%) than other types of school.

When asked about the down sides of the visualization their answers ranged from 'too little of the context was given' and 'exact years and locations would be useful' to 'did not like the symbols and/or colors used' and 'more details about characters'. But, as mentioned earlier, storyline visualization, just like other types of visualizations, can be used to present only some types of data – not all.

If we cross reference test students' answers, we find some discrepancies. For example, there are students who find the visualization to be presented clearly, to have helped them better understand the relations between characters and was helpful to have during the reading time. However, they did not find it helpful for answering the test questions. This may have something to do with their test scores. Unfortunately, since the questionnaire was anonymous, we do not have means to test this assumption. All the other combinations of their results are presented in Figure 12.

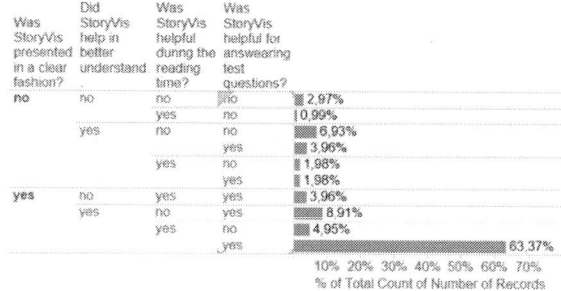

Figure 12. Distribution of Test group results

After control group was introduced to the storyline visualization 71.55% of them responded that it would have helped them during the reading time. More male students (53.01%) than female students (46.99%) were in

810

favor of this belief as well as the most gymnasium students. Their distribution regarding the school type and gender is given in Figure 13.

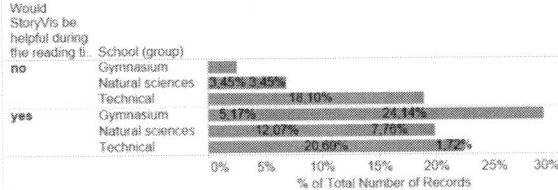

Figure 13. Distribution of Control group answers about helpfulness of storyline visualization during the reading time

Even more control group students (80.17%) believe that the storyline visualization would have helped them with understanding of relations between characters. Again, more male (56.99%) than female students (43.01%), but more technical school students than others. Their distribution regarding the school type and gender is given in Figure 14.

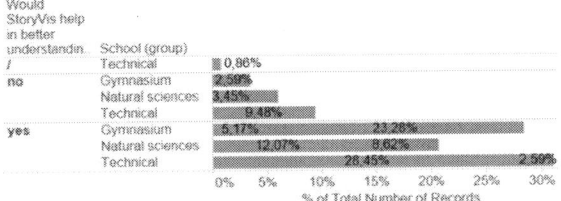

Figure 14. Distribution of Control group answers about helpfulness of storyline visualization in uderstanding of relations between characters

VI.   CONLUSION

The main purpose of this paper is to test if *storyline visualization* can be introduced as a legitimate educational tool for presenting information that includes a type of narration where the characters and their interaction over a period of time is important to convey. Our preliminary research results show that students would benefit from such presentation and would use it during the study period but also as a revision tool. Since the majority of people rely greatly on eyes as an information receiver, *storyline visualization* comes as a perfect tool that can be used not only for research and business, but in teaching and learning as well. The research has shown that students from Gymnasia and Natural Sciences Highschools that used the storyline visualization while reading the novel scored better on the exam than the students without visualization. Also, opposite to our expectations, students from Technical Highschools without visualization performed better on the test than their colleagues with visualization during the reading time. Considering only gender, our female participants using visualization while reading achieved better results than male participants with the visualization. This was also in contradiction with the stereotype and our questionnaire results where more male students declared themselves to be visual learning types that female ones. The one-tailed Fisher's exact test value (P=0.0172) show statistical significance between gender and helpfulness of storyline visualization in understanding

while χ^2 test value (P=0.0006) for the association between school type and helpfulness of storyline visualization show high statistical significance.

ACKNOWLEDGMENT

We would like to thank teachers and students of Technical Highschools *Ruđer Bošković* in Zagreb and Technical Highschool in Bjelovar, Natural Sciences Highschool *Vladimir Prelog* in Zagreb, and Gymnasium in Metković and Bjelovar without whose help this research would not have been possible.

REFERENCES

[1]   8 Technologies That Will Shape Future Classrooms. Hongkiatcom.   URL: http://www.hongkiat.com/blog/future-classroom-technologies/ (At: 24.6.2016.)

[2]   B. Fry, Visualizing Data. Beijing: O'Reilly Media, 2008.

[3]   B. Šverko, S. Szabo, M. Kolega, P. Zarevski, and T. Turudić-Čuljak Psihologija. Udžbenik za psihologiju za gimnazije. Zagreb: Školska knjiga, 2006.

[4]   I. Kostitsyna, M. Nöllenburg, V. Polishchuk, A. Schulz and D. Strash. "On Minimizing Crossings in Storyline Visualizations." Lecture Notes in Computer Science Graph Drawing and Network Visualization (2015): 192-98.

[5]   J. Heer, M. Bostock and V. Ogievetsky, "A Tour through the Visualization Zoo", Queue – Visualization, vol. 8, issue 5, 2010.

[6]   J. Wakefield. "Technology in Schosols: Future Changes in Classrooms."   BBC   News.   URL: http://www.bbc.com/news/technology-308143020(At: 24.6.2016.)

[7]   M. Ogawa and K.-L. Ma, "Software Evolution Storylines" in Proceedings of the 5th International Symposium on Software Visualization, ACM, 2010, pp 35-42.

[8]   M. Friendly, A Brief History of Data Visualization. Iz: Handbook of Data Visualization, editor Chen et al, 16-48. Berlin: Springer, 2008.

[9]   M. Sandbergs, "DataViz History: Charles Minard's Flow Map of Napoleon's Russian Campaign of 1812." Data Visualization Blog. URL: https://datavizblog.com/2013/05/26/dataviz-history-charles-minards-flow-map-of-napoleons-russian-campaign-of-1812-part-5/ (At: 22.6.2016.)

[10]   N. Yau, Visualize This: The FlowingData Guide to Design, Visualization, and Statistics. Indianapolis, Wiley Pub., 2011.

[11]   N.W. Kim, S.K. Card and J. Heer, "Tracing Genealogical Data with TimeNets" in Proceedings of the International Conference on Advanced Visual Interfaces, ACM, 2010, pp 241-248.

[12]   S. Coughlan. "Tablet Computers in '70% of Schools'" BBC News. URL: http://www.bbc.com/news/education-30216408   (At: 24.6.2016.)

[13]   S. Few, Now You See It: Simple Visualization Techniques for Quantitative Analysis, Analytic Press, 2009.

[14]   S. Liu, Y. Wu, E. Wei, M. Liu and Y. Liu, "StoryFlow: Tracking the Evolution of Stories" in Proceedings of IEEE InfoVis 2013, IEEE Transactions on Visualization and Computer Graphics, vol. 19, nb. 12, 20013, pp 2436-2445.

[15]   S. Mahoom, R. Al-Rfou and K. Mueller, "Visualizing Linguistic Shift" in arXiv preprint arXiv:1611.06478, 2016.

[16]   S. Silvia, J. Abbas, S. Huskey and C. Weaver, "Storyline Visualization with Force Directed Layout" in IEEE VIS 2015 Conference   Compendium,   Chicago,   2015.   URL: http://www.cs.ou.edu/~weaver/academic/publications/silvia-2015a/materials/silvia-2015a.pdf

[17]   T. Chen, Aidong Lu, and Shi-Min Hu. "Visual Storylines: Semantic Visualization of Movie Sequence." Computers & Graphics 36.4 (2012): 241-49.

[18]   Y. Tanahashi and M. Kwan-Liu, "Design Considerations for Optimizing Storyline Visualizations" in Proceedings of IEEE Transactions on Visualization and Computer Graphics vol. 18, nb. 12, 2012, pp 2679-2688.

Impact of ICT on Archival Practice from the 2000s Onwards and the Necessary Changes of Archival Science Curricula

Hrvoje Stančić*, Arian Rajh*, Mario Jamić*
* Faculty of Humanities and Social Sciences/Department of Information and Communication Sciences,
Zagreb, Croatia
hstancic@ffzg.hr, arajh@ffzg.hr, mjamic@ffzg.hr

At the outset authors of this article investigate the influence of ICT on archival practice. All archival processes are affected by latest developments in ICT. DMS and eSignature requirements and legislation (like MoReq and eIDAS) influenced record management practice. PAIS, OAIS and storage solutions have their impact on the archival management. Special audio and video format solutions influenced specialised archives. Graph databases as a more agile tool for organising information and the upcoming Records in Contexts (RiC) standard are influencing archival description. New preservation methods have emerged - Chain of Preservation model (COP) and the upcoming Preservation-as-a-Service for Trust (PaasT) model. PaasT is the result of growing cloud services. Semantic web influenced retrieval and usage of materials in all heritage institutions. The authors proceed to discuss how the elaborated changes should reflect the archival science academic curricula. They investigate what are new archival science core competences that should be taught to students and archival professionals by university teachers today, how should universities prepare their students of archival science for contemporary (digital) labour market etc. The authors offer a clear position what archival competencies, arising from the influence of ICT and cloud environment, should be built upon in the contemporary archival science curricula.

I. INTRODUCTION

ICT has had an indisputable impact on the reality of organisations and the reality of people's lives for decades. The significant transformation of these realities from the 2000s onwards brought about a change in archival science curricula. The authors of this article investigate the extent of this influence and the magnitude of educational adjustments. Have archival professionals incorporated enough contemporary ICT knowledge into their curricula (and their daily operations) yet? Comparable impact on archival profession happened with the appearance of new storage materials and the development of new physicochemical methods in the 20th century. Those materials and methods had a substantial influence on archival storage facilities and procedures of preservation, conservation and restoration. But, what is happening now is of far more consequence to the future of the archival profession. ICT, led by fresh approaches, trends, values and practices, has moved not just those parts of archival work that were oriented towards the management of digital records, but also parts of more

traditional archival practice, like archival description[1], because of new possibilities of computer generation and system-to-system communication of archival finding aids[2].

Digital signature and other authentication and integrity associated technologies altered management of digital records. New digital audio and video formats appeared. Electronic recordkeeping technologies related to Records Management Systems (RMS) / Enterprise Content Management Systems (ECM, ECMS) influenced organisations and supported their abilities to preserve digital records. Storage became portable, designed for massive scalability, and with archival qualities – like SDS (Software-Defined Storage), SSA (Solid State Arrays), scale-out NAS (Network-Attached Storage), and object storage. Cloud services and cloud storage have been developing for the last decade [1]. These services are today fully accepted by individuals for their personal archives and more or less accepted by corporate bodies. Concepts and standards like OAIS (Open Archival Information System), PAIS (Producer-Archive Interface Specification), PAIMAS (Producer-Archive Interface – Methodology Abstract Standard) and XFDU (XML Formatted Data Unit) are shaping the process of transfer of digital content among environments [2-6]. Semantic web and graph databases also occurred in the realm of today's records and archive management. These fresh ICT concepts and their effect on archival science are described in the second chapter of this article. The authors consider that ICT influenced archival science considerably and that archival curricula responded to these developments to a great extent. But, do these curricular changes respond enough to the progress in ICT field(s)?

The aftermath of ICT impact is visible in the courses offered to university students on the global and European

---

[1] Archival description refers to "the process of capturing, analyzing, organizing, and recording information that serves to identify, manage, locate, and explain the holdings of archives and manuscript repositories and the contexts and records systems which produced them" and "the products of the above process.", International Council on Archives, "Dictionary of Archival Terminology" (Draft Third Edition/DAT III, 1999) and Multilingual Archival Terminology, http://www.ciscra.org/mat/mat/term/51 (accessed March 2017)

[2] Finding aid – "a tool that facilitates discovery of information within a collection of records", Pearce-Moses, Richard. A Glossary of archival and Records Terminology, http://www2.archivists.org/glossary/terms/f/finding-aid (accessed March 2017)

level (analysed in 2003 and 2016). Nevertheless, it is important to constantly monitor changes in ICT and to make timely recommendations for continuing adjustment of archival science curricula.

Next, we will firstly discuss the influence of ICT on the archival science and practice. Then we will show the effects of the development of ICT on university and professional curricula by analysing the changes that were introduced between the two research activities – the first in 2003 and the second in 2016. Finally, we will conduct a SWAT analysis, draw conclusions and offer recommendations.

II. INFLUENCE OF ICT ON ARCHIVAL SCIENCE AND PRACTICE

In relation to the processes of transfer and storage, a numerous media types were developed by the ICT industry [7-8]: glass disks, PIQL film, new types of network storage etc. Some recognised media technologies got better over time (e.g. Data Tresor Disks) while the ICT industry has abandoned some of the media types in recent years, e.g. Ultra Density Optical (UDO) disks, High-Density DVD, Holographic Versatile Disks (HVD) etc.

Record Management Systems (RMS) and Enterprise Content Management Systems (ECM, ECMS) have entered the organisations and changed their record management practices and workflows. ECMS and RMS systems are practically comparable, from the pure record management perspective, but then again ECMS systems are conceptually[3] and functionally[4] broader than RMS. From the archival perspective, this broadness of functions is relevant because it shapes creator's business activities and affects archival fonds. Adoption of ECMS and change in workflows can result in business process reengineering activity and change in the business (business activities, business roles etc.). For this reason, ECM systems are not just automation tools; implementations of ECMS systems in organisations are not just trivial modifications. Archivists should be well-aware of this. Recently, the ECMS solutions have been implemented in many organisations. Examples of the most spread out technologies [9-11] are IBM FileNet, Dell EMC Documentum, Alfresco One, OpenText,

Hyland OnBase etc.[5] For illustration, the FileNet ECMS implementation in the Croatian Agency of Medicinal Products and Medical Devices (HALMED) in 2013 and 2014 led to changes in recordkeeping, archival management, IT landscape, and business processes.[6] FileNet ECMS in HALMED has been linked with XINO scanner and DPU Scan application, case management application and archival management application.[7] Outcomes of these integrations and interfaces are a) ability to use digitised records as a replacement for paper originals in the workflows, b) linked paper originals and digitised copies, and c) ability to archive new digitally-born records and cases. A threefold usage of various HALMED's ECMS content happens at this time. There are users in HALMED that use ECMS and its content through its own FileNet Navigator interface, users that prefer to use their previously developed business applications (ECMS handles the content in the background), and groups of users that work with new applications, developed on the FileNet platform. Rearrangement of business tasks was done because new applications had required different workflows (e.g. Quality Management application). Outcomes of these new workflows were new groups of records, and thus the change in archival fonds.

Storage and its diversification became very complex; therefore archivists should learn more and absorb more facts about the world of storage. There are several notions about the storage archivists should be able to consider. They have to consider storage memory feature (primary, secondary, tertiary, online, offline, near-line, volatile, non-volatile); the way storage utilises equipment (SAN, NAS, virtualisation); storage hardware dependency (SDS); the question what storage really addresses (content, location, file-addressable); the question what storage actually manages and on which level (file hierarchies, objects, blocks, and entry level, business level, enterprise level). There are numerous subtypes of storage solutions [12]. Today, for archivists it is necessary to understand the software characteristics behind the typology, characteristics like scalability (of solid state arrays, NAS, object storage), hardware-independency (of SDS) etc. Archivists should also know what to do once active storage environment is set up – to apply business continuity and disaster recovery policies as well as to develop long-term digital preservation policies together with IT colleagues and to archive what has to be archived. Therefore, archivists should be acquainted with backup procedures (traditional, snapshots, cloud), replication (replication of content after its creation, host-based, array-, network-, snapshot-based), mirroring etc.

Semantic web brought requirements to semantically enrich descriptions of existing materials in order to

---

[3] Content vs. record. ECM deals with unstructured organisational (enterprise) content. ECM is "a strategic framework and a technical architecture that supports all types of content (and format)" (Definition from Gartner MQ ECM report 2014 and 2015) and "a set of services and microservices [...] to exploit diverse content types, and serve [...] numerous use cases across an organization." (Gartner MQ ECM 2016). Gilbert, M. R.; Shegda, K. M.; Chin, K.; Tay, G.; Koehler-Kruener, H. "Magic Quadrant for Enterprise Content Management, September 2014", Gartner, Inc.; Koehler-Kruener, H.; Chin, K.; Hobert, K.A. "Magic Quadrant for Enterprise Content Management, October 2015", Gartner Inc.; Hobert, K.A.; Tay, G.; Mariano, J. "Magic Quadrant for Enterprise Content Management, October 2016", Gartner, Inc. (MQ ECM 2014-2016 v. Licenced for distribution).

[4] ECMS consists of document and records management, image management, collaboration tools, web content management, business process workflows, and vendor-specific components. It is also linked with business applications of organisations. Gartner report from 2016 predicts shifting major ECM solutions to cloud-based platforms.

---

[5] Gartner MQ ECM 2014-2016.

[6] Part of the IPA 2009 TAIB project preparations for eCTD and Implementation of Digital Archival Information System (in cooperation with Ericsson Nikola Tesla d.d. and AAM Management Information Consulting Ltd.).

[7] Imaging subsystem was developed by EMES d.o.o., and case management and archival management applications by Omega Software d.o.o.

disseminate them in a more efficient manner. Many institutions from the museums/libraries/archives (MLA) community had started with the implementation of the semantic web a long time ago and have enabled better connectivity of descriptions of their objects of interests for different communities. The emerging archival standard Records in Contexts (RiC) represents one step toward this [13-14].

Records in Contexts standard is using graph databases [15]. Graph databases appeared as more flexible and reactive technology for processing entities and their relationships than the technology of relational databases.[8] Creation of graph database (model) implies the creation of nodes (entities) with labels (entity types), relationships, and nodes' and relationships' properties. It is still too early to tell whether RiC will become a new description standard or just another schema in archival description. Nevertheless, it is important to understand the principle behind RiC and to be aware of these changes in the world of databases.

Archivists are often interpreters – they translate from the business language connected with a function to the ICT language that supports the function and vice versa. Some standards that are close to the archival domain, and records & archival management domain-specific languages [16], already possess this translated logic. OAIS standard represents a mixed language of such a sort. ICT professionals can build an OAIS compliant system. A person with a task to transfer records between two systems can conduct a structured PAIS project. But usually archivists have to familiarise ICT professionals with these standards. Also, archivists translate records management parts of business functions to ICT workflows and functionalities. Records contain business transactions metadata and remaining logic while business is driven by records to an appreciable degree.

An OAIS-based system has several functions – ingest of records being the starting one. Archive receives a submission from internal or external, human or computer producer. It has to be validated and transformed into the form suitable for long-term preservation. These standards can be applied [2-6]: PAIMAS to define producer-archive relationships in different phases[9] and conditions of their cooperation; PAIS to formally define digital objects and specify submission models[10] required for transfers into archives; XFDU to support these transfers as the precise packaging standard. Finally, there is an ISO 16363 standard [6] for evaluating the trustworthiness of repositories (i.e. digital archives). It brings to maturity OAIS, PAIMAS and PAIS standards. ISO 16363 covers ingest function, organisational infrastructure related issues and maintenance of producer-archive relationships.

Shifting the focus from the RMS systems to the digital records they are holding, it should be noted that archivists are increasingly dealing with the records having some kind of cryptographic mechanisms applied to them. Namely, the records are being signed by (advance) digital signatures which rely on (qualified) digital certificates; the digital timestamps are being added as well as digital seals. Recently implemented EU-level regulation eIDAS states that "an electronic signature shall not be denied legal effect and admissibility as evidence in legal proceedings solely on the grounds that it is in an electronic form" [21]. This means that the archivists need to know how to preserve digitally signed records in the long-term while addressing the fact that digital signatures rely on digital certificates which are valid only 2-5 years. While considering re-certification and while applying standard long-term preservation technical procedures like file format conversion, media migration, application emulation or system virtualisation they should constantly have in mind standard archival requirements for preservation of authenticity, integrity, reliability, usability and non-repudiation of digital records being preserved. All this illustrates just a fraction of ICT-connected tasks that the archivists need to proactively and successfully accomplish.

Besides the impact of ICT by itself, the authors also recognise the influence of recent professional endeavours like an education and training component of the InterPARES 3 – global, archival science research project (2007-2012) and other similar impacts on professional curricula. Projects like InterPARES laid emphasis on the need to incorporate state-of-the-art technologies into students' and trainees' programmes.

III. EFFECTS ON UNIVERSITY AND PROFESSIONAL PROGRAMMES

The effects of the development of ICT on university and professional programmes were analysed by comparing the two studies. The first one was conducted in 2003[11] and the second one in the last quarter of 2016.

The first study aimed to analyse 23 programmes of archival science globally. They were from Australia (2), Canada (2), Finland (1), Ghana (1), Israel (1), USA (14), and United Kingdom (2). Descriptions of the total of 370 courses were manually identified, collected from the official web pages and analysed. The analysis was focused on identifying if the courses found in the archival science programmes could be relevant for the library science study or museum science/heritage study (the MLA group) or social-humanistic informatics or a combination of them. Also, the level of influence of IT was determined. The results show that:

- 62% of courses are relevant for the study of library science,

- 58% of courses are relevant for the study of museum science,

---

[8] "Because of their emphasis on global queries, graph compute engines are normally optimized for scanning and processing large amounts of information in batches....",.Robinson, I.; Webber, J.; Eifrem, E. "Graph Databases". O'Reilly Media Inc., 2015, p. 7.

[9] Preliminary phase, formal definition phase or PAIS phase, transfer phase, validation phase.

[10] PAIS define types of submissions and submission containers. Actual submission implementation should be done by applying XFDU standard.

---

[11] As part of the project HERITAGE Live (Operative programme IPA – Cross-border Cooperation Slovenia-Croatia 2007-2013).

- 37% of courses are relevant for the social-humanistic informatics study,

- 1.9% of courses are exclusively relevant for the combination of archival and library science study (while not for other study combinations),

- 1.6% of courses are exclusively relevant for the combination of archival and museum science study (while not for other study combinations),

- 1% of courses are exclusively relevant for the combination of archival science and social-humanistic informatics study (while not for other study combinations),

- 56% of courses are relevant for the combination of archival, library and museum (MLA) study,

- 35% of courses are relevant for the combination of archival, library and museum (MLA) and social-humanistic informatics study,

- 7.7% of courses have information technology aspect explicitly mentioned in the course name.

The analysis showed that there are firm grounds for the organisation of the multidisciplinary studies (MLA group + social-humanistic informatics) while the combinatorial exclusiveness showed statistically insignificant overlap. This study also served as a starting point for the second research study.

The aim of the second research study was to compare the archival science programmes at the European universities taking into account the programme offered by the Faculty of Humanities and Social Sciences (FHSS) at the University of Zagreb (where the authors are from) as well. For the purpose of research, 20 emails were sent to the official addresses of different universities in Europe. Out of 20 inquiries, 4 faculties confirmed that they are offering archival science programme, 1 faculty has requested to send the inquiry to a different address, 6 faculties reported that they do not offer an archival science study, while 9 faculties did not respond to the inquiry to the date of writing this paper. The authors used faculties' official web pages to gather additional information needed for the analysis. Therefore, the analysis focused on the 5 archival science programmes at the universities that have them (Lund University, University of Amsterdam, University of Oslo, University of Stockholm, University of Zagreb). The goal of the second research was to compare the differences between the studies in Europe by looking at the amount of theory focussed courses versus practical ones as well as the courses that implement ICT. The total of 88 courses was analysed. The difference between the analysed programmes is that the archival science programme offered by FHSS officially starts at the graduate level but also have several archival courses at the undergraduate level. Other analysed programmes are mainly offered at the graduate level, so the archival science courses do not appear during the undergraduate study, the exception being the University of Oslo.

The fact that the graduate level programmes were analysed may be the main reason why the courses are mainly based on theory – the students need to learn the basics of archival science. History is not significantly present. Instead, there are courses which are more practice-oriented and which prepare students to work at the MLA institutions (usually in the second year of the study).

The courses that are IT-focussed use the technology existing in the archives of their countries, therefore giving students the opportunity to learn how to use the technology before graduating. The analysis showed that 17% of courses have information technology aspect explicitly mentioned in the course name. Although the first and the second research did not analyse the same sample so the results could not be directly correlated the increase in number of IT-focused courses may be noticed.

To enrol in the graduate program, students have to finish an undergraduate study. Most of the analysed courses (56%) are in the programme during the first year of graduate archival science studies (Figure 1). Those courses mainly serve to introduce students to the archival science. Broadening the view of the influence of ICT we have identified 27% of courses in the second year of graduate studies that involved work with information technology used in archival practice but not fully focussed on it. This shows that IT has significantly influenced the archival studies.

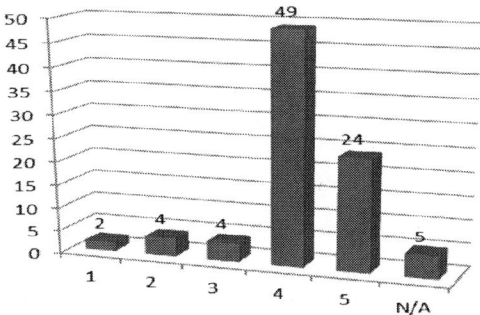

Figure 1.   Number of courses per year
(1-3 – undergraduate level; 4-5 – graduate level)

In fact, when we look at the distribution of the courses per semester, 30 courses are in programme during the seventh semester, or first semester of the graduate study (Figure 2).

While the first semester of graduate study is aimed at studying the theory of archival science, the courses which involve information technology used in practice are in the programme of the ninth semester. The tenth semester is the semester when the students have their internship in the archives, so it makes sense that they should be prepared to use the technology.

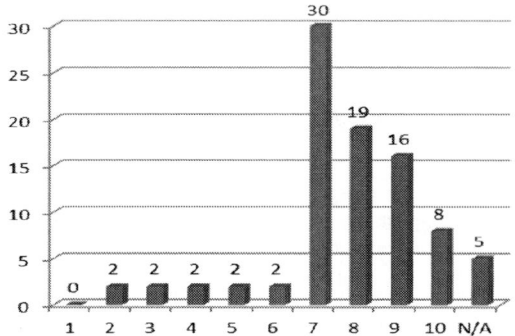

Figure 2. Number of courses per semester

Archival science courses in the analysed programmes of the second research study focus more on theory than practice, but it is clear that they follow the development of ICT. As it was detected in the first research study, here the archival science courses also do not prepare the students to work only in the archives, but in all MLA institutions as well. The analysed archival science programmes in Europe does not perceive archival science as a science that deals only with the preservation of documents, but as a science that deals with information systems, digital preservation and other activities normally conducted by other information institutions too. The analysed programmes last between one and two years, but the trend is to put less focus on courses directly linked to

---

[12] Archival arrangement – "1. The process of organizing materials with respect to their provenance and original order, to protect their context and to achieve physical or intellectual control [...] 2. The organization and sequence of items ...", Pearce-Moses, R. A Glossary of Archival and Records Terminology (accessed March 2017)

[13] "Documentary form is both physical and intellectual. The term physical form refers to the external make-up of the document, while the term intellectual form refers to its internal articulation. Therefore, the elements of the former are defined by diplomatists as external or extrinsic, while the elements of the latter are defined as internal or intrinsic." Duranti, L. "Diplomatics: New Uses for an Old Science - Part V." // *Archivaria 32* (Summer 1991)

[14] Provenance – "*Archivists* base their work on the idea that the origin of archival *documents* and *records*, or their provenance, must be known if the purposes of the archives are to be achieved [...] The application of the principle of provenance results in division of archival holdings into particular groups of records linked to their creator. "The principle of provenance" // Duranti, L.; Franks, P. C. (ed.) "Encyclopedia of Archival Science", 2015, p. 284.

[15] Computational archival science – "An interdisciplinary field concerned with the application of computational methods and resources to large-scale records/archives processing, analysis, storage, long-term preservation, and access....", Marciano, Lemieux, Hedges, Esteva, Underwood, Kurtz, Conrad. "Archival records and training in the Age of Big Data" (Draft submitted Oct. 2016) // Sarin, L. C.; Percell, J.; Jaeger, P. T.; Bertot, J. C. (ed.) Advances in Librarianship – Re-Envisioning the MLIS: Perspectives on the Future of Library and Information Science Education.

[16] "The term 'personal digital archiving' refers to how individuals manage or keep track of their digital files, where they store them, and how these files are described and organised.", Redwine, Gabriella. "Personal Digital Archiving" DCP Technology Watch Report 15-01, December 2015, Digital Preservation Coalition, http://dx.doi.org/10.7207/twr15-01 (accessed January 2017), p. 2.

---

the current archival practices and current technology, and more focus on practical work.

TABLE I. ANALYSIS OF STRENGTHS, WEAKNESSES, OPPORTUNITIES AND TREATS OF/FOR ARCHIVISTS WITH THE MISSION RELATED TO PRESERVATION IN THE DIGITAL ENVIRONMENT

STRENGTHS (KNOWLEDGE)	WEAKNESSES (LACK OF KNOWLEDGE)
Classification; arrangement[12]; organisation; document types; document forms; intrinsic and extrinsic characteristics[13]; provenance[14] and the quality of source analysis	Digital formats and standards; formatting standards; system or repository architecture and functioning; storage
OPPORTUNITIES (KNOWLEDGE)	**THREATS** (MINDSET ISSUES)
New types of databases; semantic web; machine learning software; computational archival science[15]; new types of accumulations; personal digital archiving[16]; possible usage of new interactive technologies	Holding tenaciously to workflows and procedures created for handling paper records. Choosing to keep working in closed professional ecosystems.

IV. CONCLUSION

Table 1 uses a SWOT analysis template and links strengths and opportunities with knowledge and positive aptitude of archival professionals. The proposed SWOT analysis derivative outlines today's archival corpus of knowledge related to professional ability to contribute to digital archival management and archivists' ability to absorb IT-related knowledge. Archival professionals are strong in using classifications, so their knowledge can be used in the system design phase. For a good reason the DIRKS recordkeeping methodology [17] had assigned the same role for archivists. Although their new role in the organisational environment is being recognised, the role of the archival professional in open digital social environment has not been defined yet. Archivists' ability to evaluate provenance and the quality of sources could be helpful in this new environment. That ability is also related to ICT and participation of archivists in training phases in pattern analysis and machine learning projects.

Weaknesses are related to archivists' lack of knowledge of particular technologies. Threats are related to archivists' and records managers' way of thinking allied with the paper administration and its functioning. This refers to applying records schedule in some later stage of digital document's lifecycle, sticking to workflows created for paper records, insisting on perfect custodial conditions in a dynamic digital environment etc. If we neglect the threats, they will turn into professional weaknesses.

Therefore, archivists should be abreast of the rapid ICT developments. It is important to recognise and address the needs for further improvements. Opportunity for growth lies in making archival outputs more interoperable, dynamic and usable than before, by embracing semantic web concepts, graph databases concepts, machine learning tools concepts etc. Another opportunity for archivists in the complex contemporary world appeared when they accepted personal digital archiving (PDA) conception and harnessed their theoretical knowledge to address PDA challenges. PDA represents a new type of archival fonds or accumulation because it differs from paper personal fonds.

How much knowledge we collect through the never-ending learning and practice makes us competent scientists and good professionals. If archivists will adopt these ICT-related opportunities, they will turn them to their strengths. That is why they have to be recognised and used for fine-tuning of curricula. The analysis of the archival science programmes and their courses in the two presented research studies showed the positive developments. In our opinion, the contemporary archival science programmes should continue adjusting to technological reality. Faculties and archival schools should embrace the new technologies and educate future archivists about them. Consequently, the archivists would be able to bring the needed changes to the archival practice, users' preservation capabilities and the digital meta-literacy related to the preservation in general.

LITERATURE

[1] Wells, Joyce. "10 Lessons learned from 10 years of cloud services", Database trends and applications. May 11, 2016, http://www.dbta.com/Editorial/News-Flashes/10-Lessons-Learned-from-10-Years-of-Cloud-Services-111020.aspx (accessed January 2017)

[2] ISO 14721:2012 Space data and information transfer systems -- Open archival information system -- Reference model

[3] ISO 20104:2015 Space data and information transfer systems -- Producer-Archive Interface Specification (PAIS)

[4] ISO 20652:2006 Space data and information transfer systems -- Producer-archive interface -- Methodology abstract standard (PAIMAS)

[5] ISO 13527:2010 Space data and information transfer systems -- XML formatted data unit (XFDU) structure and construction rules

[6] ISO 16363:2012 Space data and information transfer systems -- Audit and certification of trustworthy digital repositories

[7] Harris, R. "Blu-rey vs HD DVD: Game over", Storage bits, 20.6.2007, http://www.zdnet.com/article/blu-ray-vs-hd-dvd-game-over/ (accessed January 2017)

[8] Harris, R. "Holographic storage bites the dust", 18.2.2010. http://www.zdnet.com/article/holographic-storage-bites-the-dust/ (accessed January 2017)

[9] Gilbert, M. R.; Shegda, K. M.; Chin, K.; Tay, G.; Koehler-Kruener, H. "Magic quadrant for enterprise content management", September 2014, Gartner Inc.

[10] Koehler-Kruener, H.; Chin, K.; Hobert, K.A. "Magic quadrant for enterprise content management", October 2015, Gartner Inc.

[11] Hobert, K.A.; Tay, G.; Mariano, J. "Magic quadrant for enterprise content management", October 2016, Gartner Inc.

[12] Staimer, M. "Defining the software-defined storage market. Techtarget", 4.3.2016, http://searchstorage.techtarget.com/feature/Defining-the-software-defined-storage-market (accessed December 2016)

[13] International Council of Archives. Records in Contexts – A Conceptual Model for Archival Description. Consultation Draft

v.01, September 2016. http://www.ica.org/en/egad-ric-conceptual-model (accessed October 2016).

[14] Stančić, H.; Rajh, A.; Crnković, K. "Mapirani metapodaci arhivskih, knjižničarskih i muzeoloških normi i interdisciplinarne platforme", Arhivi, knjižnice, muzeji 20, presentation held 24.11.2016 (manuscript in print).

[15] Robinson, I.; Webber, J.; Eifrem, E. "Graph Databases". O'Reilly Media Inc., 2015, p. 7.

[16] Rajh, Arian; Meze, Krešimir. "A Domain-Specific Records Management and Information Governance Solution Designed to Support the Implementation of the General International Standard Archival Description" // INFuture 2013: Information Governance / Gilliland, Anne; McKemmish, Sue; Stančić, Hrvoje; Seljan, Sanja; Lasić-Lazić, Jadranka (ur.). Zagreb: Odsjek za informacijske znanosti, Filozofski fakultet Sveučilišta u Zagrebu, 2013, pp. 51-61.

[17] "DIRKS manual", NSW State Records and Archives. https://www.records.nsw.gov.au/recordkeeping/advice/dirks-manual (accessed January 2017)

[18] Yakel, Elisabeth. "Educating archival professionals in the twenty first century" // OCLC Systems & Services 20(4):152-154, December 2004

[19] Couture, Carol. "Education and research in archival science: General tendencies"// Archival Science 1(2):157-182, June 2001

[20] Lučić, Melina. "Obrazovanje arhivista i spisovoditelja za novo okruženje: praksa u svijetu i izgledi u Hrvatskoj" // Arhivski vjesnik, (44)2001, pp. 33-42. (http://hrcak.srce.hr/9307, accessed January 2017)

[21] Regulation (EU) No 910/2014 of the European Parliament and of the Council of 23 July 2014 on electronic identification and trust services for electronic transactions in the internal market and repealing Directive 1999/93/EC http://eur-lex.europa.eu/legal-content/EN/TXT/PDF/?uri=CELEX:32014R0910&from=EN (accessed January 2017).

[22] InterPARES 3 project. http://www.interpares.org/ip3/ip3_products.cfm (accessed January 2017)

[23] Duranti, Luciana "Diplomatics: New Uses for an Old Science - Part V." // Archivaria 32 (Summer 1991)

[24] Duranti, Luciana; Franks, Patricia C. (ed.) Encyclopedia of Archival Science, Rowman & Littlefield, 2015

[25] International Council on Archives, "Dictionary of Archival Terminology" (Draft Third Edition/DAT III, 1999)

[26] Multilingual Archival Terminology, http://www.ciscra.org/

[27] Pearce-Moses, Richard. "A Glossary of archival and Records Terminology", The Society of American Archivists, 2005 (http://www2.archivists.org/glossary)

[28] Marciano, Lemieux, Hedges, Esteva, Underwood, Kurtz, Conrad . "Archival records and training in the Age of Big Data" (Draft submitted Oct. 2016) // Sarin, Lindsay C.; Percell, Johanna; Jaeger, Paul T.; Bertot, John C. (ed.) Advances in Librarianship – Re-Envisioning the MLIS: Perspectives on the Future of Library and Information Science Education. http://dcicblog.umd.edu/cas/wp-content/uploads/sites/13/2016/05/submission_final_draft.pdf (accessed March 2017)

[29] Redwine, Gabriella. "Personal Digital Archiving", DCP Technology Watch Report 15-01, December 2015, Digital Preservation Coalition, http://dx.doi.org/10.7207/twr15-01 (accessed January 2017)

Publishing of Personal Information on Facebook with Regard to Gender: Comparison of Pupils and University Students

Paula Džapo* and Mirko Duić**
* National and University Library, Zagreb, Croatia
** University of Zadar, Department of Information Sciences, Zadar, Croatia
pauladzapo@gmail.com, miduic@unizd.hr

Abstract - The goal of this research was to explore and compare personal information Facebook publishing behaviour of male and female pupils and university students. One important finding is that very small percentage of all examinees publish items in most sensitive data types regarding the privacy: religious and political views, mobile phone, email, address. Additionally, it was established that a similar percentage of male and female pupils publish information in almost all data types. The exception is a Books data type where female pupils publish about two times more than male pupils. Also, on average, female pupils publish about 7.5 times more often information in Books data type than male pupils. Male pupils, on average, publish information in Sports data type about 4.5 times more often than female pupils. Considerably higher percentage of male students publish information in data types: High School, Apps and Games, Music, Friends, TV Shows, Movies. On average, male students publish about 8 times more often information in Sports data type than female students. Female students, on average, publish information considerably more frequently in Check-Ins data type: about 3.5 times more often than male students.

I. INTRODUCTION

Many young people use Facebook every day publishing various information about their interests, opinions, friends and other topics. The goal of this study is to provide insights about the Facebook publishing behaviour of young people with regard to their gender. These insights could help educators to better understand similarities and differences between the interests of female and male pupils and students. This could help them to create educational programs which are more closely connected with the young people's interests and with regard to their gender.

Also, these insights could be helpful for educators to create programs for the improvement of privacy protection behaviour. Privacy is a fundamental human right. It is „the claim of individuals, groups, or institutions to determine for themselves when, how, and to what extent information about them is communicated to others" [1, p.154]. Today, when information technology has a strong presence in our lives, privacy on the Internet is something everyone should be paying attention to. With the fast development of information technology and the creation of new devices and forms of media, such as smartphones and social networking sites (SNS), privacy of individuals becomes

an issue that needs to be addressed and protected against those who want to harm it.

Gross and Acquisti said that "privacy implications associated with social networking sites depend on the level of identifiability of the information provided, its possible recipients, and its possible uses." [2, p.73]. They stated that even if some SNS do not expose their users' identities, they still may provide information that identifies the profile's owners [2].

On social media sites, such as Facebook, people can voluntarily disclose their personal information: their names, comments, likes, posts and photos or videos can be visible to their friends as well as friends of their friends and anyone else with Facebook account. These publicly available personal traces can be seen by many unknown persons and organizations, including advertisers [1]. Personal information can be taken from social media and used without the knowledge of users who are facing risks like identity theft, physical or online stalking and even blackmailing [2].

In the next section we will present insights from the studies about young Facebook users. In the third section we will present the research methodology used to explore the Facebook publishing behaviour of young people. In the fourth section all findings will be presented in detail. In the fifth section we will emphasize the most important findings, suggest some possibilities for using these findings and suggest several ways for additional research that could be made as an extension of this research.

II. YOUNG FACEBOOK USERS

With the rise of Facebook and other social media, online privacy of children and young people becomes an especially important issue, as they are less likely to understand the possible effects of disclosing personal information which may affect their lives [3]. In a study conducted by the Pew Research Center it was confirmed that teens share a wide range of information about themselves on SNSs, as the sites are designed to encourage the sharing of private details with others [4]. Feng and Xie [1, p.156] claim that it seems hard to stop teens from disclosing personal information on SNSs. Boyd [5] points out that the notion that teens do not care about privacy is very entrenched in the public discourse. However, the truth is that teens often do care about privacy although this can be ignored by the public and the media [5]. Herring and Kapidzic [6] emphasized that there are concerns about the lack of parental control over teenage use of the Internet, but there are indications that more and more parents are trying to protect their child's online privacy. While older users may perceive privacy as protecting personal data from outside intrusion such as advertisers and the government, younger users perceive privacy as control over their personal space and social situations, as they are more worried about parents and teachers than advertisers and other third parties [7].

Today there are many types of social networks, but as Lenhart [8] pointed out, Facebook remains the most used social media site among American teens. Herring and Kapidzic cite the claim that "girls on average spend more time on social network sites and use them more actively than boys do" [6]. In several studies it was found that teenagers, from 15 to 17 years old, post more pictures and other personal information on their Facebook profiles [9]. Girls post more romantic pictures and the boys create more self-promotional pictures and comments with sexual content or alcohol references [6]. In a study of young adults' perceptions of appropriate content, it was established that they expressed "little concern about sharing posts and pictures on social network sites such as Facebook. Female participants expressed more concern about future employers seeing some of their pictures and comments" [6, p.4]. It seems that young adults predominantly utilize social media to build and strengthen relationships with their peers. They are not concerned with their more formal, professional image on social media [6].

III. RESEARCH

A. Research goal and methods

The research goal of this study was to explore and compare the extent to which personal information about male and female pupils and university students is publicly available on Facebook. The content analysis method was used to determine the extent to which male and female pupils' and university students' personal information of the various data types is publicly available on their Facebook profiles [10]. We determined the extent of publicly available items for various data types by counting the number of all items that were publicly published on Facebook profiles of examinees.

The method of comparative analysis was used to compare the extent to which various types of personal information about examinees are publicly available on their Facebook profiles [11]. We compared the percentage of examinees of different gender that published at least one item in a specific data type of personal information. We also compared the average number of items in a specific data type that were published by examinees of different gender.

B. Study participants

400 Facebook profiles were selected: 100 profiles of male university students and 100 profiles of female university students were randomly chosen from among Facebook community „Sveučilište u Zadru" which gathers students from the University of Zadar in Croatia [12]. Also, 100 profiles of female pupils were randomly chosen from among the pupils' Facebook community „OK je OK!" [13]; additionally, 100 profiles of male pupils were randomly chosen from among the Facebook friends of the members of „OK je OK!" Facebook community.

The pupils were about 10 to 13 years old and the university students were about 18 to 24 years old. The data were collected in March and April 2016. This study is a continuation of research about the Facebook publishing behaviour of young people which was presented in the year 2016 at European Conference on Information Literacy. In that presentation the same sample of Facebook profiles was used, but the data were analysed regardless of Facebook profile owners' gender [14].

IV. FINDINGS

In this section various findings from the research are presented in detail. Findings are presented according to following themes: items published by female examinees; items published by male examinees; comparison of all examinees with regard to gender; comparison between female and male pupils; comparison between female and male students.

A. Items published by female examinees

In Table 1 we can see to what extent female pupils and university students publish items in 15 most populated data types. In Part 1 of Table 1 we see that data types in which at least one item was published publicly by more than a half of female examinees are: Photos, Posts, Gender, Groups. A considerable percentage of them have published at least one item in data types about their interests: Movies, TV Shows, Apps & Games. Many of them have published about Current City (38.5%), Friends (37%) and Hometown (35%). Also, about one third of them published about their interests (Music, TV Shows, Movies, Books, Apps & Games), wrote reviews and marked their location with mobile devices (Check-Ins). It is important to note that very small number of them published at least one item in these data types: Religious Views (4%), Mobile Phone (0.5%), Email (0.5%), Political Views (0.5%), Address (0%). These most private data types of personal information are mostly unavailable on their Facebook profiles.

In Part 2 of Table 1 we see that on average a female pupil or university student has 123.1 friends, 51.4 published photos and 5.8 posts. On average, they published the largest amount of items in data types about their interests and especially about their music interests: 32 items. Other interests that have a high average of items are: Music, TV Shows and Movies. In comparison to these interests, female examinees published considerably less information about Sports, Apps & Games and Books. The following data types are also less present on their profiles: Groups, Check-Ins, Events.

TABLE I.   TO WHAT EXTENT FEMALE PUPILS AND UNIVERSITY
STUDENTS PUBLISH ITEMS IN 15 MOST POPULATED DATA TYPES

Part 1: Percentage of female pupils and univ. students that published at least one item		Part 2: Average of items published by female pupils and university students	
Data Type	%	Data Type	av. it.
Photos	98%	Friends	123.1
Posts	86.5%	Photos	51.4
Gender	81.5%	Music	31.8
Groups	71%	TV Shows	14.4
Current City	38.5%	Movies	13.4
Friends	37%	Groups	6.2
Music	37%	Posts	5.8
TV Shows	36.5%	Sports	5
Hometown	35%	Check-Ins	4.9
Movies	35%	Events	3.6
Check-Ins	30.5%	Apps & G.	2.1
Reviews	28.5%	Books	1.9
Books	28.5%	Reviews	0.9
Apps & Games	27.5%	Gender	0.8
Family Members	17.5%	Current City	0.4

B. Items published by male examinees

In Table 2 we can see to what extent male pupils and university students publish items in 15 most populated data types. In Part 1 of Table 2 we see data types in which at least one item was published publicly by more than a half of male examinees: Photos, Gender, Posts, Groups, Music, Friends.

TABLE II.   TO WHAT EXTENT MALE PUPILS AND UNIVERSITY
STUDENTS PUBLISH ITEMS IN 15 MOST POPULATED DATA TYPES

Part 1: Percentage of male pup. and univ. students that published at least one item		Part 2: Average of items published by male pupils and university students	
Data Type	%	Data Type	av. it.
Photos	95.5%	Friends	290.7
Gender	84.5%	Photos	39.7
Posts	81.5%	Sports	24.1
Groups	71%	Music	18.1
Music	54.5%	Movies	7.7
Friends	51.5%	Groups	6.4
Sports	47.5%	TV Shows	5.7
TV Shows	45%	Posts	4.3
Current city	44%	Apps & Games	3.8
Movies	44%	Check-Ins	2
Hometown	43.5%	Reviews	1.7
Apps & Games	40.5%	Events	1.6
Check-Ins	27%	Books	1.1
Books	27%	Gender	0.9
Reviews	21%	Current City	0.4

Almost half of them have published at least one item in data types about following interests: Sports, TV Shows, Movies, Apps & Games. Only 27% of male examinees published at least one book item as well as information about their locations (Check-Ins). While almost half of them published the names of the city in which they live and their hometown, nobody or very few of them published at least one item in most private data types: Religious Views (4.5%), Political Views (2.5%), Address (1%), Mobile Phone (0%), Email (0%).

C. Comparison of all examinees with regard to gender

Table 3 presents the comparison of the percentage of female and male pupils and students for data types in which at least one item was published by more than 30 percent of pupils or students. In relation to male examinees, a higher percentage of female examinees published at least one item in three data types. In Part 1 of Table 3 we see that among these data types, Check-Ins is a data type in which disparity between items published by female and male examinees is greatest. 1.13 times more female examinees than male examinees published at least one item in this data type.

In Part 2 of Table 3 we see nine data types in which male examinees published more often than female. Music, as well as Apps & Games are two data types in which disparity between items published by female and male examinees is greatest. 1.47 times more male examinees published at least one item in these data types. They also published more frequently at least one item in following data types: Friends (proportion: p:1.39), Movies (p:1.26), Hometown (p:1.24), TV Shows (p:1.23), Sports (p:1.23), Current City (p:1.14).

TABLE III.   COMPARISON OF THE PERCENTAGE OF FEMALE PUPILS
AND UNIVERSITY STUDENTS WITH MALE PUPILS AND UNI. STUDENTS

Part 1: Data types in which more items were published by female pupils and university students			
Data Type	Female	Male	Proportion
Check-Ins	30.5%	27%	1.13
Posts	86.5%	81.5%	1.06
Photos	98%	95.5%	1.03

Part 2: Data types in which more items were published by male pupils and university students			
Data type	Male	Female	Proportion
Music	54.5%	37%	1.47
Apps & G	40.5%	27.5%	1.47
Friends	51.5%	37%	1.39
Movies	44%	35%	1.26
Hometown	43.5%	35%	1.24
TV Shows	45%	36.5%	1.23
Sports	47.5%	38.5%	1.23
Curr. City	44%	38.5%	1.14
Gender	84.5%	81.5%	1.04

Table 4 presents comparison of the average of items that were published by female and male pupils and students for data types in which at least one item was published by more than 30 percent of pupils or students. A

820

higher percentage of female examinees published, on average, more items in six data types. In Part 1 of Table 4 we see that among these data types, the disparity between items published by female and male examinees is greatest for TV shows (proportion: 2.52) and Check-Ins data types (p:2.49). Also, about 1.75 times more female examinees than male examinees published, on average, more items in Movies and Music data types.

In Part 2 of Table 4 we see four data types in which male pupils and university students published more items on average than female. Sports is a data type in which there is by far the largest disparity: 4.85 times more male examinees published more frequently items in Sports data type. They also published more frequently in following data types: Friends (proportion: 2.36) and Apps & Games (p:1.77).

TABLE IV.   COMPARISON OF THE AVERAGE OF ITEMS PUBLISHED BY FEMALE PUPILS AND UNI. STUDENTS WITH MALE PUPILS AND UNIVERSITY STUDENTS

Part 1: Data types in which more items on average were published by female pupils and univ. students			
Data Type	Female	Male	Proportion
TV Shows	14.42	5.73	2.52
Check-Ins	4.85	1.95	2.49
Movies	13.42	7.65	1.75
Music	31.83	18.8	1.69
Posts	5.79	4.28	1.35
Photos	51.44	39.66	1.3

Part 2: Data types in which more items on average were published by male pupils and univ. students			
Data type	Male	Female	Proportion
Sports	24.07	4.96	4.85
Friends	290.65	123.1	2.36
Apps & G.	3.75	2.12	1.77
Groups	6.36	6.21	1.02

D. Comparison between female and male pupils

Table 5 presents the comparison of the percentage of items that were published by female and male pupils for data types in which at least one item was published by more than 30 percent of female or male pupils.

In relation to male pupils, a higher percentage of female pupils published, on average, more items in six data types that are presented in Part 1 of Table 5. Among these data types, disparity between female and male pupils is by far the largest for books (proportion: 2.06). Also, 1.28 times more female pupils than male pupils more frequently published at least one item in TV Shows (p:1.28) and Posts data types (p:1.16). In other data types there is very small difference between two sexes.

In Part 2 of Table 5 there are data types in which male pupils published more frequently at least one item. For all those data types (Apps & Games, Groups, Music) there is very small difference between two sexes because for all those data types proportion is less than 1.05.

TABLE V.   COMPARISON OF THE PERCENTAGE OF ITEMS THAT WERE PUBLISHED BY FEMALE AND MALE PUPILS

Part 1: Data types in which more items were published by female pupils			
Data Type	Female	Male	Proportion
Books	35%	17%	2.06
TV Shows	51%	40%	1.28
Posts	78%	67%	1.16
Movies	50%	46%	1.09
Curr. City	40%	38%	1.05
Friends	52%	50%	1.04
Gender	99%	97%	1.02
Sports	51%	50%	1.02
Photos	100%	99%	1.01

Part 2: Data types in which more items were published by male pupils			
Data type	Male	Female	Proportion
Apps & G.	43%	41%	1.05
Groups	72%	69%	1.04
Music	52%	51%	1.02

Table 6 presents the comparison of the average of items that were published by female and male pupils for data types in which at least one item was published by more than 30 percent of female or male pupils. In relation to male pupils, a higher percentage of female pupils published more items on average in six data types that are presented in Part 1 of Table 5. Among these data types, disparity between female and male pupils is by far the greatest for Books (proportion: 7.57). Large disparity is also found in data types TV Shows (p:4.23), Music (3.51), Movies (2.7).

In Part 2 of Table 6 there are four data types in which male pupils published more items on average. The disparity between female and male pupils is by far the greatest for Sports (p: 4.58). Large disparity is also found in data types Friends (p:1.78) and Apps & Games (1.34).

TABLE VI.   COMPARISON OF THE AVERAGE OF ITEMS THAT WERE PUBLISHED BY FEMALE AND MALE PUPILS

Part 1: Data types in which more items on average were published by female pupils			
Data Type	Female	Male	Proportion
Books	2.8	0.37	7.57
TV Shows	23.44	5.54	4.23
Music	49.83	14.19	3.51
Movies	23.36	8.66	2.7
Photos	73.42	47.49	1.55
Posts	7.32	5.03	1.46

Part 2: Data types in which more items on average were published by male pupils			
Data type	Male	Female	Proportion
Sports	41.62	9.09	4.58
Friends	315.5	177.3	1.78
Apps & G.	5.13	3.84	1.34
Groups	5.57	4.98	1.12

E. Comparison between female and male students

Table 7 presents the comparison of the percentage of items that were published by female and male university students for data types in which at least one item was published by more than 30 percent of female or male pupils. In relation to male students, a higher percentage of female students published more frequently at least one item in four data types that are presented in Part 1 of Table 7. Among these data types, disparity between female and male students is greatest for Reviews (proportion: 1.5). The much smaller disparity is found in other three data types Check-Ins (p:1.15), Photos (p:1.04), Groups (p:1.04).

In Part 2 of Table 7 there are 13 data types in which male students published more frequently at least one item. The disparity between female and male students is largest for following data types High School (p: 2.83), Apps & Games (p:2.71), Music (p:2.48), Friends (p:2.41), TV Shows (p:2.27), Movies (p:2.1). Significant disparity is also found in data types College (p:1.85), Sports (p:1.73), Books (p:1.68), Hometown (p:1.53), Current City (p:1.35).

TABLE VII.   COMPARISON OF THE PERCENTAGE OF ITEMS THAT WERE PUBLISHED BY FEMALE AND MALE UNIVERSITY STUDENTS

Part 1: Data types in which more items were published by female university students			
Data Type	Female	Male	Proportion
Reviews	33%	22%	1.5
Check-Ins	39%	34%	1.15
Photos	96%	92%	1.04
Groups	73%	70%	1.04

Part 2: Data types in which more items were published by male university students			
Data type	Male	Female	Proportion
Hi. School	34%	12%	2.83
Apps & G.	38%	14%	2.71
Music	57%	23%	2.48
Friends	53%	22%	2.41
TV Shows	50%	22%	2.27
Movies	42%	20%	2.1
College	48%	26%	1.85
Sports	45%	26%	1.73
Books	37%	22%	1.68
Hometown	49%	32%	1.53
Curr. City	50%	37%	1.35
Gender	72%	64%	1.13
Posts	96%	95%	1.01

Table 8 presents the comparison of the average of items that were published by female and male university students for data types in which at least one item was published by more than 30 percent of female or male university students. In relation to male students, a higher percentage of female students published more items on average in four data types that are presented in Part 1 of Table 8. Among these data types, disparity between female and male students is by far the greatest for Check-Ins (proportion: 3.43). Considerably smaller disparity is found in Posts data type (p:1.2).

In Part 2 of Table 8 there are eight data types in which male students published more items on average. The disparity between female and male students is by far the greatest for Sports (p:7.86), Apps & Games (p:5.9) and Friends (p:3.86). Large disparity is also found in data types Movies (p:1.91), Music (p:1.69), Books (p:1.67). In two data types: TV Shows (p: 1.1) and Photos (p:1.08), there is very small difference between two sexes.

TABLE VIII.   COMPARISON OF THE AVERAGE OF ITEMS THAT WERE PUBLISHED BY FEMALE AND MALE UNIVERSITY STUDENTS

Part 1: Data types in which more items on average were published by female university students			
Data Type	Female	Male	Proportion
Check-Ins	8.23	2.4	3.43
Posts	4.25	3.53	1.2
Reviews	0.78	0.74	1.05
Groups	7.43	7.15	1.04

Part 2: Data types in which more items on average were published by male university students			
Data type	Male	Female	Proportion
Sports	6.52	0.83	7.86
Apps & G.	2.36	0.4	5.9
Friends	265.8	68.89	3.86
Movies	6.64	3.47	1.91
Music	23.41	13.82	1.69
Books	1.77	1.06	1.67
TV Shows	5.92	5.39	1.1
Photos	31.83	29.45	1.08

V.   CONCLUSION

A.   Main insights from the research

Through this research, we acquired three types of insights about publishing behaviour of young Facebook users:

a) **Comparison of pupils' and university students' publishing data.** Very small percentage of examinees of both sexes publish items in most sensitive data types regarding the privacy: Religious Views, Mobile Phone, Email, Political Views, Address. In relation to male examinees, there is no data type in which female examinees publish information in somewhat higher (1.2<proportion<2) or considerably higher percentage (p>2). On average, female examinees publish information considerably more frequently (p>2) in following data types: TV Shows and Check-Ins. They also publish information somewhat more frequently (1.2<p<2) in data types Movies, Music, Posts, Photos. Male examinees on average publish information considerably more frequently in data types: Sports and Friends. They also publish information somewhat more frequently in Apps & Games data type. Regarding the most sensitive data types, both male and female examinees publish on average similar amount of information except in data type Political Views for which male examinees publish five times more often.

b) **Comparison of pupils' publishing data.** A similar percentage of both male and female pupils publish information in almost all data types (proportion<1.2). The only exceptions are Books data type where female pupils publish about two times more than male pupils and TV Shows data type where they publish about 1.3 times more than male pupils. On average, female pupils publish information considerably more frequently in following data types: Books, TV Shows, Music, Movies. Books data type is by far the most popular type for publishing information because female pupils publish about 7.5 times more often information in that type than male pupils. They also publish somewhat more frequently photos and posts. On average, male pupils publish information considerably more frequently in Sports data type - about 4.5 times more often than female pupils. They also publish information somewhat more frequently in data types: Friends and Apps & Games.

c) **Comparison of university students' publishing data.** In relation to male university students, somewhat higher percentage of female university students publish information in Reviews data type. Somewhat higher percentage of male students publish information in data types: College, Sports, Books, Hometown, Current City. Considerably higher percentage of male students publish information in data types: High School, Apps and Games, Music, Friends, TV Shows, Movies. On average, female students publish information considerably more frequently in Check-Ins data type where they publish about 3.5 times more often than male students. They also publish somewhat more frequently in Posts data type. On average, male students publish information considerably more frequently in data types: Sports, Apps & Games, Friends. Sports data type is by far the most popular type for publishing of information because male students publish about 8 times more often information in that type than female students. They also publish somewhat more frequently in data types: Movies, Music, Books.

B. *Some possibilities for using these findings and further research*

This study provided many insights about the characteristics of Facebook publishing behaviour of young people with regard to their gender. These insights could be helpful to create better educational programs for the improvement of privacy protection behaviour to alleviate risky information disclosure behaviour. For example, the focus of these programs could be on Facebook data types in which majority of male or female pupils and students are publishing much information (photos, posts, memberships in Facebook groups, interests: movies, music, books...). Based on insights from this study, young people educational programs could be made that are more adjusted to their gender. Various institutions (schools, universities, libraries, museums...), could use the findings to create more interesting and relevant programs with regard to their users' gender. Promotion of various educational and other types of programs could also be improved when interests

of young people of different gender (presented through their Facebook publishing behaviour) are more thoroughly known. Also, this study and its findings could be informative and inspirational for further research on the related topics. Facebook profiles of the same pupils and students explored in this study could be explored a few years from now to determine the extent of changes in their publishing behaviour. Additionally, Facebook data types that were explored in this study could be analysed by using the method of the qualitative content analysis. The research about adults' publishing behaviour on Facebook could be compared with the behaviour of the young people of different genders. The comparisons of educational and cultural differences regarding the Facebook publishing behaviour are another way to broaden the findings related to this area.

REFERENCES

[1] Y. Feng, W. Xie, „Teens' concern for privacy when using social networking sites: An analysis of socialization agents and relationships with privacy-protecting behaviors," Computers in Human Behavior, vol. 33, pp. 153–162, 2014.

[2] R. Gross, A. Acquisti, „Information revelation and privacy in online social networks," in Proceedings of the 2005 ACM Workshop on Privacy in the Electronic Society. Association for Computing Machinery, 2005, pp. 71–80.

[3] E. K. Clemons, J. Wilson, „Students' and parents' attitudes towards online privacy: An international study," in 48th Hawaii International Conference on System Sciences, IEEE, 2015, pp. 4844–4853.

[4] M. Madden, A. Lenhart, S. Cortesi, U. Gasser, M. Duggan, A. Smith, M. Beaton, Teens, social media, and privacy, Pew Research Center, 2013.

[5] D. Boyd, It's complicated: the social lives of networked teens, Yale University Press, 2014.

[6] S.C. Herring, S. Kapidzic, „Teens, gender, and self-presentation in social media," in International encyclopedia of social and behavioral sciences, 2nd ed., Oxford: Elsevier, 2015,

[7] D. Bradbury, „The kids are alright," Engineering & Technology, vol. 10, pp. 30–33, 2015.

[8] A. Lenhart, Teens, social media, and technology overview, Pew Research Center, 2015.

[9] E. Vanderhoven, T. Schellens, M. Valcke, A. Raes, „How safe do teenagers behave on Facebook? An observational study," PLoS One, vol. 9, 2014.

[10] D. O. Case, Looking for information: a survey of research on information seeking, needs, and behavior, Emerald Group Publishing Limited, 2006.

[11] M. Bray, B. Adamson, M. Mason (eds.), Comparative education research: approaches and methods, Springer Netherlands, 2007

[12] Facebook. Sveučilište u Zadru, Retrieved January 1, 2017. from https://www.facebook.com/groups/7507921803/?ref=bookmarks

[13] Facebook. OK je OK!. Retrieved January 1, 2017. from https://www.facebook.com/okjeokmagazin/

[14] M. Duić, P. Džapo, „Social Media Networking Literacy and Privacy on Facebook: Comparison of Pupils and Students Regarding the Public Availability of Their Personal Information," in press.

Web sources of literature for teachers and researchers: practices and attitudes of Croatian faculty toward legal digital libraries and shadow libraries such as Sci-Hub

Mirko Duić, Barbara Konjevod and Laura Grzunov
University of Zadar, Department of Information Sciences, Zadar, Croatia
miduic@unizd.hr, bkonjevod@student.unizd.hr, lgrzunov@student.unizd.hr

Abstract - In this paper, we have explored the extent to which faculty use different web sources of scientific papers, as well as their practices and attitudes related to availability and use of scientific papers. The questionnaire was sent to employees of the Faculty of Humanities and Social Sciences and the Faculty of Science at University of Zagreb. It was found out that majority of respondents use scientific papers on the web, several times a week. Google Scholar is the most used web portal for finding papers. Also, very popular are Croatian web portals. Pirate web portals (shadow libraries) are the least used, although almost half of respondents use them for various reasons: access to papers is expensive and not available; easy and fast access, etc. Only a small minority of respondents completely or partly agree that they are content with offer of international scientific journals and papers which are available in Croatia through subscription by scientific institutions. Through log analysis method, we also explored which scientific papers are requested in the Sci-Hub pirate web portal by users located in Croatia. Papers from Natural Sciences domain were most requested, while papers from Social Sciences and Humanities domains are not so often requested.

I. INTRODUCTION

Through the subscription of *Ministry of Science and Education*, Croatian faculty has an access to various databases of scientific journals. However, there is a problem with financing of this subscription. Therefore, the access to journals is limited. Availability of scientific journals is also problematic because higher education institutions don't have a lot of financial resources for subscribing to printed or electronic journals [1]. For example, Krajna and Markulin have established that the majority of libraries at the University of Zagreb doesn't have an adequate budget for literature procurement. For many years *Ministry of Science and Education* has co-financed their procurement, but because of the reduction of those financial subventions, libraries had to reduce the number of their journal subscriptions [2]. The additional problem is that between various university libraries there is no coordinated purchase of journals and other literature, there is no agreement about the „library fond development (...), planning and rational budget spending, more comprehensive and faster addressing of user needs, avoiding of unnecessary duplication of information sources" [2, 35]. Martek believes that the golden year of

subscription to international scientific journals and databases was 2005 when many information sources were available to the academic community. Since then, the budget cuts have had considerable negative impact on the availability of international journals [3]. However, availability of Croatian scientific journals is very good, although it was not always like that. According to a study from the year 2002, among 223 analysed Croatian scientific journals, 120 journals or 54% had a web presence. Among these 120 journals only 21 or 18% were providing full text of papers [4]. According to study from the year 2009, a study conducted on the same sample of 223 journals, 174 journals or 78% had a web presence. Among these 174 journals 61 or 36% were providing full text of papers. An important factor is that 54 of those 61 journals were providing open access to papers [5]. The considerable improvement in Croatian scientific journal availability was made when the Hrčak web portal was launched in 2006. Hrčak is online journals platform which provided to publishers a simple tool for creating online open access versions of their journals. It included 170 open access (OA) journals and had an average of about 10000 unique visits per day in 2009. Approximately 50% of visitors were from Croatia [6]. On 29 January 2017 Hrčak included 430 journals with 165141 papers. From these papers 96.7% were in open access [7]. With Hrčak platform, which is supported by governmental funds, open access movement got a stronghold in Croatia. In the same period when Hrčak was launched, many government scientific agencies in the world have accepted or started to prepare the open access policy (UK, Germany, Austria, China, Canada, France, Sweden, USA...). Many OA archives began their work, as well as important online catalogs of OA archives: *Directory of Open Access Repositories*; *ScientificCommons*; *OpenArchives.eu*. In these catalogs, including the ROAR and OAISter catalogs, the number of registered OA archives increased [8]. An important reason for the rise of OA movement was the unfavorable situation in academic publishing in which an oligopoly of large publishers was formed. For example, five big academic publishers have published more than half of all scientific papers in 2013. The largest academic publishers, such as Elsevier, Springer, Taylor & Francis, Wiley-Blackwell, Wolters, use their large market shares and influence to significantly increase subscription rates for scientific journals and databases. Academic libraries and government agencies, especially those in financially lagging countries, don't have budgets to pay subscriptions for many of these journals and databases [9]. Even universities in wealthy countries have financial problems to provide journal access for their faculty. Therefore, the scientific community began to protest against this situation as well as trying to find new ways to acquire access to scientific papers. For example, Harvard University faculty advisory council proclaimed that major scientific publishers had made scholarly communication fiscally unsustainable and academically restrictive. Therefore, the university wants its scientists to publish in open access journals [10]. One of the most recent initiatives of protest against excessive subscription prices is canceling of journal subscription of the Elsevier publisher by 60 major German research institutions. This cancellation was made at the end of 2016 and it will be continued until the acceptable subscription price is agreed [11]. Interesting idea is proposed in the analysis made within Max Planck Society in Germany. The authors of this analysis think that whole commercial, academic publishing could be replaced by open access publishing. They have calculated that if the libraries worldwide use entire budgets that they have yearly at their disposal to pay for scientific journal subscriptions, this would be sufficient financial resources to fund the work of OA journals in which all yearly production of scientific papers could be published [12].

Besides legal ways to access scientific papers, there is another way for faculty to access scientific papers. It is through pirate web portals which are called shadow libraries: „piratical text collections which have now amassed electronic copies of millions of copyrighted works and provide access to them usually free of charge to anyone around the globe" [13, 75]. One of the largest shadow libraries is the Sci-Hub, which contains more than 50 million papers. Bohannon wrote that Sci-Hub was created in 2011 by Alexandra Elbakyan, who was then 22-year-old graduate student of neuroscience in Kazakhstan. According to the server log analysis of 28 million documents which were requested from the Sci-Hub in the period from September 2015 to March 2016, it was established that this shadow library has users in the whole world, that it has users from developing and most developed countries. At the end of February 2016, Sci-Hub received 200000 paper requests per day! Also, Bohannon found out that a quarter of the Sci-Hub paper requests were made by users from 34 wealthy countries that are members of the Organization for Economic Co-operation and Development [14]. Users from wealthy countries probably have solid legal access to the same papers but their motivation for using a Sci-Hub may be speed and convenience in accessing papers as well as the comprehensiveness of its database. An online survey of attitudes toward the Sci-Hub was made by *Science* journal. Based on nearly 11000 responses, here are some of the interesting insights: Nearly 60% of respondents used Sci-Hub; 88% think that it is not wrong to download pirated papers. This opinion is held even by respondents who have never used Sci-Hub or those who are older than 51; For more than 50% of respondents primary reason for using a Sci-Hub is lack of access to journals, for about 17% it is a simple convenience and for 23% respondents

the primary reason for using a Sci-Hub is dissatisfaction with the large profits of publishers [15]. In one paper Sci-Hub usage data for Latin America was analysed. It was established that in Argentina Sci-Hub downloads represent 13.3% of downloaded papers in relation to the legal downloads made through scientific databases subscribed by scientific institutions of that country. The Sci-Hub use is considerably lower in Mexico – only 2.3% of downloaded papers in relation to the legal downloads [16]. Bodó emphasized that users from Central and Eastern European countries are among the biggest per capita users of shadow libraries. According to this author, shadow libraries are a crucial resource in the modernization of these countries [17].

II. RESEARCH

The research goal of this study was to explore the extent to which faculty use different web sources of scientific papers, as well as their practices and attitudes related to availability and use of scientific papers. In order to reach these insights, online survey was used. The information about this survey was sent in December 2016 to employees with teaching responsibilities at the *Faculty of Humanities and Social Sciences* [18], as well as at the *Faculty of Science at University of Zagreb* [19]. The survey was completed by 147 respondents who work in different domains of knowledge, primarily in Natural Sciences, Social Sciences and Humanities. Additionally, we used the log analysis method to explore which scientific papers are used in the Sci-Hub by users located in Croatia. Insights from this part of the research were compared with insights from the aforementioned survey of faculty practices and attitudes. The Sci-Hub is one of the largest shadow or pirate libraries found on the Internet. Data about the use of this library from 1 September 2015 to 29 February 2016, is publicly available on web in Sci-Hub server logs [20]. To protect the privacy of Sci-Hub users, no identifying internet protocol (IP) of users was published. Regarding the user location, only countries and approximate places of residents were published (nearest city or village) [14]. We analysed which papers from Sci-Hub were used in February 2016 by users located in Croatia. Two variables were used in the analysis: paper scientific domain and date of publishing.

III. FINDINGS

A. Survey of Croatian faculty practices and attitudes

The majority of respondents are female (60.5%). Also, the majority of respondents are 30 to 39 years old (39.5%), although there are more than 20% of respondents who are 40 to 49 years old, as well as those who are 50 to 59 years old. This means that the survey was filled up by faculty of various ages. The survey was filled up by approximately similar number of respondents from both higher education institutions. Most of survey respondents are working in domains of Natural Sciences (41.5%), Social Sciences (24.5%) and Humanities (23.8%). There is a small number of respondents from the domains of Technical Sciences (2%) and Interdisciplinary Sciences (5.4%).

In Table I we see that majority of respondents download and read scientific papers on the web, several times a week (42.2%). When we add to this group the respondents who daily use scientific papers on the web, it is evident that there is a huge share of respondents who are very active in using web for finding and reading scientific papers. Respondents who use scientific papers on the web, several times a month are also not infrequent users (25.2%).

TABLE I.   FREQUENCY OF DOWNLOADING AND READING SCIENTIFIC PAPERS ON THE WEB

Frequency	%	N
Daily	25.9%	38
Not daily, but several times a week	42.2%	62
Not every week, but several times a month	25.2%	37
Not every month, but several times a year	5.4%	8

In Table II we see that there is about 25 to 30% of respondents who never ask anyone to send them scientific papers from international journals (grade 1). About 30% of respondents almost never ask anyone to send them these papers (grade 2). Also, there are about 10% of respondents who very often ask paper authors or non-authors to send them these papers (grade 5). About 10% of respondents often ask for papers (grade 4). Respondents who are working in Natural Sciences will more often ask other people to send them papers and respondents who are working in Social Sciences are least prone to ask for papers. Also, there is a notable difference between male and female respondents with regard to asking for papers. In relation to male respondents, female respondents are about twice more often asking for papers (grades 5 and 4).

TABLE II.   ASKING SOMEBODY FOR SCIENTIFIC PAPERS

Estimate on a scale from 1 to 5 to what extent do you access scientific papers from international journals so that you ask somebody to send them to you through the Internet (1 = never; 5 = very often; n. r. = no response)

Senders	1	2	3	4	5	n. r.
Paper authors	27.2%	29.9%	17%	8.2%	12.9%	4.8%
Persons in Croatia	29.3%	29.9%	15%	10.9%	8.8%	6.1%
Persons abroad	24.5%	27.9%	17.7%	12.2%	12.2%	5.4%

In Table III we see that most popular web portal for accessing scientific papers is Google Scholar. There are about 30% of respondents who very often use Google Scholar (grade 5) and about 10% of respondents who often use that portal (grade 4). Interesting fact is that female respondents are considerably more frequent users – about 40% of them very often use Google Scholar and about 20% of male respondents very often use that portal (grade 5). Also, popular web portals are at the *Faculty of Humanities and Social Sciences* (FHSS cat. / FHSS DB.). Its library catalog and database list are used very often for accessing scientific papers by about 25% of respondents and often by about 8% of respondents. Other web portals are less used, although there are about 15% of respondents

who very often or often use DOAJ (Directory of open access journals) and NUL portal (National and university library portal). We also asked respondents to write the names of other web portals that they use. Following portals were mentioned most frequently: Research Gate (22 respondents), Science Direct (14 respondents), Hrčak (13), Academia.edu (12), Web of Science (12), Scopus (8), PubMed (7), arXiv.org (6), EBSCO (4), Google (4).

TABLE III.   ACCESSING SCIENTIFIC PAPERS THROUGH WEB PORTALS

Estimate on a scale from 1 to 5 to what extent do you access scientific papers through the specific web portal (1 = never; 5 = very often; n. r. = no resp.)

Web portal	1	2	3	4	5	n. r.
Google Scholar	17%	17.7%	15.7%	11.6%	32%	6.1%
DOAJ	44.9%	18.4%	10.2%	8.2%	6.8%	11.6%
OpenAIRE	68.7%	12.2%	3.4%	0.7%	1.4%	13.6%
NUL Portal	48.3%	12.2%	15.7%	6.8%	9.5%	7.5%
PERO	66%	8.8%	11.6%	4.1%	0.7%	8.8%
FHSS cat.	48.3%	3.4%	8.2%	8.2%	25.9%	7.5%
FHSS DB	46.3%	4.8%	4.8%	8.1%	27.2%	8.8%

In Table IV we see how often respondents access papers by using two large shadow libraries or pirate web portals. The Sci-Hub is used very often by about 18% of respondents (grade 5) and it is used often by about 7% of respondents (grade 4). Library Genesis is used very often by about 12% of respondents and it is used often by about 5% of respondents. About 50% of respondents have never used Sci-Hub and about 65% of respondents have never used Library Genesis. It seems that these portals are predominantly used pirate web portals. Namely, we asked respondents to wrote the names of the other pirate web portals they use and only four respondents indicated other portals: Bookza, booksc.org, Uz-translations, Bookfi.

TABLE IV.   ACCESSING PAPERS USING PIRATE WEB PORTALS

Estimate on a scale from 1 to 5 to what extent do you access scientific papers from international journals using the following pirate web portals (1= never; 5 = very often; n. r. = no response)

Web portal	1	2	3	4	5	n. r.
Sci-Hub	49%	10.2%	6.1%	6.8%	17.7%	10.2%
Libr. Genesis	65.3%	8.2%	6.1%	5.4%	11.6%	3.4%

In Table V we see that there are almost half of respondents who download and read scientific papers using the pirate web portals (46.3%). There are about 10% of respondents who daily or several times a week use these portals. About 15% of respondents use pirate web portals several times a month and about 18% of respondents use these portals several times a year. With regard to gender, there are about 15% of male respondents who use daily or several times a week these portals. About 40% of male respondents never use them. There are only about 10% of female respondents who use daily or several times a week these portals. About 50% of female respondents never use them. Respondents were also asked about reasons for using pirate portals. Following reasons were mentioned most frequently: papers are not available in any other way (28 respondents); legal papers are expensive (14); easy access (10); papers are available so why not to use them (8); fast access (8); Ministry or scientific institution doesn't pay subscription (7); it's a matter of principle: science should be available to all (5); relevancy and huge quantity of papers (5). Additionally, we asked respondents who don't use pirate web portals, about their reasons. Following reasons were mentioned most frequently: non-awareness about the existence of pirate web portals (11 respondents); availability of papers through legal web portals (11); it's an illegal activity / respect for the legal rights of authors (8); there is no need to use pirate web portals (7); colleagues are sending me papers (4).

TABLE V.   FREQUENCY OF DOWNLOADING AND READING SCIEN. PAPERS USING THE PIRATE WEB PORTALS

Frequency	%	N
Daily	3.4%	5
Not daily, but several times a week	8.2%	12
Not every week, but several times a month	15.7%	23
Not every month, but several times a year	18.4%	27
Never	46.3%	68
No response	8.2%	12

In Table VI we see frequency of web portals use. It was established that legal international web portals are mostly used: about 60% of respondents use these portals very often or often (grades 5 and 4). They are followed by Croatian web portals which are very often and often used by about 35% of respondents. Pirate web portals are the least used with about 25% of respondents who very often and often use these portals. It is interesting that respondents from Humanities domain are most often using pirate web portals – about 40% of them said that they are using them very often or often. About 20% of respondents from Natural Sciences and Social Sciences domains said that they are using pirate web portals very often or often.

TABLE VI.   FREQUENCY OF WEB PORTALS USAGE

Estimate on a scale from 1 to 5 how often do you use the following three types of web portals to find scientific papers (1= never; 5 = very often; n. r. = no response)

Web portals	1	2	3	4	5	n. r.
Croatian	34%	20.4%	9.5%	10.2%	23.8%	2%
International	10.9%	12.2%	15%	12.9%	46.3%	2.7%
Pirate	46.9%	15.7%	5.4%	9.5%	14.3%	8.2%

In Table VII are statements for which we wanted to get opinions from users of pirate web portals. Regarding the Statement 1 we found out that about 45% of respondents completely or partly agree that it's OK that scientists read and download scientific papers using pirate web portals if they don't have access to these papers in no other way (grades 5 and 4). This interesting finding indicates that despite the fact that it is not legal, almost half of respondents have a positive attitude to using pirate web portals in case there is no other way to access literature. Regarding the Statement 2 we found out that about 37% of respondents completely and partly agree that papers which they read and download using pirate

web portals are usually not available in any other way. Regarding the Statement 3 we found out that about 20% of respondents completely or partly agree that they use pirate web portals because it's a fast and simple way to access scientific papers.

TABLE VII.   EXTENT OF AGREEMENT WITH STATEMENTS

If you read and download scientific papers using pirate web portals, estimate on a scale from 1 to 5 to what extend do you agree with the following statements (1 = completely disagree; 5 = completely agree; n. r. = no response)

Statements	1	2	3	4	5	n. r.
Statement 1	It's OK that scientists read and download scientific papers using pirate web portals if they don't have access to these papers in no other way					
	6.1%	4.1%	7.5%	8.8%	37.4%	36.1%
Statement 2	Scientific papers which I read and download using pirate web portals are usually not available in any other way					
	6.1%	2%	9.5%	8.8%	27.9%	45.6%
Statement 3	I use pirate web portals because it's a fast and simple way to access scientific papers, in relation to legal, non-pirate web portals					
	19.1%	9.5%	6.8%	4.1%	15.7%	44.9%

In Table VIII are statements for which we wanted to get opinions from respondents, regardless if they use pirate portals.

TABLE VIII.   EXTENT OF AGREEMENT WITH STATEMENTS

Estimate on a scale from 1 to 5 to what extend do you agree with the following statements (1 = completely disagree; 5 = completely agree; n. r. = no response)

Statements	1	2	3	4	5	n. r.
Statement 1	I know very well how to use computers and internet					
	0.7%	1.4%	11.6%	23.8%	61.9%	0.7%
Statement 2	It's OK to use the internet to share for free the scientific papers of which you are author					
	1.4%	2.7%	6.8%	11.6%	74.8%	2.7%
Statement 3	It's OK to use the internet to share for free the scientific papers of which you are not an author					
	23.8%	15.7%	19.7%	8.2%	25.9%	6.8%
Statement 4	I'm contented with offer of international scientific journals and papers which are available in Croatia through subscription by scientific institutions					
	29.3%	29.9%	23.8%	9.5%	4.1%	3.4%
Statement 5	Teaching and research in Croatia could be significantly improved if availability of international scientific papers is increased through subscription of the Ministry and other institutions					
	5.4%	1.4%	10.9%	17%	63.3%	2%
Statement 6	Subscriptions for databases with international scientific journals are too expensive					
	1.4%	1.4%	16.3%	12.9%	59.2%	8.8%
Statement 7	All scientific papers should be freely available to all who are interested, without subscription payment to publishers of scientific journals					
	4.1%	3.4%	17.7%	14.3%	57.1%	3.4%

Regarding the Statement 1 and 2 we found out that about 85% of respondents completely or partly agree that they know very well how to use computers and the internet. They also think that it's OK to use the internet to share for free the scientific papers of which they are authors (grades 5 and 4). However, only about 35% of respondents completely or partly agree that it's OK to use the internet to share for free the scientific papers of which they are not authors (statement 3). Interesting fact is that female respondents are considerably less approving of this activity – about 25% of them completely or partly agree with this statement, in relation to 40% of male respondents. Regarding the Statement 4, we found out that only about 15% of respondents completely or partly agree that they are satisfied with offer of international scientific journals and papers which are available in Croatia through subscription by scientific institutions. Most of them think that teaching and research in Croatia could be significantly improved if availability of international scientific papers is increased through this subscription. About 80% of respondents gave grades 5 and 4 regarding statement 5. About 73% of respondents completely or partly agree that subscriptions for databases with international scientific journals are too expensive (statement 6). There is also very large percentage of respondents who think that all scientific papers should be freely available to all who are interested, without subscription payment to publishers of scientific journals – about 70% of respondents gave grades 5 and 4 regarding statement 7.

B. Scientific domains and publication dates of Sci-Hub papers which are requested by users located in Croatia

We used log analysis method to explore which scientific papers are used in the Sci-Hub, one of the largest shadow libraries, by users located in Croatia. We analysed 886 papers regarding their scientific domains. That is a sample of about 25% of papers requested in February 2016 by users located in Croatia. It was found out that the papers from Natural Sciences domain were most requested (41.8%). Papers from Biomedicine and Healthcare domain are at the second place with 23% of requests, followed by requests for papers in the Biotechnical Sciences domain (12.2%). It is interesting that papers from the Social Sciences domain are not so often requested (10%) and that the papers in Humanities domain are requested in only 2% of cases. Also among less requested papers are those in Technical Sciences (8.7%) and Interdisciplinary Sciences (2.7%). We compared these insights about use of SciHub, with insights from the survey. Analysis of Sci-Hub server log shows considerable differences in frequency of use of papers from different scientific domains. For example, papers from the Natural Sciences domain are by far the most frequently requested papers. In contrast to these findings, survey results indicate that the majority of respondents from domains of Social Sciences, Humanities and Natural Sciences claim that they approximately equally often use the Sci-Hub portal: about 60% of these respondents are not using this portal and about 20% of respondents are very often using it. These survey results are in contrast with findings from server log analysis of Sci-Hub portal where it was found out that the papers from Humanities were requested by only 2%

of users, papers from Social Sciences were requested by about 10% of users and papers from Natural Sciences were requested by about 40% of users.

We also analysed 1697 papers regarding their date of publishing. That is a sample of about 50% of papers requested in February 2016 by users located in Croatia We found out that most recent papers are considerably more often requested. The largest percentage of requested papers is published in the most recent period from 2010 to 2016 (54.9%). After that, the most requested papers are published from 2000 to 2009 (24%) and from 1990 to 1999 (13.2%). There are only 4.7% of requested papers published from 1980 to 1989. There is very small percentage of requested papers published before the year 1979 (3.2%).

IV. CONCLUSION

In this study, we acquired various insights about availability and use of web sources of scientific papers by teaching and research employees of the *Faculty of Humanities and Social Sciences* and *Faculty of Science at University of Zagreb*. Additionally, we explored which scientific papers are requested in the Sci-Hub, one of the largest shadow libraries, by users located in Croatia. Here are some of the main insights from the research.

Respondents are very active in using web for finding and reading scientific papers. The majority of them download and read scientific papers on web, several times a week. About two third of respondents never ask or almost never anyone to send them scientific papers. Respondents who are working in Natural Sciences will more often ask other people to send them papers and respondents who are working in Social sciences are least prone to ask for papers. In relation to male respondents, female respondents are about two times more often asking for papers. Web portals which are used most frequently by respondents for finding papers are legal international web portals such as Google Scholar (indicated by respondents as most used international legal portal). They are followed by Croatian web portals. Shadow libraries or pirate web portals are the least used. Almost half of respondents download and read scientific papers using the pirate web portals, mainly for following reasons: papers are not available in any other way; legal papers are expensive; easy access; papers are available so why not to use them; fast access; Ministry and scientific institutions don't pay database and journal subscriptions; it's a matter of principle – science should be available to all. Pirate web portal Sci-Hub is used very often and often by about one quarter of respondents. The Library Genesis is used slightly less. These portals are the most used pirate web portals among respondents. Almost half of respondents have a positive attitude to using pirate web portals in case there is no other way to access literature. About one third of respondents completely or partly agree that it's OK to use the internet to share for free the scientific papers of which they are not authors. Female respondents are considerably less approving of this activity. A small minority of respondents completely or partly agree that they are content with offer of international scientific journals and papers which are available in Croatia through subscription by scientific institutions. The vast majority of respondents think that teaching and research in Croatia could be significantly improved if availability of international scientific papers is increased through subscription.

Through log analysis method we explored which scientific papers are requested in the pirate web portal Sci-Hub by users located in Croatia. Papers from Natural Sciences domain were most requested – about 40% of all requests. Papers from Social Sciences domain are not so often requested (10%) and papers in Humanities domain are requested in only 2% of cases. These insights are in contrast with findings from the survey where respondents from all those scientific domains said that they use Sci-Hub with approximately equal frequency.

Findings from this study indicate that many respondents are very active users of legal and pirate web sources of scientific papers. Findings also indicate that there is a major problem with availability of papers from subscription journals and web portals. Although the vast majority of respondents think that the Ministry of Science and Education as well as scientific institutions don't pay enough for these subscriptions, it is evident from the findings that there are other important issues beside payment problems. The majority of respondents thinks that all scientific papers should be freely available to all who are interested, without subscription payment to publishers of scientific journals. Open access journals could help to expand the availability of scientific literature. Croatian web portal *Hrčak* is a great resource of open access papers and proof that it is possible to provide scientific papers without involvement of profit oriented publishers. Also, university faculty has various approaches for acquiring papers which are unavailable through open access and commercial databases. Many of them are exchanging papers and visiting pirate web portals. We also found out that many are using legal web portal *Research Gate* to access papers. It enables authors to upload and give access to their papers even if they have published these papers in journals closed behind subscription paywalls. Use of *Research Gate* and similar participative, scientific web portals by Croatian faculty is an important activity which should be further explored. Other findings from this study should also be further explored to better understand ways of accessing the scientific literature by Croatian faculty. For example, the impact of the differences in gender regarding the use of scientific web sources, could be more thoroughly examined. Also, other scientific institutions could be included in future surveys to compare the activities and attitudes of faculty from various scientific institutions and domains. We hope that findings from this study will enable better understanding of various issues related to practices and attitudes of Croatian faculty toward web sources of literature.

REFERENCES

[1] A. Zubac and A. Tominac, „Digitalna knjižnica kao podrška sveučilišnoj nastavi i istraživačkome radu na daljinu: elektronički izvori za elektroničko učenje na hrvatskim sveučilištima," Vjesnik bibliotekara Hrvatske, vol. 55 (2), pp. 65-82, 2012.

[2] T. Krajna and H. Markulin, „Nabava knjižnične građe u visokoškolskim knjižnicama," Vjesnik bibliotekara Hrvatske, vol. 54 (3), pp. 21-42, 2011.

[3] A. Martek, „Konzorcijska nabava u Hrvatskoj: stanje i perspektiva," Vjesnik bibliotekara Hrvatske, vol. 54 (3), pp. 79-94, 2011.

[4] S. Konjević, „Hrvatski znanstveni časopisi na Internetu," Vjesnik bibliotekara Hrvatske, vol. 46 (3), pp. 111-118, 2003.

[5] S. Konjević, „Hrvatski znanstveni i znanstveno-stručni časopisi u elektroničkome mrežnom okruženju," Vjesnik bibliotekara Hrvatske, vol. 52 (1), pp. 75-88, 2009.

[6] J. Stojanovski, J. Petrak and B. Macan, „The Croatian national open access journal platform," Learned Publishing, vol. 22 (4), pp. 263-273, 2009.

[7] Hrčak – portal of scientific journals of Croatia. Retrieved January 29, 2017. from http://hrcak.srce.hr/

[8] V. Silobrčić, „Slobodan pristup ocijenjenim znanstvenim informacijama," Vjesnik bibliotekara Hrvatske, vol. 50 (1), pp. 51-61, 2007.

[9] V. Larivière, S. Haustein and P. Mongeon, „The oligopoly of academic publishers in the digital era," PloS one, vol. 10 (6), pp. 1-15, 2015.

[10] I. Sample, „Harvard University says it can't afford journal publishers' prices," The Guardian. Retrieved January 15, 2017. from http://www.theguardian.com/science/2012/apr/24/harvard-university-journal-publishers-prices

[11] „No full-text access to Elsevier journals to be expected from 1 January 2017 on," Göttingen State and University Library. Retrieved January 15, 2017. from http://www.sub.uni-goettingen.de/en/news/details/voraussichtlich-keine-volltexte-von-zeitschriften-des-elsevier-verlags-ab-dem-112017/

[12] R. Schimmer, K. K. Geschuhn and A. Vogler, „Disrupting the subscription journals' business model for the necessary large-scale transformation to open access," A Max Planck Digital Library Open Access Policy White Paper, 2015.

[13] B. Bodó, „Libraries in the post-scarcity era," in Copyrighting Creativity: Creative values, Cultural Heritage Institutions and Systems of Intellectual Property, H. Porsdam, Eds. Ashgate, 2015, pp. 75-92.

[14] J. Bohannon, „Who's downloading pirated papers? Everyone," Science. vol. 352 (6285), pp. 508-512, 2016.

[15] J. Travis, „In survey, most give thumbs-up to pirated papers," Science. Retrieved January 15, 2017. from http://www.sciencemag.org/news/2016/05/survey-most-give-thumbs-pirated-papers

[16] J. D. Machin-Mastromatteo, A. Uribe-Tirado and M. E. Romero-Ortiz, „Piracy of scientific papers in Latin America: An analysis of Sci-Hub usage data," Information Development, vol. 32 (5), pp. 1806-1814, 2016.

[17] B. Bodó, „Eastern Europeans in the pirate library," Visegrad Insight 2015, 1 (7), pp. 98-102, 2015.

[18] Faculty of Humanities and Social Sciences, University of Zagreb (Filozofski fakultet, Sveučilište u Zagrebu). Retrieved January 15, 2017. from http://www.ffzg.unizg.hr/

[19] Faculty of Science, University of Zagreb (Prirodoslovno-matematički fakultet, Sveučilište u Zagrebu). Retrieved January 15, 2017. from http://www.pmf.unizg.hr/

[20] Sci-Hub world data set. Retrieved July 15, 2016. from http://datadryad.org/resource/doi:10.5061/dryad.q447c

Supporting mobile learning: usability of digital collections in Croatia for use on mobile devices

Radovan Vrana, Denis Gaščić, Marina Podkonjak

Department for information and communication sciences,
Faculty of humanities and social sciences, University of Zagreb, Zagreb, Croatia
rvrana@ffzg.hr, denisgascic9@gmail.com, marina.cernik777@gmail.com

Abstract - Mobile learning is growing in popularity as use of mobile devices by students worldwide increases. At the same time the number of digital information resources such as digital collections in libraries available online has also increased. Digital collections in libraries are valuable and proven resource of information in education. As students use mobile devices more frequently it is expected that use of digital information resources including digital collections in education will also increase. To achieve this goal user interfaces of digital collections must be developed in such a manner to be easily used, especially on mobile devices. This paper presents results of the research of elements of user interfaces of digital collections in libraries in Croatia available on the Web to establish the level of their mobile usability and accessibility as a prerequisite for their potential use in the learning process. To fulfill that potential, digital collections have to support different screen sizes on mobile devices and be accessible on new and old mobile devices, offer assortment of common functions to user etc. The results of this research will help librarians in planning further development of digital collections and educators in planning use of digital collections in their curricula.

I. INTRODUCTION

Mobile learning is growing more popular as value of the learning industry (especially in e-learning) is increasing [1] and as the number of advanced mobile devices owned by students with potential for use in the learning process in higher education institutions is also increasing. Mobile technology has a great influence on students: it changes "how they communicate, gather information, allocate time and attention, and potentially how they learn. The mobile platform's unique capabilities — including connectivity, cameras, sensors, and GPS — have great potential to enrich the academic experience" [2]. The success of mobile learning depends on several factors like better and cheaper devices capable of accessing the internet; improved global network speeds; changing customer expectations; social media; the ability of learners to connect to the internet; and on content available online, usually on the Web either separately as a standalone Web site or as a part of an information institution holdings such as digital collections in libraries. Digital collections in libraries are highly organized information resources with user interfaces which are somewhat different from the interfaces of other Web sites so it may not be easy for users to access and use easily

content in digital collections unless the user interfaces are carefully adjusted to user needs. Sandnes [3] suggested that content of digital libraries should be presented similarly (in all collections) as to create a holistic unified experience for users. In order to find out the current level of development of user interfaces of digital collections to the user needs, this paper will present results from the research study in mobile usability and accessibility of digital collections in libraries in Croatia accessible on the internet for the purpose of their potential use in education.

II. MOBILE LEARNING

Mobile learning presents a step forward in development of education during which educational institutions worldwide transform into true e-learning institutions. Mobile learning "is a type of learning whose learner is determined previously, is not in a specific location, or benefits the opportunities offered by mobile technologies" [5]. Behera [4] explained that "mobile learning is sometimes considered merely an extension of e-learning, but quality m-learning can only be delivered with an awareness of the special limitations and benefits of mobile devices" (mobile learning combines e-learning and mobile computing). By using mobile technologies (and the internet), mobile learning presents a learning model which allows learners to obtain learning materials anywhere and anytime [6].

Mobile learning includes "the exploitation of ubiquitous handheld hardware, wireless networking and mobile telephony to facilitate, support enhance and extend the reach of teaching and learning" [7]. By definition, this type of learning involves the use of different mobile technologies and communication technology (ICT) thus enabling earning anytime and anywhere (Mobile learning). Mobile learning implies use of mobile devices such as "laptops, netbooks, e-readers, tablets, mobile phones, smart phone MP3/ MP4 players and internet capable handheld devices" [9]. According to Choy [10], to be referred to as mobile devices, devices should have the following characteristics: telecommunication functions such as voice, email, data services (SMS); high speed access to internet; screen for viewing and interacting with multimedia content; able to run third-party software, with PC-like functionality; GPS functions and be highly portable, i.e. without being an extra burden when a person is moving about (not applicable to laptops and large e-book readers). Mobile devices have become relevant in daily activities including the learning process because they are convenient to use and are highly portable; they are ubiquitous because almost anyone owns a mobile phone; mobile technology is a fast-changing as it develops all the time and becomes more sophisticated and user and eventually everyone will have a smart phone; they are multi-function computing machine [10]. Fast-changing technologies [11] could impose problems on teachers and learners because they cannot all obtain new devices fast enough, there are no pedagogical guidelines and experiences derived from research about use of new technologies and there is no adequate content prepared for use by new technologies at the moment they appear on the market.

Seppälä and H. Alamäki [12] analyzed mobile learning in educational activities and pointed out important features of mobile learning: teaching and learning outside classroom, moving to another location while communicating via information networks; it enables learners to enter an information network at the precise moment when necessary by using a portable learning device and a wireless network. Same authors also suggested that "mobile environment integrates studies that take place on campus, at home or outside university facilities into one shared, flexible learning environment" [12]. Ozdamli and Cavus [6] suggested the following core characteristics of mobile learning: ubiquitous, portable size of mobile tools, blended, private, interactive, collaborative, and instant information. Anilkumar [13] enumerated a number of advantages of mobile learning: convenience and flexibility, relevance, learner control, good use of "dead time", fits many different learning styles, improves social, easy evidence collection, encourages reflection, supported decision making, speedier remediation, easily digestible learning, heightened engagement, direct interaction with learning, context sensitive learning etc. The learning process itself can unfold in a variety of ways under the influence of mobile technology: people can use mobile devices to access educational resources, connect with others, or create content, both inside and outside classrooms. Furthermore, mobile learning has become common and is gradually becoming ubiquitous because it does not "separate the learner from their normal activities or routines; learning can be delivered in a manner that either enhances such activities or allows them to occur in tandem with that process" [14]. Learners can learn using mobile devices anytime and anywhere and be not restricted to a particular locations or a time schedule; they can be more engaged and self-directed in learning because he/she can self-tailor his/her pace [10].

Finally, mobile learning cannot be effectively carried out without content. Content is one of basic elements of effective mobile learning which (in addition to content) also include learner, teacher, environment and assessment [6]. Quality learning material needed in education can be found in digital collections in libraries. The purpose of development of digital collections in libraries is to make their content available for use in different contexts. One such context is mobile learning.

III. DIGITAL COLLECTIONS

During the last two decades, libraries have changed significantly, from centuries old information delivery through printed materials to delivery through computers to the desktop of the user and, most recently, directly to the users' mobile device thus freeing the information consumer from his desktop and fixed location [10]. As the number of mobile devices in the general population increases, so increases the demand for digital content adapted for use on these devices. In libraries tablets and smart phones are used for browsing online content, downloading, streaming and lending digital content from libraries as well as searching a database, downloading articles, seeking catalogs, access documents, including e-books, audio-visual objects and Websites [9]. Saxena and Yadav [15] investigated library services accessible on mobile devices and found out pros and cons of such use of library resources and services. The pros were: user-friendly aid; personalized service; ability to access information; time saving; user participation; location awareness; limitless access; access to print-disabled users. The cons were: compared to wired internet service, mobile services have relatively slow transmission speed; limited computational power; inconvenient input and output interface; insufficient contents and high price.

As much as digital content could be found anywhere on the Web, in this paper special attention is paid to digital collections in libraries as a measure of the content quality because everything that is included in digital collections goes through the process of a careful selection done by library professionals. The content is then used in libraries, at universities, at home and elsewhere. Availability of access to digital collections is particularly important for distance learners interested in library collections [16]. That is the reason why libraries are very important stakeholder in the learning process because they "could provide access to resources tailored to mobile devices and services to guide users in their self-directed learning effort" [10].

To become more accessible to students (i.e. to be reachable online) on all levels of education, the access to the digital content in library collections should be also adapted for use on mobile devices. Since digital collections in libraries are accessible mostly on the Web (and some are accessible only locally in libraries on dedicated desktop computers), their use has become part of the Web content use paradigm. Therefore, the rest of the paper will focus on usability. Aldrich [17] offered some key thoughts about the relation between the Web and mobile devices by pointing out the mobile Web characteristics and ubiquity because of which the mobile Web has become an integral part of the devices used to access it "to the point that the mobile Web refers to any Web-based content or function configured for access through mobile devices" [3]. One of the biggest problems in use of the Web is it inconsistency in content organization and use. Use of digital content on mobile devices on the Web depends heavily on Web page design which must present consistency in domains of appearance, controls, and functions [18]. Furthermore, user "(…) want to control graphical quality, typography, window size, location of information within a window, and everything that is important in good design" [18].

IV. USABILITY

Use of Web based content is closely related to the concept of usability. Usability is "generally regarded as ensuring that interactive products are easy to learn, effective to use, and enjoyable from the user's perspective" [19]. The goals of usability are usually the following: effectiveness, efficiency, safety, utility, learnability and memorability [19]. The most common method for achieving usability is user-centered design which includes "user-oriented methods such as task analysis, focus groups, and user testing to understand user needs and refine designs based on user feedback" [20]. The user centered design consists of the following phases or stages: "1.) specify the context of use: identify the people who will use the product, what they will use it for, and under what conditions they will use it 2.) specify requirements: Identify any business requirements or user goals that must be met for the product to be successful 3.) create design solutions: this part of the process may be done in stages, building from a rough concept to a complete design 4.) evaluate designs: evaluation - ideally through usability testing with actual users - is as integral as quality testing is to good software development" [21]. A special attention should be given to user preferences when accessing digital content in libraries on mobile devices of users because different mobile devices have different screen sizes, orientation (portrait or landscape), users use them to access the content with different network speed, software and context of use [22] thus having different user experience. User experience "refers to the overall experience users have when using a service. User experience is a very wide ranging concept but it is often associated with aspects of usability in relation to Web based services" [23]. To improve the user experience of use of a digital collection, one needs to consider the needs of your users, how they find and navigate your collections, and the tasks they go on to complete as a result [23].

McCray and Gallagher [24] recommended that every digital library (which contains digital collections) development should follow ten principles: 1.) expect change, 2) know your content, 3) involve the right people, 4) design usable systems, 5) ensure open access 6) be(a)ware of data rights, 7) automate whenever possible, 8) adopt and adhere to standards, 9) ensure quality, 10) be concerned about persistence. As we can see, design of a usable system is one of the key points in digital library development.

V. MOBILE USABILTY RESEARCH STUDY

The goal of this part of the paper is to present results of the mobile usability test of user interfaces of digital collections in Croatian libraries accessible online.

The main idea for this research was taken from the article "Mobile Usability in Educational Contexts: What have we learnt?" written by Kukulska-Hulme [26] who suggested that mobile usability is not researched frequently and that its goal is to improve usability of the user interface or content which must adaptable to, or by, the user of the mobile device. The research hypothesis is that the user interfaces of digital collections in the Croatian libraries are not ready for use on mobile devices. The goal of this research was to detect problems in access to digital collections by use of mobile devices and to help librarians (digital collection developers) and their IT colleagues to develop digital collections highly usable by mobile devices.

During the initial phases of development of digital collections in libraries, mobile devices used today didn't exist at all. It wasn't until recently that library users started to access digital collections content more frequently by mobile devices. This triggered researches with focus on different but complementary issues in access to digital content for the purpose of learning. Wagner [26] analyzed implementation of mobile learning and selected a group of attributes related to the rich mobile internet experience in access to digital content. Since digital collections in libraries are accessible on the internet, most if not all suggested attributes are applicable to library digital collections and could be included in mobile usability testing to a degree: ubiquity, access, richness, efficiency, flexibility, security, reliability and interactivity. Authors like Saxena and Yadav [15] analyzed library services accessible via mobile technology and offered the following list of what such library services should offer: SMS notification services, formal education, distance learning and e-learning, database browsing, my library, e-resources with mobile interfaces, library guide, mobile document supply, text reference service, library virtual/ audio tours, QR codes on mobiles.

These examples served as a foundation for the mobile usability research study in the following part of the paper which aims to answer the following questions: 1.) are Web pages of digital collections developed for use on mobile devices? 3.) which additional Web page options or functions are available for easier viewing of the content of digital collections that can also contribute to the better functionality of digital collections? Seventeen digital collections accessible online in the Croatian libraries were selected for this research. The list of digital collections for research was created from information about all available digital collections at the portal of Public libraries in Croatia at http://knjiznica.hr [45]. Additional digital collections in libraries in Croatia were found by help of the Google search engine. Due to space restrictions, names of digital collection in tables were replaced by numbers from the following list: Digital library SveVid (1), Digitized Rijeka newspapers (2), ViZZ – Virtual local collection (3), Istrian newspapers online (4), Digitized Šibenik library material (5), „DELMATA" digital collection (6), Digitized local collection Spalatina (7), Split university library digital collections (8), Dubrovnik library (9), City library Poreč digital collections (10), DIKAZ – Digital library Zadar (11), Digital collection HAZU (12), Digital Zagreb heritage (13), Faculty of Humanities and Social Science Zagreb – repository (14), Exam literature digital collection at the Faculty of Humanities and Social Science Zagreb (15), National and university library Zagreb – digital collections (16), City library Vinkovci – digitized library material (17) [28-44].

Criteria for the analysis and comparison of the digital collections user interface elements were formed on the basis of previous use of digital collections by authors of this paper. Without the elements listed in tables 1 and 3, digital collections would lose some if not most of their functionality and usability. The research results in tables 1-3 show whether or not this is true for 17 digital collections. They were accessed, analyzed and compared by use of an actual mobile phone or a mobile phone emulator. JavaScript and HTML5 were tested by use the Website http://builtwith.com on a mobile phone.

TABLE 1. DIGITAL COLLECTION USER INTERFACE ELEMENTS BASIC FUNCIONALITY TEST

NUMBER	DOWNLOAD	JAVASCRIPT SUPPORT	HTML5 SUPPORT	SEARCH	ADVANCED SEARCH	INDEXING	EXISTENCE OF METADATA	SEARCHABLE METADATA
1	YES	YES	NO	YES	partially	YES	YES	partially
2	YES	NO?	NO	YES	YES	YES	YES	YES
3	NO	YES	NO	YES	YES	YES	YES	YES
4	NO	YES?	NO	YES	YES	YES	YES	YES
5	YES	YES	NO	YES	NO	NO	NO	NO
6	YES	YES	NO	NO	NO	NO	NO	NO
7	NO	YES	NO	NO	NO	YES	NO	NO
8	YES	YES	NO	YES	NO	YES	YES	YES
9	NO	YES	NO	YES	NO	NO	NO	NO
10	YES	YES	NO	YES	NO	YES	NO	NO
11	NO	YES	NO	YES	YES	YES	YES	YES
12	YES	YES	NO	YES	YES	YES	YES	YES
13	YES	YES	NO	YES	NO	YES	YES	YES
14	YES	YES	NO	YES	YES	YES	YES	YES
15	YES	NO	NO	NO	NO	YES	NO	NO
16	NO	YES	NO	YES	YES	YES	YES	YES
17	NO	NO	NO	YES	NO	YES	YES	YES

Table 1 presents results from digital collections user interface elements basic functionality test. The results indicate that one of the main functions - download of the digital collection content - is supported by 10 out of 17 (58,8%) digital collections which is very favorable for digital collections users. Web technologies such as HTML5 and Javascript were represented differently in the researched digital collections. HTML5 was not used in any of 17 digital collections while JavaScriptis was used in 13 out of 17 digital collections (76,5%). Search of the content was possible in 14 out of 17 (82,4%) digital collections and advanced search was fully supported in 7 (41,2%) digital collections, not supported in 9 (52,9%) and partially supported in 1 (5,8%) digital collection. Indexing as „a process of establishing access points to

facilitate retrieval of records and/or information" [27] was present in 14 (82,4%) digital collections while metadata were included in 11 (64,7%) digital collections. Metadata search was possible in 10 (58,8%) and not possible in 6 (35,3%), and partially possible in 1 (5,8%) digital collection. The researched digital collections also supported categorization of the content: books, magazines, picture categories were found in 11 (64,7%) digital collections.

TABLE 2. ADAPTABILITY OF USER INTERFACES OF DIGITAL COLLECTIONS TO USE ON MOBILE DEVICES WITH DIFFERENT SCREEN SIZES

No.	4,7"	5"	5,5"
1	NO	NO	NO
2	NO	NO	NO
3	NO	NO	NO
4	YES	YES	YES
5	NO	NO	NO
6	NO	NO	NO
7	NO	NO	NO
8	NO	NO	NO
9	NO	NO	NO
10	NO	NO	NO
11	NO	NO	NO
12	NO	NO	NO
13	NO	NO	NO
14	NO	NO	NO
15	NO	NO	NO
16	YES	YES	YES
17	NO	NO	NO

Table 2 represents the results of the research of adaptability of the user interfaces of the selected digital collections for use on mobile devices of different screen sizes. In addition to use of actual mobile phone for research, Google Chrome extension called Mobile/Responsive Web Design Tester was also used to emulate mobile devices with three most commonly used mobile device screen sizes: iPhone 6S (4,7"), Sony Xperia M4 Aqua (5,0") and LG G3 (5,5"). Research showed that only 2 out of 17 (11,76%) pages were prepared for use on smaller screen sizes though the same content was not displayed identically or even similarly on screens with dimensions of 4,7'',5'' and 5,5''. Ten out of 17 (58,8%) digital collections were not prepared for use on mobile devices so actual use of these digital collections was possible but required additional vertical and horizontal scrolling as if there were used on a desktop PC. Another method was use of the zooming option in the Web browser to scale the Web page for normal use. Such Web page use was difficult because with every click on the Web page of a digital collection it was mandatory to use the zoom option again. In case of 5 out of 17 (29,41%) digital collections, Web pages were extremely difficult to use because parts of their content were not displayed on mobile device screen correctly or Web page parts overlapped.

833

Picture 1. An example of the best Web-page usable on a 5,5" mobile screen

TABLE 3. DIGITAL COLLECTION USER INTERFACE FUNCTIONAL ADDITIONS

NUMBER	PAGING	SKIPPING PAGES	ZOOM	BROWSING FILTERS	REDUCED VIEW	ROTATION	PRINT	PURE TEXT
1	NO	NO	YES	NO	YES	YES	YES	NO
2	YES	YES	YES	NO	NO	YES	YES	NO
3	YES	NO	NO	NO	YES	NO	NO	NO
4	YES	YES	YES	NO	YES	YES	NO	YES
5	NO	NO	YES	NO	NO	YES	YES	NO
6	NO	NO	YES	NO	YES	YES	YES	NO
7	YES	YES	YES	NO	YES	NO	NO	NO
8	YES	YES	YES	NO	YES	YES	YES	YES
9	YES	YES	YES	NO	YES	NO	NO	NO
10	YES	YES	YES	NO	YES	YES	YES	YES
11	YES	YES	YES	NO	YES	YES	YES	YES
12	YES	NO	NO	YES	YES	NO	NO	NO
13	YES	YES	YES	NO	YES	YES	YES	YES
14	NO	NO	YES	NO	YES	YES	YES	NO
15	NO	NO	NO	NO	NO	NO	NO	YES
16	YES	YS	YES	NO	YES	YES	NO	NO
17	YES	YES	YES	NO	YES	YES	NO	YES

Digital collections user interface functions enumerated in Table 3 demonstrated the possibility of use of additional functions which would enable easier use of digital collection content. The following functions were observed: turning pages, jumping to particular page(s), zooming, rotating the digital object, objects filtering, printing, pure text viewing and reduced view of objects parts. Larger number of the researched digital collections provided the possibility of turning the pages function (most commonly represented in form of icons of left and right arrows). In one digital collection the function of turning the pages was enabled in form of paging of individual pictures which was not common but pages could still be turned, while in 4 other collections digital objects were available in pdf format where turning pages is accomplished by scrolling pages. In most digital collections jumping pages was not possible which made use of these digital collections harder (if an user needed to jump on page 345 of some book in 9 out of 17 (52,94%) digital collections this would require turning page by page until the desired page appeared). The zooming function has a big importance and impact in use of digital collections because letters on the Web page can be small and hardly legible or pictures can contain details which are impossible to see on a small screen. Three out of 17 (17,64%) collections did not offer the zooming function. Similar result was found when digital collections were searched for the rotating function. Five digital collections (29,41%) did not provide this function. Six out of 17 (35,29%) digital collections provided the function of getting the editable clear text out of digital objects which is useful. Only one digital collection (5,88%) showed clear text as a default option. Eight digital collections (47,05%) offered the printing function. The scaling function for pictures of digital objects were provided in 14 out of 17 (82,35%) observed digital collections. The worst situation by far appeared during the attempt to filter the digital collection content. This option was provided only in 1 out of 17 (5,88%) observed digital collections.

VI. CONCLUSION

As mobile learning gains more popularity, many scientists and educators put a number question about its implementation in different learning environments, scenarios and contexts. Ozdamli and Cavus [7] summed up their vision of achieving efficient results and the maximum performance from students using mobile learning in education in the following inspirational statement: "each of the elements of mobile learning should be prepared carefully, and the mobile learning characteristics should be planned and prepared with a knowledge of the teaching medium, learning environment and the learning activities. Otherwise, positive results cannot be expected from the mobile application". By having these thoughts in mind as well as the results of the mobile usability research study, we can conclude that the most of researched digital collections are not developed for use on mobile devices and that some elements of user interfaces in the observed digital collections are still underdeveloped and important options and functions are still unavailable to users which confirms the research hypothesis. Consequently, the potential of digital collections for use in education is still underused. Time waits for no-one, and libraries should increase their active involvement in education and understand their important role in providing support to mobile learning.

REFERENCES

[1] B. Little, "The Rising Popularity of Mobile Learning Southern Europe", E-learn magazine, 2011. Retrieved October 5, 2016. from http://elearnmag.acm.org/featured.cfm?aid=1966304

[2] B. Chen, R. Seilhamer, L. Bennett and S. Bauer, "Students' Mobile Learning Practices in Higher Education: A Multi-Year Study" Educause Review, 2015. Retrieved September 20; 2016. from http://er.educause.edu/articles/2015/6/students-mobile-learning-practices-in-higher-education-a-multiyear-study

[3] S.T. Sandnes, SResponsive Web Design for Digital Libraries: Accessing information anywhere, anytime. Retrieved October 5, 2016. from https://www.ntnu.no/documents/10401/1264433962/SigridArtikkel.pdf/84f3c5aa-a5e1-4587-94b4-4eab457e2f07

[4] S.K. Behera. E- and m-learning: a comparative study, International Journal on New Trends in Education and Their Implications, vol. 4, pp. 65-78, 2013.

[5] C. O'Malley, G. Vavoula, J. Glew, J. Taylor, M. Sharples and P. Lefrere, Guidelines for learning/teaching/tutoring in a mobile environment. Mobilearn project deliverable, 2003. Retrieved May 20, 2015. from http://www.mobilearn.org/download/results/guidelines.pdf

[6] F. Ozdamli, N. Cavus, "Basic elements and characteristics of mobile learning", Procedia - Social and Behavioral Sciences, vol. 28, pp. 937–942, 2011.

[7] MoLeNET. Retrieved December 11, 2016. from http://www.molenet.org.uk/about.html

[8] Mobile Learning. Retrieved December 11, 2016. from http://www.unesco.org/new/en/unesco/themes/icts/m4ed/

[9] S. Vongjaturapat and S. Chaveesuk, "Mobile technology acceptance for library information service: A theoretical model" in International Conference on Information Society (i-Society), Piscataway, NJ: IEEE, pp. 290-292, 2013.

[10] F.C. Choy, Digital Library Services: Towards mobile learning, Seminar on E-books as Learning Resources in Chinese Libraries in Asia, 2010. Retrieved October 5, 2016. from http://hdl.handle.net/10220/6286

[11] G. N. Vavoula, P. Lefrere, C. O'Malley, M. Sharples, J. Taylor. Producing guidelines for learning, teaching and tutoring in a mobile environment in Proceedings of the The 2nd IEEE International Workshop on Wireless and Mobile Technologies in Education (WMTE'04), Washington, DC : IEEE, 2004, 173-176.

[12] Seppälä, P. and Alamäki, H., "Mobile learning in teacher training", Journal of Computer Assisted Learning, vol. 19, pp. 330-335, 2003.

[13] P.D. Anilkumar. Mobile learning: a new approach in information communication technology in Libraries: traditional to modernization, Jagjit Singh Ed. Solapur : Laxmi book publication, 2015, pp. 128-144.

[14] S.J. Saravani and G. Haddow, "A theory of mobile library service delivery", Journal of Librarianship and Information Science, pp. 1–13, 2015.

[15] A. Saxena and R.D. Yadav, "Impact of mobile technology on libraries: a descriptive study", International Journal of Digital Library Services, Vol. 3, pp. 1-13, 2013.

[16] A.M. Gattan and M. Razek, "Towards Efficient Mobile Library Services" in 5th International Conference on Intelligent Systems, Modelling and Simulation (ISMS), Piscataway, NJ: IEEE, 2014, pp. 562-567.

[17] A.W. Aldrich, "Universities and Libraries Move to the Mobile Web" Educause Quarterly, vol. 33, 2010. Retrieved September 20,

2016. from http://er.educause.edu/articles/2010/6/universities-and-libraries-move-to-the-mobile-Web

[18] W.Y. Arms, Digital Libraries. Boston: Massachusetts Institute of Technology, 2000.

[19] J. Preece, Y. Rogers and H. Sharp, Interaction design: beyond human-computer interaction, New York : John Wiley & Sons, Inc., 2002.

[20] P.J. Lynch and S. Horton, Web style guide, 2009. Retrieved October 5, 2016. from http://www.Webstyleguide.com/wsg3/index.html

[21] User-Centered Design Basics. Retrieved October 5, 2016. from https://www.usability.gov/what-and-why/user-centered-design.html

[22] R. Fox, "Being responsive", OCLC Systems & Services, Vol. 28, pp. 119-125, 2012.

[23] I, Chowcat, D. Kay and O. Stephens, "Improve the user experience of your digital collection" Retrieved September 20, 2016. from https://www.jisc.ac.uk/guides/improve-the-user-experience-of-your-digital-collection

[24] A.T. McCray and M.E. Gallagher, "Principles For Digital Library Development", Communications of the ACM, vol. 5, pp. 49-54, 2001.

[25] A. Kukulska-Hulme, "Mobile Usability in Educational Contexts: What have we learnt?", The international review of research in open and distributed learning, vol 8, pp. 1-16, 2007.

[26] E.D. Wagner. Enabling mobile learning. EDUCAUSE Review, vol. 40, pp. 40-53, 2005.

[27] ISO (2016). ISO 15489-1. Information and documentation, Records management

[28] Digitalna knjižnica SveVid. Retrieved January 11, 2017. from http://svevid.locloudhosting.net

[29] Digitalizirane riječke novine. Retrieved January 11, 2017. from http://libraries.uniri.hr/liste/002n/

[30] ViZZ – Virtualna zavičajna zbirka. Retrieved January 11, 2017. from http://vizz.gkc-pula.hr

[31] Istarske novine online. Retrieved January 11, 2017. from http://www.ino.com.hr/

[32] Digitalizirana građa. Retrieved January 11, 2017. from http://www.knjiznica-sibenik.hr

[33] „DELMATA" projekt digitalizacije opusa Ljube Stipišića Delmate. Retrieved January 11, 2017. from http://www.delmata.org/o-projektu

[34] Digitalizirana zavičajna zbirka Splatina. Retrieved January 11, 2017. from http://www.gkmm.hr/digitalizirana_bastina.htm

[35] Digitalne zbirke sveučilišne knjižnice u Splitu. Retrieved January 11, 2017. from http://dalmatica.svkst.hr/?sitetext=363

[36] Dubrovačke knjižnice Dubrovnik: Digitalizacija. Retrieved January 11, 2017. from https://www.dkd.hr/index.php?option=com_content&view=article&id=355&Itemid=192

[37] Digitalna zbirka Gradske knjižnice Poreč. Retrieved January 11, 2017. from http://porec.arhivpro.hr/index.php

[38] DIKAZ – Digitalna knjižnica Zadar. Retrieved January 11, 2017. from http://dikaz.zkzd.hr

[39] Digitalna zbirka HAZU. Retrieved January 11, 2017. from http://dizbi.hazu.hr/

[40] Digitalizirana zagrebačka baština. Retrieved January 11, 2017. from http://kgzdzb.arhivpro.hr/?sitetext=317

[41] Repozitorij Filozofskog fakulteta Sveučilišta u Zagrebu. Retrieved January 11, 2017. from http://darhiv.ffzg.unizg.hr/

[42] Digitalna zbirka ispitne literature na FFZG-u. Retrieved January 11, 2017. from http://dzs.ffzg.unizg.hr/

[43] Digitalne zbirke Nacionalne i sveučilišne knjižnice u Zagrebu. Retrieved January 11, 2017. from http://digitalna.nsk.hr/

[44] Digitalizirana građa Gradske knjižnice i čitaonice Vinkovci. Retrieved January 11, 2017. from https://library.foi.hr/zbirke/vinkovci/index.php?page=oprojektu

[45] Portal narodnih knjižnica. Digitalne zbirke. Retrieved January 11, 2017. from http://www.knjiznica.hr/mods/digitalne-zbirke/

The Use of ICT in the English Language Classroom

Rebeka Lerga*, Sanja Candrlic**, Martina Holenko Dlab**

*First Croatian High School of Rijeka, Rijeka, Croatia
** Department of Informatics, University of Rijeka, Rijeka, Croatia
lerga.rebeka@skole.hr, sanjac@inf.uniri.hr, mholenko@inf.uniri.hr

Abstract – All aspects of society are constantly developing while adapting to the challenges of the modern world. Education is also included in this continuous process of transformation with the emphasis on the fast information retrieval and acquisition of knowledge from multiple resources. Young people live in a digital environment where they are constantly available on some digital device. This paper presents research on the use of information and communication technology (ICT) in education. Attitudes towards the use of technology in the English language classroom were tested among elementary students. Several surveys were conducted to compare the attitudes of students who use tablets in classrooms on the daily basis with those of students who have only experienced one lesson using tablets. The results suggest that the use of ICT in education is perceived more positively among those students who use technology in their classrooms on the daily basis.

I. INTRODUCTION

The development of the Information and Communication Technology (ICT) has affected many different aspects of the society. In the last two decades, the use of ICT has completely transformed almost all forms of social and business endeavours [1], including education. Moreover, learning and teaching has been enhanced with its dynamic and interactive contents [2]. It fosters students' skills and motivation, places them in the centre of that process, forming autonomous and ambitious students. Furthermore, by using technology in the classroom students learn how to relate to the working practices [3], and lifelong education. This all indicates a crucial need to integrate ICT in modern educational system, and numerous educational institutions have already been equipped with computers and digital devices.

Today's students mostly belong to the generation of digital natives - those who own a digital device and spend hours using it every day [4]. As digital natives, students have already developed skills they need to handle such devices. Implementation of technology in classrooms simply enhances those skills. Numerous studies were conducted exploring the impact of the use of ICT on students' motivation [6] and achievement [5], [7], and testing the attitudes of teachers and students [8] towards its application in classrooms. One of the possibilities for ICT implementation are tablet computers.

This paper presents a research on application of the ICT on foreign language teaching. It aims to increase the awareness on the importance of the use of technology in second language learning and teaching in primary schools. Since the use of tablets for learning and teaching has just recently begun in Croatia, this paper proposes a possible way of using these devices to enhance the language teaching. A lesson using exclusively ICT had been prepared for English language learning and conducted in one of the Croatian primary schools.

Moreover, the purpose of the described research was also to test elementary students' attitudes towards the use of technology in classrooms and to explore possibilities of developing more positive attitudes towards the integration of ICT in classrooms.

The paper is organized as follows: Section II gives background information focusing on the ICT implementation in Croatian schools while Section III describes the research methodology. Section IV presents and discusses results of the three surveys that had been conducted to test students' attitudes towards the benefits of ICT application in English language classroom. Section V concludes the paper.

II. BACKGROUND

There are several ongoing projects with the aim of ICT implementation in Croatian schools. The Croatian Academic and Research Network (CARNet) has initiated e-Schools: a comprehensive informatization of school operation processes and teaching processes, with the aim to create digitally mature schools for the 21st century. The e-Schools programme is carried out through several projects aimed at introducing ICT into the school system in the 2015-2022 period. The e-Schools project is part of a wider e-Schools programme. The full name of the project is "e-Schools: Establishing a System for Developing Digitally Mature Schools (pilot project)". The abbreviated name of the project is the 'e-Schools pilot project'. The specific objective of the e-Schools pilot project is to pilot organizational, technological and educational concepts of introducing ICT in the educational and operational processes in selected schools during two school years and to develop, based on the experience of the pilot project, a strategy for the implementation of a system of digitally mature schools in the entire primary and secondary education system in Croatia [9].

There are 150 elementary and high schools participating in this project. The process of equipping schools include the development of a local network, the equipping of classrooms with ICT equipment and the equipping of the teachers and administration staff with hybrid computers, laptops and tablets. Schools within the pilot project will also be equipped with two different types of classrooms, a presentation one (equipped with a PC computer, a touch screen monitor and loudspeakers) and an interactive one (equipped with 30 tablets and the presentation equipment) [10]. Also, there are educational activities and organised support for teachers and learning materials for students.

Another programme was initiated in cooperation with Apple. Since 2012 elementary school Vežica has been the part of Apple Lighthouse Schools. In this programme, the school was equipped with Apple technology (iPad tablets, etc.). Thus, Vežica became the first public elementary school in Croatia, and second public school in Europe to integrate iPads in classrooms [11]. Lessons from each subject are assisted with interactive learning contents, multimedia and Internet. Learning materials and assignments are sent to students' e-mail addresses, and tests and readings are accessed online. Moreover, students acquire digital competences they need for the future [12]. Apple is included in similar projects, equipping and transforming education in USA.

E-Schools presents a global challenge, accepted by many different countries. Through its 2008 e-Šolstvo project, the Republic of Slovenia prepared comprehensive digital content for the curricula of primary and secondary schools, which it offered to the entire educational system, and wider – as open educational resources under the Creative Commons licence. On the other hand, in 2012 the Polish Council of Ministers approved the national programme "Digital Schools" aimed at developing digital competences of students and teachers in the application of ICT with four component (e-teacher, e-student, e-textbook and e-school) [13].

III. METHODOLOGY

A. Participants

This research encompassed two groups of participants; all eight graders from the elementary school Vežica in Rijeka, Croatia. The research was conducted in 2016.

Participants in the first part of the research were students attending traditional classes. They use textbooks and notebooks. They don't use technology in classroom, except for the occasional use of computers and projector. These students filled in the survey, testing their attitudes towards the use of ICT in education, immediately before the English language lesson in which they have for the first time used only ICT in language classroom. At the end of the lesson, the same group of students has again filled in the survey (henceforth: first group). There were 16 students who filled in the first survey. At the lesson were present 14 students (8 boys and 6 girls), who immediately after the lesson filled in the second survey.

Participants in the second part of the research were students who for longer time period (few years) use tablets in classroom (henceforth: second group). These students have also filled in the survey testing their attitudes towards the use of ICT in education. There were 10 students (8 boys and 2 girls) who filled in the last survey.

B. ICT enhanced lesson

English language lesson was held in June 2016 in elementary school Vežica in Rijeka, Croatia. During the lesson, iPads, PC and projector were used in school's computers classroom. The content of the lesson was completely adjusted to the use of ICT in classroom. These students used tablets during the English language lesson for the first time.

The unit *In sync with your parents*, listed in course syllabus, was covered in this lesson. Using *Articulate Storyline 2*, lesson materials were prepared and presented on iPads. Before the class, the browser *Articulate Mobile Player* was installed on all tablets. This browser enabled teacher and students to view materials prepared using software *Articulate Storyline 2*. After the installation, materials needed for the lesson were downloaded and opened. The lesson material was prepared in digital format. Each click opened new content (including images that resemble phrases, additional descriptions, or audio and video clips). Moreover, new types of exercises, available only on digital devices were added (including pick many, drag and drop, and hotspot). This way the lesson was made more interesting and familiar to students.

At the beginning of the lesson, students were given iPads. Application Articulate Mobile Player and the material needed for the lesson were already lunched. The lesson started with a short revision. Students were given three different sentences with the same bolded phrase and different images describing it. After short class discussion, they came up with the meaning of the phrase, and then solved *multiple choice exercises*. The same pattern followed revision of all phrases covered in the previous lesson. In order to test students in their understanding of the specific words or phrases, the revision ended with two extra tasks. In the *pick many task*, students were asked to pick all the images that presented people who were in sync with others. In the *hotspot task*, students marked the picture that best described given words. They were only allowed to continue to the next slide, once they had solved the task correctly.

The main part started with the class *discussion*. Slide with question and central image opened up. Students were instructed to click on that image and then additional images appeared providing more ideas for the discussion (Fig. 1). Furthermore, images were followed by the list of adjectives describing characteristics of parents and children. Click on each adjective opened its definition (Fig. 2). After discussing all the adjectives, students were asked to sort them in two categories (positive and negative) in the *drag and drop task* (Fig. 3). Listening was accompanied with the two types of tasks: *sorting* and *matching*. In the first, students listened and sorted sentences in the correct order, and in the second they matched the expressions in the two columns. Before the video, students were presented with a family photo. Click on each member of the family opened different slide with their perspective on the family.

The lesson ended with a short video in which teenagers, native speakers, described their families. The video was accompanied with *pick many* and *drag and drop exercises*.

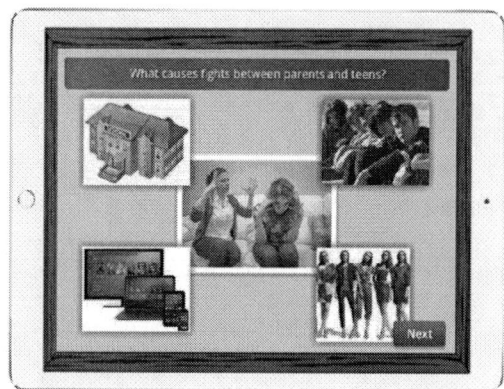

Figure 1   Discussion

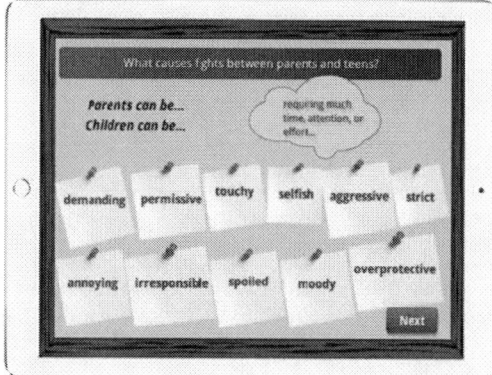

Figure 2   List of adjectives

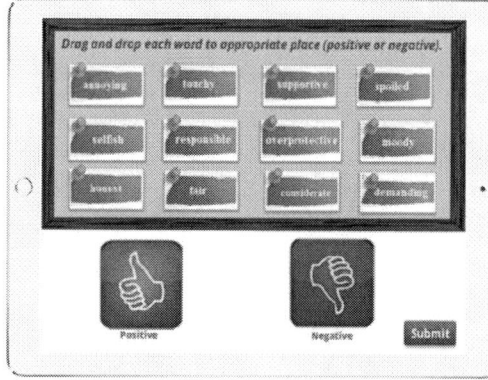

Figure 3   Drag and drop task

C. Surveys

Three surveys were conducted in order to test students' attitudes towards the integration of ICT in classrooms. Participants who have for the first time experienced the lesson using exclusively ICT filled in the first two surveys. Second group of participants, those who attend such lessons every day, filled in the third survey in order to compare their attitudes with those of the first group. The first survey was divided in two parts. The first part questioned students' acquaintance with the technology, whether they had a digital device, how much time they used it every day, and for what purposes. In the second part of the survey, using Likert scale, students expressed their attitudes towards given statements in order to test their general attitudes towards the use of ICT in classrooms. All 9 statements are shown in Table 1. In the second survey, students were asked express their attitudes towards the conducted lesson and ICT used in the classroom. All 13 statements are shown in Table 2. The last part of the survey offered the comment section where students were asked to write additional comments on the conducted lesson.

The third survey was also divided in two parts. The first part questioned students' spare time and classroom technology habits. The second part offered 23 statements on the use of ICT in language classroom. All statements are shown in Table 3.

In all three surveys, students expressed their attitudes on given statements using the 5-point Likert scale (1 – Strongly Disagree, 5 – Strongly Agree).

IV.   RESULTS AND DISCUSSION

According to the results of the first survey, which was filled in by the students who attend traditional classes without the use of ICT, all respondents own a digital device (PC, smartphone or tablet). At their own estimation, most of those students (37.5%) spend 1-2 hours each day using some of the digital devices. Moreover, none of the respondents spend less than 30 minutes using technology each day. At home, only 25% of respondents use digital devices for learning, while the remaining 75% use technology for entertainment (watching movies, social networks, etc.). At school, students who attend traditional classes only use computers in IT classes. On the other hand, students from the second group, besides using computers in IT classes, also use tablets in all other classes. On the other hand, general attitudes testing of the first group of students towards the use of ICT in classrooms offered divided results. Students mostly agree with the general advantages of the use of ICT in classrooms. However, they still prefer traditional classes. This can be seen from the results presented in the Table 1. Percentages for scale elements (1-5) are shown together with average result and standard deviation.

After the students had for the first time experienced English language class with the use of tablets, they filled in the second survey. Results of the survey are shown in the Table 2. In comment section of the survey, 78.57% of respondents had positive comments on the conducted lesson. Most of them mentioned the use of iPads, interesting exercises and easier acquisition of lesson as positive aspects of the lesson. On the other hand, students who did not enjoy the lesson, mentioned that they prefer classes without the use of iPads.

When expressing their general attitude towards the integration of ICT in classrooms, first group mostly agrees with the statements that technology makes learning easier and the information retrieval simpler. However, they express more positive attitudes towards traditional classes. Even though the students mostly agree with the statement that the technology facilitates learning, they claim that they learn better using textbooks than technology.

Most respondents from the first group agree with the statement that the technology interferes with learning and lead to frivolous education. Most of them prefer traditional classes because they know exactly what they are expected to do and how they need to learn. Therefore, before preparing the lesson, teachers should analyse their classes and students, and adapt the lesson materials to their needs. At the beginning of the lesson, students should be notified on the content and expected outcomes. Also, it is important to explain evaluation and assessment process, and to suggest adequate learning techniques and preparations for their evaluation and assessment tasks.

After the lesson was conducted in the first group of respondents, most students expressed that they enjoy clicking better than writing in their notebooks. However, they prefer learning from their textbooks. When it comes to implementation of tablets in English language classroom, students express divided attitudes. While some students prefer using iPads in classroom, and some traditional classes, other students think that the occasional use of iPads is the best solution. In fact, most students agree with the benefits of using tablets (including interesting exercises, inserting video and audio clips) but find the use of textbooks simpler because they pay more attention to the technology than the content of the lesson.

This was the first experience these students had with the use of iPads in the English language classroom. Thus, it is not uncommon that the use of technology distracted them in following the content. Therefore, students need to be educated in the use of technology before its integration in classrooms. But it is important to say that students quickly grasped the principle, and once they had correctly solved the exercises, most of them assume that they would easily solve them again.

TABLE I   FIRST SURVEY RESULTS (N=16)

No	Statement	1 (%)	2 (%)	3 (%)	4 (%)	5 (%)	AV	SD
1	Technology facilitates learning.	18,75	6,25	56,25	12,50	6,25	2,81	1,11
2	Technology makes information retrieval simpler.	12,50	0,00	25,00	25,00	37,50	3,75	1,34
3	The use of technology in classroom is advantageous.	18,75	31,25	37,50	6,25	6,25	2,50	1,10
4	I learn better using technology than textbooks.	43,75	18,75	18,75	6,25	12,50	2,25	1,44
5	I have learned more while surfing the Internet than in traditional classes.	37,50	12,50	12,50	31,25	6,25	2,56	1,46
6	In my opinion, technology interferes with learning and lead to frivolous education.	12,50	12,50	25,00	25,00	25,00	3,38	1,36
7	The use of technology interferes with the learning of the new lessons.	18,75	12,50	31,25	31,25	6,25	2,94	1,24
8	Students pay more attention to the different options technology offers than the very content they should learn.	6,25	12,50	12,50	25,00	43,75	3,88	1,31
9	I prefer traditional classes because I know exactly what I am expected to do and how I need to learn.	6,25	6,25	12,50	25,00	50,00	4,06	1,24

TABLE II   SECOND SURVEY RESULTS (N=14)

No	Statement	1 (%)	2 (%)	3 (%)	4 (%)	5 (%)	AV	SD
1	I find clicking more enjoyable than writing in notebook.	14,29	21,43	42,86	14,29	7,14	2,79	1,12
2	The lesson is more interesting when we use tablets in classroom.	21,43	14,29	28,57	28,57	7,14	2,86	1,29
3	I am used to textbooks and I learn easier using them.	7,14	14,29	14,29	28,57	35,71	3,71	1,33
4	Using textbooks is simpler than using technology.	7,14	0,00	35,71	21,43	35,71	3,79	1,19
5	I find it hard to simultaneously follow instructions where I need to click on tablet and to acquire new lesson.	14,29	35,71	28,57	7,14	14,29	2,71	1,27
6	It is easier to remember new words accompanied with many images.	0,00	28,57	28,57	35,71	7,14	3,21	0,97
7	I find it more interesting to do exercises on tablet.	7,14	7,14	42,86	28,57	14,29	3,36	1,08
8	I enjoy watching videos on tablet.	7,14	7,14	35,71	35,71	14,29	3,43	1,09
9	I prefer tasks accompanying video over tasks accompanying audio clips.	0,00	14,29	50,00	21,43	14,29	3,36	0,93
10	I learn better when while watching and listening to the speaker at the same time.	0,00	14,29	42,86	35,71	7,14	3,36	0,84
11	I would easily solve the tasks again.	0,00	7,14	28,57	35,71	28,57	3,86	0,95
12	I have learned more at this lesson.	7,14	14,29	57,14	14,29	7,14	3,00	0,96
13	I would like to use tablets in all English language classes.	28,57	21,43	28,57	14,29	7,14	2,50	1,29

Taking into account that students find new types of exercises and watching videos more interesting than traditional lesson materials, it might be that the integration of more innovative lesson materials and new activities would lead to more positive attitudes towards the use of ICT in English language classroom. Some of those activities in language classroom include digital storytelling, vocabulary games (e.g. memory game), recording and listening to different speakers, and watching documentaries. Even though most students found it hard to click and follow the lesson simultaneously, the whole lesson was held in English language and there was no need for additional instructions in Croatian language. Moreover, students had easily realized the principle how the materials were designed and by the end of the lesson they were able to do the tasks individually.

The results of the first two surveys conducted in the first group of respondents, show that positive experience in using technology in classroom influences the change of attitudes of students. Most of students who expressed positive comments on conducted lesson, also expressed more positive attitudes towards the use of tablets generally in education. In order to test if the positive attitude is related to the habit and experience in using technology in classrooms, another survey was conducted in second group of respondents. This group is comprised of eight grade students who use tablets in classroom since their fifth grade. All second group respondents confirmed that they use tablets in each English language lesson. They named several applications they use in their lessons, including *iTranslate*, *Edmodo*, *Quizlet*, *iMovie*, *Keynote*, *e-udžbenik*, and *Skitch*.

TABLE III   THIRD SURVEY RESULTS (N=10)

No	Statement	1 (%)	2 (%)	3 (%)	4 (%)	5 (%)	AV	SD
1	Technology facilitates learning.	0	10	40	30	20	3,60	0,97
2	Technology makes information retrieval simpler.	0	0	10	30	60	4,50	0,71
3	The use of technology in classroom is advantageous.	0	20	30	20	30	3,60	1,17
4	I learn better using technology than textbooks.	30	30	10	10	20	2,60	1,58
5	I have learned more while surfing the Internet than in traditional classes.	20	30	20	30	0	2,60	1,17
6	In my opinion, technology interferes with learning and lead to frivolous education.	20	30	40	10	0	2,40	0,97
7	The use of technology interferes with the learning of the new lessons.	20	10	40	30	0	2,80	1,14
8	I prefer traditional classes because I know exactly what I am expected to do and how I need to learn.	10	0	20	40	30	3,80	1,23
9	Students pay more attention to the different options technology offers than the very content they should learn.	10	10	20	50	10	3,40	1,17
10	I find clicking more enjoyable than writing in notebook and using textbooks.	10	10	10	30	40	3,80	1,40
11	The lesson is more interesting when we use tablets in classroom.	0	20	40	30	10	3,30	0,95
12	I find it easier to lean using textbooks.	0	0	20	30	50	4,30	0,82
13	I find it simpler to use notebooks and textbooks.	12,5	12,5	12,5	25	37,5	2,60	2,17
14	I find it hard to simultaneously follow instructions where I need to click on tablet and to acquire new lesson.	30	50	20	0	0	1,90	0,74
15	I enjoy using tablets in English language classroom.	0	0	30	60	10	3,80	0,63
16	During English language lesson, I lean more using tablets than traditional textbooks.	0	50	20	10	20	3,00	1,25
17	It is easier to remember new English vocabulary, when words are accompanied with many images.	0	40	0	40	20	3,40	1,26
18	I find solving English language exercises more interesting when I solve them on tablet instead of textbooks.	0	20	20	30	30	3,70	1,16
19	I enjoy watching videos on tablet.	0	0	10	60	30	4,20	0,63
20	I learn better when while watching and listening to the speaker at the same time.	0	0	40	50	10	3,70	0,67
21	Exercises that I solve during the English language lesson, I can easily solve again.	0	30	20	10	40	3,60	1,35
22	Tablet offer more possibilities for language learning than traditional textbooks.	0	0	30	50	20	3,90	0,74
23	I think that all schools should use tablets in language classrooms.	20	0	40	20	20	3,20	1,40

According to the survey results (Table 3), students who actively use ICT in classroom on daily basis have more positive attitudes towards the use of technology in education than the first group. Most of them agree that technology makes information retrieval simpler; they find clicking more enjoyable than writing in notebooks and reading textbooks. Moreover, unlike the first group, clicking does not hinder them in following the lesson. Most of them find the use of tablets in English language classroom more enjoyable, and they claim that tablets offer more possibilities than the traditional textbooks for the foreign language learning. Yet, they think that they learn better using textbooks. On the other hand, while the first group expresses completely divided attitudes, majority of the second group expresses positive attitudes towards the use of tablets in English language classroom and they mostly agree with the statement that all schools should use tablets in language classrooms. These results show the importance of experiencing such classes. Students need to become accustomed to the use of ICT in classroom, students' relationship with technology in classrooms is developing, and the benefits of ICT implementation are recognized after longer period of time. Also, after just one positive experience, many students show approval of this ways of teaching. Therefore, ICT implementation in classroom should begin with useful and interesting activities at very beginning in order to assure the first positive experience that will motivate students for future uses of the technology in education.

Moreover, it is also important to mention that implementation of technology in classrooms imply more than just purchasing devices and equipping schools. Different ways of teaching require appropriate strategies, tools and resources. Therefore, teachers need additional education to acquire knowledge and skills needed for the successful implementation of ICT in classrooms. Moreover, there needs to be a detailed plan for the use of technology in education with respect to pedagogical and didactic principles. Although the final goal of using technology is to facilitate the process of learning and teaching, it is necessary to invest time in preparation so that the use of ICT would give successful results in classroom.

V. CONCLUSION

This paper presents the results of the research that investigated the difference in attitudes towards the use of ICT in classroom, of students with different prior experience in use of ICT in classroom.

This research showed that students who use tablets in classrooms on daily basis have more positive attitudes towards the use of technology in classroom than the students who attend traditional classes. However, attitudes of those students have become somewhat more positive after their first English language lesson that was held exclusively using ICT within this research. With the implementation of these devices in classrooms, students would be presented with new and advantageous ways of using technology. Although students mostly own and use those devices every day for entertainment, they lack the experience of the use of technology in classrooms.

Since this research was conducted with the small sample, future research will include bigger sample. Furthermore, it is questionable whether the prior experience of using ICT in classroom has effects on students' achievement. In that sense, future research would investigate the achievement of students learning the same content in four groups: with/without ICT for students who attend traditional classes, and with/without ICT for students who use ICT in classroom. Moreover, the investigation should include the impact of the use of other aspects of ICT (besides tablets) on students' attitudes towards the use of ICT in classrooms. Additionally, future research will also investigate students' motivation for active interaction in the classroom, with ICT as motivating factor.

REFERENCES

[1] Noor-Ul-Amin, Syed. "An effective use of ICT for education and learning by drawing on worldwide knowledge, research, and experience: ICT as a change agent for education." Scholarly Journal of Education 2.4 (2013): 38-45.

[2] Yusuf, Mudasiru Olalere. "Information and Communication Technology and Education: Analysing the Nigerian National Policy for Information Technology." International Education Journal 6.3 (2005): 316-321.

[3] Davis, Niki. "Technology in Teacher Education in the USA: what makes for sustainable good practice?." Technology, Pedagogy and Education 12.1 (2003): 59-84.

[4] Berk, Ronald A. "Multimedia teaching with video clips: TV, movies, YouTube, and mtvU in the college classroom." International Journal of Technology in Teaching and Learning 5.1 (2009): 1-21.

[5] Kruchinina, Galina A., et al. "Information and Communication Technologies in Education as a Factor of Students Motivation." International Review of Management and Marketing 6.2S (2016).

[6] M. Domancic, J. Baksa, T. Jagust, I. Boticki, P. Seow and C. K. Looi, "A tale of two mobile learning journeys with smartphones and tablets: The interplay of technology and implementation change," 2015 IEEE Frontiers in Education Conference (FIE), El Paso, TX, 2015, pp. 1-7.

[7] Comi, Simona Lorena, et al. "Is it the way they use it? Teachers, ICT and student achievement." Economics of Education Review 56 (2017): 24-39.

[8] Chia, Steven Puay Chong. "An investigation into student and teacher perceptions of, and attitudes towards, the use of information communications technologies to support digital forms of summative performance assessment in the applied information technology and engineering studies courses in Western Australia." (2016). Retrieved from http://ro.ecu.edu.au/theses/1806

[9] CARNet, »E-Schools in the World,« 6.4.2016. [Online]. Available: www.carnet.hr/e-schools/find_out_more/e_schools_in_the_world. [Last accessted 26.1.2017].

[10] CARNet, »Adequate ICT Infrastructure in Pilot Schools,« 6.4.2016. [Online]. Available: www.carnet.hr/e-chools/results/adequate_ICT_infrastructure. [Last accessed: 26.1.2017].

[11] Vrbanus, S. »Osnovna škola Vežica je prva iŠkola,« Bug, 10.3.2015. [Online]. Available: www.bug.hr/vijesti/osnovna-skola-vezica-prva-iskola-svi-osmasi/141102.aspx. [Last accessed: 12.6.2016].

[12] Krikšić, K.»iŠkola: Tableti ušli u nastavu Osnovne škole Vežica,« Novi List, 17.11.2012. [Online] Available: http://www.novilist.hr/Vijesti/Rijeka/Ri-Servis/iSkola-Tableti-usli-u-nastavu-Osnovne-skole-Vezica [Last accessed: 27.1.2017].

[13] CARNet, »E-Schools in the World,« 6.4.2016. [Online]. Available: www.carnet.hr/e-schools/find_out_more/e_schools_in_the_world. [Last accessed: 26.1.20].

LIS students and plagiarism in the networked environment

Ivana Hebrang Grgić

Faculty of Humanities and Social Sciences, University of Zagreb / Department of Information and Communication
Sciences, Zagreb, Croatia
ihgrgic@ffzg.hr

Abstract - Plagiarism is unethical behaviour that can have negative consequences on the development of science and society. It also ruins the reputation of individuals and institutions. Plagiarism can be intended or unintended. The paper will focus on unintended plagiarism that is a result of absence of that topic in the education curricula. Teaching about plagiarism is part of so-called ethical literacy that is a subcategory of information literacy. A survey was conducted with the aim of finding out about the degree of knowledge about plagiarism among the students of Library and Information Science at the Faculty of Humanities and Social Sciences, University of Zagreb. The analysis is based on an anonymous web questionnaire with 20 questions. The presumption is that the students do not know enough about plagiarism, they have not had any experiences with plagiarism during their education and they do not know what self-plagiarism is. Another presumption is that the students get detailed instructions about how to cite sources, but do not know enough about the concept of authorship. The conclusion will be made about the inclusion of this segment of information literacy in education curricula. New surveys will be proposed (on local and national level).

I. INTRODUCTION

In the networked environment, students have access to enormous amount of information. That can be a good thing if they understand what research integrity is, so they can act ethically. However, there are many challenges in the electronic environment (easy copy-paste solution being one of them). In this paper, we will not focus on intended unethical practices, but on the unintended ones. Unintended practices in the field of research integrity are based on the lack of education about ethical issues. The aim of this research is to give a short overview of literature about students' understanding of plagiarism and authorship; to present the results of a survey of graduate students of Library and Information Science; and to conclude about their understanding of some ethical issues in academia.

Olson and Show [1] show that children at the age of 5 or 6 are capable of understanding the originality of ideas – they consider characters that copied other characters' words as "bad". Based on the study, Bailey [2] concludes that the best time to start plagiarism education is in the third grade (age 8). The education has to be well planned. Bailey also poses a question about the generation gap – how is it possible that young children understand the value of originality and students lose the understanding (as many studies and practices show)?

II. LITERATURE OVERVIEW

McCabe [3] conducted several longitudinal researches about academic dishonesty and cheating among students. A part of one of his researches was about North American undergraduate and graduate college students and their practices in written assignments. 36% of undergraduate students and 24% of graduate students have, at least once, copied or paraphrased Internet sources without giving credits to authors (i.e. without citing the sources). At the same time, 58% of undergraduate and 62% of graduate students think that such a behaviour is not appropriate. Risquez et al. [4] found out that 75% of students had high ethical awareness that copying text without citation was wrong. Armstrong and Delbridge [5] found out that 30-55% of students from the field of Information Science admitted minor plagiarism activity (e.g. copying a small part of another student's paper or inclusion of a quote without in-text citation). Major plagiarism activity is reported by 9-20% of students (e.g. buying an essay from a writing service or unacknowledged summarizing of a large amount of published work). According to the survey, the first reason why students commit plagiarism is easy access to online material and the second is that they do not know what plagiarism is. They plagiarize mostly because of pressure to complete their assignments and because of poor time management skills.

Analysing the reasons for cheating and plagiarism, Dornan et al. [6] conclude that the main reasons are misunderstanding of the assignment and running out of time. Some other reasons could be lack of research skills, problems with evaluating sources, confusions about terminology, confusions between plagiarism and paraphrasing, careless notetaking, confusions about how to cite sources etc. [7]. There are four main forms of plagiarism among students [8]:
1. intra-corpal (students cheat by copying from their colleagues in the same class);
2. collusion (student presents the paper as his own, but in fact it is the result of cooperation with another person who is not stated as a co-author)
3. extra-corpal (students cheat by copying from external source, e. g. book, journal or a web site)
4. self-plagiarism (or auto-plagiarism – students use their own works previously submitted for another assignment without acknowledgment).

Kokemuller [9] writes how plagiarism affects students. They can fail an assignment, fail a class or even be suspended. Besides destroying student's reputation, there can also be legal and monetary repercussions. Sometimes, if the students are involved in medical researches, plagiarism can cause the loss of human lives [10].

When students' papers are published, authorship should be properly assigned. There are four main authorship criteria according to ICMJE: substantial contributions to the conception or design of the work; drafting the work or revising it critically; giving final approval of the version to be published; agreement to be accountable for all aspects of the work [11]. American Psychological Association (APA) published guidelines on determining authorship credit and authorship order for students [12]. The guidelines are based on the APA code of ethics. The most important is an open discussion on authorship issues among all the students and/or researchers involved in a project. In addition, student should be listed as principal author on an article based on the students' thesis. It is possible to use authorship agreements to outline the types of contributions (responsibilities, roles, efforts etc.). Wager [13] thinks that listing contributions can make easier for editors to detect ghost authors. Ghost authors are those authors who made substantial contributions to the paper (they meet all the authorship criteria) but are not listed as authors [14]. Other forms of unacceptable authorship are honorary or guest authorship (authorship is granted out of respect) and gift authorship (offered from a sense of obligation to a person who has not contributed to the work).

Oberlander and Spencer [15] discuss how students are a special group within research community when it comes to authorship. They are in relationships with their mentors where there is a possibility of exploitation [16], because of students' inexperience and their lack of knowledge. Oberlander and Spencer think that the best solution to discourage inappropriate authorship is education of students (as possible future scientist). They also give some recommendations for students as co-authors and for their mentors. Some of them are:
1. authorship guidelines from professional organizations and journals should be consulted
2. authorship should be discussed in an early phase of the research/work.
3. roles should be clarified
4. authorship should be based on relative contribution
5. acknowledgment section should be used appropriately
6. mentors should give some time to students to be innovative [15].
Committee on Publication Ethics (COPE) gives some advice about inappropriate authorship on students' papers [17]. They give examples of some anonymized cases (e.g. editor of a journal gives an example of a submitted manuscript that is work of students, but supervisor is listed as the first author).

One of the author's moral rights is the right to be stated as an author of his/hers work. Moral rights are part of author's rights, and author's rights are basic human rights, as stated in Universal Declaration on Human rights: "Everyone has the right to the protection of the moral and material interests resulting from any scientific,

literary or artistic production of which he is the author" (article 27) [18]. Besides, violating moral rights is against the law: "Author has the right to be acknowledged and specified as the author of his/her work...author has the right to oppose to the use of his/hers work that is against his/hers reputation..." (Articles 15 and 16 of Croatian Copyright Act) [19].

III. RESEARCH

A. Aim and scope

The purpose of the research was to find out about the degree of knowledge about plagiarism and authorship issues among the Library and Information Science (LIS) students. The hypothesis were:

H1: Students know that plagiarism is unethical behaviour, but are not sure what exactly plagiarism is.

H2: There are students that commit plagiarism, but they are a minority.

H3: Students are not sure about the authorship criteria – they sometimes think that any involvement in the research and/or paper is enough for a person to be listed as an author.

H4: Students learn about ethical issues on higher education level.

The results were supposed to be a starting point for deeper study (on national and international level) that should involve students from other departments, from other faculties and universities. The findings should be the basis for planning education on plagiarism and authorship.

B. Methods

An anonymous online questionnaire was sent to 75 graduate students of Library and Information Science at the Department of Information and Communication Sciences, Faculty of Humanities and Social Sciences, University of Zagreb. The questionnaire was sent in December 2016. Response rate was 76% - there was 57 responses.

There were 20 questions in the questionnaire. The first two questions were about gender and year of study (1st or 2nd graduate year). Another 18 questions were: multiply choice questions (6); one-choice questions (3) and yes/no questions (9).

IV. FINDINGS

Question 1. Gender
There were 52 female and 5 male students in the sample. The percentage of male students is too small to make gender-related conclusions.

Question 2. Year of graduate study
In the sample are 36 (63%) 1st year students and 21 (37%) 2nd year students.

Question 3. Have you learned about plagiarism at any degree of your education?
The question was multiple-choice type and the answers were: in elementary school; in secondary school; during undergraduate study; during graduate study; in some other informal form of education; no, never. Majority of the students have learned about plagiarism at undergraduate level, but seven students have never learned about plagiarism (Figure 1).

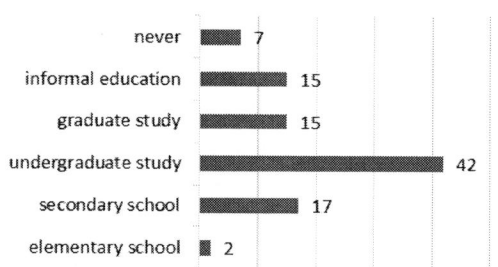

Figure 1. Degree of education when students learned about plagiarism

Question 4. If yes, what did you learn?
The question was multiple-choice type and the answers were: definition of plagiarism; ways of avoiding plagiarism; plagiarism detection; procedures in cases of detected plagiarism; definitions of authorship; authorship types; something else. Forty students (70%) have learnt about the definition of plagiarism, 36 (63%) have learnt about the definition of authorship, 28 (49%) have learnt about authorship types, 26 (46%) have learnt about avoiding plagiarism, 19 (33%) have learnt about plagiarism detection and 9 (16%) about procedures in cases of detected plagiarism. Three students (5.3%) have learnt about all the topics.

Question 5. Have you ever presented other people's words, thoughts or ideas as your own, while preparing an assignment?
Eleven answers (19%) were affirmative and 46 (81%) were negative.

Question 6. Do you consider presentation of other people's words, thoughts or ideas as your own, ethically correct?
Forty-seven students (82%) answered negative, and the other 10 (18%) answered that it depended on the situation.

Question 7. Have you ever, before preparing an assignment, been taught about proper citing practice?
Twelve students (21%) have never been taught about proper citing of sources, 22 (38%) have been partly taught, and 23 (41%) have been taught at least once about proper citing.

Question 8. Have you ever, before preparing an assignment, been taught how to make quotations?

Forty-nine answers (86%) were affirmative and eight (14%) were negative.

Question 9. Has any of your assignments ever been rejected because of wrong citation and/or quotation practice?
Nine answers (16%) were affirmative and 48 (84%) were negative.

Question 10. Have you ever been reported to an ethics committee for plagiarism?
There are no students in the sample that have been reported to an ethics committee for plagiarism.

Question 11. Have you ever been listed as author on a paper that you have not authored?
There were four affirmative answers (7%) and 53 negative answers (93%).

Question 12. Have you ever been omitted from the list of authors on a paper that you authored?
There were three affirmative answers (5%) and 54 negative answers (95%).

Question 13. Do you know about the case of wrong authorship assignment among your colleagues?
There were 21 affirmative answers (37%) and 36 negative answers (63%).

Question 14. Do you know about the case of plagiarism among your colleagues?
There were 24 affirmative answers (42%) and 33 negative answers (58%).

Question 15. Have you ever addressed an ethical committee or similar body to notify it about suspected plagiarism?
All the answers to the question were negative.

Question 16. Do you believe that self-plagiarism is possible?
8 students (14%) think that self-plagiarism is impossible, 18 students (32%) thing that it is possible, and 31 students (54%) is not sure if self-plagiarism is possible.

Question 17. Indicate what plagiarism means to you.
The question was multiple-choice type and the answers were: quoting without citing sources; quoting with proper citation of sources; paraphrasing without citing sources; not implementing the rules of proper citing; compilation of texts with proper citation of sources; not assigning authorship to someone who authored the paper; something else. Fifty-six students (98%) say that plagiarism is quotation without citing sources; 42 students (74%) say that plagiarism is paraphrasing without proper citation of the sources; 41 students (72%) think that plagiarism is when authorship is not properly assigned; 41 think (72%) that plagiarism is when the rules of proper citing are not implemented; one student (2%) thinks that quoting is plagiarism even if proper citing is

applied and one (2%) thinks that compiling texts is plagiarism even if proper citing is applied. Results are in Figure 2.

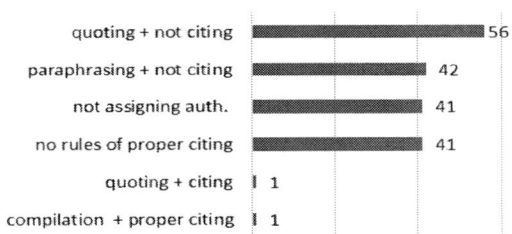

Figure 2. What is plagiarism?

Question 18. Mark all the criteria that are, in your opinion, criteria for authorship?

The question was multiple-choice type and the answers were: contribution to the concept of the work; writing the first version; financing of the research; collecting data; statistical analysis; giving final approval; technical support; something else.

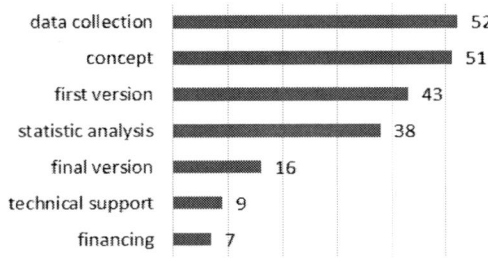

Figure 3. Authorship criteria

Fifty-two students (91%) think that collecting data is authorship criteria; 51 (89%) think that contribution to the research is authorship criteria; 43 (75%) think that writing the first version is authorship criteria; 38 (67%) think that doing statistical analysis is authorship criteria; 16 (28%) think that approving final version is authorship criteria; 9 think that giving technical support is authorship criteria and 7 (16%) think that financing is authorship criteria. Results are in Figure 3.

Question 19. Mark the applications you have heard about: Plagiarisma.net; iThenticate; Text Compare; Copy Leaks; Plagiarism Checker; MyText; Google Checker; nothing.
Thirty-nine students (68%) have not heard about any of the plagiarism detection software; 12 students (21%) have heard of GoogleChecker (the software does not exist!); 7 (12%) have heard of PlagiarismChecker; 6 (11%) have heard of Copy Leaks; 3 (5%) have heard of MyText (the software does not exist!); 2 (4%) have heard of Plagiarisma.net and one (2%) has heard of iThenticate.

Question 20. Mark the applications you have used: Plagiarisma.net; iThenticate; Text Compare; Copy Leaks; Plagiarism Checker; MyText; Google Checker; nothing.
Fifty students (88%) have never used any of the applications; four (7%) have heard of PlagiarismChecker; three (5%) have heard of GoogleChecker (the software does not exist!); two (4%) have heard of Plagiarisma.net and one (2%) has heard of MyText (the software does not exist!).

V.   DISCUSSION AND CONCLUSION

Students mostly do learn about plagiarism during higher education. Only two students have learned about it in elementary school. That is the basic problem – children should be taught about ethics and plagiarism in elementary school, as it is showed in the Bailey's research [1]. Lack of early education about ethical issues seems to be problem in the later stages of education. When they learn about plagiarism, they are mostly taught about basic definition (of plagiarism and authorship) but not enough about detecting plagiarism and about procedures in the cases of suspected or detected plagiarism.
The lack of education is the most probable reason why:
- 19% of students have at least once presented other people's words, thoughts or ideas as their own, while preparing an assignment;
- 8% of students think that sometimes (depending on situation) presenting other people's words, thoughts or ideas as their own can be ethically correct;
- 16% of students have at least once experienced rejection of their assignments because of wrong citation and/or quotation practice;
- 42% of students know at least about one case of plagiarism among their colleagues, but have never declared plagiarism;
- 37% know about at least one case of wrong authorship assignment among their colleagues, but have never declared it;
- 5% have at least once been omitted from the list of authors on papers that you authored, but have never declared it;
- 7% have at least once been listed as authors on papers they have not authored, but have never declared it;
- 86% do not know what self-plagiarism is;
- 93% do not use plagiarism detection software.
The limitation of the survey is that definition of plagiarism and/or authorship can be differently interpreted, since 30% of the students have never learnt about academic misconduct. However, 70% of the students claim that they have learned about the definition of plagiarism and high percentage of students can detect different forms of plagiarism (quoting without proper citation, paraphrasing without proper citation etc.). Sixty-three percent of the students state that they have learnt about the definition of authorship, but in fact, they are not sure what the authorship criteria are (e.g. 91% think that collecting data is important for assigning authorship and

67% think that doing statistical analysis is important for authorship).

Another possible limitation of the survey is the use of self-report methodology. It is the most commonly used methodology in similar surveys [20] – students are mostly asked to answer anonymous questionnaires and to express their attitudes and experiences. Therefore, biased opinions are possible.

Despite small sample of the survey, results can be compared to some other studies. In 2011 Schrisher et al. [21] published results of a plagiarism survey from one US university. They concluded that academic misconduct is on the rise for many reasons, and the most important reason is that students think information on the Internet is public knowledge that is no one's intellectual property. However, most of them understand it is unacceptable to submit a paper written by someone else.

In Fish and Hura [22] 2013 survey, percentage of students that have used another author's phrases or ideas without citing the source is higher (60% of students have at least once done that) than in our survey (about 20% of students). Fish and Hura found that college students consider some types of plagiarism as more serious than other types. Frequency of plagiarism is overestimated by students and therefore students are more likely to plagiarise. It is important for students to have accurate information about the frequency of plagiarism in their classes and institutions because such information could reduce the number of plagiarism incidents.

A study made by RefME service in 2016 [23] among almost 5000 US students shows that more than 50% of students have lost points for incorrect references (e.g. using the wrong style or not submitting full reference list). The survey has similar result to our survey where students were asked about lack of information on referencing (14% of Croatian and about 20% of American students report lack of the information). There is a difference in the students' usage of plagiarism detection software – 88% of Croatian students do not use it while 54% of the US students do not use the software.

Bretag [20] describes various surveys noting that the rate of plagiarism among the students vary widely (from 18% to 81%). International students and students for whom English is not their native language are more likely to commit plagiarism.

Important conclusion of our survey is that ethical issues are not part of education curricula. That is the reason of low awareness about ethical issues, namely plagiarism. Students should learn about authorship criteria and they should use plagiarism detection software more often. This survey was focused on unintended plagiarism, i.e. cases when students plagiarise without knowing they are doing something unethical or even illegal. Unintended plagiarism is undoubtedly result of lack of education.

The situation is similar in some other countries. Bretag [20] shows that 20% of postgraduate Australian students have never heard of academic integrity and 40% do not know whether their university has an academic integrity policy. As Adam et al. [24] assert – it is difficult to define plagiarism and there are no many researches that show

the differences between institutional and student understanding of plagiarism.

Future researches of plagiarism in higher education should concentrate on:

- students' understanding of plagiarism;
- finding out how, where and when students learn about academic misconduct;
- finding out which teaching methods and techniques are (and should be) used for teaching about plagiarism
- analysing ethics policy documents in academic institutions;
- investigating and analysing higher education curricula;
- finding out what tutors/teachers/professors know about plagiarism and what they do (or do not do) to avoid it in their work.

As mentioned earlier, the most commonly used methodology is self-report methodology (mostly questionnaires [22] and interviews [24]). Other methods should also be used, e.g. content analysis ([24] and [25]) where university policy documents are analysed; or experiments [26] that could help to investigate students' practices.

This research has a small sample and several upper mentioned limitations, but it can be used as a pilot study of students' plagiarism awareness in Croatia. New researches should be done with different samples – among students of other studies at the Faculty but also among students of other faculties and universities in Croatia. Comparative analysis should be conducted and it should be a basis for some changes in educational curricula.

In conclusion, ethics literacy is one important part of information literacy – it is not enough to know how to find and evaluate information without knowing how to use it in a proper way, i.e. without violating anyone's basic rights. Therefore, action should be undertaken by educators, policy makers and funders of higher education institutions in order to raise knowledge and awareness of unethical nature of plagiarism and to deploy students' skills of avoiding plagiarism.

REFERENCES

[1] K. Olson, and A. Shaw, "'No fair, copycat!': what children's response to plagiarism tells us about their understanding of ideas," Developmental Science, vol. 14, pp. 431-439, March 2010, doi: 10.1111/j.1467-7687.2010.00993.x

[2] J. Bailey, "What age do children see plagiarism as wrong?" Plagiarism Today. September 23, 2010. Retrieved December 27, 2016 from https://www.plagiarismtoday.com/2010/09/23/what-age-to-children-see-plagiarism-as-wrong/

[3] D. L. McCabe, "Cheating among college and university students: a North American perspective," International Journal of Educational Integrity, vol. 1, pp. 1-11, 2005. Retrieved December 27, 2016 from
http://www.ojs.unisa.edu.au/index.php/IJEI/article/view/14

[4] A. Risquez, M. O'Dwyer, and A. Ledwith, "'Thou shals not plagiarize': from self-reoprted views to recognition andavoidance of plagiarism," Assessment and Evaluation in Higher Education, vol. 38, pp. 34-43, 2013

[5] L. Armstrong and R. Delbridge, "Final year undergraduate student plagiarism: academic staff and student perceptions," Learning and Teaching in Action, vol. 7, 2008

[6] R. W. Dornan, L. M. Rosen, and M. Wilson, "Within and beyond the writing process in the secondary English classroom," Pearson Allyn & Bacon, 2003.

[7] Middle Georgia State University, "Plagiarism prevention guide: why students plagiarise?," Retrieved December 27, 2016 from http://www.mga.edu/student-success-center/plagiarism/why.aspx

[8] The University of Melbourne, "Why and how students plagiarise?," Retrieved December 27, 2016 from https://academichonesty.unimelb.edu.au/turnitin/students/whyhow

[9] N. Kokemuller, "How plagiarism affects students," Seattle.pi. Retrieved December 27, 2016 from http://education.seattlepi.com/plagiarism-affects-students-1023.html

[10] iThenticate, "6 consequences of plagiarism," Retrieved December 27, 2016 from http://www.ithenticate.com/resources/6-consequences-of-plagiarism

[11] ICMJE, "Defining the role of authors and contributors," Retrieved December 27, 2016 from http://www.icmje.org/recommendations/browse/roles-and-responsibilities/defining-the-role-of-authors-and-contributors.html

[12] APA Science Student Council, " A graduate student's guide to determining authorship credit and authorship order," 2006. Retrieved December 27, 2016 from http://www.apa.org/science/leadership/students/authorship-paper.pdf

[13] E. Wager, "Authors, ghosts, damned lies, and statisticians," PLoS Medicine, vol. 4, e34, 2007, doi: 10.1371/journal.pmed.0040034

[14] Washington University in St. Louis, "Washington University policy for authorship on scientific and scholarly publications," Retrieved December 27, 2016 from http://research.wustl.edu/PoliciesGuidelines/Pages/AuthorshipPolicy.aspx

[15] S. E. Oberlander and R. J. Spencer, "Graduate students and culture of athorship," Ethics and Behaviour, vol. 16, pp. 217-237, 2006, Retrieved December 27, 2016 from http://155.97.32.9/~bbenham/Phil%207570%20Website/pdfs-Authorship/Oberlander-Spencer-2006EB_GraduateAuthorship.pdf

[16] M. M. Costa M. Gatz, "Determination of authorship credit in published dissertations," Psychological Science, vol. 3, pp. 354–357, 1995.

[17] COPE, "Inappropriate authorship on students paper," Retrieved December 27, 2016 from

http://publicationethics.org/case/inappropriate-authorship-students-paper

[18] UN, "Universal declaration on human rights", Retrieved December 27 2016 from http://www.ohchr.org/EN/UDHR/Documents/UDHR_Translations/eng.pdf

[19] "Croatian Copyright act", Narodne novine, no. 167, 2003

[20] Bretag, T. "Challenges in addressing plagiarism in education," PLOS Medicine, 20, 2013, doi: 10.1371/journal.pmed.1001574

[21] R. H. Schrimsher, L. A. Northrup, and S. P. Alverson, "A survey of Samford University students regarding plagiarism and academic misconduct," International Journal for Educational Integrity,vol. 7, pp. 3-17, 2011. Retrieved February 22, 2017 from http://www.ojs.unisa.edu.au/index.php/IJEI/article/view/740

[22] R. Fish, and G. Hura. "Students' perceptionsof plagiarism," Journal of the Scholarship of Teaching and Learning, vol. 13. Pp. 33-45, 2013. Retrieved February 22, 2017 from: http://josotl.indiana.edu/article/download/4254/3862

[23] RefME. "10 things we discovered about students' attitudes towards plagiarism," July 25, 2016. Retrieved February 22, 2017 from: https://www.refme.com/blog/2016/07/25/10-things-discovered-students-attitudes-towards-plagiarism/

[24] L. Adam, V. Anderson and R. Spronken-Smith, "'It's not fair': policy discourses and students' understandings of plagiarism in a New Zealand university," Higher Education, vol. 72, pp. 1-16, 2016. doi: 10.1007/s10734-016-0025-9

[25] I. Hebrang Grgić, "IL and information ethics: how to avoid plagiarism in scientific papers?" in Information literacy: lifelong learning and digital citizenship in the 21st century: second European Conference ECIL: proceedings, pp. 217-226. Heidelberg: Springer, 2014

[26] T. S. Dee and B. A. Jacob, "Rational ignorance in education: a field experiment in student plagiarism," National Bureau of Economic Resaerch Working Paper Series, 2010, Retrieved March 6, 2017 from http://www.nber.org/papers/w15672.pdf

A platform for supporting learning process of visually impaired children

A. Kavčič*, M. Pesek* and M. Marolt*
* University of Ljubljana/Faculty of Computer and Information Science, Ljubljana, Slovenia
{alenka.kavcic, matevz.pesek, matija.marolt}@fri.uni-lj.si

Abstract - Although ICT supported tools and e-learning material are widely available in schools to support teaching and learning, there is still a lack of specific tools and material designated for children with impairments. The costs of adapting and preparing such material is often economically not justifiable due to a small number of such children, and commonly, for the best leaning outcome the material has to be adapted for each individual child and their deficits and level of impairments. Our solution to this problem is a web platform for delivering customized exercises intended for visually impaired children. There are two sorts of exercises already prepared: a tutorial for learning and practicing Braille and ten-finger typing, and various exercises for practicing vision, memory, and motor skills. For each individual impaired learner, the teacher can select appropriate type of exercise and customize it by adjusting visual aspects of the exercise, setting the specific content (e.g., words for typing, or items to sort), or selecting the level of difficulty (e.g., set timing, complexity levels, or number of shown images). A set of such customized exercises is given to a learner for practicing and their progress is constantly monitored and saved for later inspection.

I. INTRODUCTION

Various ICT supported tools and e-learning material for assisting teaching and learning are becoming commonly available in schools and are a new standard in the teaching process. Although their availability to teachers and students is nowadays indisputable, there are still groups of potential users that cannot take a full advantage of such teaching and learning approaches.

Inclusive education and access to e-learning tools and material is also an important strategic issue. "Promoting equity, social cohesion and active citizenship" is one of the strategic objectives of the EU Strategic framework for Education & Training 2020 [1]. For more than ten years, the European Agency for Special Needs and Inclusive Education[1] has been helping member states to improve inclusive policies and practices in the field of education. Even though the EU and local governments are promoting equity and inclusion of impaired learners, there is a gap between the state legislation, strategies and policies, and the classroom practice.

Beside accessibility, the main challenge with educational tools and material is the level of their adaptability. This issue is important not only to suit the individual needs of the impaired students, but also to keep the content up to date and in line with the educational process in class. Most educational materials include fixed content, which is not modifiable by the user. Such content may soon become obsolete and consequently its usefulness in the educational setting degrades.

In this paper, we introduce a novel web platform for inclusive and accessible educational games intended mostly for visually impaired children. The platform offers a variety of games for vision and memory training, learning braille, and extending the typing skills. All games are flexible and allow for modification of the content, which permits teachers to use the game in several teaching domains, and update and adapt the game to the current needs in the class. Besides, the platform and the games provide accessibility support for the blind and visually impaired. Each game can be further customized by adjusting its visual aspects, the level of difficulty, or employing the individual user preferences.

A. Related Work

Bocconi et al. [2] discuss the issue of accessibility and usability of educational tools by visually impaired students. After examining different educational products, they claim that the issue of usability is poorly addressed and needs more attention. Not only the interface features but also the product's basic functionalities have to improve their usability. Moreover, they argue that different types of visual impairments lead to different needs and therefore ask for different solutions.

Several initiatives are actively promoting braille literacy. American Foundation for the Blind has launched the Braille Bug[2], an interactive web site for teaching braille concepts through activities and games. Milne et al. [3, 4] present some examples of learning games for blind and visually impaired children on mobile devices. However, all these approaches focus only on accessibility and usability for visually impaired, yet lack the option of altering the content of the educational games.

A key skill for digital literacy is also the ability to use a standard keyboard. Typing is an important psychomotor skill, required not only for smoother inclusion of visually impaired children, but also for all digital literates as one of the basic practical skills. Ratatype typing tutor[3] is one of the many web-based educational environments that

---

[1] https://www.european-agency.org/

[2] http://braillebug.afb.org/
[3] http://www.ratatype.com/

encourage users to learn touch-typing through various educational game settings. Nevertheless, they all use a predefined content not customizable by the teacher.

Another interesting initiative is All Abilities ePlayground4, an online space offering games for children of all abilities. Although the example web-based games are all accessible, the customization and personalization opportunities are still limited.

We can see that the existing approaches focus more on the accessibility issues, but less on the usability. They do not provide sufficient support for the content customization, especially for teachers that are not IT experts.

II. PERCIEVECONCIEVE PLATFORM

In the PercieveConcieve5 project, we cooperated with the Institute for Blind and Partially Sighted Children Ljubljana, the main institution in Slovenia in the field of education of children and young people with visual impairment. The PercieveConcieve platform was designed to meet the needs of the teachers and visually impaired children and is used directly in the Institute's adapted education programs.

The main goal of our platform is to provide a variety of modifiable educational games, which include concepts that can be applied to a number of learning domains [5]. The platform provides a special interface for teachers, which enables preparing specific instances of the game suited to a particular student and their specific needs. A teacher can take a game template and define the parameters of the game, like the learning domain, shown items, or visual look. Besides, the game instance is also personalized based on the individual student preferences, adjusting colors, text fonts, text size etc. Several predefined styles have been prepared: default, blind, inverted colors, high contrast, protanopia, tritanopia, and achromatopsia. The user can select any of them in the profile settings. Moreover, the user can also define their preferred language in the settings, as the platform supports multiple languages via the use of language files.

The platform supports three different user roles: administrator, teacher and student. The administrator's view of the platform is presented in Fig. 1. Administrators manage users (e.g., adding new users, changing roles, enrolling to classes) and classes (e.g., create new classes, assigning teachers to classes), and upload new game templates and media files (e.g., images, sounds). Each class has at least one assigned teacher, who can edit the class information and create new game instances from the available templates. Students are regular users that enroll in the class and participate in the activities the class offers (i.e., the activities the teacher prepares for them).

All games communicate with the server via application programming interface (API) [5]. The API is opened and available to any developer who wishes to create new games and offer them through the platform.

---

4 http://allabilitiesplayground.net.au/
5 http://zaznajspoznaj.si

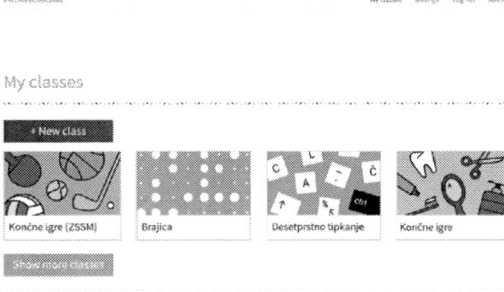

Figure 1.   Administrator's view on the PerceiveConceive platform. Administrator accesses all functionalities as a regular user, but can, in addition, administer the platform, including adding new classes

This allows for further expanding the set of game templates by the contributing community.

Although the developer of the game template has full control over the adaptability of their game and how the predefined user styles are processed in the game, the use of the entire available range of customization potentials offered by the platform, through game attributes and user preferences, is highly important for inclusion of users with different impairments.

III. PREPARED GAME TEMPLATES

In the first stage, we have prepared several game templates: a tutorial for learning and practicing Braille, a tutorial for practicing ten-finger typing, and a series of repetitive games for practicing vision, memory and fine motor skills. More games templates will be developed and added according to the wishes and demands of the teachers.

A guiding light in all games is our game character in a form of a yellow personified star, which was introduced in order to encourage students throughout the playtime. The character, shown in Fig. 2, proved to provide positive stimuli on student performance and achievements.

A. Braille Tutorial

This tutorial is intended for children and their parents that want to learn reading and writing braille. Currently, the tutorial focuses on basic characters of the Slovenian braille, but it can be extended and adjusted to the specifics

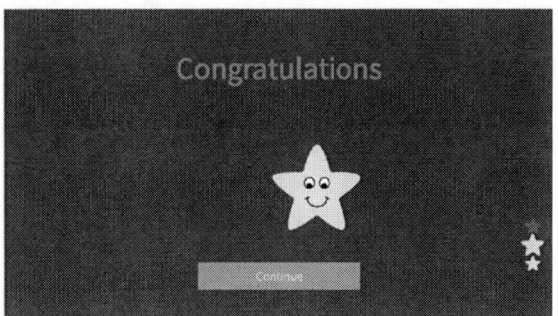

Figure 2.   A star character, which encourages players throughout the game. Players also collect stars as rewards for correct answers (shown in the column on the right)

of any language. The tutorial includes a description and explanation of this tactile writing system, including a short history, and additionally offers three types of exercises: Mirror, Hidden Character (or word), and Braille Machine for letters and words.

All exercises can be performed using the keyboard (a braille writer is not required), where the keys f, d, s and j, k, l on the keyboard (or number keys 1 to 6, consequently) are used to mark the raised dots of a braille symbol.

In the Mirror activity, the user can experiment with braille symbols. This basic interactive activity introduces braille to the user via various letter conversions. The conversion form black print to braille shows a corresponding braille symbol (an image with six dots where the raised dots are emphasized) when the user enters a letter in black print. An inverse transformation is also possible: for a given combination of raised dots (entered either using keys f, d, s, j, k, l, or number keys 1 to 6) a corresponding letter mirrors in black print. Fig. 3 shows the exercise where the user presses the key numbers 1 and 5, thus raising dots 1 and 5. Consequently, the left field displays the selected braille symbol while the right field displays the corresponding letter e in black print.

Hidden Character/Word is an activity for practicing braille by recognizing the displayed braille symbols or, in case of higher difficulty level, entire words. A user has to find corresponding letter(s) for the displayed braille symbol(s) and enter the answer within the limited time. An audio pronouncing the places of the raised dots in the braille symbol helps visually impaired users, but can be turned off any time during the exercise.

The third activity for practicing braille is Braille Machine, which is the hardest of all exercises. The user practices braille by selecting the right combination of raised dots in the symbol. A letter in black print is presented to the user together with its audio recording (which can also be turned off) and the user has to enter a braille symbol via keys f, d, s, j, k, l. If the user fails to provide the correct input in the available time, the answer is displayed to provide an immediate feedback to the user. A more challenging version of Braille Machine deals with the whole words instead of individual letters.

In all braille exercises, the teacher can select the

available time for answering as well as the letters and words for practicing, thus allowing the exercise modification to meet each individual user needs and difficulty level.

B. Ten-finger Typing Tutorial

This tutorial explains the basic concepts of two-handed touch-typing and encourages practicing through a series of customized exercises.

Each exercise uses a predefined sequence of letters or words (either default or customized and prepared by the teacher) that is displayed on the screen and provided via audio dictation. The user has to type the sequence while the time is also being measured. Immediate visual and audio feedback is given on the user performance.

One such exercise is shown in Fig. 4, where the two basic letters f and j have to be typed interchangeably using left and right index finger. The progress of the user is visually marked in yellow and the mistyped letters in red. The help below the letter sequence shows the position of the next letter on the keyboard and the finger that has to be used for typing this letter. This help can be turned off for higher difficulty levels.

Each exercise can be customized in different ways. The most obvious is the content (i.e., the sequence of letters and words) that the user uses for practicing typing. Besides, the teacher can select different ways of displaying the content to the user (e.g., all at once, line by line, word by word, or letter by letter), adjust the maximum number of characters in one line, set the colors schemes for the content (e.g., foreground and background colors, the color for marking progress and mistakes), or shuffle the displayed content by enabling randomly generated subsets of the content.

C. Vision, Memory, and Precise Movement Games

An important part of the platform are various games in the form of memory games, puzzles, matching, sorting, pair identification, image understanding and description, and object navigation, controlled with fingers or keys. These exercises support practicing vision (e.g., pattern matching, standard puzzles), memory (e.g., standard memory games, finding the correct subset of the shown items, finding exact sequences, finding reverted subsets) and precise movement (e.g., object navigation, following the path through the labyrinth), and are useful directly as a

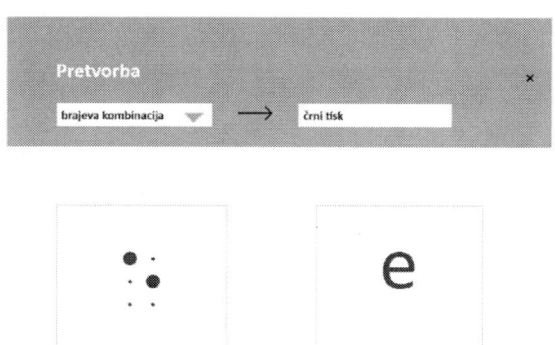

Figure 3. The Mirror exercise for practicing braille. Selecting the combination 1 to 6 for raised dots displays (mirrors) the corresponding letter in black print

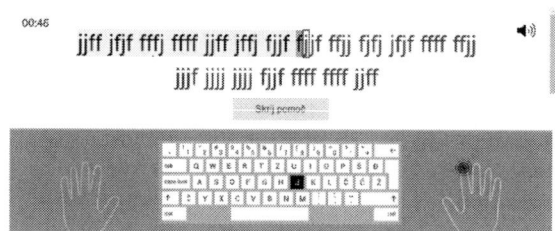

Figure 4. The Ten-finger Typing exercise for practicing touch-typing. Displayed is the sequence of letters that the user has to type, progress is visually displayed (mistyped letters are marked in red), and help (the keyboard with the marked key in question and the finger to use for pressing the marked key) is provided or hidden on demand

teaching aid for visually impaired children. Fig. 5 presents a subset of available games.

Fig. 6 shows an example of a game Sequence intended for training visual memory. The player looks at the sequence of objects on one page and, when ready, continues to the next page, where the perceived sequence has to be reconstructed from the available images. The game starts with the sequence of two images, while an additional image is added to the sequence on each next difficulty level. The player collects stars as rewards for correct answers (displayed in the bar on the right). After obtaining the third star, the player can progress to the next level.

All games are adaptable by setting various attributes, from the available topics (e.g. tools, fruit, animals, sports) and images (e.g. the selection of images can vary from color to black-and-white and line drawings) to difficulty levels (e.g. number of levels, number of images on the first level, number of additional images on the next level) and the speed of progress between levels (i.e. how many stars have to be collected before going to the next level). Therefore, the games can be adapted to each individual player, their level of difficulty, and specific deficit. This individual adaptation can be set on two levels. One level are the individual settings for the game that are prepared by the teacher, taking into account the student's impairment, performance ability, level of background knowledge, age and social background. Another level of adaptability are the student's individual settings that include pre-prepared styles according to the student deficit (e.g. changing the color scheme or contrast) and language settings, both affecting the interface presented to the user.

IV. EVALUATION

The PercieveConcieve platform is still under development, thus only one preliminary evaluation on end users has been conducted so far. We evaluated user experience, while testing the platform's games for vision, memory, and precise movement. The focus was on the games for visual perception.

Three visually impaired children, two boys and one girl, participated in the evaluation. Their average age was 10 years and all had severe low vision on both eyes. One teacher, who was an expert for teaching visually impaired children, participated in the evaluation as the leader of the testing and support for participating subjects. A solution designer had a role of an independent observer in the evaluation. The tested subjects were video recorded,

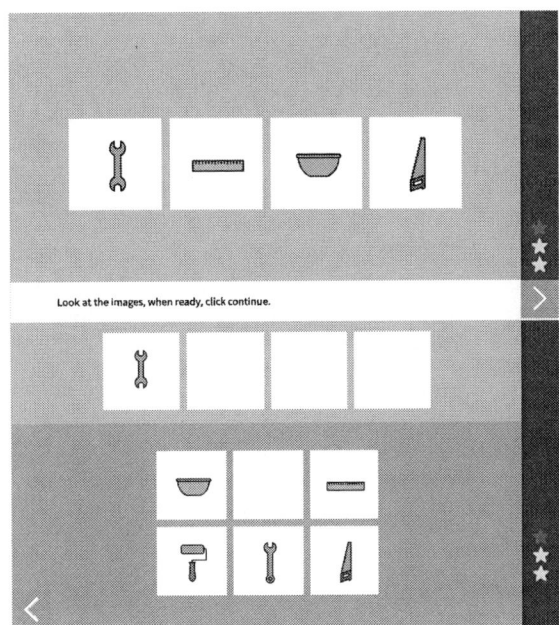

Figure 6. The Sequence Game for practicing visual memory. A sequence of images is shown to the user on the first page (top). Now the user has to reconstruct the sequence from the provided images (bottom)

observed while playing the games, and interviewed. The teacher prepared observations and notes on tested subject's performance and completed one evaluation questionnaire for each tested subject.

Evaluation tests were prepared for usability testing according to the specifics of working with and teaching the tested population. We combined three types of metrics in the tests: performance metrics (e.g. efficiency in playing the game, error frequency while playing the game, efficiency of interactions, ease of learning), issue-based metrics (e.g. why the user did not finish the task, level of satisfaction and level of frustration) and self-reported metrics (e.g. did the user like the game and wished to play again). Consequently, the evaluation questionnaire had four parts: game usability and efficiency (e.g. percentage of tasks completed, number of errors, success with performing the tasks), game effectiveness (e.g. number of help requests, number of tries before giving-up), student satisfaction (e.g. want to play more), and student dissatisfaction (e.g. student frustrations and uneasiness). The time needed to perform a task was not measured in these tests as it is not a relevant success criterion for the focus group of visually impaired children.

The main objective of the evaluation was to overcome the standard guidelines and to develop a user interface as well as to design graphic solutions that enable a visually impaired child such a user experience that is comparable to the one that children without impairments experience when interacting with didactic games.

As the tested subjects were observed while playing a specific game and performing given tasks, we could determine problematic navigational, functional, and content related elements in the application. We encountered several problems (e.g. some content elements

Figure 5. Various games for practicing memory, vision, and fine motor skills. Games are grouped in activities that can be customized for a particular learner and their impairment

were not visible or not understood, some user activities did not lead to the correct solutions, problems with magnification, problematic counting of images) and identified a number of suggestions on how to improve the games. Therefore, the results of this evaluation impacted directly the development decisions and the application was improved accordingly.

The analysis of the user experience from the viewpoint of success and difficulties in interaction with the application showed that the platform and the games were adequately designed. Main disturbing factors that in some aspects restricted tested subjects from accomplishing the tasks and caused major problems were essentially related to cognitive-behavioral characteristics of our target group (e.g. age, visual impairment, social background) rather than to the game design solution.

The evaluation results [5] show that all three children were excited to play the game and the game character (shown in Fig. 2) proved a valuable visual motivation. Nevertheless, the children still needed additional verbal support during the game, especially when playing the games for the first time. Particularly, the games for visual perception are in general more difficult for the first-time players, thus an expert teacher has to be present for additional help and guidance. Later the player is more autonomous and help is needed only occasionally.

We intend to perform a more comprehensive evaluation after the winter school break. The evaluation is planned on two levels: first with the teachers that use the platform in the educational setting, and later with the impaired children that use the customized games for exercising.

The first study will focus on the teachers and will measure user acceptance and usability of the platform. We plan to gather the data for this evaluation via a questionnaire, using a set of questions from two validated questionnaires: the technology acceptance model (TAM) questionnaire [6] and USE (Usefulness, Satisfaction and Ease) questionnaire [7].

The TAM model is valuable for explaining usage behavior, as it measures the perceived usefulness and perceived ease of use, which are two fundamental determinants of user acceptance [6]. On the other hand, the USE questionnaire will help us measure usefulness, ease of use, ease of learning, and user satisfaction with the platform [7].

The second study will focus on visually impaired children and their use of the customized games for exercising. We intend to measure accessibility of the games, student engagement, and effectiveness of the games as a teaching tool.

V. Conclusion

PerceiveConcieve is an open-source platform for accessible and inclusive educational games. The platform is designed to support accessibility, adaptability, and

customization of the games by the use of game templates. This way, even the less IT skilled teachers are able to freely adjust the games and modify their content, thus customizing the games according to the educational setting and the needs and preferences of their visually impaired students. The use of the PerceiveConcieve platform enables teachers to include motivating and accessible elements of game-based learning into inclusive learning scenarios, and consequently enrich and diversify the learning process.

The preliminary evaluation of the platform on target groups has shown positive acceptance by teachers and visually impaired children. The platform proved to be well designed, with appropriate graphic design as well as prepared content. We are preparing the next stage of evaluations, a more comprehensive study that will focus on user acceptance and usability of the platform.

Acknowledgment

The work was co-funded by the European Regional Development Fund of the European Union and the Ministry of Education, Science and Sport of the Republic of Slovenia. The operation was carried out within the Operational Programme "Strengthening Regional Development Potentials" for the period 2007-2013, Priority axis "Economic Development Infrastructure", Priority Focus "Information Society".

References

[1] "Strategic framework for European cooperation in education and training ET 2020," Official Journal of the European Union, C 119/2, 28.5.2009, available online: http://eur-lex.europa.eu/legal-content/EN/TXT/PDF/?uri=CELEX:52009XG0528(01)&from=EN

[2] S. Bocconi, S. Dini, L. Ferlino, C. Martinoli, and M. Ott, "ICT educational tools and visually impaired students: different answers to different accessibility needs," in Universal Access in Human-Computer Interaction. Applications and Services, LNCS, vol. 4556, pp. 491–500.

[3] L. R. Milne, C. L. Bennett, and R. E Ladner, "VBGhost: a braille-based educational smartphone game for children," in Proceedings of the 15th International ACM SIGACCESS Conference on Computers and Accessibility, ASSETS '13, ACM, New York, NY, USA, 2013, pp. 75:1–75:2.

[4] L. R. Milne, C. L. Bennett, R. E Ladner, and S. Azenkot, "BraillePlay: educational smartphone games for blind children," in Proceedings of the 16th International ACM SIGACCESS Conference on Computers and Accessibility, ASSETS '14, ACM, New York, NY, USA, 2014, pp. 137–144.

[5] M. Pesek, D. Kuhl, M. Baloh, and M. Marolt, "ZaznajSpoznaj – a modifiable platform for accessibility and inclusion of visually-impaired elementary school children," in Proceedings of the International Conference on Informatics in Schools – ISSEP 2015, September 28 – October 1, Ljubljana, Slovenia, 2015, pp. 173–179.

[6] F. D. Davis, "Perceived Usefulness, Perceived Ease of Use, and User Acceptance of Information Technology," MIS Quarterly, vol. 13, no. 3, 1989, pp. 319–340.

[7] A. M. Lund, "Measuring usability with the USE questionnaire," STC Usability SIG Newsletter, 8(2), 2001.

Gap in pagination due to withheld paper.

Pages 853-857

Integration of the future technologies to High schools and Colleges

Primož KURENT, BSc Engineering
ŠC Kranj, Višja strokovna šola, Kranj, Slovenia
kurent.primoz@gmail.com

Abstract

Science and technology are progressing with the speed which seems to be unfamiliar to us. What seems to be science-fiction some 30 years ago is now becoming our everyday reality. 3-D printing, O-led technology, modern communication, smart materials, mind recognition etc. How to follow the progress and all novelties and what is more, how to integrate them into the educational systems of high schools and colleges, is practically an art.

Article addresses the mind recognition system, EMOTIV Epoc technology, and its integration in the educational process.

Mind recognition device EMOTIV Epoc, (https://www.emotiv.com/the-science/) works in way to recognize brain activities in specific brain parts and sends this information to the computer. The computer processes the information and produces appropriate exit signals, whereas by strengthening the latter machines and devices can be controlled. Next to mind recognition, the device also has integrated the gyroscope and system for facial movement recognition.

In ten years such technology could thoroughly change the life of disabled people and the operational working processes in industries, as by controlling the mind, human would be able to control more machines than today by using hands.

Main obstacle for quicker integration in the practical use is the reliability – repeatability. For successful use in industry the rate must be at least 99.5 %, whereas it currently reaches only roughly 70 %.

Key words:

Mind recognition, EMOTIVE Epoc technology, brain activities, future technologies.

I. INTRODUCTION

A constant challenge professors and teacher in school are constantly facing is, how to present new technologies being developed to students, since the main teaching method, frontal method, is no longer effective. Currently prevailing methods are based on practical experiences, where students have the opportunity not only to gain theoretical knowledge, but also to test their own practical skills. We are talking about case studies, team work, and project management approach, whereas the addressed topic needs to be interesting and current.

II. INTEGRATION OF FUTURE TECHNOLOGIES TO HIGH SCHOOLS AND COLLEGES

At School Center Kranj we have thus decided to join the two concepts: we searched for a technology of future (i.e. Mind recognition device EMOTIV Epoc) and integrate it to curriculum by using modern teaching methods (i.e. project and team work), which is also what I wish to address in my article.

A. Mind Recognition Device EMOTIVE Epoc

The brain is a very complex system. The frontal cortex is the part where most of our conscious thoughts and decisions are made and it conducts much less than a tenth of the total activity in the brain.

Planning, modelling of your surroundings, interpretation of sensory inputs up to and including your perception of reality, memory processing and storage and the basic drivers of your moods and emotions all occur in many functional regions distributed around the brain, including the visual cortex at the rear, temporal cortex at the sides, parietal cortex behind the crown of your head and the limbic system deep inside the brain. The limbic system controls your basic moods and emotions, your fight/flight response and deeper long-term memory encoding as well as control of basic bodily functions such as breathing and heartbeat.

Most of these deeper functions interact intimately with different parts of the cortex, the outer layer of which is accessible to EEG measurements, however the interaction is quite complexly distributed. In order to map the true activity of the brain it is very important to measure signals from many different cortical structures located all around the brain surface. It is not possible to map these signals purely from the frontal and temporal regions. Determination of the user's complete mental state is very poorly approximated unless signals from the rear of the brain are also considered.

With proper coverage and electrode configuration, however, it is possible to reconstruct a source model of all important brain regions and to see their interplay [1].

EMOTIV Epoc is a new approach to the design and development of mobile EEG systems. It consists of hardware, which collects data, and software, which analysis the data.

B. Hardware:

The award winning EMOTIV Epoc+ is a 14 channel wireless EEG, designed for contextualized research and advanced brain computer interface (BCI) applications. The EMOTIVE EPOC+ provides access to dense array, high quality, raw EEG data using our subscription based software, Pure EEG (Picture 1).

C. Software: Detection algorithms

EMOTIVE EPOC+ offers different kinds of detection algorithms, all of them built on extensive scientific studies aimed to develop accurate machine learning algorithms to classify and grade the intensity of different conditions.

Facial Expressions – muscle artefact, which commonly get rejected in laboratory EEG studies, are diverted and classified to map the activation in different muscle groups and eye movement events. Our universal detections can be fine-tuned for each individual to indicate 12 different facial expressions or events. Individuals with partial paralysis or unusual musculature can custom-train the activations. These events can be used to animate an avatar, detect specific responses and they may be tasked to execute commands.

Performance Metrics – EMOTIV EPOC+currently measures 6 different emotional and sub-conscious dimensions in real time – Excitement (Arousal), Interest (Valence), Stress (Frustration), Engagement/Boredom, Attention (Focus) and Meditation (Relaxation). Performance Metrics algorithms are being continuously improved and upgraded.

These detections were developed based on rigorous experimental studies involving at least 20-30 volunteers for each state, where subjects were taken through experiences to elicit different levels of the desired state. They were wired up with many additional biometric measures (heart rate, respiration, blood pressure, blood volume flow, skin impedance and eye tracking), observed and recorded by a trained psychologist and also self-reported. EMOTIV Performance Metrics have been validated in many independent peer-reviewed studies [1].

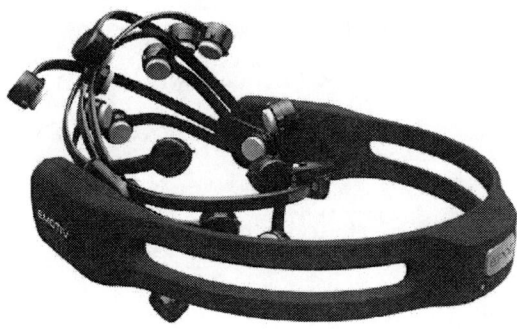

Picture 1

Mental Commands – based on unique and highly efficient methods, EMOTIV has developed a system for users to train direct mental commands where the user trains the system to recognize thought patterns related to different desired outcomes, such as moving objects or making them disappear. The system can be trained to recognize a single command in less than 20 seconds.

EmoKey – the custom software allows untrained users to be able to link their mental commands and emotional reactions directly to keystrokes, mouse operations or gestures within the host machine, allowing non-programmers to incorporate mental commands and mental state detections directly into existing applications.

Combined with the on-board motion sensor it can provide a hands-free brain-controlled mouse, adjust music volume or track skip depending in mood, and many other options.

D. Integration of the future technologies into learning process (curriculum)

Mechatronics is a new science combining engineering, electrical engineering and computer science. We wished to include the technology of the future EMOTIVE Epoc as cross-curricular connector during Mechatronics classes. Since this is a rather complex process, we integrated it through project work.

Project work, however, requires exact definition of goals, activities, sources and timeline [2]. We approached to realize our goals through the following phases:

- Task definition,
- Allocation of groups,
- Timeline,
- Conceptual design,
- Plan creation together with description of working, materials etc.
- Device making,
- Device testing,

Task definition:
Aim of the task was to create a mobile robot (car), which will be mind-controlled:

- Mobile robot (car),
- Size: length cca. 70 cm,
- Carbon made,
- Functions: forward, backward, left, right
- Additional functions: facial expressions recognition (with LED we show ☺, ☹), steering with gyroscope,
- Autonomous functioning (battery charged),
- Wireless connection with a computer,
- 4 wheels.

Allocation of groups:
We divided the project to four lots, whereas each lot is being led by one of the professors – mentors.

- 1st Lot: Computer science (Mentor Andrej Arh),
- 2nd Lot: Electrical engineering (Mentor Jožef Polak),
- 3rd Lot: Mechanics (Mentor Aljaž Rogelj),
- 4th Lot: Project Management (Primož Kurent).

Timeline

Well prepared timeline of the corresponding activities, responsibility allocation and costs management is main foundation of successful execution of a project.

Conceptual design

As aforementioned the EMOTIV Epoc senses the brain activities and only transfers to the computer the signals which require further analysis. Such program was written by students. With the help of one of the soft wares, the computer can be taught which impulse (brain activity) means which order (e.g. forward, backward etc.). Using wireless, the computer then sends the order to the vehicle. The vehicle holds a recipient, which strengthens the signal and sends it to electro engine, causing the vehicle to move. (see Picture 2: Scheme of components connection).

1st Lot: Computer science (Mentor Andrej Arh)

The computer science group studied the functioning of the mind recognition device EMOTIVE Epoc, with installed gyroscope and facial movement recognition system. The signals, which EMOTIVE Epoc transmits, were changed by a software program designed by students, to the extent that they can be transmitted to the computer holding the wireless data transfer system. This system was developed by the electrical engineering group.

2nd Lot: Electrical engineering (Mentor Jožef Polak)

The electrical engineering group developed a system for wireless data transfer. (Picture 3: Device for wireless data transfer.) [3]. The system has four different speeds, based on the number of signals. In practice this means, if we think "forward", this represents speed 1, if we think "backward", this is speed 2, etc. The same logic applies for the gyroscope: when we move head forward, this is Speed 1, if we move it forward again this is again another speed etc., whereas if we move our head backward, this reduced the speed level.

3rd Lot: Mechanics (Mentor Aljaž Rogelj)

The mechanics group had to develop a physical model of the vehicle. The vehicle is made out of carbon fibres, representing another challenge on its own Student Luka Kondič is well skilled in vehicles design and he drew a roadster sketch (Picture 4: sketch of the vehicle), after which a 3D model was designed with the help of a computer. To produce the vehicle we first needed to make a model out of Styrofoam in the scale 1:1. This model was made with the help of the School Center Celje. Later on, we applied carbon and resin to the model which gave us the vehicle model as seen on Picture 5: Vehicle made out of carbon. The physical model was made by a student with the help of company Fanell, d.o.o..

III. CONCLUSION

In the article I presented a new, innovative approach of integration of future technologies in high schools and colleges. We developed a mechatronic device steered with by human mind.

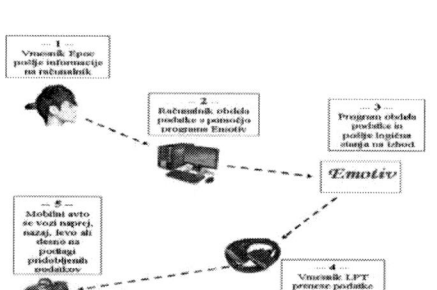

Picture 2: Scheme of components connection.

Picture 4: sketch of the vehicle.

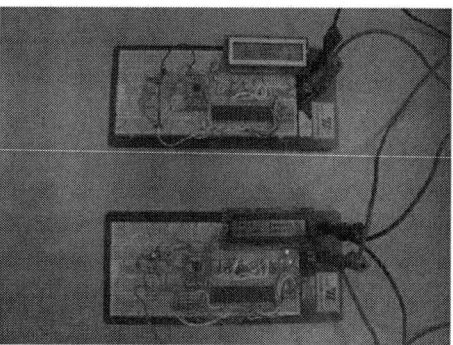

Picture 3: Device for wireless data transfer.

Picture 5: Vehicle made out of carbon.

The system entails a device which is being put to a person's head and recognizes the electrical impulses from brain when we think of a certain thing/object/feeling. These impulses can be transmitted to a computer and processed accordingly. We can form exits which manage different actuators. This gives us endless possibilities to develop new and interesting devices which are mind controlled. The device also holds a gyroscope and has the possibilities of facial activities recognition.

Knowledge and experiences which we gained with this project brought us to a level, where we practically have endless possibilities for various applications in the future; from medicine to mechanical and entertainment industries.

All involved participants learned something new in this project. Talented students, full of interests, gained new skills and insights into industry, whereas mentors gained new pedagogical and professional skills. It was a great pleasure to work with students full of enthusiasm, which also brings quality and promotion to the school.

I believe that such projects are necessary in the future and this is why we will proceed with such successful practice also in the upcoming years.

LITERATURE

[1] (https://www.emotiv.com/the-science/)

[2] A. Stare, Projektni management: teorija in praksa, Agencija poti, Ljubljana, 2011,

[3] M. Čeh, Mehatronika, Pasadena, Ljubljana, 2009

Preschool Children and Computers: Who lives in a meadow?

Barbara Strnad

Gimnazija Novo mesto, Čarunalnik - Društvo za napredno uporabo računalnikov, Novo mesto, Slovenia
barbara.strnad@gmail.com

Abstract - In recent years, most developed societies have realized that it is very important for students to acquire the skill of algorithmic thinking and the basic knowledge of computer programming. Nowadays we have numerous ways that allow us to teach programming with appropriate first steps. The paper will present one of the possibilities which we have to introduce basic programming concepts to younger children – with lego robots and a topic Who lives in a meadow?

I. INTRODUCTION

In order for children to become active creators in computer science and informatics, it is necessary to learn the algorithmic way of thinking. Knowing the basic steps of programming is one of the best ways to do it. The purpose of this paper is to present a project we did in a kindergarten with Lego WeDo, which is a great tool to start introducing the world of programming to kids [1], and is practically suitable for all age groups. Most of the activities are based on a game and co-working. Lego robots allow us to do many cross-curricular links, including topics related to life, physical science, Earth science and space technology.

II. COOPERATION WITH PRESCHOOL CHILDREN

A. Lego WeDo 2.0

Lego® Education WeDo 2.0 [2] basic set (Fig.1) is designed primarily for elementary school students, but we can also use it for working with preschool kids and high school students with no prior experience. The kit allows us to create and programme simple Lego® models. The set contains more than 280 items, including WeDo 2.0 Smarthub, one motor, a motion sensor that can detect objects at a distance of 15 cm and a tilt sensor that can detect six different positions.

Smarthub is the electronic system, which is part of the product Lego® Power Functions (LPF) 2.0, a new technology platform for the program Lego® Education.

It has a built-in low-energy bluetooth technology that connects smarthub with software. Energy can be obtained via two AA batteries or rechargeable batteries. There are two I/O outputs, through which you can connect external motors, sensors or any of the other LPF 2.0 components. Built-in RGB light area can display 10 different colors, which can be controlled by software.

It is an excellent tool for introducing the basics of robotics and enables interdisciplinary cooperation. It has the option of documenting and sharing. Programming is based on the principle of drag and drop.

B. Starting the activity

We start with an exciting polygon, where children try to lead a live robot. First, we choose a volunteer and blindfold him. Then we place a dynamic polygon made of tables and chairs in the classroom, which enables them to securely lead "a live blindfolded robot" (Fig.2). At first glance it may not be immediately apparent, but there are some programming concepts already hiding in the game.

The aim of the game is that the children learn from the experience what are the good and sufficient detailed instructions that enable successful and fast crossing of the polygon. Sometimes we can prepare a polygon in such a way that children can use the instructions in the loop or a conditional statement....

Children can see through this activity that some instructions are understood completely clear by some of them, while others may be misinterpreted and that there are also several procedures for the solution of polygons ...

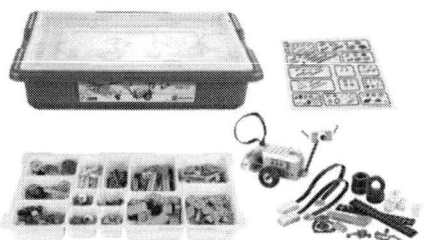

Figure 1: Lego WeDo 2.0

Figure 2: Polygon

We further discuss how we perceive space when we are blindfolded and what is the role our senses. Children usually figure out what is the connection between the robot and our game. An interesting topic is also to talk about the differences between robots and humans.

C. Robots in everyday life

We continue with a conversation about robots in everyday life. Children often say that they have not yet met a robot, but after some consideration they soon think of a robotic vacuum cleaner. They have already heard about or seen a robotic lawn mower or industrial robots... Then we talk about how we command the robots. The most thorough analysis of the robot's function can be carried out on a robotic vacuum cleaner that children know best. They know that robotic vacuum cleaner has a sensor for not falling which prevents falling down the stairs. Children can describe its moving really precisely. They know the cleaner uses a profiled rubber bumper to detect obstacles. They even know that it is necessary to lodge a border wire in areas where we do not want the robotic lawn mower to mow.

D. Creating our own robots

Our robots, which we will build, will also require sensors for better orientation in space. We introduce two sensors in the kit, a tilt sensor and a motion sensor (Fig.3).

It is time to build a robot and equip it with sensors. The focus is firstly on commands which will tell the robot how to move. We use printed command blocks on a bigger format. Children try to conclude what each command means just by observing the image (Fig.4). Most children are very good at observing. They identify a motor which rotates in the clockwise direction or in the opposite direction. They connect the engine with the hourglass and with the length of time that the motor will spin. They explain that the command block which determines the engine power shows the engine and power meter pointer.

We proceed with the assembly of lego bricks. Following the instructions first robots are made and programmed to move, and then equipped with a motion sensor and a tilt sensor.

Figure 3: Motion and tilt sensor

Figure 4: Command blocks

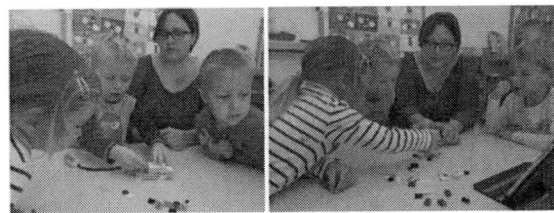

Figure 5: Building robot with preschool children

Figure 6: Our robots in their environment

III. CREATIVITY OF PRESCHOOL CHILDREN

We had quite a few meetings with preschool children (Fig.5), where we build different robots: a frog, a mantis, a spider, a caterpillar, a bee... Members of Čarunalnik were teaching children how to build correctly and follow the instructions and we did some basic programming.

First we started with the topic: Who lives in a meadow? We divided the children into six groups, each group picked one animal – bee, spider, snake, caterpillar, mantis, and firefly. There were four children in each group. Two groups worked on a robot, while the other two groups worked on modelling an animal's environment, a meadow with a pond (Fig.6). The last two groups learned about a chosen animal, explored how they live: how bees take an active role in plant reproduction, they researched different stages in the life circle of a frog, butterfly, ... Groups rotated on each session.

Before visiting the kindergarten, we elected all needed bricks for creating a specific robot, so the children didn't spend too much time searching, and they did not lose interest in searching among too many bricks. Group members that were constructing and building robots had different tasks: they had to find the right bricks, one had to put the next page of instructions on the tablet, and another had to check if pieces were joined correctly. The children did all of the tasks alternately, except the search for needed parts, which was everybody's job. They built alternately; some children had more experience following instructions and better spatial performance. They helped children with more difficulties and they were »supervisors«.

Once we constructed the robot, we wrote simple programs, where animals were moving forward and backward, they were making different sounds and the RGB light surface on Smart Hub changed colors. The children then tried to modify programs.

Because of a major interest and enthusiasm among the children, we added two more topics: jungle, and working machines and heavy-equipment vehicle (Fig. 7).

We experienced major progress in following building instructions, better spatial performance, and better motoric skills when putting bricks together. The children were excited and they happily presented their work to other kindergarten groups and their parents.

The reactions to the project we did together were extremely positive. Children were satisfied, because they really assembled their own robot, wrote a program and

Figure 7: Robots in their invironment

created an environment for the robot-animal in the meadow. Meanwhile they visited some interesting experts. They visited a beekeeper and they molded their own wax candles. Children learned a lot about animals in the meadow. They were able to use their imagination and creativity. Children enjoyed in the implementation of their own ideas and comparing their products. After completing the project, the children were excited and eager for new meetings.

IV. CONCLUSION

Panellists on computer literacy in primary schools this year agreed that the school system is in urgent need of change in the field of computer skills, because without these skills, students will not be competitive in almost any field of expertise. There is almost no activity associated with digitization, so computer skills are vital.

We introduce children to the world of programming and algorithmic thinking through attractive robotics by programming small robots. In this way there is a connection between the mind maps and the physical world. Learning about computer concepts and developing procedural way of thinking help students acquire knowledge and skills that are much more durable than the rapidly evolving technology.

REFERENCES

[1] C. R. Fisher, "Key-stage computing: Evaluating the suitability of Lego Mindstorms NXT 2.0 for use in early computer science education." Discovery, Invention & Application, University of Derby, 2014.

[2] LEGO. Lego WeDo 2.0. (2016, January 2) Retrieved from https://education.lego.com/en-us/products/wedo-2-0-core-set-software-and-get-started-project/45300

MIPRO 2017, May 22- 26, 2017, Opatija, Croatia

Web Service Model for Distance Learning Using Cloud Computing Technologies

Dragan Cvetkovic* Marko Mijatovic ** Marijan Mijatovic** Branko Medic ***

* Faculty of Education in Sombor /Department of Informatics and media, University of Novi Sad, Serbia
**Faculty of informatics/International University of Brcko, Brcko District, Bosna and Hercegovina
*** Technical College of Applied Sciences, Subotica, Serbia

dcveles@gmail.com, marko.mijatovic@hercegovina.edu.ba, mile-dj@hotmail.com, brankomedic@gmail.com

Abstract: The improvement of communication and the exchange of teaching resources between teachers and students has developed a model of web services based on cloud computing technology. Today's models and systems (in the application approach) are used according to the user's needs, while the rest of the time computing resources remain not used. Such services are characterized by extremely high level of security and synchronization of files that are continuously exchanged. The infrastructure that is based on cloud technology provides uninterrupted simultaneous work of great number of different virtual machines on which are installed different systems and application software. For the purposes of the experiment, teaching resources are offered in the form of a virtual machine. Experiment included approximately 500 respondents in the territory of the Republic of Croatia and Bosnia and Herzegovina. The experiment results show that the system for e-learning (online education) developed in cloud environment is more scalable and shows better performance from the currently used classical server oriented system.

I. INTRODUCTION

With the rapid development of information technology we see a strong ongoing need to improve the current state of information systems in education. Cloud computing or cloud technology is the answer that fulfill the above mentioned need in order to increase the capacity and provide new features on the existing infrastructure without investing in new infrastructure and personnel. This technology has enabled customers to use additional computing resources using only a web browser (e.g. Internet Explorer, Chrome or Firefox) thereby using the existing computing resources. Cloud technology is a kind of technology mainly oriented to end users and is implemented through a number of services like providing additional disk space, memory and network capacity as well as providing quality free software applications.

One of the biggest benefits of cloud technologies are new opportunities in business, economics and education. A simple exchange of information and collaborative work on documents (teaching materials) has increased flexibility and has enabled faster execution of tasks. Therefore, there is no longer need to invest in expensive IT and logistics infrastructure, we can use cloud services only when needed, which greatly reduces costs.

Based on several research papers conducted by well-known analysts, manufacturers and IT users of cloud computing and according to the National Institute of Standards and Technology (NIST) [1] , this modern cloud computing technology can be said to represent:

The shape of this cloud technology over the existing internet browsers delivers a multitude of applications to end users. Using this model, users do not have to invest in new server and network resources and licensed programs. Cost of service providers are in fact lower than in the traditional service of data storage on a separate server.

II. THE LOGICAL ARCHITECTURE OF THE SYSTEM - WEB SERVICES

Applications must manage all resources of Cloud Computing infrastructure in order to ensure quality services to end users (students). Applications are to be developed for various platforms: Android mobile platforms, Linux, windows and other. The applications would depend on existing Cloud Computing infrastructure, and would be integrated with the root computer of end users (LDAP) which contains accounts for access and relies on web services, what allows future development of other applications that have similar purposes and capabilities [2]. Through the comparable method, a similar application would be developed as well as the web applications with the same purpose.

The application would primarily benefit students chosen to conduct the research experiment, which would allow them to schedule and run predefined virtual machines with all the installed operating systems and necessary software that is required for the particular course.

The software application would use the service-oriented architecture, which allows comparative development of web and mobile applications. The mentioned approach to software development would lead to the creation desktop applications and integration with more some existing system.

Web service has the primary role in business logic and system integration. It integrates the following components: OpenLDAP user directory, Moodle LMS, and MySQL databases [3]. System architecture for execution of the application is shown below (Figure 1)

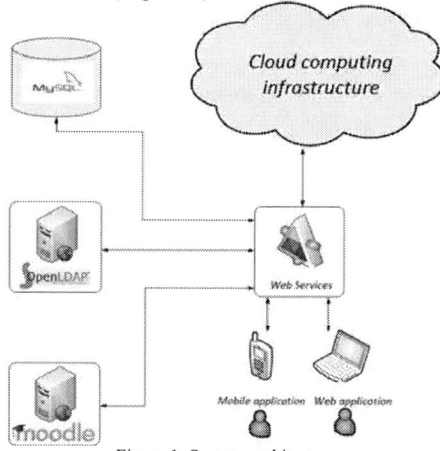

Figure 1. System architecture

The end user (student) can access the system using two channels, web applications and mobile applications. In the future, it would be possible to increase the number of channels access to the system.

The complete administration of the system would be conducted through web applications [4] [5]. One thing that would be needed is to ensure that the end user is assigned rights of system administration. Administrators would be able to define the exact DVD image of the operating system stored on Cloud Computing infrastructure and the ultimate users would be able to reserve and initiate the OS image through the application. Since the system would be integrated with Moodle LMS, Virtual machines can be grouped according to the available Moodle courses.

In order to ensure greater scalability of the system and the introduction of opportunities of new services for customers (students) the ultimate Cloud Computing infrastructure would be presented [6]. The complete software for the management of this infrastructure is package Eucalyptus. The complete software code is labeled as open source software and that is a strong advantage of the entire project as it significantly reduces the economic costs of the implementation of this software solution.

For the purposes of evaluation of the developed model of IT infrastructure for education we have conducted an experiment. The experiment covered high school students of third and fourth years as they have been divided into experimental and the control group. Students have voluntarily consented to participate in the experiment. The experiment will be carried out in the municipalities

of Stitar, Zupanja and Vinkovci (Croatia) and Samac, Orasje and Odzak (Bosnia and Herzegovina) in the academic year 2015/16. The experiment will include about 470 teachers and students.

The main goal of the experiment is eliciting answers to the following questions:

- In what measure does this online e-learning system stored in the cloud affect the results which students achieve on their final exams (main hypothesis);
- Are the teachers satisfied with the implemented infrastructure for e-education in privately developed Cloud (secondary hypothesis);
- Do we see the expected performance of the implemented system of e-education in the privately developed Cloud (secondary hypothesis).

In order to determine the degree and properly assess the performance of the infrastructure for e-education in privately developed cloud and its effects on the recently achieved results of the students, we shall compare the students' results which are grained in the environment of the privately developed cloud where the students could have used all the available educational services to those students' results gained through the system for e-education developed on classical infrastructure, in which new services had not been available (Table 1). Students of the experimental groups had the possibility of using educational resources which are offered through virtual machines. For the control group, the selected students are those who have attended the same courses as the previous generation. The students belonging to the control group have only used the classical system Moodle LMS without accessing the resources and services that our developed platform offers in the privately developed cloud.

Course title	Experimental group		Controlled group			
	N	Average	N	Average	F	Sig
Databases	461	8.01	470	7.97	0.352	0.553
Internet Technologies	256	8.07	228	8.03	0.283	0.595
Programming Languages	176	8.11	171	8.05	0.285	0.594

Table 1 Statistical summary medium rating experimental and control group students

In order to estimate the performance of the implemented system of e-education in privately developed clouds we have used a sophisticated software tool - Ganglia. This tools (Ganglia) is for gathering and representation of performances of the server resources. This particular software tool allows for deeper analysis of all the performance attributes such as the stability of the system, availability, speed and the analysis of performance of its processors (Table 2).

	The period of maximum load	The period of normal load
The maximum number of simultaneous accesses	80	32
Stability (The number of interruptions in the system)	1	0
Accessibility (Expressed in percentage)	98.75 %	100 %
Speed (System response in milliseconds)	15 ms	7 ms
Performance of CPU (Average load: CPU u %)	65 %	22 %

Table 2. Indicator of performances and the system stability

The research survey for researching the selected teachers' attitudes on electronic education system in this private cloud has been developed to contain five questions total. The questionnaire is confirmed using the well-known Cronbach alpha coefficient, which is 0.814. in this particular case. All the survey questions are formed so as to elicit answers in the following areas: teachers' productivity during the preparation phase of the course, teacher productivity during the course realization phase, feedback on technical support satisfaction, feedback on customer experiences of the e-learning system and the general opinions of the developed infrastructure for e-education. All research questions are formed with the well-known Likert five-scale (Table 3). The presentation bellow displays all the research questions of the survey:

Results:	5	4	3	2	1
P1: Systems for e-learning education based on the privately developed cloud facilitates improvements in productivity in the very process of course preparation.					
Databases	3	3	3	2	0
Internet Technologies	2	2	1	3	0
Programming Languages	3	2	0	2	0
P2: Systems for e-learning education is based on the cloud technologies and enhances productivity in the process of course realization.					
Databases	5	4	2	0	0
Internet Technologies	5	1	2	0	0
Programming Languages	5	2	0	0	0
P3: Technical support is at the right level					
Databases	3	4	2	1	1
Internet Technologies	2	3	2	1	0
Programming Languages	2	3	2	1	1
P4: The reports of the system are at the right level					
Databases	2	3	5	1	0
Internet Technologies	0	5	3	0	0
Programming Languages	0	4	2	1	0
P5: According to my opinion, the developed infrastructure based on cloud computing technologies is adequate for the process of e-learning and education.					

E-learning Education					
Databases	5	4	2	0	0
Internet Technologies	6	1	1	0	0
Programming Languages	4	2	1	0	0
P6: Systems for managing the digital identities is efficient and provides a comprehensive access to all the resources.					
Databases	4	5	2	0	0
Internet Technologies	4	3	1	0	0
Programming Languages	5	1	1	0	0

Table 3. Results of the teachers' survey

The experiment or the test covered all the students of the controlled and experimental groups. The testing approach has been used in order to present the statistical significant difference between the results results of the controlled and experimental group. The tests are made for each observed course (Databases, Internet technology, Programming languages). Cronbach alpha coefficients are 0910, 0875 and 0890, respectively. The same tests had been used during the previous five years for the observed courses. Pupils had been enabled to use the provided services in this context of the given environments of private clouds: portal service, e-mail, file management system, service for database management, services for communication and cooperation. These services provide the educational resources, and provide students with:

- Easy access to a variety of useful information and educational services;
- Open interaction between users;
- Information sharing of all the common activities;
- Learning Management system integration

The Strategy of Creation of Educational Resources is defined for needs of each course in context of the electronic education. For each course (Databases, Internet technology, programming languages) we have defined expensive virtual machines with specific hardware and software requirements. Those setup virtual machines contain the names, descriptions, roles and types of virtual machines for each of the three discussed courses.

With the development and realization of the above mentioned IT model of infrastructure for e-education, we have effectively provided contemporary services from the IT sphere and from the shpere of communication technology in order to improve the main educational work. To all the registered tutors, teaching staff and administrators we have provided and made possible a unique, fast and easy access to all the teaching resources and network services and thereby increasing the overall efficiency, safety and quality of the education process.

The procedure of integration services in electronic education in the cloud environment and conducting the electron learning:

Educational courses are created by all the teachers through Home LMS system. After that, for each course we prepare the instructional materials, activities and tasks. Also, we prepare the most specific software tools needed for each course. All the pupils' accounts are created in the Home LMS and integrated with LDAP directory linked to the educational institutions. After that, we prepare the virtual machines with the main operating system, application software and tools necessary for each individual course. Following up, we prepare virtual machines stores as DVD images deposited in the private cloud. Students use the familiar and user-friendly web interface or mobile application interface to book some courses from the predefined virtual machines. Finally, teachers and administrators of the system can see and analyze the final results of the students performance in the system and only store the results in the private clouds.

Course title	N	Average	Standadr deviation
Databases			
Exprerimetal group	461	7.98	0.961
Controlled group	470	7.72	0.949
Internet techologies			
Exprerimetal group	256	8.09	1.032
Controlled group	228	7.88	1.262
Programming languages			
Exprerimetal group	176	8.22	1.059
Controlled group	171	7.98	0.994

Table 4. Students' results achieved after the testing phase

For testing the existence of the significance difference variable between the obtained results gathered from the students of the controlled and experimental groups we have used the variable of analysis. The final results show (Table 4) that there are significant statistical differences between the achieved results of the controlled group of our students and the experimental group of our students for all three observed courses:

- Database: $F_{(1,929)} = 16.648$ ($p < 0.05$)
- Internet technology: $F_{(1,482)} = 4.130$ ($p < 0.05$)
- Programming languages: $F_{(1,345)} = 4.930$ ($p < 0.05$)

III. CONCLUSION

In the process of analyzing all the collected information from the recently conducted surveys of teachers, we can conclude the following:

- Most teachers agrees that this infrastructure of private clouds enhances the productivity in preparing for the courses intended for e-education experience. However, several teachers have had contrary opinions. Those teachers are mainly the individuals who have had to invest more efforts in preparation of certain teaching themes like setting up specific virtual machines for the given teaching themes;
- When considering the application possibilities of this infrastructure stored in private clouds for realization of electronic education, most teachers agree that the implemented infrastructure enhances productivity when conducting electronic courses;
- The technical support used in the given infrastructure of the private clouds is at an adequate level for most of our teachers. However, several teachers have not been satisfied with the quality of the provided technical support;
- The reporting channel implemented in the system of electronic education is at an adequate level for most teachers. However, there has been a significant number of teachers who have had different opinions on reporting. Those are mainly the teachers who did not need to use the reporting services;
- not was teachers which think that realization infrastructure private clouds not suitable for e education. most teachers you agreed that is infrastructure very suitable for system e-education;
- Our teachers are satisfied with the quality of the system for identity management. Most teachers agree that in order to access all the necessary resources, one can easy access them over the integrated system for management of identities.

IV. REFERENCES

[1] Peter Mell, Timothy Grance (Septembar 2011) The NIST Definition of Cloud Computing, Special Publication 800-145, Computer Security Division Information Technology Laboratory National Institute of Standards and Technology Gaithersburg,MD20899-8930:
http://nvlpubs.nist.gov/nistpubs/Legacy/SP/nistspecialpublication800-145.pdf

[2] Voorsluys, William; Broberg, James; Buyya, Rajkumar (February 2011). "Introduction to Cloud Computing". In R. Buyya, J. Broberg, A.Goscinski. *"Cloud Computing:*

Principles and Paradigms". New York, USA: Wiley Press. pp. 1–44. ISBN 978-0-470-88799-8.

[3] Amies, Alex; Sluiman, Harm; Tong, Qiang Guo; Liu, Guo Ning (July 2012). *"Infrastructure as a Service Cloud Concepts"*. *Developing and Hosting Applications on the Cloud*. IBM Press. ISBN 978-0-13-306684-5.

[4] Cvetkovic D., Rastovac D., Mandic M. (2012) . *"EyeOS – Cloud operativni system za obrayovanje"*, NIR, Internacionalni univerzitet u Brckom, Brcko, BiH (2012), ISSN: 2233-1603, pp. 83-87.

[5] B Rochwerger, J Caceres, RS Montero, D Breitgand, E Elmroth, A Galis, E Levy, IM Llorente, K Nagin, Y Wolfsthal, E Elmroth, J Caceres, M Ben-Yehuda, W Emmerich, F Galan. "The RESERVOIR Model and Architecture for Open Federated Cloud Computing", IBM Journal of Research and Development, Vol. 53, No. 4. (2009)

[6] Hsu, Wen-Hsi L., "Conceptual Framework of Cloud Computing Governance Model - An Education Perspective", IEEE Technology and Engineering Education (ITEE), Vol 7, No 2 (2012).

The Struggle with Academic Plagiarism: Approaches based on Semantic Similarity

Tedo Vrbanec[1], Ana Meštrović[2]

[1]Faculty of Teacher Education, University of Zagreb, Croatia
[2]Department of Informatics, University of Rijeka, Croatia
tedo.vrbanec@gmail.com, amestrovic@inf.uniri.hr

Abstract - Academic plagiarism is a serious problem nowadays. Due to the existence of inexhaustible sources of digital information, today it is easier to plagiarize more than ever before. The good thing is that plagiarism detection techniques have improved and are powerful enough to detect attempts of plagiarism in education. We are now witnessing efficient plagiarism detection software in action, such as Turnitin, iThenticate or SafeAssign. In the introduction we explore software that is used within the Croatian academic community for plagiarism detection in universities and/or in scientific journals. The question is - is this enough? Current software has proven to be successful, however the problem of identifying paraphrasing or obfuscation plagiarism remains unresolved. In this paper we present a report of how semantic similarity measures can be used in the plagiarism detection task.

Keywords: academic plagiarism, plagiarism detection, obfuscation plagiarism, natural language processing, semantic similarity

I. INTRODUCTION

Academic plagiarism is nowadays one of the most pressing problems of the academic community. Many successful plagiarism detection tools and software products have been developed. However, the detection of paraphrasing or obfuscation plagiarism remains a challenge because most of the existing tools are only able to detect copy-paste cases of plagiarism. Plagiarism is not just the direct copying of one's text. It can be far more complicated if it is a case of paraphrasing or obfuscation. According to [1] high-obfuscation plagiarism can be realised by modifying original text by reduction, combination, paraphrasing, summarizing, restructuring, concept specification and concept generalization.

Due to this, there are serious drawbacks of systems such as TurnitIn or SafeAssign. More precisely, these tools cannot deal with the vocabulary problems such as synonymy, homonymy, hyperonymy and hyponymy. There are many various approaches in the area of natural language processing (NLP) that may offer a solution for this task.

However, there is still lot of room for improvement. This is why we are focused on paraphrasing and obfuscation plagiarism detection. We analyse various approaches that may identify paraphrasing and obfuscation by means of semantic similarity measures. Some are based on external knowledge resources such as WordNet or ontologies, and others are based on statistical NLP techniques.

The goal of this paper is to analyse the possibilities of existing semantic similarity measures and somehow classify existing approaches. This is just a preliminary study of this wide domain with the further goal of providing an extensive overview and classification of all semantic similarity measures of text and possible approaches in paraphrasing identification. In addition to this, we give a short review of the situation with academic plagiarism in Croatia. This is important because the struggle against academic plagiarism nowadays is on-going and there are frequent clashes.

In the first part of the paper we describe the problems of academic plagiarism. Furthermore, we give a brief review of academic anti-plagiarism efforts in Croatia. In the second part of the paper we are focused on various approaches for plagiarism detection based on semantic measures. After this we give an overview of other papers which try to resolve plagiarism detection problems with NLP techniques and semantic similarity. Finally, the sixth section contains a conclusion and possible directions for future research.

II. ACADEMIC PLAGIARISM

Academic plagiarism in other words the plagiarism of digital text is most often the target of plagiarism during education and in academic papers. Academic plagiarism is a syntagma which indicates the plagiarism of a complete or part of documents of the following kinds: of programs in the source programming code, seminars, critical reviews, professional or scientific papers and non-literary books. The first tagmeme of syntagma - academically, points out that this kind of plagiarism most often appears in the academic community. At the same time it means that in the academic context plagiarism is a particularly worrying phenomenon which attracts the attention of all academic structures.

Below we provide a list of the methods of academic plagiarism and after that comment upon the anti-plagiarism efforts in the Croatian academic community.

A. Methods of Academic Plagiarism

The most famous plagiarism software manufacturer TurnitIn [2], distinguishes academic plagiarism methods and research paper plagiarism methods. As academic plagiarism methods it lists:

- the submission of someone else's work as one's own, in order to fulfil a specified teaching obligation,
- the copying of words or ideas without giving credit to the original author,
- copying the majority of the words or ideas that compromises the work,
- submitting an already submitted work (e.g. from another colleague),
- not using quotation marks when quoting,
- giving incorrect data about sources,
- the use of someone else's sentences by using substitute words,
- using someone else's ideas without referencing.

TurnitIn also lists the research plagiarism methods [2, Pt. 1, p. 5]:

- *"Claiming authorship on a paper or research that is not one's own.*
- *Citing sources that were not actually referenced or used.*
- *Reusing previous research or papers without proper attribution.*
- *Paraphrasing another's work and presenting it as one's own.*
- *Repeating data or text from a similar study with similar methodology without proper attribution.*
- *Submitting a research paper to multiple publications.*
- *Failing to cite or acknowledge the collaborative nature of a paper or study".*

B. Software solutions used in Academics

In order that the academic community could effectively fight the modern plague in science and education – plagiarism, it is necessary [3, p. 123] (1) "to form warnings and measures in the education of students and scientists" ... "at all levels of education" [4, p. 108] and (2) to develop or use existing software systems for plagiarism detection. According to [3], [4], the most widely used software products in the world are: iThenticate, SafeAssign and CrossCheck. All the papers checked by TurnitIn are stored in a database for further comparison. SafeAssign is optional in this regard. CrossCheck uses a large database of papers which have been handed over for the use of scholarly publishers affiliated to their CrossRef organisation. In return they can use the plagiarism detection system free of charge. Individuals do not generally have some cheap or free of charge choice. One of the possibilities is the Viper desktop application, which only works in the Windows environment.

It is known that with the software detection of plagiarism there still exists the insufficient detection of so-called intelligent plagiarism cases [3]: the plagiarism of ideas, complex paraphrasing, and plagiarism between multiple languages. This is trying to be resolved in existing commercial software systems with the adding of translation modules, however the solution is in a qualitative upturn: the detection of sematic similarities between documents.

C. Academic Anti Plagiarism Efforts in Croatia

In the EU and around the world, plagiarism is a very worrying phenomenon against which educational, preventative and restrictive methods are used. In 2015 The Council of Europe supported this effort by establishing the Pan-European Platform on Ethics, Transparency and Integrity in Education (ETINED), the aims of which priority activities are, amongst others, [5]: "Ethical behaviour of all actors in education" and "Academic integrity and plagiarism". In this context the European Commission led a project (2010-2013) in which Croatia did not participate: The Impact of Policies for Plagiarism in Higher Education Across Europe (IPPHEAE) "whose aim was to explore policies and systems of assuring academic integrity and deterring plagiarism in a system of higher education" [6, p. 5].

However, in Croatia as a member of the same EU, it is possible that the secretary of a parliamentary party, a rector of a university or a member of the Constitutional Court use plagiarism, yet in doing so are not sanctioned, what's more – they continue to perform their high office. And this is all despite the fact that in 2006 a Committee of Ethics in Science and Higher Education was established as a body of the highest legislative authority in Croatia (of Parliament).

TurnitIn has been used by the University of Rijeka since 2014 [3, p. 12] and immediately after by University of Osijek. The VERN Polytechnic and the Zagreb School of Management use TurnitIn, too [6, p. 12], as well as by individual faculties of the University of Zagreb, where there is unfortunately no joint financing, although the Rector announces it. In the meantime, individual faculties autonomously negotiate plagiarism checking services at their own expense. For example, the Faculty of Teacher Education in Zagreb used Ephorus for several years, however due to its high price and it crossed over to the cheaper solution: desktop only and Windows only application Plagiarism-Detector Personal.

The University of Split planned to begin using some software from 2015, but since the beginning of 2017 they have still not done so, due to the high prices. On the level of collaboration between Croatian universities, the president of the Rector's Assembly announced the collaboration of projects of mutual interest and a reduction of the burden of cost to each university.

Some unspecified plagiarism checking software is used by the School of Dental Medicine and the Faculty of Political Science [7, p. 2]. Furthermore, some scientific journals in Croatia use plagiarism detection software systems, e.g. Biochemia medica [8], the journal of the Croatian Society of Medical Biochemistry and

Laboratory Medicine which both use iThenticate since 2013.

Members of professional committees for (re)accreditation from abroad are participating in the processes of the (re)accreditation of Croatian higher education institutions. Usage of plagiarism detection tools is one of the parameters for measuring the quality of professional and scientific work in academic institutions. This has prompted many of them to start using some plagiarism checking software or to consider options of using it. Surely the best solution for all of them is to unite, as many educational institutions as possible, in a common approach to suppliers, i.e. service providers, because in this way individual institutional usage is the cheapest, and further on, the database of documents – the corpus, as the basis for software comparison, becomes more complete. The current practice is that individual universities or faculties sign contracts with the software dealers or manufacturers, usually in three years' contracts, for a price that is of much greater than if they bought together and without less limitation of usage.

III. SEMANTIC SIMILARITY

The concept of semantic similarity is fundamental and widely understood in many domains of natural language processing [9]. It can be defined as the degree of taxonomic proximity between terms [10]. The terms that can be also used instead are semantic proximity or semantic distance as an opposite concept. According to Resnik [11], semantic similarity represents a special case of semantic relatedness. He gives an example that cars and gasoline seem more closely related than cars and bicycles, but the latter pair is more similar.

Semantic similarity can be measured in terms of numerical score that quantifies similarity/proximity. In existing research's various semantic similarity measures (SSMs) have been defined and many semantic similarity computational models have been proposed. Semantic similarity measures have been widely used in many NLP and related fields such as text classification, information retrieval, information extraction, word sense disambiguation, machine translation, question answering, plagiarism detection, etc.

Semantic similarity can be calculated on different levels of granularity (between words, between paragraphs or between whole texts/documents).

In the case of plagiarism detection, semantic similarity should be expressed on the text level as the final result. The score of semantic similarity between suspected document and one or more other documents may indicate the existence of plagiarism. The whole procedure of plagiarism identification, semantic similarity can be calculated on the sentence and paragraph level.

IV. SEMANTIC SIMILARITY MEASURES

In this section we will give an overview of different measures and approaches in measuring the semantic similarity of texts. Semantic similarity measures are defined in the domain of NLP for various tasks. One of the first application of SSMs was obtained in 1970s seventies for the task of information retrieval [12].

There are various approaches in detecting semantic similarity. According to [13], there are two main approaches in measuring the semantic similarity of texts: corpora-based and knowledge-based. There is also a class of ontology-based semantic similarity measures extensively described in [10]. These measures also belong to the class of knowledge-based measures. Here we present another approach in which we classify measures it two categories: SSMs on the concept/word level and SSMs on the document/text level (Fig. 1).

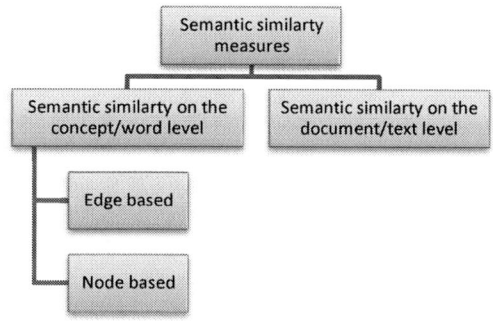

Figure 1. Simple classification of semantic similarity approaches

A. Semantic similarity on the concept/word level

Semantic similarity on the concept or word level is based on the hierarchy between concepts or words. It is usually defined for the taxonomy such as WordNet or for some more extensive ontology [10]. These measures assume as input a pair of concepts or words, and return a numeric value indicating their semantic similarity. Based on word similarities it is possible to compute the similarities of texts according to variously defined equations [13].

If we have an ontology or "IS-A" taxonomy with a set of concepts C. The similarity between two concepts can be computed by taking into account the edges (edge-based measures) or by taking into account the nodes (node-based measures).

Edge-based measures estimate the similarity of two concepts according to how near these two concepts are. The simplest way to measure semantic similarity of two concepts c1 and c2 is to estimate the distance of two concepts as the shortest-path between them [14]. There are more sophisticated approaches that take into account the possible weights of edges between two concepts. In most cases, the semantic similarity of two concepts is estimated as a function of the depth of the Least Common Ancestor (LCA) or Least Common Subsumer (LCS) [10]. An example of this approach is Wu and Palmer metrics [15] with a very simple similarity measure of two concepts:

$$sim_{WP}(c_1, c_2) = \frac{2depth(LCA_{c_1,c_2})}{depth(c_1) \cdot dept(c_2)}.$$

Node based approaches can be divided into two categories: a feature-based and one based on information theory.

In a feature based approach a concept is described with a set of features F. Thus, two concepts can be compared in terms of classical binary or distance measures. An example of this approach is the Concept-Match similarity measure defined by Maedche and Staab [16]:

$$sim_{CMatch}(c_1, c_2) = \frac{|F(c_1) \cap F(c_2)|}{|F(c_1) \cup F(c_2)|}.$$

Approaches based on information theory are based on Shannon's theory. The similarity of concepts is defined according to the amount of information they share. Resnik define a measure:

$$sim_{Resnik}(c_1, c_2) = \max_{c \in S(c_1,c_2)} [-\log p(c)],$$

where $S(c_1, c_2)$ is the set of concepts that subsume both concepts c_1 and c_2.

Pointwise Mutual Information is proposed by Turney [17]. It is based on word co-occurrence using counts collected over very large corpora. For two words w_1 and w_2, their PMI-IR is measured as:

$$sim_{PMI}(w_1, w_2) = log_2 \frac{p(w_1, w_2)}{p(w_1) \cdot p(w_2)}.$$

There are many other concept level measures, named by their authors, amongst whom the important ones are: Leacock & Chodorow, Lesk, Wu and Palmer, Resnik, Lin, Jiang and Conrath, Zhong, Nguyen and Al-Mubaid, Caviedes and Cimino, etc.

B. Semantic similarity on the document/text level

Similarity measures on the document or text level are mainly based on the approaches developed in the NLP domain. Most of these approaches have their roots in machine learning. In this section we briefly describe the most used approaches for calculating the semantic similarity between two documents.

One of the first approaches in measuring semantic similarity between documents was vector space model (VSM) originally proposed by Salton et al. [18] for the task of information retrieval domain.

In the VSM each document is a point in an *n*-dimensional space. For a given set of k documents $D = \{D_1, D_2, ..., D_k\}$, a document D_i can be represented as a vector $D_i = (w_{i1}, w_{i2}, ..., w_{in})$. In the classical word-based VSM, each dimension corresponds to one term or word from the document set. Weights may be determined by using various weighting schemes; TF-IDF is usually used in the word-based VSM. The similarity between two documents D_i and D_j can be calculated as cosine similarity:

$$sim_{cos} = \frac{D_i \cdot D_j}{\|D_i\| \cdot \|D_j\|}.$$

The main drawbacks of this model are high dimensionality, sparseness and vocabulary problems. Therefore, there are various modifications and generalisations of this classical version of the VSM.

Another approach is Latent semantic analysis (LSA) proposed by Landauer (1998). It also exploits the vector space model, but in this approach it uses a dimension reduction technique known as Singular Value Decomposition (SVD) of the initial matrix. In this way LSA overcomes the high dimensionality and sparseness of the standard SVM model. The similarity between two documents is again calculated as a cosine similarity as in equation or some other similarity measure.

One more recent approach is deep learning. Similarly with the previous methods, in deep learning documents or texts can be represented as vectors by the using document to vector technique (doc2vec). Moreover, words are also represented as vectors by using the word to vector strategy (word2vec). There are variations in learning word vector representation; one is the matrix decomposition method such as LSA; another is context-base methods such as skip-grams, a continuous bag of words. Furthermore, there is an unsupervised algorithm that learns representation for documents or smaller samples of texts (paragraphs, sentences). At the end, vectors can be compared using cosine similarity or some other similarity measure.

There are also other similar measures for text similarity such as Explicit Semantic Analysis (ESA), Salient Semantic Analysis (SSA), Distributional Similarity, Hyperspace Analogues to Language, etc.

V. RELATED WORK

There are attempts to identify plagiarism by semantic similarity measures. Researchers have approached the problem of determining plagiarism through semantic similarity in a multitude of ways. Many have used ontologies, usually WordNet with some additions like fuzzy similarity measures, a combination of WordNet with morphological and syntactic analysis, machine learning from examples or with hashing, etc. Others have turned to intrinsic methods, citation-based plagiarism detection, graphical representation, natural language processing, deep learning and even multidimensional approach.

In [19] Tsatsaronis et al. (2010) presented a semantic-based approach to text-plagiarism. Their approach is based on WordNet and the Wikipedia as knowledge bases which can resolve the problems of synonymy and hyponymy/hypernymy. Similarly, in [20] Fernando and Stevenson (2008) used WordNet as the knowledge base in the proposed algorithm for paraphrase identification.

Shenoy et al. (2012) [21, p. 59] "proposes an automatic system for semantic plagiarism detection based on ontology mapping". They invented an algorithm which is capable of learning ontology from documents using Web Ontology Language OWL and then applying it to ontology mapping to "detect correspondences between the various entities" [21, p. 60].

WordNet has been used by more than a few researchers. Al-Shamery and Gheni (2016) [22] consider that finding synonyms (over WordNet) on the same place in comparing documents is to be the proof of semantic plagiarism.

Alzahrani and Salim (2010) [23] presents "plagiarism detection method using a fuzzy semantic-based string similarity approach". They pull potential source documents using shingling and Jaccard coefficient, then compare them to the sentence granularity, simultaneously computing fuzzy degree of similarity with the help of WordNet a (different words gets 0, WordNet synonyms gets 0.5, the same word gets 1 fuzzy degrees).

Marsi and Krahmer (2010) [24] suggested the semantic similarity method for analysing comparable text that relies on a combination of morphological and syntactic analysis, WordNets, and machine learning from examples. They analyse semantic similarity between sentences "by aligning their syntax trees, where each node is matched to the most similar node in the other tree (if any)" [24].

Czerski et al. (2015) [25] proposed using an algorithm based on sentence hashing but the approach was combined with replacing some word for some representative ones using synonyms and the WordNet, thesaurus and "IS A" ontology, so the number of sentences in the document is reduced and the remaining sentences consist only from lemmas.

Eissen and Stein (2006) [26, pp. 566–567] used the intrinsic method of semantic similarity discovering: stylometry analysis, using five categories of stylometric features: "(i) text statistics, which operate at the character level, (ii) syntactic features, which measure writing style at the sentence-level, (iii) part-of-speech features to quantify the use of word classes, (iv) closed-class word sets to count special words, and (v) structural features, which reflect text organization." They also introduced a new stylometric measure: averaged word frequency class, as "the most powerful concept with respect to intrinsic plagiarism detection" [26, p. 567].

Gipp et al. (2013) [27, p. 1119] noticed that plagiarists "usually do not substitute or rearrange the citations copied from the source document", so they developed several Citation-based Plagiarism Detection algorithms using citation patterns within scientific documents "as a unique, language-independent fingerprint to identify semantic similarity" with reasonable success in detecting disguised plagiarism.

Osman et al. (2010) designed a method of detecting plagiarism based on graph representation. For two documents their method [28, p. 36] "build the graph by grouping each sentence terms in one node, the resulted nodes are connected to each other based on order of sentence within the document, all nodes in graph are also connected to top level node" which is "formed by extracting the concepts of each sentence terms and grouping them in such node".

Chong et al. (2010) [29] applied several NLP techniques on short paragraphs to analyse the structure of the text to automatically identify plagiarised texts. They proved that NLP techniques can improve the accuracy of detection tasks, although there remain challenges such as multilingual detection, synonymy generalisation (word sense disambiguation) and sentence structure generalisation.

Gharavi et al. (2016) [30, p. 1] proposed a "deep learning based method to detect plagiarism" in the Persian language. "In the proposed method, words are represented as multi-dimensional vectors, and simple aggregation methods are used to combine the word vectors for sentence representation. By comparing representations of source and suspicious sentences, pair sentences with the highest similarity are considered as the candidates for plagiarism. The decision on being plagiarism is performed using a two level evaluation method" [30, p. 1].

Mihalcea et al. (2006) presented a method that outperforms a vector-based similarity approach for measuring the semantic similarity of texts using two corpus-based and six knowledge-based measures of word similarity which they used "to derive a text-to-text similarity metric" [13, p. 775].

In [1] Kong et al. (2014) tried to detect high obfuscation plagiarism with a Logical Regression model. The proposed model integrated lexicon features, syntax features, semantics features and structure features which are extracted from suspicious documents and source documents.

VI. DISCUSSION AND CONCLUSION

In this paper we want to describe our preliminary research on semantic similarity measures and their possible usage for paraphrasing detection in the task of plagiarism identification. We analyse some existing measures for quantifying the semantic similarity of texts. There is a plethora of measures and approaches proposed and we divided them into two basic categories with some subcategories. All these measures are defined for different purposes in NLP domain. However, according to the related work, it is obvious that some of these measures can be utilized in the task of plagiarism detection.

One possible approach is based on external knowledge represented in some formalism. External knowledge can be represented in ontology or simpler taxonomies such as WordNet. However, the formalism is not limited to these classical ontologies/taxonomies; it can be any kind of graph representation of lexical relations [31]. Another approach is to use statistical methods designed in the domain of NLP. In the section about related work we present all the approaches that have been used recently for plagiarism detection.

There may be one drawback of the approach based solely on the semantic similarity measures. We need to point out that, perhaps these semantic measures are not enough, and that they can be combined with classical approaches that may identify copy-paste plagiarism.

For further research we plan to experiment with the NOK method [32] or some other graph based formalism for lexical relation representation [33]. Additionally, we would like to experiment with the approach that we try with an ontology-based information retrieval in which the classical VSM is projected onto a smaller vector space [34].

VII. REFERENCES

[1] L. Kong, Z. Lu, H. Qi, and Z. Han, "Detecting High Obfuscation Plagiarism: Exploring Multi-Features Fusion via Machine Learning," *Int. J. u-and e-Service, Sci. Technol.*, vol. 7, no. 4, pp. 385–396, 2014.

[2] Turnitin Europe, *Plagiarism in a Digital World series*. Turnitin, 2016.

[3] S. Lampret, V. Pupovac, and M. Petrovečki, "Računalni programi i programske usluge za otkrivanje plagiranja u znanosti i obrazovanju," *MEDIX*, vol. 18, no. 98/99, 2012.

[4] K. Baždarić, V. Pupovac, L. Bilić-Zulle, and M. Petrovečki, "Plagiarism as a violation of scientific and academic integrity," 2009.

[5] Council of Europe, "ETINED - The Pan-European Platform on Ethics, Transparency and Integrity in Education," 2017. [Online]. Available: http://www.coe.int/en/web/ethics-transparency-integrity-in-education. [Accessed: 22-Jan-2017].

[6] T. Birkić, D. Celjak, M. Cundeković, and S. Rako, "Izvještaj: Analiza softvera za otkrivanje plagiranja u znanosti i obrazovanju," Zagreb, Sep. 2016.

[7] S. Vuković and B. Kopić, "Plagiranje - Sveučilište u Zarebu još nije uvelo sustav provjere radova," *Global 21*, Zagreb, p. 24, Nov-2016.

[8] V. Supak Smolcic and A.-M. Simundic, "Biochemia Medica has started using the CrossCheck plagiarism detection software powered by iThenticate," *Biochem. Medica*, pp. 139–140, 2013.

[9] J. O. Shea, Z. Bandar, K. Crockett, and D. Mclean, "A Comparative Study of Two Short Text Semantic Similarity Measures," *Artif. Intell.*, vol. 4953, pp. 172–181, 2008.

[10] S. Harispe, D. Sánchez, S. Ranwez, S. Janaqi, and J. Montmain, "A framework for unifying ontology-based semantic similarity measures: A study in the biomedical domain," *J. Biomed. Inform.*, vol. 48, pp. 38–53, Apr. 2014.

[11] P. Resnik, "Semantic Similarity in a Taxonomy: An Information-Based Measure and its Application to Problems of Ambiguity in Natural Language," *J. Artif. Intell. Res.*, vol. 11, pp. 95–130, May 2011.

[12] G. Salton and M. E. Lesk, "Computer Evaluation of Indexing and Text Processing," *J. ACM*, vol. 15, no. 1, pp. 8–36, Jan. 1968.

[13] R. Mihalcea, C. Corley, and C. Strapparava, "Corpus-based and knowledge-based measures of text semantic similarity," 2006, vol. 6, pp. 775–780.

[14] R. Rada, H. Mili, E. Bicknell, and M. Blettner, "Development and Application of a Metric on Semantic Nets," *IEEE Trans. Syst. Man Cybern.*, vol. 19, no. 1, pp. 17–30, 1989.

[15] Z. Wu and M. Palmer, "Verbs semantics and lexical selection," in *Proceedings of the 32nd annual meeting on Association for Computational Linguistics -*, 1994, pp. 133–138.

[16] a Maedche and S. Staab, "Comparing ontologies-similarity measures and a comparison study," *Proc. of EKAW-2002*, no. 408, 2002.

[17] P. Turney, "Mining the Web for Synonyms: PMI-IR Versus LSA on TOEFL," 2001.

[18] G. Salton, A. Wong, and C.-S. Yang, "A vector space model for automatic indexing," *Commun. ACM*, vol. 18, no. 11, pp. 613–620, 1975.

[19] G. Tsatsaronis, I. Varlamis, A. Giannakoulopoulos, and N. Kanellopoulos, "Identifying free text plagiarism based on semantic similarity," in *Proceedings of the 4th International Plagiarism Conference*, 2010, no. i.

[20] S. Fernando and M. Stevenson, "A semantic similarity approach to paraphrase detection," in *Proceedings of the 11th Annual Research Colloquium of the UK Special Interest Group for Computational Linguistics*, 2008, pp. 45–52.

[21] M. K. Shenoy, K. C. Shet, and U. D. Acharya, "Semantic Plagiarism Detection System Using Ontology Mapping," *Adv. Comput. An Int. J.*, vol. 3, no. 3, pp. 59–62, May 2012.

[22] E. S. Al-Shamery and H. Q. Gheni, "Plagiarism Detection using Semantic Analysis," *Indian J. Sci. Technol.*, vol. 9, no. 1, Feb. 2016.

[23] S. Alzahrani and N. Salim, "Fuzzy semantic-based string similarity for extrinsic plagiarism detection," *Braschler and Harman*, 2010.

[24] E. Marsi and E. Krahmer, "Automatic analysis of semantic similarity in comparable text through syntactic tree matching," 2010, pp. 752–760.

[25] D. Czerski, P. Lozinski, A. Cacko, R. Szmit, and B. Tartanus, "Fast plagiarism detection in large scale data," 2015.

[26] S. M. Zu Eissen and B. Stein, "Intrinsic plagiarism detection," Springer, 2006, pp. 565–569.

[27] B. Gipp, N. Meuschke, C. Breitinger, M. Lipinski, and A. Nürnberger, "Demonstration of citation pattern analysis for plagiarism detection," 2013, pp. 1119–1120.

[28] A. H. Osman, N. Salim, and M. S. Binwahlan, "Plagiarism detection using graph-based representation," *J. Comput.*, vol. 2, no. 4, 2010.

[29] M. Chong, L. Speciali, and R. Mitkov, "Using natural language processing for automatic detection of plagiarism," 2010.

[30] E. Gharavi, K. Bijari, H. Veisi, and K. Zahirnia, "A Deep Learning Approach to Persian Plagiarism Detection," 2016.

[31] M. Pavlić, M. Meštrović, and A. Jakupović, "Graph-based formalisms for knowledge representation," 2013, vol. 2, pp. 200–204.

[32] A. Jakupović et al., "Comparison of the Nodes of Knowledge method with other graphical methods for knowledge representation," 2013, pp. 1004–1008.

[33] A. Meštrović and M. Čubrilo, "Monolingual Dictionary Semantic Capturing Using Concept Lattice," *Int. Rev. Comput. Softw.*, vol. 6, no. 2, pp. 173–184, 2011.

[34] A. Meštrović and A. Cali, "An Ontology-Based Approach to Information Retrieval," in *Semantic Keyword-based Search on Structured Data Sources*, 2017, pp. 150–156.

Personal Learning Environment as Support to Education

Bojan Grba*, Božidar Kovačić**
* Gymnasium Bernardin Frankopan, Ogulin, Croatia
**Department of Informatics, University of Rijeka, Rijeka, Croatia
*grbabojan@gmail.com, **bkovacic@uniri.hr

Abstract - The methodologist's cognitions achieved by experience in education resulted in the fact that the centre of education must be transferred from a teacher to a student. The result of such an approach is the development of personal environment for learning. Former generations of the learning system by which introduction of information technology is introduced into educational process points to the need for introduction of individual approach to learning by development of personal environment for learning. The results of the most important researches implemented at the European universities are presented. Current leading systems for the development of personal environment for learning and tools for creation of personal environment for learning are described. The directions of development of educational environment and guidelines for future researches are established.

I. INTRODUCTION

Transferring and acquisition of knowledge is an individual process for persons who transfer and those who acquire knowledge. It is expected that identical way of transferring the same material on the homogeneous group provides approximately the same results in learning and it is not a repeated case. Actually, individual differences among single individuals are the main reasons for that. The levels of acquisition of new knowledge are conditioned by differences in characteristics of personality, cognitive abilities, learning styles, previously acquired knowledge and skills as well as the level of motivation what is the base for further learning of new materials [1].

There are four types of learning within Croatian qualifying framework and those are:

- Lifelong Learning which denotes the all forms of learning during life and for the purpose of improving the knowledge and skills
- Formal Learning denotes an activity of authorized institution which is carried out according to the approved programs and for the purpose of improving the knowledge and skills and for which a public document is issued
- Non-formal Learning denotes the organized learning activities for the purpose of improving the knowledge and skills, but a public document isn't issued
- Informal learning denotes the unorganized activities of acquisition of knowledge and skills, as well as an adequate self-reliance and responsibility resulting from

everyday experiences and other influences and sources from an environment, for personal, social and professional needs [2].

Comparing researches on education in adult almost 70-80% of learning is acquired informally [1]. The importance of informal education regarding population is definite. Such way of acquiring knowledge, with regard to percentages isn't irrelevant, but the leading way of knowledge acquisition.

We differ three generation of Learning Management System (LMS). The first generation was only the distribution of material. It had possibility of very remote or none communication and integration way communication among attendants and teachers were developed and is still being developed in period which is very similar for all attendants without taking into consideration individual differences. From the absences of the second generation arises the third generation which allows user to create an individualized contents and transfers centralization from the position of teacher upon attendant [3, 4].

LMS systems are included into institutional education. The use of that tool initiated consideration on defects and possibilities of improving. One of more important defects is the inflexibility of the contents [5]. Development of personal environment for learning contributes considerably in elimination of such defect. The reason for the development of PLE are enormous defects of LMS what is pointed out by Comacho and Guilana in their work considering that LMS reproduces teaching where a teacher is in the center of a closed model, so attendants are joined in standardized system [6]. Dabbagh and Kitsantas pointed out the defect of the system which doesn't allow attendants to manage the education space. It is impossible to show your own works to others neither to go through other works [7].

Chapter 2 defines the term of the Personal Learning Environment (PLE) after which the approach of implementation is presented in chapter 3. Chapter 4 analyses the most important approaches to implementation of PLE. Chapter 5 contains guidelines for further work and future researches. The work is closed with chapter 6 by conclusion.

II. PERSONAL LEARNING ENVIRONMENT

PLE can be described as personal environment where somebody learns. That environment must be set and designed according to the user's needs, that is according to his style, needs and contents [3, 8]. It isn't a technological tool, but pedagogical approach which is based on technology [9].Today, PLE has become more than an approach to how people learn. It is a system which helps a student to take control over his learning. It includes setting your own goals, control of learning, communication with other participants in the learning process. Its strategy to understand and promote formal, informal and self-regulated learning of students, is of great interest. It's important to emphasize the relevance of social media in sharing learning achievements and making sense of them [7]. PLE can be compiled on one or more systems. As such it can be a desktop application or the set of systems placed on the Internet [10].

PLEM (Personal learning environment manager) is a personal learning environment that supports students in creating personalized spaces. In PLEM we distinguish four types of elements of learning: educational materials, educational services, educational experts and educational community [11].

In a personal environment a person can engage, organize, and manage formal, informal and by informal educational materials, tools and experiences.

PLE consists of a client and server parts. In the client part there are applications which communicate with web servers. Applications lead students according to increasing flexibility and effective environment. The server's part is responsible for the data that is stored locally on the PLE web server.

A. Main Function of the PLE model

Fig.1 shows a model of the main functions of the PLE, and it consists of:

- Searching- it includes searching the Internet content archives and the organization of content

- Organizing- includes labelling, reference management, archiving, either on disk or cloud.

- Creating- includes all activities that are directly related to teaching, writing, producing releases, messages, posts, blogs.

- Communicating, collaborating- interaction, discussion, debates, comments, team work.

- Publishing and sharing- The publication on the Internet when the materials and work are finished

- Project management- defines the settings of learning, rate of learning and learning goals [8].

During the time of formal and informal education the student is a part of the community which allows him to communicate with all the elements of the educational process, but in informal education student learns independently, he manages the complete the process of learning without anyone's monitoring.

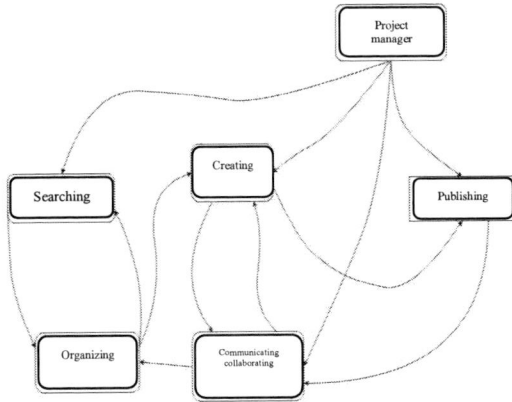

Figure 1: Main function in PLE-u [13]

PLE is neither specified nor completed system and the best is to use it in order to get sooner and easier new insights. It's the concept of using different technologies and applications in the process of independent learning, and a key determinant of the concept is that the user independently determines which applications will be used and for what purpose.

In PLE a student should use a simple set of tools, adapted to his needs and characteristics in his personal environment. The tools should enable the student to:

- Learn with other people by determining relationships with other people and other persons as well as the share of the relationship among the contacts that are not a part of his formal educational network,

- Control his resources for learning and allow them structuring, sharing, commenting on sources which were found by their side or by side of the age mates

- Manage with activities in which they participate, providing them with the opportunity to set up and join in activities such as studying groups or gathering of certain groups of people with appropriate resources,

- Integrating learning: allow them to combine learning with different institutions or allow the connection between the formal and informal education [12].

Finally, the environment has to be enough easy for use, maintenance and administration, but still has to enable the collection and linking of the resource that the user will use and store so that he could adapt the computing support to learning processes, with an emphasis on informal learning [13].

III. APPROACHES TO THE IMPLEMENTATION OF PLE

The learning process ids unique to each individual. The student's individuality is evident in creating learning plan and the selection of the environment in which he learns. If the student made the learning plan he needed the prior knowledge. Studies have shown that students most often primarily reed mentor's texts or the texts approved by the Organization, and only then resort to the individual tools. Such concept is the result of the use of classic LMS system and poor prior knowledge. Up to now, this problem is resolved in a way that a master makes a plan and an environment for what he thinks will suit students.

Linda Castenada, Jordi Adell in their work did research in the way that they wanted to meet PLE from pedagogical approach. It is important to understand the learning process that is in the background of the educational system. The author investigated how students would integrate their education process in their own environment of PLE and build PLE setting tools in it. Their idea is to understand how is the PLE organized by the students, but not by technology but with the processes that take place at students. Students who participated in the study were the first year students.

The study included 30 students who were divided into 6 different groups. All school materials were involved in the LMS SAKAI. Corse duration was limited to two weeks in which students participated three hours per week in classes in the classroom, and other obligations had to do on their own. The authors divided all processes into reading processes, creating and sharing and for each of them is made the cloud of the used technology. Examples of the clouds are shown in fig.2, 3 and 4 [9].

It is difficult to made environment that would suit all students. The solution is shown through the individual monitoring of the student movement according to classic LMS. If we keep track of the activities of attendants, we could anticipate the habits of the attendants and thus ease him to make the learning environment from which he would be allowed to return to classic LMS view. Monitoring of students and proposing the material for which it is considered to be at ease and would have helped him to create a personal environment.

There are several approaches to increasing individualization of the learning environment. PLE is considered to be the educational concept that allows the student to choose methods and tools for his education.

Figure 2: Actions reading [10]

Figure 3: Actions create [10]

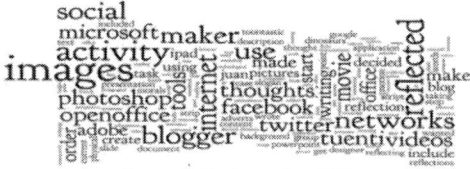

Figure 4: Sharing actions [10]

Looking from that perspective PLE looks like a tool based on the pedagogical approach rather than a technological platform.

According to Volton's attitude it is necessary to allow students to choose the learning methods and ways of learning. Looking from that attitude a personal environment is the informational-communicative technology. Thus Volton advocates PLE as a model, and not as a technological platform [14].

Castaneda and Soto consider that the personal environment should contain sources that the student uses in order to find information, relationships between information and relationships among the different sources. Their main focus is directed towards the mechanism and the tool that independently develop the environment for future learning [15].

Most authors define only-routing either as the ability necessary for the development of the personal environment or as an ability that is developed through the process of creating such environment.

Dabbagh and Kintsantes describe the correlation between self-directed learning and social networks. They develop a pedagogical framework for social networks. The aim of such a framework is to inform on the way how to make the PLE which supports self-directed learning. Their starting points were the students who were constantly viewing and sharing information using technology and thus become co-authors of the content [7].

Shaikh and Khaje consider that the process of building the PLE requires an equal participation of the students and teachers as well. Teachers don't necessarily play out all the steps at managing the learning process but the focus needs to be on collaboration with students. Also it can't be ignored competitions of teachers directed according to the nature and complexity of the tasks that students should understand [16].

Joao Paz in his research in which 20 students of the University Aberta, Portugal, took part, tried to integrate and to determine characteristics of LMS and PLE, advantages and disadvantages and ways of designing PLE by students. Total number of tools used by students was 29.

The most popular were: Skype, Youtube, Googlesearch, Side Share, Bloger, The tools that were used after its main function were used at percentage: 36% for communication, 20% for making the materials and projects, 19% for search, 11% for the Organization of the content and 9% for publishing and sharing their materials [8].

In the research that was carried out at University in Balears in Spain, students should have evaluated the use of Symbaloo EDU tools. The questions were divided in a way that it was supposed to evaluate the environment that students independently create and the environment given as a template by the University. Regarding questions on the usefulness of personal environment, about 60% of students agreed that the personal environment is useful and rated it with high grades 4 and 5. None of the respondents didn't evaluate a personal environment with the lowest grade of 1, while 5% of respondents evaluated with grade of 2. The tool also got high marks as a manager of the personal education on the course and the University. 87% of students evaluated it with the mark 4 and 5. Simplicity of using tools is evaluated with high grades in 82% of the students in a situation where they had the template made by the University, but in situation where they had to make their own personal template that percentage drops to 62% [17].

In research carried out at the University of Southampton, students use social networks to divide information 30%, 29% work in network, communication 59%, putting questions 14% and replying to questions 10% [18].

Social networks analysis is a social analysis of the internet. It is a method which attempts to understand the formal and informal connection among social networks. In research attended by 41 professors from 12 countries the tools that were used to a greater extend were Facebook 78%, Twitter 76% and LinkedIN 68%. A tiny percentage of the used tools were Classroom 2.0, Tuenti and Google+. As the main advantage are pointed out two things:

- Sharing of the content what is useful for 63% of Facebook and 81% of Twitter users

- Learning new contents which is useful for 50% of Facebook users and 81% of Twitter.

Even 70% of teachers noted that Twitter could be used for the purpose of education while the same thing was recognized by 50% of Facebook users [18].

IV. ANALIYSIS OF DIFFERENT APPROACHES TO PLE IMPLEMENTATION

Personal environment is possible to apply through a variety of tools. When using the tools it is necessary to take care of prior knowledge of the participant as well as popularity and capabilities of the tools. Setting up the PLE is a demanding job and requires a lot of planning. The teacher should be innovative and has to know to find material which will be able to use by trainees. Moreover, he should know why PLE is necessary, he should also know how is it possible to incorporate Web 2.0 tools into the plan and learning program in order to enable collaborative learning [19].

In an effort of implementation at high school students used the most familiar tolls. So for homepage they usually chose Google, for making target folders they used Mind Meister. For making the presentations they used Prezi, while for exchanging documents they used GoogleDocs, and for exchanging videos they used Youtube. The conclusions are drawn from researches [19]:

- You don't need to overestimate the digital skills of the students, they need to prepare themselves so that they could adapt web tools for their needs and learning activities

- You don't need to overboard with the number of web tools in a short period of time

- Include students in design process allowing them to make decisions on learning activities and selection of web tools

- Explain the role of teachers and students in the process.

A. ROLE

Responsive Open Learning Environment (ROLE) is European collaborative project with 16 internationally recognized research groups from 6 EU countries and China.

The main objective of this project was to support teachers in development of personal learning environment for every student ROLE offers flexibility and personality in the content and of the overall learning environment and its functionality. It uses the open source approach, and any tool created by an individual is available for all those who learn by means of internet, regardless the subject they learn, learning environment, operating system and device they use.

ROLE covers five main objectives:

- Supports individual use of available educational services, tools and resources

- For research and development of psycho-pedagogical sound framework

- For the creation of new engineering methodology

- For development and maintenance of valuation methodologies

- For the exploitation and expanding of the result of the ROLE.

For students and teachers ROLE infrastructure allows the targeting the students to PLE and so reinforces the user's journey towards lifelong learning across the institutions.

B. PLEF

Personal Learning Environment Framework (PLEF) system has been developed at the University in Achenu. The system enables the aggregation of data from multiple sources. It supports pupils in taking control over their own learning through management, tagging, commenting, and sharing their favorite materials in a personalized environment. PLEF is prominent among the following items:

- Uses Open ID

- Supports commenting and sharing all the PLE elements

- Access control is defined on each page separately

- Beside traditional approach PLEF offers a view of all the elements. Students can add tags to make the materials classified, categorized, searched and found again

• Supports lateral navigation bar where you can 'drug and drop' method drug the new element or modify the order.

Students can divide the complete content according to privacy policy and they have insight in all of their materials and those which are publicly available and which are created by other students.

C. ELGG

ELGG is free open source tool developed in New Zeland. It is based on open standards and open APIs allowing users to improve it with their own favorite tools. Courses and materials are directly displayed in ELGG. One of the major advantages is the possibility of maintenance ELGG account and when the students leave the institution. In that framework it is possible to build all kinds of social environment from the campus of the wider social networks for the university, school or college.

ELGG is free program that must be installed on your computer. It uses the GPL and MIT licenses which enable that content which is made independently by user could be sold but the client who buys that content can redistribute it.

D. Comparison of the tools

In table 1, the tools for creation of PLE are compared. The tools are compared by properties: client's technology, identity verification, possibilities of multiple identity, commenting, user's tags, learning plan, ways of communicating and the status of the tool on the basis of production.

TABLE 1: COMPARISON OF THE TOOLS

	ROLE	PLEF	ELGG
Technology client	-	Ajax	Ajax
Identity verification	project	OpenID, vlastita	OpenID, vlastita
Multiple identities	No	No	No
Commenting	Yes	Yes	Yes
Username	Yes	Yes	Yes
Learning Plan	Yes	No	Yes
Synchronous communication	Yes	Ne	Yes
Asynchronous communication	Yes	Yes	Yes
Status	Project	Under construction	Functionally

V. DIRECTIONS OF PLE DEVELOPMENT

Educational environments are becoming more flexible and more decentralized. In that development, all the processes are directed towards students and their interaction. A large number of the authors explore the tools and systems which are the most common in the student's choice. If we understand the path of construction of student's personal environment it is possible to

understand his style of learning and adjust our methods to those styles.

We can conclude how the student's learning process has the following phases:

• Students first read professors' documents

• Engage in independent creation and solving the problems

• They have need to share their work with others.

Changing educational process can be motivating and prosperous process, but at the same time even very responsible process. The result of that educational process is hardly measurable size and it is showed through a certain period of time but not through one variable.

Pascoa wonders whether we understand how quickly and dramatically the world changes and do we understand that in ten years today technology will be out of date. Furthermore, he claims that the top ten most popular jobs in 2010. didn't even exist in 2004. [20].

In consideration with the fact that education must be available 24 h a day continuously with social connections.

Downes (2010.) points out that teacher should change himself in the way to become more varied and that his action is spread in different functions. Teachers need to make the most of desirable technology and be aware of the importance of pedagogy so that it can effectively support the learning in the new context [21].

Lagarto (2012) believes that an educational strategy dramatically changes its paradigm and today offers broad opened options. He sees teachers as persons who will need to have also the good functions of the on line content manager and at the same time the environment that is most appropriate for his students [22].

Attwell (2012.) the future of e-education is seen not only in involving of that process in basic, secondary and high education (formal/informal) but as a part of the long life educational process [23].

In research of materials used by students at SAPO campus, it was shown that students chose materials and platforms which were supported by the institutions. They believe in the institutions and their selection and only when they have control over content engage in the independent selection of applications and other systems [24].

Two interesting results are shown in the CAPPLE, four-year research project. The majority of students use strategies related to work on paper much more than digital ones. Some changes are evident in the trends of use of online tools, among senior university students. One of the main changes relates to the preference for using personal instant messaging, rather than forums [19].

Guidelines for further work are directed towards the individual approach to learning and thus according to learning styles. The content that students review in classic LMS system or on the internet is necessary to divide into several groups. Division of materials is possible according to:

- Type – categorization of materials according to look: picture, video, text, animation….

- Importance – the ranking of materials by the institutions

- Successfulness – other student graded materials through which they passed.

The biggest disadvantage in the presented research, as the students evaluated, was the creation of the environment itself. If the system would follow the user and suggest him the further steps, look and materials, it would greatly assist the attendant. According to user's opinion in that way would be ironed out the biggest former disadvantage. Such categorization of information is possible through evaluation of the materials by students who spend their time on it but also the success of process. The success of the material is difficult measurable for reason that we can never be sure which material was that one which most helped the attendant to come to the finish line. However, the classification according to parameters, time he spends on certain materials, or his subjective grading information as well as the grade of authorized institutions, if you bring that into relationships, certainly has some weight.

From such a system would be possible even analyses of LMS, PLE and other tools in pedagogical researches from which it would be possible to run the methods and the ways of work of the pupils and attendants.

VI. CONCLUSION

Today, students have the possibility and opportunity to participate in education that is no longer controlled by the institutions. In that way students build and then upgrade their own identity.

PLE includes technological and methodological processes. PLE isn't a technological tool but a pedagogical approach that is based on technological approach. That means that it is necessary to develop a system that will track movements and a student's knowledge that is taking place in the monitoring of the materials by tutors, of renowned institutions and other relevant sources. Such knowledge would be of great benefit primarily the student who would structure their knowledge all in one place with further links and connections. Beside students, the system would be of great benefit to educators who could track the habits and movements of students from which could be developed concrete theories about acquiring new knowledge and upgrading the existing ones.

The system should be completely individualized but should enable to suggest some trusted paths. The attendants would be self-rated contents according to relevance and the system would suggest them the content after efficiency. Linking of learning materials with other students from different groups would set up a system without limits and restrictions.

Acknowledgement

This work has been fully supported by/supported in part by the University of Rijeka under the project number [13.13.1.2.02].

VII. REFERENCE

[1] J. Nakić, »Prilagođavanje sustava za upravljanje učenjem individualnim razlikama među korisnicima,« *Prirodoslovni - matematički fakultet, Sveučilište u Splitu*, 2011.

[2] R. Fuchs, S. Uzelac, M. Tatalović, N. Zorić, M. Dželalija, »Hrvatski kvalifikacijski okvir,« *Vlada Republike Hrvatske, Ministarstvo znanosti obrazovanja i sporta*, 2009..

[3] D. Kalapić, K. Žubrinčić, »The Web as personal learning environment,« *International Journal of Emerging Tecnologies in Learning*, 2008..

[4] A. Hernandez, A. Robles, R. Salvador, »Open Service oriented platforms for personal learning environments,« *IEEE Computer Society*, 2013..

[5] J. Mott, »Envisioning the Post-LMS Era: The open learning network,« *EDUCAUSE Quarterly vol. 33*, 2010..

[6] G. Simens, »Learning of management systems,« *Learning Tecnologies Centre*, 2006..

[7] A. Kitsantas, N. Dabbagh, »Personal learning environment, social media and self-regulated learning: A natural formula for connecting formal and informal learning,« *Internet and Higher Education 15*, 2012..

[8] J. Paz, »First time building of a PLE ina n ICT Post Graduation Course: Main Function and Tools,« *PLE Conference*, 2012..

[9] J. Adell, L. Castaneda, »Future teachers looking for their PLEs: The personalized process behind it all,« *PLE Conference*, 2012..

[10] M. Harmelen, »Personal learning environment,« *ICALT*, 2006..

[11] U. Schroeder, H. Thus, S. Dakova, M.A. Chati, »Harnesing collective inteligence in PLE,« *12th IEEE International Conferenceon Advances Learning Tecnologies*, 2012..

[12] P. Beauvoir, M. Johnson, P. Sharples, S. Wilson, O. Liberi, Colin D. Miligan, »Developing a reference model to describe the personal learning environment,« *Inovative approaches for learning and knowledge sharing volume 4227 of the series lecture Notes in Computer*, pp. 506-511.

[13] K. Žubrinčić, »Programska potpora stvaranju osobnog okoliša za učenje,« *Fakultet elektrotehnike i računalstva*, 2010..

[14] S. Hacklin, P. Dillon, M. Kukkonen, A. Hietanen, T. Valtonen, »Perspectives on personal learning environments held by vocational students,« *Computers & Education*, 2012..

[15] J. Soto, L. Castaneda, »Building personal learning environments by using and mixing ICT tools in a profesional way,« *Digital Education Review*, 2010..

[16] S. A. Khoja. Z. A. Shaikh, »Role of teacher in personal learning environment,« *Digital Education Review*, 2012..

[17] J. Salinas, B. d. Benito, V. Marin, »Using SymbalooEDU as a PLE organizer in higher education,« *PLE Conference*, 2012..

[18] G. Gosseck, C. Holotescu, M. Ivanova, »Analysis of personal learning networks in support of teachers presence optimization,« *PLE Conference*, 2012..

[19] M.Paz Prendes Espinosa, L. Castaneda, I. Gutierrez, M. del Mar Roman, »Still far from personal learning: Key aspects and emergent topics about how future professionals' PLEs are,« *digitalEDUCATION, No 29*, pp. 15-30, 2016..

[20] S. Downes, »Learning networks in practice,« *Emerging technologies for learning*, 2007..

[21] R. Pascoa, S. Lagoa, F. Brogueira, J. Mota, »Pedagogical practices, personal learning environment and the future of eLearning,,« *PLE Conference* , 2012..

[22] J. R. Lagarto, » b-learning in a Distance Learning,« *Journal of postsecondary education and Disability*, 2013..

[23] G. Attwel, »Personal learning environments - future of elearning?,« *eLearning Papers*, 2007..

[24] M. Arosta, L. Pedro, C. Santos, A. Moreira, »Building Identity ina n Institutionaly Supported Personal Learning Environment – the case of SATO Campus,« *PLE Conference*, 2012.

Free and open source software in the secondary education in Bosnia and Herzegovina

M. Pezer, N. Lazić**, M. Odak*,

*Faculty of Philosophy,Department of Information Science, Mostar, Bosnia and Herzegovina
**Faculty of Humanities and Social Sciences, Department of Phonetics, Zagreb, Croatia
mario.pezer@posteo.de; nlazic@ffzg.hr; odakmarko@yahoo.com

Abstract - Significant benefits, like an IT cost savings, security, customability, knowledge sharing, can be achieved through the introduction and use of Free/Libre and Open Source Software (FLOSS) in high schools. By using FLOSS, level of students' innovation, computer literacy, knowledge and creativity could be increased. That benefit could be important for developing countries as a chance for strengthen local IT industry. This paper shows differences regarding FLOSS and proprietary source software treatment in a computer science classroom in high schools with curriculum in Croatian language in Bosnia and Herzegovina (BiH). In this paper survey of computer course teachers' interest and their knowledge regarding use of free and open source software was conducted by on-line questionnaire. The FLOSS presence regarding operating systems and office package at computer classrooms was also explored. An additional analysis of curriculum in secondary education in schools with curriculum in Croatian language in Bosnia and Herzegovina was performed.

I. INTRODUCTION

Free and open source software (FLOSS) gives the user the freedom to use, copy, distribute, examine, change and improve the software. These rights are stipulated in licenses like the General Public License (GPL) which is created to protect access to the source code and free use in contrast to copyright that protects the property rights of software. GPL represents a copyleft – protects the freedom of copying, distribution and modification of programs [1].

FLOSS has become very popular in public sector of EU countries in last fifteen years, but its usage in schools is almost nonexistent. EU tries to encourage the FLOSS usage in its administration.

The European Commission has its strategy for internal use of Open Source Software and plan to increase the role of this type of software internally. The Commission will ensure that open source solutions and proprietary solutions will be assessed on an equal basis, being both evaluated on total cost of ownership, including exit costs [6].

FLOSS has many advantages in comparison to commercial software. Significant financial savings and the reduced share of illegal commercial proprietary software can be achieved through the introduction and use of Free and Open Source Software. Also, by using FLOSS older computers can be re-used and newer will work faster and better. FLOSS expands the horizons of knowledge. By

using FLOSS the latest information and communication technologies become available to everyone, regardless of their financial status [2].

FLOSS can provide a chance to develop a local IT industry and to develop young FLOSS specialist who can offer their services on EU job market. There is a significant growth potential for IT service providers and demand for SME use and develop FLOSS.

According to Ghosh, 54% of firms employing software developers have FLOSS in production. Worldwide, 71% of developers use FLOSS, and 68% of firms from IT-intensive sectors in Europe incorporate FLOSS in their own software-based products [4].

There is no research regarding FLOSS usage in secondary education in Bosnia and Herzegovina, but similar researches of FLOSS usage in primary education were done in Croatia.

According to Croatian results on primary education, there is no reason why pupils should not use non-commercial and even open source software for every task in their computer science courses. The problem lies in schools: each school (elementary, secondary) has a freedom to make their curricula the way they feel fit [3]. One of the main problems for ICT teachers is that syllabi are based entirely on the commercial proprietary software. In contrast to that, the primary education curriculum of the Republic of Croatia does not prescribe, order or decree any specific software for use in teaching [2].

Unfortunately, the vast majority of schools at Bosnia and Herzegovina do not have the financial possibility to invest in new ICT solutions. Having in mind FLOSS advantages for developing countries and a poor economic standard at BiH, and constant requirements for cost reduction in public sector, we wanted to analyze the usage of software in secondary education based on sample of three cantons in BiH. In addition, the paper provides an analysis of education curriculum in secondary schools in Croatian language in BiH regarding specific software for use in teaching. Moreover, the paper shows the analysis of survey on teachers' attitude and knowledge regarding FLOSS adoption in schools. These experiences are valuable for stimulating teachers and school managers to increase usage of FLOSS at schools.

Hypothesis was that use of FLOSS in secondary education could increase students ICT knowledge and reduce IT costs, which would help local economy. If schools do not have to pay for software then this money could be spent for IT innovation, student projects and IT services to local IT SMEs, which could push local economy.

II. SURVEY

A. Methodology

The data is collected through an on-line questionnaire on a representative sample on secondary school with curriculum in Croatian language (January 2017 – February 2017). The sample included secondary schools in three cantons in BiH: Herzegovina Neretva canton, West Herzegovina canton and Herceg-Bosnia canton (also known as Canton 10). Main reason for such sample selection is because in those cantons are majority of secondary schools with curriculum in Croatian language in BiH. These schools were the main target of this survey. There are 41 public secondary schools with curriculum in Croatian language in BiH, out of which 30 schools are located in aforementioned three cantons. The survey has been conducted on sample of 20 computer teachers who teach in 22 secondary schools which present a 53% of all secondary schools with curriculum in Croatian language in BiH. A 65% of respondents were from secondary schools located in Herzegovina Neretva canton, 20% were from West Herzegovina canton and 15% of respondents were from Herceg-Bosnia canton.

Online questionnaire was presented to computer course teachers with eight questions grouped in line with survey purpose: computer course teachers' attitude regarding FLOSS, FLOSS treatment in secondary education curriculum, FLOSS usage vs commercial software at classroom related to office packages and operational system, obstacles for FLOSS adoption at schools.

Computer course teachers attitude regarding FLOSS were tested with four questions (1,2,3 and 4). For first question the Lickert scale was used and second to fourth were closed answer type (for two questions multiple selection were offered). FLOSS treatment in secondary education curriculum was tested by fifth question which was closed answer type (four answers were offered plus other answer as option). FLOSS usage at classroom was tested by sixth and seventh question. For each software type (office packages and operational system) teachers were presented with several names of programs with multiple answer option. The participants were asked to select software used for computer course at their schools. Obstacles for FLOSS possible increased FLOSS adoption at schools were tested by eight questions which were single selection answer type.

B. Result and discussion

The results were then collected and each answer counted, analyzed and displayed as a chart. The most of computer course teachers are interested in FLOSS

(interested 85% and very interested 10%), but 5% of respondents have no information about FLOSS (Graph 1).

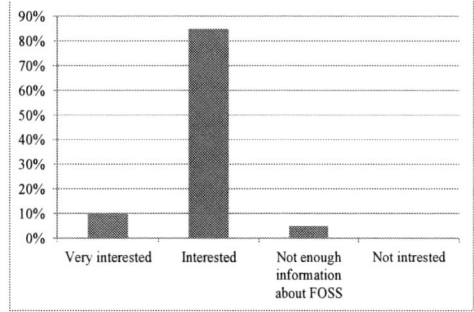

Graph 1. Number of computer course teachers interested in FLOSS

While being interested in FLOSS (95%), only 55% of computer course teachers use FLOSS in the classroom teaching (Graph 2).

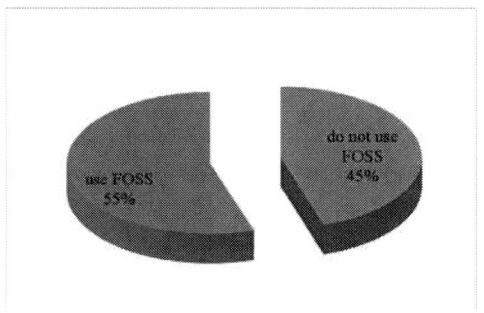

Graph 2. Computer course teachers using FLOSS at classroom

The main reasons for using FLOSS (Table 1) are the following: it is free (37%), it is available (33,3%) and quality of FLOSS software (25,9%). Reasons like openness of its source code and security have no or low importance for respondents.

Table 1. Main reasons for using of FLOSS

No.	Reason	%
1.	It is free	37,0%
2.	Available	33,3%
3.	Quality	25,9%
4.	Better than commercial software	0,0%
5.	Security	3,7%
6.	Access to source code	0,0%
7.	Other	0,0%
	TOTAL	**100,00%**

According to responses, main reasons for not using FLOSS in teaching process (Table 2) are: teachers have obligation by curriculum to teach only commercial software (25%), teachers do not need FLOSS because they have existing quality commercially software (18,8%), lack of books and training regarding FLOSS (25%) and

883

teachers do not have enough knowledge about FLOSS (12,5%). Other important reasons against using FLOSS are: a lack of customer support and IT specialist on local market (6,3%), incompatibility to commercial software formats (6,3%) and other reasons (8,3%). Other reasons are related to lack of quality hardware and computer course classes.

Table 2. Main reasons against using of FLOSS

No.	Reason	%
1.	we do not need FLOSS due to existing commercial software	18,8
2.	we have obligation by curriculum to teach only commercial software	25,0
3.	incompatibility to commercial software formats	6,3
4.	not enough knowledge about FLOSS	12,5
5.	not enough books and training regarding FLOSS	25,0
6.	not enough customer support and IT specialist on local market	6,3
7.	quality of FLOSS is lower than its commercial software	0,0
7.	other	6,3
	Total	**100,00**

Table 3 shows results related to FLOSS treatment in computer course curriculum. The respondents do not have clear opinion on this point. The majority of respondents said that FLOSS usage is not clearly defined by curriculum (40%) but the significant part of teachers consider that curriculum requires explicitly teaching on commercial software (35%). Also, 20% of respondents consider that curriculum equally treated FLOSS and commercial software and 7,1% of teachers said do not know.

Table 3. FLOSS treatment in computer course curriculum

No.	Reason	%
1.	curriculum explicitly requires teaching on commercial software	35,0%
2.	curriculum equally treated FLOSS and commercial software and give the teacher for right to choose software	20,0%
3.	FLOSS usage is not clearly defined by curriculum	40,0%
4.	I do not know	5,0%
5.	Other (please specify)	0,0%
	Total	**100,00%**

Education in BiH is decentralized and it is responsibility of regional authorities (cantons, entities). On the state level there is Agency for pre-primary, primary and secondary education (APOSO) as an advisory body. APOSO is responsible for advising, monitoring, promoting and developing the common core curricula as minimum nucleus of all existing regional curricula, syllabus and teaching plans. The Common Core Curricula in Technics & IKT [7] advice and defines a minimum of topic which has to be introduced by computer course teaching plans and local syllabuses, but regional authorities make final decisions and create syllabus and curricula.

Considering that the scope of survey was FLOSS at secondary schools with curriculum in Croatian language, we examined the syllabi and curricula for computer course

only in secondary schools with curriculum in Croatian language curriculum in BiH (based on sample of schools at Herzegovina Neretva canton, West Herzegovina canton and Herceg-Bosnia canton).

Computer course syllabus exists only for gymnasium. Other secondary schools, except gymnasiums, do not have unique syllabus and every canton and each school have its own curricula for computer course.

Analysis showed that the Syllabus for Gymnasiums with curriculum in Croatian language in BiH explicitly prefers commercial software (regarding operation system and office packages). Mentioned commercial software is property of the one foreign company. Furthermore, the Syllabus does not allow a lot of possibility to teach using FLOSS.

Although other secondary schools do not have strict rules, they also prefer commercial software in the classroom, according to survey results in Tables 4 and 5. Computer teachers have the right to choose software for teaching and they mostly prefer a commercial software.

On the other side, there is no mention what type of software should be used in The Common Core Curricula in Technics & IT (http://www.aposo.gov.ba/en/2015/10/15/development-of-the-common-core-curricula-in-technics-ikt/). This document defines that computer science courses in secondary school should, among other things, result with a following learning outcomes: students have to be enable use the software for word processing, spreadsheets, presentations, and for modeling, making and using of databases. The Common Core Curricula in Technics & IT only says what kind of outcome should be reached, so there is no reason why students should not use FLOSS in their computer science courses.

Commercial software are widespread in operating systems area, based on the data in Graph 3 (commercial software are present on 91,4% desktop at classrooms and Linux is present on 8,6% of classroom as a dual boot operating system). Students are taught Linux only on extra computer course classes at two schools while in regular computer course classes they are taught proprietary operating system.

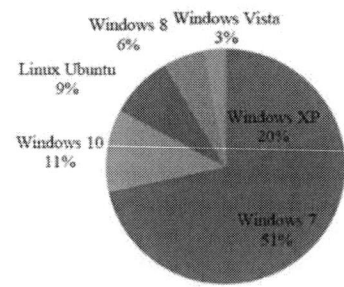

Graph 3. Operating systems on desktops at computers classroom

Almost identical result is on office packages on computers classroom in high schools (Graph 4). Commercial software is installed on 88,7% desktops and open source software packages (LibreOffice/OpenOffice) are present on 11,3% desktops (only present at two schools).

Although, a half of respondents use FLOSS at classroom its usage regarding operating systems and office package are a low.

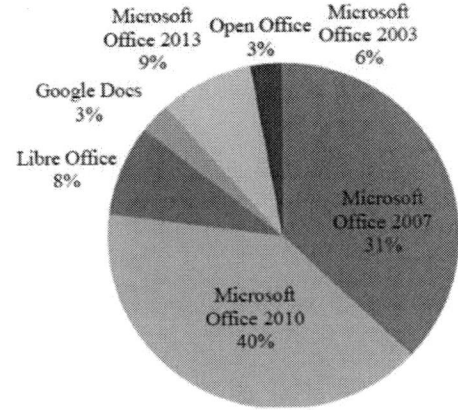

Graph 4. Office packages on desktops at classrooms

Graph 5 shows that the most important precondition for an increased adoption of FLOSS software in teaching is an increased number of computer course classes (37%). Those answers are related to fact that the most of student attend computer class during one or two of four years of high school. Other important preconditions are related to increasing of customer support for FLOSS at local level (18,5%) and the increasing popularity of FLOSS among students (18,5%). Furthermore, 14,8% of respondents consider that a decision of the local ministry of education is important to increase FLOSS presence at schools, while 11,1% of them consider that "teachers need to learn FLOSS and have further training regarding FLOSS".

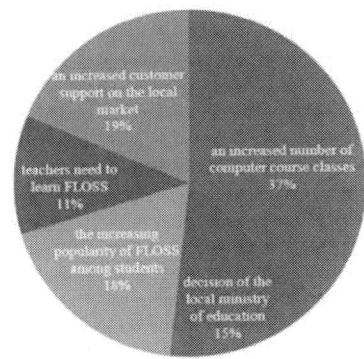

Graph 5. Precondition for an increased adoption of FLOSS software in teaching

III. CONCLUSION

Despite the fact that the computer course teachers are interested in FLOSS, its usage in classroom at secondary schools with curriculum in Croatian language in Bosnia and Herzegovina is scarce. Key obstacles for higher FLOSS usage at secondary schools are: insufficient number of computer course classes; gymnasium syllabus requires usage of commercial software; lack of knowledge and popularity of FLOSS and lack of activities by local ministries of education to introduce FLOSS in schools. As a result of obstacles mentioned above commercial software is preferred by most of teachers.

By unequal software treatment teachers and students are not given an opportunity to choose software to be used for their own learning, work and play. Consequently, a lower students' computer literacy and their IT skills will be achieved. Teaching only using commercial software results with the increased IT cost in secondary education. Also, without right to choose students are taught only for software consuming instead to encourage software developing among students through access to source code.

Considering that the most of computer course teachers are interested in FLOSS, FLOSS software should be introduced into curriculum and classroom in secondary education. But before that, number of computer course classes should be increased. An increased number of classes and teaching using FLOSS provide opportunities for developing of students advanced IT skills (like coding or system administration). Secondary education mustn't teach students to use only commercial software of only one multinational company. Today there are so many quality open source software that is used for educational purposes. These measures provide an opportunity to increase students' knowledge, ICT literacy and also reduce IT costs at schools. Savings on an IT costs could be spent for IT innovation, student projects and IT services to local IT SMEs which could push local economy.

Teachers are mostly required to use commercial software in already constrained time reserved for ICT classes in the Curricula. FLOSS is left only to teachers' enthusiasm in extra classes offered to pupils in some schools in Bosnia and Herzegovina.

REFERENCES

[1] "What is free software?" avalable on: <https://www.gnu.org/philosophy/free-sw.en.html>

[2] P. Oreški, V. Šimović, "Reasons for and against the use of free and open source software in the primary education in Croatia" , in Croatian Journal of Education, vol: 14 (1/2012), pp. 11-23.

[3] N. Lazić, J. Klindžić, M. Banek Zorica, "Open Source Software in Education" in M. Čičin-Šain, I. Uroda,I. Turčić Prstačić, I. Sluganović, (edit.). *Proceedings of the 34th International Convention MIPRO 2011: Computers in Education,* Rijeka, Croatian Society for Information and Communication Technology, Electronics and Microelectronics - MIPRO, 2011, pp. 311-314.

[4] R.A.Ghosh,,,Study On the Economic Impact of Open Source Software on Innovation and the Competiveness of the Information and Communication Technologies (ICT) Sector in the EU", UNU-

MERIT, Nederland, 2006., pp. 91–95., available on: <http://ec.europa.eu/enterprise/sectors/ict/files/2006-11-20-flossimpact_en.pdf>

[5] "Syllabus for Gymnasium on Croatian in Bosnia and Herzegovina", 2013, pp. 385-402., available on: < http://www.mozks-ksb.ba/Dokumenti/OpciDokumenti/Nastavni%20plan%20i%20program%20na%20hrvatskome%20jeziku%20za%20gimnazije%20%20u%20Bosni%20i%20Hercegovini.pdf>

[6] "Open source software strategy", available on: <http://ec.europa.eu/info/european-commissions-open-source-strategy_en>

[7] "Zajednička jezgra nastavnih planova i programa za tehniku i informacijske tehnlologije" (eng. Common Core Curricula in Technics & IKT), Agencija za predškolsko, osnovno i srednje obrazovanje (eng. Agency for pre-primary, primary and secondary education), Mostar, 2016, pp 15-23., available on: <http://aposo.gov.ba/wp-content/uploads/2016/ZJNPP%20za%20Tehniku%20i%20IT%20-%20H.pdf>

Gap in pagination due to withheld paper.

Pages 887-890

Research Methodology in the 21st Century

Dragan Cvetkovic* and Branko Medic**
* Faculty of Education in Sombor /Department of Informatics and media, University of Novi Sad, Serbia
** Technical College of Applied Sciences, Subotica, Serbia
dcveles@gmail.com, brankomedic@gmail.com

Abstract: Within this paper, we consider recent developments in methods of research and consider the context of how students are introduced to this compared with late 20th Century methodologies. We compare these strategies and consider why changes of approach have occurred and what this means for the quality of research. Finally, we suggest two further routes that will help in this analysis which are likely to provide yet more insight in to how to develop better research strategies.

1. INTRODUCTION

Work looking in to the development of research across the last few decades is an area of great interest to both students and teachers alike. From the student point of view, there is often a daunting step into their first steps of research – most often it is through a doctoral program (which we cover in predominantly in this paper). Examples of these of these kinds of trepidations might include the body of previous material to consider being excessive or there may be no simple conclusions that can be drawn in a relatively concise way which can be presented appropriately. To overcome these, universities throughout the world have developed their programmes to ensure a broader introduction. While the stereotype of academics dumping papers on a student's desk and asking for a report in 6 months are clearly overblown (but the adage that there is no smoke without fire also runs true), there is clear evidence that this style of research is less productive than that which is continuously developing in today's programmes. Hence there is a keen relationship to be fostered in a better structure to doctoral degrees between: universities, whose goal is to produce quality research; academics, whose own work can be enriched by being challenged to prepare students for the rigours of an academic career; and, students, who will gain a more fulsome education designed to get the best out of them. And so, the cycle can continue as those students go on to become the academics who supervise future students, and they are employed by the universities keen to develop a research portfolio which brings the

rewards in various guises. Within this paper, we will reflect on recent developments in the approach to doctoral studies and suggest further improvements to the work being developed.

2. A TYPICAL PhD PROGRAM

In previous years, it was common to finish a master's degree and then for those who wished to continue in their studies, the next four or five years would be spent working towards a doctoral degree. Those five years were spent learning new materials, teaching undergraduates, working as a lab assistant, and doing research for a dissertation. According to the science and communications blog Figure One [1], a decade ago it was very rare for a dissertation program to take longer than seven years. Today however, a person entering a doctoral program can expect to spend at least six or seven years earning that degree. In Humanities, as seen in [2], this can be extended to over 9 years, while for an Educational doctorate, the length of time can be just shy of 13 years. In the US, the NSF commission an annual report, in [3], which compares from 1957 to 2015 both the number of PhDs awarded and the length of time PhD completion takes. It is notable that in the 15 years of data from the year 2000, to see that while more PhDs, by mean, per institution, are being awarded, compared with the previous 43 years, that the length of time taken has gradually increased. There are 3 questions which suggest themselves to this: firstly, why are PhDs taking longer?; secondly, what conclusions can be drawn in terms of the types of graduates who complete their Bachelor's or Master's degree that continue to pursue a PhD?; and, finally, what can be done to improve the preparations for research being undertaken to ensure better quality research is produced in a quicker timeframe?. I will deal with the first of these questions last as it is important to consider overall educational changes as well as other factors once we have the answer to the other two questions.

3. CHANGES TO A PhD PROGRAM SINCE 2000

For our purposes, the systems in the Europe and the USA (which make up the majority as a percentage of PhDs produced) are the obvious places to start. In [4], the authors look at, and provide tips on, PhD research and what will be involved in the process. Depending on the student, there are various ways to enable the study of a PhD. We will concentrate on those who have completed a related subject-specific undergraduate degree who then go on (either with or without a Master's degree) to study for a PhD. There are, of course, different requirements across the spectrum of research areas: for example, most Mathematics programs require a 1st Class Bachelor's Degree and a Master's degree before consideration of a PhD program, while in some other subjects, they are less stringent with entry requirements. A broad sample of university entrance requirements booklets has given an average of an Upper Second Class Bachelor's degree as a starting point. Within this paper, this is our standard starting point. We will assume the role of a student with a 2:1 degree and consider what a PhD program entails now compared with pre-2000. Finally, we have considered 12 Universities throughout Europe and the US. These are listed in [5].

Following the standard application procedures, all of the institutions listed in [5] mentioned that eligibility criteria will be considered on a case by case basis subject to funding. When pressed, this suggested that those with private funds, to pay for the teaching element and other usual research costs, may be able to secure a PhD place despite not meeting the published eligibility requirements in usual documentation such as prospectuses.

Many European PhDs in Science are funded by EPSRC [6]. Funding is awarded based on merit of the applicant with support from supervisors and institutions will often bid for a pot of money which they can distribute through the usual procedures. The level of PhD funding has not grown as much as the growth in terms of numbers of students studying for PhDs and the likely conclusion from this is that the private funding options for a PhD will have increased over the past 15 years.

After an undergraduate degree, many academics will admit that many students are not at a level to immediately start reading relevant literature. Some institutions in [5], labelled with a *, have added additional teaching or research programmes to the PhD as part of a way to assist and provide support for students who need to bridge the gap between their knowledge at the start of a PhD and where they will need to be in sufficient time to submit by the time funding runs out.

As an example, take Mathematics at The University of Liverpool. At the end of their undergraduate program, all PhD students are required to complete a course during the first 2 years of their PhD which relates to both higher level Mathematics in their area (often this can take the form of tutorial type lessons with the supervisor and others in the general research area e.g. Number Theory) and then there are research projects to be completed which provide credits which must be passed before the PhD degree can be conferred [7]. Going forward, more universities are beginning to appreciate that it is not possible, given changes to undergraduate programmes, to jump in as would have occurred over 15-20 years ago.

Given that undergraduate studies have been using first semesters of first years to provide a high school equivalency programme, this has led to funding from sources such as [6] to increase their minimum funding to, firstly 4 years in around 2005, and now 5 years since 2012.

4. CRITICAL OBSERVATIONS AND COMPARASION – HAS OUTPUT IMPROVED?

One of the most difficult academic tasks is to determine the borderline between PhD pass and fail. Most PhDs will be very simple to deal with: in some subjects, it will involve providing research data and analysing it in a standard way, or providing the proof of some assertion either positively or negatively. Certainly, this deterministic PhDs are easily describable to a lay person in terms of why an award has been made. More difficult is those from [2], such as Humanities and Education, where the PhD will be dependent solely on subjective critical analysis. One conclusion might be that students are struggling to adapt to the rigours and demands of the PhD program in these areas. Others might suggest that it is simply more difficult to present new research in these areas. The conclusion that is proposed here is that developments of learning and research methodology are more designed and better interpreted by students and supervisors in deterministic PhD theses as opposed to those in Humanities and Educational Research.

There are 2 reasons for this conclusion. Firstly, by nature, it is certainly easier to contribute new analysis and data by running experiments or finding a counterexample to a proposition. These skills are developed through undergraduate learning in these subject and there are clear ways to compare with the current literature in the subject. Secondly, the processes for obtaining a PhD as outlined in section 2, are better developed to provide transferrable skills across disciplines which mimic the deterministic approach of those disciplines such as Economics, Mathematics, Physics and other Scientific subjects.

The timeframe to complete a PhD is not necessarily an indication as to the quality of research or the likelihood of a student to go on to complete a research career in the given subject. Many PhD students have completed within 3 years, been awarded their doctorate, and left academia, while many other students have completed their PhD in a much longer time and gone on to have a glittering academic career and body of research. The costs associated, however, are prohibitive to those from lower socio-economic backgrounds. This is addressed in section 4 below.

Finally, we return to our example of the student with a 2:1. Pre-2000, there is no doubt that this student would have been in a similar state to someone today in terms of reading a research article from the last year. With a better support infrastructure in place of recent times, there has clearly been a better realisation of the demands on a student at the end of their undergraduate studies. Making further progress in this arena, will unquestionably improve the readiness of a student, and is likely to enhance yet more a student with higher aptitudes than our exemplar student.

5. FURTHER DEVELOPMENTS – THE LIVING DOCUMENTS OD RESEARCH

The socio-economic backgrounds of a student are one important area of research into how research developments will occur over the next 15-20 years. If the continued growth of numbers continues, funding will be stretched further in the ways that will require gaps to be filled by other sources. Many private companies, such as those in [8], offer PhD grants and studentships associated to their area of research. The standard academic route for students is no longer the only option and this is a positive. PhDs are no longer just for future professional academics and many secondary school teachers and many industry workers, even politicians, have PhD degrees. The emphasis has shifted and this is to be commended.

We must additionally consider whether there are options for offering more incentives for the private route to be cheaper which will open the door to older researchers entering the field to contribute the experience they have. Looking solely at the age of students is not a taxing demand, but there must be more analysis on where people have come from to decide to choose the PhD route at a later stage of their working/professional lives. Also, whether this is linked to potential future earnings and future employment possibilities, should lead to companies being further supplied with incentive to offer PhD student sponsorship as this has only been a development of the last 15 years as opposed to 30-40 years ago.

6. CONCLUSION

In summary, we must consider the relative success in approach to the development of attacks at research problems into those for deterministic subjects and non-deterministic subjects. There is evidence, as presented, that for deterministic subject's, outcomes are easily determined. As an example, the finishing mark of the solution, by Wiles, and later in conjunction with Taylor, of the famous problem of Fermat, written in the margin of a book, made newspaper headlines around the work. Similarly, descriptions of the Higgs' boson, have received widespread commentary from many a non-expert. It seems, for these subject, the fact that amateurs can feel connected to the problem, leads to a greater interaction through arenas such as the media. No inference can be drawn as to the quality of the research, when comparing very different and diverse subjects, just because it makes the newspapers, but it is possible to conclude that an easily defined problem will lead to an easily presentable solution which can excite non-experts.

Finally, we must consider the impact on the interconnected chain of universities, academics and students and see whether we can make commentary on whether these are improved by the evidence we have. There is clear motivation from the universities to provide these programs and this development should be applauded for the wide-ranging types of programs available. In many cases, the better supervised student will be more productive in the future, and with the support of the university communities, this productivity can be nurtured for the better. The only conclusion to make must be that remain, as in Chapter 4, "living documents", persistently refined to provide a general outline with an individual tailoring to the students.

For research methodology, this nurturing of the triskelion of components is the most fundamental of change to delivery from universities in the past 50 years.

REFERENCES

[1] Jessica Stoller-Conrad
https://figureoneblog.wordpress.com/2013/01/22/the-5-year-phd-an-endangered-species/

[2] Council of Graduate School, 2010 Report
http://www.cgsnet.org/ckfinder/userfiles/files/DataSources_2010_03.pdf

[3] National Science Foundation, Data Tables, 1957-2015
https://www.nsf.gov/statistics/2017/nsf17306/data.cfm

[4] Estelle M. Phillips and Derek S. Pugh

How To Get A Phd: a handbook for students and their supervisors

[5] Contacted institutions:

Harvard University: http://www.harvard.edu/

Yale University: http://www.yale.edu/

Brown University: https://www.brown.edu/

Oxford University: http://www.ox.ac.uk

The University of Liverpool: http://www.liv.ac.uk

The University of Glasgow: http://www.gla.ac.uk

London School of Economics: http://www.lse.ac.uk

York University: http://www.york.ac.uk

Cornell University: https://www.cornell.edu/

University Paris-Sorbonne: http://www.paris-sorbonne.fr/

Vrije Universiteit Amsterdam: http://www.vu.nl/en/

Charles University Prague: https://www.cuni.cz/UKEN-1.html

[6] European Physical Sciences Research Council
https://www.epsrc.ac.uk/skills/students/

[7] Liverpool University PhD workshops
http://pcwww.liv.ac.uk/~pgro/All%20workshops.htm

[8] Industrial PhDs
https://www.newscientist.com/article/mg21328522-700-industrial-phds-exploring-the-dark-side/

A Survey and Evaluation of Free and Open Source Simulators Suitable for Teaching Courses in Wireless Sensor Networks

Dj. Pesic*, Z. Radivojevic** and M. Cvetanovic***

*University of Belgrade, School of Electrical Engineering, Belgrade, Serbia,
RT-RK Institute for Computer Based Systems, Novi Sad, Serbia (email: djordje.pesic@rt-rk.com)
** University of Belgrade, School of Electrical Engineering, Belgrade, Serbia (email: zaki@etf.bg.ac.rs)
*** University of Belgrade, School of Electrical Engineering, Belgrade, Serbia (email: cmilos@etf.bg.ac.rs)

Abstract - This paper attempts to give a survey of some free software tools for wireless sensor networks. It targets teachers and students who might have the need for education in this area. As a first stage in creating this survey, a process of software collection is presented. First collection step is search through the Shanghai list for the first 50 universities. From these universities, available information about used software for wsn-related courses is collected. Additionally, other wsn software surveys are considered. Then, only free, open source and available software are kept, because they are the easiest for students to get it. Then evaluation based on topics coverage and software features is done. Topics are selected from IEEE Curriculum guidelines, while features give emphasis on practical usage from user aspect. Selected topics are embedded systems, computer networks, operating system and system resource management. Features used in the evaluation are support for GUI and command line, programming language learning overhead, ease of installation, extensibility, and platform portability. As a result, following software tools are described: TinyOS, Prowler, Riot, Castalia, Avrora, Shawn, TRMSim-WSN, and Shox.

I. INTRODUCTION

Wireless sensor network (wsn) is a network consisted of a large number of "motes": tiny sensing and computing devices, often severely constrained with processing and communication possibilities and energy efficiency [1]. There are various sensor network applications: area monitoring, environmental sensing, industrial monitoring, and military surveillance. A number of nodes in the network can vary from few to several thousand. Node is typically built of the radio transceiver and the microcontroller with the sensors. Node is powered by batteries or it can harvest energy from the environment [2]. Wsn is not useful on its own. Its inevitable part is some system which collects and process information from the network. It can be a single computer system [1] or the cloud system [3]. From this stage, information is distributed to the users. Also, various data mining technologies can be used to process the collected information.

Wsn is a very attractive research area and a lot of students show interest in taking active role in it. So, there is a need for wsn courses in the university curricula. Wsns could be difficult for learning in practice, because of all issues related to sensors usage (e.g. need for specific hardware, price, low level programming). Therefore, wsn software tools are very good learning starting point.

This survey selects, describes, and evaluates software tools that could be used for educational purposes on courses related to wsn. However, it is up to an instructor to decide what topics from different knowledge areas related to wsn should be presented on the course, while this paper aims to elaborate in what degree a knowledge area could be covered by the selected tools. Wsn is not an easy-to-learn area, so it is expected from students to have some previous knowledge about programming and computer networks. In order to help students comprehend wsn challenges selected software tools should have reasonably small learning curve [4].

Because, there are plenty of wsn software tools (free and commercial, open and closed software), this survey will focus on free and open source software. Advantages of the free software are apparent: students and teachers can without any fee get the software and use it without limit. Some advantages of the open source software are large developer community, and available source code from which students can learn more than from bare usage of proprietary tools. One more criterion is that software tool should be wsn specific. Non-specific wsn software tools require an extra effort in order to get the software functional for usage in wsn area. For wsn related operating systems, one extra criterion is adopted: students must be able to run the software tools on a standard PC without using specific hardware with sensors. It means that a wsn operating system must support emulated hardware or that it can be compiled as a standalone application for PC. One practical limitation of software is support availability. If the software is not under active development, it can still be usable, if the available version is stable. But, for open source software, when the last version is old, it usually means that it is not compatible with new tools (compilers, linkers or other supporting software). If examined wsn software passes all criteria, but can't be used because of obsoleteness it will not be selected for further examination.

This paper is organized as follows: Chapter 2 contains a description of collecting process of the selected software and descriptions of the software themselves, together with usage examples. Then, chapter 3 contains software evaluation and chapter 4 concludes the paper.

II. SURVEY OF SELECTED WSN SOFTWARE TOOLS

This chapter describes characteristics of the selected software tools and gives the usage examples. First, it lists overall characteristics and then describes every software tool. Overall software information is summarized in Table I. It lists following information: An operating system on which the software can be used for development, a programming language used for their development, wsn support, and availability. Table I also includes software tools found in other surveys. Next step is to apply established criteria on the found software from the Table I. Available, free and open source software are kept for further analyzing. If selected software is an operating system, only operating systems with support for running without a real hardware are kept. Software which passed these filters will be described in following chapters. Selected software: TinyOS, Prowler, Contiki, Riot, Castalia, Avrora, Shawn, TRMSim-WSN, and Shox. As it can be seen from the Table I, there is a number of operating systems for wsn, but only TinyOS, Contiki, and Riot are now available at all. All other operating systems mentioned in Table I and [26] are too old and are not available anymore. SENSE simulator is not compatible with new gnu toolchain (the last version is from the year 2008) and WSNet can't be compiled, so they are excluded from further examination.

Analyzed software tools can be divided into two major groups: terminal-based and GUI-based. Terminal-based software tools are TinyOS, Riot, Castalia, Avrora and Shawn. GUI based software tools are Cooja simulator from Contiki, TRMSim-WSN, Prowler and Shox. GUI based simulators can be separated into two groups: With topology and status on GUI and without topology and status on GUI. Table II shows this division into groups.

Every selected GUI-based software tool has the different screen, but their screens can be generalized. Every screen has a part for simulation control and configuration, a part for simulation status and a part for topology, if present. Figure 1 and Figure 2 show two kinds of present GUI screens. Both figures have parts for software title and menu bar. Title and menu bar are software-specific. Configuration part of the screen contains various configuration fields, check boxes, etc. which enables concrete simulation configuration. Control part must contain simulation start and stop controls. Additionally, there can be other controls like a number of cycles. Status part often contains simulation logs. Additionally, there can be timing diagrams (like Cooja). Topology part contains current topology. Simulators can use this part for animating wsn nodes behavior or to display various information about concrete topology.

A. Software tools description

TinyOS is an operating system used in low-power wireless devices. Its main parts are task scheduler and

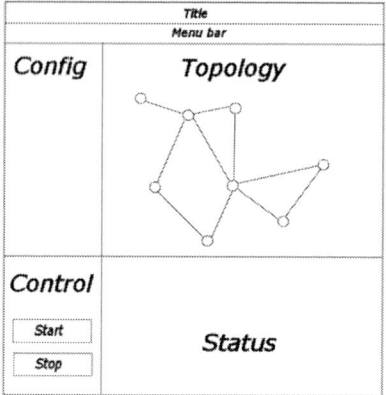

Figure 1. Screen with control, configuration status, and topology

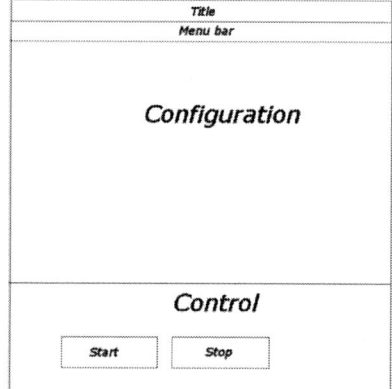

Figure 2. Screen without topology

driver collection. TinyOS is developed in NesC, the extension of C programming language. This is a component-based language with event-driven programming paradigm. TinyOS have simulator: Tossim, which simulates entire TinyOS applications. Simulation levels are very flexible: from radio communication level to component level. Tossim is a discrete event simulator and a library: a Python or C+ program for configuring and running simulation must be written.

Prowler [7] supports only Berkley Mica motes. It is event driven and has two working modes: the deterministic and the probabilistic. In the deterministic mode, it can reproduce simulation results. In the probabilistic mode, it simulates non-deterministic nature of radio communication. The simulator is developed in Matlab. Nominally, there is no limit on the number of the simulated motes.

Contiki [8] is an operating system which connects cheap low power embedded systems to the network. It completely supports IPv4, IPv6, and other low-power wireless standards. It also has a simulator, Cooja. Contiki is developed in C, and applications for Contiki are also written in C. Some characteristics of Contiki are high memory efficiency (it has own memory management, besides memory management from C), complete IP stack, power consumption tracking, wireless protocols, dynamic

linking and loading of modules, sleepy routers, efficient

TABLE I.   OVERALL SOFTWARE TOOLS CHARACTERISTICS

Software	Operating system	Language	Needs hardware	Wsn support	Availability
TinyOS [6], [12]	Linux	NesC, Python, C++	No	Yes	Open Source
Emstar [13]	C	Linux	-	Yes	No
SensorSim [14]	-	-	-	Yes	No
Prowler [7], [15]	Linux, Windows	Matlab	No	Yes	Open Source
Contiki [8][16]	Linux	C	No	Yes	Open Source
Mantis [19]	-	C	-	Yes	No
Riot [9], [17]	Linux	C/C++	No	Yes	Open Source
Nano-RK [18], [27]	Linux	C	Yes	Yes	Open Source
LiteOS [28],[29]	Linux	C	Yes	Yes	Open Source
GloMoSim [30]	Linux	Parsec	-	Yes	No
Ns2 [31]	Linux	C++, TCL	No	No	Open Source
Castalia [32], [33]	Linux	C++, NED	No	Yes	Open Source
SENSE [34]	Linux	CompC+	No	Yes	Too old
SenSim	-	-	-	-	No
J-Sim [35]	Linux, Windows	Java, TCL	No	No	Open Source
SENS [36]	-	C++	-	-	No
Avrora [37], [38]	Linux	Java	No	Yes	Open Source
Shawn [39], [40]	Linux, Windows	C++, Java	No	Yes	Open Source
ATEMU [41]	Linux	C	No	Yes	No
TRMSim-wsn [42], [43]	Linux, Windows	Java	No	Yes	Open Source
WSNet [44], [45]	Linux	C	No	Yes	Build error
Shox [46], [47]	Linux Windows	Java	No	Yes	Open Source
WSNSim	Windows	Java	No	Yes	No
SOS [48]	-	C	-	-	No
SenOS [49]	-	-	-	-	No
VM* [50]	-	-	-	-	No
MagnetOS [51]	-	-	-	-	No

TABLE II.   SOFTWARE TOOLS DIVISION INTO GROUPS

Terminal based	GUI based	
	With topology and status	*Without topology and status*
TinyOS	Contiki (Cooja simulator)	Shox
Riot	TRMSim-WSN	
Castalia	Prowler	
Avrora		
Shawn		

multithreading, file system dedicated to flash memory devices, command line shell and well developed build system.

Riot is a general purpose operating system for IoT area. It is extremely good for programmers: There is the standard set of development tools: C/C++, GCC, and GDB. An amount of hardware dependent code is minimized and particular support for POSIX are implemented. The unique characteristic is a possibility for native programming on Linux and Mac. In this usage, the whole Riot is running inside the Linux/Mac process, so development for Riot stops to be different from classic development under Linux/Mac. There is no need for virtual machines in the case of no hardware available. Riot is organized as a microkernel, maximizes energy efficiency, and supports real-time operation (extremely

low interrupt processing time and priority based scheduling). It also supports standard protocols, static and dynamic memory allocation, timers, shell, etc.

Castalia is a wsn simulator based on Omnet++ platform. It can be used for testing distributed algorithms and protocols in realistic wireless channels and radio models, with realistic node behavior. Castalia is highly parametric and can be used to evaluate different platform characteristic for specific applications. Main features of Castalia are advanced channel model based on empirically measured data, advanced radio model based on real low power radios, extended sensing modeling, node clock drift and CPU power consumption and MAC and routing protocols support. Castalia takes modules and messages from Omnet++. Module's parameters, name, structure, and interface are defined with Omnet++ NED language. Module behavior is coded in C++. Castalia cannot be used for sensor platform-specific applications. It cannot test code compiled for some specific platform, because it is designed for algorithm first-order validation, before platform-specific implementation.

Avrora is a set of simulation and analysis tools for a\ programs written for AVR microcontrollers, which are the part of Atmel and Mica2 sensor nodes. Avrora simulates and analyzes assembly programs. Main features are testing with cycle accurate execution time, monitoring, profiling and instrumenting without simulation disturbance, energy analyzer and stack checker. Avrora is a command line based simulator. It has good documentation and a rich set of options, so it is very easy to use and command outputs are simple and descriptive.

Shawn differs from other simulators with its algorithmic background instead of network stack simulation. Its main design goals are:

1) Simulate the effect caused by a phenomenon, not the phenomenon itself: For example, instead of simulating complete MAC layer, Shawn simulates only MAC layer effects on the application (packet loss, corruption, delay)

2) Scalability and support for extremely large networks: Shawn supports order of magnitudes higher number of nodes than any other available simulator.

TRMSim-WSN (Trust and Reputation Models Simulator for Wireless Sensor Networks) is a Java-based simulator aimed to test Trust and Reputation models for wsns. It allows researchers to test and compare their trust and reputation models against a wide range of wsns. They can decide whether they want static or dynamic networks, the percentage of fraudulent nodes, the percentage of nodes acting as clients or servers, etc. It has been designed to easily adapt and integrate a new model within the simulator

Shox is a discrete-event ad hoc network simulator written in Java. It is geared towards mobile wireless networks with many nodes in both indoor and outdoor scenarios. As opposed to other general-purpose simulators, which all require a tremendous learning and training effort before first results can be obtained, Shox is very straightforward to use and makes common simulation scenarios in network design like hundreds of nodes with a specific mobility generating certain traffic patterns very easy to implement. Shox has strong GUI support and offers a collection of statistical data and automatic generation of graphical representations (graphs).

III. SOFTWARE EVALUATION

This section presents the evaluation of selected software tools. Evaluation criteria are the topics coverage and the software features [4]. Because wireless sensor network is only one knowledge unit in CE Computer Networks knowledge area, all knowledge areas from curricula guidelines for computer science [10] and computer engineering [11] are searched, in order to get a broader view of topic coverage. Topics coverage for every knowledge area is calculated with the following formula: *number of related knowledge units/total number of knowledge units*. Table III shows topics coverage trough selected knowledge areas. Software tools relations to particular topics are not included in the paper for brevity. It can be noticed that CE Computer networks knowledge area has the best coverage. This is the expected result because wireless sensor networks belong to this knowledge area. It can also be noticed that Prowler simulator covers only topics from Computer networks area. This is reasonable because it simulates only radio communication. Riot has the best coverage of System resource management because of its supports real-time systems. It also has the best coverage of Operating systems area, because it supports virtualization and real-

time systems. As Castalia is platform unspecific and simulates protocols and behavior, it mostly covers Computer networks area. Avrora is code analysis tool and is dedicated to concrete platforms, so it mostly covers embedded systems area. Shawn is not platform specific and it is oriented to network applications; TRMSim runs trust & relationship models on the wireless networks so they mostly cover Computer networks area. The major drawback of Shox is the lack of documentation.

Various software features can be chosen for simulator comparison. As this survey pays attention to education, features that should primary related to education and usage had been selected. Selected features are the presence of GUI and command line support, programming languages learning overhead, ease of installation, extensibility, and platform portability. The presence of GUI is a very important feature. Most users prefer visual representation instead of the command line. Visual representation makes learning and understanding much easier, but command line gives more flexibility. The need for learning new programming languages can be unattractive for software tool because it forces the user to spend more time in preparation for usage and it can lead to less productivity. Also, when software installation is complex, users can be frustrated and they can ease give up on software usage. Extendibility gives the user flexibility for creating their own applications and models and platform portability give the commodity for using various host operating systems. Listed features are mainly related to user experience and users can make a decision which software to use, based on with what kind of listed features are they familiar. Table IV presents selected features and software tools

TinyOS is complicated for installation and requires learning of NesC programming language. Prowler requires knowledge of Matlab. Students at the department of Computer Engineering and Information Theory at School of Electrical Engineering have poor coverage of Matlab in their courses, so they probably must additionally learn it. Castalia requires learning of NED language. TRMSim-wsn is only covered software without any command line support. All covered software can be executed on Linux and they all are extensible.

Someone can tell that there is very large number of wsn software tools to be covered with one survey. A large number of them are commercial and unavailable, so contributions of this survey are:

1) A collection of free and open source software.

2) Simulators and operating systems covered in one place.

3) The inclusion of Riot operating system, which, at the best of the author's knowledge, does not exist in any other survey

IV. CONCLUSION

Wireless sensor networks are an emerging area in Computer Science and Computer engineering, so there is a need for students to be familiarized with it. The first step of education in this area probably will include usage of simulators and other software tools, so the aim of this survey is to help students with the selection of free software tools, which can help them in learning. The paper presents the process of software selection, gives the description of selected software and at last evaluates them.

TABLE III. TOPICS COVERAGE THROUGH KNOWLEDGE AREAS

Knowledge area/software	TinyOS	Contiki	Riot	Prowler	Castalia	Avrora	Shawn	TRMSim-WSN	Shox
CE Embedded systems	46.15%	46.15%	46.15%	0.00%	0.00%	23.08%	0.00%	0.00%	0.00%
CE Computer networks	55.56%	55.56%	55.56%	55.56%	55.56%	11.11%	55.56%	55.56%	33.33%
CE System resource management	28.57%	28.57%	42.86%	14.29%	14.29%	0.00%	14.29%	14.29%	14.29%
CS Operating systems	7.69%	15.38%	30.77%	0.00%	0.00%	0.00%	0.00%	7.69%	0.00%

TABLE IV. SOFTWARE FEATURES

Feature/software	TinyOS	Contiki	Riot	Prowler	Castalia	Avrora	Shawn	TRMSim-WSN	Shox
Software type	os	os	os	sim	sim	sim	sim	sim	sim
Gui and command line	CL	CL-GUI	CL	CL-GUI	CL	CL	CL-GUI	GUI	CL-GUI
Programming language learning overhead	yes	no	no	probably	yes	no	no	no	no
Ease of installation	C	E	E	M	E	E	E	E	E
Extendibility	Yes	Yes	Yes	Yes	Yes	Yes	Yes	Yes	Yes
Platform portability	L	L	L	LW	L	LW	LW	LW	LW

LEGEND:
OS: SOFTWARE IS AN OPERATING SYSTEM.
SIM: SOFTWARE IS A SIMULATOR
CL: SOFTWARE HAS A COMMAND LINE SHELL, GUI: SOFTWARE HAS GUI.
PROGRAMMING LANGUAGE OVERHEAD: YES – ALMOST ANY USER WILL HAVE TO LEARN A NEW LANGUAGE, NO – ALMOST NO ONE USER WILL HAVE TO LEARN A NEW LANGUAGE, PROBABLY: THERE IS A DECENT CHANCE THAT USER WILL HAVE TO LEARN THE NEW LANGUAGE.
EASE OF INSTALLATION: C – COMPLEX, M – MEDIUM, E – EASY
PORTABILITY: L – SOFTWARE CAN BE USED ON LINUX HOST. W – SOFTWARE CAN BE USED ON WINDOWS HOST.

Software selection is done in two phases. First, websites of the first 50 universities from the Academic Ranking of World Universities (Shanghai list) [5] are searched. Wsn related courses are searched and based on the available information, only four software tools, used at the found courses are selected. Second, other wsn software surveys [20-25] are examined and simulators and operating systems are included in this paper. The majority of them are unavailable or closed-source. At last, 3 operating systems and 6 simulators are selected for detailed description and evaluation. Every description contains basic information about software tool. Evaluation is done based on two criteria: Simulator features and Topic coverage at appropriate knowledge areas selected from Computer Engineering and Computer Science ACM Curricula recommendations.

Selected software has very different purposes and covered operating systems can serve any application planed to be executed on wsn platforms. TinyOS is the most complicated for use, while Riot seems to be the easiest. Also, TinyOS lacks a good documentation, which is the significant drawback. Command line based software proved to be more complicated for usage, but the lack of the GUI does not make it less powerful or less useful. Selected simulators are dedicated for various purposes. Castalia test first phase of algorithm development. Prowlers can quickly prototype radio communication. Shox can simulate large networks, while Avrora is dedicated for simulating and analyzing AVR assemblies

for concrete platforms. Shawn is unique with its ability for vast simulation of extremely large networks. Therefore it could be concluded that there is no one software tool or operating system that could be used for covering all topics related to wsn.

ACKNOWLEDGMENT

Work on this project was co-funded by the Ministry of Education, Science, and Technological Development of the Republic of Serbia (III44009 and TR32047). The authors gratefully acknowledge the support.

REFERENCES

[1] Sundani, Harsh, et al. "Wireless sensor network simulators a survey and comparisons." International Journal of Computer Networks 2.5 (2011): 249-265.

[2] Niyato, Dusit, et al. "Wireless sensor networks with energy harvesting technologies: A game-theoretic approach to optimal energy management." IEEE Wireless Communications 14.4 (2007): 90-96.

[3] Dash, Sanjit Kumar, et al. "Sensor-cloud: assimilation of wireless sensor network and the cloud." International Conference on Computer Science and Information Technology. Springer Berlin Heidelberg, 2012.

[4] Nikolic, Bosko, et al. "A survey and evaluation of simulators suitable for teaching courses in computer architecture and organization." IEEE Transactions on Education 52.4 (2009): 449-458.

[5] Academic Ranking of World Universities 2016. Available: http://www.shanghairanking.com/ARWU2016.html

[6] TinyOSoperating system page. Available: http://tinyos.stanford.edu/tinyos-wiki/index.php/Main_Page

[7] Simon, Gyula, et al. "Simulation-based optimization of communication protocols for large-scale wireless sensor networks." IEEE aerospace conference. Vol. 3. 2003.

[8] Contiki operating system page. Available: http://www.contiki-os.org

[9] Riot operating system page. Available: https://riot-os.org/

[10] Computer Engineering Curricula 2016. Available: http://www.acm.org/education/CS2013-final-report.pdf

[11] Computer Science Curricula 2013 Available: http://www.acm.org/binaries/content/assets/education/ce2016_final.pdf

[12] Levis, Philip, et al. "TinyOS: An operating system for sensor networks." Ambient intelligence. Springer Berlin Heidelberg, 2005. 115-148.

[13] Elson, Jeremy, et al. "Emstar: An environment for developing wireless embedded systems software." Center for Embedded Network Sensing (2003).

[14] Park, Sung, Andreas Savvides, and Mani B. Srivastava. "SensorSim: A simulation framework for sensor networks." Proceedings of the 3rd ACM international workshop on Modeling, analysis, and simulation of wireless and mobile systems. ACM, 2000.

[15] Prowler simulator. Available:http://w3.isis.vanderbilt.edu/projects/nest/downloads/PROWLER/prowler_v1_25.zip

[16] Dunkels, Adam, Bjorn Gronvall, and Thiemo Voigt. "Contiki-a lightweight and flexible operating system for tiny networked sensors." Local Computer Networks, 2004. 29th Annual IEEE International Conference on. IEEE, 2004.

[17] Baccelli, Emmanuel, et al. "RIOT OS: Towards an OS for the Internet of Things." Computer Communications Workshops (INFOCOMWKSHPS), 2013 IEEE Conference on. IEEE, 2013.

[18] A. Eswaran, A. Rowe, and R. Rajkumar, "Nano-RK: an energy-aware resource-centric RTOS for sensor networks," in 26th IEEE International Real-Time Systems Symposium, 2005. RTSS 2005., Dec 2005, pp. 10 pp.–265.

[19] Bhatti, Shah, et al. "MANTIS OS: An embedded multithreaded operating system for wireless micro sensor platforms." Mobile Networks and Applications 10.4 (2005): 563-579.

[20] Musznicki, Bartosz, and Piotr Zwierzykowski. "Survey of simulators for wireless sensor networks." International Journal of Grid and Distributed Computing 5.3 (2012): 23-50.

[21] Živković, M., Nikolić, B., Protić, J., and Popović, R. "A Survey and Classification of Wireless Sensor Networks Simulators Based on the Domain of Use." Adhoc & Sensor Wireless Networks 20 (2014).

[22] Stehlık, Martin. Comparison of simulators for wireless sensor networks. Diss. Ph. D. dissertation, Masaryk University, 2011.

[23] Phani, A. M. R. V. A., D. Janakiram Kumar, and G. Ashok Kumar. "Operating systems for wireless sensor networks: A survey technical report." (2007).

[24] Yadav, SG Shiva Prasad, and Dr. A. Chitra. "Wireless Sensor Networks-Architectures, Protocols, Simulators and Applications: a Survey." International Journal of Electronics and Computer Science Engineering (IJECSE, ISSN: 2277-1956) 1.04 (2012): 1941-1953.

[25] Farooq, Muhammad Omer, and Thomas Kunz. "Operating systems for wireless sensor networks: A survey." Sensors 11.6 (2011): 5900-5930.

[26] Eswaran, Anand, Anthony Rowe, and Raj Rajkumar. "Nano-rk: an energy-aware resource-centric rtos for sensor networks." 26th IEEE International Real-Time Systems Symposium (RTSS'05). IEEE, 2005.

[27] Nano-RK Operating system. Available: http://www.nanork.org/projects/nanork

[28] Cao, Qing, et al. "The liteos operating system: Towards Unix-like abstractions for wireless sensor networks." Information Processing in Sensor Networks, 2008. IPSN'08. International Conference On. IEEE, 2008.

[29] LiteOS: A Unix-like Operating System. Available: http://lanterns.eecs.utk.edu/software/liteos/

[30] Bajaj, Lokesh, et al. "Glomosim: A scalable network simulation environment." UCLA Computer Science Department Technical Report 990027.1999 (1999): 213.

[31] N2 network.Available:http://www.isi.edu/nsnam/ns/

[32] Castalia simulator page. Available: https://castalia.forge.nicta.com.au/index.php/en/index.html

[33] Boulis, Athanasios. "Castalia: revealing pitfalls in designing distributed algorithms in WSN." Proceedings of the 5th international conference on Embedded networked sensor systems. ACM, 2007.

[34] Chen, Gilbert, et al. "SENSE: a wireless sensor network simulator." Advances in pervasive computing and networking. Springer US, 2005. 249-267.

[35] J-Sim Official Page. Available: https://sites.google.com/site/jsimofficial/

[36] Sundresh, Sameer, Wooyoung Kim, and Gul Agha. "SENS: A sensor, environment and network simulator." Proceedings of the 37th annual symposium on Simulation. IEEE Computer Society, 2004.

[37] Titzer, Ben L., Daniel K. Lee, and Jens Palsberg. "Avrora: Scalable sensor network simulation with precise timing." Proceedings of the 4th international symposium on Information processing in sensor networks. IEEE Press, 2005.

[38] Avrora page. Available: http://compilers.cs.ucla.edu/avrora/

[39] Kröller, Alexander, et al. "Shawn: A new approach to simulating wireless sensor networks." arXiv preprint cs/0502003 (2005).

[40] Shawn project page. Available: https://github.com/itm/shawn

[41] Polley, Jonathan, et al. "ATEMU: a fine-grained sensor network simulator." Sensor and Ad Hoc Communications and Networks, 2004. IEEE SECON 2004. 2004 First Annual IEEE Communications Society Conference on. IEEE, 2004.

[42] Mármol, Félix Gómez, and Gregorio Martínez Pérez. "TRMSim-WSN, trust and reputation models simulator for wireless sensor networks." 2009 IEEE International Conference on Communications. IEEE, 2009.

[43] TRMSim-WSN, Trust & Reputation Models Simulator for Wireless Sensor Networks. Available: http://ants.inf.um.es/~felixgm/research/trmsim-wsn/index.php#relatedwork

[44] Fraboulet, Antoine, Guillaume Chelius, and Eric Fleury. "Worldsens: development and prototyping tools for application specific wireless sensors networks." 2007 6th International Symposium on Information Processing in Sensor Networks. IEEE, 2007.

[45] WSNet simulator page. Available: http://wsnet.gforge.inria.fr/index.html

[46] Shox project page. Available: http://shox.sourceforge.net/

[47] Lessmann, Johannes, Tales Heimfarth, and Peter Janacik. "Shox: An easy to use simulation platform for wireless networks." Computer Modeling and Simulation, 2008. UKSIM 2008. Tenth International Conference on. IEEE, 2008.

[48] Han, Chih-Chieh, et al. "SOS: A dynamic operating system for sensor networks." Third International Conference on Mobile Systems, Applications, And Services (Mobisys). 2005.

[49] Hong, Seongsoo, and Tae-Hyung Kim. "Senos: state-driven operating system architecture for dynamic sensor node reconfigurability." International Conference on Ubiquitous Computing. 2003.

[50] Koshy, Joel, and Raju Pandey. "VMSTAR: synthesizing scalable runtime environments for sensor networks." Proceedings of the 3rd international conference on Embedded networked sensor systems. ACM, 2005.

[51] Liu, Hangzhou, et al. "Design and implementation of a single system image operating system for ad hoc networks." Proceedings of the 3rd international conference on Mobile systems, applications, and services. ACM, 2005.

Excessive Internet Use Among Elementary School Students in Vojvodina

Silvija Tapiška*, Milena Kresoja**, Zoran Putnik**, Mirjana Ivanović**
*Pedagogical Institute of Vojvodina, Novi Sad, Serbia
** University of Novi Sad, Faculty of Sciences, Department of Mathematics and Informatics, Novi Sad, Serbia
silvija@pzv.org.rs, milena.kresoja@dmi.uns.ac.rs, zoran.putnik@dmi.uns.ac.rs, mirjana.ivanovic@dmi.uns.ac.rs

Abstract— In this paper we investigate factors that affect excessive use of Internet among elementary school students in Vojvodina. The quantitative analysis was performed using datasets from research carried out in 66 elementary schools across Autonomous Province of Vojvodina. Using a large data set on elementary school students, we have tried to describe the profile of students who excessively use Internet. Results reveal the significant impact of using Internet in non-study related purposes for differentiation between students who excessively use Internet and those who do not.

I   INTRODUCTION

It can not be overemphasized that the application of information technologies and multimedia in education is of great importance for educational system, and also for any individual in lifelong learning process. Contemporary education requires changes to traditional educational system, thus it is also essential to enable conditions for active learning and adequate improvement of teachers.

The main aim of the research presented in this article is establishing the frequency of pupil's computer use, both at home and in school, and the main objective of that use. The second aim was to research into the rate of Internet use, and checking for possible differences in its' frequency based on pupils gender, age, and place of residence.

This article uses results from a survey conducted among pupils of elementary schools in Vojvodina, in order to check what differences in use of computers and Internet exist among them. It is interesting to notice that schools usually do not allow Internet use during breaks, which has the largest effect on the differences of Internet use at home and in schools. As might be expected, the most often use of Internet is connected with social networks and game playing, easily pointing to lack of time for personal contact. Another straightforward conclusion is that excessive use of Internet and computers influences psychomotor and cognitive development of pupils [1]. More or less direct consequences of such a behavior by pupils are the lack of:

- capacity to find needed information,

- ability to quickly solve given problem,

- development of abstract and logical thinking,

- increase of knowledge and experience in synthesis of gathered data, and

- progress in reading and writing skills and understanding of read material,

which all certainly influence pupils motivation for learning. It is also noticeable that through Internet use, pupils acquire new knowledge in their own time, usually connected with their age and needs.

II   RELATED WORK

Article [2] elaborates on the results of a survey in which 291 parents of an urban elementary school (K-6) participated. Parents reported on their computer equipment at home, the type and frequency of their children's educational software and Internet use, and shared their ideas how better connections between computer use at home and school might be created. The results indicate that most of students' computer use was dedicated to game playing followed by various other software activities. Students reported more limited Internet activities. While home computer ownership is not necessarily contingent upon gender, some software and Internet use tended to be gender specific activities.

One of the common conclusions [3] is that girls tend to use computers less frequently than boys Also, boys tend to represent the vast majority of video game players, as can be proved by observations of school behavior and research into video game playing at home, so the results presented in are not unexpected. The results showed that the video games have a both positive and negative aspects. In order to maximize the good aspects of video games, parents should learn about and set rules and guidelines regarding the games their children play and how often they play them and also consider engaging in games with their children in order to better understand the games as well as promote the positive aspects of gaming. Video also games have the potential to be used as powerful teaching tools. On the negative side, use of violent video games is associated with increased aggressive behaviors. Additionally, time spent playing video games can crowd out time spent in more healthy activities, such as exercise, reading or school work.

Internet addiction and behavior concerning Internet among pupils was examined in [4]. It is recognized that the persons' manners and activities on the Internet, closely reflect their conducts in a real, everyday social life. Mentioned paper deals with the use of Internet in performing mainly illegal activities, such as molestation and abuse of other Internet users, giving false statements on behalf of others, or illegal gambling. The paper also investigates the amount of time users spend on the Internet. Difficult consequences of Internet addiction were research topic in [5]. Within this research, it has been established that the students in the USA are spending more than 25 hours per week on the Internet, or more than 5 hours per day, either using the computers, tablets or mobile phones. The paper focuses on three facts distinguishing factors of Internet usage, Internet activities, and consequences of Internet use. Achieved results show that the Internet overuse comes from various reasons. Students are not using Internet only for studying purposes, but also a lot of time they spend within social networks, searching the data, or playing games.

In study presented in [6], conclusions can be found about children in Hong Kong schools. On the daily average, children spend their time with Internet and electronic screen devices as follows (numbers go over 100%, because multiple answers were allowed):

- 44.2% for the use of TV,

- 31.6 % for the use Tablet PC, and

- 31.8% for the computer usage.

Considering the average amount of time a day spent on electronic screen devices and Internet, this research shows that 70.9 % spends less than 1 hour, 22.6% between 1-3 hours, additional 3.3 % spend more than 3 hours, while the data for the rest of 3.2% is not available. Article [7] shows analysis of the children from "European union Online", a 25-country survey regarding excessive use of the Internet by children. Children were most likely to report that they spend their time surfing (about 42% of the children), or that they had gone "without eating or sleeping" because of the Internet (about 17% of them). Without Internet about 67% of the children are bored! This paper also shows that there were significant differences between boys and girls, between young people from different age groups, and between children from households where parents had mutually diverse education.

The trends considering Internet usage by students and pupils today were examined in [8]. Paper shows that there exists a large growth of use of Internet and digital media in general, including computers, mobile phones, tablets, and video-games. Article shows that children start using in an extremely early age, at the preschool institutions. Overuse of these digital media can have very negative effect on children at such an early age, but also later, and can influence psycho-physical development, unless it is aimed to train and educate children to a proper use of those media.

In [9] autors showed the results about students Facebook users. The results show that over 90 percent of pupils were already existing Facebook users. Approximately 58 percent of the respondents were below the age of 25 and used Internet for various reasons on a daily basis. A further 10 percentage used the Internet at least once per week. 65 percentage of the pupils have access to the Internet at home and a good news is that 75 percentage of pupils had a personal computer with Internet access.

Finally, we found a paper [10] that show that the most frequent activity was playing computer games (64%responses), then working on the internet (27% responses), writing (26% responses) and others. The results showed too that boys preferred playing games whereas girls preferred writing. The use of internet was similar between two sexes. Computers at homes were used mainly for computer games, while prevalent activities in schools were related with searching for information from the internet and using e-mail. This means that the use of PCs in the school was very consistent, but the use of e-mail and information from the internet at home was quite different than the use of PC for games or MS Word.

III DATA AND METHODS

The dataset used in this paper was derived from a research carried out in 66 elementary schools across Autonomous Province of Vojvodina, Republic of Serbia. It included N=7007 elementary school students who filled in the questionnaires. The sample was formed from students attending higher grades (VII and VIII) of elementary school. The surveyed students attend monolingual, bilingual, or trilingual schools. The main aim of exploring and expanding the national community lies in the fact that Vojvodina is an intensely multiethnic community. This way, we were able to focus on general, overall state in the elementary schools in Vojvodina when it comes to ICT integration.

The survey consisted of questions regarding demographic characteristics, frequency of using computer and Internet at school and home, purposes of using ICT, integration of ICT in the classrooms, as well as students perception of ICT integration for learning purposes.

Data were analyzed using descriptive statistics, nonparametric statistical methods, and binary logistic regression. Summary statistics are presented by percentages for categorical variables. Crosstab procedure and chi square test of independence were used to examine differences in characteristics between groups for nominal variables. Binary logistic regression was employed to model outcomes of dependent variable with only two categories. It was used to assess the factors influencing excessive Internet use among elementary school pupils. Pupils covered by the research were divided into two focus groups:

- students with excessive Internet use, and

- with non-excessive, "normal/usual" Internet use.

The student was considered to use Internet excessively of he/she spent more than 25 hours/week on the Internet. The dependent variable was coded according to this classification: students were assigned the value 1 if they use Internet excessively and the value 0 otherwise. Total of 8 predictors entered the model, and all variables were categorical. Performances of the model were tested using Omnibus Tests of Model Coefficients. Model with all independent variables against a constant-only models was statistically significant (chi square= 626.389, p=.000).

This result indicated that the set of chosen variables makes reliable distinguishing between mentioned two groups of students. The regression results are presented as odds ratios. Odds ratios above 1 stand for positive associations between independent and dependent variables, while odds ratios below 1 and above 0 stand for negative associations. The level of significance in all conducted tests was 0.05.

IV   RESULTS

The survey sample offers a balance across the gender and the grade of the students. The results are given in Table 1.

TABLE I.   BASIC CHARACTERISTICS OF STUDENTS

Var.	Cath.	N	%
Grade	7th	3619	52
	8th	3374	48
Place	City	4341	62
	Countryside	2652	38
Gender	Male	3562	51
	Female	3431	49

Table 2 gives the distribution of students according to their use of internet. According to the results, exactly half of the students use internet more than 5 hours every day.

TABLE II.   INTERNET USAGE

	Percentage %
Non Excessive use	50
Excessive use	50

Table 3 presents results gained while researching the main purposes of student's use of Internet. The results gained about "game playing" and "fun" are *not* unexpected, while the results considering "study purposes", are a rather pleasant surprise.

TABLE III.   PURPOSES OF INTERNET USAGE

Study purposes	N	52	1%
	Y	6941	99%
Fun	N	2362	34%
	Y	4631	66%
News	N	3791	54%
	Y	3202	46%
Games	N	1734	25%
	Y	5259	75%
Communication	N	3657	52%
	Y	3336	48%

Further, we have investigated whether there are any significant differences between mentioned purposes of Internet use, and gender and/or place of residence. The statistical analysis revealed that there is a statistical significant difference between gender and use of Internet for study purposes (chi square = 7.011, p=.008). According to the results we reached, male are using Internet more for study related activities than females (61%). There is also a statistically significant difference between place of residence and Internet use for study purposes (chi square = 21.844, p=.000). Students from cities are using Internet more for study purposes (63%). There were no statistical significant differences between gender and place of residence and Internet use for socializing and fun. Differences in a sense of Internet use for reading news were only significant considering the place of residence (chi square = 41.592, p=.000). Students from cities use Internet more for reading news (66%). There is a statistically significant difference between gender and Internet use for games (chi square = 14.154, p=.000). As expected, and proved in many research papers, males are using Internet more for playing games (61%). Females are using Internet primarily for communication with other students, more than males (53%), and that difference is significant (x2= 27.874, p=.000).

TABLE IV.   COMPUTER AND INTERNET USAGE IN SCHOOLS

How many hours per day do you use computer at school?	0-2h	6889	99%
	3-4h	83	1%
	5+	21	0%
How many hours per day do u use Internet at school?	0-2h	6777	97%
	3-4h	222	3%
	5+	0	0%

Regression results are given in Table 4. All variables, except grade, were significant.

The rest of the results in connection with the regressions, are given in Table 5.

TABLE V.   REGRESSION RESULTS

Variables	EI	
Grade	1.026	
Gender	1.106	*
Place of residence	1.209	*
Internet use for study purposes	0.553	*
Internet use for socializing and fun	1.082	*
Internet use for reading news	2.000	*
Internet use for games	2.928	*
Internet use for communication with other students	1.627	*
*denotes significant variables		

Older students have higher odds to be in the group of students who excessively use Internet, than younger students. The results also show that girls are more prone to use Internet excessively, than boys. Finally, students from smaller places of residence are more likely to use Internet excessively, compared to students living in bigger cities. Still, there is a negative association of Internet use for study purposes, and excessive use of Internet. On the other hand, there is positive association of excessive Internet use for socializing and fun, for reading news, for games and for communication with other students. The results confirmed our main hypothesis that students who use Internet for non-study related purposes are more likely to be Internet addicted.

V CONCLUSIONS

To give a verbal explanation and discussion of presented results, we can induce the following. Based on the presented results, we can conclude that the students of 7th and 8th grade of elementary schools in Vojvodina have a rather high rate of Internet use, sometimes very extensive. Comparison also shows that Internet is used more among students of the 7th grade, compared to the students of the 8th grade (70%).

In addition, surveyed data show that boys tend to use Internet more at home (54%) than in schools. This result can be expected since, as we already mentioned, Internet use in schools is usually forbidden during breaks. As both common sense and research show, game playing is one of the most popular ways of computer use. Our research confirm to that, showing that about 86% of surveyed students mention this activity as one of their main methods of computer use. Following this activity, on the list of the most popular methods of computer use are "gaining new information" for about 71% of students, "doing their homework" for 57% of pupils, and "expanding their existing knowledge" for about 31% of

pupils. Significant and rather expected is the result showing that students living in bigger cities have better access, and use Internet more often (62.5%).

REFERENCES

[1] L. Mills Kathryn, „Effects of Internet use on the adolescent brain: despite popular claims, experimental evidence remains scarce", Institute of Cognitive Neuroscience, University College London, UK Child Psychiatry Branch, National Institute of Mental Health, Bethesda, MD, USA, 2014.

[2] B. Kafai Yasmin and S. Suton, "Elementary school students computer and Internet use at home: current trends and issuses", Journal of Educational Computing research, 1999.

[3] E. O. Jennifer Brain, W. Li, Susan M. Snyder and Matthew O. Howard, "Characteristics of Internet Addiction/Pathological Internet Use in U.S. University Students: A Qualitative-Method Investigation", PLOS one, 2015.

[4] E. Provenzo, "Videokids", Harvard University Press, Cambridge, Massachusetts, 1991.

[5] H. Keung Ma, "Internet Addiction and Antisocial Internet Behavior of Adolescents", The Scientific Worls Journal, 11, pp. 2187–2196, 2011.

[6] W. Chai, "Non-Communicable Diseases Whatch, of Internet and Electronic Screen Products among children and Adolescents", Surveillance and Epidemiology Branch, Centre for Health Protection of the Department of Health 18/F Wu Chung House, Hong Kong, 2015.

[7] D. Smahel, E. Helsper, L. Green, V. K.almus, B. Lukas and K. Ólafsson, "Excessive internet use among European children", EU Kids Online, London School of Economics & Political Science, London, UK, 2012.

[8] W. Chai, "Use of Internet and Electronic Screen Product among Children and Adolescents", Surveillance and Epidemiology Branch, Centre for Health Protection of the Department of Health 18/F Wu Chung House, Hong Kong, 2015.

[9] Lenandlar Singh, "Guided Assessment or Open Discourse: A Comparative Analisys of Students Interaction on Facebook Groups", Turkish Online Journal of Distance Education-TOJDE, ISSN 1302-6488, Volume:14, article 3, 2013.

[10] J. Fančovičová and P. Prokop, "Students Attitudes Toward Computer Use in Slovakia", Eurasia Journal of Mathematics, Science & Technology Education, 4(3), pp.255-262, 2008.

E-learning on Polytechnic Nikola Tesla – Analysis and comparison

Kovačić, Božidar; PhD[1]; Skendžić, Aleksandar; PhD[2], Devčić Kristina, univ.spec.oec.[3]

[1] University of Rijeka, Department of Informatics, Radmile Matejčić 2, 51000 Rijeka, Croatia
[2] Polytechnic Nikola Tesla in Gospic, Bana Ivana Karlovića 16, 53000 Gospić, Croatia
[3] Polytechnic Nikola Tesla in Gospic, Bana Ivana Karlovića 16, 53000 Gospić, Croatia
bkovacic@inf.uniri.hr , askendzic@velegs-nikoltesla.hr, kdevcic@velegs-nikolatesla.hr

Abstract: E-learning System has become a common constituent unit of web page of any higher education institution. At the Polytechnic Nikola Tesla in Gospic Loomen is established since 2010, a system that is designed in a role of teaching repository for students and for preparation of e-courses for the teachers and students. The aim of this paper is by using descriptive and inferential statistics, and based on a survey conducted among students, examine how, and to what purpose and how often students used Loomen E-learning System.

Keywords: e-learning system, Loomen, higher education, statistical methods.

I. E-learning and Learning Management System

The concept of E-learning can be defined in various ways. One of definitions can be E-learning is learning with the help of ICT technologies. For example, Zhang and others (2004.) defined e-learning as technology-based learning in which learning materials are delivered electronically to remote learners via computer network.

E-learning has following forms of teaching [3]:
- teaching in which the application of ICT technologies is exclusively as a supplement to traditional teaching (face-to-face, f2f),
- teaching in which you can combine traditional forms of teaching supported by ICT technology. This form of teaching is called Hybrid Learning i.e. learning mix mode. Hybrid learning uses the system for learning on distance (Learning Management System, LMS) and videoconference,
- exclusively online teaching without face to face contact.

However, the use of ICT technologies in the educational process requires some technical, and also pedagogical and organizational skills of the authors of the online content. The digital teaching content requires from author a special approach in understanding the design process of online content. In the process of creating and designing online content, it is important to follow certain pedagogical and didactic rules.

Due to development of e-learning technologies we can talk about three generations[1][4]:

- e-learning 1.0,
- e-learning 1.3,
- e-learning 2.0.

According to Kljakić (2007.),[5] the third generation of e-learning (e-learning 2.0) takes advantage of Web 2.0 technology in which the emphasis is on collaboration, knowledge creation, virtual communities, and for the first time students can participate in developing teaching materials. The advantage of e-learning 3.0 is in enabling to achieve 95% of success (as shown in Table 1).

Table 1. Achieving success in learning.

Learning technics	Success
Of what we read	10%
Of what we hear	20%
Of what we see	30%
Of what we see and hear	50%
Of what we discuss	70%
Of what we experience	80%
Of what we teach others	95%

Source: thinkexist.com/quotes/william_glasser/ (February 4th, 2017) [11]

The potentials of e-learning 2.0 are reflected in the following [6]:
- unlimited possibilities for creating, distributing, publishing, commenting of the content using Internet,
- easier use of e-learning solutions based on Web 2.0 technologies,
- focusing on students,
- the learning process becomes creation of a network.

Regardless of the generation of e-learning, all e-learning systems can complement each other, depending on the needs of the educational context and can be independently applied.

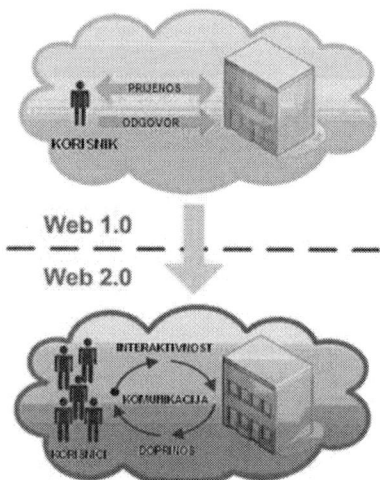

Figure 1. Web 2.0 vs Web 1.0.
Source:https://mi2web20.wikispaces.com/file/view/komu nikacija_na_webu.png/146685749/322x383/komunikacij a_na_webu.png (7.2.2017.) [10]

A. Learning Management System (LMS)

Development of systems for distance learning began in 1996 as a part of the www internet service. Using the LMS system is reflected in the technological advancement of the teaching, but maintaining the support of the classical model of learning (f2f). The advantages of using LMS are the grouping of students, starting learning at the same time, periodic distribution of new materials and expectations of the end of the learning at the same time. Today there are numerous LMS systems on the market that differ in the way how they save teaching materials and other information required in the process of learning and teaching. The main features of all LMS systems can be divided into two basic groups [7]:

- administrated group of functions,
- teaching group of functions.

Administrator functions enable the monitoring and adjustment of functions that are not related to the transfer of knowledge (of system functions). Systemic functions are related to the following processes (examples):

- creating user accounts,
- creating access permissions,
- creating user groups,
- various forms of reporting, analysis, process monitoring,
- support in the development of teaching materials,
- connectivity with other systems within the organization,
- adjustment of the system to the various program platforms,
- other system features.

Teaching functions are extremely important because the final result in learning and teaching process depends on them. If there is not enough functional support, that will be reflected in the final learning outcomes (quality) in the use of LMS by students, and teachers. Some of the most important teaching functions of the LMS system are: [8]

- the content of teaching (lessons, modules),
- navigation (management) through the elements of the system,
- self-evaluation of participants (students),
- communication between teacher and students,
- additional authoring tools for creating additional content (eg. tests, discussions, etc.).

II. Learning Management System Loomen at Polytechnic Nikola Tesla in Gospić

Loomen is a software tool used for the creation of online educational contents and teaching at distance. Loomen is an open source software, licenced as a free making it extremely popular in the implementation of the processes of education supported by ICT technologies. The Loomen system is based on Moodle [9] (Modular Object-Oriented Dynamic Learning Environment) within CARNet service designed for creating digital learning content. From the security and authentication point of view, system is enabled through an electronic identity in the AAI@EduHR system.

The Loomen system on Polytechnic Nikola Tesla in Gospić is used since 2012. In the initial period the most of the courses were introduced from all three studies (Professional Study of Road Transport, Professional Study of Economics of Entrepreneurship and Professional Administrative study) and a brief training of employees and students was carried out. The system is used primarily as a teacher repository system and for forwarding information to students in the current academic year. Over the coming years the system was updated, and became one of the primary sources of information and preparations of the exams. The next step in upgrading the Loomen system is the introduction of e-courses.

Loomen is the application for creating and maintaining online courses based on the GNU GPL[2] (GPL - General Public License) license. Originally, Loomen is written in the PHP program language and supports multiple types of databases (eg. MySQL). Also, language interface has been translated into several languages, and many users of open-source software were involved in the creation of new functionality as well as those existing, and made a new modules and test existing one and provide a customer support. Table 2 shows the basic features of the Loomen LMS system.

Table 2. Basic features of the Loomen LMS system.

---

[2] GPL licence preserves the freedom of users of softwares: the right for use, the right form making copies and the right to modify and redistribute of the modofied program. GNU GPL is copyleft licence. Software that may arise as a modification of the existing software guarantees the same freedom (ie. to be distributed under the same licence).

LMS	Functions
Loomen LMS	Development of online courses and planning
	Managing users and groups
	Work with existing educational content
	Examination and evaluation of users, self-evaluation
	Monitoring activities
	Collaboration between users
	Log records, backup, statistics
	Help system

Source: http://www.ssmb.hr/libraries/0000/2796/e_learning_LMS.pdf (12.02.2017.)

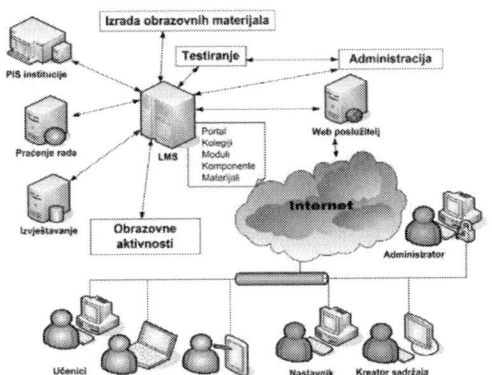

Figure 2. Example of the LMS model.
Source: http://www.ssmb.hr/libraries/0000/2796/e_learning_LMS.pdf

III. Theoretical background: about survey in Šibenik

Perišić, Goleš and Devčić (2012) conducted a research about e-learning system at the Polytechnic in Šibenik. In their study they used methods of descriptive and inferential statistics: graphical and numerical methods, estimations of the parameters and hypothesis testing procedures (χ^2 test). The level of use and students' satisfaction with e-learning system and other forms of online information and relation between using e-learning system and students' success were analyzed. The study was conducted at the Polytechnic in Šibenik on three departments: the Department of Management (professional study of Management and specialist professional graduate study of Management), Department of Administration (professional study of Administration) and Department of Transport (professional study of Transport). Analysis was based on 304 respondents. The level of use of online information sources, and especially e-learning system and the quality of existing online information sources was examined. The analysis showed that students of the Polytechnic in Šibenik in a great extent use e-learning system. Authors showed that there is statistically significant difference in the use of e-earning system between students with different study success. Students with better study success used e-learning system in a greater extent than those with lower

study success. Students are mainly satisfied with the coverage of the courses with e-learning system.

IV. Sample and Questionairre

The study was conducted on a sample of 71 students of professional studies Economics of Entrepreneurship (56% of respondents) and Road Transport (44% of respondents) at the Polytechnic of Nikola Tesla in Gospic. The average age of respondents was 20.4 years. Among the respondents was 42.3% of male and 57.7% of female respondents. 38% of respondents were the first, 18% the second, and 44% third year of the study. Mostly, full-time students were tested (90% of respondents), and only 10% of respondents were part-time students. 84% of respondents had average success E or D or C, and 16% of respondents had average success B or A. Data were collected using a questionnaire on a random sample of students of the Polytechnic. The survey was conducted in November 2016. The questionnaire contained six questions related to the demographic characteristics of respondents (age, gender, year of study, mode of study, average success and professional study) and 9 questions that gave information about forms of sources of information used, with an emphasis on the Loomen.

V. Empirical Results

Students can look for information on different ways: standard (bulletin board, Registry of Polytechnic, consultations, phone calls) or online forms of getting information (website, Loomen, social networks). According to conducted research 4% of respondents used standard forms, 59% of respondents online forms of information, and 37% both. 45% of respondents believed that the standard forms of information were sufficient to obtain information on time, and 55% think that were not sufficient. 82% of respondents believed that online sources of information are sufficient to obtain information on time, and 18% think that they are not enough. Students used online forms of information: website (37%), Loomen (22%), social networks (7%) and all of the above (34%).

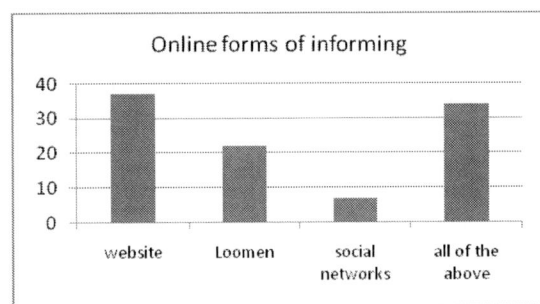

Figure 3. Online forms of informing.
Source: authors' calculation.

Students used Loomen never (4% of respondents), sometimes (72% of respondents), often (24% of respondents).

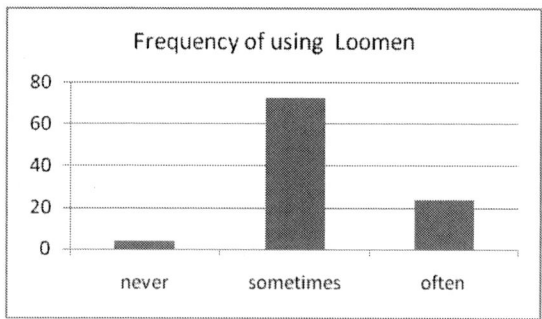

Figure 4. Frequency of using Loomen
Source: authors' calculation.

The purpose in which students most commonly used Loomen is literature and teaching materials (28% of respondents), exam results (30% of respondents), professor's notices (8% of respondents), examples of the old exams (9% of respondents), all of the above (25% of respondents).

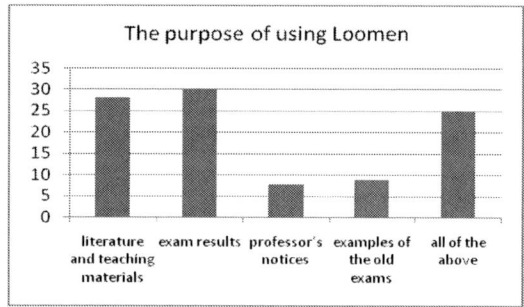

Figure 5. The purpose of using Loomen
Source: authors' calculation.

86% of respondents believed that Loomen help them while studying, and 14% think it didn't. 46% of respondents believed that there is room for improvement of Loomen, and 54% believed that there isn't. Students gave some recommendations for improving the Loomen and the most common recommendations we can extract creating and implementing the application and implementing all courses on Loomen.

The obtained data showed that there is no statistically significant relationship between year of study and frequency of using Loomen ($\chi^2 = 6.294$, p = 0.178), nor between the year of study and online forms of informing that students use ($\chi^2 = 6.908$, p = 0.329), nor between the year of study and the purpose of which students most commonly used Loomen ($\chi^2 = 13.52$, p = 0.095). Furthermore, there is no statistically significant relationship between the modes of study and online forms of informing that students use ($\chi^2 = 1.609$, p = 0.6573), nor between modes of study and frequency of using Loomen ($\chi^2 = 0.291686$, p = 0.8643), nor between modes

of study and the purpose of which students most commonly used Loomen ($\chi^2 = 0.2608$, p = 0.992). Also, there is no statistically significant relationship between average success and online forms of informing that students use ($\chi^2 = 1.609$, p = 0.6573), nor between the average success and the frequency of using Loomen ($\chi^2 = 0.291686$, p = 0.8643), nor between the average success and the purpose of which students most commonly used Loomen ($\chi^2 = 0.2608$, p = 0.992).

Also, there is no statistically significant relationship between modes of study and online forms of informing that students use ($\chi^2 = 0.9922$, p = 0.803), nor between modes of study and the frequency of using Loomen ($\chi^2 = 0.4646$, p = 0.793), nor between the modes of study and the purpose of which students most commonly used Loomen ($\chi^2 = 0.5877$, p = 0.964).

The results indicated that there is a habit among students to use online sources of information, such as Loomen. Unfortunately, so far it has not been proved that the use of the Loomen can help students achieve better results. Management of the Polytechnic should have to make further efforts to facilities at Loomen do more help in preparing exams, which would then have a positive impact on the success of students.

In subsequent studies we should try to include more part-time students to give their answers in questionnaires..

VI. Conclusion

In today's online environment, each higher education institution must take into account the availability of information, teaching materials, literature, professors' notifications, exam results etc. via online forms of informing such as web pages, Loomen or other E-learning systems, social networks and more.

Loomen e-learning system is one of the ways how teaching materials and information can be available for students and without requiring physical presence at the university/college. In this paper, the study about frequency and purposes of using Loomen was conducted on the random sample of students of the Polytechnic Nikola Tesla in Gospić. Analysis of the results showed that the students of the Polytechnic have the habit of using online forms of information (mostly Polytechnic's website and then Loomen), that students use Loomen sometimes (72% of students) and often (24% of students). The purpose in which students most commonly used Loomen is literature and teaching materials (28% of respondents), exam results (30% of respondents), professor's notices (8% of respondents), examples of the old exams (9% of respondents), all of the above (25% of respondents).

There was no statistically significant relationship between the frequency and purpose of using Loomen system and characteristics of students (year of study, student status, success of studying). In the future Polytechnic should enrich the content of the Loomen in terms of introducing new courses on Loomen (which was the proposal for the improvement of a number of students) and in the future should certainly consider the introduction of e-courses which would be an additional step in raising the quality of teaching and quality of entire contents that Polytechnic is offering to its students.

If we compare the obtained results with a similar survey about using e-learning system that was conducted in 2012 by Perišić, Goleš and Devčić we can notice certain similarities, and differences also. Thus, it was shown that, which is the case in Šibenik also, that students developed the habit of using online sources of information; 59% of students used only online sources of informing, and 37% of students used online and standard sources of informing (in Šibenik 45% of students used only online, and 47% of students both forms).

However, 24% of students used Loomen often, and 72% of students sometimes (in Šibenik 71% of students often, and 24% sometimes). In this study, as well as in Šibenik, the relationship between frequency of using Loomen and student's status was statistically insignificant. Furthermore, the relationship between frequency of using Loomen and Department of the Polytechnic was statistically insignificant, while in Šibenik this relationship appeared to be statistically significant. Unlike the students in Gospić, the relationship between the frequency of using Loomen and average student's success measured by average score in Šibenik showed statistically significant.

References:

[1] Zhang, D., Zhao, J.L., Zhou, L., Nunamaker, J.F. (2004.) "Can E-learning replace classroom learning?" Communications of the ACM, Vol. 47, No. 5, pp. 75—79.

[2] Perišić, A., Goleš, D., Devčić, K. (2012.) „Analysis of the Use of E-learning System: Example of the Polytechnic in Šibenik". 1st Internet & Business Conference, IBC 2012, 27.-28. lipnja 2012., Rovinj.

[3] http://www.ssmb.hr/libraries/0000/2796/e_learning_LMS.pdf (12.02.2017)

[4] https://www.td.org/Publications/Newsletters/Learning-Circuits/Learning-Circuits-Archives/2007/07/Understanding-E-Learning-20(07.02.2017.)

[5] Kljakić, Dušan: Evolucija elektronskog učenja: Elearning 2.0. Časopis za teoriju i praksuodgoja i obrazovanja «Naša škola» (Sarajevo), broj 42, 2007, str 3-19.

[6] https://pogledkrozprozor.wordpress.com/200929/uvod-u-e-%E2%80%93-learning-1-dio/ (7.2.2017)

[7] http://www.ssmb.hr/libraries/0000/2796/e_learning_LMS.pdf (12.02.2017.)

[8] http://www.ssmb.hr/libraries/0000/2796/e_learning_LMS.pdf (12.02.2017.)

[9] http://www.carnet.hr/loomen/o_usluzi (7.2.2017.)

[10]https://mi2web20.wikispaces.com/file/view/komunikacija_na_webu.png/146685749/322x383/komunikacija_na_webu.png (7.2.2017.)

[11] thinkexist.com/quotes/william_glasser/ (4.2.2017.)

Determination of time criteria for assessment in Learning Management Systems

Trpimir Alajbeg, Mladen Sokele, Vladimir Šimović
Zagreb University of Applied Sciences, Department of Electrical Engineering, Zagreb, Croatia
trpimir@tvz.hr, msokele@tvz.hr, vsimovic@tvz.hr

Abstract - Today's Learning Management Systems, in conjunction with knowledge and skill acquisition modules, offer the possibility for automated assessment of the aforementioned knowledge and skills. Their implementation in teaching offers a wide range of advantages: reducing teacher administrative work, reducing the possibility of errors concerning preparation, execution and evaluation of exams; elimination of teacher's subjectivity during the evaluation phase; lessening of inappropriate student actions during exams etc. A majority of tests that are done through LMS have a time limit, which opens up a problem of determining the optimal time parameters for a specific exam. The paper includes a statistical data analysis that was collected during five years via LMS Moodle on the Personal Computer Applications (PCA) course in professional study of electrical engineering at the Zagreb University of Applied Sciences. Through analysis and modelling the results of the statistical data processing, guidelines for time criteria for future tests were made.

I. INTRODUCTION

The advantages of using a Learning Management Systems (LMS) are significant. For students: the ability to prepare themselves for classes, laboratory and construction exercises with the emphasis on the ease of access to all course material, computer applications and tasks for individual preparation and work.

LMS systems are also beneficial for lecturers. They help to reduce the time spent in preparation and execution of exams, as well as to minimise errors in their evaluation and grading, especially when quick grading is required. Additionally, they considerably reduce the administrative tasks (e.g. class attendance tracking system for obligatory laboratory exercises, benefits of centralized grade database). Also, the automated knowledge and skill tests eliminate the teacher's subjectivity during the evaluation phase of the exam [1]. Another benefit is the possibility for the student to check the results of the exams in real time i.e. immediately after its end. [2]. By using the application Net Support School that is installed on all of the computers used for LMS, a reduction of inappropriate student actions during exams is achieved.

The results of automated knowledge and skills exams can be used to assess the existence of questions or tasks that are either too easy or too complex. Consequently, the questions or tasks can be then reformulated or replaced with new ones without altering the structure of the existing test. With all that in mind, setting of time criteria for assessment in LMS is challenge. The adjustment of

time criteria is in direct correlation with the number and complexity of tasks in each individual exam. The basis for it are the log data that was collected during five years of usage LMS Moodle for the Personal Computer Applications (PCA) course in professional study of electrical engineering at the Zagreb University of Applied Sciences.

The paper examines possible directions of LMS time criteria determination and/or optimization. Primarily, the optimization is not focused on saving the time (i.e. overall test duration) but on the adequate number of tasks increase in existing time period framework. The increase in the number of tasks ensures better class material coverage, enables implementation of more specific tasks as well as improvements in cognitive linkage of knowledge and skills. Therefore, the leading idea is to examine how possible time reduction and/or number of tasks increase reflects on overall successful attempts as well as identification and differentiation of better and barely enough prepared students. In order to achieve proposed idea, analysis and modelling of the results of the statistical data processing and the procedures for time criteria determination were made. The results of the aforementioned analysis and modelling could be used for developing of time criteria guidelines, noting that each test must be adapted to the specific requirements of particular lesson.

In Section II., explanatory parameters describing test solving dynamics are defined. In Section III. explanatory parameters are applied on results of attempts separated according to achieved scores as well as possible effect of test duration reduction is investigated by analysis of test solving dynamics slope near the test completion time. Based on analogy of test solving dynamics with diffusion process, in Section IV. the new model, named Declining Bass Model (DBM) is introduced. Finally, in Section V. the new model is applied in improved form (doubled DBM) that shows differentiation of better and barely enough prepared students. In addition, it is shown that new model enables determining of intensity of test solving that clearly identifies above mentioned differentiation.

II. INPUT DATA AND ITS REPRESENTATION

Automated knowledge and skills exams are used to asses acquired skills in case of laboratory exercises and acquired knowledge in case of theoretical course parts (i.e. (midterm exams and final exam). Initially, the duration of the exams was determined by having the teaching staff of the course solving exams and obtained average measured time were tripled for students.

Automated knowledge and skills exams via LMS Moodle have been used for nine years. Data for the first four years are not analysed because the tests were in constant improvement: the question database was constantly expanded and altered to optimally adapt to the course needs (Single or Multiple Choice; Short Textual or Numerical Answer, Matching, etc.). In the last five years, when automated knowledge and skills exams are in stable phase, analyses and modelling of the gathered data is reasonable.

Analysis of test solving dynamics is obtained through tests' logs processing. According to the graphical representation (see Fig. 1 – case of *Image processing test*, all attempts: passed and failed), the following **explanatory parameters** can be identified and defined:

- number of students solving test M [#] (in presented case $M = 755$)

- defined level for characteristic duration of test v [%] (in presented case v is set to 90%)

- characteristic duration of test Δt [s] i.e. time needed for v [%] of students to finish test (in presented case for $v = 90\%$, Δt is 1191 s: from $t_e - \Delta t = 609\ s$ to $t_e = 1800\ s$)

- overall test duration t_e [s], after the expiration of the time t_e the test is locked (in presented case $t_e = 1800\ s$)

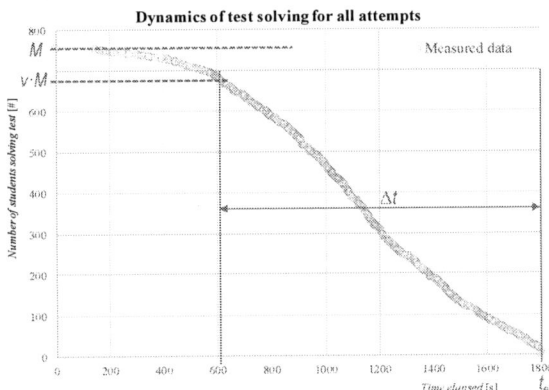

Fig. 1 Example of test solving dynamics (case: *Image processing test*), M – number of students solving test, v – defined level for characteristic duration of test, Δt - characteristic duration of test, t_e – overall test duration

The main idea is to get quantification of test completion slope. In the case of *Image processing test*, the slope is moderate: within the first 34% of available time 10% of students submitted exam, and during the remaining part of $\Delta t / t_e = \mathbf{66\%}$ of total time 90% students submitted exam.

Quite the opposite is the case of *Spreadsheet test*, where slope is steep: within the first 71% of available time (from $t_e - \Delta t = 854\ s$ to $t_e = 1200\ s$) only 10% of students submitted exam, and during the remaining part of $\Delta t / t_e = \mathbf{29\%}$ of total time even 90% students

submitted exam. Therefore, dimensionless parameter defined as:

$$g = 1 - \Delta t / t_e \qquad (1)$$

with predefined level for characteristic duration v can be used for quantification of test completion slope gradient or in other words for the level of rush at the end of test solving. Parameter g is in range [0,1] so it can be represented as percentage, where higher percentage means higher slope gradient.

Values for abovementioned parameters for all encompassed tests (all attempts, passed and failed) are given in Table 1.

Table 1. Overview of parameters for all tests attempts

	Number of students' attempts M	Characteristic duration of test Δt (for $v = 90\%$)	Overall test duration t_e	Slope gradient g
Spreadsheet test	972	347 s	1200 s	71%
Text processing test	682	1028 s	3600 s	72%
Flowchart test	901	2564 s	3600 s	29%
EDA test	928	681 s	1200 s	43%
Image processing test	755	1191 s	1800 s	34%

III. ANALYSIS OF SUCCESSFUL ATTEMPTS

Previous example illustrates statistics for all student attempts: passed and failed scores together. For proper determination of time criteria for test duration, data should be separated according to achieved scores. Intervals of scores 0%-33%, 34%-67% and 68%-100% are used for representation on Fig. 2, noting that first interval means that a student failed the test.

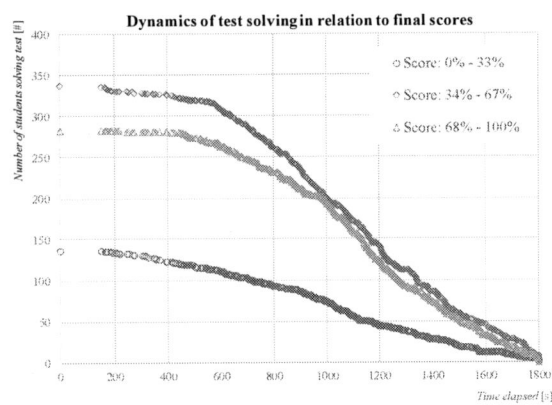

Fig. 2 Dynamics of test solving in relation to final scores (case: *Image processing test*)

Values of parameter g (test completion slope gradient) from Table 2 shows that the better prepared students that achieved higher scores are under lower time pressure (lower parameter g).

Table 2. Number of students that finished exam within the first and the remaining interval divided according the final scores (case: *Image processing test*)

Interval of scores	Δt [s]	$g = 1 - \Delta t / t_e$	Fastest* 10% students		Remaining 90% students	
			Time interval [s] $(0, t_e - \Delta t]$	# stud.	Time interval [s] $(t_e - \Delta t, t_e]$	# stud.
0%-33%	1379	76.6 %	(0, 421]	14	(421, 1800]	122
34%-67%	1184	65.8 %	(0, 616]	34	(616, 1800]	303
68%-100%	1139	63.3 %	(0, 661]	28	(661, 1800]	254

* the first time interval is defined through level for characteristic duration of test $v = 90\%$

Detailed analysis of the test solving dynamics slope near the test completion time can be obtained by its first derivative in the point $(t_e, 0)$. Obtained gradient gives the ratio of decrease in the number of successful attempts depending on the reduction in the duration of the test. With that idea in mind, Fig. 3 shows the possible effect of overall test duration reduction on successful attempts, all percentage wise.

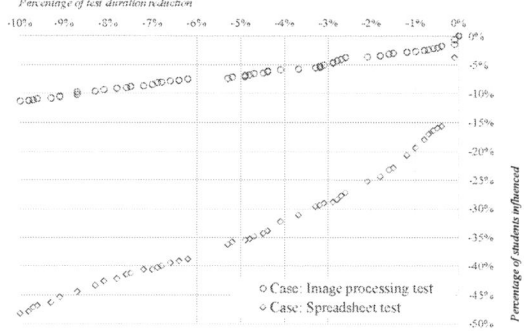

Fig. 3 Gradient of test solving dynamics in point $(t_e,0)$ for cases: *Image processing test* and *Spreadsheet test*

With appropriate scaling of the abovementioned results, the actual amount of reduction in the number of successfully completed tests can be obtained, which is the goal of future authors' research. However, in this paper is an opportunity to show that for some tests this dependence is more strongly expressed. As it is shown on Fig 3, in case of *Image processing test* 10% reduction of overall test duration results in possible reduction of successful attempts of 11 %, and even 48% in case of *Spreadsheet test*.

IV. INTRODUCTION OF THE DECLINING BASS MODEL (DBM)

By analysing the graphical representation of test solving dynamics on Fig. 1 it can be concluded that in the sense of time series methods resembles to a declining growth model [3]. More precisely, customers (students) are leaving service (test solving) and after overall test duration (t_e) there is no customers left on market (knowledge assessment).

The best known model for a full description of the genesis of the new service adoption is the Bass model [4]. The Bass diffusion model (2) is defined by four parameters: M – market capacity, $M = B(t \to \infty)$; p – coefficient of innovation, $p > 0$; q – coefficient of imitation, $q \geq 0$ and t_s – time when service is introduced, $B(t \leq t_s) = 0$:

$$B(t; M, p, q, t_s) = M \frac{1 - e^{-(p+q)(t-t_s)}}{1 + \frac{q}{p} e^{-(p+q)(t-t_s)}} \qquad (2)$$

The model $B(t)$ introduces the effect of innovators via coefficient of innovation p which corrected deficiency of simple logistic growth, i.e. considers a population of M adopters who are both innovators (with a constant propensity to adopt service) and imitators (whose propensity to adopt service is influenced by the amount of previous adoption). To emphasize model dependence of its parameters, it is convenient to indicate the model as $B(t; M, p, q, t_s)$, $t \geq t_s$.

The Bass model describes growth with positive gradient. On contrary, test solving dynamics has negative gradient i.e. it declines. Therefore, prior to use the Bass model needs to be modified in declining Bass model. Declining growth is accomplished by mirroring of time variable axis, $t \to -t$ (Fig. 4).

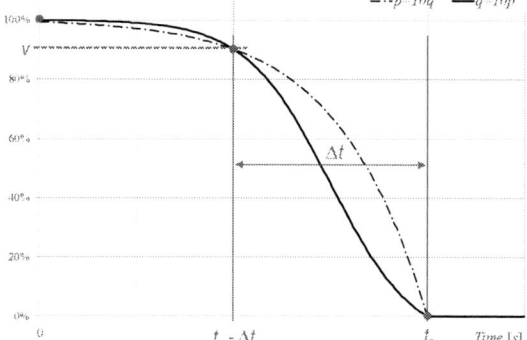

Fig. 4 The Declining Bass model (DBM): dependence on the coefficient of imitation q and coefficient of innovation p ($M = 100\%$, $v = 90\%$, Δt and t_e are fixed)

The new model, named **Declining Bass Model (DBM)** has the following form:

$$DB(t; M, p, q, t_e) = M \frac{1 - e^{-(p+q)(t_e-t)}}{1 + \frac{q}{p} e^{-(p+q)(t_e-t)}} \qquad (3)$$

Similar to (2), the DBM (3) is defined by four parameters: M – starting amount of customers; p – coefficient of innovation, $p > 0$; q – coefficient of imitation, $q \geq 0$ and t_e – time when service decease, $B(t \geq t_e) = 0$. Higher coefficient of innovation p, $(p > q)$

contributes to a steep disappearance of the service from market, and higher coefficient of imitation q, $(q > p)$ smooth disappearance of the service from market.

Due to fact that M is the asymptote of DBM, i.e. model (3) obtains value M only for $t \to \infty$, characteristic duration Δt should be defined through certain level smaller than M. (see description of v [%] in Section II – explanatory parameters definitions).

V. TEST SOLVING DYNAMICS MODELLING

For better understanding dynamics of test solving for successful attempts, data is modelled with DBM as time series.

The first step was fitting with **single DBM** where M (number of students solving test) and t_e (overall test duration) were fixed, i.e. values for them are taken directly from sample. The remaining parameters p and q are obtained by ordinary least squared method.

For case *Image processing test* - successful attempts, results are as follows (Fig. 5):

$\qquad M = 619$ students
$\qquad t_e = 1800$ s
$\qquad p = 3.38\text{E-}04$
$\qquad q = 3.48\text{E-}03$

Achieved correlation coefficient is $r = 0,9980$, and root-mean-square error is $RMSE = 11.511$.

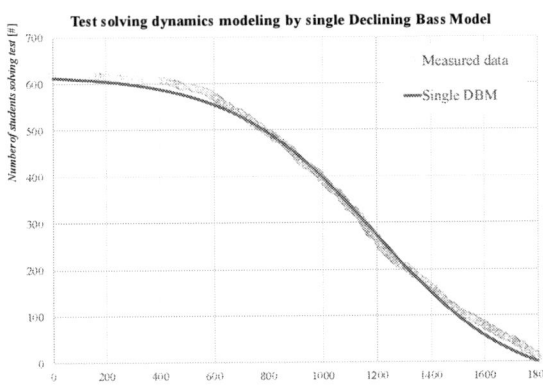

Fig. 5 Test solving dynamics modelling by a single DBM (case: *Image processing test* – successful attempts)

Students take the tests with different levels of preparation. Even when the analysis is done on the successful attempts only, the graph on Fig. 5 shows discontinuities in smoothness of experimental data. It can be assumed that this is result of inhomogeneous sample, i.e. a group of students is consisted of a better prepared ones and others. It is worth mentioning that unprepared students are not included (only successful attempts are analysed).

Following the abovementioned assumption, in **the next step** data is modelled with **doubled DBM** (4):

$$DDB(t) = DB_1(M_1, p_1, q_1, t_1; t) + DB_2(M - M_1, p_2, q_2, t_e; t) =$$

$$= M_1 \frac{1 - e^{-(p_1 + q_1)(t_1 - t)}}{1 + \frac{q_1}{p_1} e^{-(p_1 + q_1)(t_1 - t)}} + (M - M_1) \frac{1 - e^{-(p_2 + q_2)(t_e - t)}}{1 + \frac{q_1}{p_1} e^{-(p_2 + q_2)(t_e - t)}} \qquad (4)$$

The idea is to identify two groups of students – one better prepared that will successfully finish test in the first wave (parameters indexed with 1) and other one that will finish test in the second wave (parameters indexed with 2). According to that assumption, parameters of model (4) are prepared as follows:

$\qquad t_1 < t_e$
$\qquad M_1 < M$
$\qquad DDB(t \to \infty) = M$
$\qquad DB_1(t \geq t_1) = 0$
$\qquad DB_2(t \geq t_e) = 0$

In the first step **single DBM** has 2 free parameters (p and q), and **doubled DBM** has 6 (M_1, p_1, q_1, t_1, p_2, and q_2) which are obtained by ordinary least squared method, too.

For case *Image processing test* – successful attempts, result are as follows (Fig. 6):

$\qquad M_1 = 284$ students
$\qquad t_1 = 1300$ s
$\qquad p_1 = 9.20\text{E-}04$
$\qquad q_1 = 4.05\text{E-}03$
$\qquad M - M_1 = 335$ students
$\qquad t_e = 1800$ s
$\qquad p_1 = 9.55\text{E-}04$
$\qquad q_1 = 2.93\text{E-}03$

Achieved correlation coefficient is $r = 0.9995$, and root-mean-square error is $RMSE = 5.797$.

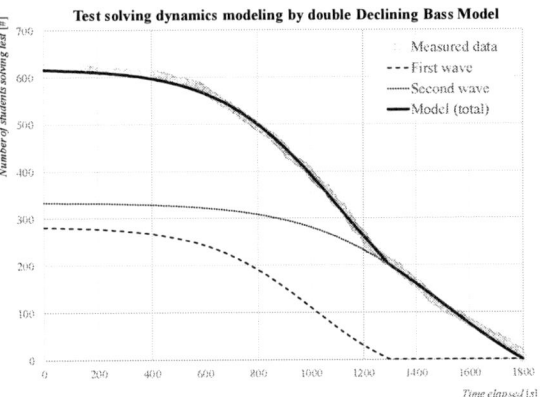

Fig. 6 Modelling by a doubled DBM showing two separate groups of students: better prepared and others (case: *Image processing test* – successful attempts)

Now when model that identifies groups of students according to their knowledge/skills is founded and it provides good fitting results, it is possible to determine intensity of test solving $I(t)$. The **intensity of test solving** is a rate of test solved with respect to the change of the time variable. It can be calculated as a negative derivative in time of doubled DBM's components (5):

$$I_i(t) = \lim_{\Delta t \to 0} \frac{DB_i(t) - DB_i(t+\Delta t)}{\Delta t} = -\frac{\partial DB_i(t)}{\partial t}, \quad i = 1, 2 \quad (5)$$

Intensity of test solving $I(t)$ for each separate group is graphically shown on Fig. 7 for case: *Image processing test* – successful attempts. Differentiation between groups are especially visible through different peak times of their intensity functions. Numerical results are as follows:

- Maximal intensity of solving for the first wave is achieved around $t_{m1} = 1005\ s$, and for second wave around $t_{m2} = 1515\ s$,
- Maximal values of $I(t)$ for both waves are almost equal $I_1(t_{m1}) \cong I_2(t_{m2}) = 0.433$.

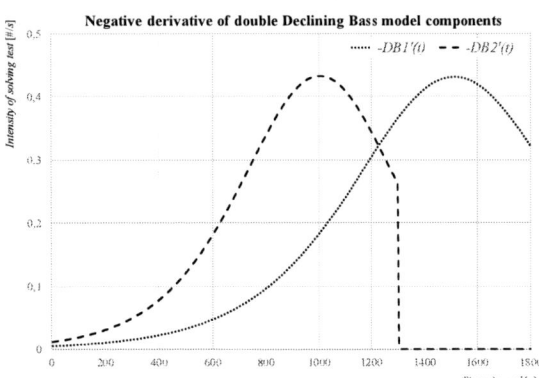

Fig. 7 Intensity of solving test according to two separate groups of students: better prepared and others (case: *Image processing test* – successful attempts)

From Fig. 7, it is evident that success of better prepared students will not be affected with appropriate test time reduction and/or number of tasks increase as it will affect the other ones.

VI. CONCLUSION

On account of the nine-years usage of the LMS Moodle on courses in professional study of electrical engineering at the Zagreb University of Applied Sciences it has been possible to perform a detailed and reliable processing of the acquired statistical data.

The data was analysed as a time series and explanatory parameters that are extension of the simple overall test duration parameter are defined: characteristic test duration, test completion slope gradient and peak of test solving intensity. These parameters are determined and evaluated for different knowledge and skills exams and could be used as time criteria optimization guidelines for similar LMS provided tests.

Based on the similarities with diffusion models, a new model named Declining Bass Model, is developed. If new model is applied in improved form (doubled DBM), the differentiation of better and barely enough prepared students can be obtained.

Further research will be focused on the quantifying abovementioned indicators for their practical applications in support of LMS Moodle usage for other courses, too.

REFERENCES

[1] T. Alajbeg; T. Horvat; T. Novak: The organization of classes and assessment system using Moodle, MIPRO, 2014

[2] T. Alajbeg; D. Ćika; T. Ražov: Using Moodle to automatically grade laboratory exercises, MIPRO, 2011

[3] Sokele, M., Growth models, Wiley Encyclopedia of Management Volume 9 (ed. N. Lee & A. M. Farrell), John Wiley and Sons, New York, 2015, (Online ISBN: 9781118785317)

[4] Sokele, M., Bass Model, The SAGE Dictionary of Quantitative Management Research (ed. L. Moutinho and G. D. Hutcheson), SAGE Publications Ltd., London, 2011, pp 18-23

Using Moodle in English for Professional Purposes (EPP) Teaching at the University North

Jadranka Lasić-Lazić*, Tomislav Ivanjko*, Iva Grubješić**

* Faculty of Humanities and Social Sciences, University of Zagreb / Department of Information and Communication Sciences, Zagreb, Croatia
** University North / Department of Civil Engineering, Varaždin, Croatia
jlazic@ffzg.hr; tivanjko@ffzg.hr; igrubjesic@unin.hr

Abstract – This paper discusses the possibilities of employing the process of the blended learning of professional English through the Moodle platform. It discusses different theoretical approaches to language learning implemented within the Moodle platform and the associated cognitive domains according to Bloom's taxonomy. The paper focuses on examining the Moodle activities within the language learning framework along with the levels of each activity within the cognitive domain. These activities are implemented at the University North within a wide range of courses covering English for Professional Purposes aimed at the engineering education. The paper can serve as a framework for the possible applications of different Moodle elements in the teaching of English for Professional Purposes.

I. INTRODUCTION

Rapid advances in technology and computer-aided teaching have offered new opportunities and enforced the application of new tools, methodologies, approaches and frameworks to promote active foreign language teaching and learning. English, as a global language, is probably the most commonly spoken language in the world when combining the native and non-native speakers, and currently the language most often taught as a foreign language in the Croatian higher education.

Creating a successful language learning environment within the higher education environment is becoming ever more complex. Information behavior and information preferences of students are also constantly changing and should be regularly examined. Faced with the new Web 2.0 services that include both the technological side by introducing new technologies and the new patterns of information behavior geared towards a more participative and socially oriented environment, new approaches based on individualization or autonomy are constantly examined.

In order to fulfill its mission of educating competent professional personnel for the needs of the real economy, University North implemented Moodle, an open-source Learning Management System (LMS), one of the currently fastest growing successful e-learning systems in the higher education environment, used in 231 countries and with nearly 100 million users worldwide [1].

In that context, Moodle has been receiving more and more popularity in higher education in the recent years, especially in the learning-teaching process applied in the English courses, offering a wide range of courses covering English for Professional Purposes. Teaching foreign languages for professional purposes is considered to be a priority in the updating of engineering education, and the ability to communicate in other languages is becoming an integral part of the professional competence of any specialist.

II. MOODLE AND LANGUAGE LEARNING APPROACHES

From the perspective of language learning approaches, Moodle implements two main approaches: the blended learning approach and the communicative approach.

The term *blended learning* originated around the year 2000 in the business world in connection to corporate training, followed by higher education, and finally it appeared in the world of language teaching and learning [2]. Blended language learning (i.e. integrating the use of technology into the classroom-based learning and teaching) is still a relatively new concept, but recent research appears to indicate that when "appropriately" implemented, blended learning can significantly improve the learning experience [3].

The communicative approach to language learning allows language teachers to follow the goal of making "the student the center of the learning experience wherever possible." Moodle supports the key features of Communicative Language Learning, some of which are learner autonomy, the social nature of learning, curricular integration, focus on meaning, diversity, alternative assessment, the role of teachers as co-learners, etc. [3].

While the blended learning approach provides the environmental setting within the Moodle platform, the communicative approach focuses on the roles of students and teachers. The connecting point between those two approaches, namely the learner and the environment, is the learning activity. For this reason, the design of blended learning needs to be centered on the activity design [4].

Along with the two approaches discussed, it is important to indicate and categorize the activities based on their learning outcomes. One of the most widely applied taxonomies of learning outcomes is Bloom's taxonomy that identifies six levels within the cognitive domain, from the simple recall or recognition of facts as the lowest level, through the increasingly more complex and abstract mental levels, to the highest order which is classified as evaluation [5]:

1. *Knowledge:* define, list, name, order, recognize, relate, recall, repeat,
2. *Comprehension:* classify, discuss, explain, identify, indicate, report, review, select,
3. *Application:* apply, choose, demonstrate, sketch, solve, use, write,
4. *Analysis:* analyze, calculate, compare, contrast, discriminate, examine, experiment,
5. *Synthesis:* assemble, construct, create, design, develop, formulate, prepare, propose, write,
6. *Evaluation:* assess, attach, choose, compare, predict, rate, select, evaluate.

In order to select the best approach in EPP teaching (using the most appropriate activities within the Moodle platform to develop targeted outcomes), it is necessary to examine the Moodle activities within the language learning framework along with the levels of each activity within the cognitive domain.

III. MOODLE ACTIVITIES IN EPP TEACHING

In the last few years, while trying to improve the management of the English language course at the University North, Moodle has emerged as an invaluable asset of that process. Considering the fact that the course is aimed at teaching a foreign language for professional purposes, a certain level of autonomy is required from students, and that is seen as an additional means of developing the students' skills needed in the professional environment. Students who participate in Moodle-based courses not only gain knowledge in the areas stipulated by the curriculum, but they also develop practical skills such as decision-making, applying critical evaluation, and problem solving [6]. As a learning platform that allows students to access the contents of the course 24/7, students are given a certain amount of autonomy in their learning process. They can access the lectures and assignments at any time and it is up to them to decide when and how fast they want to go through the course and cover the discussed topics [7]. In lieu of being pressured into learning in a traditional class environment only, they exercise individual control over the speed of their work by being given the option to learn and revise any material that they have missed or insufficiently acquired during the face-to-face lectures.

TABLE I. MOODLE USAGE STATISTICS AT THE UNIVERSITY NORTH (2016)

Moodle module	
NO. OF COURSES (ACTIVE)	647 (300)
NO. OF ENROLLED STUDENTS	2000
FORUM	724
LABEL	1557
QUESTION	465
RESOURCE	12 937
TEST	800

Moreover, Moodle offers a wide range of activities that can be used to improve the learning of a language. The activities and their advantages in EPP language teaching are presented and discussed in the next section.

Chat - Students can get together for a chat session which allows them to enter into dialogue with each other. By discussing a certain topic with each other and the teacher, they practice the use of the language and they draw each other's attention to the possible errors they might have overlooked, or they give each other additional ideas for the solving of the given task. However, it is important for the teacher to offer guidance when needed and to steer the wheel if the students go off track [8]. Levels of activity within the cognitive domain: Comprehension; Application; Synthesis; Evaluation.

Database - With the help of the Database module, teachers can help students build vocabulary lists through which students master a specific professional matter introduced during the lecture. The database can function as a personal glossary which students can consult when they revise for the exam. Levels of activity within the cognitive domain: Knowledge, Application, Analysis, Synthesis, Evaluation

Forum - Forum is an activity that allows a group of Moodle users to ask and answer questions [8]. Students can ask the teacher and each other for help by asking questions, and at the same time help others by answering their questions and giving their input on a certain matter. Moreover, students can subscribe to a certain forum and receive an e-mail when someone offers a new answer to the posed question. As with the Chat module, it is important to give students a certain level of autonomy, i.e. to make them feel as if they were in charge and thus give them the option to express their views and opinions more freely, without the sense that they are being "monitored". Levels of activity within the cognitive domain: Comprehension, Analysis, Synthesis, Evaluation.

Hot Potatoes - This module is defined as a free, easy-to-use quiz-making program ideal for setting up quick quizzes to review or test students [8]. This module can cover the testing of basically all main elements of a language, from grammar and vocabulary to reading and listening. Cloze tests, crosswords, matching exercises and multiple-choice are the most common types of Hot Potatoes tests [3]. They are quite attractive for students because they require their full attention when eliminating the incorrect answers, but also because they are allowed to use the "back" button, which in a way allows them to "cheat" and improve their results. Levels of activity within the cognitive domain: Knowledge, Comprehension, Application, Analysis, Synthesis.

Journal - This module allows students to keep a diary where they can reflect on their learning process and discuss the matters that they did not manage to master during class and have thus been left unclear. There they can also write drafts of their writing assignments. Considering the fact that it is only the student and the teacher who have access to this module, they can ask the teacher questions about a certain lecture or matter that they do not feel comfortable enough to ask in the Forum. They can then correct their assignments in accordance with the teacher's comments [7]. Levels of activity within the cognitive domain: Comprehension; Analysis.

Lesson/Lecture - The Lesson/Lecture module requires a lot of patience from teachers because it is quite complicated to set up, but at the same time it is very rewarding as it allows them to progress step by step through a lesson. For example, students are given a certain text to read, and then they are given a question accompanying that text. If they answer the question correctly, they move on to the next page where a next question or assignment awaits them. On the other hand, if they answer the question incorrectly, they can either take the question again or choose to go to an easier one. This module can be used for the presentation of cultural information, teaching grammar, reading professional texts, listening, etc. [7]. Levels of activity within the cognitive domain: Knowledge, Comprehension, Application, Analysis, Synthesis, Evaluation.

Questionnaire - The Questionnaire module is a customizable survey which can be very useful for getting feedbacks or opinions of various aspects of the course, or for getting students to write their own surveys and practice asking questions [8]. Moreover, by asking them questions about a certain text that they had to go through, the teacher can get a solid evaluation of the students' preferences and on the basis of the results, work on the improvement of the materials. Levels of activity within the cognitive domain: Analysis, Synthesis.

Dictionary - This module serves a similar purpose as the Database module. Students make vocabulary lists to which they add definitions, and that can help them master a particular matter in a more concrete, substantial way. It also allows them to find in one place the vocabulary they need in order to pass the course and that potentially shortens the time they would usually need to prepare for the exam. Levels of activity

within the cognitive domain: Knowledge, Application, Analysis, Synthesis, Evaluation

SCORM – The abbreviation stands for Shareable Content Object Reference Model. Teachers can save Hot Potatoes quizzes as a SCORM file, which bundles all the files in the activity and teachers can then import the SCORM file directly into the course home page [8]. SCORM files enable a more synoptic reading of the data due to the fact that what SCORM does for the programs is equal to what PDFs do for documents. [8] Levels of activity within the cognitive domain: Application.

Survey - A Survey is comprised of questions with slightly different answers, which then allows the teacher to collect the answers in the form of a percentage and thus get a better view of the students' opinion on a certain topic. Teachers are yet again given the opportunity to get an insight into the students' opinion of the course and on the basis of that they can implement certain changes and improve the course. Levels of activity within the cognitive domain: Analysis, Evaluation.

Test - This module allows teachers to conduct continuous assessment and/or final assessment via Moodle. Teachers can create a number of questions for the evaluation, limit the time the students have for solving the questions, and based on their performance, and decide on the final grade. Levels of activity within the cognitive domain: Knowledge, Comprehension, Application, Analysis, Synthesis, Evaluation.

Wiki - As the name discloses, this module functions as the Wikipedia webpage, meaning that its content can be edited by all users, i.e. course participants. This module is good for collaborative work such as joint writing projects and task planning [8]. This module enforces the sense of team work, while at the same time it allows the students to keep a certain level of autonomy. Levels of activity within the cognitive domain: Application, Analysis, Synthesis, Evaluation.

Assignment - The teacher gives out an Assignment and posts the instructions in the activity description. Students have to respect the deadline when they submit their Assignment as an attachment. The Assignment is then reviewed and assessed by the teacher, who is the only one who can see the submitted file. This module often goes hand in hand with the Test module in the sense that it plays a crucial role in the final assessment. Levels of activity within the cognitive domain: Comprehension, Application, Analysis, Synthesis, Evaluation.

Workshop - Workshop is similar to the Assignment module, the only difference being that as opposed to the Assignment module, it can be peer-reviewed before the file is submitted, meaning that it is not only accessible to the teacher, but also to other students who can write their own comments and contribute to the final version. Levels of activity within the cognitive domain: Application, Analysis, Synthesis, Evaluation.

As we can see, Moodle offers a wide range of learning activities that can be implemented in the EPP teaching. The activities are imbued with a number of advantages, both for the students and the teachers. Firstly, the students are given a much higher level of autonomy not only on the basis of source availability, but also on the basis of modules such as Chat and Forum, where they can ask both the teacher and their colleagues questions that can help them in mastering the course. As it is often the case, some students might find it easier to ask a question via an online platform than during a face-to-face lecture where they feel more exposed to the reaction of their environment. Such modules increase the level of student activity and investment. Moreover, students of technical studies often shift their focus away from foreign language courses towards their main interest, the technical courses [4]. Under such circumstances, it is necessary to spark their interest by engaging them in cloze tests, multiple-choice exercises, crosswords, and matching exercises offered by the Hot Potatoes module. One additional advantage offered by Moodle through the Hot Potatoes module is the involvement of multimedia tools [7]. The Hot Potatoes module not only makes language learning more interesting to students by employing a variety of game-like teaching possibilities, but it also enhances their ability to focus on particular information and it develops their skills of information processing [9]. Moreover, what is one of the most important features of Moodle is the development of student autonomy, which stays with the individual far beyond the completion of the EPP course. Throughout the years, it has been observed at the University North that with the blended learning method that is implemented in the EPP course, students have become more active and interested in the lectures, and with each assignment, the number of students who fulfill all the requirements introduced at the beginning of the course is growing.

IV. THE TEACHER'S ROLE

As is the case with face-to-face lectures, the role of the teacher is indispensable in the blended learning method. Firstly, teachers should always be on the alert when it comes to their professional expertise due to the fact that the area of ICT advances rapidly and if they wish to offer their students the best possible working environment, they have to follow all the novelties that are constantly being introduced. Teachers have for a very long time been blending face-to-face lectures with various kinds of technology-assisted methods [8]. However, with the upsurge of Web 2.0, teachers have been given a very powerful tool which can help them with the advances in the foreign language teaching. Hence, it is crucial that they be methodologically careful when choosing the materials for their students and the means through which

they will implement them. Back in 1999, Egbert and Hanson-Smith suggested eight "optimal" conditions for learning a foreign language, and one of them stresses the importance of learner autonomy [10]. Much of the learner autonomy lies in the hands of the teacher. It is on the teacher to encourage students to actively participate in the discussions, but yet again, they should not be too overwhelming in doing so [8]. One of the best modules that plays a significant role in the development of student autonomy is the Forum, which empowers and encourages class members to give answers to their classmates' questions, which not only supports camaraderie, but also gives students the sense of "being in charge", whereas the teacher is no longer seen as the only source of knowledge [8]. The students then not only feel autonomous, but they also become more motivated. However, autonomy aside, it is of extreme importance to set boundaries. Naturally, the whole point of the blended learning method is to allow students to decide by themselves when they want to study, but it is up to the teacher to ensure that they have indeed met all the requirements by the end of the course. Therefore, it is important to clearly state and explain the goals and objectives of the course and set the necessary deadlines when it comes to the enforcement of modules. Through the implementation of rules, certain managerial skills that they will need in their future professional life are developed.

Last but not least, feedback. When it comes to all types of learning, but especially language learning, it is very important for students to receive feedback as a confirmation that their learning style and strategies have been successful, or as an input on what should be modified or additionally worked on in order to reach the desired level.

V. CONCLUSION AND FUTURE WORK

The role of the blended learning method has in the last couple of years become increasingly popular, especially when it comes to the teaching of foreign languages. Moodle, an open-source learning management system, was introduced at the University North as a supplement to face-to-face lectures and it has proved to be of vital importance in the EPP teaching. Employing a variety of Moodle modules, the EPP course has progressively been working on the development of student autonomy and achieving the optimal conditions for learning a foreign language. Even though there is still room for improvement in the use of the modules, the blended learning method has been recognized as a tremendous asset in the EPP teaching, as it gives an exceptional opportunity to integrate the information advances with the advances of students both in their studies and their future professional life.

In this light, this paper can serve as a framework for the possible applications of different Moodle elements in the teaching of English for Professional Purposes. Future work in the field should be aimed at carrying out case studies of applying different Moodle activities based on the proposed theoretical framework and measuring the student competence and satisfaction level within the blended learning approach.

REFERENCES

[1] Moodle.net, Moodle statistics. Available at: https://moodle.net/stats/ (Accessed 3 February 2017)

[2] B. Tomlinson and C. Whittaker, Blended Learning in English Language Teaching. London: British Council, 2013.

[3] D. Marsh, Blended Learning: Creating Learning Opportunities for Language Learners. New York: Cambridge University Press, 2012.

[4] H. He and B. Zhu, "Blended learning of professional English for computer science based on Moodle," in Proceedings 2013 Fourth International Conference on Networking and Distributed Computing, L.

[5] J. Rutkowski, K. Moscinska, and P. Jantos, "Application of Bloom's taxonomy for increasing teaching efficiency–case study," unpublished, International Conference on Engineering Education ICEE-2010, July 2010. Available at: http://www.ineer.org/Events/ICEE2010/papers/W13A/Paper_1292_141 7.pdf (Accessed 3 February 2017)

[6] O. I. Shaykina, "Blended learning in English language teaching: open educational resources used for academic purposes in Tomsk Polytechnic University," Mediterranean Journal of Social Sciences, vol. 6, no. 3, 2015, pp. 255-260.

[7] T. Krasnova and T. Sidorenko, "Blended learning in teaching foreign languages," in Conference Proceedings: ICT for Language Learning: 6th Conference Edition, Pixel, Ed. Limena: Libreriauniversitaria.it, 2013, pp. 45-50.

[8] J. Stanford, Moodle 1.9 for Second Language Teaching. Birmingham: Packt Publishing, 2009.

[9] L. Pospíšilová, Z. Bezdíčková, and D. Ciberová, "English for science using LMS Moodle," in 2011 14th International Conference on Interactive Collaborative Learning - 11th International Conference Virtual University. New York: IEEE, 2011, pp. 169-171.

[10] J. Egbert and E. Hanson-Smith, Call Environments: Research, Practice, and Critical Issues. Alexandria, VA: TESOL, 1999.

Air Traffic Controllers' Practical Part of Basic Training on Computer Based Simulation Device

M. Pavlinović*, B. Juričić* and B. Antulov-Fantulin*

*Faculty of Transport and Traffic Sciences/Department of Aeronautics, Zagreb, Croatia
mira.pavlinovic@fpz.hr
biljana.juricic@fpz.hr
bruno.antulov@fpz.hr

Abstract - Air traffic controllers are responsible for guiding aircraft through the airspace and for ensuring timely, safe and expeditious flow of air traffic. Throughout the whole education and later work, air traffic controllers undergo extensive training that is divided into three phases – initial training (basic and rating training), unit training (transitional, pre-on-the-job and on-the-job training) and continuation training (conversion and refresher training). In all three phases of the training, the focus is placed on practical training exercises performed on computer based simulation devices. This paper focuses on basic part of the initial training and gives an overview of practical training exercises performed on a particular computer based simulation device. It also provides an insight into challenges the candidates are faced with while mastering the techniques of performance based training

I. INTRODUCTION

Air traffic controllers are responsible for safe and expeditious flow of air traffic. Their main task is to provide separation between aircraft. They perform a complex task of guiding aircraft through the airspace safely and efficiently. The goal of Air Traffic Control (ATC) is to minimize the risk of aircraft collisions while maximizing the number of aircraft that can fly safely in an airspace at the same time. To be able to perform all the mentioned duties, air traffic controller candidates have to undergo extensive training and skill acquisition process. During the training, air traffic controller candidates acquire knowledge on procedures, learn about characteristics of a certain airspace, master how to detect and solve potential conflicts between aircraft, learn how to use equipment on their working positions, etc. This paper focuses on basic part of the initial training and gives an overview of practical training exercises performed on computer based simulation devices. It is divided into five sections: 1. Introduction, 2. Air Traffic Controller's Training, 3. Air Traffic Controller's Basic Training, 4. The Simulator - BEST Simulation System, and 5. Conclusion.

II. AIR TRAFFIC CONTROLLER'S TRAINING

Since air traffic control is a highly regulated sector, air traffic controller's (ATCO) training is defined by thorough regulations that prescribe minimum training requirements. In the EU ATCO training has to comply with and meet the requirements laid down in the Commission Regulation (EU) 2015/340 that enables overall standardization of training.

Standardization of training and required competences should also reduce fragmentation and differences in licensing process and enable mutual recognition of licenses among different countries [1].

The training is divided into three phases – initial training, unit training and continuation training that are shown on Figure 1.

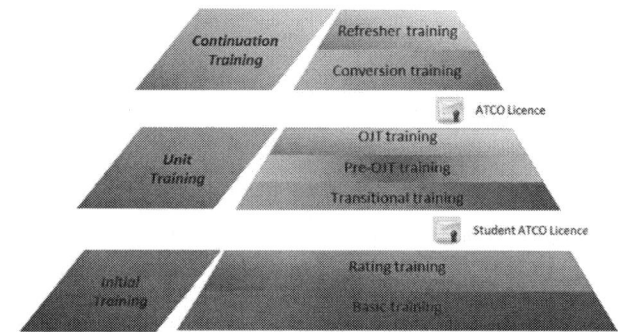

Figure 1. Progression of ATCO training [2].

Initial training consists of basic training and rating training. Basic training is defined as theoretical and practical training designed to impart fundamental knowledge and practical skills related to basic operational procedures [2].

Basic Training provides theoretical knowledge and practical skills to enable an ab initio candidate to progress to more specialized Rating Training. The Rating Training provides knowledge and skills related to a job category and appropriate to the discipline to be pursued in the ATS environment [2]. It consists of theoretical subjects and practical exercises. After successful completion of initial training candidates are awarded Student ATCO License. This license is a prerequisite for starting the following phase of training.

The Unit Training leads to the issue of an Air Traffic Controller License that enables ATCOs to work with live traffic. It is subdivided into three parts: Transitional Training, Pre-On-the-Job Training and On-the-Job Training.

Transitional Training is designed primarily to impart knowledge and understanding of site-specific operational procedures and task-specific aspects. It ensures the development of skills through the use of site-specific simulations and training [3].

Pre-On-the-Job Training (Pre-OJT) is locally based training during which extensive use of simulation using site-specific facilities will enhance the development of previously acquired routines and abilities to an exceptionally high level of achievement [2].

On-the-Job Training - OJT is the final phase of unit training during which previously acquired job-related routines and skills are integrated in practice under the supervision of a qualified on-the-job training instructor in a live traffic situation. [2].

Continuation Training is training for ATCOs with valid license that enables an upgrade or improvement of existing knowledge and skills and includes refresher and conversion training.

Refresher training is designed to review, reinforce or enhance the existing knowledge and skills of air traffic controllers to provide a safe, orderly and expeditious flow of air traffic. Conversion Training is designed to provide knowledge and skills appropriate to a change in the operational environment [2].

All the phases of training consist of theoretical and practical training. Practical training is performed on computer based simulation device, so called synthetic training device (STD). There are two different types of STDs - simulators and part-task trainers. Simulators are computer based devices that simulate important functions of the real situation of ATCO working positions, airspace, procedures, flight trajectories etc. Part-task trainers are computer based devices that enable simulation of partial ATCO functions. Both STDs are used to train candidates in gaining practical skills. Candidates take part in different practical exercises created for each segment of the training according to prescribed requirements. Each certified ATCO training organization defines the number of practical exercises to be performed for different phases of ATCO training in accordance with international standards.

III. AIR TRAFFIC CONTROLLER'S BASIC TRAINING

As previously mentioned, Basic Training is a part of Initial Training. Basic training can be provided as separate course or integrated with rating training. Basic Training is designed in a way that candidates (ab initios) acquire fundamental knowledge and practical skills related to basic operational procedures. The goal of the course is to teach candidates basic theory needed for future work and for rating training. Basic Training is developed and provided by licensed ATCO training organizations and approved by the competent authority.

In Croatia there is only one training organization approved for provision of ATCO basic training and that is Croatian Air Traffic Control Training Centre (HUSK).

HUSK is a unit established at the Faculty of Transport and Traffic Sciences of the University of Zagreb that is certified to provide basic training. HUSK provides two different basic training plans and programs: integrated program of training provided through undergraduate study of aeronautics, air traffic control module, and a separate basic training course [4].

The training program integrated in the undergraduate study of Aeronautics lasts 6 semesters. After the completion of the program, candidates are awarded bachelor's degree (bacc.ing.aeronaut.) and a Certificate on successful completion of Basic Air Traffic Controller Training [5]. The other training program is organized as a separate course lasting from 16 to 24 weeks depending on tender requirements. After the completion of the course, candidates are also awarded Certificate on successful completion of Basic Air Traffic Controller Training [6].

Both HUSK programs are compliant and harmonized with requirements of the EU REG 2015/340 and of EUROCONTROL's Specifications on the ATCO Common Core Content Initial Training.

During the basic training candidates are obliged to take lessons and successfully fulfil courses' requirements. Theoretical trainings are comprised of the following subjects as prescribed by EU REG 2015/340: Introduction to the Course, Aviation Law, Air Traffic Management, Meteorology, Navigation, Aircraft, Human Factors, Equipment and Systems and Professional Environment,

According to the EU REG 2015/340 practical training is provided within Air Traffic Management subject. During the provision of practical training on simulator, candidates develop skills of maintaining aircraft separation, monitoring aircraft movement through airspace and communicating with pilots. Candidates have to incorporate acquired theoretical knowledge into practical training skills for all three types of air traffic control (aerodrome, approach and area control).

IV. THE SIMULATOR - BEST SIMULATION SYSTEM

There are a few companies in the world involved in the development and production of ATC simulators that provide a realistic simulation of aircraft flight and human-in-the-loop real time simulation of air traffic controller's work. All simulation systems used by ATCO training organizations should be certified and approved by competent authorities for the usage in practical training. A simulation system that is very often used by ATCO training organizations in Europe is called BEST (Beginning to End for Simulation and Training) Simulator produced by Micro Nav Ltd. According to HUSK internal survey and according to the research Comparison of Radar Simulator for Air Traffic Control [7], BEST simulator is rated as the best solution to be used in ATCO training.

BEST simulator covers all levels and types of training:
- basic (ab initio)
- rating
- validation
- on-job-training support
- conversion
- refresher
- competency checks

- handling emergencies and unusual situations
- approach control – radar and non-radar
- tower control
- ground and ramp control
- tower data assistant working
- civil and military [8].

BEST is constructed in such a way that it gives a very realistic simulation of the real traffic situations. The following section describes the system functionality and the minimum hardware requirements, software information and additional system and support information.

A. Hardware Requirements

BEST runs on commercial standard PCs, networking and peripherals. The minimum hardware specifications are Pentium 4 dual core processor 2.8GHz or equivalent, 2GB RAM, 40GB hard disk drive (workstation position) or 80GB hard disk drive (system manager position), CD-ROM drive (required at the system manager position only), 100Mbps/1Gbps network capability and operating system Windows 7 Professional(workstation position or stand-alone system) and Windows 2003 Server (networked System Manager) [8].

B. BEST Software Development Platform

The BEST software is written in C++ using the Borland C++ Builder development environment and it uses Windows operating systems and networking. BEST simulation system comprises of the following ATC radar simulation and training facilities:

- ATC radar controller/student facilities
- Supervisor facilities
- Pilot facilities
- Simulation facilities for driving real radar workstations
- System manager facilities
- Data preparation facilities
- Self-teach facilities
- Voice recognition & output facilities
- Simulated audio communications facilities
- Scripting facilities
- Networking facilities
- Data management facilities [8].

C. BEST simulaton system at the Faculty of Transport and Traffic Sciences

As it was mentioned earlier, HUSK is an approved ATCO training organization that is certified for provision of basic training and is a part of the Faculty of Transport and Traffic Sciences. From 2013the Faculty owns BEST Radar Simulator that is situated in The Laboratory for Control of Air Navigation at the Department of Aeronautics and used by HUSK. It has two ATCO working stations, one pseudo-pilot working station and one supervisor working station which is also used as a pseudo-pilot working station. The user interface is similar to the real ATC workstations. As it can be seen on Figure 2, the ATCO working station consists of a radar screen, an auxiliary screen, a voice communication interface screen, a keyboard, a mouse, two sets of headphones and some communication switches.

Standard operating system has recording and playback system, which allows a recorded exercise to be played back. A replay may be paused at any time and candidates may be faced with their performance.

Figure 2.The ATCO working stations of the BEST Radar Simulator at the Faculty of Transport and Traffic Sciences

Process of training and roles of all persons involved as well as their interactions can be seen on Figure 3.

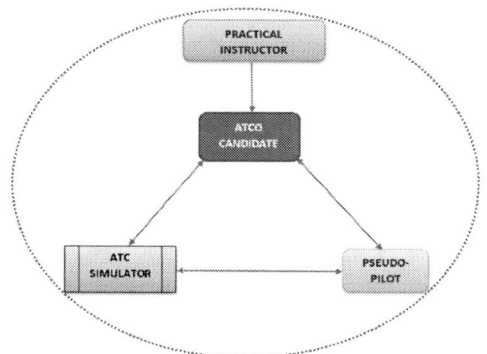

Figure 3. Simulation process during ATCO practical training.

As it can be seen, there are constant MMI (Man–Machine Interface) interactions in three processes: ATC Simulator-ATCO candidate, ATC Simulator -pseudo-pilot and ATCO candidates and pseudo-pilots when using radio communication. Practical instructor supervises the work of ATCO candidate. Roles and functions of all participants are explained in further text.

D. The ATCO working station

The ATCO working station enables the ATCO candidate to monitor the air traffic situation on the radar screen and to pass instructions through a set of headphones to a pseudo-pilot. The candidate has to perform a complex task of monitoring all the aircraft through the delegated airspace.

At the same time the candidate has to take into consideration the simulated traffic situation, decide what instruction to issue to the pilot, inform the pilot through a voice channel on the actions to be taken, observe if the pilot follows the instructions, record all the changes in the system, make an update of the current traffic situation and be familiar with the all equipment. The candidate continuously observes the simulated traffic situation on the radar screen that shows the current traffic situation. The standard radar screen display is shown on Figure4. The position of an aircraft is marked with an * sign and accompanied with the label that mandatory contains aircraft's call sign some additional information regarding aircraft type, speed and altitude.

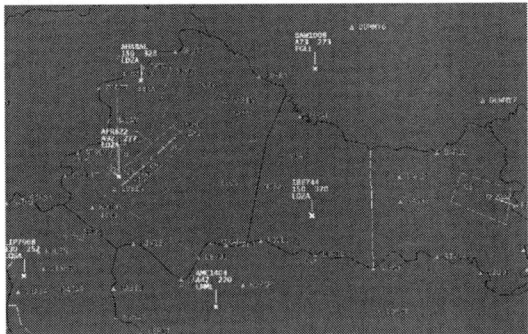

Figure4. The standard radar screen display.

The display controls are located on the toolbar at the top of the display. Different characteristics of the radar display can be altered by selection of an appropriate tool such as:

• display manipulation tools (zoom facility, recentre, drag facility, map layer selection, display range, range rings, compass rose, and text size)

• target display options (history trail, predict vector, track label position, target filters route and trail length)

• information display (exercise time and status (active/frozen), QNH and ATIS identifier

• target symbols: primary cover only, secondary cover only, combined cover, forced onto display, filtered by height or SSR code), etc. [8].

E. Pseudo- pilot working stations

The BEST pseudo-pilot working station facilities are user-friendly. The pseudo-pilot working station consists of a radar screen, an auxiliary screen, a voice communication interface screen, a keyboard, a mouse, a set of headphones and communication switches. The pseudo-pilot facilities allow the pilot to make changes in flight's behavior as instructed by the controller or instructor. This is achieved via Pilot MMI that is shown on Figure 5.

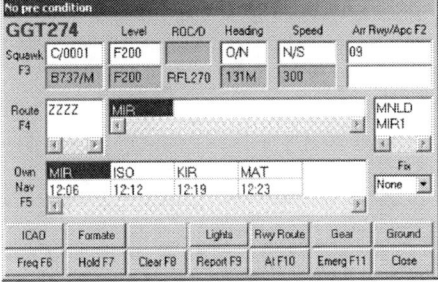

Figure 5. The Pilot Man–Machine Interface [8].

The BEST pilot interface uses standard Windows input methods working with a keyboard or a mouse, or with a combination of both. This allows complex, chained commands to be entered easily. All the inputs are optimized for fast keyboard entry [8].

All aircraft under pilot's control are listed in the *Call sign List*. The target aircraft can be selected with the keyboard or with the mouse by selecting from the radar display or the *Call sign List*.

Using Pilot MMI, the pseudo-pilot can enter a command for the currently selected aircraft or can select another aircraft and make necessary changes. When the keyboard is used to select an aircraft, pressing the letter key displayed in the first column of the *Call sign List* will display the *Pilot MMI* for the associated aircraft [8].

F. Practical Instructor role in simulation process

Practical instructor is instructor authorized to train candidates during practical exercises. Practical instructor sits next to the ATCO candidate. He/she instructs the candidate, provides advice and tuition, observes how a candidate interacts with the equipment, pays attention to candidate's behavior and attitude, monitors if a candidate takes appropriate actions and issues correct clearances. After each exercise, the instructor fills in Daily performance lists for each of the candidates giving his/her opinion on the candidate's performance.

Practical instructor can communicate with the ATCO candidate and with the pseudo-pilot (without the controller's knowledge) via voice communication facilities. The voice communication facilities support simulated radio transmissions between the candidate and pseudo-pilot, communication between the instructor and the candidate and pseudo-pilot for teaching purposes, simulated intercom communications between positions and simulated telephone communications.

G. Supervisor working stations

Supervisor working station is the central part of the complete simulation system. A supervisor has the access to the pseudo-pilot working station facilities and to the majority of functions of student-controller facilities. The supervisor is in charge of managing an exercise – he/she starts the exercise, restarts the paused exercise, changes the traffic situation adding additional flights, removes flights, saves a text file of all script commands run during the course of the exercise and edits the script files associated with the airspace, exercise and/or weather set etc. When the simulator is ready to run, the *Supervisor Controls* dialog is presented on the screen at the supervisor position.

H. Provision of Basic Training at HUSK

As it was said earlier HUSK uses BEST Radar Simulator for the provision of practical exercises during basic training. When the program integrated through undergraduate study of aeronautics is used, candidates do seven exercises of Aerodrome Control, seven of Approach Control and seven of Area Control. Each candidate does altogether 21 exercises (seven for each type of control) and spends 21 hours training on BEST radar simulator individually.

Exercises of each type of control differ in traffic, number of aircraft involved, number of conflicts, different traffic flows and in performance objectives that candidates need to fulfil. Before the beginning of practical training for each type of control, the candidates are introduced to the simulator equipment and its functionalities. Provision of every exercise is conducted in accordance to the activities given in Figure 6.

Every practical exercise on the simulator starts with group briefing. It is an introduction to the exercise where candidates are introduced with the performance objectives, particular characteristics of the exercise, separation methods and radio-telephony communication. It is provided in the classroom. Group briefing is followed by individual briefing when instructor leads candidate through an exercise, explains the goals of the exercise and emphasizes the important segments, shows flight plans. Individual briefing is as done one-on-one and provided at the BEST simulator.

The candidate does the exercise run on his/her own.

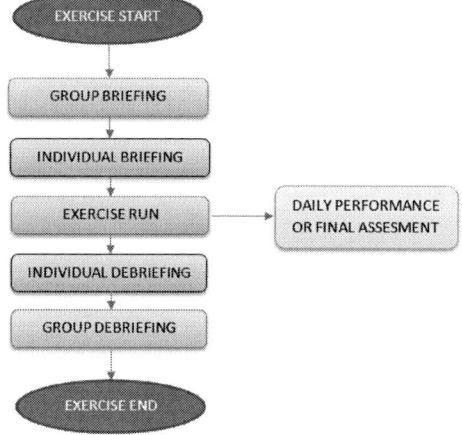

Figure 6. Activities in practical exercise provision

Exercise run lasts approximately 45 min. During the exercise run, instructors train and monitor candidate's work on BEST radar simulator. The instructor observes candidate's performance and provides guidance, if necessary. They use Daily performance lists to track and evaluate candidate's progress or do the final assessment of the candidate's performance using Simulator Assessment List in the case of the last exercise.

After the completion of the exercise run, the instructor does the individual debriefing and evaluates the candidate's work according to the set objectives of the exercise, points out candidate's strengths and weaknesses and gives the recommendations for the subsequent exercises. Individual debriefing is provided while the candidate is still at the simulator.

As it is said, the instructor uses Simulator Assessment List to make final assessment and evaluation of the candidate's knowledge, performance and attitude. A candidate needs to achieve at least 75% on the final assessment to pass.

For each of the exercises, each candidate is awarded the following performance indicators:

Excellent (90-100%) - the candidate has reached the highest passable standard; the performance was excellent but with a few minor errors.

Very good (85-89%) - the candidate has reached higher passable standard; the performance was very good but with more minor errors.

Good (80-84%) - the candidate has reached passable standard; the performance was good but with errors.

Sufficient (75-79%) - the candidate has reached the lowest passable standard; the performance was acceptable but with major errors.

Insufficient (74 % and bellow) - the candidate has failed to reach a passable standard; the performance was less than acceptable and/or erratic [9].

The last segment of each exercise is group debriefing. When all candidates finish the same exercise, a group debriefing is held. It is review of the exercise run and discussion of the outcome of the exercise and candidates' achievements and gives the recommendations for the subsequent exercises.

The instructors repeat the objectives of the exercise, summarize the key points, emphasize well-performed elements, discuss the most frequent mistakes and provide advice for the future.

After completion of the Basic Training simulation exercises on BEST radar simulator the candidate can perform the following performance objectives [3]:

- checking and using the working position equipment
- developing and maintaining situational awareness by
- monitoring traffic and identifying aircraft when
- applicable
- monitoring and updating flight data display(s)
- maintaining a continuous listening watch on the
- appropriate frequency
- issuing appropriate clearances, instructions and
- information to traffic
- using approved phraseology
- communicating effectively
- applying separation
- applying coordination as necessary
- applying the prescribed procedures for the simulated
- airspace
- detecting potential conflicts between aircraft
- appreciating priority of actions
- choosing appropriate separation methods.

I. Example of area control exercises at HUSK training organisation

To achieve smooth and performance based training process at HUSK, all candidates undergo through the set of exercises specifically designed for basic training. All exercises are designed with the assistance of certified practical instructors and are carefully constructed to achieve candidate's continuous progression in skills and are in accordance with performance objectives. As it was said earlier candidates attend three types of simulator exercises, aerodrome, area and approach control exercises and are trained and supervised a by certified practical instructors

To show progress and differences in the exercises during basic training at HUSK, exercises for area control will be explained in more detail in the terms of ATCO workload, number of aircraft and traffic complexity.

Exercise number one is the first exercise for area control. Before the beginning of this exercise, instructors introduce and demonstrate the functionalities of the computer based training device – BEST simulator. All its segments needed for performance of area control exercises are explained in detail and shown to candidates. The objectives of exercise number one are familiarization with the equipment, introduction of generic airspace and basic scenario with low complexity traffic and low ATCO workload requiring heading change. The number of aircraft in the first exercise is four.

In the second exercise, candidates work in the same generic airspace, but are faced with slightly higher workload and traffic complexity in basic scenario. The number of aircraft is increased to ten.

Croatian upper control area is simulated in exercise number three. Candidates go through scenario with overflying traffic and intermediate complexity and ATCO workload requiring heading, level, speed and frequency changes. There are eight to nine aircraft in this exercise.

The fourth exercise is similar to the third exercise, but with slightly higher workload. The scenario includes converging traffic. The complexity is intermediate requiring again some heading, level, speed and frequency changes. The number of aircraft is nine.

Exercise number five is the same as the fourth exercise. Complexity and ATCO workload are the same as in the fourth exercise, only the number of aircraft is increased to ten. This exercise enables candidate's consolidation in progress. Exercise number six is the last one before the final assessment exercise where candidates deal with the highest traffic load on crossing and opposite tracks resolving conflicts by level change or by vectoring. This is the most complex exercise with the highest ATCO workload The number of aircraft in this exercise is twelve.

Exercise number seven is the last exercise and also the assessment exercise. The final assessment and evaluation are carried out during this exercise according to the criteria said in the part H of this paper. In the course of this exercise, the candidates should show what skills and competence they have gained during the practical training. After the completion of practical exercises a candidate shall reliably and consistently apply standard coordination, approved radiotelephony, vertical, longitudinal and radar separation and control techniques [10].

If we compare candidates' behavior and attitude during the performance of exercises, it can be concluded that during the first two exercises candidates have difficulties with the adaptation to the computer based simulation device and familiarization with its functions. With the progression of exercises it is expected that candidates are adopted and familiarized with the system and that they are capable of coping with higher number of aircraft, higher ATCO workload and traffic complexity.

V. CONCLUSION

Air traffic control candidates undergo extensive training. Their training consists of three phases and incorporates theoretical and practical parts. Theoretical training includes all the subjects and topics needed to perform complex ATCO work. Practical training is provided on computer based simulation device where candidates gain practical skills in maintaining aircraft separation and guiding aircraft through airspace. HUSK, as well as every other air traffic control training organization, is obliged to use an adequate computer based simulation device in provision of practical training. BEST radar simulator used by HUSK training organization meets all the objectives and requirements prescribed for the provision of basic ATCO training.

REFERENCES

[1] B. Juričić, I. Varešak, and D. Božić, "Air traffic controller training – Regulatory phrame and practices," 14th International Conference on Transport Science: Maritime, Transport and Logistics, ICTS 2011 Conference Proceedings, Marina Zanne, Patricija Bajec (ur.). Portorož: University of Ljubljana, Faculty of Maritime Studies and Transport, pp. 1-9, 2011.

[2] "EUROCONTROL Specification for the ATCO Common core content initial training," Brussells: EUROCONTROL, 2015.

[3] "Commission regulation (EU) 2015/340 of 20 February 2015 laying down technical requirements and administrative procedures relating to air traffic controllers' licences and certificates pursuant to Regulation (EC) No 216/2008 of the European Parliament and of the Council, amending Commission Implementing Regulation (EU) No 923/2012 and repealing Commission Regulation (EU) No 805/2011 Official Journal of the European Union," Volume 58, 2015.

[4] B. Juričić, D. Novak, and E. Bazijanac, "Model of air traffic controller education at the Faculty of transport and traffic sciences," INAIR 2013 - International Conference on Air Transport, Antonin Kazda(ed.).Bratislava:Žilinska univerzita v Žiline, pp 54-58, 2013.

[5] "Basic ATCO Training Plan and Program – Undergraduate Study," HUSK, Faculty of Transport and Traffic Sciences, University of Zagreb, Rev 00, 01.01.2016., neobjavljen.

[6] Basic ATCO Training Plan and Program – Training Course, HUSK, Faculty of Transport and Traffic Sciences, University of Zagreb, Rev 03, 15.03.2017., neobjavljen.

[7] J. Vagner, E. Pappová, "Comparison of Radar Simulator for Air Traffic Control," "Naše more" 61(1-2)/2014. - Supplement, pp. 31-35.

[8] "BEST radar specification,"Bournemouth: Micro Nav Ltd, 2010.

[9] Obrazac za ocjenjivanje rada na simulatoru (Simulator Assessment List – SIM ACS), neobjavljen.

[10] "Guidance for developing ATCO basic training plans," Brussells: EUROCONTROL, 2010.

The perspective of use of digital libraries in era of e-learning

Radovan Vrana

Department for information and communication sciences,
Faculty of humanities and social sciences, University of Zagreb, Zagreb, Croatia
rvrana@ffzg.hr

Abstract – **The aim of this paper is to review digital libraries' role in supporting e-learning. Digital libraries offer technology based information resources and services to enable learners to access relevant knowledge anywhere anytime. As such, digital libraries are inseparable from the learning process because they made e-learning possible. Though highly relevant for e-learning, digital libraries haven't been used in education to their full capabilities due to their unequal development worldwide, legal and technical barriers and the fact that the industrial sector (commercial publishers) still manages access to digital content used in education on commercial basis (making it inaccessible to wide audience without paying a fee). The paper also present results from the research of digital libraries and their capabilities for inclusion in education.**

I. INTRODUCTION

During centuries, libraries have been the guardians and distributors of "books, journals, maps and other materials that are used by students in the learning process" [1]. Today, they are concerned about their sometimes diminished role in society and this role relies increasingly on information and communication technology which gradually leaves behind traditional information institutions including libraries and printed resources for learning used so far. The response of libraries to the current challenges was development of new digital information resources supplemented by new content delivery services. Presently, libraries worldwide display successfully both their traditional and digital faces and efficiently serve different purposes and different audiences. One such purpose in which libraries participate successfully is e-learning. Libraries have already been introduced to the scene of computer-based learning in the mid-20th century and they are now connected in the unity of libraries and e-learning [2]. They „serve as specific media to implement the learning process" [3]. Libraries support e-learning by offering carefully selected information resources and a variety of ICT supported services to facilitate access to their holdings which include learning material as well [4]. These services include: online bibliographic instructions, computerized library catalogs, digital libraries, distance learning services, e-databases, instant messaging services, inter library loan and document services, ready references, virtual classrooms, virtual references etc. [5]. In the era of ubiquitous learning in which learners of all age use electronic devices on daily basis to access learning material on local computer network as well as on the

internet, libraries are expected to make an extra effort to adapt and prepare content in their existing digital collections and their online library services supported by rapidly changing information and communication technologies. Research [2] confirmed existence of such a trend in which students of today have closely embraced modern technologies to enhance their learning, and libraries have become their partners in role of knowledge organizer offering various services that are ICT incorporated. Having this in mind, this paper will discuss the role and perspective of use of digital libraries in today's evolving e-learning environment.

II. LIBRARIES IN THE DIGITAL ERA AND E-LEARNING

Libraries were among pioneering institutions which introduced computer technology in their daily operations in the middle of 20th century and they continued to introduce new technological successes throughout the rest of the century to improve access to and use of library holdings. Major step forward in library development happened in the beginning of the 1990s when digital libraries were introduced. The concept of digital libraries was a major success from its very start as "it came out of closed walls of library and reached users in their home, workplace and even while traveling, with the help of laptops" [6]. Digital libraries "have emerged as a leading edge technological solution to the persistent problem of enhancing access, process of archiving and expanding the dissemination of information" [7]. Digital libraries generally speaking "consist of digital contents (which are sometimes but not necessarily text-based), interconnections (which may be simple links or complex metadata or query-based relationships), and software (which may be simple pages in HTML or complex database management systems)" [8]. More specifically, digital libraries offer "online catalogues, databases, multimedia, online journals, digital repositories, electronic books, electronic archives and online / electronic services" [9]. Digital libraries have brought digital revolution which has affected almost every aspect of library services „from the automation of internal recordkeeping systems to the digitization of physical collections, and from the acquisition of new „born-digital" works of art or library publications, to the use of technology to present collections and engage audiences" [10]. The engagement of audiences of different types proved to be crucial for the development of digital libraries because libraries have formed their services according to information needs and requests of different audiences including students and educators.

In spite of the engagement of different audiences including students [11] digital libraries didn't have any substantial role to play in the learning sector because in some cases they were not included fully in the learning process in favor of the industrial sector (mostly commercial publishers) which controls digital information resources. They suggested that digital libraries had (and still have) a great potential in learning due to the nature of resources they bring together and that can be used "to illustrate a variety of educational topics in practically all areas of knowledge, or to support individual learning" [11]. Digital libraries are indispensable for education because the offer up to date material, immediate access to a wide range of sources which do not exist physically, provide resources via an internet connection each moment from each place and because of these advantages, in digital libraries learning is independent process [12]. The role of digital libraries is clear: they function as „digital schools that offer formal packaging for specific skills and topics as well as general browsing for creative discovery and self-guided, informal learning" [13]. Same authors [13] suggested that digital libraries serve at least three roles in learning: they serve a practical role in sharing expensive resources including physical and digital resources, equipment, human resources – librarians who serve to allow instructors and students to share expensive materials and expertise; they serve a cultural role in preserving and organizing artifacts and ideas as well as ensuring access to materials through indexes, catalogs, and other aids that allow learners to locate items appropriate to their needs; and finally, they serve social and intellectual roles by bringing together people and ideas with formal, informal, and professional learning missions. Libraries "both traditional and digital one have three roles in education: place for sharing reach information, maintaining ideas, and give awareness to bring together individual with learning aims" [12]. Tanner and Deegan [14] analyzed values and benefits of use of digitized resources in teaching and learning and concluded that "The increasing availability of digitized resources allows educational institutions to provide students with more varied, more accessible and richer teaching materials than ever before. This encourages a more exploratory, research based approach to teaching and learning. Entirely new kinds of topics and courses can be studied, new modes of assessment are possible, and students are given a richer educational experience". Abbott and Cohen [15] investigated possibilities of use of large digital collections (in libraries) in education and found out that students (as library users) face a challenge when they try to find and access primary documents in digital libraries because documents (digital resources) can be found in any number of digital libraries on the internet and this presents a problem. The solution to this problem would be better integration of information about digital resources and networked systems on global level so they could be more easily located and used.

The relationship between digital libraries and e-learning is quite straightforward. Digital libraries as networked information systems are already part of the learning process [2]. The need for digital libraries in e-learning could be also found in the very definition of e-learning: „E-learning could be interpreted as electronic learning; the learning that involves the Internet; learning from a distance via the aid of the internet and, or other electronic gadgets" [2]. E-learning includes „all forms of electronic supported learning and teaching, which are procedural in character and aim to effect the construction of knowledge with reference to individual experience, practice and knowledge of the learner" [3]. E-learning is not possible without the learning material and that is where digital libraries as the aid come in convenient. The existing experiences, practices and knowledge of the student become more rich as they gain access to the quality information resources in digital libraries which have become an important stakeholder in the learning process in the last two decades. Pavani [1] analyzed whether digital libraries are suitable for inclusion in the learning process and found out the following reasons for doing so: documents in all formats are managed in a unified way: texts, animations, interactive exercises, audio files, video streams, e-books, e-journals and online tests can be stored, described and distributed through computers and networks; access control: contents can be assigned different types of access according to the classes of users that are entitled to them; content sharing: authors can make their contents available for other faculty to aggregate into their courseware; interactivity: contents stored in digital libraries can be interactive and based on multimedia; customization: some users may require special characteristics of the contents and the system; reuse: courseware can be developed with a granularity that makes it flexible to combine and support multiple syllabus; cross-institution cooperation: digital libraries are networked information resources and this allows that contents can be used from different cooperating institutions; any place and any time learning: students study in different hours of the day any day of the week. Wangila [16] shares similar thoughts: digital libraries were developed to enable higher education (and other educational) institutions provide library services to all its students and other stakeholders in and off campus to facilitate access to digital library resources from any location or work station making it easy to share digital information as well as available to everyone at any time. Huang [17] analyzed the concept of library 2.0 (best of libraries to date with community support and ICT support) and its relationship with e-learning. The author offered the following characteristics that would make stronger mutualism of libraries and e-learning: library is everywhere; no barriers (library is accessible and age-friendly and it provides access to the library's resources); library is individualized (actively providing readers with diversified content according to their needs); library is inviting participation; library promotes flexibility and best-of-breed systems (support for the e-learning to make them customized and flexible according to the needs of different subjects). Finally, digital libraries also facilitate knowledge sharing between the stakeholders in the learning process and encourage them "to work together, develop their skills, and form strong and trusting relationships" (Dhiman, 2010).

For students, digital libraries provide technology based information and services to enable them "to access relevant information and services anywhere anytime, as well as provide empowerment for innovative and life-long learning" [3]. A document "Digital libraries in education" [18] suggested existence of three chains of support for students in digital libraries. First, if a digital library provides users with profiles, they are able to form learning communities. Second, content supported by metadata enables the formation of customizable collections of educational objects and learning materials. Third, tools supported by common protocols or standards enable the development of varied application services that enhance the value of the library's content for the learner.

Librarian's job has also changes as it began to depend more intensively on information and communication technology. According to Sreenivasulu [19], the librarian's role is now oriented towards „consultancy to the users and providing digital deference services, electronic information services, navigating, searching and retrieval of digitized information through web documents that pan the universal digital library or the global digital library". For Ashikuzzaman [20], librarians must assert themselves as key players in the learning process, thereby changing their roles from the information providers to educators, they should become information gateways and advocates the librarian's involvement in teaching communities so as to meet information needs of the students. For Hermosa and Anday [4] librarians transformed their jobs into virtual or digital environments, while customizing their services and resources for e-learners (they provide remote access to, and electronic delivery of, library resources, and are using communication technologies to deliver electronic reference services and instructional support).

III. DIGITAL LIBRARY SERVICES FOR E-LEARNING

Digital libraries are opened to the wide public and as such they offer many possibilities of inclusion of their content in formal and informal learning. Calhoun [21] investigated social roles of digital libraries which also include teaching, learning and the advancement of knowledge. For formal education, digital libraries can offer the following services: specialized educational digital libraries, portals for teachers or students, integration with learning management systems and access to primary sources [21].

For progress of knowledge digital libraries offer the following services: self-archiving, deposit incentives; mandatory deposit, open access journals. libraries as publishers, digital libraries of theses and dissertations, cross-repository services. object reuse and exchange services, workflow-based content creation and management. data curation and researcher profiling services [21], digital information resources usable on different electronic devices, library services for information discovery, course materials, exhibits,

workshops etc. Wangila [16] pointed out that "the integration of the digital library technology with the educational enterprise has come at a similar time when the student requirements for access to library resources also heightened". The same author suggests the need for policies, guidelines and standards to ensure professional efficiency and quality of information service delivery in digital library.

In spite of wealth of materials and services, all digital libraries are not equal. Their differences define their possible use. This part of the paper will present digital libraries which offer support for the learning process. The focus will be on the most famous world digital libraries as there is no one global register which would list all existing digital libraries offering educational resources. Since each digital library (and institution behind digital library) has its own approach to presenting learning and teaching materials, it is rather difficult to compare them against a fixed set of criteria.

The purpose of the research is to identify major digital libraries which offer educational information resources and services as part of their Web site. The research aims to provide answers to two questions: 1.) do researched digital libraries have a section with educational content as an integral part of their Web sites that can be clearly and undoubtedly identified; 2.) is such educational section of a digital library clearly structures as to offer learning material, teacher support, lesson plans, and other aid necessary for teaching and learning? The following section of the paper will present digital libraries which have educational section.

The British Library

The British library at http://www.bl.uk [22] offers online resources, visits and workshops and teaching resources in the „Learning" section of its Web site. The teaching section presents teaching resources divided into sections „By subject", „By age", „By resource". Each of these sections is further divided into subsections „Language", „Curriculum", „Subject" and „Themes" with ready-made teaching material. The online resources section of the British Library Website offers items and expert commentary related to History, English and Citizenship with complete insight into topics in the enumerated categories.

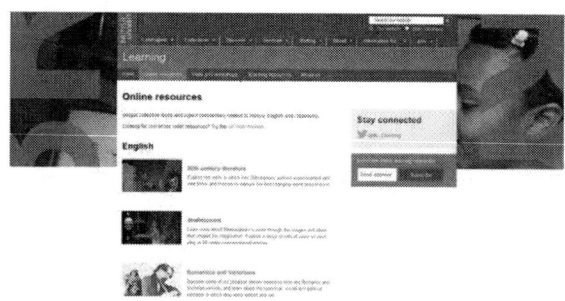

Fig. The British Library teaching resources

The Library of Congress

The Library of Congress at https://www.loc.gov/education/ [23] offers classroom materials and professional development to help teachers use primary sources from the Library's vast digital collections" (Library of Congress). The Library of Congress Web site for education is divided into two big categories „For Students & Lifelong Learners" and „For Teachers". The first category offers materials for advancement of reading, poems for high schools, American history for elementary and middle school students, fun science facts from the Library of Congress and highlights from the Library's online collections. In category for teachers, the Library of Congress offers blog with posts about use of primary sources and tools and techniques to use them, free resources to help teachers effectively use primary sources from the Library's digital collections in their teaching, teacher-created lesson plans using Library of Congress primary sources, sets of primary sources on frequently taught topics and primary source-based, ready-to-use resources for teachers and facilitators. This Website offers one of the most thoroughly prepared content among the researched libraries.

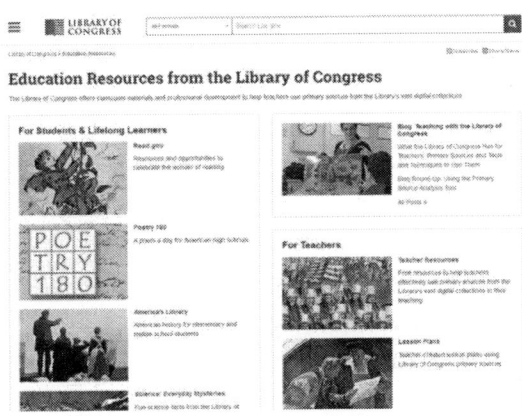

Fig 2. The Library of Congress education resources

Europeana Space

Europeana Space at http://www.europeana-space.eu/education/ [24]„is an EU funded project focused on the creative re-use of available digital cultural content" (Europeana Space: About). The Web site offers 5 examples of Educational Demonstrators, a MOOC (Massive Open Online Course), dissemination events and collection of resources and best practices as weel as 6 thematic Pilots. Europeana Space targets users from primary school to universities and engaging them with digital cultural heritage.

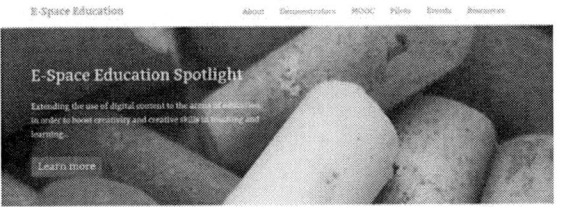

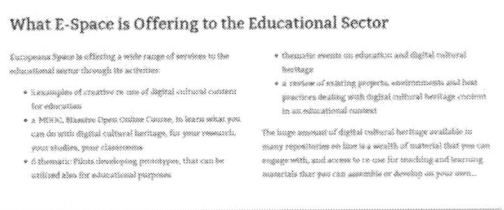

Fig 3. Europeana Space

Historiana

Historiana at http://historiana.eu/ [25]The EUROCLIO programme is an „on-line educational multimedia tool that offers students multi-perspective, cross-border and comparative historical sources to supplement their national history textbooks" (Historiana). Historiana offers three categories that can be used in teaching and learning: „Historical thinking" which introduces and individual to historical intepretation of facts, analysis of historical sources etc.; „Teaching methods" with instructions on how to teach history and „Teaching challenges" that offer content on how to improve history teaching.

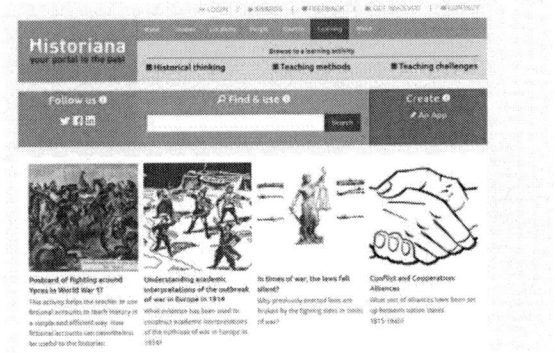

Fig 4. Historiana

National library of Australia

National library of Australila at http://www.nla.gov.au/using-the-library/learning [26] offers the learning section at its Website. The section is further divided into learning program and school and teacher program. The learning program includes learning sessions to help users use libraty collections for research, work and study and topics like getting started at the Library, using microform readers, academic eResources and Trove, online resources from Australian libraries. Additional topics include resources for family history research like searching passenger lists, finding newspapers, findmypast.com and ancestry.com and Trove for family history. The schol and teacher program include library exhibitions, collections and reading rooms for stuednts to find out more about Australian history and culture. They critically interpret texts and learn search methods to understand historical and contemporary events and issues.

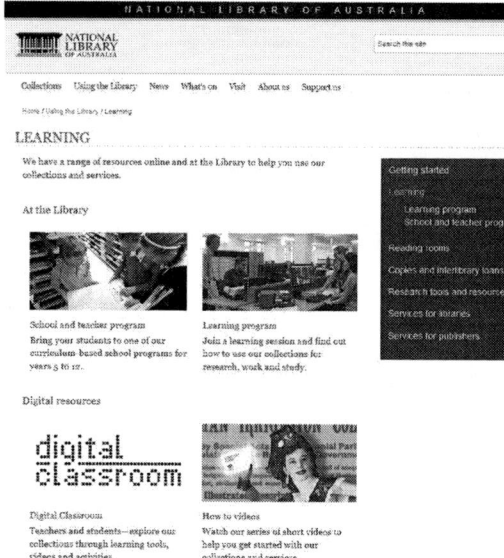

Fig 5. National library of Australia

Gallica

Gallica at http://gallica.bnf.fr [27]is the digital library of Bibliotheque nationale de France which provides free online access to millions of documents of all types (books, magazines, images, videos, partitions, maps, manuscripts, etc.). At http://gallica.bnf.fr/blog/28112016/75-epub-gallica-selectionnes-par-le-ministere-de-leducation-nationale it offers selected books from the national literature history for use in education. Actually, all Gallica is intended for research and learning but it doesn't offer a specialized section of its Website with support to teachers and students. Bibliotheque nationale de France offers a number of services to students but does not single out special collections for education.

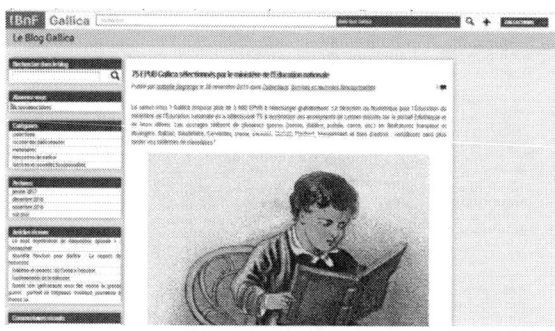

Fig 6. Gallica digital library

IV.   CONCLUSION

In theory, digital libraries have great potential and predispositions for support to the learning process. In practice, digital libraries are usually part of national libraries which are more frequently oriented towards preservation of national heritage and less to support of schools and universities. While the first part of the paper straightforwardly explained reasons for better inclusion of digital libraries in education and the second part demonstrated lack of actual sections of digital libraries prepared for inclusion in the learning process by offering review of very few digital libraries equipped for participation in education. Answers to research question woulde be: 1.) there are few digital libraries that clearly offer a section with educational content as an integral part of their Web sites that can be clearly and undoubtedly identified; 2.) these sections a digital library frequently (but not generally) have structures as to offer learning material, teacher support, lesson plans, and other aid necessary for teaching and learning. One could argue that all library holdings are to be used in the learning process and that might be true to some extent. However, education is well defined activity which relies heavily on national educational standards and pedagogical frameworks and procedures and require clearly defined approach of use of any material in the learning process. As a result, national libraries must prepare the neccesary library material together with instructions for teachers and students to become fully recognized as educational aids by those for whome they are built.

REFERENCES

[1]   A. Pavani, The Role of Digital Libraries in Higher Education. Retrieved October 22, 2016. from http://www.ineer.org/Events/ICEE2007/papers/637.pdf

[2]   H.N. Eke, "The perspective of e-learning and libraries in Africa: challenges and opportunities", Library Review, vol. 59, pp. 274-290, 2010.

[3]   A.K. Dhiman, "Evolving Roles of Library & Information Centres" in World library and information congress: 76th IFLA general conference and assembly, 2010, Retrieved October 22, 2016. from http://www.ifla.org/past-wlic/2010/107-dhiman-en.pdf

[4]   N.N. Hermosa and G. Anday, "Distance learning and digital libraries: the UP open university experience", Journal of Philippine librarianship, vol. 28, pp. 90-105, 2008.

[5]   L. Li, "Leveraging quality webbased library user services in the digital age", Library Management, vol. 27, pp. 390-400, 2006.

[6]   P. Singh, "Role of Digital Libraries in E-Learning" in 3rd International CALIBER - 2005, 2005, pp. 559-564. Retrieved October 22, 2016. from http://ir.inflibnet.ac.in/handle/1944/1564

[7]   C. Nayak, "Impact and challenges of e-learning in digital environment", Asian Journal of Library and Informationm Science, vol.5, pp. 76-80, 2013.

[8]   M. Seadle and E. Greifender, "Defining a digital library", Library Hi Tech, vol. 25, pp. 169-173, 2007.

[9]   S. Sen, "Academic Libraries in e-Teaching and e-Learning" in ICAL 2009 – Vision and roles of the future academic libraries, 2009, pp. 176-179. Retrieved October 22, 2016. http://crl.du.ac.in/ical09/papers/index_files/ical-29_46_135_1_LE.pdf

[10] WebWise 2006: Inspiring Discovery: Unlocking Collections. Washington DC: Institute of museum and library services, 2006. Retrieved October 22, 2016. from https://www.imls.gov/assets/1/AssetManager/WebWise_2006.pdf

[11] M. Dobreva, G. Angelova and G. Agre, "Bridging the Gap between Digital Libraries and e-Learning", Cybernetics and information technologies, vol. 15, pp. 92-100, 2015.

[12] F. Abbasi and S Zardary, Solmaz, "Digital libraries and its role on supporting E-Learning", AWERProcedia Information Technology & Computer Science. vol. 1, pp. 809-813, 2012.

[13] G. Marchionini and H. Maurer, "Roles of Digital Libraries In Teaching and Learning", COMMUNICATIONS OF THE ACM. vol. 38, pp. 67-75, 1995.

[14] S. Tanner and M. Deegan, Inspiring Research, Inspiring Scholarship. The Value and Benefits of Digitised Resources for Learning, Teaching, Research and Enjoyment. London: JISC, 2011. Retrieved October 22, 2016. from http://www.kdcs.kcl.ac.uk/fileadmin/documents/Inspiring_Researc h_Inspiring_Scholarship_2011_SimonTanner.pdf

[15] F. Abbott and D. Cohen, "Using Large Digital Collections" in Education: Meeting the Needs of Teachers and Students 2015, Digital Public Library of America Retrieved October 22, 2016. from http://dp.la/info/wp-content/uploads/2015/04/Using-Large-Collections-in-Education-DPLA-paper-4-9-15-2.pdf

[16] F. Wangila. "An Assessment of the Implementation of Digital Library Technologies in Institutions of Higher Learning: A Case Study of Kenyatta University", International Journal of Academic Research in Business and Social Sciences, vol. 4, 532-541, 2014.

[17] T.C. Huang, "What Library 2.0 has taught libraries in Taiwan about e-learning", The Electronic Library, vol. 33, pp. 1121-1132, 2015

[18] Digital libraries in education. Moscow: UNESCO Institute for Information Technologies in Education, 2003. Retrieved October 22, 2016. from http://iite.unesco.org/pics/publications/en/files/3214609.pdf

[19] V. Sreenivasulu. "The role of a digital librarian in the management of digital information systems (DIS)", The Electronic Library, vol. 18, pp. 12-20, 2000.

[20] Ashikuzzaman. E-Learning and Digital Library, 2016. Retrieved October 22, 2016. from http://www.lisbdnet.com/e-learning-digital-library/

[21] K. Calhoun, Social Roles of Digital Libraries, Exploring Digital Libraries: Foundations, Practice, Prospects. London : Facet Publishing, 2014.

[22] The British library. Retrieved January 5, 2017. from http://www.bl.uk

[23] The Library of Congress. Retrieved January 5, 2017. from https://www.loc.gov/education/

[24] Europeana Space: About. Retrieved January 5, 2017. from http://www.europeana-space.eu/education/about/

[25] Historiana. Retrieved January 5, 2017. from http://historiana.eu/about/

[26] National library of Australila. Retrieved January 5, 2017. from http://www.nla.gov.au/using-the-library/learning

[27] Gallica. Retrieved January 5, 2017. from http://gallica.bnf.fr

Gap in pagination due to withheld paper.

Pages 932-936

MIPRO 2017, May 22- 26, 2017, Opatija, Croatia

Introductory Physics Course for ICT Students: Computer-Programming Oriented Approach

Vladimir Krstić

College for Information Technologies, Zagreb, Croatia
vladimir.krstic@vsite.hr

Abstract - Physics is an important subject and it is good that introductory physics courses are obligatory subjects in many, if not all, ICT colleges and universities in Croatia. However, these courses often cover too many different areas of physics and are presented in a traditional way – lectures / solving problems on paper / laboratories. We think that the introductory physics courses for ICT students should be more oriented towards computer programming and solving physics problems using computers. In this work we show how we plan to use SageMath, a free open-source mathematics software system licensed under the GPL, to boost students' understanding of the electric potential and the electric field.

I. INTRODUCTION

Numerical computations followed by visualizations allow students to study and explore non-trivial physical models, develop and test their intuition and physical reasoning better than the standard paper-and-pencil approach [1, 2]. Hence, we at the College for Information Technology in Zagreb are currently in the process of restructuring our physics curriculum. The main goal of this process is to incorporate numerical computations to the current curriculum.

This structural change in the curriculum implies changes in the type and amount of educational materials that can be covered in one semester introductory physics course. Some materials in the current curriculum are more suitable to be included in the new curriculum and some must be left out because numerical computations occupy a substantial part of the course in the new curriculum [2].

In this article we present a part of our modified lesson plan on the subject of electric field and electric potential. Some of the goals of the lesson are the development of students' understanding of superposition principle and integral calculus. All programs given in this article have been written in SageMath 7.4 [3].

II. PROBLEM DESCRIPTION AND ALGORITHM

A. Physical Problem

A straight wire of length $L = 1.0$ m is uniformly charged and the total charge on the wire equals $Q = 1.0$ nC (Fig. 1). Find the electric potential and the electric field everywhere in space [4].

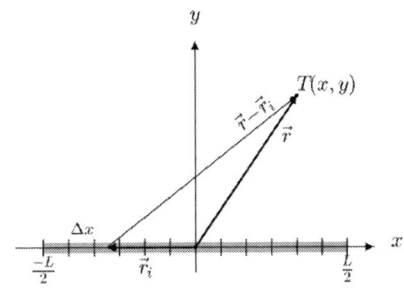

Figure 1. Graphical representation of the problem.

B. Coordinate System: Which one to choose?

Ideally, students have to decide what coordinate system best suits the problem and where to position the wire in the chosen coordinate system. We use the Cartesian coordinate system and position the wire in such a way that we can easily exploit the symmetry of the problem (Fig. 1). Because of one of the symmetries of the problem (the problem is invariant with respect to the rotation around the x-axis), it is enough to find the solution of the problem in the x-y plane.

C. Algorithm

The wire is divided into N segments of equal length: $\Delta x = \frac{L}{N}$. Since the wire is uniformly charged, the electric charge of each segment equals $\Delta Q = \frac{Q}{N}$. The electric potential and the electric field vector at the point $\vec{r} = (x, y)$ is a sum of contributions of each segment:

$$\varphi(x, y) = \sum_{i=1}^{N} \Delta \varphi_i \approx \sum_{i=1}^{N} k \frac{\Delta Q}{|\vec{r} - \vec{r}_i|}, \qquad (1)$$

$$\vec{E}(x, y) = \sum_{i=1}^{N} \Delta \vec{E}_i \approx \sum_{i=1}^{N} k \frac{\Delta Q}{|\vec{r} - \vec{r}_i|^2} \frac{\vec{r} - \vec{r}_i}{|\vec{r} - \vec{r}_i|}, \qquad (2)$$

where $k \approx 9.0 \cdot 10^9 \frac{\text{Nm}^2}{\text{C}^2}$ and $\vec{r}_i$ is a position vector of an arbitrary point within the i-th segment. In our realization of the algorithm, we will take the point in the middle of the segment.

```
reset()
k = 9.0e9
Q = 1.0e-9      # Total charge on the wire
L = 1.0         # Length of the wire

N = 100         # Number of subdivisions
dx = L / N
dQ = Q / N

def potential(x,y):
    pot = 0.0
    r = vector([x,y])
    x_i, y_i = -L/2+dx/2, 0
    for i in range(N):
        r_i = vector([x_i,y_i])
        pot += k * dQ / (r-r_i).norm()
        x_i += dx
    return pot
```

III. WRITE THE PROGRAM AND TEST IT

It is straightforward to translate the electric potential function $\varphi(x, y)$ into Python function `potential(x,y)` in SageMath. Here we use SageMath function `vector` to create vector objects so we can employ methods associated with vector objects to write more readable code. In our case we use only vector subtraction and the `norm()` method to calculate the norm of a vector.

If possible, students should write their own program, but providing them with one is quite acceptable due to the fact that some students might have no previous programming experience. However, we think that it is very important to encourage students to extensively test the program on their own. In order to test the program, they should start thinking about the physics of the problem to determine which properties the solution of the problem must have:

1. As we move away from the wire, the shape and size of the wire become less and less important and the approximation of the wire with the point charge becomes more and more acceptable. In terms of equipotential lines: the shape of the equipotential lines approaches the shape of the equipotential lines of a point charge.

2. Because of the symmetries of the problem, the electric potential function (the solution of the problem) is symmetric with respect to the x-axis and the y-axis:

$$\varphi(x, y) = \varphi(x, -y),$$
$$\varphi(x, y) = \varphi(-x, y). \qquad (3)$$

3. Sometimes, often because of the symmetries of the problem, it is not too difficult to calculate the analytical solution of the problem or to find the analytical solution on the Internet. Students should use the analytical solution to test the numerical solution. For example, in our problem the electric potential function on the y-axis is easy to calculate or find:

```
var('x','y') # Creating symbolic variables
p1 = contour_plot(potential(x,y), (x,-1,1),
    (y,-1,1), contours=[10,20,30],
    cmap=[(0,0,0)], linestyles=':',
    fill=False, labels=True,
    label_fmt="%.1f V", label_inline=True)
p1 += line([(-L/2,0),(L/2,0)],
    rgbcolor=(0,0,0))  # Draw the wire
p1.axes_labels(['$x[\\mathrm{m}]$',
               '$y[\\mathrm{m}]$'])
p1.show(aspect_ratio=1, gridlines='minor')
```

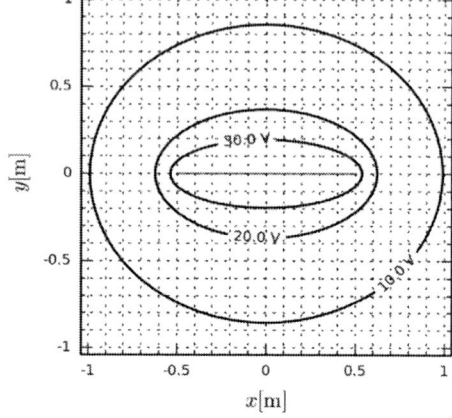

Figure 2. Several equipotential lines of prespecified values are calculated and plotted using the `contour_plot` function. Straight wire of length $L = 1.0$ m is also shown.

$$\varphi(0, y) = k\frac{Q}{L}\ln\frac{\frac{L}{2} + \sqrt{\left(\frac{L}{2}\right)^2 + y^2}}{-\frac{L}{2} + \sqrt{\left(\frac{L}{2}\right)^2 + y^2}}. \qquad (4)$$

Equipotential lines can be calculated and plotted in SageMath using the `contour_plot` function. You can evaluate `contour_plot?` in the SageMath cell to find out details about its parameters and usage. The number of parameters for this function is huge but some parameters are more important than the others: the most important is to provide the function we are examining, region where to search for equipotential lines and value of the electric potential for each equipotential line. We have calculated and plotted several equipotential lines (Fig.2). As expected, their shapes are more circular as we move away from the charge distribution. The straight line in Fig. 2 represents the straight wire. Notice in our SageMath program how we combined two graphics objects (`contour_plot` graphics and one of the SageMath graphics primitives, `line`) with the use of the `+=` operator. Axes labels are added to the combined graphics; labels are typesetted in LaTeX if given within two $ signs and LaTeX commands escaped with an extra slash.

The second property of the solution implies that equipotential lines should be symmetric with respect to the x-axis and y-axis and it seems that our solution has this property (Fig. 2). Students should check this property by calling the function `potential` for several points in the $x - y$ plane.

```
#analytical solution, Eq(4)
potential_0_y(y) = k * Q / L *
            ln( (L/2+((L/2)^2+y^2)^(1/2)) /
                (-L/2+((L/2)^2+y^2)^(1/2)) )
```

```
var('y') # Creating symbolic variable
y_min, y_max = 0.1, 1.0
p = plot(potential(0,y), (y,y_min,y_max))
p += scatter_plot([(y,potential_0_y(y)) for y
            in srange(y_min,y_max,step=0.05)])
p.axes_labels(['$y[\\mathrm{m}]$',
               '$\\varphi [\\mathrm{V}]$'])
p.show(gridlines='major')
```

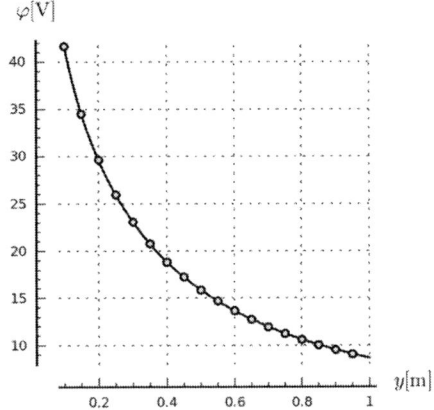

Figure 3. Numerical (full line) vs. analytical solution (circles) for the electric potential on the positive y- axis.

If analytical solutions of the problem exist (probably in some special cases of the problem), students should always compare them with numerical solutions. In Fig 3 we plotted on top of the numerical solution for the electric potential function on the y axis, several points of the analytical solution given by the Eq. (4). These points are generated using the Python list comprehension iterating over the list of numbers returned by the SageMath `srange` function and plotted using the SageMath `scatter_plot` function.

IV. IT IS TIME FOR PRACTICE

There are many problems that students should now be able to analyze and solve. For example:

- A wire can be placed along the y-axis.

- A wire can be placed at the angle $\pi/4$ with respect to the x-axis.

- Two parallel wires with the same type of charge or with the opposite types of charges.

- Two short wires with the opposite charges placed along the x-axis.

Student should be encouraged first to analyze the problem and predict the solution of the problem, then write or sketch an algorithm and numerically solve the problem by modifying the existing SageMath programs, analyze solutions and extensively test their programs.

```
def electric_field(x,y):
    el_field = vector([0.0, 0.0])
    r = vector([x,y])
    x_i, y_i = -L/2+dx/2, 0
    for i in range(N):
        r_i = vector([x_i,y_i])
        el_field += k * dQ / (r-r_i).norm()^3
                        * (r-r_i)
        x_i += dx
    return el_field
```

```
var('x','y') # Creating symbolic variables
p2 = plot_vector_field(electric_field(x,y),
    (x,-1,1), (y,-1,1))
(p1+p2).show(aspect_ratio=1)
```

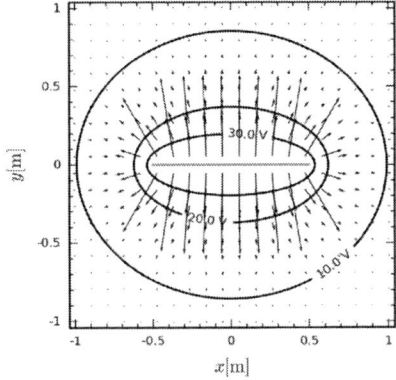

Figure 4. Electric field vectors plotted on top of the equipotential lines from Fig 1.

The electric field vector is normal to the equipotential line and students can test this claim by actually computing the electric field for a given charge distribution and plot it on top of the plot with equipotential lines. Students should notice that functions (1) and (2) are very similar from a programmer point of view except that one returns a scalar and the other returns a vector. Hence, when writing the Python function `el_field(x,y)` they can use `potential(x,y)` as a starting point.

It is instructive to plot the electric field on top of the plot with equipotential lines (Fig. 4) so students can study the relation between equipotential lines and length, direction and orientation of the electric field vectors.

V. CONCLUSION

SageMath is a great free and open-source alternative to *Mathematica* and *Matlab*. SageMath uses Python as its main language. Python is usually the first computer language that many students meet because teachers often use this language in their introductory computer programming courses. This opens an opportunity to modify our introductory physics courses for ICT students. We can now use physics to support programming courses and make physics more suitable and usable for this specific group of students. In this article, we have shown how one can use SageMath to work with students on their understanding of the electric potential and the electric field.

REFERENCES

[1] E.F. Redish and J.M. Wilson, "Student programming in the introductory physics course: M.U.P.P.E.T.," Am. J. Phys. 61, pp. 222–232, 1993.

[2] R. Chabay and B. Sherwood, "Computational physics in the introductory calculus-based course," Am. J. Phys. 76, pp. 307–313, 2008.

[3] The Sage Developers, SageMath, the Sage Mathematics Software System (Version 7.4), 2016, http://www.sagemath.org.

[4] D.J. Griffiths, Introduction to Electrodynamics, 4th ed., Pearson, 2013.

Gap in pagination due to withheld paper.

Pages 941-943

Theoretical and practical challenges of using three ammeter or tree voltmeter methods in teaching

Vladimir Šimović, Trpimir Alajbeg, Josip Ćurković
Department of Electrical Engineering
Zagreb University of Applied Sciences
vsimovic@tvz.hr, trpimir.alajbeg@tvz.hr, jcurkovic@tvz.hr

Abstract - The tree ammeter method and the three voltmeter method are used for measurements of power. More specifically, they are used to calculate the power factor of a specific load. Both methods are used on the „Fundamentals of electrical engineering" course in professional study of electrical engineering at the Zagreb University of Applied Sciences as an introduction to measurements of power and phasor arithmetic. Both methods are susceptible to accuracy problems caused by small errors in measuring devices. These accuracy problems in specific scenarios that were used during the course presented an opportunity to educate students on the difference of theory and its practical application. This paper examines our solution to the perceived accuracy problems and discusses the changes that were made in course material and teaching methods to highlight the difference between theory and practice.

I. INTRODUCTION

As an integral part of the Fundamentals of electrical engineering, course students must attend and earn a passing grade in five laboratory exercises. The exercises are used as a tool to educate students on the process of conducting experiments, interpretation and discussion of results and applying theoretical knowledge from course classes. For many of the students, the laboratory exercises present the first real opportunity to handle instruments specific to electrical engineering.

Among the five laboratory exercises, one deals in measurements of power, specifically with the measuring and calculating the power factor (cos φ) of a specific load (Z_t) in a simple AC circuit. This is done by using two methods; three ammeter and the three voltmeter method.

II. THE METHODS

A. Three voltmeter method

The three voltmeter is used in an inductive circuit to measure the value of the power factor. As seen on Figure 1, one voltmeter is used to measure the voltage of the circuit (U_i), the second one measures the voltage on the non-inductive resistance (U_R) that is connected in the series with the load branch and the third voltmeter is used to measure the voltage of the load (U_t).

The power (P) consumed by the load can be determined as the product of the voltage and current of the

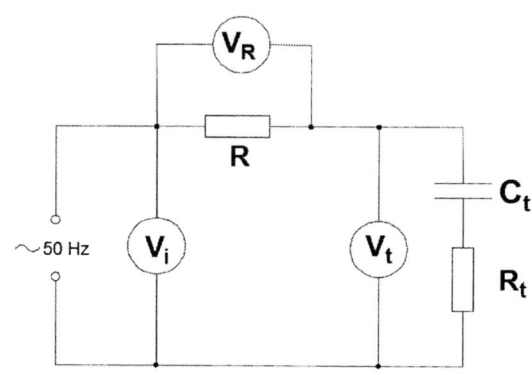

Figure 1. The three voltmeter method

load (U_t,I_t) and the cosine of the phase angle between these two (the power factor).

$$P = U_t\, I_t \cos \varphi \qquad (1)$$

Using the law of cosines, the power factor (cos φ) can be calculated.

$$\cos \varphi = (U_i^2 - U_R^2 - U_t^2) / (2\, U_R\, U_t) \qquad (2)$$

For optimal accuracy, the non-inductive resistance should be large enough so that the voltmeter (or multimeter) can measure it with satisfactory accuracy, but not too large, otherwise the voltage available to the load would be too small. Ideally, it should be close or equal to load impedance.

B. Three ammeter method

The three ammeter is also used in an inductive circuit to measure the value of the power factor, independent of source frequency and waveforms. In this method, as seen on Figure 2, across the inductive circuit load in which the power factor is to be determined, a non-inductive resistance is connected parallel with the load branch. One ammeter is used to measure total current of the circuit, the second one measures current going through the non-inductive resistance and the third ammeter measures the current of the load branch.

Current through the resistive branch (I_R) is in phase with source voltage while the current of the load (I_t) has

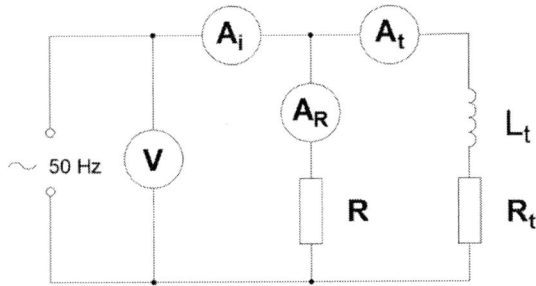

Figure 2. The three voltmeter method

its own phase, which will affect the power factor (cos φ). Total current (I_i) is a vector sum of the other two currents. Using the law of cosines, the power factor can be calculated.

$$\cos \varphi = (I_i^2 - I_R^2 - I_t^2) / (2\,I_R\,I_t) \qquad (3)$$

For optimal accuracy, ideally, the non-inductive resistance should be close or equal to load impedance as it should be for the three voltmeter method.

III. METHOD IMPLEMENTATION

A. The actual configuration

In these experiments for measuring the power factor in both three ammeter and three voltmeter methods the AC source is set to 7,5 V RMS (Root-Mean-Square), with the frequency set at 50 Hz. The voltmeter that is connected in parallel to the source, as seen on Figures 1 and 2, is used to set its output voltage. For both methods, the same electrical circuit elements were used; a non-inductive resistance value of 680 Ω and a load phase that consisted of a variable resistor of 100 Ω and a capacitor of 10µF. Three Digital Multimeters PeakTech 2010 were used to measure current or voltage, depending on the method. The used elements were connected as seen on Figures 1 and 2. For the three ammeter method a fourth Digital Multimeters PeakTech 2010 was used to set the voltage of the source.

B. Occurring problem

One of the problems that affected the results of both methods was the fact that in the original configuration the measurement devices that were used were all analog. This was mostly due to the fact that students were divided in six pairs that did the laboratory exercise simultaneously. This meant that at any given time during the class, at least eighteen instruments are needed. Since analog meters use a needle and scale to indicate values, the accuracy of the results is also affected by the operator's ability to read the readout on the meter, as seen on Figure 3.

There is also the problem of the meter's impedance. Digital voltmeters have significantly higher impedance than their analog counterparts do; they are more accurate when measuring voltage.

It is well documented [7][8][9] that for optimal accuracy the methods require that the non-inductive resistance should be close or equal to load impedance.

In the original configuration of the laboratory exercise, an adjustable resistor (rheostat to be specific) was used in the two-part load impedance. As a wire-wound resistor, a rheostat has a large inductance at higher frequencies, which affects the impedance of the load and its power factor, which would translate into further deviation from the theoretical one. Since the used frequency is 50 Hz, the effect in negligible.

There is also the matter of internal impedance of circular wires that for most low-frequency applications has no effect on the circuit or the results of the experiment and therefore can be ignored.

C. Presented opportunity

The noted problems presented an opportunity to further educate the students on the disparity between the practical and theoretical. To be more specific, to educate them about:

- risks of misreading analog instruments

- effects of measurement devices on the measurements themselves

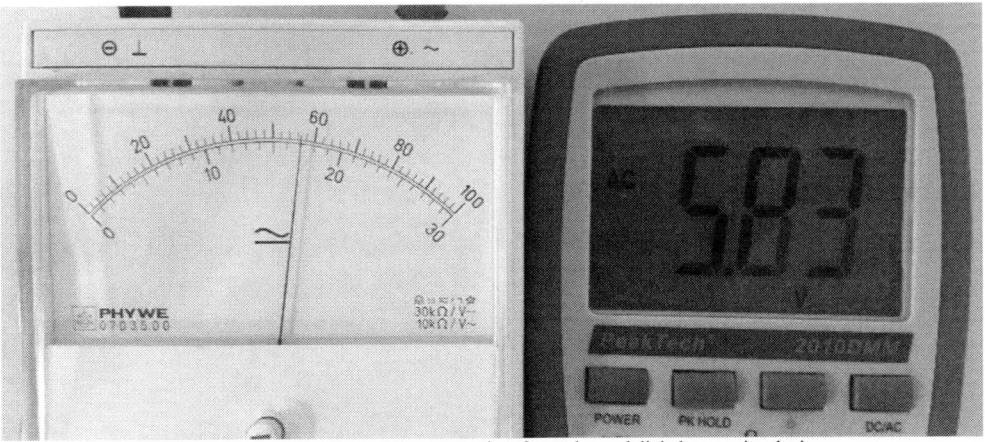

Figure 3. Example of measuring uncertainty for analog and digital measuring device

Figure 4. The actual configuration of the experiment with the three ammeter method used as described

- understanding instrument and circuit components specifications

- parasitic elements of electrical components

- effects of change in AC frequency

IV. CHANGE IN TEACHING METHODS

In the process of tackling the occurring problems different angles had to be considered. In the end, changes were made in all stages of this specific laboratory exercise and even some other exercises.

A. Pre-experiment

As part of the preparation for the laboratory exercise, students are assigned problems to solve to better prepare themselves for the exercise itself. With effects of measurement devices on the measurements themselves in mind, a specific circuit, similar to the one in the experiments, was assigned. Students were instructed to calculate the deviation in measurement devices readouts between two cases; when the instruments were ideal (no internal impedance) and when they were real (internal impedance was given). The assignments were done for a specific frequency, but can be altered and expanded for calculation for a wide range of frequencies using Microsoft Excel or some other spreadsheet program (Open Office Calc comes to mind as an open source alternative). This approach requires some knowledge of using functions in a spreadsheet program, which makes it an ideal task for a different course called Personal Computer Applications that is also taught as a part of professional study of Electrical engineering at the Zagreb University of Applied Sciences.

B. During the experiment

For the laboratory experiment itself, few key elements were changed.

Since uncertainty of reading analog instruments or human error is not a part of the intended lesson, nor the curriculum of Fundamentals of electrical engineering, analog instruments were swapped with their digital counterparts. These subjects are a part of a different course; Electrical Measurements, and therefore, because of the change, teaching materials have been altered for a different laboratory exercise on that course to educate the students on accuracy problems related to reading of analog instruments.

The most important change is the change in load

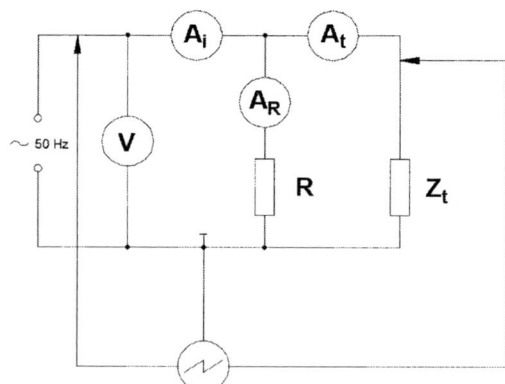

Figure 5. Updated experiment using the three ammeter method

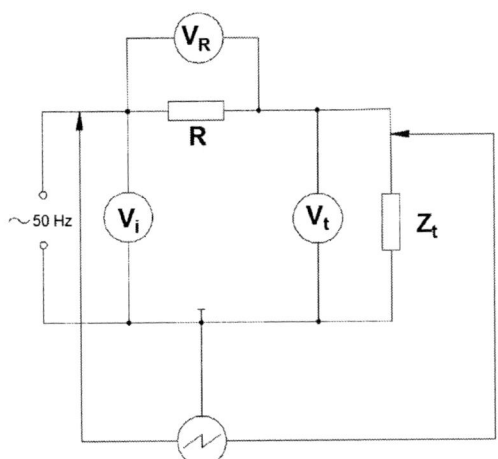

Figure 6.   Updated experiment using the three voltmeter method

impedance. Instead of the wire-wound resistor (rheostat), a typical carbon composition resistor was used. This lessened the effect of the rheostat's large inductance on the impedance of the load and its power factor but more importantly, because of its small size, the swap enabled us to construct eight different load phase impedances that were placed in 3D printed black boxes This was done to insure that all students will have different measurement results which among other things would reduce cheating. The changes are noted on Figures 5 and 6.

In addition, an oscilloscope was added to the experiment with one channel connected to the AC source and the other channel to the load phase. The oscilloscope is used for both methods but is crucial to the three voltmeter method where it is used to teach students about

phase shifts. The students are instructed to compare the value calculated from the power factor and the value of the phase shift that can be read from the oscilloscope. By doing so, they can determine the nature of the load, more specifically weather it is inductive or capacitive in nature. With the usage of a digital oscilloscope, the readout can be stored for analysis on the computer at home.

C. Post-experiment

As a tool for further education of the students on all of the above-mentioned topics, an EDA (Electronic Design Automation) program is used post-experiment. The Proteus ISIS (Intelligent Schematic Input System) package is the EDA program of choice. Students can freely and legally download a demonstration version of the program. Among several other restrictions, the main restriction, in comparison to a full version, is that the project cannot be saved or stored digitally in any way, but the readouts can be manually noted in the post-experiment documentation. These can be used later for comparison with the results of the pre-experiment assignment and the experiment itself.

The Proteus ISIS package, as seen on Figure 7, enables further education post-experiment in a way that could not be realised during the laboratory exercise because of time constraints. With Proteus, the student can simulate the behaviour of the circuit and its elements on a wider frequency spectre without the need for additional physical circuit elements. To execute that in a laboratory exercise environment, an addition of a signal generator would be needed and perhaps even a data acquisition module. This would further complicate the exercise itself for both the student and the educator.

Changing the values of electric circuit elements is an easy task in Proteus and that helps to evaluate both

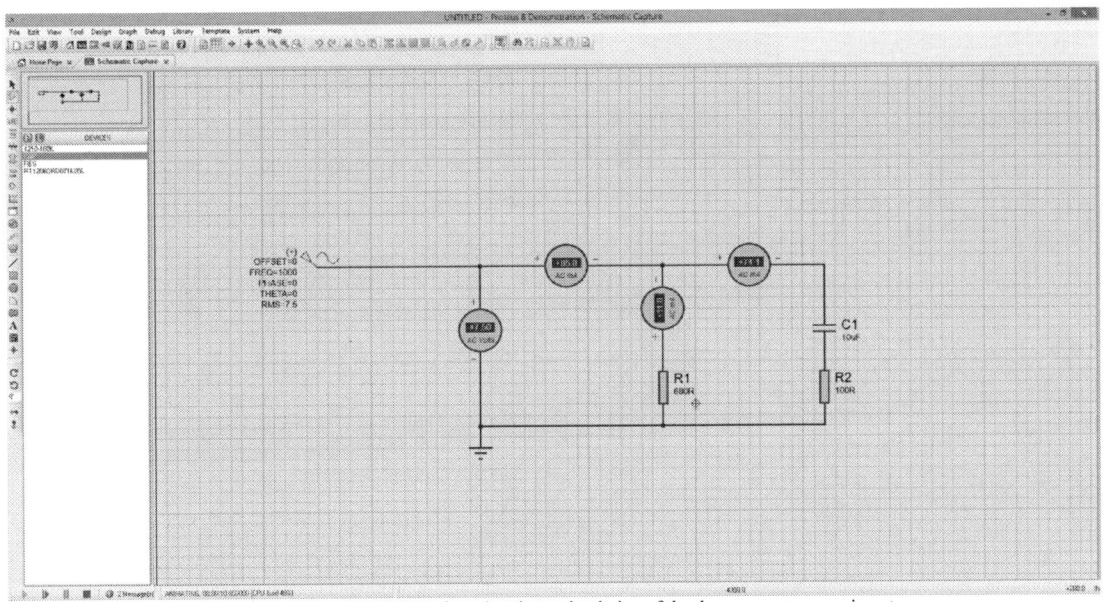

Figure 7.   Proteus ISIS interface showing a simulation of the three ammeter experiment

methods in sub-optimal configurations, specifically, when the non-inductive resistance is not close or equal to load impedance. In addition, changing the load impedance in Proteus, as with any other element in the circuit, results in an instant change of instrument readouts. This is something that yet again, is not so easy to accomplish in the laboratory environment. Not only because of time constrictions, but because it is not practical to have so many circuit elements at student's disposal. So, instead of having multiple sets of capacitors, inductors and resistors, physically available, to evaluate the behaviour of the method when the non-inductive resistance is not close or equal to load impedance, this can be achieved by simply adjusting the elements in Proteus with a few clicks of the mouse.

V.   CONCLUSION

Although power factor meters exist, even today, the tree ammeter method and the three voltmeter method are commonly used for calculation of the power factor of a specific load, both in practice and as a teaching tool. Both methods have practical and theoretical downsides and limitations, those same downsides and limitations can be used as a part of the teaching process in educating students on the divergence of the practical from the theoretical.

The addition of an EDA program proved to be an invaluable asset and a teaching tool even for a „Fundamentals of electrical engineering" course in professional study of electrical engineering at the Zagreb University of Applied Sciences, where up until recently, the usage of computers, for this specific course, was a rarity. Given the methods susceptibility to accuracy problems outside specific conditions described in section II, and physical and time constraints of working with students in a laboratory setting, the addition of an EDA program after the exercise helped us immensely to further

educate the students about the problems when the mentioned methods were applied in sub ideal conditions.

Given the satisfaction with the involvement of computer usage in the education process in the mention course, further developments and EDA program inclusions are planned.

BIBLIOGRAPHY

[1]   B. Andò, S. Baglio, and V. Marketta, "Volt-Ammeter Method introducing Principles and developing Technologies to undergraduates," Proceedings of the 8th WSEAS International Conference on Education and Educational Technology, 2009, pp.93–97

[2]   F. P. Baumgartner, R. Heule, M. Peter, "Web-Based Laboratory Training on Electrical Measurement Systems", Proc. of XVII IMEKO World Congress 2003, Dubrovnik, 2003.

[3]   B. Andò, S. Baglio, N. Pitrone, "New Trend in Laboratory Sessions", Proc. of the 12th IMEKO TC4 International Symposium, Zagreb, 2002.

[4]   A. Muciek, F. Cabiati, "Analysis of a three-voltmeter measurement method designed for low-frequency impedance comparisons", Metrology and Measurement Systems, Vol. 13, 2006, pp. 19-33

[5]   T. Alajbeg, T. Horvat, T. Novak, "The organization of classes and assessment system using Moodle", MIPRO 2014, pp 757-760

[6]   T. Horvat, T. Alajbeg, S. Predanić, "Experiences and Practices in Blended Learning Environment", MIPRO 2015, pp 1039-1043

[7]   L. A. Marzetta, "An Evaluation of the Three-Voltmeter Method for AC Power Measurement", IEEE Transactions on Instrumentation and Measurement, Vol. 21, n 4, Nov. 1972, pp.353-357

[8]   L. Callegaro, G. Galzerano, C. Svelto, "Precision impedance measurements by the three-voltage method with a novel high-stability multiphase DDS generator", IEEE Transactions on Instrumentation and Measurement,. Vol. 52, n 4, Aug. 2003, pp.1195-1199

[9]   A. Muciek, F. Cabiati, "Analysis of a three-voltmeter measurement method designed for low-frequency impedance comparisons", Metrology and Measurements Systems, vol. 13, 2006, pp.19-33

National Competition of Photography as Visual Art in Croatian Primary Schools, High Schools and Schools of Applied Arts

Zvjezdana Prohaska, BFA[*], Zdenko Prohaska, PhD[**], Ivan Uroda, PhD[***]

[*] Senior Adviser for Visual Arts, Education and Teacher Training Agency, Office Branch Rijeka, Rijeka, Croatia
[**] Full Professor, University of Rijeka, Faculty of Economics, Rijeka, Croatia
[***] Assistant Professor, University of Rijeka, Faculty of Economics, Rijeka, Croatia
zvjezdana.prohaska@azoo.hr, zdenko.prohaska@ri.t-com.hr, ivan.uroda@gmail.com

Abstract - Photography as a visual art is studied in subjects in Croatian primary schools, high schools and schools of applied arts and design.

In this article the preparation of a national competition in using photography and computers in Croatian schools at three levels mentioned above, will be presented, discussed and evaluated.

For evaluation of the photos a special and original computer program, called Photobox, was developed. It is actually a database in graphic format, where all photos can be stored, viewed and comments added.

For grading, scoring and ranking of the participants a main Excel spreadsheet with two additional Excel spreadsheets, which represent input databases, where also carefully elaborated.

I. INTRODUCTION

Under the auspices by the Ministry of Science and Education (Ministry), the Education and Teacher Training Agency (Agency) and one of the selected secondary or primary schools organize every school year, a competition - exhibition of pupils of primary and secondary schools in the field of visual arts and design called - LIK.

The competition includes three target groups: pupils from fifth to eighth grade of primary schools in the field of art, pupils from first to fourth grade in high schools, art schools in the field of visual arts, art history and theory of fine arts and pupils from first to fourth grade in secondary schools who are educated in the field of visual and applied art and design in a variety of programs and fields of fine art.

This competition - exhibition LIK has a 19 years tradition and represents the visual range of the pupils in the performance of artworks on a given topic and different art techniques (traditional and modern). Artworks by pupils selected at the three levels and assessed by a commission, at the school, county and national levels.

The artworks are selected and evaluated based on the criteria set up by the State Commission (Commission). This Commission is composed of professors from art academies, study of design and architecture, history, fine arts, and teachers of primary schools and secondary schools in the field of fine art.

The Commission prepares and creates the tasks for all three target groups of primary and secondary schools. Members of the Commission are appointed by the director of the Agency with the prior approval of the Ministry of Science and Education. From the members of the State Commission three committees in charge of the selection and evaluation of artworks (photographs) of the three above-mentioned groups of primary and secondary schools are appointed.

High-quality and objective evaluation of artworks and photos provided by the criteria for the assessment of individual elements that apply to all three levels of competition: school, county and national levels. All the artworks are coded and the results made public (e.g. the results of the national level will be published on the website of the Agency).

The most successful artworks of pupils at all three levels will be presented to the public at exhibitions

The exhibitions are regularly well-attended at schools and in public cultural and art institutions, i.e. museums and galleries (e.g. at the national level in the Mestrović Gallery in Split, the Amphitheatre in Pula, etc.).

Details of the competition - exhibition LIK are integrated in the Guidelines published on the website of the Agency. There is the data related to the organization of all three levels of competition, time schedule, category, method of rewarding pupils, a list of members of the Commission as well as a list of tasks that it performs.

Together with the instructions of the competition - exhibition LIK three documents on the tasks of the above mentioned three target groups of pupils are published. Art tasks are made at the school level with the expert assistance of teachers or teachers who are mentors of pupils. At the county level artworks are selected, and the most successful are sent to a higher level, i.e. the state level, for a new selection.

II. THE ORGANIZATION OF THE COMPETITION - GENERAL VIEW

The competition-exhibition of pupils of primary and secondary schools in the field of visual arts and design - LIK 2017 (main technique: photography) organized by the Ministry and the Agency has the following time schedule and levels: [2]

School competition-exhibition:
30th January 2017 (Monday)

County / inter-county competition-exhibition:
27th February 2017 (Monday)
Submission of the artworks by 10th February 2017

National competition-exhibition:
26th to 28th April 2017 (Wednesday, Thursday and Friday)
Submission of (all) artworks to by 10th March 2017

Note: Pupils from first to fourth grade of secondary schools and schools of fine and applied arts and design have just the school and state level.

Main topic and categories of the Competition - Exhibition are:

Topic - incentive of the competition: Sound image

Categories

1. Pupils from 5th to 8th year of primary schools / ART CULTURE
A - competition-exhibition of artworks on the theme - incentive
B - competition-exhibition of artworks on the set theme and visual problem

2. Pupils from 1st to 4th year high school / FINE ART (subjects: fine art, history of art and theory of design)
A - competition-exhibition on the theme - incentive realized artistic expression based on research work
B - competition essay on a given topic - incentive

3. Pupils from 1st to 4th year schools of applied arts and design and schools implementing programs of fine art and design
A - competition-exhibition of artworks on the theme - incentive
B - competition-exhibition of artworks set on the theme and visual problem

School competition
The state commission prepares the Tasks for the school competitions. *The school committee is* appointed by the school director.
Primary schools - visual arts category 1**A** can send from **the school to the county level no more than 10 artworks.** Secondary schools-character visual art (general education subject) in the category 2**A** can send from **the school to the county / inter-county level maximum 2 artworks.** Secondary school of fine and applied arts and design and the schools that educate pupils in programs of fine and applied arts and design (verified learning programs by the Ministry of Science, and Education / Curricula and framework programs for secondary schools of fine arts and applied arts and design, 2001) in category 3**A.**

School committees (1**A** and 2**A** category) must prepare a list of selected pupils in **Excel tables** downloaded from the website of the Agency / Competitions and festivals and forward to the county level committees (except 3**A** category directly to the state level).
After the school competition has finished, the schools have to forward electronically to the upper levels (inter-county or the county commission in addition to 3**A** category directly to the State Commission, i.e. the secretary of the State commission. Tables must be sent **by e-mail.** The organizer of each level of competition is required to consolidate the pupils registered in one table (spreadsheet), what should be done by copy and paste technique. All artworks should be evaluated according to prescribed propositions and criteria.

County competition
County commissions are appointed by the county offices. County offices determine the place and the host school, which will organize and carry out the selection of the best artworks of the competition-exhibition. It is desirable to set up an exhibition in a museum or a gallery and to prepare a simple and adequate catalog.

National competition
The State Commission has a president, secretary and members. They are appointed by the director of the Agency with the prior approval of the Ministry of Science and Education. The Evaluation committee (selected among the members of the State Commission) is split up by categories. Each category is coordinated by one of the members. The artworks in each category are evaluated by three members. The members of the evaluation committees at the national level cannot be mentors to pupils who are participating in the competition.
The evaluation committees choose up **to 13 artworks** of pupils in category 1**A**, 2**A** and 3**A,** which will participate in the exhibition (up to 39 pupils). At the state level, pupils are also invited to compete in the category 1**B**, 2**B** and 3**B.**
Pupils who win the first three places in these competitions will receive a special recognition. To the competitions and festivals on the state level, pupils will be invited according to their results in competitions held on the county level, what has to be confirmed by the Commission.

III. COMPETITION AT DIFFERENT LEVELS

In this chapter details of the competitions at different levels will be elaborated.

A. Primary schools competition

The tasks for pupils at the primary school level are defined below:
The task for pupils from 5th to 8th grade of primary schools, 1**A** category - ART CULTURE
A - competition-exhibition of artworks on the theme - incentive
Sub-theme-stimulus: Rhythm of light, the rhythm of colors
Teaching area: Flatbed design / 2D

Artistic techniques: traditional photography, digital photo. (photos must be in color /C/)

Size: size of photography should be prepared for print 30x40 or 30x45 cm (depending on the extent of filmed) and 300 dpi resolution. The artworks at the school, county and national level are reviewed and evaluated in a digital format and in printed form size 13x18 cm

Key concepts: photo, color, light, rhythm, contrast, composition, framing

Goals: Pupils will observe, learn about and evaluate the role of color, light, rhythm, contrast, composition and framing to create images on the sub-theme-stimulus *rhythm lights, the rhythm of colors.* Pupils will explore, express and evaluate different kinds of rhythms in their own work.

Task:

Pupils will:
- explore art / visual elements and key concepts for subtopic-stimulus *rhythm lights, the rhythm of colors* (separating bit of artwork pupils): photo, color, light rhythm, contrast, composition, framing
- express own idea in the medium of photography
- apply photographic procedures in making photographs (classical and digital, the ability of computer processing to the extent that does not distort and does not alter the basic meaning of the recording)
- evaluate role of color, light, rhythm, contrast, composition and framing

Evaluation:

Works will be evaluated according to the following criteria:
- divergent opinion
- art problem / key concepts: photo, color, light, rhythm, contrast, composition, framing
- understanding and creative application of art techniques and resources: the classic photography, digital photography
- connecting / interdisciplinary and correlated access

Primary schools:

Photo:

Pupils take photos by their camera or mobile phone (without limitation). Captured images can be processed in programs for photo editing to the extent that does not distort and does not alter the basic meaning of the recorded contents. Photographs must be in color /C/. Permitted file extension is JPG maximum size of 5 MB.

B. High schools competition

The tasks for pupils at the high school level are defined below:

The task for the pupils from first to fourth grade of high schools (subjects: Fine arts, art history and theory of design), 2A category - FINE ART

A - Competition - Exhibition of artworks on the theme incentive realized by artistic expression - based on research

Key concepts: photography, composition, rhythm, framing, light, contrast, composition - re-composition, definition - redefinition, recycling, texture, structure

Intersubject correlation: musical art, computer science, mathematics, Croatian language and literature and physics

Goals: The aim of the competition is to accomplish a picture using photography as a medium, taking into account its historical significance, technological capabilities, functions and aspects of photography. Selected access to traditional and contemporary photography pupils will be able to interpret the artistic manipulation of a photograph through the history of the connotation of the stylistic periods and artistic direction 20th century. Research activities will encourage the realization of artistic creations, develop creativity, the ability of divergent and convergent thinking and independence in planning and realizing the artistic task. [6], [8]

Task:

Pupils will:

Pupils prepare exclusively a collage of photos that are individually recorded. It can be done in digital or analog (classical) paper collage of printed photos. In the collage, with elements of photography, can use a collage of colored paper, can be used as a newspaper article, but not in terms of content, but in terms of its graphism.

Assignments:

Pupils will:
- apply photographic procedures in making photographs (classical and digital, the ability of computer processing to the extent that does not distort and does not alter the basic meaning of the recording)
- formulate a picture as a matrix for further consideration of the techniques of collage
- devise a collage as a redefinition, recomposition, confrontation and recycling questioning tone, rhythm, composition, structure, texture, schematisation, variation, modularity, multiplication and fragmentation
- realize their own idea as a piece of artwork in the medium of photography and in the technique of collage and photo collage
- evaluate the role of composition, rhythm, framing, light, contrast, texture, structure, chiaroscura
- conclude intermedia, multimedia, interdisciplinary, correlation creating new compositions photography

Artistic techniques: traditional photography, digital photography (photos may be color /C/, Monochrome /M/ or black and white /BW/), a collage and photo collage

Format: Minimum dimensions of the artwork is 15x15 cm, and the maximum 50x50 cm and all possible formats between the smallest and the largest dimension are allowed (e.g. 15x50, 30 x 50, etc.)

Evaluation:

The artworks will be evaluated according to the following criteria:
- visibility and correlation interdisciplinary approach to research and artistic expression and study forms of communication in different areas of art
- creative self-expression and world view and appreciation of ethical and aesthetic values in the artistic creations
- creativity and expression in artistic expression

Each criterion can be evaluated with a maximum of twenty points and the total sum amounts to sixty points.

Fine art:
Photo collage:
Pupils create exclusively collage of photos that are individually recorded. They can do digital or analog (classical) collage of photos and print them. In the collage, beside photo elements, paper and newspaper articles can be used, not in content but in terms of its graphism. Permitted file extension is JPG maximum size of 5 MB.

C.   Schools of applied arts and design competition

The tasks for pupils at the level of schools of applied art and design are defined below:
The task for the pupils from first to fourth year high school, 3**A** category - ART AND APPLIED ART AND DESIGN
A - competition-exhibition of artworks on the theme incentive
Topic-incentive: sound image / incentive: intermediality / The multimedia sound track
Intermediality - multimedia sound track:
- Intermediality defined as the process by which structures and materials are typical from one medium transferred to another; one of these media usually is artistic
- Multimedia is manifested in art as the expression of different art forms in one integrated medium. A related concept is the transmediality - an area in which identity and subjectivity changes and transforms using other media.
Note to teachers:
Research and creativity of pupils through the medium of photography and multimedia has numerous possibilities, which can easily seduce the naive observer, therefore it is necessary to explain the basic concepts of the media to the pupils.
Task:
ARTICULATION DESIGN FEATURES - intermediality - multimedia sound track
Pupils need to formulate and define systems within the composition on the concept of intermediality - multimedia sound track and show the symbolism of cooperation each element.
Goals: The pupil will be able to observe, explore and make an artistic technique / technology of his choice (in accordance with the curriculum at the department and school) on the subject of incentive-Sound Image / intermediality / Multimediality intermediality - multimedia audio clip.
During the realization of tasks pupils will be introduced to photography as a medium of expression, historical and artistic development of the same, and the role and importance throughout the 20th century. [3]
They will investigate the possibilities of photography, the achievements of modern technical / technological capabilities and to traditional or contemporary processes reinterpret obtained record (procedures redefinition / recomposition create new visual facts). [5]

Evaluation:
The artworks will be evaluated according to following criteria:
- interdisciplinary and correlated approach to research and artistic expression
- unifying themes as incentives (research access to forms of communication) and skills development in the material / medium
- the relationship of form and ideas
Each criterion can be evaluated with a maximum of twenty points and the total sum amounts to sixty points. Artworks that have less than fifty percent of the points (less than thirty points) cannot go to a higher level (state).
Schools of applied arts and design:
Pupils can realize their ideas as: photogram, chemo-gram, collage (analog and digital), photography, photomontage, photo story, photo albums, photo books, photo object, photos in textile design, jewelry, photo graphic design, photo design furniture (pupil artworks in 3D simulation), photos in interior design (pupil artworks in 3D simulation) and as a video clip, no longer than 3 minutes.
Two-dimensional artworks may have a minimum size of 10x10 cm and a maximum size of 70x70 cm and all derivative formats, both small and great.
Three-dimensional artworks may have minimum size dimensions of 15x15x15 cm and 50x50x50 cm as maximum, as well as all derivative formats, from small to the greatest format.

IV. PROGRAM PHOTOBOX AND EXCEL SPREADSHEETS

For evaluation of the photos at all three levels a special computer program called PhotoBox, actually a database in graphic format, was developed and programmed in Visual Basic for Windows. [1], [4], [7]

In this software all photos of the participants or pupils can be stored, viewed and notes could be made. Similar computer programs were already presented by the authors in articles at two earlier MIPRO Conferences. [9], [10]

The first screen of the program PhotoBox with the main menu, where photos can be loaded, saved, printed and pasted from the clipboard and to the clipboard, is shown in (Fig. 1) below.

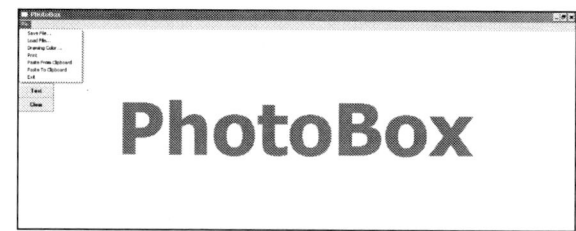

Figure 1. Program PhotoBox

For the scoring and ranking of the photos or pupils in the competition, a special Excel spreadsheet was developed, to get not only the results, but also a basic statistical analysis of the artworks submitted to this competition. (Fig. 2)

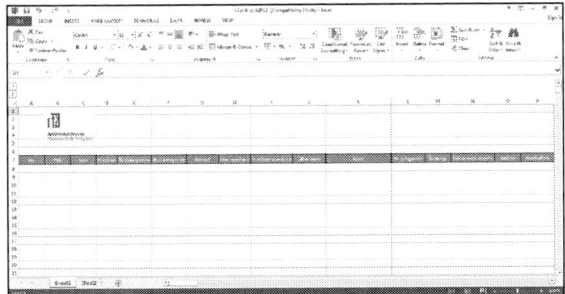

Figure 2. Spreadsheet - Results of the competition

Beside that, two additional spreadsheets containing basic data and information about the art competition itself and the names and addresses of the schools, whose pupils are participating in this competition, are downloaded from the Internet site of the Education and Teacher Training Agency and prepared for data input in the main spreadsheet presented above.

The first additional spreadsheet, containing the data about the different art competitions or school levels is presented below. (Fig. 3)

Figure 3. Spreadsheet - Art competitions

The second additional spreadsheet which contains the names, addresses and numeric labels or codes of the schools is shown in (Fig. 4).

Figure 4. Spreadsheet - Schools (basic data)

After the presentation of the program PhotoBox, which main task is to view and sort photos being sent to the competition and the three Excel spreadsheets for the evaluation and ranking, the basic installation procedure for all of them will be described in the next chapter.

V. INSTALLATION OF PROGRAM PHOTOBOX AND EXCEL SPREADSHEETS

The computer program PhotoBox consists only of two application files, i.e. the PhotoBox program file (photobox.exe) and the Visual Basic runtime module (vbrunxxx.dll). To install the program PhotoBox it is necessary to copy both files to a directory on the hard disk of a PC (for example C:\PhotoBox) and to start the program by clicking the program file or icon (photobox.exe) with a Microsoft compatible mouse.

Concerning the installation of the three Excel spreadsheets presented above, it is necessary to copy these files in the working directory of the PC where Microsoft Excel program can access them.

VI. CONCLUSION

In this article the preparation of the national competition of photography in Croatian primary schools, high schools and schools of applied arts and design was elaborated. A special computer program, PhotoBox, for viewing and sorting of the photos was developed, as well as a main Excel spreadsheet for grading, scoring and ranking of the participants and two additional Excel spreadsheets as input databases for the main spreadsheet were programmed.

REFERENCES

[1] Čičin-Šain, M., Čapko, Z., Vukmirović, S., Informatika za informatičko poslovanje, Ekonomski fakultet Sveučilišta u Rijeci, Rijeka, 2008
[2] Education and Teacher Training Agency, Guidelines and Instructions for National Competitions, Zagreb, 2017
[3] Freeman, M., The Complete Guide to Digital Photography, Thames & Hudson, London, 2006
[4] Halvorson, M., Microsoft Visual Basic Professional 6.0 Step by Step, Microsoft Press, Redmond, 1998
[5] Hedgecoe, J., Complete Guide to Photography, Sterling, London, 2004
[6] Janson, A.F., History of Art, Prentice Hall, NJ, 2004
[7] Microsoft, Microsoft Visual Basic 6.0 Programmer`s Guide, Microsoft Press, Redmond, 1998
[8] Preziosi, D., The Art of Art History, Oxford University Press, Oxford, 2009
[9] Prohaska, Z., Uroda, I., PaintArt – An Interactive Computer Software for Education of Visual Arts, in Čičin-Šain, M. et al., 36th International Convention MIPRO 2013, Mipro Proceedings, Computers in Education, Opatija, 2013
[10] Prohaska, Z., Prohaska, Z., Uroda, I., PaintArt 2.0 – Computer Program for Education of Visual Arts in Secondary Schools, in Čičin-Šain, M. et al., 37th International Convention MIPRO 2014, Mipro Proceedings, Computers in Education, Opatija, 2014

Generating large random test data table for SQL training

U. Sterle*

* Šolski center Kranj/Srednja tehniška šola, Kranj, Slovenia
uros.sterle@sckr.si

Abstract - For a quality SQL training we need many different exercises and many different tables. There are many databases along with the exercises in English. In Slovenian and other languages we have a lack of them. The idea was to create a random table from existing data. On the web we find a lot of data in our own language. When we assemble these data according to some random logic, we can obtain a large table, which is needed for practising SQL language.

We've compiled a table of people which includes the following information: name, sex, place of residence, birth date and time, the profession and a popular place. From the Web we obtained information about the most common first names by gender, family names, places in Slovenia and the professions. Birthdays were generated randomly, so that a person's age is between eighteen and eighty years. Popular places were also randomly generated as a geolocation in Slovenia.

We could easily use any other data to generate random tables.

Table was used in exercises of the module entitled "Advanced usage of databases" in the third year of secondary school program of Computer Technician.

I. INTRODUCTION

These instructions give you guidelines for creating your own large random test data table for SQL training.

All you need is find your own data, create your own tables and adjust SQL code to fit your data.

We compiled a table of people with basic information such as name, sex, place of residence, birth date and time, the profession and a popular place.

We could easily use any other data to generate random tables. This is the main advantage of this method, because it is not dependent on data we used. Disadvantage is that you need to know SQL to use this method. There is idea for web application, which would create table itself, when we choose CSV file for importing and this is the next step of development.

We used MySQL database management system (DBMS), but the code can be easily adjusted to any other DBMS such as SQL Server, Oracle, PostgreSQL etc.

II. GETTING DATA FROM THE WEB

A. First names and family names

From the web page of The Statistical Office of the Republic of Slovenia [1] we obtained data for first names and family names. Together we downloaded six csv files. One file for male first names, one for female first names and four files for family names. We adjusted files for importing in MySQL tables. We excluded the first names that appear less than five times, and so we get 3595 men and 3871 women's names. Similarly, we have done with family names and we get 28369 different family names.

B. Places of residence, settlements

We took data from another subpage of The Statistical Office of the Republic of Slovenia [2]. We downloaded one file. There are 6037 settlements from Slovenia together with number of households in individual settlements.

C. Proffesions

Webpage mojaizbira.si [3] contains a list of 496 professions and we used them for our database.

III. PREPARING DATABASE

After we retrieved data from web, we had to prepare out database for generating random data. We created several tables:

A. Table for first names

First we created a table named first_names with columns for name, number of names and gender. Here is the MySQL code for creating this table:

```
CREATE TABLE first_names (
    first_name VARCHAR(20),
    number INT,
    gender CHAR
)
DEFAULT CHARSET='utf8'
ENGINE=MyISAM;
```

TABLE I.   SOME DATA FROM TABLE FIRST_NAMES

first_name	number	gender
Franc	25083	M
Janez	21877	M
Ivan	18322	M
Anton	18246	M

Data from first_names.csv file:

```
ime;stevilo;spol
Franc;25083;M
Janez;21877;M
Ivan;18322;M
Anton;18246;M
```

We used LOAD DATA INFILE sentence to import data from file to database table.

```
LOAD DATA INFILE 'first_names.csv'
INTO TABLE first_names
CHARACTER SET 'utf8'
FIELDS TERMINATED BY ';'
LINES TERMINATED BY '\r\n'
IGNORE 1 LINES;
```

We see some imported data in Table I.

B. Table for family names

Second table was named family_names with columns for family name and number of them. Here is the MySQL code for creating this table:

```
CREATE TABLE family_names (
    family_name VARCHAR(20),
    number INT
)
DEFAULT CHARSET='utf8'
ENGINE=MyISAM;
```

Data from family_names.csv file:

```
priimek;stevilo
Novak;11124
Horvat;9648
Kovačič;5605
Krajnc;5522
```

We used LOAD DATA INFILE sentence to import data from file to database table.

```
LOAD DATA INFILE 'family_names.csv'
INTO TABLE family_names
```

TABLE II.   SOME DATA FROM TABLE FAMILY_NAMES

priimek	stevilo
Novak	11124
Horvat	9648
Kovačič	5605
Krajnc	5522

TABLE III.   SOME DATA FROM TABLE SETTLEMENTS

koda	ime	vrsta	stanovanja
001	AJDOVŠČINA	občina	7202
001001	Ajdovščina	kraj	2576
001002	Batuje	kraj	140
001003	Bela	kraj	9
001004	Brje	kraj	175

```
CHARACTER SET 'utf8'
FIELDS TERMINATED BY ';'
LINES TERMINATED BY '\r\n'
IGNORE 1 LINES;
```

C. Table for settlements

Settlements table contains code, name of settlement type (settlement or municipality) and the number of households. Here is the MySQL code for creating this table:

```
CREATE TABLE settlements (
    code varchar(6),
    name VARCHAR(35),
    type VARCHAR(10),
    households INT
)
DEFAULT CHARSET='utf8'
ENGINE=MyISAM;
```

There are some imported data in Table II.

Data from settlements.csv file:

```
001;AJDOVŠČINA;6896
001001;Ajdovščina;2609
001002;Batuje;132
001003;Bela;8
001004;Brje;136
```

We used LOAD DATA INFILE sentence to import data from file to database table.

```
LOAD DATA INFILE settlements.csv'
INTO TABLE settlements
CHARACTER SET 'utf8'
FIELDS TERMINATED BY ';'
LINES TERMINATED BY '\r\n'
IGNORE 1 LINES;
```

Imported data for settlements are located in Table III.

D. Table for proffesions

Next table was named professions with columns for id and name of proffesion. Here is the MySQL code for creating this table:

```
CREATE TABLE professions (
    ID INT,
    profession VARCHAR(120)
)
DEFAULT CHARSET='utf8'
ENGINE=MyISAM;
```

TABLE IV.   SOME DATA FROM TABLE PROFESSIONS

id	poklic
1	administrator
2	agronom
3	agronom za sadjarstvo
4	aktuar

Data from professions.csv file:

```
id;poklic
1;administrator
2;agronom
3;agronom za sadjarstvo
4;aktuar
```

We used LOAD DATA INFILE sentence to import data from file to database table.

```
LOAD DATA INFILE professions.csv'
INTO TABLE professions
CHARACTER SET 'utf8'
FIELDS TERMINATED BY ';'
LINES TERMINATED BY '\r\n'
IGNORE 1 LINES;
```

Table IV represents imported data for professions.

E. Table for random generated names

Last table was named persons with columns for first name, family name, settlement, profession, birthday and gender. Here is the MySQL code for creating this table:

```
CREATE TABLE persons (
    first_name VARCHAR(20),
    family_name VARCHAR(20),
    settlement VARCHAR(35),
    profession VARCHAR(120),
    birthday DATETIME,
    gender CHAR
)
DEFAULT CHARSET='utf8'
ENGINE=MyISAM;
```

IV. GENERATING RANDOM DATA

We created following procedure to create random person from previously generated tables.

```
DELIMITER //
CREATE  PROCEDURE  generate_random_names
(n INT, year1 INT, year2 INT)
  BEGIN
    DECLARE  fi, fa,  s,  g  VARCHAR(40)
CHARACTER SET utf8;
    DECLARE id_p INT;
    DECLARE b DATETIME;
    WHILE n > 0 DO
      SELECT first_name, gender
      INTO fi, g
       FROM names
       ORDER BY -LOG(RAND())/number
```

```
       LIMIT 1;

      SELECT family_name
      INTO fa
       FROM family_names
       ORDER BY -LOG(RAND())/number
       LIMIT 1;

      SELECT code
      INTO s
       FROM settlements
       WHERE type='kraj'
       ORDER BY RAND()
       LIMIT 1;

      SELECT id
      INTO id_p
       FROM professions
       ORDER BY RAND()
       LIMIT 1;

      SELECT CURDATE()
-       INTERVAL    (FLOOR(RAND()*(year2-
year1+1)*365)+year1*365) DAY
      - INTERVAL (FLOOR(RAND()*24*60)) MINUTE
       INTO b;

      SET n = n - 1;

      INSERT   INTO   persons(first_name,
family_name,    settlement,    profession,
birthday, gender)
        VALUES (fi, fa, s, id_p, b, g);
    END WHILE;
  END//
  DELIMITER ;
```

Finally we run the procedure:

```
CALL  generate_random_names(10000,  18,
80);
```

The meaning of the last sentence is that we generate 10000 persons whose age is between 18 and 80 years.

V. APPLICATION IN TEACHING PROCESS

We will show you just three examples of question and answers during our classes.

- Question 1: Write a query to display the names (first_name, family_name) and birthday for all people who were born in Ljubljana.

  Answer:

  ```
  SELECT first_name, family_name,
  birthday
  FROM persons
  WHERE settlement = 'Ljubljana';
  ```

- Question 2: How many female persons were born in 60's?

956

Answer:

```
SELECT COUNT(*)
FROM persons
WHERE GENDER = 'Ž' AND
YEAR(birthday) BETWEEN 1960
AND 1969;
```

- Question 3: List all data for persons who were born on Friday the 13th in alphabetical order.

Answer:

```
SELECT *
FROM persons
WHERE DAYNAME(birthday)='Friday' AND
DAYOFMONTH(birthday)=13
ORDER BY family_name, first_name;
```

VI. CONLUSION

Table was used in exercises of the module entitled "Advanced usage of databases" in the third year of secondary school program of Computer Technician. The responses of students and other teacher were excellent.

We conducted a short survey at the end. Previously our exercises were based on English tables so students were very satisfied with data in their own language. They would also like to have more exercises with data in their own language.

I also introduced this method to two other database teachers and they find it very useful and included method in their own curriculum.

REFERENCES

[1] Statistical Office of the Republic of Slovenia, First names and family names of the population. (n.d.) Retrieved from http://pxweb.stat.si/pxweb/Database/Demographics/05_population/46_names_surnames/06_05X10_names_surnames/06_05X10_names_surnames.asp.

[2] Statistical Office of the Republic of Slovenia, Households by number of members, settlements, Slovenia, multiannually. Retrieved from http://pxweb.stat.si/pxweb/Database/Demographics/05_population/17_Households/20_05F40_Households_NAS/20_05F40_Households_NAS.asp.

[3] List of proffesions. (n.d.) Retrieved from http://www.mojaizbira.si/poklici.

Programming Lego Mindstorms for First Lego League Robot Game and Technical interview

Barbara Strnad

Gimnazija Novo mesto, Novo mesto, Slovenia
Društvo Čarunalnik – društvo za napredno uporabo računalnikov Novo mesto, Slovenia
barbara.strnad@gmail.com

Abstract - **It is of great importance in the information age to equip children with the capacity to independently navigate in the flood of all available data and be responsible towards acquiring knowledge. We must provide a learning environment, where children can be enhanced to think creatively and develop innovative solutions. The article further presents the robotic part (a robot game with a technical interview) that is half of the multidisciplinary program First® Lego® League, which was brought to Slovenia by Super Glavce, the Institute for the Promotion of Knowledge. There is an overview of several months' work and the preparation for the robotic part of the competition, where children have to build and programme an autonomous robot that can complete as many tasks on the competition mat as possible in a limited amount of time. During this process a technical portfolio is made, where children record the progress, the problems and describe ways of solving the problem, explain strategies and present innovative solutions. The technical portfolio is the basis for the technical interview. The ways of working with lego robots in preparation for the First® Lego® League competition can be successfully replicated in class with the topic of algorithmic ways of thinking and programming.**

I. INTRODUCTION

In this article, we want to show that Lego robots are a very effective learning tool, which facilitates the development of logical thinking, creativity, and greatly increases children's interest in learning programming. Children's participation in the preparations for the robot game in the context of a multidisciplinary program First® LEGO® League (FLL) encourages the search for creative and innovative design and software solutions. The working methods of the FLL program, where the activity focuses on children, who learn through their own work and creation, can be replicated in class.

II. FLL PROGRAM

Since we live in an age of high technology and we are entering the era of robots, we started with robotic activities in classroom three years ago. Lego robots allow students to learn how robots work and how we can manage robots in a fun way. Likewise, we have in mind the finding [4] that on average Slovenian children do not want to learn science and technology topics, the same is true in developed countries [3].

Because children are highly motivated if they have their goals in mind, we joined the FLL program, because of the robot game and technical interview. Also, studies have shown that the FLL program increases children's interest in science and technology.

FLL is an international, multidisciplinary research program for children from 9 to 16 years old. Program promotes curiosity, creativity, collaborative learning, and teamwork. Because of a thoughtful, efficient and attractive concept young people learn about the concept of STE(A)M (science, technology, engineering, (art) and math) in a different way than what we are used to in regular classes. With the help of attractive robotics challenges, children get enthusiastic about science and technology, through the game they develop and strengthen their logical and technical way of thinking.

For the robotic part of FLL, teams plan, construct, programme and test an autonomous robot so that it can carry out a range of missions in the 2,5 minute long match on the FLL mat. Tasks on the mat represent the most typical problems of the current FLL challenge: the 2014 season was themed World's Class, in the 2015 season it was themed Trash Trek, and in the 2016 season, Animal Allies (Fig. 1).

III. PREPARING FOR THE COMPETITION

Preparations for the regional, national competition and later international, are very intense. Only in the third season did we first encounter a child who has already had some experience with Lego robots. The vast majority of children have no experience with programming, only a few have some experience with programming in Scratch.

Work starts in early September when teams receive bricks to assemble models to FLL mat. Students must follow building instructions very carefully, as imprecisely built models can influence the choice of strategies and can impair the robot's performance.

Figure 1: FLL mat from 2014, 2015 and 2016

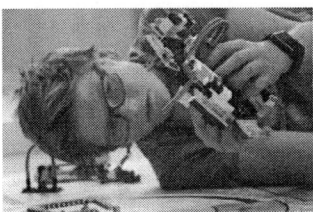

Figure 2: Constructing models

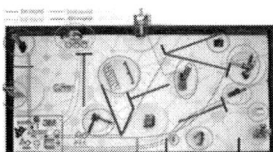

Figure 3: Plan route

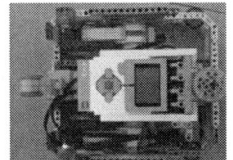

Figure 4: Base of the robot

When constructing models (Fig. 2) children can get a lot of knowledge and ideas associated with mechanical devices and mechanisms. The models include many basic mechanical principles that can be found in advanced mechanical devices, mechanisms, and structures of real life.

Students can explore the gear mechanisms when constructing models. When the FLL mat is ready, a detailed review of the missions must be done, including precise and repetitive reading instructions and watching videos with a description of the requirements. Students then determine a strategy and make a detailed plan route. (Fig. 3). It is important to indicate the points of assurance, sensors may also be used.

We can now start building the base (Fig.4) of the robot from the Mindstorms EV3 robotic kit. When constructing the base, it is very important to pay attention to the size and weight, to select the most appropriate tires, to plan the use of sensors and their positions, to think where the robotic arm will be located, etc. A lot of attention must be paid to whether the robot is well balanced and whether the center of gravity is in the right place. Improper weight distribution may cause the robot to behave erratically, for instance undesirable turning, so the location of the brick is very important. We should also not ignore the weight of the attachments and connections, which can increase the uncertainty of the robot.

When choosing tires it is necessary to carry out several test runs. Larger wheels may be faster, but less precise, with smaller wheels, the robot is slower, but you can achieve greater accuracy. Harder rubber is less deformed and less likely to withdraw from the rim. The rear wheels or sliders should move in all directions and be of the same height as the front wheels.

We can use four large motors or three large and one middle motor. The base of the robot usually has two to three light sensors, a gyro sensor, possibly a touch sensor. We want to construct the most solid and compact base with straight sides. When driving along the wall you might find a small side wheels come in handy. Changing attachments on the robot has to be as sleek, fast, and efficient as possible, even with shaky hands during the competition. The base must allows free access to the battery charging input and USB port.

Parallel to the construction of the base we start learning how to programme. We find knowledge in books and on the internet, where there is a real treasure trove of ideas and tools for learning. We begin with a variety of challenges: driving straight is required, especially when

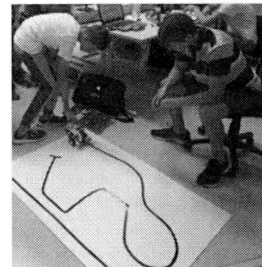

Figure 5: Line following

you want to drive fast, maintain the direction of the gyro sensor, tracking lines (Fig. 5).

When we are satisfied with the base of the robot, which is agile, and how it moves around the mat with appropriate speed and power, we can start with the construction of attachments.

Older children who have already participated in the program can transfer their experience and knowledge: their younger colleagues can learn the basics from them and get information about the barriers and traps which can occur in the competition. With peer involvement in all phases and areas of work, a positive climate can be felt in the team. Children learn from each other through conversation, discovery, research, and debate. We're trying to create a learning environment that encourages students to be more active, with more effective learning, commitment to the quality knowledge and taking responsibility for it.

Initially fun and very comfortable work with robots brings children new challenges in mechanics and programming. Usually it turns out that some children are best constructors, others are better programmers, and few are very good at both. To effectively solve tasks and eliminate problems in the robotic table, good cooperation of both constructors and programmers is crucial. Both carefully observe what is happening with the robot, analyze its mission, try to determine the cause of unwanted behavior of the robot, and remove it. Slow-motion video of the robot can be very helpful. Each ride has to be tried out as often as possible, in different conditions, in different racing fields and this way we can gain as much experience as possible to foresee several possible mistakes. We want to ensure maximum accuracy in the performance of all tasks, so we try to prevent mistakes either with the design of the robot or with programming.

That children do not lose motivation, training matches are organized in early November. Training matches are usually a new momentum for a team, because new ideas are born and usually we realize there is still quite some

Figure 6: Robot game

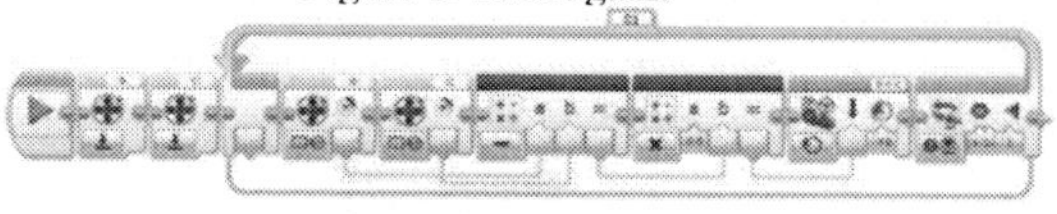

Figure 7: Example of a program

work to be done. The training is excellent preparation for the competition (Fig. 6).

While working on a robot, teams need to write a robot diary, where the process of construction and the programming process is well described and documented. A technical map is divided into three parts, three skill areas. One chapter is dedicated to mechanical design, where teams put evidence of structural integrity: durability, mechanical efficiency, and mechanical equipment of the robot. The robot must be able to withstand the rigors of competition. The second chapter is dedicated to programming (Fig. 7): the quality and efficiency of programming. To visualize the programming process and the steps in improving programs (use of loops, my blocks, math, and comments). The third chapter is devoted to strategy and innovation. Here we present a complete cycle of improvement, process development and interpretation of improvements, modifications, advantages and disadvantages of our solutions, the possible errors ... Here teams should clearly define a strategy to tackle all tasks and specifically highlight innovation and the basic features of the robot, which add a lot of value.

Robotic diary serves as a basis for the preparation for a technical interview with judges where students have to present the process of building a robot, attachments and programming. During the interview there should be a demonstration of the drive of the robot we are most proud of, and to answer the judges' questions.

The main competencies that students acquire are the development of imagination and spatial performance, conquering mathematical and technological skills, learning programming, and learning a variety of computer technology while cooperating with each other and with experts. It is experiential learning, developing

creativity and innovative problem solving. Creativity is creating new connections, which can happen only when the mind is enabled to do flexible and adaptive functioning. FLL allows and encourages all this.

IV. CONLUSION

Participating in the FLL has had a great influence on author of this paper. As a teacher, the author has changed his way of teaching a lot. Children now have a much more active role in the learning process. As a teacher, the author of this paper is usually their supervisor, assistant, and sometimes even the mediator. Teacher directs students to reflect on the achievements, thereby increasing their involvement in monitoring their own learning process and, consequently, their responsibility for their own work and performance quality.

Children who had participated and still participate in the program obtain a wide range of skills. Computer logic allows them to understand and program the robot; their technical knowledge comes in handy when they are constructing machines, because working with motors, sensors, levers, and gears do not pose any major problems. They must also be able to cooperate with other members of their group and add their piece to the mosaic.

This type of work is quite different from everyday school life because it requires a different type of work from the students, which allows them to become more independent, and the quality of their knowledge in combination with experience is sharply increased.

REFERENCES

[1] Melchior, Alan. Cutter, Tracy. Kingsley, Chris. 2013. Evaluation of FIRST LEGO League. (2016, January 2). Retrieved from: http://www.firstinspires.org/sites/default/files/uploads/resource_library/impact/impact-executive-summary-first-lego-league-impact-study-2012-2013.pdf

[2] Sjøberg, Svein. Schreiner, Camilla. 2010. The ROSE project. An overview and key findings. (2016, January 5). Retrieved from: http://roseproject.no/network/countries/norway/eng/nor-Sjoberg-Schreiner-overview-2010.pdf

[3] STRAŠEK, Rok, DOLINŠEK, Slavko and SKURJENI, Drago, 2008, Zanimanje in želje osnovnošolcev za učenje naravoslovja. [online]. 2008. Vol. 4, no. let 3, p. 363–378. (2017, April 4). Retrieved from: http://www.fm-kp.si/zalozba/ISSN/1854-4231/3_363-378.pdf

Children online safety

J. Žufić[*], T. Žajgar[**], S. Prkić[***]
[*]Juraj Dobrila University of Pula, Pula, Croatia
jzufic@unipu.hr
[**]Primary school Kaštanjer Pula, Pula, Croatia
tzajgar@gmail.com
[***]Primary school Kaštanjer Pula, Pula, Croatia
svjetlana.prkic@gmail.com

Abstract - The article provides results of a research study on safety of children on the Internet, and on use of the information-telecommunication technology (ICT) from the aspect of safety. The study was conducted on the elementary school children using paper or web-based questionnaires. It involved 1,232 students from grades one through eight of elementary school (approximately 400 students per survey), ages 7-15. The study was conducted over a period of eight years on three occasions (2008, 2011, and 2015) in elementary school Kastanjer, in Pula, Croatia. Based on the questions and answers, the findings are separated into groups: a) use of ICT among students; b) parents/teachers cooperation and provision of support to children/students in regard to cyber bullying; and c) students' understanding of what constitutes cyber bullying and their response towards unwanted harassment. The results show poor informational literacy of students in terms of safety. Parents need to invest much more effort in recognition of predation and dangers of the ICT use in order to protect their children, and teachers should spend more time and more educational contents related to the elements of cyber bullying and their prevention in order to teach their students.

I. INTRODUCTION

Use of computers, smart phones, and other devices that can connect to the Internet is spreading fast, in business and education as well as in a family circle. In families, the most frequent users of the information-telecommunication technology (ICT) are children. Most children have access to technologies with the Internet access, and learn in fun ways how to use ICT. At the same time, development of technology, its interactivity and increasing possibilities of communication represent a risk for all users, particularly children.

Today, parents are often facing the fact that their children know more than they do, and that children are growing up in a virtual world that is significantly different from the world the parents know. Therefore, parental home-upbringing becomes more complex and presents ever increasing challenges in providing support and safety for their children. Parents' worry is warranted because there are real dangers, and the worry is increased by parents' lack of understanding of dangers brought in by the technology. On one hand, parents do not want to stop progress of their children since the ICT bring many

benefits, while on the other hand they do not know enough, which represents potential danger.

Most youth and children, while playing online video games, watching videos clips, or socializing on social networks, develop their digital competences [1]. Through playing video games, whether online or not, a child develops, because video games stimulate thinking processes, associative thinking, support intuition and hypothetical thinking, improve coordination of movements, and represent unbiased teachers with unlimited patience [2]. A child releases emotions while playing video games, and video games can be useful tools in learning because they provide instantaneous reward.

European Parliament and the Council of the European Union in their recommendations on key competences for lifelong learning state that digital competency is a tool every person should have, in order to adapt to the fast-changing world [3]. The key elements of digital competency are the basic ICT skills: use of computers for exploration/browsing, assessment, storage, creation, presentation, and exchange of information, as well as creation of networks of cooperation through the Internet [4].

Children who had earlier exposure to the Internet and who grew up in that world show better knowledge and skill in computer use, mathematics, and reading, as well as in the general educational achievements. Research conducted in Australia, intended to establish the vocabulary development in children between four and eight years of age, on a sample of more than 9,000 children who had access to the Internet, has shown a developed vocabulary and well-developed verbal abilities. The exception observed by this research was in a negative relationship between use of game consoles and vocabulary development. Children-users that used only game consoles had lower linguistic abilities than other children [1]. It is important to understand the relevance of digital technology as an integrated tool for learning, which – when used reasonably – promotes linguistic, cognitive, and social development of youth and children.

The conclusion that arises is that children should be encouraged to use ICT, while it is exceptionally important to teach them how to use such devices in the correct and appropriate ways [5].

However, the fact remains that is not possible to control the contents other users will publish over the Internet, which brings into focus the other aspect of the ICT use, which is safety of children on the Internet and their exposure to disturbing contents. Children and youth are facing media challenges, primarily cyber bullying, as well as potential sexual harassment through social networks [6].

Results of the *EU Kids Online* survey conducted in 2010 in 25 EU countries indicates that it is necessary to continuously teach children and youth regarding safety standards and contents to ensure that all children have a minimal basic safety knowledge. Such an approach would prevent children from being digitally isolated and unqualified [7], [8], [9] and digitally illiterate. To be literate in a digital age involves more elements, one of which is the skill for responsible use of technology.

In the upbringing and education of children, from the point of the ICT use, teachers and parents have equal roles, teachers in school, and parents at home.

Review of the existing curriculum for elementary school [10] clearly shows that none of them require teachers to cover areas of safety of students on the Internet in any of the subjects. While use of computers, the Internet, and ICT is used in many subjects (with specific goals and tasks), from Croatian language, nature and social sciences, technical culture, music and arts, foreign language, extracurricular activities of the information-communication technology (which is a part of the teaching process only in a very small number of elementary schools outside of the regular curriculum – in Pula, for example, only in one of 10 schools) and the elective subject of informatics (taught only to those students who choose this subject), at the same time, there is only one mention of the acceptable use of e-mail (Theme 13. *Acceptable behavior when using e-mail. Educational goals: communicating over the Internet in an acceptable mode*). Since this educational content is mostly not presented in schools, most students have no chance to hear it, and it all depends on their teachers of *Informatics science* (for students who elected the subject) who may choose, but are not required to talk about safety of students on the Internet while working on some other subject. The same is true for other subjects, where the situation is even worse since teachers of other subjects have no obligation to mention the issue, and have poor understanding of ICT.

II. SAMPLES AND RESEARCH METHODOLOGY

This article describes the findings of the survey conducted in the Elementary school "Kastanjer" in Pula, with participation of students from the first to the eighth grade. The survey was conducted on three occasions over the period of eight years (2008, 2011 and 2015). Elementary school students participated in the survey by manually filling in paper forms or, alternatively, by filling in web forms. Survey questions and findings are divided into three groups: a) Use of ICT among students, b) Parents/Teachers cooperation and provision of support to Children/Students in regard to cyber bullying, c) Students' understanding of what constitutes cyber bullying and their response towards unwanted harassment.

The survey involved 1232 students with valid answers (403 students in 2008, 362 in 2011 and 467 students in 2015). Participants were students from the 1st to 8th grade of the elementary school, 7-15 years of age and gender-wise, with slightly higher participation of boys (645) than girls (587).

In the first survey, students from 2nd to 8th grade manually filled in the paper forms. The second survey applied the same method of data collection and students from the 2nd to 7th grade participated in the survey. In the third survey, students from the 1st to 4th grade manually completed paper questionnaires and from the 6th to 8th grade, students completed web questionnaires. Questionnaires were, on all three occasions, completed anonymously and teachers assisted younger students when they came across new words or phrases. First and second survey had the same questionnaire (21 questions, 19 of the closed type and 2 of the open type), while the third was more extensive (81 questions, all of them of the closed type) and was consistent with the questionnaire used in the survey "*EU Kids Online*".

III. QUESTIONNAIRE, FINDINGS AND ANALYSIS

The questionnaire contained three groups of questions. The first group contained questions about ICT use among students, the second group dealt with cooperation of parents/teachers and provided support to children/students from the aspect of cyber bullying and the third group of questions was about students' knowledge of the elements of cyber bullying and their response to the unwanted harassment. All answers were grouped by grades: 1st and 2nd, 3rd and 4th, 5th and 6th and last, but not least, 7th and 8th. The reason for such combination is that there are minor differences in students' age. Distinctions are also made in terms of the gender of participants and the year the surveys were conducted.

A) ICT use among students

Findings on the ICT use among students are shown in Table I (Findings of the posed questions). Questions are as follows: Q1 – How many times a day do you use the Internet?; What do you do most often on the Internet? Options provided were: Q2 – Play online games; Q3 – I am exploring data and school curriculum; Q4 – Use E-mail; Q5 – Watch movies and listen (and download) music; Q6 – Chat with friends; Q7 - Read news.

In the first question (Q1 – "How much time students spend on the Internet"), it is indicative that, as expected, older the students, the more time they spend on the Internet.

TABLE I. PUPIL'S USE OF ICT

Grade	Year	Gender	Q1 [min]	Q2	Q3	Q4	Q5	Q6	Q7
First and second	2008	M	45	5	20	3	7	2	0
		F	35	4	7	0	1	2	1
		Total	X	9	27	3	8	4	1
	2011	M	10	4	32	0	6	1	2
		F	45	0	17	1	4	0	0
		Total	X	4	49	1	10	1	2
	2015	M	96	45	54	15	24	6	15
		F	63	33	42	6	15	2	6
		Total	X	78	96	21	39	8	21
Third and fourth	2008	M	45	11	24	7	21	5	3
		F	50	18	33	6	20	4	1
		Total	X	29	57	13	41	9	4
	2011	M	55	9	43	4	20	8	2
		F	45	9	35	1	20	5	4
		Total	X	18	78	5	40	13	6
	2015	M=39	124	41	38	11	17	2	14
		F=30	66	32	21	9	11	2	11
		Total=69	X	73	59	20	28	4	25
Fifth and sixth	2008	M	50	29	51	23	34	23	3
		F	50	29	37	18	40	12	5
		Total	X	58	88	41	74	35	8
	2011	M	52	12	31	9	24	18	2
		F	45	14	19	3	22	11	1
		Total	X	26	50	12	46	29	3
	2015	M=43	155	23	23	7	14	4	7
		F=34	146	21	6	5	7	1	4
		Total=77	X	44	29	12	21	5	11
Seventh and eighth	2008	M	90	13	8	5	16	5	5
		F	55	21	12	15	30	8	2
		Total	X	44	20	20	46	13	7
	2011	M	70	30	31	23	40	29	6
		F	75	10	14	11	41	26	4
		Total	X	40	45	34	81	55	10
	2015	M	238	41	46	25	35	6	24
		F	175	63	24	22	46	6	23
		Total	X	104	70	47	81	12	47
Total	2008	M=220		58	103	38	78	35	11
		F=183		72	89	39	91	26	9
		Total=403		130	192	77	169	61	20
	2011	M=177	X	55	137	36	90	56	12
		F=185		33	85	16	87	42	9
		Total=362		88	222	52	177	98	21
	2015	M=248		160	161	58	90	18	60
		F=219		149	93	42	79	11	44
		Total=467		309	254	100	169	29	124

exploring educational contents, watching (and downloading) movies/listening (and downloading) music and reading news and entertaining contents. They spend less time using e-mail and chat services. Most of boys (2/3 in 2015) spend the majority of time playing online games (Q2) and exploring educational contents, while girls (also 2/3 of them) spent the majority of time playing online games. In 2008, the findings were different: 45% of boys used the Internet most often to explore educational contents and girls (50%) to play online games. Findings of the survey conducted in 2011 are quite consistent with the findings of the 2008's survey.

B) Parent/Teacher cooperation and provision of support to Children/Students in regard to cyber bullying

The second group identified four questions (Q1-Q4) that were posed in 2008 and 2011 and 10 questions (Q5-Q14) that were posed in 2015.

The questions were: Q1 – Do your parents check the Internet sites you visit?; Q2 – Have your parents talked to you about dangers on the Internet?; Q3 – Have you talked to your parents about the Internet content that bothers you?; Q4 – In the near future, the school will give a lecture for parents about the *Dangers of the Internet*. What do you think, are your parents going to attend it?; Q5 – Do your parents encourage you to explore the Internet and learn online on your own?; Q6 – Do you have some shared online activities with your parents?; Q7 – Have you ever talked to your teacher about your online activities?; Q8 – Have you ever received any help from your parents, teachers or friends when you couldn't do or find something on the Internet?; Q9 – Have your parents, teachers or friends ever explained to you why some sites are good or bad?; Q10 – Have your parents, teachers or friends ever suggested how to safely use the Internet?; Q11 – Have your parents, teachers or friends helped you when something bothered you on the Internet?; Q12 – Have your teachers set rules of what you can and cannot do on the Internet, while at school?; Are your parents using a) software for parental control or some other methods of blocking and filtering certain web sites? (13a); b) software for parental control or other tracking tools of the visited web sites? (13b); c) program, service or contract that limits your time spent on the Internet? (13c); d) software that prevents spam or junk messages and viruses? (13d); Q14 – Do your parents sometimes check a) what web sites you are visiting? (14a); b) your e-mail messages or instant messages? (14b); c) your profile on social networks or online communities? (14c); d) friends or contacts you are adding to your social profile / instant messages account? (14d).

Although it is not shown in the Table II, almost all students started using ICT before they started going to school: Average age when boys started using ICT was 5.6 years and 6 years of age for girls.

Findings of the first four questions are shown in the Table II, and the others in the Table III.

Based on the answers to the first question (Q1) parents, in general, insufficiently control online activities of their children. Among younger students, it is at around 50% (which is not so bad), among older students, from the

It was also expected to find that boys spend more time on the Internet than girls. In 2015 there is a significant increase of time spent on the Internet among all age groups. At average, boys spend 238 minutes per day on the Internet and girls, 175 minutes per day. This finding cause concern in light of the fact that many students answered that they spend approximately 6-10 hours per day on the Internet.

Q2-Q7 was focused on the answers of students' activities on the Internet. Findings show that top four activities are Q2, Q3, Q5 and Q4: Playing online games,

TABLE II. PARENT/TEACHER COOPERATION AND PROVISION OF SUPPORT TO CHILDREN/STUDENTS IN REGARD TO CYBER BULLYING (Q1-Q4)

Grade	Year	Gender	Q1	Q2	Q3	Q4
First and second	2008	M=27	10	14	3	11
		F=14	4	3	1	5
		Total=41	14	17	4	16
	2011	M=38	19	17	5	14
		F=24	8	13	2	8
		Total=62	27	30	7	22
	2015	M=76	40	50	47	
		F=60	41	41	40	
		Total=136	81	91	87	
Third and fourth	2008	M=52	23	32	3	31
		F=56	34	31	2	36
		Total=104	57	63	5	67
	2011	M=59	31	41	3	51
		F=58	36	41	1	50
		Total=117	67	82	4	101
	2015	M=75	12	36	34	
		F=69	9	29	28	
		Total=144	21	65	62	
Fifth and sixth	2008	M=65	26	35	7	39
		F=74	34	51	7	58
		Total=139	60	86	14	97
	2011	M=54	32	37	8	38
		F=37	27	33	3	33
		Total=91	59	70	11	71
	2015	M=26	3	24	19	
		F=26	3	23	21	
		Total=52	6	47	40	
Seventh and eighth	2008	M=33	8	11	1	16
		F=41	6	22	5	23
		Total=74	14	33	6	39
	2011	M=68	14	50	7	54
		F=63	24	45	10	56
		Total=131	38	95	17	110
	2015	M=70	4	47	34	
		F=74	12	63	57	
		Total=144	16	110	91	
Total	2008	M=220	67	92	14	97
		F?183	82	107	15	123
		Total=403	149	199	29	220
	2011	M=177	96	145	23	157
		F=185	95	142	16	147
		Total=362	191	287	49	304
	2015	M=248	39	157	134	
		F=219	45	156	134	
		Total=467	84	313	268	

5th grade on, that percentage is decreasing to 10-20%. Findings of the survey conducted in 2015 and later, are encouraging as more parents control online activities of their children than it was done before (in 2008).

Parents are communicating more about the dangers of the Internet (Q2) with children in 3rd and 4th grades. Recently, almost all parents talk to their children about this issue. There are similar findings with (Q3) children talk to their parents about contents that bother them on the Internet. Again, findings from the survey conducted in 2015 are quite satisfying (over 90%), in comparison to the earlier years when percentage of parents talking to their children about the dangers were at around 10%. This data primarily relates to the students from the 1st to the 4th grades.

The last question in this group was "In the near future, the school will give a lecture for parents about the *Dangers of the Internet*. What do you think, are your parents going to attend it?" Findings are diverse and they go from 30% to 90%. Lower percentage is in the lower grades and significantly higher, in the higher grades. This question was not posed in the survey conducted in 2015, therefore, comparative analysis could not be completed.

It is very important to encourage children to use the Internet (Q5), as it brings many comparative advantages. When the answer to the Q6 is positive (Do you have some shared online activities with your parents?), it means that children have parental support. Sadly, only 29%-59% of parents understand advantages of such support and encourage children to use the Internet or share activities with their children on the Internet. This will become a problem when children will have questions and won't ask parents to help them, and would instead ask either a friend or, more often, try to find answers on the Internet. It is also very important that teachers show interest and concern (Q7) for students' activities on the Internet. Findings show that at average 40% of students interacted with their teachers and got a positive response. In lower grades, the average responses were 17% and 25%, and in higher grades, the percentage was slightly higher. It is possible that teachers lack practical knowledge in this area, but it is always positive to show concern.

Answers to Q8 to Q11 are elaborating on the effort of parents, teachers and friends to ensure children's safety. Results go from 63% to 73%. It is not bad, but it could be better.

Q12 deals with rules set by teachers, what can and cannot be done on the school's Internet. The average findings are at 40%, and it is again lower in the lower grades (from 10% to 28%), and it indicates that primary school teachers in 1st to 4th grades lack knowledge of the safety on the Internet and need further education in order to be able to act responsibly.

Q13 offered four options and gives us information how much parents invest in the technical protection of their children from dangers of the Internet (protective software and/or filters). Average result is at the modest 38%. It is very low, especially in light of public campaigns to protect children from dangers in cyber space.

The last question Q14 answers how engaged parents are in the protection of their children. Findings in this segment are quite modest (21%-43%), reasons could be parents' negligence or lack of knowledge. Outcome should be better, especially because it requires only time spent with children and not any financial investments.

TABLE III. PARENT/TEACHER COOPERATION AND PROVISION OF SUPPORT TO CHILDREN/STUDENTS IN REGARD TO CYBER BULLYING (Q5-Q14)

Grade	Gender	Q5	Q6	Q7	Q8	Q9	Q10	Q11	Q12	Q13				Q14			
										13a	13b	13c	13d	14a	14b	14c	14d
First and second	M=76	26	29	14	50	50	50	46	7	20	15	0	26	28	11	16	5
	F=60	22	18	9	39	41	31	41	7	12	10	6	26	19	5	13	4
	Total=136	48	47	23	89	91	81	87	14	32	25	6	52	47	16	29	9
Third and fourth	M=75	27	22	23	46	46	36	36	21	8	5	6	27	35	23	22	17
	F=69	16	12	14	36	37	29	31	20	8	6	5	15	28	15	18	19
	Total=144	43	34	37	82	83	65	67	41	16	11	11	42	63	38	40	36
Fifth and sixth	M=26	16	9	12	22	23	24	20	19	4	4	1	11	11	10	9	6
	F=26	15	13	9	23	25	23	21	19	5	4	2	13	16	8	16	12
	Total=52	31	22	21	45	48	47	41	38	9	8	3	24	27	18	25	18
Seventh and eighth	M=70	37	30	19	49	53	47	41	41	7	6	8	29	25	6	24	17
	F=74	46	36	41	65	66	63	58	49	9	5	10	30	42	21	50	33
	Total=144	83	66	60	114	119	110	99	90	16	11	18	59	67	37	74	50
Total	M=248	106	90	68	167	172	157	143	88	39	30	15	93	99	50	71	45
	F=219	99	79	73	163	169	146	151	95	34	25	23	84	105	49	97	68
	Total=467	205	169	141	330	341	303	294	183	73	55	38	177	204	99	168	113

C) Students' understanding of the elements of cyber bullying and their response to the unwanted harassment

Third group of questions relates to the knowledge of cyber bullying. As in the second group of questions, questions that were posed in 2008 and in 2011 (Q1-Q5) are picked out as well as those posed in 2015 (Q7-Q15).

The questions were: Q1 – What bothers you the most on the Internet: a) Security – identity theft (a0); b) Advertisment – commercials (a1); c) Pornography (a2); Bad manners of the members of the Internet communities, such as insults, swearing etc (a3); Q2 – Have you ever felt bad for experiencing one the following things on the Internet a) seen photos of violence (a4); b) been insulted (a5); c) been threatened (a6); Q3 – Has anybody used your e-mail with your permission? Q4 – Has anybody used any part of your Internet identity without your permission?; Q5 – Has anybody posted your photo without your permission?; Q7 – Have you ever looked for new friends on the Internet?; Q8 – Have you added a new friend to your friend's list or on your contact list without meeting him in the real life?; Q9 – Have you ever sent your personal data to somebody you never met outside of the cyber space?; Q10 – Have you ever sent your photos or video clips with your images to someone you have never met outside of the cyber space?; Q11 – Have you ever contacted somebody online that you first met outside of the cyber space?; Q12 – Have you ever went to meet a person you met online?; Q13 – Sometime kids tell or do things that can bother you or the others, offend you, hurt you or make you or the others feel uncomfortable. Has anybody treated you that way on the Internet within the last 12 months?; Q14 – Harm, hurt, assault, anxiety or humiliation can be the outcome of the sexual images or messages of the other person on the Internet – Has anybody hurt you in that way within the last 12 months?

Findings are shown in Tables IV (Q1-Q5) and V (Q7-Q13)

First question (Q1) was testing students' sensibility to disturbing contents that may be received via e-mail or could come across on the Internet. Findings show that students have low understanding of the potential dangers. While low percentage of students in the lower grades is reasonably expected (they are still little and don't understand terminology), results for the group of students from the 5th through the 8th grade raise concern

TABLE IV STUDENTS' KNOWLEDGE OF THE ELEMENTS OF CYBER BULLYING AND THEIR RESPONSE TO UNWANTED HARASSMENT (Q1-Q5)

Grade	Year	Gender	Q1				Q2			Q3	Q4	Q5
			A0	A1	A2	A3	A4	A5	A6			
I and II	2008	M=27	12	8	5	4	9	8	3	4	0	1
		F=14	6	3	1	1	4	0	2	2	2	0
		Total=41	18	11	6	5	13	8	5	6	2	1
	2011	M=38	10	4	3	3	6	7	3	1	3	0
		F=24	9	2	1	3	4	1	1	4	3	2
		Total=62	19	6	4	6	10	8	4	5	6	2
III and IV	2008	M=52	12	7	20	14	12	8	4	12	9	0
		F=56	18	8	26	16	8	14	2	13	7	2
		Total=108	30	15	46	30	20	22	6	25	16	2
	2011	M=59	23	16	11	15	7	7	5	13	6	9
		F=58	19	19	13	15	8	2	0	3	4	6
		Total:117	42	35	24	30	15	9	5	16	10	15
V and VI	2008	M=65	28	19	19	14	18	11	10	13	4	3
		F=74	22	11	45	18	12	14	7	13	6	7
		Total=139	50	30	64	32	30	25	17	26	10	10
	2011	M=54	24	6	22	16	10	11	4	15	3	5
		F=37	16	9	15	18	5	4	2	6	2	4
		Total=91	40	15	37	34	15	15	6	21	5	9
VII and VIII	2008	M=33	10	14	4	3	5	6	2	8	6	4
		F=41	13	5	20	16	12	12	5	7	3	4
		Total=74	23	19	24	19	17	18	7	15	9	8
	2011	M=68	32	24	19	16	12	12	8	20	9	13
		F=63	27	16	22	17	8	20	11	21	13	13
		Total=131	59	40	41	33	20	32	19	41	22	26
Total	2008	M=220	62	38	48	41	44	33	19	37	19	8
		F=183	59	27	92	51	36	40	16	35	18	13
		Total=403	121	65	140	92	80	73	35	74	37	21
	2011	M=177	88	50	55	50	35	37	20	49	21	27
		F=185	71	46	51	53	25	27	14	34	22	25
		Total=362	159	96	106	103	60	64	34	83	43	52

965

Individual percentages go up to 40%-50%, but that is comparatively low, knowing what kind of situations they may come across. Parents and teachers share the responsibility for the situation.

Next question (Q2) relates to the potentially harmful experiences students' came across. There is low percentage of such experiences, up to 20%, but that is, never the less, raising concern. It is important to educate children – students. When teachers or parents don't understand dangers that ICT users can face, let them learn about it by themselves, or help them take a course.

Following three questions (Q3-Q5) relate to identity theft or fake user's profiles. Students need to be aware how to protect their e-identity. Again, percentages are not high, they go up to 30%, but that is more than enough to raise concern.

Following six questions (Q7-Q12) tested students' behavior in terms of safety. Although activities can be well intended (to make friends), caution is always recommended, as activities are occurring in the electronic media environment. This is primarily parents/teachers domain, as they need to educate children/students how to safely use the ICT and avoid potential dangers. Risky behavior in some elements is alarming, as it goes up to 30%.

The last two questions (Q13 and Q14) deal with the exposure to cyber bullying. Questions pertained not to the potential dangers, but to the actual bullying that occurred (or students experienced them as such). Average of 16% is very high as it targets the most vulnerable children/students.

Due to the limited space, this research is presenting some 20% of the obtained data. The article aims to raise awareness among professionals and parents about dangers that anyone can face, and especially to educate children how to avoid unwanted harassment on the Internet.

More information about the research and findings in Croatia, can be obtained from the first author, and for the findings of the EU survey, on the official web site of the project [11].

TABLE V STUDENTS' KNOWLEDGE OF THE ELEMENTS OF CYBER BULLYING AND THEIR RESPONSE TO UNWANTED HARASSMENT (Q7-Q14)

Grade	Gender	Q7	Q8	Q9	Q10	Q11	Q12	Q13	Q14
First and second	M=76	13	7	5	2	9	2	4	18
	F=60	3	1	0	0	3	0	0	6
	Total=136	16	8	5	2	12	2	4	24
Third and fourth	M=75	20	14	0	1	6	7	6	9
	F=69	5	4	0	1	6	1	1	3
	Total=144	25	18	0	2	12	8	7	12
Fifth and sixth	M=26	7	4	0	0	13	1	0	3
	F=26	10	2	1	2	13	2	4	6
	Total=52	17	6	1	2	26	3	4	9
Seventh and eighth	M=70	34	8	8	8	44	11	13	20
	F=74	40	7	6	13	31	14	15	10
	Total=144	74	15	14	21	75	25	28	30
Total	M=248	74	33	13	11	72	21	23	50
	F=219	58	14	7	16	53	17	20	25
	Total=467	132	47	20	27	125	38	43	75

IV. CONCLUSION

The study describes findings of the survey that was conducted over the period of eight years. The survey focused on children's safety on Internet, use of the ICT and its trends. Students from the 1st to 8th grade of elementary school participated in the survey. Findings were summarized in three groups: a) ICT use among students, b) Parent/Teacher cooperation and provision of support to Children/Students in regard to cyber bullying, and c) Students' understanding of what constitutes cyber bullying and their response to unwanted harassment. Obtained data were compared to the findings of a similar survey "EU Kids Online", conducted in 25 EU countries.

Obtained data show that students start using ICT at younger age than before, they use it more frequently and for longer periods of time, while still having low informational literacy from the aspect of safety. Parents need to invest more effort in learning advantages and dangers of using ICT in order to support and encourage children to use ICT and, on the other hand, to protect them from cyber bullying. Teachers need to invest more time and effort in teaching children the elements of cyber bullying and its prevention, even when school curriculum does not require them to do it.

REFERENCES

[1] D. Holloway, L. Green and S. Livingstone, „Zero to eight. Young children and their internet use", LSE, London: EU Kids Online, 2013.

[2] N. Laniado and P. Giaufilippe, „Naše dijete, videoigre, Internet i televizija", Studio tim, Rijeka, 2005.

[3] European Parliament and the Council „Key Competencies for Lifelong Learning, European Reference Framework", Office for Official Publications of the European Communities, Luxembourg 2006. http://eur-lex.europa.eu/legal-ontent/HR/TXT/?uri=uriserv:c11090

[4] A. Ferrari, „DIGCOMP: A Framework for Developing and Understanding Digital Competence in Europe", in JRC Scientific and Policy Reports, Eds.: Y. Punie and B.N. Brečko, European Commission Joint Research Centre Institute for Prospective Technological Studies, Seville, Spain, 2013.

[5] L. Mark and T.K. Ratliffe (2011) „Cyber worlds: New Play-grounds for Bullying", Computers in the Schools, 28:2 pp. 92–116.

[6] I. Kanižaj, V. Car and L. Kralj, (2014) „Media and Information Literacy Policies in Croatia (2013)", ANR TRANSLIT and COST "Transforming Audiences/Transforming Societies", 2014. http://ppemi.ens-cachan.fr/data/media/colloque140528/rapports/CROATIA_2014.pdf (10.12.2016).

[7] S. Livingstone, L. Haddon, A. Görzig, and K. Ólafsson, „Risks and safety on the internet: the perspective of European children: full findings and policy implications from the EU Kids Online survey of 9-16 year olds and their parents in 25 countries" London: EU Kids Online, LSE, 2011.

[8] J. Pregrad, M. Tomić Latinac, M. Mikuluć, and N. Šeparović, „Iskustva i stavovi djece, roditelja i učitelja prema elektroničkim medijima" - Izvještaj o rezultatima istraživanja provedenog među djecom, učiteljima i roditeljima u sklopu programa prevencije elektroničkog nasilja „Prekini lanac!" Ured UNICEF-a za Hrvatsku, Zagreb, 2011.

[9] L. Kralj, „Children's Safety on the Internet - Development of the School Curriculum", 37th International Convention on Information and Communication Technology, Electronics and Microelectronics (MIPRO), 2014. pp 593-596.

[10] Ministarstvo znanosti obrazovanja i sporta (2013.) „Nastavni plan i program za osnovnu školu" public.mzos.hr/fgs.axd?id=20542

[11] http://www.lse.ac.uk/media@lse/research/EUKidsOnline/EU%20Kids%20Online%20reports.aspx (15.1.2017.)

Metals Industry: Road to Digitalization.

Andrea Merluzzi*, Gianpiero Brunetti**

* Technical Coordinator – B.U. DIGI&MET - at Danieli Automation S.p.A., Buttrio (UD)
a.merluzzi@dca.it
** Executive Vice President – B.U. Steelmaking - at Danieli Automation S.p.A., Buttrio (UD)
g.brunetti@dca.it

Abstract – The challenge of the new globalized market and the current steel market outlook characterized by plant underutilization, are the elements leading metal producers to seek for low Capex investments, aiming at improving the efficiency of the production facilities, the quality of the products, the health and safety of the workers as well as the environmental sustainability. In this scenario, DANIELI has created a new cross-functional business unit named DIGI&MET whose mission consists of developing and implementing new plant design concepts, based on digital innovation, and also new business models, based on servitization and outcome economy principles. DANIELI summarizes in this article its approach for both brown and green field plants.

I. INTRODUCTION

The global economy is adjusting to a slower level of Chinese growth – what is called the "new normal". Between 2000 and 2014, global steel production doubled from approximately 800 million ton to 1.6 billion ton mainly driven by rising production capacity in China. During the same period, demand in China rose of approximately the same rate. Once the "construction" growth of China slowed down, the exceeding capacity was put on the foreign market causing a dumping of steel prices. As a result, the overall production faces a -3% in 2015 followed by a very limited growth in the following years. The steel market is then expected to enter in an inflection point with limited growth in the next years [1] [2].

There are other important trends, which are influencing the steel market: the research for newer materials with improved mechanical characteristics (high-strength steels), the greenhouse gas emissions reduction, the circular economy and the requirement for a safer working environment.

Metal producers have then to seek for low Capex investments, aiming at improving the efficiency of the production facilities, the quality of the products, the health and safety of the workers as well as the environmental sustainability.

During last decade, we have seen an incredible development and diffusion of ICT technologies: the availability and speed of mobile internet connections is increased. The possibilities of billions of people connected by mobile devices, with unprecedented calculation power is then boost by emerging technology such as artificial intelligence, robotics, autonomous vehicles and quantum computing. These changes are not only affecting our way of living but will affect the way we work and are the milestones of a new industrial revolution the fourth one where the new technologies are blurring the line between the physical and digital spheres.

In this scenario, DANIELI has created a new cross-functional business unit named DIGI&MET to provide customer digital innovation under new business models.

II. VISION, MISSION, OBJECTIVES & STRATEGY

The DIGI&MET vision can be summarized by the statement "From a Plant to a Smart Plant". The *Smart Plant* is a safe, flexible, efficient and environmentally friendly concept of manufacturing founded on the extensive digitalization of processes, the deep integration of cyber and physical worlds and the strong interconnection of intelligent systems and humans. In a *Smart Plant,* systems and equipment autonomously execute complex tasks and support humans in complex decision-making or even provide decision-automation. The below figure 1 shows the DIGI&MET vision but also its mission, which represents the path to follow in order to achieve the objectives.

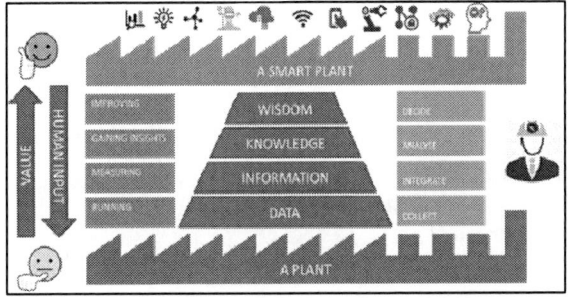

Figure 1.   DIGI&MET Vision & Mission

The mission consists of:

- Increase data collection getting data also directly from the field (e.g. installing by default required sensors and instrumentation on the equipment), creating and using for this purpose a common and certified analytics platform.

- Aggregate and integrate data, transforming it into information.

- Analyze information in order to transform it into knowledge.

Part of DIGI&MET mission is also to get closer to customers, establishing a relationship based on trust, loyalty and confidence, in order to be selected as partners in their business evolution path from the "running the business" to the "improving the business" approach. This process, which requires building a culture around filling customer needs and then engaging them every step of the way, will also transform DANIELI from a product supplier to a service/solution partner.

In light of the above, the DIGI&MET short-term objectives consist of implementing a customer-centric model that allows its customers to:

- Increase the overall plant efficiency in terms of higher productivity, higher yield, reduced lead times, increased plant availability and optimized usage of resources.

- Deliver quality products to its customers in order to establish stable and effective relationships, creating the so-called customer intimacy.

- Improve the workers' health and safety by adopting solutions aiming at avoiding accidents or reducing their effects.

- Monitor and control the plant energy and utility consumption as well as to implement efficient recovery strategies, which is currently one of the most important objectives of the metals industry and represents a first step ahead towards the so-called "Green Metals" challenging target.

Which is the DIGI&MET strategy to accomplish its objectives? The strategy consists of applying the most innovative technologies in the field of Information & Communications Technology, Process Automation and Sensors in order to implement the following levels of integration [3]:

- Vertical Integration: The hierarchical automation pyramid and its organizational and technical barriers are going to be replaced by a flat structure of intelligent, flexible and autonomous units; these units are the so-called cyber-physical systems that, through their control, computational and communication capabilities, implement the decentralized intelligence concept.

- Horizontal Integration: Product traceability and real-time quality assessment require a strong supply-chain integration, including all the internal processes that are part of the production chain as well as the external ones involving raw material suppliers and customers.

- Transversal Integration: Taking into consideration economical, technical and environmental aspects at the same time is mandatory in order to get the best decision-making support.

III. APPROACH

The process of pursuing the *Smart Plant* target, starting from plant digitalization, is justified by the large benefits expected in terms of overall plant efficiency, product quality, workers' health and safety and environmental sustainability. In order to drive its customers in this business evolution path DIGI&MET has

designed and developed a specific and effective approach. It consists of a first phase aiming at the delivery of a digital-enabled plant and then a second phase focusing on its transformation process towards the *Smart Plant* target.

The DIGI&MET approach is summarized in the following figure:

Figure 2.   DIGI&MET Approach

A key step of this approach is represented by the *Digital Maturity Assessment*, which consists of understanding if a given plant is enabled for digitalization, eventually performing the gap analysis, assessing its current digitalization degree and addressing digitalization efforts and investments considering their impact on the above-mentioned major objectives. In fact, today every function can be supported by digital systems, however a huge amount of economic resources should be invested to explore the overall spectrum of possibilities or, looking from another perspective, the risk to invest in directions with lower return could be even very high; this analysis definitely aims to support the business in selecting the best opportunities.

The assessment is performed through the following tasks:

- Estimation of a Basic Digital Index, which indicates whether a plant is enabled for digitalization and it is based on the sum of contributions of a given set of enabling technologies grouped into the following categories:

  - Connectivity & Data Sharing: It represents the capability of collecting, storing and sharing data for its further transformation into information; this capability is measured considering the presence of sensors, instrumentation, data repositories and network infrastructure.

  - Data Processing: It represents the capability of manipulating data in order to produce valuable information through validation, aggregation and analysis steps.

  - Human in the Loop: Human-in-the-Loop (HITL) is the concept based on systems and models that require human interaction inside the computation chain for accomplishing tasks. This concept includes applications where humans and machines cooperate for accomplishing tasks having the following characteristics: complexity, high cost to implement a fully automatic execution, need of human wisdom to speed up decision-making process.

- Modelling: It considers the availability of process control and optimization models, either of deterministic or statistic nature, and technological packages.

- Autonomous & Robotic Systems: Autonomous and robotic systems are foreseen to replace humans in highly repetitive and dangerous jobs, especially the ones performed in very harsh environments like steelmaking plants, in order to increase both efficiency and safety.

• Estimation of a Global Digital Index considering the implementation of a given set of smart plant enabling solutions (later on described in this article) as well as the contribution of an extended Manufacturing Execution System (or a subset of its modules).

The approach applies to both brown and green field plants with some obvious differences that are described here below.

A. Brown-Field Plant Approach

In this scenario, it is very important during the assessment phase to understand if the customer is already able to measure the performances of its business process or not. In the latter case, such monitoring solution can be designed and implemented, with the minimum impact for the plant operations, starting from the customer requirements and the DANIELI know-how and expertize in the metals industry, which has led to the creation of a large KPI library that covers supply-chain, technological and maintenance business processes.

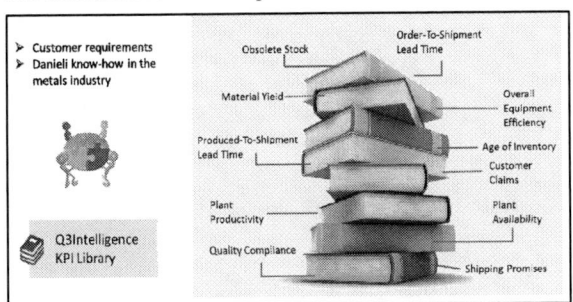

Figure 3.   KPI Library Excerpt

B. Green-Field Plant Approach

The design of a green-field smart plant starts from the layout, which should at least take into consideration possible automated logistic, robotic and material identification solutions, and then from plant engineering, which means smart sensors, instrumentation, equipment-embedded condition monitoring and control modules as well as all the required infrastructures for data collecting and sharing.

IV.   KEY-ENABLING FACTORS

The key-enabling factors that DIGI&MET applies to its innovative solutions are represented in figure 4 and include the following elements that are briefly described in the next paragraphs:

• Data-driven Approach, which represents the core of the digitalization process and the most disruptive innovation among the ones introduced by the Fourth Industrial Revolution.

• Digitalization Enabling Technologies, which represent the foundations of a digital-enabled plant.

• Smart-Plant Enabling Solutions.

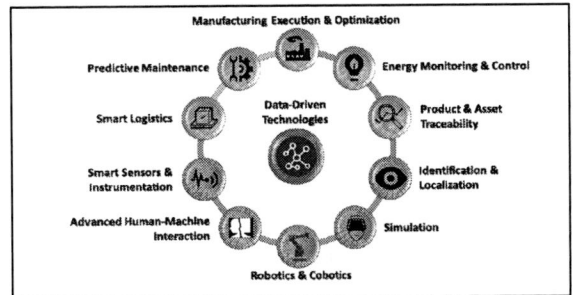

Figure 4.   Key-Enabling Factors

A. Data-Driven Approach

It represents the most innovative and disruptive element among the ones introduced by the Fourth Industrial Revolution and aims at applying innovative digital technologies such as machine learning and artificial intelligence to extract knowledge from data for supply chain and technological process optimization.

In the metals industry this approach is not supposed to supersede the deterministic one but to integrate it, thanks to its capabilities of estimating process variables that are not observable, identifying process variables correlations and providing accurate results where process is too complex or chaotic.

An excerpt of DIGI&MET business cases based on data-driven approach is the following:

• Quality Compliance Prediction: It consists of predicting the mechanical properties of the final product starting from the process conditions. This can be done both online, for the certification of the product, but also offline, to study the effects of the changes of the process conditions. The goal is to reduce the costly laboratory tests (whenever the standards allow it) and reduce the occurrence of non-prime quality thanks to fast prediction time.

• EAF Process Fingerprint: A robust strategy for EAF automatic process control based on the fingerprint concept. It allows an adaptive control of the chemical package for a smart management of the oxygen injection, considering both the slow process modifications over time and the feedback from the field in real time.

B. Digitalization Enabling Technologies

This category includes the following disciplines:

- Robotics & Cobotics

    Robotic solutions can be employed to replace humans in several shopfloor tasks, which are executed in a hot, noisy and polluted environment, are regularly heavy and demanding on the body and often subjected to process-related hazards. At the same time robot-based automated solutions can also replace workers in executing regular and repetitive tasks, which, especially on long-term and when the work environment is not friendly, may become soul-destroying and even produce psychophysiological effects. The design of these solutions, aiming at improving both plant efficiency and safety, takes of course into consideration the safety constraint represented by the fact that, in some cases, operators and robots share the same spaces.

    Some examples from the DIGI&MET robotic solutions portfolio [5] are the following:

    - Q-Robot Sample: Robotic solution designed for automatic sampling of temperature and chemical composition of a steel bath in both Electric Arc Furnace and Ladle Furnace applications.

    - Q- Robot Zinc: Robotic solution designed to skim, using a special tool, the surface of the zinc bath to remove dross, and place it in a dross container.

    - Q- Robot Antimix: Robotic solution designed for automatic anti-mixing, color-marking and labelling of special steel bars.

    The long-term challenging target of this approach, together with the automated logistic solutions that are described later on in this article, can be summarized by the sentence "No man on the shopfloor" which aims to a fully automatic plant with all the personnel sitting in control rooms and performing only supervision and control.

- Advanced Human-Machine Interface

    Pursuing efficiency and product quality often implies increasing complexity and increasing use of complex machines, processes or systems, which leads to an increase in operators' mental workload. In this scenario, the human-machine interaction is of particular relevance in order to create a comfortable and safe work environment combined with ergonomic and friendly user interfaces.

    In these regards, the DIGI&MET 3Q Digital Pulpit represents a revolutionary solution. One of the key points of the digital pulpit concept is the provision of a full "soft-desk", totally based on computer screens, through which the operator can both monitor the plant and operate it; in most cases, there is no hard-wired indication facility, with the exception sometimes of an emergency shutdown system. Simply by substituting the software of its automation system, this revolutionary pulpit is then able to drive processes like Electric Arc Furnace, Continuous Caster or Rolling Mill.

Figure 5.   3Q Digital Pulpit

The most important components of this desk are:

- The Operator Assistant (OA), a tool that offers a new approach to the control of the process, by reducing the number of commands that the operator previously had to operate with and by minimizing his intervention to a limited number of situations. There are three control modes to drive the plant: Auto Pilot, Assisted Manual and Maintenance Mode. Ergonomic and Cognitive Engineering are the key elements in the development of this system.

- The Area Performance Indicator (API), which provides a detailed view of a specific process, highlighting the main set points, the relevant KPIs (e.g. Production Rate[%], Production Yield [%] or Time Utilization [%]), the most significant trends and, where possible, the estimated time to end.

- The Plant Performance Indicator (PPI), an advanced monitoring system that allows the production manager to obtain an overall vision of the given process area and of the upstream/downstream ones. The goal of this function is to provide a graphic overview across most areas of the plant areas in order to simplify the synchronization of the production flow.

    In this field, an important role is also played by the mobile technologies, which provide several advantages such as real-time access to all information independently from location and simplicity in collecting and sharing information. These technologies have been extensively applied by DANIELI over the last years for different purposes such as:

- Q3-Mobile Intelligence: Mobile access to reporting and analytics platform.

- Q3-Mobile Shopfloor: Equipment local control (smart local control boxes).

- Q3-Mobile Recognizer: Equipment information retrieval in combination with maintenance management system.

- Q3-Mobile Safety: Passive safety support through operators' localization.

    Today the new challenge consists of applying to the metals industry the latest innovative technologies like augmented reality, taking into consideration the most critical safety-related constraints which are the following:

- The selected devices must be either certified as personal protective equipment or compatible with the personal protective equipment in use in the plant.

- The operating procedures in terms of safety must be defined considering the usage of such devices in a dangerous environment (e.g. moving equipment, extreme environmental conditions).

• Simulation

DANIELI entered this field of technology already some years ago with the development and the successfull implementation of innovative solutions such as:

- Kinematic Simulator (iStand), for personnel training as well as for virtual commissioning purposes.

- Plant Logistic Simulator, having the purpose of simulating plant operations in order to verify overall productivity, machine workload, transportation equipment utilization and sizing, storage yard sizing, potential bottlenecks and to perform relevant "what-if" analysis by changing machine/equipment characteristics and/or plant layout. Moreover, this solution is also applied for the simulation and validation of the production schedules generated by MES Production Scheduling module.

The new challenge in this field is to fully connect such virtual systems to the physical ones allowing the real plant and its digital twin to run and be maintained in parallel for off line multi-objective optimization strategy design, deployment and fast tuning. In these regards, DIGI&MET has already made a step ahead developing a process simulator called Q-Live, a tool that adopts real-time prediction technologies for simulating material, mechanics and their interactions.

• Smart Sensors & Instrumentation

The plant digitalization process starts from a wide deployment of intelligent sensors giving access to data that was not accessible or monitored before; such sensors should not be considered only as suppliers of information but as intelligent units capable of direct evaluation and processing of measured values. This approach allows to significantly improve the efficiency, availability and utilization of the plants as well as to leverage high rationalization opportunities through the integration of new systems.

In light of the above, the new DIGI&MET plant design concept includes the following:

- Install instrumentation on the equipment by default.

- Install smart sensors on the equipment.

- Embed in the equipment a smart and flexible automation module in order to perform critical real-time control tasks as well as condition monitoring.

Continuous research, in electromagnetic fields, infrared detection, lasers and x-rays, are the key factors in the development of new sensors and new instrumentation for the metals industry.

• Identification & Localization

The adoption of these technologies can provide a great contribution supporting automatic identification and localization of materials, assets and workers, therefore improving plant efficiency, product quality but also enhancing workers safety. Even if the work environment and the extreme operating conditions of the involved items often represent a restraint to their application in the metals industry, DANIELI has developed several solutions based on optical, radio frequency and barcode/QR code technologies.

C. *Smart-Plant Enabling Solutions*

The implemented solutions are the following:

• Manufacturing Execution & Optimization

An extended Manufacturing Execution System is a solution in charge of supporting, at plant-level, manufacturing coordination, execution and optimization; it includes the functionalities of a MES system plus some features which are which are normally provided by ERP or APS systems, with the purpose of supporting the entire customer order lifecycle from its acquisition up to material dispatch, through planning and execution steps. For this reason, its contribution to plant digitalization is unique for an integrated plant.

• Predictive Maintenance

The digitalization of plant maintenance processes is important in order to cut the relevant operating costs, which are normally very high. It is well known that large benefits in these regards can be gained with the implementation of predictive maintenance solutions that could be quite costly, despite its benefits.

Taking as a reference a given number of steel plants, it has been estimated that almost the 70% of maintenance activities (and therefore costs) is normally carried out on the 20% of the plant equipment. Moreover, the experience suggests also to take into consideration the following basic assumptions:

- Not all the equipment in a plant is of equal importance to either operations or safety.

- Equipment design and operation differs and different equipment will have a higher probability to undergo failures from different degradation mechanisms than others.

- A plant does not have unlimited financial and personnel resources.

In light of the above, the maintenance strategy pursued by DIGI&MET is the so-called "Reliability-Centered Maintenance" which represents a systematic approach to evaluate a facility's equipment and resources to best mate the two and result in a high degree of facility reliability and cost-effectiveness. In a few words, this strategy suggests to perform a deep analysis of plant equipment in order to:

- Identify the most critical ones (for either operations or safety) and implement predictive maintenance only for such subset of plant equipment.

- Implement preventive maintenance for most of the remaining equipment by developing relevant job guidelines with life counter-based frequency (when possible) or time-based frequency.

- Perform reactive maintenance for all the equipment that is not mission-critical.

This maintenance strategy includes also the implementation of combined Condition and Process Monitoring using a common analysis platform; it allows, when it is not possible to monitor equipment health directly, to infer a possible equipment malfunctioning from the relevant process data (e.g. a roll wear early deterioration can be inferred from a material surface quality defect).

- Energy Monitoring & Control

The capability to monitor and control the plant energy and utility consumption as well as to implement efficient recovery strategies is currently one of the most important objectives of the metals industry and represents a first step ahead towards the so-called "Green Metals" challenging target. Energy and utility consumption and recovery are not only important for what concerns the environmental aspect but also represent an important factor for the overall steel manufacturing costs, considering that actual energy market is highly variable due to local and geo-political factors.

In these regards, DIGI&MET has addressed its efforts to the implementation of an integrated Energy Management System, which allows energy control and optimization, intelligent access to power market on the basis of reliable forecast of energy demand associated with manufacturing planning and scheduling. In Europe, where the current trend is to focus on the increment of renewable sources, the most power-demanding operations are planned preferably during lower global power demand time windows such as during night shifts or weekends. This can be a regional opportunity to offer balancing services to the network in moments characterized by high level of power offered to the short-term free market, in particular for the electrical steelmaking but also for oxygen one. These opportunities ask for reliable power demand forecast and intelligent and flexible scheduling systems.

- Smart Logistics

The DIGI&MET strategy is also focused on automated logistic solutions, especially automatic yards with relevant automation and control systems that improve the plant overall efficiency through the following functionalities:

- Material Tracking & Inventory

- Yard Optimization

- Crane Movement Optimization

- Automatic Material Classification (for raw materials)

- Vehicle Identification & Tracking

Logistic operations in the metals industry are also critical in terms of safety due to the presence of heavy material items, vehicles and overhead cranes, which expose workers to several risks such as falls, being hit by falling objects and crushing injuries. Moreover, logistics often involve third-party personnel, which is not fully aware of plant specific risks and safety procedures. In these regards, automatic yard solutions can significantly improve safety conditions thanks to:

- Remote control rooms where the crane operator can comfortably and safely supervise the operations through cameras (eyes-on technology).

- Automatic execution of vehicle loading and unloading operations.

- Yard fencing and relevant automated entrance procedures designed to avoid personnel to enter the yard when cranes are in operation.

This approach applies for raw material, semi-finished and finished product handling.

- Product & Asset Traceability

Significant improvements have been achieved over recent years in terms of both product and asset tracking, thanks to a strong integration between automation and process control systems. The point of weakness of such solutions is represented by the manual identifications that are foreseen in the relevant processes and represent a source of possible mistakes.

Today the adoption of the above-mentioned innovative identification and localization technologies allows, even in harsh environments, to overcame these problems and provide product and asset visibility along the entire production chain.

This approach applies for raw materials, semi-finished and finished products as well as operational equipment with a common goal, which consists of having the right item in the right place at the right time.

V. CONCLUSION

The metals industry has historically been very conservative and generally skeptical and reluctant to embrace possible technology innovations. In this case, the large benefits of digitalization have been quickly understood but, at the same time, also some restraints have been already identified.

In these regards, a survey [4] involving more than 150 metals companies has been conducted and has produced some interesting key findings:

- Metals companies plan to invest 4% of annual revenue in digital operations solutions over the next five years. Nearly two-thirds (62%) expect to reach an advanced level of digitalization and integration within the next five years.

- The expected benefits of digitalization over the next five years are increased revenues (2.7% per annum) and cost savings (3.2% per annum).

- The two top challenges identified by metals companies are the unclear economic benefits of digital investments (49%) and the lack of a digital culture (49%). The third biggest challenge is the absence of a clear digital operations vision and leadership from top management (39%).

- Today only 11% of metals companies have advanced data analytics capabilities.

- The pace at which metals companies expect to accrue benefits from digitalization investments leads a majority (58%) to estimate a return on investment timescale of two years or less. Just over a third (37%) of companies anticipate a longer timescale of two to five years but relatively few (5%) think that it will take any longer than five years.

The above results are valuable and clearly have to be taken into consideration while approaching customers on digitalization topics.

ACKNOWLEDGMENT

Thanks to all DIGI&MET team members for the professionalism, commitment and enthusiasm that they constantly put into this challenging path.

REFERENCES

[1] World Steel Association, "World Steel in Figures 2016", 2016.

[2] World Economic Forum in collaboration with Accenture, "Digital Transformation Initiative – Mining and Metals Industry", White Paper, January 2017.

[3] ESTEP Platform, ROADMAP "Integrated Intelligent Manufacturing (I2M)", Created by the Working Group "Integrated Intelligent Manufacturing (I2M)" of ESTEP, 2016.

[4] PwC, "Industry 4.0: Building the Digital Enterprise – Metals Key Findings", 2016 Global Industry 4.0 Survey, 2016.

[5] Danieli Automation Web Site, "www.dca.it".

IoT Gateway for Smart Metering in Electrical Power Systems - Software Architecture

M. P. Shopov

Technical University of Sofia / Department of Computer systems and Technologies, Plovdiv, Bulgaria
mshopov@tu-plovdiv.bg

Abstract - The paper presents an implementation of IoT gateway for smart metering in electrical power systems. The gateway is based on Ubuntu Core operating system and connects power meters with external cloud services. Several software architecture design patterns are evaluated. Adaptation of them for the design of the software architecture of the gateway application is described in the paper. Based on the proposed software architecture an example IoT gateway implementation is suggested.

I. INTRODUCTION

With the development of sensor networks, wireless mobile communication, embedded system and cloud computing, the technologies of Internet of Things (IoT) have been widely used in areas such as smart meters, public security, smart homes and so on [1]. There are three essential components of IoT [2]: embedded devices consisting of both low cost/low power devices and high-end gateways; scalable connectivity – each embedded device should be connected; cloud-based mass device management – centralized management of distributed devices.

The support of legacy devices and devices without proper communication capabilities, will require the use of IoT gateways. These gateways could also be used for some local intelligence and value-added services [2]. Smart metering is an example of IoT application. It integrates communication capabilities with electrical power systems and delivery infrastructure to automate monitoring and control. Dynamically linking utility supply with demand could result in optimization of resource consumption [2].

The paper presents a software architecture and its implementation on an IoT gateway for smart metering in electrical power systems based on the scenario presented in [3].

II. BACKGROUND AND RELATED WORK

The Internet of Things brings new challenges for software developers. One major aspect in connecting things is that there is little value by providing just the connection. The benefit from being connected always goes hand in hand with the provision of a specific service, which is generating the benefit from the connection [4].

IoT gateway for smart metering applications has many responsibilities: communication with smart meters; communication with cloud-based services; locally processing data; caching data and synchronization; configuring devices and sensors; providing user interface. All these requirements and their dependencies bring additional complexity to the implementation of IoT gateway system and appropriate software architecture is very important.

There have been various proposals for IoT gateway architecture in the research literature. In [5] the authors propose an architecture of IoT gateway consisting of a south interface for connecting to sensors and a north interface for connection to client applications. The integration of sensors is done through M2M proxies. Architecture of a semantic gateway as service (SGS) is presented in [6]. The authors suggest the use of multi-protocol proxy as a central part of the architecture, mediating between other functional blocks. The integration with higher-level services is achieved through gateway interface.

In a closely related research, the authors of [7] suggest an architecture for open smart metering. The proposed architecture deals with the heterogeneity of smart metering equipment by the use of advanced metering infrastructure (AMI) and a central meter data management (MDM) system. The authors also suggest an UML data model that defines the objects and their attributes, exchanged between the components of the system. Although all these architectures define the functional blocks of IoT gateway and their interfaces with external systems, they do not present insights into the architecture of the application software that will run on the gateway device and implements these functional blocks.

The software architecture of a program or computing system is a depiction of the system that aids in the understanding of how the system will behave. It serves as a blueprint for both the system and the project developing it. The architecture is the primary carrier of system qualities such as performance, modifiability, and security. It is an artifact for analysis to make sure that a design approach will yield an acceptable system. By building effective architecture, one can identify design risks and mitigate them early in the development process [8].

Supported by the National Science Fund of Bulgaria – contract E02/12.

Design patterns provide approved generic solution for recurring architecture design problems, created to achieve quality-attribute goals and requirements [8-12]. Following, an overview and discussion on some of the design patterns evaluated as candidates for IoT gateway is given.

A. *Model-View-Controller*

Model-view-controller (MVC) design pattern [13] divides the software application into three interconnected components to separate internal representations of information from the presentation to and interaction with the user (Fig. 1). The *model* component is responsible for the application data, the *view* component is responsible for presentation of *model* data to the user, and the *controller* component is responsible for updating *view*, handle user interactions and updating the *model*.

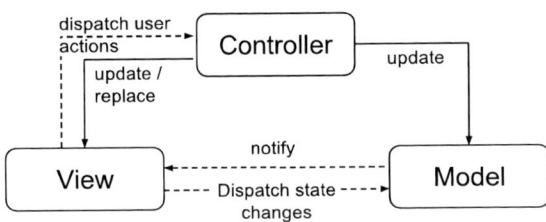

Figure 1. Model-View-Contoller

In MVC design pattern the three components are tightly coupled, which reduces their reusability and testability. Another shortcoming of MVC in the context of IoT gateway is that the *controller* component has to deal with too many responsibilities. This breaks the single responsibility principle and result in complex implementations [14, 15].

B. *Model-View-Presenter*

Model-View-Presenter (MVP) is a derivate of MVC first proposed in [16]. In MVP the *view* and *model* components are decoupled which increase their testability. The *presenter* retrieves data from the *model* and formats it for the *view* (Fig. 2).

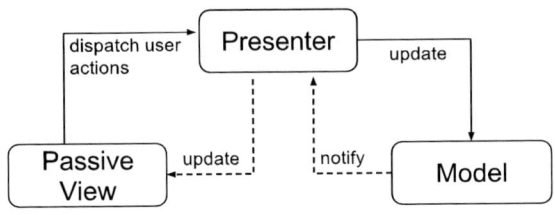

Figure 2: Model-View-Presenter

The presenter component is responsible for keeping the application synchronized [9, 11]. It improves testability of the model component in the price of doubling the amount of code compared to MVC [14].

C. *Model-View-ViewModel*

Model-View-ViewModel (MVVM) [17] is another derivate of MVC and MVP, where the

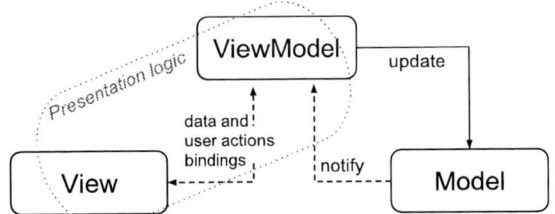

Figure 3: Model-View-ViewModel

controller/presenter component is replaced by *view model*. The *view model* is an abstraction of the view – independent representation of the View and its state – exposing public properties and commands [14]. The *view* and *view model* comprise the presentation logic of the application. The synchronization between them is mediated through software bindings (Fig. 3).

D. *Clean Architecture*

The clean architecture [18], as its author suggests is an attempt to integrate different architecture principles and practices into a single independent of technologies and frameworks idea (Fig. 4). Each circle represents a different areas of software. The outer circles are mechanisms and the inner circles are policies. The overriding rule beside this architecture model is the *dependency rule* - source code dependencies can only point inwards. *Entities* represent the business objects of the application. The *use cases* layer encapsulates and implements all of the use cases of the system. It orchestrates the flow of data to and from the *entities*. Next layer represents a set of adapters that convert data from the format most convenient for the *use cases* and *entities* - MVC, MVP, and MVVM belongs to that layer. The outermost layer is generally composed of frameworks and tools [18].

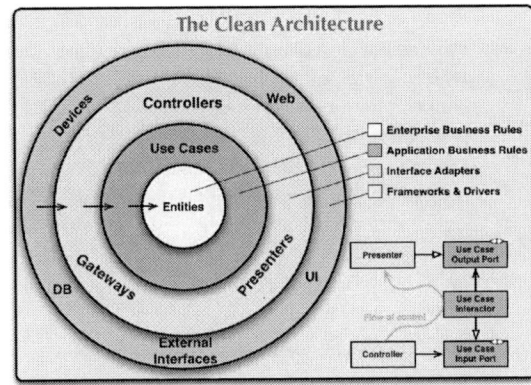

Figure 4: Clean architecture [18]

E. *Micro-services*

A micro-service is a small application that can be independently deployed, scaled, and tested and with a single responsibility. The micro-service architecture is defined as developing an application as a set of small independent services, where each of the services is running in its own independent process [19].

Three micro-services communication patterns are presented and discussed in [19]. By considering the asynchronous nature of IoT applications the authors suggests the use of service/message bus (Fig. 5). It is based on Publish/Subscribe model and allows addition of new components without changing the existing components of the system [19].

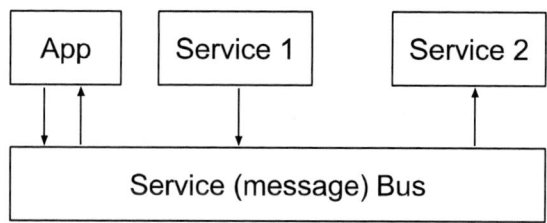

Figure 5. Message bus [11]

III. IoT Gateway software architecture

In a previous work [3] the author presents an example solution for smart metering in electrical power systems. The solution is based on a power meter, an IoT gateway, cloud services and client applications. This paper continues the work in the part constituting the applications running on IoT gateway.

By following the common functional requirements for smart metering systems (2012/148/EU Recommendation) the main features of the smart metering application can been identified – these are the *use cases* in the context of *clean architecture*:

- provide frequent readings of the consumption data from sensors;
- store consumption data locally for a reasonable time;
- provide local and remote standardized interface which provides visualized individual consumption data to the consumer, any third party designated by the consumer, and operators;
- provide a local reasoning capabilities based on on-line and past consumption data and other sensors readings and notify user to stimulate energy-savings;
- provides two-way communication for remote maintenance and control;
- provides secure data communications and data privacy.

Some of the functionality, like remote access to sensor readings and remote management and control, can be extracted to the cloud service. This reduces IoT gateway application use cases to: providing local user interface; integration of power sensor(s); two-way connection to cloud service(s); local storage of power consumption data (Fig. 6).

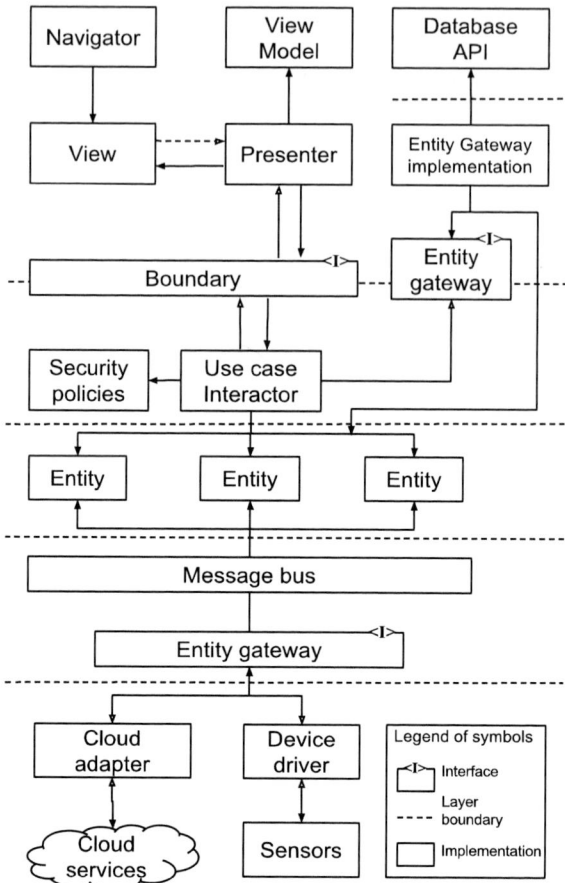

Figure 6: IoT gateway software architecture

Entities of the system encapsulate business objects of the application. There is an entity encapsulating a power reading (collection of power readings). This entity is independent from the concrete power meter sensor or cloud service being used. There are entities for encapsulation of device configurations (e.g. the gateway, a power sensor, or external service) and user accounting.

On the interface adapters layer a decision is made to design some of the functionality as micro-services – specifically the connection with the sensors and with the cloud services. The communication between IoT gateway application and those micro-services follows message bus patterns.

The design of the local user interface – the presentation logic of the application is based on MVP pattern. It could be easily changed to reflect MVC or MVVM pattern if they are better suited to the use case. The presenter contacts the use case interactor through a boundary interface. It converts the entites representation of the data to a format convenient to the view component constructing a view model. The navigation between views is extracted in a separate component – navigator.

The local storage *use case* uses an entity gateway interface and its implementation to interact with a database for persistent storage. This is used to abstract the concrete database mechanism used. Entity gateway implementation converts entities data to/from a form convenient to the persistence framewok being used. In case of a SQL database it will encapsulate all the SQL statements, in case of NoSQL – all query mechanisms.

The proposed IoT gateway software architecture is shown on Fig. 6. It follows the clean architecture *dependency rule* – inner layer knows nothing about the outer layers, and the interface segregation principle. The two-way interaction between layers is achieved through *dependency inversion principle*. Each component of the architecture also follows *single responsibility principle*.

Security is not a single component of the system, rather it is distributed in layers. There are application specific security policies that are addressed in the use case layer. There are also communication security that should be part of the adapters layer and external interfaces layer. Security policies should also be applied on the message bus securing the message bus for micro-service.

IV. IMPLEMENTATION AND TESTING

This section provides information about a concrete implementation of the proposed software architecture. The IoT gateway is based on an A20-Olinuxino embedded hardware platform. On the hardware platform an Ubuntu Snappy Core [20] operating system is installed and running. Ubuntu snappy is a stripped down version of Ubuntu, designed to run securely on autonomous machines, embedded devices and other Internet-connected devices. IoT gateway application is implemented as a snap package that runs confined under a restrictive security sandbox.

The micro-services message bus is implemented on top of the standard Linux system bus – D-Bus. It is an inter-process communication mechanism that supports two modes of interchanging messages between parties – one-to-one request-response and publish/subscribe.

The electrical power meter used is PST04. It measures the values of the main parameters of the three phase electric power system – voltage, current, frequency, active power, reactive power, and power factor [21]. The power meter sensor is integrated through a Linux device driver that implements its serial protocol.

The integration with cloud services is based on a cloud adapter that implements the DeviceHive [22] RESTful API. DeviceHive is an open-source M2M framework that contains a set of services and components for establishing a two-way communication with the remote devices using cloud technologies as a middleware. The devices can be anything connected: sensor networks, smart meters, telemetry, smart home devices and etc [22].

The document-oriented NoSQL database MongoDB [23] is selected for realization of the local persistence storage. Map-reduce algorithm can be used for batch processing of data and aggregation operations. PyMongo database driver is used as an implementation of database API component.

The implementation of the rest of the system components uses Python programming language. For implementation of MVP a wxPython is used. wxPython is is a wrapper for the cross-platform GUI API. The dependency inversion principle is implemented by passing dependency code as a parameters in constructors. Another possibility is to use broadcasters-broker implementation.

Currently, the following operations are implemented:

• configuration and management of sensors and remote service – currently only one sensor and one remote service are supported. Since the sensor does not accept any configurations those are implemented in device driver component and include naming the sensor, selecting communication interface and configuring readings period. The remote service configuration include service endpoint and credentials;

• fetching sensor readings periodically – these are stored in local persistent storage and synced with the remote storage if/when a remote connection is available;

• providing local user interface for user management and visualization of consumption data (charts, diagrams and reports).

Data collection can be initiated both locally and remotely – by default it is initiated locally. The IoT gateway implementation does not require a persistent network connection, although in current implementation it uses one. In case of no remote connection local persistent storage is used and the remote service also support queue for commands send to the gateway.

Since the software architecture comply with the dependency rule it is intrinsically testable. Each of its components can be tested separately. The Python unit testing framework is used and a number of unittests – for each component of the system and mocks – for the user interface are written. Python unittests are also written for testing the D-Bus micro-services. These kind of testing allows the testing of each component independently. All prepared unittests pass successfully, however testing with unittests is as accurate as the percentage of coverage of all possible input/output data. Finally, the system operations are tested and verified as a whole through real life tests where each operation worked as expected.

V. CONCLUSION

The paper presents the design and implementation of IoT gateway for smart metering in electrical power systems. Several software architecture design patterns and principles are envisioned and discussed in the paper. These are later mapped to the functional requirements for smart metering systems and are used for the design of the IoT gateway application software architecture. The proposed software architecture follows the clean architecture dependency rule, single responsibility and interface segregation principles for easy extendability and testability of the system.

Based on the proposed software architecture an example IoT gateway implementation is suggested. It uses an A20-Olinuxino embedded hardware platform running Ubuntu Snappy Core operating system. The system integrate a power meter, a smart energy application, local persistent storage and cloud connection. The use of cloud integration increase reliability and protection of collected data (it is stored and replicated in multiple secure, commercial-grade storage systems) and eases the software developers in management of remote devices – the actual devices are managed through web services and provided REST-based interfaces.

The future work includes continuing the process of implementing the IoT gateway application with more power meters and cloud services, developing of a local reasoning algorithms based on power consumption data and possibly other sensors readings for power-saving suggestion, and introduction of a complete security audit test suits.

ACKNOWLEDGMENT

The presented work is supported by the National Science Fund of Bulgaria project "Investigation of methods and tools for application of cloud technologies in the measurement and control in the power system" under contract E02/12 (http://dsnet.tu-plovdiv.bg/energy/).

REFERENCES

[1] Beng, L., "Sensor cloud: Towards sensor-enabled cloud services," Intelligent Systems Center Nanyang Technological University, 2009.

[2] Wu, Geng, Shilpa Talwar, Kerstin Johnsson, Nageen Himayat, and Kevin D. Johnson, "M2M: From mobile to embedded internet," IEEE Communications Magazine, vol. 49, no. 4, pp. 36-43, 2011.

[3] M. Shopov, "An M2M solution for smart metering in electrical power systems," 39th International Convention on Information and Communication Technology, Electronics and Microelectronics (MIPRO), art. no. 7522311, pp. 1141-1144, 2016.

[4] T. Schneider and A. Wolfsmantel, "Achieving cloud scalability with microservices and DevOps in the connected car domain," Software Engineering (Workshops), pp. 138-141. 2016.

[5] S. Datta, C. Bonnet, N. Nikaein, "An IoT gateway centric architecture to provide novel M2M services," IEEE World Forum on Internet of Things (WF-IoT), pp. 514-519, 2014.

[6] P. Desai, A. Sheth, P. Anantharam, "Semantic gateway as a service architecture for IoT interoperability," IEEE International Conference on Mobile Services, pp. 313-319, 2015.

[7] S. Vukmirović, A. Erdeljan, F. Kulić, S. Luković, "Software architecture for smart metering systems with virtual power plant," 15th IEEE Mediterranean Electrotechnical Conference, pp. 448-451, 2010.

[8] Software engineering institute, "Defining software architecture". [Online]. Available: http://www.sei.cmu.edu/architecture/.

[9] F. Buschmann, R. Meunier, H. Rohnert, P. Sommerlad, M. Stal, Pattern-oriented software architecture: a system of patterns, John Wiley & Sons, 2001.

[10] P. Clements and M. Shaw, "The golden age of software architecture," IEEE Software, vol. 26, no. 4, pp. 70-72, 2009.

[11] A. Syromiatnikov and D. Weyns, "A journey through the land of model-view-design patterns," IEEE/IFIP Conference on Software Architecture, Sydney, NSW, pp. 21-30, 2014.

[12] E. Gamma, R. Helm, R. Johnson, and J. Vlissides, Design patterns: elements of reusable object-oriented software, Addison-Wesley Longman Publishing, 1995.

[13] G. E. Krasner and S. T. Pope, "A cookbook for using the model-view-controller user interface paradigm in smalltalk-80," Journal Object Oriented Programming, vol. 1, no. 3, pp. 26–49, 1988.

[14] B. Orlov, "iOS architecture patterns. Demystifying MVC, MVP, MVVM and VIPER," Nov. 28, 2015. [Online]. Available: https://medium.com/ios-os-x-development/ios-architecture-patterns-ecba4c38de52

[15] R. Khandelwal, "Effective Android Architecture," 360andev conference, 21 Sept. 2016.

[16] M. Potel, "MVP: Model-View-Presenter the taligent programming model for C++ and Java," Taligent Inc, 1996.

[17] J. Smith, "WPF apps with the model-view-viewmodel design pattern," MSDN magazine, vol. 24, no. 2, 2009.

[18] Martin, R. C., "The clean architecture," Aug. 13, 2012. [Online]. Available: http://blog.8thlight.com/uncle-bob/2012/08/13/the-clean-architecture.html

[19] D. Namiot and M. Sneps-Sneppe, "On micro-services architecture," International Journal of Open Information Technologies, vol. 2, no. 9, pp. 24-27, 2014.

[20] Ubuntu Snappy Core: Ubuntu for the IoT [Online]. Available: http://www.ubuntu.com/internet-of-things

[21] P. Yakimov, S. Ovcharov, N. Tuliev, E. Balkanska, R. Ivanov, "Three phase power transduser for remote energy management system application," Annual Journal of electronics, vol. 4, no. 2, pp. 31-34, 2011.

[22] DeviceHive: IoT data platform [Online]. Available: http://www.devicehive.com/

[23] MongoDB [Online]. Available: http://mongodb.com

Application Models for Ubiquitous Systems with Sporadic Communication Availability

I. Čavrak, I. Žagar and A. Dražić[*]

[*] University of Zagreb, Faculty of Electrical Engineering and Computing, Zagreb, Croatia
{igor.cavrak, ivan.zagar, alen.drazic}@fer.hr

Ubiquitous computing systems are predominantly characterized by limited computing power, communication capabilities and energy sources. Yet novel usage scenarios shift their roles from basic data acquisition and forwarding to performing more sophisticated tasks such as data processing, local service provisioning and actuation. To meet their goals, horizontal or vertical collaboration and task distribution are often required; utilizing resources identified in the nearby communication environment or offered by remote, cloud- or edge-based services. However, different infrastructural and energy-related constraints can severely influence device's communication capabilities thus dictating the level of provided functionality. In this paper, we focus on the class of ubiquitous systems without permanent communication links, utilizing only sporadic communication with global computing infrastructure. Such systems exploit the short-term presence of third-party communication and computational resources in their environment, adapting their service level to available resource properties. Several identified communication and computation scenarios are described for such a resource constrained and time-variable environment, along with the incentive model for using and provisioning of short-term resources. A general model of adaptive distributed applications exploiting such sporadic communication and computation environment is presented, relying on the cloud- and edge-computing paradigms and microservices as its building blocks.

I. INTRODUCTION

Ubiquitous computing [1] and succeeding Internet of Things [2,3] have brought a new perspective on computing and collaboration, where computing is dispersed in space over numerous computational nodes pervading the human environment. Such pervasive computational nodes, by their sheer number and diversity, require the change in interaction paradigm, emphasizing machine-to-machine communication and collaboration over direct human-to-machine interaction.

The advent of the cloud computing [4,5], on the other hand, addressed the need for efficient computing resource usage by centralizing the computing infrastructure into large data centers and allowing for their on-demand provisioning. In addition to centralized cloud resources in data centers, fog or edge computing introduced an additional layer of computational resources, bringing the computation to smaller centers closer to service users [6].

Collaboration between application components running on mobile devices and cloud computing has been studied extensively [7,8] where complex and resource demanding application sub-functionality is off-loaded from mobile devices to resource-rich and elastic cloud computing infrastructure. With the proliferation of edge computing, additional application layering is introduced, assigning application tasks between edge and cloud layers with the aim of enhancing user experience and optimizing overall resource usage [9].

The introduction of microservice architectural style [10,11], based on independently deployable services further added to the resource elasticity of edge- and cloud-computing, enabled the application components to be deployed and orchestrated in an efficient manner using different underlying infrastructural layers with additional support of containerization technology [12,13]. Proposals of microservice usage in the context of cloud and IoT are also starting to emerge [14,15].

Our proposed ubiquitous application model strives to provide a uniform, component-based model, spanning all computational layers of the ubiquitous system; embedded, mobile, edge and cloud. In this model, application components are hosted in different layers, on platforms with diverse computational, communication and power resources. The component and application design reflects the microservice architectural style, allowing for fast deployment of components on computational nodes and using specific layer properties to achieve application flexibility such as spatial and temporal component redundancy in embedded and mobile layers, and elasticity and containerization in edge and cloud layers.

The application adaptation mechanism is based on the chain of components' Quality of Service (QoS) profiles and profile usage prices, where each component exposes a single service interface with QoS profiles and current usage price of each profile. QoS profile prices reflect the current resource state of the component's hosting platform, communication channels and QoS profile prices of components the service providing component relies on in providing its service. The proposed adaptation mechanism allows for *upstream* (application-dictated) and *downstream* (resource level-dictated) dynamic adaptation of ubiquitous application, both in *parametric space* (QoS levels of used components) and in *structural space* (application graph; the set of used components and their deployment hosts in computational layers).

In the paper, we specifically describe the ability of the proposed model to provide the application adaptation in situations of sporadic communication availability among computational layers, in particular between the embedded and the mobile layers. This application adaptation mechanism should enable preservation of application functionality, albeit with reduced QoS, efficiently utilizing computational resources within the communication range by agile adaptation of the application structure.

The rest of the paper is organized as follows. Section II presents our ubiquitous application model for component-based applications with application composition adaptable to changing component QoS parameters. Section III focuses on the usage of the model in ubiquitous applications with sporadic communication availability between computational layers. Section IV presents two example applications based on the model, forming a model testbed, and Section V concludes the paper.

II. UBIQUITOUS APPLICATION MODEL

We divide our ubiquitous computational model into two basic sub-models: *(i)* computational model representing the diversity of computational resources found in ubiquitous computing and forming the highly heterogeneous execution environment for the *(ii)* application model, comprised of heterogeneous service components, dynamically composed in space and time to perform desired application-defined functions.

A. Computational Layer

In our model of ubiquitous computing, resources and their hosting devices forming the computational substrate for ubiquitous applications are conceptually divided into four layers, each having distinct properties (Table I.):

- *Embedded layer* – computational devices statically embedded into the environment.

- *Mobile layer* – computational devices can traverse spaces at different speeds, autonomously or as part of a larger system.

- *Edge layer* – computational resources are available at network edges - fixed communication infrastructure.

- *Cloud layer* – computational resources are elastic, available within a computational cloud.

We use a simplified model of computational devices, focusing on the ones present in the first and second computational layers, assuming almost unlimited resources availability for edge- and cloud-layer devices. Each device state is defined using the following tuple:

$$Ds = <E, P, C> \quad (1)$$

where E denotes device's energy-related, P denotes processing and C denotes communication properties, each property being a compound one.

Device energy-related property E consists of the *(i)* currently available energy in the power subsystem, *(ii)* maximal power the subsystem can deliver at any time and *(iii)* predicted inflow of energy from external sources or by energy harvesting. Derived parameters, constraints and predictions can be used in higher level decision making such as energy level flow, estimated battery life, estimated energy inflow, maximal power consumption in relation to current energy levels, etc. Device processing properties P are modeled with regards to the inbuilt *(i)* computing capabilities and related parameters (number of cores, core

TABLE I.   RESOURCE AVAILABILITY PER LAYER

Layer	Computation	Communication	Energy
embedded	Low	low	low
mobile	low-medium	low-medium	low-medium
edge	medium-high	high	high
cloud	High	high	high

processing power, adjustable clock frequencies, power consumption) and *(ii)* available memory and memory types. Communication properties C portray all the available wireless and/or wired communication subsystems and their major properties (available modes of operation, throughput, latency, communication range and power consumption).

B. Component-based Application Layer

We model a distributed ubiquitous application as a directed acyclic graph (DAG)

$$App = <C, L> \quad (2)$$

where C is a set of vertices representing the required application's software components and L is a set of directed edges representing data and control flows among application components.

A software component forms a basic building block of a ubiquitous distributed application, where each component implements a single and narrowly-scoped functionality, following the microservice architectural style approach. Each component defines one or more quality of service (QoS) profiles, consisting of an internal and external sub-profile. External sub-profile declares the QoS capabilities of the component to other components - potential consumers of component's service. Internal sub-profile defines *(i)* the set of execution properties for the component (required power, computation and communication resources and their usage) as well as *(ii)* additional application components and their (external) QoS profiles required to implement the component's service. Application *initiator* components originate application activities, partially implementing application logic and using services implemented by other application components down the application graph. *Initiator* components do not need to expose their service to other components.

Data and control flows, represented as edges, connect application components forming an application graph and allowing the execution of an application on a distributed, heterogeneous computational substrate. The type of directed edge in the application model represents either one-way flow (signal) between application components or a request-reply form of communication. Active QoS profile of a component can dictate different data and control flows, thus making the application topology a function of the set of active QoS profiles of all application's components.

Components are hosted on computational devices in the computational substrate, as depicted in Fig. 1. In the lower two layers, embedded and mobile, it is expected that hosted components are of lower computational, communication and power requirements, and are deployed statically. In the upper two layers, deployment is dynamical and, considering available resources, scaled according to service demand. Multiple QoS profiles of the component can reflect different levels of the service provided by the component itself: *(i)* by mirroring the usage of resources of the hosting device depending on the service level, *(ii)* reflecting the current state of resources of the hosting device by offering all or reduced service levels, or *(iii)* by acting as a proxy towards the component implementations residing on hosts in resource-rich layers.

The model's topology, defined by the underlying DAG, allows easier and power-efficient calculation of a component's QoS profile price, dependent on the prices further down the application graph. However, it permits only partial ordering of component interactions in the application model and only a single appearance of a component instance in an application graph. The current model restricts the application state to be held only at the *initiator* components, and a tiny amount of data representing application state to be transferred to service-providing components down the application graph during *(i)* the component binding process and *(ii)* as part of the individual service invocations. Allowing robust mechanisms for the state to be preserved at each application component, even replicated in case of dynamic component rebinding, would allow for the model to be applicable to a wider class of applications, but would introduce a significant burden to component hosts' resources, especially in the embedded and mobile layers.

To form a functional distributed application, according to the proposed model, it is necessary that the application graph is connected, i.e. that all the required application components are present *(i)* in the *application neighborhood* (spatial dimension), i.e. within the communication neighborhood of the devices hosting the components, *(ii)* in the same timeframe (temporal dimension) during which the application is declared operational and that *(iii)* all interacting components are bound, forming links among application's graph nodes.

The application graph is formed and maintained using a dynamic process of local negotiations between application components. A component informs other components of its services either by broadcasting service announcements or by replying to a request for service queries. Service offer is based on the set of currently available QoS profiles for a service and associated *prices*. Components enter the process of negotiation resulting in a component binding (creating a link of an application graph) or abandoning the negotiation process. A derivative of ContractNet interaction protocol [16] has been adopted allowing the selection of one, or even more, components out of collected offers. A component can, due to changes in resource availability of the host or in changes of QoS of the used components, inform the client component of the QoS changes, where re-negotiation process can be triggered between them, possibly resulting in unbinding components and loss of application functionality.

The role of prices in local inter-component negotiations is two-fold: they reflect the required resource consumption

of the service-offering component for an offered service QoS, and they act as a resource-preserving mechanism at the application level, directing the application to use the application's QoS profile, and all the application components' QoS profiles, with the best QoS/price ratio at a given time moment. Service prices for the components in the lower two layers are expected to be considerable and fluctuating, highly dependent on the host's current resource availability. In contrast, components hosted on the edge and cloud layers are expected to have significantly lower service prices, dominated by costs of data transfer from two lower layers, encouraging their usage in the application graph.

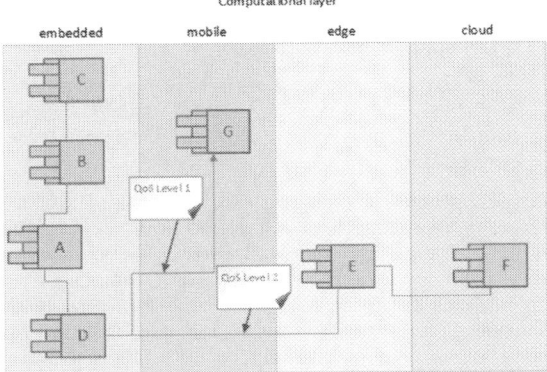

Figure 1. Ubiquitous application deployment in computational layers

III. SPORADICALLY CONNECTED APPLICATIONS

In the previously described distributed application model, for the application to be functional it is necessary for the application graph to be connected. However, in many ubiquitous application scenarios, such requirement is almost impossible to satisfy for prolonged periods of time.

Extending our previously described application model, we extend it with two probabilistic functions (Q – set of available QoS profiles for a specific application component, C – set of application components):

$$App = <C, L, p_{QoS}, p_{PRICE}> \qquad (3)$$

the function

$$p_{QoS}(q,c,t) \in [0,1], \qquad (4)$$

where $q \in Q, c \in C, t \in \mathbb{R}_{\geq 0}$ models the probability of the component c supporting the QoS profile q at the time moment t, and the function

$$p_{PRICE}(v, q, c, t) \in [0,1] \qquad (5)$$

where $v \in [0,1], q \in Q, c \in C, t \in \mathbb{R}_{\geq 0}$ models the probability of normalized price v for a given QoS profile q at time t for the component c, where the price value 0 indicates unavailability of the component under the given QoS profile.

The extended model allows for simulation and assessment of different application properties, for example *(i)* the availability of QoS profiles for different application components in time periods of interest, *(ii)* resource consumption of the application on different computational substrate nodes and in different application components and *(iii)* the availability of redundant application components residing in different computational layers.

We classify distributed applications with sporadic communication availability into two main usage scenarios:

- *Soft scenario* - applications whose functionality is preserved, but with a lowered QoS and/or increased price because of irregular communication availability; include applications employing one or more nodes from the embedded layer, with sporadic assistance of components from upper layers during (short periods of) communication availability.

- *Hard scenario* – applications whose functionality is available only during (short periods of) communication availability towards the upper layer devices.

Each application component can be realized as an *implementation* component – implementing component's service and exposing supported QoS levels, optionally using other components – and *proxy* component – implementing the same interface as the *implementation* component, but only forwarding service requests to another component instance on a host possibly in another computational layer. *Proxy* components are predominantly used in the mobile layer, where a mobile node can host the *implementation* component and provide real services to the client nodes, as on Fig. 2, or act as a dynamic link between lower-layer clients and services residing on the upper layer hosts, as shown in Fig. 3.

In the *hard* scenario, the problem of application connectivity is augmented with the problem of connection duration, highly dependent on communication technology used, with minimal duration T_{min} defined as

$$T_{min} = T_{discovery} + T_{negotiation} + T_{service} \qquad (6)$$

where $T_{discovery}$ denotes the time required for the discovery of services offered by mobile node by embedded node, $T_{negotiation}$ as the time required to finish the QoS level negotiation and $T_{service}$ as the time to forward data to the component's service, process the data and return the result to the client component. In the case of fast mobile nodes with relatively short time span within communication range with other application components, time available might not be long enough to complete the service usage transaction. Application components, with the assistance of their hosts and depending on the communication technology used, can assess the speed and available timespan using different techniques usually based on monitoring received signal strength indicators (RSSI). If supported by the *proxy* component and the *implementation* component, *asynchronous proxy* service transaction can be employed, with Fig. 4 providing an example of the service request and reply being two separate, decoupled transactions using different mobile nodes present in communication neighborhood at different times.

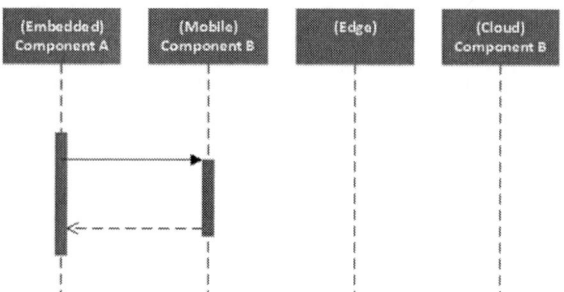

Figure 2. Component implementation hosted on the mobile node

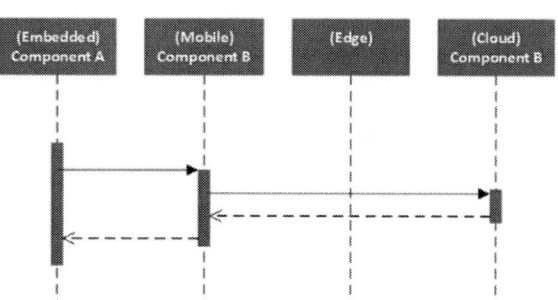

Figure 3. Component implementation hosted on the cloud node, proxy component on the mobile node

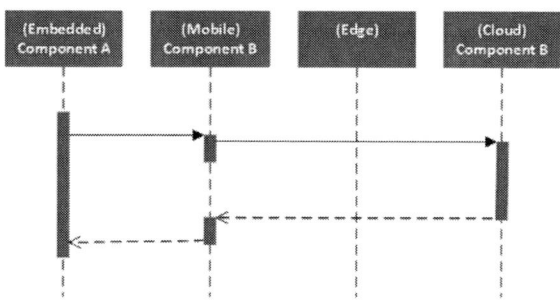

Figure 4. Component implementation hosted on the cloud node, asynchronous proxy on mobile nodes to complete the transaction

Other scenarios addressing the short available communication timespan might include:

- Assessing time span according to RSSI and selecting the component with the longest predicted availability and acceptable QoS/price ratio.

- Selecting multiple negotiation winners according to the best QoS/price ratio regardless of the predicted availability, thus increasing the cost of resources used, but increasing the success rate.

- Assessing time span according to RSSI and selecting between *proxy* (less resource spent) or *asynchronous proxy* (more resources spent) scenarios.

To perform an early evaluation of our application model in the context of *soft* and *hard* sporadic communication availability scenarios, we have conducted simulations of application behavior for a simple ubiquitous application.

The application is comprised of *initiating* application component C_I, residing on a single host in the embedded layer, and component C_S offering two QoS profiles for an exposed service, with instances hosted in multiple layers, on a single or multiple hosts at a time, and with a variable availability of those instances. All simulations have been conducted within an extended custom Java-based simulation framework, initially developed for simulating self-organizing distributed applications within the context of wireless sensor networks [17], including modeling of node mobility and stochastic communication links.

Fig. 5 presents the simulation results of the *soft* scenario by depicting the QoS profile price variability of C_S as perceived by the initiating component instance $C_I^{(1)}$, with C_S instances residing in embedded, mobile and cloud layers. In the timespan framed by points A and B, the C_S instance $C_S^{(1,QoS-A)}$ residing on the embedded layer host is used by the $C_I^{(1)}$, with price persistently rising due to host resource consumption. At the moment B, the $C_S^{(1,QoS-B)}$ QoS profile is selected by the $C_I^{(1)}$, lowering the usage price due to lower resource consumption of the selected profile. At the time moment C, the $C_S^{(1,QoS-B)}$ host resources are exhausted, and the price reaches 0 (service unavailability), re-linking shortly the application graph to redundant host and component instance $C_S^{(2,QoS-B)}$. At the time moment D, $C_S^{(2,QoS-B)}$ host resources are depleted and the application graph is incomplete due to C_S instance unavailability in the $C_I^{(1)}$ communication neighborhood until the time moment E when the component instance $C_S^{(3,QoS-A)}$ residing in the mobile layer becomes available. Implementation of C_S for the mobile layer hosts allows for service forwarding, and after negotiation with cloud-layer host and C_S instantiation, at the time moment F cloud-hosted $C_S^{(4,QoS-A)}$ with the lowest QoS/price ratio takes over the service provisioning.

Fig. 6 presents the simulation of price variability for the *hard* scenario and a single component C_S QoS profile A, where at time point A, the component instance $C_S^{(1,QoS-A)}$ becomes available for only a short period on a mobile node with low available resources and high QoS price. Point B marks the availability of instance $C_S^{(2,QoS-A)}$ on a mobile node with adequate local resources but only during a short period. At time point C there is a longer availability of instance $C_S^{(4,QoS-A)}$ in the cloud layer, with mobile layer instance $C_S^{(3,QoS-A)}$ acting as proxy. At time points D, there are two short periods of reachability of instance $C_S^{(4,QoS-A)}$ in the cloud layer, employing asynchronous proxy communication between it and $C_I^{(1)}$. Also depicted are different QoS prices reflecting resource levels of mobile nodes hosting component proxies $C_S^{(5,QoS-A)}$ and $C_S^{(6,QoS-A)}$.

IV. TESTBED APPLICATIONS

As a proof-of-concept, besides simulations, we are implementing two ubiquitous applications based on the proposed model, thus providing a testbed for developing more time– and power–efficient negotiation protocols over multiple communication channels and technologies.

The first example application uses a LoPy [18] multi-modal wireless communication node with Bluetooth LE [19] and LoRa [20] as communication channels. The application component A on the LoPy module performs periodic ambient measurements and tries to forward them to the central monitoring and storage component B using LoRa communication channel and one of the modes, depending on the channel availability. If LoRa-based channel persists to be unavailable for a prolonged period or the change rate of acquired data is high, the component seeks the assistance of nearby mobile phones equipped with Bluetooth LE and hosting the *proxy* component B to relay the data to the *implementation* monitoring and storage component B, as shown in Fig. 7. Each hardware device in this scenario hosts only one application component, where component A is the application's *initiator* component.

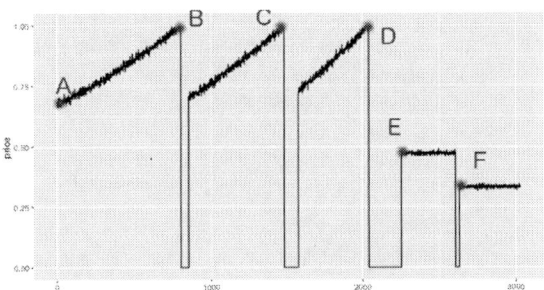

Figure 5. QoS profile price variability for a multiple-hosted component in soft scenario of sporadic communication availability

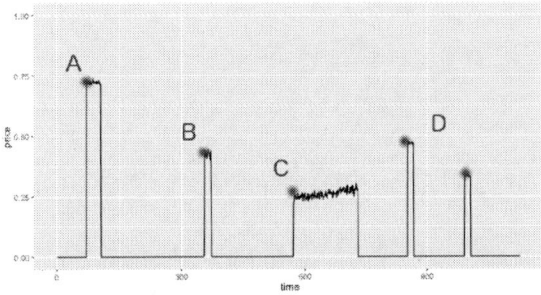

Figure 6. QoS profile price variability for a multiple-hosted component in hard scenario of sporadic communication availability

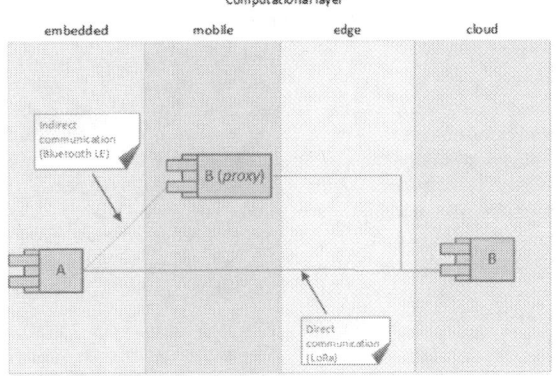

Figure 7. Test application – multimodal communication with proxy mobile component in case of direct communication failure

The second example scenario, depicted in Fig. 8, includes clusters of low-resource wireless sensor nodes using ZigBee [21] communication protocol and TinyOS platform [22] for a power-efficient collection of ambient data over large areas. In this scenario, devices or execution environments, denoted by dashed rectangles, can host multiple application components. The statically deployed data collection system is augmented with mobile nodes in the form of wearables (Intel Edison [23] wearable platform and Android-based smartphones), providing the environment with variable sensing, forwarding and processing demands reflected in variable component QoS. Data are forwarded to components in upper layers using both direct or proxy communication, subject to variable data link availability. While basic data processing is available at the cluster and mobile nodes, complex data processing and storage is reserved for application components in the edge and cloud layers utilizing OpenStack [24] as the IaaS platform and Docker [25] as the microservice container.

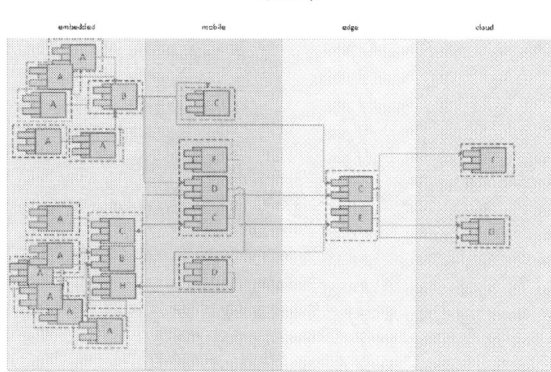

Figure 8. Test application deployment – multimodal communication, forward- and backward-adjustable component QoS

V. CONCLUSION

The ever-increasing complexity and adaptability of novel ubiquitous distributed systems present a challenge to the architects of ubiquitous applications, requiring composition and interplay of many adaptable application components spread among different computational layers; from tiny resource-constrained pervasive devices to large-volume data systems in clouds.

In this paper, we proposed a general model of adaptive ubiquitous applications based on components and application graphs defining current application configurations based on the available or required quality of service profiles. Particular attention was given to applications lacking permanent communication links towards application components in higher computational layers: edge and cloud. We proposed different interaction and computation scenarios exploiting sporadic resource availability in the mobile layer.

Our future work will, besides enhancements to the model itself and the accompanying modeling and simulation tools, include devising more time- and power-efficient negotiation protocols and communication strategies for scenarios with short periods of component availability due to high node dynamism in the mobile layer.

REFERENCES

[1] M. Weiser, "The computer for the 21st century," in SIGMOBILE Mob. Comput. Commun. Review vol. 3, no. 3, pp. 3–11, July 1999.

[2] A. Al-Fuqaha, M. Guizani, M. Mohammadi, M. Aledhari and M. Ayyash, "Internet of Things: A Survey on Enabling Technologies, Protocols and Applications," in IEEE Communications Surveys & Tutorials, vol. 17, no. 4, pp. 2347–2376, Fourthquarter 2015.

[3] J. Heuer, J. Hund and O. Pfaff, "Toward the Web of Things: Applying Web Technologies to the Physical World," in Computer, vol. 48, no. 5, pp. 34–42, May 2015.

[4] N. Antonopoulos and L. Gillam, "Cloud computing: Principles, systems and applications", Springer, 2010.

[5] M. Armbrust et al., "A view of cloud computing," Communications of the ACM, vol.53, no. 4, pp. 50–58, April 2010.

[6] F. Bonomi, R. Milito, J. Zhu and S. Addepalli, "Fog computing and its role in the internet of things," in Proc. 1st Ed. MCC Workshop Mobile Cloud Computing, pp. 13–16, 2012.

[7] IBM Corporation, "Smarter wireless networks,", North Castle, NY, USA, White Paper WSW14201USEN, February 2013. www.ibm.com/services/multimedia/Smarter_wireless_networks.pdf (2017–02–18)

[8] ETSI, "Mobile–Edge Computing – Introductory Technical White Paper," https://portal.etsi.org/tb.aspx?tbid=826&SubTB=826 (2017–02–18)

[9] W. Zhang, Y. Wen and D. O. Wu, "Collaborative Task Execution in Mobile Cloud Computing Under a Stochastic Wireless Channel," in IEEE Transactions on Wireless Communications, vol. 14, no. 1, pp. 81–93, January 2015.

[10] J. Lewis and M. Fowler, "Microservices", 2014, http://martinfowler.com/articles/microservices.html (2017–02–18)

[11] A. Sill, "The Design and Architecture of Microservices," in IEEE Cloud Computing, vol. 3, no. 5, pp. 76–80, Sept. –Oct. 2016.

[12] C. Pahl, "Containerization and the paas cloud," in IEEE Cloud Computing, vol. 2, no. 3, pp. 24–31, 2015.

[13] C. Pahl and B. Lee, "Containers and Clusters for Edge Cloud Architectures – A Technology Review," in 3rd International Conference on Future Internet of Things and Cloud, Rome, pp. 379–386, 2015.

[14] K. Vandikas and V. Tsiatsis, "Microservices in IoT clouds," Cloudification of the Internet of Things (CIoT), Paris, 2016, pp. 1–6.

[15] T. Vresk and I. Čavrak, "Architecture of an interoperable IoT platform based on microservices," 39th International Convention on Information and Communication Technology, Electronics and Microelectronics (MIPRO), Opatija, 2016, pp. 1196–1201.

[16] Foundation for Intelligent Physical Agents, "FIPA Contract Net Interaction Protocol Specification", 2001.

[17] D. Rojković, T. Crnić and I. Čavrak, "Agent-based topology control for wireless sensor network applications," 35th International Convention MIPRO, Opatija, 2012, pp. 277-282.

[18] Pycom LoPy module, https://www.pycom.io/product/lopy/ (2017–02–18)

[19] Bluetooth SIG, "Bluetooth Core Specification v5.0", 2016.

[20] LoRa Alliance, "LoRa Specification", 2015.

[21] ZigBee Aliance, http://www.zigbee.org (2017–02–18)

[22] TinyOS, https://github.com/tinyos/tinyos-main (2017–02–18)

[23] Intel Edison Compute Module, https://software.intel.com/en-us/iot/hardware/edison (2017–02–18)

[24] OpenStack, https://www.openstack.org (2017–02–18)

[25] Docker, https://www.docker.com (2017–02–18)

Towards the Utilization of Crowdsourcing in Traffic Condition Reporting

P. Rantanen*, P. Sillberg* and J. Soini*
* Tampere University of Technology, Pori, Finland
petri.rantanen@tut.fi

Abstract - The use of traffic information in map applications designed for stand-alone navigation devices as well as in mobile devices has become a common trend. This information is often governed by the various service providers with little or non-existent feedback from the users. Using a wide user base it is possible to collect information on traffic conditions faster and more efficiently. Additionally, many of the events faced on the road can be challenging to detect by automatic means, but are easily noticed by the road users – animals on the road and broken or missing road signs are only a few examples.

To better facilitate the utilization of information gathered from road users, simple and easy-to-use software solutions are required. This paper presents a prototype mobile application, which the road users can take advantage of for both following the on-going traffic conditions while driving and for collectively reporting traffic events other users might be interested in. The high-level architecture, application and data utilized in the reports are presented in addition to the preliminary findings of the on-going research. This paper will also discuss the challenges identified while developing the application.

I. INTRODUCTION

There have been several studies on how to take advantage of automated means in the traffic environment. Topics such as the detection of pavement conditions [1], winter conditions [2], vehicles [3][4] and road signs [5] have been of constant interest for the past decade. There is little doubt that the use of automated means will increase in the future. Nevertheless, events such as animals on the road, broken road signs and poor visibility are easily recognized by humans, but can be challenging use cases for automatic methods. Thus, the potential wide user base offered by the people travelling on the roads should be more tightly integrated into the road condition reporting.

Most map applications provide "layers" that show basic details (for example, road works, traffic jams) concerning the current road conditions, but generally, no option is given to the user to give feedback on the accuracy of the information. In Finland, there is an official telephone number provided by the local road agency the road users can call to report bad road conditions, and several radio stations accept calls concerning events met on the road. Unfortunately, these services are very seldom connected to the devices and the services the people use in their everyday lives. In essence, the application (or service) presented in this paper could be seen as a more modern replacement for these telephone-based services currently available.

The initiative to develop the application presented in this paper came from one of the corporate partners (a logistics and transportation company) of our project. They had a need for a very simple and easy-to-use application that the professional drivers of the company could use to notify each other about various road incidents, events and road conditions that could affect the deliveries.

This paper presents the preliminary results of the APILTA (Avoin PIlvipalvelukonsepti joukkoistetun Liikennedatan TArpeisiin, or "Open Cloud Platform for Crowdsourcing Based Services", in English) project currently on-going in Tampere University of Technology, Pori Department. The primary purpose of the research within the context of this paper is to design an easy to use application (user interface) that would not distract the driver and the users would be willing to take advantage of while travelling. Keeping this goal in mind, the following section (Section II) discusses the related studies, Section III provides a high-level look on the architecture and the use case in general, Section IV will describe the application itself, Section V offers a discussion on the challenges and opportunities, and finally, Section VI summarizes this paper.

II. RELATED STUDIES

The use of traffic data in crowdsourcing applications is not a completely new idea. For example, in [6] the focus was on evaluation of anonymized data created by an electronic travel journal to be used as study material for mobility research. If the collected data was accurate enough, it could be used in design of the traffic network. There are efforts to estimate the traffic conditions with data that are gathered from different sources and by different methods (e.g. traffic monitoring by inductive loop detectors, satellite positioning, and mobile phone positioning) [7][8]. To reliably evaluate the traffic conditions a lot of traffic monitoring is required. This is where the crowdsourcing can become a feasible source of data.

Efforts for the concept of Smart City (SC) can be useful as transportation and traffic are typically studied together with this topic. Sensor networks, data collection and data management are key components in development of Smart City applications. Literature review [9] classifies a wide range of architectures and technologies used in the field of SC. The newest trend seen was the introduction of IoT (Internet of Things) and in the future IoE (Internet of Everything) is expected to be the next important milestone. IoT enabled SC applications often employ Smart Objects (SOs) in the monitoring of traffic and environment. A large number of devices, software and services (e.g. mobile phones or even vehicles) can be categorized as a SO [10] [11].

In the use case of Smart Transportation [10], the sensor data is seen to have a crucial role in managing the transport system. The trustworthiness, robustness, timeliness, and security of the communicated data together with maintaining the privacy of users is of major importance in a SC system [10]. The security and reliability of a cloud-based city management platform have also been studied [12].

There have been studies on using mobile phones as location location sensors in a Smart City service for car sharing to improve the traffic conditions in the City of Barcelona [11]. Furthermore, there has been research on utilizing event-based architecture to conduct Machine to Machine (M2M) communication in vehicular context [13], the Internet of Vehicles (IoV) and the Social Internet of Vehicle (SIoV) [14], detecting road conditions using smartphone sensors [15], and analyzing video with a computer vision system to recognize overall road conditions and the status of traffic infrastructure [16].

There are several applications that are similar to our approach. Beat the Traffic [17] used to offer real-time traffic information partially provided by users in the United States and Canada. Waze [18] similarly offers its users to participate in gathering new traffic information. Inrix [19] provides a more complex service for traffic management including several mobile applications for tracking traffic conditions, but lesser options for user to manually create reports. In Finland, a new pilot (NordicWay Coop) was launched last year in co-operation with industry partners and the Finnish Transport Safety Agency with a goal of allowing people to report traffic conditions [20].

When it comes to reporting new alerts, each application has a slightly different approach: Beat the Traffic offered more complicated forms and a simpler "shake to report" feature; Waze uses speech recognition; and NordicWay Coop utilizes a more traditional menu-based approach. Based on the literature each controlling method (voice recognition [21][22][23], touch, tactile and gesture [24]) have their own advantages and disadvantages with perhaps no clear winner to be found amongst the options.

Similarly to Beat the Traffic, Inrix and NordicWay Coop, our application utilizes touch-based controls with the main difference being in the user interface design – the look'n feel of the application and the user interface "flow" when creating a new alert. In our case, the goal is on enabling an easy reporting of alerts where as many of the commercial applications are primarily focused on map-based features (e.g. navigation).

As mentioned before, the initiative for our application development came from a logistics and transportation company. An individual company might have the need to report alerts that are not by default included in the system, and the company might not be willing to share the reports outside the company, or the company might only trust the alerts received from a particular user group. User groups and customization would also allow regular users to team up with friends or colleagues for sharing alerts in a more social way. With the exception limited features offered by the Waze service there seems to be apparent lack of customization and user grouping options in the existing applications.

III. ARCHITECTURE

The system architecture (seen in Fig. 1) is based on receiving alerts from different clients, and transmitting the alerts to clients who are or would be affected by them. The service is implemented by utilizing commonly used open source components such as Apache Tomcat, Apache Solr and Oracle MySQL.

In the first prototype of the client, the available alert types are predefined to contain just a few examples, but the architecture and alert syntax itself are extendable. Also, the system can accept photos attached to the alerts, but this feature is not used in the application presented in this paper. City road maintenance is one use case where the additional info could be useful.

The alerts on the client are updated by regularly polling the representational state transfer (REST) Application Programming Interface (API) to retrieve the newest alerts that affect the user. Thus, the clients do not need to subscribe to listen for new alerts. The client would need to continuously report its location to the service to receive up-to-date alerts in any case, but in our approach the service will immediately respond with alerts that are in range. This would also allow anonymous or unregistered users to use the API. In our case, because of the limited computing resources available, the service is open only for users accepted to participate in our piloting phase. In practice, there could also be real business cases (e.g. pay-to-use, providing user preferences across devices based on the user account) that require the users to be authenticated before using the service.

Each alert reported by the clients includes the geographical coordinate of the event and the search method provided by the API accepts a polygon based area filter and a heading based filter. This way the client device can construct more accurate search terms that cover more area in front of the device than behind. The client can also adapt to the increasing speed by retrieving the alerts more often, or by growing the search area bigger.

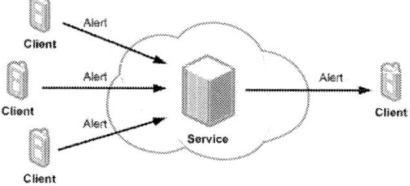

Figure 1. High level architecture.

In the first phase of the pilot tests, there are only a few alert types to be included (animals, broken street signs, slow traffic). The architecture is built to accept new, user created alert types for extending the usability of the system. Any authenticated user can create a new alert type, but by default the application presented in this paper will only utilize the built-in types. New types must be configured (chosen from a list of types known by the service) in the application settings – this is to prevent users from spamming the service with unnecessary global alerts, but the approach will still allow the utilization of custom tags with relative ease of use.

The API provides the data in Extensible Markup Language (XML) documents by default, and supports JavaScript Object Notation (JSON) format as well. The data format of the alert includes a timestamp, location (coordinate), alert type, alert description and information about how long the alert is valid. The validity period of the alert varies case by case, and at the moment there is no good approximation when or how a certain alert should become inactive. In accidents and similar events an official user such as the police or ambulance staff could be given the permission to remove the corresponding alert, but in the case of animal alerts the validity period can be challenging to predict. The default validity period for each alert can be defined when the alert type is created in the service. In our use cases, alerts will be automatically invalidated within a couple of hours. In principle, the validity period could also be set by the client though choosing the correct time value when posting a new alert would probably be just as difficult as choosing a correct pre-defined validity time.

IV. PROTOTYPE APPLICATION

As the main use case is driving a car, this poses certain requirements for the application. Most importantly, the application should not distract the driver. For this goal, the application has been designed to require as little interaction as possible with emphasis on automation both in the information gathering and in the flow of the user interface.

There are no technical limitations why the application could not be used, for example, when walking, and alternative means of transport could provide possibilities for features purposefully not implemented in the current

version. Features such as alert rating (feedback) or reporting of incorrect alerts, or zooming and panning of the map view could be easily managed by the user when walking, but could be too distracting when driving a car.

The application has two basic functions (or features): reporting new alerts to the service; and showing alerts that are near to the user. The features are shown in Fig. 2 and in Fig. 3. Additionally, the application contains a settings view (not visible in the figures), which allows the user to change the common configuration parameters such as the credentials used to authenticate the user with the service. The settings view can also be used to select the alert types the user wishes to listen for new alerts and the option to select the types available for reporting new alerts.

The screenshots are taken from a Google Nexus 7 device running the Android 7 operating system. The application should work on any relatively recent (even low-end) Android phone or tablet with a touch screen and an Internet connection. As the transferred data consists of simple Hypertext Transfer Protocol (HTTP) GET and POST requests with small XML or JSON payloads (depending on the amount of alerts, generally from ten to 50 kilobytes) even a relatively low bandwidth network connection can be utilized.

A. Feature 1: Report a New Alert

Fig. 2 illustrates the user interface flow applied when the user reports a new alert. The first picture (left side, Fig. 2) shows the default (idle) view visible when no alerts are in range. In this example case the user has configured the application so that there are two alert types available for reporting new alerts: reports for animals on road (center, Fig. 2); and reports for slow moving traffic (right, Fig. 2).

To browse the available alert types the user only needs to touch anywhere on the screen. This will automatically show the next view: from the default view to the "animal on road" view; from the "animal on road" view to the "slow traffic" view; and from the "slow traffic" back to the default view – as seen in the Fig. 2. A move from one view to another is accompanied by a sound (the spoken name of the alert type). This allows the use of custom alert type names by utilizing text-to-speech features provided by the Android platform.

Figure 2.   User interface flow when user reports a new alert.

If the user does not touch the screen within a predefined time and the active screen is one of the alert views, a new alert (with applicable details such as timestamp and user's location) will be automatically sent to the service. The time left until the alert will be sent is indicated to the user by a countdown timer visible on-screen (as can be seen in Fig. 2 middle and right, below the icons).

The time until the alert is posted can be configured from the application and a longer time will naturally give the user more time react. Touching anywhere on the screen while the counter is running will abort the operation and screen will change to show the next view. If the next view is an alert view a new counter will start, if the next view is the default view the application will return to the idle state.

The audio cues and the fact that the user can touch anywhere on the screen are designed to allow navigation without looking at the screen. On devices that do not have a touch screen the navigation could also be implemented by providing a single button to use for changing the views. The user interface is also meant to be simple enough to be usable on various screen sizes. Alternative approaches such as multiple buttons provided for navigation or a scrollable list of alerts could cause difficulties for the driver for using the application especially on devices with smaller screen sizes.

In principle, there could be any number of alert types to choose from, and if more alert types were used, new views would appear in the user interface cycle (for example, in Fig. 2, a view with a new traffic sign could be between the elk sign and the slow traffic sign). In practice the user should use the application settings to configure only the alert types he/she is planning to use to be visible in the user interface to reduce the total amount of browsing needed.

Another option could be to allow the user to speak the name of the alert type when choosing the alert to be reported, but in a location with a potential ambient noise (e.g. in a car) it can be challenging to implement reliable speech recognition. Additionally, this would require the user to remember the actual names of the alerts or implementation should be smart enough to decipher the user's meaning from alert names pronounced only partially correct. Furthermore, as the application is entirely separate from the in-car systems, the voice control could also interfere with the built-in speech recognition features of the car itself or the use of two separate systems could confuse or annoy the user.

B. Feature 2: Show Nearby Alerts to the User

The user interface flow for showing the nearby locations is can be seen in Fig. 3. As long as the visible view is the default (idle) view (Fig. 3, left) the application will automatically show the map when alerts are detected within a pre-configured range or the travel time to the alert is within a pre-defined threshold (for example, the location indicated by the alert is within 100 meters, or the location would be reached in 30 seconds). The change to the map view is also indicated by an audio cue. If the user is currently browsing for alert types and is not in the

default view, only a sound cue is played and the application will wait for the user to return to the default view before showing the map view. After the alerts go out of range, the application will automatically switch back to the default view. The user can also change to the map view by tapping the "Map" option on the top right, and the user can return from the map view to the default view (and to reporting new alerts) by simply touching anywhere on the map screen or by touching the "Home" option on top right. The "Home" option can also be used to return to the main view from any other view, and can be used at any time to cancel a reporting of a new alert.

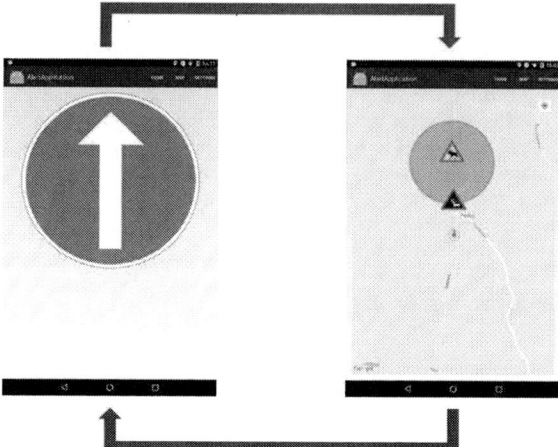

Figure 3. User interface flow when an alert becomes in range.

In the current version of the application, the map will automatically zoom to and follow the user's current location. Manual panning and zooming of the map often implemented by touch gestures are disabled. The primary purpose of this is to allow quick and easy navigation back from the map view.

The upcoming alerts are illustrated using the same icons as seen in Fig. 2. In our use case, the alerts are represented by road signs, which are easily recognized and associated by the driver to the events in question. Naturally, in many cases new icon designs are required as not all alert types can be illustrated by road signs alone. If the range of the event is known, it is illustrated by a transparent area around the icon. In Fig. 2, the "Animal on road" event contains range information, but the "Slow traffic" event does not. In the current service implementation, the area is visible only on alert types that have the range predefined (e.g. animal warning have a default range of 200 meters). The prototype application (user interface) does not support the creation of new alerts with range defined.

In the background the application will continuously poll the service for every couple of minutes requesting new alerts (i.e. alerts reported after the previous request) within a larger area (for example, within two kilometer radius) around the user's location. The retrieved alerts are cached by the application. The cache is checked whenever the user's actual location changes (within certain accuracy) and alerts that have moved too far from the user are automatically removed from the cache and alerts that are close enough are shown in the map view.

The new alerts could be requested from the service each time the user's location changes as opposed to more relaxed background polling, but this would both increase the amount of requests on the server-side as well as use more power and network bandwidth on the device itself. The downside is that the alerts do not appear on the client devices in real time, but in practice, real-time updates are seldom needed.

V. DISCUSSION

In essence, there have been two main challenges in developing the application: to make the user interface as easy to use as possible; and to validate the achieved results in the larger context of crowdsourcing.

To further aid the use of the application, automation would provide assistance for the user, and there are several promising, emerging and already implemented technologies such as speech recognition, image analysis (for example, the detection of objects or events from frames captured from a video camera), and the utilization of various sensors embedded in road modern road vehicles. Unfortunately, the most in-car systems cannot be directly accessed or extended with new features either because of security reasons or closed nature of the systems. This also means that buttons located, for example, in the steering wheel cannot be easily used to control a smart phone application.

Similarly, extending commonly used navigator devices and map applications can be a challenging task. Several companies (e.g. Google, Microsoft) provide online APIs for navigation applications, but the utilization of these APIs often require redundant work for the implementation of basic navigation features already present in the mobile device itself. Due to the fact that the user would need to use multiple applications for two closely related tasks (for example, the application presented in this paper and a proper navigation application) may decrease the user's willingness to use the provided, otherwise advantageous features. In an optimal situation, the functionalities presented in this paper would be integrated with an existing system or device, preferably in a navigation application.

As explained earlier, this paper presents only the preliminary results of an on-going study. The piloting phase utilizes people participating in the project either directly in the university or in the co-operating companies. The initial feedback on the idea has been positive, but because of the smaller user base, it is difficult to validate whether a larger audience would be willing to participate in reporting the alerts. A wider user base would be required for conclusive results, but unfortunately organizing a larger crowdsourcing effort can be challenging both for finding the applicable audience and because of the increased (computing) resources required for hosting the service online. In principle, it could also be possible to realize a working service even with a help of a smaller, but more motivated group enthusiasts.

Another issue, closely related to the widening of the target audience is the validation of the alert reports the users have submitted. Within a smaller and more controlled user base it is easier to protect the service against malicious use. By nature the reports are only "alerts" and, for example, spamming the service with incorrect alerts should not pose danger to the users (drivers), but it might render the service unusable, or at the very least, lower the users' interest in the service. Possible solutions could be to allow "rating" of the alerts or to require a certain amount of alerts of the same type to be reported within an area before the alert is shown publicly.

In the first solution, the primary problem would be how to implement the feature in the application so that it does not distract the driver. On one hand, it is entirely possible that the feature would take the user's attention away from driving causing more accidents than the application would prevent. On the other hand, if the feature would be implemented on a separate web portal, it is unknown how many of the users would bother to report the alerts after arriving at their destination.

The problem with the seconds approach is that if there are only a few users that report alerts using a road, the reported alert may never be published, or it might take a longer time before the reported alert becomes public causing potentially a larger number of road users to miss the alert. As the service does not allow anonymous access, rating alerts and users could (at least in theory) in time filter out users that post unnecessary or incorrect alerts. In the case of animals on the road, one way to verify a low number of reports could be done by making a cross-check with another database with details of animal density in that area to determine the chance for a such event to occur.

Information and guidelines for minimizing driver distraction have been released by [25][26]. Furthermore, there have been studies on user interface and system response design in the context of minimizing driver distraction [27]. However, the guidelines are only applicable to original equipment electronic devices, and do not cover aftermarket or portable devices. This means that because there is less control over mobile applications in official requirements, the applications can be organized in a more liberal way, possibly even in a distracting manner to the driver. Thus, it can be easy to overlook traffic safety aspects, especially if the recommendations would interfere with the designer's vision for the look'n feel of the application.

In conclusion, it can be said that there remain several currently unresolved and challenging tasks. Regardless, the authors of this paper believe that more could be achieved for the goals of averting accidents and reporting road conditions through better integration of road users' feedback potential into the overall road ecosystem.

VI. SUMMARY

Despite the progress in the development of automatic means to detecting road conditions, there are still cases where humans are more capable of detecting the events faced while travelling. Yet, the commonly used navigation devices and applications seldom offer any means for the user to provide feedback on the accuracy of the road information.

To better realize the potential of the road users in reporting road conditions, easy to use applications that the users can take advantage are required. This paper presented an example of an application the road users can utilize to report events faced while travelling by car. The high-level architecture utilized in the overall service was briefly introduced in combination of the data used to describe the alerts. Additionally, this paper provided a discussion on the challenges and unresolved issues identified while designing the application.

REFERENCES

[1] S. C. Radopoulou and I. Brilakis, "Improving road asset condition monitoring," Transportation Research Procedia, vol. 14, pp. 3004-3012, June 2016.

[2] R. Omer and L. Fu, "An Automatic Image Recognition System for Winter Road Surface Condition Classification," 2010 13th International IEEE Conference on Intelligent Transportation Systems (ITSC), Madeira Island, Portugal, pp. 1375-1379, September 2010.

[3] S. Sivaraman and M. M. Trivedi, "Active learning for on-road vehicle detection: a comparative study," Machine Vision and Applications, Vol. 25, Issue 3, pp. 599-611, April 2014.

[4] S. Sivaraman and M. M. Trivedi, "Looking at Vehicles on the Road: A Survey of Vision-Based Vehicle Detection, Tracking, and Behavior Analysis," IEEE Transactions on Intelligent Transportation Systems, Vol. 14, Issue 4, pp. 1773-1795, December 2013.

[5] Y.-Y. Nguwi and A.Z. Kouzani, "A Study on Automatic Recognition of Road Signs," IEEE Conference on Cybernetics and Intelligent Systems, Bangkok, Thailand, pp. 1-6, June 2006.

[6] H. Keskikiikonen, "The possibilities of crowdsourcing in travel surveys," Master's Thesis, 2014.

[7] S. Kytö, "Quantifying car traffic fluency with geospatial methods: Case Helsinki metropolitan area," Master's Thesis, 2016.

[8] H. Mattila, "Real-time estimation of traffic conditions on road sections [Linkkikohtaisen liikennetilanteen ajantasainen arviointi]," Helsinki 2003. Finnish Road Administration. Finnra Reports 61/2003. 175 p. + app. 13 p. ISSN 1459-1553, ISBN 951-803-179-7, TIEH 3200848-v.

[9] C. Kyriazopoulou, "Smart city technologies and architectures: A literature review," 2015 International Conference on Smart Cities and Green ICT Systems (SMARTGREENS), pp. 1-12, May 2015.

[10] E. Z. Tragos et al., "Enabling reliable and secure IoT-based smart city applications," 2014 IEEE International Conference on Pervasive Computing and Communication Workshops (PERCOM WORKSHOPS), Budapest, Hungary, pp. 111-116, 2014.

[11] C. Doukas and F. Antonelli, "A full end-to-end platform as a service for smart city applications," 2014 IEEE 10th International Conference on Wireless and Mobile Computing, Networking and Communications (WiMob), Larnaca, Greece, pp. 181-186, 2014.

[12] W. Li, J. Chao and Z. Ping, "Security Structure Study of City Management Platform Based on Cloud Computing under the Conception of Smart City," 2012 Fourth International Conference on Multimedia Information Networking and Security, Nanjing, China, pp. 91-94, 2012.

[13] J. Wan, D. Li, C. Zou and K. Zhou, "M2M Communications for Smart City: An Event-Based Architecture," 2012 IEEE 12th International Conference on Computer and Information Technology, Chengdu, China, pp. 895-900, 2012.

[14] K. M. Alam, M. Saini and A. E. Saddik, "Toward Social Internet of Vehicles: Concept, Architecture, and Applications," in IEEE Access, vol. 3, pp. 343-357, 2015.

[15] A. Ghose, P. Biswas, C. Bhaumik, M. Sharma, A. Pal and A. Jha, "Road condition monitoring and alert application: Using in-vehicle Smartphone as Internet-connected sensor," 2012 IEEE International Conference on Pervasive Computing and Communications Workshops, Lugano, Switzerland, pp. 489-491, 2012.

[16] Vionice Ltd, "RoadAI - Record, View, Share & Benefit," 2016. Retrieved February 6, 2017 from https://vionice.fi/en/service/2

[17] Wikipedia, "Beat the Traffic," 2017. Retrieved March 31, 2017 from https://en.wikipedia.org/wiki/Beat_the_Traffic

[18] Wace Mobile, waze, 2017. Retrieved March 31, 2017 from https://www.waze.com

[19] Inrix, 2017. Retrieved March 31, 2017 from http://inrix.com

[20] HERE, "Public pilot of HERE-powered NordicWay Coop launched in Finland". Retrieved March 31, 2017 from http://360.here.com/2016/06/02/public-pilot-of-here-powered-nordicway-coop-launched-in-finland

[21] A. Barón and P. A. Green, "Safety and usability of speech interfaces for in-vehicle tasks while driving: a brief literature review," Technical Report UMTRI-2006-5, University of Michigan Transportation Research Institute, Ann Arbor, Michigan, USA, 2006.

[22] Z. Hua and W. L. Ng, "Speech recognition interface design for in-vehicle system", Second International Conference on Automotive User Interfaces and Interactive Vehicular Applications (AutomotiveUI 2010), Pittsburgh, Pennsylvania, USA, pp. 29-33, 2010.

[23] J. C. Chang, A. Lien, B. Lathrop and H. Hees, "Usability evaluation of a Volkswagen Group in-vehicle speech system," 1st International Congerence on Automotive User Interfaces and Interactive Vehicular Applications (AutomotiveUI '09), Essen, Germany, pp. 137-144, 2009.

[24] K. M. Bach, M. G. Jæger, M. B. Skov, N. G. Thomassen, "You can touch, but you can't look: interacting with in-vehicle systems," SIGCHI Conference on Human Factors in Computing Systems (CHI '08), Florence, Italy, pp. 1139-1148, 2008.

[25] Department of Transportation, "Visual-Manual NHTSA Driver Distraction Guidelines For In-Vehicle Electronic Devices," 2012.

[26] European Transport Safery Council, "'PRAISE': Minimising In-Vehicle Distraction," 2010.

[27] J. Sonnenberg, "Service and user interface transfer from nomadic devices to car infotainment systems," In Proceedings of the 2nd International Conference on Automotive User Interfaces and Interactive Vehicular Applications (AutomotiveUI '10). ACM, New York, New York, USA, pp. 162-165, 2010.

Survey of Prototyping Solutions Utilizing Raspberry Pi

M. Saari*, A. Muzaffar bin Baharudin ** and S. Hyrynsalmi*

* Tampere University of Technology/Pori Department, Pori, Finland
** Keio University/Graduate School of Media and Governance, Yokohama, Japan
mika.saari@tut.fi

Abstract Sensor networks are a highly researched application area in the field of Internet of Things (IoT). A key cost and resource question in the development of IoT network sensor solutions is prototype implementation. In this study, the Raspberry Pi—a widely used single board computer—is investigated as it is one of the most commonly used prototyping devices available and is also widely used in scientific research. In this paper, we address which technologies, the usefulness and what kinds of issues arise when the prototyping of a sensor network solution is done with Raspberry Pi. The extant literature is studied by selecting papers with the systematic literature review method. Based on an extensive survey of the selected studies, we found several sensor-based implementations where Raspberry Pi has been used. In addition, this survey revealed subjects, such as e-health and education, which expanded the research topic in new ways. Further research opportunities have been identified in specifying the usefulness of various technologies with single board computers.

I. INTRODUCTION

The Internet of Things (IoT) is the expansion of Internet services, which connects everyday physical objects to a network. This connection between a network and physical objects makes it possible to access remote sensor data and to control the physical world from a distance. The first mention of the term 'IoT' is said to have come from Kevin Ashton in 1999. A survey of the areas of "Internet of Things" was made by Atzori, Iera and Morabito [1]. In this research, the focus has been redirected toward the Wireless Sensor Network (WSN) type of solution. The basic features of sensor networks were compiled in a survey conducted in 2002 [2].

'Sensor networks' refer to distributed autonomous sensors that are used to monitor the physical environment, e.g., temperature or pressure. Sensor networks are a widely-studied area. For example, the use of embedded Linux for sensor networks has been proposed [3] and a simple model of a sensor network has been introduced earlier [4]. There has also been research on long-range wireless sensor networks with geolocation tracking [5] and on a low energy algorithm for a sensor network [6]. Sensor networks have several development possibilities, such as the one introduced in Fog the gateway unpublished [7] study.

Single-board computers, such as Intel Galileo, BeagleBone and Raspberry Pi, are low-cost development devices for testing or educational purposes. These are fully customizable and programmable, and have the features required to implement small and low-cost IoT devices [8]. The Raspberry Pi is the most popular of these three in the field of research according to the keyword search. Single-board computers are an rising technology area in the development of prototypes. Often, developing a prototype is experienced as difficult and expensive due to the costs of hardware design, software design and developing as well as hardware manufacturing and building. However, by using single-board computers, these costs can be easily reduced. Ready-to-use hardware solutions already exist, such as Raspberry Pi, which have ready-to-use software with embedded Linux. Furthermore, there are many communities and user groups available online where a developer could ask for help and support.

Despite the potential value that single-board computers could provide, we are currently lacking a good overall picture of the studied and tested information on the real benefits and drawbacks brought about by the use of single-board computers. The objective of this study is to fill this research gap by mapping the current state of the art in the use of single-board computers in prototyping sensor networks. Thus, we seek answers to the following research questions:

RQ1: What do we know about the benefits and limitations of using the single-board computer Raspberry Pi in supporting prototyping work?

RQ2: How is the functionality in these single-board systems tested?

RQ3: Are there any specific test methods?

To answer these questions, we performed a literature study. We used the Systematic Literature Review approach (SLR) [9] to collect primary studies regarding this topic. The selected primary studies were then analyzed and categorized with the content analysis method.

The rest of this paper is structured as follows. In Section II, we introduce the research approach used in this literature study. In Section III, we present the analysis of the findings. Section IV includes a discussion and suggestions for future research on the topic and finally, the study is summarized in Section V.

II. RESEARCH APPROACH

As previously mentioned, to answer the presented research questions, we decided to perform a literature study in order to map the extant knowledge on this domain. We decided to use the SLR method for collecting relevant primary studies and followed the guidelines given by Kitchenham and Charters [9].

For the SLR, we decide to do an electronic search. The database used was IEEE Xplore Digital Library, the search engine of which was used in this study. The survey was started by selecting the following search terms: "Raspberry Pi" AND "Internet of Things" AND "Sensor Networks." We decided to use these simple keywords in order to receive good coverage of potential primary studies.

The keywords were commonly used alone: "Raspberry Pi" returned over 500 hits, "Internet of Things" returned over 12,000 hits, and the most popular was the third keyword "Sensor Network," returning over 110,000 hits. Together the keywords returned 11 (IEEE), and 1 (ACM) results. These were only the keyword search. The full text search gave too many hits, over 400, for this research. Nevertheless, this result combined with the search keyword "smart home" gave us 24 hits. The final searches were targeted to keywords and limited only to research papers.

In the first phase of the research, we selected studies based on abstracts. We used the following inclusion and exclusion criteria: Peer-reviewed studies conference and journal articles as well as book sections written in English focusing on all three aspects were included. We excluded studies written in languages other than English, posters, abstracts, and short papers, as well as studies only mentioning the keywords but not focusing on the issue.

In the second phase of the research, the selected studies were read through carefully. In this phase, we still excluded studies unless they focused on the development of sensor networks with Raspberry Pi. In the end, we selected 15 primary studies for inclusion in this study. The selected studies were read through and analyzed. We studied which technologies were used, what kinds of issues were faced, how the testing was reported, and whether there were any problems in testing or not. Finally, the notes were reviewed and the results were analyzed.

In the remaining sections of this paper, we will first present the selected key studies in the section III. This is followed by a discussion section where we answer the presented research questions.

III. RESULTS

In total, we selected 15 relevant primary studies for this paper. The primary studies found and selected were [10-24]. In the following, we will briefly summarize the key primary studies and their findings.

Baranwal, Nikita and Pateiya [10] presents a monitoring system for detecting and preventing rodents in grain stores. The system consists of a webcam, a repeller for undesirable rodents, and Raspberry Pi with a set of sensors, which were an Ultrasonic Range Detector (URD)

and a Passive Infrared sensor (PIR). The algorithm for the software was introduced. The test cases consisted of functionality tests where the hardware and software were tested. The test results showed that the observation distance was small, seven centimeters. In addition, the reliability of the system was tested. This test shows that 85 percent of system notifications were real. The 15 percent of tests that were unsuccessful were due to the connectivity of the device, data transmission, notification, and other factors such as PIR sensors being configured to generate discrete values.

Two studies [11] and [12] introduced the developed IoT-based E-learning testbed developed on the basis of a combination of five Raspberry Pis and a microwave sensor. The testbed controlled several factors: Chair Vibrator Control, Light Control, Smell Control, Sound Control, and Remote Control Socket. The purpose was to improve and stimulate the e-learner's motivation by using this testbed. The study introduced the usage of Optimized Link State Route protocol (OLSR) technology in the testbed software. The first study introduced the idea and the second study from the same authors handled the same issue more profoundly. In the second study, the testbed network communication was tested and the results were shown. The testbed usage for improving and stimulating e-learner's motivation was not tested. Also, the functionality of the testbed was not tested apart from the communication protocol.

Mahmoud and Qendri [13] introduced a sensor shield, the Sensorian platform, for Raspberry Pi. The aim was to transform Raspberry Pi into an IoT platform. This study could be categorized as hardware development. The shield consists of sensors: light, accelerometer, temperature, pressure, touchpad. It also includes a TFT display, LEDs, a real clock, and memory for software development. This shield was developed by means of crowdfunding. However, the functionality tests were not presented. It was mentioned regarding software testing that the firmware had been tested with the Raspbian operating system, but no test cases or results were mentioned.

References [14] and [15] focused on education on their research. The first study introduced the challenges and experiences of introducing IoT as an open elective course. The second focused on teaching Python programming. In these courses, Raspberry Pis and a set of sensors were used for teaching purposes. Students on this course built prototypes using the hardware mentioned. These studies did not handle testing the built prototypes and the focus was more on education than the other pieces of research in our study. However, these studies were included because of the good requirement specifications of the systems.

Maksimovic, Vujovic and Perisic introduced IoT-based e-health systems [16]. Their research also highlighted the economic impacts of IoT applications and especially the economic growth of healthcare applications. This research compares different applications: one is the e-health sensor shield V2.0 for Raspberry Pi and the other is a custom-made body sensor measuring system. Both enable data gathering and sending to the server application. This research does not have any special testing part but the research extensively points out the security issues of the gathered data.

The research by Hsiao, Liang and Sung [17] introduced a smart home system. This system uses infrared communication inside a room, Zigbee communication inside the house, and Wifi communication for data transfer to the cloud. The research focus was the combination of communication types in one system. The actual sensor or controller was not introduced. The features of these communication methods were also compared. A comparison of different means of communication was made at a general level, but no specific test cases were presented.

Hentschel, Jacob, Singer and Chalmers [18] introduced a smart campus system based on Raspberry Pis. This system uses hardware-software-service architecture, where the hardware consists of Raspberry Pis and sensors. This combination has software for collecting and sending data. The cloud has a service where the data are stored and served. This research presents sensor-to-sensor technology and delay-tolerant data transfer. This is for not-so-urgent data in the case of disrupted network connectivity. This research described several use cases of the system: Room temperature, free meeting room, room occupancy, custom event triggering, and robotic support infrastructure. These use cases are interesting but the physical test cases for them were not described. There were a few cases where the improvement of hardware design was introduced by changing the type of sensors.

As above, sensor network based systems were introduced in several studies [19-24]. These systems present the different ways to use master node - sensor node type solutions. Common to all these studies was the model of one master node and several sensor nodes. The sensor nodes collect the data and send it to the master node. The master node processes the data or sends it to the cloud service. These studies present systems from different angles or focus only one part of the system. In these studies, the test cases focused on functionality tests, communication test, or processing power tests. These tests usually support the main ideas of the studies and test cases which did not support them were dealt with.

IV. DISCUSSION

This study aims to resolve three research questions: What do we know about the benefits and limitations of using the single-board computer Raspberry Pi in supporting prototyping work (RQ1); how are the functionalities in these single-board systems tested (RQ2); and, finally, are there any specific test cases (RQ3)?

To answer the first research question, it can be said that there are a reasonable number of studies focusing on prototyping sensor networks with Raspberry Pi devices. However, most of the papers reported a single case study on the development of an interesting system. There is a clear lack of formalized approaches, methods, and tools.

The second research question dealt with the practices used in the testing of a prototyped Raspberry Pi sensor network solution. Again, only little has been reported on the testing practices utilized, problems faced, and

approaches used in the development of a sensor network or a module for it. As the testing of interconnections between the nodes in IoT networks is of utmost importance for the reliability of the system, the lack of studies is a worrying finding.

The third research question focused on whether specific test methods were used. This research shows that formal software testing was used only in the minority of research. For example, often it was only mentioned that the test cases were passed. Of course, the software parts are small, especially in sensor nodes, but if the developed system has some algorithms to process the gathered data, the software may have several functionalities. These should be tested in some way. One good test is: Will the software perform its functions within acceptable time? Data validation tests were used in a minority of studies. Validation might be important when a system uses the gathered data or the results of processed data in some way.

The results of the RQs raised several new research topic ideas. One possible future research topic that this paper does not analyze in depth is embedded operation systems. There are several types of single board computers on the market and usually each device has its own operating system. These operation systems are mostly Linux-based. An interesting topic would be the variety or modifications of these Embedded Linux components when the target is the increase of processing power.

The second research topic focuses on reliability. The Raspberry Pi based prototypes are usually connected using an experiment board and soldering of connections is not common. This came up from the studies explored. During our previous studies [3], [4] and [5], connection faults were common and it was attempted to prevent them by soldering all possible connections. However, the articles explored in this study did not commonly handle these reliability issues. Only a few of them even mentioned these problems.

Another interesting topic is recovery from faults. When we have noticed that the reliability might be an issue, so we should think about how to recover from the fault and which kinds of faults we could recover. Power failure is one common fault. However, only a few of the studies have handled this situation and the recovery process from power failure. In particular, the Raspberry Pi and its generally used Linux-based operating system are vulnerable to this kind of fault.

These selected studies tested the developed systems in various ways. This research shows the lack of systematic testing procedures of the systems of this kind. If there are systematic ways of testing Raspberry Pi based systems, those were not used. This might be also an interesting future research topic.

V. CONCLUSIONS

This survey showed that the Raspberry Pi is a widely-used device in research implementations of different kinds. The Raspberry Pi is an inexpensive, fully customizable, and programmable credit card-sized single board computer, which supports a large number of peripherals and network communication. Therefor the Raspberry Pi is suitable for small scale prototype testing. In this paper research on sensor network solutions were focused on using a literature review. This paper identified three research questions, which were answered using the systematic literature review approach. The answer of RQ1 showed a lack of formalized approaches, methods, and tools. RQ2 highlighted the minimal use of testing practices and methods, and the third RQ tried to find specific ways of using test methods. Some methods were found: software testing, software performance testing, and validation of data tests. Some further research topics were also identified. These include modifications to the embedded operating system for better performance, reliability, or fault recovery.

VI. ACKNOWLEDGMENT

This study was performed in intensive collaboration between the Tampere University of Technology (TUT) in Finland and Keio University in Japan. The survey was carried out at the TUT Pori department in December 2016.

REFERENCES

[1] Atzori, L., Iera, A., & Morabito, G. (2010). The Internet of Things: A survey. Computer Networks, 54(15), pp. 2787–2805.

[2] Akyildiz, I. F., Su, W., Sankarasubramaniam, Y., & Cayirci, E. (2002). Wireless sensor networks: a survey. Computer Networks, 38(4), pp. 393–422.

[3] Saari, M., Sillberg, P., Rantanen, P., Soini, J., & Fukai, H. (2015). Data collector service - practical approach with embedded linux. 38th International Convention on Information and Communication Technology, Electronics and Microelectronics (MIPRO) pp. 1037–1041.

[4] Saari, M., Muzaffar, A., Sillberg, P., Rantanen, P., & Soini, J. (2016). Embedded Linux controlled sensor network. 39th International Convention on Information and Communication Technology, Electronics and Microelectronics, MIPRO 2016 - Proceedings.

[5] Muzaffar, A., & Yan, W. (2016). Long-Range Wireless Sensor Networks for Geo-location Tracking : Design and Evaluation. In 18th International Electronics Symposium (IES), Bali, Indonesia, 29-30th September 2016. Bali, Indonesia.

[6] Muzaffar, A., Saari, M., Sillberg, P., Rantanen, P., Soini, J., & Kuroda, T. (2016). Low-energy algorithm for self-controlled Wireless Sensor Nodes. In the 2016 International Conference on Wireless Networks and Mobile Communications (WINCOM) pp. 42–46. IEEE.

[7] Muzaffar, A., Saari, M., Sillberg, P., Rantanen, P., Soini, J., Jaakkola, H., & Yan, W. (2017). Development of Fog Gateways for Cloud/Crowd System Back-End. Submitted

[8] Vujović, V., & Maksimović, M. (2015). Raspberry Pi as a Sensor Web node for home automation. Computers & Electrical Engineering, 44, pp. 153–171.

[9] Kitchenham, B.A., Charters, S. (2009): Guidelines for performing systematic literature reviews in software engineering. version 2.3. EBSE Technical Report EBSE-2007- 01, Keele University, Keele, Staffordshire, United Kingdom

[10] Baranwal, T., Nitika, & Pateriya, P. K. (2016). Development of IoT based smart security and monitoring devices for agriculture. 6th International Conference - Cloud System and Big Data Engineering (Confluence) pp. 597–602.

[11] Yamada, M., Oda, T., Matsuo, K., & Barolli, L. (2016). Design of an IoT-Based E-learning Testbed. In 2016 30th International Conference on Advanced Information Networking and Applications Workshops (WAINA) pp. 720–724.

[12] Yamada, M., Oda, T., Liu, Y., Matsuo, K., Ikeda, M., & Barolli, L. (2016). Performance Evaluation of an IoT-based e-Learning Testbed Considering OLSR Protocol in a NLoS Environment. In 2016 19th International Conference on Network-Based Information Systems (NBiS) pp. 451–457.

[13] Mahmoud, Q. H., & Qendri, D. (2016). The Sensorian IoT platform. In 2016 13th IEEE Annual Consumer Communications Networking Conference (CCNC) pp. 286–287.

[14] Raikar, M. M., Desai, P., & Naragund, J. G. (2016). Active Learning Explored in Open Elective Course: Internet of Things (IoT). In 2016 IEEE Eighth International Conference on Technology for Education (T4E) pp. 15–18.

[15] Guerra, H., Cardoso, A., Sousa, V., Leitão, J., Graveto, V., & Gomes, L. M. (2015). Demonstration of programming in Python using a remote lab with Raspberry Pi. In 2015 3rd Experiment International Conference (exp.at'15) pp. 101–102.

[16] Maksimović, M., Vujović, V., & Perišić, B. (2015). A custom Internet of Things healthcare system. In 2015 10th Iberian Conference on Information Systems and Technologies (CISTI) pp. 1–6.

[17] Hsiao, S. J., Lian, K. Y., & Sung, W. T. (2016). Employing Cross-Platform Smart Home Control System with IOT Technology Based. In 2016 International Symposium on Computer, Consumer and Control (IS3C) pp. 264–267.

[18] Hentschel, K., Jacob, D., Singer, J., & Chalmers, M. (2016). Supersensors : Raspberry Pi Devices for Smart Campus Infrastructure.

[19] Scott, G., & Chin, J. (2013). A DIY approach to pervasive computing for the Internet of Things: A smart alarm clock. In 2013 5th Computer Science and Electronic Engineering Conference (CEEC) pp. 57–60.

[20] Kruger, C. P., Abu-Mahfouz, A. M., & Hancke, G. P. (2015). Rapid prototyping of a wireless sensor network gateway for the internet of things using off-the-shelf components. In 2015 IEEE International Conference on Industrial Technology (ICIT) pp. 1926–1931.

[21] Osorio, F. G., Xinran, M., Liu, Y., Lusina, P., & Cretu, E. (2015). Sensor network using Power-over-Ethernet. In 2015 International Conference and Workshop on Computing and Communication (IEMCON) pp. 1–7.

[22] Lukas, Tanumihardja, W. A., & Gunawan, E. (2015). On the application of IoT: Monitoring of troughs water level using WSN. In 2015 IEEE Conference on Wireless Sensors (ICWiSe) pp. 58–62. IEEE.

[23] Thaker, T. (2016). ESP8266 based implementation of wireless sensor network with Linux based web-server. In 2016 Symposium on Colossal Data Analysis and Networking (CDAN) pp. 1–5.

[24] Griffiths, N., & Chin, J. (2016). Towards unobtrusive ambient sound monitoring for smart and assisted environments. In 2016 8th Computer Science and Electronic Engineering (CEEC) pp. 18–23.

Evaluating Robustness of Perceptual Image Hashing Algorithms

Andrea Drmic, Marin Silic, Goran Delac, Klemo Vladimir, Adrian S. Kurdija
University of Zagreb, Faculty of Electrical Engineering and Computing, Zagreb, Croatia
andrea.drmic@fer.hr, marin.silic@fer.hr, goran.delac@fer.hr, klemo.vladimir@fer.hr, adrian.kurdija@fer.hr

Abstract - In this paper we evaluate the robustness of perceptual image hashing algorithms. The image hashing algorithms are often used for various objectives, such as images search and retrieval, finding similar images, finding duplicates and near-duplicates in a large collection of images, etc. In our research, we examine the image hashing algorithms for images identification on the Internet. Hence, our goal is to evaluate the most popular perceptual image hashing algorithms with the respect to ability to track and identify images on the Internet and popular social network sites. Our basic criteria for evaluation of hashing algorithms is robustness. We find a hashing algorithm robust if it can identify the original image upon visible modifications are performed, such as resizing, color and contrast change, text insertion, swirl etc. Also, we want a robust hashing algorithm to identify and track images once they get uploaded on popular social network sites such as Instagram, Facebook or Google+. To evaluate robustness of perceptual hashing algorithms, we prepared an image database and we performed various physical image modifications. To compare robustness of hashing algorithms, we computed Precision, Recall and F1 score for each competing algorithm. The obtained evaluation results strongly confirm that P-hash is the most robust perceptual hashing algorithm.

I. INTRODUCTION

The image tracking has recently become an interesting problem since a significant number of images get uploaded on various social network sites instantly. For instance, the recent statistics show that Facebook handles over 350 million image uploads each day. Once the image gets uploaded on the social network, other users and influencers share the image, modify it and upload it as their own content. In case the image becomes very popular, it is very hard to distinguish who is the original author (source) without having an insight in social network private data.

In general, it is very hard to obtain such insider statistics since social networks have very strict policy about their data. A possible approach to track images online would be to utilize steganography [1]. However, this approach is very limited due to a fact that social network usually performs image compression prior to image upload which results in all hidden data being lost. The only promising solution to track images on various social networks from outside is to utilize image hashing algorithms.

The classic hashing algorithms used in cryptography such as MD5 and SHA1 are not suitable for this purpose since their aim is to hash entities with high dispersity. This means that similar entities are very likely to hash to quite different hashes. Even small changes in input will result in completely different hash value which is a good property of cryptographic hashing algorithm. On the other hand, we need a different family of hashing algorithms that will hash similar entities to similar hashes. Generally, the hashing algorithms having such property are called locality-sensitive hashing algorithms [2] [3]. More specifically, for image entities, *perceptual* image hashing algorithms are used, which possess the property of invariance for "perceptually similar" media objects x and x' [4].

In this paper, we analyze robustness property of perceptual image hashing algorithms. We consider the following algorithms: A-hash, D-hash, W-hash and P-hash. We conducted experiments in which we evaluated robustness property of mentioned image hashing algorithms. In our experiments, we measured robustness of image hashing algorithms regarding visible image modifications (1), and with the respect to images upload to popular social network sites: Facebook, Instagram and Google+ (2). To compare robustness of competing algorithms, we used well established measures: *Precision, Recall* and F_1 *score*. The evaluation results obtained from conducted experiments strongly suggest that P-hash algorithm is the most successful algorithm for image identification upon visible modification is applied or image is uploaded on social network.

The rest of the paper is organized as follows. Section II presents the related work. The competing hashing algorithms are described in section III, while the evaluation is presented in section IV. Section V concludes the paper.

II. RELATED WORK

The researchers have proposed a variety of image hashing algorithms that extract popular image features (i.e. HOG, DOG, SIFT, GIST etc.) as large high-dimensional vectors which are later reduced using some dimensionality reduction technique with the aim to detect near-duplicate images. For instance, in [5], the authors extract local features based on DOG for image representation, and then use locality sensitive hashing as the core indexing structure. In [6], spectral hashing is used by performing PCA to find the principal components of the data, and then the data is fit to a multidimensional rectangle. Similar as previous, the approach proposed in [7] finds the maximum variance direction using PCA, except that the original covariance matrix gets "adjusted" by another matrix arising from the labeled data. Instead of representing an image using a single feature vector, the approach proposed in [8] independently indexes large number of local descriptors derived from PCA-SIFT which results in approach highly resistant to occlusions and cropping. In [9], the authors use bag-of-words techniques for text analysis for creation of bag-of-visual-words using vector quantized local feature descriptors (SIFT), and they propose Min-Hash algorithm for retrieval of similar images. In addition, geometric image hashing [10] is proposed to improve standard Min-Hash by considering the spatial dependency among visual words. Further improvement over geometric Min-Hash is a hashing scheme for partial duplicate image discovery [11] i.e. finding groups of images in a large dataset that contain the same object. An interesting novel graph-based approach which automatically discovers the neighborhood structure inherent in the data is introduced in [12]. Kernelized locality sensitive hashing for scalable image search is proposed in [13]. More recent works propose deep learning frameworks to generate binary hash codes for fast image retrieval [14] [15].

All mentioned approaches are quite complex and computationally challenging since they usually create high dimensional image representation which is later reduced. However, our goal is to focus on perceptual image hashing algorithms that can produce fast image fingerprint and still preserve image identity.

III. PERCEPTUAL IMAGE HASHING ALGORITHMS

In this section, we briefly describe the following perceptual image hashing algorithms: A-hash, D-hash, P-Hash and W-hash.

A. A-Hash

Average hash (A-hash) is a perceptual image hashing algorithm that creates compact 64-bit image hashes by focusing on properties of image structure. As image is decomposed into its underlying harmonics, the higher frequencies represent image details, while the lower frequencies represent image structure. To make the image fingerprint as small as possible, the higher frequencies are eliminated by reducing the image in its size. Specifically, prior to computing the hash, the image is scaled down to an 8x8 block and it thus contains a total of 64 pixels. Each pixel is then converted to greyscale. Note that all perceptual image hashing algorithms employ this step since the essential semantic information is preserved in the luminance component of an image. Next, an average color for all 64 pixels is computed. Subsequently, the hash is constructed so that each bit representing a single pixel is set based on whether the color value of that pixel is below or above the computed image average.

B. D-Hash

Difference hash (D-hash) is a perceptual hashing algorithm that leverages image structure, much like in the A-hash approach. The hashing principle focuses on the image structure and it achieves so by reducing the image size, i.e. by removing higher frequencies from the image

spectrum. Unlike in the A-hash approach where the fingerprint was computed be averaging out the pixels, the D-hash approach tracks image gradient. Prior to hashing, each image is reduced to a 9x8 block and converted to gray scale, i.e. a total of 72 pixels. Then for each row, the difference between each two adjacent pixels is computed, a total of 8 differences per row. Thus, 64 differences are computed for each image and then subsequently used to construct the fingerprint. This is done by setting each bit based on the computed difference d. For instance, if d < 0, the hash bit is set to 0, and if d ≥ 0, the bit is set to 1.

C. P-Hash

The P-hash algorithm [4] is based on discrete cosine transformation (DCT). The algorithm produces a 64-bit length binary sequence as an image hash. First, the image is converted to a greyscale representation using its luminance. Next, a mean filter (i.e. smoothing, averaging or box filter) is applied to the image. To apply the filter, a kernel with dimension 7x7 is used. The kernel is applied using a special convolution function [4].

Once the convolution is applied, the image is resized to 32x32 pixels. To extract the hash, 64 low frequency coefficients are used except the lowest frequency coefficients are omitted. The low frequency coefficients are used because they are mostly stable under various image modifications. Moreover, most of the signal information is preserved in low frequency components of the DCT. The reason the lowest frequency coefficients are omitted is they tend to be quite different from others and can significantly throw off the average.

To produce the hash, DCT coefficients from (1, 1), which is the upper left corner of 64-size matrix, to DCT coefficient (8, 8), representing the lower right corner of the same matrix, are concatenated to an array of length 64. Next, the median m of coefficients array is computed. Finally, the hash is transformed into a binary form using the following procedure:

$$h_i = \begin{cases} 0, & C_i < m \\ 1, & C_i \geq m \end{cases} \tag{1}$$

where C_i is the ith coefficient in the constructed array, and h_i is the ith bit in a 64-bit length perceptual hash.

D. W-Hash

Wavelet hash (W-hash) algorithm is a perceptual image hashing algorithm that transforms the original problem into frequency domain. Similar as P-hash, W-hash utilizes frequency domain, but instead of discrete cosine transform, W-hash uses discrete wavelet transform (DWT). Wavelet transform represents a signal using wavelet functions with different locations and scales. Wavelets are particularly well suited for the representation of signals with the aggressive change in input signal, thus resulting in smaller amount of information in the frequency domain. DWT is often used to remove redundancy in a data with highly correlated neighboring values, such as pixels in images. DWT is successfully used for noise reduction [16], image compression [17], dimensionality reduction [18], EEG analysis [19] and audio signal analysis [20]. Basic W-hash algorithm used during the experiments works by first scaling the image followed by transformation of an image to frequency domain using the *Haar* wavelet (optionally removing the lowest low level frequency). Provided implementation of the W-hash algorithm is using the wavelet transform software for the Python programming language.

Due to the nature of the wavelet transformation, it is expected that W-hash will perform better on images with smaller amount of intense changes i.e. high contrast spatial data. Also, it is reasonable to assume that P-hash will perform better for images with more restrained spatial changes.

IV. EVALUATION

In this section, we evaluate the robustness of perceptual image hashing algorithms. The implementation we use is https://pypi.python.org/pypi/ImageHash. To evaluate the robustness property, we define two different experiments. In the first experiment, we evaluate the algorithms robustness with respect to various visible modifications applied to images. In the second experiment, we evaluate algorithms robustness regarding the ability to track images on the Internet. To be more specific, we uploaded, and subsequently downloaded, images on various social network sites such as: Facebook, Instagram and Google+.

A. Evaluating robustness on visible image modifications

In this experiment, our goal is to check which is the most robust algorithm when considering various visible image modifications. First, we collected an image dataset containing 1480 original images. We computed a local database containing hashes for each considered image hashing algorithm: A-hash, D-hash, W-hash and P-hash. Then, we performed various images modifications for all images in the dataset. More specifically, the following image modifications were applied: resizing, rotating, noise injecting, swirling, sharpening, adding border color, contrast-stretching, colorizing, thresholding and drawing. To introduce all mentioned modifications, we used a very popular open source tool *ImageMagick* [21] which exposes its functionalities using Python API. The modification parameters were chosen randomly under the appropriate constraints. All together, we created an amount of 8868 modified images. We introduced the appropriate naming convention for modified images so we can easily track which is the original image used to obtain each modified image. For instance, if the original name was "100201.jpg", for modified image obtained by resizing the original image was named "100201_resized_N.jpg", where N is the index of modification created from the respected original image.

To evaluate the sole robustness of each image hashing algorithm, we used well established measures for this purpose: *precision*, *recall* and *F₁ score*. Precision measure is defined as:

$$Precision = \frac{TP}{TP + FP} \qquad (2)$$

where *TP* is the number of true positives and *FP* is the number of false positives. Recall measure is defined as:

$$Recall = \frac{TP}{TP + FN} \qquad (3)$$

where *FN* is the number of false negatives. Finally, the F_1 score is defined as follows:

$$F_1 = \frac{2 \times Precision \times Recall}{Precision + Recall}. \qquad (4)$$

Once we have the number of true positives, false positives and false negatives, we can easily compute measures for each competing hashing algorithm. To

```
for mod_img in modified_images:
    mod_hash = compute_hash(mod_img, alg)
    for org_hash, org_img in hashes_db:
        d = hamming_distance(mod_hash, org_hash)
        if mod_img.name.startsWith(org_img.name):
            if d <= N:
                TP = TP + 1
            else:
                FN = FN + 1
        else:
            if d <= N:
                FP = FP + 1
            else:
                TN = TN + 1
```

Figure 1 Procedure used during the evaluation for each algorithm

compute number of true positives, false positives and false negatives in our experiment, we try to reconstruct which is the original image, for each modified image. Specifically, for each modified image, we compute its hash. Then, we search the original images hashes database to see if the computed hash matches any of hashes in the database. To determine if the two hashes *A* and *B* are considered equal, we examine hashes' in binary representation and we compute their Hamming distance as follows:

$$d(A, B) = \sum |A_i - B_i|. \qquad (5)$$

To consider two hashes A and B equal, we require:

$$d(A, B) \leq N, \qquad (6)$$

where *N* is a system parameter which can be tuned for each environment and each algorithm. During the evaluation, we perform the procedure depicted in Figure 1.

Knowing the ground truth based on a modified image name, and having a Hamming distance for each modified image and each original image hashes pair, we can easily figure out if some hash pair (i.e. original image hash, and modified image hash) is a true positive, false positive, true negative or false negative. We repeat the depicted procedure for each algorithm, varying the parameter *N* in range from 0 to 30. It should be noted that the hash size for each algorithm is 64 bits.

The precision, recall and F1 score results for algorithms A-hash, D-hash, W-hash and P-hash are depicted in Figure , Figure , Figure 5, and Figure 66, respectively. As can be seen in the figures, the most successful algorithm is P-hash, which achieves F_1 *score* value of 0.738 at *Precision* value of 0.856 and *Recall* value of 0.649 for a distance threshold $N = 14$. The second most successful algorithm is D-hash, which reaches F_1 *score* value of 0.641 at *Precision* value of

0.745 and *Recall* value of 0.563 for a distance threshold N = 10. The third place is reserved for A-hash algorithm, which gains F_1 *score* value of 0.406 at *Precision* value of 0.431 and *Recall* value of 0.385 for a distance threshold N = 1. Finally, the worst performance is obtained by W-hash algorithm, which obtains F_1 *score* value of 0.271 at *Precision* value of 0.249 and *Recall* value of 0.297 for a distance threshold $N = 0$.

As already described in the beginning of this section, the modifications introduced in this experiment setup were quite aggressive. This means that it is relatively hard to associate the original and the modified image for some examples by sole human visual inspection. Figure presents an assembled collage of some randomly chosen examples of original images (marked with yellow border), and their modifications. The modifications whose original was successfully identified using P-hash algorithm at distance threshold $N = 14$ are marked with green border, while modified images whose original could not have been discovered are marked with red border.

One could also argue the effectiveness of various algorithms depends of what the goal function is. In case the aim is finding very similar images, P-hash algorithm may be considered too tolerant to aggressive image modifications, i.e. the results will contain some images which are not similar to the original image. In this case, more appropriate choice would be A-hash algorithm which is less tolerant to visible image modifications.

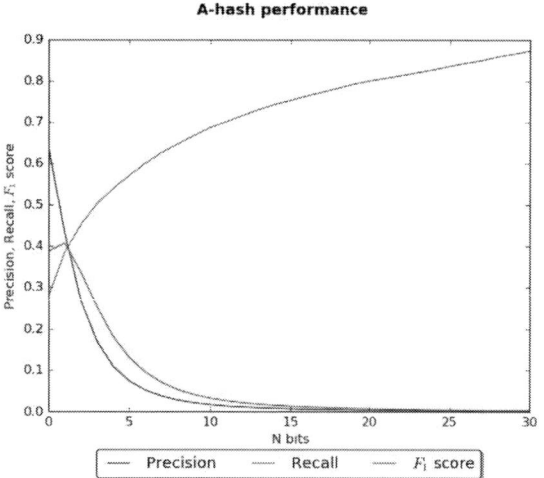

Figure 2 Examples of original images and their modifications

Figure 3 A-hash algorithm performance on modified images

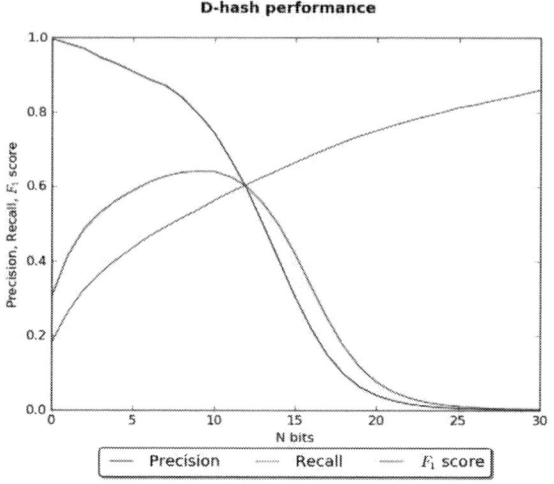

Figure 4 D-hash algorithm performance on modified images

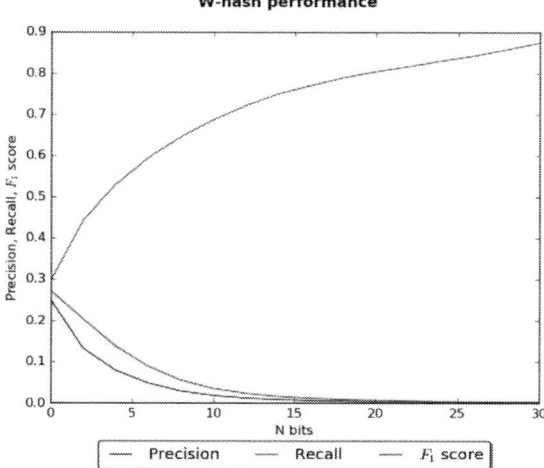

Figure 5 W-hash algorithm performance on modified images

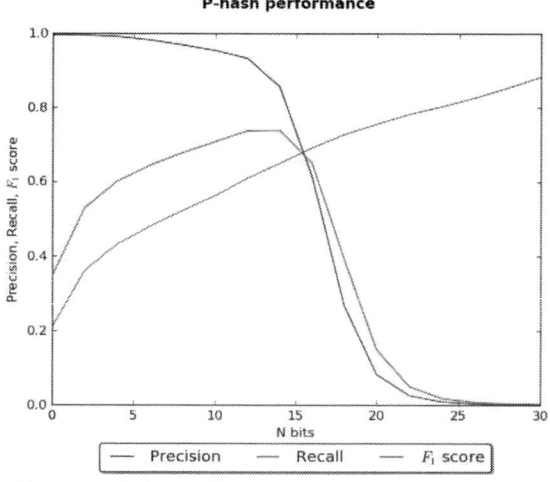

Figure 6 P-hash algorithm performance on modified images

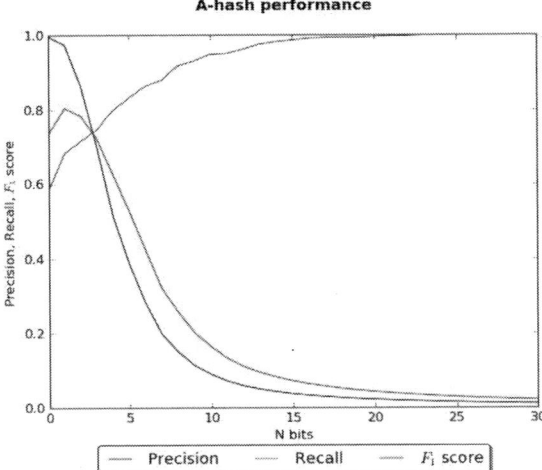

Figure 2 A-hash performance on social network dataset

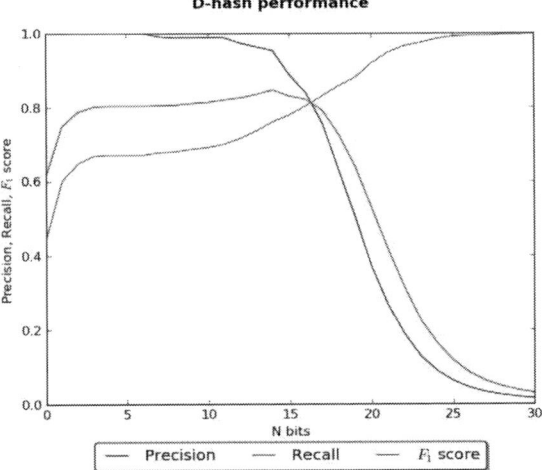

Figure 3 D-hash performance on social network dataset

B. *Evaluating robustness on images uploaded on social networks*

In this experiment, our goal is to analyze robustness on less aggressively modified images – images uploaded on social network sites. More specifically, in our experiment we included the following social network sites: *Facebook*, *Instagram* and *Google+*. It is well known that all considered sites provide users the ability to upload and share images as part of their profiles and feeds. However, while being uploaded the images are resized to dimensions dominantly used on uploading site (1), and the images are processed by *jpeg* compression engine which removes any irrelevant information (2). The resizing operation is quite likely to change the aspect ratio and the quality of the original image. Hence, it is likely that the image will change after being uploaded. Our goal is to determine the robustness of hashing algorithms with the respect to upload on social network sites.

We selected an amount of 150 original images from the dataset used in the previous experiment (see IV.A). Then, each image was uploaded to all considered social networks. After the upload, the images were downloaded from social networks and included to modified images dataset containing overall 450 images. In the same manner, as in the first experiment we created a database of hashes for each original image and each competing hashing algorithm. To measure the robustness for each competing algorithm, we used the same evaluation measures as in the first experiment: *Precision*, *Recall* and *F_1 score*. Furthermore, we used the same evaluation procedure described in pseudocode in shown Fig. 1.

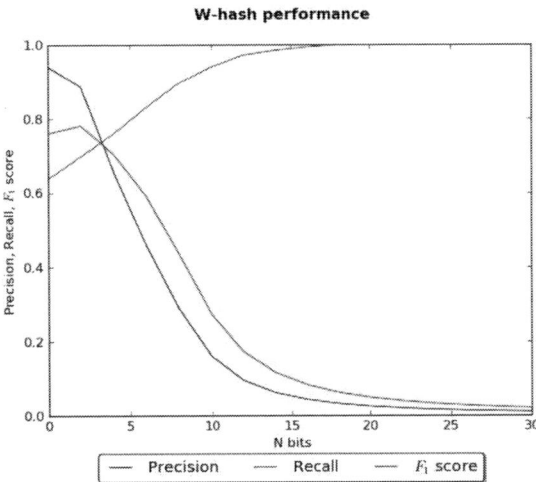

Figure 4 W-hash performance on social network dataset

The results for algorithms A-hash, D-hash, W-hash and P-hash are depicted in Figure 2, Figure 3, Figure 4, and Figure 5, respectively. Same as in the first experiment, the most successful algorithm is P-hash, which achieves *F_1 score* value of 0.864 at *Precision* value of 0.926 and *Recall* value of 0.81 for a distance threshold $N = 16$. The second most successful algorithm is D-hash, which reaches *F_1 score* value of 0.846 at *Precision* value of 0.952 and *Recall* value of 0.761 for a distance threshold $N = 14$. D-hash is followed by A-hash algorithm, which gains *F_1 score* value of 0.804 at *Precision* value of 0.974 and *Recall* value of 0.685 for a distance threshold $N = 1$. Finally, the worst performance is obtained by W-hash algorithm, which obtains *F_1 score* value of 0.781 at *Precision* value of 0.886 and *Recall* value of 0.698 for a distance threshold $N = 1$.

It is obvious that all algorithms perform significantly better on social network dataset when compared to the original image modification dataset. However, this behavior is expected since some of the introduced image modifications are quite intense, and it is hard for a human observer to associate modified image with its origin.

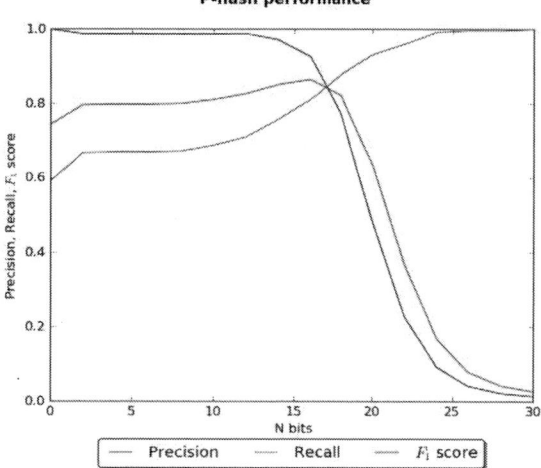

Figure 5 P-hash performance on social network dataset

V. CONCLUSION

In this paper, we studied robustness property of the following perceptual image hashing algorithms: A-hash, D-hash, W-hash and P-hash. To be precise, we analyze how robust image identification and tracking is with the respect to: visible physical image modifications (1), and image upload on social networks (2).

To assess robustness with the respect to visible image modifications, we created a dataset containing original images and their modifications. To evaluate robustness regarding upload on social network, we selected original images and uploaded them on the following social networks: Facebook, Instagram and Google+. To compare the performance of different algorithms, we used common measures: *Precision*, *Recall* and F_1 *score*.

The evaluation results, obtained on visible image modifications and both social networks uploaded dataset, strongly confirm that the most robust algorithm is P-hash which achieves F_1 *score* value of 0.738 on image modifications dataset, and F_1 *score* value of 0.864 on social networks uploaded dataset.

A detailed analysis of how a particular image hashing algorithm is robust with respect to a particular image modification is left for future work.

ACKNOWLEDGMENT

The authors acknowledge the support of the Croatian Science Foundation through the Recommender System for Service-oriented Architecture **(IP-2014-09-9606)** research project. The authors also acknowledge the support of Consumer Computing Laboratory of Faculty of Electrical Engineering and Computing at the University of Zagreb.

REFERENCES

[1] G. C. Kessler and C. Hosmer, "An overview of steganography," *Advances in Computers*, pp. 51-107, 2011.

[2] J. Leskovec, A. Rajaraman and J. D. Ullman, Mining of massive datasets, Cambridge University Press, 2014.

[3] K. Zhao, H. Lu and J. Mei, "Locality Preserving Hashing," *AAAI*, pp. 2874-2881, 2014.

[4] C. Zauner, Implementation and benchmarking of perceptual image hash functions, 2010.

[5] X. Yang, Q. Zhu and K.-T. Cheng, "MyFinder: near-duplicate detection for large image collections," in *ACM international conference on Multimedia*, 2009.

[6] Y. Weiss, A. Torralba and R. Fergus, "Spectral hashing," *Advances in neural information processing systems*, pp. 1753-1760, 2009.

[7] J. Wang, S. Kumar and S.-F. Chang, "Semi-supervised hashing for scalable image retrieval," in *Computer Vision and Pattern Recognition (CVPR)*, 2010.

[8] R. Datta, "Image retrieval: Ideas, influences, and trends of the new age," *ACM Computing Surveys*, 2008.

[9] O. Chum, "Near Duplicate Image Detection: min-Hash and tf-idf Weighting," *BMVC*, pp. 812-815, 2008.

[10] O. Chum, M. Perd'och and J. Matas, "Geometric min-hashing: Finding a (thick) needle in a haystack," in *Computer Vision and Pattern Recognition*, 2009.

[11] D. C. Lee, Q. KE and M. Isard, "Partition min-hash for partial duplicate image discovery," in *European Conference on Computer Vision*, 2010.

[12] W. Liu, "Hashing with graphs," in *International conference on machine learning*, 2011.

[13] B. Kulis and K. Grauman, "Kernelized locality-sensitive hashing for scalable image search," in *International Conference on Computer Vision*, 2009.

[14] K. Lin, "Deep learning of binary hash codes for fast image retrieval," in *Conference on computer vision and pattern recognition*, 2015.

[15] R. Xia, "Supervised Hashing for Image Retrieval via Image Representation Learning," in *AAAI*, 2014.

[16] M. Lang, H. Guo, J. E. Odegard, C. S. Burrus and R. O. Wells, "Noise reduction using an undecimated discrete wavelet transform," *IEEE Signal Processing Letters*, vol. 3, no. 1, pp. 10-12, 1996.

[17] S. Grgic, M. Grgic and B. Zovko-Cihlar, "Performance analysis of image compression using wavelets," *IEEE Transactions on industrial electronics*, pp. 682-695, 2001.

[18] L. M. Bruce, C. H. Koger and J. LI, "Dimensionality reduction of hyperspectral data using discrete wavelet transform feature extraction," *IEEE Transactions on geoscience and remote sensing*, pp. 2331-2338, 2002.

[19] H. Ocak, "Automatic detection of epileptic seizures in EEG using discrete wavelet transform and approximate entropy," *Expert Systems with Applications*, pp. 2027-2036, 2009.

[20] G. Tzanetakis, G. Essl and P. Cook, "Audio analysis using the discrete wavelet transform," in *Acoustics and Music Theory Applications*, 2001.

[21] "ImageMagick," ImageMagick Studio LLC, 2017. [Online]. Available: http://imagemagick.org/script/index.php.

[22] B. Wang, "Large-scale duplicate detection for web image search," in *International Conference on Multimedia and Expo*, 2006.

Adaptive models for security and data protection in IoT with Cloud technologies

N. Kakanakov* and M. Shopov*

* Technical University of Sofia branch Plovdiv, Dept. Computer Systems and Technologies, Bulgaria
{kakanak, mshopov}@tu-plovdiv.bg

Abstract – The paper presents an example Sensor-cloud architecture that integrates security as its native ingredient. It is based on the multi-layer client-server model with separation of physical and virtual instances of sensors, gateways, application servers and data storage. It proposes the application of virtualised sensor nodes as a prerequisite for increasing security, privacy, reliability and data protection. All main concerns in Sensor-Cloud security are addressed: from secure association, authentication and authorization to privacy and data integrity and protection. The main concept is that securing the virtual instances is easier to implement, manage and audit and the only bottleneck is the physical interaction between real sensor and its virtual reflection.

I. INTRODUCTION

The wide use of intelligent devices at home, industry, transportation, medicine and education lead to new challenges in technologies of data protection, security and privacy. Many researchers and businesses search for effective and reliable way for collection of data and statistics from internet-based objects, services, people, information without crossing the border of privacy and security of the people and society. Some of the new challenges in front of the intelligent connected object consists of: secure and reliable association of autonomous devices; secure and protected communication in uncontrolled environment; trust on services, people or devices; authentication of users and connecting them to their devices, data and services without compromising their privacy. There are some solutions on the market that can be applied as RFID (Radio Frequency Identification), NFC (Near Field Communication), SSO (Single Sing-On), OAuth (Open Authentication) but none of them is widely adopted or applicable in all areas of Sensor-Cloud application. The new model of Internet which encompasses of Internet of Thing (IoT), Internet of Data, Internet of Services and Internet of People (social networks) requires a security model that address all types of instances in the architecture. the producers of intelligent devices and providers of services for them are applying different technologies to protect and secure the information but often without success or with many trade-offs. The challenge is bigger when you need to protect personal data but making it available for research and statistics. This is crucial for electronic public services as e-Health, e-Government, Intelligent Transportation Systems (ITS), Smart Home and Smart Cities.

Development and research of adaptive models for applying security and data protection could provide a base for systematic analyses and application of best practices and security techniques in variety of use cases. This could provide environment to look for general and standardized security decisions that suit the high demand of dynamic internet-based applications and smart objects.

II. RELATED WORK

A. IoT models and virtual devices

The models for development of IoT are a subject of research of many authors. Most of proposed models are based on a layered approach with varying number and varying functionality of layers. The most commonly used layers are [1, 2, 3, 4, 5]: Hardware, Middleware and Application layers.

Hardware layer (object/perception layer) – main tasks of this layer are identification of devices, interaction with environment (sensing and actuating) and data preparation. Some authors define this layer as wireless sensor network with heterogeneous devices – various interfaces and communication media and unstructured or semi-structured data [6]. Their main characteristics are limited processing and storage resources, energy constrains and autonomous living. Due to the restriction of the objects data processing on this layer is limited.

Middle-ware layer – the main task on this layer is reliable and secure transportation of the data from hardware layer to the data processing service or unit. It includes managing the data formats, filtering and adding context (data meaning, location, local conditions, precision, units of measurement, etc.). In common case this layer includes sub-layers – e.g. object-abstraction, service management [2] and service composition [5]. Object-abstraction sub-layer [7] could be used to announce and present the functions available on hardware layer to the upper services. In [8] and [9] such layer is called Virtualized Object (VO) and store data for current and previous states of a physical devices from the hardware layer. The main advantage of using VO as a virtual copy of a physical devices are: device and services lookup; development of mash-up measurements, integration of data sources for complex analyses, energy consumption optimization, abstracting the heterogeneity of interfaces and communication protocols, scalability [10].

National Science Fund of Bulgaria, project number E02/12

Application layer – the main task on this layer is to present and provide services and functionality to the end users. It may include some data processing, data mining, filtering, and in most cases primitives and services of data presentation and statistical analyses of the data. It should use the functions of the middle-ware to present structured and processed information to the end user based on the data from multiple sources to support decision making.

Other authors [11] discuss the benefits of using virtual sensor in sensor-cloud architecture to cope with communication heterogeneity, geographical distribution and simultaneous multi-user access. Virtual sensors provide data series as the physical ones but can easily change context and omit some physical properties (e.g. manufacturer, battery status, network address).

B. Main security issues

Most of the research on IoT security propose an application of common protection methods on common security problems focusing on communication security. This addresses important threats like: network access control, device authentication, data confidentiality and integrity, but is limited on RFID, Wireless Sensor Networks (WSN) and 6LoWPAN [12], [13].

But addressing IoT and Cloud some specific threats should be taken in mind [14]:

- Secure hardware initialization, Secure boot and OS initialization;
- Secure update and reconfiguration;
- Authorized association to networks;
- Trust between peers;
- Access control;
- Secure communication;
- Internal Data protection;
- Denial of service and energy waste.
- Depersonalization of data but keeping the context.

In [11] it is suggested that the IoT to Cloud architectures should have three main pillars of protection: trust, security and privacy. The trust should be based on: lightweight Public Key Infrastructure (PKI) or decentralized self-configuring key management; metadata to control the quality of information; novel methods for assessing trust in users, devices and data; policies to control the usage of data; and methods to assure the trust in hardware and software platforms used. Security should be integrated in the data model used but need to be expanded to provide defense to IoT specific Denial-of-Service (DoS) attacks and provide reliable security monitoring and audit. The device producers should provide security reconfiguration through all device life-cycle and even provide intelligent self-configuration. Privacy could rely on encryption but it is more appropriate to keep data minimized according to the specific use (e.g. removing user/device location if not used, removing, processing and filtering information as local as possible to

reduce its leakage) and take advantage of the use of soft identities (user can use different soft identities for different applications and services) [11].

Some authors [15] suggest that IoT security should be network-based not host-based as the devices have limited resources. The proposition is to use Software defined networking (SDN) for dynamic adaptation of network security to match the application and context requirements. The reconfiguration could be managed by using hierarchical control structure with logical partitioning based on device interactions.

III. PROPOSED ARCHITECTURE

Taking in mind the benefits of using virtual devices for scalability and flexibility and considering different security issues at a different step of transferring data from devices to the Cloud and vice-versa, we propose the following architecture for integration of IoT and Cloud - "Fig. 1".

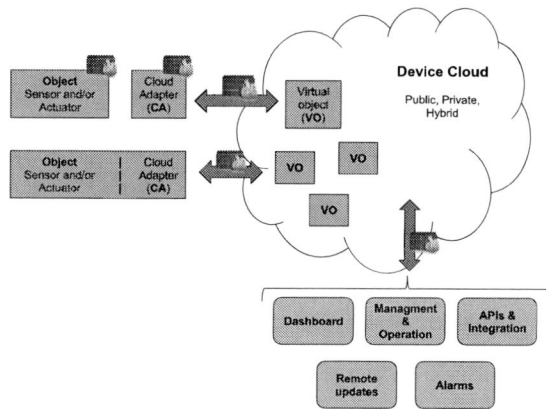

Figure 1. Device-Cloud architecture with virtual devices

The device-cloud architecture is complex and comprises of many interactions which leads to many points that should be secured – devices should be securely paired to local gateway (or cloud adapter); each physical device should be securely mapped to its virtualized entity; communication from device to gateway should be protected; communication from gateway to the cloud should be protected; data in the cloud should be kept in privacy; users should be authenticated in the cloud and their access rights should be defined. None of these elements can be underestimated as the whole system is as secure as its weakest element.

The first level of protection in such architecture is based on the separation between user-to-cloud and device-to-cloud data exchange. This will significantly reduce the low level DoS attacks on the devices which is considered as one of the main issues in battery base IoT objects. The second simple additions is the use of firewall at cloud adapter gateway and every entrance point of the cloud. The third thing is keeping the device network private and secure – authentication and encrypted messages should be

1002

used. And the last pillar of secure architecture is the simplification of communication channels – physical devices will pair only with its virtual instances. Thus, it will be easier to monitor communication and provide trust.

Nevertheless, keeping the device protected can be achieved by minimizing the communication from Cloud to it. Device can periodically request configuration and/or command from its virtual entity. In such case physical devices will not need access control as they always will be clients in communication. With one way communication the use of asymmetric encryption is possible which will increase reliability.

One specific situation must be considered – the cloud security rely on the isolation between instances but in the case of virtual devices some grouping should be considered as virtual devices often work in groups (sensor networks). If virtual devices exchange data and messages between each other, it could be done using internal sharing in the cloud. This will increase the performance but reduce isolation.

Additionally, in complex applications in the Cloud and IoT domain many services and devices from different vendors and providers could be used. In such cases chain of trust and policies for data protection should provided that spread over the whole ecosystem and allow audit and certification of security.

A.　　　　　　　Security technologies and protocols

Such complex implementation as IoT and Cloud cannot be secured with one general protocol. On every layer of the architecture different requirements should be fulfilled – secure pairing, authorization, confidentiality, integrity, reliability, low power consumption, low processing complexity, low delay and latency, transparency and ease of use. In Table 1 some of off-the-shelf protocols and technologies in IT security are shown and mapped to appropriate layer in proposed architecture.

Devices should encompass some method for secure pairing with the gateway which can be chosen from available methods on the market: Resurrecting Duckling, WPS (WiFi Protected Setup), WCN (Windows Connect Now), exchange of patterns as in Bluetooth or existing wireless security protocols can be used. But more important is to secure the mapping between the physical and virtual device. It could be done with pre-shared keys or certificates but should be reliable and error prone as it must be done once for the whole life-cycle of the device. Even emerging services at hardware layer could be applied for establishing trust – like Near Field Communication (NFC). The security can be increased with keeping the state of the device both on its physical and virtual instance – the huge difference in states could be used as an alarm for an attack. Moreover, the mirroring of configuration and data at both instances will allow faster recovery after attack. The main drawback of existing methods for secure pairing is that they are based on some user interaction and most of them work interactively (exchange of keys, entering pins, pressing buttons, preliminary setup in vicinity, etc.) The expansion and independence of IoT devices will call for finding more pervasive methods for pairing based on some

internal seamless and secure identification – something that RFID lacks.

In industrial and automotive applications the communication channel are kept private and therefore secured more easily. In IoT there is always possibility to connect through public insecure network, so data integrity, confidentiality and identification of endpoints should be provided. The public communication channels could be secured by network-based protocols as IPSec, SSL/TLS or SSH transport protocol and if it is needed the messages could be secured separately on application layer. On hardware layer this could be expanded using wireless security (WPA, 802.1X) or point to point authorization (PAP/CHAP).

For proving the identity of endpoints and defining their access rights ticket based systems could be used (Kerberos, Radius). This will allow even providing temporally access to objects and will limit the possible brute force attacks. If the chosen middle-ware is based on web services, WS-Security could be used for proving the data origin and separation of public and private parts in the messages. If the data is encoded in JSON format the endpoints could be secured using JSON Web Token (JWT). Depending on the application level protocol PGP (Pretty Goog Privacy) or S/MIME (Secure/Multipurpose Internet Mail Extensions) could be used for keeping integrity and confidentiality of messages.

TABLE I.　　PROTOCOLS FOR IoT AND CLOUD SECURITY

layer	Security		
	What to secure	Protocol/method	Confidentiality / Integrity / Authentication
HW[a]	device	Bluetooth security	yes / yes / device
HW[a]	device	WPA/802.1X	yes / yes / device
HW[a]	endpoints	PAP/CHAP	no / no / client
HW[a]	endpoints	WPS	no / no / client
HW[a]	endpoints	WCN	no / no / client
HW[a]	endpoints	NFC	no / yes / client
HW[a]	device/user	Kerberos	no / no / user
HW[a]	device/user	Radius	no / no / user
MW[b]	link	IPSec	yes / yes / host
MW[b]	link	SSL/TLS	yes / yes / server
MW[b]	link	SSH transport	yes / yes / server
App[c]	user	HTTP Digest	no / no / user
App[c]	message	SMTP S/MIME	yes / yes / message
App[c]	message	PGP/GnuPG	yes / yes / message
App[c]	data parts	WS-Security	yes / yes / data
App[c]	user	O-Auth	no / yes / user
App[c]	message	JWT	no / yes / user

a. Hardware layer

b. Middleware layer

c. Application layer

The general communication between a user and a device should be limited to the connection to the virtual instance using the appropriate protection for the purpose. If a strong encryption is needed it will be easier to implement on the virtual device in the cloud than on the low-resource physical device. Removing the shortcuts from user to physical devices will reduce significantly the security requirements for the architecture and will reduce the complexity. The use of VO could be in benefit of privacy protection as different VO could be assigned to one physical device that provide only the data that is appropriate for the application which will limit the chance of data leak and privilege stealing. Each user cloud have its personal view of the device services and parameters implementing Senor-as-a-service or Sensing-as-a-service paradigm.

No matter what security protocols and technologies are applied the best practices in complex security systems as IoT and Cloud are relying on regular monitoring and vulnerability checks [16]. The issue is that in such complex systems the monitoring of the security is a BigData task which is complex by itself.

IV. CONCLUSION

The integration of smart devices and data services to internet has lead to new security requirements that need new adaptive decisions. In such systems many aspects of security should be taken in mind from secure association via data protection and depersonalization to reliable authentication and authorization. The proposed architecture should provide application of security at each level will provide flexibility and scalability and will allow provision of security needs at design level.

The application of security at design level will fasten the integration of important new systems as intelligent transportation, public e-services, smart home, smart grid, e-health and so on.

The development of proposed model and architecture could lead to direct integration of security best practices in areas like:

- e-Health and m-Health and especially in Personal Health-care Systems and Clinical Information Systems and Ambient Assistance Living for elder people;

- Energy consumption monitoring and control in Smart Home and Smart Grid systems;

- Intelligent Transportation systems for smart railway, smart city and autonomous automobiles.

Some performance benchmarks could be made to evaluate the prefect security protocol for each practical implementation and test bed experiments could be carried out to compare performance

ACKNOWLEDGMENT

The presented work is supported by the National Science Fund of Bulgaria, project "Investigation of methods and tools for application of cloud technologies in

the measurement and control in the power system" under contract E02/12.

REFERENCES

[1] Yang, Y. Yue, Y. Yang, Y. Peng, X. Wang, and W. Liu, "Study and application on the architecture and key technologies for IoT," in IEEE Int. Conf. on Multimedia Technology (ICMT), 2011, pp. 747–751.

[2] A. Al-Fuqaha, M. Guizani, M. Mohammadi, M. Aledhari, and M. Ayyash, "Internet of things: A survey on enabling technologies, protocols, and applications," IEEE Comm. Surveys & Tutorials, vol. 17, no. 4, pp. 2347–2376, 2015.

[3] P. Porambage, M. Ylianttila, C. Schmitt, P. Kumar, A. Gurtov, and A. V. Vasilakos, "The quest for privacy in the internet of things," IEEE Cloud Computing, vol. 3, no. 2, pp. 36–45, 2016.

[4] M. Wu, T.-J. Lu, F.-Y. Ling, J. Sun, and H.-Y. Du, "Research on the architecture of internet of things," in 3rd IEEE Int. Conf. on Advanced Computer Theory and Engg. (ICACTE), vol. 5, 2010, pp. 484–487.

[5] L. Atzori, A. Iera, and G. Morabito, "The internet of things: A survey," Computer Networks, vol. 54, no. 15, pp. 2787–2805, Oct 2010.

[6] A. Botta, W. De Donato, V. Persico, and A. Pescapé, "On the integration of cloud computing and internet of things," in IEEE Int. Conf. on Future Internet of Things and Cloud (FiCloud), 2014, pp. 23–30.

[7] K. Evangelos A, T. Nikolaos D, and B. Anthony C, "Integrating RFIDs and smart objects into a unified internet of things architecture," Advances in Internet of Things, vol. 2011, pp. 5–12.

[8] C. Sarkar, A. U. Nambi, R. V. Prasad, A. Rahim, R. Neisse, and G. Baldini, "DIAT: A scalable distributed architecture for IoT," IEEE Internet of Things Journal, vol. 2, no. 3, pp. 230–239, 2015.

[9] E. Welbourne, L. Battle, G. Cole, K. Gould, K. Rector, S. Raymer, M. Balazinska, and G. Borriello, "Building the internet of things using RFID: the RFID ecosystem experience."

[10] M. Nitti, V. Pilloni, G. Colistra, and L. Atzori, "The virtual object as a major element of the internet of things: a survey," IEEE Communications Surveys & Tutorials, vol. 18, no. 2, pp. 1228–1240, 2015.

[11] O. Vermesan, P. Friess, P. Guillemin, H. Sundmaeker, M. Eisenhauer, K. Moessner, M. Arndt, M. Spirito, P. Medagliani, R. Giaffreda, S. Gusmeroli, L. Ladid, M. Serrano, M. Hauswirth, G. Baldini, "Internet of Things Strategic Research and Innovation Agenda", In: O. Vermesan, P. Friess, (Eds), *Internet of Things – From Research and Innovation to Market Deployment*, ©2014 River Publishers, Aalborg, Denmark, ISBN: 978-87-93102-95-8

[12] Jing, Q., Vasilakos, A. V., Wan, J., Lu, J., and Qiu, D. "Security of the internet of things: perspectives and challenges. Wireless Networks, 20(8):2481–2501, 2014

[13] Granjal, J., E. Monteiro, and J. Sá Silva. "Security for the Internet of things: a survey of existing protocols and open research issues." IEEE Communications Surveys & Tutorials 17.3 (2015): 1294-1312.

[14] M. Radochinski, "Security for IoT", Mentor Embedded Web seminar, 29th April 2014, online: https://www.mentor.com/embedded-software/multimedia/security-strategies-iot-systems

[15] Yu, T., Sekar, V., Seshan, S., Agarwal, Y., and Xu, C., "Handling a trillion (unfixable) flaws on a billion devices: Rethinking network security for the internet-of-things" In Proc. 14th ACM Workshop on Hot Topics in Networks, HotNets-XIV, pages 5:1–5:7, 2015.

[16] A. Sajid, H. Abbas and K. Saleem, "Cloud-Assisted IoT-Based SCADA Systems Security: A Review of the State of the Art and Future Challenges," in IEEE Access, vol. 4, no. , pp. 1375-1384, 2016.

[17] E. Oftedal, A. van der Stock, T. Chih- Hsiang, REST Security Cheat, Sheet, OWASP, Last revision (mm/dd/yy): 11/17/2016: https://www.owasp.org/index.php/REST_Security_Cheat_Sheet

Making a Smart City Even More Intelligent Using Emergent Property Methodology

A. Šarić, B. Mihaljević and K. Marasović

Rochester Institute of Technology Croatia

andrej.saric@croatia.rit.edu, branko.mihaljevic@croatia.rit.edu, kristina.marasovic@croatia.rit.edu

Abstract - With the emergence of IoT (Internet of Things) smart cities have assumed a crucial role in providing intelligent services in real time to residents and visitors, by integrating services such as smart transportation, smart parking and smart people management. By introducing emergent properties as the system's new behaviors or new structures, the system gains another higher level of intelligence that cannot be derived from its underlying substructures individually. In this paper we propose a new model for introducing emergent properties by integrating existing smart city subsystems which are already in use in the city of Dubrovnik, Croatia. The article starts with an overview of the smart city technologies and emerging property methodology. The existing smart city subsystems are explained along with their integration within emergent properties model. The paper concludes with our findings and recommendations, open issues and future research possibilities.

I. INTRODUCTION

With rapid evolution of computing and technology, and increased communication using modern devices like smartphones, people started a new way of life through constant connectivity and ubiquitous computing. Aside from enabling people to use their smart devices anywhere and anytime, most important change happened in the universal, continuous and omnipresent connectivity. Evolution of pervasive technologies enabled ubiquitous computing-based connectivity such as connecting people to various systems and accompanying information gathered from numerous heterogeneous Internet of Things (IoT) devices as data sources.

This shift to constant connectivity provides even more data for an even better analysis. Ability to use such data in real-time analysis and prediction allowed to discover correlation within the data to control the system in real-time, and avoid undesirable conditions. As with all emerging technologies, these too have found their way to the place where people mostly live – cities.

Cities are becoming overpopulated, having a negative impact on the citizens' quality of life. To provide citizens with better services, a smart city, as a new way of managing city's assets and services, could prove to be quite useful for this type of urban environment. By integrating modern technologies such as IoT into an existing municipal infrastructure and services, improvements can be made towards expansion, sustainability and budget saving. Connecting various IoT

devices to smart city background services and subsystems we can achieve a network-shaped smart city system that can later lead towards a fully integrated smart city.

In this paper we introduce emergent property methodology as the system's new behavior along with additional structures that help the system achieve a higher level of intelligence. This new approach enables the smart city an ease of integration with other services such as smart transportation, smart parking system and smart people management. Through predefined protocols, procedures and standards, data exchange across all integrated subsystems is being made possible. This architecture is flexible and allows for any future extension which may include integrating additional subsystems with new services and features.

The remainder of this paper is organized as follows. Section 2 introduces smart cities, and explains the driving force behind this concept. Section 3 describes what emergent property is and how it is applied to different software solutions, in particular to smart city subsystems. Section 4 presents an overview of existing smart city services and subsystems found in the city of Dubrovnik, Croatia, along with the technical specifications, concerns and possibilities for improvement. Section 5 discusses possible changes and adaptions of the existing smart city subsystems to introduce and apply the emergent property methodology. Section 6 concludes the paper with our findings, recommendations, open issues and future research possibilities.

II. SMART CITIES

With more than half of the world's population living in cities and constant growth there is a need to find new ways of improving cities to become more efficient while providing citizens with better services in the areas of mobility, energy, building and home infrastructure, transparency, public services and others. The popular term "smart city" appeared almost two decades ago [1], however, its definition, meaning and understanding varies according to context, problem domains and used technologies. A more recent survey of smart city research and literature could be found in [2][3].

To make necessary changes to implement smart city concept older cities need to be redesigned starting with their infrastructure to allow easy integration of IoT and various sensors, as well as new information and communications technology (ICT) based services. Newer cities need to be designed with smart city concept in mind from the start, which consequently allows easier integration of overall ICT infrastructure and IoT devices.

By making people conscious of this smart city implementation options and by introducing them to overall smart city concept there will be valuable feedback that will help in building and designing new services that will improve governance, transparency and efficiency. To achieve all of this investments need to be made in public city infrastructure, ICT services and their integration, as well as increasing public awareness about advantages and benefits of living in a smart city.

With the emergence of the new IoT market which has enabled smart cities to easily integrate new services using these production-ready platforms and associated IoT devices [4]. Smart cities can also aim towards building completely new smart city systems which will be based on selected IoT platforms, subsystems and technologies. To build smart city IoT system a city needs to invest into infrastructure and hardware, which is usually consisting of various sensors and other control smart devices integrated together and supported with backend servers or cloud, which will both collect data from the environment and consequently trigger appropriate actions. Since any such system will highly rely on connectivity, which is usually smart devices and sensors sending and receiving data over the cloud, second step should be implementation of standardized protocols and format for data transfer and storage. Final step is achieved through implementation of various software, which will analyze the data collected from the sensors and use internal decision making subsystem to instruct other smart devices to initiate actions based on decisions made. Software should be also responsible for providing monitoring dashboards and control interfaces with integrated statistics and analytics, where administrators will have access to real-time graphically presented data and easy to use management controls, while public might have access to client apps mostly used for information dissemination.

III. Emergent Properties

Concept of emergent properties occurs in many different areas; most prominent ones would be philosophy and biology but it can also be applied on software system designs. There are many definitions about what emergent properties are and more about it can be found in [5]. In a nutshell, emergent properties are properties that some complex system has but its individual components such as underlying modules and subsystems do not have. Emergent properties concept can be easily extended to software development [6] and implemented in software design and architecture. In this paper we are introducing emergent properties as new behaviors that add value to the existing smart city systems and overall whole smart city platform.

IV. Smart City Subsystems

Smart city is composed of multiple services which are in fact specific subsystems. These subsystems provide services which can be directly consumed by citizens or

used by city officials. These systems focus on domains such as public services, mobility, energy, buildings, etc. In this section we will present some of the smart city subsystems already implemented in the city of Dubrovnik.

A. Smart Parking System

Smart Parking System [7] (Fig. 1) is a complete smart city solution developed a year ago with the goal to be easily replicable and installable to any location that has GSM network coverage. Solution is cloud based which makes it fully scalable and it makes no difference if the system implementation happens on one parking lot in a small city or many parking lots in a metropolis. Hardware part of this solution is composed of battery-powered infrared wireless sensors that are designed to be easily placed on each parking spot. Their installation is simple and fast, which is an advantage compared to some other implemented solutions which require extensive construction work as well as power and communication lines. These sensors communicate and send data to the router using standard HTTP [8] and MQTT protocol [9]. They come with multiple useful features such as:

- Network reconfiguration in case of router failure
- Automatic registration and network configuration of newly added sensors
- Error states and malfunction reports are sent using same channel to simplify maintenance and diagnostics

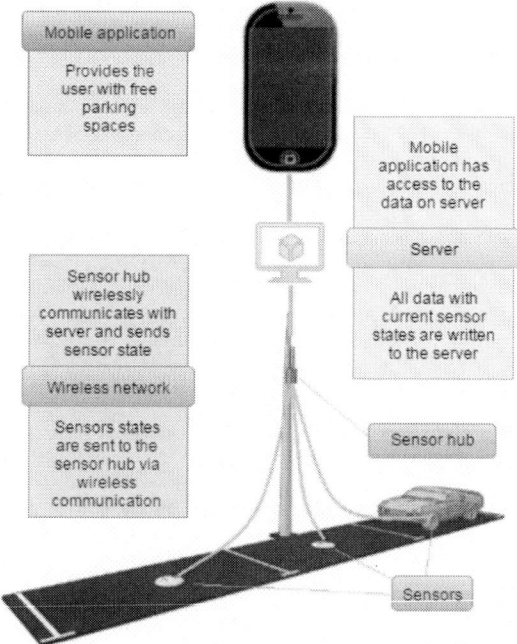

Figure 1.   Smart Parking System overview

Technological stack used in the system is PHP for server and Angular (Version 2) in TypeScript with Bootstrap 4 for web interface.

1) Sensor Node for Smart Parking System

Sensor node (Fig. 2) is a device which main task is car detection so that it can provide information whether the parking spot is vacant or not. Every single sensor contains the information about current status of the parking spot. The main goal of every sensor node is to send information to the router as fast as possible over HTTP using MQTT. HTTP usage is based on common REST API and MQTT is using standard publish/subscribe pattern which enables other clients to subscribe to publishers. All sensor nodes are wireless, which allows to easily install them onto parking spots, and battery-powered with battery life expectancy estimated to three or more years.

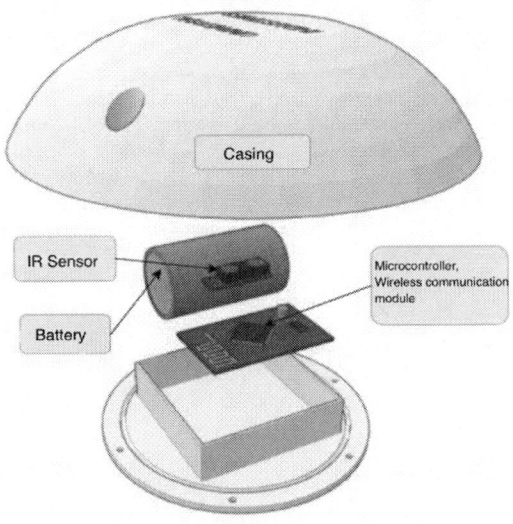

Figure 2.   Sensor node for Smart Parking System

2) Routers and Server for Smart Parking System

Router (Fig. 3) uses two communication methods; one to get information about parking spot vacancy, which is retrieved from sensor nodes, and second method using wireless network to forward information to the server in the cloud.

Server gathers information from routers over wireless network for every single sensor node using its id, GPS location and other available information, and it is responsible for saving their current state in the database. All states are stored in the relational database so that we could use that data for statistics, analytics and overviews.

3) Software and User Interface

Solution comes with a publicly available web site that has an integrated interactive map which presents current state of a parking lots in terms of vacant and non-vacant spots. Mobile applications are implemented for both iOS and Android devices (Fig. 4) that show to end users the information about current state of the parking lot. Mobile applications are also used to gather usage data for analytics.

An administrator's web-based dashboard is used for monitoring the system and its components, and it provides important control functionalities such as remotely resetting or enabling and disabling sensor nodes.

Figure 3.   Router for Smart Parking System

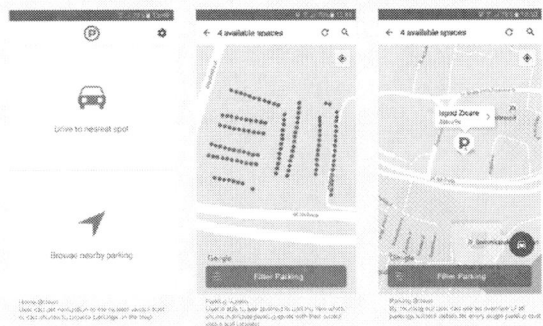

Figure 4.   Smart Parking Mobile Application user interface

4) Possibilities for improvement

This system relies on network reliability to send and receive data so going towards edge and fog computing might be one of the largest improvements. More about fog and edge computing can be found in [10]. Other improvements can be made in the area of features towards end users and data collection. Suggested improvements include but are not limited to:

- Integrating fog and edge computing into the system.
- Providing a hotspot with internet connection to end users which would allow more information gathering about end users and application usage.
- Parking spot reservation which would allow users to block entrance to a parking spot using mobile application.
- Payment system that will be fully integrated with mobile applications.

B. People Counter System

People Counter System (Fig. 5) is a smart city solution that was motivated by the need of solving the problem of too many people visiting the Old City of Dubrovnik and the main street – Stradun. To fully resolve the problem this solution must be dependent on full integration with smart city platform and other smart city subsystems. Solution is composed of both hardware and software parts; hardware part of this solution are Hikvision cameras that are installed on five locations around the Old City of Dubrovnik area. All designated camera locations are on entrances to the Old City and the primary goal of this solution is being able to count the number of people (visitors) in that area in real-time. Technology stack used by the system is as follows; server is developed using ASP.NET Core in C# and web interface using Angular (Version 2) in TypeScript with Bootstrap 4. Solution is deployed on Windows Azure Web service using three slots: development, staging and production.

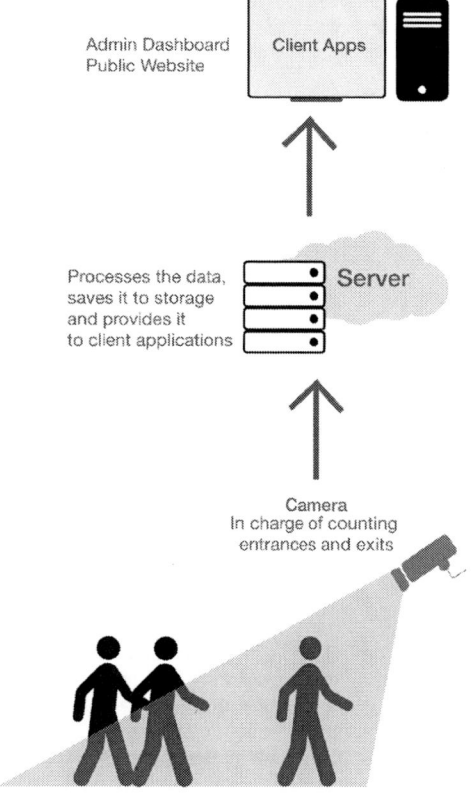

Figure 5. People Counter System overview

1) Cameras for People Counter System

Since all cameras used are network-based cameras, once they are installed they need constant and reliable network access used to transfer the data. Providing cameras with the network and power cables needs construction work upon the installation and can extend the development process. These cameras have already implemented people counting capabilities which are currently used to count how many people enter and exit the area of Old City. All cameras are located at all possible entrances to this area at the position and angle that allows them to give precise data about entrances and exits of visitors.

2) Server for People Counter System

Server gathers information from the cameras and stores it into a relational database. Information that is gathered provides an overview of real time situation in that area, as well as statistics and analytics for different time ranges.

3) Software and User Interface

Solution is initially integrated with public website that provides information about current status and number of visitors in the Old City of Dubrovnik. Public website also provides more detailed view for every single camera and statistics dating back to one week. Aside from public website, solution provides administrator dashboard which contains more information about current status in the Old City area, including single camera information and management with exact count of entrances and exits for that area (Fig. 6), "heat maps" showing current situation and which of the entrances is most commonly used to enter or to exit the Old City area(Fig. 7), and general statistics for overall readings.

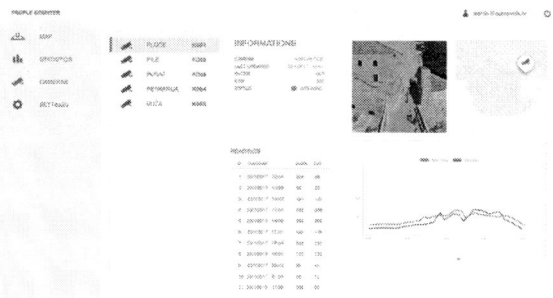

Figure 6. Administator Dashboard for Single Camera Information and Management in People Counter System

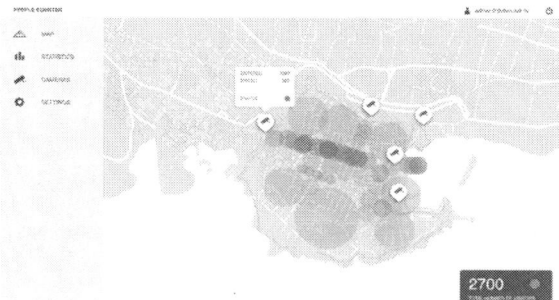

Figure 7. "Heat map" in Administrator Dashboard in People Counter System

4) Possibilities for improvement

Improvements that can be done in this subsystem focus on hardware and data gathering which would enable achieving better and more precise data, which can later be used in other subsystems of smart city. Some of the improvements include but are not limited to:

- Using wireless cameras so that maintenance is easier and with less construction work on the Old City walls when installing .

- More cameras on "hot spot" locations in the city which would help in tracking movements and finding movement patterns.

- Integration with other smart city subsystems to provide better pattern recognition and possible traffic and movements predictions.

V. ADOPTION OF EMERGENT PROPERTY METHODOLOGY

In the previous sections already implemented services and smart city subsystems were discussed and we propose how to adopt emergent property methodology through already existing subsystems and integrate it into new overall smart city system.

Emergent property methodology is derived from the concept of emergent properties and its integration in software design and architecture [6], [11]. This methodology enables better testability of subsystems as individual components of a larger system, exploration of new possibilities and features that can be achieved through the emergence.

To achieve this, we propose a smart city framework and platform that allows multi-subsystem integration. The smart city platform is a top-level system into which all other subsystems integrate and where we have emergent properties developed as new services, which weren't possible to implement before. This smart city platform system implementation is in test in the city of Dubrovnik, using its already implemented smart city subsystems, as discussed in previous section. Those are used to explain integration in smart city framework and introduction of emergent properties.

Our smart city platform proposal is using as a cloud-based servers that have two ways of providing integration for subsystems via Deep linking or Superficial linking. Deep linking assumes that platform hosts the database, server and backend of a whole subsystem, while Superficial linking presumes that subsystems will be deployed externally.

Every smart city subsystem needs to be able to collect and store data, share that data with the platform and have control commands. Smart city framework defines generic standardized inputs and outputs needed for every subsystem integration, together with protocols and procedures that subsystems need to adhere to.

Smart city framework's core components are:

- Schema that allows edge computing on devices and parsing libraries, and

- Support for various data interchange formats and transfer protocols.

Schema brings the logic down to the smart IoT devices and defines device's sensors, readings and controls, which enables easy device-client configuration. Client apps are able to read schema and request data from the device, thus knowing which sensors and mapping for data is used. This

enables flexibility and allows clients to connect and parse readings from any device that adheres to this standard without previously knowing device's sensor and data mapping.

Data interchange formats that are supported by the framework are JSON and XML based; JSON is supported because it is widely used today and XML because of legacy systems that still use it and want to integrate with the platform. Transfer protocols supported are HTTP [8] and MQTT [9]; we use REST-enabled services for platform-subsystem communication and lightweight MQTT [12] to leverage almost instantaneous message publish and topic subscribe, which is crucial for IoT systems that have critical performance and need to be deployed on any kind of network.

First step in adopting smart city framework is having all devices use schema to map their controls, sensors and readings. The next step is using framework parsers on server or in client apps to read schema and data from devices. Once the subsystem is able to communicate with devices it needs to adopt REST service that will enable smart city platform connection and data transfer. Upon successful integration of subsystems smart city platform achieves emergent property methodology environment. Because of multiple integrations and different data sources smart city platform provides three data categories:

- Paid Data – data that can be used if paid

- Private Data – internal data, only accessible to administrator and city officials

- Open Data – data that is free to use for public

These data categories are crucial for achieving emerging properties, and depending on integrations within the platform we can develop different systems and services emerging from it.

A. Smart City use case example

Smart City use case example is based on previously implemented subsystems in city of Dubrovnik and we are presenting how each of this subsystems is able to adopt the smart city framework by adhering to its standards.

Smart parking subsystem implements schema to map different sensors and controls needed for smart devices to operate. This enables server to parse the schema and receive the data from devices. In the long run this will prove crucial if changes to the smart devices are made or additional IoT sensors are added, because by updating the schema server is able to parse new data from devices without any additional mediation needed. Smart devices implement MQTT to send data to the router, and router implements HTTP to forward the data to the server in the cloud. Server implements REST API needed for integration to the smart city platform and gives access to subsystem's data.

People counter subsystem's communication protocol for cameras is HTTP because performance is not critical. Subsystem integrates with the smart city platform through REST API and shares its data with it. However, this subsystem varies from some standards needed to fully adopt the smart city framework. Since this subsystem uses commercial cameras unable to adopt dynamic schema mapping, subsystem implements static schema saved on the server. For example, if changes are made to just a couple of devices, server will still use local schema that will not be retrieved and updated for every individual device.

Both of these systems collect different kinds of data from accompanying applications. Smart parking subsystem data includes the location and time of the day, collected from users when they using the mobile application to find available parking spot. Additionally, parking spot related data includes timeframes when with most frequent changes in vacancy, which can be used to find patterns and predict traffic increase. Data collected by people counter subsystem includes current number of people in the Old City area, as well as which entrances are mostly used to enter and exit. Based on this information system can limit the number of people entering the Old City area and predict which locations will have big influx of people.

B. New emergent services

Before abovementioned integration to smart city system all of this data was confined within its own subsystem, therefore having limited use. Once the subsystems integration will be totally completed and data will be collected and analyzed, system will be able to create data categories from which we can easily create new emerging services. For example, emerging services that we can develop are People management subsystem and Traffic management subsystem.

People management system comes from data collected by cameras in People Counter System and mobile applications from Smart Parking System, which can be used to direct people to different streets, exits and entrances to avoid overcrowding and traffic jams.

Traffic management system includes public transportation such as buses and taxies and allows control over possible paths, i.e. when to enable passage for taxies and buses towards the Old City. This system will use parking spot data about vacancy of public transportation parking spots. It will be used to inform when there are vacant spots for taxies and buses to stop. It will be used in conjunction with data from People Counter System to know when there is a lot of people leaving the Old City so that number of buses and taxies can be increased without creating traffic jams.

VI. CONCLUSION

It is important to stress that this research and smart city platform and framework proposal would not be able without previous implementation of smart city subsystems in the city of Dubrovnik, Croatia. This research has

proved to be successful based on results achieved through smart subsystems integration into the smart city platform. Authors believe that they have proved the achievability of Emergent Property Methodology in a smart city system using smart city platform and framework to easily integrate its subsystems. Future research will focus on how new technologies influence on already implemented architecture and possible improvements in areas of security and reliability in smart city subsystems.

ACKNOWLEDGMENT

This research was supported by City of Dubrovnik and DURA City of Dubrovnik Development Agency.

REFERENCES

[1] B. van Bastelaer, "Digital cities and transferability of results," in *Proc. of 4th EDC conference on digital cities*, Salzburg, Oct 29-30, 1998, pp. 61-70.

[2] A. Cocchia, "Smart and Digital City: A systematic Literature Review," in *Smart City: How to Create Public and Economic Value with High Technology in Urban Space?*, Springer International Publishing, 2014, p. 13-43.

[3] L. G. Anthopoulos, "Understanding the smart city domain: A literature review," in *Transforming City Governments for Successful Smart Cities*, Springer International Publishing, vol. 8, 2015, pp. 9–21.

[4] R. R. Harmon, E. G. Castro-Leon, and S. Bhide, "Smart Cities and the Internet of Things," in Proceedings of PICMET '15: Management of the Technology Age, 2015, pp. 485–494.

[5] T. O'Connor and H. Y. Wong, "Emergent Properties," in *The Stanford Encyclopedia of Philosophy*, Summer 2015., E. N. Zalta, Ed. Metaphysics Research Lab, Stanford University, 2015.

[6] O. Pomorova and T. Hovorushchenko, "The way to detection of software emergent properties," in Proc. of the 2015 IEEE 8th International Conference on Intelligent Data Acquisition and Advanced Computing Systems: Technology and Applications, IDAACS 2015, 2015, vol. 2, no. September, pp. 779–784.

[7] K. Hassoune, W. Dachry, F. Moutaouakkil, and H. Medromi, "Smart parking Systems: A Survey," in 11th International Conference on Intelligent Systems: Theories and Applications (SITA), 2016.

[8] R. Fielding, J. Reschke, "Hypertext Transfer Protocol (HTTP/1.1): Message Syntax and Routing," RFC 7230, IETF June 2014. [Online]. Available: https://tools.ietf.org/html/rfc7230. [Accessed: 06-Feb-2017]

[9] "MQTT Version 3.1.1 Plus Errata 01, OASIS Standard Incorporating Approved Errata 01", OASIS, 2015, [Online]. Available: http://docs.oasis-open.org/mqtt/mqtt/v3.1.1/mqtt-v3.1.1.html. [Accessed: 06-Feb-2017]

[10] F. Ai-Doghmant, Z. Chaczko2, A. R. Ajayan3, and R. Klempous, "A Review on Fog Computing Technology," 2016.

[11] M. Patrascu, M. Dragoicea, and A. Ion, "Emergent intelligence in agents: A scalable architecture for smart cities," 18th International Conference on System Theory, Control and Computing - ICSTCC 2014, pp. 181–186, 2014.

[12] T. Yokotani, "Transfer Protocols of Tiny Data Blocks in IoT and their Performance Evaluation," pp. 54–57.

MIPRO 2017, May 22- 26, 2017, Opatija, Croatia

Digital Chess Board based on Array of Hall-Effect Sensors

Filip Sušac, Ivan Aleksi, Željko Hocenski

Faculty of Electrical Engineering, Computer Science and Information Technology Osijek
Josip Juraj Strossmayer University of Osijek, Osijek, Croatia
filip.susac@etfos.hr, ivan.aleksi@etfos.hr, zeljko.hocenski@etfos.hr

Abstract-This paper addresses a novel concept of digital chess board realized with an array of Hall-Effect sensors. Main task of digital chess board is autonomous detection of player's moves. Therefore, players do not have to write down their own moves on a sheet of paper. Non-invasive piece moves detection is considered, with minimum required installations in a game room. The digital chees board presented in this paper has a Hall-Effect sensor placed under the middle of each field on the chess board, while each chess piece has a permanent magnet placed on its bottom. Sensors, total of 64, are organized as 8x8 array. Microcontroller reads sensor data and sends them to a remote PC for storage in a database and board visualization. Rows are multiplexed in time, i.e. one row is active in a certain time instance. States in the column of the currently active row are stored in a corresponding matrix column. Con to this approach is in its ability to detect piece presence on the field, without the information about the piece type. The proposed approach is low cost, non-invasive and requires only power and communication cable.

I. INTRODUCTION

Development of computer science and technology advances are followed by integration of new technologies into social games, and chess is one of them. First aspirations were focused on creating a computer system that can beat the chess Grandmaster. One of the most famous chess games, which marked a milestone in the pursuit of creating a superior chess computer system is a chess game between, at that time the world's best chess player, Garry Kasparov and IBM supercomputer Deep Blue in 1997. Deep Blue defeated Kasparov and became the first computer which defeated world chess champion [1]. With further development of computer technology, there is a remaining tendency of creating new algorithms for chess playing and chess learning. On the other hand, there is a need for automation of chess game itself and developing possibilities of remote access and control. A novel approach is an implementation of Hall-effect sensors for detecting the presence of chess pieces and getting access to a current position on a chess board, cf. Fig. 1.

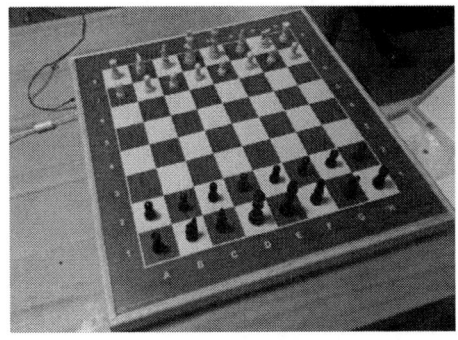

Figure 1. Digital chess board with: 64 Hall-effect sensors placed under each field, PC communication and power cables.

II. RELATED WORK

In recent years, many papers and projects are published about computers and chess. Chess playing combine elements of artificial intelligence for player's move decision, computer vision and robotics for position detection and performing a movement of pieces. Considering developments in computer vision, numerous papers are published in the field of recognition and detection of chess pieces. Despite that, some of the papers are published with the aim to improve automation of the game. This allows playing against computer in real world, playing from remote location and improving playing possibilities for people with disabilities.

Usually, camera-based systems for visual inspection require adequate and constant level of light source without shadows from surrounding objects. Therefore, overhead camera is used in [2] with ambient lightning to detect player's moves. Human and chess piece shadows caused difficulties in moves detection. Lightning source was adjusted so that each white piece does not have a black shadow. Specular reflection from pieces is removed with image filtering. Like a chess game, Janggi chess board detection under severe conditions in camera point of view is described in [3]. Using Hough transform, Harris corner detection and Canny edge detection, Janggi chess board and pieces are detected. Piece detection problem is relaxed in Janggi chess due to smaller piece heights and less lines hidden with pieces. Another camera-based approach is presented in [4] where authors created multimodal interface for making chess moves. Using a stereo camera and difference in image entropy, hand gestures are detected and translated to chess moves.

One approach in detection of chess position is detection based on player's hand movement (not piece's). Authors in [5] realized their project named "Hand-Motion Chess" using gloves with sensors and a microcontroller. Hand gestures are detected with sensors and are translated into chess move. This approach can be implemented in creating universal equipment (glove) which could detect chess moves regardless of variations in size and color of pieces and chess board.

Due to rapid development of commercial robotic arms, accessibility to this technology is increasing. Numerous projects were made with educational purposes on moving chess pieces around the chess board using robotic arm. Authors in [6] presents autonomous chess playing system based on robotic manipulator arm. Fully autonomous chess playing robot "MarineBlue" is presented in [7]. It combines elements of computer vision, chess engine and robot control. Authors in [8] implemented detection of chess pieces using reed sensors that are triggered by the magnetic field from permanent magnet placed on the bottom of each piece. Computer Numerical Control (CNC) machine with 2D Cartesian coordinate system is used for moving chess pieces around the board. Interactive chess board is presented in [9]. It detects player's move based on 8x8 membrane keypad, and makes its own move using 2D CNC machine. The idea behind proposed approach is to enable players to play chess over the internet on a real chess board. Thus, when player makes a move on his board, chess piece moves on another board, keeping synchronized positions between boards.

A commercial product commonly used in FIDE chess tournaments is digital chess board named "DGT e-Board" [10]. The board is using passive LC-circuits with ferit core in pieces with active LC-circuit in each chess field [11]. Passive LC-circuits are induced with active LC-circuit and resonant frequency is measured. However, development of digital chess board using Hall-effect sensors is an approach for educational purposes because of relatively low price and simplicity of assembling.

III. DEVELOPMENT OF CHESS BOARD

This section describes the main components of the digital chess board proposed in this work. The board consists of embedded microcontroller system for detection of chess pieces, chess pieces with integrated permanent magnets and acrylic board with printed wooden chess board image. System for detection of chess pieces is connected to a PC with a USB cable.

A. Chess Board and Chess Pieces

Chess is a game that is usually played on wooden chess board with corresponding pieces. That is not an explicit rule, but more a part of tradition of a chess game. Chess is a two-player game, where each player has 16 chess pieces: 8 pawns, 2 rooks, 2 bishops, 2 knights, 1 queen and 1 king. The game is played on board with 8x8 (64) chess fields (32 black and 32 white) [12]. According to the World Chess Federation (FIDE) official width of each chess field is ranging from 5 to 6 cm [13]. To make a digital chess board applicable for real tournaments, a width of 5 cm is used in this work. The total dimensions of the digital chessboard are 52x52 cm, of which 40x40

cm goes for chess fields and 6 cm for borders. In the bottom part of each chess piece two permanent are placed. Magnets diameter is 10 mm, with height 2 mm. Experimental testing of 3 and 5 mm acrylic board thickness showed that both are good for detection of magnetic field. Due to better mechanical properties, 5 mm acrylic board is used.

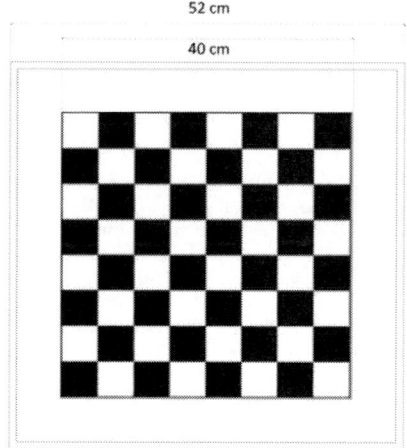

Figure 2. Dimensions of the board

B. Microcontroller System for Chess Position Detection

The system for detection of chess pieces is consisted with platform Arduino Nano platform, a set of 64 A3144 Hall-effect sensors and 5V power supply. Additional power supply is involved since the overall current consumption is 100 mA per pin, while the maximum is 40 mA. Connecting a single Hall-effect sensor A3144 requires one 10kΩ pull up resistor per pin, cf. Fig 3.

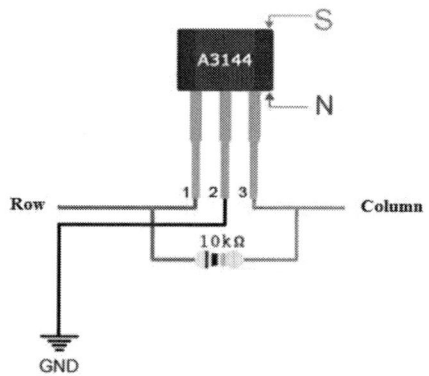

Figure 3. Connecting Hall-effect sensor to row and column of an array.

Printed Circuit Board (PCB) is made for detection of chess position method proposed in this work. Electronic scheme of designed PCB is presented, cf. Fig 4, as well as the final look of the board, cf. Fig. 5.

Internal pull-up resistors, which ATmega328p microcontroller has, can't be used since 8 sensors in a row require more power per pin than available. P-channel MOSFET transistors are used to activate each row with

5V power supply. Total of 4 dual-transistor ICs APM4953 are used with 10kΩ pull up resistors, cf. Fig 4.

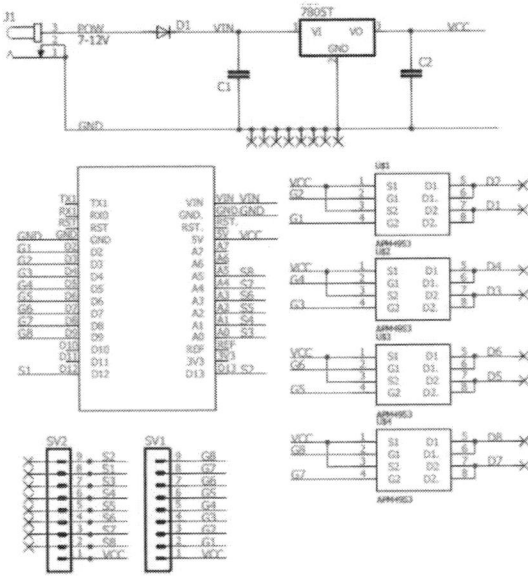

Figure 4. Electronic scheme of proposed system for chess position detection.

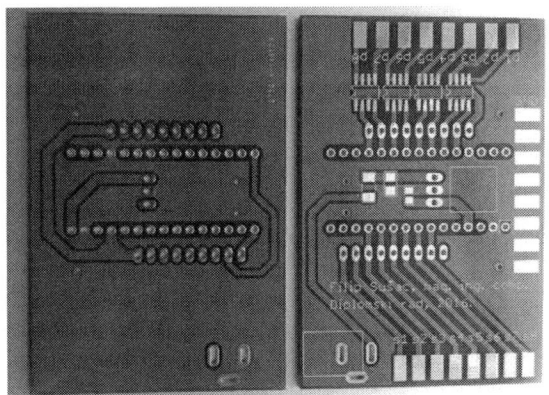

Figure 5. Designed Printed Circuit Board (PCB): bottom layer (left); top layer (right).

C. Programming Microcontroller

Serial communication (UART) is used as a communication protocol between a microcontroller and a PC. Communication is done in two directions, in a master (PC) and slave (microcontroller) fashion. Firstly, master is requesting data from slave by sending request character "R". Subsequently, slave collects data from sensors and responds with 8 bytes of data, where each bit corresponds to an occupied (1) or free (0) field, cf. Fig. 6.

Microcontroller (slave) is programmed to receive states from 64 sensors and send data to a PC. Due to a limited number of input and output pins, an array of 64 sensors is multiplexed [14]. Multiplexing is done row-wise, meaning that at one time instance only one row is active. Columns are active all the time and states from entire row can be

read. In this way, 8x8 binary matrix is reconstructed, where 1 (0) corresponds to an occupied (free) field.

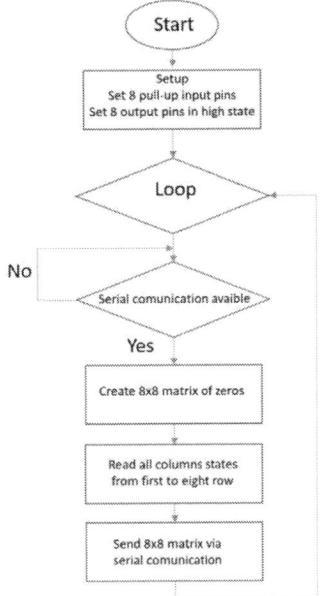

Figure 6. Flowchart of Arduino code

D. Player's Move Detection

Algorithm for player's move detection is done on a PC. It is based on a difference between two subsequently received binary matrices. Proposed method compares the currently received and previously received matrix. Difference between the two is calculated with

$$\Delta M(x,y) = M(x,y,n) - M(x,y,n-1), \quad (1)$$

where x and y are denoting chess field index, ranging from 1 to 8, respectively, n is the current time instance and $n-1$ is the previous time instance.

In the case when there are no changes on the chess board (no moves), 8x8 matrix ΔM is a zero matrix, cf. eq. (1). On the other side, when a player starts a move, one must take a chess piece and lift it up and away from the chess board field. Moving the piece from the field will be detected at a corresponding x, y location, and a value -1 is assigned to $\Delta M(x,y)$. The first occurrence of value -1 in $\Delta M(x,y)$ denotes either a *moving chess piece* or *taking chess piece*. Distinction between the two is done by knowing the player's color. If a piece's color corresponds to player's color, it is a *moving chess piece*, otherwise it is a *taken chess piece*. *Taken chess piece* is removed from the board. Finally, when a value +1 appears in $\Delta M(x,y)$ it denotes a *destination location* of a *moving chess piece*. Commonly used algebraic chess notation is used for tracking chess moves [12].

E. Visualizing the Game of Chess

Instead of using an open-source chess Graphical User Interface (GUI) (Winboard, Stofkfich, GNU Chess, etc.), our own GUI is made for educational purpose and further development of chess programs. GUI is done on a PC and is integrated with the logic described in Section III. D. When the chess board position is set, GUI requires an information about it. Therefore, a user can set-up an initial position by a button click, cf. Fig. 7. Chess positions, beside initial position, must be manually synchronized in a GUI to start from a desired position.

Figure 7. Chess game Graphical User Interface (GUI).

F. Performance analysis

Proposed approach is incrementally reconstructing a chess game by recognizing each move independently. In this paper, a chess move is recognized by detecting presence or absence of a chess piece. Limitations of proposed approach are in terms of inability to reconstruct starting chess position without a user interaction. Except starting position, the user interacts in pawn promotions. One can choose to automatically promote to queen, or to manually set each promotion. However, even if the chess board position could detect piece's type and color, it would still require the following inputs from the user: possibilities of the chess position: side to move, allowed castling moves, allowed *en-passant* move, piece move counter without taking (required for a 50-move rule) [12].

If the program fails to recognize one move, following moves will also be incorrect. Therefore, a user interaction is required to correct a false move detection. An experiment is done to objectively express. Total of 50 trials for every type of chees move is tested and illustrated in Table 1. Moves are made on an empty and occupied, respectively.

TABLE 1. SUCCESSFULLY DETECTED CHESS MOVES.

Number of experiments N=50	Successful detection	
Move a chess piece on an empty field	48 / 50	96 %
Capture an opponent's chess piece	43 / 50	86 %

IV. CONCLUSION AND FUTURE WORK

Chess remains a popular domain for experimenting in the field of robotics, computer vision and artificial intelligence. This paper gives an overview of the design and implementation of an autonomous digital chess board. Proposed approach combines low-cost elements with an array of Hall-effect sensors controlled by a microcontroller. Con to proposed approach is in possibility to detect only piece presence and not its type and color. Also, false move detection requires a user interaction to correct misdetections.

In the future, we intend to improve robustness of proposed system by considering better wiring and improved noise resistance in hardware and software, respectively.

REFERENCES

[1] Feng-Hsiung Hsu, "Behind Deep Blue: Building the Computer that Defeated the World Chess Champion," Princeton University Press, 2002.

[2] V. Wang and R. Green, "Chess move tracking using overhead RGB webcam," 2013 28th International Conference on Image and Vision Computing New Zealand (IVCNZ 2013), Wellington, 2013, pp. 299-304.

[3] VQ Nhat, G Lee, "Chessboard and Pieces Detection for Janggi Chess Playing Robot," International Journal of Contents, vol. 4, no. 4, pp. 16-21, Oct 2013.

[4] Doe-Hyung Lee and Kwang-Seok Hong, "Game interface using hand gesture recognition," 5th International Conference on Computer Sciences and Convergence Information Technology, Seoul, 2010, pp. 1092-1097.

[5] Cornell University Electrical & Computer Engineering, [Online]: https://people.ece.cornell.edu/land/courses/ece4760/FinalProjects/f 2012/oaq3_cig23_rk447/oaq3_cig23_rk447/, time of access Januar 2017.

[6] C. Matuszek et al., "Gambit: An autonomous chess-playing robotic system," 2011 IEEE International Conference on Robotics and Automation, Shanghai, 2011, pp. 4291-4297.

[7] David Urting, Yolande Berbers, "MarineBlue: A Low-Cost Chess Robot," IASTED International Conference Robotics and Applications, RA 2003, pp. 76-81, June 25-27, 2003, Salzburg, Austria,

[8] Sunandita Saker, "Wizard Chess: An Autonomous Chess Playing Robot," IEEE International WIE Conference on Electrical and Computer Engineering, p.p. 476 -478., December 2015.

[9] Allen R. Mendes, Atur M. Mehta, Bhavya H. Gohil, "Implementation of the Automatic and Interactive Chess Board," ISOR Jurnal of Electrical and Electronics Engineering, Volume 9, pp. 1-4, December 2014.

[10] DGT- Digital Game Technology, Januar 2017. http://www.digitalgametechnology.com/

[11] Bernard Johan Bulsink, "Device for detecting playing pieces on a board",US6168158 B1, January 2, 2001.

[12] John Emms, "Concise Chess: The Compact Guide for Beginners," Everyman Chess, 2013.

[13] Standards of Chess Equipment and tournament venue for FIDE Tournaments, 2014.

[14] James Feher, "Introduction to Digital Logic with Laboratory Exericises," Creative Commons Attribution, 2009.

Influence of Human-Computer Interface Elements on Performance of Teleoperated Mobile Robot

Stanko Kružić[1], Josip Musić[1], Ivo Stančić[1]

[1]University of Split, Faculty of Electrical Engineering, Mechanical Engineering and Naval Architecture
E-mail: {skruzic,jmusic,istancic}@fesb.hr

Abstract—**Mobile robots are becoming ubiquitous, with applications which usually include a degree of autonomy. However, due to uncertain and dynamic nature of operational environment, algorithms for autonomous operation might fail. In order to assist the robot, the human operator might need to take control over the robot from remote location. In order to efficiently and safely teleoperate the robot, the operator has to have high degree of situational awareness. This can be achieved with appropriate human-computer interface (HCI), so that the remote environment model constructed with sensor data is presented at appropriate time, and that robot commands can be issued intuitively and easily. In the research, influence of HCI elements on performance of teleoperated mobile robot was studied for several tasks and with several HCI setups. The user study was performed, in which accuracy and speed of completion of given tasks were measured on a real robot. Statistical analysis was performed in order to identify possible setup dependencies. It showed that, in majority of analysed cases and based on introduced metrics, there is no significant difference between the setups, and between the visual control and teleoperation. Finally, conclusions were drawn with emphasis on benefits of information technology in particular case.**

I. INTRODUCTION

Mobile robots are becoming more popular for various applications in everyday life [1], [2], [3]. The robots usually have a certain degree of autonomy while performing their tasks. However, occasionally it is necessary that the operator takes control over the robot, usually from remote location, either because robot's algorithms for autonomous operation failed or in order to complete a task that robot cannot complete during autonomous operation.

For efficient teleoperation of the robot, the operator must be well aware of robot's state and state of its environment. This is often referred to as situational awareness [4]. Situational awareness is understanding of the conditions of dynamic environment and is defined as *"The perception of the elements in the environment within a volume of time and space, the comprehension of their meaning, and the projection of their status in the near future."* [5]. When interacting with a robot, it is also required that the operator is aware of the consequences of the robot's actions both for the robot and for the operational environment. Hence, high degree of situational awareness is of great importance because it is a prerequisite for the efficient and safe teleoperation. This in turn can be achieved with appropriate and efficient design of HCI which should present all necessary information about the robot, it's state and the

state of the environment. The interface should be designed in a way that it offloads some of the workload from the operator and makes robot control easy and intuitive.

For the purpose of this research, safe navigation is defined as one with minimal (preferably zero) collisions with objects from the robot's operational environment. This is not always a trivial task due to dynamic and uncertain nature of the robot's environment, which is made more difficult by implementation of teleoperation control. Given only live video information from a camera mounted on the robot, operator is usually not sufficiently aware of the state of the robot's environment, especially robot's distance to obstacles (lack of depth perception), and possible nearby obstacles (camera's limited field-of-view). Therefore, some additional data must be provided to improve the operator's awareness. This can be achieved by using different sensing modalities like infrared, ultrasound or laser-based proximity sensors, RGB-D cameras, panospheric cameras, or usage of multiple cameras mounted on the robot. Additional information that is often included in navigation tasks, which provides limited amount of data, is the environment map. It can be a helpful aid, since the operator may determine robot's pose within environment at any time during navigation, as well as provide approximate information about proximity of (static) obstacles, and can be used as a reference.

The manner in which information from sensing devices is presented to the operator also affects performance of teleoperation [6]. Graphical interfaces containing multiple information fused in one frame (a form of augmented reality) were shown to be more efficient than arrangement when they were presented in multiple frames, side-by-side. Additional element that can influence operator's performance in teleoperation tasks is choice of HCI, which is considered as means for communication and interaction between the human and computer [7]. Choosing efficient HCI is a challenging task and may not be unique for different operators, because it may depend on operators' level of training [8]. Studies showed that virtual reality (VR) approach to teleoperation of robots is possible as a way to enhance both display and situation awareness [9]. Additional benefit of virtual reality is that safe training of operator in teleoperation can be achieved.

Another important variable that needs to be addressed is time (network) delay. It is well known that teleoperation is affected by latency, that may have a significant impact on teleoperation performance [10]. High time delay can make teleoperation impossible regardless of quality of HCI, while variable time delay (which is frequent in practical applications) can contribute further to degradation of performance in teleoperation. The sources of delay are various, and include both network delays and processing and sensing delays. The impact of time delay on teleoperation performance is not studied in this paper, but is only recorded and reported. This means that it changed freely, but due to dedicated network equipment used in the experiment is kept as low as possible.

Teleoperation of a mobile robots has been extensively studied topic and various and innovative approaches have been proposed recently. In [11], various aspects of newly proposed ecological 3D interfaces are compared to traditional 2D interfaces, while in [12] different control interfaces (including virtual reality based ones) are compared in order to assess impact on operator's performance. In [13], a haptic interface is proposed for teleoperation of mobile robots. However, to our best knowledge, majority of research on teleoperation was done in simulations [6], [11], [10]. In the paper we present a study of teleoperation that was performed on a real mobile robot, and draw conclusions based on it.

The rest of the paper is structured as follows. In Section 2 experimental design and setup are explained and used analysis tools are briefly presented. Section 3 presents obtained results as well as associated discussion, while in Section 4 conclusions are drawn and possible future research directions discussed.

II. MATERIALS AND METHODS

A. Experimental Design

The experiment was designed as a between-subject user study where each participant had to complete set of four tasks using one of four setup designs. One variable was input method which had two levels: gamepad (joystick) and steering wheel. Second variable was graphical interface elements, again with two levels: with and without depth information. This resulted in 2x2 design setup. Besides depth information, both graphical interfaces included live video from robot-mounted camera and map of the location with robot model localized within it.

The user study was performed with 32 test subjects, 8 in each of four setups: gamepad + GUI (Graphical User Interface) without depth, gamepad + GUI with depth, steering wheel + GUI without depth and steering wheel + GUI with depth. Participants were all volunteers recruited from Faculty staff and students which gave informed consent for participation in the study. Summary of participants' demographics are presented in Table I.

The experiments were performed using *Turtlebot 2*, differential drive mobile robot, depicted in Fig. 1. Developed software support was based on Robot Operating System (ROS) [14]. The robot was equipped with a laptop, running Ubuntu 14.04 with ROS Indigo (which was used as a ROS master), live camera and 3D Kinect sensor with 640 px × 480 px resolution and 30 fps for both camera and depth streams.

TABLE I
PARTICIPANTS' DEMOGRAPHIC DATA

(F)emale and (M)ale.

Setup	No.	F/M	Age [years]	
			μ	σ
Gamepad + GUI without depth	8	5/3	30.63	5.04
Gamepad + GUI with depth	8	2/6	31.63	6.23
Wheel + GUI without depth	8	1/7	33.13	10.08
Wheel + GUI with depth	8	4/4	32.38	6.28
OVERALL	32	12/18	31.94	6.87

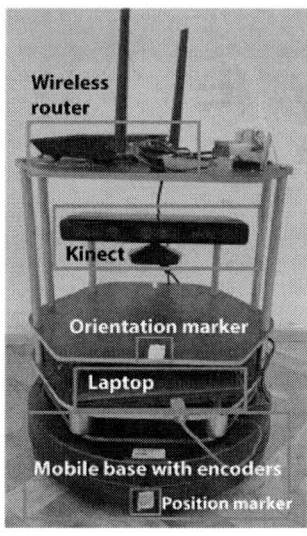

Fig. 1. An overview of Turtlebot 2 robot used in the experiments

The communication between the robot and the PC on which experiment was conducted was based on dedicated IEEE 802.11n wireless network. On robot's side, Asus RT-N12+ wireless router was used, while on PC side, running Ubuntu 16.04 and ROS Kinetic, TP-Link TP-WN722N USB wireless antenna was used. This setup ensured robot range of about 25 m radius (in non-ideal, out-of-line-of-sight conditions). The setup is depicted in Fig. 1. Robot's base ROS node (used for robot control) and Kinect driver for capturing both live camera and depth data were running on robot's laptop, while control interface driver, map (localization) and depth information interpreter (if used) were running on a PC.

The map of the environment (Fig. 2) in which the experiments were conducted was also provided to the test subjects, with robot's location visible on it in the real time. The map itself was built prior to the experiments using LIDAR sensor mounted on another (in-house built) robot and ROS package *gmapping*, which is based on [15]. More advanced robot was used for mapping due to better odometry and depth sensors, which resulted in more precise environment map.

All test subjects were given 10 minutes before the experiments to get familiar with the system and to try controlling the robot (both visually and in teleoperation mode).

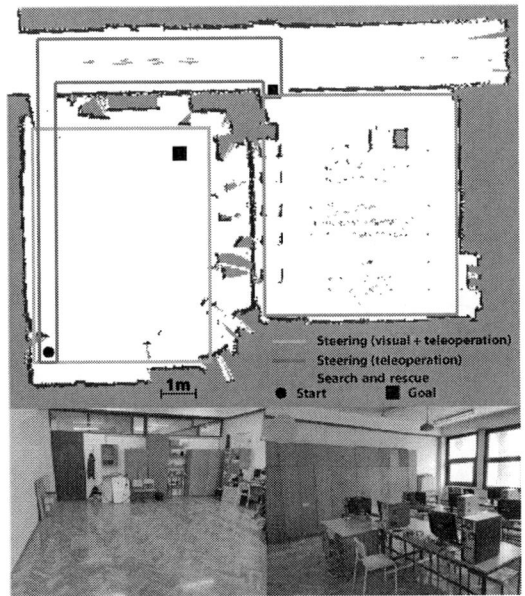

Fig. 2. Map of the entire experimental environment. Please note areas used for particular task. Tasks T1 and T2 were conducted in the green marked area, T3 in the red marked area, and T4 in the blue marked area.

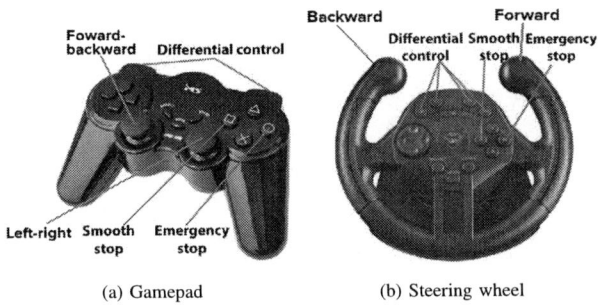

(a) Gamepad   (b) Steering wheel

Fig. 3. Control interfaces used in the experiment

1) Control Interfaces: All control interfaces used in the experiment were programmed so that the operator could use holonomic or differential control mode and could easily switch between them (if needed). Also, all interfaces have keys for both smooth and emergency stop. Control interfaces that were used in the experiment were MS Industrial Console 6in1 wireless gamepad which communicates on 2.4 GHz frequency with 6 m range, and Trust GXT 570 gaming steering wheel (which was used without floor pedals).

Both control interfaces (gamepad and steering wheel) used the same command interpreter that generated robot's velocity commands based on input commands given by the test subject. The maximum velocity of the robot was set to 0.45 m/s for linear motion and 1 rad/s for angular motion, with constraint that the velocity of each robot wheel cannot exceed 0.5 m/s in case of combined linear and angular motion. This

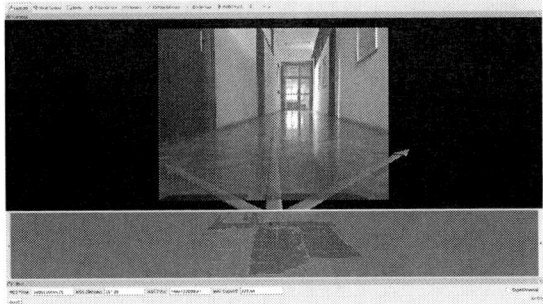

Fig. 4. GUI with depth information (green arrows) used in the experiments. When not using depth information, GUI is identical with exception of arrows.

configuration enabled cornering while driving at maximum allowed linear velocity.

2) Graphical User Interface: The intuitive and efficient GUI is one of the key factors for efficient and safe navigation of the robot from remote location. Our aim was to assess if performance of the operator improved when (s)he was presented with depth information as opposed to only being presented with camera (2D) information. Depth information from Kinect sensor was overlaid on top of the live image from camera, reducing number of windows which test subject had to observe. Depth information was provided in a form of three transparent green arrows: one pointing directly forward and two at 25 degree angles to the left and to the right from central arrow (see Fig. 4). The length of the arrows was proportional to the distance to the nearest obstacle in direction of the particular arrow. Additionally, over each arrow a number representing distance in meters was added so to provide full information with minimal effort of the operator. In cases when obstacle distance was out of the range of sensing device, the arrow turned red and in that case last good measurement distance was displayed. Please note that this type of interface can easily be upgraded with additional arrows covering wider field of view since Kinect 1.0 has 57° horizontal field of view and angular resolution of 0.09°.

3) Operator tasks: The operators, regardless of used design setup, had to complete four tasks. In the first two tasks, the operators had to guide the robot to a desired position and orientation as accurately and as fast as possible (without emphasising either), both visually, without any feedback from the robot (T1), and in teleoperation mode (T2). The starting point, as well as desired position and orientation were the same in both tasks, and were marked on the floor. Both locations were known to test subjects. While teleoperating the robot, the test subjects could not see or hear the robot due to physical barriers (cardboard screen) and the fact they wore earmuffs. Time of completion, distance between desired and actual position (in direction of x and y axes) and orientation (angle) were measured, and number of collisions was counted (if any). It should be noted that there were no obstacles in the direct path of the robot (from start to goal position) but there were obstacles outside it, and some of the subjects collided

1017

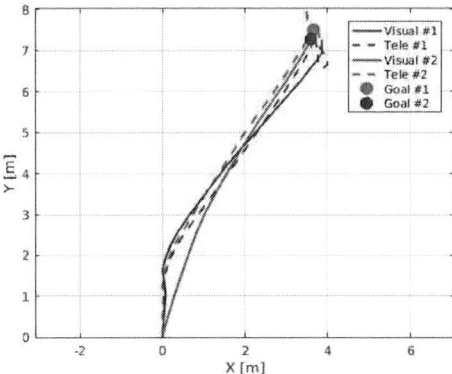

Fig. 5. Example of comparison of motion trajectories obtained using visual and teleoperaton conditions for two test subjects

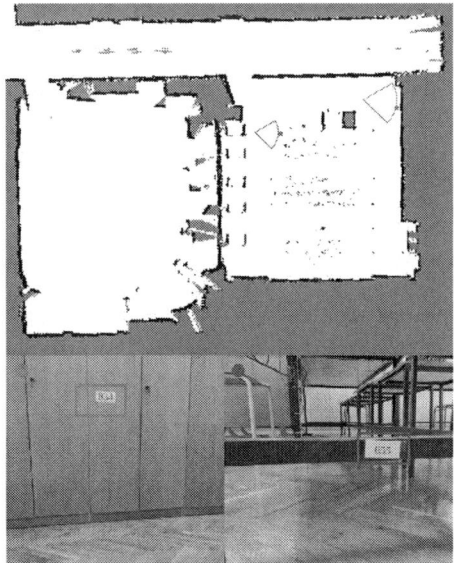

Fig. 6. Example of labels used for search and rescue task and their location on the map (marked in red and blue along with their locations on the map)

with them while trying to position the robot more accurately. Example of resulting robot motion trajectories can bee seen in Fig. 5 for two test subjects for tasks T1 and T2.

Another, more challenging, steering task in teleoperation mode (T3) was also performed. The goal of this task was the same as previous one, i.e. to steer the robot to given position and orientation as accurately and as fast as possible, but this time the route the operator had to take was significantly longer than in previous tasks (please see Fig. 2). Time of completion, distance between desired and actual position (in direction of x and y axes), and orientation (angle) were measured, as well as number of collisions.

The final task was a simulation of a simple search and rescue mission (T4) where the goal was to find as many objects as

possible in a dislocated environment in a limited time frame of five minutes. In the experiment ten white labels (examples shown in Fig. 6) with unique random three-digit codes were printed in different font sizes (from 20 pt to 200 pt) and were attached to objects of interest that the test subjects had to find. Test subjects had to identify the codes. The labels were positioned so that they were concealed and that they were visible only from some (relatively) small part of the environment. This in turn resulted in the requirement that the operator had to steer the robot around the environment to a specific position and orientation to be able to identify it (just like trying to find victims in urban search and rescue operation). The time of completion of the task was measured (in case test subject successfully found all the objects before time limit), number of collisions, as well as number of identified labels (and which ones).

Tasks T1, T2, T3, and T4 were given to each test subject in predefined random order, so to compensate for possible learning effects.

B. Statistics

1) Two-Sample Student's t-test: This test was used to test the null hypothesis that means and variances of two population samples are equal.

2) Kruskal-Wallis test: This test is a non-parametric test which aims to determine if medians of multiple groups are different (similar to ANOVA but without underlying assumptions). In case of our experiments, these groups are experiment setups. The null hypothesis for Kruskal-Wallis test is that population medians are equal. Rejecting the null hypothesis provides the information that there is significant difference between the populations, but will not indicate which ones are different and thus post-hoc testing needs to be performed (Tukey Honestly Significant Difference test in our case).

All statistics calculations were done at $\alpha = 0.05$ significance level. For more details about statistical tests, please refer to [16].

III. RESULTS AND DISCUSSION

A. Comparison of visual and teleoperated steering

The analysis was performed on the results of steering tasks where test subjects had to steer the robot from initial to final position and orientation using both visual feedback and teleoperation. Summary of obtained results are presented in Fig. 7. They show that steering tasks were completed, on average, 20% slower in teleoperation, as well as that positioning error was almost doubled in teleoperation (per individual setup). However, both for T1 and T2, considering time of completion there is no statistically significant difference between the setups, according to Kruskal-Wallis test. This is a bit surprising, but might be due to relatively simple tasks which required short amount of time. Similar results were obtained for positioning error (Euclidean distance from the desired position to the measured position). The two-sample t-test was used to test if means of visual and teleoperation controls (for all three dependent variables) were different with

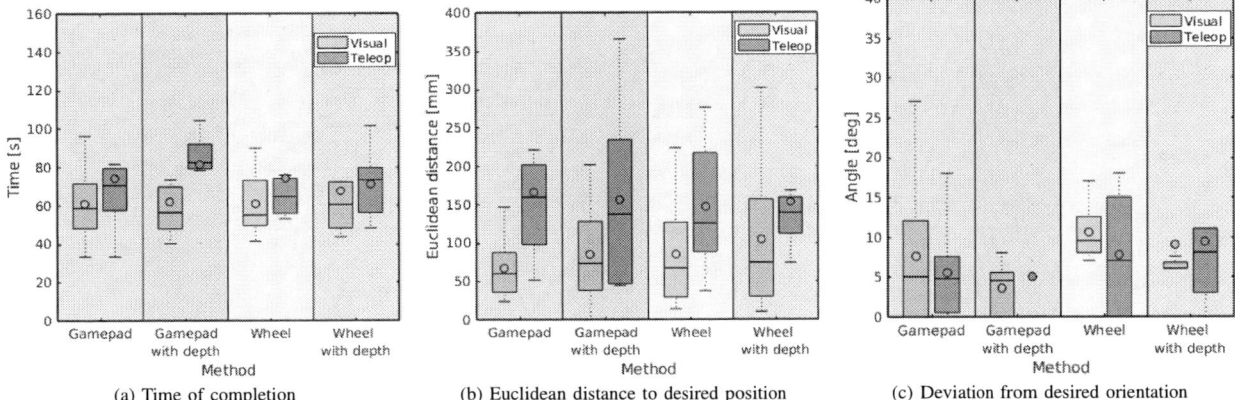

(a) Time of completion   (b) Euclidean distance to desired position   (c) Deviation from desired orientation

Fig. 7. Comparison of performance between visual feedback and teleoperation with different setups

statistical significance. Tests for all the setups, after applying Bonferroni correction for multiple comparisons, showed that samples do not differ, which might suggest that either tasks were relatively simple so differences did not emerged or that operators' situational awareness was good in both cases. However, when considering positioning errors along x and y axes separately, the two-sample t-test significant difference was observed only for steering wheel with depth test setup and only along y axis ($p = 0.0343$ after applying Bonferroni correction), but not along x axis. We believe this might be due to test setup configuration (i.e. distances between start and goal point in x and y direction), but further testing is required.

Considering number of collisions, a very low number of such events occurred, a total of 6 in T1 and 9 in T2. In T1 number of collisions was almost equal for each of the setups, while in T2 8 out of 9 collisions occurred in setups with depth, which might suggest that users were distracted with additional information which was presented to them.

B. Assessment of teleoperation steering setup

In third task (teleoperation in more demanding conditions), there was also no statistically significant differences between the setups, in time of completion, deviation from desired final position and absolute deviation from desired final orientation. Results obtained in this task are presented in Table II, which demonstrates that, on average, subjects that used steering wheel completed their task faster than subjects using gamepad.

C. Search and rescue analysis

For search and rescue task, summarized results for all test setups are presented in Fig. 8. They demonstrate that for setup with steering wheel without depth information operators achieved less collisions (in total) than in other setups. However, no statistical significance was detected both for number of identified labels and for number of collisions (using Kruskal-Wallis test).

The maximum number of identified labels in this task was seven, while only one of the labels was not been found by any

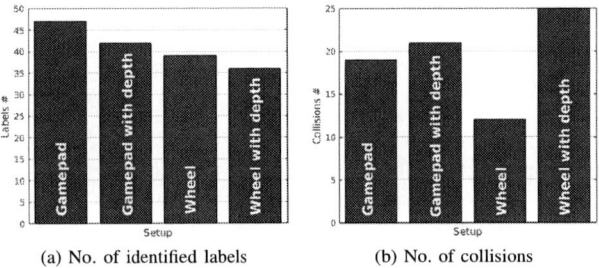

(a) No. of identified labels   (b) No. of collisions

Fig. 8. Search and rescue task results

of the test subjects. The average number of identified labels per setup was 5.875 for gamepad, 5.25 for gamepad with depth, 4.875 for steering wheel and 4.5 for steering wheel with depth. The average number of identified labels across all setups was 5.125.

D. Time delay

The time delay was measured for live camera feed and for depth feed. The summary of delay data, which was recorded with a resolution of up to 1 ns, is shown in Table III. From the table it can be seen than latency was on average similar across all test conditions.

IV. CONCLUSIONS

In the paper, the impact of user interface elements on teleoperation of mobile robot was studied with two control interfaces and two GUIs for a total of four setups. The experiment was designed as a between-subject user study in which a subject had to complete three steering tasks and a search and rescue task. Obtained results showed, as was expected, that all performance parameters were better for visual feedback control than for teleoperation mode (regardless of used setup). It was also observed that although various parameters degraded for teleoperation setups in most cases they did not have statistically significant difference, which was

TABLE II
RESULTS OF TELEOPERATION TASK IN T3

Setup	Time of compl.		Pos. error		Orient. error	
	μ [s]	σ [s]	μ [mm]	σ [mm]	μ [°]	σ [°]
Gamepad + GUI without depth	96.55	22.1	171.73	33.87	6.13	6.42
Gamepad + GUI with depth	106.50	22.27	138.43	75.38	7.37	7.00
Wheel + GUI without depth	92.10	30.02	138.65	81.48	4.25	4.53
Wheel + GUI with depth	92.94	16.32	172.56	101.11	10.50	9.13
TOTAL (mean)	97.02	22.8	155.34	75.1	7.06	7.01

TABLE III
TIME DELAY (IN SECONDS)

Setup	Live image		Depth	
	μ	σ	μ	σ
Gamepad + GUI without depth	0.0440	0.1700	0.0188	0.2049
Gamepad + GUI with depth	0.0432	0.1600	0.0179	0.1925
Wheel + GUI without depth	0.0445	0.1652	0.0186	0.1966
Wheel + GUI with depth	0.0455	0.2016	0.0209	0.2344
TOTAL (mean)	0.0443	0.1753	0.0191	0.2080

surprising. We believe that this lack of statistical significance might be in part due to the fact that sample size per condition was somewhat small and tasks were rather simple. Thus we plan to perform additional measurements in the future to obtain more general results. It is also interesting to note that addition of depth information did not improve the performance. In fact this might be due to the fact that users could (based on their previous experience and knowledge of relative object sizes) estimate depth information from live camera stream.

For the search and rescue task additional observations were made. While there were again no statistically significant results (with similar note as before) obtained result seem to suggest that gamepad interface resulted in more labels being found. This might suggest that robot was more dexterous with that particular control interface. Numbers of collisions were also more consistent across subjects for gamepad control interface than for steering wheel. However, steering wheel had lowest cumulative number of collisions (interestingly, for the case of no depth information, which might suggest that used depth information mode overloaded and distracted test subjects). While majority of our analyses showed no significant statistical difference between the setups, use of more advanced control devices and more complicated user tasks might produce statistically significant differences as was the case in [12].

In the future we plan to conduct similar research with more advanced HCIs (keyboard, VR, haptic feedback devices, etc.) and provide a map overlaid on top of live camera image, as was done in [11] instead of in separate window, to study the impact on teleoperation. Furthermore, we also plan to build a driver model based on control commands given by the operators, and study the impact of time delay on teleoperation of real mobile robots.

ACKNOWLEDGEMENT

The authors would like to express gratitude to all test subjects who participated in the user study.

REFERENCES

[1] A. Birk, K. Pathak, S. Schwertfeger, and W. Chonnaparamutt, "The iub rugbot: an intelligent, rugged mobile robot for search and rescue operations," in *IEEE International Workshop on Safety, Security, and Rescue Robotics (SSRR)*. IEEE Press, vol. 10, 2006.

[2] H. Durrant-Whyte and T. Bailey, "Simultaneous localization and mapping: part i," *IEEE robotics & automation magazine*, vol. 13, no. 2, pp. 99–110, 2006.

[3] V. Čelan, I. Stančić, and J. Musić, "Cleaning up smart citieslocalization of semi-autonomous floor scrubber," in *International Multidisciplinary Conference on Computer and Energy Science (SpliTech)*. IEEE, 2016, pp. 1–6.

[4] J. M. Riley, D. B. Kaber, and J. V. Draper, "Situation awareness and attention allocation measures for quantifying telepresence experiences in teleoperation," *Human Factors and Ergonomics in Manufacturing & Service Industries*, vol. 14, no. 1, pp. 51–67, 2004.

[5] M. R. Endsley, "Design and evaluation for situation awareness enhancement," in *Proceedings of the human factors and ergonomics society annual meeting*, vol. 32, no. 2. SAGE Publications, 1988, pp. 97–101.

[6] C. W. Nielsen and M. A. Goodrich, "Comparing the usefulness of video and map information in navigation tasks," in *Proceedings of the 1st ACM SIGCHI/SIGART conference on Human-robot interaction*. ACM, 2006, pp. 95–101.

[7] S. Smith-Atakan, *Human-computer interaction*. Cengage Learning EMEA, 2006.

[8] T. W. Fong, F. Conti, S. Grange, and C. Baur, "Novel interfaces for remote driving: gesture, haptic, and pda," in *Intelligent Systems and Smart Manufacturing*. International Society for Optics and Photonics, 2001, pp. 300–311.

[9] L. A. Nguyen, M. Bualat, L. J. Edwards, L. Flueckiger, C. Neveu, K. Schwehr, M. D. Wagner, and E. Zbinden, "Virtual reality interfaces for visualization and control of remote vehicles," *Autonomous Robots*, vol. 11, no. 1, pp. 59–68, 2001.

[10] S. Vozar and D. M. Tilbury, "Driver modeling for teleoperation with time delay," *IFAC Proceedings Volumes*, vol. 47, no. 3, pp. 3551–3556, 2014.

[11] C. W. Nielsen, M. A. Goodrich, and R. W. Ricks, "Ecological interfaces for improving mobile robot teleoperation," *IEEE Transactions on Robotics*, vol. 23, no. 5, pp. 927–941, 2007.

[12] J. Jankowski and A. Grabowski, "Usability evaluation of vr interface for mobile robot teleoperation," *International Journal of Human-Computer Interaction*, vol. 31, no. 12, pp. 882–889, 2015.

[13] W. Li, Z. Liu, H. Gao, X. Zhang, and M. Tavakoli, "Stable kinematic teleoperation of wheeled mobile robots with slippage using time-domain passivity control," *Mechatronics*, vol. 39, pp. 196–203, 2016.

[14] M. Quigley, K. Conley, B. Gerkey, J. Faust, T. Foote, J. Leibs, R. Wheeler, and A. Y. Ng, "ROS: an open-source Robot Operating System," in *ICRA workshop on open source software*, vol. 3, no. 3.2. Kobe, 2009, p. 5.

[15] G. Grisetti, C. Stachniss, and W. Burgard, "Improved techniques for grid mapping with rao-blackwellized particle filters," *IEEE Transactions on Robotics*, vol. 23, no. 1, pp. 34–46, 2007.

[16] S. M. Ross, *Introductory Statistics*. Academic Press, 2010.

The Challenge of Measuring Distance to Obstacles for The Purpose of Generating a 2-D Indoor Map Using an Autonomous Robot Equipped with an Ultrasonic Sensor

R. Rodin and I. Štajduhar

University of Rijeka, Faculty of Engineering, Department of Computer Engineering, Rijeka, Croatia
rino.rodin@hotmail.com

Abstract – In this paper, we deal with some preliminaries concerning the problem of creating a reasonably accurate 2-D indoor map, using an autonomous mobile robot equipped with an ultrasonic sensor. For this purpose, we utilize the robotic frame DiddyBorg, coupled with the latest Raspberry Pi model. The accuracy of the used ultrasonic sensor is tested in a simulated and a real-world environment. Ultrasonic sensor flaws cause errors in the area mapping process, in the form of precision decline and unwanted ghost points, usually around the corners of an indoor environment. We present a method for filtering these unwanted occurrences. The method has been tested in a controlled test box environment and in the real-world indoor environment. We have used a square, a triangle, and a hexagon as shapes for the testing environment. As a basis for the mapping process, ordinary trigonometry was used, whereas cluster analysis was used for the removal of errors. Experimental results suggest that the proposed method is reasonably accurate in determining correct position and distance of measured points from single-point scans.

I. INTRODUCTION

One of highly important features of autonomous robots is their ability to move independently (i.e. without human intervention) in different environments. This feature can find its application in many different areas. For example, a robot capable of moving autonomously can be used in human rescue missions [1], as a means for daily transportation [2], for firefighting [3], or for security and military-related applications [4]. In order to move autonomously, i.e. to calculate a path from its source to its destination, a robot must be able to position itself with regard to its environment. The best way for a robot to obtain information about its current position is to have a genuine map of its current surroundings stored inside its memory. This problem is generally known as SLAM or Simultaneous localization and mapping [5]. To acquire a spatial model of its surroundings, a robot should have a sensor that helps it perceive its environment. Seeing that there exists a need for reducing the cost of building an autonomous robot [1], we opted for using a low-cost ultrasonic sensor instead of a significantly more expensive laser rangefinder. A short comparison of sensor prices is

This work has been supported in part by the University of Rijeka under the project number 16.09.2.2.05.

shown in Table I. Furthermore, the measuring accuracy of ultrasonic sensors is not affected by any changes in type, color or shade of the reflecting surface. However, ultrasonic sensors also have some drawbacks. In order to obtain the highest possible accuracy of an indoor map using limited hardware resources, we have primarily focused on improving the accuracy of the data obtained from the ultrasonic sensor. Seeing that our ultimate goal is to create an autonomous robot capable of mapping and navigating in an indoor environment entirely by itself, we have also described the robotic platform used to achieve that goal. However, the focus remains on the challenges concerning the ultrasonic sensor and their plausible solutions.

TABLE I. ULTRASONIC AND LASER SENSOR PRICE COMPARISON

Sensor	Store	Price
HC-SR04 Ultrasonic Range Finder	robotshop.com	USD $2.5
HC-SR04 Ultrasonic Range Finder	ebay.com	USD $1.3
Parallax 15-122cm Laser Rangefinder	robotshop.com	USD $99
LIDAR-Lite v3	sparkfun.com	USD $150

II. HARDWARE USED

In this section we describe the hardware necessary for assembling a robot which can be utilized for the purpose of indoor mapping.

A. Raspberry Pi 3 Model B

We decided to use the latest Raspberry Pi 3 model B as a control unit for our robot. Raspberry Pi is a COTS (Commercial of-the-shelf) small-size computer that runs Linux Raspbian operating system. It is powered by Broadcom 1.2GHz 64-bit quad-core ARMv8 CPU, having 1GB RAM. Seeing that the Raspberry Pi 3 offers a relatively high computing power for a low price, and has rather small dimensions (85x56mm), we concluded that it should meet our needs. Furthermore, the chosen model has a built-in 802.11n Wireless LAN module, which allowed us to access the robot remotely and, therefore, made parameter changing during our experimenting a lot easier.

B. Ultrasonic sensor HC-SR04

The ultrasonic sensor uses sonar to determine distance to an object. There are two main pins on HC-SR04, trig pin and echo pin. By connecting the high level signal for at least 10μs to the trig pin, the module transmits a 40kHz sound wave and we store the start time into a variable. As the sound wave reflects and returns to the module, Echo pin comes to a state of a logical 1. At this point, we store an end time. We get the measured distance *dist* using the following expression:

$$\text{dist} = (\text{EndTime-StartTime}) / 2 * 340 \text{ m/s} \quad (1)$$

As stated in the datasheet [6], the ultrasonic ranging module HC-SR04 provides 2cm-400cm non-contact measurement functionality with ranging accuracy up to 3mm. However, our tests show different results. Measuring short distances matches the data in the datasheet [6], whereas measuring long distances produces repeatable errors. While driving or rotating the robot, the CPU usage is usually around 25%. Seeing that the measuring is performed before any computationally intensive tasks, the percentage of the CPU usage does not affect measuring accuracy. Regardless of that fact, the repeatable errors are still present. This is an important factor because it means that we have to define the optimal distance range at which our sensor works well enough for creating an accurate map. Therefore, the plane surface had been placed at a distance of 2 cm in front of the robot and 10 measurements have been made. We compared each distance measurement acquired using the ultrasonic sensor against the real distance measured by a meter, and noted the number of accurate measurements and the number of inaccurate measurements. A measurement was considered inaccurate if it differed by more than 2cm from the one measured by a meter. After every 10th measuring, the distance was incremented by 2cm. At the end, the number of correct measurements was divided by the total number of measurements for each group of distances, expressing the accuracy as a percentage. The results of the measuring experiment are shown in Table II. The results suggest that the distance obtained using a sensor at a distance from anywhere between 2cm and 80cm has a high probability of matching the real distance. This means that our sensor has to be at most 80cm away from an obstacle to be certain how far the obstacle really is. If a sensor is more than 80cm away from the obstacle, the obstacle will not be represented on a map. The obstacle will be represented on a map as soon as the robot moves close enough to the obstacle. Also, the minimum distance is 2cm, meaning that the robot should never approach the obstacle in a way that puts an obstacle closer than 2cm to the sensor. Otherwise, the sensor could be damaged.

TABLE II. Sensor accuracy decline with respect to the increase of the measured distance

Distance in cm	Accuracy
2-40	100,00%
40-60	99,50%
60-80	99,33%
>80	56,14%

C. DiddyBorg frame

As a backbone for our robot, we decided to use DiddyBorg. Produced by Freeburn Robotics Limited, DiddyBorg is a COTS bundle of chassis, motors with motor controller, wheels, power converter and a battery holder. Chassis is laser-cut 3mm cast acrylic perspex of robust build quality, having 5kg maximum carrying load. Robot characteristics are shown in Figure 1 and Figure 2. The bundle includes six 1:71 (60rpm) 6V 220mA motors. Each motor has the output power of 1.3W and produces 0.35 kg cm of torque. For a power converter, we have used the BattBorg. The BattBorg is also produced by Freeburn Robotics Limited and is based on OKI78SR DCDC converter with maximum 1.5A output. It allowed us to power the Raspberry Pi using 1.2V 2500mAh Panasonic Eneloop batteries.

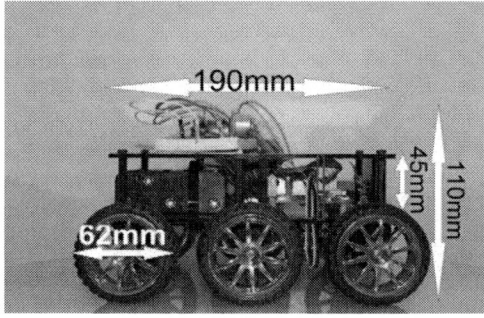

Figure 1. Robot side measurements.

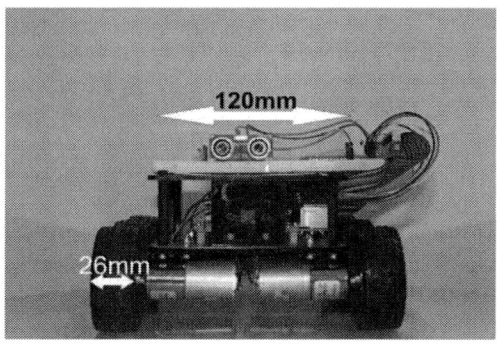

Figure 2. Robot front measurements.

Figure 3. General overview of the robot. Ultrasonic sensor is visible on the top of the robot.

III. GENERATING THE MAP

First attempts of creating a space (obstacle) map were executed in a controlled environment. We constructed three small wooden boxes and placed the robot in the center. For the shape of the box we chose a square, a triangle and a hexagon. Those shapes were chosen because we believe them to be representative for closed indoor environments. The polygons with the measurements are shown in Figures 4-6. By rotating the robot for 360° in the direction shown by the blue arrow on Figures 4-6, we scanned the entire environment. Clockwise direction of rotation is, otherwise, not important. To calculate the x, y coordinates of the points measured inside the boxes, the following expressions were used:

$$x = len * \cos(\alpha * \pi/180) \qquad (2)$$

$$y = len * \sin(\alpha * \pi/180) \qquad (3)$$

Len variable represents the measured length received from the sensor, whereas α is the angle of rotation relative to the origin converted to radians. Every 5° or $\pi/36$ radians a new point was obtained. This angle was optimal for the type of surface under the robot because the robot changes the speed of the wheels to match the given angle of rotation. A wider angle requires a higher speed of rotation. A higher speed causes the robot to lose traction and finish the rotation after the point from which it started. An angle of rotation of 6° on the same surface would result in the robot not stopping at 360° but rather at 384°. The results of the mapping are shown in Figures 7-9.

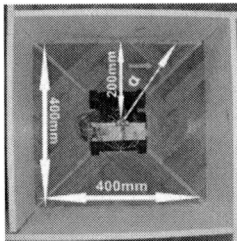

Figure 4.   Square test box. Dimensions: 400x400mm. Direction of rotation: right. Angle of rotation: 5°.

Figure 5.   Triangle test box. Dimensions: 700x700mm. Direction of rotation: right. Angle of rotation: 5°.

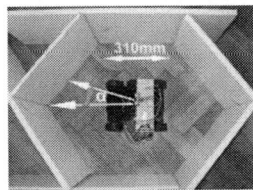

Figure 6.   Hexagon test box. Each edge is 310mm long. Direction of rotation: right. Angle of rotation: 5°.

IV. SONAR LIMITATIONS

Because of the limitations of the used ultrasonic sensor, raw maps are not accurate, primarily due to the phenomenon known as "ghost points" [7]. When the robot rotates 35° from the initial orientation, the emitted sound wave does not reflect directly back to the sensor. On its way back to the sensor, the wave reflects from other surfaces of the box as well, causing an unwanted time delay. This creates an illusion that the corners of the box are further away from the edges of the box, as shown in Figures 7-9. The ghost points are visible in the square and in the triangle, but not in the hexagon. This only seems so because of the angle between the edges of the hexagon, but the phenomenon is still present. Blue line in Figures 7-9 represents the real outline of the testing boxes used, whereas the red points represent the locations obtained using the ultrasonic sensor. The ghost points are in the corners far away from the outline of the boxes. When performing a real indoor environment mapping, this problem is also expected to appear. Therefore, in order to obtain an accurate map, it has to be solved. We discuss a possible solution next.

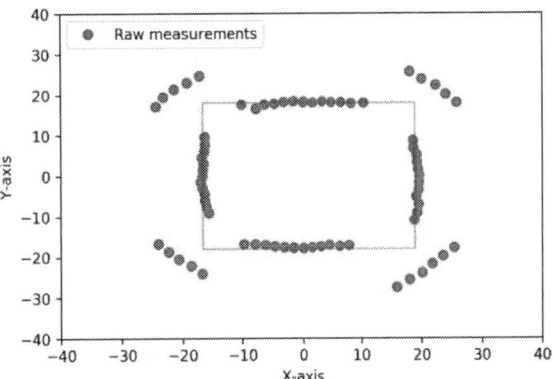

Figure 7.   Raw map of the square test box. The blue line represents the real edges of the test box.

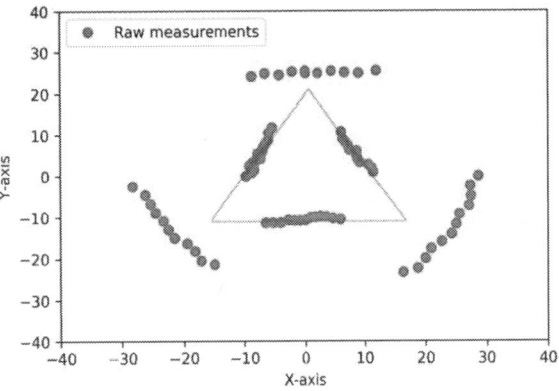

Figure 8.   Raw map of the triangle test box. The blue line represents the real edges of the test box.

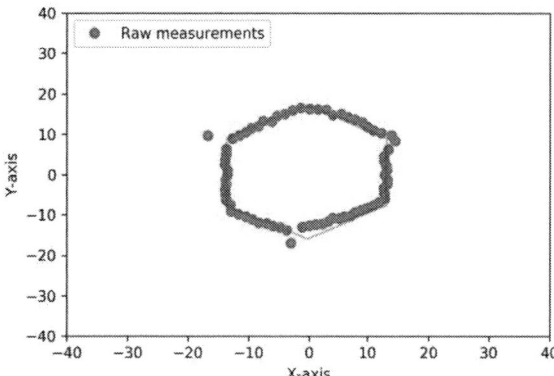

Figure 9. Raw map of the hexagon test box. The blue line represents the real edges of the test box.

V. ELIMINATING MEASUREMENT ERRORS

For solving the problem of ghost points, we decided to use cluster analysis. Cluster analysis or clustering is a technique that groups individual measurements in a population together, based on their similarity. The idea is to group the nearest values obtained from the ultrasonic sensor into clusters. Once the clusters are formed, the robot has to decide which ones consist of ghost points and which ones are populated with accurate measurements. The overview of the proposed method is given in Figure 10.

A. Cluster analysis

In order to cluster points, K-means algorithm was used [8]. K-means is an iterative algorithm which operates in three steps. First, it randomly scatters a predetermined number of points called centroids among the data points. Second, it assigns each data point to the closest centroid, thus forming a cluster. Finally, it calculates the average of all data points in a cluster, and moves the centroid to that average location. Cluster analysis was performed using the Python library *scikit-learn*.

K-means algorithm requires setting the number of clusters to work properly. Each of our controlled test-box environments required a different number of clusters. For example, the triangle required 3 clusters for the accurate measurements, and 3 for the ghost points. The square required 4 clusters for the accurate measurements, and 4 for the ghost points. Finally, the hexagon required 6 clusters for the accurate measurements, and 0 for the ghost points. To determine the optimal number of clusters for each individual setup, a function for estimating the efficiency-of-clustering [9] was used. This score, called *silhouette* score, is determined using the mean intra-cluster distance x and the mean nearest-cluster distance y for every cluster. The expression is given in (4).

$$\text{silhouette score} = (y-x) / \max(x, y) \quad (4)$$

Silhouette score is calculated independently for different numbers of clusters, ranging from 2 to 10. The upper bound was chosen as such because the optimal number of clusters in our experimental environments has never surpassed 10. This, however, can be adapted to the complexity of an indoor environment under inspection. The number of clusters having the highest score will be set as the optimal number of clusters for that specific set of data points.

B. Discarding the ghost points

After the robot has successfully estimated the correct amount of clusters, it will start discarding the clusters populated with the ghost points. Since the ghost points clusters are always smaller than the clusters populated with the accurate measurements, our method will find the biggest cluster and discard all the clusters that are smaller than the size of the biggest cluster, minus the tolerance. The tolerance represents the permissible limit of variation in the size of the biggest clusters, i.e. the clusters populated with the accurate measurements. In our experimental environment, the tolerance of -3 points was sufficient. However, the tolerance should be adjusted depending on the angle of rotation because any change in the angle of rotation affects the overall number of data points on the graph. The clusters generated by the method for every test-box environment are shown in the Figures 11-13. The final elimination of ghost points is shown in the Figures 14-16. Figure 17 shows the polygon and the result of applying the method to the real-world environment.

```
1) FOR i IN RANGE(2,10):
      (silhouette_score[i], noClusters[i])
        <- K-means(noClusters=i)

2) optimal_i <- argmax(silhouette_score[j])

3) largestClusterSize <- MAXIMUM CLUSTER
      SIZE FOR optimal_i CLUSTERING

4) SHOW ALL CLUSTERS HAVING SIZE(i) >=
      largestClusterSize - tolerance
```

Algorithm 1. Pseudocode for the proposed method of eliminating measurement errors.

VI. RESULTS

Experimental results are presented in this section. Figures 11-13 are depicting clusters found by the K-means algorithm. The accurate measurements that remain after filtering the ghost points are shown in Figures 14-16.

By observing the results, we can state that the method is highly efficient in the removal of ghost points. However there are also some imperfections of the method to be aware of. For example, it can be noticed that in Figure 12 and in Figure 13, the number of clusters does not match the number of clusters that we would intuitively assign to that set of data points. The 3 sets of ghost points in Figure 12 are divided into 4 clusters, but only 3 are necessary to distinguish the ghost points from the accurate data points. Furthermore, in Figure 13 it is possible to see that the number of clusters is 5, but we

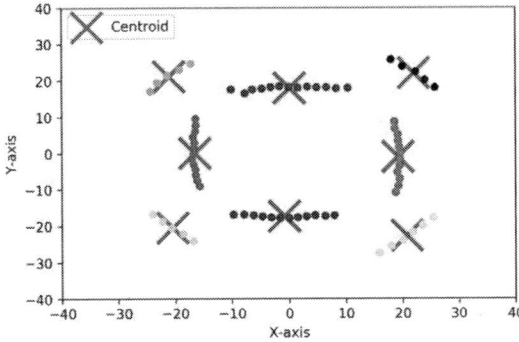

Figure 11. Cluster analysis of the square. Every color represents a different cluster. Centroids are in the middle of each cluster.

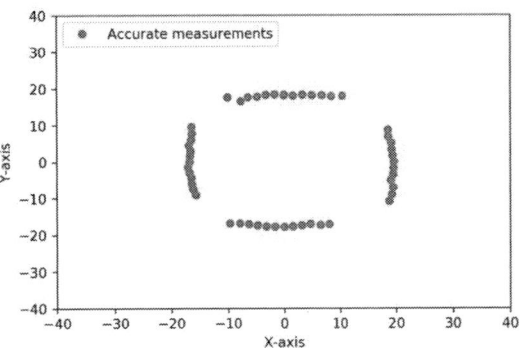

Figure 14. Accurate measurements of the square after the removal of ghost points.

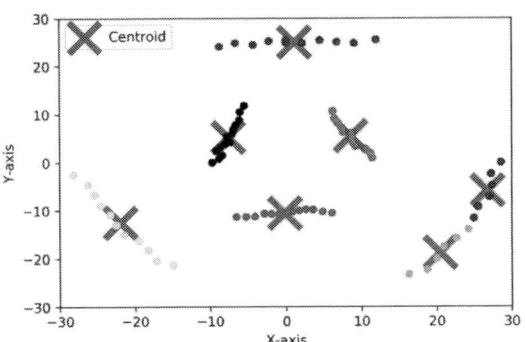

Figure 12. Cluster analysis of the triangle. Every color represents a different cluster. Centroids are in the middle of each cluster

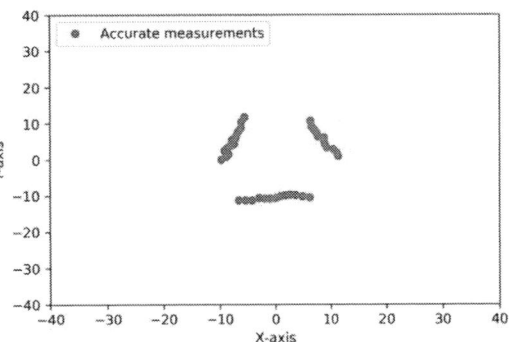

Figure 15. Accurate measurements of the triangle after the removal of ghost points.

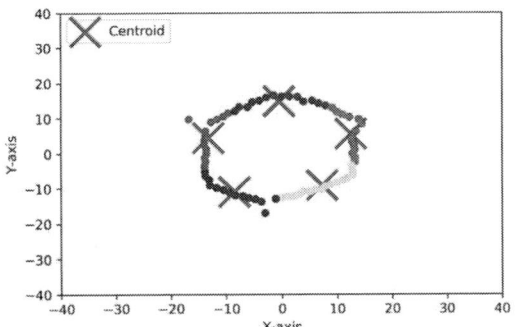

Figure 13. Cluster analysis of the hexagon. Every color represents a different cluster. Centroids are in the middle of each cluster

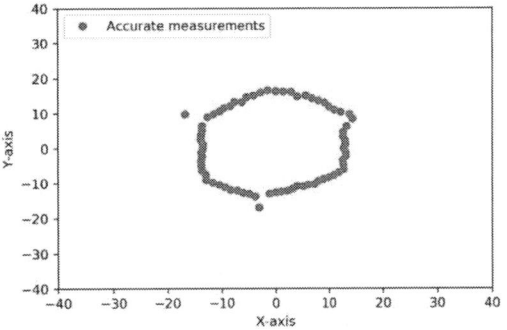

Figure 16. Accurate measurements of the hexagon after the removal of ghost points.

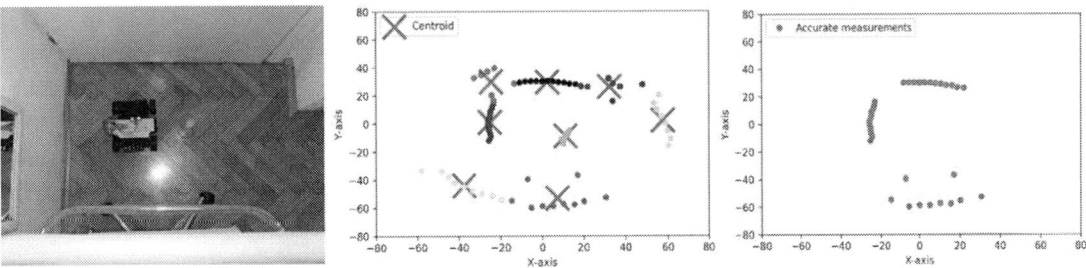

Figure 17. The real-world environment with the office chair and the bottle next to the robot (left), cluster analysis with each cluster in different color (middle) and accurate measurements with the ghost points removed (right).

1025

would expect the number of clusters to be 6 in a hexagon. This is an imperfection inherited from the K-means algorithm. The K-means algorithm assigns the clusters randomly, therefore causing minor deviations in the total number of clusters. This does not pose a great problem, since the method will usually assign more clusters than necessary, therefore reducing the size of each cluster. If cluster size is reduced, then there exists a higher probability that the cluster will be discarded, since we tend to keep only the biggest clusters. If a cluster is discarded, it only means that the area where that cluster would be is now empty. An empty area suggests that there is a part of an indoor environment that is yet to be discovered, and the robot should approach it and examine it more thoroughly.

Considering that the robot will ultimately operate in a real-world indoor environment we decided to test our method in one. We placed the robot between the walls in different sizes with the sensor facing the closed doors. An office chair and a bottle were added since we believe them to be the objects that the robot will often meet during the process of mapping. The raw map was then acquired by rotating the robot for 360° and measuring the distance every 5°.

Figure 17 shows the real-world indoor environment, the cluster analysis for the acquired set of data points and the final removal of ghost points. This experiment suggests that rejecting small clusters and keeping only the biggest ones also comes with a price. A small obstacle, bottle, visible in a real-world environment in Figure 17, represented with a gray cluster in cluster analysis, is not visible on the graph showing the accurate measurements. This happens because a bottle is represented by a small cluster and is therefore mistaken for a set of ghost points. This is an imperfection of the proposed method that has to be approached in a different way. One of the possible ways to detect such small objects would be to equip the robot with a tactile sensor that would additionally complement the accuracy of the map. However, this is a topic for our future work.

VII. CONCLUSION

The main focus of this paper was on solving the problem of ghost points caused by the imperfections of the ultrasonic sensor. We presented a method for elimination of ghost points, based on cluster analysis. An important task concerning the described method, was automatically determining an optimal number of clusters for any given environment. This was successfully solved using the *silhouette* scoring function. Another important task was the correct detection and removal of the ghost-point clusters. This is done with respect to the cluster sizes, i.e. smaller ones were discarded, whereas the larger ones were kept. It is important to note that the tolerance in the size of the clusters, not to be rejected, should be set with respect to the angle of rotation. The proposed method has shown some imperfections dependent on the K-means algorithm, but it has also successfully eliminated the ghost points from the test box

environments as well as from the larger part of the used real-world indoor environment.

Our research can be viewed as an extension to the research performed by Ilias et al. [10]. In their paper, the authors have presented a method for eliminating ghost points from data collected using an ultrasonic sensor bank attached to a mobile robot. However, there are some considerable differences. The method presented in [10] uses a trigonometric approach to eliminating the ghost points, whereas our method uses cluster analysis based on the K-means algorithm. As it can be observed in Figure 6 in [10], the method proposed by the authors leaves some fragments in the corners of the scanned environment, whereas our method completely removes all the ghost points. Furthermore, the authors have used 16 ultrasonic sensors while we managed to reduce the number of sensors to only one. The contribution of our paper is expressed in the use of the K-means algorithm for the purpose of eliminating the ghost points.

REFERENCES

[1] A. Davids, "Urban Search and Rescue Robots: From Tragedy to Technology", IEEE Intelligent systems 17.2, 2002, pp. 81-83.

[2] T. Rakib, and M.A. Rashid Sarkar, "Design and fabrication of an autonomous fire fighting robot with multisensor fire detection using PID controller.", 5th International Conference on IEEE Informatics, Electronics and Vision (ICIEV), 2016.

[3] T. Lozano-Perez, "Autonomous robot vehicles", Eds. Ingemar J. Cox, and Gordon T. Wilfong, Springer Science & Business Media, 2012.

[4] G. Song, Y. Kaijian, Z. Yaoxin, and C. Xiuzhen, "A surveillance robot with hopping capabilities for home security", IEEE Transactions on Consumer Electronics 55.4, 2009.

[5] H. Durrant-Whyte, and T. Bailey, "Simultaneous localization and mapping: part I.", IEEE robotics & automation magazine 13.2, 2006, pp. 99-110.

[6] HCSR04 datasheet: http://www.micropik.com/PDF/HCSR04.pdf

[7] H. Choset, K. Nagatani, and N.A. Lazar, "The arc-transversal median algorithm: a geometric approach to increasing ultrasonic sensor azimuth accuracy", IEEE Transactions on Robotics and Automation 19.3, 2003, pp. 513-521.

[8] R.O. Duda, P.E. Hart, and D.G. Stork, "Pattern classification", John Wiley & Sons, 2012.

[9] R.C. de Amorim, and C. Hennig, "Recovering the number of clusters in data sets with noise features using feature rescaling factors", Information Sciences 324, 2015, pp. 126-145.

[10] B. Ilias, S. A. A. Shukor, A. H. Adom, M. F. Ibrahim and S. Yaacob, "A novel indoor mobile robot mapping with USB-16 ultrasonic sensor bank and NWA optimization algorithm," *2016 IEEE Symposium on Computer Applications & Industrial Electronics (ISCAIE)*, Penang, 2016, pp. 189-194.

Automated Posture Assessment for Construction Workers

R.J. Dzeng*, H.H. Hsueh*, and C. W. Ho*

* National Chiao-Tung University/Department of Civil Engineering, Hsinchu, Taiwan

Abstract - Construction workers often suffer various kinds of musculoskeletal disorders (MSDs), which are injuries in the human musculoskeletal system. MSDs are often aroused due to sudden exertion such as lifting heavy equipment, or repeated and cumulative stressed motions. OWAS (Ovako Working posture Assessment System) may be used to evaluate the exposure of MSDs risk by sampling the snapshots of a worker's postures and categorize the postures. However, tracking and categorizing postures of different body parts by human eyes are tedious work with limited accuracy and easy to make mistakes even with facilitation of video recording. This research develops an automatic tracking and categorizing system, named Posture Assessment System for MSD (PAS-MSD) for OWAS using Microsoft Kinect. The PAS-MSD captures human postures during his/her movement, recognizes human skeleton, and assesses the risk of MSD. An experiment with typical construction activities such as handling and moving of materials, hammering, and tiling was conducted. Except for the hammering activity where the subjects' body parts were easily blocked by the target hammered box and could not be detected by Kinect, the posture identification accuracies for all other activities exceed 90% (i.e., 91.6%-93.9%). The OWAS categorization accuracies are also satisfactory, ranging from 85.4%-88.5%.

I. INTRODUCTION

Construction industry has been a major contributor to work injuries. Majority of the research and workplace guidelines has been focusing on mitigation of specific event-based or accidental injuries. Cumulative injuries caused due to prolonged adoption of stressful positions, occur in considerable measure and warrant an equal attention on development of solutions for helping in reduction of their occurrences. For this purpose, a tool has been developed that employs the body tracking abilities of Kinect to identify the adopted postures and evaluates them based on a predefined method specified in OWAS [1] to obtain an improved understanding of the long-term hazards of the common work postures of construction workers. OWAS was developed in the 1970s for identification of work postures that cause discomfort and are detrimental to workers' health. The developed system can be used as an assisting tool for health professionals and occupational health experts for identifying poor working postures and diagnosing possible musculoskeletal injuries. It can also lead to improved work health regulations and reduced instances of occupational injuries in the future.

Musculoskeletal disorders (MSDs) refer to injuries and disorders that affect the human body movement or musculoskeletal system such as muscles and tendons. Since repetitive motion or stress is one major risk factor contributing the MSDs, many other similar terms such as repetitive motion injury, cumulative trauma disorder, and repetitive stress disorder were also used in the literature.

According to NIOSH report [2] based on annual surveys from 1982 to 1996 conducted by Labor of Statistics of U.S Department of Labor, while the number of cases and the rates increased slightly for most of the illnesses over the time period, disorders associated with repeated trauma increased nearly 10-fold. In 1996, repeated-trauma disorders accounted for close to two-thirds of reported occupational illnesses. The proportion of occupational illnesses due to repeated trauma was nearly twice as high as in 1986. Later on the rate declined and was speculated to have occurred as a result of a better recognition of MSDs and the implementation of industrial health and safety programs [3].

The prevention of the development of MSD for workers requires the efforts on both company and each worker. Some researchers have worked on improving the work environment. For example, Hisao and Stanevich [4] investigated various tasks pertaining to scaffolding to identify the risks of overexertion injury and also develop an assistive device to help lifting scaffold end frames.

In a review of epidemiological studies on low-back, neck, shoulder, and upper extremity disorders, several physical load factors were identified as risk factors for the disorders. Many of these factors have been repeatedly identified, and for different types of outcomes of an anatomical area (e.g. pain, disc herniation, disc degeneration of the low-back or neck) [5].

Many postural observational methods, e.g., Joseph et al. [5] and Pehkonen et al. [6], have been developed to evaluate the posture exposed to the MSD risk. The OWAS (Ovako working posture analysis system) is also a postural observational method commonly used to identify poor postures at a worksite. OWAS has been widely used in several industries for postural analysis (e.g., [8], [9], [10], and [11]), and also in the construction industry (e.g., [12], [13], [14], [15], [16]). The OWAS evaluation is based on sampling from typical working postures of major body parts including back, forearms, and legs, and the information about the force exerted or load carried during work upon the observed subject. This research adopts OWAS as the postural evaluation method. Other related measurement methods that were not used in this research include the subjective Nordic Musculoskeletal Questionnaire developed by Nordic Council of Ministries [17], RULA (Rapid Upper Limb Assessment) [18], REBA (Rapid Entire Body Assessment, 2000) [19], and LUBA (Loading on the Upper Body Assessment) [20]. Both RULA and LUBA focused only on upper limbs, and RULA requires the measurement of muscular forces and frequency of muscular force exertions. REBA's approach is also based on RULA.

OWAS uses a four-digit code to describe various postures and weight combinations. As shown in Fig. 1, the codes include four back postures, three forearm postures, seven leg postures, and three levels of exerted force. For example, three forearm postures are (1) both below elbow joint, (2) one above elbow joints, and (3) both above elbow joint. The OWAS then categorizes the total number of 252 possible combinations of the four digits into four levels of actions according to their risk of injuries.

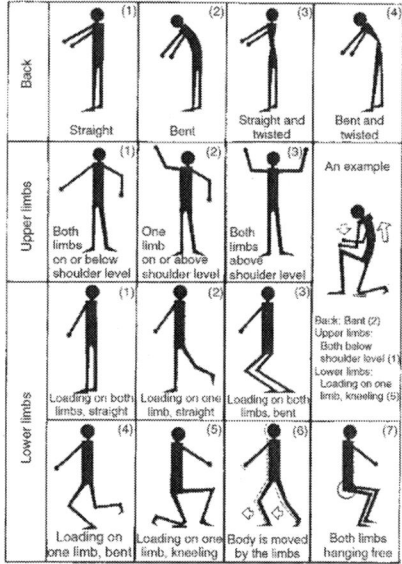

Figure 1. OWAS postures [1]

AC1: postures are normal and natural with no particular harmful effect on the musculoskeletal system, no action is required;

AC 2: postures have some harmful effect on the musculoskeletal system, corrective actions are required in the near future;

AC3: postures have a distinctly harmful effect on the musculoskeletal system, corrective actions should be done as soon as possible;

AC4: postures have an extremely harmful effect on the musculoskeletal system, immediate corrective actions for improvement are required.

Although OWAS has broken down common human postures contributing to MSDs into precise categorization of postures of body parts, the utilization of OWAS still remains somehow subjectively and imprecise due to the inconvenience of statically measuring angles of body parts in a dynamic continuous motion. Motion capture has been

widely used in a variety of industries such as cinema, entertainment, and computer games. Sharma et al. [21] reviewed several motion capture methodologies including marker-based motion capture (e.g., acoustical system, mechanical system, magnetic system, optical system) and marker-less motion capture such as Microsoft's Kinect solution, which is low-cost and does not require any special equipment attached to the observed subject. This research used KINECT as a tool for tracking the movement of human skeleton to implement the proposed system for its low cost and availability of SDK (software development kit) [22].

II. SYSTEM DESIGN

This study develops an automated system, named Posture Assessment System for Construction Workers (PAS-CW) for MSDs for assessing postures of construction workers and identifying high-risk postures pertaining to MSDs. The following describes first the OWAS categorization, and how the system is used with the user interface, and then the design of the system so that the text is easier to follow.

Figure 1 shows the main control user interface of the system. Once the Kinect is setup and the observed subject is ready to perform his/her regular work activity, the ergonomics expert or physiotherapist starts the posture capture and evaluation according to the following steps (from top to bottom on the right hand side of Fig.2).

1. Click the appropriate weight carried by the subject. This is necessary because the OWAS evaluation requires the carried weight information which cannot be detected by the Kinect camera.
2. Adjust the angle of camera so that it covers the activity area.
3. Press the red recording button to start capturing and recording the subject's skeleton motion.
4. The result window (shown in the bottom of Fig. 1) instantaneously displays the categorization of the postures and the subject's body parts including back, forearm, leg, and the previously input carried weight, and also the injury risk level assessed based on the OWAS.

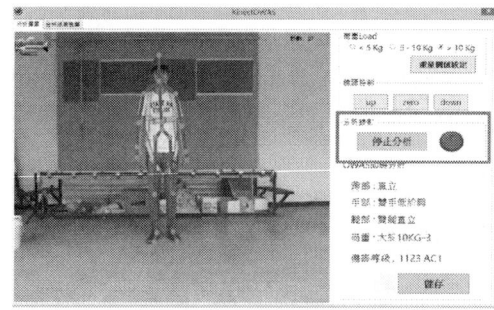

Figure 2. Main control user interface for PAS-CW

The system was developed using Microsoft Visual Studio 2010 in C# language, and Kinect for Windows SDK [22]. The system also uses SQLite Database, a popular public-domain database engine for local data storage, and EmguCV, a cross platform that allows Intel's OpenCV's image processing functions to be called from .NET compatible languages such as C# and Visual Basic.

The system includes several modules as shown in Figure 3. The Kinect video cameras record the observed subject. For each video image frame, the skeletal tracking module tracks the posture of the subject at a frame rate of 2 Hz, and transforms into skeleton with information of twenty joint positions. The posture categorization module categorizes the posture type for each body part (e.g, back, arm). Given the categorization of each body part, the general posture of the subject in the frame can be determined, and the corresponding OWAS's AC levels can be determined.

The user interface module controls the main display windows of the system, and includes the video display of the subject's skeleton (window left), the input area (window top right), and the analysis result area (window bottom right). With the estimated exerted force input by the user, for each video frame, the analysis module receives the data of the skeleton and the joint positions and determines the postural categorization of back, forearms, and legs. The output of the analysis includes the categorization of each body part and the resulting AC levels according to OWAS. The database module stores the video images, and the data of the skeleton and joint positions with the determined AC levels. The display module displays the statistics of the data visually to provide the expert a holistic view of the subject posture in a continuous fashion instead of a single frame. Examples are bar chart graphics showing the AC level at intervals of 0.5 seconds, or the average or peak AC levels for the choice of intervals of 5, 10, 15, and 30 seconds.

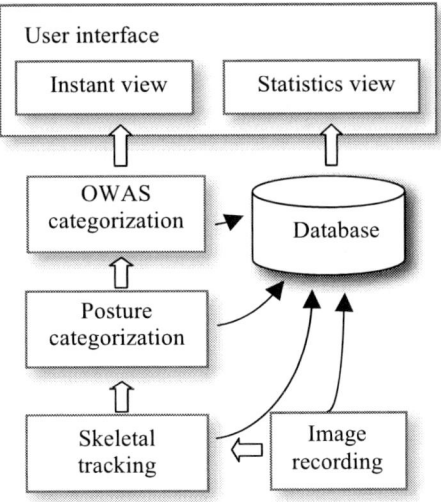

Figure 3. PAS-CW's implementation modules

The Kinect cameras can provide the positions for 20 joint of a human skeleton (e.g., head, hand right, wrist right, elbow left, hip center, knee left, ankle left, and foot right) in a three-dimension coordinate system. With this information, the relative distances, positions of joints, angles of the associated limbs or other body parts can be derived. The individual posture of back, arms, and legs need to be categorized first in order to categorize the entire body posture of each sampled video frame based on the OWAS.

For example, the OWAS categorizes the back posture into 4 types, e.g., *Straight-back*, *Bent-back*, *Twisted-back*, and *Bent&twisted-back*. The back posture will be identified as *Straight-back* if the smaller angle between Vector1 passing through the spine and the shoulder center and Vector2 passing through left hip and right hip is greater than or equal to 75°. The back posture will be identified as *Bent-back* if the smaller angle between Vector1 passing through the spine and the shoulder center and Vector2 passing through left hip and right hip is smaller than 75°. The back posture will be identified as *Twisted-back* if the angle measured in the X-Z plane, between the vectors normal to Vector1 passing through left shoulder and right shoulder and Vector2 passing through left hip and right hip is smaller than 15°. The back posture will be identified as *Bent&twisted-back* if both *Bent-back* and *Twisted-back* conditions are met.

III. SYSTEM EVALUATION

The evaluation of the PAS-MSD focuses on two aspects: skeleton identification accuracy and OWAS categorization accuracy. The skeleton identification accuracy represents the rate of frames in which all necessary skeletons and joint positions required for the OWAS assessment are successfully identified. Skeleton identification may be incorrect or even fail in some frames due to obstructed view, extreme twisting or bending, or sudden movement. The OWAS categorization accuracy determines the rate of the correct assignment of AC level for each frame by the PAS-MSD system compared to the participating experts.

Fifteen graduate students with or without construction working experience were recruited to participate the experiment. The participants were asked to perform four tasks: (1) materials moving, (2) materials handling, (3) nail hammering, and (4) tiling.

Taking task-1 as an example, it required the participant to lift a bucket of materials (19 kg for male, and 9 kg for female), move the bucket, and place at the designated area. The task requires 4 repetitions. The MSD risk factors [23] included in the task are unnatural work posture, repetitive work, and exerted force. The task was designed to include common risk factors to MSDs but not actually result in MSDs to the participants.

All the tasks were performed on the same position except for Task-1, for which the subjects were required to walk 2 to 3 steps. During the tasks, the subjects were asked to perform the tasks naturally without any special constraints on body rotation or blocking camera view from body parts. Because the participants may turn around their bodies in all directions while performing the tasks, we set up three Kinect cameras in the face angle and side angle horizontally, and another from a 45° birds eye angle.

Three experts participated in the evaluation of the experimental result, and include a rehabilitation medical physician, a professor in physiotherapy, and a professor in medical engineering.

Table 1 shows the identification accuracy and categorization accuracy results for each work task. Except for Task 3, all other tasks show similar results. First, the identification accuracies all exceed 90% (i.e., 91.6%-93.9%), and all categorization accuracies exceed 85% (i.e., 85.4%-88.5%). Such performances make the proposed PAS-CW feasible to be used to help analyze the posture of construction workers for MSDs. Note that the calculation of the identification and categorization accuracies was first prepared by the authors, and then examined by the participating professionals because the process was too time consuming due to the large amount of data involved.

Task 1 required subjects to lift heavy bucket, moved it, and placed it at the designated place. The motions involved large and slow movement, which result in the highest accuracy in terms of identification and categorization.

Task 2 required subjects to take materials from the

TABLE 1. IDENTIFICATION AND CATEGORIZATION ACCURACY RESULT

	Task 1	Task 2	Task 3	Task 4
Identification accuracy	93.9%	92.2%	69.4%	91.6%
Categorization accuracy	88.5%	88.2%	48.4%	85.4%

bucket and put them in place piece by piece. The data showed that the PAS-CW successfully identified the skeleton of the upper body most of the time, but often failed in identifying the skeleton of the legs due to the squat posture and view blocked by the bucket. Nevertheless, since legs were successfully identified when subjects were standing and the legs seldom make large movement, the lack of updating on the leg part did not affect the accuracy significantly. Task 4 requires subjects to take tiles at squad position and place the magnetic tiles, which was customized designed for this experiment, on the wall. Even the PAS-CW's camera was facing the subjects from behind, the detection accuracies were still satisfactory.

Task 3 is the one that resulted in poorer identification accuracies compared to all the other tasks. The subjects were required to hammer nails on the designated wood box (Fig. 4), which was specifically designed to protect the subjects from accidental injuries. As shown in Fig. 5, the wood box blocked the Kinect camera's view of the subject's legs most of the time during the task and resulted in the poor identification accuracies.

Figure 4. Wood box for Task 3

Figure 5. Subject nailing at the box for Task 3

The experts took much longer time in evaluating the captured images without precise constant sampling frequency. After long hour of evaluation, the experts may also give inconsistent categorizations within themselves or among them. The inconsistency usually occurred when the subjects posed bending legs or waists, and turning backs. In conclusion, all participating experts agreed the accuracy rates of the machine identification and categorization are good enough to be helpful to their diagnosis.

IV. CONLUSION

Construction workers often suffer various kinds of MSDs due to sudden exertion such as lifting heavy equipment, or repeated and cumulative stressed motions. This research uses Microsoft Kinect tool kit to develop an automatic tracking, named Posture Assessment System for MSD. The PAS-MSD captures a worker's postures during his/her movement, recognizes human skeleton, and produce and assesses the risk of MSD based on OWAS.

An experiment of four types construction tasks was designed to evaluate the PAS-MSD. Except for the nailing task, in which the target box easily block Kinect's view of subjects' legs and which had poor identification performance, both the identification and OWAS categorization accuracies were satisfactory (i.e., from 85% to 93.9%). The experiment also provided the following lessons learned.

- The PAS-MSD has satisfactory identification and OWAS categorization as long as the cameras' view of any worker's body part is not blocked by surrounding environment, work object, self's body, or other people.

- The PAS-MSD is able to provide an evaluation of much higher sampling rate of captured images and consistent categorization compared to human experts. This is particularly beneficial when longer observation time is required for determining the postural causes of MSDs.

- In addition to the one facing the observed worker, we set up additional two Kinect cameras in the side angle horizontally and from a 45° birds eye angle to capture the participants' gestures in all directions especially when they turned their bodies around during the task. However, while the gestures were captured in video, the successful rates for identifying the skeletons were too low to be useful. This is because the current skeleton capturing algorithm provided by Microsoft work poorly for side view and top view. New sets of capturing algorithms need to be developed if additional cameras are really needed to avoid blocking the camera's view. The categorization rules are also needed to develop to transform the side view or top view body postures into the normal view of the conventional OWAS.

Due to the limitation of the Kinect-based PAS-CW, we also learned that the system is not suitable to be directly used on a normal-daily-setting construction site to monitor the postures of construction workers. Other workers, temporary work, and equipment may easily affect the image capturing ability of Kinect and the setup and the clearance of the area between the cameras and the observed worker may also affect the normal construction operation. A temporary cleared and safe area such as a room inside the jobsite office needs to be setup to use the PAS-CW directly on a construction site. This research only tested four types of construction operations, but the experiment already indicates that some works may not be applicable if the postures of the observed body part or potentially high-risk posture may be blocked by the working elements, temporary works, or self body parts.

ACKNOWLEDGMENT

The authors like to thank the Ministry of Science and Technology of Taiwan, which financially supported this research under contract 105-2221-E-009 -022 -MY3.

REFERENCES

[1] O. Karhu, P. Kansi, and I. Kuorinka (1977) "Correcting working postures in industry: A practical method for analysis." Applied Ergonomics, vol. 8(4): 199-201, 1977.

[2] NIOSH, A NIOSH Look at Data from the Bureau of Labor Statistics: Worker Health by Industry and Occupation. National Institute for Occupational Safety and Health, DHHS (NIOSH) Publication No. 2001-120. 2011.

[3] H. Conway and J. Svenson, Occupational injury and illness rates, 1992– 96: Why they fell. Monthly Labor Review, vol. 121 (11), pp. 36– 58, 1998.

[4] H. Hisao and R. L. Stanevich, "Biomechanical evaluation of scaffolding tasks", International J. of Industrial Ergonomics, vol. 18, pp. 407-415, 1996.

[5] E. Viikari-Juntura, "The scientific basis for making guidelines and standards to prevent work-related musculoskeletal disorders, Ergonomics, vol. 40(10), pp. 1097-1117, 1997.

[6] C. Joseph, D. Imbeau, and I. Nastasia, "Measurement consistency among observational job analysis methods during an intervention study," International J. of Occupational Safety and Ergonomics, vol. 17(2), pp. 139–46, 2011.

[7] I. Pehkonen, R. Ketola, R. Ranta, E. P. Takala, "A video-based observation method to assess musculoskeletal load in kitchen work," International J. of Occupational Safety and Ergonomics. 15(1):75–88, 2009.

[8] J. A. Engels, J. A. Landeweerd, and Y. Kant, "An OWAS-based analysis of nurses' working postures," Ergonomics," vol. 37(5), pp. 909–19, 1994.

[9] E. J. Wright and R. A. Haslam, "Manual handling risks and controls in a soft drinks distribution centre," Applied Ergonomics, vol. 30(4), pp. 311–8, 1999.

[10] G. B. Scott and N. R. Lambe, "Working practices in a perchery system, using the OVAKO working posture analysis system (OWAS)," Applied Ergonomics, vol. 27(4), pp. 281–4, 1996.

[11] D. P. Gilkey, T. J. Keefe, P. L. Bigelow, R. E. Herron, K. Duvall, and J. E. Hautaluoma, "Low back pain among residential carpenters: ergonomic evaluation using OWAS and 2D compression estimation," International J. of Occupational Safety and Ergonomics, vol. 13(3), pp. 305–21, 2007.

[12] A. Burdorf, G. Govaert, and L. Elders, "Postural loads and back pain of workers in the manufacturing of prefabricated concrete elements," Ergonomics, vol. 34 (7), pp. 909–918, 1991.

[13] P. Kivi. and M. Mattila, "Analysis and improvement of work postures in the building industry: application of the computerised OWAS method", Applied Ergonomics, vol. 22 (1), pp. 43-48, 1991.

[14] M. Mattila, W. Karwowski, and M. Vilkki. "Analysis of working postures in hammering tasks on building construction sites using the computerized OWAS method" Applied Ergonomics, vol. 24(6), pp. 405-412, 1993.

[15] B. Buchholz, V. Paguet, L. Punnett, D. Lee, and S. Moir, "PATH: A work sampling-based approach to ergonomic job analysis for construction and other non-repetitive work," Applied Ergonomics, vol. 27 (3), pp. 177–187, 1996.

[16] T. H. Lee and C.S. Han, "Analysis of working postures at a construction site using the OWAS Method", International J. of Occupational Safety and Ergonomics. Vol. 19(2), pp. 245–250, 2013.

[17] I. Kuorinka, B. Jonsson, A. Kilbom, H. Vinterberg, F. Biering-Sorensen, G. Andersson, and K. Jorgensen, "Standardised Nordic questionnaires for the analysis of musculoskeletal symptoms," Applied Ergonomics, vol. 18(3), pp. 233-237, 1987.

[18] L. McAtamney and E. N. Corlett, "RULA: a survey method for the investigation of world-related upper limb disorders," Applied Ergonomics, Butterworth-Heinemann Ltd., 24(2), pp. 91-99, 1993.

[19] S. Hignetta and L. McAtamney, "Rapid Entire Body Assessment (REBA)," Applied Ergonomics, vol. 3(2), 2000, pp. 201–205.

[20] D. Kee and W. Karwowski, "LUBA: an assessment technique for postural loading on the upper body based on joint motion discomfort and maximum holding time," Applied Ergonomics vol. 32, pp. 357-366, 2001.

[21] A. Sharma, M. Agarwal, A. Sharma, and P. Dhuria "Motion capture process, techniques and applications" International Journal on Recent and Innovation Trends in Computing and Communication, vol. 1(4) pp. 251-257, 2013.

[22] Microsoft, Kinect for Windows SDK 2.0, Microsoft Corporation, retrieved Feb. 4 , 2017, from: https://www.microsoft.com/en-us/download/details.aspx?id=44561

[23] CCOHS (Canadian Centre for Occupational Health and Safety) "Work-related Musculoskeletal Disorders (WMSDs) - Risk Factors," retrieved Feb. 2 2017, from: https://www.ccohs.ca/oshanswers/ergonomics/risk.html

MIPRO 2017, May 22- 26, 2017, Opatija, Croatia

Software Toolbox for Analysis and Design of Nonlinear Control Systems and Its Application to Multi-AUV Path-Following Control

Sergey Ul'yanov and Nikolay Maksimkin
Institute for System Dynamics and Control Theory, Irkutsk, Russia
Email: sau@icc.ru

Abstract—In the paper, we present a software toolbox for rigorous analysis and design of nonlinear continuous and discrete-continuous (digital) control systems based on the reduction method and sublinear vector Lyapunov functions. For these systems, the toolbox provides solution of the following problems: verification of dynamic properties of dissipativity, asymptotic and practical stability; computation of estimates of basic direct indicators of dynamic quality (accuracy, domains of attraction and dissipativity, settling time and others); synthesis of parameters that ensure desirable or optimal quality of the system. In addition, we show how the toolbox could be applied to solve the path-following control problem for a single autonomous underwater vehicle (AUV) as well as for multi-AUV formations. In our statements of the path-following control problem, we take into account together uncertainties of AUVs model, inaccuracy of measurements, and constraints on control actions.

I. INTRODUCTION

Nowadays the classical and modern control theories provide a wide range of methods for analysis and design of control systems, most of which are efficiently implemented in software and used in practice. However, these techniques are mainly applicable to linear systems and do not give satisfactory results when dealing with complex and highly nonlinear models. Existing tools for analysis and design of nonlinear systems lack of constructiveness and therefor the development of new techniques that have the potential to be developed up to explicit computing algorithms as well as their efficient software implementation is a high-priority problem in control theory.

In this paper, we present a software toolbox VLF-REDUCTOR for rigorous analysis and design of nonlinear continuous and continuous-discrete (hybrid) control systems designed in MATLAB environment. The toolbox implements a numerical technology for analysis of nonlinear systems based on the reduction method [1], [2] and sublinear vector Lyapunov functions (VLF) [3], [4]. The main features of the technology are the possibility to formulate analysis and design problems in terms of system requirements, which are convenient for engineers, and its constructiveness, which implies that after reduction of an investigated model to some standard form, subsequent computations are performed, for the most part, automatically. The main aim of the paper is to demonstrate how the developed software can be applied to

solve challenging control problems arising in practice. As an example, the path-following control problem is addressed.

Steering a vehicle to converge to and follow a desired spatial path is a traditional problem in robotics. With regard to autonomous underwater vehicles (AUV), the problem is studied in [5] where a nonlinear path-following control algorithm, which takes into account the AUV dynamics and exploits the idea of tracking the motion of a virtual target along the path by controlling explicitly its progression rate, is proposed. This approach is extended then to deal with uncertainties [6]. The path-following problem for AUV with actuator saturations is also considered in literature [7]. This paper concerns the path-following problem under parameter uncertainties, measurement errors, and saturations of control signals. We consider the case when the discrete (sampled-data) control of AUV is used. It proves to be reasonable for formation control of multi-AUV systems, as will be shown below.

The formation control problem for multi-AUV systems is being actively studied by researchers from all over the world due to a wide range of its applications: exploring and monitoring underwater environment, seafloor mapping, and search-and-rescue operations. To get an insight about advances in this area, we refer the reader to survey papers [8], [9].

Most of the known multi-AUV formation control algorithms are based on the leader-follower approach [10]–[12], according to which all the vehicles in the formation are connected in pairs by the leader-follower relation so that there is a vehicle, being not a follower for others and called the formation leader, that determines the motion of the group, whereas other vehicles as followers try to maintain a desired position with respect to their leaders.

Since only slow and unreliable acoustic communications are available under the water, the issue of delays and packet dropouts is also addressed [12], [13]. The path-following control scheme for multi-AUV systems is proposed in [14]. It assumes that all vehicles in formation have to follow their own predefined paths and communications are used to synchronize their speeds. In our study, we admit that only the formation leader is aware of the path to be followed. This condition arises when the path is constructed or reconstructed during the motion in order, for example, to avoid obstacles.

The rest of the paper is organized as follows. Section II

gives a brief description of the designed software toolbox for analysis and design of nonlinear systems. Particularly, it introduces classes of control system models and formulations of analysis and design problems captured by the toolbox. Path-following digital controllers for a single AUV and multi-AUV systems are built with the use of the toolbox in Section III. Section IV provides simulation results for the designed controllers. Section V contains the conclusions and expiations of future work.

II. Software toolbox for analysis and synthesis of nonlinear systems

A. Control System Models

The software toolbox is designed to deal with a wide range of continuous and continuous-discrete (digital) control systems. It assumes that a system under consideration is represented in some typical form. Such a representation for continuous systems has the form

$$(J + \Delta J(t, \mathbf{x}))\dot{\mathbf{x}} = (A + \Delta A(t, \mathbf{x}))\mathbf{x} + (B + \Delta B(t, \mathbf{x}, \mathbf{u}))\mathbf{u} + GF(t, \mathbf{x}, \mathbf{u}) + H\Phi(t, \mathbf{x}, \mathbf{u}), \quad (1)$$

$$\mathbf{u} = \varphi(\cdot, \sigma), \quad \sigma = C\psi(\cdot, \eta), \quad \eta = D\mathbf{x}, \quad \mathbf{x} = \mathbf{x}(t), \quad (2)$$

where $\mathbf{x} \in \mathbb{R}^n$ is the state vector of the system formed by stacking state variables of the plant as well as other elements of the system (state observers, dynamic filters and so on), $\mathbf{u} \in \mathbb{R}^m$ is the control input, $\eta \in \mathbb{R}^l$ is the output vector; $\Delta J(\cdot)$, $\Delta A(\cdot)$, $\Delta B(\cdot)$, $F(\cdot)$, $\Phi(\cdot)$ are functions that specify uncertainties and nonlinearities of the system, external disturbances and satisfy inequalities: $|J^{-1}\Delta J(\cdot)| \leq J^0$, $|\Delta A(\cdot)| \leq A^0$, $\Delta B(\cdot) \leq B^0$, $|F(\cdot)| \leq U|\theta|$, $\theta = \Theta\mathbf{x} \in \mathbb{R}^q$, $|\Phi(\cdot)| \leq \Phi^0 \in \mathbb{R}^r$ (henceforward, inequalities between two matrices (vectors) and modulus of a vector are understood as the component-wise ones); A, B, C, D, G, H, J_0, A_0, Φ^0, B_0, U, Θ are constant matrices and vectors of appropriate dimensions. Functions $\varphi(\cdot)$, $\psi(\cdot)$ describe various characteristics of actuators and sensors with their nonlinearities, noise and quantization errors, saturation of signals and belong to a special class of functions called $\mathcal{SN}$ [3], [4]. For a function $\theta(\xi)$ of variable $\xi \in \mathbb{R}^k$, we will write $\theta(\xi) \in \mathcal{SN}^k(K_\xi, K_\xi^0, \xi^0, \bar{\xi})$ if

$$\min\left\{K_\xi\left(\xi - \xi^0\right) - K_\xi^0|\xi|; K_\xi\bar{\xi}\right\} \leq \theta(\cdot, \xi) \leq$$
$$\leq \max\left\{K_\xi\left(\xi + \xi^0\right) + K_\xi^0|\xi|; -K_\xi\bar{\xi}\right\},$$

where $0 \leq \xi^0 < \bar{\xi} \leq \infty$, $\xi^0 \in \mathbb{R}_+^k$, $\bar{\xi} \in \mathbb{R}_+^k$, K_ξ, K_ξ^0 are diagonal matrices. Parameters K_ξ, K_ξ^0 specify the nominal values and admissible uncertainties of the gain coefficients, respectively; ξ^0 characterize noise and quantization errors, dead zones; $\bar{\xi}$ defines saturation levels of actuators or the operating ranges of sensors. Using the notation above, assume that $\varphi(\cdot) \in \mathcal{SN}^m(K_\sigma, K_\sigma^0, \sigma^0, \bar{\sigma})$, $\psi(\cdot) \in \mathcal{SN}^l(K_\eta, 0, \eta^0, \bar{\eta})$.

In the case of continuous-discrete systems, the model also includes continuous part (1), but now control signal $\mathbf{u}(t)$ is generated by a zero-order hold function with a sequence of hold times $t_k = t_0 + kh$, $k = 0, 1, 2, \ldots$ as $\mathbf{u}(t) = \mathbf{u}(t_k) \equiv \mathbf{u}_k$,

$t \in T_k \equiv [t_k, t_{k+1})$ and feedback control law, instead of (2), is defined as follows:

$$\mathbf{u}_k = \varphi(\cdot, \sigma_k), \quad \sigma_k = C_\eta \psi(\cdot, \eta_k) + C_z z_k, \quad (3)$$

where $\eta_k = D\mathbf{x}(t_k) \in \mathbb{R}^l$ is the output vector, $z_k \in \mathbb{R}^{n_d}$ is the state of a discrete controller, C_η, C_z are some constant matrices.

The discrete controller is described by difference equations

$$z_{k+1} = A_d z_k + B_d \mathbf{u}_{dk} + \sum_{i=1}^{\mu} C_d^i \psi_i(\cdot, \eta_{ik}) + H_d \Phi_{dk}(\cdot), \quad (4)$$

$$\mathbf{u}_{dk} = \varphi_d(\cdot, \sigma_{dk}), \quad \sigma_{dk} = C_{d\eta}\psi_d(\cdot, \eta_{dk}) + C_{dz} z_k, \quad (5)$$

$$\eta_{dk} = D_d \mathbf{x}(t_k), \quad \eta_{ik} = D_i \mathbf{x}(t_i), \quad k = 0, 1, 2, \ldots \quad (6)$$

where $\mathbf{u}_{dk} \in \mathbb{R}^{m_d}$ is the control input of the controller, $\eta_{dk} \in \mathbb{R}^{l_d}$, $\eta_{ik} \in \mathbb{R}^{l_i}$ are output vectors sampled at time moments t_k and $t_{k_i} = t_k + \tau_i$, $i = \overline{1, \mu}$, $\tau_i \leq h$, $\mu \geq 1$ is the number of samples available on each control step (memory depth); $\Phi_{dk}(\cdot)$ is a vector function satisfying $|\Phi_{dk}(\cdot)| \leq \Phi_d^0 \in \mathbb{R}^{r_d}$, $\varphi_d(\cdot) \in \mathcal{SN}^{m_d}(K_{d\sigma}, K_{d\sigma}^0, \sigma_d^0, \bar{\tau}_d)$, $\psi_d(\cdot) \in \mathcal{SN}^{l_d}(K_{d\eta}, 0, \eta_d^0, \bar{\eta}_d)$, $\psi_i(\cdot) \in \mathcal{SN}^{l_i}(K_{\eta i}, 0, \eta_i^0, \bar{\eta}_i)$; A_d, B_d, H_d, $C_{d\eta}$, C_{dz}, C_d^i, D_d, D_i, Φ_d^0 are constant matrices and vectors of appropriate dimensions.

To specify a control system for further investigation using the software toolbox, one needs to assign values to all or a part of elements of the matrices included in the definition of the model. In case of partial specification, undefined parameters are assigned default values: $J^0 = O_{n \times n}$, $A^0 = O_{n \times n}$, $B^0 = O_{n \times m}$, $G = O_{n \times p}$, $U = O_{p \times q}$, $H = O_{n \times r}$, $\Phi^0 = 0_r$, $H_d = O_{n_d \times r_d}$, $\Phi_d^0 = 0_{r_d}$; for any class $\mathcal{SN}^k(K_\xi, K_\xi^0, \xi^0, \bar{\xi})$, the default values are $K_\xi = E_k$, $K_\xi^0 = O_{k \times k}$, $\xi^0 = 0_k$, $\bar{\xi} = \infty_k$. Here, $O_{k \times l}$ denotes the zero matrix of dimensions $k \times l$, E_k denotes the identity matrix of size k, 0_k is the all-zeros column vector of size k, ∞_k is a column vector of size k filled with infinite values.

Using the toolbox, the following problems arising in practice can be solved for both continuous model (1), (2) and continuous-discrete model (1), (3)-(6): verification of dynamic properties of dissipativity, asymptotic and practical stability; estimation of basic indicators of dynamic quality[1]:

- stabilization accuracy γ^0 (steady state error) with respect to the output vector $\gamma = \Gamma\begin{bmatrix}\mathbf{x} & \mathbf{z}\end{bmatrix}^T \in \mathbb{R}^\nu$, that is, a vector γ^0 that satisfies the following inequality:

$$\varlimsup_{t \to \infty}|\gamma(\mathbf{x}(t), z_{k(t)})| \leq \gamma^0, \quad k(t) = \lfloor(t - t_0)/h\rfloor; \quad (7)$$

- domain of attraction (dissipativity)

$$\Omega_0 = \{(\mathbf{x}_0, \mathbf{z}_0) : |\gamma_0(\mathbf{x}_0, \mathbf{z}_0)| \leq c\bar{\gamma}_0\}$$

with respect to the vector $\gamma_0 = \Gamma_0\begin{bmatrix}\mathbf{x}_0 & \mathbf{z}_0\end{bmatrix}^T \in \mathbb{R}^{\nu_0}$ for the given direction $\bar{\gamma}_0 \in \mathbb{R}_+^{\nu_0}$, $c > 0$;

- maximum deviations $\max_t|\gamma(\mathbf{x}(t), z_k(t))|$ with respect to the vector γ from a given set of initial states of the form $|\gamma_0(\mathbf{x}_0, \mathbf{z}_0)| \leq \bar{\gamma}_0$;

- control time up to a desirable accuracy $\bar{\gamma}^0 \geq \gamma^0$ from the given initial state $|\gamma_0(\mathbf{x}_0, \mathbf{z}_0)| \leq \bar{\gamma}_0$;

---

[1]For brevity, we list here only indicators for the continuous-discrete model.

• the Lyapunov exponent (exponential decay rate).

Solution of the problems above is based on the construction of an auxiliary (comparison) monotonous system related to the initial one by sublinear VLF and subsequent analysis of the auxiliary system using a special theory of monotonous comparison systems. For details, we refer the reader to [3], [4], [15], [16].

One of the key features of the toolbox is the possibility to design control systems with desired or optimal dynamic quality. Typically, the design problem is formulated as follows: find unknown parameters of the system (e.g. controller feedback coefficients) that provide its dissipativeness for all admissible uncertainties and disturbances and minimize the steady-state error of the system.

More formally for the continuous-discrete system, it can be stated as follows. Let $\beta \in \mathfrak{B}$ be the vector of parameters to be determined and $\mathfrak{B}$ be the admissible domain for β. Define the domain of dissipativity Ω_0 as a set of initial states containing the neighborhood of the origin, for which the relation (7) holds. The control design problem consists in finding $\beta \in \mathfrak{B}$ that minimize criteria $J_f = \alpha^T \gamma^0$ ($\alpha \in R_+^\nu$ is a given vector of weighting coefficients) provided that closed system (1), (3)-(6) is dissipative, i.e., $\Omega_0 \neq \emptyset$ (domain of dissipativity is not empty).

The posed problem cannot be rigorously solved, since, in the general case, it is impossible neither to get the exact value of the criteria, nor to test the feasibility of the restrictions. To deal with it, we adopt an approach, in which, instead of exact values of performance indicators included in criteria and constraints, their estimates computed on the basis of sublinear VLF, are exploited. In the toolbox, the given optimization problem with computable criteria and constraints is efficiently solved by use of the penalty function method and non-gradient methods of optimization.

The advantages of the technology implemented in the toolbox are, on the one hand, the strictness of results obtained and guaranteed nature of the judgments and estimates, and, on the other hand, its constructiveness. It is for the accuracy of numerical results VLF method is advantageous over other known methods. It is important that analysis problems are formulated directly in terms of system requirements, which are convenient for engineers; the results are also obtained in this form. After reduction of an investigated model to some standard form, subsequent computations are performed, for the most part, automatically.

In the following section, we will demonstrate the application of the designed software to path-following problem for multi-AUV systems.

III. Path-following Control of Multi-AUV formation

A. Dynamic Model of AUV

In this paper, we use the full dynamic model of the INFANTE AUV borrowed from [5]. The kinematic and dynamics equations of the vehicle can be defined using a global coordinate frame $\{U\}$ and a body-fixed coordinate frame $\{B\}$ (see Fig. (1)).

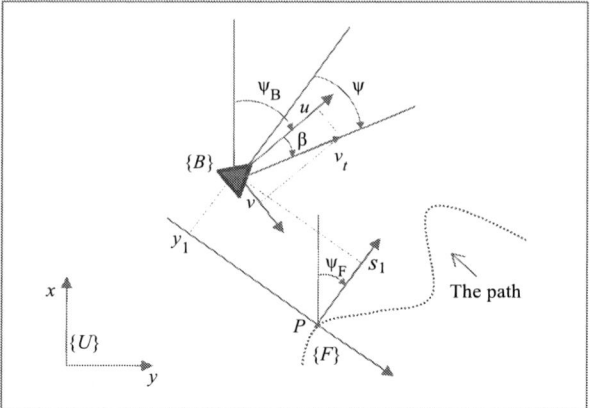

Fig. 1. Reference frames.

The kinematic equation of the AUV can be written

$$\dot{x} = u\cos(\psi_B) - v\sin(\psi_B),$$
$$\dot{y} = u\sin(\psi_B) + v\cos(\psi_B),$$
$$\dot{\psi}_B = r, \qquad (8)$$

where x, y are the coordinates of the center of mass of the vehicle, ψ_B denotes the yaw angle, u and v are respectively the surge and sway velocities expressed in $\{B\}$ and r is the angular yaw rate.

Neglecting the equations in heave, roll and pitch, the equations for surge, sway and yaw can be presented as

$$\mathcal{F} = m_u \dot{u} + d_u,$$
$$0 = m_v \dot{v} + m_{ur} ur + d_v,$$
$$\mathcal{G} = m_r \dot{r}, \qquad (9)$$

where

$$m_u = m - X_{\dot{u}}, \quad d_u = -X_{uu}u^2 - X_{vv}v^2,$$
$$m_v = m - Y_{\dot{v}}, \quad d_v = -Y_v uv - Y_{v|v|}v|v|,$$
$$m_r = I_z - N_{\dot{r}}, \quad d_r - N_v uv - N_{v|v|} - N_r ur,$$
$$m_{ur} = m - Y_r,$$

m is the mass of the AUV, I_z is the moment of inertia around AUV's vertical axis, $X_{(\cdot)}$, $Y_{(\cdot)}$, $N_{(\cdot)}$ are classical hydrodynamic derivatives, and $[\mathcal{F} \ \mathcal{G}]^T$ is the vector of force and torque applied to AUV.

B. Path-following Controller Design for Single AUV

The conception of the virtual target is used here to design path-following controller for a single AUV. Define a virtual target as a point P moving along the path to be followed by the AUV. Associated with P, consider the corresponding Serret-

Frenet frame $\{F\}$ (see Fig. 1). As shown in [5], the dynamics of the virtual target in $\{F\}$ can be described by

$$\dot{s}_1 = v_t \cos\psi - \dot{s} + \dot{\psi}_F y_1,$$
$$\dot{y}_1 = v_t \sin\psi - \dot{\psi}_F s_1,$$
$$\dot{\psi} = r + \dot{\beta} - \dot{\psi}_F, \tag{10}$$

where s_1, y_1 are the coordinates of the target in $\{F\}$, s is the signed curvilinear abscissa of P along the path, $\beta = \arctan(v/u)$ is the side-slip angle, $v_t = (u^2 + v^2)^{1/2}$ is the absolute value of the total vehicle velocity vector; ψ_F is the angle that defines the orientation of F with respect to B, $\dot{\psi}_F \equiv c_c(s)\dot{s}$, c_c is the path curvature, $\psi \equiv \psi_B + \beta - \psi_F$. In this study, we suppose that the virtual target move along the path with constant velocity $\dot{s} = u_d$ and the path curvature is constrained ($|c_c| \le \bar{c}_c$).

Now, the path-following control problem can be formulated as followers. Given the AUV model (8) and a path to be followed, derive feedback control laws for the force $\mathcal{F}$ and torque $\mathcal{G}$ that minimize the steady-state values of y_1, s_1, and ψ.

To solve the problem, the following discrete control law is proposed:

$$\mathcal{F}(t) = \mathcal{F}_c + \mathcal{F}_s, \quad \mathcal{G}(t) = \mathcal{G}_c + \mathcal{G}_s, \quad t \in T_k,$$
$$\mathcal{F}_c = d_u(t_k), \quad \mathcal{G}_c = d_r(t_k) + m_r(\ddot{\psi}_F^k - \hat{\ddot{\beta}}^k),$$
$$\mathcal{F}_s = \text{sat}(k_1 \hat{s}_1^k + k_2 \Delta \hat{u}^k, \overline{\mathcal{F}}_s),$$
$$\mathcal{G}_s = \text{sat}(k_3 \hat{y}_1^k + k_4 \hat{\psi}^k + k_5 \Delta \hat{r}^k, \overline{\mathcal{G}}_s), \tag{11}$$

where $T_k \equiv [t_k, t_{k+1})$, $t_k = kh$, $k = 0, 1, 2 \ldots$, h is the control step; $\mathcal{F}_c$, $\mathcal{G}_c$ are feedforward control terms aimed to cancel terms d_u, d_r in equations (9) and terms $\ddot{\beta}$, $\ddot{\psi}_F$ in the equation for $\Delta r \equiv r + \dot{\beta} - \dot{\psi}$, $\hat{\ddot{\beta}}^k$ is an estimate of acceleration $\ddot{\beta}$ at t_k obtained using the dynamic model of the AUV (see [5] for details), $\ddot{\psi}_F = \dot{c}_c \dot{s} + c_c \ddot{s}$; $\mathcal{F}_s$, $\mathcal{G}_s$ are feedback control terms, $\overline{\mathcal{F}}_s$, $\overline{\mathcal{G}}_s$ are the shares of control resources reserved for stabilization, $\text{sat}(\sigma, \bar{\sigma}) = \text{sign}(\sigma) \min(|\sigma|, \bar{\sigma})$ is the saturation function; $\hat{s}_1^k$, $\hat{y}_1^k$, $\hat{\psi}^k$, $\Delta \hat{u}^k$, $\Delta \hat{r}^k$ are measurements of s_1, y_1, ψ, $\Delta u \equiv u - u_d$, Δr at time moment t_k that contain some additive errors:

$$\hat{s}_1^k = s_1(t_k) + \tilde{s}_1(t_k), \quad |\tilde{s}_1(t_k)| \le s_1^0,$$
$$\hat{y}_1^k = y_1(t_k) + \tilde{y}_1(t_k), \quad |\tilde{y}_1(t_k)| \le y_1^0,$$
$$\hat{\psi}^k = \psi(t_k) + \tilde{\psi}(t_k), \quad |\tilde{\psi}(t_k)| \le \psi^0,$$
$$\Delta \hat{u}^k = \Delta u(t_k) + \Delta \tilde{u}(t_k), \quad |\Delta \tilde{u}(t_k)| \le \Delta u^0,$$
$$\Delta \hat{r}^k = \Delta r(t_k) + \Delta \tilde{r}(t_k), \quad |\Delta \tilde{r}(t_k)| \le \Delta r^0,$$

s_1^0, y_1^0, ψ^0, Δu^0, Δr^0 are positive constants; k_i are feedback coefficients to be determined ($i = \overline{1, 5}$).

For synthesis of the feedback coefficients with the use of toolbox VLF-REDUCTOR, the dynamic model of the vehicle given by equations (9), (10) should be simplified. Using the Taylor expansion, neglecting the equation in sway and keeping

in mind feedforward control terms, model (9), (10) can be rewritten as

$$\dot{s}_1 = \Delta u + c_c u_d y_1,$$
$$\dot{y}_1 = u_d \psi - c_c u_d s_1,$$
$$\dot{\psi} = \Delta r,$$
$$m_u \Delta \dot{u} = \mathcal{F}_s,$$
$$m_r \Delta \dot{r} = \mathcal{G}_s. \tag{12}$$

Model (12) combined with feedback control law (11) can be presented in terms of model (1), (3)-(6) of toolbox VLF-REDUCTOR:

$$\mathbf{x} = \begin{bmatrix} s_1 & y_1 & \psi & \Delta u & \Delta r \end{bmatrix}^T, \quad \mathbf{u} = \begin{bmatrix} \mathcal{F}_s & \mathcal{G}_s \end{bmatrix}^T,$$

$$J = \begin{bmatrix} 0 & 0 & 0 & 0 & 0 \\ 0 & 0 & 0 & 0 & 0 \\ 0 & 0 & 0 & 0 & 0 \\ 0 & 0 & 0 & m_u & 0 \\ 0 & 0 & 0 & 0 & m_r \end{bmatrix}, \quad A = \begin{bmatrix} 0 & 0 & 0 & 1 & 0 \\ 0 & 0 & u_d & 0 & 0 \\ 0 & 0 & 0 & 0 & 1 \\ 0 & 0 & 0 & 0 & 0 \\ 0 & 0 & 0 & 0 & 0 \end{bmatrix},$$

$$B = \begin{bmatrix} 0 & 0 \\ 0 & 0 \\ 0 & 0 \\ 1 & 0 \\ 0 & 1 \end{bmatrix}, \quad G = \begin{bmatrix} 1 & 0 \\ 0 & -1 \\ 0 & 0 \\ 0 & 0 \\ 0 & 0 \end{bmatrix},$$

$$\Theta = \begin{bmatrix} 0 & 1 & 0 & 0 & 0 \\ 1 & 0 & 0 & 0 & 0 \end{bmatrix}, \quad U = \begin{bmatrix} u_d \bar{c}_c & 0 \\ 0 & u_d \bar{c}_c \end{bmatrix},$$

$$D = E_5, \quad \bar{\sigma} = \begin{bmatrix} \overline{\mathcal{F}}_s & \overline{\mathcal{G}}_s \end{bmatrix}^T, \quad \eta_0 = \begin{bmatrix} s_1^0 & y_1^0 & \psi^0 & \Delta u^0 & \Delta r^0 \end{bmatrix}^T.$$

The other matrices and vectors in model (1), (3)-(6) are assigned default values. Now, specifying the criteria for synthesis as

$$J_f = \omega_1 \limsup_{t \to \infty} |s_1(t')| + \omega_2 \limsup_{t \to \infty} |y_1(t')| + \omega_3 \limsup_{t \to \infty} |\psi(t')|$$

(ω_i are the wighted coefficients), the toolbox can compute feedback parameters k_i.

C. Path-following Controller Design for Multi-AUV Formation

As mentioned above, the formation control strategy considered in the paper is based on the leader-follower approach. Assume that each leader can accurately estimate the value of its state variables (x, y, ψ_B, $\cdots$) at time instant t_{k+1} on the basis of its dynamic model and measurements of the state vector obtained at t_k ($t_{k+1} = t_k + h$). Let the leader sends these estimates at every time moment t_k and its followers reliably receive them at t_{k+1}. Moreover, we assume that the AUVs have their clocks synchronized and update their control signals at the same time moment t_k, $t_k = kh$, $k = 0, 1, 2 \ldots$. Under these assumptions, the problem of communication delays can be left out.

In what follows, focusing on a given leader-follower pair, denote the leader as l and the follower as f. Considering the leader as a virtual target for its follower, one can derive a kinematic model of the leader-follower pair in coordinates (s_1, y_1) as

$$\dot{s}_1 = v_{tf}\cos(\psi_{Wf} - \psi_{Wl}) - v_{tl} + \dot{\psi}_{Wl}y_1,$$
$$\dot{y}_1 = v_{tf}\sin(\psi_{Wf} - \psi_{Wl}) - \dot{\psi}_{Wl}s_1,$$
$$\dot{\psi}_{Wf} = r_f + \dot{\beta}, \qquad (13)$$

where $\psi_W = \psi_B + \beta$.

Let a desired position of the follower with respect to the leader in coordinates (s_1, y_1) be defined by vector $[s_1^*, y_1^*]^T$. The control law that provides stabilization of the desired position of the follower with respect to the leader can be defined as in (11) except for the feedback control terms given by

$$\mathcal{F}_s = \mathrm{sat}(k_1 \triangle \hat{s}_1^k + k_2 \triangle \hat{u}_{lf}^k, \overline{\mathcal{F}}_s),$$
$$\mathcal{G}_s = \mathrm{sat}(k_3 \triangle \hat{y}_1^k + k_4 \triangle \hat{\psi}_{lf}^k + k_5 \triangle \hat{r}_{lf}^k, \overline{\mathcal{G}}_s), \qquad (14)$$

where $\triangle \hat{s}_1^k$, $\triangle \hat{u}_{lf}^k$, $\triangle \hat{y}_1^k$, $\triangle \hat{\psi}_{lf}^k$, $\triangle \hat{r}_{lf}^k$ are estimates of $\triangle s_1$, $\triangle u_{lf} \equiv u_l - u_f$, $\triangle y_1$, $\triangle \psi_{lf} \equiv \psi_{Wf} - \psi_{Wl}$, $\triangle r_{lf} \equiv \dot{\psi}_{Wf} - \dot{\psi}_{Wl}$ at time moment t_k computed with some errors using measurements of the state vector of the follower and estimates of the state variables of the leader at t_k, which are sent by the leader at t_{k-1} and received by the follower during time interval T_{k-1}. As before, we need to design feedback coefficients k_i, $i = \overline{1,5}$.

Linearizing equations in (13) and (9) in the neighborhood of "unperturbed motion" ($s_1 = s_1^*$, $y_1 = y_1^*$, $\psi_{Wl} = \psi_{Wf} = \psi_F$, $u_l = u_f = u_d$, $v_l = v_f = 0$, $r_f = \dot{\psi}_{Wl} = \dot{\psi}_F$) and having in mind (14), one can derive the following error model for feedback control design:

$$\triangle \dot{s}_1 = \triangle u_{lf} + \dot{\psi}_F \triangle y_1 + y_1^* \dot{\psi}_l,$$
$$\triangle \dot{y}_1 = u_d \triangle \psi_{lf} - \dot{\psi}_F \triangle s_1 - s_1^* \dot{\psi}_l,$$
$$\dot{\psi}_{lf} = \triangle r_{lf},$$
$$m_u \triangle \dot{u}_{lf} = \mathcal{F}_{sf} + m_u \dot{u}_l,$$
$$m_r \triangle \dot{r}_{lf} = \mathcal{G}_{sf} + m_r \dot{r}_l. \qquad (15)$$

There are two ways to synthesize the feedback control parameters for the follower AUV using toolbox VLF-REDUCTOR. The first way implies that the velocities of the leader $\triangle v_l$, $\dot{\psi}_l$ and its accelerations $\dot{u}_l$, $\dot{r}_l$ are constrained and terms in equations (15) that contain these variables are regarded as bounded uncertainties, which can be accommodated in the toolbox model. In the second one, a full model of the multi-AUV system is built by aggregating error models (either (12) for the formation leader or (15) for the other AUVs) of all AUVs and then control design is performed for the full system. The first way gives more conservative results but it is less time-consuming compared to the second one. In practice, both approaches are used simultaneously.

IV. NUMERICAL COMPUTATION AND SIMULATION RESULTS

Numerical computations and simulations were conducted for a formation of four INFANTE underwater vehicles uniformly distributed around a circle of radius 5 m. We borrowed the AUV model parameters from [5], including $m = 2234.5$ kg and $I_z = 2000$ N m^2. The formation parameters are presented in Table I. The other parameters of the model used in control

AUV ID	Leader ID	s_1^*	s_1^*
0	–	–	–
1	0	-7	7
2	0	-7	-7
3	1	-7	-7

TABLE I
FORMATION PARAMETERS.

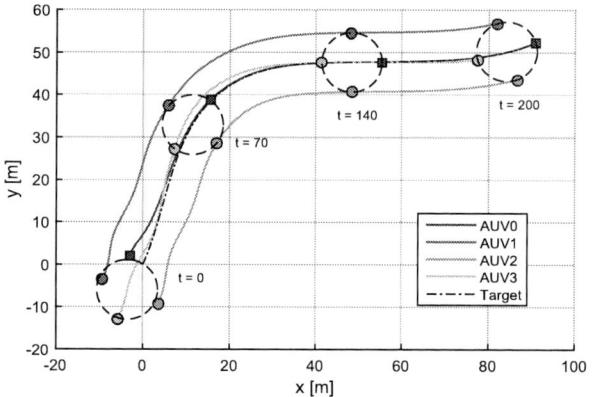

Fig. 2. Formation trajectories.

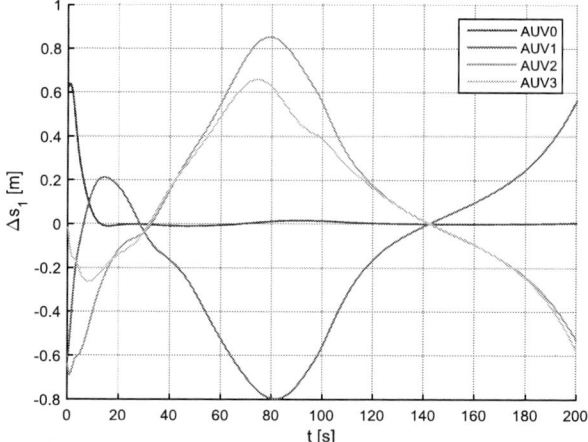

Fig. 3. Errors in the course direction.

design and simulations are: v_d=0.6 m s^{-1}, h=0.2 s (control step), $\overline{\mathcal{F}}_s$=200 N and $\overline{\mathcal{G}}_s$=100 N m (constraints on the feedback control signals), $\bar{c}_c$=0.12 (constraints on the path curvature), s_1^0=y_1^0=0.1 m, ψ^0=0.01 rad, $\triangle u^0$=0.1 m s^{-1}, $\triangle r^0$=0.01 rad s^{-1} (estimation errors). Simulation results of the designed control system are depicted in Fig. 2-4.

Simulation results and analysis of model (15) show that the greater reference linear velocity u_d and the larger values s_1^*, y_1^* the more difficult for the follower to keep a desired position with respect to the leader. If these parameters are large enough, the proposed formation control algorithm does not

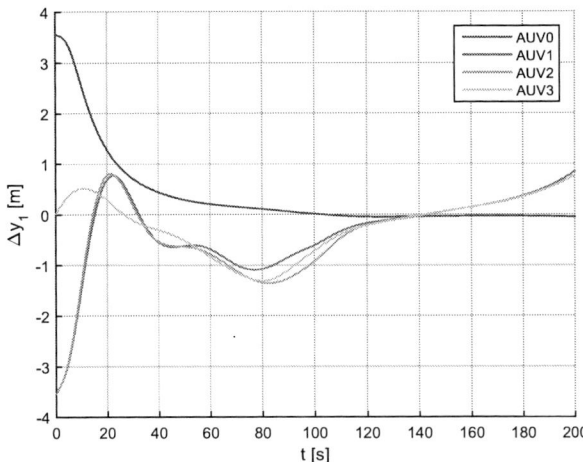

Fig. 4. Errors in the direction perpendicular to the course.

solve the path-following problem and other control schemes, for example, based on measurements of distance and bearing angles (see e.g. [17]) are preferable. However, the appearance of the follower's trajectory generated by these algorithms may significantly differs from the appearance of the path to be followed, which is not acceptable in many applications.

It should be noted that the controller is designed for precise stabilization of the formation and works well only if the initial deviations from a desired configuration are small enough. However, for the case of large deviations, other formalizations of the desired formation behavior [18], [19] supported by VLF-REDUCTOR can be used.

V. CONCLUSION

In this paper, we presented a MATLAB-based software toolbox for analysis and synthesis of nonlinear control systems. Using the toolbox, a path-following digital controller for multi-AUV formations was designed taking account measurement errors, constraints on control actions, and the fact that only the formation leader holds information about the path to be followed. Simulation results illustrated the performance of the control system, which turns out to be acceptable for many cases, however, has the potential to be improved. The future work intension is in improving performance of the control system by including integral actions in the controller to reject constant disturbances and introducing extra controls for the virtual target to adjust its speed to the current errors of the followers.

ACKNOWLEDGMENT

The software toolbox designed for analysis and synthesis of continuous-discrete (digital) control systems using VLFs and the proposed multi-AUV formation model have been created under support of RSF (grant №16-11-00053). The reported study was also funded by RFBR according to the research projects №16-29-04238.

REFERENCES

[1] S. N. Vassilyev, "Method of reduction and qualitative analysis of dynamic systems:i, ii," *Journal of Computer and Systems Sciences International*, vol. 45, no. 1, 2, pp. 17–25, 167–179, 2006. [Online]. Available: http://dx.doi.org/10.1134/S1064230706010023

[2] S. N. Vassilyev, A. E. Druzhinin, and N. Y. Morozov, "Derivation of preservation conditions for properties of mathematical models," *Doklady Mathematics*, vol. 92, no. 3, pp. 658–663, 2015. [Online]. Available: http://dx.doi.org/10.1134/S1064562415060058

[3] R. Z. Abdullin, L. J. Anapolski, R. I. Kozlov, A. I. Malikov, V. M. Matrosov, A. A. Voronov, and A. S. Zemljakov, *Vector Lyapunov functions in stability theory*. University of Michigan: World Federation Publisher Company, 1996.

[4] R. I. Kozlov and O. R. Kozlova, "Investigation of stability of nonlinear continuous-discrete models of economic dynamics using vector lyapunov function. i," *Journal of Computer and Systems Sciences International*, vol. 48, no. 2, pp. 262–271, 2009. [Online]. Available: http://dx.doi.org/10.1134/S1064230709020105

[5] L. Lapierre and D. Soetanto, "Nonlinear path-following control of an {AUV}," *Ocean Engineering*, vol. 34, no. 1112, pp. 1734 – 1744, 2007.

[6] L. Lapierre and B. Jouvencel, "Robust nonlinear path-following control of an auv," *IEEE Journal of Oceanic Engineering*, vol. 33, no. 2, pp. 89–102, April 2008.

[7] D. W. Kim, "Tracking of {REMUS} autonomous underwater vehicles with actuator saturations," *Automatica*, vol. 58, pp. 15 – 21, 2015.

[8] B. Das, B. Subudhi, and B. B. Pati, "Cooperative formation control of autonomous underwater vehicles: An overview," *International Journal of Automation and Computing*, vol. 13, no. 3, pp. 199–225, 2016. [Online]. Available: http://dx.doi.org/10.1007/s11633-016-1004-4

[9] X. Li, D. Zhu, and Y. Qiun, "A survey on formation control algorithms for multi-auv system," *Unmanned Systems*, vol. 2, no. 4, pp. 351–359, 2014.

[10] R. Cui, S. S. Ge, B. V. E. How, and Y. S. Choo, "Leader-follower formation control of underactuated autonomous underwater vehicles," *Ocean Eng.*, vol. 37, pp. 1491–1502, 2010.

[11] N. Burlutskiy, Y. Touahmi, and B. H. Lee, "Power efficient formation configuration for centralized leader–follower auvs control," *Journal of Marine Science and Technology*, vol. 17, no. 3, pp. 315–329, 2012. [Online]. Available: http://dx.doi.org/10.1007/s00773-012-0167-0

[12] P. Millan, L. Orihuela, I. Jurado, and F. R. Rubio, "Formation control of autonomous underwater vehicles subject to communication delays," *IEEE Transactions on Control Systems Technology*, vol. 22, no. 2, pp. 770–777, 2014.

[13] H. Yang, C. Wang, and F. Zhang, "Brief paper: a decoupled controller design approach for formation control of autonomous underwater vehicles with time delays," *IET Control Theory Applications*, vol. 7, no. 15, pp. 1950–1958, October 2013.

[14] X. Xiang, C. Liu, L. Lapierre, and B. Jouvencel, "Synchronized path following control of multiple homogenous underactuated auvs," *Journal of Systems Science and Complexity*, vol. 25, no. 1, pp. 71–89, 2012.

[15] R. I. Kozlov, E. I. Druzhynin, S. A. Ulyanov, V. A. Voronov, and B. B. Belyaev, "Synthesis of a combined precision stabilization system for a space telescope," *AIP Conference Proceedings*, vol. 1493, no. 1, pp. 1059–1065, 2012. [Online]. Available: http://scitation.aip.org/content/aip/proceeding/aipcp/10.1063/1.4765619

[16] I. V. Bychkov, V. A. Voronov, E. I. Druzhinin, R. I. Kozlov, S. A. Ul'yanov, B. B. Belyaev, P. P. Telepnev, and A. I. Ul'yashin, "Synthesis of a combined system for precise stabilization of the spektr-uf observatory: Ii," *Cosmic Research*, vol. 52, no. 2, pp. 145–152, 2014. [Online]. Available: http://dx.doi.org/10.1134/S0010952514020014

[17] S. A. Ul'yanov and N. N. Maximkin, "Stabilization of multi-auv formation with digital control," in *2016 39th International Convention on Information and Communication Technology, Electronics and Microelectronics (MIPRO)*, May 2016, pp. 1108–1113.

[18] S. N. Vassilyev, R. I. Kozlov, and S. A. Ul'yanov, "Analysis of coordinate and other transformations of models of dynamical systems by the reduction method," *Proceedings of the Steklov Institute of Mathematics*, vol. 268, no. 1, pp. 264–282, 2010. [Online]. Available: http://dx.doi.org/10.1134/S0081543810050184

[19] ——, "Multimode formation stability," *Doklady Mathematics*, vol. 89, no. 2, pp. 257–262, 2014. [Online]. Available: http://dx.doi.org/10.1134/S1064562414020173

Remote alarm reporting system responsive to stoppage of ballast water management operation on ships

Goran Bakalar, Ph. D.*, Myriam Beatriz Baggini, B. Sc.**, Sebastian Gabriel Bakalar***

*Bonum Mare Consulting, President & General Manager, 148 Young Ave, Cocoa Beach 32931, FL, USA
** Bonum Mare Consulting, Operations Manager, 148 Young Ave, Cocoa Beach 32931, FL, USA
*** Bonum Mare Consulting, Board Member, 148 Young Ave, Cocoa Beach 32931, FL, USA
Goran.Bakalar@xnet.hr

Abstract - **This paper reviews ballast water treatment systems regarding functional and operational problems. Ballast water treatment systems (BWTS) could be in failure or under repair. New invented sensors based on flow measurement methodologies should report malfunctions utilizing satellites. This invention is the solution concerning the quality control of ship's ballast water, and this is the system which would timely and with a higher degree of certainty establish the quality of ballast water treatment operations on ships. The system according to this invention is autonomous in relation to the ship's crew and eliminates a possibility that the members of the ship's crew influence the control of the ballast water, and it excludes the possibility of concealing a possible defect or irregular operation of the device for ballast water management.**

Key words: Ballast water management; Flow measurement; Sensor; Communication; Evaluation; Innovation; Invention

I. INTRODUCTION

The technical problem being addressed in this article is the efficiency of BWTS and the necessity of supervising the results. There are various systems for the purification or neutralization of unwanted ingredients contained in ship's ballast water [1][2]. The purification of ballast water implies the harmonization with the international laws on the allowed quantity of certain ingredients contained in ballast water [3]. The existing technologies do not provide a certification that a ship having entered a port has clean ballast water [4]. This problem is still tended to be solved by means of laboratory analyses of the samples taken on board a ship after it has entered a port. The solution concerning the quality control of ship's ballast water should be found, and the system which would timely and with a higher degree of certainty establish the quality of ballast water on board a ship should be created [5]. Then, it could be established with certainty that certain ship would not create potential problems in the port of arrival from the legal or time aspects. In that way, any additional control by the sampling and laboratory analyses of the ship's ballast water would not be necessary. A technical problem to be solved by this

innovation relates to the question how to carry out the quality control of ballast water, as regards the compliance with international standards, before the ship enters a port and present such evidence to the port authorities [6]. The overall problem is that the ability of modern ships to cover long distances in a short amount of time, provides a means for non-native marine species, including algae, invertebrates, and pathogens to arrive in ports via ballast water [7]. The International Convention for the Control and Management of Ships' Ballast Water & Sediments was adopted by consensus at a Diplomatic Conference at IMO in London on Friday 13 February 2004 [8].

The today's potential control of ship's ballast water is provided for by the International Maritime Organization Convention according to which the port authorities are given a possibility to carry out laboratory analysis of ballast water, where the inspector working in that port orders so [8]. The ballast water samples are taken on board a ship, after the ship enters the port, and they are submitted to a long-term analysis in a laboratory. If the analysis of the ballast water shows deviation from the values specified by the Convention, the ship shall leave the port without the cargo operations being performed, and the total analysis cost shall be paid by the shipping company. If the results of ballast water laboratory analysis show to be negative, the total amount of analysis cost shall be paid by the port authorities located in the port in which the laboratory analysis has been carried out and the ship shall be granted permission to load the cargo and to discharge ballast water [9]. Ballast water treatment systems could have clogged filters (UV ballast water management systems), burned UV bulbs, improper function of de-chlorination unit (chlorination ballast water management system) and many other possibilities [10]. The operational efficiency of a BWTS is questionable considering the long number of years that these systems have to operate once installed on a ship. There were more than 60 existing certified treatment systems listed in year 2014 [11] and more new systems have been invented at this year 2016. Different technologies were suggested for ballast water treatment [12][13], such as disinfection with chlorine [14], adding biocides to ballast water for harmful microorganisms neutralisation [15], sterilisation with ozone [16], filtration with ultraviolet light [17], sonication [18], hydrodynamic cavitation [19], and many other technologies such as electroionization [20] and treatment by heat [21]. Contribution of this study is in proposed monitoring system which helps to exclude the possibility of concealing a possible unreported defect or irregular operation of a ballast water treatment system.

II. FAILURES AND MAINTENANCE OF BWTS

There are many ships already equipped with ballast water treatment systems today. Not all of them have a good operation performance of their certified treatment systems. Unsuitable function of automated ballast water management systems on board the ships could cause invasion to domestic waters. The ship's movement in the storms combined with the salt exposure brings up a very big risk of damages or improper functionality of any device or parts of the systems of ballast water treatment [22]. Certain scientists have tried to prove the effectiveness and efficiency of a particular BWTS. A scientific paper and study results have proved the efficiency of BWTS on a very large passenger ship [23]. Most of the attention on these very large ships has been on passengers having the most stable journey possible. Everything possible is done so that the impact of waves and winds on the ship is kept to a minimum. These almost ideal conditions cannot be completely accepted as a measure of the quality of the operations and state of certain technologies for the treatment of ballast water. The operation and results of the BWTS after these conditions could be very different to those on mid-sized tankers that travel in difficult weather conditions of very high waves, as well as devastating and life threatening winds [22]. These working conditions for the BWTS are, from spring until autumn for example, quite frequent in certain areas when choosing the shortest possible route. These areas include the North Sea, the western coast of Algeria, the Bay of Lyon, Biscayne, the coast of Greece, the Adriatic Sea, the Aegean Sea, as well many other areas that are very difficult to travel in. Ships that transport liquid cargo, and usually travel in these areas the most, have a total load of between 25,000 and 50,000 tons DWT. These ships are far smaller than many large passenger ships. Due to the difficult weather conditions that these boats have to go through and after which their BWTS has to function perfectly, it is inappropriate to take into relevant consideration solely the results of BWTS operations in the best possible conditions, such as is the case in the largest passenger ships or cargo ships that travel in ideal conditions. Other scientists make no mention in their experiments of how long the BWTS, which was already installed on the boat, was used before their research, whether the water taken for analysis was during winter or in difficult weather conditions, and whether the installed system had already been under the influence of extreme weather and sea conditions.

A scientist has conducted a study related to consistency of BWTS operation. In certain study, the total calculated reliability of a particular ballast water treatment system on board ships which consists of the two computing subsystems as part of units of which one is software redundant, was 0.916 or 91.6% [22]. This means that for 8.4% of the operational time, any of the mentioned systems could be in failure or under repair. This is a significant risk for the operation of ballast water treatment systems. In same study, survey of operational experiences from ships has proven that more time would be needed for the maintenance of ballast water management systems. It can be clearly concluded from that survey result of 7.3% from all participants and interviews that ship officers by-pass ballast water treatment systems due to lack of time for rectifying malfunctions, failures, maintenance or spare parts replacement in ballast water treatment systems. [22]. This implies an unclear future for port environment protection management. Malfunctions occur and the reactions of the crew members can be loyal towards the shipping company [24][25]. When viewed from the aspect of making a profit, completing cargo operations and ballasting and then sailing towards the next port is of the utmost importance to a shipping company [26]. Ballast water treatment systems are not always in optimal operational condition and ships' crew members by-pass them, as often occurred with automatic oil discharge content monitors in the past [25] [27]. Also, sensors of a system would inhibit the operation of the ballast water treatment system due to the TRO value being too high. It can take from three to three and a half hours to replace a UV bulb in a BWTS with a UV generator. Bulbs burn out often [10].

III. METHODS AND TOOLS FOR EVALUATION AND MONITORING

Efficacy of BWTSs was evaluated during certification tests [28] [29]. Evaluation of the treatment efficacy is based on the determination of living organisms after the treatment. Different methodologies were suggested for monitoring content of treated ballast water. Some methods that have been proposed regarding the supervision of ballast water have been the use of PAM fluorometry [30], flow cytometry [31] [32] [33] [34], FDA and ATP content [35] [36], as well as classic laboratory methods [37] [38].

Those methodologies use their detection tools. Significant difference of detection tools is coming from the aspect of time needed for ships' ballast water sample analyses [4]. All the detection methods regarding the quality of ship ballast water have been proposed with the purpose of analysing the samples upon neutralisation. As such, depending on the analysis results of the treated ballast water samples, questions are raised regarding responsibility should the analysis results of this ballast water be such that it is not acceptable for unloading [9]. If the BWTS was monitored and if action was taken

1039

immediately when treatment problems occurred, it would be easier to designate who is responsible for the poor operation of the BWTS. The existing technologies do not provide a certification that a ship having entered a port has clean ballast water. This problem is still tended to be solved by means of laboratory analyses of the samples taken on board a ship after it has entered a port. The system which would timely and with a higher degree of certainty establish the quality of ballast water on board a ship should be created. Then, it could be established with certainty that the detected ship would not create potential problems in the port of arrival from the legal or time aspects. In that way, any additional control by the sampling and laboratory analyses of the ship's ballast water would not be necessary [39] [40]. The conclusion that can be drawn upon completing one of the best ballast water projects ever, the NSBWO (North Sea Ballast Water Opportunity) project, is that there should be no doubts regarding the capabilities of BWTS that has received all necessary working permits. These conclusions have also been drawn by many scientists in their studies. Biological scientists do not contest the obligatory installation of BWTS on ships, but confirm the repeated growth of numerous microorganisms upon treatment. Certain samples analyses after suitable BWTS function, showed re-growth of phytoplankton which represents risk of aquatic invasion and contamination [41] [42]. Growth of viable organisms can be assumed to occur cases where at least one cell keeps its viability after the treatment, and then, it is exposed to favorable conditions [43]. Fraction of organisms is expected to keep their viability after the treatment application [44]. Effects of the combination of UV irradiation and dark storage in ship's ballast water tank provide a framework for studies and research [30] [45] [46] [47]. However, the comments are always that the introduction of BWTS is good as it reduces the risk of the invasion of microorganisms in new areas. These interpretations leave too much room for discrepancies regarding treated ballast water compared to the prescribed standards. Overall, not one suggestion for compliance monitoring has been considered, let alone accepted, despite the efforts and lobbying of scientists. Drawing an indirect conclusion whether the treated ballast water has reached the required standard, which allows a certain number of microorganisms to remain in the water after treatment, is a debatable method. This is why not one suggestion has been accepted at any level. Despite the lobbying efforts of scientists from renowned institutes and universities worldwide that their monitoring method be accepted, the fact remains that lawyers can bail out the incriminated ship and company should any analysis show unacceptable results. Up to the present, not one suggested method has been seriously considered for the indicative monitoring of the quality and contents of ballast water. It is needed to have an information immediately after some failure happen in

a ballast water treatment system during the operation [5].

The invention, as a result of this study, excludes the possibility of concealing a possible unreported defect or irregular operation of the ballast water treatment system, because the follow-up system results indicate this on time to the authorities of the port which the ship leaves. Figure 1 shows a block scheme of the system of the automatic measurement of ship's ballast water, via satellite communication, which includes the assembly A of the system that is located on land, and the assembly B of the system that is located on board a ship. The function of this innovation is explained in further text.

IV. THE INNOVATION

Solutions that are suggested in this article reveal irregularities in the operations of BWTS immediately after they occur. In this way, the need for sampling and analysing ballast water is avoided as this type of supervision is not approved by the IMO in any case, even though it was the subject of analysis of many scientists. In this monitoring system, shown in Figure 1, the on duty operator is located at the ship terminal where the ship unloads its cargo and loads ballast water. He has contact with the ship via an internet connection or a cable and has immediate insight into any BWTS malfunctions. The indication is the amount of water that has gone through the treatment system. It is monitored and compared with the total amount of ballast water after loading the ballast. The method used in this system is flow measurement. There are several types of products and technologies that can be used to measure flow. The options for measuring flow include the use of magnetic and ultrasonic flow meters, and pressure transmitters. Magnetic flow meters can be affected by air in the pipe and can give false readings of the pipe if it is not kept full. They can also be affected by stray. Ultrasonic flow devices can be affected by solids in the flow, and if the pipe is not maintained full, they will not be able to read and report the flow. There are level measurement options in technology, such as radar. In USA, radar level technology has been partially approved. Meters are classified into two basic types: positive displacement and velocity. Meters that feature both positive displacement and velocity are known as compound meters. The unit of measurement could be in cubic meters. Positive displacement meters operate by filling and emptying a small compartment with a known volume of liquid which moves with the flow of water. The flow rate is calculated based on the number of times these compartments are filled and emptied. The movement of a disc or a piston drives an arrangement of gears that registers and records the volume of liquid. Velocity meters operate on the principle that water passing through a known cross-sectional area with a measured velocity can be equated into a volume of flow. Velocity meters are good for high flow applications. In this invention, two different meters could be used: one for main ballast water pipeline flow measurement and the second one for treated ballast water flow measurement. The effectiveness of the proposed method has been proved by certain scientists **[48] [49] [50] [51]**. The tool and computed program measures flow of ballast water through main ballast water pipeline and flow just before and after the treatment **[52]**.

The performance control of ballast water treatment system using the automatic alarm reporting system responsive to stoppage of ballast water treatment operation, starts in the port in which cargo will be loaded. The electronic devices 17, from the office 1 in the assembly A of the system that is located on land, start follow-up computed program, remotely, via satellite communication 2, through electronic devices 17 in the assembly B of the system that is located on board a ship 3, through the lines 16, and both flow meters 12. The ballast water is driven from the sea, through the pipes 15, via the valve 4, and pressure gauge 13 into the ballast water treatment system TS. By the automatic flow measurement of ballast water, the results of operation are obtained.

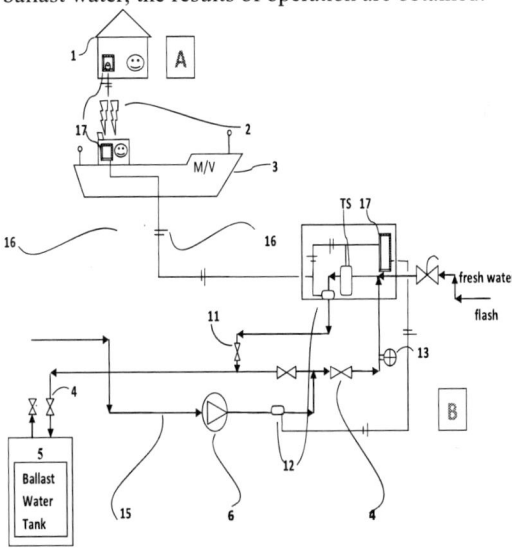

Fig. 1. Block scheme of the invented system

The data are collected automatically in the electronic device 17, and are, after the filtration and archiving, sent as messages, via electronic devices 17 on board a ship and satellite communication 2, to the assembly A of the system that is located on land. It can be utilized hybrid communications that combine satellite connectivity with links to terrestrial mobile networks. The procedure is completed when the staff from the assembly A of the system that is located on land gives its statement on the performance and quality of the ballast water treatment system operation of the ship under control. That statement includes exact amount of ballast water loaded on board a ship

and exact amount of treated ballast water driven through a treatment unit TS and measured on flow meter 12. The statement will be sent by email, using internet protocol (IP). This statement given by the staff from the assembly A of the system that is located on land is manifested in the issue of the Certificate of the Quality and Performance of the Observed Ballast Water that is filed to the port authorities. It will contain a comment of proper or improper ballast water treatment on a ship according to statement data sent by terminal operator. On the basis of such Certificate the port authorities decide to grant or to refuse to grant the ship's permission to enter the next port, where the observed results indicated irregular operation of the ship's ballast water neutralization system. In this monitoring system, the on duty operator is located at the ship terminal where the ship unloads its cargo and loads ballast water. He has contact with the ship via an internet and has immediate insight into any BWTS malfunctions. The tool and computed program measures flow of ballast water through main ballast water pipeline and flow just before and after the treatment. Measured data is transmitted through router, utilizing satellites. If there is a difference in the amount that had been treated and the total amount, this would then indicate the need to analyse the samples of treated ballast water. Should an alarm go off on his monitor regarding malfunctions (lack of flow) in the treatment of ballast water, the on duty operator has to contact the ship from the terminal and the ship has to be inspected. Records and notes of the event have to be made on the ship and it has to be photographed. Ballasting must not continue until the ship's BWTS malfunction is fixed and the system works properly. The method of determining proper operations using a flow meter is rapid, preventative and the cheapest method for shipping companies owners. Pressure in the ballast system pipes is compared with pressure at the opening of the BWTS. In this way, the possibility of bypassing the BWTS when loading the ballast water is avoided. In addition, crew members cannot solve this problem in inventive ways so that the ship can continue its journey with untreated ballast. On the basis of the ship terminal operator's report, the ship is either given permission or not given permission to sail to the next planned port for cargo operations. If a failure in the operation of the BWTS occurs that is difficult to fix, the ship would need to terminate cargo operations and anchor in front of the port. The next planned ship enters and moors itself to the port terminal and undertakes loading or unloading cargo operations and water ballast. Responsibility and expenses are determined according to the current law that is related to these situations. The advantages of this system above others in the quality detection of ship ballast are that it is a timely, economical and simple procedure, accurate and the cost of the flow meter device is very low, as is its installation and maintenance. The essence of this invention is the system of automatic detection of ship's ballast water operation stoppage. The system is

1041

composed of the following assemblies: an assembly A of the system that is located on land and an assembly B of the system that is located on board a ship. The assembly A of the system that is located on land and the assembly B of the system that is located on board a ship mutually communicate remotely, via satellite communication. The aim of the invention is to establish the quality and purity of ship's ballast water, via satellite communication, before the ship leaves the port, and to deliver a certificate containing the received information to the port authorities that grant the ship permission to enter the next port. In this way, an additional sampling of the ballast water after the ship enters the next port is avoided, at the same time avoiding the possibility of having to send the ship back from the local waters due to unwanted ingredients contained in the ballast water. The system is autonomous and does not depend on the ship's crew influence. This invention also eliminates the possibility that the members of the ship's crew influence, in any manner whatsoever, the ballast water control [53]. The invention excludes the possibility of concealing a possible unreported defect or irregular operation of the device for the purification and neutralization of ballast water, because the results indicate this on time to the authorities of the port which the ship leaves [54]. It can be implemented through maritime remote maintenance protocols [55]. The system which is presented by this innovation is proposed to be implemented in the world maritime legislation, since it provides for a frequent taking of observation of the ship's ballast water management operation order that it could be asserted with high credibility whether the ballast water is acceptable for discharge in the port of arrival or not. The system presented by this invention may be applied in the control of the water used for washing cargo tanks after unloading chemical substances. The system according to this invention may be applied in the remote control and management of the water supply and sewage systems.

V. CONCLUSION

The suggested methods of monitoring the operations and functionality of a particular BWTS unify the analysis results of previous research and suggestions for monitoring and controlling the operations of BWTS. The methods suggested in previous research were based on controlling the contents of treated ballast water. Not one of these methods has been accepted at a worldwide level nor implemented in the shipping industry worldwide. The reasons for this are, among others, due to the interests and profits of large shipping companies who would not like their ship to be stopped in future due to malfunctions in the BWTS. As such, there is a large risk of the invasion of sea microorganisms. One could say that the 2004 Convention has not fulfilled it task. Automatic alarm reporting system responsive to stoppage of ballast water treatment operation on ships continues the fight to protect the marine environment

and endemic species in local seas. The advantages of this system are that it is simple, economic, harmless technology and there is no need for additional training of the ship's crew as would be the case in using other methods. In this system, the port terminal has immediate insight into the proper operation of the BWTS and should its monitoring and supervision program malfunction – an alarm is activated. In this way, should the flow of the ballast water (treated or untreated) not be shown on the sensors that are monitored at the port terminal, they have to personally go to the ship and record the reasons for this. The best characterestic of this system is its timeliness as the ship should not continue cargo operations until the ballast water can be unloaded as per the regulations of the 2004 Convention.

REFERENCES

[1] E. Tsolaki, and E. Diamadopoulos, "Technologies for ballast water treatment: a review," J. Chem. Technol. Biot., vol. 85 (1), pp. 19-32, 2010.

[2] J. Mouawad, "Ballast water treatment technologies - technical considerations, experiences and concerns in retrofitting and new - builds," IMarEST Shipbuilders Forum, Istanbul, 2011.

[3] C. Hallers-Tjabbes, "Ballast Water Technologies and International and Regional Policy Developments - Knowns and Unknowns," Marine Tech Summit 2011, Busan, Korea, 23–25. September 2011.

[4] G. Bakalar, "Review of interdisciplinary devices for detecting the quality of ship ballast water," SpringerPlus, vol. 1(3), pp. 468. DOI: 10.1186/2193-1801-3-468., 2014.

[5] G. Bakalar, and M. B. Baggini, "Automatic communication system ship to shipping terminal, for reporting potential malfunctions of a ballast water treatment system operation," 39th International Convention MIPRO, Proceedings, pp. 836-840, IEEE Xplore, DOI: 10.1109/MIPRO.2016.7522236, http://ieeexplore.ieee.org/document/7522236/, 2016. Accessed: 10 November 2016.

[6] G. Bakalar, "Automatic Control System for Ship Ballast Water Treatment by Using Flow Cytometry and Satellite Communications Technologies," Dissertation, Rijeka University Croatia. 170 pp., 2013.

[7] http://hawaii.gov/dlnr/dar/ballast.html., Accessed: 8 April 2016.

[8] http://globalast.imo.org/the-bwmc-and-its-guidelines/, Accessed : 9 April 2016.

[9] G. Bakalar, "A thesis on a remotely controlled BWTS," The Ballast Water Times II, Journal of Institute NIOZ, project NSBWO (North Sea Ballast Water Opportunity)

[10] G. Bakalar, "Comparisons of interdisciplinary ballast water treatment systems and operational experiences from ships," SpringerPlus, vol. 5(1), pp. 240, 2016. DOI: 10.1186/s40064-016-1916-z.

[11] Lloyd's, "Ballast water treatment technology - current status," http://globalast.imo.org/wp-content/uploads/2015/01/BW-Treatment-Technology-Sept.-2012.pdf Accessed: 28 March 2017.

[12] G. Greensmith, "Ballast water technology availability-an update," IMO-WMU R&D Forum, Malmo, Sweden, 26–29. January 2010.

[13] A. Cangelosi, "Overview of current ballast water technologies: meeting the compliance criteria," Global R&D Forum on Ballast Water Management, Istanbul, Turkey, 26–28. October 2011.

[14] G. Simpson, "Ballast water disinfection with ClO_2," Proceedings of the 20th International Conference on Marine Biology, pp. 131, 2001.

[15] E. Chelossi, and M. Faimali, "Comparative assessment of antimicrobial efficacy of new potential biocides for treatment of cooling and ballast waters," Sci Total Environ 356, pp. 1–10, 2006.

[16] J. C. Perrins, W. J.Cooper, J. H. van Leeuwen, and R. P. Herwing, "Ozonation of seawater from different locations:

formation and decay of total residual oxidant implications for ballast water treatment," Mar. Pollut. Bull. 52, pp. 1023–1033, 2006.

[17] T. F. Sutherland, C. D. Levings, S. Petersen, and W.W. Hesse, "Mortality of zooplankton and invertebrate larvae exposed to cyclonic pre-treatment and ultraviolet radiation," Mar. Technol. Soc. J. 37, pp. 3–12, 2013.

[18] M. R. Gavand, B. M. McClintock, C. D. Amsler, R. W. Peters, and R.A. Angus, "Effects of sonication and advanced chemical oxidants on the unicellular green alga Dunaliella tertiolecta and cysts, larvae and adults of the brine shrimp Artemia salina: a prospective treatment to eradicate invasive organisms from ballast water," Mar. Pollut. Bull. 54 (11), pp. 1777–1788, 2007.

[19] M. Cvetković, M. Grego, and V. Turk, "The efficiency of a new hydrodynamic cavitation pilot system on Artemia salina cysts and natural population of copepods and bacteria under controlled mesocosm conditions," Mar. Pollut. Bull., doi: 10.1016/j.marpolbul.2016.01.030, 2016.

[20] J. Aliotta, A. Rogerson, C.B. Campbell, and M. Yonge, "Ballast Water Treatment by Electro-Ionization," 1st International Ballast Water Treatment R&D Symposium, Proceedings, pp 61-69, 2001.

[21] D. Mountfort, T. Dodgshun, and M. Taylor, "Ballast water treatment by Heat-New Zealand Laboratory $ Shipboard trials," 1st International Ballast Water Treatment R&D Symposium, Proceedings, pp 45-50, 2001.

[22] G. Bakalar, and M. B. Baggini, " A response to pollution of hazardous materials from ships," International Journal for Traffic and Transport engineering, 6(4), pp. 406-415, 2016. http://ijtte.com/article/100/Papers_Accepted_for_Publication.html, Accessed: 7 Novemberl 2016.

[23] D. A. Wright, N. A. Welschmeyer, and L. Paperzak, "Alternative, indirect measures of ballast water treatment efficacy during a shipboard trial: a case study," J. Mar. Eng. Technol., 14, pp. 1-8, 2015.

[24] G. Bakalar, "Efforts to develop a ballast water detecting device," Global IMO R&D Forum on Compliance Monitoring and Enforcement, Proceedings, Istanbul, Turkey, pp. 117-126.,2011.

[25] G. Bakalar, V. Tomas, and Z. Sesar, "Remote Monitoring of Ballast Water Treatment System Quality by Using Flow Cytometry and Satellite Communication Technologies," 54th International Symposium ELMAR 2012, Procceedings, pp. 290-295, http://IEEExplore.ieee.org/document/6338520/, 2012., Accessed: 9 September 2016.

[26] G. Bakalar, and V.Tomas, "Possibility of using satellite communication technologies for remote maintenance in marine industry," 5th GNSS Vulnerabilities and Solutions Conference, 2011. https://bib.irb.hr/prikazi-rad?lang=EN&rad=520610, Accessed: 7 Septeomber 2016.

[27] C. McLaughlin, D. Falatko, R. Danesi, and R. Albert, "Characterizing shipboard bilgewater effluent before and after treatment," Environ. Sci. Pollut. Res. 21(8):5637-5652, 2014.

[28] N. Dobroski, L. Takata, C. Scianni, and M. Falkner, "Assessment of the Efficacy, Availability and Environmental Impacts of Ballast Water Treatment Systems for Use in California Waters," California State Lands Commission Marine Facilities Division, 2007.

[29] MEPC, g/2012/Individual Guidelines for reference/G4.pdf, 2005., Accessed: 1 April 2016.

[30] M. R. First, and L. A. Drake, "Life after treatment: detecting living microorganisms following exposure to UV light and chlorine dioxide," J. Appl. Phycol. 26, pp. 227-235, 2014.

[31] G. Bakalar, and V. Tomas, "Possibility of Using Flow Cytometry in the Treated Ballast Water Quality Detection," Pomorski zbornik, vol. 51(1), pp. 43-55, 2016.

[32] G. Bakalar, and M. B. Baggini, "A verification of a remote monitoring of results of ballast water management system," ICTTE 2016 Belgrade, Proceedings, pp. 322-327 2016.

[33] I. van de Star, F. Fuhr, E. Brutel, and C. ten Hallers, "A matter of life and dead," IMO ICBWM, 2012.

[34] K. Peterson, "Updated experiments using a portable imaging instrument as a rapid diagnostic tool for IMO indicative monitoring compliance for ballast water treatment systems," Global R&D Forum on Ballast Water Management, 2011.

[35] C. van Slooten, T. Wijers, A. Buma, and L. Peperzak, "Development and testing of a rapid, sensitive ATP assay to detect living organisms in ballast water," Journal of Applied Phycology, DOI:10.1007/s10811-014-0518-9., 2014.

[36] C. van Slooten, "Quick compliance tools: ATP & FDA," The Ballast Water Times II, Journal of Institute NIOZ, project NSBWO (North Sea Ballast Water Opportunity) http://www.northsearegion.eu/files/repository/201210051742 24_TheBallastWaterTimes-Vol.2LR.pdf., 2012., Accessed 1. April 2016.

[37] S. Golasch, and M. David, "Sampling methodologies and approaches for ballast water management compliance monitoring," Promet, 5(23), pp. 397-405, 2011.

[38] S. Gollasch, "What is the overall error of different sampling and analyses methods which are proposed to verify compliance with the BWM Convention D-2 Standard?," ICBWM2012, Singapore, 2012.

[39] G. Bakalar, "The System of Remote Control of the Automatic Detection of Ship's Ballast Water via Satellite Communication from Land," Patent app., HR P20150144A, 2015.

[40] G. Bakalar, "Possibility of Remote Monitoring BWTS Quality," IMO ICBWM 2012 Globallast R&D Forum, Singapore, 2012.

[41] P. P. Stehouwer, A. Buma, and L. Peperzak, "A comparison of six different ballast water treatment systems based on UV radiation, electrochlorination and chlorine dioxide," Environmental Technology, DOI:10.1080/09593330.2015.1021858, 2015.

[42] V. Liebich, P. P. Stehouwer, and M. Veldhuis, "Re-growth of potential invasive phytoplankton following UV based ballast water treatment," Aquat. Invas. 7, pp. 29-36, 2012.

[43] L. F. Martinez, M. M. Mahamud, A. G. Lavin, and J. L. Bueno, "Evolution of phytoplankton cultures after ultraviolet light treatment," Mar. Poll. Bull. 64, pp. 556-562, 2012.

[44] J. Kain, and G. Fogg, "Studies of the growth of marine phytoplankton," J. Mar. Biol. Ass. UK. 37, pp. 781-788, 1958.

[45] P. P. Stehouwer, F. Fuhr, and M. Veldhuis, "A novel approach to determine ballast water vitality and viability after treatment," IMO-WMU R&D Forum, 2010.

[46] A. Sneekes, "Ecological risk of treated ballast water: a mesocosm experiment," Global R&D Forum on Ballast Water Management, Istanbul, Turkey, 26–28 October 2011.

[47] H. Litved, "Enumeration by staining techniques may overestimate viable cells in UV treated ballast water," Global R&D Forum on Ballast Water Management, 2011.

[48] G. Dinardo, L. Fabbiano, and G. Vacca, "How to directly measure the mean flow velocity in square cross-section pipes," Flow Measurement and Instrumentation, 49, pp. 1-7, 2016.

[49] T. Ihara, N. Tsuzuki, and H. Kikura, "Application of ultrasonic Doppler velocimetry to molten glass by using broadband phasedifference method," Flow Measurement and Instrumentation, 48, pp. 90-96, 2016.

[50] G. Bakalar, and M. B. Baggini, "Automated remote method and system for monitoring performance of ballast water treatment system operation on ships," 58th International Symposium ELMAR 2016, IEEE Xplore, Proceedings, pp. 229-232, 2016.

[51] T. Ziegenhein, and D. Lucas, "On sampling bias in multiphase flows; Particle image velocimetry in bubbly flows," Flow Measurement and Instruments, 48, pp. 36-41, 2016.

[52] G. Bakalar, "The system of automatic and autonomous flow and quantity measurement for detection of stoppage of ballast water treatment system operation," Patent app. HR P20160144A, 2016.

[53] G. Bakalar, and M. B. Baggini, "Bridge officers' operational experiences with electronic chart display and information systems on ships," Pomorski zbornik, vol. 52(1), 2016

[54] G. Bakalar, "Intensifying efforts for furthering safety culture in shipping - training aspects," International conference IMLA 19, Proceedings, pp 63-70.

[55] G. Bakalar, V. Tomas, and A. Buksa, "Monitoring of chemical pollution from the ships in coastal areas," International Conference on Climate Friendly Transport "Shaping Climate Friendly Transport in Europe, Key Findings & Future Directions", 2011.

Parameters for Condition Assessment of the High Voltage Circuit Breakers Arcing Contacts using Dynamic Resistance Measurement

Kerim Obarčanin*, Radenko Ostojić ** and Samir Džuzdanović***
DV Power, Stockholm, Sweden
kerim@dv-power.com*, radenko@dv-power.com**, samir@dv-power.com***

Abstract — Circuit Breakers are one of the most important elements in the transmission and distribution of electrical energy, and assessing their condition is of the upmost importance regarding the health and safety of the transmission and distribution system. Numerous procedures and parameters exist that directly reflect the state of the circuit breaker, but this paper focuses on a relatively new procedure that measures and estimates the states of the arcing contacts of the circuit breaker. The state of the arcing contacts is in a direct correlation with the health of the circuit breaker and it is assessed by applying test called the Dynamic resistance measurement. The important parameters during this Dynamic Resistance test will be explained. The results obtained using developed algorithm are demonstrated in the field on a real circuit breaker, completing thus all necessary steps for the assessment of the arcing contacts.

Keywords — High Voltage Circuit Breaker; Dynamic Resistance Measurement; Arcing Contacts;

I. INTRODUCTION

The design of modern high-voltage SF6 gas circuit breakers (HVCB) is based on the switching of two parallel contact sets. The first ones are the low-resistance contacts – main contacts which are specifically designed to carry the load current without excessive temperature rise. The second set, arcing contacts with tungsten – copper tip operate (as standalone contacts in the circuit for a few milliseconds) at the breaker opening following the main contact part. The electrical arc starts after the separation of the arcing contacts. The tungsten-copper material is designed to carry the arc until it is cleared at the next current zero-crossing.

Measurement of the dynamic resistance of the contact system (DRM) was presented in 1993 [1][2] as a tool to diagnose the condition of the arcing contacts. The method consists of applying constant current through the circuit breaker contact system, measuring the voltage drop over the contacts and plotting the resistance waveform during open or close operation of the HVCB. Since arcing contacts have higher resistance than the main contacts, during opening operation increasing of the voltage drop will appear at the moment of the main contacts separation. Relevant information related to contact condition, wear and/or misalignment can be obtained by analyzing resistance waveform together with the waveform of the motion of the contact system.

Beside the waveform shape, there must be numerical values that are comparable and that determine the condition of the arcing contacts - if they are in a good or bad condition or fall into the "gray zone". Arcing contacts degradation due to thermal stress can be determined using two parameters:

- Arcing contact overlapping time

- Arcing contact wipe (overlapping distance) length

Those two parameters might be calculated manually, or automatically extracted from the resistance and motion waveforms using an appropriate algorithm.

II. IMPORTANCE OF PARAMETERS FOR ARCING CONTACT CONDITION ASSESSMENT

The arcing contacts are the most important part of the high voltage circuit breaker. During the trip operation the main contacts open first and arcing contacts open after a few milliseconds. The electrical arc starts after the separation of the arcing contacts, and is cleared at the next current zero – crossing. During the closing operation the arcing contacts close first and main contacts close after a few milliseconds. The electric arc starts before arcing contacts are closed. Therefore, any electrical arc formed during breaker operation will appear on the arcing contacts.

During each opening and closing operation, a fraction of the arcing contact material burns away since temperature at the center of the arc is around 25,000° C [4]. This temperature is four times higher than the temperature of the surface of the Sun, there is no material which can withstand such a high temperature. The standard operation for circuit breakers requires a test in a high power laboratory under rated short circuit current condition and there are three opening operations O – t1 – CO – t2 – CO. The standard operation practically represents the life span of a circuit breaker and is directly connected to the shape of the arcing contact. The life span of a circuit breaker, and the life span of the arcing contacts, is represented graphically (Figure 1).

Figure 1 show the number of interruptions versus short-circuits current that the arcing contacts in a circuit breaker can withstand before they must be replaced.

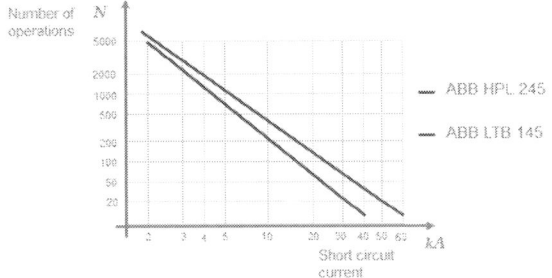

Figure 1. Service Life of Arcing Contacts depending on Short-Circuit Current Value [4]

Each circuit breaker type has its own permissible electrical wear curve. In most designs, a minimal number of circuit breaker operations under rated short circuit conditions are from 3 to 10 [4]

One of the consequences of material loss at the arcing contacts is the shortening of the arcing contacts and thus shortening time difference between the separation of the main contacts and the separation of arcing contacts. This time difference is called the contact overlapping time. In accordance with this, arcing contact wipe is shorter. Other characteristics of material loss is the change the arcing contact surface (Figure 2.), which will result in the changing of the dielectric strength

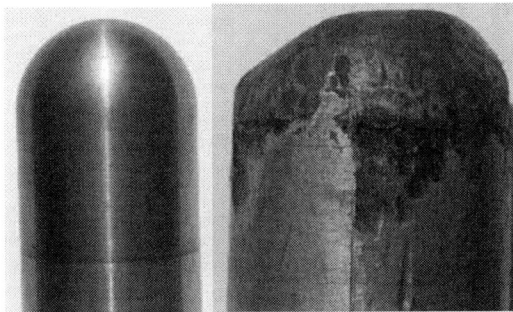

Figure 2. New (left) and worn (right) arcing contacts [2]

of space between the arcing contacts.

The shape of the arcing contacts is probably their most important characteristic. At some point during an arc the alternating current which forms the arc will pass through zero; this is the instant at which current is zero. At that moment a transient voltage will be present across arcing contacts. According to circuit breaker design, the distance between the arcing contacts has to be sufficient to withstand the transient voltage to perform the opening operation successfully. If the voltage is greater, the arc will "re-strike" and alternating current begins to flow again.

Assuming a sufficient distance is reached at the zero crossing to withstand the transient voltage presumes the shape of the arcing contacts is unchanged, according to the design, the contacts are undamaged and the breaker separates the contacts at a given velocity. In the case where arcing contact shape has changed, through wear from previous arcs for example, dielectric strength has decreased and it is probable that this planned distance will not be sufficient to withstand transient voltages. Breaking time will increase as the contacts must now travel further at a finite velocity to reach next current zero-crossing. If the opening velocity of the contacts is not increased the breaker will not clear the current within the specified time; this leads to a longer arcing time and consequently more material wear. Clearly, the performance of the arcing contacts is critical to the performance of the breaker.

Change of the arcing contact shape will most likely affect the main parameter values used for arcing contact assessment (Contact overlapping time, arcing contact wipe)

III. MEASUREMENT PRINCIPLES

Dynamic resistance measurement is performed using a Circuit Breaker Analyzer and Timer instruments (CAT) and software solution for the acquisition and analysis of the obtained results [6]. The device has the ability to generate true DC ripple free current up to 500 A and it has the channel for recording motion of the contact system.

Measurement and calculation of the resistance of the current path is based on the 4 wire Kelvin method (Figure 3).

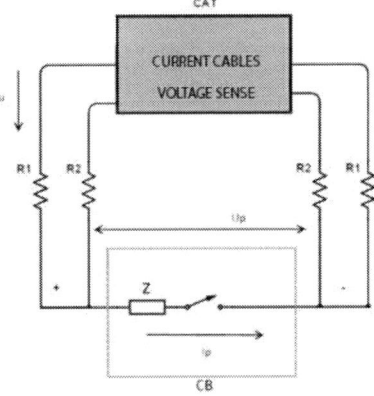

Figure 3. Resistance calculation using 4 wire Kelvin method

This principle of resistance calculation has been chosen since the cables have the resistance in the range of mΩ and the contact resistance is in the range of μΩ thus the measurement error using classical two wires method would be significant.

IV. MEASURED VALUES AND PRESENTATION

The result of the DRM consists of:

- Resistance waveform
- Motion waveform
- Voltage drop waveform
- Numerical results

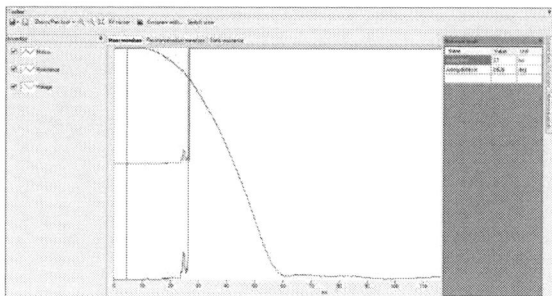

Figure 4 DRM result presentation

The resistance change of CB main circuit during opening operation can be seen the best in the resistance waveform (blue plot). For the interpretation of the result, the most interesting part is the moment in time where the arcing contacts are engaged.

Prior to point A in the Figure 5 the contact system is still in idle state. At point A the contacts starts moving, at point B main contacts separates, at point C the arcing contacts separates and the HVCB changes its role from ideal conductor to ideal insulator. The time difference between points C and B on the time axis represents the arcing contacts overlapping time. Arcing contact overlapping length is extracted from the motion plot, correlating it with previously mentioned points on the time axis.

V. ALGORITHM

Even if points B and C can be obtained manually using

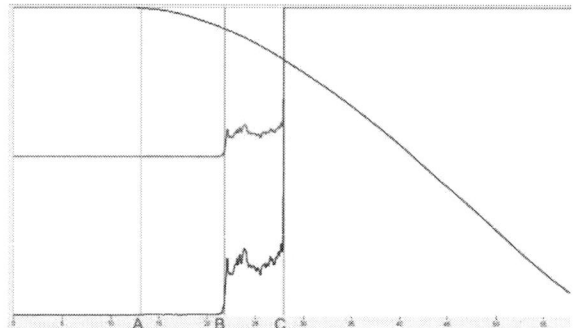

Figure 5 Period in time (B-C) of the arcing contacts engagement

cursors, it is highly important to provide the test engineer with the fully automated measurement system and shorten the time of the testing and analysis.

Resistance waveform in an ideal case does not have any superposed noise and points B and C are clearly emphasized. However, depending on a large number of factors, the waveform of the real results contains noise. Also, depending on the state of the contact system, waveform shape in the arcing area varies for different HVCB. All those factors influence the algorithm for determining points B and C on the time scale.

The assumed premise is that in point B and C the waveform has its highest gradient. However, using only this premise to determine points B and C is not enough due to noise and waveform shape differences caused by contact system itself. Because of this, the resistance waveform is filtered.

Figure 6 shows the interesting section of a typical Dynamic Resistance operation and a comparison between the filtered and the raw signal. The purpose of the filter is to "smooth out", to a reasonable amount, the aggressive rate of change of the signal but still keep the qualitative form of the signal intact. This will result in a signal more suitable for certain numerical analysis procedures required to obtain the arcing contact overlapping time. For the smoothing of the signal it has been used classical moving average filter of the order 5.

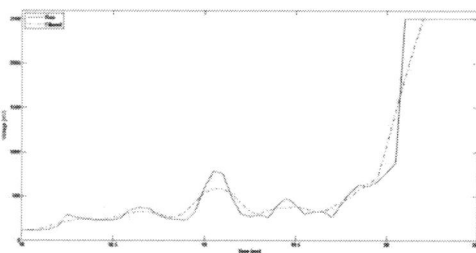

Figure 6 - Comparison between the filtered (dashed red line) and the raw signal (blue line).

As previously explained, the arcing contact overlapping time from the signal point of view, can be defined as the time difference between the point in time where the signal shows its first "significant" change and the point where the signal displays a final jump. The former point signifies the initial opening of the main contacts and the latter signalizes the completion of the opening process (arcing contacts separation). In other words, we are actually interested in the rates of change of the signal, and by straight forward numerical differentiation of the filtered signal, the key points such as in Figure 7 are obtained:

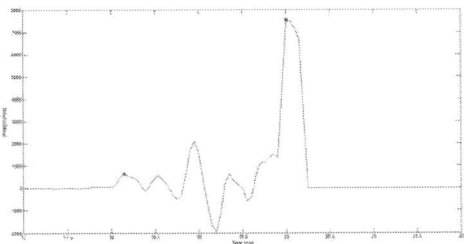

Figure 7 – The key points obtained from filtered signal

As one can see, from the plot in the Figure 7, interesting points can be found and thus the calculation of the arcing contact overlapping time as their time difference, as well.

VI.   CASE STUDY

To validate the method and developed algorithm, DRM has been performed on 21 different circuit breaker within period of 6 months. One of the test objects used for the validation of the method has been Energoinvest SFE 11/ 18 – G 123 kV HVCB.

Table 1. – HVCB data sheet

Type	SFE 11/ 18 – G
Manufacturer	Energoinvest
Number of breaks per phase	1
Rated voltage	123 kV
Rated short circuit current	31.5 kA
Rated short circuit duration	1 s
Operating mechanism	Pneumatic
Control type	Three pole control
Medium	SF6

Obtained result is shown in the Figure 8.

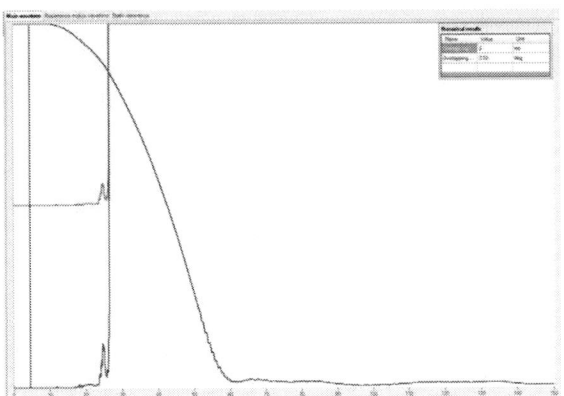

Figure 8. - DRM result for Energoinvest SFE 11/ 18 – G

Using the previously described algorithm, numerical results calculated are shown in the following table:

Table 2. – Arcing time and distance

Parameter	Value
Overlaping time	3 ms
Overlaping length	$2.52° \approx 8.4\ mm$

To verify the validity of the algorithm, a manual calculation is performed using the waveforms and cursors, as shown in the Figure 9.

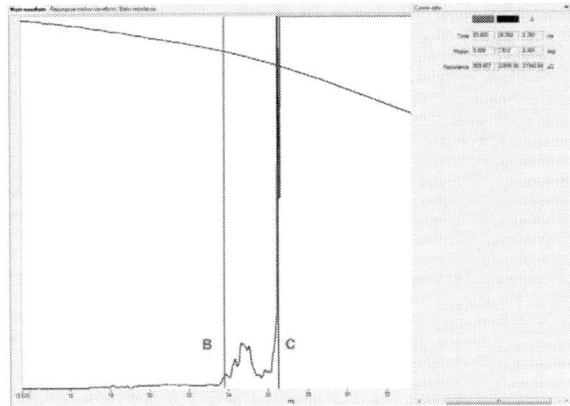

Figure 9. – Overlaping time and length calculation using waveform and cursors

As can be seen in the Figure 9, the moment when main contacts separate is the moment when the resistance changes (significantly increases) for the first time (23. 6 ms). The moment when the arcing contacts separate finally is the moment when the resistance goes (theoretically) to infinity (26.6 ms). The time difference (3 ms) between these two moments is the arcing contact overlapping time.

Based on these two moments and motion curve the arcing contact wipe is calculated (2.52°). Since motion is measured with rotary transducer, it is expressed in degrees. When this value is converted into contact motion (60° corresponds to 200 mm), it corresponds to 8.4 mm.

Arcing contacts overlapping distance should not be less than 12 mm for this circuit breaker, so calculated parameters indicate that arcing contacts are worn and should be considered to be investigated or replaced.

Validation of this method will be demonstrated on one more example. This time it is shown on GIS type of circuit breaker. Data about circuit breaker used for this example are given in the table 3.

Table 3. – HVCB data sheet

Type	SFT 11 I K (GIS)
Manufacturer	Energoinvest
Number of breaks per phase	1
Rated voltage	123 kV
Rated short circuit current	31.5 kA
Rated short circuit duration	3 s
Operating mechanism	Pneumatic
Control type	Three pole control
Medium	SF6

Obtained result is shown in the Figure 10.

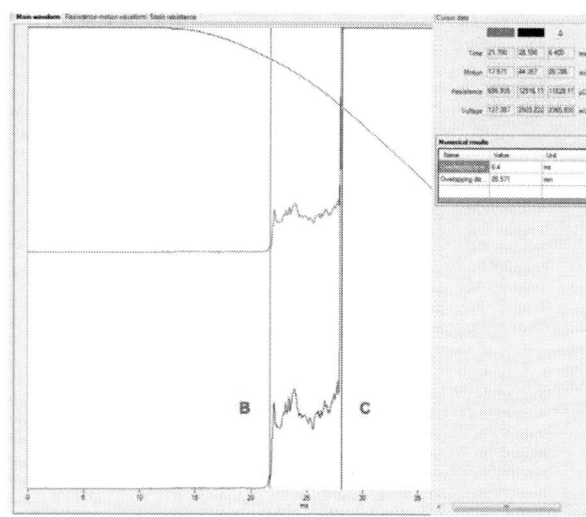

Figure 10. - DRM result for Energoinvest SFT 11 I K (GIS)

Using the proposed method and algorithm, calculated numerical results are shown in the Table 4.

Table 4. – Arcing time and distance

Parameter	Value
Overlaping time	6.4 ms
Overlaping length	26.6 mm

Furthermore, using the waveforms and cursors a manual calculation is performed, as shown in the Figure 10.

As can be seen in the Figure 10, the moment when main contacts separate is 21.7 ms. The moment when the arcing contacts finally separate is 28.1 ms. The time difference (6.4 ms) between these two moments is the arcing contact overlapping time.

Based on these two moments and motion curve the arcing contact wipe is calculated (26.6 mm). Calculated parameters indicate that arcing contacts are in good conditions since minimum required overlapping distance (contact wipe) of the arcing contacts for this type of CB is 20 mm.

VII. CONCLUSION

Measuring dynamic resistance of the CB main circuit is highly important to assess the condition of the arcing contacts. Also, the automation of the testing procedures is crucial for saving test engineers valuable time. The important parameters of Dynamic Resistance testing have been explained in this paper, and the algorithms and procedures for the extraction of key information about the arcing time and distance have been demonstrated.

REFERENCES

[1] F.Salamanaca, F.Borras, H.Eggert and W.Steingaber, "Preventive diagnosis on high voltage circuit breakers", in Proc. Cigre Symp., Berlin, Germany, 1993, pp.120-02

[2] M.Ohlen, B.Dueck and H.Wernli in Proc. "Dynamic Resistance Measurement – A Tool for Circuit Breaker Diagnostics", Stockholm Power Tech Int. Symp. Electric Power Engineering, Sweden, June 18-22 1995, vol 6,pp.108-113

[3] A. Secic, N. Hadzimejlic i M. Brorsson, "Safety Improvements in Testing Procedures for High and Medium Voltage Circuit Breakers with Both Sides Grounded," u The 11th Annual Euro TechCon 2013 Conference and Expo, Glasgow, 2013.

[4] R.Ostojic, J.Levi, "Use of Micro ohm meter as a Power Source for DRM testing on Dead Tank Circuit Breakres", 80th Annual International Double Client Conference, 2013.

[5] J.R. Arias, "Application of the Dynamic Resistance Measurement Technique in the Measurement of Operating Times for High Voltage Circuit Breakers", Doble Europen Colloquium, Seville, 1995

[6] K. Obarcanin, A. Secic, N. Hadzimejlic, "Design and Development of the Software Solution for Analysis and Acquisition of the High Voltage Circuit Breakers Dynamic Resistance Measurement Results", 38th International Convention on Informatics and Communication Technology, Electronics and Microelectronics Mipro 2015, Opatija, Croatia, 2015.

Measurement Noise Propagation in Distribution-System State Estimation

Urban Kuhar
Jozef Stefan Institute,
Department of Communication Systems,
and International Postgraduate School,
Ljubljana, Slovenia
Email: urban.kuhar@ijs.si

Gregor Kosec
Jozef Stefan Institute,
Department of Communication Systems,
Ljubljana, Slovenia
Email: gregor.kosec@ijs.si

Aleš Švigelj
Jozef Stefan Institute,
Department of Communication Systems,
Ljubljana, Slovenia
Email: ales.svigelj@ijs.si

Abstract—The distribution power networks are growing in complexity due to the increasing penetration of the distributed power generation such as from wind and solar plants, and distributed resources such as electric vehicles and batteries, which are making the operation of the system more and more challenging. The first step towards more efficient operational capabilities is to introduce an observability of the system. Power system is considered observable, when all its system variables are known, in other words, system variables are chosen in such a way that when known, they allow computation of all other physical quantities in the system. In this paper a State Estimation (SE) software that considers bus voltage magnitudes and phase angles as state variables is presented. To deduce the state of the system from measurements a non-linear relation between measurements and state variables is used. The system is also overdetermined, i.e. there are more measurements then state variables. A non-linear Weighted Least Square (WLS) approach is used to solve the problem at hand with the final goal to demonstrate how quality of measurements affects the quality of power system state awareness. The results are presented in terms of statistical information about the state variables estimated with presented state estimator using generated measurements with added zero-mean Gaussian noise. We show that noise propagation through the SE algorithm is greatly influenced by the selection and placement of measurement devices.

I. INTRODUCTION

A. Motivation

In recent decade it became apparent that global climate change is a man-caused phenomena and consequently a vast societal consensus for shift to low carbon energy sources and transportation appeared [1]. One aspect of this shift are increased investments into the solar and wind power plants, and increased adoption of electric vehicles. Since the nominal powers of these are typically less than 1 MVA, they are normally connected to distribution networks. The distribution networks are thus growing in complexity and their operation is more and more challenging [2], [3]). It is expected that in less than 10 years the complexity of the system will overgrow current operational capabilities and therefore make safe and stable operation of the system impossible without a real-time regulation of assets in the system. The first step towards more efficient operational capabilities is to introduce the observability of the system, which can be achieved in

various ways. The most economically feasible is a combination of special measurement devices (Phasor Measurement Units - PMUs) and real-time SE algorithms [2].

The SE tool has been used in transmission systems for decades, but it's straightforward application to distribution systems is not possible due to some inherent differences between the two systems. This differences are:

- Load imbalance - distribution systems have many single phase loads and therefore all three phases of the system need to be modelled
- Low line X/R ratio - since the lines are smaller in diameter and shorter in length, their inductance is smaller, which in turn means that phase displacement is smaller and harder to measure.
- High number of nodes - this poses a demand for high number of measurements and increases computational burden.
- Network model uncertainty - network parameters such as series impedances or shunt admittances may be incorrect as a result of inaccurate measurement campaigns. Measurement devices can become inaccurate also due to ageing. All these errors impact the results of the SE.

In this work we present a WLS based SE and attempt to provide an answer about how the measurement errors propagate through state estimation algorithm into its results.

B. Literature review

The existing literature describes several different methods for evaluation of noise propagation. In [4] authors presented a method that compares the covariance matrices of the true state vector based on the error free model and a state vector that is based on a model with a known error. The drawback of this method is that the complete calculation has to be repeated for every measurement in the system. Comparison of two statistical measures namely bias and variance gives a good indication for evaluation of measurement errors. In [5] Macii et al. propose a similar technique that allows computation of noise propagation properties for the complete measurement set. In [6] a two-step method for noise propagation evaluation is proposed. In the first step a WLS estimate is calculated and the lower and upper bounds are computed for the specified range of the measurement error using linear programming. In the second step, the method specifies an uncertainty range within which we can be certain that the true value lies.

In [7] authors use the Monte Carlo (MC) method for the evaluation of noise propagation properties in a one phase model. This paper extends [7] with application on a novel unified three-phase branch model.

C. Paper organization

This paper is organized as follows. Section II describes the underlying state-estimation problem and its WLS solution. Section III presents the three-phase branch model and relationships between SE variables and measurements. Section IV describes the simulation framework and demonstrates the results. Section V concludes the paper.

II. STATE ESTIMATOR

Power system is considered observable, when all system variables are known, in other words, system variables are chosen in such a way that when known, they allow computation of all other physical quantities in the system. Bus voltage magnitudes and phase angles (commonly referred to as voltage phasors) or branch current magnitudes and phase angles, i.e. current phasors, are most commonly chosen as state variables.

State of the system can be deduced from sufficient number of measurements deployed in the system. Generally we divide measurements into two groups, namely real measurements, i.e. the measurements from the actual devices in the network, and pseudo measurements that are derived from other, typically less accurate, data about the system, e.g. customer load profiles. In order to deduce the state of the system from measurements, we need to establish a relation between the measurements and state variables. In general this relation is non-linear and can be written as

$$
\begin{bmatrix} f_1 \\ f_2 \\ \vdots \\ f_m \end{bmatrix} = \begin{bmatrix} h_1(x_1, x_2, \ldots, x_n) \\ h_2(x_1, x_2, \ldots, x_n) \\ \vdots \\ h_m(x_1, x_2, \ldots, x_n) \end{bmatrix} + \begin{bmatrix} w_1 \\ w_2 \\ \vdots \\ w_m \end{bmatrix} \quad (1)
$$

$$
\mathbf{f} = \mathbf{h}(\mathbf{x}) + \mathbf{w} \quad (2)
$$

where f_k is a measurement of the system's output, $h_k(x_1, x_2, \ldots, x_n)$ are the nonlinear functions relating measurement k to state variables, and $w \sim \mathcal{N}(\mathbf{0}, \mathbf{R})$ is a Gaussian, zero mean, with covariance matrix $\mathbf{R}(= diag\{\sigma_1^2, \sigma_2^2, \sigma_3^2, \ldots, \sigma_m^2\})$ noise contribution to measurement k. To construct a robust method that tolerates failure or inaccuracy of certain number of measurements, the system (1) needs to be overdetermined, i.e. $m > n$ where n is the number of nodes in the system, and m is the number of real and pseudo measurements; typically $m \approx 2n$ should hold. There are various ways to solve such overdetermined systems, i.e. Weighted Least Squares (WLS) [8], variations of Kalman filter (KF) [9], and Bayesian Estimation (BLSE) [10].

Since the problem is overdetermined ($m > n$), we define the residual function:

$$
r_i = \frac{f_i - h_i(\mathbf{x})}{\sigma_i} \quad (3)
$$

and write the cost function as a minimization of the weighted least squares (WLS):

$$
J(x) = \sum_{i=1}^{m} R_{ii}^{-1} r_i^2 = (\mathbf{f} - \mathbf{h}(\mathbf{x}))^T \mathbf{R}^{-1} (\mathbf{f} - \mathbf{h}(\mathbf{x})). \quad (4)
$$

It can be shown that the WLS approach is a minimum-variance unbiased estimator (MVUE) if the stated noise assumptions hold true [11].

Since the problem has no direct solution, the Newton-Gauss method can be used to solve it iteratively. Starting with an initial guess $\mathbf{x}^{(0)}$, the method proceeds with:

$$
\mathbf{x}^{(k+1)} = \mathbf{x}^{(k)} - \left(\mathbf{J}^{(k)T} \mathbf{J}^{(k)} \right)^{-1} \mathbf{J}^{(k)T} \mathbf{R}^{-1} \mathbf{r}(\mathbf{x}^{(k)}) \quad (5)
$$

where $\mathbf{J}^{(k)}$ is a Jacobian matrix with the elements:

$$
J_{ij}^{(k)} = \frac{\partial R_{ii}^{-1} \left(f_i - h_i(\mathbf{x}^{(k)}) \right)}{\partial x_j}. \quad (6)
$$

The method terminates when the difference between two successive approximations falls below a selected threshold:

$$
|\mathbf{x}^{(k+1)} - \mathbf{x}^{(k)}| < \epsilon. \quad (7)
$$

The first partial derivatives of the measurement functions with respect to the state variables (Eqn. 6) required by the Gauss-Newton method can be found in [12].

III. THREE-PHASE BRANCH MODEL

A three-Phase Branch Model [13], considered in this paper (Fig. 1), is symmetrical and establishes a relationship between the voltages and currents between the nodes k and m. The

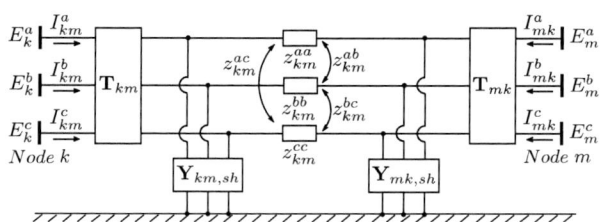

Fig. 1. Three-phase unified branch model

node voltage phasors are denoted as:

$$
E_m^a = |V_m^a| e^{j\phi_m^a}, \quad (8)
$$

where $|V_m^a|$ represents the phasor magnitude, and ϕ_m^a represents a phase angle at the node m on phase a. The self and mutual series impedances are denoted as z_{km}^{xy}, and $\mathbf{Y}_{km,sh}$ is the matrix of self and mutual shunt admittances. The elements $\mathbf{T}_{km}$ and $\mathbf{T}_{mk}$ are presented as matrices and are used for the modeling of voltage regulators or tap changers.

Line current equations that follow from the model are:

$$
\begin{aligned}
I_{km}^a = \sum_{j \in \Omega_p} \sum_{n \in \Omega_p} \sum_{i \in \Omega_p} & \Big[(y_{km}^{jn} + y_{km,sh}^{jn}) t_{km}^{ni} t_{km}^{aj} E_k^i \\
& - t_{mk}^{ni} t_{km}^{aj} y_{km}^{jn} E_m^i \Big] \\
I_{mk}^a = \sum_{j \in \Omega_p} \sum_{n \in \Omega_p} \sum_{i \in \Omega_p} & \Big[(y_{km}^{jn} + y_{mk,sh}^{jn}) t_{mk}^{ni} t_{mk}^{aj} E_m^i \\
& - t_{km}^{ni} t_{mk}^{aj} y_{km}^{jn} E_k^i \Big],
\end{aligned} \quad (9)
$$

where $\Omega_p = \{a, b, c\}$ denotes the phases. A matrix form of the equations can be obtained as shown in [13] which is more convenient for computer implementation.

Formulation of the state-estimation problem usually includes active and reactive line power-flow and active and reactive node power injection measurements, as well. Expressions for them are provided in Eqn. 11, Eqn. 12, Eqn. 13, and Eqn. 14 respectively, where $y_{km}^{aa} = g_{km}^{aa} + jb_{km}^{aa}$ is the branch series admittance, and G_{km}^{ap} and B_{km}^{ap}, respectively, represent the real and imaginary components of the (a, p) element of the (k, m) system-admittance submatrix, and κ is the set of buses adjacent to bus k, including bus k.

IV. UNCERTAINTY PROPAGATION

The quality of estimated state variables is reflected from the noise in the measurement. We are interested in how the is noise propagated through the SE algorithm into the state variables, i.e. we attempt to evaluate how the uncertainty of different types of measurements in different measurement configurations in the network is transmitted into the estimated state variables. This is done by means of MC simulations on several different measurement configurations. In each particular experiment noise is added to a single selected measurement, while other measurements are assumed to be perfectly accurate, without added noise. The measurements are generated from the load-flow results for the reference feeder loading provided by the IEEE [14], more precisely the methodology was tested on the 13-bus IEEE distribution test feeder [14] presented in Fig. 2 (labels on branches represent phasing sequence, labels near spot loads represent load type and connection). Measurement that has added noise is calculated as follows:

$$
X_{meas} = \mathcal{N}(X_{refVal}, X_{refVal}X_{acc}/2). \quad (10)
$$

For each experiment 500 MC runs were performed. The result sets were then processed to evaluate their statistical behaviour.

Uncertainty propagation was evaluated for several measurement configurations, namely:

1) All buses were equipped with injection active and reactive power measurements. No other measurements were installed.
2) All buses were equipped with voltage PMU measurements. No other measurements were installed.
3) All branches were equipped with active and reactive power flow measurements. No other measurements were installed.

4) All buses but the selected one were equipped with voltage PMU measurements. The selected bus was equipped with active and reactive power injection measurement.

Measurement device accuracies were selected as 2 sigma values for the Gaussian random generator, and the values for different devices were selected as follows:

- 10% of the reference value for active and reactive power flow measurements.
- 50% of the reference value for active and reactive power injection measurements.
- 1% of the reference value for the PMU magnitude.
- 0.005° for the PMU angle.

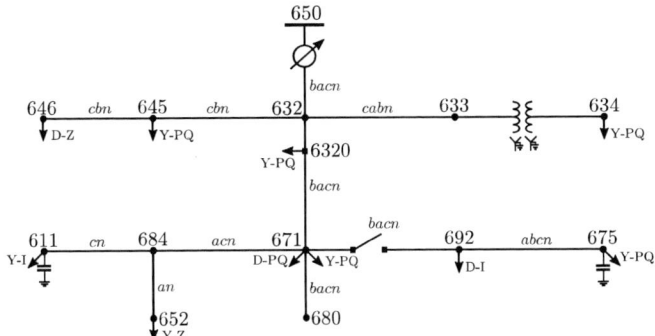

Fig. 2. The 13-bus IEEE test feeder

1) Measurement configuration 1: In this measurement configuration noise is added to the measurement on the first phase of active power measurement on bus 671. All other measurements have no noise added as shown in Fig. 3. The results of SE for buses 634 and 675 are shown in Fig. 4 and Fig. 5 respectively. It can be observed that the distribution of the results is also Gaussian, and that the noise spreads to other two phases. Bias and variance for all buses for both experiments are shown in Table I. It can be observed that added noise in single measurement spread across all buses in this measurement configuration.

2) Measurement configuration 2: In this measurement configuration, noise was added to the phase measurement on the bus 632 as shown in Fig. 6. Since the SE problem for this measurement configuration is linear (the configuration consists solely of PMU measurements), the noise does not spread across other buses in the SE results. The results for bus 632 are shown in Fig. 7, since there is no influence on the other buses the results for them are not presented.

3) Measurement configuration 3: In this measurement configuration, noise was added to the flow measurement between buses 671 and 684. Similarly to measurement configuration 1, the noise due to the non-linear formulation of state estimation problem spreads across all buses. Results for all buses are presented in Table II.

4) Measurement configuration 4: In this measurement configuration all the buses, except bus 671, were equipped with the voltage PMU measurements. Bus 671 was equipped with a real/reactive power injection measurement. Noise was added

$$P_{km}^a = |V_k^a| \sum_{j \in \Omega_p} \sum_{n \in \Omega_p} \sum_{i \in \Omega_p} \left\{ t_{km}^{ni} t_{km}^{aj} |V_k^i| \left[(g_{km}^{jn} + g_{km,sh}^{jn}) cos(\phi_k^a - \phi_k^i) + (b_{km}^{jn} + b_{km,sh}^{jn}) sin(\phi_k^a - \phi_k^i) \right] \right.$$
$$\left. - t_{mk}^{ni} t_{km}^{aj} |V_m^i| \left[g_{km}^{jn} cos(\phi_k^a - \phi_m^i) + b_{km}^{jn} sin(\phi_k^a - \phi_m^i) \right] \right\} \tag{11}$$

$$Q_{km}^a = |V_k^a| \sum_{j \in \Omega_p} \sum_{n \in \Omega_p} \sum_{i \in \Omega_p} \left\{ t_{km}^{ni} t_{km}^{aj} |V_k^i| \left[(g_{km}^{jn} + g_{km,sh}^{jn}) sin(\phi_k^a - \phi_k^i) - (b_{km}^{jn} + b_{km,sh}^{jn}) cos(\phi_k^a - \phi_k^i) \right] \right.$$
$$\left. - t_{mk}^{ni} t_{km}^{aj} |V_m^i| \left[g_{km}^{jn} sin(\phi_k^a - \phi_m^i) - b_{km}^{jn} cos(\phi_k^a - \phi_m^i) \right] \right\} \tag{12}$$

$$P_k^a = |V_k^a| \sum_{m \in \kappa} \sum_{p \in \Omega_p} |V_m^p| (G_{km}^{ap} cos(\phi_k^a - \phi_m^p) + B_{km}^{ap} sin(\phi_k^a - \phi_m^p)) \tag{13}$$

$$Q_k^a = |V_k^a| \sum_{m \in \kappa} \sum_{p \in \Omega_p} |V_m^p| (G_{km}^{ap} sin(\phi_k^a - \phi_m^p) - B_{km}^{ap} cos(\phi_k^a - \phi_m^p)) \tag{14}$$

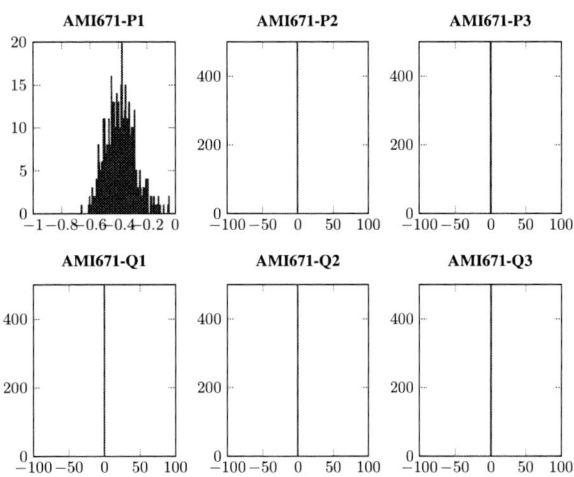

Fig. 3. Noise distribution in active power injection measurement

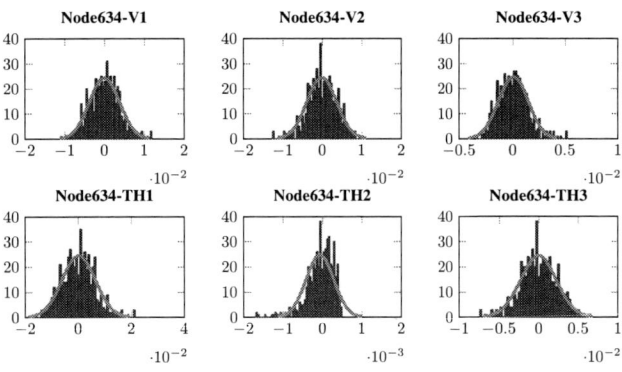

Fig. 4. Noise distribution in SE results on bus 634

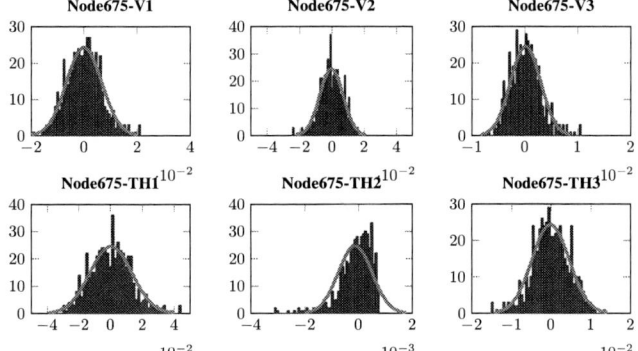

Fig. 5. Noise distribution in SE results on bus 675

V. CONCLUSION

In this paper measurement noise propagation in distribution-system SE was evaluated through means of MC simulation. It has been shown that configuration of measurements has a large influence on noise spread in the SE variables. It has also been shown that in the configuration consisting solely of PMU measurements, the noise stays localized to a particular node, as a result of linear formulation of SE problem. However, using other measurement types with non-linear measurement functions, the noise of a particular measurement spreads into the adjacent phases and into other buses. This effect is most severe in the configuration with power injection measurements.

The main drawback of presented analysis is the computational inefficiency, which renders it useless for larger networks. The future work will be therefore focused on the formulation of more computational effective techniques for sensitivity analysis of SE algorithms.

ACKNOWLEDGMENT

Work was supported within FP7 SUNSEED project, contract number 619437, http://sunseed-fp7.eu. The authors also

to the real measurement on phase one. Here, the influence of the noise stays isolated to the 671 bus, but it spreads across the two other phases as shown in Fig. 8.

TABLE I
BIAS AND VARIANCE FOR MEASUREMENT CONFIGURATION 1

Bus	Bias			Variance		
	Ph.1	Ph.2	Ph.3	Ph.1	Ph.2	Ph.3
611_ϕ	0	0	2.02	0	0	4.72e-3
$611_{|V|}$	0	0	9.74e-1	0	0	2.89e-3
632_ϕ	-4.33e-2	4.16	2.06	6.22e-3	3.00e-4	2.31e-3
$632_{|V|}$	1.02	1.04	1.02	3.66e-3	3.68e-3	1.38e-3
633_ϕ	-4.44e-2	4.16	2.06	6.22e-3	2.98e-4	2.32e-3
$633_{|V|}$	1.02	1.04	1.01	3.68e-3	3.69e-3	1.38e-3
634_ϕ	-5.62e-2	4.15	2.05	6.31e-3	3.54e-4	2.29e-3
$634_{|V|}$	9.94e-1	1.02	9.96e-1	3.77e-3	3.76e-3	1.40e-3
645_ϕ	0	4.16	2.06	0	3.28e-4	2.31e-3
$645_{|V|}$	0	1.03	1.02	0	3.72e-3	1.40e-3
646_ϕ	0	4.15	2.06	0	3.42e-4	2.31e-3
$646_{|V|}$	0	1.03	1.01	0	3.72e-3	1.41e-3
652_ϕ	-9.13e-2	0	0	0	0	1.31e-2
$652_{|V|}$	9.83e-1	0	0	0	0	6.71e-3
671_ϕ	-9.22e-2	4.15	2.02	1.31e-2	6.18e-4	4.74e-3
$671_{|V|}$	9.90e-1	1.05	9.78e-1	6.65e-3	7.27e-3	2.89e-3
675_ϕ	-9.65e-2	4.14	2.03	1.32e-2	6.29e-4	4.72e-3
$675_{|V|}$	9.83e-1	1.06	9.76e-1	6.68e-3	7.24e-3	2.87e-3
680_ϕ	-9.22e-2	4.15	2.02	1.31e-2	6.18e-4	4.74e-3
$680_{|V|}$	9.9e-1	1.05	9.78e-1	6.65e-3	7.27e-3	2.89e-3
684_ϕ	-9.26e-2	0	2.02	1.31e-2	0	4.73e-3
$684_{|V|}$	9.88e-1	0	9.76e-1	6.67e-3	0	2.89e-3
692_ϕ	-9.22e-2	4.15	2.02	1.31e-2	6.18e-4	4.74e-3
$692_{|V|}$	9.90e-1	1.05	9.78e-1	6.65e-3	7.27e-3	2.89e-3
320_ϕ	-5.52e-2	4.16	2.05	7.88e-3	3.75e-4	2.91e-3
$320_{|V|}$	1.01	1.04	1.01	4.47e-3	4.58e-3	1.75e-3

TABLE II
BIAS AND VARIANCE FOR MEASUREMENT CONFIGURATION 3

Bus	Bias			Variance		
	Ph.1	Ph.2	Ph.3	Ph.1	Ph.2	Ph.3
611_ϕ	0	0	2.02	0	0	2.70e-5
$611_{|V|}$	0	0	9.74e-1	0	0	1.83e-5
632_ϕ	-4.35e-2	4.16	2.06	0	0	0
$632_{|V|}$	1.02	1.04	1.02	0	0	0
633_ϕ	-4.46e-2	4.16	2.06	0	0	0
$633_{|V|}$	1.02	1.04	1.01	0	0	0
634_ϕ	-5.64e-2	4.15	2.05	0	0	0
$634_{|V|}$	9.94e-1	1.02	9.96e-1	0	0	0
645_ϕ	0	4.16	2.06	0	0	0
$645_{|V|}$	0	1.03	1.02	0	0	0
646_ϕ	0	4.16	2.06	0	0	0
$646_{|V|}$	0	1.03	1.01	0	0	0
652_ϕ	-9.17e-2	0	0	8.75e-5	0	0
$652_{|V|}$	9.82e-1	0	0	8.50e-5	0	0
671_ϕ	-9.26e-2	4.15	2.02	0	0	0
$671_{|V|}$	9.90e-1	1.05	9.78e-1	0	0	0
675_ϕ	-9.69e-2	4.14	2.03	0	0	0
$675_{|V|}$	9.83e-1	1.06	9.76e-1	0	0	0
680_ϕ	-9.26e-2	4.15	2.02	0	0	0
$680_{|V|}$	9.90e-1	1.05	9.78e-1	0	0	0
684_ϕ	-9.30e-2	0	2.02	8.78e-5	0	2.71e-5
$684_{|V|}$	9.88e-1	0	9.76e-1	8.45e-5	0	1.82e-5
692_ϕ	-9.25e-2	4.15	2.02	0	0	0
$692_{|V|}$	9.90e-1	1.05	9.78e-1	0	0	0
320_ϕ	-5.55e-2	4.16	2.05	0	0	0
$320_{|V|}$	1.01	1.04	1.01	0	0	0

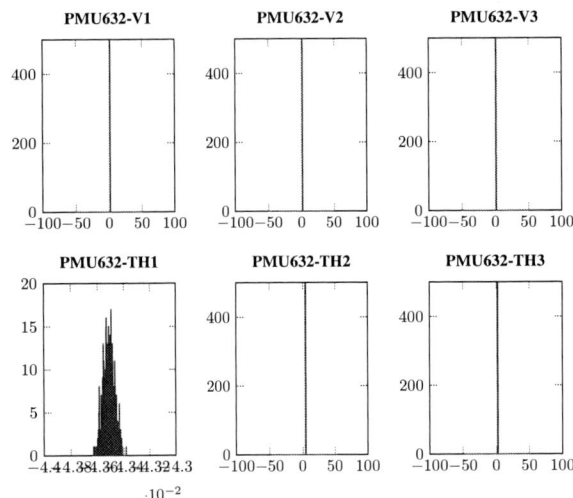

Fig. 6. Noise distribution in phase measurement on bus 632

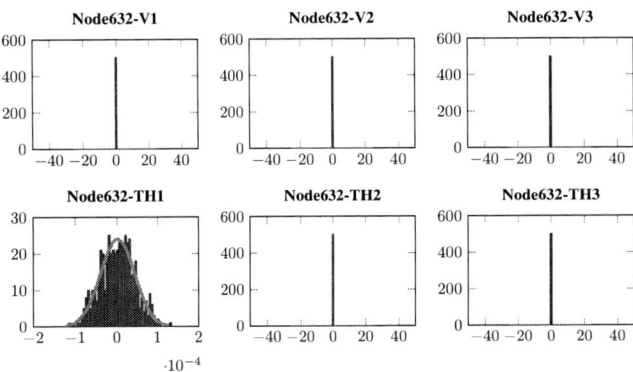

Fig. 7. Noise distribution in SE results on bus 632

acknowledge the financial support from the Slovenian Research Agency (research core funding No. P2-0095).

REFERENCES

[1] D. Yergin, *The Quest: Energy, Security, and the Remaking of the Modern World.* Penguin Publishing Group, 2011.

[2] Y.-F. Huang, S. Werner, J. Huang, N. Kashyap, and V. Gupta, "State estimation in electric power grids: Meeting new challenges presented by the requirements of the future grid," *IEEE Signal Processing Magazine*, vol. 29, no. 5, pp. 33–43, Sep. 2012.

[3] D. Della Giustina, M. Pau, P. A. Pegoraro, F. Ponci, and S. Sulis, "Electrical distribution system state estimation: Measurement issues and challenges," *IEEE Instrumentation and Measurement Magazine*, vol. 1094, no. 6969/14, 2014.

[4] T. A. Stuart and C. J. Herczet, "A sensitivity analysis of weighted least squares state estimation for power systems," *IEEE Transactions on Power Apparatus and Systems*, no. 5, pp. 1696–1701, 1973.

[5] D. Macii, G. Barchi, and D. Petri, "Uncertainty sensitivity analysis of wls-based grid state estimators," in *IEEE International Workshop on Applied Measurements for Power Systems (AMPS) Proceedings, 2014.* IEEE, 2014, pp. 1–6.

[6] A. Al-Othman and M. Irving, "Uncertainty modelling in power system state estimation," *IEE Proceedings - Generation, Transmission and Distribution*, vol. 152, no. 2, p. 233, 2005.

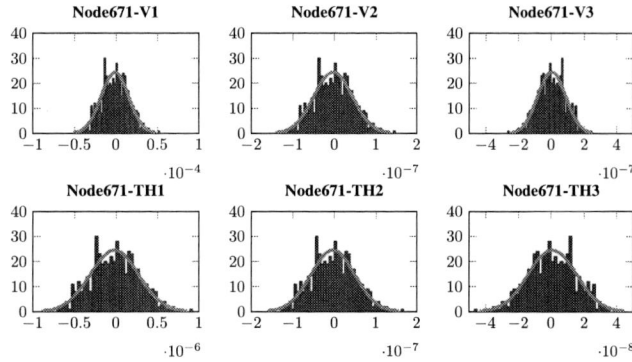

Fig. 8. Noise distribution in SE results on bus 671

[7] C. Muscas, S. Sulis, A. Angioni, F. Ponci, and A. Monti, "Impact of different uncertainty sources on a three-phase state estimator for distribution networks," *IEEE Transactions on Instrumentation and Measurement*, vol. 63, no. 9, pp. 2200–2209, Sep. 2014.

[8] M. E. Baran and A. W. Kelley, "State estimation for real-time monitoring of distribution systems," *IEEE Transactions on Power Systems*, vol. 9, no. 3, pp. 1601–1609, 1994.

[9] S. Sarri, M. Paolone, R. Cherkaoui, A. Borghetti, F. Napolitano, and C. A. Nucci, "State estimation of active distribution networks: comparison between WLS and iterated Kalman-filter algorithm integrating PMUs," in *3rd IEEE PES International Conference and Exhibition on Innovative Smart Grid Technologies (ISGT Europe), 2012*. IEEE, 2012, pp. 1–8.

[10] L. Schenato, G. Barchi, D. Macii, R. Arghandeh, K. Poolla, and A. Von Meier, "Bayesian linear state estimation using smart meters and pmus measurements in distribution grids," in *IEEE International Conference on Smart Grid Communications (SmartGridComm), 2014*. IEEE, 2014, pp. 572–577.

[11] S. M. Kay, *Fundamentals of Statistical Signal Processing: Estimation Theory*. Upper Saddle River, NJ, USA: Prentice-Hall, Inc., 1993.

[12] U. Kuhar. Power flow and power injection expressions for a unified three-phase branch model. [Online]. Available: http://link.com

[13] U. Kuhar, J. Jurse, K. Alic, G. Kandus, and A. Svigelj, "A unified three-phase branch model for a distribution-system state estimation," in *IEEE PES innovative smart grid technologies, Europe, Ljubljana*, Oct. 2016, p. 6.

[14] W. H. Kersting, "Radial distribution test feeders," in *Power Engineering Society Winter Meeting, 2001. IEEE*, vol. 2. IEEE, 2001, pp. 908–912.

Gap in pagination due to withheld paper.

Pages 1055-1060

MIPRO 2017, May 22- 26, 2017, Opatija, Croatia

Brief Review of Self-Organizing Maps

Dubravko Miljković
Hrvatska elektroprivreda, Zagreb, Croatia
dubravko.miljkovic@hep.hr

Abstract - As a particular type of artificial neural networks, self-organizing maps (SOMs) are trained using an unsupervised, competitive learning to produce a low-dimensional, discretized representation of the input space of the training samples, called a feature map. Such a map retains principle features of the input data. Self-organizing maps are known for its clustering, visualization and classification capabilities. In this brief review paper basic tenets, including motivation, architecture, math description and applications are reviewed.

I. INTRODUCTION

Among numerous neural network architectures, particularly interesting architecture was introduced by Finish Professor Teuvo Kohonen in the 1980s, [1,2]. Self-organizing map (SOM), sometimes also called a Kohonen map use unsupervised, competitive learning to produce low dimensional, discretized representation of presented high dimensional data, while simultaneously preserving similarity relations between the presented data items. Such low dimensional representation is called a feature map, hence map in the name. This brief review paper attempts to introduce a reader to SOMs, covering in short basic tenets, underlying biological motivation, its architecture, math description and various applications, [3-10].

II. NEURAL NETWORKS

Human and animal brains are highly complex, nonlinear and parallel systems, consisting of billions of neurons integrated into numerous neural networks, [3]. A neural networks within a brain are massively parallel distributed processing system suitable for storing knowledge in forms of past experiences and making it available for future use. They are particularly suitable for the class of problems where it is difficult to propose an analytical solution convenient for algorithmic implementation.

A. Biological Motivation

After millions of years of evolution, brain in animals and humans has evolved into the massive parallel stack of computing power capable of dealing with the tremendous varieties of situations it can encounter. The biological neural networks are natural intelligent information processors. Artificial neural networks (ANN) constitute computing paradigm motivated by the neural structure of biological systems, [6]. ANNs employ a computational approach based on a large collection of artificial neurons that are much simplified representation of biological neurons. Synapses that ensure communication among biological neurons are replaced with neuron input weights. Adjustment of connection weights is performed by some of numerous learning algorithms. ANNs have very simple principles, but their behavior can be very complex. They have a capability to learn, generalize, associate data and are fault tolerant. The history of the ANNs begins in the 1940s, but the first significant step came in 1957 with the introduction of Rosenblatt's perceptron. The evolution of the most popular ANN paradigms is shown in Fig. 1, [10].

B. Basic Architectures

An artificial neural network is an interconnected assembly of simple processing elements, called artificial neurons (also called units or nodes), whose functionality mimics that of a biological neuron, [4]. Individual neurons can be combined into layers, and there are single and multi-layer networks, with or without feedback. The most common types of ANNs are shown in Fig. 2, [11]. Among training algorithms the most popular is backpropagation and its variants. ANNs can be used for solving a wide variety of problems. Before the use they have to be trained. During the training, network adjusts its weights. In supervised training, input/output pairs are presented to the network by an external teacher and network tries to learn desired input output mapping. Some neural architectures (like SOM) can learn without supervision (unsupervised) from the training data without specified input/output pairs.

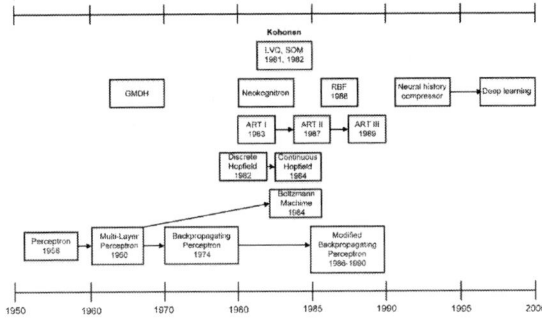

Figure 1. Evolution of artificial neural network paradigms, based on [10]

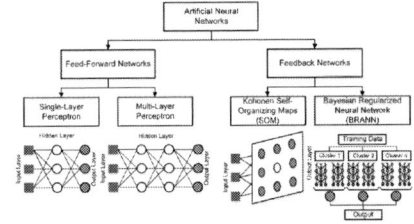

Figure 2. Most common artificial neural networks, according to [11]

III. SELF-ORGANIZING MAPS

Self-organized map (SOM), as a particular neural network paradigm has found its inspiration in self-organizing and biological systems.

A. Self-Organized Systems

Self-organizing systems are types of systems that can change their internal structure and function in response to external circumstances and stimuli, [12-15]. Elements of such a system can influence or organize other elements within the same system, resulting in a greater stability of structure or function of the whole against external fluctuations, [12]. The main aspects of self-organizing systems are increase of complexity, emergence of new phenomena (the whole is more than the sum of its parts) and internal regulation by positive and negative feedback loops. In 1952 Turing published a paper regarding the mathematical theory of pattern formation in biology, and found that global order in a system can arise from local interactions, [13]. This often produces a system with new, emergent properties that differ qualitatively from those of components without interactions, [16]. Self-organizing systems exist in nature, including non-living as well as living world, they exist in man-made systems, but also in the world of abstract ideas, [12].

B. Self-Organizing Map

Neural networks of neurons with lateral communication of neurons topologically organized as self-organizing maps are common in neurobiology. Various neural functions are mapped onto identifiable regions of the brain, Fig. 3, [17]. In such topographic maps neighborhood relation is preserved. Brain mostly does not have desired input-output pairs available and has to learn in unsupervised mode.

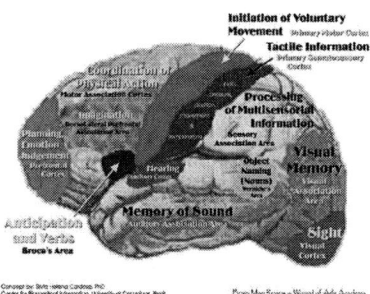

Figure 3. Maps in brain, [17]

A SOM is a single layer neural network with units set along an *n*-dimensional grid. Most applications use two-dimensional and rectangular grid, although many applications also use hexagonal grids, and some one, three, or more dimensional spaces. SOMs produce low-dimensional projection images of high-dimensional data distributions, in which the similarity relations between the data items are preserved, [18],

C. Principles of Self-Organization in SOMs

Following three processes are common to self-organization in SOMs, [7,19,20]:

1. Competitive Process

For each input pattern vector presented to the map, all neurons calculate values of a discriminant function. The neuron that is most similar to the input pattern vector is the winner (best matching unit, BMU).

2. Cooperative Process

The winner neuron (BMU) finds the spatial location of a topological neighborhood of excited neurons. Neurons from this neighborhood may then cooperate.

3. Synaptic Adaptation

Provides that excited neurons can modify their values of the discriminant function related to the presented input pattern vector by the process of weight adjustments.

D. Common Topologies

SOM topologies can be in one, two (most common) or even three dimensions, [2-10]. Two most used two dimensional grids in SOMs are rectangular and hexagonal grid. Three dimensional topologies can be in form of a cylinder or toroid shapes. 1-D (linear) and 2-D grids are illustrated in Fig. 4, with corresponding SOMs in Fig. 5 and Fig. 6, according to [19].

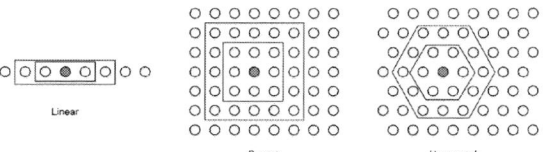

Figure 4. Most common grids and neuron neighborhoods

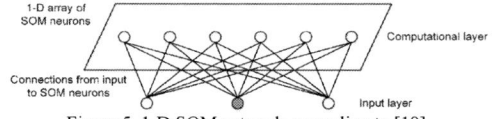

Figure 5. 1-D SOM network, according to [19].

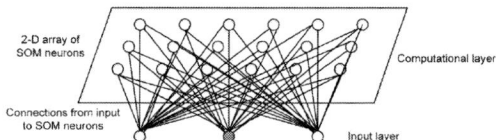

Figure 6. 2-D SOM network, according to [19].

IV. LEARNING ALGORITHM

In 1982 Professor Kohonen presented his SOM algorithm, [1]. Further advancement in a field came with the Second edition of his book "Self-Organization and Associative Memory" in 1988, [2].

A. Measures of Distance and Similarity

To determined similarity between the input vector and neurons measures of distance are used. Some popular distances among input pattern and SOM units are, [21]:

- Euclidian
- Correlation
- Direction cosine
- Block distance

In a real application most often squared Euclidean distance is used, (1):

$$dj = \sum_i \left(x_i - w_{ij} \right)^2 \qquad (1)$$

B. Neighborhood Functions

Neurons within a grid interact among themselves using a neighbor function. Neighborhood functions most often assume the form of the Mexican hat, (2), Fig. 7, that has biological motivation behind (rejects some neurons in the vicinity to the winning neuron) although other functions (Gaussian, cone and cylinder) are also possible, [22]. Ordering algorithm is robust to the choice of function type if the neighbor radius and learning rate decrease to zero. The popular choice is the exponential decay.

$$h_g\left(w_{ij}, w_{mn}, r\right) = e^{-\frac{1}{2}\left(\frac{\sqrt{(i-n)^2+(j-m)^2}}{r}\right)^2} \qquad (2)$$

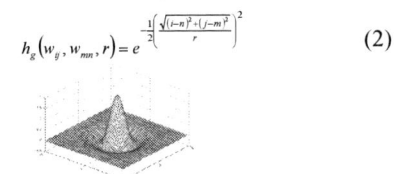

Figure 7. Mexican hat function

C. Initialization of Self-Organizing Maps

Before training SOM, units (i.e. its weights) should be initialized. Common approaches are, [2,23]:
1. Use of random values, completely independent of the training data set
2. Use of random samples from the input training data
3. Initialization that tries to reflect the distribution of the data (Principal Components)

D. Training

Self-organizing maps use the most popular algorithm of the unsupervised learning category, [2]. The criterion D, that is minimized, is the sum of distances between all input vectors x_n and their respective winning neuron weights w_i calculated at the end of each epoch, (3), [21]:

$$D = \sum_{i=1}^{k} \sum_{n \in c_i} (x_n - w_i)^2 \qquad (3)$$

Training of self-organizing maps, [2,18], can be accomplished in two ways: as sequential or batch training.
1. Sequential training
   - single vector at a time is presented to the map
   - adjustment of neuron weights is made after a presentation of each vector
   - suitable for on-line learning
2. Batch training
   - whole dataset is before any adjustment to the neuron weights is made
   - suitable for off-line learning

Here are steps for the sequential training, [3,7,19,22]:
1. Initialization
   - Initialize the neuron weights (iteration step n=0)
2. Sampling
   - Randomly sample a vector $x(n)$ from the dataset
3. Similarity Matching
   - Find the best matching unit (BMU), c, with weights w_{bmu}=w_c, (4):

$$c = \arg\min_{i} \left(\|x(n) - w_i(n)\|\right) \qquad (4)$$

4. Updating
   - Update each unit i with the following rule:

$$w_i(n+1) = w_i(n) + \alpha(n)h(w_{bmu}(n), w_i(n), r(n))\|x(n) - w_i(n)\| \qquad (5)$$

5. Continuation
   - Increment n. Repeat steps 2-4 until a stopping criterion is met (e.g. the fixed number of iterations or the map has reached a stable state).

For the convergence and stability to be guaranteed, the learning rate $\alpha(n)$ and neighborhood radius $r(n)$ are decreasing with each iteration towards the zero, [22].

SOM Sample Hits, Fig. 8, show the number of input vectors that each unit in SOM classifies, [24].

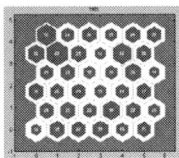

Figure 8. SOM Sample Hits, [24]

During the training process two phases may be distinguished, [7,18]:
1. Self-organizing (ordering) phase:
   Topological ordering in the map takes place (roughly first 1000 iterations). The learning rate $\alpha(n)$ and neighborhood radius $r(n)$ are decreasing.
2. Convergence (fine tuning) phase:
   This is fine tuning that provides an accurate statistical representation of the input space. It typically lasts at least (500 x number of neuron) iterations. The smaller learning rate $\alpha(n)$ and neighborhood radius $r(n)$ may be kept fixed (e.g. last values from the previous phase).

After the training of the SOM is completed, neurons may be labeled if labeled pattern vectors are available.

E. Classification

Find the best matching unit (BMU), c, (5):

$$c = \arg\min_{i} \left(\|x - w_i\|\right) \qquad (5)$$

Test pattern x belongs to the class represented by the best matching unit c.

V. PROPERTIES OF SOM

After the convergence of SOM algorithm, resulting feature map displays important statistical characteristics of the input space. They are also able to discover relevant patterns or features present in the input data.

A. Important Properties of SOMs

SOMs have four important properties, [3,7]:

1. Approximation of the Input Space

The resulting mapping provides a good approximation to the input space. SOM also performs dimensionality reduction by mapping multidimensional data on SOM grid.

2. Topological Ordering

Spatial locations of the neurons in the SOM lattice are topologically related to the features of the input space.

3. Density Matching

The density of the output neurons in the map approximates the statistical distribution of the input space. Regions of the input space that contain more training vectors are represented with more output neurons.

4. Feature Selection

Map extracts the principal features of the input space. It is capable of selecting the best features for approximation of the underlying statistical distribution of the input space.

B. Representing the Input Space with SOMs of Various Topologies

1. 1-D

2D input data points are uniformly distributed in a triangle, 1D SOM ordering process shown in Fig. 9, [2].

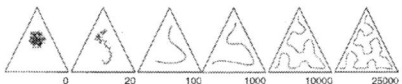

Figure 9. 2D to 1D mapping by a SOM (ordering process), [2]

2. 2-D

2D input data points are uniformly distributed in a square, 2D SOM ordering process shown in Fig. 10, [3].

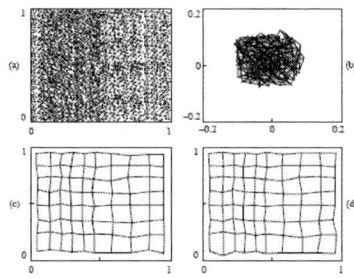

Figure 10. 2D to 2D mapping by a SOM (ordering process), [3]

3. Torus SOMs

In conventional SOM, the size of neighborhood set is not always constant because the map has its edges. This problem can be mitigated by use of torus SOM that has no edges, [25]. However torus SOM, Fig. 11, is not easy to visualize as there are now missing edges.

Figure 11. Torus SOM

4. Hierarchical SOMs

After previous topologies, hierarchical SOMs should also be mentioned. Hierarchical neural networks are composed of multiple loosely connected neural networks that form an acyclic graph. The outputs of the lower level SOMs can be used as the input for the higher level SOM, Fig. 12, [10]. Such input can be formed of several vectors from Best Matching Units (BMUs) of many SOMs.

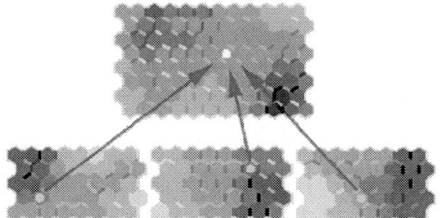

Figure 12. Hierarchical SOM, [10]

VI. APPLICATIONS

Despite its simplicity, SOMs can be used for a various classes of applications, [2,26,27]. This in a broad sense includes visualizations, generation of feature maps, pattern recognition and classification. Kohonen in [2] came with the following categories of applications: machine vision and image analysis, optical character recognition and script reading, speech analysis and recognition, acoustic and musical studies, signal processing and radar measurements, telecommunications, industrial and other real world measurements, process control, robotics, chemistry, physics, design of electronic circuits, medical applications without image processing, data processing linguistic and AI problems, mathematical problems and neurophysiological research. With such an exhaustive list provided here, as space permits, it is possible only to mention some of them that are interesting and popular.

A. Speech Recognition

The neural phonetic typewriter for Finnish and Japanese speech was developed by Kohonen in 1988, [28]. The signal from the microphone proceeds to acoustic preprocessing, shown in more detail in Fig. 13, forming 15-component pattern vector (values in 15 frequency beans taken every 10 ms), containing a short time spectral description of speech. These vectors are presented to a SOM with the hexagonal lattice of the size 8 x 12.

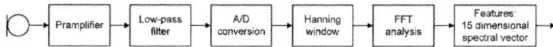

Figure 13. Acoustic preprocessing

After training resulting phonotopic map is shown in Fig. 14, [7]. During speech recognition new pattern vectors are assigned category belonging to a closest prototype in the map.

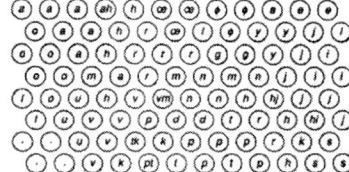

Figure 14. Phonotopic map, [7]

B. Text Clustering

Text clustering is the technology of processing a large number of texts that gives their partition. Preparation of text for SOM analysis is shown in Fig. 15, [29], and

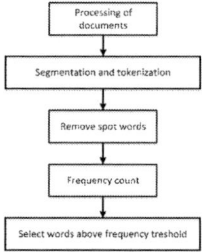

Figure 15. Preparation of text for SOM analysis, according to [29]

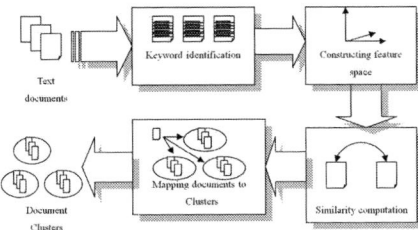

Figure 16. Framework for text clustering, [29]

complete framework in Fig. 16, [29]. Massive document collections can be organized using a SOM. It can be optimized to map large document collections while preserving much of the classification accuracy Clustering of scientific articles is illustrated in Fig. 17, [30].

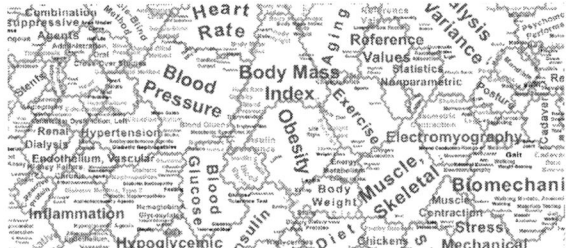

Figure 17. Clustering of scientific articles, [30]

C. Application in Chemistry

SOMs have found applications in chemistry. Illustration of the output layer of the SOM model using a hexagonal grid for the combinatorial design of cannabinoid compounds is shown in Fig. 18, [11].

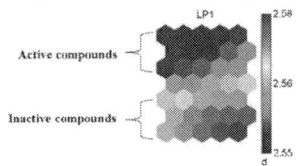

Figure 18. Application of SOM in chemistry, [11]

D. Medical Imaging and Analysis

Recognition of diseases from medical images (ECG, CAT scans, ultrasonic scans, etc.) can be performed by SOMs, [21].This includes image segmentation, Fig. 19, [31], to discover region of interest and help diagnostics.

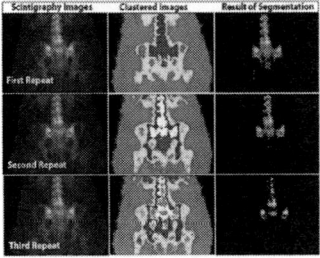

Figure 19. Segmentation of hip image using SOM, [31]

E. Maritime Applications

SOMs have been widely for maritime applications, [22]. One example is analysis of passive sonar recordings. Also SOMs have been used for planning ship trajectories.

F. Robotics

Some applications of SOMs are control of robot arm, learning the motion map and solving traveling salesman problem (multi-goal path planning problem), Fig. 20, [32].

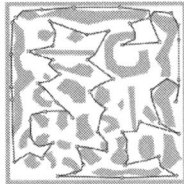

Figure 20. Traveling Salesman Problem, [32]

G. Classification of Satellite Images

SOMs can be used for interpreting satellite imagery like land cover classification. Dust sources can also be spotted in images using the SOM as shown in Fig. 21, [33].

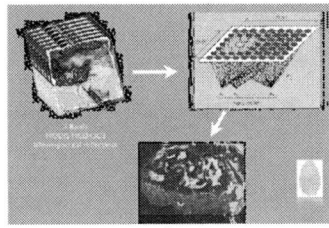

Figure 21. Detecting dust sources using SOMs, [33]

H. Psycholinguistic Studies

One example is the categorization of words by their local context in three word sentences of the type subject-predicate-object or subject-predicate-predicative that were constructed artificially. The words become clustered by SOM according to their linguistic roles in an orderly fashion, Fig. 22, [18].

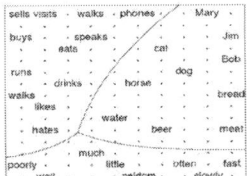

Figure 22. SOM in psycholinguistic studies, [18]

I. Exploring Music Collections

Similarity of music recordings may be determined by analyzing the lyrics, instrumentation, melody, rhythm, artists, or emotions they invoke, Fig. 23, [34].

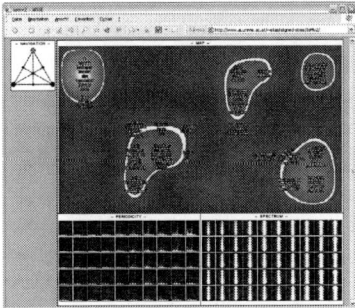

Figure 23. Exploring music collections, [34]

J. Business Applications

Customer segmentation of the international tourist market is illustrated in Fig. 24, [35]. Another example is classifying world poverty (welfare map), [36]. Ordering of items with the respect to 39 features describing various quality-of-life factors, such as state of health, nutrition, and educational services is shown in Fig. 25. Countries with similar quality of life factors clustered together on a map.

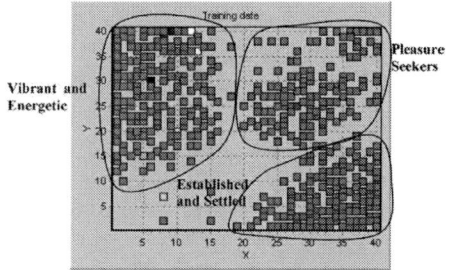

Figure 24. Customer segmentation of the international tourist market, [35]

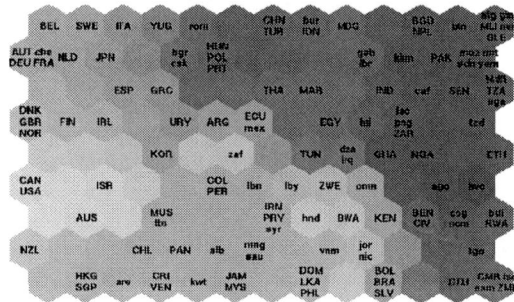

Figure 25. Poverty map based on 39 indicators from World Bank statistics (1992), [36]

VII. CONCLUSION

Self-organizing maps (SOMs) are neural network architecture inspired by the biological structure of human and animal brains. They become one of most popular neural network architecture. SOMs learn without external teacher, i.e. employ unsupervised learning. Topologically SOMs most often use a two-dimensional grid, although one-dimensional, higher-dimensional and irregular grids are also possible. SOM maps higher dimensional input onto the lower dimensional grid while preserving topological ordering present in the input space. During competitive learning SOM use lateral interactions among the neurons to form a semantic map where similar patterns are mapped closer together than dissimilar ones. They can be used for broad type of applications like visualizations, generation of feature maps, pattern recognition and classification. Humans can't visualize high-dimensional data, hence SOMs by mapping such data to a two-dimensional grid are widely used for data visualization. SOMs are also suitable for generation of feature maps. Because they can detect clusters of similar patterns without supervision SOMs are a powerful tool for identification and classification of spatio-temporal patterns. SOMs can be used as an analytical tool, but also in myriad of real world applications including science, medicine, satellite imaging and industry.

REFERENCES

[1] T. Kohonen, "Self-organized formation of topologically correct feature maps", Biol. Cybern. 43, pp. 59-69, 1982
[2] T. Kohonen, Self-Organizing Maps, 2nd ed., Springer 1997
[3] S. Haykin, Neural Networks: A Comprehensive Foundation, 2nd ed., Prentice Hall PTR Upper Saddle River, NJ, USA, 1998
[4] K. Gurney, An introduction to neural network, UCL Press Limited, London, UK, 1997
[5] D. Kriese, A Brief Introduction to Neural Networks, http://www.dkriesel.com
[6] R. Rojas: Neural Networks, A Systematic Introduction, Springer-Verlag, Berlin, 1996
[7] J. A. Bullinaria, Introduction to Neural Networks - Course Material and Useful Links, http://www.cs.bham.ac.uk/~jxb/NN/
[8] C. M. Bishop, Neural Networks for Pattern Recognition, Clarendon Press, Oxford, 1997
[9] R. Eckmiller, C. Malsburg, Neural Computers, NATO ASI Series, Computer and Systems Sciences, 1988
[10] P. Hodju and J. Halme, Neural Networks Information Homepage, http://phodju.mbnet.fi/nenet/SiteMap/SiteMap.html
[11] Káthia Maria Honório and A. B. F. da Silva, "Applications of artificial neural networks in chemical problems", in Artificial Neural Networks - Architectures and Applications, InTech, 2013
[12] W. Banzhafl, "Self-organizing systems", in Encyclopedia of Complexity and Systems Science, 2009, Springer, Heidelberg, 2009
[13] A. M. Turing, "The chemical basis of morphogenesis", Philosophical Transactions of the Royal Society of London. Series B, Biological Sciences, Vol. 237, No.641. pp. 37-72, Aug. 14, 1952
[14] W. R. Ashby, "Principles of the self-organizing system", E:CO Special Double Issue Vol. 6, No. 1-2, pp. 102-126, 2004
[15] C. Fuchs, "Self-organizing system", in Encyclopedia of Governance, Vol. 2, SAGE Publications, 2006, pp. 863-864
[16] J. Howard, "Self-organisation in biology", in Research Perspectives 2010+ of the Max Planck Society, 2010, pp. 28-29
[17] The Wizard of Ads Brain Map - Wernicke and Broca, https://www.wizardofads.com.au/brain-map-brocas-area/
[18] T. Kohonen, MATLAB Implementations and Applications of the Self-Organizing Map, Unigrafia, Helsinki, Finland, 2014
[19] Bill Wilson, Self-organisation Notes, 2010, www.cse.unsw.edu.au/~billw/cs9444/selforganising-10-4up.pdf
[20] J. Boedecker, Self-Organizing Map (SOM), .ppt, Machine Learning, Summer 2015, Machine Learning Lab, Univ. of Freiburg
[21] L. Grajciarova, J. Mares, P. Dvorak and A. Prochazka, Biomedical image analysis using self-organizing maps, Matlab Conference 2012
[22] V. J. A. S. Lobo, "Application of Self-Organizing Maps to the Maritime Environment", Proc. IF&GIS 2009, 20 May 2009, St. Petersburg, Russia, pp. 19-36
[23] A. A. Akinduko and E. M. Mirkes, "Initialization of self-organizing maps: principal components versus random initialization. A case study", Information Sciences, Vol. 364, Is. C, pp. 213-221, Oct. 2016
[24] MathWorks, Self-Organizing Maps, https://www.mathworks.com/help/nnet/ug/cluster-with-self-organizing-map-neural-network.html
[25] M. Ito, T. Miyoshi, and H. Masuyama, "The characteristics of the torus self organizing map", Proc. 6th Int. Conf. on Soft Computing (IIZUKA'2000), Iizuka, Fukuoka, Japan, Oct. 1-4, 2000, pp. 239-44
[26] M. Johnsson ed., Applications of Self-Organizing Maps, InTech, November 21, 2012
[27] J. I. Mwasiagi (ed.), Self Organizing Maps - Applications and Novel Algorithm Design, InTech, 2011
[28] T. Kohonen, "The 'neural' phonetic typewriter", IEEE Computer 21(3), pp. 11–22, 1988
[29] Yuan-Chao Liu, Ming Liu and Xiao-Long Wang, "Application of self-organizing maps in text clustering: a review", in "Self Organizing Maps - Applications and Novel Algorithm Design", InTech, 2012
[30] K. W. Boyacka et. all., Supplementary information on data and methods for "Clustering more than two million biomedical publications: comparing the accuracies of nine text-based similarity approaches", PLoS ONE 6(3): e18029, 2011
[31] A. Aslantas, D. Emre and M. Çakiroğlu, "Comparison of segmentation algorithms for detection of hotspots in bone scintigraphy images and effects on CAD systems", Biomedical Research, 28 (2), pp. 676-683, 2017
[32] J. Faigl, "Multi-goal path planning for cooperative sensing", PhD Thesis, Czech Technical University of Prague, February 2010
[33] D. Lairy, Machine Learning for Scientific Applications, slides, https://www.slideshare.net/davidlarv/machine-learning-for-scientific-applications
[34] E. Pampalk, S. Dixon and G. Widmer, "Exploring music collections by browsing different views", Computer Musical Journal, Vol. 28, No. 2, pp. 49-62, Summer 2004
[35] J. Z. Bloom, "Market segmentation - a neural network application", Annals of Tourism Research, Vol. 32, No. 1, pp. 93–111, 2005
[36] World Poverty Map, SOM research page, Univ. of Helsinki, http://www.cis.hut.fi/research/som-research/worldmap.html

MIPRO 2017, May 22- 26, 2017, Opatija, Croatia

Fault Detection for Aircraft Piston Engine Using Self-Organizing Map

Dubravko Miljković
Hrvatska elektroprivreda, Zagreb, Croatia
dubravko.miljkovic@hep.hr

Abstract - Aircraft piston engine can be monitored using an advanced graphic engine monitor. Such engine monitor can supply a large amount of data containing evolution of engine parameters through the time. Analysis of a vast amount of multidimensional temporal data by a self-organizing map may aid in data visualization, but also in detection of engine parameter deviations from normality indicating potential problems in operation of aircraft piston engine. For determination of engine parameter space that corresponds to normal engine operation quantization error of the self-organizing map is used.

I. INTRODUCTION

Piston engine is a heat engine designed to convert energy into rotational mechanical motion. It uses reciprocating pistons to convert pressure into a rotating motion, [1]. The chances for an engine failure are pretty remote, but do happen. Piston engine reliability depends on the complexity of the engine (number of cylinders, turbocharging), use (private or club aircraft) and maintenance. Piston engine has a lot of moving parts and in comparison to a turbine engine its reliability is significantly lower (about seven times), yet low cost of such engines makes them a popular choice among most general aviation airplanes. An engine failure is a serious situation, both in single and twin engine aircrafts. Single engine aircraft can attempt to glide to a nearest airport (if altitude and winds permit) or perform an off airport landing (challenging situation if over inhospitable terrain, at night, low cloud ceiling and low visibility). Twin engine (particularly piston engine) aircrafts encounter an asymmetric thrust situation that pose increased risk during take-off, initial climb and possible go-around (that should be avoided altogether). Single engine climb rate is severely reduced and is only about 20% of climb rate available when operating on both engines (not 50% as one would expect). Failures further complicate the fact that it can be partial power loss instead of full power loss.

II. ENGINE MONITOR

Engine monitors (also called engine analyzers or engine management system) provide monitoring of vital engine parameters, [1]. These parameters are measured, by engine probes, recorded and presented on a graphic display with parameters usually shown as vertical bars. Cylinder head temperatures (CHTs), Exhaust gas Temperatures (EGTs) for each cylinder and Turbine inlet temperatures (TIT) are shown graphically as bars on the display of an engine monitor, [2,3], as shown in Fig. 1. Some additional engine parameters like engine rotational speed (RPM), calculated % of maximal horsepower (% HP), etc. are also shown. All parameters are logged and can later be analyzed on the ground. Such an advanced piston engine-monitoring instrument helps pilots to better manage engine operation and detect engine problems in real time (while the engine is running). It is also of great value to maintenance personnel. Logged data are available for post-flight analysis helping to detect impeding problems and suggest appropriate preventive actions. Engine parameters that are most commonly monitored in engine monitors are listed in Table I, [2,3].

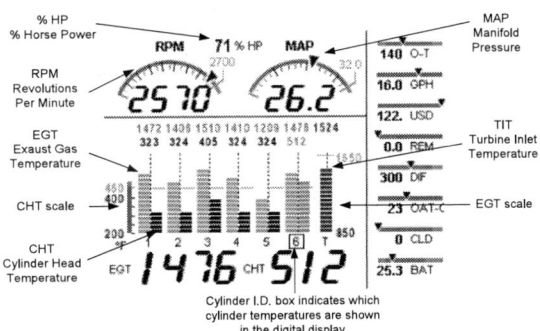

Figure 1. Engine monitor with separate bars for EGT and CHT
(JPI EDM 830)

TABLE I. MONITORED ENGINE PARAMETERS

Parameter	Description
EGT	Exhaust Gas Temperature
CHT	Cylinder Head temperature
OIL TEMP	Oil Temperature [1]
OIL PRES	Oil Pressure [1]
TIT1	Turbine Inlet Temperature 1[1]
TIT2	Turbine Inlet Temperature 2[1]
OAT	Outside Air Temperature
CDT	Compressor Discharge Temperature [1]
IAT	Intercooler Air Temperature [1]
CRB	Carburetor Air Temperature [1]
CDT - IAT	Intercooler cooling
RPM	Rotations Per Minute
MAP	Manifold Pressure
%HP	% Horse Power
CLD	CHT cooling rate [2]
DIF	EGT span [3]
FF	Fuel Flow [1]

[1]optional, [2]fastest cooling cylinder, [3]difference between the hottest and coolest EGT

An engine monitor typically has two modes: monitoring and lean operation mode (for accurate adjustment of fuel mixture). Various engine problems can be detected using an engine monitor by spotting characteristic EGT/CHT bar patterns, [4]. Such patterns are catalogued and included in a technical documentation accompanying an engine monitor, [2].

III. VISUALIZATION OF ENGINE PARAMETERS

Engine parameters recorded during each flight can later be visually presented using specialized plotting software, e.g. the EzTrends2, [5]. Graphical representation of engine parameters (multidimensional data) through the duration of the whole flight as presented by JPI EZTrends2 is shown in Fig. 2 (upper curves represent EGTs and lower CHTs, inverted colors and background).

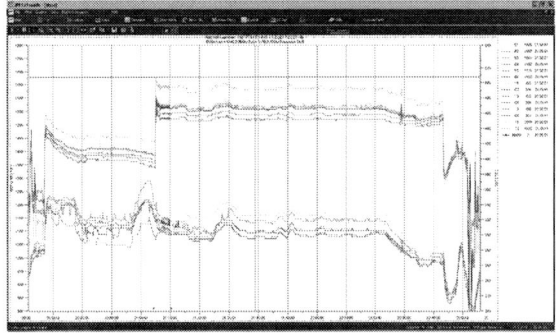

Figure 2. Engine parameters plot

As can be seen form the figure, EGT and CHT curves from the engine with no faults present group together. Another, simpler, representation of engine operation that doesn't include evolution of parameters during a time is use of flight summaries, Fig. 3. Temperature ranges and average temperatures are shown in summary table. Discrepancies from the symmetry between temperatures of various cylinders can also be easily spotted in accompanying graphical representation. Engine monitor data can also be analyzed statistically, one such analysis is given in [6].

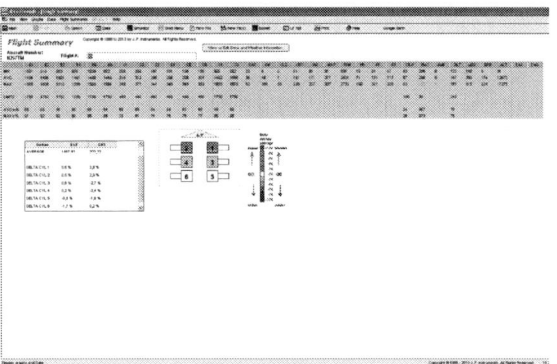

Figure 3. Flight summary

Despite interesting graphical representations of engine parameters, both in real-time on engine monitor display during a flight and later on computer using specialized accompanying software, more subtle engine problems are not so easy to discern and process require trained maintenance personnel with the experience in engine diagnostics from available recorded engine parameters. Catalogued patterns of various engine problems are commonly supplied with the engine monitor documentation, [2].

IV. SELF-ORGANIZNIG MAPS

Self-Organizing Map (SOM) is a type of neural network architecture that is trained using unsupervised training and produces a low-dimensional (most common 2D) discretized representation of the input space of the presented training samples. It provides dimensionality reduction of available multidimensional training data. The map is used in two modes: training mode and mapping mode. During the training mode a two-dimensional discrete representation of input space is formed. In the mapping mode input sample is assigned to the closest member of a map, thus classifying newly presented input vector. The most common two dimensional SOM lattice with neuron neighborhood is shown in Fig. 4. SOM uses competitive learning with lateral inhibition function. The most often used neighborhood function is the Mexican hat. An engine monitor with its parameters can form an input vector that is simultaneously fed to all SOM neuronal units. The input vector is mapped to the winner node (node with highest activation). More on SOM and learning algorithm could be found in [7,8].

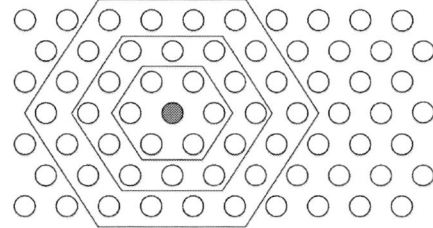

Figure 4. Self-organizing map (SOM) with hexagonal lattice and the local neighborhood of the one particular neuron

V. FAULT DETECTION USING SOM

SOMs could be used for the visualization of engine and equipment parameters, [9,10]. Sometimes further step is taken with the application of the SOM to fault detection and condition monitoring, [11,12]. Fault detection using SOM is based on the assumption that a SOM (or its parameters like quantization error or SOMs hits map) that belongs to the normal engine operation differs from one that belongs to a faulty engine. The main problem is that SOMs trained on similar data may not always give quite similar results (e.g. occurrence of data shift in a map), [13]. Resulting self-organization is influenced by the various initial conditions: SOM initial

1068

weights, [14], the choice of the neighborhood function, data normalization, the learning rate and the order of presentation sequence of training vectors. One analysis of statistical measures to assess the stability of the results of SOM training is given in [15].

A. Methods for Fault Detection

Some ideas how to compare SOMs are listed below:

1) Comparison of SOM maps

Comparison of SOM maps (test and reference) is done by comparing weights belonging to SOM planes, Fig. 5. These maps can be compared visually, an analytical approach is, however, more difficult, [13,16].

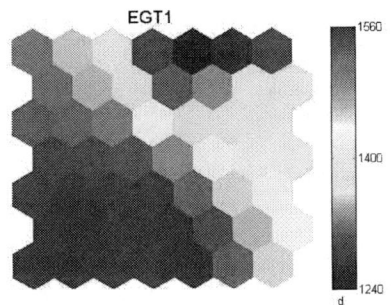

Figure 5. SOM plane for EGT1 (7x7 lattice example)

2) Comparison of Hit Maps

Comparison of SOM maps is done by comparing hits count belonging to SOM (depicted as size of hexagon line), Fig. 6. The method is based on the assumption that particular region of the map belongs to normal engine operation. Any visit to SOM units outside this region, particularly if held during a longer period, should be regarded as a suspicious engine condition.

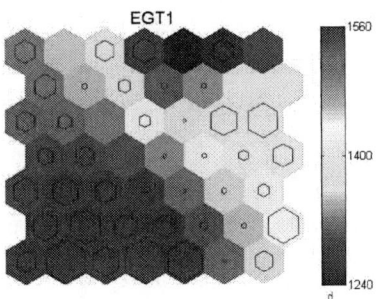

Figure 6. SOM plane for EGT1 with hits count

3) Use of Mean Quantization Error (MQE)

Quantization error e_{QE} is the distance between the input pattern vector z and the weight vector w_{bmu} of the Best Matching Unit (BMU), w_k is the weight vector of kth SOM unit and K is the number of SOM units, (1), (2):

$$e_{QE} = \left\| z - w_{bmu} \right\| \tag{1}$$

$$e_{QE} = \min_{1 \le k \le K} \left\| z - w_k \right\| \tag{2}$$

Mean Quantization Error (MQE), e_{MQE} is given by (3):

$$e_{MQE} = \frac{1}{N} \left(\sum_{i=1}^{N} \min_{1 \le k \le K} \left\| z_i - w_k \right\| \right) \tag{3}$$

where N is the number of patterns in a dataset, and z_i is the ith pattern vector in a dataset.

The accuracy of a SOM in representing its inputs can be validated using the mean quantization error (MQE). It quantifies how well a previously trained SOM approximates the presented data items. Low values for MQE show that the data set is well represented by the SOM.

B. Method Used in This Paper

Method with MQE is applied in this paper. MQE is a more robust measure than a comparison of SOMs as it is resistant to data shift in a map.

1) Training Phase

During the training phase, SOM captures the statistical distribution of the data from a non-faulty engine, Fig. 7.

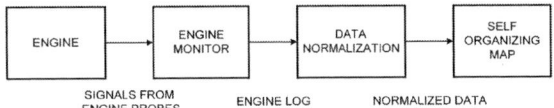

Figure 7. Training phase

2) Monitoring Phase

During the monitoring phase, input vectors are classified to the Best Matching Unit (BMU) and Mean Quantization Error (MQE) is determined, Fig. 8.

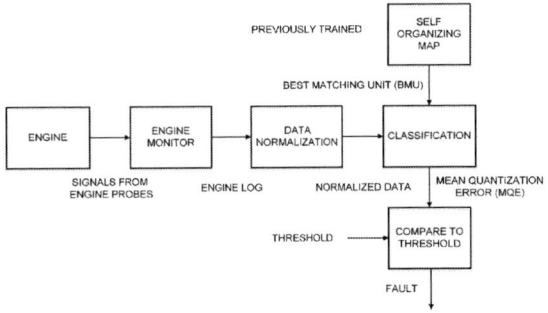

Figure 8. Monitoring phase

C. Off-Line and On-Line Detection

Method for fault detection can be performed as off-line detection (compares complete engine log with the SOM after the flight) and on-line detection (compares each new vector with the SOM).

1) Off-Line Detection

- Normal condition

The new engine log is classified as normal if e_{MQE} (mean quantization error) is less than the threshold L, (4):

$$e_{MQE} < L \tag{4}$$

- Suspicious condition

The new engine log is classified as suspicious if e_{MQE} is greater than the threshold L, (5):

$$e_{MQE} \geq L \qquad (5)$$

2) On-Line Detection

- Normal condition

The new vector is classified as normal if e_{QE} (quantization error) is less than the threshold L, (6).

$$e_{QE} < L \qquad (6)$$

- Suspicious condition

The new vector is classified as suspicious if e_{QE} is greater than the threshold L, (7).

$$e_{QE} \geq L \qquad (7)$$

In real applications additional filtering will be required (e.g. minimal number of threshold violations in a specific period of time).

VI. EXPERIMENT

The SOM was trained using engine parameters from available engine monitor logs.

A. Available Data

An experiment was performed using engine logs accompanying the EzTrends2 software, belonging to a six cylinder engine, with no known faults present, [5]. Available experimental data were flight logs Flt#192 (duration 1.49h) and Flt#193 (duration 1.23h). As these files belong to twin engine aircraft, only one (left) engine was selected for analysis. Engine parameters are logged every 6 seconds. Part of engine log Flt#192 is shown in Fig. 9, (left engine EGTs: LE1-6 and CHTs: LC1-6).

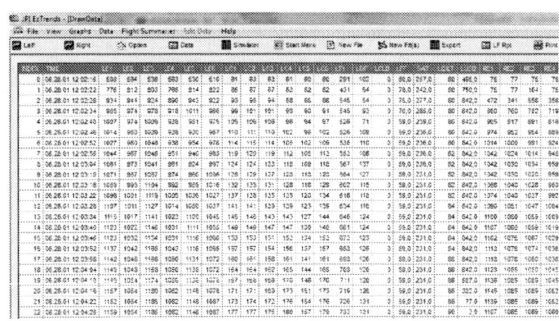

Figure 9. Example of engine log

B. Synthetic Faults

Because it is difficult to obtain data logs from faulty engines (as they are still quite reliable) and one has rarely access to large maintenance facility, testing datasets were created artificially by modifying existing engine data log (Flt#193) according to common descriptions of engine problems that include EGT and CHT temperature deviations. Examples of two patterns corresponding to problems in engine operation are shown in Fig. 10 and Fig. 11. Such patterns are often catalogued in documentation belonging to an engine monitor. Detailed description of fault indications as EGT/CHT bars on the engine monitor with temperature differences is given in [2]. This gives following datasets, as shown in Table II.

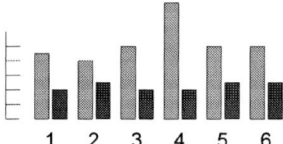

Figure 10. Example of fault pattern, failure 1, EGT rise for one cylinder

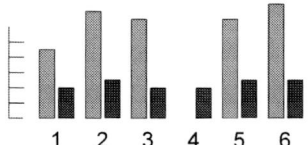

Figure 11. Another example of fault pattern, failure 2, loss of EGT for one cylinder

TABLE II. USED FILES

File	Duration	Frames	Description
Training Flt#192	1.49h	896	Training data
Test (normal) Flt#193	1.23h	736	Test (normal) data
Test (Failure 1) [1]	1.23h	736	Fouling, faulty plug, wire or distributor
Test (Failure 2) [1]	1.23h	736	Stuck valve
Test (Failure 3) [1]	1.23h	736	Faulty valve lifter
Test (Failure 4) [1]	1.23h	736	Dirty fuel injectors or fouled plugs
Test (Failure 5) [1]	1.23h	736	Burned exhaust valve
Test (Failure 6) [1]	1.23h	736	Detonation
Test (Failure 7) [1]	1.23h	736	Pre-ignition
Test (Failure 8) [1]	1.23h	736	Leaking exhaust gasket

[1]Flt#193 modified according to the particular fault description

C. Selection of Input Parameters

Engine monitors are primarily designed to monitor engine temperatures EGTs and CHTs as they reflect a combustion process that is happening in engine cylinders. The other parameters were added to most monitors later as additional information about engine operation. Parameters, not directly related to a combustion process, are not considered in this experiment. Selected input parameters are:

- Exhaust Gas Temperatures (cylinders 1-6):
  EGT1, EGT2, EGT3, EGT4, EGT5 and EGT6
- Cylinder Head Temperatures (cylinders 1-6):
  CHT1, CHT2, CHT3, CHT4, CHT5 and CHT6

Input vector z to SOM that consists of EGT and CHT temperatures is given in (8), (static, current state, neighborhood for time evolution analysis is not included):

$$z = [\, T_{EGT,1}, \ldots, T_{EGT,6}, T_{CHT,1}, \ldots, T_{CHT,6}\,] \qquad (8)$$

Software used for analysis is the SOM Toolbox for MATLAB 5, [17]. Initialization was linear. Training was performed using a batch algorithm.

D. Sizing of SOM

The size of the SOM was determined according to (9), where M denotes number of SOM units and N is the number of samples, [12]:

$$M \approx 5\sqrt{N} \qquad (9)$$

Most flights last about an hour. One hour flight consists of 600 frames. For 600 frames $M \approx 123$. In case of SOM lattice with equal number of rows and columns it is a 11x11 lattice.

E. Normalization of Parameters

Due to various ranges of values for EGT and CHT parameters should be normalized (as parameters are features it is also feature scaling) before applying to SOM. One interesting analysis of data normalization applied to SOMs is given in [18].

Two methods for feature scaling are popular:

- Rescaling (normalization): rescales the values into to a [0, 1] range. This is useful in cases where all parameters must belong to the same range. The disadvantage of this method is that the outliers from the data set are lost, (10):

$$x' = \frac{x - x_{\min}}{x_{\max} - x_{\min}} \qquad (10)$$

- Standardization (variance method): rescales data to have a zero mean (μ) and standard deviation (σ) of 1 (unit variance), (11):

$$x' = \frac{x - \bar{x}}{\sigma} \qquad (11)$$

Standardization method is recommended for most applications. It was also chosen for SOM visualizations, partially because it was already included in the SOM Toolbox.

Previous normalization methods are single-variable operations suitable for visualization of isolated parameter planes. However, for preserving multivariate anomalies and comparisons of MQEs of test datasets classified by a SOM produced by a normal dataset with the threshold for the purpose of fault detection, following normalization is applied to the CHT data instead, (12). Considering that the average EGT value is approximately 3.8 times greater than the average CHT value this puts normalized CHT values in the same range with EGT for each cylinder i:

$$T_{N,CHT,i} = 3.8 T_{CHT,i} \qquad i = 1, \dots, 6 \qquad (12)$$

F. Data Visualization

Following are three examples of visualization of engine logs using SOMs. The training process is accomplished with normalized vectors, however legend corresponds to real temperatures. SOM corresponding to

training data from a normally operating engine is shown in Fig. 12. In Fig. 13 there is another example of a normally operating engine, similar to one in Fig. 12, and in Fig. 14 is an example of faulty engine operation (see EGT4 plane).

- SOM Training data (normal engine operation, Flt#192)

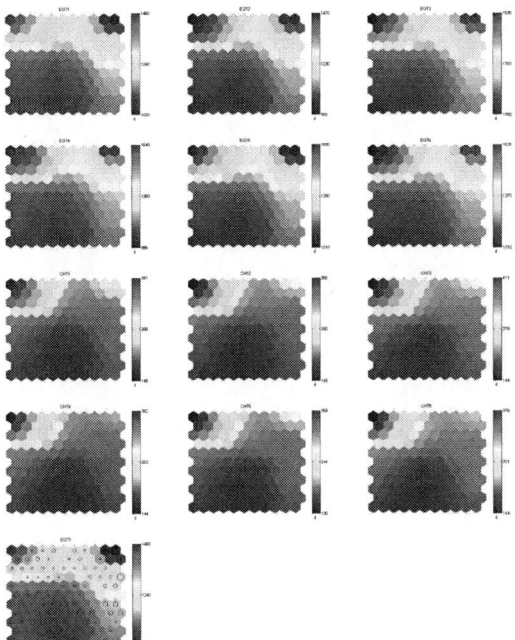

Figure 12. Visualization of training data in SOM (normal engine operation), planes EGT1-6 and CHT1-6 plus hit map

- SOM Test data (normal engine operation, Flt#193)

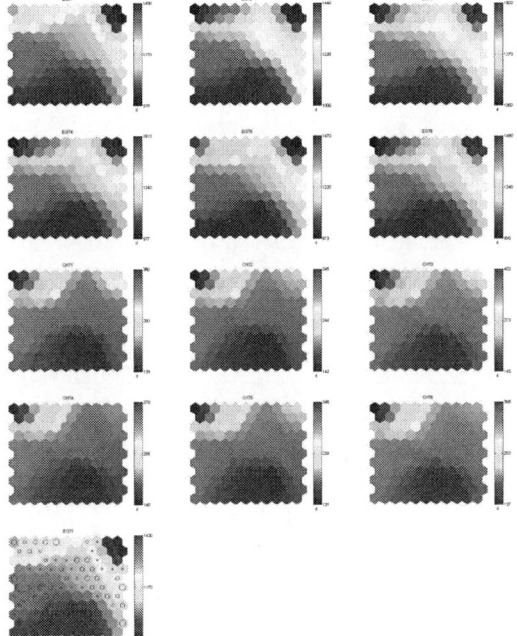

Figure 13. Visualization of test data in SOM (normal engine operation), planes EGT1-6 and CHT1-6 plus hit map.

- SOM Test data (example of problematic engine operation, artificial data: Failure 3)

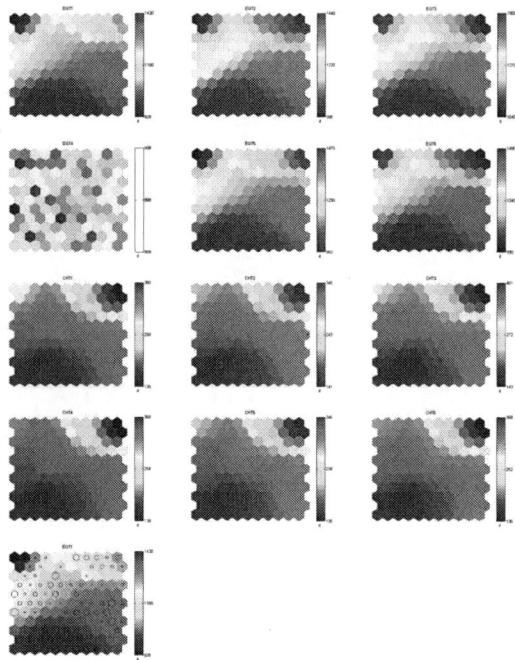

Figure 14. Visualization of test data in SOM (abnormal engine operation, Failure 3), planes EGT1-6 and CHT1-6 plus hit map

G. Fault Detection of Failure Datasets

Artificial data and MQEs for abnormal engine operation (eight different failures) were also analyzed. Examples of eight different engine problems were included in eight test datasets. Example of visualization of abnormal engine operation for one failure (Failure 3) is shown in Fig. 14. MQEs (e_{MQE}) were calculated for different data sets and results are shown in Table III, including the original, non-normalized data and data normalized according to (12). These values could be used for determination of the threshold L by multiplying with a constant α for the safety margin (e.g. α=0.8), (13). More failure examples are needed for more precise value of L.

TABLE III. MQE FOR SOM AND VARIOUS DATASETS

Dataset	e_{MQE}[1]	e_{MQE}[2]
Training data (Flt#192)	19.6937	32.756
Normal data (Flt#193)	33.412	58.8373
Failure 1	389.6016	605.1097
Failure 2	721.1847	747.0054
Failure 3	360.6092	374.6447
Failure 4	500.2829	694.2098
Failure 5	342.6307	500.4933
Failure 6	333.4395	419.8564
Failure 7	466.7379	522.2226
Failure 8	229.1554	825.1329
Min (for Failures 1-8)	229.1554	374.6447

[1]non-normalized data, [2]normalized data according to (12)

$$L = \alpha \min_{k}(e_{MQE,k}) \quad k=1, 2, \dots, 8 \qquad (13)$$

In previous experiment one SOM is used for all engine operating regimes. Better accuracy could be achieved if separate SOM is prepared for each particular operating regime. Methods for regime selections are given in [3].

VII. CONCLUSION

Engine monitors for aircraft piston engines record large amount of data that may be difficult to visualize and analyze. SOMs with its unsupervised learning algorithm are able to produce the two-dimensional mapping of multivariate data. This provides dimensionality reduction while topologically preserving similarities in the input space. Such two-dimensional representation is suitable for visualization. Resulting SOM can be further analyzed and used for the fault detection. Use of SOM for monitoring of aircraft piston engine operation is proposed. By comparing MQE of engine logs from engines under the test with the threshold it is possible to detect suspicious engine operation. On-line detection is also possible by comparing input vector to best matching unit (BMU) and comparing quantization error with the predetermined threshold.

REFERENCES

[1] D. Miljković, "Engine Monitors for General Aviation Piston Engines Condition Monitoring", CrSNDT Info Journal, Vol. 3, No. 2 (10), pp. 19-23, 2013

[2] Pilot's Guide - Engine Data Management, EDM-730, EDM-830, EDM-740, J. P. Instruments, Inc., 2010

[3] D. Miljković, "Regime Dependent Aircraft Piston Engine Monitoring", Proc. MIPRO 2014, Opatija, Croatia, 26-30 May 2014, pp. 1311-1316

[4] J. P. Instruments, Technical Support - Video Tutorials, https://www.jpinstruments.com/technical-support/video-tutorials/

[5] EzTrends (Download and Plotting Software), JPI, California, USA, Available: http://www.jpitech.com/softwaredownload.php

[6] D. Miljković, "Statistical Properties of Aircraft Piston Engine Monitor CM Data", Proc. MATEST 2013, Zagreb, 9 October 2013

[7] T. Kohonen, Self-Organizing Maps, 2nd ed., Springer 1997

[8] T. Kohonen, MATLAB Implementations and Applications of the Self-Organizing Map, Unigrafia, Helsinki, Finland, 2014

[9] R. Ji-Hong, C. Jiang-Cheng and W. Nan, "Visual Analysis of SOM Network in Fault Diagnosis", Proc. 2011 International Conference on Physics Science and Technology (ICPST 2011), Hong Kong, 12 – 13 December 2011, pp. 333-338

[10] E. Come, M. Cottrell, M. Verleysen and J. Lacaille, "Aircraft engine health monitoring using Self-Organizing Maps", Proc. 10th Industrial Conference, ICDM 2010, Berlin, Germany, 12-14 July 2010, pp. 405-417

[11] A. Zachrison, "Fluid Power Applications Using Self-Organising Maps in Condition Monitoring", PhD Thesis, Linköping University, Linköping, Sweden, 2008

[12] J. Tian, M. H. Azarian. and M. Pecht, "Anomaly Detection Using Self-Organizing Maps-Based K-Nearest Neighbor Algorithm", Proc. European Conference of the Prognostics and Health Management Society 2014, Nantes, France, 8-10 July 2014, pp. 110-118

[13] R. Mayer, D. Baum, R. Neumayer, A. Rauber: "Analytic Comparison of Self-Organising Maps", Proc. 7th International Workshop on Self-Organizing Maps (WSOM 2009), St. Augustine, Florida, USA, 8-10 June 2009, pp. 182-190

[14] A. A. Akinduko and E. M. Mirkes, "Initialization of Self-Organizing Maps: Principal Components Versus Random Initialization. A Case Study", CoRR abs/1210.5873, 2012

[15] M. Cottrell, E. de Bodt and M. Verleysen, "A Statistical Tool to Assess the Reliability of Self-Organising Maps". In: Advances in Self-Organising Maps, Springer, London, 2001

[16] S. Kaski and K. Lagus, "Comparing self-organizing maps", Proc. ICANN'96, Bochum, Germany, July 16 - 19, 1996, pp. 809-814

[17] J. Vesanto, J. Himberg, E. Alhoniemi and J. Parhakangas, SOM Toolbox for MATLAB 5, Helsinki University of Technology, April 2000

[18] P. Demartinesand and F. Blayo, "Kohonen Self-Organizing Maps: Is the normalization necessary?", Journal of Complex Systems, Vol. 6, Iss. 2, pp. 105-123, 1992

PicoAgri. Realization of a Low-cost, Remote Sensing Environment for Monitoring Agricultural Fields Through Small Satellites and Drones

S. Marsi *, A. Gregorio **, M. Maris *** and M. Puligheddu **

* University of Trieste, Department of Engineering and Architecture, Trieste, Italy
** University of Trieste, Department of Physics, Trieste, Italy
*** INAF – Istituto Nazionale di AstroFisica, Osservatorio Astronomico di Trieste, Trieste, Italy
marsi@units.it

Abstract - In this paper, we analyze the accomplishment of a study for a low cost system, named PicoAgri, to monitor the status of agricultural fields. We here deal with the small detection system we are developing representing the first element to build the final PicoAgri concept: a system consisting of two elements, an array of small satellites coupled to drones. The array of CubeSat small satellites will provide data for an initial multispectral analysis with a resolution at ground of 30 m; when required a well localized area will be explored employing suitable drones capable of analyzing the territory with a resolution at least two orders of magnitude higher.

I. INTRODUCTION

Small satellites represent a new generation of satellite platforms that are opening their own market niche expanding extremely quickly. Particularly interesting is the Earth Observation and remote sensing field with practical applications for agriculture.

The *"PicoAgri"* concept will consist of two elements, an array of small satellites coupled to drones. The array of small satellites will be orbiting at a height of 400 km to provide data for an initial multispectral analysis with a resolution at ground of 30 m. When required a well-localized area will be explored employing suitable drones capable of analyzing the territory with a resolution at least two orders of magnitude higher.

In this context, the University of Trieste is developing a new low cost detector system dedicated to agriculture monitoring based on an array of CubeSats [1]. The PicoAgri system is mostly based on COTS (Commercial Off The Shelf) devices that are being adapted to the requirements of the full system aimed at monitoring the status of agricultural fields.

In this paper we briefly presents a feasibility study of the PicoAgri system and in particular we will focus the attention on the detection system, composed by the optical environment and by a suitable sensor.

II. SMALL SATELLITES

From the late 90s, with the construction of the first small satellites with very compact mass and size, space technology, prerogative of large companies specialised in the aerospace field, wants to become a technology available to small companies, facilities and research institutes, historically smaller and less rich. The design can be complex, given the environment in which such systems have to operate, but it is clear that the CubeSats small satellite systems [1], initially designed for educational purposes [2,3], are being brought to the daily technological and scientific use [4,5,6,7]. Over the last decade, more than one hundred CubeSats, modular small satellites whose basic element is an Aluminium cubic structure of 10 cm side, have been launched. This number will double very soon and in recent years interest for small satellites has grown considerably also for commercial applications further expanding the potential use of these systems.

Particularly interesting is the Earth Observation [8] and remote sensing field segment, that has large potential applications for safety issues or even more practical applications for agriculture. In this context the University of Trieste, with the two departments of Physics and Engineering and Architecture, in Collaboration with the Astronomical Observatory of Trieste, is developing the PicoAgri detection system dedicated to agriculture monitoring.

Figure 1. A picture of the CubeSat mechanical prototype produced by PicoSaTs

The University of Trieste started a CubeSat program that finally converged in the creation of a spin-off company, PicoSaTs S.R.L. [9], that is developing its own CubeSat satellite to be used as platform for this system. Figure 1 shows a picture of the mechanical prototype.

A. Remote Sensing and Earth observations

Remote sensing from Low Earth Orbit (LEO) satellites can significantly contribute to provide a timely and accurate picture of the agricultural sector [10,11], as it is suitable to gather information over large areas with high revisit frequency. The increase in agriculture production, required to cope with the foreseen increased worldwide population, must be achieved while minimizing the environmental impact of agriculture. Agriculture must cope with climate change and compete with land users not involved in food production (e.g., biofuel production, urban expansion). The necessary changes and transitions have to be monitored closely to provide decision makers with feedback on their policies and investments. In particular remote sensing studies of vegetation, adopting spectral radiance data from three bands has been proved to be suitable to gather properties related to pigment absorption, leaf density, and the canopy leaf water content[12,13,14] and eventually to the vegetation health.

The rapid wide coverage capability of satellite platforms allows monitoring of rapidly changing phenomena. Sensors on board satellites [15] can provide prompt information about global patterns on Earth (surface vegetation cover and its time and seasonal variations, surface morphologic structures, ocean surface temperature, near-surface wind), and in the atmosphere (dynamics of clouds).

B. The market view

There are a number of scientific satellites whose data are available to scientists but typically not suitable for commercial use. On Europe side, the European Space Agency (ESA) GMES (Global Monitoring for Environment and Security) Sentinel family and in particular Sentinel-2 [16], a multispectral high-resolution imaging mission for land monitoring to provide, for example, imagery of vegetation, soil and water cover, inland waterways and coastal areas, represents the status of the art of Earth Observation from space.

On business side, according to Euroconsult [17] more

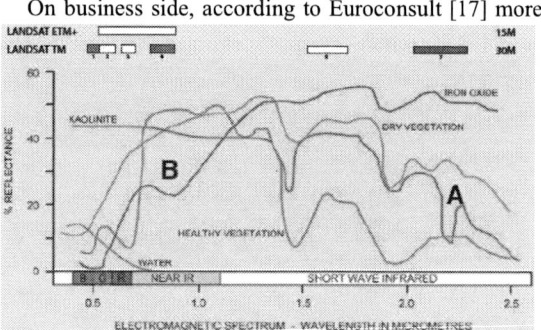

Figure 3. Vegetation reflectance.
(Courtesy of [20])

than 150 satellites (class larger than 50 kg) were launched providing ad hoc services for civil and Earth observation over 2006-2015. According to similar analysis by Pixalytic Ltd [18], the fleet is even larger, almost 200 satellites overall, of which about 6% are commercial satellites. About 15% of the fleet belongs to Europe,

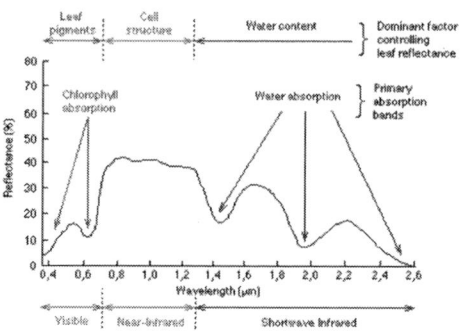

Figure 2:Typical spectral reflectance curve of green leaf
(Courtesy of [19])

demonstrating once more that Earth Observation is one of the most promising commercial space applications. A number of market and economic studies can be analysed [19,20,21], but it is clear that services varies according to the type of system and on board sensors. In this context we think that state of the art technology, as hyper-spectral imaging, is now ready for miniaturization. High resolution can be achieved also by small satellite systems, well below 50 kg, by using dedicated detection systems as the one proposed here, definitely at lower costs with respect to actual large space Earth Observation missions.

III. VEGETATION REMOTE SENSING

Remote sensing studies of vegetation use specific wavelength bands that enhance leaf optical properties (scattering and absorption) providing spectral contrast with respect to background. Figure 2 gives the typical reflectance curve of green leaves, enhancing the dominant reflectance factors. Five primary regions are typically identified in the 0.4÷2.5 µm band:

B1. 0.4÷0.5 µm. Strong spectral absorption by chlorophylls and carotenoids;

B2. 0.6÷0.7 µm. Strong chlorophyll absorption;

B3. 0.7÷1.1 µm. Minimum absorption, leaf scattering mechanisms resulting in high levels of spectral reflectance, especially for dense canopies;

B4. 1.1÷1.3 µm. Increase of water absorption coefficients;

B5. 1.3÷2.5 µm. Absorption by liquid water.

The amount of green biomass also affects the reflectance signature of biologic materials. As the vegetation grows, vegetation spectral signature becomes dominant; this signature can be observed in Figure 3. In principle, the biomass can be measured by comparing the reflectance in the 0.8÷1.1 µm region to the reflectance near 0.4 µm.

IV. THE PICOAGRI SYSTEM

A. The satellite system

Very simple considerations are given here for what regards satellite requirements since the full system

1074

concept is still preliminary; requirements are related on one side to the system size and on the other to the orbital characteristics. For what regards the attitude control system, it is assumed the satellite is controlled on three axes and the instrument will be pointing in the nadir direction.

For what regards the size, the system shall fit within a so-called 3U CubeSat, a parallelepiped satellite of 10 cm side and 30 cm length and maximum power consumption of the order of a few tens W. The optical system will be aligned with long side of the parallelepiped satellite whose size limits the optical system diameter to 10 cm and length to less than 30 cm if one consider that there should be enough room to host the PicoAgri detection system and all the other satellite systems.

Satellites will be in Sun-Synchronous[1] (SS) orbits at 400 km altitude corresponding to an inclination of 97° (polar regressive orbits): the surface illumination angle will be nearly the same every time the satellite is overhead, being almost in constant sunlight, useful for imaging, and for Earth Observation in particular.

About the data acquired by the satellite, we have to consider that all the acquired data can be downloaded, of course, only when the satellite orbit passes near the ground station. Since the time during which the satellite and the ground station can communicate is very limited (few minutes) and moreover it depends by the available communication bandwidth, the information provided by satellite should thus be quite limited. However, it could be noted that even if the satellite will scan the entire earth surface, the data we are interested to acquire is typically limited to some predefined targets, where every target would have the dimension of one or more agricultural field. Thus, the strategy we suggest to adopt is the following: When the satellite is near to the target, the acquisition system will be turned on and all the data acquired will be stored inside a proper memory device, on the opposite if the target is outside the field of view, the satellite acquisition system remains in stand-by and no further acquisition is performed. A suitable on board control system will control both the acquisition system and the transmission procedure according to the data provided by the GPS system and to the current time. The acquired data will eventually be analized by the ground station which would provide several processing steps : Mapping, mosaicking, cloud and cloud shadow detection, lens distortion correction, atmospheric correction, compositing procedure, etc. [26].

B. Mission Analysis

With the aid of the STK commercial software [24] we analysed possible CubeSaT constellations that use the SS orbit. Five test constellations were analysed:

1. One satellite;
2. Two satellites on the same orbit plane;

3. 12 satellites orbiting on two planes (six on each plane). The angle between the orbital planes is 90°;
4. 24 satellites orbiting on two planes (12 each orbit). The angle between the orbital planes is 90°;
5. 36 satellites orbiting in three planes (12 each orbit). The angle between the orbital planes is 45°.

The parameter that best ranks the different configurations is the revisit time, defined as the time elapsed between two consecutive observations of the same point on Earth. To calculate it, we used a grid of 33 target points placed at different latitudes (but with the same longitude) and run the simulation over three months. Figure 4 shows the result. One or two satellites do not provide a good revisit time, from 100 to 1000 hour to see the same point on Earth (according to their latitude). To improve this result the overall number of satellites should be increased, a constellation made of 12 satellites distributed into two planes can provide good coverage, with a worst case of about one week at the equator. The constellation performance grows almost linearly at the beginning as a function of the number of satellites in the constellation, but after about 30 satellites the performance reaches a plateau.

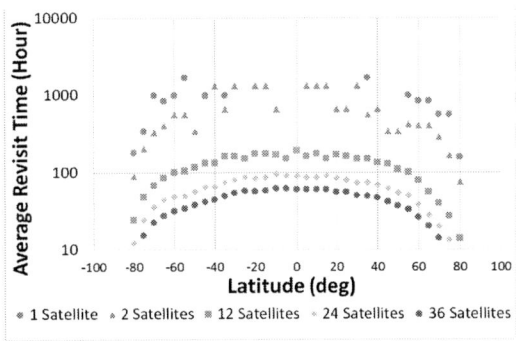

Figure 4.   Average Revisit Time simulated at different latitudine

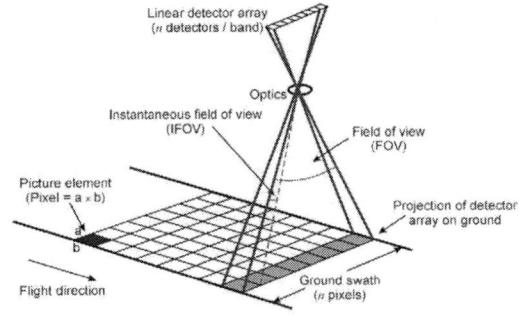

Figure 5.   Scheme of an along track scanner

C. Communication Time

In a Low Earth Orbit (LEO), the time during which the satellite and the ground station can communicate is very

---

[1] A Sun-Synchronous orbit profits by the Earth non-uniform gravitational field and combines orbit altitude and inclination so that the satellite passes over any given point of the Earth surface at the same local solar time.

limited. With STK we analysed the satellite connections with a ground station placed in Trieste (Italy). On average two links per day are available, the mean duration is 236 seconds (minimum elevation angle of 15º), ranging from a minimum of 25 s to 302 s.

D. Detection System

The detection system is made of an optical system and a sensor, the processing unit used by now is an Arduino 1, that have been already demonstrated to work in flight [25] although we are analysing the possibility of using a more powerful unit.

For what regards both the optical and sensor systems, we analysed possible COTS optical and detection systems and linear CCD systems.

1) Working Principle

The acquisition of the satellite image is performed via a *multispectral sensor* that cover three spectral bands simultaneously:

1. Blue   459÷479 nm

2. Red   620÷670 nm

3. Near Infra-Red (NIR)   841÷875 nm

The satellite acts as an *along track scanner* (*push-broom scanner*) (Fig.5) to obtain the spectroscopic image. A push-broom camera consists of an optical system projecting an image onto a linear array of sensors, typically a CCD array, arranged perpendicular to the flight direction of the spacecraft. At any time only those points are imaged that lie in the plane defined by the optical center and the line containing the sensor array. This plane is called the *instantaneous view plane* or simply *view plane*. The push-broom sensor is mounted on a moving platform, a satellite in our case, and as the platform moves, the view plane sweeps out a region of space. At regular intervals of time, 1-dimensional images of the view plane are captured. The ensemble of these 1-dimensional images constitutes a 2-dimensional image.

PicoAgri uses three different linear sensors, one for

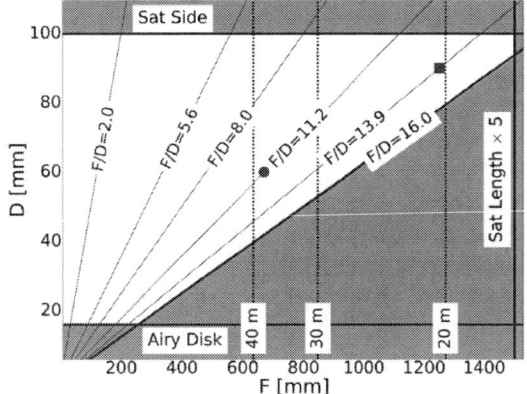

Figure 6. Optical instruments constrains

TABLE I. REQUIREMENTS

Requirement		Nominal
Orbital:		
Altitude	h	400 km
Inclination	i	97°
CubeSat size:		
Optical system diameter	D	< 100 mm
Length		< 300 mm
Detection system:		
Resolution on ground	r	20 - 40 m
Swath on Earth	w	> 25 km
Corresponding FoV	fov	> ± 2.1°
Wavelength	λ	0.4 – 1. μm

each spectral band, to reconstruct an approximate spectral image of the soil. The three sensors lay parallel on the focal plane perpendicular to the satellite flight direction so that they are scanning quasi simultaneously the *view plane* on Earth. In such a device the size of detectors (in our case the physical dimension of pixels) determines the size of each ground resolution cell. Another parameter that determines the instrument resolution is the acquisition time: the satellite is moving at a speed of approximately 8 km.s^{-1} with respect to the ground, this has an impact on the image resolution along the satellite moving direction.

A desired resolution of 30 m corresponds to a maximum sampling time of 4 ms, a value easily achievable with Arduino or with any other acquisition systems. The main advantage of this type of sensors is the lack of mechanical components, preventing from malfunctions and extending satellite lifetime. The requirements of the detection system are summarized in Table I.

2) Feasibility Study

Requirements for optics and optoelectronics can be derived from analysis of mission requirements. In this section we want to determine the most important parameters for the optical system: the effective focal length F, the plate scale S, the diameter D or the F/D ratio. For what may concern the optoelectronic part of the device, relevant parameters are: the sampling frequency f_s, the pixels size P_s and the detector number of pixels N_{samp}.

With the figures given in Table I, the angular resolution r_{ang} is:

$$r_{ang} = r/h = 7.5 \cdot 10^{-5} \text{ rad} \quad (1)$$

The diameter of the lens has to be large enough to assure proper resolution for a diffraction limited system. Since the typical resolution r_{obj} is:

$$r_{obj} = 1.22 \, \lambda / D \quad (2)$$

The constrain becomes:

$$D > 1.22 \, \lambda / r_{ang} = 16 \text{ mm} \quad (3)$$

The *fov* required to accommodate w is:

$$fov = w/h > 6.2 \cdot 10^{-2} \text{ rad} \quad (4)$$

This puts a limit for the F/D ratio:

$$F/D < h/w \ < 16 \quad (5)$$

Figure 6 depicts the allowed F and D for the optical instrument. Suitable F and D combinations are those included in the white area. Tilted lines represent the geometric locus with a constant F/D. The D is limited on the lower side by the Airy disk as reported by eq. (3) and on the upper side by the satellite dimensions.

For a single lens or mirror optics, the upper limit for F would be the satellite size itself which is 300 mm, however more sophisticated optical configurations allows a compression of the telescope size till a factor of five, leading to a maximum F of about $1.5 \cdot 10^{-3}$ mm. The F/D ratio is limited on the right side by eq. (5) while on the left side there is not a real physical limit, but COTS objectives have typically $F/D > 2$.

The pixel size P_s of the detector fixes the scale S of the optics. If O is the oversampling factor i.e. the number of pixels corresponding to a single element of image with an r size on ground, the scale is:

$$S = r_{ang}/(O \times P_s) = 1/F \qquad (6)$$

In our simulation we assumed two possible pixel sizes: 7 μm and 63.5 μm as representative of typical commercial sensors, and we also assumed $O = 1$. We then obtain $S = 1.1 \cdot 10^{-2}$ rad/mm in the first case and $1.2 \cdot 10^{-3}$ rad/mm in the second case. These values correspond to a focal length F respectively of 93 mm and 847 mm. While the first can be easily accommodated in the satellite, the second actually exceeds the satellite size, but it can be accommodated using a two lens objective, one for instance with an objective 180 mm lens, with a scale $5.5 \cdot 10^{-3}$ rad/mm and a 36 mm camera lens suitable to magnify a portion of the image of about a factor 5.

The minimum sampling rate for the sensor is determined by the scanning speed of the satellite to limit motion blurring. The *fov* scans the Earth surface at a linear rate given by:

$$rate\_scan = 2 \pi R_{earth}/P_{sat} \qquad (7)$$

with R_{earth} the Earth radius and P_{sat} the orbital period of the satellite (~90 min), the *rate_scan* is ~ 7.4 km/s.
The crossing time for resolution element r is

$$t_{res} = P_{sat} r / 2 \pi R_{earth} \sim 4 \cdot 10^{-3} \qquad (8)$$

Assuming we accept a motion blurring of at most 50% of an image element, the maximum integration time for the detector is:

$$t_{exp} < t_{res}/2 = 8 \cdot 10^{-3} \text{ s} \qquad (9)$$

The w is equivalent to $N_{samp} = w/r > 833$ samples the minimal sampling rate for a linear detector is then:

$$f_s > N_{samp} \, 2/t_{res} \sim 500 \text{ kHz} \qquad (10)$$

Commercially available linear detectors allows sampling rates larger than few MHz, with frame formats equivalent to 256, 512, 1024 pixels per line or even larger, and pixel sizes in the range 7–63.5 μm. In order to reduce the size of the camera, small pixel sizes should be preferred. However low cost linear detectors are more easily available with large pixel sizes. So we consider two basic figures schemes for the camera as summarized in Table II.

Scheme 1 has been the first developed since it is based on on-the-shelf optics available, as describe in the next

TABLE II. PROPOSED SCHEME

	Scheme 1	Scheme 2
Ps	63.5 μm	7 μm
N pixels	1024	1024
Detector width	65 mm	7 mm
Sampling Rate	500 kHz	500 kHz
F	900 mm (180 mm × 5)	100 mm
D	90 mm	60mm

section while Scheme 2 is under development since it could be a good solution for a more compact optical system, suitable to be installed on smaller satellite.

V. HARDWARE DEVICES

A. Optical System

We acquired two telescopes for amateur astronomical observations, namely "Skywatcher Maksutov MC 90/1250 SkyMax OTA" and "Omegon Maksutov MightyMak 60". Their characteristics are summarized in Table III together with the minimal characteristics of the detection system to be placed on the focal plane. Both the telescopes have a

TABLE III. CHARACTERISTICS OF THE CHOSEN TELESCOPES

Parameter	Maksutov MC 90/1250	Maksutov MightyMak 60
Aperture (mm)	90	60
Length	250	200
Focal length	1250	670
f/#	13.9	11.2
Field of View	± 2.1°	± 2.4°
Resolution power	1.28	2.3
Max magnification	180	120
Sensor		
N_{pixels}	960	1142
Pixel size	93 μm	52 μm

Maksutov design, a catadioptric telescope design that combines a spherical mirror with a weakly negative lens placed at the entrance pupil of the telescope.

In figure 6 the parameters of the two telescopes under test are marked with a circle for the 60 mm and a square for the 90 mm. It can be noted in particular that vertical dashed lines represent the required focal lengths for a pixel size of 63.5 μm to have a resolution at ground of respectively 20, 30 and 40 m.

It could be interesting to compare how the two proposed instruments copes with the required resolution at ground. Assuming a 63.5 μm pixel size, the 90 mm focal length allows a resolution at ground of about 20 m. On the other side the 60 mm focal length allows a resolution at ground of about 37 m. While the first case represents a slight oversampling, the second represents a slight degradation of performance. A better match can be obtained either using a sensor with different pixel sizes, or adding a camera lens to properly match the required scale.

B. Sensor

The first sensor that is being proposed is a linear CCD sensor, the TSL 1410R. The TSL 1410 linear sensor array consists of two sections of 640 photodiodes each and associated with acharge amplifier circuitry, aligned to form a contiguous 1280×1 pixel array. The pixels measure 63.5 μm by 55.5 μm, with 63.5 μm center-to-center spacing and 8 μm spacing between pixels. This sensor reasonably fits on the two optical systems by coupling three sensors.

VI. CONCLUSION

We have proved the feasibility of a hyper-spectral system to be placed on board a small satellite, the first element of a low cost system, named PicoAgri, to monitor the status of agricultural fields at low costs.

At this level we verified the existence of COTS devices that are suitable for the system and acquired them, we designed and produced a box to contain a single detection sensor interfaced with an Arduino processing unit for the readout. The black box allows simulating the detection system since light comes from a slit and can be mounted on an optical bench that is used for all the measurements. We are calibrating the system by performing flat field, dark current and sensitivity of linear device exposures to properly characterize the system.

As a second step in the next future we are going to simulate and produce a prototype of the full scanning system, coupling the detection to the optical system, first as a monochromatic device, followed by a set of three linear cameras with ad hoc filters.

ACKNOWLEDGMENT

We acknowledge support by the University of Trieste through FRA (*Fondi Ricerca Ateneo*) 2014 funded activity "Osservazione e monitoraggio dei terreni: sistema di rivelazione per nano-satelliti".

REFERENCES

[1] R. Hevner, W. Holemans, J. Puig-Suari, and R. Twiggs, "An advanced standard for CubeSats," in Proc. AIAA/USU Conf. Small Satellites, 2011.

[2] Eber Huanca Cayo; Reginaldo Duarte; Thiago Da Silva "*Low cost yaw controller for CubeSat oriented to education and entertainment*" E International Symposium on Consumer Electronics (ISCE),Year: 2016,Pages: 107 - 108, DOI: 10.1109/ISCE.2016.7797393

[3] Jeremy Straub "Consideration of the versatility of the Open Prototype for Educational NanoSats CubeSat design" 2016 IEEE International Conference on Electro Information Technology (EIT) Year: 2016 Pages: 0586 - 0591, DOI: 10.1109/EIT.2016.7535305

[4] Keith Menezes; Tremayne Gomes "Obtaining infrared spectral imagery of the upper atmosphere using a cubesat" IEEE

Communications Magazine Year: 2015, Volume: 53, Issue: 5 Pages: 205 - 207, DOI: 10.1109/MCOM.2015.7105664

[5] Chantelle Dubois; Pawel Glowacki; Ahmad Byagowi "Biological investigations using a triple-cubesat" IEEE Communications MagazineYear: 2015, Volume: 53, Issue: 5 Pages: 211 - 213, DOI: 10.1109/MCOM.2015.7105666

[6] Constance Fodé; Jacopo Panerati; Prescilia Desroches; Marcello Valdatta; Giovanni Beltrame "Monitoring glaciers from space using a cubesat" IEEE Communications Magazine Year: 2015, Volume: 53, Issue: 5Pages: 208 - 210, DOI: 10.1109/MCOM.2015.7105665

[7] Therese Moretto "CubeSat mission to investigate ionospheric irregularities Space Weather" Year: 2008, Volume: 6, Issue: 11Pages: 1 - 2, DOI: 10.1029/2008SW000441

[8] Blocker, A and Litton, C and Hall, J and Romano, M, TINYSCOPE – "The Feasibility of a 3-Axis Stabilized Earth Imaging CubeSat from LEO", AIAA/USU Conference on Small Satellites 2008, https://calhoun.nps.edu/handle/10945/37323

[9] http://picosats.eu/ PicoSaTs SRL Padriciano, I- c/o Area Science Park 34149 TRIESTE ITALY. tel: +39 040 375 5445 e-mail: info@picosats.eu

[10] Chunming Peng; Meixia Deng; Liping Di "Relationships Between Remote-Sensing-Based Agricultural Drought Indicators and Root Zone Soil Moisture: A Comparative Study of Iowa" IEEE Journal of Selected Topics in Applied Earth Observations and Remote Sensing Year: 2014, Volume: 7, Issue: 11 Pages: 4572 - 4580, DOI: 10.1109/JSTARS.2014.2344115

[11] Frederic Jacob; Marie Weiss "Mapping Biophysical Variables From Solar and Thermal Infrared Remote Sensing: Focus on Agricultural Landscapes With Spatial Heterogeneity" IEEE Geoscience and Remote Sensing Letters Year: 2014, Volume: 11, Issue: 10 Pages: 1844 - 1848, DOI: 10.1109/LGRS.2014.2313592

[12] Clement Atzberger. "Advances in remote sensing of agriculture: Context description, existing operational monitoring systems and major information needs". In: Remote Sensing 5.2 (2013), pp. 949–981.

[13] Pietro Ceccato et al. "Detecting vegetation leaf water content using reflectance in the optical domain". In: Remote Sensing of Environment 77.1 (2001), pp. 22–33.

[14] Compton J. Tucker. "Remote sensing of leaf water content in the near infrared". In: Remote Sensing of Environment 10.1 (1980), pp. 23–32.

[15] John G. Nellist "Satellite Communications" Understanding Telecommunications and Lightwave Systems:An Entry-Level Guide Year: 2002 Pages: 63 - 77, DOI: 10.1002/0471722855.ch10 Wiley-IEEE Press eBook Chapters

[16] M. Drusch et al. , "Sentinel-2: ESA's Optical High-Resolution Mission for GMES Operational Services", Remote Sensing of Environment Volume 120, 15 May 2012, Pages 25–36 The Sentinel Missions - New Opportunities for Science

[17] Euroconsult www.euroconsult-ec.com/earthobservation

[18] Pixalytics www.pixalytics.com/how-many-eo-space/

[19] ESA Earth observation Market Development www.eomd.esa.int

[20] European Association of Remote Sensing Companies www.earsc.org

[21] EOVOX Study www.eovox.org

[22] https://de.wikiversity.org/wiki/Projekt:FE_Auswerteverfahren_1/ Vegetationsindizes/Strahlungstheoretische_Grundlagen

[23] http://www.geoimage.com.au/media/satellite_pdfs/Landsat_summ ary.pdf

[24] STK – System Tool Kit by AGI https://www.agi.com/

[25] Ardusat https://www.ardusat.com/

[26] E.Wolters, W.Dierckx, J.Dries, E.Swinnen " PROBA-V Products User Manual v1.1" Date 7/10/2014 http://proba-v.vgt.vito.be/sites/default/files/Product_User_Manual.pdf

Gap in pagination due to withheld paper.

Pages 1079-1083

Feeding a DNN for Face Verification in Video Data acquired by a Visually Impaired User

Jhilik Bhattacharya[a,b], Stefano Marsi[b], Sergio Carrato[b], Herbert Frey[c], and Giovanni Ramponi[b]

[a]Thapar University, India
[b]University of Trieste, Italy
[c]Ulm University of Applied Sciences, Germany

Abstract—Some experiments on a face verification tool based on FaceNet are presented in this paper. The task of the system is to perform face verification in a real-time assistive system aiming at facilitating the approach between a blind person and a preselected acquaintance of his/her who enters the field of view. Face detection is made easier by the fact that an almost frontal view of the face is highly probable; verification on the contrary is difficult due to the poor quality of the acquired images and to the necessity of achieving a very low error rate.
A custom database consisting of subjects required for verification is populated with face images provided by a suitable detection tool. The cascade of FaceNet and a Bayesian Classifier proves to be an effective tool for this unconstrained face verification task.

Index Terms—face detection, face verification, convolution neural network, face recognition

I. INTRODUCTION

We are developing a system for facilitating a blind person to interact with other people in a way similar to the one of a person with normal vision [1], [2]. The scenario we have agreed upon with the users is the one of a blind person who needs to meet one of his/her acquaintances in a public place, and is not willing to wait for the acquaintance to engage interaction e.g. by speaking: the users prefer to autonomously recognize the person they are meeting, in order to be able to behave consequently. In computer vision, this is a problem of face verification. To enable such a scenario, the system has to access the visual information of the surrounding environment and process it to extract information which gives an understanding of different non-verbal communication cues. Some of them may include the number of people in the scene, distance and position of identified people, physical appearance, gesture and expression of known people. This research devises the various steps needed to verify the presence or absence of particular people in the scene captured by the blind user, i.e. video acquisition, face detection, preprocessing, feature extraction and finally classification for end use.

The scene is simultaneously recorded by two commercial devices: one camera is mounted on the bridge of a pair of sunglasses, another is held by a short necklace on a light support. The glasses-mounted camera has a resolution of 1280 × 720 pixel and an angle of view of 135 deg.; the resolution and angle of view of the necklace-mounted camera are 1920 × 1080 pixel and 124 deg. respectively. In order to keep the prototype system close to its final goal,

several test videos used for experimentations and validations are actually recorded by users who are fully blind from birth. Consequently, the acquired video data suffer from geometrical distortion due to wide angle camera optics, back-lighting, and disturbances due to fast and wide movement of the blind person. As the user of course lacks any feedback about the subjects in the field of view of the cameras, faces can be partially occluded or partially outside the frame. Moreover, the field of view of both cameras may be partially occluded, typically by a tuft of hair or by a lapel of the dress.

The video sequences are acquired in different indoor and outdoor environments where it may be required to identify subjects. These include a university library, a coffee shop, the hall of a public building, the neighborhood of a bus stop. These scenes reflect some typical scenarios in terms of natural and artificial lighting and crowd where the user may have to find and approach his/her acquaintance [3]. Our research focuses on preprocessing detected faces from these video sequences and feeding a feature extractor which provides face representation embeddings for classification.

The performances of face recognition tools have gradually increased even in unconstrained situations with the application of biologically-inspired Deep Neural Networks (DNN), which have been shown to largely outperform shallow nets. The literature reports both the existence of standard DNNs trained with millions of face images and the continuous evolution in layer architecture and patch selection [4], [5], [6], [7], [8], [9], [10], [11].

The performance of recognition or verification is greatly influenced by the preceding face detection and preprocessing steps. In our system, faces are detected using PICO [16]; they are validated based on a quality parameter, preprocessed and then passed on to pretrained networks which are variants of FaceNet, developed by Google for feature extraction. The features extracted from the second last layer of the network are fed into a Bayesian classifier for the face verification tasks. The method is hence able to exploit the deep layered feature extraction of FaceNet and adapt it for recognition or verification with a classifier-training phase which uses a customized dataset. Moreover, we also analyse verification results of Euclidean distance classifiers on the two different FaceNet versions [11] and [4].

Novel contributions of this paper are related to the usage of truly-in-the-wild video data acquired by a blind user, a refined method for region of interest (RoI) preprocessing, and a study of intensity preprocessing methods and their effects. The rest of the paper is organized as follows. Section II discusses face detection and RoI processing; the feature extraction models are discussed in III, while classification results are given in Section IV; section V provides the conclusions and future directions of the current research.

II. FACE DETECTION AND RoI PROCESSING

A lot of datasets have been reported in the literature for face detection. These differ in their level of annotation detail, which may vary from a simple bounding box to few or more facial landmarks such as eyes, nose, lips etc. The use cases considered in our project require to work on video sequences and hence the need to cope with larger amounts of data. Moreover, the final deliverable in this respect includes eye-related detections for gaze estimation and pose modifications between successive frames as a suggestion of an intention to communicate; although this counts as a future direction and is not in the scope of the work discussed here, the currently used face detector is chosen to accommodate this scope as an add-on without major modifications.

Popular face detectors like Viola Jones [12], Visage [13], NPD [14], FaceID [15], PICO [16], and GMS Vision [17] were tested to get an idea of which works best for the current dataset. Indeed, as elaborated in [18], all the face detectors performed poorly in the considered sequences. PICO and NPD, however, provide a confidence value which can be utilized to successfully refine the outcome by discarding the detected regions for which a low confidence is obtained. PICO is chosen in our experiments as it gives a higher average precision compared to NPD for the data under consideration.

The PICO software reports a rectangular RoI and a score for each object found. When a face is actually present, PICO reports several RoIs in slightly different positions for the same face in each frame; these RoIs are reported in subsequent frames as long as the face is inside the scene. In turn, for false positive results PICO often reports only one or two RoIs in isolated frames. To get rid of false positive results we implement a filter that uses the RoIs and their scores as well as their occurrence in subsequent frames. For each RoI reported by the detector we test if this RoI belongs to an already existing face object of our filter by calculating the distance of the center of the region from the center of all the face objects. It the distance is below a given threshold, the score of the RoI is added to the score of this face object, but only up to a given maximum. If a RoI does not fit any of the already existing face objects, a new face object with the data of this RoI is generated. All face objects that were not hit by a new RoI in the current frame are penalized by subtracting a given value from their score; if the score is below zero the face object is deleted. Finally, all face objects with a score above a suitable threshold are reported as a positive result of the face detector for the current frame, as depicted in Figure 1.

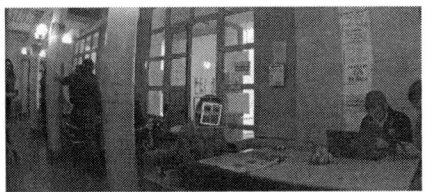

Fig. 1. Example of Face Detection.

III. FACE FEATURE REPRESENTATION

Deep convolutional networks have recently become the core of face recognition and verification tools even in unconstrained situations. The various networks reported in the literature differ in patch selection and network architecture. The models consist of multiple interleaved layers of convolutions, non-linear activations, local response normalizations, and max pooling layers. $1 \times 1 \times d$ convolution layers and inception models [4], [7] are two variants of using a large number of wide kernel sized filters for convolution in some deep layers; the latter consists in parallel mixed layers of convolutional and pooling layers concatenated together, and have been reported to provide almost twenty times reduction in time complexity and an improved feature representation.

In general, to use these networks for feature representation the CNN bottleneck layer output is further processed using PCA for dimensionality reduction and an SVM or Bayesian classification tool [19], [20], [21].

The performances of these different networks are improved in various ways:

- The input to the network is an aligned or frontalized face. DeepFace uses 3D frontalization to align the face, whereas OpenFace utilizes a 2D affine transform to align and get a tight crop of the face
- Different networks on different face patches or alignments are computed and their responses are combined. [21] combines the responses of 25 networks and predicts the distance using PCA and Joint Bayesian Model; [20] uses SVM to combine the predictions of three networks using different face alignments
- The network is trained with a combination of classification and verification loss [4], [22]. This also avoids the extra dimension reduction and the nonlinear classification tasks.

According to an analysis of popular networks in [23] the following facts are highlighted:

- Large fully connected layers are inefficient for small batches of images as the operations are better optimized over large matrices rather than small ones, hence utilizing resources more efficiently. For example AlexNet takes 84% of its inference time for batch size 1 and 33% for batch size 16

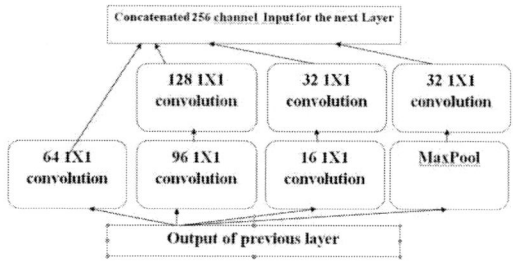

Fig. 2. Inception Layer 1:$Inception_1$.

Fig. 3. Face Detection without a tight face crop.

Layers	no.of filters,filter size
Spatial Convolution	48,11
Maxpool	3,2
ReLU	
Spatial Convolution	128,5
Maxpool	3,2
ReLU	
Spatial Convolution	192,3
ReLU	
Spatial Convolution	192,3
ReLU	
Spatial Convolution	256,3
Maxpool	3,2
ReLU	
View	9216
Linear	1024
ReLU	
Linear	1024
ReLU	
Linear	128
Normalize	128

TABLE I
NETWORK 1.

Layers	no.of filters,filter size
SpatialConvolution	64,7
Batch Norm	
ReLU	
MaxPooling	3,2
CrossMapLRN	
SpatialConvolution	64,1
Batch Norm	
ReLU	
SpatialConvolution	192,3
Batch Norm	
ReLU	
CrossMapLRN	
MaxPooling	3,2
$Inception_1$	64,1:(96,1|128,3):(16,1|32,5):32p
Inception	64,1:(96,1|128,3):(32,1|64,5):64p
Inception	(128,1|256,3):(32,1|64,5)
Inception	256,1:(96,1|192,3):(32,1|64,5):128p
Inception	224,1:(112,1|224,3):(32,1|64,5):128p
Inception	(160,1|256,3):(64,1|128,5)
Inception	384,1:(192,1|384,3):(48,1|128,5):128p
Inception	384,1:(192,1|384,3):(48,1|128,5):128p
AveragePool	1024
View	896
Linear	
Normalize	128

TABLE II
NETWORK 2.

- Accuracy and inference time are in a hyperbolic relationship: in general, model averaging is carried out for a better accuracy thus increasing the inference time. Consequently, the accuracy vs. inference time graph shows a steep slope which eventually flattens when cost complexity outgains accuracy
- Energy constraints are an upper bound on the maximum achievable accuracy and model complexity as it is obvious that in order to achieve a greater accuracy resource usage, power-consumption, and latency increase to a large extent
- The number of operations is a reliable estimate of the inference time.

In real-time situations, it may be meaningless to consider combined networks (for example the concatenation of 25 network outputs which extracts features from 25 different patches of a face) for performance elevation. Although alignment plays a crucial role, also the face detection phase has to be optimized in order to get good performances. Consequently, we need to balance the cost of the overhead (in terms of both time and computational effort) with improvements in accuracy.

The current work utilizes two models as shown in Table I and II. Network 1 is a modified version of FaceNet, which was kindly provided by the e-lab laboratory at Purdue University; Network 2 is the OpenFace network nn4.small2.v1 [11]. In Table I, we provide number of filters and filter sizes for the Spatial Convolution layers, window size and stride for the Maxpool layers, and output feature size for the View and Linear layers. The Inception Layer in Table II is a concatenation (indicated with the symbol ":") of two or four operations: $(f1, n1|f2, n2)$ denotes a layer with $f1$ and $f2$ filters of size $n1$ and $n2$. This is further shown in Figure 2. Both networks provide a 128-dimensional feature representation.

We feed the networks with the images of the detected faces; they are not aligned and may not contain a tight face crop (e.g. in case of a miss of the PICO preprocessing), as shown in Figure 3.

IV. CLASSIFICATION RESULTS

Even if the final goal of our study is face verification, the results we show are for top-1 face recognition in a set of 1700 faces; in this phase we found this approach more informative, since it does not require to set a threshold to determine the reliability of the system. The results are analysed with three different classifiers; the effect of some basic preprocessing on the images fed to the network is also evaluated. It should be noted that an initial preprocessing is first carried out to obtain the best possible crop of the face RoI; this is followed by histogram and other normalizations of the cropped faces.

The ground truth for the various classifiers consists in 11 classes having 32 images each. 11 different sets of scene sequences are used for testing. After face detection, the presence or absence of a particular person is searched for in each scene; then, the detected face images are preprocessed, and finally they feed the network. The network requires a 3-channel input and is tested with (a) a gray image on all the three channels and (b) an RGB color image; it may be observed (Table III) that RGB color images provide a better average performance with a much lower standard deviation between different test sets. Even though grayscale images show some improvement for a few sets, their performances on others (2, 7, 10) are poor, so that the use of RGB images seems to be a better and stable solution.

As already mentioned, three kinds of classifiers are used, namely Bayesian (B), Euclidean Distance (E), Euclidean Distance with the mean face of each class (EM). E and EM are computed as follows:

$$E = \arg\min_{i=1}^{n}(\overline{f_{test_x} - f_{train_i}}) \quad (1)$$

$$EM = \arg\min_{i=1}^{c}(\overline{f_{test_x} - f_{train_i}}) \quad (2)$$

where f_{train_i} is the feature vector for each sample in Eqn. 1 and the mean feature vector of each class in Eqn. 2, respectively for n samples and c classes.

The results are shown in Table IV. As shown in Equation 2, Euclidean mean provides better results with respect to standard Euclidean distance 1) for most cases. The performances of Bayesian and Euclidean mean are almost the same with the exception of Sets 2 and 7. In Figure 4) it may be seen that Set 2 has little or no variations in terms of pose and illumination but faces are not tightly cropped in most cases, whereas Set 7 has huge variations in terms of scale and illumination. Consequently, we suppose that the Bayesian classifier provides the best verification results on using tight image crops, while the Euclidean mean distance classifier performs better with aligned images.

The different preprocessing variants used in this work include normalized histogram equalization (HN) and normalized average histogram equalization (HAN) (mean of original and equalized image). Two normalizations have been tested, namely global normalization of the face according to neural network training data (NT), and classifier training set mean and standard deviation (CT). As shown in Figure 5-a,-b, the

Set	Gray	RGB
1	100	100
2	74.68	87.34
3	98.27	86.20
4	94.54	92.59
5	96.77	100
6	98.12	95.78
7	47.82	94.88
8	88.88	87.30
9	96.55	94.31
10	84.37	94.89
11	97.82	95.58
mean	88.89	93.53
std	15.61	4.78

TABLE III
PERFORMANCE (ACCURACY IN %) OF BAYESIAN CLASSIFIER FOR FACE RECOGNITION ON NORMALIZED GRAYSCALE AND COLOR IMAGE.

Set	Bayes	Euclidean	Euclidean mean
1	100	100	96.87
2	79.74	92.40	92.40
3	91.37	100	98.27
4	94.44	96.29	96.29
5	100	100	100
6	97.19	97.54	97.89
7	96.69	72.77	89.43
8	87.30	88.88	88.88
9	96.59	97.72	96.59
10	97.95	96.93	96.93
11	91.17	100	92.64
mean	93.85	94.77	95.11
std	6.12	8.10	3.68

TABLE IV
PERFORMANCE (ACCURACY IN %) OF BAYESIAN, EUCLIDEAN AND EUCLIDEAN MEAN CLASSIFIER FOR FACE RECOGNITION ON NORMALIZED COLOR IMAGE.

global histogram equalization and global average histogram normalization are computed over three different kinds of data referred to as RGB-RGB, RGB-HSV, RGB-YUV, where the first part of the name denotes the color channel data that feeds the network whereas the second part denotes the color channel used to perform the histogram equalization. For example, in case of RGB-HSV the RGB image is converted to the HSV color space and the V channel is normalized before converting the image back to RGB and feeding the network. Using the Bayesian classifier, we verified that HN,NT on RGB-RGB and HAN,NT on RGB-YUV provide better performances with lower deviation between sets. The best performing categories are also evaluated using the two other classifiers as shown in Table V. The results show that histogram normalized images with NT data give the best results with a Bayesian classifier. Normalization (N) results on RGB images using the different classifiers are depicted in Figure 6. It may be seen that all the classifiers give the same average performance; however, the Euclidean mean has a lower standard deviation across the different datasets and can be considered the best one. It may be also observed that the deviation in verification accuracy results among the different test set is larger when using CT normalization instead of NT normalization for E and EM; it is

(a) Snapshots from Set 2

(b) Candidate 1 for identification

(c) Cropped face by face detector

Fig. 4.   Snapshots and their RoI detections.

Set	HN NT-E	HAN NT-B	HN NT-B	HN NT-EM	HAN NT-EM	HAN NT-E
1	100	100	96.87	93.75	96.87	100
2	96.20	93.67	92.40	97.46	97.46	96.20
3	96.55	81.03	89.65	96.55	86.2	94.82
4	98.14	98.14	90.74	90.74	100	100
5	100	100	100	100	100	100
6	97.54	98.94	97.54	97.19	98.94	98.24
7	66	94.71	93.23	86.30	89.43	79.53
8	93.65	85.71	96.82	92.06	85.71	87.30
9	92.04	94.31	89.77	90.90	90.90	93.18
10	97.95	94.89	95.91	96.93	91.83	96.93
11	95.58	95.88	92.64	89.70	92.64	98.52
mean	93.97	94.27	94.14	93.78	93.64	94.97
std	9.58	5.94	3.47	4.17	5.3	6.36

TABLE V
PERFORMANCE (ACCURACY IN %) OF BAYESIAN CLASSIFIER FOR FACE
RECOGNITION ON COLOR IMAGE INPUTS AFTER APPLYING DIFFERENT
KINDS OF PREPROCESSING TECHNIQUES.

just the opposite for the Bayesian classifier. Consequently, we
decided to use the Bayesian classifier with CT normalization
for the comparison of the two networks, shown in Table III.

In Figure 7 it may be seen that the OpenFace network
performs poorly when compared to Network 1. This may be
due to underlying network differences, the fact that the images
are not aligned as expected by the network, and the use of 96
X 96 patches of already very poor quality images, whereas
the other network utilizes 240 X 240 patches. We are aware
of the fact that the small amount of experiments we have
performed does not permit to draw final conclusions about
the performances of the system we are building. However, we

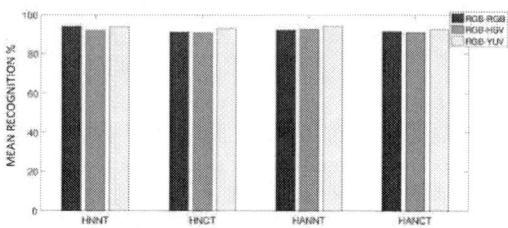

(a) Mean of recognition accuracies (in %) using different histogram equalizations

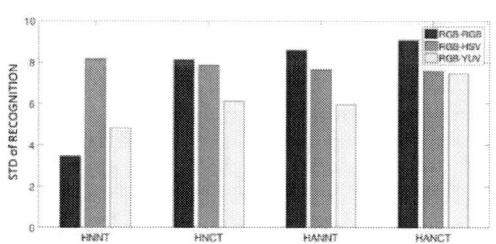

(b) STD of recognition accuracies using different histogram equalizations

Fig. 5.   Preprocessing results.

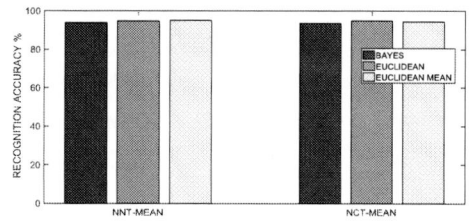

(a) Mean of recognition accuracies (in %) using different kinds of
normalization

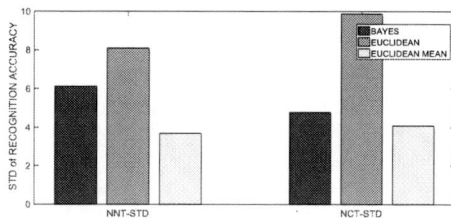

(b) STD of recognition accuracies across sets using different kinds
of normalization

Fig. 6.   Effect of different kinds of normalization on CNN Input.

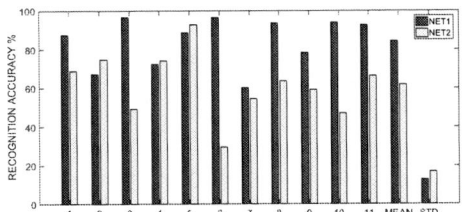

Fig. 7. Face recognition performance with two different models using the Euclidean classifier.

think that useful indications are already present (e.g. the larger or smaller standard deviations of the performances) that can guide a reader interested in the design of a system in this field of application.

V. CONCLUSION

The identification of faces in video sequences captured by devices worn by a blind person has been studied in this paper. This includes face detection using PICO filter, preprocessing, and face representation using deep convolution filters. The already challenging task of face recognition/verification from an unconstrained environment is further aggravated here by the fact that the faces may be partially or completely occluded and have to be detected from videos subject to backlighting, low resolution, distortion due to the use of wideangle cameras, and fast movements. Although the overall performance of the face detectors in such scenarios is quite poor, the current task is motivated by the fact that the user mainly has to recognize or verify faces who are approaching, interacting or at least looking directly at him/her; this favors the condition to a certain extent. The current work analyses the performance of two convolution models without face alignment (2D or 3D). The obtained results are satisfactory considering the single unaligned patch approach and promise considerable improvement subject to alignment operations, even if realtime implementation issues have to be taken into account. Indeed, for the same reason the authors do not plan to consider a multi-patch feature extraction for performance enhancement. Work is in progress towards testing the effect of 3D frontalization and 2D affine transformation on the detected faces before feature extraction. As the OpenFace network is supposed to be trained with aligned faces, the former step may significantly enhance the performance on our dataset. Moreover, the effect of finetuning on the networks instead of training a classifier will also be analyzed.

VI. ACKNOWLEDGEMENT

This work has been supported by the University of Trieste - Finanziamento di Ateneo per progetti di ricerca scientifica - FRA 2014, and by a private donation in memory of Angelo Soranzo (1939-2012).
The authors also thank Eugenio Culurciello and Alfredo Canziani for kindly providing Network 1.

REFERENCES

[1] S. Carrato, G. Fenu, E. Medvet, E. Mumolo, F.A. Pellegrino, G. Ramponi, "Towards More Natural Social Interactions of Visually Impaired Persons", Int. Conf. on Advanced Concepts for Intelligent Vision Systems, ACIVS 2015, Catania, Italy, Oct. 26-29, 2015

[2] S. Carrato and G. Ramponi, "Assistive technologies based on image processing for people with visual impairments: a tool to help social interaction of the blind," in Proc. UNIversal Inclusion Rights and Opportunities for Persons with Disabilities in the Academic Context, (Torino (Italy)), May 2016

[3] S. Carrato, S. Marsi, E. Medvet, F. A. Pellegrino, G. Ramponi, and M. Vittori, "Computer vision for the blind: a dataset for experiments on face detection and recognition," in Proc. MIPRO 2016 - 39th International Convention on ICT, Electronics and Microelectronics, (Opatija (HR)), 30 May - 3 June 2016

[4] Florian Schroff, Dmitry Kalenichenko, James Philbin, "FaceNet: A Unified Embedding for Face Recognition and Clustering", Proceedings of the IEEE Computer Society Conference on Computer Vision and Pattern Recognition 2015

[5] K. Kavukcuoglu, P. Sermanet, Y. Boureau, K. Gregor, M. Mathieu, and Y. LeCun, "Learning convolutional feature hierarchies for visual recognition," in Proc. NIPS, 2010, pp. 1090-1098

[6] Tsung-Han Chan, Kui Jia, Shenghua Gao, Jiwen Lu, Zinan Zeng, and Yi MaPCANet: "A Simple Deep Learning Baseline for Image Classification", IEEE Trans. on Image Processing, vol. 24, no. 12, Dec. 2015.

[7] Christian Szegedy, Vincent Vanhoucke, Sergey Ioffe, Jonathon Shlens, "Rethinking the Inception Architecture for Computer Vision", arXiv:1512.00567v3 [cs.CV] 11 Dec 2015

[8] Y. Taigman, M. Yang, M. Ranzato, and L. Wolf "Deepface: Closing the gap to human-level performance in face verification". In CVPR, pages 1701-1708, 2014.

[9] M. D. Zeiler and R. Fergus. "Visualizing and understanding convolutional networks". CoRR, abs/1311.2901, 2013. 2, 4, 6

[10] C. Szegedy, W. Liu, Y. Jia, P. Sermanet, S. Reed, D. Anguelov, D. Erhan, V. Vanhoucke, and A. Rabinovich. "Going deeper with convolutions". CoRR, abs/1409.4842, 2014. 2, 4, 5, 6, 9

[11] B. Amos, B. Ludwiczuk, M. Satyanarayanan, "Openface: A general-purpose face recognition library with mobile applications," CMU-CS-16-118, CMU School of Computer Science, Tech. Rep., 2016.

[12] Viola, P., Jones, M.J., "Robust real-time face detection". International journal of computer vision 57(2) (2004) 137154

[13] Visage Technologies- Face Tracking and Analysis, https://visagetechnologies.com/products-and-services/visagesd

[14] Liao, S., Jain, A.K., Li, S.Z., "A fast and accurate unconstrained face detector". IEEE Transactions on Pattern Analysis and Machine Intelligence 38(2) (2016)

[15] Dundar, A., Jin, J., Martini, B., Culurciello, E., "Embedded streaming deep neural networks accelerator with applications". IEEE Transactions on Neural Networks and Learning Systems (2016)

[16] Markus, N., Frljak, M., Pandzic, I.S., Ahlberg, J., Forchheimer, R., "Object detec- tion with pixel intensity comparisons organized in decision trees". arXiv preprint arXiv:1305.4537 (2013)

[17] Google Developers, https://developers.google.com

[18] M. De Marco, G. Fenu, E. Medvet, and F.A. Pellegrino, "Computer Vision for the Blind: a Comparison of Face Detectors in a Relevant Scenario," in Proc. GoodTechs 2016 - 2nd EAI International Conference on Smart Objects and Technologies for Social Good, Venice, Italy, November 30 - December 1, 2016.

[19] Z. Zhu, P. Luo, X. Wang, and X. Tang. "Recover canonicalview faces in the wild with deep neural networks". CoRR, abs/1404.3543, 2014. 2

[20] Y. Taigman, M. Yang, M. Ranzato, and L. Wolf. "Deepface: Closing the gap to human-level performance in face verification". In IEEE Conf. on CVPR, 2014. 1, 2, 5, 8

[21] Y. Sun, X. Wang, and X. Tang. "Deeply learned face representations are sparse, selective, and robust". CoRR, abs/1412.1265, 2014. 1, 2, 5, 8

[22] K. Q.Weinberger, J. Blitzer, and L. K. Saul. "Distance metric learning for large margin nearest neighbor classification". In NIPS. MIT Press, 2006. 2, 3

[23] Alfredo Canziani & Eugenio Culurciello, "An Analysis of Deep Neural Network Models for Practical Applications", arXiv:1605.07678v2 [cs.CV] 30 May 2016.

Acquiring ISAR Images using Measurement Instruments

Hyeon-Cheol Lee*, Sang Gye Lee *, Seung Hoon Lee*, and Chul Ho Jung**
* Payload Electronics Team, Korea Aerospace Research Institute, Daejeon, 34133, Rep. of Korea
** HyperSensing, Daejeon, 34133, Rep. of Korea
hlee@kari.re.kr, sglee@kari.re.kr, shlee@kari.re.kr, chjung@hypersensing.net

Abstract - We introduce the image acquisition from X-band Inverse Synthetic Aperture Radar (ISAR) system using measurement instruments for chirp generation and signal modulation in this paper. We design and develop a power amplifier and a signal demodulator, but we use an Arbitrary Waveform Generator (N8241A, Keysight) for chirp generation and a Vector Signal Generator (E8267D, Keysight) for signal modulation. Moreover, 'Andale', the Commercial-Off-The-Shelf (COTS) for data recording and Matlab for image post-processing are used. Then, we integrate all together for ISAR image acquisition and acquire vivid ISAR images. The final resolutions of acquired ISAR images are 0.41m range, 0.14m azimuth, the PSLRs are -22.8dB range, -22.3dB azimuth, the ISLRs are -20.2dB range, -17.1dB azimuth, better than the target specification.

I. INTRODUCTION

There are many devices to observe subjects on the earth from a satellite, such as an optical camera, an infrared camera, a Synthetic Aperture Radar (SAR) [1, 2], etc. The optical camera can show targets clearly, but it has limitation when it is rainy, cloudy, snowy, and dark. The Infrared camera can observe targets in the night though, it also cannot distinguish objects when it is rainy, cloudy, and snowy. The SAR, however can observe and monitor the targets regardless of rain, cloud, snow, and dark night. In addition, it can detect underground objects at certain depth, such as ground tunnels. But, the weak point of the SAR image is that the target image is not clear as much as the optical camera image.

The SAR payload of Korea Multi-Purpose Satellite (KOMSAT)-5 made by Thales Alenia Space Italy was launched and is working successfully and then, the development of SAR payload of KOMSAT-6 are driven by Korean industry so that the interest and demands of SAR developments are increasing now in Korea.

Pieraccini [3] et al. suggested ground based radar techniques. Pieraccini [4, 5] and Ji [6] also suggested the Ground-based SAR (GB-SAR) using RailSAR for topographic mapping application. Nico et al. [7] and Leva et al. [8] suggested the interferometry test with GB-SAR systems. Cho [9] had designed and developed GB-SAR mounted on an automobile. In this paper, we construct the Inverse Synthetic Aperture Radar (ISAR) system using

measurement instruments for its transmitter and perform the [1, 2, 10, 11] ISAR image acquisition test.

The ISAR equipment consists of a chirp generator, a signal modulator, a power amplifier, antenna, a signal demodulator, a signal processor, and post-processing by software for images. In this test, we use measurement instruments for the functions of chirp generation and modulation. Using measurement instruments saves monetary budget and manufacturing time.

This paper organizes as follows. Section II introduces ISAR equipment and test scheme, section III shows the test concept of ISAR and describes test results. Section IV describes resolution measurement of the system. Section V concludes.

II. DESCRIPTION OF TEST SCHEME

The equipment (see Figure 1) of ISAR consists of the chirp generator that produces linear FM chirp pulses, the signal modulator that converts its signal to X-band center frequency, the power amplifier to amplify the magnitude of the signal, the antenna to emit its energy, and the signal demodulator that demodulates and detects the reflected signals, the signal processor that stores in order and processes to be sent to the ground by datalink. In addition, the post-processing software manages the ISAR image with the given raw data.

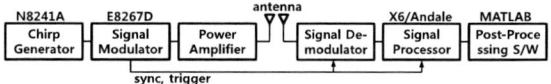

Figure 1: Block diagram of ISAR system

For chirp generation and modulation, we use measurement instruments, an Arbitrary Waveform Generator (N8241A, Keysight) that has an option of linear FM chirp pulse generation and a Vector Signal Generator (E8267D, Keysight) to convert baseband signal to X-band. We make use of COTS, 'Andale' (high speed customized 24TByte) with X6-GSPS FPGA for its signal processing and data recording. Using these measurement instruments and COTS has given us some advantages, such as stable triggers and changing bandwidth, carrier frequency, PRF, etc. easily.

Finally, we employ range compression, range cell migration compensation, and azimuth compression from the range doppler algorithm [1, 2] by Matlab for post-processing to produce ISAR images.

Hence, we design and develop a power amplifier and an X-band signal demodulator, subsequently interface them with each equipment. We apply IF sampling and Digital Down Conversion (DDC) scheme here(the sampling rate is over gigahertz.). System parameters are summarized in Table 1.

TABLE I.   PARAMETERS OF THE ISAR SYSTEM

	Method #1	Method #2	Method #3
Center Freq.	X-band		
Bandwidth	500MHz		
IF Freq.	xxxMHz		
PRF	1000Hz		
Pulse Width	xxxus		
Max. Range	4~5km	7~8km	60~70km
Power Amp.	10W	70W	70W
Ant. Gain	25dBi	25dBi	45dBi
Beamwidth	20deg	20deg	1deg

III.   ISAR TEST

Figure 2: Configuration of ISAR test with two 25dBi antenna for acquiring a ship image on the bridge

We carry every equipment inside van like Figure 2. We mount two antenna for transmitting and receiving respectively that means we do not use a circulator, so this scheme reduces RF losses. Because ISAR needs the doppler effect of the target, we position antenna along with the horizontal direction of the moving target. Several ISAR images are obtained.

A.   Method #1 (10W, 25dBi)

Figure 3 is an ISAR image of a container ship. The 100m-length, 20m-width ship we take pictures on the bridge, approximately 380m away is shown clearly by

radar pulses. The velocity that we estimate is 21.5 km/h. The flagpole of the ship is dimly visible, because we believe the size of flagpole is relatively smaller than body size of the ship itself.

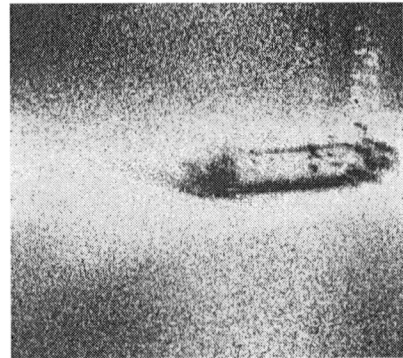

Figure 3: ISAR image of ship (velocity 21.5km/h, distance 380m)

The Figure 4 is ISAR images of an airplane. Target images is 380m away, especially about 4m-long engines and 34m-long wigs are shown.

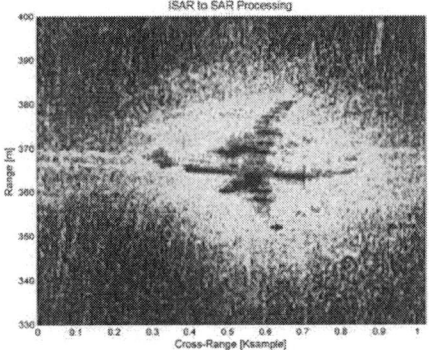

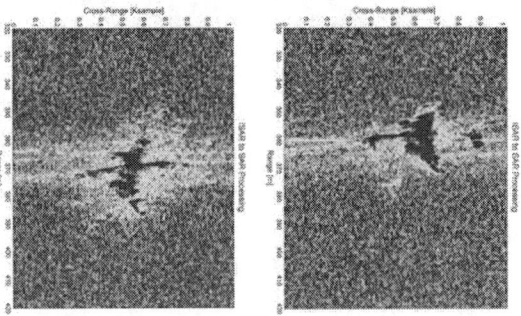

Figure 4: ISAR images of Aircrafts (length 35m, velocity 250km/h, distance 390m)

B.   Method #2 (70W, 25dBi)

In this test, we use the 70W power amplifier instead of 10W for a longer distance. At this time, the target is 6.5km away like Figure 5.

1091

Figure 5: Actual target image (red circle, length 60m, wingspan 60m)

The right picture of Figure 6 is an ISAR image out of focus, because analyzing velocity 342km/h we used is 10km/h off from nominal velocity. The left image, velocity matched one shows clear the figure of aircraft.

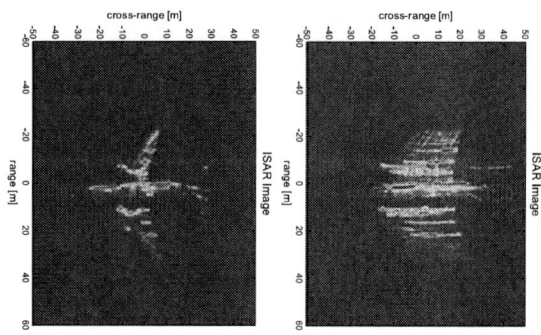

Figure 6: ISAR image of an Aircraft (velocity 352km/h, distance 6.5km) (left), Velocity unmatched image (right)

C. Method #3 (70W, 45dBi)

In this test, instead of 25dBi antenna, we use the 2.4m-diameter antenna whose gain is 45dBi in the right picture of Figure 7. At this time, the target is a Cessna 208, smaller than the previous airplane.

Figure 7: 2.4m antenna (left) (45dBi) and actual target image (right) (Cessna 208, length 12.6m, wingspan 15m)

Figure 8 shows an ISAR image of the Cessna. Due to the small sized target and narrow beamwidth of high gain antenna, this ISAR image is not much clearer than the image of Figure 6.

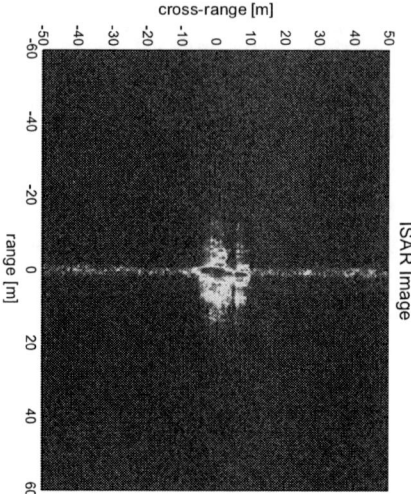

Figure 8: ISAR image of Cessna208 (velocity 240km/h, distance 3.5km)

IV. RESOLUTION MEASUREMENT

We measure the resolution of this system with Impulse Response Function (IRF), the range resolution is 0.41m, the azimuth resolution is 0.14m in Table 2. These results are better than the target resolution specification, we believe signal processing with windowing and filtering gives positive effects to the result.

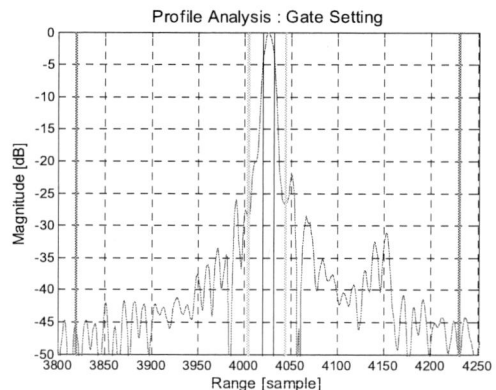

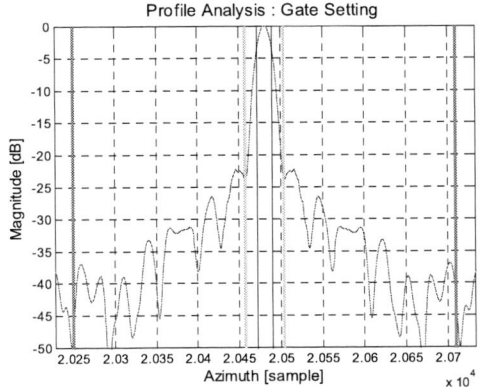

Figure 9: IRFs of Range (left), Azimuth (right)

TABLE II.   SYSTEM RESOLUTIONS OF THE ISAR TEST

parameter	Range		Azimuth	
	target	measure	target	measure
Resolution	0.5m	0.41m	0.4m	0.14m
PSLR	-12.8dB	-22.1dB	-12.8dB	-22.3dB
ISLR	-8.8dB	-20.2dB	-8.8dB	-17.1dB

V. CONLUSION

We design and develop ISAR system and try to get images from the moving target. We simply compose X-band ISAR system of the measurement instruments (Keysight) for chirp generation and modulation, and COTS. We, furthermore use the range doppler algorithm by Matlab for the post-processing of ISAR images so that those cost effectively and save manufacturing time. After measuring IRF, we finally get 0.41m range resolution, 0.14m azimuth resolution, better than our target resolution (0.5m(range), 0.4m(azimuth)) specification.

ACKNOWLEDGMENT

This paper was performed for the IRND (Core Technology Development of Next Generation SAR Payload) of KARI, Korea Aerospace Research Institute.

REFERENCES

[1] I. Cumming and F. Wong. Digital Processing of Synthetic Aperture Radar Data, Artech House, Boston, 2004.

[2] J. Curlander and R. McDonough. Synthetic Aperture Radar Systems and Signal Processing, John Wiley & Sons, Inc., New York, 1991.

[3] M. Pieraccini, G. Luzi, D. Mecatti, and C. Atzeni. "A ground based remote sensing radar technique for dynamic testing of large structures", Int'l Geosci. and Remote Sens. Symposium(IGARSS) 2003, Toulouse, France, pp. 4326-4328, July 2003.

[4] M. Pieraccini, G. Luzi, and C. Atzeni. "Ground-based interferometric SAR for terrain elevation mapping", IEEE Electronics Letters, vol. 36, no. 16, pp. 1416-1417, Aug. 2000.

[5] M. Pieraccini, G. Luzi, and C. Atzeni. "Terrain mapping by ground-based interferometric radar", IEEE Trans. Geosci. Remote Sens., vol. 39, no. 10, pp. 2176-2181, Oct. 2001.

[6] Y. Ji, H. Han, and H. Lee. "Construction and application of tomographic SAR system based on GB-SAR system", Int'l Geosci. and Remote Sens. Symposium(IGARSS) 2014, Quebec, Canada, pp. 1891-1894, July 2014.

[7] G. Nico, D. Leva, G. Antonello, and D. Tarchi. "Ground-based SAR interferometry for terrain mapping: Theory and sensitivity analysis", IEEE Trans. Geosci. Remote Sens., vol. 42, no. 6, pp. 1344-1350, June 2004.

[8] D. Leva, G. Nico, D. Tarchi, J. Fortuny-Guasch, and A. Sieber. "Temporal analysis of a landslide by means of a ground-based SAR interferometer", IEEE Trans. Geosci. Remote Sens., vol. 41, no. 4, pp. 745-752, April 2003.

[9] Byung-Lae Cho, Young-Kyun Kong, Hyung-Geun Park, and Young-Soo Kim. "Automobile-based SAR/InSAR system for ground experiments", IEEE Geosci. Remote Sens. Letters, vol. 3, no. 3, pp. 401-405, July 2006.

[10] P. Spudis, S. Nozette, B. Sussey, K. Raney, H. Winter, C. Lichtenberg, W. Marinelli, J. Crusan, and M. Gates. "Mini-SAR, An imaging radar experiment for the Chandrayaan-1 mission to the Moon", Current Sci., vol. 96, no. 4, Feb. 2009.

[11] N. Agrawal and K. Venugopalan. "Saturation adaptive quantizer design for synthetic aperture radar data compression", Int'l Journal on Comp. Sci. and Eng., vol. 02, no. 01S, pp. 76-81, 2001.

Segmentation of Kidneys and Abdominal Images in Mobile Devices with the Android Operating System by Using the Connected Component Labeling Method

Seda Arslan Tuncer*, Ahmet Alkan**

* Firat University, Department of Software Engineering, 23119 Elazig, Turkey
** Kahramanmaras Sutcu Imam University, Department of Electrical and Electronic Engineering,
Kahramanmaras, Turkey
satuncer@firat.edu.tr, aalkan@ksu.edu.tr

Abstract - The purpose of this study was the segmentation of kidneys and abdominal images to assist the diagnosis and to focus on the required area. Kidney segmentation from abdominal images is not an easy task due to the proximity of those organs in the image, the similarity of organ tissues and the occurrence of different properties of the image in each cross-section. In this study, a fully automatic approach was suggested for the kidney segmentation in abdominal computed tomography (CT) images. Both the success of the suggested approach was tested and the performance of the process was evaluated. Area Error Rate (AER) criteria were used to reveal the accuracy of the segmentation operation. Because the vertebral column was used as the reference in the suggested approach, the coordinates of the vertebral column were determined by applying pre-processing to the images. In the second step, the kidney areas were obtained using the Connected Component Labeling (CCL) method. The final step of the study included transferring the operations performed on a PC to a mobile platform. The results obtained reveal that the suggested methodology is a kidney segmentation process that experts can use.

I. INTRODUCTION

Imaging for diagnosis has become a significant subject because detailed information about anatomy is now possible by using various medical imaging methods. Data presented by imaging devices can be easily processed by separating the desired tissues from the rest of the data. Segmentation and imaging are two very close fields that are used in numerous radiological applications. Side effects caused by the treatment are diminished if there is an early diagnosis of the illnesses at the onset, which increases the rate of recovery from the illness. Moreover, early diagnosis decreases the cost of the treatment, and the time and effort spent on treatment.

There are some instances where the data obtained from imaging devices used in radiology are insufficient for medical doctors to make a diagnosis. Therefore, such data must be processed to make it more understandable. The segmentation method of computed tomography (CT) images obtained from different phases is an effective method to find lesions in kidneys and to characterize them.

Segmentation is defined as the separation of any pattern in the image, or any part of the image, from the remaining parts [1]. Segmentation in the medical image-processing field has been widely used in studies of anatomical structures, treatment planning and computer-assisted surgery [2]. Literature shows that the methods developed are currently semi-automated and user-interactive methods. The most significant disadvantage of these is the introduction of a delay in diagnosis. Thus, automatic segmentation is usually preferred by experts.

In this study, unlike others, anatomical characteristics of the abdominal region were utilized. Right and left kidneys were separated by using the coordinates of the vertebral column obtained from its segmentation. The Connected Component Labeling (CCL) method was used for automated kidney segmentation. The performance of this method was evaluated by using an Area Error Rate (AER) method. The following studies were then transferred to mobile media for use on other larger platforms. The materials and the method used in the study are explained in the second section. Next, the performance of the method was evaluated, and experimental results are given. In the third section, the infrastructure of the mobile platform is described and the use of the system on mobile media is explained using schematics. In the last section, the results obtained are discussed.

II. MATERIALS AND METHOD

Thirty patient images were obtained from the image archival storage system of the Department of Medical Faculty Radio-diagnosis, Firat University. CT images were taken in the portal phase, following transfusing using an opaque substance. The image matrices have dimensions of 512 x 512 pixel and are in color. They are in the DICOM format. The kidney images obtained were manually segmented by an expert radiologist to obtain references images.

In the pre-processing step of the study, morphological operations were applied to the images. The purpose of these operations was to ensure the success in determining the vertebral column in the kidney segmentation. The segmentation steps are shown in Figure 1.

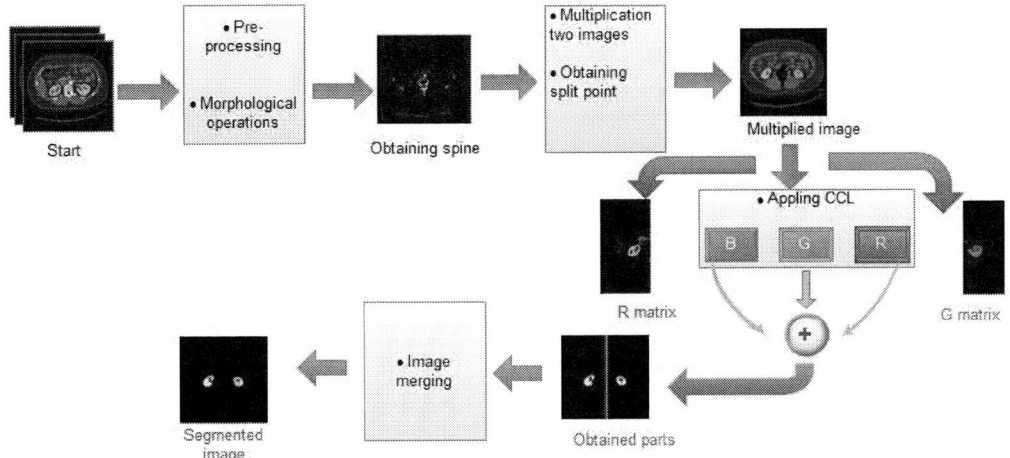

Figure 1. Segmentation steps of kidneys from abdominal CT images

A. Connected Component Labeling Algorithm

In the second step, the CCL algorithm was used for kidney segmentation. CCL is an algorithm that first labels combined objects in the image, then collects neighboring pixels in a group and finally makes them distinguishable. After grouping, each group on the image is numbered to represent an object. Therefore, the desired result is possible by distinguishing the desired group-numbered object from the others. The CCL algorithm is divided into two groups: 4-neighborhood and 8-neighborhood. Adjoining pixels are included in the operation as neighbors because 8-neighborhood is preferred in many applications [3, 4, 5]. The algorithm gives the pseudo code for the object-numbering for 1 to 8 neighborhoods. The flowchart of the segmentation using the CCL method is shown in Figure 2.

Algorithm.1. Numbering objects by CCL

Step 1: Scan all pixels. If the pixel is not equal to black,
Step 2: Scan all the neighbors of the pixel,
Step 3: If all neighbors are either black or white, that is a new pixel; assign a new label to the pixel
and go to another pixel,
Step 4: If at least one neighbor is labeled (not black or white); assign the least labeled neighbor to
that pixel and save the other labels as the same
Step 5: Go back to step 2

R, G, and B values are obtained for each part after using the CCL algorithm. Segmented kidney images are obtained by assuming that the largest component in certain channels is the kidney. The results obtained are again combined from the split point used in the first step with the original images until dimensions of 512 x 512 pixel are obtained.

B. Area Error Rate

Area error rate (AER) is a method indicating the percentage rate of change of the difference between the segmented areas of the image to be compared [6,9]. The method uses the area manually selected by a medical doctor and the area automatically found by the suggested method. AER is calculated based on equation.1 [7]. In this equation, M indicates the area manually segmented, while A indicates the area obtained by the suggested method.

$$AER = \frac{(M \cup A) - (M \cap A)}{M} * 100 \quad (1)$$

Table 1 shows the performance evaluations of both right and left kidneys. An error rate for 30 right kidneys was 9.74%, while for 30 left kidneys it was found to be 13.1%.

Table 1. AER average values for the right and left kidney

	Right Kidney	**Left Kidney**
	CCL	CCL
AER	9.74	13.1

The discrepancy between the error rate of right and left kidneys is due to the diversity of organs surrounding both kidneys. This value indicates approximately a 15% error rate for both kidneys, but it also reveals an 85% success rate.

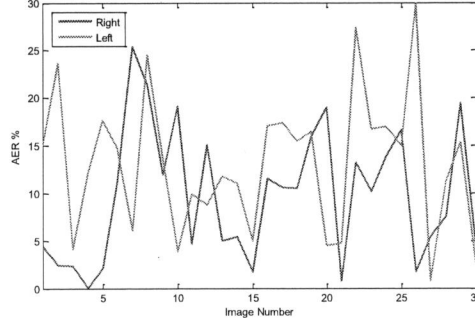

Figure 2. AER values obtained for each image

III. ANDROID APPLICATION

Transfer of the study to the mobile platform is a bilateral process. Eclipse IDE and Android SDK software was used for the client application, while MATLAB, PHP and MySQL software was used for the server. Figure 3 show the server and client structure that was developed.

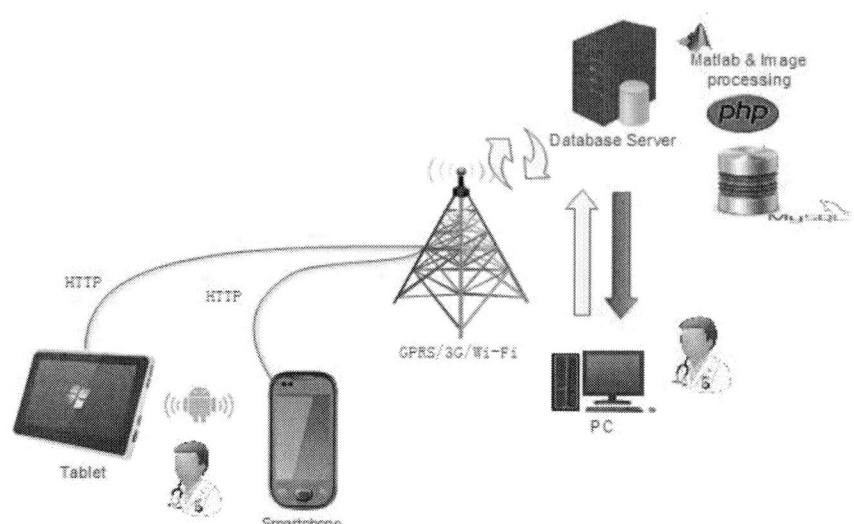

Figure 3. The server and client structure

The first step in the application is to select patient information and display the corresponding CT data. A medical doctor can send the data to the system over the mobile device or he/she can connect to the system over the mobile device by sending the patient information. Then, the application is run via a command issued by the mobile device. Many image processing steps are used in the application.

In this study, segmentation of the right and left kidneys was conducted on axial abdominal CT images. Next, the segmented image was displayed on the mobile device. Figure 4 shows the original image on the mobile device. After taking the original image, the system performs the segmentation operation without needing a separate starting point for each piece of data. Figure 5 indicates the segmented image on the mobile device.

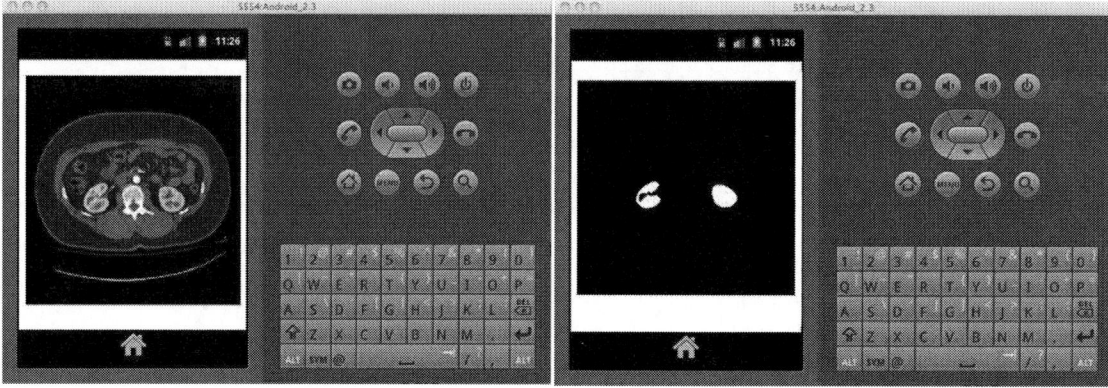

Figure 4. Original image

Figure 5. Output of mobile device

IV. CONCLUSION

Lesions on kidneys can appear without warning because they do not show any signs for a long time. It is known that 1 one person out of approximately 10,000 people will get kidney cancer, and 1 person out of 30,000 dies due to this illness [8]. Because the onset of the illness is rarely noticed by the patient, early diagnosis on abdominal images is of vital importance. The purpose of the study was to detect any cyst in the kidneys before it reaches an advanced stage by segmenting the abdominal images and perform this operation on mobile media independent of any platform. The CCL algorithm was used for kidney segmentation in CT images whose contrasts were increased. AER success criteria were used

to evaluate the performance of the operation. When the segmentation results obtained by the manual and suggested methods are compared, an 85% success rate was achieved. To use the application in a wider environment, a mobile platform was selected that is widely used in daily lives. It was possible to focus on a desired area in the images obtained by segmenting the abdominal images; thus, a successful diagnosis could be made. The present study constitutes the first step in studying kidneys to diagnose the presence of any kidney cysts. Moreover, medical doctors can access data independently from hospital systems and their working areas, which provides a convenient way for doctors to work outside the hospital. The system studied will serve many purposes and provide a basis for further studies.

REFERENCES

[1] R. C. Gonzales, R. E. Woods, Digital Image Prossesing, PrenticeHall, 2001.

[2] Dzung L. Pham, Chen yang Xu, and Jerry L. Prince. "Current methods in medical image segmentation", Vol. 2: 315-337 (Volume publication date August 2000)Plan Z.A., Lin R.T. and Richer J.A. Nanotechnology Devices. in "The World of Nanotechnology," G.E. Goodfellow and A.T. Mann, Eds., Butterworth Publishers, Boston, MA (1989), pp. 61–67.

[3] Michael B. Dillencourt and Hannan Samet and Markku Tamminen "A general approach to connected-component labeling for arbitrary image representations". J. ACM. 1992.

[4] http://www.cse.unr.edu/~bebis/CS791E/Notes/ConnectedCompon ents.pdf

[5]   Burn, D. H., Zrinji, Z., and Kowalchulk, M., Regionalization of Catchments for Regional Flood Frequency Analysis. Journal of Hydrologic Engineering, 2(2), 76–82, 1997.

[6] Alkan A., Arslan Tuncer S, Gunay M., "Comparative MR image analysis for thyroid nodule detection and quantification, Measurement, Volume 47, Pages 861-868, January 2014.

[7] Selver A.,Kocaoğlua.,Doğanh.,Demir K. ,Dicle O.,Güzeliş C. Nakil Öncesi Verici Değerlendirmeleri İçin Otomatik Karaciğer Bölütleme Yordamı, Hastane ve Yaşam , Sayı 29, 80-87, 2008.

[8] Health system. virginia.edu. ,Department of Radiology and Medical Imaging, School of Medicine at The University of Virginia", 2012.

[9] S. A. Tuncer and A. Alkan, "Segmentation of thyroid nodules with K-means algorithm on mobile devices," Computational Intelligence and Informatics (CINTI), 2015 16th IEEE International Symposium on, Budapest, 2015, pp. 345-348.

MIPRO 2017, May 22- 26, 2017, Opatija, Croatia

An Overview of Action Recognition in Videos

M. Burić, M. Pobar, M. Ivašić Kos

University of Rijeka/ Department of Informatics, Rijeka, Croatia
matija.buric@hep.hr, mpobar@inf.uniri.hr, marinai@uniri.hr

Abstract - Action recognition in videos is currently in the focus of scientific research due to improvements made in automatic analysis of static images and greater availability of processing power. The paper provides an overview of the key models and methods for action recognition that comprise human models and methods based on estimation of joint trajectories, silhouettes and template matching and spatio-temporal local descriptors. To deal with compound actions and activities, action semantic models are proposed with help of expert knowledge. Since the action recognition task is domain dependent, the methods and models are built and tested on domain specific databases. The paper provides an overview and description of recent video datasets that were created for developing action recognition methods, with an emphasis on datasets with additional modalities such as depth images or accelerometer data.

I. INTRODUCTION

Automatic analysis of videos is nowadays in the focus of research interest. This is due to their ubiquity in ordinary life and usability in many different fields like security surveillance, automation of processes based on visual references, detection of all sorts of abnormalities, and anything that is connected to visual observation.

The key part of automatic analysis of videos is the recognition of individual actions, such as falling down, or jumping. The recognized actions can be used per se, for example for interacting with electronic devices, or can serve as a starting point towards recognizing activities that require further interpretation, such as distinguishing between a normal encounter or a threatening approach.

Over the years, many techniques have been proposed for solving the task of human action recognition in video. Most of them use some kind of supervised machine learning approach, where the models of actions that should be recognized are constructed using the training data consisting of features extracted from videos containing the appropriate actions. Fig. 1. shows the typical workflow of the supervised learning approach, where a prediction model is created in the first workflow and then is used in the second workflow to make a prediction.

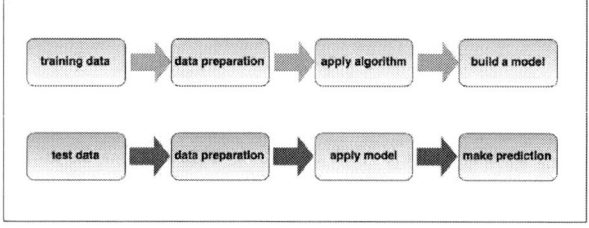

Figure 1.   Traditional model of action recognition

In the other, knowledge-based approach, the models of actions used for recognition are constructed mainly by encoding by hand the expert knowledge using a descriptive language and logic operators.

Lately, a newer kind of machine learning approaches for action recognition using deep neural networks have emerged. This approach has been used for automatic extraction of features that will be used for recognition instead of handcrafted features in a classic ML style, as well as an integrated framework of feature extraction and classification where both processes are completely handled in the deep network.

The purpose of this paper is to give an overview of the most used models and methods for activity recognition with their strengths and weaknesses. A list of datasets that are appropriate for this purpose will also be presented.

II. MODELS AND METHODS

Raw video sequences are inappropriate for modeling and classification due to their extremely high dimensionality (proportional to image dimensions and time), where most dimensions do not actually contain any information that is relevant for action recognition. On the other hand, machines perform classification based solely on primary elements like pixels or features. That is why the first step after acquiring videos is typically the extraction of features. Simply put, features are the interesting parts of the image altered in a way that machine can use. Some of the common low-level features are edges (divides between lighter and darker part of the image), corners (points where edges intersect), blobs (regions of image too smooth for an edge to detect), etc.

In action recognition, the features can be roughly divided into those that encode positions or trajectories of different body parts, into those that aim to track the whole figure and into local features that operate on interest points in videos without regard if they belong to a person or not. Both the body part based features and figure based features require additional video processing in order to detect, segment and track the persons in video, which is a non-trivial task. Local features, on the other hand, have an advantage of not requiring any kind of person localization, so they have lately became increasingly interesting to researchers in the field.

A. Body based models

Feature representations for action recognition that are based on human body parts use 2D and 3D features like stick figures (Fig. 3), silhouettes (Fig. 4) or volumes (Fig. 5) to represent information about position and movement of different body parts [1].

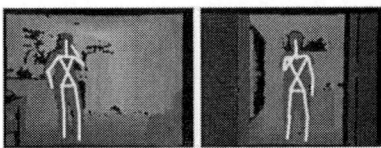

Figure 2. Stick figure from the Cornell Activity Datasets database [2]

Figure 3. A video frame depicting a person walking (left) and the corresponding silhouette mask (right) [3]

Figure 4. Volumes formed by stacking the silhouettes of persons while performing actions [4]

Most approaches use an explicit model of human body, such as a stick figure model, and strive to optimize the match between the model projections and an observed image frame while simultaneously keeping a correspondence of joints between frames. The resulting representation is a set of joint trajectories in 2D-time space or 3D-time space, as shown in Fig. 5.

 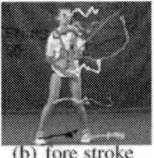 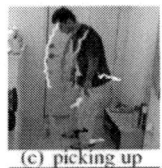

(a) running   (b) fore stroke   (c) picking up

Figure 5. Tracked trajectories of joints generated by performing different actions [4]

The drawback of 2D-time representations is that they are view-dependent, meaning that the features for the same action will be very different depending on the relative orientation of the camera and the person performing the action. On the other hand, the 3D pose estimation from a single view requires models that are more complex. If the single camera requirement is relaxed, solutions using multiple cameras to reconstruct the depth information have been proposed, as in [4]. Lately, affordable depth cameras have emerged that capture 3D depth information using an infrared sensor along with RGB video, which can aid greatly in estimation of joint positions in 3D, but are still limited in use due to modest range of depth sensing. An additional consideration is whether camera view is stationary or mobile [4], whether a single camera is used or there are multiple views available [3] to either reconstruct a 3D representation of the world or to mitigate the effect of occlusions and expand the field of view.

Precise full pose reconstruction is a difficult task, and the results can be unreliable due to noise and occlusion, yet it turns out that is not necessary for action recognition. Thus, a different direction of research focused on representation of actions using body silhouettes, without attempting to segment or match individual body parts to a model. This has been used in an early work [6] where a vector quantization scheme was used to turn the silhouette image sequence into a sequence of symbols. The symbol sequences are used to train Hidden Markov Models (HMMs) of actions, which are applied in the recognition phase to find the model that best matches the observed symbol sequence, thus incorporating a flexibility in the duration of actions.

In a more recent work [4], the actions are represented as templates that are volumes formed by stacking the silhouettes of persons performing actions along the time axis (Fig. 4), and the matching is done by template matching. The drawback of silhouette and volume based representations is the fact that silhouette segmentation can be difficult in noisy environments, and view dependence, which can be mitigated by forming different models for different views.

B. Bag of visual words models

Recognition of actions using local features and bag of visual words have an advantage over the approaches that rely on body models, since the extraction of local features doesn't require any kind of human model or person localization. The local features are extracted by first using an interest point detector and then extracting a local descriptor of that interest point. The descriptors are clustered into *visual words*. Each image is then represented by a bag of visual words (BOW) which is then used for learning, Fig.6.

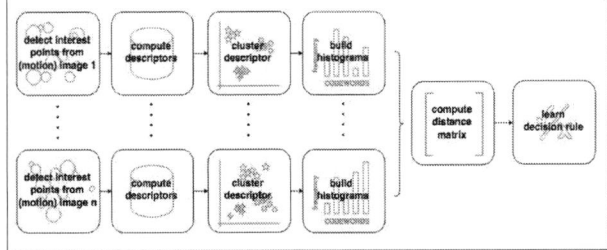

Figure 6. Bag-of-features learning diagram

Following the division in [7], interest point detectors can be grouped into 3 groups – contour based, intensity based and parametric based. Contour based detectors extract contours in order to find edges and corners of most interest. Intensity based detectors compute interest points directly from the grey values. Parametric based identify corners in images using analytic approximation of grayed-value structures [8]. Harris 3D detector [9] computes spatio-temporal corners at each video point and determines final locations of space-time interest points by local maximum. Cuboid detector based on Gabor filters [10] detects regions with spatially distinguishing characteristics undergoing a complex motion. Hessian [11] detects spatio-temporal blobs and dense sampling [12] is extracting 3D patches at regular positions with variation in scale.

After the points of interest or trajectories information about shape and movement in local surrounding area are successfully detected, they are presented using interest point descriptors. Feature trajectories are typically extracted using the Kanade–Lucas–Tomasi (KLT) [13] tracker or by matching SIFT descriptors [14] between frames.

Majority of descriptors fall under the group of spectra descriptors which are based on computed quantities like color and light intensity, local area gradients, statistical features and moments, surface normals and sorted data like 2D or 3D histograms of any spectral type.

3D, 4D, volumetric and multimodal descriptors are becoming more interesting because of the development of affordable 3D sensors and accelerometers built into mobile devices. Since the field of 3D feature description is early in the development, it is not yet clear which methods will be widely adopted. The most notable at this point are 3D HOG [15], 3D SIFT [16] and HON 4D [17] which are based on familiar 2D methods that are extended into a spatio-temporal 3D space.

The extracted features are then encoded using BOW or Fisher Vector approach [18].

Common classification methods are used in conjunction with the above features for action recognition such as Multi-Layer Perceptron (MLP) in [19], and Support Vector Machine (SVM) [10,9,20,21].

C. Deep learning approaches

Motivated by the success that has been achieved with deep learning methods such as convolutional neural networks (CNNs) [22], used in image and video classification [23], there is an increasing interest to apply that approach in the action recognition field as well. CNNs can learn to extract features automatically from a large number of labeled images and have outperformed the classical ML approach using handcrafted features for image classification task. In [24] the image classification CNN method is extended to handle the temporal dimension of videos. The proposed architecture used several layers of 3D convolution starting from the initial 7-frames deep cube, that generates multiple channels of information that are analogous to handcrafted features, and layers of subsampling that reduces the dimension of the feature vector. The classification itself is also performed by the last layer of the neural network.

An architecture where two parallel networks capture spatial and temporal information were proposed in [25]. One network operates on individual video frames, performing action recognition from still images, and the other operates on the optical flow explicitly describing the motion between frames and forms the temporal recognition stream. The output of the two networks is fused into a final decision score using a SVM classifier. The results obtained with CNNs in action recognition show the similar performance as classical methods [24, 25].

III. ACTIVITY UNDERSTANDING

The previously mentioned methods, due to intensive research that has been done are very accurate in recognition of simple actions but they lack the ability to deal with more complex and hierarchically related actions and activities.

Therefore, more descriptive model and logical operators should be used for presenting that kind of activities with help of expert knowledge.

Authors of [26] give an overview of the current research in the field of activity understanding and an overall model of activity understanding task, comprising abstraction and action modeling, Fig.7. Abstraction deals with a problem of translating motion image input into form understandable by action models to determine if an interesting action has occurred. Final output can be a particular activity or summary of actions. Abstraction is performed using either pixel features, objects and their properties or logical facts of knowledge. Action modeling includes traditional classification methods for action recognition, state models for knowledge representation in space-time domain and semantic models for reasoning about sequential actions. State modeling formalisms include: finite state machines (FSMs), Bayesian networks (BN), HMM, etc. Semantic models use an interesting subset of actions defined by the semantic relationships between their sub-actions. Semantic model depends on knowledge of domain expert to classify an activity. It is usually applied in case of more complex actions that vary in their appearance [27, 28, 29]. Semantic models include grammars, Petri nets, constraint satisfaction, etc. The deterministic nature of semantic models makes them sensitive to inaccuracies in lower-level recognition, therefore mechanism of fuzzy reasoning to handle uncertainty in observation and interpretation is generally desired [30]. Authors of [31, 32] used a fuzzy knowledge representation scheme that enables modeling of uncertain knowledge about relations between entities that could be used for indistinct interpretation of borders between actions in motion image sequence.

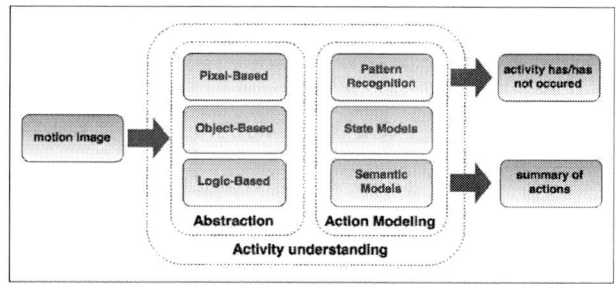

Figure 7.   Activity understanding model after [26]

IV. DATASETS

To start building models for activity recognition it's essential to acquire good dataset of video samples which reflect desired activities. There are many datasets publicly available for scientific research including recently growing datasets with additional information acquired using RGB-D sensors, accelerometers and position markers that are placed directly on a model that is being observed, multiple sources, etc. In this paper the emphasis will be on these datasets.

Princeton Tracking Benchmark [33] introduced 2013 consists of 100 RGB-D tracking datasets with tracking software and online submission script. Datasets include real world footage of variety of actions performed by humans, pets and object presentations in form of RGB (8 bit PNG format) and accompanying depth (16 bit PNG format) images (Fig. 8). Annotations are per-frame in a form of bounding box covering target object only. These datasets are more related to tracking than to action recognition but could serve as a starting point in segmentation of scenery.

Figure 8. Princeton Tracking Benchmark [33]

Cornell Activity Datasets: CAD-60 & CAD-120 [2]. CAD-60 comes with 60 RGB-D videos of 2 male and 2 female persons in real world closed environment: office, kitchen, bedroom, bathroom, and living room describing 12 activities: rinsing mouth, brushing teeth, wearing contact lens, talking on the phone, drinking water, opening pill container, cooking (chopping), cooking (stirring), talking on couch, relaxing on couch, writing on whiteboard, working on computer, Fig. 9. CAD-120 consists of 120 RGB-D videos with same number of people in similar environment. Activities are divided into 10 high-level activities (making cereal, taking medicine, stacking objects, unstacking objects, microwaving food, picking objects, cleaning objects, taking food, arranging objects, having a meal) and 10 sub-activity labels (reaching, moving, pouring, eating, drinking, opening, placing, closing, scrubbing, null) with object affordance labels like reachable, movable, pourable, closable. Skeleton joint position and orientation is labelled on each frame. RGBD data has resolution of 240 by 320. RGB is saved as three-channel 8-bit PNG file and depth is saved as single-channel 16-bit PNG file.

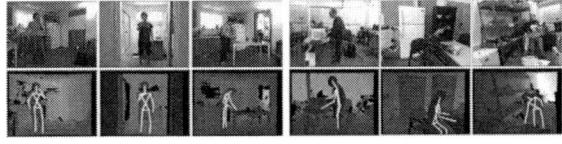

Figure 9. Cornell Activity Datasets: CAD-60 & CAD-120 [2]

Northwestern-UCLA Multiview Action 3D Dataset [34] contains RGB, depth and human skeleton data captured simultaneously by three Kinect cameras. This dataset include 10 action categories: pick up with one hand, pick up with two hands, drop trash, walk around, sit down, stand up, donning, doffing, throw, carry (Fig. 10). Each action is performed by 10 actors in a library from a variety of viewpoints.

Figure 10. Northwestern-UCLA Multiview Action 3D Dataset [34]

RGB-D People Dataset [35] was gathered by a three vertically mounted Kinect sensors on a tower at approximately 1.50 m height. It contains 3000+ RGB-D frames acquired in a university hall and contains mostly upright walking and standing persons seen from different orientations and with different levels of occlusions, Fig. 11. Annotations are made in a form of a square box. Depth images are saved as 16 bits, 1 channel PGM images - 640 by 480. They contain the raw data content from the Kinect sensor. Namely, each pixel has value between [0, 1084]. RGB images are saved as 8 bits, 3 channels PPM images - 640 by 480. Dataset doesn't provide activity annotations but offers material for an art gallery research.

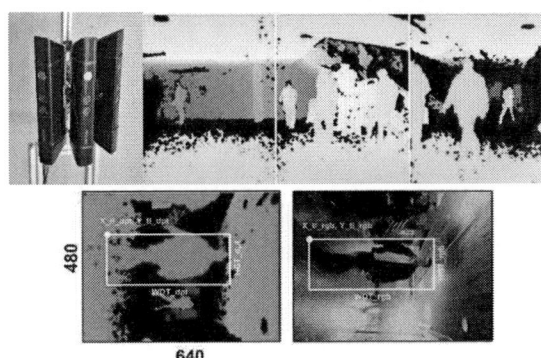

Figure 11. RGB-D People Dataset [35]

UTD Multimodal Human Action Dataset (UTD-MHAD) [36] was collected using a Kinect sensor and a wearable inertial sensor in an indoor environment. The dataset contains 27 actions performed by 8 subjects (4 females and 4 males). Each subject repeated each action 4 times. The dataset includes 861 data sequences. Four data modalities of RGB videos, depth videos, skeleton joint positions, and the inertial sensor signals were recorded in three channels or threads (Fig. 12). One channel was used for simultaneous capture of depth videos and skeleton positions, one channel for RGB videos, and one channel for the inertial sensor signals (3-axis acceleration and 3-axis rotation signals). For data synchronization, a time stamp for each sample was recorded. The inertial sensor was worn on the subject's right wrist or the right thigh (see the figure below) depending on whether the action was mostly an arm or a leg type of action.

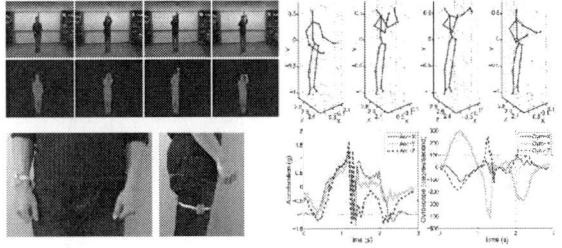

Figure 12. UTD Multimodal Human Action Dataset (UTD-MHAD)

Berkeley Multimodal Human Action Database (MHAD) [37] contains 11 actions performed by 7 male and 5 female subjects in the range 23-30 years of age except for one elderly subject. All the subjects performed 5 repetitions of each action, yielding about 660 action sequences which correspond to about 82 minutes of total recording time. In addition, a T-pose for each subject was recorded which can be used for the skeleton extraction along with the background data (with and without the chair used in some of the activities). The specified set of actions comprises of the actions with movement in both upper and lower extremities, actions with high dynamics in upper extremities and actions with high dynamics in lower extremities. Each action was simultaneously captured by five different systems: optical motion capture system, four multi-view stereo vision camera arrays, two Microsoft Kinect cameras, six wireless accelerometers and four microphones (Fig. 13).

Figure 13. Berkeley Multimodal Human Action Database (MHAD) [37]

Dataset of a human performing daily life activities in a scene with occlusions [38] consists of 12 RGB-D video sequences of a person moving in front of a Kinect in a scene with obstacles, Fig.14. In addition to the depth and RGB image, each sequence contains the synchronized ground truth data obtained from a Qualisys motion capture system with 8 infrared cameras. 3D representation of a human model is achieved by using 15 position markers: one for a head, neck and torso and 2 for shoulders, elbows, wrists, hips and knees.

Figure 14. Dataset of a human performing daily life activities in a scene with occlusions [38]

Recording video is another way of acquiring dataset. Using private dataset allows customization to adopt to a desired method, however, it is time consuming and it requires significant resources to collect desired footages.

When building private dataset one of the things that needs consideration is what approach regarding camera number would be most suitable. When one camera is used depth perception is lost and observed object can be completely or partly hidden. Multiple cameras concentrated on an object of observation from different positions give much more information. The position of cameras can be calculated like in [39] for an optimal solution. Multiple cameras can be set up two ways. Camera fields can overlap, which is more suitable for detail action examination, or they can be put side by side, the so-called "art gallery". This way is used more in surveillance [40]. Art gallery, also, allows greater filed coverage with a same resources but shares the problem as one camera approach, only partial image is visible.

Overlapping cameras provide info about the object from different perspectives but there is an occlusion problem. This can be avoided by using algorithms like in [41] where input from different cameras are used sequentially which require less samples and computational power.

Art gallery and one camera approach generally suffer from absence of depth perception. Fortunately, latest development in game industry brought affordable RGB-D sensors like Microsoft Kinect and Asus Xtion which give depth based on two cameras and widespread use by using infrared spectrum. Still, even though these sensors perform well they are still inferior compared to marker based systems [39].

V. CONLUSION

In this paper, an overview of models, methods for action recognition is presented. Human models and methods that are based on trajectories, silhouettes and template matching, spatio-temporal local descriptors and descriptor-based methods are presented along with their advantages and disadvantages. Recognition of actions using local features have an advantage over the classical approaches, since the extraction of local features doesn't require any kind of human model or person localization. There is in an increased interest in using convolutional neural networks that can automatically learn to extract features, and show state-of-the art performance.

In case of recognition of actions and activities that are more complex, semantic models that use expert knowledge are commonly proposed.

In addition, paper presents, in detail, available image databases appropriate for the tasks of action recognition with additional information acquired using RGB-D sensors, accelerometers and position markers.

ACKNOWLEDGMENT

This research was fully supported by Croatian Science Foundation under the project Automatic recognition of actions and activities in multimedia content from the sports domain (RAASS).

REFERENCES

[1] J.K. Aggarwal, Cai, Q., (1999) "Human motion analysis: A review," Computer vision and image understanding 73, 428 – 440

[2] Koppula, H. S., Gupta, R., & Saxena, A. (2013). Learning human activities and object affordances from rgb-d videos. The International Journal of Robotics Research, 32(8), 951-970.

[3] Ivašić-Kos, Marina; Iosifidis, Alexandros; Tefas, Anastasios; Pitas, Ioannis. Person De-Identification in Activity Videos, BiForD, Opatija, Hrvatska - MIPRO, 2014. 75-80

[4] Gorelick, L., Blank, M., Shechtman, E., Irani, M., & Basri, R. (2007). Actions as space-time shapes. *IEEE transactions on pattern analysis and machine intelligence*, 29(12), 2247-2253.

[5] Yilmaz, A., & Shah, M. (2005, October). Recognizing human actions in videos acquired by uncalibrated moving cameras. In Computer Vision, 2005. ICCV 2005. Tenth IEEE International Conference on (Vol. 1, pp. 150-157). IEEE.

[6] J. Yamato, J. Ohya, and K. Ishi, "Recognizing human action in time sequential images using hidden markov model," Computer Vision and Pattern Recognition, 1992. Proceedings CVPR'92., 1992 IEEE Computer

Society Conference on, IEEE. pp. 379-385.

[7] C. Schmid, R. Mohr, and C. Bauckhage, "Evaluation of Interest Point Detectors," International Journal of Computer Vision, 2000, Volume 37, Number 2, Page 151

[8] K. Rohr, "Recognizing Corners by Fitting Parametric Models," Arbeitsbereich Kognitive Systeme, FB Informatik, Universitat Hamburg, International Journal of Computer Vision, 1992, 9:3, pp 213-230

[9] I. Laptev and T. Lindeberg, "Space-time interest points," In ICCV, 2003.

[10] P. Dollar, V. Rabaud, G. Cottrell, and S. Belongie "Behavior recognition via sparse spatio-temporal features," In VS-PETS, 2005.

[11] G. Willems, T. Tuytelaars, and L. Van Gool, "An efficient dense and scale-invariant spatio-temporal interest point detector," In ECCV, 2008.

[12] H. Nakayama, T. Harada, and Y. Kuniyoshi, "Dense Sampling Low-Level Statistics of Local Features," CIVR '09, July 8-10, 2009, Santorini, GR

[13] B. D. Lucas and T. Kanade "An iterative image registration technique with an application to stereo vision," In International Joint Conference on Artificial Intelligence, 1981.

[14] D. Lowe, "Method and apparatus for identifying scale invariant features in an image and use of same for locating an object in an image", U.S. Patent 6,711,293, March 23, 2004

[15] A. Klaser, M. Marszalek, and C. Schmid. "A Spatio-temporal Descriptor Based on 3d-gradients," British Machine Vision Conference, 2008.

[16] P. Scovanner, S. Ali and M. Shah. "A 3-dimensional SIFT Descriptor and its Application to Action Recognition," ACM Proceedings of the 15th International Conference on Multimedia, pages 357–360., 2007.

[17] O. Oreifej and Z. Liu., "HON4D: Histogram of Oriented 4D Normals for Activity Recognition from Depth Sequences," Conference on Computer Vision and Pattern Recognition, 2013.

[18] F. Perronnin, J. S´anchez, and T. Mensink, "Improving the Fisher Kernel for Large-Scale Image Classification," Xerox Research Centre Europe (XRCE), 2010.

[19] A. Iosifidis and A. Tefas, "View-invariant action recognition based on Artificial Neural Networks," Neural Networks and Learning Systems, IEEE Transactions on 23, 412-424

[20] A. Karpathy, G. Toderici, S. Shetty, T. Leung R. Sukthankar, and L. Fei-Fei, "Large-scale Video Classification with Convolutional Neural Networks," in: Computer Vision and Pattern Recognition (CVPR), 2014 IEEE Conference on, IEEE. pp. 1725-1732.1

[21] I. Laptev, M. Marszałek, C. Schmid, and B. Rozenfeld, "Learning realistic human actions from movies," in: Computer Vision and Pattern Recognition, 2008. CVPR 2008. IEEE Conference on, IEEE. pp. 1-8.

[22] H. Wang, A. Klaser, C. Schmid, L. Cheng-Lin, "Action Recognition by Dense Trajectories," CVPR 2011 - IEEE Conference on Computer Vision & Pattern Recognition, Jun 2011, Colorado Springs, United States. IEEE, pp.3169-3176, 2011

[23] LeCun, Yann, Yoshua Bengio, and Geoffrey Hinton. "Deep learning." Nature 521.7553 (2015): 436-444.

[24] Ji, S., Xu, W., Yang, M., & Yu, K. (2013). 3D convolutional neural networks for human action recognition. IEEE transactions on pattern analysis and machine intelligence, 35(1), 221-231.

[25] Simonyan, K., & Zisserman, A. (2014). Two-stream convolutional networks for action recognition in videos. In Advances in neural information processing systems (pp. 568-576).

[26] G. Lavee, E. Rivlin and M. Rudzsky, "Understanding Video Events: A Survey of Methods for Automatic Interpretation of Semantic Occurrences in Video," Technion - Computer Science Department - Tehnical Report CIS-2009-06 - 2009

[27] V. T. Vu, F. Bremond, and M. Thonnat, "Automatic video interpretation: A recognition algorithm for temporal scenarios based on pre-compiled scenario models," in International Conference on Computer Vision Systems, 2003, pp. 523–533.

[28] A. Borzin, E. Rivlin and M. Rudzsky, "Surveillance interpretation using Generalized Stochastic Petri Nets," in The International Workshop on Image Analysis for Multimedia Interactive Services (WIAMIS), 2007.

[29] A. Bobick and J. Davis, "The recognition of human movement using temporal templates," IEEE Transactions on Pattern Analysis and Machine Intelligence, vol. 23, no. 3, pp. 257–267, March 2001.

[30] M.S. Ryoo and J.K. Aggarwal, "Semantic Representation and Recognition of Continued and Recursive Human Activities," Springer Science+Business Media, Int J Comput Vis (2009) 82: 1–24

[31] M. Ivašić-Kos, I. Ipšić, and S. Ribarić, "A knowledge-based multi-layered image annotation system," Expert systems with applications. 42 (2015) , 2015; 9539-9553.

[32] M. Ivašić-Kos, M. Pobar, and S. Ribarić, "Two-tier image annotation model based on a multi-label classifier and fuzzy-knowledge representation scheme," Pattern recognition. 52 (2016); 287-305

[33] S. Song and J. Xiao, "Tracking Revisited using RGBD Camera: Unified Benchmark and Baselines,"Princeton University Proceedings of 14th IEEE International Conference on Computer Vision (ICCV2013)

[34] UCLA database http://users.eecs.northwestern.edu/~jwa368/

[35] L. Spinello, K. O. Arras. People Detection in RGB-D Data. IEEE Int. Conf. on Intelligent Robots and Systems (IROS), 2011.

[36] C. Chen, R. Jafari, and N. Kehtarnavaz, "UTD-MHAD: A Multimodal Dataset for Human Action Recognition Utilizing a Depth Camera and a Wearable Inertial Sensor", Proceedings of IEEE International Conference on Image Processing, Canada, September 2015.

[37] F. Ofli, R. Chaudhry, G. Kurillo, R. Vidal and R. Bajcsy. Berkeley MHAD: A Comprehensive Multimodal Human Action Database. In Proceedings of the IEEE Workshop on Applications on Computer Vision (WACV), 2013.

[38] A. Dib, F. Charpillet, "Pose Estimation For A Partially Observable Human Body From RGB-D Cameras," Proc. of the International Conference on Intelligent Robot Systems (IROS), 2015.

[39] E. J. Almazan and G. A. Jones, "Tracking People across Multiple Non-Overlapping RGB-D Sensors," The IEEE Conference on Computer Vision and Pattern Recognition (CVPR) Workshops, 2013, pp. 831-837

[40] F. Faion, S. Friedberger, A. Zea, and U. D. Hanebeck, "Intelligent Sensor-Scheduling for Multi-Kinect-Tracking," Intelligent Robots and Systems (IROS), 2012 IEEE/RSJ International Conference

[41] S. Han, M. Achar, S. Lee and F. Peña-Mora, "Empirical assessment of a RGB-D sensor on motion capture and action recognition for construction worker monitoring," SpringerOpen Journal, Visualization in Engineering 2013, 1:6

Pothole Detection: An Efficient Vision Based Method Using RGB Color Space Image Segmentation

Amila Akagic, Emir Buza, Samir Omanovic
University of Sarajevo
Faculty of Electrical Engineering, Department for Computer Science and Informatics
Zmaja od Bosne bb, Kampus Univerziteta, 71000 Sarajevo
Email: {amila.akagic, emir.buza, samir.omanovic}@etf.unsa.ba

Abstract—The proper planning of repairs and rehabilitation of the asphalt pavement is one of the important tasks for safe driving. The most common form of distress on asphalt pavements are potholes, which can compromise safety, and result in vehicle damage. Timely repairing potholes is crucial in ensuring the safety, quality of driving, and reducing the cost of vehicle maintenance. Many of the existing methods for pothole detection often use sophisticated equipment and algorithms, which require substantial amount of data for filtering and training. Consequently, as a result of intensive computational processing, this can lead to long execution time and increased power consumption. In this paper, we propose an efficient unsupervised vision-based method for pothole detection without the process of training and filtering. Our method first extracts asphalt pavements by analysing RGB color space and performing image segmentation. When the asphalt pavement is detected, the search continues in detected region only. The method is tested on online image data set captured from different cameras and angles, with different irregular shapes and number of potholes. The results indicate that the method is suitable as a pre-processing step for other supervised methods.

Keywords: Pothole detection, Unsupervised method, Image processing, Image segmentation, Computer Vision

I. INTRODUCTION

The continuous inspection and assessment of physical and functional condition of civil infrastructure is cruital for safety, serviceability and rideability [1]. Currently, the manual visual inspection at regular intervals is the main form of conditions assesment in most countries. Unfortunately, due to insufficient inspection and condition assessment this method had some tragic consequences in countries such as Japan [2] and USA [3]. The other form of inspection is automatic visual inspection which acounts for only 0.4% of all inspections. Automatic inspection uses specialized vehicles that are mounted with laser scanners, pavement profilers, accelerometers, image and video cameras, and positioning systems. Current automatic inspection methods have several restrictions, such as the number of defects they can detect, the price of sensor equipment and the level of detail they can attain.

The most common form of distresses on asphalt pavements are *potholes* - small, bowl-shaped depressions in the pavement surface. The repair of potholes is necessary in situations where

potholes compromise safety and pavement rideability. The automatic pothole detection systems are based on 3D reconstruction, vibration and vision methods. The 3D reconstruction requires high cost laser scanners, while vibration based approach is unreliable on a surface such as bridge expansion joints due to surface vibrations. The vision-based methods use various image processing techniques [4] [5] to detect potholes on 2D images or video data. Among existing methods many use supervision, which requires a training data set and some form of filtering. Unfortunately, supervision methods suffer from the problem of overfitting, where characteristics of a training set are memorized, instead of capturing the desired pattern.

In this paper, we address the problem of automatic pothole detection by automatic analysis of selected 2D images of asphalt pavement. We propose a new unsupervised vision-based method to detect potholes. First, we extract the asphalt pavement (region of interest) and we limit the search for potholes in that region only. The region of interest is detected through manipulation of B component in RGB color space, and two-level dynamic selection of alphalt pixels (seed points) based on a standard deviation of an image. Once the asphalt pavement is found, we proceed to detect potholes by comparing two cropped images, the idea based on the method [6] from our previous research.

The effectiveness of our method is verified on a newly formed image data set from selected images from Google search engine. The data set contains highly unstructured images taken from different cameras and shooting angles, with different irregualr shapes and number of potholes. Our method is designed for use under daytime fair weather conditions, which is consistent with the current practice. The method is low-cost and efficient since it does not require expensive equipment, filtering nor training data. Also, the image quality does not significantly affect the accuracy. The results show that the method detected potholes with reasonable accuracy. Since the method uses manipulation of B component in RGB space and image segmentaion it can be easily and widely adopted for hardware implementation.

This paper is organized as follows: related work is briefly

reviewed in section II. In section III, we proposed a new method for pothole detection. The implementation details, data set and experimental results are presented in section IV. We concluded the paper with some remarks in section V.

II. RELATED WORK

The existing approaches for pothole detection can be classified into three groups [7]: 3D reconstruction-based, vibration-based and vision-based. A 3D reconstruction of the pavement surface can be acquired by laser scanners [8] [9], stereo-vision algorithms by using a pair of video cameras [10], [11], [12] and visualization using Microsoft Kinect sensor [13], [14], [15]. These methods visualize 3D pavement surface data by using sophisticated, high cost equipment, and exhibit high computational cost. A vibration-based approaches [16], [17], [18], [19] use accelerometers to assess pavement condition based on the mechanical responses of the vehicle which carries equipment. The main problem of this approach is that it cannot be used at bridge expansion joints and cannot detect pothole in the center of a lane.

A vision-based approach uses various image processing techniques to detect potholes on 2D images or video input. In [20], Karuppuswamy et al. proposed a new method based on integrating vision and motion system for detection of simulated potholes. The application of the method is limited due to the simulation limitations (potholes smaller than 2 feet in diameter, and the color of pothole). In [21], Koch et al. proposed a novel supervised approach for automated pothole detection based on asphalt pavement images. This approach is based on pothole texture extraction and comparison, where the surface texture inside a pothole candidate has to be described and compared with the texture of the surrounding region. This implies existence of a number of pothole texture samples from which the system is trained. System can detect a pothole based on this training results. Additionally, four spot filters are required to be applied to original gray-level image in order to emphasize structural texture characteristics.

In our previous work [6], we proposed new unsupervised vision-based method which does not require expensive equipment, additional filtering nor training phase. Our method deployed image processing and spectral clustering for identification and rough estimation of potholes. The effectiveness has been verified on image data selected from Google image collection. The accuracy of a pothole surface area was about 81%, thus we concluded that the method can be used for rough estimation for repairs and rehabilitation of pavements.

In regards to detection of potholes over a sequence of frames, Koch et al. [22] extended their original method with video processing [23]. Radopoulou et al. [24] proposed the use of Semantic Texton Forests (supervised learning algorithm) for detection of several defects occurring in video frames.

III. METHODLOGY

The overall design of our method is depicted in Figure 1. The method consists of three major steps:

(A) Image Pre-processing

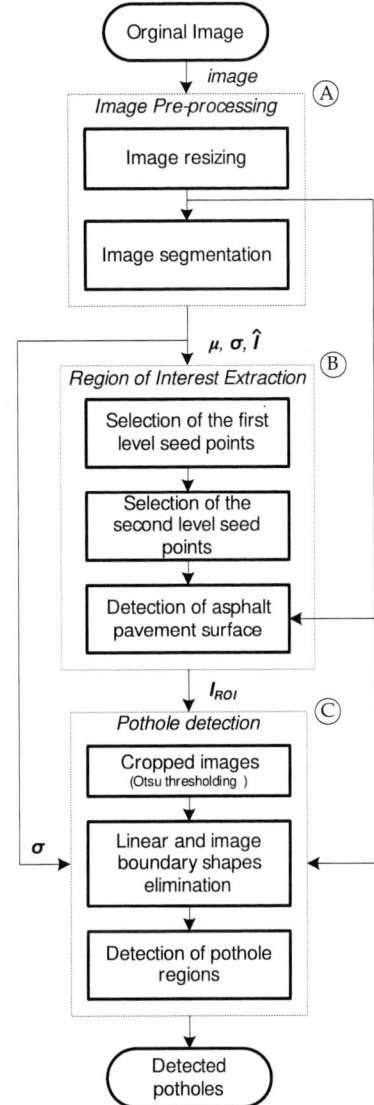

Fig. 1: Block diagram of proposed method for pothole detection.

(B) Region of Interest Extraction
(C) Pothole detection

A. Image Pre-processing

The first step consists of two phases: image resizing and image segmentation. The size of an image does not effect the quality of pothole detection. Thus, in order to descrease the number of computations, we first decrease the size of an image by α times (1).

$$\alpha = \begin{cases} 400/\text{Columns}, & \text{for Rows} > 301 \\ & || \text{ Columns} > 401 \\ 400/\text{Columns}, & \text{for Columns} < 181 \\ 1, & \text{otherwise} \end{cases} \quad (1)$$

In the image segmentation phase, the arithmetic mean μ and the standard deviation σ are caluculated by (2) and (3).

$$\mu = \frac{1}{N \times M \times 3} \sum_x \sum_y \sum_{[R,G,B]} p(x,y)_{[R,G,B]} \quad (2)$$

$$\sigma = \sqrt{\frac{1}{3NM-1} \sum_x \sum_y \sum_{[R,G,B]} [p(x,y)_{[R,G,B]} - \mu]^2} \quad (3)$$

The parameters x, y, N, M and $[R,G,B]$ are row and column indices, image height and weight, and RGB color components of an image I, respectively. The $p(x,y)_{[R,G,B]}$ represents pixel values on the point (x,y) for the color components: red, green and blue.

We replace the original image pixels of R, G and B components as shown by criterias in the equations (4), (5) and (6). The Fig. 2. depicts results of mentioned equations (a,b,c), respectively, and the final $\hat{I}$ image (d), which is passed to the next step together with parameters μ and σ.

$$I'_{[R,G,B](x,y)} = \begin{cases} (0,0,0), & I_{G(x,y)} - \frac{I_{R(x,y)}+I_{B(x,y)}}{2} \\ & +15 > 0 \\ (0,0,0), & I_{R(x,y)} - I_{B(x,y)} > 20 \\ (0,0,0), & I_{G(x,y)} - I_{R(x,y)} > 15 \end{cases} \quad (4)$$

$$I''_{[R,G,B](x,y)} = \begin{cases} (0,0,0), & I'_{[R,G,B](x,y)} > \\ & (255 - \sigma) \\ (0,0,0), & I'_{[R,G,B](x,y)} < \sigma \end{cases} \quad (5)$$

$$\hat{I}_{[R,G,B](x,y)} = \begin{cases} (255,255,255), & I''_{[R,G,B](x,y)} > 0 \\ (0,0,0), & I''_{[R,G,B](x,y)} = 0 \end{cases} \quad (6)$$

B. Region of Interest Extraction

The region of interest is the area on which potholes can be found, i.e. asphalt pavement. The input for this step is the B component from RGB color space ($\hat{I}_{[B]}$) from the previous step, which is a black-white image (0,255). First, we form new matrix which contains pixel positions of all white pixels in an image. Second, we select a dynamic number of pixels, based on the standard deviation of an image ($\sigma/2$), to form *the first level seed points*. The algorithm for finding the first level seed points is shown in Algorithm 1. The result of Algorithm 1 is presented in the figure Fig. 2a). The red circles represent the first level seed points.

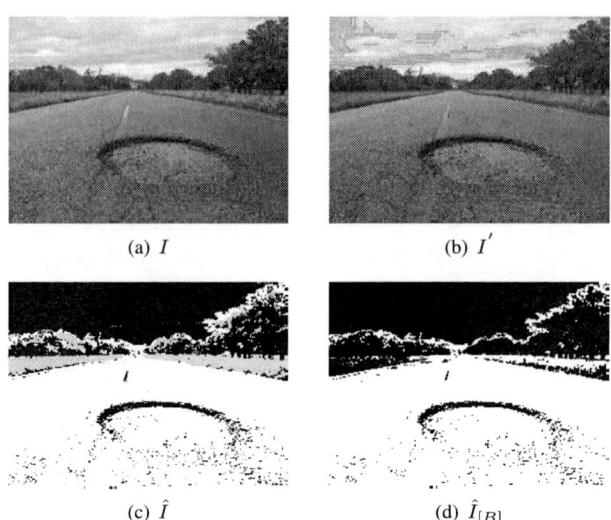

Fig. 2: Image Pre-processing: a) The orginal image I; b) The resulting image I' after (4); c) The resulting image $\hat{I}$ after (5) and (6) and d) The blue component $\hat{I}_{[B]}$ of the image $\hat{I}$.

Algorithm 1 The finding of first level seed points

Require: $\hat{I}_{[B]}$; σ - standard deviation of the original image I

$(N,M) \leftarrow$ size of $\hat{I}_{[B]}$
for $i \le N; i \leftarrow i+1$ **do** ▷ N is the number of rows of the image $\hat{I}_{[B]}$
    for $j \le M; j \leftarrow j+1$ **do** ▷ M is the number of columns of the image $\hat{I}_{[B]}$
        if $p_{i,j} == 255$ **then**
            $points \leftarrow [points; p_{i,j}]$
        end if
    end for
end for

$n \leftarrow \frac{length(points)}{\frac{\sigma}{2}}$ ▷ n is the step for seed points selection

for $i \le length(points); i \leftarrow i \times n$ **do** ▷ Seed points selection on the first level
    $seedPoints \leftarrow [seedPoints; points_i]$
end for
return $(seedPoints)$ ▷ Selection of seed points at the first level

---

$$C = \begin{cases} \frac{\sigma}{5}, & \sigma > 60 \\ \frac{\sigma}{3}, & \text{otherwise} \end{cases} \quad (7)$$

Third, we introduce *the second level seed points* in order to detect asphalt pavement more accurately. The second level is formed based on the values of neighbouring pixels from the first level. For each first level seed point (red circle), we inspect four neighbouring windows of size $C \times C$, where C is formed by (7). The windows are illustrated in Fig. 3b). Then,

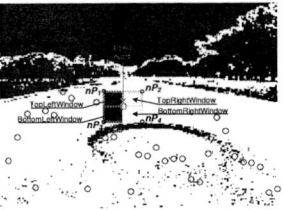

(a) The first level seed points (red circles).
(b) The process of determination of the second level seed points.

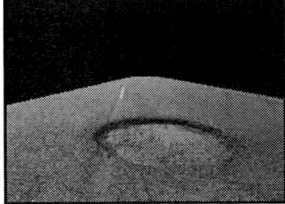

(c) The second level seed points (blue points).
(d) Detected asphalt pavement (I_{ROI}).

Fig. 3: Region of Interest Extraction.

for each selected window (*TopLeftWindow, TopRightWindow, BottomLeftWindow, BottomRightWindow*), we caluclate average pixel value. The second level seed points are selected by comparing average pixel value to 250. If the value is larger or eqaul to 250, the selected second level seed point is diagonally opposite pixel from the first level seed point (for example nP_1, nP_2, nP_3 and/or nP_4). This step is repeated n times in order to cover as much of the region of interest. The number of iterations $n = 5$ is emirically selected. The second level seed points are illustrated in Fig. 3c) as blue points. The region of interest is extracted based on the position of these points, while the values of pixels from the rest of the region is changed to 0. The final result is presented in the figure Fig. 3d).

C. Pothole detection

The third and final step consists of three phases: the method of cropped images and Otsu thresholding, linear and image boundary shapes elimination and detection of pothole regions.

The first phase is based on our previous work [6], where we proposed a new unsupervised method based on image processing and spectral clustering for identification and estimation of potholes on asphalt pavements. In this paper, we use only the first and a small portion of the second step from our previous work for the purpose of pothole detection. The majority of second and third steps were based on spectral clustering, and they are now not necessary due to a newly proposed preprocessing and region of interest extraction steps. There are additional differences when compared to the previous method. The first difference is the input image into the first step, which is now the result of Region of Interest Extraction step (I_{ROI}). The second difference is that we use only the step for elimination of all linear and image boundary shapes from the second step of the previous method, and disregard the rest.

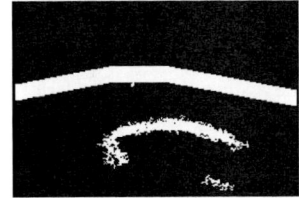
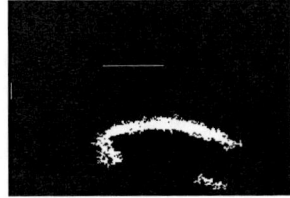

(a) The result of Cropped images and Otsu thresholding method.
(b) All linear and image boundary connected shapes elimination.

Fig. 4: Detection of shapes.

Fig. 5: Detected potholes.

The measure for removing all linear regions is the same and it is defined by eccentricity ϵ of shapes [25].

The resulting image after the first and second steps is shown in Fig. 4a) and b), respectively. Based on the shapes detected, we draw rectangles over detected potholes in the final step, as shown in Fig. 5.

IV. IMPLEMENTATION AND RESULTS

A. Implementation

Our method has been implemented in MATLAB version 7.11.0 (R2010b) with Image Processing Toolbox. The image processing was performed on Intel Core2 2.80 GHz CPU with 8GB of RAM.

B. Dataset

The common practice for testing an pothole detection method is to generate a a data set with in-expensive and omnipresent equipment mounted on a passenger vehicle, i.e. off the shelf digital cameras for video and photo acquisition. Such a data set usually captures images taken from only one angle (front or back camera), which enables extraction of parameters (such as vanishing point, pavement lane and its surroundings, etc) for narrowing the scope of search. Unfortunately, no publically available data set exists for detection and classification of any defect type [26]. This prevents true verification of results and makes comparison between different methods very difficult.

The effectiveness of our method is verified on an image data set[1] formed from selected images from Google search engine (keyword "pothole"). The selected images are collected with different types of cameras. Images are taken from various angles with different irregualar shapes and number of potholes. We believe that this unstructured data set can be used to examine the true potential of any pothole detection method.

C. Results

The method was tested on more than 80 different pothole images captured from various cameras, shooting angles, different irregualar shapes and number of potholes, different sizes and backgrounds. The capabilities of our method on selected images are shown in Fig. 6. The execution time varies significantly on the complexity of an image. The average execution time on a given data set is 465.71 ms with average mean error of 7%.

The detection accuracy is calculated by comparing the detected number of potholes and an actual number of potholes visiable on an image (calculated manually). We considered detection succesful if the method detected more than half of potholes on an image. On a given data set the accuracy was 82%. The method shows promising results on pothole images from simple to complicated background, and high illumination. Since our method does not require any training stage, it has low computational cost when compared with supervised methods. The results indicate that the method is suitable as a pre-processing step for other supervised methods.

V. CONCLUSION

In this paper we proposed a new unsupervised method for automatic detection of potholes based on RGB color space image segmentation. Potholes are bowl-shaped depressions in the asphalt pavement, thus the region of interest was defined as asphalt pavement. Hence, we narrowed the search for potholes in this region only. The effectiveness of the method depends on the extraction of ROI accuracy. Once ROI is extracted, the potholes are detected by comparing two cropped images and performing Otsu thresholding method. Once all linear and image boundary shapes are eliminated, the remaining regions are potential potholes.

The method was tested on 80 different pothole images and the detection accuracy is calculated manually. On a given data set, our method detected all potholes and the surface estimation was 82% accurate. Thus, the method is suitable for rough estimation of potholes. The results show that the method detected potholes with reasonable accuracy.

REFERENCES

[1] T.P. Wilson and A.R. Romine, *Materials and Procedures for Repair of Potholes in Asphalt-Surfaced Pavements*, Report No. FHWA-RD-99-168, 1999

[2] Asakura, Toshihiro, and Yoshiyuki Kojima. "Tunnel maintenance in Japan." Tunnelling and Underground Space Technology 18.2 (2003): 161-169.

[3] National Transportation Safety Board, "Collapse of I-35W Highway Bridge, Minneapolis, Minnesota, August 1, 2007," 2008. [Online]. Available: "http://www.dot.state.mn.us/i35wbridge/ntsb/finalreport.pdf"

[4] J.R. Parker *Algorithms for Image Processing and Computer Vision*, Second Edition, Wiley Publishing, Inc., 2011

[5] N. Paragios, G. Tziritas, *Adaptive detection and localization of moving objects in image sequences*, Signal Processing: Image Communication, Vol. 14, pp. 277-296, 1999

[6] Buza, Emir, Samir Omanovic, and Alvin Huseinovic. "Pothole detection with image processing and spectral clustering." Proceedings of the 2nd International Conference on Information Technology and Computer Networks. 2013.

[7] Kim, Taehyeong, and Seung-Ki Ryu. "Review and analysis of pothole detection methods." Journal of Emerging Trends in Computing and Information Sciences 5.8 (2014): 603-608.

[8] K.T. Chang, J.R. Chang, J.K. Liu, *Detection of pavement distresses using 3D laser scanning technology*, Computing in Civil Engineering, pp. 1-11, 2005

[9] Li, Qingguang, et al. "A real-time 3D scanning system for pavement distortion inspection." Measurement Science and Technology 21.1 (2009): 015702.

[10] Wang, Kelvin CP. "Challenges and feasibility for comprehensive automated survey of pavement conditions." Applications of Advanced Technologies in Transportation Engineering (2004). 2004. 531-536.

[11] Hou, Zhiqiong, Kelvin CP Wang, and Weiguo Gong. "Experimentation of 3D pavement imaging through stereovision." International Conference on Transportation Engineering 2007. 2007.

[12] Staniek, Marcin. "Stereo vision techniques in the road pavement evaluation." XXVIII International Baltic Road Conference. 2013.

[13] Joubert, Deon, et al. "Pothole tagging system." (2011).

[14] Jahanshahi, Mohammad R., et al. "Unsupervised approach for autonomous pavement-defect detection and quantification using an inexpensive depth sensor." Journal of Computing in Civil Engineering 27.6 (2012): 743-754.

[15] Moazzam, I., et al. "Metrology and visualization of potholes using the microsoft kinect sensor." Intelligent Transportation Systems-(ITSC), 2013 16th International IEEE Conference on. IEEE, 2013.

[16] A. Mednis, G. Strazdins, R. Zviedris, G. Kanonirs, L. Selavo, *Real time pothole detection using Android smartphones with accelerometers*, Distributed Computing in Sensor Systems and Workshops (DCOSS), pp. 1-6, 2011

[17] Yu, Bill X., and Xinbao Yu. "Vibration-based system for pavement condition evaluation." Applications of Advanced Technology in Transportation. 2006. 183-189.

[18] De Zoysa, Kasun, et al. "A public transport system based sensor network for road surface condition monitoring." Proceedings of the 2007 workshop on Networked systems for developing regions. ACM, 2007.

[19] J. Eriksson, L. Girod, B. Hull, R. Newton, S. Madden, H. Balakrishnan, *The pothole patrol: using a mobile sensor network for road surface monitoring*, Proceeding of the 6th international conference on Mobile systems, applications, and services, pp. 29-39, 2008

[20] J. Karuppuswamy, V. Selvaraj, M.M. Ganesh, E.L. Hall, *Detection and avoidance of simulated potholes in autonomous vehicle navigation in an unstructured environment*, Intelligent Robots and Computer Vision XIX: Algorithms, Techniques, and Active Vision, vol. 4197, pp. 70-80, 2000

[21] C. Koch, I. Brilakis, *Pothole detection in asphalt pavement image*, Advanced Engineering Informatics, Vol. 25(3), pp. 507-515, 2011

[22] C. Koch, G. Jog and I. Brilakis, "Towards Automated Pothole Distress Assessment Using Asphalt Pavement Video Data," Journal of Computing in Civil Engineering, 2012.

[23] Huidrom, Lokeshwor, Lalit Kumar Das, and S. K. Sud. "Method for automated assessment of potholes, cracks and patches from road surface video clips." Procedia-Social and Behavioral Sciences 104 (2013): 312-321.

[24] Radopoulou, Stefania C., and Ioannis Brilakis. "Automated Detection of Multiple Pavement Defects." Journal of Computing in Civil Engineering (2016): 04016057.

[25] Matlab Toolbox for Image Processing, Image Analysis, Region and Image Properties, *http://www.mathworks.co.uk/help/images/ref/regionprops.html#bqkf8id*

[26] Koch, Christian, et al. "A review on computer vision based defect detection and condition assessment of concrete and asphalt civil infrastructure." Advanced Engineering Informatics 29.2 (2015): 196-210.

---

[1] For the purpose of verification and comparison of future methods, the data set is available at http://people.etf.unsa.ba/~aakagic/pothole_detection/.

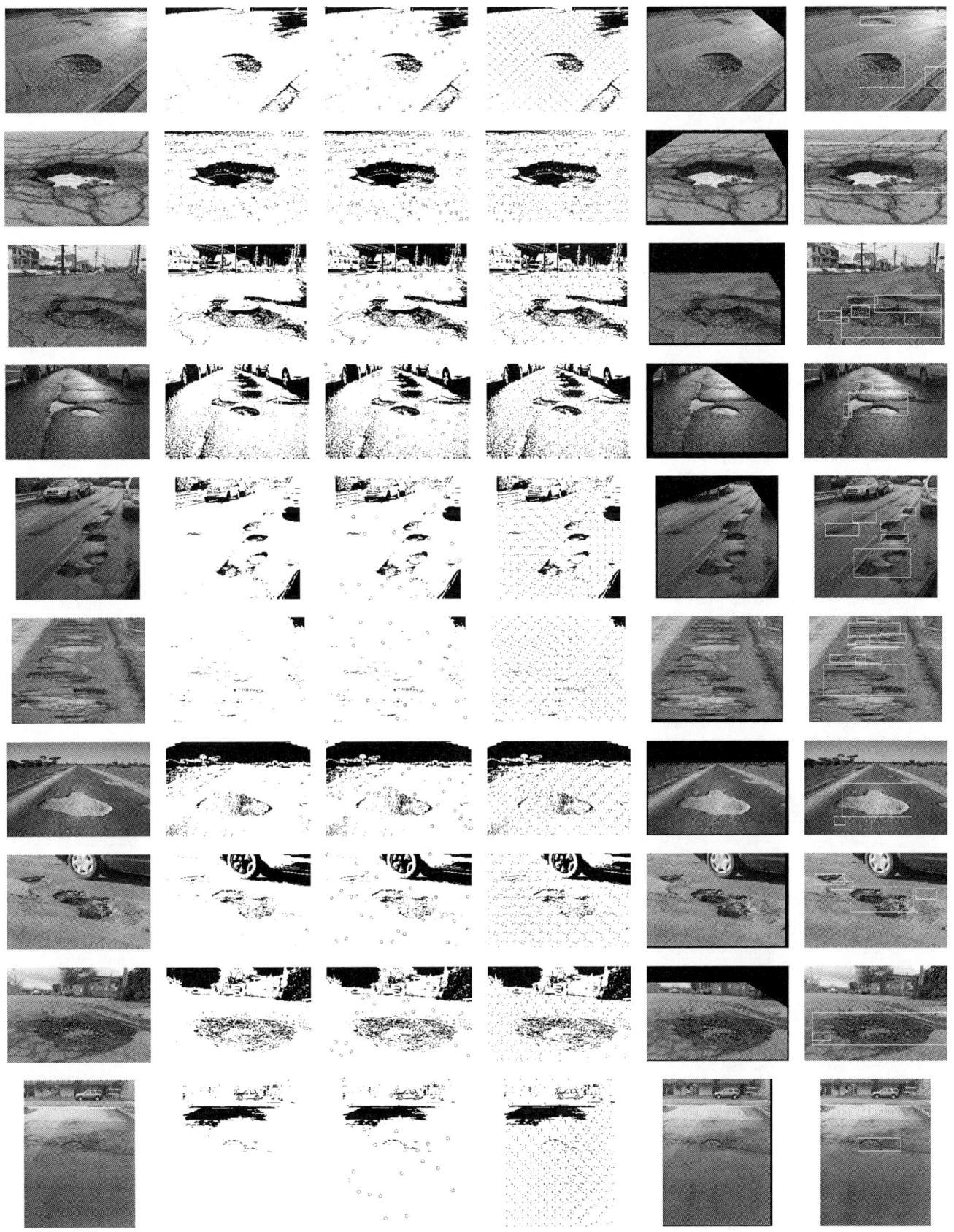

Fig. 6: Pothole detection capabilities of our method: (1) Orginal image, (2) Segmented image, (3) The first level seed points, (4) The secound level seed points, (5) Original image with removed non-asphalt area, (6) Pothole detected.

Classification of the Hand-Printed and Printed Medieval Glagolitic Documents using Differentiation in Orthography

D. Brodić*, and A. Amelio**

* University of Belgrade, Technical Faculty in Bor, Bor, Serbia
**DIMES University of Calabria, Rende (CS), Italy
dbrodic@tfbor.bg.ac.rs, aamelio@dimes.unical.it

Abstract - The angular Glagolitic documents dated from the XIII to XV century in the hand-printed form and from the XV up to XIX century in the printed form. During such a wide period of time, the way of writing documents was changed. It is particularly true for so-called old and new Glagolitic orthography. The old Glagolitic orthography was used in all documents whose origin is Croatia as well as when the editors were Croatians. This was spread in the period from the XIII to the first half of XIV century. After that period, under the influence of the Latin script, the Glagolitic orthography changed to so-called new one. The paper proposes a new automatic methodology for the classification and differentiation of the Glagolitic documents written by old and new orthography style. It consists of initial document transformation to coded text according to the horizontal energy levels in the text line. This coded text is then seen as an image. The image is subjected to the texture analysis in order to extract a feature vector. The extracted feature vector is used as an input to the classification tool. As a final result, the classification tool differentiates Glagolitic documents written in different orthography styles. The experiment is based on an excerpt of the original Glagolitic documents dated from the XIV to XIX century. The obtained results regarding differentiation of Glagolitic documents are very promising.

I. INTRODUCTION

Documents in a given language can be characterized by different orthography rules. Orthography is an important aspect of the natural language processing which is mainly connected to the written language. In particular, it defines the set of rules of "correct writing", realizing a certain capitalization, emphasis, hyphenation, punctuation and word breaks style [1].

A language in a given historical period is subjected to evolution due to multiple influences. Consequently, its orthography is subjected to variation, too. Sometimes, these variations in orthography are so visible that they are connected to a new development of the language. In this case, specific orthography rules may characterize the origin of the historical document. For some languages, dating and printing origin of the documents may be easily identified by the differences in orthography style.

Some work has been introduced in the literature for orthography recognition. In particular, it analyzed the connection between the distinction of some tokens during the evolution of the language [2] and in multiple languages [3]. This was accomplished by employing traditional linguistic methods, such as bi-grams, tri-grams, probabilistic mapping and changes in tokens and vocabulary [4]. The main limitation of the proposed approaches is their lack of generality. In particular, they have been introduced for orthography differentiation in the same language or among different languages, but they have not been or cannot be applied in both the contexts.

To overcome the limitations of the previous works, we propose a new more general method for orthography differentiation and recognition. It may capture the changes in orthography style determining the evolution of the script in the same language or identify the orthography changes tracing the evolution in a different language. In this paper, the method is specifically employed for recognition of documents in angular Glagolitic script of the Croatian language as written in old or new orthography style. To the best of our knowledge, it is the first time that a similar approach has been introduced in the literature. The angular Glagolitic documents dated from the XIII to XV century mainly in their hand-printed form, and from XV up to XIX century in their printed form. The old Glagolitic orthography, whose spreading advanced in the period from the XIII to the first half of XIV century, was used in hand-printed and printed documents from the Croatian region or where editors were Croatians [5]. After that period, under the influence of the Latin language, the orthography was changed into the new one. It followed the change between hand-printed and printed documents [6]. Because the old and new orthography styles are mainly related to two different historical periods, the method is implicitly able to identify the origin of the Medieval Croatian documents. In particular, old and new orthography styles of Glagolitic documents represent the same combination of letters in spite of different ways of writing capitals. Although a local approach has a clear advantage over a global one, the orthography differentiation by extracting certain letters is superfluous. The only solution to solve such a problem is a transformation of a given information into another space, which can differentiate given orthography styles. This has been established by coding each letter according to its energy projection profile. After a coding phase, transforming the document into a codified image, texture analysis is applied to the image for feature extraction. Finally, features of the document are classified for orthography recognition.

The paper is organized as follows. Section II describes the main concepts underlying the method, including script coding, texture analysis and classification. Section III presents the experiment. Section IV makes a discussion about the experiment. Finally, V makes the conclusions and outlines future work directions.

II. THE METHOD

The proposed method has the aim to differentiate and recognize the angular Glagolitic documents as written in old or new orthography style.

It is composed of the following steps: (i) script coding, (ii) texture analysis, and (iii) classification. First, the horizontal projection profile is applied to segment text lines in text establishing a central line in each text line. Then, each blob is framed by the bounding box. Furthermore, the distribution of the blob heights and its center point can be extracted. These features are used in a classification according to typographical features [7]. It is illustrated in Fig. 1. Then, script coding classifies the letters of the document into four types according to their horizontal energy profile in the text line. It provides a mapping of each letter to a certain code representing the class of the letter. In this way, the document is transformed into a sequence of codes. This determines linear patters of codes inside the document which are able to characterize the script of the document. Then, coded text is subjected to transformation into an image. Disposition, location and length of the linear patterns are quite dissimilar in documents written by different scripts. For this reason, application of texture analysis on the image corresponding to the coded text determines features which can differentiate documents as given in the different scripts [8]. Because orthography style of the document is related to script characteristics, this procedure directly realizes the differentiation of the orthography styles. Finally, classification employed on the obtained document features is able to recognize the orthography style of the document (old or new one). Fig. 2 shows the main steps of the proposed method.

A. Script Coding

Script coding is the first and essential step of the proposed method for orthography recognition. Its aim is to classify each letter in the text into a certain script type, based on the horizontal energy profile of the letter determining its position in the text line area. Fig. 3 shows such a classification of the letters based on their location in the text line area. According to that, letters are classified as: (i) base letters, (ii) ascender letters, (iii) descendent letters, and (iv) full letters [9]. Each class is associated to a numerical code in the set of 4 codes {0, 1, 2, 3}. In particular, we have that: (i) base letter corresponds to 0, (ii) ascender letter corresponds to 1, (iii)

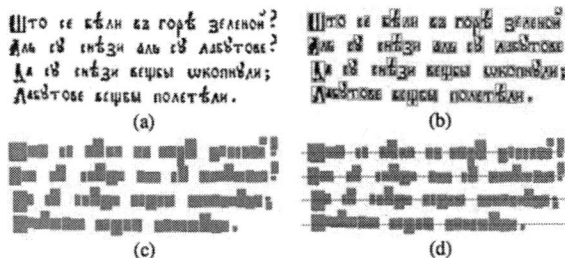

Figure 1. Pre-mapping steps: (a) initial text, (b) bounding box extraction, (c) bounding box filling, and (d) center line [7]

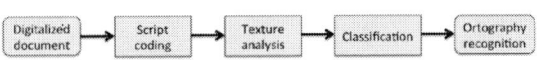

Figure 2. The main steps of the proposed method: (i) script coding, (ii) texture analysis, and (iii) classification

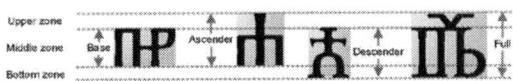

Figure 3. Classification of the letters according to their position in the text line area

descendent letter is associated to 2, and (iv) full letter is associated to 3 [10]. Hence, each letter of the document is mapped into a numerical code based on its classification.

Hence, the document is transformed into a sequence of numerical codes which may be considered as amplitude values of a signal. Consequently, each numerical code is represented as a gray level of an image, and the document is treated as an image I with four gray levels. Fig. 4 illustrates the association between the four numerical codes and the gray levels of the image. This transformation determines a reduction of the variables from the size of the alphabet to only four ones, which provides a noticeable advantage in space and time for the

Figure 4. Association between numerical codes and gray levels of the image

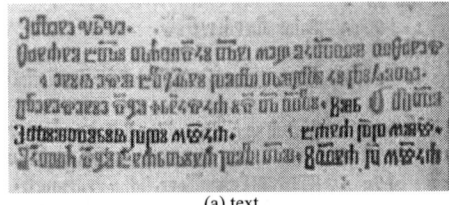

(a) text

```
31000 0100
30100 010 010100100 00200100 0300
0 0000 000 012100 2010 0030 00 301000
300000000 120 0010001011110 200 3 310
310000000 220 0100 0101 22 000
30001 1210100001201010 3101 0 0101
```
(b) coding

Figure 5. Printed text excerpt in old orthography style (1493) [11]

1111

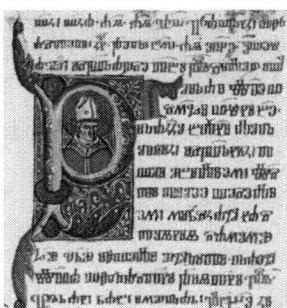

```
000 001 10 10 30 21102 03
100000 3 3000200 10 222 2000
100 0201200 000 3 20101 01
0010 0 100
0022 00000
01000 010 100
00000 0020000000
00 00000 000010
000 0000013010
000000 010000
30 000 30010 030000 0103
101 03010000 300000 3
20010 0100000010 3020 00
```

(a) text   (b) coding

Figure 6. Hand-printed text excerpt in old orthography style (1368)

ⰗⰊⰈ, ⱃⰔⰡⰀⱁⰠⰴⰑⰅ ⰒⰊⰍⰓⰍⰈⰴⱑⰔⰈ ⰖⰗⰈⰴⰊⰟⰈ, * ⰵⰝⰊⰭⰭⰟⰊ-ⱇⰎⰗⰊ ⰈⰋⰕ ⰝⰅⰠⰀⰛⰵⰔⱃⰟⱓⰵⰝⰊ ⰲⰢⱃⰠⰋⰕⰝⰭⰭⰟⰕ ⰓⰈⰡⰈⰵⰀⰊⰓⰑⰖⰂ.

ⰗⰊⰈ Ⰸ ⱍⰎⰈⰵⰊⰓⰀⰗ ⰥⰈⰟⰢ ⰵⰈⰎⰔⰟⰡ * ⰝⰊ ⰵⰴⱚⰡⰝⰭⰭⰀⰰ ⰵⰝⰊⰴⰰⰈⰕⰈ.

ⰛⰡⰍⰎⰡⰡ * ⰝⰊⰰⰦⰡⰈⰴⰊⰟⰈ ⰈⰕ ⱃⰎⰡⰡⰈⰗⰊⰈ.

(a) text

```
100 20010000 0000001000 1010 000000
310 000 010020000 000100000 000001000
100 0 0001002 1000 000000 00 01001
00100
10000 0100000 00 200000
```

(b) coding

Figure 7. Printed text excerpt in new orthography style (1862) [12]

feature extraction process.

Next, we show that script coding is a very useful task for orthography differentiation in Glagolitic documents. In particular, old orthography style represents the capital letters as descendent ones, while new orthography style writes the capital letters as ascender ones, like in the Latin alphabet. This characteristic allows to make some differentiation based on script types. As an example, we report the coding of a printed text excerpt in Glagolitic script and of a hand-printed text excerpt in Glagolitic script in the old orthography style, and of a printed text excerpt in Glagolitic script in the new orthography style. Fig. 5 depicts the printed text excerpt in the old orthography style and its corresponding coding. Fig. 6 shows the hand-printed text excerpt in the old orthography style. The printed text excerpt in the new orthography style is depicted in Fig. 7. We may intuitively observe that some differences occur at the level of linear patterns between old and new orthography (see Figs. 5-6 (b) and Fig. 7 (b)). Table I quantifies the distribution of the script types for the three text excerpts in Figs. 5, 6 and 7. We can observe that some distinction between old and new orthography is clearly visible in terms of descendent and full letters. In fact, we have around 3.5% more of descendent letters and around 5% more of full letters in the new orthography style. Differently, hand-printed and printed texts in old orthography style exhibit a similar percentage of descendent and full letters. This confirms the efficacy of the proposed script coding.

TABLE I. DISTRIBUTION OF THE SCRIPT TYPES IN THE TEXT EXCERPTS IN OLD AND NEW ORTHOGRAPHY

Letter	Base	Ascender	Descendent	Full
Old Handprinted				
No.	145	29	15	14
%	71.42	14.29	7.39	6.90
Old printed				
No.	98	35	11	11
%	63.22	22.58	7.10	7.10
New printed				
No.	95	17	4	1
%	81.20	14.53	3.42	0.85

B. Texture Analysis

Image I is subjected to texture analysis for the features extraction. Texture brings information about the disposition of the intensities in the image, in order to generate the features. Run-length statistics method for texture extraction is in the focus.

Run-length statistics are characterized by the concept of run. It is a sequence of consecutive pixels of the same intensity along a given direction of the texture. If image I has M intensity levels and maximum length of a run N, the run-length matrix p can be computed by considering the runs in I along a given direction. An element at position (i,j) of p contains the number of runs of intensity level i and length j. Fig. 8 shows an example of run-length matrix generated from an image pattern where each gray level is represented by the numerical code.

Starting from p, eleven features can be generated: (i) Short run emphasis (SRE), (ii) Long run emphasis (LRE), (iii) Gray-level non-uniformity (GLN), (iv) Run length non-uniformity (RLN), and (v) Run percentage (RP) [13]; (vi) Low gray-level run emphasis (LGRE), and (vii) High gray-level run emphasis (HGRE) [14]; (viii) Short run low gray-level emphasis (SRLGE), (ix) Short run high gray-level emphasis (SRHGE), (x) Long run low gray-level emphasis (LRLGE), and (xi) Long run high gray-level emphasis (LRHGE) [15].

Table II reports the 11 run-length features. N_r represents the total number of runs, while n_p is the number of pixels of image I.

In particular, SRE quantifies the distribution of short runs and it is sensitive to their occurrence. Accordingly, it will take higher values for finer textures. LRE evaluates the distribution of long runs. Hence, it will have higher values for coarser textures. GLN quantifies the similarity of gray levels over the image. If image presents some differences in gray levels, GLN will take low values. RLN measures the similarity of the length of the runs over the image. Hence, it will have low values if runs exhibit different length. RP quantifies the homogeneity and distribution of runs along a given direction. If the length of the runs is 1 for all gray levels along a given direction, RP will take the highest value.

Furthermore, SRLGE, SRHGE, LRLGE and LRHGE are based on the concept of joint statistical measure of gray level and run-length.

TABLE II. THE ELEVEN RUN-LENGTH FEATURES

Feature	Formula
Short run emphasis (SRE)	$\dfrac{1}{n_r}\displaystyle\sum_{i=1}^{M}\sum_{j=1}^{N}\dfrac{p(i,j)}{j^2}.$
Long run emphasis (LRE)	$\dfrac{1}{n_r}\displaystyle\sum_{i=1}^{M}\sum_{j=1}^{N}p(i,j)\cdot j^2.$
Gray-level non-uniformity (GLN)	$\dfrac{1}{n_r}\displaystyle\sum_{i=1}^{M}\left(\sum_{j=1}^{N}p(i,j)\right)^2.$
Run length non-uniformity (RLN)	$\dfrac{1}{n_r}\displaystyle\sum_{j=1}^{N}\left(\sum_{i=1}^{M}p(i,j)\right)^2.$
Run percentage (RP)	$n_r / n_p.$
Low gray-level run emphasis (LGRE)	$\dfrac{1}{n_r}\displaystyle\sum_{i=1}^{M}\sum_{j=1}^{N}\dfrac{p(i,j)}{i^2}.$
High gray-level run emphasis (HGRE)	$\dfrac{1}{n_r}\displaystyle\sum_{i=1}^{M}\sum_{j=1}^{N}p(i,j)\cdot i^2.$
Short run low gray-level emphasis (SRLGE)	$\dfrac{1}{n_r}\displaystyle\sum_{i=1}^{M}\sum_{j=1}^{N}\dfrac{p(i,j)}{i^2\cdot j^2}.$
Short run high gray-level emphasis (SRHGE)	$\dfrac{1}{n_r}\displaystyle\sum_{i=1}^{M}\sum_{j=1}^{N}p(i,j)\cdot i^2 / j^2.$
Long run low gray-level emphasis (LRLGE)	$\dfrac{1}{n_r}\displaystyle\sum_{i=1}^{M}\sum_{j=1}^{N}p(i,j)\cdot j^2 / i^2.$
Long run high gray-level emphasis (LRHGE)	$\dfrac{1}{n_r}\displaystyle\sum_{i=1}^{M}\sum_{j=1}^{N}p(i,j)\cdot i^2\cdot j^2.$

C. Classification

Feature vector representing the document composed of 11 run-length features is classified by Naive Bayes (NB) method [16] for recognition of its orthography style. We adopted NB method because it is quite robust to noise and extensively used for text document classification [17], for which it obtained good accuracy results [18]. After a training phase, the test feature vector is the input of the classification process, recognizing the script of the corresponding document as given in the old or new orthography style.

NB method is based on the assumption of *conditional independence* between the features, which means that all features are mutually independent, considering the class

label value. In particular, given a binary classification problem with the class label given as 1 or 0, let n be the number of features (in our case we have $n=11$ features) and $x_i=\{x_i^1,...,x_i^n\}$ be a feature vector with the associated class label y_i. Considering the assumption of *conditional independence*, the NB classifier is defined as follows:

$$f_{nb}(x_i)=\frac{p(y_i=1)}{p(y_i=0)}\prod_{h=1}^{n}\frac{p(x_i^h\mid y_i=1)}{p(x_i^h\mid y_i=0)}, \qquad (1)$$

where $p(y_i=1)$ is the probability of occurrence of class label $y_i=1$, and $p(x_i^h|y_i=1)$ is the probability of occurrence of the feature value x_i^h given the class label $y_i=1$. In order to classify a test feature vector x_t, the probability terms in (1) will be computed for its feature values according to the training set. In the end, x_t will take the class label 1 if and only if $f_{nb}(x_t)>=1$, otherwise it will take the class label 0. When the features take numerical values, which is specifically our case, the probability term is computed by the normal function as follows:

$$p(x_i^h\mid y_i)=f(x_i^h,\mu_{y_i},\sigma_{y_i})=\frac{1}{\sqrt{2\pi}\sigma}e^{\frac{(x_i^h-\mu_{y_i})^2}{\sigma_{y_i}^2}}, \qquad (2)$$

where μ_{yi} and σ_{yi} are respectively the mean and the standard deviation of the values of h-th feature when class label is y_i.

III. EXPERIMENT

An experiment is conducted for evaluating the accuracy of the proposed method in old and new orthography recognition of the angular Glagolitic documents. Accordingly, the method is tested on a custom-oriented database of thirteen printed text excerpts in Glagolitic script. Five out of thirteen texts are written in the old Glagolitic orthography style, while eight out of thirteen texts are given in the new Glagolitic orthography style. Texts in the old orthography style have been extracted from the well-known book entitled *Missale Romanum Glagolitice* dated from 1483 [7], [19]. It is the first printed book in Glagolitic script and represents a wonderful and important contribution to Croatian history. About the texts in the new orthography style, six out of eight ones have been extracted from the book *The Confession and Knowledge of the True Christian Faith* dated from 1564 [6]. Finally, two out of eight texts have

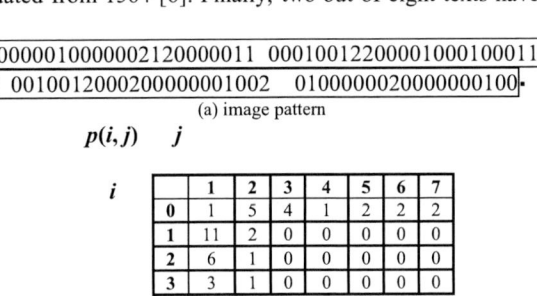

Figure 8. Run-length matrix generated from the image (• marks the end of the text

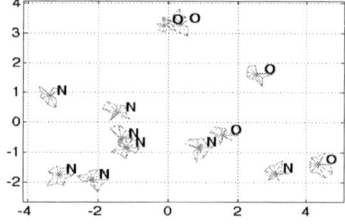

(a)

(b)

Figure 9. Glagolitic text excerpts from selected books in (a) old orthography style, and (b) new orthography style

been extracted from the book *Foundations of the Old Slavic language* dated from 1862 [12]. All historical Glagolitic books have been found at the National Library of Zagreb in their digitalized format. Texts excerpts have been randomly selected from the books, extracted from them and processed in their digital format. Fig. 9 shows an example of two text excerpts from selected books in the old and new Glagolitic orthography style.

IV. RESULTS AND DISCUSSION

The proposed method has been implemented in MATLAB R2015a on a laptop computer Quad-Core 2.2 GHz with 8 GB RAM and UNIX operating system.

A test involving different feature representations, such as co-occurrence-based and ALBP-based features [20], has been performed. In particular, image patterns have been processed by run-length, co-occurrence and ALBP methods in order to obtain three different representations. In the end, we observed that run-length statistics received the best accuracy results in this context. For this reason, they have been definitively employed for this analysis.

Because run-length statistics determine numerical feature values, NB method uses (2) to compute the probability terms by the normal function. Furthermore, classification of the texts for orthography recognition has been performed by other two different methods: Support Vector Machine (SVM) [21], and K-Nearest Neighbors (K-NN) [22]. Specifically, a linear kernel function for SVM demonstrated to be enough for our purposes. For K-NN, the cosine similarity was used, which performed considerably better than the other measures. The number of neighbors has been fixed to 5, because different values did not obtain considerable improvements in classification accuracy. Obtained results demonstrated that NB is the most accurate method for orthography recognition. According to that, it has been employed as the classifier inside the system.

Because SVM and K-NN methods are based on the concept of distance which is sensitive to large differences in the values, the feature vectors have been normalized before the application of the classification task. In particular, we adopted the following normalization

formula for every value inside each feature vector of the database:

$$x'^h_i = (x^h_i - min_h)/(max_h - min_h). \qquad (3)$$

where x^h_i is the value of h-th feature inside the x_i feature vector, min_h is the minimum value of h-th feature in the database, and max_h is the maximum value of h-th feature in the database.

Our task is a binary classification problem, where the orthography of the text excerpt is identified as the new or old one, according to the script characteristics. In order to evaluate the classification accuracy, the confusion matrix is generated between the classification result of the method and the ground truth labeling of the texts in old and new orthography [23]. From the confusion matrix, precision, recall, and f-measure are computed. *Precision* is the ratio between retrieved and relevant documents and all retrieved documents. *Recall* is the ratio between relevant retrieved documents and all relevant documents. *F-Measure* is the harmonic mean of precision and recall.

In order to make the evaluation independent from the adopted training and test sets, the K-fold cross validation is employed on the database [24]. In particular, the database is randomly divided into K folds. Then, each fold is alternately used as the test set, while the remaining K-1 folds are employed as the training set. Every time, the performance measures are computed on the found classification solution. In the end, the average precision, recall and f-measure, together with the standard deviation values, are reported as the final result. In this case, we fixed K=10 for cross validation, which is a typical value.

Fig. 10 shows the glyphplot of the feature vectors in the database. It realizes a representation and analysis of the multivariate data in the database. Each feature vector is represented as a "star", called glyph, where i-th spoke has a length which is proportional to i-th feature value. Each feature (column of the database) is shifted and scaled separately in the interval [0,1], and glyphs are centered inside a rectangular grid for visualization. Finally, glyphs are labeled as 'O' (old orthography style) or 'N' (new orthography style), according to the class label of the corresponding feature vector in the database. We can clearly observe that proposed feature representation has a good discriminatory capability. In fact, most of the glyphs in the old orthography style are easily separable inside the grid from glyphs in the new orthography style.

Figure 10. Glyphplot of the feature vectors in the database

Table III reports the average precision, recall and f-measure, together with the corresponding standard deviation (in parenthesis), when NB method is employed inside the system. Results are computed for each class of old and new orthography. Furthermore, the results of NB method are compared with those obtained by SVM and K-NN methods. They confirm that NB method receives the best classification results, with the lowest values of standard deviation. In fact, it obtains an f-measure value of 0.97 for new orthography and of 0.90 for old orthography, which are high values of the performance measure. Differently, SVM does not overcome 0.87 for old orthography and 0.80 for new orthography. Finally, K-NN obtains 0.93 for new orthography and 0.80 for old orthography, which are lower than values obtained by NB method. Similar results are clearly visible also in terms of precision and recall. We may observe that results obtained by SVM and K-NN never overcome the results obtained by NB for each orthography class. In conclusion, it demonstrates the efficacy of our system in solving the orthography recognition problem.

V. CONCLUSION

This paper proposed a new method for recognition of the old and new orthography style in hand-printed and printed angular Glagolitic documents. Because differentiation in orthography style is related to different historical periods of the Glagolitic script, this method is implicitly able to identify the historical period of the document. It consists of coding the document into a long sequence of numerical codes, transforming it into an image and extracting textural features from image by run-length statistics. Finally, classification by NB method is employed on the feature vector of the document for orthography recognition. Results obtained on a custom-oriented database of Glagolitic text excerpts in the old and new orthography style show positive and encouraging results. Future work will provide an experiment on a larger database of text excerpts extracted from a huge variety of Glagolitic books.

ACKNOWLEDGMENT

This work was supported by the Ministry of Education, Science and Technological Development of the Republic Serbia [TR33037].

TABLE III. CLASSIFICATION RESULTS

Method	Precision		Recall		F-measure	
	New	Old	New	Old	New	Old
NB	**0.95** (0.16)	**0.90** (0.32)	**1.00** (0.00)	**0.90** (0.32)	**0.97** (0.10)	**0.90** (0.32)
SVM	0.80 (0.42)	0.85 (0.34)	0.80 (0.42)	0.90 (0.32)	0.80 (0.42)	0.87 (0.32)
K-NN	0.90 (0.21)	0.80 (0.42)	1.00 (0.00)	0.80 (0.42)	0.93 (0.14)	0.80 (0.42)

REFERENCES

[1] F. Coulmas, "The Blackwell Encyclopedia of Writing Systems", Oxford: Blackwell, p. 379, 1996.

[2] D. Garrette, and H. Alpert-Abrams, "An Unsupervised Model of Orthographic Variation for Historical Document Transcription", 15th Annual Conference of the North American Chapter of the Association for Computational Linguistics: Human Language Technologies, San Diego, USA, pp. 467-472, 2016.

[3] O. Biller, J. El-Sana, and K. Kedem, "The influence of language orthographic characteristics on digital word recognition", 11th IAPR International Workshop on Document Analysis Systems, Tours, France, pp. 131-135, 2014.

[4] U. Reffle, and C. Ringlstetter, "Unsupervised profiling of OCRed historical documents", Pattern Recogn., Vol. 46, pp. 1346-1357, 2013.

[5] L. Febvre, and H.J. Martin, "The Coming of the Book: The Impact of Printing 1450- 1800", Verso, 1976.

[6] S. Lipovčan, "Discovering the Glagolitic script of Croatia", Zagreb, Croatia: Erasmus Publisher, 2000.

[7] D. Brodić, A. Amelio, and Z.N. Milivojević, " Classification of the Scripts in Medieval Documents from Balkan Region by Run-Length Texture Analysis", 22th Int. Conf. on Neural Information Processsing (ICONIP), Istanbul, Turkey, pp. 442-450, 2015.

[8] D. Brodić, A. Amelio, and Z.N. Milivojević, "Identification of Fraktur and Latin Scripts in German Historical Documents Using Image Texture Analysis", Appl. Art. Int., Vol. 30, No. 5, pp. 379-395, 2016.

[9] A. Zramdini, and R. Ingold, "Optical font recognition using typographical features", IEEE Trans. Pattern Analysis and Machine Intelligence, Vol. 8, No. 20, pp. 877-882, 1998.

[10] D. Brodić, A. Amelio, and Z.N. Milivojević, "Language Discrimination by Texture Analysis of the Image Corresponding to the Text", Neural Computing and Applications, pp. 1-21, 2016.

[11] "Baromic's Breviary", Venice, 1493.

[12] I. Berčić, "Foundations of the Old Slavic language written by Glagolitic scripts to read the church books", Prague,1862.

[13] M.M. Galloway, "Texture analysis using gray level run lengths", Comp. Graph. Im. Proc., Vol. 4, No. 2, pp. 172-179, 1975.

[14] A. Chu, C.M. Sehgal, and J.F. Greenleaf, "Use of gray value distribution of run lengths for texture analysis", Pattern Recogn. Lett., Vol. 11, No. 6, pp. 415-419, 1990.

[15] B. R. Dasarathy, and E.B. Holder, "Image characterizations based on joint gray-level run-length distributions", Pattern Recogn. Lett., Vol. 12, No. 8, pp. 497-502, 1991.

[16] S. Russell, and P. Norvig, "Artificial Intelligence: A Modern Approach (2nd ed.)", Prentice Hall, 2003 [1995].

[17] S. Raschka, "Naive Bayes and Text Classification: Introduction and Theory", Ithaca, USA: Cornell University Library, 2014.

[18] M. Shahid, S.S. Hassan, and M. Rafi, "Comparing SVM and naive Bayes classifiers for text categorization with Wikitology as knowledge enrichment", IEEE 14th International Multi-Topic Conference, Karachi, Pakistan, 2011.

[19] "Missale Romanum Glagolitice", Venice, 1483.

[20] D. Brodić, A. Amelio, and Z.N. Milivojević, "Characterization and Distinction Between Closely Related South Slavic Languages on the Example of Serbian and Croatian", 16th Int. Conf. on Computer Analysis of Images and Patterns (CAIP),Valletta, Malta, pp. 654-666, 2015.

[21] C. Cortes, and V. Vapnik, "Support-vector networks", Mach. Learn., Vol. 20, No. 3, pp. 273-297, 1995.

[22] N.S. Altman, "An introduction to kernel and nearest-neighbor nonparametric regression", The American Stat., Vol. 46, No. 3, pp. 175-185, 1992.

[23] Confusion Matrix, http://www2.cs.uregina.ca/dbd/cs831/notes/ confusion_matrix/confusion_matrix.html

[24] Cross Validation, https://www.cs.cmu.edu/ schneide/tut5/node42. html, 1997.

An evolutionary approach to route the heterogeneous groups of underwater robots

M.Yu. Kenzin, I.V. Bychkov and N.N. Maksimkin

Matrosov Institute for System Dynamics and Control Theory SB RAS, Irkutsk, Russia
gorthauers@gmail.com

Abstract - An evolutionary approach to solve the dynamic routing problem for the heterogeneous group of robots is presented. Since robots in the group may differ by their speed and, more importantly, by their functionality, each robot is able of performing only a specific subset of tasks among all tasks of the mission. The routing problem is to find a feasible group route ensuring well-timed accomplishment of all tasks. We propose a variation of the evolutionary algorithm to effectively solve the problem described. The heterogeneity factor implies some specific constraints on the genetic operators, thus we have developed both a new multi-mode mutation and crossover operators as well as the adapted algorithm structure to answer these changes. A software modeling system implementing all the necessary computational procedures has been developed; the results of computations are given.

I. INTRODUCTION

The rapid evolution of the subsea technologies in recent years has significantly expanded the scope of the autonomous underwater vehicles (AUV) implementation, which includes the usage of distributed groups of underwater robots to perform stand-alone multi-objective missions of long duration. Such large-scale missions demand handling of a range of different underwater works and operations in different areas of the underwater space. The effective coordination of AUVs is required not only during group movement while following preplanned trajectories [1], but, in the first place, on the upper level of the control system that is responsible for both task allocation and path planning. In this regard, it is a problem of considerable practical interest to effectively route the group of functionally heterogeneous vehicles.

A significant number of papers examining different approaches to solving the vehicle routing problems (VRPs) address classical statements of the problem such as capacitated routing, routing with time windows, etc. In recent years, the research community has turned to more advanced variants of the VRPs, which are aimed at a more accurate simulation of real-world problems and were considered too difficult to handle [2] before. These variants include new complex requirements and restrictions, such as the heterogeneity of the vehicles in the group, new types of restrictions on length and duration of the routs, multiple visits requirement, etc.

In this paper, we propose an approach to solving the

The research is partially supported by RSF (project № 16-11-00053: problem statement and evolutionary algorithm) and by RFBR (project 16-29-04238-офи_м: simulation framework).

problem of heterogeneous robots group control as a new variation of a vehicle routing problem, which is to find a feasible and efficient route for the group under specific spatio-temporal constraints. To do this, we combine the genetic algorithms with new specialized constructed and improvement heuristics.

II. GENERAL MODEL

In general, the multi-objective mission of the AUVs group is to visit and inspect (perform some underwater works) the set of waypoints (objectives) under certain requirements [3]. The vehicles in the group have different sets of on-board equipment, enabling each AUV to inspect only a certain subset of mission objectives. Thus, the routing problem here is to find a feasible group route ensuring, as far as possible, the most effective inspection of the objectives. It is formally defined as follows.

Assume there is a set of waypoints (objectives) $N = \{1,...,n\}$ within the given water area. These objectives are defined not only by their location in space, but also by the time interval s_i and the type of equipment $w_i \in \{1,...,w\}$ required for its inspection. In addition, each objective $i \in N$ receives its periodicity value p_i. It means that the duration of the time interval between its two successive inspections by robots of the group should preferably be equal to p_i. In that way, in case of arriving to the objective ahead of time, an AUV should loiter before starting the inspection. The inspection of waypoints in equal time periods provides an efficient way to study the dynamics of various underwater processes with AUVs taking samples and measurement, photo and video shoots.

The group of robots performing the mission consists of M vehicles with cruising speed v^k, $k = 1,...,M$ and a set of on-board equipment $u^k = \langle u_1^k, u_2^k,...,u_w^k \rangle$, $u_l^k \in \{0,1\}$, $l = 1,...,w$. Thus, the AUV $k \in M$ is able to inspect i-th objective if and only if $u_{w_i}^k = 1$ (Fig. 1).

We define the dummy nodes n_0^k for the AUVs initial positions. Let $V = N \cup \{n_0^k\}$ denote the set of all waypoints and AUVs locations. Let the $\varepsilon = \{(i,j): i,j \in V, i \neq j\}$ define the set of edges. The connected graph $G = (V, \varepsilon)$ represents the "roadmap" for the current mission of the group.

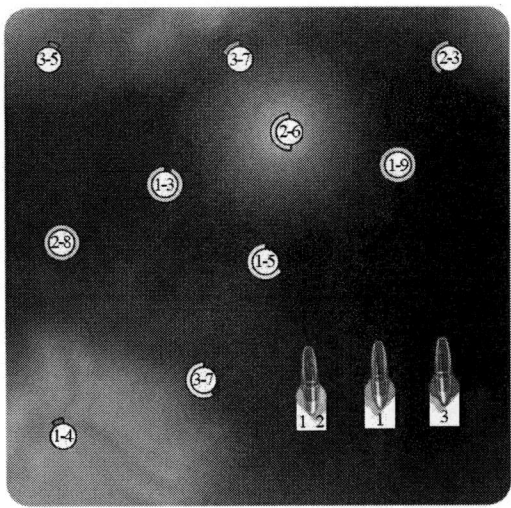

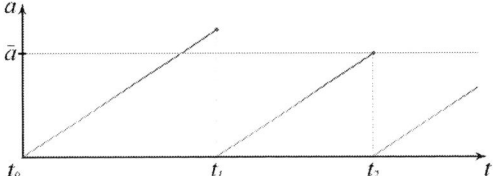

Figure 2. Graphical representation of the "hotness" function

$\bigcirc$ - the objective of the mission with its inner timer

(a-b) - equipment **a** is required to inspect each **b** time units

$\bigcirc$ - the objective with delayed inspection

(c d) - the vehicle with both **c** and **d** on-board equipment

Figure 1. The schematic representation of the group mission

The route of a single vehicle $r = \langle V_1(r), V_2(r), ..., V_h(r) \rangle$ is a list of vertices (waypoints numbers) in the consecutive order of their planned visit, and h is the route length. It should be noted that any vertex could be included more than once into the route of a single robot. The group route $R = \{r_1, ..., r_M\}$ is a combination of routes of all single AUVs. We suggest to use here the group movement scheme based on the planning periods [4] of the known and limited horizon (time duration).

III. OBJECTIVE FUNCTION

The effectiveness of the group work is defined by regularity of well-timed inspections of all objectives. Situations, when AUV arrives too late and delays the inspection of the objective, are undesirable and should be penalized via efficiency criteria (objective function).

In order to construct reasonable efficiency criterion for the group route, we define an additional function $a_i(t)$ corresponding to each objective i of the mission that defines the "hotness" of corresponding objective at specified moments of time:

$$a_i(t) = \begin{cases} \overline{a}(t - (t_{ik} + s_i)) / p_i, & t \in [t_{ik} + s_i, t_{ik+1}] \\ 0, & t \in [t_{ik}, t_{ik} + s_i] \end{cases}, \quad (1)$$
$$k = 1, 2, ...,$$

where $t_{i1}, t_{i2}, ...$ is a sequence of moments when the objective i is expected to be visited by AUVs according to the current route, and $\overline{a}$ is a constant threshold value

that is equal for all objectives. This way, the inspection of the objective i resets its "hotness" to a zero value, following that the function (1) begins to increase until next inspection, while reaching the threshold value $\overline{a}$ in the period of p_i (Fig. 2).

Now we define a function to calculate a penalty for delaying the inspection of the objective:

$$\varphi(a_i, t) = \begin{cases} a_i(t) - \overline{a}, & a_i(t) > \overline{a} \\ 0, & a_i(t) \le \overline{a} \end{cases}. \quad (2)$$

Hence, the total penalty for the group route R is a sum of penalties (2) for each inspection made by all AUVs within the route:

$$\Phi(R) = \sum_{i=1}^{M} \sum_{j=1}^{h} \varphi(a_m, t_m), \quad (3)$$

where m stands for $V_j(r_i)$ and t_m values represent the time moments of expected inspections of corresponding objectives.

However, function (3) considers only those waypoints that are being included in the group route R and only in the moments of their inspections. For this reason, it is needed to consider an additional function, which would neglect the well-timed inspection, but would estimate the level of "completeness" of all objectives at the end of group movement:

$$\Psi(R) = \sum_{i=1}^{N} a_i(t_0 + t_R), \quad (4)$$

where t_R is the moment of the mission end. The function (4) also allows us to instantly provide the group routes with two positive features: it encourages the group to inspect each waypoint of the mission and it indirectly normalizes durations of routes of all vehicles in the group.

Hence, the final efficiency criterion for the group route is as follows:

$$f(R) = M \cdot \Phi(R) + \Psi(R), R \in Z, \quad (5)$$

where Z is the set of all possible routes. It is also should be noted that function (1) is constructed in such a way that "hotness" of objectives with lesser periodicity grows faster. In that way, delaying of inspections would happen preferably with waypoints of the biggest periodicity.

Since we have the final criterion in the explicit form, we can use it as the objective function to compare different group routes on the graph G.

IV. EVOLUTIONARY APPROACH

For a broad combinatorial class of vehicle routing problems there are no algorithms solving it in polynomial time, which leads us to the class of approximation algorithms that allow obtaining rational sub-optimal solutions in low computational time. Another layer of complexity is implied by a "bad" neighborhood structure of the described patrol routing problem, making it difficult to allocate and find qualitative and feasible solutions, as they may not be in the neighborhood of other feasible high-quality solutions in the search space.

A comprehensive survey of various approaches to solving the VRP concludes that evolutionary algorithms (EA) generally outperform any other heuristics or metaheuristics [5]. Their main advantage is the ability to find solutions to poorly structured problems and problems with complex constraints, because EAs require a relatively small amount of information about the nature of the problem. The main drawback here is that algorithm's speed and efficiency strongly depend on the construction heuristics and improvement heuristics.

We propose a hybrid evolutionary approach featuring specialized genetic operators and solutions improvement heuristics addressing both the expectable large-dimension of the problem and spatio-temporal constraints that primarily arise from the AUVs heterogeneity.

The construction of the initial population of solutions is the first and crucial step to achieve good rational solutions rapidly. The aim of this step is to construct a set of only feasible chromosomes that ensures coverage of a significant portion of the search space and contains a variety of good solutions. We propose to achieve this requirement by means of three different construction heuristics: sequential insertion and two parallel insertions.

The first heuristic is a "*random insertion with no duplicates*". It builds AUVs routes sequentially: at first, the route for the one single vehicle is fully generated, then for the next vehicle, etc. The starting objective for each vehicle is randomly selected from the entire set of permissible objectives for the current AUV. All subsequent objectives are selected in the same manner but with the additional requirement to be different, if possible, from the previous objective of this vehicle.

The next two constructive heuristics are based on the time-oriented nearest-neighbor insertion heuristic by Solomon [6] and are parallel, i.e. they build routes for all vehicles in the group simultaneously. Both heuristics require the construction of an auxiliary vector $Q = (q_1, q_2, ..., q_{M \times h})$ of the mission objectives, ordered by the time of their required visits. Each element q_i of this vector is a pair $\langle e_i, t_i \rangle$, where e_i stands for the index (number) of the objective and t_i corresponds to the time moment of its scheduled inspection. The vector Q is formed as follows:

1. Construct Q from the pairs $\langle e, t \rangle$ corresponding only to the first objective. Thus, we obtain a sequence of time moments when this objective should be inspected by the vehicles of the group.

2. Perform the same procedure for the next objective with the only difference being that the temporary vector $\hat{Q}$ of the same structure is used instead of Q.

3. Merge two created vectors together by "attaching" vector $\hat{Q}$ to the vector Q and then reorder the elements q of the new extended vector Q in the ascending order of values t_i in pairs $\langle e_i, t_i \rangle$.

4. Reduce vector Q to the first $M \times h$ elements, which returns Q to its initial length.

5. Repeat 2-4 for the rest of objectives of the mission. The resulting sequence of indexes e in the pairs $\langle e, t \rangle$ will match the order in which the objectives of the mission should be visited and inspected according to their periodicity.

Both parallel heuristics use the vector Q as a source of objectives to pick and insert into routes of different vehicles. For the first heuristic of "*time-greedy insertion*", the probability to select each AUV for the current objective is inversely proportional to the number of objectives already assigned to this vehicle, while for the second "*quality-greedy insertion*" this probability depends also on the distance between the current objective and the last objective in the route of each AUV. The use of two parallel heuristics here provides the population diversity not only among the solutions built with the random insertion, but also within the set of more qualitative solutions.

It should be noted that at each iteration only those AUVs are considered, which have all the required on-board equipment. This restriction guarantees us the feasibility of each solution obtained by the heuristics proposed. The population constructed is then evaluated with the objective function (5). According to the results of ranking, the tournament selection chooses a set of solutions for procreation and mutations.

We suggest using a number of specialized genetic operators that guarantee the feasibility of the offspring solution: two different variants of crossover and the multimode mutation.

The multimode mutations proposed consists of four operators: add new objective into the route, remove the objective from the route, change the objective within the rote to another objective, swap two objectives within the route of the same AUV. When mutation operators are implied, all changes are also verified on admissibility.

The first crossover is called "*one-plus-one-point*" and it is a modification of the standard two-point crossover with the additional rule to choose randomly only the first point. The second one is then defined automatically within the route of the same vehicle as the first point.

The second crosover is "*focused on inheritance*" and it is based on the "*adaptive memory*" crossover proposed in [7]. The proposed crossover identifies the common characteristics of the parental individuals and copies them to the offspring-solution, which is a kind of adaptive memory procedure. Initially, the two solutions are selected. If there are common parts between the solutions, then these parts are inherited to the offspring; all parts left empty inherit the values from the best solutions in the adaptive memory or from some random solutions in the population. At each iteration, the adaptive memory is updated based on the best solutions found. The structure of such crossover also ensures the feasibility of resulting chromosomes.

The mechanisms of parallel populations with migration (island model) and elitism provide faster algorithm convergence rate while preventing premature convergence to a local optima. The procedure of clone removal preserves diversity of new populations on each iteration of evolutionary algorithm.

To improve the efficiency of the population creation procedure, all probabilistic genetic parameters are constantly changed at the end of each iteration of the algorithm. Being determined by the current efficiency of the genetic operators these changes allow algorithm to adapt to the different situations on the different steps of processing. The implementation of the adaptation mechanism may significantly increase the speed of computing in those cases when some genetic operators begin to work significantly better than others [3].

V. CONCLUSION

The approach proposed has been implemented in our modeling framework «AUV Mission Planner» and has demonstrated the high efficiency and reliability even under the strongest requirements and restrictions during a series of simulation studies. The suggested modification of the evolutionary algorithms has the ability to rapidly generate a set of rational and feasible solutions of high quality and to effectively improve the populations of solutions to meet the required conditions.

We have evaluated the impact of each embedded heuristic on the rate and quality of the algorithm convergence (Fig. 3). The average deviation of the resulting solutions from the optimal ones is 2,8%.

The proposed algorithm structure and new heuristics can be used not only to efficiently route the heterogeneous robot groups during multi-objective missions, but also to address other statements of VRPs and its modern variations.

The routing problem statement described by itself can be singled out as a separate subclass of VRP that combines the features of a periodic VRP and VRP with time windows and requires the original approach to address it effectively.

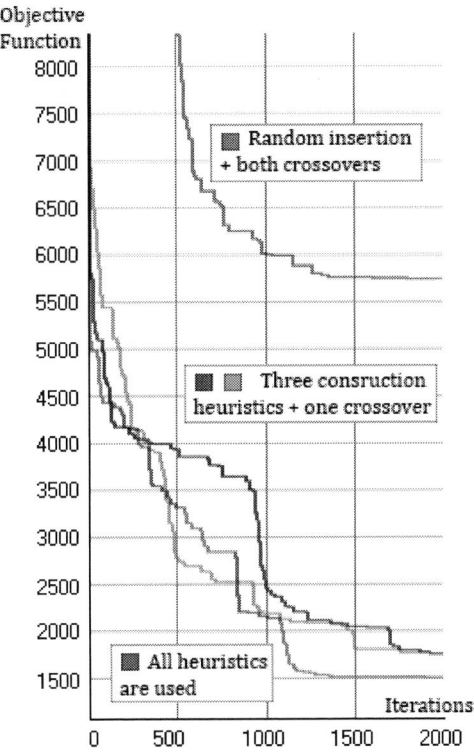

Figure 3. Rate of the algorithm convergence with different heuristics in use

REFERENCES

[1] S.N. Vassilyev, R.I. Kozlov, S.A. Ul'yanov, "Multimode formation stability", Doklady Mathematics, Vol. 89, Issue 2., pp. 257-262, 2014.

[2] R. Hartl, G. Hastle, G.E. Jansens, "Special issue on rich vehicle routing problems", Central European Journal of Operations Research, Vol. 13(2), pp. 103-104, 2006.

[3] M.Yu. Kenzin, I.V. Bychkov, and N.N. Maksimkin, "Hybrid evolutionary approach to multi-objective mission planning for group of underwater robots", Communications in Computer and Information Science, Vol. 549: Mathematical Modeling of Technological Processes, pp. 73-84, 2015.

[4] M.Yu. Kenzin, I.V. Bychkov, N.N. Maksimkin, "A hybrid approach to solve the dynamic patrol routing problem for group of underwater robots", 39th International Convention on Information and Communication Technology, Electronics and Microelectronics (MIPRO), pp. 1114-1119, 2016.

[5] O. Braysy, and M. Gendreau, "Vehicle routing problem with time windows, part I: route construction and local search algorithms," Transportation science, Vol. 39, No.1, pp. 104–118, 2005.

[6] M.M. Solomon, "Algorithms for the vehicle routing and scheduling problems with time windows constraints", Operations Research, Vol. 35(2), pp. 254–265, 1987.

[7] M. Yong, "Solving vehicle routing problem with time windows with hybrid evolutionary algorithm", In proceeding of Intelligent Systems (GCIS), 2010 Second WRI Global Congress on, Vol. 1, pp. 335-339, 2010.

MIPRO 2017, May 22- 26, 2017, Opatija, Croatia

Low Cost Robot Arm with Visual Guided Positioning

Petra Đurović, Ratko Grbić, Robert Cupec and Damir Filko

Josip Juraj Strossmayer University of Osijek, Faculty of Electrical Engineering, Computer Science and Information Technology Osijek, Osijek, Croatia

pdurovic@etfos.hr, rgrbic@etfos.hr, rcupec@etfos.hr, dfilko@etfos.hr

Abstract – Low cost robotic solutions are of great importance for improvement and development of robotics. In this paper, two visually guided low cost robot arms are proposed. The proposed system performs automatic hand-eye calibration and, after the calibration, positions its end effector above the object of interest using visual servoing based on off the shelf marker tracker. The presented experiments demonstrate positioning accuracy of the proposed setup.

I. INTRODUCTION

Robot arms are today widely used in industrial automation. Their accuracy, reliability and flexibility make them unavoidable components of many production processes. However, high cost of industrial robot arms hinders their wide application in households and education. The prices of the smallest industrial robot arms with sub-millimeter repeatability are incomparably higher than prices of average household machines. Therefore, low cost robotic solutions are of the great importance for the further development of robotics, since they broaden the population of robot users and also expand the base of researchers in this field.

In this paper, development of low cost robot arms is considered. The proposed solution consists of only basic components, which allow implementation of a vision-based robot manipulation system. The presented research is biologically inspired in the sense that hand-eye coordination, typical for living beings, is used to achieve positioning accuracy instead of relying on proprioceptive sensors in the robot joints, common in a vast majority of industrial robots. The goal of this investigation is to find out what accuracy can be achieved with such a simple configuration.

The paper is structured as follows. Section 2 provides a short overview of related research. Section 3 explains components of the proposed vision-based robot manipulation system and implementation of the calibration and visual servoing algorithms, while Section 4 presents the performed experiments and their results. Finally, Conclusion comments on the results and offers ideas for future work.

II. RELATED RESEARCH

There exist many researches where visual servoing is applied for control of a robotic arm. In some of them, industrial robots [1] are being used and eye-in-hand [2], or hybrid eye-to-hand/eye-in-hand [3] configuration is applied. Also, variety of hand-eye calibration methods are applied, such as convex linear matrix inequality [4]. Our setup consists of a low-cost robotic arm and an off-the-shelf camera in eye-to-hand configuration. This setup, which consists of significantly less expensive components in comparison to the hardware used in the aforementioned related research, together with an evaluation of the absolute positioning accuracy achieved by the proposed system represents the contribution of this paper.

III. VISION BASED ROBOT MANIPULATION SYSTEM

A. Components

Since the main goal of this paper is to provide an insight in the positioning accuracy which can be achieved by low cost vision guided robot manipulators, all

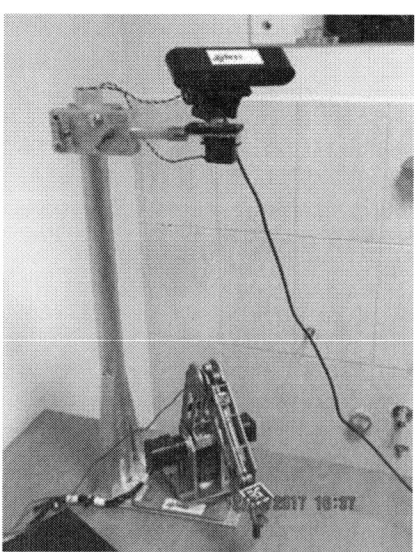

Figure 1. Dobot Arm mounted on the stand

components are acquired from the lower price specter. The robot manipulation system shown in Figure 1 consists of a robot arm, a camera mounted on a stand, a marker used for visual servoing installed on the robot's end effector and a vision-based robot control software.

Two robots are considered: Dobot Arm V1.0 and a custom made robot arm in SCARA configuration, named VICRA (Vision Controlled Robot Arm). The price of the Dobot Arm was 900$ and the manufacturing of the mechanical construction which connects the robot arm with a camera on a pan-tilt head costed 320 EUR. The development and manufacturing of VICRA prototype costed approximately 3000 EUR.

The Dobot package includes Dobot Arm, Dobot controller, 5 pieces of effectors, power adaptor, USB cable, toolkit, extension cable and base. Dobot Arm has 4 axes and 0.5 kg payload with maximum reach of 320 mm and +/- 0.2 mm position repeatability. Dobot Arm has three joints with stepper motors: base, rear arm and forearm, and the fourth joint with rotation servo. It weighs 3 kg and it is made of aluminum alloy. Dobot Arm is controlled by Dobot Controller which is based on Arduino Mega 2560 and FPGA board placed inside a control box. It can communicate with a PC via USB-UART. A Dobot Arm is shown in Figure 2. The Dobot arm and the RGB-D camera are mounted on a stand thereby forming a compact vision-based robot manipulation unit. The stand was designed in SketchUp [13] and built in a local workshop. The stand shown in Figure 1 was designed to provide secure camera holding in adequate position w.r.t. the robotic arm. The camera is positioned in such a way that the robot arm workspace is completely contained inside the camera's field of view. It is located sufficiently close to the robot's workspace in order to achieve precise measurement of objects which the robot should manipulate with. In the same time, the distance to the closest object must be greater than the lower bound of the camera range. Furthermore, the marker used for visual servoing must always be clearly

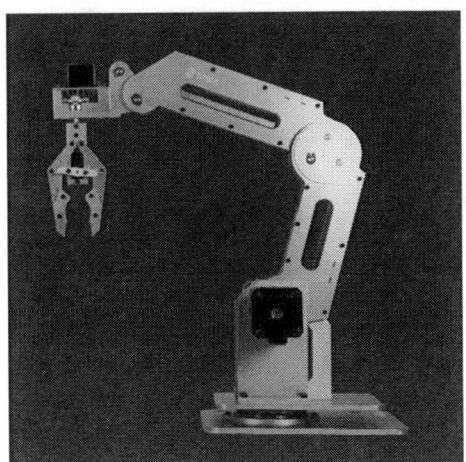

Figure 2. Dobot Arm V1.0

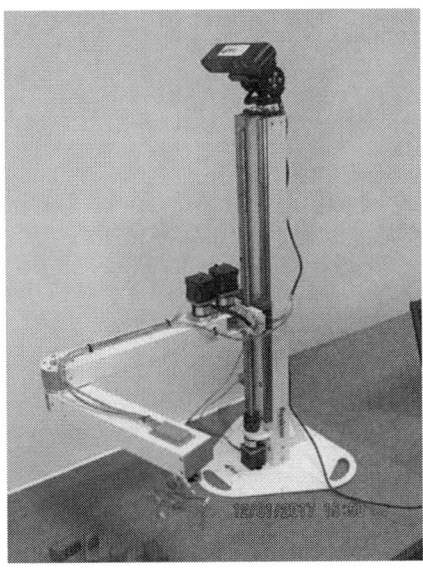

Figure 3. Robot manipulation system with VICRA

visible.

VICRA, shown in Figure 3, is a custom made robot arm in SCARA configuration. It has one translational and three rotational joints. The first three joints, which position the tool, are driven by stepper motors, while the fourth joint, which defines the tool orientation, is driven by a DC servo motor. The currently used tool is a gripper driven by a servo motor, but it can easily be replaced by alternative tools. The first translational joint enables vertical reach of approximately 0.6 m and the two rotational joints enable horizontal reach of approximately 0.6 m. The weight of the robot is approximately 8 kg, which makes it suitable for mounting on a mobile platform. The robot is controlled by Arduino Mega 2560 with RAMPS 1.4 shield which communicates with a PC via USB. The robot is designed for a pan-tilt head with a camera to be mounted on the top of the first axis, in order to support computer vision-based control. Also, inexpensive micro-switches are used as end-stop for all three axes.

Each of the two considered robot arms has its advantages and drawbacks. VICRA is more expensive than Dobot, but it has greater payload and wider operative range. Also, because of its configuration, in specific positions, Dobot's joints interfere with camera's view of the marker used for visual feedback. On the other hand, VICRA provides undisturbed view of the marker in each position.

Both considered robot configurations are four axes robot arms, whose absolute positioning is achieved by a 3D perception sensor being the only sensor applied. The first three axes of the robot are driven by stepper motors without encoders and the fourth axis is driven by a DC servo motor. The robots are controlled by an Arduino-based controller and the applied 3D perception sensor is an affordable off-the-shelf RGB-D camera, Orbbec Astra S, costing 150 US$.

Although a common RGB-camera can be used for implementation of visual feedback, a RGB-D camera Orbbec Astra S is used in the considered system because we wanted to build a system with 3D vision capability, since results of recent computer vision research clearly demonstrate advantages of 3D vision systems [6]–[10]. Orbbec Astra S is a camera optimized for short-range use cases, from 0.35 m to 2.5 m. It is compatible with existing OpenNI [14] applications and provided with Astra SDK software. Astra SDK is used for acquiring image from the camera which is then processed by algorithms presented in Sections III.B and III.C. Even though presented manipulation module currently doesn't use depth obtained from the RGB-D camera, it is planned to use this information in future versions.

Both robot manipulators are based on stepper motors and no absolute encoders or any other proprioceptive sensor are used in this research for measuring absolute position. Stepper motors are controlled by series of impulses defining relative changes in joint angles. Hence, in order to achieve absolute positioning, the initial position must be known. Furthermore, stepper motors could lose impulses, which could lead to wrong positioning. In order to achieve precise absolute positioning, visual servoing using RGB-D camera is applied.

Visual servoing is implemented by detection and localization of a marker mounted on the robot arm close to the end-effector. Marker detection and pose estimation is implemented using ArUco library for augmented reality [11] based on Open CV [12]. Since an object of interest is not always captured in the current camera field of view, the camera is mounted on a pan-tilt head. This way, the camera is able to rotate up and down, left and right and find the right angle to successfully capture objects of interest. The pan-tilt head also allows navigation of mobile robot manipulator, which can be created by mounting of the presented vision guided robot arm on a mobile platform. In the experiments reported in this paper, the capabilities of the pan-tilt head are not used, i.e. the camera orientation remains fixed during the experiment.

The control software consists of functions for controlling robot's joints [15], built in Python, and functions for visual servoing written in C++.

B. Hand-Eye Calibration

Systems where a robotic "hand" is guided by a camera, which represents "eyes", are called Hand-eye systems. There are two main types of these systems: (i) the eye-in-hand configuration, where the camera is mounted on the arm, close to its end effector, and moves along with the end-effector, and (ii) the eye-to-hand configuration, where the camera is stationary and overlooks the robot's movements. In this paper, the

second configuration is used, primarily because the camera was too big to be mounted on the applied robot arms. In order to be able to coordinate the robot movement with the information obtained from the camera images, the pose of the robot reference frame S_R w.r.t. the camera reference frame S_C must be determined. A calibration procedure used to compute this pose is known as hand-eye calibration.

To perform a hand-eye calibration, a set of images of a calibration object is acquired. The calibration object used in the research presented in this paper is a marker mounted close to the end effector of the robot arm, which is also used for visual servoing.

First, let's introduce the notation used in the rest of this paper. In this paper, ${}^{B}\mathbf{T}_A$ denotes the homogeneous transformation matrix representing the pose of a reference frame S_A w.r.t. a reference frame S_B. Analogously, ${}^{B}\mathbf{R}_A$ and ${}^{B}\mathbf{t}_A$ denote rotation matrix and translation vector defining the orientation and position of S_A w.r.t. S_B respectively. A rotation matrix and a translation vector are included in a homogeneous transformation matrix as follows

$$ {}^{B}\mathbf{T}_A = \left[\begin{array}{ccc|c} & {}^{B}\mathbf{R}_A & & {}^{B}\mathbf{t}_A \\ \hline 0 & 0 & 0 & 1 \end{array} \right]. $$

Furthermore, the following notation is used for reference frames: R represents robot, C camera, G tool, and M marker reference frame.

The pose of S_R w.r.t. S_C can be computed from three marker positions, referred to in this section as position 0, 1 and 2. Assuming that the marker is clearly visible, the applied vision software provides us with 3D position of the marker denoted by ${}^{C}\mathbf{t}_M(k)$, for each robot position $k = 0, 1, 2$.

If the robot moves from position 0 to position 1 along the x-axis of its reference frame, then the marker positions ${}^{C}\mathbf{t}_M(0)$ and ${}^{C}\mathbf{t}_M(1)$ define the x-axis of the robot reference frame viewed from the camera reference frame, i.e. unit vector

$$ {}^{C}\mathbf{x}_R = \frac{{}^{C}\mathbf{t}_M(1) - {}^{C}\mathbf{t}_M(0)}{\left\| {}^{C}\mathbf{t}_M(1) - {}^{C}\mathbf{t}_M(0) \right\|}, \qquad (1) $$

represents the x-axis of S_R w.r.t. S_C.

Furthermore, if the robot moves from position 1 to position 2, where the both positions have the same z-coordinate in the robot reference frame, then the marker positions ${}^{C}\mathbf{t}_M(1)$ and ${}^{C}\mathbf{t}_M(2)$ define a vector perpendicular to the z-axis of the robot reference frame. Consequently, unit vector

$$ {}^{C}\mathbf{z}_R = \frac{{}^{C}\mathbf{x}_R \times \mathbf{b}}{\left\| {}^{C}\mathbf{x}_R \times \mathbf{b} \right\|}, \qquad (2) $$

where

$$\mathbf{b} = {}^{C}\mathbf{t}_{M}(2) - {}^{C}\mathbf{t}_{M}(1), \tag{3}$$

represents the z-axis of S_R w.r.t. S_C. Vectors ${}^{C}\mathbf{x}_R$ and ${}^{C}\mathbf{z}_R$ completely define the orientation of S_R w.r.t. S_C, which can be represented by rotation matrix

$$ {}^{C}\mathbf{R}_{R} = \left[{}^{C}\mathbf{x}_{R} \mid {}^{C}\mathbf{y}_{R} \mid {}^{C}\mathbf{z}_{R} \right] \tag{4}$$

where

$$ {}^{C}\mathbf{y}_{R} = {}^{C}\mathbf{z}_{R} \times {}^{C}\mathbf{x}_{R}. \tag{5}$$

The position of the origin of S_R w.r.t. S_C can be computed from ${}^{C}\mathbf{t}_M(0)$, the previously determined rotation matrix ${}^{C}\mathbf{R}_R$ and the position of the marker w.r.t. the robot end effector ${}^{G}\mathbf{t}_M$, which is expected to be known in advance. The relation between ${}^{C}\mathbf{t}_M(0)$ and the corresponding position of the end effector w.r.t. robot reference frame can be described by the following equation

$$ \begin{bmatrix} {}^{C}\mathbf{t}_{M}(0) \\ \hline 1 \end{bmatrix} = {}^{C}\mathbf{T}_{R} \, {}^{R}\mathbf{T}_{G} \begin{bmatrix} {}^{G}\mathbf{t}_{M} \\ \hline 1 \end{bmatrix}. \tag{6}$$

The both considered robots are constructed in such a way that the z-axis of the tool reference frame S_G is always parallel to the z-axis of S_R. Hence, if position 0 is selected so that the x-axis of S_G is parallel to the x-axis of S_R, then ${}^{R}\mathbf{R}_G$ is the identity matrix and (6) can be written as

$$ {}^{C}\mathbf{t}_{M}(0) = {}^{C}\mathbf{R}_{R} \cdot \left({}^{G}\mathbf{t}_{M} + {}^{R}\mathbf{t}_{G}(0) \right) + {}^{C}\mathbf{t}_{R}. $$

Consequently, vector ${}^{C}\mathbf{t}_R$ can be computed by

$$ {}^{C}\mathbf{t}_{R} = {}^{C}\mathbf{t}_{M}(0) - {}^{C}\mathbf{R}_{R} \cdot \left({}^{G}\mathbf{t}_{M} + {}^{R}\mathbf{t}_{G}(0) \right). \tag{7}$$

The eye-to-hand calibration is performed in the following steps.

Step 1. Position the robot tool at an arbitrary position ${}^{R}\mathbf{t}_G(0)$ on the x-axis of the robot RF.

Step 2. Capture an image and measure ${}^{C}\mathbf{t}_M(0)$.

Step 3. Move the robot tool in the x-direction for distance a. The resulting position is ${}^{R}\mathbf{t}_G(1)$.

Step 4. Capture an image and measure ${}^{C}\mathbf{t}_M(1)$ using the tracking tool.

Step 5. Move the robot tool in the y-direction for distance b. The resulting position is ${}^{R}\mathbf{t}_G(2)$.

Step 6. Capture an image and measure ${}^{C}\mathbf{t}_M(2)$ using the tracking tool.

Step 7. Compute ${}^{C}\mathbf{R}_R$ using equations (1) – (5) and ${}^{C}\mathbf{t}_R$ by (7).

C. Visual Servoing

The visual servoing algorithm presented in this section computes changes of joints variables required to move the marker in a wanted position based on the image acquired by the camera, a desired marker position ${}^{C}\mathbf{t}_{Mr}$ w.r.t. the camera reference frame and the pose of the robot reference frame w.r.t. the camera reference frame ${}^{C}\mathbf{T}_R$, obtained by calibration. In a practical application, the desired marker position ${}^{C}\mathbf{t}_{M,r}$ is computed from a pose of an object of interest obtained by an appropriate computer vision software, which is not a subject of this paper. The task of the presented visual servoing algorithm is to position the marker at ${}^{C}\mathbf{t}_{M,r}$ within a given tolerance. The visual servoing algorithm consists of the following steps.

Step 1. Based on a given image, the current marker position ${}^{C}\mathbf{t}_M$ w.r.t. the camera is computed using a tracking tool (in the current implementation, this is ArUco software).

Step 2. Compute the current marker position it the robot reference frame by ${}^{R}\mathbf{t}_M = {}^{R}\mathbf{T}_C \cdot {}^{C}\mathbf{t}_M$.

Step 3. Compute joint angles $\mathbf{q}$ from ${}^{R}\mathbf{t}_M$ by inverse kinematics.

Step 4. Compute the reference marker position in the robot reference frame by ${}^{R}\mathbf{t}_{Mr} = {}^{R}\mathbf{T}_C \cdot {}^{C}\mathbf{t}_{Mr}$.

Step 5. Compute new joint angles $\mathbf{q}_r$ from ${}^{R}\mathbf{t}_{Mr}$ by inverse kinematics.

Step 6. Compute the required changes in joint angles by $\Delta\mathbf{q} = \mathbf{q}_r - \mathbf{q}$.

Step 7. Send $\Delta\mathbf{q}$ to the robot controller.

These steps are repeated iteratively until the marker reaches the desired position within a given tolerance.

IV. EXPERIMENTAL EVALUATION

In order to test the developed system, two test case scenarios were performed. In the first experiment, the accuracy of the camera calibration was evaluated by comparing the estimation of the tool position measured by vision, the tool position given to the robot and the actual tool position measured manually. In the second experiment, the goal was to test the accuracy of the implemented visual servoing algorithm. The experiments were performed using Dobot Arm. Since the positioning accuracy depends on visual feedback and not on the mechanical construction of the robot, similar results are expected to be obtained by VICRA. Nevertheless, experiments with VICRA are not included in this paper, since its development is still in progress.

In order to facilitate manual measurements, in the first experiment, a laser pointer is mounted instead of a gripping tool. The manually measured positions are considered in this analysis as the ground truth positions. The tool positions given to the robot are referred to in this section as robot positions and the positions measured by

TABLE I.   STATISTICAL DATA ANALYSIS OF TOOL COORDINATES MEASURED BY VISION.

	BEFORE CORRECTION	AFTER CORRECTION
Δx_{max} [mm]	5.94	4.87
Δy_{max} [mm]	14.61	2.30
Δd_{max} [mm]	8.93	4.64
σ_x [mm]	1.80	1.74
σ_y [mm]	4.17	0.93
σ_d [mm]	2.61	1.76

TABLE II.   STATISTICAL DATA ANALYSIS OF TOOL POSITIONS GIVEN TO THE ROBOT.

	BEFORE CORRECTION	AFTER CORRECTION
Δx_{max} [mm]	7.00	1.86
Δy_{max} [mm]	2.00	1.25
Δd_{max} [mm]	7.23	1.75
σ_x [mm]	1.95	0.77
σ_y [mm]	0.61	0.44
σ_d [mm]	1.95	0.78

vision are referred to as vision measurements. After camera calibration, a square matrix of 10×10 points is given to the robot as reference positions. The robot's task was to move to these points, one by one, wait for the vision system to calculate the tool pose and a human operator to measure the tool position. In order to allow comparison of the measurement obtained by vision to the open-loop robot positioning and the manual measurements, the pose estimated by the vision system is transformed in the robot reference frame. Since the robot has a laser as the end effector, the point projected by the laser on a millimeter paper allowed an easy measurement in the robot reference frame. After all 100 points were processed, a statistical analysis is performed. The results of this analysis are shown in the middle column of Table I and Table II. In Table I, the statistics of the difference between the ground truth positions and the vision measurements is shown, where Δx_{max} and Δy_{max} represent the maximal absolute difference in the x- and y-coordinates, Δd_{max} represents the maximal distance, while σ_x, σ_y, and σ_d represent the standard deviations of the x-coordinates, y-coordinates and distances. The analogous statistical analysis of the difference between the ground truth positions and the tool positions given to the robot is shown in Table II.

The presented analysis shows that the calibration procedure proposed in Section III.B, although fast and simple, provides the transformation ${}^{C}\mathbf{T}_R$ with significant inaccuracy. We noticed that both the measurements obtained by vision and the robot open-loop positioning have significant bias, which could possibly be reduced by an additional calibration step. In order to estimate to which extent the considered positioning errors represent random noise and to which extent these errors contain a bias, which could be reduced by an additional correction step or a more thorough calibration procedure, we performed the following analysis. We computed values a_x, b_x, a_y and b_y which minimize the following error functions

$$\mathfrak{I}_x\left(a_x, b_x\right) = \sum_i \left(x_{r,i} - x_i'\right)^2,$$

$$\mathfrak{I}_y\left(a_y, b_y\right) = \sum_i \left(y_{r,i} - y_i'\right)^2,$$

where $x_{r,i}$ and $y_{r,i}$ are reference values obtained by manual measurement, while x_i' and y_i' are values obtained by linear mapping of the measured values x_i and y_i defined by

$$x_i' = a_x x_i + b_x, \qquad y_i' = a_y y_i + b_y. \qquad (8)$$

The statistics of the vision error and robot error for the values obtained by mapping (8) is shown in the last column of Table I and Table II. This analysis indicates that the positioning error can be significantly reduced by an additional correction step, such as the linear mapping proposed in this section. If the robot is required to have a good positioning accuracy within an approximately planar workspace, then the presented linear mapping can be used as the final calibration step, resulting in a more accurate positioning. If the robot is required to have accurate positioning in a wider workspace including significant changes in z-direction, then a more thorough calibration procedure should be applied.

In the second experiment, visual servoing accuracy was evaluated. A marker which represented an object on the scene, was put in 30 different positions, one by one, in the robot's working region. This marker is referred to in this paper as *target marker*. After capturing an image, the center of the target marker representing the desired position was determined.

The robot initially moved to this position by setting its joints to the values computed by inverse kinematics. If the difference between the robot's position and the target marker after this movement was greater than a given threshold, which was 3 mm in this case, the visual servoing algorithm described in Section III.C was applied. This process is repeated until the positioning error falls below the given threshold. After the robot stopped moving, the distance between the laser point on the millimeter paper, which represents the position of the end effector, and the center of the target marker was measured by the human operator. The results of the

TABLE III. STATISTICAL DATA ANALYSIS OF DISTANCES BETWEEN TARGET AND OBTAINED POSITIONS BASED ON VISUAL SERVOING ALGORITHM

Δd_{max} [mm]	6
σ_d [mm]	1.42

described experiment are given in Table III, where Δd_{max} represents the maximum distance between the center of the target marker and the laser position and σ_d represents the standard deviation of this distance.

V. CONLUSION AND FUTURE WORK

In this paper, a simple vision guided robot manipulation system consisting of a low cost robot arm and an off-the-shelf RGB-D camera is presented. As the main contribution, we emphasize the development of a low-cost solution which can be used by a wide population of developers. Furthermore, an analysis of the positioning accuracy of the developed system achieved by visual servoing without proprioceptive sensors is presented. From the results of this analysis it can be concluded that the positioning error is within 6 mm with standard deviation of 1.42 mm.

The main error source in the proposed positioning system comes from inaccurate determination of the marker distance w.r.t. the camera reference frame. We expect that a significant improvement to the given system can be achieved by fusing the measurement obtained by ArUco software with depth information provided by a RGB-D camera. This will be the topic of our future research.

ACKNOWLEDGMENT

This work has been fully supported by the Croatian Science Foundation under the project number IP-2014-09-3155.

REFERENCES

[1] S. van Delden and F. Hardy, "Robotic Eye-in-hand Calibration in an Uncalibrated Environment."

[2] K. H. Strobl and G. Hirzinger, "Optimal hand-eye calibration," in *Intelligent Robots and Systems, 2006 IEEE/RSJ International Conference on*, 2006, pp. 4647–4653.

[3] V. Lippiello, B. Siciliano, and L. Villani, "Eye-in-hand/eye-to-hand multi-camera visual servoing," in *Decision and Control, 2005 and 2005 European Control Conference. CDC-ECC'05. 44th IEEE Conference on*, 2005, pp. 5354–5359.

[4] J. Heller, D. Henrion, and T. Pajdla, "Hand-eye and robot-world calibration by global polynomial optimization," in *2014 IEEE International Conference on Robotics and Automation (ICRA)*, 2014, pp. 3157–3164.

[5] A. Aldoma, F. Tombari F, L. Di Stefano and M. Vincze, "A global hypothesis verification method for 3d object recognition," European Conference on Computer Vision (ECCV), 2012.

[6] C. Papazov and D. Burschka, "An Efficient RANSAC for 3D Object Recognition in Noisy and Occluded Scenes," *Asian Conference on Computer Vision (ACCV)*, Part I, 2010, pp. 135–148.

[7] S. Hinterstoisser, C. Cagniart, S. Ilic, P. Sturm, N. Navab, P. Fua and V. Lepetit, "Gradient Response Maps for Real-Time Detection of Textureless Objects," *IEEE Transactions on Pattern Analysis and Machine Intelligence*, vol. 34, no. 5, 2012, pp. 876–888.

[8] C. A. Mueller, K. Pathak and A. Birk, "Object Shape Categorization in RGBD Images using Hierarchical Graph Constellation Models based on Unsupervisedly Learned Shape Parts described by a Set of Shape Specificity Levels," *IEEE/RSJ International Conference on Intelligent Robots and Systems (IROS)*, Chicago, IL, USA, 2014, pp. 3053–3060.

[9] R. Detry, C. H. Ek, M. Madry and D. Kragić, "Learning a Dictionary of Prototypical Grasp-predicting Parts from Grasping Experience," in *2014 IEEE International Conference on Robotics and Automation (ICRA)*, 2014, pp. 3157–3164.

[10] L. Twardon and H. Ritter, Interaction skills for a coat-check robot: identifying and handling the boundary components of clothes, *IEEE International Conference on Robotics and Automation (ICRA)*, 2015, pp. 3682–3688.

[11] ArUco: a minimal library for Augmented Reality applications based on OpenCV, *https://www.uco.es/investiga/grupos/ava/node/26*, February 2017.

[12] A. Kaehler and G. Bradski, Learning OpenCV 3, O'Reilly Media, December 2016.

[13] SketchUp, *https://www.sketchup.com/*, March 2017.

[14] OpenNI, The standard framework for 3D sensoring, http://www.openni.ru/, March 2017.

[15] open-dobot , *https://github.com/maxosprojects/open-dobot*, March 2017

Gap in pagination due to withheld papers.

Pages 1126-1137

Knowledge Elicitation in Multi-Agent System for Distributed Computing Management

Alexander Feoktistov*, Andrey Tchernykh**, Sergey Gorsky* and Roman Kostromin*

* Matrosov Institute for System Dynamics and Control Theory of SB RAS, Irkutsk, Russia
** Computer Science Department, CICESE Research Center, Ensenada, B.C., Mexico
agf65@yandex.ru, chernykh@cicese.mx, gorsky@icc.ru, romang70055@gmail.com

Abstract - The effective management of scalable applications for solving large problems in a heterogeneous distributed computing environment is the non-trivial problem. Scalable applications generate competitive job flows that have be executed with the help of shared resources of the environment. The promising approach to solve this problem is to use multi-agent technologies. To this end, we develop a multi-agent system for the management of scalable applications. In contrast to known multi-agent systems, our system is based on applying a special conceptual model of the environment. It includes several components of a comprehensive knowledge about both the environment and subject domains of solved problems. We propose a new approach to an elicitation of these knowledge components through an integrated use of the conceptual modelling of distributed computing, classification of jobs and resources, and parameters adjustment for agent algorithms. With this approach, specialists in various fields of distributed computing considered as users of the environment, can apply their own knowledge at different levels of the problem solving process. This flexibility and algorithm adaption are benefits of our approach. Extensive modeling and practical experiments, with variation of important parameters of applications execution, show the efficiency of our management under developed multi-agent system for scalable applications.

I. INTRODUCTION

High-performance computing (HPC) is an important component for effective solving large problems and mathematical modeling in various subject domains. HPC is based on the parallel execution of computational processes that are characterized different degrees of an interrelationship between them.

A rapid increase of the computational elements in systems for HPC leads to need for scalable applications. A computations scalability means that a problem solving time decreases with increasing a number of the nodes used by an application. Thus, the solved problem has to have an ability to be decomposed into sub problems, which can be solved as much independently as possible.

HPC supercomputers and clusters provide solutions within various distributed computing environments that include heterogeneous, autonomous and geographically distributed nodes that can have complex hybrid structures, special computational characteristics and administrative policies [1]. The nodes can have different owners with

own criteria of the usage (efficiency, profit, load balancing, power consumption, etc.).

Grids [2] and Clouds [3] are typically form such an environment.

Resources management system (RMS) is used for computations management at the node level. The known traditional RMSs are Grid Engine [4], PBS Torque [5], HTCondor [6], and SLURM [7].

The specialized middleware is used for computations management of distributed nodes through an interaction with RMSs and their substitution. For example, the Globus Toolkit [8] supports various the standards and compatibility with various DRMSs, and GridWay [9].

Usually, the environment supports the collective usage of its resources by users for the large range of problem classes. Problem specification includes information about the required computational resources, executable programs, input and output data, quality criteria (time, cost, reliability, etc.), and other required issues. The specification describes a computational job for the traditional RMSs and meta-schedulers used in environment nodes.

In general, we assume that the job includes the set of interconnected programs. The job specifications are generated by applications in the form a workflow [10, 11, 12]. It can be represented by Directed Acyclic Graph (DAG) according to logical and information relations between the programs.

Workflow management systems (WMSs) largely used for solving scientific problems are known. There are the following systems: Condor DAGMan, Pegasus, Taverna, UNICORE, etc. [13]. Nevertheless, the workflow management problems in heterogeneous distributed computing environment is not completely solved [14].

One of problem is a concordance of all solving processes. WMS operates at the application level. Its objective is to obtain best resources for workflows.

Unlike WMSs, RMSs and meta-schedulers operate at the environment level. Node administrators can configure RMSs and meta-schedulers to determine common rules of a resources usage and policies for the allocation of quotas and rights for users and their jobs. However, the effective coordination of users and owners of the resources in processes of their allocation for solved problems, taking into account the specificity of subject domains, often remains beyond their capabilities.

A promising approach for solving this non-trivial problem is applying a multi-agent technologies based on the self-organization of agent communities [15]. Within such an approach, a Multi-Agent System (MAS) implements the computations management. Agents can represent users and owners of the resources, and interact themselves with aim to meet their interests.

We know the range of MASs, successfully used for distributed computing management in practice [16,17]. Nevertheless, there are additional components of knowledge that do not used by these systems. They do not include the knowledge about the specific subject domain, job properties, agent algorithm parameters, etc.

To this end, we develop a multi-agent system for the management of scalable applications that allows specialists in various fields of distributed computing can apply their own knowledge at different levels of the problem solving process. The special attention is on a job classification system.

II. MULTI-AGENT SYSTEM

The developed MAS has a hierarchical structure, which includes two or more functional levels, and operates on the base of self-organization. At each level, agents play a variety of roles, and perform different functions. The roles may be permanent or temporal. Their changes occur at discrete times when agents need to solve new problems. Each level is related with the conceptual model knowledge layers.

Agents are autonomous, and computing management is based on their interaction. They can be organized in virtual communities. In these communities, agents interact using a cooperation or competition. The formation of virtual communities allows the MAS to adapt the management process to the new challenges.

The MAS applies the algorithm for the static resource allocation based on economic mechanisms for regulation of resources demand and supply. It uses a model of the Vickrey auction with one-round bidding [18].

The users formulate problems and define criteria of their solving. User agents create and classify jobs based on the problem formulations. Then, these agents submit the jobs to agents responsible for resource allocation.

A virtual community of agents that most suitable for the problem solving is created. Agents use the classification system of resources and jobs for self-organization into the virtual community.

Each agent of the virtual community sets bids for all jobs that reflect the genuine value of these jobs and maximal utility of the bids for this agent. The steady state of the virtual community is achieved at the end of the auction. This state is similar to the equilibrium that has been defined by Nash [19] in the game theory. Rules of the Vickrey auction provides simplicity and satisfactory rate of the decision-making by agents.

According to results of bidding, the virtual community manage the job execution. In the process of computing, agents may reallocate its own computational load among other agents through interaction with their neighbors.

Agents can also provide dynamic decomposition of a problem into sub-problems, when additional free resources are available. The decomposition is carried out in order to accelerate the problem solving and make it scalable.

There are large variety of knowledge elicitation methods [20]. Conceptual and simulation modeling are major methods used by agents. Our implementation of these methods is based on conceptual programming [21, 22].

III. CONCEPTUAL MODEL

With high-level languages and methods, conceptual modeling becomes a powerful tool for complex systems design. A conceptual model has to allow the system designers and developers to represent and structure the results of their expert analysis of a subject domain.

In this regard, the conceptual model of a subject-oriented distributed computing environment includes the following components of comprehensive knowledge:

- Computational knowledge about software modules (programs) for solving problems in subject domains and working with the environment objects, including planning of their actions;

- Schematic knowledge about the modular structure of models and algorithms, problem formulations and workflows;

- Production knowledge to support decision-making for selecting optimal algorithms, depending on the workflow and environment states;

- Knowledge about the software and hardware infrastructure of the environment and administrative policies in its nodes, including an information about users, their problems and jobs.

We develop the specialized high-level toolkit named SIRIUS II for designing the conceptual models. This toolkit provides elicitation the subject-oriented knowledge of application developers. Detailed information about the conceptual model can be found in [23, 24].

IV. JOB CLASSIFICATION SYSTEM

We developed the job classification system to improve the resource allocation. This system allows to elicit knowledge of the administrators about a conformity of jobs and resources. Such a knowledge provides a possibility to determine a job class and select conformable resources.

Requirements to a computer system for a problem solving included in jobs represent explicit characteristics such as the problem solving time, RAM, disk memory, number of nodes, processors and cores, program libraries, compilers, their keys, etc.

1139

Our job classification system allows the administrators to use these characteristics without changes, and develop new characteristics based on the value ranges for original characteristics. For example, they can create the new special characteristics with different ranges for the requested core number, problem solving time, and define job classes based on these characteristics.

Additionally, the administrators can develop implicit job characteristics using the conceptual model knowledge. These characteristics are based on the knowledge about methods, algorithms, programs execution conditions, computational history, etc.

A. Classification Model

A classification model has a set of explicit and implicit job characteristics $h_1, h_2, ..., h_k$. Each characteristic h_i for $i \in \overline{1,k}$ is defined by its rank $r_i \geq 1$, weight $w_i \geq 0$, and range R_i of values including the symbol θ of an uncertainty [25, 26]. Characteristics are partially descending ordered in accordance with their ranks.

We can define m job classes based on the set of characteristics. Each new class c_j is defined using the two disjoint subsets of characteristics: mandatory and optional subsets. The characteristic h_i included in one of these subsets for the class c_j has the specialized range $R_{ij}^{spec} \subseteq R_i \setminus \{\theta\}$ of values.

Let us denote by x and y the Boolean vectors of dimensions k and m, respectively.

Value of the element x_i is defined as follows:

$$x_i = \begin{cases} 0, & if \ R_i^{job} \equiv \{\theta\}, \\ 1, & if \ R_i^{job} \subseteq R_i \setminus \{\theta\}, \end{cases}$$

where R_i^{job} is the range of values, requested for the characteristic h_i. The value $x_i = 1$ ($x_i = 0$) shows that the characteristic h_i is used (not used) in the job specification.

The vector x satisfies $\overset{k}{\underset{i=1}{\wedge}} \overline{x_i} = 0$.

Define the following characteristic function:

$$\chi_j(x) = \begin{cases} 0, if \ \exists i : (x_i = 1) \wedge (R_i^{job} \not\subseteq R_{ij}^{spec}), \\ 1 \ otherwise, \end{cases}$$

where $i \in \overline{1,k}$, $j \in \overline{1,m}$. The value $\chi_j(x) = 1$ ($\chi_j(x) = 0$) means that the job is matched (mismatched) to the class c_j.

We implement a primary job classification applying the function $\chi_j(x)$ for m classes: $y_j = \chi_j(x)$, where $y_j = 1$ ($y_j = 0$), means that the job match (mismatch) to the class c_j, $j = \overline{1,m}$. It is obvious that the job can match to several classes.

To this end, we use the additional characteristic function $\varphi_{jl}(x,y)$ to determine when the job match ($\varphi_{jl}(x,y) = 1$) or mismatch ($\varphi_{jl}(x,y) = 0$) to the class c_j taking into account the four properties of job characteristics:

1. Probabilistic measure based on the number of optional characteristics matching to the class ($l = 1$);

2. Aggregated rank ($l = 2$);

3. Summary weight ($l = 3$);

4. Probabilistic measure based on the computational history of jobs ($l = 4$).

These functions allow implementing various variants of the additional job classification based on the primary job classification. The job classification intended for the primary filtration of the resources set. In general, it provides forming the residual set V of resources for the job execution.

B. Job Classification Algoritm

In the job classification system, the administrator creates the sets of characteristics and classes, and defines conformity of jobs and resources using own expert knowledge. The MAS uses the sorting method for the knowledge elicitation.

During the system operation, the administrator can modify and add resources and job classes including inheritance mechanisms usage.

When the job submitted in the MAS, agents apply the job classification algorithm described below.

Majority stages of the algorithm:

1. Initialization of the vector y: $y_j = 0$, $j = \overline{1,m}$.

2. Primary job classification: $y_j = \chi_j(x)$, $j = \overline{1,m}$.

3. If the vector y satisfies $\overset{m}{\underset{j=1}{\vee}} y_j = 0$, then the job classification is not possible. Go to the stage 6.

4. Otherwise, if the vector y satisfies $\left(\overset{k-1}{\underset{l=1}{\vee}} \overset{k}{\underset{q=l+1}{\vee}} (y_l \wedge y_q) \right) \vee \left(\overset{k}{\underset{l=1}{\wedge}} \overline{y_l} \right) = 0$, then the single element $y_j = 1$ exists and the job matches to the single class c_j. Go to the stage 6.

5. Otherwise, $y_j = \varphi_{jl}(x,y)$, $\forall j : y_j = 1, l = \overline{1,4}$.

6. Completion of the algorithm.

After the job classification, the MAS specializes this job by adding the directive how to use the selected resources. The MAS creates the virtual community of agents corresponding to selected resources. Then, it submits the specialized job to these agents. They perform the final selection and allocation of resources.

1140

V. AGENT ALGORITHM ADJUSTMENT

Algorithms for the resource allocation have the following control parameters:

- Limits of the node components load;
- Bonuses for limits satisfying;
- Penalties for limits exceeding;
- Priorities of job classes;
- Degree of desire to carry out the jobs of defined classes.

Administrators as experts set the initial values of these parameters in configuring the MAS. These parameters are automatically adjusted based on simulation modeling and meta-monitoring in a process of the MAS operation.

Figure 1 shows the block diagram of the parameter adjustment of agent algorithms.

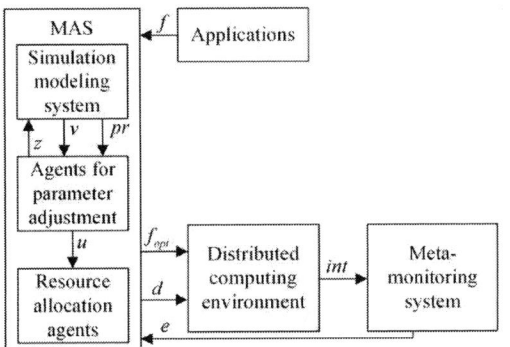

Figure 1.   Block diagram of the parameter adjustment

Table I describes the parameters used on the block diagram.

TABLE I.   PARAMETERS DESCRIPTION

Notation	Description
f	Job flow
f_{opt}	Optimized job flow
d	Job distribution
inf	Information about the environment state
e	Evaluation of the environment state
z	Vector of input variables of the simulation model
v	Vector of observed variables of the simulation model
pr	Prediction of the environment state changes
u	Vector of the control parameters

Agents for parameter adjustment use the simulation modeling to solve the following problem:

$$v_i(z) \to \min(\max) , \qquad (1)$$

$$v_i^{\min} \le v_i \le v_i^{\max} , \ i = \overline{1,n} . \qquad (2)$$

We create the simulation model as the parameter sweep application that generates the large number of model copies [27]. Each independent copy runs with the single values variant of its input variables. A meta-monitoring system provides the actual evaluation of the environment state. After execution of all copies, the environment services provide the collection of simulation results represented by variants of the vector v. We apply multi-criteria selection of the results to form the vector u with evaluation of components of the vector v using the conditions represented in (1) and (2) [28, 29]. Administrators assign an extremum for each component of the vector v before the simulation modeling.

The parameter adjustment provides for agents the efficiency of their bids for jobs at the Vickrey auction.

We consider the parameter adjustment as the element of agent learning. A feedback for learning process is implemented using the bonuses and penalties. Agents correct own decision-making taking into account this feedback.

VI. EXPERIMENTAL ANALISYS

In this section, we provides results of extensive modelling and practical experiments with the developed MAS.

We obtained and analyzed the computational history about 80000 real jobs executed in one day on three computer clusters with different computational resources (Table II).

TABLE II.   COMPUTER CLUSTERS

Cluster	Peak performance	Processor	Nodes/ processors/ cores numbers	RMS
Cluster 1	1.50 TFlops	Intel Xeon E5345	20/40/160	Cleo [30]
Cluster 2	0.77 TFlops	AMD Athlon II X4	16/16/64	HTCondor
Cluster 3	0.17 TFlops	Intel Xeon	16/32/32	SUPPZ [31]

Clusters 1 and Cluster 3 include dedicated nodes that are used within cluster only. Non-dedicated nodes of the Cluster 2 are used by owners and cluster users. In the Cleo, all cores of a node are allocated for a job request, regardless of the number of requested cores. Often, it leads to inefficient resource loading. We solve this problem using classification system.

In the classification system, we define six parental classes of jobs based on the analysis of computational history (Table III). These classes differ by the number of requested cores and execution time. Then, we created additional classes of jobs based on the parental classes taking into account the required RAM, disk space, system, applied software, etc. Each additional class is related to one of the clusters.

1141

TABLE III.   CLASSIFICATION RESULTS

Parental class	n_1	t	n_2	n_3
c_1	1	1 - 5	10221	4675
c_2	1	5 - 60	7927	5861
c_3	1	≥ 60	1211	972
c_4	≥ 2	1 - 5	33453	492
c_5	≥ 2	5 - 60	6125	50
c_6	≥ 2	≥ 60	3312	95

We generated the job flow in accordance to the obtained computational history. Each job is classified and submitted by the MAS to the matched cluster.

Table III presents the experimental results for 6 classes showing the following parameters:

- Number n_1 of requested cores,

- Requested execution time t in minutes,

- Number n_2 of jobs for the given class,

- Number n_3 of reallocation jobs.

When the classification system is used, about 12000 jobs have been relocated for better resource utilization in comparison to initial processing. The relocated jobs retained all characteristics of the classes defined for them. Figure 2 and Figure 3 illustrate a jobs distribution change for hours of the day without job classification and with it.

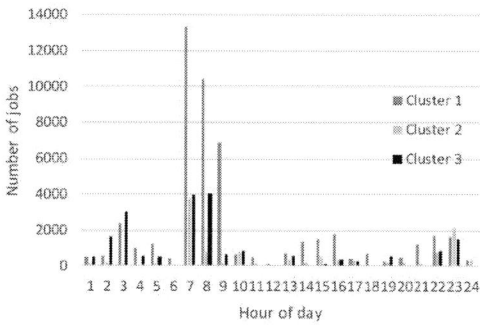

Figure 2.   Jobs distribution with classification

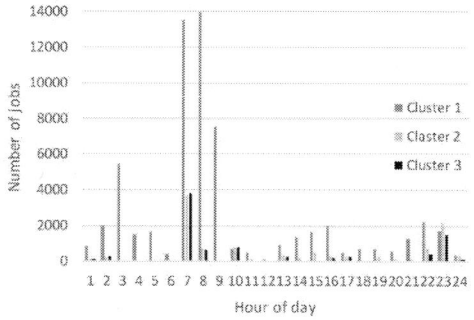

Figure 3.   Jobs distribution without classification

The average processor load for three clusters during the job flow processing is improved by 18 %, 25 % and 30 % respectively (Figure 4).

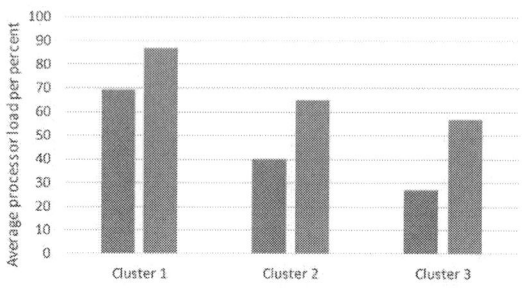

Figure 4.   Jobs distribution with their classification

In the following experiment, the synthetic job flow is processed in the distributed computing environment including three clusters. The jobs represent a copy of the same workflow.

We compare the efficiency of processors load with management under the meta-scheduler of HTCondor version 6.6.8, GridWay version 5.12 and developed MAS, which as middleware interact with clusters RMSs.

We execute an application for warehouse logistics problem solving. Figure 5 illustrates the advantages of the MAS in comparison with target systems for this example. These advantages are achieved due to the control parameters adjustment represented in the Section V.

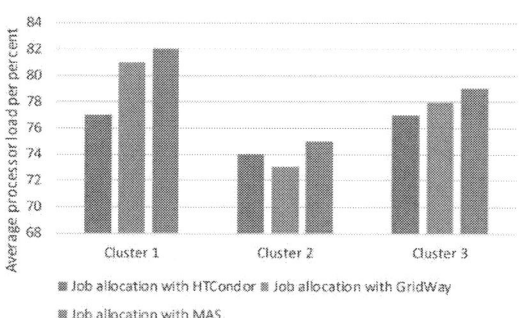

Figure 5.   Cluster load with three management systems

Experiments is carried out in the distributed computing environment based on resources of the Irkutsk state university and Center of collective usage "Irkutsk Supercomputer Center of SB RAS" [32].

VII. CONLUSION

In this paper, we address a problem of the multi-agent management in distributed computing. We showed that its efficiency significantly depends on ability to use the subject-oriented knowledge.

We propose the approach to elicit, represent and use the comprehensive knowledge about both the distributed computing environment and subject domains of solved problems. Specialists in various fields of distributed computing can generate this knowledge at different levels of the problem solving process based on own skills and experience. We also developed the MAS for the management of scalable applications that generate job flows. The MAS operates based on the self-organization.

Comprehensive modelling and practical experiments, with variation of important parameters, show the efficiency of our management under the developed MAS due to elicitation and using the conceptual knowledge.

As future work, we will continue the MAS evolution, expanding the subject-oriented knowledge usage.

ACKNOWLEDGMENT

The study is partially supported by Russian Foundation of Basic Research, projects no. 15-29-07955 and no. 16-07-00931, and Program 1.33P of fundamental research of Presidium RAS, project "Development of new approaches to creation and study of complex models of information-computational and dynamic systems with applications".

REFERENCES

[1] L. Zeng, L. Xu, Z. Shi, M. Wang, W. Wu, "Distributed Computing Environment: Approaches and Applications," IEEE International Conference on Systems, Man and Cybernetics, IEEE Publisher, pp. 3240-3244, October 2007.

[2] I. Foster, C. Kesselman, "Computational Grids," in The Grid: Blueprint for a New Computing Infrastructure, I. Foster, C. Kesselman, Eds. San Francisco: Morgan Kaufmann, pp. 2-48, 1999.

[3] R. Buyya, C. Vecchiola, S. T. Selvi, "Mastering Cloud Computing," Burlington, Massachusetts: Morgan Kaufmann, 2013.

[4] "Oracle Grid Engine," http://www.oracle.com/technetwork/oem/grid-engine-166852.html [online, accessed: Jan 31 2017].

[5] "Torque Resource manager," http://www.adaptivecomputing.com/products/open-source/torque/ [online, accessed Jan 31 2017].

[6] "HTCondor," http://research.cs.wisc.edu/htcondor/ [online, accessed: Jan 31 2017].

[7] "Slurm Workload Manager," http://slurm.net/ [online, accessed: Jan 31 2017].

[8] "Globus Toolkit Homepage," http://www.globus.org/toolkit/ [online, accessed Jan 31 2017].

[9] "GridWay Metascheduler," http://www.gridway.org [online, accessed Jan 31 2017].

[10] A. Rodriguez, A. Tchernykh, K. Ecker, "Algorithms for Dynamic Scheduling of Unit Execution Time Tasks," European Journal Operation Research, vol. 146, no. 2, pp. 403-416, April 2003.

[11] D. Kliazovich, J. E. Pecero, A. Tchernykh, P. Bouvry, S. U. Khan, A. Y. Zomaya, "CA-DAG: Modeling Communication-Aware Applications for Scheduling in Cloud Computing," Journal of Grid Computing, vol. 14, no. 1, pp. 22-39, March 2016.

[12] A. Hirales-Carbajal, A. Tchernykh, T. Roblitz, R. Yahyapour, "A Grid simulation framework to study advance scheduling strategies for complex workflow applications," IEEE International Symposium on Parallel and Distributed Processing, Workshops and Phd Forum (IPDPSW), IEEE Publisher, pp. 1-8, May 2010.

[13] J. Yu, R. Buyya, "A Taxonomy of Workflow Management Systems for Grid Computing," Journal of Grid Computing, vol. 3, no. 3, pp. 171-200, September 2005.

[14] D. Talia, "Workflow Systems for Science: Concepts and Tools," ISRN Software Engineering, vol. 2013, 2013, http://dx.doi.org/10.1155/2013/404525 [online, accessed: January 31 2017].

[15] G. Di Marzo Serugendo, M.-P. Gleizes, A. Karageorgos, "Self-organization in Multi-agent Systems," The Knowledge Engineering Review, vol. 20, no. 2, pp. 165-189, June 2005.

[16] D. Talia, "Cloud Computing and Software Agents: Towards Cloud Intelligent Services," 12th Workshop on Objects and Agent, CEUR Workshop Proceedings, vol. 741, pp. 2-6, June 2011.

[17] P. Leitao, U. Inden, C.-P. Ruckemann, "Parallelising Multi-agent Systems for High Performance Computing," 3rd International Conference on Advanced Communications and Computation, Red Hook, NY: IARIA, pp. 1-6, June 2014

[18] W. Vickrey, "Counterspeculation, Auctions, and Competitive Sealed Tenders," Journal of Finance, vol. 16, no. 1, pp. 8-37, March 1961.

[19] J. Nash, "Equilibrium points in n-person games," Proceedings of the National Academy of Sciences of the United States of America, vol. 36, no. 1, pp. 48-49, 1950.

[20] N. J. Cooke "Varieties of knowledge elicitation techniques," International Journal of Human-Computer Studies, vol. 41, no. 6, pp. 801-849, December 1994.

[21] J. Sowa, "Conceptual Structures – Information Processing in Mind and Machine," in The Systems Programming Series, Addison-Wesley, 1984.

[22] E. Tyugu, "Knowledge-Based Programming (Turing Institute Press Knowledge Engineering Tutorial Series)," Boston: Addison-Wesley, 1988.

[23] A. G. Feoktistov, I. A. Sidorov, "Logical-Probabilistic Analysis of Distributed Computing Reliability," 39th International Convention on information and communication technology, electronics and microelectronics, Riejka: MIPRO, pp. 247-252, May 2016.

[24] I. Bychkov, G. Oparin, A. Tchernykh, A. Feoktistov, V. Bogdanova, S. Gorsky, "Conceptual Model of Problem-Oriented Heterogeneous Distributed Computing Environment with Multi-Agent Management," Procedia Computer Science, vol. 103, pp. 162-167, January 2017.

[25] A. Tchernykh, U. Schwiegelsohn, V. Alexandrov, E. Talbi, "Towards Understanding Uncertainty in Cloud Computing Resource Provisioning," Procedia Computer Science, vol. 51, pp. 1772–1781, May 2015.

[26] A. Tchernykh, U. Schwiegelsohn, E. Talbi, M. Babenko, "Towards Understanding Uncertainty in Cloud Computing with risks of Confidentiality, Integrity, and Availability," Journal of Computational Science, http://www.sciencedirect.com/science/article/pii/S1877750316303878 [online, accessed: January 31 2017].

[27] H. Casanova, F. Berman, G. Obertelli, R. Wolski, "The apples parameter sweep template: User-level middleware for the grid," ACM/IEEE conference on Supercomputing, Washington: IEEE Press, pp. 111-126, November 2000.

[28] I. V. Bychkov, G. A. Oparin, A. G. Feoktistov, V. G. Bogdanova, A. A. Pashinin, "Service-oriented multiagent control of distributed computations," Automation and Remote Control, vol. 76, no. 11, pp. 2000-2010, November 2015.

[29] I. V. Bychkov, G. A. Oparin, A. G. Feoktistov, I. A. Sidorov, V. G. Bogdanova, S. A. Gorsky, "Multiagent Control of Computational Systems on the Basis of Meta-monitoring and Imitational Simulation," Optoelectronics, Instrumentation and Data Processing, vol. 52, no. 2, pp. 107-112, June 2016.

[30] "Cleo Cluster Batch System," https://sourceforge.net/projects/cleo-bs.html [online, accessed: January 31 2017].

[31] SUPPZ http://suppz.jscc.ru/ [online, accessed: January 31 2017].

"Irkutsk Supercomputer Centre of SB RAS," http://hpc.icc.ru [online, accessed: January 31 2017].

Improving a Distributed Agent-Based Ant Colony Optimization for Solving Traveling Salesman Problem

Aleksandar Kaplar*, Milan Vidaković*, Nikola Luburić* and Mirjana Ivanović**

* Faculty of Technical Sciences, University of Novi Sad, Novi Sad, Serbia
** Faculty of Sciences, University of Novi Sad, Novi Sad, Serbia
* {aleksandar.kaplar, minja, nikola.luburic}@uns.ac.rs
** mira@dmi.uns.ac.rs

Abstract - Optimization of a large-scale Traveling Salesman Problem, which is a well-known NP-hard problem in combinatorial optimization, is a time-consuming problem. A modern approach to dealing with such time-consuming problems is with the use of distributed computing, which can significantly improve the speed of the problem-solving algorithm. In this paper, we discuss the design approaches for an agent-based distributed algorithm and their benefits. Based on further analysis and experiments, we have improved our previous agent-based Ant Colony Optimization algorithm for Solving Traveling Salesman Problem using Siebog multiagent middleware.

I. INTRODUCTION

Distributed system can be defined as a network of independent components that communicate and coordinate their actions only by passing messages [1]. The motivation behind the use of distributed systems is the fact that distributed data computing can lower the cost of data processing and increase the system's robustness with data replication. When used properly, distributed computing can achieve a computational result much more quickly than a single computer can.

One area inherently amenable to parallelization and distributed computing is swarm intelligence. [5]

Ant colony optimization (ACO) [2] is one of the swarm based optimization algorithms used for solving wide range combinatorial optimization problems. It is inspired by the behavior exhibited by ant colonies while searching for a food source. One of the characteristics of ant colonies is that they have the ability to find the shortest path to a food source, by using a large number of independent ants that use only pheromones as a means of communication. In ACO each ant can be seen as an independent processing agent searching through an environment.

In this paper, we describe our intention to improve the execution speed of our previous agent-based distributed ACO algorithm for Traveling Salesman Problem (TSP). The algorithm was initially developed to showcase Siebogs distributive capabilities [6, 7]. Although successful, the algorithm didn't fully utilize all the capabilities offered by the Siebog design.

Siebog is our FIPA compliant multiagent middleware built using the Java Platform, Enterprise Edition (Java EE) [8, 9, 10]. Siebog supports both client-side and server-side agents. Client-side agents execute their code on browsers, as JavaScript code, while server-side agents are being executed as EJB (Enterprise Java Beans) components. Siebog utilizes the standards and technologies readily available in Java EE, in order to provide automatic agent load balancing and fault tolerance on the server side [7]. It also uses Java EE built-in features to implement core agent functionalities:

- it uses JMS (Java Message Service) to exchange messages between agents,

- it uses JNDI (Java Naming and Directory Service) as agent discovery service (agent "yellow pages"),

- it uses JAAS (Java Authentication and Authorization Service) to implement security concepts.

This paper is organized as follows. Section 2 provides an overview of relevant literature. The formulas of ACO for TSP are presented in Section 3. Section 4 presents our proposed model. The results of the model are presented in Section 4. Finally, Section 5 outlines general conclusion and further work.

II. RELATED WORK

Over the years there have been many attempts at developing optimal parallel and distributed algorithms for ACO. A survey on parallel ant colony optimization was reported by [11]. Authors of that work proposed a new taxonomy for classifying parallel ACO. Their classification includes: master-slave model, cellular model, parallel independent runs model, multi-colony model, and hybrid model. Classification best suited for this work is the multi-colony model.

In a standard multi-colony model configuration, introduced by [12, 13], different colonies of ants work on the same problem independently. Pheromones aren't shared between the colonies, but after a number of iterations colonies exchange their best solution in order to influence each other via elitist strategy.

The standard multi-colony configuration was successfully applied to various optimization problems, with improvements over the sequential ACO. Notably, authors of [14], developed a parallel ant colony systems (PACS). Their preliminary test on three datasets showed that PACS outperforms sequential algorithms, especially with larger datasets.

Another notable distributed agent-based algorithm for solving TSP, named ACODA, was developed by [3]. Their experimental results show execution time improvement, and that the system, by partitioning the map between agents, supports scalability.

III. ACO FOR TSP

ACO algorithm is inspired by the behavior exhibited by ants while searching for a food source. Ants, while searching for a food source, secrete pheromones on their way back to the anthill. Other ants can detect paths with secreted pheromones and they may become attracted to them. The paths become more attractive when more pheromone is deposited on them. Another property of the pheromones is that they evaporate over time. Evaporation makes the longer paths less interesting, which decreases the probability that an ant will choose it. However, evaporation will have less influence on shorter paths, due to the fact that they are refreshed more frequently. In time, ants will converge towards the shortest path due to largest concentration of pheromones [2][3].

Mathematical model of ACO for TSP we used in our work is fully described in [2][3], briefly summarized the formulas are the following:

An ant located at a city (node) i chooses to go to city j with the probability:

$$p_{i,j} = \frac{(\tau_{i,j})^{\alpha}(\eta_{i,j})^{\beta}}{\sum_j (\tau_{i,j})^{\alpha}(\eta_{i,j})^{\beta}}, \qquad (1)$$

where $\tau_{i,j}$ - is the amount of the pheromone deposited on edge (i,j), $\eta_{i,j}$ - is the inverse of the weight of edge (i,j), α - is a parameter to control the influence of parameter $\tau_{i,j}$, β - is a parameter to control the influence of $\eta_{i,j}$, j - represents a city reachable from city i that was not yet visited by that ant. When an ant determines a new tour of a cost L it will increase every edge of the tour with the value $\Delta\tau_{i,j}$ which is inversely proportional to the cost of tour. Pheromone update and evaporation, are done while ant traces back it's steps, with the formula:

$$\tau_{i,j} = (1 - \rho)\tau_{i,j} + \rho\Delta\tau_{i,j}, \qquad (2)$$

where ρ is the evaporation rate in the range $0 \leq \rho < 1$. In order to increase the exploration rate an ant can apply local evaporation of to a node with the formula:

$$\tau_{i,j} = (1 - \xi)\tau_{i,j} + \xi\tau_0, \qquad (3)$$

where ξ is the local evaporation rate in the range $0 \leq \xi < 1$ and $\tau_0 = 1/nC$ in which n represents the number of nodes and C represents the tour cost approximation (C usually equals to product of number of nodes and average edge cost between them).

IV. PROPOSED MODEL

When designing an agent based ACO algorithm, we considered two distinct approaches. Approaches are characterized by the representation of the ants. That is, ants can be represented as either an agent in the system or as a message exchanged in the system.

Initially, to display Siebogs capabilities, an ACO algorithm was developed in which ants were represented as agents, and the problem space was represented by a single map agent. Although not optimized, that algorithm showed that our multi-agent middleware had all the necessary capabilities required for distributed computing, but it had a major flaw brought on by the nature of distributed systems. As mentioned before, distributed systems communicate over a physical network which introduces a delay to the system. That delay combined with a bottleneck which is a single map agent and synchronous access to it resulted in a significant increase in time required for algorithm execution (for a simple map of 16 nodes, the time required was about 1h55m on commodity hardware and about 10m on high-end hardware). In order to improve the execution time we have redesigned the ant itself. In fact, we have implemented it as a message instead of an agent.

In the first approach, ants are agents that search through the problem space. The problem space is represented and maintained by one or more map agents. Ant agents are the algorithm executors, they start the search, use the problem space data to perform calculations, and notify map agents about necessary updates to the problem space. The behavior of map agents can be described as purely responsive because all of their actions are responses to queries generated by ant agents. There are three distinct messaging types, illustrated in Figure 1A, in this approach:

1. Initialization messaging – with which an ant agent request a list of nodes in the problem space,

2. Local move messaging – when choosing the next node to visit an ant queries the map agent for pheromone levels in path from current node to next nodes,

3. Update tour – after finding a possible solution, an ant notifies all map agents about it.

To decrease the number of messages in our distributed system, for the second approach, we represented ants as messages exchanged by map agents. To accommodate such changes, we extended map agents by adding them algorithm execution functionality. Algorithm execution functionality is represented as an infinite loop which takes and processes an ant from the ant queue. After processing (choosing the next node that ants will visit, and updating the problem space), the ant is either sent to another map agent or added to the end of the ant queue, depending on the node to which the ant will go to next. In this approach, there are two distinct messaging types (illustrated in Figure 1B):

1. Send ant – this message happens when an ant needs to visit a node managed by another map agent,

1145

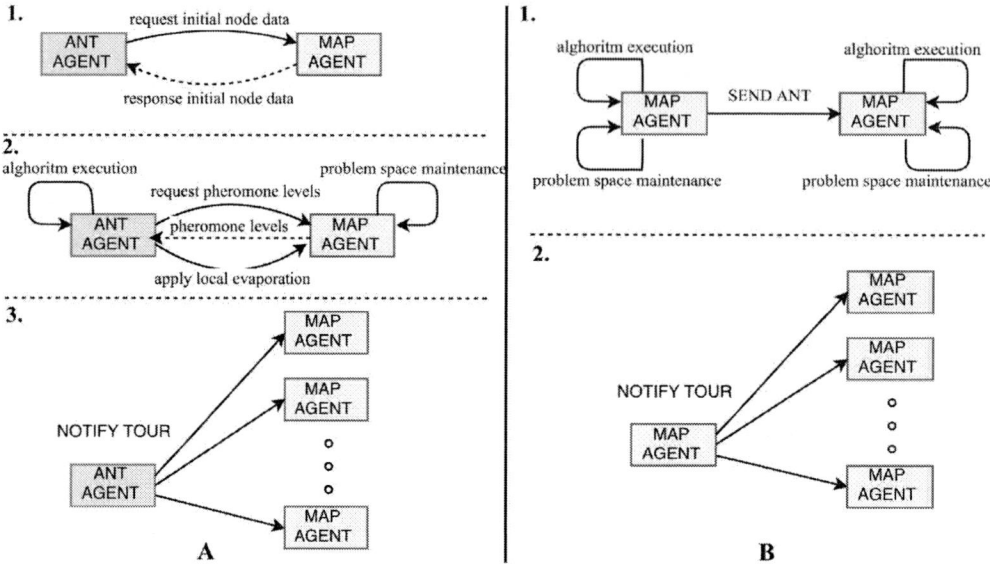

Figure 1. Typical messaging exchange between the agents in: A – ant as an agent approach, B – ant as a message approach.

2. Update tour – when the ant returns to its initial map agent with a possible solution, that map agents notifies other map agents about it.

To fully optimize the algorithm execution in this approach, we have used Siebogs flexibility to separate the agents' message processing from the algorithm execution. The separation is achieved by using a separate thread for the algorithm execution. With that, each map agent is able to process messages, update the problem space, and execute the algorithm steps in parallel.

Inspired by [2], our agent-based algorithm, given in table 1., consists of several steps regardless of the chosen approach.

TABLE 1. STEPS OF AN ACO ALGORITHM

```
SEARCH_STEP(ant)
    GET_MAP_DATA()
    if COMPLETED_TOUR(ant):
        NOTIFY_TOUR(ant)
        REINITIALIZE(ant)
    else:
        ANT_MOVE(map_data, ant)

ANT_MOVE(map_data, ant):
    node = perform random choice per equation 1
    UPDATE_ANT_PATH(ant,node)
    apply local evaporation using equation 4
    SEND_ANT(ant, node)

SEND_ANT(ant,node):
    if node.agent != current_map_agent:
        SEARCH_STEP(ant)
    else:
        SEND(ant, node.agent)

NOTIFY_TOUR(ant):
    for edge in ant.tour:
        update pheromone levels using equation 3
    if STOP_CRITERIA_MET():
        STOP_SEARCH()
```

When the client initiates the search, the problem space is loaded and distributed among a number of map agents. In order to decrease the initial workload of each map agent, ants are evenly assigned to. The client defines the number of ants used in the search. Each ant begins its search of the problem space from a randomly chosen node managed by their assigned map agent. After the initial pick of a start node, ants repeatedly execute the SEARCH_STEP functionality until they return to their start node. In the ant as an agent approach, to complete each step, ant first requests the pheromone levels of each possible path it can take from the current node. When the ant gets a response from a map agent, it executes appropriate calculations to choose the next node it will visit. After choosing the node, it notifies the map agent to apply the calculated local pheromone evaporation and repeats this process again with the next node. On the other hand, in ant as a message approach all calculation are done by map agents. The step of the search process is represented by taking the ant from the ant queue, calculating the next node, and putting the ant at the back of the ant queue (or sending it to another map agent).

During a single ANT_MOVE the ant uses pheromone data, of all the paths it can take, to choose a path that is a city, it will visit next. When the ant chooses the next city, it updates its tour cost by adding the cost of the taken path. With the new data generated by ant, appropriate map agents apply local evaporation to the taken path. In the ant as a message approach, the last step of the ant move is to send it to next map agent (if the next city belongs to another agent), or to add him to the end of the ant queue (if the next city belongs to the map agent). On the other hand, in the ant as an agent approach, the ant agents receive the agent id of the appropriate map agent it needs to query for the map data.

When the ant finishes its tour, it returns to the ant hill (starting city) while depositing the pheromone on the taken path. Ants return is implemented via NOTIFY_TOUR message that is used to send tour data to all map agents. With the tour data, the map agents update their part of the map and update the best tour if needed. After the updates, map agents check if the stop criteria are met and if it is they send out stop messages to terminate the algorithm execution and return the best tour to the client.

V. EVALUATION AND RESULTS

In order to properly represent the results and compare them with the execution speed of initial implementation, we used a single Siebog node for testing. The machine used for Siebog node had a 4 virtual CPUs and 8 GB of RAM, running a 64-bit version of Windows 7.

Execution speed was tested by measuring the time required for each ant to complete 200 tours of the search space. It happens under requirement that the number of ants is equal to the number of cities on the map. We ran the test on benchmark TSP maps selected from TSPLIB [3], the results are presented in table 2.

The results show that ant as a message approach is significantly faster than ant as an agent approach. When compared to the initial implementation which required 1h55m for ulysses16, both approaches show considerable improvement in the execution speed.

The speed discrepancy between two approaches can be attributed to several factors. For the smallest map of 16 nodes, the major factor is the delay introduced by messages subsystem, which can introduce the delay from 5ms up to 25ms per message. On the other hand, with larger maps in which more ants are required to properly conduct the search, another major contributing factor to execution time is the number of available threads. Ant represented as an agent requires a separate thread for each agent to fully parallelize the execution of the algorithm. When working with large maps, that number can exceed the number of threads available in the appropriate thread pool (in the conducted test, Siebog had 50 threads available in the thread pool).

From the given results we can conclude that the better approach for this type of problem is the ant as a message approach. Advantage of the ant as a message is twofold. Firstly, it separates message processing from the algorithm execution, which parallelizes the task done by

TABLE 2. RESULTS OF TESTING

Map	Algorithm	Average execution time (seconds)
ulysses16	ant as an agent	957
	ant as a message	4
st70	ant as an agent	14600
	ant as a message	32
ch150	ant as an agent	67000 (estimated)
	ant as a message	101

single map agent. Secondly, the number of messages exchanged in this approach is greatly reduced, which in turn improves the execution speed. On the other hand, ant as an agent approach can be considered as an underutilization of processing power. The processing requirements done by ant agents are not complex and require less time than it's required by message exchange. Benefits of ant as an agent approach could only be achieved with the problems where the performed tasks, done by agents, are computationally intensive.

VI. CONCLUSION

In this paper, we have presented our attempts at improving the execution speed of a distributed agent-based ACO algorithm for TSP, designed for the Siebog multiagent middleware. During the design phase, we considered two distinct approaches in the design of our agent-based algorithm: *ant as an agent* and *ant as a message*.

Evaluation of those approaches has shown that optimizing and reducing the number of messages exchanged greatly improves the execution speed of the algorithm. Achieved results show that, with the utilization of the Siebog flexibility, the better solution for this type of problem is the realization of ant as a message.

For the future work, we intend to test and optimize the fully distributed algorithm on a Siebog cluster and compare results with other similar approaches. We believe that with further optimizations of the algorithm, combined with Siebog capabilities, we can produce competitive results.

REFERENCES

[1] Coulouris, George; Jean Dollimore; Tim Kindberg; Gordon Blair (2011). Distributed Systems: Concepts and Design (5th Edition). Boston: Addison-Wesley. ISBN 0-132-14301-1.

[2] M. Dorigo and T. Stutzle. Ant Colony Optimization. MIT Press, 2004.

[3] Ilie, S., Bǎdicǎ, A., Bǎdicǎ, C.: Distributed agent-based ant colony optimization for solving traveling salesman problem on a partitioned map. In: Proceedings of the International Conference on Web Intelligence, Mining and Semantics, WIMS 2011, pp. 23:1–23:9. ACM (2011)

[4] G. Reinelt. Tsplib - a traveling salesman problem library. ORSA Journal on Computing, 3(4):376-384, 1991.

[5] Marco Dorigo and Mauro Birattari (2007) Swarm intelligence. Scholarpedia, 2(9):1462.

[6] Mitrović, D., Ivanović, M., Vidaković, M., Budimac, Z.: Extensible Java EE-based agent framework in clustered environments. In: Müller, J.P., Weyrich, M., Bazzan, A.L.C. (eds.) MATES 2014. LNCS, vol. 8732, pp. 202–215. Springer, Heidelberg (2014)

[7] Mitrović, D., Ivanović, M., Vidaković, M., Budimac, Z.: A scalable distributed architecture for web-based software agents. In: 7'th International Conference on Computational Collective Intelligence (ICCCI), pp. 67–76 (2015)

[8] Mitrović, D., Ivanović, M., Budimac, Z., Vidaković, M.: Supporting heterogeneous agent mobility with ALAS. Computer Science and Information Systems 9(3), 1203-1229 (2012)

[9] Vidaković, M., Ivanović, M., Mitrović, D., Budimac, Z.: Extensible Java EE-based agent framework - past, present, future. In: Ganzha, M., Jain, L.C. (eds.) Multia-gent Systems and

Applications, Intelligent Systems Reference Library, vol. 45, pp. 55-88. Springer Berlin Heidelberg (2013)

[10] Vidaković, M., Milosavljević, B., Konjović, Z., Sladić, G.: EXtensible Java EE-based agent framework and its application on distributed library catalogues. Com-puter Science and Information Systems, ComSIS 6(2), 1-16 (2009)

[11] M. Pedemonte, S. Nesmachnow, H. Cancela, A survey on parallel ant colony optimization, Applied Soft Computing 11 (8) (2011) 5181–5197.

[12] R. Michel, M. Middendorf, An island model based ant system with lookahead for the shortest supersequence problem, in:

Proceedings of the 5th International Conference on Parallel Problem Solving from Nature, Lecture Notes in Computer Science 1498 (1998) 692–701.

[13] R. Michel, M. Middendorf, An ACO algorithm for the shortest common supersequence problem, in: D. Corne, M. Dorigo, F. Glover, D. Dasgupta, P. Moscato, R. Poli, K. Price (Eds.), New Ideas in Optimization, McGraw-Hill, 1999, pp. 51–62.

[14] S. Chu, J. Roddick, J. Pan, Ant colony system with communication strategies, Information Sciences 167 (1–4) (2004) 63–76.

Towards an Agent-Based Automated Testing Environment for Massively Multi-Player Role Playing Games

Markus Schatten, Igor Tomičić, Bogdan Okreša Đurić, Nikola Ivković
Artificial Intelligence Laboratory
Faculty of Organization and Informatics
University of Zagreb
Pavlinska 2, 42000 Varaždin, Croatia
Email: (markus.schatten, igor.tomicic, dokresa, nikola.ivkovic)@foi.hr

Abstract— **Automated testing in massively multi-player on-line role playing games (MMORPG) is a challenging task due to the complexity of such games and their large numbers of mutually distributed but interacting components. Large-scale multi-agent systems (LSMAS) provide us with a suitable formalism to address such complex problems. Herein a first step towards an automated game testing environment, built for the open source Mana World MMORPG, will be presented that allows the implementation of software agents and agent organizations and provide the developers with valuable game-play and testing data.**

Keywords— *MMORPG, LSMAS, organization, automated testing, games, software*

I. INTRODUCTION

Automated game testing in general, presents a challenging problem most game developers have to face eventually due to the emerging complexity of contemporary video games. Especially massively multi-player on-line games (MMO) which allow for thousands and sometimes even hundreds of thousands players playing simultaneously, are an additional challenge due to the possible mutual interactions between players, especially in the form of organizing their behaviour to perform certain game-related tasks. Additionally MMO role-playing games (MMORPG) which, beside massiveness of simultaneous players, feature also complex logical tasks or quests for the players, are even more complex for automated testing, since implemented player bots, have to perform complex reasoning and plan their actions to solve certain tasks.

While most of the literature deals with techniques of preventing automated players from playing MMO games (like [1], [2], [3]) there were only few attempts to create automated testing environments for large-scale settings. For example in [4] Jung et al. developed the VENUS simulator and present an efficient method for simulating large numbers of virtual clients in on-line games in order to ensure game stability. In [5], [6] the VENUS II system is proposed, which allows for blackbox

and scenario-based testing by generating user packages based on game grammar. In this way various game scenarios can be generated and tested in the simulation environment. In [7] Mellon provides us with lessons learned from automated testing of The Sims On-line MMO game. Therein the development team used a constrained test client, based on the actual game client application, that was controlled using various scripting approaches.

Herein we will use a game logic oriented approach that, besides load and stability testing, should also allow for the testing of actual quests in a given MMORPG. All the previously outlined approaches are not developed for role-playing games (RPGs) which include complex logical quests that often foster various forms of organizing between players. Large-scale multi-agent systems (LSMAS) provide us with the necessary foundations to address such complex scenarios [8], [9], [10], [11]. For our purpose we have used the open source MMORPG called the Mana World[1] in order to develop an agent based testing framework. The testing framework is part of a larger context of building a graphical modelling tool for the semi-automated development of LSMASs [12], [13], as part of the ModelMMORPG[2] project. This work in progress paper shows the initial efforts in building the various components of the to be established framework.

The rest of this paper is organized as follows: firstly in section II we provide a short overview of the Mana World MMORPG game. In section III a motivation for using an agent based approach to automated testing is given. In section IV an overview of the implementation of the framework is provided. In the end in section V we draw our conclusions and give guidelines for future work in this area.

II. THE MANA WORLD

The Mana World is a free open source MMORPG which is implemented with a 2D graphics interface (see figure 1), similar to classic RPG games like Zelda or Final Fantasy.

---

[1]https://www.themanaworld.org/
[2]http://ai.foi.hr/modelmmorpg

Fig. 1: The Mana World – an open source MMORPG

There are several official game servers available for players to connect to, but being an open source game developed under the GPL license, it allows players to implement their own local servers by using the server source code freely available at the GitHub service.

The game itself facilitates character personalization and development, mutual interaction by using several in-game mechanisms such as chat, trade, fights, social network creation, tagging (friends/enemies), party creation, story development, quest solving, etc.

The important aspect of solving quests is to encourage players to form parties as in-game organizations within which there are internal rules, task delegations, responsibilities, interrelations and other aspects of such goal-oriented communities which all are scientifically engaging for analytical observations and represent a critical input for modeling and simulations of artificial players.

The game server used in this work is called TMWAthena, and is composed of three distinct servers:

- **Character server** – dedicated to managing characters and connecting them to the map server.
- **Login server** – dedicated to managing accounts and connecting to the character server.
- **Map server** – dedicated to managing the game content (such as monsters, items, maps, etc.) and the interaction of such game contents with the players.

Understanding the structure and the communication protocols of the client/server interaction formed the basic prerequisite for further game development and manipulation.

III. AGENT-BASED APPROACH TO AUTOMATED MMORPG TESTING

Herein we have chosen to approach the problem of automated testing of MMORPGs using LSMAS. There are several reasons for that: (1) MMORPG players are by definition distributed entities that interact, compete, collaborate and organize – a definition that closely matches the one of (artificial) agents, (2) MMORPG players have to solve complex tasks in forms of puzzles and quests indicating reasoning capabilities as of intelligent agents, (3) MMORPG avatars are situated in a predefined but complex environment in which they have to act both reactively and proactively in order to achieve game objectives, as do agents in (more or less) realistic environments.

In the mentioned ModelMMORPG project we have partially developed an organizational ontology and metamodel for graphical modelling of LSMAS (see for example [14], [15], [8]) that allows for modelling a wide range of various LSMAS scenarios. Automated testing of MMORPGs is one such scenario, as we will show further.

IV. IMPLEMENTATION

The testing framework consists of three logical layers: (1) a lower-level interface (dealing with the technical implementation of the Mana World client network protocol), (2) a higher-level interface (implementing actual agent classes, agent behaviours, roles, agent communication, agent reasoning, knowledge bases etc.), and (3) a modelling tool (allowing for graphical modelling of tests and generating agent implementation stubs to be developed further for individual test).

These three layers are described in more detail in the following subsections.

Our approach in developing tests is loosely a top down approach. Firstly, agent organizations, which represent the various players' forms of organizing (e.g. guilds, parties, etc.) are modelled using a graphical modelling language that we have developed. Then the model is translated into a concrete implementation of agent classes facilitated through and agent-based platform called SPADE (Smart Python Agent Development Environement) [16]. These agent classes, are then extended with low-level game client methods that allow various game related tasks like walking around, picking up items, fighting etc., to be performed by agents.

A. Lower-level Interface

The lower-level interface was loosely based on an old Python script used to implement bots for the Mana World. The script has been rewritten in full and extended with multiple details of client implementation in order to allow agents to connect to the game server, create avatars, and perform all relevant tasks to be able to solve quests.

figure 2 shows a sample session of the lower-level interface. The provided screenshots illustrate changing Python player coordinates by using the "setDestination" function of the "Connection" class.

The following listing shows the actual implementation of the Python function which enables the basic navigation of the character on the loaded map by using two-dimensional coordinate values.

```
def setDestination( self, x, y, direction
    ):
        '''Set destination (walk to given
            x, y coordinates with
            orientation direction like in
            setDirection)'''
        debug( "SET_DESTINATION" )
        debug( "X: _%d" %x )
        debug( "Y: _%d" %y )

        data = bin(x)[-10:].replace( 'b',
            '0' ).rjust(10).replace( '_',
            '0' ) + bin(y)[-10:].replace(
            'b', '0' ).rjust(10).replace(
            '_', '0' ) + bin(direction)
            [-4:].replace( 'b', '0' ).
            rjust(4).replace( '_', '0' )

        data = data[ :8 ], data[ 8:16 ],
            data[ 16: ]
        data = [ int( '0b' + i, 2 ) for i
            in data ]

        b1, b2, b3 = data
```

```
self.srv.sendall( "\x85\0%s" %
    struct.pack( "<BBB", b1, b2,
    b3 ) )
```

B. Higher-level Interface

The higher-level interface is being developed using the mentioned SPADE platform in Python. It currently features basic agent classes, agent behaviours and communication facilities to allow agents to organize mutually. The agent classes closely resemble a BDI (belief-desire-intention) architecture in which agents act based on their beliefs (knowledge about the world they live in), desires (objectives they want to pursue) and intentions (plans and commitments on achieving objectives through acting upon the environment).

The knowledge about the Mana World is stored in a specially developed [10] SWI Prolog knowledge base that can be directly accessed by the agent through its knowledge base interface.[3].

The knowledge base also featured an automated planner based on the STRIPS algorithm [17]. In the planner all quests are coded in form of STRIPS rules comprising three parts: (1) preconditions – statements that have to be true about the world in order for the rule to be applicable, (2) deletions – statements that won't be true about the world when the rule is executed, and (3) insertions – statements that will become true about the world only after the rule has been executed. In this way game related quests can be modelled quite easily, for example to solve a quest in which the player has to buy a certain item from some NPC (non-playing character) for a certain amount of money a rule would include the following statements:

- **Preconditions**: player has enough money, player is within reach of the NPC
- **Deletions**: player hasn't the payed money any more
- **Insertions**: player has acquired the wanted item.

The planner also features a list of quests to be solved, which is queried by the player agent every time the environment is updated in order to test if the preconditions of a certain quest to be solved are met. If this is the case, the agent tries to solve the quest, else the quest is put back into the list for a later time. In case the agent doesn't have any quests to be solved it does a random walk in the environment to find an NPC and get an actual quest.

C. Graphical Modeling Tool

The basis of the system of agents using the mentioned interfaces, allowing them to communicate with the game engine, is, in the context of this paper and the accompanying research, constructed using a customized graphical modelling tool[4].

The modeling tool, in its present work-in-progress version, provides the user with functionalities that allow them to model

---

[3]SPADE allows for using SWI, XSB, eCLiPSe, Flora-2, SPARQL and integrated first-order logic knowledge-bases

[4]The model is available at https://github.com/Balannen/LSMASOMM

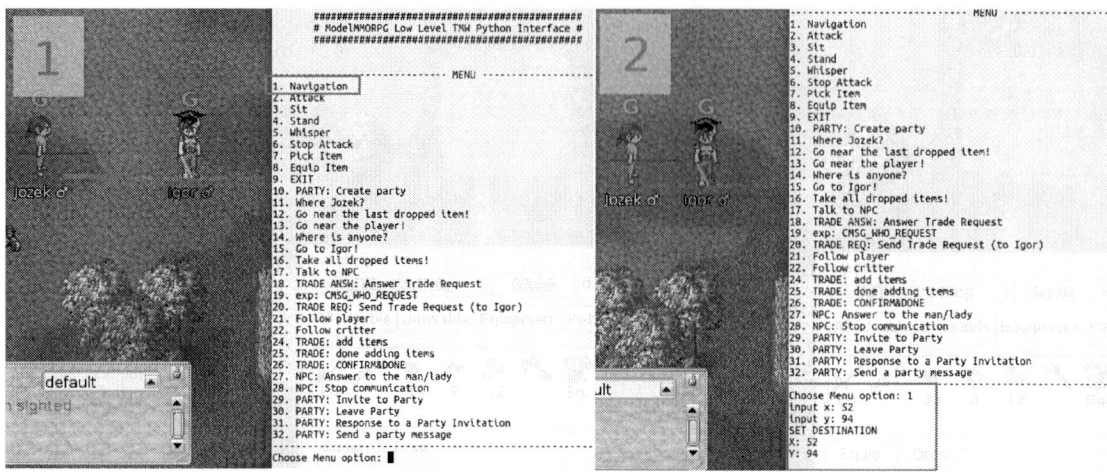

Fig. 2: Manipulating player coordinates with the Python client

a system comprising agents, groups of agents, roles as special types of grouped constraints, actions granted to individual agents upon playing a specific role, sets of actions designated as processes that are poised to fulfill specific goals, and quests (in-game goals) that consist of a series of subgoals or tasks that are reachable by a specific combination of processes, i.e. action sets.

The elementary concept of such a model, applicable to the example shown in this paper, is an organisational unit representing an individual agent. Since the idea of roles is integral to the organizational approach which, in turn, is in the basis of the models developed for the mentioned modelling tool, roles are a *de facto* central elements of a model. Figure 3 shows a situation that is possible within The Mana World game, detailed as follows. Individual agents, shown yellow in Fig. 3, are modeled following the *tabula rasa* idea, i.e. the only behavior they possess at this stage of the metamodel is intended for changing roles, thus gaining new actions. It is useful to note here that the element named *Player* in Fig. 3 represents a class of individual players, i.e. individual agent-players will be instantiated at runtime. Those individual players can form groups (most commonly called parties in an MMORPG) that utilize cooperative power of players. Each player can play a number of roles (e.g. Herbalist, Scout, Warrior), shown in blue in Fig. 3, at the moment constrained to one at any given point in time. The roles shown in Fig. 3 provide players with actions needed to successfully complete tasks (shown in red in Fig. 3) that form the quest called *The Quest for the Dragon Egg* (the topmost red element in Fig. 3.

One of the novel features in the context of LSMAS models and modeling tools, that this particular tool provides, even though only in initial stages at the moment, is SPADE code generating functionality. This functionality, still in a work-in-progress state, generates basic code, i.e. a code skeleton, for the modeled system, thus giving the model users basic overview of their system, and advanced system functionalities

not explicitly visible in the model, e.g. system and services for agents' role changing actions and utilization of available knowledge on ways of solving quests based on their respective wanted tasks.

V. CONCLUSION & FUTURE WORK

In this work-in-progress paper we have presented the current state of implementation of an agent-based automated MMORPG testing system using a game-oriented approach. As opposed to most other approaches, the proposed system will allow for testing game-logic by implementing agents that use a specially created knowledge-base to be able to solve quests in the selected MMORPG called the Mana World. We believe that this approach can be extended to any MMORPG, given that the various quests or rules of the game can be coded into a new knowledge base.

Currently the proposed system comprises three components: (1) a lower-level interface, (2) a higher-level interface, and (3) a graphical modelling tool. Whilst the lower-level interface is fully developed at the time of writing this paper, the other two components are still being tested and developed further. Especially, in the higher-level interface, the connection between an agent's perception of the environment and the knowledge base has to be improved, so that agents can rely on up-to-date information from the environment. The modelling tool features necessary concepts for successful description of a system, but even slightly complex systems cause a cluttered and unfriendly interface. Therefore a multi-perspective modeling approach has to be enabled, allowing users to develop a model on various levels presented by different models. The code generating feature has to be upgraded appropriately to conform to the idea of multi-model modeling. Lastly, one of the foreseen improvements of the modelling tool is introduction of concepts for modelling additional selected organizational features of the system. These and similar tasks are part of our future work.

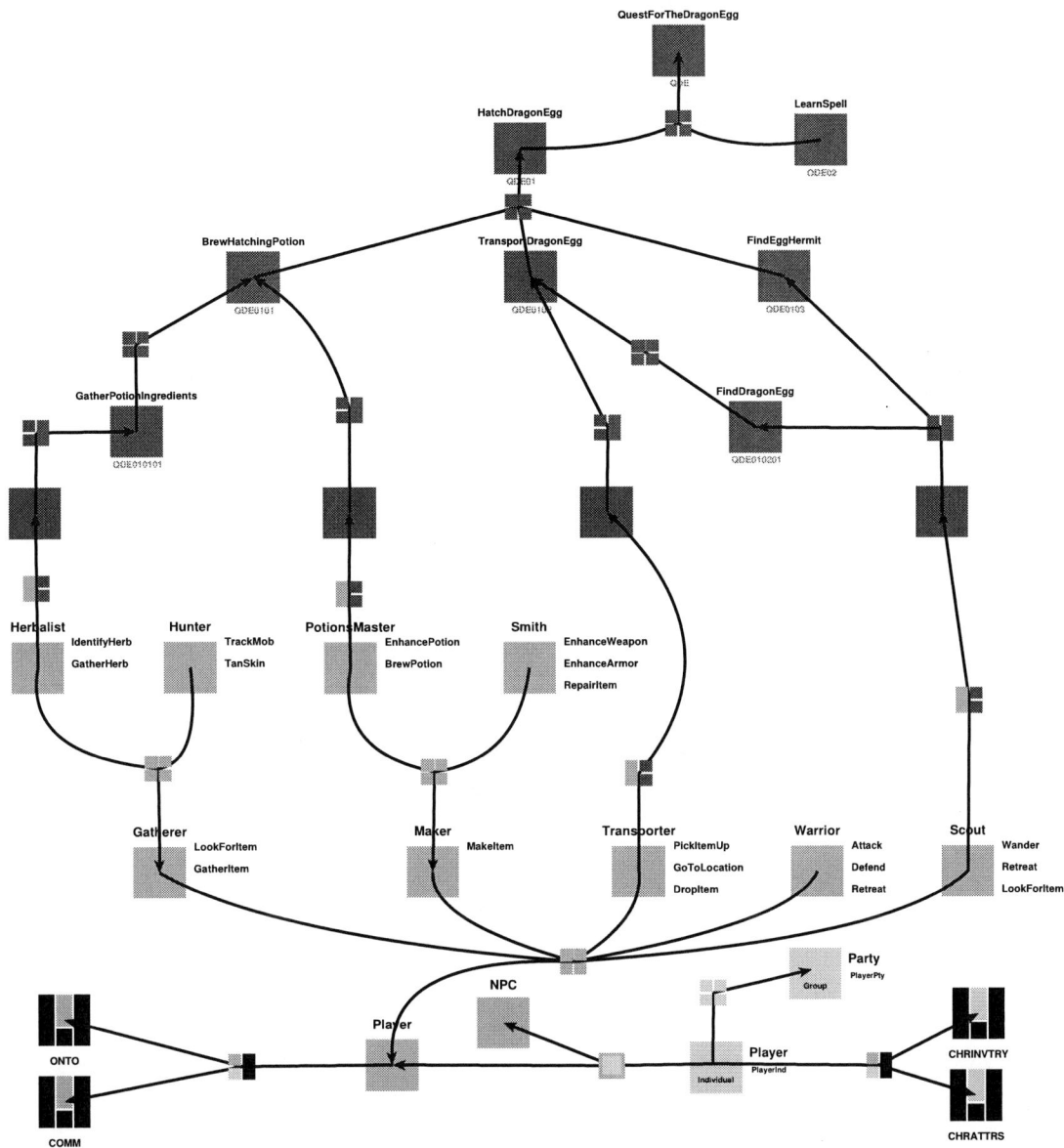

Fig. 3: Modeled selected concepts of The Mana World

ACKNOWLEDGMENT

This work has been supported in full by the Croatian Science Foundation under the project number 8537. We would also like to acknowledge the Mana World development team which often helped us with various implementation specific details.

REFERENCES

[1] J. Yan, "Bot, cyborg and automated turing test," in *International Workshop on Security Protocols.* Springer, 2006, pp. 190–197.

[2] P. Golle and N. Ducheneaut, "Preventing bots from playing online games," *Computers in Entertainment (CIE)*, vol. 3, no. 3, pp. 3–3, 2005.

[3] M. van Kesteren, J. Langevoort, and F. Grootjen, "A step in the right direction: Botdetection in mmorpgs using movement analysis," in *Proc. of the 21st Belgian-Dutch Conference on Artificial Intelligence (BNAIC 2009)*, 2009, pp. 129–136.

[4] Y. Jung, B.-H. Lim, K.-H. Sim, H. Lee, I. Park, J. Chung, and J. Lee, "Venus: The online game simulator using massively virtual clients," in *Asian Simulation Conference.* Springer, 2004, pp. 589–596.

[5] C.-S. Cho, K.-M. Sohn, C.-J. Park, and J.-H. Kang, "Online game testing using scenario-based control of massive virtual users," in *Advanced Communication Technology (ICACT), 2010 The 12th International Conference on*, vol. 2. IEEE, 2010, pp. 1676–1680.

[6] C.-S. Cho, D.-C. Lee, K.-M. Sohn, C.-J. Park, and J.-H. Kang,

"Scenario-based approach for blackbox load testing of online game servers," in *Cyber-Enabled Distributed Computing and Knowledge Discovery (CyberC), 2010 International Conference on.* IEEE, 2010, pp. 259–265.

[7] L. Mellon, "Automated testing of massively multi-player systems: Lessons learned from the sims online," *GDC 2003*, 2003.

[8] M. Schatten, J. Ševa, and I. Tomičić, "A roadmap for scalable agent organizations in the internet of everything," *Journal of Systems and Software*, vol. in press, 2016.

[9] M. Schatten, I. Tomičić, and B. O. Đurić, "Multi-agent modeling methods for massivley multi-player on-line role-playing games," in *Information and Communication Technology, Electronics and Microelectronics (MIPRO), 2015 38th International Convention on.* IEEE, 2015, pp. 1256–1261.

[10] M. Maliković and M. Schatten, "Artificial intelligent player's planning in massively multi-player on-line role-playing games," in *26th Central European Conference on Information and Intelligent Systems*, 2015.

[11] B. O. Đuric, I. Tomicic, and M. Schatten, "Towards agent-based simulation of emerging and large-scale social networks. examples of the migrant crisis and mmorpgs," *European Quarterly of Political Attitudes and Mentalities*, vol. 5, no. 4, p. 1, 2016.

[12] M. Schatten and P. Terna, "From players to agents via the project modelmmorpg and the slapp simulation shell," in *SWARMFEST 2016: 20th Annual Meeting on Agent Based Modeling & Simulation*, 2016.

[13] B. O. Đurić and M. Schatten, "Defining ontology combining concepts of massive multi-player online role playing games and organization of large-scale multi-agent systems," in *Information and Communication Technology, Electronics and Microelectronics (MIPRO), 2016 39th International Convention on.* IEEE, 2016, pp. 1330–1335.

[14] B. O. Đurić, "Organizational metamodel for large-scale multi-agent systems," in *Trends in Practical Applications of Scalable Multi-Agent Systems, the PAAMS Collection.* Springer, 2016, pp. 387–390.

[15] B. O. Đurić and M. Konecki, "Specific owl-based rpg ontology," in *Central European Conference on Information and Intelligent Systems 26th Inernational Conference*, 2015.

[16] M. E. Gregori, J. P. Cámara, and G. A. Bada, "A jabber-based multi-agent system platform," in *Proceedings of the fifth international joint conference on Autonomous agents and multiagent systems.* ACM, 2006, pp. 1282–1284.

[17] R. E. Fikes and N. J. Nilsson, "Strips: A new approach to the application of theorem proving to problem solving," *Artificial intelligence*, vol. 2, no. 3, pp. 189–208, 1972.

On the Properties of Discrete-Event Systems with Observable States

Nadezhda Nagul

Matrosov Institute for System Dynamics and Control Theory, Siberian Branch of the Russian Academy of Science,
Lermontov str., 134, 664033 Irkutsk, Russia
Email: sapling@icc.ru

Abstract—We address the issue of study the properties of discrete-event systems (DES) with partially observed events and observable states. Two ways of introducing state observations in the system are considered. To show that properties of languages generated by DES, which are essential for constructing supervisors to control DES behavior, such as controllability and observability, are preserved after this transformation, the method of logical-algebraic equations (LAE-method) is exploited. The LAE-method allows one to examine preservation of property of system under morphism-like mapping to another system that is usually more complex than the original system. Thus, once the property is proved to be valid in the original system, it is immediately valid in another.

I. INTRODUCTION

This paper is a continuation of the previous work [1] presented an application of the method of logical-algebraic equations (LAE-method) to the field of discrete-event systems (DES), in particular, for studying properties of supervisors for controlled partially observed DES. Widely used, DES describe system evolution by considering the occurrence of some event sequences. The development of DES theory is driven by the rapid progress of manufacturing systems and communication networks, technological processes, transportation networks, automated and robotic systems, and others, primarily man-made systems. To regulate DES behavior, the Ramadge-Wonham framework of supervisory control is commonly used, adopting ideas from logic, language and automaton theory.

Since in practice only a part of system behavior may usually be observed and used to generate proper control action, the concept of observability plays an important role in supervisory control theory. Initially RW-supervisory control was language based, so the concept of observability was applied to events only and information on states was not used to form control patterns. However, it is often the case that sensors monitor system states as well as changes of those states, usually interpreted as events. In robotics applications observations made by sensory measurements monitoring environment or observations of effects of control actions on system state are presented as observable uncontrollable events [2]. So taking state observations into account seems useful. Even so, there are various ways to take state observations in consideration, and two of them are suggested in [3] and [4]. While the latter uses state observations for the purpose of control and demonstrates that supervisor existence and constructing problem for DES with event and state observation may be reduced to the same

problem in the context of event observation only, the former embraces them for fault events diagnosis. In [1] supervisor non-rejecting property preservation for the model from [4] has been considered. In what follows we examine both models using the the method of logical-algebraic equations (LAE-method).

Lying at the intersection of system dynamics, algebra and logic, the LAE-method is a method of mathematical systems theory which serves to synthesize criteria for preservation properties of systems connected by special mappings called *morphisms*. One of the main applications of the preserving criteria obtained is the reduction of studying some complex system, say S, to studying a much simpler one, say S'. To exploit the LAE-method [5], a property of the system under consideration is treated as a property of an algebraic system, name it $\mathfrak{A}$. Due to the complex nature of dynamical systems, the process of algebraizing of their models usually leads to many-sorted algebraic systems (MAS) where the basic sets have the meaning of a state space, a time scale, etc. Moreover, in [5] the notion of a *general many-sorted algebraic system of finite type* (GMAS) $\mathfrak{A} = \langle A, \Omega_F, \Omega_P, \Omega_E \rangle$ was introduced, where $A = \{A_\lambda | \lambda = \overline{1,k}\}$ is a family of basic sets, $\Omega_F = \{\mathbf{F}_\beta^{n_\beta} | \mathbf{F}_\beta^{n_\beta} : S_{1\beta}[A] \times S_{2\beta}[A] \times \ldots \times S_{n_\beta\beta}[A] \to S_{n_\beta+1,\beta}[A], \beta = \overline{1,k_F}\}$ a set of functions, $\Omega_P = \{\mathbf{P}_\gamma^{n_\gamma} | \mathbf{P}_\gamma^{n_\gamma} \subseteq T_{1\gamma}[A] \times \ldots \times T_{n_\gamma\gamma}[A], \gamma = \overline{1,k_P}\}$ a set of relations, $\Omega_E = \{\mathbf{E}_\delta | \mathbf{E}_\delta \in U_\delta[A], \delta = \overline{1,k_E}\}$ a set of distinguished elements. The elements of the set $\Omega_F \cup \Omega_P \cup \Omega_E$ are defined on extended Bourbaki *steps* $S[A]$ over the family A. Omitting detailed description of the extended step notion, note that step is a set formed from the basic sets A_λ with the help of operations of cartesian product, boolean and sequence forming, called *schemes* [1], [5].

Let a family of mappings

$$\varphi = \{\varphi_\lambda | \varphi_\lambda : A_\lambda \to A'_\lambda, \lambda = \overline{1,k}\}, \tag{1}$$

maps the basic sets of GMAS $\mathfrak{A}$ to the basic sets $A' = \{A'_\lambda | \lambda = \overline{1,k}\}$ of the GMAS $\mathfrak{A}' = \langle A', \Omega'_F, \Omega'_P, \Omega'_E \rangle$ which is of the same type as $\mathfrak{A}$. The "same type" means that the powers of the sets A and A', Ω_F and Ω'_F, and so on accordingly match, and the step $S'_{1\beta}[A'] (S'_{2\beta}[A'], T'_\gamma[A'], U'_\delta[A']$ respectively) is formed from the sets A'_λ with the same scheme as $S_{1\beta}[A]$ ($S_{2\beta}[A], T_\gamma[A], U_\delta[A]$ respectively) from the sets A_λ.

The LAE-method considers either a logical-algebraic equation $\mathcal{X}\&\mathcal{F} \Rightarrow \mathcal{F}'$ or $\mathcal{X}\&\mathcal{F}' \Rightarrow \mathcal{F}$. Here $\mathcal{F}$ is a formula predicate which describes the studied property of the system $\mathfrak{A}$, $\mathcal{F}'$ is a property of another algebraic system $\mathfrak{A}'$ corresponding to the system S', and $\mathcal{X}$ is the subject for searching. The second equation corresponds to the case of preservation of the property in the direction which is opposite to the direction of mappings (1). We describe the considered system property with a formula predicate $\mathcal{F}(\overline{x}) \overset{df}{=} \mathcal{F}(x_1, \ldots, x_p)$ of the signature σ of the chosen GMAS $\mathfrak{A}$, where x_μ is a free variable, $\mu = \overline{1,p}$, $p \geq 0$. Without loss of generality, the formula $\mathcal{F}$ is considered to be formed of *literals* (*concluding statements*, or *c-formulas*) $\mathcal{F}^\nu$, that is, atomic formulas $\mathcal{F}^\nu_+$ or their negations $\mathcal{F}^\nu_-$, with the help of connectives $\&$, $\vee$, and type quantifiers $\hat{\omega}_\alpha \overset{df}{=} \forall z_\alpha : Z_\alpha \overset{df}{=} \forall z_\alpha(Z_\alpha \Rightarrow \sqcup)$ (universal type quantifier), $\check{\omega}_\alpha \overset{df}{=} \exists z_\alpha : Z_\alpha \overset{df}{=} \exists z_\alpha(Z_\alpha \& \sqcup)$ (existential type quantifier), $\alpha = \overline{1,n}$, $\nu = \overline{1,M}$. Such formulas, unlike positive formulas used, for example, in general algebraic system theory, are referred to as *generalized positive formulas*.

A solution, name it $\mathcal{R}$, of the chosen LAE is constructed algorithmically. Such a solution in place of $\mathcal{X}$ guarantees preserving the truth values of formula predicate $\mathcal{F}$ under the mappings of many-sorted algebraic system $\mathfrak{A}$ to $\mathfrak{A}'$. $\mathcal{R}$ is constructed in the form of traditional morphisms, i.e. it is of the meaning of preservation operations and relations only. Morphisms of dynamical systems proved to be especially useful, for example, for those procedures of studying stability and other dynamic properties that require changing variables, since we should ensure that the property under consideration in old variables is equivalent to that one in new variables, or at least guarantee its unidirectional preservation. In [6] classes of properties which are preserved by the morphisms of the same class were defined. Exploiting these results, in what follows a connection between properties which are basic for supervisory control implementation, of specification languages for DES with event and state observation and specification languages for a common DES, will be investigated.

II. THE NOTION OF CONTROLLED DES

Let $\mathcal{G} = (Q, \Sigma, \delta, q_0, Q_m)$ be a discrete event system modeled as a generator of a formal language [7]. Here Q is the set of states q; Σ the set of events; $\delta: \Sigma \times Q \to Q$ the transition function; $q_0 \in Q$ the initial state; $Q_m \subset Q$ the set of marker states. As usual, Σ^* denote the set of all strings over Σ, including the empty string ε. Function δ is naturally extended on strings. Language generated by $\mathcal{G}$ is $L(\mathcal{G}) = \{w : w \in \Sigma^*$ and $\delta(w, q_0)$ is defined$\}$, while language marked by $\mathcal{G}$ is $L_m(\mathcal{G}) = \{w : w \in L(\mathcal{G})$ and $\delta(w, q_0) \in Q_m\}$.

The Ramadge–Wonham supervisory control framework assumes the existence of a means of control $\mathcal{G}$ presented by a *supervisor* [7]. Let Σ_c be a controllable event set, $\Sigma_{uc} = \Sigma \setminus \Sigma_c$, $\Sigma_c \cap \Sigma_{uc} = \emptyset$. The supervisor switches control patterns so that the supervised discrete event systems achieve a control objective described by some regular language K. Let $\mathcal{G}$ be partially observable, i.e. a set Σ_o of observable events is distinguished from all events, $\Sigma_{uo} = \Sigma \setminus \Sigma_o$, $\Sigma_c \cap \Sigma_{uo} = \emptyset$. The observation function is usually defined

as the natural projection $P : \Sigma^* \to \Sigma_o^*$ which just erases unobservable events. The supervisor observes only events from the Σ_o and, basing on this information, disables events in Σ_c. Formally, a supervisor is a pair $\mathcal{J} = (\mathcal{S}, \phi)$ where $\mathcal{S} = (X, \Sigma_o, \xi, x_0, X_m)$ is a deterministic automaton with input alphabet Σ_o. $\mathcal{S}$ is considered to be driven externally by the stream of observable event symbols (words) generated by $\mathcal{G}$ (i.e. words from $P(L(\mathcal{G}))$), while $\phi : X \to \Gamma$ is a function that maps supervisor states x into control patterns $\gamma \in \Gamma \subseteq 2^\Sigma$. If $\sigma \in \gamma = \phi(x)$, then σ is enabled, while if $\sigma \notin \phi(x)$ then σ is disabled (prohibited from occuring). Note that, unlike DES models with *forced events*, enabled events should not necessary occur. Because uncontrollable events cannot be disabled, it is required $\Sigma_{uc} \subseteq \gamma = \phi(x)$.

Extend the function δ to the function $\delta_c : \Gamma \times \Sigma \times Q \to Q$ accounting control patterns as

$$\delta_c(\gamma, \sigma, q) = \begin{cases} \delta(\sigma, q), & \text{if } \delta(\sigma, q) \text{ is defined and } \sigma \in \gamma; \\ \text{undefined}, & \text{otherwise.} \end{cases}$$

Construct the function $\xi \times \delta_c : \Sigma \times X \times Q \to X \times Q$, where $(\xi \times \delta_c)(\sigma, x, q) = (\xi(\sigma, x), \delta_c(\phi(x), \sigma, q))$ is defined iff $\delta(\sigma, q)$ is defined, $\sigma \in \phi(x)$ and $\xi(\sigma, x)$ is defined. Denote $L(\mathcal{J}/\mathcal{G})$ a language generated by the closed-looped behavior of the plant and the supervisor: $L(\mathcal{J}/\mathcal{G}) = \{w : w \in \Sigma^*$ and $(\xi \times \delta_c)(w, x, q)$ is defined$\}$. Let $L_m(\mathcal{J}/\mathcal{G})$ denote the language marked by the supervisor: $L_m(\mathcal{J}/\mathcal{G}) = \{w : w \in L(\mathcal{J}/\mathcal{G})$ and $(\xi \times \delta_c)(w, x_0, q_0) \in X_m \times Q_m\}$.

Supervisory control and observation problem. Given a plant $\mathcal{G}$ over an alphabet Σ, a language $L_A \subseteq L(\mathcal{G})$, a language $L_E \subseteq L(\mathcal{G})$, and sets Σ_o, $\Sigma_c \subseteq \Sigma$, construct a supervisor $\mathcal{J}$ for $\mathcal{G}$ such that $L_A \subseteq L(\mathcal{J}/\mathcal{G}) \subseteq L_E$.

Since the above problem face high computational complexity, it is often substituted by the less complex problem of finding such control patterns that the language marked by the supervisor is equal to some required language. Thus, the special case of the supervisory control and observation problem is to construct such supervisor that $L_m(\mathcal{J}/\mathcal{G}) = K$ where K is called a *a specification language*. The notions of controllable and observable languages are essential in solving this problem.

Let $L \subset \Sigma^*$. The *closure* of L is the set of all strings that are prefixes of words of L, i.e. $\overline{L} = \{s | s \in \Sigma^*$ and $\exists t \in \Sigma^* : s \cdot t \in L\}$. Symbol $\cdot$ denotes string concatenaton and is often omitted. A language L is *closed* if $L = \overline{L}$. If $\mathcal{G}$ is any generator then $L(\mathcal{G})$ is closed.

Definition 1 ([7]): K is controllable (with respect to $L(\mathcal{G})$ and Σ_{uc}) if

$$\overline{K}\Sigma_{uc} \cap L(\mathcal{G}) \subseteq \overline{K}.$$

We have that K, thinking of it as the admissible behavior of the system, is controllable if occurring of any uncontrolled event after prefix of the word from K leads to a word from K, i.e. still admissible.

Definition 2 ([4]): K is observable (with respect to $L(\mathcal{G})$ and P) if

$$\forall s, t \in \Sigma^* \, (P(s) = P(t) \to consis(s, t)),$$

where $consis(s,t)$ is true if and only if

$$(\forall \sigma \in \Sigma)(s\sigma \in \overline{K} \,\&\, t\sigma \in L(\mathcal{G}) \,\&\, t \in \overline{K} \to t\sigma \in \overline{K}).$$

Definition of observability means that if two strings look the same, then they must be consistent in the sense that no conflict of one event continuable after one string but not continuable after the other should occur. *Supervisor existence criterion* sounds as follows: given $K \subseteq L(\mathcal{G})$, there exists supervisor $\mathcal{J}$ such that $K = L_m(\mathcal{J}/\mathcal{G})$ iff K is controllable and observable w.r.t. $L(\mathcal{G})$ and P. Note that it is also required K to be $L_m(\mathcal{G}) - closed$, i.e. $K = \overline{K} \cap L_m(\mathcal{G})$ [7].

III. DES WITH STATE OBSERVATION

Assume that in addition to event observation, supervisor also possesses some information on the states of $\mathcal{G}$, provided by some sensors installed in the system of interest. For a given string generated by $\mathcal{G}$, there is a unique sequence of states visited by this string. Define [3] the sensor maps $h_j : Q \to Y_j$, $j = \overline{1,M}$, where M is the number of sensors, Y_j is a set of outputs of j-th sensor, and a global sensor map defined as $h(q) = (h_1(q), h_2(q), \ldots, h_M(q))$. To use sensor information $\mathcal{G}$ may be transformed to a new system $\tilde{\mathcal{G}}$ where sensor outputs are explicitly exploited. In [3] this transformation is as follows. Let $q' = \delta(q, \sigma)$ with $q, q' \in Q$, $\sigma \in \Sigma$, $h(q) = y$, $h(q') = y'$. Then

1) if $\sigma \in \Sigma_o$ then σ is replaced with the new event $\langle \sigma, y' \rangle$ and let $\tilde{\delta}_1(q, \langle \sigma, y' \rangle) = q'$;
2) if $\sigma \in \Sigma_{uo}$ and $y = y'$, then σ is left unchanged in $\tilde{\mathcal{G}}$ and $\tilde{\delta}_1(q, \sigma) = q'$;
3) if $\sigma \in \Sigma_{uo}$ and $y \neq y'$, then new event $\langle y \to y' \rangle$ and new state q_{new} are introduced and let $\tilde{\delta}_1(q, \sigma) = q_{new}$ and $\tilde{\delta}_1(q_{new}, \langle y \to y' \rangle) = q'$.

Define $Q_{new} = \bigcup\{q_{new}\}$, $|Q_{new}| \leq |Q|$. Let $\tilde{\mathcal{G}}_1 = (\tilde{Q}_1, \tilde{\Sigma}_1, \tilde{\delta}_1, \tilde{q}_{0,1}, \tilde{Q}_{m,1})$ with $\tilde{Q}_1 = Q \cup Q_{new}$, $\tilde{\Sigma}_{c,1} = \Sigma_c$, $\tilde{\Sigma}_{uo,1} = \Sigma_{uo}$ and $\tilde{\Sigma}_{o,1} = \{\langle \sigma, y' \rangle, \langle y \to y' \rangle\}$ where $\sigma \in \Sigma_o$, $y \neq y'$, $\tilde{q}_{0,1} = q_0$, $\tilde{Q}_{m,1} = Q_m$, and $\tilde{\delta}_1$ is constructed according to rules 1) − 3) above.

In [4] for the supervisor construction problem $\mathcal{G}$ is modified in the other way. For each $q \in Q$ new state q_{new} is defined. Let $\tilde{\Sigma}_2 = \Sigma \cup Y$, $\tilde{Q}_2 = Q \cup Q_{new}$, $|Q| = |Q_{new}|$ and given $\delta(\sigma, q) = q'$, $h(q) = y$, define $\tilde{\delta}_2$ according to rules:

1') $\tilde{\delta}_2(\sigma, q) = q'_{new}$ for $\sigma \in \Sigma, q \in Q$, whenever σ is observable or not;
2') $\tilde{\delta}_2(y, q) = q$ for $q \in Q$;
3') $\tilde{\delta}_2(y, q_{new}) = q$ for $q_{new} \in Q_{new}$;

and everything else is undefined. The artificial events in Y are considered to be observable and uncontrollable. Denote $\tilde{\mathcal{G}}_2 = (\tilde{Q}_2, \tilde{\Sigma}_2, \tilde{\delta}_2, \tilde{q}_{0,2}, \tilde{Q}_{m,2})$ a DES with $\tilde{\Sigma}_2 = \tilde{\Sigma}_o \cup \tilde{\Sigma}_{uo}$, $\tilde{\Sigma}_{o,2} = \Sigma_o \cup Y$, $\tilde{\Sigma}_{uo} = \Sigma_{uo}$, $\tilde{\Sigma}_{uc,2} = \Sigma_{uc} \cup Y$, $\tilde{q}_{0,2} = q_{0,new}$, $\tilde{Q}_{m,2} = Q_m$, and $\tilde{\delta}_2 : (\Sigma \cup Y) \times (Q \cup Q_{new}) \to Q \cup Q_{new}$ is constructed according to rules 1') − 3') above.

To compare these two constructions consider Fig.1, there a part of an automaton for $\mathcal{G}$ is presented. Fig. 2 presents modifications $\tilde{\mathcal{G}}_1$ and $\tilde{\mathcal{G}}_2$ for $\mathcal{G}$.

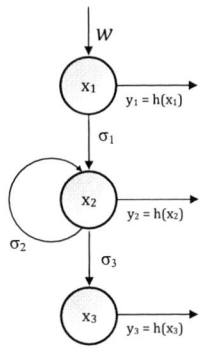

Figure 1. Generator $\mathcal{G}$

Here only σ_1 is observable, $y_1 = y_2$, $y_2 \neq y_3$. Note that structure of $\tilde{\mathcal{G}}_2$ is not effected by observational properties of events or equality of outputs. If a word $v = w\sigma_1\sigma_2\sigma_3$ is generated by $\mathcal{G}$ which gives observation $P(v) = P(w)\sigma_1 y_1 y_3$, then $\tilde{\mathcal{G}}_1$ gives the string $v_1 = w\langle \sigma_1, y_2 \rangle \sigma_2 \sigma_3 \langle y_2 \to y_3 \rangle$, or projection $P(v_1) = P(w)\sigma_1 y_2 y_3$. It should be noted that appearance of y_1 in v_1 depends on previous string w. For the same string $\tilde{\mathcal{G}}_2$ gives $v_2 = wy_1 y_1^* \sigma_1 y_2 \{y_2^* \{\sigma_2 y_2\}^* y_2^*\}^* \sigma_3 y_3 y_3^*$ and a projection $P(v_2) = P(w)y_1 \sigma_1 y_2 y_3$ if we unable to recognize multiple occurrence of the same outputs.

The example shows that approaches providing $\tilde{\mathcal{G}}_1$ and $\tilde{\mathcal{G}}_2$ are quite close. Since the latter one is more stable in the sense of generator structure, in what follows we focus on it, leaving the former one for future investigation.

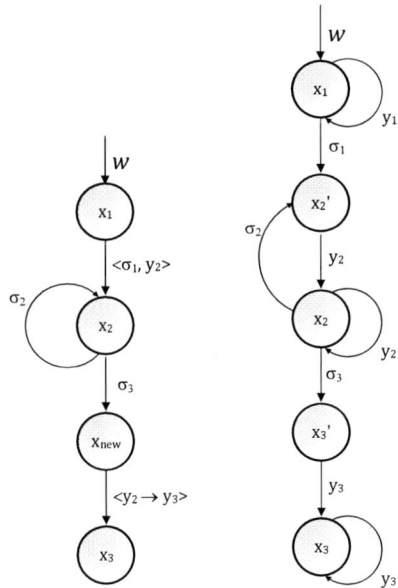

Figure 2. Generators $\tilde{\mathcal{G}}_1$ and $\tilde{\mathcal{G}}_2$

Supervisor for $\tilde{\mathcal{G}}_2$ is able to immediately take the information about the output $y = h(q)$ into consideration since an output y is inserted before each occurrence of event. Such supervisor realizes a mapping $\theta_s : P(L(\tilde{\mathcal{G}}_2)) \to 2^{\Sigma_c \cup Y}$. Note that the language generated and marked by $\tilde{\mathcal{G}}_2$ may be defined via the language generated and marked by $\mathcal{G}$ [4]. For this define $e : \Sigma^* \to (\Sigma \cup Y)^*$ as follows:

$$e(\varepsilon) = h(q_0),$$
$$e(w\sigma) = e(w)\sigma h(\delta(w\sigma, q_0)),$$

which is different from [4] but we believe is more correct. Here, unlike [4], we do not repeat insertion of the output, since it is redundant to form control pattern. The following lemma may be easily proved in the same way as in [4].

Lemma 1: If $L(\mathcal{G})$ is not empty, then

1) $L(\tilde{\mathcal{G}}_2) = e(L(\mathcal{G}))$,
2) $T(L(\tilde{\mathcal{G}}_2)) = L(\mathcal{G})$,
3) $L_m(\tilde{\mathcal{G}}_2) = e(L_m(\mathcal{G}))$,

where $T : (\Sigma \cup Y)^* \to \Sigma^*$ is the projection.
Let K be a nonempty language that describes, for instance, the control objective for the system modeled by $\mathcal{G}$. In [4] it is suggested to extend K till language K_s, such that

$$K_s = T^{-1}(K) \cap L_m(\mathcal{G}).$$

Lemma 2: [4] Assume that K is $L_m(\mathcal{G}) - closed$. Then

1) $\overline{K}_s = T^{-1}(\overline{K}) \cap L(\tilde{\mathcal{G}}_2)$,
2) $K_s = \overline{K}_s \cap L_m(\tilde{\mathcal{G}}_2)$,
3) $T(\overline{K}_s) = \overline{K}$.

In order to include state observation in $\mathcal{G}$ directly, the observation mapping may be presented by the extended projection $\tilde{P} : L(\mathcal{G}) \to ((\Sigma_o \cup \{\varepsilon\}) \times Y)^*$ recursively defined as

$$\tilde{P}(\varepsilon) = (\varepsilon, h(q_0)),$$
$$\tilde{P}(w\sigma) = \tilde{P}(w)(P(\sigma), h(\delta(w\sigma, q_0))).$$

Again, unlike [4], multiple occurring of the artificial events corresponding to the output is not included in this definition. Only one pair $(\sigma, y) \in \Sigma \times Y$ is added to the observation sequence, since adding more than one output is redundant for control purposes. Since S is now driven by the words of $L(\mathcal{G})$ after observation provided by extended projection, a supervisor realizes a mapping $\theta_e : \tilde{P}(L(\mathcal{G})) \to 2^{\Sigma_c}$. Here $\theta_e(\tilde{P}(w))$ is interpreted as a set of events enabled by θ_e after observing $\tilde{P}(w)$, $w \in L(\mathcal{G})$. θ_e should guarantee that $L_m(\theta_e/\mathcal{G}) = K$. In [4] it was proved that non-blocking θ_s such that $L_m(\theta_s/\tilde{\mathcal{G}}_2) = K_s$ exists iff θ_e such that $L_m(\theta_e/\mathcal{G}) = K$ exists and an existence condition for θ_s is expressed in terms of controllability and observability of K_s. In the next section we use the LAE-method to show the connection between these two properties of K and K_s because they are basic for supervisor existence.

IV. LANGUAGE PROPERTIES PRESERVATION

A. Controllability

The definition of controllability plays a key role in characterizing those languages that can be generated by the closed-loop structure plant–supervisor. Controllability property (Def-

inition 1) may be expressed with the formula of the first-order predicate calculus language

$$\forall w \in \overline{K} \ \forall \sigma \in \Sigma_{uc} (w \cdot \sigma \in L(\mathcal{G}) \to w \cdot \sigma \in \overline{K}). \quad (2)$$

Transform (2) to a generalized positive form and in the notation of the language $L(\mathcal{G}_s)$ because we are going to consider LAE $\mathcal{X} \& \mathcal{F}' \Rightarrow \mathcal{F}$ with $\mathcal{F}$ describing controllability of the language K_s w.r.t. $L(\tilde{\mathcal{G}}_2)$ which will be denoted from now on as $L(\mathcal{G}_s)$ for simplicity. This leads to the formula

$$\mathcal{F}_1 = \forall w_s : \overline{K}_s(w_s) \forall \sigma_s : \Sigma_{uc}^s(\sigma_s) \, (\neg L(\mathcal{G}_s)(\cdot(w_s, \sigma_s)) \vee \\ \vee \overline{K}_s(\cdot(w_s, \sigma_s))). \quad (3)$$

In what follows we do not distinguish relations and predicate symbols, i.e. $\overline{K}_s(w_s)$ is true iff $w_s \in \overline{K}_s$. While in general DES is treated as a many-sorted algebraic system with the sets of events, states and so on as basic sets, the property of controllability may be dealt with as the property of the single-sorted algebraic system $\mathfrak{A} = \langle A, \Omega_F, \Omega_P, \Omega_E \rangle$, where $A = \{\Sigma \cup Y\}$, $\Omega_F = \{\cdot\}$, $\Omega_P = \{\overline{K}_s, \Sigma_{uc}^s, L(\mathcal{G}_s)\}$, $\Omega_E = \emptyset$. But nevertheless it remains general. Indeed, since $\overline{K}_s, L(\mathcal{G}_s) \subseteq (\Sigma \cup Y)^{I\!N}$, the scheme $\mathcal{N}(a)$ is associated with the predicates $\overline{K}_s$ and $L(\mathcal{G}_s)$. Therefore the operation of a sequence forming, corresponding to the scheme $\mathcal{N}(a)$ [5], is used to build $\mathfrak{A}$, which is a specific feature of GMAS.

Let $\mathfrak{A}' = \langle A', \Omega_F', \Omega_P', \Omega_E' \rangle$ be the GMAS of the same type as $\mathfrak{A}$ with $A' = \{\Sigma\}$, $\Omega_F' = \{\cdot\}$, $\Omega_P' = \{\overline{K}, \Sigma_{uc}, L(\mathcal{G})\}$, $\Omega_E' = \emptyset$, i.e. $\mathcal{F}'$ describe controllability of the language K w.r.t. $L(\mathcal{G})$. The family of mappings (1) then consists of a single function

$$\varphi : \Sigma \cup Y \to \Sigma. \quad (4)$$

To omit bulky formal manipulations of the algorithm of preservation criteria synthesis [5], results obtained in [6] will be used. For any formula $\mathcal{F}(\overline{x})$ let $ex(\mathcal{F})$ $(all(\mathcal{F})$, respectively) denote a set of variables of existential type quantifiers $\tilde{w}_\alpha$ of $\mathcal{F}$ (universal $\hat{w}_\alpha$, respectively). Let Q_{ex} $(Q_{all}$, respectively) denote a set of predicates which form type conditions of quantifiers for the variables from $ex(\mathcal{F})$ $(all(\mathcal{F})$, respectively). In case of (3) $Q_{ex} = \emptyset$, $Q_{all} = \{\overline{K}_s, \Sigma_{uc}^s\}$. A set of predicate symbols which form c-statements $\mathcal{F}_+^\nu$, we denote as $pos(\mathcal{F})$, while a set of predicate symbols which form c-statements $\mathcal{F}_-^\nu$ we denote as $neg(\mathcal{F})$. In (3) $pos(\mathcal{F}_1) = \{\overline{K}_s\}$, $neg(\mathcal{F}_1) = \{L(\mathcal{G}_s)\}$. Next we form the sets

$$\Omega_{P+} = Q_{ex} \cup pos(\mathcal{F}) = \{\overline{K}_s\},$$
$$\Omega_{P-} = Q_{all} \cup neg(\mathcal{F}) = \{\overline{K}_s, \Sigma_{uc}^s, L(\mathcal{G}_s)\},$$
$$\Omega_{P_-^+} = \Omega_{P+} \cap \Omega_{P-} = \{\overline{K}_s\}, \Omega_{P+} \setminus \Omega_{P_-^+} = \emptyset.$$

Let $\mathcal{I} \subseteq \{1, \ldots, k\}$ be a subset of the index set of the family (1). Let $\mathcal{I}_{S[A]}$ denote the index set of basic sets A_λ which are used to construct the step $S[A]$ with the scheme S.

Definition 3: $\mathcal{MI}_\leftarrow^-$ class is the set of those general positive formulas $\mathcal{F}(\overline{x})$ which satisfy the following conditions:

1) predicates in type conditions of quantifiers do not contain functional symbols;

2) $\mathcal{I} = \bigcup_{S[A]} \mathcal{I}_{S[A]}$, where $S[A] = |v|$, v are all variables z_β or x_μ that enter the equalities $z_\alpha \equiv v$, which are included in c-statements $=_-$ or type conditions of the universal quantifiers $\hat{z}_\alpha$, and z_α is a defining variable;

3) $\Omega_{P+} \setminus \Omega_{P^+_-} = \varnothing$.

Recall that z_α is called a *defining variable*, if in $\mathcal{F}(\overline{x})$ the domain of the type quantifier $\tilde{z}_\alpha$ is larger than the domain of the type quantifier of the variable v. If $v = x_\mu$ than defining variable is z_α. The $t_1 \equiv t_2$ denote any of the equalities $t_1 = t_2$ or $t_2 = t_1$, while symbol $=_-$ (resp. $=_+$) denote entrance of the symbol $=$ in positive (resp. negative) literal (c-formula) of the initial formula $\mathcal{F}(\overline{x})$. The formula (3) obviously belongs to $\mathcal{MI}_\leftarrow^-$ class. It does not contain equalities so the set $\mathcal{I}_{S[A]}$ in this case is empty.

Definition 4: Let Θ be a set of relation symbols, $\Theta \subseteq \Omega_P$. The family of mappings (1) is said to be a $\mathcal{I}$-injective Θ-morphism GMAS $\mathfrak{A} = \langle A, \Omega_F, \Omega_P, \Omega_E \rangle$ to $\mathfrak{A}' = \langle A', \Omega'_F, \Omega'_P, \Omega'_E \rangle$, if

1) $\langle \varphi \rangle^{S_{n_\beta+1,\beta}[A]}(\mathbf{F}_\beta^{n_\beta}(z_1, \ldots, z_{n_\beta})) = (\mathbf{F}_\beta^{n_\beta})'(\langle \varphi \rangle^{S_{1\beta}[A]}(z_1), \ldots, \langle \varphi \rangle^{S_{n_\beta\beta}[A]}(z_{n_\beta}))$, $z_1 \in S_{1\beta}[A]$, ..., $z_{n_\beta} \in S_{n_\beta\beta}[A]$, $\beta = \overline{1, k_F}$;

2) $\langle \varphi \rangle^{U_\delta[A]}(\mathbf{E}_\delta) = \mathbf{E}'_\delta$, $\delta = \overline{1, k_E}$;

and for all $\gamma = \overline{1, k_P}$

$$\langle \varphi \rangle^{T_{1\gamma}[A] \times \ldots \times T_{n_\gamma \gamma}[A]}(\mathbf{P}_\gamma^{n_\gamma}) \subseteq (\mathbf{P}_\gamma^{n_\gamma})'$$

if $\mathbf{P}_\gamma^{n_\gamma}$ correspond to a predicate symbol $P_\gamma \in \Theta$,

$$\langle \varphi \rangle^{T_{1\gamma}[A] \times \ldots \times T_{n_\gamma \gamma}[A]}(\mathbf{P}_\gamma^{n_\gamma}) = (\mathbf{P}_\gamma^{n_\gamma})',$$

if $\mathbf{P}_\gamma^{n_\gamma}$ correspond to a predicate symbol $P_\gamma \in \Omega_P \setminus \Theta$, and mappings φ_λ are injective for all $\lambda \in \mathcal{I}$.

Theorem 1 ([6]): Let $\mathcal{F}(x_1, \ldots, x_q)$ be a formula of the class $\mathcal{MI}_\leftarrow^-$ with free variables $x_1, \ldots, x_q$, $q \geqslant 0$. Then satisfiability of $\mathcal{F}'(f^{|x_1|}(x_1), \ldots, f^{|x_p|}(x_p))$ implies satisfiability of $\mathcal{F}$ under $\mathcal{I}$-injective $\Omega_{P-} \setminus \Omega_{P^+_-}$-morphism of GMAS $\mathfrak{A}$ to GMAS $\mathfrak{A}'$.

Symbol $\varphi^{|x_i|}(x_i)$ denotes *the canonical expansion of mappings (CEM)* φ on the step $S[A] = |x_i|$ to which variable x_i belong [5]. To use Theorem 1 for obtaining conditions for the controllability property preservation it is necessary to show that (4) may be specified as $\Omega_{P-} \setminus \Omega_{P^+_-}$-morphism. Let us start with condition 1) of Definition 4. The functional symbol "$\cdot$" is associated with triple occurring of the scheme $\mathcal{N}(a)$, and for all $v_s, w_s \in (\Sigma \cup Y)^{\mathbb{N}}$ an equality

$$\varphi|^N(v_s \cdot w_s) = \varphi|^N(v_s) \cdot \varphi|^N(w_s).$$

should be valid, since the operation of concatenation is the same in both $\mathfrak{A}$ and $\mathfrak{A}'$. It actually has a place if we define φ as follows:

$$\varphi(\sigma_s) = \begin{cases} \sigma, & \text{if } \sigma_s \in \Sigma; \\ \varepsilon, & \text{if } \sigma_s \in Y, \end{cases}$$

where ε is an empty string, or null event. This definition obviously corresponds to the projection $T : (\Sigma \cup Y)^* \to \Sigma^*$.

Since for (3) $\Omega_E = \varnothing$, we skip the condition 2) and consider elements of the set $\Omega_{P-} \setminus \Omega_{P^+} = \{\Sigma_{uc}^s, L(\mathcal{G}_s)\}$. To satisfy Definition 4, it is necessary to guarantee

$$\langle \varphi \rangle^{(A^{\mathbb{N}})}(L(\mathcal{G}_s)) \subseteq L(\mathcal{G})$$

since the scheme $\mathcal{N}(a)$ is associated with the predicate $L(\mathcal{G}_s)$, a correspond to a basic set, symbol $\mathcal{N}(X)$ corresponds to forming all sequences of elements of the set X, i.e. $\mathcal{N}(X) = X^{\mathbb{N}}$. So far as $\langle \varphi \rangle^{(A^{\mathbb{N}})}(L(\mathcal{G}_s)) = \varphi|^N(L(\mathcal{G}_s))$ then $\varphi|^N(L(\mathcal{G}_s)) \subseteq L(\mathcal{G})$ should be valid where $\varphi|^N(L(\mathcal{G}_s))$ mean application of φ on each letter of the words from $L(\mathcal{G}_s)$ [5]. It actually has place because elimination of symbols of the set Y from the words of $L(\mathcal{G}_s)$ leads to the words from $L(\mathcal{G})$. For the predicate Σ_{uc}^s it is necessary to guarantee

$$\varphi(\Sigma_{uc}^s) \subseteq \Sigma_{uc},$$

where $\Sigma_{uc}^s = \Sigma_{uc} \cup Y$. If we suppose $\varepsilon \in \Sigma_{uc}$, this inclusion is satisfied.

It remains to prove that $\varphi|^N(\overline{K}_s) = \overline{K}$. It is easy to do using the same reasoning as in the proof of Lemma 2 [4]. Thus, it is shown that (4) is a $\Omega_{P-} \setminus \Omega_{P^+}$-morphism and Theorem 1 implies the following proposition, taking into account that $\mathcal{F}_1$ do not contain free variables.

Proposition 1: If K is controllable then K_s is controllable.

Definition 5: $\mathcal{MI}_\rightarrow^-$ is a class of general positive formulas $\mathcal{F}(x_1, \ldots, x_q)$ which satisfy the following conditions:

1) predicates in type conditions of quantifiers do not contain functional symbols;

2) $\mathcal{I} = \bigcup_{S[A]} \mathcal{I}_{S[A]}$, where $S[A] = |v|$, and v are all variables z_β or x_μ, which enter to equalities $z_\alpha \equiv v$, which are included in c-statements $=_-$ or type conditions of the universal quantifiers $\hat{z}_\alpha$, and z_α is a defining variable;

3) $\Omega_{P+} \setminus \Omega_{P^+_-} = \varnothing$.

Since the formula (3) does not contain equalities, it belongs to $\mathcal{MI}_\leftarrow^-$ class as well as $\mathcal{MI}_\rightarrow^-$, with an empty set $\mathcal{I}_{S[A]}$.

Definition 6: Let Θ be a set of relation symbols, $\Theta \subseteq \Omega_P$. The family of mappings (1) is said to be a mighty $\mathcal{I}$-injective Θ-morphism GMAS $\mathfrak{A} = \langle A, \Omega_F, \Omega_P, \Omega_E \rangle$ to $\mathfrak{A}' = \langle A', \Omega'_F, \Omega'_P, \Omega'_E \rangle$, if 1) and 2) of Definition 4 is satisfied and for all $\gamma = \overline{1, k_P}$

$$(\mathbf{P}_\gamma^{n_\gamma})' \subseteq \langle \varphi \rangle^{T_{1\gamma}[A] \times \ldots \times T_{n_\gamma \gamma}[A]}(\mathbf{P}_\gamma^{n_\gamma})$$

if $\mathbf{P}_\gamma^{n_\gamma}$ correspond to a predicate symbol $P_\gamma \in \Theta$,

$$\langle \varphi \rangle^{T_{1\gamma}[A] \times \ldots \times T_{n_\gamma \gamma}[A]}(\mathbf{P}_\gamma^{n_\gamma}) = (\mathbf{P}_\gamma^{n_\gamma})'$$

if $\mathbf{P}_\gamma^{n_\gamma}$ correspond to a predicate symbol $P_\gamma \in \Omega_P \setminus \Theta$, and mappings φ_λ are injective for all $\lambda \in \mathcal{I}$.

Theorem 2 ([6]): Formulas of the class $\mathcal{MI}_\rightarrow^-$ are preserved under mighty $\mathcal{I}$-injective $\Omega_{P-} \setminus \Omega_{P^+}$-morphism of GMAS $\mathfrak{A}$ to GMAS $\mathfrak{A}'$.

To prove that (4) is also a mighty $\Omega_{P-} \setminus \Omega_{P^+}$-morphism it is sufficient to show that $L(\mathcal{G}) \subseteq \varphi|^N(L(\mathcal{G}_s))$ and $\Sigma_{uc} \subseteq \varphi(\Sigma_{uc}^s)$. Since these inclusions are obviously satisfied, combination of Theorem 1 and Theorem 2 implies

1159

Proposition 2: K is controllable if and only if K_s is controllable.

The proposition states that controllability is not affected by the state observation. This fact was previously demonstrated in [4] but its proof considerably differs from the one presented here.

B. Observability

The observability property (Definition 2) of the language $\overline{K}_s$ may be described by the generalized positive formula

$$
\begin{aligned}
\mathcal{F}_2 = &\forall v_s : (\Sigma \cup Y)^*(v) \quad \forall w_s : (\Sigma \cup Y)^*(w_s) \\
&\forall \sigma_s : \Sigma \cup Y(\sigma_s) \quad (\neg(P_s(v_s) = P_s(w_s)) \vee \\
&\vee \neg \overline{K}_s(\cdot(v_s, \sigma_s)) \vee \neg L(\mathcal{G}_s)(\cdot(w_s, \sigma_s)) \vee \\
&\vee \neg \overline{K}_s(w_s) \vee \overline{K}_s(\cdot(w_s, \sigma_s))).
\end{aligned} \tag{5}
$$

In this case consider $\mathfrak{B} = \langle B, \Omega_F, \Omega_P, \Omega_E \rangle$ with $B = \{\Sigma \cup Y\}$, $\Omega_F = \{\cdot, P_s\}$, $\Omega_P = \{(\Sigma \cup Y)^*, \Sigma \cup Y, \overline{K}_s, L(\mathcal{G}_s), =\}$, $\Omega_E = \emptyset$. Note that predicates $\Sigma \cup Y$ and $(\Sigma \cup Y)^*$ here are artificial elements which introduced to embody the formula in the language of GMAS. Let $\mathfrak{B}' = \langle B', \Omega'_F, \Omega'_P, \Omega'_E \rangle$ be the GMAS of the same type as $\mathfrak{B}$ with $B' = \{\Sigma\}$, $\Omega'_F = \{\cdot, P\}$, $\Omega'_P = \{\Sigma^*, \Sigma, \overline{K}, L(\mathcal{G}), =\}$, $\Omega'_E = \emptyset$. In signature of $\mathfrak{B}'$ (5) describe the observability property of the language K. Here $P_s : \Sigma \cup Y \to (\Sigma_o \cup Y)^*$ and $P : \Sigma \to \Sigma_o^*$. Again, the family of mappings (1) includes the single function $\varphi : \Sigma \cup Y \to \Sigma$.

In case of (5) and $\mathfrak{B}$ $Q_{ex} = \emptyset$, $Q_{all} = \{(\Sigma \cup Y)^*, \Sigma \cup Y\}$, $pos(\mathcal{F}_2) = \{\overline{K}_s\}$, $neg(\mathcal{F}_2) = \{=, \overline{K}_s, L(\mathcal{G}_s)\}$,

$$\Omega_{P+} = Q_{ex} \cup pos(\mathcal{F}) = \{\overline{K}_s\},$$

$$\Omega_{P-} = Q_{all} \cup neg(\mathcal{F}) = \{(\Sigma \cup Y)^*, \Sigma \cup Y, =, \overline{K}_s, L(\mathcal{G}_s)\},$$

$$\Omega_{P^+_-} = \Omega_{P+} \cap \Omega_{P-} = \{\overline{K}_s\}.$$

Though the formula (5) contains an equality, the set $\mathcal{I}_{S[A]}$ is still empty. As in the case of (3), $\mathcal{F}_2$ belong to both $\mathcal{MI}^-_\rightarrow$ and $\mathcal{MI}^-_\leftarrow$ classes. Therefore we proceed to check if φ is the morphism we need. In company with the condition on the symbol $\cdot$, previously obtained, for the functional symbols P and P_s the equality

$$\varphi|^N(P_s(w_s)) = P(\varphi|^N(w_s))$$

is valid, what may be easily checked. We skip the trivial equalities $\varphi((\Sigma \cup Y)^*) = \Sigma^*$, $\varphi(\Sigma \cup Y) = \Sigma$, and consider predicate "=". We now state that

$$[P_s(v_s) = P_s(w_s)] \Rightarrow [P(\varphi|^N(v_s)) = P(\varphi|^N(w_s))].$$

Indeed, as far as the output symbols $y \in Y$ are inserted regardless of observability of the symbols from Σ, this implication is true.

Thus, all conditions of Definition 4 are satisfied and (4) is a $\Omega_{P-} \setminus \Omega_{P^+_-}$-morphism. Therefore, according to Theorem 1, if K is observable then K_s is also observable. However, the property is not preserved in the opposite direction. According to Theorem 2, it should be valid

$$[P(\varphi|^N(v_s)) = P(\varphi|^N(w_s))] \Rightarrow [P_s(v_s) = P_s(w_s)].$$

But this is not the case. Indeed, let two words $v_s = y_1 a y_2 b_1 y_3 c \in (\Sigma \cup Y)^*$ and $w_s = y_1 a y_4 b_2 y_3 c \in (\Sigma \cup Y)^*$ are given, where b_1 and b_2 are unobservable. Then $\varphi(v_s) = ab_1c$, $\varphi(w_s) = ab_2c$ and $P(\varphi(v_s)) = P(ab_1c) = ac$, $P(\varphi(w_s)) = P(ab_2c) = ac$. But $P_s(v_s) = y_1 a y_2 y_3 c$ while $P_s(w_s) = y_1 a y_4 y_3 c$. So we formulate

Proposition 3: If K is observable then K_s is observable while the opposite is not the case.

V. Conclusion

In this paper an application of the LAE-method to prove system properties preservation under its mapping into another system has been demonstrated. This method combines algebraic and logical approaches and may be of considerable interest to the audience. The theorems concerning the classes of the properties which are preserved under the similar type morphisms allow one to easily obtain preservation criteria, basing just on the structure of the formula and omitting some numerous formal manipulations. Although the presented results are quite simple, they illustrate the power of the LAE-method. The LAE-method allows to generate the properties preservation conditions for different dynamical systems and the only stipulation of its applicability is the issue of algebraization of a dynamical system model.

Note that there are still a lot of open problems in the theory of DES, especially partially observed and decentralized DES, therefore new approaches and methods are needed. For solving some problems of DES supervisory control theory seems perspective to apply the calculus of positively constructed formulas [8] which proved to be effective means of solving problems of dynamic systems control.

Acknowledgment

The research is supported by the Russian Science Foundation (project no. 16-11-00053).

References

[1] N. Nagul, "Logical-algebraic equations application in discrete-event systems studying," in *Proc. of the 39th Intern. Convention MIPRO, 2016. CIS - Intelligent Systems*, P. Biljanovic, Ed. Rijeka, Croatia: MIPRO Croatian Society, 2016, pp. 1566–1572.

[2] J. Kosecka and R. Bajcsy, "Discrete event systems for autonomous mobile agents," *Robotics and Autonomous Systems*, vol. 12, no. 3, pp. 187–198, 1994.

[3] M. Sampath, R. Sengupta, S. Lafortune, K. Sinnamohideen, and D. C. Teneketzis, "Failure diagnosis using discrete-event models," *IEEE Transactions on Control Systems Technology*, vol. 4, pp. 105–124, 1996.

[4] C. Cao, F. Lin, and Z.-H. Lin, "Why event observation: Observability revisited," *Discrete Event Dynamic Systems*, vol. 7, no. 2, pp. 127–149, 1997. [Online]. Available: http://dx.doi.org/10.1023/A:1008226632478

[5] N. V. Nagul, "The logic-algebraic equations method in system dynamics," *St. Petersburg Math. J.*, vol. 4(24), pp. 645–662, 2013.

[6] ——, "Classes of properties preserved under morphisms of generalizations of many-sorted algebraic systems in studying dynamics (in Russian)," *Trudy Instituta Matematiki i Mekhaniki UrO RAN*, vol. 20(1), pp. 185–200, 2014.

[7] P. Ramadge and W. Wonham, "Supervisory control of class of discrete event processes," *SIAM J. Control and Optimisation*, vol. 25(1), pp. 206–230, 1987.

[8] I. Bychkov, E. Cherkashin, A. Davydov, and A. Larionov, "Software for automated theorem proving based on the calculus of positively constructed formulas," in *2016 IEEE 11th Conference on Industrial Electronics and Applications (ICIEA)*, June 2016, pp. 940–945.

The Formal Description of Discrete-Event Systems Using Positively Constructed Formulas

Artem Davydov, Aleksandr Larionov, and Nadezhda Nagul

Matrosov Institute for System Dynamics and Control Theory at Siberian Branch of Russian Academy of Sciences

134, Lermontov str., Irkutsk, Russia

E-mails: artem@icc.ru, bootfrost@zoho.com, sapling@icc.ru

Abstract—The approach to the first-order logic formalization of discrete-event systems based on the positively constructed formulas calculus is presented. Main concepts of the calculus are given and an example of logical deductions which model the behavior of a discrete-event system under supervisory control is presented.

I. Introduction

The calculus of positively constructed formulas (PCF) was originally developed by Russian scientists S.N. Vassilyev and A.K. Zherlov [1], [2] while describing and solving control theory problems. In [2] the proof of soundness and completeness of the PCF calculus as first-order logical formalism is presented (further development may be found in [3]).

Being both machine-oriented with interactive capabilities for a proof search, the PCF calculus is naturally aimed at solving problems of dynamic systems control due to its features, such as modifiability of semantics (constructive, nonmonotonic, temporal, etc.) and an ability to construct intuitionistic inferences of some non-Horn formulas while explicit usage of ∀– and ∃–quantifiers, since the skolemization procedure is not required. Constructive semantics is important since we need to extract some knowledge (for example, action plans) from the proofs, nonmonotonicity and a treat of time help to construct plans in dynamically changing subject areas. Interactive properties of the PCF calculus, its application for dynamic systems control and action planning are described in [2] with the examples of elevator group control, mobile robot action planning and telescope guidance. Problems of automatic theorem proving software (called *provers*) design and implementation are briefly considered in [4].

The current paper presents PCF calculus application in the field of supervisory control of discrete-event systems (DES). During the last 40 years, to investigate poorly structured and poorly formalized control systems discrete-event approach is widely used. Among other things, DES are extensively exploited in group control, for example, to describe the behavior of groups of autonomous mobile robots [5]. In hierarchical control systems DES may model lower level of control to ensure safe movement of each unit [6] as well as upper levels, which are responsible for planning actions of a unit or a whole group [7].

Almost all existing methods of formalizing DES (automata models, Petri nets, Markov chains, queuing systems, generalized semi-Markov processes, minimax algebras and so on)

have the possibility of influencing the behavior of the system. To control DES presented in form of automata, the method of supervisory control [8] is used, proven itself in many applications and widely recognized.

Computational complexity of the problem of supervisor construction, usually based on system model and behavior specification model, especially for partially observed DES [9], and NP-hardness of supervisor minimization problem [10] are well known. New methods are needed to check systems properties essential for supervisor existence, to construct supervisors, and to simplify them. For example, the method of logical-algebraic equations may be successfully applied in this field [11], [12]. In what follows a new logical approach basing on PCFs to solve problems of DES supervisory control theory is suggested.

II. Preliminaries

Let us consider the basic ideas behind PCFs and their calculus. The language of PCFs is a restricted variant of the language of first-order logic (FOL), which consists of first–order formulas (FOFs) built out of atomic formulas with operators $\&, \vee, \neg, \rightarrow, \leftrightarrow$, quantifier symbols $\forall$ and $\exists$, and constants **True** and **False**. The concepts of *atom*, *literal*, and *term* are defined in the usual way.

Let $X = \{x_1, \ldots, x_k\}$ be a set of variables, $A = \{A_1, \ldots, A_m\}$ a set of atomic formulas, and $F = \{F_1, \ldots, F_n\}$ a set of subformulas. Then formulas $((\forall x_1) \ldots (\forall x_k)(A_1 \& \ldots \& A_m \rightarrow (F_1 \vee \ldots \vee F_n)))$ and $((\exists x_1) \ldots (\exists x_k)(A_1 \& \ldots \& A_m \& (F_1 \& \ldots \& F_n)))$ are denoted as $\forall_X A \colon F$ and $\exists_X A \colon F$ respectively, keeping in mind that the $\forall$–quantifier corresponds to the disjunction of all subformulas, and $\exists$–quantifier corresponds to the conjunction. If $F = \varnothing$ then above formulas turn to the form $\forall_X A \colon \varnothing \equiv \forall_X A \rightarrow$ **False** and $\exists_X A \colon \varnothing \equiv \exists_X A \&$ **True**, since the empty disjunction is understood as **False**, as usual whereas the empty conjunction is understood as **True**. Let $\forall_X A$ and $\exists_X A$ be abbreviations of such formulas. If $X = \varnothing$ then $\forall A \colon F$ and $\exists A \colon F$ are analogous abbreviations.

The set of atoms A is called *conjunct*. Variables from X, bound by corresponding quantifiers, are called $\forall$–variables and $\exists$–variables, respectively. In $\forall_X A$, a variable from X that does not appear in conjunct A is called *unconfined* variable. Note that $\forall \varnothing \equiv \forall \varnothing \colon \varnothing \equiv \forall$**True** $\rightarrow$ **False** $\equiv$ **False**.

1161

Constructions $\forall_X A$ and $\exists_X A$ are called *positive type quantifiers* (TQ), because A is a conjunction of positive atoms only and referred to as *type condition* for X. In practice, these constructions denote phrases such as "for all X satisfying A there is...", "there exist X satisfying property A such that...", and so on; for example, "for all integer x, y, z and $n > 2$ there is $x^n + y^n \neq z^n$". Originally, the term "type quantifier" was introduced by N. Bourbaki [13] as a part of notation for formalization of mathematics.

A. PCF Language Explicit Definition

Definition 1 (Positively constructed formulas (PCF)). Let X be a set of variables and A a conjunct. Then

1) $\exists_X A$ and $\forall_X A$ are $\exists$–PCF and $\forall$–PCF respectively.
2) If $F = \{F_1, \ldots, F_n\}$ is a set of $\forall$–PCFs, then $\exists_X A \colon F$ is a $\exists$–PCF.
3) If $F = \{F_1, \ldots, F_n\}$ is a set of $\exists$–PCFs, then $\forall_X A \colon F$ is a $\forall$–PCF.
4) Any $\exists$–PCF or $\forall$–PCF is a PCF.

This form of logical formulas is referred to as positively constructed formulas (PCFs), as they are written with only positive type quantifiers. The formulas contain no explicit logic negation sign. Without loss of generality only closed formulas will be considered. Any FOF can be represented as PCF [2].

A PCF starting with $\forall\varnothing$ is called a PCF in the *canonical form*. Any PCF can be represented in the canonical form. If F is a non–canonical $\exists$–PCF then $\forall\varnothing \colon F$ is the canonical PCF since $\forall\varnothing \colon F \equiv \mathbf{True} \to F \equiv F$. If F is a non–canonical $\forall$–PCF then the canonical PCF is $\forall\varnothing \colon \{\exists\varnothing \colon F\} \equiv \mathbf{True} \to \mathbf{True} \& F \equiv F$. Type quantifiers $\forall\varnothing$ and $\exists\varnothing$ are called *fictitious*, since they do not influence truth value of an original PCF and do not bind any variables. They are used to regularize PCFs, i.e, transform them to canonical ones.

PCFs are usually represented as trees for easier reading, i.e. $Q_X A \colon \{F_1, \ldots, F_n\}$ is represented as

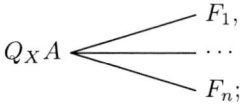

where Q is a quantifier. Tree elements have conventional names: *node, root, leaf, branch*, etc. As the quantifier $\forall$ corresponds to a disjunction of formulas $\{F_1, \ldots, F_n\}$ (quantifier $\exists$ corresponds to conjunction), then each $\forall$–node is considered as *disjunctive branching*, and each $\exists$–node corresponds to *conjunctive branching*.

Parts of canonical PCF are named as follows:

1) The root of a PCF's tree-view $\forall\varnothing$ is called a PCF *root*.
2) Each PCF root child $\exists_X A$ is called a PCF *base*, conjunct A is called *base of facts*, and a PCF rooted from base is called *a base subformula*.
3) PCF base children $\forall_Y B$ are called *questions* to the parent base. If a question is a leaf of a tree then it is called *a goal question*.

4) Subtrees of questions are called *consequents*. If a question has no consequent then the question is referred to as *goal question*, and it is identical to **False**.

Example 1. Consider a PCF representation of a FOF

$$\mathcal{F} = \neg\big(\forall x \exists y P(x, y) \to \exists z P(z, z)\big).$$

An image $\mathcal{F}'$ of $\mathcal{F}$ in the PCF language is $\mathcal{F}' = \forall \colon \varnothing\{\exists \colon \varnothing\{\forall x \colon \varnothing\{\exists y \colon P(x, y)\}, \forall z \colon P(z, z)\{\exists \colon \mathbf{False}\}\}\}$. The tree–like form of the latter is as follows:

$$\forall \colon \varnothing - \exists \colon \varnothing \Big\langle \begin{array}{l} \forall x \colon \varnothing \text{——} \exists y \colon P(x, y) \\ \forall z \colon P(z, z) - \exists \colon \mathbf{False} \end{array}$$

B. The Inference rule

Definition 2 (Answer). A question $\forall_Y D \colon \Upsilon$ to a base $\exists_X A$ has an answer θ if and only if θ is a substitution $Y \to H^\infty \cup X$ and $D\theta \subseteq A$, where H^∞ is Herbrand universe based on constant and function symbols that occur in corresponding base subformula.

Definition 3 (Splitting). Let $B = \exists_X A \colon \Psi$, and $Q = \forall_Y D \colon \Upsilon$, where $\Upsilon = \{\exists_{Z_1} C_1 \colon \Gamma_1, \ldots, \exists_{Z_n} C_n \colon \Gamma_n\}$ then $split(B, Q) = \{\exists_{X \cup Z_1'} A \cup C_1' \colon \Psi \cup \Gamma_1', \ldots, \exists_{X \cup Z_n'} A \cup C_n' \colon \Psi \cup \Gamma_n'\}$, where $'$ is a variable renaming operator. We say that B is split by Q. Obviously, $split(B, \forall_Y D) = split(B, \forall_Y D \colon \varnothing) = \varnothing$.

Definition 4 (Inference rule ω). Consider some canonical PCF $F = \forall\varnothing \colon \Phi$. Let there exists a question Q that has an answer θ to appropriate base $B \in \Phi$, then $\omega F = \forall\varnothing \colon \Phi \setminus \{B\} \cup split(B, Q\theta)$.

In other words, if a question has an answer to its base, then the base subformula is split by this question. In the case of a goal question, we say that the basic subformula is refuted because $split(B, \forall_Y D) = \varnothing$. The refuted base subformula B removed from the set of base subformulas Φ, since $\Phi \setminus \{S\} \cup \varnothing = \Phi \setminus \{S\}$

As soon as all the bases subformulas from Φ have been refuted, the formula F is also refuted, since $\forall\varnothing \colon \varnothing \equiv \mathbf{False}$. The PCF language and the inference rule ω form the calculus oriented to refutation of a negation of an original formula. The only axiom of PCF calculus is $\forall\varnothing \colon \varnothing$, i.e., **False**.

III. Supervisory Control of Discrete Event Systems

Let $\mathcal{G} = (Q, \Sigma, \delta, q_0, Q_m)$ be a discrete event system modeled as a generator of a formal language [8], also called a *plant*. Here Q is the set of states q; Σ the set of events; $\delta \colon \Sigma \times Q \to Q$ the transition function; $q_0 \in Q$ the initial state; $Q_m \subset Q$ the set of marker states. As usual, Σ^* denote the set of all strings over Σ, including the empty string ε. δ is easily extended on strings from Σ^*. Language generated by $\mathcal{G}$ is $L(\mathcal{G}) = \{w \colon w \in \Sigma^* \text{ and } \delta(w, q_0) \text{ is defined}\}$, while language marked by $\mathcal{G}$ is $L_m(\mathcal{G}) = \{w \colon w \in L(\mathcal{G}) \text{ and } \delta(w, q_0) \in Q_m\}$.

The Ramadge–Wonham supervisory control framework assumes the existence of a means of control $\mathcal{G}$ presented

1162

by a *supervisor* [8]. Let Σ_c be a controllable event set, $\Sigma_{uc} = \Sigma \setminus \Sigma_c$, $\Sigma_c \cap \Sigma_{uc} = \varnothing$. Let K be a nonempty language that describes the control objective for the system modeled by $\mathcal{G}$. A supervisor should switch control patterns so that the supervised DES generates exactly K. Formally, a supervisor is a pair $\mathcal{J} = (\mathcal{S}, \phi)$ where $\mathcal{S} = (X, \Sigma, \xi, x_0, X_m)$ is a deterministic automaton with input alphabet Σ. $\mathcal{S}$ is considered to be driven externally by the words from $L(\mathcal{G})$, while $\phi : X \to \Gamma$ is a function that maps supervisor states x into control patterns $\gamma \in \Gamma \subseteq 2^{\Sigma}$. If $\sigma \in \gamma = \phi(x)$, then σ is enabled, while if $\sigma \notin \gamma$ then σ is disabled (prohibited from occuring). Because uncontrollable events cannot be disabled, it is required $\Sigma_{uc} \subseteq \gamma = \phi(x)$.

Extend the function δ to the function $\delta_c : \Gamma \times \Sigma \times Q \to Q$ accounting control patterns as

$$\delta_c(\gamma, \sigma, q) = \begin{cases} \delta(\sigma, q), & \text{if } \delta(\sigma, q) \text{ is defined and } \sigma \in \gamma; \\ \text{undefined}, & \text{otherwise.} \end{cases}$$

Construct the function $\xi \times \delta_c : \Sigma \times X \times Q \to X \times Q$, where $(\xi \times \delta_c)(\sigma, x, q) = (\xi(\sigma, x), \delta_c(\phi(x), \sigma, q))$ is defined iff $\delta(\sigma, q)$ is defined, $\sigma \in \phi(x)$ and $\xi(\sigma, x)$ is defined. Denote $L(\mathcal{J}/\mathcal{G})$ a language generated by the closed-looped behavior of the plant and the supervisor: $L(\mathcal{J}/\mathcal{G}) = \{s : s \in \Sigma^*$ and $(\xi \times \delta_c)(s, x, q)$ is defined$\}$. Let $L_m(\mathcal{J}/\mathcal{G})$ denote the language marked by the supervisor: $L_m(\mathcal{J}/\mathcal{G}) = \{s : s \in L(\mathcal{J}/\mathcal{G})$ and $(\xi \times \delta_c)(s, x_0, q_0) \in X_m \times Q_m\}$. The main goal of supervisory control is to construct such supervisor that $L_m(\mathcal{J}/\mathcal{G}) = K$.

The definition of controllability plays a key role in characterizing those languages that can be generated by the closed-loop structure plant–supervisor. Let $L \subset \Sigma^*$. The *closure* of L is the set of all strings that are prefixes of words of L, i.e. $\overline{L} = \{s | s \in \Sigma^*$ and $\exists t \in \Sigma^* : s \cdot t \in L\}$. Symbol $\cdot$ denotes string concatenation and is often omitted.

Definition 5. K is controllable (with respect to $L(\mathcal{G})$ and Σ_{uc}) if

$$\overline{K}\Sigma_{uc} \cap L(\mathcal{G}) \subseteq \overline{K}.$$

If K represents the admissible behavior of the system, K is controllable if occurring of any uncontrolled event after prefix of the word from K leads to a word from K, i.e. still admissible. Only controllable languages may be exactly achieved by the joint behavior of the plant and supervisor.

Supervisor existence criterion sounds as follows: given $K \subseteq L(\mathcal{G})$, there exists supervisor $\mathcal{J}$ such that $K = L_m(\mathcal{J}/\mathcal{G})$ iff K is controllable and $L_m(\mathcal{G}) - closed$, i.e. $K = \overline{K} \cap L_m(\mathcal{G})$ [8].

Example 2 (DES model of AUV mission). Consider a simplified general DES model of autonomous underwater vehicle (AUV) functioning during some mission implementation (Fig. 1). For this DES $Q = \{ready, check, complete\}$, corresponding to AUV readiness for a mission, hardware and software checking, and mission completion, respectively, $\Sigma = \{start, tuning, perfom, end\}$, corresponding to a start of AUV functioning, AUV systems'

checking, mission performing and operation quitting, respectively, $q_0 = \{ready\}$, $Q_m = Q$. Suppose $\Sigma_c = \{start, tuning, perfom\}$.

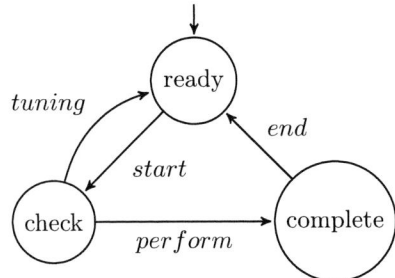

Fig. 1. Generator G

Suppose that specification on AUV behavior requires that after first AUV systems' checking procedure AUV will return for tuning or sensors calibrating before the main mission implementation. The automaton generating the specification language K is shown on Fig. 2.

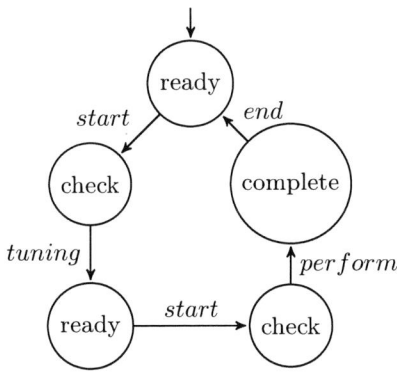

Fig. 2. Specification K

In this case it is easy to check that K is controllable and $L_m(\mathcal{G})$-closed therefore supervisor $\mathcal{J} = (\mathcal{S}, \phi)$ such that $K = L_m(\mathcal{J}/\mathcal{G})$ exists and may be constructed on the base of the automaton for K [14]. Non-reduced $\mathcal{S}$ will have five states $X = \{x_0, x_1, x_2, x_3, x_4\}$, all marked, and its structure coincides with the structure of the automaton for K. The function ϕ is shown in Table I. Dashes there denote that it does not matter if event is enabled or disabled.

TABLE I
Mapping $\phi : X \to \Gamma$

	x_0	x_1	x_2	x_3	x_4
start	enable	—	enable	—	—
tuning	—	enable	—	disable	—
perform	—	disable	—	enable	—

IV. PCF Presentation of Supervised DES

In this section it will be shown how above example of DES may be formalized using a PCF. The following predicates are needed. $E(x)$ will be interpreted as "x is an event", $S(x,y)$ will denote "x is a current sequence of events in a state y". The function symbol "$\cdot$" will denote the strings concatenation, and the "ϵ" symbol will be for an empty string. To save the space event names will be contracted to its first letters and denote states *ready, check, complete* as s_1, s_2 and s_3, correspondingly. The generator G (corresponding to Fig. 1) is described by the following single base of PCF.

$$\exists \mathcal{B} \begin{cases} \forall \sigma \ E(s), S(\sigma, s_1) \text{——} \exists S(\sigma \cdot s, s_2) \\ \forall \sigma \ E(t), S(\sigma, s_2) \text{——} \exists S(\sigma \cdot t, s_1) \\ \forall \sigma \ E(p), S(\sigma, s_2) \text{——} \exists S(\sigma \cdot p, s_3) \\ \forall \sigma \ E(e), S(\sigma, s_3) \text{——} \exists S(\sigma \cdot e, s_1) \end{cases}$$

Here, $\mathcal{B}$ is the conjunct $\{E(t), E(s), E(p), E(e), S(\epsilon, s_1)\}$.

Let's go through first steps of this formula's inference to demonstrate the strategy that will generate the language $L(\mathcal{G})$ in the base conjunct $\mathcal{B}$ as first arguments of atoms S. At the beginning of the inference search, there is the only one possible answer $\{\sigma \to \epsilon, x \to s_1\}$ to the first question, that will add $S(s, s_2)$ to the $\mathcal{B}$. Next steps include answers to the second and the third questions, that will add $S(st, s_1)$, $S(sp, s_3)$ to the $\mathcal{B}$. Note that after these last additions, there will be the new possible answer to the first question, and an extra answer to last question. But we will not use those until the next cycle of questions bypassing, so the inference rule is applied one time to each question in one cycle. The first cycle of question bypassing will end with the answer to the last question, adding the $S(spe, s_1)$ to the $\mathcal{B}$. The next cycle starts with three possible answers to the first question as there are three S atoms, with the constant argument s_1 in the base. One of those answers is the same as the very first one used, generating $S(s, s_2)$ that will be consumed by existing one in the base, thus, our strategy will not use answers that was already used. Going on further, the base will be updated with atoms $S(spes, s_2)$, $S(sts, s_2)$, then $S(stst, s_1)$, $S(spesp, s_3)$, $S(spespe, s_1)$ and so on.

The language constructing PCF of the supervisor (Fig. 2) is the following, where $\mathcal{B}_x$ is $\{E_x(t), E_x(s), E_x(p), E_x(e), S_x(\epsilon, x_1)\}$.

$$\exists \mathcal{B}_x \begin{cases} \forall \sigma \ E_x(s), S_x(\sigma, x_0) \text{—} \exists S_x(\sigma \cdot s, x_1) \\ \forall \sigma \ E_x(t), S_x(\sigma, x_1) \text{—} \exists S_x(\sigma \cdot t, x_2) \\ \forall \sigma \ E_x(s), S_x(\sigma, x_2) \text{—} \exists S_x(\sigma \cdot s, x_3) \\ \forall \sigma \ E_x(p), S_x(\sigma, x_3) \text{—} \exists S_x(\sigma \cdot p, x_4) \\ \forall \sigma \ E_x(e), S_x(\sigma, x_4) \text{—} \exists S_x(\sigma \cdot e, x_0) \end{cases}$$

Junction of the supervisor and the system guarantees that the specification language is constructed. The PCF imitating the concurrent work of the supervisor and the system will be as follows.

$$\exists \mathcal{B}, \mathcal{B}_x \begin{cases} \forall \sigma \ E_x(s), S_x(\sigma, x_0) \text{————} \exists S_x(\sigma \cdot s, x_1) \\ \forall \sigma \ E_x(t), S_x(\sigma, x_1) \text{————} \exists S_x(\sigma \cdot t, x_2) \\ \forall \sigma \ E_x(s), S_x(\sigma, x_2) \text{————} \exists S_x(\sigma \cdot s, x_3) \\ \forall \sigma \ E_x(p), S_x(\sigma, x_3) \text{————} \exists S_x(\sigma \cdot p, x_4) \\ \forall \sigma \ E_x(e), S_x(\sigma, x_4) \text{————} \exists S_x(\sigma \cdot e, x_0) \\ \forall \sigma, x \ E(s), S(\sigma, s_1), S_x(\sigma \cdot s, x) \cdot \exists S(\sigma \cdot s, s_2) \\ \forall \sigma, x \ E(t), S(\sigma, s_2), S_x(\sigma \cdot t, x) \text{-} \exists S(\sigma \cdot t, s_1) \\ \forall \sigma, x \ E(p), S(\sigma, s_2), S_x(\sigma \cdot p, x) \cdot \exists S(\sigma \cdot p, s_3) \\ \forall \sigma, x \ E(e), S(\sigma, s_3), S_x(\sigma \cdot e, x) \cdot \exists S(\sigma \cdot e, s_1) \end{cases}$$

Let's discuss the inference of the above formula. It consists of two groups of questions, and the first one corresponds to the supervisor, and the second to the initial system with the supervisor atoms added. Latter do not allow incorrect answers that could have added those atoms to the base that contain sequences of events that do not belong to the specification language K.

V. Conclusion

This paper is the starting point in developing a novel approach to formalizing and solving various control problems for important class of dynamic systems known as DES. With the simple example of AUV mission control basic concepts of PCF calculus implementation was shown.

Computational complexity of inference search of PCFs is rely strongly on the complexity of the inference rule application. The application of the PCFs inference rule for the formulas presenting DES has the polynomial, close to linear complexity, since the obtained structures are quite simple and the answers search procedure (consuming the most of time) is also not so hard.

Further investigations will include, among other issues, subsets of controllable and observable events choosing for uncontrollable languages, supervisor reduction problem solving, sensors activation policies investigation. Obtained results will be exploited in group control, for instance, for supporting AUV missions.

Acknowledgment

The research is supported by the Russian Science Foundation (project no. 16-11-00053).

References

[1] S.N. Vassilyev, "Machine Synthesis of Mathematical Theorems", The Journal of Logic programming, Vol. 9, no.2–3, pp. 235–266, 1990.
[2] S.N. Vassilyev, A.K. Zherlov, E.A. Fedunov, B.E. Fedosov, "Intelligent Control of Dynamic Systems", Moscow, Russia: Fizmatlit, 2000. (in Russian)

[3] A.V. Davydov, A.A. Larionov, E.A. Cherkashin, "On the calculus of positively constructed formulas for automated theorem proving", Automatic Control and Computer Sciences (AC&CS), Vol. 45, no. 7, pp. 402–407, 2011.

[4] E.A. Cherkashin, A.A. Larionov, A.V. Davydov, "Calculus of Positively Constructed Formulas, its Features, Strategies and Implementation", MIPRO 2013/CIS (36-th international convention on information and communication technology, electronics and microelectronics), 20–24 May 2013, Chroatia, Opatija, 2013. pp. 1289–1295.

[5] B. Moore and K. Passino, "Distributed coordination strategies for wide-area patrol", Journal of Intelligent and Robotic Systems, vol. 56, pp. 23-45, 2009.

[6] J. Kosecka and R. Bajcsy, "Discrete Event Systems for autonomous mobile agents", Robotics and Autonomous Systems, vol. 12, no. 3, pp. 187–198, 1994.

[7] R. Huq, G. K. I. Mann, and R. G. Gosine, "Behavior-modulation technique in mobile robotics using fuzzy discrete event system", IEEE Transactions on Robotics, vol. 22, pp. 903–916, 2006.

[8] P.J. Ramadge, and W.M. Wonham, "Supervisory control of class of discrete event processes", SIAM J. Control and Optimisation, vol. 25(1), pp. 206–230, 1987.

[9] J. N. Tsitsiklis, "On the control of discrete-event dynamical systems", Mathematics of Control, Signals, and Systems, vol. 2, pp. 95-107, 1989.

[10] R.Su, and W.M. Wonham, "Supervisor Reduction for Discrete-Event Systems", Discrete Event Dynamic Systems, vol. 14, no. 1, pp. 31–53, 2004.

[11] N.V. Nagul, "Generating Conditions for Preserving the Properties of Controllable Discrete Event Systems", Autom. Remote Control. 2016, vol. 77, no. 4, pp. 637–652, DOI: 10.1134/S0005117916040111.

[12] N. Nagul and I. Bychkov, "On the problem of discrete-event systems properties preservation", AIP Conference Proceedings, vol. 1798, no. 1, p. 020107, 2017, doi: 10.1063/1.4972699.

[13] N. Bourbaki, "Theory of Sets", Paris: Hermann, 1968.

[14] C. Seatzu, M. Silva, and J. H. Schuppen, "Control of discrete-event systems: Automata and petri net perspectives", London: Springer, 2013.

Runtime Estimation for Enumerating all Mutually Orthogonal Diagonal Latin Squares of Order 10

Stepan Kochemazov, Oleg Zaikin, Alexander Semenov

Matrosov Institute for System Dynamics and Control Theory SB RAS, Irkutsk, Russia

Email: veinamond@gmail.com, zaikin.icc@gmail.com, biclop.rambler@yandex.ru

Abstract—In the present paper we estimate how long it will take state-of-the-art combinatorial algorithms to enumerate all possible mutually orthogonal diagonal Latin squares of order 10. For this purpose we first evaluate the performance of DLX algorithm and contemporary algorithms for solving SAT in application to finding orthogonal mates of an arbitrary diagonal Latin square of order 10. Then we estimate the number of diagonal Latin squares of order 10 and use this information in combination with some techniques for exploiting symmetries and equivalences to approximate the amount of time it would take to process them using existing hardware.

I. Introduction

Latin square of order n is a square table $n \times n$ filled with elements from the set $\{0, \ldots, n-1\}$ in such a way that all elements within a single row or single column are distinct [1]. A Latin square is called diagonal if both its main diagonal and main antidiagonal contain all elements from 0 to $n-1$. Two Latin squares $A = (a_{ij})$ and $B = (b_{ij})$ of the same order are called orthogonal if all ordered pairs (a_{ij}, b_{ij}), $i, j \in \{0, \ldots, n-1\}$ are distinct. A set of at least two Latin squares that are mutually orthogonal is often called a set of MOLS.

Latin squares represent one of the most widely known, well studied and easy to understand combinatorial designs. However, even for relatively small values of order n there remain exceptionally hard open problems. Probably, the most famous of them is to prove the existence or non-existence of a triple of MOLS of order 10.

In the present paper we focus on diagonal Latin squares of order 10. The existence of orthogonal pairs of these squares was determined in [2]. They present an interesting case of Latin squares because mutually orthogonal diagonal Latin squares (MODLS) are quite rare compared to MOLS. That is why we can hope that their number is relatively small. Nevertheless the problem outlined in the title appears to be very hard. We want to estimate how hard it is from the point of view of state-of-the-art combinatorial algorithms. In particular, we estimate the runtime of the simple algorithm, which generates all diagonal Latin squares of order 10 and for each square determines if it has orthogonal diagonal mates. From empirical evaluation it is clear that the second part of the algorithm is much harder than the first, so we consider it in detail and analyze two approaches to the problem of finding orthogonal mates of a specific diagonal Latin square: the approach based on reducing this problem to Boolean satisfiability problem (SAT) [3], and the approach that employs the DLX algorithm proposed by D. Knuth [4].

Let us present the brief outline of the paper. In the second section we give necessary definitions and outline in more details the general algorithm for enumerating all possible MODLS of order 10. In the third section we consider SAT-based and DLX-based approaches in the context of the general Euler-Parker approach [5] to finding orthogonal mates and analyze their effectiveness. In the fourth section we apply the Monte Carlo method [6] to estimate the number of diagonal Latin squares of order 10 and construct our runtime estimation. After this we review related works and discuss how our research fits into general picture. In the last section we make final conclusions regarding the obtained results.

II. Preliminary information

There are several terms and definitions that we will use throughout the paper. Let us introduce them first. A *transversal* of a Latin square of order n is a set of n of its entries, in which there are representatives of each row, column and value [1], [7]. In practice for a specific Latin square A of order n by transversal we mean an n-tuple $T = (t_0, \ldots, t_{n-1})$, where pairs (i, t_i), $i = 0, \ldots, n-1$ mark the coordinates of corresponding elements in A. It is known that a Latin square of order n has an orthogonal mate if and only if it can be partitioned into n disjoint transversals (we can effectively construct the corresponding orthogonal Latin square based on this set of transversals). This fact is often used to find MOLS [5], [8], [9]. Diagonal Latin squares impose an additional constraint in this context: for a diagonal Latin square to have diagonal orthogonal mate, all of its disjoint transversals must intersect exactly once with its main diagonal and exactly once with its main antidiagonal [2], [10]. Let us refer to transversals that satisfy this condition as to *diagonal transversals*. It is easy to see that the set of all possible transversals of a Latin square of order n actually coincides with the set of all possible permutations of n elements $\{0, 1, \ldots, n-1\}$. Meanwhile, the number of diagonal transversals is significantly smaller than that. For $n = 10$ the number of transversals is 3 628 800 among which only 494 080 are diagonal. This fact is the main reason why MODLS are rare: for example, the second square from the first pair found in [2] has 5 504 transversals, of which only 866 are diagonal.

For the purposes explained later it is convenient to split the space of all possible diagonal Latin squares into relatively large chunks. It can be done in various ways. In the context of the present paper we follow a simple path and consider the sets of Latin squares which have the same values of elements in the first several rows. In particular, let us denote by LS_n^k (DLS_n^k), $k < n^2$ an incomplete Latin square (diagonal Latin square) of order n in which only first k elements (traversing a Latin square from left to right, from top to bottom) are known. It is then natural to consider all Latin squares (diagonal Latin squares) constructed by filling the remaining $n^2 - k$ elements.

For convenience we consider only normalized diagonal Latin squares, i.e. Latin squares in which the elements of the first row are in ascending order. An arbitrary diagonal Latin square can be represented in such form using simple transformations that do not violate any conditions. It also reduces the search space by the factor of $n!$. In more detail we will consider the ways to reduce the search space in the end of Section 3.

A. General outline of the algorithm for finding MODLS of order 10

We consider a basic algorithm for solving the considered problem. It looks as follows.

1) Generate all possible diagonal Latin squares of order 10.
2) For each generated diagonal Latin square check if it has diagonal orthogonal mates.

From a practical point of view this basic algorithm can be viewed as a computational scheme involving two main parts:

- algorithm for generating diagonal Latin squares of order 10;
- algorithm for finding orthogonal mates of a specific Latin square (or for proving that there are none).

It means that to construct the runtime estimation for this algorithm we need, first, to estimate the number of diagonal Latin squares of order 10, and, second, to estimate the speed with which we will be able to process them.

To estimate the number of diagonal Latin squares of order 10 we can use the Monte-Carlo method [6] in the following form: first for a specific $0 < k < n^2$ we compute (or estimate) the number of all possible DLS_{10}^k. Then using a random sample of DLS_{10}^k we compute an estimation of the expected value representing the average number of diagonal Latin squares that share the same DLS_{10}^k. Then by multiplying these two numbers we will obtain our estimation. In detail we will address this problem in Section 4.

The two stages of outlined computational scheme can be viewed as independent (of course in practice it is reasonable to generate and analyze Latin squares in batches of relatively small size). Since there are known algorithms for generating diagonal Latin squares that make it possible to construct about 10^6 squares per second on one core of mainstream PC [11], in the remainder of the paper we assume that the bottleneck of the general algorithm lies in finding orthogonal mates. Our empirical evaluation showed that this point of view is justified. That is why we want to consider it in more detail in the following section.

III. COMPUTATIONAL APPROACHES FOR FINDING ORTHOGONAL MATES

The problem of finding orthogonal mates for a specific Latin square of order n has a history of its own. As it is outlined in Chapter 7 of [12], two main approaches have been formed for solving this problem. To the first one we can informally refer as to Paige-Tompkins approach. In it we fill the cells of the second (potentially orthogonal) square in some order to satisfy all the constraints (imposed by Latin square conditions, orthogonality condition, etc.).

The second one is usually referred to as Euler-Parker approach [5]. It implies that we first construct all transversals of a considered Latin square and then search for subsets of n disjoint transversals in a constructed set, considering it as an instance of *exact cover* problem. After thorough analysis D. Knuth makes a conclusion that Euler-Parker approach is much more effective, therefore we will employ it in the remainder of the paper. Since it represents a kind of a general framework and does not specify any particular algorithms, we implemented several state-of-the-art combinatorial algorithms within its context and compared their effectiveness. In particular, we used for this purpose state-of-the-art SAT-solving algorithms and DLX. Let us first address the *exact cover* problem and then describe the algorithms employed in more detail.

A. Finding orthogonal mates of a Latin square as exact cover problem

In the context of Euler-Parker approach we can consider both the problem of finding transversals of a Latin square of order n and the problem of finding sets of n disjoint transversals as instances of *exact cover* problem. This problem is formulated as follows: given a set X and the collection of its subsets $S = \{S_0, \ldots, S_k\}$, $S_i \subseteq X$, $i \in \{0, \ldots, k\}$, $k < 2^{|X|} - 1$ to find such subcollection S^* of sets from S that each element from X is contained exactly in one subset of S^*. Note that when searching for transversals we need to find all possible solutions of this problem.

To reduce the search for transversals to *exact cover* we represent the problem in the following form. Let $A = \{a_{ij}\}$, $a_{ij}, i, j \in \{0, \ldots, n-1\}$ be an arbitrary Latin square of order n. Let the set X contain $3 \times n$ elements $\{x_0, \ldots, x_{3n-1}\}$. The elements $x_0, \ldots, x_{n-1}$ correspond to rows, $x_n, \ldots, x_{2n-1}$ to columns, and $x_{2n}, \ldots, x_{3n-1}$ to values of Latin square cells. Then the collection of subsets S will have exactly $n \times n$ subsets, that correspond to Latin square cells in the following way: with the cell a_{ij} we associate the subset $S_{i \times n + j} = \{x_i, x_{n+j}, x_{2n+a_{ij}}\}$. If we want to search only for diagonal transversals we need to introduce slight modifications. Remind that we call a transversal diagonal if and only if it has exactly one entry from the main diagonal and exactly one entry from the main antidiagonal. The modifications are as follows: first, we add to X two more elements x_{3n}, x_{3n+1} that correspond to main diagonal and main antidiagonal, respectively. After this we add x_{3n} to subsets corresponding to Latin square cells on the main diagonal, and add x_{3n+1} to subsets corresponding to cells on the main antidiagonal. It is easy to see that each solution of this problem will yield a transversal (diagonal transversal) of a considered Latin square.

The representation of the problem of finding n disjoint transversals in a set of transversals $T^A = \{T_1, \ldots, T_k\}$ is similarly straightforward. In this case the set X contains $n \times n$ elements $x_0, \ldots, x_{n^2-1}$. Here an element $x_{i \times n + j}$, $i, j \in \{0, \ldots, n-1\}$ corresponds to a Latin square cell with coordinates (i, j). The subset corresponding to transversal $T_k = \{t_k^0, \ldots, t_k^{n-1}\}$ is formed by elements $x_{t_k^0}, x_{n+t_k^1}, \ldots, x_{n \times (n-1) + t_k^{n-1}}$. To find diagonal orthogonal mates we need to find *exact cover* solution containing only subsets corresponding to diagonal transversals.

It is interesting that we can specify an *exact cover* instance, all solutions of which will yield all possible normalized diagonal Latin squares. Moreover, the corresponding instance can be adapted to produce only Diagonal Latin squares that share some particular LS_n^k (DLS_n^k). Let us now consider the algorithms that we employ.

B. SAT approach

Boolean satisfiability problem (SAT) is the historically first NP-hard problem. Despite this fact there are many areas in computer science in which state-of-the-art SAT solving algorithms, usually referred to as SAT solvers, are successfully applied [3]. SAT approach was used to solve problems related to Latin squares, for example, in [13]. We, too, carried out a computational experiment in the volunteer computing project SAT@home[1] aimed at finding MODLS of order 10 [14]. However, in all these works the SAT encoding used was in a way implementing the Paige-Tompkins approach. Possibly because of this reason we managed to find only about 50 pairs of MODLS of order 10 in several months. In the present paper we apply SAT approach in the context of Euler-Parker approach.

For this purpose we need to reduce *exact cover* instances to SAT. By SAT instance it is usually meant a Boolean formula in Conjunctive Normal Form (CNF). CNF is essentially a conjunction of clauses, where clause is a disjunction of literals. By literal we mean either Boolean variable or its negation. Thankfully, the transition from exact cover instance to SAT is very simple. First we associate with each subset S_k of a set S a Boolean variable b_k. Then for each $x_i \in X$ we form a set X_i that contains all b_j such that $x_i \in S_j$. A SAT instance we need will be formed by the set of constraints specifying that in each set X_i exactly one variable can take the value of $True$. This constraint can be written in CNF in several different ways. In our experiments we used pairwise scheme. Note that given a CNF C a SAT solver either finds its satisfying solution or reports that a CNF is unsatisfiable. It is easy to see that from a satisfying assignment we can effectively extract the solution to *exact cover* problem. Note that to find all solutions we need to implement an iterative process in which we restrict

[1] http://sat.isa.ru/pdsat/

each found solution and search for different ones until a CNF becomes unsatisfiable, meaning that we found all solutions.

C. DLX approach

The DLX algorithm proposed in [4] is the most widely known algorithm for solving *exact cover* problem. It is a recursive depth-first backtracking algorithm that heavily relies on the use of special technique called 'Dancing Links' to represent data in the computer memory. It takes as an input a description of *exact cover* problem that does not need any adjustments. While there are available implementations of DLX[2], we implemented our own simple version. Since it almost directly corresponds to the algorithm described in [4] we believe that its performance is comparable to that of other existing implementations.

D. Comparison and performance evaluation

All the experiments described below were carried out on one core of Intel Core i7-3770k with 16 Gb RAM (Windows 10). All considered algorithms were implemented in C++ (Microsoft Visual Studio 2015). To solve SAT instances we used the MINISAT 2.2 solver [15]. We made a wrapper for MINISAT and introduced small changes in standard SAT solving procedure in order to find all satisfying assignments without having to reload SAT instance each time. The source code of the application can be found online[3].

An important stage of experiments consists in generating diagonal Latin squares of order 10. Existing implementations, for example the one described in [11] make it possible to generate about 1 million of them per second on one processor core. We found out that by applying DLX to a set cover instance for this problem we get the generation speed of about 300 000 diagonal Latin squares of order 10 per second. This speed was satisfactory for our purposes.

Apart from SAT and DLX-based algorithms for finding transversals and orthogonal mates we also implemented a straightforward depth-first backtrack algorithm for both purposes to evaluate their performance better. In Table I we show how the algorithms fared against each other. The total time required to process the sample is averaged for 10 samples of size 100 000 diagonal Latin squares of order 10 each. The samples were produced by generating first 100 000 diagonal Latin squares for a randomly selected DLS_{10}^{40}. In 'Finding Transversals' we measured total time required only to find all diagonal transversals for each square from the sample. In 'Finding orthogonal mates' we measured total time spent on both finding transversals and subsequent finding disjoint sets of them. The entries SAT and DLX imply that we employed the corresponding approaches for both finding transversals and searching for their disjoint sets. In $SAT + DLX$ we find transversals using DLX and search for orthogonal mates using SAT.

Let us comment the obtained results. It is clear that DLX definitely wins in all categories, with SAT being from 4 to

[2] http://koti.kapsi.fi/pottonen/libexact.html
[3] https://github.com/veinamond/LS_search/

TABLE I
EVALUATION OF PERFORMANCE OF DIFFERENT ALGORITHMS FOR
FINDING MODLS, AVERAGED FOR 10 SAMPLES OF SIZE 100 000
DIAGONAL LATIN SQUARES EACH, TIME IN SECONDS

Finding Tranversals	Time
Backtrack search	578.88
SAT	3 937.95
DLX	97.54
Finding orthogonal mates	
Backtrack search	784.18
SAT	4 238.61
SAT + DLX	393.67
DLX	111.22

40 times worse. When finding transversals, SAT loses even to simple backtrack search. The explanation here is simple: to find, say, 100 transversals using SAT approach we need to launch SAT solver 101 times, each time restricting it from encountering already found solutions. Meanwhile DLX (and Backtrack search) simply traverses the search space without stops and restarts. We can modify the SAT solver to imitate such behavior by incorporating the so-called warm restarts technique, but judging by the overall picture, it won't make much difference. Another interesting observation is that more than 90% of time spent on checking orthogonal mates the algorithms spend to search for transversals. It is surprising because in both papers [12], [9] the corresponding process takes much less time. One possible reason for this is the fact that most diagonal Latin squares of order 10 in these random samples have relatively small number of diagonal transversals (about 100) and do not have orthogonal mates.

Overall, it means that for the purpose of our runtime estimation we will use DLX results. In particular we will assume that it can process roughly 900 diagonal Latin squares per second on one mainstream processor core. It is more or less similar to the performance achieved in [9], where the authors used special algorithms for enumerating all MOLS of order 9.

E. Correctness of results

To safeguard from possible errors, we cross-checked the results of our computational experiments. In particular, we compared on random samples of diagonal Latin squares the output of transversal finding procedures. We even went one more step in this direction and made a fourth implementation of this procedure based on lookup. We did the same in application to procedures for finding orthogonal mates. It was less trivial due to the fact that the majority of diagonal Latin squares of order 10 from random samples do not have orthogonal mates. That is why for this purpose we tested the implemented methods on diagonal Latin squares found in SAT@home[4] and in [2].

F. Symmetries and equivalence classes

Taking into account various symmetries and equivalence classes can greatly impact the performance of any algorithm

[4]http://sat.isa.ru/pdsat/

operating with combinatorial designs. A good example of how it is used in computational experiments can be found in [16]. In the case of Latin squares there are several variants of constructing equivalence classes, covered in detail for example in [9]. However, diagonal Latin squares present a special case. While they can be normalized effectively (transformed to a form where the first row is in ascending order), many simple transformations used to produce isotopic/isomorphic Latin squares lead to violation of a diagonality constraint. Surprisingly, the question of constructing isomorphic diagonal Latin squares is well studied in the research area related to magic squares. A magic square is an $n \times n$ table filled with integer numbers in such a way that the sums of numbers in each row, each column and also in the main diagonal and main antidiagonal are all equal to the same so-called 'magic constant'. It is clear that each diagonal Latin square is a Magic square. Unfortunately, we could not trace the original references to the ideas on constructing isomorphic Latin squares presented below, despite the fact that they became a sort of common knowledge. Let us briefly describe them.

So, we are interested only in transformations of a diagonal Latin square of order n that make it possible to produce other diagonal Latin squares of order n. Note that when we change the order of rows or order of columns in a Latin square, the square remains Latin and has the same number of orthogonal mates as an original Latin square. However, as a result we can violate the diagonality constraint. This is not true for the case of renaming elements. However, since we consider only normalized diagonal Latin squares (i.e. with the first row in ascending order), we can not use it to produce isomorphic squares. It means that we need to find such combinations of rows/columns permutations that preserve the contents of main diagonal and main antidiagonal. In total there are three kinds of such transformations of which the second and third are the so-called M-transformations.

- The first class of such transformations is formed by 4 transformations that are produced by mirroring a diagonal Latin square with respect to its main diagonal or main antidiagonal and also horizontally or vertically.
- The first kind of M-transformations contains all transpositions of two columns that are positioned symmetrically with respect to the middle with simultaneous transposition of two symmetrically positioned rows, for example, transposition of 0-th and $(n-1)$-th columns with simultaneous transposition of 0-th and $(n-1)$-th row. The number of such transformations for $n = 10$ is 2^5 (equal to the number of all subsets of a set with 5 elements).
- The second kind of M-transformations consists of all transpositions where we choose two columns in the left half of a Latin square and transpose them and simultaneously transpose the two columns positioned symmetrically with respect to the middle in the right half of a square with simultaneous similar transposition of

rows, for example transposition of 0-th column with 1-th column, $(n-2)$-th column with $(n-1)$-th column, 0-th row with 1-th row and $(n-2)$-th row with $(n-1)$-th row. The number of these transformations for $n = 10$ is $5!$ (equal to the number of permutations of a set with 5 elements because by transposing pairs we form all possible permutations).

It is easy to see that each of the above transformations preserves the contents of main diagonal and main antidiagonal. Note that we can normalize (by renaming elements) each diagonal Latin square produced as a result of the above transformations. Thus, we can produce for $n = 10$ exactly $5! \times 2^5 \times 4 = 15\,360$ normalized isomorphic squares. Also if we sort all isomorphic squares lexicographically and choose the first one – we can use it as a representative associated with this isomorphism class (the corresponding square is usually referred to as *canonical form* of an original square).

It is currently unclear if it is feasible to construct canonical form for each diagonal Latin square to be checked, but our preliminary evaluations show that it should be possible to employ the outlined technique to significantly reduce the search space.

IV. ESTIMATING THE NUMBER OF DIAGONAL LATIN SQUARES OF ORDER 10

Since we have evaluated the performance of state-of-the-art combinatorial algorithms in application to the considered problem, the only thing we need to do in order to construct our runtime estimation is to estimate the number of diagonal Latin squares of order 10. Basically it means that enumerating all possible diagonal Latin squares of order 10 in reasonable time is out of question. Thus we will have to resort to the Monte Carlo method. As we mentioned in the introduction, we can compute the number of possible variants of incomplete diagonal Latin squares of order 10 comprised of the first k elements (we refer to them as DLS_{10}^k) and use probabilistic experiment to estimate the expected value of a number of diagonal Latin squares produced by completing arbitrary DLS_{10}^k.

Since we consider only normalized diagonal Latin squares, we can safely fix the first row to 0123456789. An important part of the Monte Carlo method is that we have to choose species representatives randomly according to the uniform distribution. Assume that we need to randomly choose DLS_{10}^k, $10 < k < 100$ and that we have a procedure that can enumerate all possible DLS_{10}^k in a fixed order in a reasonable time. We first count the number of corresponding DLS_{10}^k, let us refer to it as N_{10}^k. Then for a fixed random sample size u choose according to the uniform distribution over the set $[0, \ldots, N_{10}^k]$, u values $\alpha_0, \ldots, \alpha_{u-1}$. After this we can repeat the DLS_{10}^k generation procedure to output the instances of DLS_{10}^k with numbers α_i, $i = 0, \ldots, N_{10}^k$.

In our experiments we used the modified version of the algorithm proposed in [11] to enumerate DLS_{10}^k. To produce proper estimation we need to choose k in such a way that N_{10}^k can be computed fast, and at the same time the number of squares that share a specific DLS_{10}^k on average can be

enumerated in reasonable time. Computing the number of possible DLS_{10}^{30} requires quite significant effort, and the corresponding number N_{10}^{30} is 284 086 571 712. However, to compute the number of Latin Squares produced by completing arbitrary DLS_{10}^{30} it takes several days on one core of state-of-the-art processor. Therefore we chose $k = 32$ and computed our estimation using this value. The number of DLS_{10}^{32} is 12 611 543 636 160. We generated a random sample of size 10 000 DLS_{10}^{32} instances and used it to estimate the expected value of the number of diagonal Latin squares of order 10 with fixed DLS_{10}^{32}. The corresponding expected value was equal to 11 931 268 344.

By multiplying the two values we can estimate that the number of normalized diagonal Latin squares of order 10 is 150 471 711 355 090 461 719 040 or about 1.5×10^{23}. Following the optimistic scenario that we can exploit the features outlined above we can divide this number by 15 360 thus having to process about 9.8×10^{18} distinct diagonal Latin squares. Taking into account the fact that we can process about 900 squares per second it means that the runtime estimation on one mainstream processor core is about 1×10^{16} seconds or 345.15 million years. If we employ the second best supercomputer (*Tianhe-2*) in Top500[5], that has 3 120 000 cores, it will take us about 110.6 years to enumerate all MODLS of order 10.

V. RELATED WORKS

There is a significant interest to enumerating various combinatorial designs, reflected in the existence and popularity of OEIS (The On-Line Encyclopedia of Integer Sequences)[6]. It has over 40 entries directly related to Latin squares, 5 of which are about diagonal Latin squares, and also about 100 more entries that concern related combinatorial designs. We would like to mention the most relevant of them. They are A266166 (Number of reduced pairs of orthogonal Latin squares), which was obtained as a result of research described in [9], A000315 (Number of reduced Latin squares of order n) that is a result of cumulative effort of many research groups throughout the world, and A274171 (Number of diagonal Latin squares of order n with first row $1..n$), the authors of which recently finished a large scale computational experiment on enumerating all diagonal Latin squares of order 9 in the volunteer computing project Gerasim@home[7]. Note that there is little to no available information regarding enumeration of (diagonal) Latin squares or MODLS of order 10.

Ideologically, one of the most closely related works is [9] in which authors managed to enumerate and classify all MOLS of order 9. Note, that they used completely different approach based on special combinatorial properties of specific objects. The performance of their computational algorithm seems to be more or less similar to that of our DLX implementation since they comment that they process about 1200 Latin squares of order 9 per second on a mainstream PC (supposedly, on

---

[5]https://www.top500.org/lists/2016/11/

[6]http://oeis.org

[7]http://gerasim.boinc.ru/

one core). They also consider some problems related to Latin squares of order 10, in particular to construct the triple of Latin squares closest to being a triple of MOLS of order 10, however, they do not study specifically diagonal Latin squares of order 10.

Constructing runtime estimations for solving hard problems is performed quite often, especially for problems found in cryptography. We are not aware of other attempts at constructing runtime estimations for finding specific configurations of MOLS or MODLS of order 10.

Despite the fact that both SAT and *exact cover* problems are NP-complete, there exist state-of-the-art combinatorial algorithms that manage to solve many instances arising in practice relatively fast. DLX algorithm proposed by Donald Knuth in [4] is the de facto algorithm for solving *exact cover* problem. In case of SAT while there are many different directions of development, the majority of algorithms are usually based on the Conflict-Driven Clause Learning concept [17]. SAT approach was successfully applied in [13] to solve several algebraic problems. Also the author of [13] noted that he searched for a triple of MOLS of order 10 using SAT solvers for several years without any success.

Usually the problems related to Latin squares are encoded to SAT using the so-called naive encoding, which represents Latin square as a set of its cells with specific constraints over them. It is described, for example, in [14]. We applied this encoding to search for MODLS of order 10 in the volunteer computing project SAT@home, and managed to find about 50 previously unknown pairs. Currently, we believe, that while SAT approach using naive encoding might be good for finding specific cases of Latin squares which satisfy a lot of additional constraints, it is inferior to the one employed in the present paper for a systematic search or enumeration purposes. In our recent work [10] we applied SAT approach using naive encoding to prove that there is no triple of MODLS of order 10 with fixed values of the first 45 elements. It took about 38 minutes of multithreaded SAT solver on 32 cores to solve the corresponding SAT instance. Meanwhile using the DLX implementation from the present paper we can do it in about 42 seconds on one processor core.

Conclusions and future works

The constructed runtime estimation for enumerating MODLS of order 10 means that, unless there are some theoretical constructions that make it possible to significantly reduce the search space or to drastically improve the effectiveness of the algorithms for traversing it, it is completely unrealistic and pointless at the present moment to launch the corresponding computational experiment or expect that it will be finished at least in the course of one lifetime. In particular it means that the problem of finding triples of MOLS of order 10 will most likely require a different approach. Nevertheless, the technical progress, especially in supercomputing, is evident and it is possible that in a decade or two new algorithms and hardware will make it possible to reconsider the current evaluation. In the nearest future we plan to apply DLX approach to finding

triples of diagonal Latin squares of order 10 that are closest to being triples of MODLS of order 10.

Acknowledgment

The research was funded by Russian Science Foundation (project No. 16-11-10046).

The authors would like to thank Eduard Vatutin, Alexey Zhuravlev, Maxim Manzyuk, and citerra (Russia team at Boinc.ru) for helpful discussions. We are also grateful to Sergey Belayev for sharing his implementation of straightforward backtrack search implementing Euler-Parker approach and to Alexey Belishev for information about M-transformations and canonical forms of diagonal Latin squares.

References

[1] C. J. Colbourn, J. H. Dinitz, and I. M. Wanless, "Latin squares," in *Handbook of Combinatorial Designs, Second Edition (Discrete Mathematics and Its Applications)*, C. J. Colbourn and J. H. Dinitz, Eds. Chapman & Hall/CRC, 2006, ch. 3.1, pp. 135–152.

[2] J. Brown, F. Cherry, L. Most, E. Parker, and W. Wallis, "Completion of the spectrum of orthogonal diagonal Latin squares," *Lecture notes in pure and applied mathematics*, vol. 139, pp. 43–49, 1992.

[3] A. Biere, M. Heule, H. van Maaren, and T. Walsh, Eds., *Handbook of Satisfiability*, ser. Frontiers in Artificial Intelligence and Applications, vol. 185. IOS Press, 2009.

[4] D. Knuth, "Dancing links," *Millennial Perspectives in Computer Science*, pp. 187–214, 2000.

[5] E. T. Parker, "Computer investigation of orthogonal Latin squares of order ten," *Proc. Symp. Appl. Math.*, vol. 15, pp. 73–81, 1963.

[6] N. Metropolis and S. Ulam, "The Monte Carlo Method," *J. Amer. statistical assoc.*, vol. 44, no. 247, pp. 335–341, 1949.

[7] R. J. R. Abel, C. J. Colbourn, and J. H. Dinitz, "Mutually orthogonal Latin squares (MOLS)," in *Handbook of Combinatorial Designs, Second Edition (Discrete Mathematics and Its Applications)*, C. J. Colbourn and J. H. Dinitz, Eds. Chapman & Hall/CRC, 2006, ch. 3.3, pp. 160–193.

[8] B. D. McKay, A. Meynert, and W. Myrvold, "Small Latin squares, quasigroups, and loops," *Journal of Combinatorial Designs*, vol. 15, no. 2, pp. 98–119, 2007.

[9] J. Egan and I. M. Wanless, "Enumeration of MOLS of small order," *Math. Comput.*, vol. 85, no. 298, 2016.

[10] O. Zaikin, A. Zhuravlev, S. Kochemazov, and E. Vatutin, "On the construction of triples of diagonal Latin squares of order 10," *Electronic Notes in Discrete Mathematics*, vol. 54, pp. 307–312, 2016.

[11] E. Vatutin, O. Zaikin, A. Zhuravlev, M. Manzyuk, S. Kochemazov, and V. Titov, "Using grid systems for enumerating combinatorial objects on example of diagonal Latin squares," in *Proceedings of GRID 2016 conference*, ser. CEUR Workshop Proceedings, vol. 1787. CEUR-WS.org, 2016, pp. 486–490.

[12] D. E. Knuth, *The Art of Computer Programming, Volume 4, Fascicle 0: Introduction to Combinatorial Algorithms and Boolean Functions (Art of Computer Programming)*, 1st ed. Addison-Wesley Professional, 2008.

[13] H. Zhang, "Combinatorial Designs by SAT Solvers," in *Handbook of Satisfiability*, ser. Frontiers in Artificial Intelligence and Applications, A. Biere, M. Heule, H. van Maaren, and T. Walsh, Eds. IOS Press, 2009, vol. 185, pp. 533–568.

[14] O. Zaikin, S. Kochemazov, and A. Semenov, "SAT-based search for systems of diagonal Latin squares in volunteer computing project SAT@home," in *39th International Convention on Information and Communication Technology, Electronics and Microelectronics, MIPRO 2016, Opatija, Croatia, May 30 - June 3, 2016*. IEEE, 2016, pp. 277–281.

[15] N. Eén and N. Sörensson, "An extensible SAT-solver," in *Proc. of SAT 2003*, ser. Lecture Notes in Computer Science, vol. 2919. Springer, 2003, pp. 502–518.

[16] C. Hartman, M. Heule, K. Kwekkeboom, and A. Noels, "Symmetry in gardens of Eden," *Electr. J. Comb.*, vol. 20, no. 3, p. P16, 2013.

[17] J. Marques-Silva, I. Lynce, and S. Malik, "Conflict-driven clause learning SAT solvers," in *Handbook of Satisfiability*, ser. Frontiers in Artificial Intelligence and Applications, A. Biere, M. Heule, H. van Maaren, and T. Walsh, Eds. IOS Press, 2009, vol. 185, pp. 131–153.

Improving the Effectiveness of SAT Approach in Application to Analysis of Several Discrete Models of Collective Behavior

Stepan Kochemazov, Oleg Zaikin, Alexander Semenov
Matrosov Institute for System Dynamics and Control Theory SB RAS, Irkutsk, Russia
Email: veinamond@gmail.com, zaikin.icc@gmail.com, biclop.rambler@yandex.ru

Abstract—In this paper we study several discrete models of collective behavior based on Synchronous Boolean Networks. For these models we consider a number of related problems that we solve by reducing them to Boolean satisfiability problem (SAT) and applying state-of-the-art parameterized SAT solving algorithms. We describe a greedy algorithm that exploits the features of functions used to recalculate network nodes weights and interconnections between neighborhoods of network nodes. This algorithm in combination with several specific propositional encoding techniques, makes it possible to significantly reduce the size of propositional encoding. We compare the effectiveness of SAT solving algorithms on the new encodings with that on previously employed encodings and evaluate the total performance gain achieved by using state-of-the-art tools for finding efficient SAT solver parameters.

I. INTRODUCTION

Many hard combinatorial problems arising in different areas of science can be reduced to Boolean satisfiability problem (SAT) and solved using state-of-the art SAT solving algorithms called SAT solvers. Since SAT is NP-hard, it means that these algorithms heavily employ different heuristics and have many parameters. In the recent years there appeared several tools that make it possible for specific classes of problems to find effective values of these parameters that significantly improve SAT solvers performance on these problems compared to that of SAT solvers with default parameters values. Another relatively active direction of research in the area of SAT consists in development of new techniques for propositional encoding — methods used to reduce original combinatorial problems to SAT.

In the present paper we apply both propositional encoding techniques and parameterization algorithms to improve the effectiveness of SAT approach in application to analysis of discrete models of collective behavior. In the corresponding models we observe a system of a number of agents in discrete time steps. A system of agents is specified by simple graph, where vertices (associated with agents) have Boolean weights. All weights are recalculated and refreshed synchronously using Boolean functions associated with each agent (vertex). We consider one special class of these functions, that can be used to model the so-called conforming behavior. It means that for each graph vertex we need to count the number of vertices with weight '1' in its neighborhood. In propositional encoding we use sorting networks for this purpose. We propose two

improvements to the previously used encoding. In the first one we analyze the interconnections between neighborhoods of network vertices in order to process their 'intersections' once (instead of multiple times) and thus reduce the redundancy of the encoding. For this purpose we employ a specially designed greedy algorithm. The second improvement consists in using the recent findings related to encoding of cardinality constraints to significantly reduce the number of comparators in encoded sorting networks. In total, it is possible to reduce the size of propositional encodings up to several times depending on the graph structure. After this we figure out if the improvements of encodings translate into the improvements of effectiveness of solving. In particular, on this stage we apply state-of-the-art parameterization algorithms in order to find good values of parameters for each of constructed encodings and evaluate the total gain.

Let us give a brief outline of the paper. In the next section we formally define the studied discrete models of collective behavior and describe the previously used propositional encoding. In the third section we consider sorting networks in somewhat more detail, discuss their specific features from the SAT perspective, and then propose two improvements to the existing scheme. In the fourth section we use state-of-the-art parameterization tools to optimize SAT solvers parameters and improve their effectiveness on the constructed encodings, thus comparing the encodings between each other. In the two last sections we consider related works and draw conclusions.

II. SAT APPROACH TO ANALYSIS OF DISCRETE MODELS OF COLLECTIVE BEHAVIOR

The discrete models of collective behavior that we study in this paper have been considered in detail in [1], [2]. Let us describe the general idea on which we built them. It was first expressed in the classical paper [3].

A. Discrete model of conforming behavior

Assume that we have n distinct agents. We want them to show conforming behavior – meaning that the decisions of each agent should conform to the decisions of other agents from its neighborhood. It is interesting to view the spread of decisions of agents in a system in the course of several discrete time moments, assuming that all agents re-evaluate their decisions synchronously. In the simple case each agent can decide either to act (1) or not to act (0). Then we can assign each agent a special parameter c called *conformity level* and say that this agent decides to act at the next time step if and only if at least c agents from its neighborhood act at the current time step.

Formally, it is convenient to define the network of agents by a simple directed graph $G = (V, E)$, where a set of vertices V, $|V|$=n, corresponds to a set of agents, and a set of edges E reflects the interconnections between agents. Consider discrete time moments $t = 0, 1, \ldots, k$. At each time moment associate with each vertex v_i its weight $w_i^t \in \{0, 1\}$. We define the neighborhood of vertex v_i as the set of vertices N_i, from each of which there is an edge to v_i ($N_i = \{v_u \in V | (v_u, v_i) \in E\}$. Assume that with each vertex v_i we also associate the conformity level c_i, $c_i \in \{1, \ldots, |N_i|\}$. Then we can define the function for recalculating the value of each vertex as follows:

$$
w_i^{t+1} = \begin{cases} 1, & \sum_{v_j \in N_i} w_j^t \geq c_i \\ 0, & \sum_{v_j \in N_i} w_j^t < c_i \end{cases} \quad (1)
$$

The model defined this way despite its simplicity makes it possible to study a lot of different phenomena in a way it is done in agent-based modeling, when we define simple rules for each agent and then see how a collective of such agents develop in different conditions. In the present paper (similar to [2]) we focus on the following combinatorial problem. First we introduce special kind of agents called *instigators*. Their distinctive feature consists in the fact that they are always active (i.e. the corresponding $w_i^t = 1$ for each possible value of t). Assume that at initial time moment ($t = 0$) only instigators are active in a network. Then for an arbitrary graph $G = (V, E)$, $|V| = n$, with specified conformity levels $C = (c_1, \ldots, c_n)$ we consider the following problem: to find a disposition of at most S, $S << n$ instigators (assuming that we replace some agents by instigators) in such a way that after k time steps at least F, $F >> S$, agents in a network are active.

On the one hand, it is easy to see that in the general case the proposed problem is very hard. On the other hand, essentially Boolean weights of graph vertices and simple nature of the function used to recalculate them allow for the natural reduction of this problem to SAT [4]. For practical purposes, SAT is formulated as follows: for a system of logical equations to find its solution or to prove that there is none. Usually, the system of logical equations is represented in the conjunctive normal form (CNF). Let us now consider how can we reduce the problem of analysis of the discrete model of collective behavior outlined above to SAT.

B. General Outline of Propositional Encoding

Assume that we have a system of n agents. Without the loss of generality, let us consider only the transition from one time step to the next, in particular, from $t = 0$ to $t = 1$. It means that essentially we need to write a system of logical equations that specifies how we move from Boolean variables $w_1^0, \ldots, w_n^0$ to $w_1^1, \ldots, w_n^1$ according to (1).

Again, without the loss of generality let us focus on encoding the transition from w_i^0 to w_i^1. From the (1) it is clear that we need to somehow count the number of 1's among Boolean variables $W^0(N_i) = \{w_j^0 | v_j \in N_i\}$. It can be done in several different ways: for example, we can write logical equations that encode computing the sum of corresponding variables viewed as binary numbers. Or we can construct for $W^0(N_i)$ its sorted version $S(W^0(N_i)) = (t_1, \ldots, t_{|N_i|})$, where $t_1, \ldots, t_{|N_i|}$ are auxiliary Boolean variables, $\#\{t_j = 1 | t_j \in S(W^0(N_i))\} = \#\{w_j^0 = 1 | w_j^0 \in W^0(N_i)\}$ and $t_u > t_{u+1}$, $u = 1, \ldots, |N_i| - 1$. We believe that the second path (with sorting) is more convenient and effective due to several reasons, arising from how state-of-the-art SAT solving algorithms work. We will discuss them in more detail in the next section. Once we have $S(W^0(N_i)) = (t_1, \ldots, t_{|N_i|})$, it is clear that, $w_i^1 \equiv t_{c_i}$. If there are at least c_i ones in $W^0(N_i)$ then $t_{c_i} = 1$ and according to (1) it follows that $w_i^1 = 1$. If there are less than c_i ones in $W^0(N_i)$ then $t_{c_i} = 0$ and $w_i^1 = 0$ too. Thus the problem is to construct for an arbitrary $W^0(N_i) = \{w_j^0 | v_j \in N_i\}$ the sorted array $S(W^0(N_i)) = (t_1, \ldots, t_{|N_i|})$. It can be done using several methods, developed in recent years for encoding so-called *cardinality constraints* to SAT [5], [6]. In [1], [2] we used sorting networks for this purpose, and we continue to use them in the present paper, albeit in a more efficient manner.

III. IMPROVING PROPOSITIONAL ENCODING

From the layout of the function (1) it is easy to see, that for most graphs G there exist $i, j \in \{1, \ldots, n\}$, $i \neq j$ that $N_i \cap N_j \neq \emptyset$. It means, that when counting ones from $W^u(N_i)$ and $W^u(N_j)$, we can actually count ones in $W^u(N_i \cap N_j)$ once, and thus reduce the redundancy of the procedure. Of course, the benefit is the larger the more common elements we find, however, there is no effective algorithm for this purpose. Nevertheless, having in mind several special constraints imposed by the use of sorting networks, it is possible to propose a greedy algorithm specifically for this purpose. To progress further let us give some details about sorting networks.

A. Sorting Networks from SAT Perspective

In computer science, sorting networks are constructions, that are built to sort fixed number of values. They comprise of so-called *wires*. Modules called *comparators* connect pairs of wires. When a pair of values corresponding to connected wires move through a comparator, they switch wires if their order is not desirable, and remain on their wires otherwise. The example of sorting network for 8 inputs is shown in the left lower part of Fig. 1. On most depictions it is assumed that input values are on the left and output values are on the right.

The advantages of sorting networks from SAT perspective compared to other variants of counting ones in an array of Boolean values were considered in detail in [5]. Let us mention those that we see as the most important. First is that sorting networks are easy to encode. In the context of propositional encoding, by $comparator(a, b, c, d)$ we mean two logical equations: $c \equiv a \lor b$, $d \equiv a \land b$. Second, propositional encodings of sorting networks allow propagation of partial knowledge, i.e. if the values of several input variables are already known, it makes it possible to derive values of several output variables (depending on several additional factors). The same can not be said regarding, for example, the variant when we compute the sum of Boolean variables.

There are several algorithms that can be used to construct sorting networks for $n = 2^k$ inputs [7], [8]. In the present paper, as in [1], [2] we use Batcher's odd-even merge algorithm [7]. Essentially, the algorithm presents the construction, that makes it possible to merge two sorted sequences of $n/2$ numbers into one sorted sequence of n numbers. Since a single comparator can be used to sort a sequence of length 2, it means that by recursive application of merging procedure we can sort any sequence of length $n = 2^k$. The merging procedure, to which we will refer as *Merge* is defined as follows (the description is taken from [6]).

$$Merge(\langle a_1 \rangle, \langle b_1 \rangle, \langle c_1, c_2 \rangle) \leftrightarrow comparator(a_1, b_1, c_1, c_2).$$
$$Merge(\langle a_1 ... a_n \rangle, \langle b_1 ... b_n \rangle, \langle d_1, c_2 ... c_{2n-1}, e_n \rangle) \leftrightarrow$$
$$Merge(\langle a_1, a_3 ... a_{n-1} \rangle, \langle b_1, b_3 ... b_{n-1} \rangle, \langle d_1 ... d_n \rangle) \bigwedge$$
$$Merge(\langle a_2, a_4 ... a_n \rangle, \langle b_2, b_4 ... b_n \rangle, \langle e_1 ... e_n \rangle) \bigwedge$$
$$\bigwedge_{i=1}^{n-1} comparator(e_i, d_{i+1}, c_{2i}, c_{2i+1}).$$

There are at least two papers (see [6], [5]), which study in detail how to encode sorting networks to SAT, however, focused on working with *cardinality constraints* (constraints of the kind $a_1 + a_2 + ... + a_k \Delta h$, $\Delta \in \{\leq, \geq, =\}$, $h \leq k$, where $a_i \in \{0, 1\}$). Nevertheless, a lot of methods from these papers can be adapted to our needs.

B. Improvements in Encoding Sorting Networks

In previous works [1], [2] when we needed to encode sorting network to sort l inputs, we constructed a sorting network for $2^{\lceil log(l) \rceil}$ inputs, encoded it to SAT and fed inputs with number $> l$ the Boolean variable, assigned with the False (0) value. It means, that, for example, if we needed to sort only 33 values — we nevertheless encoded the full sorting network with 64 inputs and 64 outputs. In the present work we decided to adapt the algorithm for constructing sorting network to our means. In particular, it is convenient for our purposes to group comparators of the sorting network into *mergers*. By *merger* of size l we mean the construction, that takes as input two sorted sequences of size $l/2$, merges them into one sorted sequence of size l and outputs it. Each merger directly corresponds to the application of *Merge* procedure. It is easy to see that mergers naturally form layers, and the number of layers coincides with $\lceil log(l) \rceil$: in the first layer there are $\lceil l/2 \rceil$ mergers of size 2 (with 2 inputs and 2 outputs, i.e. comparators), on the second layer $\lceil l/4 \rceil$ mergers of size 4, etc. The last layer always consists of one merger of size $2^{\lceil log(l) \rceil}$. On the first glance it may seem that this construction provides no benefits compared to the one when we generate the whole sorting network at once. But let us consider as an example the network of mergers for 5 inputs, that is shown in the right part of Fig. 1. Dashed lines correspond to void

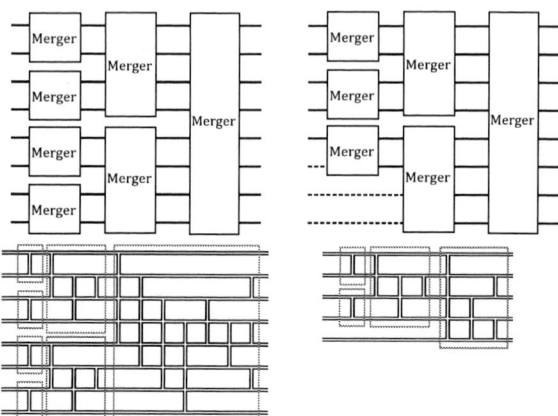

Fig. 1. Network of mergers with 8 inputs (left) and with 5 inputs (right)

values. The third merger of size 2 in the first layer has only one input — it means that to produce its output we do not need to introduce new comparators at all: the output of the merger will be the non-void value and void value. Now let us proceed to the second merger in the second layer. Since it has only output of one merger as an input — it does not need to introduce new comparators as well, only to add to the sorted sequence of one nonempty value and one void value two more void values. By combining these two techniques we now can construct sorting networks for an arbitrary length of input sequence, which do not have comparators that are never used. Also, we modified the previously used procedure for introducing comparators in such a way, that if one of comparator input values is void, it does not add new comparator at all, just puts non-void value (if any) as first output and void value as second.

Now let us return to the idea, where we want to reduce the redundancy of our encoding by careful processing of neighborhood intersections. The use of networks of mergers imposes one welcome additional constraint that we will use to our advantage: indeed we are interested only in neighborhood intersections with sizes equal to powers of 2.

C. Using Greedy Algorithm to Determine Neighborhood Intersections

At first we need to determine all neighborhood intersections of size 2 that should be processed. In order to do it we use the following greedy algorithm. Assume that we have n agents. Then as an input our algorithm takes n neighborhoods N_i, $N_i \subseteq \{1, 2, ..., n\}$, $i = 1, ..., n$. It outputs two sets. One is the ordered and numbered set of pairs R. Second is the set of sets of pairs NU. It has size n components, and each component $NU[j]$ is the set that contains pairs from R included in the corresponding N_j. Then we perform the following algorithm.

1) Introduce auxiliary sets $WS_i = N_i$, $i = 1, ..., n$.
2) Construct all possible pairs of elements from WS_j: $P = \{(a, b) | \exists j : (a, b) \in WS_j\}$.

3) For each pair $(a, b) \in P$ count how many WS_j contain it: $W[(a, b)] = \#\{WS_j : (a, b) \in WS_j\}$.
4) Choose pair $(a^*, b^*) \in P$ for which $W[(a^*, b^*)]$ is maximal.
5) If $Max_{(a,b) \in P}(W[(a, b)]) = 1$ then return R, NU.
6) Remove (a^*, b^*) from P and add it to resulting set R and NU.
7) For each $j : (a^*, b^*) \in WS_j$
   a) Add (a^*, b^*) to $NU[j]$.
   b) Remove a^*, b^* from WS_j.
8) Go to 3.

Let us refer to the set of obtained intersections of size 2 as R^2, and the set detailing which pair is included in which neighborhood as NU^2. In order to obtain intersections of size 4 we can use the same greedy algorithm. Indeed, assume that we order all pairs in R^2 and assign them numbers from 1 to $|R^2|$. Then for each N_i we construct set $N_i^2 = \{j | (a, b)^j \in NU[i]\}$. Then we give the constructed n sets $\{N_i^2\}$ as an input to our greedy algorithm. It is easy to see, that from its outputs R^4, NU^4 using information about numbers assigned to pairs from R^2 we can extract all intersections of size 4 constructed from intersections of size 2. We can repeat this process until the resulting $R^{2^l} = \emptyset$.

Of course, by means of this greedy algorithm we do not obtain the optimal solution for the problem. But it works relatively fast (with proper implementation), gives us the result that is suitable for our needs (i.e. only intersections of sizes equal to powers of 2) and is easy to implement.

Now we can use the obtained sets $R^2, R^4, \ldots$ and $NU^2, NU^4, \ldots$ to reduce the redundancy of networks of mergers. Assume that we first introduce the mergers for all R^2, then for all R^4, etc. Note that, for example, each set of size 4 in R^4 is comprised of two pairs from R^2 by definition. It means that we use outputs of mergers for corresponding pairs as inputs for merger of size 4, etc. Of course, in general case not all mergers of smaller size are used to construct mergers of larger size. Assume that we now construct network of mergers for an agent j. Then we fill its network of mergers by already constructed mergers corresponding to entries in the corresponding $NU^2[j], NU^4[j], \ldots$ (put in the proper order). After this we introduce mergers for all elements that are not covered by existing ones, etc.

Let us consider one more improvement which consists in adapting the sorting network generation procedure to specific value of conformity level parameter.

D. Using Advanced Sorting Network Construction Procedures

There are several ways, how we can take into account the value of conformity level to reduce the size of propositional encoding. The most drastic effect we can achieve by noticing that conformity levels $c_i = 1$ and $c_i = |N_i|$ do not require to construct any sorting network at all. Indeed, $c_i = 1$ means that if there is at least one agent v_j with w_j^t in the neighborhood of agent v_i at time moment t, then $w_i^{t+1} = 1$. It is easy to express it by simple logical equation $w_i^{t+1} \equiv \bigvee_{v_j \in N_i} w_j^t$. Following

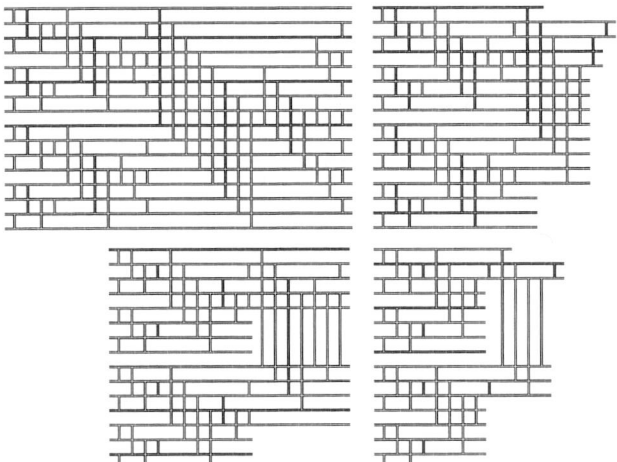

Fig. 2. Network of mergers with 16 inputs and 16 outputs (top left), network with 16 inputs and 4 outputs constructed using simplified mergers (bottom left), and their variants with removed redundant comparators (assuming that $c_i = 3$, top right and bottom right, respectively).

the similar reasoning, for $c_i = |N_i|$ the logical equation looks as follows: $w_i^{t+1} \equiv \bigwedge_{v_j \in N_i} w_j^t$. It also means that we do not need to process the corresponding neighborhoods by our greedy algorithm described earlier.

Now, for the cases when $1 < c_i < |N_i|$ we can use the information about conformity level value to a lesser, but quite significant extent. For this purpose we combine one simple heuristic with a method for reducing the size of sorting network outlined in [5]. It is used in that paper to construct *cardinality constraints*. Essentially, it proposes that if for a sorting network that sorts 2^l values, we are interested only in the first 2^h outputs, $h < l$, then we can use a simplified merging procedure to merge sorted sequences of size 2^h into sequences of size 2^h. It is defined as follows:

$$SMerge(\langle a \rangle, \langle b \rangle, \langle c_1, c_2 \rangle) \leftrightarrow comparator(a, b, c_1, c_2)$$
$$SMerge(\langle a_1 \ldots a_n \rangle, \langle b_1 \ldots b_n \rangle, \langle d_1, c_2 \ldots c_{n+1} \rangle) \leftrightarrow$$
$$SMerge(\langle a_1, a_3 \ldots a_{n-1} \rangle, \langle b_1, b_3 \ldots b_{n-1} \rangle, \langle d_1 \ldots d_{\frac{n}{2}+1} \rangle) \bigwedge$$
$$SMerge(\langle a_2, a_4 \ldots a_n \rangle, \langle b_2, b_4 \ldots b_n \rangle, \langle e_1 \ldots e_{\frac{n}{2}+1} \rangle) \bigwedge$$
$$\bigwedge_{i=1}^{\frac{n}{2}} comparator(e_i, d_{i+1}, c_{2i}, c_{2i+1})$$

We use it as follows. Consider an agent v_i with conformity level c_i, $|N_i| = k_i$, $c_i < k_i$. We construct a network of mergers, and if $\lceil log(c_i) \rceil < \lceil log(k_i) \rceil$ then we replace each merger which takes as input sequences of length $2^{\lceil log(c_i) \rceil}$ by simplified merger. The resulting sorting network has much less comparators than original. At the left part of Fig. 2 we show full sorting network with 16 inputs and 16 outputs at the top, and sorting network with 16 inputs and 4 outputs, constructed using simplified mergers at the bottom.

Another small improvement consists in the following: in fact, we only need one output of the sorting network for our purposes – the c_i-th one. It means that we can discard all comparators that do not influence the value on the c_i-th wire.

1175

Interesting and relatively counter-intuitive observation consists in the fact, that for Batcher sorting networks the application of this technique gives different results depending on whether we use simplified mergers or not. As an example, we assume that $|N_i| = 16$ and $c_i = 3$. The results of application of the proposed technique to the sorting networks with 16 inputs constructed using mergers and simplified mergers are shown in the right part of Fig. 2: the top variant corresponds to removing redundant comparators from the standard network constructed using mergers, and the bottom variant corresponds to processed sorting network constructed using simplified mergers.

E. Comparison of Encodings on Random Graphs

To measure the effect of proposed encoding techniques, we need to specify the problem and define graphs specified above. Assume that we consider a graph with n vertices. We encode to SAT the following problem: to determine the disposition of at most $\lfloor 0.15 \times n \rfloor$ instigators at the initial time moment (remind, that only instigators are active at the initial time moment), so that after 15 time steps at least $\lceil 0.8 \times n \rceil$ agents in the network are active. We compare the encoding techniques in application to two series of random graphs generated according to different random graphs model: Erdős-Rényi model [9] and Barabáshi-Albert model [10]. We will not specify how the graphs are constructed since it was covered in detail in many sources, including [1]. In Table I we show the sizes of propositional encodings for outlined problems for graphs with $n = 250$ vertices. Graphs corresponding to Erdős-Rényi model are referred to as 'GNP' and were constructed using parameter $p = 0.05$. Barabási-Albert model does not have explicit parameters per se, and the corresponding graphs are referred to as 'BAR'. As for encodings, 'OLD' stands for the encoding method employed in [2], [1], 'BASE' – to the new encoding method without employing greedy algorithm and techniques for removing redundant comparators, 'GR' – to the encoding produced using greedy algorithm, 'RED' - to the encodings produced using techniques for removing redundant comparators, 'GR-RED' – to the encodings fully benefeting from all techniques proposed in this paper. For each random graph model we generated 100 instances and show the average size in megabytes. Also to show that greedy algorithm works best when the graph is dense, we show the statistics for a complete graph with 250 vertices (referred to as 'Complete250' in the Table I). For each graph, conformity level of each agent v_i was generated according to the uniform distribution specified on $[1, |N_i|]$. It is easy to see that our

old scheme used in previous works was very redundant. It is also worth mentioning that the size gap in the transition from 'OLD' to 'BASE' should not theoretically translate into easier problems for SAT solvers (unlike that for other cases) because it is achieved mainly by techniques for constructing sorting networks of specified size instead of that of size equal to closest power of two (see subsection III.B). Overall, it should be noted that the results are quite remarkable: proposed techniques make it possible to significantly reduce the size of encodings. The impact of greedy algorithm is the larger the denser is the graph, as it is illustrated by 3 times difference between the sizes of 'GR' and 'BASE' encodings for complete graph. As for techniques that 'prune' sorting networks based on the value of conformity level – they generally work quite well, however, it is possible that their effect will be worse the higher (on average) the conformity levels are.

IV. USING STATE-OF-THE-ART TOOLS TO OPTIMIZE SAT SOLVERS PARAMETERS

Now let us compare the effectiveness of SAT solving algorithms on the constructed encodings. In particular, for this purpose we use the LINGELING SAT solver [11] (version bbc-9230380-160707) and employ the SMAC tool [12] (version 2.10.03) for optimizing solver parameters. As a computing platform we use cluster nodes of the computing cluster 'Academician V.M. Matrosov' of Irkutsk supercomputing center SB RAS [1], one node equipped by two 16-core AMD Opteron 6276 CPUs and 64 Gb RAM. We disable one of 331 LINGELING parameters ('mem-lim'), thus SMAC has to vary the remaining 330. The LINGELING solver is a single-threaded application, while SMAC is multi-threaded. For each test series when SMAC had to optimize parameters values it was launched for two days with time limit of 5000 seconds on several nodes, using 8 CPU cores within a node (so that solvers do not run out of memory) with enabled option –shared-model-mode, allowing SMAC instances to exchange data.

In the first series of experiments we applied SMAC+LINGELING to 8 series of 100 tests corresponding to 4 encodings for 'BAR250' and 'GNP250', discussed in the previous section. For each series we randomly chosen 10 instances as *training set*, for which SMAC optimized parameters, and the remaining 90 as *test set*. It means that we performed 8 separate experiments with SMAC. In Table II we show the results of LINGELING on the *test set* with default values of parameters and with tuned parameters. It is clear that the benefit is very substantial and the 'GR-RED' encoding is the best or very close to the best.

After this we (following [2]) applied the parameter values found for graphs with 250 vertices to the corresponding problems for graphs (of the same model) with 500, 1000 and 2000 vertices. We show the results for 'BASE' and 'GR-RED' encodings (see Table III). Note that for the case of 2000 vertices we generated only 10 SAT instances with 'GR-RED' since their size is very large. Also, because Erdős-Rényi model

TABLE I
COMPARISON OF SIZE OF PROPOSITIONAL ENCODINGS FOR DETERMINING DISPOSITIONS OF INSTIGATORS. MODELS SPECIFIED BY RANDOM GRAPHS WITH 250 VERTICES. SIZE OF ENCODINGS IN MEGABYTES.

	OLD	BASE	GR	RED	GR-RED
BAR250	56.514	31.067	29.171	22.269	20.368
GNP250	34.163	18.294	17.411	13.365	12.477
Complete250	273.502	219.317	74.336	169.393	31.763

[1] http://www.hpc.icc.ru

TABLE II
RESULTS FOR RANDOM GRAPHS WITH 250 VERTICES

	Default		Tuned	
BAR				
Encoding	Solved	Mean	Solved	Mean
BASE-250	90	196.61	90	5.46
GR-250	90	155.22	90	5.33
RED-250	90	188.89	**90**	**4.60**
GR-RED-250	90	106.81	90	4.88
GNP				
Encoding	Solved	Mean	Solved	Mean
BASE-250	90	36.84	90	3.53
GR-250	90	32.08	90	4.94
RED-250	90	34.28	89	34.33
GR-RED-250	90	29.11	**90**	**2.63**

TABLE III
RESULTS FOR RANDOM GRAPHS WITH 500 – 2000 VERTICES

	Default		Tuned	
BAR				
Encoding	Solved	Mean	Solved	Mean
BASE-500	100	423.20	100	13.71
GR-RED-500	99	426.40	**100**	**10.16**
Encoding	Solved	Mean	Solved	Mean
BASE-1k	95	1809.73	100	32.92
GR-RED-1k	96	1670.54	**100**	**24.93**
Encoding	Solved	Mean	Solved	Mean
GR-RED-2k	0	-	**3**	**2328.87**
GNP				
Encoding	Solved	Mean	Solved	Mean
BASE-500	97	674.27	**100**	**154.79**
GR-RED-500	93	577.20	72	770.22
Encoding	Solved	Mean	Solved	Mean
BASE-1k	67	1466.18	**74**	**1444.64**
GR-RED-1k	64	1507.10	25	1326.40
Encoding	Solved	Mean	Solved	Mean
GR-RED-2k	**6**	**1204.56**	3	2020.4

has a parameter p which directly influences the number of arcs in a graph, we had to scale it down with the increase of dimension to maintain balance between the dimension and the size of encoding. Thus for 500 we used $p = 0.025$, for 1000 $p = 0.0125$, and for 2000 $p = 0.00625$. It is clear from Table III that the parameters found by SMAC for GNP250 do not work well for GNP graphs of higher dimension. We believe that it can be explained by relatively chaotic nature of GNP graphs. On the other hand in the case of graphs constructed according to the Barabási-Albert model the effect from the corresponding tuning is very impressive.

V. RELATED WORKS

The model of conforming behavior in quite similar formulation was first proposed in [3]. In [1] we extended it, formulated several combinatorial problems aimed at its study and used SAT approach to solve them. In [2] we applied it to the corresponding problems parameterized SAT solving algorithms and showed that they make it possible to improve the effectiveness of SAT approach quite significantly. In the present paper we completely reworked the propositional encoding procedures used to encode the combinatorial problems described in [1]

to SAT. New procedures are heavily optimized in a sense that we reduced the amount of redundant repeated actions to minimum, thus making encodings more compact and less difficult for SAT solvers. The optimizations of the algorithm use both new (however, relatively simple and problem-specific) methods, such as greedy algorithm to figure out neighborhood intersections and removing redundant comparators that do not influence specific outputs of a sorting network, and the results from known works, such as Simplified Merge procedure [5].

VI. CONCLUSIONS

In this paper we significantly improved the effectiveness of procedures that can be used to study several discrete models of conforming behavior using SAT approach. We tested it on two families of random graphs, generated according to Erdős-Rényi and Barabási-Albert models. The decrease in size turned out to be from 30 % to 700 %. The corresponding effect on SAT solving algorithms was good as well.

ACKNOWLEDGMENTS

The research was funded by Russian Science Foundation (project No. 16-11-10046). Stepan Kochemazov and Oleg Zaikin were additionally supported by Council for Grants of the President of the Russian Federation (stipends SP-1829.2016.5 and SP-1184.2015.5, respectively).

The authors would like to thank Alexey Hmelnov for the general idea of proposed greedy algorithm.

REFERENCES

[1] S. Kochemazov and A. Semenov, "Using synchronous boolean networks to model several phenomena of collective behavior," *PLOS ONE*, vol. 9, no. 12, pp. 1–28, 12 2014.
[2] S. Kochemazov, A. Semenov, and O. Zaikin, "The application of parameterized algorithms for solving SAT to the study of several discrete models of collective behavior," in *39th International Convention on Information and Communication Technology, Electronics and Microelectronics, MIPRO 2016*, 2016, pp. 1288–1292.
[3] M. Granovetter, "Threshold models of collective behavior," *American Journal of Sociology*, vol. 83, no. 6, pp. 1420–1443, 1978.
[4] A. Biere, M. Heule, H. van Maaren, and T. Walsh, Eds., *Handbook of Satisfiability*, ser. Frontiers in Artificial Intelligence and Applications, vol. 185. IOS Press, 2009.
[5] R. Asín, R. Nieuwenhuis, A. Oliveras, and E. Rodríguez-Carbonell, "Cardinality networks: a theoretical and empirical study," *Constraints*, vol. 16, no. 2, pp. 195–221, 2011.
[6] M. Codish and M. Zazon-Ivry, "Pairwise cardinality networks," in *Logic for Programming, Artificial Intelligence, and Reasoning - 16th International Conference, LPAR-16*, 2010, pp. 154–172.
[7] K. E. Batcher, "Sorting networks and their applications," in *American Federation of Information Processing Societies: AFIPS Conference Proceedings: 1968 Spring Joint Computer Conference*, 1968, pp. 307–314.
[8] I. Parberry, "The pairwise sorting network," *Parallel Processing Letters*, vol. 2, pp. 205–211, 1992.
[9] P. Erdős and A. Rényi, "On random graphs i." *Publicationes Mathematicae (Debrecen)*, vol. 6, pp. 290–297, 1959 1959.
[10] A.-L. Barabási and R. Albert, "Emergence of scaling in random networks," *Science*, vol. 286, no. 5439, pp. 509–512, 1999.
[11] A. Biere, "Splatz, Lingeling, Plingeling, Treengeling, YalSAT Entering the SAT Competition 2016," in *Proc. of SAT Competition 2016 – Solver and Benchmark Descriptions*, vol. B-2016-1, 2016, pp. 44–45.
[12] F. Hutter, H. H. Hoos, and K. Leyton-Brown, "Sequential model-based optimization for general algorithm configuration," in *Proc. of LION-5*, 2011, pp. 507–23.

Effects of the distribution of the values of condition attribute on the quality of decision rules

V. Ognjenović*, E. Brtka*, V. Brtka*, I. Berković*

* University of Novi Sad/Technical faculty "Mihajlo Pupin", Zrenjanin, Serbia
visnjao@tfzr.uns.ac.rs, norab@ tfzr.uns.ac.rs, brtkav@ tfzr.uns.ac.rs, berkovic@tfzr.uns.ac.rs

Abstract – **The table-organized data can be analyzed by various algorithms; some of them are capable of generating IF THEN decision rules which comprises of condition attributes and decision attributes. However, it is possible to reduce the set of condition attributes but without information loss. By analysis of the condition attributes set and cuts histogram obtained by discretization and rule consistency, it is possible to choose condition attributes. This paper gives some directions and the practical example.**

I. INTRODUCTION

This paper presents the process of attribute selection in order to generate IF THEN rules from table-organized data. The problem is recognized by the CRISP-DM standard [1], which clearly separates Data Understanding from Data Preparation. Data Understanding is an activity that uncovers the very essence of the data, while the Data Preparation includes more activities dealing with lost information, simpler and better data processing. The problem of choosing data for research is especially present in the field of Big Data.

There are situations when the result of data classification obtained by IF THEN rules set is good. However, if we analyze the consistency of this set, we can conclude that there are situations when the rules are not precise enough. If the OR logical operator is included in the THEN part of the rule, the rule is inconsistent and imprecise, while classification is often good. The possibilities of further detailed selection of condition attributes will be considered based on Rough Set Theory [2], and by attribute histogram analysis as well.

II. ROUGH SET THEORY

The Rough Set Theory has been developed having in mind data analysis and information systems [2]. The basic purpose of these sets is the approximation of unfamiliar knowledge using the familiar one [3]. Based on the principle of indiscernibility relation of objects and the concept of approximation, this theory enables recognition of inter-dependability between the decision attributes and condition attributes [4].

In the rough set theory, an information table [5], consisting of ordered quadruple $S = \langle U, Q, V, f \rangle$, is defined, where:

- U is the finite set of objects - universe;

- $Q = \{q_1, q_2, ..., q_m\}$ is a finite set of attributes;

- $V = \bigcup_{q \in Q} V_q$, where V_q is the domain of attribute q (attribute values);

- $f = U \times Q \rightarrow V$ is the total function, such that $f(x,q) \in V_q$ for each $q \in Q, x \in U$ and is called information function.

Each object $x \in U$ is described by the vector:

$$\inf_q(x) = [f(x,q_1), f(x,q_2), ..., f(x,q_m)] \qquad (1)$$

which defines the values of object x attributes.

If P is a non-empty subset of attribute set Q, then the relation I_P is defined on the objects from the universe U, in the following way:

$$I_P = \{(x,y) \in U \times U : f(x,q) = f(y,q), \forall q \in P\} \quad (2)$$

Relation (2) is called indiscernibility relation. If $(x,y) \in I_P$, we say that the objects x and y are P-indiscernible. The indiscernibility relation is the equivalence relation and it generates the partitions – equivalence classes. [6]. The family of equivalence classes, generated by I_P , are marked with $U|I_P$.

Equivalence classes generated by relation I_P are called P-elementary sets, and the equivalence class containing the object $x \in U$ is marked with $I_P(x)$ or $[x]_P$.

For the set X, a non-empty subset of U, called a rough set, and for $\varnothing \neq P \subseteq Q$, the following are defined:

$$\underline{P}(X) = \{x \in U : I_P(x) \subseteq X\} \qquad (3)$$

$$\overline{P}(X) = \bigcup_{x \in X} I_P(x) \qquad (4)$$

$\underline{P}(X)$ is the P-lower approximation, which means that the objects certainly belong to X, and $\overline{P}(X)$ is the P-

upper approximation, which means that the objects may belong to the set X [7].

P-boundary of subset X in S is defined as follows:

$$Bn_P(X) = \overline{P}(X) - \underline{P}(X) \qquad (5)$$

Graphical interpretation of P-boundary is shown in Fig.1. It cannot be said with certainty that the elements from the boundary region belong to the set X – so it is said that they are the elements of the set X approximation [8].

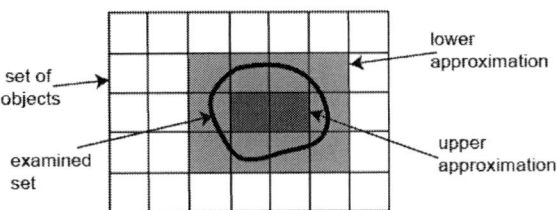

Figure 1.   P-boundary of set

This research will use simple calculation of consistency and reduct sets, which are used for IF THEN rules generation. Consistency is defined on the basis of generalized decision function) ∂_P (x_i) in the rough set theory. Moreover, a direct implication of inconsistent table can be observed: it is the decision rules which in IF part have the same conditions, and in THEN part various decisions. Information reduct, or just reduct is intuitively recognized as a minimal subset of condition attributes which keep the discernibility between the objects [9]. The reduct set includes only these condition attributes which are sufficient for evaluation of the decision attribute value.

The variation of the Johnson's algorithm [10] will be used for calculating the minimal simple implicants of Boolean function, so that IF THEN classification rules will be generated by single reduct set. Johnson's algorithm is a simple greedy heuristic algorithm, often applied to discernibility functions to find a single reduct. The algorithm first sets the current reduct candidate, to the empty set. Then, it evaluates each conditional attribute appearing in the discernibility function according to the heuristic measure. This measure is usually a count of the number of appearances an attribute makes within clauses. The algorithm adds the attribute with the highest heuristic value, removing all clauses in the discernibility function containing this attribute at the same time [11].

After reduct set calculation, it is possible to determine rules of the **IF a THEN b** form. Here **a** denotes a conjunction of attributes from some reduct set and their corresponding values: attribute value or subinterval obtained through discretization. The **b** consists of decision attribute and corresponding value.

III.   HISTOGRAM

Histogram is a well-known mathematical tool for graphical presentation of object distribution, i.e. frequency of certain data values. Its advantage in presenting data is firstly in the fact that the graphical presentation can show

the center of distribution, the spread of distribution, shape of data distribution, as well as noticing of some unusual data characteristics. If a distribution has only one peak, then the middle of such a distribution is usually at the peak. The shape of distribution is usually used in recognizing the data patterns [12]. Unusual characteristics often refer to certain breaks in histogram, as well as in isolated values (Fig. 2). Some distributions can be similar to a mathematical distribution, while a significant data set often comes in the form of an irregular multimodal distribution (Fig. 3).

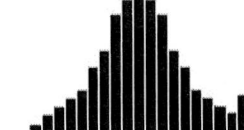

Figure 2.   Histogram edge peak data distribution

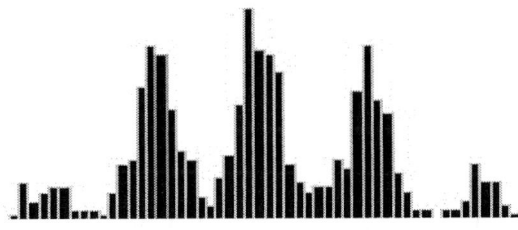

Figure 3.   Histogram multimodal data distribution

Previous to rough set theory application, the data values need to be prepared so that all values are discrete and the consequent histogram is a suitable for data analyses.

IV.   THE RELATION OF HISTOGRAM AND DATA DISCRETIZATION

In the domain of Data Mining, numerous Machine Learning methods can work only with the discrete attribute values. This is the reason why it is necessary to transform the continuous attribute values into the discrete ones, before machine learning process. This process of data discretization is an essential task in preprocessing of data. The result of discretization is the set of cuts by which the data are classified into intervals [13].

A.   The relation of data distribution and discretization algorithm

The relation of data distribution and large databases has been observed within the algorithm of maximal discernibility, by using median [5]. Paper [14] suggests the usage of the incremental algorithm for data streams, in order to modify the data distribution. The algorithm for estimating a histogram density based on the MDL principle (MDL Histogram Density Estimation) uses the entropy to generate the histograms for data regularity. The consequent analysis results enable the discretization [15]. Therefore, the MDL based algorithm discovers patterns in the data histogram.

B. Histogram segmentation

Histogram segmentation definition is related to recognition of the thresholds of multimodal distribution. Firstly, the segmentation of multimodal distribution is done by 'smoothing out' the existing histogram – so that the distribution function becomes a smooth curve, as shown in Fig. 4 (the figure was taken from Paper [16]). Secondly, based on the cross-section of these smooth curves the threshold point is obtained, dividing the histogram into two parts.

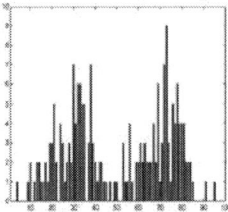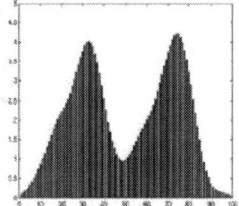

Figure 4. Original histogram (left) i smooth histogram (right)

Appropriate histogram segmentation example is described in [16]: Fig. 5 shows the blood cells and the histogram with two thresholds (the figure was taken from Paper [16]). In detail, Fig. 5 shows the picture of blood cells, histogram, smooth curve obtained through the histogram and the picture obtained through histogram segmentation. The first part of segmentation of the interval (0, 128) shows the picture of blood cells. The second part of the interval segmentation (128, 184) shows the blood plasma. Finally, the segmentation of the interval (184, 250) shows the cell membranes.

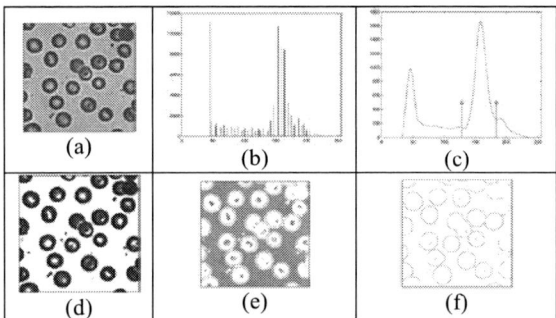

Figure 5. The picture of blood (a), histogram (b), smooth histogram (c) and the results obtained through histogram segmentation, blood cells (d), blood plasma (e) and cell membranes (f)

Based on the histogram analysis, the unimodal parts of multimodal distribution can be recognized. In the case of no-peak histograms or the ones with only one distinct peak, the segmentation would not be possible.

C. The idea of choosing the condition attributes on the basis of histogram and cuts

In [13] the significance of histogram segmentation in the processes of discernibility-based discretization is described. The cuts obtained through the algorithm of maximal discernibility, which are closest to the histogram segmentation thresholds, are very important for the preservation of discernibility. Based on this, the algorithm of approximate discretization APPROX MD was developed [13]. It uses the cuts obtained through the maximal discernibility algorithm, which are closest to the multimodal histogram segmentation thresholds. The aim is to define the most significant cuts for all condition attributes.

Consideration of process of choosing the condition attributes for generating IF THEN rules is the extension of the process described in [13]. Since the reduct of the condition attribute set distinguishes those attributes which can describe the entire set, the subject of the research is the analysis of the histogram of the attributes which make up the reduct.

This paper will give the basic directives for the analysis of histogram and cuts obtained from reduct set. The directives are:

- the existence of the histogram thresholds,
- the position of cuts on the multimodal distribution histograms,
- the position of cuts on the unimodal distribution histograms,
- the comparison of the reduct of the entire information table with the reduct of the modified information table obtained through the reduction of certain condition attributes set,
- the comparison of the reduct of the entire information table with the reduct of the modified information table obtained through the enlargement of certain condition attributes set.

Mandatory part of the reduct analysis is the observation of consistency of the IF THEN rules set.

V. EXAMPLE

The histograms of condition attributes will be shown on the example of Iris database [17]. The database contains 3 classes, 150 instances while each class refers to a type of the flower iris. As a result, a discretization has been done by the maximal discernibility algorithm and a reduct was calculated by the variation of the Johnson's algorithm. The reduct consists of third and fourth attribute, which satisfies multimodal histogram distribution. First two condition attributes do not satisfy typical multimodal distribution. The histograms of all the condition attributes are presented on Fig. 6 – Fig 9. The cuts are shown on the ordinates by wide black vertical lines.

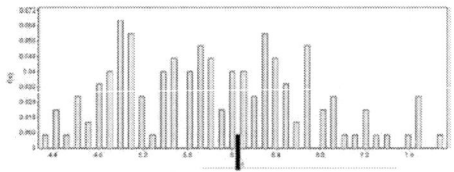

Figure 6. Data distribution of the 1. attribute sepal length, with the cut obtained by the maximal discernibility algorithm

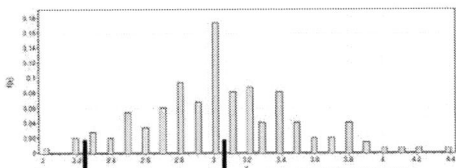

Figure 7. Data distribution of the 2. attribute sepal width, with the cut obtained by the maximal discernibility algorithm

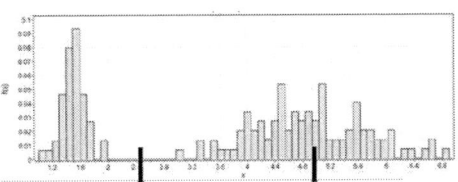

Figure 8. Data distribution of the 3. attribute petal length, with the cut obtained by the maximal discernibility algorithm

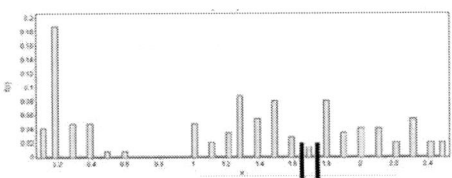

Figure 9. Data distribution of the 4. attribute petal width, with the cut obtained by the maximal discernibility algorithm

For previous cuts and for reduct $\{a_3, a_4\}$, obtained IF THEN rules are consistent. If any cut from Fig. 8 or Fig. 9 is omitted, a significantly larger reduct set with much less precise IF THEN rules will be obtained.

Concretely, if the first cut of the third attribute (petal length) is omitted, the reduct $\{a_1, a_2, a_3, a_4\}$ is obtained, while the inconsistency increases. In the case of omitting the third attribute, the inconsistency of the IF THEN rules increases even more. It is also noticeable that the first cut of the third attribute overlaps with the multimodal distribution threshold.

VI. CONCLUSION

For multimodal distribution of attribute values, the existence of thresholds describes the natural line between data values, as was shown by the authors who have researched the picture histogram segmentation. The question whether reduct attributes have a more specific multimodal distribution in comparison to the other attributes, would answer the question on influence of data distribution on quality of decision rules. The correlation between the rule consistency and attribute distribution would be the feedback on certain thresholds' and cuts' importance. The comparison of various reducts obtained through a range of attributes on the basis of data distribution would also contribute to the research of relations between the attribute value data distribution and the reducts.

Condition attributes histograms have certain hidden patterns which can be discovered by investigation of the position of the cuts in relation to the thresholds of segmentation of multimodal distribution. The influence of

multimodal distribution on the reduct set should be researched on the bigger number of the databases.

ACKNOWLEDGMENTS

This paper has been supported by the Ministry of Education and Science of Republic of Serbia, within the project TR32044 "The Development of Software Tools for Business Process Analysis and Improvement", 2011-2017.

REFERENCES

[1] C. Shearer, The CRISP-DM model: the new blueprint for data mining, J Data Warehousing, 5: pp 13—22, 2000.

[2] Z. Pawlak Rough sets, International Journal of Computer and Information Sciences, 11, 341-356, 1982.

[3] A. Skowron, H.S. Nguyen, Boolean reasoning scheme with some applications in data mining, Principles of Data Mining and Knowledge Discovery, 107-115, 1999.

[4] A. Øhrn, J. Komorowski, A. Skowron, and P. Synak, The ROSETTA, software system. In L. Polkowski and A. Skowron, editors, Rough Sets in Knowledge Discovery 2. Applications, Case Studies and Software Systems, number 19 in Studies in Fuzziness and Soft Computing, pages 572–576, 1998., Physica-Verlag, Heidelberg, Germany.

[5] H.S. Nguyen, Approximate boolean reasoning: foundations and applications in data mining, Transactions on rough sets V, 2006., pp. 334-506.

[6] D. Sajter, Metode predviđanja poslovnih poteškoća, Ekonomski fakultet u Osijeku, http://oliver.efos.hr/~dsajter/PDF/3.)%20Metode%20pred.%20pos l.%20pot..pdf

[7] N. Dragun, Granularno predstavljanje znanja zasnovano na aspektnom pristupu, Sveučilište u Zagrebu Fakultet elektrotehnike i računarstva, https://www.fer.unizg.hr/_download/repository/Dragun_Granularn o_predstavljanje_znanja_zasnovano_na_aspektnom_pristupu.pdf

[8] V. Brtka, Automatska sinteza baze pravila u inferentnim sistemima, doktorska disertacija, Univerzitet u Novom Sadu, Tehnički fakultet „Mihajlo Pupin" Zrenjanin, 2008.

[9] E. Brtka, V. Ognjenovic, V. Brtka, The evaluation of the overall knowledge of the students by usage Dynamic Reducts, Technics Technologies Education Management, Vol7No4, ISSN: 1840-1503, pp. 1672-1680, 2012.

[10] D. S. Johnson, Approximation algorithms for combinatorial problems. Journal of Computer and System Sciences, 9:256–278, 1974.

[11] R. Jensen, Q. Shen, Rough set-based feature selection: A review, in A. E. Hassanien , Z. Suraj , D. Slezak and P. Lingras, eds., Rough Computing: Theories, Technologies and Applications, Information Science Reference, IGI Global Press pp. 70-107, 2007.

[12] H. Berman, Stat Trek, How to Describe Data Patterns in Statistics, http://stattrek.com/statistics/charts/data-patterns.aspx

[13] V. Ognjenovic, Approximative Discretization of Table-Organized Data (Doctoral dissertation), University of Novi Sad, Technical faculty "Mihajlo Pupin", Zrenjanin, 2016.

[14] J. Gama, C. Pinto, Discretization from Data Streams: Applications to Histograms and Data Mining, SAC '06 Proceedings of the 2006 ACM symposium on Applied computing, Pages 662-667, 2006.

[15] P. Kontkanen, P. Myllymaki, MDL Histogram Density Estimation, Proceedings of the Eleventh International Conference on Artificial Intelligence and Statistics March 21-24, 2007, San Juan, Puerto Rico, 2:219-226, 2007.

[16] J.H. Chang, C. Fan Kuo, Y.L. Chang, Multi-modal gray-level histogram modeling and decomposition, Image and Vision Computing 20, pp. 203-216, 2002.

[17] Iris, Iris Data Set, 2006, https://archive.ics.uci.edu/ml/datasets/Iris

Complexity comparison of integer programming and genetic algorithms for resource constrained scheduling problems

Rebeka Čorić, Mateja Đumić
Department of Mathematics, University of Osijek
Osijek, Croatia
Email: rcoric/mdjumic@mathos.hr

Domagoj Jakobović
Faculty of Electrical Engineering and Computing
Zagreb, Croatia
Email: domagoj.jakobovic@fer.hr

Abstract—Resource constrained project scheduling problem (RCPSP) is one of the most intractable combinatorial optimization problems. RCPSP belongs to the class of NP hard problems. Integer Programming (IP) is one of the exact solving methods that can be used for solving RCPSP. IP formulation uses binary decision variables for generating a feasible solution and with different boundaries eliminates some of solutions to reduce the solution space size. All exact methods, including IP, search through entire solution space so they are impractical for very large problem instances. Due to the fact that exact methods are not applicable to all problem instances, many heuristic approaches are developed, such as genetic algorithms. In this paper we compare the time complexity of IP formulations and genetic algorithms when solving the RCPSP. We present two different solution representations for genetic algorithms, permutation vector and vector of floating point numbers. Two formulations of IP and and their time and convergence results are compared for the aforementioned approaches.

I. INTRODUCTION

Scheduling is process which all of us use on everyday basis. In scheduling the goal is to allocate resources to tasks over given time periods. Usually the resources that we have have some limitations. For example we have limited project budget, limited number of employees, vehicle capacity etc.

This work will focus on one of the problems with constraints known as resource constrained project scheduling problem (RCPSP). RCPSP is a problem in which one has to deal with two kinds of constraints - precedence and resource. In this scheduling process we define starting point of some activity taking care of its precedence and amount of resources that are available.

RCPSP belongs to class of NP hard problems [1] and that, with fact that it is common problem in everyday world, results with numerous methods for solving it. Which method to use is a decision that has to be made upon the problem characteristics. In literature two different groups of solving methods can be found - exact methods and heuristics [2].

Exact methods search the entire space of feasible solutions and guarantee the optimality [3], but because of size of solution space they are impractical and almost useless for problems with large instance size [4]. Because of that many heuristic methods were developed. Heuristics do

not search the entire solution space and do not guarantee the optimality but they are fast and can result with good enough solutions for the specific problem.

Heuristics can be divided into two groups: priority rule-based methods (or constructive heuristics) and metaheuristic-based approaches (or improvement heuristics). Methods in the first group start with empty schedule, and iteratively put in each step one activity based on it's priority, until unscheduled activities exists. The second group starts with initial complete solution and in each iteration tries to improve it.

In this work we will present two IP formulations as representatives of exact methods, and Genetic algorithm (GA) as representative of heuristics methods. In GA approach two different solution representations will be used. Result comparison and time execution comparison will be given for these methods.

This paper is organized as follows: in section 2 a short overview of solving methods for RCPSP is given. The definition of RCPSP can be found in section 3. In section 4 two different IP formulations are given, and GA adaption for RCPSP is explained in section 5. Experimental results together with implementaion details of this methods are given in section 6. Finally in section 7 a short conclusion is given.

II. RELATED WORK

Many researchers investigate RCPSP and a lot of work for exact and heuristic methods can be found in literature.

Exact methods can be divided in four different groups:
1) Integer Programming - using large number of binary variables [5], [6]
2) Implicit Enumeration - using enumeration tree and bounds to reduce the space of feasible solutions [7], [8]
3) Branch-and-bound methods - using trees and lower bounds to eliminate nodes in tree that cannot lead to optimal solution [9], [10]
4) Dynamic programming - divide problem on problems with smaller size and solve that problems, and combine the solution [11].

Some of heuristic methods that can be found in literature are: Genetic algorithms [12], [13], Simulated annealing

[14], Ant colony [15]. For achieving better results there are attempts of combining different solution methods like hybrid algorithms, memetic algorithms etc. Usually these approaches combine some evolutionary algorithm and local search methods. One of newer work shows that combining local search method with GA in Genetic-Local Search Algorithm gives good results [16].

III. RCPSP

RCPSP is a problem in which we need to schedule activities which have their duration and resource demands, while all resources have limited amount that can be used in one time period. Beside of the limited resource amount, there are precedence constraints that have to be satisfied, meaning that every activity has predecessors that need to be finished before it starts. The goal of scheduling is to find a feasible schedule that minimizes some objective, by assigning start times to each activity that satisfies the precedence and resource constraints.

Formally, RCPSP can be defined as a combinatorial optimization problem as tuple [17]:

$$RCPSP = (A, E, p, R, B, D, c). \quad (1)$$

In this tuple A denotes the set of all activities $A = \{A_0, A_1, \ldots, A_{n+1}\}$, where activities A_0 and A_{n+1} are dummy activities that represent the start and the end of a schedule. Vector $p \in \mathbb{N}_0^{n+2}$ is vector of durations, in which component p_i is duration of activity A_i. Dummy activities have duration 0.

The precedence relations between activities are given by set E. Set E consists of pairs of activities (A_i, A_j) which mean that activity A_i precedes activity A_j. The assumption for dummy activities is that activity A_0 precedes all other activities and that A_{n+1} succeeds all other activities. So, by determining the start point of A_{n+1} the total project duration is found.

Resources that are needed for activities are given by set $R = \{R_1, \ldots, R_r\}$, and amount of each resource that is available in all time period is given by vector $B \in \mathbb{N}^r$. Demands on resources for each activity are given by matrix $D \in \mathbb{N}_0^{(n+1) \times r}$.

The objective function is $c : \chi \subseteq \mathbb{R}^{n+2} \to \mathbb{R}$, where χ is a feasible schedule. Sometimes this tuple member is omitted, in which case the objective function which needs to be minimized is project makespan (total project duration). The solution is a schedule $S \in \chi$, where component S_i of vector S represents the start time of activity A_i. When vector S is known it is easy to find vector of completion times of activities denoted by C, by computing component $C_i = S_i + p_i$.

If we mark the set of all activities which are active at a time t with $A_t = \{A_i \in A : S_i \leq t \leq S_i + p_i\}$ RCPSP can be defined as a problem of finding a feasible schedule S with minimal makespan S_{n+1} for which following constraints are satisfied:

$$S_j - S_i \geq p_i \quad \forall (A_i, A_j) \in E \quad (2)$$

$$\sum_{A_i \in A_t} D_{ij} \leq B_j \quad \forall t \geq 0, \forall R_j \in R. \quad (3)$$

For further problem analysis we also need to define set $E(t)$ - the set of eligible activities at time t. For definition of that set we need to define the following properties of each activity:

- ES_j - the earliest start time of activity A_j
- EF_j - the earliest finish time of activity A_j
- LS_j - the latest start time of activity A_j
- LF_j - the latest finish time of activity A_j.

To calculate mentioned properties the horizon (the total project time) T, must be known. If it is not given, it can be calculated by some heuristic or using the formula

$$T = \sum_{j=1}^{n} p_j. \quad (4)$$

With setting $ES_0 = EF_0 = 0$ for calculating ES_j and EF_j we can use the following formulas:

$$ES_j = max\{EF_i : i \in P_j\} \quad (5)$$

$$EF_j = ES_j + p_j, j \in \{1, \ldots, n\}, \quad (6)$$

where P_j is set of activities that directly precede the activity A_j.

Similarly, for calculating LS_j and LF_j we set $LF_n = LS_n = T$ and use formulas:

$$LF_j = min\{LS_i : i \in F_j\} \quad (7)$$

$$LS_j = LF_j - p_j, j \in \{n - 1, \ldots, 1\}, \quad (8)$$

where F_j is set of activities that directly succeed the activity A_j.

Now we can define $E(t)$ as:

$$E(t) = \{A_j : A_j \in A, ES_j + 1 \leq t \leq LF_j\}. \quad (9)$$

IV. IP FORMULATION OF RCPSP

IP is a method in which number of binary variables are introduced and used for creating a schedule. In this section two different IP formulations of RCPSP will be given.

A. Formulation 1

The first IP formulation was given by Pritisker et al. [6] in 1969. In this formulation, the solution of RCPSP is represented with series of zero-one variables x_{jt}, $j \in J$ (J is set of activity indexes), $t \in [EF_j, LF_j]$. Variables are defined as:

$$x_{jt} = \begin{cases} 1, & \text{if activity } A_j \text{ is completed at the end of} \\ & \text{period t} \\ 0, & \text{otherwise.} \end{cases}$$

The variable x_{jt} can have the value 1 just in a time period in which the activity A_j can be completed so it is introduced just for $t \in [EF_j, LF_j]$. Time in which activity A_j finished is given by:

$$\sum_{t=EF_j}^{LF_j} t \cdot x_{jt} \quad (10)$$

1183

IP formulation for RCPSP can be now given as:

$$\sum_{t=EF_n}^{LF_n} t \cdot x_{nt} \to \min \qquad (11)$$

subject to:

$$\sum_{t=EF_j}^{LF_j} x_{jt} = 1, \ j \in J \qquad (12)$$

$$\sum_{t=EF_j}^{LF_j} (t-p_j)\, x_{jt} - \sum_{t=EF_i}^{LF_i} t \cdot x_{it} \ge 0, \ j \in J, i \in P_j \qquad (13)$$

$$\sum_{j \in E(t)} u_{jr} \cdot \sum_{q=\max\{t,EF_j\}}^{\min\{t+p_j-1,LF_j\}} x_{jq} \le a_r, \ r \in R, t \in [1,\dots,T] \qquad (14)$$

$$x_{jt} \in \{0,1\}, \ j \in J, t \in [EF_j, LF_j] \qquad (15)$$

From (11) it is clear that in this formulation one minimizes the completion time of the last activity (makespan). If the completion time of the last activity is minimal then the completion time for all activities is minimal too. Constraints (12) and (15) ensure that the activity can be completed exactly in one time period. Constraint (13) ensures that all predecessors of activity A_j are completed before it starts with execution, and constraint (14) ensures that in each time period the amount of resources being used is smaller or equal to the available amount of resources.

B. Formulation 2

RCPSP can be formulated as IP using variables y_{jt}, which can also only become 0 or 1 for $j \in J$, $t \in [ES_j + 1 - p_j, LF_j]$ and are defined in the following way:

$$y_{jt} = \begin{cases} 1, & \text{if activity } A_j \text{ is completed at the beginning} \\ & \text{of period t or earlier} \\ 0, & \text{otherwise.} \end{cases}$$

For activity A_j there is vector with $LF_j - ES_j - 1 + p_j$ components with two blocks: the first consists of zeroes, and the second consists of ones. Time period in which change between these two blocks happens is the time point in which activity A_j started. In accordance with that, the time point in which activity A_j starts is given with:

$$LF_j - \sum_{t=ES_j+1}^{LF_j} y_{jt} \qquad (16)$$

IP formulation of RCPSP with introduced variables is:

$$T - \sum_{t=ES_n+1}^{LS_n} y_{nt} \to \min \qquad (17)$$

subject to:

$$y_{j,t+1} - y_{jt} \ge 0, \ j \in J, t \in \{ES_j + 1, \dots LS_j - 1\} \qquad (18)$$

$$y_{i,t-p_i} - y_{jt} \ge 0, \ j \in J, i \in P_j, t \in \{ES_j + 1, \dots LF_i\} \qquad (19)$$

$$\sum_{j \in E(t)} u_{jr} \cdot \left(y_{jt} - y_{j,t-p_j}\right) \le a_r, \ r \in R, t \in [1,\dots,T] \qquad (20)$$

$$y_{jt} = 0, \ j \in J, t \in [ES_j + 1 - p_j, ES_j] \qquad (21)$$

$$y_{jt} = 1, \ j \in J, t \in [LS_j + 1, LF_j] \qquad (22)$$

$$y_{jt} \in \{0,1\}, \ j \in J, t \in [EF_j, LF_j] \qquad (23)$$

Objective function is given with (17) and minimizes makespan. Constraints (18) and (21-23) ensure that activity A_j starts in one time point only. Constraint (19) ensures that all predecessors $A_i \in P_j$ of activity A_j start at least p_i time units earlier. Constraint (20) ensures that resource usage in every time period satisfies the amount of resource which can be used.

V. GA FOR RCPSP

In this section, genetic algorithms and how they work will briefly be explained.

The idea for genetic algorithms (GAs) came from the principles of natural selection and genetics. Because of that, certain terms in GAs are equal to the terms one can see in genetics. In GAs decision variables are coded as finite dimensional vectors which come from an alphabet of some cardinality. Those vectors are called chromosomes, parts of the alphabet are called genes, and the values of the genes are called alleles. E.g. for the Traveling Salesman Problem (TSP), one chromosome represents a route and one gene represents a city in a route [18]. In GAs solutions can be represented in various ways. Some of them are [19]: array of bits, permutation array, matrix representation, floating point vector, integer vector etc.

In order to get better solutions in the process of natural selection, there should exist some measure which will determine which solutions are better than the others and then lead to better solutions in GAs. Genetic algorithms use population of solutions rather than only one solution to find optimum of some problem. It is necessary to determine the optimal size of population, which is an optimization problem by itself. After determining the size of population and chromosome coding as well as some fitness measure for solution candidates, algorithm can start to evolve solutions for a given problem using the following steps [18]:

1) Initialization: initial population of solutions is generated randomly, or if some additional information about solution space is known, that knowledge is incorporated in generating better initial population
2) Evaluation: after initialization, the fitness of each population member is evaluated

3) Selection: selection chooses few of the better solutions and that way one can secure survival of the fittest
4) Crossover: crossover operators combine parts of two or more parents in order to create one or more children solutions
5) Mutation: mutation modifies existing solution in order to keep diversity in population
6) Replacement: population of children obtained in selection, crossover and mutation, replaces old population of parents

After that, evaluation, selection, mutation and replacement are repeated until some of stopping criteria is met.

GAs can be sequential and parallel [19]. Sequential GAs can be steady-state GA or generational GA. Steady-state GA chooses two parents from population, makes one child, mutates it and evaluates it. Child can then be inserted in population (in that case some other solution candidate gets thrown out) or rejected. In generational GA in every step whole new population of children is created which replaces old parent generation. If in that process one does not want to lose best found solution, elitism can be introduced, i.e. the best current known solution is always put in child population. In GA implementation used for this work, a sequential steady-state GA with elitism is used.

VI. EXPERIMENTAL RESULTS

A. Integer Programming

For solving IP problems it's common to use cutting plane methods and variants of the branch and bound method. In this work Gurobi [20] was used for solving IPs. The solvers in the Gurobi Optimizer use parallelism while solving problems. The process of solving can be divided in two phases: presolving and solving. Presolving phase is a phase in which problem is simplified which can significantly reduce the time needed for solving problem. Methods that are used in this phase are: removal of empty rows and columns, finding and removing rows dominated by linear combination of other rows etc. In the solving phase Gurobi uses branch and bound and cutting plane methods for solving new reduced problem created in presolve phase.

B. Genetic Algorithm

In GA implementation for this problem two solution representations were used: permutation representation (as proposed in [21]) and floating point vector representation. In permutation representation, every candidate solution is represented as integer vector which contains numbers from 0 to $n+1$, where n denotes number of jobs. E.g. for $n = 5$, $\{0, 2, 3, 5, 1, 4, 6\}$ means that job 2 is executed first, after that job 3 is executed etc. In floating point vector representation, every candidate solution is presented as vector which contains numbers between 0 and 1. The higher the number, the more important it is to start the job earlier. E.g. $(0.3, 0.95, 0.17, 0.82, 0.5)$ means that jobs should be executed in the following order: $(2, 4, 5, 1, 3)$. In order to

get feasible solutions in the initial population (solutions that meet precedence constraints and resource constraints) for both representations, serial schedule generation scheme has been used which schedules a job in earliest possible time in which both precedence constraints and resource constraints for given job are met. Due to the lack of space, serial schedule generation scheme pseudocode is omitted but can be found in [22].

For GA tests, Evolutionary Computation Framework[1] (ECF) was used with the following setup: 3-tournament selection was used, where in each iteration three random individuals are chosen and the worst of the three is eliminated. A new individual is created using crossover from the remaining two, and is subjected to mutation with a given probability. In the permutation encoding, order-based crossover operators OBX and OX2 as well as cycle crossover where used as crossover operators. The mutation operator randomly chooses two genes in an individual and swaps their positions. For the floating point vector a discrete crossover was used and a simple mutation that randomly recreates a single vector element. More information on the genetic operators is available at the framework website.

A short tuning phase was conducted for both representations, which resulted in using the population size of 1000 individuals and a mutation rate of 0.7 (i.e. on average 7 out of 10 new individuals will get mutated). Rather than using a maximum number of generations or evaluations, the stopping criterion is set to a common time limit (of 8 hours) for the purpose of comparison with the IP solver.

C. Benchmark

Implementation of methods mentioned in this work were tested on problems from PSBLIB Library. This library consists of different sets of RCPSP with their so far known optimal or heuristic results. Problems in PSBLIB Library are divided in 4 groups based on number of jobs. The first group consists of problems with 30 activities, the second with 60 activities, the third one has problems with 90 activities and last group problems with 120 activities. More about this library and generation of problem instances can be found in [23].

D. Result and execution time comparison

Tests were conducted on 48 problem instances from group of 30 activities and on 48 problem instances from group of 90 activities. Results are given in tables I and II.

For IP formulations tables show execution times if solution was found in under 8 hours, otherwise the x is put in corresponding row. To make the comparison fair, for the genetic algorithm the stopping criterion was set at the same time limit or achieving the optimal solution (if it is known in advance). Furthermore, the GA was automatically restarted (under the time limit) if no improvement was achieved in the last 500 generations. The GA results in Tables I and II present the best found solution in the

---

[1]http://ecf.zemris.fer.hr/

TABLE I
PROBLEMS WITH 30 ACTIVITIES

Instance #	Opt. solution	Formulation 1		Formulation 2		GA - permutation		GA - floating point	
		Presolve phase (in s)	Solve phase (in s)	Presolve phase (in s)	Solve phase (in s)	Solution	Time (in s)	Solution	Time (in s)
1	43	0.07	0.47	0.14	0.31	43	0.41	43	1
2	51	0.08	0.37	0.34	0.63	51	0.48	51	1
3	57	0.10	0.13	0.21	0.31	57	0.44	57	1
4	45	0.03	0.04	0.16	0.24	45	0.41	45	1
5	82	0.16	176.50	0.62	144.45	82	0.44	82	1
6	67	0.19	2.81	0.62	5.01	67	0.45	67	1
7	44	0.08	0.33	0.21	0.33	44	0.42	44	1
8	53	0.11	0.12	0.27	0.37	53	0.43	53	1
9	63	0.15	2470.00	0.48	2155.83	63	5.65	63	20
10	58	0.21	3.99	0.18	3.22	58	2.90	**59**	**125**
11	62	0.26	0.64	0.21	0.90	62	0.45	62	1
12	53	0.25	0.26	0.18	0.32	53	0.42	53	1
13	77	x		x		77	5.43	77	1
14	50	0.19	12.22	0.32	32.43	50	0.69	50	1
15	47	0.24	0.27	0.36	0.48	47	0.45	47	1
16	51	0.32	0.33	0.39	0.45	51	0.44	51	1
17	47	0.07	2.17	0.35	1.92	47	0.44	47	1
18	56	0.08	0.24	0.53	0.65	56	0.45	56	1
19	49	0.04	0.10	0.41	0.57	49	0.41	49	1
20	41	0.03	0.03	0.17	0.21	41	0.41	41	1
21	69	0.14	99.00	0.50	172.20	69	0.38	69	1
22	42	0.06	0.65	0.28	0.54	42	0.75	42	1
23	65	0.12	0.85	0.73	1.01	65	1.14	65	2
24	58	0.15	0.16	0.34	0.60	58	0.45	58	1
25	72	0.16	2159.31	1542.38	0.66	72	1.33	72	3
26	58	0.15	0.27	0.43	1.59	58	0.41	58	1
27	64	0.21	0.50	0.49	0.88	64	0.40	64	1
28	62	0.32	0.33	0.47	0.75	62	0.46	62	1
29	78	x		x		78	3.47	78	18
30	53	0.25	3.46	0.43	15.09	53	1.73	53	13
31	63	0.26	0.34	0.54	0.71	63	0.43	63	1
32	54	0.36	0.37	0.46	0.61	54	0.43	54	1
33	60	0.12	0.66	0.37	0.73	60	0.43	60	1
34	44	0.05	0.15	0.25	0.46	44	0.40	44	1
35	57	0.08	0.16	0.29	0.49	57	0.45	57	1
36	46	0.04	0.05	0.22	0.30	46	0.42	46	1
37	81	0.17	62.41	0.95	229.26	81	0.40	81	1
38	63	0.13	0.70	0.42	1.59	63	0.44	63	1
39	54	0.10	0.17	0.35	0.56	54	0.44	54	1
40	51	0.14	0.15	0.30	0.48	51	0.42	51	1
41	88	0.23	1954.62	1.04	2385.91	88	1.41	88	9
42	66	0.24	6.13	0.54	6.24	66	1.48	66	8
43	43	0.12	0.17	0.30	0.42	43	1.10	43	1
44	57	0.29	0.30	0.46	0.59	57	0.44	57	1
45	92	0.26	274.07	1.44	3819.38	92	2.46	92	15
46	67	0.38	4.75	1.14	117.24	67	0.77	67	2
47	49	0.13	1.00	0.44	0.99	49	0.42	49	2
48	50	0.28	0.29	0.46	0.63	50	0.44	50	1

8 hours alloted to the algorithm, regardless of the number of restarts.

It can be seen that when using IP for problems with 90 activities the number of instances that are not solved in under 8 hours is bigger than in problems with 30 activities, which supports the fact that instances of larger size are more difficult to solve. If we compare formulation 1 and formulation 2, we can see that formulation 1 gives results in less time than formulation 2. However, it is worth noticing that there is no problem instance which is solved by one formulation and not by the other.

The tables also show results and execution times for both solution representations in GA. The GA was always able to reach the known optimum solution for 30 activity problems, whereas it did not reach the optimum in ten of

the problem instances with 90 activities.

If we compare the two solution representations we can see that the permutation representation obtains slightly better results. This is evident from the deviation of the obtained results from the optimum solutions: for the cases where the optimum was not obtained, the permutation encoding exhibits an average deviation of 7.16%, while for the floating point it averages 8.8%.

To better illustrate the average performance of the two GA representations, Table III shows the results of both variants on the test instances from Table II in which the GA did not manage to obtain the optimal solution. The data in the table indicates the standard statistical properties obtained with running both variants in 20 repetitions with the stopping criterion of 10^6 evaluations.

TABLE II
PROBLEMS WITH 90 ACTIVITIES

Instance#	Opt. solution	Formulation 1		Formulation 2		GA - permutation		GA - floating point	
		Presolve phase (in s)	Solve phase (in s)	Presolve phase (in s)	Solve phase (in s)	Solution	Time (in s)	Solution	Time (in s)
1	92	0.36	5.84	1.05	19.38	92	16	92	12
2	70	0.35	1.01	0.64	1.85	70	5	70	13
3	87	0.45	2.36	0.85	1.85	87	6	87	13
4	78	0.35	0.39	0.84	1.09	78	5	78	13
5	105		x		x	**110**	382	**110**	2008
6	71	0.34	2.58	1.22	6.69	71	5	71	12
7	90	0.64	1.14	1.56	2.83	90	5	90	13
8	70	0.95	0.99	1.24	1.90	70	5	70	13
9	110		x		x	**123**	10140	**124**	1812
10	95	1.41	8.27	1.95	54.73	95	22s	95	81
11	99	1.36	2.43	2.12	12.00	99	5	99	13
12	83	1.47	1.65	1.66	2.54	83	5	83	14
13	112		x		x	**121**	10710	**124**	1924
14	94	1.61	2.32	2.24	10.28	94	5	94	13
15	92	1.73	2.67	2.06	4.45	92	6	92	14
16	71	1.66	1.97	1.55	3.69	71	5	71	14
17	94	0.43	45.71	1.20	21.73	94	5	94	150
18	90	0.41	0.99	1.08	13.91	90	5	90	13
19	66	0.18	0.51	0.78	1.76	66	5	66	11
20	83	0.36	0.41	1.03	1.78	83	6	83	14
21	124		x		x	**127**	10560	**131**	2085
22	83	0.57	6.72	1.56	51.72	83	121	**84**	2019
23	116	0.85	1.75	2.33	4.43	116	6	116	14
24	92	1.25	1.35	1.68	3.13	92	6	92	14
25	114		x		x	**129**	8410	**130**	1878
26	108	1.54	2.93	2.51	4.33	108	17	108	16
27	81	0.80	2.96	1.75	4.25	81	12	81	14
28	80	1.29	1.44	1.74	2.59	80	11	80	13
29	141		x		x	**152**	19570s	**157**	1844
30	102	2.16	15.2	2.57	13.72	102	68	102	218
31	106	2.16	2.84	2.65	4.49	106	11	106	13
32	104	2.37	2.87	2.55	4.89	104	12	104	14
33	112	0.56	19.89	1.61	87.08	112	6	112	14
34	83	0.37	0.86	1.17	3.36	83	6	83	14
35	98	0.44	2.15	1.26	2.96	98	6	98	15
36	79	0.69	0.74	1.02	1.83	79	5	79	13
37	123		x		x	**124**	21770	**128**	1971
38	78	0.57	7.94	1.54	6.10	78	52	78	179
39	102	0.91	4.21	2.13	4.40	102	6	102	14
40	91	1.18	1.38	1.91	2.27	91	6	91	14
41	158		x		x	**172**	21130	**179**	1893
42	102	1.78	8.51	2.63	41.06	102	6	102	39
43	92	0.96	3.62	2.16	4.70	92	11	92	14
44	110	1.85	2.11	2.71	4.21	110	5	110	14
45	136		x		x	**152**	20540	**156**	1866
46	93*		x		x	**95**	384	**97**	2248
47	90*	1.66	2.93	2.32	5.31	90	5	90	2002
48	114*	2.58	2.95	3.08	4.71	114	6	114	2331

If we compare IP and GA we can see that for instances in which IP found a solution, GA also found the optimal solution with execution time that is approximately the same as the execution time of IP. For some instances, IP (especially for instances with 90 activities) found solution in less time, but for problems where IP did not found a solution, GA always managed to obtain a solution of an acceptable quality. One can conclude that when dealing with problems with less activities, it would be better to use an IP solver because then one can be sure the that optimal solution is achieved. But when dealing with problems with more activities, GA shows its strength because in a reasonable amount of time one can get a good enough solution for most applications.

VII. CONCLUSION

In this paper two different formulations of IP and GA using two different representations for RCPSP were compared. The results indicate that RCPSP belongs to NP hard problems for which the exact methods are applicable only on smaller size problems and even for that problems there is no guarantee that solution will be found in reasonable time. On the other hand GA shows better results than IP, but for some instances the solving process lasts long considering the fact that there was only 30 or 90 activities and real problems are usual larger than that. Also the downside of both approaches is the fact that these methods are generally not applicable in dynamic environments.

Due to these facts, as future research it is planned to try out other heuristic methods that could give better results,

TABLE III
GENETIC ALGORITHM PERFORMANCE COMPARISON

Instance#	Permutation GA			Floating point GA		
	Best (min)	Median	Worst (max)	Best (min)	Median	Worst (max)
5	110	111.5	113	112	115	118
9	123	124	126	127	130.5	135
13	121	123	125	125	128	131
21	127	130	131	132	135	138
22	83	83.5	85	84	85	86
25	129	131	133	132	136	140
29	152	156	158	159	161	164
37	124	127	129	128	133	137
41	172	175	178	179	183.5	189
45	152	156	158	156	160	164
46	95	96	97	97	100.5	103

especially hybrid optimization algorithms in continuous domain, which could be applied to the floating point encoding. Also it is planned to find some methods that will be applicable to scheduling in dynamic conditions.

REFERENCES

[1] J. Blazewicz, J. K. Lenstra, A. R. Kan, Scheduling subject to resource constraints: classification and complexity, Discrete Applied Mathematics 5 (1) (1983) 11–24.

[2] K. S. Hindi, H. Yang, K. Fleszar, An evolutionary algorithm for resource-constrained project scheduling, Evolutionary Computation, IEEE Transactions on 6 (5) (2002) 512–518.

[3] M. Abdolshah, A review of resource-constrained project scheduling problems (rcpsp) approaches and solutions, International Transaction Journal of Engineering, Management, Applied Sciences and Technologies.

[4] P. Brucker, A. Schoo, O. Thiele, A branch and bound algorithm for the resource constrained project scheduling problem, European Journal of Operation Research 17 (1998) 143–158.

[5] O. Icmeli, W. O. Rom, Solving the resource-constrained project scheduling problem with optimization subroutine library, Computers and Operation Research 23 (1996) 801–817.

[6] A. A. B. Pritsker, L. Watters, P. Wolfe, Multiproject scheduling with limited resources: A zero-one programming approach, Management Science 16 (1969) 93–108.

[7] J. Patterson, F. Talbot, R. Slowinski, J. Weglarz, Computational experience with a backtracking alogrithm for solving a general class of precedence and resource-constrained project scheduling problems, European Journal of Operational Research 49 (1990) 68–79.

[8] L. Schrage, Solving resource-constrained network problems by implicit enumeration - nonpreemptive case, Operations Research 18 (1970) 263–278.

[9] E. Demeulemeester, W. Herroelen, A branch-and-bound procedure for the multiple resource-constrained project scheduling problem, Management Science 38 (1992) 1083–1818.

[10] J. P. Stinson, E. W. Davis, B. M. Khumawala, Multiple resource-constrained scheduling using branch and bound, AIIE Transactions 10 (1978) 252–259.

[11] R. Klein, Bidirectional planning: improving priority rule-based heuristics for scheduling resource-constrained projects, European Journal of Operational Research 127 (3) (2000) 619–638.

[12] S. Kadam, N. Kadam, Solving resource-constrained project scheduling problem by genetic algorithms, Business and Information Management (ICBIM) (2014) 159–164.

[13] H. Ouerfelli, A. Dammak, The genetic algorithm with two point crossover to solve the resource-constrained project scheduling problems, Modeling, Simulation and Applied Optimization (ICMSAO) (2013) 1–4.

[14] V. Valls, F. Ballestn, Population-based approach to the resource-constrained project scheduling problem, Annals of Operations Research 131 (2004) 305–324.

[15] D. Merkle, M. Middendorf, H. Schmeck., Ant colony optimization for resource-constrained project scheduling, IEEE Transactions on Evolutionary Computation 6 (2002) 333346.

[16] S. Kadam, S. Mane, Genetic-local search algorithm approach for resource constrained project scheduling problem, Computing Communication Control and Automation (ICCUBEA) (2015) 841–846.

[17] C. Artigues, S. Demassey, E. Neron, Resource-constrained project scheduling: models, algorithms, extensions and applications, John Wiley & Sons, 2013.

[18] K. Sastry, D. Goldberg, G. Kendall, Search methodologies, Springer, 2014, Ch. Genetic algorithms.

[19] M. Čupić, Prirodom inspirirani optimizacijski algoritmi. Meta-heuristike., Fakultet elektrotehnike i računarstva, Zagreb, 2013.

[20] I. Gurobi Optimization, Gurobi Optimizer Reference Manual., http://www.gurobi.com/documentation/.

[21] S. Hartmann, A competitive genetic algorithm for resource-constrained project scheduling, Manuskripte aus den Instituten fr Betriebswirtschaftslehre der Universitt Kiel 451.

[22] R. Kolisch, S. Hartmann, Handbook on recent advances in project scheduling, Christian-Albrechts-Universitt, Kiel, 1999.

[23] R. Kolisch, C. Schwindt, A. Sprecher, Benchmark instances for project scheduling problems, in: Project Scheduling, Springer, 1999, pp. 197–212.

Balancing academic curricula by using a mutation-only genetic algorithm

Kadri Sylejmani, Arbnor Halili, Arbnor Rexhepi

University of Prishtina/Department of Computer Engineering, Prishtina, Kosovo
(kadri.sylejmani, arbnor.halili, arbnor.rexhepi)@uni-pr.edu

Abstract - In universities, the academic programs are organized in a number of periods, usually in six or ten semesters, for a bachelor or a master degree, respectively. It usually happens that a given semester is much loaded with courses than the others. This makes it hard for the students to comprehend and deal with a high volume of learning material per certain semesters. This problem is difficult, because some courses have prerequisites (e.g. Math2 should be taught after Math1), and this means that course correlation mast be taken into account. Therefore, in this paper, we present an intelligent method that is based on genetic algorithms to optimize the academic curricula of a given program, by trying to dispatch the courses over the available semesters, so that the load of individual semesters, in terms of course credits, is balanced as much as possible. The proposed genetic algorithm explores the search space by means of two mutation operators, which swap or shift courses between the semesters. The algorithm performance is fine-tuned and evaluated by using three state of art instances from the literature. The results show that the proposed algorithm is comparable with the state of the art solutions for the problem at hand.

I. INTRODUCTION

A study curricula (program) taught at some university is commonly structured based on some professional academic domain (e.g., computer science, economics, medicine, etc.). Such study curricula consist of several courses, which are usually dispatched, as even as possible, into several study periods (semesters). Since, some courses are more difficult to comprehend than the others, each course is assigned a number of credits based on learning hours needed to grasp the required material. In addition, some courses might not be taught before some other courses, due to perquisites a course might have (e.g. in an economic study program the microeconomics course is taught before the macroeconomic course). An intuitive intention in such situations would be to have the courses distributed into semesters so that the load of the students (in terms of learning hours) over different semesters is as balanced as possible. This problem in the literature is known as Balanced Academic Curriculum Problem (BACP) [1]. Another extension is the Generalized BACP (GBACP) [2] problem, where two new concepts are introduced, explicitly allowing multiple study programs (curricula) and consideration of professor's preferences about the terms (winter or spring) they want to teach the courses. In this paper, the earlier version of BACP problem is considered.

The remaining of this paper is organized as in the following. Section II presents a literature review of approaches solving the BACP problem; Section III presents the mathematical definition of the problem; Section IV presents the proposed approach; Section V shows the computational experiments; and Section VI concludes the paper.

II. LITERATURE REVIEW

In the literature, there is a plethora of approaches that tackle and solve the BACP problem by means of deterministic and non-deterministic algorithms. Castro & Manzano [1] solve the BACP problem by using constraint programming techniques, where parameter values and variables are ordered based on some specific heuristics. Lambert et al. [3] present a hybrid approach consisting of genetic algorithms and constraint propagation techniques, by employing some elementary functions that can be parameterized and used by different search strategies. Castro et al. [4] present a genetic local search approach that uses an operator that is based on the simulated annealing algorithm. Further, in [4], with the aim of making the BACP problem more realistic, new diverse evaluation functions are introduced by considering different mathematical models. The Generalized BACP problem is tackled by Chiarandini et al. [2] through applying several local search heuristics, where also a new alternative objective function is introduced. Rosas-Téllez et al. [5] present a tabu search metaheuristic for the BACP problem, by utilizing short-term memories to escape from local optima. Rubio et al. [6] solve the BACP problem by means of Ant Colony Optimization approach, in particular by employing the version of Best-Worst Ant System.

III. MATHEMATICAL MODELLING

The mathematical formulation for balancing the academic curricula problem is done based on [1], which is an Integer Linear Programming (ILP) model that uses a range of parameters and three decision variables, as described below:

Parameters:

m – Number of courses

n – Number of academic periods

α_i – Number of credits of course i, $\forall i=1, ..., m$

β – Minimum credit load allowed per period

γ – Maximum credit load allowed per period

δ – Minimum number of courses per period

ε – Maximum number of courses per period

Decision variables:

x_{ij} -equals 1, if course i is set in period j,
otherwise it is 0, $\forall i=1, ..., m, \forall j=1, ..., n$

c_j – Credit load of period j, $\forall j=1, ..., n$

c – Maximum credit load of all periods

Objective function:

$$Min\ c = Max\{c_j\},\ where\ c_j = \sum_{i=1}^{n} \alpha_i x_{ij}, \quad (1)$$
$$\forall j=1, ..., n$$

Constraints:

$$\sum_{j=1}^{n} x_{ij} = 1, \forall i=1, ..., m \quad (2)$$

$$x_{bj} \leq \sum_{r=1}^{j-1} x_{ar} = 1, \ \forall j=2, ..., n \quad (3)$$

$$c_j \geq \beta, \ \forall j=1, ..., n \quad (4)$$

$$c_j \leq \gamma, \ \forall j=1, ..., n \quad (5)$$

$$\sum_{i=1}^{m} x_{ij} \geq \delta, \ \forall j=1, ..., n \quad (6)$$

$$\sum_{i=1}^{m} x_{ij} \leq \varepsilon, \ \forall j=1, ..., n \quad (7)$$

In the above mathematical formulation, Equation (1) represents the objective of optimizing the problem, which is minimizing the credit load of the most loaded period, in terms of the academic credits assigned to it. Constraint (2) makes sure that all courses are assigned to some period, whereas Constraint (3) ensures that the perquisites of individual courses are complied. Constraints (4) and (5) together guarantee that the minimum and the maximum credit load is respected, whilst constraints (6) and (7) confirm that the rule for the minimum and the maximum number of courses per period is enforced.

IV. SOLUTION APPROACH

In order to obtain a real time response when balancing the academic curricula, it would be required to design an algorithm that computes the results in period of less than one second, while ensuring that the quality of the solutions stays at a high level. Genetic Algorithms (GAs) [7] have been successfully applied in different domains, to optimize problems like flow shop scheduling [8], determining multiple routes [9], aircraft sequencing [10], etc. Hence, in this paper, we solve the problem of balancing the academic curricular by using a genetic algorithm that generates new individuals only by utilizing mutation operators. At each iteration, GAs make the three basic steps, namely selection of the parents for constructing the new population, mutating the parents to bread new children, and updating the population for the next generation.

A. Representation

The representation of a candidate solution (i.e. a member from the population P) is done by using a one dimensional array, where each member of the array is a list. The length of the array equals the number of periods (n), whereas the length of each list (that represents a single member of the array), is determined by the number of assigned courses at a certain period. In the following, we present a sample member representation of a scenario of an academic curriculum with 10 courses and 4 periods, where it can be seen that that the first and the second period contain three courses each, whereas the other two periods contain only two courses each.

$I = \{\{1, 3, 4\}, \{2, 6, 10\}, \{7, 5\}, \{8, 9\}\}$

B. Mutation operators

The mutation mechanism of the algorithm consists of two operators, namely *swap* and *shift*, where the earlier swaps two courses belonging to distinctive periods, while the later shifts a course from some period to some other period.

As presented in Algorithm 1, in order to apply a number of swaps between different courses of a given individual, the *swap* operator iterates through a loop for a number of iterations (as specified by *sw* parameter). During the course of a single iteration, initially, two distinct periods (i.e. p and q) are selected randomly, and then, their respective feasible courses (i.e. courses that satisfy the perquisite constraint) are enlisted (i.e. CL_p and CL_q). Afterwards, the first pair of courses, which is obtained by combing courses from CL_p and CL_q lists and that produces a feasible mutation (i.e. courses that meet other hard constraints), is used to mutate the current individual I_m by swapping the places of the selected courses. In case, no feasible swap can be achieved, the operator continues with next iteration.

Algorithm 1 Swap operator procedure

Input: individual I, swap mutate intensity sw
Output: mutated individual I_m

1:   $I_m = I$
2:   $i = 1$
3:   **repeat**
4:      p = Random period from I_m
5:      q = Random period from I_m different from p
6:      CL_p = Get courses that can be swept from p
7:      CL_q = Get courses that can be swept from q
8:      **repeat**
9:         I_m = Mutate I_m by randomly swapping a course from list CL_p with a course from list CL_q
10:      **until** a feasible swap cannot be found
11:      $i = i+1$
12: **until** $i > sw$

Algorithm 2 Shift operator procedure

Input: individual I, shift mutate intensity sh
Output: mutated individual I_m

1:   $I_m = I$
2:   $i = 1$
3:   **repeat**
4:      **repeat**
5:         p = Random period from I_m
6:         **repeat**
7:            c = Get a random course from p
8:         **until** no course provides a feasible shift
9:      **until** no period provides a feasible shift
10:      PL_q = Get periods that can accept course c
11:      **if** PL_q is not empty **then**
12:         q = Random period from PL_q
13:         I_m = Mutate I_m by shifting course c from period p to period q
14:      $i = i+1$
15: **until** $i > sh$

The shift operator (see Algorithm 2) is also executed for several iterations, as specified by *sh* parameter. During the progress of a given iteration, initially, a random period is selected, and then, one of its courses (i.e. course *c*) is picked as a course that will be shifted in some other period, while leaving the current period in a feasible state. Next, the periods that can accept course *c* (i.e. PL_q list) are initially outlined, and then, one of them is selected randomly as the period that will accept the envisioned course. In case there are no periods that can feasibly accept the selected course, the operator proceeds with the next iteration.

C. Initialization

In a typical instance of the academic curricula balancing problem, courses are (or can be) ordered based on the number of prerequisites they have, where courses with no prerequisites are placed first then courses with one prerequisite and so on.

At the beginning (see Algorithm 3), in order to start with a balanced curricula, in terms of the number of courses, each period is set (estimated) to accept equal number of courses, unless there is an odd number of courses, which requires that some of the periods have an additional course. Next, in the first phase of the algorithm (lines 3 to 17), by following the ascending order of the number of perquisites, the courses are dispatched into the

---

Algorithm 3 Initial solution procedure

Input: academic curricula $C(m, n, \alpha_i, \beta, \gamma, \delta, \varepsilon)$, initial solution randomness intensity *is*
Output: random individual *I*
1:   Initialize periods of individual *I* to an empty state
2:   *IL* = Get estimated number of courses for each period
3:   *sp* = 1
4:   *c* = 1
5:   **repeat**
6:      *RC* = Get required courses of course *c*
7:      *p* = *sp*
8:      **repeat**
9:        **if** number of courses in period *p* less then *IL(c)* **then**
10:           **if** course *c* satisfies the constraints of min/max credit load **and** none of the courses in *RC* is placed in period *p* **then**
11:             Add course *c* to period *p* of individual *I*
12:        **else**
13:          *sp* = *sp* + 1
14:        *p* = *p* + 1
15:      **until** *p* > *n* **or** *c* cannot be settled to period *p*
16:      *c* = *c*+1
17:   **until** *c* > *m*
18:   *i* = 1
19:   **repeat**
20:      *p* = Random period
21:      *q* = Random period that is different from *p*
22:      *I* = Mutate *I* by randomly swapping a course from period *p* to period *q*
23:      *i* = *i* + 1
24:   **until** *i* > *is*

---

available periods, subject to the constraints of min/max credit load and course perquisites. If during this phase, a given course cannot be settled at some particular period, due to the hard constraints, it gets shifted into one of the next available periods. As a result, the end of this phase will produce an individual that satisfies all the hard constraints foreseen by the problem modelling.

Further, in the second phase (lines 18 to 24), with the aim of incorporating some randomness on the initial solution, subject to the parameter of *randomness intensity*, a couple of feasible swaps are made between any two selected courses belonging to distinct periods.

D. Evaluation and selection

The evaluation of a given candidate solution is done by using Equation (1), which returns the credit sum of the courses within the period that has the maximum load. In the case of the running scenario with 10 courses and 4 periods, for the specific example presented in the previous subsection, the load of the individual periods, in terms of course credits, is 4, 11, 5 and 8, respectively, hence the evaluation of the solution is 11.

The process of the selection of the parents that will take part in breading the next population is completed by using the tournament selection algorithm. This algorithm initially selects a number of individuals (as specified by the tournament size parameter) at random from the population, and then, from those that are selected, it finds the best fit individual.

E. Genetic algorithm

The implementation of the proposed algorithm has a generative nature, meaning that its population will be totally renewed at each generation. As shown in Algorithm 4, the envisioned approach has 7 parameters, which can be used for fine tuning the performance for different problem complexities and sizes. Besides the default genetic algorithm parameters, such as population size (*ps*), maximum generations (*mg*) and tournament size (*ts*), the implementation at hand uses three so called "intensity" parameters, namely initial solution randomness intensity (*is*), swap mutate intensity (*sw*) and shift mutate intensity (*sh*), for specifying the number of times a certain operator (i.e. swap or shift) will be applied when called upon. In addition, the algorithm uses a special parameter (i.e. operator alternation frequency – *af*) to change the mutation operator from swapping to shifting and vice versa every *af* number of generations.

At the very start of the algorithm, a list of two possible operators to be used is defined, which is made of *swap* and *shift* operators. Then, a population *P* of *ps* individuals is created by using the procedure for creating the initial solution explained above. Next, in the repetitive phase of the algorithm, at each iteration, the following steps are undertaken:

- Evaluation of all individuals. In case a new best individuals is found, it is saved as the best solution found so far,

- Formation of the new population by selecting the parents based on tournament selection and by

1191

Algorithm 4 Genetic algorithm

Input: academic curricula $C(m, n, \alpha_i, \beta, \gamma, \delta, \varepsilon)$, population size ps, maximum generations mg, tournament size ts, initial solution randomness intensity is, swap mutate intensity sw, shift mutate intensity sh, operator alternation frequency af

Output: S_b

```
 1:   Operators = {Swap, Shift}
 2:   P = {}
 3:   i = 1
 4:   repeat
 5:       P = P ∪ Random Individual (C, is)
 6:       i = i+1
 7:   until i > ps
 8:   Sb = Best solution from P
 9:   k = 1
10:   j = 1
11:   repeat
12:       Sc = Best solution from P
13:       if Sc better Sb then
14:           Sb = Sc
15:       Q = {}
16:       i = 1
17:       repeat
18:           I = Select Individual (P, ts)
19:           Mutate = Operators (k)
20:           Q = Q ∪ Apply Mutate (I, sw, sh)
21:           i = i+1
22:       until i > ps
23:       P = Q
24:       k = (i / af) % 2
25:       j = j+1
26:   until j > mg
```

mutating them through the operator (i.e. swap or shift) used in the running iteration

- Before the next generation commences, based on the *af* parameter, one of the two operators is picked for acting as a mutation operator in the next stage.

The algorithm terminates when the maximum number of foreseen generations is achieved.

V. COMPUTATIONAL EXPERIMENTS

In this section, we initially present the test set used for experimentation, and then, we show the experimental results for fine tuning algorithm parameters, as well as presentation of the comparative results of the proposed approach against the known optimal solutions.

TABLE 1 DETAILS OF TEST INSTANCES

Instance	Number of periods	Number of courses	Total credits	Maximal number of prerequisites per course
BACP8	8	46	133	4
BACP10	10	42	134	3
BACP12	12	66	204	5

TABLE 2 PARAMETER FINE TUNING

Parameter		Fitness		
Abbreviation	Value	Best	Average	Worst
ps	50	17	17.5	17.7
	100	17.3	17.5	17.7
	200	16.7	17.3	17.3
	500	16.6	**17.1**	17.3
mg	20	16.7	17.2	17.3
	50	16.7	17	17.3
	100	17	17.1	17.3
	200	16.7	**16.9**	17
ts	10	16.3	16.9	17
	15	16.7	16.9	17.3
	20	16.3	**16.5**	17
	25	16.3	**16.5**	16.7
is	10	16.3	16.5	17
	20	16.3	**16.4**	17
	30	16.3	**16.4**	16.7
	40	16.3	**16.4**	16.7
sw	1	16.3	16.5	16.7
	2	16.3	**16.4**	16.7
	3	16.3	**16.4**	16.7
	4	16.3	**16.4**	16.7
sh	**5**	16.3	**16.4**	16.7
	10	16.3	**16.4**	16.7
	15	16.3	16.5	16.7
	20	16.3	16.5	16.7
af	5	16.3	16.5	16.7
	10	16.3	**16.3**	16.3
	15	16.3	**16.3**	16.3
	20	16.3	**16.3**	16.3

The algorithm is developed by using C# programming language that is part of Microsoft Visual Studio 2012 development environment. The experiments are done by means of a machine with CPU of type Intel Core i5 2.2GHz with 16GB of RAM memory, while running on a Windows 10 Operating System. Since, the proposed approach is a non-deterministic one, every experiment constitutes of 10 independent executions of the algorithm.

A. Test set

The evaluation of the proposed approach is done by using a widely used test set from the literature that consists of three instances, which are taken from the CSPlib online library [11]. This basic details of individual instances are given in Table 1, where it can be noticed that the range of periods varies from 8 to 12, whereas the number of courses and credits goes from 46 to 66 and 133 to 204, respectively. In terms of course interrelationship, the courses that are the most constrained in Instances 8, 10 and 12 have 4, 3 and 5 prerequisites, respectively.

B. Parameter settings

Based on some preliminary algorithm execution, for each of the seven algorithm parameters (i.e. *ps, mg, ts, is, sw, sh,*and *ah*), four best performing values are chosen (as presented in second the column of Table 2). Then, a systematic computation, by using the three above described test instances, is performed, with the aim of finding out which value of the range, for a given parameter, yields to better algorithm results.

In Table 2, we present the computation outcome for different parameters, where the results are averaged over the complete test set. The best performing values for specific parameters are emphasized in bold. In overall, the results show that the algorithm is more sensitive for the general parameters (i.e. *ps*, *mg* and *ts*) of the genetic algorithms, than for the specific parameters (i.e. *is*, *sw*, *sh*, and *af*) introduced here for this particular problem. As a result, in the forthcoming experiments, the best parameter value sets are utilized.

With the aim of illustrating the level of sensitivity of algorithm performance for different runs, in Fig. 1, we show the results of Instance BACP8, for 10 different runs, by using the best parameter values. As it can be seen, in each and every run of the algorithm, the optimal solution is found, while the running time at most goes to 500ms. In all of the cases of the algorithm execution, a solution with fitness 18 (which is one more than the fitness of the optimal solution), is found within 100ms of computation time.

C. Comperative resutls

In this section, we compare the results of the proposed approach, denoted as the Mutation only Genetic Algorithm (MGA), against the known approaches from the literature, namely Variable and Value Ordering (VVO) [1], Hybridization of Genetic Algorithms and Constraint Propagation (HGACP) [3], Tabu Search (TS) [5] and Ant Colony Optimization (ACO) [6]. For the three data sets presented above, the optimal solutions, obtained from some of the state of the art solutions described above, read as in the following: BACP8=17, BACP10=14 and BACP12=17.

Table 3 presents a comparison, in terms of the quality of solutions, of the results of the proposed approach against the state of the art approaches in the given test set. The results are averaged over 10 different algorithm executions. The computational results show that the performance of the proposed approach is equal (for Instance BACAP8) or

better than the performance of VVO approach, which does not find the optimal solutions for instances BACAP10 and BACAP12. While, the proposed approach produces the same results as HGACP, TS and ACO approaches for instances BACAP8 and BACAP10, it can be noticed that, for instance BACAP12, the results are for 6% worse than the optimal solution, which is obtained by ACO approach. In overall, it can be stated that the proposed approach can be used to balance complex academic curricula (such as BACAP12 instance with 12 periods), in average, in less than one second. Furthermore, the results outline that for more complex problems (i.e. BACP12 instance), the algorithm can provide good quality solutions, although not always the optimal ones.

In addition, Table 4 presents the results of the computation time of all state of the art approaches, along with computation time of the proposed approach. Nonetheless, the results should be taken more as informative rather than comparative, since all of the envisioned approaches utilise distinct computation machines when doing the experiments, hence no direct comparison can be made between specific approaches in this respect. However, the last column of Table 4 shows that the proposed approach is able to solve any of the described instances within a period of less than one second.

VI. CONLUSIONS

In this paper, we presented an approach for balancing the academic curricula based on genetic algorithms, where two mutation operators are employed alternatively for exploring the search space. The swap operator exchanges two courses between two distinct periods, whereas the shift operator shifts a single course from one period to another period.

The computation results showed that the algorithm is not sensitive to the set of values of the new parameters introduced for this particular problem. When the parameters are fine-tuned, the algorithm is always able to find the optimal solution for instances BACP8 and BACP10, whereas for instance BACP12, the algorithm

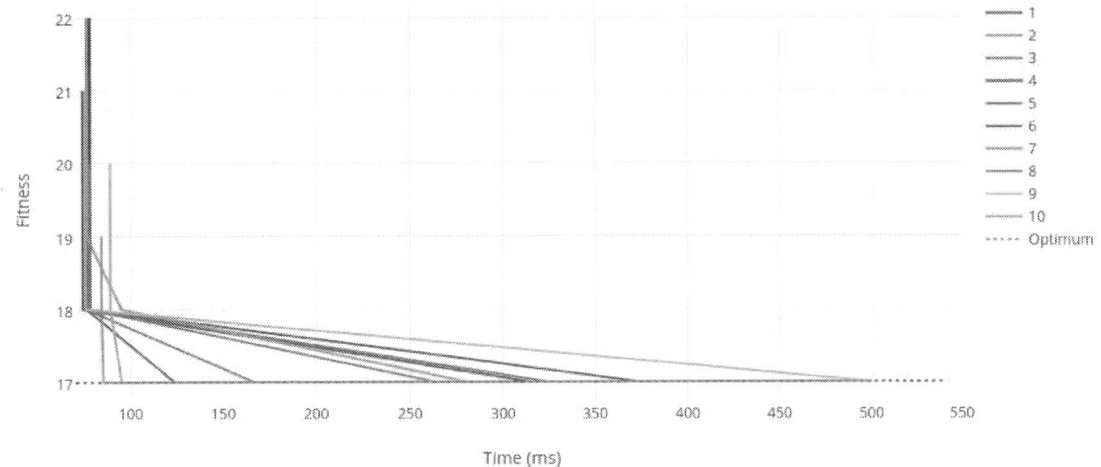

Figure 1 Solution improvement over time for different algorithm executions in the case of BACP8

TABLE 3 COMPARISION OF QUALITY OF SOLUTIONS AGAINST THE OPTIMAL SOLUTIONS (EXPRESSED AS MAXIMUM NUMBER OF CREDITS PER PERIOD)

Instance	Optimal solution	Gap from optimal solution (%)				
		VVO	HGA-CP	TS	ACO	MGA
BACP8	17	0	0	0	0	0
BACP10	14	NA	0	0	0	0
BACP12	17	NA	6	0	6	6

TABLE 4 COMPARISION OF COMPUTATION TIME (SECONDS)

Instance	VVO	HGACP	TS	ACO	MGA
BACP8	1459.73	15.05	5	1.25	0.25
BACP10	NA	34.84	4	1.92	0.09
BACP12	NA	35.20	5	6.37	0.88

finds solutions that are for 6% worse than the optimal solution.

Since for more complex problems, such as instance BACP12, the algorithm seems to get caught in local optima, as part of future work, it would be interesting to investigate other variants of the swap and shift operators, such as for example swapping one course with a set of other courses, or shifting a set of courses from one period to another one. In addition, it could be worthwhile developing new mutation operators, such as for example a rotation operator, which could for instance, select a number of courses from distinct periods and rotate them based on the their actual placements in the periods.

REFERENCES

[1] C. Castro and S. Manzano, "Variable and value ordering when solving balanced academic curriculum problems," in *Proceedings of the ERCIM Working Group on Constraints*, Prague, Czech Republic, 2001.

[2] M. Chiarandini, L. Di Gaspero, S. Gualandi and A. Schaerf, "The balanced academic curriculum problem revisited," *Journal of Heuristics,* vol. 18, no. 1, pp. 119-148, 2012.

[3] T. Lambert, C. Castro, E. Monfroy and F. Saubion, "Solving the balanced academic curriculum problem with an hybridization of genetic algorithm and constraint propagation," in *International Conference on Artificial Intelligence and Soft Computing*, Zakopane, 2006.

[4] C. Castro, B. Crawford and E. Monfroy, "A genetic local search algorithm for the multiple optimisation of the balanced academic curriculum problem," in *Cutting-Edge Research Topics on Multiple Criteria Decision Making*, Springer Berlin Heidelberg, 2009, pp. 824-832.

[5] L. V. Rosas-Téllez, J. L. Martínez-Flores and V. Zanella-Palacios, "Solution to the Balanced Academic Curriculum Problem Using Tabu Search," *Computer Technology and Application,* vol. 3, no. 9, pp. 630-635, 2012.

[6] J.-M. Rubio, W. Palma, N. Rodriguez, R. Soto, B. Crawford, F. Paredes and G. Cabrera, "Solving the balanced academic curriculum problem using the ACO metaheuristic," *Mathematical Problems in Engineering ,* 2013.

[7] J. Holland, Adaptation in Natural and Artificial Systems, University of Michigan Press, 1975.

[8] T. Murata, H. Ishibuchi and H. Tanaka, "Multi-objective genetic algorithm and its applications to flowshop scheduling," *Computers \& Industrial Engineering,* vol. 30, no. 4, pp. 957--968, 1996.

[9] J. Inagaki, M. Haseyama and H. Kitajima, "A genetic algorithm for determining multiple routes and its applications," in *Circuits and Systems, 1999. ISCAS'99. Proceedings of the 1999 IEEE International Symposium on*, 1999.

[10] K. Sylejmani, E. Bytyçi and A. Dika, "Solving Aircraft Sequencing Problem by using Genetic Algorithms," *International Journal of Intelligent Decision Technologies,* in press.

[11] B. Hnich, Z. Kiziltan and T. Walsh, "CSPLib Problem 030: Balanced Academic Curriculum Problem (BACP)," {CSPLib}: A problem library for constraints, [Online]. Available: http://www.csplib.org/Problems/prob030.

A Hybrid Method for Prediction of Protein Secondary Structure Based on Multiple Artificial Neural Networks

Haris Hasic, Emir Buza, Amila Akagic
Faculty of Electrical Engineering
Department for Computer Science and Informatics
University of Sarajevo, Bosnia and Herzegovina
Email: {haris.hasic, ebuza, aakagic}@etf.unsa.ba

Abstract—The prediction of protein secondary structure is the method of finding the way in which an amino acid sequence causes the protein structure to fold and bend into *alpha helices*, *beta strands* and other shapes. Until today, the problem of finding protein secondary structure is not fully resolved. Classification or clusterization based methods have an accuracy rate of circa 80 percent and they mainly work on a reduced set of shapes and folds. It is very difficult to predict how a local sequence of amino acids is going to behave and in which way it is going to affect the future of protein structure. Based upon the predicted secondary structure of the protein, the tertiary and quaternary predictions show the real nature and function of the protein as a whole. In this paper, we address the problem of the secondary structure prediction of protein and propose a new hybrid method based on the usage of multiple neural networks with the use of a consensus function and compare our approach with other efficient methods.

Keywords—**Bioinformatics, Protein Secondary Structure Prediction, Hybrid Method, Neural Networks, Machine Learning**

I. INTRODUCTION

Bioinformatics is an important interdisciplinary field in which information technologies are used to successfully solve existing biological problems in the world today. Usually, the most accurate ways to solve these kinds of problems are through experimental methods. Some of those approaches are described in [1]–[3]. The main obstacle, which undermines the significant progress of experimental methods, is the high cost of the entire process. One of the reasons for the high cost is the high amount of pure protein required to perform experiments on, not to mention the required amount of computational power. On the other hand, information technologies are rapidly improving and spreading over various fields, which opens up the space for new approaches in solving biology problems. These kinds of approaches are typically referred to as *ab initio* approaches since they usually focus on solving problems "from scratch" rather than using existing structures obtained through experimental methods. According to [4], some of those methods can achieve up to 75% to 80% accurate results, while the theoretically highest possible accuracy lies at 90%.

The problem that has proven difficult to solve efficiently is the prediction of protein structures. Predicting the protein structure and function does not solely include predicting

classes and structures, but also predicting the environmental and other potential influences, protein-protein interactions etc. The problem is rather complicated to be precisely determined only with machine learning and other mathematical-based methods. However, a combination of different algorithms merged together and combined with some biology field knowledge might just give good results. This fusion of different experimental and mathematical approaches represents the domain of hybrid algorithms and it is proven to be an efficient way to improve existing algorithms.

In this paper, we focus on the improvement of the machine learning algorithms which are typically referred to as *in silico* methods, especially neural network approaches. The accuracy of neural network based methods is around 60% [5] and the results largely depend on the protein that is being analyzed. We try to increase this accuracy and also overcome the large oscillations that can occur if the input datasets which contain a large number of differently structured proteins. Our focus is the identification and exact classification of the two most common secondary structural classes, *alpha helices* and *beta sheets* (Fig.1). Other structures that can form during the process are aggregated in the *coil* class. We present a new hybrid method based on multiple neural networks combined together through a census function. The networks that compose the ensemble are trained with different parameters which are determined empirically, through analysis of benchmarking results. The local results obtained from the neural networks are analyzed and, through a majority voting process, combined into a global ensemble result. This ensures the higher consistency and accuracy of the method in comparison to the single network approach, regardless of the diversity of the input dataset. The accuracy of our method is 65% for all the datasets used.

This paper is organized as follows: Related work regarding different machine learning approaches is briefly reviewed in section II. In section III a new, neural network based, hybrid method for protein secondary structure prediction is proposed. The implementation of the proposed method and discussion of the achieved results are stated in section IV. We conclude the paper in section V with appropriate remarks.

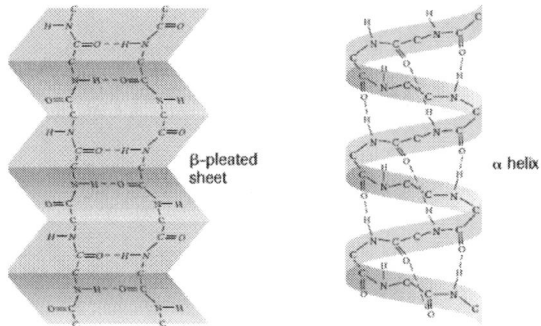

Fig. 1: Two of the most common protein secondary structure elements: *alpha helices* and *beta sheets*.

II. RELATED WORK

In the early development stages of secondary structure prediction methods, amino acids had been mostly observed statistically, one residue at a time. Those methods were also constrained by the small amount of predetermined protein structures available at the time. As more secondary structures were acquired through experimental methods, the *in silico* methods also advanced in consistency and accuracy as they had much more example data to work with. One of the first fairly consistent methods for secondary structure prediction was the *Chou-Fasman* method [6] which combined different statistical and heuristic rules. The main problem with this method was the mentioned inspection of isolated amino acids in the chain which couldn't exactly reflect the real state of the protein as a whole. This issue was resolved in the *GOR* [7] [8] method where the surrounding of the amino acids was also included in the secondary structure prediction.

After the initial simple approaches, the algorithms began to improve drastically as higher degree interactions between elements were observed. More advanced statistical approaches were implemented and elements of machine learning were integrated in the methods. Advanced methods make use of nearest neighbor approaches and fuzzy logic [9]–[11], hidden Markov models [12] [13], support vector machines [14] [15], neural networks etc. Today, approaches that only predict protein structure from a single organism are getting more popular since they avoid the need for generalization and therefore offer higher accuracy in prediction. One example is the specified structure protein interaction for the yeast species *Saccharomyces Cerevisiae* is described in [16].

One of the more interesting approaches is the usage of neural networks. The network is trained with a primary structure and the corresponding secondary structure later predicts secondary structural classes for protein with unknown secondary structure. There are many different ways to tackle the defined problem using different types of neural networks, sometimes in a combination with other algorithms as described in [17]–[20]. The deep learning algorithms are also gaining popularity as described in [21]–[24]. They emphasize the learning process of the networks and achieve more accurate results.

III. NEW HYBRID METHOD

The problem of the secondary structure prediction is one of the most challenging problems in bioinformatics. The neural networks "style" of problem solving is a one way of solving this problem. We formalize the main steps of the modeling process for our method as:

(A) Window length selection
(B) Binarization of inputs and outputs
(C) Construction of the neural network as a classifier
(D) Ensemble construction

The first two steps offer a detailed description of the data preparation process with focus on the inclusion of the immediate surroundings of each residue. After the input and output data format is established, available datasets are pre-processed and divided into appropriate training, validation and test sets. The main contribution of our method is in the next two steps, where we introduce network selection and ensemble construction. Based on achieved results for different parameters, the best performing networks are chosen and a diversified ensemble is constructed. The voting process which unifies single neural network results is implemented. In the end, a series of tests containing data from different datasets than the ones used in the training process are carried out to measure performance of the algorithm. The next subsections offer detailed description of all individual steps of the process.

A. Window Length Selection

Looking at a single amino acids individually does not get good results. The interaction between residues in the chain needs to be preserved in some way. For example, if the window size is 11 and an amino acid at the nth position in the chain is in focus, elements at positions $\{n-5, ..., n, ..., n+5\}$ also need to be taken into consideration as depicted in Fig. 2. In [25], it is shown that there are many factors in successfully determining the optimal window size, but it largely depends on the protein in focus. Many protein secondary structures depend on factors such as hydrophobicity, motifs, b-factors etc. and that makes it difficult to find the general optimal sliding window size for all the protein in existence. For example, the transmembrane proteins have the average of one transmembrane *alpha helix* spanning through the membrane so the optimal size should be around 20 residues. The only way to determine the optimal window size is empirical and thus multiple neural networks with different windows must be constructed, trained and evaluated.

Fig. 2: Sliding window size.

B. Binarization of Inputs and Outputs

The string representation of the amino acid chain is simple for people to understand, but difficult for machines to process. Because of that, we need a fitting transformation of the input in order to model a neural network and gain efficiency in terms of processing speed. Generally speaking, one of the easier types of data for machines to process is the binary format. If the amino acid chain is composed from a total of 20 different amino acid types, the matching binary form will contain 20 positions, where only one position is set to 1 and the other to 0 to represent the type of amino acid as annotated in Table I.

TABLE I: Binary codes for each of the 20 amino acids.

Ala	[10000 .. 0 .. 0 .. 0]	Met	[0 .. 0 .. 10000 .. 0]
Arg	[01000 .. 0 .. 0 .. 0]	Phe	[0 .. 0 .. 01000 .. 0]
Asn	[00100 .. 0 .. 0 .. 0]	Pro	[0 .. 0 .. 00100 .. 0]
Asp	[00010 .. 0 .. 0 .. 0]	Ser	[0 .. 0 .. 00010 .. 0]
Cys	[00001 .. 0 .. 0 .. 0]	Thr	[0 .. 0 .. 00001 .. 0]
Gln	[0 .. 10000 .. 0 .. 0]	Trp	[0 .. 0 .. 0 .. 10000]
Glu	[0 .. 01000 .. 0 .. 0]	Tyr	[0 .. 0 .. 0 .. 01000]
Gly	[0 .. 00100 .. 0 .. 0]	Val	[0 .. 0 .. 0 .. 00100]
His	[0 .. 00010 .. 0 .. 0]	Asx	[0 .. 0 .. 0 .. 00010]
Ile	[0 .. 00001 .. 0 .. 0]	Glx	[0 .. 0 .. 0 .. 00001]

This indicates that the input needs to be at least a 20xm matrix where m is the number of amino acids in the chain. Because of the sliding window size that needs to be incorporated into the input, the matrix also needs to have n residues on each side of the current amino acids. Therefore, the final input matrix needs to be a (20*ws)xm matrix where ws represents the sliding window size. For example, if we use a window with a size of 17, the final input matrix will be 340x($n-16$). The 16 elements that are subtracted are the first and last 8 elements of the amino acid sequence, that are ignored because the window size can be applied only at the 9th position. That leads to the conclusion that some of the elements of the primary structure will not be included in the prediction of the secondary structure which is one of the disadvantages of this surrounding-inclusive approach. The input data has grown in dimensionality since it went from a simple string to a fairly big matrix but in this format it is much easier for the computers to process and also an excellent fit for the input of a neural network.

The same principle of binarization can also be applied for the output. The main difference is that there are only three classes to represent. The *alpha helix* labeled A, the *beta sheet* labeled B and the *coil* labeled C.

$$bout(c) = \begin{cases} [1 \quad 0 \quad 0]^T & \text{if } c \text{ is } alpha\ helix \\ [0 \quad 1 \quad 0]^T & \text{if } c \text{ is } beta\ strand \\ [0 \quad 0 \quad 1]^T & \text{if } c \text{ is } coil \end{cases}$$

The c represents the resulting structural class, and *bout(c)* represents the output binarization function which translates the three structural classes from the string into the binary format.

C. Construction of the Neural Network

Multilayer feed-forward neural networks with backpropagation learning algorithm are suitable for advanced classification problems [26] as the one that is being treated in this paper.

1) Neural Network Architecture: In a feed-forward network the output of a node y is described as a function of the input x. The input to a given node is a sum of previous nodes and their associated weights:

$$X = \sum_{i=1}^{n} x_i w_i \tag{1}$$

where n is the number of neurons and w_i is the associated weight. This value is then passed through a sigmoid activation function:

$$Y^{sigmoid} = \frac{1}{1+e^{-X}} \tag{2}$$

which guarantees that the neuron output is bounded between 0 and 1. If we consider equation (1), the activation of the nodes y_i can be defined as:

$$y_i = f_i(X) = f_i(\sum_{i=1}^{n} x_i w_i) \tag{3}$$

For any dataset given as input and the corresponding weights, there is a certain error measured by an error function. Since the backpropagation learning rule is applied, there are two contrary directions of information flowing across the network. Input signals $(x_1, x_2, ..., x_n)$ are propagated from the left to right and the error signals $(e_1, e_2, ..., e_n)$ from right to left. Error signals are calculated for the output of each neuron and the general error function for one epoch is defined as the sum of the squares of the differences between all target node outputs and actual node outputs:

$$E_p = \frac{1}{2} \sum_n (t_{jn} - w_{jn})^2 \tag{4}$$

where t_{jn} is the target activation value for the node n and p marks the current epoch. Given the equation (4), the networks overall error is simply calculated by summing all of the E_p values for a given set of training patterns. The respective formula and the standardized version, the MSE (abbr. *Mean Squared Error*) equation are shown below.

$$E = \sum_p E_p = \sum_p \sum_n (t_{jn} - w_{jn})^2 \tag{5}$$

$$MSE = \frac{1}{2PN} \sum_p \sum_n (t_{jn} - w_{jn})^2 \tag{6}$$

The MSE shows the difference between the correct output and what's estimated. Since the algorithm uses the backpropagation learning rule, which is based on the *Widrow-Hoff delta learning* rule, the main goal is to adjust the neural network parameters in a way that the MSE is minimized below a certain threshold. As that is not always bound to happen, an additional maximum epoch number is given after which the algorithm terminates regardless of the current MSE value.

2) Important Parameters of the Neural Network: One of the important factors in this algorithm is the configuration of the parameters that suits the secondary structure prediction problem. Three neurons make up the initial hidden layer. By increasing this number, more accurate results can be achieved, but there is also the risk of *overfitting*. The neural network can get too adapted to the training data and try to memorize the previous examples instead of learning how to generalize and adapt its structure to successfully solve unknown structures that show up as the input after training ends. All the different layers and the previously described backpropagation learning flow of data inside a neural network is depicted in Fig 3.

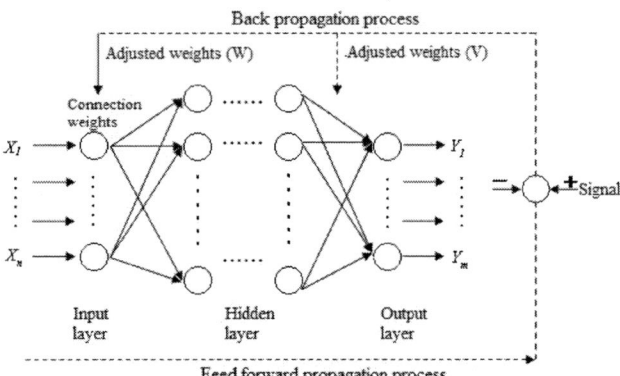

Fig. 3: Neural network architecture and data flow.

For the training of the networks in the ensemble the Rost-Sander RS121 [27] and the FC699 datasets [28] are used, since they contain a wide range of different protein primary structures and their corresponding secondary structures. This makes the selection of datasets justifiable for the initial training. Through the *k-fold cross-validation* method of determining training, validation and test sets, the individual datasets are split and translated into the correct input matrix format with a split ratio of 70-15-15 for all the sets, respectively. Other parameters that have a significant impact on the networks performance are also: the training algorithm, number of neurons in the hidden layer, number of input vectors etc.

D. Construction of the Ensemble

The idea of creating a learning ensemble is relatively simple. Since the neural network largely depends on the quality and diversity of the input data, it cannot provide the correct prediction all the time. This is especially true within the structure prediction problem since a wide range of protein families exists. That is why multiple networks are created and their results are combined through one of the consensus methods. Individual networks can have different architectures, different number of neurons in their layers, different window sizes etc. The important aspect is that the output is one of the structural classes. In this way, if H_n is the hypothesis space of one of the ensemble members, multiple hypothesis spaces narrow down the possible solution space with their intersections as depicted in Fig. 4. This

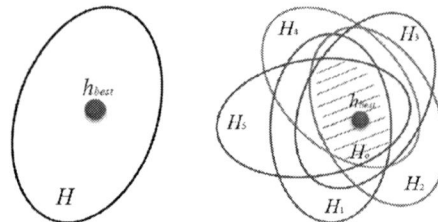

Fig. 4: Multiple hypothesis space intersection.

limits the search space for the optimal solution marked as h_{best}. Thus, the best classification rule is constructed through approximation of multiple classification rules. One important factor is the aggregation of multiple outputs. There are many different approaches, however in this paper we used majority voting method to determine the final output. If one or two networks fail to correctly classify a structural class, other networks with correct predictions can override the bad result. It all depends on the way other networks are constructed and trained. Therefore, the entire ensemble will give incorrect results only if the majority of the networks fails to identify the correct structural class. The ensemble approach in our method can add some additional security to the classification that is sometimes needed to provide satisfactory results.

The complete pseudo code for our hybrid method, based on multiple neural networks working together within an ensemble, can be formulated as in Algorithm 1. The method requires certain parameters which are usually determined empirically as described in the previous sections. The first two lines of the for loop represent on of the advantages of the method as pre-processed datasets with diversified data are given as inputs to neural networks to ensure the stability of the ensemble.

Algorithm 1 Hybrid Method based on an Ensemble of Multiple Artificial Neural Networks

---

function PREDICTPROTEINSECONDARYSTRUCTURE
  Require:
  size - Size of the ensemble;
  ws - Sliding window size;
  datasets[] - Datasets used for training;
  annParameters[] - Network parameters;
  inputSequence - Amino acid chain;
  k - k-fold cross-validation parameter;

  for $i < size; i \leftarrow i + 1$ **do**
    binIn = binarizeInputs(*datasets*[n], *ws*);
    binOut = binarizeOutputs(*datasets*[n]);
    dataset = combine(*binIn, binOut*);
    [*tr,te,val*] = crossValidation(*dataset, k*);
    ann = constructANN([*tr,te,val*], *annParameters*[i]);
    results[i] = ann.predict(*inputSequence*);
  end for

  secondaryStructure = consenusMethod(*results*);
  return *secondaryStructure*;

IV. RESULTS

The algorithm described in previous chapters was implemented in MATLAB version 8.5.0 (R2015a) using the Neural Network Toolbox. The datasets used for training were the RS121 and FC699. The dataset used for testing was a combined dataset containing parts from the 25PDB [28] and CB513 [29] datasets and other proteins and their respective secondary structures which did not occur in the training sets. The standard *Q3* or *Average Percentage Accuracy* method was used as the quality measurement of the results. The results that were achieved for a single neural network with different parameter combinations are listed in Table II. Throughout all runs, the highest percentage achieved is at about 61% with 10 neurons in the hidden layer and a window size of 17. It is also visible that, with these configuration parameters, the network has big oscillations, since the accuracy for window size of 3 lies little below 35%. That makes these results not trustworthy.

TABLE II: Q3 accuracy results for a single neural network.

Train	10	15	20	[3 5 3]	[5 10 5]	[10 20 10]
3	35,0%	57,1%	54,3%	55,3%	53,4%	53,3%
5	58,4%	60,1%	59,7%	56,3%	54,9%	59,0%
9	60,2%	61,6%	62,3%	58,3%	52,8%	62,0%
17	61,6%	65,1%	64,1%	47,1%	61,4%	58,9%
21	64,8%	58,5%	65,6%	46,6%	46,7%	59,0%
Test	**10**	**15**	**20**	**[3 5 3]**	**[5 10 5]**	**[10 20 10]**
3	34,8%	55,8%	52,8%	52,9%	52,3%	54,5%
5	56,9%	58,8%	59,2%	53,9%	52,6%	55,6%
9	57,1%	59,5%	60,1%	55,4%	55,0%	60,2%
17	61,0%	59,8%	60,3%	47,9%	56,3%	59,1%
21	59,8%	57,1%	59,1%	46,8%	48,9%	56,2%

According to the measurements, the best and most consistent results are achieved with a window size of 17 and 20 neurons in the hidden layer. If Table II is translated into a percentage bar chart for the most successful parameter configurations, the correlation between the window size and the prediction accuracy becomes visible. That leads to the conclusion that the optimal empirical values for the window size are between 17 and 20, depending on the protein structure.

These results show that the isolated neural network performance possibilities lie at around 60% as shown in Fig. 5. Of course, these numbers can be increased by implementing some advanced network improvements as mentioned in Section II, but in this paper we focus is on the ensemble and integration of methods. The described ensemble method is tested by executing 20 runs with 20 different test sets than the ones used to train the neural networks. The results achieved through the whole process of testing the proposed method are as shown in Table III. The highest and lowest accuracy is also marked.

The average accuracy lies at approximately 65,3% which shows the improvement made by simply combining differentiated neural networks together. If the datasets that caused the two best, two worst and a near-average performance in the ensemble are given as input in a single neural network and *Naive Bayes* classificator, a good accuracy comparison can

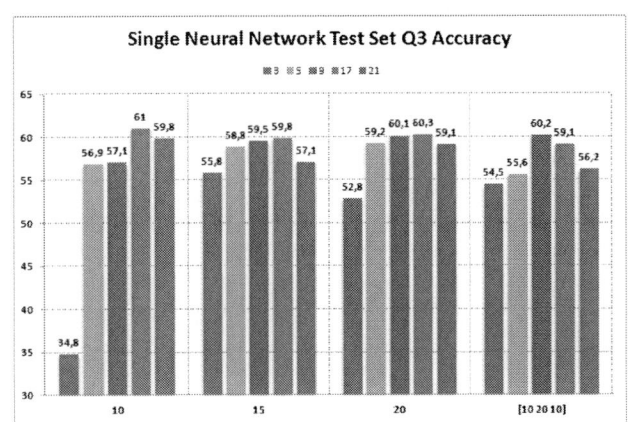

Fig. 5: Neural network Q3 accuracy bar chart.

be made. As depicted in Fig. 6, the proposed hybrid method solves the structural classification much more efficiently than the single neural network approaches and the common classification methods such as the *Naive Bayes* classificator [30]. That is to be expected because of the lack of diversity in network training and the previously described problems with purely statistical approaches cannot cover all the processes within the secondary structure formation process.

TABLE III: Q3 accuracy results for a network ensemble.

	1	2	3	4	5
Q3 Accuracy	64,5%	68,9%	64,9%	64,8%	65,4%
	6	**7**	**8**	**9**	**10**
Q3 Accuracy	68,7%	65,2%	65,0%	65,5%	64,2%
	11	**12**	**13**	**14**	**15**
Q3 Accuracy	64,2%	64,0%	66,1%	64,4%	64,3%
	16	**17**	**18**	**19**	**20**
Q3 Accuracy	65,1%	65,0%	64,6%	66,5%	64,9%

It is also worth noting that the isolated neural networks work well under the additional pressure of differentiating datasets. That means that, for all the individual neural networks, a good parameter configuration is chosen and that the networks are capable of good generalization. The common problems that can arise, such as *overfitting* and *underfitting*, are thereby successfully avoided.

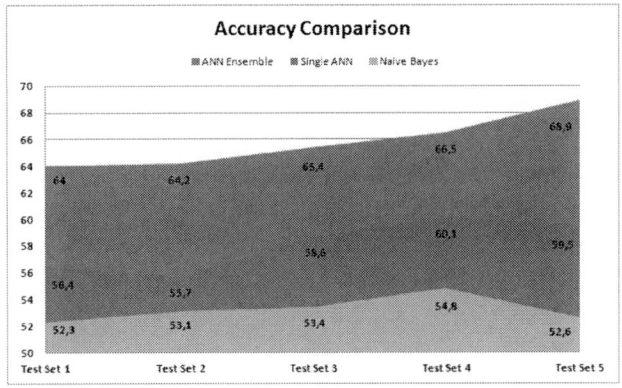

Fig. 6: Accuracy comparison for different approaches.

V. CONCLUSION

The search for a universal algorithm for the protein secondary structure prediction is not an easy problem to solve. However, in this paper, a hybrid, multiple neural network ensemble approach is proposed which shows promising results of improving the accuracy of existing algorithms. Through a simple aggregation of different prediction methods, this approach narrows down the possible hypothesis space in which the optimal solution is located and therefore increases the time that is needed to find the optimal solution. That also makes it more likely that the optimal solution will be found, i.e. that the MSE parameter will drop down below the given accuracy threshold within the set number of epochs.

The proposed method, based on multiple differently trained neural networks achieves around 65% accuracy of successfully predicted secondary structures. The final evaluation was based upon creations of different smaller datasets partially derived from the 25PDB and CB513 datasets and other protein structures gathered for the purpose of testing. The input data was formed by combining data from different sources, which proves that our method, along with the accuracy increase, is stable in prediction of diverse protein structures. Since the accuracy of methods based solely on neural networks lies around 60% [5], and those methods can have oscillations for differently structured protein than those used for training, our proposed method is suitable for classification of protein secondary structures. Also, it offers a good example on how to combine different methods and, more importantly, how to properly train and incorporate these elements into a bigger, more advanced algorithm for secondary structure prediction.

REFERENCES

[1] M. Pukáncsik, Á. Orbán, K. Nagy, K. Matsuo, K. Gekko, D. Maurin, et al. (2016). *Secondary Structure Prediction of Protein Constructs Using Random Incremental Truncation and Vacuum-Ultraviolet CD Spectroscopy.* PLoS ONE 11(6): e0156238. doi:10.1371/journal.pone.0156238

[2] L. Whitmore, B.A. Wallace (2008). *Protein Secondary Structure Analyses from Circular Dichroism Spectroscopy: Methods and Reference Databases.* Biopolymers 89: 392–400. PMID: 17896349

[3] K. Matsuo, K. Sakai, Y. Matsushima, T. Fukuyama, K. Gekko (2003). *Optical Cell With a Temperature-Control Unit for a Vacuum-Ultraviolet Circular Dichroism Spectrophotometer.* Analytical Sciences 19: 129–132. PMID: 12558036.

[4] O. Dor, Y. Zhou; Zhou (2006). *Achieving 80% Tenfold Cross-validated Accuracy for Secondary Structure Prediction by Large-scale Training.* Proteins. 66 (4): 838–45. doi:10.1002/prot.21298. PMID 17177203.

[5] J. Chandonia, M. Karplus (1994). *Neural Networks for Secondary Structure and Structural Class Predictions.* Protein Science (1995), 4: 275-285.

[6] P. Privilege Jr., G.D. Fasman (1989). *Chou-Fasman Prediction of the Secondary Structure of Proteins.* MIT.

[7] J. Garnier, B. Robson (1989). *Prediciton of Protein Structure and the Principles od Protein Conformation.* Plenum Press, New York.

[8] J. Garnier, J.F. Gibrat, B. Robson (1996). *GOR Method for Predicting Protein Secondary Structure from Amino Acid Sequence.*

[9] J. Sim, S.Y. Kim2, J. Lee (2005). *Prediction of Protein Solvent Accessibility using Fuzzy k-nearest Neighbor Method.* Vol. 21 no. 12 2005, pages 2844–2849. doi:10.1093/bioinformatics/bti423

[10] E.G. Mansoori, M.J. Zolghadri, S.D. Katebi (2009). *Protein Superfamily Classification Using Fuzzy Rule-Based Classifier.* IEEE Transactions on Nanobioscience, Vol. 8, No. 1.

[11] H. Shen, J. Yang, X. Liu, K. Chou (2005). *Using Supervised Fuzzy Clustering to Predict Protein Structural Classes.* Biochemical and Biophysical Research Communications 334 (2005) 577–581.

[12] J. Martin, J.F. Gibrat, F. Rodolphe (2005). *Choosing the Optimal Hidden Markov Model for Secondary-Structure Prediction.* IEEE Intelligent Systems, vol. 20, no. , pp. 19-25, November/December 2005, doi:10.1109/MIS.2005.102.

[13] N. Nguyen, M. Nute, S. Mirarab, T. Warnow (2016). *HIPPI - Highly Accurate Protein Family Classification with Wnsembles of HMMs.* 14th Annual Research in Computational Molecular Biology (RECOMB) Comparative Genomics Satellite Workshop, Montreal, Canada. BMC Genomics 2016, 17(Suppl 10):765. doic 10.1186/s12864-016-3097-0.

[14] Y. Dong, X. Liu, G. Zhou (2001). *Support Vector Machines for Predicting Protein Structural Class.*doi:10.1186/1471-2105-2-3.

[15] B. Bhushan, M.K. Singh (). *Protein Structure Prediction using Neural Networks and Support Vector Machines.* International Journal of Engineering Science and Advanced Technology, Volume-3, Issue-3, 145-156.

[16] J. Zubek et al. (2015). *Multi-level Machine Learning Prediction of Protein–Protein Interactions in Saccharomyces Cerevisiae.* PeerJ 3:e1041; DOI 10.7717/peerj.1041.

[17] Z. Sun, X. Rao, L. Peng, D. Xu (1997). *Prediction of Protein Supersecondary Structures Based on the Artificial Neural Network Method.* vol.10 no.7 pp.763–769, 1997.

[18] K. Lin, V.A. Simossis, W.R. Taylor, J. Heringa (2005). *A Simple and Fast Secondary Structure Prediction Method Using Hidden Neural Networks.* Vol. 21 no. 2 2005, pages 152–159. doi:10.1093/bioinformatics/bth487.

[19] Z. Li, Y. Yu (2016). *Protein Secondary Structure Prediction Using Cascaded Convolutional and Recurrent Neural Networks.* International Joint Conferences on Artificial Intelligence. arXiv:1604.07176v1[q-bio.BM].

[20] C. Mirabello, A. Adelfio, G. Pollastri (2014). *Reconstructing Protein Structures by Neural Network Pairwise Interaction Fields and Iterative Decoy Set Construction.* Biomolecules 2014, 4, 160-180; doi:10.3390/biom4010160.

[21] K. Paliwal, J. Lyons, R. Heffernan (2015). *A Short Review of Deep Learning Neural Networks in Protein Structure Prediction Problems.* Adv Tech Biol Med 3:139. doi: 10.4172/2379-1764.1000139.

[22] S. Wang et al. (2016). *Protein Secondary Structure Prediction Using Deep Convolutional Neural Fields.* Sci. Rep. 6, 18962; doi: 10.1038/srep18962 (2016).

[23] Z. Lin, J. Lanchantin, Y. Qi (2016). *MUST-CNN: A Multilayer Shift-and-Stitch Deep Convolutional Architecture for Sequence-based Protein Structure Prediction.* AAAI 2016. arXiv:1605.03004v1 [cs.LG].

[24] A. Busiay, J. Collins, N. Jaitly (2016). *Protein Secondary Structure Prediction Using Deep Multi-scale Convolutional Neural Networks and Next-Step Conditioning.* RECOMB 2017. arXiv:1611.01503v1 [cs.LG].

[25] K. Chen, L. Kurgan, J.Ruan (2006). *Optimization of the Sliding Window Size for Protein Structure Prediction.* CIBCB '06: 2006 IEEE Symposium on Computational Intelligence and Bioinformatics and Computational Biology, pp.1-7, 28-29, doi:10.1109/CIBCB.2006.330959.

[26] G.P. Zhang (2000). *Neural Networks for Classification: A Survey.* IEEE Transactions on Systems, Man and Cybernetics-Part C: Applicationss and Reviews, vol. 30, No. 4, November 2000. Publisher Item Identifier S 1094-6977(00)11206-4.

[27] B. Rost, C. Sander, R. Schneider (1994). *Redefining the Goals of Protein Secondary Structure Prediction.* J. Mol. Biology. 235:13–26. [PubMed].

[28] L. Kurgan, K. Cios, K. Chen (2008). *SCPRED: Accurate Prediction of Protein Structural Class for Sequences of Twilight-zone Similarity with Predicting Sequences.* BMC Bioinformatics, 9:226.

[29] J. Cuff, G. Barton (1999). *Evaluation and Improvement of Multiple Sequence Methods for Protein Secondary Structure Prediction.* PROTEINS: Structure, Function, and Genetics 1999; 34508–519.519. [PubMed].

[30] Q .Li, D. Dahl, M. Vannucci, J. Hyun, J. Tsai (2014). *Bayesian Model of Protein Primary Sequence for Secondary Structure Prediction.* PLoS ONE 9(10): e109832. doi:10.1371/journal.pone.0109832.

Hardware-in-the-Loop Architecture with MATLAB/Simulink and QuaRC for Rapid Prototyping of CMAC Neural Network Controller for Ball-and-Beam Plant

Vjosa Shatri*, Lavdim Kurtaj*✉ and Ilir Limani*

* Faculty of Electrical and Computer Engineering, University of Prishtina "Hasan Prishtina", Prishtina, Kosova
✉ lavdim.kurtaj@uni-pr.edu

Abstract - Cerebellar Model Articulation Controller (CMAC) is type of neural network inspired from part of the brain named cerebellum. It has long history in control applications, in simulated or in real-time implementations. MATLAB/Simulink is used for rapid building of CMAC neural network controller prototypes than can run in real-time. To run the controller in the same personal computer, QuaRC is used as real-time operating system that will execute controller task along Windows operating system. Hardware-in-the-loop architecture was used to test CMAC controller in real ball-and-beam physical plant from Quanser. Library of developed Simulink functional blocks enable easy exploration of different structural aspects of CMAC neural network controller, including selection of receptive field types, receptive field widths, Albus overlays, and fully interconnected multi-dimensional receptive fields. Same CMAC controller can be used also for controlling model of the plant, which can serve as means for controller to acquire knowledge before operating with real plant. CMAC controller learned to control the plant, and progressively became main control signal generator, while control signal from primary proportional-derivative controller almost vanished. Developed library can serve as easy startup for working with CMAC networks also for other types of applications.

I. INTRODUCTION

All technical systems that are built to serve in fulfillment of some intended functional behavior that can be described with desired output quantities. Some of these quantities can be requested to maintain constant value, while for others may be requested to follow some time trajectories. To achieve acceptable performances plant must be driven with proper command inputs. But the trivial cases, complexity of these commands necessitates appending the plants with other functional blocks that in a concerted ways will generate them in an automated fashion without permanent human intervention. These functional blocks are generally named as controller [1]. Designing and building proper controller for desired certain plant and desired reference behavior is main task of automatic control.

Most widespread type of controller in use is PID (proportional-integral-derivative) controller [2], where not

all of three types of action must be present, resulting in different versions of the controller, like P, PI, PD. For plants with more complex dynamics PID controller will not be able to achieve satisfying performance. Many other types of the controllers have been devised to aid in this task. One family of these controllers tries to emulate some working principles of animal's nervous system, named neural network controllers [3]. In general, even they were inspired from real neurons and neural systems, they evolved as a separate family of controllers with many functional variations, not necessarily biologically based or not yet proven they exist. An attractive feature of neural network controllers is ability of learning [4] control task from past and present information they get from inputs of the plant to be controlled and from outputs they generate to act on that system. This contrasts with rigid structure of classical PID controller that can not give long run satisfactory performance, without retuning [5], when system parameters or working conditions change.

One practical and safe way for testing new controllers is by simulations. Usual simulations will require also the model for plant to be controlled [6]. Accuracy of plant model will influence the final results of the controller, when it goes to real implementation and gets installed to work with real plant. To have more accurate and realistic data, especially the one related to timing aspects of controller action, simulation can be made to interact with real plant. In this case controller will be "built" in the simulation environment, with all commodities that it offers, but simulation runs will happen in real-time interacting with real plant to better emulate real working conditions. In cases when using real plant could be very costly, or dangerous to use, hardware plant simulators [7], [8] can be used. These would provide similar interface as the real plant has, and from controller point of view it would behave like real one. Since most control structures form a closed loop including controller and plant, this technique of simulation is termed hardware-in-the-loop (HIL) [9], and is used for rapid prototyping of some functional part of bigger system [10], [11], controller in this case, like in [11]–[13].

Neural network controller that has long history in control applications, in simulated or in real-time implementations is the one inspired from part of the brain named cerebellum [14], with resulting Cerebellar Model Articulation Controller (CMAC) [15]. Processing of information on layers of the cerebellum and development of corresponding Simulink models are shortly described in Section II. Results of simulations with ball-and-beam plant model and with HIL simulations with real plant are contained in Section III. Paper in the end concludes with Section IV.

II. CMAC SIMULINK MODEL

Since it was introduced CMAC [15] was attractive for real time applications, as a consequence of simple logical operations needed for implementation. It enabled very modest microcomputer systems to handle all computations for real time operation [16]. To experiment with different CMAC configurations and performance on different plants, availability of the model for more general tools would ease analysis and design process. MATLAB is powerful tool for numerical computing that enables graphical programming in common form of block diagram through tightly integrated Simulink environment. It enables automatic generation of C source code that can be targeted to wide range of hardware platforms for operation in real time. With additional real-time operating system, like QuaRC from Quanser, models can be run in real-time on platforms with Windows operating systems, including the same platform where MATLAB/Simulink is running.

Simulink CMAC model generation can follow several approaches. Authors of [17] used MATLAB S-function implementation that can be included in models developed in Simulink. Approach of this paper is to preserve visual layered structure, and in same time to leave open possibilities of customizing final structure, in creating averaged cerebellar structures, like standard CMAC [15] or Cerebellar elementary Processing Unit (CePU) [18].

A. CMAC Mossy Fiber Input

Information carried by mossy fibers, during most of the simulations, usually is in form of receptive fields (population code). On the other side, standard form of information in other parts of the system is in form of continuous values lying in a specified range (rate code). Blocks of this layer will convert rate coded signals to population coded signals, with specified bases function for receptive fields. First three rows of Simulink block library, Fig. 1, show blocks for generation of mossy fibers with square, triangular, and radial basis functions. Blocks have corresponding masks for customizing functionality, like number and width of receptive fields. If looked from perspective of ANFIS artificial neural network [19] this layer would correspond to layer 1 of the ANFIS, whereas looked from cerebellums perspective layer corresponds to pre-cerebellar structures.

B. CMAC Granule Cell Layer

This layer will create multidimensional receptive fields from (assumed) one-dimensional receptive fields of several dimensions. Most of the granule cells have 4 dendrites that receive information from mossy fibers. We assume that idealized elementary granule cell-Golgi cell

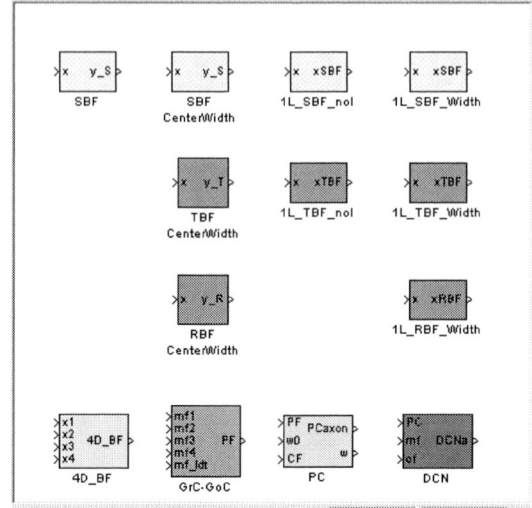

Figure 1. Simulink library of processing blocks for building different models of Cerebellum. First three rows receive standard rate coded input and generate population coded mossy fiber output, with square, triangular, and radial basis functions. Next two blocks of bottom row, 4D_BF and GrC-GoC blocks, correspond to processing of granule cell layer. First one will create standard up to four dimensional receptive fields, with multiplication operation, from inputs x1 to x4. Unused input dimensions should be tied to constant 1. Standard CMAC [15] will use this block. Second one corresponds to CePU model [18], and will create multidimensional receptive fields, from inputs mf1 to mf4, modulated with mf_ldt rate coded input. PC block models Purkinje cell with synapses if makes with paralell fibers, representet with weights. CF input represents climbing fiber input, and guides learning. It is assumed that CF input carries error information. DCN block models deep cerebellar nuclei as simple summation of inputs.

processing circuit (GrC-GoC) [18] is a group of several thousand granule cells, inhibited from same Golgi cell, [14] that receive information from up to four groups of population coded mossy fibers, corresponding up to 4-dimensional input space. Bottom row of library contains two blocks that perform function of this layer, 4D_BF and GrC-GoC blocks. First one will create standard up to four dimensional receptive fields, with multiplication operation, from inputs x1 to x4. Unused input dimensions should be tied to constant 1. Standard CMAC [15] will use this block. Second one corresponds to CePU model [18], and will create multidimensional receptive fields, from inputs on mf1 to mf4, modulated with mf_ldt rate coded input.

C. CMAC Purkinje Cell Layer

Single Purkinje cell with synapses it makes with parallel fibers is part of this layer, PC block of library in Fig. 1. Purkinje cells in CMAC model are modeled as simple perceptron with adjustable weights. Block has PF input that mimics Purkinje cell inputs from parallel fibers. Other input, CF, matches climbing fiber input to Purkinje cell and will guide learning. Usual assumption is that CF input carries error information. Initial vales for weights are set with input w0, whereas w output will show current weight values. Output PCaxon of this block is in the same time output of standard CMAC model Mask of the block can be used to set learning rate and update period.

1202

D. CMAC Deep Cerebellar Nuclei

In standard model of the CMAC deep cerebellar nuclei are not modeled. Every output would have one Purkinje cell model. In current version, block DCN of library in Fig. 1, model will be simple summation of all signals in the incoming vector, usually outputs from several Purkinje cells, inputs from some set of mossy fibers, and from some climbing fibers, inputs PC, mf, and cf respectively.

III. SIMULATIONS AND RAPID PROTOTYPING WITH CMAC

Inclusion of the neural controllers as part of the general control system usually goes jointly with some form of conventional controller [20], but it can be used also as a standalone controller [21]. Neural controller will acquire knowledge about the system to control during learning or training phase. This phase can be accomplished off-line or on-line. When on-line mode is used, it can last some time and then stop, or if can be active all the time adapting itself to possible changes of plant behavior. Different types of neural network manifest quite different behavior at different phases of their operation, during initial, during, or after learning [22], i.e. influence that "fresh" network has on system behavior, how fast it gains knowledge during learning phase where learned and adaptive behavior are intertwined, and final quality of learning when learning stops. Conventional controller used along neural one is some version of PID family of controllers [2].

Figure 2 show diagram for building of full overlaid CMAC and for CMAC with Albus overlays from library blocks. When building Albus CMAC, receptive fields of the same layer should not overlap, achieved by proper parameters of corresponding blocks, blocks with names starting with 1L_TBF_Width. According to signal dimensions from figure, number of receptive fields for each dimension of each layer is 26. Number of two-dimensional receptive fields on each layer is 26·26=676. Total number of two-dimensional 4-quanta wide receptive fields for this 4-layer CMAC is 4·676=2704. Same resolution with full overlaid CMAC would have (4·26)·(4·26)=104·104=10816 two-dimensional receptive fields. Block implementation is meant to correspond directly to defining equations, for easy understanding and building different models that mimic connections in real biological version of the cerebellum. This has consequences in calculation time, and directly influencing size of CMAC that can be used during HIL simulation. QuaRC will report error if calculation time exceeds fundamental sample time selected for simulation. Size of the CMAC is not limited for offline simulations, but it will directly influence simulation speed.

CMAC was used as secondary controller along PD primary controller for position of the ball, in cascade control structure of Quanser the SRV02 Ball and Beam plant. Used CMAC had two dimensions, ball position error and ball speed error. These same signals usually serve as input to standalone fuzzy controllers [21]. Output of PD controller was used by Purkinje cell during learning. Fig. 3a shows connection of the CMAC controller. Iside this block is subsystem from Fig. 2a.

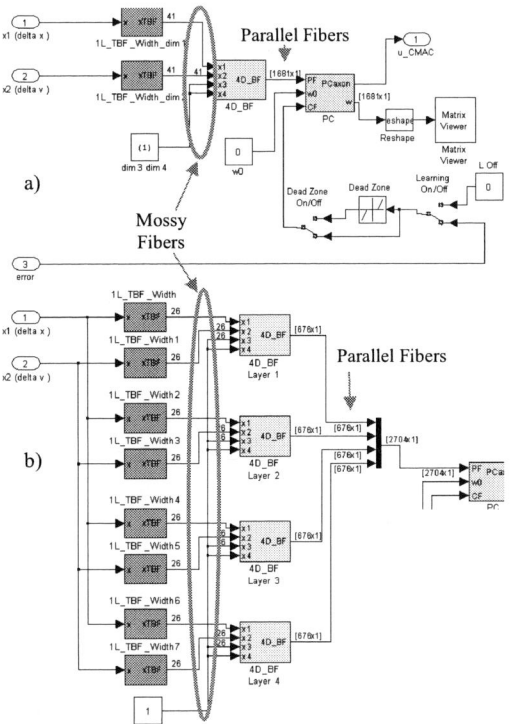

Figure 2. Diagrams for realization of Simulink CMAC models from library blocks: a) full overlaid model, b) Albus overlaid with 4 layers. Only part from rate coded inputs up to parallel fibrs is shown. Model around Purkinje cell is the same for both models. Part with matrix viewer and reshape blocks is used only for offline simulations. Switches are used to select mode of operation: Learning On/Off and Dead Zone On/Off, respectively.

Testing was done with sinusoidal reference signal. First 50 seconds are controlled with PD controller alone. From Fig. 3c we see that control signal from CMAC is zero (red, thick solid line). When CMAC starts learning error drops considerably, traces on Fig. 3b, caused by CMAC taking main control, followed by reduction of control from PD, Fig. 3c. Color map of the weight space for this simulation and one with step reference are given in Fig. 4a and Fig. 4b, correspondingly. Number of the receptive fields for second case was 101 for each dimension was 8-quanta wide.

Any simulation CMAC model that is build can be compiled and run in real-time with QuaRC, if hardware platform can handle all processing timely. All we have to do is change simulated plant with interface to physically real plant (or physical model of real plant), under condition that interface signal are of the same type and range. Quanser provides simulated plant and interface to physically real SRV02 Ball and Beam plant. Behavior of physical plant under sinusoidal or triangle reference signal was very different, with unpredictable sticks [23] and jumps. This can be caused by not modeled dynamic effects, and by limits of control components. To control sinusoidal 5 cm reference signal, amplitude of reference angle to inner control loop is only ±2°. Gear backlash under this condition can easily deform final behavior. Looking to components of control part of the problem, simulated control voltage that will drive motor for this

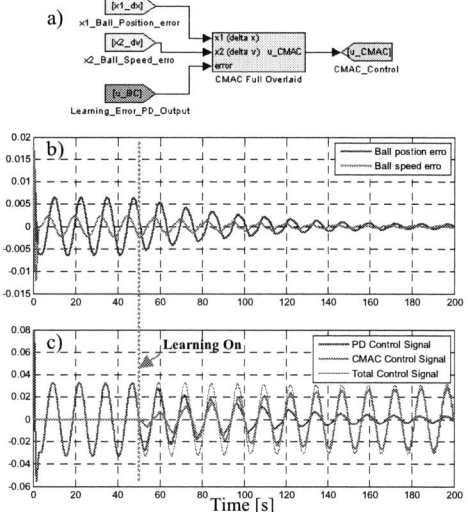

Figure 3. Tracking ball position on Quanser SRV02 Ball and Beam plant with PD and full overlaid CMAC and results. a) Secondary full overlaid CMAC controler, b) Ball position and speed error, c) PD, CMAC, and total (PD+CMAC) controll signal. First 50 seconds only PD controler is activ, afterward CMAC learning is switchen on.

sinusoidal steady state condition is only ±0.12 mV, where if only 1 quantum is used for both positive and negative side, number of quanta would come close to that provided by 12-bit digital-to-analog converter resolution of control board, 3333 vs. 4096. Under this conditions CMAC controller in most of cases, if some dead-zone for error was not set, it came to the solution with high oscillatory motion around reference path, which in some form solves stiction problem, thought not the desirable one. Same can happen with simulated plant with no dead-zone in learning error. This dead-zone block is present at CMAC models in Fig. 2, with a switch for possibility to bypass it.

Situation is more favorable with square wave reference. At jump discontinuities position reference for servo will saturate to selected saturation point, ±π in this case, and continue with rich dynamic behavior. Motor control voltage swings between about ±4 volts. Much

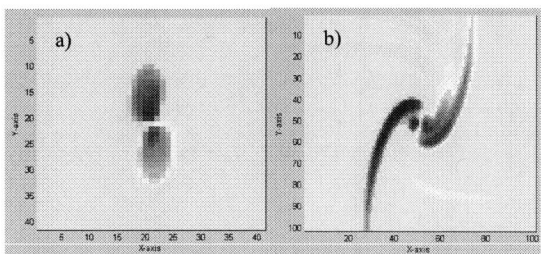

Figure 4. Color map of CMAC weight space: a) Results at the end of simulation run with conditions as for Fig. 3. CMAC has 41 receptive fields for each dimension 4-quanta wide. b) CMAC with 101 receptive fields for each dimension 8-quanta wide and wider input range. Learning was on from beginning of the simulation. Reference signal was square wave with amplitudes 20 cm, 10 cm, and 5 cm. Most of the green space belongs to weights not used by corresponding simulation, i.e. parts of the input space that were never visited during simulation and gained no knowledge. Generalization is local and equals width of the receptive field.

more part of the input space will be traveled, resulting in more receptive fields and corresponding weight been used, as can be seen in Fig. 4b, where step of selected amplitude will travel path with two arc routs, one for rising and one for falling edge. It will be traveling fast through these arc routes, with longer stay at central part of input space. This type of behavior favors use of receptive fields with nonuniform widths, as result of the design, or as part of learned process. Fig. 5 shows an example of real-time control of Quanser SERV02 Ball and Beam plant with full overlaid CMAC, with 41 receptive fields on each dimension. CMAC was placed as secondary controller at inner loop position controller for servo sub-plant SRV02 that drive beam sub-plant. Range of input values was read from one run with PD control only. What happens when input is outside of the working range depends on type of receptive fields used. Usual solution in fuzzy control is to hold at constant level activities of edge receptive fields. Receptive fields of same type were used through this paper, where fall of total activity on active receptive fields can be used for purpose of position and width adaption of same. As desirable value of total activity would be normalized value equal to 1. For generation of Fig. 5 scope block provided by Quanser (Time Figure) was used, that can show whole range in same plot during real-time HIL simulation. Two first periods of the simulation in Fig. 5 are controlled with PD controller alone. For time after 40 second CMAC learning is on and contributes to control process. We notice dampening of oscillation after jump change in the reference input. Initial overshoots did not change considerably. Most of behavior is preserved after switching learning off, signifying that CMAC has acquired knowledge for controlled plant. For lower values of learning rate, set from mask of Purkinje cell block, speed of learning is proportional learning rate parameter value. When values become larger control action of CMAC can contain considerable amount of adaptive behavior and less learned one. At these conditions, control can be with small errors during learning on phase, but it can increase significantly when learning is switched off. CMAC neural controllers are in good position of fast learning and preserving knowledge [22], if high vales of learning rates are avoided and with structure of design that suits to the problem it will control for best generalization capabilities [18]. Not every problem can be solved efficiently with same uniform type of network; even thought approximation capabilities for that specific type of

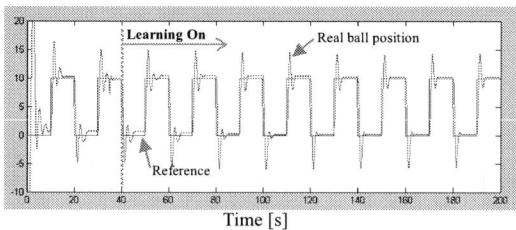

Figure 5. Hardware-in-the-Loop (HIL) simulation of cascade controlled Quanser SRV02 Ball and Beam, with square wave as reference signal (blue trace) and real ball position (green trace), for CMAC as a secondary controller in inner position controller. First two periods of square wave were tracked with PD controller alone. At t=40 s learning was turned on.

neural network are proved to be general. Specific structure can lead to much simpler network, with much better generalization. Brain that served as inspiration and still inspires can serve as an example, by having different structures at specific part of it, presumably serving different purposes.

Same library can be used for any type of target selected for development, and for more intensive computations stand-alone target can be used, where pure hardware based implementations with complex programmable devices, like field programmable generic arrays (FPGA), would offer superior performance. Initial tries for preserving block based building, but by performing operations on specific limited group that would correspond to active receptive fields did not provide significant improvement in performance. It may be that selection process takes about same time as for operation being performed uniformly on all receptive fields. The reason behind can be to optimized structure of MATLAB and Simulink for working with vectors and matrixes. At this stage simplicity of structure was favored, with possibility of building general neural networks with biological similarity, and more specifically cerebellar structures, but with non-spiking averaged signals.

IV. CONLUSION

Cerebellum was an inspiration from the time that its neuronal circuit was constructed [24], [25]. Initially thought uniform structure inspired theories of cerebellar functioning, and models that would be used for solving problem of coordinated motion, as one of main functions that was thought to be carried by cerebellum. Modern view of possible functions where cerebellum is involved is very broad, up to cognition and emotion.

To be able to develop behavioral models of the cerebellum that would follow some functionally specific group of neuronal circuit, library of functional blocks with models of specific cerebellar structures was developed. Library includes blocks that provide interface to real signals, and converts them in appropriate form to provide compatible inputs to the cerebellar model. Development had in mind to use same development environment for practical control applications, with simulated and real physical plants. This objective is achieved with library of block developed in Simulink, and same can be used in real-time applications through QuaRC. Library blocks were used to build Cerebellar Model Articulation Controllers (CMAC) with full or with Albus overlays. They were used in simulation with simulated and real Quanser SRV02 Ball and Beam plant. Results with simulated plant proved effectiveness and flexibility of the different CMAC models built with library blocks, and proved good learning abilities. During simulations with physical plant, nonlinearities that were not modeled in simulated plant, like static friction and backlash, generated very "fuzzy" signals that necessitate different approach for interfacing them to CMAC. Parts where signals are of comparable quality to simulated one, like in the inner position control loop of cascaded control strategy, results were similar.

REFERENCES

[1] J. J. D'Azzo and C. H. Houpis, Linear Control System Analysis and Design: Conventional and Modern, 3rd Edition, McGraw Hill, 1988, pp. 139–184.

[2] K. J. Åström and T. Hägglund, Advanced PID Control, ISA - Instrumentation, Systems, and Automation Societ, 2006, pp. 64–94.

[3] A. G. Barto, "Connectionist learning for control: an overview," in Neural Networks for Control, W. Thomas Miller, Richard S. Sutton, and Paul J. Werbos, Eds. 5th Printing, Cambridge, The MIT Press, 1996, pp. 5–58.

[4] D. H. Nguyen and B. Widrow, "Neural networks for self-learning control systems," IEEE Control Systems Magazine, vol. 10, no. 3, pp. 18–23, April 1990.

[5] V. M. Alfaro and R. Vilanova, Model-Reference Robust Tuning of PID Controllers, Switzerland, Springer International Publishing, 2016.

[6] D. C. Karnopp, D. L. Margolis, and R. C. Rosenberg, System Dynamics: Modeling, Simulation, and Control of Mechatronic Systems, Fifth Edition, New Jersey, John Wiley & Sons, Inc., 2012.

[7] M. Saez, F. Maturana, K. Barton, and D. Tilbury, "Real-time hybrid simulation of manufacturing systems for performance analysis and control," 2015 IEEE International Conference on Automation Science and Engineering (CASE), 2015, pp. 526–531.

[8] C. E. Agüero, N. Koenig, I. Chen, H. Boyer, S. Peters, J. Hsu, et al., "Inside the virtual robotics challenge: Simulating real-time robotic disaster response," IEEE Transactions on Automation Science and Engineering, vol. 12, no. 2, pp. 494–506, 2015.

[9] J. A. Ledin, "Hardware-in-the-loop simulation," Embedded Systems Programming, pp. 42–60, February 1999.

[10] R. T. Ogan, "Hardware-in-the-loop simulation," Chapter 14 in Modeling and Simulation in the Systems Engineering Life Cycle Core Concepts and Accompanying Lectures, Margaret L. Loper Ed. London, Springer-Verlag, 2015, pp. 167–174.

[11] H. K. Fathy, Z. S. Filipi, J. Hagena and J. L. Stein, "Review of hardware-in-the-loop simulation and its prospects in the automotive area," in Modeling and Simulation for Military Applications, Kevin Schum and Alex F. Sisti, Eds. Proc. of SPIE, Vol. 6228, pp. 62280E-1–62280E-20, 2006.

[12] J. Toman, Z. Ančík, and V. Singule, "Hardware-in-the-loop testing of control algorithms for brushless DC motor," in Mechatronics: Recent Technological and Scientific Advances, Ryszard Jabloński and Tomaš Březina, Eds. Berlin Heidelberg, Springer-Verlag, 2011, pp. 165–173.

[13] M. Ram´ırez-Neria, H. Sira-Ram´ırez, R. Garrido-Moctezuma and A. Luviano-Juarez, "Linear robust generalized proportional integral control of a ball and beam system for trajectory tracking tasks," 2016 American Control Conference (ACC), pp. 4719–4724, July 6-8, 2016.

[14] Masao Ito, The Cerebellum: Brain for an Implicit Self, New Jersey, Pearson Education, Inc., 2012.

[15] J. S. Albus, "New approach to manipulator control: the cerebellar model articulation controller (CMAC)," Transactions of the ASME Journal of Dynamic Systems, Measurement, and Control, vol. 97, no. 3, pp. 220 - 227, September 1975.

[16] L. Kurtaj, I. Limani, V. Shatri and A. Skeja, "The cerebellum: new computational model that reveals its primary function to calculate multibody dynamics conform to Lagrange-Euler formulation," IJCSI International Journal of Computer Science Issues, vol. 10, issue 5, no 2, pp. 14–24, September 2013.

[17] K. Herrup and B. Kuemerle, "The compartmentalization of the cerebellum," Annual Review of Neuroscience, vol. 20, pp. 61–90, March 1997.

[18] N. L. Cerminara, E. J. Lang, R. V. Sillitoe, and R. Apps, "Redefining the cerebellar cortex as an assembly of non-uniform Purkinje cell microcircuits," Nature Reviews Neuroscience, vol. 16, no. 2, pp. 79–93, February 2015.

[19] J.-S. R. Jang, "ANFIS: adaptive-network-based fuzzy inference system," IEEE Transactions on Systems, Man, and Cybernetics, vol. 23, no. 3, pp. 665–685, May/Jun 1993.

[20] D. T. Pham and L. Xing, Neural Networks for Identification, Prediction and Control, London, Springer-Verlag, 1995.

[21] J. H. Lilly, Fuzzy control and identification, New Jersey, John Wiley & Sons, Inc., 2010.

[22] L. Kurtaj, I. Limani, V. Shatri, and A. Skeja, "Dependence of CMAC neural network properties at initial, during, and after learning phase from input mapping function," Proceedings of the 12th WSEAS International Conference on Systems Theory and Scientific Computation (ISTASC'12), pp. 187–192, August 2012.

[23] P. H. Eaton; D. V. Prokhorov; and D. C. Wunsch, "Neuro-controller alternatives for "fuzzy" ball-and-beam systems with nonuniform nonlinear friction," IEEE Transactions on Neural Networks, vol. 11, no. 2, pp. 423–435, March 2000.

[24] David Marr, "A Theory of Cerebellar Cortex," The Journal of Physiology, vol. 202, no. 2, pp. 437–470, Jun 1969.

[25] J. S. Albus, "Theory of cerebellar function," Mathematical Biosciences, vol. 10, no. 1/2, pp. 25–61, February 1971.

Primary and Secondary Experience in Developing Adaptive Information Systems Supporting Knowledge Transfer

S. Lugović*, I. Dunđer ** and M. Horvat*

* Zagreb University of Applied Sciences, Department of Computer Science and Information Technology, Vrbik 8, Zagreb, Croatia

** University of Zagreb, Faculty of Humanities and Social Sciences, Department of Information and Communication Sciences, Ivana Lučića 3, Zagreb, Croatia

slugovic@tvz.hr, ivandunder@gmail.com, mhorvat1@tvz.hr

Abstract - One can evidence an increased and complex dynamic in the today's society in terms of technology that is being used, amounts of data that are being collected and processed by computers, and the resulting information overloads. The main challenge is how to address such a change with regard to the design of information systems and how those information systems are aligned with the change in human behaviour. This paper aims to address the issue of designing information systems that support Science 2.0 – a new phenomenon of interrelated sociotechnical interactions in which communication is the heart of science and the environment that enables critiquing, suggesting and sharing ideas and data in real time with almost no costs. This paper underlines problems related to the institutionalisation of knowledge transfer and its weaknesses. It also pinpoints John Dewey's primary and secondary experience as a point of departure which provides theories, theoretical frameworks and models such as information behaviour, documents and communities of action, the quadruple Helix model, activity theory, evolutionary learning and knowledge sharing communities. All these can be utilised for designing adaptive information systems that support better knowledge transfer.

Keywords - adaptive information systems, science 2.0, information behaviour, socio-technical systems, open innovation, information and communication sciences

I. INTRODUCTION

Humanity evidences major social, technological, economic and cultural transformations producing a new kind of society: network society [1]. Such an environment is described as turbulent, and is more complex, with higher uncertainty and with more interdependence.

In such contextual turbulent environment, technocratic bureaucracies, with its mechanical authoritarian control structure of the organisational form, cannot absorb or reduce such environmental turbulence. The absorption and reduction are necessary, opening the way to a viable human future [2].

Information systems can be understood as the "extension of meaning engagement practice through mediating and organising social interactions" [3].

Empirical evidence of such a proposition can be found in a recent massive-scale experiment on Facebook users in which the emotional state of the user changed accordingly to the amount of positive or negative content in their news feed placed without their acknowledgment [4]. Besides emotions, patterns of the information system use can configure cognition and behaviour of a user in the process of accomplishing work-related tasks [5].

It is also important to discuss how new principles of information system design will succeed in overcoming separation between technology, work and organization.

If an information system consists of social, technological and informational components, which are not separate but interrelated [6], and the social component of such a system changes according to patterns of behaviour, whereas there is an inherent inseparability between the technical and the social [7], can we search for causality between those patterns and adaptiveness of the information technology?

Human and material agencies are shared building blocks of routines and technologies, but by being isolated, neither of them (human or material agencies) are important. Namely, what is essential is the moment when they become imbricated, i.e. interlocked in a particular sequence, and as a whole they produce, sustain, or change routines and technologies [8]. To observe this phenomenon and to find an answer to the aforementioned question, the particular sequence of the relationship between human and material agencies, the inherent inseparability between the technical and the social, and the complexity of real situations should be examined, rather than analysing separate aspects [9].

II. SCIENCE 2.0 AS A SOCIO-TECHNICAL SYSTEM

In a recent proposal for development of science based on socio-technical progress, the term Science 2.0 emerged as a new phenomenon of interrelated socio-technical interactions, claiming that socio-technical systems are best studied at scale, in the real world, by rigorous observation, carefully chosen interventions and ambitious data collections [10]. In such an environment, which is fruitful for critiquing, suggesting, sharing ideas and data, communication is the heart of science, the most powerful tool ever invented for correcting errors, building on colleagues' work and fashioning new knowledge [11].

To understand technology in society, we have to treat it as an action system, where its subfunctions could be performed by humans or technical objects (human or material agencies) acting as subsystems. This allows us to transform the abstract action system into a socio-technical system by conceiving an object for every suitable acting function and by integrating them into the human acting or working relations [9].

Software to be run on such a socio-technical system must be able to sense, interpret and respond [12] to patterns of system behaviour that emerge according to internal system properties or reflections to the environment.

III. SECONDARY EXPERIENCE RESEARCH

The aim of this research is to investigate and eventually enable the exchange and (re)use of scientific papers created on the universities in the Danube region with their wider external environment including public, private and non-governmental organisations.

Scientists working at the universities publish scientific papers and get the papers' reflection according to the usage of the outside environment including public, private and non-governmental organisations. The main research question is, can we build an artefact in form of an information system that supports such an exchange and reflection?

In a critical review of the literature related to university governance of knowledge transfer, institutionalisation of linkage between universities and industry is defined as a new phenomenon, underlying various forms of knowledge transfer activities, ranging from collaborative research projects involving universities and companies (e.g. research contracts), intellectual property rights and spinoffs, labour and student mobility, consultancy etc., as well as "soft" forms of knowledge transfer, such as attendance at conferences and creation of electronic networks.

Universities' governance of knowledge transfer applies only to research contracts, intellectual property rights and spin-offs, but most university knowledge is transferred via traditional channels such as personnel exchanges, publishing, consulting and conferences. However, these types of knowledge transfer activities have not been institutionalised and little attention has been paid to their management and governance [13]. We believe that an intervention into this area by designing an adaptive information system could extend the capabilities of human and material agencies.

As a starting point for the conceptualisation of our research, we use primary and secondary experience proposed by John Dewey. The primary experience is the one with "minimum of incidental reflection", while secondary experience is described as "what is experienced in consequence of continued and regulated reflective inquiry... experienced only because of the intervention of systematic thinking". Dewey contrasted two different kinds of experience, primary and secondary, proposing that objects in secondary experience "get the meaning contained in a whole system of related objects; they are rendered continuous with the rest of nature and take on the import of the things they are now seen to be continuous with" [14].

To successfully transfer knowledge an information system is needed that supports better exchange of scientific papers between the academia and the environment, but which also enables acquiring feedback based on reflections from the environment.

In our view, the primary object in designing an information system is the one in which the observed object is excluded from the context with other objects, while secondary objects are those objects which are observed as a part of the higher-level system, consisting of the object itself and its relationships and behaviour in interaction with other related objects. Such a higher-level system includes an information system itself, but also its users and their information behaviour observed as a whole.

To design such an information system, we have to understand the information behaviour in socio-technical systems consisting of technologies that support the interaction between scientists, organisations they are working for, and published papers. The environment consists of public, private and non-governmental organisations.

Those three sectors together with the academic actors create a Quad Model [15] or Quadruple Helix [16] creating a framework for EU Digital Agenda for Europe in which government, industry, academia and civil participants work together [17].

To do so we have to extend our research not only to the design of the information system, but also towards the information behaviour research in such a socio-technical system.

We have to research what type of information resource (e.g. abstract, full paper etc.), and what type of media (e.g. scientific journal, conference proceedings, web pages etc.), are being utilised, but also what are the patterns of information seeking behaviour in the process of accessing information resources.

Those three research variables (type of information resources, type of communication channels and information seeking patterns) will provide us with insight into the phenomena of impact and usage of already published scientific papers by their environment (public, private and NGO). Such an insight is essential for the design of such an artefact, i.e. information system.

Another research inquiry is the area of interaction, or precisely speaking, what are the motivation drivers and factors that influence the interaction between scientists and their environment. If we understand the motivation drivers and factors that influence interaction, we can implement them into the design of an information system.

But we cannot know the effect of such functions in the information system, unless we incorporate them and put them into use.

IV. THEORETICAL BACKGROUND OF RESEARCH

Main theoretical background of this research is in the Activity theory [18], describing the three-way relationship between a person (the subject), an object, to which an activity is directed, and the tools or instruments used in the activity. A further theoretical extension is in different models of information behaviour [19], which will provide us with the framework for collecting data about the usage of already published scientific papers existing in the area of interaction between universities, public, private and NGO organisations.

In our research we will use the Documents of Action concept [20], which gives us an analytical framework to analyse usage, interaction and co-operations around already published scientific papers. Another theoretical concept used in this research is Evolutionary Learning [21], suggesting that sustainability requires collaboration among governments, businesses and civil society.

A clear distinction was made between growth, development and evolution, where growth is the increase in size or quantity, development is an amelioration of conditions or quality, and evolution is a tendency towards greater structural complexity and organisational simplicity, more efficient modes of operation and greater dynamic harmony.

Another theoretical contribution to this research is based on knowledge sharing communities [22] and communities of action [23] providing us with detailed frameworks for an information system functionality that supports both, social and technical, aspects of information behaviour.

Also different social cybernetic concepts about self-organisation, self-reference, self-steering, autocatalysis and cross-catalysis and autopoiesis [24] will be used in researching feedback between the information system and its users. This proposed research also contributes to the discipline of information system design science [25-27] and contributes to the extended definition of the information system, seen as an artefact which consists of information, social and technical elements, creating a whole which is greater than the sum of its parts [6].

Two papers have already been published with regard to this very particular research – one paper dealing with the possibilities of developing an information system based on the recognition of usage patterns [28]. In another paper, the authors provided an overview of the scientific communications models that are present for the last 50 years and suggested a synthesised model [29].

V. CONCLUSION

We evidence a trend of blurring the line between technological and social in information system research, moving the focus from deterministic to more casual logic in their design. Main aim of our research is to search for feedback from users' socio-cognitive behaviour that could be used as a signal that triggers information system adaption.

One of the theoretical fields we are currently exploring is information behaviour, which results in patterns of that behaviour. As the dynamic of patterns is observable by a machine, we believe that there is a possibility to use this signal to automatically (or semi-automatically) trigger restructuring of the information system, to generate new functions to support existing and create new information system goals.

We perceive an information system as a system consisting of informational, social and technological components acting as a whole, and that is aligned with findings from related theoretical and empirical studies presented in this paper.

Those components interact between each other and such an interaction, which is not only deterministic but casual, could provide fundaments for adaptive information systems which evolve along their usage. Researching interaction around scientific papers by universities, public, private and non-governmental organisations could provide us with valuable information on where and how to intervene in such a system.

Building an artefact in form of an information system for the purpose of the research could provide us with empirical insights which of the interventions and interactions give optimal results in terms of information system performance.

For example, if we knew what types of documents have the most impact on the environment and trigger the cognitive, communicative and co-operation processes [30] (with public, private, non-governmental organisations), we could further design amplification towards this area of the system which could then produce change in dynamics of information behaviour and related patterns.

New patterns will open up new areas of research interests, which then again could be amplified or attenuated. In that way we could design feedback loops in the information system which could enable deterministic but also casual properties.

REFERENCES

[1] M. Castells, The rise of the network society: The information age: Economy, society, and culture. John Wiley & Sons, NJ: 2nd ed.; vol. 1, p. 17, 2011.

[2] E. L. Trist, The Evolution of Socio-technical Systems: A Conceptual Framework and an Action Research Program. Ontario Quality of Working Life Centre, Toronto: p. 39, 1981.

[3] M. Aakhus, P. J. Agerfalk, K. Lyytinen and D. Te'eni, "Symbolic Action Research in Information Systems: Introduction to the Special Issue," MIS Quarterly 2014, vol. 38, no. 4, pp. 1187–1200, 2014.

[4] A. D. Kramer, J. E. Guillory and J. T. Hancock, "Experimental evidence of massive-scale emotional contagion through social networks," Proceedings of the National Academy of Sciences of the United States of America, vol. 111, no. 24, pp. 8788–8790, 2014.

[5] A. O. de Guinea and J. Webster, "An Investigation of Information Systems Use Patterns: Technological Events as Triggers, the Effect of Time, and Consequences for Performance," MIS Quarterly, vol. 37, no. 4, pp. 1165–1188, 2013.

[6] A. S. Lee, M. A. Thomas and R. L. Baskerville, "Going back to basics in design: From the IT artifact to the IS artifact," Proceedings of the Nineteenth Americas Conference on Information Systems, pp. 1757–1763, 2013.

[7] W. J. Orlikowski and S. V. Scott, "Sociomateriality: challenging the separation of technology, work and organization," The

Academy of Management Annals, vol. 2, no. 1, pp. 433–474, 2008.

[8] P. M. Leonardi, "When flexible routines meet flexible technologies: Affordance, constraint, and the imbrication of human and material agencies," MIS Quarterly, vol. 35, no. 1, pp. 147–167, 2011.

[9] G. Ropohl, "Philosophy of Socio-Technical Systems," Techné: Research in Philosophy and Technology, vol. 4, no. 3, pp. 186–194, 1999.

[10] B. Shneiderman, "Computer science. Science 2.0," Science, vol. 319, no. 5868, pp. 1349–1350, 2008.

[11] M. M. Waldrop, "Science 2.0," Scientific American, vol. 298, no. 5, pp. 68–73, 2008.

[12] G. Stock, Metaman: The merging of humans and machines into a global superorganism. Simon and Schuster, New York, 1993. Recited from W. Hofkirchner, "A critical social systems view of the internet," Philosophy of the Social Sciences, vol. 37, no. 4, pp. 471–500, 2007.

[13] A. Geuna and A. Muscio, "The governance of university knowledge transfer: A critical review of the literature", Minerva, vol. 47, no. 1, pp. 93–114, 2009.

[14] J. Dewey, Experience and Nature. George Allen & Unwin Ltd., London, 1929.

[15] E. J. Wilson III, "How to Make a Region Innovative. strategy+business", no. 66, 2012, available at: http://www.strategy-business.com/article/12103?pg=all

[16] E. G. Carayannis and D. F. J. Campbell, "Mode 3" and "Quadruple Helix": toward a 21st century fractal innovation ecosystem," International Journal of Technology Management, vol. 46, no. 3/4, pp. 201–234, 2009.

[17] European Commission, "Open Innovation 2.0", 2015, available at: http://ec.europa.eu/digital-agenda/en/open-innovation-20

[18] T. D. Wilson, "Activity theory," chapter 1 in "Theory in information behaviour research," T. D. Wilson (Ed.), Eiconics Ltd., Sheffield, 2013.

[19] T. D Wilson, "Models in information behaviour research," Journal of documentation, vol. 55, no. 3, pp. 249–270, 1999.

[20] M. Zacklad, "Documents for action (Dofa): infrastructures for Distributed collective Practices", Actes du workshop "Distributed

Collective Practice: Building new Directions for Infrastructural Studies CSCW", p. 6, 2004.

[21] K. C. Laszlo and A. Laszlo, "Fostering a sustainable learning society through knowledge-based development," Systems Research and Behavioral Science, vol. 24, no. 5, pp. 493–503, 2007.

[22] S. Nousala, A. Miles, B. Kilpatrick and W. P. Hall, "Building knowledge sharing communities using team expertise access maps (TEAM)," Proceedings of the KMAP05 Knowledge Management in Asia Pacific, p. 30, 2005.

[23] M. Zacklad, "Communities of action: a cognitive and social approach to the design of CSCW systems," Proceedings of the 2003 international ACM SIGGROUP conference on supporting group work, pp. 190–197, 2003.

[24] F. Geyer, "The challenge of Sociocybernetics," Symposium VI: "Challenges to Sociological Knowledge", Session 04: "Challenges from Other Disciplines", as part of the 13th World Congress of Sociology, 1994, available at: http://www.critcrim.org/redfeather/chaos/006challenges.html

[25] A. R. Hevner, S. T. March, J. Park and S. Ram, "Design science in information systems research," MIS Quarterly, vo. 28, no. 1, pp. 75–105, 2004.

[26] S. T. March and G. F. Smith, "Design and natural science research on information technology," Decision support systems, vol. 15, no. 4, pp. 251–266, 1995.

[27] P. Järvinen, "Action research is similar to design science," Quality & Quantity, vol. 41, no. 1, pp. 37–54, 2007.

[28] S. Lugović, I. Dunđer and M. Horvat, "Patterns-based information systems organization," Proceedings of the 5th International Conference: The Future of Information Sciences (INFuture), pp. 163–174, 2015.

[29] S. Lugović, I. Dunđer and M. Horvat, "Secondary Experience of an Information System Enabling Scientific Communication," Proceedigs of the 1st International Communication Management Forum (CMF2015), pp. 562–587, 2015.

[30] W. Hofkirchner, "How to Achieve a Unified Theory of Information. TripleC: Communication, Capitalism & Critique," Open Access Journal for a Global Sustainable Information Society, vol. 7, no. 2, pp. 357–368, 2009.

The Captology of Intelligent Systems

Zdenko Balaž
Zagreb University of Applied Sciences, (TVZ)
Department of electrical engineering
Zagreb, Croatia
zbalaz@tvz.hr

Davor Predavec
Zagreb University of Applied Sciences, (TVZ)
Department of electrical engineering
Zagreb, Croatia
dpredavec@tvz.hr

Abstract — Captology is an acronym, derived from Computer As Persuasive TechnOLOGY, where the instance persuasive, (lat. *persuasibilibus - enticing*), refers to the convincing persuasion caused by computer technology. Transitive and interactive technologies as intelligent systems, have imposed, by their persuasivity, "the cult of information", after which the information became type of goods that as utilitarian resource need to be quickly and efficiently exploited. Such widely accepted fact resulted as hype, presenting perspective that approach to large amount of information and faster "digestion" of their content and sense will enable users to quickly get desired knowledge. Intelligent systems are ubiquitous in almost all segments of society and life, although there are no relevant researches to confirm the claims of all their acceptability. The paper presents the results of research and testing processes of computer persuasion to show that its success is primarily dependent on its factors. The design for "cloud work", interactive computer programs, web, desktop and other factors directly affect the user, and its attitudes, beliefs, learning and behavior. That impact can be positive or negative.

Keywords — *Artificial intelligence, AI, inteligent technology, ILLG, intuigence, persuasion, permission, expert systems.*

I. INTRODUCTION

Studies have shown that artificial intelligence, (AI), contributes to development of the human mind, through its products, "smart" creations, [1]. Researches of AI from the very beginning of its recognition, (the end of 40-ies of the last century), had a much more complex relationship to philosophical researches related to the science, all to find the methodological guidelines and practical tools for self-analysis and synthesis, simultaneously providing a comprehensive theory of human nature, [2-9]. AI seen as a "new philosophy", but as "anti-philosophy", as well, defined the human being and its place in the universe, with a radically new empirical approach, (Herbert Simon, Alan Turing and Margaret Boden), [10-19]:

- AI with its content, goals and methods produces "smart" creations that simulate and reproduce quasi-cognitive characteristics, (quasi-abilities), of people to reduce symbol manipulation by captological access through the application of specific formal rules.

- By symbol processing, AI emphasizes the key role of cognitive processes calculation and their criteria as crucial intellectual foundations.

- Recognition of AI within technologically-intelligent multitude related to informatics, ensured the required

start for intelligence assessment beyond the conceptual point of view.

- The link established between AI and informatics is extended to engineers of knowledge, designers of creations whose successes must be judged based on their performance, regardless these processes are used by AI and whether any similarity with the corresponding human processes is found.

- "Smart" creations of AI, whether a computer program, computer itself or the intelligent system, are assessed through their activities, which are compared with the human mind.

A. The concept of expert system for processing

In technical sciences, field of computing, there is a branch of artificial intelligence. One of its products is an expert system. Model of such system is simply displayed with three components, (database, base of knowledge and inference mechanism). When the expert system is being compared with information processing type, problem solving and decision making related to human cognition and mental potential, a parallel with intuition and intelligence can be drawn, Fig. 1.

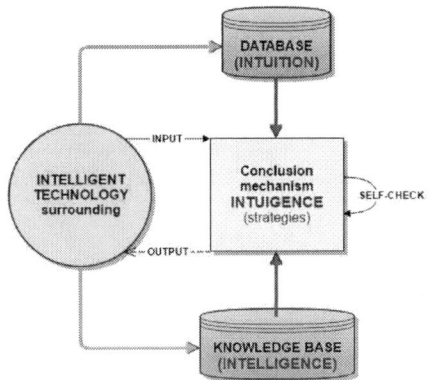

Fig. 1 - Mental competences - model of expert system

The base of knowledge is "cognitive" dependent, based on the symbolic representation and is identical to the intelligence. For the sake of purposefulness, it is necessary to undergo both databases and knowledge to the mechanism of reasoning. In the biological sense of identification with the functioning of the human brain, (way of thinking), latter suggests that the purposeful use of intuition and intelligence need some sort of trained strategy, (*intuigence*). That was confirmed through detailed studies results conducted since 2008 till today, and these findings are groundwork for this paper, [20].

II. CAPTOLOGIES - CONDUCTED RESEARCHES

A. 3P-Model of intelligent technologies and empathy

The strategy development of world computer production and sales, confirmed the promoted 3P-model, [20], (*Persuasive / Permissive / Pervasive*). Within this model, **persuasive**, (lat. *persuasibilibus* - enticing), is associated with convincible and suggestible influence by the intelligent system, while **permissive**, (compliant), is linked to the educational approach that does not require the implementation of almost any control, (the effort of meeting all wishes and requirements); and lastly, **pervasive**, (permeate), is associated with indicators of developmental disorders, (PDD). The common product of latter results as impact on empathy loss, in psychology described as a process of direct involvement to emotional states, opinions and behavior, as well as the ability to understand the emotions of other people, (the way they are reacting to perceived emotions). As a basic emotion, empathy is extremely important for healthy emotional and moral development. Two levels exist therein: emotional and rational, (it is possible to feel what others feel, but also understand rationally).

Apart from being partly inherited through genetics, empathy may be encouraged and learned as any other skill much needed in society. As a skill that develops, it depends on the environment in which lives. It is well known that empathetic people are more successful and happier and have more friends. Empathy is a positive emotion and it's highly advised to develop. By strengthening empathy, the connection between people strengthens and the ability of emotional tolerance is achieved.

A prerequisite for a positive impact on empathy is primarily interest in other people and sincere desire to overcome internal borders. The most difficult task herein is learning to listen others without judgment. Since the characteristic of human mind is that it constantly classifies something, defines, shares, asks - developing empathy is a very demanding task, although is one of the ways of self-development. By listening the other person, it's needed to involve parallelly within its inner world and truthfully understand its point of view. If alongside listening comes words reasoning, then the subject of conversation becomes susceptible for approval or disapproval, understanding and misunderstanding, which leads to the conclusion that much other is made but mere hearing. Speech and hearing are greatly suppressed with the presence of intelligent technology. Examples of studies conducted show that speech is lost in the flood of images and messages that quickly disappear because they cannot be memorized over created archetypes, [21].

B. Persuasive addiction and the human mind

The perfidy of persuasion in the context of information treatment and understanding, comes to the fore when the relative value of each information on network, via intelligent systems, including "cloud-work", (e.g. the ubiquity of web-information), and is evaluated by two factors:

- the number of incoming links information, (page), attracted,

- the authority of the pages from which these links originate.

Studies have confirmed that this is about "third-party tools for experiments on users", [22]. Software algorithms evaluate and rank the value of all information on networks from these databases. Ever since the beginning of the new millennium so called A/B testing have been introduced. By that case, millions of users were informed about 6,000 conducted experiments annually on the search mode with 10, 20 and 30 links. Within five years, more than 450 changes are made in search algorithms and page layout. Therefore, it is not surprising fact that over 300 trials are conducted daily and about 200 signals are continuously carried out therein, all to have effect on the maintenance of "freshness" and speed of collecting, dissecting and transferring data, which is directly related to provider's profit of information services, [23]. Although rarely mentioned, but persuasively successful, captological arsenal was discovered as:

- Cognitive biases, (from more than 170 of cognitive biases, stresses the "risk aversion in time deficit", and which is led by persuasive message),

- The weapon of persuasion, (this is a persuasive technique that rely on the "principle of scarcity" and the "theory of authority acceptance"),

- Productive laziness, (confirmed by the fact of the of the human mind laziness and skipping preferences or careless reading, which, because of persuasive guidance result in unwanted consequences),

- Fatigue of the decision, (associated with the depletion of user by persuasive excessive offers that encourage the application of the least resistance line), [24].

The human mind evolved thanks to the unfailing desire to try and change, while the desire to follow only "reliable and truthful" answers is not the case.

Brain stimulation usually happens while new situations are about to occur, such as: solving new problems, visiting new places, seeing new shows, meeting new people. A larger beneficial of eclectic energy is created due to the novelty of mental or motor stimulus, rather than when it comes as already familiar knowledge.

Although interesting insight as to when learning occurs at the cellular level of the brain, cellular learning and behavior of individuals differ significantly. It is possible to learn good learning processes from books, and to not produce any outward signs of acceptance and application of acquired knowledge. Behavior change that is the result of learning is dependent on numerous factors such as emotional state, previous knowledge, daily fluctuations in brain chemistry, the amount of peptide hormone and captology.

The result of learning should be a human intelligence. The brain is a structure, and mind is what the brain does, (function). The mind is a process. Nowadays, psychometric studies confirmed that the brain can continually create new connections to increase mind learning. The learning capacity of the brain is huge and designed for more demanding intellectual activity than those in which it is usually involved. The human brain has two independently acting systems, for receiving and for thinking. The leading one is in the left half of the brain and it analyzes, writes, reads, speaks and receives logic. The main center for emotions is in the right half of the brain and acts more completely, emotive, and associative - intuitive. Selected activities of the left and right hemispheres of the brain helps the brain learn by pairs model, (challenges and feedback).

Right side of the brain uses for intuitive problem processing based on two components:

- simple provisional rules used by provisional order to recognize the most important information, but ignore everything else, which enables quick response.

- Predispositions of brain evolved abilities - skills acquired genetically or by learning, all to make these predispositions into competences.

The real question seeks how human mind is progressing thanks to the unfailing desire to try and change, and not through desire to follow only "reliable" answers. It also includes the question if the brain development can be achieved when captologies are not new but uniform, deadening situation.

C. Example of the study

In support to abovementioned persuasive addictions, the results of the "Tests / Exercises" can be joined, as educational material originally developed through course "Intelligent Systems" at the Electrical Department of Zagreb University of Applied Sciences, (TVZ). It is basically about learning and reasoning components through the knowledge base and databases of expert and agent system, [25]. The conclusion is carried out by applying the acquired knowledge of distributed artificial intelligence in correlation with human intelligence and social predetermination. The instruction accurately describes the task which each student receives in envelope. The teacher reads instructions out loud before the beginning of the task, in front of the students at the initial location of the implementation. Two groups are solving the task shortly after given instructions, at two different locations. Students are reminded that they have the ability to cooperate with each other, where this is exactly the object of exercise. Five simple questions are conceived that must be answered in a way that all students participate and create a common, identical answer. The guideline text is divided in three parts, (description of the assignment, approach to the resolution and guidelines for the solution). Through four educational periods in three academic years, 148 students took part in this task and in any case the task has not been solved.

The reason such a simple task is not solved lies in the data, (information), hyper-offer. Students are taught that with the help of information technology comes fast data browsing. Therefore, task understanding comes down to short view and is very shallow, which makes the easy way to create oversight in the understanding of approaches to solve task. Too much data distract from deep and lasting engaging in any single argument, fact or idea. Too much details and characters nullify each other. Hence, students are prone to shorten sometimes exhausting search and move toward each connector, (link): Thusly piled but

seemingly interesting data are the start of the interruption in current potential of available human concentration. Thence, the externally generated interference of human attention and lack of cooperation is performed. As in browsing and search within vastness of data the case is to cause users to quickly come and go, the same effect is achieved within the messy text. Rapid collection and transmission of data causes cognitive efficiency loss. Furthermore, a multitude of confusing information and parameters that draw attention, fragment, impede and impair the flow of thoughts finnally end up as distractions. Self-efficacy is heavily influenced by intelligent technology and therefore under the pressure. Part of the human mind, in charge for critical thinking, analysis of reality and which should be used daily is not evident and has a lack of understanding. It should be used as a form of thinking when solving problems, but also in social interaction when used to connect to the outside world. That element has been lost in captology for it does not follow the sequence of evaluation, Fig. 2., which is the process where mind through stages of gathering information and understanding them acquires the higher forms of evaluation.

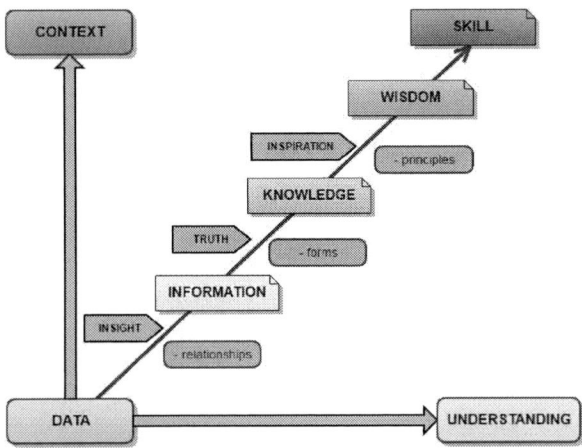

Fig. 2 - The levels and forms of understanding evaluation

D. Example of understanding while meeting other people

One example of understanding related to learning to understand while meeting other people was carried out by William Outhwaite, who observed the situation meeting other people through four categories:

1. Category of physical facts that can be observed, (how people look like and where that result come from). These are understandable phenomena's in the relevant sense when seen as a mark of something not physical from last time or something related with the mental state.

2. Category of mind states at others that may be identified immediately, or can be concluded based on previous findings.

3. Category of others actions, (detection of what others do and what they consider they're doing), is described through three subcategories:

1213

- Action identification accessed from distance,
- Acceptance of what is said about on the same elemental level,
- Sorted judgment on the actual significance of what it is or what is really meant by what is said about it.

4. Category of questions about why certain things work, and what motivates people to do so.

The difference is tried to be made through abovementioned categories, (especially between third and fourth category, i.e. the difference between motives and "hermeneutic understanding"). The psychological understanding of human mind mental states can be observed based on three conclusions about people motives and their intentions, and they are: a), visible signs, b), explicit statements and c), knowledge of the "situation facts", which is in fact confirmed by the psychology of the situation, [26]. From latter examples, the complexity of understanding knowledge can be seen, which is theoretically processed, and mere social process of knowing people, is incorporated as algorithm on a subconscious level, [27].

E. Example of understanding while reading out loud

For confirmation of this research is useful to consider the adoption of understanding through example of literacy development research. This case leads to the conclusion that central process in learning to read actually to understand the text and active search for meaning and purpose, as Whitehurst & Lonigan determined. The process of reading, i.e. reading comprehension, comes down to understanding encryption and decryption, which is nowadays included in media environment, Fig. 3.

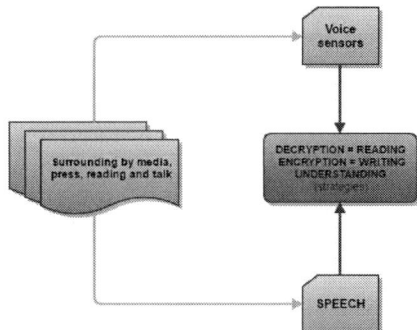

Fig. 3 - The process of reading - modern developed understanding of reading, (encryption/decryption and understanding)

In fact, the presented is mere concept of expert system functioning, groundwork for explanation and processing.

Researches on the reading impact has set the foundations for mechanism of reading explanation, which are:

- comparison,
- memory recall,
- connecting what is seen and what is heard, (or being heard),
- identifying the relevant data and discarding redundant.

On that basis, data processing parallel to reading takes place with five concurrent methods, known as the "Model of 5 babushkas", developed by R. J. Marzano and D. E. Paynter, which includes:

1. Understanding the reading goal,
2. Assessment read data conformity with the goal,
3. The meaning of smallest thought units, (assertions),
4. Recognizing the meanings of words,
5. Recognizing sense of what was read.

Reading, listening, watching and pronunciation are learning modes, and each one has a different contribution to what is learned. According to the learning mode, left and right cerebral hemisphere is activated, but the emotional center of the brain as well, Table I, [20].

#	Percentage	Learning mode	Activation
1	10 %	Reading	L[a]
2	20 %	Listening	R[b]
3	30 %	Watching and reading	L+R
4	50 %	Watching and listening	L+R
5	70 %	Pronunciation, (aloud)	L+R+EC[c]
6	90 %	Pronunciation and learning	L+R

a. Left brain hemisphere; b. Right brain hemisphere; c. Emotional brain center

Understanding the learning by reading, (reading aloud), is a strategy that along with understanding involves understanding development, voice sensitivity development and motivation development. These examples, in particular data processing operation when reading, is an important basis for different types of reading analysis, such as:

- matured reading,
- linear text reading - printed book,
- linear text reading - e-books,
- reading by text search - computer,
- hypertext reading – computer.

It is necessary to distinguish the reading from search, which being influenced by captology is turned into an addictive need, because they held worthless power over digital information flow. Imposed "Internet Ethics" as the media, results in persuasive addiction that manifests through:

- stimulation and "thirst" quench for small segments of short-lived information,
- curious dependence on uninterrupted progress which is updated in real time, "updated status" that loses concreteness already few moments after the announcement and
- desire to speed up the flow of information.

These events do not have any contributions for the storage of learned, for nothing has been taught. There was a disturbance and activation of the brain hemisphere nor emotional center of the brain hadn't been recognized, which nowadays can be confidently confirmed by psychometric tests, [28].

F. Examples of hermeneutic understanding

Accepting the presented facts, human as a rational being can observe its understanding on two levels. The first is a subjective level or psychological level of participation, and the other is an objective level, or the level of intellectual participation. Thusly, an objective understanding of the meaning and subjective understanding of the motives and intentions can be distinguished. Next to it, basic and higher forms of understanding can be recognized, (graph in Fig. 2). The elementary forms of understanding are regulated by the concept of objective mind or spirit, which guarantees the possibility of inter-communication at the basic level.

"Higher forms of understanding" act confrontative with an internal difficulty or contradiction with what is already known. Therefore, recall and input of all living structures is necessary, (establishment of relations between life and expression of its mental content), whose foundation can be found in the hermeneutic understanding, [29].

Hence, it is possible to conclude that the hermeneutics is universal theory of understanding where is no misperception for hermeneutic understanding is not an interpretation or explanation. The ancient interpreters of hermeneutics confirm latter by saying: *"sensus non est inferendus, sed eferendus"*, i.e. meaning must be parsed from the outside, not from the text, [30].

Studies that processed captological components of intelligent interactive technology are confirmed through examples of human thinking simulation by computer in the field of cognitive science. Model of expert system has processed numerous examples of research studies. The research results show that style of thinking and information processing develops and modifies over long period. On a slightly higher level of flexibility, mental skills which process information intuitively and analytically can be acquired, [31]. These findings, along with persuasive context and information technology close the "hermeneutic circle of artificial intelligence", which requires certain psychoculture, Fig. 4.

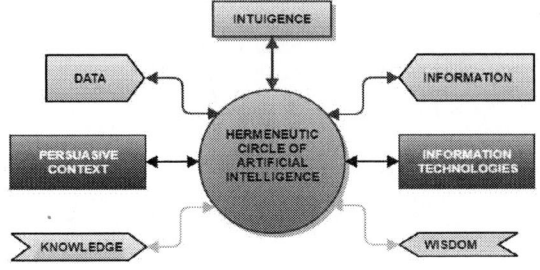

Fig. 4 - Hermeneutic circle of artificial intelligence

G. Psychoculture and captologies

The term "psychoculture", according to [32], calls for culture, (lat. *colere* – to breed, nurture; *cultos* - cultivation), which represents an integrated system of attitudes, beliefs and behavior patterns typical for members of society, and which are not the result of biological heritage, but the social product, transmitted and maintained through communication and learning. The function and meaning of culture are assistance, extension and development of human society. Culture arises from the people's need to master the nature and to control the animal part, hence not only external nature is cultivated but its own, which realizes the humanization process. Under various influences and thusly captological ones, culture is the subject of collective mind programming that differs members of one community from others. Psychoculture, which is associated with the cultivation of the soul, is mentioned in ancient times, e.g. by Cicero, (lat. *cultura anima*). Developing and training noble qualities, but also the creation of a collective mentality in society can be achieved by dominant ideology that directs the social and political processes. Every individual has own ideology or value system and environmental landmarks that use as directions in life, but also determine the behavior that can be influenced by captology. Socio-cultural neuroscience enables better insight into the complex interactions between the socio-cultural and neural structures and processes, and thus the biological mechanisms that underlie social and cultural phenomena and people behavior, including all components of existentialism.

From these findings is useful to point out, according to [33], the structure of neural networks of the cerebral cortex is the biological basis of our mental abilities. In adulthood, new knowledge and skills are simply acquired, without changing the structure of the neural network. Education, psychological, social and emotional environment during development change structure of the neural network, especially in parts of the brain essential for adoption of the most complex brain functions. This includes the affective modulation of emotional expression, conceptualization of the own mind, mentalization, cognitive flexibility and working memory. Captological impact is confirmed on the mental and cognitive abilities of people, at a sufficient level of psychometric tests, but the results of these impacts are not transparently recognized, therefore initiated research continues, [34-37].

III. CONCLUSION

When education, psychological, social and emotional environment connect with the presence and intimacy of intelligent technology, it leads to confirmation of their multiple influence hypothesis, as in a positive way by raising level of integrated development knowledge of the human mind, thus in the negative way through insights about origin of the neuropsychiatric diseases development. Biomedical and social concept is one of the main subjects of continuing debate between advocates of selective, as opposed to design theory of mental processing, unfortunately, only present at the expert level of medical profession. This topic arises an important question: how to use the mind and intelligent technology in modern times? One part of the human mind is responsible for critical thinking, analysis of reality, and as a way of thinking, it is used daily to solve problems, but also in social interaction, when used to connect to the outside world. The second part of the mind is used for imagination and can be operated when the analytical thinking is excluded. The realization that imagination and analytical thinking are not two opposing forms of thought, but two kinds of thinking with different functions, is a result of intelligent technology contribution, whose use must be decided by human intelligence.

REFERENCES

[1] B. J. Fogg, Persuasive Technology: Using Computers to Change What We Think and Do (Interactive Technologies), publisher: Morgan Kaufmann; 1st edition (December 30, 2002), ISBN-13: 978-1558606432

[2] K. W. Heisenberg, "Promjene u osnovama prirodne znanosti", six lectures, translation by Mladen Klepac, Kruzak, press: Librikon, ISBN 953-6463-16-4, Zagreb, 1998.

[3] K. W. Heisenberg, "Fizika i filozofija", translation by Stipe Kutleša, Kruzak, press: S. Brusina– Donja Lomnica, ISBN 953-96477-3-8, Zagreb, 1997.

[4] H. Gardner, M. L. Kornhaber and W. K. Wake: Intelligence: Multiple Perspectives, Hercourt Brace College Publishers, Fort Worth, Philadelphia, San Diego, New York, Orlando, Austin, San Antonio, Toronto, Montreal, London, Sydney, Tokyo, 1996. Translation by Inteligencija različita gledišta, Gordana Keresteš, Vlasta Vizek Vidović, press: Slap, editor: Krunoslav Matešić, library: PARNAS, press: Slap, ISBN 953-191-094-4, Zagreb, 1999.

[5] J. M. P. Teilhard de Chardin: L'Energie Humaine, Editions du Seuil, Paris, 1962., translation by M. Dobrović, Ljudska snaga, IPT Naprijed, Zagreb, 1991.

[6] T. Matulić, "Metamorfoze kulture", Tertium mullenium, ISBN 9789-5324-1161-14, CIP 86343, Zagreb, 2009.

[7] F. M. Arouet - Voltaire, "Rasprava o toleranciji", translation by Bosiljka Brlečić, MH, editor: Jelena Hekman, library: PARNAS, press: Targa, ISBN 953-150-082-7, Zagreb, 1997.

[8] P. Vranicki, "Filozofija historije", the second book, Press: Naprijed, editor: Srđan Dvornik, library: Enciklopedija filozofskih disciplina, press: Hrvatska tiskara, ISBN 953-178-012-9, Zagreb, 1994.

[9] V. Hoesle, "Filozofija ekološke krize", Moskov lectures, Matica Hrvatska, Urednica Jelena Hekman, Press: Targa, , ISBN 978-953-150-051-7, Zagreb, 1996.

[10] A. Newell and H. A. Simon, "Theleologic theory machine", IRE Transactions on Informations Theory, Vol.2(3), pp 61-67, 1956.

[11] A. Newell, J. C. Shaw and H. A. Simon, "Elements of a theory of human problem solving", Psychological Review, Vol. 65, pp 151-166, 1958.

[12] H. A. Simon, "Experimentswith a heuristic compiler", Journal of the Association for Computing Machinery, Vol. 10, pp 493-506, 1963.

[13] H. A. Simon and A. Newell, "Information processing in computer and man", American Scientist, Vol. 52, pp 281-300, 1964.

[14] H. A. Simon, "Motivational and Emotional Controls of Cognition", Psychological Review, Vol. 74, No 1, pp 29-39, 1967.

[15] H. A. Simon, "The structure of structured problems", Artificial Intelligence, Vol. 4, pp 181-202, 1973.

[16] H. A. Simon, "The design of large computing systems as an organisational problem", Organisation, pp. 163-180, 1976.

[17] H. A. Simon, "Artificial intelligence systems that understand", Proceedings of the Fifth International Joint Conference on Artificial Intelligence, Vol. 2, pp 1059-1073, 1977.

[18] H. A. Simon, "Search and reasoning in problem solving", Artificial Intelligence, Vol. 21, pp 7-29, 1983.

[19] H. A. Simon and A. Newell, "Information processing language V on the IBM 650", Annals of the History of Computing, Vol. 8, pp 47-49, 1986.

[20] Z. Balaž, "Intuigencija KAPtološke ZAmke UMjetne InteliGENcije", The manuscript of the book is to be reviewed, Zagreb, September, 2016.

[21] R. Lachmann, "Phantasia Memoria Rhetorica, translation by Davor Beganović, ISBN 953-150-150-560-8, Matica Hrvatska, Zagreb, 2002.

[22] N. Carr, "Plitko: Što Internet čini našem mozgu", Jesenski i Turk, translation by Ognjen Strpić, Zagreb, 2011.

[23] O. Parmy, "Ekonomija zamoraca", FORBES, Vol XIV, No 3-4, UDK 167.7:316, pp. 63-68, III. 2015.

[24] H. Gardner, M. L. Kornhaber and W. K. Wake," Inteligencija različita gledišta", editor: Krunoslav Matešić, press: PARNAS and Naklada slap, ISBN 953-191-094-4, Zagreb, 1999.

[25] Z. Balaž and K. Meštrović, "Učenje i poučavanje iz umjetne inteligencije", Zagreb University of Applied Sciences, Polytechnic & Design, Vol. 2, No 1, pp 9-14, ISSN 1849-1995, Zagreb, 2014.

[26] U. V. Detlev, "Psychologie und Welt", Verlag W. Kohlhammer, Stuttgart, Berlin; Koeln, Meinz, 1972.

[27] B. H. Lipton, "Biologija vjerovanja – Znanstveni dokazi o nadmoći uma nad materijom", editor: Darko Imenjak, translation by Goran Bosnić, press: Denona d.o.o., graphic design: Teledisk, ISBN 978-953-7039-55-2, CIP 642337, NSB Zagreb, 2007.

[28] N. Birbaumer, "Der Weg zum Super Hirn", Hoerzu Wissen, pp 21-25, IV. 2014.

[29] V. Afrić, "Teorijske osnove hermeneutike", sociological revue, Vol XIV, No 3-4, UDK 167.7:316, pp. 191-200, 1984.

[30] J. Grondin, "Smisao za hermeneutiku", Matica Hrvatska, editor: Jelena Hekman, press: Targa, , ISBN 953-150-512-8, Zagreb, 1999.

[31] Z. Balaž, "Intuigencija i kaptologije inteligentnih tehnologija - izazovi i strahovi", article sent to the newspaper Polytechnic & Design, ISBN 1849-1995, Zagreb University of Applied Sciences, September, 2016.

[32] M. Jakovljević, "Mozak, duša i kultura u zdravlju i bolesti - Izazovi pred psihijatrijom 21. Stoljeća", HAZU: Symposium on research and diseases of the brain, Zagreb, 16th March, 2016.

[33] Z. Petanjek, "Sinaptički izazovi u djetinjstvu i adolescenciji: Izazovi i smjerovi suvremenog istraživanja mozga", HAZU: Symposium on research and diseases of the brain, Zagreb, 16th March, 2016.

[34] Z. Balaž, "Promatranje vjere kroz kaptologije inteligentnih tehnologija", The work sent to the chief editor inspired by New Evangelization article – "From interpersonal contact to the new media", GK, nr. 51, (2217), 18th December, 2016.

[35] R. Rohr, "Golo sada – Vidjeti kao što mistici vide", Synopsis, Zagreb – Sarajevo, ISBN 978-953-7968-32-6, 2016.

[36] D. Goleman, "Emocionalna inteligencija, Zašto je važnija od kvocijenta inteligencije", 2nd edition, editor: Ivanka Borovac, Translation by Damir Biličić, Mozaik knjiga, Zagreb, 1997.

[37] D. Predavec, "Kaptologija inteligentnih sustava", Thesis nr. E697, Zagreb University of Applied Sciences, Polytechnic specialist graduate studies - Electrical Engineering department, Intelligent Systems course, mentor: Ph.D. Zdenko Balaž, mag.ing.el., Zagreb, March, 2017.

Automatic Information Behaviour Recognition

S. Lugović* and I. Dunđer**
* Zagreb University of Applied Sciences, Department of Computer Science and Information Technology, Vrbik 8,
Zagreb, Croatia
** University of Zagreb, Faculty of Humanities and Social Sciences, Department of Information and Communication
Sciences, Ivana Lučića 3, Zagreb, Croatia
slugovic@tvz.hr, ivandunder@gmail.com

Abstract - This paper aims to propose and discuss concepts of how users can recognise information seeking behaviour automatically and what implications such an automatic recognition can have. The authors develop the discussion around variables proposed in Wilson's second model of information behaviour and state how they can collect data necessary to recognise information behaviour automatically. The authors give an overview of different parts of Wilson's second information behaviour model such as activating mechanisms (stress/coping, risk/reward, and self-efficacy), intervening variables, different stages in the information acquisition process, and types of information seeking. They also discuss streams of data that can be collected and processed automatically. From a set of thought experiments, the authors propose that by processing computer log files and recognising user affective states, which employ affective computing techniques, it is possible to recognise information seeking behaviour. Analysis also shows that Wilson's second model of information behaviour can be used as a reference model in designing automatic information behaviour systems. By quantitatively describing the information behaviour sequence, such a description could be implemented into the computer algorithm, and that process of information searching could be replicated reducing the time needed to satisfy information needs. Modern society is one step closer to making the trails Vannevar Bush proposed a reality.

Keywords - data mining, information behaviour, systems modelling, adaptive information systems, information and communication sciences

I. INTRODUCTION

This paper aims to propose conceptual ideas related to automatic information behaviour recognition. As a fundamental block of this very analysis, the authors use T. D. Wilson's second model of information behaviour [1] and discuss different sets of variables that it proposes in terms of how they could be used to recognise behaviour automatically.

The authors discuss two main techniques used in collecting data necessary to recognise information behaviour. One is computer log file processing, and another is affective computing. In their view, instead of observing the user or sources of the information separately, it is necessary to observe "patterns" of the interaction of the user with the sources of information (stored in an information system).

The authors define "patterns" as a time sequence happening between the moment a user experiences information, i.e. when structured data become information and the information seeking process starts, until the moment the user stops interacting with the information source as the user satisfies his information needs. Such a process can occur in social and technical domains [2].

This is a working definition describing the scope of the analysis. Such a proposition can connect with the concept of trails proposed by [3]. The authors would also like to propose that every interaction of the user, described as a "pattern", and the source of information is observable as an open system of interest.

Patterns, as defined above, are a dependent variable of system behaviour, and the user's affective states accompany them.

The authors believe in the importance of such a view, as system behaviour is not quantitative, which makes it hard to collect and process data automatically. But by observing the user's log files created while interacting with the information resources and using affective computing techniques, one can describe "patterns" with rich quantitative data sets.

The main question the authors would like to raise and discuss is – how can different variables from Wilson's second model be used to facilitate automatic information behaviour recognition?

II. INFORMATION BEHAVIOUR

Information behaviour is "the totality of human behaviour in relation to sources and channels of information, including both active and passive information seeking, and information use" [4].

Different people may engage in different types of information behaviour even when faced with similar tasks and circumstances [5]. Information behaviour can be understood as an adaptive mechanism [5] that responds to changes in the environment, creating complex system behaviour.

There are two descriptions that can help to understand complex systems: state descriptions and process descriptions. The former characterises the world as sensed, and the latter characterises the world as acted upon [6].

Reference [6] indicates that this distinction is an essential condition for the survival of the adaptive organism – in this case, an adaptive system of the user and the information source – and it should develop correlations between goals in the sensed world and actions in the world of process. When such correlations become conscious and verbalised, it is possible to find differences between them.

The task is to discover information behaviour sequences resulting in changes in state and process descriptions, and this changes are goal-driven. For example, if a person starts seeking for information by using an information system, then the person's affective states will be different at the moment when the person starts to seek for information (one state description) and at the moment when the person has found the information (another state description) [7].

Observing the changes in the process description enables one to recognise when the information need is satisfied.

By observing how log files in the information system change while a user is performing information seeking, it is possible to determine when the process is finished, as there are no new log files being produced.

Central parts of Wilson's second model are two activating mechanisms and intervening variables. An activating mechanism is what prompts someone to engage in activities, and intervening variables are factors that influence those activities [5].

The activating mechanisms and intervening variables can help in describing the sequence of a process description. State descriptions tell about the information need, i.e. the initial state, and about satisfying this need, i.e. the goal state.

There are correlations between goals in the sensed world (information need and information need satisfaction) and actions in the world of process (activating mechanism and intervening variables).

But if there is no awareness of those correlations – i.e. the satisfaction of the information need and the corresponding satisfying actions – and the difference between the two isn't recognisable, then the information system does not produce much of value to the user in terms of adaptiveness to particular user information needs.

For instance, let us assume that there is a new librarian working in a library and he does not know on which shelf a particular book is. A new visitor enters the library and he is looking for this very book and finds it by himself. The librarian recognises only that the visitor found the book and that his information need is therefore satisfied. But the librarian does not know the steps the visitor took to find the book he was looking for. Consequently, when another user comes to look for the same book a few hours later, the librarian cannot help. But if the librarian knew the process the first visitor followed to find the book, then he could easily help a second user to find it.

How the process description, i.e. the sequence of events or steps the first visitor used to find a book, is verbalised will influence the quality of its transition to the state description, i.e. the instructions on how to find a particular book.

So if one understands information behaviour as a process description and if one is able to recognise it automatically, then it is clearly possible to translate it into the state description.

In other words, the more one knows about information behaviour, the easier it gets to move users from process (actions) to state (goals) descriptions.

According to Wilson's second model, when an information need arises, a system, which consists of a user and a source of information, starts with the process of acquiring information.

There are two activating mechanisms. The theory of stress and coping explains the first activating mechanism [1]. Stress is a relationship between the person and the environment, and the person appraises such a relationship through his reflection on the resources available to him and his wellbeing [1]. Coping is cognitive and produces behavioural effects to master, reduce, or tolerate the internal and external demands that stressful situations create [1].

Reference [1] indicates that various states will exist as a result of intolerance to uncertainty and arousal. If one looks at the user and the source of the information as one open system, uncertainty is important as the information that enters the system reduces it. It is possible to understand the arousal as an intention of the system to move into the steady state, and it is done by inputting new information into the system.

Reference [8] states that a living organism maintains a disequilibrium called the "steady state of an open system". Characteristics of an open system in the steady states are that it can do work [8], and such a state is reachable by equifinality, that is, a process specific to open systems in which a system may reach the same state from different initial conditions, and only the system parameters determine it. In this case, such parameters will depend on properties of the user and the information system that stores information resources.

Such changes may occur since the living system initially is in an unstable state and tends towards a steady state, such as, roughly speaking, the phenomena of growth and development [8].

In terms of information behaviour, when a user experiences an information need and starts to interact with the information sources, and if one perceives these two entities, i.e. the user and the information source, as an open system, in the beginning, it is in an unstable state, but as information needs become satisfied, such a system moves towards a steady state. And the closer it is to the steady-state system the more it develops and the more it is capable of doing better work.

For example, when a student has to pass an exam, the information need begins, as he starts the process of seeking and searching for information. And the more relevant papers he finds and reads, the more he knows about a specific subject, whereas his chances of passing the exam increase, being more capable of doing work.

At the same time, if the information system – that stores those papers and acts as an information source for the student – could recognise student information behaviour and adapt accordingly to the student's information needs, it would also develop and would be capable of performing better. And if one understands this distinct joint environment, which consists of a student and an information system storing relevant papers, as an open system, such a system will evolve and develop too.

III. AUTOMATIC INFORMATION BEHAVIOUR RECOGNITION

An important question emerges – how can an information system that acts as an information source to the user automatically recognise user information behaviour?

The authors believe that doing this automatically is possible if two sets of techniques are applied. The first is by collecting, analysing, and interpreting computer log files, and the other is affective computing, which is the study and development of systems and devices that can recognise, interpret, process and simulate human affects [9].

Those thoughts derive from recent researches that are based on behavioural computing and which are pointing out the distinct differences between demographic and behavioural data, and which show how they are structured for the purpose of conducting analyses. Such a data structure is essential for developing specific algorithms on top of the data which represent the user behaviour [10-11].

Any change in the emotional states of the user and change in the information system's interaction patterns could signal that there are changes in the information need [7].

If stress/coping theory explains the first activating mechanism in Wilson's second model [1] and involves user affective states, recognising user emotions in the process of seeking and searching for information is important and could be done automatically.

The next group of variables proposed by Wilson relate to the psychological, demographic, role-related, environmental, and source characteristics of the user who is interacting with the information system while seeking and searching for information [1].

Data describing those variables can be found in many sources. If seeking and searching happen in an organisation that has an information system with functions that support human resources management, then most of the data needed is already in those systems.

Other sources are social media platforms, but it is also possible to collect relevant data through sensors (e.g. biofeedback, ubiquitous computing). In today's communication environments, one can gather lots of data from the interaction between users (user to user, user to system, user to system to user). It is also possible to collect data related to time spent while interacting with the system and, for instance, geo-locations of the system users. By collecting those data and comparing them with the log files that describe patterns and affective states, one

can develop various computing techniques that automatically recognise an information need.

The types of information seeking [1] proposes are passive attention, passive search, active search and ongoing search. The authors hold that those four types of search are easily recognisable by observing log files collected from users interacting with the information source, since different log files probably correlate to different types of searches.

In Wilson's second model, the second activating mechanism is described as risk/reward and self-efficacy [1]. The risk/reward model is essential for understanding information seeking behaviour.

User behaviour will depend on the risk/reward ratio, and such a risk is not only evaluated through financial resources, but also trough psychological and physical resources.

But how can one recognise and measure the risk/reward ratio from the signal, i.e. the stream of data, one can capture from the user-information system interaction?

Components that are included in risk are performance risk, financial risk, physical risk, social risk, ego risk, safety risks and time-convenience loss [1]. If one recognises the type of information seeking behaviour from the log files, it is possible to build models that can help one to automatically recognise and connect different types of risks with different seeking behaviours.

For example, if a student is looking for exam-related information at the information kiosk (i.e. information system) in the university hall, and if the system is able to recognise that particular exam within a certain timeframe (e.g. 10 minutes from the moment a student starts to interact with the information kiosk), then the main goal of the information system is to reduce the performance risk by providing the student with basic data (e.g. in plain text format) for fast information retrieval.

But if the system knows that the student, that is looking for exam-related information, had already failed the exam a few times, and has to pay additional exam fees if he fails again, then the system should reduce the student's financial risk by offering him more extensive and relevant data, such as books to read, in order to decrease the chances of failing the exam once more.

And let us assume that the information kiosk has sensors that can detect a student's stress level, and if such a level is high, then the system might play relaxing music to reduce the student's physical risk.

By observing log files while a student is searching for information, the system can detect if the student is not focused and therefore cannot find information, and adapt accordingly. Also, by comparing log files of a student's search process with the log files of his colleagues, the social risk will possibly be reduced.

By using trails [3] that are already stored in the information kiosk, the system can reduce time/convenience ratios by employing implemented

1219

recommendation techniques, which can improve the person's state of happiness, i.e. can reduce ego risk.

Another part of the second activating mechanism is related to the self-efficacy, as an "individual may be aware that use of an information source may produce useful information, but doubt his or her capacity properly to access the source" [1].

There is a link between self-efficacy and coping strategies, as people are more likely to act, or in this case to search for information, if they are convinced in their own effectiveness. It will also influence the amount of effort invested in the process.

According to [12], efficacy expectations are based on performance accomplishments, i.e. carrying out the actions oneself; vicarious experience, i.e. learning from others; verbal persuasion, i.e. which may include self-instruction; and physiological states, particularly emotional arousal.

The authors believe that it is possible to recognise from log files whether the student is looking for information in order to learn by himself (e.g. by searching for papers), from others (e.g. by searching for students that already passed an exam and by looking up their contact information), or from postings to find the date of the next exam.

By using affective computing techniques, one can recognise different levels of emotional arousal. Reference [1] proposed the following stages in human computer interaction for the purpose of the information acquisition process: intelligence, intention or goal formation, design, choice or selection, information extraction (and integration), and review or evaluation.

Breaking down and observing the process of student interaction with the system (for example the information kiosk) throughout these stages, one can hypothesise that it is possible to recognise behaviour by observing interaction log files and data describing the student's affective states.

IV. CONLUSION

The authors briefly demonstrated that Wilson's second information behaviour model can be very useful as a rich source of variables that describe the process of satisfying information needs.

Following Wilson's second model the authors describe the sequence from the initial state to the goal state. The

sequence can be described quantitatively by implementing computing techniques that collect and process data related to the user's stress/coping, intervening variables that influence the behaviour, different types of associated risks, and different stages in information acquisition.

Such a description could be implemented into an algorithm, reducing the time needed to go from the initial state to the goal state. By doing so, the "recipe" for how to find information (or trails) is stored in the system.

Through observing the user and the information system he interacts with as one open system, one can compare and analyse different systems and test how different "recipes" (or trails) perform. And the ones that perform well can be implemented into the design of the information system.

REFERENCES

[1] T. D. Wilson, "Information behaviour: an interdisciplinary perspective," Information Processing & Management, vol. 33, no. 4, pp. 551–572, 1997.

[2] S. Lugović, I. Dunđer and M. Horvat, "Patterns-based information systems organization," Proceedings of the 5th International Conference: The Future of Information Sciences (INFuture), pp. 163–174, 2015.

[3] V. Bush, "As we may think," The Atlantic Monthly, vol. 176, no. 1, pp. 101–108, 1945, available at: http://www.theatlantic.com/unbound/flashbks/computer/bushf.htm

[4] T. D. Wilson, "Human information behavior," Informing Science, vol. 3, no. 2, pp. 49–55, 2000.

[5] N. Ford, Introduction to Information Behaviour. Facet Publishing, London, 2015.

[6] H. A. Simon, "The architecture of complexity," Proceedings of the American Philosophical Society, vol. 106, no. 6, pp. 467–482, 1962.

[7] C. C. Kuhlthau, "Inside the search process: Information seeking from the user's perspective." Journal of the American Society for information Science, vol. 42, no. 5, pp. 361–371, 1991.

[8] L. von Bertalanffy, General System Theory: Foundations, Development, Applications. George Braziller, New York, 1969.

[9] J. Tao and T. Tan, "Affective computing: A review," Proceedings of the First International Conference on Affective Computing and Intelligent Interaction (ACII), pp. 981–995, 2005.

[10] L. Cao and S. Y. Philip, "Behavior Informatics: An Informatics Perspective for Behavior Studies," IEEE Intelligent Informatics Bulletin, vol. 10, no. 1, pp. 6–11. 2009.

[11] L. Cao, "In-depth behavior understanding and use: the behavior informatics approach," Information Sciences, vol. 180, no. 17, pp. 3067–3085, 2010.

[12] A. Bandura, "Self-efficacy: toward a unifying theory of behavioral change," Psychological Review, vol. 84, no. 2, pp. 191–215, 1977.

Adjective Representation with the Method Nodes of Knowledge

M. Pavlic[*], Z. Dovedan Han[**], A. Jakupovic [***] M. Asenbrener Katic[*] and S. Candrlic[*]
[*] Department of Informatics, University of Rijeka, Rijeka, Croatia
[**] Faculty of Humanities and Social Sciences, University of Zagreb, Zagreb, Croatia
[***] Polytechnic of Rijeka, Rijeka, Croatia
mile.pavlic@ris.hr, zdovedan@hotmail.com, ajakupovic@veleri.hr, masenbrener@inf.uniri.hr, sanjac@inf.uniri.hr

Abstract - The paper analyzes the semantics of adjectives in sentences in English language. The goal is, based on the analysis of individual sentences and the method of induction, to show the semantic relationship between the adjective and other word types by using the method Nodes of Knowledge (NOK). The NOK method is used for formal knowledge representation expressed in text, i.e. to represent the knowledge network. The method incorporates the enrichment of representation with hidden semantics contained in the context of an individual sentence. The paper analyzes and graphically represents the location of the adjective in a sentence using the NOK method diagram (DNOK). DNOK graphically shows the knowledge network, which can be computer-implemented into the knowledge base as a part of the construction of intelligent information systems. The construction of intelligent information system is not possible without a complete metamodel of language and this work is a contribution to the research.

I. INTRODUCTION

One of the unsolved problems in the field of the development of database and programming tools is the input of language and knowledge of the language in a computer system which could understand the language and communicate intelligently. In addition, there is also the problem of machine translation of languages. Both problems are being addressed through the construction of intelligent software product and a corresponding knowledge base [1].

From a series of necessary properties, from the standpoint of communication with human, an intelligent system is any system that has the following properties [2], [3]:

- the system must communicate with humans and other intelligent systems in a friendly manner – therefore, it uses natural language and speech.

- such communication involves handling ambiguities and grammatically incorrect sentences.

Knowledge representation (KR) began to develop in the area of artificial intelligence in 1970-ies and remained one of the strongest sub-areas of artificial intelligence [4], [5].

There are numerous ways to display knowledge [6]. Schemes for knowledge representation can be divided into four basic approaches [7], [8]: network schemes, logical schemes, procedural schemes and framework theories. There are also other approaches to represent knowledge such as: object-oriented approach, descriptive logic, taxonomies, ontologies and others.

Semantic network [9] is a network scheme for knowledge representation that represents knowledge in the form of a graph. Nodes mark objects or concepts, their properties and corresponding values. The arcs mark the relationships between nodes. Nodes and arcs generally have names (arcs have weight). Nodes can represent concepts, objects, events, features, time and so on [5], [8], [10]. The NOK method belongs to the group of semantic networks. A comparison of the NOK method to other methods is given in [11].

Here we can provide a short comparison of the NOK method to two other methods for knowledge representation: Quillian's model of semantic memory [12], [13] and Minsky's frame theory [14].

In Quillian's model of semantic memory, each word is stored along with the configuration of the pointer to other words, and this configuration represents the meaning of the word. For example, if we want to store the following statement: "Canary is a yellow bird.", then the word "canary" is stored in memory, which has the pointer to the word "bird" (the category name of canary) and pointer to "is yellow" (representing a property of canary). The NOK method also has pointers to other words but these pointers have a semantic identifier which indicates the reason for the existence of the pointer (or link). Semantic identifier is represented by a question. In the example of the statement "Canary is a yellow bird.", the NOK method would record the words "canary", "is", "yellow", "bird", the links between them, as well as questions that represent the reason for the existence of links ("canary" - (who?) - "is" - (what?) - "bird" - (what kind?) - "yellow").

In Minsky's frame theory, the frame is a data structure (which is stored in slots) that stores information such as: Facts or Data, Procedures, Default Values and Other Frames or Subframes. Table 1. shows the statement "Boy is a male person under 12 years.". It can be seen that this type of representation results in the loss of the semantic relationship between words that exists in the sentence, which does not occur with the NOK method.

TABLE I. REPRESENTATION OF THE STATEMENT "BOY IS A MALE PERSON UNDER 12 YEARS." USING THE MINSKY'S FRAME THEORY

Slot	Value	Type
BOY	_	(This Frame)
ISA	Person	(parent frame)
SEX	Male	(instance value)
AGE	Under 12 yrs.	(procedural attachment - sets constraint)

The NOK method starts from verbalized knowledge, which is composed of parts and has structure and content, and can be presented as such in the form of knowledge network. In the process of the formation of the network, the network is explicitly enriched with hidden implicit contextual knowledge contained in sentences [15].

The NOK method uses two elements for the graphical representation, node and link. Links between nodes allow grouping into complex expressions. A detailed description of the NOK method is presented in [15]–[17].

In the beginning of the development of the NOK method, verbs were selected, as separate word type, and a special symbol in the form of an ellipse was introduced for their modelling. Other word types are treated equally. This paper analyzes adjectives and suggests their modelling in the context of the NOK method.

Based on the NOK method, a special formalized language for knowledge representation was developed (FNOK) [18] by which knowledge can be stored and processed using a computer program [19], [20].

The application of the NOK method can proceed towards the development of QA system, Chatterbot system (Elize, ALICE, Evi, Wolfram Alpha) and more recently digital assistant system (such as Google Assistant, Apple Siri, Amazon Echo) [19], [21]–[26]. All these systems have a myriad of problems and will require further research and development.

II. THE NOK METHOD

This paper deals with knowledge representation formalisms and with problems of natural language semantic representation. It introduces a new method for knowledge representation named Nodes of Knowledge (NOK). The NOK method has a graphic representation form, a diagram (DNOK). The NOK method uses two elements for graphical representation, nodes and links. Different kinds of nodes are used for terms representation. Links between nodes enable grouping terms into more complex expressions. Process node is introduced as an aggregation point for representation of knowledge described in sentences. Furthermore, an array of interconnected process nodes can represent knowledge expressed in a sequence of sentences.

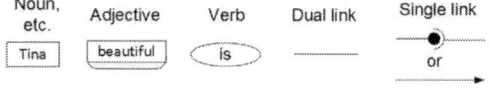

Figure 1. Graphical concepts in the NOK method

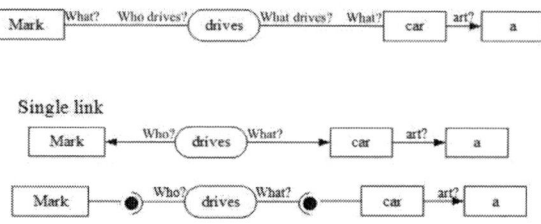

Figure 2. Example of NOK diagrams

Our idea was to define a method that can capture natural language semantics, but simple enough to allow efficient reasoning and learning process implementation. The goal of the NOK method is to represent a knowledge network of knowledge written in a textual form. Furthermore, the NOK method can capture different kinds of knowledge: knowledge from dictionaries and encyclopaedias, knowledge from existing databases, knowledge embedded in business processes, knowledge stored in business documents, etc.

Concepts of the NOK method are used for modelling knowledge. In the NOK method, nouns are independent nodes, verbs are independent processing nodes and adjectives are a special type of nodes that belong to nouns [15]. Basic graphic symbols are visible in Fig.1. Verbs and adjectives are separately marked, while other word types are displayed as a rectangle.

Let us observe the sentence "Mark drives a car." With the proposed symbols of the NOK method, the semantics of the sentence is represented using a diagram as shown in Fig. 2. The following discussion will use the version of the NOK method with a single link.

III. ADJECTIVES AS ENTITY ATTRIBUTES

This paper will briefly present basic facts about adjectives. It should be noted that the detailed descriptions are given in other papers. Adjectives fall into the category of open word classes, that is, the category which can acquire new members (either by borrowing from other languages or by coining new words).

Persons, places, things and ideas are collectively called entities. A noun is a word type (term), which appoints entities (persons, places, things or ideas). For example, thing "house" in objective reality is named with the word "house". Entities may have some properties (or attributes). The properties of entities are also appointed by nouns. The house has the following properties: number of rooms, colours, etc.

In English language, an adjective is "a word that describes a noun or pronoun" [27]. In Croatian language, adjectives are defined as "words which express the characteristics of persons, places, things and ideas" [28]. From the viewpoint of database modelling, in regard to the definition of adjectives in both languages, adjectives specify the values of entity (noun) attributes or relationships between persons, places, things and ideas. The adjective is the value of a certain property of the entity. For example, for the noun "house", we can ask the following: what size it is and what colour it is. The answers

1222

are adjectives: small and blue. Therefore, adjectives additionally describe a noun or highlight some of the possible values of a chosen property.

The most common inherent adjective category is the comparison (lat. comparatio) [29]. From the viewpoint of comparison, adjectives are changeable word types in both languages. One of the inherent properties of adjectives is also Motion (lat. motio, mobility), according to which the number of adjectives depends on the number of grammatical genders of the adjective (Croatian, Latin and German have three grammatical genders, French two, English none) [30]. In the Croatian language, as well as some other languages, adjectives are additionally changeable according to: number and case. In addition, the Croatian language also differentiates adjectives for describing living and non-living things in masculine gender. In some grammatical cases there is also a difference between definite and indefinite adjectives. Both languages have three degrees of comparison.

According to function, adjectives can be classified as [31]:

- Adjectives which can occur in attributive function, i.e. they can premodify a noun, appearing between the determiner and the head of a noun phrase (e.g. a round table).

- Adjectives which can occur in predicative function, i.e. they can function as subject complement (e.g. he is diligent) or object complement (e.g. I consider him diligent).

According to the semantic features of adjectives, they can be classified as:

- Stative/dynamic adjectives. Adjectives are characteristically stative, but some can be seen as dynamic. In particular, adjectives that are susceptible to subjective measurement can be dynamic. For example, funny can be static (He is funny) or dynamic (He is being funny).

- Gradable/nongradable adjectives. Gradability of adjectives is manifested through comparison, and most adjectives are gradable. However, certain adjectives are nongradable, such as sheer, British, etc.

- Inherent/noninherent adjectives. An inherent adjective applies the meaning directly. For example, a wooden cross implies that the cross is made of wood. Noninherent, on the other hand, implies an allegorical meaning (a wooden actor is not made of wood).

Therefore, adjectives can fall into two categories at the same time. For example, in the phrase "a perfect stranger", perfect is an attributive and noniherent adjective.

Adjectives can be compared. Comparison of adjectives is a grammatical category that has three degrees (three states) as follows: positive, comparative and superlative.

What follows is an overview of representing adjectives in sentences using the NOK method.

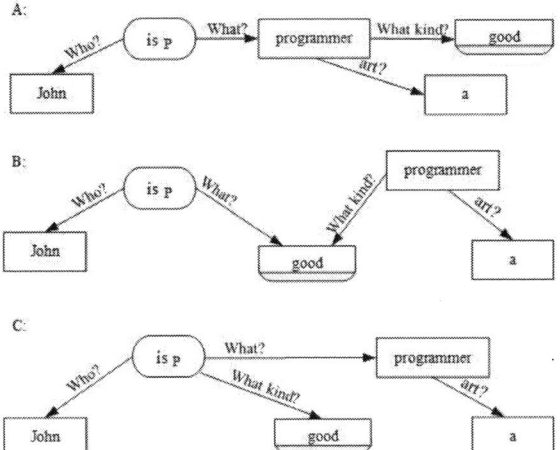

Figure 3.   Different representation of a sentence using the NOK method

IV.   CREATING DNOK FOR ADJECTIVES

Let us analyze the semantics of the sentence "John is a good programmer" and propose a corresponding NOK model. This sentence is represented in three versions A, B and C in Fig. 3.

Let us analyze version A. Nouns "John" and "Programmer" are independent nodes (rectangle symbol). The verb "is" brings together the two nouns into a thought "John is a programmer", because the verb is transformed into a process node named "is P", which is linked with single links towards nodes. The thought John and thought programmer are included into the thought "is P" and form a unique whole, a new thought. John is, therefore, a programmer, and then we can ask: what kind of a programmer? The adjective Good is associated with Programmer through a single relationship "What kind". This means that the adjective is passive and that it gives its characteristic of "good" to the noun, and the noun "Programmer" is active in this relationship because it receives the property from the adjective.

The adjective "good" is linked to the noun "programmer", and thus we attribute "good" to the noun programmer. Then we link the noun which is characterized by the adjective to the person John. Thus, we say that John is a programmer, a programmer who is good.

Other possible forms of DNOK are shown in versions B and C. The B version emphasizes that John is good, and only then that he is a good programmer. In version C dual links emphasize that John is a programmer and the adjective is added as an attribute of the process relationship, thus "good" is transferred to both John and programmer. Therefore, versions B and C are poor solutions since they do not reflect the semantics of the starting sentence.

A. General Rules of Writing and Reading Knowledge on DNOK

Let us point out the differences and similarities of the three ways to store knowledge: in mind, in textual records and in the DNOK. In the text, the terms are written and all

1223

the interpretation is given in the mind of the person reading it. In DNOK terms are written in a network of related terms in an effort to represent their relationship. The meaning of certain terms in the network is given by the reader, and the relationships between them are shown with the structure of DNOK.

Reading the DNOK network cannot directly lead to same sentences from which the network was created, but it leads to the same meaning. The situation is similar in the mind of a man listening to a story; he/she remembers the elements and relationships between elements. When asked to retell the story told in his/her own language, new sentences are created, that convey the same knowledge in a new way. An even clearer example is a teacher who can successfully lecture the same lesson in completely different ways.

In the text, the same noun (or verb, adjective) referring to the same concept can be repeated several times. Noun John can be written twice in two sentences, as is natural for a language made up of sentences. In DNOK, the noun does not have to be re-written, and it can be seen that the same John is included in a set of skills that apply to him. DNOK has no limitations which are typical for the textual record. In DNOK, a noun representing the same concept (e.g. John) is drawn only once. We repeatedly draw the same concept only for clarity of figures and put a label (*), but it is understood that this is the same concept. John is a genuine entity in reality and in thoughts, and in DNOK, it represents a unique implementation of the concept from mental reality.

The question is how to show relationships of nouns, verbs and adjectives in DNOK? For example, let us observe a simple sentence "John has a big house." The verb Has and adjective Big, unlike nouns John and House, are not objects in reality, but merely mental concepts. Verb and adjective used in this sentence are a unique implementation (application) of general terms in the language of the particular sentence and they represent original terms from mental reality in DNOK and therefore unique concepts in DNOK. Thus, every adjective, e.g. Big, is an original word in all sentences, and it can therefore be said that no two sizes are the same. They have the umbrella term adjective Big that connects all of them, but when that word is used in a particular sentence, its original meaning is of a special "size". This is also true for verbs and other types of words. Thus, "having" a house by John is a completely original "having", different from other "having" of the house of all other people.

If new sentences express new knowledge related to John, it is added to DNOK by expanding the network. For example, for the sentence "John is diligent, fast and wise programmer", we would add three new adjectives for "programmer" in Fig. 3.A.

The sentence has a beginning (initial capital letter of the first word), duration and end (a full stop at the end of the sentence). DNOK does not have the beginning of the sentence, word order or the end of the sentence. In DNOK there is a relationship between words that has a unique meaning and no homonymous meaning.

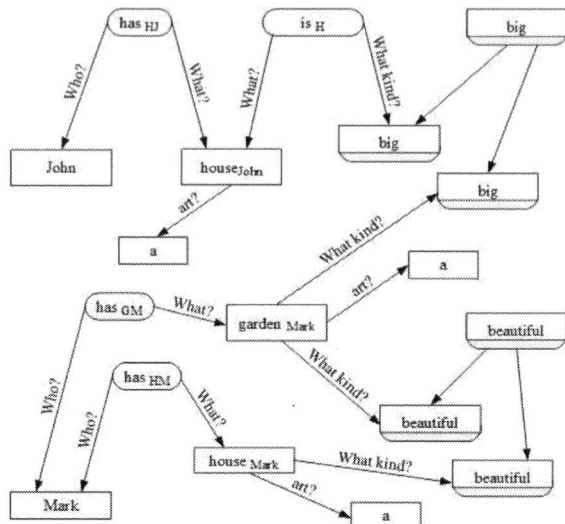

Figure 4.   Sentences represented in NOK method

To clarify the above statements, let us observe the following sentences:

- John has a house. The house is big.
- Mark has a beautiful and big garden and a beautiful house.

Let us analyze the sentences and their representation on DNOK (Fig. 4). Mark and John's house is not the same house, so two nodes are generated in the diagram. The verb "has" results in two process nodes for having a house and one process node for having a garden, because it involves a different possession, of another house belonging to another person.

In human language and in text, it is implied that the reader understands it out of the context. To build a network of knowledge, this is not enough, and it is necessary to represent every verb as a different concept on the model.

Texts in language abound in such "implicit" homonyms and they do not present a problem because human intelligence interprets it in the context of the sentence. For a computer system, it is necessary to resolve the homonyms by introducing different process nodes for verbs.

It is similar for adjectives. The adjective "big" (and "beautiful") itself outside the sentence context is the same umbrella term, but in every sentence it is an original term associated with something specific. Every word in the dictionary, therefore every adjective represents one umbrella term. And each used adjective in a sentence represents the implementation of the term from the dictionary in a sentence and is drawn using an arrowed line from the contextual umbrella term to its implementation.

B. Variants of Modelling Adjectives

There are two basic forms of DNOK model for adjectives: adjective "subordinate" to the noun and the adjective "nonsubordinate" to the noun. Mostly, attributive adjectives are subordinate, while predicative are nonsubordinated forms in the DNOK diagram, with few

1224

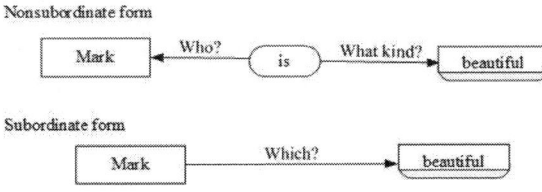

Figure 5. DNOK model for adjective (positive)

exceptions. For subordinate form, an analysis of the diagram shape in the example of "John is a good programmer" was conducted. DNOK form is shown in Fig. 3A. Let us take two simple sentences, the first representing the subordinate and the second nonsubordinate form:

- Mark is beautiful.

- Beautiful Mark.

These sentences are represented with DNOK in Fig. 5. Since "beautiful" is an attributive adjective, an adjective that is semantically associated with and subordinate to the noun, then the question in Fig. 5 is "Which?" instead of "what kind?".

The adjective from the viewpoint of comparison has three forms, namely: positive, comparative and superlative. Fig. 5 shows a form of DNOK for positive of the adjective. Let us take the example of a sentence for comparative of adjective "Mark is more beautiful than Noah" and example of a sentence for superlative of adjective "Mark is the most beautiful." DNOK form for this sentence is shown in Fig. 6.

Fig. 6 shows the way of translating sentences in DNOK that have the comparison of adjectives. For comparative, it can be seen that every word in the sentence represents a separate node. Links between nodes Mark and Noah are single links and go from the verb "is".

The model with the comparative is an aggregate of three relationships: 1. Who? - the one talked about is Mark, 2. What kind? - that particular somebody is beautiful and 3. Than who? - This particular somebody is something more than Noah.

For comparative of adjectives, the verb "to be" is superior to three nodes: the first noun which the adjective belongs to, the adjective describing the first noun and the second noun that is compared with the first noun. At the same time, the proposition "than" is linked as a subordinate node of another noun with which it makes a unity.

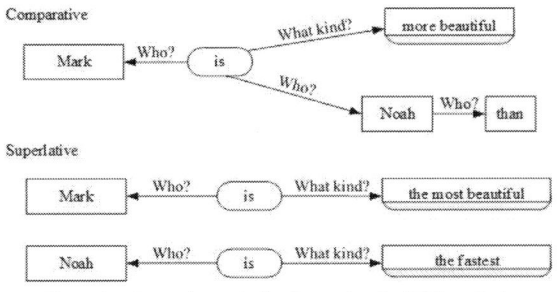

Figure 6. Comparison of adjectives using the NOK method

There may be other variant solutions for the comparative, which would follow the implementation and only practice can show the advantages of the solution, such as:

- Link node "Noah" with "more beautiful" using the question "Than who". The argument for this version is the ability to ask the question "Who is Mark more beautiful than?"

- Introduce a comparative process node "than" linked to three nodes that correspond to the question of who?, what kind? and than who?.

The superlative is resolved in the same manner as the positive, i.e. as an adjective nonsubordinate to the noun.

V. CONLUSION

The paper graphically presents modelling adjectival sentences using the NOK method. Based on the analysis of individual sentences, the semantic relationship of adjectives to nouns and verbs is identified.

For subordinated adjectives, there is a special relationship between an adjective and a noun where the adjective is "subordinate" to the noun and there is no verb between them. For nonsubordinated forms of adjectives, there is the verb "to be" which is superior to the noun described by the adjective, as well as to the adjective itself.

Comparative is presented by a process node that links the three nodes, where the two words "than Noah" represent a semantic whole.

The NOK method, in addition to the nodes that contain words in a sentence, adds links that represent semantic relationships between words. Each link is named with questions that examine the relationship between words and thus enriches the graph in relation to the starting sentence.

Based on the DNOK, it is possible to build a formalized language to record sentences into the knowledge base, the so-called FNOK, required for the construction of intelligent information systems.

After the development of the system prototype, an initial test was conducted using 42 sentences and 88 questions. The obtained results (82% correct answers) indicate that the direction in which the development of the prototype is headed is correct. The authors are currently upgrading the prototype by installing knowledge of modeling adjectives and other relationships in a sentence. The success of the new prototype will be tested using a larger sample of sentences and authors will try to find ways to increase the established success.

Further expansion of the NOK method requires resolving two important types of words: pronouns and conjunctions. Automation of the proposed transformation process of sentences into a new record is a particular problem that the authors are addressing and it is important for the massive input of articles into the knowledge network.

The construction of an intelligent information system is not possible without the knowledge base and precise knowledge representation in accordance with the selected method, and the proposed research is only part of the process.

REFERENCES

[1] S. Russell and P. Norvig, Artificial Intelligence: A Modern Approach, Third edition. Prentice Hall, 2010.

[2] R. Reddy, "The challenge of artificial intelligence," Computer (Long. Beach. Calif)., vol. 10, no. 10, pp. 86–98, 1996.

[3] R. Lujic, T. Saric, and G. Simunovic, "Applying of expert system for determination of work order priority rule class in the single production (Primjena ekspertnog sustava pri određivanju klase prioriteta radnog naloga u pojedinačnoj proizvodnji)," Teh. Vjesn., vol. 1, no. 2, pp. 65–75, 2007.

[4] L. Morgenstern and R. H. Thomason, "Teaching Knowledge representation: Challenges and Proposals," in KR2000, 2000.

[5] A. Mestrovic, "Knowledge representation, teaching materials (Prikaz znanja, nastavni materijali)," Odjel za informatiku, Rijeka, 2013.

[6] M. Chein and M.-L. Mugnier, Graph-based Knowledge Representation, Computational Foundations of Conceptual Graphs. Springer, 2009.

[7] L. Budin, B. Dalbelo Basic, N. Pavesic, and S. Ribaric, Inteligent systems (Inteligentni sustavi). Zagreb: Inteligentni informatički sustavi, 2001.

[8] J. Grundspenkis, "Knowledge representation and networked schemes," 2013. [Online, last access 25.01.2017.]. Available: stpk.cs.rtu.lv/sites/all/files/stpk/lecture_7.pdf.

[9] M. R. Quillian, "Semantic Memory," in Semantic Information Processing, M. Minsky, Ed. MIT Press, 1968, pp. 227–270.

[10] A. Gomez-Perez and V. R. Benjamins, "Knowledge Engineering and Knowledge Management, Ontologies and the Semantic Web," in Proceedings of 13th International Conference, EKAW 2002, 2002.

[11] A. Jakupovic, M. Pavlic, A. Mestrovic, and V. Jovanovic, "Comparison of the Nodes of Knowledge method with other graphical methods for knowledge representation," in Proceedings of the 36th international convention /CIS/, MIPRO 2013, 2013, pp. 1276–1280.

[12] M. R. Quillian, "Word concepts: A theory and simulation of some basic semantic capabilities," Behav. Sci., vol. 12, pp. 410–430, 1967.

[13] M. R. Quillian, "The Teachable Language Comprehender: A simulation program and theory of language," Commun. Assn. Comp. Mach., vol. 12, no. 8, p. 459–476., 1969.

[14] M. A. Minsky, "Framework for Representing Knowledge." [Online, Accessed: 10-Mar-2017]. Available: http://web.media.mit.edu/~minsky/papers/Frames/frames.html.

[15] M. Pavlic, A. Jakupovic, and A. Mestrovic, "Nodes of knowledge method for knowledge representation," Informatologia, vol. 46, no. 3, pp. 206–214, 2013.

[16] M. Asenbrener Katic, M. Pavlic, and S. Candrlic, "The representation of database content and structure using the NOK method," Procedia Eng., vol. 100, pp. 1075–1081, 2015.

[17] M. Rauker Koch, M. Pavlic, and M. Asenbrener Katic, "Homonyms and Synonyms in NOK Method," Procedia Eng., vol. 100, pp. 1055–1061, 2015.

[18] A. Jakupovic, M. Pavlic, and Z. Dovedan Han, "Formalisation method for the text expressed knowledge," Expert Syst. Appl., vol. 41, no. 11, pp. 5308–5322, 2014.

[19] M. Pavlic, Z. Dovedan Han, and A. Jakupovic, "Question answering with a conceptual framework for knowledge-based system development 'Node of Knowledge,'" Expert Syst. Appl., vol. 42, no. 12, pp. 5264–5286, 2015.

[20] M. Asenbrener Katic, S. Candrlic, and M. Pavlic, "Comparison of two versions of formalization method for text expressed knowledge," in Beyond Databases, Architectures and Structures (BDAS) 2017, CCIS, 2017, p. accepted for publishing.

[21] M. Ueno and N. Mori, "Novel chatterbot system utilizing web information," Distrib. Comput. Artif. Intel. Springer, Berlin Heidelb., 2010.

[22] D. Deutsch, The beginning of infinity. Penguin Books, Ltd., 2011.

[23] E. Pariser, The filter bubble. The Penguin Press, 2011.

[24] R. Kurzweil, How to create a mind: the secret of human thought revealed. Viking Penguin, 2012.

[25] M. Rauker Koch, M. Pavlic, and A. Jakupovic, "Application of the NOK method in sentence modelling," in Proceedings of the 37th Internation Convention MIPRO 2014, 2014, pp. 1426–1431.

[26] J. Tomljanovic, M. Pavlic, and M. A. Katic, "Intelligent question - Answering systems: Review of research," in 2014 37th International Convention on Information and Communication Technology, Electronics and Microelectronics, MIPRO 2014 - Proceedings, 2014, pp. 1228–1233.

[27] Longman Dictionary of Contemporary English. Barcelona, 1995.

[28] S. Babic and S. Tezak, Croatian language grammar: a manual for elementary language education (Gramatika hrvatskoga jezika: Prirucnik za osnovno jezicno obrazovanje). Zagreb: Skolska knjiga, 2005.

[29] S. R. Anderson, "Inflectional morphology," in Language typology and syntactic description, Cambridge University Press, 1985, pp. 150–201.

[30] I. Markovic, An Introduction to Linguistic Morphology (Uvod u jezičnu morfologiju). Zagreb: Disput, 2012.

[31] R. Quirk and S. Greenbaum, A student's grammar of the English language. Addison Wesley Longman, 1998.

Identification of image source using serial-number-based watermarking under Compressive Sensing conditions

Andjela Draganić*, Milan Marić**, Irena Orović* and Srdjan Stanković*

* University of Montenegro, Faculty of Electrical Engineering, Podgorica, Montenegro
** S&T Crna Gora d.o.o, Podgorica, Montenegro

Abstract – Although the protection of ownership and the prevention of unauthorized manipulation of digital images becomes an important concern, there is also a big issue of image source origin authentication. This paper proposes a procedure for the identification of the image source and content by using the Public Key Cryptography Signature (PKCS). The procedure is based on the PKCS watermarking of the images captured with numerous automatic observing cameras in the Trap View cloud system. Watermark is created based on 32-bit PKCS serial number and embedded into the captured image. Watermark detection on the receiver side extracts the serial number and indicates the camera which captured the image by comparing the original and the extracted serial numbers. The watermarking procedure is designed to provide robustness to image optimization based on the Compressive Sensing approach. Also, the procedure is tested under various attacks and shows successful identification of ownership.

Keywords – digital watermarking, digital signature, public key, TrapView, watermarking

I. INTRODUCTION

Development of new ICT technologies improved the ease of creation and access to digital information. However, ease access to digital information increases the doubt about who is the creator or owner of the digital content, as well as its authenticity. Namely, by using various digital tools, digital media copies can be distributed with changed information about source, author or owner. The public key digital signature (PKCS), embedded into the digital media, can provide services for strong origin authentication and reliable content integrity of digital image. PKCS technology is reliable and strong in terms of protection on hacker attacks, but it is fragile for use in environments where digital images are subject to modification or exposed to communication noise during transport to receiver. Therefore, this paper proposes the watermarking procedure that aims in preserving the embedded part of the PKCS, even if the digital image is exposed to the various attacks.

Watermarking techniques provides indication of ownership, and/or indication of the identity of a licensed user, by embedding the information in the digital object [1]-[12]. This information may be visible or hidden, but the security properties of this technology are limited due to possible malicious attacks on the original image. Image watermarking techniques usually embed the security information throughout the digital pixels in a manner that does not impact its normal use [1],[11]-[13]. There is always the requirement that the watermark should be capable of surviving routine transformations to the image, such as blurring, cutting or compression. Adding digital signature in cover work of the watermarking techniques could highly improve safeguarding against malicious source origin change and unauthorized image modification.

In the proposed procedure, part of the PKCS – a serial number (SN), is used for embedding. The SN, that is available in the hexadecimal form, is transformed into the 32-bit binary form. Based on the binary form, logo is created and embedded into the image as a watermark. Watermarked image is transmitted through the network and may be exposed to the attacks. At the receiver side, the watermark detection extracts 32-bit binary sequence. This sequence is then compared with the original sequence, i.e. with the SN corresponding to the camera device that captured the observed image. This original SN is available through the Certification Authority system.

The procedure of watermarking and camera identification is tested on the images from the TrapView pest monitoring system [14],[15]. The TrapView system is consisted of traps distributed through the fields/orchards with a purpose of monitoring pests captured in those traps. The images of caught insects are sent regularly to the TrapView cloud that later provides pest recognition and pest occurrence statistics. When the number of insect specimens becomes too large, trap has to be renewed. The SN-based watermarking procedure helps to locate the trap that has to be renewed, by revealing the SN of the camera device.

Having in mind that the images has to be regularly uploaded to the cloud, and that they are of high resolution, the image size optimization has to be done prior it is sent over mobile network. The optimization is done by using the Compressive Sensing (CS) approach [16]-[23]. By randomly selecting only small number of image samples from its frequency domain, the image can be successfully reconstructed at the receiver size by using an optimization algorithms. The SN has to be preserved after the random samples selection and optimization. In other words, the proposed watermarking procedure should be robust to the CS attack.

The paper is organized as follows. The TrapView system and image optimization by using the CS is described in the Section 1. Theoretical background on the image watermarking is provided in the Section 2. Section 3 describes the proposed watermarking procedure, while in Section 4 the experiments are given. Conclusion is in the Section 5.

II. THEORETICAL BACKGROUND

TrapView pest monitoring system is a platform that provides information about occupancy of traps distributed through the fields/orchards. The system captures and uploads the images to the TrapView cloud, at daily basis. An automated pest monitoring is useful in cases when there is need for monitoring large areas [14], [15]. Each TrapView camera has its own PKCS, and part of the PKCS is the SN. System is illustrated in Figure 1.

In order to be able to detect which camera captured the observed image, the SN is embedded into the image. The SN embedding can be added to the standard TrapView system after image capturing, as it is shown in the Figure 1. As the TrapView captured image is around 1MB large, our goal is to reduce its size without loss of quality and by preserving the SN embedded in each captured image.

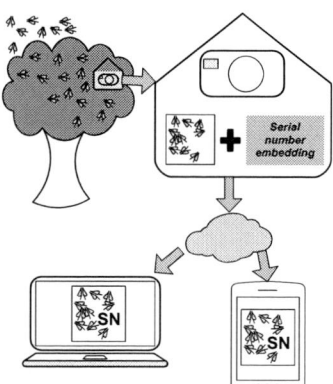

Figure 1. TrapView system with SN embedding

Let us firstly describe the size optimization approach. The optimization is done by applying the CS method. The CS aims at recovering data from the small set of available samples [11]-[13], [16]. Signal can be intentionally under-sampled, or samples corrupted by noise or some other environmental factors can be considered as missing. Random selection of the signal coefficients, in the domain where signal has dense representation, assures successful

reconstruction from the small number of acquired samples. If an N-dimensional signal $\mathbf{x}$ has a sparse representation in the certain transform domain $\mathbf{\Psi}$:

$$\mathbf{x} = \sum_{i=1}^{N} \mathbf{X}_i \psi_i = \mathbf{\Psi X}, \quad (1)$$

where $\mathbf{X}_i$ is a transform domain coefficient and ψ_i is a basis vector, then the vector of acquired samples $\mathbf{y}$ is defined as [11]:

$$\mathbf{y} = \mathbf{\Phi x} = \mathbf{\Phi \Psi X} = \mathbf{A X}. \quad (2)$$

The matrix $\mathbf{A}$ denotes the measurement (CS) matrix, while the matrix $\mathbf{\Phi}$ models random selection of the coefficients. The signal is reconstructed by solving the set of linear equations (2), using an optimization algorithms, i.e. finding the sparsest solution of the system. In the case of 2D data, commonly applied method for the optimization problem solving is the TV optimization, based on minimization of the image gradient [11]-[13],[15],[23].

For solving the system (2), the minimization over $\mathbf{X}$ of the regularization function $J(\mathbf{X})$ is performed [12],[23]:

$$J(\mathbf{X}) = \frac{\mu}{2}\|\mathbf{y} - \mathbf{A X}\|^2 + \lambda R(\mathbf{X}), \quad (3)$$

where $\lambda \in (0,\infty)$, $R(\mathbf{X})$ is the TV of the signal $\mathbf{X}$:

$$R(\mathbf{X}) = \|d\mathbf{X}\|_{\ell_1}, $$
$$d_{i,j}\mathbf{X} = \begin{bmatrix} \mathbf{X}(i+1,j) - \mathbf{X}(i,j) \\ \mathbf{X}(i,j+1) - \mathbf{X}(i,j) \end{bmatrix}, \quad (4)$$

and d is a gradient operator.

The TV optimization is applied on the TrapView images. The TrapView camera, placed in the field, takes pictures of insects captured in the trap and sends them to the cloud. Before sending, the image is transformed in the discrete cosine transform (DCT) domain and under-sampled. Therefore, the vector of measurements $\mathbf{y}$ is consisted of the DCT coefficients. The optimization problem is defined as:

$$\min_{\mathbf{X}} \text{TV}(\mathbf{X}) \text{ subject to } \mathbf{y} = \mathbf{A X}, \quad (5)$$

Or, in the discrete form:

$$\text{TV}(\mathbf{X}) = \sum_{i,j} \sqrt{(\mathbf{X}_{i+1,j} - \mathbf{X}_{i,j})^2 + (\mathbf{X}_{i,j+1} - \mathbf{X}_{i,j})^2} . \quad (6)$$

III. THE PROCEDURE FOR TRAPVIEW IMAGES WATERMARKING AND CS OPTIMIZATION

A. TrapView image watermarking

The watermarking procedure is based on embedding of the SN, part of the digital certificate, that corresponds to certain camera device. Digital certificates are issued from the Certification Authority (CA), and each issued certificate has its own and unique SN, validity period and a subject to whom is issued. The SN provides the name or source origin of the image, stored in register of digital certificates issued by CA, and consequently identifies the camera which is the source of the captured image.

The SN consists of 32 bits grouped in 8 hex numbers, thus forming unique digital identifier for a subject. In the proposed procedure, the group of 8 hex numbers is firstly modified from its original form and binary sequence is created, as it is shown in the Figure 2.

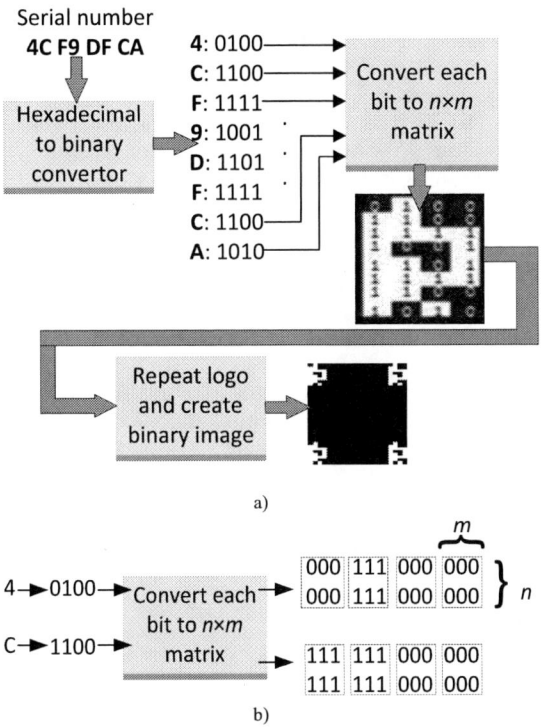

a)

b)

Figure 2: a) Logo creation for the SN example "4CF9DFCA"; b) simplified illustration of the bit-to-matrix conversion

Each symbol is represented with 4 bits, which results in 32-bit sequence. After hex-to-binary conversion, each bit from the sequence is represented in the $n \times m$ matrix form. Therefore, one hex symbol forms an $n \times 4m$ matrix (see Figure 2). After all of the 8 symbols are converted into the matrix form, the binary logo image is created.

In order to increase robustness to different watermark attacks, the logo image is repeated and new binary image with repeated logos is created. This binary image is of the same size as the original image. The starting logo is chosen to be embedded 4 times, in the image corners, as it is shown in Figure 2.

The next step is logo embedding. The original logo is spread into the several bit planes, and only part of the logo is embedded into each plane. If we assume that the image coefficient is represented with B-bits, then B bit planes are available. We chose $L=4$ bit planes for embedding and four $n \times 4m$ matrices in the image corners are used. Middle bit

planes are considered [9], in order to provide robustness and, at the same time, to avoid the influence of the watermark to the image quality. The logo is divided into the L parts (layers), by using a unique security key, i.e. a unique random matrix A [9]. The matrix is separated on L layers, and each layer contains values from the certain interval - there are L non-overlapping intervals. Therefore, the random matrix A and the threshold values, together with selection of the bit planes, form a security key. The complete logo is obtained after summation of the L layers (Figure 3).

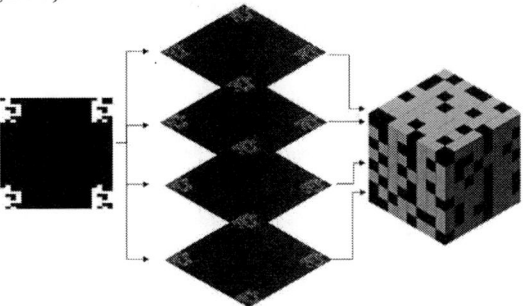

Figure 3: Logo separation and embedding into the bit planes

If W_k denotes the k-th layer of the logo W, B number of bit planes, M and N are image sizes and ω_k are the threshold values for the layers, then logo creation procedure can be defined within the Table 1.

Table 1: Procedure for logo creation

$k \in \{1, \dots, B\}$
- for $i=1:M$
- for $j=1:N$
- if $\omega_{k-1} < A(i,j) < \omega_{k-1}$ then
- $W_k(i,j) = W(i,j)$
- else
- $W_k(i,j) = 0$
- end if
- end for
- end for

The threshold values are chosen by using an equidistant rule, which is one way of secrecy providing. In our case, the elements of the random matrix A take values from the interval $(0,1)$ and the threshold values for the corresponding layers are $(0, 1/4)$, $(1/4, 1/2)$, $(1/2, 3/4)$ and $(3/4, 1)$. If we denote the observed image with $\mathbf{x}$, the embedding can be described within the Table 2.

The watermark extraction is done on the receiving side. Depending on the attack to which image is exposed during transmission, the logo will be degraded at certain degree. The SN is extracted from the received logo. Since the logo is consisted of 4 small logos at the image corners, the 4 SN will be extracted. If we find at least one SN corresponding to the embedded original SN, and if we have successful comparison between original and received SN, we are proving the source origin. The details for the original SN are requested from the Montenegro Public Certification Authority – Post CG CA.

Table 2: Procedure for logo embedding

$k \in \{1,...,B\}$
. for $i=1:M$
.   for $j=1:N$
.   if $W_k(i,j)=1$ then
. $x(i,j)=W_k(i,j)$
.   else
. $x(i,j)=x(i,j)$
.   end if
.   end for
. end for

The procedure for logo extraction starts by observing the bit planes where watermark is added, and is described in the Table 3:

Table 3: Procedure for logo extraction

$k \in \{1,...,B\}$
. for $i=1:M$
.   for $j=1:N$
.   if $W(i,j)=1$ and $w_{k-1} < A(i,j) < w_k$ then
. $W_k^{ext}(i,j) = W_k(i,j)$
.   else
. $W_k^{ext}(i,j) = 0$
.   end if
.   end for
. end for

Logo extraction cannot be done unless the random matrix A , together with the thresholds and the bit plane order numbers, are known.

B. CS as a watermark attack

A special attention is devoted to the CS attack of the watermarked image [12], [13]. After the SN embedding, the CS is applied to the captured image, with an aim to decrease the amount of transmitted information. Certain percent of the pixels from the watermarked image is selected. Selection is done at a random manner [11],[12],[15],[16]. After that, the selected pixels are transmitted and the image is reconstructed at the receiver side. High image quality after the CS reconstruction is important for:

- image post-processing;
- SN extraction.

Post-processing of the images means counting the number of captured insect specimens, based on counting algorithms [14],[15]. Therefore, it is of great interest to save the quality of the reconstructed image as better as it is possible, in order to minimize the possibility of error occurrence during counting. As it was mentioned previously, an accurate SN extraction is important for the camera device location.

IV.   EXPERIMENTAL RESULTS

Let us consider the SN-based watermarking of the TrapView images. The overview of digital certificate for certain user/device is shown in Figure 4.

The observed SN is in the hexadecimal form: 4C F9 DF CA. The first step of the proposed procedure is hexadecimal to binary conversion. Therefore, after the conversion, the following sequence is obtained: 0100 1100 1111 1001 1101 1111 1100 1010. The next step assumes forming matrix from each bit in the sequence. This part of the procedure results in binary logo formation, as it is shown in Figure 2a. The logo is created as a binary image of equal size as the original image. The sizes of the logos that are embedded into the image corners are 50×48.

The logo is repeated 4 times (see Figure 2b), prior it is embedded into the 4 bit planes (the coefficients of the image are represented with 8 bits). In order to determine the positions of sub-images pixels used in watermarking procedure, the random matrix A is exploited and it has to be known at the receiver side.

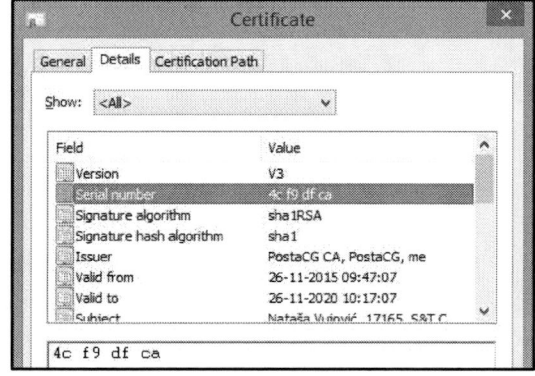

Figure 4: Overview of digital certificate with serial number details

The original TrapView image is of 2MB size. In order to reduce the procedure complexity, we have observed only part of the original image – part of 256×256 size. The 4 logos of size 50×48 are embedded into the observed image part. The same watermarking procedure can be applied to the whole image, or only to the selected image part.

The original and the watermarked images are shown in Figure 5. As it can be seen, the watermark does not degrade the image quality. In order to decrease the number of transmitted samples per image, as well as to increase transmission and upload speed, only 21% of randomly

1230

selected samples per image are chosen (14000 samples out of the total number of 256×256 samples).

TV-based optimization is done at the receiver side and the image is reconstructed. After the optimization, the logo and SN are extracted. The CS reconstructed image is shown in Figure 6a, while the extracted logo is shown in Figure 6b. The obtained peak signal to noise ratio (PSNR) is 30.0082 dB, which numerically proves satisfactory reconstruction quality. Therefore, we can conclude that the CS approach will not affect the process of specimens counting.

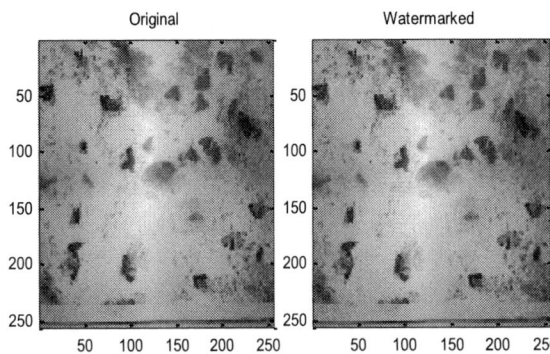

Figure 5: Original and watermarked image

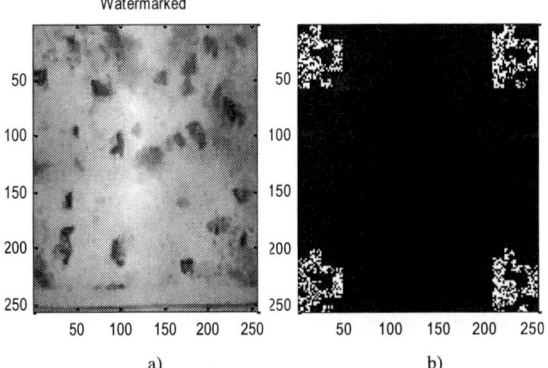

Figure 6: a) Image reconstructed using 21% of the total number of image samples; b) the extracted logo from the image shown in a)

The SN is detected by using the extracted logo. Based on counting the "1" and the "0" in each $n\times4m$ matrix and taking the value that correspond to the greater number of occurrences, the decision on serial number's bit is made and the extracted SN appears in binary form.

The SN extraction is done from all 4 logos, because if only one bit from 32 sequence is modified, the identification fails. Therefore, we check the extracted SN 4 times, and if match with the original SN is obtained just once, we can say that the camera device identification is successful. In this case, all 4 extracted logos provide the exact SN. Beside the CS, the robustness of the watermarking procedure is tested under other commonly appeared attacks in real applications. The success of the SN extraction is observed, and the results are given in the Table 4.

Table 4: Logo and SN extraction results for various attacks

Attack	Gaussian noise	Impulse noise
Extracted logo		
Extracted SN	All 4 SN extracted	All 4 SN extracted
	PSNR=10.1167 Db	40% of the image samples corrupted
Attack	JPEG compression, quality = 10	CS with 21% available samples
Extracted logo		
Extracted SN	All 4 SN extracted	All 4 SN extracted
Attack	Image brightening (80%)	Image darkening
Extracted logo		
Extracted SN	All 4 SN extracted	All 4 SN extracted
Attack	Median filtering	Image blurring
Extracted logo		
Extracted SN	All 4 SN extracted	All 4 SN extracted

V. CONCLUSION

The procedure for image watermarking, based on the Public Key Cryptography Signature and serial number embedding, is proposed in the paper. The images captured with an automatic observing cameras in the Trap View pest monitoring system, are used. Each camera has its own serial number, based on which the device identification is done. The images are watermarked in order to ease the camera identification that captured the observed image. Based on 32-bit PKCS serial number, the binary logo is created and embedded into the captured image. Detection of the watermark is done at the receiver side, and indicates the camera which captured the image by comparing the extracted serial number with an original one. The watermarking procedure is defined in a way that provides robustness to CS-based image optimization. The 80% of the image samples can be avoided and image can still be reconstructed, preserving the embedded serial number. The robustness of the procedure is also tested under other common watermarking attacks and shows successful serial number extraction in all considered cases.

ACKNOWLEDGMENT

This work is supported by the Montenegrin Ministry of Science, project grant funded by the World Bank loan: CS-ICT "New ICT Compressive sensing based trends applied to: multimedia, biomedicine and communications". The authors are thankful to Mr. Matej Štefančič, the director of EFOS, for providing us test images.

REFERENCES

[1] G. V. Mane, G. G. Chiddarwar , "Review Paper on Video Watermarking Techniques," *International Journal of Scientific and Research Publications (IJSRP)*, Volume 3, Issue 4, April 2013 Edition.

[2] I. Orović, S. Stanković, "Time-frequency-based speech regions characterization and eigenvalue decomposition applied to speech watermarking", *EURASIP Journal on Advances in Signal Processing* 2010 (1), 572748.

[3] D. K. Tsolis, S. Sioutas, T. S. Papatheodorou, "A multimedia application for watermarking digital images based on a content based image retrieval technique," *Multimedia Tools and Applications*, 47 (3), 581-597, (2010).

[4] S. Stanković, "Time-Frequency Analysis and its Application in Digital Watermarking," Review paper, *EURASIP Journal on Advances in Signal Processing, Special Issue on Time-Frequency Analysis and its Application to Multimedia signals*, Vol. 2010, Article ID 579295, 20 pages, 2010.

[5] I. Cox, M. Miller, J. Bloom, J. Fridrich, T. Kalker "Digital Watermarking and Steganography", 2nd Edition, Morgan Kaufmann, 624 pages, November 2007, ISBN: 978012372585.

[6] S. Stanković, I. Djurović, R. Herpers, LJ. Stanković, "An approach to the optimal watermark detection," *AEUE International Journal of Electronics and Communications*, Vol. 57, No. 5, pp. 355-357, 2003.

[7] A. Nikolaidis, I. Pitas, "Asymptotically optimal detection for additive watermarking in the DCT and DWT domains," *IEEE Transactions on Image Processing*, vol.12, no.5, pp.563,571, May 2003.

[8] H. B. Golestani, M. Joneidi, M. Ghanbari. "Logo watermarking with unequall strength for improved robustness against attacks" *7th International Symposium on Telecommunications (IST)*, 2014.

[9] I. Orović, P. Zogović, N. Žarić, S. Stanković, "Speech Signals Protection via Logo Watermarking based on the Time-Frequency Analysis," *Annals of Telecommunication*, Vol. 63, No. 5-6, pp. 276-284, 2008.

[10] S. Bansal, P. Pandey, "On the security of robust reference logo watermarking scheme in Fractional Fourier Transform domain" *2013 International Conference on Information Systems and Computer Networks*, ISCON, Mathura, 2013, pp. 200-205.

[11] S. Stanković, I. Orović, and E. Sejdić, "Multimedia Signals and Systems," Springer-Verlag, New York, 2012.

[12] I. Orović, A. Draganić, S. Stanković, "Compressive Sensing as a Watermarking Attack," *21st Telecommunications Forum TELFOR 2013*, Novembar, 2013.

[13] M. Orovic, T. Pejakovic, A. Draganic, S. Stankovic, "MRI watermarking in the Compressive Sensing context," *57th International Symposium ELMAR-2015*, Zadar, Croatia, 2015.

[14] http://www.trapview.com.

[15] M. Marić, I. Orović, S. Stanković, "Compressive Sensing based image processing in TrapView pest monitoring system," *39th International Convention on Information and Communication Technology, Electronics and Microelectronics*, MIPRO 2016.

[16] E. J. Candes, M. B. Wakin, "An Introduction to Compressive Sampling," *IEEE Signal Processing Magazine*, vol. 25, no. 2, pp. 21, 30, March 2008.

[17] T. Blumensath, M. E. Davies, "Gradient Pursuits," *IEEE Transactions on Signal Processing*, vol.56, no.6, pp.2370-2382, June 2008.

[18] M. A. Davenport, M. B. Wakin, "Analysis of Orthogonal Matching Pursuit Using the Restricted Isometry Property," *IEEE Transactions on Information Theory*, vol.56, no.9, pp. 4395-4401, September 2010.

[19] S. Stanković, I. Orović, LJ. Stanković, "An Automated Signal Reconstruction Method based on Analysis of Compressive Sensed Signals in Noisy Environment," *Signal Processing*, vol. 104, pp. 43 - 50, 2014.

[20] E. J. Candes, T. Tao, "Near optimal signal recovery from random projections: universal encoding strategies?," *IEEE Transactions on Information Theory* 52(12), 5406-5425 (2006).

[21] M. A. T. Figueiredo, R. D. Nowak, and S. J. Wright, "Gradient Projection for Sparse Reconstruction: Application to Compressed Sensing and Other Inverse Problems," *IEEE Journal of Selected Topics in Signal Processing*, vol.1, no.4, pp.586-597, Dec. 2007.

[22] S. Stanković, I. Orović, "An Approach to 2D Signals Recovering in Compressive Sensing Context," *Circuits Systems and Signal Processing*, pp. 1-14, ISSN 0278-081X, 2016.

[23] J. Romberg, "Imaging via Compressive Sampling," *IEEE Signal Processing Magazine*, March 2008.

Anti-computer forensics

K. Hausknecht[1]
S. Gruičić[1]
[1]INsig2 d.o.o., Zagreb, Croatia
Kresimir.Hausknecht@insig2.eu
Savina.Gruicic@insig2.eu

Abstract - Generally speaking, anti-computer forensics is a set of techniques used as countermeasures to digital forensic analysis. When put into information and data perspective, it is a practice of making it hard to understand or find. Typical example being when programming code is often encoded to protect intellectual property and prevent an attacker from reverse engineering a proprietary software program.

Through this paper the focus will be on anti-forensics methods which in sense is how information obfuscation is affecting digital forensic investigation. The paper will describe some of the many anti-forensics methods used under the broad classifications of data hiding, artefact wiping, trail obfuscation and finally attacks on the forensic tools themselves.

With any modern-day investigation relying more and more on digital forensics, investigators are required to deal with anti-forensics methods on a daily basis. This paper will explore the challenges investigators and forensic practitioners are facing when conducting investigations. The methods used will be separated into low-tech and high-tech techniques, how they are being used, how they are affecting digital forensic investigation and what the mitigation possibilities are. Focus will be on high-tech techniques that will not stop the investigation but rather prolong or make the process extremely time consuming and therefore not possible to complete in a timely manner or be cost effective.

Index Terms - information, obfuscation, artefacts, anti-forensics, digital forensics

I. DIGITAL FORENSICS

Digital forensics (sometimes known as digital forensic science) is a branch of forensic science encompassing the recovery and investigation of material found in digital devices, often in relation to computer or mobile device crime. The term digital forensics was originally used as a synonym for computer forensics but has expanded to cover investigation of all devices capable of storing digital data. With roots in the personal computing revolution of the late 1970s and early 1980s, the discipline evolved in a haphazard manner during the 1990s, and it was not until the early 21st century that national policies emerged.

Digital forensics investigations have a variety of applications. The most common is to support or refute a hypothesis before criminal or civil (as part of the electronic discovery process) courts. Forensics may also feature in the private sector; such as during internal corporate investigations or intrusion investigation (a specialist probe into the nature and extent of an unauthorized network intrusion) [1].

The technical aspect of an investigation is divided into several sub-branches, relating to the type of digital devices involved; computer forensics, network forensics, forensic data analysis, mobile device forensics and new emerging cloud/internet or cyber forensics. The typical forensic process encompasses the seizure, forensic imaging (acquisition) and analysis of digital media and the production of a report of collected evidence. Later it will be shown how obfuscation directly affects each of these steps.

As well as identifying direct evidence of a crime, digital forensics can be used to attribute evidence to specific suspects, confirm alibis or statements, determine intent, identify sources (for example, in copyright cases), or authenticate documents. Investigations are much broader in scope than other areas of forensic analysis (where the usual aim is to provide answers to a series of simpler questions) often involving complex time-lines or hypotheses. To make this harder or impossible to do, information obfuscation is used.

Obfuscation is method used for obscuring intended meaning in communication, making the message or content confusing, wilfully ambiguous, or harder to understand. It may be intentional or unintentional (although the former is usually connoted) and may result from circumlocution (yielding wordiness) or from use of jargon or even argot (yielding economy of words but excluding outsiders from the communicative value). Unintended obfuscation in expository writing is usually a natural trait of early drafts in the writing process, when the composition is not yet advanced, and it can be improved with critical thinking and revising, either by the writer or by another person with sufficient reading comprehension and editing skills. Conventionally, obfuscation is commonly tied to encryption since it is the main way of making any type of information unreadable unless the cypher is known. Nevertheless, information can be obscured in many other ways that will be described later on [1][2][3].

The combination of information obfuscation methods and digital forensics form anti-forensics techniques.

II. ANTI-FORENSICS

Anti-forensics was first defined by Ryan Harris in 2006 in his paper "Arriving at an Anti-forensics Consensus: Examining How to Define and Control the Anti-forensics Problem" as "Attempts to negatively affect the existence, amount, and/or quality of evidence from a crime scene, or make the examination of evidence difficult or impossible to conduct". He was also one of the first to produce classification of common anti-forensics methods as described in the table below [4]:

Classification of common anti-forensic methods				
Name	Destroying	Hiding	Eliminating source	Counterfeiting
MACE alterations	Erasing MACE information or overwriting with useless data			Overwriting with data which provides misleading information to investigators
Removing/wiping files	Overwriting contents with useless data	Deleting file (overwriting pointer to content)		
Data encapsulation		Hiding by placing files inside other files		
Account hijacking				Evidence is created to make it appear as if another person did the "bad act"
Archive/image bombs				Evidence is created to attempt to compromise the analysis of an image
Disabling logs			Information about activities is never recorded	

TABLE 1 Classification of common anti-forensics methods

In his paper Ryan reflected only on several methods but in today's world many new have emerged and the ways how data is being hidden or obfuscated has changed. This paper will go much further than this. Still, the goal of anti-forensics remains the same, avoid detection of the true meaning of data, disrupting information collection, increasing time needed to conduct the investigation, trail obfuscation, information modification, data hiding, data saturation and general casting doubt on the forensics report or testimony.

Anti-forensic techniques aimed towards digital forensics tools are especially of interest, since they exploit shortcoming of tools that are the main means of giving broader meaning to data. It is important to mention that in the period when this paper was written an average storage capacity of a computer is about 1TB of data and personal computers hold about 8GB

or Random Accesses Memory (RAM). On the other hand, mobile devices have at least 16GB of data while the top capacity phones now stretch up to 256GB of data. The amount of data that must be examined is truly vast and the only way of putting any meaning to it, in a reasonable amount of time, is to use digital forensic tools that accelerate the process. Manually going through such high capacity storages would be too time consuming for police officers and the back log (number of cases waiting to be processed) would be vast, while in the perspective of corporate investigations, it would be financially unprofitable.

As mentioned earlier, there are many ways of making information hard or impossible to interpret. To build on previously mentioned classification, means of obfuscation will be separated into two main classes, low tech and high tech. As their names suggest, low tech requires basic knowledge of computing and electronics while high tech requires excellent conversance of computing/programming and electronics.

III. LOW TECH ANTI-FORENSICS TECHNIQUES

A. *Physical data destruction*

The simplest method of them all that doesn't require any special knowledge is complete physical data destruction. Typically, this means taking data holding media such as computer hard drive or USB thumb drive and smashing it with a hammer. In some cases, this procedure can be used as an official procedure for media destruction when certain hardware becomes obsolete or is due for replacement/upgrade. Besides using a hammer, criminals will do some of the following:

- Use a power drill and bore through the hard drive plates or through memory chips
- Throw the media in water – nearby pond, toilet, pour water over electronic parts
- Use power press to completely crush the media
- Use strong magnets to demagnetize the media
- Pour acid over the media

In today's modern world, these very primitive methods are not used very often since if applied, the media will become unusable and what is even more troublesome is the fact that data will be destroyed even for the criminals to use again. Today even the media can be somewhat expensive to replace but the data can be irreplaceable. Because of these facts, criminals will tend to use other techniques that will allow them to access the data after it was investigated by the legal authorities.

B. *Hard drive scrubbing*

Besides physical data destruction, next low tech method that is commonly used is Hard Drive (HDD) scrubbing or wiping. This is easy to implement since modern HDD's and file system will do this automatically when data is being deleted from a digital source. Basically, a user will delete everything he doesn't want others to find. This can be a full HDD wiping – writing zeros to the whole HDD or just deleting important files. Since this is also a destructive method on a logical form where data is being completely destroyed, persons usually go for less destructive method where they can still retrieve the data for the same reason described in the section above. The most common and fastest

1234

method is quick formatting. By doing a quick format, only index that contains information where the data is located is deleted, but data still resides on the media. If untrained investigator performs a preview of the media he will not see any data and disregard it when in reality, everything is still there but "hidden" from the operating system.

Data deletion is commonly used by low level criminals but even an average investigator will know that soft deleted data can be retrieved by all digital forensic tools and how to do it. What is also important to state is that data deletion often implies guilt and intention to destroy evidence. Furthermore, often absence of information can be evidence itself. To explain on an example, it would be very strange to see a mobile phone that doesn't have any pictures on it since everybody takes pictures and it would be anticipated to find at least some. Another example would be to find an older computer that has no user files on it while there are traces of past user activities. Same would be if someone is buying a 5 year old car that has very low mileage – for experienced buyer a definitive sign of mileage count manipulation.

C. *Artefact wiping*

Continuing with the previous section, besides actual information deletion or partial deletion which is very destructive mechanism persons can delete artefacts that make digital forensics analysis more difficult. This can be considered as removing metadata – information that describes other information. To accomplish this, users can utilize various free programs used for fixing or clearing up space on computers as CCleaner, Clean Master, BC Wipe or Eraser. These programs are advertised as PC optimization tools, speeding up systems, clearing up space, safe browsing, privacy protection etc [5]. In reality what they do is remove browsing history, delete cache files, delete operation system files, wipe slack and unallocated space and "clean" registry files. All this information is used to create a better picture of what the person was doing on the system, what was their intention and if they were trying to hide their true goal. For example, by wiping artefacts investigators can find an incriminating picture but they don't have the information on who was the original author, to whom was it sent to or how did it get on the system.

Mitigating this method is relatively easy. No tool is perfect and therefore "anti-forensic" tools are not perfect and will leave something behind that could be used during the investigation. As before, the mere presence of the tool will raise suspicion with the investigator and make him dig deeper into analysis.

D. *Steganography*

A bit of history… Roots in hiding data/information begin with steganography - the practice of concealing messages or information within other non-secret text or data [5]. When put into digital information perspective, steganography can be used on computers and networks through steganography applications that allow for someone to hide any type of binary file in any other binary file, where image and audio files are today's most common carriers. To put it simple, hide pictures in MS Office PowerPoint or word document or hide a message in a spam email. Persons can go even one step further by covering picture, table or text block under a white block so if the document is quickly reviewed it would be hard to spot. Most modern digital forensic tools will detect most of these anti-forensics methods since they unpack and index documents with their metadata [6][7]. Of course, this can also be a high-tech technique if data is being embedded in an audio file or inside the picture, null cyphers can be used to select a pre-determined pattern of letters from a sequence of words or similar. These methods are rarely used since they are hard to implement, take time to do and also require lot of time to decode. Steganography is especially hard do implement on data that is accessed often.

E. *Cryptography*

Cryptography is a very easy and commonly used technique to hide data. Sometimes it is also described as an ultimate anti-forensic tool since if properly implemented, will put the digital forensic investigation to a complete halt. It is very easy to implement since there are variety of tools, both paid and free that have excellent encryption algorithms and are simple to install and use. Most commonly known programs widely used all over the globe are Truecrypt, Veracrypt and Bitlocker for Windows and Linux machines while Apple devices use proprietary encryption FileVault. They can be implemented in two ways – full disk encryption or file/container encryption.

If full disk encryption is used, whole media (hard drive, memory card, USB thumb drive) will be fully encrypted, first to last byte and without a password no data is accessible or retrievable. If less known product is used for encryption, even with knows password it can be troublesome to retrieve the data.

Other type is file or container encryption. This method will create an encrypted file or encrypted container that can store other files and acts as a vault. Everything stored in this container will be fully encrypted but rest of the files on a system or media will be readable. Digital forensics tools will often detect encryption but don't have any means of decrypting the data without the proper password. There are tools that will attempt to break the encryption by trying to guess the password, but this method of decryption is a very time consuming process and with no guarantee of success. The table below represents the number of combinations per number of letters in a password.

Number of Letters	Possible Combinations
1	94
2	8836
3	830584
4	78074896
5	7339040224
6	689869781056
7	6.4847759e+13
8	6.0956894e+15

TABLE 2 Password combinations

This table is only for English dictionary and does consider capital letters, numbers or special characters which make up total of 94 different characters:

- numbers (10 different ones: 0-9)
- letters (52 different ones: A-Z and a-z)
- special characters (32 different ones)

Usual password length today is 6-8 characters with requirement for at least 1 small and 1 capital, and 1 number. As seen from the table, even with today's powerful computers that can compute 100.000 passwords per second, for 6 character password it would take several years to guess the password [8][9].

Encryption can be implemented not only on stored data but also to hide data/information that is being transmitted over the network. This can be implement through various ways such as PGP, encrypted VPN, TOR networks and similar. Again, if an investigator is collecting network traffic, all data will be fully encrypted and not possible to interpret.

Similarly to data deletion, encryption in respect to the circumstances, can be interpreted as intention to hide data and raise suspicion which for suspects is an undesirable effect since it will cause investigators to dig deeper into the case. On the other hand, encryption it widely used as a necessity, corporate or government rule or is turned on by the manufacturer by default. Usage of encryption can be debated but it not the subject of this paper.

IV. HIGH TECH ANTI-FORENSICS TECHNIQUES

Most of low tech methods can be utilized with various free tools and do not require any special interaction from the user except running the tool and making few selections. The main issue they have in common is that they raise suspicion from the investigator side, since they leave noticeable traces and it's very easy to spot something is wrong, missing or hidden. On the other hand, high tech methods are not destructive and are focused more on hiding data, breaking digital forensics tools and process, or causing prolongation of the whole investigation. These methods will try to confuse the automated process of evidence discovery and basically make the whole investigation last longer and therefore making it not financially profitable. To better understand how this is achievable, the regular forensic process must be defined. These are common steps that are part of any digital forensic investigation:

1. Collection – gathering all relevant evidence from the crime scene
2. Preservation – respecting the chain of custody. Marking and transporting the digital evidence from the scene to the lab
3. Identification – reviewing and identifying gathered evidence and determining what and how it should be processed
4. Analysis - preforming digital forensic analysis of collected evidence
5. Presentation – creating a report on the findings

Each of this basic steps will be challenged with the goal not to prevent the forensics from happening but rather just slowing down the examination process down until the data loses its value or intelligence or cost.

There are many high tech anti-forensics methods and this paper will mention just some of them since they are very technical and as mentioned before, require a lot of advance knowledge of computing.

A. Data saturation

To start simple, it is very easy to create problems in collection phase of investigation - own a lot of media. Simply, persons will never throw out old hard drives, USB drives, memory cards, phones, laptops or any storage media. When investigators come to the scene, they will find all this media that they cannot neglect and must either review on the scene, image or take with them to the laboratory. This will prolong the time needed to image all storage media before it can be processed and examined.

To mitigate this, investigators must parallelize the acquisition process – utilize multiple duplicators and machines. Also, investigators can use the suspects hardware against them to preview the evidence prior to imaging.

In the later investigation phase, data saturation can also be applied by creating or owning a lot of false data or senseless data. This method will again prolong the investigation since the investigator must divide false from real data and later on come to some concrete conclusion. The same is often used when one is trying to conceal its public information by creating vast number of false information.

Mitigation technique involves having as much as possible data about the case so that through analysis investigator can easily pinpoint desired information. Never the less, there will always be cases where data crosscheck will be required.

B. Hiding data

Similar to steganography where binary data is hidden in other binary data, the same effect can be accomplished in numerous ways. This section will address only few that are very common and appear in many digital forensics investigations.

Most common would be hiding data in virtual machines. In computing, a virtual machine (VM) is an emulation of a computer system. Virtual machines are based on computer architectures and provide functionality of a full physical computer. Their implementations may involve specialized hardware, software or a combination [10]. Even an average personal computer today is capable of running virtual machines and software for creating them is very easy to use. Persons that want to hide their activity create VM's and use them for malicious or secret work while the host machine is used for everyday work. These machines can also be fully encrypted or password protected for strengthen security.

Next possibility of hiding data can be accomplished by storing data in other user disk spaces, closed sessions on compact discs and public or shared servers. By storing data away from the original machine where it is actually being used, if not all data storage spaces are inspected it can easily be omitted. Furthermore, analysis of such data can be tedious work since it can be difficult to determine the owner of the information. This poses a serious problem for digital forensic investigator since data today can very easily be transferred and received over the network to and from a remote location. In cases where data is being stored on remote servers outside the person's country, police officers usually don't have any jurisdiction over that kind of data and is often inaccessible.

To mitigate this, investigators must perform live machine analysis and dawn raids where persons of interest don't have time to delete, disconnect or close their sessions to remote data. When data is stored in such way, there is always some data leftovers that will provide clues to investigators on what was going on and how the user was operating.

C. Hiding data in slack and unallocated space

Over 80% of computers today are running Windows OS which implies that the most used file system is Windows New Technology File System (NTFS) [11]. When formatting a hard drive to NTFS users have a choice of how many partitions they want to create, e.g. if they want to format the whole drive or create several partitions. After partition is created, certain portion of hard drive space will be reserved for file system files and there will be partitions that will remain unused by the file and operating systems. Some of these protected and unused areas are Master Boot Record (MBR), Host Protected Area (HPA), Device Configuration Overlay (DCO), unallocated hard drive space. These areas are never used by regular user and are by default inaccessible, but for skilled user they can be used for storing sensitive data or for data exchange between users. If storage media is not properly examined, it is very easy to omit these areas.

Main stream digital forensic tools will always examine most of these areas for data but some, such as HPA and DCO, must be explicitly reviewed if any trace of hidden data exists. If investigator is performing a manual analysis, it is very easy to overlook.

D. Nonstandard RAID configurations

RAID or redundant array of independent disks provide a way of storing the same data in different places (thus, redundantly) on multiple hard disks (though not all RAID levels provide redundancy). By placing data on multiple disks, input/output (I/O) operations can overlap in a balanced way, improving performance. Since multiple disks increase the mean time between failures (MTBF), storing data redundantly also increases fault tolerance [12]. In order to create a RAID array, users must define share stripe patterns, block sizes and various other parameters. By using nonstandard parameters, if investigators image or confiscate the drives, before they start the analysis they must rebuild the RAID array. If parameters are not known this is not possible. RAID arrays commonly need not only proper parameters to function but also dedicated hardware. If nonstandard hardware is used, this will also cause issues during investigation.

To mitigate, investigators can de-RAID volumes on suspects machines, create images on suspect machine or preview the data. It would be recommended to image volumes rather than physical drives and therefore worry only about copying data to destination drive. Recording configuration of the suspects machines is a necessity.

E. File signature masking

All digital forensic tools work with file signature rather than file extensions. This basically means that all files will be analysed and handled by their binary header and/or footer (signature) rather than the file name extension which is easily changed just by renaming the file with a single click. In the past, this method was widely used to hide data on the system, but today it would be detected automatically by forensic tools. To make it harder, users can "hollow out" a file and store wanted data inside it. To go even one step further, data that is pasted in the middle of the other file can be encrypted. This method is sometimes also called transmogrify since the original file is being significantly changed. Other method would be to change the file header/footer rather than the file extension to confuse the digital forensic tool. For an example, all *.jpg picture files have header FF D8 in hexadecimal representation and by changing it to e.g. 00 00 00 14 66 74 79 70 would mask it as Quick Time movie file. If somebody tried to open such modified file, they would get an error.

Mitigating file signature masking can be accomplished with usage of fuzzy hashing. In order to understand this, general file hashing must be explained. A hash function is any function that can be used to map data of arbitrary size to data of fixed size. The values returned by a hash function are called hash values, hash codes, digests, or simply hashes. Put simple, hash function is a mathematical formula that takes any value as an input and returns a fixed size result. If the input changes even by one digit or one byte, the resulting output will be a completely different number. In forensics, most common hash functions used are MD5 which returns 128-bit value and SHA-1 which returns 160-bit value. Hashing in computing is mostly used to find and/or compare data. This will be discussed later. Going back to fuzzy hashing, it is a concept which involves the ability to compare two distinctly different items and determine a fundamental level of similarity (expressed as a percentage) between the two. Chances are that the person or suspect chose a file from his own system and copied it or hollowed it out. By analysing recent files investigators can easily spot suspicious activity such as opening system file as rundll.dll with Excel. Also during analysis when fuzzy hashing produces results that specific file is very similar to notepad.exe. One other mean of mitigation is to use National Software Reference Library (NSRL) which is a large database of known files with their hash values. NSRL database contains large physical collection of commercial software packages (e.g., operating systems, off-the-shelf application software as MS Office, Adobe etc.), detailed information, or metadata, about each file that makes up each of those software packages and smaller public dataset containing the most widely used metadata for each file in the collection that is published and updated quarterly [13] Investigators use NSRL libraries to filter out "typical" files from the evidence so that only unknow remains. This process is sometime called De-NISTing.

F. NSRL Scrubbing

In previous section NSRL database was introduced. As investigators use NSRL on regular basis to reduce the amount of data needed to be analysed, suspect will try to disable the usage of this database by modifying system, program and other files in general. As mentioned before, if only a fraction of input data is changed, the resulting hash value will be completely different and therefore it will not match the record in the database. This is accomplished by modifying strings or insignificant part of files. Furthermore, by turning off Data Executing Prevention (DEP) which is a set of hardware and software technologies that perform additional checks on memory to help prevent malicious code from running on a system, hashes will again be changed and no longer match with the NSRL databases.

There is no easy way of mitigation for this technique of masking files and therefore obfuscating information they hold. There are some guidelines investigators can follow not to omit important information. Instead of using blacklisting, whitelisting approach is preferable – approach that looks for things that match. Investigators are encouraged to identify useful files rather than eliminating them.

G. Scrambled MACE Times

Timing is everything. Same can be said for digital forensics investigations. It is of utmost importance to establish proper timeline of events. Investigators will also utilize dedicated tools for creating histograms in addition to timelines of events in order to create a better picture of what and when something happened. All files on computer store multiple timestamps:

- Modified - the last time file was modified or written to
- Accessed - the last time file was read
- Created - the file's creation date
- Entry - the last time the Master File Table (MFT) entry was updated

As it can be seen above, these four times form MACE acronym. By changing these times either manually or by randomizing them, investigators will have hard time in determining the proper timeline of events. Time changes can be also accomplished by changing BIOS time, turning off "Last Access" update in Windows or by changing time zone of the system.

Mitigation process involves ignoring MACE times and creating new timeline by determining proper ones. As an example, suspects will rarely go to such lengths of changing all possible times on the system and since majority of systems are connected to the internet, investigators only need to find times that are correct by comparing them to real time. Log files will usually be sequential and therefore anything that is off can be considered to be changed intentionally. Also, identifying small sets of similar times can prove to be helpful since investigator only needs to determine the time offset to the real time. All times that are completely scrambled can be ignored. Scrambled MACE times is one of the most difficult anti-forensic techniques to mitigate since it can be very hard to determine when something happened to the second, since that information can be crucial to any case.

H. Restricted filenames

Since this paper is addressing information that is stored in a digital form, in order for that to be possible a storage media is needed. Every storage media must be formatted to a specific file system to hold data. In computing, a file system or filesystem is used to control how data is stored and retrieved. Without a file system, information placed in a storage medium would be one large body of data with no way to tell where one piece of information stops and the next one begins. By separating the data into pieces and giving each piece a name, the information is easily isolated and identified [14]. A filename (or file name) is used to identify a storage location in the file system. Most file systems have restrictions on the length of filenames. In some file systems, filenames are not case sensitive (i.e., filenames such as FOO and foo refer to the same file); in others, filenames are case sensitive (i.e., the names FOO, Foo and foo refer to three separate files). Most modern file systems allow filenames to contain a wide range of characters from the Unicode character set. However, they may have restrictions on the use of certain special characters, disallowing them within filenames; those characters might be used to indicate a device, device type, directory prefix, file path separator, or file type (File system, 2016). As mentioned earlier, most widely spread file system is Windows NTFS. This file system has specific restricted file names such as:

- CON
- PRN
- AUX
- NUL
- COM1, COM2, COM3, COM#
- LPT1, LPT2, LPT#

Furthermore, file names cannot have nonstandard or hidden characters as 0xFF 'e, ~u or similar. If file has a restricted file name or file name contains nonstandard or hidden characters, it will not be accessible through the operating system and will confuse the digital forensic tools. Also, some functions in tools will not work properly such as exporting files. This again will not stop the investigation but will cause prolongation of the data analysis since live preview and live analysis will not be possible until the data is properly imaged and analysed in a forensic tool.

Mitigation is relatively easy since the only requirement is to process all acquired data in a digital forensic tool. Files with such modified file names must be exported with different files names or with file ID number as a name.

I. Circular references

As defined in the previous chapter, file system not only has file name restriction but also file location restrictions. Users can also exploit this restriction for hiding data. Folders have a limit of 255 characters for the full file path. If file is stored with a file path longer than 255 it will become inaccessible to the operating system and therefore to the investigator. File path can also contain "junctions" or "symbolic links" that change the actual file location. File can be represented to the operating system in way that the actual file location is in a completely different location than reported. Users can also use circular references in a way that the file is being stored in file path as: C:\Parent\Child\Parent\Child etc. to confuse the investigator or they can be stored in multiple nested folders to cause the tool to run into an infinite loop or throw errors during acquisition or analysis [15].

As in previous case with restricted filename, mitigation is relatively easy. Just knowing about this method will help investigator circumvent it. Investigators should always work from digital forensic images and be mindful about this anti-forensic method when dealing with a live system.

J. *Broken log files*

In addition to file MACE times, logs can also be vital evidence. Logs will show important information such as when a user logged in to the system, when did he log out, what programs were used, when did the system restart, how application was used and in general will lay out an audit trail. To hide their activity and make the information hard to retrieve, users will make changes to log files as described in artefact wiping, file signature masking or restricted filenames sections of the paper. This will confuse certain data parsers and make them throw errors [16].

To mitigate broken log file method, investigators first must ask themselves whether they actually need the log file. Usually there should be sufficient data to gather all necessary information from evidence files – try to prove a point without logs. If there is a necessity for examination of the log file, they can potentially be parsed in portions (parts that are needed), create custom small scripts that will perform automatic parsing or zero in on the specific records that are of interest rather than parsing the whole log.

K. *Portable systems and programs*

With the introduction of restricted user profiles that don't have administrator privileges and no possibility to modify the system (eg. install new programs or change settings), many portable software's have emerged ranging from simple programs as file browser to full portable operating systems. These applications and systems can be simply copied to a portable USB thumb drive and used on whatever system is available. Most common usage would be in public internet shops or libraries. Person just plugs the USB thumb drive into computer and boots into his own operating system, usually Linux. There he has a full control over all hardware that is available. After he is done, he simple takes his USB and boots back to the original operating system leaving almost no trace on the host machine. Similar can be done with programs. Portable version of internet browser can be run directly from the USB. In this way majority of data will be stored on USB drive itself and therefore very little will be left on the machine hard drive for investigators to go through.

Mitigation method requires thorough preparation before going to field. On the scene, machines must be imaged live, computer memory must be captured and all media storages must be confiscated. This procedure will ensure that investigators have all possible data sources that will enable them to perform full analysis.

L. *Non standard program usage*

The last method that this paper will address is based on usage of nonstandard tools. As mentioned multiple times throughout this paper, what was addressed were the most common operating systems, file systems, programs etc. Digital forensic tools parse best "common" data that is coming from common programs or sources. To lay out an example, there is excellent support for parsing artefacts coming from Microsoft Office but few coming from LibreOffice. All Microsoft Windows operating system versions are supported but not all Linux operation systems are supported. The same can be said for smaller types of software such as internet browser, encryption programs, communication programs etc. It is easy to spot an emerging pattern that is exploited by users. By using non-popular programs, it is very easy to hide activity, information or intentions since tools will not parse data coming from non-popular/standard sources.

Mitigation of this method can only be achieved by constant education and by following current trends. Investigators should train themselves to spot suspicious applications and educate on how they work. Unfortunately, investigation of such data should be performed manually and will definitely prolong the overall investigation time.

V. CONCLUSION

In today's modern world where almost all information is exchanged in a digital from, digital forensics plays a big role. Conceiving true form of information, person's intention, information destination or source are just some of the ways data and therefore information is being manipulated and therefore making discovery and data analysis hard or impossible to perform. In digital forensics, anti-forensics methods are trying to confuse investigators and their tools in accomplishing their task. Over the years, goal of anti-forensics changed from trying to completely deny access to information, to just making it too hard to obtain or being cost effective. The main reason for this change is that the lack of evidence is evidence itself and intentional (obvious) hiding of information, raises unwanted suspicion. It is also necessary to mention that complete data destruction is not an option since in today's world, information is money.

Taking all methods that were explained throughout this paper into account and considering that there are many more available, it can be concluded that it is very easy to disrupt digital forensic investigation. There are also other factors that make investigation hard to perform, one of them being data stored in the cloud that takes it out of current investigators jurisdiction and making it hard to obtain and investigate. The future with everyday development of hardware and encryption possibilities will bring even more obstacles in conducting investigations. The only method that can help investigators to come on top of this never-ending battle is education. Knowing about these and other methods and their mitigation is to only way investigators will be successful in their jobs.

REFERENCES

[1]. K. Conlan, I. Baggili, F. Breitinger, Anti-forensics: Furthering digital forensic science through a new extended, granular taxonomy, Digital Investigation, Volume 18, Supplement, 7 August 2016, Pages S66–S75

[2]. Obfuscation. (2016, December 12). Retrieved from Wikipedia: https://en.wikipedia.org/wiki/Obfuscation

[3]. Strickland, J. (2016). How Computer Forensics Works. Retrieved 2016, from How Stuff Works Tech: http://computer.howstuffworks.com/computer-forensic3.htm

[4]. Harris, R. (2006). Arriving at an Anti-forensics Consensus: Examining How to Define and Control the Anti-forensics Problem. DIGITAL FORENSIC RESEARCH CONFERENCE.

[5]. Piriform. (2016, December). Piriform. Retrieved from CCleaner: https://www.piriform.com/ccleaner

[6]. Steganography. (2016, December 14). Retrieved from Wikipedia : https://en.wikipedia.org/wiki/Steganography

[7]. Kessler, G. C. (2015, February). Gary Kessler Associates. Retrieved from An Overview of Steganography for: http://www.garykessler.net/library/fsc_stego.html

[8]. Brute forcing passwords (2012), Retrieved from Extreme Tech: https://www.extremetech.com/computing/133110-are-fpgas-the-future-of-password-cracking-and-supercomputing

[9]. Password combinations (2016), Retrieved form AceBit: https://www.password-depot.com/know-how/brute-force-attacks.htm

[10]. Virtual machine. (2016, December). Retrieved from Virtual machine: https://en.wikipedia.org/wiki/Virtual_machine

[11]. Net market share. (2016, December 14). Retrieved from Desktop Operating System Market Share: https://www.netmarketshare.com/operating-system-market-share.aspx?qprid=10&qpcustomd=0&qpob=ColumnName

[12]. TechTarget. (2015, April). Retrieved from RAID (redundant array of independent disks): http://searchstorage.techtarget.com/definition/RAID

[13]. National Software Reference Library. (2016, February). Retrieved from National Institute of Standards and Technology : http://www.nsrl.nist.gov/

[14]. File system. (2016). Retrieved from Wikipedia: https://en.wikipedia.org/wiki/File_system

[15]. Chhabra, G. S. (2014). Anti-Forensic Techniques: An Anlytical Review. Thepar University.

[16]. Palmer, C., Newsham, T., Stamos, A., & Ridder, C. (2007). Breaking Forensics Software: Weaknesses in Critical Evidence Collection. Retrieved 2016, from Black Hat USA 2007: http://www.blackhat.com/html/bh-usa-07-bh-usa-07-speakers.html#Palme

Analysis of Mobile Phones in Digital Forensics

Sengul Dogan* and Erhan Akbal*
* Department of Digital Forensics Engineering, Firat University, Elazig, Turkey
sdogan@firat.edu.tr

Abstract – Nowadays, the need to tackle rapidly increased crimes is increasing day by day to help ensuring justice. Digital forensics can be defined as the process of collecting, examining, analyzing and reporting of digital evidence without any damage. Digital forensics requires a detailed examination of devices such as computers, mobile phones, sim cards, tablets that contain digital evidence regardless of whether the crime is large or small. Among these devices, mobile phones take an important place in digital forensics because of their widely usages by every individual. The importance of examining of the data called as evidence in mobile phones has increased with advances in technology, operation capacity, storage capacity and functionality. In a forensics case, mobile phones must be examined by authorized persons and the data obtained from the device must be brought to standards that can be used forensically. In this study, examination and analysis of mobile phones in terms of digital forensics is evaluated. At the same time, data that can be obtained from mobile phones through a sample application has been investigated.

Keywords: Mobile Phone, Digital Evidence, Digital Forensics.

I. INTRODUCTION

Digital forensics is defined as the analysis of data, such as audio, video, video, etc., obtained after the examination of electronic devices, to help the legal process [1,2]. Today, with the advancement of technology, electronic devices are diversified such as tablet, flash memory, memory cards. At the same time, the storage capacities of devices are increasing day by day. People use these devices widely in many areas such as facilitating their work and following social environments [3]. It is an important issue to properly store and analyze this increased data in the electronic environment. Digital forensics aims to examine these devices and data to help the legal process. When forensics analysis is performed, the data on these devices must be evaluated unchanged and not destroyed [4-6]. The obtained results can be used in the judicial process with this condition. Digital forensics is divided into sub-disciplines as given below [3,7,8].

- Computer Forensics
- Mobile Forensics
- Memory Forensics
- Network Forensics
- Malware Forensics
- OS Forensics

Digital forensics has become a discipline that needs to be examined in detail at sub-disciplines because of different operating conditions of each device, progress of technology, storage structure.

After a crime occurred, digital forensics consists of four important processes: collection, examination, analysis and reporting as shown in Figure 1. The detailed planning of these steps and use of effective software and hardware tools ensure quickly and accurately assessment of the case [2].

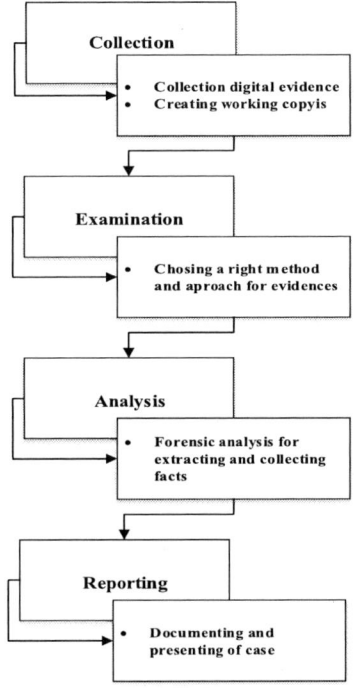

Figure 1. The process of digital forensics

1. Collection: The evidence is collected and the image is taken.

2. Examination: The method is selected for the evidence to be examined.

3. Analysis: The stage of analysis is to obtain findings from digital evidence in accordance with the information required by judicial authorities.

4. Reporting: The reporting phase is the preparation of the documentation to be submitted to the judicial authorities.

Among available devices, mobile phones are one of the most used devices. Mobile phones meet the general needs of people like a computer with the cheaper and widespread use of the Internet. In terms of digital forensics, mobile phones that people do not separate from themselves can contain important evidences [3].

In this study, the evidence that can be obtained by examining mobile phones containing important data in digital forensics is presented.

II. MOBILE FORENSICS

Mobile devices have the ability to store and process data in a variety of methods. Mobile forensics, which is called examination of mobile devices, is a difficult area for the judicial process because it includes devices such as different brand-model mobile phones, smartphones, tablets [9,10]. Different models, hardware, memory structure, operating systems of each of these devices are available. These structures are removed from the standard when every device is examined in the light of digital forensics. However, just as digital forensics, mobile forensics also consist of basic steps as shown in Figure 2 [11].

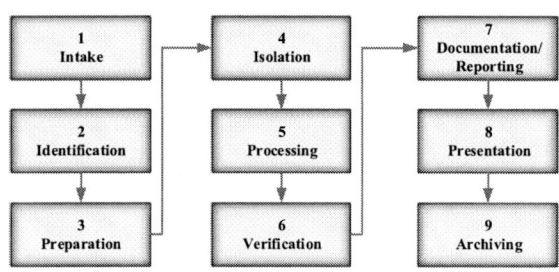

Figure 2. The processing steps for mobile phone examination

The basic information obtained from mobile phones is given in Figure 3, despite the device differences [9,11].

Phone Harddisk Data

Communication Network

SIM Cards Information

Multimedia Data

Messages

Call Records

Location

Social Media Data

Figure 3. The basic data obtained by mobile phone examination

The threats detected on mobile devices can be classified as direct attacks, malware, data interception, exploitation and social engineering. At the same time, it is among the most important security issues that users have insufficient knowledge of technology. There are many software and hardware tools developed for use in judicial cases in the detection of all these problems. The basic software and hardware tools used in mobile phone examinations are given in below [12,15]:

- Cellebrite
- Paraben's Device Seizure
- XRY
- EnCase Neutrino
- Oxygen Forensic
- MOBILedit
- Faraday
- Tarantula

On the basis of all these programs, evidence has to be obtained without harming the data in the mobile phone to help the judicial process [3,15].

III. FINDING AND RESULTS

In this study, the data that can be obtained by a mobile phone examination are presented. For this purpose, the phone having android operating system examined for the computer with the features given in Table 1 is analyzed with Oxygen Forensic and MOBILedit programs.

TABLE I. THE SYSTEM FEATURES

Operation System	Windows 8.1 Professional 64 Bit
System Producer	Sony Vaio
RAM	8 GB
CPU	Intel Core i5 – 3317U 1.7 GHz

Screenshots for the examination software tools are given in Figures 4 and 5.

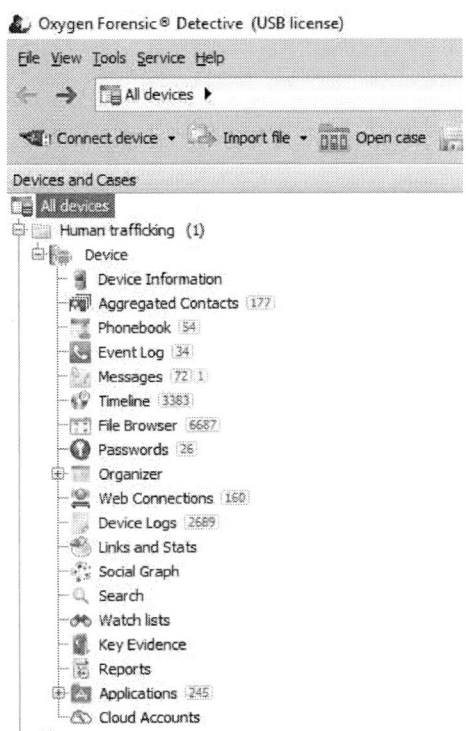

Figure 4. Screenshot for oxygen forensic

Figure 5. Screenshot for MOBILedit

Findings obtained after the analysis are presented in Table 2.

TABLE II. THE COMPRASION RESULTS OF SELECTED SOFTWARES

	Oxygen Forensics	MOBILedit
Version	8.4.0.99	8.7
Processing Start Time	17.20 pm	15.05 pm
Processing Finish Time	17.38 pm	15.16 pm
Total Time	18 minutes	11 minutes
Live Acquisition	✓	✓
Phone Image Analysis	✓	-
Social Graph	✓	-
Simcard Serial Number	-	✓
SHA-2 Hash (Wireless Network)	✓	✓

Wireless Security Protocol (WPA / WPA2 / WEP)	✓	-
Frequently contacted person support	✓	-
Interface	Advanced	Simple
Disable Screen Lock	✓	-
Import Credentials Package Support	✓	-
Telephone Brand / Model / Serial No	✓	✓
Position Finding	✓	-
Phone Book	371	371
Event Log	35	35
SMS	7	7
Calendar	2	3
Application	31	27
Cloud	0	0
Timeline	63	-

At the same time, the findings with mobile phone examination can be listed as follows:

- Device Information

  - Retail Name
  - Manufacturer
  - Platform
  - IMEI
  - Software Revision
  - Network Information
  - Device Information, etc.

- Sim card information

  - IMSI

  - PIN

  - PUK

  - ICCID

  - LAI, etc.

- Contacts

  - Phonebook

  - Event Log (Incoming, missed call, outgoing)

  - Messages

  - Applications

  - E-mail

  - Groups, etc.

- Timeline of events made during phone use

- List of web browsers entered from the phone

- The available files on the phone

- Passwords stored on the phone

- Phone calendar events
  - Meetings
  - Appointments
  - Anniversaries, etc.
- Applications installed on the phone
- Cloud account information and data
- Device log information
- User Data
  - Photo
  - Videos
  - Audio
  - Sound recordings,
  - Document, etc
- Deleted data
- Location information
- Wi-Fi connection logs

According to the information obtained above;

- It is necessary to act on the dump in order for the case to be able to help the legal process.

- Live analysis can damage the evidence. For this reason, live analysis is not recommended in the judicial process. Choosing a program to perform on the dump is important for judicial cases.

- If the number of evidences is too high, the program should be selected according to the need for fast processing.

- Mobile forensics programs have advantages and disadvantages. For this reason, the correctness and diversity of the findings must be revealed by using different programs

IV. CONLUSION

Almost everyone in today's world has a mobile phone. In the past, mobile phones were only used for voice communication, but now people are using a lot of fields such as writing messages, data transferring, socializing, data storing. The widespread use of mobile phones has made these devices the most important factor applied in judicial events. For this reason, mobile forensics, a subdivision of digital forensics, has emerged. At the same time, the development of hardware and software to provide mobile phone investigations with mobile forensics has been achieved. In this study, the analysis of information obtained from the investigation of mobile phones are evaluated. At the same time, evidences from mobile phones are presented by the investigation. When evaluated in terms of running time, it was observed that MOBILedit program performed faster for this analysis. In addition, Oxygen Forensics advantages over MOBILedit program can be listed as dump analysis, Social graph, Wireless Security Protocol. It has been shown that the results of the analysis can be obtained from a mobile device in terms of forensic information of the user's criminal data.

REFERENCES

[1] A. C. Popescu, and H. Farid, "Statistical tools for digital forensics." International Workshop on Information Hiding. Springer Berlin Heidelberg, 2004.

[2] S. L. Garfinkel, "Digital forensics research: The next 10 years." digital investigation 7 (2010): S64-S73.

[3] E. Casey, Digital evidence and computer crime: Forensic science, computers, and the internet. Academic press, 2011.

[4] E. Delp, N. Memon, and M. Wu. "Digital forensics." IEEE Signal Processing Magazine 26.2 (2009): 14-15.

[5] M. Karyda, and L. Mitrou. "Internet forensics: Legal and technical issues." Digital Forensics and Incident Analysis, 2007. WDFIA 2007. Second International Workshop on. IEEE, 2007.

[6] M. K. Rogers, and K. Seigfried. "The future of computer forensics: a needs analysis survey." Computers & Security 23.1 (2004): 12-16.

[7] R. W. Taylor, E. J. Fritsch, and J. Liederbach. Digital crime and digital terrorism. Prentice Hall Press, 2014.

[8] R. Leigland, and A. W. Krings. "A formalization of digital forensics." International Journal of Digital Evidence 3.2 (2004): 1-32.

[9] K. K. R. Choo, A. Dehghantanha, eds. Contemporary Digital Forensic Investigations of Cloud and Mobile Applications. Syngress, 2016.

[10] R. Ayers, "Mobile Device Forensics-Tool Testing." National Institute of Standards and Technology May 6 (2009): 1-23.

[11] S. G. Punja, and R. P. Mislan. "Mobile device analysis." Small scale digital device forensics journal 2.1 (2008): 1-16.

[12] R. Ayers, "Mobile Device Forensics." NIST Mobile Forensics Workshop and. 2014.

[13] S. Saleem, O. Popov, and I. Baggili. "A method and a case study for the selection of the best available tool for mobile device forensics using decision analysis." Digital Investigation 16 (2016): S55-S64.

[14] J. Brunty, "Mobile device forensics: threats, challenges, and future trends." Digital Forensics (2015): 69-84.

[15] R. Ahmed, and R. V. Dharaskar. "Mobile forensics: an overview, tools, future Trends And Challenges From Law enforcement perspective." 6th International Conference on E-Governance, ICEG, Emerging Technologies in E-Government, M-Government. 2008.

Security Analysis of Open Home Automation Bus System

Milan Ramljak

University of Zagreb, Faculty of Electrical Engineering and Computing
Unska 3, 10000 Zagreb, Croatia
ramljak.milan@gmail.com

Abstract - Today's modern homes are becoming complex live systems in which virtually all functionality, from lighting and heating control to security and occupancy simulation, is mediated by computerized controllers leading to IoT future. The smart nature of these homes raises obvious security concerns and history has shown that a vulnerability in only one component may provide the means to compromise the system as a whole. Thus, the addition of every new component, and especially new components with external networking capability, increases risks that must be carefully considered. In this paper we examine one of the most active open source home automation framework, Open Home Automation Bus (openHAB) which is used as platform for many other IoT supported devices. First, we go through openHAB security architecture and supported features following the challenge of a static source code analysis of several most used openHAB packages (called bindings) and carefully crafted test cases that revealed many undocumented features of the platform. Next, we exploited security flaws by constructing two proof-of-concept attacks that: (1) openHAB system denial of service; (2) inject custom binding for message bus monitoring and control; We conclude the paper with security best practices for the design of custom openHAB bindings.

I. INTRODUCTION

With the unstoppable advance of technology, most research projects in field of smart homes have been designed to help human and increase their quality of everyday living experience. A home, which is smart, is the technology result used to make all electronic equipment around the home act "intelligent" or more smart. Smart home has advanced automatic rules for lighting, temperature control, security and many other functions. Each and every new home appliance are having embedded intelligence with automation, monitoring and controlling possibilities. Exposing the data outside smart home system, except security concerns, provides wide range of potential benefits for users and companies building those systems. There are many smart home solution, but of interest in this paper is Open Home Automation Bus (openHAB) system with main focus on its security aspects. We chose openHAB for several reasons. First, openHAB has a growing community driven approach. Second, connection to 153 different devices and protocols such as Samsung TV, Z-Wave, Asterisk and many other. And third, source code is available giving easy way for analysis.

OpenHAB is a software for integrating different home automation systems and technologies into one single solution that allows reusable automation rules and offers uniform user experience. OpenHAB does not try to replace existing solutions, but rather wants to enhance them - it can thus be considered as a system of systems. A core concept for openHAB is the notion of an "item". An item is a data-centric functional atomic building block. [1] Under the hood, openHAB does not directly communicate with a physical device. Instead, it communicates with an instance of an "item" that encapsulates a physical device. This approach makes it even easier to replace device communication protocol by another without doing any changes to rules and UIs.

A most advanced aspect of openHAB's architecture is its modular design giving simple way to extend system at runtime without stopping it. Each feature is developed as completely standalone Java OSGi library also known as openHAB binding. Modular binding approach has been a huge enabler for the active community around openHAB with many engaged contributors [2]. Looking from security side this could lead to many problems in security implementation as different bindings are developed by different users implying that static code analysis is important. Our security analysis explores the above security-oriented aspects of the openHAB framework. Performing the security analysis of an open-source system revealed devastating result of binding documentation. Some bindings are well document but some without any documentation at all, referring to code documentation. For an experienced Java developer is not that hard to dig into the code to get more understanding but some more complex bindings do need significant amount of time. To overcome these challenges, we used a combination of existing static analysis tools and manual analysis on a dataset of 153 bindings that we downloaded in source code form. As openHAB is based on Java programming language for this purpose we used Java enabled analysis tool Find Security Bugs.

II. RELATED WORK

Although the openHAB system analyzed in this article is widely used within the open source community, to day of writing this article, no information of existing security assessments has been found. Still many researchers touched the field of smart home systems and their security implications. Except the analyzed system, there are few other open source solutions such as openHAB [12]:

- Domoticz – a home automation system with native HTML5 support, developed in C/C++ programming language
- OpenMotics – open source system designed to deliver full wired solution, including software and hardware
- Calaos – full-stack home automation platform, including preconfigured Linux operating system

Search for any structured security analysis documents related to any of above mentioned resulted without success. Most of the security analysis articles focus only to smart home devices and their protocols, building blocks, and barely on the complete system utilizing them. All automation systems support set of widely used smart devices such as switches, thermostats, pin locks. In following part, we show some of security flaws of most popular smart devices. We start with security assessment of smart thermostat, Google Nest, where researchers have revealed how it can be exploited and then used as smart spy device. Vulnerability required physical access to thermostat to load custom software over underlying Linux system. The compromised Nest thermostat acts as an entry point to attack other nodes within the local network exposing the data to the attacker [15].

Another device, where more realistic security issues were identified, is based on Insteon Hub, a network enabled device that can control light bulbs, wall switches, power outlets, thermostats, cameras and more. During a first Insteon Hub setup, user is asked to set up port forwarding from the Internet to the device. System documentation basically led user to expose access to anybody from the Internet. Furthermore, there was no option to enable authentication for the Web service running on the Insteon Hub that receives commands [16].

Smart home system with the most detailed security analysis is the Samsung propriety automation system called SmartThings [13]. Results revealed in SmartThings security analysis were trigger for this paper, and comparing the results similar set of issues is identified although the system is propriety [14]:

- SmartThings application can read all events a device generates if the application is granted at least one capability the device supports
- 42% of applications are granted capabilities that were not explicitly requested or used
- exploited framework design flaws to construct proof-of-concept attacks

On the protocol front, researchers demonstrated flaws in the ZigBee and ZWave protocol implementations for smart home devices [17]. Exploiting these bugs requires proximity to the target home but later can be used in attacks that do not require physical access to the home. Researchers also discussed the presence of some well-known vulnerabilities in home automation systems, such as Cross Site Scripting (XSS). Attacker could embed

persistent Javascript in the log pages of one of the products. The researchers also observed that in some home automation systems, every communication between the homeowner and home automation system, both from within the home network and over the Internet, is done in clear text allowing an attacker to eavesdrop on the communication and gather legitimate login credentials [18].

III. OPENHAB SECURITY ANALYSIS

OpenHAB is designed in multilayer ecosystem with OSGi as baseline for its runtime engine showed on Fig 1. In this chapter we put focus on following components:

- openHAB Add-on Libraries
- openHAB REST service
- openHAB HTTP service

The Add-on libraries layer represents most vivid component of architecture also known as openHAB bindings. There are many bindings available, from house heating control to network binding allowing user to check, whether a device is currently available on the network. All bindings communicate with openHAB core through event based message bus. Next architectural component is HTTP service exposed as main management interface for openHAB configuration and administration. Every openHAB binding can be monitored, installed, or removed from this point. Last but not least important component is REST service. Most of the logic of REST service implements JavaScript Object Notation Application Programming Interface (JSON API) providing HTTP endpoints for third-party applications to interface with openHAB. This interface is main point for openHAB Android and iOS mobile applications to communicate

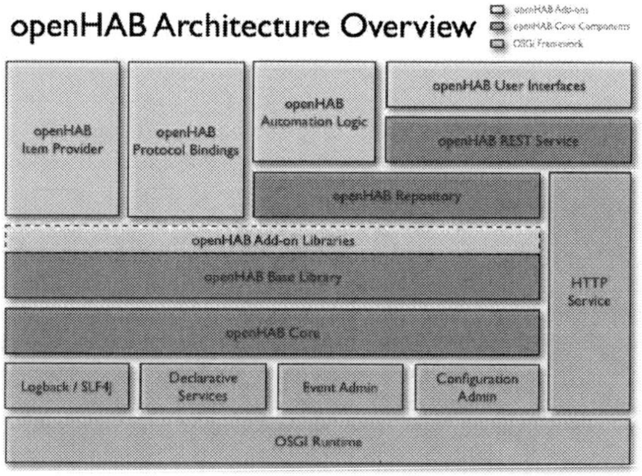

Figure 1. openHAB architecture [3]

towards openHAB event bus.

A. Security analysis of core components

To understand how the community based binding development manifest in practice from security point, we downloaded 153 [4] bindings from the openHAB official web page and performed a static code analysis. We first present the number of bindings that are potentially vulnerable and then drill down to determine the extent to which apps are over privileged due to design flaws.

Table I shows the result overview of our dataset and most important that not all of these bindings are vulnerable. To show example, we pick few of these bindings to show actual vulnerability. Almost all bindings have some sort of Java coding bugs with resulting on its performance issues. Attaching higher number of bindings to openHAB runtime engine puts more focus on mentioned performance issues leading to process crash. After systems is crashed all openHAB functionality is terminated leading to smart home system which needs to be manually restarted. The same goes for openHAB REST service and mobile applications using it.

Adding more and more bindings to running system is very simple, but having in mind less code quality of some of bindings, could lead to openHAB process crash. From security perspective this is denial of service (DoS). However, this bug can be minimized by configuring openHAB in high availability mode (HA) where two instances are used. If one openHAB instance fails, other will take over and system will remain functional.

Next we show how we can utilize over privilege bindings to perform operations on system without permissions. One of widely used binding is openHAB Exec binding used to execute system scripts or commands directly from binding. Once started on system, this binding lets user to perform a command without any access control. The same goes for remote user using REST interface or using mobile application. Binding requires command in following structure:

```
exec="<[/bin/sh@@-c@@command]"
```

where command can be any shell command. To make a DoS showcase this can do remote system shutdown.

```
exec="OFF:ssh pi@192.168.1.4 shutdown -p
now"}
```

Exec binding is not installed by default, but is usually first choice for developers to do quick progress. This exact vulnerability is perfect example of user input validation problem we found by scanning bindings. Most of 153 bindings provide a user to control them by providing input arguments without any validation. Input from untrusted sources must be validated before use. Maliciously crafted inputs may cause problems, whether coming through binding arguments or external sources.

Another result analysis revealed is that some bindings uses direct communication with openHAB message bus. This is not an issue but shows that an attacker can create such binding which can listen for all events on openHAB message bus. Having complete access to all messages exchanged on bus can provide much information, such as security pins and passwords used in bindings connecting to external services such as Gmail Calendar. This kind of attack is not hard to achieve as all openHAB bindings are possible to download from web store where no cryptographic hash integrity functions are used by time of this analysis.

TABLE I.   BINDINGS STATIC CODE ANALYSIS

Bindings count	153
Performance bugs	3019
Bindings with event bus access	9
User input not validated	103

OpenHAB exposes HTTP based web service using the Representational state transfer (REST) or RESTful protocol. REST-compliant Web services allow requesting systems to access and manipulate textual representations of Web resources using a uniform and predefined set of stateless operations [5].

This kind of service is widely used for many IoT systems as usage is very simple and does not require special programs to be used. RESTful service exposes several predefined operations that can be utilized to integrate openHAB with other systems as it allows read access to items and item states as well as status updates or the sending of commands for items. To access REST service of the running openHAB users simply browse link:

```
http://ip_address:8080/rest
```

where `ip_address` is IP address of machine running openHAB process. Complete REST schema is based on Swagger specification [6] supporting different media types. The REST service furthermore supports server-push, so one can subscribe client on change notification for certain resources. Reaching REST endpoint returns structured openHAB data such as:

- Bindings
- Binding Discovery
- Items
- Data persistence

Default configuration starts openHAB without any security activated at all. In general, it is advised to use HTTPS communication over HTTP. On the very first start, openHAB generates a personal (self-signed, 256-bit ECC) SSL certificate and stores it in the Jetty Web service keystore. This process makes sure that every installation has an individual certificate, so that nobody else can falsely mimic your server.

Maybe the worst thing regarding openHAB REST interface is that it does not support restricting access through HTTPS for certain users. There is no authentication in place, nor is there a limitation of functionality or information that different users can access. So, once someone has access to openHAB service it has access to all of its functionality. This looks like very bad approach, but the answer why relies on the fact that openHAB uses Eclipse Smart Home code as baseline [7]. Complete HTTP service part exposing REST interface is used from that code so the expected solution needs to be there.

1247

Fortunately, there are few recommended workarounds to overcome this security vulnerability. First is to limit access and only allow REST requests coming from local loopback interface. The default value allows access from all interfaces:

```
web.listening.addresses = 0.0.0.0
```

and needs to be changed to:

```
web.listening.addresses = 127.0.0.1
```

to allow requests only from local machine.

Second solution is to run openHAB behind reverse proxy. A reverse proxy simply redirects client requests to the appropriate server. This means we can proxy openHAB request connections to other web service. Running openHAB behind a reverse proxy allows to access openHAB runtime via port 80 (HTTP) and 443 (HTTPS). It also provides a simple way of protecting server with authentication and secure certificates.

The last option is to setup account for openHAB Cloud service which enabled that openHAB runtime tunnels to it hiding all the communication. The main core features of openHAB Cloud are a user-management frontend, secure remote access, remote proxy-access, device registry & management, messaging services and data management & persistence. The openHAB Cloud also serves as core backend integration point for cloud-based features (e.g. IFTTT) and provides an OAuth2 application enablement [4]. Cloud openHAB will proxy home sitemap over HTTPS so the communication to UI is TLS encrypted.

To see how openHAB REST interface is vulnerable we used penetration testing tool OWASP Zed Attack Proxy (ZAP), an open-source web application security scanner intended to be used by both those new to application security as well as professional penetration testers [9]. When used as a proxy server it allows the user to manipulate all the traffic that passes through it, including traffic using https. It can also run in a 'daemon' mode which is then controlled via a REST Application programming interface.

Scan discovered four type of issues, two with medium severity (orange) and two with low (yellow) showed in Table II. First issue is related to application error disclosure. Some responses contain an error/warning message that may disclose sensitive information like the location of the file that produced the unhandled exception. This information can be used to launch further attacks against the web application. To solve this bug, review the source code of this page by implementing custom error pages. Consider implementing a mechanism to provide a unique error reference/identifier to the client (browser) while logging the details on the server side and not exposing them to the user. For example:

```
"error": {
  "message": "HTTP 404 Not Found",
  "http-code": 404,
  "exception": {
    "class":
"javax.ws.rs.NotFoundException",
    "message": "HTTP 404 Not Found",
    "localized-message": "HTTP 404 Not
Found"
  }
}
```

Second detected medium severity vulnerability is that X-Frame-Options header is not included in the HTTP response to protect against 'Click Jacking' attacks. Most modern Web browsers support the X-Frame-Options HTTP header. Ensure it's set on all web pages returned by your site (if you expect the page to be framed only by pages on your server (e.g. it's part of a frameset) then you'll want to use deny, otherwise if you never expect the page to be framed.

Going forward to low severity vulnerabilities we start with XSS protection not being enabled, or is disabled by the configuration of the 'X-XSS-Protection' HTTP response header on the web server. Posting data to REST API could make an attacker to post script which can lead to leakage of some user data. To minimize this kind of vulnerability openHAB web browser's XSS filter needs to be enabled, by setting the X-XSS-Protection HTTP response header to '1'.

The last but not the least important vulnerability is the X-Content-Type-Options header was not set to 'nosniff'. This allows older versions of Internet Explorer and Chrome to perform MIME-sniffing on the response body, potentially causing the response body to be interpreted and displayed as a content type other than the declared content type. As a solution administrator needs to ensure that the application/web server sets the Content-Type header appropriately, and that it sets the X-Content-Type-Options header to 'nosniff' for all web pages.

B. Exploiting system weaknesses

OpenHAB HTTP server is based on Java Jetty web service. Looking inside Jetty configuration it is noticeable that default configuration is used with no request filtering policy. Default HTTPS connection timeout is set to 30 seconds, giving easy way to produce high load request flooding and cause DoS. To do this test in this paper we

TABLE II.   OWASP ZAP SCAN RESULTS

Vulnerability Description	Count
Application error disclosure	8
X-Frame-Options HTTP header not set	829
Web browser XSSprotection not enabled	657
X-Content-Type-Options not set	829

used JMeter load generator software build by utilizing Java HttpClient object [8]. This object gives plain implementation of HTTP protocol without overhead resulting in high performance. Used generator is written in asynchronous programming style, so that limited threads do not limit the maximum number of users that can be simulated. Generator is configured to start thread pool of 1000 parallel HTTP requests over the network as it gives more realistic performance and latency. Each thread represents unique user. The results revealed that 1000 parallel requests decreased openHAB availability by 30 %. Increasing number of parallel requests up to 10000 decreases availability to an extent that openHAB rest service cannot be used where 80 % of sent requests failed. To overcome this vulnerability default Jetty configuration needs to be updated by adding request filter. Filter places throttled requests in a priority queue, giving priority first to authenticated users and users with an HTTP session, then to connections identified by their IP addresses. The DoS filter limits exposure to request flooding, whether malicious, or as a result of a misconfigured client. The DoS filter keeps track of the number of requests from a connection per second. If the requests exceed the limit, Jetty rejects, delays, or throttles the request, and sends a warning message.

Proposed filter is shown here:

```
<filter
  <filter-name>DoSFilter</filter-name>
  <filter-class>DoSFilter /filter-class>
  <init-param>
    <param-name>maxRequestsPerSec</param-name>
    <param-value>30</param-value>
  </init-param>
</filter>
```

where filter limits maximum number of requests per second to 30. The filter works on the assumption that the attacker might be written in simple blocking style, so by suspending requests you are hopefully consuming the attacker's resources. For a high reliability system, it should reject the excess requests immediately (fail fast) by using a queue with a bounded capability. The DoS filter is used to avoid Jetty thread starvation what is of great importance if openHAB runs on embedded system such as Raspberry Pi.

Earlier is showed that delivering openHAB bindings doesn't incorporate integrity validations, so all bindings downloaded and used directly could be potentially harmful. In this chapter we show how is easy to build an openHAB binding which could be used to expose system protected data over the REST interface. The purpose of a binding is to translate between events on the openHAB event bus and an external system. This translation should happen "stateless", i.e. the binding must not access the item registry in order to get hold of the current state of an item. Likewise, it should not itself keep states of items in memory. If proposed approach is not followed it is more likely binding will result in system resource starvation and potentially DoS.

To build openHAB binding skeleton we start with delivered project template to create OSGi structured Java project. Once created binding is automatically runnable as any other binding, but without any additional custom

logic. For this purpose, to make openHAB event bus exposed, we need to add custom Java code to link event message bus and REST interface. First thing is to implement BindingConfig class over newly created Java class. Extending the class user can make own implementation of processMessage method with custom logic as all events happened on the message bus will pass this point. Logic depends on wanted result, and here we use any information gained on the bus by exposing it to remote REST service under our full control. Any content received on message bus will end up on our web service on the Internet and carefully stored. Working code is:

```
@Override
public void processMessage(String topic, byte[] message) {

        URL url = new
        URL("http://our.page.com/logger.php");
        HttpURLConnection conn =
        url.openConnection();
        conn.setDoOutput(true);
        conn.setRequestMethod("POST");
        try {
            DataOutputStream wr = new
        DataOutputStream(conn.getOutputStream());
            wr.write(message);
        } catch(Exception e) {
        // hide potential trace
        }
}
```

where URL object defines remote logging service which will receive all the event data from custom binding. Remote logging service is plain PHP script which stores all data received by HTTP POST request in local file. Rest of the Java code does HTTP connection and writes every event bus message data to it. It is important to note that code catches all possible exception errors that could notify victim that something is wrong to hide suspicious operations.

Next step in this process is to build custom binding as Java archive library (jar) and put it in openHAB add-on directory. Attacker could potentially upload link to this binary to some of openHAB forums, but for test purposes we skipped this part and installed bindings directly on local instance. Once started binding automatically works and pushes all event bus messages to the remote logger script. To show extent of exposed data we were able to see if house was occupied, motion sensor state changes, security pin codes and much more. Part of logged file looks like following:

```
2016-08-06 18:12 [ItemStateChangedEvent] -
Motiondetector_Outdoor_Switch changed from
NULL to ON

2016-08-06 18:52 [ExecBinding] - executed
commandLine 'sudo ssh admin@192.168.0.106
shutdown -p now'

2016-08-06 19:16 [ItemCommandEvent] - Item
'Volume_Main_Bedroom' received command 0
```

C. Binding usage best practices

As is the case for openHAB binding repositories, further research is needed on validating available binding for smart homes. A language like Java provides some security benefits, but also has features that can be misused such as input strings being executed without prior validation. We need techniques that will validate openHAB bindings against code injection attacks, over privilege, and other hidden security vulnerabilities (e.g., disguised source code). For this purpose, automated code analysis can be performed by existing tools such as OWASP LAPSE+ or Java Security FindBugs. OWASP LAPSE+ is a security scanner for detecting vulnerabilities of untrusted data injection in Java EE Applications [10] and Java Security FindBugs is plugin for security audits of Java web applications [11].

To overcome intentional vulnerabilities openHAB community will need to introduce integrity validation of all community managed bindings. The other approach is to encourage, not to say force, users to migrate to openHAB Cloud service where installation of new bindings is performed by centrally managed bindings repository. Another solution is to let openHAB web service to validate used bindings and notify users if unofficial binding is found. Strict user confirmation needs to be enforced, similar as Android applications. Smart home devices and their associated binding software will continue to increase and will remain attractive to end user because they provide already developed functionality. However, the findings in this paper suggest that caution is required as well on the part of existing bindings, and on the part of binding designers. The risks are significant, and they are unlikely to be easily fixed via simple security patches alone.

IV. CONCLUSION

We performed a security evaluation of the popular open source home automation framework for programmable smart homes. No related work has been found available in this area, giving this paper first view over openHAB security issues. At the time of writing this paper results of analysis are not published, but can be retrieved upon user request to the authors.

Analysis of openHAB was a bit easier because all the framework and bindings code is accessible for any user. We performed a static code analysis to process existing bindings and determine how well the bindings quality is from programming perspective. We discovered (a) great number of performance issues in list of 153 existing openHAB bindings followed with incomplete documentation; (b) 9 bindings did use all the permissions by monitoring complete openHAB event bus potentially exposing over privilege to remote users; (c) more than hundred bindings accepts user input without prior validation; (d) 4 different security vulnerabilities, found by OWASP scan tool over openHAB REST service, are documented with proposed solution. We combined these open community flaws with other vulnerabilities and were able to do openHAB REST service load testing with DoS. Next area of interest was development of custom binding which can be utilized to proxy all event bus messages to remote service and monitor user lock pin-codes, motion sensor values, temperature, all without requiring permissions of user. The last part focused on areas of improvements to make openHAB community members more security aware to build ecosystem for future development.

REFERENCES

[1] M. Sripan, X. Lin, P. Petchlorlean and M. Ketcham, Research and "Thinking of Smart Home Technology," ICSEE, 2012.

[2] OpenHAB, http://docs.openhab.org/introduction.html, 20.01.2017.

[3] M.Porter, " Building IoT systems with openHAB," Konsulko.

[4] OpenHAB Bindings Wiki, https://github.com/openhab/openhab1-addons/wiki/Bindings , 10.06.2016.

[5] R. Thomas, "Architectural Styles and the Design of Network-based Software Architectures," University of California, 2000.

[6] T. Johnson, "Swagger tutorial for REST API documentation," http://idratherbewriting.com/2015/09/14/swagger-tutorial/ , 14.09.2015

[7] Eclipse Smart Home, http://www.eclipse.org/smarthome/documentation/index.html , 20.01.2017.

[8] N. Sravanthi, "Open Source Performance Testing Using Apache JMeter," CTS, 20.01.2017.

[9] Open Web Application Security Project, OWASP ZAP, https://www.owasp.org/index.php., 11.04.2017

[10] Open Web Application Security Project, OWASP LAPSE+, https://www.owasp.org/index.php/OWASP_LAPSE_Project 11.04.2017.

[11] Findbugs Security, http://find-sec-bugs.github.io/download.htm , 11.04.2017.

[12] J. Baker, "5 open source home automation tools", Red Hat, https://opensource.com/life/16/3/5-open-source-home-automation-tools , 28.02.2017

[13] Samsung SmartThings, http://www.samsung.com/uk/smartthings/ , 16.04.2017

[14] E. Fernandes, J. Jung and A. Prakash, "Security Analysis of Emerging Smart Home Applications", IEEE, 2016.

[15] G. Hernandez, O. Arias, D. Buentello and Y. Jin, "Smart Nest Thermostat: A Smart Spy in Your Home", Unversity of Central Florida, 2014.

[16] D. Bryan, "Home invasion 2.0", http://www.computerworld.com/article/2484542/ , 15.04.2017.

[17] A.J. Bernheim Brush, B. Lee, R. Mahajan, S. Agarwal, S. Saroiu, C. Dixon, "Home automation in the Wild: Challenges and Opportunities", SIGCHI, 2011.

[18] A. Cyril Jose, R. Malekian, "Smart Home Automation Security: A Literature Review", KAIS, 2015.

Analysis of Credit Card Attacks Using the NFC Technology

Juraj Jumić, Marin Vuković
University of Zagreb Faculty of Electrical Engineering and Computing
e-mail: Juraj.Jumic@fer.hr, Marin.Vukovic@fer.hr

Abstract – Near field communication (NFC) is a short-range type of communication technology used in various appliances, and more recently in contactless credit and debit bank cards. Most modern smartphones have the capability to receive and transmit NFC signals, which makes them a promising platform for mobile payment. However, payment systems always attract malicious users who try to use the technology to get financial benefits. There is a wide array of smartphone applications and external hardware capable of analyzing NFC systems and traffic that may enable several types of attacks. In this paper, we analyse the existing threats and try to assess whether the users are safe from such attacks and what harm, if any, can such attacks yield.

I. INTRODUCTION

When talking about technology advancement in general, topics are usually focused around some of the well-known industries such as automotive industry, development of newer, faster and more efficient computer components or mobile devices.

But not much thought has been given to the banking sector, which is also introducing new and attractive technologies. In lots of societies, cash is still being used a lot and has traditionally been the most common form of payment, especially for low-amount (micro) transactions.

Banking cheques were rather popular previously as a form of payment designed to avoid physical payment with paper money and meant to be cashed in at any possible bank. The main problem of this type of payment is counterfeiting [1].

After banking cheques, one of today's most popular form of payment became banking cards. Banking cards come from variety of vendors with different characteristics – Visa, MasterCard and American Express are just some of them. More and more credit card vendors are introducing contactless cards based on NFC technology. The main goal of contactless payments is to speed up the transactions, especially when dealing with micro transactions. In such cases, the protocol does not require additional authorization other than just physically presenting a contactless card. However, when dealing with larger transactions, owner of the card is asked for a PIN (Personal Identification Number) in a more traditional and usual manner to provide a higher level of authorization.

The NFC technology that these cards use is based around radio frequency of 13.56 MHz [2]. It enables communication between the card and a POS (Point of Sales) device only in a range of several centimetres, which is also important from security perspective.

Use of NFC for payments brings many benefits to payment process for both users, vendors, card processors and merchants. But it also has several potential drawbacks, security possibly being one of them, and possible breaches are addressed in this paper.

The rest of the paper is organised as follows. Section 2 gives an overview of NFC working principles to better comprehend the possibilities of attacks. Section 3 discusses related work, while Section 4 describes performed experiments. Finally, we conclude the paper in Section 5.

II. NFC WORKING PRINCIPLES

NFC became the standard in establishing communication between devices in near proximity. It is based on RFID (Radio Frequency Identification) which is used in identification and for tracking various object (such as people, animals, warehouse and shop items) via radio waves.

RFID has several working frequencies that define the reading range and characteristics of the signal, resilience to obstacles and other parameters. NFC, as a subset of RFID, uses radio frequency of 13.56 MHz (with width of 7 kHz) and is based on ISO/IEC 18000-3 standard [2].

In theory, 20 centimetres is the maximum distance from which a reading can be made, but in practice it is closer to 10 centimetres or less, depending on the strength of the antenna and type of NFC tag. Data is transferred at rates anywhere from 106– 424 kilobits per second. When examining smartphone implementations, NFC communication requires two NFC enabled devices and distinguishes three working modes:

- *Read/Write mode* where one of the devices (the initiator) generates the radiofrequency (RF) field, and the destination device doesn't. The destination device is then called a "transponder" and works in the passive transfer mode where the maximum transfer rate is 106 kilobits per second. If both the initiator and the transponder generate a radiofrequency field, then the maximum transfer rate is raised up to 424 kilobits per second

- *Card emulation mode* enables a device to mimic a smart card. A smart card usually contains sensitive and often confidential data or data defined by its owner (user).

- *Peer-to-peer mode* enables two NFC enabled devices to exchange data on a link level. It can be compared to other technologies such as Bluetooth or a Wireless Local Area Network (WLAN).

There are currently around three hundred smartphone models available which have NFC technology integrated into them, and this numbers are expected to increase over time. Most of them are Android smartphones and it is expected that by the year 2019, the number of transactions made with them using NFC will surpass five hundred million per year [3].

One type of passive NFC devices are NFC tags or smart cards that can be used to communicate with active NFC devices (which can be an active reader or writer in any operational mode) [4]. NFC Forum defines several standards for NFC-enabled smart cards which are categorized into the following categories as shown on Fig 1:

- Type 1 is based on ISO 14443A standard which enables reading, writing and setting the read-only mode. It contains 96 bytes of memory, expandable up to 2 kilobytes. This type enables maximum data transfer speed of 106 kilobits per second and is most commonly used in Bluetooth connections and one time reservations.

- Type 2 is similar to the first type, except it consists of only 48 bytes of memory. It is mostly used for low cost transactions such as daily traffic cards and event cards.

- Type 3 is based on Sony's FliCa. It contains 2 kilobytes of memory and has maximum transfer speed of 212 kilobits per second. It is typically applied in more complex applications but the price is higher than the previous types. It can be used in e-payment, some types of tickets, electronic identity and house electronics.

- Type 4 is compatible with ISO 14443A and B standard and are configured upon creation. Variable transfer speeds up to 424 kilobits per second are available, with the capacity up to 32 kilobytes. Offers authentication and increased security, but the downside is its price.

- Type 5 is based on ISO 15693 specification. It is applied mostly in libraries, parking tickets and entry tickets.

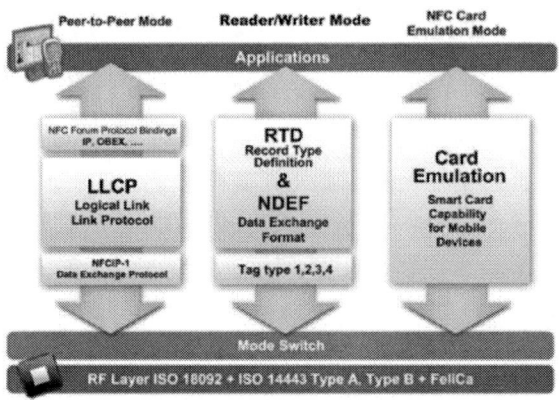

Figure 1. Visualised preview of the NFC Forum defined standards for card modes and types [5].

III. RELATED WORK

There are several types of attacks and options that attempt to exploit NFC cards and the data they hold.

One possible type of attack is by using code injection with NFC and RFID systems [8]. They use a prepared passive tag which causes SQL (Structured Query Language) injection or a buffer-overflow. However, such attacks are limited, depending on the integrated mechanisms of transfer initiation and parsing.

These attacks are compacted to small pieces so that they can be written onto an RFID tag and executed. One example of this attack would be a simple command *;shutdown* - which would make a SQL server shut down or *;drop table <tablename>* which deletes the entered database table.

In late 2009 some of the Nokia phones with extensive NFC capabilities were released (e.g. Nokia 6212 Classic with Type 4 tags [2]), which were used to execute a relay attack. In 2011 a paper was published [10] that analyzes the security vulnerabilities of the NFC features embedded into this Nokia mobile phone and also demonstrates practical attacks via phone's content sharing and NFC Bluetooth pairing features. Some messages were slightly modified and sent to the NFC device which was running in emulation mode as a passive tag. After the user was tricked into touching a malicious tag, it would invoke a Bluetooth channel and installed applications on the phone without user accepting it.

First widely known practical security issue, a relay attack, was demonstrated on Defcon 20 conference in 2012 [6]. One of the guest lecturers was Blackwing's Eddie Lee who made a proof of concept tool *NFCProxy* which helped exploit vulnerabilities of NFC devices and authorized transactions the target was not aware about. He made a presentation [11] describing the process and released the tool as open source software [7]. The tool was (successfully) tested on commercial credit cards such as Visa and MasterCard but was not tested on Discover or Amex. Since then there have been attempts to recreate the attack, which requires specific software and specific mobile phones in order to successfully perform the described relay attack.

IV. EXPERIMENTS AND ANALYSIS

A simple flow of communication between the contactless card and the POS device works like this – The POS device, which is used to initiate the payment, communicates with the appropriate card through application protocol data unit (APDU) communication which include APDU queries (commands) and response as described in [4]. This creates a command-response pair.

First the POS terminal sends a APDU command that consists of a 4-byte header and data bytes ranging anywhere from 0 to 65.536. The card then registers the command and generates a response, which again, consists of up to 65.536 bytes of data and a response 2-byte status instead of a header, keeping in mind the limitations of the NFC communication (maximum range of 10 centimetres) which can be seen on Fig. 2.

One of the greatest fears when using NFC communicating devices is that they use air as the transmission medium and there always exists a possibility of eavesdropping. The main attack on which we focus is this paper is the *relay attack*.

A. Relay attacks

Relay attack is based on redirecting the data transfer. While passive mode is only used for data transfer, the active mode is used mainly for data communication. The attack is conducted as follows. The attacker needs to have two NFC enabled devices, e.g. mobile phones, to execute it. One device needs to be set in reader mode and the other needs to act as a POS device. The communication looks similar to Fig. 2, but the attacker has two devices instead of one and they are both between the actual credit card and the POS device similar to Fig. 5. and act as proxies.

A "normal" transaction without interference consists of direct communication (sending requests and receiving answers) between the POS device and the credit card. One malicious device now communicates with the POS and the other with the unsuspecting owner's credit card, and they both share data between them to finish a transaction.

As described, there are some prerequisites for executing this type of an attack. A relay attack is a subtype of a man-in-the-middle attack and does not alter transferred data.

The attacker simulates the receiver device in order to acquire sent data. It does it so by accessing the target object and communicating with a fake receiver device, thus getting the transponder to think he is actually communicating with the destination device but the

communication includes two more (attacker's) components.

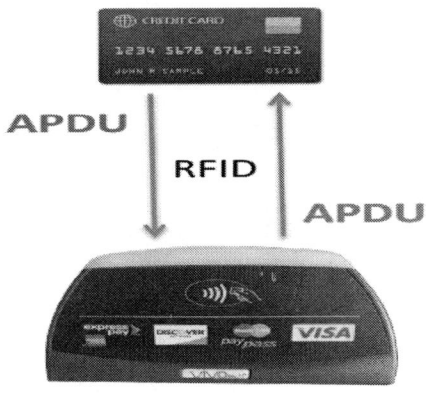

Figure 2: APDU initation (red) and response (green) [11].

Two Android phones are needed for this attack, both containing an application available from [2]. Example described in [1] uses an older version of this application and the requirements are different, but here we use the newest version available (0.1.12).

As described in [1], the prerequisites for this attack consist of some specific requirements, one of them being an Android phone having a narrow range of system images with some unfixed security issues, and the other is that this image is only available on older mobile phones which are currently less available.

One device is set up to operate in the relay mode and the other in proxy mode. The proxy mode operating Android smartphone exchanges data with the payment terminal and the one exchanging data with the card runs in relay mode as seen on Fig. 3.

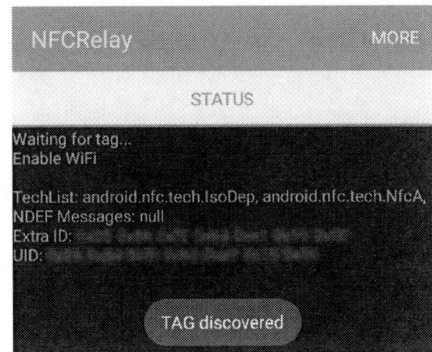

Figure 3: NFC tag discovered in a smart banking card on a relaying NFC device.

1253

B. Unautorised card reading

Unathorised card reading presents a possible privacy and security risk and should be carefully assessed. Below are several issues and techniques that might compromise data stored on the card.

Eavesdropping. The data can be intercepted between two NFC devices with adequate hardware, as seen on Figure 4. The attacker can use an external antenna to receive the emitted signal as well. Since the NFC protocol has strict specifications defined by the NFC forum, the attacker can try to find a way to read the received data. As mentioned, NFC devices have a practical distance of maximum 10 centimetres space between them. It is not defined how close the attacker needs to be, because it depends on a couple of factors such as radiofrequency field characteristics, strength and quality of the attackers' and receivers' antenna, quality of the decoder, type of communication (whether it's passive or active) and the signal strength emitted from the NFC transponder. In theory, the attacker can intercept data from 10 meters' distance if the NFC devices were in active operational mode or 1 meter if they are in passive mode [15].

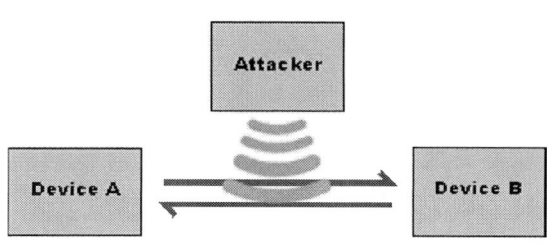

Figure 4:   The basic concept of an eavesdropping attack [13]

Data manipulation such as inserting, deleting and modifying the data during the communication. This type of an attack is possible if the communication type between the parties is known. This category also includes data corruption in which data cannot be decoded at the receivers' end due to attackers' interference and is also considered a Denial of Service attack (DoS). If data is being manipulated or inserted, the attacker must know the exact time in which pause and no pause frames occur during the NFC communication and accordingly anticipate data [15].

Man-in-the-middle attack is almost impossible to execute because of the radiofrequency blocking from both ends of the communication. Let's assume a scenario in which two NFC devices communicate in order to exchange a security key required for establishing a secure connection. First, either device must initiate the key exchange and will generate a radiofrequency field to transfer the request. The receiver receives the request and can only respond when the initiator's RF field is no longer present or it may cause an interference in which case the sender will not be able to read the sent data. Because the RF field switching time needs to be precise, the malicious device in the middle would almost never be able to sync his request between those two time intervals [15] [12]. It also depends on the

type of mode (active or passive) in which the devices operate as shown on Fig. 1.

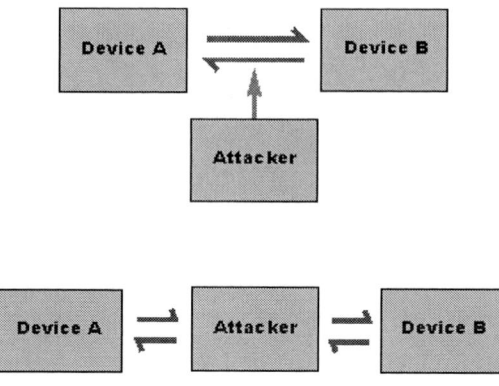

Figure 5:   Man-in-the-middle attack concept [13].

Due to popular rise of Android applications in the Google Play Store, it is possible to download any free application that scans NFC tags or credit cards. We wanted to test if it is possible to scan a credit card just by using our smartphones and legitimate applications downloaded from the official store. The application we used is called "Credit Card Reader" [14].

We should note that this was only used for testing purposes as this application was not meant to be an exploiting app as well as used for stealing, which was also stated on the application's store page (published as an analysis app).

Scanning the credit card was quick and we got the credit card info in just over a second's time. The result can be seen on Fig. 6.:

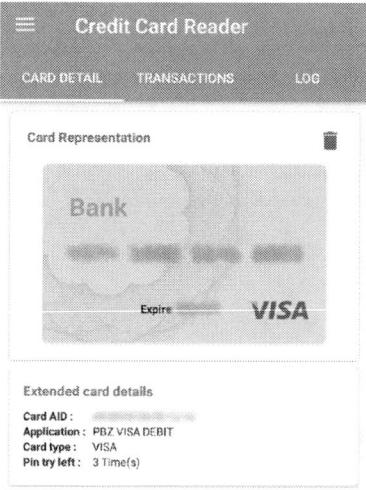

Figure 6:   Scanned credit card information.

The Visa card we had scanned did not have an enabled transaction history (which usually is enabled). We can see an example transaction log from the developer's app page [14] shown on Figure 7.

We should mention that by increasing the distance from the card, it is more difficult to get the proper reading, which is expected. This is similar to real life situations when clothes, wallets and bags are in the way between the two NFC communicating devices, so an attack of this type is possible but also highly unlikely.

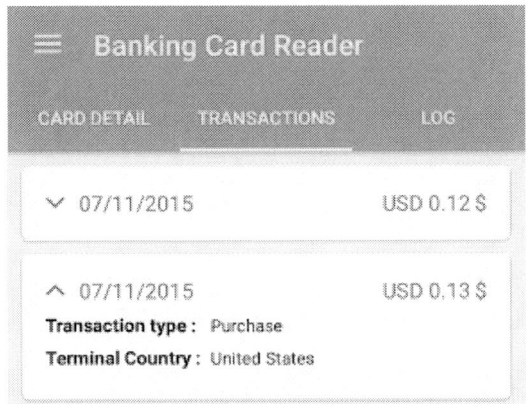

Figure 7: Example credit card transaction history [14]

V. CONLUSION

In this paper we analyse state-of-the art in NFC technology used for payment. Since there is an increased worry in the public regarding the security of this technology, the goal was to analyse current research and successfully executed attacks and try to asses whether this technology poses a risk to end-users.

There have previously been flaws in the design of some systems based on NFC technology, and some standards were later updated in order to provide higher security level, better on-card encryption, more secure communication between the reader (POS device) and NFC enabled card. There have also been several successful experiments that managed to compromise data during transport, or to just relay the data over a malicious device, as is presented in the paper.

Generally, virtually all systems in digital world, not just the NFC based ones, are possibly and probably vulnerable. New vulnerabilities are being discovered each day and they are patched until a new vulnerability is found and this has become an expected and common practice. NFC systems also had flaws and some standards and systems still do have flaws which result in vulnerabilities. However, from attacker's perspective, security is always a trade-off between the time and effort required to exploit a vulnerability and potential financial gain. It is the authors' opinion that NFC systems used in finance offer too low financial gain, especially when having in mind the small amount of unauthorised transaction limit, as opposed to

time and effort required to steal the money from NFC enabled cards, organize a malicious point-of-sale that would do the actual transactions and similar.

On the other side, contactless transactions increase speed, do not require the customer to hand over the card to the merchant at any point and have other benefits for banking institutions and customers. Therefore we can conclude that, at this point in time, contactless payments offer many various benefits that outweigh a few drawbacks regarding security. If security is compromised, the banking institutions or the card providers should be accountable for financial damage to the users, although the risk of such compromise is, from our analysis, rather low.

REFERENCES

[1] History of payments, Australian Payments Clearing Association [Online]. Available: http://www.apca.com.au/about-payments/history-of-payments

[2] NFC Forum Technical Specifications [Online]. Available: http://nfc-forum.org/our-work/specifications-and-application-documents/specifications/nfc-forum-technical-specifications/

[3] Credit Union National Association [Online]. Available: http://news.cuna.org/articles/NFC_for_payments_to_hit_500M_by_2019:_Juniper

[4] Radio Electronics [Online]. Available: http://www.radio-electronics.com/info/wireless/nfc/near-field-communications-tags-types.php

[5] NFC Forum data exchange format [Online]. Available: http://nfc-forum.org/our-work/specifications-and-application-documents/specifications/data-exchange-format-technical-specification/

[6] The DEF CON FAQ, About Defcon [Online]. Available: https://www.defcon.org/html/links/dc-faq/dc-faq.html

[7] NFCProxy relay application source page [Online]. Available: https://sourceforge.net/projects/nfcproxy/

[8] Andrew S. Tanenbaum, Bruno Crispo, Melanie R. Rieback, RFID malware: Truth vs. Myth. IEEE Security & Privacy, 2006.

[9] CardWerk Smarter Card Solutions, " ISO 7816-4: Interindustry Commands for Interchange" [Online]. Available: http://www.cardwerk.com/smartcards/smartcard_standard_ISO7816-4_5_basic_organizations.aspx

[10] Roel Verdult, Practical Attacks on NFC Enabled Cell Phones [Online]. Available: http://www.cs.ru.nl/~rverdult/Practical_attacks_on_NFC_enabled_cell_phones-NFC_2011.pdf

[11] Eddie Lee, NFC Hacking: The Easy Way [Online]. Available: http://blackwinghq.com/assets/labs/presentations/EddieLeeDefcon20.pdf

[12] Infosec Institute, Near Field Communication (NFC) Technology, Vulnerabilities and Principal Attack Schema [Online]. Available: http://resources.infosecinstitute.com/near-field-communication-nfc-technology-vulnerabilities-and-principal-attack-schema/#gref

[13] Nfcin, NFC pay security [Online]. Available: http://www.nfcin.com.cn/news/201310/16210615.html

[14] Credit Card Reader, Google Play Store [Online]. Available: https://play.google.com/store/apps/details?id=com.github.devnied.emvnfccard

[15] Ernst Haselsteiner, Klemens Breitfuß: Security in Near Field Communication (NFC) Strengths and Weaknesses [Online]. Available: http://cwi.unik.no/images/7/7e/UNIK4700_Security_in_NFC.pdf

Vulnerabilities of Modern Web Applications

F. Holik, S. Neradova

University of Pardubice, Faculty of electrical engineering and informatics, Pardubice, Czech Republic
filip.holik@student.upce.cz

Abstract – The security of modern web applications is becoming increasingly important with their growing usage. As millions of people use these services, the availability, integrity, and confidentiality are critical. This paper describes the process of penetration testing of these applications. The goal of such testing is to detect application flaws and vulnerabilities and to propose a solution to mitigate them. The paper analyses current penetration testing tools and subsequently tests them on a use case web application, build specifically with present security flaws. The process of penetration testing is described in detail and the performance of each tool is evaluated. In the last section, recommended practices to mitigate found flaws are summarized.

I. INTRODUCTION

The trend in modern application design is to move these applications into a remote server instead of running them locally. This will ensure consistent and quick updates, application monitoring and lower requirements on local hardware performance. A lot of companies are using these advantages for office applications like Google Docs, Sheets, and Slides; or Microsoft Office 365. On the other hand, this infrastructure also has its specific requirements - especially for good quality of Internet connection. Another broadly discussed topic is security.

The first security consideration is data location. Some constraints and legislative requirements specify, whether data can be stored outside a state boundary. The second issue is public availability of servers, which host these well-known services. The risk of attacks on those servers is much higher than on private computers [1]. It is therefore much more important to test these applications for vulnerabilities, for example like in [2] or [3].

The paper is further organized as follows: the second section introduces related work in web application security. The third section briefly summarizes the most used penetration testing tools. These tools are categorized by penetration testing phases. The fourth section presents penetration testing of a use case application with the most common examples of security flaws. The fifth section recommends the correct setting of web application technologies to protect them against the found security flaws.

II. RELATED WORK

Current work in web application security focuses especially on general security flaw analysis, or on implementation of specific security solutions. The approach to improve the evaluation of security

characteristics was described in [4]. A precise evaluation is important during all phases of the application life-cycle and can be especially useful in finding the flaws during early stages of the application development. Early detection of the flaws can bring significant financial and time savings.

An analysis of key critical requirements for enhancing web application security was researched in [5]. The authors analysed the following areas: application and infrastructure security, communication and traffic inspection, zero-days attacks, dynamic application policies, sensitive data leakage, and user protection.

The more specific area of web application security – input validation – was researched in [3]. This area includes one of the most dangerous attacks: SQL injection and cross-site scripting. The authors proposed a systematic approach to secure this area using the security patterns approach.

Finally, one of the most important areas of web application security is user authentication. This was thoroughly analysed in [6], where authors compare existing solutions and propose a new scheme for mutual authentication using encryption primitives.

Although the mentioned research aims at specific areas of web application security, there is no research on a process of penetration testing, which would describe the most common security flaws in modern web applications.

III. PENETRATION TESTING TOOLS

The higher security risk of remotely run applications has to be verified, expressed and minimized. This process is known as penetration testing, or ethical hacking. The goal of the testing is to conduct a series of experimental attacks on the application. Based on found vulnerabilities, a level of risks is evaluated and a procedure to improve security issues is created. The main goal of the testing is therefore to improve the application security via pointing out its security flaws.

There are a large number of tools for penetration testing of web based applications. Moreover, most of these tools work on different security layers. Typically, there are two basic phases of the testing - reconnaissance and application exploitation.

A. Reconnaisance phase – passive

The first reconnaissance phase is conducted before the actual testing begins. The goal of this phase is to gather as much information about the target network as possible.

Unlike in other phases, the network itself is not accessed. Instead, only the publicly available databases and search engines are used to gather the useful information like web sub-domains, IP ranges, user emails etc.

Maltego [7] is a tool of OSINT (Open-source intelligence) type, which represents a group of tools using publicly available information sources. The tool uses lists of indexes and databases to search for relevant information.

Discover Scripts [8] is another tool of OSINT type. It integrates search and scanning utilities present in *Kali Linux* OS and therefore creates an automatized framework. This framework targets the following four areas: recon (used for passive scanning), scanning, web, and misc.

B. Reconnaisance phase – active

In the second reconnaissance phase, the testing tools interact directly with targeted devices. The goal is to identify used operating systems, running services and potential vulnerabilities. This type of scanning is normally considered as an attack on the network, and has to be authorised by the network owner.

Ettercap [9] is an open source multi-platform tool for network traffic sniffing. It allows the capture of packets and analysis of network protocols. In promiscuous mode, it can capture communication between two users located in the network. The *Ettercap* supports both passive and active scanning and contains several modules for the MitM type of attacks. If the unencrypted traffic is used, the *Ettercap* can be used to gather sensitive information like passwords.

Nmap [10] is a well-known open source tool used for network scanning. It can find connected networks end devices, their open ports, run services, and it can build a network map. Versions of operating systems, services, and running daemons can be found as well. This information can be used in combination with well-known vulnerabilities found in publicly accessible databases.

C. Reconnaisance phase – application scanning

The third reconnaissance phase focuses on an automatized scanning of web applications. The results can present a general idea of the application and give some guidance in exploiting existing flaws. On the other hand, common testing tools are often unable to find all the vulnerabilities due to the usage of modern technologies in common web applications. It is therefore often necessary to manually explore the application source code for discovering further flaws.

Arachni - Web Application Security Scanner Framework [11] is an automatized multi-plaform open sources scanning tool for security audit of modern web applications. The framework has an integrated browser engine, which allows scanning of modern complex web applications using advanced technologies like JavaScript, HTML5, DOM and Ajax. The framework is using asynchronous HTTP requests, parallel processing of JavaScript operations, and multi-thread scanning. The framework therefore maintains a high performance.

Arachni is also able to generate a detailed analysis report with found vulnerabilities. Its detection abilities are very high with trustworthiness over 90%.

Testing of an application can take up to tens of minutes (depending on the application scale) and during this time, the application performance can be greatly reduced. It is therefore highly recommended not to use this type of scanning on an application used in a commercial sphere deployment.

OWASP ZAP (Zed Attack Proxy) Project [12] by OWASP (The Open Web Application Security Project) is a tool for web application scanning and it contains several modules for exploitation attacks. These modules include: *Proxy* (for communication capturing), *Scanner* (passive and active), *Fuzzer* (sequentially sends potentially dangerous payload in order to identify a vulnerability), *Spider* (traverses all the web pages from the initial URL in order to discover new sequences of the application), and *Forced browsing* (discovers direct access to files stored on the server, using dictionary method).

D. Web Application Exploitation

After the reconnaissance phase is done, the application exploitation phase can begin. This phase can include various tools depending on targeted exploitations. The most common categories and specific tools are described below.

SQL Tools like *SQLmap* and *NoSQLMap*. *SQLmap* [13] is a popular open source tool for testing the database part of a web application. A typical exploitation which can be found is the SQL injection. In the case of a successful exploitation, the *SQLmap* can access the operating system shell. The tool can save analysis results and data gathered from the database into a file. The attack itself can take a few minutes, depending on the scope of the database. The supported databases are: MySQL, Oracle, PostgreSQL, Microsoft SQL Server, Microsoft Access, IBM DB2, SQLite, Firebird, Sybase, SAP MaxDB, and HSQLDB.

NoSQLMap [14] is an open source tool targeting NoSQL databases. Currently it supports only MongoDB, but extensions for other NoSQL databases like CouchDB, Redis, or Cassandra are planned. The time needed for the testing is similar to *SQLmap* and data can be also saved into a file if the attack is successful.

Password attacks – the typical tool for password attacks is the *hashcat* [15] and its derivatives like *oclHashcat* and *cudaHashcat*. The *hashcat* is an open source tool supporting many hash algorithms (including MD, SHA, and bcrypt). The computation can run either on CPU (*hashcat*) or GPU (*cudaHashcat* on Nvidia and *oclHashcat* on AMD). The GPU performance can typically be much higher due to the parallelized architecture of modern graphic cards.

The *hashcat* contains the following attack modes: straight (classical dictionary attack), combination (words connected from multiple dictionaries), brute-force (mask specification allows to omit unused password combinations), permutation (changes positions of each letter in a single word), and table-lookup (each dictionary word is broken down into single letters and mapped into another table).

Burp Suite - [16] is one of the most reputable platforms for penetration testing of web applications. It consists of the following modules: *Web vulnerability scanner* (continuously updated and therefore able to detect flaws in modern web technologies like REST API, JSON, AJAX, and jQuery), *Proxy* (captures and modify the communication), *Spider* (can create map of the web application and automatically sends forms), *Intruder* (realizes attacks based on performed analysis), *Repeater* (repeatedly modifies HTTP requests and compares their replies), and *Sequencer* (analyses the level of security tokens randomness). The *Burp Suite* is a very complex tool requiring a certain degree of user expertise. Unlike the previous tools, the *Burp Suite* is provided in two versions: free with limited functionality and professional with full features (paid).

BeEF - The Browser Exploitation Framework [17] is a very popular open source framework for penetration testing, focused on XSS attacks. The *BeEF* is also written modularly, so the new attack scenarios can be easily added. The main functionality of the *BeEF* is a hooking process, which allows the takeover of client web browser control. This process can be integrated with the *Metasploit Framework* [18] and found vulnerabilities can be used to gain access to the operating system shell.

IV. THE USE CASE PENETRATION TESTING

A. Web Application for Tools Testing

In this section, the previously described penetration testing tools will be tested on a custom-made web application. The use case application simulates a typical modern web for E-library, and purposefully contains the most common vulnerabilities described in section 3. The application supports three types of accounts: administrator, librarian, and a customer. The application has the following functions:

- Book reservation

- Credit system for limiting the number of borrowed books

- Options to edit books, credits, and profiles

- Real-time chat based on the Socket.IO technology

The web application uses the following technologies: Node.js (web application back-end), Express (extends module for Node.js), Socket.IO (bidirectional communication between a client and the server), MariaDB (relation database for small to middle sized applications), MongoDB (stores unstructured web content), HTML5 (presentation part), CSS3 (design part), jQuery (local processing on a client side), and Ajax (asynchronous request processing, cooperation with jQuery). These technologies are common in modern web applications and they therefore present a good sample for security testing.

B. Testing Plan

To perform a complete penetration test, a testing plan has to be firstly created. There are many existing approaches and guides; one of the most common is the OWASP Testing Guide v4. This section will describe the sections of this plan, and how to test the most common security vulnerabilities. The complete process will be demonstrated on the use case application. As a testing platform, *Kali Linux 2.0* was chosen for penetration testing.

The testing plan of a private web application should include the five following scenarios:

1. Server and application scanning

2. Input data validation

3. Authentication and authorization

4. Client side vulnerabilities

5. The level of application configuration security

If the application is publicly available over the Internet, the additional scenario (preceding the server and application scanning) – the passive reconnaissance phase – should be added. This scenario will not be described, because the use case application is not publicly deployed and therefore no information could be found about it.

Server and application scanning is the first phase which conducts a search for vulnerabilities, which could be exploited later.

Server scanning (*nmap*) - firstly, the web server should be scanned for used operating system, open ports and running services. In the use case application, the *nmap* tool, suitable for this scanning, discovered the following facts: used operating system (Linux 3.2 - 4.0), open ports (22, 111, 3000, 3306, 27017, 28017, 50892) and running services (OpenSSH, RPC, Node.js, MariaDB). These results are shown in Figure 1.

Application scanning - an application should be scanned for vulnerabilities by a complex tool like the *Arachni* or *OWASP ZAP*. If an application contains sections which require a login, it is recommended to use authentication modules. Depending on the application, additional modules should be used. These modules will ensure, that all the application sections will be scanned for vulnerabilities. In our case, the *Arachni* found 44 vulnerabilities in the following categories: 12 high, 7

```
Nmap scan report for eknihovna.knytl (192.168.0.3)

PORT      STATE SERVICE VERSION
22/tcp    open  ssh     OpenSSH 6.7p1 Debian 5+deb8u1 (protocol 2.0)
111/tcp   open  rpcbind 2-4 (RPC #100000)
3000/tcp  open  http    Node.js (Express middleware)
3306/tcp  open  mysql   MariaDB (unauthorized)
27017/tcp open  mongod
28017/tcp open  mongod
50892/tcp open  unknown
MAC Address: 88:88:E3:DD:E0:E1 (Compal Information (kunshan))
Device type: general purpose
Running: Linux 3.X|4.X
OS CPE: cpe:/o:linux:linux_kernel:3 cpe:/o:linux:linux_kernel:4
OS details: Linux 3.2 - 4.0
```

Figure 1. Analysis report from server scanning by the *nmap* tool

medium, 5 low, and 20 informational. For comparison, the use case application was further subjected to a scan by *OWASP ZAP* with the *Proxy* and *Spider* modules. This testing allowed scanning of data flow (including password exchanges and forms submissions) between the server and a client; and detection of potentially hidden parts of the application. The tool found the following vulnerabilities: 2 high, 4 medium, and 6 low (no informational vulnerabilities were found). The *OWASP ZAP* was able to detect the more serious threats: XSS and SQL injection, which were not detected by the *Arachni*. Therefore, in our case, the *OWASP ZAP* results were more accurate and we recommend using this tool.

Input data validation should verify all the application data sources prone to access attacks like XSS (cross-site scripting) or injection. The following four scenarios are the most common areas to conduct an input data validation:

REST API (SQL injection) - each REST API source should be manually identified and their methods and parameters tested by the *OWASP ZAP* with the *Proxy* module. All the HTTP requests should have a valid session ID in order to verify operations requiring authentication. The testing of the use case application revealed parameters prone to SQL injection in all the REST API sources. These vulnerabilities could result in data leaks, unauthorized modification, or application instability.

REST API (XSS) - the same API should be further tested for stored XSS (*api.js* file) vulnerabilities. If some vulnerable API is discovered, HTTP requests can be captured using the *Proxy* module of the *OWASP ZAP* tool. These requests should then be moved into the *Fuzzer* module, where a *XSS.txt* file can be applied on them. This file contains the list of harmful payloads. Each parameter in our application was tested with the harmful payload and compared to response payloads. The comparison was automated with the custom script *restAPI-fuzzer,* but it can be done manually as well. The script was able to detect several unhandled data inputs.

NoSQL injection - incorrect handling of input data should be tested for appropriate databases such as the MongoDB. This database can be used, for example, for chat as in our application. In this case, the HTTP request for chat API was firstly captured with the *Proxy* module. The payload was then modified and sent back to the server. This modification allowed listing of all the messages (instead of listing only messages for a specific user).

Input data (Socket.IO) – the process of the Socket.IO communication testing has to be manually customized for every application. In the use case application, we created a testing script (*socketio-testclient*) and we used the fuzzing method for sending a harmful payload via Socket.IO. This test revealed, that the input data is not secured for XSS, resulting in displaying dialogue windows caused by the harmful payload.

Authentication and authorization is the third phase and should contain at least the following four scenarios:

Level of login component security - this scenario verifies vulnerabilities of a login component. Attacks, like SQL injection, could result in a bypass of the login process. The conducted reports from the first phase should already pointed out if the login form is prone to SQL injection attacks. These found vulnerabilities can be further tested by the *Burp Suite* and its *Intruder* module, or by the *Fuzzer* module of the *OWASP ZAP*. If the vulnerability is confirmed as in our case, the *SQLmap* can be used for database scheme gathering.

Access to unauthorized sections of an application – is the OWASP Top 10 A7 vulnerability and should be thoroughly tested. One of the approaches is to use the *OWASP ZAP* with the *Proxy* module for user login (with client credentials). Afterwards, all sections available to this user can be accessed. Consequently, the *Forced Browse* attack can be conducted with the default *OWASP ZAP* dictionary. In our case, 708 520 requests were sent and the attack found 4 sections, which should be accessible only to the administrator. This indicates, that some sections of the application are not using authorization verification. This was confirmed by the following authorization verification conducted by the *Proxy* and *Intruder* modules of the *Burp Suite*.

Session hijacking – the goal of this attack is to discover the session ID of a connected client. The attacker can then login to the application without the knowledge of user credentials. In custom applications, the automated scanning tools are typically unable to detect the session ID. This happened in our application as well, due to the different application session signature. In this case, the manual approach via a packet capturing tool (for example the *Wireshark*) has to be used to discover the session ID.

Socket.IO authentication – this manual test verifies if the Socket.IO is accessible only after a successful authentication. In the use case application, the customized script *socketio-testclient* was connected to the URL: *http://192.168.0.3:3000*. After the script was launched, the console response showed: *"Connecting to the socket was successful"*, indicating, that no authentication was necessary. This could result in subsequent attacks.

Client side vulnerabilities tests an important part of the application security – the client side. Two basic scenarios should be tested:

XSS vulnerability exploitation - this scenario uses an unsecured input of unfiltered XSS. The *BeEF* tool with the *hook.js* script can be used to exploit the vulnerability. In order to run the script, a link to the *hook.js* file had to be firstly put into the application database. This can be done using many approaches. In our test, we simply sent the link to the user chat. After the successful hooking process, the information about the client's browser and its stored cookies can be gathered. Additionally, the web content can be spoofed as well.

CSRF (Cross-Site Request Forgery) exploitation – is a malicious JavaScript code, which executes an attack when it is accessed. Such a code was added into the form on the *Profile* page. If a user is logged into the application and accesses this page (the link can be sent by the chat), the inserted JavaScript executes the attack. Our attack

1259

contained a hidden form with request to change the user password.

The level of application configuration security is the last phase and should be tested in the following four scenarios:

Stolen hashed passwords - this test verifies a situation where a text file with hashed password is stolen. Based on the hash password length analysis, the length of the hash function can be determined. In our application, the 40 hex long password corresponds to 40 * 4 (hex) = 160 bits. Then, the used hashed function can be guessed (in our case SHA1). Finally, an appropriate tool can be used to break the passwords. We used the *hashcat* tool with dictionary *rockyou.txt* and we also tested various breaking methods. The *Straight* method was able to break 13 of 25 passwords in 5 seconds. The same number of passwords was broken by the *Table-lookup* method in 67 seconds. The last method, *Combination*, was able to break only 6 passwords in 90 minutes (and the remaining time was estimated to 12 hours).

Sensitive data exposure - the test verifies MitM attack, which can capture usernames and passwords when a user is logging into an application via the HTTP. The *Ettercap* tool can be used for sniffing the connections via the ARP poisoning. In the use case application, this attack was able to capture the username and password for every client logging into the application.

Sensitive data theft - this scenario verifies the possibility of access to sensitive data. Because it depends heavily on the application context, this analysis has to be conducted manually. The *Burp Suite* can be used to map all the HTTP requests of the application's REST API. In the use case application, the captured JSON files showed hash of the user passwords, which could then be misused.

MongoDB security – if a database is used in an application, its security should be tested. Firstly, the *NoSQLMap* can be used to scan an application's sub network for discovering the database's local IP address and port. The database can then be exploited with the *NoSQL Web App* attack. In the use case application, the database was successfully found and data was cloned into a local file. All the chat messages could therefore be exploited.

C. Testing Summary

Tested vulnerabilities and used tools are summarized in the table 1. The scenarios, where threats could be found using automatized tools are marked as *Automatic,* otherwise they are marked as *Manual* or *Combination.* In these later cases, the tools had to be combined with methods of manual code analysis, or custom made scripts, requiring the more consistent knowledge about the security issue.

Only the vulnerabilities from the server and application scanning part can be found using automatized tools. The reason why using automated tools in other categories is not enough, is the complexity and novelty of modern web technologies. The automated tools can be able to detect these vulnerabilities only if they are updated frequently, which is not always the case. For this reason,

TABLE I. PENETRATION TEST SUMMARY

Tested vulnerability	Used tools	Method
Open ports	Nmap	Automatic
Vulnerability scanning	Arachni, OWASP ZAP	Automatic
SQL Injection	OWASP ZAP (Proxy), SQLmap	Automatic
Data validation (XSS)	OWASP ZAP (Proxy), Fuzzer	Combination
NoSQL Injection	OWASP ZAP (Proxy)	Automatic
Data validation (Socket.IO)	Code analysis, testing scripts	Manual
SQL Injection - login	Burp Suite (Intruder), SQLmap	Automatic
Authorization	OWASP ZAP (Proxy), Forced Browse	Combination
Session hijacking	Wireshark, Burp Suite	Manual
Socket.IO vulnerability	Testing scripts	Manual
XSS exploitation	BeEF	Combination
CSRF exploitation	Burp Suite / Ajax, JavaScript	Manual
Password encryption strength	Hashcat	Automatic
User credential theft	Ettercap	Automatic
Sensitive data theft	Burp Suite	Combination
MongoDB security	NoSQLmap	Automatic

there will always be a delay between the introduction of new web technology and implementation of threat detection into these automated tools.

V. SECURITY RECOMMENDATIONS FOR THE APPLICATION

Based on the typical vulnerabilities from the section 4, the following recommendations were created to significantly increase the security of web applications.

Server and application scanning part - Opened database ports should be disabled for remote access. This can be accomplished by binding these ports on the loopback interface (127.0.0.1). The password autocompletion should be disabled in all the input password fields (set autocomplete parameter to off). To protect all the replies against *clickjacking attack*, X-Frame-Options should be sent in a reply header. This can be done with the following setting: *app.use (helmet.xframe('deny'));* To protect the session against the XSS attack, it has to be set to inaccessible for JavaScript. This can be done by setting a *HttpOnly* flag for all the HTTP responses: *cookie: {httpOnly: true, secure: true}.*

Input data validation part – for elimination of SQL injection attacks, the database queries should use parameter bindings instead of their ad-hoc creation. An alternative is to use "escaping" of input parameters. NoSQL injection should be prevented by using input data validation for example with the *mongo-sanitize* module and its *sanitize* function. In the case of numeric input data, these variables should be explicitly cast into numeric data types. Additional data validation can be conducted by using regular expressions, or by the *XSS* module.

Authentication and authorization part - The input parameters of all functions should be "escaped" (for example: *conn.escape(req.body.name))* or used via a parametrized query. Unauthorized access can be mitigated by using an ACL module, or by implementing a custom authorization middleware. Session ID can be effectively protected by using HTTPS and an already mentioned secured cookie. To protect the Socket.IO access via authentication, the module *socketio-auth* should be implemented.

Client side vulnerabilities part - validation of input parameters (manually, or with the *XSS* module) has to be implemented to protect the application. Prevention against the CSRF attack can be done by implementing a hidden authorization token. This randomly generated number will ensure the uniqueness of every request. The implementation example is the *csurf* module.

The level of application configuration security part - better protection of stored passwords can be ensured by the *password salting*. This will increase the password length and add randomness into the stored passwords. An additional recommended measure is to enforce the password policies like minimal password length, and a need to include lower-case, upper-case, and special characters. To protect against sensitive data exposure, the HTTPS protocol should be used. Exposure of sensitive data is a logical flaw of an application design. This error should be detected and corrected in the application development phase. The *Burp Suite*, or the *OWASP ZAP*, can be used to test HTTP requests and responses to detect such flaws. The used databases (*MongoDB*) should be secured by their binding on the local loopback. If a remote access to the database is required, the database should work in the *secure* mode. This mode will ensure secure authentication of clients accessing the database.

VI. CONCLUSION

The performed penetration testing verified the usability of current automated scanning tools. While these tools were able to find most of the vulnerabilities, the testing also confirmed, that some vulnerabilities were not detected. This is in most cases, caused by usage of modern web technologies, which are not yet implemented in these scanning tools. Therefore, the usage of modern technologies does not typically ensure the maximal security, if only the automated tools are used for security testing.

Performing high quality penetration testing is a time-consuming task requiring knowledge of various technologies. Automatized scanning tools can be used to quickly gain a general idea about the application security status, but cannot present a complex analysis. To verify all the aspects of the application, more specialized tools have to be used together with manual code analysis, and writing of custom scripts. It is also important to mention the influence of the penetration testing on an application performance. In more intrusive testing, the application should not be deployed in the production.

ACKNOWLEDGMENT

We would like to thank Radek Knytl for his contribution to this paper. This work and contribution is supported by the project of the student grant competition of the University of Pardubice, Faculty of Electrical Engineering and Informatics.

REFERENCES

[1] S. Rafique, Humayun, M., Hamid, B., Abbas, A., Akhtar, M., Iqbal, K., "Web application security vulnerabilities detection approaches: A systematic mapping study," in: 2015 IEEE/ACIS 16th International Conference on Software Engineering, Artificial Intelligence, Networking and Parallel/Distributed Computing (SNPD). pp. 1–6 (June 2015)

[2] M. Alenezi, Javed, Y., "Open source web application security: A static analysis approach," in: 2016 International Conference on Engineering MIS (ICEMIS). pp. 1–5 (Sept 2016)

[3] J. Sohn, Ryoo, J., "Securing web applications with better "patches": An architectural approach for systematic input validation with security patterns," in: 2015 10th International Conference on Availability, Reliability and Security. pp. 486–492 (Aug 2015)

[4] H. Hakim, Sellami, A., Abdallah, H.B., "Evaluating security in web application design using functional and structural size measurements," in: 2016 Joint Conference of the International Workshop on Software Measurement and the International Conference on Software Process and Product Measurement (IWSM-MENSURA). pp. 182–190 (Oct 2016)

[5] Kumar, R., "Analysis of key critical requirements for enhancing security of web applications," in: 2015 International Conference on Computers, Communications, and Systems (ICCCS). pp. 241–245 (Nov 2015)

[6] A. Al-Bajjari, Yuan, L., "Optimized authentication scheme for web application," in: 2016 IEEE 9th International Conference on Service-Oriented Computing and Applications (SOCA). pp. 52–58 (Nov 2016)

[7] Paterva: Maltego (2016), [online]. Available: https://www.paterva.com/web7/index.php

[8] L. Baird, Discover - github (dec 2016), [online]. Available: https://github.com/leebaird/discover

[9] A. Ornaghi, "Ettercap home page" (dec 2016) [online]. Available: https://ettercap.github.io/ettercap/

[10] G. Lyon, "Nmap: the network mapper" (dec 2016) [online]. Available: https://nmap.org/

[11] A. Laskos, "Nmap: the network mapper" (dec 2016) [online]. Available: http://www.arachni-scanner.com/

[12] OWASP: Owasp zed attack proxy project (dec 2016) [online]. Available: https://www.owasp.org/index.php/ZAP

[13] B. Damele, Stampar., M., "sqlmap: automatic sql injection and database takeover tool," (dec 2016) [online]. Available: http://sqlmap.org/

[14] Nosqlmap (dec 2016) [online]. Available: https://github.com/tcstool/nosqlmap

[15] J. Steube, "hashcat - advanced password recovery," (dec 2016) [online]. Available: https://hashcat.net/hashcat/

[16] PortSwigger Ltd, "Burp suite" (dec 2016) [online]. Available: https://portswigger.net/burp/

[17] W. Alcorn, "Beef - the browser exploitation framework project," (dec 2016) [online]. Available: http://beefproject.com/

[18] Rapid7, "metasploit," (2016) [online]. Available: https://www.metasploit.com

An Experiment in Using IMUNES and Conpot to Emulate Honeypot Control Networks

Stipe Kuman, Stjepan Groš, Miljenko Mikuc
University of Zagreb
Faculty of Electrical Engineering and Computing
Unska 3, 10000 Zagreb, Croatia
E-Mail: stipe.kuman@gmail.com, {stjepan.gros, miljenko.mikuc}@fer.hr

Abstract—**Honeypots are used as a security measure both to divert the attention of a potential attackers intentions and to reveal the attacker since the only reason someone would interact with honeypots is if they are looking for a vulnerable target. Honeypots emulate only a part of the machine they are supposed to represent and contain no valuable data. ICS (Industrial Control System) is a term that is used for a system that monitors industrial plants, distributed control systems or other systems that mostly contain PLCs (Programmable Logic Controllers). Conpot is an open source honeypot that emulates PLC devices so it can be used in ICSs. However, Conpot can not emulate complex honeypot networks. The aim of this project is to make a tool that can be used to design a honeypot network which emulates an ICS. A network designed with that tool will be simulated as a part of this project and the data collected during the simulation will be analyzed.**

Index Terms—**ICS, honeypot, Conpot, IMUNES**

I. INTRODUCTION

Devices in Industrial Control Systems (ICS) often use protocols that aren't secure because, at the time when those protocols were developed, security wasn't an issue for control systems. Among other, those systems were not connected in any way to the Internet, or accessible via the Internet or any other network outside the premises. But today, things are different. Those old protocols are retrofitted into new ones, ICSs are connected to business networks, and sometimes directly to the Internet. All this means that hackers can access control systems and use flaws in their original protocols to do harm, e.g. disable certain devices or steal data from them. This is a big issue, as evidenced by multiple incidents that happened [1], and the problem is very complex mainly because of the specifics of ICS and historical development that led to the present situation.

Multiple mechanisms are experimented with and applied in order to make ICS more secure. Although honeypots don't enhance security of the individual ICS directly, they can be used as an indicator of malignant activities on the network, thus protecting the whole network. *Honeypots* [2] are systems that emulate different devices, operating systems, or services and are used to protect systems from hackers and/or to study hacker's behavior. Honeypots are placed within the network that should be protected where they emulate vulnerable devices in order to attract the attention of potential attackers and to alert owners of the network that it is being attacked. Conpot is currently the most popular honeypot specialized for ICSs that emulates PLCs. It uses templates to define PLCs that will be emulated, thus making Conpot adjustable to a specific situations. Users can define and use their own custom templates. *Honeynets* [3] are a form of honeypots that emulate a network of devices instead of a single device, i.e. they can be viewed as a network of honeypots. They have the potential to further slow down the attackers and give the owners of the network more time to react to the attack. Conpot by itself doesn't provide the functionality to emulate honeynets, but it could be used for such a purpose by having multiple real or virtual machines mutually interconnected with each running a copy of Conpot. But this is something that has to be done manually, requires a lot of time and resources, and is error prone.

In this paper we describe how the IMUNES network simulator/emulator [4] can be expanded to create a honeynet which emulates an Industrial Control System. IMUNES is a network simulator that allows users to define and simulate/emulate networks of almost arbitrary complexity on a single physical or virtual machine. The nodes of the emulated/simulated network are made of virtual nodes. It allows users to customize the virtual nodes and provides them with the tools needed to connect the virtual network to the internet. The Conpot honeypot is used to emulate PLCs in the simulated network and the OSSEC tool is used to monitor all activities on the honeypots and alert the owner of the honeynet when something interesting is happening.

The paper is structured as follows. In Section II of the paper, the tools used for this project are described, as well as the purpose whey they were used. An overview of related work is given in Section III. In Section IV we describe how we built Conpot honeynet. Few experiments done with Conpot honeynet are described in Section V. The paper finishes with conclusions and future work in Section VI.

II. TOOLS USED TO BUILD ICS HONEYNET

ICS is a term that describes systems used in industrial plants and distributed control systems that are composed mostly of PLCs [5]. As stated before, protocols for PLCs rarely provide any security protections so systems that use those devices need to use additional protection mechanisms. Devices in those systems can, and do, exchange important data that the owners of the system don't want to be interrupted, modified, delayed, faked, etc. Furthermore, because of the processes ICSs are connected to that are controlled by PLCs, any downtime on ICSs could be very dangerous, up to the point of losses of human lives. That is why those systems need to be protected from alternations or any other action by the attacker that would impact its availability and/or integrity.

In the following subsections we will briefly describe three components used to implement honeynet, namely Conpot honeypot, OSSEC host intrusion detection system (HIDS) and IMUNES network simulator/emulator.

A. Conpot

Conpot is a honeypot for Linux that emulates devices commonly used in Industrial Control Systems [6]. It comes with templates that can be used to emulate Siemens S7-200, Guardian AST and Kamstrup 382 PLCs. All of the templates can be customized by the users to suit their needs. Users can also make their own templates to emulate the devices they need. We will be using this property to diversify emulated devices in the network making the honeynet more realistic, specifically we emulate Siemens S7-300 PLCs due to a certain vulnerabilities which would make that device more tempting to potential attackers [7].

Conpot supports emulation of SNMP [8] and Modbus [9] protocols which are commonly used in PLCs. Like with all other honeypots, Conpot has no other function in the network except for luring potential attackers. Therefore, any outside access to it could be considered an attack on the network. That is why, along with emulating a device, Conpot also records all the network accesses to the honeypot and stores it into a *conpot.log* file. However, Conpot does not provide a means of instantly notifying the user that the honeypot is being attacked. Which will be solved using OSSEC HIDS.

B. OSSEC Host Intrusion Detection System

The problem with monitoring changes on the honeynet nodes is solved by employing OSSEC Host Intrusion Detection System (HIDS) [10]. OSSEC has several functions, one of which is to monitor changes in specified files. We'll monitor changes in the log file produced by Conpot. Furthermore, OSSEC allows central collection of logs and log monitoring with customizable regular expressions. But, we are not using those features in this version of the ICS honeynet. Users are notified of changes to Conpot honeypot by email.

C. IMUNES

IMUNES is a network simulator and emulator that runs on FreeBSD and Linux [4] and allows configuration and simulation/emulation of number of nodes interconnected with links. For this work we used Linux version of IMUNES and in the following text it is implicitly assumed so. IMUNES is a "lightweight" virtualization system in which all nodes are using the same kernel. File systems are mounted using overlay, union mount file system and the simplest solution is to mount the same file system on all nodes (read-only) and use copy-on-write mechanism for changes. The virtual nodes can emulate different roles, like routers, switches and Linux hosts and servers. Nodes with those roles are shipped with IMUNES, but the system is configurable and it is possible to add new nodes by customizing existing ones or creating a new ones. The reason is that each node is actually a Docker [11] image. The possibility to create new nodes is used in this project, i.e. PC nodes are modified to run Conpot and OSSEC so that they can be used as honeypots in the honeynet.

IMUNES provides an easy to use GUI that enables the users to easily create networks by drawing them on a canvas. All the network settings for nodes, such as protocols being used by the nodes, can also be configured from the GUI.

III. RELATED WORK

The project *A Virtual Honeypot Framework* (VHF) [12] has some similarities with our project. The VHF

1263

created Honeyd [13], a low-interaction honeypot which can be used to run a network of honeypots on a single machine. When used to run multiple honeypots, Honeyd emulates an arbitrary network. It also provides features that make that network seem more realistic such as latency and packet loss.

There are several differences between Honeyd and the tool described in this paper. The first is that Honeyd isn't meant to be used for PLC or ICS emulation. Second difference is that Honeyd isn't focused on defining specific network topologies so it doesn't provide the users with a tool to design their own topologies. By using IMUNES to design and run topologies, this tool provides more options when it comes to network emulation.

IV. CONPOT HONEYNET

The process of using IMUNES to emulate a Conpot honeynet can be divided into three steps. The first step is modifying the IMUNES PC nodes by installing tools like Conpot and OSSEC on them. This will enable the users to use the virtual nodes as honeypots in a honeynet. The second step is connecting the virtual network to the Internet so that it can be scanned and accessed by potential attackers. Linux provides the tools needed to forward all the network traffic between the internet and the virtual network. The last part is starting all the services, like Conpot and OSSEC, on the virtual nodes and running the honeynet. The whole process of starting the honeynet and connecting it to the internet was scripted to make it easier for users to start their own honeynets.

A. Modifying IMUNES Nodes

In order to add Conpot to the virtual nodes in IMUNES, the *vroot-linux* git repository was forked and modified [14]. The repository contains all the files needed to build a Docker image that can be used by IMUNES. The *utilities.sh* script in *vroot-linux* is used to install tools and all their dependencies on IMUNES virtual nodes so it was modified to install additional tools on virtual nodes. Along with Conpot, the OSSEC monitoring tool was installed on virtual nodes to enable email alerts. Before installing Conpot and OSSEC, dependencies for those tools were added to the *Often used tools* part of the *utilities.sh* file. Since the normal OSSEC installation requires user inputs, OSSEC was installed with preloaded variables by using the *preloaded-var.conf* file so that the user doesn't have to manually configure the tool during installation. This file is used to define which services will be used, the email address that will

receive OSSEC notifications and where the tool will be installed. The *utilities.sh* file was also used to make any needed modifications to the newly installed tools. After the *utilities.sh* script was modified, all the specified tools were installed on the virtual nodes using the following command:

```
sudo docker build \
    -t IMUNES/vroot vroot-linux/image/
```

Virtual nodes can now execute Conpot and thus act as honeypots. They also have OSSEC to monitor Conpot log file which will send email notifications to users.

B. Connecting the Virtual Network to the Internet

In order to connect the simulated network to the internet an *external connection* node was used. That type of node connects the virtual network to the computer that runs the simulation. Figure 1 shows an example network with 3 PLCs connected to a router that is connected to an *external connection* node.

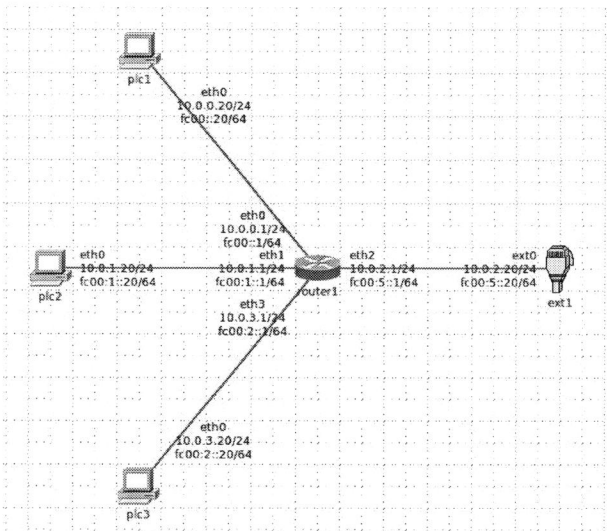

Fig. 1.   Simple Network With 3 PLCs

The *ext0* node creates an interface on the user's computer that can be used to interact with the virtual network. The *ext0* node configuration window enables the user to define the IP of the newly created interface on the host machine. The virtual router connected to the external interface was also configured, setting the external interface's IP address as a next hop for a default route on the router. That way, when the router receives a packet with an unknown destination address, it will send it to the host machine which will forward it accordingly.

1264

In order for the host machine to route properly packets received from the virtual router, the host OS was configured to forward packets between the virtual network and the internet. This was done by setting appropriate routes on the host, enabling forwarding using sysctl parameter, and using the `iptables` command to setup source and destination NAT as it is necessary to map host's IP address to IP addresses in virtual network. Additionally, host was configured to forward TCP segments reaching certain ports on the host's IP address to ports on virtual honeypots, i.e. destination NAT was used. This enables attackers to communicate with the honeypots in the virtual network even though honeypots don't have public IP addresses.

For example, any TCP segment with destination port 502 and destination IP address of host was forwarded to port 502 of one of the virtual honeypots in the virtual network. Port 502 is used for the Modbus protocol [15]. The same forwarding of ports was done for other ports used by Conpot. This means that one virtual Conpot honeypot would be easily found by port scanning the host's IP address. Other virtual honeypots weren't connected to the host machine's ports but should instead be accessed from machines inside the virtual network.

C. Starting Services on Virtual Nodes

IMUNES provides a few commands which enable the user to interact with the nodes in the virtual network from shell running on the host. In order to execute a command on a virtual node `himage` command is provided. It enables the user to run a command in the virtual node's shell. For example, Conpot can be executed on the *plc1* node by running the following command:

```
himage plc1 conpot
```

`himage` could also be used to open a shell on some virtual node. That could be acomplished using `himage <node name>`. IMUNES also provides the `hcp` command, which can be used to copy files between the virtual nodes and the host. This command is useful for fetching *conpot.log* files from virtual honeypots, among others.

In order to make the honeynet more user friendly for users two scripts were added to the forked *vroot-linux* git repository. The *start.sh* script is used to initialize a honeynet defined by a topology file with `sudo bash start.sh -b <topology file>`. The `-e` option can also be used in order to define the ID of the simulation that is executed, otherwise the simulation will be assigned a random ID by IMUNES.

The honeynet can be further customized with the *conpotimunes.conf* file which is used to define which template will be used on which node, as well as the ports on the honeypot nodes that will be forwarded to the host machine. The *conpotimunes.conf* file also defines the IP of the virtual *external connection* node and the interface on the host machine that is used to connect to the internet. The *start.sh* script will initialize the virtual network and then start both Conpot and OSSEC on the virtual nodes. OSSEC will be configured to send email notifications when a change occurs in the *conpot.log* file of a virtual node. The second script, *collectLogs.sh*, is used to fetch *conpot.log* files from honeypot nodes with the `sudo bash collectLogs.sh <node name>` command. This script will use the `hcp` IMUNES command copy the log file to a directory on the host machine. The name of the directory indicates the time and date when the log was fetched in order to make data analysis easier.

V. EXPERIMENTAL WORK

As a part of this project, an ICS was modelled and then initialized. The network used the default Conpot template to emulate Siemens S7 devices. Additionally, the default Conpot template was modified to more closely resemble a Siemens S7-200 PLC. That template was later used to make a new template which could be used to emulate a Siemens S7-300 PLC because those devices had a vulnerability which could easily be reproduced [7].

A. Modifications Made to the Conpot Templates

The default Conpot template can be used to emulate a Siemens S7-200 PLC. In order to do that it has to be modified since the default template will easily be spotted by attackers [16]. This is mostly because it contains hardcoded values for things like the serial number, PLC name, etc. To see some examples of more realistic PLC devices, the Shodan [17] search engine was used. The default template was then modified to resemble a real PLC device. After comparing Siemens S7-200 devices that were classified as honeypots and devices that were classified as ICSs, modifications were made to the default template which resulted in a honeypot that was also classified as an ICS on Shodan. Listing 1 shows parts of the changed *template.xml* file which is used to define basic honeypot information. Original values for the changed entries were either too specific or weren't in the right format.

1265

Listing 1. Modified template.xml file for S7-200

```
...
<key name="SystemName">
  <value type="value">"SIMATIC 200"
</value>
</key>
...
<key name="s7_id">
  <value type="value">"S C-Z1DH98170013"
</value>
</key>
...
```

Some entries, like *FacilityName*, were changed to an empty string because most of ICS examples had that value undefined. Changes were also made to other XML files that define the default template such as *s7comm.xml* and *http.xml*. This of course does not mean that the honeypot could fool every hacker, but it is still an improvement to the standard honeypot. Additionally, by using the same method, the default template was further modified in order to emulate a Siemens S7-300 and Siemens S7-400 devices. The changes were mostly made in the *s7comm.xml* file along with the *template.xml* file. Listing 2 shows some of the changes in the *template.xml* file which were made in order to emulate a Siemens S7-300.

Listing 2. Modified template.xml file for S7-300

```
...
<key name="SystemName">
  <value type="value">"SIMATIC 300(1)"
</value>
</key>
...
<key name="basicFirmware">
  <value type="value">"v.2.6.11"</value>
</key>
...
<key name="s7_id">
  <value type="value">"S C-Y3TT99128012"
</value>
</key>
...
```

The new devices were useful because some Siemens S7-300 devices had a vulnerability which could easily be exploited. The vulnerability was in the hardcoded `basisk` [18] username and password which enabled anyone to gain control over the PLC. There is also a Metaspolit exploit for that vulnerability which increases the chances that a potential attacker would use that vulnerability to attack the virtual network. The S7-300 templates http protocol file *http.xml* was modified

to partially emulate the vulnerability in question. The modification allowed the honeypot to respond to the `basisk` login attempt and to the memory dump request. The memory dump consisted mostly of random data but it also had IP addresses and ports from other devices inserted into that random data in hopes that it would help the attackers find the other devices in the honeynet.

B. Designing a Control Network

In this section, an example of an Siemens PLC network was partially modelled in IMUNES. Figure 2 shows the selected network which consists mostly of Siemens S7 PLCs [19].

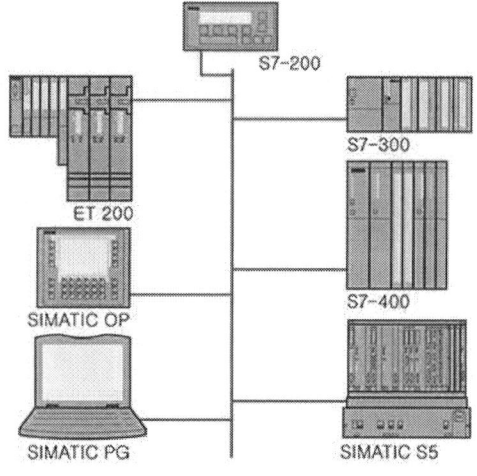

Fig. 2. An example of a Siemens ICS

The modified IMUNES tool can emulate some of the devices in this network, but not all of them so some modifications were made to the network. Siemens ET 200, Simatic S5, SIMATIC OP and SIMATIC PG are the devices that couldn't be emulated. Simatic PC is an industrial laptop used for programming PLC devices so it was replaced by a standard *pc* node. Other devices that can't be emulated were replaced with Siemens S7-200 PLC. The resulting IMUNES topology is shown in Figure 3.

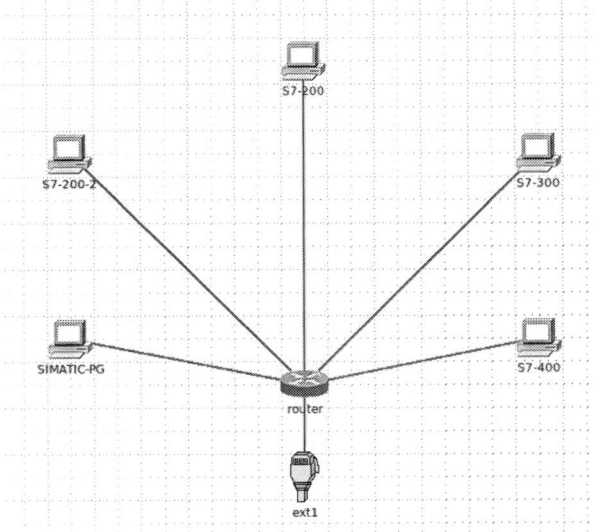

Fig. 3. ICS modelled in IMUNES

The Siemens S7-300 ports were connected to the standard ports on the host machine while other devices were connected to random ports on the host machine. The reason for this is the basisk vulnerability added to the S7-300 honeypot which would allow the attacker to request a memory dump from that device. The memory dump of that device contains the private IP addresses of other devices which could be used to access those devices from inside the network.

C. Analysis of Obtained Results

In this section the results of running the honeynet are analysed. The experiment lasted for two weeks with a couple of short breaks to adjust honeypot settings. Once the honeynet was initialized, PLCScan [20] was used to check that all the honeypots are working properly.

During the simulation of the network described in the previous section some activity on the honeypots was recorded. Every time a honeypot was accessed OSSEC notified the owner of the network that the *conpot.log* has changed. Listing 3 shows one example of an OSSEC email.

Listing 3. OSSEC notification about changes in the conpot.log file

```
OSSEC HIDS Notification .
2017 Feb 09 14:07:50

Received From: plc1 ->syscheck
Rule: 550 fired (level 7) -> "Integrity
checksum changed."
```

Portion of the log(s):

```
Integrity checksum changed for:
'/etc/conpot.log'
Size changed from '4918' to '9973'
```

Conpot honeypots recorded all their activity in *conpot.log*. Parts of the resulting *conpot.log* can be seen in Listing 4. That specific part is a record of a port scan directed at the modbus 502 port on the hosts IP address.

Listing 4. Activity recorded in the conpot.log file

```
2017-02-19 02:16:30,279 New modbus session
from 185.35.63.142
(e5ac81d6-dad5-4abc-8f42-2e8170738b38)

2017-02-19 02:16:30,280 New Modbus
connection from 185.35.63.142:37443.
(e5ac81d6-dad5-4abc-8f42-2e8170738b38)

2017-02-19 02:16:30,280 Modbus traffic
from 185.35.63.142: {'function_code': 43,
'slave_id': 0, 'request':
'000000000005002b0e0100', 'response': ''}
(e5ac81d6-dad5-4abc-8f42-2e8170738b38)

2017-02-19 02:16:30,281 Modbus connection
terminated with client 185.35.63.142.
```

The recorded activity, however, consisted mostly of port scans targeting the modbus 502 port and the http 8080 port of the Siemens S7-300 honeypot. Port scans on the modbus 502 port mostly requested some basic information from the devices and then didn't explore the device further. Port scans on port 8080 also didn't result in any further activity, but that is to be expected seeing how that port is not specific to control devices and was meant to be used by attackers that have previously scanned the 502 port or 102 port and found a Siemens device that is potentially vulnerable.

None of the port scans resulted in attacks on the virtual network so there was no activity recorded on the other honeypots in the virtual network. There were also no attempts to exploit the Siemens S7-300 vulnerability. Since the time frame of the experiment was only two weeks, more testing should be done in order to see how successful this particular honeynet is at attracting hackers. However, as stated before, Shodan has classified the simulated network as a valid ICS and the Siemens S7-300 honeypot was successfully modified to emulate the hardcoded password vulnerability so a longer experiment could have resulted in more interesting activity.

VI. Conclusions and Future Work

This article shows how the IMUNES network simulator can be used to simulate a honeynet. The steps used in this article can be used to emulate any kind of honeynet with IMUNES. Because of the IMUNES GUI it is easy for users to model a network that suits their needs and the addition of the OSSEC email alerts makes supervising the honeynet much easier.

In future work, the tool presented in this article will be improved by adding more templates to Conpot and by adding other honeypot tools to the virtual nodes. That would allow the users to emulate honeynets that better resemble production networks. Additionally, better monitoring capabilities are necessary.

Honeynets emulated by this tool provide owners of ICSs with more options for securing their systems. Along with potentially improving the security of an ICS, this tool could serve as a template to build other kinds of honeynets.

References

[1] "The repository of industrial security incidents," http://www.risidata.com/, 2017.

[2] F. Zhang, S. Zhou, Z. Qin, and J. Liu, "Honeypot: a supplemented active defense system for network security," in *Parallel and Distributed Computing, Applications and Technologies, 2003. PDCAT'2003. Proceedings of the Fourth International Conference on.* IEEE, 2003, pp. 231–235.

[3] L. Spitzner, "The honeynet project: Trapping the hackers," *IEEE Security & Privacy*, vol. 99, no. 2, pp. 15–23, 2003.

[4] M. Zec and M. Mikuc, "Operating system support for integrated network emulation in imunes," 2004.

[5] K. Stouffer, J. Falco, and K. Scarfone, "Guide to industrial control systems (ics) security," *NIST special publication*, vol. 800, no. 82, pp. 16–16, 2011.

[6] L. Rist, "Introducing conpot," *The Honeynet Project*, vol. 11, 2013.

[7] E. Byres, "Siemens PLC Security Vulnerabilities – It Just Gets Worse," Tofino Security, August 2011. [Online]. Available: https://www.tofinosecurity.com/blog/siemens-plc-security-vulnerabilities---it-just-gets-worse

[8] J. D. Case, M. Fedor, M. L. Schoffstall, and J. Davin, "Simple network management protocol (snmp)," Tech. Rep., 1990.

[9] D.-g. Peng, H. Zhang, L. Yang, and H. Li, "Design and realization of modbus protocol based on embedded linux system," pp. 275–280, 2008.

[10] R. Bray, D. Cid, and A. Hay, *OSSEC host-based intrusion detection guide.* Syngress, 2008.

[11] J. Fink, "Docker: a software as a service, operating system-level virtualization framework," *Code4Lib Journal*, vol. 25, 2014.

[12] N. Provos *et al.*, "A virtual honeypot framework." in *USENIX Security Symposium*, vol. 173, 2004, pp. 1–14.

[13] N. Provos, "Honeyd-a virtual honeypot daemon," in *10th DFN-CERT Workshop, Hamburg, Germany*, vol. 2, 2003, p. 4.

[14] S. Kuman, "Imunes with conpot," https://github.com/Qmaan/vroot-linux/, 2017.

[15] N. Goldenberg and A. Wool, "Accurate modeling of modbus/tcp for intrusion detection in scada systems," *International Journal of Critical Infrastructure Protection*, vol. 6, no. 2, pp. 63–75, 2013.

[16] D. Martyn, "OPSEC For Honeypots," Xiphos Research Labs, December 2015. [Online]. Available: http://xiphosresearch.com/2015/12/09/OPSEC-For-Honeypots.html

[17] R. C. Bodenheim, "Impact of the shodan computer search engine on internet-facing industrial control system devices," DTIC Document, Tech. Rep., 2014.

[18] D. Beresford, "Exploiting siemens simatic s7 plcs," pp. 723–733, 2011.

[19] "Siemens S7-200 – Getting Started with Siemens S7 PLC's," PLC Academy, June 2015. [Online]. Available: http://www.plcacademy.com/siemens-s7-200/

[20] L. Yan, "Plcscan," https://github.com/yanlinlin82/plcscan, 2012.

A wireless propagation analysis for the frequency of the pseudonym changes to support privacy in VANETs

Eduardo Cano Pons, Gianmarco Baldini, Dimitrios Geneiatakis

European Commission
Joint Research Centre
Ispra, Italy
E-mail: eduardo.cano-pons@ec.europa.eu, gianmarco.baldini@ec.europa.eu, dimitrios.geneiatakis@ec.europa.eu

Abstract — Vehicle Ad Hoc Networks (VANETs) in Cooperative Intelligent Transport Systems (C-ITS) are based on the exchanges of messages among ITS-Stations (e.g., vehicles and roadside infrastructure) using the wireless G5 Dedicated Short Rate Communication (DSRC) standard to support safety-critical applications. VANETs require the authentication of ITS-stations and messages but the privacy of the drivers of the vehicles must be supported. In recent years, researchers have proposed solutions to mitigate privacy risks based on the use of pseudonyms. A key design decision is related to the frequency of the change of pseudonyms. The activity of a vehicle under one pseudonym can be linked to another thus providing traceability of the vehicle and a privacy risk for the driver. To prevent link-ability of actions, the vehicle must change pseudonyms over time. In this paper, the authors propose a radio frequency physical layer analysis to determine the frequency of the pseudonym changes. The rationale is that different wireless propagation conditions will impact the capability of the privacy attacker to trace the vehicle, thus reducing the need to frequently change the pseudonyms. The analysis has been performed in different channel fading conditions and for different relative speed values.

Keywords—wireless propagation, privacy, VANET,

I. INTRODUCTION

Cooperative Intelligent Transport Systems (C-ITS) is a set of technologies, which will allow road users, service providers in road transportation and traffic managers to share information and use it to coordinate their actions. One of the main elements is the wireless connectivity among vehicles (V2V) and between vehicles and transport infrastructure (V2I), collectively called V2X. Wireless connectivity is mainly proposed by standardization bodies [1] (e.g., ETSI) with the Dedicated Short Range Communications (DSRC) wireless standard based on the IEEE 802.11p standard, which belongs to the 802.11 family of standards and it is described in the subsequent section of this paper. While other authors have recently proposed the application of different standards like LTE [2], the authors of this paper have focused their study only on the IEEE 802.11p standard. In this context, vehicles and road infrastructures nodes (called ITS-Stations in the ETSI standards [1]) will exchange messages through wireless communications, thus

creating dynamic Vehicle Ad Hoc Networks (VANET). The ITS-Stations will broadcast messages like the Cooperative Awareness Messages (CAM) to support the creation of dynamic cooperative awareness in the ITS stations [1]. The broadcast of CAMs from the vehicle can potentially endangers the privacy of drivers, as an eavesdropper could identify the mobility patterns of individual drivers and track the vehicles.

To mitigate these privacy risks, various papers have proposed pseudonymity schemes for VANETs, where the ITS-stations use pseudonyms rather than the real identity when generating and broadcasting the CAMs. A recent and very detailed survey identifies and describes the main pseudonymity schemes for VANETs [3]. As presented in [3] and [4], a static pseudonym may not be sufficient to mitigate privacy risks in VANETs, because a single vehicle could still be identified and tracked based on a time series of eavesdropped messages. In other words, a malicious privacy attacker could be equipped with a RF receiver to eavesdrop messages broadcasted from a vehicle and still correlate specific vehicles or even drivers based on reoccurring travel patterns. As a consequence, there is the need to change the pseudonyms generated by the mobile ITS-station, but it is still an open problem on what could be the frequency of pseudonyms changes [3]. This problem is common to most of the proposed pseudonym schemes, as discussed in [7]. Note that there are trade-offs in the definition of the frequency of the pseudonyms. If the frequency is too high, it can have a significant impact on the storage of the pseudonyms or the computing power, which is needed to generate them. In addition, it increases the frequency of pseudonym refills in the vehicle, which impacts the communication channel used to distribute or activate pseudonyms [5]. Conversely, if the frequency is too low, privacy attackers can easily track the vehicle. To identify the point of equilibrium in this, the objective is to calculate the required maximum frequency of changes of pseudonyms.

In this paper, we propose an analysis for the definition of the frequency of pseudonym changes based on wireless propagation conditions. The rationale is that there is no need to frequently change pseudonyms if a receiver is not able to collect and process, with an acceptable Bit Error Rate (BER), targeted values of the CAM messages broadcasted by the mobile ITS-station. If the wireless propagation conditions are not optimal, as in an urban environment, the frequency of change of pseudonyms can be kept to a low value but it should be raised to a higher value in different propagation conditions or depending on the speed of the vehicle. In other words, the frequency of change of the pseudonyms depends on the context where the mobile ITS-station is travelling or even the types of modulation used to transmit the CAM messages. While there can be different contexts which could drive the frequency of changes of the pseudonyms, in this paper, we specifically focus on the radio frequency physical layer context and the capability of the privacy attacker to collect and process the CAM messages based on the 802.11p standard used in the mobile ITS-station.

The structure of this paper is the following: Section II describes the system model with a presentation of the scenario, a brief description of the 802.11p standard and the conditions (attenuation and fading), which can impact the capability of the privacy attacker to collect and process the CAM messages (e.g., if a BER threshold is overcome). Section III provides the simulations results, which gives an indication on the needed change of frequency pseudonyms. Section IV concludes this paper.

II. RELATED WORK

The issue of privacy in VANET is directly linked to the broadcasting of messages as described in the introduction section. The periodic broadcasting of messages from a vehicle can be used to track the vehicle and the driver/passenger and it can be used to attack his/her privacy. In literature, the privacy risk related to the tracking of the position of an individual are part of computational location privacy, meaning computation-based privacy mechanisms that treat location data as geometric information [6]. While the authors in [6] have investigated computation-based privacy in general, recent papers have investigate the risks and related mitigation techniques for privacy in VANET. In VANETS, security is of paramount importance because the integrity of the exchanged messages (e.g., CAM) must be ensured to avoid malicious attacks tampering with the messages and then creating safety hazards. The authors in [7] have proposed digital signatures to ensure the integrity of the messages, which is the approach based on [1]. On the other side, the certificates used for the digital signatures can be uniquely linked to the owner of the vehicle, does creating a privacy threat. Then, pseudonyms were proposed in VANETs to break this linkability [8]. A pseudonym allows authentication of a specific entity without knowing the holders's real identity. On the other side, an eavesdropper collecting signed messages, could still track a vehicle because it will use the same certificate. To prevent this privacy threats, authors have investigated the possibility to change pseudonyms [3]. The issue is how frequently the pseudonyms should be changed.

The authors in [4] have used a Multi Hypothesis Tracker (MHT), which relies on Kalman filters to calculate the frequency of changes of pseudonyms. The evaluation was based on the definition of a discrete event simulator and a vehicular mobility model to generate mobility traces. Then, the

anonymized position samples are processed by the MHT. The evaluation metric the maximum period of time the tracker was able to correctly reproduce the trace of each vehicle, averaged over all traces in the simulation.

An alternative approach to calculate the frequency of pseudonyms changes was proposed in [9], where the authors use the context information (such as the number of neighbors, their direction and speed) for initiating a pseudonym change. For example, vehicles can cooperatively identify good opportunities to blend in a number of vehicles and hence increase their anonymity. Then, the simulations were executed using the vehicular mobility model provided with STRAW.

Both [4] and [9] do not address the aspect that the CAM messages transmitted through wireless communications are subject to wireless propagation errors due to the distance of the vehicle from the eavesdropper and the surrounding environment. This paper addresses this gap in literature, which can be a significant factor to evaluate the practical feasibility to implement the privacy threat.

III. SYSTEM MODEL

A. Scenario

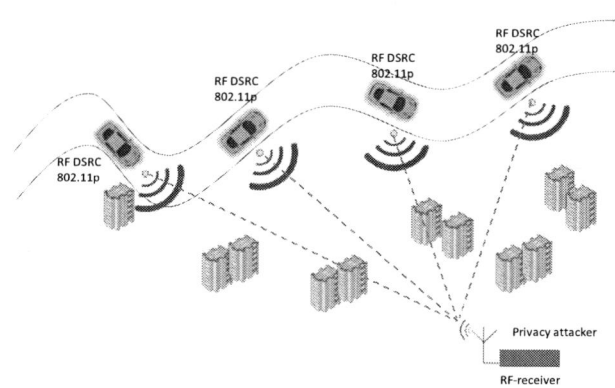

Fig. 1. Privacy threat scenario

The privacy threat scenario is shown in figure 1. A Privacy attacker is equipped with an RF receiver to collect the radio frequency signal based on the 802.11p standard. In this scenario, we make the following assumptions:

1. The privacy attacker is located in a fixed position.
2. The pseudonym scheme is not based on "silent" periods, where the vehicle stops transmitting for pre-defined times to avoid being tracked. While a "silent" period could mitigate privacy risks, it also negates or reduces the benefits of V2X [3].
3. The pseudonym scheme is based on traditional public key cryptography schemes, where the vehicle is equipped with a set of public key certificates and

corresponding key pairs. This is the most common approach proposed by industry and standardization bodies ([10],[11]). The public key certificates are stripped of any identifying information and used as unlinkable pseudonyms. Vehicles sign messages with the secret key of the currently active pseudonym.

As described before, the privacy attacker can be successful only if the receiver is able to collect and process the CAM messages based on the 802.11p signal standard, which depends on the wireless propagation conditions.

It is considered that a vehicle is moving at constant velocity, with a single eavesdropper present within the communications range R_{max}. If the location of the eavesdropper is randomly chosen within a circle defined by the transmission range, the average distance between the vehicle and the eavesdropper is exactly R_{max}. Thus, the average time in which the eavesdropper can listen the transmission from the transmitter is obtained by applying the following relationship $T_{max} = R_{max}/v$, being v the speed of the vehicle. Then, the objective is to calculate the value or the range of values of T_{max}, because it is inversely correlated with the maximum frequency of change of pseudonyms.

B. A description of the 802.11p standard

IEEE 802.11p is a consolidated standard used for DSRC between vehicles [12]. This standard provides both the physical layer (PHY) specifications for the DSRC and the Medium Access Control (MAC) of vehicular communication operating at the 5.9 GHz band. The physical layer consists of two sublayers, the Physical Medium Dependent (PMD) sublayer and the Physical Layer Convergence Protocol (PLCP) sublayer. The first sublayer defines signal parameters, such as modulation and coding rates, and performs the digital-to-analog conversion. The PLCP sublayer is responsible for communicating with the MAC layer. It is also a convergence process that transforms the PLCP Service Data Unit (PSDU) arriving from the MAC layer into PLCP Protocol Data Unit (PPDU). During this conversion, the PSDU is appended with preamble and header. The PLCP preamble is composed of ten repetitions of a short training sequence and two repetitions of a long training sequences preceded by a guard interval in order to combat inter-symbol interference. The PLCP header contains information regarding the transmission rate, tail and padding bits and, also, how many octets are used in the PSDU.

The IEEE 802.11p physical layer is based on an orthogonal frequency division multiplexing (OFDM) technique. The OFDM technique transmits data on spaced orthogonal subcarriers, allowing spectrum efficiency improvement and coping with severe channel conditions. A total number of 64 subcarriers are defined, but only the inner 52 subcarriers are employed. Among the 52 subcarrier, 48 of them contain modulated-coded data and the other 4, called pilot subcarriers, convey a fixed pattern used for channel and synchronisation

purposes at the receiver [13]. Each of the 48 subcarriers can be modulated by using BPSK, QPSK, 16-QAM or 64-QAM schemes. In addition, different convolutional coding rated can be used, leading to nominal data rate values of 3 to 27 Mb/s when using a 10 MHz bandwidth. The main parameters that define the IEEE 802.11p standard are defined in Table 1.

TABLE I
802.11P PARAMETERS

Parameter	Value
Bandwidth (MHz)	10
Bit Rate (Mbps)	3, 4.5, 6, 9, 12, 18, 24, 27
Modulation Schemes	BPSK, QPSK, 16-QAM, 64-QAM
Code Rate	1/2, 2/3, 3/4
Data subcarriers	48
Pilot subcarriers	4
Fast Fourier Transform (FFT) size	64
FFT Period (μs)	6.4
GI duration (μs)	1.6
OFDM Symbol Duration (μs)	8
Subcarrier frequency spacing (MHz)	1 / 6.4
Error Correcting Code	Convolutional Codes

The transmitter chain is shown in figure 2, where the first block is the bit source unit, which generates random bits following a uniform distribution. Then, the serial number of bits are going through to an encoder and modulator unit. The raw bits are firstly scrambled in order to randomise the data pattern, which may contain large strings of 0s or 1s. Then, a convolutional encoder is applied to the bit stream in order to introduce redundancy. This redundancy is used for error correcting coding, which allows the receiver to combat the channel propagation effects. Subsequently, an interleaver process is applied to the coded bits in order to mitigate correlation channel noise effects, such as burst errors or fading. After the interleaver, the modulation mapper is applied to convert the bit stream into symbols. The following block is the OFDM symbol assembler, which constructs the OFDM symbols by means of mapping the modulated data into the 48 data subcarriers, the pilot sequence of bits into the 4 designated subcarriers and the 12 resting subcarriers are nulled. Once the OFDM symbols are built, an Inverse Fast Fourier Transform (IFFT) process is applied to obtain the temporal representation of the symbols. Finally, the transmitted data is generated by assembling the preamble, signal field and the IFFT modulated data. Note that the receiver performs the opposite processes with respect to the transmitter (i.e. demodulation, decoding, etc.).

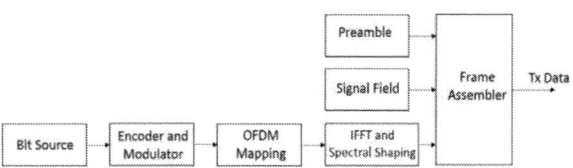

Fig. 2. Transmitter block diagram

C. Rician Fading Channel Model

In wireless communications, the Rician distribution is commonly used to describe the statistical variations in time of the envelope of the received signal. This signal can present a unique component or several components as a result of reflections and scattering phenomena, also known as multipath signal. In the situation that there is a dominant stationary signal component, i.e. line-of-sight communications, the small-scale fading envelope follows a Rician distribution. The Rician fading channel is characterised by mainly two parameters, K and Ω, where K represents the ratio between the power amplitude of the strongest received path with respect to the other secondary multipath components. The other parameter, Ω, is the result of the total available power from all the multipath components. Additional details on the Rician distribution and the K factor are available in [14].

Two particular cases can arise from the Rician fading channel definition. The first one is when the ratio between Line-of-Sight (LoS) components and multipath rays is very large ($K \rightarrow \infty$). In that case the amplitude of the received signal is constant and the mobile channel acts like an Additive White Gaussian Noise (AWGN) channel. The other case corresponds to $K=0$. In this situation, the envelope amplitude follows a Rayleigh distribution, which can be modelled as the sum of two quadrature Gaussian noise signals. In Rayleigh fading, the signal weakening can cause that the main component is not noticed among the multipath components

In addition, another important factor that needs to be taken into account in vehicular communications is the Doppler effect associated to the relative velocity among transmitter and receiver. The Doppler effect is represented by frequency variations of the received signal. In this context, the frequency of the received signal is given by $f = (1+\Delta v/c)f_0$, where f_0 is the carrier frequency of the transmitted signal, c is the speed of light and Δv is the relative velocity between transmitter and receiver. The Doppler spectrum used in this work follows the Jakes spectrum model [15].

D. Maximum Communication Distance

The maximum reachable communications distance is calculated based on the SNR_{min} values for a targeted BER. Following the Friis transmission equation, which is commonly used in Telecommunications to calculate the received power

level at a given distance, and the expression of the thermal noise power [16], the maximum reachable distance between transmitter and receiver is given by

$$R_{\max} = \frac{c}{4\pi f} \sqrt{\frac{P_{Tx} G_T G_R}{SNR_{\min} K_B TWN_F}}, \qquad (1)$$

where f is the centre frequency of the system, P_{Tx} is the transmitted power, W is the noise bandwidth, SNR_{min} is the minimum Signal to Noise Ratio, T is the ambient temperature and K_B is the Boltzmann constant. The variables G_T and G_R in (1) are the transmit and receive antenna gains respectively. Also, N_F is the noise figure of the receiver. Note that free-space propagation losses conditions are considered in (1). The values of parameters of (1), which are used to calculate the results of this paper are provided in section IV.B.

IV. RESULTS

A. Bit Error Rate Simulations

The Bit Error Rate (BER) performance of a radio communications systems is a unitless metric used for representing the quality of the system in terms of bit reception. This percentage metric computes the relationship between erroneous received bits and transmitted bits. The BER performance of a wireless system is commonly represented as a function of the received Signal-to-Noise Ratio (SNR).

The first step is to fix the value of the minimum required BER value (BER_{min}), which allows the receiver to demodulate and obtain the transmitted message with a determined quality of service (QoS). In our situation, the goal is to establish this BER value for which the eavesdropper will be able to properly detect the message. In this context, a $BER_{min} = 0.001$ value is specified by the IEEE 802.11p standard [13]. Correspondingly, the threshold established by the BER_{min} value will provide the minimum required SNR value for correct detection, SNR_{min}.

In this analysis, two modulation-coding schemes, QPSK with $R_c = 1/2$ and 16-QAM with $R_c = 1/2$, have been chosen for BER evaluation in different channel conditions. Initially, the BER performances as a function of the received SNR are evaluated by considering four relative speeds between transmitter and receiver. These speed values are $v = 0$ (static), $v = 30$ km/h, $v = 70$ km/h and $v = 120$ km/h. Note that a perfect Channel State Information (CSI) and no Doppler estimation mechanism are applied for these Monte Carlo simulations. BER curves as a function of the received SNR are plotted in Figures 3 and 4 for both modulation schemes. Results show the performance degradation caused by the Doppler effect and the low values of K-factor of the Rician channel. It is observed that, in some scenarios, the targeted BER is never reached by increasing the SNR values. In particular, this situation is more pronounced for larger values of the modulation scheme, lower values of K and higher relative speed values.

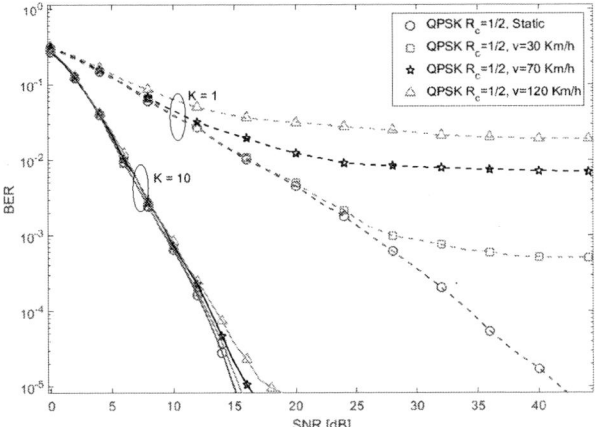

Fig. 3. BER as a function of the received SNR for a QPSK $Rc=1/2$ system with different relative speed values. Two K-factor values of the Rician channel are considered, K=10 (continuous lines) and K=1 (dashed lines).

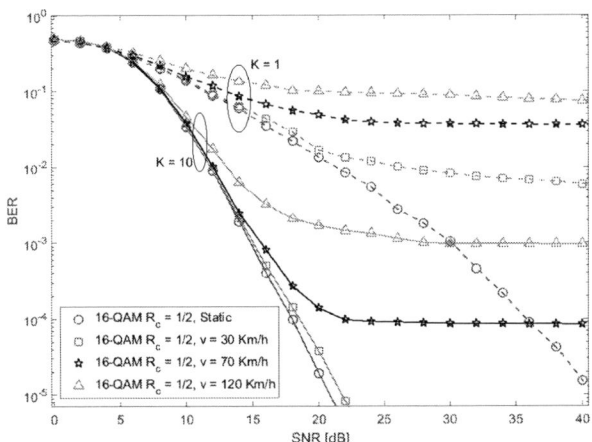

Fig. 4. BER as a function of the received SNR for a 16-QAM Rc=1/2 system with different relative speed values. Two K-factor values of the Rician channel are considered, K=10 (continuous lines) and K=1 (dashed lines).

B. Eavesdropper Visibility Time

The time in which an eavesdropper (receiver) has in view the transmitter (vehicle in motion) based on the simulated SNR_{min} values and the maximum achievable distance in (1), is computed for different channel conditions and modulation schemes. Thus, the frequency rate for changing the pseudonym cryptographic key can be obtained straightforward as the inverse of these eavesdropper visibility times. Initially, the simulation scenario and the hypothesis used in this work are defined as follows. As described in section III.A, it is

considered that a vehicle is moving at constant velocity and driving in a straight direction with only a single eavesdropper, which is present within the communications range calculated in equation (1). The straight direction is chosen because it is the worst case scenario for the calculation of the frequency of the pseudonyms. As pointed out before, the average distance between the vehicle and the eavesdropper is exactly R_{max}. Thus, the average time in which the eavesdropper can listen the transmission from the transmitter is obtained by applying the following relationship $T_{max} = R_{max}/v$.

The average eavesdropper visibility time as a function of the relative speed between vehicle and eavesdropper are plotted in Figures 5,6, and 7 for the three modulation-coding schemes (QPSK, 16-QAM and 64-QAM) and different K-factor values of the Rician fading channel. For this simulation analysis, the values of the parameter defined in equation (1) are following: $G_T=G_R=0$ dBi, $P_{Tx} = 20$ dBm, $N_F=15$ dB, f= 5.9 GHz and $T= 290$K are used. Simulation results showed that the visibility time is minimised by increasing the relative speed due to the Doppler effect, and/or using a large level of modulation-coding scheme and/or transmitting information in urban environments (low values of the K-factor of the Rician fading channel). From these results, it is clear that the maximum frequency of the pseudonyms depends on the context where the mobile ITS-station is operating and the transmission parameters used to transmit the CAM messages. Higher modulation schemes like 64-QAM allows a lower frequency of change of pseudonyms because the visibility time of the eavesdropper drops drastically in comparison to the QPSK modulation.

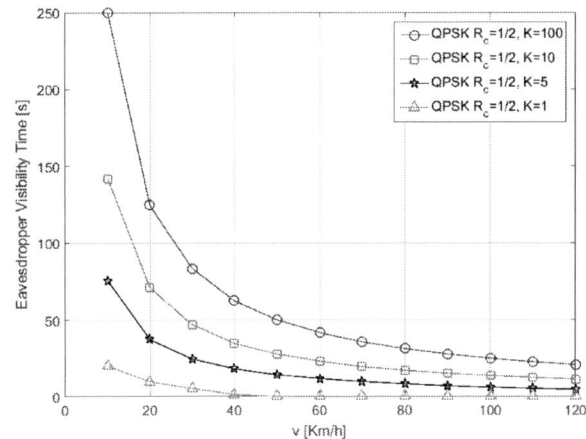

Fig. 5. Eavesdropper visibility time as a function of the relative speed for QPSK Rc = 1/2 and K-Rician fading channel.

1273

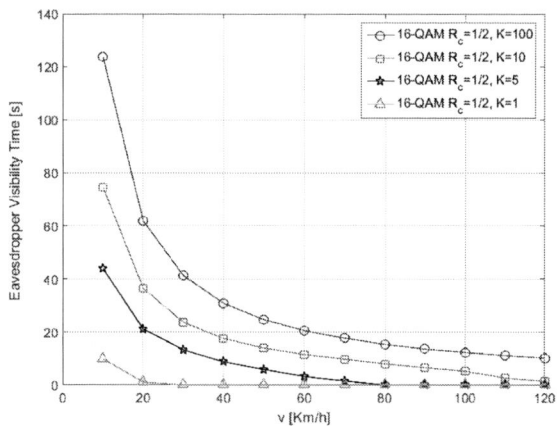

Fig. 6. Eavesdropper visibility time as a function of the relative speed for 16-QAM R_c=1/2 and K- Rician fading channel

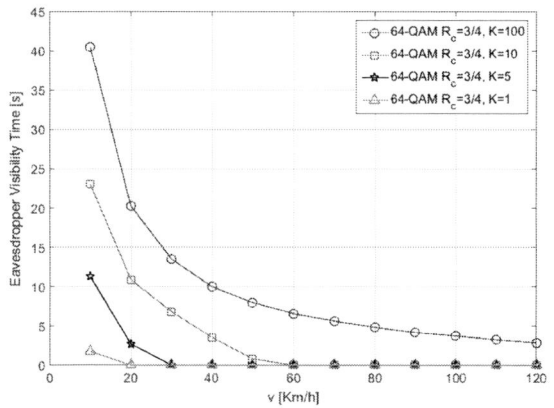

Fig. 7. Eavesdropper visibility time as a function of the relative speed for 64-QAM Rc = 3/4 and K-Rician fading channel.

V. CONCLUSIONS

The maximum visibility time available for the eavesdropper to obtain the DSRC transmitted data has been obtained by means of Monte Carlo simulations. This analysis has been performed considering different mobile channel conditions, different relative speeds and three modulation-coding schemes. The pseudonym frequency rate used to decide how many times the pseudonym cryptographic key must be changed is inversely proportional to these eavesdropper visibility times. Simulation results showed that, as expected, that the visibility time is minimised by increasing the relative speed due to the Doppler effect, and/or using a large level of modulation-coding scheme and/or transmitting information in urban environments (low values of the K-factor of the Rician fading channel). From the results it can be

concluded that using conventional power control mechanism, directional antennas and/or applying dynamic modulation-coding and multiple access coding schemes helps to mitigate the eavesdropping impact. Future developments will focus on more complex distributions of eavesdroppers and mobile ITS-stations.

ACKNOWLEDGMENT

We acknowledge the Cooperative ITS platform coordinated by DG MOVE Unit C.3 - Intelligent Transport Systems, which has driven the need for the study presented in this paper.

REFERENCES

[1] ETSI, TS. 102 637-2: Intelligent Transport Systems (ITS). Vehicular communications, 2014-06.

[2] G. Araniti, C. Campolo, M. Condoluci, A. Iera and A. Molinaro, "LTE for vehicular networking: a survey," in IEEE Communications Magazine, vol. 51, no. 5, pp. 148-157, May 2013.

[3] J. Petit, F. Schaub, M. Feiri and F. Kargl, "Pseudonym Schemes in Vehicular Networks: A Survey," in IEEE Communications Surveys & Tutorials, vol. 17, no. 1, pp. 228-255, Firstquarter 2015.

[4] B. Wiedersheim, Z. Ma, F. Kargl and P. Papadimitratos, "Privacy in inter-vehicular networks: Why simple pseudonym change is not enough," 2010 Seventh International Conference on Wireless On-demand Network Systems and Services (WONS), Kranjska Gora, 2010, pp. 176-183S.

[5] Z. Ma, F. Kargl, and M. Weber, "Pseudonym-on-demand: A new pseudonym refill strategy for vehicular communication," in Proc. 2nd IEEE Int. Symp. WiVec, Sep. 2008, pp. 1–5.

[6] J. Krumm, "A survey of computational location privacy," Pers. Ubiquitous Comput., vol. 13, no. 6, pp. 391–399, Aug. 2009.

[7] L. Gollan and C. Meinel, "Digital signatures for automobiles," in Proc. 6th World Multiconf. SCI, Jul. 2002, pp. 1–5.

[8] M. Gerlach, "Assessing and improving privacy in VANETs," in Proc. 4th Workshop ESCAR, Nov. 2006, pp. 1–9.

[9] M. Gerlach and F. Guttler, "Privacy in VANETs using changing pseudonyms—Ideal and real," in Proc. IEEE 65th VTC-Spring, Apr. 2007, pp. 2521–2525.

[10] IEEE Standard for Wireless Access in Vehicular Environments (wave)—Security Services for Applications and Management Messages, IEEE 1609.2, 2013Ss

[11] "Public key infrastructure memo," Car 2 Car Commun. Consortium, 2010.

[12] IEEE Std 802.11TM-2012, "IEEE Standard for Information technology – Telecommunications and information exchange between systems – Local and metropolitan area networks – Specific requirements – Part 11: Wireless LAN Medium Access Control (MAC) and Physical Layer (PHY) Specifications," New York, 29 March 2012.

[13] M.S.A. Abdeldime and W. Lenan, "The Physical Layer of the IEEE 802.11p WAVE Communication Standard: The Specifications and Challenges," In proceedings of the World Congress on Engineering and Computer Science (WCECS 2014), vol.2, pp.22-24, October, 2014.

[14] Abdi, A., Tepedelenlioglu, C., Kaveh, M. and Giannakis, G., "On the estimation of the K parameter for the Rice fading distribution," IEEE Communications Letters, pp. 92 -94, March 2001.

[15] J. G. Proakis, "Digital Communications," McGraw–Hill Book Co. pp. 767–768. ISBN 0-07-113814-5, 1995.

[16] D. Layne, "Receiver sensitivity and equivalent noise bandwidth," High Frequency Electronics Magazine, pp.22, June 2014.

Resilience of students' passwords against attacks

Boštjan Brumen* and Tadej Makari*
* University of Maribor, Faculty of Electrical Engineering and Computer Science, Smetanova 17, Si-2000 Maribor,
Slovenia
bostjan.brumen@uni-mb.si

Abstract - Passwords are still the predominant mode of authentication in contemporary information systems, despite a long list of problems associated with their insecurity. Their primary advantage is the ease of use and the price of implementation, compared to other systems of authentication (e.g. two-factor, biometry, …). In this paper we present an analysis of passwords used by students of one of universities and their resilience against brute force and dictionary attacks. The passwords were obtained from a university's computing center in plaintext format for a very long period – first passwords were created before 1980. The results show that early passwords are extremely easy to crack: the percentage of cracked passwords is above 95 % for those created before 2006. Surprisingly, more than 40 % of passwords created in 2014 were easily broken within a few hours. The results show that users – in our case students, despite positive trends, still choose easy to break passwords. This work contributes to loud warnings that a shift from traditional password schemes to more elaborate systems is needed.

I. INTRODUCTION

The computers of pioneering era needed no passwords. Their security relied on restricted access to the physical machines. Even if an adversary would have gained access to these machines, the lack of knowledge to manipulate the computers was preventing the attacks. Besides, these early computers barely stored anything interesting and/or valuable.

Today, it has become commonplace to note that computer systems protected by authentication systems are either mathematically robust or user-friendly, but not both [1]. Predominant authentication scheme today is the use of the username:password combination. The implementation of this scheme is straightforward, simple, easy to implement, has low cost of maintenance, and is overall accepted by the users.

However, since passwords are at users' responsibility, they cause many problems. Interestingly, these problems became evident already with first operating systems that implemented the time sharing functionality, all the way back in 1960s. When a PhD student Allan Scherr became fed up with time restrictions imposed on the system, he found a way to print the password file and use other people's accounts to use the system [2]. In 1978, Thompson and Morris wrote their seminal paper on password security, identifying numerous issues and proposing several countermeasures, which (still) have not been implemented in modern operating systems decades later [3], but only after a breach has occurred.

Badly enough, passwords are not used only on the operating system security level, but also on the application level, ranging from simple stand-alone desktop few-user applications, to multi-million user web applications.

The password issues should have been taken seriously. Breaches are not sporadic and by chance, they are frequent and systematic. They cause several millions of damage, and even take lives – as was the case of a preacher being listed as one of the ashleymadison.com (adult dating website) users [4].

Sadly, the computer community has not made a very much-needed shift in password management for more than 35 years [5].

The present study aims to shed some light on how a young population, students aged 18-24, manage and create their passwords.

In the next section, we present related studies that deal with the issue of users' passwords and their properties.

II. RELATED WORK

The related work reviews the works by others about similar researches on passwords.

The authors of study [6] have found that users tend to form their own mental models of security and what makes good passwords and they create insecure but easily recalled password by favoring memorability over security.

To help users select better (stronger and/or easy to remember passwords), several solutions have been proposed, such as mnemonic passwords [7, 8], cognitive passwords [9], associative passwords [10] and password checkers and meters [11-14].

A body of research was conducted on the user-centered approach, such as [15-19]. Here, the authors have identified the end user as the weakest link in the security chain and special attention needs to be given to user-related issues, such as memorability.

The education of users and its influence on the password (and overall system) security was researched in several works, e.g. in [20-23].

The impact of password policies was investigated in works of Summers et al [8], Inglesant & Sasse [24], Komanduri et al [25], Farrell [26] and Duggan [27].

The studies show that the average length of passwords is below 8 characters [11, 28, 29], yet these passwords are not completely random, thus of inadequate length, which should be at least 11 characters (mixed upper and lower, case, digits, symbols) [30].

The analysis of the related research has shown that most researchers collected their data, related to textual passwords, through laboratory experiments and/or surveys. In these settings the participants were aware of the setting, and this very fact may lead to false, fake, less accurate or biased data. Participants may have created or used different password than those used in their daily life in order to protect the latter.

A. Aims, scope, and organization of the paper

In this paper, our research goal is to analyze the strength of real users' passwords and their resilience against attacks.

Our research question is as follows: how resilient are real users' passwords against two most common attacks, a dictionary and a brute force attack?

The paper is organized as follows: in the next section, we present the methods of our research, and in the section IV we present the results. Our contribution is concluded in Section V with final remarks.

III. METHODS

A. Data collection

To answer the research question we obtained and analyzed real students' passwords. We have obtained 185.643 plaintext passwords of students of one of universities located in the EU. The passwords were used by students to log-in into their university accounts where they accessed to various functionalities. The university's student account system was developed in the 1980s and all passwords were stored in plaintext ever since; fortunately, with no known security incident.

The passwords were made available for research purposes only based on a strict non-disclosure agreement between the researchers and the university management; this agreement prevents the researchers from disclosing the name of the university and/or its location. The passwords resided on a computer that was locked in a separate room without internet or any other network connection, and access to the room and the computer was restricted and logged. Passwords are encrypted using a contemporary encryption system and only when analyzed they are being decrypted to a sandboxed environment and deleted after the analyses.

B. Ethical and legal considerations

Before the research commenced we considered whether it would break any ethical rules. Since the passwords were in plain text, their leakage could have been misused for attacks on other accounts. However, no personally identifiable information (e.g. username, student's number, etc.) were stored alongside the passwords. For the purpose of the research, other students' data were available alongside the password, such as sex, year of entry to university, study field, and few others. The passwords (and the corresponding accounts) are no longer being used by the university computing system as the students' accounts have been upgraded; each student was assigned a new identity and a new default password. Due to password reuse there is still a possibility that a user re-uses her password in any other application or system. However, we believe the chance of a password being matched to a single user is very low. Additionally, the passwords available and the results of the studies based on these passwords will help the professional community to understand better the patterns and the behavior of users and their passwords. We believe the positive aspects of using passwords overweight the potential perils. Last, but not least, the study is in line with local laws and EU regulations as in Directive 95/46/EC [31]. The Article 11/§2 allows the personal data to be used for research purposes.

C. Hardware and software

The passwords were hashed using MD5 algorithm, and the hashes were fed into the Hashcat v3.10. The properties of the hardware used are listed in Table I.

TABLE I. HARDWARE CONFIGURATION

Item	Description
CPU	Intel Core i5 6600 @ 3.30GHz
GPU	NVIDIA GeForce GTX 960 4GB
RAM	Crucial DDR4 16GB @ 2400MHz
HDD	Samsung SSD 250GB 850 EVO
motherboard	ASUS Z170-P
OS	Windows 10 Pro x64

The upper time limit of running the Hashcat software was set to 72 hours to mimic an attacker that has limited resources; from Table 1 one can note that the cost of hardware used was below 1.000 EUR / US$.

IV. RESULTS

A. Brute force attack

For the first part of the brute force attack, the software was configured to use lower case English alphabet, plus special characters from Slovenian alphabet ('č', 'š', and 'ž'), and digits (0..9). The search space was set to maximum 9 characters (calculations for 10 characters were estimated to last a bit less than 44 days).

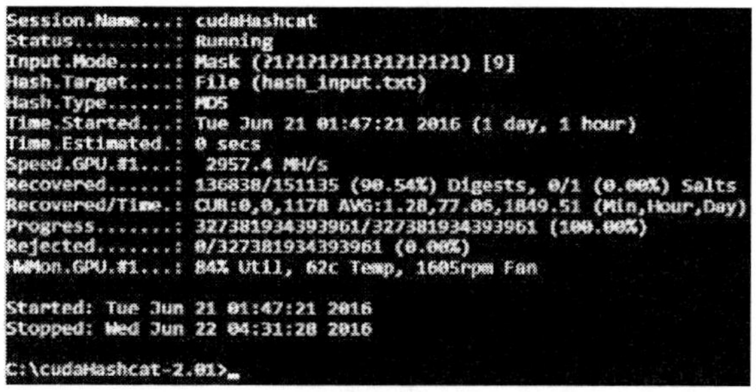

```
Session.Name...: cudaHashcat
Status.........: Running
Input.Mode.....: Mask (?1?1?1?1?1?1?1?1?1) [9]
Hash.Target....: File (hash_input.txt)
Hash.Type......: MD5
Time.Started...: Tue Jun 21 01:47:21 2016 (1 day, 1 hour)
Time.Estimated.: 0 secs
Speed.GPU.#1...: 2957.4 MH/s
Recovered......: 136838/151135 (90.54%) Digests, 0/1 (0.00%) Salts
Recovered/Time.: CUR:0,0,1178 AVG:1.28,77.06,1849.51 (Min,Hour,Day)
Progress.......: 327381934393961/327381934393961 (100.00%)
Rejected.......: 0/327381934393961 (0.00%)
HWMon.GPU.#1...: 84% Util, 62c Temp, 1605rpm Fan

Started: Tue Jun 21 01:47:21 2016
Stopped: Wed Jun 22 04:31:28 2016

C:\cudaHashcat-2.01>_
```

Figure 1. Command line output of Hashcat

The program ran for 1 day, 3 hours and 45 minutes and was able to crack 168.081 or 90,5 % of all passwords.

In the next part we lowered the number of characters from 9 to 8 and added the following special characters: !"#$%&/()=?*+'-_.:,;<>. With this setting we cracked 4,30 % of all passwords, but they partially overlapped with the previous results. Actually there were only 66 passwords with special characters, or 0,04 % of all. This part took 7 hours and 25 minutes.

The last part was configured to check only special characters and digits, with lengths of up to 9 characters. This way additional 10 passwords were cracked, for the cost of 7 hours and 10 minutes.

B. Dictionary attack

For dictionary attack we used a publicly available word lists from http://hashcrack.blogspot.si/p/wordlist-downloads_29.html. The size of all the dictionaries was 6 GB. The software ran for 2 minutes and 40 seconds and revealed 19,7 % (36.497) of passwords.

Next, we configured the software to add up to 5 digits at the end of each word. In 22 hours and 46 minutes we cracked 89,9 % of passwords.

Figure 2 shows an excerpt from the output file with examples of cracked passwords.

```
252    62d096fd1d286b23c7d231e963233475:031mobitel
253    c59c73c8fe7f2160fbbbfd2b13fa6499:0505idol
254    4a89f4288fe1415a0ea74626e8778d45:112gasilci
255    833f1e2bf72cc5734bd6d5e847e968bb:12975stronger
256    c66d652001a2163342cf8a08365e3fc4:2233geslo
```

Figure 2.   An example of cracked passwords (white: hash value, yellow: front digits, cyan: dictionary word)

Finally, the software added up to 5 digits in front of each word. In 10 hours and 5 minutes we additionally cracked 4,0 % of passwords.

At the end of combined dictionary attack only 11,853 passwords (6,4 %) remained intact.

C. Combined attacks

After combining the results from both types of attacks we were able to crack 95,3 % of all passwords (see Table II).

TABLE II.   RESULTS OF DIFFERENT ATTACKS

Item	Count	Percent	Time
Complete list of passwords	185.643	100,0 %	
Brute force attack list	168.157	90,6 %	1d 18h 20m
Dictionary attack list	173.790	93,6 %	1d 8h 54m
Combined attack list	176.919	95,3 %	3d 3h 14m

The total time to complete the attacks was 3 days, 3 hours and 14 minutes (Haschcat runtime only), where brute force attack took 1 day and 18 hours and dictionary attack lasted 1 day and 9 hours. Combined time, with necessary preparations, was thus less than four days.

It is interesting to observe the length of the cracked passwords (see Table III). Almost 95 % of passwords are of length 9 characters or less, with majority of passwords of length 6. The average length was 6,7 characters.

TABLE III.   PASSWORD LENGTHS AND THEIR COUNTS

Password length	Count	%
3 or less	472	0,3
4	1.631	0,9
5	1.711	1,0
6	156.670	88,6
7	3.451	2,0
8	9.227	5,2
9	2.231	1,3
10	811	0,5
11	350	0,2
12 or more	365	0,2
Total	176919	100

This result shows the effect of the hint given to the users that "passwords should be 6 characters or longer", but this policy was not enforced.

We also checked whether male students have had their passwords' lengths different than their female colleagues.

The difference is slight, 0,21 characters. Namely, male students have their average password lengths of 6,8, and females 6,6.

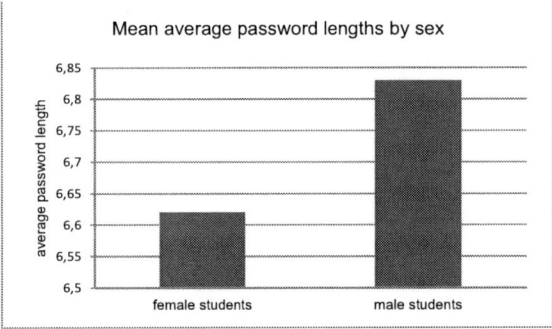

Figure 3.   Mean average password lengths by students

We checked if the difference is statistically significant. The independent t-test has indicated that the difference (0,21 characters) is not statistically significant (P=0,079) at α=0,05 level.

Another interesting finding is the speed of cracking. It was variable in different parts of attacks, but highest values were obtained with plain brute force attack. The cracking speed was close to 3 billion hashes per second (see Figure 1, parameter "Speed.GPU.#1…") on a single low-cost machine. One can imagine that a serious brute force attacker would have 1 million equivalent machines available, e.g. by means of a botnet. This means that a password should consist of 15-16 totally random characters from upper and lower case letters, digits and special symbols [30]!

Next, we checked how passwords differed by the year when they were generated.

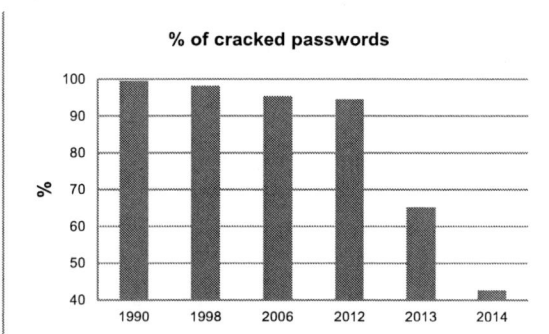

Figure 4. % of cracked passwords, by years

From Figure 4 one can notice that until 2012, the percentage of cracked passwords was in the 95 % range (1990: 99,5 %; 1998: 98,2 %; 2006: 95,4 %), with no statistically significant difference (ANOVA test, P<0,000). In 2013, a new policy was introduced for the default, system-assigned passwords, which changed from "XY9999" template (where X and Y were students' first and last name initials) to randomly generated 8-character password. For 2013 and 2014, the percentage of successfully cracked passwords fell to 65,2 % and 42,7 %, respectively.

V.   DISCUSSION AND CONCLUSION

The work presents an analysis to resilience of real students' passwords against two most common attacks, brute force and dictionary attacks. The passwords were obtained from a university's student account software upon its migration to a new system. Ethical and legal considerations were taken into the account before the analysis and research was conducted. The passwords were obtained from a single university from a single central European country, from a young body of population. The results are invertedly influenced by these facts, as localization, culture and level of technology all influence the choice of a password. Hence, the results of this study cannot be freely generalized to other types of users, geographical and cultural areas.

The analysis has shown that students' passwords are extremely vulnerable against these attacks. Astonishing 95,3 % of passwords were broken in a bit more than three days' time on a single low-cost machine with very simple, publicly available tools and methods. The cracking speed approached 3 giga-hashes, or passwords, per second.

At the beginning we have asked the following research question: how resilient are real users' passwords against two most common attacks, a dictionary and a brute force attack? The answer is simple and worrying: they are not resilient at all. The password lengths are inadequate and do not come close to the required "safe" password lengths of 15-16 *random* characters.

The results, in light of the above statement, may well indicate that the authentication systems using username:password are becoming obsolete and urgently need upgrading to more secure (e.g. two-way authentication) schemes. As Haque et al observe, users reuse high level passwords, such as those for banking applications, with little or no modifications for low level services, leading to a possibility for attackers to use unsafely shared or leaked passwords to crack higher-level passwords [32].

The passwords already have several alternatives, but they are costly, harder to implement and may require additional hardware. Two or three way authentication requires additional devices and/or services to display the second authentication token; biometrics requires additional scanning equipment and raises privacy issues.

The drawbacks of this study are the limitations posed by the data themselves. Results cannot be freely generalized due to the sample of users from one university, using a specific account for specific purposes. The future work will address the questions if different groups of students, such as technical vs. non-technical, have different password lengths; if the grade of a student correlates to the student's password length and strength, and alike.

ACKNOWLEDGMENT

The authors acknowledge the financial support from the Slovenian Research Agency (research core funding No. P2-0057).

REFERENCES

1.   McLennan CT, Manning P and Tuft SE. An evaluation of the Game Changer Password System: A new approach to password security. International Journal of Human-Computer Studies 2017; 100: 1-17. DOI: http://dx.doi.org/10.1016/j.ijhcs.2016.12.003.

2.   Corbató F, Daggett M, Daley R, Denning P, Grier DA, Mills R, Roach R and Scherr A. The Compatible Time Sharing System (1961–1973), Fiftieth Anniversary Commemorative Overview. Washington DC, U.S.A.: IEEE Computer Society; 2011. ISBN:

3.   DiBona C and Ockman S. Open sources: Voices from the open source revolution. O'Reilly Media, Inc.; 1999. ISBN: 0596553900

4.   Segall L. Pastor outed on Ashley Madison commits suicide. CNNMoney. p. September 8, 2015.

5.   Taneski V, Heričko M and Brumen B. Password security — No change in 35 years? 2014 37th International Convention on Information and Communication Technology, Electronics and Microelectronics (MIPRO); 2014. Opatija, Croatia; DOI: 10.1109/MIPRO.2014.6859779.

6.   Forget A, Chiasson S and Biddle R. Helping Users Create Better Passwords: Is This the Right Approach? 3rd Symposium on Usable Privacy and Security; 2007

7.   Yan J, Blackwell A, Anderson R and Grant A. Password Memorability and Security: Empirical Results. IEEE SECURITY & PRIVACY 2004; 2(5): 25-31. DOI: DOI: 10.1109/MSP.2004.81.

8.   Summers WC and Bosworth E. Password Policy: The Good, the Bad, and the Ugly. Winter International Synposium on Information and Communication Technologies; 2004.

9.   Zviran M and Haga WJ. Cognitive passwords: the key to easy access control. Computers & Security 1990; 9(8): 723-736. DOI: DOI: 10.1016/0167-4048(90)90115-A.

10. Helkala K, Svendsen N, Thorsheim P and Wiehe A. Cracking Associative Passwords. Secure IT Systems; 2012.

11. Egelman S, Sotirakopoulos A, Muslukhov I, Beznosov K and Herley C. Does my password go up to eleven?: the impact of password meters on password selection. Proceedings of the 2013 SIGCHI Conference on Human Factors in Computing Systems; 2013. April 27 - May 2, Paris, France: ACM.

12. Bishop M and Klein DV. Improving system security via proactive password checking. Computers & Security 1995; 14(3): 233-249. DOI: DOI: 10.1016/0167-4048(95)00003-Q.

13. Zviran M and Haga WJ. Password security: an empirical study. Journal of Management Information Systems 1999; 15: 161-186.

14. Yan J, Blackwell A, Anderson R and Grant A. The memorability and security of passwords: some empirical results. in Technical report # UCAM-CL-TR-500.Cambridge, UK: University Of Cambridge; 2000.

15. Barton BF and Barton MS. User-friendly Password Methods for Computer-mediated Information Systems. Computers & Security 1984; 3(3): 186-195.

16. Adams A and Sasse MA. Users are not the enemy. Communications of the ACM 1999; 42(12): 40-46. DOI: DOI: 10.1145/322796.322806.

17. Fidas CA, Voyiatzis AG and Avouris NM. When Security Meets Usability: A User-Centric Approach on a Crossroads Priority Problem. 14th Panhellenic Conference on Informatics (PCI); 2010.

18. Adams A, Sasse MA and Lunt P. Making passwords secure and usable. People and Computers XII, chapter 1. . London: Springer London; 1997. p. 1-19. ISBN: 3540761721

19. Sasse MA, Brostoff S and Weirich D. Transforming the 'weakest link'—a human/computer interaction approach to usable and effective security. BT technology journal 2001; 19(3): 122-131. DOI: DOI: 10.1023/A:1011902718709.

20. Horcher A-M and Tejay GP. Building a better password: The role of cognitive load in information security training. IEEE International Conference on Intelligence and Security Informatics, 2009. ISI'09. ; 2009. IEEE.

21. Grawemeyer B and Johnson H. Using and Managing Multiple Passwords: A Week to a View. Interact. Comput. 2011; 23(3): 256–267.

22. Vidyaraman S, Chandrasekaran M and Upadhyaya S. Position: The User is the Enemy. 2007 Workshop on New Security Paradigms; 2007.

23. Lorenz B, Kikkas K and Klooster A. The Four Most-Used Passwords Are Love, Sex, Secret, and God: Password Security and Training in Dierent User Groups. Human Aspects of Information Security, Privacy, and Trust; 2013.

24. Inglesant PG and Sasse MA. The True Cost of Unusable Password Policies: Password Use in the Wild. SIGCHI Conf. on Human Factors in Computing Systems; 2010.

25. Komanduri S, Shay R, Kelley PG, Mazurek ML, Bauer L, Christin N, Cranor LF and Egelman S. Of Passwords and People: Measuring the Efect of Password-composition Policies. Proc. of the SIGCHI Conf. on Human Factors in Computing Systems; 2011.

26. Farrell S. Password policy purgatory. IEEE Internet Computing 2008; 12(5): 84-87.

27. Duggan GB, Johnson H and Grawemeyer B. Rational Security: Modelling Everyday Password Use. Int. J. Hum.-Comput. Stud. 2012; 70(6): 415–431.

28. Voyiatzis AG, Fidas CA, Serpanos DN and Avouris NM. An Empirical Study on the Web Password Strength in Greece. 15th Panhellenic Conference on Informatics (PCI); 2011.

29. Jakobsson M and Dhiman M. The Benefits of Understanding Passwords. Mobile Authentication 2013: 5–24.

30. Brumen B and Černezel A. Brute force analysis of PsychoPass-generated Passwords. in 37th International Convention on Information and Communication Technology, Electronics and Microelectronics (MIPRO)Opatija, Croatia, 26-30 May 2014; 2014. p. 1366-1371.

31. Directive 95/46/EC of the European Parliament and of the Council of 24 October 1995 on the protection of individuals with regard to the processing of personal data and on the free movement of such data. Rule Number: 95/46/EC. Bruxelles: Official Journal L 281; 1995.

32. Taiabul Haque SM, Wright M and Scielzo S. Hierarchy of users' web passwords: Perceptions, practices and susceptibilities. International Journal of Human-Computer Studies 2014; 72(12): 860-874. DOI: http://dx.doi.org/10.1016/j.ijhcs.2014.07.007.

Empirical study on the risky behavior and security awareness among secondary school pupils - validation and preliminary results

T. Velki[1], K. Solic[2], V. Gorjanac[3] and K. Nenadic[4]

[1] J.J. Strossmayer University of Osijek, Faculty of Education, Croatia
[2] J.J. Strossmayer University of Osijek, Faculty of Medicine, Croatia
[3] Center for Missing and Exploited Children Osijek, Croatia
[4] J.J. Strossmayer University of Osijek, Faculty of Electrical Engineering, Computer Science and Information Technology, Croatia

tena.velki@gmail.com, kresimir.solic@mefos.hr, vanjagorjanac@gmail.com, kresimir.nenadic@etfos.hr

Abstract - This study was based on the validated Users' Information Security Awareness Questionnaire (UISAQ). Authors gathered information on risky behavior and security awareness among 355 pupils from three secondary schools: General program secondary school (Gymnasium), Business and administrative high school and Trade school.
Aim of the study was to test adapted version of UISAQ questionnaire on secondary school student population.
Questionnaire was slightly modified and validated to meet new requirements regarding young pupils' behavior. Analysis outcome represent results of secondary school pupils regarding 6 subareas: Users' usual behavior (x_1=4.12), Personal computer maintenance (x_2=3.33), Borrowing access data (x_3=4.77), Awareness about security in communications (x_4=2.67), Belief into securing data (x_5=2.27) and Importance of backup quality (x_6=3.86). Also some detailed comparisons regarding demographic variables and correlation with students and employed users were given. Additionally, correlations with revealing password and Internet abuse are given. The percentage of young pupils that have revealed their password for their e-mail system access is much higher than usual (77.7%). This information should alert government institutions and schools teachers.
Main results of this study have shown that slightly modified UISAQ can be used on the secondary school population. Additional results have shown that there is a great reason to be concerned about pupils' risky behavior on the Internet and their low awareness about information security and privacy issues.

I. INTRODUCTION

The importance of knowledge, behavior and awareness about information security and privacy issues among Internet users is firstly recognized among network administrators and security experts. Only later scientists

This work is financed by the Croatian Government Office for Cooperation with NGOs and co-financed by the European Union's Connecting Europe Facility, under project named "Safer Internet Centre Croatia: Making internet a good and safe place", Agreement Number: INEA/CEF/ICT/A2015/115320
The sole responsibility of this publication lies with the authors. The European Union is not responsible for any use that may be made of the information contained therein.

acknowledged that average information system's user is the weakest link in the information security and privacy chain [1, 2]. Much research was made later on this subject [3-16], but systematic research by using validated questionnaire as the scientific measurement instrument was first made three years ago in Croatia named UISAQ [17, 20] then in the USA named SeBIS [18]. By now only Turkish scientists with their Four Measurements Scales [19] and Australian scientist with HAIS-Q [20] have followed. Main reason for this situation is probably because security experts are mostly technicians or partly managers and are not behavioral scientists. Today research on this subject is usually carried out by a team of scientists covering different scientific areas, both computer science and behavioristic science.

The main aim of this study was to adapt existing UISAQ questionnaire for usage among secondary school population and to validate this new version. Secondary school students, teenagers, mainly differ in some basic characteristics like cognitive, social and emotional stage of development from adult population as adults and students. The risky behavior in the teenager population is proven to be on highest level, so existing questionnaire has to be adapted and revalidated on this population.

The secondary aim was to gain some new conclusions based on the preliminary results regarding potentially risky behavior and security awareness among secondary school pupils, by carrying out an empirical study with this new version of UISAQ questionnaire.

II. METHOD

A. Participants

Participants in this study were secondary school pupils (N=355) from three different schools: Trade school, Business and administrative high school and General program secondary school (Gymnasium). In this manner sample is representative and stratified because students were chosen from both high school, three and four year vocational schools.

Proportion of male students was 36.3% while proportion of female students was 63.1%. The average age of participants was 16.39 +/- 1.15 (arithmetic mean +/- SD).

B. Procedure

During regular classes students were asked to voluntarily give some general information about themselves (age and gender) and to fill out the UISAQ. Filling out the questionnaire lasted for approximately 30 minutes. Survey was carried out in all three schools during one week period.

C. Instruments

For the purpose of this research authors used adapted version of UISAQ [21, 22] for secondary school population. UISAQ consists of two parts with total of 33 questions. Some questions (k=9) were changed and adjusted for secondary school pupils (i.e. *Borrow your personal login data to colleague from work* into *Borrow your personal login data to friend from your classroom*). Adaptation was made with minor changes, as it was important to have a parallel form of UISAQ questionnaire.

First part of UISAQ consisted of 17 questions measuring computer users' potentially risky behavior. Second part of questionnaire consisted of 16 questions measuring the level of user's information security knowledge and awareness.

III. RESULTS

Using descriptive statistics, factor analysis and reliability analysis authors tested if adjusted version of

TABLE I. STRUCTURE MATRIX FOR THE FIRST PART OF UISAQ EXTRACTION

Items	Factors		
	1	*2*	*3*
sc1	.412		
sc2	.383		
sc3	.527		
sc4	.859		
sc5	.776		
sc6		.496	
sc7		.735	
sc8		.739	
sc9			.484
sc10			.588
sc11			.613
sc12			.588
sc13			.412
sc14		.474	
sc15			-.499
sc16		.461	
sc17		.366	

a. Rotation Method: Oblimin

TABLE II. RELIABILITY ANALYSIS: ITEM – TOTAL STATISTICS FOR THE FIRST PART OF UISAQ EXTRACTION

Items	Analysis results			
	Scale Mean if Item Deleted	*Scale Variance if Item Deleted*	*Corrected Item - Total Correlation*	*Cronbach's Alpha if Item Deleted*
1. factor (Subscale of access data lending/ borrowing)				
sc1	4.96	2.420	.339	.475
sc2	4.49	1.635	.288	.600
sc3	5.00	2.436	.436	.434
sc4	5.11	2.653	.468	.453
sc5	5.09	2.830	.247	.527
2. factor (Subscale of (personal) computer systems' maintenance)				
sc6	16.36	18.542	.308	.521
sc7	16.51	17.691	.467	.457
sc8	16.79	17.909	.435	.469
sc14	16.95	17.834	.279	.538
sc16	15.65	19.941	.227	.554
sc17	17.15	19.073	.184	.584
3. factor (Subscale of computer users' usual behavior)				
sc9	9.11	9.359	.246	.567
sc10	9.48	8.727	.413	.491
sc11	9.96	9.904	.396	.518
sc12	9.68	9.144	.402	.501
sc13	9.89	10.090	.285	.548
sc15	8.07	7.935	.269	.584

UISAQ has good psychometrics characteristics.

For the first part of UISAQ Confirmatory factor analysis with 3 factors (method principal components, Oblimin rotation) was used. Analyses have shown

TABLE III. MEASURES OF SENSITIVITY FOR THE FIRST PART OF UISAQ EXTRACTION

Items	Analysis results					
	Min	*Max*	*Range*	*Mean*	*Std. Deviation*	*Test of normality*
sc1	1.00	5.00	4.00	1.20	.574	.504[a]
sc2	1.00	5.00	4.00	1.67	.993	.360[a]
sc3	1.00	5.00	4.00	1.16	.497	.509[a]
sc4	1.00	5.00	4.00	1.06	.373	.529[a]
sc5	1.00	5.00	4.00	1.08	.422	.530[a]
sc6	1.00	5.00	4.00	3.56	1.451	.238[a]
sc7	1.00	5.00	4.00	3.38	1.301	.169[a]
sc8	1.00	5.00	4.00	3.10	1.317	.178[a]
sc9	1.00	5.00	4.00	2.14	1.100	.239[a]
sc10	1.00	5.00	4.00	1.76	1.014	.325[a]
sc11	1.00	5.00	4.00	1.29	.742	.486[a]
sc12	1.00	5.00	4.00	1.56	.913	.375[a]
sc13	1.00	5.00	4.00	1.35	.809	.463[a]
sc14	1.00	5.00	4.00	2.94	1.670	.202[a]
sc15	1.00	5.00	4.00	3.16	1.447	.182[a]
sc16	1.00	5.00	4.00	4.22	1.356	.422[a]
sc17	1.00	5.00	4.00	2.75	1.668	.227[a]

a. p < 0.01

explanation of 37.28% of overall variance. First factor explained 16.07% of overall variance, second factor explained 12.96% of overall variance and third factor explained 8.25% of overall variance. For all 3 factors the saturation (factor loadings) was larger than 0.3 and had exactly the same distribution of questions as the original validated version of UISAQ. The factor structure of the first part of UISAQ is shown in Table 1. Then reliability analysis was done for three subscales (Table 2). All subscales had satisfactory internal consistency (first k=5; Cronbach α=0.63; second k=6; Cronbach α=0.59; third k=6; Cronbach α=0.62), and all questions contributed significantly to good internal consistency which implies that this form of subscale should be kept as final one. Results of sensitivity test of new formed questionnaire are shown in Table 3. All items have full range of response which implies good sensitivity. Distribution of results was not normal (Kolmogorov-Smirnov Statistic was significant for all items), which was expected. For the first and the third subscale means were at lower part of subscale (positive asymmetry) meaning less risky behavior of computer users and for the second subscale means were at higher part of subscale (negative asymmetry) meaning more risky behavior of computer users, e.g. low level of users' maintenance of personal computer systems.

The Confirmatory factor analysis with 3 factors (method principal components) was also used for the second part of UISAQ, which have shown explanation of 54.02% of overall variance. In Table 4. factor structure of the second part of UISAQ is shown and has exactly the same distribution of questions as the original validated version of UISAQ and a satisfactory factor loadings. First

TABLE IV.   STRUCTURE MATRIX FOR THE SECOND PART OF UISAQ EXTRACTION

Items	Factors		
	1	*2*	*3*
sa1			.667
sa2			.746
sa3			.811
sa4			.704
sa5			.709
sa6	.684		
sa7	.770		
sa8	.780		
sa9	.750		
sa10	.763		
sa11		.566	
sa12		.720	
sa13		.747	
sa14		.758	
sa15		.657	
sa16		.783	

a. Rotation Method: Oblimin

TABLE V.   RELIABILITY ANALYSIS: ITEM – TOTAL STATISTICS FOR THE SECOND PART OF UISAQ EXTRACTION

Items	Analysis results			
	Scale Mean if Item Deleted	*Scale Variance if Item Deleted*	*Corrected Item - Total Correlation*	*Cronbach's Alpha if Item Deleted*
1. factor (Subscale of level of awareness about security in communications)				
sa1	13.62	12.020	.481	.764
sa2	13.76	10.588	.591	.728
sa3	12.86	10.570	.664	.705
sa4	12.96	10.910	.508	.758
sa5	13.46	10.786	.547	.744
2. factor (Subscale of belief into securing data)				
sa6	8.89	14.314	.517	.789
sa7	9.26	13.912	.612	.760
sa8	9.08	13.776	.621	.757
sa9	9.29	14.196	.590	.767
sa10	8.83	13.493	.608	.761
3. factor (Subscale of backup quality importance)				
sa11	19.59	15.813	.409	.799
sa12	19.33	14.691	.571	.759
sa13	18.86	15.278	.591	.755
sa14	19.41	14.593	.611	.749
sa15	19.65	15.506	.496	.777
sa16	19.00	14.901	.640	.744

factor explained 19.25% of overall variance, second factor explained 14.07% of overall variance and third factor explained 20.07% of overall variance. Reliability analysis (Table 5) had shown high internal consistency (first k=5; Cronbach α=0.78; second k=6; Cronbach α=0.81; third k=6; Cronbach α=0.80). All items had full range of

TABLE VI.   MEASURES OF SENSITIVITY FOR THE SECOND PART OF UISAQ EXTRACTION

Items	Analysis results					
	Min	*Max*	*Range*	*Mean*	*Std. Deviation*	*Test of normality*
sa1	1.00	5.00	4.00	3.04	.982	.207[a]
sa2	1.00	5.00	4.00	2.90	1.142	.239[a]
sa3	1.00	5.00	4.00	3.80	1.057	.290[a]
sa4	1.00	5.00	4.00	3.71	1.175	.258[a]
sa5	1.00	5.00	4.00	3.20	1.156	.174[a]
sa6	1.00	5.00	4.00	2.45	1.251	.242[a]
sa7	1.00	5.00	4.00	2.08	1.199	.255[a]
sa8	1.00	5.00	4.00	2.27	1.209	.226[a]
sa9	1.00	5.00	4.00	2.04	1.167	.263[a]
sa10	1.00	5.00	4.00	2.51	1.269	.215[a]
sa11	1.00	5.00	4.00	3.58	1.160	.274[a]
sa12	1.00	5.00	4.00	3.84	1.133	.250[a]
sa13	1.00	5.00	4.00	4.31	1.005	.336[a]
sa14	1.00	5.00	4.00	3.76	1.098	.231[a]
sa15	1.00	5.00	4.00	3.52	1.087	.220[a]
sa16	1.00	5.00	4.00	4.17	1.013	.269[a]

a. p < 0.01

TABLE VII. COMPARISON AMONG SECONDARY SCHOOL PUPILS AND GROWN-UPS FROM PREVIOUS STUDY [22]

Subscales of UISAQ	Pupils (n=355)	Adults (n=701)	P
UB/ x±SD	4.12±0.58	4.52±0.43	<0.001[a]
PCM/ x±SD	3.33±0.82	3.18±0.91	0.009[a]
BAD/ x±SD	4.77±0.37	4.74±0.39	0.230[a]
SC/ x±SD	2.67±0.81	3.48±0.83	<0.001[a]
SD/ x±SD	2.27±0.92	2.06±0.79	<0.001[a]
BQ/ x±SD	3.86±0.76	4.18±0.68	<0.001[a]
Password Revealing	77.7%	28.8 %	<0.001[b]

[a] Student T Test

[b] Chi-Square Test

response which implies good sensitivity (Table 6). Distribution of results was not normal (Kolmogorov-Smirnov Statistic was significant for all items), which was expected. Means for level of security in communications and backup importance were at higher part of distribution (negative asymmetry) meaning low level of user's information security awareness.

Additional results present mean values for each subscale among secondary school pupils on 5-point Likert-type scale where five means excellent from the aspect of information security and privacy (Table 7). In the same table, comparison of tested secondary school pupils with results of adults gathered in the previous study [22] is shown. Statistical significance was found between mean values in all subscales except in the subscale regarding Borrowing Access Data. In comparison to adults, secondary school pupils were worse in subscales regarding Usual behavior, Criticism on security in communications and Importance of backup quality. Examined pupils were better in subscales regarding Personal computer maintenance and Belief into securing data.

Among tested secondary school pupils there were 77.7% of them that had revealed password for currently used e-mail system, which is significantly more than adults. Also only 20.0% of tested pupils had made backup during last month period. While 7.6% said that more than two other persons know their password for currently used e-mail system, 5.7% said the same for their Facebook account. Percentage of pupils that have ever experienced cyber violence is 28.8% while percentage of pupils that behaved cyber violently was 22.9%.

Regarding age there was no statistically significant difference in any of subscales for pupils and regarding gender, there was a statistical difference only in subscale about backup quality (p<0.001; Student T Test). Female pupils are making backup more often than male ones.

Pearson's correlation coefficients are either "low correlation", "very low correlation" or even "negligible correlation", as their value is closer to zero than to positive or negative value one (Table 8). Prominent correlation coefficients, with statistical significance, are found between subscale describing Users' usual risky behavior and Being violent by doing Internet abuse actions; than between subscale describing Borrowing access data and number of persons that know access data for e-mail or Facebook and being violent or experiencing abuse on the Internet; also between subscale describing Personal computer maintenance and time of making last backup. Subscale describing Importance of backup quality is in low or very low correlation with all five external variables that present controlling questions. Subscale regarding Awareness on security in communications is negative, but very low correlation with external variable measuring being violent on the Internet.

IV. CONCLUSION

Main aim is achieved as results show that modified UISAQ questionnaire has the same structure as version for adults and students, satisfying consistency of subscales and excellent sensibility. The new adapted parallel form of UISAQ questionnaire can be used to trace development of computer risky behavior and information security knowledge from teenage to adulthood. Also it can be used to compare data about risky behavior and information security knowledge between secondary school pupils and both their teachers and parents who are primarily modeling pupils' behavior.

Validation results have proven that questionnaire can be used in future research applied on the secondary school

TABLE VIII. PEARSON'S CORRELATION COEFFICIENTS REGARDING EXTERNAL VARIABLES

Subscales of UISAQ	Time of the last backup	Number of persons that know access data for e-mail	Number of persons that know access data for Facebook	Cyber violence	Cyber victimization
UB	-0.024	0.077	0.074	0.306[a]	0.194[a]
PCM	0.260[a]	0.148[a]	0.079	0.011	0.003
BAD	0.011	0.286[a]	0.173[a]	0.282[a]	0.219[a]
SC	-0.003	-0.097	-0.100	-0.095	-0.157[a]
SD	0.025	0.006	0.073	-0.194[a]	-0.174[a]
BQ	0.266[a]	0.229[a]	0.178[a]	0.266[a]	0.164[a]

a. p < 0.01

population.

Additional results had shown that percentage of the secondary school pupils that have revealed their password for their e-mail system access is much higher than usual. In comparison to adults, they are worse in subscales regarding Usual behavior, Criticism on security in communications and Importance of backup quality. Secondary school pupils are better in subscales regarding Personal computer maintenance and Belief into securing data. In addition, there is relatively high percentage of young pupils that were cyber victims, and also a certain percentage of young pupils that were cyber violent.

Regarding result on correlation analysis there are some expected conclusions. More risky behavior is in positive correlation with cyber violence, either being violent or being victim. Also, borrowing access data is in positive correlation with higher number of other persons knowing access data for e-mail and Facebook. It is also expected that better PC maintenance is in positive correlation with more often backup making.

It is unexpected positive correlation between high awareness about issues regarding importance of backup quality and all other external variables regarding risky and insecure behavior, measured in number of other persons knowing access data or cyber violence. This finding is similar to some previous findings using UISAQ questionnaire among adults, and may mean that persons who have high level of awareness are acting more careless on the Internet.

Also similar to some previous studies, female users are more careful in their behavior regarding information security and privacy issues.

Results of this study have shown that there is a reason to be concerned about pupils' risky behavior on the Internet, and this is conclusion regardless of pupils' gender or age. This information should alert government institutions and school teachers and should result in better education and prevention programs in order to increase safe behavior when online and information security and privacy issues.

Main future work is going to be measurement and definition of norms regarding behavior on the Internet among Croatians by carrying out empirical study using the UISAQ including secondary school population, students and adults. Other future plans are to develop international scientific questionnaire, possibly in collaboration with teams that also have some results on this subject e.g. Australian team and Turkish team.

REFERENCES

[1] K. D. Mitnick, The Art of Deception - Controlling the Human Element of Security, John Wilwy & Sons, 2002.

[2] M. A. Sasse, S. Brostoffand, and D. Weirich, "Transforming the 'weakest link' - a human/ computer interaction approach to usable and effective security" BT Technology Journal, vol. 19, pp. 122–131, July 2001.

[3] A. Tsohou, S. Kokolakis, M. Karyda and E. Kiountouzis, "Process-variance models in information security awareness research", Information Management & Computer Security, vol. 16, pp. 271-287, July 2008.

[4] S. Williams and S. Akanmu, "Relationship between Information Security Awareness and Information Security Threats", IJRCM, vol.3, pp. 115-119, August 2013.

[5] P. Tasevski, "Methodological approach to security awareness", CyberSecurity for the Next Generation. (Politechnico di Milano, Italy), 11-12 December 2013.

[6] P. Puhakainen and M. Siponen, "Improving Employees' Compliance through Information Systems Security Training: An Action Research Study", MIS Quarterly, vol. 34, pp. 757-778, December 2010.

[7] K. Solic, D. Sebo, F. Jovic and V. Ilakovac, "Possible Decrease of Spam in the Email Communication", Proceedings IEEE MIPRO, (Opatia), pp. 170-173, May 2011.

[8] K. Beckers, L. Krautsevich, and A. Yautsiukhin, „Analysis of Social Engineering Threats with Attack Graphs", Proceedings of the 3rd International QASA - Affiliated workshop with ESORICS, (Wroclow, Poland), September 2014.

[9] K. Solic and V. Ilakovac, "Security perception of a portable PC user (The difference between medical doctors and engineers): a pilot study", Medicinski glasnik Dobojsko-Tuzlanskog kantona, vol. 6, pp. 261-264, August 2009.

[10] R. E. Crossler, A. C. Johnston, P. B. Lowry, Qing Hu, M. Warkentin and R. Baskerville, "Future directions for behavioral information security research", Computers&Security, vol. 32, pp. 90–101, June 2013.

[11] A. Keszthelyi, "About Passwords", Acta Polytechnica Hungarica, vol.10, pp. 99-118, September 2013.

[12] K. Solic, H. Ocevcic and D. Blazevic, "Survey on Password Quality and Confidentiality", Automatika, vol. 56, Juny 2015.

[13] A.G. Voyiatzis, C.A. Fidas, D.N. Serpanos and N.M. Avouris, "An Empirical Study on the Web Password Strength in Greece", 15th Panhellenic Conference on Informatics, (Kastonija Greece), pp. 212-216, September-October 2011.

[14] M. Dell'Amico, P. Michiardi and Y. Roudier, "Password Strength: An Empirical Analysis", Proceedings IEEE INFOCOM, (San Diego, CA) pp. 1-9, March 2010.

[15] Ma Wanli, J. Campbell, D. Tran and D. Kleeman, "Password Entropy and Password Quality", 4th International Conference on Network and System Security, (Melbourne, VIC), pp. 583-587, 1-3, September 2010.

[16] P.G. Kelley, S. Komanduri, M.L. Mazurek, R. Shay, T. Vidas, L. Bauer, N. Christin, L.F. Cranor and J. Lopez, "Guess Again (and Again and Again): Measuring Password Strength by Simulating Password-Cracking Algorithms", IEEE Symposium on Security and Privacy, (San Francisco, CA), pp. 523-537, May 2012.

[17] T. Velki, K. Solic and H. Ocevcic, "Development of Users' Information Security Awareness Questionaire (UISAQ) - Ongoing Work", Proceedings IEEE MIPRO, (Opatia), pp. 1417-1421, May 2014.

[18] S. Egelman, M. Harbach and E. Peer, „Behavior ever follows intention? A validation of the Security Behavior Intentions Scale (SeBIS)", Proceedings of Annual ACM Conference on Human Factors in Computing Systems, (San Jose, CA, USA), pp. 7-12, May 2016.

[19] G. Öğütçü, Ö.M. Testik and O. Chouseinoglou, „Analysis of personal information security behavior and awareness", Computers&Security, vol. 56, pp. 83–93, February 2016.

[20] K. Parsons et. al, „The Human Aspects of Information Security Questionnaire (HAIS-Q): Two further validation studies", Computers&Security, vol. 66, pp. 40–51, May 2017.

[21] T. Velki, K. Solic and K. Nenadic, „Razvoj i validacija Upitnika znanja i rizičnog ponašanja korisnika informacijskog sustava (UZRPKIS)", Psihologijske teme, vol. 24, pp. 401-424, December, 2015.

[22] T. Velki, K. Solic and T. Galba, „Empirical study on ICT system's users' risky behavior and security awareness", Proceedings IEEE MIPRO, (Opatia), pp. 1356-1359, May 2015.

Interoperability and Lightweight Security for Simple IoT Devices

Darko Andročec, Boris Tomaš, Tonimir Kišasondi
{darko.androcec,boris.tomas,tonimir.kisasondi}@foi.hr
Faculty of Organization and Informatics,
University of Zagreb
Pavlinska 2, Varaždin

Abstract - The Semantic Web can be used to enable the interoperability of IoT devices and to annotate their functional and nonfunctional properties, including security and privacy. In this paper, we will show how to use the ontology and JSON-LD to annotate connectivity, security and privacy properties of IoT devices. Out of that, we will present our prototype for a lightweight, secure application level protocol wrapper that ensures communication consistency, secrecy and integrity for low cost IoT devices like the ESP8266 and Photon particle.

Keywords: interoperability, IoT security, Semantic Web, IoT device

I. INTRODUCTION

This paper shows our efforts in creating a method for interoperability for IoT devices and creation of a lightweight communication protocol for cheap MCUs (microcontroller unit), that can be used on almost all devices.

Internet of things potential is still underexploited. One of the main reasons is that there is no interoperability among IoT devices and services, because of many differences among IoT devices and IoT services. Each IoT protocol is suitable for different types of applications, and IoT devices support only a small subset of available IoT protocols. The interoperability problems can be solved or its effects can be reduced by using ontology to semantically annotate things and their services. There are many existing IoT ontologies, but none describes well the connectivity, security and privacy properties of IoT devices, so we have developed and presented here briefly a new ontology. JSON for Linking Data (JSON-LD) is a lightweight Linked Data format that provides a way to help JSON-data interoperate at Web-scale. Ontology and JSON-LD enable us to semantically annotate things and their capabilities, together with the mentioned non-functional properties.

One of the more interesting use-cases for lightweight communication and security is the usage of low cost MCU (Micro Controller Units) like ESP8266 [1] or ESP32. One of the main benefits of those MCUs is their low cost. An ESP8266 board costs about 5 USD at the time this paper is published. With a 32-bit processor, 16 GPIO pins and support for connecting to 802.11 based WiFi networks,

This work has been fully supported by the Croatian Science Foundation under the project IP-2014-09-3877.

ESP8266 and other MCUs became quickly popular as an important building block in many IoT devices. One of the reasons for their increased popularity is that with classic SDK support for writing applications in C, open source developers started enabling support and started porting languages like MicroPython [2], Lua [3] or Javascript to the ESP8266 and similar cheap MCU's.

One of the benefits of using high level languages such as MicroPython to develop applications for MCUs is the same reason why high level languages are used in modern software development: rapid prototyping and the use of high level abstractions and libraries enables faster and easier development. Coupled with the advancements in the design and availability of various embedded components like sensors and corresponding libraries that drastically simplify development to the point that a developer needs to correctly connect the sensor and write some glue code to collect data from the sensor, enables software developers to venture into the embedded systems domain, for better or for worse.

The background for development of this protocol is one specific professional implementation we needed to develop. The project required a system that would have the following features:

1. Rapidly prototype a system on ESP8266 platform and MicroPython

2. System contains multiple nodes collecting sensor data and/or communicating with other nodes (mesh network scenario)

3. System must enable data exchange and communication with enhanced security

4. Only some nodes have access to a broker / internet, and the network needs to communicate in a best effort scenario.

5. System should support light and deep sleep states for the MCU

6. Develop security in *userspace* and don't rely on any specifics like crypto acceleration from the MCU

Since languages like MicroPython, Lua and others when implemented on MCUs lack some features (like raw packet crafting or support for various low level protocols), we wanted to develop a scheme that we could use on all of our preferred development platforms which would not be restricted to a specific C language only SDK like [4]. Therefore, an application level scheme and approach that doesn't need low level packet access needed to be devised to fulfill the requirements.

Another of our requirements was to avoid the usage of various protocols such as XBee, ZigBee or LoRA, since those protocols solve the problem of scalable mesh communication, but their price is restrictive if we want to explore the possibilities of using sub 5$ MCUs that have basic 802.11 WiFi networking capability.

II. RELATED WORK

Boman et al. [5] used several sensors to demonstrate sensor networks interconnected with systems like GSN and Firebase. They conclude that having XML to define sensor node and its capabilities, is complicated and they suggest researching for different solutions. Cassar et al. propose a new semantic service for IoT device matchmaking method that is based on probabilistic service matchmaking. They state that their hybrid method can overcome most cases of semantic synonymy in semantic service description [6]. Souza et al. propose vocabulary for IoT service discovery. To identify thing in IoT they propose use of URI (Uniform Resource Identifier) the same way it is used in Web and Semantic web. Syntax of this URI is according to XML standard [7]. They also define WSDL (Web Services Description Language) for IOT device that is accessible using provided URI. URI is used in *Event driven* SOA (Service Oriented Architecture) [8], a concept defined by Zhang et al. In their work, they focus on IoT as a service (publish and subscribe model), however to achieve that level, IOT device must be identified. To achieve that they use custom XML service description.

Hur et al. identify interoperability as an issue between things and platforms and it is one of major challenges in achieving the vision of the Internet of Things (IoT). They propose concept of SSD (Semantic Service Description) ontology that semantically represent heterogeneous things across various platforms [9]. They use JSON-LD to store semantic metadata. To retrieve such data, they propose use of Web Linking [10] and POWDER *describedby* property. SSD ontology contains three main concepts: Property, Capability and Server Profile. Each of those is designed to support interoperability between platforms and physical objects. Server Profile defines connectivity for device configuration, for example: HTTP methods, server URI, API keys, etc. It is notable that they assume that IoT device is equipped with full ISO/OSI stack, and Internet connection. All of services rely on TCP/IP protocol.

In their next paper, same authors emphasize use of existing, conventional Web technologies to enable interaction between things. They recognize that each IoT platform may have different data structure, data representation format and APIs. Furthermore, this diversity may cause problems caused by challenging interoperability and data inconsistency. They investigated Automated Deployment, more specific TOSCA (Topology and Orchestration Specification for Cloud Applications) that is designed for describing IT services. TOSCA might be used for service deployment in IoT

environments [11]. After that they conducted an experiment with five IoT devices that used their SSD processor that automatically publishes semantic metadata to the Internet. Experiment results showed that their approach can be applied to many IoT devices on various platforms without significant effort.

Nordahl and Magnusson present PON (PalCom Object Notation). It is a lightweight format for data interchange that they developed to be used in IoT applications. PalCom is a middleware JAVA framework that simplifies creation of dynamic networks between devices in IoT environments. PON defines a format for structured data that combines the structure and compactness of JSON with efficient data handling defined by FORTRAN's string format: <Length><Type><Data> for example: 5sIOTIF. PON is purely textual data format and authors present translation scheme between PON and JSON. They demonstrated successful implementation in healthcare environment where PON is used to communicate with HTTP server that marshals all the devices that communicate with managing Database also using PON. In evaluation phase, they compared PON scheme with JSON, conclusion is that PON is 33% shorter than JSON and 70% faster than Gson (Java implementation) [12]. Ngu et al. have classified different architecture types of IoT middleware, they also analyzed composition, adaptability and security of IoT middleware. They emphasize that IoT service discovery is an issue, and propose non-ontological approach: search engine for heterogeneous IoT device and services: ServiceXplorer [13]. Their engine uses WSDL as service descriptor and searches for similarities in descriptions. They also suggest another non-ontological approach that is based on data analytics of IoT device usage log. After analysis, they try to reconstruct relationships between devices and services. This relationship map help with load balancing between similar IoT devices. Ngu et al. have also analyzed security and privacy issues, and they propose lightweight device authentication that uses low bandwidth and low CPU and other resources. It is based on public key cryptographic techniques like NTRU, ECC and AE. They conclude that IoT middleware needs to be designed with security, privacy and trust in the first place. Those concerns should be propagated throughout all of IoT layers. Also, they state that existing approach: using existing high level concepts is not good and that IoT ecosystem needs a dedicated and from-the-scratch approach.

III. DEVELOPING AN IoT ONTOLOGY

We have developed the open IoT ontology using Ontology Development 101 methodology [14] that answers to the following questions: What are the concepts to describe IoT devices and things as a service? How to support mappings of data types among the heterogeneous things and existing cloud services? There are many IoT ontologies in the current literature, but no one was suitable for our case and our implementation, so we have developed a new open IoT ontology. The existing IoT ontologies are shown in Table 1. The ontology is publicly available at:

https://github.com/dandrocec/IoTOntology.

As a basis for sensor descriptions in our ontology, we have used concepts defined in the W3C defined Semantic Sensor Network (SSN) ontology [15]. We have also used the actuator concepts from Semantic Actuator Network Ontology developed by Spalazzi et al. [16]. A total of 173 classes were defined that are organized in 20 top level classes (Figure 1). The developed ontology will be used to annotate things. The ontology is richer than any existing IoT ontology, because it contains concepts to annotate privacy, security and supported protocols as non-functional properties of things and their services. Main IoT security problems are listed as subclasses of *IoTSecurityProblems* OWL class and are derived from OWAPS IoT security guidelines [17]. IoT acceptance factors and service privacy factors are also listed in the ontology, as subclasses of the OWL classes *IoTAcceptanceFactors* and *ServicePrivacyFactors*. IoT protocols are defined as subclasses of the following OWL classes: *IoTDataProtocols*, *IoTDiscoveryProtocols*, *IoTInfrastructureProtocols*, and *IoTTransportProtocols*.

Table 1: Existing IoT ontologies

Authors	Short ontology description
Mrissa et al. [18]	The main classes of their ontology are Appliance, Capability, and Functionality. Each functionality is described in terms of composing functionalities and implementing capabilities in a common shared ontology
Wang et al. [19]	The main concepts in their ontology are sensor, physical capability (location, battery, platform, size, weight, working status, etc.), observation value (value, its precision, frequency, response model, sensing range), and measurement value range (measurement unit, quality, sampling method).
W3C SSN [15]	The Semantic Sensor Network (SSN) ontology describes the capabilities and properties of sensors, the act of sensing and the resulting observations.
Mathew et al. [20]	They classified things into four fundamental dimensions: identity (use of an appropriate identification systems), processing (functions which allow things to be controlled or managed), communication (thing's communication interface to enable interactions among things), and storage (the type and amount of information that a thing retains).
Nambi et al. [21]	Resource ontology represents an entity (sensors, actuators, physical objects, composite objects) in IoT, and it was developed as extension of the SSN ontology.
Spalazzi et al. [16]	They extended the SSN ontology with concepts and roles that describe actuators.

Table 2: Samples of subclasses of IoT discovery protocols

Class	Superclass	Description
HyperCat	IoTDiscovery Protocols	An open, lightweight JSON-based hypermedia catalogue format for exposing collections of URIs.
mDNS	IoTDiscovery Protocols	mDNS (multicast Domain Name System) - Resolves host names to IP addresses within small networks that do not include a local name server.

PhysicalWeb	IoTDiscovery Protocols	The Physical Web enables you to see a list of URLs being broadcast by objects in the environment around you with a Bluetooth Low Energy (BLE) beacon.
UPnP	IoTDiscovery Protocols	Universal Plug and Play

Table 3: Samples of some identified security problems described in the ontology

Class	Superclass	Description
UnnecessaryPortsAreOpen	InsecureNetworkServices	Unnecessary ports are opened
InsecureSoftwareFirmware	IotSecurityProblems	Insecure software firmware
DeviceUpdatesNotSigned	InsecureSoftwareFirmware	Device updates are not digitally signed
DeviceUpdatesTransmittedWithoutEncryption	InsecureSoftwareFirmware	Device updates transmitted without encrpytion
UpdateServersAreNotSecured	InsecureSoftwareFirmware	Update servers are not secured
InsecureWebInterface	IotSecurityProblems	Insecure web interface

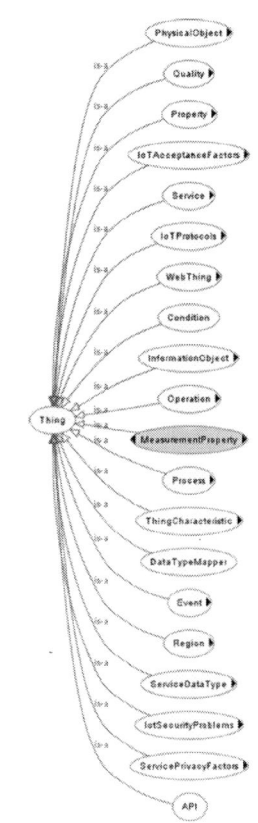

Figure 1: The main hierarchy of the IoT ontology

1287

IV. SEMANTIC ANNOTATION OF THINGS

Thing will be annotated using concepts from the ontology by means of JSON-LD. JSON-LD [22] is a simple method to add semantics to existing JSON documents. Semantic data is serialized in a way that is often indistinguishable from traditional JSON [18]. "The basic idea of JSON-LD is to create a description of the data in the form of a so-called context. It links objects and their properties in a JSON document to concepts in an ontology. A context can either be directly embedded in a JSON-LD document or put into a separate file and referenced from different documents" [22]. We list one JSON-LD example below:

```
{
 "@context":
"http://www.foi.unizg.hr/ontologies/ThingAsAServ
iceOntology.owl",
 "@type": "Sensor",
 "name": "TMP36",
 "type": "Analog Temperature Sensor",
}
```

Other interesting related work related to our use case is W3C Member Submission: Web Thing Model [23]. The mentioned document describes a model and API for things (physical objects) that will be connected to the Web of Things. They also recommend use of JSON-LD to support semantic extensions and to semantically describe things. Sample extract of one possible Web Thing model [23]:

```
{
 "id": "pi",
 "name": "My WoT Raspberry PI",
 "description": "A simple WoT-connected
Raspberry PI",
 (…)
    "properties": {
      "link": "/properties",
      "title": "List of Properties",
      "resources": {
        "temperature": {
          "name": "Temperature Sensor",
          "description": "An ambient temperature
sensor.",
          "values": {
            "temp": {
              "name": "Temperature sensor",
              "description": "The temperature in
celsius",
              "unit": "celsius",
              "customFields": {
"gpio": 21
              }
           }
         }
 (..)
}
```

V. DEVELOPING A PROTOCOL

After developing the possible concept for the interoperability part for our use case, we wanted to develop a protocol that enabled simple and secure communication between multiple MCUs. We started our research by analyzing the lowest set of common features that we have available on popular ESP8266 [1] based MCUs. One of the problems we encountered is that a lot of MCUs and their firmware, where we mean the development scaffolding - bootloader and SDK or the bootloader and the high-level language base implementation on the MCU, don't support ad-hoc or infrastructure mode for 802.11 wireless communication. The only two modes that are usually available are station mode (STA) and Access point mode (AP). Other limitations that we found include:

1. No specific support for multithreading or multiprocessing. The lack of support for multiprocessing or multithreading makes any parallel tasks difficult. There are attempts to implement MT for the last 4 years into Micropython [2], but unfortunately such attempts are mostly unsuccessful and result in very buggy implementations [24], Where there is one implementation from March 2017 which looks promising [25], [26]. One possible approach was to write a MCU specific task scheduler, like [27] which was scrapped because we wanted a minimal, portable solution that could work without any specific bootloader/language combination or any other low level modifications. There is an available Micropython support for uasyncio [28] but it's not available on ESP8266 [29]. This was our biggest issue which mostly guided our design. One interesting case is that the 802.11 radio of the device can be in both STA and AP mode at the same time, but can't parallelize the operation of both modes. The MCU will send 802.11 AP regardless if any other code is running on the device. This is because the 802.11 subsystem has one dedicated MCU core that will handle such low-level communication operations but any other code like *userspace* code that isn't in the 802.11 PHY layer must be run on the second core.

2. There is no specific supported implementation for services like multicast DNS, LLDP (Link level discovery protocol), SSDP (Simple Service Discovery Protocol), UPnP or any other. Usually such implementations are way too heavy since most MCUs have limited storage space.

3. Multicast support and reliability depends on the firmware image used. In mixed mode environments, this can be a limitation, but we created a simple bypass to avoid implementing a DHCP like protocol, and still be able to use the default TCP/IP stack. If the desired MCU/Firmware combination supports reliable multicast, multicast is the preferred and simplest way to solve the addressing problem on such point to point connection with least hassle. If not, the WiFi interface on the MCU has an interface MAC address, which is broadcast with the SSID name and other information in AP beacon frames and can be seen by every other STA node in range. The first 3 octets of the MAC address are the organizationally unique identifier, while the last 3 octets are network interface controller (NIC) specific part. Each NIC suffix octet has the decimal value between 0 and 255,

same as an IP octet. This means we can easily map the NIC part to a RFC1918 24 bit private IP address by mapping (10.(NIC0).(NIC1).(NIC2)) and thus creating a set of addresses in the range of 10.0.0.0 to 10.255.255.255 with a /8 (255.0.0.0) netmask, which enables us to calculate and both IP addresses at the time the STA is connecting to the AP without resorting to DHCP. If two nodes have the same NIC suffix, both nodes can identify such a collision by comparing the NIC suffixes of both nodes at the time of connection. The resolution method for this problem is by adding +1 mod 255 to the last octet of the NIC suffix. This eliminates the need to use DHCP in an already constrained environment that only focuses on a point to point configuration at a single time.

To facilitate the protocol each node (N) needs to be pre-programmed with some information. One interesting piece of research is from [30] called "Flashing displays: user-friendly solution for bootstrapping secure associations between multiple constrained wireless devices." which shows we can use a cell phone or other device to securely program and key a device with only an added photodiode to the device. Each node has to contain the following information:

1. MeshID - A random 16 character Base58 string. MeshID is used to identify the mesh network. All nodes that want to communicate need to share the same MeshID. The AP interface SSID of the MCU is set to the value of MeshID and the network is set to Hidden. The reason of hiding the SSID name is not for security reasons, but just to remove the network name from a casual observer (like someone connecting to a normal wifi network).

2. MeshPW - A WPA2 pre-shared key, all nodes need to share the same key in order to connect to each other, ideally at least a 24 to 63 character random Base58 string.

3. ListenInterval - A pair of two values [LiMin, LiMax] defining the lowest and highest time a node will spend serving data.

4. TransmitInterval - A pair of two values [TiMin, TiMax] defining the lowest and highest time a node will spend trying to receive data.

5. Any cryptographic material used for secure communication phases.

Both values ListenInterval and TransmitInterval need to be chosen to account for several characteristics. Depending on the needed communication and data transfer frequency between nodes. For instance, one sensor network needs to send the measurement data to the broker twice daily, while a requirement for another network might be that the data is transmitted approximately every 30 minutes.

It depends on the system and sleep states (modem, low clock, light or deep sleep). In the case that light sleep or deep sleep states are used, all phases of the protocol need to be synchronized in all nodes of the network. Usually by selecting a time interval when the communication will

occur. This requires an RTC + battery and code for timekeeping. The needed time for the data exchange which usually needs to account for:

- The time needed for the MCU to discover and for the STA to connect to the AP (other MCU)

- The time needed for the data transfer to be initiated

- The amount of data that needs to be transferred over a low bandwidth channel.

When using the ListenInterval or TransmitInterval, each node stays in the listen or transmit state for a random time that won't be lower than the minimum bound or greater than the maximum bound. The reason for this choice is trying to remove possible overlaps where two nodes are in the same state at the same time and never exchange data. Our scheme tries to create a simple mesh based communication protocol with the following three main phases:

1. Node discovery and communication:
   a. Each node tries to discover other nodes sharing the same MeshID (SSID).
   b. For each discovered node with the same MeshID add the BSSID values of the node to a table (SeenIDs) sorted by RSSI.
   c. Each node also keeps the track of all seen nodes in a LastSeen table.
2. Data transfer between nodes:
   a. Data is exchanged with the help of an embedded web server running on the MCU. Communication is done with the help of a RESTful service since json encoding/decoding, HTTP requests and a simple HTTP server is available in the standard library for most implementations of alternate firmware on MCUs.
3. Processing phase:
   a. Arbitrary data processing phase like collecting data from sensors, processing data or running any other arbitrary code.

Protocol description:

The protocol can be described with the following pseudocode. Each node in the network runs the following algorithm:

```
Run the network discovery phase
Randomly pick if the node will be an AP or STA
    if STA mode:
        Stay in STA mode for a random time
        between TransmitInterval
        if there is an available AP in SeenIDs:
            Try to connect to AP, if connected,
            access the webserver on the other node
            Remove AP from SeenIDs table
            Update the LastSeen table
        else:
            Run network discovery phase
            Try to connect to any other device
    If AP mode:
        Stay in AP mode for a random value inside
        RecieveInterval
        Enable the local webserver and accept
        connections
```

The core logic behind this method is that in a long enough timeframe every pair of nodes will stay in both AP and STA mode and at some time. The random intervals are used to make sure that each two nodes can't overlap enough to miss connecting to each other. This enables the system to converge on a long enough timeframe.

Protocol security

The network security part of the system is reduced down to the security of the WPA2 protocol implemented in the MCUs and the strength of the WPA2 passphrase. The problem with WPA2 connections is that each MCU and firmware stack is implementing the WiFi security protocols on its specific way and isn't reusing a common security library like wpa_supplicant, so an audit of security protocol implementations for common MCU/firmware pairings might be interesting future research path for a professional paper. The current state of SSL/TLS implementations in MCUs is not in a stable phase and since getting TLS right is hard even in standard systems as we can see with the numerous problems of vulnerabilities in standard, wide use libraries like OpenSSL [31], the whole security of the TLS stacks in embedded MCUs is questionable. The integrity and security of communications in reduced to the security of WPA2, and we can't guarantee reliable communications in such lightweight system. One problem is always present in IoT components: what if an attacker obtains physical access to a device, for example if an attacker steals one sensor node or one device? Such an option opens a large attack surface for an attacker to exploit and enables a large number of attacks that can lead to full compromise of the device [32]. In our approach, we tried to set a tradeoff between known security, and ease of implementation. If additional amount of communications security is needed, approaches such as [13], [33] and others can be used on top of our implementation.

VI. CONCLUSION

In this paper, we have explored some possibilities for interoperability and communication between cheap IoT devices. We have developed an open IoT ontology richer than any existing IoT ontology, because it contains concepts to annotate privacy, security and supported protocols as non-functional properties of things and their services. To semantically annotate things using concepts from defined ontology, we propose usage of JSON-LD to extend the Web Thing model [19]. In this way, we can semantically annotate things, their services, and their functional and non-functional properties. Next, we have presented a use-case for lightweight communication and security of low cost MCUs (MicroController Units). Additionally, the lightweight communication protocol for cheap MCUs that can be used on almost all devices, was developed in this work. We plan to further work on IoT security and interoperability problems. Specifically, we plan to use and extend our ontology, semantic annotations approach, and lightweight communication protocol in a number of use cases and on different devices (things).

REFERENCES

[1] Espressif, ESP8266: https://espressif.com/en/products/hardware/esp8266ex/overview Accessed 30 Jan 2017

[2] Micropython: https://micropython.org/ Accessed 30 Jan 2017

[3] NodeMCU: http://nodemcu.com/index_en.html Accessed 30 Jan 2017

[4] Espressif, ESP-MESH: http://espressif.com/products/software/esp-mesh/overview Accessed 30 Jan 2017

[5] J. Boman, J. Taylor, and A. Ngu, "Flexible IoT Middleware for Integration of Things and Applications," Proc. 10th IEEE Int. Conf. Collab. Comput. Networking, Appl. Work., no. CollaborateCom, pp. 481–488, 2014.

[6] G. Cassar, P. Barnaghi, W. Wang, and K. Moessner, "A hybrid semantic matchmaker for IoT services," in Proceedings - 2012 IEEE Int. Conf. on Green Computing and Communications, GreenCom 2012, Conf. on Internet of Things, iThings 2012 and Conf. on Cyber, Physical and Social Computing, CPSCom 2012, 2012, pp. 210–216.

[7] M. de Souza Lima, A. de Ribamar L. Riberio, and E. David Moreno, "Proposal of a Standard Vocabulary for Services Discovery on the Internet of Things," Proc. 11th Int. Conf. Web Inf. Syst. Technol., pp. 129–134, 2015.

[8] Y. Zhang, L. Duan, and J. L. Chen, "Event-driven SOA for IoT services," in Proceedings - 2014 IEEE International Conference on Services Computing, SCC 2014, 2014, pp. 629–636.

[9] K. Hur, S. Chun, X. Jin, and K. H. Lee, "Towards a Semantic Model for Automated Deployment of IoT Services across Platforms," in Proceedings - 2015 IEEE World Congress on Services, SERVICES 2015, 2015, pp. 17–20.

[10] RFC5988, Available online 28.3.2017 at https://tools.ietf.org/html/rfc5988

[11] K. Hur, X. Jin, and K. H. Lee, "Automated deployment of IoT services based on semantic description," in IEEE World Forum on Internet of Things, WF-IoT 2015 - Proceedings, 2016, pp. 40–45.

[12] M. Nordahl and B. Magnusson, "A lightweight data interchange format for internet of things with applications in the PalCom middleware framework," J. Ambient Intell. Humaniz. Comput., vol. 7, no. 4, pp. 523–532, 2016.

[13] A. H. H. Ngu, M. Gutierrez, V. Metsis, S. Nepal, and M. Z. Sheng, "IoT Middleware: A Survey on Issues and Enabling technologies," IEEE Internet Things J., vol. X, no. X, pp. 1–1, 2016.

[14] Noy NF, McGuinness DL. Ontology Development 101: A Guide to Creating Your First Ontology [Internet]. Stanford University; 2001 [cited 2013 Jun 30]. Available from: http://www-ksl.stanford.edu/people/dlm/papers/ontology-tutorial-noy-mcguinness.pdf

[15] M. Compton, P. Barnaghi, L. Bermudez, R. García-Castro, O. Corcho, S. Cox, J. Graybeal, M. Hauswirth, C. Henson, A. Herzog, V. Huang, K. Janowicz, W. D. Kelsey, D. Le Phuoc, L. Lefort, M. Leggieri, H. Neuhaus, A. Nikolov, K. Page, A. Passant, A. Sheth, and K. Taylor, "The SSN ontology of the W3C semantic sensor network incubator group," Web Semant. Sci. Serv. Agents World Wide Web, vol. 17, pp. 25–32, Dec. 2012.

[16] L. Spalazzi, G. Taccari, and A. Bernardini, "An Internet of Things ontology for earthquake emergency evaluation and response," 2014, pp. 528–534.

[17] OWASP (2016) IoT Security Guidance. In: IoT Secur. Guid. https://www.owasp.org/index.php/IoT_Security_Guidance. Accessed 30 Jan 2017

[18] M. Mrissa, L. Medini, and J.-P. Jamont, "Semantic Discovery and Invocation of Functionalities for the Web of Things," in 2014 IEEE 23rd International WETICE Conference, 2014, pp. 281–286.

[19] X. Wang, H. An, Y. Xu, and S. Wang, "Sensing Network Element Ontology Description Model for Internet of Things," in 2015 2nd International Conference on Information Science and Control Engineering (ICISCE), Shanghai, 2015, pp. 471–475.

[20] S. S. Mathew, Y. Atif, Q. Z. Sheng, and Z. Maamar, "Web of Things: Description, Discovery and Integration," in 2011 International conference on Internet of Things and 4th International Conference on Cyber, Physical and Social Computing, Dalian, 2011, pp. 9–15.

[21] S. N. A. U. Nambi, C. Sarkar, R. V. Prasad, and A. Rahim, "A unified semantic knowledge base for IoT," in 2014 IEEE World Forum on Internet of Things (WF-IoT), Seoul, 2014, pp. 575–580.

[22] M. Lanthaler and C. Gütl, "On using JSON-LD to create evolvable RESTful services," 2012, p. 25.

[23] V. Trifa, D. Guinard, and Da. Carrera, "Web Thing Model," W3C, W3C Member Submission, Aug. 2015.

[24] https://github.com/micropython/micropython/issues/595 Accessed 30 Jan 2017

[25] https://www.pycom.io/news/qa-micropython-multi-threading-garbage-collector/ Accessed 30 Jan 2017

[26] https://forum.micropython.org/viewtopic.php?t=1864 Accessed 30 Jan 2017

[27] https://github.com/peterhinch/Micropython-scheduler Accessed 30 Jan 2017

[28] https://github.com/peterhinch/micropython-async Accessed 30 Jan 2017

[29] https://docs.micropython.org/en/latest/esp8266/library/index.html Accessed 30 Jan 2017

[30] Tonko Kovačević and Toni Perković and Mario Čagalj. "Flashing displays: user ‐ friendly solution for bootstrapping secure associations between multiple constrained wireless devices." Security and Communication Networks (2015)

[31] https://www.openssl.org/news/vulnerabilities.html Accessed 30 Jan 2017

[32] https://www.owasp.org/index.php/OWASP_Internet_of_Things_Project#tab=IoT_Vulnerabilities Accessed 30 Jan 2017

[33] https://tools.ietf.org/html/rfc7539 Accessed 30 Jan 2017

Security and Privacy Issues for an IoT based Smart Home

Dimitris Geneiatakis, Ioannis Kounelis, Ricardo Neisse, Igor Nai-Fovino
Gary Steri, and Gianmarco Baldini
European Commission, Joint Research Centre (JRC)
Cyber and Digital Citizens' Security Unit
Via Enrico Fermi 2749, 21027 Ispra, Italy
Email: {firstname.surname}@ec.europa.eu

Abstract—Internet of Things (IoT) can support numerous applications and services in various domains, such as smart cities and smart homes. IoT smart objects interact with other components *e.g.,* proxies, mobile devices, and data collectors, for management, data sharing and other activities in the context of the provided service. Though such components contribute to address various societal challenges and provide new advanced services for users, their limited processing capabilities make them vulnerable to well-known security and privacy threats. Until now various research works have studied security and privacy in IoT, validating this claim. However, to the best of our knowledge literature lacks research focusing on security and privacy flaws introduced in IoT through interactions among different devices supporting a smart home architecture. In particular, we set up the scene for a security and privacy threat analysis for a typical smart home architecture using off the shelf components. To do so, we employ a smart home IoT architecture that enables users to interact with it through various devices that support smart house management, and we analyze different scenarios to identify possible security and privacy issues for users.

I. INTRODUCTION

The development of new type of sensors and actuators combined with the deployment of increasingly powerful and pervasive network connectivities is shaping the concept of the Internet of Things (IoT). Several factors are contributing to the evolution of the current Internet into IoT including the lower market price of IoT devices and the higher demand of customers for new services.

Manufactures are now able to provide mobile, wearable or embedded devices with more memory, processing power, and more diverse sensing technology. As a consequence, this increased capability of IoT devices also increases the amount of data available to services and their value to end users. However, even if IoT is capable of supporting new business models, increasing the efficiency of many applications, and enriching the life of citizens with new services, the risks are also significantly higher. The collection of even larger amount of data and merging of the cyber and physical world implies a higher number of privacy and safety issues than the cyber-only Internet.

More specifically our focus in this paper is on a *Smart Home* scenario. In this scenario, the potential for privacy breaches is limited if we consider the direct and explicit collection of data regarding the individuals living in the house. However, the activities of these individuals can be indirectly tracked through the observation of the cyber and physical activities of their connected domestic devices, assisted living systems, or smart meters. The protection of privacy in these complex scenarios where different entities and IoT technologies coexist and work together requires new approaches and solutions. Even if various Privacy Enabling Technologies (PET) have been proposed in literature, their market adoption is still relatively weak and many concrete threats still persist.

In this paper we set up the scene for a security and privacy threat analysis for a typical smart home architecture that relies on existing and readily available market IoT devices and platforms. In contrast to existing security and threat analysis of IoT scenarios, we target a real IoT smart home environment deployed in our testbed focusing on the interactions among the different IoT components. In this architecture, we identify points of interest that an adversary might manipulate either to gain access to unauthorized information or to cause a denial of service. Our contribution, in addition to a concrete threat analysis, is a practical feasibility evaluation of the identified vulnerabilities showing how exploits can be implemented in practice.

The rest of the paper is structured as follows. In Section II we overview a smart home's architecture and in Section III we analyse its threat model. In Section IV we study the realization of the threat model in a test-bed architecture, and analyze possible consequences to end-users in terms of security and privacy. In Section V we provide guidelines and protection measures for eliminating the threats presence, while in Section VI we overview the related work. Finally, in Section VII we conclude this paper and present some pointers for future work.

II. IoT BASED SMART HOME ARCHITECTURE

Smart home can be defined as the symbiosis of different elements *i.e.,* sensors, connections, and applications that build a dynamic heterogeneous architecture with the aim of efficiently managing home devices, and providing to users advanced services.

Due to a still missing generic interoperability standard among IoT devices, in this architecture, without loss of generality, the IoT devices organised in islands are connected to a corresponding hub and are not directly accessed by other devices. Moreover, the majority of commercial sensors do not provide direct Internet connectivity; instead the intermediate

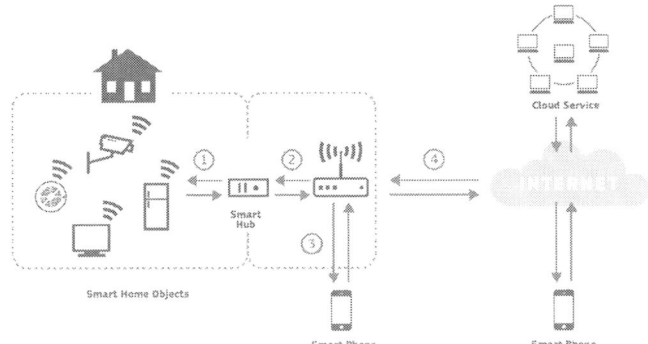

Fig. 1.   Architecture of a Smart Home

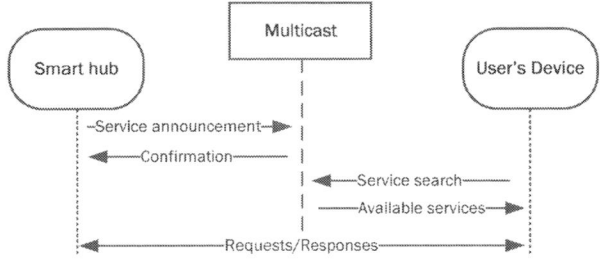

Fig. 2.   Plug and play architecture for smart house

hub is the component responsible of providing such connectivity.

The communication between the IoT devices and the hub is usually wireless, based on different protocols depending on the device's manufacturer. The most popular are:

- Zigbee[1];
- Z-wave[2].

The hub is then connected to the smart home's router either via an Ethernet or a Wi-Fi interface, depending on its capabilities in order to connect the IoT devices with the outside world.

Users can interact with IoT devices and manage their smart home through different platforms such as PCs, smart phones, and tablets. The interaction modalities are in general two:

1) directly interacting with them using the connectivity and services provided by the hub;
2) accessing Internet cloud services which interact with the IoT hub and the connected IoT devices.

These two scenarios are quite often present at the same time and mixed together to support local and remote interactions with IoT objects. In case of remote management all the information is forwarded to the smart hub through the cloud service, while if the user is acting from the same network where the smart hub is installed the traffic is routed directly to it and thus no Internet access is needed.

However, in order for users to enable IoT devices management, regardless of their location, they must first follow a procedure for correlating their devices with the corresponding hub. In most cases, for an off the shelf based solution to successfully complete this procedure user's physical action is involved, e.g., pressing a button on the smart hub.

Furthermore, IoT manufacturers support protocols such as the Simple Service Discovery Protocol (SSDP) [1] to enable the transparent configuration of the smart home devices in a plug and play mode that requires minimum user interactions. In this case, the smart hub generates a presence announcement

[1]http://www.zigbee.org/
[2]http://www.z-wave.com/

to the multicast channel, i.e. the default IoT devices' gateway. Any device that searches for available services sends to the multicast channel a discovery request and receives as a response the available requested services. Then the device can communicate directly with the newly discovered services. This procedure is illustrated in Figure 2.

So consider an example in which a user would like to control the smart home's lights status using his mobile phone, while he is outside the home. To do so, firstly he must have already successfully completed the correlation procedure, otherwise it is not possible to have remote access to the smart home's devices. Once setup, he can launch the mobile application and request the lights' status. This request reaches the cloud service which forwards it on behalf of the user to the hub that is responsible for controlling the lights, using a reverse communication channel kept open, through the house router, by the hub itself. As soon as the hub receives such a request, it sends the corresponding command, i.e., status, to the lights in order to receive back their response, and forwards it to the user via the cloud service. All the interactions between the different components of a smart house are illustrated in Figure 1.

III. SMART HOME THREAT MODEL

The formulation of a smart home's cyber threat model should consider two types of adversaries; internal and external entities that can act maliciously on a passive or an active way depending on their goal. On one hand, the former category consists of malicious entities that are located close or inside the smart home's premises. On the other hand, external adversaries can interact only through an Internet connection. In both cases adversaries target either the smart home's infrastructure, or the information stored in the related cloud services. Note that in this work we do not consider adversaries having physical access to IoT components. This is because, cyber-sphere entities have only virtual access to the components of a provided service.

In this context, similar to any other IP based service, adversaries acting passively will try to eavesdrop available communications in order to acquire information that can either be used to monitor users' behaviour or can be accumulated and exploited in a later step of an active attack. Adversaries, to access this type of information, will try to capture the traffic in different points of the smart home architecture depending on their capabilities and goals. In this way, adversaries could impact users' confidentiality and privacy as they can collect information related to the smart home's status. For instance,

1293

taking as example the architecture shown in Figure 1, if the adversary monitors the communication link (1) between the smart hub and the router in the smart house premises, he could identify which entities the smart hub communicates with, while if he monitors the communication link (4) he can deduce the users daily habits *e.g.,* when the lights turn on and off that might correspond to user's absence from home.

On the other side, an active adversary will interact actively with the IoT components, instead of only eavesdropping the underlying communication. He could identify the existence of components by generating the appropriate probes through different network devices. Moreover, the adversary could try to impersonate a legitimate user in order to gain access to the smart home devices. Then he could be able to control them, use them or even extract sensitive information from them. An active adversary could impact not only users privacy and the provided service's confidentiality but also affect data integrity, gain unauthorized access and ultimately disrupt the proper function of provided services.

Obviously in more complex scenarios an adversary could combine a passive and active attack. Consider, for instance the case where a smart socket provides electricity to a health device. If the adversary knows its unique identifiers by eavesdropping the communication traffic, he could cause a denial of service to the IoT that could have an immediate impact on the users safety. For that reason this type of information should be considered of high importance.

Besides the passive and active network layer threats, software exploitation is another aspect that an active adversary will capitalize in order to gain access to otherwise private domains. This is because, IoT relies on light weight versions of well known operating systems that adversaries look to exploit with very few resources. Moreover, most of the largely used IoT devices have a corresponding mobile application that acts as a controller. The mobile application's execution environment could be used as an attack vector for an adversary as he could be in position to exploit well-known software vulnerabilities of the underlying operating system.

Table I overviews the threats and the possible consequences that they can have to a smart home's infrastructure.

IV. IoT BASED SMART HOME CYBER-FLAWS

To study the feasibility of the generic threats reported in Section III we deployed a test-bed architecture similar to the one illustrated in Figure 1. This architecture is based on commercial products that provide connectivity to IoT sensors through a smart hub that is connected to a network and to

TABLE I.   IoT SMART HOME THREATS OVERVIEW

Type	Threat	Impact	Target
Passive	Eavesdropping	Confidentiality	User-IoT
Passive	Eavesdropping	Privacy	User
Active	DoS	Availability	User-IoT
Active	Impersonation	Integrity	User
Active	Impersonation	Availability	User-IoT
Active	Impersonation	Unauthorized access	User
Passive	Software exploitation	Confidentiality	User-IoT
Passive	Software exploitation	Privacy	User
Active	Software exploitation	Integrity	User
Active	Software exploitation	Availability	IoT

the Internet via a traditional wireless router. Specifically, a computer (A) supported with a WiFi network card and Internet connection is configured as the access point to which the mobile device with the IoT management apps is connected to. In a second computer (B), which is connected to computer's A WiFi for having Internet, the IoT device or Hub (depending on the device type) is attached. This way, computer A monitors the traffic shown in point 3 and 4 of Figure 1, while computer B monitors the traffic of point 2. By running wireshark on both computers we are able to capture all packets passing through the specific points.

As our goal is to illustrate the security and privacy issues of IoT in general, and not to criticize a specific product, we do not provide any related information for it. Note that, as we are interested in studying the interaction between the different components, we assume that a powerful adversary [2] can get access to the underlying communication *e.g.,* by exploiting a specific device or protocol vulnerability, using default or common WiFi and router passwords, cracking insecure passwords, social engineering, *etc.,* however, such an analysis is out of scope of this work.

In the following subsections we discuss the implementation of the threats described in Section III considering information that we gather from our test-bed architecture as well as other related research reports.

A. Eavesdropping

An adversary might use different tools and techniques for capturing the traffic among the different components of an IoT based smart home infrastructure, considering its heterogeneous architecture. These techniques are highly related with the attacker's location and capabilities.

If the attacker manages to connect to smart home network components, *e.g.,* adsl router, he is able to capture all the traffic between the smart hub and the local or the remote users; that corresponds to reference points 2, 3 and 4 of Figure 1. In that case the adversary relies on well known tools such as tcpdump[3], wireshark[4], *etc.,* to gain access to the data. In case the communication is wireless, the adversary might use specific hardware equipment, for instance the WiFi Pineapple[5], that can spoof access points and intercept the underlying communication; this corresponds to reference points 1, 2, 4 of Figure 1.

So taking into consideration an adversary that intercepts the traffic among the reference points 1,2 and 3, he can identify:

1) whether or not the smart hub communicates with a cloud supported service *e.g.,* (cloud.iot.com:80)
2) the user device's type (*e.g.,* Linux, Android 7.1.1; Nexus 5X Build/NMF26F)
3) unique identifiers for user's access to the smart hub services
4) the device's status through traffic analysis
5) the smart hub's operating system (*e.g.,* unix like OS)
6) methods that can be used to send commands to the smart hub (*e.g.,* POST, GET, DELETE, *etc.*)

---

[3]http://www.tcpdump.org/
[4]https://www.wireshark.org/
[5]https://wifipineapple.com/

```
NOTIFY * HTTP/1.1
HOST: 239.255.255.250:1900
CACHE-CONTROL: max-age=100
LOCATION: http://192.168.1.24:80/description.xml
SERVER: Ubuntu UPnP/1.0
NTS: ssdp:alive
BridgeId: 40285567791489150
```

Listing 1. An SSDP NOTIFY message example for service announcement

Listing 1 illustrates the example of an eavesdropped message during a smart hub service announcement in the local network. In this case the adversary can deduce the operating system that the smart hub runs, its IP address, and its unique ID. In this point it should be mentioned that this information might be slightly changed depending on the IoT manufacturer.

B. Impersonation

An adversary could try to impersonate and act on behalf of a legitimate user. To do so, he requires either access to users' credentials or to any other information that provide access to the IoT resources. The former case is used in IoT architectures requiring access to the IoT devices remotely, while the latter is normally used to access the IoT resources from the local network (an example of key information needed in this case is for instance the unique identifier that the smart hub generates during device registration for enabling local access). Note that the smart hub often recognizes the location of the user based on his IP address. So if the adversary captures such a message, he can impersonate the user and can consequently interact with the smart hub on behalf of the user. This can be achieved by simply crafting the appropriate request towards the smart hub's resources with the appropriate parameters, *e.g.,*: `http://ip-address/api/unique-id/rsrc`

Recall that the adversary can intercept the unique identifier during an eavesdropping attack. In this category we also classify replay attacks as the adversary reuses previous requests either towards the hub or the cloud based service so that the user does not have the proper information about the status of his devices. Though this type of attack assumes that the adversary acts on the same network that the smart hub belongs to, it introduces a vulnerability in the smart home architecture in case that the adversary can reach the smart home router from outside world.

C. DoS

Similarly, to other IP based services, the adversary using different techniques might try to cause a Denial of Service (DoS) either to the hub or to the sensors themselves. As the adversary knows the smart hub's IP through eavesdropping, he can easily launch a single DoS or a Distributed DoS (DDoS) attack against it by simply sending numerous requests to it. Note that as IoT devices rely on low capabilities processing hardware they are susceptible to low rate DoS [3] as well.

Alternatively, the adversary might try to craft specific messages *e.g.,* malformed messages, so that the provided service cannot process them properly and cause either a DoS or provide unauthorized access.

DoS can also take place directly at the IoT devices, without passing through the smart hub. An adversary having the appropriate hardware that enables him to use the IoT devices' protocols can send directly messages to them and attempt to interfere with their proper functioning.

Finally, a DoS attack can also take place at the router or the cloud service. This may be a general attack, not linked to IoT, but the consequences for the end user would be the same. Without a working router or a could service, he will not be able to access his smart home's IoT devices through the Internet.

D. Software exploitation

Malicious software (malware) can affect the IoT services and devices. As IoT devices run autonomously light weight versions of well known operating systems adversaries will search for software vulnerabilities and exploit them to gain access to otherwise private information.

However, currently IoT is becoming an attack target for executing DoS in order to increase the amplification factor of the generated traffic to break down a target. For instance, IoT devices have been exploited for launching a DoS attack [4] against DNS servers in order to paralyze Internet access. In such cases the adversaries exploit the fact that these devices are running over the Internet with default configurations, *e.g.,* default passwords, while most of these devices are not patched in most of the cases against security flaws.

The malware can reach the IoT devices through different channels:

1) Device Acquisition: When the device is bought by an end user there is a risk of buying malware infected devices. For instance, an adversary could purchase many new devices, infect them, and sell them to users through online auction sites (*e.g.,* eBay[6]).

2) Firmware Upgrade and Trusted Boot: Orthogonal to device infection during acquisition, adversaries might be able to upgrade IoT firmware with a malicious acting version. For instance, this was a channel that adversaries exploited in the case of Mirai malware [5].

3) Apps and services: Users control their IoT devices through corresponding applications and online services. Lately, the most common way to manage one's devices is to use a mobile application on the user's mobile device. Almost all manufacturers provide such applications that can be downloaded directly from the mobile operating system's application market. Since these applications are executed on the user's personal device, they can be infected by malware that is already present on the device, or exploited directly by an adversary that takes advantage of either the mobile application's or the operating system's vulnerabilities.

Moreover, several IoT manufacturers assume that they can delegate security on the smart home's underlying architecture, such as the router's firewall. However, this assumption should not be taken for granted. As Sivaraman *et al.* demonstrate with their work [6], the security measures deployed at one's

---

[6]http://www.ebay.com/

smart home could easily be bypassed with a malware mobile application.

Similarly to mobile applications, online services that interact directly with the smart hub could pose a weak point in the IoT architecture chain. If one manages to access them through standard web services attack methods, he could then easily manipulate all connected IoT devices.

V. Discussion

Considering the above mentioned discussion in current smart home architectures, adversaries can gain access to underlying infrastructure and exploit it. As a result, smart homes should deal with security "flaws" in the same manner other IP based services with advanced resources do.

So in the case of smart home, adversaries can eavesdrop the underlying communication and extract different information due to the lack of an end-to-end encryption between the different components of IoT. This is also a flaw for the existing protocols that IoT builds on; for instance, the SSDP protocol does not use any encryption and thus the adversary could exploit this fact and identify available smart hubs and their capabilities.

Furthermore, the current access control approaches that smart home deployments follow, *e.g.,* generating unique identifiers during correlation of user's device with the smart hub, expose IoT services to impersonation attacks as the adversary can eavesdrop the unique identifier, and use it for future attack attempts. However, these types of attacks can be mitigated if the appropriate integrity and authentication mechanisms are deployed.

Protection against DoS and their distributed counterpart (DDoS) is a challenging task especially for IoT architecture considering its limited capabilities, while currently we even lack effective solutions for IP based services that are supported by high power security infrastructures. To the best of our knowledge, only research related solutions such as [7] have been proposed for the protection of IoT against DoS attacks. However, such approaches do not focus on the application layer, but are mainly dealing with network layer protection.

As the majority of low cost IoT manufacturers do not usually consider mechanisms for validating firmware integrity during installations, upgrades and on execution, for instance using a trusted boot, IoT devices are exposed to possible software flaws. To eliminate software exploitation users should also use applications and services provided through well known channels, as unknown third party applications can manipulate the existing infrastructure introducing backdoors for future attacks.

One major issue is the possibility to deploy IoT installations using the default configurations. As explained in the previous section this is what made the DNS attack [4] possible. A recommendation in this could be to force users to properly configure the devices, otherwise the services cannot be started (*i.e.,* routers cannot have ports open by default, remote login can be enabled only if you set a strong password, IoT services can be started and accessed only if a good password was set.) Next to this, for what concerns wireless communications, open connections should not come as default configuration, the router can be started only if the default password was changed, vulnerable protocols should be deprecated and removed from configuration options.

VI. Related Work

Jacobsson *et al.* in [8], [9] presents the results of a risk analysis of a smart home automation system developed in a research project involving leading industrial actors. Their architecture was identical to the one discussed in this paper including sensors/devices, in-house gateway, cloud server, mobile devices and apps. The risk analysis was performed during collaborative workshop sessions with nine persons including security experts, domain experts, and smart home system developers. The discussion was organized using an open information security risk assessment questionnaire used to reason, identify, analyze, and evaluate threats. The identified threats were linked to the respective system vulnerabilities and the corresponding probability, likeliness of occurrence, and potential impact associated with each threat was estimated by each participant using a five level scale (1-5) from unlikely/negligible to likely/disastrous. The risk analysis results were organized in five categories relating to: software, hardware, information, communication, and human-related risks. The higher ranking risks in each category were related respectively to the software security in apps and APIs, inadequate physical security, inadequate access control policy/mechanisms, inadequate authentication and confidentiality, and poor password management. The results of their risk analysis are presented in a high-level and are in-line with our findings in this paper. Furthermore, in their work the a main observation was the need of empirically based methods that support the evaluation of risks in smart home environments, which is precisely the focus of this paper.

Kozlov *et al.* in [10] describe a threat analysis for an overall IoT architecture including security, privacy, and trust issues. Their threat analysis is mostly a high-level selection of threats discussed in the literature considering many scenarios and application domains (e.g., smart home, road transportation, smart energy meters, and mobile apps). The scenarios discussed illustrate threats to personal data privacy, availability, and also safety, for example, when an exploited vulnerability in a road traffic system could cause an accident. In contrast to their work, the analysis performed in this paper is more concrete with respect to the threats and vulnerabilities identified, and is also focused specifically on a deployed smart home scenario. We show not only the high-level issues but also demonstrate how they can be realized by attackers in real IoT devices.

Perera et al. propose in [11] a privacy-by-design framework for assessing IoT applications and platforms, which is proposed as a systematic method to guide privacy analysis and design in IoT based on a set of 30 guidelines. Each guideline should be applied during different phases of the data lifecycle including consent and data acquisition, data pre-processing, data processing and analysis, data storage, and data redistribution. The major privacy risks addressed by the guidelines are unauthorized access and secondary usage of information, meaning the use of already collected data for purposes not initially allowed by the data owners. The authors show the application of their framework in two open source IoT middleware platforms: OpenIoT and Eclipse SmartHome. For each middleware a table was constructed showing if the guideline is supported, extendible, or not supported considering each of the phases in the data lifecycle. An extendible support means the middleware provides a plug-in mechanism that could make it straightforward to implement the functionality. By comparing the support for each guideline it is possible to compare the middleware with respect to their privacy-by-design features and gaps; in a sense more compliance to the guidelines implies a lower privacy risk. In contrast to the technical contributions in this paper the proposed guidelines are more abstract (*e.g.,* data anonymization, encryption, *etc.*) and can be mapped to the threats and vulnerabilities detailed in this paper.

Ziegeldorf *et al.* in [12] classify and examine RFID privacy threats in a broad sense with the goal of presenting relevant challenges that should be overcome in future deployments. In their reference model, they consider the collection of information by devices in the user environment, the processing and dissemination of the information by services that exploit the RFID technology. In their analysis they list the main threats to privacy, namely: identification, localization, tracking, profiling, privacy-violating interaction and presentation, decommissioning of devices, inventory attacks, and linkage of RFID related components. As a result of their analysis profiling is considered the most severe threat. In contrast to our approach this work focuses on RFID based IoT systems, while we concentrate on smart home components threat analysis.

VII. CONCLUSIONS & FUTURE WORK

IoT architectures will be an important component of future Internet as it closes the gap between physical and virtual objects. Among others, smart home is one of the main developments of IoT environments as it enhances the user's experience when using home devices.

Albeit the advantages that IoT offers to smart home users do not only expose homes to well known attacks but also the (IoT) sensors should deal with flaws that have not been previously considered. This is due to the fact that such devices are of limited processing power, and rely on heterogeneous network architectures that increase the attack surface of the provided service.

In this paper, we introduced a smart home threat model, and analysed it in our test bed architecture considering off the shelf components. Our initial analysis demonstrates that existing smart home IoT infrastructure could be vulnerable to eavesdropping, impersonation, DoS and software exploitation attack vectors under specific conditions *e.g.,* considering an attacker that manages to get access to the underlying network.

Currently, we are planning to extend our analysis demonstrating in detail the consequences and the impact of the different threats on users and the IoT infrastructure, as well as introduce the appropriate countermeasures for enhancing smart home security. In this direction, we envisage a framework that is able to automatically identify vulnerable points in a smart home architecture.

REFERENCES

[1] U. forum Members, "UPnP Device Architecture 1.1." [Online]. Available: http://upnp.org/specs/arch/UPnP-arch-DeviceArchitecture-v1.1.pdf

[2] J. King, K. Lakkaraju, and A. Slagell, "A taxonomy and adversarial model for attacks against network log anonymization," in *Proceedings of the 2009 ACM Symposium on Applied Computing*, ser. SAC '09. New York, NY, USA: ACM, 2009, pp. 1286–1293. [Online]. Available: http://doi.acm.org/10.1145/1529282.1529572

[3] "LOIC." [Online]. Available: https://sourceforge.net/projects/loic/

[4] M. Smith, "IoT botnets used in unprecedented DDoS against Dyn DNS; FBI, DHS investigating," Oct. 2016. [Online]. Available: http://www.networkworld.com/article/3134093/security/iot-botnets-used-in-unprecedented-ddos-against-dyn-dns-fbi-dhs-investigating.html

[5] "Hacker Claims To Push Malicious Firmware Update to 3.2 Million Home Routers." [Online]. Available: https://motherboard.vice.com/en_us/article/hacker-claims-to-push-malicious-firmware-update-to-32-million-home-routers

[6] V. Sivaraman, D. Chan, D. Earl, and R. Boreli, "Smart-phones attacking smart-homes," in *Proceedings of the 9th ACM Conference on Security & Privacy in Wireless and Mobile Networks*, ser. WiSec '16. New York, NY, USA: ACM, 2016, pp. 195–200. [Online]. Available: http://doi.acm.org/10.1145/2939918.2939925

[7] P. Kasinathan, C. Pastrone, M. A. Spirito, and M. Vinkovits, "Denial-of-service detection in 6lowpan based internet of things," in *2013 IEEE 9th International Conference on Wireless and Mobile Computing, Networking and Communications (WiMob)*, Oct 2013, pp. 600–607.

[8] A. Jacobsson, M. Boldt, and B. Carlsson, "A risk analysis of a smart home automation system," *Future Generation Computer Systems*, vol. 56, pp. 719 – 733, 2016. [Online]. Available: http://www.sciencedirect.com/science/article/pii/S0167739X15002812

[9] A. Jacobsson and P. Davidsson, "Towards a model of privacy and security for smart homes," in *2015 IEEE 2nd World Forum on Internet of Things (WF-IoT)*, Dec 2015, pp. 727–732.

[10] D. Kozlov, J. Veijalainen, and Y. Ali, "Security and Privacy Threats in IoT Architectures," in *Proceedings of the 7th International Conference on Body Area Networks*, ser. BodyNets '12. ICST, Brussels, Belgium, Belgium: ICST (Institute for Computer Sciences, Social-Informatics and Telecommunications Engineering), 2012, pp. 256–262. [Online]. Available: http://dl.acm.org/citation.cfm?id=2442691.2442750

[11] C. Perera, C. McCormick, A. K. Bandara, B. A. Price, and B. Nuseibeh, "Privacy-by-Design Framework for Assessing Internet of Things Applications and Platforms," in *Proceedings of the 6th International Conference on the Internet of Things*, ser. IoT'16. New York, NY, USA: ACM, 2016, pp. 83–92. [Online]. Available: http://doi.acm.org/10.1145/2991561.2991566

[12] J. H. Ziegeldorf, O. G. Morchon, and K. Wehrle, "Privacy in the internet of things: threats andchallenges," *Security and Communication Networks*, vol. 7, no. 12, pp. 2728–2742, 2014. [Online]. Available: http://dx.doi.org/10.1002/sec.795

Towards Overall Information Security and Privacy (IS&P) Taxonomy

K. Solic[1], H. Ocevcic[2], I. Fosic[3], I. Horvat[4], M. Vukovic[*5] and T. Ramljak[6]

[1] J.J. Strossmayer University of Osijek, Faculty of Medicine, Osijek, Croatia
[2] Addiko Bank, Zagreb, Croatia
[3] HEP Telekomunikacije d.o.o. PS Osijek, Osijek, Croatia
[4] OTIS d.o.o., Osijek, Croatia
[5] University of Zagreb, Faculty of Electrical Engineering and Computing, Zagreb, Croatia
[6] Center for Missing and Exploited Children, Osijek, Croatia

kresimir@mefos.hr, ocevcic@gmail.com, igor.fosic@hep.hr, ivan@otis-os.hr, marin.vukovic@fer.hr, tomislav@cnzd.org

Abstract - Although the importance of information security in every day's business and private life is obvious, there is a constant increase in number of security breaches. A wide array of all kinds of issues have either direct or indirect impact on the overall information security and privacy level.

The aim of this work is to collect and describe these issues in a standard manner. For this purpose, a taxonomy modeling was used, since it is a commonly used technique for classification.

Building taxonomy scheme means classifying elements on some subject with parent-child relations between entities. After defining main parent node as the Overall Information Security and Privacy, its main child nodes are defined to represent main subareas in the scheme: Network Protection, Software Protection, Physical Protection, Security Procedures, Web Site, System's Paper Elements, Security Legislations and Users' Influence. Each subarea is further divided into smaller information security and privacy subareas, and further into even smaller subareas, until it was no longer possible to divide.

This taxonomy can be used as a basis for security ontology and various analytical and decision support systems used for security evaluation and risk assessment.

I. INTRODUCTION

Today, there is no special need to highlight the importance of information security in every day's business processes and private life. Value of the information in today's world is becoming higher than the value of time and even money, since the important piece of information can save time and bring profit. Generally, a wide range of all kinds of issues have either direct or indirect impact on the overall level of information security and privacy

This work is financed by the Croatian Government Office for Cooperation with NGOs and co-financed by the European Union's Connecting Europe Facility, under project named "Safer Internet Centre Croatia: Making internet a good and safe place", Agreement Number: INEA/CEF/ICT/A2015/115320

The sole responsibility of this publication lies with the authors. The European Union is not responsible for any use that may be made of the information contained therein.

(IS&P) and cause constant increase of number of security breaches.

Most of the actual information security and privacy projects do not include all possible segments of information security, but are focused on one or several security segments, e.g. only privacy, or only password quality [1]. Conceptually, they are more often organizationally than technically oriented. So, there is a need for integration of separate security areas in the overall security solutions and new information security and privacy tools. Those new solutions should combine risk management, infrastructure safety, hardware solutions, security protocols and especially users' education and control [2].

Taxonomies of security concepts and security ontologies are common methods for modeling and sharing knowledge on information security and privacy. Although OWL (W3C Web Ontology Language) based ontology is more often used to describe information security concepts and protocols than taxonomy, former is simpler, has wider range of usage, is more flexible in modeling and is easier for humans to read [3].

Existing security ontologies and security taxonomies found in literature lack generalization. Each rare overall taxonomy on information security and privacy have different aim and scope than IS&P Taxonomy or they are rather old [3-5].

Development of overall information security and privacy taxonomy is based on previous work [6, 7] and on cooperation with other security experts. Different security standards like ISO 27000 family; security analysis tools like CORAS method, GSTool, EBIOS methodology and COBIT method; and different security recommendations published by international and national security offices like ENISA, ANSI, BSI, CSEC are also included as basis for taxonomy modeling.

The aim of this work was to collect and describe a wide range of security and privacy issues in standard manner, from perspective of both ICT security technicians and security consultants and managers, by using taxonomy modeling procedure. This work is still at its beginnings and security taxonomy is conceived as ever-changing model. The security taxonomy model should be basis for different security ontologies and various analytical decision support systems.

II. IS&P TAXONOMY

Constructing taxonomic scheme is commonly used technique of classification modeling by classifying elements on some subject of interest. It is not general requirement for taxonomy to have hierarchical structure, but in this work it is defined in order to meet requirements for future work. So taxonomy as knowledge scheme in the area of information technology should be defined with entities and corresponding parent-child relations among them. The hierarchical tree structure starts with main parent node named "Overall Information Security and

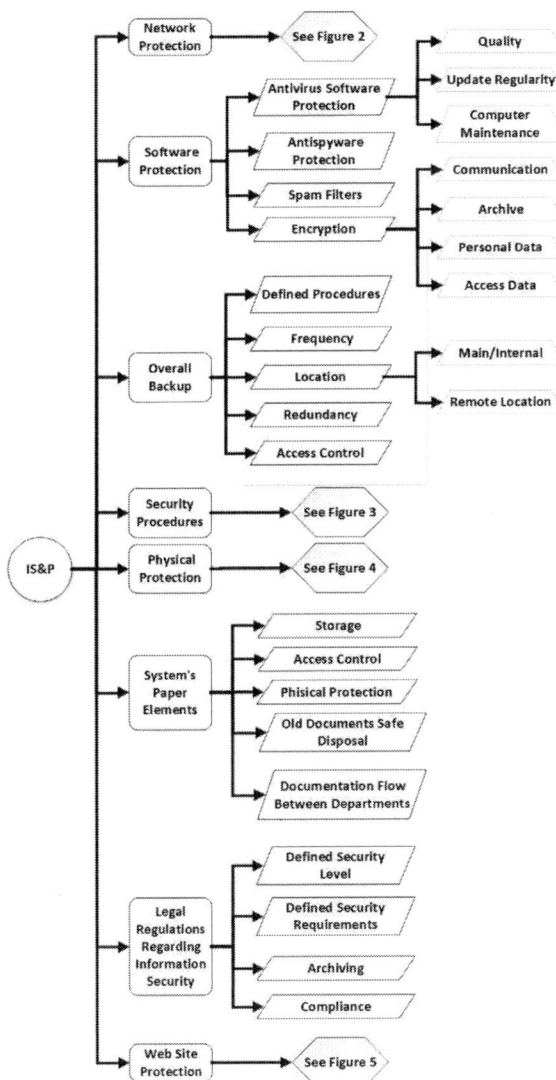

Figure 1. General IS&P Taxonomy scheme

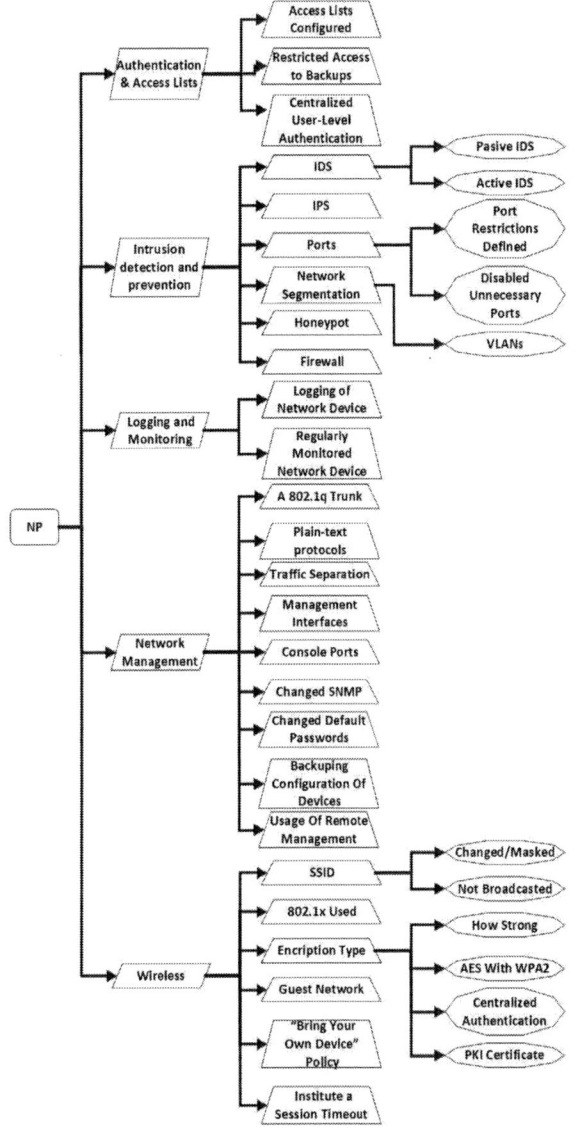

Figure 2. The Network Protection sub-taxonomy scheme

Privacy".

In the (current) IS&P Taxonomy main parent node has eight child nodes presenting main information security and privacy elements in the scheme. They define eight main subareas on the first level of structure complexity (Fig.1). Further steps in modeling taxonomy are defining child nodes of those eight parent nodes at the second level of complexity. Eight child nodes of first level of complexity are now parent nodes for the second level of complexity. This modeling steps were applied until the smallest elements in the scheme were defined as ending child nodes in the hierarchical tree structure. Current version of the IS&P Taxonomy has four levels of complexity in depth.

1299

Four of eight main subareas that have most of the constructing elements are presented in the separate and more detailed schemes. They can be used as a sort of

sub-taxonomies and are named as follows: „Network Protection" (Fig.2), „Security Procedures" (Fig.3) "Physical Protection" (Fig.4), and "Web Site Protection" (Fig.5).

The „Network Protection" sub-taxonomy describes all the hardware elements with belonging software that can be used to build network infrastructure. It encompasses administration with previous configuration of network hardware; functionality and optimization regarding business needs; and usage of network infrastructure with compliance towards system requirements (Fig.2).

Second main sub-taxonomy named "Security Procedures" covers all the security procedures and protocols that should be defined in every company's IS&P documents (Fig.3).

The "Physical Protection" sub-taxonomy covers all kind of possible methods of physical protection regarding data and ICT system's elements, both hardware and software. Physical protection encompasses hardware equipment that is not necessary computational, software

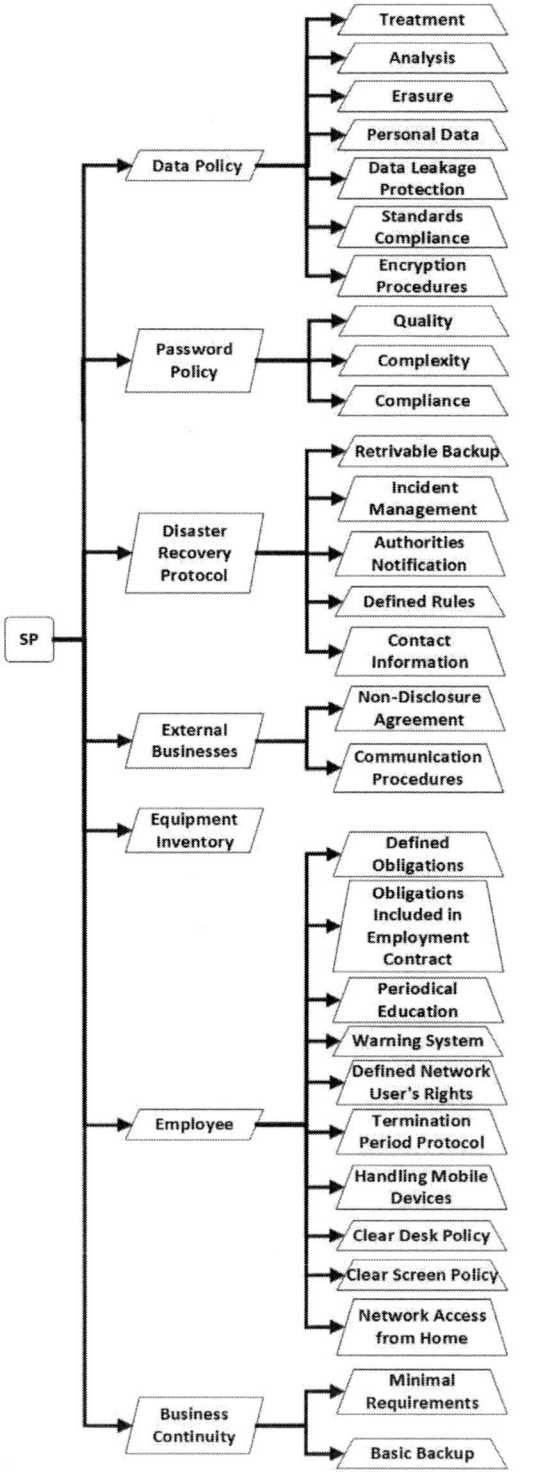

Figure 3.   The Security Procedures sub-taxonomy scheme

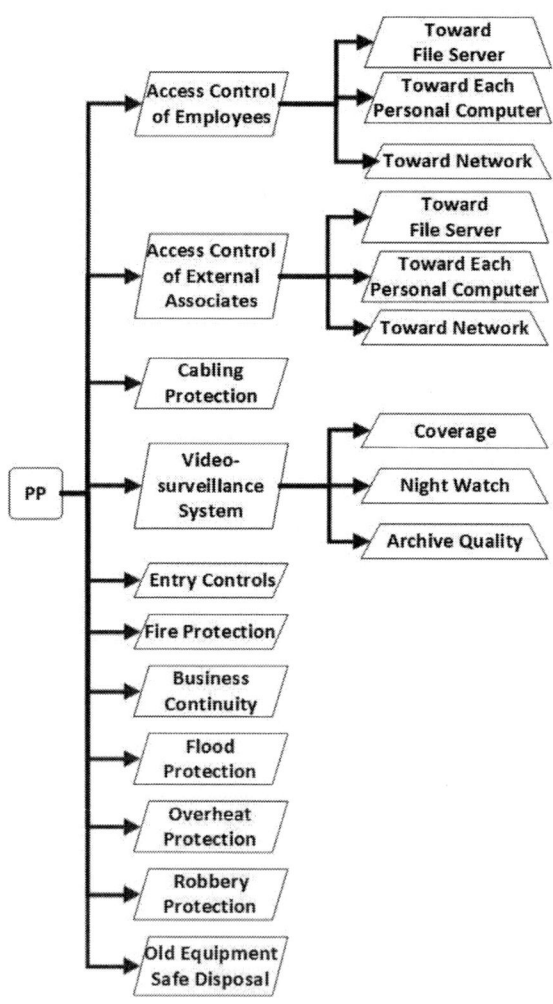

Figure 4.   The Physical Protection sub-taxonomy scheme

1300

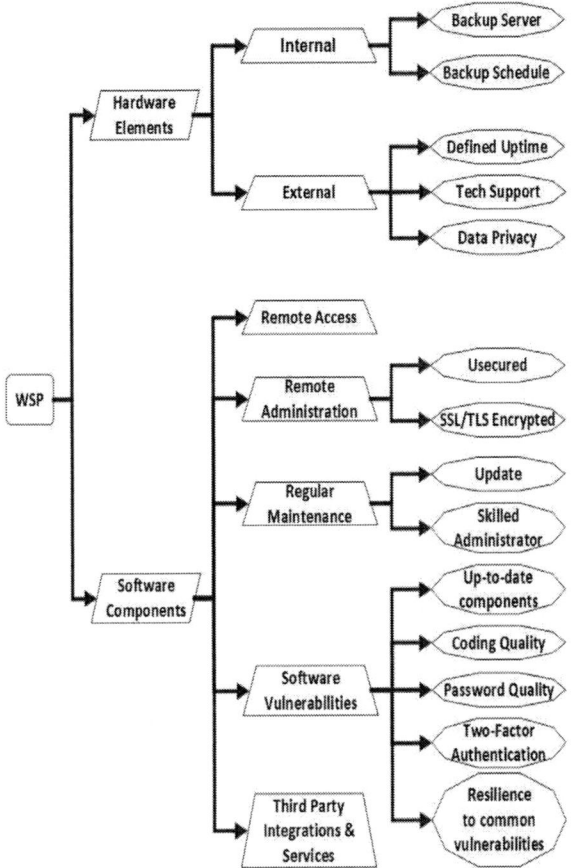

Figure 5. Web Site Protection sub-taxonomy scheme

modules as well as some security processes that involve humans and external business (Fig.4).

Forth main sub-taxonomy "Web Site Protection" describes protection of both hardware and software elements in details (Fig.5).

The other five sub-taxonomies are described in details in IS&P Taxonomy scheme (Fig.1) and are covering protection by software solutions, overall backup quality, paper elements of the information system and legislative with legal regulations.

As it was more important to cover all considered security issues and ICT system's elements there are partial overlaps between sub-taxonomies, but it is not considered as problem in the proposed solution. When using this taxonomy for e.g. security evaluation, it is probably better to double-check than to omit some important security and privacy issue.

III. CONCLUSION

The proposed information security and privacy taxonomy is the first proposal and should be considered as a live construct. It will be further developed in the future, along with new issues that may, and surely will, arise.

Main information security and privacy areas that are covered with this taxonomy are: network protection, software protection, overall backup quality, physical protection, security protocols including employees' influence, system's paper elements protection, legislative and legal regulations and web site protection.

Presented IS&P Taxonomy can be used as basis for security ontology and all kinds of analytical and decision support systems used for security evaluation and risk assessment. Plans for future work are to validate this taxonomy and to develop the decision support system as information security and privacy self-assessment software tool based on taxonomy presented in this work.

LITERATURE

[1] A. Shameli-Sendi, R. Aghababaei-Barzegar and M. Cheriet, „Taxonomy of Information Security Risk Assessment (ISRA)", Computer & Security, vol. 57, pp. 14–30, March 2016.

[2] R.E. Crossler, A.C. Johnston, P.B. Lowry, Q. Hu, M. Warkentin and R.Baskerville, "Future Directions for Behavioral Information Security Research", Computer & Security, vol. 32, pp. 90–101, February 2013.

[3] J. Heurix, P. Zimmermann, T. Neubauer and S. Fenz, "A Taxonomy for Privacy Enhancing Technologies", Computer & Security, vol. 53, pp. 1–17, September 2015.

[4] A. Souag, C. Salinesi and I. Wattiau, „Ontologies for Security Requirements: A Literature Survey and Classification" CAiSE, 2012.

[5] H. Gomes, A. Zúquete and G.P. Dias, „An Overview of Security Ontologies", Proc. of the 9th Conferência da Associação Portuguesa de Sistemas de Informação (CAPSI 2009), Viseu, 28-30 October, 2009.

[6] K. Solic, H. Ocevcic and M. Golub, "The Information System's Security Level Assessment Model Based on an Ontology and Evidential Reasoning Approach", Computer & Security, vol. 55, pp. 100–112, November 2015.

[7] T. Galba, K. Solic and I. Lukic, "An Information Security and Privacy Self-Assessment (ISPSA) Tool for Internet Users", Acta Polytechnica Hungarica, vol. 12, pp. 149–162, 2015.

Trends in IoT Security

M. Radovan[*], B.Golub [**]

[*] Daimler AG, Stuttgart, Germany
mihael.radovan@daimler.com
[**] Adnet d.o.o., Zagreb, Croatia
boris.golub@adnet.hr

Abstract – The incredible rapid development of internet technologies, primarily thanks to omnipresent access of high speed broadband internet access and supporting technologies like Big Data, Cloud Computing, REST/Web services as well as cheap electronic equipment that use new wireless communications standards, lead to equal rapid growth of number of smart devices – "things" - connected to the internet. Increased number of connected smart devices results with huge daily data traffic on the internet and data volume stored and available on the internet. IoT becomes part of our homes and our companies, and security in these systems is very important. What does that outlook hold for the next 5, 10 or 20 years, will mostly depend on the development of security standards, user behavior and education in a next few years. This paper is trying to summary and analyze trends in IoT standards regarding security.

Keywords - Internet of Things (IoT); Cybersecurity; SCADA;

I. INTRODUCTION

Each new technology potentially involves some risks. The worst of which are security threats. This is especially the case with a fast-growing technology where it is difficult to predict all possible risks involved. On the other hand, the number of skilled malignant hackers, highly motivated to abuse new technology inadequacies is rising.

Quoting Kevin Ashton, the probable author of the term "Internet of Things" [1], from 1999, the size of the Internet in 1999 was about 50 petabytes of data, and the prediction for 2020 is that the size of the Internet will be about 40 zettabytes of data.

Although the industrial giants are for security reasons very conservative with beginning to utilize new things, IoT is already there. Smart devices are everywhere around – smart wearable, smart medical devices, smart homes, smart buildings, smart cars, smart cities, smart grids, smart agriculture and many other aspects of life [2] (see Fig. 1). The challenge grows further as IoT devices have the option to control important industry infrastructure. This will certainly increase society exposure to cyberthreats. In today's approximations, the damage from cybercrime is assessed at about 400 billion dollars per year.

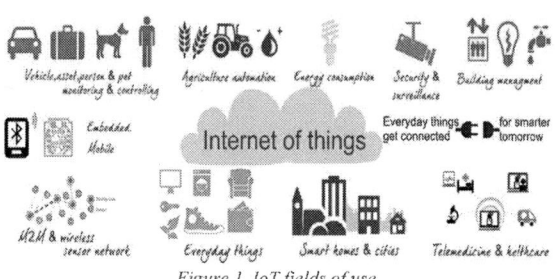

Figure 1. IoT fields of use

The smart future has already arrived. Are we ready for it? Do we behave responsibly in order to have all around us "smart"? Is it smart to be surrounded with smart things, can we be sure that our privacy is not compromised? These smart devices are smarter and smaller, but are they reliable and secure enough? Especially when we know that parts of IoT have not been well defined and designed, or still use parts of old architecture from "wired" era.

The main aim of this paper is to summarize the current state of security trends in IoT technologies and to analyze what needs to be considered to shape the security standards for the new generation of connected systems.

This paper is organized as follows: Section II gives a short timeline of IoT, the present state and predictions for the future. Section III shows security and privacy trends in the development of IoT and new protocols which cover these trends, and Section IV impact of IoT to industrial security. Section V gives a conclusion of the paper.

II. HISTORY, PRESENT AND FUTURE OF IoT

At the beginning of computer era (1950's until late 1970's), there were dominant mainframe computers with terminal-server architecture. During 1980's, the computer market takes over minicomputers and personal computers. 1990's comes internet era, first with analog modems (2400, 9600, 14400, 28800 bps) through analog telephone wire, then from 2000's with a digital connection (ISDN, DSL, VDSL) and finally with the broadband and wireless internet connectivity. Late 1990's comes mobile era, and a few years later, in the third millennium, the mobile device meets computer – first in shape of PDA's, and later as tablets and smart phones. Although the internet for commercial use was available first in the late 1980's, the first internet appliance was a Coke machine at Carnegie Melon University in 1982. The commercial internet began to exist in late 1980's as the "Internet of Computers" [3], providing global services such the World Wide Web. Broadband internet connectivity is becoming fast, cheap and omnipresent. Last 10 years Internet has changed to the Internet of People. This "network of humans" covers more than 1 billion people. According to the statistics portal Statista (https://www.statista.com), in 4th quarter 2016, just Facebook had 1.86 billion active users. A number of active e-mail accounts worldwide in 2016 was 4.626 billion, and prediction for 2019 is about 5.6 billion (see Fig. 2).

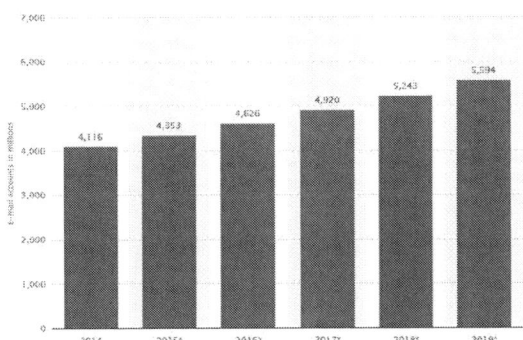

Figure 2. Number of active e-mail accounts in mil.

Fast broadband internet, Web services architecture, Big Data and Cloud Computing technologies, together with small and cheap microprocessors, led to "Internet of Things". In this case, "things" are any beings (humans, animals, plants) or items (cars, buildings, cities, factories, devices) with unique ID, internet connection and set of sensors or actuators. At any moment in time this unique items carry its status and position and provide remote access to read data from its sensors or control its actuators. It was initially proposed to use RFID to uniquely identify, track and monitor any object with RFID tag [3]. RFID is a foundational technology for IoT, while it allows microchips to transmit the ID information to a reader through wireless communication [4], widely used in industry since 1980's. Another foundational technology for IoT is wireless sensor network (WSNs), from 1990's – set of a wirelessly interconnected smart sensor attached to devices.

The newest trends in IT have established a way to Big Data era – everything goes to the Cloud. All data will be stored somewhere on the internet, or will be transferred to the storage destination through the internet. This trend can easily be seen by observing the internet size since 2000. The amount of all data published on the internet in 2000 was about 50 PB. In 2012, this size was 4 ZB, today is about 8 ZB, and expected size in 2020 is about 40 ZB (see Table 1.).

Label	Title	Size in minor unit	Size in bytes
1GB	Gigabyte	1024 MB	1024^3 byte
1TB	Terabyte	1024 GB	1024^4 byte
1PB	Petabyte	1024 TB	1024^5 byte
1EB	Exabyte	1024 PB	1024^6 byte
1ZB	Zettabyte	1024 EB	1024^7 byte
1YB	Yottabyte	1024 YB	1024^8 byte

Table 1.: Orders of magnitude of data

Just 25 years ago, in 1992 was less than 100.000 devices connected to the internet. In 2012 number of connected devices exceeded human population in the world. Predictions of Cisco [5] for 2020 is about 50 billion devices [6] (See Fig. 3).

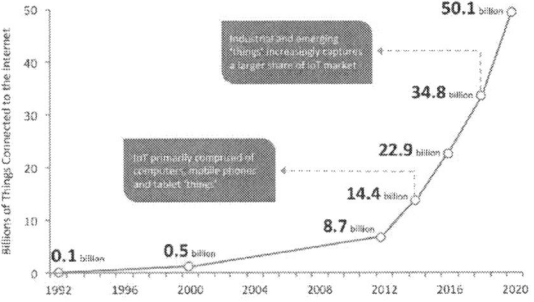

Figure 3. Number of connected devices

The numbers in the timeline – number of people connected to the internet, number of active e-mail accounts, number of connected devices, amount of published or transferred data – are just presenters of development of technology in last years, and show how fast this development is. Such a rapid development of technology enables a range of new opportunities, but also demands a development of new communications and security standards.

III. SECURITY TRENDS IN IoT

Security in IT includes availability, integrity and confidentiality. Availability means that the remote system in online and can offer the service to the customer. In the moment when system is not available, either system is down or communication is broken, the security is compromised. Integrity means that all data are the same on the source as on the destination, in communication process between client and server, or between two or more nodes in the network, and can be accessed or modified just by those authorized to do that. If during the communication process there is an unauthorized data modification the security is compromised. Confidentiality refers to the data protection on both sides in communication (client/server, sender/recipient) as well as on the network side, during data transportation. The data must be encrypted and protected, so that just sender and recipient can read it.

One of IoT technology main security challenges is in its fast growth in development of electronic devices. Security standards are often not implemented, not proven enough, or even do not exist. This is particularly the case with

1303

cheaper devices intended for home use made by no name producers. IoT technology is today still based on the standard communication architecture, which was defined for stationary clients or mobile clients under owner's control. These clients are usually located in some protected environment, and under security standards set to protect this environment from malicious interventions in every way. IoT architecture has three basic parts: Perception, Network and Application [7] (see Fig. 4).

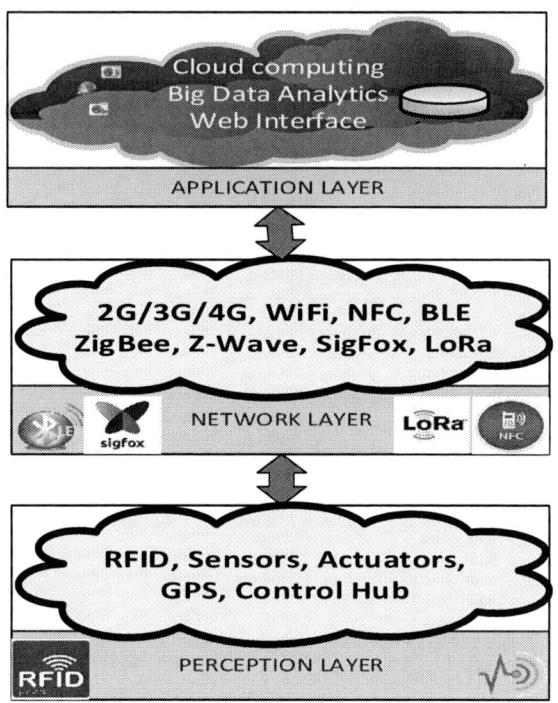

Figure 4. Architecture of IoT

Perception layer contains hardware – sensors, actuators, RFID, programmable control unit with processor and memory. These are all necessary to collect and send data or receive data. These devices are connected to the Network layer – whether standalone with independent wireless connection to ISP, or wireless grouped by hub or internet access point and through it connected to ISP. They are placed everywhere, and not always protected enough from unauthorized access, so they can easily be damaged or stolen. In this case, by unauthorized access to data or just by breaking connection between devices and the cloud, the security system will be compromised. Also, these devices are often powered by batteries or solar cells, which raises another important issue – autonomy time.

Network layer serves for the communication between sensing devices and data storage in cloud. There is a wide spectrum of technologies and standards used to exchange data. New standards are in continuous development, based on existing LAN, WLAN and WAN standards. They differ from each other by radio frequency spectrum, data exchange rate, range and power consumption. Being easily exposed target to malicious attempts security is especially important on this layer.

Highest layer is Application layer – Cloud architecture and data management systems as well as analytic and presentation software. At this point data must be protected from unauthorized access and anonymized when used for public purpose.

Together with good known WiFi, 2G/3G/4G Cellular, NFC and Bluetooth communication technologies, for industrial and commercial use are developed some new low-power and wide-range protocols: Bluetooth Low-Energy (BLE)[1], ZigBee[2], Z-Wave[3], 6LowPAN, Sigfox[4], Thread[5], LoRaWAN[6].

NFC (Near Field Communication - http://nearfieldcommunication.org/) is RFID based technology for contactless data exchange using electromagnetic induction in range of 10 cm at a rates of 106, 212 or 424 kbps [7].

BLE is Bluetooth 4.2 version designed for small chunks of data. Works at 2.4 GHz, has 50-150m range by 1Mbps speed [8].

Zigbee with last version 3.0, is based on the IEEE802.15.4 industry protocol at 2.4 GHz that require low data rates of 250 kbps within 100m range. This protocol is robust, scalable and highly secure.

Z-Wave is a low-power RF simple protocol made by Sigma Designs for home use products. Works at 900 MHz frequency within 30 m range at rates of 9.6/40/100 kbps.

6LowPAN is an acronym Ipv6 Low-Power Wireless Personal Area Networks, formed by Internet Engineering Task Force (IETF)[7]. The standard can use number of frequency band across multiple communications platforms. Its specialty is using IPv6, which offers $5*10^{28}$ addresses – practically each connected device can have its own unique IP address [9].

Sigfox is technology designed to handle low rate up to 1 kpbs, but consumes only 50 μW. for communication at range of 30-50 km in rural and 3-10 km in urban environment. For example, the solution COPERNIC (https://partners.sigfox.com/products/copernic) for monitoring of thousands of fire hydrants sending status and timestamp by an email or SMS, runs on lithium battery with estimated lifetime of 10 years. Startup Sigfox was developed in 2009. Today covers 31 countries with more than 7 million registered devices.

Thread is newer protocol based on 6LowPAN, primarily designed as a complement to WiFi on existing IEEE802.15.4. wireless chips as Freescale or Silicon Labs.

---

1 Bluetooth Special Interest Group (SIG) – Ericsson, IBM, Intel, Nokia, Toshiba Alliance, https://www.bluetooth.com/
2 Zigbee Alliance, http://www.zigbee.org/
3 Z-Wave Protocol, http://www.z-wave.com/
4 Sigfox, www.sigfox.com
5 Thread Group, http://www.threadgroup.org/
6 LoRa Alliance, https://www.lora-alliance.org/
7 Internet Engineering Task Forces, www.ietf.org

LoRaWAN is another concept similar to Sigfox. It works over WAN area and is designed for low-power bi-directional communication for industrial purpose and smart city solutions. It uses unlicensed public spectrum called the ISM 8 (Industrial Scientific and Medical) frequency band. Its range is 2-5 km in urban and 15 km in suburban environment, at rates of up to 50 kbps. As well as Sigfox, LoRaWan is world-wide present, with more than 150 on-going trials and city deployments and more than 400 members in the Alliance. Korea's ICT giant Samsung, together with Korea's Telecom is building nationwide LoRaWAN IoT network, covering 99% of the population in South Korea.

Proposed IoT strategy to operators is to offer hybrid technologies (See Table 2). For applications that need long range and high data rates, there are existing networks such a cellular, LTE or WiFi (15% of IoT market). Short range technologies as ZigBee, NFC, Z-Wave, BLE should be used for short range systems such a smart meter systems, buildings, parking systems etc., and in this cases, it is possible to use hubs to aggregate metering devices and send data to the cloud (40% of IoT market). Nevertheless, to meet long range coverage it is necessary to use low-power WAN devices with low data rate and low cost (45% of IoT market).

Range	Long	Short	Long
Data rates	High	Low/High	Low
Market share	15%	40%	45%
Technology	Cellular, LTE, WiFi	ZigBee, NFC, Z-Wave, BLE	LoRa, Sigfox, 6LowPAN

Table 2: Technology usage

Here listed technologies are all based on the RF communication, and hacking a radio connection can be performed without physical access to the device. The attacker has unlimited time to try to perform malicious operation. A good protected radio communication would be a first line of defense against hackers. Second line of defense is to keep data signed with encrypted fingerprint, using hashing mechanisms as well as asymmetric mechanisms for encryption. Asymmetric mechanisms can also be used for authorization and authentication. Firmware and internal programs should be stored in the internal memory rather than to external memory and also protected with some locking mechanisms, similar to hard drive encryption software. In this way, the attacker has limited access to the communication interface through hardware using for example oscilloscope or logic analyzer. This keeps the data stored in internal memory safe and protected from unauthorized access.

---

8 ITU – United Nations specialized agency for ICT, http://www.itu.int

IV. IMPACT OF IOT TO THE INDUSTRIAL SECURITY

Traditionally, cyber security for Information Technology (IT) focuses on the protection required to ensure the confidentiality, integrity, and availability of the electronic information communication systems. IoT imposes that cyber security must include a balance of both cyber system technologies and processes in IT and industrial system operations and governance.

For example, in the power industry, the focus has been on the implementation of equipment that could improve power system reliability [9]. Until recently, communications and IT equipment were typically seen only as supporting service. However, the distinction is getting less and less visible (OT – Operational Technology and IT Network Integration). For example, electricity transmission system operators face the challenges of adopting large amounts of energy coming from renewable sources. Many of renewables are inherently volatile and thus difficult to predict and plan. New methods and algorithms of planning had to be adopted to maintain the grid reliability. All this could not be done without the support of evolving IT technologies.

Supervisory control and data acquisition (SCADA) is an information system architecture that uses computers, networked data communications and graphical user interfaces for high-level industrial, infrastructure or facility processes supervisory management.

SCADA systems are used to control and monitor physical processes, examples of which are a transmission of electricity, transportation of gas and oil in pipelines, water distribution, traffic lights etc. The security of these SCADA systems is important because compromise or destruction of these systems can have an impact on multiple areas of society. Other systems for automatic meter reading or planning usually extend basic SCADA functionalities.

We can roughly speak of three phases of SCADA system evolution. At the beginning, SCADA systems were standalone with proprietary protocols and completely isolated thus making everything intrinsically secure.

The next phase included increased number of connections between SCADA systems, office networks and the Internet which made them more vulnerable to types of network attacks that are relatively common in IT world.

Today with the increasing availability of cloud computing, SCADA systems have adopted Internet of things technology to significantly reduce infrastructure costs and improve ease of maintenance and integration. As a result, SCADA systems can now report state in near real-time and use the horizontal scale available in cloud environments to implement more complex control algorithms than are practically feasible to implement on traditional programmable logic controllers. All these changes imply many new threat vectors to a modern SCADA system.

1305

SCADA systems may differ. But from a security point of view, they can be broken down into the components that are present in every system in some form. A typical SCADA system consists of four elements.

Data acquisition includes sensors, meters and field devices, such as photo sensors, pressure sensors, temperature sensors and flow sensors. For example, in sensor accepting modifications without sufficient checks may cause the system to default to a failsafe condition. Another example could be unsecure energy meter reading devices allowing the unauthorized users to change the internal data.

Conversion and control include devices like remote terminal units (RTU), intelligent electronic devices (IEDs) and programmable logic controllers (PLC). Remote terminal units connect to sensors and actuators in the process, and are networked to the supervisory computer system. Programmable logic controllers are connected to sensors and actuators in the process, and are networked to the supervisory system in the same way as remote terminal units but have more sophisticated embedded control capabilities. Unauthenticated ports could allow modification of memory and logging. This can allow attackers to change system configuration and furthermore remove log records that indicate system change to hide malicious activity.

Communication infrastructure connects the supervisory computer system to the remote terminal units and programmable logic controllers, and may use industry standard or manufacturer proprietary protocols. Failure of the communications network does not necessarily stop the process controls, and on the resumption of communications, the operator can continue with monitoring and control. Some critical systems will have dual redundant data highways, often cabled via diverse routes. A theoretical attacker could (if gained access to the unencrypted network) retrieve sensible industry data.

The supervisory computer is the core system component responsible for gathering process data and sending control commands to the connected devices. To increase the integrity of the system the multiple servers it can usually be configured in dual configurations.

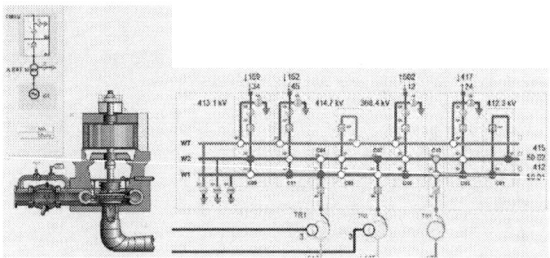

Figure 5. Screenshots from NetVision SCADA HMI

The human-machine interface presents process information to the operating personnel graphically in the form of mimic diagrams, which are a schematic representation of the process being controlled, and alarm and event logging pages (See Fig. 5).

The interface is linked to the SCADA supervisory computer to provide live data to drive the mimic diagrams, alarm displays and trending graphs. Typically, it also includes a drawing program that the operators or system maintenance personnel use to change the way these points are represented in the interface as well as historian database which accumulates time series, events, and alarms data which can then be used to populate graphic trends etc.

Majority of all SCADA vulnerabilities fall in this category. Most SCADA vendors are shifting to web based HMIs. As a result, a lot of web related vulnerabilities could affect this component. Also, traditionally IT system administrators have unrestricted access to all the system databases allowing them the possibility to change or share the data outside the organizational security perimeter.

	Traditional	Modern
Communication	Proprietary protocols and networks	Standard protocols and Public Networks
Systems	Custom HW and SW Centralized Architectures Physical Access	Commercial Off-The-Shelf Distributed Architectures Remote Access
Human resources	Employee	Outsourcing
Risk level	Lower	Higher

Table 3.: SCADA evolution

Traditional isolated SCADA system have security implicitly applied to it by keeping the very tight border between who has access to what particular SCADA function or data together with the dedicated secure communication channels (see Table 3.). The focus was on keeping thing tightly coupled and locked as much as possible. On the other hand, in modern systems number of different types of connections within and across network boundaries continue to grow at an accelerating rate, opening new points of exposure. The introduction of IP-based technology and general-purpose computing devices into operational environments is introducing new vulnerabilities along with their benefits. The focus is now on large scale IoT data integration and sharing information as much as possible. Now it is not only the protection of the core SCADA functionalities which has to be secured from cyber-attacks we also have to deal with security at the data layer. For example, settlement data is the pillar of modern energy market structure and can be affected with targeted subtle security attacks affecting metered data [11].

1306

Technologies (tools, strategies) available to protect the control system:

- Access controls – authentication and authorizations at functional and data level, including industry topology segmentations

- Bidirectional gap (e.g., firewall) between control systems and rest of network

- Application whitelisting - to control which applications are permitted to install or execute on a specific host

- Anomaly detection tools

- Patch management / upgrade services

- Vulnerability scanning

- Security awareness training for staff, contractors and vendors

- Asset identification - visibility of components within the control system equipment and network activity

- Antimalware/Antivirus

- Assessment and audit

- Continuous Monitoring and log analysis

- Data forensics

The energy sector benefits from a head start. From the very beginning, cybersecurity was a key requirement of the Smart Grid initiative and significant efforts have been invested in defining requirements that have been formalized in reports such as the NIST 7628 Guidelines for Smart Grid Security. While these guidelines do not address the IoT per se, they do define the cybersecurity requirement for applications in the energy sector, from generation plants to customer premises. Furthermore, many utilities are required to meet the NERC CIP cybersecurity standards and have thus acquired valuable experience in protecting their assets.

The most common vision of IoT is thus for a loosely coupled network of devices and sensors that publish their data through a messaging architecture using web services and messaging protocols such as Message Queuing Telemetry Transport (MQTT), Constrained Application Protocol (CoAP), Data Distribution Service (DDS), and Advanced Message Queuing Protocol (AMQP). Besides AMQP, most of these protocols are not yet widely used. The AMQP protocol is used in the financial industry and supports a transactional model, making it more complex and not appropriate for edge devices. To our knowledge, none of these protocols are used in energy sector automation systems and devices.

IoT evolution will certainly force the industrial system architectures to change significantly. The idea is to abandon the monolithic and monolithic like designs and replace it with the more natural and simpler design paradigms which can more easily be formally checked for security deficiencies. We can think of sensors and actuators as a primary service providers in a networked environment. By using the data which those primary services provide we can add additional services to the network (secondary services) which can then be further combined or exploited by adding new services. Formally we designed a graph of interconnected nodes. Nodes represent services and edges represent the data flows between them (service dependencies). If we further exploit that idea we can easily connect the IoT with the newest information system architectural patterns of today. First major pattern is using microservices as an architectural model for the deployment of lightweight and loosely-coupled services and the second is the adoption of containers instead of virtual machines for certain types of workloads and thus benefit from the more efficient resources usage as well as easier deployment operations. Both introduce more visibility into the services topology, versions tracking, logging errors etc. This implicitly improves security model by increasing visibility, resiliency and elasticity of the system. What we described is implementation of a more agile design for failure concept where the system is structured in a way that in case of service failure the effect of the compromised service is minimized.

Another benefit of that kind of approach is the side effect of reducing the dependency on the network itself. The service nodes take on a greater role in terms of processing and behavior without hard-coding it in the communication layer (as was the case in Enterprise Service Bus component of the Service Oriented Architecture).

V. CONCLUSION

As IoT brings numerous new possibilities it is expected to be widely applied to industry and home solutions. As it integrates various internet technologies – powerful devices capable of sensing, processing and exchanging data, new low-energy communication protocols, big data and cloud systems it is obviously very important to use existing security technologies for dealing with security issues. But in the future, we will have to address this problem further by moving to a more systematic approach to IoT security. More work will have to be done in the area of communication standards of consumer IoT products as well as enforcing them. Also, new methods for testing the formal security in communication protocols will probably emerge as well. Data forensics as a tool to pinpoint and understand the subtle changes (potentially unauthorized) in the sensor data received from the smart devices is already widely used through the industry.

Systems today are reusing and repurposing existing technology in conjunction with newer, more IoT aware solutions. This is obviously only the first step towards fully interconnected IoT network.

VI. REFERENCES

[1] K. Ashton, "Internet of Things", RFID Journal 06/2009, online available:

[2] N. Hughes, "The Internet of Things Explained", 2014, Picture source online available: https://www.linkedin.com/pulse/20140804163105-98377657-the-internet-of-things-explained

[3] R. van Kranenburg, "The Internet of Things: A Critique of Ambient Technology and the All-Seeing Network of RFID", The Netherlands Institute of Network Cultures, 2007

[4] L. Coetzee, J. Eksteen, "The Internet of Things – Promise for the future", IST-Africa 2011 Conference Proceedings

[5] D. Evans, "The Internet of Things", 2011, online available: http://blogs.cisco.com/news/the-internet-of-things-infographic

[6] CompTIA, "Sizing Up the Internet of Things", 2015, picture source online available: https://www.comptia.org/resources/sizing-up-the-internet-of-things

[7] X. Jia, T. Fan, Q. Lei, "RFID Technology and its application in IoT", 2nd IEEE Int. Conf. Consum. Electron., CECNet China, 04/2012, pp. 1282 – 1285

[8] J. S. Kumar, D.R. Patel, "A Survey on Internet of Things: Security and Privacy Issues", International Journal of Computer Applications Vol. 90 – No 11, 03/2014

[9] "Telecommunications and information exchange between systems – Near Field Communication – Interface and Protocol NFCIP-1", ISO/IEC 18092:2004 Information technology, 11/2011, online available: http://www.iso.org/iso/catalogue_detail.htm?csnumber=38578

[10] "Bluetooth low energy wireless technology backgrounder", Nordic Semiconductor, 03/2011

[11] IETF Technical Documentation for 6LowPAN, online available: https://datatracker.ietf.org/wg/6lowpan/documents/

[12] Introduction to NISTIR 7628 Guidelines for Smart Grid Cyber Security, 09/2010

[13] White Paper WP152016EN, The Internet of Things and the energy sector: myth or opportunity, Power and Energy Automation Conference Spokane, WA March 8–10, 2016

[14] Saša Radomirović, "Towards a Model for Security and Privacy in the Internet of Things"

International cyber security challenges

I. Duić*, V. Cvrtila**, T. Ivanjko***
*Croatian Radiotelevision, Zagreb, Croatia
** University of Applied Sciences Vern, Zagreb, Croatia
***Faculty of Humanities and Social Sciences/Information and Communication Sciences, Zagreb, Croatia
igor.duic@gmail.com

Summary - The opportunities provided by the information and communications technology, with a special emphasis on the Internet, have become an integral part of life. However, are we sufficiently aware and prepared as individuals, nations or the international community for the threats coming from cyberspace or for the denial of the use of that dimension of communication, commerce and even warfare? Namely, despite the growing number of users, the Internet is still beyond or below minimum regulation. Those are precisely the conditions for the organization and realization of hostile action in cyberspace. There are security issues within the cyberspace that represent a security risk and challenge of modern times.

The development and application of the information and communications technology has created a new battleground. As a special challenge to international security, cyber terrorism arises. Cyber security will significantly affect international relations in the 21st century. This paper gives an overview of the concepts and principles of cyber threats that affect the safety and security in an international context.

Keywords: cyberspace, cyber-attack, cyber terrorism and crime, international security.

I. INTRODUCTION

Cyber warfare and terrorism do not know borders. Action in cyberspace requires the rejection of the common assumptions related to time and space because such attacks, by means of modern information and communications networks, can be performed from anywhere in a very short time. The processes of globalization did not have the impact only on the achievements of civilization, but also on the development of new threats to the civilization. It is a fact that terrorism and national threats changed under the influence of the globalization process and the Internet information revolution. Strategic advantage no longer lies in the fighting power or geographical location, but in the information and knowledge. International cooperation and intelligence sharing are essential for an effective prevention of cyber threats. Even though cyber threats have in the recent years been specifically emphasized in the modern military doctrines of great powers and NATO, they are still shrouded in secrecy.

The purpose of this paper is to draw attention to cyber threats, which endanger the safety of modern states, organizations and international relations. By combining

the principles of a review and professional research paper, this paper aims to show cyberspace, in terms of security challenges, as a dimension in which international relations unfold. It is necessary to distinguish the main subjects of the international cyber security environment, analyze their intentions and set a paradigm of the multipolarity of cyberspace and analyze its uniqueness and principles.

NATO's Strategic Concept, adopted at the end of 2010 at the Lisbon summit, determines that cyber-attacks have become more frequent, more organized and more expensive, causing damage to the government administration, the business sector, economies, and potentially to the transport and supply. It also states that cyber-attacks can reach the level that threatens the national and Euro-Atlantic prosperity, security, and stability. Foreign military and intelligence services, organized criminals, terrorists and extremist groups are the potential sources of such attacks. What is also emphasized in the conclusions of the Lisbon summit is the need to further develop the skills of prevention, recognition, defense and recovery from cyber-attacks, including the use of the NATO planning process for the advancement and coordination of the national abilities of cyber protection, assembling all NATO bodies under a centralized cyber protection, and a better integration of the NATO cyber awareness, warnings and common response of the member states [1].

It should be borne in mind that the rapid development and adoption of technologies through their use in everyday life opens up many opportunities for the attackers, whether they are in the form of states, terrorists or criminals, because they are always at an advantage in cyberspace. We can therefore conclude that, a new concept of cyber security in which prevention represents a significant portion is being created.

The initial hypothesis is that cyberspace is a growing security risk and challenge of modern times. Moreover, cyber security will significantly affect international relations in the 21st century, while the threats and challenges will exponentially increase.

The goal of this paper is the synthesis and analysis of knowledge based on a review of recent literature and professional and scientific articles that problematize the challenges of international security in cyberspace. The scientific work seeks to show cyberspace as an operational dimension of international relations in terms of the cyber security challenges. With the systematization of the cyber warfare strategy and the very methods of attack, links with the planned action will be set up through the application of technical, computing and network systems.

II. INTERNATIONAL CYBER SECURITY

The cyber domain has a great influence on the transformation of the international security and the very concept of security. Many authors highlight the necessity of the duly understanding and setting up of cyber doctrines.

The new, cyber dimension of international relations is a major challenge for the theories of the preservation of power and intimidation. Cyber threats are serious, destabilizing and on the increase. The theories and strategies of intimidation designed and implemented during the Cold War cannot be implemented in the cyber domain. Many scientists are working on the understanding of the cyber revolution in international relations. Authorities have also taken certain steps in cooperation, especially in the area of crime and the establishment of CERTs (Computer Emergency Response Teams) [2]. Tatalović, Grizold and Cvrtila write that the processes of internationalization and globalization have brought a greater cohesion and efforts for a unified regulation of the world order, more than it was in the system of sovereign states during the Cold War. This is reflected in the core of the states' security policies. In that context, a new concept – human security concept – emerged in theory and political practice. In contrast to the traditional concept of national security, it primarily emphasizes the security of an individual, not the state [3]. Lin theorizes [4] about cyber security. The concept of intimidation was the basic idea of the nuclear strategy. However, the question is whether the dissemination of the principles of intimidation on cyberspace is a viable strategy. Even though nuclear and cyber weapons share a key feature – the superiority of the attack in comparison with the defense – they differ in many ways. Only a few countries possess nuclear weapons and the number of possible enemies is limited, as is then the application of intimidation. The situation is completely different when it comes to cyberspace. Unlike nuclear weapons, each state has access to cyber "weapons", and such attacks cannot be firmly linked to state action. The protection of national infrastructure against attack could become another common interest of states. Experts and analysts estimate that the efforts of Russia and China to dominate cyberspace have over the past few years intensified so much that any delay in this area could present a big problem for the modern West.

Cyber-attack, whether it happens as a conflict between states, a terrorist or a criminal act, is an attack in cyberspace with the aim of compromising a computer system or network, but also of compromising physical systems as it was the case with the Stuxnet worm. In layman's, popular terms, most often mentioned in the media, it is called a hacker attack. Identical methods of a hacker attack are applied for both military and terrorist purposes.

Janczewski and Colarik [5] divided cyber-attacks into phases, which they consider to be basically the same as the phases of conventional criminal offenses:

- the first phase of the attack is the scouting of potential victims. By observing the implementation of the normal operations of targets, useful information that are accumulated and determined through the used applications and hardware;

- the second phase of the attack is intrusion. Until the attacker gets into the system, there is not much that can be done against the target apart from disrupting the availability or access to certain services provided by the target;

- the next phase is the identification and dissemination of internal opportunities by examining the resources and the right to access the restricted and important parts of the system;

- in the fourth phase the intruder does damage to the system or steals certain data;

Furthermore, they indicate that today cyber-attacks consist primarily of:

- malware via attachments in the Internet browser, e-mail or other system vulnerabilities;

- denial of service (DoS) to prevent the use of computer systems and networks;

- deletion or transformation (leaving a message) to government and commercial websites for propaganda purposes or in order to disrupt the informing;

- unauthorized intrusion into systems for the theft of confidential and/or proprietary information, compromising of data or using the system in order to launch attacks against other systems.

In such circumstances of transformation and different views and understandings of security in general and international security, cyber threats certainly redefine those terms. In line with the efforts to ensure security on one hand and specificities of cyber threats and motives of the actors who initiate them on the other, it will be necessary to set up a new international security paradigm of the cyber age.

III. MULTIPOLARITY OF CYBERSPACE

The USA, Russia and China are nations known for their skilled military cyber units. In addition to the aforementioned states, France and Israel are working on the development of cyber capabilities. American intelligence officers believe that there are 20 to 30 armies with respectful capabilities for cyber-war, including Taiwan, Iran, Australia, South Korea, India, Pakistan and several NATO countries. The United States Cyber Command, along with the agencies they work with, has some of the most intelligent, patriotic-minded civil servants, both military and civilian, who create plans and capabilities for the domination in cyberspace with the goal of preserving the national security and peace [9].

Strategic domination in cyberspace has not yet been achieved by any of the entities of international relations. That is undoubtedly the goal of many nations such as the

1310

USA, China and Russia. However, as much as they might invest in their defense system and offensive capabilities, the system of power has not been set up. As opposed to the bloc division of the world into two centers of power during the Cold War, intimidation based on offensive capabilities is not crucial in cyberspace and there are many centers of power. The strength of those nations will mostly depend on the possibility of establishing an adequate defense system which is also influenced by their dependence on the information infrastructure. The dependence on information infrastructure is in correlation with the level of vulnerability of the economically and militarily developed digitized countries.

Due to the specificity of cyberspace, especially the asymmetry with the actual time and space and the geostrategic factors, a new security challenge that requires new military concepts is put before states and organizations. Namely, it is necessary to develop specific defense doctrines, but also offensive plans for action in cyberspace.

The dependence on networked computers and computer communication leaves the USA vulnerable to possible attacks, which made the cyber world a major source of uncertainty [6]. The vulnerability to attacks and the possibility of action is defined by Clarke and Knake [8] as the national cyber power. They state that the national cyber power is the net estimate of the ability of a nation to wage a cyber-war. National cyber power takes into account three factors: offensive cyber capabilities, national dependency on cyber networks and the nation's ability to control and defend its own cyberspace by implementing measures such as stopping the traffic outside the state. Based on these three factors, the authors provide an assessment of the overall cyber power of the United States, Russia, China, Iran and North Korea. To facilitate the comparison and analysis, the results of the assessment are systematized in the following table. The measurement scale goes from 1 to 10, with the smaller value representing a worse assessment and the higher value representing a better assessment.

Nation	USA	Russia	China	Iran	North Korea
Offensive capabilities	8	7	5	4	2
Dependency on the cyber network	2	5	4	5	9
Defensive capabilities	1	4	6	3	7

Table 1. Assessment of the national cyber power

They further explain why the USA, according to the assessment, is not the dominant power of cyberspace. If the total national cyber power was observed only on the basis of the offensive capabilities, the USA would occupy the first place. However, the outcome of a cyber-war does not depend only on the offensive capabilities. The important part is the dependence of a nation on the systems in cyberspace. Unlike the USA, China is developing its offensive cyber capabilities, but it is also oriented on the defense. Cyber warriors of the Chinese military have both offensive and defensive tasks in cyberspace and in contrast to the military of the USA, when talking about the defense, they also refer to the defense of the nation, i.e. the civil networks, not just the military networks. In China, the networks that make up their Internet infrastructure are under the control of the government. The Chinese government has the power and means to shut down the Chinese portion of the Internet from the rest of the world, which it would very likely do in case of a conflict with the USA. On the other hand, the USA has no plans or the capacity to do so, because their cyber connections are largely privately owned. China may limit the use of cyberspace in a crisis, refusing access to certain users. The USA cannot do it. North Korea has high scores when it comes to the defense and low dependence on the network infrastructure. Namely, that country may terminate its limited connections with cyberspace in an easier and more effective way than China. North Korea has few systems that are dependent on cyberspace that a large cyber-attack on its systems would have a minimal effect. The authors warn that one should bear in mind that cyber dependency is not the percentage of households with a broadband connection or the number of people who have smartphones, but the degree to which the critical infrastructure (electricity, railways, supply chains) dependent on the network systems. Thus, a state which is largely dependent on the systems in cyberspace has greater challenges in the creation of a national cyber defense. This is why the USA is more vulnerable to cyber-war than Russia or China. It is certainly more risky for the USA to engage in cyber-war than it is for a small country such as the North Korea. With three large entities of international relations (the USA, China and Russia) and the balance of power in cyberspace, the overall cyber power of two states that pose a threat to the world because of their totalitarianism and nuclear problems has been analyzed. Clarke and Knake estimate that they do not have great offensive capabilities, but have participated in the abuse of cyberspace.

The Iranian presidential election of 2009 sparked a huge public protest against election fraud. Social media platforms, mostly the two most popular, Twitter and Facebook, served for the organization, rebellion and spreading of anti-regime news. The Iranian government responded by introducing harsh police actions against the demonstrators, by shutting down media channels, and disabling Internet access within the country. Some members of the

1311

opposition launched DDoS attacks (distributed denial of service) against the websites of the Iranian government. Due to the speed and ease of communication, they used Twitter to organize and recruit cyber activists. They also used it to exchange links on an software that facilitated the inclusion of participants in the DDoS attack [7]. It is clear from the available data that this is not an international, but intrastate conflict. This is by no means a cybercrime because the attackers were politically motivated.

Because of its nuclear program, Iran was a target of an attack by the computer worm Stuxnet in June 2010. The worm was created to infect the industrial systems, and it proved its danger by sabotaging the Iran's nuclear program. In addition to the Iran's nuclear program, it also infected thousands of computers and industrial facilities worldwide. The Stuxnet worm can hide in cyberspace for a longer period. Analysts disclosed that the complex worm was written specifically for the breaching and taking control of the computer systems of Natanz nuclear facility in Iran. The worm takes very good care of itself for a longer period in cyberspace. Experts describe Stuxnet as a sophisticated piece of software with half a million program lines of code. For such a complex malware, it is necessary to have knowledge of the certain types of industrial control systems that are being attacked, and it seems that the code was written by an expert team, and not just one person [11]. Therefore, there was a suspicion that it was done by American or Israeli programmers. In an article published in the New York Times, Sanger [12] writes that the American President Obama ordered the cyber-attack on Iran, i.e. on the centrifuges used for the uranium enrichment.

North Korea, due to its poor technological development, is not very dependent on the systems in cyberspace. That is also the reason behind the very good assessment of their defense capabilities. Even though it has no developed offensive capabilities, it is obvious that it has recognized the importance of playing an active role in cyberspace. In fact, in July 2009, a few dozen American websites, including the website of the White House, were under a DDoS attack (denial of service). The main suspect was North Korea. That status was confirmed after the attacks spread to South Korea. The South Korean media and government officials publicly accused its northern neighbor, and the officials of the USA advocated a cyber-counterattack "in order to send a strong message" [7]. In November 2014. a group which calls itself GOP or The Guardians Of Peace, hacked its way into Sony Pictures and stole the data that included personal information about the Sony Pictures employees and their families, e-mails between the employees, information about the executive salaries at the company, copies of the then-unreleased Sony films, and other information [9]. The purpose of the attack, attributed to North Korea, was to deter Sony Pictures from releasing a movie which was (correctly) understood as ridiculing that country's dictator

and portraying the North Korean regime and its leader, Kim Jong-Un, with sarcasm and mockery [10].

IV. CONCLUSION

The topic of the paper, cyber threats to international security, stands out merely by its title as an interesting and challenging area of research. The explanation for it is first and foremost that the area has not yet been sufficiently explored, especially not in the Croatian context. Due to the intensive development of international relations in cyberspace, conditioned and supported by the speed of the development of technologies and their implementation in the relations of states, organizations and individuals, this area will always be interesting and challenging. That conclusion arises from the constant change of attitudes and technology. It is precisely that instability which indicates that from that specific, interdisciplinary field of research, in 5 or 10 years, it will be possible to draw some new conclusions, and according to them, set some new paradigms and doctrines. Carr [7] states that cyber-warfare has been present for about a decade, but that it is still not well defined. There is no valid international agreement which would establish a legal definition of an act of cyber aggression. In fact, the entire area of international cyber law is still unclear.

The development and availability of information and communications technologies and the ever-present tensions between politically and ideologically different states have conditioned the international relations in cyberspace. Strategic domination in cyberspace has not yet been achieved by any of the entities of international relations. A large number of international entities demonstrated their presence and willingness to act in cyberspace. That demonstrates a multipolar dimension of cyberspace in which it is very unlikely that domination or bloc division will occur. The reasons lie in the mutual mistrust and fear of espionage in the case of linking the defense systems. However, the nations that are the most influential are the ones that are the most powerful, economically and militarily, and at the same time are the most dependent of the cyber-infrastructure – the USA, Russia and China. NATO also plays an active role. We can conclude that in the recent years, a new concept of cyber security that can be defined as a paradigm of the multipolarity of cyberspace is being created.

Most authors predict an escalation of conflicts and intelligence activities in cyberspace, which supports the confirmation of the initial hypothesis of this paper. They state that cyber-attacks are among the biggest threats to the international security. Unlike conventional conflicts, such attacks will become increasingly common, and they could, as a conventional attack, cause large-scale destruction, even with fatal consequences. It is therefore essential to establish an effective defense in which the key role is that of prevention, international cooperation and the adoption of the internationally recognized, legally binding norms.

Due to the increase in cyber-terrorism and crime, it is necessary to organize systematic education and to strengthen operational military, intelligence, police and civil centers for the defense from cyber-attacks.

If we take into consideration all that has been stated in the elaboration, and the confirmation of the initial hypothesis, we can conclude that cyber security has become one of the prerequisites of the democratic concept of life in the modern society.

REFERENCES

[1] NATO, "Strategic concept for the defence and security of the members of North Atlantic Treaty Organization," 2010, Available: http://www.nato.int/nato_static/assets/pdf/pdf_publications/20120214_strategic-concept-2010-eng.pdf (3.2.2017.)

[2] N. Choucri and D. Goldsmith, "Lost in cyberspace: harnessing the Internet, international relations, and global security," Bulletin of the Atomic Scientists, vol. 68, no. 2, 2012, pp. 70-77.

[3] S. Tatalović, A. Grizold, and V. Cvrtila, Suvremene sigurnosne politike. Zagreb: Golden marketing-Tehnička knjiga, 2008.

[4] H. Lin, "A virtual necessity: some modest steps toward greater cybersecurity," Bulletin of the Atomic Scientists, vol. 68, no. 5, 2012, pp. 75-87.

[5] L. J. Janczewski and A. M. Colarik, Cyber warfare and cyber terrorism. Hershey: Information Science Reference, 2008.

[6] J. S. Nye, "Cyber war and peace," 2012, Available: http://www.project-syndicate.org/commentary/cyber-war-and-peace (3.2.2017.)

[7] J. Carr, Inside cyber warfare, 1st ed. Sebastopol, CA: O'Reilly Media, 2010.

[8] R. A. Clarke and R. K. Knake, Cyber war: the next threat to national security and what to do about it. New York: Ecco, 2010.

[9] Risk Based Security, "A Breakdown and Analysis of the Sony Hack," 2014, Available:
https://www.riskbasedsecurity.com/2014/12/a-breakdown-and-analysis-of-the-december-2014-sony-hack/#thebeginning (9.4.2017.)

[10] G. Siboni and D. Siman-Tov, "Cyberspace Extortion: North Korea versus the United States," INSS Insight No. 646, 2014, Available: http://www.inss.org.il/uploadImages/systemFiles/No.%20646%20-%20Gabi%20and%20Dudi%20for%20web.pdf (9.4.2017.)

[11] M. Petrović, "Obrana od cyber-napada", Hrvatski vojnik, vol. 9, no. 385, 2012, pp. 26-29.

[12] D. E. Sanger, "Obama order sped up wave of cyberattacks against Iran," The New York Times, 2012, Available: www.nytimes.com/2012/06/01/world/middleeast/obama-ordered-wave-of-cyberattacks-against-iran.html?pagewanted=1 (3.2.2017.)

Cloud Computing Threats Classification Model Based on the Detection Feasibility of Machine Learning Algorithms

Z. Masetic*, K. Hajdarevic**, N. Dogru*

* International Burch University, Sarajevo, Bosnia and Herzegovina
** Faculty of Electrical Engineering, University of Sarajevo,
Sarajevo, Bosnia and Herzegovina

zerina.masetic@ibu.edu.ba, khajdarevic@etf.unsa.ba, nejdet.dogru@ibu.edu.ba

Abstract – Cloud computing became very popular in past few years, and most of the business and home users rely on its services. Because of its wide usage, cloud computing services became a common target of different cyber-attacks executed by insiders and outsiders. Therefore, cloud computing vendors and providers need to implement strong information security protection mechanisms on their cloud infrastructures. One approach that has been taken for successful threat detection that will lead to the successful attack prevention in cloud computing infrastructures is the application of machine learning algorithms. To understand how machine learning algorithms can be applied for cloud computing threat detection, we propose the cloud computing threat classification model based on the feasibility of machine learning algorithms to detect them. In this paper, we addressed three different criteria types, where we considered three types of classification: a) type of learning algorithm, b) input features and c) cloud computing level. Results proposed in this paper can contribute to further studies in the field of cloud threat detection with machine learning algorithms. More specifically, it will help in selecting appropriate input features, or machine learning algorithms, to obtain higher classification accuracy.

I. INTRODUCTION

Cloud computing is a technology that delivers IT resources and applications as a service, over the Internet, with pay-as-you-go pricing. It consists of number of individual computing nodes with corresponding networking and storage subsystems [1]. Cloud technologies and services have been used by individuals or companies daily, from storage services using Google drive to chatting services using Skype. Skype, which was known as peer to peer architecture (P2P), has been transitioning from P2P to cloud based architecture, with improved existing features and the new ones that are launched, such as mobile group video calling, Skype Translator, etc. [2].

Most of the services such as emails, search engines, social networks are hosted in the cloud [3]. Adopting cloud computing over traditional data center shows several benefits, and National Institute of Standards and Technology (NIST) defined it in following way [4]:

"Cloud computing is a model for enabling ubiquitous, convenient, on-demand network access to a shared pool of configurable computing resources (e.g. networks, servers, storage, applications, and services) that can be rapidly provisioned and released with minimal management effort or service provider interaction."

A. Cloud Computing Security Overview

However, the biggest concerns of cloud users and main hurdles to move their in-house business to a public cloud is security. The issue might be that cloud vendors do not provide security of the data while being transported to the cloud [1]. Moreover, remote access is under the client's responsibility and clients need to have a strong knowledge of security aspects related to the networking and Internet, before obtaining access to a cloud. Nothing connected to the Internet is perfectly secured. Therefore, many data breaches happen [1, 5].

Stealing information is the biggest concern when it comes to a cloud computing. According to the Breach Level Index website [6], totally 3 989 114 567[1] records were lost since 2013. In 2016, around 554 million records were stolen, with highest peak in October (around 472 million stolen records). Even the technology giants were not left out, namely Microsoft, with the 8 recorded data breaches in period 2013 – 2015 [6].

Other popular companies in the field of technology attacked by data breach are shown in Table 1.

TABLE 1. STATISTICS ON SOME DATA BREACHES IN TECHNOLOGY SINCE 2013 [6]

Organization Breached	Records Breached	Year
Evon Gaming Company	33 million	2016
Apple Inc.	10 000	2016
Minecraft (Microsoft)	2.8 million	2015
CyberVor	1.2 billion	2014
Korea Telecom	12 million	2014
Gmail	5 million	2014
iCloud Apple	2 million	2014

[1] Number of records stolen has been growing montly. This number is obtained on June, 2nd, 2016.

SalesForce	2 million	2014
Adobe Systems	152 million	2013
Facebook	6 million	2013

Source of the breach for each of them was malicious outsider, and all of them are with severe and catastrophic risk score [6].

Many cloud attacks such as phishing and fraud, unauthorized access, malicious insider, insecure interfaces, etc., lead to the data breach [7, 8] .

This paper presents the model for classification of cloud security threats based on the feasibility of machine learning algorithms to detect them.

Machine learning algorithms are powerful tools, being used in financial services, government, health care, transportation, information security, etc. for different prediction problems. These algorithms get computers to act without being explicitly told or programmed how to do something. Moreover, they make computers learn from the past experiences and solve some future problems. Considering their wide usage, we decided to see whether we can apply machine learning algorithms for cloud computing threat detection. Therefore, this paper provides general overview of the detection model with machine learning algorithms, in the field of cloud computing security.

The work presented in this paper is the part of the larger study and PhD work of the first author who recognized, in the process of the literature review research in cloud computing security, that there is no such available classification model. Therefore, decision was made to contribute by making this classification model.

This paper is organized as follows: Section II outlines the review of the literature on cloud security threats. Section III introduces our proposed model for threat classification based on the feasibility of machine learning methods to detect them. Section IV concludes and ends the paper.

II. RELATED WORK

Cloud computing security threats and their classification is the subject of research in many articles [9, 10, 11, 12, 13, 14].

A. Overview of Cloud Computing Threats Classifications

Ahmed, Litchfield and Ahmed [9] provided the generalized threat taxonomy in a cloud computing environment. In their taxonomy, they involved human and technological factors as root categories, from which security challenges in cloud computing emerge. Chraibi, Harroud and Maach [14] provided taxonomy for cloud computing security issues according to the level to which they belong. They identified security issues on hardware level, virtual machine manager (VMM) level, guest operating system level, application level, network level and governance level. Jouini, Rabai and Aissa [10] proposed the model for threat classification, based on the following threat classification criteria: source, agent, motivation, intention; and their potential impacts: destruction of information, corruption of information, theft

or loss of information, disclosure of information, denial of use, elevation of privilege and illegal usage.

Srinivasan, Sarukesi and Rodrigues [13] categorized security challenges based on architectural & technological aspects: logical storage segregation & multitenancy security issues, identity management issues, insider attacks, virtualization issues, cryptography and key management, and process & regulatory related aspects: governance and regulatory compliance gaps, insecure APIs, cloud & CSP migration issues, SLA & trust management issues. Gupta and Kumar [11] provided an exhaustive cloud security classification, by location of security attack and based on the cloud layer. Each category is further divided into different subcategories. Hashemi and Ardakani [12] provided taxonomy of security threats in cloud computing, based on the service models of cloud computing, as well as the administration.

However, it is important to know what are the most important cloud computing threats in recent years. Therefore, Cloud Security Alliance [15], a not-for-profit organization that promotes the use of best practices to provide security assurance within cloud computing, identified twelve cloud computing threats specifically related to the shared, on-demand nature of cloud computing, and ranked them by the severity of each threat.

These threats are: a) data breach, b) insufficient identity, credential and access management, c) insecure interfaces and APIs d) system vulnerabilities, e) account hijacking, f) malicious insiders, g) advanced persistent threats, h) data loss, i) insufficient due diligence, j) abuse and nefarious use of cloud services, k) denial of service, l) shared technology issues [16].

B. Threat Classification Principles

The most important thing when it comes to the cloud computing security and cloud computing threats is to successfully detect those threats, and efficiently protect the system from attacks.

One of the approaches that has been taken for threat detection and classification is the application of machine learning algorithms. Many research articles focused on cloud computing threat classification, based on the system performance information, or based on the information within the packet that has been exchanged over the network, when some attack occurs. They are summarized in the Table 2.

Classification results which are shown in the Table 2 show high classification accuracy of the machine learning algorithms that have been used in previous work. Furthermore, sensitivity or true positive rate (TPR) which shows the rate of correctly classified attack cases, and false positive rate (FPR) which shows the rate of non-attack cases, incorrectly classified as attack cases, are included in the table.

Therefore, it would be beneficial to identify what cloud computing security threat types could be detected using machine learning algorithms. The aim of this paper is to make cloud computing threats classification model, which will help in identification and organization of security threats into the classes, finding out the best possible way to detect them and protect system and data from the attacks.

Based on the previous research works in the field of cloud threat detection with machine learning algorithms, we have identified criteria to our threat classification:

- type of machine learning algorithm
- input data features
- cloud computing level

TABLE 2. PREVIOUS STUDIES ON APPLICATION OF MACHINE LEARNING ALGORITHMS ON CLOUD COMPUTING THREAT DETECTION

Author (s)	Meth.	Data	Threat	Class. acc.	TPR	FPR
Parveen, Weger, Thuraisingham, Hamlen & Khan [17]	Support Vector Machine (SVM)	System logs data	Insider threat	87%	-	13%
Chiu, Yeh & Lee [18]	Machine learning algorithms for mining the frequent patterns	System logs data	DDoS, backdoor	99.4%	100%	2.6%
Kumar, Lal & Sharma [19]	One-class SVM	Network traffic features	DoS attacks	68% - 100%	43%-100%	-
Bhat, Patra & Jena [20]	Naïve Bayes, Naïve Bayes + Random Forest	Network traffic features	Probe, DoS, U2R, R2L	99.5% 99%	99.7% 99.1%	2% 2%
Chen et al. [21]	k-means, Naïve Bayes	Network traffic features	DDoS, Port scanning	90% 90%	-	0.5% 1.8%
Joshi, Vijayan & Joshi [22]	Neural Networks	Network traffic features	DoS attacks	75%	76%-81%	-
Modi, Patel, Patel & Muttukrishnan [23]	Bayesian Classifier	Network traffic features	Intrusions	Avg: 96%	Avg: 94%	Avg: 0.94%
Khorsed, Ali & Wasimi [24]	C4.5 Decision Tree	System performance data	DoS attacks	93.47%	-	-
Khorsed et al. [25]	Random Forest	System performance data	DoS attacks	93.9%	93.9%	1.3%
Win, Tianfield & Mair [26]	SVM	System call data	Malware actions	-	-	-
Prwez & Chatterje [27]	AdaBoost	Network traffic data	U2R, DoS, port scanning	-	-	-
Mishra, Pilli, Varadharajan &	Gradient Boosting, Random Trees	Network traffic features	DoS, Conficker	86.67%-94.54%	-	2.7%-6.23%
Tupakula [28]	Embedding, Random Forest, Logistic Regression					
Chou & Wang [29]	Decision Tree	Network traffic features	Malicious activities	-	94.79%	4.28%
Li, Sun & Wang [30]	Neural Networks	Network traffic features	Intrusion	99.7%	-	-

III. THE PROPOSED CLASSIFICATION MODEL

In our paper, we propose threats classification based on three criteria:

- Criterion 1: Machine Learning Algorithm Type
- Criterion 2: Input Data Features
- Criterion 3: Cloud Computing Level

A. Criterion 1: Machine Learning Algorithm Type

We classified cloud computing security threats firstly by machine learning algorithm, or categories of learning, these algorithms belong to: supervised and unsupervised learning. Supervised learning is the classification learning where the actual outcome for each training example is provided, whereas in unsupervised learning, training example outcome is not specified. Classification learning success can be judged by trying out what has been learned on the independent set, with known, true classification outcomes. When there is no specified outcomes/class, items are grouped based on the features that denote similar items [31]. Popular supervised learning algorithms are support vector machine (SVM), random forest (RF), artificial neural networks (ANN), decision tree (DT), Naïve Bayes (NB), etc., and clustering (k-means, expectation-maximization) belongs to unsupervised learning algorithms [32]. Selecting appropriate machine learning algorithm is a critical step. Algorithm is selected based on different criteria, but usually it is the prediction accuracy (or percentage of correctly predicted items over the total number of items). Furthermore, computational time, handling high dimensional values or overfitting are also important factors to consider, when selecting machine learning algorithm [33].

B. Criterion 2: Input Data Features

Each instance that provides an input to machine learning algorithms is described by the number of attributes where each attribute can have number of values. Therefore, cloud attacks have attributes that describe them and are input to machine learning algorithms. It is important to determine appropriate input data to the machine learning algorithms, as classification performance depends much on it. However, determining input features is not easy and it requires considerable effort, as it is hard to determine what features are important and what not, and in most cases, it contains noise and missing data, which affects the classification performance and its classification accuracy. In our research, we analyzed two categories of input data: system performance data [16, 17, 23, 24, 25], such as: CPU, network, memory, disk, hypervisor, guest OS performance when the attack starts and during its execution, and network traffic features, or packet data being exchanged over the network while executing an attack [18, 19, 20, 21, 22, 26, 27, 28, 29]: source and destination IP address, protocol, packet length, port numbers, flags, etc. While considering system performance data, it is important to detect activity pattern by looking at the change of parameters in the computer. For network packet data, packet sniffer is used to catch all packets being sent during the attack execution. This step also involves features selection process, where irrelevant and redundant features are removed. This enables machine learning algorithms to operate faster and more efficiently.

C. Criterion 3: Cloud Computing Level

Finally, attack categories to which cloud security attacks belong are added to the whole schema. We classified cloud security threats into two broad categories, considering the level of cloud on which attack might happen: network specific threats and cloud computing specific threats. When developing cloud solutions, important factor to consider is network security, as network is a vulnerable spot for different attacks. The purpose of network attacks is one of the following: to monitor the network and obtain useful information, to change the content with the intent to destroy or corrupt the data, or to consume network's resources. Popular network attacks in the cloud are account or service hijacking that happens when the attacker eavesdrops on the user's activities and transactions, or denial of service (DoS) attacks that prevent user from a normal use of resources. Cloud computing specific threats are major threats in cloud computing too. Cloud computing specific threats include insecure interfaces and APIs, which are offered to clients for the interaction with cloud services, shared technology vulnerabilities, or malicious insiders who has an authorized access to organization's network, system or other resources and intentionally misuse it for some malicious activities [16].

Fig. 1 represents the diagram of the cloud threat classification, considering all three criteria that we proposed above.

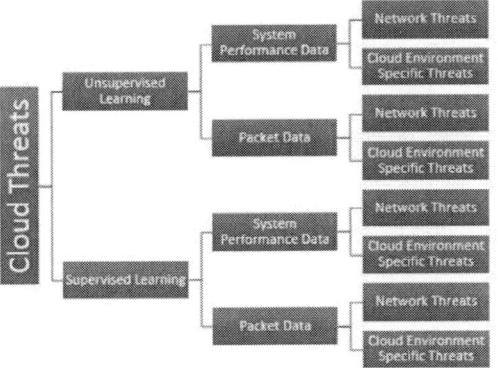

Figure 1. Diagram of the cloud computing threats classification model based on the detection on feasibility of machine learning algorithms.

Fig. 1 shows that network specific threats and cloud computing specific threats can be described by both, system performance and network traffic data. The choice depends on the aim of the research, tools that will be used in the experimentation phase, interest of the researcher and what a researcher is trying to find out or prove.

Furthermore, system performance data and network traffic data are input to both types of learning, supervised and unsupervised. The researcher decides on the machine learning algorithm by considering different criteria, such as complexity of the algorithm, computational time, handling missing values, overfitting, etc.

IV. CONCLUSIONS

Popularity of the cloud computing has been increasing over the years, which made the cloud computing services a target for many cloud specific attacks. Security of the cloud computing is a critical factor for both, cloud providers and cloud users. Therefore, cloud providers need to implement strong protection system on their cloud infrastructure. In this paper, we have developed a threat classification model that uses machine learning algorithms as tools for threats detection. In our model, we considered three criteria: a) type of machine learning algorithm, b) input data features, and c) cloud computing level. First criterion considers two types of learning: supervised and unsupervised. Second criterion considers system performance data and network traffic data, which are input to both types of learning. Finally, last criterion considers network specific and cloud environment specific threats, which are described by system performance and network traffic data. Results proposed in this paper can contribute in further studies in the field of cloud computing threat detection with machine learning algorithms in a way that it can help researchers to select the appropriate learning algorithm for their research, or appropriate input features to the selected algorithm. Proposed model can be optimized, by extending it with additional criteria, such as: threat motivation (malicious, or non-malicious), or threat intention (accidental, or intentional).

REFERENCES

[1] N. Antonopoulos and L. Gillam, Cloud Computing: Principles, Systems and Applications, Springer, 2010.

[2] Skype Team, "Skype – the journey we've been on," 20 July 2016. [Online]. Available: https://blogs.skype.com/news/2016/07/20/skype-the-journey-weve-been-on/.

[3] S. Singh, B. K. Pandey, R. Srivastava, N. Rawat, P. Rawat and Awantika, "Cloud Computing Attacks: A Discussion With Solutions," *Open Journal of Mobile Computing and Cloud Computing,* vol. 1, no. 1, 2014.

[4] P. Mell and T. Grance, "The NIST Definition of Cloud Computing," National Institute of Standards and Technology, 2011.

[5] D. M. Kroenke, Using MIS, Pearson, 2014.

[6] "Breach Level Index," [Online]. Available: http://breachlevelindex.com/#!home. [Accessed 15 March 2016].

[7] H. Kuchler, "10% of phishing scams lead to data breach," 26 April 2016. [Online]. Available: https://www.ft.com/content/9ef7aebc-0b1a-11e6-9456-444ab5211a2f.

[8] A. Collar, "Most Common Causes of Data Breaches," March 2015. [Online]. Available: http://www.oriontech.com/most-common-causes-of-data-breaches/.

[9] M. Ahmed, A. T. Litchfield and S. Ahmed, "A Generalized Threat Taxonomy for Cloud Computing," in *25th Australasian Conference on Information Systems*, 2014.

[10] M. Jouini, L. B. A. Rabai and A. B. Aissa, "Classification of security threats in information systems," in *5th International Conference on Ambient Systems, Networks and Technologies (ANT-2014)*, 2014.

[11] S. Gupta and P. Kumar, "Taxonomy of Cloud Security," in *International Journal of Computer Science, Engineering and Applications (IJCSEA)* , 2013.

[12] S. M. Hashemi and M. R. M. Ardakani, "Taxonomy of the Security Aspects of Cloud Computing Systems - A Survey," *International Journal of Applied Information Systems (IJAIS)*, 2012.

[13] M. K. Srinivasan, K. Sarukesi and P. Rodrigues, "State-of-the-art Cloud Computing Security Taxonomies - A classification of security challenges in the present cloud computing environment," in *ICACCI '12 Proceedings of the International Conference on Advances in Computing, Communications and Informatics*, 2012.

[14] M. Chraibi, H. Harroud and A. Maach, "Classification of Security Issues and Solutions in Cloud Environments," in *IIWAS '13 Proceedings of International Conference on Information Integration and Web-based Applications & Services*, 2013.

[15] "About CSA," [Online]. Available: https://blog.cloudsecurityalliance.org/about-cloud-security-alliance/. [Accessed 15 March 2016].

[16] C. S. A. Top Threats Working Group, "The Treacherous 12 Cloud Computing Top Threats in 2016," Cloud Security Alliance, 2016.

[17] P. Parveen, Z. R. Weger, B. Thuraisingham, K. Hamlen and L. Khan, "Supervised Learning for Insider Threat Detection Using Stream Mining," in *2011 IEEE 23rd International Conference on Tools with Artificial Intelligence*, 2011.

[18] C.-Y. Chiu, C.-T. Yeh and Y.-Y. Lee, "Frequent Pattern based User Behavior Anomaly Detection for Cloud System," in *2013 Conference on Technologies and Applications of Artificial Intelligence*, 2013.

[19] R. Kumar, S. P. Lal and A. Sharma, "Detecting Denial of Service Attacks in the Cloud," in *2016 IEEE 14th Intl Conf on Dependable, Autonomic and Secure Computing*, 2016.

[20] A. H. Bhat, S. Patra and D. Jena, "Machine Learning Approach for Intrusion Detection on Cloud Virtual Machines," *International Journal of Application or Innovation in Engineering & Management (IJAIEM)*, vol. 2, no. 6, 2013.

[21] Z. Chen, G. Xu, V. Mahalingam, L. Ge, J. Nguyen, W. Yu and C. Lu, "A Cloud Computing Based Network Monitoring and Threat Detection System for Critical Infrastructures," *Big Data Research*, 2015.

[22] B. Joshi, A. S. Vijayan and B. K. Joshi, "Securing Cloud Computing Environment Against DDoS Attacks," in *International Conference on Computer Communication and Informatics (ICCCI-2012)*, 2012.

[23] C. N. Modi, D. R. Patel, A. Patel and R. Muttukrishnan, "Bayesian Classifier and Snort based Network Intrusion Detection System in Cloud Computing," in *Third International Conference on Computing, Communication and Networking Technologies (ICCCNT'12)*, 2012.

[24] T. M. Khorsed, S. A. B. M. Ali and S. A. Wasimi, "Classifying different denial-of-service attacks in cloud computing using rule-based learning," *Security and Communication Networks*, vol. 5, pp. 1235-1247, 2012.

[25] M. T. Khorshed, N. A. Sharma, K. Kumar, M. Prasad, A. B. M. S. Ali and Y. Xiang, "Integrating Internet-of-Things with the power of Cloud Computing and the intelligence of Big Data analytics — A three layered approach," in *2nd Asia-Pacific World Congress on Computer Science and Engineering (APWC on CSE)*, 2015.

[26] T. Y. Win, H. Tianfield and Q. Mair, "Detection of Malware and Kernel-level Rootkits in Cloud Computing Environment," in *IEEE 2nd International Conference on Cyber Security and Cloud Computing*, 2015.

[27] T. M. Prwez and K. Chatterjee, "A framework for Network Intrusion Detection in Cloud," in *6th International Advanced Computing Conference*, 2016.

[28] P. Mishra, E. S. Pilli, V. Varadharajan and U. Tupakula, "NvCloudIDS: A Security Architecture to Detect Intrusions at Network and Virtualization Layer in Cloud Environment," in *2016 Intl. Conference on Advances in Computing, Communications and Informatics (ICACCI)*, 2016.

[29] H.-H. Chou and S.-D. Wang, "An Adaptive Network Intrusion Detection Approach for the Cloud Environment," in *The 49th Annual International Carnahan Conference on Security Technology (ICCST)*, 2015.

[30] Z. Li, W. Sun and L. Wang, "A Neural Network based Distributed Intrusion Detection System on Cloud Platform," in *IEEE CCIS2012*, 2012.

[31] I. H. Witten, E. Frank and M. A. Hall, Data Mining: Practical Machine Learning Tools and Techniques, Elsevier Inc., 2011.

[32] I. H. Witten, E. Frank and M. A. Hall, Data Mining - Practical Machine Learning Tools and Techniques, Elsevier , 2011.

[33] I. Maglogiannis, K. Karpouzis , B. A. Wallace and J. Soldatos, Emerging Artificial Intelligence Applications in Computer Engineering, IOS Press, 2007.

Legal Framework Issues Managing Confidential Business Information in the Republic of Croatia

Goran Vojković, Ph.D.
Melita Milenković, LL.M.

University of Zagreb
Faculty of Transport and Traffic Sciences
Vukelićeva 4, Zagreb, Croatia
E-mail: goran.vojkovic@fpz.hr
melita.milenkovic@fpz.hr

Abstract – Confidential business information in the Republic of Croatia is regulated in an insufficient mode - the Data Confidentiality Act provisions are still on force for more than 20 years, and entirely inadequate for today's time. Considering that information is now greatly kept in electronic form, it is necessary by the internal documents, and in accordance with the existing outdated legislation, to regulate in detail the issues of confidential business information. In case of exchange of business information which are confidential - it is necessary to sign the appropriate contract between the parties. This paper is about existing of legal shortcomings and how to overcome them in practice.

I. INTRODUCTION

On 13 of July 2007, that is nearly 10 years ago, Croatian Parliament passed the Data Confidentiality Act [in Croatian: Zakon o tajnosti podataka, further: ZTP] [1] which was in the matter of data confidentiality arranged on a fundamentally different level than the old Act on Protection of Data Confidentiality [in Croatian: Zakon o zaštiti tajnosti podataka, further: ZZTP] [2], back from the 1996.

The scope of the ZTP is defined significantly narrower than the scope of the old Act because the purpose of the ZTP is something essentially different. In fact, in the old Act from 1996 it has been stated that it "stipulates the concept, types and degrees of confidentiality and the measures and the procedures for determining, use and protection of confidential business information", and it regulates the state, military, official, business and professional confidentiality.

Unlike the above, Art. 1, para. 2 of the ZTP, which determines: "This Act shall apply to the state authorities, local and regional government, legal persons with public authorities and legal and natural persons, who in accordance with this Act, gain access or handle classified and unclassified information." [3]

The definition of classified and unclassified business information, in the art. 2 of ZTP, according to which, classified information is the one that the competent authority, (in the prescribed procedure), marked and determined by the degree of confidentiality. As well as the fact that the Republic of Croatia was handed by data from another state, an international organization or institution that the Republic of Croatia cooperates with. Furthermore,

unclassified data is data without the determined level of classification, which is used for official purposes, as well as the fact that the Republic of Croatia was also handed by data from another state, an international organization or institution that the Republic of Croatia cooperates with. ZTP harmonized categories of data confidentiality, in accordance with international safety standards (top secret, secret, confidential or restricted). [4]

For our presentation, it is important to emphasize the fact that the scope of the ZTP is limited on the information whose confidentiality is of public interest and the information without an established level of classification, used for official purposes. ZTP does not regulate the question of confidential business information.

In the ZTP's Final Proposal [5] it states: "This Act does not prescribe the terms of business and professional confidentiality, unlike the ZZTP (Croatian Official Gazette 108/96), since it is not common in the legislation of the most EU and NATO's member states. Therefore, the provisions from the Head 8 and the Head 9 will remain in force until this matter is not regulated by the other relevant laws."

ZTP's First Proposal had an interesting observation: "The confidentiality of data of legal and natural persons outside the state system (including private companies) is regulated by their own internal rules, especially within those companies who have their headquarters in the other countries and had already established a proper system of confidentiality of their data". [6]

However, except for certain adjustments of the Croatian Criminal Act [7], the new regulations have not been enacted. Therefore, the question of confidential business information in the Republic of Croatia is solved by the transitional provisions of ZTP and by the two heads of the old ZTTP's, from 1996.

In more than 20 years, there has been a remarkable development of the high-speed Internet, as well as completely new business models, which include engagement of many business subjects, on one or several projects. Also, the business innovation and the 'know-how' is becoming increasingly important in today's modern business practice.

In short, questions of regulation and storage of confidential business information are getting more important regardless the fact that Croatian legal framework is based on several out-dated legal provisions. This deficiency is partly possible to be solved in practice – by the quality agreements between business partners but it is necessary to be aware of this deficiency. Foreign business partners can come across this problem, during the fact that they are not familiar with the Croatian legal practice.

II. CONFIDENTIAL BUSINESS INFORMATION BY THE CURRENT REGULATIONS

ZZTP regulates confidential business information in the head VIII, in the articles 19 - 26.

According to this Act, the confidential business information is represented by the information which is defined by the Act of law, also other regulation or by the general company Act, or by the other legal entities, which represent confidential business information, results of research or construction work and other information whose disclosure to an unauthorized person could have harmful effects on its economic interests (ZZTP Art. 19, para. 1). Further, it states that the general Act cannot determine that all the information relating to the business of legal persons is considered for confidential business information nor can such information be in the contrary to the legal person's interests. (ZTTP Art. 19, para. 2).

In the Article 19, para 1, the legislator wanted to restrict confidential business information to the information that makes sense for declaring it confidential, and not for confidential business information to become a way of introducing police state for the economic subjects. [8]

Here we come to the first important provision related to the confidential business information – which must be stipulated. At the same time, the legislation adopted by the state is an exception - usually, confidential business information should be stipulated by the internal act of the organization (institution, company or association). If the organization fails to adopt such an act and if it is not systematic and appropriate to business organizations – then the whole concept of confidential business information cannot be applied.

By doing so, we must avoid the general rulebooks where the only guidance for confidential business information is: "everything that a responsible person declares as a confidential business information", because there is a high probability that at some changes of the extent and the type of business of the organization, part of the data remains unprotected.

The legal person is obliged to keep as a secret the information, which has been learned from the other legal entities (ZZTP's Art. 20, p. 1). The provision is obviously ambiguous: what's happening with the natural persons - sole traders and natural persons who are self-employed or who are employed by the service contracts? [8]

The question is: how can a legal entity know if the information that it received from other entities is indeed confidential. The simplest answer is - "it says on the paper" which would hardly be enough in the era before computers. If the relationship between the two parties includes access, processing of data which is confidential business information, it is necessary to sign an appropriate non-disclosure agreement (NDA). (see below)

ZZTP does not speak a lot about NDA, but because confidential business information is extensively described; it is giving a wrong message that this issue is resolved through legislation. As we have said, this is referred only to the legal persons, but contemporary business often includes external teams of experts, freelancers and similar categories of non-classical legal persons. It is a failure of the legislator, however, originated in another time, because the complex forms of work from today were not common.

The ZZTP in its provisions if there is a proper general act, in which any legal person; (the Act forgot to add craftsmen and freelancers!); but more closely it regulates the issue of confidential business information in accordance with its business and needs.

General Act elaborates questions of confidential business information and in the Art. 21 of ZTTP it states, where it provides a framework for handling with classified information. The information that is considered confidential business information under The General Act can be transmitted to the other persons by the persons authorized and determined by The General Act (Art. 21, para. 1). Also, The General Act describes in detail the cases and the manner of protection when the above information may be disclosed to the other persons (Art. 21 para. 2).

The general "guidelines" are given on the information which is supposed to be described by the regulations of the legal person. The legal person will be determined by the authorized person or by the special body that has access to confidential business information, the task of their storage, and deciding which of the persons employed in that legal person could be authorized to keep those confidential business information, and to which persons' confidential business information could be told (Art. 21, para. 3).

The data that are considered as confidential business information should not be disclosed or made accessible to unauthorized persons, unless a special act provides differently (Art. 22, para. 1). The employees are obliged to keep the confidential business information if they find out an information that is the confidential business information (Art. 22, para. 2). That limits the ability for disclosure of the confidential business information by an informal communication.

The 24th Article of the ZZTP's consists of exceptions to the provisions which provide obligation of keeping the confidential business information. It is not considered a violation of the disclosure of the confidential business information to the natural or legal persons, to which, such information can or should be disclosed: 1) under the basis of this Act and other regulations, 2) under the authority which is resulting from the duties they perform, the position or the workplace where they are employed (Art. 24, para. 1).

Also, it is not considered as a violation of confidential business information the disclosure of information that are considered as confidential business information in the sessions, if for example, the communication as such is necessary to perform the work (Art. 24 para. 2). The authorized person (at the meeting), who disclosures confidential business information should inform those present at the meeting that the information is considered confidential, and that they are obliged to keep the information confidential (Art. 24, para. 3).

The 25th Article of the ZZTP's stipulates that as a violation of confidential business information, it won't be considered as a statement that a person familiar with the confidential business information gives while reporting the crime, (also an economic offense or misdemeanour), to the competent body for the exercise of their labour rights.

The 26th Article of ZZTP's contains a general provision according to which a general act of a legal person specifies the method of administration and storage of information that is considered confidential business information, measures, procedures and other circumstances of keeping the confidential business information.

III. MARKING OF THE CONFIDENTIAL BUSINESS INFORMATION IN THE CROATIAN LEGISLATION

As already stated, ZTP's Article 34 has determined that by its coming into force, provisions of the ZZTP ceased to be valid, except of the provisions stated in the Head 8 and 9, provisions about business confidential information.

However, the manner of marking the business confidential information is not stated in the Head 8 or 9! How to mark that some information is confidential? In our opinion, it is necessary to apply the ZZTP's provisions which ceased to be valid, but the ones that are writing about marking the business confidential information.

According to the ZZTP's business confidential information has its own type and level. The 3rd Article of the ZZTP has determined, to the type of confidentiality, the confidential business information can be state, military, official, business or professional, and according to the level of confidentiality – state's confidentiality, top-secret, secret and confidential.

In accordance with the ZZTP's Article 5, para. 2, information representing the confidential business information can have level of business confidential information as 'top-secret', 'secret' and 'confidential' and they are marked by the appropriate mark, type or level of confidentiality. Therefore, to the ZTTP, the proper way of marking was: 'business confidential information – as top secret', 'business confidential information – as secret' or 'business confidential information – as confidential'. [8]

We can certainly recommend that such marking could be used today – for, it is simple and understandable, even if that part of the ZTTP is not in force. What can prove to be a problem - is that this is the authors' opinion, and as such the court in the case of a dispute, does not have to accept it.

IV. THE CONFIDENTIAL BUSINESS INFORMATION IN THE CRIMINAL LEGISLATION

The Republic of Croatia regulates the protection of confidential business information in the criminal legislation by very strict (long) prison sentences. In the Article 262 of the Criminal Act, it states that whoever, without an authorization, delivers or otherwise makes available any information that is confidential business information, or whoever collects such information with an aim of delivering it to an unauthorized person, will be punished by the imprisonment of up to three years. There is also much worse form of this criminal act - if the offender acquired considerable material gain for himself or for another, or caused a considerable damage, will be punished by imprisonment from six months to five years. [7]

In the same article, there is one exception – there is no criminal offense of disclosure and unauthorized acquisition of the confidential business information, if the offense is committed, predominantly in the public interest.

The stated criminal offense is prosecuted by the proposal. This means that the State's Attorney will not initiate proceedings *ex officio*, but in concern with the proposal of the injured party.

Naturally, the Criminal Act does not define what the confidential business information is - it only states that it should be defined by the other already mentioned acts, (by the separate act or the general act of the legal person's).

V. COMPARATIVE VIEW

Among other things, the ZTP's Final Proposal [5] states: "According to the criteria of comparative legislation, business confidential information isn't regulated together with the secrets of the state bodies sphere, but separately, and mostly within the framework of regulations on market operations. If we consider the practice of several relevant countries, it is only partially true. Austria has a dual model of protection of business confidential information.: "The Austrian Law provides for the protection of trade secrets prevalently in the Act against Unfair Competition, specifically in sections 11 to 13 but also in the Penal Code. These provisions are supported by the general clause according to section 1 of the Austrian Unfair Competition Act. Also, Sections 11 and 12 are provisions of criminal law. The general nature of these provisions is the disclosure or exploitation of trade or business secrets, that have been entrusted to the offender during a professional occupation, or which have been obtained by espionage. The penalties in question are imprisonment or fines. The legal values protected under this framework are the legitimate interest in the confidentiality of trade and business secrets and to punish infringements of such legitimate interests." [9]

A similar solution is offered by the United States: "In contrast to other types of intellectual property (trademarks, patents, and copyrights) that are governed primarily by federal law, trade secret protection is primarily a matter of state law. Thus, trade secret owners have more limited legal recourse when their rights are violated. State law provides trade secret owners with the power to file civil lawsuits against misappropriations. A federal criminal statute, the Economic Espionage Act (EEA), allows U.S. Attorneys to prosecute anyone who engages in "economic espionage" or the "theft of trade secrets." The EEA's "economic espionage" provision punishes those who misappropriate trade secrets with the intent or knowledge that the offense will benefit a foreign government, instrumentality, or agent." [10]

The Federal Republic of Germany has solved this issue through several regulations, primarily through the Unfair Competition Regulations but also through the labour regulations, which does not exclude criminal liability. It is interesting that there is a special obligation of keeping the information of the other side, in case of business negotiations. [11]

It is obvious that also in the other countries there is a considerable disagreement over the legal protection of business confidential information. It is regulated by the Regulations of Competition's Protection, by the Civil Code, but also by the system of Criminal law. Considering that numerous business ventures are supranational, it represents pre-regulation. It should be taken into consideration that the perpetrators (especially those for the first time) were condemned on parole or short prison sentences, which in practice is proven to be troublesome. [12] *De lege ferenda*, it should be advocated that the business confidential information issue remains within the civil liability for damage.

.

VI. A NON-DISCLOSURE AGREEMENT

A non-disclosure agreement (further: NDA), also known as a confidentiality agreement is a private contract, which is used for keeping and saving valuable information. It can be of a great benefit for researchers and organizations that are involved in research and development projects. NDAs' are legally binding, for determining the conditions under which one party is (the 'discloser' of information), the one who reveals confidential business information to another person (the 'recipient' of the information). In doing so, one party provides information, and the other party receives the information. Depending on the number of parties who share information, NDA can be unilateral, bilateral or multilateral.

For an NDA to become valid, in Croatian legal system, it is not required to be verified by a notary public, but the signature of the contracting parties is sufficient, (in case of organizations, the agreement should be concluded by the authorized representative/s). Other than mentioned, the ZZTP (in the Head 8), is familiar with the protection of confidential business information, but there is no NDAs' in Croatian legislation.

Any information which may not be in public domain, or generally known, may be protected by a non-disclosure agreement. It can protect any sort of confidential information like: a concept for a new restaurant, a new business venture, or any sort of confidential business information that could be of value to others if it was disclosed. [13] Additionally, confidentiality agreements should contain a provision stating that no implied license to the technology or information is to be granted to the recipient and that all tangible embodiments of the information (e.g., models, data, and drawings) should be returned upon request and in no event later than the end of the agreement term, and that no copies shall be retained by the recipient. [14]

In conclusion, there are several situations where a confidentiality agreement is appropriate and may be proposed. Knowing a few basic points concerning confidentiality agreements can ensure that the important purposes they serve will not be defeated by ambiguities or ignorance of the meaning of terms used in the agreement. [14]

VII. CONCLUSION

The confidential business information in The Republic of Croatia is, therefore, defined by the 1996 ZTTP's, and by the provisions of the Criminal Act. At first glance, one might conclude that it is extensively regulated by the acts of public authorities.

However, after a review we have given in this paper, it can be concluded that the entire state regulation concerning the confidential business information in Croatia is deficient. It is based on the over 20 years old provisions of the ZTTP, which are applied based on the transitional provisions of the

ZTP from 2007, and partly by an outdated criminal legal liability (the issue of confidential business information in the developed countries, mainly lays within the scope of civil liability for damage).

This regulation gives only an illusion of protection of the confidential business information - so it could easily be said that it is even harmful, in practice. Therefore, for all, that are doing business in Croatia and want to protect their information, we recommend that they primarily do:

a) their own quality act on the protection of business information;

b) the signing of an appropriate NDA agreement with their partners (standardized, or if necessary, custom),

c) Signing NDA agreements with external partners (individuals, craftsmen, freelancers).

Also, it is necessary to inform its own staff about confidentiality and ask them confirmation and familiarity with the same.

Only, if the organization does its own measures then it can partly rely on the national system of protection of business information.

In the end, we emphasize that this legal system of protection of confidential business information, literally represents a rarity in the world, so it is important to familiarize foreign business partners in businesses which include confidential business information.

REFERENCES

[1] Zakon o tajnosti podataka, Narodne novine*, br. 79/07

[2] Zakon o zaštiti tajnosti podataka, Narodne novine, br. 108/96.

[3] Što donosi Nacrt prijedloga novog Zakona o tajnosti podataka u RH, http://www.media.ba/bs/etikaregulativa-novinarstvo-etika/sta-donosi-nacrt-prijedloga-novog-zakona-o-tajnosti-podataka-u-rh

Retrieved on 20/02/2017

[4] B. Peran, M. Goreta, K. Vukošić, "Pojam i vrste tajni", Zbornik radova Veleučilišta u Šibeniku, 3-4/2015

[5] PZ 618, Konačni prijedlog Zakona o tajnosti podataka, Hrvatski Sabor, 2007.

[6] Nacrt Zakona o tajnosti podataka, Vlada Republike Hrvatske, 2007.

[7] Kazneni zakon, Narodne novine, br. 125/11, 144/12, 56/15, 61/15.

[8] G. Vojković, "Poslovna tajna", Pravo i porezi. XVIII. (2009), 4; 3-8

[9] MMag. Juliane Messner, Dr. Max W. Mosing, LL.M. (IT-Law), LL.M. (Strathclyde), Mag. Rainer Schultes: LIDC 2015, Question B: The protection of trade secrets and know-how, Austrian Report, International League of Competition Law

[10] Brian T. Yeh, Protection of Trade Secrets: Overview of Current Law and Legislation, Congressional Research Service, April 22, 2016

[11] Protecting Trade Secrets in Germany, http://blogs.orrick.com/trade-secrets-watch/trade-secrets-laws/protecting-trade-secrets-in-germany

Retrieved on 16/04/2017

[12] B. Tot: Alternativa kazni zatvora – rad za opće dobro na slobodi, Polic. sigur. (Zagreb), godina 16. (2007), broj 1-2, str. 21-39

[13] What is Non-Disclosure Agreement? http://blog.ipleaders.in/what-is-non-disclosure-agreement/ Retrieved on 19/02/2017

[14] D. V. Radack: Understanding Confidentiality Agreements http://www.tms.org/pubs/journals/JOM/matters/matters-9405.html, Retrieved on 19/02/2017

* Narodne novine is Official Gazette of the Republic of Croatia.

Combinatorial Optimization in Cryptography

Karlo Knežević
University of Zagreb
Faculty of Electrical Engineering and Computing
Unska 3, 10000 Zagreb, Croatia
Email: karlo.knezevic@fer.hr

Abstract—The known attacks on different cryptosystems lead to a number of criteria that the implemented cryptographic algorithms (ciphers) must satisfy. The design of cryptographic systems needs to consider various characteristics simultaneously, which can be regarded as a multi-objective combinatorial optimization problem. Evolutionary computation present a range of problem-solving techniques based on the principles of biological evolution. Evolutionary algorithms can quickly offer satisfactory solution to combinatorial optimization problems. Evolutionary computation can be also used in evolving pseudorandom number generators which play important role as a countermeasure against side channel attacks. The purpose of this paper is to give a state-of-the-art overview of the evolutionary computation area in symmetric and asymmetric cryptography, as well as for the evolving pseudorandom number generators. In symmetric cryptosystem, one of the important components is the substitution box which can be successfully built by evolutionary algorithm. In asymmetric cryptosystem, evolutionary algorithms can be used to speed-up some discrete mathematic operations, like modular exponentiation.

I. INTRODUCTION

Combinatorial optimization is a topic that consists of finding an optimal object from a finite set of objects. In many such problems, exhaustive search is not feasible. It operates on the domain of those optimization problems, in which the set of feasible solutions is discrete or can be reduced to discrete, and in which the goal is to find the best solution.

Cryptography is the study of mathematical techniques for all aspects of information security. The security of information encompasses the fundamental aspects: confidentiality, data integrity, authentication, and non-repudiation. To ensure all the previous aspects, cryptographic algorithms and modes should satisfy a number of criteria, which can be regarded as a multi-objective combinatorial optimization problem.

Evolutionary computation algorithms represent a range of problem-solving techniques based on principles of biological evolution. Such algorithms can be used to solve a variety of difficult problems, among which are those from area of cryptography and which can be represented as combinatorial optimization problem. It is justified to mention several other methods like cellular automata or addition chains, which in cooperation with evolutionary algorithms can result with better results than conventional algorithms.

This paper is not intended as a detailed survey of possible applications of evolutionary algorithms in cryptography nor detailed explanation of mathematical background in cryptography. Rather, the purpose of this paper is to give a state-of-the-art overview of evolutionary computation methods applied to cryptographic systems design.

The paper begins by discussing evolutionary computation algorithms in Section II. Section III describes relevant theory in cryptography and the criteria that cryptographic algorithms should satisfy. In Section IV the state-of-the-art of different cryptographic problems and the applications of evolutionary computation methods used in cryptography is given. Finally, Section V draws a conclusion.

II. EVOLUTIONARY COMPUTATION

Evolutionary computation is a family of algorithms inspired by biological evolution, and the subfield of artificial intelligence and soft computing. It can be roughly divided to three groups: evolutionary algorithms, swarm algorithms, and other algorithms. Evolutionary computation uses iterative progress, like growth or development in population. This population is then selected in a guided random search to achieve the desired goal. In following paragraphs, briefly is explained each of these three groups of evolutionary computation.

A. Evolutionary Algorithms

Evolutionary algorithms could be broadly divided into 4 different approaches: genetic algorithms [1], genetic programming [2], evolutionary strategies [3], and evolutionary programming [4]. All of these algorithms have been successfully applied to the variety of problems in the field of combinatorial optimization [5]. Evolutionary algorithms are based on the Darwinian theory of evolution.

The population of individuals which represent possible solutions to an optimization problem evolve toward a better solution. To measure the quality of a solution, the fitness function is defined, which is always problem dependent. The evolution usually starts from a pool of randomly selected individuals and then, by utilizing evolutionary operators, new and better population is generated. Main evolutionary operators are selection and recombination. Selection is a process of selecting individuals that will produce a new generation. Recombination consists of two operators: crossover and mutation. Crossover combines two or more parent individuals to form one or more child individuals, that inherit parents' good characteristics. Mutation is a random change of individual alleles in an individual [6].

Main difference between all evolutionary algorithms is the way of individual encoding and evolutionary operators implementation. The genotype in genetic algorithm is usually represented as a string of bits, array of floating point numbers or a permutation vector. In genetic programming, the genotype is encoded as a tree structure. That tree structure is built by functional nodes and terminal nodes. The result of genetic programming is called a program. One of possible problems in genetic programming is bloat and there are numerous methods how to avoid it. After some generations, the search for better programs halts as the programs become too large. Bloat can be caused by inefficient code, e.g. two times $x = x + 1$, instead of $x = x + 2$, or a large calculation that is multiplied by zero in the end.

Cartesian genetic programming [7] is a type of genetic programming where a genotype is encoded as a graph structure. It is called Cartesian because functional nodes are organized in rows and columns, which reminds to a Cartesian system. This type of genetic programming solves bloat problem as well as the number of individuals in population. The genotype length is constant size and because of that the bloat problem is inherently solved. Typically, evolution strategy ES-(1+4) is used in Cartesian genetic programming. For more details on evolutionary algorithms, we refer readers to [6].

B. Swarm Algorithms

Swarm algorithms consist of ant algorithms [8], particle swarm algorithm [9], bee algorithms [10], and others. Some swarm algorithms are inspired by experiments made by biologists or by software imitation of swarms in computer graphics. Common to the all algorithms is a swarm population consisting of individuals of different characteristics. For more details on ant algorithms refer to [11] and bee algorithms refer to [12]. In the next paragraphs the particle swarm algorithm will be described.

Particle swarm optimization algorithm is created to simulate a flock of flying birds. Flock of birds is represented as a set of particles and each particle must satisfy several rules according to the neighbor particles: avoiding collision, speed harmonization, and low distance. Those rules determine social aspect of particles. Each particle is described by velocity and position. Position is a possible solution of a problem, a velocity vector changes the particle position according to the previous rules. For more information on particle swarm optimization algorithm refer to [13].

C. Other Algorithms

Other algorithms relevant for this paper are artificial immune system algorithms and cellular evolutionary algorithm. Each of these algorithms are specific and have different individual encoding or solution finding method.

Artificial immune system algorithms is a family of algorithms inspired by human immune system. Clonal Selection Algorithm (CLONALG) is the most used algorithm from this family. The algorithm begins by creating initial population of antibodies. The antibodies are made by some random mechanism. Then, affinity of each antibody is measured. By selection operator, according to affinity, antibodies are selected for cloning. The operator of hyper-mutation is applied to clones in order to improve fitness between antibody and antigen. This procedure is repeated until termination condition

is reached. To read more about CLONALG, we refer interested readers to [14].

A cellular evolutionary algorithm is a type of evolutionary algorithm in which individuals cannot mate arbitrarily, but every one interacts with its close neighbors on which a basic evolutionary algorithm is applied. A cellular evolutionary algorithm usually evolves a structured bi-dimensional grid of individuals, although other topologies are also possible. In this grid, cluster of similar individuals are naturally created during evolution, promoting exploration in their boundaries, while exploitation is mainly performed by direct competition and merging inside them. This reproductive cycle is executed inside the neighborhood of each individual and consists in selecting two parents among its neighbors according to a certain criterion, applying the variation of evolutionary operators. For further information about cellular evolutionary algorithm, we refer readers to [15].

III. MODERN CRYPTOGRAPHY

Modern cryptography designs cryptographic algorithms that are assumed to be hard to break by an adversary. It is divided into symmetric key cryptography, asymmetric key cryptography, and hash functions [16]. Pseudo-random number generators can be singled out as separate algorithm that have relevance to this paper.

A. Asymmetric Key Cryptography

This type cryptography is also known as public key cryptography. In public key cryptography two keys are used, one for encryption and other for decryption. The encryption key is a public key and the decryption key is private. Public key cryptography can be divided into three groups: cryptosystems based on factorization problem, cryptosystems based on discrete logarithm problem and others.

The most famous public key cryptosystem is RSA, called after its creators [17].The security of the RSA cryptosystem is based on the difficulty of the integer factorization problem. In the process of encryption, modular exponentiation is used. Let p and q be prime numbers. Then $n = pq$, where $p, q \in \mathcal{Z}$, $de \equiv 1 \bmod \varphi(n)$ and $\varphi(n)$ is Euler function. Public key is then defined as (n, e) and private key as (p, q, d). The RSA algorithm is believed to be secured as long as large p and q are used.

Famous cryptosystems or protocols based on discrete logarithm problem are Diffie-Hellman [18] or ElGamal [19]. Another large family cryptosystems are elliptic curve discrete logarithm problems (ECDLP). One representative of that family is Menzes-Vanstone cryptsystem [20]. Other public key cryptosystems are Merkle-Hellman, McEliece and NTRU.

B. Symmetric Key Cryptography

Symmetric key cryptography uses the same key for encryption and decryption, therefore some mechanism to keep those keys secret is needed. This type of cryptography is called symmetric because communication participants have the same secret key. Symmetric key cryptography's main advantage over

1325

asymmetric key cryptography is the fact that it is much faster. Symmetric key cryptography can be divided into block and stream ciphers.

Block ciphers take as an input block of plaintext and a key, and the output is a block of the ciphertext of the same size as the plaintext. Stream ciphers create an arbitrary long stream of key material, which is combined with the plaintext bit by bit. In a stream cipher, the output is created based on a hidden internal state which changes as the cipher changes.

The design of block ciphers relies on two fundamental principles: confusion and diffusion. Confusion means that each digit of ciphertext should depend on several parts of the key. Diffusion means that if a single bit of a plaintext is changed, then, statistically, half of the bits in the ciphertext should change and vice versa.

An S-box or a substitution box is a basic component of symmetric key algorithms. The main purpose of using an S-box is to introduce a confusion. The nonlinearity property of S-boxes is one of the most important cryptographic criteria because cryptographic algorithm should be resistant to linear cryptanalysis [21]. S-box (n, m) consists of n input variables and m output variables. These m outputs can be viewed like m Boolean functions. If nonlinearity of a Boolean function is equal to maximum then it is called bent function. The nonlinearity bound, show in 1, is called the covering radius bound and is strict for bent Boolean functions.

$$N_f \leq 2^{n-1} - 2^{\frac{n}{2}-1}. \qquad (1)$$

Other criteria which is considered in Boolean or vectorial functions are algebraic degree, balancedness, resiliency, algebraic immunity, and others [21]. If all criteria are used simultaneously then the problem of finding the best Boolean function or an S-box is a multi-objective problem. To inform more about definitions or mathematics background see [22], [23].

C. Pseudo-random Number Generator

A pseudorandom number generator (PRNG) is a deterministic algorithm for generating a sequence of numbers that have the properties of random numbers. The sequence obtained by the pseudorandom number generator is not truly random because it is determined by small set of initial values, called the state. Pseudorandom number generators play important role in the area of cryptography. General security requirements of PRNGs are listed in below [24]:

- the random numbers should not show any statistical weaknesses,
- the knowledge of subsequences of random numbers shall not allow to compute predecessors or successors practically or to guess them with non-negligibly larger probability than without knowledge of these subsequences,
- it shall not be practically feasible to compute preceding random numbers from the internal state or to guess them with non-negligibly larger probability than without knowledge of the internal state and
- it shall not be practically feasible to compute future random numbers from the internal state or to guess

them with non-negligibly larger probability than without knowledge of the internal state.

The most famous and the simplest PRNG is the linear congruential generator (LCPRNG) given in equation 2. The s_0 is called seed. Another PRNGs are RSA PRNG and Blum-Blum-Shub PRNG.

$$s_{i+1} = (as_i + b) \ mod \ M, a, b \in \{1, ..., M-1\}. \qquad (2)$$

IV. EVOLUTIONARY COMPUTATION IN CRYPTOGRAPHY

In next subsections the state-of-the-art overview of the evolutionary computation algorithms will be given.

A. Evolutionary Computation in Asymmetric Key Cryptography

Two problems in asymmetric key cryptography chosen to be described are finding short addition chains and cracking of the Merkle-Hellman cryptosystem using genetic algorithm.

1) Addition Chain: Field of the modular exponentiation has several important applications in cryptography. Well-known public key cryptosystem such as RSA adopt modular exponentiation. Modular exponentiation is defined as follows:

$$B = A^c \ mod \ p. \qquad (3)$$

where A is an integer in the range $[1, ..., p-1]$ and c and p are an arbitrary positive integers. In cryptosystems, p is a large positive prime number defined in III-A. One possible way of reducing the computational load of Eq. (3) is to minimize the total number of multiplications required to compute the exponentiation. Since the exponent in Eq. (3) is additive, the problem of computing powers of the base A can be formulated as an addition calculation, for which so-called addition chains are used. For example, if A^{50} wants to be calculated, the traditional procedure would require 50 multiplications. If the addition chain is used, then only 7 multiplications are required: $[1 \rightarrow 2 \rightarrow 4 \rightarrow 8 \rightarrow 16 \rightarrow 32 \rightarrow 48 \rightarrow 50]$.

Usual deterministic algorithm is the binary method. The exponent is written in binary representation (of length d) and then each occurrence of digit 1 is replaced with the letters "SM" and each digit 0 with letter "S", where "S" stands for squaring and "M" stands for multiplication. After all digits are replaced, the first "SM" occurrence on the left is crossed. This method is deterministic but it does not give the optimal solution. There are two version of binary method: left-to-right and right-to-left. Both methods have same number of operations: d times squaring and multiplications as much as there is number of digit 1.

In [25] experiment with the Particle Swarm Optimization algorithm in order to find optimal short addition chains. Authors in [26] implement the Ant Colony Optimization algorithm on a System on s Chip (SoC) in order to speed up the modular exponentiation in cryptographic applications. In [27] use a GA to find minimal Brauer chains where a Brauer chain is an addition chain in which each member uses the previous member as a summand. In [28] authors investigate the usage of evolutionary programming for minimizing the length of addition chains.

In [29] authors proposed a genetic algorithm to find short addition chains for a given exponent. The main paper contributions were a new representation of solutions, design several mutation and crossover operators designed to improve convergence and design a special part of algorithm called *repair heuristics* that is believed to be an integral part of the algorithm.

Internally the alleles were pair of positions (i_1, i_2) that hold values (n_1, n_2) which form value on given position. That type of encoding is called encoding with *summand positions*. *Repair heuristics* is strategy needed to recreate invalid individual. The genetic algorithm in that paper is described as follows. First, set zeroth element to 1 and the first element to 2. Then uniformly at random select between all minimal subchains consisting of three elements and a random choice of the second element. With a probability equal to $3/5$ double the elements until they reach half of the exponent size. Check weather the current element and any previous element sum up to the exponent value. Uniformly at random choose mechanisms given in paper to obtain the next value in the chain, under the constraint that it needs to be smaller than the exponent value. The used fitness function was the number of elements in the chain.

Authors showed that with proposed genetic algorithm is possible to find better solution than with the binary method.

2) Merkle-Hellman Cryptosystem: Merkle-Hellman cryptosystem is a public key cryptosystem based on a concrete case of the knapsack problem [30]. The knapsack problem is defined as follows: for a given set $\{v_1, v_2, ..., v_n\}$ of n integers and an integer V, find an array $m = (\epsilon_1, \epsilon_2, ..., \epsilon_n)$, where $\epsilon_i \in \{0, 1\}$, so that $\epsilon_1 v_1 + ... + \epsilon_n v_n = V$, if such m exists. This problem is NP-complete, which means that there is no polynomial algorithm to solve that problem. However, the special case of the knapsack problem is the super-increasing sequence, which can be solved in polynomial time. Super-increasing sequence is satisfied if $v_j > v_1 + v_2 + ... v_{j-1}$ for $j = 2, 3, ..., n$. One example of the super-increasing sequence is $v_i = 2^{i-1}$. The main idea of the Merkle-Hellman cryptosystem is to mask the super-increasing sequence to look like a random sequence. The message receiver should know how to remove the mask, and than solve the super-increasing problem. Masking is done by modular multiplication.

Shamir [31] showed existence of polynomial algorithm for breaking Merkle-Hellman cryptosystem. However, there is a genetic algorithm twice faster. In [32] authors proposed a genetic algorithm as a method to crack the cryptosystem. The authors proposed a new selection and crossover method.

The selection function chooses the selection candidates, which the most satisfy the fitness function, i.e., their fitness values are higher than those of others. In case the population size is L only $L/5$ solution candidates is chosen. Exactly these solution candidates form new generations.

The crossover function receives the population of the solution candidates. From this population solution candidates with $t1$ and $t2$ numbers are chosen in pairs by means of a random generator taking into account that $t1$ and $t2$ do not coincide

with each other and the used pair is not repeated. Each solution candidate is divided in two parts (at the mid point). To explore more about given algorithm refer to [32].

B. Evolutionary Computation in Symmetric Key Cryptography

In next subsections the state-of-the-art overview of the evolutionary computation algorithms applied to Boolean and vectorial functions construction will be shown.

1) Cryptographic Boolean Functions: Boolean functions represent an important primitive in the design of various cryptographic algorithms. Three main approaches for generating Boolean functions for cryptographic usages are distinguished: algebraic constructions, random generation, and heuristic constructions.

There exists a number of works that examine Boolean functions in cryptography and their generation with evolutionary computation techniques. Here will be enumerated only relevant papers.

[33] compare the effectiveness of Cartesian genetic programming and genetic programming approach when looking for highly nonlinear balanced Boolean functions of eight inputs. Boolean function is balanced if there is same number of ones and zeros. Nonlinearity is minimal Hamming distance between Boolean function and its all affine functions. [34] investigate several evolutionary algorithms in order to evolve Boolean functions with different values of the correlation immunity property. In the same paper, the authors also discuss the problem of finding correlation immune functions with minimal Hamming weight, but they experiment with only one size of Boolean functions. More extensive investigation on finding correlation immune Boolean functions with minimal Hamming weight and different sizes is conducted by [35].

[36] use Particle Swarm Optimization to find Boolean functions with good trade-offs of cryptographic properties for dimensions up to 12. The same authors use genetic algorithms where the genotype consists of the Walsh-Hadamard values in order to evolve semibent (plateaued) Boolean functions [37]. Semibent Boolean function has all values in Walsh-Hadamard spectrum in $\{0, 2^{\frac{n+2}{2}}, -2^{\frac{n+2}{2}}\}$, where n is input space dimensionality.

[38] conduct a detailed analysis of the efficiency of a number of evolutionary algorithms and fitness functions for Boolean functions with 8 inputs.

In [39] authors used several evolutionary algorithms to evolve Boolean functions with several cryptographic criteria. Authors used genetic algorithm, evolutionary strategy, genetic programming and Cartesian genetic programming. Individuals were, according to the algorithm, encoded by string of bits whose values define a truth table of th Boolean function, trees of Boolean primitives which are then evaluated according to the truth table they produce or directed graphs that are also evaluated in accordance to the truth table they produce.

Authors used two objective functions. First function was the nonlinearity where the goal is maximization. In the work, truth table was transformed by using Walsh-Hadamard transformation as follows:

1327

$$W_f(\vec{a}) = \sum_{\vec{x} \in F_2^n} (-1)^{f(\vec{x}) \oplus \vec{a} \cdot \vec{x}} \qquad (4)$$

where $\vec{a} \cdot \vec{x}$ is the inner product of two vectors. With that objective function the goal was to find the bent function. Although bent functions are not appropriate as parts of filter and combiner generators, there is nevertheless motivation in their generation. Filter and combiner generators are generators of random sequence which uses linear feedback shift registers (LFSR). The first reason is that this problem can be used as a benchmark to test various evolutionary algorithms. The second reason lies in the fact that using bent functions it is possible to build highly nonlinear functions that are balanced [22]. The second objective was combination of balancedness and nonlinearity. The goal was to find highly nonlinear function that is balanced.

In that paper authors showed that genetic programming and Cartesian genetic programming performed the best, which indicates that the truth table representation in not the most appropriate one when evolving cryptographically suitable Boolean functions.

In [40] authors investigate the efficiency of two immunological algorithms: CLONALG and optimization immune algorithm with elitism. For all algorithms is experimented with two different representations for encoding a Boolean function: string of bits and floating point representation. Same objective functions were used as in [40].

By using those algorithms and individual encodings, it is impossible to conclude whether the string of bits or floating point encoding should be used. Furthermore, the results show that the optimization immune algorithm performs better than the CLOANLG, but both perform slightly worse than the genetic algorithm and evolutionary strategy.

2) Evolution of S-boxes: Substitution boxes (S-boxes) play an important role in many modern-day cryptographic algorithms, more commonly known as ciphers. Without carefully chosen S-boxes, such ciphers would be easier to break. In the process of the design of S-boxes (similarly as in the design of Boolean functions), one can roughly follow three directions, namely, algebraic constructions, random search, and heuristics.

There are several unique S-box representations. An (n, m)-function F can be represented as a list of values (lookup table - LUT), with each value ranging from 0 to $2^m - 1$. Alternatively said, an (n, m)-function can be implemented as a lookup table with 2^n words of m bits each. When $n = m$ it is usual that the S-box is bijective, i.e., that each value in the output appears exactly once.

A Boolean function f on $\mathbb{F}_2^n$ is represented by a truth table (TT), which is a vector $(f(\vec{0}), \ldots, f(\vec{1}))$ that contains the function values of f, ordered lexicographically, i.e., $\vec{a} \leq \vec{b}$, where $\vec{a}$ and $\vec{b}$ are two input entries for the truth table [22]. An S-box can be represented in the truth table form as a matrix of dimension $2^n \times m$ where each column m represents one Boolean function (i.e., one coordinate function).

The Walsh-Hadamard transform of an (n, m)-function F equals [23]:

$$W_F(\vec{a}, \vec{v}) = \sum_{\vec{x} \in \mathbb{F}_2^n} (-1)^{\vec{v} \cdot F(\vec{x}) \oplus \vec{a} \cdot \vec{x}}, \qquad (5)$$

where $\vec{a} \in \mathbb{F}_2^n$ and $\vec{v} \in \mathbb{F}_2^{m*}$.

An (n, m)-function F is called *balanced* if it takes every value of $\mathbb{F}_2^m$ the same number 2^{n-m} of times. Balanced (n, n)-functions are permutations on $\mathbb{F}_2^n$ [23].

[41] experiment with a GA that works in a reverse way in order to generate bijective S-boxes of dimensions from 8×8 to 16×16. They seed the initial S-boxes population with solutions based on the finite field inversion method and then evolve them to find new solutions. The same authors use a modified immune algorithm to generate 8×8 S-boxes that are balanced, with high nonlinearity, and low δ-uniformity [42].

[43] explore how to generate S-boxes of size 8×8 with a better resistance against side-channel attacks (SCA) as measured with the transparency order property. Next, [44], [45] investigate side-channel resilience of 4×4 S-boxes as well as when considering the confusion coefficient property. [46] use genetic algorithms to evolve S-boxes with better SCA resilience and they implement such S-boxes in both software and hardware settings in order to properly evaluate them. Finally, [47] experiment with the modified transparency order property in order to achieve S-boxes with better SCA resistivity.

[48] use Cartesian genetic programming (CGP) and genetic programming (GP) to evolve S-boxes where they discuss how to obtain permutation-based encoding with those algorithms.

In [49] authors experimented with three different S-box representations: truth table, Walsh-Hadamard transform coefficients and the lookup table representation where in the case that $n = m$ an S-box can be represented as a permutation.

The goal was to evolve highly nonlinear S-box with different dimensions. The authors proposed a new cost function that is faster than those in given related work. This cost function is defined as a two-component vector with the nonlinearity as the first component and a cost associated with a part of the Walsh-Hadamard spectrum as the second component; when using in a single criterion optimizer, to determine better solutions, a hierarchical comparison should be performed.

C. Evolving Cryptographic Pseudorandom Number Generators

Random number generators are used in a range of applications spanning from producing simple values and adding randomness to programs, over online betting to various cryptographic applications. One real-world application of pseudorandom number generators in cryptography is to use them for masking as a countermeasure against side channel attacks [50]. When used for such a purpose, those generators needs to be extremely fast and small when implemented in hardware. To obtain a generator with such characteristics, "expensive" operations like multiplication or addition should be avoided as much as possible.

John Koza used genetic programming (GP) to evolve programs that output random numbers [51]. As a fitness function he used the notion of information entropy as defined by Shannon and the end result was a program that was able to accept a sequence of consecutive integer values and transform it into random binary digits.

Hernandez, Seznec, and Isasi used GP to evolve random number generators where they used the strict avalanche criterion (SAC) as a fitness function [52]. Martinez et al. designed a pseudorandom number generator suitable for cryptographic usage by means of GP. The obtained generator, Lamar was tested with a number of tests where the input values were obtained via a counter function.

In [50] authors used Cartesian genetic programming to evolve a pseudorandom number generator. Functional nodes consisted of logical functions like AND, NOT, XOR, operations RR/RL (rotate right or rotate left) and SR/SL (shift right or shift left), and a function P which is shuffle function. The test for randomness was the bias of the output sequence, which is defined as a ratio of zeros and ones in the sequence.

As the fitness function the entropy is used, where the goal is to maximize entropy in order to achieve randomness. With that fitness function and by using Cartesian genetic programming authors evolved pseudorandom number generators that are extremely small and fast in hardware implementation because they don't rely on expensive operations like multiplication or addition.

V. CONCLUSION

In this paper, we showed how evolutionary computation can be successfully used in cryptography. We present different approaches in asymmetric key cryptography, symmetric key cryptography and pseudo-random number generator. In asymmetric key cryptography, evolutionary computation can speed-up exponentiation calculation or be used for breaking Merkle-Hellman cryptosystem. In symmetric key cryptography, evolutionary computation can build Boolean functions or S-boxes which satisfy various characteristics simultaneously. Finally, evolutionary computation can be successfully used in evolving pseudorandom number generator.

ACKNOWLEDGMENTS

This work has been supported in part by Croatian Science Foundation under the project IP-2014-09-4882.

REFERENCES

[1] D. E. Goldberg and J. H. Holland, "Genetic algorithms and machine learning," *Machine learning*, vol. 3, no. 2, pp. 95–99, 1988.

[2] J. R. Koza, *Genetic programming: on the programming of computers by means of natural selection*. MIT press, 1992, vol. 1.

[3] H.-G. Beyer and H.-P. Schwefel, "Evolution strategies–a comprehensive introduction," *Natural computing*, vol. 1, no. 1, pp. 3–52, 2002.

[4] D. B. Fogel, *Evolutionary computation: toward a new philosophy of machine intelligence*. John Wiley & Sons, 2006, vol. 1.

[5] M. Affenzeller, S. Wagner, S. Winkler, and A. Beham, *Genetic algorithms and genetic programming: modern concepts and practical applications*. Crc Press, 2009.

[6] A. E. Eiben and J. E. Smith, *Introduction to Evolutionary Computing*. Springer-Verlag, Berlin Heidelberg New York, USA, 2003.

[7] J. F. Miller, "Cartesian genetic programming," in *Cartesian Genetic Programming*. Springer, 2011, pp. 17–34.

[8] O. Cordón García, F. Herrera Triguero, and T. Stützle, "A review on the ant colony optimization metaheuristic: Basis, models and new trends," *Mathware & soft computing. 2002 Vol. 9 Núm. 2 [-3]*, 2002.

[9] A. Banks, J. Vincent, and C. Anyakoha, "A review of particle swarm optimization. part i: background and development," *Natural Computing*, vol. 6, no. 4, pp. 467–484, 2007.

[10] D. Karaboga and B. Basturk, "On the performance of artificial bee colony (abc) algorithm," *Applied soft computing*, vol. 8, no. 1, pp. 687–697, 2008.

[11] M. Dorigo, M. Birattari, and T. Stutzle, "Ant colony optimization," *IEEE computational intelligence magazine*, vol. 1, no. 4, pp. 28–39, 2006.

[12] X.-S. Yang, "Engineering optimizations via nature-inspired virtual bee algorithms," in *International Work-Conference on the Interplay Between Natural and Artificial Computation*. Springer, 2005, pp. 317–323.

[13] J. Kennedy, "Particle swarm optimization," in *Encyclopedia of machine learning*. Springer, 2011, pp. 760–766.

[14] D. DasGupta, "An overview of artificial immune systems and their applications," in *Artificial immune systems and their applications*. Springer, 1993, pp. 3–21.

[15] M. Tomassini, "Cellular evolutionary algorithms," in *Simulating Complex Systems by Cellular Automata*. Springer, 2010, pp. 167–191.

[16] J. Katz and Y. Lindell, *Introduction to modern cryptography*. CRC press, 2014.

[17] R. L. Rivest, A. Shamir, and L. Adleman, "A method for obtaining digital signatures and public-key cryptosystems," *Communications of the ACM*, vol. 21, no. 2, pp. 120–126, 1978.

[18] W. Diffie and M. Hellman, "New directions in cryptography," *IEEE transactions on Information Theory*, vol. 22, no. 6, pp. 644–654, 1976.

[19] T. ElGamal, "A public key cryptosystem and a signature scheme based on discrete logarithms," *IEEE transactions on information theory*, vol. 31, no. 4, pp. 469–472, 1985.

[20] N. Koblitz, A. Menezes, and S. Vanstone, "The state of elliptic curve cryptography," in *Towards a quarter-century of public key cryptography*. Springer, 2000, pp. 103–123.

[21] C. Carlet, "Boolean functions for cryptography and error correcting codes," *Boolean models and methods in mathematics, computer science, and engineering*, vol. 2, pp. 257–397, 2010.

[22] ——, "Boolean Functions for Cryptography and Error Correcting Codes," in *Boolean Models and Methods in Mathematics, Computer Science, and Engineering*, 1st ed., Y. Crama and P. L. Hammer, Eds. New York, NY, USA: Cambridge University Press, 2010, pp. 257–397.

[23] ——, "Vectorial Boolean Functions for Cryptography," in *Boolean Models and Methods in Mathematics, Computer Science, and Engineering*, 1st ed., Y. Crama and P. L. Hammer, Eds. New York, NY, USA: Cambridge University Press, 2010, pp. 398–469.

[24] F. Özkaynak, "Cryptographically secure random number generator with chaotic additional input," *Nonlinear Dynamics*, vol. 78, no. 3, pp. 2015–2020, 2014.

[25] A. León-Javier, N. Cruz-Cortés, M. A. Moreno-Armendáriz, and S. Orantes-Jiménez, "Finding minimal addition chains with a particle swarm optimization algorithm," in *Mexican International Conference on Artificial Intelligence*. Springer, 2009, pp. 680–691.

[26] N. Nedjah and L. de Macedo Mourelle, "High-performance soc-based implementation of modular exponentiation using evolutionary addition chains for efficient cryptography," *Applied Soft Computing*, vol. 11, no. 7, pp. 4302–4311, 2011.

[27] A. Rodriguez-Cristerna and J. Torres-Jimenez, "A genetic algorithm for the problem of minimal brauer chains," in *Recent Advances on Hybrid Intelligent Systems*. Springer, 2013, pp. 481–500.

[28] S. Domínguez-Isidro, E. Mezura-Montes, and L.-G. Osorio-Hernández, "Evolutionary programming for the length minimization of addition chains," *Engineering Applications of Artificial Intelligence*, vol. 37, pp. 125–134, 2015.

[29] S. Picek, C. A. C. Coello, D. Jakobovic, and N. Mentens, "Evolutionary algorithms for finding short addition chains: Going the distance," in *European Conference on Evolutionary Computation in Combinatorial Optimization*. Springer, 2016, pp. 121–137.

[30] S. Martello and P. Toth, *Knapsack Problems: Algorithms and Computer Implementations*. New York, NY, USA: John Wiley & Sons, Inc., 1990.

[31] A. Shamir, "A polynomial-time algorithm for breaking the basic merkle-hellman cryptosystem," *IEEE transactions on information theory*, vol. 30, no. 5, pp. 699–704, 1984.

[32] Z. Kochladze and L. Beselia, "Cracking of the merkle–hellman cryptosystem using genetic algorithm," 2016.

[33] S. Picek, D. Jakobovic, J. F. Miller, E. Marchiori, and L. Batina, "Evolutionary Methods for the Construction of Cryptographic Boolean Functions," in *Genetic Programming - 18th European Conference, EuroGP 2015, Copenhagen, Denmark, April 8-10, 2015, Proceedings*, 2015, pp. 192–204.

[34] S. Picek, C. Carlet, D. Jakobovic, J. F. Miller, and L. Batina, "Correlation Immunity of Boolean Functions: An Evolutionary Algorithms Perspective," in *Proceedings of the Genetic and Evolutionary Computation Conference, GECCO 2015, Madrid, Spain, July 11-15, 2015*, 2015, pp. 1095–1102.

[35] S. Picek, S. Guilley, C. Carlet, D. Jakobovic, and J. F. Miller, "Evolutionary Approach for Finding Correlation Immune Boolean Functions of Order t with Minimal Hamming Weight," in *Theory and Practice of Natural Computing - Fourth International Conference, TPNC 2015, Mieres, Spain, December 15-16, 2015. Proceedings*, 2015, pp. 71–82.

[36] L. Mariot and A. Leporati, "Heuristic Search by Particle Swarm Optimization of Boolean Functions for Cryptographic Applications," in *Genetic and Evolutionary Computation Conference, GECCO 2015, Madrid, Spain, July 11-15, 2015, Companion Material Proceedings*, 2015, pp. 1425–1426.

[37] ——, "A Genetic Algorithm for Evolving Plateaued Cryptographic Boolean Functions," in *Theory and Practice of Natural Computing - Fourth International Conference, TPNC 2015, Mieres, Spain, December 15-16, 2015. Proceedings*, 2015, pp. 33–45.

[38] S. Picek, D. Jakobovic, J. F. Miller, L. Batina, and M. Cupic, "Cryptographic Boolean functions: One output, many design criteria," *Applied Soft Computing*, vol. 40, pp. 635–653, 2016.

[39] ——, "Cryptographic boolean functions: One output, many design criteria," *Applied Soft Computing*, vol. 40, pp. 635–653, 2016.

[40] S. Picek, D. Sisejkovic, and D. Jakobovic, "Immunological algorithms paradigm for construction of boolean functions with good cryptographic properties," *Engineering Applications of Artificial Intelligence*, 2016.

[41] G. Ivanov, N. Nikolov, and S. Nikova, "Reversed genetic algorithms for generation of bijective s-boxes with good cryptographic properties," *Cryptography and Communications*, vol. 8, no. 2, pp. 247–276, 2016.

[42] ——, "Cryptographically Strong S-Boxes Generated by Modified Immune Algorithm," in *Cryptography and Information Security in the Balkans - Second International Conference, BalkanCryptSec 2015, Koper, Slovenia, September 3-4, 2015, Revised Selected Papers*, 2015, pp. 31–42.

[43] S. Picek, B. Ege, L. Batina, D. Jakobovic, L. Chmielewski, and M. Golub, "On Using Genetic Algorithms for Intrinsic Side-channel Resistance: The Case of AES S-box," in *Proceedings of the First Workshop on Cryptography and Security in Computing Systems*, ser.

CS2 '14. New York, NY, USA: ACM, 2014, pp. 13–18. [Online]. Available: http://doi.acm.org/10.1145/2556315.2556319

[44] S. Picek, B. Ege, K. Papagiannopoulos, L. Batina, and D. Jakobovic, "Optimality and beyond: The case of 4x4 S-boxes," in *2014 IEEE International Symposium on Hardware-Oriented Security and Trust, HOST 2014, Arlington, VA, USA, May 6-7, 2014*, 2014, pp. 80–83. [Online]. Available: http://dx.doi.org/10.1109/HST.2014.6855573

[45] S. Picek, K. Papagiannopoulos, B. Ege, L. Batina, and D. Jakobovic, "Confused by Confusion: Systematic Evaluation of DPA Resistance of Various S-boxes," in *Progress in Cryptology - INDOCRYPT 2014 - 15th International Conference on Cryptology in India, New Delhi, India, December 14-17, 2014, Proceedings*, 2014, pp. 374–390.

[46] B. Ege, K. Papagiannopoulos, L. Batina, and S. Picek, "Improving DPA resistance of S-boxes: How far can we go?" in *2015 IEEE International Symposium on Circuits and Systems (ISCAS)*, May 2015, pp. 2013–2016.

[47] S. Picek, B. Mazumdar, D. Mukhopadhyay, and L. Batina, "Modified Transparency Order Property: Solution or Just Another Attempt," in *Security, Privacy, and Applied Cryptography Engineering - 5th International Conference, SPACE 2015, Jaipur, India, October 3-7, 2015, Proceedings*, 2015, pp. 210–227.

[48] S. Picek, J. F. Miller, D. Jakobovic, and L. Batina, "Cartesian Genetic Programming Approach for Generating Substitution Boxes of Different Sizes," in *Proceedings of the Companion Publication of the 2015 Annual Conference on Genetic and Evolutionary Computation*, ser. GECCO Companion '15. New York, NY, USA: ACM, 2015, pp. 1457–1458. [Online]. Available: http://doi.acm.org/10.1145/2739482.2764698

[49] S. Picek, M. Cupic, and L. Rotim, "A new cost function for evolution of s-boxes," *Evolutionary Computation*, vol. 24, no. 4, pp. 695–718, 2016.

[50] S. Picek, D. Sisejkovic, V. Rozic, B. Yang, D. Jakobovic, and N. Mentens, "Evolving cryptographic pseudorandom number generators," in *International Conference on Parallel Problem Solving from Nature*. Springer, 2016, pp. 613–622.

[51] J. R. Koza, "Evolving a computer program to generate random numbers using the genetic programming paradigm." in *ICGA*. Citeseer, 1991, pp. 37–44.

[52] J. C. Hernandez, A. Seznec, and P. Isasi, "On the design of state-of-the-art pseudorandom number generators by means of genetic programming," in *Evolutionary Computation, 2004. CEC2004. Congress on*, vol. 2. IEEE, 2004, pp. 1510–1516.

Gap in pagination due to withheld paper.

Pages 1331-1336

Multidimensional Mining of Big Social Data for Supporting Advanced Big Data Analytics

Alfredo Cuzzocrea

University of Trieste and ICAR-CNR, Trieste, Italy

alfredo.cuzzocrea@dia.units.it

Abstract - *Big social data* are now everywhere. They constitute a rich source of knowledge that is prone to be explored and mined in order to support *advanced big data analytics*. *Multidimensional mining* identifies a promising collection of tools to this end. Following this recent trend, in this paper, we provide an overview on two state-of-the-art proposals that show how big data analytics over big social data work in practice.

I. INTRODUCTION

Big data (e.g., [2,30]) and *big data analytics* (e.g., [13,14,15,16,17,18]) are now top keywords in actual database and data mining research. Among the plethora of possible big data settings, *big social data* (e.g., [31,32,40]) play a leading role. *Twitter* is a typical case of big social data. Indeed, tweets are a rich source of knowledge that is prone to be explored and mined in order to support advanced big data analytics.

Multidimensional mining identifies a promising collection of tools to this end. These tools allow us to extract interesting knowledge from big social data via fortunate *multidimensional metaphors*, among which OLAP (e.g., [7]) plays the leading role. Indeed, the majority of tweets already identifies multidimensional spaces, mining that Twitter contents usually incorporate multiple variables. For instance, consider the case of political elections. Here, tweets related to this topic are used to contain, for instance, the following dimensions: (*i*) *region*, which models the place where the election will take place (including the respective hierarchy); (*ii*) *time*, which models the time when the election will occur (including the respective hierarchy); (*iii*) *elector*, which models the electors that will express their vote during the elections (including the respective hierarchy). This rich multidimensional knowledge base is prone to support a variety of big data analytics tools, thanks to which decision makers can perform effective decision-making processes.

Figure 1 shows a reference architecture for supporting big data analytics from tweet data, which is multi-layer in nature. As shown in Figure 1, the layers of the reference architecture are the following:

- *Big Tweet Data*, where the input data are located;

- *Big Data Storage Layer*, where tweet data are collected and stored – here, technologies like *MongoDB* [33] and *Hadoop* [34] are adopted;

- *Big Data Analytics Layer*, where multidimensional analytics tools extract interesting knowledge from tweet data – here, technologies like *Hive* [35] and *Pentaho* [36] are adopted;

- *Big Data Knowledge Layer*, where the final knowledge from tweet data is delivered to knowledge makers.

Following this recent trend, in this paper we provide an overview on two state-of-the-art proposals that show how big data analytics over big social data work in practice. In particular, the first one is represented by *a framework for supporting OLAP analysis over multidimensional tweet streams* [1]. The second one is represented by *a framework for supporting advanced big data analytics over geo-tagged tweets via intelligent multidimensional clustering methodologies* [19].

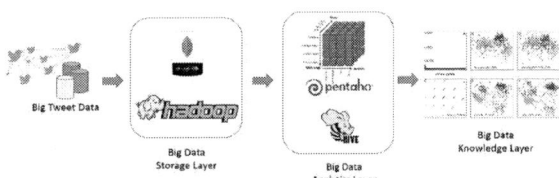

Figure 1. Big Social Data Analytics Reference Architecture

II. OLAP ANALYSIS OVER MULTIDIMENSIONAL TWEET STREAMS

[1] introduces a framework for supporting OLAP analysis over multidimensional tweet streams. Nowadays, people extensively use social networks (e.g., *Facebook*, Twitter, etc.) and multimedia aggregators (e.g., *YouTube*, *Wikipedia*, etc.) to share with their friends across the world: tastes, opinions, ideas, and so on. Millions of tweets about brands, news and so forth, hundreds of thousands of Facebook "likes" and check-ins on *Foursquare*, happen every day. So, we are heading into a social media data explosion. Collecting these data and performing analytics on social media data streams is one of the main challenges of big data because very interesting business goals could be achieved, such as: addressing marketing strategies, profiling people tastes, targeting advertisements, and so forth (e.g., [2]).

Specifically, [1] is focused on the metadata available in Twitter and we point out that there is a need to investigate and define suitable *knowledge mining approaches* to go beyond explicitly available metadata analyzing unstructured data in order to provide intelligent analytics services to support decision-making.

Twitter is the most popular microblogging service with more than half billion of users. Recent statistics pointed out that the average number of tweets per day is more than 100 million, and thousands of them happen every minute. So, tweets exchanged over the Internet represent an important source of information. Tweets are associated with meta-information that cannot be included in messages (e.g., date, location, etc.) or included in the message in the form of tags having a special meaning. Tweets can be represented in a multidimensional way by taking into account all this meta-information as well as associated temporal relations. Twitter's API enables us to acquire public tweet metadata (e.g., *Twitter's Search API* and *Streaming API*).

[1] focuses on the definition of *a multidimensional data model for the storage of tweet data streams for enabling OLAP analysis*. Furthermore, we exploit the implicit information that could be derived or discovered in the tweets by investigating beyond explicit the metadata. In particular, we propose an aggregation operator for the text embedded in the tweets' content based on the *Formal Concept Analysis* (FCA) theory [3,8].

The data warehouse model enables us to manipulate a set of indicators (measures) according to different dimensions that may be provided with one or more hierarchies. Associated operators (e.g., roll-up, drill-down, etc.) allow an intuitive navigation on different levels of the hierarchy. Indeed, the collected tweets will be represented by means of fuzzy mathematical model in order to extract timed fuzzy lattice through fuzzy extension of FCA [4]. Taking into account the resulting lattice of tweets, a *microblog summarization algorithm* is introduced to provide subset of the tweets that best represents the values of the chosen dimensions in a specific region of the OLAP cube.

In order to prove the benefits of our proposed innovative multidimensional tweet stream OLAP aggregation and analysis model, we provide a real-life case study focusing on tweets associated to the 2015 Italian election campaign.

In our model [1], the main research result consists in the definition of the so-called *OLAP Tweet Cube*, along with its multidimensional modeling. A data cube can model an analysis subject called fact F defined on the schema $D = \{T_1, T_2, \ldots, T_n, M\}$ where T_i $(i = 1, \ldots, n)$ are several dimensions and M stands for a measure. Every dimension defined on a domain D includes attributes $A = \langle a_1, a_2, \ldots, a_m \rangle$ that can be organized in several hierarchical levels $\langle l_1, l_2, \ldots, l_m \rangle$. In the proposed cube of tweets, the tweet represents the fact and some of its metadata explicit or implicit will be selected as hierarchical or flat dimensions.

OLAP dimensions can be classified in the following two types.

Semantic Dimension. In our model, this dimension is extracted from Wikipedia knowledge base and it makes use of the titles of Wikipedia articles and of the Wikipedia category graph. Wikipedia provides a knowledge base for computing word relatedness in a more structured fashion than a search engine and with more coverage than WordNet. Obviously, semantic dimension could be extracted from other domain ontologies or vocabularies related to the dimension area as external knowledge source (e.g., *GeoNames*, *WordNet*, etc.). Each level $l_i = \langle c_1, c_2, \ldots, c_n \rangle$ includes a set of concepts c_j $(j \in [1; n])$ extracted from Wikipedia category graph.

Metadata Dimension. Metadata refers to the information about the tweet that can be derived from its metadata, such as: timestamp, user, hashtag, location, etc. So, we have to create a metadata dimension to represent each metadata information that we want consider. In this case we focus on time and location that are both hierarchical dimensions.

OLAP measures definition starts from considering that tweets are significant source of evidence when extracting information related to the reputation of a particular entity (e.g., a particular politician, singer or company) or, more general, a topic. In order to analyze the textual content of tweets, there is a need of an automated methods to disambiguate tweets with respect to entity or topic names in their content. To address this issue we propose a measure which exploit wikification service, wikify, that is the practice of representing a sentence with a set of Wikipedia concepts [5,6]. Wikification enables us to recognize sense of main concepts and named entity mentioned in the tweet associating a Wikipedia link and a corresponding weight representing uncertainty degree of the disambiguation results.

Specifically, the tweet content is wikified to extract a set of $\langle topic, rd \rangle$ pairs corresponding to Wikipedia articles that are related to the tweet content itself with a specific relevance degree (rd) [5]. Let us report an example by considering the following i-th tweet in the tweet stream:

$tweet_i$ = *"Hillary Clinton is running for president to be a champion for everyday Americans. Join Hillary for America today."*

The wikification process extracts from the above text a set of $\langle topic, relevance \rangle$ pairs. These pairs are features characterizing meaning of the input text. Taking into account the example above, the extracted topic are:

$\langle Hillary\ Rodham\ Clinton, 0.94 \rangle$, $\langle Bill\ Clinton, 0.24 \rangle$

Hence, let S be the vector space defined by the set of *topics* :

$$S \langle topic_1, topic_2, \ldots, topic_n \rangle$$

A *tweet*$_i$ in a multidimensional space is represented by a weights vector as follows:

$$tweet_i \left\langle (topic_{i_1}, rd_{i_1}), (topic_{i_2}, rd_{i_2}), (topic_{i_m}, rd_{i_n}) \right\rangle$$

where $topic_{i_k}$ is one of the topic associated to $tweet_i$, rd_{i_k} is the *relevance degree* associated to $topic_{i_k}$, n is the number of topics detected by sentence wikification of the $tweet_i$.

Our innovative aggregation model for tweet data lies along classical aggregation mechanisms [7] with the novelty of being target to such kind of data that are inherently textual in nature, hence achieving the proposed methodology for microblog summarization. This methodology essentially makes use of a fuzzy extension of *Fuzzy Formal Concept Analysis* (FFCA) [3,8] as fundamental theoretical tool. In more detail, based on the lattice theory, the FCA deals with concepts (objects and their attributes) and their hierarchical relationships. It supplies a basis for conceptual data analysis, knowledge processing and extraction. To this end, [4] combines fuzzy logic into FCA representing the uncertainty through membership values in the range [0, 1].

Aggregation is the fundamental step for computing the target OLAP cube over tweet data. Before that, microblog summarization is applied. Our summarization algorithm over multidimensional tweet streams has been defined walking across concepts of the timed fuzzy lattice structure resulting from temporal extention of FFCA. The general idea behind is to explore fuzzy formal concepts according to the chronological order of the peak areas. The algorithm incrementally selects the *best tweet*, that is the tweet with highest degree of membership belonging to the most representative concept C, at each exploration stage. The most representative concept is one that has highest weight $w(C)$. Formally, the weight $w(C)$ of fuzzy formal concept C will be evaluated as follows:

$$w(C) = \frac{\sum_{m \in M'} \mu_m}{|M'|}$$

where $|M'|$ is the number of attributes in C and the membership μ_m is defined as follows:

$$\mu_m = max_{g \in I_{G'}} \mu(g, m)$$

where $\mu(g, m)$ is the membership value between object g and attribute m.

In the following, we describe our case study that follows on the 2015 Italian election campaing. Nowadays, Twitter plays an important role in the political agenda at national and international level. So, the definition of advanced analytics services on Twitter is assuming much interest. Therefore, a motivating example of application of the methodology proposed in this work is about political elections. For the sake of clarity, here we illustrate a case study of OLAP and timed FFCA integration applied to tweet streams collected during Italian regional election campaign held in May 2015, thus showing how our microblog summarization works. A large round of

regional elections were held in Italy in seven of the twenty regions composing the country, including four of the ten largest ones: Campania, Veneto, Apulia and Tuscany. The other three regions holding elections were Liguria, Marche, Umbria, along with more than 700 of Italy's municipalities went to the polls. An estimated 23 million Italians are eligible to vote.

The target case study clearly demonstrates how OLAP analysis methodologies are perfectly suitable of supporting advanced analytics over tweet data, similarly to other recent and correlated research experience (e.g., [9,10]). In fact, summarizing the information content of massive amounts of data like streaming tweet data plays a leading role with respect to the goal of extracting useful knowledge from so abundant data. Indeed, even semantics issues must be considered, as summarization is not only data reduction but, better, *knowledge synthesis* (e.g., [11,12]), which is now becoming more and more relevant in the emerging context of big data analytics.

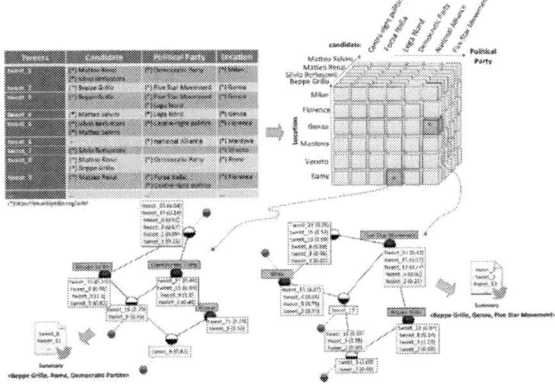

Figure 2. OLAP over Tweet Data: Case Study

Figure 2 shows the three-dimensional OLAP data cube defined on a set of collected tweets, having as measure the set of tweets belonging to the dimension *candidate* (e.g., Beppe Grillo), *location* (e.g., Rome) and *Political Party* (e.g., Democratic Party). The detail of a data cube cell (i.e., the lattice computed via microblog summarization) is also shown.

III. CLUSTERING-BASED BIG DATA ANALYTICS OVER GEO-TAGGED TWEETS

[19] proposes a framework for supporting advanced big data analytics over geo-tagged tweets via intelligent multidimensional clustering methodologies. Nowadays, we are entering the so called *big geo-data era* (e.g., [20,21]), i.e. we are in the condition of having a huge amount of information that could be exploited to answer the above question, whenever appropriately filtered and analyzed. This aspect plays a critical role, especially when dealing with big data analytics (e.g., [22,23,24]). In this respect, the widespread diffusion of smart devices has favored a new way of sharing impressions about places visited by travelers on social networks; people can post geo-tagged (i.e., geo-localized) and time-stamped (i.e., with a date) messages and pictures; these meta-data give a ready to use key to know where and when the messages

were sent. Also, *adaptive methapors* (e.g., [39]) could be integrated in this respect.

Among all social networks, Twitter (as well as other social networks that adopt the same approach) is particularly attractive to the purpose of searching messages that have something to do with travels: in fact, the limitation on the length of tweeter messages makes them a suitable means to exchange impressions in a couple of words, when people are moving around and are not keen to write long posts. Furthermore, they represent a kind of voluntary contribution, because users voluntarily install the (Twitter) app and voluntarily post messages (tweets) so that every user can see messages by other users without limitations. Even if only 20% of the tweets are geo-tagged, this posts' information, are hardly acquirable with traditional survey methods.

The above considerations motivate the research that we describe in this paper. It is necessary to design an architecture of tools that: 1) are able to continuously gather geo-localized posts from social networks; 2) provide novel methods to analyze travelers' trips, on the basis of gathered posts; 3) enable to visualize results in an integrated fashion, i.e., exploiting open geo-data sources in an integrated way; 4) make results available to other geographic services. This objective is very relevant for emerging *big data analytics tools* (e.g., [25,26,27,28]).

In order to be compliant with modern standards and to enable interoperability with open data sources and other services, we defined a *web service architecture*, in which standard components and external sources are integrated with services and suites of tools specifically developed for this research. One suite (called the *FollowMe Suite* has been described in a previous work (see [29]), queries social networks to discover posts sent by travelers from specific geographic areas, identifies the possible tourists and follow their tweets to reconstructs their time lines, i.e., the history of posted messages. A new suite (called the *Trip Analysis Suite*) proposes and implements an original method for knowledge-based trip clustering. Finally, a geo-portal (named *Tourist Tracker Geo-portal*) enables to analyze most popular tours (i.e., clustered trips) contextualizing them with the characteristics of the territory.

A main characteristic of the approach is that it is a knowledge-based approach. In facts, our goal is to semantically analyze the tracked tourist trips, by correlating them with information regarding POIs such as commercial, social and natural attractions of a specific territory of interest. The final and ambitious goal is to be able to guess which are the most plausible resources the tourists have visited.

Different knowledge of the territorial resources can be exploited, in order to perform distinct analysis of the tourists' trip. For example, in our experiments we represented tours as sequences of zip areas they crossed; however, it could be possible to represent trips as sequences of the closest POIs they came across. This is the knowledge-based geo-clustering method, developed as part of the *Trip Analysis Suite*.

Furthermore, by publishing the popular tours by an interoperable Web GIS enables their mapping and analysis contextually to other multi-source open geo-spatial data themes relative to the territory resources such as historical and world heritage, naturalistic places, shopping centers. This is provided by the *Tourist Tracked Geo-portal*.

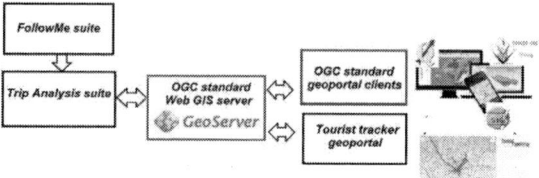

Figure 3. Architecture for Supporting Big Data Analytics over Geo-Tagged Tweets

The interoperable framework we devised for tracking and analyzing trips is depicted in Figure 3. It is constituted by two main suites of tools: the *FollowMe Suite* [29] for trips identification and the *Trips Analysis Suite* for its geographic analysis.

A third component of the architecture is the *Geoserver* open source Web GIS, that is used to publish on the Web the analyzed trips in order to share then as open data, so that any OGC compliant geo-portal client, such as QGIS, can visualize them and spatially analyze trips with respect to other geo-spatial data.

A forth component of the framework is the *Tourists Tracker Portal*, which provides users with specific functionality to analyze popular tours, getting them from *Geoserver*.

The communication is based on the web service paradigm and related protocols and formats. The joint adoption of the web service architecture and of the OGC standard for open geographic data enables the easy extension with new dedicated services; this choice guarantees flexibility in the future development of the project. This way, the framework is actually interoperable, both internally (new services and components can be easily added) and externally (easy bidirectional exchange of data with external source and services).

In the rest of this Section, we describe the tasks performed by each service.

The *FollowMe Suite* implements the first service developed during the project. The interested reader can find details in [29]. Here, we briefly describe its three main functionalities.

1. The *FollowMe Suite* queries Twitter API to find hang tweet, i.e., tweets posted in the area of the monitored airports.

2. For each user identified by means of hang tweets, the *FollowMe Suite* queries (through Twitter API) his/her *Timeline*, i.e., the history of tweets posted by the user, to get *tracked tweets*. These are geo-localized tweets posted in the next 8 days after the date of the hang tweet. Both *hang tweets* and *tracked tweets* are stored in a local storage area.

1340

3. Given an area of interest, trips that occur in that area are reconstructed and extracted, by querying hang tweets and tracked tweets previously stored in the local storage area. Reconstructed trips are exposed and exported through the web service interface.

The *Trip Analysis Suite* performs the geographical analysis and tour discovery on trips tracked by the *FollowMe Suite*. These are the main functionalities in our framework.

Basically, the *Trip Analysis Suite* performs the activities of knowledge discovery on trips collected by the *FollowMe Suite*. To this end, we propose a knowledge-based clustering method, where semantics is given by geo-slots in which punctual coordinates of tweets fall. Different geo-slots partitions give different results. Analysts can import geographic description of geo-slots from external interoperable sources. For example, it is possible to analyze tours with respect to municipalities, regions, countries, neighborhoods, etc.

This means that row trips must be pre-processed before discovering tours. The tasks hereafter described constitute the knowledge discovery process performed by the *Trip Analysis Suite*:

1. fetching of the *Geographic Slots* descriptions;

2. fetching the gathered trips w.r.t. an area of interest;

3. geo-partitioning of tweets in trips, in order to semantically labeling tweets with geographic slots;

4. clustering trips in order to identity *tours*, i.e., groups of trips that mostly visited the same geographic slots.

Results of the tour discovery tasks are deployed to *Geoserver*.

To illustrate the rationale behind the clustering technique, in Figure 4 we report five sample trips, labeled from *A* to *E* (left upper corner). Matrices in the right side of Figure 4 reports the similarity measures at each step, and the dendogram in the bottom left corner shows the hierarchy of generated clusters.

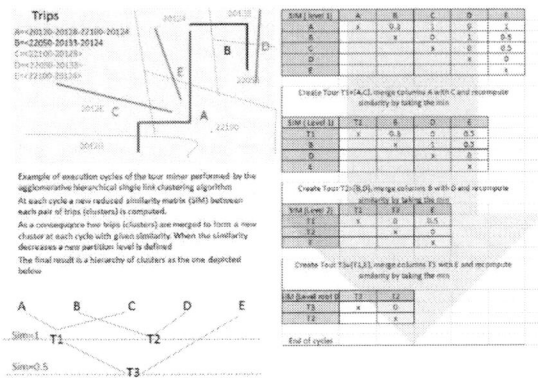

Figure 4. Tour Clustering Example

Finally, the *Tourists Tracker Geo-Portal* is the last service in our architecture. It is an OGC compliant geo-portal that provides the end-user interface to visualize tours, by getting them from *Geoserver*.

Mainly, trips belonging to the same tour are depicted with the same color, while each tour is visualized with a different color; this way, for users it is easier to analyze different tours.

IV. CONLUSIONS AND FUTURE WORK

In this paper, we have provided an overview on two state-of-the-art proposals focusing on big data analytics over big social data. Future work comprises advanced topics such as *privacy of big social data management* (e.g., [37]) and *performance of big social data processing* (e.g., [38]).

ACKNOWLEDGMENT

The author is grateful to colleagues who contributed to early versions of referred state-of-the-art works: C. De Maio, G. Fenza, V. Loia, M. Parente, G. Bordogna, L. Frigerio, G. Psaila.

REFERENCES

[1] A. Cuzzocrea, C. De Maio, G. Fenza, V. Loia, M. Parente, "OLAP analysis of multidimensional tweet streams for supporting advanced analytics". *Proceedings of ACM SAC 2016*, pp. 992–999, 2016

[2] A. Cuzzocrea, D. Saccà, J.D. Ullman, "Big data: a research agenda". *Proceedings of ACM IDEAS 2013*, pp. 198–203, 2013

[3] R. Wille, "Restructuring lattice theory: An approach based on hierarchies of concepts". *Proceedings of ICFCA 2009*, p. 314, 2009

[4] C. De Maio, G. Fenza, V. Loia, and S. Senatore, "Hierarchical web resources retrieval by exploiting fuzzy formal concept analysis". *Information Processing and Management 48(3)*, pp. 399–418, 2012

[5] R. Mihalcea, A. Csomai, "Wikify!: linking documents to encyclopedic knowledge". *Proceedings of ACM CIKM 2007*, pp. 233–242, 2007

[6] Y. Miao, C. Li, "Enhancing query-oriented summarization based on sentence wikification". *Proceedings of FGSIR 2010*, p. 32, 2010

[7] J. Gray, S. Chaudhuri, A. Bosworth, A. Layman, D. Reichart, M. Venkatrao, F. Pellow, H. Pirahesh, "Data Cube: A Relational Aggregation Operator Generalizing Group-by, Cross-Tab, and Sub Totals", *Data Mining and Knowledge Discovery 1(1)*, pp. 29–53, 1997

[8] B. Ganter, R. Wille, *"Formal Concept Analysis: Mathematical Foundations"*, 1st Edition, Springer-Verlag New York, Inc., Secaucus, NJ, USA, 1997

[9] T. Mühlbauer, W. Rödiger, A. Reiser, A. Kemper, T. Neumann, "ScyPer: A Hybrid OLTP&OLAP Distributed Main Memory Database System for Scalable Real-Time Analytics". *Proceedings of BTW 2013*, pp. 499–502, 2013

[10] A. Cuzzocrea, "Analytics over Big Data: Exploring the Convergence of Data Warehousing, OLAP and Data-Intensive Cloud Infrastructures". *Proceedings of IEEE COMPSAC 2013*, pp. 481–483, 2013

[11] L.V.S. Lakshmanan, J. Pei, Y. Zhao, "QC-Trees: An Efficient Summary Structure for Semantic OLAP". *Proceedings of ACM SIGMOD Conference 2003*, pp. 64–75, 2003

[12] T. Munger, S. Desa, and C. Wong, "The Use of Domain Knowledge Models for Effective Data Mining of Unstructured Customer Service Data in Engineering Applications". *Proceedings of IEEE BigDataService 2015*, pp. 427–438, 2015

[13] A. Cuzzocrea, I.-Y. Song, K.C. Davis, "Analytics over large-scale multidimensional data: the big data revolution!". *Proceedings of ACM DOLAP 2011*, pp. 101–104, 2011

[14] A. Cuzzocrea, I.-Y. Song, "Big Graph Analytics: The State of the Art and Future Research Agenda". *Proceedings of ACM DOLAP 2014*, pp. 99–101, 2014

[15] M. Franklin, "Making Sense of Big Data with the Berkeley Data Analytics Stack". *Proceedings of ACM WSDM 2014*, pp. 1–2, 2014

[16] K. Kolomvatsos, C. Anagnostopoulos, S. Hadjiefthymiades, "An Efficient Time Optimized Scheme for Progressive Analytics in Big Data". *Big Data Research 2(4)*, pp. 155–165, 2015

[17] V. Kantere, M. Filatov, "A Workflow Model for Adaptive Analytics on Big Data". *Proceedings of IEEE BigData Congress 2015*, pp. 673–676, 2015

[18] F. Zhang, M. Liu, F. Gui, W. Shen, A. Shami, Y. Ma, "A distributed frequent itemset mining algorithm using Spark for Big Data analytics", *Cluster Computing, 18(4)*, pp. 1493–1501, 2015

[19] G. Bordogna, L. Frigerio, A. Cuzzocrea, G. Psaila, "Clustering Geo-tagged Tweets for Advanced Big Data Analytics". *Proceedings of IEEE BigData Congress 2016*, pp. 42–51, 2016

[20] L. Zhao, L. Chen, R. Ranjan, K. R. Choo, J. He, "Geographical information system parallelization for spatial big data processing: a review". *Cluster Computing 19(1)*, pp. 139–152, 2016

[21] P. Baumann, P. Mazzetti, J. Ungar, R. Barbera, D. Barboni et al., "Big data analytics for earth sciences: the earthserver approach". *Int. J. Digital Earth 9(1)*, pp. 3–29, 2016

[22] G. L. Andrienko, N. V. Andrienko, H. Bosch, T. Ertl, G. Fuchs, P. Jankowski, D. Thom, "Thematic patterns in georeferenced tweets through space-time visual analytics". *Computing in Science and Engineering 15(3)*, pp. 72–82, 2013

[23] S. Kumar, X. Hu, H. Liu, "A behavior analytics approach to identifying tweets from crisis regions". *Proceedings of ACM HT 2014*, pp. 255–260, 2014

[24] Y. Lu, X. Hu, F. Wang, S. Kumar, H. Liu, R. Maciejewski, "Visualizing social media sentiment in disaster scenarios.". *Proceedings of ACM WWW 2015*, pp. 1211–1215, 2015

[25] S. Yang, A.L. Kavanaugh, "Collecting, analyzing and visualizing tweets using open source tools". *Proceedings of DG.O 2011*, pp. 374–375, 2011

[26] D. Cameron, A. Finlayson, R. Wotzko, "Visualising social computing output: Mapping student blogs and tweets.". *Social Media Tools and Platforms in Learning Environments*, pp. 337–350, 2011

[27] L. Wang, R. Ranjan, J. Kolodziej, A. Y. Zomaya, L. Alem, "Software tools and techniques for big data computing in

healthcare clouds". *Future Generation Computer Systems 43-44*, pp. 38–39, 2015

[28] K. Slavakis, G.B. Giannakis, G. Mateos, "Modeling and optimization for big data analytics: (statistical) learning tools for our era of data deluge". *IEEE Signal Process. Mag. 31(5)*, pp. 18–31, 2014

[29] A. Cuzzocrea, G. Psaila, and M. Toccu, "Knowledge discovery from geo-located tweets for supporting advanced big data analytics: A real-life experience". *Proceedings of MEDI 2015*, pp. 285–294, 2015

[30] J. Manyika, M. Chui, B. Brown, J. Bughin, R. Dobbs, C. Roxburgh, A.H. Byers. *"Big data: The next frontier for innovation, competition, and productivity"*, 2011

[31] L. Oneto, F. Bisio, E. Cambria, D. Anguita, "Statistical Learning Theory and ELM for Big Social Data Analysis". *IEEE Computational Intelligence Mag. 11(3)*, pp. 45–55, 2016

[32] X. Zhang, Z. Yi, Z. Yan, G. Min, W. Wang, A. Elmokashfi, S. Maharjan, Y. Zhang, "Social Computing for Mobile Big Data". *IEEE Computer 49(9)*, pp. 86–90, 2016

[33] K. Chodorow, *"MongoDB: the definitive guide"*. O'Reilly Media, Inc., 2013

[34] K. Shvachko, H. Kuang, S. Radia, R. Chansler, "The Hadoop Distributed File System". *Proceedings of MSST 2010*, pp. 1–10, 2010

[35] A. Thusoo, J.S. Sarma, N. Jain, Z. Shao, P. Chakka, N. Zhang, S. Anthony, H. Liu, R. Murthy, "Hive - a petabyte scale data warehouse using Hadoop". *Proceedings of IEEE ICDE 2010*, pp. 996–1005, 2010

[36] Pentaho, *Pentaho: Data Integration, Business Analytics and Big Data*, http://www.pentaho.com/

[37] A. Cuzzocrea, "Privacy and Security of Big Data: Current Challenges and Future Research Perspectives". *Proceedings of ACM PSBD 2014*, pp. 45–47, 2014

[38] A. Cuzzocrea, U. Matrangolo, "Analytical Synopses for Approximate Query Answering in OLAP Environments". *Proceedings of DEXA 2004*, pp. 359–370, 2004

[39] M Cannataro, A Cuzzocrea, A Pugliese, "A probabilistic approach to model adaptive hypermedia systems". *Proceedings of WebDyn 2001*, 2001

[40] N. Makrynioti, A. Grivas, C. Sardianos, N. Tsirakis, I. Varlamis, V. Vassalos, V. Poulopoulos, P. Tsantilas, "PaloPro: a platform for knowledge extraction from big social data and the news". *International Journal of Big Data Intelligence 4(1)*, pp. 3–22, 2017

Accelerating Dynamic Itemset Counting on Intel Many-core Systems

Mikhail Zymbler
South Ural State University, Chelyabinsk, Russia
mzym@susu.ru

Abstract—The paper presents a parallel implementation of a Dynamic Itemset Counting (DIC) algorithm for many-core systems, where DIC is a variation of the classical Apriori algorithm. We propose a bit-based internal layout for transactions and itemsets with the assumption that such a representation of the transaction database fits in main memory. This technique reduces the memory space for storing the transaction database and also simplifies support counting and candidate itemsets generation via logical bitwise operations. Implementation uses OpenMP technology and thread-level parallelism. Experimental evaluation on the platforms of Intel Xeon CPU and Intel Xeon Phi coprocessor with large synthetic database showed good performance and scalability of the proposed algorithm.

Index Terms—frequent itemset mining, dynamic itemset counting, bitmap, OpenMP, many-core, Intel Xeon Phi

INTRODUCTION

Association rule mining is one of the important problems in data mining [1]. The task is to discover the strong associations among the items from a transaction database such that the presence of one item in a transaction implies the presence of another. Association rule mining is decomposed into two subtasks [1]. The first one is to find all frequent itemsets that consist of items which often occur together in transactions. The second one is to generate all the association rules from the frequent itemsets found.

In this paper, we address the task of frequent itemset mining which can be formally described as follows. Let $\mathcal{I} = (i_1, \ldots, i_m)$ be a set of literals, called *items*. Let $\mathcal{D} = (T_1, \ldots, T_n)$ be a database of *transactions*, where each transaction $T_i \subseteq \mathcal{I}$ consists of a set of items (*itemset*). An itemset that contains k items is called a k-itemset. The *support* of an itemset $I \subseteq \mathcal{I}$ denotes the percentage of transactions in $\mathcal{D}$ that contain the itemset I. If support of an itemset $I \subseteq \mathcal{I}$ satisfies the user-specified minimum support threshold (called $minsup$) then I is *frequent itemset*. Let the set of frequent k-itemsets be denoted by $\mathcal{L}_k$ and $\mathcal{L} = \cup_{k=1}^{k_{max}} \mathcal{L}_k$ denotes a set of all frequent itemsets, where k_{max} is number of items in the longest frequent itemset. Given the transaction database $\mathcal{D}$ and minimum support threshold $minsup$ the goal of frequent itemset mining is to find the set of all frequent itemsets $\mathcal{L}$.

There is a wide spectrum of algorithms for frequent itemset mining and none of them outperforms all others for all possible transaction databases and values of $minsup$ threshold [7]. *Apriori* [1] is one of the most popular itemset mining algorithms for which many refinements and parallel implementations for various platforms were proposed. Dynamic Itemset Counting (DIC) [2] is a variation of Apriori, which tries to reduce number of passes made over a transaction database while keeping the number of itemsets counted in a pass relatively low. Despite the fact that DIC has good potential of parallelization [2] it still has not been implemented for modern many-core CPU and accelerators, to the best of our knowledge.

In this paper we propose parallel implementation of the DIC algorithm for Intel Xeon and Intel Xeon Phi (Knights Landing) many-core platforms. Intel Xeon Phi device is an x86 many-core coprocessor of 61 cores, connected by a high-performance on-die bidirectional interconnect where each core supports $4\times$ hyperthreading and contains 512-bit wide vector processor unit. Knights Landing [13] is a second generation MIC (Many Integrated Core) architecture product from Intel. As opposed to predecessor it is an independent (bootable) device, which runs applications only in native mode.

We suggest a bit-based internal layout for transactions and itemsets assuming that such a representation of a transaction database fits in main memory. This technique has a few major merits. It reduces memory space of storing the transaction database and simplifies support counting and generation of candidate (potentially frequent) itemsets via logical bitwise operations. We parallelize the algorithm through OpenMP technology and thread-level parallelism. We conduct experiments on large synthetic database to evaluate performance and scalability of our algorithm.

The rest of the paper is organized as follows. Section I provides a brief description of an original DIC algorithm. The proposed parallel algorithm is presented in section II. In section III related work is discussed. The results of experimental evaluation of the algorithm are described in section IV. The conclusion contains summarizing remarks and directions for future research.

I. SERIAL DIC ALGORITHM

Dynamic Itemset Counting (DIC) [2] is a variation of the most well-known Apriori algorithm [1]. Apriori is an iterative, level-wise algorithm, which uses a bottom-up search. At the

---

This work was financially supported by the Russian Foundation for Basic Research (grant No. 17-07-00463), by Act 211 Government of the Russian Federation (contract No. 02.A03.21.0011) and by the Ministry of education and science of Russian Federation (government order 2.7905.2017).

first pass over transaction database it processes 1-itemsets and finds $\mathcal{L}_1$ set. A subsequent pass k consists of two steps, namely candidate generation and pruning. At the *candidate generation* step Apriori combines elements of $\mathcal{L}_{k-1}$ set to form candidate (potentially frequent) k-itemsets. At the *pruning* step it gets rid of infrequent candidates using the *a priori* principle, which states that any infrequent $(k-1)$-itemset cannot be a subset of a frequent k-itemset. Apriori counts support of candidates which have not been pruned and proceeds with such passes so forth until no candidates remain after pruning.

Algorithm 1. DIC(in $\mathcal{D}$, in $minsup$, in M, out $\mathcal{L}$)

$\qquad\qquad\qquad\qquad\qquad$ ▷ Initialize sets of itemsets
2: SolidBox $\leftarrow \varnothing$; SolidCircle $\leftarrow \varnothing$; DashedBox $\leftarrow \varnothing$
$\quad$ DashedCircle $\leftarrow \mathcal{I}$
4: **while** DashedCircle $\cup$ DashedBox $\neq \varnothing$ **do**
$\qquad\qquad\qquad$ ▷ Scan database and rewind if necessary
6: $\quad$ Read($\mathcal{D}, M, Chunk$)
$\quad$ **if** EOF($\mathcal{D}$) **then**
8: $\qquad$ Rewind($\mathcal{D}$)
$\quad$ **for all** $T \in Chunk$ **do**
10: $\qquad\qquad\qquad\qquad$ ▷ Count support of itemsets
$\qquad$ **for all** $I \in$ DashedCircle $\cup$ DashedBox **do**
12: $\qquad\quad$ **if** $I \subseteq T$ **then**
$\qquad\qquad$ support(I) $\leftarrow$ support(I) $+ 1$
14: $\qquad\qquad\qquad\qquad$ ▷ Generate candidate itemsets
$\qquad$ **for all** $I \in$ DashedCircle **do**
16: $\qquad\quad$ **if** support(I) $\geq minsup$ **then**
$\qquad\qquad$ MoveItemset(I, DashedBox)
18: $\qquad\qquad$ **for all** $i \in \mathcal{I}$ **do**
$\qquad\qquad\qquad$ $C \leftarrow I \cup i$
20: $\qquad\qquad\qquad$ **if** $\forall s \subseteq C \; s \in$ SolidBox$\cup$DashedBox **then**
$\qquad\qquad\qquad\qquad$ MoveItemset(C, DashedCircle)
22: $\qquad\qquad\qquad$ ▷ Check full pass completion for itemsets
$\qquad$ **for all** $I \in$ DashedCircle **do**
24: $\qquad\quad$ **if** IsPassCompleted(I) **then**
$\qquad\qquad$ MoveItemset(I, DashedBox)
26: $\qquad$ **for all** $I \in$ DashedBox **do**
$\qquad\quad$ **if** IsPassCompleted(I) **then**
28: $\qquad\qquad$ MoveItemset(I, SolidBox)
$\quad \mathcal{L} \leftarrow$ SolidBox

The DIC algorithm tries to reduce the number of passes made over the transaction database while keeping the number of itemsets counted in a pass relatively low. Alg. 1 depicts pseudo-code of the DIC algorithm. DIC processes database with stops at equal-length intervals between transactions (parameter M of the algorithm). At the end of the transaction database it is necessary to rewind to its beginning.

DIC maintains four sets of itemsets, namely *Dashed Circle*, *Dashed Box*, *Solid Circle* and *Solid Box*. Itemsets in the "dashed" sets are subjects for support counting while itemsets in the "solid" sets do not need to be counted. "Circles" contain infrequent itemsets while "boxes" contain frequent itemsets.

Thus, *Dashed Circle* and *Dashed Box* contain itemsets that are suspected infrequent and are suspected frequent respectively while *Solid Circle* and *Solid Box* contain itemsets

that are confirmed infrequent and are confirmed frequent respectively. At start *Dashed Box*, *Solid Circle* and *Solid Box* are assumed to be empty and *Dashed Circle* contains all the 1-itemsets.

Before the stop, DIC counts support of itemsets from "dashed" sets for each transaction. At any stop DIC performs as follows. Itemsets whose support exceeds $minsup$ are moved from *Dashed Circle* to *Dashed Box*. New itemsets are added into *Dashed Circle*, they are immediate supersets of those itemsets from *Dashed Box* with all of its subsets from "box" lists. Itemsets that have completed one full pass over the transaction database are moved from the "dashed" set to "solid" set. DIC proceeds if any itemset in "dashed" sets remains.

II. PARALLEL DIC ALGORITHM

A. Internal Data Layout

In this work we suggest *direct bit representation* for both transactions and itemsets. For a transaction $T \subseteq \mathcal{D}$ (for an itemset $I \subseteq \mathcal{I}$, respectively) this means that it is represented by a word where each p-th bit is set to one if an item $i_p \in T$ ($i_p \in I$, respectively) and all other bits are set to zero. The word's length W in bytes depends on system environment and it is calculated as $W = \lceil \frac{m}{sizeof(\text{byte})} \rceil$. In our implementation we use C++ and `unsigned long long int` data type, so we have $W = 8$ and $m = 64$. This could be extended through an open-source library for arbitrary precision arithmetic, for instance, GNU Bignum Library[1].

Let us denote by $BitMask$ a function that returns direct bit representation of a given itemset or transaction as a word, i.e. $BitMask : \mathcal{I} \to \mathbb{Z}_+$. Then direct bit representation of transaction database $\mathcal{D}$ is an n-element array $\mathcal{B}$, where $\forall j, 1 \leq j \leq n \; \mathcal{B}[j] = BitMask(T_j)$.

Direct bit representation has several major merits. It often requires less space than byte-based representation for dense transaction database with long transactions. In fact, $\mathcal{B}$ requires $n \cdot W$ bytes to store and allows $\mathcal{B}$ to fit in main memory. For instance, *netflix*[2], one of the most referenced datasets, contains $n = 17,771$ transactions consisting of $m = 480,189$ distinct items. Hence, direct bit representation of the *netflix* dataset takes about 1 Gb. Thus, in what follows we assume that $\mathcal{B}$ has been preliminary produced from $\mathcal{D}$ and it is available in main memory.

Direct bit representation simplifies support counting as well. The fact of $I \subseteq T$ can be checked by the predicate with one logical bitwise operation, that is $BitMask(I)$ AND $BitMask(T) = BitMask(I)$.

Thereby, we implement an itemset as a record structure with the following basic fields, namely $mask$ to provide direct bit representation, k as number of items in the itemset, $stop$ as counter to determine when full pass for the given itemset is completed, and $supp$ to store support count.

---

[1] The GNU Multiple Precision Arithmetic Library
[2] http://www.netflixprize.com

To implement a set of itemsets, we use vector, which represents an array of elements belonging to the same type and provides random access to its elements with an ability to automatically resize when appending elements. Such a data structure is implemented in C++ Standard Template Library as a class with iterator and methods for inserting an element and removing an element with complexity of $O(1)$ and $O(s)$ respectively, where s is the current size of a vector.

To reduce costs of moving elements across vectors, we establish a *DASHED* vector for "dashed box" and "dashed circle" itemsets and a *SOLID* vector for "solid box" and "solid circle" itemsets and provide the itemset's record structure with fig field to indicate an appropriate set the given itemset belongs to.

B. Parallelization of the Algorithm

The proposed parallel version of DIC algorithm is presented in Alg. 2 and basic sub-algorithms are depicted in Alg. 3–5.

Algorithm 2. ParalDIC(in $\mathcal{B}$, in $minsup$, in M, out $\mathcal{L}$)

                                  ▷ Initialize sets of itemsets
2: SOLID.$init()$; DASHED.$init()$
     $k \leftarrow 1$
4: **for all** $i \in 0..m - 1$ **do**
         $I.fig \leftarrow$ NIL; $I.bitmask \leftarrow 0$
6:       $I.mask \leftarrow$ SetBit($I.mask, i$)
         $I.stop \leftarrow 0$; $I.supp \leftarrow 0$; $I.k \leftarrow k$
8:       SOLID.$push\_back(I)$
     $stop_{max} \leftarrow \lceil \frac{n}{M} \rceil$; $stop \leftarrow 0$
10: FirstPass(SOLID, DASHED)
     while not DASHED.$empty()$ **do**
12:                ▷ Scan database and rewind if necessary
        $stop \leftarrow stop + 1$
14:     **if** $stop > stop_{max}$ **then**
           $stop \leftarrow 1$
16:     $first \leftarrow (stop - 1) \cdot M$; $last \leftarrow stop \cdot M - 1$
        $k \leftarrow k + 1$
18:     CountSupport(DASHED)
        CutDashedCircle(DASHED)
20:     GenCandidates(DASHED)
        CheckFullPass(DASHED)
22: $\mathcal{L} \leftarrow \{I \in$ SOLID, $I.fig =$ BOX$\}$

We enhance the classical DIC algorithm by adding two more stages, namely *FirstPass* and *CutDashedCircle* where each of them is aimed to reducing the number of itemsets to perform support counting of.

We parallelize the following stages of the algorithm, namely support counting (cf. Alg. 3), reduction of *Dashed Circle* set (cf. Alg. 4) and checking full pass completion for itemsets (cf. Alg. 5) through OpenMP technology and thread-level parallelism.

In the classical DIC algorithm, the *Dashed Circle* set is initialized by all the 1-itemsets (cf. Alg. 1, line 3). In contrast with classical DIC, we use the technique of full first pass [4]. This means that we initially perform one full pass over $\mathcal{D}$ to find $\mathcal{L}_1$, the set of frequent 1-itemsets (this done similarly

to Alg. 3). Then candidate 2-itemsets are computed from $\mathcal{L}_1$ through the Apriori join procedure [1]. This done via logical bitwise OR operation on each pair of frequent 1-itemsets and candidates are inserted in the *Dashed Circle* set. This technique helps to reduce cardinality of the *Dashed Circle* set in further computations because infrequent 1-itemsets and their supersets have been pruned according to the *a priori* principle.

Algorithm 3. CountSupport(in out $DASHED$)

     if DASHED.$size() \geq num\_of\_threads$ **then**
2:     #pragma omp parallel for
        for all $I \in$ DASHED **do**
4:         $I.stop \leftarrow I.stop + 1$
          for all $T \in \mathcal{B}[first] .. \mathcal{B}[last]$ **do**
6:            **if** $I.mask$ AND $T = I.mask$ **then**
              $I.supp \leftarrow I.supp + 1$
8: **else**
       omp_set_nested(true)
10:    #pragma omp parallel for
       num_threads(DASHED.$size()$)
12:    **for all** $I \in$ DASHED **do**
        $I.stop \leftarrow I.stop + 1$
14:    #pragma omp parallel for reduction(+:$I.supp$)
       num_threads($\lceil \frac{num\_of\_threads}{\text{DASHED}.size()} \rceil$)
16:       **for all** $T \in \mathcal{B}[first] .. \mathcal{B}[last]$ **do**
         if $I.mask$ AND $T = I.mask$ **then**
18:           $I.supp \leftarrow I.supp + 1$

In the original algorithm support counting is performed through two nested loops (cf. Alg. 1, lines 9–13) where the outer loop takes transactions and the inner loop takes the "dashed" itemsets. As opposed to the classical DIC algorithm we change the order of these loops to parallelize outer loop through omp parallel for pragma (cf. Alg. 3). This shuffle avoids data races when threads process different transactions but need to change support count of the same itemsets simultaneously.

Additionally, our algorithm balances the load of threads depending on the current total number of elements in both *Dashed Circle* and *Dashed Box* sets. If the number of available threads does not exceed current total number of "dashed" itemsets, we parallelize the outer loop (along itemsets) using all the threads. Otherwise, we enable nested parallelism and parallelize the outer loop using a number of threads equal to the current total number of "dashed" itemsets. Then we parallelize the inner loop (along transactions) so that each outer thread forks an equal-sized set of descendant threads where descendants perform counting through reduction of summing operation. This balancing technique allows to processing data effectively in the final stage of counting when the number of candidate itemsets tends to zero and increases overall performance of the algorithm.

After the support counting, in addition to moving appropriate itemsets from *Dashed Circle* set to *Dashed Box* set as in classical DIC (cf. Alg. 1, line 17), we reduce *Dashed Circle* set pruning clearly infrequent itemsets as follows [9]. We compute an itemset's highest possible support by adding

Algorithm 4. CutDashedCircle(in out $DASHED$)

    #pragma omp parallel for
2: **for all** $I \in$ DASHED **and** $I.fig =$ CIRCLE **do**
    if $I.supp \geq minsup$ **then**
4:              ▷ Move appropriate itemsets to Dashed Box set
        $I.fig \leftarrow$ BOX
6:     **else**
                            ▷ Prune clearly infrequent itemset
8:     $supp_{max} \leftarrow I.supp + M \cdot (stop_{max} - I.stop)$
        if $supp_{max} < minsup$ **then**
10:         $I.fig \leftarrow$ NIL
                         ▷ Prune supersets of infrequent itemset
12:         **for all** $J \in$ DASHED **and** $J.fig =$ CIRCLE **do**
            if $I.mask$ AND $J.mask = I.mask$ **then**
14:             $J.fig \leftarrow$ NIL
    DASHED$.erase(\forall I, I.fig =$ NIL$)$

its current support to the number of transactions have not been processed yet (cf. Alg. 4). If the value of the itemset's highest possible support is less than *minsup* threshold, then the itemset is pruned and after that we prune all its supersets according to the *a priori* principle.

After the reduction of *Dashed Circle* set we generate afresh itemsets to be inserted in that set performing Apriori join procedure [1] via logical bitwise OR operation between all the itemsets marked as "boxes".

Algorithm 5. CheckFullPass(in out $DASHED$)

    #pragma omp parallel for
2: **for all** $I \in$ DASHED **do**
    if $I.stop = stop_{max}$ **then**
4:     **if** $I.supp \geq minsup$ **then**
        $I.fig \leftarrow$ BOX
6:     SOLID$.push\_back(I)$
    $I.fig \leftarrow$ NIL
8: DASHED$.erase(\forall I, I.fig =$ NIL$)$

Finally, for all itemsets in the *Dashed Circle* set we check if an itemset has been counted through all the transactions and if yes, we make the itemset "solid" and stop counting it (cf. Alg. 5). This activity is also parallelized along itemsets through omp parallel for pragma.

In the end *DASHED* vector contains "box" itemsets as a result of the algorithm.

III. Related Work

The Original DIC algorithm was presented by Brin et al. in [2], where the authors briefly discuss a way to parallelize DIC using the distribution of the transaction database among the nodes so that each node counts all the itemsets for its own data segment. The authors noticed that it is unnecessary to perform synchronization and load balancing in parallel version of DIC.

Paranjape-Voditel et al. proposed *DIC-OPT* [10], a parallel version of DIC for distributed memory systems. The key idea is that each node sends messages to other nodes after every M

transactions have been read regarding the counts of potentially frequent itemsets. This initiates the early counting of the itemsets on other nodes without waiting for synchronization with other nodes. Authors carried out experiments on up to 12 nodes where their implementation showed sub-linear speedup.

Cheung et al. suggested *APM* [4], a DIC-based parallel algorithm for SMP systems. APM is an adaptive parallel mining algorithm, where all CPUs generate candidates dynamically and count itemset supports independently without synchronization. The transaction database is partitioned across CPUs with a highly homogeneous itemset distributions. This technique addresses to the problem of a large number of candidates because of the low homogeneous itemset distribution in most cases. The experiments on the Sun Enterprise 4000 server with up to 12 nodes showed that APM outperforms Apriori-like parallel algorithms. However, APM's speedup gradually drops down to 4 when the number of nodes is grater than 4. This is because APM suffers from the SMP's inherent problem of I/O contention when the number of nodes is large.

Schlegel et al. proposed *mcEclat* [12], a parallel version of the well-known mining algorithm Eclat [14] for the Intel Xeon Phi coprocessor. mcEclat converts a dataset being mined into a set of tid-bitmaps, which are repeatedly intersected to obtain the frequent itemsets. *Tid-bitmap* maps the IDs of transactions in which an itemset occurs to bits in a bitmap at certain positions. For instance, if the itemset i exists in 4-th and 7-th transactions then the respective bits of i's tid-bitmap are set to one while all its other bits are set to zero. Tid-bitmaps are intersected via logical bitwise AND operation and then support of an itemset is obtained by counting the one bits in its respective tid-bitmap. Experiments showed up to $100\times$ speedup of mcEclat on the Intel Xeon Phi. However, the algorithm's performance on the Intel Xeon Phi coprocessor is similar or slightly worse (for smaller values of $minsup$) than on system with two Intel Xeon CPUs when the maximum number of threads is employed on both systems. The reason is that mcEclat does not fully exploit the Intel Xeon Phi's powerful vector processing capabilities.

Kumar et al. presented *Bitwise DIC* [9], a serial version of the DIC algorithm based upon tid-bitmap technique mentioned above. Bitwise DIC outperforms the original DIC. Unfortunately, the authors poorly supported their study by experiments and discussion of the results (only five runs of the algorithms on one dataset with 5,000 transactions for fixed value of $minsup$ were conducted and only runtime was presented).

In serial algorithms *MAFIA* [3] and *BitTableFI* [6] Burdick et al. and Dong et al., respectively, used vertical bitmap to compress the transaction database for quick candidate itemsets generation and support count. *Vertical bitmap* is a set of integer in which every bit represents an item. If an item i appears in a transaction j, then bit j of the bitmap for item i is set to one; otherwise, the bit is set to zero. This idea is applied to transactions and itemsets. In cases where itemsets appear in a significant number of transactions, the vertical bitmap is the smallest representation of the information. However, the weakness of a vertical representation is the sparseness of the bitmaps, especially at the lower support levels.

In this paper we suggested a parallel version of the DIC algorithm for Intel Xeon and Xeon Phi many-core systems (which was done for the first time, to the best of our knowledge) where we use direct bit representation of both transaction database and itemsets.

IV. EXPERIMENTAL EVALUATION

To evaluate the developed algorithm, we performed experiments on the Tornado SUSU [8] supercomputer's node (cf. Tab. I for its specifications).

TABLE I: Specifications of hardware

Specifications	CPU	Coprocessor
Model, Intel Xeon	X5680	Phi SE10X
Cores	6	61
Frequency, GHz	3.33	1.1
Threads per core	2	4
Peak performance, TFLOPS	0.371	1.076
Memory, Gb	24	8
Cache, Mb	12	30.5

We compiled source code using Intel `icpc` compiler (version 15.0.3). Experiments have been performed on realistic and synthetic[3] datasets summarized in Tab. II.

TABLE II: Specifications of datasets

Dataset	Category	# transactions	# items
SKIN [5]	Real	245,057	11
RECORDLINK [11]	Real	574,913	29
20M	Synthetic	$2 \cdot 10^7$	64

In the experiments, we studied the following aspects of the developed algorithm. We compared the performance of parallel DIC with serial implementations of DIC[4] and Apriori[5] algorithms. We also evaluated the scalability of our algorithm depending on the value of M (the number of transactions that should be processed before stop) and on $minsup$ threshold.

Fig. 1 illustrates the results of the first set of experiments where we compare the performance of parallel DIC with serial DIC and Apriori on CPU. As was seen, serial DIC performs the worst for all the datasets we have tested on[6] and this is in accordance with testing results of B. Goethals. For datasets with relatively small number of short transactions, serial Apriori performs best whereas parallel DIC demonstrates degradation of the performance. However, in case of a large dataset, parallel DIC outperforms serial Apriori. Hence, our algorithm behaves the best way when the transaction database provides sufficient amount of work in support counting, which

---

[3]We use IBM Quest Synthetic Data Generator similar to original paper [2].
[4]Frequent Pattern Mining Implementations by Bart Goethals
[5]Apriori – Frequent Item Set Mining by Christian Borgelt
[6]For 20M dataset Serial DIC was stopped after 30 hours without output.

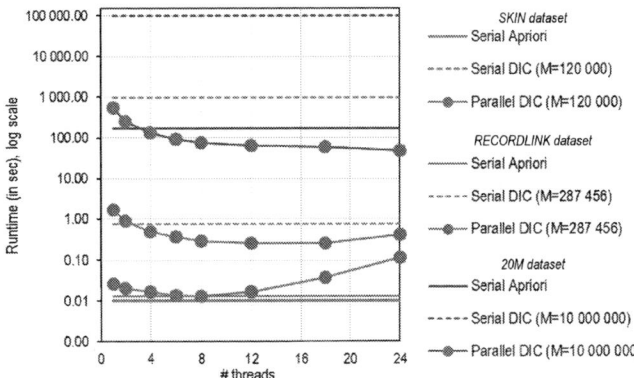

Fig. 1: Comparison of performance ($minsup = 0.1$)

is the heaviest part of the algorithm. This is why we use 20M dataset in the next set of experiments.

Fig. 2 depicts the results of the second set of experiments where we studied the scalability of parallel DIC w.r.t. M parameter on both platforms using 20M dataset. Experimental results show that parallel DIC outperforms serial Apriori much more often than not. A greater value of M results in less runtime and greater speedup. On both platforms at greater value of M our algorithm shows speedup closer to linear when the number of threads matches the number of physical cores the algorithm is running on and speedup becomes sub-linear when the algorithm uses more than one thread per physical core. Parallel DIC achieves up to $12\times$ and $90\times$ speedup on CPU and Xeon Phi, respectively.

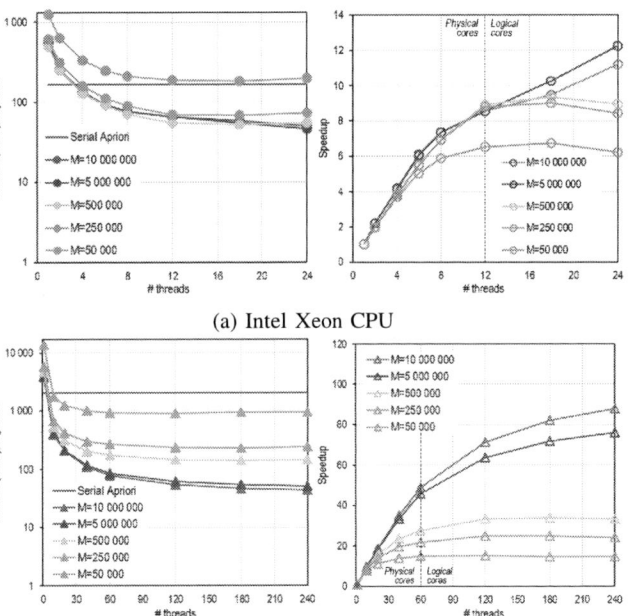

(a) Intel Xeon CPU

(b) Intel Xeon Phi coprocessor

Fig. 2: Scalability w.r.t. M (20M dataset, $minsup = 0.1$)

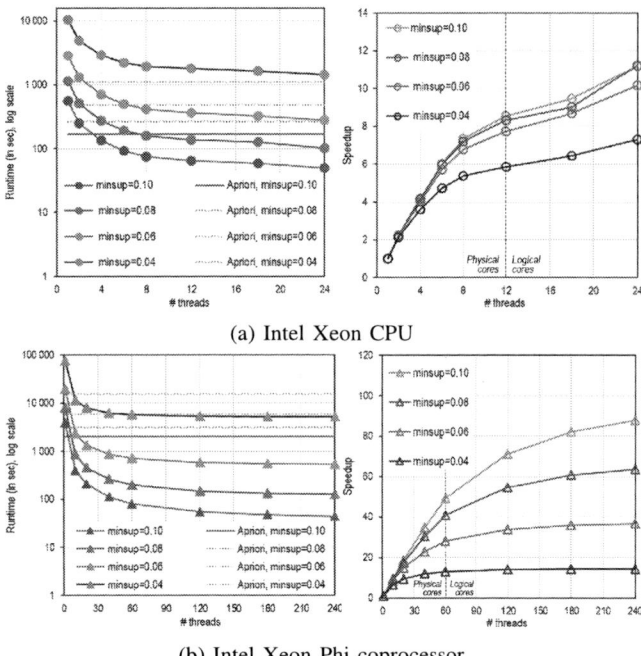

(a) Intel Xeon CPU

(b) Intel Xeon Phi coprocessor

Fig. 3: Scalability w.r.t. $minsup$ (20M dataset, $M = 10^7$)

Fig. 3 presents the results of the third set of experiments where we evaluated scalability of parallel DIC w.r.t. $minsup$ threshold on both platforms. Our algorithm still shows better speedup when only physical cores are involved. Runtime and speedup of the algorithm expectedly suffer from decreasing of $minsup$ value. Parallel DIC outperforms serial Apriori with the exception of hard case $minsup = 0.02$ on CPU where our algorithm shows almost the same runtime as Apriori when the maximum number of threads is employed. On the Intel Xeon Phi platform it is enough for our algorithm to use 10 threads to overtake serial Apriori.

Summing up, parallel DIC demonstrates good performance and scalability on both platforms.

CONCLUSION

In this paper we presented a parallel implementation of Dynamic Itemset Counting (DIC) algorithm for Intel many-core systems, where DIC is a variation of classical Apriori algorithm for frequent itemset mining.

We enhance the DIC algorithm by adding two more stages, which are devoted to reducing the number of itemsets to perform support counting of. We also propose direct bit representation for transactions and itemsets with the assumption that such a representation of the transaction database fits in main memory. This technique reduces memory space for storing the transaction database and also simplifies support counting and candidate itemsets generation via logical bitwise operations. We parallelize the DIC algorithm through of OpenMP technology and thread-level parallelism. Our algorithm balances support counting between threads depending on the current

total number of candidate itemsets. We performed experimental evaluation on the platforms of the Intel Xeon CPU and the Intel Xeon Phi coprocessor with large synthetic database, showing the good performance and scalability of the proposed algorithm.

In continuation of the presented research, we plan to implement the developed parallel algorithm for the case of cluster systems based on nodes with the Intel Xeon Phi many-core coprocessor on-board.

REFERENCES

[1] R. Agrawal and R. Srikant, "Fast algorithms for mining association rules in large databases," in *VLDB'94, Proceedings of 20th International Conference on Very Large Data Bases, September 12-15, 1994, Santiago de Chile, Chile*, J. B. Bocca, M. Jarke, and C. Zaniolo, Eds. Morgan Kaufmann, 1994, pp. 487–499.

[2] S. Brin, R. Motwani, J. D. Ullman, and S. Tsur, "Dynamic itemset counting and implication rules for market basket data," in *SIGMOD 1997, Proceedings ACM SIGMOD International Conference on Management of Data, May 13-15, 1997, Tucson, Arizona, USA.*, J. Peckham, Ed. ACM Press, 1997, pp. 255–264.

[3] D. Burdick, M. Calimlim, J. Flannick, J. Gehrke, and T. Yiu, "MAFIA: A maximal frequent itemset algorithm," *IEEE Trans. Knowl. Data Eng.*, vol. 17, no. 11, pp. 1490–1504, 2005.

[4] D. W. Cheung, K. Hu, and S. Xia, "An adaptive algorithm for mining association rules on shared-memory parallel machines," *Distributed and Parallel Databases*, vol. 9, no. 2, pp. 99–132, 2001.

[5] A. Dhall, G. Sharma, R. Bhatt, and G. M. Khan, "Adaptive digital makeup," in *Advances in Visual Computing, 5th International Symposium, ISVC 2009, Las Vegas, NV, USA, November 30 - December 2, 2009, Proceedings, Part II*, ser. Lecture Notes in Computer Science, G. Bebis, R. D. Boyle, B. Parvin, D. Koracin, Y. Kuno, J. Wang, R. Pajarola, P. Lindstrom, A. Hinkenjann, L. M. Encarnação, C. T. Silva, and D. S. Coming, Eds., vol. 5876. Springer, 2009, pp. 728–736.

[6] J. Dong and M. Han, "Bittablefi: An efficient mining frequent itemsets algorithm," *Knowl.-Based Syst.*, vol. 20, no. 4, pp. 329–335, 2007.

[7] M. HooshSadat, H. W. Samuel, S. Patel, and O. R. Zaïane, "Fastest association rule mining algorithm predictor (FARM-AP)," in *Fourth International C* Conference on Computer Science & Software Engineering, C3S2E 2011, Montreal, Quebec, Canada, May 16-18, 2011, Proceedings*, B. C. Desai, A. Abran, and S. P. Mudur, Eds. ACM, 2011, pp. 43–50.

[8] P. Kostenetskiy and A. Safonov, "Susu supercomputer resources," in *PCT'2016, International Scientific Conference on Parallel Computational Technologies, Arkhangelsk, Russia, March 29–31, 2016*, L. Sokolinsky and I. Starodubov, Eds. CEUR Workshop Proceedings. vol. 1576, 2016, pp. 561–573.

[9] P. Kumar, P. Bhatt, and R. Choudhury, "Bitwise dynamic itemset counting algorithm," in *Proceedings of the ICCIC 2015, IEEE International Conference on Computational Intelligence and Computing Research, December 10–12, 2015, Madurai, India*, N. Krishnan and M. Karthikeyan, Eds. IEEE, 2015, pp. 1–4.

[10] P. Paranjape-Voditel and U. Deshpande, "A dic-based distributed algorithm for frequent itemset generation," *JSW*, vol. 6, no. 2, pp. 306–313, 2011.

[11] M. Sariyar, A. Borg, and K. Pommerening, "Controlling false match rates in record linkage using extreme value theory," *Journal of Biomedical Informatics*, vol. 44, no. 4, pp. 648–654, 2011.

[12] B. Schlegel, T. Karnagel, T. Kiefer, and W. Lehner, "Scalable frequent itemset mining on many-core processors," in *Proceedings of the Ninth International Workshop on Data Management on New Hardware, DaMoN 1013, New York, NY, USA, June 24, 2013*, R. Johnson and A. Kemper, Eds. ACM, 2013, p. 3.

[13] A. Sodani, R. Gramunt, J. Corbal, H. Kim, K. Vinod, S. Chinthamani, S. Hutsell, R. Agarwal, and Y. Liu, "Knights landing: Second-generation intel xeon phi product," *IEEE Micro*, vol. 36, no. 2, pp. 34–46, 2016.

[14] M. J. Zaki, S. Parthasarathy, M. Ogihara, and W. Li, "New algorithms for fast discovery of association rules," in *Proceedings of the Third International Conference on Knowledge Discovery and Data Mining (KDD-97), Newport Beach, California, USA, August 14-17, 1997*, D. Heckerman, H. Mannila, and D. Pregibon, Eds. AAAI Press, 1997, pp. 283–286.

ETLator – a scripting ETL framework

Miran Radonić[*1], Igor Mekterović[*2]
[*] Faculty of Electrical Engineering and Computing, University of Zagreb, Zagreb, Croatia
miran.radonic@gmail.com[1], igor.mekterovic@fer.hr[2]

ETL (Extract Transform Load) process is the industry standard term for data extraction, transformation and loading into the Data Warehouse (DW). ETL process is the most resource demanding process in DW implementation and typically has to be evolved and maintained for the duration of the DW. To facilitate the development and maintenance of ETL processes many ETL tools have been developed featuring Graphical User Interfaces and various built-in functionalities (parallelism, logging, rich transformation libraries, documentation generation, etc.). The downside of such GUI ETL tools is that development is carried out heavily using mouse operations and less by writing programming code, which feels unnatural for some developers, especially with many similar, repetitive tasks. In this paper we present an alternative approach – an ETL framework "ETLator" based on Python scripting language where ETL tasks are defined by writing Python code. ETLator implements various typical ETL transformations and allows the user to simply and efficiently define complex ETL tasks with multiple sources and parallel tasks whilst leveraging full flexibility of Python. ETLator also provides logging and can document ETL tasks by generating data flow images. On a test case we show that ETLator simplifies ETL development and rivals the GUI approach.

I. INTRODUCTION

Data processing is the purpose of most computer programs. For the purposes of storage and manipulation, the data is usually structured and stored in relational databases. However, relational model (normalized to the third normal form) used in relational databases is not the optimal solution for analytical purposes. For that part, data is restructured and stored in a Data Warehouse (DW) - a special database used for storing analytical data. To populate the DW, data is typically transferred from (often multiple) relational databases and potentially additional data sources (text files, Excel files, etc.) by a process known as ETL (Extract Transform Load). ETL process is a complex and most resource intensive part of the DW project. To reduce development and maintenance costs ETL tools have been developed. ETL tools are program solutions intended to transfer and transform data. They gather data from various sources, transform and integrate the data, and, finally, store it in the destination system. Since every ETL process is specific, users need the ability to configure the tool to suite their needs. Most of the ETL tools today (e.g. Informatica, Microsoft SSIS, Talend [3]) feature Graphical User Interface (GUI) which makes them more user-friendly for non-programmers, but can limit their flexibility. In a different approach, ETL frameworks enable defining ETL processes using programming code, but without having to write the whole program from scratch, because they provide most of the code leaving only the specifics of the process to the user. Scripting

programming languages are especially suited for this task processes because of their simplicity and efficiency. The question of text based versus GUI based modelling/interface is neither new nor decided (e.g. [1]), in software development in general, and thusly in the ETL domain. For instance, in [2] authors conclude that "that text-based modelling constitutes a noteworthy alternative to graphical modelling because of its simple usage, scalability and easy development and reuse of tool support". This is very much true in the ETL process definition where GUI approach implies a lot of point and click mouse operations, property settings editing, drawing, etc. which can be tedious and tiresome, especially with many similar tasks, often found in the ETL environment. Text is much more easily reused, or even generated when appropriate. Of course, GUI tools have their advantages, namely when documentation is concerned, and also when introducing new developers in the existing projects. Without going into further debate over this issue, in this paper we present a text based ETL tool named "ETLator". It is an ETL framework written in Python scripting language that allows the user to simply and efficiently define ETL processes with advanced and automated features while maintaining the flexibility and extensibility of Python. ETLator provides standard ETL tool features like logging, parallelism, batch loading, caching, support for slowly changing dimensions [14], and can even generate documentation featuring images of data flows.

II. RELATED WORK

ETL tools fall under a more generic category of data integration tools, which are heavily used in the industry, not only in the DW domain. There is a number of commercial and a few open-source data integration tools available. Gartner Magic Quadrant brings a yearly overview of these tools [3] where companies like Informatica, IBM, SAP, Oracle, Microsoft and, most notable open-source company – Talend, dominate the market. All of these tools employ a GUI approach to ETL process design, as opposed to the approach presented here. A survey paper [4] brings an overview of academic papers in the ETL domain, most of them using UML or graphs to model an ETL workflow [5].
Scriptella [6] is a Java based scripting ETL framework where XML is used to define tasks. It is also possible to mix Java code and XML. Scriptella provides no support for DW specific constructs, like Slowly Changing Dimensions (SCDs). Also, there is no support for parallelism. Petl [7] is a general purpose Python package for ETL supporting a number of different data sources (relational databases, text files, XML, HTML, JSON, Excel). ETL tasks are designed via ETL pipelines

representing data flow. Petl does not support SCDs or parallelism. Bubbles [8] is an open-source Python framework for data processing and data quality measurement. ETL tasks are represented as data processing pipeline graphs. Bubbles does not support SCDs or parallelism. Most notably, Pygrametl [9] is an open source ETL framework written in Python first released in 2009 and continually upgraded ever since. Version 2.4 was released in 2015 with extended support for Slowly Changing Dimensions (SCD) and batch-load of data. Pygrametl consists of multiple classes that represent fact tables, dimensions and data sources. Data is accessed by iterating through a data source. Each row is represented by a Python dictionary with keys being the column names. Pygrametl supports parallelism and SCD types 0, 1 and 2. Several concepts from Pygrametl were adopted in ETLator including table representation with table name and attribute names, representing a row with a dictionary, a class for processing connections with databases and a special class for handling SCD type 2. As an added value, and compared to Pygrametl, ETLator has more automated advanced features such as *logging and documentation generation*. Those features require user programming in Pygrametl. Also, while Pygrametl offers parallel execution, it uses threads to achieve it, which makes it run sequentially in CPython, the most common Python interpreter, because of Global Interpreter Lock, so it forces users to use JPython to run parallel ETL tasks. ETLator takes a different approach (described later), which does not even require programming to achieve parallelism, resulting in *simpler development and maintenance*.

III. BASIC FEATURES OF ETL TOOLS

ETL tools have to be able to connect to a data source, process the data and store the processed data to the destination. Sources are usually different databases and various files, and destinations are in most cases DWs. The ability to connect to databases of various vendors is a necessary feature of all ETL tools, as is the ability to parse various files (CSV, XLS, JSON, etc.). ETL tools also have to be able to validate and integrate the data from multiple sources and lookup the data on the destination to reference it from new data. For example when adding a row into the fact table it has to reference existing data in the dimension tables.

Most ETL tools (e.g. Informatica, Microsoft SSIS, Oracle Data Integrator [3]) also provide some more advanced features that speed up the ETL process and make maintenance of ETL and the data itself easier. Some of the more advanced features of ETL tools are:

- Batch loading the data to the destination
- Caching the data when reading from source
- Parallelism of ETL jobs within a process
- Reusability of jobs or parts of jobs
- SCD support
- Data transformations libraries
- Support for snowflake schema
- Documentation generation

- Logging

IV. ETLATOR ETL FRAMEWORK

ETLator is an open source framework for ETL processes definition in Python programming language. The main goal of ETLator is simplicity and productivity - to provide means to the user to define an ETL process with the least amount of written code, leveraging Python features to the full extent, in the process. ETLator can work with various relational databases, CSV files and MongoDB NoSQL database as a source or a destination. Tables are, like in Pygrametl, iterators and rows are represented by dictionaries. Parallel task execution is realized without writing any additional code (addressed later). ETLator automatically generates logs and documentation in form of flow charts and text files with table information and data changes within the table.

A. Main classes

Fig. 1 shows a partial class diagram of ETLator, where attributes are omitted for the sake of simplicity.

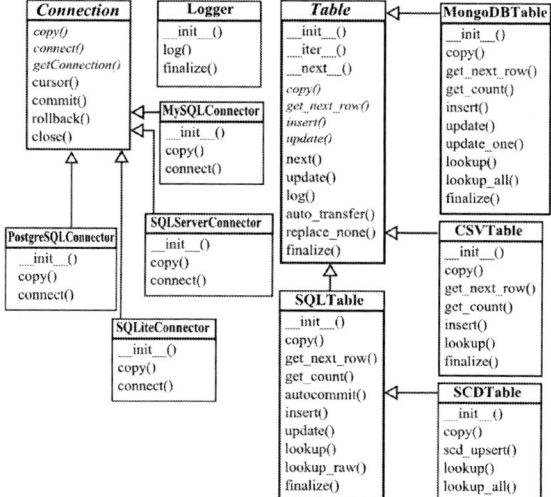

Figure 1. ETLator class diagram

ETLator contains three main classes:

- Connector – used to connect to various relational databases
- Logger – enables logging feature
- Table – represents tables in databases

1) Connector Class

Connector class is an abstract class providing common functionalities for connecting to relational databases. It is extended by concrete classes representing various database systems. Connections are executed using Python modules that comply with PEP 249 specification [5?]. All methods of Connector class correspond with PEP 249 methods of the same name.

Subclasses of the Connector class must contain their own implementation of the methods "copy", "connect" and the constructor. Loading of modules to access various databases is done dynamically in the constructor to allow

1350

the use of ETLator on systems that do not have the modules and database drivers for all of the supported databases, but only the ones that are used. Currently supported databases are: PostgreSQL, SQLite, MySQL and Microsoft SQL Server, but additional connector classes can be easily added. File "connector.py" in the "lib" directory contains all class definitions that extend the Connector class for connecting to supported databases. Users can write their own subclasses of the Connector class to connect to originally unsupported databases.

2) Table Class

Class Table is an abstract class from which classes used to define tables of data are extended. Name and attribute names must be defined for all tables. Location where local logs are stored, master log, operation counters and source table can also be defined. Table class does not have to be based on physical table, it can also be defined via SQL query, e.g. a query that joins and filters multiple tables.

Each class that extends Table class must implement "copy", "get_next_row", "insert" and "lookup" methods. Tables can be in a relational database, CSV file or MongoDB database. Subclasses of the Table class are: SQLTable, CSVTable, and MongoDBTable.

3) SQLTable Class

SQLTable class represents tables in relational databases. It can be used as a source or a destination of data, but not as both at the same time. SQLTable class automatically generates SQL queries for all the operations and executes them against the database.

Caching is used to store accessed or previously loaded data for faster lookup. It allows lookup without sending a query to the database. Cache size is by default set to 1000 records, but it is recommended to change this parameter depending on the table size and the system specifications. Caching is very useful in the DW ETL context, because, in accordance with best design practices, production tables' keys are replaced with surrogate keys in the dimensional model typically used in the DW. This implies looking up dimension members when loading fact tables to replace production foreign keys with corresponding dimension table surrogate keys. Issuing lookup SQL statement for every fact table row is expensive in terms of performance, hence leveraging ETLator's client computer RAM by caching dimension table's production key, surrogate key pairs can greatly improve performance. Another performance booster is batch loading. Batch loading is not enabled by default, but enabling it can substantially improve performance because, again, the number of queries sent to the database is greatly reduced. Batch loading by default can be confusing for novice developers since one erroneous row in the batch (e.g. violated referential integrity) can cancel the entire batch, which can be hard to spot in the development phase.

In dimensional modelling, type 2 slowly changing dimensions are a classic technique for tracking historical changes to the dimension tables. With SCD's a change in the tracked attribute causes a new record in the dimension table, consequently a SCD can have multiple records with the same production key. SCDTable class is defined to simplify SCD type 2 operations. SCDTable class extends SQLTable class and contains new implementations of abstract methods and defines new methods: "scd_update" and "lookup_all".

The "scd_update" method adds a new record into the table, sets it as active and sets the old record as inactive and changes the timestamp indicating the end of its activity. The "lookup_all" method returns all versions of the record, not only the active one as the "lookup" method does.

4) Other Subclasses of the Table Class

CSVTable class is used to define tables stored as CSV files. It can be used both as data source and as a destination. CSVTable's constructor accepts a number of parameters used to define CSV format (delimiter, quote char, header definition, etc.). There is also a concept of "dialect" where well-known CSV formats are represented with a dialect (e.g. Microsoft Excel dialect) so that one can easily define a CSV format via dialect.

MongoDB class is used to define tables stored as single-depth JSON objects in MongoDB NoSQL database.

5) Logger Class

Every instance of the Logger class generates its own text file where log messages are stored. In accordance with most major logging frameworks, messages are ranked in five levels:

1. "DEBUG" – Notifications for user who is tasked with ETL process development
2. "INFO" – General notifications for the user
3. "WARNING" – Alerts that show anomalies in execution
4. "ERROR" – Errors in execution
5. "CRITICAL" – Critical error that prevents ETLator from proceeding with work

It is also possible to define a Master logger – a parent instance of Logger class to which all local logger messages are forwarded ("bubble up").

Each log message entry comprises of log level, timestamp and message text. By default, ETLator generates master logger object and logger objects for all tables defined in the "flow" directory, but developers can also define their own, additional logger objects. ETLator automatically generates messages of "INFO" and "ERROR" levels. By default, master logger dismisses all messages below "ERROR" level.

B. Project structure

File and directory structure of ETLator is predetermined and is shown in Fig. 2.

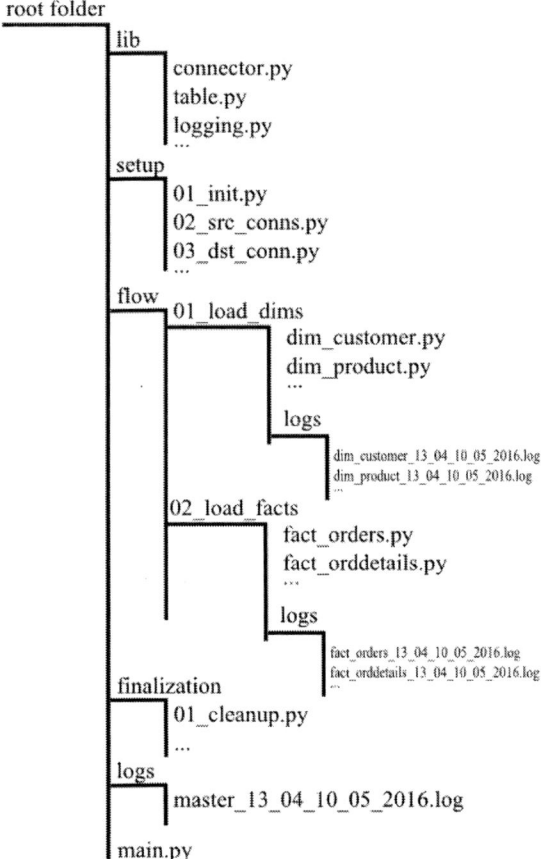

Figure 2. ETLator project structure

Directory "lib" contains classes necessary for ETLator to work and the contents of the directory should not be changed by the user. Changes are possible, but only in case the user wants to expand the features of existing classes.

Remaining three folders are used to store user defined scripts that will be executed in the following order: setup, flow, and finalization.

Directory "setup" should be populated with scripts written by user that are executed at the beginning of ETL run. Typically, scripts in "setup" directory are used for setting up initial values and defining global parameters and connections.

Directory "flow" contains user-defined directories and subdirectories that contain scripts representing the main part of ETL process – various data flows. ETLator used a very simple and intuitive way to define execution behaviour:

- In all folders, scripts are executed in the lexicographic order
- In the flow folder, scripts can be placed in subdirectories. Subdirectories are accessed sequentially, in the lexicographic order, and scripts within them are executed in parallel

With these two simple rules it is possible to define arbitrary data flows in terms or ordering and parallelism by leveraging the well-known paradigm - file system. For instance, Fig. 3 shows data flow as it would look like in a typical GUI ETL tool corresponding to the example in the Fig 2.

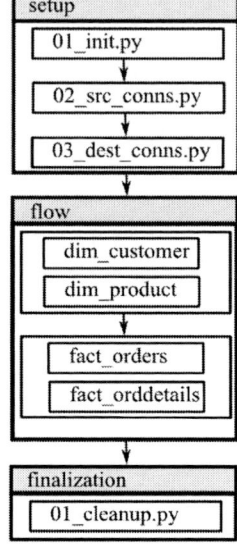

Figure 3. Data flow corresponding to the project in Fig 2.

We argue that it is even easier to define flows using file system operations (make directory, rename, copy, move) than it is to draw them, especially since these file system actions can be easily scripted, e.g. if we want to generate a flow for each table in the database.

Within each subdirectory of the directory "flow" ETLator automatically generates directory "logs" where logs and documentation files are stored.

Directory "finalization" contains user-defined scripts that will be executed at the end of the run.

Root directory "logs" is automatically generated by ETLator and is used to store summary logs.

Script "main.py" is used to run ETLator and should not be changed by the users unless there is need to change the behaviour or add functionality. All the parameters that are set by default in "main.py" can be overridden on the global level using scripts in the "setup" directory or locally in the scripts within subdirectories of the "flow" directory, without changing "main.py".

Note that this project setup makes it portable even between different OS platforms, provided of course, that Python interpreter is installed. A simple zip file is all it takes to deploy a project.

C. Parallelism

Parallel execution of scripts within subdirectories of the "flow" directory can significantly speed up the execution on multiprocessor systems, especially when multiple data sources and destinations are involved.

For each script in the directory a process is created that evaluates the script. Maximum number of processes can be set in scripts in the "setup" directory. Processes are used instead of threads because memory management of Global Interpreter Lock (GIL) in CPython, the most commonly used implementation of Python interpreter, is not thread safe. GIL prevents multiple native threads to execute Python bytecode at once [13]. Consequently, multithreaded Python programs work slower than equivalent single threaded programs because GIL incurs additional costs for thread synchronization are. GIL can also cause starvation of important threads while unimportant threads are using all processor cycles because thread priority cannot be defined. Jthon, an implementation of Python interpreter in Java programming language doesn't have GIL so multithreaded programs can work in parallel, but is rarely used because of lower performance and module incompatibility problems.

By using separate processes we circumvent GIL so that the execution of operations that do not require often synchronization is faster than a multithreaded program, despite the slower initial operation of starting a new processes.

D. Running example

A well-known Northwind database is taken as a running example. Due to space considerations, we consider only a subset of Northwind database shown in Fig. 4:

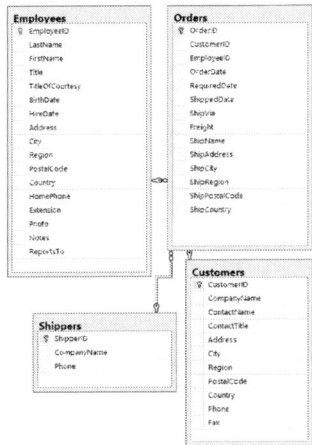

Figure 4.   Source database schema.

1) Directory "setup"

Setup directory contains the definitions of Connector objects that will be used multiple times and definitions of destination tables that will be the target of lookup operations from other tables. These scripts can set ETLator parameters such as turning off documentation generation and set the maximum number of processes used. In this example the "setup" directory contains files "connSrc.py" and "connDest" that contain the definition of connections to the source database and the DW, and files "dCity.py", "dCustomers.py", "dDate.py", "dEmployees.py" and "dShippers.py" that contain definitions of the tables that have to be accessible from every process, as can be seen in Fig 5. Order of execution is usually not important in the setup directory, which

holds true also for this example, otherwise file names would be prefixed with a number.

2) Directory "flow"

Three directories are added into the "flow" directory: "1", "2" and "3" with scripts for populating dimension tables (directories 1 and 2), and the fact table (directory 3), as shown in Fig 5. Scripts in directories 1 and 2 will be evaluated in parallel.

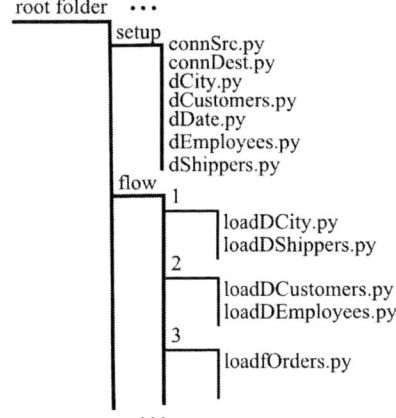

Figure 5.   Project structure subset for the running example

City source table is actually not a physical table, but is defined via SQL query that is extracting attributes used in dCity dimension in the DW. The following snippet is used in loadDCity.py to populate dCity dimension:

```
columns = ['PostalCode', 'CityName',
           'Region', 'Country']
connectorSrc = connectorSrc.getConnection()
tabCity = lib.table.SQLTable(
    connector=connectorSrc
    , name='City'
    , column_names=columns
    , default_args=default_args)
connectorDest = connectorSrc.copy()
tabDCity = lib.table.SQLTable(
    connector=connectorDest
    , name='dCity'
    , column_names=columns
    , src_table=tabCity
    , default_args=default_args)
tabDCity.auto_transfer()
```

To fetch the connection getConnection() method is used which returns a clone of the Connector objects as to avoid potential collisions during parallel execution. Primary key definition is omitted above since dCity uses autogenerated surrogate key. "default_args" argument is generated by "main.py" and is typically reused (as above) to confirm default documentation generation, logging and automatic table finalization settings.

The following snippet shows the script for populating SCD dimension dEmployees:

```
tabdCustomers = lib.table.SCDTable(
    connector=connectorDest
    , name='dCustomers'
    , column_names=['CustomerID',
'CustomerIDDB', 'CompanyName', 'ContactName',
'ContactTitle', 'Address', 'CityID', 'CityIDDB',
'CityPostalCode', 'CityName', 'CityRegion',
```

1353

```
'CityCountry', 'Phone', 'Fax', 'DateValidFrom',
'DateValidTo']
    , scd_lookup_columns=['CustomerIDDB']
    , commit_count=1
    , date_from_cloumn='DateValidFrom'
    , date_to_column='DateValidTo'
    , pkey='CustomerID'
    , default_args=default_args)
## for loop with additional adjustments
tabdCustomers.scd_upsert(row,)
```

Finally, the following snippet shows part of the script used to populate the fOrders fact table:

```
for row in tabOrders:
    row['OrderIDDB'] = row['OrderID']
    custLkp = tabdCustomers.lookup(row,
                            ['CustomerIDDB'])
    if custLkp is not None:
     · row['CustomerID'] = custLkp['CustomerID']
    else:
        row['CustomerID'] = 0
    ## other lookups, and field transformations
    tabfOrders.insert(row);
```

Data transfer is achieved by iterating over rows of source table, calculating each row of the destination table and executing "insert" method of the destination table (actual insert to the database occurs depends on the batch load settings). To fetch data from other tables (e.g. to fetch dimension keys in the fact table), lookup method of the target table is used. Lookup method takes a row that contains attributes it will try to match and, optionally, a list of attribute names to match. The return value of "lookup" method is either the found row or a "None" value if a matching row is not found.

3) Directory "finalization"

Simple ETL processes usually do not need any finalization operations. In this example the "finalization" directory is left empty, and is not shown in the Fig 5, as well as log folder, lib, etc.

4) Running

Upon starting "main.py", all scripts in "setup", "flow" and "finalization" directories are executed in described order, and logs and documentation is generated. Fig. 6 shows data flow diagram generated as a part of the documentation process.

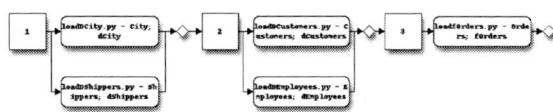

Figure 6. Data flow diagram generated for the running example

E. Future development

Future development of ETLator will include support for additional connections, from relational databases to various NoSQL systems (e.g. Hadoop). Also, our plan is to develop rich data transformation and data quality library.

V. CONCLUSION

In this paper we present ETLator - a scripting ETL framework. ETLator is an open-source, platform independent ETL framework written in Python programming language, with support for various data sources, parallel task execution, slowly changing dimensions, etc. ETLator prescribes the project's file and folder structure, and enables users to define data task's ordering and parallel execution simply by using conventions for file and directory naming and nesting. In comparison with other frameworks ETLator is simpler to use, while retaining typical ETL features. Also, unlike similar tools, ETLator features built-in logging and documentation generation features (data flow charts included) which are traditionally features found only in GUI based ETL tools.

Definition of ETL process in ETLator consists of writing short scripts in Python, with the ETLator library and all other Python libraries at one's disposal, which results in terse and expressive code, and overall an easier and faster experience of developing and maintaining ETL projects.

REFERENCES

[1] S. Melia, C. Cachero, J.M. Hermida, E. Aparicio, "Comparison of a textual versus a graphical notation for the maintainability of mde domain models: an empirical pilot study", Software Quality Journal:1–27, 2015

[2] H. Gronniger, H. Krahn, B. Rumpe, M. Schindler, and S. Volkel. Text-based Modeling. In Proc. of ATEM, 2007

[3] 2016 Gartner Magic Quadrant for Data Integration Tools, https://www.informatica.com/data-integration-magic-quadrant.html, as of Jan 30, 2017

[4] P. Vassiliadis: A Survey of Extract–Transform–Load Technology. IJDWM 5(3):1–27, 2009.

[5] C. Thomsen and T. Bach Pedersen, "pygrametl: A Powerful Programming Framework for Extract–Transform–Load Programmers - Technical Report", http://pygrametl.org/

[6] Scriptella, http://scriptella.org/, as of Jan 30, 2017.

[7] Petl, https://petl.readthedocs.io, as of Jan 30, 2017.

[8] Bubbles, http://bubbles.databrewery.org/, as of Jan 30, 2017.

[9] C. Thomsen and T. Bach Pedersen, "Easy and effective parallel programmable ETL", Proceedings of the ACM 14th international workshop on Data Warehousing and OLAP, p37-44, 2011

[10] R. Kimball, J. Caserta, "The Data Warehouse ETL Toolkit", ed. 1, Indianapolis, IN, USA, Wiley, 2004, pp. 21.

[11] W. H. Inmon, "Building the Data Warehouse", ed. 3, Indianapolis, IN, USA, Wiley, 2002., pp.

[12] "PEP 249 - Python Database API Specification v2.0", https://www.python.org/dev/peps/pep-0249/, Accessed on January 28 2017.

[13] "Global Interpreter Lock", https://wiki.python.org/moin/GlobalInterpreterLock, Accessed on January 29 2017.

[14] R. Kimball and M. Ross. The Data Warehouse Toolkit, 2nd Edition., Wiley, 2002.

The role of Alignment for the Impact of Business Intelligence Maturity on Business Process Performance in Croatian and Slovenian Companies

V. Bosilj Vukšić*, M. Pejić Bach*, T. Grublješič**, J. Jaklič **, A.M. Stjepić*

* University of Zagreb, Faculty of Economics & Business, Zagreb, Croatia
** University of Ljubljana, Faculty of Economics, Ljubljana, Slovenia
vbosilj@efzg.hr, mpejic@efzg.hr, tanja.grubljesic@ef.uni-lj.si, jurij.jaklic@ef.uni-lj.si, astjepic@efzg.hr

Abstract - Business intelligence (BI) allows companies to analyze business information in order to support successful decision making. Currently, the research on the level of BI maturity in Croatian and Slovenian companies is limited. In addition, several BI maturity models have been developed, but most of them are not comprehensive. In order to shed some light to this issue, this paper is focused on two goals: (1) to investigate the impact of BI maturity on business process performance and (2) to explore the requirements for the alignment of two concepts, BI and business process management (BPM) within the organization. Paper presents the following: (i) investigation of BI and BI systems in general, (ii) adaption of the BI maturity model (called biMM) for the purpose of this research, (iii) results of the primary research on the sample of Croatian and Slovenian companies which has been conducted as one of the activities of the project financed by the Croatian Science Foundation: IP-2014-09-3729 Process and Business Intelligence for Business Excellence, (iv) level of BI maturity and the role of BI and business process alignment for the impact of BI maturity on business process performance in investigated companies.

I. INTRODUCTION

The vast amount of daily generated data in organization is a result of a growing number of business transactions. It becomes crucial for organizations to transform collected transaction data into valuable information using information technology (IT). According to Dinter [6] and Forrester [38] business intelligence (BI) has the essential role in this part since it encompasses all processes and systems that transform raw data into meaningful and useful information and enable effective, systematic and purposeful analysis of an organization and its competitive environment. BI can be defined as the acquisition of skills and abilities to adapt the organization to different business conditions ([23]; [24]; [16]; [22]). Agile, real-time BI becomes a prerequisite in the environments of constant change in which organizations

operate [39]. In general, the role of BI in improving business process performance has gained a lot of attention because of its ability to provide a detailed insight into business operations and to enhance operational intelligence ([13]; [7]; [8]; [20]).

However, the empirical evidence of how BI impact organizational processes with improving business processes is still lacking. Accordingly, the aim of this paper is to explore if and how the higher BI maturity increases process performance.

This papers presents following sections: a theoretical background about BI and BI maturity model introduction, a discussion about BI and BPM alignment, the detailed empirical research methodology, the results analysis and a discussion on the results and, finally, the conclusions and implications for further research.

II. BI AND ORGANIZATIONAL PERFORMANCE

BI uses different tools and applications for collecting and analyzing business data. Business intelligence systems (BISs) can be defined as software platforms that provide users of the system with relevant information, which enable them to make better decisions. Today, BISs combine methodology, applications, technologies and platforms for storing information such as data warehouses, data marts, analytical tools such as reporting tools, ad hoc analytics (OLAP), in-memory analytics, planning, alerts, forecasts, scorecards, data mining and online analytical mining (OLAM) ([7]). The results of the researches claim that BI became the driver of organizational success and process performance, so business practitioners and academics search for strengths and weaknesses of existing BI structures in order to make these more effective [18]. For this purpose, BI maturity models are applied.

In general, maturity models usually consist of several (4-6) dimensions of maturity and enable organizational assessment and organizational development [21]. Though a large number of BI maturity models exist, the majority of these models still indicate certain shortcoming such as the lack of transparency, comprehensiveness, systematization, appropriate assessment tools and the lack of empirical data ([6]; [26]). Dinter [6] developed a BI maturity model (called biMM) that was structured in 3

This study has been partly founded by the Croatian Science Foundation under the project PROSPER – Process and Business Intelligence for Business Excellence (IP-2014-09-3729) and by the Slovenian Research Agency under the research programme No. P2-0037 - Future internet technologies: concepts, architectures, services and socio-economic issues and under the project No. J5-7287 - Big Data Analytics: From Insights to Business Process Agility

dimensions (functionality, technology and organizational dimension), within each dimension several categories were introduced and the design objects for each category were defined. According to Dinter [6] functionality dimension includes aspects of the use and impact of BI within the organization, as well as other content-related issues. Technology dimension highlights the system and the architecture of the data and the BI tools along with their associated functionality while organizational dimension refers to separate organizational structures, processes, profitability and the strategy of BI in an organization [6]. For the purpose of this research an aggregated BI maturity model based on Dinter's biMM is designed. This model distinguishes ten categories comprising the most important elements of Dinter's biMM.

III.   BI AND BPM ALIGNMENT FOR PROCESS PERFORMANCE MEASUREMENT

The emergence of a large number of activities in the process, possible automation and a large number of errors result in the need for business process management (BPM). One of the most widely used definitions of BPM is Harmon's [11]: "BPM is a management discipline focused on improving corporate performance by managing a company's business processes". BPM enables the alignment of business processes and the organization's strategic goals [4] and comprises different phases, from analysis and design, through implementation and automation, to process monitoring and measurement phase [2]. Both of these concepts, BI and BPM, are considered key drivers for process and organizational performance improvement.

Business process performance is measured by the business data on the costs and time required to perform the activities so as by the costs and resource capacities involved in the activities, keeping in mind that the focus should be on the quality of products or services [3]. For this purpose, business process management systems (BPMSs), whose functionalities support the documentation, automation and tracking of business processes, real-time monitoring and collecting historical data on the business processes performance and optimization of processes, are used [19]. In terms of internal processes, an organization is directed towards measuring product quality and production costs, while the focus of external processes is aimed at measuring customers' satisfaction. According to empirical data by Huang [12] collaboration of process performance measurement together with the reduction of costs and increased speed of the internal processes, as well as an improved quality of the external processes, result in the improvement of the organizational performance.

Kueng [17] defines process performance measurement systems (PPMSs) as information systems that integrate relevant information of the performance of one or more business processes, compare historical and desired future values with achieved business process value and report the results to participants of the process. Also, data on business processes can be collected from ISs systems, which are focused on the collection and storage of financial and time-oriented data [17]. Similarly, Hammer

[9] emphasizes the need for "match between organization's information and management systems and the process's needs, and the quality of the metrics that the company uses to measure process performance". Since the BISs also serve to collect data relevant to the assessment of process performance this study discusses the effects of connecting BISs and BPMSs.

The results of the researches from business practice show that the full potential of BISs and BPMs integration and information exchange is still not recognized [3]. According to the authors the major obstacles are systemized as follows: "(1) the main goal of BPM initiatives is to improve business processes, while BI initiatives are usually focused on achieving marketing, customers and sales objectives; and (2) although management starts both initiatives, the results evidence lack of the strong commitment towards coordinated usage of BISs and BPMs as tools for supporting performance management" [3].

Hammer et al. [10] specify seven obstacles of performance measurement success in organizations. Some of these imply the need for BPM and BI alignment. According to Hammer et al. [10] "Provincialism" refers to the existence of conflicting goals among departments that result in defining process metrics and KPIs within the functional boundaries. Besides, "Pettines" or "the policy to measure only a small component of what matters" must be discussed [10]. "Inanity" is concerned with the organization's approach "to implement metrics without giving any thought to the consequences of these metrics on the performance on organizational level" [10]. Such situations could be avoided by setting up and measuring the organizational KPIs on the end-to-end process level and by implementation of performance management systems in line with cross functional processes ([15]; [1]). This approach could help organizations to increase the relevance and usefulness of their performance measurement systems. The authors suggest to coordinate BPM and BI initiatives in a company and to provide very intensive communication between BPM and BI experts and mangers ([27]; [3]). According to Nenadal [34] the main goal of process performance system is to identify if processes meet strategic goals and this goal is achieved through "the monitoring of agreed performance indicators". For this reason, it is very important to establish the role of the process owner and to ensure that process owners monitor business process execution and KPIs achievement through BISs and BPMs [25]. It's very important to achieve a common understanding of business process terminology on the level of entire organization ([43]; [9]). Customer focus, collaboration, teamwork, and a willingness to reach process KPIs must be shared and accepted by employees across the organization.

Some authors emphasize the need to apply BI techniques in BPM, so the new term "business process intelligence" is introduced in literature [41]. According to van der Aalst [42] the combination of BPM and data mining, called "process mining" is applied in business practice with the aim to extract and discover knowledge from process events (logs) data. Schifer, Jeng and Chowdhary [40] present a solution for merging business data with typical workflow data by adding process metrics

1356

(such as throughput time, utilization and cost of resources, process volume and frequency) to existing data warehouse technologies. Bucher and Gericke [43] specify several reasons for the integration of BI and BPM platforms, such as: "(1) a lot of operational processes require "data analytics" in real-time or near-real time; (2) operational processes provide the context for data analysis and decision making, hence BI techniques can be used to merge and consolidate raw data about process execution into KPIs; (3) once operational decisions have been made based on KPIs, useful knowledge about KPIs, corresponding decisions and related consequences can be added into a dedicated data store, so developing the rules for future decision making."

Since the ultimate goal of both, BI and BPM, is to increase process performance, this paper aims to investigate if the need for alignment of these concepts is recognized within the organizations and to evaluate its impact to process performance.

IV. EMPIRICAL RESEARCH METHODOLOGY

Based on the previous theoretical discussion, it is reasonable to expect, that BI and BPM will result in higher process performance improvement when they are implemented as coordinated and aligned initiatives. Thus, we put forward the following hypotheses:

H1: Business intelligence maturity positively influences Process performance.
H2: BPM/BI alignment positively influences Process performance.
H3: Business intelligence maturity positively influences BPM/BI alignment.
H4: The impact of Business intelligence maturity on Process performance is mediated by BI/BPM alignment.

Our questionnaire is presented in appendix, and was developed based on the previous theoretical knowledge, as presented in parts II and III of this paper, in order to assure content validity. We used a structured questionnaire with five-point Likert scales, with anchors ranging from totally disagree (1) to totally agree (5), for all items used in our study. The data were collected through a survey of medium- and large-sized business organizations in Croatia and Slovenia. Questionnaires were addressed to top management in the contacted organizations. The two rounds of call-up were conducted yielding altogether a sample of 65 completed surveys in Croatia and 118 in Slovenia.

V. RESEARCH RESULTS: ANALYSIS AND DISCUSSION

To conduct the data analysis, partial least squares (PLS), a component-based structural equation modeling (SEM) technique, was used. This is a widely used methodology in the IT and IS field as it is suitable for predicting and theory-building because it examines the significance of the relationships between the research constructs and the predictive power of the dependent variables ([28]; [32]). The estimation and data manipulation were performed using SmartPLS [36] and SPSS.

We have examined the reliability and validity measures for our reflective measurement model. In the

model, all Cronbach's alphas by far exceeded the 0.7 threshold [35]. Without exception, the latent variables composite reliabilities were higher than 0.8 and in all cases even higher than 0.9, showing the high internal consistency of indicators measuring each construct and thus confirming construct reliability ([35]; [32]). The Average Variance Extracted (AVE) was generally around 0.7 or higher, thus exceeding the threshold of 0.5, demonstrating the convergent validity of the constructs [29]. The reliability and convergent validity of the measurement model were also confirmed by computing standardized loadings for the indicators and Bootstrap t-statistics for their significance. All standardized loadings of the indicators in the model exceeded the 0.7 threshold and they were found without exception to be significant at the 0.001 significance level, thus confirming the high indicator reliability and convergent validity ([33]; [32]). Based on the discriminant validity tests, that "two conceptually different concepts should exhibit sufficient difference" [32], BI3 and BI9 indicators were excluded due to too high cross loadings with BI/BPM ALIGNMENT indicators. Discarding the mentioned items substantially and sufficiently improved the discriminant difference between the items of the two constructs.

The assessment of the indicator loadings on their corresponding constructs indicated that manifest variable correlations with their theoretically assigned latent variables are an order of magnitude larger than other loadings to other constructs [30]. Therefore, all the item loadings met the criteria. The square roots of AVE for constructs were significantly higher (and also substantially larger than the threshold of 0.5) than the correlations between the constructs, thus confirming that they are sufficiently discriminable ([28]; [29]).

We further estimated the inner path model. We tested the significance of the hypothesized relationships between the constructs by bootstrapping with 1,000 replicates. The structural model was then assessed (see Figure 1) by examining the coefficient of determination (R2) of the endogenous latent variable, the estimates for the path coefficients of relationships in the structural model and their significance levels (via bootstrapping) [28].

The path and indicator loadings are shown in Figure 1. The influence of BI MAT and BI/BPM ALIGN explain 20.3 % of the variance in PP. Further the influence of BI MAT on BI/BPM ALIGN explains 48.3% of the variance in BI/BPM ALIGN. Since the exogenous variables explain a moderate to high proportion of the variance of the endogenous variable, we may conclude that the model holds sufficient explanatory power and is capable of explaining the constructed endogenous latent variable [32]. The direct impact of BI MAT on PP is not statistically significant ($\hat{\beta}$ =0.200; p=1.660) rejecting H1. The direct impact of BI/BPM ALIGN on PP is statistically significant ($\hat{\beta}$ =0.289; p=2.592) and positive thus our H2 is supported. The positive impact of BI MAT on BI/BPM ALIGN is statistically significant ($\hat{\beta}$ =0.694; p=13.189) and supporting our H3. To test that the impact of BI MAT on PP is mediated by BI/BPM alignment we followed the procedures explained in Kenny [14] and Rucker et al. [37].

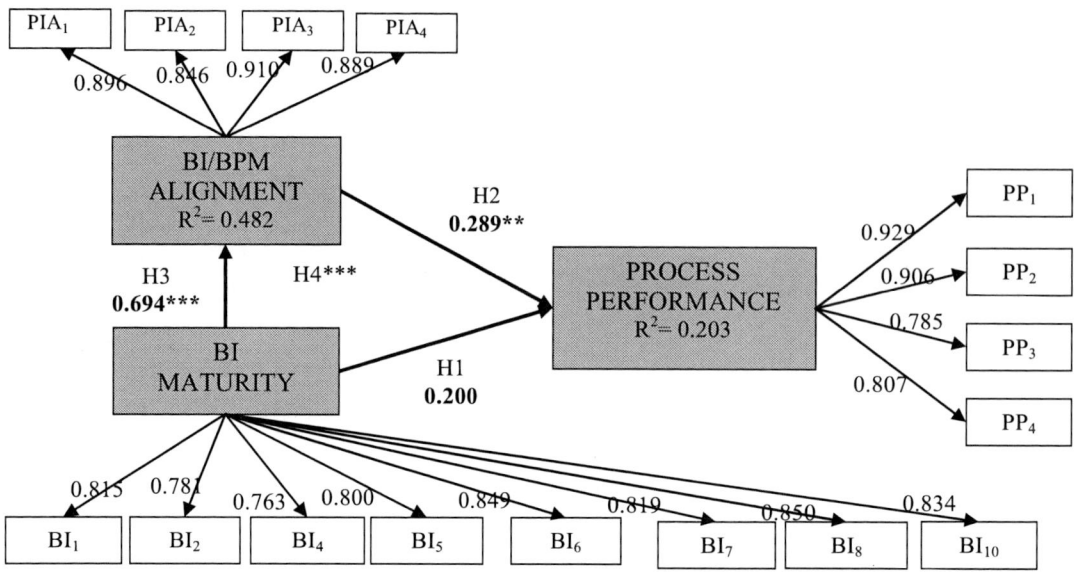

Figure 1: The final structural and measurement model

Notes (Figure legend): The structural model (the inner path model): statistical significance (no stars) - non-significant; * significant at the 0.05 level (two-tailed test); ** significant at the 0.01 level (two-tailed test); *** significant at the 0.001 level (two-tailed test); grey squares – constructs; R^2 – explanatory power of the constructs; $\hat{\beta}$ - written in bold – indicating path loadings.

The (reflective) measurement model: white squares – indicators of the constructs; loadings of the measurement model presented on the (thin) black arrows.

First, the direct effect of BI MAT on use intentions was tested. The direct impact of BI MAT on PP is not statistically significant, thus our hypothesis H1 is rejected. However, the causal variable (BI MAT) is correlated (0.400) with the outcome (PP), thus there is an effect that may be mediated [14]. In the second step, the mediator (BI/BPM ALIGN) needs to be treated as the outcome variable and BI MAT as the causal variable [14]. The path is statistically significant at 0.001% significance level, where BI MAT have direct positive influence on BI/BPM ALIGN. The causal variable is also correlated with the mediator (0.693). The third step involves showing that the mediator affects the outcome variable [14], which was tested with H2 and proven to be statistically significant. Since, the direct impact of BI MAT on PP is not statistically significant, BI/BPM ALIGN completely mediates the impact of performance perceptions on use intentions [37], proving our H4.

VI. CONLUSION

In this paper we have presented preliminary results of the comprehensive research conducted in Croatian and Slovenian companies. Structural equation model was developed in order to test hypothesis. First, the impact of BI maturity to business process performance (H1) was not confirmed. Second, the positive impact of BPM/BI alignment to the process performance (H2) was confirmed. Third, positive impact of business intelligence maturity to BPM/BI alignment (H3) was confirmed. Fourth, the mediating effect of BI/BPM alignment to the

impact of BI maturity on Process performance (H4) was confirmed, which is an important finding, since the direct impact was not confirmed. The contribution of this model is thus to indicated a path towards the future research on a interrelation of BPM/BI alignment and Process performance. Practical implication of this research stems in the greater need of the synchronized efforts in the implementation and development of BI and BPM.

REFERENCES

[1] J. Blasini and S. Leist, "Success factors in process performance management," *Business Process Management Journal*, vol. 19, no. 3, pp. 477-495, 2013.

[2] V. Bosilj Vukšić, M. Indihar Štemberger and A. Kovačič., "Business Process Management and Business Intelligence as Performance Measurement Drivers," *The Business Review*, Cambridge, vol. 10, no. 1, pp. 338-343., 2008.

[3] V. Bosilj Vukšić, A. Popovič and M. Pejić Bach, "Supporting Performance Management with Business Process Management and Business Intelligence: a case analysis of integration and orchestration," *International Journal of Information Management*, vol. 33, no. 4, pp. 613-619, 2013.

[4] R.T. Burlton, Business Process Management: Profiting form Processes, Sams Publishing, 2003.

[5] D. Chaffey, and S. Wood, Business Information Management: Improving Performance Using Information Systems, Prentice Hall, 2005.

[6] B. Dinter, "The Maturing of a Business Intelligence Maturity Model", AMCIS 2012 Proceedings. Paper 37., July 2012.

[7] M.Z. Elbashir., P.A. Collier and M.J. Davern, "Measuring the effects of business intelligence systems: The realitonship between business process and organizational performance," *International Journal of Accounting Information Systems*, vol. 9, no. 3, pp. 135-153, September 2008 [Eighth International Research Symposium on Accounting Information Systems (IRSAIS)].

[8] M. Golfarelli, S. Rizzi, and I. Cella, "Beyond data warehousing: what's next in business intelligence?," *Proceedings of the 7th ACM international workshop on Data warehousing and OLAP*, Washington, DC, USA, pp. 1-6, December 2004 [CIKM Conference on Information and Knowledge Management].

[9] M. Hammer, The process audit. Harvard business review, Harvard business school publishing, vol. 85, no. 4, pp. 111-123, April 2007.

[10] M. Hammer, C. J. Haney, A. Wester, P. Gaffney and R. Ciccone, "The 7 deadly sins of performance measurement and how to avoid them," *MIT Sloan Management Review*, vol.48, no. 3, pp. 19-28, 2007.

[11] P. Harmon, Business Process Change: a Guide for Business Managers and BPM and Six Sigma Professionals. Waltham, Morgan Kaufmann Publishers, 2007.

[12] S.Y. Huang, CH. Lee, AA. Chiu and DC. Yen, "How business process reengineering affects information technology investment and employee performance under different performance measurement," *Information Systems Frontiers*, vol. 17, no. 5, pp. 1133–1144, October 2015.

[13] R. Kaula, "Performance trends for operational inetlligence throuhg value chain model", *Journal of Advanced Computer Science & Technology*, vol. 5, no. 2, pp. 34-41, 2016.

[14] D. A. Kenny, (2016), Mediation, Available at: http://davidakenny.net/cm/mediate.htm.

[15] M. Kohlbacher and S. Gruenwald, "Process ownership, process performance measurement and firm performance," *International Journal of Productivity and Performance Management*, vol. 60, no. 7, pp. 709-720, September 2011.

[16] M.N. Koupaei, M. Mohammadi, B. Naderi, "An Integrated Enterprise Resources Panning (ERP) Framewrok for Flexible Manifacturing Systems Using Business Intelligence (BI) Tools," *International Journal of Suplly and Operationg Managemenet*, vol. 3, no. 1, pp. 1112-1125, May 2016.

[17] P. Kueng, "Process performance measurement system: a tool to support process based organizations," *Total Quality Management*, vol. 11, no. 1, pp. 67-85, August 2000.

[18] G. Lahrmann, F. Marx, R. Winter, and F. Wortmann, "Business Intelligence Maturity Models: An Overview," in*: VII Conference of the Italian Chapter of AIS (itAIS 2010)*, Naples, Italy, 2010.

[19] A. Margherita, "Business process management system and activities," *Business Process Management Journal*, vol. 20, no. 5, pp. 642 – 662, January 2014.

[20] O. Marjanovic, "Business Value Creation through Business Processes Management and Operational Business Intelligence Integration," *IEEE, System Sciences (HICSS)*, Honolulu, HI, USA, pp. 1-10, 2010 [43rd Hawaii International Conference on System Sciences (HICSS)].

[21] T. Mettler and P. Rohner, "Situational Maturity Models as Instrumental Artifacts for Organizational Design," *Proceedings of the 4th International Conference on Design Science Research in Information Systems and Technology Article No. 22*, May 2009 [4th International Conference on Design Science Research in Information Systems and Technology (DESRIST 2009)].

[22] G. Muang Sang, L. Xu and P. de Vrieze, "Implementing a Business Intelligence System for Small and Medium-sized Enterprises," *SQM 2016*, March 2016, [24th International Software Quality Management Conference Bournemouth, UK, 2016].

[23] P. Patil, "Survey of business intelligence system," *International Journal of research in computer applications and robotics*, vol.4, no. 9, pp. 13-17, September 2016.

[24] S. Pavkov, P. Poščić and D. Jakšić, "Business intelligence systems yesterday, today and tomorrow – an overview," *Proceedings of the University in Rijeka*, vol.4, no. 1, pp. 97-108, May 2016.

[25] M. Schläfke, R. Silvi and K. Möller, "A framework for business analytics in performance management," *International Journal of Productivity and Performance Management*, vol. 62, no. 1, pp. 110-122, January 2013.

[26] K.D. Schulze, U. Besbak, B. Dinter, A. Overmeyer, C. Schulz-Sacharow and E. Stenzel, Business Intelligence Study, Steria Mummert Consulting AG, 2009.

[27] P. A. Smart, H. Maddern, R.S. Maull, "Understanding Business Process Management: Implications for Theory and Practice," *British Journal of Management*, vol. 20, no. 4, pp. 491–507, December 2009.

[28] W.W. Chin, "Issues and opinions on structure equation modeling", *MIS Quarterly*, vol. 22, no. 1, pp. 7-16, March 1998.

[29] C. Fornell and D. F. Larcker, "Evaluating structural equation models with unobservable variables and measurement error," *Journal of Marketing Research*, vol. 18, no.1, pp. 39-50, February 1981.

[30] D. Gefen and D. Straub, "A practical guide to factorial validity using PLS-graph: Tutorial and annotated example," *Communications of the Association for Information Systems*, vol. 16, no. 5, pp. 91-109, July 2005.

[31] J. Henseler, and G. Fassott, "Testing moderating effects in PLS path models: AN illustration of available procedures" in Handbook of Partial Least Squares, V. Esposito Vinzi , W.W. Chin, J. Henseler and H. Wang, Eds. Springer Handbooks of Computational Statistics, Springer-Verlag Berlin Heidelberg, 2010, pp. 713-735.

[32] J. Henseler, C. M. Ringle and R. R. Sinkovics, "The use of partial least squares path modeling in international marketing," in New Challenges to International Marketing (Advances in International Marketing), vol. 20, R.R. Sinkovics, P.N. Ghauri, Eds. Bingley: Emerald, 2009, pp. 227-319.

[33] J. Hulland, "Use of partial least squares (PLS) in strategic management: A review of four recent studies," *Strategic Management Journal*, vol. 20, no. 2, pp. 195-204, February 1999.

[34] J. Nenadal, "Process performance measurement in manufacturing organizations," *International Journal of Productivity and Performance Management*, vol. 57, no. 6, pp. 460–467, July 2008.

[35] J. C. Nunnally and I. H. Bernstein, Psychometric theory, 3rd ed., New York, NY: McGraw-Hill, 1994.

[36] C. M. Ringle, S. Wende and S. Will, SmartPLS 2.0 (M3), University of Hamburg, 2007., URL: http://www.smartpls.de.

[37] D. D. Rucker, K. J. Preacher, Z. L. Tormala and R.E. Petty, "Mediation analysis in social psychology: Current practices and recommendations," *Social and Personality Psychology Compass*, vol.5, no. 6, pp. 359-371, June 2011.

[38] Forrester (2015), Business intelligence, Available at: https://www.forrester.com/Business-Intelligence (February 12, 2017).

[39] B. Azvine, Z. Cui, D. D. Nauck and B. Majeed, "Real time business intelligence for the adaptive enterprise," *E-Commerce Technology*, pp. 29-29, February 2006 [The 8th IEEE International Conference on and Enterprise Computing, E-Commerce, and E-Services, The 3rd IEEE International Conference on].

[40] J. Schiefer, J. Jeng and P. Chowdhary, "Process information factory: a data management approach for enhancing business process intelligence," *Proceedings CEC*, San Diego, pp. 162-169, July 2004 [IEEE International Conference on e-Commerce Technology].

[41] I. Linden, "Proposals for the integration of interactive dashboards in business process monitoring to support resources allocation decisions," *Journal of Decision Systems*, vol. 23, no. 3, pp. 318-332, September 2013.

[42] W.M.P. Van der Aalst, Process Mining – Discovery, Conformance and Enhancement of Business Processes, Springer-Verlag, Heidelberg, Berlin, 2011.

[43] T. Bucher, A. Gericke and S. Sigg, "Process-centric business intelligence," *Business Process Management Journal*, vol. 15, no.3, pp. 408-429, 2009.

APPENNDIX I BUSINESS INTELLIGENCE MATURITY

BI	BUSINESS INTELLIGENCE MATURITY		
Please, indicate how you would rate BI maturity in your organization along the following dimensions (X = don't know, can't judge).			
Statement A			Statement B
BI-1	What is the scope of business intelligence systems use in your organization?		
BI is used in isolated manner by individuals.	1 2 3 4 5 X		BI is used in all (wherever needed) organizational units, hierarchical levels, and application areas.
BI-2	What is the level data architecture maturity in your organization?		
Business data management is not addressed in organization. There is non-existing or heterogeneous semantics.	1 2 3 4 5 X		Internal (both structured and unstructured) and external data are fully integrated, and requirements (e.g. data quality) are met.
BI-3	What is the impact of business intelligence in your organization?		
Impact of BI is not considered as relevant.	1 2 3 4 5 X		Decision-making is based on BI and BI is perceived as having a critical impact on organizational performance.
BI-4	What is the level of technical architecture maturity of BI in your organization?		
There is no dedicated BI storage used.	1 2 3 4 5 X		Enterprise-wide data warehouse is used.
BI-5	What is the level of data management maturity in your organization?		
Data integration is manual.	1 2 3 4 5 X		Data integration is automated, dedicated tools for data management and integration are used.
BI-6	What kind of BI tools is used your organization?		
We don't use any specific BI tool, manual analysis is performed.	1 2 3 4 5 X		A broad range of BI tools and techniques is used, such as reporting tools, ad hoc analytics (OLAP), in-memory analytics, planning, alerts, forecasts, scorecards, mobile BI, data mining, predictive analytics, and other advanced techniques of analysis and visualization.
BI-7	What is the the organizational structure related to business intelligence in your organization?		
There are no specifically defined roles and organizational units for BI.	1 2 3 4 5 X		BI (business data analytics or similar) competence center with a comprehensive spectrum of tasks and competences exists.
BI-8	What is the level of maturity of BI processes (e.g. requirements engineering and service management) in your organization?		
BI is used in isolated manner by individuals.	1 2 3 4 5 X		BI is used in all (wherever needed) organizational units, hierarchical levels, and application areas.
BI-9	What is the level of the profitability assessment of business intelligence in your organization?		
There is no profitability assessment of BI.	1 2 3 4 5 X		Cross-project and benefit oriented profitability assessment of BI takes place.
BI-10	What is the level of BI strategy in your organization?		
No BI strategy exists in our organization.	1 2 3 4 5 X		There exists a dedicated BI strategy that clearly reflects business/IT alignment.

APPENNDIX II BPM/BI ALIGNMENT

PIA	BPM/BI ALIGNMENT	
Please, indicate to what extent you agree / disagree with the following statements.		1 = completely disagree; 5 = completely agree; X = don't know, can't judge
PIA-1	BPM initiative is coordinated with BI initiative in a company. Very intensive communication between BPM and BI experts and mangers exists.	1 2 3 4 5 X
PIA -2	BPM and BI terminology is aligned. BI and BPM use common terms, a glossary of BPM&BI terms exists.	1 2 3 4 5 X
PIA -3	BI system enables performance measurement and management of cross-functional processes.	1 2 3 4 5 X
PIA -4	BI system is regularly used by process owners and other process actors for monitoring business process execution.	1 2 3 4 5 X

APPENNDIX III PROCESS PERFORMANCE

PP	PROCESS PERFORMANCE	
Please, indicate to what extent you agree / disagree with the following statements.		1 = completely disagree; 5 = completely agree; X = don't know, can't judge
PP-1	The efficiency of our processes is high above the average of the industry.	1 2 3 4 5 X
PP-2	The quality of our processes is high above the average of the industry.	1 2 3 4 5 X
PP-3	The flexibility of our processes is high above the average of the industry.	1 2 3 4 5 X
PP-4	The quality of our products/services is high above the average of the industry.	1 2 3 4 5 X

MapReduce Research on Warehousing
of Big Data

M. Ptiček and B. Vrdoljak
University of Zagreb, Faculty of Electrical Engineering and Computing
Department for Applied Computing, Zagreb, Croatia
{marina.pticek, boris.vrdoljak}@fer.hr

Abstract - The growth of social networks and affordability of various sensing devices has lead to a huge increase of both human and non-human entities that are interconnected via various networks, mostly Internet. All of these entities generate large amounts of various data, and BI analysts have realized that such data contain knowledge that can no longer be ignored. However, traditional support for extraction of knowledge from mostly transactional data - data warehouse - can no longer cope with large amounts of fast incoming various, unstructured data - big data - and is facing a paradigm shift. Big data analytics has become a very active research area in the last few years, as well as the research of underlying data organization that would enhance it, which could be addressed as *big data warehousing*. One research direction is enhancing data warehouse with new paradigms that have proven to be successful at handling big data. Most popular of them is the MapReduce paradigm. This paper provides an overview on research and attempts to incorporate MapReduce with data warehouse in order to empower it for handling of big data.

I. INTRODUCTION

Data-oriented systems have recently been faced with overwhelming amount of various-typed data. Data warehouses, core of all traditional decision-support systems, also have to face these challenges. It has become crucial to analyze almost all *possible* data to keep up with the competition by following the shift in data warehousing trends. Transactional data do not provide sufficient support anymore and other emerging data sources must be embraced. These new data sources are part of *big data* phenomenon. Traditional data warehouses and RDBMSs are not an adequate storage for big data, and that needs to change because the quantity of unstructured data exceeds the quantity of structured data by four to five times [1]. In big data analytics, MapReduce programming framework has gained popularity because of its ability to effectively process large databases, as well as big data. For that reason, researchers have turned to application of MapReduce to overcome the gap between data warehouses and big data.

Research approaches include integration of data storage systems, changing relational paradigm to non-relational, and new algorithms and approaches to creating and storing data cubes. This paper presents an overview of some of the interesting work done in this field. The paper is organized as follows: section II presents the current data trends, features, storages and paradigms; section III

presents main issues of data warehouses with big data and unsuitability of Hadoop alone for solving them; section IV presents two research approaches to solving problems described in previous section through describing representative work that has been done; section V discusses the presented approaches; and the section VI concludes this paper.

II. DATA TRENDS

A. Big data

The term *big data* was invented to describe new types of data that emerged with popularization of social networks and data-generating smart or simple devices. There are quite a few big data issues from analytical point of view: they may be unstructured and/or semi-structured and lack formal model and metadata; they arrive fast and in large quantities; they may be incorrect to a certain level. Most importantly, it is questionable if a certain set of big data holds any value at all, but that can only be known through data exploration. Big data features are usually described by *4Vs – volume, variety, velocity,* and *veracity* – but, three more *V* features have been identified [2]: *variability* (in ways to interpret same data), *value* (actual usefulness of extracted information), and *visualization* (need for illustrative visualization of big data sets to understand them).

B. Shared-nothing architecture

Shared-nothing architecture is a distributed computing architecture that features a collection of independent and self-sufficient nodes. These nodes may be various (virtual) machines, where each one has a local disc and main memory. The nodes are connected via network and there is no contention point in it. The phrase *shared-nothing* came from the fact that nodes do not share any kind of memory. The most popular paradigm based on this concept is the MapReduce.

C. MapReduce

MapReduce (MR) has recently become one of the most popular frameworks for processing large amounts of data. It is specially designed for scalability of data processing in the cloud environments. It hides the implementation of its main features from the programmer, e.g. fault tolerance, load balancing, parallelization and data distribution.

D. NoSQL

The term NoSQL is interpreted as "Not Only SQL". In this case, having *SQL* as a synonym for relational databases, *NoSQL* does not imply the complete absence of *relational*. NoSQL databases are usually open-source and their main features are: data and processing distribution, horizontal scalability, fault tolerance, fast data access, simple APIs, relaxed consistency and being schema-free [3]. However, they also have their downsides. The current issue with NoSQL databases is that they do not offer a declarative query language similar to SQL and there is no single, unified model of NoSQL databases.

There are multiple NoSQL database families, which differ significantly in their features. Main NoSQL database families (sometimes referred to as 4 *true* NoSQL database families) are: column-oriented, document-oriented, key-value stores, and graph. Also, there are less commonly mentioned families that had been present long before NoSQL term was used: object-oriented, XML-native, and other database families like multimodel, multivalue, and multidimensional databases.

In attempts of big data warehousing, mostly used NoSQL systems were based on Hadoop's implementation of MR, running on top of it or along with it, e.g. HBase, Hive, Pig, Spark, Sqoop, Impala, etc., but attempts have also been made without using a MR-based system. Those that best served for research purposes are briefly described below.

• **HBase** is a column-oriented distributed database that runs on top of Hadoop. It was modelled after Google BigTable. It features data compression and Bloom filters on column-level. Its tables can be used as input and output for MR. One of the most popular data-driven websites that uses HBase is Facebook, which uses it for its messaging platform.

• **Hive** [4] is a data warehouse infrastructure running on top of Hadoop. In Hive, data can be stored in various databases and file systems, all interconnected with Hadoop. It supports SQL-like commands in its language, *HiveQL*, which shifts the focus from low-level programming to a much more declarative level. Given queries are translated into MR jobs that are executed on Hadoop. Hive contains a system catalogue called *metastore* that stores metadata about tables, and that makes it a *data warehouse*.

The idea behind Hive was to bring in well-known concepts of *structuredness* (e.g. tables, columns, partitions and subset of SQL) into unstructured Hadoop environment, but still keeping the flexibility and extensibility Hadoop has. Organization-wise, table is a one logical unit of data and is stored as a directory in HDFS; table partitions are stored as a subdirectory within table's directory; buckets are stored as files within partition's or table's directory, depending on table being partitioned or not.

Hive adopted one of the main changes in approaches to data (warehouse) modelling: *schema-on-read* approach rather than *schema-on-write*, meaning that the data are loaded more quickly because there is no validation against a pre-defined schema. The validation is performed during run-time, which impacts the query time to be slower.

One of the main downsides of Hive's design is its naive simple rule-based optimizer. The authors predicted optimizer improvements as the future research work, as well as exploring other data placement ways, e.g. column-oriented to improve data reading performance.

• **Pig** is a high-level platform that runs on top of Hadoop and serves for creating programs running on it. Unlike Hive, which is meant for exploring datasets, Pig is more appropriate for processing data flows [4]. Its language *Pig Latin* is more declarative comparing to MR programming in Java and also supports user-defined functions, written in Python, Java, etc. Pig Hadoop jobs cannot only be run in MR, but also in other frameworks, e.g. Spark.

• **Google BigTable** [5] is a NoSQL database for unstructured data. It was built on a few Google technologies, including Google File System. In BigTable, data are mapped by triplets *<row key, column key, timestamp>* and each stored value is an uninterpreted array of bytes, while the timestamps are used for versioning. There is a false perception that BigTable is a column-store database, but its connection to column-oriented approaches lies in usage of *column families*. Column families are groups of column keys that are grouped into sets, i.e. columns are clustered by key. Usually, a column family contains data of the same type.

• **Apache Spark** is a cluster-computing framework featuring data parallelism and fault tolerance. It was made to overcome MapReduce's dataflow limitations and has proved to be faster than MR framework. For that reason it should be considered at the time of choosing the right technologies to deal with big data in general, as well as in OLAP.

III. DATA WAREHOUSE'S *BIG DATA ISSUE*

There is a misconception that standard data warehouse architecture is not completely compatible with big data. One of the main arguments of this misconception is *the notion that data warehouses are primarily focused on "stock" or fixed data supply oriented on the past and not suited for real time streaming analytics* [2], but real-time BI, which is supported by real-time data warehouses is its counterargument. The problem lies in the intensity of the data streaming, which has become too high with the emergence of big data. The hype around big data has even caused some vendors and analysts to foretell the death of data warehousing and even relational databases. Most popular BI platforms do not support integration with NoSQL databases, but some provide native access to them (e.g. Hive).

MapReduce is considered "appropriate to serve long running queries in batch mode over raw data"[1], which is exactly what ETL process does. However, integrating or replacing traditional systems with Hadoop is a challenging research topic. Despite Hadoop's ability to provide instant access to large amounts of high quality data, the implementation of ETL processes in Hadoop is not simple. The users need to manually write custom programs, which are sometimes not effective and cannot compare to ETL tools for data cleaning, transformation and coordination. Authors in [6] suggest that using ETL tools is still suitable for small amounts of data, while design of custom programs in Hadoop is more feasible for larger amounts. Another important feature Hadoop lacks is adding dynamic indexes to the framework, which is still an open issue.

IV. MAPREDUCE APPROACHES IN RESEARCH

Research of MapReduce approaches in data warehousing can be grouped into two categories:

i. system-level research: focused on system-level improvements and/or integration;

ii. data cube research: focused on an OLAP data cube component of data warehouse.

A. System-level research

1) Simple data warehouse augmentation

The simplest way of data warehouse's coexistence with *new* big data solutions was presented in [2], where big data are treated as an additional data source for the data warehouse. This idea does not include integration of data warehouse and big data system on any level, but the data sequentially going through processing systems based on their kind, and in the end - ending up in the data warehouse. Before loading, big data are processed (analysed) in a MR framework, while the rest go through standard cleaning and transformation processes. In this case, the resulting data from the MR processing enrich dimension tables by adding more attributes to them and thus enrich the overall analytical power of the data contained in the data warehouse. Authors pointed out that the good approach would be to leave the original data source intact to enable different types of analyses because the same data processed in different ways can yield different results. However, the issue with this solution is that the programmer or analyst needs skills and knowledge to write map and reduce functions, which may have a big span of complexity – from simple counting to using data mining algorithms and machine learning.

2) HadoopDB

HadoopDB [7] is a hybrid of parallel databases and MR. It is an open-source solution made of other open-source technologies like Hive, Hadoop and PostgreSQL, all organized in a shared-nothing architecture. The idea behind such integration was to benefit from scalability advantages of MR, and from performance and efficiency featured in parallel databases. Comparing its used storage systems with those used in Hive, HadoopDB uses relational databases in each of the nodes instead of distributed file systems. MR is used as communication layer; Hive serves as a translation layer (queries expressed in SQL are translated into MR jobs); on the database layer there can be multiple nodes running DBMS. The main research contribution was improving query optimization compared to optimization in Hive by pushing query processing logic maximally down to the databases (because Hive did not consider table data collocation in a single node, while HadoopBP does), and modifying Hive by adding four more components to it – *database connector* (for connecting to databases, executing queries and returning the result), *catalog* (for maintaining information about the databases), *data loader* (for

partitioning and re-partitioning of data and data loading), and *SQL-to-MapReduce-to-SQL planner* (for extending Hive query planner). So far, HadoopDB implemented filter, projection and aggregation operators, while the authors expect that this system will benefit from future Hadoop improvements, despite of parallel databases currently still being faster.

3) Starfish

Starfish [8] is a self-tuning system for Hadoop for big data analytics: it tunes Hadoop parameters, but it does not improve its peak performance. Starfish is placed between Hadoop and MR job-submitting clients like Pig, Hive, command line interface, etc. MR jobs, workflows, or collections of workflows are expressed in language *Lastword* (Language for Starfish Workloads and Data). The main focus of this system is its adaptation to user's and system's needs. Reasoning behind self-tuning approach lies in Hadoop's suboptimal usage of resources and time. The problem with Hadoop is that there might be around 190 configuration parameters controlling its behaviour, and most of them need to be set manually, e.g. actions like adding and removing a node based on the needs, or rebalance of data. Even for a single MR job, multiple parameters need to be set, e.g. number of mappers and reducers, controls for network usage, memory allocation settings. Higher-level languages for defining MR jobs – like HiveQL and PigLatin – do not consider cost-based optimization during the query compilation into MR jobs, but only rely on naive hard-coded rules or hints specified by the users. The basic endeavour is to *move the computation to the data* as much as possible to achieve minimal data movement. Starfish provides automatic system tuning throughout the whole data lifecycle: it is able to perform tuning at different levels of granularity – workload, workflow, and job-level – and the interaction of these components provides Starfish's tuning in whole.

4) Olap4cloud

Olap4cloud[1] is an open-source experimental OLAP engine built on top of Hadoop and HBase, designed to efficiently perform queries containing aggregations and grouping on large data sets, and to be horizontally linearly scalable. Its main features include data defragmentation, indexing, and preaggregations. Olap4cloud stores rows with the same dimensions close to one another. For preaggregations, aggregation cuboids are constructed using classic lattice modes, which help to achieve lower latency during execution of queries that use already calculated aggregations. For now, Olap4cloud is still in an early experimental stage: it is able to manage and query cubes (fact tables containing dimensions and measures), but it does not fully support dimension tables, hierarchies and measures: dimensions can only be Java data type *long*, and measures must be in Java data type *double*.

5) HBase Lattice

HBase Lattice [9] is an open-source solution that attempted to provide BI "OLAP-ish" solution based on HBase. Author argues that, in discussion of *"Cassandra is for OLTP, HBase is OLAP* mantra", this system, based on

---

[1] https://github.com/akolyadenko/olap4cloud/blob/wiki/UserGuide.md

HBase, has limited support for OLAP features: it does not directly support cube models and has no concepts of dimensions, hierarchies, measures and facts, nor has its own query language. It implements incremental OLAP-like cubes: by pre-building cuboids in a cube lattice model, it is able to cope with aggregate queries.

6) HaoLap

HaoLap [10] (Hadoop based OLAP) is an OLAP system designed for big data that has some common features with MOLAP tools. It uses simplified multidimensional model for mapping of dimensions and measures, and most importantly - it supports dimension operators (e.g. roll-up). In this system, optimization and speed-up are not achieved using indexes nor pre-computations, but by sharding (horizontal partitioning) and chunk selection. Instead of storing large multidimensional arrays, *calculations* are stored and that simplifies the dimension by having low storage cost and makes this system's OLAP efficient.

Dimension and cube metadata are stored in XML format in a metadata server, while measures are stored in HDFS. Data cubes do not have to be instantiated in memory - OLAP algorithm and MR are used for that. MR jobs are consisted of four parts (input-formatter, mapper, reducer and output-formatter) and are specified as such quadruplets. MR job operates on the level of chunks, and at the end of the processing chunk is stored in the created chunk files, which are then logically combined into the result – a cube.

7) Octopus

Octopus [11] is a hybrid data processing engine. It is a middleware engine that integrates various backend systems and serves as a one access point (Figure 1.) between the user and them. The reasoning behind building such integration system is that neither traditional warehousing systems, nor recent big data systems are able to fully leverage the power of the other – neither can fully take over the other's role.

Data are distributed in multiple locations of the system, while the system itself bases its optimization on minimizing the amount of data movement by making a query push-down, which is based on the data's residence. Octopus takes a given query task in an SQL-like form from the user and divides it into multiple subqueries. The

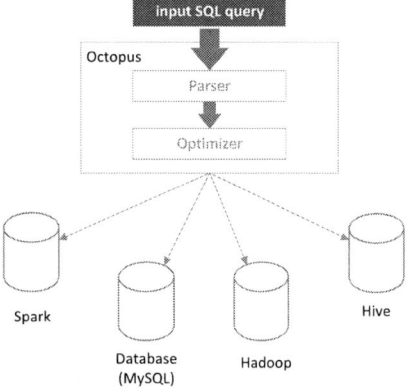

Figure 1. Octopus system overview [11]

initial query is processed with system components, *parser* and *optimizer*; parser produces a logical operator tree and optimizer produces an optimal set of query plans. Subqueries derived from the initial query are pushed down to the backend systems and processed inside of them. By sending the queries to systems where the data are physically stored and optimally processed, data movement costs are reduced.

Authors have compared Octopus to Spark, because Spark also supports integration of external data sources and implements unnecessary data filtering for reducing data migration. Experiments proved Octopus to be faster than Spark in processing aggregation queries. Possible future development of Octopus lies in adding support for more types of NoSQL databases, as well as integrating more Spark modules, e.g. machine learning.

B. Data cube research

1) Approach by Abello, Ferrarons and Romero

An approach to building data cubes on BigTable was presented in [1]. BigTable is used for storing all the data together temporarily - data are just partially cleaned and integrated before they are stored - without having a concrete schema, so the schema may evolve easily over time. Also, the authors argue that column storage is good enough for data marts because they have well-defined columns, but for the entire data warehouse more flexibility is required because of unclear data structure and continuous schema evolution.

The basic idea is to have a non-formatted chunk of data just stored and associated to the key, which is later, once it is needed, retrieved and used according to the specific needs. That is the opposite from putting a lot of resources and effort into formatting and structuring the data according to the strict star schema, which after data import may not be used for a couple of months, or may even not be used at all. Upon having an analysis to perform, data quality is specified, suitable technology is selected (e.g. ROLAP or MOLAP), and then ETL is used to prepare the data according to needs. On demand, specific and customized data cubes are extracted and stored in a data mart, which may or may not reside in cloud.

For building cubes three typical table access approaches (assuming denormalized data) have been applied in a way they could benefit from MR and BigTable: full scan, index access, and range index scan.

2) MRDataCube

A new cube computation algorithm *MRDataCube* (MapReduce Data Cube) was presented in [12]. MRDataCube is a large-scale data cube computation algorithm that uses MR framework and uses concepts of cuboids and cells. *Cuboid* is considered an aggregated result of aggregate operations upon a cube with the goal of extracting each combination of attributes. *Cells* construct cuboids, and they represent the values stored in them and can be represented as m-membered tuple, where m is aggregation measurement of the cell.

This algorithm computes the cube in two phases, *MRSpread* and *MRAssemble*. Its main feature is continuous data reduction through these two phases. The first phase produces partial cuboids cells, where one cuboid cell values may spread through outputs of several reduce functions. In the second phase, complete cuboids are generated by computing all partial cells, i.e. merging partial cuboids into one cell. Between two phases, result is stored in HDFS. This algorithm showed 2-16 times better performance compared to previous work in that area, algorithms *MRNaive*, *MR2D*, *MRGBLP* and *MRPipeSort*.

3) CN-CUBE

An operator *CN-CUBE* (Columnar NoSQL cube) for computing OLAP cube from column-oriented data warehouse was presented in [13], and it proved to be faster than Hive. The main technology they used was HBase engine. By implementing CN-CUBE operator, authors have attempted to solve one of the most common obstacles of column-oriented databases – the lack of OLAP operators. The CN-CUBE reduces disk access by using a result on an extraction query as a view – based on attributed needed for the cube computation requested by analysts - and avoids re-accessing the data warehouse for performing different aggregations. Using this approach, authors have reduced input and output flows substantially.

This operator performs in three phases. During the first phase, attributes are identified and necessary columns (most recent values) are extracted into an intermediate result. In second phase, intermediate relation columns with the values are hashed to produce the lists of values' positions. These lists allow some columns to be aggregated separately by each dimension. In the final phase, lists of dimensions' positions of intermediate results are associated through the logical *AND* operator.

4) MC-CUBE

Similarly to [13], [14] presents an aggregation operator MC-CUBE (MapReduce Columnar cube) that uses HBase as main technology, and introduces a different approach to building OLAP cubes from columnar "big data warehouse". In this approach, data are taken at a single, least aggregated level, and other aggregates – at other, higher granularity levels – are computed from it. To perform aggregation, this operator performs *invisible join*, typical in column stores, and extends the aggregation to cover "all possible aggregations at different levels of granularity of the cube" [14].

MC-CUBE operator performs seven MR jobs during four phases. During the first phase, aggregations are produced from the data warehouse at lowest and highest granularity levels – joins between fact and dimension tables are materialized and aggregates are calculated. In the second phase, aggregations are performed of each dimension separately, and in the third, rest of the aggregates composing the cube are computed. The final phase replaces dimensions' identifiers with dimensions' attributes, and completes the cube.

V. DISCUSSION

For big data warehousing solution, system integration approaches usually include integration of one or more MR-based technologies like Hive and Hadoop, and combine them with either NoSQL and/or traditional relational storage. Particularly interesting combination is the integration of MapReduce and parallel databases, where the best features are taken from both of them and combined complementarily because one does well what the other does not.

One of the suggested scenarios [3] for big data analysis is to use Hive in Hadoop environment for OLTP, and ROLAP tool for data analysis. However, problems may occur in case when data arrive too quickly and are stored in real-time in Hive, because the existing aggregates in RDBMSs used for OLAP could in a very short time become inaccurate. Other problems are related to the volume of the data; when using partial aggregates, it still may affect the speed OLAP browsing in a negative way. The solution may lie in using MapReduce processing on both OLTP and OLAP side.

In a few works [8], [15], [16] an acronym *MAD* appeared, denoting features expected from big data analytics systems: *magnetism* means that system attracts all kinds of data, regardless of possibly unknown structure, schema, or even partial values; *agility* stands for system's ability to adapt to data evolutions; *depth* means that the system supports analyses that go to bigger depths than just roll-ups and drill-downs. Authors in [8] even went further and described three more features that would *MADDER* analytics systems need to have: *data-lifecycle-awareness*, *elasticity*, and *robustness*. In future data warehouse modifications and research, these features may prove to be a good guiding principle. It is noticeable that some of these features are found in NoSQL databases, but never all of them. This shows that NoSQL technologies are a good candidate for a data warehouse asset, but chances are small that a NoSQL database would ever completely replace data warehouse's relational basis.

Another potential problem that may affect the course of big data analytics and warehousing development is a large portfolio of technologies and implementations, even among open-source products, and even industry has pointed out this as potential problem [17]. The vendors have approached this matter each in their own way, making new products most compatible with their existing ones. Analyzing the general situation from the perspective of data warehousing future, the existing products from which data warehouses may eventually benefit - NoSQL databases and Hadoop implementations and tools - still have many downsides in terms of lacking concepts commonly used in data warehouses and relational databases (e.g. indexing, profound optimizer, ability of having OLAP operators, dimensions and hierarchies, etc.). All of these downsides combined together create obstacles that are being worked on by the current research.

Through the research overview presented in this paper, we can see various system integration approaches [2], [7], [8], [9], [10], [11] to overcoming these issues. System-level integration raises questions of middleware design, data localization, computation locality i.e. amount data movement, etc. Much work has also been done on optimization issues: it included designing special tuning middleware; building better query and MapReduce job plans; preaggregations – most often in forms of cuboids that are stored in a lattice model; using views to access fewer data – but, one of the most applied approaches is the

1365

query push-down technique, i.e. moving the computation to the data, which is guided by the thought that the data should be stored and processed in the system that does suits their nature the best, whether that are RDBMSs, NoSQL databases or MapReduce environment. Possible progress is seen in designing new middleware, either from scratch or by modifying existing systems like Hive.

Presented research overview from the perspective of building data cubes [1], [12], [13], [14], describes representative approaches that have been made with various ways of data storage, especially with column-oriented [13], [14]. Approaches included: storing measures in HDFS while the metadata reside in simple format on server; storing whole data marts in column-oriented storage; collocating data by their type (rows) in column-oriented storage, etc. The diversity is significant, and further research of this matter is expected.

However, one of the most noticeable downsides in some of the presented approaches is considerable simplification – dimensional model is simplified to the extent that the system does not support dimensions and hierarchies, and sometimes even does not differ dimensions from facts. Simplification is an interesting approach, e.g. solving explosive dimension cardinality issue by using some aggregative calculations, but attention should be paid to not over-simplifying the model to the level of its inapplicability for OLAP. To bring big data into the data warehouse, ETL process may need to be loosened to ensure the flexibility for the warehouse and enable the schema to evolve. Still, the main focus remains on creating the much needed OLAP operators.

VI. CONCLUSION

This paper gives an overview of latest research on bringing big data into data warehouses and organization of big data for OLAP analysis using the MapReduce framework. Big data were described by their main features, which are the main reason they cannot be easily let into the strictly structured and highly organized data warehouse environment. However, they are potentially greatly valuable and their share in available data pool is rapidly growing. Thus, they should not be treated as a distinct issue and completely separated from other data, but should be dealt with and made supportive for BI analytics along with traditional transactional data.

NoSQL databases show great potential for handling big data analytics, especially those that implement MapReduce framework. But, there are many issues with these two concepts. The industry has overwhelmed data analytics field with too many implementations that are still not mature enough, mostly in terms of all the underlying supporting features that RDBMSs offer. NoSQL database families differ a lot and may not handle all types of data equally well, and having no unified database model nor a clear data structure leads to great difficulties in design of analytical systems.

The most realistic future scenario is that the data warehouses will not be replaced by some completely different innovative systems. The more likely scenario is that they will continue to coexist in a symbiotic relationship with the new, more suitable solutions for big data, where big data would serve as an additional analytical asset.

REFERENCES

[1] A. Abelló, J. Ferrarons, and O. Romero, "Building cubes with MapReduce," in *Proceedings of the ACM 14th international workshop on Data Warehousing and OLAP (DOLAP '11)*, Glasgow, Scotland, UK, 2011, pp. 17 – 24.

[2] N. Jukić, A. Sharma, S. Nestorov, and B. Jukić, "Augmenting Data Warehouses with Big Data," *Information Systems Management*, vol. 32, no. 3, pp. 200–209, 2015.

[3] J. Duda, "Business Intelligence and NoSQL Databases," *Information Systems in Management*, vol. Vol. 1, no. 1, pp. 25–37, 2012.

[4] A. Thusoo *et al.*, "Hive - a petabyte scale data warehouse using Hadoop," presented at the 2010 IEEE 26th International Conference on Data Engineering (ICDE), 2010, pp. 996–1005.

[5] F. Chang *et al.*, "Bigtable: A Distributed Storage System for Structured Data," *ACM Transactions on Computer Systems*, vol. 26, no. 2, pp. 1–26, Jun. 2008.

[6] T. Šubić, P. Poščić, and D. Jakšić, "Big data in data warehouses," presented at the 4th International Conference The Future of Information Sciences (INFuture2015), Zagreb, Croatia, 2015, pp. 235-244.

[7] A. Abouzeid, K. Bajda-Pawlikowski, D. Abadi, A. Silberschatz, and A. Rasin, "HadoopDB: an architectural hybrid of MapReduce and DBMS technologies for analytical workloads," *Proceedings of the VLDB Endowment*, vol. 2, no. 1, pp. 922–933, 2009.

[8] H. Herodotou *et al.*, "Starfish: A Self-tuning System for Big Data Analytics," presented at the 5th Biennial Conference on Innovative Data Systems Research (CIDR 2011), Asilomar, CA, USA, 2011, pp. 261–272.

[9] D. Lyubimov, "HBase Lattice Quick Start", software manual, source: https://github.com/dlyubimov/HBase-Lattice/

[10] J. Song, C. Guo, Z. Wang, Y. Zhang, G. Yu, and J.-M. Pierson, "HaoLap: A Hadoop based OLAP system for big data," *Journal of Systems and Software*, vol. 102, pp. 167–181, Apr. 2015.

[11] Y. Chen, C. Xu, W. Rao, H. Min, and G. Su, "Octopus: Hybrid Big Data Integration Engine," presented at the 2015 IEEE 7th International Conference on Cloud Computing Technology and Science, 2015, pp. 462–466.

[12] S. Lee, S. Jo, and J. Kim, "MRDataCube: Data cube computation using MapReduce," presented at the 2015 International Conference on Big Data and Smart Computing (BigComp), 2015, pp. 95–102.

[13] K. Dehdouh, F. Bentayeb, O. Boussaid, and N. Kabachi, "Columnar NoSQL CUBE: Agregation operator for columnar NoSQL data warehouse," presented at the *2014 IEEE International Conference on Systems, Man and Cybernetics (SMC)*, 2014, pp. 3828–3833..

[14] K. Dehdouh, "Building OLAP Cubes from Columnar NoSQL Data Warehouses," in *Model and Data Engineering*, vol. 9893, L. Bellatreche, Ó. Pastor, J. M. Almendros Jiménez, and Y. Aït-Ameur, Eds. Cham: Springer International Publishing, 2016, pp. 166–179.

[15] J. Cohen, B. Dolan, M. Dunlap, J. M. Hellerstein, and C. Welton, "MAD skills: new analysis practices for big data," *Proceedings of the VLDB Endowment*, vol. 2, no. 2, pp. 1481–1492, 2009.

[16] A. Cuzzocrea, I.-Y. Song, and K. C. Davis, "Analytics over large-scale multidimensional data: the big data revolution!," in *DOLAP '11 Proceedings of the ACM 14th international workshop on Data Warehousing and OLAP*, 2011, pp. 101–104.

[17] J. Vaughan, "EMC, Intel unveil new Hadoop distributions, but how many is too many?," *TechTarget*. Internet: http://searchdatamanagement.techtarget.com/news/2240179304/EMC-Intel-unveil-new-Hadoop-distributions-but-how-many-is-too-many, Mar. 8, 2013 [Oct. 26, 2016].

1366

Selection of Variables for Credit Risk Data Mining Models: Preliminary research

M. Pejić Bach*, J. Zoroja*, B. Jaković*, N. Šarlija**

* University of Zagreb, Faculty of Economics & Business, Zagreb, Croatia
** University of Osijek, Faculty of Economics, Osijek, Croatia
mpejic@efzg.hr, jzoroja@efzg.hr, bjakovic@efzg.hr, natasa@efos.hr

Abstract - Credit risk is related to the risk of the borrower that the lender will not be able to return their debt including interest. Numerous researches have been conducted in the area of credit risk, both using classical models such as Altman Z-score and using machine learning methodology. However, the research using the data from Croatian financial institutions is scarce, especially research focused on the selection of the demographic and/or behavior variables. In addition, it is important to develop robust models that estimate credit risk as accurately as possible. The goal of this research is to develop a data mining model for prediction of credit risk, using the data from Croatian financial institutions on defaulted clients (demographic and behavior data). Decision tree models are constructed for the prediction of credit risk. Different algorithms for the variable selection are evaluated based on the classification accuracy of the decision trees developed based on the selected variables.

I. INTRODUCTION

Data mining methods are used for finding undiscovered valuable information from large databases. In other words, the main goal of data mining techniques is to extract knowledge in order to make successful management decisions [1]. Applications of data mining methods are used in almost every industry: banking, marketing, finance, manufacturing, medicine, education, trade, supply [2, 3, 4]. Each industry has its own characteristics, which implies usage of different data mining methodologies. Therefore, in the banking industry, characterized with a high level of fraud and risks, which requires successful prediction of credit default, scoring and applicants, usage of data mining techniques is very common [5].Data mining usage is one of the most common techniques used in the financial analysis, especially in the banking industry. Prediction of credit risk, mostly prediction of credit default, presents an important activity of the banking industry [6]. There are several different data mining techniques that can be used for financial data analysis because of their high level of success. However, their success also depends on on the data available, its cleaning, and transformation. Therefore, decision trees are one of the most commonly used methods [7, 8]. Decision trees are one of the classification methods which group variables into one or more categories of the target variables [9]. When using the decision trees process it is important to follow three main steps: (i) determine the sample, (ii) choose variables, and (iii) select an appropriate algorithm. In this paper we have used the algorithm C4.5, as one of the ten

most popular variables. The goal of the paper is to compare the classification of banking clients according to the credit default, with the C4.5 decision tree algorithm, using different sets of the variables: Entrepreneurial idea; Growth plan; Marketing plan, Personal characteristics of entrepreneurs, Characteristics of SME, Characteristics of credit program, and Relationship between the entrepreneur and a financial institution. The variables are selected by the usage of three different algorithms, provided in the Weka software: Class CfsSubsetEva algorithm, ChiSquaredVariableEval algorithm, and ConsistencySubsetEval algorithm. Previous research that tested the efficiency of algorithms for the selection of variables was mostly conducted on the retail credit risk datasets, e.g. Oreski et al. [10]. The scientific contribution of our paper is that the algorithms for the selection of variables are tested on the real-world dataset of credit risk of Croatian financial institution's business clients (entrepreneurs from Eastern Croatia). The paper consists of six sections. After the Introduction, as the first section, there is Literature review. In Literature review, data mining methodology and its usage for predicting credit default are presented. Decision trees, as one of the data mining methods, are described as well as variables and techniques selection approaches used in this research. In the third section named Methodology, data, decision trees techniques and the variable selection process are discussed. Research results are provided in the fourth and the fifth section. The fourth section elaborates on results of the different variable selection strategies, while the fifth section of the paper discusses results regarding classification efficiency measures, classification matrices and falsely predicted good and bad debtors with different variable selection approaches. The last section is Conclusion. *This work has been fully supported by the Croatian Science Foundation under the project "Process and Business Intelligence for Business Performance" - PROSPER (IP-2014-09-3729).*

II. LITERATURE REVIEW

A. Data mining methodology

The amount of data has been constantly increasing, which creates difficulties for managers and successful decision making. A high growth of valuable as well as invaluable data in databases has created a need for the use of different methodologies which can help finding, extracting and analyzing data important for decision makers [11].

Data mining technology combines different approaches, e.g. machine learning, statistics and database management, which are used for finding valuable patterns in data for further prediction and decision making. In addition, data mining techniques can also be used for determining relationships among data in order to create knowledge [12]. The main purpose of data mining is to find and analyse disorganized information with the goal of improving business knowledge and activities.

The most commonly used data mining methods are: classification, regression, clustering, visualization, decision trees, association rules, neural networks, support vector machine [5, 13]. In our research we conducted a decision tree analysis, which is mainly used for classification, in order to predict the credit default.

The main purpose of the decision tree analysis is to predict behavior of the target variable using different algorithms to get the best outcome [14]. Some of possible algorithms are: C4.5, ID3, CART, CHAID, and MARS. One of the disadvantages of the decision tree analysis is that, in the process, analysts should pay more attention to a high variance. Other disadvantage of the decision tree is the overfitting, which occurs when model is excessively complex, and it has a poor predictive efficiency. However, the most important advantage of the decision trees is the possibility of interpretation, which together with the simple usage and implementation make the decision tree method appropriate for a wide range of research.

B. Predicting credit default with data mining approach

Countries, especially their economies and financial institutions, have been facing a strong financial crisis in the last years. Therefore, in many countries, governments have brought many saving measures in order to decrease costs and to restart economy development. In addition, credit default has increased and nowadays banks pay much more attention to credit risk assessment and to prediction of credit default with the goal to reduce risk [15].

Financial institutions and banks are using different intelligent techniques, e.g. mathematical models, statistics analysis, data mining methods with the goal to make efficient credit decisions. A detailed analysis of data on the characteristics of current and previous credit users plays an important factor in forecasting the future credit default of new clients [16].

C. Variables and techniques selection approaches

In order to predict credit default, financial institutions and banks mostly use behavioral and demographic variables of previous and current clients, e. g. monthly income, marital status, real-estate owner, employment, age, gender [17].

The main purpose of our research is to classify banking clients regarding credit default with a decision tree analysis, using different variables related to entrepreneurship activity. In addition, financial institutions and banks, when approving credits to clients,

strive to select those clients who will be able to repay it in the given period of time [18]. In other words, they are focused on good clients.

There are also studies about methods used in credit scoring. One of the examples is a research which used demographic and behavioral data and three data mining methods: credit scorecard, logistic regression and a decision tree model [9]. The results of the research showed that all three methods are appropriate for use but the scorecards method is the easiest to apply.

There is also a study in which authors have investigated recent researches conducted in the field of credit risk assessment regarding clients and their ability to repay credits [19]. Research results showed that logistic regression is the most commonly used method to group clients into good or bad debtors.

Recent studies showed that intelligence methods used for discovering credit scoring are mostly non-parametric methods and computational intelligence techniques, e.g. decision trees, artificial neural networks, support vector machines and evolutionary algorithms [20, 21, 17].

III. METHODOLOGY

A. Data

Data used in this research were collected from an entrepreneurship credit dataset. Data were collected randomly from the database of clients (entrepreneurs from Eastern Croatia) of the financial institution, that is focused on financing small and medium enterprises, mostly start-ups. There are 200 applicants in the sample.

There are two main reasons for a small dataset: (i) a quite low level of business activity regarding entrepreneurship in Croatia (Total Early State Entrepreneurial Activity (TEA) index for the year 2015 = 7.69; TEA index for the year 2004 = 7.97) which means that a low number of people is taking a credit to start a business, and (ii) financial institutions are rejecting too risky start-ups applications for a credit. Therefore, collecting a larger sample will be possible in the next few years, when the entrepreneurial climate and perception of entrepreneurship activity improves.

The following variables were used for the development of the credit scoring model: Entrepreneurial idea; Growth plan; Marketing plan, Personal characteristics of entrepreneurs, Characteristics of SME, Characteristics of credit program, Relationship between the entrepreneur and a financial institution, and Creditworthiness. Most of the variables are nominal, while two of them are numeric (Entrepreneurs' age, Number of employees, and Credit amount).

Variables related to the future plans for the SME were estimated by a banking clerk (Table I).

First, it was estimated whether the entrepreneur has a clear vision of the business development (for newly established SMEs), or it is already an established business (Variable Vision).

Second, the variable Better estimated what the main competitive advantage of the SME (better quality, technology, price, or expertise of employees) is.

1368

Third, it was estimated what the main market for SME's products/services is: local, national, wider region, or narrow targeted customers (Variable Market). Entrepreneurs stated which percentage of the profit is planned to be reinvested in the business operations (Variable Reinvest), and what the plans for the promotion of products/services are (Variable Ad). Also, it was estimated whether the entrepreneur can identify who SME's main competitors are.

TABLE I VARIABLES RELATED TO THE FUTURE PLANS FOR THE SME

Variable	Question asked	Answers
I Entrepreneurial idea		
Vision	Does the entrepreneur have a clear vision of the business?	1 – no clear vision (for newly established SMEs) 2 – clear vision (for newly established SMEs) 3– established business
Better	Advantages of products/services	1 – better quality 2 - better technology 3 – good price 4 – expertise of employees 100 – no answer
Market	Market for products/services	1 – local 2 – narrow targeted customers 3 – wider region 4 – Croatia 100 – no answer
II Growth plan		
Reinvest	Projected percentage of the invested profit (reinvested profit/profit*100)	1 - 0 to 30% 2 - 30.01 to 50% 3 - 50.01 to 70% 4 –70.01 to 100% 100 – missing value
III Marketing plan		
Ad	Promotion of products/services	1 – without promotion 2 – no need for promotion 3 – all media 5 – personal selling, presentation 6 – posters, leaflets, internet 100 – missing value
Compet	Can the entrepreneur identify competition?	1 – no competition 2 – not defined 3 – defined competition 100 – no answer

Source: Authors, using Entrepreneurship credit dataset

Table II presents the variables related to the characteristics of the entrepreneur and SME. Entrepreneurs' occupations are grouped into 5 main groups: 1 - farmer, veterinarian; 2 - trader, restaurateur; 3 - construction worker; 4 - engineer, physician, and pharmacist, 5 - Technologist, chemist. Entrepreneurs' ages are expressed as a numeric variable. Entrepreneurs' locations refer to 4 geographic areas in Croatia.

Table III represents the variables related to the credit program and the bank: interest repayment frequency

(monthly, quarterly, half-yearly), grace period, principal repayment, repayment period (expressed in months), interest rates, and amount of credit (expressed in local currency). Also, the variable Client measures whether an entrepreneur has applied for a credit before.

Table IV represents the classification variable that was used for the credit scoring (variable Default), and groups clients as "bad" or "good" based on the regularity of their payment.

TABLE II VARIABLES RELATED TO THE CHARACTERISTICS OF ENTREPRENEUR AND SME

Variable	Question asked	Answers
IV Personal characteristics of entrepreneurs		
Occup	Entrepreneurs' occupations	1 - farmer, veterinarian 2 - trader, restaurateur 3 - construction worker 4 - engineer, physician, pharmacist 5 - technologist, chemist
Age	Entrepreneurs' ages	numeric
Location	Entrepreneurs' locations	1 – Baranja, Osijek 2 - S.Brod, Požega 3 - Đakovo, Našice 4 - Vinkovci, Vukovar
V Characteristics of SME		
Ind	Industry	1 - plastics, textiles 2 - car service 3 - food production 4 - health and intellectual services 5 - agriculture 6 - construction 13 - tourism
Start	Is this a new business venture	1 – yes 2 – no
Equip	Does the entrepreneur have some equipment?	1 – yes 0 – no
Emp	Number of employees	numeric

Source: Authors, using Entrepreneurship credit dataset

TABLE III VARIABLES RELATED TO THE CREDIT PROGRAM AND THE BANK

Variable	Question asked	Answers
VI Characteristics of credit program		
Int	Interest repayment frequency	1- monthly 2 - quarterly 4 - half-yearly
Grace	Grace period	1 – yes 0 – no
Prin	Principal repayment	1- monthly 5 – yearly
Period	Repayment period (months)	numeric
I_rate	Interest rate	4,9% 6,9% 8,9%
Amount	Credit amount (local currency)	numeric
VII Relationship between the entrepreneur and a financial institution		
Client	Is this the first time the entrepreneur is applying for a credit?	1 – yes 2 – no

Source: Authors, using Entrepreneurship credit dataset

TABLE IV GOAL VARIABLE USED FOR THE CREDIT SCORING

Variable	Question asked	Answers
Default	"bad" clients are defined as those who have been late with their payments for more than 45 days at least once. Other clients who have not been late for more than 45 days are labeled as "good".	1-bad 0-good

Source: Authors, using Entrepreneurship credit dataset

B. Decision trees

Table V represents the Weka Description of the used algorithm. The C4.5 (weka.classifiers.trees.J48-C0.25-M5) algorithm was used for developing models for classification of debtors as good or bad. There are 200 instances in the model and the used test mode is 10-fold cross-validation.

TABLE V WEKA DESCRIPTION OF THE USED ALGORITHM

Weka feature	Used feature
Scheme:	weka.classifiers.trees.J48 -C 0.25 -M 5
Relation:	credit_scoring.arff
Instances:	200
Test mode:	10-fold cross-validation

Source: Authors, using Entrepreneurship credit dataset

C. Variable selection

Three approaches to the variable selection were applied: (1) selection of the variables using the Class CfsSubsetEval algorithm (searching approach BestFirst), (2) selection of the variables using the ChiSquaredVariableEval algorithm (searching approach Ranker), and (3) selection of the variables using the ConsistencySubsetEval (searching approach Greedy Stepwise).

There are differences in definition and usage of the three mentioned approaches to the variable selection which were applied. The Class CfsSubsetEval algorithm is based on the individual estimation of the variables that are highly correlated with the class variables but are not highly mutually correlated. The ChiSquaredVariableEval calculates the value of a variable regarding the value of the chi-squared statistic with respect to the class. The ConsistencySubsetEval calculates the value of a subset of variables by the level of reliability in the class values [22].

Table VI presents the variables used for a different approach to the variable selection. The Class CfsSubsetEval algorithm selected only three variables: Variable Vision - Does the entrepreneur have a clear vision of the business?; Variable Ad - Promotion of products/services, and Variable Reinvest - Projected percentage of the invested profit.

The ChiSquaredVariableEval algorithm selected all of the variables except the Variable Location. The ConsistencySubsetEval selected all of the variables related to the Entrepreneurial idea, Growth plan, and Marketing plan. However, only a few algorithms were selected that were related to the Personal characteristics of entrepreneurs, Characteristics of SME, Characteristics of a credit program, and Relationship between the entrepreneur and a financial institution.

TABLE VI VARIABLES SELECTED BY DIFFERENT ALGORITHMS

Variable	Class CfsSubsetEval	ChiSquaredVaria bleEval	ConsistencySubs etEval
I Entrepreneurial idea			
Vision	✓	✓	✓
Better	∅	✓	✓
Market	∅	✓	✓
II Growth plan			
Reinvest	✓	✓	✓
III Marketing plan			
Ad	✓	✓	✓
Compet	∅	✓	✓
IV Personal characteristics of entrepreneurs			
Occup	∅	✓	✓
Age	∅	✓	∅
Location	∅	∅	∅
V Characteristics of SME			
Ind	∅	✓	✓
Start	∅	✓	∅
Equip	∅	✓	∅
Emp	∅	✓	∅
VI Characteristics of credit program			
Int	∅	✓	✓
Grace	∅	✓	✓
Prin	∅	✓	∅
Period	∅	✓	∅
I_rate	∅	✓	∅
Amount	∅	✓	∅
VII Relationship between the entrepreneur and a financial institution			
Client	∅	✓	∅

Source: Authors, using Entrepreneurship credit dataset

IV. RESULTS

Table VII presents the results of different strategies for variable selection. The largest tree is produced by the ChiSquaredVariableEval (34 leaves and 48 nodes), which could be expected since this algorithm generated the largest number of independent variables. The next is the ConsistencySubsetEval algorithm with 24 leaves and 32 nodes. The smallest tree is produced using the Class CfsSubsetEval algorithm (3 leaves and 4 nodes).

TABLE VII CHARACTERITICS OF THE TREES DEVELOPED WITH DIFFERENT VARIABLE SELECTION APPROACHES

Variable selection	Number of Leaves	Size of the tree (number of nodes)
Class CfsSubsetEval	3	4
ChiSquaredVariableEval	34	48
ConsistencySubsetEval	24	32

Source: Authors, using Entrepreneurship credit dataset

Figure 1 represents the complete decision tree generated with the C4.5 algorithm, using the algorithms selected by the Class CfsSubsetEval algorithm. However, only one variable was selected for the tree development. The first number in the bracket is the total number of cases classified in the leaf. The second number is the number of those cases that are misclassified. For example, among 153 entrepreneurs that had a vision of established business, 104 were correctly classified as good, while 49 were incorrectly classified as bad.

```
vision = established_business: good (153.0/49.0)
vision = no_clear_vision: bad (14.0/2.0)
vision = clear_vision: good (28.0/5.0)
```

Figure 1 Decision tree developed using variables selected by the Class CfsSubsetEval algorithm (Source: Authors, using Entrepreneurship credit dataset)

Figure 2 represents a part of the decision tree generated with the C4.5 algorithm, using the algorithms selected by the ChiSquaredVariableEval algorithm.

```
vision = established_business
|  ad = missing_value
|  |  grace = yes
|  |  |  period = 12_months
|  |  |  |  occup = farmer_veterinarian: good (17.0/2.0)
|  |  |  |  occup = construction_worker: bad (1.0)
|  |  |  |  occup = trader_restaurateur: bad (2.0)
|  |  |  |  occup = engineer_physician_pharmacist: bad
                                                  (1.0)
|  |  |  |  occup = technologist_chemist: good (0.0)
|  |  |  period = 24_months
|  |  |  |  occup = farmer_veterinarian
|  |  |  |  |  reinvest = missing_value
|  |  |  |  |  |  age <= 38: bad (9.0/1.0)
|  |  |  |  |  |  age > 38
|  |  |  |  |  |  |  amount <= 22227.6: bad (2.0)
|  |  |  |  |  |  |  amount > 22227.6: good (6.0)
|  |  |  |  |  reinvest = 30.1.1950: bad (4.0/1.0)
|  |  |  |  |  reinvest = 70.01-100: good (1.0)
|  |  |  |  |  reinvest = 0-30: bad (2.0)
|  |  |  |  occup = construction_worker: good (1.0)
|  |  |  |  occup = trader_restaurateur: bad (0.0)
|  |  |  occup = engineer_physician_pharmacist: good
                                                  (4.0/1.0)
```

Figure 2 Decision tree developed using variables selected by the ChiSquaredVariableEval algorithm (Source: Authors, using Entrepreneurship credit dataset)

Figure 3 represents a part of the decision tree generated with the C4.5 algorithm, using the algorithms selected by the ConsistencySubsetEval algorithm.

```
vision = established_business
|  ad = missing_value
|  |  int = quarterly
|  |  |  occup = farmer_veterinarian: good (15.0/1.0)
|  |  |  occup = construction_worker: bad (1.0)
|  |  |  occup = trader_restaurateur: bad (2.0)
|  |  occup = engineer_physician_pharmacist: good (0.0)
|  |  |  occup = technologist_chemist: good (0.0)
|  |  |  int = monthly: good (36.0/10.0)
|  |  |  int = half-yearly: bad (14.0/4.0)
|  ad = personal_selling_presentation: good (20.0/1.0)
|  |  ad = all_media
|  |  |  compet = not_defined
|  |  |  |  reinvest = missing_value
|  |  |  |  |  grace = yes: good (3.0)
|  |  |  |  |  grace = no: bad (7.0/1.0)
|  |  |  |  reinvest = 30.1.1950: bad (2.0)
|  |  |  |  reinvest = 70.01-100: good (6.0/1.0)
|  |  |  |  reinvest = 0-30: good (0.0)
```

Figure 3 Decision tree developed using variables selected by the ConsistencySubsetEval algorithm (Source: Authors, using Entrepreneurship credit dataset)

V. DISCUSSION

The following section elaborates on research results regarding classification efficiency measures, classification matrices and falsely predicted good and bad debtors with different variable selection approaches.

Table VI presents classification efficiency measures. According to the percentage of correctly classified instances and the root mean squared error the best approach was to use the ConsistencySubsetEval algorithm. However, according to the Kappa statistics, the best approach was to use the Class CfsSubsetEval algorithm.

Table VIII Classification Efficiency Measures

Variable selection	Correctly classified instances	Root mean squared error	Kappa statistic
Class CfsSubsetEval	71,28 % (2)	0.4556 (2)	0.2059 (1)
ChiSquaredVariableEval	65.64 % (3)	0.5188 (3)	0.1747 (3)
ConsistencySubsetEval	70.77 % (1)	0.4548 (1)	0.1954 (2)

Note: Rank of the measure in parenthesis

Source: Authors, using Entrepreneurship credit dataset

Table VII presents classification matrices for the decision trees generated by the different sets of variables selected by the three algorithms. The decision tree generated using the variables selected by the Class CfsSubsetEval falsely predicted 27% of good clients as bad clients, but it falsely predicted only 1% of bad clients as good clients. The decision tree generated using the variables selected by the ChiSquaredVariableEval falsely predicted 22% of good clients as bad clients, but it falsely predicted 12% of bad clients as good clients. Finally, the decision tree generated using the variables selected by the ConsistencySubsetEval falsely predicted 27% of good clients as bad clients, but it falsely predicted only 2% of bad clients as good clients, thus producing the similar results as the Class CfsSubsetEval.

TABLE VII CLASSIFICATION MATRICES

	Predicted - good	Predicted - bad
Class CfsSubsetEval		
Observed - good	127 (64%)	2 (1%)
Observed - bad	54 (27%)	12 (6%)
ChiSquaredVariableEval		
Observed - good	105 (53%)	24 (12%)
Observed - bad	43 (22%)	23 (12%)
ConsistencySubsetEval		
Observed - good	126 (63%)	3 (2%)
Observed - bad	54 (27%)	12 (6%)

Source: Authors, using Entrepreneurship credit dataset

Figure 4 presents falsely predicted good and bad debtors with different variable selection approaches.

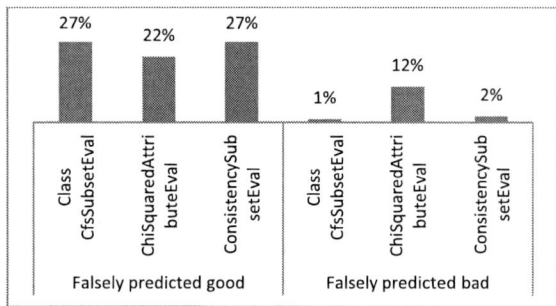

Figure 4 Falsely predicted good and bad debtors with different variable selection approaches (Source: Authors, using Entrepreneurship credit dataset)

VI. CONCLUSION

The main purpose of the paper is to compare the classification of banking clients regarding credit default through the analysis of different entrepreneurial variables using decision tree algorithms. Research results showed that variables selected by the algorithm Class CfsSubsetEval have the best results regarding the percentage of correctly classified instances. On the other hand, according to the percentage of bad debtors falsely predicted as the good ones, the decision tree generated using the variables selected by the ChiSquaredVariableEval is the worse. According to the criteria of the minimal percentage of falsely predicted bad debtors as good, the best approach was to use the decision tree generated using the variables selected by the Class CfsSubsetEval or the decision tree generated using the variables selected by the ConsistencySubsetEval. In addition, for financial institutions, especially for banks, the most valuable data are the data on prediction of bad debtors, and in our case two mentioned algorithms should be used. However, since the Class CfsSubsetEval generates a decision tree that is based only on the Variable Vision, it is prone to subjective mistakes, since this variable was estimated by a banking clerk. The ConsistencySubsetEval could be considered as more reliable, since it produces similar results as the Class CfsSubsetEval, and it is based on a larger number of variables. Most of the variables are related to 'what has been done' instead of 'who is doing it'. In other words, variables related to Entrepreneurial idea, Growth plan, and Marketing plan were more relevant than variables related to Personal characteristics of entrepreneurs, Characteristics of SME, Characteristics of credit program, and Relationship between the entrepreneur and a financial institution.

REFERENCES

[1] R.-S. Wu, C.S. Ou, H.-Y. Lin, S.-I. Chang, D.C. Yen, "Using Data Mining Technique to Enhance Tax Evasion Detection Performance," Expert Systems with Applications, vol. 39, no. 10, pp. 8769-8777, 2012.

[2] J.-T. Wei, M.-C. Lee, H.-K. Chen, H.-H. Wu, „Customer Relationship Management in the Hairdressing Industry: An Application of Data Mining Techniques," Expert Systems with Applications, vol. 40, no. 18, pp. 7513-7518, 2013.

[3] M.A.P.M. Lejeune, "Measuring the Impact of Data Mining on Churn Management," Internet Research: Electronic Networking Applications and Policy, vol. 11, no. 5, pp. 375-387, 2001.

[4] A.K. Choudhary, J.A. Harding, M.K. Tiwari, "Data Mining in Manufacturing: A Review Based on the Kind of Knowledge," Journal of Intelligent Manufacturing, vol. 20, no. 5, pp. 501-521, 2008.

[5] E.W.T. Ngai, Y. Hu, Y.H. Wong, Y. Chen, X. Sun, "The Application of Data Mining Techniques in Financial Fraud Detection: A Classification Framework and an Academic Review of Literature," Decision Support Systems, vol. 50, no. 3, pp. 559-569, 2011.

[6] L.C. Thomas, R.W. Oliver, D.J. Hand, "A Survey of the Issues in Consumer Credit Modelling Research", Journal of the Operational Research Society, vol. 56, no. 9, pp. 1006-1015, 2005.

[7] J. Quinlan, „C4.5: Programs for Machine Learning," San Francisco, Calif.: Morgan Kaufmann, 1992.

[8] L. Breiman, J. H. Friedman, R. A. Olshen, C. J. Stone, „Classification and Regression Trees," Belmont, Calif.: Wadsworth, 1984.

[9] B.W. Yap, S.H. Ong, N.H.M. Husain, "Using Data Mining to Improve Assessment of Credit Worthiness via Credit Scoring Models", Expert Systems with Applications, vol. 38, no. 10, pp. 13274-13283, 2011.

[10] S. Oreski, D. Oreski, G., Oreski, "Hybrid system with genetic algorithm and artificial neural networks and its application to retail credit risk assessment", Expert systems with applications, vol. 39, no. 16, pp. 12605-12617.

[11] P. I. Priya, D. K. Ghosh, "A Survey on Different Clustering Algorithms in Data Mining Technique" International Journal of Modern Engineering Research (IJMER), vol. 3, no. 1, pp. 267-274, 2013.

[12] E.W.T. Ngai, L. Xiu, D.C.K. Chau, "Application of Data Mining Techniques in Customer Relationship Management: A Literature Review and Classification," Expert Systems with Applications, vol. 36, no. 10, pp. 2592-2602, 2009.

[13] S. Strohmeier, F. Piazza, "Domain Driven Data Mining in Human Resources Management: A Review of Current Research," Expert Systems with Applications, vol. 40, no. 7, pp. 2410-2420, 2013.

[14] H.G. Patel, K. Sarvakar, "Research Challenges and Comparative Study of Various Classification Technique Using Data Mining", International Journal of Latest Technology in Engineering, Management & Applied Science, vol. 3, no. 9, pp. 170-176, 2014.

[15] Y. Marinakis, M. Marinaki, M. Doumpos, N. Matsatsinis, C. Zopounidis, "Optimization of nearest neighbor classifiers via metaheuristic algorithms for credit risk assessment," Journal of Global Optimization, vol. 42, no. 2, pp. 279–293, 2008.

[16] L.C. Thomas, "A Survey of Credit and Behavioral Scoring: Forecasting Financial Risk of Lending to Consumers," International Journal of Forecasting, vol. 16, no. 2, pp. 149-172, 2000.

[17] A. Lucas, "Statistical challenges in credit card issuing, "Applied Stochastic Models in Business and Industry, vol. 17, no. 1, pp. 83–92, 2001.

[18] X. Wu, V. Kumar, J.R. Quinlan, J. Ghosh, Q. Yang, H. Motoda, G.J. McLachlan, A.F.M. Ng, B. Liu, P.S. Yu, Z.-H. Zhou, M. Steinbach, D.J. Hand, D. Steinberg, "Top 10 Algorithms in Data Mining", Knowledge and Information Systems, vol. 14, no. 1, pp. 1-37, 2008.

[19] J.N. Crook, D.B. Edelman, L.C. Thomas, "Recent Developments in Consumer Credit Risk Assessment", European Journal of Operational Research, vol. 183, no. 3, pp. 1447-1465, 2007.

[20] D.F. Zhang, S. Leung, Z.M. Ye, "A decision tree scoring model based on genetic algorithm and K-means algorithm," In: Proceedings of the 3rd International Conference on Convergence and Hybrid Information Technology, Daejeon, Korea, pp. 1043–1047, 2008.

[21] A. Mahmoud, M. Pourzandi, K. Babaei, "Using genetic algorithm in optimizing decision trees for credit scoring of banks customers," Journal of Information Technology Management, vol. 2, no. 4, pp. 23–38, 2010.

[22] M.A. Hall, "Correlation-Based Feature Subset Selection for Machine Learning", Hamilton, New Zealand, 1998.

Brand communication in social media: the use of image colours in popular posts

L. Zailskaitė-Jakštė[*], A. Ostreika[#], A. Jakštas[†], E. Stanevičienė[#], R. Damaševičius[‡]

[*] Department of Marketing
[#] Department of Multimedia Engineering
[†] Department of Mechanical Engineering
[‡] Department of Software Engineering
Kaunas University of Technology, Kaunas, Lithuania
robertas.damasevicius@ktu.lt

Abstract - Recent scientific and theoretical studies defining brand communication in social media emphasize the significance consumer engagement in brand-related content. Popularity of brand messages and the reach of target audiences depends on consumer engagement in social media. Therefore, many business companies are seeking to increase an impact on consumers using social media analysis and consumer engagement technics. Usually, consumer actions such as likes, comments and shares in social media channels are used to estimate the popularity of brand posts. One of the factors, which has not widely analyzed before, is the impact of colors for popularity of visual brand-related posts. In this paper, we analyze the effect of colors for popularity of brand-related posts in social media. We analyze our own dataset of images collected from 35 most popular brand Facebook groups. Our results show that black, gray and brown colors were more often used in images of more popular brand-related posts.

I. INTRODUCTION

Social media has become an influential communication channel, where business companies can foster relationships and interact with customers. In 2016, about a half of social media users followed brands on different social media channels such as Facebook, Instagram, Twitter, etc. According to eMarketer [1], 91% of retail brands use two or more social media channels, and by 2017, social network ad spending will reach $36B, representing 16% of all digital ad spending globally.

The importance of visual perception in marketing and advertisement, including shapes, forms, colors, materials, textures, graphical elements and logos has long been acknowledged [2, 3]. One of the main factors of consumer attraction in the modern world is the color [4]. Psychologically, colors may attract attention, carry specific meanings and communicate information [5], evoke certain emotions or motivations [6], and stimulate our memory, thought and experiences [7]. Colors operate via sensory and cognitive mechanisms. In the sensory mechanism, color helps distinguishing an object from its background or a set of other similar objects. In the cognitive mechanism, color helps assign a specific meaning to the object [8].

Colors impact consumer behavior directly by stimulating visual cortex and the effects of colors

determine attitude, feelings, affects, emotional reactions and behavioral intentions of consumers of consumers towards colorful items, e.g., packaging, ads, websites [9]. Targeting of user groups by careful deployment of colors can promote attitudes, feelings and moods and thus differentiate products. For similar products and services, their visual images (logos, posters, ads) have become an important factor in a consumer's shopping decision-making process [10, 11]. The decision is strongly influenced by color components of imagery used [12]. For example, Lindgaard *et al.* [13] showed that users' first impressions are constructed in about 50 ms and about 62-90% of the assessment is based on colors alone [14].

One way of enhancing the engagement and popularity of corporate brand pages such as Facebook brand pages is to include images or animation. Previous research has demonstrated that vivid colors have been effective at improving attitudes to web-based ads [15] and increasing popularity of brand fan pages [16]. Images are the most successful in terms of received likes in Facebook [17]. However, there is still a lack of studies evaluating brand post popularity in relationship with colors. Management-oriented studies about brand post popularity are mainly descriptive as they have no theoretical background and do not formally analyze which activities actually may influence brand post popularity [16].

In this paper, we study the effect of colors and their combinations in images attached to brand-related posts on the popularity of posts. The aim is to identify colors that social network users may find appealing and to determine whether some colors favor popularity of social network posts as well as the engagement of users. The contribution of the paper is a method for identifying the association between image colors and image popularity. To support our claims we have carried out an experimental study of color use in images of corporate Facebook posts, and have analyzed whether the frequency of the use of colors in images posted in social media is related to post popularity (i.e., the number of comments, shares and likes).

The structure of the remaining parts of the paper is as follows. The methodological backgrounds of color research are discussed in Section II. The proposed method is presented in Section III. The results of experiments are presented in Section IV and discussed in Section V. Finally, the conclusions are given in Section VI.

II. METHODOLOGICAL BACKGROUND OF COLOR RESEARCH

A. Colors and their psychology

People experience psychological change when they are in contact with different colors [5]. Colors can stimulate, excite, and invoke different emotions. Each color may result in a different psychological reaction. For example, a color with a long wavelength (such as red) has a stimulus effect, and a color with a short wavelength (such as blue) has a comforting effect [9]. Colors also can be mapped to higher social qualities, e.g., orange indicates friendliness, and grey specifies professionalism [18]. Producing desired emotions through the target user oriented use of color is a key challenge in marketing. But first, one needs to understand how the colors transmit specific traits and cause emotional reactions, which have impact on how we judge the environment and take decisions.

In the first color theory, Goethe back in 1808 proposed the concept of a color circle that separates colors into positive and negative parts. Positive colors (yellow, orange, and red) represent activity and ambition, while negative colors (blue and purple) represent passiveness and obedience [18]. The theories of colors has been summarized by Shirley Willett's color codification [18], which has interpreted the colors in terms of positive traits, emotions and negative traits. Wright and Rainwater [19] categorized 48 color-emotion scales into six factors: happiness, showiness, forcefulness, warmth, elegance, and calmness. The ability of colors to invoke different emotions can be used to create a variety of experiences for consumers, and these can be described as flow, presence, immersion, and fun. People also take advantage of color terms to strengthen their messages and convey emotions in natural interactions [20]. However, little research has been done on the impact of colors in social networking and only a handful of researchers have conducted studies on this topic in recent years (see, e.g., [21]).

B. Consumer engagement and colors in brand communication

There is a theory in design that people will respect and care for surroundings and objects that they find beautiful, and will disdain and neglect, even damage, those that are unappealing [22]. The success of brand post in attracting readers depends in large part on the appeal of its design and colors. Consumers rate color as one of three most important factor in making a decision to purchase [23]. In some cases a good visual design of a website, with respect to colors, resulted in trust, loyalty and satisfaction [24] as well as raised positive attitudes towards products in terms of their perceived efficiency and usability [25]. Colors also contribute to understandability and efficient use of software tools [26].

With expansion of self-service stores, the color of goods and their packaging has become one of the main factors of customer attraction in the modern world [4]. The effects of colors applied in packaging design similarly could be applied to increase brand post popularity as well. Unique icons, distinctive graphics and easily recognizable color palette can be combined to create instantly recognizable brand that subconsciously prompts the "buy"

response from consumers. Because of psychological implications of color, companies often deliberately choose colors that are associated with certain domains or qualities to elicit a desirable emotional response. Examples include organic products marketed using green; gold or silver used for luxury goods, children toys having bright primary colors of red, blue and yellow, and high tech products using black color. Some studies claimed that color impresses consumers to recognize corporate brands and, consequently, improve purchase and profit [18].

In a virtual environment, colors may influence the amount of time spent on the site and items memorized and retrieved [24]. The latter is extremely important for business content in social media, which is steadily increasing from year to year, but is facing the decreased engagement of consumers [27]. Therefore, business companies, which use social media to advertise their products and attract customers, face a challenge to engage consumers into interaction with brand message and to receive consumers reaction (likes, comments, shares and other actions), while at the same time compete with other social media channels. According to empirical studies visual content especially images make an impact on consumer engagement as compared to other forms of posts [28], can stimulate the interest and increase purchase intentions, and determine product choices [29].

C. Color representation in images and analysis

In color image processing, various color models are used. RGB is a non-uniform color space that is universally accepted by the image processing community for color representation. RGB uses three additive primary colors, red, green and blue, which are added together to reproduce other colors. RGB color model refers to the biological processing of colors in the human visual system [30].

However, perceptually uniform color spaces, such as HSV, better correspond to the human perception of color similarity as the measured color differences are proportional to the human perception of such differences. In the HSV model, the luminous component (brightness) has been separated from color-only components (hue and saturation). Hue controls the perceived qualitative experience of; saturation is an expression of the relative purity or the degree to which a pure color is diluted by white light, and value is defined as the largest component of a color [29]. Furthermore, HSV color space is better suited for extracting features that can be used both for image segmentation and color histogram generation, which are important in content-based image retrieval, segmentation and clustering.

III. METHOD

To identify the most popular colors in social media post images we propose the following algorithm:

Inputs: let I be a set of images, and $L_i, i \in I$ be the popularity value of an image (the number of likes, shares or comments), and N be the number of desired colors.

Outputs: colors C_j, their probabilites p_j, consistency (p-value), and agreement (Jaccard distance).

1) Resize all images to 128x128 resolution.

2) From each image, sample colors of random pixel with probability of p, here

$$p = \frac{\max(L)}{\sum_I L_i},$$

3) Cluster the colors of random pixels in 3D space (representing the red, green and blue channel values) into N clusters.

4) Calculate the centroids of each cluster as $C_j, j = 1..N$

5) Calculate the number of pixels belonging to each cluster as n_j

6) Let p_j be the probability of a color C_j defined as follows:

$$p_j = \frac{n_j}{\sum_C n_j}$$

7) To evaluate the consistency of our result, we perform a random permutation test, where the popularity vector is bootstrapped (randomly permuted) and the results compared. The *p*-value of the test is computed by comparing color probabilities of the original probability vectors with color probabilities obtained using the permuted popularity vector. The *p*-value > 0.95 indicates the statistical significance of the result.

8) To evaluate the agreement of the results obtained using the like, share and comment data, we have used the Jaccard distance between two color vectors C_1 and C_2:

$$J(C_1, C_2) = \frac{\|C_1 - C_2\|}{\|C_1 + C_2\|}$$

The Jaccard distance between three color vectors C_1, C_2 and C_3 is calculated as the mean of Jaccard distances:

$$J(C_1, C_2) = \frac{1}{3}(J(C_1, C_2) + J(C_1, C_3) + J(C_2, C_3))$$

IV. DATA AND RESULTS

A. Data

We have selected 35 most popular globally brands on Facebook (according to statistics provided by website www.socialbakers.com). These brands represent different domains such as technologies (BlackBerry, Intel, Microsoft Lumia, Play Station, Samsung Mobile, Samsung Mobile USA, Skype, Huawei Mobile, KitKat, Sony Mobile, Windows), food and drinks (KFC, McDonalds, Nescafe, Pizza Hut, Pringles, Starbucks, Oreo, Subway, Pepsi, Red Bull), clothing and shoes (Zara, adidas, adidas Originals, Converse, H&M, Nike Football, Victoria's Secret), beauty (L'Oreal Paris, Dove), retail and e-commerce (Amazon, Walmart), entertainment (iTunes, Netflix) and automobiles (Volkswagen).

We have used a custom script written in Perl language, which used Facebook Graph API to collect images from the brand posts and their popularity information (likes, shares, comments) automatically.

In total, we have collected 801 images from posts made from 2010/11/29 to 2017/02/05. To reduce the influence of outliers, we have performed the removal of outliers, i.e., removed images, whose popularity was more than 3 standard deviations off the mean of the corresponding popularity metric. This procedure has resulted in 785 images. To normalize the popularity data, we have divided the number of likes, shares and comments per post by the total number of likes, because the total number of comments and shares per page is not provided by Facebook API. The statistical characteristics of analyzed images and their popularity characteristics are presented in Table I and Table II, respectively.

TABLE I. STATISTICAL CHARACTERISTICS OF DATA

Feature	Min	Max	Mean	Std	Skewness	Kurtosis
Total likes	24,242,400	68,822,434	34,159,000	9,609,200	1.51	5.87
Likes per post	3	95528	2281	6451	8.63	102.0
Shares per post	0	31725	409.4	2315	10.2	119.2
Comments per post	0	25	17.0	9.66	-0.61	1.654

TABLE II. STATISTICAL CHARACTERISTICS OF IMAGE DATA (GRAND MEANS)

Feature	Mean	Std	Skewness	Kurtosis	Color of mean
Red	125.9	57.6	-0.007	5.494	
Green	120.6	56.1	0.088	5.639	
Blue	118.8	55.9	0.137	5.760	
Hue	0.352	0.235	0.641	16.05	
Saturation	0.323	0.202	1.224	12.72	
Value	0.556	0.215	-0.212	5.649	

The mean values of all images in RGB and HSV are very similar, both representing light blue color. The distribution of values was evaluated using normality based on skewness and kurtosis [31]. Skewness is a measure of the lack of symmetry. Kurtosis is a measure of outlier. Note that skewness and kurtosis is much higher in the HSV space than in the RGB space. For sample sizes greater than 300, an absolute skew value larger than 2 or an absolute kurtosis larger than 7 may be used as reference values for determining substantial non-normality. In our case, the popularity data is not normal, the RGB data of images can be considered as normally distributed, and the HSV data can not be considered as normally distributed. Image representations in the HSV space are very asymmetric and contain many outliers. Therefore, hereinafter we use the RGB color space for color representation and analysis.

B. Procedure

We use MATLAB Version 8.6.0.267246 (R2015b) running on Microsoft Windows 10 Enterprise Version

10.0 with Java 1.7.0_60-b19 64-Bit Server VM. For color clustering, we used the k-means clustering algorithm with 10 replicates. For bootstrapping, we used 1000 repetitions.

C. Results

We have performed the clusterization of image colors into 5 clusters. The number of clusters was established heuristically. A smaller number of clusters leads to mixed colors of cluster centroids, which are not present in the original images. The results are summarized in Table III. Color names are given according to Bang V2 tool (http://www.perbang.dk/). The name of most similar color as well as the values of similarity to the named color are given.

TABLE III.   RESULTS OF COLOR CLUSTERIZATION

Cluster	Centroid			Color name	Similarity
	Red	Green	Blue		
Likes					
1	31	26	27	20 Reddish brownish black	98%
2	239	238	238	1 Pinkish white	96%
3	173	166	157	125 Yellowish gray	96%
4	65	97	107	385 Dark grayish cerulean	94%
5	154	88	59	44 Moderate vermilion	95%
Shares					
1	177	163	145	69 Grayish brown	95%
2	40	96	135	396 Strong cornflower blue	95%
3	27	23	27	585 Magentaish black	99%
4	237	237	237	1 Pinkish white	95%
5	120	81	65	58 Dark tangelo	91%
Comments					
1	175	162	146	125 Yellowish gray	95%
2	45	108	134	384 Strong cerulean	92%
3	29	25	27	585 Magentaish black	99%
4	235	235	234	1 Pinkish white	95%
5	124	82	68	58 Dark tangelo	90%

Figure 1 shows the probability that a random pixel, sampled from an image at the normalized frequency of its likes will belong to a cluster. Each cluster is represented by a color-filled bar with the color of its centroid (sorted by decreasing probability value).

The respective probability calculated by sampling pixels with respect to the number of shares and number of comments of a post is given in Figures 3 and 5, respectively.

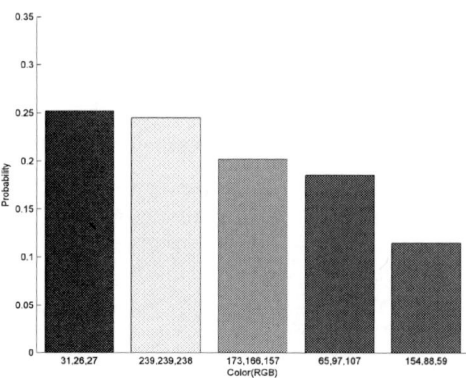

Figure 1.   Probability of top-5 colors in images with respect to the number of post likes

Figure 2 shows the p-value that represents the confidence of our results. The p-value was calculated by applying bootstrapping on randomly permuted like values and comparing the obtained probability with the original probability value. The p-value evaluates probability that the obtained result has not been obtained by chance, while p-value > 0.95 is considered as statistically significant. Again, the color of each bar represents the color of the centroid of each respective cluster (sorted by the decreasing p-value).

The p-values calculated with respect to the number of shares and the number of comments of posts are given in Figures 4 and 6, respectively.

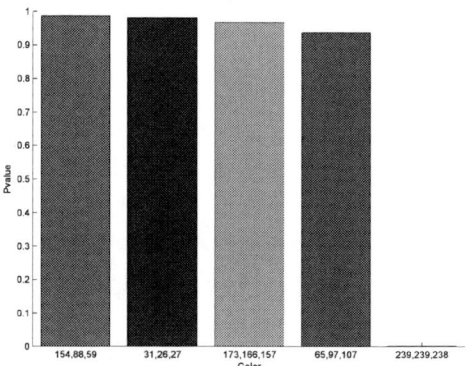

Figure 2.   P-value of top-5 colors in images with respect to to the number of post likes

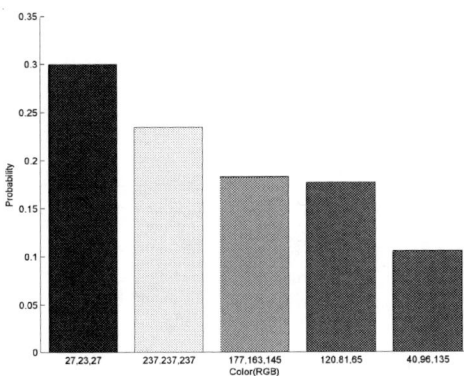

Figure 3. Probability of top-5 colors in images with respect to to the number of post shares

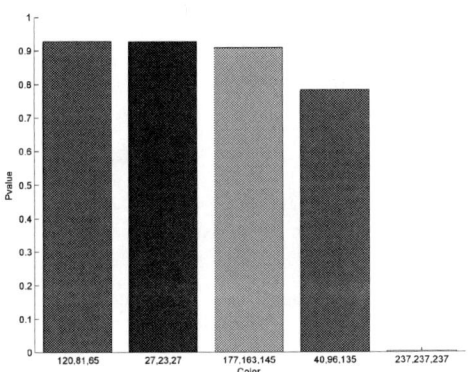

Figure 4. P-value of top-5 colors in images with respect to to the number of post shares

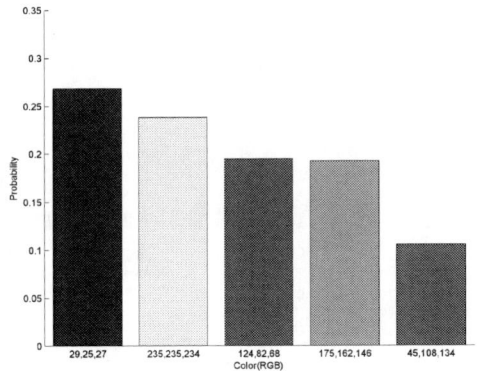

Figure 5. Probability of top-5 colors in images with respect to to the number of post comments

The mean Jaccard distance between the color centroids of color clusters obtained using the like, comment and share data was only 0.03, which demonstrates an excellent agreement.

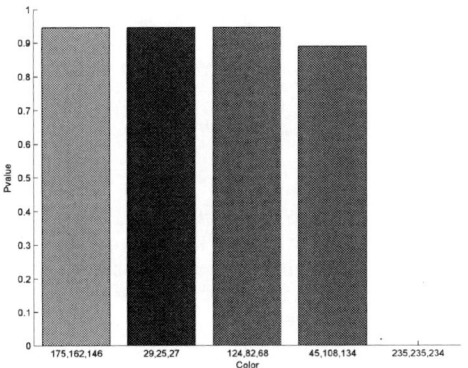

Figure 6. P-value of top-5 colors in images with respect to to the number of post comments

V. EVALUATION AND DISCUSSION

Our study has identified a set of colors, which are more likely to be used in images posted in social media. After analyzing 785 images from 35 brand Facebook groups, we have identified the statistically significant prevalence of four colors, which are shades of black, brown, gray and blue (navy).

We have evaluated the probability of colors found in the popular (in terms of the number of likes, shares and comments received) posts with images. The results show that more liked post images contain colors that are close to black ($p = 0.25$; $p\text{-}value = 0.98$), dark gray ($p = 0.20$; $p\text{-}value = 0.96$), blue ($p = 0.19$; $p\text{-}value = 0.94$), brown ($p = 0.11$; $p\text{-}value = 0.98$); most shared images contain colors that are close to black ($p = 0.30$; $p\text{-}value = 0.93$), light brown ($p = 0.23$; $p\text{-}value = 0.91$); brown ($p = 0.18$; $p\text{-}value = 0.93$), and most commented contain colors that are close to black ($p = 0.26$; $p\text{-}value = 0.95$), dark brown ($p = 0.19$; $p\text{-}value = 0.95$), light brown ($p = 0.19$; $p\text{-}value = 0.95$). Note that light gray (or white) in all cases is not statistically supported (has low p-value), maybe due to its common use as background color in images.

The results are consistent with several studies, which, e.g., claim that blue and black are among the most preferred colors amongst Generation Y (18-24 years) who are also the most active group of social [32]; blue is also the color most favored by both males and females [33]. Identified colors also have a strong achromatic component (black, gray), which psychologically are associated with style, wealth and elegance [34]. Blue color is associated with trustworthiness [35]. Cooler colors, such as blue or gray, are generally considered to have a calming effect [36]. While brown is not a popular color, it is often used as the color of a popular product (e.g., drinks or clothes).

Nethertheless we do not claim that the intentional use of colors identified by our research in posted images would increase the popularity of social media posts themselves. That hypothesis would require a separate research of its own, including the analysis of context of the use of color, which has not been performed here. Also note the limitations of our study as most of the brands analyzed here have their roots in Western countries and also mainly target Western audience, therefore, the results of the study may be biased towards colors preferred in Western (European and North American) culture.

VI. CONCLUSIONS AND FUTURE WORK

In this paper, we have empirically investigated the use of colors in brand post images and identified colors, which are used more often in popular brand posts. The identified colors (black, gray, brown, blue) are consistent with colors, which have been identified by other cultural and marketing studies. We also have demonstrated statistical validity using bootstrapping and good agreement between the results, obtained using like, share and comment data.

Further research may enrich our initial findings about the factors that determine the popularity of brand posts with a wider investigation of image descriptors (including those, which describe the textures and shapes) that may influence the behavior of social media users. We also plan to expand our dataset to include posted images from other cultural environments (such as Asian, African) to study possible cultural differences between brand image popularity. Furthermore, we have gathered data from the brand pages of only one social networking site. It could be interesting to replicate this research for other social networking sites such as Twitter or Instagram.

REFERENCES

[1] eMarketer, "Social Network Ad Spending to Hit $23.68 Billion Worldwide in 2015", 2015 http://www.emarketer.com/Article/Social-Network-Ad-Spending-Hit-2368-Billion-Worldwide-2015/1012357

[2] T. M. Karjalainen, "It Looks Like a Toyota: Educational Approaches to Designing for Visual Brand Recognition", International Journal of Design, 1(1), 67-80, 2007.

[3] K. L. Keller, Strategic brand management. Building, measuring and managing brand equity (3rd ed.), New Jersey: Pearson, 2008.

[4] M. Babolhavaeji, M.A. Vakilian, and A. Slambolchi, "Color Preferences Based On Gender As a New Approach In Marketing", Advanced Social Humanities and Management 2(1):35-44, 2015.

[5] A. J. Elliot, "Color and psychological functioning: a review of theoretical and empirical work," Front Psychol. 6:368, 2015.

[6] P. Sorokowski, and M. Wrembel, "Color studies in applied psychology and social sciences: An overview," Polish Journal of Applied Psychology, 12(2), 9–26, 2015.

[7] A. E. Crowley, "The Two-Dimensional Impact of Color on Shopping," Marketing Letters, 4(1), 59-69, 1993.

[8] M. Ghaderi, F.J. Ruiz, and N. Agell, "Understanding the impact of brand colour on brand image: A preference disaggregation approach," Pattern Recognition Letters 67: 11-18, 2015.

[9] P. Valdez, and A. Mehrabian, "Effects of color on emotion," Journal of Experimental Psychology: General 123, 394–409, 1994.

[10] B.J. Babin, D.M., T. Hardesty, A. Suter, "Color and shopping intentions: The intervening effect of price fairness and perceived affect", Journal of Business Research, 56(7), 541-551, 2003.

[11] L.I. Labrecque, and G.R. Milne, "Exciting red and competent blue: the importance of color in marketing," Journal of the Academy of Marketing Science, 40(5), 711-727, 2012.

[12] W. Kang, S. Qin, and Q. Zhang, "Computer-Aided Color Aesthetic Evaluation Method Based on the Combination of Form and Color," Mathematical Problems in Engineering, vol. 2015, Article ID 153103, 8 p., 2015.

[13] G. Lindgaard, G.J. Fernandes, C. Dudek, and J. Brownet, "Attention web designers: you have 50 ms to make a good first

impression!," Behaviour and Information Technology 25, 115–126, 2006.

[14] S. Singh, "Impact of color on marketing", Management Decision, 44(6), 783 – 789, 2006.

[15] D.R. Fortin, and R.R. Dholakia, "Interactivity and Vividness Effects on Social Presence and Involvement with a Web-Based Advertisement," Journal of Business Research, 58(3), 387-396, 2005.

[16] L. de Vries, S. Gensler, and P.S.H. Leeflang, "Popularity of Brand Posts on Brand Fan Pages: An Investigation of the Effects of Social Media Marketing," Journal of Interactive Marketing, 26(2), 83-91, 2012.

[17] T. Trefzger, C. Baccarella, C.W. Scheiner, and K.-I. Voigt, "Hold the Line! The Challenge of Being a Premium Brand in the Social Media Era", LNCS 9742, 461-471, 2016.

[18] W.-L. Chang and H.-L. Lin, "The impact of color traits on corporate branding," African Journal of Business Management, 4(15), 3344-3355, 2010.

[19] B. Wright, and L. Rainwater, "The meanings of color," J Gen Psychol, 67:89 –99, 1962.

[20] S. Volkova, W. B. Dolan, and T. Wilson, "Clex: A Lexicon for Exploring Color, Concept and Emotion Associations in Language", Proc. of the 13th Conference of the European Chapter of the Association for Computational Linguistics, 306-314, 2012.

[21] M.N.A. Corvette, Consuming Colour: A Critical Theory of Colour Concerning the Legality and Implications of Colour in Public Space. Doctoral thesis, Goldsmiths, University of London, 2016.

[22] L. Holtzschue, Understanding Color: An Introduction for Designers, 4th Edition, 2011.

[23] S.C. Mason, T.W. Starman, R.D. Lineberger, and B.K. Behe, "Consumer preferences for price, color harmony and information level of container gardens," HortScience 42(4):892, 2007.

[24] N. Bonnardel, A. Piolat, L. Le Bigot, 'The impact of colour on website appeal and users' cognitive processes," Disp. J. 32(2), 69–80, 2011.

[25] G. Nordeborn, The Effect of Color in Website Design: Searching for Medical Information Online. MSc thesis, University of Lund, 2013.

[26] I. Plauska, and R. Damaševičius, "Usability Analysis of Visual Programming Languages Using Computational Metrics," IADIS Interfaces and Human Computer Interaction (IHCI), 63-70, 2013.

[27] TrackMaven, "Social Media Impact Report: B2B Industry Edition," 2016. http://pages.trackmaven.com/rs/251-LXF-778/images/TrackMaven-B2BIndustry-Report.pdf

[28] D.-H. Kim, L. Spiller, M. Hettche, "Analyzing media types and content orientations in Facebook for global brands", Journal of Research in Interactive Marketing, 9(1), 4 – 30, 2015.

[29] O. Akcay, "Marketing to Teenagers: The influence of Color, Ethnicity and Gender," International Journal of Business and Social Science, 3(22), 10-18, 2012.

[30] M. Loesdau, S. Chabrier, and A. Gabillon, "Hue and Saturation in the RGB Color Space," Proc. of the 6th International Conference, ICISP 2014, LNCS 8509, 203-212, 2014.

[31] H.-Y. Kim, "Statistical notes for clinical researchers: assessing normal distribution (2) using skewness and kurtosis," Restor Dent Endod. 2013 Feb; 38(1): 52–54, 2013.

[32] R. Müller, A. L. Bevan-Dye, W. P. Viljoen, "Generation Y Students' Product Colour Preferences," Mediterranean Journal of Social Sciences, 5(21), 2014.

[33] Y. Ling, and A.C. Hurlbert, "A new model for color preference: Universality and individuality," In Color and Imaging Conference 2007(1), 8-11, 2007.

[34] G. Keskar, "Color Psychology and its Effect on Human Behavior. Paintindia," 60 (5), 61, 2010.

[35] S. Lee, and V.S Rao, "Color and store choice in electronic commerce: The explanatory role of trust," Journal of Electronic Commerce Research, 11, 110-126, 2010.

[36] J. Eckstut, and A. Eckstut. The secret language of color: Science, nature, history, culture, beauty of red, orange, yellow, green, blue & violet. Black Dog & Leventhal, 2013.

Recommender System Based on the Analysis of Publicly Available Data

Goran Antolić*, Ljiljana Brkić*

*University of Zagreb, Faculty of Electrical Engineering and Computing, Zagreb, Croatia
goran.antolic@fer.hr
ljiljana.brkic@fer.hr

Abstract - A recommender system is a software system aimed to make recommendations. To be able to do that, recommender system feature several components, such as: data collection and processing, recommender model, recommendation post-processing and a user interface. Recommender systems apply one or the combination of few of the recommendation techniques. In this paper we present recommender system developed to provide users with recommendations in accordance with their interests in different domains. We deduce user interests based on his activities and posts in social network. Social network used as a source of information on user (Facebook) provides Open API allowing access to the information about the user collected on the social network. Thanks to this data we are overcoming the so-called "cold start" problem and building user profile. A recommender system is commonly associated with only one domain, while the recommender system described in this paper is able to generate recommendations from different domains (movies and music). In addition to recommendations related with the specific domain, our system is able to recommend the web articles (unstructured text), relevant to the user that may belong to more than one category of interest.

I. INTRODUCTION

The unrestrainable growth in the amount of available digital data and the increasing number of Internet user hinders timely access to items of interest on the Internet. Information retrieval systems, such as search engines, have somewhat solved this problem but issues such as personalization and item prioritisation are yet to solve. In an attempt to solve these issues recommender systems were developed. Recommender systems are software systems that filter relevant information out of large amount of available information according to user's preferences and interest. They play important role in business on the World Wide Web. Recommender system's goal is to provide the particular user with recommendations based on his profile. To be able to make personal recommendations, recommender systems maintain users' profiles that take into account their preferences. The profile serves to filter items that will be presented as recommendations assuming that they match user's preferences. Items can be news, movies, music, books, restaurants, scientific publications, webpages, etc. Service providers on the Internet and customers both benefit from recommender system. They are widely used by online shopping sites [1] [2] and internet streaming media services [3] [4] [5]. Social

networking services may suggest future connections with people who you may know or would possibly like to become your friend, groups you might like to follow, jobs or companies you might be interested in [6] [7]. There are also interests-based social networking websites that provide users with recommendations of movies, music, restaurants, food, articles and other items based on users' ratings [8] [9]. When it comes to the problem of a document recommendation in a large digital library like e.g. in research-paper library, recommender system can be of great help. In a higher education and research institutions, students, academic staff and other researchers need to find the papers most relevant to their research topics. Hence, finding the right papers to read becomes a relevant part of their academic lives. A research paper recommender system will benefit these users by recommending newly published papers that may be of interest to them or papers related to their previous research affinities. Despite the great need, only a few recommendation systems exist for scientific papers [10] [11] [12].

Taking all this into consideration the need for effective recommender systems exist in many aspects of our lives. Efficient recommender system have to implement effective and accurate recommender technique. Three main categories of recommender systems techniques are: collaborative filtering (CF), content-based filtering (CBF) and hybrid approach which combines the previous two methods [13].

The collaborative filtering and the content-based filtering essentially differ in their concept. The basic idea of CBF is that users will be interested in items that are similar to items they previously liked. While, the idea of CF is that users might like items that are liked by his peers. Yet, these ideas are rather fuzzy and are implemented using different approaches.

A. Collaborative filtering

A recommender system using collaborative filtering, recommends items by finding other users with similar taste. The fundamental assumption is that similar users (peers) have similar interests on similar items. An active user is recommended with the items that are most liked by his peers. The correlation between users is usually measured on the basis of common items they have rated. Consequently, the accuracy of collaborative filtering-based systems depend on user willingness to rate the items.

Providing ratings requires extra effort and users tend to avoid that, so in many systems many related ratings will not be available for the prediction. This missing user-item ratings cause two types of problems: the data sparsity problem and the cold start problem.

The cold start problem may occur in three situations [14]: a new users, a new items, and a new communities or disciplines. A new user has not rated any item, so the system cannot find his peers and hence cannot provide recommendations. A new item has not been rated by a single user, so it cannot be recommended. In a new community, there are no user-item ratings, so no recommendations can be made and consequently, the users are not motivated to rate items. Although cold start problem affects CF more, it is present in the CBF recommender systems too. A possible approach to overcome the cold start problem is proposed and described later in this paper.

An algorithms for collaborative filtering measure the similarity between users (user-based CF) or the similarity between items (items-based CF) or both. To generate recommendations, different methods for calculating similarity between items or users are used. Some of widely used measures are Cosine similarity, Pearson correlation and Jaccard index [15].

The collaborative filtering technique is implemented in many recommender systems developed in academia and industry. Just two examples are the book recommender system from Amazon [2] and the Jester system that recommends jokes [16].

B. Content-based filtering

In the content-based filtering (CBF) the interests of users are deduced from the items that users interacted with in the past. These items are commonly textual like webpages, scientific publications, books etc. User's preferences are concluded from the items user downloaded, bought or tagged. A systems that deal with textual documents, commonly use structured representation of the documents that originated with text search systems [17]. This structured representation includes words contained in text. Instead of a words in their original form, the root forms of words, created through a process called stemming [18], are used. The goal of stemming is to represent all words having the common meaning (e.g. organize, organizes, and organizing) with a single term reflecting that common meaning. The relevance or importance of the term in the text is presented with the real number usually called the tf-idf weight (term-frequency-inverse document frequency) [17]. Some recommender systems implementing the content-based filtering use non-textual characteristics like a writing style [19] or a layout information [20].

While developing the recommender system presented in this paper we were guided by the idea of using data provided by the public services to collect information about: (1) the users of the system and (2) on the possible recommendations. We use these data as a starting point in building our own system of recommendations and evaluation of offered recommendations. As a source of information on users, ubiquitous and extremely popular

social networks, have imposed for two reasons: (1) based on users comments, tags, posts, likes and expressions of emotions regarding other people's publications, a lot about users preferences can be concluded and (2) almost all social networks (Facebook, Twitter, Instagram etc.) provide an application programming interface (API) allowing access to the data about the user collected on that particular social network. From all the data we can get from the social network API we take over only the data approved by the user. Also, we are interested only in the data from those categories for which our system is able to generate recommendations.

Which publicly available data sources, in the context of generating a recommendations, make sense to use, depend on the domain for which the recommender system makes recommendations. For now, we focus on recommendations for a movie and music domain, so that's why we chose The Movie Database API [21] and Last.Fm API [22] as a starting point for recommendations.

The main contributions of this paper are: a) the model of building user's profile based on (i) Facebook API data to overcome cold-start issue and on (ii) user's ratings of recommended items for tweaking and improving the profile; b) the prototype for recommending text-based content based on matching concepts between user profile and unstructured text.

II. RELATED WORK

Using a social media information to feed recommender system with the data is present in some previous papers. In [23] authors calculate similarities between social network users based on data obtained from Facebook API. They used acquired data to recommend a tourist attractions to users they consider to have similar taste. The authors of [24] also used data extracted from a real online social network to overcome the data sparsity and the cold-start issues.

A recommender systems for the movie and music domain are subject of many research papers. Some dealing with a movies recommendation are based on movie genre correlations [25], while others generates recommendations using various types of data - about users, the available items, and previous transactions [26]. Our recommender system leverages the existing publicly available recommendations APIs for the movie and music domain to get initial set of recommendations for the given genre or author. We further enrich that data based on users feedback and thereby improve subsequent recommendations.

III. THE SYSTEM ARCHITECTURE AND TECHNLOGIES USED

Figure 1 presents the overall architecture of the software system developed. Rectangles represent program modules, ellipses data components and cloud callouts serve to introduce and describe either outer services or input/output data. Using data gathered from social networks, the user profile generator, creates the initial user profile for each category for which the user has expressed interest. We support several types of sources for recommendation: an application programming interfaces (API), a web links from RSS feeds and a web links found in process called web crawling.

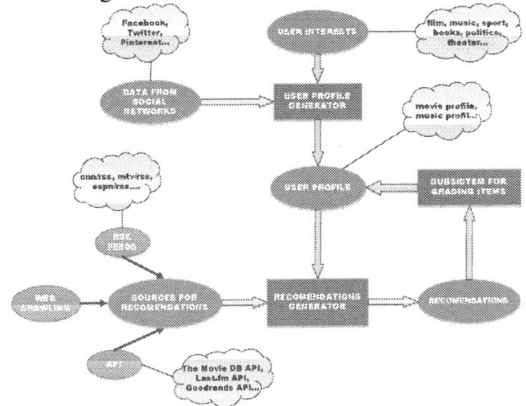

Figure 1 Architecture of the software system

Consulting these sources, a recommendations generator finds potential items for recommendation and calculate score using the expression presented later in the paper. The score is a numeric value that represents the relevance of the individual item for a specific user profile. The higher the score the greater the likelihood that the item will be of interest to the user. The subsystem for items grading enhances the system with the active learning component.

The user profile is updated based on the grade user has given to the recommended item. Updating a user profile affects the future recommendations. If the user rated an item high, the more likely similar items will be recommended to him in the future. For example, if the user has given the high grade to film starring Viggo Mortensen, the bigger will be the probability that the system will, in the future, to the same user recommend content (movies or articles) that includes Viggo Mortensen. And vice-versa.

All data pertaining to system configuration are stored in a relational database, Microsoft SQL Server, while the data on user profiles and recommendations are stored in the NoSQL database RavenDB. The relational database proved to be a good choice for the system's parametric data because the data occurring is structured, of simple type and the amount of data is not overwhelming. Data about user profiles and recommendations we store in NoSQL system because of the expected large number of users and improper scheme that is hard to describe and manage using the relational data model.

Figure 2 presents the relational data model holding data about: the interest categories (*category*); the web sources (*webSource*) which can be of different type (*webSourceType*) and can have different parameters (*endPoint*) for a different categories and response types (*responseType*); the users (*user, userRole*), the categories they are interested in (*userCategory*) and their social network accounts (*socialAccount*) along with permissions to retrieve data associated with their social network account (*socialAccountPermission*).

The system collects information about the user or the items for a recommendation by sending HTTP requests to a specific URL. The **endPoint** holds the exact location-URL (**path**) on the data source. The answer to the request

can be in various forms eg. json, xml or txt (*responseType*).

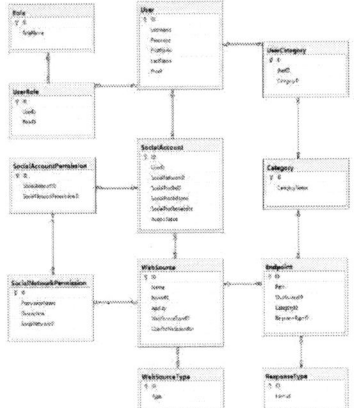

Figure 2 System configuration data - data model

The repository stored in a RavenDB contains several collections: t*User* (the basic data about user along with the relationships to category profiles), *CategoryProfile* (each user have separate profile for each category he is interested in), *DbMovies* (data about movies), *DBArtists, Recommendations* (for a specific user), *UserRatings* (user's grades for the recommendations).

IV. CREATING USER PROFILE

To create the initial user profile, our recommender system uses data taken from the Facebook API. Upon the analysis of the most popular social networking APIs (Facebook, Twitter, Instagram, LinkedIn) we came to the conclusion that the Facebook offers best quality data for user profile construction. The Instagram and LinkedIn APIs offer only the basic user information such as name, surname and email. As such, they are not sufficient for the construction of a general user profiles that will present user in different domains (categories). The recommender system presented in this paper, out of all the data that can be acquired from Facebook profile, takes over only the data explicitly permitted by the user and relevant to the domains of recommendations. Therefore, we use Facebook Login for authentication. Data about the user is retrieved in form of JSON document gathered as a result to HTTP GET request to a particular endpoint. The most valuable data about the user's preferences are the Facebook pages marked as "Like". Each page belongs to a category designated by the Facebook page administrator who is, also, in charge for definition of the fields (attributes) that describe the particular category. Data from a different categories have different attributes that describe the page. For example, some of the attributes describing movies (page category "Movie") are: title, genre, list of actors, producer, director, brief description and so on.
Example: Suppose that user with id equal to 575480316 marked with "Like" movie having title "Captain Fantastic". The following HTTP request is sent to Facebook API:

```
/{575480316 }?fields=
        likes{id, category, name, genre, starring,
            directed_by, id}
```

1381

Result will be JSON file containing the following segment:
```
{ ...
    "category":        "Movie",
    "name":            "Captain Fantastic",
    "genre":           "Comedy, Drama",
    "starring":        "Viggo Mortensen,George ...",
    "directed_by":     "Matt Ross",
    "id":              "74089565764"
}
```

For each category of interest that the user follows (table *userCategory*), the system generates a separate category profile because different domains/categories are described with different concepts. Each category profile contains a number of list of classes (representing concepts) describing that specific domain. For the movie domain it may be list of movie titles, actors, genres, directors etc. Each concept in a category profile is described by three attributes: *Name*, *ExternalId* and *Rank*. *ExternalId* is external unique identifier that serves as a link to the item in outer system. *ExternalId* for the movie is a unique movie identifier in The Movie Database, where more detailed information about the movie can be found. For music artist, album or song attribute *ExternalId* is a reference to an item in an open music encyclopedia MusicBrainz. Attribute *Rank* is a numeric value that represents the relevance of the observed concept for a specific user profile. The initial value for the *Rank* is set to value 1 for each concept. Later the *Rank* will be updated as a consequence of user's grading on recommended items. Grading recommendations is described in section VI.

The following is the example of JSON document from the collection *CategoryProfile* stored in RavenDB. It presents movie profile for the user "Sk8trGal". The meaning of the field *Alchemy* is discussed in section VI.
```
{ "Username": " Sk8trGal",
  "CategoryName": "Movie"
  "Titles": [ {"Name": "Captain Fantastic",
               "ExternalID": null,
               "Rank": 1
             },   ...],
  "Actors": [ {"Name": "Viggo Mortensen",
               "ExternalID": null,
               "Rank": 1
             },   ...],
  "Genres":     ...
  "Directors":  ...
  "Alchemy": { "Entities": [] }
}
```

V. GENERATING RECOMMENDATIONS

A. Recommendations for a specific domain

Out of a recommendations related to a particular domain, we have implemented recommendations for the movie and music domain. The methods are essentially the same, except that they use different external data sources, to retrieve additional information (The Movie Database API for the movie and the Last.Fm API for the music domain). We were guided by the idea to connect concepts from the user profiles with items returned by this public APIs, fetch and store additional information about the found items, and then look for similar items that will be recommended to the user. The Movie Database API provides a method that will, for the given movie ID, return a list of IDs of similar movies. Similar movies are good candidates for the recommendation. To rank

recommendations according to the user profile, we present the matching between recommendation and profile with numerical value. The expression that calculates matching between the movie item and the user profile, for example, takes into account the actors who appear in the movie, genre, film director and producer. Assuming that the user's profile and the movie description both contain N concepts, they contribute to the total score according to:

$$\sum_{i=1}^{n} \text{Weight}(\text{concept}_i) * \text{Rank}(\text{concept}_i)$$

Different concepts have different importance meaning different weight. In the movie domain eg. it is natural to assign bigger weight to the movie title or genre than to the word that appears in a short summary of the movie. For example, the weight for the movie title can be 20, for the genre 15 and for the actor 12. Weights are parameters in our system and are determined using the method of trial and error by the administrator. In the present implementation the weight of the particular concept is the same for each user, but in the future implementation different weights of concept for different users can be foreseen.

B. Recommendations based on semantic features

The recommender system introduced in this paper additionally supports recommending web articles that are considered to be of interest to the user. The system is guided by the interests of the user stored in users' profile. Recommending web articles, as unstructured text, is suitable for CBF method. Analysis of content-based recommender systems showed that the representation of items (potential recommendations) and user profiles using keywords is efficient when a lot of text documents describing user's interests is available. In that case it is possible to find the matching words in web documents and users profile. However, the approach using keywords exclusively has its limitations. For example, if the user likes the "French impressionism" keyword-based system will find only documents containing the words "French" and "impressionism", but not the documents containing words "Claude Monet", the originator of this style [27]. To be semantically intelligent, the content-based system needs more advanced strategies for the presentation of items and the user profile, than using the list of keywords. The semantic analysis implies including a knowledge base, such as a lexicon or ontology, for labeling items. Keywords need to be connected to external data sources such as Linked Data sources (Semantic Web). The area of an artificial intelligence - natural language processing deals with assigning a semantic meaning to unstructured text. To decide whether the web article is a suitable recommendation for a user, it is necessary to semantically process the article i.e. to find concepts and entities mentioned in the text. To be suitable as a reference, article and user profile must contain matching concepts. For example, if the user shows the interest for movies, and graded with high grades movies starring Viggo Mortensen, concepts "movie" and "actor" and entity "Viggo Mortensen" appear in his profile. It is expected that this

1382

particular user will be interested in web articles that contain same concepts and entity. Since the semantic analysis of the text is outside of the scope of this paper, we therefore use the free version of Alchemy API [28] tool to perform semantic analysis of text in web articles.

Links (URLs) to web articles are periodically programmatically fetched from RSS feeds and forwarded to AlchemyAPI for processing. After the AlchemyAPI processes article on a given URL, as a result, among other things, a list of concepts and entities that appear in the text is obtained. For each entity found, AlchemyAPI returns its name, type and relevance in the article. After that, entities are compared with those contained in the user profile to calculate the suitability of the article as a recommendation for a particular category.

The comparison is done by the name of the entity. We try to pair each entity found by AlchemyAPI with entities contained in the user profile. Each matching contributes to the overall result according to the following expression:

```
Weight(concept) * Rank(concept) *
                Relevance(concept, article)
Weight("actor") * Rank("Viggo Mortensen") *
            Relevance("Viggo Mortensen", articleId)
```

VI. GRADING RECOMMENDATIONS

Recommender systems usually rely on a users' grading to assess the effectiveness and the degree of user's satisfaction. Users are commonly asked to quantify their overall satisfaction with the recommendations. This general principle is implemented in our recommender system too. With regard to the type of recommendation, user grades (movie, music artist, a web article) are used to update the user category profile.

Figure 3 illustrates the process of updating the movie profile upon grading the recommended movie "Captain Fantastic". On a scale of 1-5 user graded movie with grade 4. This grade is normalized to the value 0.5 on the interval [-1, 1]. The grade applies to all concepts/entities related to the movie (title, all actors, genres and director) because there is no way to find out what is actually graded. We used this simplified model since we don't want to burden user with the detailed assessment. Each concept (entity) describing graded movie is either found and updated or added in user's movie profile. The grade affects the value of attribute *Rank*. The normalized grade is added to existing *Rank* or if the concept does not exist in the profile, a new record with an initial value of attribute *Rank* equal to normalized grade is added. The procedure for updating the music profile after grading musician is the same.

Remark: The concept that was first found in the data retrieved from the social network is added to the user's profile with default value for the Rank equal to 1 (eg. if the user liked page of the actor Viggo Mortensen and if this concept does not exist in the user profile yet, it is added to the user's profile with Rank equals to 1). The reason for this is the belief that if the user liked the page, that is strong evidence of his high opinion on the actor (concept).

Figure 3 Updating the user profile upon grading

Upon assessing web articles different logic is applied. AlchemyAPI, for each entity found in the text, returns relevance of that particular entity (floating point number

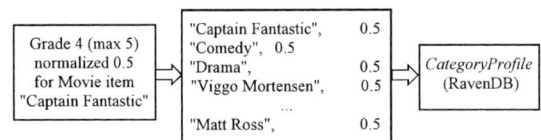

between 0 and 1). When a user grades web article, attribute *Rating* in his profile is updated, for each entity AlchemyAPI had found in the article, with a value equal to the normalized grade multiplied by the relevance of the entity. *Example*: The user graded with grade 2 article in which the Alchemy API found entity Viggo Mortensen with relevance 0.6. Supposing that the entity Viggo Mortensen exists in the movie profile for the user, the value of that attribute will be modified as follows:

```
Rank("Viggo Mortensen") += normalizedGrade(article)*
            relevance("Viggo Mortensen", article)
                += -0.5*0.6
```

In case that concept/entity found in the article does not exist in the user's profile, it is added to the list of "Entities" under the class "Alchemy" (see JSON presenting user's profile). In assessing the film or music items we could not determine the relevance of certain concepts/entities so we applied same grade to all concepts.

VII. EVALUATION

For the evaluation we used user studies [29]. A user study is carried out by recruiting a set of test users, asking them to interact with the recommender system and additionally to give answers to questions aimed to collect data that is not directly observable.

We evaluated only the recommendations from the movie domain with plans to conduct evaluation on other domains in the near future. The evaluation involved 5 users. After taking over data on users from Facebook API, the initial user profiles were created and the first recommendations from the movie domain were generated. The users were asked to state the number of previously watched movies, out of all the recommended. This information we consider relevant because the system recommending movies that user had already watched is in accordance with user's taste. The columns *Recomm.* and *watched* under the joint column "*1ˢᵗ cycle*" in Table 1 present results for recommendations users received for the initial profiles. Out of a total of 125 recommended movies users have already watched 30 of them or 24%. Given the size of the movie database, we found this result satisfactory.

The users are then asked to grade each recommendation regardless they already watched the movie or not. These grades reflect their satisfaction with the recommendation, ie. expresses a willingness of the user to watch the movie he hasn't seen yet. The average score of all users is 3.28. After users graded movies recommended in the first cycle, new recommendations, were generated (2ⁿᵈ cycle). Recommendations from the second cycle take into account the user's feedback from the first cycle. Results show that the percent of already watched movies (41%) increased which implies that the users' grades embedded in the user profile closer recommendations to users' tastes.

	1st cycle		2nd cycle		AVG(grade)
	Recomm.	watched	Recomm.	watched	(1-5)
user1	25	4	30	10	2.4
user2	29	9	-	-	3.3
user3	14	0	23	14	3.4
user4	27	6	30	6	3.1
user5	30	11	30	16	4.2

Table 1 Results of evaluation

Sometimes this is not a good feature of the recommender system because it is expected that the recommendations bring novelty not obvious, already consumed items. On the other hand, the evaluation showed that the user's grades improves recommendations.

VIII. CONCLUSION

Recommender systems are a powerful technology for entertainment industry and social networks. We presented recommender system that exploits existing publicly available services to gather data needed to build user profile and to generate initial recommendations. User's interests are deduced from his activities and posts in social network. Using this data for building user profile we bridge the cold start problem. Currently, we have implemented recommendations for items related to two domains: movies and music. We connected concepts from user profile with external items obtained through Open API calls. These and similar items are subsequently recommended to the user. In generating content based recommendations of web articles we use AlchemyAPI to perform semantic analysis of the text. To get feedback from users we implemented recommendations grading. That way users affect future recommendations because grades change and tune user profiles.

Recommender system evaluation showed that generated recommendations for the movie domain are in line with user's taste.

REFERENCES

[1] T. Pinckney, "eBay NYC - Graph-based Recommendation Engine on Cassandra," 30 July 2013. [Online]. Available: https://www.hakkalabs.co/articles/tom-pickney-discusses-ebays-recommendation-engine.

[2] G. Linden, B. Smith i J. York, »Amazon.com Recommendations Item-to-Item Collaborative Filtering,« 2003.

[3] C. A. Gomez i N. Hunt, »The Netflix Recommender System: Algorithms, Business Value, and Innovation,« *ACM Trans. Manage. Inf. Syst.*, 2016.

[4] C. Mims, "MIT Technology Review - How iTunes Genius Really Works," 2010. [Online]. Available: https://www.technologyreview.com/s/419198/how-itunes-genius-really-works/.

[5] L. Xiang, "Hulu Tech blog - Hulu's Recommendation System," 19 September 2011. [Online]. Available: http://tech.hulu.com/blog/2011/09/19/recommendation-system.html.

[6] L. Wu, "Browsemap: Collaborative Filtering At LinkedIn," 23 October 2014. [Online]. Available: https://engineering.linkedin.com/recommender-systems/browsemap-collaborative-filtering-linkedin.

[7] P. Gupta, A. Goel, J. Lin, A. Sharma, D. Wang i R. Zadeh, »WTF: The Who to Follow Service at Twitter,« u *Proceedings of the 22nd international conference on World Wide Web*, Rio de Janeiro, Brazil, 2013.

[8] S. K. Rogers , »Item-to-item Recommendations at Pinterest,« u *Proceedings of the 10th ACM Conference on Recommender Systems*, Boston, Massachusetts, USA, 2016.

[9] J. Moore, "Building a recommendation engine, foursquare style," 22 March 2011. [Online]. Available: https://engineering.foursquare.com/building-a-recommendation-engine-foursquare-style-4df6dc23ea15#.tk20r8sd0.

[10] "Mendeley," [Online]. Available: https://www.mendeley.com/.

[11] "Google Scholar," [Online]. Available: https://scholar.google.ca/.

[12] "PubChase," [Online]. Available: https://www.pubchase.com/.

[13] G. Adomavicius and A. Tuzhilin, "Toward the Next Generation of Recommender Systems: A Survey of the State-of-the-Art and Possible Extensions," *IEEE TRANSACTIONS ON KNOWLEDGE AND DATA ENGINEERING*, 2005.

[14] B. Schafer, D. Frankowski, J. Herlocker and S. Sen, "Collaborative Filtering Recommender Systems," *Lecture Notes In Computer Science*, vol. 4321, pp. 291-324, 2007.

[15] "Cosine similarity, Pearson correlation, and OLS coefficients," 13 march 2010. [Online]. Available: https://brenocon.com/blog/2012/03/cosine-similarity-pearson-correlation-and-ols-coefficients/.

[16] D. Gupta, M. Digiovanni , H. Narita and K. Goldberg, "Jester 2.0 (demonstration abstract): collaborative filtering to retrieve jokes," in *Proceedings of the 22nd annual international ACM SIGIR conference on Research and development in information retrieval*, Berkeley, California, USA, 1999.

[17] T. Joachims, "Text Categorization With Support Vector Machines: Learning with Many Relevant Features," in *European Conference on Machine Learning*, Chemnitz, Germany , 1998.

[18] M. Porter, *An Algorithm for Suffix Stripping, Program*, 1980, pp. 130-137.

[19] Y. Seroussi, I. Zukerman and F. Bohnert, "Collaborative Inference of Sentiments from Texts," in *User Modeling, Adaptation, and Personalization, 18th International Conference, UMAP 2010*, Big Island, HI, USA, 2010.

[20] F. Esposito, S. Ferilli, T. Basile and N. Di Mauro, "Machine Learning for Digital Document Processing: from Layout Analysis to Metadata Extraction,," in *Machine Learning in Document Analysis and Recognition*, 2008, pp. 105-138.

[21] "The Movie Database API," The Movie DB, [Online]. Available: https://www.themoviedb.org/documentation/api.

[22] "Audioscrobbler The Social Music Technology Playground," Audioscrobbler Ltd. , [Online]. Available: http://www.audioscrobbler.net/.

[23] C.-C. Chang and K.-H. Chu, "A Recommender System Combining Social Networks for Tourist Attractions," in *Fifth International Conference on Computational Intelligence, Communication Systems and Networks*, 2013.

[24] W. W. Chu, "A social network-based recommender system (SNRS)," in *Data Mining for Social Network Data*, 2010, pp. 47-74.

[25] S.-K. Ko, S.-M. Choi, H.-S. Eom, J.-W. Cha, H. Cho, L. Kim and Y.-S. Han, "A Smart Movie Recommendation System," in *Lecture Notes in Computer Science*, vol. 6771 , 2011, pp. 558-566.

[26] M. Kumar, D. K. Yadav, A. Singh i V. K. Gupta, »A Movie Recommender System: MOVREC,« *International Journal of Computer Applications*, 2015.

[27] P. Lops, M. de Gemmis and G. Semeraro, "Content-based Recommender Systems: State of," in *Recommender Systems Handbook*, 2010, pp. 73-105.

[28] " http://www.alchemyapi.com," IBM, [Online]. Available: http://www.alchemyapi.com/..

[29] G. Shani and A. Gunawardana, "Evaluating Recommendation Systems," in *Recommender Systems Handbook*, Springer US, 2010, pp. 257-297.

Alternative Business Intelligence Engines

Ivan Kovačević, Igor Mekterović
FER, Zagreb, Croatia
ivan.kovacevic@fer.hr, igor.mekterovic@fer.hr

Business intelligence (BI) engines enable companies and researchers to perform fast and efficient data analysis and extract relevant information from their data. At the present, it involves the usage of reporting tools and OLAP on top of traditional relational database systems. This paper gives an overview of alternative BI systems, based on various NoSQL and NewSQL databases. NoSQL (Not-only SQL) systems were developed to address the scalability issues of traditional relational databases when exposed to huge amounts of data and large read and write capacities. In addition to fulfilling those requirements, NewSQL systems aimed to preserve the useful properties of relational databases as well, such as ACID transactions, SQL and a relational schema. The various systems bear significant differences in terms of architecture, data representation, query capabilities and optimization techniques, and because of that, they offer diverse means of data analysis. The paper comments on key differences, presents an overview of various NoSQL and NewSQL systems that might be considered in the context of BI engine, and brings an overall feature comparison of those systems.

I. INTRODUCTION

At the time of writing, the amounts of available data are larger and produced at greater velocities than in any earlier period in history. For example, as stated in [1], in 2014, 90% of data on Facebook and Twitter has been created within the preceding two years, with new data adding up 7 to 10 terabytes every day. Most of the recently created data is unstructured, coming from sources such as web logs, social networks, radio-frequency identification tags (RFID), sensor networks, etc. As such, it requires solutions capable of managing and analyzing massive amounts of data while maintaining minimal latency and adaptability to previously unpredicted usage scenarios. According to [2], the crucial requirements for big data processing are the following: fast data loading, fast query processing, highly efficient space utilization, and strong adaptability to dynamic workload patterns. In contrast, traditional BI usually involves relational database systems and On-Line Analytical Processing systems (OLAP), which show serious drawbacks when confronted with big data. In this paper, we explore the use of newly emerging database technologies (many of them prompted by the big data movement) in the context of BI. In the following chapters, we comment on key differences of alternative engines and traditional RDBMSs with regards to storage medium, data model and query processing capabilities, present most notable database systems in this context, and bring a detailed feature comparison of those systems.

II. STORAGE SYSTEMS

Both NoSQL and NewSQL systems can be split into two subgroups based on the type of the underlying storage medium. These are *Disk-Based Database Management Systems* (DBMS) and *In-Memory DBMS*.

A. Disk-Based DBMS

Disk based DBMSs use disks as the main storage medium, and use the main memory primarily for buffering. Their data structures and algorithms are optimized to minimize the amount of the I/O operations required to perform given queries and data management operations. As the latency of the disks is several orders of magnitude larger than the latency of the main memory, these systems can afford fairly complex and lengthy algorithms to optimize disk access and buffering. Data structures, such as indices, are optimized for the physical structure of disks, since each disk I/O operation can only read or write complete blocks of data. An example of a disk optimized data structure is the B+ tree index, which matches its node size to the pages of the disk [3]. Most NoSQL and traditional relational DBMS are disk based.

B. In-Memory DBMS

One of the goals of research in big data systems is to support ultra-low latency and enable real-time data analytics. Existing disk-based systems are limited by the high access latency of the hard disk, so they cannot offer timely response required by real-time services, such as real-time bidding, social gaming, and advertising [4]. During the last decade, multi-core processors and the availability of large amounts of main memory at decreasing costs are creating new breakthroughs, making it possible to build database systems in which a significant part or the whole of the database can be situated inside the main memory. Note that mere relocation of data to RAM (e.g. using *Ram disk*) would bring limited performance gains, since the disk-based DBMSs are built to optimize disk access, rather than main memory performances. In-Memory DBMS solve these problems by using main memory specific data structures and algorithms. For example, most such systems use T-trees of Trie structures for indexing [5], and instead of optimizing disk access, aim to optimize the utilization of the cache hierarchy, adapting to *Non-Uniform Memory Access* (NUMA), and providing efficient in-memory data management. Furthermore, in order to leverage the low latency of the CPU caches (L1, L2, and L3), applications must ensure data locality, so as to reduce the amount of cache misses requiring the data to be fetched from the lower and slower memory levels. An additional concept called hardware transactional memory support, which is important to consider in in-memory DBMS design, is explained in [4]. Transactional memory allows for an optimistic approach to concurrency, where concurrent transactions are performed without locking by using L1 cache lines as their local isolated copies of data. If a transaction fails, e.g. by trying to commit to a cache line already updated by another transaction, its local copy in L1 cache is dropped and the transaction is either rolled back and started anew, or its locking employing variant is started. Ensuring data durability poses an important challenge for in-memory systems. Since the main memory is predominantly volatile, a power loss could result in loss of data. One potential solution is the introduction of the newly-emerging *non-volatile main memory* (NVRAM), with properties similar to those of solid state drives (SSD). Such memory is still quite expensive, and suffers from a limited number of successful write operations. Various in-memory DBMS take different strategies to ensure durability. Some systems, like e.g. VoltDb, incorporate checkpoints and logs stored in non-volatile storage to reconstruct the data after a power loss. Furthermore, in-memory systems supporting transactions do not have to log all operations, but just the committed transactions, since a power loss would invalidate the need to undo uncommitted operations.

III. DATA MODELS

In relational systems, such as the NewSQL DBMS, data warehouses are usually built with a star schema, which consists of a single fact table containing measures, and several dimension tables in which the aspects of analysis, such as e.g. time and product are stored. NewSQL databases support the same principles of data warehousing employed in traditional relational systems, since they support SQL and relational schema. On the other hand, with regards to data model, NoSQL systems can be classified in four categories [6] [7]: *Key-value stores*, which store data as key-value pairs and are useful for caching, *Column family oriented databases*, adept at storing data in sparse tables, *Document oriented databases*, organized as collections of documents with variable structure and dynamic number of fields, and *Graph oriented databases*, which store graph data, such as data related to social networks. Most surveyed sources are focused on using column and document oriented DBMS for BI applications.

A. Column Oriented DMBS in BI

Column based DBMS store data column-wise, and support grouping columns in column-families. The authors in [1] and [8] describe a desirable implementation of the star schema in a column based DBMS, in which the fact and dimension tables are organized as column families, with their attributes being stored in individual columns. In case of building an OLAP cube, an additional cube identifier column is required. Hierarchies are implemented using columns inside the hierarchical dimension's column families.

B. Document Oriented DBMS in BI

Since document stores, such as MongoDb, often have limited join support, the authors in [6] present a method of using embedded array structures for associating dimensions with documents. Each document represents a fact and contains measures and associated dimensions in its attributes. In this manner, documents satisfying a given predicate can be identified and aggregated using simple filtering.

IV. DATA PROCESSING AND STORAGE CAPABILITIES

Most distributed storage systems implement the Map-Reduce processing framework. MapReduce enables users to run large scale parallel operations based on the "divide and conquer" model. The model consists of two user-defined functions which are sent to all nodes, namely map, which reads data entities and generates key-value pairs, and reduce, which aggregates the key-value pairs and returns results. An additional intermediate shuffle function can be defined to optimize grouping. Several NoSQL systems are based on the Hadoop ecosystem, which offers a complete distributed architecture, including a distributed file system called Hadoop File System (HDFS) and implementations of several management and data processing functionalities, such as the aforementioned MapReduce. Hadoop is usually bundled with two high level languages based on Map-Reduce, Hive and Pig Latin [1]. Hive provides a SQL dialect called HiveQL, which enables data definition and access in a manner similar to relational systems. Map-Reduce is useful for fast aggregation operations required for data analysis, since it enables them to run in parallel on all nodes [2]. Some in-memory DBMS support *Main Memory MapReduce* (M3R), which mimics the concept of MapReduce parallel processing inside the main memory. As column oriented systems are storing data in a column-wise fashion, a significant increase in query performance can be achieved by using bitmap indexes. Distributed systems often support additional operators for data analysis purposes. Examples are the Hive's cube and rollup operators, which automatically build an OLAP cube based on a given table. An improved version of the cube operator, called cn-cube, has been proposed in [9]. As the authors of [10] point out, another important aspect of aggregation to consider is the materialization strategy, which defines how intermediate aggregation results are processed. In early materialization, which is the classical approach, row-based tuples are materialized by reading all columns, just to be filtered and aggregated. Late materialization avoids materializing the row-based tuples, and employs bitmap indexes to perform column-based aggregation, resulting in much better performance. In order to satisfy the big data analytics requirements, systems must remain constantly available, without a refreshing time window. This requires a new approach to the data warehouse extraction, transformation and loading (ETL) process. Unlike traditional systems, NoSQL and NewSQL based data warehouses can cope with increased loading frequencies, up to near real-time ETL [2].

V. NOSQL SYSTEMS

Unlike classical high-consistency RDBMSs, NoSQL systems take on a more relaxed approach, informally called BASE (*Basically Available, Soft state* and *Eventually Consistent)* which fits very well with BI requirements, where data is typically more or less stale. By using the BASE paradigm, NoSQL systems can achieve scalability vastly superior to traditional relational systems, which makes them suitable for working with large volumes of data. Furthermore, most NoSQL systems are schema-less, which means that the processed data does not have to conform to a predefined structure. Such systems possess numerous advantages over relational systems when working with unstructured data or data with frequent schema changes [6]. Most importantly in this context, NoSQL systems fall short in terms of query languages, especially when standards are concerned. A lack of schema or SQL certainly complicates data analysis. For analytical purposes, most authors present solutions based on column and document oriented databases [1] [2] [6] [8] [9] [10], which are also considered in this paper.

1) C-Store

C-Store was a proof of concept pioneering column oriented storage system released by the associated MIT research group in 2005. The system is optimized to support fast reading patterns required for data analysis, such as customer relationship management (CRM) systems, and data warehousing applications [11]. Furthermore, the system demonstrated the previously mentioned improvements in data compression compared to traditional row-oriented storage systems, and introduces heavy usage of bitmap indexing [11]. Bitmap indexes are binary vectors which enable filtering the data one column at a time [12].

2) BigTable

As described by its author Google, BigTable is a high-performance NoSQL database service intended for large analytical and operational workloads [13]. It is a column oriented storage system which, in addition to its native *Remote Procedure Call* (RPC) based API, provides an interface compatible with Hadoop based technologies, such as HBase.

3) HBase/Hadoop

Apache HBase is a highly specialized CF (Column Family) oriented database focused on operating with huge amounts of data. It is not intended for usage in small deployments, since it requires a minimum of five distributed nodes for normal operation. Queries are supported through a REST API and a query language called Thrift. In addition to those, the database supports a JDBC/ODBC driver. In CF databases, each column family consists of one or more columns (or attributes) which are stored in sequential order on the storage medium. This manner of storage enables fast column-wise access, with a minimal number of I/O operations, and a high data compression ratio, since the data inside a single column is usually more uniform and repetitive than data contained in several columns stored row-wise. Since columns are not mandatory, column based storage systems enable individual rows to possess different attributes, which allows users to store highly variable data together. HBase is built on top of the Hadoop infrastructure. It offers limited transaction support based on row-wise locks, and thus limits availability in order to satisfy consistency and partition tolerance. The system was initially inspired by Google BigTable.

4) Cassandra

Apache Cassandra is a CF database system which bears a lot of similarities to the aforementioned HBase system. In addition to CFs, Cassandra supports another layer of column hierarchy called *supercolumns*. *Supercolumns* are groups of columns which can be contained inside column families like individual columns.

5) MongoDb

According to [14], MongoDb is the most commonly used NoSQL database system. It is a highly scalable document oriented database, used in a variety of projects, such as Foursquare, bit.ly, and the CERN Large Hadron Collider. Document oriented storage systems store data in collections of documents. Each document is a data structure consisting of an arbitrary number of key-value pairs, or fields. Fields can contain nested structures. Documents do not possess an explicit schema. They can be identified and accessed using associated unique identifiers or one or more keys. MongoDB stores documents in BSON (Binary JSON) format. In contrast to relational databases, where related data is stored in several tables, in document stores most related data is usually contained within the same document. Since documents bear a significant similarity to classes in object oriented programming languages, integration is usually straightforward and simple. Up until recently, a drawback of document stores was a lack of support for joining operations. In the recent versions of MongoDb and CouchDb, operations similar to joins are supported through special operators. MongoDb supports automatic partitioning and master-slave replication. It supports a query language represented as JSON-like structures, and exposes a CRUD REST API. For more complex operations, it supports an implementation of the Map-Reduce paradigm. With regards to the Brewer's CAP theorem, the common configuration supports consistency and partition tolerance while offering a limited availability.

6) CouchDb

Like MongoDb, CouchDb is a highly scalable document oriented database system. Unlike MongoDb, the common configuration supports availability and partition tolerance while offering a reduced level of consistency through MVCC. It supports *design* documents, or *views*, which are defined through MapReduce queries and similar to views in relational database systems. Design documents show higher performance when compared to ad-hoc MapReduce queries [15]. Apart from views and Map-Reduce, queries are supported through the exposed CRUD REST API. Data conflicts caused by concurrent write operations on different nodes have to be resolved manually.

7) Apache Spark

Apache Spark is an engine (framework) for large-scale data processing on a cluster. Spark presents a data abstraction for big data analytics, called *resilient distributed dataset* (RDD). It is a coarse-grained deterministic immutable data structure, on top of which Spark SQL, Spark Streaming, MLib and GraphX are built to support SQL-based manipulation, stream processing, machine learning and graph processing, respectively [4]. Spark's data processing performance comes from its in-memory architecture, as well as the useful properties of RDDs and automatic job scheduling using the DAG scheduler. RDDs support coarse-grained operations called trans-formations, such as *map, filter,* and *join*, which are applied to the entire dataset and result in new RDDs. The main two features of RDDs are a persistence model, which can persist datasets in memory, disks or both, and a light-weight fault-tolerance mechanism which makes it possible to recreate RDDs from their dependent RDDs by using transformation logs (i.e. *lineage*). This makes RDDs suitable for iterative data analytics jobs, because data doesn't have to be shuffled on disks at each stage like in Hadoop [4]. Calculations over RDDs can be performed using special operations called *actions*, such as *count* or *reduce*. Operations in Spark are run through jobs, where the DAG scheduler captures the complex job dependencies and relieves programmers from managing them for manual scheduling [16].

8) Apache Drill

Apache Drill stands out in this list, as it is not a DBMS per se, but a distributed SQL engine with support for Hadoop and several other storage systems. It fully supports ANSI SQL:2003, which enables users to write queries like in a traditional DBMS [17]. In addition to that, it incorporates support for advanced JSON data operations through SQL extensions. Within Drill, all data is internally represented as a JSON structure, whose schema can be discovered automatically. This flexible data model makes Drill useful in operating unstructured and semi-structured data from several sources, without writing additional middleware. Drill can be used by most existing BI tools, like Tableau, Qlik, Excel, etc., to interface NoSQL engines, supported through its plugins, such as HBase, MongoDB, Hive and Apache Spark. That simplifies integration or even migration in existing systems. Finally, it can be used as a complete SQL shell over various NoSQL storage systems.

VI. NEWSQL SYSTEMS

Nowadays, and probably for many years to come, most existing systems work with relational databases, and require a complex migration procedure in order to switch to a NoSQL solution. An example of one such migration is presented in [6]. For OLTP applications, a major drawback of NoSQL systems is the lack of strong consistency. NewSQL refers to various recently developed relational systems aiming to achieve high consistency at scale, where the attribute "New" stands for the predominantly new companies developing such systems. Most of them are in-memory DBMS. All support SQL queries and ACID transactions.

1) Vertica

Vertica is a distributed massively parallel relational DMBS, based on the aforementioned C-Store research project. Although it supports SQL data management operations, it is optimized for supporting analytical query workloads, since the data is organized in columns, rather than rows. The data is stored in a hybrid in-memory/on-disk architecture, where frequently used data and buffer are contained within main memory and only passive data is stored on disk. The column structure enables users to define *projections*, groups similar to column families in column based DBMS. Data is split into two parts, the *read-optimized store* (ROS), which holds most of the data stored in index-value pairs, and *write-optimized store* (WOS), which allows fast data editing. Vertica is not appropriate for usage in update-intensive applications.

2) SAP HANA

HANA is the SAP's approach to BI based on two main components, the SAP HANA database, which is a hybrid in-memory DBMS with row-based, column-based, and object-based technologies, and the SAP HANA appliance, a combination of hardware and software which enables the analysis of large volumes of data without the need for aggregation materialization. Data is stored into two relational databases, one column oriented, and the other row-oriented. Both support SQL and Multidimensional Expressions (MDX). The column oriented database supports real-time analytics, while the row-oriented database is intended for data which is frequently inserted and updated.

3) Oracle Times Ten

Oracle Times Ten is an in-memory DBMS which supports ACID transactions and can be either used as a stand-alone system or a data subset caching provider for traditional DBMS, used by existing applications using an Oracle database to reduce query latency. The database incorporates a data aging scheme, which periodically removes data that are no longer needed. Similar to SAP HANA, Oracle has commercialized a complete appliance based on Times Ten, called *Oracle Exalytics In-Memory Machine*. Exalytics comes with a complete MOLAP server which contains a high performance MDX query engine.

4) SQL Server Column indexes

Microsoft decided to take a different approach to column-store technology, by keeping SQL Server as the central database product and incorporating the column oriented store as an optional index. These additions significantly improve the performance of analytical query processing, while maintaining a well-known database system for transactional processing. On the other hand, as the column-based query processing is optimized for smaller result sets typical of data warehouses, such as performing star joins and aggregating, queries returning large result sets suffer from low performance, since batch processing cannot be successfully executed in those cases [1]. Also, up until SQL Server 2016, tables for which a column index had been defined could not have been updated. However, as of version 2016, both *Clustered Column Store* Indexes and *Non-Clustered Column Store* indexes can be updated [18].

5) VoltDb/H-Store

H-Store, and its commercial counterpart, VoltDb, are distributed in-memory relational DBMS, which provide automatic and transparent fragmentation and replication support. In VoltDb, queries are performed using stored procedures, which are run sequentially and possess the properties of ACID transactions. Ad-hoc SQL queries are supported as well, but such queries have limited optimization support and should be avoided in production. Consistency of replicas is enforced by running the same stored procedures on all nodes containing target replicas.

6) NuoDb

NuoDb is a highly scalable in-memory distributed relational DBMS, based on a *Distributed Durable Cache* (DDC) peer-to-peer architecture. DDC allows peers to be assigned to specific tasks, such as managing durability or orchestrating transactions. Transaction requests are received by transaction orchestrating peers, which forward them to durability peers. Once durability peers have stored the transaction's effects, the updated data is synchronized on all peers containing it.

	MongoDB	C-Store	HBase/Hadoop	Cassandra	BigTable	CouchDB
data model	document (JSON)	logically relational	column	column/key-value	column	document (JSON)
architecture[1]	distributed, MS	distributed, MS	distributed, MS	distributed, P2P	distributed	distributed, MM
storage type	BSON + binary data storage	column-oriented	Hadoop, column-oriented	column-oriented	cloud storage	JSON + Hadoop
query language	proprietary, Js syntax	SQL	proprietary	CQL	proprietary	proprietary, Js syntax
QL features[2]	J, P, S, A	J, P, S, A	P, S, A	P, S, A	P, S, A	J, P, S, A
MapReduce	+	-	+	+	+	+
APIs	language bindings for all major languages	?	Java, C/C++, Thrift, Scala, Jython, REST	plugins for all major languages	RPC API, Hadoop-compatible plugin	libraries for all major languages
SQL/ACID	-/-	+/-	-/-	-/-	-/-	-/-
BI references	[4] [29] [30] [35]	-	[30] [34] [35]	[30] [35]	[13]	[37]
BI tools available	compatible with Pentaho, Informatica, MicroStrategy, QlikView, etc.	-	Pentaho, QlikView, MicroStrategy, etc.	Pentaho, MicroStrategy, etc.	Google Cloud BI Tools, HBase compatible BI tools	Pentaho, Tableau, etc.
OS[3]	W, X, M, S	X	(java) W, X, M	(java) W, X, M	Google Cloud P.	W, X, M
license	AGPL v3.0	BSD	Apache 2.0	Apache 2.0	proprietary	Apache 2.0
version	3.4.2	0.2	1.2.4 / 1.3.0	3.10	-	2.0
maturity	10 years, mature, act. developed	12 years, not actively used	10 years, mature, act. developed	10 years, mature, act. developed	2 years, actively developed	12 years, mature, act. developed
typical use cases	document data store, web app storage, etc.	column-oriented storage research	sparse schema-less data storage	product catalogs, IoT, messaging apps, etc.	big data analysis	document data store, web app storage, etc.
community	large	very small	large	large	large	small
support	yes, commercial	-	(community)	(community)	yes, commercial	(community)
popularity[4]	high	-	medium	high	small	medium

Table 1. Overview of covered storage engines, part 1

	Apache Spark	Apache Drill	Vertica	SAP Hana	Oracle Times Ten	SQL Server Column indexes	VoltDb	NuoDb
data model	RDD	relational	relational	relational	relational	relational	relational	relational
architecture[1]	distributed, MW	distributed	distributed, MM	hybrid IM	distributed, IM	IM	distributed, IM, MS	distributed, IM, P2P, MMS
storage type	in-memory, disk	plugin dependent	in-memory-disk hybrid	(column oriented) memory, disk	in-memory, disk	(column oriented) in-memory, disk	in-memory	in-memory
query language	proprietary, SQL (subset)	SQL	SQL (subset)	SQL	SQL	SQL	SQL, stored procedures	SQL
QL features[2]	J, P, S, A	J, P, S, A	J, P, S, A	J, P, S, A	J, P, S, A	J, P, S, A	J, P, S, A	J, P, S, A
MapReduce	+	+	+	-	-	-	-	-
APIs	Scala, Python, R, Java, JDBC, ODBC	REST, ODBC, JDBC, C++	plugins for all major languages	JDBC, ODBC	ODP.NET, OCI, JDBC, ODBC, C, C++, Java	support for all major languages	REST, plugins for all major languages	JDBC, ODBC, ADO.NET
SQL/ACID	+/-	+/storage dependent	+/+	+/+	+/+	+/+	+/+	+/+
BI references	[4] [33] [35] [37]	[33] [32] [35]	[1] [31] [33] [34] [35]	[4] [33] [35]	[28]	[31] [33] [34]	[36]	[23]
BI tools available	Tableau, MicroStrategy, Pentaho, etc.	Tableau, Qlik, SAS, Excel, MicroStrategy, Spotfire, etc.	QlikView, Pentaho, Tableau, MicroStrategy, etc.	SAP HANA tools, Tableau, MicroStrategy, etc.	Oracle Exalitics, QlikView, etc.	Microsoft BI Tools, Tableau, QlikView, Pentaho, etc.	MicroStrategy, Tableau, etc. (Progress DataDirect)	Pentaho, Japersoft, etc.
OS[3]	W, X, M	W, X, M	X	cloud / A	W, X, M, A	W	X, M	W, X, M
license	Apache 2.0	Apache 2.0	proprietary	proprietary	proprietary	proprietary	proprietary	proprietary
version	2.1.0	1.9.0	8.0.x	2.0	11.2.2	2016	7.0	2.6
maturity	3 years, act. developed	2 years, act. developed	12 y, mature, developed	10 y, mature, developed	20 y, mature, developed	5 y, mature, developed	7 y, mature, developed	2 years, act. developed
typical use cases	real-t. stream analytics (logs, transactions, signals, etc.)	SQL data analytics over NoSQL data stores	data warehousing and analysis	real-time analytics, appl. & warehousing data store	relational storage caching, BI	data warehousing and analytics	low-latency relat. storage for appl., data and stream BI	low-latency relational storage for appl., real-t. BI
community	large	small	medium	small	medium	medium	small	small
support	(community)	(community)	yes	yes, commercial	yes, commercial	yes, commercial	yes, commercial	yes
popularity[4]	medium	small	medium	medium	small	-	small	small

Table 2. Overview of covered storage engines, part 2

---

[1] MS = master-slave, MMS = multi-master-slave, P2P = peer-to-peer, MM = multi-master, MW = master-worker

[2] J = joins, P = projection, S = selection, A = aggregation

[3] W = Windows, X = Unix/Linux, M = Max OS, S = Solaris, F = FreeBSD, A = appliance

[4] Popularity measures are taken from db-engines [14] and stack overflow measures [26]

VII. CONCLUSION

This paper has given an introduction to the key architectural differences between classic DW technologies and emerging alternative DBMS engines, followed by a basic overview of various NoSQL and NewSQL systems. Tables 1 and 2 bring a feature comparison of presented DMBSs. In general, NoSQL and NewSQL engines show superior aggregation performance and much lower latencies than traditional BI engines for large amounts of data. However, they significantly fall behind in terms of query languages and their analytical capabilities, and consequently – data analysis tools. However, as the requirements of data analysis are changing towards larger data volumes and larger numbers of concurrent users, NoSQL and NewSQL systems are gaining popularity in BI applications, and will most likely be more and more included as an accompanying DBMSs to the existing BI platforms, in the spirit of polyglot database architecture. Along those lines, we believe that integration of existing BI systems and alternative engines is a promising area for future research and practice.

REFERENCES

[1] A. Vaisman and E. Zimányi, Data Warehouse Systems, Heidelberg New York Dordrecht London: Springer, 2014.

[2] G. S. Sureshrao and A. H. P., "MapReduce-Based Warehouse Systems: A Survey," Unnao, Indija, 2014.

[3] G. Graefe, "Modern B-Tree Techniques," *Foundations and Trends® in Databases*, vol. 4, no. 3, pp. 203-402, 2011.

[4] H. Zhang, G. Chen, B. C. Ooi, K.-L. Tan and M. Zhang, "In-Memory Big Data Management and Processing: A Survey," IEEE, 2015.

[5] T. J. Lehman and M. J. Carey, "A Study of Index Structures for Main Memory Database Management Systems," in *Twelfth International Conference on Very Large Data Bases*, Kyoto, 1986.

[6] L. Bonnet, A. Laurent, M. Sala, B. Laurent and N. Sicard, "REDUCE, YOU SAY: What NoSQL can do for Data Aggregation and BI in Large Repositories," 22nd International Workshop on Database and Expert Systems Applications, 2011.

[7] F. M. Sadalage PJ, "NoSQL distilled: a brief guide to the emerging world of polyglot persistence," Addison-Wesley, Upper Saddle River, NJ, 2013.

[8] J. You, J. Xi, C. Zhang and G. Guo, "HDW: A High Performance Large Scale Data Warehouse," Guangdong, 2008.

[9] K. Dehdouh, F. Bentayeb, O. Boussaid and N. Kabachi, "Columnar NoSQL CUBE: Agregation operator for columnar NoSQL data warehouse," San Diego, 2014.

[10] X. Dai and C. Li, "The Application of Materialization Strategies on OLAP in Column Oriented Database Systems," Tianjin Polytechnic University, Tianjin, 2011.

[11] M. Stonebraker, D. J. Abadi, A. Batkin, X. Chen, M. Cherniack, M. Ferreira, E. Lau, A. Lin, S. Madden, E. O'Neil, P. O'Neil, A. Rasin, N. Tran and S. Zdonik, "C-Store: A Column-oriented DBMS," in *Proceedings of the 31st VLDB Conference*, Trondheim, 2005..

[12] C. Y. Chan and Y. E. Ioannidis, "Proceedings of the 1998 ACM SIGMOD international conference on Management of data," in *SIGMOD '98*, Seattle, 1998.

[13] Google, "Google Cloud Platform," Google, 2 2017. [Online]. Available: https://cloud.google.com/bigtable/. [Accessed 2 2017].

[14] Solid IT, "DB-Engines Ranking," 5 5 2016. [Online]. Available: http://db-engines.com/en/ranking.

[15] E. Redmond and J. R. Wilson, Seven Databases In Seven Weeks, Dallas: Pragmatic Programmers, LLC., 2012.

[16] J. Wei, J. K. Kim and G. A. Gibson, "Benchmarking Apache Spark with Machine Learning Applications," Carnegie Mellon University, Pittsburgh, 2016.

[17] J. Scott, "MAPR," 7 12 2015. [Online]. Available: https://www.mapr.com/blog/apache-spark-vs-apache-drill. [Accessed 2 2017].

[18] J. d. Bruijn, S. Agarwal, D. Komo, B. Bordia and V. Soni, "SQL Server In-Memory OLTP and Columnstore Feature Comparison," Microsoft, 2016.

[19] R. Cattell, "Scalable SQL and NoSQL Data Stores," 2011.

[20] M. Aslett, "What we talk about when we talk about NewSQL," 6 Travanj 2011. [Online]. Available: https://blogs.the451group.com/information_management/2011/0 4/06/what-we-talk-about-when-we-talk-about-newsql/.

[21] VoltDb, "VoltDb Documentation," 2016. [Online]. Available: https://docs.voltdb.com/UsingVoltDB/ChapSecurity.php. [Accessed 1 2017].

[22] Microsoft, "Overview of SQL Server Security," 1 2017. [Online]. Available: https://msdn.microsoft.com/en-us/library/bb669078(v=vs.110).aspx. [Accessed 1 2017].

[23] NuoDB, "NuoDB," 10 5 2016. [Online]. Available: http://www.nuodb.com/product/distributed-cloud-database-architecture.

[24] M. Chevalier, M. E. Malki, A. Kopliku, O. Teste and R. Tournier, "Benchmark for OLAP on NoSQL Technologies," Toulouse.

[25] W. Romsaiyud, "Applying MVC Data Model on Hadoop for Delivering the Business Intelligence," Bangkok, 2014.

[26] Stack Exchange, "Stack Overflow Developer Survey," [Online]. Available: http://stackoverflow.com/research/developer-survey-2016#technology. [Accessed 2 2017].

[27] Oracle, "Oracle TimesTen In-Memory Database Architectural Overview," Oracle, 2006.

[28] V. Murthy, P. Deshpande, A. Lee, D. Granholm and S. Cheung, "Oracle Exalytics In-Memory Machine: A Brief Introduction," Oracle Corporation, Redwood, 2014.

[29] MongoDB, Inc., "MongoDB: Bringing Online Big Data to Business Intelligence & Analytics," MongoDB, Inc., 2016.

[30] Pentaho Corporation, "Pentaho Instaview," Pentaho Corporation, 2013.

[31] Pentaho, "Pentaho Components Reference," 2016. [Online]. Available: https://help.pentaho.com/Documentation/5.2/0D0/160/000. [Accessed 2 2017].

[32] Apache Drill, "Using Drill with BI Tools," Apache Drill, 2014. [Online]. Available: https://drill.apache.org/docs/using-drill-with-bi-tools/. [Accessed 2 2017].

[33] Tableau, "Tableau Technical Specifications," 2017. [Online]. Available: https://www.tableau.com/products/techspecs. [Accessed 2 2017].

[34] Qlik, "QlikView Data Sources," Qlik, 2017. [Online]. Available: http://global.qlik.com/tw/explore/solutions/data-source/qlikview-data-sources. [Accessed 2 2017].

[35] MicroStrategy, "MicroStrategy - Drivers and Connectors," 2017. [Online]. Available: https://www.microstrategy.com/us/services/drivers-and-connectors. [Accessed 2 2017].

[36] K. Leadley, "VoltDB Selects Progress DataDirect to Advance Rapid Time to Value for Data Connectivity for Its Customers," Progress, 2015.

[37] Pentaho, "Pentaho - Data Sources for Business Analytics," 2017. [Online]. Available: http://www.pentaho.com/data-sources-for-business-analytics. [Accessed 2 2017].

Insights into BPM maturity in Croatian and Slovenian companies

V. Bosilj Vukšić*, M. Indihar Štemberger ** and D. Suša Vugec*
* Faculty of Economics, University of Zagreb / Department of Informatics, Zagreb, Croatia
** Faculty of Economics / Academic Unit for Business Informatics and Logistics, University of Ljubljana, Ljubljana, Slovenia
vbosilj@efzg.hr

Abstract - In recent period, business process management (BPM) has been increasingly a matter of interest for numerous authors as well as numerous organizations, due to the understanding of business processes as the core part of every organization. Significant efforts have been put into researching and implementing BPM within organizations. This paper's goal is to present the current state of BPM maturity and usage of social BPM within the companies operating in Croatia and Slovenia by analysing data collected by the PROSPER research group through questionnaires. Moreover, organizational culture is included in the analysis as well. Results indicate higher BPM maturity level and higher usage of social BPM within Slovenian companies than within Croatian ones and some other differences regarding dominant organizational cultures have also been found.

I. INTRODUCTION

Business Process Management (BPM) concept has been explored by numerous researchers for about two decades. The academics define BPM as a holistic and a lifecycle approach that covers different phases; from process documentation and modelling towards process execution, monitoring and optimization, and focuses on different issues that range from organizational structures, management positions and roles to strategy alignment and usage of IT for BPM [7][27][8]. Many of the researchers agree that BPM has a significant role in the organizational performance and a lot of effort was put into investigations related to this area. Accordingly, different maturity models are used to evaluate how the higher BPM maturity drives organizations towards higher performance [26].

Through years many definitions of organizational culture are accepted among researchers, but for the purpose of this paper a one developed by [9] and cited by [34] is used. Organizational culture implies "a system of values, beliefs and customs in an organization which produces certain norms of behaviour when interacting with the formal organizational structure" [9][34]. Though many different instruments for measuring organizational

culture exist [14], a lack of researches about the importance of organizational culture type for the success of BPM implementation was evident in BPM literature through many years [36]. Nowadays, the academics and business practitioners intensify the efforts to explore this field [25]. Therefore, this study aims to examine the link between the organizational culture type and the level of BPM maturity. Recently, "the social BPM", a new term related to BPM was coined out. This concept refers to a collaboratively designed process of BPM implementation [18] and "aims to integrate social aspect throughout the different stages of BPM" [24]. The brief literature overview shows that more research needs to be carried out on the role of social BPM in BPM maturity.

The main goal of this paper is to investigate the current state of BPM maturity in Croatian and Slovenian companies. Besides, the focus is put on two factors that impact the implementation of BPM concept in organizations; these are (1) organizational culture and (2) social BPM. The paper is structured in two parts. First, a theoretical framework is given: concepts of BPM, social BPM and organizational culture are presented. Next, the methodology and the results of empirical research about BPM maturity and social BPM through different organizational culture types are depicted and the final conclusions are given.

II. BUSINESS PROCESS MANAGEMENT MATURITY

Numerous models are used by researchers to asses BPM maturity. The origin of the majority of these models is the Capability Maturity Model (CMM) which was developed to measure the maturity of software development processes [35]. In general, BPM maturity models usually consist of 4-6 dimensions that are used to assess the stage of organizational maturity. Besides, maturity models can serve business practitioners as a guidance to reach higher maturity levels, or for a comparison or benchmarking purposes. Röglinger, Pöppelbuß and Becker [26] identified and analysed ten BPM maturity models, among which Process Performance Index (PPI) - originally founded by [28]. According to their findings, PPI is "a descriptive model that defines statements for ten BPM critical success factors" [26]. It refers to different areas linked to BPM implementation in organization: (1) alignment with strategy; (2) holistic

---

This study has been partly founded by the Croatian Science Foundation under the project PROSPER – Process and Business Intelligence for Business Excellence (IP-2014-09-3729) and by the Slovenian Research Agency under the research programme No. P2-0037 - Future internet technologies: concepts, architectures, services and socio-economic issues and under the project No. J5-7287 - Big Data Analytics: From Insights to Business Process Agility

approach; (3) process awareness by management and employees; (4) portfolio of process management initiatives; (5) process improvement methodology; (6) process metrics; (7) customer focus; (8) process management; (9) information systems and (10) change management.

This model distinguishes three levels of maturity: (1) process management initiation; (2) process management evolution; and (3) process management mastery. Each of these levels and its characteristics are described by [29] and interpreted by [3], as it is presented in Table I.

TABLE I. PPI MODEL MATURITY LEVELS

Maturity level	Maturity level characteristics
process management initiation	Although organizations are „neophytes" to BPM, a strong aspiration to learn about it exists; by starting to focus systematically and formally on business processes, significant benefits could be achieved [3][29].
process management evolution	Although there are formal process improvement programs and organizations are "process-aware" there is room for BPM improvements; process roles and jobs are identified, process performance indicators and metrics defined and in some cases process measurement systems are implemented [3].
process management mastery	For the organizations BPM is fully integrated into their functioning and general performance management system; process owners are rewarded on process performance and every employee understands the processes [3].

III. THE ROLE OF ORGANIZATIONAL CULTURE IN BPM

According to [30] and [12] organizational culture represents "values, beliefs, attitudes and behaviours" of the employees and reflects their behaviour. Typically, 4 types of organizational culture are recognized [6]: clan, adhocracy, market and hierarchy culture. In business practice, different culture types are usually adopted within an organization, thus forming the combination of culture types that is aligned with the specific internal conditions (e.g. size, industry and company ownership) and with the environmental influences (e.g. legal regulations, governmental policy, market conditions). According to [22] and [23], "a classification of organizational culture called the competing values framework is depicted with 4 quadrants that are determined by a horizontal and vertical axis". The horizontal axis emphasizes the organizational internal or external orientation, while the vertical axis reflects the criteria of flexibility and control [37]. The position of organizational culture types within the classification proposed by [22] and its main characteristics are shown in Table II.

Through the years many organizational culture assessment models have been developed. Among these, Organizational Culture Assessment Instrument (OCAI) is one of the most often used by researchers. OCAI model is developed by [6] to assess the organizational culture type based on the respondent perception of current organizational culture and preferred one. The model defines six groups of statements: (1) dominant characteristics, (2) organizational leadership, (3) management of employees, (4) organizational glue, (5) strategic emphasis and, (6) criteria for success [6]. Further, each group of statements is structured out of four statements, each of these representing one of the four culture types.

TABLE II. MAIN CHARACTERISTICS OF ORGANIZATIONAL CULTURE TYPES

Culture type	Competing values framework [22]	Culture type characteristics
Clan	Internal focus, Flexibility	A workplace is flexible, friendly and people oriented [21]. A loyalty, tradition, collaboration, participation and teamwork are in focus [6][2].
Adhocracy	External focus, Flexibility	Creativity, agility and innovation are encouraged within a creative working environment and are supported by the employees willing to take risk and to experiment [21][37].
Market	External focus, Control	The main characteristics are: very competitive environment strongly oriented towards performance, productivity and achievement of business goals [21][37].
Hierarchy	Internal focus, Control	A workplace is formal, organizational structures are strong, with very deep pyramid of decision-making levels [21]. A control and stability is achieved by formal rules, regulations and policies [2].

Still, a challenging question for the researchers is if the success of BPM implementation and adoption relies on the dominant type of culture within the organization? The results of the different surveys showed that employees will support the implementation of new concept if it is compatible with the culture that prevails in organization, so it can be concluded that organizational culture strongly impacts BPM implementation success [1][2][7]. Several authors share the opinion that "a change in the organizational culture is an important factor to increase a level of business process maturity" [33][11]. These conclusions promote a need to deepen a research on the role of organizational culture in BPM maturity. Moreover, the focus of researches on the impact of organizational characteristics on the success of BPM is notable [27][36][2][32]. According to [5] and [13], clan and adhocracy culture seem to have a positive role in BPM success.

IV. SOCIAL BPM

BPM is a holistic concept where different disciplines, such as management, organizational theory and IT are linked. Through the years the attempts to integrate social elements in BPM have evolved [24], so the term social BPM has been introduced. Social BPM comprises two areas: (1) communication and collaboration aspects during all stages of BPM lifecycle and (2) implementation of IT platform to support social user behaviour. According to [20] "social BPM is the practice of actively involving all relevant stakeholders into BPM through the use of social software and its underlying principles". This approach should help to avoid the limitations of traditional BPM, but still little is known how social BPM can contribute to the enhancement of BPM implementation and governance.

As a result of brief literature overview conducted for the purpose of this research, four main principles of social BPM concept are systemized and presented in Table III. Similarly, [16] developed an a-priory model to identify the appropriate social technology for inclusion within the BPM lifecycle.

TABLE III. MAIN PRINCIPLES OF SOCIAL BPM

SOCIAL BPM FACTORS	SOURCES
Egalitarianism	
Our BPM approach relies highly on the idea of giving all participants the same rights to contribute to business process design and change.	[4][10][20][31]
Collective intelligence	
Business processes are designed and modified based on the ideas and knowledge of a group (collective) rather than individual experts or external influence.	[4][10][20][31]
Self-organization	
Employees are self-organized and interactively design and change business processes in bottom-up rather than top-down fashion.	[20][31]
Social production	
Stakeholders use social software and Enterprise 2.0 tools (e.g. blogs, wikis, social networks, Lync, Yammer) to suggest and create process content and context.	[10][15][20][31]

V. METHODOLOGY

A. Research questions

The research presented in this paper is based on the following research questions: (1) What is the current state of BPM maturity and the usage of social BPM within the Croatian and Slovenian companies?; (2) Are there any statistically significant differences between the results obtained in Croatia and those from Slovenia regarding BPM maturity and the usage of social BPM?; (3) Are there any statistically significant differences between organizations with different organizational cultures regarding BPM maturity? and, (4) Are there any statistically significant differences between organizations with different organizational cultures regarding the usage of social BPM?

B. Research instrument

For the purpose of the research conducted under the PROSPER project, a questionnaire has been developed based on the comprehensive literature review. The final version of the questionnaire contained 12 sections, being: BPM, social BPM, business intelligence (BI), corporate performance management (CPM), BPM/CPM alignment, BPM/BI alignment, CPM/BI alignment, process performance, organizational performance, organizational culture, characteristics of organization and demographic characteristics of the respondent. This paper is based on the three sections of that questionnaire: (1) BPM, (2) social BPM and (3) organizational culture.

BPM section of the questionnaire represents the Process performance index (PPI) developed by [29], containing ten statements as described earlier in this paper. For each statement, the respondents state their level of agreement on the scale from 1 to 5, with 1 being "totally disagree" and 5 being "totally agree". The cumulative score gained through these ten statements represents organization's PPI. If the PPI score is in the range from 10 to 25 it means the organization is at the lowest level of BPM maturity, PPI from 26 to 40 points puts organization into the second level of BPM maturity while 41 to 50 points are included in the third, highest (mastery) level.

Social BPM section of the questionnaire has been developed by the PROSPER research group based on the state of the art literature review [31][10][4][20][15]. It contains four statements which refer to the principles of social BPM: (1) egalitarianism, (2) collective intelligence, (3) self-organization and (4) social production, as shown in Table III. The respondents state their level of agreement with the each statement on the Likert scale from 1 to 5, with 1 representing total disagreement while 5 represented total agreement. Higher average of the stated grades represents a higher level of usage of social BPM within the observed company.

Organizational culture section of the questionnaire is the Organizational culture assessment instrument (OCAI) developed by [6] and described earlier in the paper. The respondents are supposed to split 100 points for each of the six groups of statements over a total of four descriptions of culture types in each group, according to the state within the organization. For the purpose of this research, OCAI is used to assess only the current dominant organizational culture types.

C. Data collection and sample description

The data collection for the purpose of this research has been carried out in two phases. Firstly, a preliminary research has been conducted with the purpose of testing the clarity and the design of the questionnaire draft. In February 2016, a series of interviews were conducted in order to test the questionnaire, after which slight modifications of the draft were made and the questionnaire was prepared for the main stage of the research.

The main stage of the data collection started in March 2016 and ended in December 2016. During this period, the questionnaires were distributed to the middle-sized and large companies in Slovenia and Croatia through post or e-mails. Besides the paper versions, an online version of the survey has been made in both countries. In Croatia, the questionnaires have been sent to 500 randomly selected organizations out of 1765 active middle-sized and large organizations in the Register of Business Entities. With total of 101 responses received, the response rate for Croatia was 20,2%. In Slovenia, the questionnaires have been sent to 1394 organizations, which is a whole population of registered middle-sized and large organizations. The response rate for the Slovenian data collection was 12,27%, with 171 responses received. In both countries, the questionnaires were addressed to top

1393

management or person in charge of business process management within the organization. After the overall data cleansing based on the survey fulfilment, a total of 211 answers were taken into consideration. Table IV presents the sample characteristics used in this research.

TABLE IV. SAMPLE CHARACTERISTICS

Criteria	Characteristic	N	%
Country	Slovenia	132	62,56%
	Croatia	79	37,44%
Number of employees	50-249	126	59,72%
	250-1000	49	23,22%
	more than 1000	36	17,06%
Sales revenue in 2015	up to and including 10 million €	46	21,80%
	more than 10 million and up to and including 50 million €	81	38,39%
	more than 50 million €	66	31,28%
	no answer	18	8,53%

Our final sample consists of 132 answers from Slovenia, being 62,56% of the sample, and 79 answers from Croatia, being 37,44% of the sample. Majority of total surveyed organizations have between 50 and 249 employees (59,79%), while there is 23,22% those which have between 250 and 1000 employees and 17,06% of those employing more than 1000 employees. When looking at the sales revenue for the year 2015, there is 21,80% of the total surveyed organizations with the sales revenue in 2015 being less than 10 million euros. Majority of the surveyed organizations had sales revenue in 2015 between 10 and 50 million euros (38,39%), while 31,28% of them had more than 50 million euros of sales revenue in 2015 (31,28%). 8,53% of the respondents preferred not to give the answer to this question.

D. Statistical methods

In order to check the reliability of the research instrument, a reliability analysis using Cronbach's alpha coefficients has been conducted. Since all calculated Cronbach's alpha coefficients were above the cut-off value of 0.70 recommended by [19], we concluded that the scales used in this research have good overall reliability. We based the validity of used research instruments on the fact that both PPI and OCAI were broadly used in previous researches (e.g. [5]). The validity of a social BPM part of the research instrument was tested during the preliminary phase of the survey execution and was based on the opinions of experts from both academic and practice population.

With the purpose of testing the assumption of the normality of distributions, a Kolmogorov-Smirnov (K-S) test has been used. The results indicated the data distribution is normal for both BPM and social BPM in groups with clan, adhocracy and market organizational cultures. However, the results indicate that the data distribution is not normal in either case in the group with

hierarchy organizational culture, nor if calculated on overall data. On the other hand, with the purpose of testing the assumption of the homogeneity of variance, a Levene's test has been used. The results indicated the variances are not significantly different for social BPM, but are significantly different for BPM. Since neither the assumption of normality of the data distribution nor the assumption of the homogeneity of variance were tenable, the nonparametric tests have been used in the further data analysis. In order to test if the differences in BPM maturity levels and the usage of social BPM across the different organizational culture types are statistically significant, a Kruskal-Wallis test has been used. For the purpose of testing if the differences between Slovenia and Croatia in BPM maturity level and the usage of social BPM within the organizations are statistically significant, a Mann-Whitney U Test has been used.

VI. RESULTS AND DISCUSSION

One of the aims and first research question of this study was to examine the current state of BPM maturity and the usage of social BPM within the companies from Croatia and Slovenia. The results of the BPM maturity research were based on the calculated PPI score. PPI showed that majority of surveyed Croatian companies are currently in the middle BPM maturity phase, being process management initiation (58,23%). These results are in line with previous research of BPM maturity in Croatia (e.g. [17]) and indicate some kind of stagnation in the progress of Croatian companies in achieving the highest level of BPM maturity. On the other hand, majority of surveyed Slovenian companies are in the highest BPM maturity phase, with 53,79% of them being in the process management mastery phase. The average PPI for Croatia is 36,16, while the average PPI for Slovenia is 40,14. The results through levels of BPM maturity are presented in Table V. Regarding the social BPM, the results indicate slightly higher usage of social BPM in companies from Slovenia than those from Croatia. In Slovenia, the usage of social BPM score is above 4 (out of maximum 5) in 46,21% cases, while in Croatia there were 41,77% of them.

TABLE V. BPM MATURITY LEVEL RESULTS

BPM maturity level	Total (N=211)	Slovenia (N=132)	Croatia (N=79)
1-process management initiation	4,27%	1,52%	8,86%
2-process management evolution	49,76%	44,70%	58,23%
3-process management mastery	45,97%	53,79%	32,91%

The second research question concentrates on the existence of statistically significant differences between the PPI and social BPM results regarding the respondents' country. In order to answer the stated question, a Mann-Whitney U Test has been used and the results indicated there are statistically significant differences between Croatia and Slovenia regarding both PPI results ($p < 0,01$) for the BPM maturity and social BPM results ($p < 0,05$).

With the intention of answering the third research question, a Kruskal-Wallis test has been performed to investigate whether the differences between BPM maturity results between organizational culture groups are statistically significant in overall sample. With the purpose of testing group differences based on the dominant organizational culture type, the sample size has been taken into consideration. Since there was an unequal ratio of responses between groups, in some groups there would be a very small number of responses if looking the samples from each country. Instead, the combined sample for Slovenia and Croatia has been used in order to mitigate this ratio. Table VI presents the results of the Kruskal-Wallis test for BPM. The results indicate that BPM maturity is significantly different between the organizations with different dominant organizational cultures ($H(3)=15,31$, $p<0,01$). When looking at the results, it is evident that BPM maturity had higher scores in organizations where dominant organizational culture is adhocracy or clan than in those where hierarchy or market organizational culture is the dominant one. These results are to some extent in line with the similar previous research conducted by [5] in Croatia and Slovenia as well. In her work, [5] reported the highest BPM maturity score in clan organizational culture.

TABLE VI.   BPM MATURITY THROUGH DIFFERENT ORGANIZATIONAL CULTURES

Kruskal-Wallis ANOVA by Ranks; BPM Independent (grouping) variable: OC Kruskal-Wallis test: H (3, N=211) =15,31213 p =0,0016			
Organizational culture	Valid N	Sum of Ranks	Mean Rank
Clan	70	8813,500	125,9071
Adhocracy	15	1890,500	126,0333
Market	54	4891,500	90,5833
Hierarchy	72	6770,500	94,0347

Finally, same statistical method has been used in order to answer the fourth research question, examining existence of statistically significant differences between organizations with different dominant organizational cultures regarding the usage of social BPM. The results of the Krusakal-Wallis test for social BPM are shown in Table VII. As in the previous case, the results showed there are statistically significant differences in the levels of social BPM usage in the organizations from different dominant organizational culture groups ($H(3)=22,95$, $p<0,01$). Surveyed organizations with dominant adhocracy and clan organizational culture had better scores regarding the usage of social BPM than those in the market organizational culture group, while the lowest scores were obtained from the organizations with hierarchy organizational culture as dominant. These results could be explained if looking the organizational culture types' characteristics as presented in Table II and social BPM principles as presented in Table III. The concept of social BPM is based on the principles of social software, mainly user engagement, collaboration and real-time reaction. These principles are similar to the characteristics of adhocracy and clan culture types, where flexibility, creativity and teamwork are the main factors. Following that, the higher usage of social BPM score within the

companies with adhocracy and clan organizational culture does not come as a surprise.

TABLE VII.   SOCIAL BPM THROUGH DIFFERENT ORGANIZATIONAL CULTURES

Kruskal-Wallis ANOVA by Ranks; social BPM Independent (grouping) variable: OC Kruskal-Wallis test: H (3, N= 211) =22,95403 p =0,0000			
Organizational culture	Valid N	Sum of Ranks	Mean Rank
Clan	70	9120,500	130,2929
Adhocracy	15	1958,000	130,5333
Market	54	4949,000	91,6481
Hierarchy	72	6338,500	88,0347

VII. CONCLUSION

This study has presented the results of the BPM maturity and usage of social BPM research conducted in Slovenia and Croatia in 2016. It has been shown that Slovenian companies are at the higher level of BPM maturity and higher level of usage of social BPM than Croatian companies. While majority of Slovenian companies are at the upmost BPM maturity level, majority of Croatian ones are still at the middle level. However, the limitations of this study include unequal ratio of Slovenian and Croatian companies included in the research.

Regarding the organizational culture part of the research, the results indicated there are statistically significant differences between groups of companies with different dominant organizational culture type when investigating BPM maturity and the usage of social BPM. For the future research it is planned to further examine the role that organizational culture type has on the usage of social BPM and on achieving higher level of BPM maturity.

REFERENCES

[1] H. F. A. Ahmad, and M. Zairi, "Business process reengineering: critical success factors in higher education," Business Process Management Journal, vol. 13, no.3, pp. 451-469, 2007.

[2] A. Alibabaei, M. Aghdasi, B. Zarei, and G. Stewart, „The Role of Culture in Business Process Management Initiatives," Australian Journal of Basic and Applied Sciences, vol. 4, no. 7, pp. 2143-2154, 2010.

[3] V. Bosilj Vukšić, Lj. Milanović Glavan, and Z. Merkaš, "The Success Factors of Business Process Management: a Case Study of Croatian Company," in Proceedings of the 8th International Conference "An Enterprise Odyssey: Saving the Sinking Ship Through Human Capital", June 08 – 11 2016, L. Galetić, I. Načinović Braje and B. Jaković, Eds. Zagreb, Croatia: Faculty of Economics and Business, 2016, pp. 697-706.

[4] G. Bruno, F. Dengler, B. Jennings, R. Khalaf, S. Nurcan, M. Prilla, M. Sarini, R. Schmidt, and R. Silva, "Key challenges for enabling agile BPM with social software," Journal of Software Maintenance and Evolution: Research and Practice, vol. 23, no. 4, pp. 297-326, 2011.

[5] B. Buh, Approaches towards business process management adoption under different organizational cultures, doctoral dissertation. Ljubjana, Slovenia: Faculty of Economics, 2016.

[6] K. S. Cameron, and R. E. Quinn, Diagnosing and changing organizational culture: Based on the competing values framework. Reading, MA: Addison-Wesley, 2006.

[7] T. de Bruin, and G. Doebeli, "An organizational approach to BPM: the experience of an Australian transport provider," in Handbook on Business Process Management 2, International Handbooks on Information Systems, J. vom Brocke and M. Rosemann, Eds. Berlin: Springer, 2010, pp. 559-577.

[8] M. Dumas, M. La Rosa, J. Mendling, and H. A. Reijers, Fundamentals of Business Process Management. Heidelberg: Springer, 2013.

[9] Economic lexicon, Organizational culture, 2nd ed, Zagreb, Croatia: Leksikografski zavod Miroslav Krleža & Masmedia, 2011.

[10] S. Erol, M. Granitzer, S. Happ, S. Jantunen, B. Jennings, P. Johannesson, A. Koschmider, S. Nurcan, D. Rossi, and R. Schmidt, R. "Combining BPM and social software: contradiction or chance?," Journal of software maintenance and evolution: research and practice, vol. 22, no. 6/7, pp. 449-476, 2010.

[11] C. Grau, and J. Moormann, "Investigating the Relationship between Process Management and Organizatinal Culture: Literature Review and Research Agenga," Management and Organizational Studies, vol.1, no.2, pp. 1-17, 2014.

[12] G. Hofstede, "Culture constraints in management theories.," Academy of management executive, vol. 7, no. 1, pp. 81-94, 1993.

[13] B. Hribar, and J. Mendling, "The correlation of organizational culture and success of BPM adoption," in Proceedings of the European Conference on Information Systems (ECIS), June 9-11 2014, Tel Aviv, Israel, 2014.

[14] T. Jung, T. Scott, H. T. O. Davies, P. Bower, D. Whalley, R. McNally, and R. Mannion "Instruments for the Exploration of Organisational Culture: A Review of the Literature," Public Administration Review, vol. 69, no. 6, pp. 1087 – 1096, 2009.

[15] M. Kocbek, G. Jošt, and G. Polančič, "Introduction to Social Business Process Management," in Knowledge Management in Organizations, L. Uden, M. Heričko, and I. H. Ting, Eds. Switzerland: Springer International Publishing, 2015, pp. 425-437.

[16] P. Mathiesen, W. Bandara, and J. Watson, "The affordances of social technology: a BPM perspective," in Proceedings of the 34th International Conference on Information Systems (ICIS), December 15-18 2013, Milan, Italy, 2013.

[17] Lj. Milanović Glavan, Conceptual model of process performance measurement system, doctoral dissertation. Zagreb: Faculty of Economics and Business, 2014.

[18] B. Niehavens, and J. Henser, "Business Process Management beyond Boundaries? – A Multiple Case Study Exploration of Obstacles to Collaborative BPM," 44th Hawaii International Conference on System Sciences, January 4-7 2011, Hawaii, USA, 2011.

[19] J. C. Nunnally, and I. H. Bernstein, Psychometric Theory, 3rd ed. New York: McGraw-Hill, 1994.

[20] N. Pflanzl, and G. Vossen, "Human-Oriented Challenges of Social BPM: An Overview," in Enterprise Modelling and Information Systems Architectures, Lecture Notes in Informatics, vol. P-222, R. Jung, and M. Reichert, Eds. Bonn, Germany: Köllen Druck and Verlag GmbH, 2013, pp. 163–176.

[21] D. I. Prajogo, and C. M. McDermott, "The relationship between total quality management practices and organizational culture," International Journal of Operations & Production Management, vo. 25, no. 11, pp. 1101-1122, 2005.

[22] R. E. Quinn, Beyond Rational Management. San Francisco. CA: Jossey-Bass, 1988.

[23] R. E. Quinn, and J. Rohrbaugh, "A spatial model of effectiveness criteria: towards a competing values approach to organizational analysis," Management science, vol. 29, no. 3, pp. 363-377, 1983.

[24] M. E. Rangiha, and B. Karakostas, "A Socially Driven, Goal-Oriented Approach to Business Process Management," International Journal of Advanced Computer Science and Applications, Special Issue on Extended Papers from Science and Information Conference, pp. 8-13, 2013.

[25] S. Reiter, G. Stewart, and C. & Bruce, "Integrating Qualitative and Quantitative Approaches," in Cross-cultural Research, Proceedings of the Sixteenth Americas Conference on Information Systems, August 12-15 2010, Lima, Peru, 2010.

[26] M. Röglinger, J. Pöppelbuß, and J. Becker, "Maturity models in business process management," Business Process Management Journal, vol. 18, no. 2, pp. 328 – 346, 2012.

[27] M. Rosemann, and J. vom Brocke, "The Six Core Elements of Business Process Management," in Handbook on Business Process Management 1: Introduction, Methods and Information Systems, J. vom Brocke and M. Rosemann, Eds. Berlin: Springer, 2010, pp. 107-122.

[28] G. Rummler, and A. Brache, Improving Performance: How to Manage the White Space on the Organization Chart. San Francisco, USA: Jossey-Bass Publishers, 1990.

[29] Rummler-Brache Group (2004), Business process management in US firms today, available at: http://rummler-brache.com/upload/files/PPI_Research_Results.pdf (accessed 22 February 2016)

[30] E. H. Schein, "Organizational Culture," American Psychologist, vol. 45, no. 2, pp. 109-119, 1990.

[31] R. Schmidt, and S. Nurcan, "BPM and social software," in Business Process Management Workshops, Berlin Heidelberg: Springer, 2009, pp. 649-658.

[32] T. Schmiedel, J. vom Brocke, and J. Recker, "Which cultural values matter to business process management? Results from a global Delphi study," Business Process Management Journal, vol. 19, no. 2, pp. 292-317, 2013.

[33] R. Skrinjar, V. Bosilj Vuksic, and M. Indihar Stemberger, "The impact of business process orientation on financial and non-financial performance," Business Process Management Journal, vol. 14, no. 5, pp. 338-350, 2008.

[34] D. Suša, V. Bosilj Vukšić, and D. Ivandić Vidović, "A Role of Organizational Culture in Business Process Management: a Case Study," in Proceedings of the 8th International Conference "An Enterprise Odyssey: Saving the Sinking Ship Through Human Capital", June 08 – 11 2016, L. Galetić, I. Načinović Braje and B. Jaković, Eds. Zagreb, Croatia: Faculty of Economics and Business, 2016, pp. 697-706.

[35] A. Van Looy, M. De Backer, G. Poels, and M. Snoeck, "Choosing the right business process maturity model," Information & Management, vol. 50, no. 7, pp. 466-488, 2013.

[36] J. vom Brocke, and T. Sinnl, "Culture in business process management: a literature review," Business Process Management Journal, vol. 17, no.2, pp.357-378, 2011.

[37] Y. Yang, and J. Hsu, "Organizational process alignment, culture and innovation," African Journal of Business Management, Vol. 4, no. 11, pp. 2231-2240, 2010.

Efficient Social Network Analysis in Big Data Architectures

Iva Sorić, Davor Dinjar, Marko Štajcer and Dražen Oreščanin
Poslovna Inteligencija d.o.o., Zagreb, Croatia
{iva.soric, marko.stajcer, davor.dinjar, drazen.orescanin}@inteligencija.com

Abstract - Social network analysis (SNA) is the application of graph theory to understand, categorize and quantify relationships in a social network. It can be a great tool to improve analytic capabilities in any field, for example marketing analytics, churn prediction, health care, etc. In terms of SNA, network structure is defined by nodes, edges and metrics which quantify the importance or influence of certain nodes in the network or relationship strength between nodes. Algorithms for network metrics calculation are complex and that makes SNA difficult to implement in big data environments on large datasets with many nodes and edges. In this paper we will elaborate how to efficiently and performance wise perform SNA and visualize results of the analysis on large datasets using increasingly popular GraphX and JavaScript libraries.

I. INTRODUCTION

Although the concept of social network analysis is not new, with the widespread availability of data and the progress made in recent years in computer science, social network analysis was found valuable in many fields. The increasing expansion of Internet and the growing connectivity have brought a higher level of attention to graph analytics. The benefits of analyzing and visualizing the network type data from a different perspective became clearer, which is something specific for SNA. Instead of putting the focus on the entities and their properties, SNA puts the focus on the links between them - the relations and the structure of the graph they form.

The 'network approach' means studying individuals as part of a network structure, in terms of the relations to other individuals, not just entities for themselves. This approach brings new insights, but for some time its computational complexity represented an obstacle to applying these algorithms to ubiquitous large networks. However, with the recent development of new data processing tools and techniques, new tools suited to analyzing large graphs were also developed. One of them is Spark GraphX – a library developed for large scale graph analytics.

II. CHALLENGES OF EFFICIENT SNA

A. SNA basic concepts

Social network analysis applies graph theory to study the relationships (edges or links) between entities (nodes or vertices) in order to better understand and evaluate the network and its actors. It includes social media data, but also any kind of social structure where there is information sharing and a system of connections that can be captured in a graph. Such data can be found in various fields and domains beyond social sciences, e.g. in financial industries for transaction analysis, in telecommunication for fraud detection, in any organization for diffusion of information analysis, Internet traffic, or even in health care for analyzing the spread of contagious diseases.

Nodes and edges, which compose a graph, are the main abstractions in SNA. Nodes are entities – persons, organization or items, and the links describe the relationship between them. Both the nodes and the edges can have properties that describe them, but the main focus of the analysis are the relationships and the structure, not the individual properties. SNA focuses on uncovering the patterns in the links, i.e. interactions between nodes, analyzing the communication flow in the network structure, identifying the individuals and groups playing central roles, or identifying isolated individuals. Some of the usually asked questions are: Who are the key players (by some criteria)? Who is isolated? Which connections act as bridges between groups? Are there clusters and how are they formed? Is there a hub? Etc.

To answer those questions, SNA defines a number of metrics that reveal certain characteristics of the network or specific nodes. On the network level there are metrics designed for better understanding of the overall network structure like density – the ratio of the number of existing connections and the number of all possible connections, diameter – the longest geodesic (shortest path between two nodes), and many others. On the node level, most common are centrality measures, which include: degree, strength, closeness, betweenness, eigenvector centrality, and PageRank. Their goal is to identify the most central and most important nodes in the network. There are also clustering algorithms to detect groups of nodes that communicate most frequently, etc.

The analysis includes both a mathematical and a visual component. In the next few chapters we describe the benefits of using GraphX library for the mathematical analysis. Additionally, we show how visualization using JavaScript can be used to effectively present graphs, to allow a user to comprehend the overall network structure, and a detailed view of the analysis results with the ability to filter and more closely examine certain nodes of interest. Fig. 1 describes the proposed architecture.

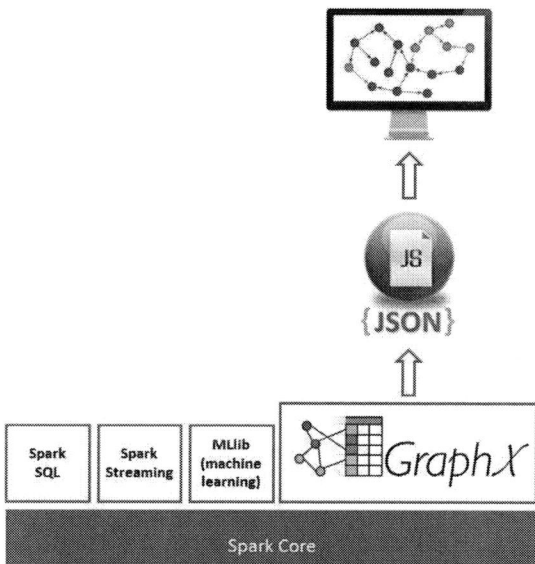

Figure 1. Schema of the proposed architecture

B. Tools development

Recently there was a lot of advances in developing general-purpose distributed processing tools, i.e. data-parallel systems like Hadoop MapReduce and Spark, developed as a response for many big data problems. Unfortunately, directly applying specific graph algorithms in those environments turned out to be a poor solution. On the other hand, developers turned to graph-parallel systems for specialized graph processing problems. These systems perform a lot better than general-purpose tools, but their big disadvantage remains that they can only be used for graph specific problems. For a larger analytics pipeline, which is often the case, they would have to be integrated with other systems, which would require unnecessary data movement and duplication [1].

In this paper we present a tool that combines the advantages of both approaches - the graph-parallel approach and the general distributed data processing tools, and thus addresses graph analytics problems in an efficient and yet generally applicable way.

C. Graph-parallel approach

Most of the SNA metrics require implementing complex and iterative algorithms, whose execution time is often not satisfying, especially for very large graphs. For example, PageRank, famously developed by Google for calculating the importance of web pages, now also a popular algorithm for analyzing other types of networks, is implemented through an iterative algorithm to calculate an approximation of the node's importance.

PageRank of a node depends on the PageRank value of its incoming nodes. Also, the number of node's outgoing links affects the PageRank value of the node it connects to: the more outgoing links it has, the lower the value it propagates to the connected node. Thus, important

nodes are the ones who are connected to many other nodes, but also the ones who are connected to fewer, but more important nodes. PageRank values are approximated by running an iterative algorithm. In the start the algorithm assigns an initial value to each vertex, and then in each step upgrades the value based on the values propagated from its incoming links [2].

Therefore, the algorithm is based on iterative local changes; vertex properties (in this case PageRank values) are transformed in each step recursively based on the properties of its direct neighbors. Graph-parallel systems are based on executing these changes in parallel and they are optimized for that kind of iterative graph algorithms. However, they are only suited to computation on static graph structure; their 'vertex-centric model' does not work well when there is a need to modify the graph structure like graph coarsening [3]. Furthermore, it is not designed for ETL kind of tasks like graph construction or manipulating the computation results.

III. GRAPHX AS A UNIFYING SOLUTION

A. Graph processing inside a larger platform

GraphX is Apache Spark's built-in library for graph analytics and graph-parallel computation. GraphX leverages graph-parallel computation for the best performance in graph-specific algorithms, but also keeps the ability to interoperate with other tools and modules for general data processing inside the Spark framework. That leaves the possibility to make graph computation a part of a larger, more general analytics pipeline, which is often a requirement of an extensive analytics project. For example there can be multiple data sources, the data needs to be joined, cleaned and prepared for the graph computation, afterwards the results need to be analyzed further and joined with other information, etc. Furthermore, GraphX inherits all Spark's good characteristics [4] like:

- Scalability

GraphX is designed for scalable graph computation. It is suitable for performing complex computation on very large graphs.

- Fault-tolerance

Unlike many specialized graph processing tools, GraphX as part of a general distributed data processing tool – Spark, does not sacrifice fault-tolerance in favor of latency.

- Integration with the rest of the Spark platform

Specialized graph tools are restricted to a limited range of graph-centric tasks, everything else is beyond their scope. For example, they require the data to be in a specified format, whereas with GraphX the user can easily transform, filter and clean the data the way the analysis needs it. It enables flexible graph construction (ETL tasks) and other data processing tasks after the graph computation part.

- Interactive computation (Spark shell)

All Spark modules can be used and tested interactively from the Spark shell for better developer productivity.

B. GraphX data model

Spark GraphX is based on some key data types that are actually extensions of Spark's main abstraction – RDD but optimized for graphs: VertexRDD, EgdeRDD, and Graph. Like all RDD-s, they are distributed, immutable, and fault-tolerant. VertexRDD is composed of a unique numeric vertex id and the properties related to the vertex. EdgeRDD contains id-s of the source and target vertex, and edge properties. Both the VertexRDD and the EdgeRDD have internal indices for fast joins. Graph class abstracts a property graph and provides methods for graph-oriented calculations. It is created using the VertexRDD and EdgeRDD i.e. it is essentially a pair of a partitioned collection of vertices and a partitioned collection of edges. Consequently, all RDD methods are available. [5] [6]

GraphX allows the Graph to be viewed and operated with in both a table-oriented and graph-oriented way. There are methods for analyzing and modifying nodes and edges properties, and also performing graph specific algorithms or changing the graph structure. Furthermore, since GraphX is a part of the Spark platform, it can be used together with other modules, including SparkSQL and DataFrames API, MLib, etc, which means that the whole analytics process, with data computation and graph computation steps, can be implemented within a single application, without sacrificing the performance of the graph specific tasks. For example a user can read raw data from various sources, clean and filter the data, extract the nodes and edges to perform graph analytics tasks, analyze the results, and optionally repeat steps for a different piece of data, or a subgraph. The biggest advantage of GraphX is that it allows graph computation to be performed as a part of a bigger analytics workflow inside a single platform.

C. Best of the graph-parallel and data-parallel world

GraphX is an embedded part of a data-parallel framework and works in that environment, but achieves similar performance as specialized graph systems by using multiple distributed join optimizations. Its graph data model implements graph-parallel abstraction as a specific pattern of join stages and group-by stages, with intervening map operations [3]. Specific graph computation patterns are reconstructed as dataflow optimizations to achieve similar performance without sacrificing the computational flexibility.

The optimizations include the way graphs are implemented as a pair of distributed collections – VertexRDD and EdgeRDD, and the way they are processed to recast the graph-parallel model by using common dataflow operators – join, group by, map. To achieve efficient distributed execution, GraphX enables vertex-cut partitioning scheme; graphs are split along vertices, which has been shown to minimize the communication and movement of data [7]. By leveraging these optimizations, GraphX performance does not fall behind specialized graph systems, and still keeps the generality and applicability.

IV. VISUALIZATION USING JAVASCRIPT

JavaScript is very commonly used for various types of visualizations across many web projects. The most common type of visualization is chart visualization which shows many different aspects of data, which can be interactive.

There are several libraries for SNA analysis which support render of edges and nodes as a network which can also be interactive. Most commonly used are SigmaJS, Linkurios and JSNetworkX [8][9][10].

JSON is the most common data format used for loading data in an SNA graph. It is a lightweight data-interchange format. It is easy for humans to read and write and is also easy for machines to parse and generate. GEXF is a language for describing complex networks structures, their associated data and dynamics.

Most of these libraries support parsing JSON or GEXF data format, but there is a limit to how much nodes and edges JavaScript can render before the page script stops being responsive, so JavaScript developers must be careful not to load too many nodes and edges in the graph network.

One way to reduce the number of nodes and edges is to provide filters on the page, which can help the user to pre-filter the data. Interactivity provides users with ability to move nodes and edges, and show only the ones which are interesting for any given business aspect. The whole process of rendering SNA graph is mainly done by the library, and the user's part is only to provide the data which needs to be loaded in graph. This data is most commonly consisted of two arrays. First array represents nodes with common properties (size, color), but extended properties can also be wrapped in nodes collection.

Other array is the edges array that describes the relations between nodes which can also be extended with additional properties that can later be rendered in network graph. Extended properties can be added to allow better visualization of network with custom icons, and more information in it, but it's important to remember that JavaScript execution becomes slower over large collections with many attributes, and the code complexity also increases with each new attribute.

SNA graph visualization is preferred to be used with smaller number of nodes and edges, where it is easy to identify crucial information needed to understand the relation between nodes, and render it smoothly in the user's browser. As new JavaScript libraries are introduced, a trend is noticed that there is a need to construct more complex networks with large number of nodes for big data analysis, so some newer libraries support loading large collections of arrays, but there is still an issue with rendering them in the web browser, as it is not designed to handle large number of DOM elements. WebGL rendering engine is used for that, but it requires users to have a better graphics card, which is then used to render large networks in one canvas element in 3D. However big data SNA analytics is still a concept, while currently big data is separated into sections, filtered and then loaded into smaller chunks with any given SNA JavaScript library such as SigmaJS or Linkurios.

JavaScript events can help introduce better user experience while browsing through graph, since each zoom or mouse click can be handled with JQuery or a similar library. Some SNA libraries already support event handling so the user is not required to write any additional code except extend logic with only specific behavior. Coloring is also commonly used to separate specific graph sectors, and visually distinct them. Colors can be injected in JSON as hex codes or string values and then rendered. For most SNA graph libraries coloring is natively supported for nodes and edges.

Searching through SNA network graph is done in JavaScript by iterating nodes or edges array and setting visibility attribute of filtered elements. After each search SNA graph is reduced to size of only the filtered result, which increases performance of whole network render in browser and also provides more detail result to user, which can then focus only on the part of the network which is interesting for business his business case. Also additional properties provided in data can allow the render of graph to support custom icons, images, and additional data which can then be displayed in filtered output with more detail keeping in mind that each of this attributes increases the dataset size. Custom attributes can also be used to provide better filtering features in graph, so user can filter data with more specific information. There is no limitation in type of custom attributes. Custom attributes can be of any type, so it is possible to define icons or images with custom URL string, dimension values as decimal numbers, or even Date or Boolean values.

Most of this JavaScript libraries also support zooming into data, which allows user to see specific part of SNA network with more detail. Usually the initial render of the graph is so out of scope that the user must zoom in to a specific part, so this comes naturally to the user and provides better user experience with higher level of data insight. Nodes are usually represented as dots, and edges are represented by lines, which can be straight or curved depending on the library used. Clicking on each node or edge, the user can get more focus to selected data area, and view only nodes connected to the one he selected, but also only edges related to that node. The way this is achieved through JavaScript is similar to searching; first step is to find selected node in provided nodes array, and then all the connecting nodes are easily identified, and all other nodes are just programmatically hidden. Usually library redraws the whole graph with simple Draw() or Refresh() method so this must be called after the filtering is done. An important thing to mention here is that this works smoothly with smaller data arrays, but as arrays get bigger, you might get unresponsive behavior. Best practice for rendering SNA graph is to always provide

filters which reduce the size of initial render to only a portion of the dataset, which can then be loaded with any provided JavaScript library, and then further drilled in more detail using mouse zoom. By drilling into data with mouse zoom, the user can better focus on a particular segment of graph.

JavaScript as a programming language for integration with SNA graph analysis proves to be a good solution if you need fast insight into your data, but for larger data collections it is not recommended since HTML document object model isn't made to be used for complex graph analysis, so this is a limitation of using JavaScript in social network analysis.

V. CONLUSION

This article presents a big prototype platform for SNA analysis in Big Data environment, and gives an overview of the architectural integration in combination with multiple technologies, frameworks and techniques through Big Data architecture. It is shown how the Spark GraphX framework in combination with JavaScript provides a platform for efficient social network analysis with different types of powerful and interactive visualizations. Since SNA can find its application in different industries for different use cases, the presented architecture has a great potential in today's big data world.

REFERENCES

[1] Joseph E. Gonzalez, Reynold S. Xin, Ankur Dave, Daniel Crankshaw, Michael J. Franklin, and Ion Stoica, "GraphX: Unifying data-parallel and graph-parallel analytics", arXiv:1402.2394v1 [cs.DB], 2014.

[2] Sergey Brin and Lawrence Page, "The anatomy of a large-scale hypertextual Web search engine", April 1998.

[3] Joseph E. Gonzalez, Reynold S. Xin, Ankur Dave, Daniel Crankshaw, Michael J. Franklin, and Ion Stoica, "GraphX: Graph processing in a distributed dataflow framework", Proceedings of the 11th USENIX Symposium on Operating Systems Design and Implementation, October 2014.

[4] Spark GraphX documentation, retrieved February 2017 from http://spark.apache.org/graphx/

[5] Sean Owen, Sandy Ryza, Uri Laserson, and Josh Wills, "Advanced analytics with Spark: Patterns for learning from data at scale", O'Reilly Media, 2015.

[6] Mohammed Guller, "Big data analytics with Spark: A practitioner's guide to using Spark for large scale data analysis", Apress, 2015.

[7] Reynold S. Xin, Joseph E. Gonzalez, Michael J. Franklin, and Ion Stoica, "GraphX: A Resilient Distributed Graph System on Spark", Proceeding GRADES '13 First International Workshop on Graph Data Management Experiences and Systems, June 2013.

[8] Sigma.js documentation, retrieved February 2017 from http://sigmajs.org/

[9] Linkurious documentation, retrieved February 2017 from https://linkurio.us/

[10] NetworkX documentation, retrieved February 2017 from https://networkx.github.io/

Integrating evolving MDM and EDW systems by Data Vault based System Catalog

D. Jakšić*, V. Jovanović** and P. Poščić*
* Department of informatics-University of Rijeka/ Rijeka, Croatia
** Georgia Southern University/ Statesboro, GA, USA
dsubotic@inf.uniri.hr, vladan@georgiasouthern.edu, patrizia@inf.uniri.hr

Abstract - The paper presents results of a research on integration of enterprise data warehouse (EDW) and a master data management (MDM) system. The primary goal was solving a schema evolution problem, and the corner stone of our approach was utilization of a data vault modeling of an integrated meta-model of EDW and MDM as an expansion of a traditional relational database system catalog. The main contributions of this paper are: a) common integration architecture, b) new system catalog based on a meta-model for DW and MDM integration, and c) research prototype used for empirical validation of the effectiveness of the proposed solution.

I. INTRODUCTION

A data warehouse is "a subject-oriented, integrated, time-variant and non-volatile collection of data in support of management's decision making process" [15]. This means that a data warehouse (DW) can be used to analyze a particular subject area (such as sales, marketing, finance, etc.), it integrates data from multiple heterogeneous data sources, it keeps the history of data, and it never alters the data once it enters a DW. A simpler form of a DW is a data mart (DM). DM is focused on a single subject area and it draws its data not from all the DW data sources, but from a limited number of them (such as retail sales applications with daily transactions). For the logical representation of a DW (or DM), we traditionally use a denormalized dimensional model [12][20] in a combination with a 3NF model for modelling a central and integrated enterprise data warehouse (also called operational data store) [12][15][16]. DW environment (namely its data sources) nowadays is in a state of constant structural (schema) change. Master data management (MDM) comprises the processes, governance, policies, standards and tools that consistently define and manage the critical data of an organization to provide a single point of reference [3][23]. Some of the fundamental tasks of MDM system are duplicate removal, data standardization, and rule implementation and incorporation - all with a goal to eliminate incorrect data from entering the system and to create an authoritative source of master data for further distribution. It is a method of enabling a business organization to link all of its critical data to a common point of reference, which is further shared throughout all the departments and relevant employees. This way, a data quality of an organization greatly improves and the organization can better serve its clients, as well as improve their business by running a

more accurate and efficient business analysis and reporting (based on a "single version of the truth" data pool [15]). Dimensional model can also be used here - for the logical representation of MDM data (master data are represented as dimensions here). Master data are the key business entities and their descriptive attributes (e.g., the customer has a name, address, etc., the product has a name, color, weight, category, etc.), which are used by multiple systems, applications, and business processes of the organization as a unique source of data. Reference data can be internal or external and are used for the validation of other data. MDM environment is also in a state of constant change – data sources nowadays often change their structure and content. In this work our focus was on an enterprise size data warehouse (DW) and a master data management (MDM) system integration. We approached the problem through a development of a common system catalog meta-model, with the goal of solving a common schema evolution problem. Schema evolution occurs in both of these systems and is traditionally resolved separately. DW integrates current and historical data from many data sources and serves for business reporting and data analysis. MDM is traditionally used (by other business systems, applications, databases and data warehouses) as a physically independent database of master and reference data (also collected from many data sources). Data sources are often the same ones for DW and MDM and they often change their structure. These changes have to be implemented into both systems, so that they could accurately reflect the current (and historical) state of the real world - so that the DW could provide for effective business analysis and the MDM could achieve the optimal data quality throughout all the systems involved. We aim to resolve this problem on a common level - we state that the schema evolution problem can be viewed as a double issue: at a DW level and at the MDM level. From the DW perspective every event (fact) that is associated with dimensions is monitored, but from the MDM perspective the master data (dimensions) are monitored and the events (facts) give them context. And in both of these cases schema evolution problem to solve exists. The paper is organized as follows: section II gives a brief overview of a related work, section III describes our DW/MDM integration research (general research idea, our common integration architecture and integration part of a system catalog meta-model), section IV presents our research prototype and some relevant test results, and in section V we conclude our work with a brief summary and some plans and guidelines for the future.

II. RELATED WORK

With respect to the literature, the DW evolution can generally be traced through three approaches - schema evolution [5][10][29], schema versioning [1][9][25] and view maintenance [2][7][14]. The first two approaches are more interesting to our research because they are based on a DW defined as a multidimensional schema, and the third is based on a DW that is defined as a set of materialized views. Our extensive and comprehensive state-of-the-art on this problem can be found in [28], but from analyzing the related work we can generally conclude that the process of schema evolution and versioning is still demanding in terms of invested time and resources. It is necessary to balance the resource requirements and the quality of schema evolution process. Perhaps the biggest problem here is the preservation of schema consistency and data integrity (there is still a lack of an integrated system-of-records), as well as the simultaneous performance of temporal queries against multiple versions of the schema. Also, migration and transformation of data is still slow and expensive, the loss of information during these processes is still present and there is a lack of effective integration, organization and management of metadata. On the other hand, the view maintenance process can cause network saturation, depending on the amount of updated views and the amount of information they contain. The problems of anomalies and inconsistent changes in the views are still unresolved, the proposed approaches for view maintenance are still limited in terms of efficiency and performance, and the ETL processes, which are an integral part of most of today's DWs are completely ignored. Different approaches to solving the DW schema evolution problem are presented in literature (including a variety of techniques, algorithms, algebras, models, prototypes, methodologies and frameworks), but there is still no widely accepted solution and general framework for managing DW schema changes. More importantly, previous research does not emphasize the fact that the DW requirements, in this day and age, are increasing in the data and meta-data scope and structure (the growing number of data sources and more new and different types of data), which additionally requires developing some new approaches and solutions to the schema evolution problem.

III. MDM AND DW INTEGRATION RESEARCH

A. General research idea

As we already mentioned, the DW needs to preserve the history of data and metadata changes, as well as the history of schema and scope changes, for a very long time period [19][26]. On the other hand, the MDM needs to preserve a data quality of an organization by achieving and maintaining a "single version of the truth" data pool as a basis for accurate and efficient business analysis [3][23]. Seeing that the DW and MDM both integrate basically the same data sources and have the same schema evolution problem, we will integrate those two systems into one and address the evolution problem at the common level. Our main research question was: *"Can our new system catalog model serve to successfully integrate DW and MDM systems?"*. In order to answer this question, we developed a new, integrated architecture for these systems, as well as a system catalog model based on a Data Vault modelling methodology [21][22]. The data vault (DV) is a data modeling method designed for supporting the long-term storage of historical data collected from various data sources and tracking the origin of data contained in the database [21][22]. The DV model (due to structural separation, usage of empiric meta-data and addition-only policy, which relation model does not implement) is able to track the data source values and the history of changes, which is a vital function of a DW system-of-records [19][26][27]. Furthermore, we developed and tested a research prototype based on said architecture and system catalog model.

B. Common integration architecture

Fig. 1 shows our common integration architecture. Central and integrated enterprise DW/MDM (purple section in Fig. 1) is in focus of our research and consists of two parts - the raw copy of the source data (SDV) and the synchronized master and business data (PMDV). Both parts are based on the Data Vault (DV) method, in contrast to traditional approaches based on the relational model [12][15][20]. SDV is focused on obtaining and preserving the original and unchanged copy of data sources for the purpose of governance and auditing, and PMDV is focused on obtaining and preserving the data that has been modified according to business and master rules – the data that later feeds DMs and master dimensions (MDMs) and is oriented to user requirements. The integration of SDV and PMDV is carried out over an extended DBMS system catalog (i.e. our new meta-data repository, MDV), based on the DV model. MDV repository serves to integrate the two parts and to monitor history of changes of meta-data and their schemas, of the business rules and transformations and of mappings between the two systems. MDV contains the history of data sources' meta-data (their domains and schemas), of central DW/MDM integration and schema changes, of DMs and MDMs schema changes and of security schema changes (user access rights). We can say that MDV repository represents the DW on DW and is a key part of our new architecture. In materialized DMs and MDMs (blue section in Fig. 1) master data and business analysis and reporting data is stored. Data is stored according to the user requirements and a dimensional model is used for representation (data are summarized, aggregated and calculated). A traditional DMs are integrated here with the traditional MDMs – the PMDV forwards the data to MDMs so the MDMs could then forward the data to DMs and data sources. The end-user can then access the DM/MDMs and analyze the data through the selected data analysis and reporting tools. Additionally, the master data is returned from enterprise DW/MDM to the data sources in order to harmonize and refine the data within business organization. Master data collected from PMDV is maintained in one central MDM location and all source systems and applications, as well as DMs, are using this data. This directly affects the quality of data, which later re-enters the raw SDV and passes again through the described layers of the architecture. Also, by reducing the amount of data over which was necessary to make "heavy" transformations we thus speed up the process of

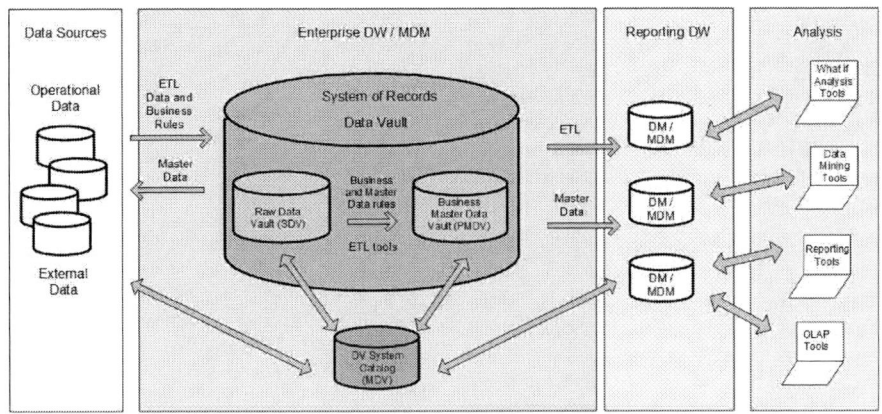

Figure 1.   Common integration DW/MDM system architecture

data integration [19]. However, these issues are out of the scope of this paper.

C.   Meta Data Vault (MDV) model for SDV/PMDV integration

Fig. 2 shows a simplified meta-data-vault (MDV) model for DW/MDM integration. Due to complexity and size of the full MDV system catalog model and the scope of this work, we will show here one small, but relevant part of the MDV model – the one for SDV/PMDV integration. The whole model is much more extensive because it serves for schema evolution purpose - it incorporates data sources, data marts, materialized views and issues of security - but those are out of scope of this paper. The model (as well as other models in this work) is made in IDEF1X method [17] with the use of the CA Erwin Data Modeler 9.5 modelling software [8]. In our MDV model, SDV/PMDV integration is managed through the same-as-links on four main hubs (H_HUB, H_LINK, H_SATELLITE, and H_ATTRIBUTE) which store meta-data about three main DV concepts in a data model (hubs, links and satellites respectively – additionally H_ATTRIBUTE stores data about satellite attributes; column meta-data). For example, if we have two different relational data sources which both have the table CUSTOMER, we integrate these sources into a DV based central EDW (EDV). This means that the relational table CUSTOMER from both sources becomes one hub H_CUSTOMER (stores business keys for CUSTOMER) with its satellites S_CUSTOMER1 (stores columns from CUSTOMER in Source1) and S_CUSTOMER2 (stores columns from CUSTOMER in Source2). This is considered as a copy of the original data sources (SDV),

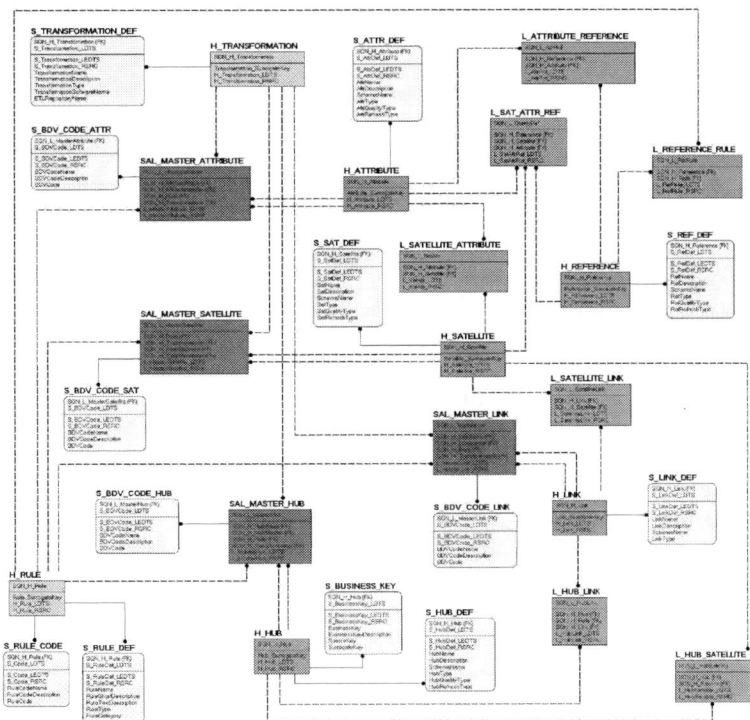

Figure 2.   MDV system catalog model for SDV/PMDV integration

1403

which is then copied into a PMDV where the business and master rules are applied. For this example it means that hub H_CUSTOMER gets only one satellite S_CUSTOMER – by the rules those two satellites are integrated into common one and the data is cleansed. Also, data in the H_CUSTOMER is integrated and de-duplicated. The corresponding meta-data is then stored in our MDV system catalog from Fig. 2. In the H_HUB and its satellites S_BUSINESS_KEY and S_HUB_DEF the meta-data about hub H_CUSTOMER (and generally about all the hubs in a enterprise DW/MDM – from both SDV and PMDV) is stored, including the data about names of the hubs, names of their keys, hub types and their general descriptions, as well as their load dates and record sources. Also, in the H_SATELLITE and its satellite S_SAT_DEF the meta-data about satellites S_CUSTOMER1, S_CUSTOMER2 and S_CUSTOMER is stored. The same is true for all other MDV structures (links, attributes, …). However, at this point the DW and MDM (SDV and PMDV) hubs are not yet integrated (list of hubs with their descriptive meta-data is just stored in a corresponding meta-hub and its meta-satellites - H_HUB, S_BUSINESS_KEY, S_HUB_DEF). For the integration we use the same-as link SAL_MASTER_HUB which relates a single hub from a list of hubs in H_HUB to another hub from that list, in a base-master relation (SDV hub is base one, PMDV hub is master). In the example it would mean that the hub H_CUSTOMER form PMDV becomes master hub to hub H_CUSTOMER in SDV. For the satellites, their integration is stored in the SAL_MASTER_SATELLITE link – for example, the S_CUSTOMER1 and S_CUSTOMER2 meta-data records stored in H_SATELLITE can have (separate) base-master relationship with S_CUSTOMER record. This relationship is stored as a record in SAL_MASTER_SATELLITE where S_CUSTOMER is master structure on S_CUSTOMER1 and S_CUSTOMER2. Additionally, SAL_MASTER_HUB (as well as other master links) relates to H_RULE and H_TRANSFORMATION, the two hub meta-concepts that store the data about business rules that can be applied to data and (ETL) transformations needed to apply those rules, respectively. This way, we get the complete history of SDV/PMDV integration – we can keep track of all the business rules and transformations applied to the specific integration, with SAL_MASTER_HUB storing the load date and record source data about that specific integration. In that way we can create and manage master hubs and store the history of their changes. SAL_MASTER_HUB has its satellite S_BDV_CODE_HUB, with descriptive data about the code (ETL or other) used for that specific integration. This satellite will later serve as a basis for automating the process of integration. The same can be applied to the other three main hubs in a MDV model (SAL_MASTER_LINK for H_LINK, SAL_MASTER_SATELLITE for H_SATELLITE and SAL_MASTER_ATTRIBUTE for H_ATTRIBUTE), so we can create complete and historicized "golden copy" of business and master data in PMDV, from the simple SDV raw copy of data. Regarding the reference table concept, as it exists only in a data model (it represents external or internal reference data and as such it is already consolidated and organized by internal business rules or external data standards), we have no need to further make

the master concepts in the meta-model. We will simply store its meta-data in a H_REFERENCE concept in MDV.

IV. RESEARCH PROTOTYPE AND RESULTS

A. Business case example

For the purpose of theoretical validation of the proposed architecture and DW/MDM integration, as well as building a prototype, we developed a simple business case (in a similar way as in [18]), which monitors the work of the employees on projects and their participation in job trainings and educations. We have two data sources (JobDB and TrainingDB) which we integrate into a common SDV and PMDV database. This integrated DW/MDM further feeds the MDM and local DMs with relevant data. JobDB data source schema is describing employees, their bosses and projects they are currently working. TrainingDB data source schema also describes employees, but this time in the context of business trainings in which they participate and competences they achieve through them. SdvDB and PmdvDB databases integrate these two data sources into a single schema, through the use of Data Vault model. However, due to size restrictions, business case data models will not be shown in this paper.

B. Prototype description

Research prototype has been developed according to the business case examples and has been tested by running a set of queries against SDV, PMDV and MDV databases, as well as the original relational system catalog. Fig. 3 shows the architecture of a prototype. Corresponding to business case models shown in the previous section, there are two source databases (JobDB and TrainingDB) which are then integrated into a raw copy of the source data (SdvDB). A business and master PmdvDB is built and loaded from the SdvDB and it keeps "purified" data. The MdmDB is loaded from PmdvDB and it practically represents the MDM system which keeps the master data. In MdmDB the same data as in PmdvDB is stored, but organized according to a dimensional model - in the dimensional database. MdmDB is then the basis for loading the DmDB, which is also based on a dimensional model and represents a local DM. This way, DW and MDM are integrated - MdmDB serves as the employee master data and the basis of loading the local DMs, and

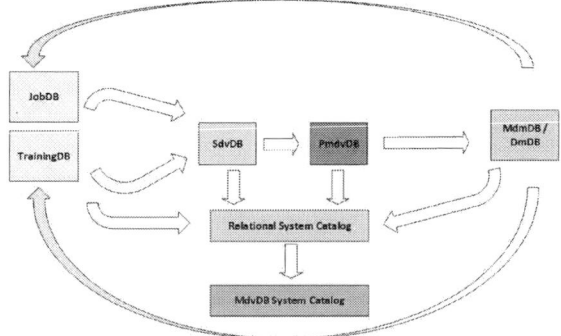

Figure 3. Prototype's architecture diagram

has the ability to return 'golden copy' of employee master data back to data sources. The prototype (with all of its separate databases) is developed and implemented on the same Windows 10 Education x64 operating system and is made in Microsoft SQL Server 2012 [24] database management system using Microsoft SQL Server 2012 Integration Services (SSIS) and SQL Server data Tool for Visual Studio 2012 for extracting, loading and transforming the data between specific databases. To generate the data with which we initially loaded the source databases JobDB and TrainingDB, we used the web tool FreeDataGenerator [11].

C. Testing the sustainability of the integration

As we already stated, our main research question was "Can our new system catalog model serve to successfully integrate DW and MDM systems?". We presume that the successful integration is a sustainabile one – if a MDV model for integration is developed (which it is, shown in Fig. 2) and if: a) a new system catalog (MdvDB in the prototype) collects and stores historical meta-data about SdvDB and PmdvDB schema mappings, and b) queries defined over MdmDB and DmDB return the same results. In order to prove the integration was successful we developed and conducted two simple tests on a prototype: a) run a query on MdvDB system catalog that can return information about SdvDB and PmdvDB schema mappings, and b) run equivalent queries over MdmDB and DmDB that return the same results.

For the test A we developed 8 queries on the new MdvDB system catalog that return information on SdvDB and PmdvDB structures (hubs, links, satellites and attributes) and their mappings, as well as the mappings between data sources and SdvDB structures, or mappings between PmdvDB and MdmDB structures. Fig. 4 shows one of those queries – a query which serves to monitor the mappings and integration between base hubs in SdvDB and their corresponding master hubs in PmdvDB. This way we can monitor the transformations and the origin of master data, as well as structural changes (Change Type and LEDTS columns in Fig. 4). Fig. 5 shows the query that returns a similar information on the history of mappings between data sources' tables and SdvDB hubs. All the queries that we have created and run by MdvDB system catalog successfully returned needed information on creation of PmdvDB structures from SdvDB structures, as well as the historical state of integration – thus we conclude the test A was successfully conducted.

For the test B we made a data change in the JobDB

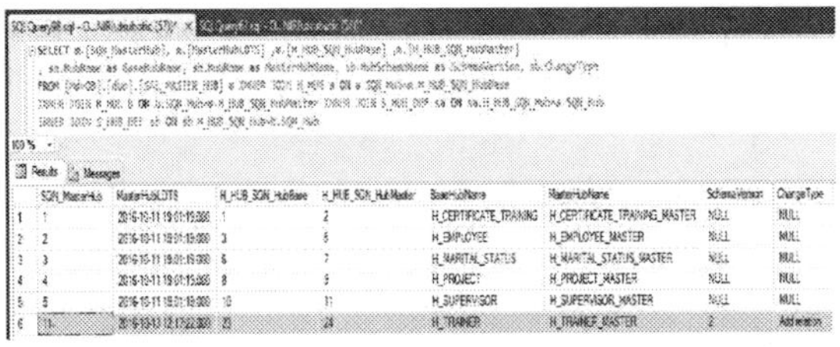

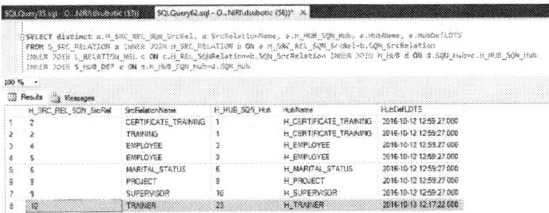

Figure 4. SdvDB and PmdvDB hub mappings stored in MdvDB system catalog

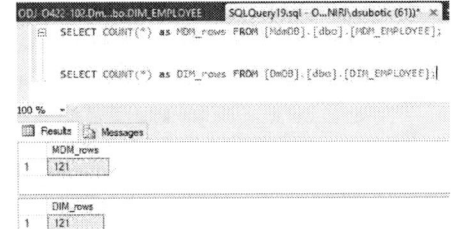

Figure 5. SdvDB and data sources hub mappings stored in MdvDB system catalog

data source, but only after loading with start data all the databases in the prototype (SdvDB, PmdvDB, MdmDB and DmDB). An employee Desiree Farmer became Desiree Farmer-Matthews – we changed her last name. Then we propagated this data change to all the levels in the prototype architecture. Fig. 6 shows a pair of equivalent queries on both MdmDB and DmDB – queries that return the number of rows in Employee master table after the data change. We can see they have returned the same results. Fig. 7 shows the data state in the MdmDB and DmDB after the change. Two equivalent queries have been made on those databases and we can also see that the results are identical. From the results from test B (as well as test A) we have gotten the answer to our research

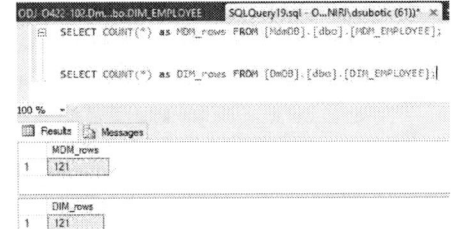

Figure 6. Number of rows in DmDB and MdmDB after a data change in data sources

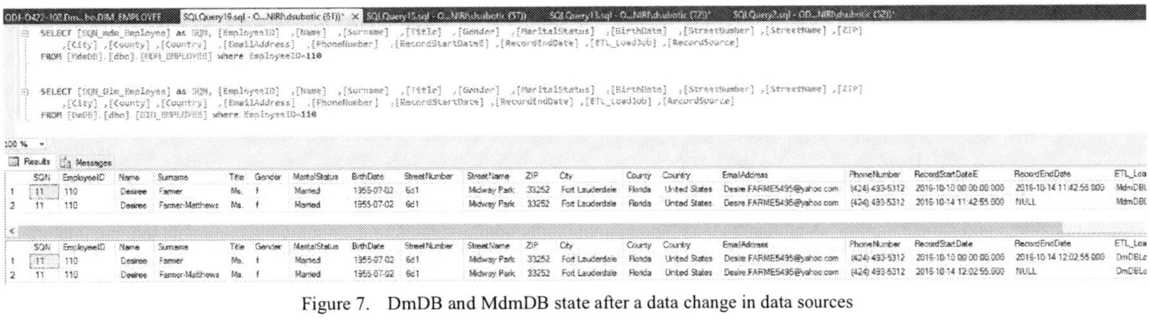

Figure 7. DmDB and MdmDB state after a data change in data sources

1405

question – yes, it is possible to successfully integrate DW and MDM systems through the usage of a new system catalog and new DW/MDM system architecture.

V. CONCLUSION AND FUTURE WORK

In this paper we presented our research on a DW/MDM integration into one common business intelligence system. Seeing that the DW and MDM both integrate basically the same data sources and have the same evolution problem (preserving the history of data and structure), our goal was to integrate those two systems into one so we could in the future address the evolution problem at the common level. We have presented our MDV meta-model which serves to integrate those two systems and to provide the means for managing their evolution at a mutual level. In our proposed common solution, the data repository preserves the history of raw (DW) and master (MDM) data, as well as their schemas. Meta-data repository (i.e. system catalog) preserves the history of meta-data for the data repository. This way, the problem of the DW and the MDM evolution could be addressed at the general level and a permanent general solution on a meta-level could be developed. The end result could be a flexible, modular solution which will be able to track and manage changes in both data and metadata, as well as their schemas. In the paper we described our common integration architecture, which includes the new system catalog based on a Data Vault modelling method. We further described the integration part of underlying data model for the new system catalog and we presented our research prototype used for empirical validation of the sustainability of the proposed solution. Also, we described a couple of tests that we conducted in order to prove this kind of integration is possible and can be successfully deployed. These are also the main contributions of this paper. The benefits of this approach could be various, the least of all a development of a simpler common solution which can effectively manage data and schema evolution in both of those systems. The next step of our research is the main one - proving that the schema evolution problem can be solved more effortlessly and efficiently through the usage of our solution. In order to do so, we plan to define a final set of structural change cases, together with evolution operations and test them against the prototype, as well as the traditional RDBMS system catalog.

REFERENCES

[1] B. Bebel, J. Eder, C. Koncilia, T. Morzy and R. Wrembel, "Creation and Management of Versions in Multiversion Data Warehouse", 19th ACM Symposium on Applied Computing, Nicosia, Cyprus, 2004.

[2] Z. Bellahsene, "Schema Evolution in Data Warehouses", Knowledge and Information Systems, pp. 283–304, 2002.

[3] A. Berson and L. Dubbov, „Master data management and data governance", 2nd ed. New York: McGraw Hill, 2011.

[4] P. Chen, "The Entity-Relationship Model - Toward a Unified View of Data", ACM Transactions on Database Systems, vol. 1 pp. 9–36, 1976.

[5] J. Chen, S. Chen and E. Rundensteiner, „A transactional model for data warehouse maintenance", In: Spaccapietra, S., March, S.T., Kambayashi, Y. (eds.) ER 2002. LNCS, vol. 2503, pp. 247–262. Springer, Heidelberg, 2002.

[6] E. F. Codd, "Relational database: a practical foundation for productivity", Communications of the ACM, vol.25, pp. 109-117, 1982.

[7] Y. Cui and J. Widom, "Practical Lineage Tracing in Data Warehouses", Proceedings of the 16th International Conference on Data Engineering, San Diego, California, 2000.

[8] (2015) The CA ERwin Data Modeler Site [Online]. Available: http://erwin.com/products/data-modeler

[9] J. Eder and C. Koncilla, „Evolution of Dimension Data in Temporal Data Warehouses", Technical Report, 2000.

[10] H. Fan and A. Poulovassilis, "Schema Evolution in Data Warehousing Environments – A Schema Transformation-based Approach", Proceedings of 23rd International Conference on Conceptual Modeling, Shanghai, China 2004.

[11] FreeDataGenerator (June 2016). Available at: http://www.freedatagenerator.com/.

[12] M. Golfarelli and S. Rizzi, „Data warehouse design", New York: McGraw Hill, 2009.

[13] M. Golfarelli, J. Lechtenbörger,S. Rizzi and G.Vossen, "Schema Versioning in Data Warehouses", In: ER Workshops, LNCS Springer, vol. 3289, pp. 415–428., 2004.

[14] A. Gupta and I. Mumick, „Maintenance of Materialized Views: Problems, Techniques, and Applications", Data Engineering Bulletin, 1995.

[15] W. H. Inmon, „Building the data warehouse, 4th ed.", Indianapolis: Wiley Publishing, 2005.

[16] W. H. Inmon, D. Strauss and G. Neushloss, „DW 2.0: The Architecture for the Next Generation of Data Warehousing", Burlington: Morgan Kaufmann Publishers, 2008.

[17] Integration Definition for Information Modeling (IDEF1X) Standard, FIPS Publication 184, Computer Systems Laboratory of the National Institute of Standards and Technology (NIST), 1993.

[18] D. Jakšić and P. Poščić, "Data Warehouse Models in Higher Education Courses", International Conference on Advanced Technology & Sciences, Antalya, Turkey, 2015.

[19] V. Jovanovic, S. Bojicic, C. Knowles and M. Pavlic, "Persistent staging area models for data warehouses", Issues in Information Systems, vol.13, pp. 121-132, 2012.

[20] R. Kimball and M. Ross, „The data warehouse toolkit, 3rd ed.", Indianapolis: Wiley Publishing, 2013.

[21] D. Linstedt, „SuperCharge Your Data Warehouse: Invaluable Data Modeling Rules to Implement Your Data Vault", USA: CreateSpace Independent Publishing Platform, 2011.

[22] D. Linstedt, and M. Olschimke, "Building a Scalable Data Warehouse with Data Vault 2.0: Implementation Guide for Microsoft SQL Server 2014". Morgan Kaufmann, 2015.

[23] D. Loshin, „Master Data Management", San Francisco: Morgan Kaufmann, 2010.

[24] Microsoft SQL Server 2012 (October 2016). Available at: https://www.microsoft.com/en-us/download/details.aspx?id=43351

[25] T. Morzy and R. Wrembel, „On Querying Versions of Multiversion Data Warehouse", In Proceedings of the International Workshop on Data Warehousing and OLAP, DOLAP"04, Washington, USA, 2004.

[26] D. Subotić, "Data Warehouse Schema Evolution Perspectives", Advances in Intelligent Systems and Computing, Springer, vol. 312, pp. 333-338, 2014.

[27] D. Subotić, V. Jovanović, and P. Poščić, "Data Warehouse and Master Data Management Evolution-A Meta-Data-Vault Approach", Issues in Information Systems, vol.15, pp. 14-23, 2014.

[28] D. Subotić, V. Jovanovic, P. Poščić, "Data Warehouse Schema Evolution: State of the Art". 25th Central European Conference on Information and Intelligent Systems CECIIS, Varaždin, Croatia, 2014.

[29] C. Quix, "Repository Support for Data Warehouse Evolution", Proceedings of the Workshop DMDW, Germany, 2004.

Gap in pagination due to withheld paper.

Pages 1407-1411

Using Public Private Partnership models in smart cities– proposal for Croatia

M. Milenković, LL.M[*], M.Rašić, LL.M[**] and G. Vojković, Ph. D.[***]

[*] University of Zagreb, Faculty of Transport and Traffic Sciences/Chair of Transport Law and Economics, Zagreb, Croatia
[**] Zagreb School of Economics and Management/Department of Law, Zagreb, Croatia
[***] University of Zagreb, Faculty of Transport and Traffic Sciences/Chair of Transport Law and Economics, Zagreb, Croatia
mmilenkovic@fpz.hr
mrasic@zsem.hr
gvojkovic@fpz.hr

Abstract - How smart is a "smart city"? According to Asian examples of newly growing smart cities, usage of technology can improve life standard and reduce cost of living, improve operational efficiency, environmental sustainability, eco-friendly infrastructure, smart technology Internet of Things (IoT), smart living and direct citizen participation in decision making process. Learning, adaptation and innovation could be the future for Croatian cities by using all mentioned, which will improve social, regulatory and safety indicators for designing a better living environment for Croatian citizens. While using Public Private Partnership (PPP) models, crowdsourcing and democratic ecologies which provide better and more efficient public services by taking advantage of private sector's "know-how", cities will create long-term investment opportunities and sustain real time optimization strategies by providing safe and reliable place to invest. The intention of this paper is to show how the (local) government's role in PPP projects is to evaluate and approve detailed execution plans of the concessionaire while the private partner's role is to design, build, finance, and operate the facilities. *In futuro*, digital technologies offer numerous possibilities for citizen participation in decision making at local and regional government level.

The key words: Internet of Things, smart city, Public Private Partnership, concession models, public procurement, decision making.

I. INTRODUCTION

A common definition of a 'smart city' has not yet been determined. Many authors use various definitions to clarify what is a 'smart city'. Authors define it as a city that bets a lot on the quality of living and where the citizens are involved as main actors in decision processes [1]. At the beginning of an attempt to define a smart city, we should start by pondering upon cities and systems which operate in them. Cities have a duty to fulfill the needs of their citizens through various systems. The types of systems are by no means exhaustive, but certainly include public services such as light management, traffic and transport organization, waste and water management, administration policies, security, energy sustainability and information services. Regular cities operate and supervise every system as a separate unit, which in return produces

more costs for taxpayers with slight to no improvements in the quality of living. On the contrary, 'smart cities' use Information and Communications Technologies (ICT) with Internet of Things (IoT), to create connections and interactions between some or all the systems, cutting expenses and improving the quality of life for citizens during the process.

We believe that a 'smart city' is a city where investments are focused towards smart citizens who use renewable energy resources wisely and widespread technological networks to combine sustainable economic growth whilst improving the quality of life, through the open government model by the interaction of all stakeholders. Regarding future, Croatian 'smart cities' use of renewable sources of energy, ICT and sustainability combined with usage of Public Private Partnerships (PPP) is a crucial model which could easily adapt to Croatian legislation and Croatian way of living.

In terms of innovation and quality of living a set of indicators, such as: air quality, water, greens, waste handling, energy consumption health issues, urban mobility and logistics should be analyzed. Also, the key features of the 'smart city' are: smart governance, smart mobility and smart energy as an educational and communicational tool for opinions of the citizens, the technology supporting the Social Networks like YouTube, Facebook, Twitter and others can be employed when influencing political decisions including functions in cities, urban spaces, water, waste, in one word living [2].

The aim of this paper is to set forth in which direction should Croatian cities advance in the near future, according to the successful models of smart cities worldwide, and implementation of PPP models for building smart cities in Croatia. Furthermore, the paper will recommend improvements of current legislation towards an open government.

II. DEVELOPING SMART CITIES

Today is the age of technological revolutions! Technology and science have brought many conveniences to the citizens of the world, most of which we are not even aware of. The idea of a 'smart city' has drawn governments' attention, especially in South Korea, Japan, India and China. They believe that 'smart cities' could bring more benefits to their citizens.

Cities around the world are facing the impact of series of emerging megatrends like accelerating urbanization, new technological developments, increased connectivity, demographic shifts, climate change, scarcity of natural resources and many more changes. But so, has the need for cities and regions to collaborate on global issues like climate change and public safety. In this context, it has become important for a city to understand its national, regional and global position. This enables the city and its stakeholders to set appropriate policies, develop an effective strategy and plan for actions accordingly, all aimed at the sustainable development of the city [3].

According to the research conducted by the United Nations in 2014 more people live in urban areas than in rural areas globally. As stated in Figure 1 below, in 1950, 70% of people worldwide lived in rural settlements and less than 30% in urban settlements. In 2014, 54% of the world's population is urban. The urban population is expected to continually grow, so by 2050, the world will be 1/3 rural and 2/3 urban, roughly the population will reverse in comparison to what we had during the 1950s. Europe already consists of 73% of its population living in urban areas. The United Nations' research predicts that over 80% of Europe's population will be living in urban areas by 2050. [4]

Development of 'smart cities' requires continuous and joint effort of all participants in the process. Urban stakeholders include governments, scientific institutes, companies, citizens and NGOs who innovate together to up the quality of life of the 'smart cities'. However, the connection between citizens, technologists, and urbanists is not strong enough. In order to improve the current urban settings Croatia needs to design a pattern to develop a universal platform between stakeholders on a local level.

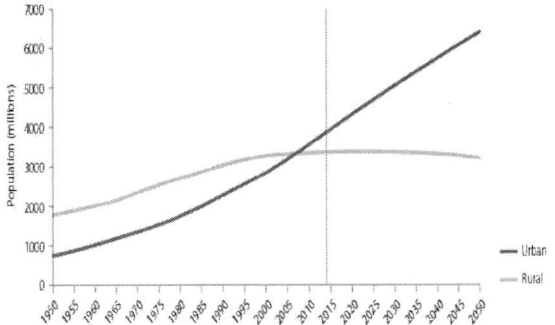

Figure 1: Urban and rural population of the world, 1950–2050. [4]

Many corporations and investors assume that fixing cities is the purview of government, and that the government will proceed in that manner. But governments around the world are stuck—financially, politically, or even both. They can't rely to single-handedly address the

problems of urbanization or to start solutions, such as efficient electrification and reliable public transport which will instantly drive economic growth. By implementing those solutions large amounts of capital, exceptional managerial skill, and significant alignment of interests— all of which are often in short supply in city governments but abound in the private sector are required. That is the main reason that South Korean 'smart cities' like Songdo (the city built on reclaimed land) [5] used PPPs to define infrastructure types, and the roles of public and private parties when they were looking and arranging the means and ways for financing 'smart cities'.

III. THE HOWs AND THE WHYs TO BUILD A 'SMART CITY'

A. Financial support

The Smart City industry is growing constantly, and it's predicted to be worth more than 20 billion dollars by 2020, while the annual global Smart City revenue is expected to reach 88.7 billion dollars by 2025. [6] In the Asian countries, funding of 'smart cities' have been conducted through government incentives, local subsidies and private entrepreneurships projects.[1]

As for Croatia, the possibilities are more diverse; other than national and local subsidies, there are European Structural and Cohesion Funds. The European Union (EU) is encouraging Member States to develop smart cities by allocating 365 million euros for this purpose. [7] Also the EU brought new financial instruments supporting environmental and climate action projects from which, cities can withdraw the funds. They include: the Financial Instrument for the Environment and Climate Action (LIFE) Programme, Horizon 2020 and Intelligent Energy Europe (IEE).

Barcelona and Amsterdam developed systems by which citizens and companies can interact on solving key city issues with 'smart' solutions. Barcelona's project "BCN Open Challenge" set out six challenges for businesses and entrepreneurs to provide solutions for transforming public spaces and services. The city government sought to procure innovative solutions, support winning companies and validate projects. Winning solutions were provided with public service contracts to fulfill their solutions. [8]

B. Government's role in the PPP projects in the 'smart cities' development

In the local governments sequence of events in the development of 'smart cities' largely depends on the mayors. If they are willing to promote sustainability, then it becomes the priority, and some of these categories of sustainability include: transportation, utilities, electricity, thermal energy, renewable energy and others. The point is; after the establishment of the 'smart grid', infrastructure will significantly reduce the environmental impact of the whole electricity supply system. The goal is a holistic

---

[1] For example, IBM provides cities around the world with grants of IBM expertise and technology which will aid cities with their strategic challenges. To find out more: https://www.smartercitieschallenge.org/

approach in which all the processes are conducted by applying IoT (Internet of things) technology and application of VTV (Vehicle to Vehicle), intelligent transport systems. It's expected a real-time optimization of traffic routes, traffic on the roads and to allow easy selection of different modes of transport and measure the effectiveness of the current system of distribution of energy available at any time on any smartphone app. E.g., in how much time does the tram or bus arrive or in which street is lesser traffic, how much moisture is in the air, what is the current outdoor or indoor temperature, CO_2 and other pollutants level? Possibility for a new clean mobility solutions that complement bicycling and public transport and the application of VTV intelligent transport systems; communications are expected within real-time optimization of traffic routes, traffic on the roads, by simply choosing between different modes of transport.

The possibility of cooperation between the public and private sector (PPP) - (following the example of South Korea's Songdo and other smart cities of Asia), which opens the possibility for withdrawal of additional funds from the EU Structural and Cohesion funds. When speaking of Croatia, investing in innovative strategies improve the efficiency, savings and promote the development of the real sector. For such operations, it's necessary for local governments to reduce utility fees in order to encourage the construction of smart infrastructure and significant energy savings, reduction of municipal fees and increase incentives for small and medium-sized enterprises (SMEs).

As one of Europe's examples of a proactive and 'smart city' oriented government - Amsterdam appointed a Chief Technology Officer (CTO) for the city. Also, they founded the Amsterdam Metropolitan Solutions (AMS) as an institute focused on applied technology. It's built by a consortium of public and private partners. AMS aims to attract and retain talent in the field of applied technology, create sustainable connections, drive a positive economic impact for Amsterdam by innovating, developing and marketing metropolitan solutions in urban themes such as water, energy, waste, food, data and mobility. In addition, their Amsterdam Smart City [9] is an online platform which connects all interested parties in one goal: dealing with 'smart city' problems and solutions. [10]

C. Benefits

Smart cities aim to improve the functioning infrastructure, access to resources, and safety and security for the population [11]. The European Commission tells us that: in 'smart cities', digital technologies translate into better public services for citizens, better use of resources and less impact on the environment [14]. Figure 2 shows an independent and symbiotic energy management system in which the value is expanded by the next generation energy use, where the residents and city itself cooperate in conservation of renewable energy resources.

There are some small changes that can save huge amounts of energy, and it can be obtained when for instance active measures (building automation systems) and passive measure (low energy bulbs) are used constantly, furthermore, some European countries like Italy have already implemented mentioned measures and the savings in thermal consumptions for heating and cooling can bring up to 26% [2].

Sustainable strategies application should already start today since those will increase energy efficiency in the next half century. Some of the most effective models of combining the PPP in order to develop smart cities in Asia (Songdo, Seoul, Fujisawa) and the EU (Barcelona, Amsterdam, Vienna) have already benefit from the implementation of 'smart city' strategies. Barcelona saved $58 million annually using smart water technology, and the city has increased parking-fee revenues by $50 million annually utilizing smart parking technology. The government also stated that Barcelona has created 47,000 new jobs through its Smart City efforts. [13]

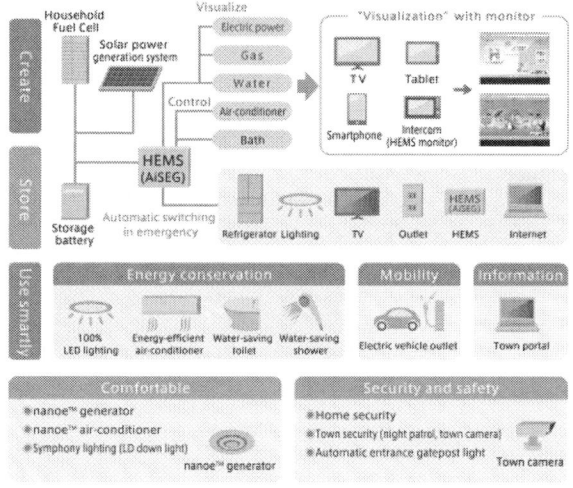

Figure 2: From smart management to power savings. [15]

Cities that implement 'smart' solutions for usage of energy resources, manage water supply and have a waste management system reduce pollution and use less energy-according to Cisco, their energy efficiency will increase by up to 30 percent within the period of 20 years. [16]

In the city of Songdo, this is obvious throughout the city. The city's garbage collection is so automated that it only takes seven employees to serve the current 35,000 residents. [17] The kind of efficiency which is unimaginable in European cities.

A Croatian example of implementing IoT into cities include Dubrovnik's praiseworthy efforts. As a part of the comprehensive Dubrovnik Smart City project, in 2016 Dubrovnik opened a Smart street which features public lighting with multifunctional sensory network, alongside a variety of access technologies, from optical links and 4G network to the Wi-Fi network, cameras to identify traffic violations, parking technology which recognizes vehicles and performs contactless charging of parking fees and offers real-time information on the parking vacancy status in the Smart Street, but also all over Dubrovnik. [18] In addition to this, Dubrovnik also held conferences and competitions which resulted in other IoT solutions towards a smart city. [19]

Other measures for improving transport sustainability in Croatian cities are the inter-modal mobility as well as carpooling and car sharing and support to the soft mobility (on foot and bikes). Cities 'green areas' such as: parks, gardens vertical gardens, green roofs and facades show not only that people prefer to live in those cities where they can breathe and see plants, trees and flowers but also contribute to climate protection (improved air quality and cooling down air temperature while producing its own microclimate and help save on heating, cooling and increasing comfort inside the building), [2] which can easily be connected and implemented by IoT (e.g. Songdo's remote energy monitoring systems which bring the power to IoT and save the energy), especially in the Croatian cities by the sea.

IV. SUCCESSFULNESS OF IMPLEMENTATION OF PPP MODELS TO THE CROATIAN ECONOMY

The promotion of PPP projects is expected to ripple effects on the national economy through three channels:

- economic growth resulting from the inflow of private capital

- increased social welfare resulting from the timely delivery of social services and the early realization of social benefits

- reduction in the government fiscal burdens through better VFM (value for money) [12].

A. Institutional settings

In accordance with the Croatian PPP Act[2], article 2, paragraph 7 the public partner may allow the establishment of the right of construction in favor of the private partner and free of charge. All issues related to the establishment or transfer the right of construction and awarding of concessions, including the question of fees, public and private partner, regulate contractually. The basic principles in the implementation of PPP projects are the principles of public procurement, the principle of protecting the public interest and the principle of economy. The PPP Act is proposed and authorized only by the public authority, and that is the reason why; primarily mayors should try to establish a setting where entrepreneurs can create solutions to improve quality of life — without added any of government expense. In order to facilitate PPP implementation, the PPP grants land expropriation rights to the concessionaire. The concessionaire may entrust the competent authority or the local government with the following responsibilities of execution of the land purchase, compensation for loss and more [12].

PPP Act, The Concession Act[3] and the Public Procurement Act[4] regulate the procurement procedure designed in a way to ensure value for money (VFM) of PPP projects. Korean study analyzed the efficiency of PPP project from 3 different perspectives: users, concessionaires and the government and it showed both cost savings and efficiency gains. The main problem that the study discovered was the level of user fees between government financed and privately invested projects, which decreased over the time in proportion to accumulated experience in PPP projects. For PPP projects to be carried out efficiently, one of the most important issues is prompting competition among private participants bidding for the project [12].

Agency for Investments and Competitiveness is a Croatian Agency set up by the PPP Act whose main tasks are to give investors full view of services to invest and implement in projects for the improvement of the economic growth and business environment and to promote Croatian PPP model as competitive. [20] In 2016 only one PPP project worth 4.6 million Croatian kuna has been contracted. But in 2014 and 2015 no PPP contracts have been signed. [21]

On the EU level, in 2014 new procurement directives have been adopted, replacing the 2004 directives and covering the award of concessions. The new directives open several opportunities for 'smart city' investments, while maintaining the basic requirements of competition, transparency, equal treatment and compliance with EU state aid rules.

The key directives are:

1. *Directive 2014/24/EU on public procurement[5],*
2. *Directive 2014/23/EU on the award of concession contracts[6],*
3. *Directive 2014/25/EU on procurement by entities operating in the water, energy, transport and postal services sectors[7].*

B. Open government

The *E-government* is changing the way in which politics interacts with citizens and the democratic processes can be enhanced but the main goal is to turn the government tools from an 'office-centric' mode to a 'citizen-centric' mode [2]. These new and innovative forms of governance have been included into the term 'smart governance' under which the government manages and implements policies toward the improvement in quality of life of citizens by leveraging ICTs and institutions and by actively involving and collaborating with stakeholders [22].

While it may be true that ICTs are introducing a range of new capacities for the design and planning of human urbanization, questions remain about the network

---

[2] The PPP Act, Official Gazette [in Croatian] Zakon o javno-privatnom partnerstvu, Narodne novine, nb. 152/14, (Narodne novine are an official Gazette of Republic of Croatia)

[3] The Concessions Act, Official Gazette [in Croatian] Zakon o koncesijama, Narodne novine, nb. 143/12
[4] Public Procurement Act, The Concessions Act, Official Gazette [in Croatian] Zakon o koncesijama, Narodne novine, nb. 120/16
[5] OJ L 094, 2014, implemented by Croatia
[6] OJ L 094, 2014 currently not implemented by Croatia
[7] OJ L 094, 2014 currently not implemented by Croatia

capabilities of smart cities to remake systems of government. Government systems are now increasingly under pressure to develop institutional frameworks that support tools and resources for empowering citizen agency.

The overarching idea is that smart city planning should – at the very least – incorporate experimentation with new forms of community and citizenship [23]. Citizens have a leading role in designing, building and maintaining 'smart cities'. This calls for a fundamental shift when we think of our cities and about urban development in the near future. Citizen participation reduces government corruption by expanding public insight and decentralizing government power, [23] by giving them the ultimate decision in the adoption of the city's services and the creation and management of public value out of them. Even large-scale projects can be enhanced by crowdsourcing, while cities can collect data, give answers to simple questions e.g. about how they want to use their public spaces or acquire services for bringing innovation and 'smart' thinking to acquire powerful results. [24]

In the USA, the Open Government was introduced by Ex-president Obama in 2009 in legislations such as: The Freedom of Information Act, E-government Act and The Paperwork Reduction Act. Open Government could improve and advance citizen participation in the process of building democratic ecologies of 21st century in The Republic of Croatia.

Croatian *Central country portal*[8] and within it an E-citizen system is an example of enabling citizens' various options within a single platform by merging and connecting IT platforms. That example should be followed on decentralized levels of cities and regions, which should include a 'smart city' sector for all stakeholders to interact.

C. Building 'smart cities' through the concession agreements and public procurement models

Decision-making within contracting authorities (especially on the local level) is generally carried out by a small number of government officials who, we believe, do not have in-depth ICT and IoT knowledge in solving a public issue. Furthermore, the process of procurement leads to nothing more than a race to the lowest price point between a small number of large suppliers. Therefore, new models offer alternatives for building 'smart cities'.

Preconditions for the application of the new Croatian Concessions Act (currently in its second reading in the Croatian Parliament) promote use of concessions through the high-quality medium- and long-term sectoral strategy, establishes safe, secure and clear rules in cases of awarding a concession in a way to protect the public interest. Also, it strengthens the competences of operators responsible for the development and maintenance of infrastructure and introduces administrative contracts into the practice which enables administrative protection during the administrative procedure.

The new Croatian Public Procurement Act introduced two new procedures which are likely to be particularly

---

[8] http://gov.hr

relevant as models for authorities who wish to purchase innovative goods, services or works:

- the innovation partnership - enables a public authority to enter a structured partnership with a supplier with the objective of developing an innovative product, service or works, with the subsequent purchase of the outcome and which will enable public authorities to select partners on a competitive basis and allow them to develop an innovative solution tailored to their requirements;

- the competitive procedure with negotiation - enables the procurement of goods, services or works that includes an element of adaptation, design or innovation, or other features which make the award of a contract without prior negotiations unsuitable. Unlike the competitive dialogue, it requires that the authority may specify the required characteristics of the goods or services in advance of the competition. [25]

Other important changes in the new Public Procurement Act include: increased flexibility and simplification on the procedures to follow, negotiations and time limits, use of 'most economically advantageous tender' as default criteria within the public procurement process, use of life cycle costing (LCC) as a method for assessing tender costs and the competitive dialogue procedure which has been simplified particularly for technically and financially complex projects.

This will in turn enable to de-risk the most promising innovations step-by-step via solution design, prototyping, development and first product testing. [26]

V. CONLUSION

Many of the challenges faced by smart cities surpass the capacities, capabilities, and reaches of their traditional institutions and their classical processes of governing, and therefore require new and innovative forms of governance [22].

Alongside 'top-down' master-planning, we need to enable 'bottom-up' innovation and collaborative ways of developing systems. The notion of the 'smart citizen' as a co-creator draws on a rich intellectual backdrop in both technology design and urban design. In practice, engaging citizens in these processes is immensely challenging. [27]

PPPs are the most effective way to make Croatian cities 'smart cities'. Certainly, it's unreasonable to expect Croatian cities to become 'smart' in a short period but by implementing the measures and strategies like creating walkable localities – reduce congestion, air pollution and resource depletion, boost local economy, promote interactions and ensure security bring Croatian cities step closer to the 'smart cities'.

Cities should ensure the visibility of procurement and PPP opportunities through a single portal and use problem-based methods for solving key issues. However, it should be noted that cities cannot simply copy the best practices from successful 'smart cities', hence must develop approaches that fit their own mindset, organization and culture in terms of broader strategies, human resource policies and demographics. Forming E-groups to listen to people and obtain feedback and use online monitoring of programs and activities with the aid of cyber tour of worksites. For example, making areas less vulnerable to disasters, using fewer resources, and providing cheaper services; green buildings and pool-sharing. [28]

That is the main reason why Croatian PPP Act should intend to simplify use of PPPs and give equal possibilities to both private and public sector companies. In that manner, we believe Croatian cities will be able to develop and have similar appearance as the cities from South Korea, Japan, China, India and other countries in the world who detected and adopted wise 'smart city' policy and achieved advanced economic conditions and improved life of its citizens.

REFERENCES

[1] R. Riva Sanseverino (ed.), Competitive Urban Models, in Smart Rules for Smart Cities, vol. 12, Springer, 2014, pp.1-14.

[2] A. Ahuja, Integration of Nature and Technology for Smart Cities, 3rd ed, Springer, 2016, pp. 200 – 210.

[3] PwC, Amsterdam, City of Opportunity, 2013, pp.3.

[4] United Nations, Department of Economic and Social Affairs, Population Division (2014). World Urbanization Prospects: The 2014 Revision, Highlights (ST/ESA/SER.A/352), pp.7.

[5] Smart cities market in Asia projected to grow rapidly through 2023, http://geospatial.blogs.com/geospatial/2014/03/smart-cities-market-in-asia-projected-to-grow-rapidly-through-2023.html, Retrieved 4/02/2017

[6] Pike Research Report, Smart Cities (2013): www.navigantresearch.com/research/smart-cities, Retrieved 15/02/2017

[7] How smart cities & IoT will change our communities, http://www.businessinsider.com/internet-of-things-smart-cities-2016-10, Retrieved 15/02/2017

[8] Barcelona City Council, https://bcnopenchallenge.org/, Retrieved 15/02/2017

[9] Amstardam Smart City, https://amsterdamsmartcity.com/, Retrieved 15/02/2017

[10] PwC, Amsterdam, City of Opportunity, 2013, pp.19-30

[11] C. L. Stimmel,Building Smart Cities, CRC Press, 2016, pp. 144-146.

[12] J-H. Kim, J. Kim, S. Shin, S-y. Lee, 'Public-Private Partnership Infrastructure Projects: Case Studies from the Republic of Korea, vol. 1, Institutional Arrangements and Performance, pp. XIX-XXI, Asian Development Bank, 2011.

[13] Cisco, IoE-Driven Smart City Barcelona Initiative Cuts Water Bills, Boosts Parking Revenues, Creates Jobs & More, 2014, pp. 6.

[14] Smart Cities | Digital Single Market - European Commission, http://ec.europa.eu/digital-agenda/en/smart-cities, Retrieved 16/02/2017.

[15] Energy | Town Services | Fujisawa SST, http://fujisawasst.com/EN/service/energy.html, Retrieved 16/02/2017

[16] Smart City Infographic, http://postcapes.com/anatomy-of-a-smart-city/, Retrieved 16/02/2017

[17] When Smart Cities are Stupid, http://newtowninstitute.org/spip.php?article1078, Retrieved 16/02/2017

[18] Hrvatski Telekom opens first Smart Street in Croatia, http://www.t.ht.hr/en/Press/press-releases/3175/Hrvatski-Telekom-opens-first-Smart-Street-in-Croatia.html, Retrieved 16/02/2017

[19] Smart City Dubrovnik, https://cityos.io/Dubrovnik, Retrieved 16/02/2017

[20] Agencija za investicije i konkurentnost: AIK, http://aik-invest.hr, Retrieved 16/02/2017

[21] T. Bašić, Kako se uključiti u novi val JPP-a, Lider, vol. 13, nb. 591, 2016, pp.27

[22] A. J. Meijer, J.R Gil-Garcia, M. P. Rodriguez Bolivar, Smart City Research: Contextual Conditions, Governance Models, and Public Value Assessment, Social Science Computer Review, Vol 34, December 2015, pp. 647 - 656.

[23] D. Araya, Smart Cities as Democratic Ecologies, Palgrave Macmillan, 2015, pp. 19-20.

[24] C. L. Stimmel,Building Smart Cities, CRC Press, 2016, pp. 182.

[25] A. Semple, Guidance for public authorities on Public Procurement of Innovation, Procurement of Innovation Platform, 2016. pp.24

[26] Innovation Procurement | Digital Single Market - European Commission, https://ec.europa.eu/digital-single-market/en/pre-commercial-procurement, Retrieved 16/02/2017

[27] D. Hemment, A. Townsend, Smart Citizens Publication, FutureEverything, 2013, pp.1-3

[28] Here is how Modi govt performed in one year, http://indianexpress.com/article/india/india-news-india/india-20-smart-cities-list/, Retrieved 16/02/2017

The platform for the content exchange between Internet music streaming services and discographers

Miodrag Sretenović* Božidar Kovačić** Aleksandar Skendžić***
* MI&DA ltd, Karlovac, Croatia, m.sretenovic@mida.hr
** Department of Informatics, University of Rijeka, Croatia, bkovacic@inf.uniri.hr
*** Polytechnic Nikola Tesla, Gospic, Croatia, askendzic@velegs-nikolatesla.hr

Abstract - Internet streaming is the type of technology that allows simultaneous reception of multimedia content and its simultaneous playback through a computer network. Global development of the Internet leads to growing demand for multimedia content. Discographers deal in distribution and sale of rights to play audio recordings (tracks), adapting their business to new technologies and modern ways of selling their products and services in order to survive in the market impacted by a large drop in demand and sales of CDs and DVDs. In the process of rationalization and automation of business processes, new standards in the electronic exchange of data are set between discographers and Internet streaming services with the goal to increase efficiency and economy. This paper describes a platform for the exchange of data and content, presents the concept and the basic methods used in the development of an automated system for the exchange of data and content between discographers and Internet streaming services and offers guidelines for future development.

I. INTRODUCTION

The process of modernisation, rationalisation and automation of business processes aimed at increasing productivity and cost-efficiency is unfeasible without continuous improvement and investment into new information technologies [1]. New technologies are forcing business entities to actively adjust themselves and follow the latest IT trends so they could survive in the market.

Multimedia (Latin *multus*: many, *medium*: means, medium). The media are: the newspapers and other press, radio and television programmes, news agency programmes, electronic publications, teletext and other forms of daily or periodic publications of editorial content by broadcasting recordings, voice, sound or picture [2].

Discography (Greek *diskos* disk, Greek *grafo* list) is a systematic list of music editions (tracks) of a certain provider-artist. Business entities engaged in distribution and sale of rights to play audio recordings (tracks) are called discographers. Availability and channels for selling multimedia contents have thoroughly changed with the

development of Internet. The age of vinyl (gramophone records) as an era of recordings, playback and sale is over. It has been succeeded by magnetic recordings on CD and DVD media, which are also slowly passing into history as audio formats. The sale of rights to play audio recordings (tracks) as a type of multimedia content is gradually moving to a new digital era.

Contemporary electronic business or e-business requires the use of information and Internet technologies.

Electronic business or e-business is a form of business which, in addition to applying information and communications technologies, exchanges information as support in all business activities [3].

Business intelligence is a process of collecting the data and information from internal and external business environment and transforming them into business knowledge based on which business decisions can be made [4].

Recording Industry Association of America (RIAA) recorded an increasing trend in using streaming services, and strong growth in sale of digital music, which accounted for 77% of the total music industry revenue in the first half of 2016 [5,6]. In the same period, streaming accounted for 47% of the US music industry revenue [5,6]. These indicators provide firm grounds for the development of a platform model for exchanging electronic data between discographers and streaming services.

The precondition for defining an electronic format for the exchange of electronic data is to define standards and rules. The rules of e-business and exchange of electronic data between discographers and streaming services are defined by standards and rules of streaming services themselves.

The aims of e-business, which also includes platforms for the exchange of data between Internet streaming services and discographers, are as follows:
- greater productivity of business processes,
- faster exchange of information and content,
- reduction in operating expenses,
- higher data integrity,
- reduced risk of error,
- faster document search and archiving.

When implementing e-business, it is important to apply the latest technological methods of business intelligence in order to achieve higher dynamics of business processes. The abstract of this paper explains the need for applying platforms for the exchange of data between Internet streaming services and discographers in business processes, and the concept will offer the answers to the following questions:

- What does the applicable solution of platforms for the exchange of data between Internet streaming services and discographers look like and how does it work?
- What benefits does the application of platforms for the exchange of data between Internet streaming services and discographers offer?

The second chapter of this paper shows the development of the platform, and the third chapter presents the conceptual platform model. The fourth chapter describes the implementation process, the fifth chapter describes the control of data import and export, as well as security, and the sixth chapter presents the conclusion.

II. PLATFORM DEVELOPMENT

The platform for the exchange of content and data between discographers and streaming services is aimed at business entities-discographers (B2B) who engage in distribution and sale of rights to play audio recordings (tracks). (Figure 1).

The platform for the exchange of content and data can be defined as business intelligence communicating between two data warehouses according to set rules.

The platform is the result of individual approach in the development of the conceptual design for exchange content and data between Internet music streaming services and discographers.

Data are exchanged in a standardised format, **EDI** (Electronic Data Interchange) – the direct transfer of structured business data and messages between computers electronically [7]. The exchange of content and data is possible between [8]:

- B2B (business to business) – electronic business between two business entities,
- B2C (business to consumer) – electronic business between a business entity and a final consumer,
- B2G (business to government) – electronic business between a business entity and government bodies,
- C2C Electronic business between final consumers,
- G2B Electronic business between government bodies and business entities,
- G2C Electronic business between government bodies and final consumers.

The platform is designed as a highly-automated business process. In addition to the exchange of electronic documents as a modern method of exchanging various types of documents between companies, the platform also allows for automatic electronic exchange of other digital contents accompanying electronic documents. At the same time, the platform can be seen as business intelligence consisting of a set of methodologies and concepts transforming the data into knowledge.

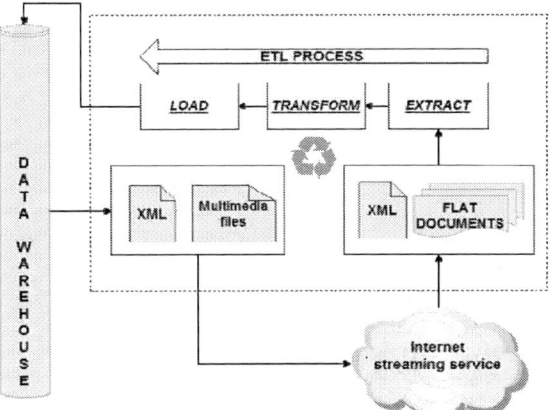

Figure 1 Presentation of the process of building a platform for the exchange of data between Internet streaming services and discographers

The advantages of selling multimedia content through streaming services are as follows:
- availability of rights to play audio recordings to a large number of potential customers,
- lower investment into audio recordings (tracks) launched on the market,
- higher earnings and profit,
- new business model,
- significant reduction in operating expenses, and increase in productivity.

In spite of evident benefits of this form of sale, the decision as to which rights to play audio recordings will be marketed through an individual streaming service is up to the business entity's management. In order to be able to assess which rights to play audio recordings to launch into individual streaming sales channel, the management needs high quality information that will give them advantage over competition. The present-day information systems are expected to provide information whose content is necessary in the decision-making process. Business intelligence methods used in the platform development are as follows:

A. Data mining

Data mining is a process of searching for new and potentially useful knowledge from the data [9]. This process reveals relations, structures and logic between data, which facilitates decision making processes. In the process of data mining, it is decided which rights to play will be offered to which market, under what market conditions and prices, and when.

B. ETL process modeling

The process of reporting on the amount of sold rights to play audio recordings and generated traffic can be described using ETL (extract-transform-load) process [10]. This means that the data on the traffic generated from rights to play audio recordings are first brought to a staging area, and afterwards transformed and transferred into a data warehouse. (Figure 1)
Therefore, modeling requires:

- Conceptual modeling of data,
- Business process modeling according to ETL (Extraction, Transformation, Loading):
  - data extraction process,
  - data transformation process,
  - data cleansing process.

The process part consists of extracting the data from the source (from the streaming service), integrating and loading the data, and finally transforming the data into a data warehouse.

- The first step includes extracting the data from the source (the data from the streaming service). In the process of data extraction, it is crucial to trace the source of data, date and time of their entry and data on entry operators, in order to enable complete insight to auditors. Data Vault concept emphasises the need to leave a trace of the source and time of storing data in the database [11].
- Transform means the transformation of data towards the data warehouse.
- Load is a process of loading.

In the Internet community, there is a large number of streaming services, and each service has its own rules and reporting methods, as well as rules for importing and exporting data. Due to various structures and formats of data from external sources and towards external sources, export and import need to be adjusted, which is usually performed by IT experts or data warehouse administrators. For this reason, the business process includes a large number of subjects, who are dependent on IT support. The result of such practice is a delay in the implementation of data both from the source and to the source, as well as the inability to timely analyse the results. The platform for the exchange of content and data between Internet streaming services and discographers provides business users with autonomy in controlling their business processes, so they do not depend on an implementer or IT sector.

III. CONCEPT OF PLATFORM

Conceptual modeling of data is a highly abstracted process of data modeling. The structure of data within the information system for tracking must reflect the structure of user's needs. The conceptual data model is a complete, consistent and concise description of data within an information system [12]. Figures 2 and 3 show Entity-Relationship Diagram (ERD), according to Martin Entity-Relationship Diagram [13]. Entity is anything we wish to collect and store information about, entity-relationship method is a diagram of mutually connected groups of data within a system in question [14,15].

A. Track

Track is multimedia content of sound and text or only sound. The data on tracks, as well as corresponding digitalized documents, are located within the data warehouse. Therefore, when exporting them into the streaming partner's system, the following control process actions need to be performed: extraction, loading, transformation and data export verification.

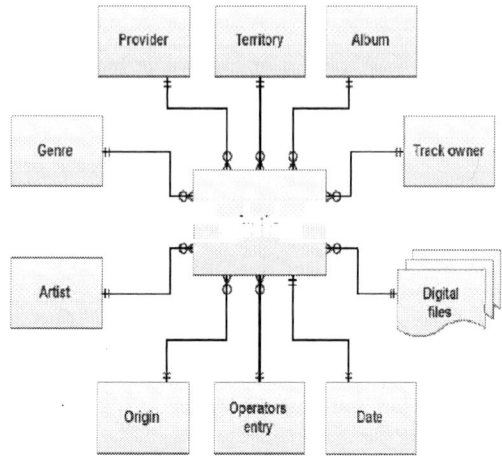

Figure 2 Entity relationships of tracks

From the entity-relationship diagram for tracks (Figure 2), entity relationships with the following entities can be seen:

- *Digital files* – digitalised documents on the basis of tracks,
- *Date* – date of change that has occurred,
- *Operators entry* – data on persons/operators recording the document,
- *Origin* – data on the track origin,
- *Artist* – data on the track artist,
- *Genre* – data on the track genre,
- *Provider* – data on the track provider,
- *Territory* – territory of broadcasting and selling the track,
- *Album* – collection of tracks to which the track belongs,
- *Track owner* – copyright holder.

B. Report on Internet music streaming services

Business relations also include the segment of reporting on the results of streaming service sales, which in essence has similar attributes. The ultimate aim of the platform is to archive the generated sale into a data warehouse, so the management could make decisions based on relevant information.

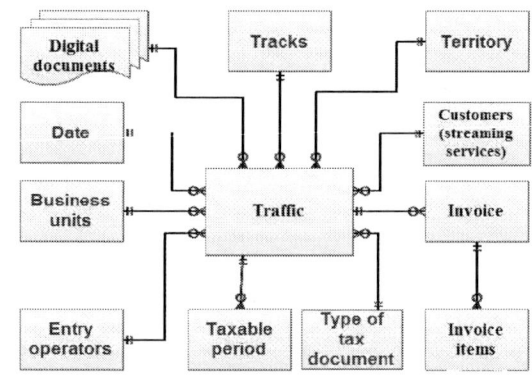

Figure 3 Entity relationships of streaming service reports

From the entity-relationship diagram for traffic (Figure 3), entity relationships with the following entities can be seen:

- *Digital documents* – digitalised documents on the basis of traffic,
- *Date* – date of change that has occurred,
- *Business units* – business unit of the business entity,
- *Entry operators* – data on persons/operators recording the document,
- *Taxable period* – period of mandatory tax return on the basis of traffic,
- *Type of tax documents* – type of traffic in the sense of taxes,
- *Invoice* – invoice issued based on traffic,
- *Invoice items* – items of an invoice,
- *Customers* – streaming service acting as customer,
- *Tracks* – track for which traffic was made,
- *Territory* – territory in which traffic was made.

The reporting form for all services is related to the track structure. In order for reports to be stored, the user's data warehouse needs to be well designed, as well as dimension tables and their attributes within the warehouse.

IV. IMPLEMENTATION

The platform for the exchange of content and data between Internet streaming services and discographers has been created in Embarcadero studio architect technology using programming language Delphi XE5, and Embarcadero InterBase XE5 database technology. Delphi development tool belongs to Embarcadero development tools, which have been among the best development tools for desktop applications for several years. The development of applications in Embarcadero programming environment enabled the development of mobile applications on the platform of the same program code. The development of applications in Embarcadero, which in addition to Delphi has C++ Builder in its portfolio, enables the same application to be delivered for various operating systems, namely for Windows, MAC, IOS Android environments, as well as for mobile devices. This makes the Embarcadero platform for the development of applications a multi-platform.
The main functionality of the platform:

- retrieval of tracks from the database,
- export of data and content to Internet music streaming services,
- import of data from Internet music streaming services.

A. Track entities

Figure 4 shows the form for entering and updating track entities, which is crucial for the functioning of the system. Within the software solution, all attributes necessary for the process of exporting the data into streaming services are designed. ISRC (International Standard Recording Code) is a code for international identification of sound recordings and video recordings,

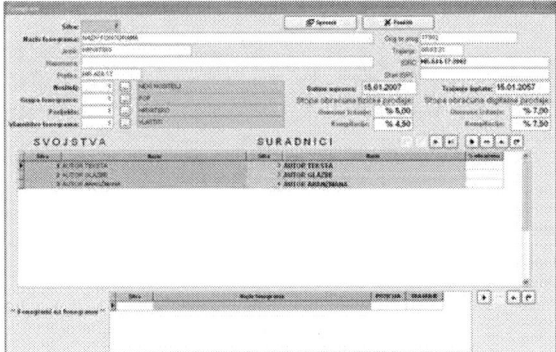

Figure 4 Data entry forms for tracks

and is a mandatory track attribute. System for the automatic exchange of data between discographers and streaming services is based on the track entity and ISRC attribute. There is a large number of Internet streaming music services, some of the most visited ones, with the largest number of subscribers, being: Spotify, Pandora, Deezer, iTunes, YouTube.

B. Data export process

The process of exporting data on tracks and transferring them into the streaming service information system is supported by defined templates. Templates enable a faster and more autonomous work on an unlimited number of digital platforms. Sample template for the choice of attributes and their formats for exporting the data to Internet music streaming services. (Figure 5)

The structure of columns and formats for export can be adjusted to the needs of automated data transfer. Through the user interface, the user can choose an entity and attributes, as well as their sequence in the export file. (Figure 5)

The structure is defined by the user instruction for each streaming service, and all required attributes, as well as corresponding music and other files need to be entered. In the communication process with streaming services, contractual relations are an important component, as well as the process of proving the rights to play tracks protected by copyright. Due to a large number of plagiarisms, there is a special component of streaming services, a programming interface that challenges the right to play a track without authorization.

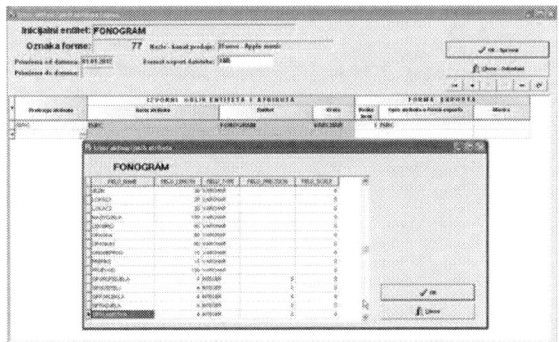

Figure 5 Template forms for the export of tracks

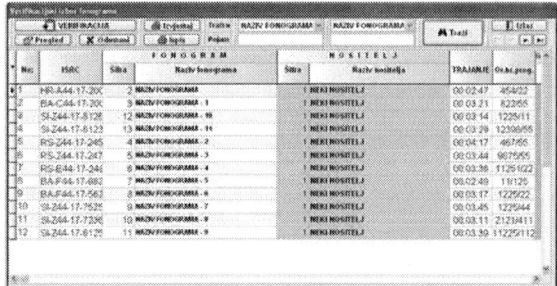

Figure 6 Selection of tracks for export

In the process of choosing the tracks that will be put into use, the management makes its decisions based on relevant assessments. Figure 6 shows the form for the selection of tracks to be put into use in the streaming service. Based on the template (Figure 5), the data will be exported according to the defined conditions from the template with attributes and their formats.

Figure 7 shows the XML file for importing the data into the streaming service platform, where all attributes of the streaming service are visible. By putting a track into use in the streaming service, the process of exporting data from the user's warehouse is completed.

The process of tracking all statistical information on exploitation and use of tracks can be found below. Depending on contractual relations, the official statistical report is compiled monthly by automated data transfer.

```
- <Album id="id_album">
   <TITLE>Title name</TITLE>
   <ARTIST>Artist name</ARTIST>
   <YEAR>Year album</YEAR>
   <TYPE>Type album</TYPE>
   <CATALOG-NUMBER>Catalog number</CATALOG-
NUMBER>
   <BARCODE>Barcode album</BARCODE>
- <EDITOR id="ID EDITORS">
   <FIRST-NAME>FIRST NAME EDITOR</FIRST-
NAME>
   <LAST-NAME>LAST NAME EDITOR</LAST-NAME>
   <EMAIL>EMAIL NAME EDITOR</EMAIL>
   <PHONE>PHONE NAME EDITOR</PHONE>
   <MOBILE-PHONE>MOBILE PHONE NAME
EDITOR</MOBILE-PHONE>
   </EDITOR>
- <RECORDING id="RECORDING">
   <ISRC>ISRC</ISRC>
   <TITLE>TITLE</TITLE>
   <ARTIST>ARTIST</ARTIST>
   <YEAR>YEAR</YEAR>
   <DURATION>DURATION</DURATION>
+ <OWNERS>
+ <COMPOSERS>
+ <LYRICS-AUTORS>
+ <ARRANGMENT-AUTORS>
   </RECORDING>
   </Album>
```

Figure 7 The selection of tracks for export to XML file

C. Data import process

Each sales report consists of track attributes, which are an important element for recognizing data in the data warehouse. In general, this is an ISRC track attribute.

The process of importing the data starts by loading the file. This is performed by choosing columns and rows from the table implemented in the data warehouse. The original file is also saved in the archive of digitalized imported files.

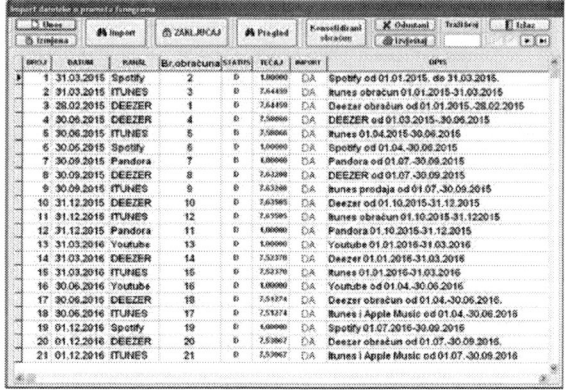

Figure 8 Overview of imported streaming service reports

If an ISRC attribute is not identified from the original file, the recording with an unidentified ISRC code will be allocated to the table of unrelated data. This functionality makes the system more automated. Figure 8 presents a form showing which data are imported into the data warehouse from which channels.

The data that follow the analytical track overview for each entry are shown in Figure 9. In addition to the leading attribute, ISRC, each track contains other attributes that are partly imported from the streaming service report, and partly from the entities dependable on the track entity.

The data loaded from the streaming source will not be visible to other data users until the verifier completes the verification and validation of received data, and checks accuracy of all transactions performed.

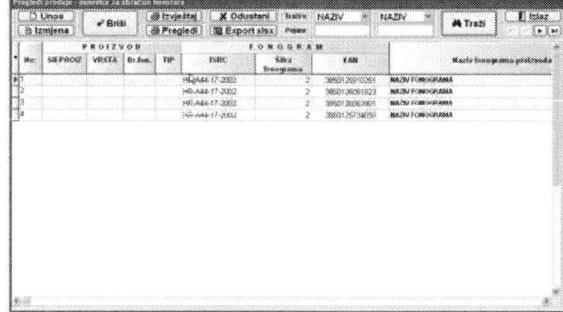

Figure 9 Data on realized transactions

1422

This way, the verifier can review all loaded data, as well as interventions on the data before the data is available to other authorized persons in an application segment.

The overview of traffic opens the possibility to further analyse the data, by connecting multiple entities dependable on traffic, as well as their attributes.

D. Control of data import and export and security

Saving files in their original form is becoming imperative in business, since it enables insight into and control of file authenticity. It also enables comparisons and controls with the data registered into the user's information system.

The application of Data Vault concept emphasises the need to track the source and moment of storing the data into database. This is a sort of security and control aspect performed through the entity of entry operators. The export of files from the user's information system with tracks, before being loaded into the streaming system and realized, is performed with verification by a supervisor. Figures 6 and 8 show the form in which verification in performed. Only users with special privileges are authorized to use it.

On the other hand, the files loaded from streaming services are not visible to other operators before they are verified by a financial supervisor. Only after the verification and comparison of monetary claims and realization of sold rights to play on the basis of tracks has been performed, the data becomes visible to other operators. This way the data is protected from any errors, and track providers and management are protected from making wrong decisions.

In the process of recording the collection of fees based on copyright, financial supervisors are included, and they have insight into and overview of all data, which prevents any illegal actions by the operator. This enables multiple controls of business processes, both in the process of recording and collecting the fees for rights to play tracks.

There are no ready-made solutions for the exchange of data and content between Internet music services and discographers that would allow the level of functionality in the exchange of data and content comparable with the described platform. The disadvantages of using platforms: their use requires knowledge of advanced techniques of artificial intelligence and a high level of education.

V. CONCLUSION

A reliable archive of electronic business documents and multimedia content, and storage of exchanged electronic content archived into data warehouses are important elements of information system functionality.

The platform for the exchange of content and data between Internet streaming services and discographers provides business users with autonomy in controlling their business processes, in the segment of import and export of data from a data warehouse towards streaming services. Such user autonomy increases efficiency and productivity,

and reduces the risk of error. The ability of the platform to adapt and memorise templates allows for connecting streaming services and their rules without any additional software interventions.

The use of templates from individual streaming services enables faster and easier data manipulation. The verification of data is enabled through the system administrator, and it ensures the correct export and import of data. The aim of this paper is to help users decide on the implementation of the platform for the exchange of content and data between Internet streaming services and discographers, and to suggest guidelines and propose a model for building a platform to designers, by applying the proposed model.

The architecture and design of such platform entails multiple benefits, from more cost-effective maintenance, centralised system of tracking costs, to the overview of all reports influencing the decision making by the business management.

By implementing the platform and adopting a development model that separates the transactional part of information business system from the data warehouse itself, we increase security. We also reduce pressure on operational databases in the sense of operation, amount of data and transaction conditions.

REFERENCES

[1] Sretenović, Miodrag, Božidar Kovačić, and Vladan Jovanović. "Organization of tax data warehouse for legal entities." *Information and Communication Technology, Electronics and Microelectronics (MIPRO), 2016 39th International Convention on.* IEEE, 2016

[2] Zakon o medijima Republike Hrvatske (The Media Act of the Republic of Croatia), Official Gazette no. 163/2003, http://narodne_novine.nn.hr/clanci/sluzbeni/2003_10_163_2338.html on 18/01/2017

[3] Dave Chaffey, E-business and E-commerce Management, Pearson Education, 2007

[4] Javorović, B., Bilandžić, M.: Poslovne informacije i business intelligence, Golden marketing – Tehnička knjiga, Zagreb 2007

[5] http://www.riaa.com/wp-content/uploads/2016/09/RIAA_Midyear_2016Final.pdf 17/01/2017

[6] https://www.riaa.com/facts/ 17/01/2017

[7] Matić Tin, Pravo virtualnih pravnih odnosa: Eelktronička trgovina, Narodne novine , 2012

[8] Chimay J. Anumba, Kirti Ruikar, e-Business in Construction, John Wiley & Sons, 2009

[9] Han, Jiawei, Jian Pei, and Micheline Kamber. *Data mining: concepts and techniques.* Elsevier, 2011

[10] Ralph Kimball, Joe Caserta, „The Data Warehouse ETL Toolkit: Practical Techniques for Extracting, Cleaning, Conforming, and Delivering Data", Wiley, 2004

[11] D.Linstedt, Olschimke, Building a Scalable Data Warehouse with Data Vault 2.0, MK 2015

[12] Mladen Varga, Baze podataka, Društvo za razvoj informacijske pismenosti (DRIP) 1994

[13] Martin, J. Principles of Object Oriented Analysis and Design Prentice Hall, Englewood Cliffs, 1993

[14] Mile Pavlić, Oblikovanje baza podataka, Rijeka: Odjel za informatiku Sveučilišta u Rijeci 2011

[15] Mile Pavlić, Razvoj informacijskih sustava: projektiranje, praktična iskustva, metodologija, Zagreb, Znak, 1996

Law and Technology in Data Processing:

Risk-Based Approach in EU Data Protection Law and Implementation Challenges in Croatia

Nina Gumzej
University of Zagreb Faculty of Law, Zagreb, Croatia
E-mail: nina.gumzej@pravo.hr

Abstract – Recently reformed EU personal data protection rules (General Data Protection Regulation) provide mechanisms to ensure that under strict legal responsibility risks inherent in modern data processing operations are appropriately and timely identified and managed. Analysis in the paper focuses on its risk-oriented rules aiming to ensure proper data security and data protection compliance from earliest data processing stages, and envisaged impact assessment procedure as an important tool toward identifying and mitigating risk in certain high-risked operations. Assessment of impacts that processing activities, as enabled by further technological advancements, may have on individuals' rights and freedoms requires continuous consideration, in particular as some possibly cannot yet be perceived and/or assessed whether in terms of nature or severity (scope). This adds to analysed open issues requiring further clarification in the area, in connection with initiatives towards the more specific, harmonized EU procedure. Examined risk assessment solutions currently envisaged under the Data Protection Directive, as implemented under Croatian law, and lacking local practice in the area point to likely local implementation challenges of the soon directly applicable, more comprehensive and strict new EU rules, as a result of which proposals are made toward urgent and intensive awareness-raising measures.

I. INTRODUCTION

In today's era of Big Data in progress is evolution from the information economy, which focuses on content and services for consumers, to the data economy where the data mainly flow between a set of non-human actors processing such data, often in real time. This evolution further expands with the development of the Internet of Things. The data economy is built on establishing data value chains and depending on possibilities of using, aggregating and generally processing the data from different sources in an automated form. As such the data economy is strongly affected by the legal framework affecting the flow of such data. In that context Lammerant and De Hert rightly establish that visions on how technology changes economical processes are important drivers of legal change [1]. Those drivers are evident in particular through development of new EU legislation in the area of personal data. Following a four year long complicated legislative procedure, on 24 May 2016 a breakthrough piece of EU legislation in this legal area entered into force: the *General Data Protection Regulation* (hereinafter referred to as: GDPR) [2]. The GDPR will apply directly in Member States as of 25 May

2018 and at that time it will repeal the currently valid EU general data protection law basis - the *Data Protection Directive 95/46/EC* (hereinafter referred to as the DPD) [3]. Technologically neutral rules signify a quantum leap in this legal area as they deliver on the long needed adaptation of the law to modern digital processing conditions especially in light of rapid technological developments in the ICT area and the ever-reaching, perplexed, extensive and in general more and more complex procedures for processing personal data ("processing" signifying any operation performed on such data), in particular online. In addition to this the often fragmented domestic frameworks based on the DPD required better harmonization and more appropriate conditions to ensure legal certainty and a levelled playing field for competition, thereby strengthening the EU internal market, as well as EU-wide harmonized protection of rights of individuals.

In addition to personal data processing activities by all relevant EU-based actors (controllers and processors) the GDPR seems to apply world-wide where non-EU based in particular digital actors are processing personal data of individuals in the EU by offering them goods or services (with or without payment) or by monitoring their behaviour in the EU, which may include online monitoring activities in the context of profiling [4]. Adding to this is also its wide material scope of application as evidenced by the expressly expanded concept of personal data to also include digital data such as online identifiers (IP addresses, cookie identifiers, RFID tags, and other) and location data [5]. Additionally to expanded duties and reinforced accountability of actors processing personal data, particular severity of the new legal framework is evidenced by stronger harmonized authorities of supervisory data protection bodies enforcing it. A clear example thereof are envisaged administrative fines, so far unprecedented in EU data protection legislation, which may rise to up to 20 million Euro, or where undertakings are concerned to up to 4% of total worldwide annual turnover, whichever is higher [6]. Monitoring relevant developments, insofar as they have an impact on personal data protection, especially development of ICT technologies and commercial practices represents a new legal task facing those same authorities, the relevance of which should not be overlooked [7]. All this adds to the general appraisal of the strict, detailed and complex nature of the new legal frame, which demands due regard for relevant technological developments and which is as such soon directly applicable in all EU Member States.

Assessment and overall management of risks that prompts legal responsibility and accountability in data protection is a crucial area in new EU data protection legislation that the author chose to examine and critically appraise in this paper. Analysis begins with an overview of current legal solutions in risk assessment according to the DPD and the selected local solutions. Data protection legislation and practice in the youngest EU Member State Republic of Croatia provides a case study for this analysis. This will allow an evaluation of a local implementation level of DPD's risk assessment solutions and of possible challenges in the uptake of a more comprehensive risk-based approach to data protection under the GDPR (which itself is examined in fourth part of the paper). Third part of this paper provides an overview of selected discussions and initiatives relating to methodology and procedures identifying and assessing risks and impacts thereof on individuals' rights and freedoms, especially in the context of data processing activities in a networked environment and with the use of newer technologies and applications. This discussion provides a basis for the ensuing examination of GDPR solutions in the risk-assessment area, which reinforce and add on the earlier analysed provisions of the currently still valid DPD.

II. RISK ASSESSMENT IN THE DPD AND ONE LOCAL IMPLEMENTATION CASE STUDY

The concept of risk appraisal and management in the DPD is evident in particular from its rules on security measures and the prior checking procedure, i.e., examination of data processing activities prior to their initiation [8]. As regards security measures, the DPD requires those measures to ensure (having regard to state of the art and implementation costs) a level of security *appropriate to risks represented by the processing and the nature of data to be protected*. With respect to prior checking, the DPD requires Member States to establish data processing activities that are likely to pose specific risks to individuals' rights and freedoms. A prior check of those activities is a duty conferred on supervisory authorities upon notification by controllers (main responsible agents that determine the purposes and means of processing personal data). Data protection officers (where appointed by controllers to help them ensure data protection compliance) can also conduct such prior checks and their duty is to consult the relevant data protection body in cases of doubt. It should be noted that important aspects of the procedure, such as concrete consequences of prior checks and determination of risks are not regulated by the DPD, i.e., they are left to regulatory discretion of Member States under their respective laws [9]. To be precise, the DPD does not list examples of processing activities likely to present specific risks in any of its operative provisions (Articles), although some indicators and examples are provided in its recitals (elements to consider risk: nature, scope and purpose of processing; examples of risks: processing that may result in excluding individuals from a right, benefit or a contract, processing with specific use of new technologies). Exact consequences of prior checking procedure are also not

specified by the DPD, although it is clarified in its recitals that the supervisory authorities may following the prior check give their opinion or provide approval of the processing under their national law. It is assumed that they should also be able to prohibit intended processing unless, for example, appropriate risk-mitigating measures are implemented. A more general view on this issue points to the scope of intervention powers that was granted to these authorities. Namely, according to the DPD Member States must ensure that the supervisory bodies have effective intervention powers, such as delivering opinions before processing operations are carried out (prior checking) and ensuring their appropriate publication, imposing bans on processing, warning the controller, and other [10].

The Republic of Croatia implemented the DPD in its *Personal Data Protection Act* (further: PDPA), which was adopted in 2003 and thereon in process of alignment with the DPD up until its latest amendment in 2011 [11]. Although in line with the earlier examined DPD rule on security measures the PDPA lays down a duty to protect the data (from any accidental or deliberate abuse, destruction, loss, unauthorized alteration or access), it was not before its last amendment in 2011 that for the first time a method of establishing security measures at least somewhat in line with the DPD was introduced. However, the adopted solution cannot be considered satisfactory [12] as it only requires security measures to be *proportionate to the nature of controller's or recipient's activities, and to contents of the data filing system.* Assuming that "contents of the data filing system" correspond to the nature of data to be protected (e.g. stronger security measures such as encryption should apply where the more sensitive data are processed), the PDPA still fails to specify that the measures need also to ensure a level of security *appropriate to risks represented by the processing*. It can therefore be concluded that the solution adopted in 2011 is a missed opportunity to appropriately introduce the risk-based approach in local data protection legislation and practice. This concern does not, however, relate to security requirements for sensitive data (i.e., special personal data categories such as health data) that have been prescribed in secondary legislation already as of 2004, and which rely on ISO standards 27001 and 17799 (27002) [13].

Where prior checking procedures are concerned, the adopted solution is somewhat detached from the prevailing practice in the EU [14]. One reason for this are unspecified risks for which prior checks are necessary, which, although required by the DPD are neither foreseen in the PDPA nor in any publicly available guidance of the Data Protection Agency. Secondly, prior checking is not independently regulated by any operative provision, i.e., Article of the PDPA. Instead, the process is only implied from the provision elaborating Agency's tasks, i.e., its duty to provide preliminary opinions if the processing presents specific risks to individuals' rights and freedoms. Furthermore, controllers' duty to seek Agency's opinion on the need for a prior check in case of doubt is also awkwardly placed in the rule specifying Agency's tasks, which therefore adds on the previous remark [15]. Though required by the DPD the PDPA does not regulate the duty to publicize specifically its prior checking opinions,

1425

although some such opinions can be tracked on its internet pages [16]. Finally to be noted is that the PDPA failed to regulate in line with the DPD also the data protection officers' duty to carry out prior checks. Combined with earlier remarks this is particularly troubling also in light of mandatory appointment of such officers for all controllers with 20+ employees (not required by the DPD) and the thereby enabled development of an entire data protection profession on one hand, and on the other failed use also of that opportunity to promote best practices in the vital risk assessment area [17]. This contributes to an overall impression of the so far undervalued risk-based approach to personal data protection in Croatia.

III. RISK ASSESSMENT CONCERNS AND INITIATIVES IN LIGHT OF TECHNOLOGICAL ADVANCEMENT

In the context of data protection legislation there is a lacking consensus in scientific literature not only on what could potentially represent risks for individuals' rights and freedoms where personal data processing operations are concerned, especially in light of rapidly advancing technologies and applications in the online environment, but also on methods of their appraisal (e.g. objective and/or subjective) [18]. Examination of the object of protection, i.e., what rights and freedoms a certain processing activity affects, is likewise under a discussion. In that sense the inevitable evolution of privacy as the object of protection, i.e., consideration for new conceptualisations of that right in light of advances in technology affecting it should here also be mentioned. Finn and others, for example, expand on previous Clarke's typology of privacy and propose consideration for seven types (privacy of the person, privacy of behaviour and action, privacy of personal communication, privacy of data and image, privacy of thoughts and feelings, privacy of location and space and privacy of association), all in view of various examples in current technological development (e.g. whole-body scanners, RFID-tagged identity documents, drones, next generation biometry and DNA sequencing, and other) [19].

In the context of privacy and data protection legislation the procedures to identify, assess and minimize risks, especially those embodied through the use of newer technologies and enabling various surveillance operations over individuals have up to today mainly focused on impacts on the right to individual privacy (*privacy impact assessment - PIA*) [20]. Certain authors even suggest that in addition to this impact (of newer technologies and applications) on ethical issues should be assessed (*ethical impact assessment*) [21]. This viewpoint closely relates to discussions on current inadequacies of PIAs where surveillance technologies are concerned, both in the public and private sectors, since PIAs focus on individual privacy and not on the wider scope of affected rights, freedoms and interests [22]. Certain projects co-funded by the European Commission elaborate even further on the issue, adjusting the PIA to the *security impact assessment* frame, which addition to privacy and personal data protection considers also ethical, social, economic and political issues during the design and implementation of smart surveillance or monitoring systems, technologies, processes and policies [23]. A need for a broader

viewpoint of affected rights and freedoms and risks in the context of EU personal data protection legislation was affirmed also by independent advisory body established under Article 29 of the DPD for the protection of personal data and privacy – the Article 29 Data Protection Working Party. According to it the risk-based approach to data protection should go further than the narrow approach based on harm and it should take into account any potential and real negative effect, which can be evidenced by effect on the individual up to the general effect on society (e.g. loss of social trust). Also, the scope of rights and freedoms that could be negatively affected by data processing should not always focus exclusively on the right to privacy, since other fundamental rights (e.g. freedom of speech, thought and movement, prohibition of discrimination, right to liberty, conscience and religion) may thereby also be negatively affected [24]. A discourse should here also be added on limitations in risk assessment methodologies *inter alia* due to limitations of current scientific knowledge on impact (of data processing activities in the context of newer technologies and applications) on individuals' rights and freedoms, as well as on shortcomings of potential "managerisation" of related risk assessment procedures. Thus for example certain authors point to the need to expressly include the *precautionary principle* (accepted in areas such as environmental law) [25] into the EU data protection law risk management frame [26] so that in case of uncertainty on existence and/or extent of risk precautionary (preventive) measures are applied without waiting for those risks to materialize [27].

Particularly in light of ongoing technological progress and data processing operations related to the use of newer technologies and applications (notably in the ICT area), all of the above noted concerns seem to justify the need to institutionalize the element of *transparency* into impact assessment procedures, i.e., including all relevant stakeholders as well as the general public therein [28]. Also to be noted are expert proposals nonexclusively related to the data protection field, in light of their assessment of European Court of Human Rights' case law as pointing to development of the *right to have technologies assessed before their launch*. In that context and in light of recognized significance of impact assessment procedures it is proposed that not only experts and industry but also individuals are included (with public consultations to be held already in the stage of planning or shaping new technologies), impact assessment results published (without prejudice to special treatment of sensitive information) and regular checks ensured [29].

Voluntary initiatives to devise impact assessment procedures on the basis of the currently valid DPD (they are not themselves envisaged in the DPD) that should here be noted are the *Privacy and Data Protection Impact Assessment Framework for RFID Applications* at EU level [30], impact assessment procedures at individual Member State level [31] and industry-based proposals on impact assessments tailored for specific sectors and/or special needs of small and medium-sized enterprises [32]. Where new (GDPR) and current legal EU frame (DPD) are concerned it can be observed that risk assessment solutions in the GDPR reinforce and expand on earlier in this paper examined DPD solutions and this is evident *inter alia* from the newly introduced provisions on impact assessments. Namely, as it will be elaborated in next part of this paper, the GDPR expressly adopts for the first time in EU data protection legislation a legal duty to conduct the so-called *data protection impact assessments* in certain cases of high-risked personal data processing. Especially in light of future technological developments affecting the data protection legal area, the earlier noted concerns on identification and assessment of risk, scope and object of protection and limitations in risk assessment procedures should also currently be appraised in view of endeavours toward EU-wide harmonization of the procedure to identify high risks and frame for impact assessments under the GDPR. The vital role of harmonized procedures in this area is not only that of enabling EU-wide consistency in supervisory practices, but also that of reducing associated compliance burdens both for the supervisory authorities and the controllers [33].

IV. RISK-BASED APPROACH AND DATA PROTECTION IMPACT ASSESSMENTS IN THE GDPR

The overall risk-based approach to personal data protection in the GDPR is based on identification of *likelihood and severity* of risks that a personal data processing activity has on individuals' rights and freedoms with respect to its nature, scope, context and purposes. Main GDPR provision on security of processing establishes the duty of both the controllers and processors (who process data on behalf of controllers) to implement appropriate technical and organisational measures to ensure a level of security appropriate to risk, taking into account state of the art, implementation costs and the nature, scope, context and purposes of processing as well as risk of varying likelihood and severity for individuals' rights and freedoms. In assessing data security risks consideration should be given to risks presented by data processing such as accidental or unlawful destruction, loss, alteration, unauthorised disclosure of, or access to, personal data transmitted, stored or otherwise processed (and which may in particular lead to physical, material or non-material damage, in line with the relevant recital) [34]. In fact, implementation of measures to ensure appropriate data security is now established as one of the fundamental data processing principles (*integrity and confidentiality principle*) and the also newly introduced *accountability* principle imposes a duty on the controllers to (be able to) demonstrate compliance of their processing activities with this and other fundamental principles [35].

Furthermore, a risk-based approach is the decisive element of the newly introduced principle of *data protection by design*. According to this principle appropriate technical and organisational measures must be implemented both at the time of determining the means for processing and at the time of processing itself, which must be designed to implement data-protection principles effectively and with safeguards integrated into processing, in order to meet GDPR requirements and protect rights of data subjects. Those measures must be chosen taking into account state of the art, implementation costs, nature, scope, context and purposes of processing as well as risks

of varying likelihood and severity for individuals' rights and freedoms [36].

GDPR's strong emphasis on a risk-based approach is evidenced also from introduced provisions on data protection officers, in particular their tasks, related appointment conditions and the duty to have in performance of their tasks due regard to risks associated with processing operations, taking into account the nature, scope, context and purposes of processing [37].

Finally to be noted are explanations of risks of varying likelihood and severity that may result from processing activities potentially leading to physical, material or non-material damage. Those activities include in particular:

1) processing activities that may lead to discrimination, identity theft or fraud, financial loss, damage to reputation, loss of confidentiality of personal data protected by professional secrecy, unauthorised reversal of pseudonymisation, or any other significant economic or social disadvantage;

2) where data subjects might be deprived of their rights and freedoms or prevented from exercising control over their personal data;

3) where special categories of personal data, data relating to criminal convictions and offences and data of vulnerable persons, especially children, are processed;

4) where personal aspects are evaluated, in particular analysing or predicting aspects concerning performance at work, economic situation, health, personal preferences or interests, reliability or behaviour, location or movements, in order to create or use personal profiles; or

5) where processing involves a large amount of personal data and affects a large number of data subjects.

In reference to earlier noted discussions on objective and/or subjective evaluation, according to explanations in the GDPR assessments of existence of risk or high risk in relation to processing should in fact be *objective* [38].

The GDPR abolishes a general duty of notifying data processing to supervisory authorities under the DPD and introduces a risk-based system involving supervisory authorities where certain procedures show that the processing would result in a *high risk* to individuals' rights and freedoms on account of its scope, nature, context and purpose [39]. In this respect it introduces *data protection impact assessments* (further also as: DPIA), which are mandatory only if it is likely that a type of processing, in particular using new technologies, and taking into account nature, scope, context and purposes of processing, is likely to result in a high risk to individuals' rights and freedoms [40]. As clarified in relevant recital DPIAs serve to evaluate especially the origin, nature, particularity and severity of that risk. In order to reduce unnecessary burden it is allowed to use a single DPIA to cover more similar processing operations presenting similar high risk. Relevant high-risked processing activities include:

(1) systematic and extensive evaluation of personal aspects relating to natural persons which is based on automated processing, including profiling, and on which

decisions are based that produce legal effects concerning or similarly significantly affecting him or her;

(2) processing on a large scale of special categories of data or of data relating to criminal convictions and offences, and

3) systematic monitoring of a publicly accessible area on a large scale (as clarified this is especially required where optic-electronic devices are used for monitoring).

National supervisory authorities may establish further examples, and as explained in that process they are to evaluate if the processing activity is likely to result in a high risk in particular because it prevents individuals from exercising a right or using a service or a contract, or because it is carried out systematically on a large scale.

Finally to be noted are prescribed minimal contents of a DPIA: systematic description of processing operations and purposes thereof, assessment of necessity and proportionality of processing operations in relation to their purposes, assessment of risks to rights and freedoms and intended measures to mitigate risks and demonstrate compliance with the GDPR.

With the exception of publishing results of impact assessments the GDPR foresees key elements pointed to in previous part of this paper resulting from discussions on optimal impact assessment procedures where the processing entails use of new technologies (transparency – public consultations, publishing results and regular checks of impact assessments). However, the effect of those elements of DPIAs as introduced by the GDPR in reality appears minimal, since their implementation is left to controllers' discretion. Concretely, it is prescribed that *where appropriate* controllers shall seek views of data subjects or their representatives on intended processing (without prejudice to protection of commercial or public interests or security of processing operations). In connection with this the ensuing process of publishing DPIA results is regrettably not at all envisaged. As far as subsequent DPIA checks are concerned, the GDPR does envisage them, however, only where necessary, i.e., the controller shall *where necessary* carry out a review (to assess if the processing complies with the DPIA) *at least when there is a change of risk represented by processing operations*. It is difficult to sensibly apply this rule without acknowledging the implied need to continually review the DPIA to be able to establish possible changes to the level of risk that brought on a mandatory DPIA in the first place. On the other hand the adopted solution "as is" fails to specify such checks and as a result it is not unlikely that they could be well understated in practice.

Taking into account that the scope of this paper does not allow a more detailed analysis of the DPIA area, the following is a brief discussion on *prior consultations* as another vital GDPR procedure, pursuant to which the controllers must consult the supervisory authority before processing, where the DPIA showed that the processing would lead to a high risk *in the absence of measures mitigating it* [41]. A concern to which it could here be pointed is coupled with an explanation in the relevant recital, pursuant to which prior consultations should only take place where the controller *cannot mitigate* identified

high risks by appropriate measures in terms of available technology and implementation costs [42]. This would appear to allow a more narrow scope of circumstances instigating control by supervisory bodies, and possible misuses where a decision on adequacy of proposed risk mitigating measures is left to controller's sole discretion. In any case, the recital can be considered as not entirely consistent also with the prescribed task of authorities to issue an opinion and provide advice where it considers that the controller insufficiently identified or mitigated the risk, and all this in fact takes place during the already initiated consultation procedure.

Where affected rights and freedoms are concerned, the very title but also goal of the DPIA ("assessment of the impact of the envisaged processing operations on the protection of personal data") suggest that impact assessments as required under the GDPR would have a narrow object of protection, i.e., it appears their focus would be strictly on the right to personal data protection. This solution is unsatisfactory especially in light of subject-matter and objectives of modern EU data protection legislation that is generally interwoven with outmost consideration for future technological developments. Namely, the GDPR lays down *inter alia* rules relating to protection of natural persons with regard to personal data processing and though prevailingly, guaranteed rights and freedoms with respect to such data processing are nonetheless not strictly limited to the data protection right [43]. It is expected that this and other open issues analysed in the paper will be resolved in the scope of work to ensure a harmonized risk assessment methodology on the basis of the GDPR and possibly specified via adopted codes of conduct at EU level or approved certification mechanisms and/or guidelines issued by the European Data Protection Board [44]. In respect of the latter, in fact, the Article 29 Data Protection Working Party (which will be instituted into the European Data Protection Board on the basis of the GDPR) announced publication of its guidelines on high-risked processing and DPIAs during 2017 [45]. In connection with this to be followed is also development of the *DPIA Template for Smart Grid and Smart Metering Systems* [46], individual DPIA initiatives as already proposed by some data protection authorities [47] and the industry-based proposals (*ISO/IEC 29134 Information technology - Security techniques - Guidelines for privacy impact assessment*; under development) [48].

V. CONCLUSION

As a result of an ever-close connection between law and technology the recently reformed EU personal data protection rules (GDPR) provides mechanisms to ensure that under strict legal responsibility risks inherent in modern personal data processing operations are timely identified and appropriately managed. Relevant risk-oriented rules toward ensuring and maintaining proper data security and data protection compliance already from earliest stages of data processing operations, and impact assessment procedures as an indispensable tool for recognizing and mitigating risks in certain high-risked processing operations represent a giant step in development of modern data protection legislation and

1428

should be applauded, particularly when taking into account that the road toward adoption of this piece of legislation was itself quite long, and quite difficult. The future-proof stance of these technologically neutral rules demands ongoing consideration of impacts that the various personal data processing activities as enabled by further technological advancement may have on individuals' rights and freedoms and out of which some possibly cannot yet be perceived and/or assessed, whether in terms of nature or severity (scope). Where the envisaged data protection impact assessment procedure is concerned and taking into account high-risked processing activities mandating them, such as systematic monitoring of publicly accessible areas on a large scale, analysis in this paper pointed to issues that should further be clarified and specified, and hopefully approached in a generally harmonized EU procedure. These include *inter alia* nature and scope of rights and freedoms that should thereby be protected, methodology to identify and evaluate risks (high risks) and other procedural matters (e.g. public consultations, checks) that significantly affect the overall effectiveness of impact assessments as an introduced legal duty. Taking into account that these procedures can utilize significant resources, all efforts toward EU-wide harmonized procedures should be seized as they could result in a significant reduction of costs of compliance, and reduced resources otherwise required in this area also for supervisory bodies. Furthermore, a harmonized methodology enables supervisory consistency, which is in any case a goal that the GDPR intends to achieve.

Examined inadequacy of (non)implemented risk assessment solutions under the DPD (which will soon be replaced by the analysed more comprehensive and stricter, directly applicable GDPR rules) in Croatian law and practice could compromise intended harmonization aims as well as cause panic for locally responsible actors due to the generally lacking focus on this vital data protection area. Even today it can be considered to be too late for some organizations to start making appropriate adjustments to their data processing operations and allocating necessary resources in order to ensure timely compliance with GDPR requirements. It is therefore strongly recommended that as a minimum measure the Data Protection Agency as well as local experts in the area carry out intensive awareness-raising activities in the risk-based approach to data protection. Utilizing data protection officers currently appointed on the basis of the PDPA is a practical opportunity to work on this goal, which should not be overlooked. Building skills in this area combined with the duty to keep a close eye on relevant developments in technology is in any case an upcoming requirement for all.

LITERATURE

[1] H. Lammerant and P. De Hert, "Visions of Technology - Big Data Lessons Understood by EU Policy Makers in Their Review of the Legal Frameworks on Intellectual Property Rights, Access to and Re-use of PSI and the Protection of Personal Data", in: Data Protection on the Move - Current Developments in ICT and Privacy/Data Protection, Law, Governance and Technology Series - Issues in Privacy and Data Protection, vol. 24, S. Gutwirth, R. Leenes and P. De Hert, Eds. Springer, 2016, pp. 163-194 at pp. 164, 166.

[2] Regulation (EU) 2016/679 of 27 April 2016 on the protection of natural persons with regard to the processing of personal data and on the free movement of such data, and repealing Directive 95/46/EC, Official Journal of the European Union (further: OJ) L 119, 4.5.2016, pp. 1–88.

[3] Directive 95/46/EC of 24 October 1995 on the protection of individuals with regard to the processing of personal data and on the free movement of such data, OJ L 281, 23.11.1995, pp. 31–50.

[4] Article 3 and recital 24 of the GDPR.

[5] Article 4, point 1 and recitals 26 and 30 of the GDPR.

[6] Article 83 and paragraphs 5-6 of the GDPR.

[7] Article 57, paragraph 1 (i) of the GDPR.

[8] Article 29 Data Protection Working Party, Statement on the role of a risk-based approach in data protection legal frameworks, 14/EN, WP 218, 30.5.2014, p. 2. See Article 17, paragraph 1 and recital 46; Article 20 and recitals 53-54 of the DPD, respectively.

[9] D. Korff, "Data protection laws in the EU: The difficulties in meeting the challenges posed by global social and technical developments", Study for the European Commission, 2010, pp. 103-104, http://ec.europa.eu/justice/policies/privacy/docs/studies/new_privacy_challenges/final_report_working_paper_2_en.pdf.

[10] Article 28, paragraph 3 of the DPD.

[11] Official Gazette of Croatia no. 103/03, 118/06, 41/08, 130/11 and 106/12 - consolidated text.

[12] Article 18 of the PDPA.

[13] Decree on the procedure for storage and special measures relating to the technical protection of special categories of personal data, Official Gazette of Croatia no. 139/04. Special categories of personal data are regulated in Article 8 of the PDPA.

[14] A good comparative overview of implementing prior checks in accordance with the DPD in majority of EU Member States is provided in: G. Le Grand and E. Barrau, "Prior Checking, a Forerunner to Privacy Impact Assessments", in: Privacy Impact Assessment, Law, Governance and Technology Series, D. Wright and P. De Hert, Eds. Vol. 6, Springer, 2012, pp. 97-116.

[15] Article 33, point 9 of the PDPA.

[16] Personal Data Protection Agency, Opinions, Control of working time by biometric fingerprint, http://azop.hr/misljenja-agencije/detaljnije/kontrola-radnog-vremena-biometrijom-prst.

[17] Article 18, paragraph 2 and Article 20, paragraph 2 and recitals 49, 54 of the DPD; Article 18a of the PDPA, respectively.

[18] See e.g. C. Kuner, F. H. Cate, C. Millard, D. J. B. Svantesson and O. Lynskey, "Risk management in data protection", International Data Privacy Law, vol. 5, no. 2, 2015, pp. 95-98 at p. 96-97.

[19] R. L. Finn, D. Wright and M. Friedewald, "Seven Types of Privacy", in: European Data Protection: Coming of Age, S. Gutwirth, R. Leenes, P. de Hert and Y. Poullet, Eds. Springer, 2013, pp. 3-32. For Clarke's typology, see *ibid* at pp. 6-7 and in author's original work: R. Clarke, "Introduction to Dataveillance and Information Privacy, and Definitions of Terms", 15.8.1997 http://www.rogerclarke.com/DV/Intro.html.

[20] For further details see: Privacy Impact Assessment, Law, Governance and Technology Series, D. Wright and P. De Hert, Eds. Vol. 6, Springer, 2012. Also see outcome of the project „Privacy Impact Assessment Framework for data protection and privacy rights", available at: http://www.piafproject.eu; K. Wadhwa and R. Rodrigues, "Evaluating privacy impact assessments, Innovation", The European Journal of Social Science Research, vol. 26, no. 1-2, 2013, pp. 161-180.

[21] D. Wright and M. Friedewald, "Integrating privacy and ethical impact assessments", Science and Public Policy, vol. 40, no. 6, 2013, pp. 755–766; D. Wright and E. Mordini, "Privacy and Ethical Impact Assessment", in: Privacy Impact Assessment, Law, Governance and Technology Series, D. Wright and P. De Hert, Eds. Vol. 6, Springer, 2012, pp. 397-418.

[22] See e.g. Raab's and Wright's discussion on the four circles (P1-P4) of privacy impact assessments: C. Raab and D. Wright, "Surveillance: Extending the Limits of Privacy Impact Assessment", in: Privacy Impact Assessment, Law, Governance and Technology Series, D. Wright and P. De Hert, Eds. Vol. 6, Springer, 2012, pp. 363-383 at pp. 376-382.

[23] SAPIENT project, http://www.sapientproject.eu/; D. Wright, M. Friedewald and R. Gellert, "Developing and testing a surveillance impact assessment methodology", International Data Privacy Law, vol. 5, no. 1, 2015, pp. 40-53.

[24] Article 29 Data Protection Working Party, Statement on the role of a risk-based approach in data protection legal frameworks, 14/EN, WP 218, 30.5.2014.

[25] The principle encompasses specific circumstances where scientific evidence is insufficient, inconclusive or uncertain and where there are indications through preliminary objective scientific evaluation that there are reasonable grounds for concern that the potentially dangerous effects (on the environment, human, animal or plant health) may be inconsistent with the chosen level of protection. Communication from the Commission on the precautionary principle (COM(2000)) 1 final, 02.2.2000.

[26] See e.g. R. Gellert, "Data protection: a risk regulation? Between the risk management of everything and the precautionary alternative", International Data Privacy Law, vol. 5, no. 1, 2015, pp. 3-19; L. Costa, "Privacy and the precautionary principle", Computer Law & Security Review, vol. 28, no. 1, 2012, pp 14-24.

[27] L. Costa, "Privacy and the precautionary principle", Computer Law & Security Review, vol. 28, no. 1, 2012, pp. 14-24 at p. 16.

[28] See e.g. R. Clarke, "Privacy Impact Assessment Guidelines", Xamax Consultancy Pty Ltd, February 1998, http://www.xamax.com.au/DV/PIA.html, point 5; R. H. Weber, "Privacy management practices in the proposed EU regulation", International Data Privacy Law, vol. 4, no. 4, 2014, pp. 290-297 at pp. 295-296; A. Warren and A. Charlesworth, "Privacy Impact Assessment in the UK", in: Privacy Impact Assessment, Law, Governance and Technology Series, D. Wright and P. De Hert, Eds. Vol. 6, Springer, 2012, pp. 205-224 at p. 224; D.Wright and E. Mordini, "Privacy and Ethical Impact Assessment", in: Privacy Impact Assessment, Law, Governance and Technology Series, D. Wright and P. De Hert, Eds. Vol. 6, Springer, 2012, pp. 397-418 at pp. 402-404; K. A. Bamberger and D. K. Mulligan, "PIA Requirements and Privacy Decision-Making in US Government Agencies", in: Privacy Impact Assessment, Law, Governance and Technology Series, D. Wright and P. De Hert, Eds. Vol. 6, Springer, 2012, pp. 225-250 at p. 233; R.Gellert, "Data protection: a risk regulation? Between the risk management of everything and the precautionary alternative", International Data Privacy Law, vol. 5, no. 1, 2015, pp. 3-19.

[29] P. De Hert, "A Human Rights Perspective on Privacy and Data Protection Impact Assessments", in: Privacy Impact Assessment, Law, Governance and Technology Series, D. Wright and P. De Hert, Eds. Vol. 6, Springer, 2012, pp. 33-76 at pp. 75-76.

[30] Privacy and Data Protection Impact Assessment Framework for RFID Applications, 12.1.2011., http://ec.europa.eu/justice/policies/privacy/docs/wpdocs/2011/wp1 80_annex_en.pdf. Also see: Article 29 Working Party, Opinion 9/2011 on the revised Industry Proposal for a Privacy and Data Protection Impact Assessment Framework for RFID Applications, 00327/11/EN, WP 180, 11.2.2011; Commission Recommendation of 12.5.2009 on the implementation of privacy and data protection principles in applications supported by radio-frequency identification, OJ L 122, 16.5.2009, pp. 47–51.

[31] Information Commissioner's Office, "PIA Code of Practice", https://ico.org.uk/media/for-organisations/documents/1595/pia-code-of-practice.pdf; Commission nationale de l'informatique et

des libertés, "Privacy Impact Assessments: the CNIL publishes its PIA manual", 10.7.2015, https://www.cnil.fr/fr/node/15798.

[32] D. Wright and M. Friedewald, "Integrating privacy and ethical impact assessments", Science and Public Policy, vol. 40, no. 6., 2013, pp. 755–766; D. Wright and E. Mordini, "Privacy and Ethical Impact Assessment", in: Privacy Impact Assessment, Law, Governance and Technology Series, D. Wright and P. De Hert, Eds. Vol. 6, Springer, 2012, pp. 397-418.

[33] R. H.Weber, "Privacy management practices in the proposed EU regulation", International Data Privacy Law, 2014, vol. 4, no. 4, p. 290-297 at p. 296.

[34] Article 32, recitals 76 and 83 of the GDPR, respectively.

[35] Article 5 of the GDPR.

[36] Article 25, paragraph 1 of the GDPR.

[37] Article 37, paragraph 5 and Article 39 of the GDPR.

[38] Recitals 75 – 76 of the GDPR.

[39] Commission Staff Working Paper, Impact Assessment Accompanying the General Data Protection Regulation, 25.1.2012, SEC(2012) 72 final, pp. 14-16.

[40] For further details see Article 35 and Article 57, paragraph 1 (k) as well as recitals 84-93 of the GDPR.

[41] For further details, see Article 36 and recitals 94-96 of the GDPR.

[42] Recital 84 of the GDPR.

[43] Article 1, paragraphs 1-2 and recital 75 of the GDPR.

[44] Recital 77 of the GDPR. For details on codes of conducts, certification mechanisms and tasks of the European Data Protection Board, see Articles 40-43 and Article 70, respectively.

[45] Article 29 Working Party, "Adoption of 2017 GDPR Action Plan – Press Release", 16.1.2017, http://ec.europa.eu/newsroom/document.cfm?doc_id=41387.

[46] Smart Grid Task Force 2012-14, Expert Group 2: Regulatory Recommendations for Privacy, Data Protection and Cyber-Security in the Smart Grid Environment, "Data Protection Impact Assessment Template for Smart Grid and Smart Metering Systems", 18.3.2014, https://ec.europa.eu/energy/sites/ener/files/documents/2014_dpia_smart_grids_forces.pdf. In connection with this see Commission Recommendation of 10.10.2014 on Data Protection Impact Assessment Template for Smart Grid and Smart Metering Systems (2014/724/EU), OJ L 300, 18.10.2014, pp. 63–68 and the related opinion of the Article 29 Working Party to which the Commission (inter alia) refers: Opinion 07/2013 on the Data Protection Impact Assessment Template for Smart Grid and Smart Metering Systems, 2064/13/EN, WP209, 4.12.2013. More information is available at: http://ec.europa.eu/energy/node/1748; also see Communication from the Commission, Delivering a New Deal for Energy Consumers COM/2015/0339 final, point 2.1.3.

[47] Commission de la protection de la vie privée, "Projet de recommandation d'initiative concernant l'analyse d'impact relative à la protection des données et la consultation préalable soumis à la consultation publique (CO-AR-2016-004)", https://www.privacycommission.be/sites/privacycommission/files/documents/CO-AR-2016-004_FR.pdf.

[48] ISO/IEC 29134 Information technology -- Security techniques -- Guidelines for privacy impact assessment (under development; target publication date: 30.5.2017, http://www.iso.org/iso/catalogue_detail.htm?csnumber=62289.

Social and Economic Effects of Investments in Primorsko-goranska County Broadband Network

Saša Vojvodić*, Saša Čegar**, Damir Medved*

* Ericsson Nikola Tesla d.d., Zagreb, Croatia
,** Faculty of Economics in Rijeka, Croatia
sasa.vojvodic@ericsson.com; sasa.cegar@efri.hr; damir.medved@ericsson.com

Abstract - The paper deals with the methodology applied to data collection, as well as social, economic, and financial analysis of broadband investments' effects related to the planning of broadband infrastructure development of the Primorsko-goranska County (PGC), with the City of Rijeka in focus. The analysis of the collected data and derived indicators has yielded results which can be significant for local and regional policy makers. It has been shown that investments in broadband infrastructure have a relatively high social return on investment and produce multiple socio-economic benefits that can not only accelerate the development of cities and municipalities in PGC, but also contribute to inhibiting or turning negative demographic trends around.

I.   INTRODUCTION

In modern times ICT technologies have become an important prerequisite to high quality of life, and an essential tool for efficient production and business in all economic sectors. Investments in the development and implementation of broadband infrastructure do not only have a positive impact on the overall society and business, but they are also an accelerator of growth and development of local economies, thus putting their huge development potential to a concrete use. Considering a broadband network as a factor of development mainly depends on the environmental, demographic, and economic specificities of a certain local unit. Furthermore, to choose an optimal broadband network technology, type, and model of financing of its implementation or modernization, it is important to consider a broader socio-economic context of such investment projects. In doing so, decisions makers must give priority to solutions that can potentially generate the biggest multiplier benefits for the local population and economy. In accordance to this, the authors have devoted special attention to integrated analysis of demographic and economic trends, spatial characteristics, and commercial potential of broadband services to determine how investment in broadband infrastructure can contribute to responding to key challenges of future development of cities and municipalities in PGC.

Taking into consideration the intended format and volume of this paper, the authors decided that presenting the obtained results and conclusions for all the cities and municipalities that were included in their study of broadband infrastructure development plans would be too extensive. Therefore, this paper deals with the assessment and interpretation of the socio-economic effects of investment in the broadband networks implementation in the City of Rijeka - the administrative center of Primorsko-goranska County (PGC).

II.   BROADBAND NETWORK - A PREREQUISITE FOR THE IMPLEMENTATION OF THE SMART CITY CONCEPT

The City of Rijeka has an extremely favorable geographical position, good transport links and an abundance of natural resources in its surroundings, of which the most significant are marine, forest and water resources. The City of Rijeka is also an important educational, cultural, health, trade, and economic center. In fact, it is the regional metropolis whose broader gravitation area extends beyond the territorial boundaries of PGC. However, considering its developmental needs, we must note that the City of Rijeka has a relatively small territory: 44 square kilometers. This considerably complicates the application of integrated urban approach in solving the key development challenges, which are primarily related to optimal planning, coordination, and allocation of development factors in accordance with the overall social and economic needs.

Considering the complex social, economic, and cultural perspective of Rijeka as the center of the county and its growing problem of lack of space, the paradigm of Rijeka makes an ideal case for the application of the so-called smart city concept, a systematic and integrated approach to the urban development management, using innovative ICT solutions and intelligent systems in integrated networking, planning, optimization and monitoring of all aspects of urban development. It aspires to socially, economically, and environmentally efficient and sustainable cities and urban areas. This enables the complementarity of investments in human and social capital and the overall city infrastructure, which provides conditions for a long-term sustainable economic development, a high standard of living, as well as an effective and efficient environmental protection and natural resources use. For these reasons, in the *City of Rijeka Development Strategy for the period 2014-2020* it was emphasized that the implementation of the smart city concept is an important determinant of its future development [1]. In this regard, the broadband infrastructure projects have a strategic priority for the City of Rijeka because the realization of such projects is the basic prerequisite for the formulation and implementation of smart city applications.

III. THE ROLE OF BROADBAND NETWORKS IN STOPPING NEGATIVE DEMOGRAPHIC TRENDS

In accordance with the last Census (2011), the City of Rijeka has 128,624 inhabitants and an average population density of 2,923 inhabitants per km², which makes it one of the most densely populated cities in the region. As the data gathered in the last two censuses show, the City of Rijeka underwent depopulation, following population trends characteristic of the higher territorial levels. However, it is indicative that the depopulation process in Rijeka is much more dynamic than the average depopulation processes at the national and county level.

GRAPH I. INTERCENSUS RATE OF POPULATION CHANGE (%) IN THE CITY OF RIJEKA, PGC AND THE REPUBLIC OF CROATIA IN 2001 AND 2011 (SOURCE: AUTHOR'S CALCULATIONS BASED ON CROATIAN CENSUS 2001 [2] AND CENSUS 2011 [3] DATA)

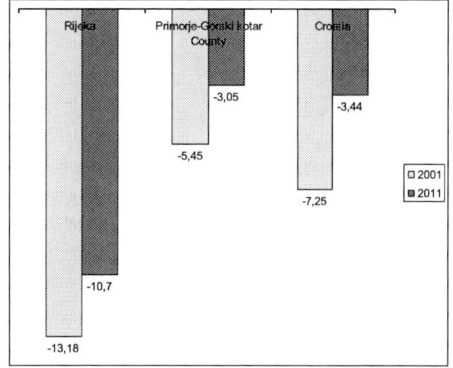

TABLE I. MECHANICAL AND NATURAL POPULATION CHANGE IN THE CITY OF RIJEKA, PGC AND THE REPUBLIC OF CROATIA IN THE INTER-CENSUS PERIOD 2001/2011 (SOURCE: AUTHOR'S CALCULATIONS BASED ON THE OFFICIAL VITAL STATISTICS [4] AND CROATIAN CENSUSES DATA [2][3]).

Territorial units	Population change in the inter-census period 2001/2011		
	Total change	Natural change	Migration balance
Rijeka	-15.419	-5.440	-9.979
Primorje-Gorski kotar County	-9.310	-10.211	901
Croatia	-152.571	-96.310	-56.261

Since over the observed period both components of general population trend are negative, the City of Rijeka suffered the most unfavorable outcome – a decrease in population, with an especially strong influence of the negative factors. In the context of discussing the unfavorable population trends in the City of Rijeka, it is important to point out that much of the problem is created by the redistribution of population within the Rijeka circle, i.e. moving of Rijeka's population to neighboring cities and municipalities. Such demographic trend is primarily the consequence of lower cost of living in the neighboring local units, as well as good traffic connection of these places with Rijeka. People can therefore relatively easily and on a lower budget enjoy all the conveniences of communal, sport, health, educational, cultural, business and entertainment infrastructure of the City of Rijeka, as well as its other urban contents.

The decrease of Rijeka's population incurred the decrease of the number of private households and the average size of households, which is typical for suburbanization processes where commutation of young and working-age population leads to the depopulation of the central city and demographic growth of its surroundings. At the same time, a part of housing capacity in the central city becomes available and the share of elderly households is increased. Another negative aspect of suburbanization process with the young population leaving the central city is a damaged reproductive potential and worsened biodynamics of population. This is also confirmed by the vital index data for the City of Rijeka in the period from 2011 to 2015. The vital index of Rijeka decreased during this period by 22,4%, reaching a value of 50,9 in 2015. That means that in Rijeka in 2015 on every new born, the city had two deceased inhabitants.

GRAPH II. RIJEKA'S VITAL INDEX IN THE PERIOD FROM 2011 TO 2015 (SOURCE: AUTHOR'S CALCULATIONS BASED ON THE OFFICIAL VITAL STATISTICS [4])

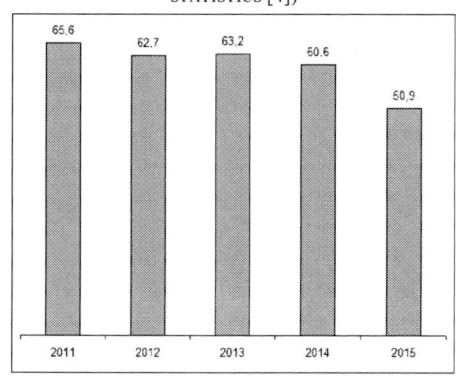

In comparison to the structural characteristics of the county and state population, it is evident that Rijeka has more retired citizens and fewer young and working-age citizens.

TABLE II. AGE STRUCTURE OF THE POPULATION OF THE CITY OF RIJEKA, PGC AND CROATIA IN 2011 (SOURCE: AUTHORS' CALCULATIONS BASED ON THE CROATIAN CENSUS 2011 DATA [3])

Age groups	Rijeka	Primorje-Gorski kotar County	Croatia
youth contingent (0-14 years)	11,63	12,48	15,22
working contingent (15-64 years)	68,63	68,62	67,1
retirement contingent (65 years and over)	19,74	18,91	17,7

The age structure of a population has a vital economic importance. It is primarily the young and the people of working-age who opt for migration. In the case of Rijeka, the analysis has shown that the age structure of the local population is not favorable for future economic development of the city. Most of Rijeka's working-age population is close to retirement. On the other hand, due to a relatively small ratio of the young constituting a pool of fresh work force, the city will have to face a serious work force problem over the next decade, which poses a threat to the future city's development.

In comparison to the county and state educational structure, the City of Rijeka is in a somewhat better position, with higher percentages of secondary and high education, and lower percentages of people with no school.

TABLE III. EDUCATIONAL STRUCTURE OF THE POPULATION OF THE CITY OF RIJEKA, PGC AND CROATIA IN 2011 (SOURCE: CENSUS 2011 [3])

Territorial units	No schooling	Secondary education	Higher education
Rijeka	0,68%	55,69%	23,79%
Primorje-Gorski kotar County	0,67%	57,74%	20,07%
Croatia	1,71%	52,63%	16,39%

The over-average share of highly educated people in the local population surely contains a large development potential and is one of the most significant comparative advantages of the City of Rijeka. The problem is, however, that the active entrepreneurial capacities in Rijeka's economy are predominantly concentrated in those activities which do not offer enough possibilities for the employment of highly educated population, with the consequence that the local economy leaves much to be desired regarding the employment of highly educated work force in recent years. This indicates that the economic orientation of the City of Rijeka has so far been in disproportion with its available demographic resources, but also with spatial limitation of the city, which has led to the structural maladjustments of the local economy (cf. chapter III.).

Due to the interdependence between demographic and economic trends, opening of new working places and increasing economic inclusion of the working population are distinctive factors of the economic growth and higher standard of living in the City of Rijeka. Investments in building and development of high speed broadband networks will definitely play an important role in achieving this. High speed broadband networks contribute to the development of high value added activities and more intensive investments into local economy. Consequently, this leads to the creation of new jobs, increase of employment, youth retention and immigration.

IV. ECONOMIC IMPLICATIONS OF INVESTING IN THE DEVELOPMENT OF HIGH SPEED BROADBAND NETWORKS

When considering local economic trends in small countries such as Croatia it is important to bear in mind that conjuncture in the national economy largely determines those trends. In this regard, Rijeka's economy was strongly influenced by the effects of the negative economic cycle in the Croatian economy between 2009 and 2015, which was considerably longer and more pronounced than in the rest of the EU. This can best be illustrated by the sharp decline in the financial results of enterprises located in the City of Rijeka (cf. Graph III).

GRAPH III. INCOME AND EXPENSES OF ENTERPRISES LOCATED IN RIJEKA IN 2008 AND 2014 IN HRK (SOURCE: SUMMARY ANNUAL FINANCIAL STATEMENTS OF CROATIAN FINANCIAL AGENCY FOR 2008 AND 2014)

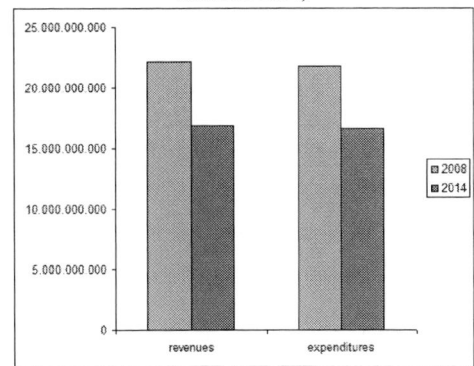

According to Croatian Financial Agency data, in 2014 the total revenue of enterprises located in Rijeka was 16.9 billion HRK, which is 23.8% lower level of total revenue than in 2008, the pre-recession year. Total expenditures of different businesses over the same period decreased, following the total revenue trends. Compared to 2008, the total expenditures of businesses located in the City of Rijeka fell to 5.2 billion (23.7%) in 2014. This volume of business and economic activity in Rijeka is lower than in pre-recession years. As there was no disproportion of income and expenditure trends, local entrepreneurs have retained profitability achieving a positive consolidated financial result, i.e. a profit of 275.1 million kuna in 2014 (31.4% less than in 2008).

It should be noted that more than 40% percent of the revenue in the City of Rijeka is generated by retail and wholesale trade along with the car repair services, which is why the monolithic economic structure of Rijeka was extremely sensitive to the general recessionary conditions in the national economy. Trade largely employs lower-educated and lower-skilled workforce, so by the nature of its role in the social division of labor it has a relatively small contribution to the creation of added value in local economies. Also, urban area trade is primarily focused on distribution of consumer goods to local people, which is why it is characterized by an enhanced pro-cyclicality, i.e. rapid growth of turnover and new jobs in periods of positive economic cycle and *vice versa*, rapid decline of turnover and the loss of jobs in periods of economic crisis and reduced citizens' purchasing power. Therefore, trade cannot be the main driver of economic development of Rijeka. This is also confirmed by the data provided by the Croatian Financial Agency on the number of employees at enterprises located in the City of Rijeka. In 2014 the enterprises of Rijeka employed 30.135 employees, which is a decrease of 17.36% compared to 2008. Over the period between 2008 and 2014 Rijeka lost as many as 6,332 jobs in the entire enterprise sector, out of which 2.295 jobs were lost in the trade sector.

Furthermore, the economic orientation of Rijeka towards trade and related service activities has led to high vulnerability of its economy to fluctuations in the national economy. Such economic orientation has also significantly limited the options for activating the unused local human potential. This is especially true in the case of highly educated people who have fewer and fewer opportunities to find adequate jobs in Rijeka.

GRAPH IV. SHARE (%) OF HIGHLY EDUCATED POPULATION IN REGISTERED UNEMPLOYMENT IN RIJEKA, PGC AND CROATIA IN 2008 AND 2015. (SOURCE: CROATIAN EMPLOYMENT SERVICE)

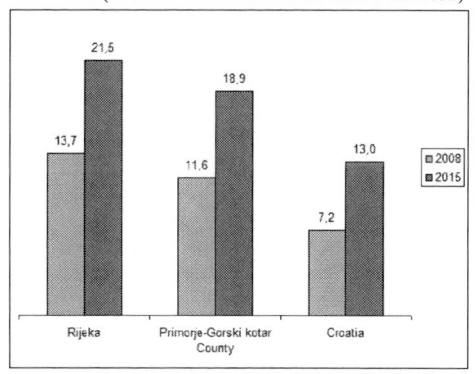

Similar trends are present on the national and county level, but they are particularly strong in the City of Rijeka. Such developments on the local labor market further indicate the existence of increased employment problems for young and educated population, which is primarily caused by structural maladjustment of Rijeka's economy. In this sense, it is opportune that the diversification of the Rijeka's economic structure, as a distinguished university center that creates high-quality professionals, focuses on the development of knowledge-intensive activities within which the highly educated workforce can apply its expertise, competencies, and entrepreneurial innovation.

Following the obtained findings on economic and demographic characteristics of Rijeka, it can be argued that setting up an advanced broadband network can certainly intensify growth and diversification of the local economy, but also positively affect the stability and quality of its future development. Numerous European experiences have shown that the multiple economic effects of high speed broadband Internet access have a strong impact on economic growth and development of local government units [5]. Therefore, it is important to emphasize multilayered, multiplicative and long-term nature of contribution of the construction and development of broadband infrastructure to economic growth:

- The primary economic effect of building broadband infrastructure in local units is usually identical to the construction of any type of infrastructure and it is reflected in the growth of income and employment in the local construction sector and production and supply chains that support all phases of broadband infrastructure life cycle.

- The introduction of broadband technology greatly simplifies the deployment of innovative business processes, products and solutions for the existing

entrepreneurs, which has a positive effect on the growth of their profitability and contributes to further development of their business.

- The introduction of broadband networks contributes to the development of new activities in local economies, the creation of new jobs and restructuring of the local economy towards activities that create higher added value. This particularly applies to professional, scientific, and technical activities and to activities within ICT sector. This happens because the creation of conditions for greater use of high speed information and communication technologies opens opportunities for teleworking, which significantly stimulates self-employment of highly educated young individuals and those with a secondary school. The availability of high speed broadband allows young people autonomy in terms of the commercialization of their competence, expertise and entrepreneurial innovation because they can develop and offer their services and applications on domestic, external and overseas markets and find partners and investors for their business projects by working from home. In macroeconomic terms, the main advantage of unimpeded access to high speed broadband in the local economy is the development of services that are completely independent of cyclical trends in the rest of the local economy.

- It is well known that investors are more inclined to invest in those areas where there is already a well-developed communication and information infrastructure. The main reason for this is that companies in such areas enjoy advantages such as the following: lower communication and operating costs; easier access to international markets; increased efficiency of sales, promotion, and customer relationships; provided external requirements for the development of a successful and secure e-business and for sustained training of employees through online educational programs. In this sense, the construction of high speed broadband infrastructure contributes significantly to attracting investments in the local economy.

- All previously presented multiplier economic effects of the construction of broadband infrastructure in the end lead to a rise of employment in local units, strengthening the purchasing power of the local population and an increase of their private consumption, which can generally initiate a new upward conjuncture in the local economies.

Summarizing the above direct and indirect social and economic benefits of broadband networks, and taking into account the spatial, demographic and economic specificities of the city of Rijeka, it can be concluded that investments in the implementation, development and modernization of the broadband infrastructure have an immeasurable importance for the economic prospects of Rijeka. Even more so considering their role in enabling an environment for innovative, creative and knowledge-intensive activities based on better evaluation and more efficient use of overall development potential of the city.

V. ASSESSMENT OF MULTIPLIER ECONOMIC EFFECTS OF INVESTMENTS IN THE CONSTRUCTION OF BROADBAND NETWORKS

The authors of this paper have developed a simulation model for the preparation of local broadband infrastructure development plans, and by using it they have estimated multiplier effects to the entire project life-cycle of building a high speed broadband network for the benefit of local consumers, production, employment, and the city budget.

The assessment of the increased consumers' well-being was based on the consumer surplus concept. Consumer surplus is the total benefit or value that consumers receive beyond what they pay for a certain product or service, or the difference between the price of the product / service that consumers were willing to pay versus the price they actually paid. The development of broadband networks contributes to the consumer surplus due to the growth of competitiveness in the tele-communications market, which leads to an increase of quality and a reduction of price of ICT services. Additional benefits for users of broadband networks are created both in their private and professional lives (e.g., faster access to information, public and health services and entertainment facilities), which leads to significant time and cost savings in the performance of business and private activities, increases the quality of life and leisure, and consequently raises the general level of satisfaction of users of high speed broadband connections. A special feature of this indicator is that it measures the positive externality of economic goods, i.e. the part of the social benefits that has not been internalized in pricing, so the same cannot be identified and expressed by means of statistics and indicators of GDP. In this respect, the calculation of the discounted cumulative value of the consumer surplus for users of high speed broadband connections at the level of the local economy, and increase of the discounted value of total production in the local economy, which is directly or indirectly associated with the development of high speed broadband networks, can give a full picture of the overall socio-economic benefits of investments in the construction and development of the broadband infrastructure in the City of Rijeka.

The multiplier effects of investment in the broadband network on total output and employment in the local economy were estimated using the input-output multipliers. These multipliers are commonly calculated using specific input-output models that describe the direct and indirect links between producers in the processes of production and consumption of economic goods and services in a specific economic area. Thus, using input-output models it is possible to simulate and quantify the direct, indirect and induced economic effects of certain economic changes on total employment and production in a given economy (e.g. investment in a particular sector, the increase in demand for a particular product, price changes, etc.).

Data on the surtax and the average wage in the City of Rijeka were an input for calculating discounted value of the fiscal effects of the increase in employment in the local economy during the presumed life of the project on the budget of the City of Rijeka.

Due to the lack of quality data, necessary for the full application of methodology to assess consumers' surplus and input-output analysis of the multiplier effects of broadband infrastructure in the example of Rijeka, the authors have used ponders, parameters and multipliers from the European Commission study on the socio-economic impact of broadband infrastructure in European countries, including Croatia [6]. Aware of the methodological shortcomings and limitations of such an approach in the assessment of overall economic benefits of the development of broadband infrastructure in the city of Rijeka, the authors' goal was not to precisely quantify future economic effects of developing different types of broadband connection, but to consider the current demographic and economic trends in the City to simulate the direction, structure and magnitude of these effects.

The results obtained from the evaluation of socio-economic benefits of the project clearly indicate substantial differences in generating multiplier effects between different network technologies. While technological solutions based on fiber-to-the-home (FTTH) concept initially require some increased capital expenditures, network infrastructure based on such solutions has significantly greater multiplier effect on the growth of production, GDP, employment, and the city budget. This ultimately means that technologies with higher average speeds generate a considerably higher net economic benefit to the local economy. For this reason, it is not advisable to take decisions on the development of broadband infrastructure based on a conservative approach, i.e. to take implementation costs and direct financial flows of the project as the sole criteria for selection of technological solutions on which to base future broadband network. In fact, those policies that give greater importance to implementation costs rather than technological features, speed and capacity of the broadband network, have immeasurably high social opportunity costs. By doing so, decision makers reduce future production potential and possibilities of production in a local economy, as well as the ability of future generations to achieve higher standards of living in local communities.

Nowadays, a broadband network from the aspect of private and business users has, in fact, no adequate substitute because it represents the basic requirement for the use of advanced ICT tools, applications and other products and services whose performance and efficiency of use are directly dependent on network speed. Therefore, it is important to emphasize that because of its ability to intensify the impact of modern technologies and innovations that are used in all sectors on the overall economic and social development and growth, broadband has all the characteristics of merit goods. This means that the implementation and development of high speed broadband networks in a given area should not entirely rely on the operators' market decisions and business plans, since they are typically profit-oriented. In a sense, the implementation of open, high-speed broadband networks has an even greater social significance in the context of overall development of PGC.

VI. CHOICE OF FINANCING MODEL

To ensure the maximum value for money, i.e. the achievement of the highest possible social rate of return, before deciding on the financing model, it is necessary to determine the ability to pay long-term liabilities of the public entity and make a preliminary calculation of the value for money.

Basic guidelines for calculating the capacity to assume long-term commitments are given in the manual of the Agency for Investment and Competitiveness, *Determining the affordability of the local and regional self-government (LRS) in public-private partnership* [7]. The process of determining the capacity to undertake long-term obligations of public institutions such as local and regional self-government (LRS) and ministries leads to consideration and analysis of their long-term debt-paying ability or undertakings of financial liabilities for long term lease, as it is the case for a contractual public-private partnership. Taking over long-term liabilities implies determining the borrowing and lease payments capacity in a way which will not jeopardize the long-term financial stability of the public sector institutions and the whole financial system.

After the analysis of the ability to pay long-term liabilities, the preliminary calculation of the value for money must be made. The main objective of this procedure is to determine which financial model provides the best value for money and the highest social rates of return. The basic steps of this procedure and general guidelines to planning public investment projects are also given in the manual of the Agency for Investment and Competitiveness *Preparation and implementation of public investment (Public-Private Partnership Projects)* [8]. The detailed procedures of the value for money calculation are given in the manual *The meaning and process of calculating the value for money for public-private partnership projects* [9].

It should be noted that if the analysis of the ability to pay long-term liabilities proves that local/regional self-government is not able to bear the liability in the context of a financial model, even if the choice of such a model provides greater value for money, the model cannot be accepted since it would be such a choice which would jeopardize the realization of the project.

Therefore, when choosing the sources of financing of investment projects, it is certainly important to give priority to financing models with private capital that consider the statistical rules and fiscal treatment prescribed by the system of national accounts ESA 2010, i.e. financing models in which the distribution of long-term financial risks between the public and private sector is such that the investments do not negatively affect the Croatian public debt.

VII. CONCLUSION

The development of high speed broadband network infrastructure in the Croatian municipalities increases the level of social standard in local communities, primarily through the practical application of smart city concept, which leads to the improved quality of public services. Investments in high speed broadband infrastructure therefore considerably contribute to stopping negative demographic trends on local, regional and state levels and enable the introduction of ICT technologies for more efficient planning and implementation of development policies for leveraging the human potential.

The results obtained from the evaluation of socio-economic benefits of the project clearly indicate substantial differences of generating multiplier effects by implementation of different broadband network technologies. While technological solutions based on FTTH concept initially require some increased capital expenditures, network infrastructure based on this technology generate significantly larger multiplier effects in local economy. This is valid both for remote rural areas in Gorski kotar and for the City of Rijeka as a regional center. Therefore, it is important to emphasize that due to its capacity to intensify the impact of modern technologies and innovations that are used in all sectors, broadband has all the characteristics of the merit good. This means that the implementation and development of high speed broadband networks in local areas cannot entirely be based on market decisions, since investments in broadband generate large socio-economic returns from which all members of society gain direct or indirect benefits.

LITERATURE

[1] Development Strategy of the City of Rijeka for the period 2014th-2020th <http://www.rijeka.hr/lgs.axd?t=16&id=70941>

[2] Census 2001 <http://www.dzs.hr/Eng/censuses/ Census2001/Popis/Edefault.html>

[3] Census 2011 <http://www.dzs.hr/Eng/censuses/census2011/results/censustabshtm.htm>

[4] Natural Change in Population in the Republic of Croatia <http://www.dzs.hr/Eng/Publication/subjects.htm>

[5] Impact Assessment accompanying the document proposal for a Regulation of the European Parliament and of the Council on a series of guidelines for trans-European telecommunications networks <http://register.consilium.europa.eu/doc/srv?l=EN&t=PDF&f=ST+16006+2011+ADD+2>

[6] Socio-economic impact of bandwidth <https://ec.europa.eu/digital-single-market/en/news/study-socio-economic-impact-bandwidth-smart-20100033>

[7] Agency for Investments and Competitiveness, <http://www.aik-invest.hr/wp-content/uploads/2015/11/m5-local-and-regional-solvency_final_iva.pdf>

[8] Agency for Investments and Competitiveness, <http://www.aik-invest.hr/wp-content/uploads/2015/10/p9_preparation-and-implementation-of-public-investments-ppp-projects.pdf>
Agency for Investments and Competitiveness, <http://www.aik-invest.hr/wp-content/uploads/2015/11/p6-v2-vfm-finalno-za-objavudoc.pdf>

Gap in pagination due to withheld paper.

Pages 1437-1441

MIPRO 2017, May 22- 26, 2017, Opatija, Croatia

The Role of Applications and their Vendors in Evolution of Software Ecosystems

Sami Hyrynsalmi and Petri Linna
Tampere University of Technology, Pori, Finland
{sami.hyrynsalmi, petri.linna}@tut.fi

Abstract - The most recent trends in the electronic commerce research have suggested that forming an ecosystem around a platform would create a winning solution. The ecosystem, consisting of vendors and external actors, would create competitive advantage for the platform owner. Furthermore, the sheer number of the actors has been used as the measure of the ecosystem's well-being against competing ecosystems. Whereas a number of studies has been devoted to analyse the well-being indicators or structures of software ecosystems and the importance of complementors and complements are acknowledged, there is lack of studies addressing how the complementors affect into the evolution of ecosystems. This conceptual analysis aims to open discussion on this topic by using the mobile application ecosystems—such as Google Play or Apple's iOS—as the case subject. While the results suggest some implications for the platform owners and complementors, more work is needed

I. INTRODUCTION

The summer of 2008 and the launch of App Store for smart devices using Apple's iOS operating system will likely remain a remarkable milestone in the history of the mobile industry. Although similar application stores by different vendors had been available for several years before the launch of App Store, Apple's marketplace—together with the new series of smart devices—was able to revolutionize the business and dethrone the old kings of the castle such as Nokia and BlackBerry [1]. Eventually, the previous market leaders were driven out from business, and Apple's new innovation was copied to several different industrial segments [2].

While several analyses for the reasons of Apple's iOs devices and marketplaces success have been written, e.g. [1] [3] [4], it seems to be clear that millions of applications by over hundred thousand application developers had also an important role in the outcome of the competition. The application developers and their offerings together with the platform provider and customers form a software ecosystem [1]. The concept of 'software ecosystem' is a descendent of Moore's [5] business ecosystem with focus on the software industry and its special characteristics [6].

There are several definitions for software ecosystems; however, the one by Jansen et al. [7] summarizes the concept well: a software ecosystem *"consists of the set of businesses functioning as a unit and interacting with a shared market for software and services, together with the relationships among them. These relationships are fre-* *quently underpinned by a common technological platform or market and operate through the exchange of information, resources and artefacts."*

Both the definition of a software ecosystem as well as the history of App Store emphasize the presence of third parties—no company alone can run an ecosystem. However, the implications of involving third-parties have been discussed only a little in the extant literature of software ecosystems.

Some previous studies have considered the third parties role in the war of competing mobile application ecosystems, a sub-type of software ecosystems [6]. On one hand, there has been an argument that the sheer number of applications and their developers would eventually be the most important factor for the success of a mobile application ecosystem [8] [9]. On the other hand, there has been argument that instead of the sheer number, it is quality of the content offered [10]. Nevertheless, both views emphasizes the application offering and suggest that, at least in this domain, the applications are holding the highest bargaining power in the ecosystem.

This conceptual paper discusses on the implications of this assumption. The mentioned mobile application ecosystems are used an example case and limitations of generalization from the case to the general type of software ecosystems are addressed. That is, while we focus on the mobile applications and their role in the ecosystems, they are only a case study subject and we aim to generalize the result to other types of software ecosystems. As a result, this study calls for further inquiries assessing in strategical management of evolving software ecosystems.

The remaining of the study is structured as follows. The following section presents the central concepts and reviews related work. The third section presents the competition of mobile applications ecosystems in the 2010s as a case and the fourth section discusses on the findings. The final section concludes the study with some suggestion for future work.

II. BACKGROUND AND MOTIVATION

The hype of different kinds of artificial ecosystems started in the 1990s when Moore [5] published his seminal article on the ecology of competition. He defined a business ecosystem as a set of companies that evolve around a shared innovation. The companies work together, both cooperatively and competitively, to satisfy customers. Moore describes that the ecosystems evolve through four

1442

different stages: Birth,Expansion, Fight of leadership, and Self-renewal or death.

The concept of 'software ecosystem' is a derivative of a business ecosystem. It was first used by Messerschmitt and Szyperski [11] in their book published in 2003. Since then, the number of studies assessing different kinds of software ecosystems has been growing steadily [12]. However, due to the popularity of the new conceptualization, there are lots of definitions and views what constitutes of, and what are differences and similarities between the concepts of 'ecosystem', 'platform', 'community', and 'two-sided market'. This study follows the view that a software ecosystem is formed around a platform and it consists of different kinds of actors [1] [7].

The software ecosystem conceptualization has become important in the field electronic commerce due to the popularity of platformization, i.e., the process of establishing a platform [13], in business. Platforms and ecosystems are nowadays seen as a winning solution in the new era business [14] and the whole field has been started to call as 'platform economy' [15]. Classic examples of platform economies—or software ecosystems—in the field of electronic commerce, are Apple App Store, Google Play as well as Valve's Steam [1].

The actors are important part of an ecosystem. In their literature review, Manikas and Hansen [16] categorized the presented roles of actors (i.e., an independent person, a team, an entity or an organization) associated with software ecosystems into five main groups: 1) *Ecosystem orchestrator*; 2) *Niche player*; 3) *External actor*; 4) *Vendor*, i.e., independent software vendor (ISV) or value-added reseller (VAR); and 5) *Customer*. The first of these is the main actor being responsible for keeping the ecosystem functioning whereas the last one is the buying customer. The remaining three are complementors, i.e., they are offering their complementing services and products to the ecosystem [17].

The differences between three remaining groups are little and one actor can serve in several roles for the ecosystem at the same time. A niche player is often developing and adding components to the platform and thus producing value to customers. External actors use the possibilities provided by an ecosystem and create, thus, indirect value to the ecosystem. External actors can, e.g., promote the ecosystem and its auxiliaries, serve as an external tester or do parallel developing to the ecosystem platform. Finally, a vendor is an actor who makes profit by selling the products of the ecosystem. A vendor can sell, e.g., integration services, components, support agreements or licenses to the main product. [16] Altogether, the actors belonging to these groups are the complementors for the main ecosystem and the remaining of this study focuses on them.

Complementors' ability to freely choose to what ecosystem being a part with [18] [19] or even to rethink its position in the ecosystem [5], makes software ecosystems interesting study subjects. A complementor can decide to be a part of several competing ecosystems at the same time, a strategy called as multi-homing whereas the opposite decision is called as single-homing [20] [21] [22]. Furthermore, in his seminal paper Moore [5] describes,

that as a part of a healthy business ecosystem's evolution, complementors will challenge the ecosystem orchestrator for the leadership of an ecosystem. As an example, Microsoft and Intel challenged and won IBM for the supremacy of a personal computer ecosystem's leadership in the 1980s [5].

In the field of software ecosystems, a remarkable number of literature studies have been published, e.g. [6] [12] [16] [23] [24] [25]. These studies were looked through for this study in order to map whether there are existing discussions on the evolution or not. So far, there seems to be no previous discussion on the implications of the complementors' roles in the evolution of software ecosystems. Therefore, this study aims to open discussion on the issue by analyzing a case and discussing research avenues that it opens. In the following section, we will present the case and it is followed by analysis in Section IV.

III. EVOLUTION OF MOBILE APP ECOSYSTEMS

The mobile application stores—such as Google Play, Apple's App Store and Microsoft's Windows Phone Store—are frequently assessed software ecosystems [12]. In these kinds of ecosystems, there are three major actor groups: *the orchestrator* (i.e., Google, Apple and Microsoft, respectively), *the customers* (i.e., the end-users of smart devices) and *the application developers* (e.g., Supercell, King Digital Entertainment) [1]. Whereas there are, e.g., niche players contributing to the core platform and external actors (e.g., Samsung and HTC) adding value to the ecosystem, they are infrequently discussed in this domain.

The three mentioned mobile application ecosystems were competing for the customers as well as from the application developers at the beginning of the 2010s. In addition to the big three, also smaller ecosystems and orchestrators such as Nokia with Ovi and RIM with Black-Berry World marketplaces were involved in the war of smart devices' supremacy. [1]

Most of the orchestrators were most likely looking for, so called, the *virtuous cycle* effect [8]. In virtuous cycle, a high number of potential applications lure more customers to use the smart device platforms. More customers mean more sales in the marketplace, which in turn tempt more developers to join into the ecosystem. Finally, more developers mean more potential applications for the customers which start the cycle again.

Due to the virtuous cycle, it was not a surprise that the sheer numbers of application developers joined and applications offered in the marketplace have been seen as the measure of success of an ecosystem. This has often been presented in the extant literature [8] [9] [26] as well as in the news analysis and marketing[1][2]. However, there are some critics of using the number of applications as the measure of well-being of an ecosystem [10] [27] and also practitioners have argued for content over quality[3].

---

[1] http://www.wired.com/gadgetlab/2010/10/app-for-that/
[2] http://www.reuters.com/article/2012/06/11/apple-developersidUSL1E8HB4Z820120611
[3] https://www.cnet.com/news/does-an-app-stores-size-matter-ifcontent-is-the-killer-app/

1443

Nevertheless, after assessing the success of an ecosystem with the sheer number of applications, arguments have been moved to claim that either the best content [22] or the killer applications [1] would define the success of a mobile application ecosystem. A case in point was a sequel of a popular mobile game that was announced to skip Windows Phone platform. Market analysts quickly judged that the lack of a blockbuster game would be a significant hit against the ecosystem and endangers its future[4].

While Windows Phone ecosystem still exists, it is currently silently dying out. Similarly, most of the other old challengers have given up and only the two of the largest application ecosystems survived: Apple's App Store for iOS devices and Google Play for smart devices with Android operating system. Often, the lack of specific applications—together with insufficient devices—is credited as the source of downfall for at least Microsoft's solution[5]. For example, official Facebook and Instagram applications did not offer the same set of features that a user could get with Android or iOS devices.

However, the app economy has also demonstrated that good ideas are swiftly copied [1]. For example, the Flappy bird game, published in 2013, was replicated by different developers to other ecosystems in a few days. After the withdrawal from the market, the number of copies was still growing[67]. Similarly, the same kinds of applications are occupying the top lists of all major mobile application ecosystems even though the applications are not necessarily produced by the same developers [22].

IV. ANALYSIS AND DISCUSSION

In the following, we will discuss on the importance of complementors for mobile application ecosystems and address shifts in relative bargaining powers. This section ends with discussion on the limitations and suggestion for future work.

A. The importance of complements

Based on the presented discussion from the mobile application ecosystems, it seems that the sheer number of applications is one of the most important measures of success in the beginning. After a certain point, adding new applications does not seem to bring as much value to the customers as previously. In this phase, content of applications seems to be more important. In other words, lacking of certain key applications such as WhatsApp, Facebook or Instagram can be a major disadvantage for an ecosystem.

This chain of thoughts leads easily to the question presented in the title of this study: *Are applications holding the highest bargaining power in the ecosystem?* Whereas this is, in the context of smart devices, a clear simplification of several complex phenomena occurring—e.g., phys-

ical devices, network operators are not considered here—at the same in the market, complements (i.e. applications) seem to be crucial for the ecosystems.

While mobile application ecosystems have some specific features such as the remarkable dependency on the physical devices and ubiquitous nature of smart devices to every aspect of people's life, they still share also remarkable similarities with general type of software ecosystems. For example, the 'app store' approach is spread in numerous different areas [2] and several, if not all, software ecosystems can be characterized as a two-sided platform connecting complementors to customers.

Therefore, an easy deduction is to argue that complementors and their offerings are important also for general type of software ecosystems. Furthermore, the importance of complementors to platforms of all kind of and their ecosystems has also been emphasized [14].

B. Shifts in bargaining powers

An important but still mainly uncovered, to the best of author's knowledge, question arises: *If content and complements have the greatest bargaining power, do they still need the basis ecosystem?* That is, when a complement has come into such a position of power that customers make decision based on availability or absence of certain services, its relative bargaining power would be higher than the ecosystem orchestrator. Thus, the complementor could even abandon the ecosystem and form a new one when it is more valuable to the ecosystem orchestrator than the ecosystem is for it. With a quick glance, one can argue that complements cannot bypass the basis ecosystem, but the recent development has hinted that this can actually be a reasonable threat to an ecosystem.

For example, if Facebook's project Spartan[8] is considered, that would have added another layer into the top of mobile operating systems. After that, application developers would have been able to pass over the mobile operating system vendor's marketplace and rules by producing for and distributing content by the Spartan platform. However, the rumoured project got eventually cancelled and this kind of a revolution did not happened.

The cancelled project Spartan was not the only option for reducing the power of the platform owner. In the mobile application domain, the number of new cross-platform development tools and techniques has been rising [28]. With these kinds of tools, a developer can program an application once and it will run on several different technological platforms. While these tools have some remarkable weaknesses [28], the technology is developing constantly and the cross-platform development methods are constantly improving. In the near future, these might be a reasonable alternative for native development tools.

When the cross-platform development tools have gained enough maturity, the application developers can be expected to use them to publish the same application instantly for several platforms. With these kinds of tools, a developer can achieve reasonable benefits from being first in several markets to cost savings in development work

---

[4] http://www.bloomberg.com/news/2012-03-22/-angry-birds-space-edition-skips-windows-phone-in-blow-to-nokia.html

[5] http://www.theverge.com/2015/10/23/9602350/microsoft-windows-phoneapp-removal-windows-store

[6] https://www.cnet.com/news/the-search-for-an-awesome-flappybird-replacement/

[7] https://techcrunch.com/2014/03/24/clones-clones-everywhere-1024-2048-and-other-copies-of-popular-paid-game-threes-fill-the-app-stores/

[8] https://techcrunch.com/2011/09/28/this-sure-looks-a-lot-like-facebooksproject-spartan-screenshots/

[19] [28]. At the same time, these kinds of tools cause that the ecosystem where a complement is published and offered becomes less relevant — a developer can publish it to almost all alternatives. This makes the platform providers' role less important and the platform can turn out to be 'just distribution channels' for the content.

C. Struggle for leadership

In his seminal work, Moore [5] already addressed the evolution stages of an ecosystem. While this aspect seems to be mostly forgotten by, at least, software ecosystem researchers, the evolution model is even more topical nowadays as the software ecosystems are coming of age.

According to Moore [5], there are two conditions that must be fulfilled that the leadership struggles would occur at the third stage of the ecosystem's life-cycle model. First, the ecosystem must be strong and profitable enough to be worth fighting for. Second, the central value-adding components of an ecosystem should be reasonable stable. According to Moore, the latter condition allows contenders to attack those components and diminish the dependence to the original ecosystem orchestrator.

For example, the mobile applications ecosystems seem to fulfil both conditions. The survived ecosystems are profitable and the components that add value to the customers are stable. Thus, according to the original theory of the business ecosystem, the fight for leadership inside the mobile application ecosystems should be expected to start. Some elements of this can be seen in Android ecosystem as the mobile phone manufacturers and Amazon has founded their own application stores and distribute the content through them.

What makes software ecosystem interesting in the light of the ecosystem evolution model is the relatively easiness of multi-homing. The same application can be offered with relatively cheap cost to several competing ecosystems [19]. When compared with, e.g., the personal computer ecosystems' fight against each other's and struggle for leadership, this would have mean that a vendor would have steadily worked for both IBM's and Apple's ecosystems. For a software vendor, this is easier than for a hardware vendor due to the intangibility, changeability and portability of software. Thus, in the software industry, it seems that vendors can challenge more easily the ecosystem orchestrator for the battle of leadership.

D. Implications and future work

To summarize the above chain of thoughts, the argument presented by Lemstra et al.'s [29] for mobile network operators is followed: Will the mobile application ecosystems become just another distribution channel when a complementor takes over the ecosystems? Based on the original theory of the business ecosystem, a struggle of leadership is expected as the preconditions seem to be fulfilled.

The conceptual analysis presented in previous sections has certain implications for practitioners. First, if the presented hypothesis, that in software ecosystems battle for leadership is more probable holds true, the ecosystem orchestrators should carefully follow their position in the market as well as in their own ecosystem. While giving more power to complementors might be a good tool in the war against other ecosystems, it can cause that the orchestrator loses its own bargaining power against its cooperators. In this case, the initial platform can turn to be only just another distribution channel.

Second, if the presented argument holds true, it questions some of the hyped platform economy arguments. By 'platforming' company's old product and opening them for cooperation, a company might also accidentally weaken its own position. However, based on the presented conceptual analysis, this seems to be only a case in software ecosystems and in the field of electronic commerce, where the role of a physical device is a smaller. Nevertheless, companies should also pay attention to this aspect when they are deciding to go or not to go in the platform economy.

Finally, to the best of author's knowledge, not much has been studied in the evolution of software ecosystems. Therefore, this study calls for further work on analysing and theorizing 1) an evolution model of software ecosystems whether they follow the same pattern and conditions that business ecosystems; 2) assessing the role of complementors and complements in the evolution of the ecosystem; and 3) investigating counter-measures for ecosystems' orchestrators to mitigate contenders' actions.

E. Limits of generalization

There are a few major questions related to the presented ideas in this paper. The first is related to software ecosystem studies itself. The software ecosystem conceptualization has been used in a wide array of different context ranging from World of Warcraft to SAP [12]. Thus, it is not a surprise that the software ecosystem literature seems to be started to diverge into different sub-communities [23].

Two large sub-communities are rather easily identifiable when the results by Suominen et al.'s [23] and Manikas' [12] bibliographical studies are combined: On one hand, a stream of literature is devoted to study large-scale software, often open-source, projects consisting of hundreds if not thousands of auxiliary projects, such as R and Python programming languages, and their packages. On the other hand, another literature stream is devoted to understand marketplace-centered ecosystems, such as Google Play and Apple App Store.

This paper contributes mainly on the latter literature stream and the division between these two literature streams is meaningful to this study: The application developers belonging to the former group are often motivated with a different set of reasons ranging from meritocracy to fame and improving the CV or just contributing for the society. Whereas these reasons are also available in more business-oriented ecosystems [30], often financial benefits are the main reason.

In the open-source related software ecosystems, the first condition presented by Moore [5] for the struggle of leadership might not be fulfilled: while the ecosystem is healthy according to its own indicators, the ecosystem might not be interesting to fight over. Thus, software ecosystems should be selected with a care for empirical studies as well as generalizability of results should be well justified.

Second, the argument presented in this conceptual study is deduced from only one case. It might be that the case is not representative enough that general rules of an ecosystem's lifecycle could be identified. It can be, for example, that there are certain specific features of mobile application ecosystems that cause the seen shifts in bargaining powers. Therefore, more case studies about different ecosystems fields are needed to verify the found observations.

V. CONCLUSION

This paper presented and analysed a case of mobile application ecosystems. Based on analysis, it can be argued that applications are likely to increase their relative bargaining power in the mobile industry due to their impact even on the sales of different phones. This conceptual analysis, however, raised the question whether complementors and complements—i.e., the applications—can obtain such position that they start to threat the initial ecosystem orchestrator for the leadership of the ecosystem. While this analysis hinted that such a phenomenon might occur in the software ecosystems due to the improvements in cross-platform development tools, this analysis also emphasized that not much is understood about the evolution of business or software ecosystems. Therefore, this study calls for further work to analyse and clarify the role of complements in the evolution of artificial ecosystems.

REFERENCES

[1] S. Hyrynsalmi, "Letters from the war of ecosystems — an analysis of independent software vendors in mobile application marketplaces," Doctoral dissertation, University of Turku, Turku, Finland, December 2014, TUCS Dissertations No 188. [Online]. Available: http: //urn.fi/URN:ISBN:978-952-12-3144-5

[2] S. Jansen and E. Bloemendal, "Defining app stores: The role of curated marketplaces in software ecosystems," in Software Business. From Physical Products to Software Services and Solutions, ser. Lecture Notes in Business Information Processing, G. Herzwurm and T. Margaria, Eds., vol. 150. Berlin, Germany: Springer Berlin Heidelberg, 2013, pp. 195–206.

[3] J. West and M. Mace, "Browsing as the killer app: Explaining the rapid success of Apple's iPhone," Telecommunications Policy, vol. 34, no. 5–6, pp. 270–286, June 2010.

[4] R. C. Basole and J. Karla, "On the evolution of mobile platform ecosystem structure and strategy," Business & Information Systems Engineering, vol. 3, pp. 313–322, 2011.

[5] J. F. Moore, "Predators and prey: A new ecology of competition," Harvard Business Review, vol. 71, no. 3, pp. 75–86, May-June 1993.

[6] S. Hyrynsalmi, M. Sepp¨anen, T. Nokkala, A. Suominen, and A. J¨arvi, "Wealthy, healthy and/or happy — what does 'ecosystem health' stand for?" in Software Business — 6th International Conference, ICSOB 2015, Braga, Portugal, June 10-12, 2015, Proceedings, ser. Lecture Notes in Business Information Processing, M. J. a. Fernandes, R. J. Machado, and K. Wnuk, Eds., vol. 210. Springer International Publishing, 2015, pp. 272–287.

[7] S. Jansen, A. Finkelstein, and S. Brinkkemper, "A sense of community: A research agenda for software ecosystems," in 31st International Conference on Software Engineering — Companion Volume, ICSE-Companion 2009. IEEE, May 2009, pp. 187–190.

[8] A. Holzer and J. Ondrus, "Mobile application market: A developer's perspective," Telematics and Informatics, vol. 28, no. 1, pp. 22–31, February 2011.

[9] T. Yamakami, "A mobile digital ecosystem framework: Lessons from the evolution of mobile data services," in Proceedings of the 2010 13th International Conference on Network-Based Information Systems, ser. NBIS '10. Washington, DC, USA: IEEE Computer Society, September 2010, pp. 516–520.

[10] S. Hyrynsalmi, A. Suominen, and M. Mäntymäki, "The influence of application developer multi-homing and keystone developers on competition between mobile application ecosystems," The Journal of Systems and Software, vol. 111, pp. 119–127, January 2016.

[11] D. G. Messerschmitt and C. Szyperski, Software Ecosystem: Understanding an Indispensable Technology and Industry. Cambridge, MA, USA: The MIT Press, 2003.

[12] K. Manikas, "Revisiting software ecosystems research: A longitudinal literature study," Journal of Systems and Software, vol. 117, pp. 84–103, July 2016.

[13] A. S. Islind, T. Lindroth, U. L. Snis, and C. Sørensen, "Co-creation and fine-tuning of boundary resources in small-scale platformization," in Nordic Contributions in IS Research: 7th Scandinavian Conference on Information Systems, SCIS 2016 and IFIP8.6 2016, Ljungskile, Sweden, August 7-10, 2016, Proceedings, U. Lundh Snis, Ed. Cham: Springer International Publishing, 2016, pp. 149–162. [Online]. Available: http://dx.doi.org/10.1007/978-3-319-43597-8 11

[14] M. Van Alstyne, G. G. Parker, and S. P. Choudary, "Pipelines, platforms, and the new rules of strategy," Harvard Business Review, vol. 94, pp. 54–62, April 2016.

[15] M. Kenney and J. Zysman, "The rise of platform economy," Issues in Science and Technology, vol. XXXII, Spring. [Online]. Available: http://issues.org/32-3/the-rise-of-the-platform-economy/

[16] K. Manikas and K. M. Hansen, "Software ecosystems — A systematic literature review," Journal of Systems and Software, vol. 86, no. 5, pp. 1294–1306, May 2013.

[17] A. M. Brandenburger and B. J. Nalebuff, Co-opetition, 1st ed. New York, NY, USA: Currency Doubleday, 1997.

[18] R. M. Stallman, Free Software, Free Society: Selected Essays of Richard M. Stallman, 2nd ed. Boston, MA, USA: Free Software Foundation, Inc., 2010.

[19] S. Hyrynsalmi, A. Suominen, S. Jansen, and K. Yrjönkoski, "Multi-homing in ecosystems and firm performance: Does it improve software companies' ROA?" in Proceedings of the International Workshop on Software Ecosystems 2016, 2016, pp. 1–14.

[20] B. Caillaud and B. Jullien, "Chicken & egg: Competition among intermediation service providers," The RAND Journal of Economics, vol. 34, no. 2, pp. 309–328, Summer 2003. [Online]. Available: http://www.jstor.org/stable/1593720

[21] J. C. Rochet and J. Tirole, "Platform competition in two-sided markets," Journal of the European Economic Association, vol. 1, no. 4, pp. 990–1029, June 2003.

[22] S. Hyrynsalmi, "To redefine ecosystem health, or not to redefine? a view of scientific knowledge on the "software ecosystem health" concept," in Proceedings of the European Workshop on Software Ecosystems 2015, K. M. Popp, P. Buxmann, P. Aidan Curran, G. Eichler, S. Jansen, and T. Kude, Eds., Synomic Academy. Books on Demand, 2016, pp. 47–51.

[23] A. Suominen, S. Hyrynsalmi, and M. Seppänen, "Ecosystems here, there, and everywhere — a barometrical analysis of the roots of 'software ecosystem'," in Software Business: 7th International Conference, ICSOB 2016, Ljubljana, Slovenia, June 13-14, 2016, Proceedings, A. Maglyas and A.-L. Lamprecht, Eds. Springer International Publishing, 2016, pp. 32–46.

[24] O. Barbosa and C. Alves, "A systematic mapping study on software ecosystems," in Proceedings of the Workshop on Software Ecosystems 2011, ser. CEUR Workshop Proceedings, S. Jansen, J. Bosch, P. Campell, and F. Ahmed, Eds., vol. 746. CEUR-WS, June 2011, pp. 15–26. [Online]. Available: http://ceur-ws.org/Vol-746/IWSECO2011-2-BarbosaAlves.pdf

[25] O. Barbosa, R. P. dos Santos, C. Alves, C. Werner, and S. Jansen, "A systematic mapping study on software ecosystems from a three-dimensional perspective," in Software Ecosystems: Analyzing and Managing Business Networks in the Software Industry, S. Jansen, S. Brinkkemper, and M. A. Cusumano, Eds. Northampton, MA, USA: Edward Elgar Publisher Inc., 2013, ch. 4, pp. 59–81.

[26] N. Schultz, R. Zarnekow, J. Wulf, and Q.-T. Nguyen, "The new role of developers in the mobile ecosystem: An Apple and Google case study," in 15th International Conference on Intelligence in Next Generation Networks (ICIN), October 2011, pp. 103–108.

[27] K. Boudreau, "Too many complementors? Evidence on software developers," HEC-Paris School of Management, Tech. Rep., January 2007, working Paper Series, Social Science Research Network. [Online]. Available: http://hal-hec.archives-ouvertes.fr/hal-00597766

[28] V. Ahti, S. Hyrynsalmi, and O. Nevalainen, "An evaluation framework for cross-platform mobile app development tools: A case analysis of adobe phonegap framework," in Compsystech'16, B. Rachev, D. Tegolo, Y. Kalmukov, S. Smrikarova, and A. Smrikarov, Eds., 2016.

[29] W. Lemstra, G.-J. Leeuw, E. Kar, and P. Brand, " 'Just another distribution channel?'," in Mobile Wireless Middleware, Operating Systems, and Applications — Workshops, ser. Lecture Notes of the Institute for Computer Sciences, Social Informatics and Telecommunications Engineering, C. Hesselman and C. Giannelli, Eds. Berlin, Germany: Springer Berlin Heidelberg, 2009, vol. 12, pp. 1–12.

[30] S. Hyrynsalmi, A. Suominen, T. Mäkilä, A. Järvi, and T. Knuutila, "Revenue models of application developers in Android Market ecosystem," in ICSOB 2012, ser. Lecture Notes in Business Information Processing, M. Cusumano, B. Iyer, and N. Venkatraman, Eds., no. 114. Berlin, Germany: Springer Berlin Heidelberg, 2012, pp. 209–222.

Open Data Based Value Networks: Finnish examples of public events and agriculture

P. Linna, T. Mäkinen and K. Yrjönkoski
Tampere University of Technology/Pervasive Computing, Pori, Finland
petri.linna@tut.fi

Abstract - In recent years, several countries have placed strong emphasis on openness, especially open data, which can be shared and further processed into various applications. Based on studies, the majority of open data providers are government organizations. This study presents two cases in which the data providers are companies. The cases are analyzed using a framework for open data based business models derived from the literature and several case studies. The analysis focuses on the beginning of the data value chain. As a result, the study highlights the role of data producers in the ecosystem, which has not been the focus in current frameworks.

I. INTRODUCTION

Openness and open data have been appearing as development trends in recent years [1-5]. The transition to openness in international agendas and agreements has been agreed. [6-7] As can be observed currently, these decisions have an effect at both national and local levels. Some of the indicators of openness are a list of application programming interfaces [8], open data datasets [9], and Census services [10]. At this moment, data releases mainly come from municipalities or government organizations.

Recently, some companies have started to take on board the ideology of openness. They have begun to pick up on this innovation and accelerate their R & D by opening up some of their production processes, and organizing so-called industrial hackathon competitions for coders. Individual data openers can also be found in the corporate world, of which perhaps the best known is Google and its maps. No cases are known where many companies in the same sector are opening up the same kind of data.

This paper presents two different cases, which at the same time represent two different starting projects. The aim is to promote openness in these projects. A single data opening would have no business meaning, but when hundreds of companies' open data via a common data format, it will generate business benefits for many different parties. Thematically, these two cases are completely different. In the first case, the focus is on the openness of public event organizations and in the second case, the focus is on the openness of agriculture.

In the second section of this paper, the business ecosystem and value network are examined from the open data point of view. Projects and cases are introduced in the following section. The concluding section contains a summary and discussion of the presented open data roles and some thoughts on the future.

II. OPEN DATA BUSINESS ECOSYSTEM AND VALUE NETWORK

A business ecosystem is a dynamic structure of organizations utilizing business potential influencing and influenced by each other, in a certain business environment. It consists of a heterogeneous set of firms that are interconnected through a complex, global network of relationships. A data-based business ecosystem is formed by organizations where each has their own part and know-how in the data-based business. The ecosystem's actors affect and are affected by the creation and delivery of the offerings of the other actors. Each actor also has a role in the flows of information, material, money, and influence the relationships between one another. [11-13]

Today, value creation is more and more seen to happen at the level of networks and alliances, instead of relationships or dyadic chains. A value network is a structure formed by all the actors that could benefit from a certain product or service offering, and add value to it. In the value network, different economic factors – supplier, partners, allies, and customers – work together to reinvent and co-produce value. [14-17] In addition, in an ecosystem, value is not created in a chain but more in a network of actors inside a certain ecosystem [13].

Business model literature has been developed in many different domains. Therefore, many different insights exist - but in general, the business model is a single-company-level concept that takes a stand on three matters: what does the customer get? (value proposition, products and service offering); how does the customer get the value? (value network, delivery channel), and how does the company get the money (cost structure, revenue models). [18-19]. The Business Model Canvas [20] is one of the most stable, most referred to and most generic business model tool. Immonen et al. [13] have identified it as suitable for use in the open data domain.

As part of their open data research program, Deloitte has published a review, which also introduces (Fig 1) five archetypes of open data business models [4]:

- Suppliers: organizations that publish their data via an open interface to allow others to use it

- Aggregators: organizations that collect and aggregate open data, typically on a particular sectoral theme

- **Developers:** organizations and software entrepreneurs that design, build, and sell applications

- **Enrichers:** organizations that use open data to enhance their existing product and services through better insight

- **Enablers:** organizations that facilitate the supply or use of open data, but are not themselves users of open data

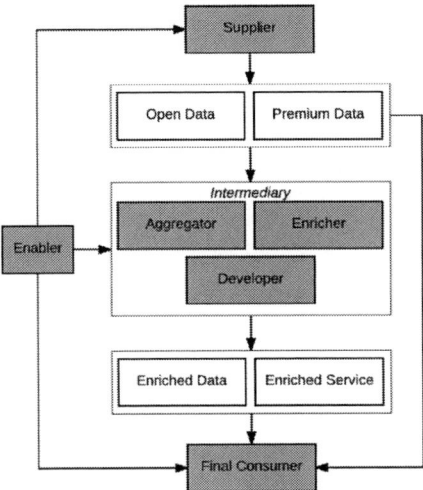

Figure 1. Deloitte's archetypes for a business model related to open data.

The open data marketplace is growing fast and is populated by an increasing number of heterogeneous organizations. Value chains between different actors are quite short and therefore simple - but on the other hand, there may be links between an entity and any other entity, and key roles can be played by any combination of individuals, companies, and public sector bodies. [4].

A variety of ecosystem models have multiple as Immonen et al. introduced [13]. In this paper Deloitte's model has been used, because it is already familiar from previous studies [1-2].

III. CASE A: THE PUBLIC EVENT DATABASE

An event organization may be, for instance, established purely for a specific music event, or the event organization may be a small unit of a bigger organization, which arranges various events, such as in municipalities and universities. Event organizations normally promote their coming events (e.g. music festivals, concerts, markets, races, seminars) via multiple channels, utilizing, for example, newspapers, radio, newsletters, and social media services. The aim is to reach potential customers, but in practice, advertising reaches many people who have little or no interest at all in the event. At the same time, event organizations compete for the attention and free time of potential customers. From the customer's point of view, newsletters in particular are challenging, because customers receive a lot of them and usually they are irrelevant.

The event information is already open data, but usually it is not in machine-readable format, and information is on the event organization's website or social media accounts. The main point would be to capture the complete attention of potential customers. The event organizations spend a lot of money on marketing and resources to get event information out via various channels.

In the near future, event data will be digitized, structured, filled automatically on an event database, and divided automatically into different applications via an open interface. At the same time, event organizations need to modify work practices and transfer their workforce to other tasks. Event data can be managed with the centralized model, where one trusted organization maintains a service where event organizations can transfer the event information. The event information can be delivered via an open interface to websites, services, or mobile apps.

In Finland, an open source event interface was developed by the City of Helsinki, called Linked Events [21]. This interface will be used in the events case in the region of Satakunta in the west of Finland. In the spring of 2017, a project will begin, called "the digital leap of events" (in Finnish, tapahtumien digiloikka), where this case will be implemented. The database and open events interface will be built by the City of Pori. The project supports the export of data to the database and aims to find solutions do this automatically. The second part of the project will be piloted in the form of a variety of ideas that take advantage of the event interface. The project is the first step of digitalization to support tourism in the region of Satakunta. [22]

The project promotes the realization of the below-mentioned structure in the pilot area in Satakunta. The project includes a new kind of funding mechanism, which seeks solutions for high-speed pilots that are aware of problems. In this case, the problem is the poor availability of transaction data from the perspective of the customer, and the event organization's waste of resources in event information management.

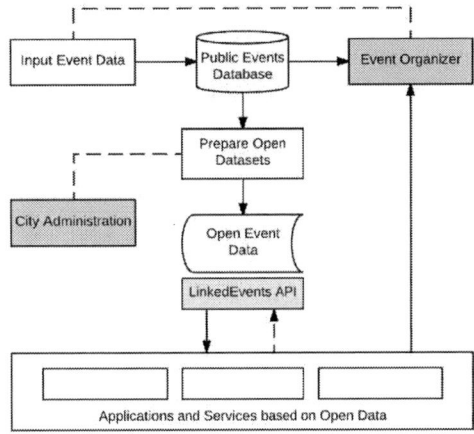

Figure 2. Project structure of digital leap of events and data flow.

This case highlights the role of data producer. Events organizations, usually companies, transfer their data to a

1449

centralized events database. In this case, the event organizations are data producers.

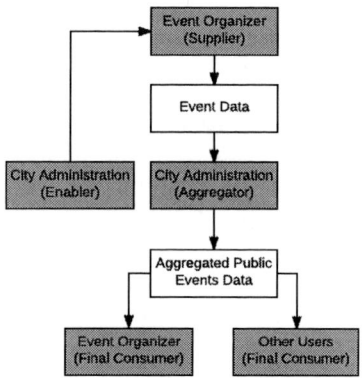

Figure 3. Case A structured in Deloitte's archetypes.

Figure 3 depicts Case A structured in Deloitte's archetypes (Fig. 1). In the figure, *Event Organizer* has been interpreted as a *Supplier* opening up data related to a public event, while *City Administration* aggregates the data of several events, thus acting as an *Aggregator*. However, in this case, the data of a single event is not openly available before it has been aggregated with the data of other events. Hence, it seems that the *Event Organizer* might not fulfill the criteria of *Supplier*. Therefore, it appears that the *City Administration* is the *Supplier,* which is preceded by some other archetype in the value network: *Event Organizer* produces event information that the City Administration supplies. Figure 4 suggests a separate archetype: *Producer* before *Supplier*.

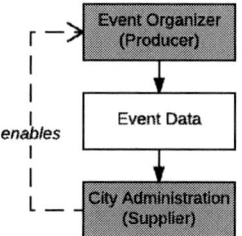

Figure 4. Case A re-structured using archetype *Producer*.

Below are listed a few of the potential motives for event organizers to open up their data via an events database.

- Resources can be moved away from the updating of event information

- New one of kind application and ecosystems opportunities arise

- Event organization does not have to develop mobile applications itself

- Events get wider visibility through a variety of applications

- Language support is broad, allowing a wider customer base

- Event data is connected with location information so event data can be retrieved / displayed on a map

- Event information can be profiled

- More information on the event can be included

- Event organizations may collaborate with each other more

To explain the last point further, collaboration means, for example, easier opportunities to hire manpower or facilities when more precise information is known about each other event organizer's needs and services.

IV. CASE B: DATA-BASED AGRICULTURE

Agriculture is one of the oldest business sectors in the world. Throughout history, the harvest yield has increased per hectare. Now agriculture is in a new stage of development, where it seems to be difficult to find significant ways to improve production via mechanical improvements. Instead of developing mechanical implements, farmers should focus on data. Data sources can be satellites, drones, harvest sensors, field sensors, and working machines with special equipment like hyperspectral cameras and many kinds of sensors. The various sources of data make it possible to see the details of the fields and find the trouble spots, and so farmers can adjust the actions to be taken to an accuracy of almost one meter.

In Finland, the development of agriculture is still more mechanically oriented than data oriented. For example, it has been evaluated in the Satakunta region that there are less than ten harvest sensors, even though there are about 3500 working farms. On average, farming has shown a loss for the fourth consecutive year in Finland. However, the average distorts the truth, because some of the farmers do not invest in cultivation of crops, instead their focus is for example on cattle. For farmers, whose main focus is on cultivation, the situation is better. However, from the business point of view, there are several kinds of threats like intensification of extreme weather events, global competition, and data ownership.

Naturally, data management is not the core skill of farmers. For this reason, data should be supplied to a trusted organization, which knows how to collect data together, to make the necessary algorithms and analyze them, and give this processed data back to the farmers. In this case, we call this trusted organization the data operator. The data operator could be a commercial company, the government, or another organization. The data operator should also provide the processed data where appropriate as open data. Opening the data allows precise monitoring of the origin of grain, such as the revision of the criteria related to grain quality. It also allows for more accurate assessment of the condition of the fields, which affects the rental and sales prices for the fields. The quality of grain increases its sale value, especially if it is possible to verify the purity of the soil and the purity of the crop. Ensuring the quality of the crop also affects products made from it, for example, specialty bread products.

1450

In this paper, the second case is based on the project, "Agricultural business development with intelligent data analytics" (the Finnish abbreviation, MIKÄ DATA), which is developing a data collection of fields, and the consolidation and display of intelligent data analysis server data items to the user. The project promotes the use of data in farming. The project also educates farmers about the introduction of technology and shows the benefits of data management. [23] In this case, the project is extended hypothetically with the Select Open Datasets process, which is not directly an original project aim.

The project highlights the value chain where a data operator (Rural Expert Organization) is required, who will take over the management of the data and its analysis. The data of individual farms is not useful, but when many of the farmers open their data to a data operator, the data operator receives a significant mass of data to be analyzed. For example, if cultivation plans are exported to the service early, the plans can still be changed if it seems that the world market is full of a particular variety of crop.

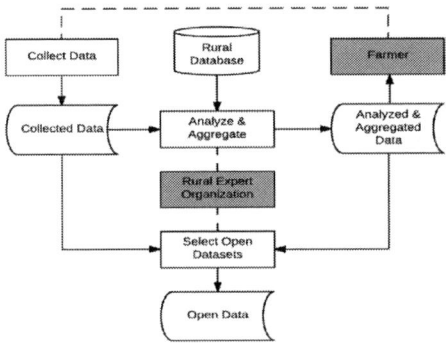

Figure 5. Project structure extended with *Select open datasets* process.

Farmers in this case are companies, which could open up their data via the data operator. Their role is to be data producers, but there are considerations that a single opener is not relevant. In the Finland case, there are about 50 000 farms, which could open their data to a data operator. It is essential that farmers also own the right to the data after they have given it to the data operator. There is needed to define mydata use in agriculture.

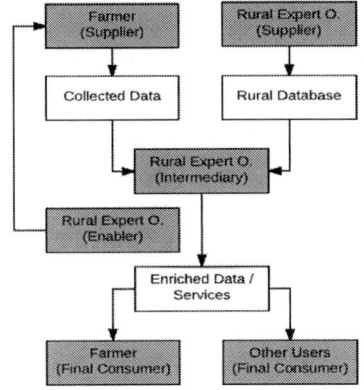

Figure 6. Case B structured in Deloitte's archetypes.

Figure 6 depicts Case B structured in Deloitte's archetypes (Fig. 1). In this figure, the *Farmer* has been interpreted as a *Supplier,* opening data related to the harvest yield of the farm. The *Rural Expert Organization* is acting as an intermediary between the *Farmer* and *Final Consumer,* e.g., by *aggregating* data from several farmers and *enriching* it with data from the Rural Database. However, by analogy with Case A, data from a single farmer is not openly available before it has been aggregated with other farmers' data. Thus, the *Rural Expert Organization* seems to be the *Supplier* rather than the *Farmer*. Such an interpretation is depicted in Figure 7, in which the *Farmer* represents the archetype *Producer,* which is not in Deloitte's model (Fig. 1).

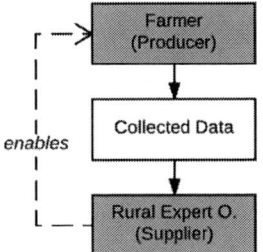

Figure 7. Case B re-structured using archetype *Producer*.

Below are listed a few of the potential motives of farmers to open up their data.

- Traceability of the origin of food is improved

- Productivity is improved when the farmer can better respond to the world market situation

- Farmers can get a wider variety of analyses from various providers

- Reliance on any particular service provider is reduced

- New products may emerge that take advantage of open data on agriculture

- Farmers do not need to specialize as data experts

- Knowledge of the actual conditions of the fields will help the sale and lease of fields

To clarify the last point, larger regeneration activities can be done for fields when it is known that this would also increase the value of the field.

V. CONCLUSION

This paper presented two cases, where data openers were companies. One individual data opener is not relevant in these cases, but several dozen openers start to become relevant. The number of openers will create new types of ecosystems. These two cases present very different sectors. The public event database case focused on event information. The data-based agriculture case focused on data on fields. There are not so many data openers that are companies and no other cases like the ones in this paper are known. Data openers are normally municipalities or various government organizations.

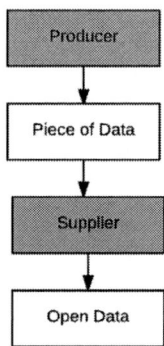

Figure 8. The archetype *Producer* placed at the beginning of the value network of open data.

Figure 8 presents a generalization of Figures 4 and 7. It suggests *Producer* as an additional archetype for Deloitte's model. While *Supplier* is an organization that publishes data via an open interface, *Producer* delivers a piece of data to the Supplier in order to make it openly available.

The data producers have a stronger role in these two cases than in Deloitte's value chain model, which is based on government open data. The data producer's role needs to be separated from the data supplier role. The structure is then a little different, because in these cases there are many data producers, while in Deloitte's open data model there is only one data provider, i.e., a municipality. In practice, this chain model has not yet been tested, but in this paper, we have briefly outlined the events organization and agriculture development projects, which will contribute to the development of these value chain models.

In both cases a trusted third party is needed to build and provide a service between the data producers and end-users. In the event case, the trusted organization is the City of Pori. In the agriculture case, the entity has not yet been resolved. This service will be built during the project, but, at the end of the project, one of the trusted organizations should take over the service.

The reasons and motivation for opening up data are quite clear when government or municipalities open data. Openness is required in international agreements. Of course, there are also other motivational aspects, such as municipalities hoping to create new businesses via open data. In business cases, motives are not based on international agreements on openness. The reasons for opening up data will probably have to be related to the profitability of the business. In both our cases, the fact was emphasized that there should be numerous data openers in order to generate sufficient data to be analyzed. For these two cases, it is still not clear which matters motivate companies to open up their data. During these two projects, the real motivations might be seen. In the report on a data ecosystem for agriculture and food, made by Godan, the motives of other organizations to open their data for use in agriculture have also been discussed [24]. In other words, when building an ecosystem many players are needed, but each has their own motives for opening up their data.

The rising value of the network can also be considered through the Virtuous Cycle theory. When there are sufficiently large data openers, it generates interest for service developers. Before a critical mass can be reached, a lot of work is needed to achieve it. The benchmark is the mobile phone sector, where there are different competing platforms.

In the future, there is a need to examine how the business model canvas would be suitable for testing the value chain for each business case. The canvas model could provide a better detailed understanding of how each company operates and can find success of open data ecosystems.

The public event case also has a significant cultural purpose. The project is part of a larger objective to promote tourism in the area. Tourists can receive comprehensive and experienced regional services. Tourists can be made increasingly aware of events in the region and its diverse culture. Digital services like Airbnb is one of this kind, which enables closer experiences of the local culture.

REFERENCES

[1] H. Jaakkola, T. Mäkinen, and A. Eteläaho, "Open data: opportunities and challenges," In Proceedings of the 15th International Conference on Computer Systems and Technologies (CompSysTech '14), Boris Rachev and Angel Smrikarov (Eds.). ACM, New York, NY, USA, pp. 25-39. 2014.

[2] H. Jaakkola, T. Mäkinen, J. Henno and J. Mäkelä, "Openn," 37th International Convention on Information and Communication Technology, Electronics and Microelectronics (MIPRO), Opatija, pp. 608-615. 2014.

[3] Open data, Driving growth, ingenuity and innovation. 2012. https://www2.deloitte.com/content/dam/Deloitte/uk/Documents/deloitte-analytics/open-data-driving-growth-ingenuity-and-innovation.pdf. Accessed 20 February 2017.

[4] Deloitte: Open growth. Stimulating demand for open data in the UK. 2012. https://www2.deloitte.com/content/dam/Deloitte/uk/Documents/deloitte-analytics/open-growth.pdf . Accessed 20 February 2017.

[5] Open data communication. Open data, an engine for innovation, growth and trasparent governance. European Commission. 2011.

[6] European Union, Digital Agenda for Europe: A Europe 2020 Initiative.

[7] Open data charter, G8. Published by the Cabinet Office of the United Kindom. 2013. Available online at https://www.gov.uk/government/publications/open-data-charter

[8] A Comprehensive List of Open Data Portals from Around the World. http://dataportals.org/. Accessed 20 February 2017.

[9] ProgrammableWeb, API directory. http://www.programmableweb.com/. Accessed 20 February 2017.

[10] Open data Census. http://census.okfn.org/. Accessed 20 February 2017.

[11] RC. Basole, MG. Russell, J. Huhtamäki, N. Rubens, K. Still, H. Park. "Understanding Business Ecosystem Dynamics: A Data-Driven Approach," ACM Transactions on Management Information System, Vol. 6, No. 2, Article 6, 2015.

[12] M. Iansiti, R. Levien, "Creating Value in Your Business Ecosystem," Boston, MA, USA: Harvard Bus. School Press, March 2004.

[13] A. Immonen, M. Palviainen, E. Ovaska, "Requirements of an Open Data Based Business Ecosystem". 2014.

[14] N. Helander, "Value-creating Networks: an Analysis of the Software Component Business," Doctoral thesis. University of

Oulu. Faculty of Economics and Business Administration, Department of Marketing, 2004.

[15] A. Ojala, N. Helander, "Value creation and evolution of a value network: A longitudinal case study on a Platform-as-a-Service provider," 2014 47th Hawaii International Conference on System Science, 2014.

[16] J. Peppard, A. Rylander, "From Value Chain to Value Network: Insights for Mobile Operators," European Management Journal, volume 24, Issue 2, 2006.

[17] R. Normann, R. Ramirez, "From value chain to value constellation: designing interactive strategy," Harvard Business Review 1993; 71 (4), pp. 65–77

[18] C. Zott, R. Amitt, L. Massa, "Business model: Recent developments and future research," Journal of Management 2011; 37(4): 1019–1042. DOI: 10.1177/0149206311406265.

[19] R. Rajala, M. Rossi, VK. Tuunainen, "A framework for analyzing software business models," Proceedings of 11th European Conference on Information Systems, IEEE, 2003; 58: pp. 1614–1627.

[20] A. Osterwalder, C. Parent, Y. Pigneur, "Setting up an ontology of business models," Proceedings of 16th International Conference on Advanced Information Systems Engineering (CAiSE02), 2004, pp. 319–324.

[21] Linked Events, developers' Portal. https://dev.hel.fi/projects/linked-events. Accessed 20 February 2017.

[22] Digital leap of events project pages. https://www.avoinsatakunta.fi/tapahtumiendigiloikkahanke/. Accessed 20 February 2017.

[23] Agricultural business development with intelligent data analytics – project pages. https://www.avoinsatakunta.fi/mikadata-hanke/. Accessed 20 February 2017.

[24] A global Data Ecosystem for Agriculture and Food. Godan, DOI: 10.1079/CABICOMM-79-12. 2016. Availlable online: http://www.godan.info/documents/data-ecosystem-agriculture-and-food

Financial impact of forensic proceedings in ICT

Saša Aksentijević[1], Edvard Tijan[2], Alen Jugović[3]
[1] Aksentijević Forensics and Consulting, Ltd.
Gornji Sroki 125a, Viškovo, Croatia
Tel: +385 51 65 17 00 Fax: +385 51 65 17 81 E-mail: sasa.aksentijevic@gmail.com
[2,3] University of Rijeka, Faculty of Maritime Studies
Studentska 2, 51000 Rijeka, Croatia
Tel: +385 51 33 84 11 Fax: +385 51 33 67 55 E-mail: etijan@pfri.hr, ajugovic@pfri.hr

Abstract – Application of scientific forensic methods in ICT has become a mainstream methodology not only in criminal and civil proceedings, but also in preventive maintenance of various aspects of ICT systems used by corporations, governmental and other institutions. However, despite efforts of solution providers to create forensic hardware, software and procedures that are purported to be easy to use even by those that are not forensic experts, in most cases forensic proceedings are connected with high utilization of financial and temporal resources. Accelerated changes in information technology and architecture also require additional regulation that will pre-emptively ensure adequate amount and form of forensic trail left for possible future investigations. This paper is an attempt to describe current state of affairs of forensic proceedings, the latest trends and to provide comment on their financial impact and consequential real-world feasibility.

Key words: forensic proceedings, forensic investigations, financial impact, ICT

I   INTRODUCTION

Generally, forensic science or forensics is the application of scientific principles and techniques to matters of criminal justice especially as relating to the collection, examination and analysis of physical evidence [1]. The main characteristic of forensics is that it produces results and reports suitable to be used in courts or judicature, and for public discussion or debate. Therefore, forensics is indubitably tied to the legal context and explaining complex facts and their mutual connection in simple, straightforward language and layman terms to those who are not subject experts, but have to reach certain conclusions or make decisions based on the presented facts.

Digital forensics is one of the latest branches of general forensic science dealing with digital traces and artefacts. One of possible definitions of digital forensics is given by the following: "The application of computer science and investigative procedures for a legal purpose involving the analysis of digital evidence after proper search authority, chain of custody, validation with mathematics, use of validated tools, repeatability, reporting, and possible expert presentation" [2]. As it is clearly visible from comparison of definitions of "general" forensics and the digital one, digital forensics deviates from general definition because it does not analyse material (physical) evidence. Material nature of evidence was initially included in the definition of forensics because reliance on physical properties of the evidence

clearly demonstrated validity of the used method. While one can argue that evidence digital forensics is dealing with is material, in a sense that it is usually inextricably tied to material media that contains them, their nature is clearly not material. This demonstrates how technical developments of the present and near future will test and challenge some definitions that seemed to be solid in the past.

Over the past two decades, dematerialization of the IT world has caused detachment of the data, media that contains them and physical locations where such media is stored. Internet became a new transport layer for such evidence, crossing geographical barriers and further pushing the envelope of well-established legal framework whose jurisdiction is usually defined by physical national borders. While crime and other events are always one step ahead of codified legislation, described course of events made it very difficult even for those very best in the legal profession of digital forensics, and those who use their services (district attorneys, lawyers and courts) even to apply existing legislature, let alone understand the presented facts that are usually dealing with very complex technical architecture, facts and conclusions.

Most authors agree that until events of 9/11 [3], separation of non-material world of digital networks and systems according to national and geographical borders was one of the main obstacles in front of experts in the field of digital forensics and users of their services. After 9/11, new legislative acts have been passed all over the world to support the agenda of "fight against terrorism" [4], further facilitating cooperation between national bodies in charge of ICT infrastructure and legislative bodies of various countries, in provision of digital evidence during investigations. However, facilitation of this process has caused some major concerns among privacy advocates. Civil liberties implications of counterterrorism policies are a hot topic of debate in the European Union whose directives still protect the privacy of its countries' citizen. Findings of Julian Assange's WikiLeaks organization and those of Edward Snowden have clearly demonstrated how pre-emptive acting on behalf of the governments creates a myriad of more or less coordinated global surveillance programs with cooperation of telecommunication companies and European governments [5] that are clearly creating breaches on behalf of privacy of their own citizens.

It is clear that there are several concerns to be addressed by experts in digital forensics and legislative branches outlining the operational framework. Privacy of citizens and security of corporate business information has to be leveraged against national and international security; digital evidence is extremely volatile in nature, therefore it has to be carefully collected and examined, and it has to be done quickly, in order to avoid data expiration. Requirements related to digital data handling have to be respected. Furthermore, international cooperation is often the main prerequisite for most forensic investigations.

The described situation requires regular engagement of significant resources: time needed to perform analysis and create reports, sophisticated software and technology to analyse media and networks, and skilled experts who are able to explain digital findings. Therefore, both the required technology and time of experts are major constraints in forensic examination of digital evidence and if improperly utilized, they could results in sunk cost or even misinterpretation of findings.

II DIGITAL FORENSICS PRINCIPLES AND LAWS

Modern forensics has its roots in ancient China, with Song Ci being the most famous forensic medial expert during Southern Song Dynasty whose book "Collected Cases of Injustice Rectified" is still regarded as a seminal book of forensic science in China. Developments in the field of forensics in Europe became rapid only in the 19th century and especially in the beginning of the 20th century with wide adoption of fingerprint analysis invented by Juan Vucetic in 1891. Widespread adoption of forensic and scientific methods and introduction of expert witnesses in most legislative systems had as a consequence the dissemination of forensics in almost all fields of human life (and related legal proceedings).

Digital forensics is only a logical development of the above mentioned and became a discipline with the introduction of computers first used as mainframes for mass data processing. In the past two decades and with further development of information society, computers, smartphones, networks, servers and Internet usage are very often the usual and even expected part of many other legal proceedings, in the areas of both the criminal and civil law.

Digital forensics adheres to several classical laws of forensics sciences that are also used in other areas of science. While different sources quote different versions of these laws, they can be summarily explained in the following way when applied to the digital world:

1. Law of individuality, stating that every digital artefact has the characteristics that are not duplicated in any other object.

2. Principle of exchange (otherwise known as Locard's Exchange Principle, as a *hommage* to its founder, professor Edmond Locard) according to which when the perpetrator or the instrument (s)he uses comes in contact with the victim or surrounding objects, they leave traces, but they also pick up graces from them. This principle is extremely important in digital forensics.

3. Law of progressive change applied to digital forensics means that every digital trace changes with the passage of time. The impact of this principle in digital forensics is immense because the passage of time logarithmically alters it. Digital

data evidence is among the most volatile evidence forms [9].

4. Law of comparison states that only the likes can be compared, meaning that only like samples and specimens can be compared.

5. Law of probability claims that all definite or indefinite identifications made consciously or unconsciously are based on probability.

6. Law of circumstantial facts, otherwise known as "facts do not lie, men can and do" requires reliance on digital data and evidence and not oral evidence, power of observation or suggestion.

Except basic forensic laws that are more or less common for all forensics disciplines, there are a few more principles of digital forensics that are specific for the field of digital investigations [10].

1. No action taken during the forensic investigation should change data which may have to be relied upon in court

2. In case that an investigator has to access original digital evidence, he has to be competent to do so and give evidence explaining the relevance and implications of their actions. Considering that operating systems and programs alter the content of digital evidence without the user being aware of such change, forensic analysis is usually performed using relevant media images.

3. An audit trail has to be created and preserved demonstrating that the chain of custody over digital evidence is maintained. An independent expert should be able to examine the utilized methodology and achieve the same result.

4. There has to be a single instance in charge of investigation and ensuring that the forensic laws and principles are being adhered to.

The golden standard in digital forensics nowadays is the Abstract Digital Forensics Model, created as a generic, technology-independent model, composed of nine different phases (Figure 1).

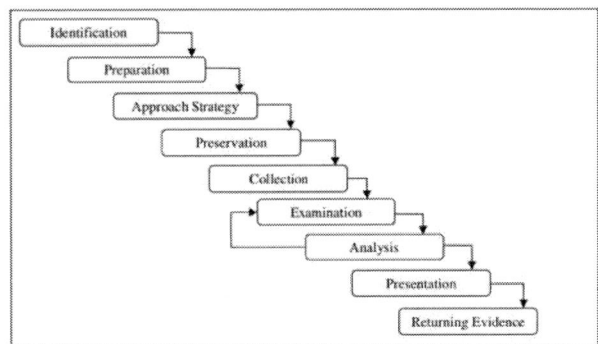

Figure 1. Abstract Digital Forensic Model [11]

This model assumes that the incident type is well recognized and determined. In comparison to previous models, this model consists of detailed pre- and post- investigation procedures.

III FINANCIAL IMPACT OF DIGITAL FORENSICS

Digital forensics trends are largely dictated by rapid advances in information technology over the past decades. In the latest period, we saw rapid growth and development of concept of Internet of Things (IoT), where ubiquitous computing principles are applied to a variety of devices and sensors in media, manufacturing, energy management, medical and healthcare, transportation, building and home automation, environmental monitoring and personal use. Each of these devices may serve as a source of digital data stream to be analysed as a part of forensic process. With the introduction of IoT, literally almost *anything* can become an object of digital forensic investigation, from wearable technologies and cars, to sensor grids. An interesting analysis of diverse topics in journals covering different areas is shown in Figure 2.

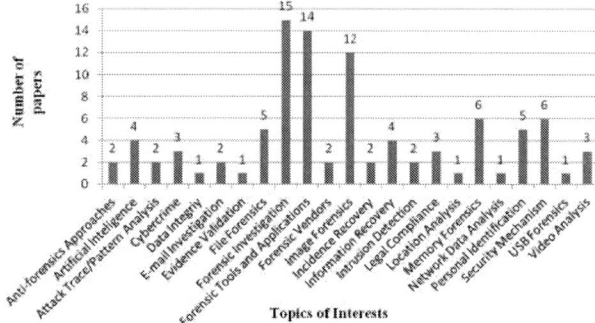

Figure 2. Coverage of digital forensics topics in journal papers [12]

While the number of overall papers shown in Figure 2. is low and its sum is not statistically significant, it shows a variety of topics covered by digital forensics. Some other papers show that 77,8 % of all cases deal with single user computers, 44,4 % with network forensics, and 55,60 % with mobile forensics [13]. It is worth noticing that the sum is above 100 % because in some investigations, there are multiple objects. So, despite variety of various objects of digital forensic investigation, some less complex or "traditional" objects still make up the majority of all investigations. This can partially be explained by the latency that is still prevalent in this field: there is a significant passage of time since the moment when certain digital evidence is created until the time it is fixed for analysis. So, forensic investigators are still working with delayed data.

On the other hand, there is an entire industry made around log data analysis and even predictive algorithms, aimed especially towards large enterprise server and network systems that are creating automated environment for large-scale data acquisition, analysis and alerting of administrative and other personnel in relation to potentially occurring anomalous events. These systems are in fact forensic in nature: they create forensically viable environment for analysis of various events. In most cases it is even possible to program various actions that will be triggered by events. These systems are primarily used to facilitate day-to-day

operations and administration of various ICT systems, and to provide audit trail for compliance purposes, but they can also be used in forensic analysis and provide a valuable source of information, especially if their usage and data storage follows forensic principles and laws.

Legislative branch is also placing forensics-motivated requests in form of various laws and requirements, especially aimed towards telecommunications and IT service providers. In most telecommunication acts, there are articles and provisions requiring service providers to install and maintain systems and software (often at their own cost) that tracks its usage and provide full access to the police and investigators. This trend is widespread in the United States and more and more present in the European Union. Privacy is still one of the main concerns and only communication meta-data and not its content is preserved, unless measures of wiretapping or surveillance are ordered by the court. It is reasonable to expect that in the future there will be more and more implementations of legislation-driven systems that will monitor patterns of usage of various information systems in order to collect data for later forensic analysis.

Anticipative inclusion of data logging in laws does not only provide audit trail and basis for further forensic investigations, it can also lower the cost of forensic analysis because it contains data that would otherwise have to be extracted using other, more costly methods, or it would not be available at all. The cost of forensic investigations in the USA is typically in the range of 10,000 US$ - 100,000 US$ with hourly rates in the range between 125 US$ and 650 US$ [14]. These costs can be significantly lowered with greater inclusion of logging tools, some of which can also be obtained as a open source data loggers and maintained as such. It is worth noticing that there is significant overlapping between solutions that are behind implemented controls in systems of information security and forensic logging tools. Organizations that have higher achieved levels of information security will also have less security breaches, and probably even less those that will result in serious consequences that might be a matter of forensic investigation. Even if that occurs, forensic investigation might be faster or easier, thus incurring less related cost.

Inclusion of legislative requirements seems to be especially important in the case of cloud computing forensics. The National Institute of Standards and Technology of the U.S. Department of Commerce has recognized this importance and has included forensic science challenges in the draft of its NISTIR 8006 standard. This draft anticipates almost all steps of forensic process described in this paper and recognizes that cloud forensics possesses certain specific traits and challenges arising from the distinctive nature of the computer cloud [15]. It further defines cloud computing stakeholders and their roles, and collection and aggregation of challenges, along with additional observations. The most distinctive are:

1. Time, either in terms of consistency or data volatility in time,

2. Location, where even locating an evidence may be a major hurdle in forensic investigation, and

3. Data sensitivity, where pervasive use of cloud computing environments by users and employees could elevate the risk of incidents that might end up as forensic investigations.

Additional requirements for laboratories performing forensic investigations in ICT and especially data acquisition are arising from some applicable ISO standards, and especially the new edition of ISO /IEC DIS 17025: General requirements for the competence and testing and calibration laboratories, slated for the next revision issue in May 2017 [16]. In United Kingdom, there will be a mandatory required certification of ICT forensic laboratories according to this standard [17] – something that was until now reserved for "wet evidence" laboratories dealing with DNA testing and organic evidence. This procedure will have far reaching consequences for all involved parties because, at this moment, four year long backlogs for analysis of seized computer equipment are not unheard of in the industry [18]. Currently, the market of digital forensics field consists of a small number of large players, and a large number of one-man forensic investigators who are very important in the process of provisioning forensic assistance to district attorneys, police and judicial system, and decrease of current forensic backlog. The anticipated certification of the process will both increase the cost of service due to less competition and lengthen (at least initially) the process of forensic investigation until the market is fully consolidated.

VI   CONCLUSION

Digital forensics deviates from the general definition of forensics because its procedures are done over a set of non-material evidence, which is one of the prerequisites of the traditional forensics. However, media where such evidence resides is still material in nature (hard drives, memory cards, network storages, volatile memory).

There are several other characteristics that separate digital forensics from other, more traditional forensic disciplines, some of them being evidence volatility, remote geographical placement of evidence, transition over several legal jurisdictions, reliance on legal framework and implemented measures of information security to obtain digital evidence, and constantly changing technology.

Digital forensics follows the same well-established laws and principles known in the other forensic fields. However, the amount of analysed material, the inability to make data acquisition automatic, high level of required skills, knowledge, specialized hardware and software and time required to perform the analysis are the main obstacles placed in front of forensic investigators. Modern digital forensics also has to perform the analysis of the new systems, like data acquired from sensor arrays and grids connected to the Internet of Things, personal devices like mobile phones or cloud computing systems. Such forensic analysis is often very complex and costly.

One possibility of control of possible forensic cost and facilitating forensic analysis is the implementation of various controls aimed to elevate the achieved level of information security. These measures often provide a good level of audit trail that has to be kept under forensically sound conditions and it could be later used as such. Therefore, there is a significant level of overlapping between the information security management systems and tentative subsequent forensic usage.

There is a noticeable global effort to standardize practices in digital forensics by using standards already applicable to other forensic fields and especially DNA and crime trace analysis. The introduction of these standards will further increase the cost of forensic services, and might render small forensic investigation teams unable to compete with large accredited laboratories, who possess capabilities to forensically analyse large volumes of ICT equipment and data.

REFERENCES

[1]   Meriam-Webster,   https://www.merriam-webster.com/medical/forensic%20science   (accessed   17th December 2016)

[2] Lynch, V.A., Duval J.B. "Forensic Nursing Science", Elsevier Health Sciences, 2010, p. 97

[3] Kean, T.H. et al. "The 9/11 Commission Report", The National Commission on Terrorist Attacks Upon the United States, August 21, 2014

[4] http://www.un.org/en/counterterrorism/ (accessed 17th December 2016)

[5] https://www.theguardian.com/world/2013/jun/09/edward-snowden-nsa-whistleblower-surveillance   (accessed   17th December 2016)

[6] Sung, T. et al. "The Washing Away of Wrongs: Forensic Medicine in Thirteenth-Century China (Science, Medicine, and Technology in East Asia)", Center for Chinese Studies, University of Michigan, 1981

[7] Vucetich, J. "*Dactiloscopia comparada – el nuevo sistem Argentino*", Establecimento Tipografico Jacobo Pkuser, La Plata, 1904

[8]   Locard's   Exchange   Principle,   http://vjestak-informatika.com/2016/12/13/locardov-princip-razmjene-u-forenzici/ (accessed 17th December 2016)

[9] DOJ National Institute of Justice, "Volatility of digital evidence",https://www.policeone.com/Officer-Safety/tips/1655664-Volatility-of-digital-evidence/, June 10 2008 (accessed 17th December 2016)

[10] http://www.computerforensicsspecialists.co.uk/blog/the-principles-of-digital-evidence (accessed 17th December 2016)

[11] Reith, M. et al. "An Examination of Digital Forensic Model", International Journal of Digital Evidence, Volume 1, Issue 3, fall 2002

[12] Dezfoli, F.N. et al. "Digital Forensics Trends and Future", International Journal of Cyber-Security and Digital Forensics (IJCSDF) 2(2), 2013, p. 50

[13] M. Tu, K. Cronin, D.Xu, S.Wira,"On the Development of Digital Forensics Curriculum", http://www.dsu.edu/research/ia/documents/ [6]-On-the-development-of-Digital Forensics-Curriculum (accessed 17th December 2016)

[14]http://blog.securitymetrics.com/2016/08/what-do-forensic-investigations-do.html1 (accessed 17th December 2016)

[15]http://csrc.nist.gov/publications/drafts/nistir-8006/draft_nistir_8006.pdf (accessed 17th December 2016)

[16]http://www.iso.org/iso/home/store/catalogue_ics/catalogue_detail_ics.htm?csnumber=66912 (accessed 6th January 2017)

[17]http://www.forensicmag.com/article/2012/02/isoiec-170252005-accreditation-digital-forensics-discipline (accessed 6th January 2017)

[18] D. Lillis, B. Becker, T. O'Sullivan, and M. Scanlon, "Current Challenges and Future Research Areas for Digital Forensic Investigation," 05 2016

Reliability, Availability and Security of Computer Systems Supported by RFID Technology

Pančo Ristov[*], Toni Mišković[*], Ante Mrvica[*], Zvonko Markić[*]
*Faculty of Maritime Studies, Split, Croatia
panco.ristov@pfst.hr, tmiskovic@pfst.hr

Abstract - The implementation of RFID technology in computer systems gives access to quality information on the location or object tracking in real time, thereby improving workflow and lead to safer, faster and better business decisions. This paper discusses the quantitative indicators of the quality of the computer system supported by RFID technology applied in monitoring facilities (pallets, packages and people) marked with RFID tag. Results of analysis of quantitative indicators of quality compute system supported by RFID technology are presented in tables.

I. INTRODUCTION

The development of elements of RFID technology leads to an improvement of its application in many sectors of the economy, such as: maritime (tracking of containers, etc.), auto industry, production processes, identification and/or tracking animals, implantation in humans, etc. [1,2,3]. According to research by IDTechEx, during 2015 8.9 billion tags were sold, while in 2016 year around 10.4 billion will be sold [4]. Most selling tags are passive UHF tags. In addition to this rapid growth in the application of RFID technology, the problems with the reliability and availability of the system elements are still present thus leading to restrictions apply in many applications in the economy. Therefore, this paper analyzes the reliability, availability and security of critical elements of RFID technology (RFID reader) and a computer system.

Reliability and availability of RFID technology depends on the reliability and availability of the reader, or moreover, depends on the reliability of reading tag's content. Reliability of tag is probability that content of tag will be read accurately for a particular operating environment, while the reliability of the reader is probability that reader will successfully read all the data from the tag in time t (reader is in working condition at the time of observation t) for a given work environment, and, as an indicator, mean time between failures is usually given.

Reliability of computer system supported by RFID technology (CSSRFIDT) refers to the ability of the system to maintain its operational state without delay or the ability to prevent the degradation of characteristics for a certain period of time, and as an indication, most often the mean time between failures is given. The total availability of CSSRFIDT system is an integral part of

the availability of all system elements. Security of CSSRFIDT system is probability that system will either function properly or will terminate in a safe manner and is calculated so that the calculation takes all work and correct states as well the state of safe failure.

By inspecting the scientific and professional papers and practices can be observed that a lot more attention is given to privacy and data security while the reliability of the elements of RFID technology and the reliability of the entire computer system supported by RFID technology is rarely analyzed.

The low level of confidence lead to compromising the security of data and information, dissatisfaction of subjects in the business process because they cannot perform all the functions CSSRFIDT system, and due to higher maintenance costs and so on. It is obvious that the availability of 99% (the system is in failure three days and 15 hours and 40 minutes in one year) or 99,95% (the system is in the cancellation of 1 day, 19 hours, 48 minutes in one year) is not permissible but values of 99.99% (the system is in the cancellation of 52 min and 36 sec. in one year) or 99.999% (the system is in failure 5 min., 15 sec. in a year) are the values to be pursued in the exploitation of the system.

Increase of reliability CSSRFIDT system has its price and economic justification. Correction after the occurrence of failure must be as short as possible. It is necessary to implement new maintenance strategies or a combination of the new strategies supported by information system for maintenance management. The application of new strategies leads to increased reliability and availability, failure prevention and reduction of maintenance costs.

The new strategy of maintaining integrates RFID technology and CMMS (Computerized Maintenance Management System) system. Elements of RFID technology enable the storage and transmission of data for maintenance in the CMMS system [5].

II. RELIABILITY OF ELEMENTS CSSRFIDT SYSTEM

Achieving the target reliability of CSSRFIDT system requires undertaking certain activities at all stages of the life cycle of the system. During exploitation of the system reliability can be increased by applying the techniques of fault tolerance and dodge failures.

By incorporating redundancy system availability is increased so price of system with partial or complete redundancy is quite understandable issue. However, if the analysis starts from the criteria of safety of object that is tracked, additional cost on basic redundant structure, and the price difference from the simple structure of the system and the complex structure that is tolerant to failures, remains to an extent that justifies the goal.

The planned failures are caused by planned and regular maintenance of elements CSSRFIDT system and therefore it is required that certain elements of the system to be inactive for other elements and users (installing new or upgrading existing systemic functions, update the operating system or systems for database management, etc.). Unplanned failures are the result of human errors, bugs in software, hardware failure (replacing the hard drive, replacing network cards, etc.) and environmental conditions (dust, humidity, vibration, high/low temperature and voltage fluctuations).

A.   Reliability of elements of RFID technology

Tags differ in shape, size, power, capacity, protocol, the radio frequency wavelengths and the method of data storing [6, 7].

The tag can store n bits of useful information of tracked object. The probability that the content from tag will be properly read depends on the probability that each bit won't be accurately read (P_{nr}) and can be represented by the equation (1) [8]:

$$P_r = \left(1 - P_{nr}\right)^n \qquad (1)$$

Readers are different in complexity depending on the type of tag with which they communicate, and the functions to be fulfilled, such as error checking and correction. The reader can be fixed or mobile. Fixed readers are mounted on a fixed structure and most commonly they use external antenna (longer range), while mobile readers are handheld readers and antenna is implemented in the reader (shorter range). The position of the antenna significantly influences the characteristics of the antenna and thus the reliability of the reader. Readers can be distinguished between the implemented protocols, the possibilities of networking multiple readers, the operating frequency, the management capabilities with multiple antennas, etc.

The reliability of the elements of RFID technology is influenced by several factors: speed, environment, distance, orientation, encryption techniques, strength, sensitivity and debugging. Many of these factors are the result of signal attenuation, which reduces the signal to noise ratio [9, 10].

The efficiency of reading content from tag depends on the reader's radio frequency beam which shape is elliptical. Therefore, as the distance from a reader increases it is possible that within the zone of reading

more than one tag exists. Consequently, as result of an increase in the number of tags there might be an issue about collision of tags. Reliability of reading depends on the distance between the tag and the reader, and the distance between the reader and antenna. Maximum read range is controlled by the power and sensitivity of reader. The choice of power levels and sensitivity reader depends on the application.

Environment has large influence on the reliability of readers, because the presence of metal, water or objects that absorb or reflect the radio frequency signal may affect the accuracy of reading the content from tag. In addition, the presence of wireless networks and other sources of noise also affects on the reliability of the reader. Furthermore, the accuracy of reading also depends on the type of matter in which tag is packed especially for tags operating in the UHF and MF frequency range. It is necessary to take into account the number of people in the work environment of reader because people like silencers signals have an effect on the accuracy of the reading.

Orientation of tag refers to a position of tag to reader. Tags that are not sensitive to the orientation can work regardless of the orientation. RFID systems that operate at higher frequencies are sensitive to the orientation of the tag. Best reading reliability is achieved when the tag is set toward the reader, while the lowest reading reliability is achieved when the tag is placed on top of the object being monitored [11]. The sensitivity on the orientation is visible with a linear polarized antenna and the best reading is achieved when the tag is parallel to its axis and then provides maximum read range.

Encryption involves the use of codes for changing the original data in a different format for storage or transmission. To store data on the tag unipolar and polar techniques are used. From polar technique being used there are irreversibly to zero (NRZ-Non-Return-to-Zero Level) and Manchester method. Unipolar encryption is the simplest. Generally, one voltage level is assigned to a binary zero, and the second level is assigned to binary unit. Polarity refers to whether the impulses are positive or negative. In methods irreversibly to zero, the signal is always positive or negative. There are two types of this type of encryption: NRZ-L (Level) where the signal level depends on the state of the bit, and NRZ-I (Inverted) where the signal turns when it encounters a binary first. Manchester methods are two-phase encryption type in which signal is changed to opposite pole in the middle of one bit interval.

Emitted power of reader has a big impact on the reliability of the reader. If the energy level falls below a certain threshold (the sensitivity of the reader) the reader will not read correctly the content of the tag and thus affect the reliability of the reader.

In order to increase the reliability of readings two or more tags are implemented to the tracking object. By increasing the number of tags the reliability of reading is also increased. In such applications position and distance between two or more tags must be taken into account [11, 12].

B. *Reliability of computer system*

Reliability, availability and security of the computer system are directly linked with the emergence of failures in hardware, software modules or errors in the database. Indicators of reliability of hardware are usually linked to a period of normal operation, which is the longest. Reliability of software subsystems increases over time because its errors are detected and corrected so their numbers are decreasing. In addition, changes in the operating profile or environment may cause a change of reliability and then decreases the reliability of software. Therefore, the measure of reliability software must be linked to the operating profile and the environment in which the system works [13].

A computer system that tolerates failures must be designed so that operates properly at all times, regardless of the defects in the hardware or software, or there must be a module that will dynamically check for accuracy and proper function. Research has shown that the module for fault detection does not detect or does not detects 100% of all failures resulting in operational work, but it only "covers," or detects all possible defects (the definition of this concept have made Bouricius, Carter, and Schneider [1969]) with a coverage factor C ($0 < C < 1$) [14].

In order to increase reliability, availability and security of computer systems in hardware and software modules appropriate diagnostic program (self-diagnosis) is incorporated and it generates a warning at the time of detecting errors. System failure occurs by failure of hardware or software components or it is caused by human error.

III. MODELING RELIABILITY, AVAILABILITY AND SECURITY OF CSSRFIDT SYSTEM

Analysis of quality indicators of CSSRFIDT system is made by the simulation methodology used to describe the behavior of the system using the Markov method. To use the Markov method two basic conditions are met, namely: the exponential character of the probability density function of failure, i.e., a constant function of frequency of failure and the exponential character of the probability density function of the correction, i.e., a constant function of frequency of repairs [15, 16].

Modern implementation of the power supply system use uninterruptible power supplies. Therefore, with the simulation considering the quality indicators of critical elements 100% reliability of the power supply system will be assumed.

Due to the specific functions that CSSRFIDT system executes; special attention is paid to the reliability, availability and security of critical elements. The changes of the critical elements as well as the frequency of transition from one state to another are described. Elements, in order to perform their function, each state that passes from the state of the work in another state,

must restore back to the state of work. Disturbing the static balance between the states, changes the reliability, availability and security of the system, but in every element of the event dynamic processes occurs.

Analyze the transitions that can take place to elements of the system under certain assumptions:

1. Failure rate λ and repair rates μ of modules are constant and not dependent on time. Failure rate of a computer system is a total failure rate of hardware (λ_h), failure rate software (λ_s), failure rate of operators (λ_o), and failure rate of database, i.e. hardware $\lambda_{PC} = \lambda_h + \lambda_s + \lambda_o + \Lambda_{db}$ and frequency of RFID reader λ_{RFID}.
2. All failures are mutually independent, i.e. any failure is independent from other failures.
3. Probability of occurrence of two or more failures in the time interval Δt is negligible.
4. The system starts in a completely proper operating condition where all system modules operate properly.

Failure of CSSRFIDT system can be in two basic conditions. One condition is defined as "safe" and another as "precarious" condition. Safe state (SO) of CSSRFIDT system means that the operator, after receiving a warning from the diagnostic program, performs a sequence of actions to close the database and system shutdown. Uncertain state (NO) of CSSRFIDT system means that diagnostic program has not detected the error and computer system simply "freeze" their work. In addition to these two states, in the analysis of quality indicators following states are also examined: SP state (initial state where all elements of the system are correct), SR state (primary RFID reader is in the failure but reserve RFID reader is active) and state SPC (primary computer is in failure but backup computer is active). If the diagnostic program detects a failure on standby RFID reader, and the recovery of the primary computer is not completed, it is possible the transition from state SR to state SO or transition from a state SR to a state NO. The transition from state SR to state SO is safe transition, and that means that diagnostic program generates a warning on the mandatory exclusion of reserve RFID reader and the operator performs the procedure of closing the database and system shutdown. The transition from a state of SR in the state of NO will occur when a diagnostic program does not detect a fault on standby RFID reader so the reserve reader "freezes" work. In such situations may arise malfunctioning hardware (RFID reader/antenna, it is impossible to program the tags or microchips) or software of reader (RFID reader cannot read the contents of a tag). Other passes are exactly the same as in the previous analysis so there is no need to repeat.

Equations for Markov model of reliability, availability and security may be written in matrix form:

$$P_{CSSRFIDT}(t + \Delta t) = P_{CSSRFIDT}(t) \cdot T_{CSSRFIDT} \qquad (2)$$

where every element of $P_{CSSRFIDT}$ *(t)* is the probability of elements of CSSRFIDT system with hot standby in a particular state at time t. $T_{CSSRFIDT}$ is the transition matrix from state to state, while $P_{CSSRFIDT}(t+\Delta t)$ is the probability that each element of the system is in the proper state at time $t + \Delta t$. Consequently, we can write:

$$P_{CSSRFIDT}(t+\Delta t) = \begin{bmatrix} P_{SP}(t+\Delta t) \\ P_{SPC}(t+\Delta t) \\ P_{SR}(t+\Delta t) \\ P_{SO}(t+\Delta t) \\ P_{NO}(t+\Delta t) \end{bmatrix} \qquad P_{CSSRFIDT}(t) = \begin{bmatrix} P_{SP}(t) \\ P_{SPC}(t) \\ P_{SR}(t) \\ P_{SO}(t) \\ P_{NO}(t) \end{bmatrix}$$

$$T=\begin{bmatrix} [-(\lambda_{SPSPC}+\lambda_{SPSR}+\lambda_{SPSO}+\lambda_{SPNO})+(\mu_{SPCSP}+\mu_{SRSP}+\mu_{SOSP}+\mu_{NOSP})]\Delta t & \mu_{SPCSP}\Delta t & \mu_{srsp}\Delta t & 0 & 0 \\ \lambda_{SPSPC}\Delta t & -(\lambda_{SPCSO}+\lambda_{SPCNO}+\mu_{CSPSP})\Delta t & 0 & 0 & 0 \\ \lambda_{SPSR}\Delta t & 0 & -(\lambda_{SRSO}+\lambda_{SRNO}+\mu_{SRSP}) & 0 & 0 \\ \lambda_{SPSO}\Delta t & \lambda_{SPCSO}\Delta t & \lambda_{SRSO}\Delta t & 1 & 0 \\ \lambda_{SPNO}\Delta t & \lambda_{SPCNO}\Delta t & \lambda_{SRNO}\Delta t & 0 & 1 \end{bmatrix} \quad (3)$$

The initial conditions at $t = 0$ are:

$$P_{SP}(t) = 1, P_{SPC}(t) = 0; P_{SR}(t)=0, P_{SO}(t)=0, P_{NO}(t)=0. \quad (4)$$

Substituting the appropriate values for the frequency of cancellation and correction, and solving matrix equation of Markov model, gives the value of $P(\Delta t)$ by multiplying the vector of initial value $P(0)$ and the matrix T, $P(2\Delta t)$ multiplying the $P(\Delta t)$ and transition matrix TRSS. General solution of equation of Markov model is given as:

$$P_{CSSRFIDT}(n\Delta t) = T^n \cdot P(t). \quad (5)$$

In accordance with Markov model reliability of redundant CSSRFIDT system is the probability that the system is in a state of SP, SR and SPC, which are the only states when a redundant system is operating correctly, i.e. the reliability of a redundant system with hot standby can be written as:

$$R_{CSSRFIDT}(t) = P_{SP}(t) + P_{SR}(t) + P_{SPC}(t). \quad (6)$$

CSSRFIDT system is safe as long as it is in one of three states: the state of the SR, state SPC and state SO. Security of redundant system with hot standby can be written as:

$$S_{CSSRFIDT}(t) = P_{SP}(t) + P_{SR}(t) + P_{SPC}(t) + P_{SO}(t). \quad (7)$$

TABLE 1: INDICATORS OF QUALITY OF CSSRFIDT SYSTEM

$\lambda_{RFID}= 5,3\ 10^{-4}$				$\lambda_{PC}=2,20\ 10^{-4}$		
Time of observation 8760 h						
μ (h^{-1})	Reliability	Availability	Security	System in failure (h)	System in operation (h)	Redundancy
0	0,999072	0,999536	0,999982	4,07	8755,93	No
2	0,999072	1	0,999982	0,00046	8759,996	No
0,25	0,999072	0,999999	0,999982	0,00371	8759,999	No
0,125	0,999072	0,999999	0,999982	0,0074	8759,993	No
0	0,995551	0,997774	0,999906	19,499	8740,501	RFID reader
2	0,995551	1	0,999906	0,0022	8759,998	RFID reader
0,25	0,995551	0,999998	0,999906	0,019	8759,981	RFID reader
0,125	0,995551	0,999998	0,999906	0,036	8759,964	RFID reader
0	0,992409	0.9962	0,999159	33,29	8726,71	RFID reader +PC
2	0,992409	1	1	0,0038	8759,9962	RFID reader +PC
0,25	0,992409	0,999997	1	0,03	8759,97	RFID reader +PC
0,125	0,992409	0,999993	1	0,06	8759,94	RFID reader +PC

Using software packages Isograph Reliability Workbench 11.0 - Markov model was created and simulated and performed experiments to data from MIL HDBK 217[1]: catalog and operational data for the real system and obtained numerical data for the relevant indicators of quality of CSSRFIDT system.

Development of the simulation model is made in several steps. In the first step a model of the critical elements of the system (RFID reader and personal computer) was created. In the second step was created a model where was added a reserve RFID reader and in the third step one spare computer is added to the model.

Table 1 shows the results of quality indicators of CSSRFIDT system inputs for the frequency of cancellation of RFID reader and a personal account or $\lambda_{RFID} = 5.3\ 10^{-4}$, $\lambda_{PC} = 2.2\ 10^{-4}$ from MIL HDBK 217 catalog and manufacturers, or elements of moderate reliability [17, 18, 19].

Based on the analysis of several maintenance contracts of CSSRFIDT system (from few companies who have contract of maintenance of CSSRFIDT and similar systems) frequency of repair ranges from 2 h^{-1} (repair executed by staff of organization with a telephone or Internet customer service help), 0.25 h^{-1} and 0,125h^{-1} (repair executed by customer service staff).

---

[1] MIL HDBK 217 is a military standard for assessing the inherent reliability of electronic equipment and systems based on the intensity of the cancellation of components.

Maintenance of CSSRFIDT system is combined, e.g. preventive and some simple repairs can be done by user, and corrective, adaptive and perfective maintenance executed by the manufacturer or service technician with certificates for repairs or maintenance. Repair of defective element is performed by one serviceman (if repair is performed by more than one serviceman than $\mu=k\cdot\mu$, where k is number of servicers). The values of availability, reliability, security, uptime time and time in dismissal of CSSRFIDT system for a year of work are shown in Table 1.

Looking at table 1 it can be observed that as the shorter the mean time to repair failure is, availability and security of CSSRFIDT system grows, and that during the cancellation time is reduced, while the reliability remains same. In order to increase the quality indicators of CSSRFIDT system redundant reader and personal computer are installed. Best availability and system security are achieved when the frequency of repairs is 2 h^{-1} and MTTR (mean time to repair) is 30 min. By increasing the mean time to repair, the value of quality indicators of system are being reduced. Such a mean time to repair can be achieved only if the user has the IT knowledge that can perform certain operations of corrective and perfective maintenance, particularly when replacing the hardware module (RFID reader), installation of system and application software with the telephone or Internet support from customer service. It can also be noted that without maintenance and with

increase number of elements the indicator "time to failure" increases.

Tables 2 and 3 shows the results of the quality for the input value of the operating frequency of failure of the RFID reader and the cancellation of the personal account or $\lambda_{RFID} = 9.6\ 10^{-4}$ and $\lambda_{PC} = 2.7\ 10^{-4}$ for the system consisting of a 10 readers deployed at different locations and the master server in the operative work. In this experiment the indicators of the quality of the system are observed after 8760h and 17520h.

Looking at tables 2 and 3, it can be observed that the availability, security and system in the cancellation are increasing, and the system in operation is reduced by increasing MTTR or by reducing the frequency of repairs. By increasing the number of electronic elements (redundancy) the quality indicators are being reduced. The worst results are obtained when redundancy of critical elements is not integrated and maintenance is not organized. Satisfactory indicators of the quality system are obtained by adding redundant system elements. Best system availability (99.9999%) and safety (99.9603%) for work time 8760h is obtained when both critical elements are redundant and the intensity repairs is $2h^{-1}$ and MTTR is 30 minutes. If the problem cannot be solved by the staff of the organization, intervention of authorized service is required where MTTR ranges from 4 hours to 8 hours depending on the type of fault and location of customer service.

TABLE 2: INDICATORS OF QUALITY OF CSSRFIDT SYSTEM

λ_{RFID}=9,6 10^{-4}		λ_{PC}= 2,7 10^{-4}				
Time of observation 8760 h						
μ (h^{-1})	**Reliability**	**Availability**	**Security**	**System in failure (h)**	**System in operation (h)**	**Redundancy**
0	0,97732	0,998866	0,999955	9,94	8750,06	No
2	0,97732	1	0,999955	0,01	8759,99	
0,25	0,97732	0,999999	0,999955	0,009	8759,99	
0,125	0,97732	0,999998	0,999955	0,018	8759,98	
0	0,987638	0,993807	0,999754	54,25	8705,75	RFID reader
2	0,987638	0,999999	0,999754	0,006	8759,999	
0,25	0,987638	0,999994	0,999754	0,049	8759,95	
0,125	0,987638	0,999989	0,999754	0,099	8759,9	
0	0,925542	0,962231	0,999603	330,86	84229,14	RFID reader +PC
2	0,925542	0,999996	0,999603	0,0309	8759,69	
0,25	0,925542	0,999965	0,999603	0,309	8759,69	
0,125	0,925542	0,999929	0,999603	0,619	8759,38	

A major impact on quality indicators of CSSRFIDT system has and the quality of the used tag. Corrupted, damaged and poorly executed tags directly affect the quality indicators of CSSRFIDT system. By increasing the time to use the system (table 3), all indicators of quality systems fall indicating that the organization needs to carry out all maintenance activities and continuous IT educate their workers. Performance of tag depend on several factors, including the matter of which is made

until the environment in which they are used (temperature, humidity, physical limitations, etc.). From the practice and research conducted at AT&T laboratories it is known that low quality tags can reduce the quality indicators of CSSRFIDT system from 20% to 30%. In order to maintain the value of quality indicators of system shown in Tables 2 and 3, it is necessary to take into account the quality of the tag. Therefore, it is necessary to use cheap redundant tags or high functional tags.

TABLE 3: INDICATORS OF QUALITY OF CSSRFIDT SYSTEM

λ_{RFID}=9,6 10^{-4}			λ_{PC}= 2,7 10^{-4}			
Time of observation 17520 h						
μ (h^{-1})	Reliability	Availability	Security	System in failure (h)	System in operation (h)	Redundancy
0	0,97732	0,998866	0,999955	9,94	8750,06	No
2	0,97732	1	0,999955	0,01	8759,99	
0,25	0,97732	0,999999	0,999955	0,009	8759,99	
0,125	0,97732	0,999998	0,999955	0,018	8759,98	
0	0,987638	0,993807	0,999754	54,25	8705,75	RFID reader
2	0,987638	0,999999	0,999754	0,006	8759,999	
0,25	0,987638	0,999994	0,999754	0,049	8759,95	
0,125	0,987638	0,999989	0,999754	0,099	8759,9	
0	0,925542	0,962231	0,999603	330,86	84229,14	RFID reader +PC
2	0,925542	0,999996	0,999603	0,0309	8759,69	
0,25	0,925542	0,999965	0,999603	0,309	8759,69	
0,125	0,925542	0,999929	0,999603	0,619	8759,38	

III. CONCLUSION

The quality and level of use CSSRFIDT system greatly depend on the quality of critical elements of the system, diagnostic programs, staff and system maintenance.

Looking at the value of quality indicators of CSSRFIDT system can note the following: availability, security of system in the cancellation increase, and the system in operation decreases with increase mean time to repair, or by reduce the frequency of repairs.

The analysis has shown that CSSRFIDT system configured with elements of moderate reliability and warm standby is highly available in the tracking systems. Availability of 99.9999% and 99.9996 achieved when both critical elements are redundant and intensity of repairs is 2 h^{-1} and MTTR is 30 minutes.

These results can be achieved and maintained only if CSSRFIDT system is maintained by professional staff that will, at all times, respond to any hardware and/or software problem at work. Furthermore, in the system of CSSRFIDT it is necessary to use highly functional tags whose price go to 6$ or cheap redundant tags whose target price may range from 20 cents to 1 $.

For availability of CSSRFIDT system maintenance strategies, and quality of the organization and management of preventive, corrective, adaptive and perfective maintenance has a big impact.

REFERENCES

[1] Narsoo, J., Muslun, W. and Sunhaloo, M. S., (2009) A Radio Frequency Identification (RFID) Container Tracking System for Port Louis Harbor: The Case of Mauritius, Issues in Informing Science and Information Technology, available at: http://iisit.org/Vol6/IISITv6p127-142Narsoo691.pdf [accessed 9.5.16.].

[2] Günther, O., Kletti, W. and Kubach, U., (2008), RFID in Manufacturing, Springer-Verlag, Berlin Heidelberg.

[3] Hansen, W-R., Gillert, F., Cox, K. and Schmid, V., (2006), RFID for the Optimization of Business Processes, JohnWiley&Sons Ltd, pp.

[4] www.IDTechEx.com/forecast - (9.5.2016.)

[5] Adgar, A., Addison, D., Chi-Yung (Alan) Yau, „Applications of rfid technology in maintenance systems", University of Sunderland, School of Computing & Technology, Edinburgh Building, Chester Road, Sunderland, SR1 3SD, U.K.

[6] Kaur. M.,Sandhu, M., Mohan. N., Sandhu, N.S.,„RFID Technology Principles, Advantages, Limitations & Its Applications", International Journal of Computer and Electrical Engineering, Vol.3, No.1, February, 2011.

[7] Ristov, P. and Mrvica, A., (2011), Primjena RFID tehnologije u pomorstvu, Proc. of 3th International Maritime Science Conference (ISSN 1847 1498), Split, pp. 247-259.

[8] Ristov, P., Mišković, T. and Mrvica, A., (2015), RFID based access control system, Proc. of 38th International Convention, MIPRO, pp. 1793-1798.

[9] Ng Ling Siew, „A reliability study of the RFID technology", Naval postgraduate school Monterey, California, 2006.

[10] Hiilltunen, M., „RFIID Reliability - Research Report", AT&T Labs – Research.

[11] Bolotnyy, L., „New Directions in Reliability, Security and Privacy in Radio Frequency Identification Systems", A Dissertation, the faculty of the School of Engineering and Applied Science, University of Virginia, 2008.

[12] Bolotnyy, L., Robins, G., Multi-Tag RFID Systems, Department of Computer Science, University of Virginia, Charlottesville, Virginia

[13] Ristov, P., Komadina, P., Tomas, V., „Pouzdanost i raspoloživost računalnih sustava koji se koriste u nadzoru, praćenju i organizaciji pomorskog prometa", Naše more, Znanstveno-stručni časopis za more i pomorstvo, Vol 60, ", broj 1-2, 2013.

[14] Shooman, M. L., "Reliability of Computer Systems and Networks: Fault Tolerance, Analysis, and Design", J. Wiley & Sons, 2002.

[15] Pukite, P., Pukite, J.: "Modeling for reliability analysis: Markov modeling for reliability, maintainability, safety, and supportability of complex systems", IEEE Press, Inc., New York, 1998.

[16] Ramović, R. M.: „Pouzdanost sistema elektronskih, telekomunikacionih i informacionih", Beograd,2005.

[17] Robert Brumnik, R., Balantič, Z., „Reliability and Efficacy of Identification Systems and Supply Chain Management1", Strojniški vestnik - Journal of Mechanical Engineering 54 (2008)11.

[18] MIL-HDBK-217, Reliability Prediction of Electronic Equipment. U.S. Department of Defense www.itemuk.com/milhdbk217.html (10.09.2016)

[19] http://w3.siemens.com/mcms/identification-systems/en/rfid-systems /Pages/rfid-systems.aspx (10.5.2016.)

MIPRO 2017, May 22- 26, 2017, Opatija, Croatia

Transportation and Power System Interdependency for Urban Fast Charging and Battery Swapping Stations in Croatia

Ivan Pavić*, Ninoslav Holjevac*, Matija Zidar *,Igor Kuzle *
Aleksandar Nešković**,

* University of Zagreb, Faculty of Electrical Engineering and Computing, Unska 3, 10000 Zagreb, Croatia
** School of Electrical Engineering, University of Belgrade, Bulevar kralja Aleksandra 73, 11120 Belgrade, Serbia
ivan.pavic@fer.hr, ninoslav.holjevac@fer.hr, matija.zidar@fer.hr, igor.kuzle@fer.hr
neshko@etf.bg.ac.rs

Abstract – An increasing penetration of electric vehicles in recent years has been driven by government and municipal subsidies, tax exemptions, parking access priority, as well as by the citizens' increased environmental awareness. Electric vehicles indisputably bring benefits to their drivers and society in general, indirectly through global warming and greenhouse gas emissions mitigation and directly through financial savings and cleaner microclimate. However, integration of electric vehicle charging spots at home or work, especially fast charging stations and battery swapping stations, without prior analysis can have a negative effect on power system. In order to predict and eliminate power grid issues before they occur, a detailed analyses should be made through a common understanding of both transportation and recharging needs of electric vehicles' users and of the power grid constraints. Power and transportation system interdependency becomes of high value for correct placement and sizing of charging stations and for overall increase of social welfare. This paper analyzes electric vehicles charging needs at the basic level, through both the power system and the transportation system. An urban transmission and power grid in the vicinity of the Croatian capital Zagreb is used as a study case. Driving and electricity consumption curves are compared, locations for charging infrastructure are selected (fast charging spots and battery swapping station) and power grid's available capacity is defined.

I. INTRODUCTION

Electric vehicles (EVs), over the past few years, made a breakthrough into personal and commuting vehicles market worldwide and they are starting to hold a meaningful share. Their rapid market sales increase at the global level can be seen from Figure 1 [1]. The highest market share occurred last year with percentage of more than 0.86% of total vehicles sold. When plug-in EVs market shares are observed in Europe by the country, Norway is "off the chart" since no other country comes even close to 24% share as it can be seen on Figure 2 [1]. Figure 3 displays estimated number of electric vehicles in use in five countries with highest EV share in total vehicles fleet [2]. Norway, even as a small country with population somewhat more than five million, is listed fourth. Such high penetration of EVs is a highly probable

future scenario for all European countries, including Croatia, especially as touristic country.

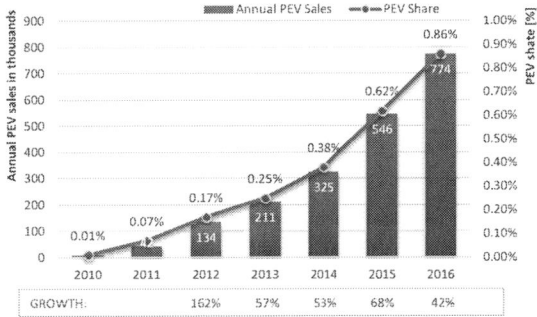

Figure 1 Plug-in EVs global market share from 2010.-2016. [1]

EVs are a powerful tool for greenhouse gas emissions' mitigation and reduction as it is recognized through European regulations and directives such as Directive 2009/33/EC on the promotion of clean and energy-efficient road transport vehicles [3]. The greenhouse gas emissions' cuts to at least 40% (from 1990 levels) are one of the main key targets within 2030 climate and energy framework [4].

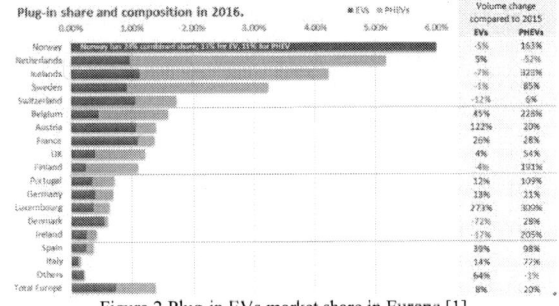

Figure 2 Plug-in EVs market share in Europe [1]

Globally, EVs decrease greenhouse gas emissions only when electricity is generated in carbon-free power plants. Therefore, impact on urban microclimate is far more beneficial EVs' feature since it does not depend on energy mix of power system, i.e. EVs can be observed as a new participant in the war against smog and urban pollution

1465

since they can reallocate emission sources from local transport sector to global electricity generation sector.

Charging of electric vehicles in urban areas should be considered from two different aspects. First one is from the transportation system viewpoint, i.e. observing usage of road vehicles, and the second one is from the power system perspective, which is related to the utilization of power system capacity [5], [6]. The interdependency of the two systems, transportation and power, is important when dealing with fast charging and battery swapping stations placement and sizing problem [7].

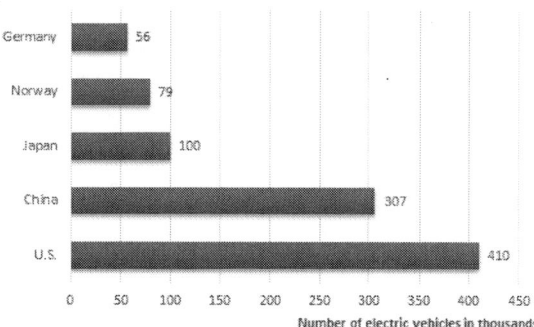

Figure 3 Estimated electric vehicles in use in selected countries as of 2015 (in 1000 units) [2]

Furthermore, one of the main aspects that should be considered is the proper selection of the location. The location needs to be chosen in way that it enables profitable use of the charging infrastructure and alleviates the problem of range anxiety connected with limited range of electric vehicles [8]. There are several other problems associated concerning planning and operation of the fast charging stations and battery swapping stations such as optimal dispatch between two charging types [9] or provision of additional ancillary services [10], [11] but they are outside the scope of this paper. This paper will deal with initial analysis of selection of appropriate charging infrastructure capacity from both transportation and power system perspective and test mentioned interdependency on case of an urban area between Croatian cities of Zagreb and Velika Gorica.

II. TRANSPORTATION AND POWER SYSTEM INTERDEPENDENCY

Transformation from fossil-fueled internal combustion vehicles (ICE) to EVs, in urban environment, should not impact transportation habits of their user. The range anxiety, fear form insufficient EV's battery capacity, is not so profound in urban centers where short range commuting is the main transportation feature as it is in long-rage traveling. EVs, in regards to ICE, offer more flexible way of refueling/recharging. ICE can only be refueled at gas stations by stopping during their trip. An EV, on the other hand, can be recharged on fast charging stations (FCS) or battery swapping stations (BSS) by stopping during their trip analogous to an ICE, but also when parked at home or work. The difference is frequency of the charging, where ICE vehicles in general require much less frequent refueling. In accordance, urban centers

can be divided into different zones regarding transportation and recharging possibilities:

- Roads:
  o Main roads with high transportation density;
  o Secondary roads with medium transportation density – streets interconnecting specific neighborhoods;
  o Local roads with low transportation density – streets within specific neighborhoods;

- Parking lots:
  o Parking lots with short vehicle's stoppage times (up to max two hours), frequent exchange of vehicles and everyday usage – parking lots designated for consumers next to commercial buildings and next to public buildings;
  o Parking lots with short/medium/long vehicle's stoppage times (up to few hours), not so frequent exchange of vehicles and periodic usage – parking lots designated for consumers next to buildings with periodic activities;
  o Parking lots with medium vehicle's stoppage times (up to 8 hours), not so frequent exchange of vehicles and everyday usage – parking lots designated for employees;
  o Parking lots with long vehicle's stoppage times (whole day, night), infrequent exchange of vehicles and everyday usage – parking lots designated for residents (residential buildings, student dormitories).

Refueling of conventional ICE vehicles requires short charging time and therefore gas stations are strategically located next to the main roads with high visibility and high transportation density. It provides both profitability to station owners and satisfaction of station users without fear of fuel shortage. The existing gas stations are also the most suitable locations for rapid DC FCS with ultra-high installed power (above 50 kW, charging up to 20 minutes) and BSS since they are already next to highly frequent road segments. Other benefit of placing FCS and BSS on existing gas stations is the availability of power grid, which entails investment cost reduction since no larger investments are needed.

Power grid is highly branched and additional infrastructure installation is easier compared to other forms of energy, e.g. heating system or gas distribution system. As mentioned before, electrification of road transport brings additional benefits for the vehicle's owners, because vehicles can be charged when they are parked on different locations. Depending on parking lot characteristics electric vehicles charging stations of different characteristics can be profitable and beneficial. Therefore, three-phase AC charging stations with high installed power (more than 20 kW, charging up to two hours) can be installed on parking lots with high frequency, fast one phase AC charging stations of medium installed power (around 10 kW, charging up to 8 hours) can be installed on parking lots with medium frequency, while slow charging stations of low installed power can be

1466

installed on parking lots with low frequency. In order to be competitive to refueling of the ICE vehicles, battery swapping and fast charging infrastructure is required. The two technologies complement each other since they deal with the different business cases and can together offer more flexibility to the drivers.

Beside the transportation aspect, electric vehicles charging station integration should be observed from the viewpoint of the power grid. The fast charging stations infrastructure require high power needs at the same time when the EVs are recharging, while battery swapping stations can recharge its battery stocks in periods when the electricity price is lower or when power system constraints are not bounding. Distribution network in Croatia is supplied through substations 110/x kV or 35/x kV to 10 or 20 kV voltage level. Most of the distribution substations under peak demand have loading of less than 50% of their rated power. The same thing applies to the most of 10 and 20 kV lines and cables supplied by distribution substations with their loading also less than 50 % of their rated power (most of them even less than 25%). In other words, for Croatian case, distribution grid usually has sufficient redundancy for integration of new demand like electric vehicles charging stations. If congestions on some locations occur during peak periods caused by FCS there is a possibility to use battery storage in combination with FCS to improve distribution network conditions [12]

III. CASE STUDY

A. Considered Transportation System

As a test case for this paper road next to north entrance of city Velika Gorica has been chosen. Reason for such selection is highly urbanized area next to city of Zagreb with high transportation density. It can be found in [13] that traffic counting place number 2014 named "Velika Mlaka" is the third transportation density counting place by average density in whole of Croatia, and first in the vicinity of the Croatia capital Zagreb. Since counting place (CP) number 2035 named "Velika Gorica Northern Bypass" is geographically close to CP 2014 it is also taken into consideration. Locations of observed CPs are shown on Figure 4. It can be seen that differences between summer and annual peak are not significant.

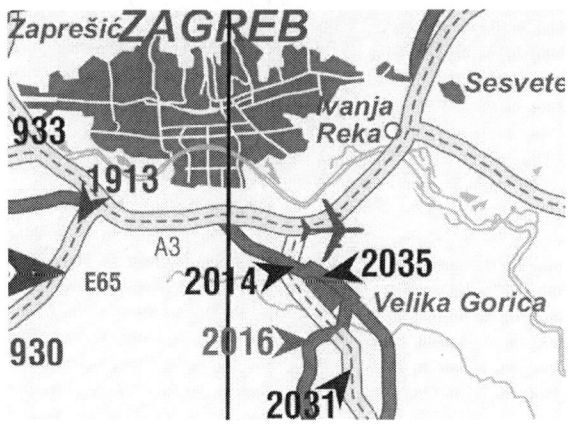

Figure 4 Locations of observed counting places [13]

General information about average traffic on analyzed CSs is presented in Table 1. Acronyms AADT and ASDT are referring to Average Annual Daily Traffic and Average Summer Daily Traffic, respectively.

Table 1 Counting place average daily traffic (AADT - Average Annual Daily Traffic and ASDT - Average Summer Daily Traffic)

CP number	CP name	AADT	ASDT
2014	Velika Mlaka	37.260	35.160
2035	Velika Gorica	7.958	7.119

Figure 5 and Figure 6 display average daily vehicle driving profiles at observed counting places. It is clear that daily driving profiles are matching typical daily behavior. At night hours, there are no activities and very few vehicles are on road. On both CPs two peaks exist, the morning peak – around 08:00 AM, and the afternoon peak – around 17:00 PM. It can be considered as "go-to-work" peak, and "back-to-home" peak.

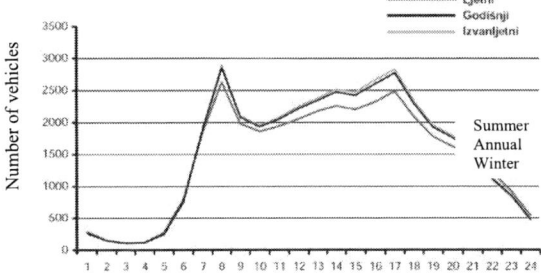

Figure 5 Charging place number 2014 "Velika Mlaka" - daily profile average hourly traffic [13]

The daily period between peak hours is full of activities and the driving profile curve is very high. After the afternoon peak, driving activities are decreasing together with other daily activities. The minimum of driving profile curve can be found around 3:00 AM.

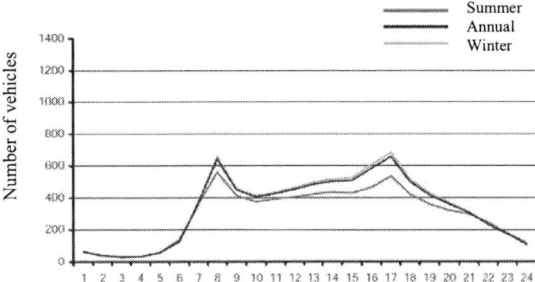

Figure 6 Charging place number 2035 "Velika Gorica Northern Bypass" - daily profile average hourly traffic [13]

Figure 7 and Figure 8 are taken from Google maps application where positions of the existing locations of gas stations are pointed out. As the existing gas stations are situated on the main road with two separate driving directions two gas station are commonly located on the same location on the opposite sides of the road. Traffic conditions at peak hours are colored on the same figures, morning and evening peaks, respectively. Main roads are not suffering from heavy congestion unless unexpected maintenance, car accidents or other unexpected events occur.

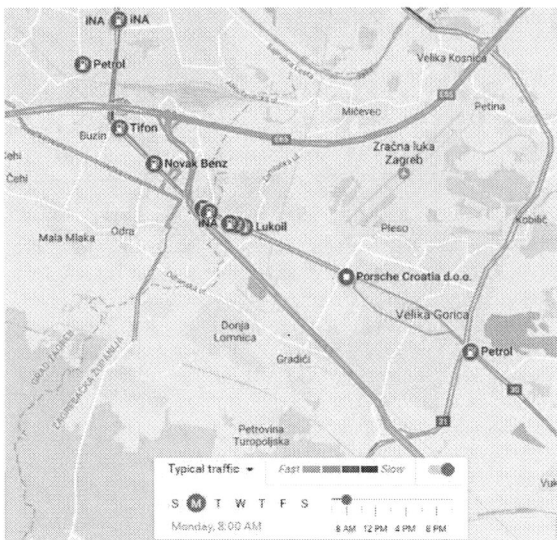

Figure 7 Traffic congestion and gas stations on observed area – morning peak 08:00 AM [14]

For the selection of appropriate charging infrastructure and taking into consideration future increased shares of electric vehicles, it can be assumed that 1% to 2% of all vehicles on the road will be in need for recharging. This means approximately 10 vehicles at each hour would be stopping to fast charge their vehicles or swap the batteries depending on the user preferences of the charging mode philosophy.

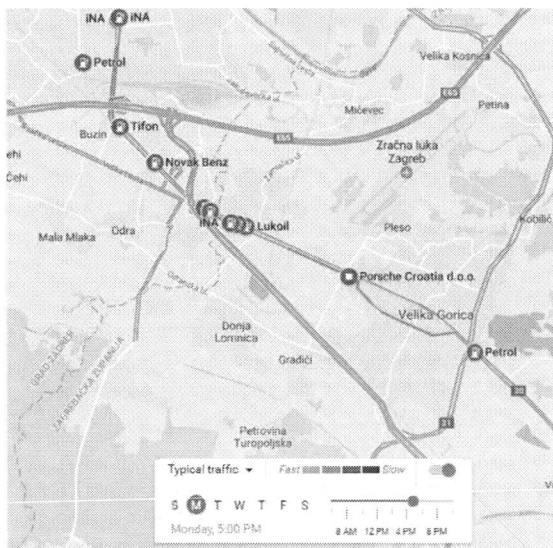

Figure 8 Traffic congestion and gas stations on observed area – afternoon peak 17:00 AM [14]

B. Considered Power System

As it was mentioned before, power system infrastructure in the considered part of the distribution system and used in the test case contains one main 110/20 kV feeder transformer substation TS 110/20 kV Botinec

and associate set of smaller distribution substations 20/0.4 kV as it can be seen on Figure 9. Only the main supply route is depicted, while all the area has an alternative route of supply (less efficient one) from the direction of TS 110/20 kV Velika Gorica. For this analysis, only the main supply route is relevant. The area of the interest next to the main traffic route is supplied through two 20 kV feeders marked with different colors. The peak loading of all distribution elements does not exceed 50%. There is one 2 km long weak overhead line in the distribution network that would require reconstruction if the demand requirement increase. If a congestion occurs, it can be easily located with this kind of analysis. The transformer capacity for the transformation form 20 kV to 0.4 kV does not represent great share of the investment and can be easily, if needed, replaced on the location of possible fast charging and battery swapping stations.

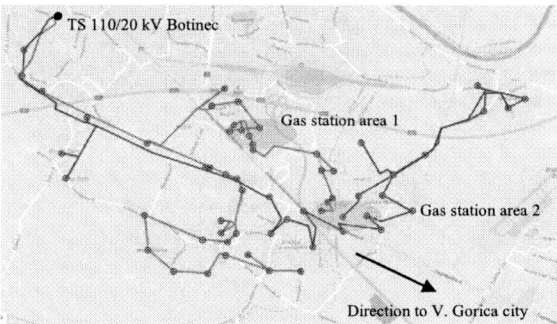

Figure 9 Distribution power grid supplied from feeder substation TS 110/20 kV Botinec

Daily loading curve of the relevant transformer is shown on Figure 10. Total installed capacity is 2x40 MVA. Maximum loading is 28.86 MW, or approximately 35 MVA (~89%). At every moment n-1 reliability requirements must be held, which means that all demand can be supplied through either of the existing 40 MVA power transformers. Considering that fact, there is an additional capacity of approximately 10 MW available (green circle) and the reliability margin is preserved. When comparing daily curves from transportation system to the daily curves in the power system it can be seen that the shape coincides but the peaks (morning and afternoon) for the power curve lag for about two hours behind the transportation curve peaks circled in red (8 AM and 5 PM).

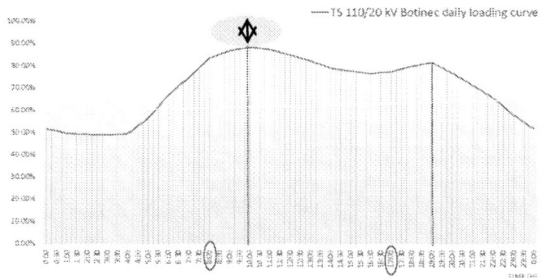

Figure 10 Daily loading profile for TS 110/20 kV Botinec

Power flow for conditions for maximum historic demand with additional 2.25 MW as the highest amount of possible demand increase (where no network investment is required) are visualized on Figure 11. In other words, additional capacity of 2.25 MW is the borderline capacity that can be integrated in the area of potential charging station locations marked with blue circle. The red circle identifies the weak spot in the network.

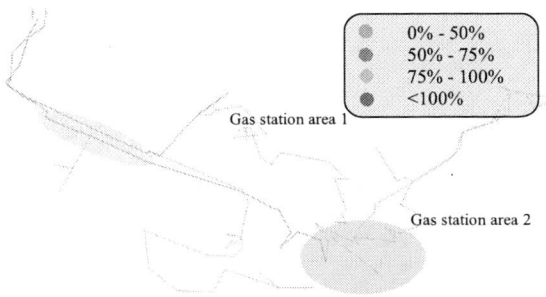

●	0% - 50%
●	50% - 75%
●	75% - 100%
●	<100%

Figure 11 Power flow analysis shows the loading of elements and overloaded segment (overloaded 35 mm^2 cross-section AlFe over-head line)

The maximum number of charging spots that can integrated into the power grid without additional network investments is approximately 18 charging spots and a battery swapping station with projected peak demand of 150 kW. The analysis conducted is conservative and is done under the assumption of coinciding loads.

C. Comparison

In the planning phase both the average and maximum expected loading/traffic should be considered. As it can be seen from figures above (Figure 6, Figure 10) the variability of the daily curve is not so distinguished. For example, the maximum hourly traffic is around 650 vehicles while on average during the day the number is around 500. The increase in share of electric vehicles on the road is projected to be as high as 50%. As mentioned before, the assumption of max 2% of all on road vehicles will be charged each hour, therefore the number of required charging spots is 6. This accounts for the fact that charging on the supercharger still takes approximately 45-60 minutes (5-10 times longer that gas refueling) and the process of battery swapping takes up to 10 minutes (competitive to gas refueling). Additionally, what should be considered is that the road Zagreb-Velika Gorica has two directions and preferably if the charging needs increase dramatically both directions should be covered to avoid the few kilometers detouring that is else required. Two peak hours that are presented occur in different directions. With proposed 6 fast charging spots and with battery swapping infrastructure enough redundancy in charging capacity would be achieved.

Table 2 Required number of charging spots on the locations of gas stations for the "V. Gorica Northern Bypass" road

Traffic hourly peak	Average hourly traffic	Number of charging spots + BSS	Power system available capacity
650 vehicles with increase potential	500 vehicles	6x120 kW + 150 kW	2250 kW

The assumed peak power of a charging spot is 120 kW (Tesla superchargers used as a reference [15]) which means that total additional capacity to be integrated into the power grid is approximately 720+150 kW. The typical transformer capacity of 1000 kVA is appropriate to be selected and added. As the analysis shown above, the power grid with minimum investment can withhold approximately 18 charging spots and a battery swapping station under the assumption of full range simultaneous charging. It is expected that the needs and power requirements would be lower.

The decision on number of the charging stations is dependent on additional factors that mostly deal with the general selection of the wider area of influence to locate the charging station and decide on the number of charging spots. The Tesla experience, as the biggest pool of fast charging stations, shows that on average 6 to 7 charging spots (810 stations with 5195 charging spots) are being installed per location. The ones that are built or under construction in Croatia consist of 4 and 6 charging spots.

Additionally, the test case for the Zagreb-Velika Gorica road concurs well with the requirements of the large populations that frequents this road daily and do not have available plug-in slow charger due to configuration of large residential blocks they live in. Furthermore, the gas station infrastructure is abundant which makes the integration process of superchargers a lot easier.

IV. CONCLUSION

The paper presents the basic analysis of both transportation system and power system with the aim of integrating fast charging stations and battery swapping stations for electric vehicles. The case study on a road connecting Zagreb and Velika Gorica shows the interdependency between two systems. The requirements set by the transportation sector need to take into consideration limitations stemming from the power system infrastructure capabilities and configuration.

To enable further integration of electric vehicles urban designers, transportation and electric power engineers should cooperate. The long term interdependent transportation and electric power system analyses is prerequisite for effective and efficient transformation to electrically driven transportation. The future work will be focused on detecting vehicle traffic distribution and density using GSM/UMTS networks [16] in the cases when traffic counters are not available on some urban locations, and exploring the characteristics (throughputs, delays) of public mobile GSM/UMTS networks for advanced control of electric power network in order to avoid power disruptions, avoid congestions in the network, reduce losses, and improve overall stability of the power network.

ACKNOWLEDGMENT

This work has been supported by the Croatia-Serbia Cooperation in Science and Technology project (2016.-2017.) under the name "Influence of an Electrical Vehicles Charging Station on Urban Electric Power Distribution Network - Smart Cities Solution" and by the Croatian Science Foundation under the project Electric

Vehicles Battery Swapping Station (EVBASS) - IP-2014-09-3517.

REFERENCES

[1] "EV-Volumes - The Electric Vehicle World Sales Database." [Online]. Available: http://www.ev-volumes.com/country/total-world-plug-in-vehicle-volumes/. [Accessed: 08-Mar-2017].

[2] "Estimated electric vehicles in use in selected countries 2015." [Online]. Available: https://www.statista.com/statistics/244292/number-of-electric-vehicles-by-country/. [Accessed: 08-Mar-2017].

[3] EUROPEAN PARLIAMENT, *DIRECTIVE 2009/33/EC OF THE EUROPEAN PARLIAMENT AND OF THE COUNCIL of 23 April 2009 on the promotion of clean and energy-efficient road transport vehicles.* 2009.

[4] "2030 climate & energy framework | Klimatska politika." [Online]. Available: https://ec.europa.eu/clima/policies/strategies/2030_hr. [Accessed: 08-Mar-2017].

[5] T. Martinsen *et al.*, "Improved grid operation through power smoothing control strategies utilizing dedicated energy storage at an electric vehicle charging station," in *CIRED Workshop 2016*, 2016, p. 154 (4 .)-154 (4 .).

[6] I. Pavić, T. Capuder, N. Holjevac, and I. Kuzle, "Role and impact of coordinated EV charging on flexibility in low carbon power systems," in *2014 IEEE International Electric Vehicle Conference (IEVC)*, 2014, pp. 1–8.

[7] B. Liao *et al.*, "Load modeling for electric taxi battery charging and swapping stations: Comparison studies," in *2016 IEEE 2nd Annual Southern Power Electronics Conference (SPEC)*, 2016, pp. 1–6.

[8] M. R. Sarker, H. Pandžić, and M. A. Ortega-Vazquez, "Optimal operation and services scheduling for an electric vehicle battery swapping station," *IEEE Trans. Power Syst.*, vol. 30, no. 2, pp. 901–910, 2015.

[9] Xian Zhang and G. Wang, "Optimal dispatch of electric vehicle batteries between battery swapping stations and charging stations," in *2016 IEEE Power and Energy Society General Meeting (PESGM)*, 2016, pp. 1–5.

[10] P. Xie, Y. Li, L. Zhu, D. Shi, and X. Duan, "Supplementary automatic generation control using controllable energy storage in electric vehicle battery swapping stations," *IET Gener. Transm. Distrib.*, vol. 10, no. 4, pp. 1107–1116, Mar. 2016.

[11] I. Pavić, T. Capuder, and I. Kuzle, "Value of flexible electric vehicles in providing spinning reserve services," *Appl. Energy*, vol. 157, pp. 60–74, Nov. 2015.

[12] N. D. Zidar, Matija, Georgilakis, Pavlos S., Hatziargyriou, D. Škrlec, and T. Capuder, "Review of energy storage allocation in power distribution networks: applications, methods and future research," *IET Gener. Transm. Distrib.*, vol. 10, no. 3, pp. 645–652, 2016.

[13] M. Božić, D. Kopić, and F. Mihoci, "Brojanje prometa na cestama Republike Hrvatske godine 2013," p. 468, 2014.

[14] "benzinske crpke – Google karte." [Online]. Available: https://www.google.hr/maps/search/benzinske+crpke/@45.7227273,15.9931155,12.5z/data=!5m1!1e1?hl=hr. [Accessed: 08-Mar-2017].

[15] "Supercharger | Tesla." [Online]. Available: https://www.tesla.com/supercharger?redirect=no. [Accessed: 15-Mar-2017].

[16] M. Borenovic, A. Neskovic, and N. Neskovic, "Vehicle positioning using GSM and cascade-connected ANN structures," *IEEE Trans. Intell. Transp. Syst.*, vol. 14, no. 1, pp. 34–46, 2013.

MIPRO 2017, May 22- 26, 2017, Opatija, Croatia

Croatian Qualification Framework – Data Model and Software Implementation in higher education

Željko Kovačević, Mladen Mauher and Miroslav Slamić
Zagreb University of Applied Sciences
Vrbik 8, 10000 Zagreb, Croatia
e-mail: zeljko.kovacevic@tvz.hr

Abstract - Project „Politehnika 2025" was started by Tehničko veleučilište u Zagrebu in cooperation with APIS IT, IN2 GRUPA and SPAN companies. The goal was to analyze future market needs until year 2025 and to upgrade existing and to develop new study programs that would modernize polytechnic studies. Project success depended on collecting and analyzing a large amount of data, and to do that it was required to develop appropriate information system. That included to create a conceptual database model and later to write applications that would allow all project participants to work with data. The goal of this paper is to describe database model and its parts, all applications used in the process of gathering and reviewing data and their features. Information system was one of the key elements of the project since without it the project would not be able to complete in desired period and its results may be prone to errors.

I. INTRODUCTION

The Croatian Qualifications Framework is a reform instrument for regulating the system of qualifications at all levels in the Republic of Croatia through qualifications standards based on learning outcomes and following the needs of the labour market, individuals and the society [1]. Using that idea project "Politehnika 2025" was started at Zagreb University of Applied Sciences to prepare our future students for incoming needs and demands of the market until year 2025 [2]. Project included many participants, from faculty staff to employers trying together to find the best solution for future study programs and educational standards.

One of the first goals was to collect data from all participants of the project in a single centralized repository (database) using appropriate application. To create a database that can contain all required data first it was necessary to identify all entities used in the project and determine their relationships. That was a basis to create a conceptual and later a physical database model.

Primary goal of the application was to enable data input for all project participants. Also, application provides many other features like various types of data analysis, reporting, class syllabus generator etc. Application is connected to a previously created central database and allows multiple users to input, review and analyze data at the same time.

II. DATABASE MODEL

When considering all required entities to develop conceptual database model first we analyzed The European e-Competence Framework (e-CF). It provides a reference of 40 competences as applied at the Information and Communication Technology (ICT) workplace, using a common language for competences, skills, knowledge, and proficiency levels that can be understood across Europe [3].

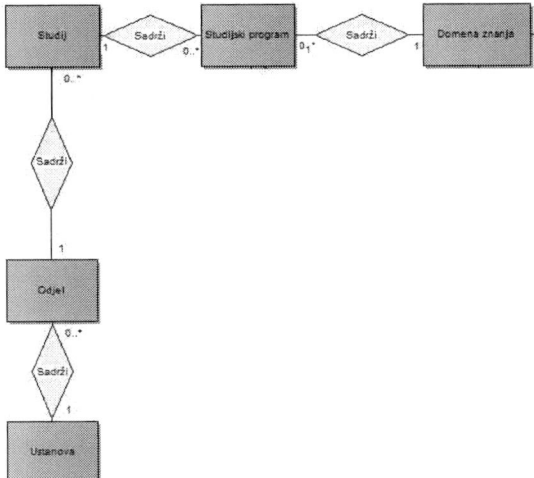

Figure 1. Basic part of the conceptual model

Entire conceptual model consists of 50 entities and basic part of the conceptual model (figure 1) displays key model entities and their relationships. Model supports

1471

having data from multiple higher-education institutions each having multiple study programs. Each of those institutions can input their own data which would also be useful in analysis when comparing same or similar study programs on different higher-education institutions.

Figure 2. Knowledge domain, branch, field and area

In the continuation of the basic part of the model knowledge domains are related to branch, field and area of expertise (figure 2). For example, one area of expertise can have multiple fields, while one field can have multiple branches. Each branch can be a part of only one knowledge domain.

During the project "Politehnika 2025" Tehničko veleučilište u Zagrebu as a higher-education institution enumerated over 330 knowledge domains for 23 study programs. All these knowledge domains were part of 116 branches and 57 fields.

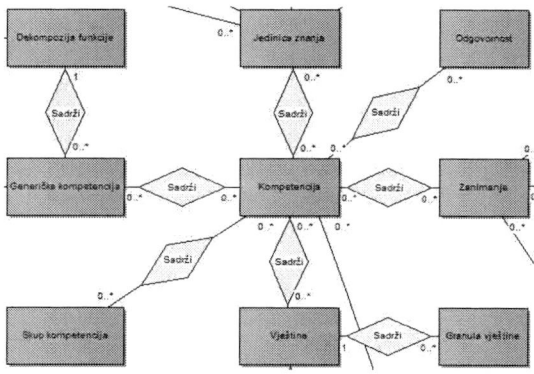

Figure 3. Competence and its relationships

Competence is in the middle of the entire model (figure 3) which shows its importance. Competence is related to generic competences, like ones from e-CF Framework and can be mapped to multiple knowledge units, responsibilities, professions and skills. Also, each competence can be a part of one or more set of competences. This part of the model is also used when creating qualification standards where each qualification standard is related to competences through set of learning outcomes.

Project participants defined 1405 competencies connected with 29 professions, 1583 knowledge units, 826 skills and 80 set of competences. This data made it possible to analyze market needs until year 2025 and what higher-education institutions currently offer and need to change to satisfy those needs.

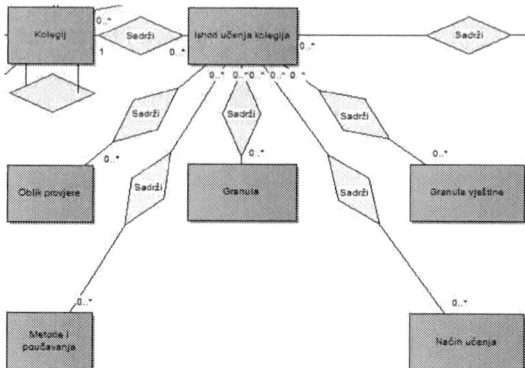

Figure 4. Classes and their learning outcomes

Sub model in figure 4 contains all entities required to describe any class and its learning outcomes. As a result we are able to create a class syllabus containing learning outcomes and its granules, skill granules, methods of learning etc. Class syllabuses for entire study programs are generated based on this part of the model.

For total of 23 study programs there are 552 classes containing 2092 learning outcomes. All learning outcomes are matched with appropriate granules (4161) and skill granules (2180). Learning outcome is in the center of this part of the model and class syllabuses are generated based on their description.

After creating conceptual model it was necessary to create physical database model that would replace entities with tables and describe them with appropriate attributes.

Physical database model is created using Microsoft SQL Server 2012 and Microsoft SQL Server Management Studio (SSMS). Microsoft SQL Server 2012 is Microsoft's first cloud-ready information platform. It gives organizations effective tools to protect, unlock, and scale the power of their data, and it works across a variety of devices and data sources, from desktops, phones, and tablets, to datacenters and both private and public clouds [4].

Resulting physical model consists of 87 database tables and 603 attributes. Physical database also contained 40 stored procedures which provided most of the features in HKO desktop and web application. Physical database also contained user groups and permissions, database and application roles, custom schemas, triggers and aggregate functions.

III. "HKO" DESKTOP APPLICATION

All participants of the project "Politehnika 2025" use desktop HKO application to input, review and analyze the data. Due to complexity of the conceptual database model main user application interface is the picture of the model itself.

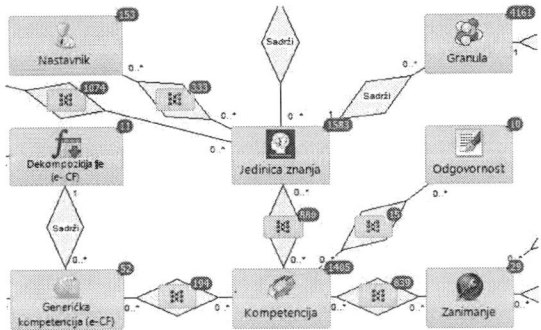

Figure 5. Part of the main application interface

Each entity is represented with a button where in the upper-right corner is the current number of existing records (figure 5). Upon clicking a button a dialog appears enabling a user to review, add, edit or delete records.

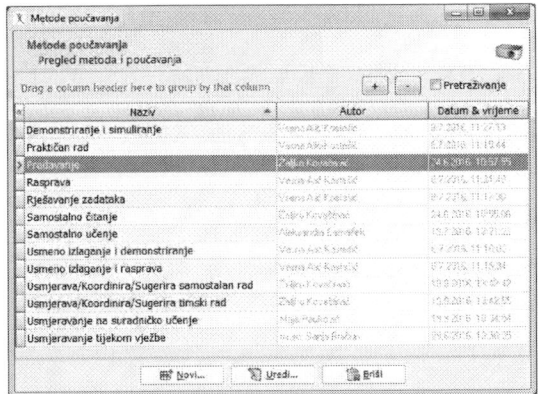

Figure 6. Data dialog

Since there are many users working with data at the same time application forces some usage rules. For example, user can only delete or edit its own records or records of other users that allowed him to do so. Also, adding or changing any data is automatically recorded by application by writing the author's full name and time of the change. That information is later used for user's activity analysis and statistics. Currently, application's statistics show over 53.000 existing records from over 50 project participants.

Application contains main menu that enables users to see a variety of different data analysis. For example, it is possible to see all skills for some profession, professions per institutions, study programs for some job type etc.

TABLE I. Data analytics example

Potreba za radnim mjestima po instituciji
Institucija : APIS IT d.o.o.
BI SPECIJALIST
DB ADMINISTRATOR
IT ARHITEKT
MULTIMEDIJALNI DIZAJNER
Poslovni analitičar
RAZVOJNI INŽENJER
SISTEM INŽENJER - MREŽA
SISTEM INŽENJER - SERVERI
SOFTVER ARHITEKT
TESTNI INŽENJER
Video producent
Voditelj projekta
Voditelj sustava informacijske sigurnosti
Institucija : IN2 d.o.o.
Arhitekt sustava
Funkcionalni arhitekt
Konzultant
Sistem inženjer za Microsoft tehnologiju
Sistem inženjer za mrežne tehnologije
Sistem inženjer za Oracle tehnologiju
Softverski inženjer
Suradnik na poslovima podrške/implementacije
Suradnik za aplikativnu podršku korisnicima i testiranje
Suradnik za podršku korisnicima
TESTNI INŽENJER
Voditelj podrške/implementacije
Voditelj projekta
Voditelj razvoja
Voditelj testiranja
Institucija : SPAN d.o.o.
Poslovni analitičar
RAZVOJNI INŽENJER
SISTEM ARHITEKT
SISTEM INŽENJER
SOFTVER ARHITEKT
TESTNI INŽENJER
UX/UI DIZAJNER

One of the main usage of the application is to create syllabuses for study programs. Application can create a single syllabus for a specific class or a complete set of syllabuses for entire study program. For example,

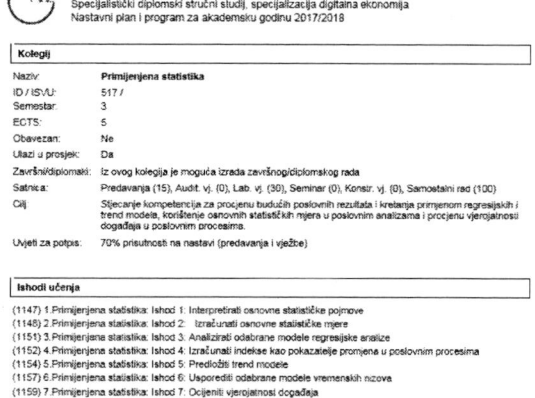

Figure 7. A part of class syllabus

Syllabus is automatically generated document that contains several sections describing the class and its learning outcomes (figure 7). Each learning outcome is described by its granules, skill granules, methods of learning and type of examination.

Application is written using C++ programming language and C++ Builder XE6 integrated development environment (IDE) from Embarcadero.

C++ programming language is more flexible than other languages because you can use it to create a wide range of applications —from fun and exciting games, to high-performance scientific software, to device drivers, embedded programs, and Windows client applications [5].

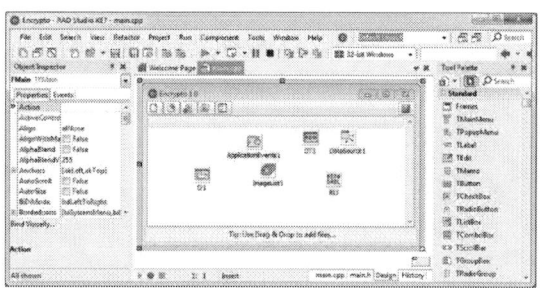

Figure 8. C++ Builder – VCL Forms application

C++ Builder is in development from year 1997. At the time Borland was developing Delphi and C++ Builder as RAD (Rapid Application Development) tools which supported Windows 16 bit platform. Today, C++ Builder is C++ application development toolset for native Windows, Mac, and Mobile development, with broad Cloud and IoT support. [6].

IV. "HKO" WEB APPLICATION

Existing database also enabled the creation of HKO web page titled "Studijski programi" (eng. Study programs)

[7]. Visitors can browse through study programs created in desktop HKO application, listing its classes, syllabuses, skills, competences, professions etc.

Web page targets three types of visitors: new students, students who graduated looking for work and employers seeking employees.

Figure 9. Browsing through study program classes

New students could be informed about study programs which support their desired professions (figure 9). They could see the classes and their syllabuses, job and job types, competencies, skills etc. Also, students who graduated and employers seeking employees could connect each other automatically. Each student can store its profile and would automatically see employers job offers while employers could search employees by searching through student profiles.

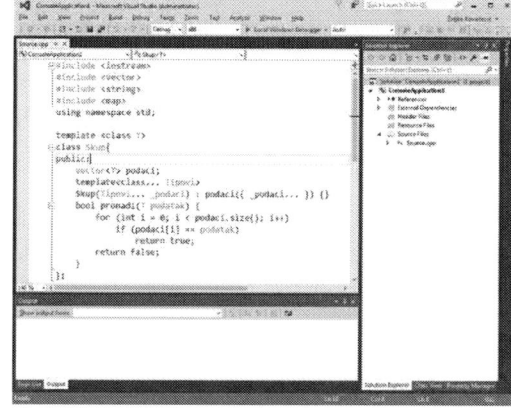

Figure 10. Visual Studio 2015 Community Edition

Web page is created using Microsoft Visual Studio 2015 (figure 10). Microsoft Visual Studio is an integrated development environment (IDE) from Microsoft. It is used to develop applications for Windows platform, web sites, web applications and web services. Also, it supports development of mobile applications. Visual Studio uses Windows API, Windows Forms, Windows Presentation Foundation, Windows Store and Microsoft

Silverlight, and can produce both native and managed code.

HKO web page is created using ASP.NET web application framework. It is an open source web framework for building modern web apps and services with .NET. ASP.NET creates websites based on HTML5, CSS, and JavaScript that are simple, fast, and can scale to millions of users [8].

This modern framework provided all required features to build the HKO web page and all its functionalities.

Project also uses another web page "Politehnika 2025" [2]. It is a project's main web page containing all relevant info about the project and its results. Visitors can find information about all project participants, partners, project results, events and press releases.

V. CONCLUSION

During 15 months of project "Politehnika 2025" existing database and corresponding desktop application helped to analyze over 53.000 records and generate thousands of documents and reports. This helped to improve three existing qualification standards for polytechnic specialist, develop 4 new educational programs in the specialist polytechnic fields of prosthetics and orthotics, information security, forensics and mechatronics and 16 new comprehensive qualification standards for polytechnic specialist. Also, 9 study programs in the specialist polytechnic fields of polytechnics have been improved – computer science, information systems, energetics and electronics, graphic design and technology in printing and civil engineering.

One of the results of the Project is a study of demand for interdisciplinary and unidisciplinary competences and professions on polytechnic job market in 2020 and 2025. It is based on anticipated technologic trends, platforms as well as transformations of real and public sector.

Project helped to improve educational skills by development of innovative teaching methods and the application of the principles of the Croatian Qualification Framework. By putting these results into practice, the Project contributes to development of the educational system in the fields of high education, improvement of human resources in education, research and development of polytechnics, creation of propositions for sustainable employability at the end of educational as well as comprehensive social inclusion at the job market. Application of the acquired competences contributes to the quality of everyday life of a modern person and reminds of permanently needed and present „human face of polytechnics". [9]

Although the project is officially complete the existing database and applications will still be used for the future improvement of existing and for the development of new study programs. All data will be available to public in interest of better communication between employers and future employees thus trying to improve employment of our students.

References

[1] Ministry of Science, Education and Sports, 19 2 2017. [Online]. Available: http://www.kvalifikacije.hr/hko-en.

[2] Tehničko veleučilište u Zagrebu, "Politehnika 2025," 19 2 2017. [Online]. Available: http://politehnika2025.tvz.hr/.

[3] European e-Competence Framework, 19 2 2017. [Online]. Available: http://www.ecompetences.eu/.

[4] R. M. a. S. Misner, Introducing Microsoft® SQL Server 2012, Redmond, Washington 98052-6399: Microsoft Press, 2012.

[5] Mircosoft, "Welcome Back to C++ (Modern C++)," 20 2 2017. [Online]. Available: https://msdn.microsoft.com/en-us/library/hh279654.aspx.

[6] Embarcadero, "C++Builder Frequently Asked Questions," 20 2 2017. [Online]. Available: https://www.embarcadero.com/products/cbuilder/faq.

[7] Tehničko veleučilište u Zagrebu, "Studijski programi," 19 2 2017. [Online].

[8] Microsoft, "ASP.NET," 20 2 2017. [Online]. Available: https://www.asp.net/.

[9] Tehničko veleučilište u Zagrebu, "Politehnika 2025," 27 3 2017. [Online]. Available: http://politehnika2025.tvz.hr/rezultati/.

Internet as a Purchasing Information Source in Children's Products Retailing in Croatia

Blaženka Knežević*, Zdravka Pavlić Šipek **, Božidar Jaković*
* University of Zagreb, Faculty of Economics and Business, Zagreb, Croatia
** Postgraduate student at University of Zagreb, Faculty of Economics and Business, Zagreb, Croatia
bknezevic@efzg.hr, bjakovic@efzg.hr

Abstract - The intensive development of technology, the global availability of the Internet, the overall network coverage and increasing of computer literacy within the population, are growth factors that contribute to popularization and importance of electronic commerce. Moreover, electronic commerce is described as an effective way of communication with consumers in terms of money and time expenditures, and, therefore, in recent decades its role is growing rapidly both in general and specialized retailing. There are numerous research studies that describe and analyze the growth and general potentials of e-commerce. However, the level of development and potential in the field of special commodity groups are not yet sufficiently explained nor practically explored. The aim of this paper is to fill this gap through explanation of Internet as a purchasing source in a particular field of children's products retailing.

I. INTRODUCTION

Electronic commerce as a term denotes using the Internet or other computer networks to conduct purchase transactions, or facilitating the process of buying or selling products or services via computer networks. In retailing, the process of purchasing consists of several phases: (1) seeking information about products in stores or via Internet, (2) buying products that can be done online or in traditional retail outlets, then (3) payment for products or services by cash or electronic money transfer and (4) collecting of product within the store or delivery of the purchased product to the consumer in case of online retailing.

In addition, the fact that the significant proportion of powerful consumers today are the part of so called "information society" generation that, from its early primary school days, intensively use information and telecommunications technology and have no fear nor puts it into a question in any segment of its application, contributes to the belief that electronic commerce has great potential for further rapid development in near future. Moreover, taking into account that this generation is the generation of today's young and future parents, we can also conclude that it is realistic to expect an increase in the application of electronic commerce in specialized retail of children's products as well.

In recent literature it is recognized that e-commerce and the use of the Internet as a source of purchasing information is growing rapidly in all product groups,

without exception, however, the level of development and potential in the field of special commodity groups are not yet sufficiently explained nor practically explored.

Therefore, this paper aims to show the potential of the Internet as a source of purchasing information in the field of one specific commodity groups, such as baby and children's equipment. Also, the paper aims to emphasize the obstacles of Internet usage in this particular product group.

The primary research was conducted on a sample of 217 buyers of children's equipment between 20 and 44 years of age. The paper analyzes and discusses research results in terms of: the length of Internet usage, frequency of Internet usage for the purpose of finding information on the products and for the purpose of children's products purchasing together with the tendency to use the Internet for these purposes in the future. For purchases made via the Internet information on the type of products purchased and features web stores used to purchase products are described. And finally, there is and critical overview on attitudes of the respondents on benefits and obstacles of seeking information and purchases via Internet for this particular kind of products.

The paper starts with an overview of contemporary streams in e-commerce research and then, based on the results of the primary research; it switches to a particular field of e-commerce.

II. AN INSIGHT INTO E-COMMERCE RESEARCH

E-commerce as one of the major fragments of e-business concept is very well researched in many scientific papers recently. E-commerce is defined as buying and selling of product, services or information via the Internet and other digital networks. When discussing forms of e-commerce, B2C e-commerce is dominant form measured by number of transactions done in some period and B2B is dominant form measured by generated revenues. Both forms of e-commerce are growing rapidly even in time in deep economic crisis. However in this paper we will focus on B2C e-commerce or electronic retailing or e-tailing and its narrow form which uses Internet as a basis for transactions with consumers.

The Internet economy presents many benefits. It offers the ability to create an environment where buyers and sellers meet to exchange goods and services without experiencing any geographical boundaries. [1].

The key to achievement in the Internet business lies in how you carry out the combination between virtual and physical channels [2]. Furthermore, e-commerce offers functionality and new ways of doing business that no company can afford to ignore [3]. Confidence that electronic markets have the potential to be more efficient in developing new information-based goods and services, finding global customers and trading partners to conduct business is the basis for moving to an e-commerce. Therefore, we can claim that e-commerce and Internet presence in various industries is a contemporary imperative.

An important part of everyday life for consumers during the twenty-first century is e-commerce. Furthermore, the variety of services in e-commerce has increased in recent years and consumers have adopted those services as part of their everyday lives. An important factor in attracting potential consumers, encouraging first-time purchases and retaining repeat purchases is the quality of e-commerce Websites. The quality of e-commerce Websites is an important piece for consumers in selecting the most favored Website that ultimately results in more revenue for the service suppliers [4].

Yen [5] explored how perceived risk affects customer loyalty in e-commerce and how switching costs mediate in the relationship between perceived risk and customer loyalty and he found that the relationship between perceived risk and customer loyalty is not directly significant in e-commerce. On the other hand, e-service quality and logistics service quality are strongly linked to customer satisfaction [6]. Moreover, Yen [7] examines quality satisfaction between transactional and relational customers in e-commerce, and it also explores the moderating effect of perceived control and perceived enjoyment on quality satisfaction and his results show that system quality satisfaction is more significant for transactional customers, but information quality and service quality satisfactions are more important for relational customers.

Trust and information sharing are very important for doing e-commerce business, in order to adopt e-commerce, managers should spend some energy on building up a trust based corporate culture and a trust based transaction relationship, both within the firm and when cooperating with external partners [8].

Purchase intention and website satisfaction positively influence conversion rates in B2C e-commerce and website satisfaction positively influences purchase intention. [9]. In addition, nowadays, ratings and reviews of products in e-commerce websites usually exert strong influences over the purchase decision of other customers. [10].

E-commerce is growing fast, and the use of the Internet has accelerated value innovation in service dimensions of speed, convenience, personalization and price [11]. The exponential growth of e-commerce has favored the upsurge of codes of conduct pointing at bringing down the barriers of consumer distrust when it comes to online purchasing. Customers are using the Internet for their purchases no longer need to be physically present when their transactions occur, they can buy products from their homes or offices. Also, consumers are able to explore goods and products on their computer monitors and view information about how the products are manufactured. Therefore, e-commerce has created major changes in the retail and service industries [12].

Despite the fact that online sales as a percentage of overall retail sales is still below 15 percent in developed economies, its absolute value and average growth per year are increasing rapidly. If we measure growth in sales over the last five years, clothing and consumer electronics have become the most important industries in the B2C e-commerce scenario [13].

As numerous research studies show, e-commerce is changing the way in which consumers are searching for purchasing information and the way of buying, no matter which specialized assortment is involved. However, research studies focusing on particular product categories or specialized retail sub-industries are very scarce. In order to fill in this gap, in advance we will focus on a particular commodity group or product category in order to explain how Internet is being used in retailing in the field of children's products.

III. RESEARCH METHODOLOGY AND SAMPLE

The survey was conducted in Croatia during 2015 through a questionnaire containing 25 questions. The survey questionnaire included questions on demographic information such as age, gender and education, then questions about Internet purchase, the reasons why respondents are or are not inclined to purchase products via the Internet or the reasons for which prefer to purchase over the Internet in comparison to traditional retail stores, together with their preference to shop online in the future. In addition, respondents are questioned about brands bought through the Internet and about web preferred web pages for Internet purchases.

The survey results were collected through Google Forms and responses were collected in one spreadsheet. The link to the survey was placed on Facebook page, the official groups within some post-graduate courses at University of Zagreb, but also distributed by other Facebook users to obtain as soon as a larger sample of respondents. Statistical analysis and visualization of data was made in MS-Excel and SPSS for Windows, version 17.0.

TABLE I. SOURCE OF INFORMATION ON NEW WEB STORES OR NEW WEB OFFER

The source of information	Number of respondents	Relative frequency
Independent own research	168	77.40%
Social networks	118	54.40%
Mass media	22	10.10%
Friends or family	95	43.80%
Blogs, forums and similar sources	45	20.70%

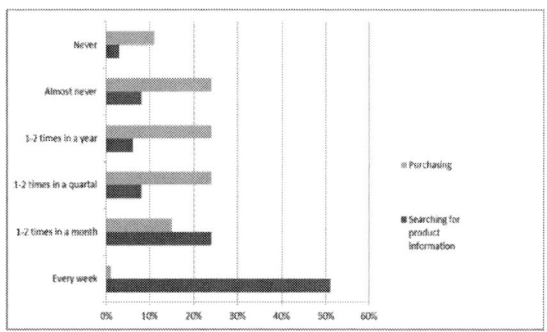

Figure 1.   Online information search and online purchase of childrens' products

The designed questionnaire was based on previously conducted research in the field of general e-commerce [14], [15], [16], [17], [18], [19], but is partially modified according to specific issues related to retail of children's equipment in order to collect data relevant for this product group. The questionnaire consisted of one-choice, multiple choice and open-ended questions, while the Likert scale questions were used to measure the satisfaction of respondents who buy through the Internet.

The questionnaire consisted 25 questions, of which the first five were focused on demographic characteristics of the sample, the following three concerned the experience in using the Internet and frequency of purchase on the Internet, following 6 questions examined the purchasing preferences of consumers on the Internet, types of online-stores and products which are preferred, as well as other issues including the strengths, weaknesses and future potential of Internet retailing of children's products. Ten questions were with one-choice questions, ten with a multiple choice, one question was open-ended, and the rest were in the form of Likert scale.

The research was conducted on a representative sample of 217 customers of children's products. The buying habits of consumers differ from customer to customer depending on their demographic characteristics, particularly by gender. Previous studies have shown that customers in retail of baby and children equipment are mostly women and they are the object in marketing activity of most retail companies in this field [20], [21], [22]. Also, purchasing decisions depend on the purchasing power of customers and their education that affects their level of IT knowledge required for purchases via the Internet. Therefore, the research included questions about these demographic characteristics of customers. The survey was conducted in 2015, and the target group was buyers of children's products between 20 and 44 years of age.

Of the 217 respondents, 189 (87%) were women, and 28 (13%) men. The most represented age group in the sample is 30-34 years, which includes 103 respondents, i.e. 47% of the respondents. Following group by frequency is 35-39 years old and it includes 74

respondents or 34% of the respondents. None of the respondents was younger than 20 years, and only one of the respondents was older than 45 years. Age of majority of respondents in the sample (81%) is between 30 and 39 years. The most common group according to education is the group that had completed high school, which includes 96 respondents or 45%. Followed by the group with secondary education and includes 58 respondents as 27% of the total sample. Majority of the respondents have one child, 111 of them or 51%, of which, 6 of them are expecting a new baby. Second group is with two children, including 75 respondents or 35% of the sample, while those with three children include 17 respondents (8%). In the sample there are no respondents with four or more children.

IV.   DISCUSSION OF RESULTS

The most frequent way in which respondents are informed about new web stores or the new offer is an independent search 77.4%, followed by social networks 54.4% and through friends / family 43.8% of respondents, as shown in Table I.

The data shown in Figure 1 demonstrate popularity of Internet as channel for researching information about children's products before making a purchasing decision.

According to the frequency of online information search, those who seek such information every week, 110 of them or 51%, are the leading group. Second group consists of 53 respondents (24%) who search for information on children's products 1-2 times a month. So we can conclude that ROBO principle is the common way of behavior in the field of children's products as well as in general e-commerce. The acronym ROBO stands for "Research Online and Buy Offline" and it describes the importance of online presence and availability of online published information which is used by consumers before actual purchasing in a traditional retail store. It includes actions as searching for information on products and services, information on company, information on stores location, but also exhaustive action of information comparison regarding product characteristics and product prices. ROBO way of behavior causes the effect of fully educated and completely informed consumer and its consequences on retail processes is discussed frequently in contemporary literature in the field of retail management and marketing.

On the other hand, the frequency of online purchases of children's products is somewhat less frequent. Equal number of respondents purchase via the Internet 1-2 times in 3 months, as well as 1-2 times per year, 53 respondents or 24% (see Figure 1). Only 1% of respondents state that they purchase every week. However, there is a significant proportion of those who purchase rarely or never (total of 35%). This represents a potential for future development of e-commerce in this product category, as the share of these respondents use the Internet to search for information about children's products.

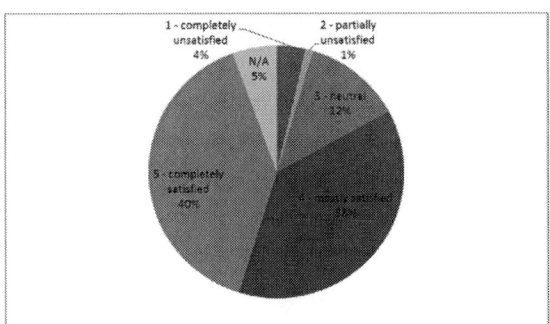

Figure 3. Satisfaction with completed online purchasing transactions in field of children's products

TABLE II. INTENTION TO PURCHASE CHILDREN'S PRODUCT ONLINE

Assesment of future e-commerce usage in field of products for children	Number of respondents	Relative frequency
I will use it more than nowadays	142	65.44%
I will use it less than nowadays	1	0.46%
I will use it similar as nowadays	74	34.10%
Total	**217**	**100,0%**

In Figure 2 product categories within children's products are ranked. We can observe that the most frequent category is "new children's clothes" which is bought by 51.4% of respondents, then follows category "toys" bought by 49.5% of respondents. The third category ic "Gifts for other children" (31.8%) and so on (see Figure 2). Among other data, it is interesting that specific children's cosmetic products were bought by 10% of respondents, while only 1.8% of respondents bought children's food online.

Figure 3 shows the general satisfaction of the purchase of children's products on the Internet. On a scale of 1 – completely unsatisfied and 5 – completely satisfied, the majority of respondents (39.1%) assessed their satisfaction with a score of 5, while 37.3% rated satisfaction with grade 4, while only 3.6% respondents declared as completely unsatisfied. Therefore, we can draw conclusion that the respondents are generally satisfied with the completed purchases of children's products via the Internet.

In addition, according to our findings, there is great potential for growth of this form of retailing for this particular product group. Specifically (see Table II), 34.10% of respondents believe that they will continue to use electronic commerce on the same level as today; while 65.44% think that they will intensify their usage of e-commerce as a purchasing channel for children's products. Thus, specialized retailers have to get more engaged into development of online sales channel.

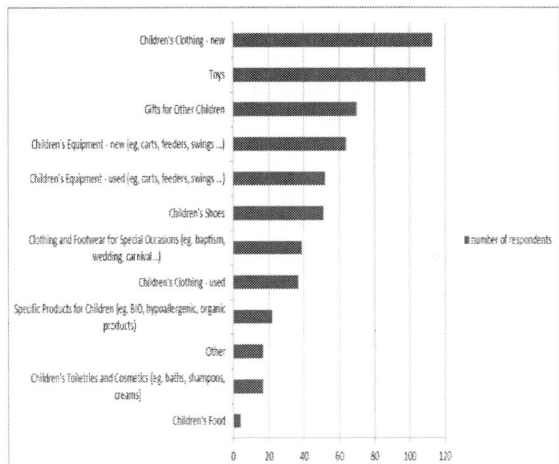

Figure 2. Categories of children's products purhased online

Out of eleven offered options regarding benefits of online purchase, three main benefits of buying children's products through the Internet were recognized by our respondents are: (1) lower prices and / or discounts (63.2%), (2) the possibility to shop 24/7 (55.9%) and (3) solution to the lack of time via home delivery (51.8%). These findings do not correspond to findings delivered in research on general e-commerce [23] where three major benefits were as follows: "I can purchase products which are not available at my city/country", "It is easier to compare prices and product information" and "I can shop abroad without traveling costs". Therefore, we can conclude that for particular retail product category (in our case children's product) perception on benefit differs in comparison to usual and common perception on benefits of e-commerce in general. Thus, for future e-commerce research we could recommend focusing on e-commerce sub-industries (in accordance to online offered product categories).

On the other hand, out of ten offered options, respondents outlined the three major disadvantages of online purchasing of children's products: (1) I am afraid that the delivered product will not be of appropriate quality (55.8%), (2) I like to see and test the product before purchasing (49.3%) and (3) the return process is more complex than in traditional stores (37.8%). This finding is to somewhat in line with previous research conducted for general e-commerce [23] where it was shown that three main obstacles perceived by young consumers in Croatia are: "return of a product is more complicated", "I prefer shopping in traditional store because I can touch product" and "It is difficult to check out validity of offer and online shop". Therefore, in near future retailer should particularly pay more attention to return policies and creating virtual environments in which consumers will feel the same or similar shopping experience as in traditional stores.

V. CONCLUSIONS

Contemporary consumers in the field of children's products are digitally literate young people, open to the Internet as a channel of communication, sales and distribution. In the primary research it was observed that majority of them Internet as a channel for collecting information before purchasing (according so called ROBO principle), but also as a purchasing place to buy children's products.

According to the primary research, three sub-categories that are most prevalent in e-commerce of children's products are: (1) the new children's clothes, (2) toys and (3) gifts for other children.

A high degree of consumer satisfaction with the completed purchasing transactions of children's products is observed. A large number of respondents believe that they will use e-commerce as a source of supply of children's products even more frequently. Therefore, we can conclude that Internet as an additional sales channel has good prospects for greater implementation and application in retail market of children's' products. Therefore, we could recommend specialized retailers to integrate this channel into their overall business strategy more strongly as an imperative goal and to pay more attention on changed information needs and information behaviour of their future consumers.

Perception on e-commerce benefits for particular product groups varies in comparison to perception on e-commerce in general. Therefore, in future researchers should focus on particular retail product categories offered online, rather than general e-commerce attitudes.

However, there are some limitations regarding the primary research that reduces the possibility of generalizing the results. Here we can outline: given limited time of the survey and a relatively small sample of respondents, consisting mostly of respondents in the same age group and similar education level. Moreover, used methodology is based on pseudo-random sample due to the intensive usage of Google forms and social networks for data collection which negatively influences the possibility of generalization of conclusions for whole population of young families with children in Croatia. Therefore, in future research should be broadened and should include more randomized approach of sample selection. Also, the future survey by using the written form of questionnaire could improve input data as well. In addition, the research could be broadened geographically to other EU countries in order to observe similarities and differences between developed and developing economies and in order to make results more robust regardless to the geographic origin of the sample.

LITERATURE

[1] T. Mayayise and O. Osunmakinde, "E-commerce assurance models and trustworthiness issues: an empirical study", Information Management & Computer Security, vol. 22 no. 1 pp.76 – 96, 2014.

[2] I.-L. Wu and S-M Wu,"A strategy-based model for implementing channel integration in e-commerce", Internet Research, vol. 25, no. 2, pp. 239 – 261, 2015.

[3] H. J. Wen, H-G. Chen and H-G. Hwang, (2001),"E-commerce Web site design: strategies and models", Information Management & Computer Security, vol. 9, no.1, pp. 5 – 12, 2001.

[4] G. Sharma and W. Lijuan, "The effects of online service quality of e-commerce Websites on user satisfaction", The Electronic Library, vol. 33, no. 3, pp. 468 – 485, 2015.

[5] Y-S. Yen, (2015),"Managing perceived risk for customer retention in e-commerce", Information & Computer Security, vol. 23, no. 2, pp. 145 – 160, 2015.

[6] Y. Lin, J. Luo, S. C. Shihua and M. K. Rong, (2016),"Exploring the service quality in the ecommerce context: a triadic view", Industrial Management & Data Systems, vol. 116, no. 3, pp. 388 -415

[7] Y-S. Yen, "A comparison of quality satisfaction between transactional and relational customers in e-commerce", The TQM Journal, vol. 26, no.6, pp. 577 – 593, 2014.

[8] P. Li and W. Xie, (2012),"A strategic framework for determining e-commerce adoption", Journal of Technology Management in China, vol. 7, no.1, pp. 22 – 35, 2015.

[9] N. Gudigantala, P. Bicen and M. Eom , "An examination of antecedents of conversion rates of e-commerce retailers", Management Research Review, vol. 39, no. 1 pp. 82 – 114, 2016.

[10] Y. Wang, Z. Wu, Z. Bu, J. Cao and D. Yang, "Discovering shilling groups in a real e-commerce platform", Online Information Review, vol. 40 no. 1, pp. 62 – 78, 2016.

[11] P. M. Torres, J. Veríssimo and M. M. Yasin, "E-commerce strategies and corporate performance: an empirical investigation", Competitiveness Review, vol. 24, no. 5, pp. 463 – 481, 2014.

[12] A. A. Jahanshahi, S. X. Zhang and A. Brem, "E-commerce for SMEs: empirical insights from three countries", Journal of Small Business and Enterprise Development, vol. 20, no. 4, pp. 849-865, 2013.

[13] R. Mangiaracina, G. Marchet, S. Perotti and A. Tumino,"A review of the environmental implications of B2C e-commerce: a logistics perspective", International Journal of Physical Distribution & Logistics Management, vol. 45, no.6, pp. 565 – 591, 2015.

[14] B. Anckar, "Adoption drivers and intents in the mobile electronic marketplace: Survey findings", Journal of Systems and Information Technology, vol. 6, no. 2, pp. 1 – 18, 2002.

[15] E. Patokorpi and K.K. Kimppa, "Dynamics of the key elements of consumer trust building online", Journal of Information, Communication and Ethics in Society, vol. 4, no. 1, pp. 17 – 26, 2006.

[16] J. M-S., Cheng, E.S.-T. Wang, J. Ying-Chao Lin and S. D. Vivek, (2009), "Why do customers utilize the internet as a retailing platform?: A view from consumer perceived value", Asia Pacific Journal of Marketing and Logistics, vol. 21, no. 1, pp. 144 – 160, 2009.

[17] R., Thakur and M. Srivastava, "Customer usage intention of mobile commerce in India: an empirical study", Journal of Indian Business Research, vol. 5, no. 1, pp.52 – 72, 2013.

[18] G. Petkovic, D. Stojkovic, D. and B. Knezevic, "Security and Privacy Issues in Shopping Through Smartphones – Comparison of Shoppers in Croatia and Serbia", in: N. Knego, S. Renko and B. Knežević, (Eds.): TRADE PERSPECTIVES 2016 - Safety, security, privacy and loyalty, University of Zagreb, [available at: http://web.efzg.hr/tp-proc], pp. 126-138, 2016.

[19] M. Stefanska, M. and T. Wanat (2017), "Benefits from using mobile applications by Millennials – a gender and economic status comparative analysis", in J. C. Andrean,. And U. Collesei. (Eds.): Proceedings of 16th International Marketing Trends Conference, Paris-Venice Marketing Trends Asociation, [available at: www.marketing-trends-congress.com/paper], pp. 1-15, 2017.

[20] L. Carlson, S. Grossbart, A. Walsh, (1990), "Mothers' Communication Orientation and Consumer-Socialization Tendencies", Journal of Advertising, vol. 19, no. 3, pp 27-38, 1990.

[21] Cook, D.T. (1995), "The Mother as Consumer: Insights from the Children's Wear Industry, 1917-1929", Sociological Quarterly, vol. 36, no. 3, pp. 505-522, 1995.

[22] L. Martens, D. Southerton, and S. Scott, (2004), "Bringing Children (and Parents) into the Sociology of Consumption", Journal of Consumer Culture, vol. 4, no. 2, pp. 155 – 182, 2004.

[23] B. Knežević, B. Jaković and I. Strugar, "Potentials and Problems of Internet as a Source of Purchasing Information – Experiences and Attitudes of University Students in Croatia", Business, Management and Education, Vol 12. No. 1, pp. 138-158, 2014.

Valuation of Common Stocks
Using the Dividend Valuation Approach and Excel

Zdenko Prohaska, PhD*, Ivan Uroda, PhD**, Anita Radman Peša, PhD***

* Full Professor, University of Rijeka, Faculty of Economics, Rijeka, Croatia
** Assistant Professor, University of Rijeka, Faculty of Economics, Rijeka, Croatia
*** Assistant Professor, University of Zadar, Department of Economics, Zadar, Croatia
zdenko.prohaska@ri.t-com.hr, ivan.uroda@gmail.com, apesa@unizd.hr

Abstract - The valuation of common stocks using the dividend valuation method is useful for stocks in so called no-growth or constant-growth situations.

In the case of constant-growth the stock price should be equal to the discounted present value of all future dividends, but in the case of non-constant growth the case becomes more complicated.

In both cases the use of Excel is a straightforward method for solving such calculations efficiently.

I. INTRODUCTION

Educational software is very important in finance courses at universities for students to better understand equity securities, like stocks, which represent a share in the ownership of a joint stock company.

To evaluate stock prices and dividends with no growth, constant growth and non-constant or variable growth rates in the future, complex calculations are to be done to get acceptable and realistic results.

Since students at universities, especially faculties of economics are usually trained in using spreadsheet programs, the authors decided to use Excel and develop spreadsheets for evaluating stocks in cases of different dividend payouts mentioned above.

To get better compatibility with future versions of Excel and increased portability on other computers and operating systems, so called *plain vanilla* Excel spreadsheets, containing only authors' custom-made formulas and common financial and statistical functions, were programmed like in two other articles at earlier MIPRO Conferences. [6], [7]

An additional reason to apply Excel as the preferred spreadsheet program is that a lot of excellent textbooks exist, so the time period for learning and applying such programs should not be a long one. [1], [4], [10]

II. THE BASICS OF INVESTMENTS IN STOCKS

Stocks are securities that represent a share in the assets of a joint stock company.

In this respect, they differ completely from bonds which represent a claim against the issuer or the debtor.

Beside that, there are four basic differences between stocks and bonds and the markets in which they are traded: [5]

a) stocks are securities without maturity

b) due to the lack of maturity most of the turnover of stocks is happening on the secondary market

c) turnover in the secondary market is mostly in the form *face to face* (over the counter - OTC) and on the securities markets, while bonds are generally traded on the OTC market or on the phone

d) while bonds are characterized by fixed interest rates and value on maturity, in case of stocks no pre-determined rate of return does exist.

In addition to ownership, stocks in a joint stock company provide certain additional rights, including the rights: [3]

- to get dividends

- of priority purchase of new issued stocks

- to vote at the general meeting of the joint stock company

- to be informed and

- to get a share of the liquidation mass if the joint stock company goes bankrupt.

Depending on the type, there are essentially common stocks (ordinary shares) and preferred stocks (preference shares).

While the rights of ordinary or common stockholders are determined by laws of a country and certain internal regulations of joint stock companies, if it comes up to preference stocks best known forms are, ones which give priority to:

a) in case of dividend payment (e.g. in the form of increased or priority dividend payments)

b) in the case of liquidation (e.g. covering its rights before common stockholders)

Given these characteristics, common stocks are or should be the rule and preferred stocks should be an exception.

There are also advantages and disadvantages of financing with common stocks. Financing companies with common stocks has two major advantages. First of all, this form of financing requires minimum restrictions in the operations of a company. The company is contractually not in the obligation to pay dividends, nor at any time obliged to pay the principal.

In addition, financing by issuing additional common stocks increases the future ability of companies to take loans. For a company, which has a higher proportion of equity, it should be much easier to apply for additional loan when needed. Thus, it comes to a reduction of possible insolvency - inability to pay due liabilities - and an increased possibility of profitable investments.

However, financing by common stocks has also some disadvantages. Issuance of new stocks usually affects the reduction in earnings per stock and the reduction in the price of the stock. Therefore, companies issue common stocks especially when market demand is increasing and common stocks and their prices are rising, because only then they can obtain the necessary amount of new capital with a relatively small volume of new stocks.

Issue of new common stocks further causes that existing ordinary stockholders have to share the control over the company with new stockholders. If control is an important right of ordinary stockholders, the gradual loss of control could become so important that a company would not issue new common stocks. In addition, the costs of issuing new common stocks and their required rate of return is higher than in the case of preferred stocks or debt capital, i.e. by issuing bonds.

Normally, to a certain degree financing companies by issuing common stocks has more advantages than disadvantages, but after a certain level the disadvantages rise and further financing of the company in the form of common stocks is no longer meaningful. [8]

III. VALUATION OF STOCKS AND THE DIVIDEND GROWTH MODEL

Concerning the value of stocks it is necessary to distinguish their nominal value, book value and market value. [8]

The nominal value (par value) is printed on the stock and it defines the share of the joint stock company, especially in the case of liquidation of the joint stock company, the owner of the stocks is entitled to share of the liquidation mass only up to the nominal value of the stocks regardless how much he had paid for it.

Most corporations assign a nominal or par value to their common stock. Today it is not very important, but in the past it was very common for companies to determine a value to each stock when the company went public. Today, this value is used only for bookkeeping and it is not in any way correlated with the book value and market value of a stock.

The book value of the stock, in case that the nominal value exists, is calculated by adding the reserve per stock on the nominal value or, if it does not exist, as sometimes in Anglo-Saxon countries, by dividing the total capital and reserves of the company by the number of stocks.

So, the book value could be defined as the accounting value of a share of common stock, i.e. common stock plus paid-in capital plus retained earnings divided by the number of stocks outstanding.

Market value of the stock or the stock price is determined on the secondary capital market (stock exchange), in the first place according to the expected future income of the respective joint-stock company.

In addition to this main factor that has a key influence on the movement of stock prices there are also other factors which are important, like a general upturn in the stock market, inflation, profitability of the company, etc.

Regarding stock valuation it has to be pointed out, generally speaking, that stock does not give ownership to any specific physical asset of a joint stock company, rather it gives a claim on the company's future flow of earnings.

Therefore, one can look at the value of the stock as the present value of future earnings of the company that the stockholders will receive. If the future flow of earnings of the company is known as well as the adequate discount rate for converting those future earnings into today`s value, the stock could be easily evaluated. Or to put it in another way, it could be said, that the value of a stock or the real stock price should be the present value of the future dividends due to the stockholders.

The very first explanations of the stock price or market price of the stocks can be found in the publications of J. B. Williams. Based on the theoretical foundations set up by A. Marshall, E. von Böhm-Bawerk and I. Fisher, J. B. Williams has defined the market value, or as he called it the present value of the stocks, as the sum of all dividends discounted at an appropriate interest rate.

While J. B. Williams claimed that the dividend is the only basis for determining the value of stocks, others primarily B. Graham and D. L. Dodd, argued that instead of the dividend the category of income (profit) can be used, because the change of the market value or stock price will affect the income or profit and therefore also potential future dividends.

Although most authors supported J. B. Williams' approach, i.e. the dividend growth model or the so-called "Growth-stock school" as it was called by B. G. Malkiel, they still thought that income is important, especially in the case of companies that have a high growth rate.

Based on that, the intrinsic value of stocks can be presented, provided that there is no future dividend growth expected, in the form of: [2]

$$P_0 = \sum_{t=1}^{\infty} \frac{D_t}{(1+r)^t} \tag{1}$$

P_0 - stock price at period 0

D_t - expected dividends at the end of period t

r - expected rate of return

According to equation (1) the price of stocks depends on the expected future dividends discounted by an appropriate discount rate or rate of return.

However, if it is expected that dividends in the future will grow at a constant rate g, then the expected dividends can be calculated as:

$$D_t = D_0 \cdot (1+g)^t \tag{2}$$

D_t - expected dividends at the end of period t

D_0 - amount of the dividends at period 0

g - growth rate of dividends

Substituting equation (2) into equation (1) the following equation is obtained:

$$P_0 = \sum_{t=0}^{\infty} D_0 \cdot \frac{(1+g)^t}{(1+r)^t} \tag{3}$$

P_0 - stock price at period 0

D_0 - amount of the dividends at period 0

r - expected rate of return

g - growth rate of dividends

Provided that r is greater than g, equation (3) can be presented as: [9]

$$P_0 = \frac{D_1}{r-g} \tag{4}$$

P_0 - stock price at period 0

D_1 - expected annual dividends at the end of period 1

r - expected rate of return

g - growth rate of dividends

Although the model given in equation (3) and modified according to the equation (4), or the so called Gordon Model, can explain the market value of stocks, it also has certain restrictions. It cannot be applied to stocks of companies that currently do not pay dividends; the result becomes infinite if r is less than g and requires prediction of future dividend growth rates.

These difficulties and the fact that, in practice, the growth rates of dividends do not increase at a constant rate of growth, but that this rate can again decline, alternative models for determining the value of stocks were developed.

In the case that the dividends grow on the basis of two different growth rates such a model valuing stocks can be presented as: [2]

$$P_0 = \sum_{t=1}^{\infty} \frac{D_0 (1+g_x)^t}{(1+r)^t} + \sum_{t=n+1}^{N} \frac{D_N (1+g_y)^{t-N}}{(1+r)^t} \tag{5}$$

where:

P_0 - stock price at period 0

D_0 - amount of the dividends at period 0

g_x - first growth rate

g_y - second growth rate

r - expected rate of return

These equations will be used for the calculations in three subsequent Excel spreadsheets.

IV. EXCEL SPREADSHEETS

For the calculations of the dividend growth models presented above Excel spreadsheets with custom-made formulas developed by the authors and containing only standard financial and statistical functions will be applied. [11]

The basic dividend growth model presented in equation (1) and its modifications explained in equations (4) and (5) will be implemented to take into account the fact that common stock dividends could grow at different rates, therefore three different cases will be considered: [12]

a) zero dividend growth model

b) constant dividend growth model

c) non-constant or variable dividend growth model

Every model will be explained by an example and calculated in Excel spreadsheets.

A. Zero dividend growth model

In this case the common stock dividend remains the same and by applying equation (4) the growth rate of dividends has to be set to 0.00%.

Example: The common stock dividend of company XYZ is 2.50 USD and remains the same in the future. What is the value of this stock if the expected rate of return is 12.00%?

1483

After typing all the necessary data in the "input data" field of the spreadsheet model (Stock-Dcg.xls), the result is calculated and displayed in the "output data" field. (Fig. 1)

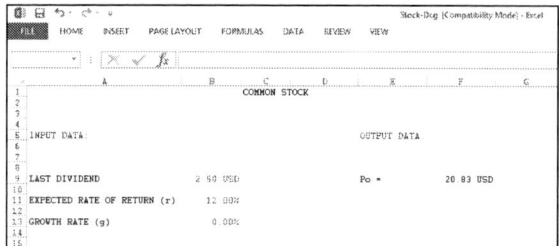

Figure 1. Common stock - zero dividend growth

The result shows that the value of the stock, if the expected rate of return is 12.00% and the dividend is constant 2.50 USD, is 20.83 USD.

B. Constant dividend growth model

In this model, also equation (4), the dividend grows at a constant rate, where the rate of growth is smaller than the expected rate of return.

Example: Calculate the value of common stock of the ABC company if it expects to increase its common stock dividend, being 2.00 USD last year, by 6.00% annually. Assume an expected rate of return of 8.00% is given.

After typing all the necessary data in the "input data" field of the spreadsheet model (Stock-Dcg.xls), the result is calculated and displayed in the "output data" field. (Fig. 2)

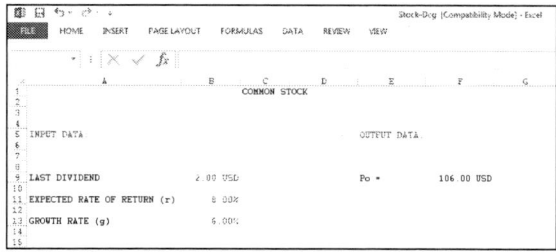

Figure 2. Common stock - constant dividend growth

The result shows that the value of the stock, in case that last dividends of company ABC were 2.00 USD, with a growth rate of 6.00% annually and an expected rate of return of 8.00%, is 106.00 USD.

C. Non-constant or variable dividend growth model

To overcome the disadvantage of the constant growth dividend model, which does not allow any changes in the growth rate, a more elaborated (up to 5 stages) variable dividend model discussed in equation (5) will be presented in an Excel spreadsheet (Stock-Dncg.xls).

Based on the current market price, the projected cash dividends will grow for the first two years and after that, they will drop. An example of a four stage model is presented below.

Example: In a four year period the company USX has paid dividends in USD (1.00 D1, 1.25 D2, 1.50 D3 and 1.75 D4) with an estimated stock price in year 4 of 48.00 USD and estimated rate of return of 17.00%. What is the value of the common stock?

After typing all the necessary data in the "input data" field of the spreadsheet model (Stock-Dncg.xls), the result is calculated and displayed in the "output data" field. (Fig. 3)

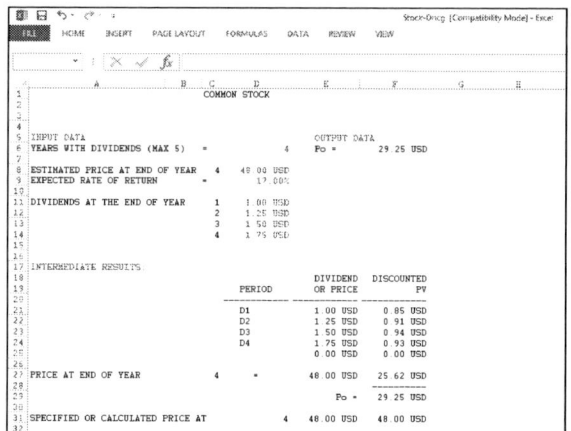

Figure 3. Common stock - non-constant or variable dividend growth

The result shows that the value of the stock, in case that dividends of company USX were first growing faster from 1.00 USD to 1.25 USD, and then slower from 1.50 USD to 1.75 USD with an expected stock price in year 4 of 48.00 USD and an expected rate of return of 17.00%, is 29.25 USD.

V. INSTALLATION OF THE SPREADSHEETS

Since all procedures and calculations are implemented in two Excel spreadsheets (Stock-Dcg.xls) and (Stock-Dncg.xls), it is only necessary to copy these files in the working directory of the PC where Microsoft Excel program can access them. In addition, the installation of the Excel Analysis Toolpak Add-In could help to improve the programming of some specific functions, but is not necessary.

VI. CONCLUSION

In this article the valuation of stocks using the dividend growth model was elaborated for so called no-growth, constant-growth and non-constant or variable growth situations.

For this purpose two spreadsheets were developed using only *plain vanilla* Excel, i.e. by implementing only custom-made formulas that were elaborated by authors and standard financial and statistical functions already built in basic Excel.

The examples and results show, that by using Excel even more sophisticated cases in business finance can be solved quickly and efficiently.

1484

REFERENCES

[1] Benninga, S., Financial Modeling, MIT Press, Cambridge, 2014

[2] Bodie, Z., Kane, A., Marcus, A.J., Essentials of Investments, McGraw Hill, New York, 2013

[3] Fabozzi, F., Investment Management, Prentice Hall, Englewood Cliffs, New Jersey, 1995

[4] Holden, C.W., Excel Modeling in Investments, Pearson, New York, 2014

[5] Mayo, H.B., Introduction to Investments, South-Western, Cengage Learning, New Jersey, 2011

[6] Prohaska, Z., Olgić Draženović, B., Šarić, V., Calculating Duration and Convexity of Bonds Using Excel, in Biljanović, P., Butković, Ž., Skala, K., Golubić, S., Čičin-Šain, M., Sruk, V., Ribarić, S., Hutinski, Ž., Baranović, M., Tijan, E., Mauher, M., Bombek, I., 36th International Convention MIPRO 2013, Mipro Proceedings, Computers in Education, Opatija, 2013

[7] Prohaska, Z., Olgić Draženović, B., Uroda, I., Valuation of Options by Using Excel, in Biljanović, P., Butković, Ž., Skala, K., Mikac, B., Čičin-Šain, M., Sruk, V., Ribarić, S., Groš, S., Vrdoljak, B., Mauher, M., Sokolić, A., 38th International Convention MIPRO 2015, Mipro Proceedings, Computers in Education, Opatija, 2015

[8] Prohaska, Z., Uvod v finančne trge – metode analize in instrumenti, Univerza v Ljubljani, Ekonomska fakulteta, Ljubljana, 1994

[9] Reilly, F.K., Investment Analysis and Portfolio Management, South-Western College Pub, Cincinnati, 2011

[10] Underdahl, B., Excel Expert Solutions, Que Corporation, Indianapolis, 1996

[11] Walkenbach, J., Excel 2013 Formulas, Wiley, New Jersey, 2013

[12] Wilmott, P., Paul Wilmott Introduces Quantitative Finance, Wiley, New York, 2007

The interconnection between investment in software and financial performance – The case of Republic of Croatia

Marija Boban, PhD*, Toni Šušak, M.Econ.**
*University of Split, Faculty of Law
Department for Economic and Financial Sciences, Split, Croatia
marija.boban@pravst.hr

**University of Split, Faculty of Economics
Department of Finance
toni.susak@efst.hr
**University Department of Professional Studies Split
Department of Finance and Accounting
toni.susak@oss.unist.hr

Abstract – Nowadays, information technology is conditio sine qua non of every successful business in information economy surroundings. The implementation of daily tasks is aggravated or impossible without use of information techologies (IT). In this sense, many questions emerge – can companies use information technology as their competitive advantage? Is IT potentially the generator of increasing the financial result of a company? Specifically, the object of consideration in this paper is software, which is indispensable in terms of giving functionality to company's assets – hardware. In terms of finance and accounting, software is considered as intangible fixed asset which will be used for period longer than one year. The era of companies mainly focused on tangible assets has passed, and that implies that the average intangible assets to total assets ratio is rapidly increasing nowadays. At the same time, there is also a growth of industries which are greatly dependent on this category of assets. Bearing in mind that software is developed by the programmers, its quality can vary significantly. When developing specialized business software, it is not only important to have skilled programmers, but also to include experts who have the knowledge of certain business processes. If we take into consideration abovementioned facts, it is evident that the quality of software is inherent to successfully established business process that could potentially lead to optimal use of resources, i.e. save time and money. Statistically, the main aim of this paper is to examine the relationship between investment in software and financial performance of a company. In other words, determine if investment in information technology is significant stimulus in terms of improving financial performance of a company. The research sample was formed using companies listed on Zagreb Stock Exchange (ZSE). The data used in the statistical analysis was gathered from the financial statements publicly available at ZSE's official website. Key words: financial perfomance, information economy, information technologies, investment, statistical research, software

I. INTRODUCTION

From ancient times the people have been trying to find ways to improve business, increase productivity, expand the market and simultaneously reduce costs and gain more profits. Going in that respect for centuries was extremely small because of the linguistic and spatial limitations and the low level of technological development that did not allow these limits to go beyond. Only the industrial revolution and the development of information technology (with the emphasis on the development of high technology) have enabled the development of new forms of business communication, organization and cooperation as well as the emergence of new types of services and thus the improvement of traditional business. [1].

A new, interactive way of communication who doesn't know spatial and time constraints, has enabled the emergence and development of new forms of communication and business and their widespread introduction into everyday work in all areas of action – as well in finance and accounting sector which is special interest of this paper. Such business does not exclude traditional forms and ways of work and business, on the contrary, it complements and improves them. Thus, the development and application of high technologies become the strategic goal and the commitment of every advanced society, and their implementation in business processes is a condition of economic and any other progress that defines the modern information economy.

Few decades ago, intangible assets were not important as they are today. Nowadays, traditional reliance on fixed assets is replaced with "*high – tech paradigm*" that implies high share of IT intangible assets. Some companies mainly rely on intangible assets (e.g. tech start-ups), and other companies with significant share of fixed assets are forced to invest in intangible assets in order to keep their competitive edge.

II. INFORMATION ECONOMY

The new, information economy has emerged in the last decades of the 20th century at the global level. Castells has defined that it is called information (or global) to establish its fundamental distinctive features and further emphasizing of their perception. It is informative because the productivity and competitiveness of units or factors (whether they are companies, regions or nations) basically depends on their ability to efficiently create, process and apply knowledge-based information. It is global because it is the core of production, consumption and circulation activities, as well as their components (capital, workforce, raw materials, management, information, technology, markets) organized globally, either directly or through the network of links between economic factors. It is information and global because, in new historical conditions, productivity is generated and competition exploited in the global network of interactions. And the new economy emerged in the last quarter of the twentieth century because the revolution of information technology creates the necessary, material basis for such a new kind of economy. It is the historical link between knowledge-based and information-based economy, its global reach, and the revolution in information technology that creates a new, distinct economic system.[2]

Thus, the foundation of productivity and competitiveness of companies in the information economy depends on their ability to create, process and apply knowledge-based information, and relies on the intensive use of information technology in today's digital single market.[15]

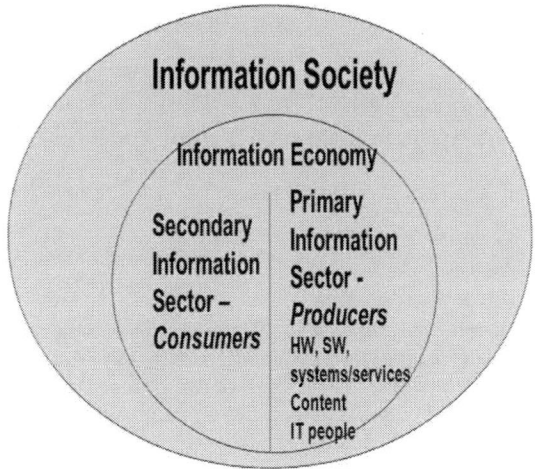

Figure 1. Information economy and information society (Source: Trauth, E. M., The Pennsylvania State University (2008), http://slideplayer.com/slide/5198224/ (last viewed 15. 03. 2017.)

III. PREVIOUS RESEARCHES

Constantly increasing importance of intangible assets made the research about investment in that part of company's assets very frequent topic among scholars all over the world. But, it must be highlighted that scientific papers which are focused exclusively on software as part of intangible assets are not so frequent. One of underlying reasons is that researches in this field are very particular due to numerous forms of intangible assets. *Apropos*, researches in this field can be categorized into two large groups: researches which analyse intangible assets as aggregate category and researches which focus on certain part of intangible assets. Some of researches are presented in this chapter chronologically.

Sheng and *Mykytyn* (2002.) have stressed the importance of the „quality of the output of the implemented IT systems", because systems don't have to function optimally. [3].
Although *Kim* (2004.) has found positive relationship between investment in IT and productivity of a company, he pinpoints some debatable issues („utilization of IT is closely related with firm-specific factors" (e.g. management ability), biases in results due to samples which are not comprehensive enough, „technological differences among industries" etc.).[4]
Doms et al. (2003.) have founded "significant relationship between IT investment intensity and productivity growth". [5]
Lim et al. (2004.) highlight conflict between some previous findings in terms of profitability and investments in IT. Besides that, they have founded „more positive returns of IT investment in information-intensive industries than non information-intensive industries" and that „the small size of the firm tends to have statistically more positive returns of IT investment than the large size of the firm".[6]
Stewart et al. (2007.) suggested that „profitability and productivity have not always emerged as a result of IT investments", and pinpointed several key factors such as firm's strategic advantage and innovative technologies. [7]
Mithas et al. (2012.) have founded that „IT has a positive impact on profitability" and that „firms have had greater success in achieving higher profitability through IT-enabled revenue growth than through IT-enabled cost reduction".[8]
Guzić (2014.) founded statistically significant positive relationship between intangible assets and profitability and also has proven that their importance in financial statements of Croatian companies is increasing. [9]
Bubić and *Šušak* (2015.) analysed relationship between investment in intangible assets as aggregately expressed value (research and development, patents, software, goodwill, advance payments for purchase of intangible asset and other intangible assets) and have founded that "companies which invested in intangible assets were less likely to open the bankruptcy proceeding" and that "there isn't strong evidence in favour of positive relationship

between investment in intangible assets and profitability". [10]

Despite the fact that logical reasoning and majority of previous researches indicate that there should be positive significant relationship between investment in IT and profitability, there is significant number of papers opposing aforementioned relationship. In process of hypotheses forming, authors of this research have accepted the prevailing group of findings and presumed positive relationship between those variables. [11]

IV. HYPOTHESES

Having in mind previous theoretical findings and practical experiences, hypotheses for this research were formed as following:

- *Hypothesis 1 – there is statistically significant positive relationship between company's financial performance and investment in software,*
- *Hypothesis 2 – there is statistically significant difference between investments in software compared to the previous financial year (increase of investment in intangible assets).*

Hypotheses have been formed using findings from previous researches, but one must be cautious when analysing these relationships. IT can be (and in most situations is) irreplaceable support which contributes to optimalization of business procedures. But, on the other hand, inadequately implemented IT (e.g. inadequate or poorly designed software) can cause situations with serious repercussions on financial performance of a company (e.g. delays, loss of goodwill, insufficient inventories etc.).

V. DATA, STATISTICAL METHODOLOGY, FINANCIAL RATIOS AND RESULTS OF RESEARCH

The research sample was formed using 125 non – financial companies listed on *Zagreb Stock Exchange (ZSE)* in financial year 2013. (commercial banks, insurance companies etc. were excluded from the sample due to the difference in financial reporting and the nature of their operations in comparison to other, non – financial companies). Approximately one half of companies included in sample (53 companies) have disclosed the structure of their intangible assets in detail. The data used in the statistical analysis was collected from the financial statements publicly available at *ZSE's* official website [12] and *Croatian Financial Agency* official website.[13] Correlational statistics was used to determine the relationship between variables – methods used were correlation and regression. Also, t – test for dependent samples was used as method which is part of group statistics.

Financial ratios included in research can be classified into two groups:

- Assets Ratios (*table 1.*),
- Profitability Ratios (*table 2.*).

TABLE 1. ASSET RATIOS INCLUDED IN RESEARCH

Abbrev.	Ratio	Formula
STA	Software to Total Assets	Software / Total Assets
SLTA	Software to Long – term Assets	Software / Long – term Assets

TABLE 2. PROFITABILITY RATIOS INCLUDED IN RESEARCH

Abbrev.	Ratio	Formula
ROA	Return on Assets	Net Profit / Total Assets
ROE	Return on Equity	Net Profit / Total Equity
NPM	Net Profit Margin	Net Profit / Total Revenues

TABLE 1. CORRELATION TABLE

		S/TA	S/LTA	ROE	ROA	NPM
S/TA	**P.C.***	1	0,975**	0,017	0,066	0,057
	Sig.		0,0001	0,902	0,644	0,690
	N	53	53	53	52	52
S/LTA	**P.C.***	0,975**	1	-0,012	0,068	0,059
	Sig.	0,0001		0,930	0,630	0,678
	N	53	53	53	52	52
ROE	**P.C.***	0,017	-0,012	1	-0,037	-0,049
	Sig.	0,902	0,930		0,794	0,728
	N	53	53	54	53	53
ROA	**P.C.***	0,066	0,068	-0,037	1	0,996**
	Sig.	0,644	0,630	0,794		0,0001
	N	52	52	53	53	52
NPM	**P.C.***	0,057	0,059	-0,049	0,996**	1
	Sig.	0,690	0,678	0,728	0,0001	
	N	52	52	53	52	53

*P.C. – Pearson Correlation
** Sig. – Sig. (2-tailed)
Software used in analysis: IBM Corp.: „*IBM SPSS Statistics for Windows*", Version 22.0., Armonk: NY: IBM Corp., 2013. [14]

Correlation analysis (*table 3.*) indicates that there isn't statistically significant relationship between software investment ratios (*Software to Total Assets and Software to Long – term Assets*) and profitability ratios (*Return on Assets, Return on Equity and Net Profit Margin*) at the 5% significance level. Intensity of correlation is mostly positive but extremely weak.

TABLE 4. INDEPENDENT VARIABLES COEFFICIENTS

Model		Unstandardized Coefficients		Standardized Coefficients	t	Sig.
		B	Std. Error	Beta		
Dependent Variable: S/TA						
1	(Constant)	0,003	0,001		2,795	0,007
	ROE_13	0,0002	0,001	0,017	0,124	0,902
1	(Constant)	0,003	0,001		2,737	0,009
	ROA_13	0,0003	0,001	0,066	0,465	0,644
1	(Constant)	0,003	0,001		2,797	0,007
	NPM_13	0,00001	0,00002	0,057	0,401	0,690
Dependent Variable: S/LTA						
1	(Constant)	0,004	0,002		2,845	0,006
	ROE_13	0,0001	0,002	-0,012	-0,089	0,930
1	(Constant)	0,005	0,002		2,881	0,006
	ROA_13	0,001	0,001	0,068	0,485	0,630
1	(Constant)	0,005	0,002		2,916	0,005
	NPM_13	0,00001	0,0001	0,059	0,418	0,678

Software used in analysis: IBM Corp.: „*IBM SPSS Statistics for Windows*", Version 22.0., Armonk: NY: IBM Corp., 2013.

The independent variables coefficients of linear regression models are positive for all analysed financial ratios (ROA, ROE, NPM) and it means that increase in one of these ratios will be accompanied by an increase in one of ratios used as dependent variables (S/TA, S/LTA) (*table 4*). But, coefficients for all analysed ratios are not statistically significant at the 5% significance level.

TABLE 5. PAIRED SAMPLES TEST

Paired Differences					t	df	Sig.
Mn	SD	SEM	CED				
			L	U			
-0,0002	0,002	0,0002	-0,0006	0,0002	-1,11	52	0,2704

Mn – Mean; **SD** – Std. Deviation; **SEM** – Std. Error Mean; **CED** – 95% Confidence Interval of the Difference; **L** – Lower; **U** – Upper; **Sig.** – Sig. (2-tailed)
Software used in analysis: IBM Corp.: „*IBM SPSS Statistics for Windows*", Version 22.0., Armonk: NY: IBM Corp., 2013.

Table 5. shows Paired Samples Test which indicates that there is no statistically significant difference between investment in software in financial year 2013. and 2012. The value of software intangible assets in year before was even slightly higher than in financial year 2013., i.e.

value of software decreased in comparison to the year before.

VI. HYPOTHESES ACCEPTANCE

According to the results of research, both initially established hypotheses were rejected (*table 6.*).

TABLICA I. TABLE 6. ACCEPTANCE AND REJECTION OF RESEARCH HYPOTHESES

Hypothesis	Content	Acceptance / Rejection
H₁	*Hypothesis 1 – there is statistically significant positive relationship between company's financial performance and investment in software,*	Rejected
H₂	*Hypothesis 2 – there is statistically significant difference between investments in software compared to the previous financial year (increase of investment in intangible assets).*	Rejected

VII. RESEARCH LIMITATIONS

Firstly, when gathering information about company's intangible assets, significant part of companies haven't disclosed the structure of their intangible assets in detail. That is the reason why the share of software in their intangible assets could not be determined. Consequentially, those companies couldn't be included in research sample.

Secondly, information was gathered from financial statements which are an end product of accounting. Accounting information is focused on past events and susceptible to manipulation.

Thirdly, economic events are very complex and sometimes it is very hard to resolve what influenced financial performance of companies because there are many macroeconomic and microeconomic variables that can possibly be connected with analysed variable.

Second and third limitation are inherent to every economic analysis of this kind.

VIII. CONCLUSION

The main aim of this paper was to statistically try to examine the relationship between investment in software and financial performance of a company. In other words, to determine if investment in information technology is significant stimulus in terms of improving financial performance of a company. The research sample, which was formed using companies listed on Zagreb Stock Exchange (ZSE) and used in the statistical analysis was gathered from the financial statements publicly available at ZSE's official website showed that both of our hypotheses: *Hypothesis 1* which said that there is

1489

statistically significant positive relationship between company's financial performance and investment in software and *Hypothesis 2-* that there is statistically significant difference between investments in software compared to the previous financial year (increase of investment in intangible assets) and according to the results of research, both initially established hypotheses were rejected meaning that by data used in research showed there is no (direct) interconnection between investment in software and financial performance considering given research limitations. But one fact still remains, information technologies have become the most important part of companies assets and business infrastructure on whom the financial performance relies on so many levels so even though we can's statistically prove the direct interconnection between software investment and financial performance they present the groundings of every business in information economy surroundings.

LITERATURE

[1] Gates, B., "Business @ the Speed of Thought: Succeeding in the Digital Economy", Grand Central Publishing (1999)

[2] Castells, M., The information age: economy, society and culture, Vol. I: The rise of the network society . Cambridge: Blackwell Publishers, 2000.

[3] Sheng, Yi Hua, and Peter P. Mykytyn Jr. "Information Technology Investment and Firm Performance: A Perspective of Data Quality." IQ. 2002.

[4] Kim, Jong-Il. "Information technology and firm performance in Korea." Growth and Productivity in East Asia, NBER-East Asia Seminar on Economics, Volume 13. University Of Chicago Press, 2004.

[5] Doms, Mark E. and Klimek, Shawn D. and Jarmin, Ron S., "IT Investment and Firm Performance in U.S. Retail Trade", FRB of San Francisco Working Paper No. 2003-19; US Census Bureau Center for Economic Studies Paper No. CES-WP- 02-14., 2013., Available at SSRN: https://ssrn.com/abstract=550442 or http://dx.doi.org/10.2139/ssrn.550442

[6] Lim, Jee Hae, Vermon J. Richardson, Tom L. Roberts, 2004, "Information Technology Investment and Firm Performance: A Meta-Analysis", Proceedings of the 37th Annual Hawaii International Conference on System Sciences

[7] Stewart, Walter; Coulson, Sheri; and Wilson, Robert (2007) "Information Technology: When is it Worth the Investment?," Communications of the IIMA: Vol. 7: Iss. 3, Article 11. Available at: http://scholarworks.lib.csusb.edu/ciima/vol7/iss3/11

[8] Mithas, Sunil and Tafti, Ali R. and Bardhan, Indranil and Goh, Jie Mein, "Information Technology and Firm Profitability: Mechanisms and Empirical Evidence", MIS Quarterly, Vol. 36, No. 1, pp. 205-224, 2012. Available at SSRN: https://ssrn.com/abstract=1000732

[9] Guzić, Š. (2014). Analysis of Correlation between Intangible Assets and Successful Performance of Companies. Journal of Accounting and Management, IV(1), 25-42. Available at: http://hrcak.srce.hr/143428

[10] Bubić J., Šušak T.: „The Impact of Intangible Assets on Financial Performance of Croatian Companies", 9th International Scientific Conference "Economic and Social Development", Istanbul, 2015.

[11] Belak, V., „Analiza poslovne izvrsnosti", RRIF plus d.o.o., 2014.

[12] The Zagreb Stock Exchange, www.zse.hr

[13] Croatian Financial Agency – FINA: „Registar godišnjih financijskih izvještaja - javna objava", available at website: http://rgfi.fina.hr/JavnaObjava-web/jsp/prijavaKorisnika.jsp

[14] IBM Corp., "IBM SPSS Statistics for Windows", Version 22.0., Armonk: NY: IBM Corp., 2013.

[15] Boban, M., Digital single market and EU data protection reform with regard to the processing of personal data as the challenge of the modern world", Proceedings of 16th International Scientific Conference ESD 2016., „The legal challenges of modern world", Split, 2016., str. 191 - 201

Data Warehouse Architecture Classification

G. Blažić, P. Poščić and D. Jakšić
Department of informatics-University of Rijeka/ Rijeka, Croatia
gordana.blazic@student.uniri.hr, patrizia@inf.uniri.hr, dsubotic@inf.uniri.hr

Abstract - The purpose of this study is to give an overlook and comparison of best known data warehouse architectures. Single-layer, two-layer, and three-layer architectures are structure-oriented one that are depending on the number of layers used by the architecture. In independent data marts architecture, bus, hub-and-spoke, centralized and distributed architectures, the main layers are differently combined. Listed data warehouse architectures are compared based on organizational structures, with its similarities and differences. The second comparison gives a look into information quality (consistency, completeness, accuracy) and system quality (integration, flexibility, scalability). Bus, hub-and-spoke and centralized data warehouse architectures got the highest scores in information and system quality assessment.

Key words: data warehouse architecture classification, organization structure, information quality, system quality

I. INTRODUCTION

None of the business organization is the same. They all have their own unique processes and actions. In the databases of this organizations, the data has to be in every moment updated and available for processing. To be able to perform analytical processing for business decision-making, the organization needs to have a data warehouse where certain data are firstly being processed, and then permanently stored for future actions. The data warehouses provide support to managers in the management and decision-making. Before the implementation of a data warehouse, it is necessary to determine the criteria to be satisfied, and on the basis of determined, select the appropriate data warehouse architecture that satisfies the needs of organization.

The purpose of this paper is to give an overlook and comparison of best known data warehouse architectures. The main features for each architecture will be presented, and architectures will be compared based on organizational structures, as well as on system and information quality.

II. DATA WAREHOUSE ARCHITECTURE CLASSIFICATION

The authors [1] gave the best overlook and explanation of various data warehouse architectures. They put data warehouse architectures into two categories. The first one includes single-layer, two-layer and three-layer

architectures that are structure-oriented one and are depending on the number of layers used by the architecture. The second classification consists of independent data marts architecture, bus, hub-and-spoke, centralized and distributed architecture where the main layers are differently combined.

A. Single-Layer Architecture

Single-layer architecture is not frequently used in practice. The goal is to minimize the amount of stored data by removing data redundancies. Data warehouse is virtual and implemented as a multidimensional view of operational data. The main weakness of a single-layer architecture is its failure to separate the analytical and transactional data processing.

B. Two-Layer Architecture

In the two-layer architecture, there is a separation between two layers: a layer of data sources and data warehouse layer. Although it is called a two-layer architecture to emphasize the separation of the two layers, it is actually consisted of four data flow stages: source layer, data staging, data warehouse layer and analysis. In contrast to single-layer architecture, in this one there is a separation between analytical and transactional data

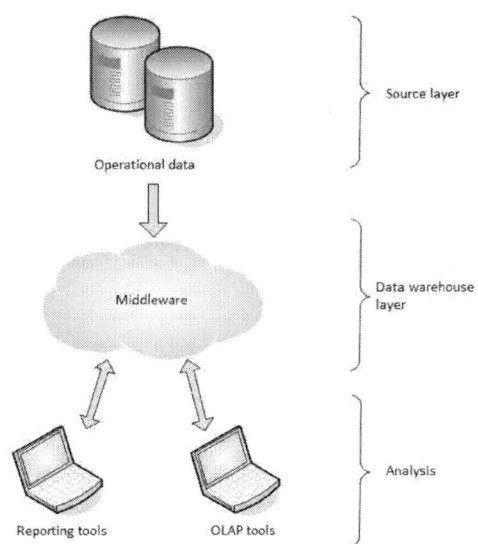

Figure 1 Single-Layer Architecture [1]

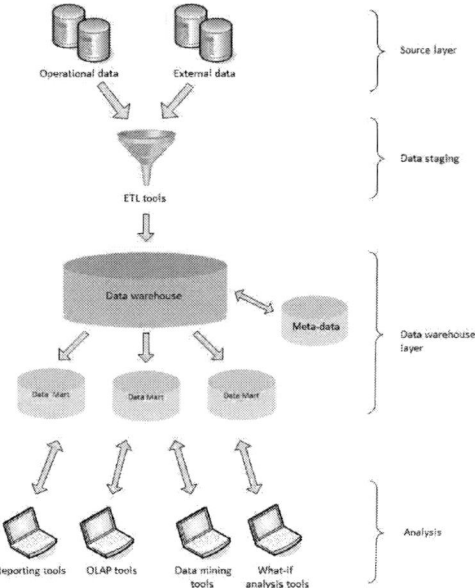

Figure 2 Two-Layer Architecture [1]

processing.

C. Three-Layer Architecture

Three-layer architecture is characterized by the fact that three layers are physically implemented: source layer, reconciled layer and data warehouse layer. Reconciled data layer materializes operational data after integrating and cleansing source data. As a result of passing through the reconciled layer, data is integrated, consistent,

accurate, correct, and detailed.

D. Independent Data Marts Architecture

As the name implies, independent data marts architecture consists of various data marts being separately designed and implemented, in another words, they are not integrated. Data marts usually have inconsistent data definitions and use different dimensions and measures, making data analysis difficult. This architecture is usually replaced by another one in order to achieve better data integration and cross-reporting.

E. Bus Architecture

The bus architecture is recommended by Ralph Kimball and is similar to independent data marts architecture with significant difference – data marts are logically integrated and there is an enterprise-wide view of information.

F. Hub-And-Spoke Architecture

The hub-and-spoke architecture consists of data sources, reconciled data and data marts. Enterprise data warehouse, called the hub, is created with a set of data marts, called spokes. Atomic, normalized data is stored in a reconciled layer that feeds a set of data marts made of summarized data in multidimensional form. In hub-and-spoke architecture the emphasis is on scalability and extensibility, as well as on retrieving large amounts of information.

G. Centralized Architecture

The centralized architecture is recommended by Bill Inmon. This architecture can be seen as a specific implementation of hub-and-spoke architecture, where in contrast to hub-and-spoke, there is no dependent data marts. It consists of one centralized data warehouse which contains integrated data and data marts.

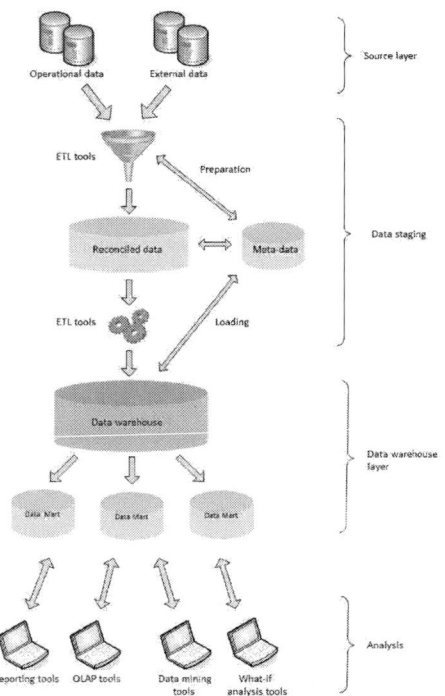

Figure 3 Three-Layer Architecture [1]

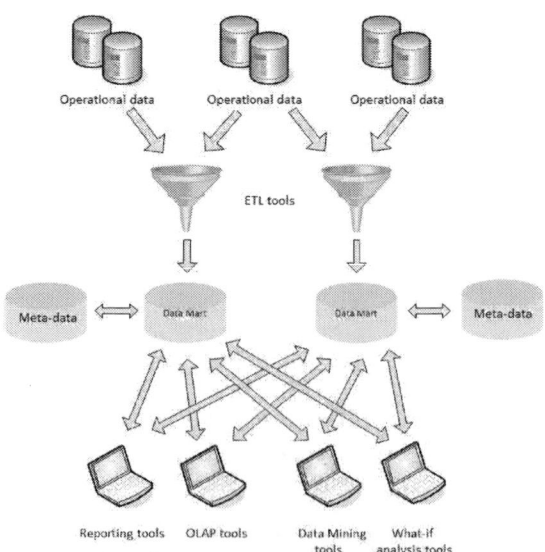

Figure 4 Independent Data Marts Architecture [1]

1492

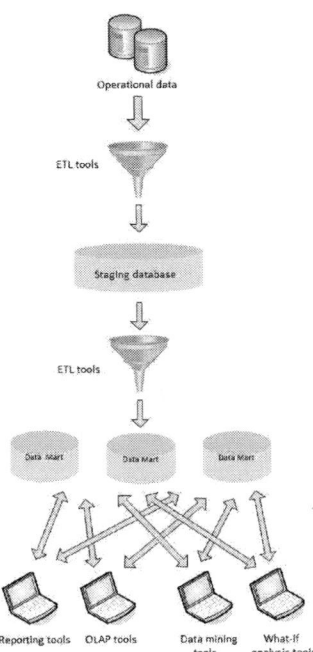

Figure 5 Bus Architecture [2]

H. Distributed Architecture

The distributed architecture is sometimes adopted in dynamic contexts where existing data warehouses should be integrated to provide a single solution. Each data warehouse/data mart is in this architecture logically or physically integrated using joint keys, global metadata, distributed queries, and other methods.

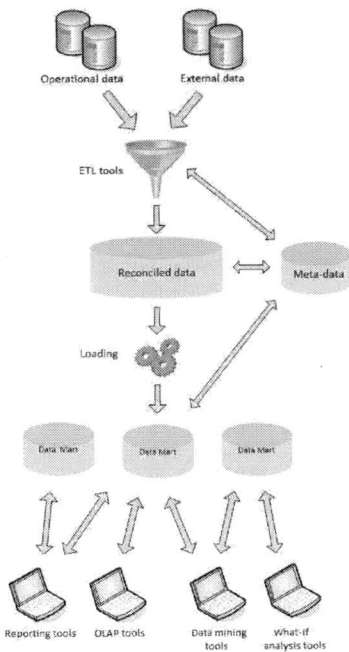

Figure 6 Hub-And-Spoke Architecture [1]

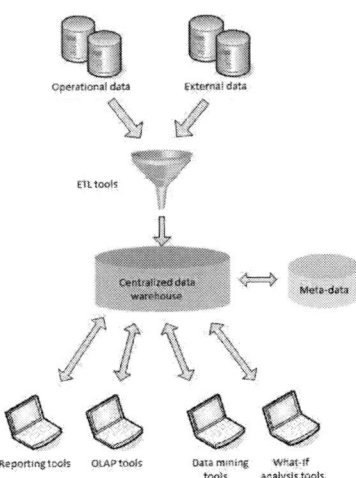

Figure 7 Centralized architecture [2]

III. DATA WAREHOUSE ARCHITECTURE COMPARISON

We start with comparison based on organizational structure. Visible features in the organizational structures of architectures were taken as comparison criteria. Is the source data being processed with ETL tools before being stored? Are the users getting data directly from data warehouse, or through data marts? Do the data marts depend on data warehouse?

In Table 1 the results are presented. In all architectures, except in single-layer architecture, data is stored and retrieved only after it has been processed by ETL tools. In single-layer and centralized architecture, data is accessed directly from data warehouse. In all other architectures, data for specific business area is being retrieved from data marts. If there is a data mart after data warehouse in the organizational structure, we can say that data mart depends on data warehouse. We can found independent data marts in independent data marts

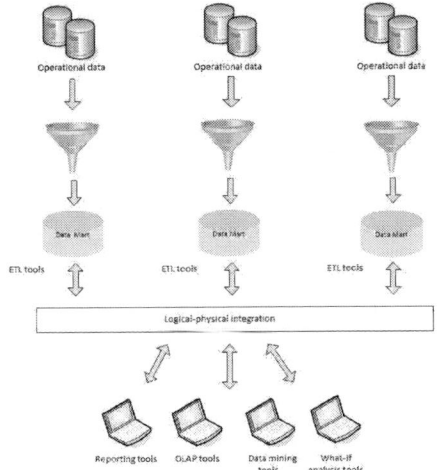

Figure 8 Distributed Architecture [1]

1493

TABLE 1 ORGANIZATION STRUCTURE COMPARISON

	Single-layer architecture	Two-layer architecture	Three-layer architecture	Independent data marts architecture	Bus architecture	Hub-and-spoke architecture	Centralized architecture	Distributed architecture
ETL tools are used in processing data.	NO	YES	YES	YES	YES	YES	YES	YES
The users are getting data directly from data warehouse.	YES	NO	NO	NO	NO	NO	YES	NE
Data marts are used in accessing data.	NO	YES	YES	YES	YES	YES	NO	YES
Data marts depend are depending on data warehouse.	-	YES	YES	NO	NO	YES/NO	-	NO

architecture, bus, centralized, and distributed architecture. Hub-and-spoke architecture has a data warehouse made of set of data marts.

For the second comparison we used numeric tables [3] to take a look into consistency, completeness, and accuracy of information, and integration, flexibility, and scalability of system. Based on these tables, Chart 1 and Chart 2 were created, representing results for independent data marts architecture, bus, hub-and-spoke, centralized and distributed architecture.

In information quality assessment on Chart 1, independent data mart architecture has the lowest results. The next one is distributed architecture. For bus, hub-and-spoke and centralized architecture, there are similar results. Independent data marts architecture has relatively small result for information completeness, what confirms the fact that this architecture is particularly weak in providing all data needed to make decisions. Hub-and-spoke and centralized architectures have higher values for information completeness. These architectures provide more comprehensive data source in decision support. The bus architecture has proved to be the best when it comes to the information accuracy.

In system quality assessment on Chart 2, the average results are analog to results in information quality assessment. Independent data marts architecture got also in this case the lowest results, and that is why this architecture is soon being replaced by another architecture. The distributed architecture has some better results. Bus, hub-and-spoke and centralized architectures also achieved similar results. Although centralized architecture got the highest result for scalability, bus and hub-and-spoke architectures have proven to be the best when it comes to all three measures of system quality.

In [4], [5] and [6] we can find more features that were used to assess the success of the architectures. The given conclusions are similar to ours.

IV. CONLUSION

This paper gave an overlook of best-known data warehouse architectures: single-layer, two-layer, and three-layer architecture, independent data marts architecture, bus, hub-and-spoke, centralized and

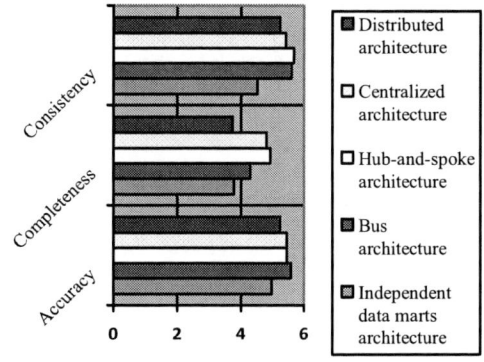

Chart 1 Information quality assessment

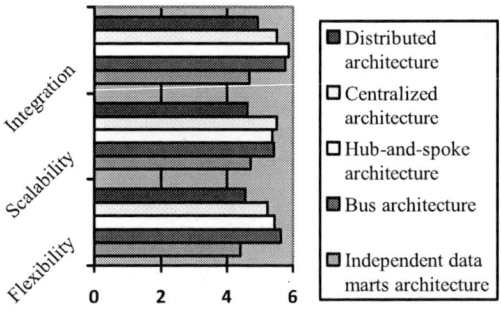

Chart 2 System quality assessment

distributed architecture. Looking at the organizational structures of these architectures, we can conclude that in the beginning there is source data. Data is then being processed by ETL tools (extraction, transformation, load) and then stored in data warehouse. Data marts can follow data warehouse for retrieving data required for analysis of particular business area. Using different tools, end users receive the required information.

The contribution of this paper is in comparison of best known data warehouse architectures. Bus, hub-and-spoke, and centralized architectures have shown the best results in assessing information quality, as well as in system quality assessment. The reason for that is that these architectures have developed over time and have become similar. We believe that architectures in the future will be very similar. Over time, there will be less disadvantages and every architecture will successfully satisfy needs of their clients.

For further work on this paper, one can pay attention to the future of data warehouse, due to the increasing popularity of NoSQL databases.

REFERENCES

[1] M. Golfarelli, S. Rizzi, Data Warehouse Design: Modern Principles and Methodologies, McGraw Hill, 2009, pp. 7-14.

[2] Il-Yeol Song, „Data warehousing systems: foundations and architecture", Drexel University, available: http://www.cis.drexel.edu/faculty/song/courses/info%20607/DWS -Foundations-Architecture.pdf, 6.5.2016.

[3] H. J. Watson, T. Ariyachandra, "Data Warehouse Architectures: Factors in the Selection, Decision and the Success of the Architectures", July 2005, pp. 39-40.

[4] T. Ariyachandra, H. J. Watson, "Which Data Warehouse Architecture Is Most Successful?", Business Intelligence Journal Vol. 11, No. 1 2006, pp. 4-6.

[5] T. Ariyachandra, H. J. Watson, "Which Data Warehouse Architecture is Best?", Communications of the ACM October 2008, Vol. 51 No. 10, pp. 146-147.

[6] Moh'd Alsqour, K. Matouk, M. L. Owoc, "A survey of data warehouse architectures – preliminary results", Computer Scuence and Information Systems (FedCSIS), 2012 Federated Conference, pp. 1121-1126. IEEE, 2012.

Comparative Analysis of the Selected Relational Database Management Systems

R. Poljak, P. Poščić and D. Jakšić

Department of informatics-University of Rijeka/ Rijeka, Croatia
rpoljak@student.uniri.hr, patrizia@inf.uniri.hr, dsubotic@inf.uniri.hr

Abstract - **The database management system is a software that enables easier work with databases i.e. to define database structure, retrieve stored data, enter data into the database and process the previously stored data in the database. In this article we have compared 3 relational database management systems (RDBMS) - Oracle 11g, MySQL and PostgreSQL. They are compared according to the simple criteria that we defined, such as the comparison of basic data, syntax, data types and speed performance. The main contribution of the article is a comparison of 3 different RDBMSs by our own score criteria.**

I. INTRODUCTION

Databases today are widely used in the field of information systems. Databases allow working with multiple levels of data, unlike conventional programming languages. They are made with the purpose to eliminate the weak points of automatic processing of data that have been most used during the 60s and 70s. Growing usage of databases has enabled the development of quality and reliable applications. The definition of a database according to [8] is a collection of interrelated data and stored in computers external memory. According to [8] system for database management is a server for a database, and with it you can create a database to display the required logical structure. It is also used to perform operations on data that its client tasks. In this paper we were working with database management systems that are based on the relational model. The relational model was designed by E. F. Codd in the 60s. The model uses mathematical concept of a relation and data and links between data are shown in table form.

II. SPECIFICATIONS

A. MySQL

According to [1] MySQL is the most popular open source database in the world, because it has cost-effective delivery of data, with high-performance, stable e-commerce and online transaction processing. According to [1] MySQL is an integrated, safe for transactions, an ACID compliant database with full support for a return to the old data, the recovery after the destruction of the base and the possibility of row-level locking. MySQL provides ease of use, high performance, scalability, as well as packet drivers of the database and visual tools to help developers (database administration). MySQL database has the following characteristics:

- High performance and scalability - to satisfy the exponential growth of data loads and a number of users.

- Self-healing cluster replication - to improve the performance, stability and availability.

- Online modification of the scheme - in order to satisfy changing business requirements.

B. PostgreSQL

According to [2] PostgreSQL is an object-relational system for database management and is based on PostgreSQL version 4.2 that was developed at the University of California at Berkeley IT department. PostgreSQL has developed many concepts that have become available only in some commercial database systems much later. PostgreSQL is an open-source successor of the original Berkeley code. It supports a large part of the SQL standard and offers many modern features, such as:

- Complex queries.

- External keys.

- Triggers.

- Transaction Integrity.

- Multi-version concurrency control.

Because of its liberal license PostgreSQL can be used, modified and distributed completely free for any purpose (private, academic or commercial).

C. Oracle database 11g

According to [3] Oracle Database 11g is a full-featured database manager for the needs of medium-sized companies. Within Oracle 11g is an *Oracle Application Express*, a unique online development environment that enables rapid construction of a database using a web browser. It also supports *Oracle Developer* graphical tool for database development, which allows the developer to browse database objects by running SQL queries, SQL scripts and editing PL / SQL statements. Oracle Database 11g is well integrated with Visual Studio through Oracle Data Access components that allow the developer to easily build .NET applications and Web services based on Oracle 11g database.

III. COMPARISON OF A DBMS

A. Comparison of basic data of DBMS

The first comparison that we will make is the comparison of basic data for each RDBMS. According to [7] below in the text is an overview of Table 1 which contains a comparison of basic data RDBMSs.

From the attached Table 1 we can see that MySQL and Oracle 11g have the same developer, Oracle 11g is the only commercial RDBMS, PostgreSQL is supported on the variety number of operating systems, even 9. Oracle supports most programming languages. In the table we mentioned some of the most popular programming languages, although there are lesser-known languages that Oracle supports. All RDBMSs have ACID.

B. Comparison of syntax

The second comparison is the syntax between MySQL, Oracle 11g, and PostgreSQL which is shown in Table 2. The commands are creating two tables Manufacturer and Tires, the input of data in tables and queries that run against created tables and which are written according to the syntax of the corresponding RDBMS. From Table 2, we can see that the difference in the syntax for each RDBMS is negligible. All that is changed is just one word, and that is NUMBER, which reads NUMERIC for MySQL and PostgreSQL. We tested the examples only with two data types (VARCHAR and NUMERIC) in this paper – in the future work we will broaden the scope.

TABLE I. COMPARISON OF BASIC RDBMS DATA

	RDBMS		
	MySQL	*Oracle 11g*	*PostgreSQL*
Database model	Relational DBMS	Relational DBMS	Relational DBMS
Developer	Oracle	Oracle	PostgreSQL Global Development Group
Licence	Open source	Commercial	Open source
Supported OS	Windows, Linux, OSX, Solaris, FreeBSD	Windows, Linux, OSX, Solaris, HP-UX, AIX	Windows, Linux, OSX, Solaris, HP-UX, Unix, FreeBSD, NetBSD, OpenBSD
Supported programming languages	C, C#, C++, D, Haskell, Perl, PHP, Java, Python, Ruby, Objectiv-C	C, C#, C++, Cobol, Fortran, Haskell, Java, JavaScript, Lisp, Perl, PHP, Python, Ruby, Visual Basic, Objectiv-C	.Net, C, C++, Java, Perl, Python, Tcl
Transactions	ACID	ACID	ACID

Source: [7]

C. Data types

The third comparison is the difference between the data types that a particular RDBMS is using. For easier and more user-friendly presentation we will use Table 3. which shows a comparison of data types. Simillar (or corresponding) data types through all three RDBMSs are shown in the same color.

According to sources [4], [5] and [6] we pulled out all the essential information for the comparison of data types. Below we will describe the advantages and disadvantages of different RDBMSs for certain types of data which they have or do not have. Data type ARRAY MySQL doesn't have, but the alternative that users of MySQL are using is the temporary table. All three RDBMSs have CHAR - it is used because it is fast, space efficient and it is practical to use when the rows are the same or approximately the same length.

The CHARACTER in MySQL is a longer version of CHAR, there is also a variant of CHARACTER SET binary for character data type. It causes the column to be made from the appropriate binary data types so that the char becomes a binary char, varchar becomes varbinary while the character in PostgreSQL is a string of fixed size.

DATE is supported in all three RDBMS and it is used to save the date in the form of strings or numbers. CHARACTER VARYING is only supported in PostgreSQL and its advantage over regular character is variable length.

DECIMAL is supported in PostgreSQL and MySQL, and MySQL also supports DEC. Oracle solves the lack of decimals in a way that converts decimal to numeric data type. In PostgreSQL user precisely defines the decimal numbers ranging from 131072 figures before the decimal point and 16383 numbers after the decimal point. In MySQL decimal determines the exact number of digits before the decimal point and the number of digits after the decimal point. The maximum number of decimal digits is 65 while the number of supported decimals is 30. It also has an option that prevents unsigned negative values. DEC is synonymous for decimal. DOUBLE is supported in MySQL and PostgreSQL. Alternative for the DOUBLE which is used by users of Oracle is BINARY DOUBLE, and it is 64-bit floating point NUMBER or BINARY FLOAT. In PostgreSQL is used DOUBLE PRECISION, and it consists of 15 decimal numbers. In MySQL DOUBLE is the number with the moving point of normal size, a range can be less than that is specified in the documentation, but it all depends on the hardware and operating system.

ENUM is supported in MySQL and PostgreSQL, but in Oracle it isn't. MySQL ENUM is a string object with a selected list of allowed values that are listed in the column with specifications at the time of creating the table. PostgreSQL ENUM is a data type that includes statically ordered set of values, and they are equivalent to ENUM type that is supported in many programming languages. In Oracle using PL / SQL users can make their own type that has a similar logic to Enum.

TABLE II. COMPARISON OF SYNTAX RDBMS

RDBMS	Commands
	CREATE TABLE 1
MySQL	CREATE TABLE MANUFACTURER (ID_PRO VARCHAR (2) NOT NULL, PRO_NAME VARCHAR (20) NOT NULL , HEADQUARTERS VARCHAR(20));
Oracle 11g	CREATE TABLE MANUFACTURER(ID_PRO VARCHAR (2) NOT NULL, PRO_NAME VARCHAR (20) NOT NULL , HEADQUARTERS VARCHAR(20));
PostgreSQL	CREATE TABLE MANUFACTURER(ID_PRO VARCHAR (2) NOT NULL, PRO_NAME VARCHAR (20) NOT NULL, HEADQUARTERS VARCHAR(20));
	INSERT INTO TABLE 1
MySQL	INSERT INTO MANUFACTURER VALUES ('GY', 'GOODYEAR', 'BUFFALO');
Oracle 11g	INSERT INTO MANUFACTURER VALUES ('GY', 'GOODYEAR', 'BUFFALO');
PostgreSQL	INSERT INTO MANUFACTURER VALUES ('GY', 'GOODYEA', 'BUFFALO');
	CREATE TABLE 2
MySQL	CREATE TABLE TIRES (ID_TIRES VARCHAR(6) NOT NULL, TIRES_NAMEVARCHAR(30) NOT NULL, ID_PRO VARCHAR(2) NOT NULL, PURPOSE VARCHAR(15) NOT NULL, PRICE NUMBER(6,2) NOT NULL, STOCK NUMBER NOT NULL, SOLD NUMBER NOT NULL);
Oracle 11g	CREATE TABLE TIRES (ID_TIRES VARCHAR(6) NOT NULL, TIRES_NAMEVARCHAR(30) NOT NULL, ID_PRO VARCHAR(2) NOT NULL, PURPOSE VARCHAR(15) NOT NULL, PRICE NUMERIC(6,2), STOCK NUMERIC NOT NULL, SOLD NUMERIC NOT NULL);
PostgreSQL	CREATE TABLE TIRES (ID_TIRES VARCHAR(6) NOT NULL, TIRES_NAMEVARCHAR(30) NOT NULL, ID_PRO VARCHAR(2) NOT NULL, PURPOSE VARCHAR(15) NOT NULL, PRICE NUMERIC(6,2), STOCK NUMERIC NOT NULL, SOLD NUMERIC NOT NULL);
	INSERT INTO TABLE 2
MySQL	INSERT INTO TIRES VALUES ('GY3455', 'DURAGRIP', 'GY', 'LJETNE', 528.66, 12, 1567);
Oracle 11g	INSERT INTO TIRES VALUES ('GY3455', 'DURAGRIP', 'GY', 'LJETNE', 528.66, 12, 1567);
PostgreSQL	INSERT INTO TIRES VALUES ('GY3455', 'DURAGRIP', 'GY', 'LJETNE', 528.66, 12, 1567);
	QUERY 1
MySQL	SELECT NAZIV_TIRES, ID_PRO FROM TIRES WHERE ID_PRO IN (SELECT ID_PRO FROM MANUFACTURER WHERE HEADQUARTERS = 'MILANO');
Oracle 11g	SELECT NAZIV_TIRES, ID_PRO FROM TIRES WHERE ID_PRO IN (SELECT ID_PRO FROM MANUFACTURER WHERE HEADQUARTERS = 'MILANO');
PostgreSQL	SELECT NAZIV_TIRES, ID_PRO FROM TIRES WHERE ID_PRO IN (SELECT ID_PRO FROM MANUFACTURER WHERE HEADQUARTERS = 'MILANO');
	QUERY 2
MySQL	SELECT AVG(SOLD/22*PRICE)"Average daily earnings" FROM TIRES WHERE ID_PRO = 'DL';
Oracle 11g	SELECT AVG(SOLD/22*PRICE) "Average daily earnings" FROM TIRES WHERE ID_PRO = 'DL';
PostgreSQL	SELECT AVG(SOLD/22*PRICE) "Average daily earnings" FROM TIRES WHERE ID_PRO = 'DL';

FLOAT is not supported in PostgreSQL. The alternative to FLOAT in PostgreSQL is a REAL, DOUBLE PRECISION or NUMERIC. In MySQL FLOAT is the value of the approximate numerical value data with a movable section, and his precision is from 0 to 23 results in a 4-byte one-precision float column. FLOAT is a subtype of number data type and it has a precision in the range from 1 to 126 binary numbers. INT is supported only in MySQL.

Oracle solves this problem so that INT converts into NUMBER data type. PostgreSQL uses instead of INT use an INTEGER for values whose size of storage is 4 bytes and BIG INT for values that require storage size of 8 bytes. MySQL supports SQL standard integer types and the size of its storage is 4 bytes. LONG is supported only in Oracle. The columns that we declare as LONG can be stored as signs of adjustable length and a length is up to 2 gigabytes of information. MySQL alternative for LONG would be text that can store up to 4 gigabytes of character, and the effective maximum length is less if the value contains multibyte signs. PostgreSQL also used TEXT data type as a replacement for LONG, and the size is undefined.

TABLE III. COMPARISON OF DATA TYPES RDBMS

Data Type	RDBMS		
	MySQL	*Oracle 11g*	*PostgreSQL*
ARRAY	-	+	+
CHAR	+	+	+
CHARACTER	+	-	+
DATE	+	+	+
CHARACTER VARYING	-	-	+
DECIMAL, DEC	+	-	+
DOUBLE	+	-	+
ENUM	+	-	+
FLOAT	+	+	-
INT	+	-	-
LONG	-	+	-
NUMBER	-	+	-
NUMERIC	+	-	+
RAW	-	+	-
REAL	+	-	+
TEXT	+	-	+
TIME	+	-	+
TIMESTAMP	+	+	+
UUID	-	-	+
VARCHAR	+	+	+
VARCHAR2	-	+	-
YEAR	+	-	-

NUMBER is supported only in Oracle. NUMBER is a data type that stores the numbers with a fixed and floating point, and their storage size is almost unlimited and works with 38 digits of precision. In PostgreSQL alternatives that are used instead of the number are SMALLINT, BIGINT, DECIMAL, DOUBLE PRECISION, and NUMERIC. MySQL equivalent is the NUMERIC data type that allows the use of unsigned attributes that represent negative values, and signed is always given so there is no effect. NUMERIC is supported in MySQL and PostgreSQL, its equivalent in Oracle is NUMBER. RAW is supported only in Oracle. RAW is variable-length binary or byte string. Full size is 2000 bytes and there must be given the value of the RAW size. MySQL alternative is a BIT. BIT is a kind of a bit field where M signifies the value of bits from 1 to 64, and the default value is 1 if M is not specified. PostgreSQL alternative is BYTES. It allows storage of binary strings, and the size of the storage is from 1 to 4 bytes. MySQL and PostgreSQL support the REAL. Alternative for REAL is a FLOAT in Oracle. REAL in MySQL works almost the same as FLOAT or DOUBLE PRECISION. REAL in PostgreSQL has 6 decimal digits of precision, and storage size is 4 bytes.

TEXT is not supported in Oracle, but the alternative in Oracle is long. TEXT in PostgreSQL, in the beginning, had a TEXT data type, but when was transferred to SQL language for compatibility, instead of being renamed they added a new type VARCHAR, but both types use the same C routines internally.

TIME is supported in MySQL and PostgreSQL while in Oracle as an alternative is used a DATE. TIME in MySQL is a data type that can be used not only to represent the time of the day but also the elapsed time or the time interval between two events. In PostgreSQL TIME data type saves the time of day. Storage size is 8 bytes. TIMESTAMP is supported in all three RDBMS. TIMESTAMP in MySQL is a data type and is used for values that contain both DATE and TIME intervals. TIMESTAMP in PostgreSQL has two variants, one is without time zone, and one is with the time zone, and both of them store date and time. TIMESTAMP in Oracle has three variants, the ordinary one, with the time zone and with local time of time zone.

UUID is only supported in PostgreSQL. UUID is written as a series of small hexadecimal digits through several groups separated by dashes, a separate group of eight digits followed by three groups of four digits followed by a group of 12 digits, and all that together is the thirty-two digits that represent one hundred and twenty-eight bits. In Oracle, the UUID is obtained using a function in PL / SQL, while in MySQL is used the UUID() function that returns the value of hundred and twenty-eight bit number as utf8 string of five hexadecimal numbers in the form of aaaaaaaa-bbbb-cccc-dddd-eeeeeeeeeee.

VARCHAR is supported in all three RDBMS. VARCHAR in MySQL is very similar to char, but the difference is in the storage and retrieval of data. VARCHAR is a string of variable length, and its length is

from 0 to 65.535 while its maximum effective length is limited to a maximum line length which is 65,535 bytes. VARCHAR in PostgreSQL is the character type of general purpose variable length with a limit. VARCHAR in Oracle is the same as varchar, but it is recommended to use VARCHAR2 because the VARCHAR is designed to redefine in a separate data type. It will be used for variable length strings compared with different comparison semantics. VARCHAR2 is supported only in Oracle. VARCHAR2 is a data type that specifies the character string of variable length, once when the VARCHAR2 is created column has an available maximum number of characters or bytes of data that can be kept. The equivalent of VARCHAR2 in MySQL and PostgreSQL is VARCHAR.

YEAR is supported only in MySQL. YEAR is the data type of 1 byte and it is used to represent the values in the form yyyy, and the range is from 1901 to 2155. Oracle and PostgreSQL use DATE data type instead.

D. Speed performance

In the last comparison we will compare the speed of execution of certain commands in each of the RDBMSs. The hardware resources that were used during testing are: CPU: Intel (R) i5 2400K, RAM: 8GB DDR3 RAM, HDD: WD 1T, OS: Windows 7 64-bit. To view the runtime we used commands within the RDBMSs. For Oracle, the commands are *set timing on* and *set autotrace traceonly*, and for PostgreSQL is *\timing*, while for MySQL command was not necessary, because it already has that option within itself as part of the MySQL Workbench. We must mention that the speed comparison done in this paper is pretty simple (with simple statements and simple database with small number of tuples). Table 4. shows our comparison of the speed performance of the simple RDBMS queries. More comprehensive speed performance will be done in the future (with larger and more complex database, as well as more complex queries).

In Table 4. we can see the results of comparing the speeds which will be described in more detail below. The first command *create table* was performed fastest by Oracle for only 0.03 ms. After Oracle was MySQL and it performed a query for 0.22 ms and at the end is PostgreSQL. The second command *insert into* was performed fastest by Oracle for 0.001 ms, followed by PostgreSQL for 0.006 ms and at the end is MySQL with 0.05 ms. The third query *select* was performed fastest by MySQL for 0.001 ms, followed by Oracle with 0.01 ms and at the end is PostgreSQL with 0.02 ms. The fourth query *select* was executed equally fast by both Oracle and MySQL for 0.001 ms, while the PostgreSQL performed it for 0.009 ms. The fifth query *join* was performed equally fast by both MySQL and Oracle in a time of 0.001 ms, while the PostgreSQL performed it in 0.006 ms. The sixth command *update* was performed fastest by Oracle for 0.001 ms, followed by PostgreSQL with 0.004 ms and at the end is MySQL with 0.06 ms. The seventh and final command *drop table* was performed fastest by PostgreSQL for 0.08 ms, followed by MySQL with 0.16 ms and at the end is Oracle with 0.44 ms.

1499

TABLE IV.	COMPARISON OF SPEED PERFORMANCE OF QUERIES IN RDBMS		

Commands and queries	Execution time in ms		
	MySQL	Oracle 11g	PostgreSQL
CREATE TABLE TIRES(ID_TIRES VARCHAR(6) NOT NULL,TIRES_NAME VARCHAR(30) NOT NULL,ID_PRO VARCHAR(2) NOT NULL,ABOVE_CLASS VARCHAR(6), PURPOSE VARCHAR(15) NOT NULL,PRICE NUMBER(6,2) NOT NULL,STOCK NUMBER NOT NULL,SOLD NUMBER NOT NULL);	0.22	0.03	0.49
INSERT INTO TIRES VALUES ('GY3455', 'DURAGRIP', 'GY', NULL, 'LJETNE', 528.66, 12, 1567);	0.05	0.001	0.006
SELECT NAZIV_TIRES, PURPOSE, PRICE FROM TIRES WHERE ID_PRO = 'MN' AND PURPOSE IN (SELECT PURPOSE FROM TIRES WHERE ID_PRO IN (SELECT ID_PRO FROM MANUFACTURER WHERE PRO_NAME='PIRELLI'));	0.001	0.01	0.02
SELECT PRO_NAME, HEADQUARTERS FROM MANUFACTURER WHERE ID_PRO IN (SELECT ID_PRO FROM TIRES WHERE PURPOSE = 'LJETNE' GROUP BY ID_PRO);	0.001	0.001	0.009
SELECT MANUFACTURER.ID_PRO, MANUFACTURER.PRO_NAME,TIRES.TIRES_NAMEFROM MANUFACTURER INNER JOIN TIRES ON MANUFACTURER.ID_PRO=TIRES.ID_PRO;	0.001	0.001	0.006
UPDATE TIRES SET NAZIV_TIRES='DURAGRIP1', STOCK=13 WHERE NAZIV_TIRES='DURAGRIP';	0.06	0.001	0.004
DROP TABLE TIRES;	0.16	0.44	0.08

IV. CONCLUSION

In this article we have compared 3 relational database management systems (RDBMS) - Oracle 11g, MySQL and PostgreSQL. They are compared according to the criteria that we defined, such as the comparison of basic data, syntax, data types and speed performance. At the very end, we will analyze each of the criteria by which we compared the systems for managing databases. The first criterion for comparing was the basic data of each of the RDBMS's. Each of them uses the relational model, MySQL and PostgreSQL have an open source license, while Oracle license is commercial. Oracle databases are designed to be used when the large budget is available and they served to solve very complex problems in contrast to the PostgreSQL and MySQL that are free to use. The number of supported operating systems for MySQL is five, for Oracle is 6, while PostgreSQL supports up to 9 operating systems. The number of supported programming languages for MySQL is 11, for PostgreSQL is 7, and Oracle supports 15 of them. In the first comparison Oracle and PostgreSQL are dominating. The second criterion is the comparison between the syntax of RDBMS's. The syntax is the same in terms of create, insert, select, drop, join for all three RDBMS, and the only difference appears in the writings of certain types of data and how the values within them are written. The third criterion of comparison of RDBMSs is data types. We compared 22 types of data for each RDBMS. For each of the systems, if there is no particular type of data supported, another type of data is supported, that performs multiple functions or has an alternative solution using a different data type or the combination of several. So on the basis of that, there is no best RDBMS, it all depends on what a person who works with database management prefers to use. The fourth and last criterion of comparison is the speed performance of queries in RDBMSs. According to Table 4. it can be seen that the speed of execution is prevailed by Oracle which is the best to use in large and complex systems, while immediately behind it is MySQL intended for free users, education and business, to the fact that MySQL is the most widely used open source system database management. PostgreSQL is also very popular open source system for managing databases and it applies the same as MySQL, only it is much less prevalent in contrast to MySQL. After all the comparisons, Oracle would be the best option for RDBMS, if to the user who uses it the speed and performance of complex operations is the most important thing and is willing to pay, while MySQL is our recommendation for the open source version which is slightly slower than the Oracle by the tests we made. However, we must state that our conclusions are based on a very simple databace and benchmark - more comprehensive syntax and data type comparison, as well as speed performance will be done in the future (with larger and more complex database, as well as more complex queries).

REFERENCES

[1] MySQL, *MySQL Enterprise Edition.*: MySQL, 2016. [Online]. https://www.mysql.com/products/enterprise/mysql-datasheet.en.pdf [04-2016]

[2] PostgreSQL, *PostgreSQL 9.5.2 Documentation.*: PostgreSQL, 1996-2016. [Online]. http://www.postgresql.org/docs/9.5/interactive/intro-whatis.html [04-2016]

[3] Westcon, *Oracle Database 11g Standard Edition with real application clusters.*: Westcon, 2009. [Online]. http://africa.westcon.com/documents/41863/datasheet-oracle-database-419233.pdf. [04-2016]

[4] MySQL, *MySQL 5.7 Reference Manual: Data Types.*: MySQL, 2016. [Online]. http://dev.mysql.com/doc/refman/5.7/en/data-types.html [04-2016]

[5] Oracle Help Center, *Database SQL Language Reference: Datatypes.*: Oracle Help Center, 2016. [Online]. https://docs.oracle.com/cd/B28359_01/server.111/b28286/sql_elements001.htm [04-2016]

[6] TutorialsPoint, *PostgreSQL – Data Types.*: TutorialsPoint, 2016. [Online]. http://www.tutorialspoint.com/postgresql/postgresql_data_types.htm [04-2016]

[7] DB-Engines, *System Properties Comparison MySQL vs. Oracle SQL vs. PostgreSQL.*: DB-Engines, 2016. [Online]. http://db-engines.com/en/system/MySQL%3BOracle%3BPostgreSQL [05-2016]

[8] R. Manger, "Baze podataka." Zagreb, *Element* [2012]. [Online]. http://jadran.izor.hr/~dadic/EKO/baze-podataka.pdf [05-2016]

A Tool for Simplifying Automatic Categorization of Scientific Paper Using Watson API

Luka Cvetković, Boris Milašinović and Krešimir Fertalj
University of Zagreb, Faculty of Electrical Engineering and Computing, Zagreb, Croatia
{luka.cvetkovic, boris.milasinovic, kresimir.fertalj}@fer.hr

Abstract – More than a decade many scientific papers have been dealing with automatic categorization of documents. Until recently there were no wide spread tools or libraries that could be integrated in a custom application. Recent cloud solutions for cognitive computing eased such integration, but integration still requires a lot of work. An idea of extendible library for simplifying document parsing and categorization using IBM Watson API is developed and presented in this paper. As a proof of concept nearly one hundred of scientific papers in a scientific journal were parsed and analysed using Watson API to detect key topics and concepts of each paper which were then compared with keywords provided by the author.

I. INTRODUCTION

Significant data growth in past few decades has been the main driver for increased interest in text analysis. Many scientific papers (e.g. [3], [4], [5], [7]) have dealt with automatic categorization of documents in different areas. Various types of documents have been analysed, from scientific papers to newspaper texts. Motivation for document analysis varies: to detect plagiarism [26], help librarians to categorize scientific papers by providing additional information beyond authors' keywords [6], assigning disease code in medical documents [5], etc.

Until recently there were no widespread tools or libraries that could be easily integrated in a custom application. Recent cloud solutions for cognitive computing eased such integration, but it still requires a lot of work for adaptation and integration in proprietary application. Furthermore, introduction of any new technology can be a misleading hype and what looks appealing at the first sight can cause many problems during development or in a production environment. Notwithstanding the fact that cloud solutions offer monthly payment plans (or some other pay per usage method) switching technology is only theoretically easy and cheap. Change of technology requires many modifications in existing applications, which requires a lot of additional work. An idea presented in this paper is to utilize one of cloud services for cognitive computing and to develop an extendible library that would encapsulate interaction with chosen service, thus easing possible switch of technology.

In the next section, we reflect on the related work followed by short description of present cloud technologies for cognitive computing. In the Selection IV architecture of a proposed solution is elaborated, followed by the description of developed prototype and presentation of obtained results. As a proof of concept nearly one hundred of scientific papers from a scientific journal [1] was parsed and analysed using IBM Watson API [2] in order to detect key topics and concepts of each paper. Comparison of obtained results and author keywords was a first step of documents categorization (i.e. clustering based on key topics). The paper ends with conclusion elaborating future work.

II. RELATED WORK

For more than a decade many scientific papers have dealt with automatic analysis of various documents but an accent was mostly on well-structured documents like scientific papers. Scientific papers are suitable for analysis due to similar formatting However, any document with semantic tags could be a good source for analysis (e.g. META tags in HTML documents [3]).

A typical application is automatically categorizing papers in one of predefined categories or assigning one or more tags to documents. Examples are [4], having predefined keywords for each category and [5] trying to automatically assign international disease codes to medical documents. An idea to observe not only abstract but rather complete text to extract information about document was a motivation for [6]. X. L. X. Lu et al. [7] went further trying to categorize figures inside a document and add additional information to the document in the categorization process.

Although most of the work is related to documents in English language which was found more suitable for text analysis, there is much ongoing research for other languages. Authors introduced new approach [8] for Croatian language that extracts keywords from news articles.

During document indexing process, scientific papers are usually paired with predefined descriptors. Authors of [9] analysed keywords given by authors with the descriptors assigned to the papers. However, it is reasonable to question author keywords because it can be misleading in some cases. Authors in [10] compared keywords with automatically generated keywords and differences were not negligible. Various techniques of keywords extraction and text classification are described in [11], [12], and [13].

Automatic extraction of key concepts and categorization is not only limited to scientific papers and motivated by ease of later information retrieval but it was used to detect religious and political sentiment of a document [14].

Emergence of cloud solutions for cognitive computing opened a whole new perspective of appliance. Thus authors of [15] and [16] developed an interactive application that can answer user questions by analysing the question and finding related facts, I. Wesley-Smith et al [17] used cognitive services to rank scientific papers.

III. TOOLS AND LIBRARIES FOR COGNITIVE COMPUTING

Cognitive services and computing widely gained popularity over the past few years. Most popular services are provided by Microsoft, Google and IBM. Those services are available to developers for making their applications more intelligent, engaging and discoverable. Services are used not only for text analysis but also for picture and video analysis. The main idea behind these services is to make computer experiences more personal and to increase productivity aided by systems that progressively can see, hear, speak, understand and even begin to reason.

A. Microsoft cognitive services

Microsoft cognitive services [18] work across devices and platforms such as iOS, Android, and Windows. API methods are categorized over few categories: Vision, Speech, Language, Knowledge and Search. Each category contains additional API methods providing services like: Computer Vision, Content Moderator, Speaker Recognition, Bing Spell Check, Language Understanding and more. Overview of Academic Knowledge API is given in [19].

B. Google Cloud Platform

Google Cloud Platform [20] provides multiple categories to work with and improves application with new features. Categories that Google provides are: Compute, Storage & Databases, Networking, Big Data, Machine Learning and Identity & Security. Machine learning is one of the most popular API which provides pre-trained models and services to generate own models. In machine learning Powerful Text Analysis reveals the structure and meaning of the text. It is used to extract information about people, places, events and much more, mentioned in text.

C. IBM Watson

IBM Watson [2] offers a variety of services for developing cognitive applications with Alchemy Language application [27] as one of applications. It is a collection of text analysis functions that derives semantic information from context and it works with input text, HTML, public URL or leverage sophisticated natural language processing [28] (NLP) techniques to get a quick high-level understanding of content and obtain detailed insights such as keywords, sentiment, concepts, authors and more.

D. Open Calais

Thomson Reuters Open Calais [21] is a web service that provides intelligent tags to text and thus provides text analysis. It is a natural processing engine that automatically analyse and tags input text so that relevant data from the text can be used in further analysis. Open Calais depends on statistics, machine-learning, big data and custom pattern-based methods so that analysed text is analysed in detail. It searches a text for things like companies, people, cities, industries, products, details and more.

Brief overview of input and output types for previously mentioned cognitive libraries can be found in [22].

IV. SOLUTION CONCEPT

Control flow of a proposed solution consists of four main components: input text (e.g. scientific article), text parser, text analysing component and an analysed data that can be represented in multiple ways (e.g. output file) as shown in Figure 1.

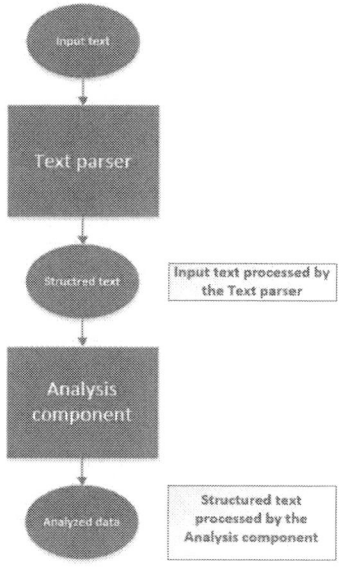

Figure 1. Control flow of the proposed solution

Main idea is independence and possibility of replacement with different implementations due to inheritance of provided interfaces allowing ease of replacement with recent versions and different implementations without changing other components.

Input text is not predefined by any format. For each text format and document type there is a parser compatible with that type of document (e.g. pdf, txt, doc) and document formatting (e.g. standard scientific paper format) must be predefined.

The parser is used for extracting text from a file and preparing text for further analysis. Also, the parser is used

1502

to determine if given text has written keywords provided by the author of the paper. Parser output is structured text that is machine processable and list of author keywords. Parser is implemented using the external library *iTextSharp* [23] that has methods to extract text from documents like pdf, doc, txt.

For a given machine processable (structured text) analysis component can determine text entities, keywords, concepts, language and more. This is the main part of the architecture and it is designed to be open for changes. It is replaceable so that text analysis component can be replaced with custom one. Provided implementation is created using IBM Watson cognitive service that can detect keywords from text. For a given text API call is made to retrieve information about the text. Result from analysis component is analysed data retrieved in JSON format and structured for further analysis

Analysed data is a custom result type. It is structured with same structure regardless of the analysis component implementation. This data can be represented in multiple ways. One of the most used is file representation (e.g. csv files with found keywords). Processed data can be used for further analysis.

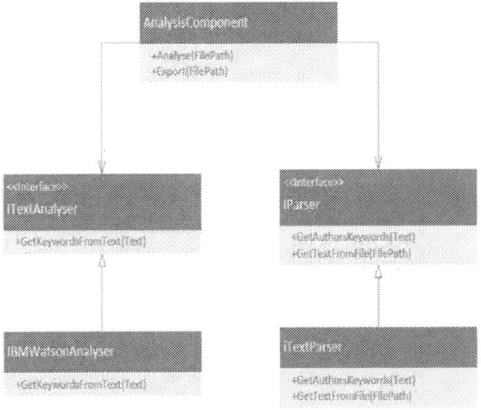

Figure 2. Class diagram

Figure 2 shows main interfaces of the proposed solution. A complete source code can be downloaded from [24]. Idea is that *AnalysisComponent* contains implementation of *IParser* and *ITextAnalyser* so it can perform actions on a given text. *AnalysisComponent* handles all the work, calls *IParser* methods to get text represented as a String and analyses it with *ITextAnalyser* component. After getting information about the text it structures given data so that result is uniformed and further analysis can be done or data can be exported.

V. EXPERIMENTAL EVALUATION

An experiment was done by examining a hundred documents from ComSIS journal [4]. ComSIS has been chosen due its relevance (the journal is indexed in Science Citation Index Expanded) and availability of well-structured documents (PDFs of each paper for each journal volume).

Each document was parsed and analysed using Watson API. Keywords returned from Watson API were compared with author keywords. Resulted accuracy defined as number of hits divided by number of words in total was varying between 0% and 100%. The 0% results were examined and it was discovered hat some of the keywords were not mentioned later in text or abbreviation was used instead of the whole keyword.

An example:

Given text have 4 author keywords: *clustering, requirements engineering, survey, tools*. The same text was parsed and analysed with Watson API and result was 30 keywords from the same text: *analysis, category, classification framework, cluster analysis, current re tools, data, data analysis itself, descriptive analysis, differences, features, global software development, globally distributed development, gsd, gsd features, hierarchical cluster analysis, information inter-rater reliability analysis, iso/iec tr, new features, number, process, rational doors, requirements scenario plus, study, survey, tool, tool capabilities, tool vendors, tools*. Even though Watson API found significantly more keywords and keywords are more detailed not all author keywords were found. Out of four given author keywords only three was found by the API.

Average number of author keywords	4,43
Average number of found keywords	29,73
Average accuracy score	0,622449
Standard deviation of accuracy	0,2393

Table 1.   Number of keywords

Table 2.   Number of keywords

The table shows statistics the program got when papers were examined. Even though system found a lot more keywords (Average number of found keywords row in table) than authors provided (Average number of author keywords row in table) themselves, results are not as good as expected. Average accuracy score calculated as sum of every accuracy divided by number of papers is 0,622449 whereas standard deviation of average accuracy is only 0,2933 which means that there are a lot of papers having low accuracy. The reason for that is keyword comparison method. The system is using exact string matching for comparison. Analysis showed that when there are too few keywords the system is not as accurate as when there are more keywords. When having too few keywords the authors tend to put multiple redundant keywords as one (e.g. one of the keywords was: crawling systems and search engines). If comparison was replaced with approximate method like Jaccard similarity [25] results could be much better. Other options for comparison are yet to be implemented and further analysis is required.

VI. CONCLUSION

Occurrence of cloud solutions enabled wide spread usage of cognitive features in custom applications. An example of such application was shown in this paper to detect difference between keywords authors have provided in papers and keywords detected by cloud cognitive services. As there are several cognitive cloud solutions with different input and output formats and idea of presented work was to develop such system that can ease switching between various cloud solutions. The same approach is utilized for parsing various documents.

A future work would be oriented in simplifying integration of this tool in custom applications, defining plugin interfaces to create an extendable tool simultaneously providing several usage scenarios (e.g. detecting keywords, clustering text, etc.).

A potential risk in future work and significant obstacle could be related to language use. Although there are many natural language processing (NLP) tools for language, current cloud solutions are primary oriented on English. This means that various forms of the same word could be recognized as different words. One of the possible ideas to work around this problem could be to use some domestic NLP processor to parse text and instead of sending raw text to analysis, use a new text combined from words without grammar cases.

REFERENCES

[1] "ComSIS | Computer Science and Information Systems." [Online]. Available: http://www.comsis.org/. [Accessed: 25-Jan-2017].

[2] "IBM Watson - Build Your Cognitive Business with IBM." [Online]. Available: http://www.ibm.com/watson/. [Accessed: 25-Jan-2017].

[3] K. He and C. Li, "Structure-based classification of web documents using Support Vector Machine," *2016 4th Int. Conf. Cloud Comput. Intell. Syst.*, pp. 215–219, 2016.

[4] Y. Ko and J. Seo, "Automatic Text Categorization by Unsupervised Learning," *Conf. Comput. Linguist.*, vol. 1, pp. 453–459, 2000.

[5] B. Ribeiro-Neto, A. H. F. Laender, and L. R. S. De Lima, "An experimental study in automatically categorizing medical documents," *J. Am. Soc. Inf. Sci. Technol.*, vol. 52, no. 5, pp. 391–401, 2001.

[6] P. K. Shah, C. Perez-Iratxeta, P. Bork, M. A. Andrade, and C. P. Iratxeta, "Information extraction from full text scientific articles: Where are the keywords?," *BMC Bioinformatics*, vol. 4, no. 1, p. 20, 2003.

[7] X. L. X. Lu, P. Mitra, J. Z. Wang, and C. L. Giles, "Automatic categorization of figures in scientific documents," *Proc. 6th ACM/IEEE-CS Jt. Conf. Digit. Libr. (JCDL '06)*, pp. 129–138, 2006.

[8] S. Beliga, A. Meštrović and S. Martinčić-Ipšić, "Toward selectivity based keyword extraction for Croatian news,"

arXiv Prepr. arXiv1407.4723, 2014.

[9] I. Gil-Leiva and A. Alonso-Arroyo, "Keywords given by authors of scientific articles in database descriptors," *J. Am. Soc. Inf. Sci. Technol.*, vol. 58, no. 8, pp. 1175–1187, 2007.

[10] C. D. Hurt, "Automatically generated keywords: A comparison to author generated keywords in the sciences," *J. Inf. Organ. Sci.*, vol. 34, no. 1, pp. 81–88, 2010.

[11] A. Onan, S. Korukoglu, and H. Bulut, "Ensemble of keyword extraction methods and classifiers in text classification," *Expert Syst. Appl.*, vol. 57, pp. 232–247, 2016.

[12] M. S and R. N, "Text Classification using Keyword Extraction Technique," *Int. J. Adv. Res. Comput. Sci. Softw. Eng.*, vol. 3, no. 12, pp. 734–740, 2013.

[13] S. Beliga, A. Meštrović, and S. Martinčić-Ipšić, "An overview of graph-based keyword extraction methods and approaches," *J. Inf. Organ. Sci.*, vol. 39, no. 1, pp. 1–20, 2015.

[14] M. Koppel, N. Akiva, E. Alshech, and K. Bar, "Automatically classifying documents by ideological and organizational affiliation," *2009 IEEE Int. Conf. Intell. Secur. Informatics, ISI 2009*, pp. 176–178, 2009.

[15] A. Goel, B. Creeden, M. Kumble, S. Salunke, A. Shetty, and B. Wiltgen, "Using Watson for Enhancing Human-Computer Co-Creativity," *Adv. Artif. Intell.*, no. 4, pp. 22–29, 2015.

[16] A. Goel et al., "Using Watson for Constructing Cognitive Assistants," *Adv. Cogn. Syst.*, vol. 4, 2016.

[17] I. Wesley-Smith, C. T. Bergstrom, and J. D. West, "Static Ranking of Scholarly Papers using Article-Level Eigenfactor (ALEF)," *WSDM Conf. Entity Rank. Chall. Work.*, 2016.

[18] "Microsoft Cognitive Services." [Online]. Available: https://www.microsoft.com/cognitive-services. [Accessed: 25-Jan-2015].

[19] A. Sinha et al., "An Overview of Microsoft Academic Service (MAS) and Applications," *Proc. 24th Int. Conf. World Wide Web Companion (WWW 2015 Companion)*, pp. 243–246, 2015.

[20] "Google Cloud Platform." [Online]. Available: https://cloud.google.com/. [Accessed: 25-Jan-2017].

[21] "Thomson Reuters Open Calais." [Online]. Available: http://www.opencalais.com/. [Accessed: 04-Feb-2017].

[22] A. Schmidt, "Cloud-Based AI for Pervasive Applications," *IEEE Pervasive Comput.*, vol. 15, no. 1, pp. 14–18, 2016.

[23] "iTextSharp" [Online]. Available: https://github.com/itext/itextsharp. [Accessed: 04-Feb-2017].

[24] L. Cvetković, [Online]. Available: https://github.com/lukacvetkovic/Dipl.Project. [Accessed: 04-Feb-2017].

[25] Wen-tau Yih and Cristopher Meek, "Improving Similarity Measures for Short Segment of Text" Available:

https://www.microsoft.com/en-us/research/publication/improving-similarity-measures-for-short-segments-of-text/ [Accessed: 20-Apr-2017].

[26]   Antonio Si, Hong Va Lewong and Rynson W. H. Lau, "CHECK: A Document Plagirism Detection System" Available:
http://www.cs.cityu.edu.hk/~rynson/papers/sac97.pdf
[Accessed: 20-Apr-2017].

[27]   Alchemy API [Online]. Available:
https://www.ibm.com/watson/developercloud/alchemy-language.html [Accessed: 04-Feb-2017].

[28]   Natural Language Processing (NLP) [Online]. Available:
http://onlinelibrary.wiley.com/doi/10.1002/aris.1440370103/full [Accessed: 04-Feb-2017].

The Role of Redundancy and Sexual Reproduction in the Conservation of the Genetic Information Tested on a Cellular Automaton

Viktor Kovács, Valéria Póser
Óbuda University / John von Neumann Faculty of Informatics, Budapest, Hungary
viktorhoz@gmail.com, poser.valeria@nik.uni-obuda.hu

Abstract— **Pixel evolution is a framework created to study various evolutionary mechanisms. It consists of a cellular automaton with programmable and highly mutable cells and a customizable dynamic environment. We modeled the basic aspects of sexual reproduction to test our hypothesis about its role in information integrity and in the curse of evolution in general. Our findings also suggest that if there is a reasonable cost of reproduction, a selective pressure emerges in favor of short genome sizes. We examine the implications of our findings in a bottom-up evolutionary model.**

I. INTRODUCTION

In our simulation we tested the adaptiveness of organisms on different levels of complexity. We were particularly interested in the accumulation/deletion of genetic information in case of sexually and asexually reproducing entities. We then compared our findings with the results from other fields of science, and examine how these results fit in bottom up, or a top down evolutionary model.

II. THE STRUCTURE OF THE CELLULAR AUTOMATON

A. Design goals

Our main goal was to create a model which is sufficiently complex to simulate higher level evolutionary phenomena (Such as resource management and optimization, comparing the fitness of organisms on different levels of complexity, role of sexual reproduction and the effect of these factors on adaptability), while its computing and memory requirements are low enough to make evolutionarily significant simulations on decent population sizes. It was also important to make our system flexible, mutable, and not to rely on hard coded mechanisms. Thus there should be evolutionary path between asexual and sexual reproduction and between simple and sophisticated resource management.

B. Evolution model

The basic unit of our model is "Matter", which could be either a resource or an organism. Organisms can transform resources to other organism through reproduction and they themselves become resources upon their death. Since the amount of resources is fixed during initialization, we have a closed system in which there is a competition for resources.

The resources are implemented as pixels, which also helps in the visual demonstration of our environment. These resources are determined by the 3 color channel of the corresponding pixel (Red, Green, and Blue) and by the amount of matter they represent. An organism can consume a resource if it meets three criteria:

- It has a corresponding "enzyme", which means there is a gene in its genome, which is coding for the RGB channels for the given pixel.
- If its size allow to consume the given amount of resource.
- If the organism occupies the same cell as the resource.

Genes carry 3 bytes of information. This 3 byte can code for an enzyme, can serve as a memory unit, or code for an instruction. In the case of enzymes, as we suggested, the bytes represent the 3 color channel of a pixel.

Every organism carries a virtual processor, which is capable to process the instructions encoded in its genes. This means, that the sequence of genes represent a Turing complete machine. Our organisms therefor can use sophisticated decision making algorithms. However, there is a limited amount of instructions each organism is allowed to process in a given cycle. This means processor time is another factor which plays a role in optimization and adaptation.

There are three ways an organism can contact its environment:

- Movement
- Sensation
- Consummation
- Reproduction

The cost of movement is dependent on the size of the organism, and the size is determined by the number of genes it carries. Sensation allows for the scanning of the information about neighboring cells, and it is independent of the size. This means that it is more efficient for a bigger entity to scan all the neighboring cells and with the help of a decision making algorithm to chose a cost efficient path.

There is no predation in the system yet, therefore to avoid the depletion of resources and reaching stasis, each organism has a specific lifetime determined by the number of its genes. When an entity dies the resources it possesses are redistributed in the environment in the form of spawning new pixels. In summary, the large genomsize leads to more expensive reproduction and movement, but it is compensated by a longer lifetime.

Environmental changes and climates can be simulated if the resources are not distributed randomly. If in a given area a certain color dominates, then in that area a way smaller but specific genepool can be sufficient for survival. (Figure 1.)

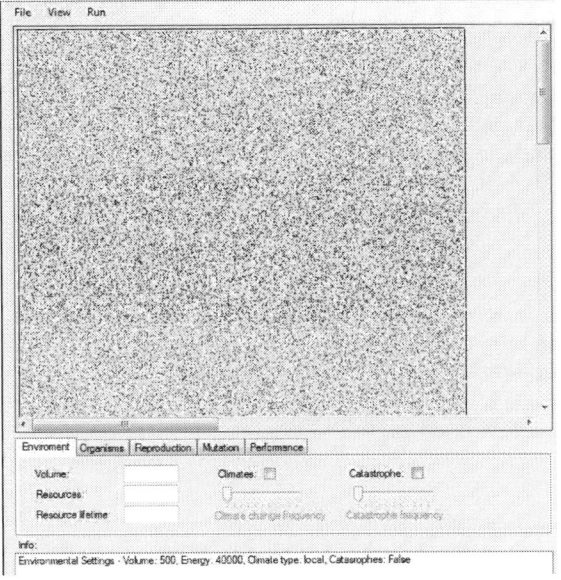

Figure 1. Climates by the non random distribution of colors

The byte base genes are easily mutable. Because of the small interval, there is a smaller mutation space for each gene. We can use point mutations, gene and chromosome duplications and deletions.

Every organism that has an even number of chromosomes is recognized as a sexually reproducing species by the system and upon reproduction it is paired up another entity which has the same number of genes and chromosomes.

C.   The architecture of the framework

To achieve programmability we had to design a processor which utilizes the 3 byte based genes. This processor has the following instruction set:

Invert, Write, Add, Subtract, Compare, And, Or, If, JumpIf, PushByAddress, PopByAddress, Push, Pop, RegA, RegB, RegC, ProgramPointer, ChromosomePointer, Move, Sense, Check, Consume, Divide.

The first byte of a genes represents the memory address, the second byte can be used for storing data or to pass a parameter, and the third byte is used for instruction codes.

In every organism the first six genes are reserved for registers and pointers which store data in the second byte, the third byte of these genes are never used for instructions, therefore we can further utilize these genes by using their third byte as a stack instead.

Addressing was another issue to solve. Addressing by bytes allow only very limited programs to work, so we had to implement some kind of paging system. This is

implemented by using chromosomes. Each chromosome can hold 256 genes, and there can be only one procedure on a chromosome.

Furthermore, we had to separate the enzymes from the instructions. To do so, we use a quite arbitrary constraint. If the last byte of a genes has a value greater then 100, then it is recognized as an enzyme. The last byte otherwise would code for instructions, but the instruction set requires values only from 0 to 21, therefore a value beyond 100 would be unknown for the processor anyway. In addition we wanted to reduce the variability of pixels, because to cover all possible color combinations we would require 256^3 (~16 million) genes. The chance of an organism to carry a pixel specific enzyme would be negligible. Therefore we reduced to problem space by three magnitudes, by the following method: We neglect the hundred decimals on each byte, then we use modulo 25 on the remaining value. Of course the values of the resources also had to obey this role. This results in a much smaller problem space (~16 thousand). In order to display more distinguishable colors, the display method multiplies the 25 based color values by 10 (as seen on figure 1).

To optimize performance each organism also carries a lookup table for its enzymes. When its processor executes the Check instruction, it uses this lookup table to decide whether a resource can be consumed or not.

One of the main features of the Pixelevolution framework is the ability to use a higher level script language to program its entities. Writing code by using bytes, would be increasingly inconvenient as the complexity of a given problem rises, so we created the 8script language as a higher level mediator.

```
block =
 ["var" ident {"," ident}]
 {"procedure" ident ":" block } statement.

statement =
 variable ":=" expression
 | "call" ident
 | "begin" statement { statement } "end"
 | "if" condition "then" statement
 | "while" condition "do" statement.

condition =
 expression("=="|"<"|">"|"!=") expression.

expression = ["+"|"-"]
 term {("+"|"-") term} .

term =
 variable
 | number
 | "(" expression ")" .

 variable =
 ident
 | indexed-ident

 indexed-ident =
 ident "[" expression "]
```

Figure 2. Extended Backus–Naur form (EBNF) of the 8Script

The framework contains a parser and a compiler to transform 8script code into a sequence of genes. The parser implemented by a recursive decent parser algorithm.

III. Results from Running Simulations

A. Selective pressure toward shorter genomsizes

While running the simulation the most remarkable phenomenon was the inevitable selective pressure towards shorter genomes. This pressure can be reduced by different parameters but seemingly cannot be eliminated. We also observed that less complex initial state results stronger pressure towards short genomes. The question naturally arises whether this phenomenon is a particularity of our model or is it a more general trend? To investigate this issue, we examined the documentation of one of the most successful evolutionary simulations, the Avida project. [1] The following remarks can be found in the configuration documentation:

"SIZE_MERIT_METHOD: This setting determines the base value of an organism's merit. Merit is typically proportional to genome length otherwise there is a strong selective pressure for shorter genomes (shorter genome => less to copy => reduced copying time => replicative advantage). Unfortunately, organisms will cheat if merit is proportional to the full genome length -- they will add on unexecuted and uncopied code to their genomes creating a code bloat. This isn't the most elegant fix, but it works." [2]

In the case of Avida, the program compensate the selective pressure in question with this parameter. But it seems in every system in which there is a real cost to reproduction this pressure also arises. The next question is if this pressure exist in the actual living biosphere as well.

B. The role of sexual reproduction

In the case of sexual reproduction, the pressure in favor of the shorter genomes was much less prevalent. At the same time the sexually reproducing population was more prone to extinction. Our hypothesis is that the main evolutionary aspect of sexual reproduction could be its role in conserving genes as it seems.

IV. The Biological Significance of the Results

It seems the selective pressure in favor of shorter genomes indeed exist in actual living systems as well. [3]

"When analyzed in a molecular phylogenetic perspective, every clade of bacteria with genome sizes of <2 Mb was derived from ancestors with substantially larger genomes"

The article also confirms that the selective pressure strongly biased in favor of gene deletion. [3]

From the standpoint of resource management this result is actually not at all surprising. The organism which uses its resources, time and energy (ATP) to copy pseudo genes and unnecessary ones is in a disadvantage against those, which copy only the essential set for survival. This resource cost is particularly significant amongst simple prokaryotes, because here the cost of DNA replication represent a large proportion of the cost of the whole reproduction process. In other words, the reduction of DNA-replication-cost significantly effects the overall cost of reproduction.

Ocman remarks "This pattern dispels the long-held notion that bacteria evolved by the successive doubling of small-genomed progenitors, and raises numerous questions about an evolutionary process that seems to affect all bacterial lineages" [3]

This statement has implications to the whole bottom up evolutionary paradigm. There is much more at stake then the evolution of certain bacterial lineages. The primary proposed mechanism of DNA accumulation in th early stages of evolution is in question. If the selective pressure towards short genomes is prevalent amongst simple organism, the gene and genome duplications can't account for the accumulation of genetic information.

These results cast a fresh light on the very basics of genetics as well, such as the optimization of the universal codon table.

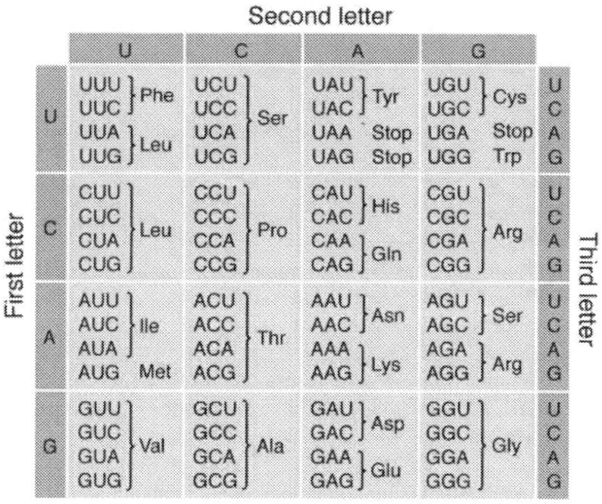

First letter	U	C	A	G	Third letter
U	UUU Phe	UCU Ser	UAU Tyr	UGU Cys	U
	UUC Phe	UCC Ser	UAC Tyr	UGC Cys	C
	UUA Leu	UCA Ser	UAA Stop	UGA Stop	A
	UUG Leu	UCG Ser	UAG Stop	UGG Trp	G
C	CUU Leu	CCU Pro	CAU His	CGU Arg	U
	CUC Leu	CCC Pro	CAC His	CGC Arg	C
	CUA Leu	CCA Pro	CAA Gln	CGA Arg	A
	CUG Leu	CCG Pro	CAG Gln	CGG Arg	G
A	AUU Ile	ACU Thr	AAU Asn	AGU Ser	U
	AUC Ile	ACC Thr	AAC Asn	AGC Ser	C
	AUA Ile	ACA Thr	AAA Lys	AGA Arg	A
	AUG Met	ACG Thr	AAG Lys	AGG Arg	G
G	GUU Val	GCU Ala	GAU Asp	GGU Gly	U
	GUC Val	GCC Ala	GAC Asp	GGC Gly	C
	GUA Val	GCA Ala	GAA Glu	GGA Gly	A
	GUG Val	GCG Ala	GAG Glu	GGG Gly	G

Figure 3. Universal codontable
source: http://www.lucasbrouwers.nl/blog/wp-content/uploads/2010/11/genetic-code.jpg

From the stand point of mutation resilience the codon table is optimized to such a degree, that it is said to be at least one of a million. [4] This means that there is a good chance that a potential point mutation doesn't effect the function of the synthetized protein. This resilience is achieved by optimizing the redundancy found in the coding system. A codon provides 4^3=64 possible state, while it has to code only for ~20 amino acids. The question is how did this system evolve? The possibility of chemical or physical necessity is out of the question, so we have to rely on natural selection. However a selection based optimization would gain millions of equally resilient but completely different codon tables, it doesn't answer its uniformity. To answer uniformity we have to assume, that there was a strong pressure for optimization of the coding system for a while, then after a genetic bottle neck, only one version remained. Then this optimization basically stopped. This account seems a bit too specific, but other than that, it assumes that the pressure is in favor of mutation resilience, while based on our results we can argue that in fact there would be a pressure for shorter genome sizes, with a less redundant coding system.

There are experiments for creating non natural base pairs since the 90s. In 2014 a team successfully expanded the DNA of E coli bacteria with artificial base pairs, which could reproduce and carry the new base pairs for successive generations. [5]

If codons would use 6 different bases instead of 4 (A, C, T, G) then 2 base would be sufficient instead of 3 to code for all the amino acids and for all the syntactic function. This would result in by 33% shorter genomes.

The organism which would use this less redundant coding would be in a huge advantage against its more redundant peers. Probably it would effect negatively only the way more complex forms of life, where the much more intricate systems of gene expressions are more sensitive to genetic errors. During the human ontogenesis for example, in the first trimester there is a 20% chance for miscarriage due to genetic deficiencies. Us, humans seems to need to mutation resilient, redundant coding, but this should be irrelevant in a bottom-up evolutionary process.

A similar case could be made for the inefficiency of sexual reproduction. It appeared more than 1,2 billion years ago, way before the first multicellular life, [6] and it introduced a huge level of redundancy, and increased the cost of reproduction by a magnitude. The supposed advantage was the possibility of new gene combinations in a given population. New gene-combinations provide selective advantage only if it increases the level of adaptability. However sexual reproduction is not optimized for adaptability, but quite the contrary as it is recognized in the form of recombination load. [7]

New gene combinations could be achieved by horizontal gene transfer, without the increased cost of reproduction and the recombination load. We would rather expect this method to evolve as the main source for new gene combinations in the light of the pressure against larger genomes.

V. CONCLUSION

Our findings, particularly the selective pressure in favor of short genomes seem to suggest, that the accumulation of genetic information in the early stages of evolution in a bottom up process is strictly impossible. The existence of the universal codon table and the mechanism of sexual reproduction are also inconsistent with such a process and more in accordance with a genetically top-down approach.

REFERENCES

[1] Avida project, avaible at: http://avida.devosoft.org/

[2] https://homes.cs.washington.edu/~weise/Avida%20Docs/genesis.html

[3] Genomes on the shrink, Howard Ochman, 2005

[4] The Genetic Code Is One in a Million Stephen J. Freeland,1 Laurence D. Hurst2

[5] Life engineered with expanded genetic code". San Diego Union Tribune. Retrieved 8 May 2014

[6] Bangiomorpha pubescens n. gen., n. sp.: implications for the evolution of sex, multicellularity, and the Mesoproterozoic/Neoproterozoic radiation of eukaryotes

[7] Sexual Reproduction and the Evolution of Sex Sarah P. Otto (2008).

Developing MOBA games using the Unity game engine

D. Polančec, I. Mekterović

University of Zagreb, Faculty of Electrical Engineering and Computing, Zagreb, Croatia
domagoj.polancec@fer.hr

Abstract - MOBA (Multiplayer Online Battle Arena) games are currently one of the most popular online video game genres. This paper discusses implementation of a typical MOBA game prototype for Windows platform in a popular game engine Unity 5. The focus is put on using the built-in Unity components in a MOBA setting, developing additional behaviours using Unity's Scripting API for C# and integrating third party components such as the networking engine, 3D models, and particle systems created for use with Unity and available through the Unity Asset Store. A brief overview of useful programming design patterns as well as design patterns already used in Unity is given. Various game state synchronization mechanisms available in the chosen networking engine, Photon Unity Networking, and their usage when synchronizing different types of game information over multiple clients are also discussed. The implemented game retains most of the main features of the modern MOBA games such as heroes with different play styles, skills, team versus team competition, resource collection and consumption, varied maps and defensive structures. The paper concludes with comments on Unity 5 as a MOBA game development environment and execution engine.

I. INTRODUCTION

MOBA games are match based multiplayer online games in which a team of players competes against another team in fulfilling one or more objectives in closed games environment usually referred to as maps. What differentiates MOBA games from similar games such as FPS (First Person Shooter) games is usually the heightened strategic aspect borrowed from RTS (Real Time Strategy) games from which MOBA games evolved. This includes resource collection and isometric game view. Most often, the primary objective of a MOBA game match is the destruction of the opposing team's defensive structure. However as more and more MOBA games are developed, alternative objectives can be found. [1]

Today MOBA games are one of the most popular online game genres. Riot Games, the publisher of arguably the most popular MOBA game League of Legends, claims that the game had over one hundred million active players monthly in 2016 [2].

This paper discusses the implementation of a prototype of a MOBA game in the Unity game engine. Game engines are software tools for game development. They offer complete implementations of software components commonly used in game development such as the physics engine, the rendering engine, the sound engine, network support and other. They also allow the game developers to

develop components specific to the game that is being developed on top of or within these systems.

The Unity game engine was chosen because its most basic version is free to use and because there are many learning resources online available for it. Unity is also one of the three currently most popular commercially available game engines, the other two being Unreal Engine and CryEngine. Unity is developed by Unity Technologies based in Denmark [3].

The implemented game is called Moghi Battles and retains most of the MOBA features which will be mentioned in Section III.

II. RELATED WORK

There hasn't been much work on the topic of MOBA game development in particular except for a couple of online video tutorials and a recorded presentation given by the lead architect of League of Legends [4]. The work on online games in general, however, is more easily found ranging from papers discussing architectures and algorithms used in online games to online video tutorials offering implementation suggestions. For example, the work by Prabha [5] discusses architectures and login methods used in online games implemented in Unity. The work also discusses implementing authentication in the Photon Unity Networking engine [5] which is also used in the implementation discussed in this paper. The work by Bharambe, Pang, Seshan and Colyseus [6] discusses a different, distributed, architecture for online games and ways of synchronizing the game state in such architectures. Another work that discusses distributed architecture in online games is work by Assiotis and Tzanov [7] which focuses on MMORPGs (Massively Multiplayer Online Role Playing Games) with large virtual worlds. The work by Hsu, Ling, Li and Kuo [8] compares the traditional client-server server architecture, the peer-to-per architecture and the mirrored-server architecture concluding that the client-server architecture is the best fit for latency sensitive online games and proposing a clustered-server architecture to improve scalability. The work also discusses the two main approaches to game state synchronization: the conservative and the optimistic approach [8]. The former attempts to prevent inconsistencies in the game state over multiple clients completely, while the latter tolerates small amount of inconsistencies [8].

There are a number of papers covering Unity engine itself. Some of these papers discuss using Unity as a tool for developing applications which aren't games. An example of such work is a paper by Craighead, Burke and Murphy [3] which discusses implementation of a robot simulator in Unity. Official Unity page also offers plenty of tutorials aimed at helping new Unity developers to get started and to understand basic Unity concepts [17]. Finally, there are plenty of video tutorials available on popular video sharing sites such as YouTube which explain how to use Unity to develop certain types of games. One very useful such video tutorial is a series of videos uploaded by user quill18creates that shows how to make an FPS game in Unity using Photon Unity Networking [12].

It is not known, to the best of our knowledge, whether there are commercially successful MOBA games developed in Unity.

III. BASIC MOBA CONCEPTS

It is difficult to precisely define the MOBA genre due to an increasing number of diverse titles on the market, however the basic concepts found in most MOBA games are: heroes, teams, different skills for different heroes, resource collection and consumption, bases with defensive structures, creeps and lanes [1].

Heroes are player controlled units. Each player usually controls only one unit, their chosen hero, and there is no army management aspect which is present in RTS games [1]. MOBA games usually allow the players to choose from a range of different heroes with different play styles, skill sets and roles. Heroes can have different resources at their disposal of which the most common is health. Health is a resource which, when depleted, causes the hero to die. In most MOBA games, heroes can *respawn* at specified locations called *respawn points*.

Teams in MOBA games are usually two to six players in size and there are usually two teams in a match [1]. Teams are competing against one another in achieving the main objective of the match which can vary from having to destroy the opposing team's defensive structures to having the highest number of *kills*. The team which achieves the main objective is the winning team.

Skills are hero-specific actions available to players which are usually the only way to interact with other player's heroes, as well as to provide boosts to one's own hero. In most MOBA games skills can be activated by a specified keyboard shortcut. Activating a skill usually costs a certain amount of non-health resources. Skills also have a *cooldown time* which starts immediately after the skill has been activated and during which that same skill can't be activated again. Skills can vary greatly between different heroes and they define a hero's play style.

Resources other than health are used to perform certain game actions such as activating a skill, resurrecting a hero or purchasing other benefits. Resources are usually automatically accumulated as the time passes, but can also be gained by defeating other players or destroying secondary defensive structures [1].

Bases are areas where players start the match and where main defensive structures are located. Each team has their own base. Defensive structures in the bases can also have health and some of them can also damage the enemy team's heroes. The main objective of a MOBA match is most often to infiltrate the enemy team's base and destroy the main defensive structure.

Lanes are paths that lead from one base to the other. The number of lanes can vary but there are usually three: the top lane, the bottom lane and the middle lane. The middle lane is usually the shortest path between the two bases.

Creeps are units controlled by artificial intelligence and which also have health. Creeps usually spawn in the base of the team that they belong to and travel across lanes to the enemy base. Creeps can damage the enemy team's heroes or other creeps. [1]

The MOBA prototype implementation discussed in this paper implements all these concepts except creeps as artificial intelligence controlled units were out of the scope of the implementation requirements.

IV. IMPLEMENTATION ARCHITECTURE

The typical components of the client-server architecture in network games are: the game client, the session server, the user database, the game server and the game database [5] (Figure 1).

Game client is an application running on the player's computer which enables the player to observe the state of the game which is simulated elsewhere or to simulate the game themselves. [5]

Session servers (also called lobby servers) are servers that open a session between a client and a game server [5]. This is usually done only after the player provided valid authentication credentials through their game client or a special launcher application by comparing the given credentials with data stored in user database.

Game server is a server which either simulates a game and/or facilitates communication between clients [5]. Commercial MOBA games usually have more than one game server to improve scalability by distributing matches over servers [4]. Game servers may have access to game databases if the game needs to store persistent

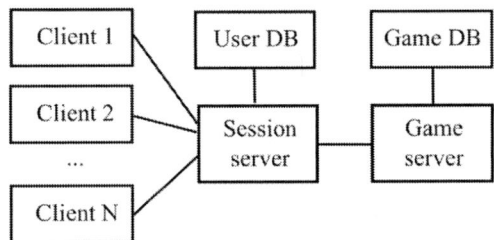

Figure 1. **Traditional client-server architecture for online games**

information [5] (for example storing persistent world data in MMORPGs).

The architecture of the implemented game differs slightly from the described client-server architecture. The game and session servers aren't part of the implementation, instead, the Photon Cloud service is used (Figure 2). Photon Cloud [9] service acts as both a session and a game server when working with PUN (Photon Unity Networking) framework. PUN will be discussed in more detail in Section VI. Furthermore, due to the simplicity of the implemented game, there is only one database which stores user and match participation data. This database is accessed using the JDBC (Java Database Connectivity) API through the data server which is implemented in Spring. Spring is a popular Java web development framework. Part of the League of Legends server stack is also implemented in Spring [4]. The data server offers HTTP endpoints for registering a new user, obtaining authentication tokens and reporting game results. Both the client and the Photon Cloud service access the Data Server. As the Photon Cloud service is generic and game-agnostic, it doesn't implement any form of authentication on its own, which means that it must delegate that task to a specialized authentication service that implements the PUN authentication interface [10]. The data server's HTTP endpoints for authentication implement that interface and the Photon Cloud is configured to use it as an authentication service. After successful authentication, the data server sends the authentication token to the Photon Cloud which relays it to clients. The clients communicate with Photon Cloud either through TCP or UDP protocol. Section VI goes into more detail on which of these protocols is used in which situations. Since data server's implementation is relatively simple, this paper will henceforth focus on client implementation.

V. BASIC UNITY CONCEPTS

This section explains the most important concepts and features in Unity. Since the implemented game is a 3D game, this section focuses on Unity features for 3D game development. The Unity Editor is a desktop application used to create games based on Unity engine [11]. The Editor user interface is composed of a scene view in the centre of the screen, a file explorer on the bottom of the screen, a hierarchy view on the left of the screen, a game object or an asset inspector on the right of the screen and a menu bar on the top of the screen (Figure 3).

A. Projects

All work in Unity is stored within Unity Projects. When a new project is created, a folder with the same name is also created in the directory specified by the user. The newly created project will contain the Assets folder which contains all game-specific data such as 3D models, sounds, scripts, etc. The Assets folder is initially empty but all assets which are accessed dynamically during runtime must be put in the Resources subfolder within the Assets folder.

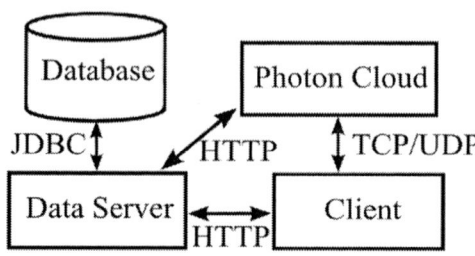

Figure 2. **Implementation architecture**

B. Scenes and Game Objects

A scene is a 2D or 3D space that contains game objects [13]. Scenes can be menus, game levels or anything that is being shown to the player [13]. Creating a new project automatically creates the initial scene. New scenes can be created through the menu bar. Scenes have a *.unity* file extension [13].

Game objects are objects that exist within a scene. When a new scene is created, it will already contain two game objects: a camera and a light source [13]. Having at least one camera in a scene is necessary for the scene to be shown to the player. Users can add additional objects to the scene through the hierarchy view, the main menu or by dragging-and-dropping objects from the file explorer on to the scene view. Another game object can be attached to a parent game object.

C. Components

By themselves, game objects don't do anything and they need to have components attached to them to have a function. This means that Unity by design enforces using the **component pattern**. The component pattern is a decoupling pattern that is used when a class interacts with many domains and it is necessary to keep domain-specific functionalities decoupled from each other or when different objects share same functionalities, but one doesn't strictly inherit another [15]. Unity provides an easy way to facilitate inter-component communication either through the Editor or programmatically.

In Unity, game objects can be activated and deactivated and components can be enabled and disabled. The components attached to deactivated game objects are disabled. This provides an easy way of implementing **the state design** pattern. The state design pattern is an often-

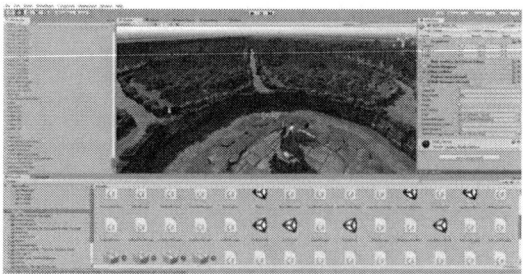

Figure 3. **The Unity Editor**

1512

used design pattern in games that allows implementation of state-machine like behaviour in object-oriented paradigms [16] (each state in the state machine is modelled as an object). In Unity, this pattern can be implemented by simply activating an object that models the current state, and deactivating other state objects.

D. The Scripting API

Although Unity ships with many built-in components, most games require additional game-specific behaviours. These behaviours can be implemented programmatically in special components called Scripts by using Unity's Scripting API. Scripts can be written in either C#, JavaScript or Boo. In this implementation, all Scripts were written in C#. The Scripting API is an IoC (Inversion of Control) framework, which means that the Scripts written by developers can't control the execution of the game. Rather, the developers have to implement certain methods in the Scripts and Unity will call those methods under certain conditions [18]. The most common methods found in Scripts are Start and Update methods. Update method is called before a frame is rendered and as such the bulk of the game logic is usually executed within Update methods. The Start method is called before the first time an Update method is called on the Script [18].

E. Unity Asset Store

The UAS (Unity Asset Store) is a web marketplace that allows Unity users to buy and sell custom assets that can be used in Unity. The UAS is available through the Editor. The Editor offers an easy way of importing assets obtained through the UAS. The newly downloaded assets are placed in their own folder within the Assets folder. Assets such as 3D models often come attached to game objects which can even have some simple scripts or other components attached to them. Other assets come in formats compatible with Unity.

VI. PHOTON UNITY NETWORKING

PUN (Photon Unity Networking) is a framework for online game development in Unity. PUN is available through the UAS in two versions, the free one and the more advanced paid one. Moghi Battles uses the free version. PUN is generally very easy to use as most of the framework's functionalities are exposed through static members of a single class that acts both as a façade in the façade pattern and a singleton object in the singleton pattern.

There are two main concepts in PUN: rooms and lobbies. Connecting to a lobby conceptually corresponds to connecting to a session server, while connecting to a room corresponds to connecting to a game server. This means that a player must connect to a lobby before they can connect to a room. After successful authentication, the player joins a lobby from which they can join or create a room. Photon Servers do not simulate the game. They only provide a means of sending data from one client to the other that fully simulate the game and use PUN to synchronize the game state between themselves. This communication is achieved through the following synchronization mechanisms: object synchronization, PUN RPCs (Remote Procedure Calls) and custom room properties [19].

Object synchronization [19] is used to synchronize simple game object states (such as current health status or location) and to instantiate game objects over the network. This is done by attaching a specialized script to the game object [19]. PUN periodically synchronizes this data between clients. PUN can be configured to either use UDP or TCP for game object synchronization. Most often, UDP protocol is used because responsiveness is often more important than, for example, the exact position of the object. This is also the case with Moghi Battles.

PUN RPCs [19] allow calling script methods on other clients. This provides an easy way to synchronize events between clients. PUN RPCs always use the TCP protocol to ensure events are synchronized correctly. PUN RPCs can be either buffered or unbuffered. Buffered RPCs will be called on all clients that are currently in the room and the clients that join a room in the future. Unbuffered RPCs are only called on clients currently in the room.

Custom room properties are key-value pairs used to synchronize values that do not belong to any objects [19].

VII. MAPS

In the implemented game, players can choose to play on one of two maps with different main objectives. Maps are implemented as Unity scenes. These scenes contain game objects that can be conceptually divided into three categories: controllers, active objects and passive objects. Controllers are empty game objects that contain scripts that implement game-specific logic which isn't a part of any other object's behaviour. Both active and passive objects are objects that have components which enable them to interact with other objects in the scene or make them visible to the players. In this case the physics related components are all built-in Unity components, while all 3D models, textures and particle effects on these objects are third-party assets obtained through the UAS except for the terrain which was sculpted with the built-in terrain editor. Active objects also contain scripts that implement game-specific behaviours for those objects. For example, active objects are the camera and the defensive structures while passive objects are terrain and non-damageable environment objects. The background music playing on

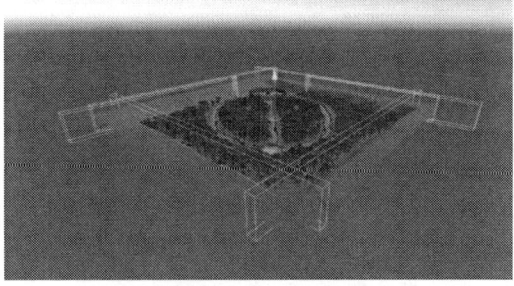

Figure 4.   **Scene view in the Editor showing the Sacred Forest map with invisible walls highlighted**

both maps is also a third-party asset obtained through the UAS.

Players are kept within the boundaries of a map by invisible walls. The invisible walls are game objects without textures but with components that enable collision (Figure 4). Players can only join matches that are still in *waiting phase* or *preparation phase*. Every newly-created match begins in the *waiting phase* and the player that created it joins a new room and sets the current server time as a custom room property and sends out a buffered RPC call that increments the player counter on all clients. When other players join a map, they join the same room and increment the player count via buffered RPCs as well. As soon as both teams have a minimum number of players, the match moves to the *preparation phase* which lasts one minute. After the *preparation phase* the match moves into the *combat phase* and teams can begin competing. If a Hero dies during the *combat phase*, they are respawned at their base.

The maximum number of players in a team on both maps is three, while minimum is one. The current number of players in the team, as well as the information about the current phase of the game is displayed to players at the top of their screens. The players within the same team can also send messages to each other via a chat function located in the bottom left part of the. The chat function also uses RPCs to synchronize messages.

A. The Sacred Forest map

The Sacred Forest map has three lanes and is set in a forest setting which is a common theme among MOBA games. The main objectives on this map are different for each team, which is unusual for MOBA games. The offensive team's objective is to destroy the defensive team's main structure within ten minutes. The structure that needs to be destroyed is located at the heart of the defensive team's base while the defensive team's main objective is to hold off the offensive team's efforts. The defensive team is helped by defensive structures in form of walls which surround the defensive team's base and rotating fire towers that damage nearby offensive team players. The fire damage effect on the fire towers is implemented as a rotating non-collideable child object that notifies offensive team players to take damage through the RPC mechanism. This map contains a relatively large number of objects compared to the other map (823 game objects in total, many of which were manually placed).

B. The Desert map

The Desert map is set in a desert setting and features only one lane. The main objective of both teams is to be the team with highest number of *kills*. Whenever a hero deals enough damage to another hero to cause that hero to die, that counts as a *kill* for the team of the player who controlled the hero that dealt that damage. This map also includes power ups strewn across the map which can be picked up. These power-ups are also game objects with collider components set in trigger mode. This map is much smaller than the other map to provide more action-packed and less strategic experience.

VIII. HEROES AND SKILLS

Players can choose between two heroes implemented in Moghi Battles. These heroes have different attributes and sets of skills which give them somewhat different playstyles. Hero attributes are starting health, attack, defence and speed. Heroes are composed of the main game object and two types of child objects. The first type defines static hero appearance (3D models, and textures). The second type are objects that define area of effects of skills. The 3D models, textures and animations are third party assets obtained through the UAS. The built-in physics components are attached to both the main object and the second type of child objects. The main object also contains the PUN script required for synchronization and the built-in component which is used to control animation transitions. Initially, this component contains no transitions and they should be defined by game developers. Other scripts attached to the main object implement game-specific logic such as movement, resource and attribute tracking and skill-related logic. All skill icons, skill particle effects and hero sound effects are third-party assets obtained through the UAS.

Hero movement is eight-directional and is achieved by pressing W (forward), A (left), S (backward) or D (right) keys. The same key combinations for movement are commonly found in many games. Pressing each of these keys results in a corresponding direction vector applied to the Hero main object. If more than one key is pressed, the direction vectors are all summed up and normalized. The final movement vector is a result of multiplying the total normalized direction vector with the speed attribute.

Activating skills costs gold. Gold is an expendable resource in Moghi Battles which is auto-generated with the passage of time. Players can gain additional gold by destroying enemy defensive structures and killing enemy heroes. Both heroes have three skills, two damage skills and one buff skill. One damage skill is weak with very short cooldown and very cheap gold cost. The other damage skill is much stronger with much longer cooldown and much higher cost. The buff skill is very powerful on both heroes and as such it has the longest cooldown time and the highest gold cost. Both types of skills are represented by scripts that extend an abstract class that implements everything that is required by both skill types such as listening to player input and managing particle effects. The extending classes need only implement three methods which are called at different points in the skill lifecycle. This is an implementation of a **template method design pattern** [20]. Skill bar at the bottom of the screen indicates to players which skills can be activated and which skills are either on cooldown time or the player doesn't have enough gold to pay their costs. The current amount of gold the player has is displayed in the top right corner.

Damage skills are skills that are used to defeat other players and destroy enemy defensive structures. These skills have an array of damage effects with defined starting times. The starting times are defined as offsets from the time when the skill is activated. Each damage effect has one area of effect associated with it. When the start time for a damage effect passes, the game looks up all damageable objects (opposing team's defensive structures or opposing team's players) currently within the area of effect and uses the RPC mechanism to notify those objects that they should take damage. Each damageable object has a health bar that indicates how much damage that object has taken (indicated by red colour) and how much more it can take (indicated by red colour).

Buff skills alter the current player attributes when they are activated and restore them to the previous state when they are done. In Moghi Battles this behaviour is implemented in accordance with **the command design pattern** [14].

IX. TESTING WITH OTHER PLAYERS

The game was tested by a group of four people. Two testers were familiar with MOBA games and had experience playing them while the other two weren't. There were two tests in total and tests consisted of playing out a full match. The second test was also recorded (Figure 5) [21].

A. The first test

During the first test, several issues were discovered. The game sometimes didn't place players in teams properly and the transition between the waiting phase and preparation phase happened before the necessary prerequisites were met. This issue was caused by a bug in the code that synchronized the number of players among clients. Another issue that was discovered was that one of the heroes was too strong compared to the other one.

B. The second test

The second test was conducted after the issues discovered in the first test were fixed. No further significant issues were discovered, only minor performance related issues due to insufficient optimization.

X. CONCLUSION

Although there are many online tutorials for working with Unity and the Unity Scripting API is very well documented the general impression is that Unity has a very steep learning curve which means that many concepts must be learned by absolute beginners before any more advanced projects can be undertaken. Performance wise, the games built with Unity run reasonably well even without optimizations. Some movement choppiness was observed with V-sync on, however, the problem was resolved by turning V-sync off. The Scripting API is very intuitive and easy to use as well, though the fact that it is an Inversion of Control framework limits what kind of algorithms can be implemented. What makes choosing Unity worthwhile is the UAS which has a very large number of third party components, both paid and free, readily available to developers. In total, 59 classes were implemented and 7 unity scenes were created (two maps and five menus).

Figure 5. **Screen capture of Moghi Battles game play from the recording of the second test.**

REFERENCES

[1] Multiplayer Online Battle Arena, December 12th, 2015, http://www.giantbomb.com/multiplayer-online-battle-arena/3015-6598/

[2] Our Games | Riot Games, http://www.riotgames.com/our-games

[3] Craighead, J., Burke, J. and Murphy, R., Using the Unity Game Engine to Develop SARGE: A Case Study (2008)

[4] Delap, S., Stafford, R., League of Legends: Scaling to Millions of Ninjas, Yordles, and Wizards, 7th April, 2011., https://www.infoq.com/presentations/League-of-Legends

[5] Prabha, C., Login and Networking Services for Online Multiplayer Games, Master's Thesis, Helsinki Metropolia University of Applied Sciences (2015)

[6] Bharambe, A., Pang, J., Seshan, S., Colyseus: A Distributed Architecture for Online Multiplayer Games, 3rd Symposium on Networked Systems Design & Implementation, San Jose, California, United States of America (2006)

[7] Assiotis, M., Tzanov, V., A Distributed Architecture for MMORPG, The 5th Workshop on Network & System Support for Games, Singapore, (2006)

[8] Hsu, C., Ling, J., Li, Q., Kuo, C.C.J., On the Design of Multiplayer Online Video Game Systems, University of Southern California (2003)

[9] Photon Unity 3D Networking Framework SDKs and Game Backend | Photon: Multiplayer Made Simple, https://www.photonengine.com/en-US/PUN

[10] Custom Authentication | Exit Games, https://doc.photonengine.com/en/realtime/current/reference/custom-authentication

[11] Unity – Editor, https://unity3d.com/unity/editor

[12] Unity 3d: Simple First-Person Shooter Tutorial, 28th February, 2013, https://www.youtube.com/watch?v=mbm9lPB5GPw

[13] Scenes, http://docs.unity3d.com/Manual/CreatingScenes.html

[14] Command Pattern | Object Oriented Design, http://www.oodesign.com/command-pattern.html

[15] Component, Decoupling Patterns, Game Programming Patterns, http://gameprogrammingpatterns.com/component.html

[16] Design Patterns - State Pattern, https://www.tutorialspoint.com/design_pattern/state_pattern.htm

[17] Unity – Manual: Unity User Manual (5.5) https://docs.unity3d.com/Manual/UnityManual.html

[18] Unity – Manual: Execution Order of Event Functions https://docs.unity3d.com/Manual/ExecutionOrder.html

[19] Synchronization and State | Exit Games, https://doc.photonengine.com/en-us/pun/current/manuals-and-demos/synchronization-and-state

[20] Design Patterns - Template Pattern, https://www.tutorialspoint.com/design_pattern/template_pattern.htm

[21] moghibattles, 15th July, 2016, https://youtu.be/zMOADfahpu8

Comparative analysis of tools for development of native and hybrid mobile applications

Tena Vilček*, Tomislav Jakopec**

* Faculty of Humanities and Social Sciences, Osijek, Croatia
tvilcek@ffos.hr
** Faculty of Humanities and Social Sciences, Osijek, Croatia
tjakopec@ffos.hr

Summary – One of the main reasons for wide acceptance of smartphones are mobile applications that offer a wide variety of features. This paper deals with types of mobile applications: hybrid and native, as well as tools that enable their development. For purposes of this paper, a total of eight simple, identical applications were made for Android, iOS and Windows Phone. Native applications for Android, iOS and Windows Phone were made in integrated development environment, or so-called IDEs: Android Studio, Xcode and Visual Studio. Hybrid applications for Android and iOS were made in Ionic, PhoneGap and NativeScript framework. The paper compares tools used for development through following criteria: supported computer operating system, supported mobile platforms, programming languages, official documentations and community of programmers, installation and development. Goal is to research the advantages and disadvantages of the tools used for development of native and hybrid mobile applications and to find out which applications are the most profitable.

I. INTRODUCTION

Features and options offered by smartphones through mobile applications are immense. Purpose of the application is to resemble a standalone application, not a web site, and it would be preferred for it to be interactive. At first, mobile applications were created for information purposes and productivity, such as electronic mail, calculator, weather forecast, calendar and similar, but their diversity and number have increased enormously in the past couple of years [1].

There are three categories of mobile applications: native, web and hybrid. Choice of the appropriate one depends on the requirements that need to be met [2].

Native mobile applications are specific to a certain mobile platform and they are developed in an "official" programming language of that platform, for instance, native Android applications are developed in Java programming language. Such applications are installed directly on a device and they communicate with its operating system. For that reason, they can utilize the features of the device to a maximum [2]. One of the main characteristics, as well as advantages, is application's speed of performance on the device, and a disadvantage is that users of such applications must update them

regularly [2]. Considering the fact that in application development on different platforms one must use different programming languages, expenses of developing it are often higher. However, despite the aforementioned disadvantages, native applications offer by far the best usability, features and overall user experience [3].

The second category is comprised of web mobile applications based on technologies and languages such as HTML5, CSS and JavaScript [3]. These applications are actually web sites adapted to smaller screens that can be opened in any browser [3]. Therefore, the only condition, which can also be a disadvantage, is the fact that Internet connection is necessary for starting such applications. Furthermore, they are independent of mobile platforms, which means that development programmer simultaneously creates an application for several operating systems or mobile platforms and that the duration of the process and expenses are smaller [3]. Web applications could not, until recently, access native features of mobile devices, but with HTML5, there is a feature of accessing, for instance, a camera or a microphone.[1] Web applications perform slowly and they do not appear in official application stores [2].

As an excellent blend of native and web applications, the third category represents hybrid mobile applications. Hybrid applications use frameworks which enable development programmers to create applications that behave like native ones, using technologies such as HTML5, CSS and JavaScript [3]. These applications are installed on a device just like native ones and they have access to most features of a device: camera, GPS, and similar [2]. Their main characteristic and a great advantage is that they can be started on several different platforms, so the price and time necessary for development are significantly decreased. However, considering the fact that hybrid applications are developed using the aforementioned technologies, they do not perform completely native-like, which results in somewhat slower performance [2].

---

[1] HTML 5: camera access,
http://stackoverflow.com/questions/9431475/html5-camera-access
(2017-02-02)

In creating applications, development programmers can use various IDEs[2] and frameworks. During the development of native mobile applications, it makes most sense to use the official IDE for a certain platform. On the other hand, for development of hybrid applications, there are different frameworks.

For needs of developing native applications, the following IDEs were used: Android Studio[3] for developing Android applications, Xcode[4] for developing iOS applications and Visual Studio[5] for developing Windows Phone application. In developing hybrid applications for Android and iOS, Ionic[6], NativeScript[7] and PhoneGap[8] frameworks were used. Out of all the aforementioned frameworks, Ionic was largely focused on the appearance of the application and user interface, and is considered a front-end work environment, so it needs a so-called wrapper, such as PhoneGap, in order for the application to get native characteristics – in other words, to access a microphone, phonebook, phone (calls) and similar. These technologies or tools are compared and analyzed according to six criteria: supported computer operating systems, supported mobile platforms, programming languages, official documentation and community of programmers, installation and development. Goal of this paper is to answer the questions: what are the advantages and disadvantages of tools used for development of native and hybrid mobile applications and which ones are the most affordable.

A. Related works

Khandeparkar, Gupta and Sindhya compared frameworks for development of hybrid mobile applications: Ionic, Famo.us/Angular and OnSenUI according to four criteria: documentation, performance, community and learning curve. On average, Ionic had the best rating. According to criteria interface design, user experience and performance, safety, tools and debugger, native applications are a better choice, while according to criteria expenses, time necessary for development, maintenance and independence of platforms, hybrid applications represent a better solution [4]. Y. A. Redda in his paper states the criteria according to which one could compare tools for development of hybrid applications, or criteria that would help choosing the appropriate tool. Criteria are: supported platforms, features of the device of access, programming languages and license [5]. D. Markov states 7 frameworks, and for each one states advantages and disadvantages, but he compares them according to several criteria: native appearance and feeling, prerequisite, community,

documentation, tools. Ionic and NativeScript, which are analyzed in this paper as well, are according to his criteria two most powerful frameworks, and the only criterion in which NativeScript got weaker rating is community, while in others it was even better than Ionic. However, at the end he concludes that there is no ideal framework because each one has its advantages and disadvantages [6].

On a webpage, *Mobile frameworks: comparison chart* there is a list with more than 20 frameworks for development of mobile applications. Also, there are criteria: platform, type of application, programming language, device's options, user interface, license and other, and they offer a feature of choice of the appropriate demand. Based on the elements chosen, if there is work environment that corresponds to all demands, a person who cannot decide while choosing a framework, this way can find a solution easily [7]. Comparisons of IDEs are mostly done according to criteria of license, computer operating system and according to various features they have, such as debugger, GUI[9] builder, autocomplete and similar [8]. It is stated how the best IDEs are determined according to customer satisfaction and various measures, based on which they are put into four categories on the grid: Leaders, High Performers, Contenders and Niche products. Examples of leading IDEs are Visual Studio, Xcode, NetBeans, IntelliJ IDEA and such, while examples of niche products are Aptana Studio and Code::Blocks [9].

B. Comparison according to criteria from works related

World wide web has many articles that compare Android and iOS applications according to various criteria, and less of those dealing with Windows Phone applications. Several articles states how development of Android applications can take 2 or 3 times longer than development of iOS applications, according to which number of lines in code is larger, which is often a result of Android fragmentation. It means that Android devices come in various shapes and sizes with large differences in performance. Furthermore, it includes a multitude of various Android versions that are active at the same time. Taking the aforementioned into consideration, it turns out that development of Android applications is more expensive than developing iOS applications [10, 11]. Tomislav Car presents statistical data proving the aforementioned. He analyzed 6 applications meant for both platforms and measured working hours and number of lines in code. In the end it was concluded how Android applications have about 40% less code than iOS application. It is interesting how in some application that percentage amounted up to 189%, while one of the projects had a difference of 4%, which was not as significant. Comparison also indicated how about 30% more working hours are necessary for developing Android applications [12].

---

[2] IDE – Integrated Development Environmnet
[3] Android Studio, https://developer.android.com/studio/index.html (2017-02-02)
[4] Xcode, https://developer.apple.com/xcode/ (2017-02-02)
[5] Visual Studio, https://www.visualstudio.com/ (2017-02-02)
[6] Ionic, http://ionicframework.com/ (2017-02-02)
[7] NativeScript, http://docs.nativescript.org/ (2017-02-02)
[8] PhoneGap, http://phonegap.com/ (2017-02-02)

---

[9] GUI – graphical user interface

Adam Sinicki and several other authors of different articles state how iOS simulator is better and faster for debugging than Android emulators, which enhances the speed of development [11, 13]. Furthermore, emulators for Windows Phone applications are considered better and faster than Android's [14].

When it comes to publishing applications in official stores, situations differ significantly in different platforms. Firstly, it is necessary to register on platform's official stores. Registering on Google Play for publishing Android applications costs 25$ and it is permanent, which means that it does not have to be paid on the annual basis [13, 15]. Apples for annual fee in App Store requires 99$ [13], while registration in Windows Store for individuals amounts 19$, and 99$ for a company, and it is also permanent [16]. After the application is sent, i.e. uploaded, it goes through a phase of testing. In Google Play Store, the process between testing and accepting takes several hours to several days [13, 17]. Situation is similar with publishing Windows Phone applications, where process takes a couple of hours up to maximum 3 working days [18]. Process of testing iOS applications can take up to two or three weeks, after which there is a great chance of rejecting the application because of Apple's high standards [13].

Nic Raboy analyzed Ionic and NativeScript. NativeScript enables access to any component of a devide or API directly with JavaScript, which means that if a necessary plugin does not exist, development programmer can independently create his own plugin with knowledge of JavaScript. Ionic is different, because it must access API through Apache Cordova, which is the same in PhoneGap. While for NativeScript it is possible to create plugins using only JavaScript, in Apache Cordova, Ionic and PhoneGap, it is necessary to create plugins in Java or Objective-C programming languages. Raboy also gives advantage to NativeScript when it comes to curve of learning, and the reason is that there are Ionic 1 that uses Angular 1, and Ionic 2 that uses Angular 2, which are drastically different. It is a problem for development programmers who have to learn everything from the beginning, which is very demanding [19]. Prantik Vaghela also compares these two frameworks, but he, unlike Raboy, gives advantage to Ionic when it comes to curve of learning. He considers Ionic easier to learn, especially if programmers have experience in web development. Furthermore, he states that debugging experience is easier and faster in Ionic, considering the fact that it is possible to debug parts of application that do not use native device features in a browser. When it comes to speed of performance, NativeScript is faster. However, one of important factors for application users is size of the application, which is a lot smaller in Ionic [20].

II. RESEARCH

A. Technical circumstances

For needs of this paper, a practical part was done first by developing a simple mobile application that enables simple mathematical operation of adding two numbers. The goal was to show the principle of functioning of six different development tools: IDEs and frameworks. The following IDEs were used: Android Studio (version 2.1), Xcode (version 6.4) and Visual Studio (version Community 2015). In development of hybrid applications for Android and iOS, Ionic (version 1.7.14) and PhoneGap (version 6.3.0) were used, while the application in NativeScript (version 2.2.1) framework was, due to technical limitations, developed only for Android.

Due to the fact that technical characteristics of a personal computer affect greatly the quality and speed of development, installation of IDEs and framework as well, it is necessary to state the characteristics of computers used. Native application and all hybrid applications for Android were created on a laptop computer with installed memory (RAM) of 4 GB and Windows 7 Professional operating system. Furthermore, personal computer with 8 GB of installed memory and OS X Yosemite operating system (version 10.10.5) was used to design all iOS applications, one native in Xcode and two hybrids in PhoneGap and Ionic frameworks. Native application for Windows Phone was developed on a laptop computer with installed memory of 4 GB and Windows 10 (it is important to state that it is not Professional).

B. Criteria and results observed

- Supported computer operating systems – operating systems of computers used for development of applications.

- Supported mobile platforms – operating systems of mobile devices on which it is possible to install and start an application.

- Programming languages – languages in which the main code of the application is written. They differ depending on platforms for which the application is developed.

- Official documentation and community of programmers – official documentation is available on web sites of the IDEs and frameworks mentioned. Detailed documentation is extremely important for programmers, especially for those who are beginning to work with certain technology. Alongside documentation, one of the key elements is the community of programmers who use the same technologies. In evaluation of official documentation and community, mark 1 means that documentation is not transparent, detailed and that it is difficult to be found, so the community itself is not wide and it is almost impossible to find an answer to any question posed. Mark 5 means that documentation is

TABLE 1. REPRESENTATION OF IDES AND WORK ENVIRONMENTS ACCORDING TO CRITERIA ANALYSED

	Native Android application – Android Studio	Native iOS application - Xcode	Native WP application – VS Community	Hybrid applications- Ionic	Hybrid applications - PhoneGap	Hybrid applications – NativeSript
Supported computer operating systems	Windows, Linux, Mac OS X	Mac OS X	Windows	Windows, Linux, Mac OS X	Windows, Mac OS X	Windows, Linux, Mac OS X
Supported mobile platforms	Android	iOS	Windows Phone	Android, iOS	Android, others[a]	Android, iOS
Programming languages	Java	Swift, Objective-C	C#, C++	JavaScript	JavaScript	JavaScript
Official documentation and community (1-5)	5	4	5	5	5	3
Speed and complexity of installation (1-5)	4	5	1	5	5	5
Complexity of development (1-5)	4	4	3	5	5	3

a. iOS, BlackBerry 10, Windows and Windows Phone, Firefox OS, Ubuntu, Tizen, Amazon Fire OS

detailed, well organized and available on web sites, the community is very widespread and for all concerns and problems, one can find an answer. It is necessary to take into consideration that marks are assigned from a subjective point of view.

- Installation – complexity itself is observed, as well as time necessary for installation of a certain IDE or framework. Prerequisites and requirements necessary to fulfill before the installation are observed as well. In analyzing and assigning marks in this criterion, mark 1 means that installation is complex and takes a lot of time, while mark 5 means that installation is simple and short.

- Development – refers to the main part in which the application is developed. In other words, complexity of making an application in certain technology is compared from a subjective point of view. Mark 1 means that technologies used are non-intuitive, non-organized and that the programming language is very difficult and incomprehensible, and mark 5 means that technologies are very intuitive, organized and that programming language is simple and comprehensible.

C. Analysis

According to the first criterion, supported computer operating systems, Android Studio is one of the IDEs that gave the best results because it can be installed on Windows, Mac OS X and Linux operating systems. It means that native Android mobile applications can be developed regardless of the operating system of a computer. While analyzing frameworks, Ionic and NativeScript can be installed on all operating systems, while development of hybrid applications in PhoneGap

framework is still impossible on Linux operating system. Possibility of installing certain IDEs and frameworks and development of mobile applications on different computer operating systems significantly contributes to lowering expenses, so it is not necessary to invest in operating systems the company does not already own – it is possible to work with the ones already available at the company.

According to the criterion supported mobile platforms, it is sensible to compare and analyze only those frameworks that are used for hybrid applications, because, as it was stated before, native ones are developed separately for every platform. PhoneGap offers the widest spectrum of mobile platforms: Android, iOS, BlackBerry 10, Windows and Windows Phone, Firefox OS, Ubuntu, Tizen, Amazon Fire OS, while Ionic and NativeScript support Android and iOS (although newer Ionic 2 supports Windows 10 Universal App as well). Taking into consideration this criterion and the analysis results, it is evident that it is more profitable to develop hybrid applications, especially using PhoneGap framework that enables development for more than five mobile platforms simultaneously. Native applications are developed in "official" programming languages for individual platforms. Applications for Android are developed in Java programming language, while C# or C++ are used for native Windows Phone applications. Objective-C or recently Swift, are programming languages for development of native iOS applications. For needs of this paper in creating native mobile applications, Java, C# and Swift programming languages were used. The introduction stated how hybrid applications are developed using HTML, CSS and JavaScript technologies, which facilitates their development. Therefore, all hybrid applications are developed using JavaScript programming language, and some frameworks require certain JavaScript frameworks. For instance, Ionic requires use of Angular JS. NativeScript recommends, but does not require, most recent Angular 2 framework. Advantages of hybrid applications are in those technologies or languages used for their development. Developers do not need several different programming languages to develop applications for different platforms, as it is the case with native applications. Also, if it is necessary, developers who are familiar with JavaScript can easily transfer from one framework to another because of similar principle of work. From all IDEs used according to criterion of official documentation and community of programmers, the most intuitive and most quality rated were documentations of Android Studio and Visual Studio. That means that all steps were described in detail starting from installation, development, up to publishing, structure of application, starting the emulator, debugging and much more. Furthermore, advice for improvement of code writing are offered. While making the application, of great help were communities of programmers that are very widespread, which is significant when a developer has a problem and when he knows that there is someone who has certainly already had the same problem, and even if he has not, there is someone willing to help. Based on the analysis of frameworks used, it is concluded that they all have excellent intuitive and detailed official documentations. However, concerning the fact that NativeScript is a relatively young framework, the community itself has not been developed and spread yet, and it is difficult to find a solution and an answer when it is necessary. Installations function as regular installations of standalone applications. Of course, it is necessary to take into consideration the speed of Internet of which depends the speed of downloading. While installing Android Studio and Visual Studio, it is possible to choose what tools and elements are to be installed, while by installing Xcode, everything is done automatically. Time of installation of analyzed IDEs drastically differs. Also, it should be taken into consideration that installations were done on different computers with different random access memory and characteristics, and that different elements to be installed were selected (Android Studio and Visual Studio). Installing Android Studio on Toshiba laptop took about 20 minutes, while the Xcode installation on a Mac computer took about 7 minutes. The longest and the largest installation was the installation of Visual Studio on Asus ZenBook. It took between 4 and 5 hours, which is a lot compared to the previous two. Taking into consideration extreme differences in duration of installation, and the complexity as well, Xcode was assigned the best mark, and Visual Studio the worst. Installation of frameworks were done using CLI[10] on Windows computer and Terminal on Mac computer. Duration of all installations is very short and simple. All installations – Ionic, PhoneGap and NativeScript, demand preinstalled Node.js. Furthermore, other prerequisites necessary to be fulfilled are stated in official documentations. For instance, versions of Android SDK,

Git and similar, which depends of the computer on which the application is made, or of what is perhaps already installed. Installations of all frameworks were assigned the best mark because they do not require too much effort and they do not take up too much time. Marks in this criterion do not necessarily mean that the IDE that requires more time for installing is bad. On the contrary, it can have excellent features, as it is case with Visual Studio. It is only one minute disadvantage of the analyzed IDE. It is difficult to make an objective statement on the process of development itself and on its complexity because it all depends on the developer and his previous knowledge. This paper, in a relatively objective way, is trying to conclude about the complexity of development in a certain IDE and framework, which depends on the programming language, intuitiveness, features of the IDE and so on. Developing a native Android application went without greater problems and Android Studio turned out to be an excellent IDE because it has all characteristics of a quality IDE: it is free, it has syntax and error coloring, autocomplete, code generator, debugging, WYSIWYG[11] editor, simplicity of starting the application in an emulator and on a device, simplicity of adding an icon, and much more. Mark 4 was assigned due to complicated Java programming language. Xcode was assigned the same mark as Android Studio because it also turned out as a quality and well organized IDE. However, the problem in developing these simple applications appeared during changing the icon, which was a bit more complicated. Designing a native Windows Phone application also turned out to be a more complicated. The first problem appeared when Visual Studio was installed on Toshiba laptop. Then, it was completely impossible to create a Windows Phone application and the problem was incompatibility of the operating system and necessary SDKs. A virtual machine was started, onto which Windows 10 was installed, but despite the fact that it was then possible to create a mobile application, it was not possible to start the emulator since it was a virtual machine. At the end, Visual Studio was installed on Asus laptop with operating system Windows 10, but despite all that, it was still not possible to start the emulator due to limitations of working with this operating system. As an alternative solution, a Universal Windows Platform application was designed, that was able to start on different devices: mobile devices, computers, Xbox and similar. Regardless of the limitations and strict requirements, integrated developmental environment has shown equally good as the rest. Problems prior to development and impossibility of starting the emulator are disadvantages due to which Visual Studio was graded worse than other IDEs. During the development of applications in frameworks analyzed, CLI were often used because if one wants to create an application, add a platform, start an application, sign it and such, it is necessary to input commands for certain actions. Code was written in Sublime Text 3 application for text editing. Structures of project in PhoneGap and Ionic are very similar. Files are logically sorted, which makes browsing

---

[10] CLI – command line interface

[11] WYSIWYG - what you see is what you get

through the directory of the project much easier. Structure of NativeScript project differs from previous two, and another difference is that XML markup language was used, not HTML. PhoneGap and NativeScript applications were written in "clear" JavaScript, while for Ionic, AngularJS was used. Since applications for Android were first made on Windows computer, in making iOS applications it was only necessary to transfer directory of the application on Mac computer. Of course, it was necessary to first install a framework, after which it was only necessary to make a position in the directory of the application, add iOS platform and build an application. Application developed in NativeScript could not be built for iOS due to the limitations of the computer itself for now, so it was, unlike Ionic and PhoneGap, assigned a lower grade.

III. CONCLUSION

From all IDEs analyzed, Android Studio made the best impression. During the installation and development there were no bigger problems and difficulties, and even if there were, solutions could be found relatively quickly, which is connected to a widespread and active community of programmers. One of great advantages of Android Studio is that it can be installed, and native Android applications can be developed on all computer operating systems. When talking about frameworks for creating hybrid applications, and taking into consideration the analyzed criteria, they all turned out to be powerful tools. However, PhoneGap is a tool with the greatest comparative advantages, mostly due to a wide spectrum of supported mobile platforms. Although native applications offer by far the best user experience and can use all features and functionalities of devices to a maximum, nowadays tools for creating hybrid applications are continuously developed. One of the reasons is that "only" knowledge of JavaScript programming language is necessary, and not of all programming languages that one needs to know for developing native applications: Java, Objective-C, C# and similar. Also, time necessary and expenses for developing the application are becoming shorter, so it is more profitable to develop hybrid applications. The discrepancy between hybrid and native applications is smaller every day, and hybrid applications are starting to resemble native ones, so the difference will surely disappear completely very soon.

Comparative analysis of tools for development of native and hybrid applications has shown only some advantages and disadvantages of tools used for development of mobile applications. Analyses through different criteria would certainly bring different results, but it is without a doubt that such analyses can be of great use to manufacturers in order to develop the aforementioned tools even further, and to make them more powerful and better on one hand, but also help users in choosing tools for work on the other.

REFERENCES

[1] V. N Inukollu, D. D Keshamoni, T. Kang and M. Inukollu, "Factors influencing quality of mobile apps: role of mobile app development life cycle," International Journal of Software Engineering & Applications, vol. 5, pp. 15-34, 2014.

[2] N. Litayem, B. Dhupia and S. Ruhab, "Review of cross-platforms for mobile learning application development," International Journal of Advanced Computer Science and Application, vol. 6, pp. 31-39, 2015.

[3] Z. Ćović, D. Babić, "Development and implementation of location based native mobile application," International journal of electrical and computer engineering systems, vol. 5, pp. 27-31, 2014.

[4] A, Khanderparkar, R. Gupta and B. Sindhya, „An introduction to hybrid platform mobile application development, "International Journal of Computer Applications, vol. 118, pp. 31-33, 2015.

[5] Y. A. Redda, Cross platform mobile applications development. Trondheim: Norwegian University of Science and Technology, 2012.

[6] D. Markov, „Comparing the top frameworks for building hybrid mobile apps," http://tutorialzine.com/2015/10/comparing-the-top-frameworks-for-building-hybrid-mobile-apps/ (2017-02-02)

[7] Mobile frameworks: comparison chart, http://mobile-frameworks-comparison-chart.com/ (2017-02-02)

[8] Comparison of integrated development environments, https://sites.google.com/site/richardgennaro/home/development/ides/comparison-of-integrated-development-environments (2017-02-02)

[9] Best integrated development environment (IDE) software, https://www.g2crowd.com/categories/integrated-development-environment-ide (2017-02-02)

[10] iOS vs Android development, https://theappsolutions.com/blog/development/ios-vs-android/ (2017-03-23)

[11] P. Halabuda, „Mobile Application Development: iOS vs Android vs Windows Phone," http://whallalabs.com/mobile-application-development-ios-vs-android-vs-windows-phone/ (2017-03-23)

[12] T. Car, „Android development is 30% more expensive than iOS. And we have the numbers to prove it!," https://infinum.co/the-capsized-eight/android-development-is-30-percent-more-expensive-than-ios (2017-03-23)

[13] A. Sinicki, „Developing for Android vs developing for iOS – 5 rounds," http://www.androidauthority.com/developing-for-android-vs-ios-697304/ (2017-03-23)

[14] Android vs Windows Phone from a developer's scope, http://blog.bugsense.com/post/26918984231/android-vs-windows-phone-from-a-developers-scope (2017-03-23)

[15] Get started with publishing, https://developer.android.com/distribute/google-play/start.html (2017-03-23)

[16] Register as an app developer, https://developer.microsoft.com/en-us/store/register (2017-03-23)

[17] 9 differences between iOS and Android app development, https://www.cleveroad.com/blog/9-differences-between-ios-and-android-app-development (2017-03-23)

[18] The app certification process, https://docs.microsoft.com/en-us/windows/uwp/publish/the-app-certification-process (2017-03-23)

[19] N. Raboy, „NativeScript vs Ionic Framework, should you switch?," https://www.thepolyglotdeveloper.com/2015/11/nativescript-vs-ionic-framework-should-you-switch/ (2017-03-23)

[20] P. Vaghela, „NativeScript vs Ionic Framework pros and cons," http://pointdeveloper.com/nativescript-vs-ionic-framework-pros-cons/(2017-03-23)

Exploring HTTP/2 Advantages and Performance Analysis using Java 9

L. M. Bach, B. Mihaljević, A. Radovan

Rochester Institute of Technology Croatia

leo.bach@mail.rit.edu, branko.mihaljevic@croatia.rit.edu, aleksander.radovan@croatia.rit.edu

Abstract - As websites have become bulkier and serve more larger files to end users, it has become harder for the HTTP/1.1 protocol to provide adequately short page load times. The HTTP protocol, in its previous versions, is a generic stateless application protocol most often used for distributed, collaborative, hypermedia information systems. The newest version, HTTP/2, introduced an updated protocol feature set based on Google's SPDY/2 protocol, aimed at improving the speed at which modern web resources load, reducing network congestion and additionally enforcing security standards. With header compression, new server push responses, and fully multiplexed connections, HTTP/2 solves the major issues previous versions were unable to overcome. This paper presents performance analysis results comparing both protocol versions in the upcoming release of Java 9 programming language, as well as describes key differences in development between versions. Analysis was run on the open source Jetty web server and benchmark code was compiled using JDK 9 Early Access. This paper shows benchmark results using different scenarios and typical usage performance gains. It outlines the major improvements and recommendations switching to the newer protocol version provides, specifically in general use object-oriented programming languages such as the new version of Java language.

I. INTRODUCTION

The Hypertext Transfer Protocol (HTTP) was introduced as an application protocol for distributed, collaborative, and hypermedia information systems [1]. It is currently the most common way to send and receive data across the Internet, making it essential to the World Wide Web as we know it.

Originally, when the Internet was a much newer technology, web pages were primarily text-based with perhaps a few embedded still images to make the web page more attractive. The HTTP/1.1 protocol, the latest version of the 1.x series, was finalized in 1999 as RFC 2616 [1] and was designed to optimally service web communications of its era. Later, in June 2014, numerous improvements to the protocol were introduced in the form of six RFCs 7230-7235 [2], which collectively update and obsolete RFC 2616. In total, these updates to the original protocol addressed 553 issues, which can be found at [3].

The Internet has steadily grown with time, however. Today, the average web site is not a few kilobytes in size but, rather, a few megabytes. Specifically, as of December 2016, the average website is roughly two and a half

megabytes in size [4]. Most of this page size increase can be attributed to web pages incorporating more images into their design; however, there is also a noticeable increase in the sizes of stylings (CSS) and scripting code (JavaScript) being sent to the end client.

TABLE I. SIZE INCREASE OF WEB PAGE COMPONENTS FOR ALL WEBSITES [4]

Technology	Size end 2014 (kB)	Size end 2015 (kB)	Increase in 2014-2015	Size Dec 2016 (kB)	Increase in 2015-2016
HTML	60	67	11.7%	52	-22.4%
Stylesheets	58	76	31.0%	79	3.9%
Scripts	295	357	21.0%	415	16.2%
Images	1248	1426	14.3%	1623	13.8%
Video	n/a	174	n/a	197	13.2%
Other	...	...	...	...	...
Total	1931	2232	15.6%	2479	11.1%

As can be seen in Table I., from the end of 2014 to the end of 2016, websites have increased from 1931 kB to 2479 kB: an increase of 28.4%. Not included in Table I is HttpArchive's data from the end of 2010, when the top websites were, on average, only 716 kB in size. From those numbers, websites have increased in size by 246% in the last 6 years.

This increase in file size and quantity is something that HTTP/1.1 cannot overcome and is the main reason for creation of the new version of HTTP protocol. HTTP/2 was designed in RFC 7540 [5] with the modern web in mind and aims at removing the need for web developers to work around HTTP/1.1's shortcomings in today's age.

With this paper, we compared the improved performance one can expect to see when migrating Java web servers from HTTP/1.1 to HTTP/2 and also to present the current state of HTTP/2 implementation within the Java ecosystem.

II. HTTP/1.1 LIMITATIONS

The HTTP/1.1 protocol is able to adequately service small size, relatively simple HTML pages. When used to serve modern web sites, which can be significantly larger than the original specification was designed for, HTTP/1.1's deficiencies become far more noticeable. Among these are: head-of-line blocking, uncompressed headers, and the Slow Start mechanism.

A. Head-of-Line Blocking

Head-of-Line blocking is the phenomena which occurs when multiple packets are held up due to needing to wait for the transmission of another packet [6]. This can occur in various scenarios, due to:

- Packets being lost, requiring that further packets wait until the lost packet is resent

- Port contention, where two packets are being sent to the same port and one must go first while the other is blocked until the next cycle

- Ordered packets being sent out of order, causing out of order packets to be delayed until the required packet is transmitted.

B. Uncompressed Headers and Slow Start Mechanism

The HTTP/1.1 standard imposes no limits as to the size of each header field contained in a request; however, most servers and clients impose some sort of limit for both practical and security reasons. Per research conducted by Google [7], header sizes range from approximately 200-2000 bytes with an average size of 750 bytes. Considering that these numbers were gathered in 2009, and with Table I and Fig. 1 showing more than an 11% increase in web site size for 2016, as well as Fig. 2 presenting an average web page size of 2479 kB, we can give a conservative estimate of typical header sizes being approx. 1500 bytes or more.

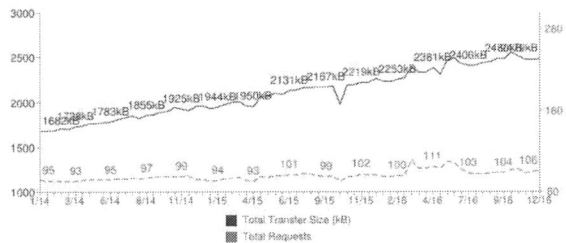

Figure 1. Total Transfer Size and Total Requests for all web pages (URLs) Jan 2014 – Dec 2016 according to httparchive.org [8]

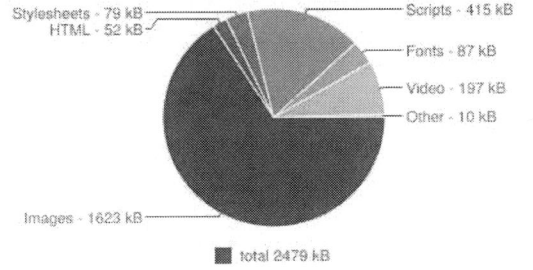

Figure 2. Average bytes per page by content type for the all web pages (URLs) as of 02-Dec-2016 according to httparchive.org [4]

The size of these headers is important due to the fact that large headers may take multiple round trips from server to client before being fully transmitted. If a website has, for example, 100 assets to download (e.g. images,

script files, stylesheets, embedded videos, and text content), each request will require headers to be sent between the server and client. Using a conservative estimate of 50 round trips [9] to transfer the contents of a web site, this amounts to 75 kB of header data being transferred, most of which is repeated data.

HTTP/1.1's Slow Start mechanism, defined in RFC 5681 [10], is a congestion-control feature built into the TCP protocol. With Slow Start, every connection starts with an "initial congestion window," which defines how many packets may be sent during the first round trip over the network. With each successful ACK response from the client, the window is increased, allowing more packets to be sent at once until, ultimately, the peak transfer rate of the connection is found [11]. This feature, specifically designed to limit data transfer in order to prevent router overload, causes problems as it introduces more round trips than would normally be necessary over the course of a connection. Each round trip is affected by the connection's latency, meaning that a 100 ms latency in the previous example would result in a total wait time of 800 ms just to open and close all of the congestion windows; the time to transfer the data contained carried within the packet payloads is not included. In addition, each window requires that headers be sent over the network, the end result being that the client must wait not only due to the latency of the connection, but it must also transfer additional, redundant data with each request.

The initial window, as specified in RFC 3390 [12], is between two and four segments, with a maximum segment size of 1460 bytes. This strict limit bottlenecks modern connections, and, as a result, many connections are never able to reach their maximum speeds. Google petitioned that the initial congestion window be increased to 10 segments based on their research that showed user-visible latencies could be reduced by up to 10% with no noticeable increase in congestion problems across the net [13], which was later accepted in the HTTP/1.1 standard.

III. MODERN WORKAROUNDS

As of writing this paper, HTTP/2 usage statistics [14] show that only 11.7% of all the websites support it as the new standard (Fig. 3).

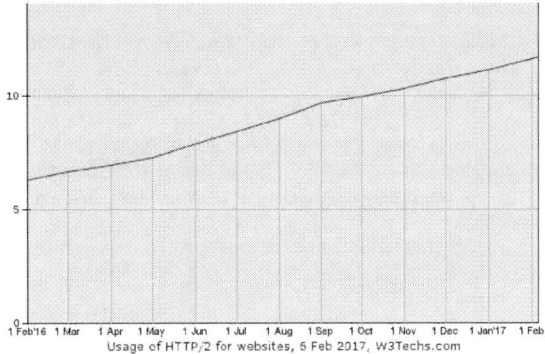

Figure 3. Adoption rate of HTTP/2 for websites as of 06-Feb-2017 according to W3Techs.com [14]

Ergo, web developers today still must find ways to improve their website performance while remaining in the confines of the HTTP/1.1 protocol. Most of these solutions are considered "hacky" as these techniques work around the protocol's limitations rather than overcoming them. It should be noted that all of them are aimed at circumventing head-of-line blocking and the Slow Start initial congestion window.

A. Domain Sharding

Instead of hosting all of a site's content on one domain, domain sharding is a method of hosting content on multiple domains in order to allow the end user's client to open more connections and download content in parallel [15]. With this method, should one domain slow down due to head-of-line blocking, the other domains will still transmit their data to the client. For example, a site may have an additional domain that starts with "img" to host its image files and yet another domain that starts with "vid" to host its video files.

B. CSS/JavaScript Concatenation and Asset Inlining

CSS/JavaScript concatenation involves merging all of a site's CSS files into one, singular file and doing the same for all of a site's JavaScript files. By reducing the number of files served to the client, this method alleviates some of the problems of the HTTP/1.1 Slow Start feature.

In a similar vein, asset inlining calls for all CSS and JavaScript to be embedded directly into the web page's HTML code. This can be automated by tools such as packify [16] or asset-inliner [17]. Much like file concatenation, this method helps alleviate Slow Start issues. A more extreme version of this is image embedding, where a desired image is embedded into HTML encoded in Base64. Combining image and CSS/JavaScript inlining could potentially greatly reduce the number of HTTP requests a client needs to send; however, it is not always possible to inline CSS styles and JavaScript code. Additionally, embedding images causes both latency and caching issues on the client side as the client browser must decode every embedded image and it is not possible to cache such images as they become part of the HTML file, itself.

C. Image Spriting

Image spriting involves making a single image mosaic of many, smaller images [15]. For example, on a music streaming site, instead of downloading four separate image files for "previous," "stop," "start," and "next," a single image containing all of these sprites would be downloaded. Using JavaScript, the client then processes the image and displays them were necessary. As with image embedding, this method puts more computational load on the client side and must be repeated every time an image mosaic is loaded.

IV. NEW FEATURES IN HTTP/2

A. Binary Protocol

HTTP/1.1 and its predecessors are based on plaintext communication. However, HTTP/2 has made a shift towards using binary, which is more condensed when compared to plaintext, and is easier to parse. It should be noted that, due to this change, HTTP/2 is not usable through Telnet.

B. HPACK Header Compression

As discussed previously in this paper, HTTP/1.1 has the problem of sending redundant data contained in its headers over the course of a connection. HTTP/2 mitigates this problem through the use of its HPACK mechanism [18] which utilizes Huffman encoding to compress the binary header values. In addition, the HPACK mechanism uses two tables to further reduce bandwidth usage: a static table which contains a cache of commonly-used headers, and a dynamic table which references other headers used in the current session. This dynamic table reduces header size by sending pointers to pre-existing values instead of the values themselves.

C. Multiplexing

HTTP/2 introduces streams [5]: independent, bidirectional sequences of frames between the client and server within an HTTP/2 connection. This technique both reduces the number of connections a server needs to handle and solves the issue of head-of-line blocking.

D. Flow Control

Streams can be further prioritized using HTTP/2's flow control algorithms [5]. Flow control enables the receiver to impose flow control limits that the sender must respect. In practice, flow control is designed to protect endpoints under resource constraints: such as when a server is under heavy load, which causes it to have a slow upstream connection while maintaining a fast downstream connection. In this scenario, flow control can allow the server to process data on the less-burdened stream while the other stream is busy.

E. Server Push

Typically, upon receiving a request for a web page, a server can predict that the client will make further requests for all the page's external resources including style sheets, scripts, and images. In HTTP/1.1, the server cannot send these dependencies without the client explicitly requesting each resource. HTTP/2 enables the server to bundle page dependencies it knows the client will need into a "push", which it can proactively send to the client [5]. Afterwards, the client has the ability to refuse or accept the data.

V. JAVA AND HTTP/2

The maintainers behind the community-based open source version of Java Development Kit (JDK) aka OpenJDK [19] have specifically targeted HTTP/2 support for JDK 9 as one of their stated goals in JEP 110 [20]. This is not only due to the desire to support the new protocol, but also to improve the existing application programming interface (API) which currently implements HTTP/1.1. These classes currently suffer from multiple problems such as: undocumented behavior, difficult maintainability, and a base connection API that was designed with multiple, now defunct, protocols in mind during its creation.

Although the Apache Software Foundation and Eclipse Foundation provide HTTP/2 implementations in the form of the Apache HttpClient and the Web server and Servlet container Jetty, respectively [21] [22], the native HTTP/2 implementation within the JDK 9 is designed with a smaller footprint in mind [20]. Furthermore, Java maintainers have expressly stated that the internal implementation of their HTTP/2 client API must be "friendly" towards newer language features, such as lambda expressions.

VI. TEST ENVIRONMENT

Out test environment consist of two different Jetty servers version 9.4.0.v20161208: one configured for the HTTP/1.1 protocol, and the other configured for the HTTP/2 protocol. Source code for the Jetty server was developed using the IntelliJ Idea integrated development environment (IDE) early access version 2017.1. The operating system used for hosting the servers was Windows 10 x64 Anniversary Update. Chrome version 57.0.2987 run on Ubuntu Linux version 16.04 was used as the browser of choice for the following benchmarks. As of writing, no major web browser supports HTTP/2 without encryption. As such, in the interest of fairness, the HTTP/1.1 server was also configured with TLS encryption in order to more accurately portray the latencies between protocols.

To obtain page and resource load times, Chrome's built-in network logger was used. In order to more accurately simulate a real-world scenario, Chrome was configured to simulate a typical 4G mobile connection. This added a 20ms delay to requests and limited downloads and uploads to 4 Mb/s and 3 Mb/s, respectively. Page load benchmarks were conducted in three different setups:

- Over the localhost interface
- Over a Local Area Network (LAN) between a Windows host machine and Ubuntu virtual machine guest

In the first setup, both Jetty servers were hosted on the same machine as the Chrome browser. Web pages were loaded 10 times each using both the HTTP/1.1 and HTTP/2 protocol and the resulting load times were then averaged.

The second setup kept the Jetty server on the same windows machine, but moved the client onto a local virtual machine created by VMWare Workstation Pro 12 running Ubuntu Linux. This test is conducted in order to quantify the performance loss when transmitting packets over a virtual Ethernet interface in addition to a real interface. As with the first setup, each web site was loaded 10 times using each protocol and the results were averaged.

A previous study conducted in 2015 [23] cloned 15 websites in order to provide a more accurate depiction of the performance differences between the protocols in real world scenarios. Using this method as a framework, this study also cloned 15 different websites, 5 of which were benchmarked previously in the aforementioned study.

TABLE II.   RESOURCE STATISTICS OF WEBSITES

	Website	HTML / PHP	CSS	JS	Images	Size (MB)
1	Amazon	4	7	12	33	4.73
2	BBC	4	8	23	19	2.47
3	eBay	4	1	4	149	1.85
4	GitHub	1	3	2	6	1.45
5	IEEE	2	13	39	37	2.41
6	IMDB	14	10	28	57	4.41
7	Microsoft	7	2	18	10	1.86
8	Skype	2	3	16	0	1.06
9	Stack Overflow	2	1	5	8	1.20
10	The Economist	5	3	29	8	5.05
11	Wikipedia	5	0	0	19	0.479
12	Wordpress	1	3	3	8	1.61
13	Yahoo	4	1	20	24	4.51
14	Yelp	2	2	10	64	1.92
15	Youtube	8	6	11	37	2.34

Note: Websites in gray are sites which were also included in [23]. All data is taken from the United States domains of each website. i.e. Amazon.com, not Amazon.de.

Table II lists every site included in benchmarking. Websites were obtained by saving them directly within the Google Chrome web browser. This method of locally saving the websites comes with the drawback of not fully downloading dynamic content and content located on external content distribution networks (CDNs); however, the purpose of this study is to showcase the differences between HTTP/1.1 and HTTP/2, not to accurately clone commercial web pages.

A. Notable Issues

At the time of writing this paper, the JDK 9 was in early access Build 153 [24] with a slated release date estimated at July 2017. As it is still in active development, certain features don't yet function as expected. One such feature is built-in Application Layer Protocol Negotiation (ALPN) support [25]. The build of the JDK used in this paper does not have an automatically loaded ALPN library, which is used for establishing connections in HTTP/2. In order to circumvent this, a Java Virtual Machine (JVM) argument must be passed with the path to a suitable ALPN JAR file, as shown in Fig. 4:

```
-Xbootclasspath/p:/path/to/alpn/jar
```

Figure 4.   The template for the JVM argument required to load the Jetty ALPN jar file before JDK 9.

This argument instructs the JVM to prepend the supplied JAR file before the standard library classes. It is important to note that the `/p:` section of the command is the section that specifies to prepend the JAR file. Omitting this section will cause the JVM to overwrite all standard library files with the provided JAR. As no standard library classes will have been loaded in this case, the JVM will force close.

With JDK 9 and the modularity provided by the project Jigsaw, the above argument is deprecated. Attempting to start the JVM with said flag will result in the JVM refusing to start, with an explicit error stating the deprecation of the `Xbootclasspath` flag. In the final release version of JDK 9, the ALPN libraries should be included with the standard library.

B. Limitations

Although the goal of these experiments is to simulate real world situations as closely as possible, certain things could not be achieved. Namely, the following limitations are present:

- All websites are held on a single server, thereby removing data related to domain name sharding

- The server is reached through a plain IP address, without the need for domain name resolution

- Cloned websites were not explicitly designed to be run on the local configuration provided, possibly decreasing site performance

VII. RESULTS

Preliminary results for cloned websites transmitted over the localhost interface showed a mildly negative performance trend for HTTP/2. On average, web sites gained 2.7% load time when migrating from HTTP/1.1 to HTTP/2 with no other changes.

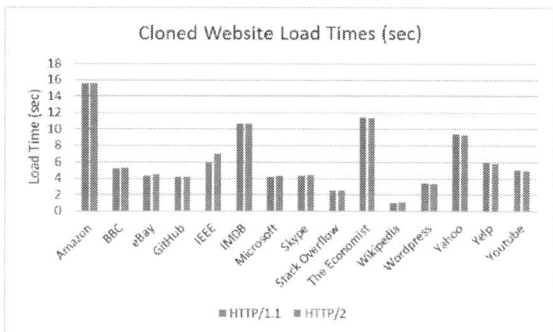

Figure 5.   Load times for locally hosted websites

The above results were run while the host computer was still connected to the internet, meaning that each websites' script files could still perform remote content requests, which may not necessarily be transmitted with the HTTP/2 protocol.

In Fig. 6, the host's Ethernet adapter was disabled within the Windows operating system in order to prevent the mentioned scripts from requesting additional data.

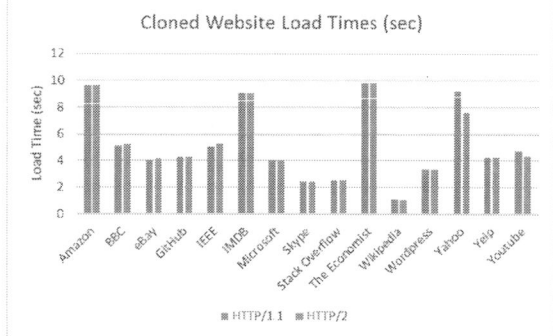

Figure 6.   Load times with the host machine's ethernet adapter disabled

With remote content disabled, the load times for HTTP/2 were able to improve by a small margin. On average, websites loaded 1.4% faster when transmitted through the HTTP/2 protocol. Most notably, Yahoo.com gained a 20% performance improvement in this test, reducing its load time from 9.187 to 7.638 seconds. YouTube.com also gained a significant performance boost and lowered its load times by 8.2%: from 4.716 to 4.356 seconds.

Unfortunately, the host machine's Ethernet adapter could not be disabled for tests involving a guest OS: disabling the adapter also prevented the operating systems from being able to communicate with each other. Results were less favorable than the first set of tests when the client browser was moved to a locally-run Ubuntu Linux virtual machine, as shown on Fig. 6:

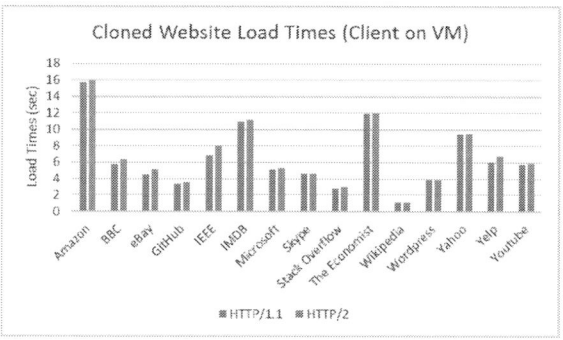

Figure 7.   Load times for locally hosted websites accessed through a web browser running on a Ubuntu Linux virtual machine

Load times in this scenario were, on average, 4.7% slower for sites accessed through HTTP/2 in this scenario.

VIII. IMPLICATIONS

Given that the tests did not show an all-around improvement when switching to HTTP/2, it can be seen that a simple switch to the new protocol does not provide significant performance gains. Although there was a slight improvement in load times when remote content was disabled, it was only a substantial increase in the case of Yahoo.com and Youtube.com. It should be noted that both Yahoo and YouTube support HTTP/2 on their production websites.

Sites with the worst load times were those that created many asynchronous requests through JavaScript during the page loads. In the case of The Economist, there was one specific script file that was dedicated to loading advertisements which slowed load times by up 5 seconds, regardless of whether the host was disconnected from the Internet or not. Given the delay introduced by this file, it is obvious that HTTP/2 is not a replacement for optimized web page design.

IX. FUTURE WORK

The current research was of limited scope and leaves much room for future study. Namely, while Chrome's simulation of slower connections is useful, it is still not as useful as a truly slow connection on a slow device: such as an older smart phone. Additionally, as only a server was created, extra data could be collected through the use of a Java HTTP client instead of using the Chrome browser. This would have the benefit of allowing more granular benchmarks through the use of low-level HTTP programming APIs.

Moving the Jetty web server along the test web sites onto an externally hosted virtual private server would also be a desirable change for future work as it would root results more in real-world situations. If a service such as Amazon Web Services is used, then there is the possibility to hosting the website in multiple different locations around the world and benchmarking performance differences.

X. CONCLUSION

HTTP/2, at this current point in time, is not enough on its own to improve web site load times. Web sites must be designed from the beginning with HTTP/2 features in mind, such as server push, in order to fully take advantage of the benefits that the newer protocol can provide.

As shown in the results, switching to the newer protocol without any other changes may actually prove detrimental to performance, particularly when accessing web sites through a virtual Ethernet interface. Given that most of the technologies shown in this paper were in early access or beta at the time of benchmarking, it is possible that, given a few years' time or less, default HTTP/2 server configurations will be enough to provide noticeable performance gains with little work on the part of developers. Frameworks or analysis tools could particularly be of use should they be able to intelligently and automatically determine which files should be pushed together.

REFERENCES

[1] R. Fielding et al., "Hypertext Transfer Protocol -- HTTP/1.1," RFC 2616, IETF, June 1999. [Online]. Available: https://tools.ietf.org/html/rfc2616. [Accessed: 06-Feb-2017].

[2] R. Fielding, J. Reschke, "Hypertext Transfer Protocol (HTTP/1.1): Message Syntax and Routing," RFC 7230, IETF June 2014. Available: https://tools.ietf.org/html/rfc7230. [Accessed: 06-Feb-2017].

[3] IETF, "All Tickets by Component," IETF, 18-Feb-2017. [Online]. Available: https://trac.ietf.org/trac/httpbis/report/19. [Accessed: 06-Feb-2017].

[4] "HTTP Archive - Interesting facts," HTTP Archive, httparchive.org, 06-Feb-2017. [Online]. Available: http://httparchive.org/interesting.php. [Accessed: 06-Feb-2017].

[5] M. Belshe, R. Peon, M. Thomson, Ed., "Hypertext Transfer Protocol Version 2 (HTTP/2)", RFC 7540, IETF, May 2015. [Online]. Available: https://tools.ietf.org/html/rfc7540. [Accessed: 06-Feb-2017].

[6] M. Scharf, S. Kiesel, "Quantifying Head-of-Line Blocking in TCP and SCTP," IETF, 15-Jul-2013. [Online] Available: https://tools.ietf.org/id/draft-scharf-tcpm-reordering-00.html. [Accessed 05-Feb-2017].

[7] Google, "SPDY: An experimental protocol for a faster web," The Chromium Projects, 12-Nov-2009. [Online]. Available: https://sites.google.com/a/chromium.org/dev/spdy/spdy-whitepaper. [Accessed: 06-Feb-2017].

[8] "HTTP Archive - Trends," HTTP Archive, httparchive.org, 06-Feb-2017. [Online]. Available: http://httparchive.org/interesting.php. [Accessed: 06-Feb-2017].

[9] M. Belshe, R. Peon. "SPDY Protocol – Draft 3," [Online.] Available: https://www.chromium.org/spdy/spdy-protocol/spdy-protocol-draft3. [Accessed: 06-Feb-2017]

[10] M. Allman, V. Paxson, E. Blanton, "TCP Congestion Control," RFC 5681, IETF, September 2009. [Online]. Available: https://tools.ietf.org/html/rfc5681. [Accessed: 06-Feb-2017].

[11] N. Dukkipati, et al. "An Argument for Increasing TCP's Initial Congestion Window," in *ACM SIGCOMM Computer Communication Review*, 2010, pp. 26-33.

[12] M. Allman, S. Floyd, C. Partridge. "Increasing TCP's Initial Window," RFC 3390, IETF, October 2002. [Online]. Available: https://tools.ietf.org/html/rfc3390. [Accessed: 06-Feb-2017].

[13] J. Chu, N. Dukkipati, Y. Cheng, M. Mathis. "Increasing TCP's Initial Window," Internet Draft, IETF, 12-Jul-2010. [Online]. Available: https://tools.ietf.org/html/draft-hkchu-tcpm-initcwnd-01. [Accessed: 05-Feb-2017].

[14] "Usage of HTTP/2 for websites," W3Techs, 5-Feb-2017. [Online]. Available: https://w3techs.com/technologies/details/ce-http2/all/all. [Accessed: 06-Feb-2017].

[15] Nginx. "HTTP/2 for Web Application Developers," 15-Sep-2015. [Online]. Available: https://cdn.wp.nginx.com/wp-content/uploads/2015/09/NGINX_HTTP2_White_Paper_v4.pdf [Accessed: 06-Feb-2016].

[16] M. Ogden. "packify," 2014, GitHub repository. [Online]. Available: https://github.com/maxogden/packify/commits/master. [Accessed: 06-Feb-2017].

[17] J. Bellamy, "asset-inliner," 2014, GitHub respository. [Online]. Available: https://github.com/jasonbellamy/asset-inliner/commits/master. [Accessed: 05-Feb-2017].

[18] R. Peon, H. Ruellan. "HPACK: Header Compression for HTTP/2," RFC 7541, IETF, May 2015. [Online]. Available: https://tools.ietf.org/html/rfc7541. [Accessed: 06-Feb-2017].

[19] "OpenJDK FAQ," OpenJDK. 10-Dec-2010. [Online]. Available: http://openjdk.java.net/faq/. [Accessed: 06-Feb-2017].

[20] "JEP 110: HTTP/2 Client (Incubator)," OpenJDK. 06-Feb-2017. [Online]. Available: http://openjdk.java.net/jeps/110. [Accessed: 06-Feb-2017].

[21] "HttpClient Overview," Apache Software Foundation. [Online]. Available: https://hc.apache.org/httpcomponents-client-ga/. [Accessed: 06-Feb-2017].

[22] "Jetty," The Eclipse Foundation. [Online]. Available: http://www.eclipse.org/jetty/. [Accessed: 06-Feb-2017].

[23] H. De Saxce, I. Oprescu, and Y. Chen, "Is HTTP/2 really faster than HTTP/1.1?," in *Proc. - 2015 IEEE Conference INFOCOM Computer Communications Workshops (INFOCOM WKSHPS)*, 26 April - 1 May 2015, pp. 293–299, 2015.

[24] "JDK™ 9 Early Access Releases", Java.net, [Online]. https://jdk9.java.net/download/. [Accessed: 20-Jan-2017].

[25] S. Friedl, A. Popov, A. Langley, E. Stephan, "Transport Layer Security (TLS) Application-Layer Protocol Negotiation Extension," RFC 7301, IETF, July 2014. [Online]. Available: https://tools.ietf.org/html/rfc7301. [Accessed: 06-Feb-2017].

Software Supporting International Student Exchange Program in Higher Education

Zvonimir Gračak*, Ljiljana Brkić*
*University of Zagreb, Faculty of Electrical Engineering and Computing, Zagreb, Croatia
zvonimir.gracak@fer.hr
ljiljana.brkic@fer.hr

Abstract - International student exchange programs allow students to study in various countries and experience challenges in new academic environment. International student exchange processes are complex and accompanied with extensive administrative tasks involving many higher education institutions (HEI). These administrative processes can be facilitated with the help of an appropriate software. This paper presents a software designed for managing data required for various processes in student exchange programme. The software, allows the user to manage data about HEI, bilateral agreements signed between institutions in European Higher Education Area (EHEA), applications of the outgoing students for an annual tender and the outcome of the competition. It supports automatic procedures for student ranking considering their choices, achievements and propositions of the tender. The main outcome of the software is the ranking list of the students with associated outgoing HEI complied with the quotas agreed in bilateral agreements and student's preferences.

I. INTRODUCTION

International exchange of students allows students from all levels and all subject areas to study abroad for a semester or an academic year and gain credits that will subsequently be recognized at their home institution. Students can participate in mobility at any time during their degree. In 1987 the European Commission established a programme for student mobility in Europe called Erasmus. This programme is constantly being improved and enhanced especially with the advent of the Bologna process which brings the advantage of the use of the ECTS credit transfer system. The most prominent student mobility program, active since January 2014, at EU level is the Erasmus+ Programme designed to, among other, support higher education and targeting EU Member State countries (Programme Countries). Programme Countries are those countries participating fully in the Erasmus+ Programme [1]. Each country participating in Erasmus+ set up a National Agency and contribute financially to the programme.

Erasmus+ Programme functions under a number of inter-institutional agreements signed between higher education institutions, through which the partners agree upon exchange quotas and the duration of mobility for students and staff. Quotas are set up on the basis of a specific number of students (and staff) to be exchanged each academic year within a specific subject area. By

signing the agreement, the partner institutions commit themselves to respect the principles of the Erasmus Charter for Higher Education (ECHE) [2] in terms of the organisation and management of mobility.

The European Commission, through the Croatian National Agency for Erasmus+ (Agency for Mobility and EU Programmes), provides Erasmus+ students with grants to help them financially while studying abroad. The University (e.g. University of Zagreb - UNIZG) receives the Erasmus+ funding from the Agency for Mobility and EU Programmes, based on the number of Erasmus students sent out in the previous academic year. The University further allocates funding to its constituents. The actual grant amount each student will receive (in case he/she receives it) is dependent on where they are going and how long they will stay.

Since the early beginnings of the institutional exchange of students in the year 2006, when a total of 5 students went on exchange, the number of exchange students is constantly growing. The number of bilateral agreements between FER and prestigious higher education institutions (HEI) in European Higher Education Area is constantly growing too. Currently, the number of agreed places for exchange students at all levels of study (244 available places) in a total duration of 1545 months surpasses the needs.

II. OUTGOING MOBILITY IN ERASMUS+ FOR PROGRAMME COUNTRIES - THE PROCESS FLOW

The most massive student exchange, incoming and outgoing, takes place through the Erasmus+ Programme for Programme Countries. In this paper we set focus on outgoing mobility process since it currently involves more students than the incoming mobility and requires much more administrative and organizational engagement.

The tender for the selection of students intending to participate in the exchange as outgoing takes place annually, usually at the beginning of a calendar year for the next academic year. An important part of processing the tender is awarding Erasmus student mobility grants to financially facilitate students studying abroad. These grants are given to Universities by the National Agency also on an annual basis. The tender processing is coordinated work of International Relations Office (IRO) of the University and corresponding offices of its constituents. The tender sets common framework applying to all university students who will submit an application and is announced by the UNIZG

IRO. Some segments of the tender processing are the responsibility of constituents and university leaves them to prescribe rules and implement those parts of the process. Students submit applications online, using web application of the UNIZG. They apply for certain institution, study level, subject area, beginning and duration of stay. Since quotas at institutions in certain degree level and subject areas are limited, sometimes it is not possible to send all students to their first choice of institution. For that reason students are allowed to submit application for up to three HEI. After the application deadline, HEI obtain data on student's applications from IRO and carries out the ranking of applicants. The ranking of students - as well as the procedure for awarding them a grant - must be transparent and documented, and also available to all parties involved in the selection process.

Each HEI determines the ranking criteria for its students. The ranking criteria are set out in a fair and transparent way before participants submit their applications. HEI usually ranks students considering the academic performance of the candidate (cumulative GPA), English skills, previous mobility experience, motivation, experience in the receiving country, etc. Once ranking procedure is completed, the list of selected students will be sent to the IRO to consolidate partial lists of individual constituents and assign funds to students. Due to the finite funding only limited number of students can receive Erasmus scholarship. Students who will not be awarded Erasmus scholarship will be offered zero-grant placements. Zero-grant is a possibility for students to participate in Erasmus exchange program on the same terms (pay no tuition fee, benefit from mentorship program, receive academic recognition) but without Erasmus scholarship.

Many parts of the described processes are performed manually or semi-automatically. There is no unique criteria for ranking students at the institutions and majority of constituents assess applications manually. Also, the data exchange between the IRO and the mobility offices at the constituents is carried out via an email. Given the increasing number of applications, there is the need for process automation in order to accelerate and facilitate the entire process of application processing. In the rest of the paper the architecture and implementation of system supporting the application ranking is described.

III. SYSTEM ARCHITECTURE AND DATA MODEL

A. System architecture

In the architecture of the system three main components stand out: client application, web server and a Database Management System (DBMS) as a repository for a database (Figure 1).

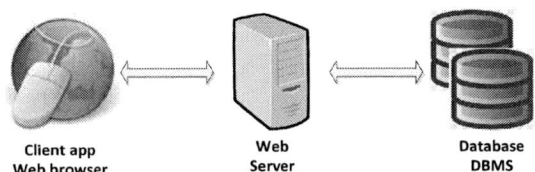

Figure 1 System architecture

End users access and manipulate data via a web browser using web application developed following the MVC architecture [3]. A web server via a HTTP protocol responds to users' requests, which are forwarded by their computers' HTTP clients. To be able to answer the clients' requests, web server, when needed, communicates with DBMS to gather and change data stored in the database. As a web server Microsoft's Internet Information Server (IIS) [4] is used, while Microsoft SQL Server [5] serves as a relational DBMS.

The client application is developed using ASP.NET programming framework, and programming languages C#, JavaScript, HTML and css. Entity Framework [6] is used as an object-relational mapping framework.

B. Data model

The database has been designed following the principles of the relational data model. We believe that the relational data model is the most suitable for this business process because the data is structured and the amount of data is small. Figure 2 shows tables used to store data occurring in the process of the Erasmus+ outgoing mobility.

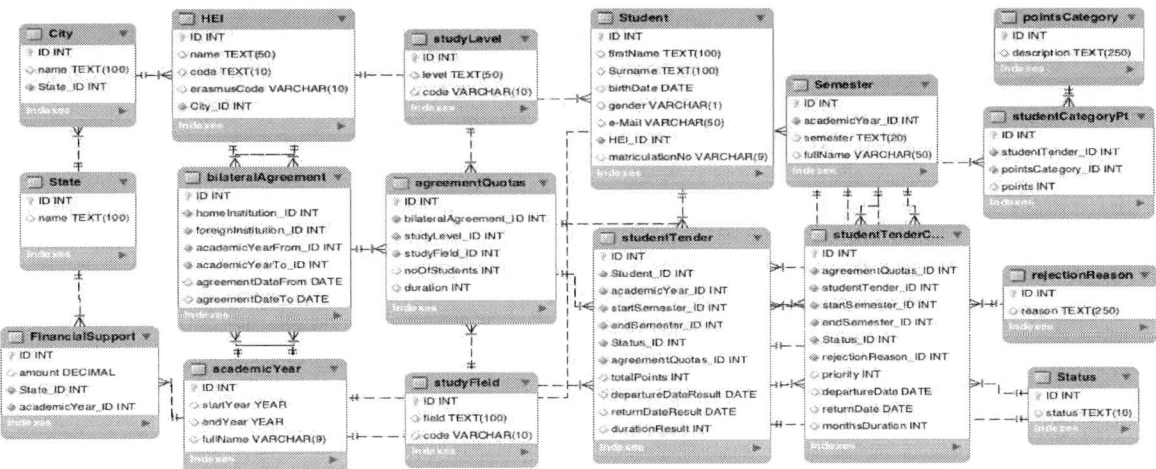

Figure 2. Relational data model

Data about HE institutions and the details of bilateral agreements signed between them are kept in tables **HEI**, **studyLevel**, **studyField**, **bilateralAgreement** and **agreementQuotas**. Data on students, their applications as well as the results of the ranking process are stored in tables **student**, **studentTender**, **studentTenderChoice** and **studentCategoryPt**.

The database design allows to store data from different HEIs in one database. A centralized database containing data from all members of the university allows the uniform processing of data and easy distribution of results. With proper software, besides members of UNIZG, database can be used by the University and thus the exchange of data via an e-mail can be avoided. Of course, in simpler variant data of only one HEI will be stored in the database, and that is currently the case.

IV. BASIC TYPES OF SCREENS IN THE WEB-APPLICATION

Ensuring the ergonomic quality of user interface (UI) is one of the major concerns in developing interactive application. While developing web application we followed guidelines for good and ergonomic software interface design [7] with an emphasis on the content of screens and the way in which typical data operations take place. Special attention was paid to ensure that the flow of data from screen to screen is consistent and corresponds to user expectations. By keeping the application consistent, users learn more quickly since they recognize and re-apply the usage patterns. For the sake of consistency and simplicity, all screens designed to enable same operations on different data (rows from different tables) look exactly the same and support the same set of functionalities. In the application it is possible to see two types of screens: a) screens designed for performing CRUD operations on data stored in a single table and b) master-detail screens. Master-detail screens allow performing CRUD operations on master row and also on detail (slave) rows related to master row in a one-to-many type relationship.

The acronym CRUD refers to create, read, update, and delete operations - the major operations that should be implemented over data stored in data repository. In our case data is stored in relational DBMS but that need not be the case. Each, of the above mentioned, operation can be mapped to a standard SQL statement or HTTP method.

A. Typical screen layout for performing CRUD operations in a single database table

Typical screen layout that serves for browsing and updating data stored in a single database table can be seen on *Figure 3*. In application described in this paper such screens exist for tables usually called catalogs such as: **state**, **city**, **academicYear**, **studyLevel**, **studyField** etc. Data is displayed in a form of a list and initially only a certain number of rows is visible.

Besides the possibility of viewing data, each page contains additional functionality, such as adding new or editing and deleting the existing rows, sorting rows, filtering rows and viewing a certain number of items per page (paging).

On all screens, equivalent operations are activated in the same way - either by clicking on the option (eg. Add) or by clicking on the appropriate icon. The Table 1 shows the icons that appear on each screen.

Table 1 Icons for typical operations

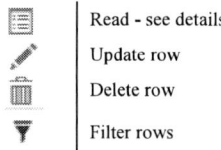

	Read - see details
	Update row
	Delete row
	Filter rows

Number of rows displayed on each page is a parameter and can be easily changed by selecting value from drop-down list. If there are more rows than the selected parameter (for display per single page), the rest is displayed on other pages that can be accessed by clicking on the page number or on the icon for the next page (>>).

Data filtering is enabled by clicking the funnel icon, after which the editable textboxes below the columns appear (Figure 4).

Figure 3. Typical screen layout for a single table

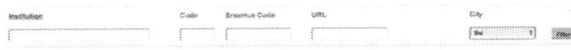

Figure 4. Filtering rows

Textboxes serve for entering the selection criteria. After applying the selection conditions by pressing the button Filter, only the records that meet the given criteria will be displayed.

B. Master-detail screen layout and functionality

In master-detail relationship, for one master tuple always exist N detail tuples. On the master-detail screens, in the simplest case, two logical parts can be noticed: 1) master data in the upper part of the screen and 2) detail data on the lower part. This type of screen is more complex than previously described screens as it has to allow the CRUD operations on both - master and detail data. Typical master-detail screen can be seen on Figure 5.

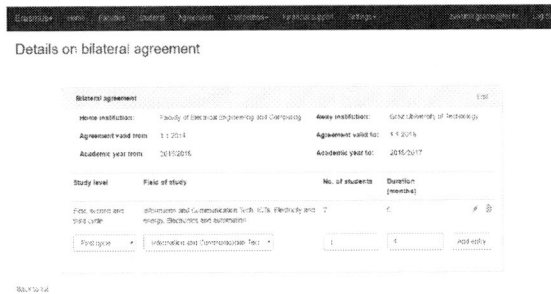

Figure 5. Example of master-detail screen

This screen enables manipulation on data regarding bilateral agreements (contracts) and the agreed quotas for student exchange. In this case, the master data are general data on the contract (partners in the contract - HEIs, the date of signing the agreement, the valid period – in dates and academic years, etc.). The detail data (tuples) presents quotas agreed with the contract for the study levels and subject area. Specific for master-detail screens is that master data can't be deleted while there are detailed data referencing master data. Besides in application, this rule, is also implemented at the database level by defining integrity constraint - foreign key.

V. WEB-APPLICATION FUNCTIONALITY

Web application is, currently, designed solely for HEI's employees participating in the process of receiving and ranking students' applications for mobility at a foreign institution. For keeping data private and preventing non-authorized access, a login system via account name and password for each user, given by the application administrator, has been implemented. User has the possibility of viewing data about HEI and students, bilateral agreements signed between sending and receiving institutions, exchange quotas per study level and subject area, student applications and the outcome of the tender. Besides viewing, users have the possibility of inserting and editing data.

Application menu allowing access to the listed data can be seen on Figure 6.

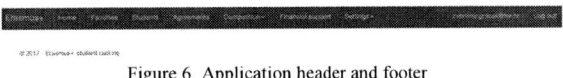

Figure 6. Application header and footer

Unlike foreign higher education institutions, majority of the members at UNIZG don't have mobility offices and this process is supported by one or more ECTS coordinators who are the targeted users of this software. They commonly have no data about the student's achievements, which are necessary for the applications ranking. These data must be delivered from other HEI's department and thereafter be imported into database via software. Besides department in charge for students' grades, process supported with this software involves UNIZG IRO so the exchange of information between all parties involved must be easy. For easier data manipulation, user has the possibility of importing student applications and their scoring via an Excel file.

Figure 7 shows the screen for importing students' applications from excel file that faculty gets from UNIZG IRO. All persons' names on Figure 7,

Figure 8, Figure 9 and Figure 10 are fictitious due to protection of data privacy.

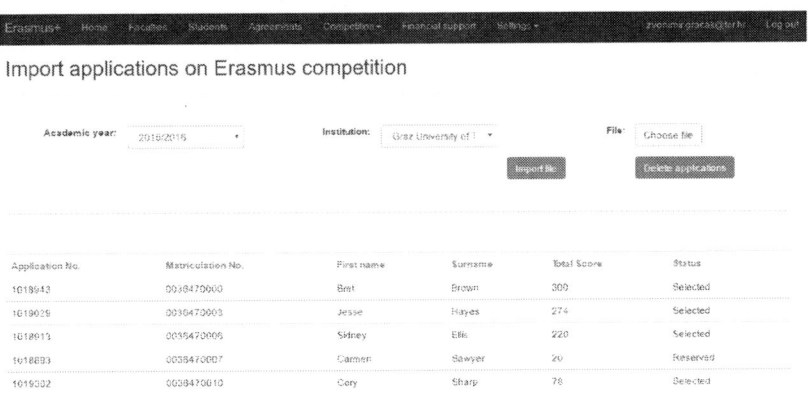

Figure 7 Importing students' applications via an Excel file

Figure 8 shows the screen for importing different categories of points students gained. The screen shows importing points won on the basis of motivational letters that are read and evaluated by ECTS coordinators.

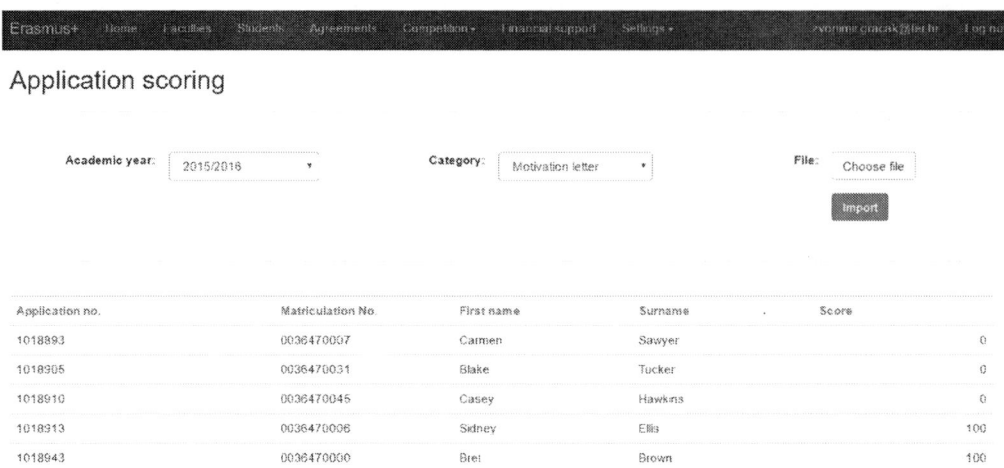

Figure 8 Importing application scoring via an Excel file

We are aware that there exists a much better solution for data exchange than importing from Excel files, e.g. the use of web services. We haven't used advanced technologies for data exchange because we could not reach an agreement with our partners in the process to develop the necessary software on their side. In our case, it would be a complete failure to develop a web services because there would be no compatible software with which they can communicate.

Main part of the application refers to processing all student applications, in terms of ranking applications considering students' scoring and assigning corresponding selections to students. After all students' applications and their scoring are imported into the system, the user has the ability to run a procedure to process all applications for a specific academic year. The applications are ranked based on the achieved scores and in accordance with calculated rank, applications are assigned their achieved selection (foreign HEI according to student's choice). The process starts with sorting all applications by their scoring and then by their priorities. After that, the system selects applications in descending order. Each application has one or more desired foreign HEI ordered by priorities. For each desired institution on the application, system is checking if there is any space left to fulfill on the selected institution. The first institution on the application that has available space will be automatically selected and application's status for that institution will change to *Selected*. If there isn't any space available on any of the institutions in the application, application's status will change to *Reserved*. After all the applications have been processed, user is presented with the results (Figure 9).

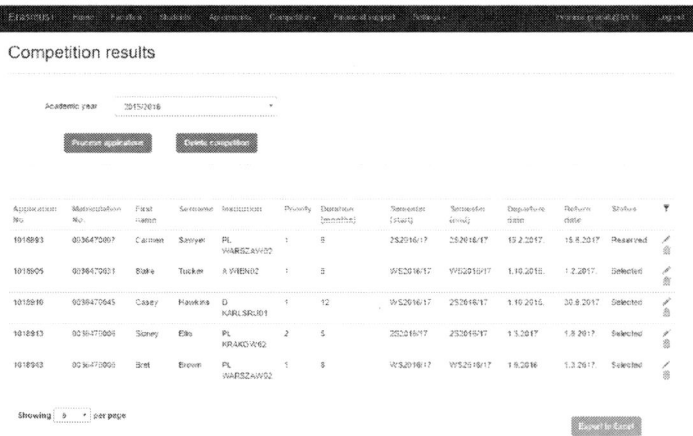

Figure 9 Competition results after processing applications

Besides *Selected* and *Reserved*, student's application's status can also be *Declined* for some specific reasons (eg. due to lack of needed documentation). Also, there's always a possibility that someone whose application was selected for a specific institution will pass the selection, which means that the best scored Reserved application will automatically be *Selected*. That part has to be manually operated by user, so besides processing the applications,

user has the possibility to modify the final results by changing applications' status to either *Selected, Reserved* or *Declined*.

After processing the applications and generating the final results, user can export results to an Excel file to enable easier data manipulation and exchange with the UNIZG IRO (Figure 10).

Matriculatio	Surname	First Nam	Birth Date	Application	Subject area	Study Level	Semester (Start	Duration (month)	Departure Date	Return Date	Foreign institution - name	Foreign institution - Erasmus	Country	Total score	SELECTED/RE
0036470000	Brown	Bret	28.05.1993	1018943	061	S	WS2016/17	6	01.09.2016	01.03.2017	Warsaw University of Technology	PL WARSZAW02	Poland	300	Selected
0036470039	Suarez	Blake	08.06.1994	1019405	0714	S	WS2016/17	5	22.08.2016	22.01.2017	Chalmers University of Technology	S GOTEBOR02	Sweeder	300	Selected
0036470038	Newton	Sam	10.09.1994	1019462	0714	S	WS2016/17	12	19.09.2016	19.09.2017	KU Leuven	B LEUVEN01	Belgium	291	Selected
0036470009	Hayes	Jesse	16.05.1994	1019029	061	S	WS2016/17	12	01.10.2016	01.10.2017	Warsaw University of Technology	PL WARSZAW02	Poland	274	Selected
0036470025	Pearson	Drew	24.08.1995	1019475	061	S	WS2016/17	5	25.08.2016	25.01.2017	Chalmers University of Technology	S GOTEBOR02	Sweeder	260	Selected
0036470035	Cox	Blake	28.02.1996	1019594	0215	F	WS2016/17	5	04.10.2016	04.03.2017	Hochschule für Musik Detmold	D DETMOLD01	Germany	236	Selected
0036470006	Ellis	Sidney	26.07.1994	1018913	0714	S	2S2016/17	5	01.03.2017	01.08.2017	AGH University of Science and Techno	PL KRAKOW02	Poland	220	Selected
0036470028	Cook	Will	02.08.1993	1019608	0713	S	2S2016/17	6	01.04.2017	01.09.2017	Technische Universitat München	D MUNCHEN02	Germany	208	Selected
0036470045	Hawkins	Casey	24.05.1994	1018910	0714	S	WS2016/17	12	01.10.2016	30.09.2017	Karlsruher Institut für Technologie	D KARLSRU01	Germany	175	Selected
0036470031	Tucker	Blake	15.04.1993	1018905	0714	S	WS2016/17	5	01.10.2016	01.02.2017	Vienna University of Technology	A WIEN02	Austria	90	Selected
0036470007	Sawyer	Carmen	03.08.1996	1018893	0714	F	2S2016/17	6	15.02.2017	15.08.2017	Warsaw University of Technology	PL WARSZAW02	Poland	20	Reserved

Figure 10. Competition results exported to an Excel file

VI. CONLUSION

In this paper we have presented the software developed to support processing students' applications in the outgoing student mobility in Erasmus+ Programme for Program Countries. This software facilitates the work of HEI's staff responsible for the ranking of students' applications and assignment of foreign HEI chosen by the student intending to study abroad.

Relational data model proved to be a good choice for this type of process because the data occurring in the process is structured, of simple type and the amount of data is not overwhelming. Also, consistency and durability of data is a necessary requirement and RDBMS is the best database system for ensuring the ACID properties of transactions. A web application, compared to a desktop application, is a better choice because of the easier installation, deployment and maintenance.

The process involves participants from different institutions that have to exchange data at different stages of the process. The data is currently exchanged semi-automatically via excel files. The plan is to change this part in the future, but to succeed we must reach an agreement with the other participants who will have to adapt their software.

REFERENCES

[1] Europaen Commission, "Erasmus+ Programme Guide," 20 01 2017. [Online]. Available: https://ec.europa.eu/programmes/erasmus-plus/sites/erasmusplus/files/files/resources/erasmus-plus-programme-guide_en.pdf. [Accessed 2017].

[2] European Commission, "Erasmus+ - Key Action 3 - Erasmus Charter for Higher Education 2014-2020," [Online]. Available: https://eacea.ec.europa.eu/erasmus-plus/actions/erasmus-charter_en.

[3] M. Fowler, D. Rice, M. Foemmel, E. Hieatt, R. Mee and R. Stafford, "Chapter 14: Web Presentation Patterns," in *Patterns of Enterprise Application Architecture*, Addison-Wesley, 2003.

[4] Microsoft, "nternet Information Services (IIS)," [Online]. Available: https://www.iis.net/.

[5] Microsoft, "Microsoft SQL Server," [Online]. Available: https://msdn.microsoft.com/en-us/library/mt590198(v=sql.1).aspx.

[6] P. Krill, "Microsoft open-sources Entity Framework," InfoWorld, 20 July 2012. [Online]. Available: http://www.infoworld.com/article/2617690/microsoft-net/microsoft-open-sources-entity-framework.html.

[7] Cornell University Ergonomics Web, "ERGONOMIC GUIDELINES FOR USER-INTERFACE DESIGN," [Online]. Available: http://ergo.human.cornell.edu/ahtutorials/interface.html.

Blood Vessel Segmentation using Multiscale Hessian and Tensor Voting

Andrea Lukač
Faculty of Electrical Engineering
and Computing
University of Zagreb
Zagreb, Croatia
Email: andrea.lukac@fer.hr

Marko Subašić
Faculty of Electrical Engineering
and Computing
University of Zagreb
Zagreb, Croatia
Email: marko.subasic@fer.hr

Abstract—We present a new method for segmentation of retinal blood vessels in color fundus images. The method utilizes Hessian eigenvalues and eigenvectors calculated for each pixel to obtain a measure of vesselness. The measure is similar to the Frangi vesselness but we use a different combination of eigenvalues, while keeping entire Hessians for further processing. Hessians are calculated at different scales, where each scale targets a certain width range of blood vessels. We perform a scale selection procedure to pick the most appropriate scale for each pixel. Hessians from the selected scales are treated as tensors and used as input for tensor voting procedure to further reduce the influence of noise and highlight the blood vessels. Green channel of color retinal photographs is used as input to the method due to better blood vessel contrast. We also present promising experimental results obtained on a test database with manually segmented retinal blood vessels of healthy patients only.

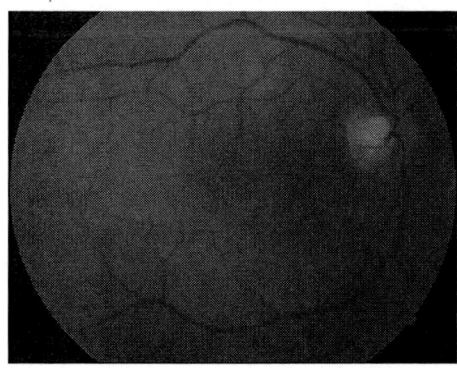

Fig. 1. Example of retinal image.

I. INTRODUCTION

Retinal image analysis provides important information for detection of several diseases with most common being Diabetic Retinopathy (DR). In order to detect the disease symptoms successfully, they must not be confused with normal retinal structures. Hence it is helpful to obtain segmentation results of all normal retinal structures prior to symptom detection. This paper presents a segmentation method for one such normal retinal structure, the blood vessel tree. Figure 1 shows an example of retinal image in which the blood vessel tree can be seen.

Analysis of a retinal image can be performed by an experienced ophthalmologist, but the process is time consuming. On the other hand, diabetic patients, which are at risk of DR, need to perform regular retinal checkups, and their number is constantly increasing. This indicates a clear need for computer vision-based retinal image analysis and a lot of algorithms have been and still are being developed. Besides providing full diagnosis, computer vision-based analysis could serve as an aid to ophthalmologists and optometrists and reduce the time required to interpret retinal images and overcome possible biases that may skew the clinical assessments [1]. Automated algorithms for retinal blood vessel segmentation have specific issues due to blood vessel tree structure with a lot of branches and some very thin vessels. Also, images can have low resolution, poor contrast and significant amount of noise.

One other issue is that other structures can be misinterpreted as vessels because of similar characteristics. Those structures can be other parts of retina, fovea and optical disc, but also disease side effects, like hemorrhages, lesions or exudates.

In this paper, we segment blood vessels in retinal images of healthy patients with no central light reflex within vessels. We use a procedure inspired by Frangi vesselness [2]. We are using a stickness feature from tensor voting framework [3] as a measure of vesselness. The obtained measure resembles Frangi vesselness but is not limited to [0,1] interval. Hessian matrices are calculated at different scales and we use this data to perform tenesor voting [3]. Two different steps of tensor voting are applied to supress the influence of noise and other retinal structures, and to highlight retinal blood vessels.

II. RELATED WORK

There are a lot of different methods that have been developed for segmentation of blood vessels in fundus images [4], [5], [6]. They mostly use DRIVE and STARE databases, which are both publicly available [5]. Vessel segmentation techniques can be supervised or unsupervised. Supervised methods exploit some prior labeling information to decide whether a pixel belongs to a vessel or not, while unsupervised methods perform the vessel segmentation without any prior labeling knowledge [5]. In this section we give a short overview of such techniques. In [7] an unsupervised approach has been used for vessel segmentation. Authors use 2-D Gabor wavelet transform for image enhancement followed by sharpening filter to reduce the blurring side effect. Vessels are segmented using Canny Edge Detector with pixel dilation to fill the gaps. Accuracy of $0.9439 \pm .0253$ was recorded when tested on STARE database. Unsupervised method that uses gradient orientation was exploited in [8]. They use Sobel operator at three different scales and the results are integrated into final image. Morphological operations and thresholding are used to segment blood vessels. This method is not affected by low contrast or poor illumination and it performs with accuracy of 0.9358 on DRIVE database and 0.9423 on STARE database. A supervised approach based on a multilayered feed forward neural network has been proposed in [9]. Tested on DRIVE database it has scored average accuracy of 0.9452, while tested on STARE it has achieved accuracy of 0.9526.

Linear combination of line detectors at various scales are used to segment retinal blood vessels and to reduce noise in [10]. Our method uses linear combination of Hessians at various scales. In [11] authors also use tensor voting procedure along with other techniques to reconstruct only small blood vessels in high resolution images. Both tensor voting stages are used while we use only the second stage. Larger blood vessels are tracked using a multi-scale line detection from [10]. In [12] authors calculate their vesselness measure based on standard deviation of Hessian eigenvalue's orientations across several scales. Then they apply clustering procedure to obtain vessel segmentation. Our approach also uses multi-scale Hessians, but we combine scales in a different way.

III. METHODS

Figure 2 shows the procedure we use for blood vessel segmentation. We use green channel from input RGB images, as it provides the best contrast between vessels and background [9]. We use multi-scale Hessians to estimate a measure of vesselness at each pixel and thus detect blood vessels. We perform a simple form of tensor voting to reduce the influence of the noise, followed by a scale selection. Hessians of the appropriate scale, and the scale are used to perform a full stick tensor voting to further reduce the noise, fill the gaps and increase our vesselness measure for pixels belonging to blood vessels. The measure is thresholded, and simple morphological opening is used to further enhance the results. We continue with descriptions of specific steps.

A. Vessel Detection

Vessels, and other tubular structures can be detected using gradients of the image, which can be estimated by convolving an image with partial derivatives of Gaussian. The second order partial derivative of a Gaussian kernel at scale σ_g generates a probe kernel that measures the contrast between the regions inside and outside the range $(-\sigma_g, \sigma_g)$ in the direction of the derivative [2]. Gaussian kernel is defined by

$$G(x,y,\sigma_g) = \frac{1}{2\pi\sigma_g^2} e^{-\frac{(x^2+y^2)}{2\sigma_g^2}}, \qquad (1)$$

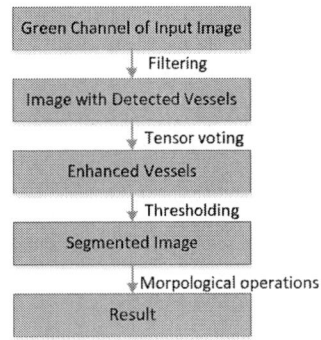

Fig. 2. Stages of proposed procedure.

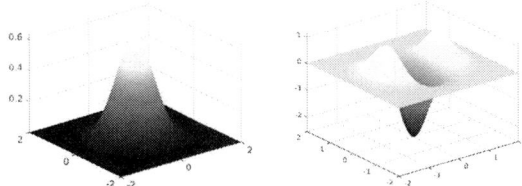

Fig. 3. 2D Gaussian kernel (left) and its second order partial derivate (right) with x in the range [-2,2] and $\sigma_g = 0.5$.

where x and y represent pixel coordinates. It's second order partial derivative by x variable is described by the equation

$$G_{xx}(x,y,\sigma_g) = \frac{1}{2\pi\sigma_g^4}\left(-1 + \frac{x^2}{\sigma_g^2}\right)e^{-\frac{(x^2+y^2)}{2\sigma_g^2}}. \qquad (2)$$

The shape of Gaussian and its second order partial derivate are shown in Figure 3. In our approach, we apply a range of scales at every pixel, as different scales are better suited for different vessel widths. For each pixels and for each scale, a 2x2 Hessian matrix H is calculated:

$$H(x,y,\sigma_g) = \begin{pmatrix} r_{xx} & r_{xy} \\ r_{xy} & r_{yy} \end{pmatrix}. \qquad (3)$$

Each element of the Hessian matrix is an estimate of the second order partial derivative of image I at specific scale, according to the following equations:

$$r_{xx} = G_{xx}(x,y,\sigma_g) * I(x,y) \qquad (4)$$

$$r_{xy} = G_{xy}(x,y,\sigma_g) * I(x,y) \qquad (5)$$

$$r_{yy} = G_{yy}(x,y,\sigma_g) * I(x,y). \qquad (6)$$

For each Hessian matrix, eigen decomposition is performed to obtain two eigenvalues λ_+ and λ_-, and two eigenvectors $\vec{e}_+$ and $\vec{e}_-$. As Hessian matrix is symmetric, two eigenvectors will always be perpendicular. The eigenvalue with the largest absolute value is denoted λ_+, and the most significant eigenvector as $\vec{e}_+$. According to the geometric interpretation of the Hessian matrix, $\vec{e}_+$ will point in the direction where intensity changes the most. Hence, $\vec{e}_+$ will be oriented perpendicular to the vessel direction. Using eigenvalues and eigenvectors we calculate a measure of vesselness and an estimated orientation of the vessel normal. As our measure of vesselness, we use stickness from the tensor voting framework [3], which is defined by following equation:

$$s = \lambda_+ - \lambda_-. \qquad (7)$$

The orientation β that we use is the orientation of $\vec{e}_+$.

$$\beta = angle(\vec{e}_+). \qquad (8)$$

Pixels belonging to linear structures will have large absolute values of λ_+ and lover absolute values of λ_-. Blob-like structures will produce simmilar absolute values for λ_+ and λ_-. Frangi vesselness uses eigenvaluess of the Hessian in a slightly different way than the stickness does, but both of them will exhibit higher absolute values for linear structures, and lower absolute values for blob-like structures. If the image compositing is such that vessels are dark on brighter background, the larger eigenvalue λ_+ will be positive.

Stickness orientation β of pixels inside blood vessels tend to change only slightly in between scales, while noise generated stickness of the background pixels tend to change orientation significantly in between scales. We perform an addition of colocated Hessian matrices (a simple form of tensor voting) from three consecutive neighboring scales. Stickness of resulting Hessians in blood vessel pixels will be amplified by this procedure, while noise induced stickness will not get amplified as much. This results in better highlighted blood vessels.

Then, for every pixel we find the scale σ_G with the highest stickness value. In [13] authors have demonstrated that such maximum along the scale axis can be expected, and corresponding scale can be linked to the width of the linear structure (blood vessel), at least for simple profiles of linear structures.

We use these maximum sticknesses and corresponding orientations as input for final tensor voting procedure.

B. Tensor Voting

Tensor voting is a method of information propagation where tokens convey their orientation preferences to their neighbors in the form of votes. Each vote is an estimate of orientation of a perceptual structure consisting of just two tokens: the voter and the receiver [3]. We can describe a tensor with a symetric 2x2 matrix (rank 2 tensor) which can be decomposed into its eigenvectors and eigenvalues as shown in:

$$T = \lambda_+ \vec{e}_+ \vec{e}_+^T + \lambda_- \vec{e}_- \vec{e}_-^T, \qquad (9)$$

Every tensor has three properties: orientation (β), stickness (s) and ballness (b). Stickness and orientation are defined by equations (4) and (5). Ballness is defined with:

$$b = \lambda_-. \qquad (10)$$

The stickness s is interpreted as a measure for the orientation certainty or a measure of anisotropy of the ellipse in orientation β. The ballness b is interpreted as a measure

for the orientation uncertainty or isotropy [3]. Since we are considering tubular structures, we are mostly interested in stickness property. In Figure 4, a rank 2 tensor is illustrated with its eigenvalues, stickness, ballness and orientation. A tensor with ballness value of zero is called a stick tensor, and a tensor with zero stickness is called a ball tensor. For vessel segmentation purposes, we use voting fields that consist only of stick tensors. Figure 5 shows an example of a stick voting field amplitude. Here, pixel intensities are interpreted as a measure of likelihood that a feature at some position x belongs to the same curvilinear structure as the feature positioned in the center of the voting field, i.e. the voting tensor. The orientation of tensor vote at position x is the presumed orientation of the curvilinear structure at that position. In [14] authors assume that the curvilinear structure should be a circular arc. The size of a voting field is determined by a scale parameter σ_{ctx}, which controls the amount of affected neighboring pixels. Resulting stickness will be calculated considering all of the received votes.

For each voting pixel, it's orientation β has to be determined prior to vote casting, to orient the voting field properly. This can be computationally expensive, so we use steerable filters combied with bandlimited voting field, as proposed in [15]. The voting field is defined by:

$$\widetilde{V}(r, \phi) = \frac{1}{G} \epsilon^{e^{\frac{-r^2}{2\sigma_{ctx}^2}}} cos^{2n}(\phi) \begin{pmatrix} 1 + cos(4\phi) & sin(4\phi) \\ sin(4\phi) & 1 - cos(4\phi) \end{pmatrix}. \qquad (11)$$

Gaussian decay function is used to penalize distance, and $cos^{2n}(\phi)$ to penalize high-curvature arcs. Parameters r and ϕ represent relative euclidean distance and angle of corresponding neighboring pixel. Speed of decay as a function of ϕ is controlled by parameter n, and G is a normalization constant. Appropriate scale σ_{ctx} of the voting field can be selected in many ways. In the most simple case, a fixed value can be chosen, but it is reasonable to presume that dynamically adaptable scale should provide better voting results. It is also resonable to presume that voting scale should change with blood vessel width, i.e scale of the largest stickness. Wider blood vessels are expected to be less curved as opposed to thinner blood vessels so we want wider vessels to have larger voting field that will favor longer, less curved structures. This relation can be described in different ways, including linear relation, but based on our experiments the following relation provides best results:

$$\sigma_{ctx} = \sigma_G^2. \qquad (12)$$

For speed of decay, we used fixed value.

IV. RESULTS

In order to test the given algorithm, we have used a small subset of DRiDB [16] database containing only 13 images of healthy patients. Currently we use only images of healthy patients to demonstrate effectiveness of our blood vessel segmentation method. An example of such image is shown in Figure 6a with corresponding manually segmented image

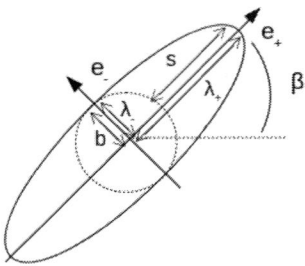

Fig. 4. Illustration of tensor with rank 2.

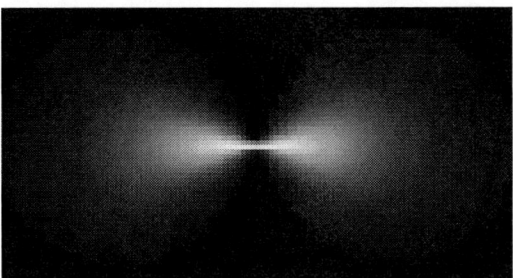

Fig. 5. Example of tensor voting field.

shown in Figure 6b. Resolution of images is 576 × 720. For Gaussian kernel, we have used σ_g in the range [1, 3] with step 0.1 and for speed of decay $n = 2$. For the thresholding step we are using threshold intensity value of 3. In post-processing step, we first erode the image with a disk structuring element with radius of 1 pixel, after which a dilatation is performed with the same structuring element.

In evaluation of our algorithm, we used accuracy and f-measure as effectiveness indicators. They are defined with equations:

$$F = 2 * \frac{precision * recall}{precision + recall} = 2 * \frac{TP}{2TP + FN + FP} \quad (13)$$

and

$$accuracy = \frac{TP + TN}{TP + FP + TN + FN}. \quad (14)$$

Where precision and recall are defined with:

$$precision = \frac{TP}{TP + FP} \quad (15)$$

and

$$recall = \frac{TP}{TP + FN}. \quad (16)$$

When tested on all healthy images, our algorithm performed with accuracy of 0.9400 and f-measure of 0.6319. Table I shows accuracy and f-measure for each individual image used for testing.

Figure 7c is an example of image segmented with our algorithm that scored well, with accuracy of 0.9380 and f-measure of 0.7183. Input image that gave this result is shown in Figure 7a and corresponding manually segmented image is shown in Figure 7d. This image had some noise left, which hasn't been removed by morphological operations. Also, there are some smaller vessels, or parts of them, that remained undetected, but even though the accuracy for this image is lower then the accuracy scored on whole database, we consider it to be a well segmented image because of higher f-measure, since f-measure is a better measure for successfulness of vessel pixels detection. This is due to the fact that, unlike accuracy which gives equal importance to negative and positive results, f-measure gives higher importance to positive results. This context can be deduced from their formulas defined in (13) and (14).

It can be seen that wider vessels have been successfully detected as well as the most of smaller vessels. To detect higher percentage of small vessels, a lower threshold can be applied, but with side effect of weaker noise suppression. Our choice for the threshold value is based on experimenting with different thresholds and evaluating their accuracy and f-score. This effect is shown in Figures 8b, 8c, 8d, 8e and 8f where results with different thresholds applied are shown. With lower thresholds, some of the wider vessel's endings haven't been detected or they contain gaps. It can also be seen that a significant part of the optic disc has been falsely detected as a part of the vessel tree.

Figure 9c shows an example of lower quality result with input and manually segmented image in Figures 9a and 9d. Accuracy for this image is 0.9343 and f-measure is 0.5367. Aside from small vessels being poorly detected, there are some bigger vessels on which the algorithm hasn't performed well. This is caused by low contrast between the vessels and the background. It can be seen in the input image that the regions where bigger vessels haven't been successfully segmented, have darker background.

Figure 10 is an example where our algorithm provides good detection of vessel, even the thin ones, but a lot of noise is present. On this particular image, a higher threshold could give better results in term of noise cancellation. Circular edge of retinal images would also produce high stickness values, but the edge is easily detected and appropriately masked out from results.

In future, additional post-processing steps should be applied to alleviate the problem of false segmentation of the optic disc. The low contrast issue should also be addressed and contrast enhancement preprocessing techniques should be explored. Also, other noise removal techniques should be considered to remove remaining noise. Various retinal disorders do interfere with blood vessel segmentation, so further post-processing steps are planned for future work, to enable successful segmentation in presence of such disorders.

TABLE I
A TABLE WITH RESULTS FOR EACH IMAGE.

	Accuracy	F-measure
1	0.9315	0.5241
2	0.9422	0.6880
3	0.9355	0.6856
4	0.9400	0.5330
5	0.9460	0.5898
6	0.9504	0.6710
7	0.9380	0.7183
8	0.9404	0.6995
9	0.9515	0.6632
10	0.9337	0.5913
11	0.9343	0.5367
12	0.9420	0.5771
13	0.9349	0.6142

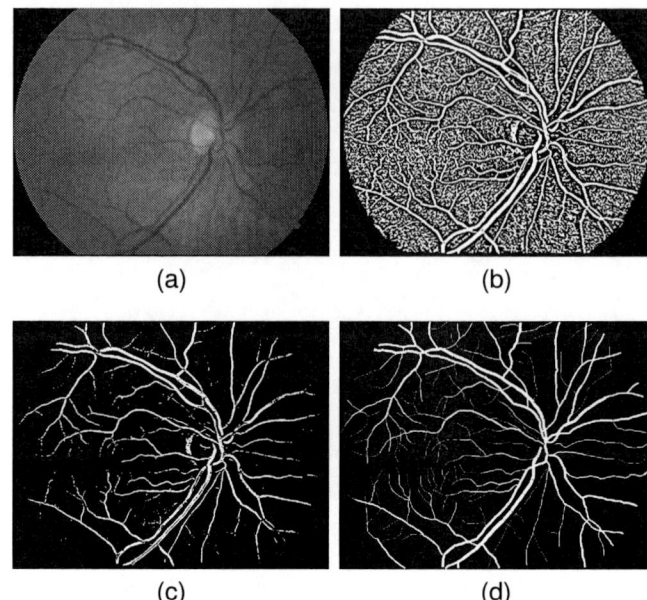

(a) (b)

(c) (d)

Fig. 7. Input image (a) on which the algorithm performed well with resulting stickness (b), result of segmentation with proposed method (c) and manually segmented image (d).

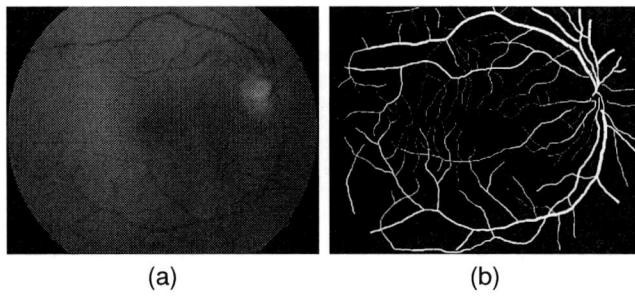

(a) (b)

Fig. 6. Example of input image (a) and corresponding manually segmented image (b).

V. CONCLUSION

Retinal vessel segmentation is important in ophthalmological diagnostics and there are a lot of different automated methods proposed. In this paper we have used Hessian matrix and tensor voting to detect the vessels. Our preliminary results on a limited database indicate that our algorithm performs well, with results comparable to other published methods. However, care must be taken when comparing the results, as they are obtained with different databases. In the future we plan to expand the method to enable good segmentation result in problematic images and in the presence of diabetic retinopathy symptoms.

REFERENCES

[1] I. Kaur and L. M. Singh, "A method of disease detection and segmentation of retinal blood vessels using fuzzy c-means and neutrosophic approach," *Imperial Journal of Interdisciplinary Research*, vol. 2, no. 6, 2016. [Online]. Available: http://www.imperialjournals.com/index.php/IJIR/article/view/870

[2] R. F. Frangi, W. J. Niessen, K. L. Vincken, and M. A. Viergever, "Multiscale vessel enhancement filtering." Springer-Verlag, 1998, pp. 130–137.

[3] G. Medioni and S. B. Kang, *Emerging topics in computer vision*, ser. IMSC Press multimedia series. Upper Saddle River, N.J. Prentice Hall PTR, 2005. [Online]. Available: http://opac.inria.fr/record=b1126919

[4] N. Singh and L. Kaur, "A survey on blood vessel segmentation methods in retinal images," in *2015 International Conference on Electronic Design, Computer Networks Automated Verification (EDCAV)*, Jan 2015, pp. 23–28.

[5] M. Fraz, P. Remagnino, A. Hoppe, B. Uyyanonvara, A. Rudnicka, C. Owen, and S. Barman, "Blood vessel segmentation methodologies in retinal images - a survey," *Comput. Methods Prog. Biomed.*, vol. 108, no. 1, pp. 407–433, Oct. 2012. [Online]. Available: http://dx.doi.org/10.1016/j.cmpb.2012.03.009

[6] S. Sharma and W. V., "Retinal blood vessel segmentation a review," *International Journal for Scientific Research and Development*, vol. 3, pp. 952 – 955, 2015.

[7] M. U. Akram, A. Atzaz, S. F. Aneeque, and S. A. Khan, "Blood vessel enhancement and segmentation using wavelet transform," in *Proceedings of the International Conference on Digital Image Processing*, ser. ICDIP '09. Washington, DC, USA: IEEE Computer Society, 2009, pp. 34–38. [Online]. Available: http://dx.doi.org/10.1109/ICDIP.2009.70

[8] D. Onkaew, R. Turior, B. Uyyanonvara, and T. Kondo, "Automatic extraction of retinal vessels based on gradient orientation analysis," in *2011 Eighth International Joint Conference on Computer Science and Software Engineering (JCSSE)*, May 2011, pp. 102–107.

[9] D. Marin, A. Aquino, M. E. Gegúndez-Arias, and J. M. Bravo, "A new supervised method for blood vessel segmentation in retinal images by using gray-level and moment invariants-based features," *IEEE Trans. Med. Imaging*, vol. 30, no. 1, pp. 146–158, 2011. [Online]. Available: http://dx.doi.org/10.1109/TMI.2010.2064333

[10] U. T. V. Nguyen, A. Bhuiyan, L. A. F. Park, and K. Ramamohanarao, "An effective retinal blood vessel segmentation method using multi-scale line detection," *Pattern Recogn.*, vol. 46, no. 3, pp. 703–715, Mar. 2013. [Online]. Available: http://dx.doi.org/10.1016/j.patcog.2012.08.009

[11] A. Christodoulidis, T. Hurtut, H. B. Tahar, and F. Cheriet, "A multi-scale tensor voting approach for small retinal vessel segmentation in high resolution fundus images," *Comp. Med. Imag. and Graph.*, vol. 52, pp. 28–43, 2016. [Online]. Available: http://dx.doi.org/10.1016/j.compmedimag.2016.06.001

[12] N. M. Salem, S. A. Salem, and A. K. Nandi, "Segmentation of retinal blood vessels based on analysis of the hessian matrix and clustering

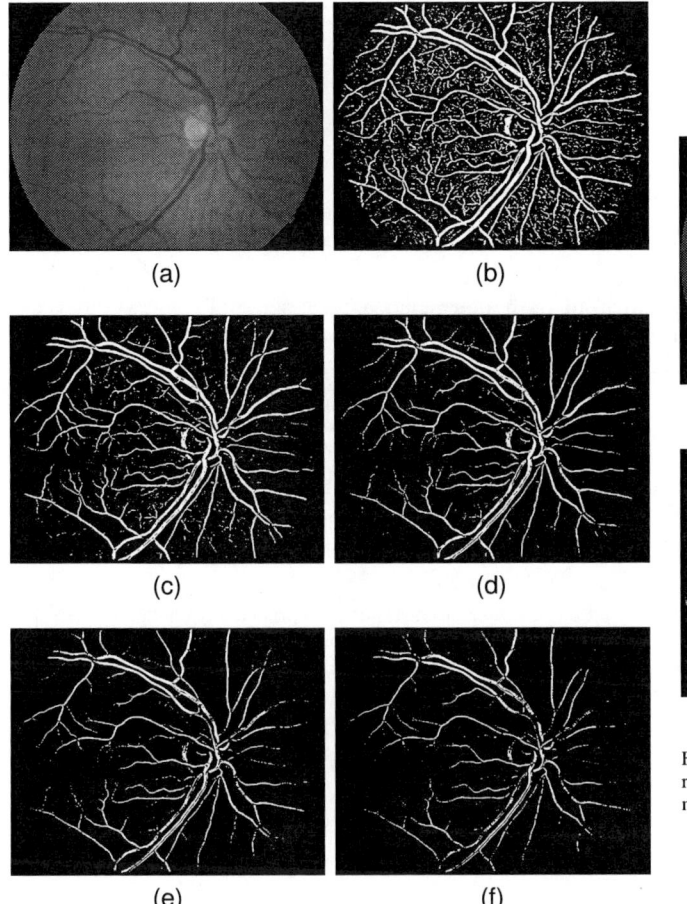

Fig. 8. Results for input image (a) with different threshold values of 1 (b), 2 (c), 3 (d), 4 (e), 5 (f).

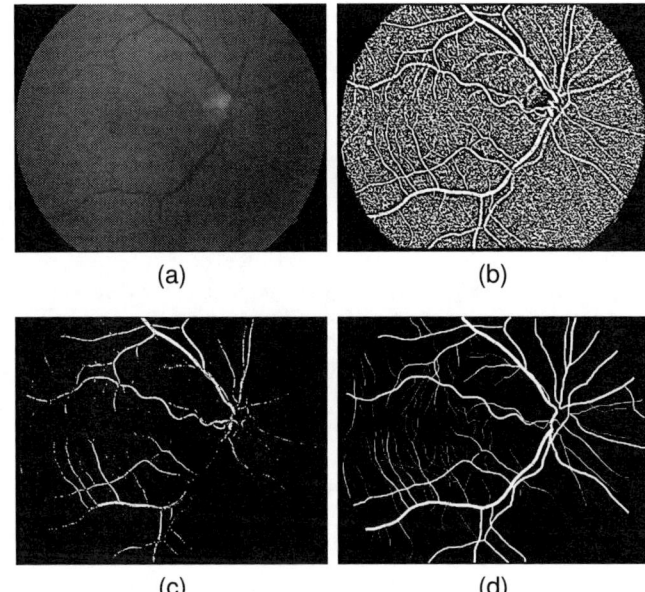

Fig. 9. Input image (a) on which the algorithm performed poorly with resulting stickness (b), result of segmentation with proposed method (c) and manually segmented image (d).

algorithm," in *2007 15th European Signal Processing Conference*, Sept 2007, pp. 428–432.

[13] C. Steger, "An unbiased detector of curvilinear structuresr," Technische Universitat Munchen, Orleansstrae 34, 81667 Munchen, Germany, Tech. Rep. FGBV9603, Jul. 1996.

[14] G. Medioni, M.-S. Lee, and C.-K. Tang, *Computational Framework for Segmentation and Grouping*. New York, NY, USA: Elsevier Science Inc., 2000.

[15] E. Franken, M. van Almsick, P. Rongen, L. Florack, and B. ter Haar Romeny, *An Efficient Method for Tensor Voting Using Steerable Filters*. Berlin, Heidelberg: Springer Berlin Heidelberg, 2006, pp. 228–240.

[16] P. Prentašić, S. Lončarić, G. Vatavuk, Zoran; Benčić, M. Subašić, T. Petković, L. Dujmović, M. Malenica-Ravlić, N. Budimlija, and R. Tadić, "Diabetic retinopathy image database(DRiDB): A new database for diabetic retinopathy screening programs research." in *Proceedings of the 8th International Symposium on Image and Signal Processing and Analysis (ISPA 2013)*, Trieste, 2013, pp. 704–709.

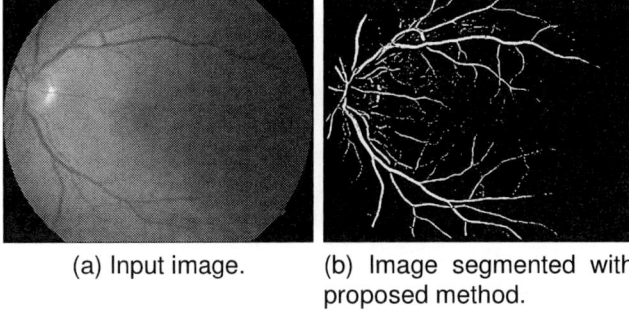

(a) Input image. (b) Image segmented with proposed method.

Fig. 10. Example of image that gave a noisy result with input (a) and segmentation result (b).

1539

Storytelling in Web Design: A Case Study

Marijana Pivac
e-mail: marijana.pivac000@gmail.com

Andrina Granić
University of Split, Faculty of Science, Department of Computer Science, Split, Croatia
email: andrina.granic@pmfst.hr

Abstract - **Developing technologies and meeting the demands and needs of a wide range of different users had rather important influence on design of user interfaces of web applications. In such a dynamic context, storytelling in web interfaces design represents a new approach which aims to realize a more attractive interfaces that also ensure and encourage the participation of the users themselves. This paper examines the pros and cons of such an approach as well as its effectiveness and impact on the development of interfaces designed for a good user experience. Developed theoretical aspects are applied on design and implementation of selected web application, whose usability as well as user experience was tested with future users by applying relevant methodology.**

Keywords: web design, user interfaces, storytelling, usability, user experience

I. INTRODUCTION

Website design has changed over the years. As technology evolved, design always followed by changing and adapting to the new requirements. As information became widely approachable it has become a real challenge to easily find valuable and accurate information on the Internet. Now web designers must find other ways to attract users' attention. In general, users are attracted by interesting and creative websites that offer different search experience. Users value experience, rather than facts. There comes an important element, which is also the topic of this paper - the story. Namely, *"stories are important cognitive events, for they encapsulate, into one compact package, information, knowledge, context, and emotion"* [1]. In this paper we developed a web application in which we've put the story as a main element of getting users attention. We've noted down the idea and the whole process of design, implementation and finally evaluation with potential end users.

II. PSYCHOLOGY BEHIND A STORY

A. The history of stories

The story plays a big role in our existence. More than before 27,000 years, since the discovery of the first pictures in the caves, storytelling was one of the most basic communication methods. People were gathering around the fire and telling their life stories and teaching the youngsters. To leave a mark for the next generations they were also drawing pictures on the walls, trying to mark down their knowledge. Thanks to those pictures, we know about their occupations and customs even today after all these years.

After a while, humans started writing, and then books appeared. Years later, humans invented radio, after that television, and finally the Internet. It all came from the need of communication which was always in some form of a story.

B. How human brain reacts to a story

People strive to understand, to process, connect and learn from stories. The human brain reacts instinctively to information seeking it, trying to simplify the whole complex so that it is simpler to interpret the experience and construct systematized view of the world. If we listen to a PowerPoint presentation with boring bullet points, certain parts in the brain get activated. Scientists call these Broca's area and Wernicke's area [2]. Overall, it hits our language processing parts in the brain, where we decode words into meaning. And that's it, nothing else happens.

When we are being told a story, though, things change dramatically. Not only are the language processing parts in our brain activated, but any other area in our brain that we would use when experiencing the events of the story are too. A story, if broken down into the simplest form is a connection of cause and effect. And that is exactly how we think. We think in narratives all day long, no matter if it is about buying groceries, whether we think about work or our spouse at home. We make up (short) stories in our heads for every action and conversation. Whenever we hear a story, we want to relate it to one of our existing experiences. That's why metaphors work so well with us. Whilst we are busy searching for a similar experience in our brains, we activate a part called insula, which helps us relate to that same experience of pain, joy, disgust and the like (see Figure 1).

Figure 1 Connecting experience with an emotion [3]

In this way we connect current events with metaphors, all of our brain associate reasons and the effects of the previous experiences [3].

C. The power of a persuasive story

Philip Pullman once said that "after nourishment, shelter and companionship, stories are the thing we need most in the world" [4]. There are few mediums more captivating than a well told story. From "what happened next?" to personal connections we make through characters and events, everyone loves them. Stories are also a very integral part of being persuasive. Those in sales and marketing have known for a long time that stories trump data when it comes to persuasion because stories are easier to understand and relate to. How you say something is just as important as what you are saying. Refusal to recognize this places you at risk of having your good information become lost in a sea of less-worthy content. You also miss out on the connections to be made via a strong narrative. Here's the thing: While we are all often resistant to the idea of being told what to do, we are very susceptible to agreeing with the "moral of the story" due to how it is presented to us.

According to research by psychologists Green & Brock [5], stories work so well on us because we are susceptible to getting "swept up" in both their message and in the manner of their telling. Quite literally, stories are able to transport our mind to another place, and in this place we may embrace things we'd likely scoff at in the "harsh, real world" [6]. Consequently, if we include story in our websites, no matter how, we can be almost absolutely certain it will get better impact than if we didn't. No matter the purpose of the website, we want it to be persuasive. That's how political campaign works, that's how commercials work and that's how we should approach the process of creating a website.

III. How to use a story in the web design

A. Basic elements of a story

Every story has a number of basic elements that makes it a whole [7, 8]:

The content – as a set of ideas and events in the story, usually formed around the conflict of the main and the side character, but sometimes even as an internal conflict of the main character. In a context of the website, content is the goal which we want the user to achieve and the process that gets him there.

The theme – the central topic, subject or concept of the story, one of the basic components.

The characters – comes in a form of a person, a place or an object that have human characteristics and function in the story. Actions and thoughts of the character lead the story forward. A well-developed characters help the audience to connect with the story. The characters in a website may be organizations, groups or companies, team members, users, mascots and similar entities.

The genre – the type of the story.

The atmosphere – well-designed atmosphere creates a greater connection between the audience and the story. In web design this element is achieved through the tone, style and colours of the design.

Imagery – the author's attempt to create a mental picture (or reference point) in the mind of the reader. Visual, strong and effective imagery can be used to invoke an emotional, sensational (taste, touch, smell etc.) or even physical response. If we talk about the website, this is a very important part, because it includes photography, illustrations, typography and all the other visual elements. It complements the atmosphere.

The highlight – the central part of the story, which is the most dramatic. It mostly causes a turning point or a change in the characters.

The outcome – comes after the highlight and gives a conclusion. In this part, the story goes back to normal by decreasing the tension and excitement of the audience.

To tell a story through a website, we need to keep in mind all aforementioned elements of the story. Firstly, create the content and the theme, then create characters and bring in the right atmosphere and imagery. Finally, we should bring users to a "call-to-action" – the goal that we wanted her/him to reach. By adding one additional element – an animation, we create a whole new palette of ways to tell our story. It allows us to include motion, which allows us to include gestures and create deeper emotions, which leads to a deeper connection with the audience, which, at the end creates a much better user experience. And that is precisely what we want to achieve.

B. Animation in a story

Initially it is important to plan well the whole animation. This section is probably the most important, because if we have to subsequently make changes, there may be situations in which the need to change such a big part of the animation so that we, mind as well, better start again. Therefore, this step is necessary for it to work out well. We should proceed only when we are certain that we have it planned exactly as we want. Accordingly, (i) we describe briefly the whole story, create the characters and determine the goal which we want to achieve, (ii) we create storyboard or a sequence of frames that will interchange during the animation and lastly (iii) we develop the atmosphere by creating well styled ambient [9]. After we have a basic parts created, we need to determine the rhythm of the animation. It's a key to find the right speed and variations. If it's too fast, the viewer can feel lost, but if it's to slow it may be boring. We need to decide which scenes are more important or require more time to process. To be sure that the user can follow up with the story, we can simply let him decide the rhythm of the scenes interchanging. That's why we decide to use a new trend in web design and development – the parallax animation.

C. Parallax

Animation activated by scrolling is a relatively new trend in web animation. In 2014. web designers started giving a little more "feel" to websites by adding subtitle movement effects to elements on the sites using jQuery programming language. It easily gave them fresh, modern and appealing look [10] although it was just a start. In rather short time, websites started looking more and more like animated movies, but controlled by the user.

The idea of parallax design is simply an awareness of movement. More specifically, the word "parallax" is used to describe the perception of distance between objects while moving along a line of sight. For example, the objects nearest to you in space will seem to move quicker and more dynamically than "background" objects. So parallax is just a measurement of the difference in position between two points from the viewer's perspective. Since the objects aren't really moving, it's just a type of illusion like 2D animation [11]. The feeling of involvement in the content of the websites improves the user experience. It makes browsing through the web a whole new experience. If it looks eye-pleasing and also tells a good story, letting them feel as they can move the elements, makes it more fun and entertaining. If it stimulates a whole package of positive emotions, it will sure be an experience they won't forget. There are all kinds of great examples of those effects implemented on the web.

Bizbrain is an advanced long one-page website that skilfully demonstrates statistical data via amazing colourful infographic. Each section includes its own piece of information that is wonderfully bolstered by activated animation (Figure 2).

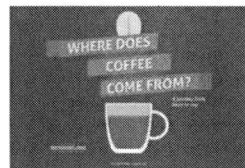

Figure 2 Bizbrain - www.bizbrain.org/coffee

Merry Christmallax, like the previous example, is creative website aimed to give its loyal customers and regular users a warm and hearty congratulation. The story begins with a clean sheet of paper and ends up with a vibrant fully-illustrated scene; the whole transformation is pulled by a scrolling technique (Figure 3).

Figure 3 Merry Christmallax - ihatetomatoes.net/merry-christmallax

Make Your Money Matters is a promotional website that graphically showcases benefits of joining a credit union. You will become an active participant of its journey. The technique allows holding the theme alive and brings the better user experience (Figure 4).

Figure 4 Make Your Money Matters - makeyourmoneymatter.org

Artem & Julia Wedding is a wedding invitation made by a designer and a programmer. Page tells their story and invites viewers to their wedding. This is an example of how websites are fully involved in our lives and can be used for all kinds of purposes (Figure 5).

Figure 5 Artem & Julia wedding - artemjuliawedding.com

IV. WEB APPLICATION DEVELOPMENT: CASE STUDY

The aim was to create a web application for the HCI (Human-Computer Interaction) group at the Faculty of Science, University of Split, Croatia; to present the group itself, its members, their activities, projects and collaborations (refer to *http://hci.marijanapivac.com/*). Since the HCI is still rather unfamiliar concept to people who are not related to the design and technology, the idea was to design a page that will introduce the concept through the parallax animation. Although, this kind of web application is not really the best scene for parallax animation website, we were eager to try and implement it anyway and hopefully get a good feedback for this approach when the consequences are not ideal.

Hence we created the "Intro page" which tells a narrative story about the HCI. Furthermore, besides the HCI story, additional relevant group information was provided in additional webpages. To ensure high usability of the web application, the target users were firstly considered. For this case, a target user is a person who wants to be informed about the HCI group, find appropriate information and be potentially engaged in collaboration. So, as a homepage we set the "About page", while the "Intro page" was set as the first element in a menu, specifically highlighted so that it draws attention but still doesn't interfere with users' browsing. For the main story we decided to illustrate few examples of an everyday frustration caused by a bad design, and present a whole discipline dedicated to exploring the little things that make our lives easier (see Figure 6).

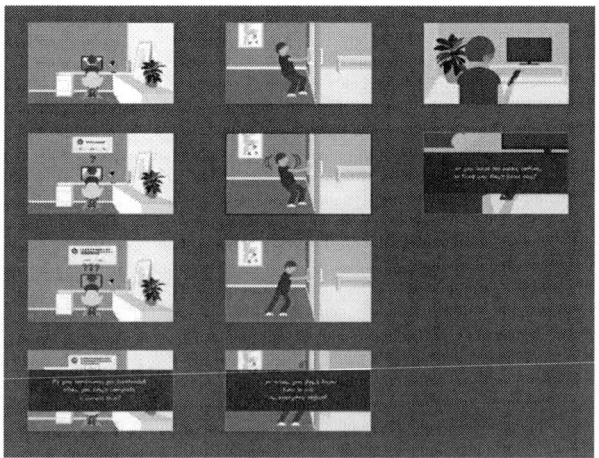

Figure 6 Storyboard in our web application

A. Design: atmosphere, style and characters

The design was cosy, relaxed, little childish, with illustrated and animated elements. The main character is a ordinary, common user, drawn in plain clothes, without facial features, in order to clearly shown the "generality". The style which was used is called a "flat design", which relies on a 2D look, minimalism and symbolism, avoids excessive use of overflow colour, texture, shadowing. This style has its roots in Bauhaus, modernism and so-called Swiss style which became popular back in the 1950's and 1960's. "Flat design" earned general popularity in the 2013 when Apple started using it [12]. Since the character we made didn't have facial features, we have paid special attention in creating an emotion in the story (see Figure 7).

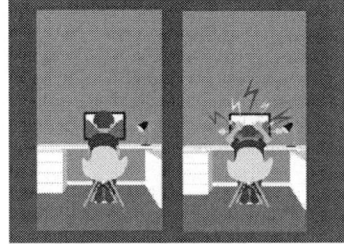

Figure 7 "Flat design" and visualization of emotions

B. Implementation

Given that today the use of a wide range of devices that are viewing the website, in the design is necessary to take account of all sizes and proportions of the screen of the device. Since it is important to provide the same experience across all screens, different techniques have been developed to cope with that issue. The most commonly used is responsive design (Figure 8). It is an approach aimed at allowing desktop webpages to be viewed in response to the size of the screen or web browser one is viewing with. A site designed with responsive design adapts the layout to the viewing environment by using fluid, proportion-based grids, flexible images and CSS3 media queries.

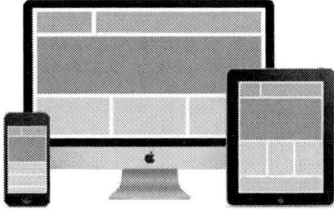

Figure 8 Responsive principles

The main downside of parallax concept is often the behaviour on mobile phones. It is considered really heavy and slow for this kind of devices since the JavaScript is used for calculating the sizes and position of elements which slows down the loading of the website [13]. It is also hard to scale the design down so it is responsive to different sizes, and it is now a common practice to make a whole other solution for those devices.

In our example we've used Skrollr, a JavaScript library made especially for parallax, which made the developing part so much quicker and easier [14]. Also, we made the whole design adjust to different screens using CSS. Since we needed an ability for the group members to be able to insert new content, we decided to use WordPress CMS (Content Management System) to implement the idea. So, after we

have adjusted the design to multiple screen sizes, we created a WordPress theme that will give us desired abilities.

V. EVALUATION OF THE USER INTERFACE

In general, human-centred design is an approach to system (application, product or service) development that focuses specifically on making systems usable. Consequently, for the success of a website in particular is necessary to raise the level of its ease of use i.e. usability what will automatically have an influence on positive user experience. Keeping in mind users' needs, experience and limitations, we need to create user interface that is transparent, intuitive and easy to use. Since the visual experience is closely linked to emotions, it is something we should pay particular attention to.

Emotions provide us with instant information about our environment: where's the risk, where a potential pleasure; what is beautiful, what is bad. The emotional system, which is changing the way the cognitive system operates, is closely related to behaviour, preparing the body to a suitable reaction to a given situation [15]. Given that all humans are different, what might work for one person may not for another. We can do our best to predict their reactions, but can't really predict the real experience. Additionally, we can't determine the effectiveness of the design based on statistics, visit rate or bounce rate; we can only create assumptions. Nevertheless, in order to get the best measurement for user experience, we must ask the users themselves.

A. Evaluation methods

When evaluating website usability, it is important to determine whether the users of the website achieve their goals and whether they have a positive experience while using the site. These parameters cannot be determined with only one evaluation method. In line with these reasoning, we conducted two evaluation studies – one related to the usability and the other of the user experience.

1) SUS questionnaire

SUS (System Usability Scale) questionnaire was developed as a simple way of assessing the usability of the system, which uses a Likert scale. It is represented by ten items on which the user completes the degree of agreement or disagreement with the statement [16]. Number of stages is always an odd number, so that the mean value is always indefinite, and extreme values are always complete agreement and complete disagreement with the statement.

2) AttrakDiff questionnaire

AttrakDiff questionnaire serves as a method to assess the user experience, as subjective user satisfaction with the service system [17]. The model separates the four essential aspects:

- The product quality intended by the designer.

- The subjective perception of quality and subjective evaluation of quality.

- The independent pragmatic and hedonic qualities.

- Behavioural and emotional consequences.

To measure the attractiveness, we applied an instrument of measurement in the format of semantic differentials. It consists of 28 seven-step items whose poles are opposite adjectives. Each set of adjective items is ordered into a scale of intensity, while every middle value of an item group creates a scale value for:

- Pragmatic Quality (PQ),
- Hedonic Quality (HQ),
- Attractiveness (ATT).

At the end of measurement, the average value of each group claims is taken to get a clearer picture of the user experience. This theoretical model is researched and tested in a number of studies which showed that the hedonic and pragmatic quality independent of each other, and both equally affect the level of attractiveness.

B. Evaluation results

The preliminary evaluation was conducted on a sample of 12 participants (six males and six females). Given that the target audience and future users are younger people interested in HCI issues and possibly collaboration and future work in the field, the participants were also mostly students. For the task they had to visit and browse through web application and then fill-in SUS and AttrakDiff questionnaires. All respondents completed both questionnaires without difficulties. Results of the SUS questionnaire showed a biggest diversity of responses for two claims: (i) "I found the website unnecessarily complex" and (ii) "I found the various functions in the website were well integrated" (see Figure 9). Other claims mostly got similar, positive results.

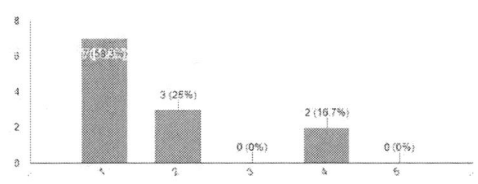

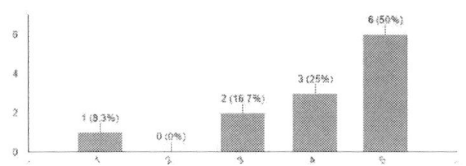

Figure 9 SUS diversities

We conclude that the use of this kind of storytelling in the selected theme could be unclear because some participants found it unnecessary. The reason for that might be in the fact that most of the users were familiar with the HCI term and didn't see the importance of "Intro page". Also some of the functionalities could be better implemented, but that part is completely of a technical nature which wasn't a

main focus of conducted study. Future development will take those results into the consideration.

The AttrakDiff questionnaire generally got better results. When measuring the pragmatic and hedonic quality results are entirely contained within the region of desired outcome (see Figure 10).

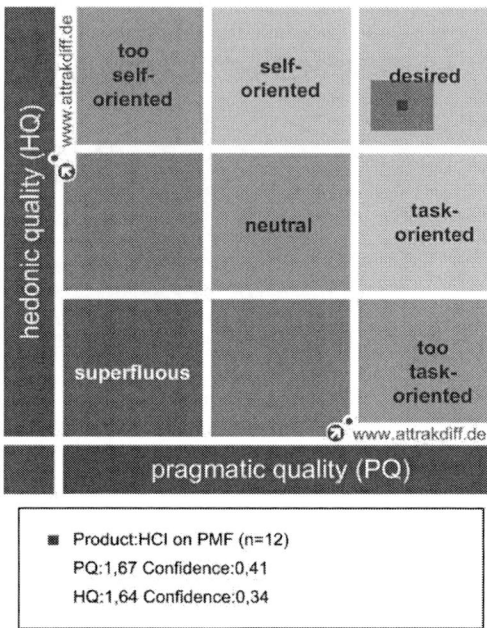

Figure 10 Pragmatic and hedonic quality results

In the "Diagram of average values", we see that the attractiveness (ATT) reached the highest level and pragmatic quality, as the two aspects of hedonic quality, (stimulation and identity) reached equal results (see Figure 11).

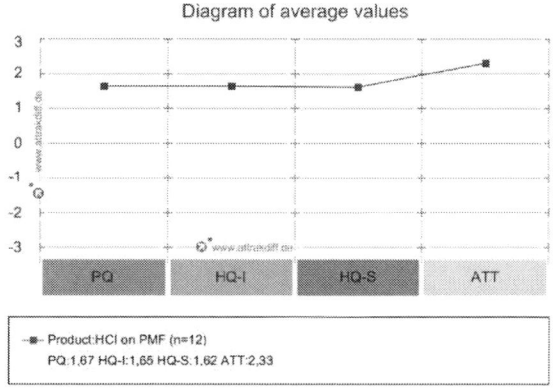

Figure 11 Diagram of average values

In the "Description of word-pairs" diagram the values are presented in claims with two extremes. All the values were in the right half of the diagram, which is the desirable one (see Figure 12).

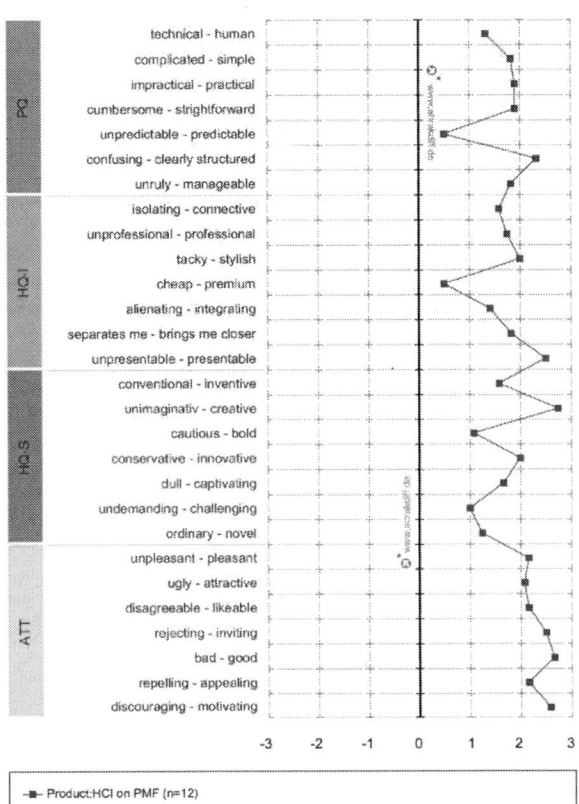

Figure 12 Description of word-pairs diagram

The lowest value reached terms of unpredictability and price value. The highest value reached concepts of clear structure, particularity, creativity, attractiveness, motivation and attractiveness.

VI. CONCLUSION

In web design creating something that differ from similar projects can make a better impact on users if it's done by following well-known principles. Experimenting with new elements demands smaller steps applied gradually, so that users can adapt. It is crucial to know the right moment and way to implement new things that might challenge users' behaviour. In such a process it's important to evaluate the outcomes by testing users' impressions.

By adding animated story we've tried to find a new way to keep users attention, which we, by our results, have accomplished. Given that this was a preliminary evaluation participants were all students as they were target audience. However, for further evaluation studies it would be more accurate if we include users of all ages so that we have a clearer picture of the real user experience.

Although there are believes that it is difficult to integrate the storytelling in all types of sites, this web application proves that the story is the one element that can (and should) be included if we are to attract user's attention.

However, when planning the design and development is necessary to pay attention to the way the story is included. Whether by using the parallax animation, or any other animation, and perhaps without it, the visual elements, as the textual, always tell a story.

REFERENCES

[1] Norman, D. 1994. Things that make us smart: Defending human attributes in the age of the machine. New York : Basic Books; Revised ed. edition (April 21, 1994), 1994.

[2] Concepts, Psychology. Broca's and Wernicke's Areas. Psychology Concepts. [Online] [Cited: 6 February 2017.] http://www.psychologyconcepts.com/brocas-and-wernickes-areas/.

[3] Widrich, L. The Science Of Storytelling: What Listening To A Story Does To Our Brain. Buffer Blog. [Online] [Cited: 6 February 2017.] https://blog.bufferapp.com/science-of-storytelling-why-telling-a-story-is-the-most-powerful-way-to-activate-our-brains.

[4] Clubs, Scholastic Book. Interview with Philip Pullman. Scholastic Book Clubs. [Online] [Cited: 6 February 2017.] https://clubs-kids.scholastic.co.uk/clubs_content/7922.

[5] The role of transportation in the persuasiveness of public narratives. Green, Melanie C. and Brock, Timothy C. Nov 2000. Nov 2000, Journal of Personality and Social Psychology, Vol 79(5), pp. 701-721.

[6] Ciotti, G. The Psychology Of Storytelling. Sparring Mind. [Online] [Cited: 6 Februaruy 2017.] http://www.sparringmind.com/story-psychology/ .

[7] Jacobs, D. 2012. Smashing Book #3 1/3: The Extension, The Science Of Storytelling: What Listening To A Story Does To Our Brains. Freiburg : U.S. Magazine, 2012.

[8] Roane State Comunity College. Literary Analysis: Using Elements of Literature. Online writing lab. [Online] [Cited: 30 8 2016.] http://www.roanestate.edu/owl/ElementsLit.html.

[9] About tech. 2016. 5 Things To Do Before You Ever Start Animating. About tech. [Online] 29 8 2016. http://animation.about.com/od/otherusefultutorials/a/5-Things-To-Do-Before-You-Ever-Start-Animating.htm

[10] Birch, Nataly. Scroll Activated Animations – New Trend in Web Design. Designmodo. [Online] [Cited: 30 8 2016.] http://designmodo.com/scroll-animations-web-design/.

[11] Rocheleau, Jake. The Ultimate Guide to Parallax Scrolling: Best Practices, Examples and Tutorials. Vandelay Design. [Online] [Cited: 30 8 2016.] http://www.vandelaydesign.com/parallax-scrolling-best-practices-examples-and-tutorials/.

[12] Interaction Design Foundation. Flat Design – An Introduction. Interaction design. [Online] [Cited: 30 8 2016.] https://www.interaction-design.org/literature/article/flat-design-an-introduction?utm_source=facebook&utm_medium=sm&utm_content=everything_you_need_to_know_about_flat_design&utm_campaign=post-

[13] Sandu, B., How the web design trend of parallx scrolling has faded, Design your way, 2017. [Online] [Cited: 314 April 2017.] http://www.designyourway.net/drb/how-the-web-design-trend-of-parallax-scrolling-has-faded/

[14] Prinzhorn. Skrollr. [Online] [Cited: 6 February 2017.] http://prinzhorn.github.io/skrollr/.

[15] Norman, Donald A. 2004.. Emotional design. New York : s.n., 2004.

[16] Brooke, J. SUS - A quick and dirty usability scale. UX for the masses. [Online] [Cited: 9 9 2016.] www.uxforthemasses.com/.../SUS-System-Usability-Scale.doc.

[17] AttrakDiff. Science behind AttrakDiff. AttrakDiff. [Online] [Cited: 9 9 2016.] http://www.attrakdiff.de/sience-en.html

Idioms in state-of-the-art Croatian-English and English-Croatian SMT systems

Maja Manojlović*, Luka Dajak*, Marija Brkić Bakarić**
*Faculty of Humanities and Social Sciences, Rijeka, Croatia
**Department of Informatics, Rijeka, Croatia
*mmanojlovic@ffri.hr, ldajak@ffri.hr
**mbrkic@inf.uniri.hr

Abstract - Idioms are well known for posing problems to non-native speakers, let alone machines. A failure to identify idioms often leads to unnatural, even hilarious outputs. This paper investigates the treatment of idioms in state-of-the art SMT systems involving English and Croatian. First we introduce the concept of idioms. Then we construct three short stories abundant with idioms per each language, and translate them into the other language by two state-of-the-art SMT systems. Next we manually inspect the outputs and present results. For the purpose of conducting analysis, we devise an error taxonomy for handling idioms.

I. INTRODUCTION

The quality of translation depends on the language pair, the type of text being translated, and the system used for translation [1]. Idioms are fixed expressions that do not convey literal meaning of their constituents. Words in their structure lose their original meaning and, together with other words within the idiom, convey new meaning. Although words within an idiom do not lose their meaning completely, they get a new context and background. These words must often be in the same order and they cannot be omitted nor replaced with other words. Idioms are comprised of at least two constituent words that are characterized by their common use, integrity and relatively fixed structure. They do not develop in spoken communication, but are integrated in discourse as a whole. This way, they either become a part of a sentence structure or they function as an independent unit [2].

Idioms will always pose a problem in translating, even to humans. Therefore, it is plausible that machines will come across even greater obstacles.

The aim of the research is to compare the way machines translate texts, more precisely idioms, in English-Croatian and Croatian-English language directions. Two online translation systems are used for the purpose of this research: Google Translate (hereinafter GT) and Asistent [3], both phrase-based statistical machine translation (PBSMT) systems. Croatian-English translations are expected to be of higher quality since English, unlike Croatian, is well resourced [4].

Section two clarifies the specificities of idioms. Section three gives an overview of related work. Section four describes our experimental study. The results are presented in section five. Section six highlights the main findings of our work.

II. IDIOMS

According to their structure, idioms can be divided into three categories. First category is comprised of sequences of words that consist of at least two independent words (e.g. *gold mine*). Second category are phonetic words. These sets of words consist of only one independent unit (e.g. *out of sight*). Finally, third category is comprised of idioms that acquired a form of a sentence (e.g. *the place is getting too hot*).

Idioms can also be categorized in accordance to their origin. First category, thus, would consist of biblical idioms. Considering the importance of the Bible and its tremendous influence since the beginning of the Middle ages, it is not surprising that biblical idioms found their equivalents in various languages. Examples of this category of idioms include: *Judas kiss*, *the lost sheep*, *to cast pearls before swine*. Second category consists of literature idioms. These are idioms that have origin in literature or mythology and are commonly used in language (e.g. *to be or not to be*, *Sisyphean task*, *Tantalus' agony*). There are also idioms that originated from various human professions (e.g. mathematics, music, theatre, nautics) and are used in everyday communication. Some of the examples include: *bring sth under the common denominator*, *have an ear for sth*, *behind the scenes*.

In addition to previously mentioned categories, there are also historical idioms. These idioms are related to historical people and events which left a strong impact in the society where they are still used in everyday communication. Some of the examples include: *die is cast, like sheep without a shepherd, meet your Waterloo* [5]. The last idiom has a national background and is unknown among other cultures. The translation cannot be found outside the culture in which this idiom has occurred. Even if one tried to translate it, the message would never be truly transmitted.

Consequently, idioms can also be differentiated according to the time and place of their emergence. Phraseology divides its constituents on national and international idioms. While former are almost impossible to translate, the latter can be transferred between languages [5].

Some idioms have both identical meaning and identical components (e.g. *swallow the bait* which is translated into Croatian as "progutati mamac"). In some instances, it is possible to come across idioms that have the same meaning but slightly different form (e.g. *get out of the bed on the wrong side* which is translated as "ustati na lijevu nogu"). Sometimes it is possible to find an idiom in the target language that has equivalent meaning in the source language, but a completely different form (e.g. *piece of cake* which is translated as "mačji kašalj"). Example translations listed are in Croatian.

One of the biggest mistakes when dealing with idioms is translating its components literally, word by word. Translator must be familiar with culture and language of both source and target language in order to recognize idioms and transfer them to the target language appropriately [5].

III. RELATED WORK

A study in [6] illustrates differing patterns between human and machine translations, but also between two different machine translation (MT) systems. Error analysis, as [7] puts it, gives a qualitative view on the MT system and should be an integral part of MT development. It can point to strengths and problem areas for a certain MT system, which is not possible using automatic evaluation metrics [8]. Automatic metrics, as well as some forms of human evaluation such as fluency and adequacy scoring or system ranking, provide quantitative system evaluation [7]. Research community would like to get answers to what kind of errors the system makes most often, whether a particular modification improves some aspect of translation, although the overall score is intact, whether one system is superior in all aspects of translation or just in some, etc. [9]. Idioms are usually explicitly tackled in MT error taxonomies, i.e. as expression in [10, 11], and as idioms in [12, 13, 14].

Standard statistical machine translation (SMT) systems do not model idioms explicitly [15]. The term phrase in PBSMT does not refer to a linguistic unit, but to a sequence of words. Although phrasal translations might indirectly capture multiword expressions (MWE), which is a more general concept, they are not distinguished from any other *n*-gram [15]. A rising interest has been detected in explicitly modelling MWEs within the SMT framework. The authors in [16] note that highly fixed expressions (e.g. *by and large*) can be represented as words-with-spaces in Natural Language Processing (NLP) applications. This, however, does not hold for semi-fixed and syntactically flexible expressions. Semi-fixed expressions adhere to strict constraints on word order and composition, but undergo some degree of lexical variation, e.g. inflection, variation in reflexive form, or determiner selection [16].

The authors in [17] show a simple approach to extract domain bilingual multiword expressions (MWE) and three methods to integrate them to Moses, the state-of-the-art PBSMT. The first method adds MWEs to the training corpus, the second adds one feature which has the value of 1 if the source language phrase contains a MWE and the target language phrase contains its translation, or the value of 0 otherwise, and the third method includes an additional phrase table containing automatically extracted MWEs.

A study in [15] proposes two different integration strategies for MWEs in SMT. In the first strategy they identify MWEs and turn them into a single unit. Therefore, from the perspective of SMT, all MWEs are considered frozen. In their second strategy they add a count feature which represents the number of MWEs in the input language phrase and integrate MWE knowledge as a feature in the translation lexicon. In that way the system is biased towards using phrases that do not break MWEs. This is sort of a generalization of a binary strategy from [17].

MWEs are integrated into the phrase table in two different ways in [18]. After identifying MWEs, the translation pairs are extracted from the corpus. In the first approach the aligner probability is kept, while in the second it is set to 1 in both translation directions. The factored model is used, which enables using lemmas beside surface forms of words.

A study in [19] shows that different MWE types require different integration methods in the SMT framework. Beside the two approaches proposed in [15], they also take zone integration approach in which they define reordering zones for all MWEs found in the test data. In that way the decoder is forced to respect the boundaries while constructing the hypothesis. However, although it is not allowed to translate out of zone phrases unless it fully finished translating the words in the zone, it is allowed to divide the zone into any combination of phrases and translate them individually and in any order. That is why the approach does not help to increase automatic scores.

The impact of idioms on SMT is evaluated in [20]. Unlike in this study, the authors focus on idiomatic expressions formed from the combination of a verb and a noun as its direct object (e.g. *lose head*). They show that even in that limited scenario idiomatic expressions pose a challenge to PBSMT systems, as witnessed by a drop in the BLEU score (BLEU score is proposed and presented in [21]).

A substitution based technique for improving SMT on idiomatic MWEs is evaluated in [22]. The method first performs substitution on the original idiom with its literal meaning before translation, and then replaces literal meanings with idioms following translation. Although a statistically significant improvement is reported, the authors conclude that there is still a lot of work to be done to solve the problems posed by idioms to SMT.

IV. EXPERIMENTAL SETUP

We compare the way that two online translation systems, i.e. GT[1] and Asistent[2], manipulate texts, more precisely idioms, in English-Croatian and Croatian-English language directions. The texts used in this analysis are constructed by hand and care was taken to ensure that the texts are abundant with idioms but make up coherent stories. More details on the test sets are given in Table I.

In the analysis of generated translations, we ignore punctuation errors, inappropriate upper case and lower case letters and discordance in gender, number and case. The focus is primarily on the translations of idioms[3]. For the purpose of analysis, we divide machine translations of idioms into the following categories: Equivalent, Appropriate meaning, Literal translation, Anomaly in form, Untranslated or partially translated, Wrong idiom, and Other.

V. RESULTS

The results are presented in Table II. For the purpose of comparison of translation directions, the distribution of error categories in percentages is given in Figure 1.

The results consist mostly of literal translations of idioms, i.e. word by word translations (*stretch a dollar* is translated into Croatian as "produžim dolar").

Some of the idioms are not translated into their equivalents in the target language, but their translation conveys appropriate meaning, which is in accordance with the context (*cold hearted* is translated into Croatian as "hladnog srca", instead of "mrtav hladan"). For some of the idioms in this category it is not possible to use an equivalent idiom because it does not exist in the target language (*Jack of all trades* is translated into Croatian as "dobar u svemu").

To a smaller extent, we identify idioms that are translated using an equivalent idiom in the target language (*armed to the teeth* is translated into Croatian as

TABLE I.   DESCRIPTION OF THE TEST SET

	text	# of words	# of sentences	# idioms	Avg sent. len
HR→EN	Text 1	318	22	23	14.4
	Text 2	347	24	19	14.4
	Text 3	339	23	22	14.7
EN→HR	Text 4	387	24	26	16.1
	Text 5	386	26	27	14.8
	Text 6	383	29	33	13.2

"naoružan do zuba"), as well as idioms with some sort of anomaly in their form. The latter are idioms that are recognizable and translated appropriately, but have some sort of anomaly (an extra word, inappropriate preposition, wrong word). For example, idiom *crying like the rain* is translated as "plakala kao kiša s", where the preposition s is unnecessary, but the rest of the idiom is translated appropriately.

The translations contain several idioms that are translated using an inappropriate idiom (*is down for the count* is translated into Croatian as "manji od makova zrna"). Additionally, some idioms are not even translated, i.e. they remain in their original form ("pomrsio konce" is translated into English as *pomrsio strings*), because of training data sparsity, which is mainly caused by rich morphology. We group these idioms into separate categories.

VI. CONCLUSION

Although one would expect Croatian-English translations to be better, as English is well resourced, the opposite proves to be true. We attribute this to the fact that English texts are more prevalent on the Internet than texts in Croatian, and therefore, translations from English into Croatian are much more accessible. Moreover, in our analysis we ignore discordance in gender, number and

TABLE II.   EVALUATION OF GT AND ASISTENT TRANSLATIONS OF IDIOMS

	CRO-EN		EN-CRO		Total	
	Asistent	*GT*	*Asistent*	*GT*	*Asistent*	*GT*
Equivalent	16	13	9	8	25	21
Appropriate meaning	6	2	21	10	27	12
Literal translation	28	35	43	60	71	95
Anomaly in form	7	3	9	7	16	10
Untranslated or partially translated	3	3	0	1	3	4
Wrong idiom	0	2	0	0	0	2
Other	4	6	4	0	8	6

[1] https://translate.google.hr/

[2] http://server1.nlp.insight-centre.org/asistent/

[3] Translations performed on January 19, 2017

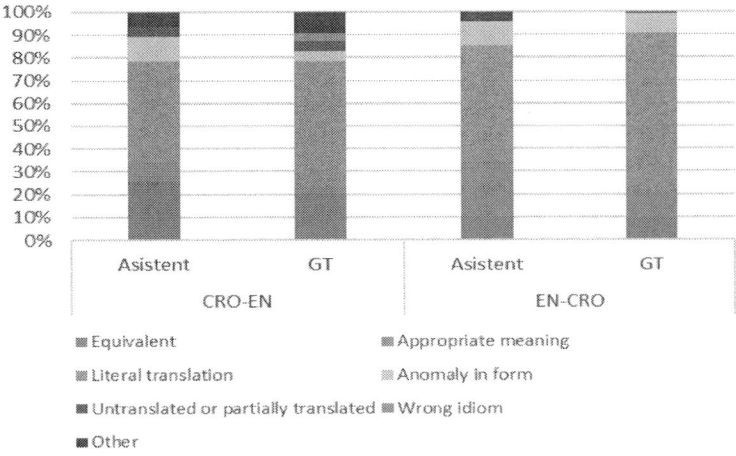

Figure 1. GT and Asistent performance with regard to idioms per both language directions

case, which could also greatly affect results when translating from morphologically poor to morphologically rich language. Furthermore, due to a great number of existing idioms in both the source and target language, it was impossible to include them all in our corpus. This could have affected our final analysis.

The majority of generated translations are literal translations of idioms and other elements of the texts. This leads to the conclusion that the systems involved do not have special treatment for MWEs. The last three categories, namely Untranslated or partially translated, Wrong idiom and Other are so poorly represented that they can be dismissed from further investigations.

In some instances, GT achieves better results, while in others Asistent offers much better translations. Both give similar translations although we observe slight superiority of Asistent especially in English-Croatian translations.

The main conclusion that can be deducted from this study is that PBSMT systems do not capture MWEs to a sufficient degree and special attention needs to be paid first to their detection, and afterwards to finding appropriate integration method. Since literal translations predominate, back translations would yield similar if not even worse results.

ACKNOWLEDGMENT

This work has been fully supported by the University of Rijeka under the project number 16.13.2.2.01.

REFERENCES

[1] S. Seljan, M. Brkić, V. Kučiš, "Evaluation of free online machine translations for Croatian-English and English-Croatian language pairs," Proceedings of the 3rd International Conference on the Future of Information Sciences: INFuture2011-Information Sciences and e-Society, pp. 331–344, 2011.

[2] A. Menac, Z. Fink-Arsovski, R. Venturin, " Hrvatski frazeološki rječnik, Nakl. Ljevak, 2003.

[3] M. Arcan, M. Popovic, P. Buitelaar, P., "Asistent–a machine translation system for Slovene, Serbian and Croatian,"

Proceedings of the 10th Conference on Language Technologies and Digital Humanities, Ljubljana, Slovenia, 2016.

[4] M. Federico, M. Negri, L. Bentivogli, M. Turchi, F. F. Kessler, "Assessing the Impact of Translation Errors on Machine Translation Quality with Mixed-effects Models," EMNLP, pp. 1643–1653, 2014.

[5] J. Forko, "Prevođenje frazema-Sizifov posao," in Hrvatistika, pp. 93–98, 2009.

[6] M. Koponen, "Assessing machine translation quality with error analysis," Electronic proceeding of the KaTu symposium on translation and interpreting studies, 2010.

[7] S. Stymne, "Blast: A tool for error analysis of machine translation output," Proceedings of the 49th Annual Meeting of the Association for Computational Linguistics: Human Language Technologies: Systems Demonstrations, pp. 56–61, 2011.

[8] S. Stymne, L. Ahrenberg, "On the practice of error analysis for machine translation evaluation," LREC, pp. 1785–1790, 2012.

[9] M. Popovic, A. Burchardt, A, "From human to automatic error classification for machine translation output," 15th International Conference of the European Association for Machine Translation (EAMT 11), 2011.

[10] M. Flanagan, "Error classification for MT evaluation," Technology Partnerships for Crossing the Language Barrier: Proceedings of the First Conference of the Association for Machine Translation in the Americas, 1994.

[11] M. Farrús, M. R. Costa-Jussa, J. B. Marino, M. Poch, A. Hernández, C. Henríquez, J. A. Fonollosa, "Overcoming statistical machine translation limitations: error analysis and proposed solutions for the Catalan--Spanish language pair," in Language resources and evaluation, pp. 181–208, 2011.

[12] A. Font-Llitjós, J. G. Carbonell, A. Lavie, "A framework for interactive and automatic refinement of transfer-based machine translation," Tenth workshop of the European Association for Machine Translation (EAMT), 2005.

[13] D. Vilar, J. Xu, J., L. F. d'Haro, H. Ney, "Error analysis of statistical machine translation output," Proceedings of LREC, pp. 697–702, 2006.

[14] M. Carpuat, M. Diab, "Task-based evaluation of multiword expressions: a pilot study in statistical machine translation," Human Language Technologies: The 2010 Annual Conference of the North American Chapter of the Association for Computational Linguistics, pp. 242–245, 2010.

[15] Â. Costa, W. Ling, T. Luís, R. Correia, L. Coheur, "A linguistically motivated taxonomy for Machine Translation error analysis" in Machine Translation, pp. 127–161, 2015.

[16] I. A. Sag, T. Baldwin, F. Bond, A. Copestake, D. Flickinger, "Multiword expressions: A pain in the neck for NLP,".

International Conference on Intelligent Text Processing and Computational Linguistics, pp. 1–15, 2002.

[17] Z. Ren, Y. Lü, J. Cao, Q. Liu, Y. Huang, "Improving statistical machine translation using domain bilingual multiword expressions," Proceedings of the Workshop on Multiword Expressions: Identification, Interpretation, Disambiguation and Applications, pp. 47–54, 2009.

[18] D. Bouamor, N. Semmar, P. Zweigenbaum, "Improved statistical machine translation using multiword expressions," Proceedings of the International Workshop on Using Linguistic Information for Hybrid Machine Translation (LIHMT 2011), pp. 15–20, 2011.

[19] M. Ghoneim, M. T. Diab, "Multiword Expressions in the Context of Statistical Machine Translation," IJCNLP, pp. 1181–1187, 2013.

[20] G. Salton, R. Ross, J. Kelleher, "An Empirical Study of the Impact of Idioms on Phrase Based Statistical Machine Translation of English to Brazilian-Portuguese," Third Workshop on Hybrid Approaches to Translation (HyTra) at 14th Conference of the European Chapter of the Association for Computational Linguistics, 2014.

[21] K. Papineni, S. Roukos, T. Ward, W. J. Zhu, "BLEU: a method for automatic evaluation of machine translation," ACL-2002: 40th Annual meeting of the Association for Computational Linguistics. pp. 311–318, 2002.

[22] G. Salton, R. Ross, J. Kelleher, "Evaluation of a Substitution Method for Idiom Transformation in Statistical Machine Translation," 10th Workshop on Multiword Expressions at 14th Conference of the European Chapter of the Association for Computational Linguistics, 2014a.

Contactless control of sanitary water flow and temperature

Dominik Gečević, Toni Bjažić

Tehničko veleučilište u Zagrebu

Vrbik 8, 10000 Zagreb, Croatia

e-mail: dgecevic@tvz.hr

Abstract—**An innovative embedded computer system for contactless control of sanitary water flow and temperature is described in this paper. The innovative component of this embedded system is in its ability to control the flow and temperature of sanitary water, without the need for physical contact with a tap. In this way, the transfer of bacteria from the surface of the tap to a user is avoided, which significantly contributes to the preservation of health and cleanliness, i.e. hygiene. A hardware and software components required for the functioning of this embedded computer system are described in this paper. A design and production of printed circuit boards (PCBs), for control of servo valves using an analog infrared proximity sensors, a digital infrared proximity sensor and signal LEDs, are included as the hardware part of the paper. The software part is consisted of description of a class for linearization characteristics of the analog infrared proximity sensor, and description of a class for noise filtering, as essential software components of the program, which ensure comfortable use of the system for the end user.**

Keywords: Nucleo-F303RE, STMicroelectronics, mbed, SN74HC595N, shift register, Texas Instruments, Circuit-Maker, Sharp, GP2Y0A21YK0F, GP2Y0D815Z0F, 42STH38-1304AF, SolidWorks, SunCor Motor, C++, Matlab.

I. INTRODUCTION

One of the most basic hygiene procedure is hand washing. Proverb that said: "Hygiene is a half way to health" is old, but it is true! Therefore, in this paper, special attention will be dedicated to health care, and how it is possible to improve the existing mixing water system and health care for the purpose of general living improvement and health standards. Classic taps requires physical contact to regulate flow control and temperature of sanitary water. The biggest drawback of this system is hygiene. Eventually a newer system that detects the presence of hand have appeared. The system does not require hand contact to start the flow of sanitary water, which prevents the transfer of bacteria from the tap surface such as Streptococcus, Staphylococcus, E. coli, shigella, hepatitis A virus, cold virus, etc. This revolutionary system was implemented around the world in public toilets, but has major drawbacks are the inability to control the flow rate of water, temperature change and excessive (unnecessary) use of water.

The second chapter describes all the components that are used in the printed circuit board (PCB). It includes working principle and how to connect the components on the PCB.

The third chapter presents and explains all the devices used in the system. It includes working principle and how to connect the input and output components on the PCB.

The fourth chapter presents the first prototype testing photos.

The fifth chapter explains the programming code used to operate this system.

Concluding remarks are given in the sixth chapter.

II. DESIGNING THE PRINTED CIRCUIT BOARD

In this section are shown and explained components that are built on the PCB, as it follows: built-in microcontroller Nucleo-F303RE and shift register SN74HC595N.

A. Embedded microcontroller Nucleo-F303RE

Nucleo-F303RE the development platform from the manufacturer ST^{TM} [1] is based on 32-bit ARM® microcontroller Cortex®-M4. Nucleo-F303RE (See table I) communicates via USB (2.0 full speed), SPI, I2C and CAN. Programming is done through the USB interface with its own decoder. The programming uses online mbed developer [2] and the code is written in object oriented programming language C ++.

B. Shift register SN74HC595N

Shift registers are integrated circuits that consist of flip flop bistables. Bistable has only two states, which means that it can only remember one bit of data (0 or 1).

Flip flops in an integrated circuit can be connected in series and parallel entering and outputting data. Shift register SN74HC595N (See table II) from the manufacturer Texas Instruments [3] in 16 PDIP (Plastic Dual Inline Package) version with serial and parallel output of entering data.

1) Principle of operation: A minimum of three pins at Nucleo development platform are needed to successfully use the shift registers: SER, RCLK and SRCLK. At the beginning SRCLK and RCLK input initiate to a low logic state, and the input SER brings (the first bit) data. After the SRCLK make the state change from low to a high

TABLE I. Characteristics of Nucleo-F303RE.

Characteristics	Min	Max	Unit
Power supply	3.3	12	V
Output voltage	0	3.3	V
Processor frequency	-	72	MHz
Flash memory	-	512	kB
SRAM memory	-	80	kB
Arduino terminals	-	32	pcs
Morpho terminals	-	76	pcs
ADC (12-bit)	-	4	pcs
DAC (12-bit)	-	2	pcs

logic state, i.e. on the rising edge, data that was previously brought is saved in the first bit register as shown in Fig. 1.

Since the SER was in a high logical state, the high logic state was brought to the register. Now, if the SER is put in a low logic state and the state of SRCLK is changed from a low logic state to high, the data from the first register will move to another register, and to its previous place comes a logical zero as shown in Fig. 2.

This process can be repeated as many times as necessary until the desired combination of data within flip flops is achieved as shown in Fig. 3.

If the state of RCLK is changed from low to high logic state it will parallely send data from flip flops to output, and every odd LED will light up as shown in Fig. 4.

The big advantage is the serial growing number of registry (to a maximum of 40) can be connected and operated with a much larger number of digital outputs. Shift register 74HC595 is compatible with low voltage Schottky TTL (Transistor - Transistor Logic).

The maximum total output current must not be higher than 70 mA and each output can be set up to a maximum of 35 mA. The Dependence of frequency to voltage is shown in Tab. III.

C. Printed circuit board

The printed circuit board (See figure 5) is designed using CircuitMaker software [4], and is shaped as a 100x100 mm square. It is made by the ShenZhen2U [5] service from a 1.6 mm thick standard double-sided FR4 epoxy (fiberglass-reinforced) material with 0.036 mm thick copper layers.

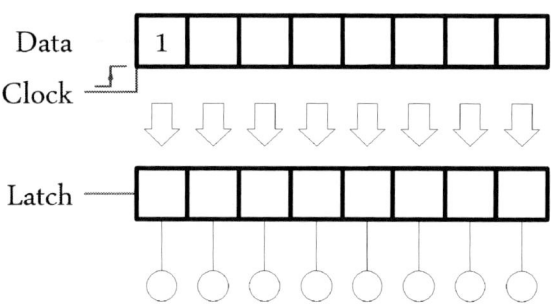

Fig. 1: The process of sending the first bit.

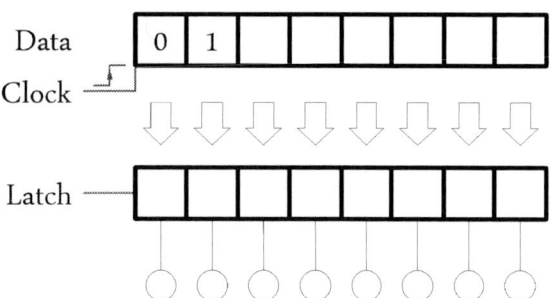

Fig. 2: The process of sending second bit.

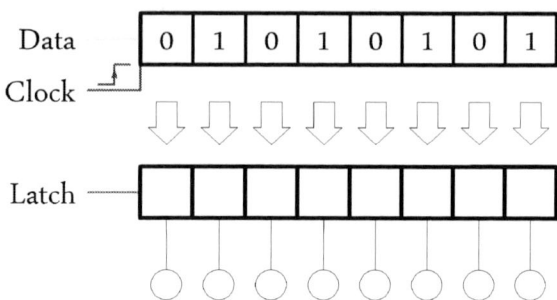

Fig. 3: The desired combination of data within flip flops.

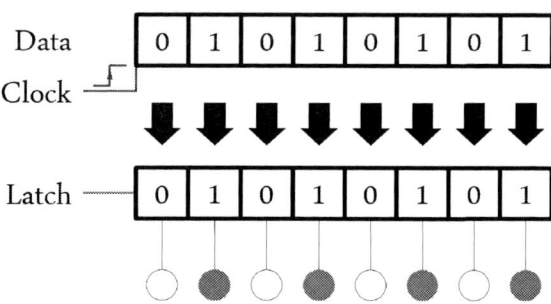

Fig. 4: Changing the status of the outputs after the activated latch pin.

Printed circuit board is covered with a green protective thermal resistant varnish (solder mask) and all copper surfaces are surface treated using HASL method (hot air solder leveling), which is an addition to the protection and makes it easier to solder SMD components. Printed circuit board meets the requirements by European Directive 2002/95/EC - the RoHS standard (Restriction of Hazardous Substances), which means that during its production no mercury, lead, chromium (VI), cadmium and other harmful substances were used. The width of the power trace is an average of 0.5 mm for signal traces and

Fig. 5: 3D view of printed circuit board.

TABLE II. Terminal functions.

Index	Meaning	Tip	Description
Q_A	-	Digital output	
Q_B	-	Digital output	They must not
Q_C	-	Digital output	be connected to
Q_D	-	Digital output	a higher voltage
Q_E	-	Digital output	than shift
Q_F	-	Digital output	register power
Q_G	-	Digital output	supply (V_{CC})
Q_H	-	Digital output	
$Q_{H'}$	-	Digital output	Parallel writing data
V_{CC}	-	-	See table III
GND	Ground	-	Ground
SER	Serial	Digital input	Serial writing data
$\overline{OE}$	Output Enable	Digital input	If it is set on low logic state outputs of shift register will be enabled, otherwise if we set on high logic state outputs will be disabled.
RCLK	Register clock	Digital input	If it is set to high logic state all data will be switched from shift register to output.
SRCLK	Serial clock	Digital input	If it is set to high logic state on each rising edge new data will be entered in shift register, depending on the state of SER.
$\overline{SRCLR}$	Serial clear	Digital input	If it is set to low logic state all data will be erased within the shift register, therefore it is necessary to be in high logic state if we want to control the outputs.

TABLE III. Dependence of frequency to voltage.

Characteristics	Power supply	Maximum frequency
Frequency	$V_{CC} = 2V$	5 MHz
	$V_{CC} = 4.5V$	25 MHz
	$V_{CC} = 6V$	29 MHz

1.0 mm for the control outputs of bipolar stepper motor. Three holes with open contact ring for mounting and grounding screw were added to prevent electromagnetic interference and parasitic capacity. All electrical traces have been carefully checked and are chamfered at $45°$ to eliminate "hot spots" (sharp turn edges where electrons travel the shortest route for which generate heat and damping) as shown in Fig. 6.

All grounding traces and unused space are covered in the Polygon Pour feature so that the maximum area of the circuit board absorbs all electromagnetic interference and parasitic capacity that could harm Nucleo development platform during program execution.

III. INPUTS AND OUTPUTS ON PCB

In this section are shown and explained all the devices used in the system, as it follows: two phase stepper motor driver, analog infrared proximity sensor, digital infrared proximity sensor and bipolar stepper motor.

A. Two phase stepper motor driver HY-DIV268N-5A

The HY-DIV268N-5A (See table IV) subdivision type is a PWM chopper type integrated driver which is used for controlling two phase hybrid stepping motor up to 4 A. This driver is using a current loop subdivision control. The motor torque noise is very small. The big advantage of this driver is high positioning accuracy.

The micro stepping (See table V) mode can be selected from the following eight operating modes using the S1, S2 and S3 inputs. The operation mode is set to standby mode if the S1=S2=S3=OFF. The power consumption in standby mode is minimized, while only protection operation is active.

The current regulation (See table VI) mode can be selected from the following eight operating modes using the S4, S5 and S6 inputs.

1) Function: To start the bipolar stepper motor it is necessary to configure the ENABLE pin high, to turn off the motor and configure the ENABLE pin low. The ENABLE pin is turning on or off all FET outputs. When the ENABLE pin is set high and the PULSE is on the

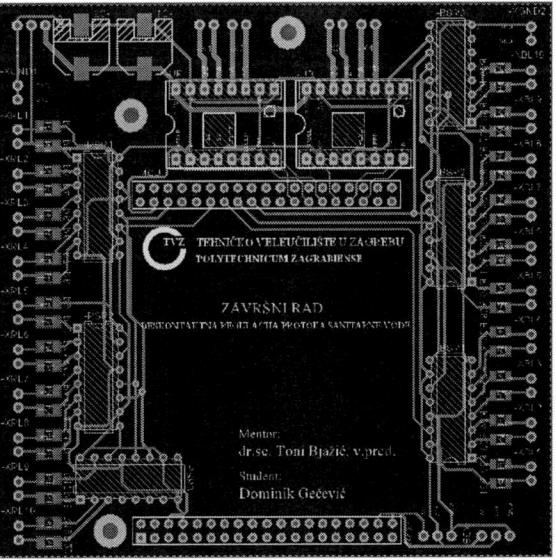

Fig. 6: Arrangement power lines and components.

TABLE IV. Characteristics of HY-DIV268N-5A.

Index	Characteristics	Min	Max	Unit
V_{CC}	Input signal voltage	3	5.5	V
V_{MOT}	Input motor voltage	12	48	V
T_A	Operative temperature	-10	45	°C
T_S	Storage temperature	-40	70	°C
I_{MAX}	Maximum drive current	-	4	A
f	Chopping frequency	-	20	kHz
m	Mass	-	200	g

1553

rising edge, the system will rotate the motor by one increment.

B. Analog infrared proximity sensor Sharp GP2Y0A21YK0F

The proximity sensor is an optical sensor that works without the use of any external force, reveals the presence of nearby objects from which reflected light was reflected.

1) Principle of operation: Diffuse optical proximity sensor is a sensor in which the transmitter and receiver are integrated in a single housing. If an object comes close to the sensor its transmitted light will be reflected back to the receiver that will detect it. Such sensors are used for short distances, providing a distance reading from objects that have a good reflecting surface. The Sharp GP2Y0A21YK0F [6] sensor has an integrated electronic circuit for filtering noise and signal processing and analog output. The characteristics of the analog infrared sensor are shown in Tab. VII.

Sharp GP2Y0A21YK0F [6] the first measurement is performed in the range from 28.7 to 47.9 milliseconds. While the sensor performs the first measurement it is necessary to add a maximum of 5 milliseconds that would allow the signal to be measured and sent to output. Infrared proximity sensor is designed using paired infrared diode and photo-sensitive diodes. The principle of operation of the infrared sensor is relatively simple. The infrared diode emits infrared light waves that is recived by a photo diode after reflection from obstacles (body).

2) Linearization of the sensor characteristics: Characteristic of the Sharp GP2Y0A21YK0F [6] sensor gives a

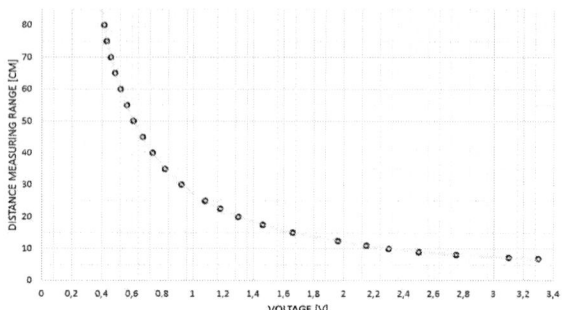

Fig. 7: Nonlinear characteristics of the analog infrared sensor.

higher voltage to the reaction at a short distance, and at a lower voltage at a longer distance so in order to have a more accurate reading the characteristic needs to be linearized. The first step is to manually type in a meter reading from the image to Excel [7] table. Then choose the "scatter plot" that would replace the axle. The "Add Trendline" will allow the mathematical function of non-linear characteristics to select the "Power" function as obtained in the enclosed Figure 7

$$y = 27,57 * x^{1,192}. \tag{1}$$

C. Digital infrared proximity sensor Sharp GP2Y0D815Z0F

Sharp GP2Y0D815Z0F [8] is a digital infrared proximity sensor used for measuring distances in the range from 0.5 to 15 centimeters. It consists of an integrated combination of the PSD (Position Sensitive Detector), IRED (InfraRed Emitting Diode) and processor unit for signal processing. The characteristics of the digital infrared sensor are shown in Tab. VIII.

D. Bipolar stepper motor 42STH38-1304AF

Stepper motors are electromechanical transducers of energy. Converted pulsed of electrical excitation are transferred to the rotational mechanical movement. The stepper motor may have 4, 5, 6, or 8 wires. Bipolar stepper 42STH38-1304AF produced by Suncor Engine [9] have only 4 wires. The characteristics of a bipolar stepper motor are shown in Tab. IX.

Normal mode to be used on a stepper motor is a full step mode. Half-step is a dividing method, which is more precise than full steps. The principle of operation is that the current flows through two motor windings. This way

TABLE V. Microstepping table.

Micro step	Pulse/rev	S1	S2	S3
-	-	ON	ON	ON
1	200	ON	ON	OFF
2/A	400	ON	OFF	ON
2/B	400	OFF	ON	ON
4	800	ON	OFF	OFF
8	1600	OFF	ON	OFF
16	3200	OFF	OFF	ON
32	6400	OFF	OFF	OFF

TABLE VI. Current regulation mode.

Current	Peak current	S4	S5	S6
0.5	0.7	ON	ON	ON
1.0	1.2	ON	OFF	ON
1.5	1.7	ON	ON	OFF
2.0	2.2	ON	OFF	OFF
2.5	2.7	OFF	ON	ON
2.8	2.9	OFF	OFF	ON
3.0	3.2	OFF	ON	OFF
3.5	4.0	OFF	OFF	OFF

TABLE VII. Characteristics of the analog infrared sensor.

Index	Characteristics	Min	Max	Unit
V_{CC}	Supply voltage	4.5	5.5	V
I	Consumption current	30	40	mA
ΔL	Distance measuring range	10	80	cm
T_A	Operating temperature	-10	60	°C
T_{TSD}	Storage temperature	-	165	°C
m	Mass	-	3.6	g

TABLE VIII. Characteristics of the digital infrared sensor.

Index	Characteristics	Min	Max	Unit
V_{CC}	Supply voltage	2.7	6.2	V
V_{MOT}	Consumption current	-	5	mA
ΔL	Distance measuring range	0.5	15	cm
T_{OPR}	Operating temperature	-10	60	°C
f	Frequency	-	400	kHz
m	Mass	-	0.8	g

1554

TABLE IX. Characteristics of a bipolar stepper motor.

Index	Characteristics	Nominal	Unit
V_{CC}	Supply voltage	3.25	V
I	Consumption current	1.3	A
R	Resistance	2.5	Ω
L	Inductance	5.0	mH
M	Holding torque	0.39	Nm
m	Mass	0.28	kg
-	Step angle	1.8	°
-	Steps per revolution	200	-

half of the rotor is held in position between two stator poles. The principle of operation is similar to the half-step microstepping, but the only difference is that the value of the current in the coil is adjustable because half of the rotor can be in a number of discrete positions between the stator [10].

IV. DESIGN AND TESTING RESULT

Total power dissipation is shown in Tab. X, flowchart is shown in Fig. 8 and first prototype testing photos are shown in Fig. 9, 10, 11, 12,

V. IN GENERAL OF WRITING CODE

The code is written in C ++ object-oriented programming language for the Nucleo-F303RE [1] development platform which is the main part of the processing electronics for signal processing in this final work. Programming the microcontroller was done from an mbed [2] integrated development environment. The code is translated using the online compiler [2] that is available for Linux, Mac and Windows operating system in the object-oriented programming language C ++. Microcontroller runs a program that is stored in its flash memory.

1) Linearization of analog infrared proximity sensor: Mbed developer is made to easily share codes (open source). The linearization signal class used in this project was written by Thomas Johansen [11], and is protected under the license of MIT Institute. For the proper use of this class a non-linear function of the infrared analog proximity sensor should be described as we did at III-B2. The resulting function (1) must be described within the class to determine the minimum and maximum voltage value of the sensor. Class SHARPIR can return values in centimeters or inches, depending on the user's needs.

2) Noise filtering using PT 1 element: The used class is written by a member of the TVZ Mechatronics Team [12]. For proper use of this class is necessary to determine the gain K, time constant T_1 and sampling time T_d. Low Pass Filter PT 1 is used to filter the noise of the analog infrared proximity sensor using the given transfer function. Transfer function PT 1 element is simulated in the software package Simulink, MATLAB from publisher MathWorks® [13].

TABLE X. Total power dissipation.

Power supply [V]	Total power dissipation [W]
5	1.4
12	72

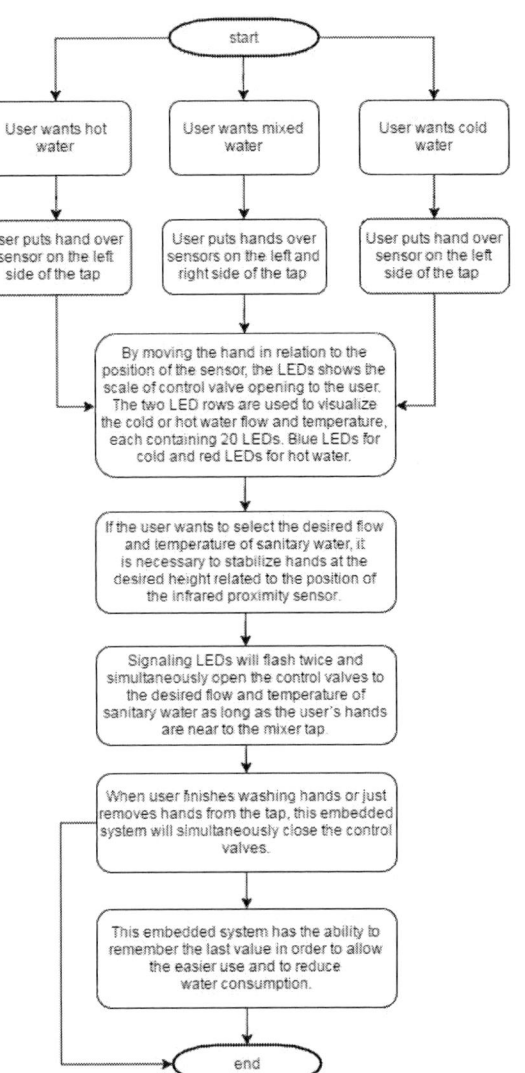

Fig. 8: Flowchart of the system.

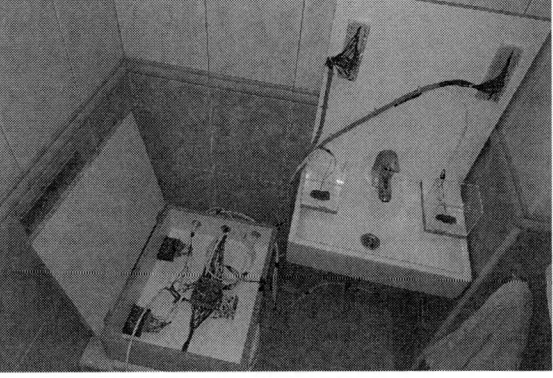

Fig. 9: First prototype.

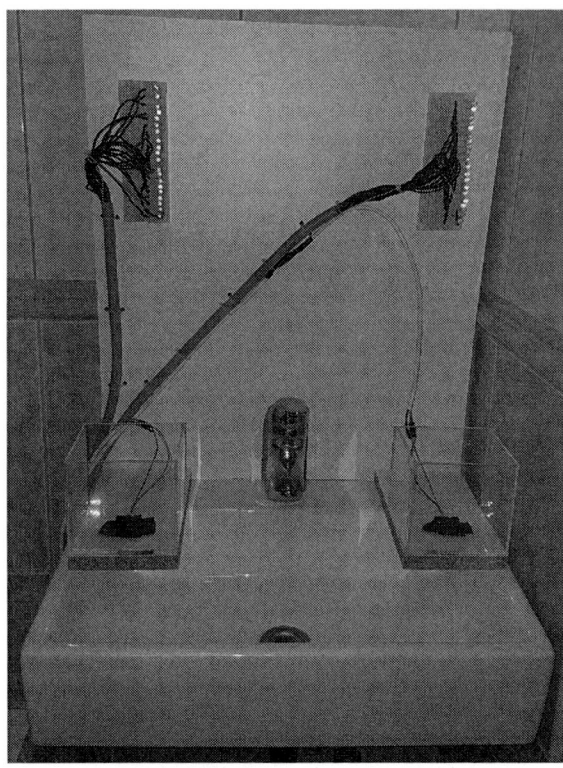

Fig. 10: Sink with built-in IR sensors and LEDs.

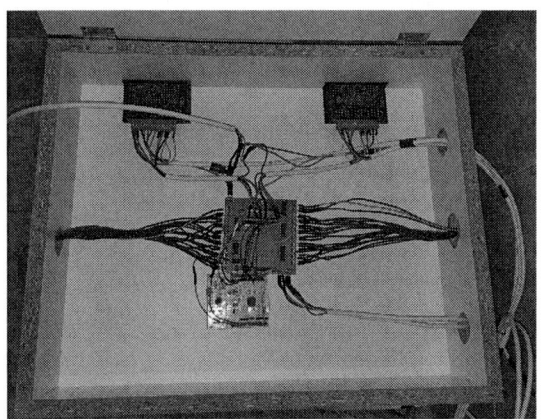

Fig. 11: Control box.

VI. CONCLUSION

An innovative embedded computer system enables the user to contactless control sanitary water flow and temperature depending on the distance of a hand to the position of the infrared proximity sensor. The offset position of the hand is shown by signaling LEDs. The use of this embedded system is very simple. If the user wants to select the desired flow and temperature of sanitary water, it is

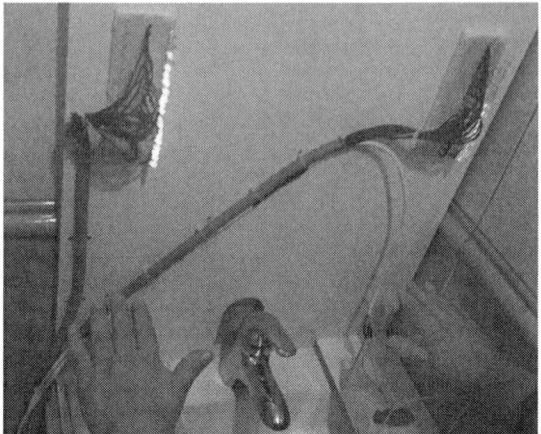

Fig. 12: Testing the prototype.

necessary to stabilize hands at the desired height to the infrared proximity sensor and then the signaling LEDs will flash twice. After setting the desired value, control valves will open to the desired flow and temperature of sanitary water as long as the user's hands are near to the mixer tap. This embedded system has the ability to remember the last value in order to allow the easier use and to reduce water consumption. Two LED rows are used to visualize the cold or hot water flow and temperature, each containing 20 LEDs. For example by activating 20 red LEDs, control valves are fully opening, the same logic hold true for cold water.

REFERENCES

[1] "Nucleo-F303RE," https://developer.mbed.org/platforms/ST-Nucleo-F303RE/, 2017, [Online; accessed 20-February-2017].

[2] "MBED compiler," https://developer.mbed.org/compiler/, 2017, [Online; accessed 20-February-2017].

[3] "SN74HC595N," http://www.ti.com/lit/ds/symlink/sn74hc595.pdf, 2017, [Online; accessed 20-February-2017].

[4] "PCB," http://circuitmaker.com/, 2017, [Online; accessed 20-February-2017].

[5] "ShenZhen2U," http://www.shenzhen2u.com/, 2017, [Online; accessed 20-February-2017].

[6] "GP2Y0A21YK0F," http://www.sharpsma.com/webfm_send/1489, 2017, [Online; accessed 20-February-2017].

[7] "Excel," https://www.microsoft.com/hr-hr/, 2017, [Online; accessed 20-February-2017].

[8] "GP2Y0D815Z0F," https://www.pololu.com/file/0J813/gp2y0d815z_e.pdf, 2017, [Online; accessed 20-February-2017].

[9] "42STH38-1304AF," http://www.suncormotor.com/Products/1468325853.html, 2017, [Online; accessed 20-February-2017].

[10] "Stepper motor," https://en.wikipedia.org/wiki/Stepper_motor, 2017, [Online; accessed 20-February-2017].

[11] "MBED, SHARPIR," https://developer.mbed.org/users/Tomas/code/SHARPIR/, 2017, [Online; accessed 20-February-2017].

[12] "TVZ Mechatronics Team, AutomationElements," https://developer.mbed.org/teams/TVZ-MechatronicsTeam/code/AutomationElements/rev/b9e11da0f2eb, 2017, [Online; accessed 20-February-2017].

[13] "MATLAB," https://www.mathworks.com/, 2017, [Online; accessed 20-February-2017].

PI controller for DC motor speed realized with Arduino and Simulink

Mario Gavran*, Mato Fruk** and Goran Vujisić**

* Faculty of Electrical Engineering and Computer Science, Maribor, Slovenia
**University of Applied Sciences - Department of Electrical Engineering, Zagreb, Croatia
mario.gavran@student.um.si, mato.fruk@tvz.hr, gvujisic@tvz.hr

ABSTRACT – In this paper we describe a technical system for DC motor speed control. The speed of DC motor is controlled using Arduino programming platform and MATLAB's Simulink coder. This paper contains introduction to using an Arduino board and Simulink PI controller in closed loop system. It will be described how to program Arduino with Simulink coder and in the end we present the results of PI controller for DC motor speed will be given.

I. INTRODUCTION

Short settling time and minimized steady state error are desired in technical system of speed controlled DC motor. To accomplish these goals, closed control loop must contain a PI controller, DC-DC power converter and a negative feedback/speed sensor. Testing was done on laboratory model of small DC motor coupled with DC generator that is used as a load on the motor. Model is shown in Fig. 1.

Figure 1 Laboratory model of DC motor

That model also includes the tachogenerator that was used as negative feedback/speed sensor. Arduino Uno board was used as controller in this closed loop. To accomplish short settling time and a small steady-state error regulator have to be adjusted to the rest of the closed loop.

An Arduino board is used due to low cost, simplicity and flexibility. MATLAB Simulink is used to create control algorithm and convert that algorithm in C++ code that can be uploaded in Arduino board. PC computer with Windows OS is required to make the model of controller in Simulink and to upload that model on Arduino board.

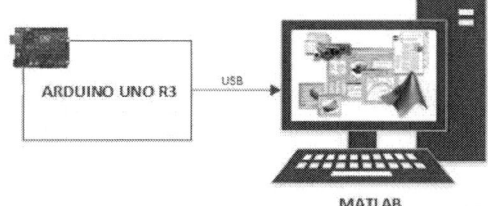

Figure 2 Aduino Uno and Simulink

II. ARDUINO UNO

Arduino Uno is microcontroller board with Atmega328p that has been used as a digital PI controller. This board is used to test the controller. It has 14 digital input/output pins, and 6 of them can be used as a PWM output pins. It also has 6 analog inputs, 16MHz clock crystal, a USB connection that is used to connect it to the PC and it operates at 5 Volts. Various Arduino boards can be used instead of Arduino Uno board.

Arduino Board	Shield Support	Interactive Tuning and monitoring	Comments
Arduino Due*	Y	Y	DAC and CAN channels not currently supported
Arduino Uno*	Y	N	
Arduino Leonardo*	Y	Y	
Arduino Mega 2560*	Y	Y	
Arduino Mega ADK*	Y	Y	
Arduino Micro*	N	N	
Arduino LilyPad USB	N	N	
Arduino Esplora	N	N	Additional IO supported via analog multiplexer
Arduino Robot	N	N	Additional IO supported via analog multiplexer
Arduino Mini* (ATmega328)	N	N	Mini with ATmega168 not supported
Arduino Nano 3.X* (ATmega328)	N	N	Nano 2.X with ATmega168 not supported
Arduino Pro* (ATmega328)	N	N	Pro with ATmega168 not supported
Arduino Fio	N	N	
Arduino Ethernet Shield			See Shield Support column for compatbility
Arduino WiFi Shield			See Shield Support column for compatbility

Figure 3 Arduino boards compatible with Simulink

III. ARDUINO AND SIMULINK

Arduino boards are usually programed by writing C/C++ code in Arduino IDE window, but in this example it will be programmed using MATLAB Simulink package for Arduino. Simulink is a block diagram based environment for simulating mathematical models and it can be used to program some of the Arduino boards. To use Simulink with Arduino a support package is needed. It can be downloaded from Mathworks website with instructions how to install it and start using it. After installation is done correctly, a new library is now ready for use in Simulink library browser. In this paper analog input, digital input and PWM output blocks will be explained and shown.

IV. TESTING ARDUINO AND SIMULINK

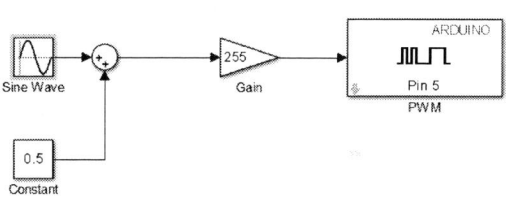

Figure 4 Block diagram of sine wave generator

This section explains how to use Simulink blocks to read analog signals from real world and how to use PWM output signals to control some kind of electric device. Figure 4 shows simple block diagram made in Simulink. Its purpose is to generate a sine wave signal from Arduino using only Simulink. First block is a sine wave block. This block outputs sine wave signal of amplitude 0.5 and frequency of 10Hz. In sum block that signal is added to the constant of 0.5. Sum block now outputs sine wave that has "DC" component of 0.5, and amplitude of 0.5. Because Arduino Uno does not have analog output, the PWM output is used with corresponding RC filter. That signal is connected to the gain block of 255 value of gain, and outputs sine signal that has amplitude of 255/2.

In the PWM block, pin number needs to be entered. The PWM block will generate a PWM signal on the specified pin. The duty cycle of that PWM signal depends on input signal. In this case pin 5 is used and generated sine wave signal with amplitude of 255 is on input. When that block diagram is compiled and uploaded in Arduino Uno board by clicking on "build model" icon (normal mode), on pin 5 of the Arduino board you can measure PWM signal and by using RC filter that PWM signal can be filtered and that is shown on Fig 5.

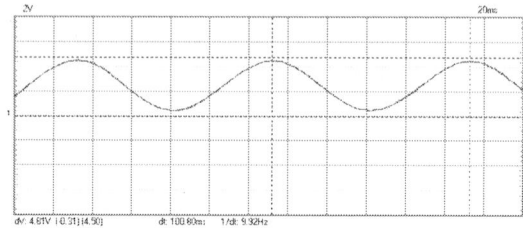

Figure 5 Sine wave generated with Arduino

For demonstration how to use analog input block and Arduino Uno and Simulink in real time, another block diagram is uploaded on Arduino board (Fig. 6) but this time in "external mode". "External mode" is used when you have to monitor any of the signals in block diagram.

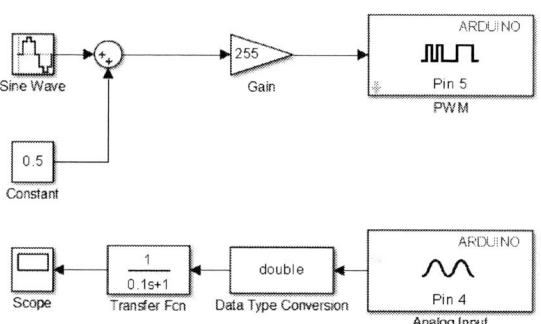

Figure 6 Block diagram of sine wave generator and analog read

Top part of this diagram is the same as on Figure 4. This time analog input block is defined by selecting analog pin 4 as an input pin. For demonstration PWM pin 5 is connected with analog input pin 4 via the RC filter. In this example Arduino generates sine wave signal and reads that same signal. Transfer Fcn block simulates the same RC filter but represented as first order filter with time constant of 0.1 s. When diagram is uploaded using external mode on scope block generated sine wave signal can be monitored as shown in Fig 7. Input block converts input voltage in uint16 type number in range from 0 to 1023.

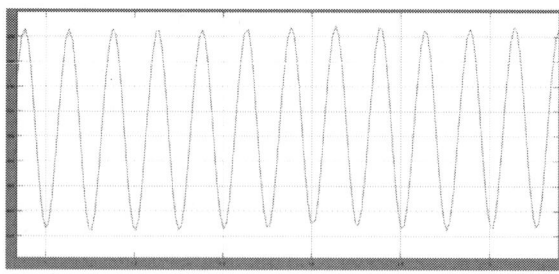

Figure 7 Sine wave read by Arduino

V. TESTING THE CONTROLLER WITH DC MOTOR

In order to apply the controller on DC motor model, the parameters of the controller need to be set. The parameters of the controller depend on the rest of closed loop components, so transfer functions must be known. Transfer functions of the motor, DC-DC power converter and speed sensor characteristics must be known to determine their transfer functions.

The motor is tested on load and data is collected. On Fig. 8 load characteristics is shown. Moment of inertia is determined with motor stopping experiment.

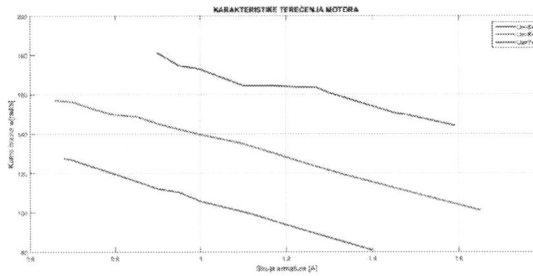

Figure 8 Load characteristics

Resistance of armature coil is determined as well as motor constant "k". In Fig. 9 time response of armature circuit is shown and inductance of armature coil is determined.

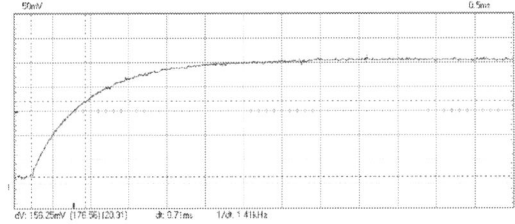

Figure 9 Time response of armature coil

Motor now can be simulated in Simulink using block diagram shown in Fig. 10.

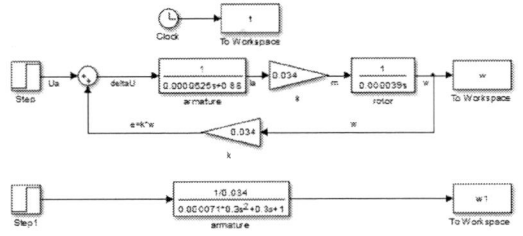

Figure 10 Simulation model of used motor

Because Arduino Uno does not have analog output, the PWM output is used and therefore chopper is used as DC-DC power converter. This is convenient because an RC filter is not needed to convert PWM to analog signal. DC-DC converter i.e. the chopper is shown in Fig. 11.

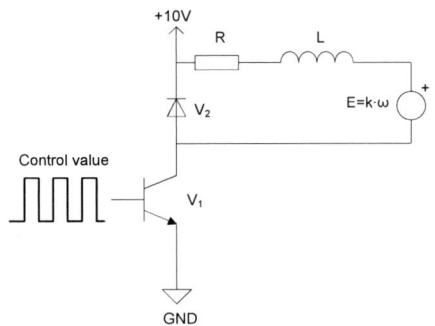

Figure 11 chopper used to power the motor

Transfer function of the speed sensor will be approximated by the first order filter.

Control block diagram is shown in Fig. 13. PI control algorithms will be tested and PID block is used. In PID block window various parameters can be changed. In "controller" drop menu, type of controller can be selected, and in the main window proportional and integral constants can be changed. In "form" drop menu window ideal or parallel algorithms can be selected. In Fig. 12 the function block parameters for PID block are shown.

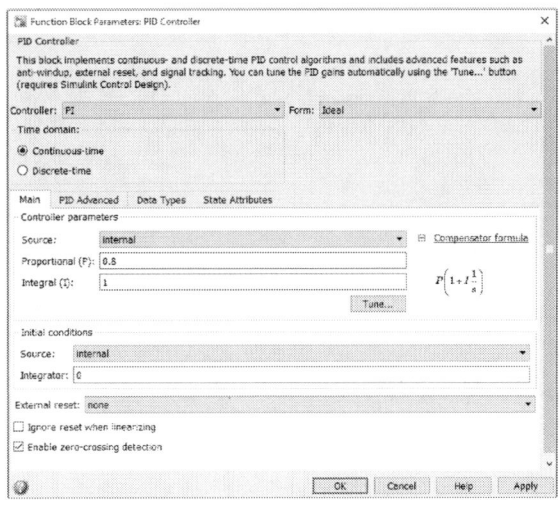

Figure 12 PID controller block parameters

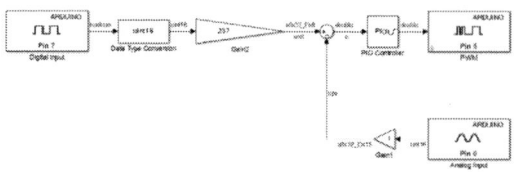

Figure 13 Block diagram of the controller

Controller diagram (Fig. 13) represents how reference and negative feedback are accomplished and used as input signals to the controller itself. Speed sensor is connected to the analog input pin 0. 5V pin on Arduino board is connected to the digital input pin 7 via the button and resistor. When button is pressed on pin 7, 5 volts appears. Speed sensor that is mounted on the motor shaft gives 1,14V for the speed of 1000 RPM. Analog input block convert voltage that is connected to the pin 0 in to number, so data conversion block is used to adapt number type. In this case it is 237. In order to initialize the motor speed of 1000 RPM the reference value has to be 237 so Gain2 block has value of 237.

When button is not pressed motor is not spinning. When button is pressed digital input block outputs 1 and multiplies it by 237. Still motor is not spinning so speed sensor connected to analog input gives 0V and value that analog input block gives is also 0. Now PID controller block has 237 on its input and starts to give signal to the PWM block. PWM output pin is connected to the DC-DC power converter and motor starts to spin. Depending on the controller parameters the motor will get to 1000 RPM speed and input in the PID block now is 0 and PI controller will maintain that speed of the motor. If motor gets more than enough energy from DC-DC converter it will overshoot wanted speed and input in the controller will be negative, so the controller will slow the motor down.

Fig. 14 shows the time response of the motor speed. Speed is measured with the speed sensor that is already mounted on the motor. For example, wanted speed will be 1000 RPM, which means the reference gain must be set to 237. Speed will be recorded using the speed sensor. Because of that speed will be represented with voltage. Speed sensor that is used for these tests has gain of 1,14 mV/RPM.

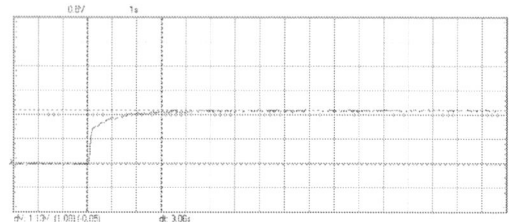

Figure 14 Motor speed time response

Steady state speed is around 1000 RPM as shown in Fig. 14. The time response graph is recorded with unloaded motor and using PI controller that is set to parallel form, proportional gain (P) set to 1 and integral gain (I) set to 1. In Fig. 15 parallel model of controller is shown.

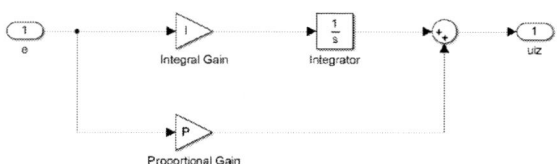

Figure 15 parallel model of controller

In this example, where P=1 and I=1, the motor will get to 1000 RPM in around 3 seconds. To test the speed response on load disturbance a generator coupled with motor is used. Motor will be spinning at 1000 RPM and the load will be switched on. When controller compensates that load disturbance and gets up to 1000 RPM again, the load will be switched off, as shown in Fig. 16.

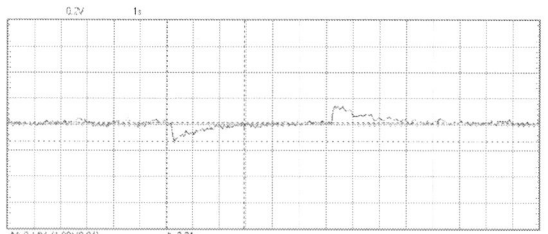

Figure 16 Motor speed response with load disturbance

It takes 3 seconds to get back to wanted motor speed. To get more dynamic response of the motor controller parameters must be changed. Speed and load disturbance time response is shown in Fig. 17. In that case the proportional gain is set to 1 and integral gain to 20.

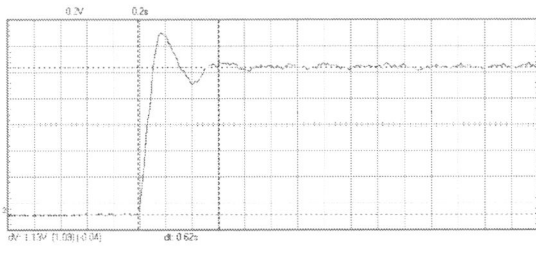

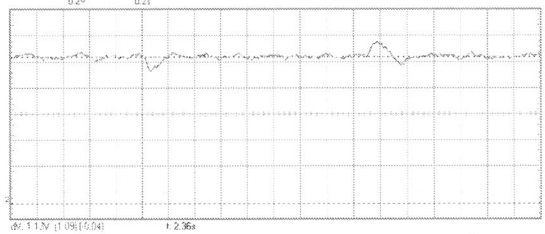

Figure 17 Speed and load disturbance response

With this set up of controller the speed response is underdamped but much faster. In this case motor spins up to 1000 RPM in 0,62 seconds and time needed for the controller to compensate the load is 0,36 seconds.

All tests of the controller were done in normal simulation mode and all sample times were the same. Sample time is a very important parameter of the model and must be set at 0.0001 s for examples shown in this paper. This model of speed controlled motor will be used in laboratory exercises, and so it is convenient to use external simulation mode. In this mode user can monitor any of the signals in the model. For example in Fig. 18 and Fig. 19 two signals are being monitored in real time.

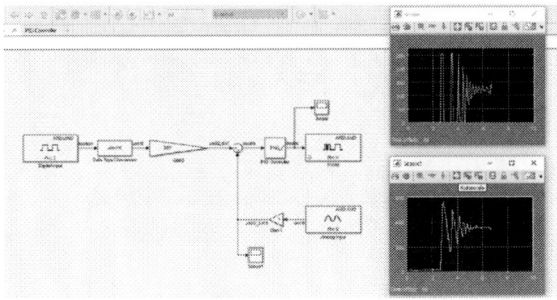

Figure 18 External mode with controller

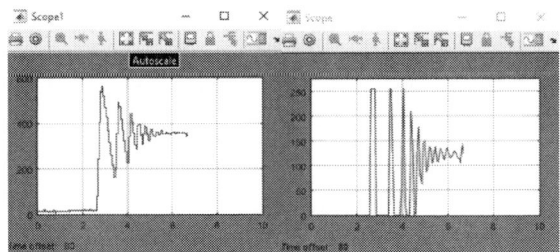

Figure 19 Scopes in real time working in external mode

VI. CONCLUSION

Using Simulink and Arduino is one of the simplest ways to make satisfying controller for many purposes in variety of technical systems. Tests that were done in this paper show that Arduino can be very flexible for programming a control algorithms using MATLAB's Simulink. One purpose of this kind of controller is using it in laboratory exercises where students can easily modify controller's parameters and see what is going on with output value of the system that they are studying. Sample time of this controller can be changed and adapted to the system requirements. Using external mode, parameters of the controller can be changed in real time and user can see what has changed in the output signal almost instantly. The model was already tested on laboratory exercises and it has yielded good results. The students were very interested how to program the controller with Simulink.

REFERENCES

1. Arduino website:
   https://www.arduino.cc/en/Main/ArduinoBoardUno
2. http://www.mathworks.com/, 29.7.2016.
3. Perić, Nedjeljko.
   Automatsko upravljanje-predavanja, FER, Zagreb, 2006.
4. https://www.fer.unizg.hr/_download/repository/SIMULINK_SKRIPTA.pdf, 10.1.2015.
5. Fruk, Mato; Vujisić, Goran; Tikvić, Ivan.
   CONTROL OF THERMAL PROCESS WITH SIMULINK AND NI USB-6211 IN REAL TIME
   // MIPRO 2013, Proceedings, Computers in Education – CE, Opatija, 2013, 697-702
6. Fruk, Mato; Vujisić, Goran; Špoljarić, Tomislav.
   Parameter Identification of Transfer Function Using MATLAB // MIPRO 2013, Proceedings, Computers in Education – CE, Opatija, 2013, 697-702

Laboratory Model of the Elevator Controlled by ARDUINO Platform

Marijo Andrija Balug*, Tomislav Špoljarić** and Goran Vujisić**

* Faculty of Electrical Engineering and Computer Science, Maribor, Slovenia

**University of Applied Sciences - Department of Electrical Engineering, Zagreb, Croatia

marioandrija.balug@student.um.si, tomislav.spoljaric@tvz.hr, gvujisic@tvz.hr

ABSTRACT - This paper presents methods for regulating the elevator's cabin speed and position for purposes of traffic control. It contains four stations. On every station there are two position sensors and cabin call button. Cabin speed regulation is realized with microcontroller, incremental encoder and chopper, where chopper controls DC motor speed. A simple traffic control is written for ARDUINO platform.

I. INTRODUCTION

This paper will show how to regulate elevator's cabin speed and position for purpose of traffic control. It is necessary to achieve speed regulation due to speed limitations, acceleration and deceleration. In that way a cabin snatch can be solved. Position regulation is needed because the cabin needs to stop on preferred position. All functions are realized with two ARDUINO development boards (Pro Mini and Mega board) they are easy to program and they easily control another elements of control systems. The elevator model was made with the aim of acquiring knowledge of automation and regulation. The picture of model is given in Fig 1.

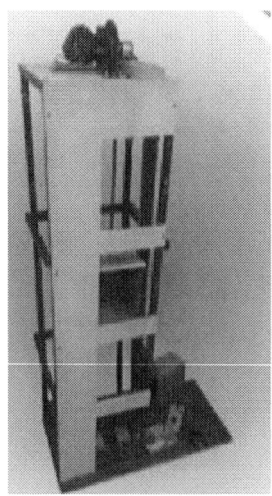

Figure 1. Laboratory model of the Elevator

II. DC MOTOR SPEED REGULATION

The speed control loop contains microcontroller ATmega328p, which is a part of ARDUINO Pro Mini development board, chopper and small DC motor with permanent magnets.

A. Controller

ARDUINO Pro Mini (Fig 2.) development board is used as a PI controller. PI controller is implemented using the PID library. The process has only one dominant pole so it can be controlled quite well with a PI controller. PI controller parameters are obtained with experimental method, and uses difference between reference value which is set by another microcontroller and feedback value as an input. First the proportional gain is increased to get a little overshoot and then the intagral constant is deacreased from large value to small to obtain the response with given overshoot. Output is a PWM signal with modifiable duty-cycle.

Figure 2. Overview of an ARDUINO Pro Mini board

TABLE I. ARDUINO PRO MINI TECHNICAL CHARACTERISTICS

Microntroller type	ATmega328p
Operating voltage	5V
Power supply	5-12V
Digital I/O	14 (6 PWM outputs)
Analog inputs	6
Flash memory	32 KB
SRAM	2 KB
EEPROM	1 KB
Clock	16 MHz

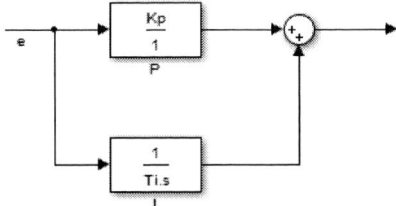

Figure 3. Parallel PI regulator block scheme [5]

PI transfer function is as follows [3, 5]:

$$F(s) = K * \frac{1 + sT}{sT} = 0.5 * \frac{1 + 0.25s}{0.25s} =$$

$$0.5 * \left(\frac{1}{0.25s} + 1\right) = 0.5 + \frac{2}{s} \qquad (1.)$$

B. H-Bridge

H-bridge is used to control direction and speed of DC motor. Inputs of H-bridge are connected to ARDUINO Pro Mini which controls direction and PWM duty-cycle. Output of an H-bridge is a voltage between 0 V and 12 V, and which depends on input's duty-cycle. It is connected to a DC motor armature.

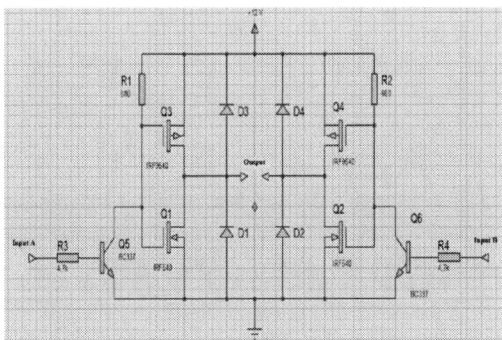

Figure 4. Electrical scheme of an H-bridge [2]

Transfer function of a H-bridge is as follows:

$$F(s) = \frac{U_{OUT}(s)}{U_{IN}(s)} = K = 2.43 \qquad (2.)$$

and its characteristic is shown on Fig.5.

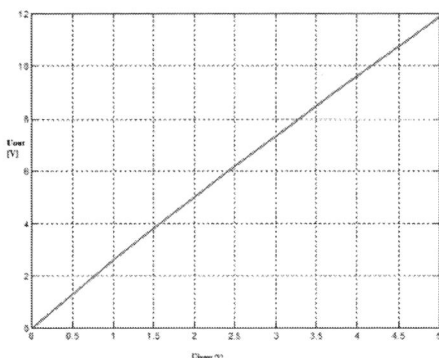

Figure 5. Steady state characteristics characteristic of an H-bridge

C. DC Motor

DC motor with permanent magnets is used as an operating machine. The only known data about the motor are armature voltage which is 12 V and rotation speed which is 120 rpm. To get the transfer function of this motor, this data was insufficient. Dependence between rotation speed and armature voltage was measured with tachogenerator, and transfer function was obtained with MATLAB's System Identification Toolbox [4]:

$$F(s) = \frac{\omega(s)}{U_a(s)} = \frac{K_p}{(1 + sT_{p1})(1 + sT_{p2})} =$$

$$\frac{K_p}{1 + s(T_{p2} + T_{p1}) + s^2 T_{p1} T_{p2}} = \frac{0.9124}{1 + 0.321s + 0.011s^2} \qquad (3.)$$

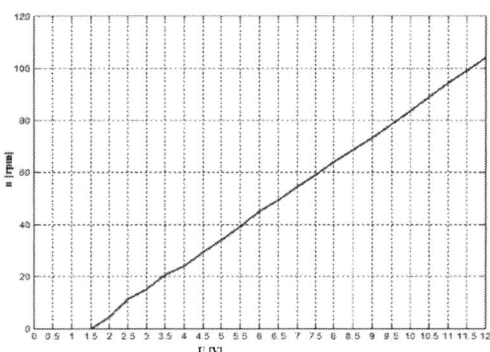

Figure 6. Input-output characteristic of a DC motor obtained from measurements

D. Incremental Encooder

DC motor rotation speed is measured with incremental encoder (Fig 7.). Incremental encoder gives 600 impulses per rotation. Output is connected to ARDUINO Pro Mini which counts impulses in time and is used as a regulation circuit feedback. Maximum input in encoder is 13.51 rad/s and maximum output is 5 V.

1563

Figure 7.   Incremental encoder

Teherefore the transfer function of incremental encoder is:

$$F(s) = \frac{y(s)}{\omega(s)} = \frac{5}{13,51} = 0.37 \qquad (4.)$$

III.   SPEED REGULATION FEEDBACK LOOP

Reference value is voltage between 0 V and 5 V. It is set on analog input of microcontroller. If the reference value is set to 5 V, then DC motor rotation speed is 13.5 rad/s. Output of measuring element is value between 0 and 5. Regulator input uses difference between referent value and output of measuring element. Depending on the difference, on the output of the regulator 500 Hz PWM signal with modifiable duty-cycle appears. PWM signal is used as an input value to H-bridge. DC motor rotation direction is chosen by Arduino Mega which sends information of direction to ARDUINO Pro Mini. Arduino Pro Mini generates PWM signal with which controls DC motor speed via H-bridge. DC motor is connected on the output of the H-bridge. Motor speed depends on PWM's duty cycle.

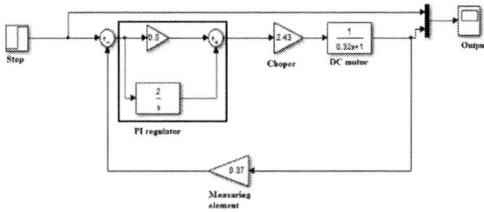

Figure 8.   Simulink model of speed control feedback loop

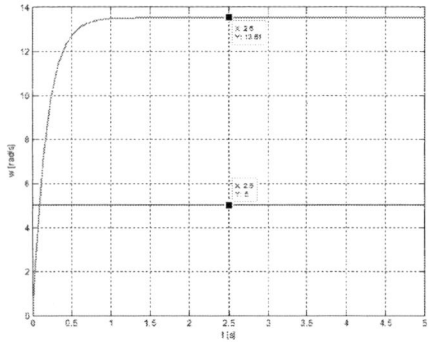

Figure 9.   Speed control time response

IV.   POSITION AND TRAFFIC REGULATION

A. ARDUINO Mega

ARDUINO Mega development board (Fig 10.) is chosen for position and traffic control. This board is chosen because it has enough digital inputs and outputs to connect outputs of 8 position sensors, 8 buttons for cabin calling and few pins to communicate with ARDUINO Pro Mini.

Figure 10.   Overview of an ARDUINO Mega board

TABLE II.   ARDUINO MEGA TECHNICAL CHARACTERISTICS

Microntroller type	ATmega2560
Operating voltage	5V
Power supply	7-12V
Digital I/O	54 (15 PWM outputs)
Analog inputs	16
I/O current	20 mA
Flash memory	256 KB
SRAM	8 KB
EEPROM	4 KB
Clock	16 MHz

B. Cabin position sensor

Cabin position sensor is an optocoupler. Paper flag is mounted on the cabin, and it goes through sensors. When the flag is between IR diode and photo transistor, sensor output is 0 V, else output is 5 V. Depending on sensors outputs microcontroller determines cabin position. Elevator model has eight sensors, two on every station.

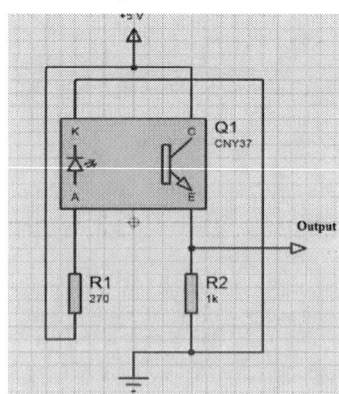

Figure 11.   Electrical sheme of an optocoupler

C. Traffic control

There is a call button (Fig. 12.) on every of four stations. Four buttons are mounted in the cabin. They are connected to digital inputs of microcontroller. Algorithms in microcontroller save every button call and send cabin to the first saved value.

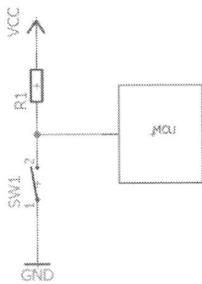

Figure 12.   Electrical scheme of a call button

An elevator has four stations and the cabin speed depends on difference between current position and position where cabin needs to go. If the cabin needs to go from the first to second station, referent speed needs to be minimal because the distance is small. Also, acceleration and deceleration depends on difference between stations. Algorithm in microcontroller calculates the difference and constantly sends parameters to second microcontroller. It sends reference value of speed and cabin direction. Reference value has the shape of PWM signal, because microcontroller does not have analog outputs. PWM signal is connected to RC filter (Fig 13). The output is approximately equal to average value of PWM signal and it is connected to ARDUINO Pro Mini analog input. Fig. 14 shows acceleration and deceleration speed reference ramps.

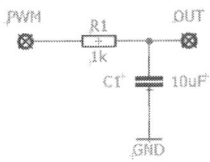

Figure 13.   Electrical scheme of an RC filter

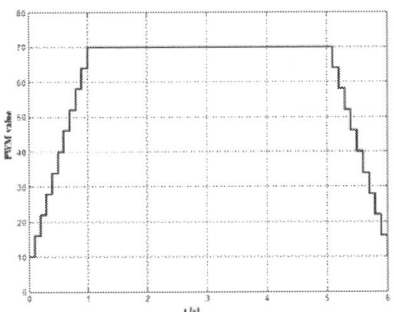

Figure 14.   Change of a reference value

D. Principles of operation of the elevator system

If the system is turned on for the first time, microcontroller checks cabin position. If the cabin is not on the first station, microcontroller positions cabin on first station. After that, microcontroller constantly checks call button and position sensors change of state. When the call is active, microcontroller calculates difference between stations and sets reference value of speed and direction of moving cabin to the speed control circuit. When the cabin is on the wanted position, microcontroller sends zero as speed reference and the cabin stops.

V.   CONCLUSION

The elevator model is made with the aim of acquiring knowledge of automation and regulation. It is useful for laboratory exercises because the parameters of regulator are easy to change. If the parameters are changed, then the speed of cabin will be different.

Benefits in using such system are:

- ARDUINO platform is easy to program and is compatible with many operating systems,
- it is easy to change elements and their parameters in code,
- it is possible to measure all signals between microcontrollers and other elements.

However, there are certain disadvantages. They include:

- low operating voltage and possible disturbances,
- ARDUINO Mini Pro and Mega have not enough interrupt pins, so microcontroller needs to have additional algorithm to constantly check changes of states on digital inputs,
- optocouplers are not resistant to dust. Electromagnetic sensors would be better.

In addition, some improvements to this system can be done:

- adding the cabin door,
- changing DC motor with gear unit with DC motor without gear unit because when the cabin goes down the motor would be in braking mode. In that way real elevators work [6].

REFERENCES

[1]   "ARDUINO Tutorials", www.arduino.cc, 2016.

[2]   "The H-bridge", www.talkingelectronics.com, 2012.

[3]   N. Perić, "Automatsko Upravljanje", FER, Zagreb, 1998.

[4]   M. Fruk, G. Vujisić, T. Špoljarić, "Parameter Identification of Transfer Functions Using MATLAB", 35th International Convention, MIPRO, Opatija, 2012.

[5]   Fruk M., "Bilješke s predavanja iz kolegija Automatsko Upravljanje", interna skripta, TVZ, Zagreb, 2015.

[6]   A. Jozić. T. Špoljarić, D. Gadže, "Laboratory Model of an Elevator: Control with Three Speed Profiles", 39th International Convention, MIPRO, Opatija, 2016.

System for acquisition and processing of pressure data around body in airflow

D. Mežnarić, K. Krajček Nikolić, D. Franjković
Faculty of Transport and Traffic Sciences / Department of Aeronautics, Zagreb Hrvatska
domagoj.meznaric@gmail.com, kkrajcek@fpz.hr, dfranjkovic@fpz.hr

The article deals with the system and methods for determination of the pressure distribution around aerodynamically shaped body immersed in airflow and further calculations of aerodynamic characteristics. Measurements are conducted in the subsonic closed-loop wind tunnel AT-1, in Aerodynamics Laboratory of Department of Aeronautics at Faculty of Transport and Traffic Sciences. Pressure distribution around airfoil NACA 2421 is sensed by the system for acquisition of pressure data Intelligent Pressure Scanner 9016, produced by the Pressure Systems Company. Data are digitalized and transferred to the computer through the Ethernet link. Data are processed by NUSS and LabVIEW software. Measurement results are displayed and compared to those obtained from piezometric harp. Results of experiments are commented and recommendations for further research are given.

I. INTRODUCTION

Wind tunnels are devices or facilities for experimental measurements in aerodynamics and they are used from the early beginning of aviation. The tunnels are used to obtain high-quality experimental data, especially when certain data can't be obtained by theoretical calculations. In the wind tunnel, a controlled flow of air is produced which is acting on the model - subject of research and physical quantities are felt and measured and physical phenomena arising from the interaction of the body and the air flow are observed and recorded.

The measurements were performed in the subsonic wind tunnel AT-1 on the model of wing made of airfoil NACA 2421. The aim of the experiment was the construction of the measuring system from the measuring model through the acquisition of pressure system to the computer with appropriate programs.

Intelligent pressure transducer of Pressure Systems company uses electrical resistance elements by which mechanical displacement caused by the pressure is converted into an electrical signal. It consists of two module of code 9016 each containing 16 measurement points, each with input for reference pressure which is equal to the atmospheric pressure of the ambient air. In addition to the two modules, there is pressure calibrator of code 9034. It is used to calibrate the measuring instrument to zero and for the range. The modules convert analog signal to digital and send it through the Ethernet connection to the computer where the signal is further processed.

The results are compared with the results of hydrostatic pressure measurements. Two softwares are used for data processing: NUSS and LabView. NUSS is a basic program used to calibrate and adjust the system, and LabView is used to obtain pressure distribution on the upper and lower wing surface and to calculate the lift and the drag force due to the pressure.

II. WIND TUNNELS

Wind tunnels are complex installations which in its test section simulate flow conditions similar to those around the actual object or model. Wind tunnels are divided in regard to the velocity in the test section (subsonic, transsonic, supersonic) and the shape of the airstream line (open- or closed-circuit). According to their purpose or operating mode, tunnels are divided into: tunnels with a controlled pressure, tunnels with variable density of the working fluid, tunnels for testing prototypes in full size, tunnels for flow visualization, tunnels for testing free flight (the model is not fixed on the sting), tunnels for testing of spiral maneuvers, testing the stability of the flight, testing icing conditions on aircraft or other vehicles, for testing V/STOL aircraft, for testing aerodynamic characteristics of cars and boats, as well as for other uses in civil engineering (wind load on the building), ecology (the spread of pollution by natural air flow, the boundary layer to the surface of the Earth), sports (car racing, sailing, ski jumping, cycling ...) and for many other purposes.

Wind tunnel AT-1 (Figure 1.) at the Faculty of Transport and Traffic Sciences is a closed-circuit tunnel with a single return line. The test section has elliptical cross-section and partly open. Other components of the wind tunnel are converging nozzle, corner sections with air routers, honeycomb and screens, diffuser, the electric motor and fan, and the return line. Scheme of the wind tunnel AT-1 is shown in Figure 2.

Figure 1. Wind tunnel AT-1

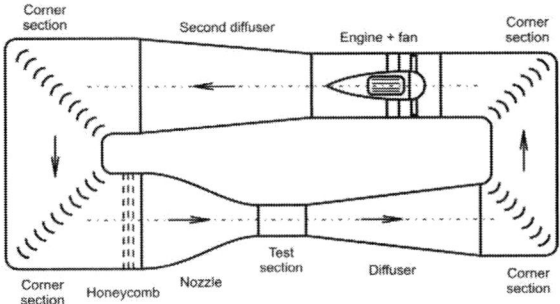

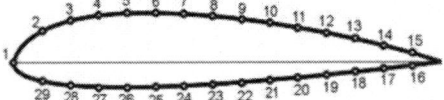

Figure 2. Scheme of the wind tunnel AT-1

The test section is the most important part of the wind tunnel, the required form of flow is obtained there, model and measurement equipment is located inside and measurements carried out. Through convergent nozzle, air is accelerated to the required speed due to the narrowing of the cross section of nozzle. Honeycomb and screens have to give enough laminar flow.

Angle sections are fitted with blades for deflecting the air stream to minimize losses due to turbulence. The purpose of the diffuser is to decrease the flow velocity of the working fluid, and thus the minimize power losses which are proportional to the third power of the speed. The electric motor drives fan which causes the airflow. Electric motor's rotation speed can be controlled and thus determines the flow velocity in the test section. Speeds attained in the test section of the tunnel AT-1 are up to 50 m/s and Reynolds numbers of flow up to 10^6. [1]

III. BODY MODEL AND PRESSURE DISTRIBUTION

The aerodynamic wind tunnel AT-1 at the Faculty of Transport and Traffic Sciences in Zagreb uses a wing model with 29 pressure measurement points. A model of "infinite" wing with constant length of chord line of 150 mm and standard airfoil shape NACA 2421 is used in this experiment. A model of the wing located in the test section of the wind tunnel is shown on Figure 3.

Figure 3. Wing model in the test section of the wind tunnel AT-1

Static pressure on the surface of airfoil is measured on measurement points allocated on upper and lower surface of the wing. Figure 4. shows 15 measurement points on the upper surface of an airfoil from leading to trailing edge. The rest of measurement points are set over the lower surface from trailing to leading edge. Measurement

points on airfoil are holes of small diameter (1 mm) which are placed perpendicular to the contour of the airfoil surface. From that place canals are installed inside the wing which are connected to the plastic pneumatic hoses.

Figure 4. Measurement points on the wing model

Pneumatic hoses are connected on a two pressure measuring devices. Each hose from one measuring point is divided into two hoses which are connected on both U-tube manometer and Intelligent Pressure Scanner.

U-tube manometer, also called piezometric harp, measures the pressure on the surface of the wing by a set of U-tube gauges. (Figure 5.). Piezometer harp consists of as many gauges as measuring points on the test model, plus one measuring tube for the determination of a reference pressure. All measuring tubes have one end connected to a common reference pressure (usually the pressure of the surrounding atmosphere). Reading height of fluid in the tubes is done using a measuring tape, which is set in the immediate vicinity of the tubes.

That is one of the oldest methods for normal pressure measurement from measuring point on the contour of the airfoil through pneumatic lines to the hydrostatic pressure gauge. Such a method of pressure reading is outdated and very time-consuming due to slow processing of each measurement point on the contour of the airfoil. Newer methods use electro-mechanical pressure transducers described in chapter IV of this article. These electromechanical transducers are integrated into digital systems for the acquisition of pressure around the airfoil. Intelligent Pressure Scanner uses silicon electro resistive pressure transmitter.

Figure 5. Piezometric harp

Pressure distribution around an airfoil at some angle of attack is shown on Figure 6. Pressure is expressed as relative to the pressure at infinity and drawn as a vertical length above each elementary surface of the airfoil. All peaks are connected and that represents the pressure distribution around the airfoil. If the difference $p - p_\infty$ has a positive value in the observed point, pressure forces acts towards aerofoil and arrow is directed towards airfoil.

That happened on the lower surface of airfoil where airflow is slower than in infinity. On the upper surface due to increased speed static pressure will decrease and will be lower than p_∞. That difference $p - p_\infty$ has a negative value which creates a vacuum that pulls the airfoil up. So overpressure on the lower surface and vacuum on the upper surface are lifting the model of the wing. [2]

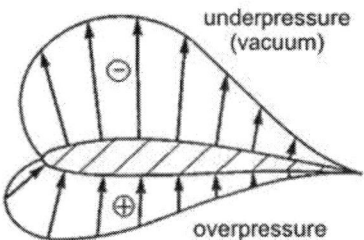

Figure 6. Pressure distribution around an airfoil

IV. INTELLIGENT PRESSURE SCANNER

The system for data acquisition is used to collect data from various sensors and to convert that data into digital numerical value used by the computer. It converts analog signals to digital. Programs for systems acquisitions are made in various programming languages and the program that is used here is LabVIEW.

Measurement begins with pressure sensing. Pressure gauges used by this system are electro resistive elements also called piezo-resistive pressure transmitters. These devices are measuring the elongation of piezo-resistive element which stretches under the influence of the diaphragm (Figure 7.). If the thin electro-resistive element is loaded by force attributable to the action of pressure, it causes a change in the geometry of electrical conductor and therefore the electrical resistance of the wire.

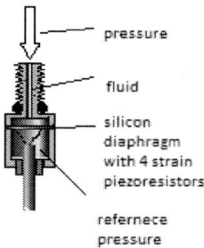

Figure 7. Piezo-resistive element

Bending (tension and compression) of piezo-element generates very small changes in resistance. Deformation and fracture of strips are the limit for greater elongation. Therefore, extremely small changes in resistance have to be measured with great accuracy. Such a need for precision resistance measurement requires bridge circuit, a Wheatstone bridge. [3]

The Model 9016 Pneumatic Intelligent Pressure Scanner is a pressure acquisition module for multiple measurements of dry, non-corrosive gases. The scanner integrates 16 silicon piezo-resistive pressure sensors of large pressure range. This scanner has scanning speed up to 100 measurements per second on a single measurement channel. Each has its own reference pressure input, in this case atmospheric. Each pressure sensor is individually separated and incorporates a temperature sensor and EEPROM memory which stores complete information on digital temperature compensation. Housing incorporates also calibration valve, 16 bit A/D converter and 32 bit processor (Figure 8.). [4]

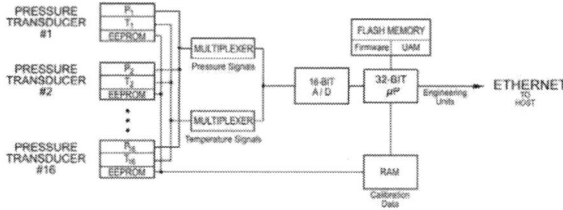

Figure 8. Piezometer harpasad[4]

Scanner modules are connected with the computer via power supply unit and Ethernet interface (Figure 9). The computer communicates with measuring modules using 10 MBit TCP/IP and UDP/IP protocols. Each measuring module and the computer have a unique IP address which is used for identification. In this network, computer represents a customer who receives information about the data of pressure and the measurement modules are servers that generate information about the data pressure. In order to connect with the modules computer must have a proper setting of a static IP address.

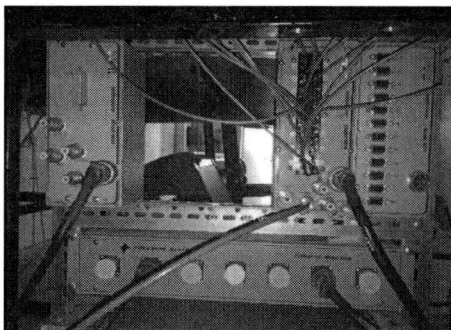

Figure 9. Intelligent Pressure Scanner connections

In order to reduce the length of the hoses and improve dynamic measurement characteristics, model 9016 is designed for installation close to the measuring point. The distance between the object which is used for pressure measurement and electronic pressure scanner can range from a meter to thirty meters. Larger hoses have an impact on dynamic characteristics of transferred measuring air pressure signal.

Electronics are located within a housing which is hermetically sealed against the entry of liquids and dirt. Housing and internal electronics are designed to withstand very high vibrations that are moving in vibrational envelope with peak acceleration of 30 G.

The effect of temperature on the measurement error may be significant if the module is exposed to temperatures outside the range of 0 °C to 60 °C. In this temperature range algorithm acts to correct pressure, and any excess of the work area temperature means uncontrolled increase of errors. [5]

V. DATA ANALYSIS

After setting the correct network properties the Netscanner Unified Startup Software can be started. This program allows system calibration, testing, cleaning and changing the IP address. There is no equipment in the laboratory to calibrate system, only the accuracy can be checked. After checking and obtaining IP address, the program LabVIEW can be started.

Manufacturer designed an application inside LabVIEW that connects to module device and reads the pressure. That application was modified for wind tunnel purpose. There is an interface of the application used for pressure acquisition for Netscanner system. To start the application it is necessary to enter correct IP address.

Air characteristics:

- temperature: $t = 26,2°C$, $T = 299,35$ K

- pressure: $p = 99580\ Pa$

- density: $\rho = 1,1589$ kg/m^3

- viscosity: $\mu = 1,8415e - 05$.

Measuring of surrounding flow speed around airfoil for motor frequency of 35 Hz:

- fluid level difference in U-tube: $\Delta h = 62$ mm

- pressure difference: $\Delta p = \rho_{H2O} \cdot g \cdot \Delta h = 1000 \cdot 9,81 \cdot 0,062 = 608,22$ Pa

- airspeed: $v = \sqrt{2 \cdot \Delta p / \rho} = 32,403$ m/s.

Using the sum of all pressure around measuring points, and lift formula we can get the normal component of the lift:

- $F_N = 71,39$ N

Difference between piezometric harp and scanner is +/-5% because of errors in reading height of harp and inability to accurately measure a harp slope.

The Figures 10. and 11. show a graphs that represents the pressure distribution around airfoil and it is very similar to graph obtained by manual calculating using piezometric harp. Y-axis shows pressure value in PSI and X-axis shows measurement point number. On the right side pressure is shown in decimal number.

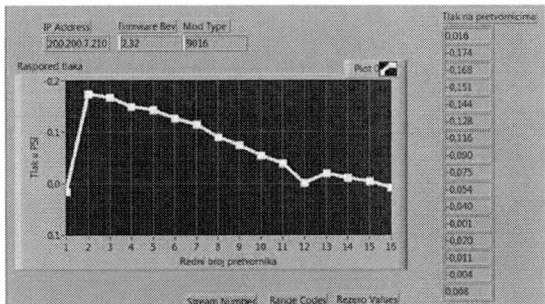

Figure 10. Pressure distribution – upper surface of the wing

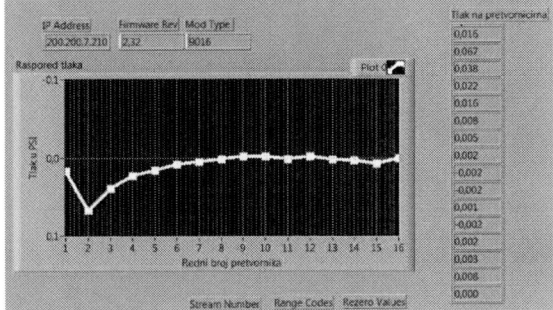

Figure 11. Pressure distribution – lower surface of the wing

VI. CONCLUSION

Use of the system for pressure data acquisition significantly shorten and simplify the process of pressure distribution determination. Measurement results obtained from piezometric harp and those from pressure acquisition module vary in acceptable extent. Intelligent pressure scanner is more accurate device and allows programming in various languages. There is no more need for manual calculations. Application for sensing the pressure created in the LabVIEW program could be upgraded so all pressure around the airfoil could be presented in one coordinate system. It is also possible to make the application to sense all the pressure around airfoil and, with given speed and conditions of the atmosphere, to calculate coefficient of pressure, lift and drag for given angle of attack. This system is independent of AT-1 wind tunnel and airfoil within it. In future experiments other airfoils could be used and in different wind tunnels for various testing.

LITERATURE

[1] D. Franjković, Razvoj sustava za eksperimentalno određivanje aerodinamičkih karakteristika s posebnim osvrtom na zrakoplove, Zagreb: Fakultet prometnih znanosti, 2003.

[2] P. Kesić, Osnove aerodinamike, Zagreb: Fakultet strojarstva i brdogradnje, 2005.

[3] J. A. Aschetz and A. E. Fuhs, Handbook of Fluid Dynamics and Fluid Machinery, John Wiley & Sons, Inc., 1996.

[4] Pressure Systems, NetScanner 9016 user manual, Hampton, 2000.

[5] Pressure Systems, Pressure calibrator user's manual, Hampton, 2001.

Author index

Abazi-Bexheti, L.	655	Berkovic, I.	1132, 1178
Adamovic, N.	146	Bernik, A.	711
Adzinets, D.	1126	Besedin, K.Y.	221
Aglic Cuvic, M.	794	Bhattacharya, J.	1084
Akagic, A.	1104, 1195	Bikov, T.D.	497
Akbal, A.	506	Bjazic, T.	1551
Akbal, E.	506, 1241	Blazic, G.	1491
Aksentijevic, S.	1454	Blech, A.	57
Alajbeg, T.	910, 944	Boban, M.	1486
Aleksi, I.	1011	Bodrusic, I.	472
Alexandrova, M.I.	125	Bogdanova, V.G.	353
Alic, D.	527	Bohak, C.	259
Alienin, O.	359	Bokan, D.	773
Aljancic, U.	336	Bosilj Vuksic, V.	1355, 1391
Alkan, A.	1094	Boukhebouze, M.	381
Allred, P.	43	Brajovic, M.	482
Amelio, A.	1110	Brcic, M.	197
Anderson, N.	624	Brenner, W.	146
Andjelkovic, M.	887	Brezany, P.	365
Androcec, D.	1285	Brezovec, I.	88, 93
Antolic, G.	1379	Brkic Bakaric, M.	1546
Antulov-Fantulin, B.	920	Brkic, K.	7
Apostolova, M.	655	Brkic, L.	1379, 1528
Arslan Tuncer, S.	1094	Brloznik, M.	303
Artemkina, S.B.	48	Brodic, D.	1110
Artetxe González, E.	359	Brodjanac, P.	740
Asenbrener Katic, M.	1221	Brscic, D.	564
Astrova, I.	215	Brtka, E.	1178
Avaroğlu, E.	171	Brtka, V.	1178
Avbelj, V.	289, 303	Brumen, B.	1275
Babic, M.	188	Brunetti, G.	967
Babic, S.	717	Brzeźniak, M.	233
Babic, Z.	192	Bubas, G.	711
Bach, L.M.	1522	Bucko, J.	750
Baggini, M.B.	1038	Budimac, Z.	570, 613
Bakalar, G.	1038	Bundalo, D.	103
Bakalar, S.G.	1038	Bundalo, Z.	103
Balaz, Z.	1211	Bunja, D.	1437
Baldini, G.	1269, 1292	Burdukovskii, V.F.	27
Balic, K.	244	Buric, M.	1098
Balota, A.	593	Busch, J.	624
Balug, M.A.	1562	Busch, K.	57
Banovic-Curguz, N.	445, 492	Buza, E.	1104, 1195
Barakovic Husic, J.	428, 434	Bychkov, I.V.	1116
Barakovic, S.	428, 434	Bziuk, W.	521
Baric, A.	13, 88, 93	Candrlic, S.	836, 1221
Batistic, L.	420	Cano Pons, E.	1269
Begicevic Redjep, N.	705	Capeska Bogatinoska, D.	411, 515
Berdinsky, A.S. 27,	48	Car, S.	330

Car, Z.	209	Drira, K.	555
Carapina, M.	853	Drmic, A.	995
Caria, M.	152	Dudarin, A.	119
Carpio, F.	521	Duic, I.	1309
Carrato, S.	1084	Duic, M.	818, 824
Cavrak, I.	979	Dundjer, I.	1207, 1217
Cegar, S.	1431	Dzapo, P.	818
Chiasera, A.	21	Dzeng, R.J.	1027
Chiussi, S.	37	Dzuzdanovic, S.	1044
Chydzinski, A.	460	Eickemeyer, C.	215
Cifrek, M.	324	Emanovic, E.	141
Cizmesija, A.	734	Esztelecki, P.	778
Cobic, A.	853	Etinger, D.	717
Colston, G.	43	Fedorov, V.E.	27, 48
Coric, R.	1182	Feilhauer, T.	365
Crnko, N.	371	Feoktistov, A.	1138
Culjak, I.	324	Ferrari, M.	21
Cupec, R.	1120	Fertalj, K.	1501
Cupic, M.	7	Filic, M.	424
Curkovic, J.	944	Filipovic Tretinjak, M.	746
Cuzzocrea, A.	1337	Filipovic-Grcic, B.	598
Cvejic, R.	887	Filjar, R.	424
Cvetanovic, M.	895	Filko, D.	1120
Cvetko, M.	393	Fischer, I.A.	37, 57
Cvetkovic, D.	865, 891	Fonseca-Pinto, R.	276, 279
Cvetkovic, L.	1501	Fosic, I.	1298
Cvijic, B.	103	Franjkovic, D.	1566
Cvrtila, V.	1309	Frankovic, I.	728
Dadic, M.	166	Franulovic, M.	608
Dajak, L.	1546	Frey, H.	1084
Dakovic, M.	478, 482	Frieiro, J.L.	37
Damasevicius, R.	1373	Friganovic, K.	330
Davydov, A.	1161	Frincu, M.	399
Debevc, M.	393	Fruk, M.	1557
Dedic, V.	887	Fulford, C.P.	649
Delac, G.	995	Funk, H.S.	31
Depolli, M.	270, 292	Galzerano, G.	21
Desic, S.	424	Gascic, D.	30
Devcic, K.	905	Gavran, M.	1557
Dinjar, D.	1397	Gecevic, D.	1551
Divjak, B.	705	Geneiatakis, D.	1269, 1292
Djedovic, A.	527	Giedrimas, V.	588
Djumic, M.	1182	Goeminne, M.	381
Djurovic, P.	1120	Golodov, V.	225
Dogan, S.	1241	Golub, B.	1302
Dogru, N.	1314	Gordienko, Y.	359
Domazet, E.	318	Gorgan, D.	253
Dorosz, D.	21	Gorjanac, V.	1280
Dovedan Han, Z.	1221	Gorsky, S.	353, 1138
Doychinov, Y.I.	407	Gracak, Z.	1528
Draganic, A.	1227	Granic, A.	1540
Drazic, A.	979	Grba, B.	876

Grbic, R.	1120	Jahandar, P.	43
Greer, D.	624	Jaklic, J.	1355
Gregorio, A.	1073	Jakobovic, D.	1182
Gros, S.	1262	Jakopec, T.	1516
Grubjesic, I.	915	Jakovic, B.	1367, 1476
Grubljesic, T.	1355	Jaksic, D.	1401, 1491, 1496
Gruicic, S.	1233	Jakstas, A.	1373
Grzunov, L.	824	Jakupovic, A.	1221
Gumzej, N.	1424	Jamic, M.	812
Gusev, M.	308, 318, 387	Jan, M.	297
Gütl, C.	619	Jazbec, D.	393
Hajdarevic, K.	1314	Jerman-Blazic, B.	188
Halili, A.	1189	Jervan, G.	359
Hanna, P.	624	Jovanovic, V.	1401
Hashad, Y.	57	Jovic, A.	330
Hasic, H.	1195	Jozic, K.	330
Hausknecht, K.	1233	Jugovic, A.	1454
Havasi, F.	755, 778	Juhasz, Z.	340
Hebrang Grgic, I.	842	Jukan, A.	152, 521
Hedji, I.	455, 501	Jukic, O.	455
Henno, J.	635, 660, 694	Jumic, J.	1251
Henriques, V.	515	Jung, C.H.	1090
Herceg, M.	109	Juric, M.B.	264
Herynek, B.	393	Juricic, B.	920
Hivziefendic, J.	603	Jurisic, D.	141
Hlupic, N.	197	Kadoic, N.	705
Ho, C.W.	1027	Kadriu, A.	655
Hocenski, Z.	1011	Kakalejcík, L.	750
Höfler, M.	619	Kakanakov, N.	205, 1001
Hoic-Bozic, N.	728	Kalpic, D.	440
Holenko Dlab, M.	672, 836	Kamimura, N.	31
Holik, F.	450, 1256	Kanda, T.	564
Holjevac, N.	1465	Kaplar, A.	1144
Horalek, J.	450	Kavcic, A.	848
Horvat, I.	1298	Kaya, D.	314
Horvat, M.	1207	Kaya, T.	202, 314
Hsueh, H.H.	1027	Kazi, L.	1132
Hurtova, V.	450	Kazi, Z.	1132
Hyrynsalmi, S.	991, 1442	Kemper, N.	152
Iliev, T.B.	407, 416, 497	Kenzin, M.Y.	1116
Ilisevic, D.	445, 492	Kersten, J.	215
Indihar Stemberger, M.	1391	Kevric, J.	603
Ivanda, M.	21	Kholkhoev, B.C.	27
Ivanjko, T.	915, 1309	Kisasondi, T.	1285
Ivanova, E.P.	416, 497	Klasinc, J.	1407
Ivanova, O.N.	685	Klemo, V.	995
Ivanovic, M.	613, 901, 1144	Knezevic, B.	1476
Ivanovski, D.	582	Knezevic, K.	1324
Ivasic Kos, M.	1098	Knezevic, T.	72
Ivkovic, N.	1149	Kochemazov, S.	1166, 1172
Ivosic, I.	533	Kocijan, K.	806
Jaakkola, H.	635, 660, 694	Koerner, R.	57

Komen, V.	375	Llorente Coto, A.	359
Konecki, M.	723	López Benito, J.R.	359
Konjevod, B.	824	Luburic, N.	1144
Koren, A.	510	Ludescher, T.	365
Koricic, M.	77, 83	Lugovic, S.	1207, 1217
Kőrösi, G.	755	Lukac, A.	1534
Koschel, A.	215	Lukac, D.	689
Kosec, G.	1049	Lukac, Z.	773
Kostecki, K.	57	Lukovac, B.	1079
Kostenetskiy, P.S.	221, 229	Lukowiak, A.	21
Kostromin, R.	1138	Lushchyk, U.	359
Kounelis, I.	1292	Machado, M.	276, 279
Kovacevic, I.	1385	Madhale Jadav, G.	420
Kovacevic, Z.	1471	Magerl, M.	88, 93
Kovacic, B.	876, 905, 1418	Makari, T.	1275
Kovács, V.	1506	Mäkelä, J.	635, 660, 694
Kozlova, M.N.	48	Mäkinen, T.	1448
Kozuh, I.	393	Makotchenko, V.G.	27
Krajcek Nikolic, K.	1566	Makovec, M.	336
Kramaric, I.	1331	Maksimkin, N.	1032, 1116
Kranjac, M.	238	Malaric, R.	162, 166
Krasna, M.	666, 678	Malcic, G.	1055
Kresoja, M.	901	Malekian, R.	411, 515
Krivec, S.	66	Mandic, T.	13
Krpan, D.	800	Manojlovic, M.	1546
Krstic, V.	937	Maras, J.	794
Kruglov, A.	57	Marasovic, K.	1005
Kruzic, S.	1015	Marcelic, M.	162
Kuhar, U.	1049	Mareva, D.D.	130
Kukolja, D.	330	Maric, M.	1227
Kuman, S.	1262	Maris, M.	1073
Kunic, L.	247	Markic, Z.	1459
Kurdija, A.S.	995	Markovic, K.	608
Kurent, P.	858	Marolt, M.	259, 848
Kurtaj, L.	1201	Marsi, S.	1073, 1084
Kuzle, I.	1465	Marsic, D.	1055
Kuznetsov, V.A.	27, 48	Márton, M.	538
Lackovic, D.	466	Martucci, A.	21
Larionov, A.	1161	Masetic, Z.	1314
Lasic-Lazic, J.	915	Matic, T.	109
Lavric, P.	259	Mauher, M.	1471
Lazic, N.	882	McGowan, A.	624
Ledneva, A.Y.	48	Medic, B.	865, 891
Lee, H.-C.	1090	Medved, D.	1431
Lee, S.G.	1090	Megill, N.D.	182
Lee, S.H.	1090	Mekterovic, I.	1349, 1385, 1510
Leitgeb, E.	404	Merluzzi, A.	967
Lekic, V.	192	Mestrovic, A.	870
Lerga, R.	836	Meznaric, D.	1566
Limani, I.	1201	Mihajlovic, Z.	7, 247
Linna, P.	1442, 1448	Mihaljevic, B.	1005, 1522
Lisek, J.	932	Mihaylov, G.Y.	416, 497

Mijatovic, M.	865
Mijatovic, M.	865
Mikuc, M.	1262
Mikulic, J.	88, 93
Milasinovic, B.	1501
Milenkovic, M.	1319, 1412
Milic, L.	723
Miljkovic, D.	1061, 1067
Milos, A.	113
Miskovic, T.	1459
Mladenovic, S.	794, 800
Mlinac, F.	1437
Mohorcic, M.	270
Molnar, G.	113, 119
Moloisane, N.R.	411
Mouetsi, S.	53
Mrvica, A.	1459
Muminovic, S.	434
Music, J.	1015
Muzaffar bin Baharudin, A.	991
Myronov, M.	43
Nagul, N.	1155, 1161
Nai-Fovino, I.	1292
Nanver, L.K.	72
Neisse, R.	1292
Nemec, G.	678
Nenadic, K.	1280
Neradová, S.	644, 1256
Neskovic, A.	1465
Ng, J.	31
Nikolov, G.T.	98
Nikolov, N.N.	125
Novak, N.M.	767
Obarcanin, K.	1044
Ocevcic, H.	1298
Ocovaj, S.	773
Odak, M.	882
Offel, N.	215
Ognjenovic, V.	1178
Ogrizovic, D.	209
Okanovic, V.	244
Okresa Djuric, B.	1149
Omanovic, S.	527, 1104
Oparin, G.A.	353
Opiła, J.	283
Orehovacki, T.	717
Orescanin, D.	1397
Oreski, D.	723
Orovic, I.	478, 1227
Osmakcic, K.	806
Ostojic, R.	1044
Ostreika, A.	1373
Ovseník, Ľ.	538
Ozdemir, M.T.	202
Paek, S.	649
Pale, P.	17
Paľová, D.	767
Papic, A.	700
Pasalic, D.	103
Pashinin, A.A.	353
Pavelin, G.	1437
Pavic, I.	1465
Pavicic, M.	182
Pavkov, S.	728
Pavlic Sipek, Z.	1476
Pavlic, M.	1221
Pavlinic, A.	375
Pavlinovic, M.	920
Pejcinovic, B.	1
Pejic Bach, M.	1355, 1367
Pełech-Pilichowski, T.	283
Pereira, J.	276
Pesek, M.	848
Pesic, D.	895
Pesut, D.	941
Petrovic, J.	17
Petrovic, K.	166
Pezer, M.	882
Pilicic, S.	608
Pita Costa, J.	558
Pivac, M.	1540
Pobar, M.	1098
Podbojec, D.	393
Podkonjak, M.	830
Polancec, D.	1510
Poljak, M.	66
Poljak, R.	1496
Poscic, P.	1401, 1491, 1496
Póser, V.	1506
Prazina, I.	244
Prcic, V.L.	440
Predavec, D.	1211
Prkic, S.	961
Prohaska, Z.	949
Prohaska, Z.	949, 1481
Pticek, M.	1361
Puligheddu, M.	1073
Pürcher, P.	619
Pusnik, M.	576, 582
Putnik, Z.	613, 901
Radev, D.I.	497
Radivojevic, Z.	895
Radman Pesa, A.	1481
Radonic, M.	1349

Radosevic, D.	711	Shopov, M. 974,	1001
Radovan, A.	1522	Sikimic, U.	238
Radovan, M.	1302	Silic, M.	995
Radovan, M.	630	Siljak, H.	603
Radulovic, B.	1132	Silkina, N.S.	685
Rajh, A.	812	Sillberg, P.	985
Rakic, G.	570	Silva, M.	276
Rakic, N.	570	Simovic, V.	910, 944
Ramljak, D.	542	Simunic, D.	440, 510, 1079
Ramljak, M.	1245	Skala, K.	347, 359
Ramljak, T.	1298	Skendzic, A.	905, 1418
Ramponi, G.	1084	Skliarova, I.	176
Ramponi, R.	21	Sklyarov, V.	176
Rantanen, P.	985	Skracic, K.	472
Rashkovska, A.	289	Slamic, M.	1471
Rasic, M.	1412	Smrikarov, A.	613
Rechem, D.	53	Smrikarova, S.	613
Repnik, R.	678	Sneler, L.	109
Révészová, L.	761	Soini, J.	985
Rexhepi, A.	1189	Sojat, Z.	347, 359
Righini, G.C.	21	Sokele, M.	910
Ristic, D.	21	Solic, K.	1280, 1298
Ristov, P.	1459	Sorgo, A.	666
Ristovski, A.	308	Soric, I.	1397
Rizvic, S.	244	Speh, I.	455, 501
Rjabov, A.	176	Speranza, G.	21
Rodin, R.	1021	Spes, M.	538
Rojbi, A.	359	Spindler, M.	576
Rolseth, E.G.	57	Spoljaric, T.	1562
Romanenko, A.I.	27, 48	Sprager, S.	264
Rybicki, J.	233	Sretenovic, M.	1418
Saari, M.	991	St. Vieth, B.	233
Sabou, A.	253	Stajcer, M.	1397
Sadinov, S.M.	547	Stajduhar, I.	1021
Sajn, L.	359	Stanchev, O.P.	98, 125
Salom, J.	238	Stancic, H.	812
Samociuk, D.	460	Stancic, I.	1015
Sarabok, A.	501	Staneviciene, E.	1373
Saric, A.	1005	Stanic, J.	1055
Sarlija, N.	1367	Stankovic, I.	478, 482
Savic, M.	613	Stankovic, L.	482
Schatten, M.	1149	Stankovic, S.	1227
Schatzberger, G.	88, 93	Stefanyuk, A.Y.	27
Schlipf, J.	37	Steri, G.	1292
Schudrowitz, J.	152	Sterle, U.	954
Schulze, J.	31, 37, 43, 57	Stirenko, S.	359
Scotognella, F.	21	Stjepic, A.M.	1355
Semenov, A.	1166, 1172	Stoyanov, I.S.	407, 416, 497
Senthil Srinivasan, V.S.	57	Stoyanov, R.S.	130
Serbet, F.	202	Stoyanov, S.	205
Serra, C.	37	Strnad, B.	862, 958
Shatri, V.	1201	Subasic, M.	1534

Sudnitson, A.	176	Vidakovic, M.	1144
Sukur, N.	570	Vilcek, T.	1516
Suligoj, T.	66, 72, 77, 83	Vilhar, A.	292
Sumak, B.	576, 582	Vinko, D.	158
Susa Vugec, D.	1391	Vlahinic, S.	420
Susac, F.	1011	Vojkovic, G.	1319, 1412
Susak, T.	1486	Vojvodic, S.	1431
Svigelj, A.	1049	Vrana, R.	830, 926
Sylejmani, K.	1189	Vrankic, M.	420
Taccheo, S.	21	Vrbanec, T.	870
Tanovic, A.	527	Vrdoljak, B.	1361
Tapiska, S.	901	Vrtacnik, D.	336
Tchernykh, A.	1138	Vucic, M.	113, 119
Temerinac, M.	773	Vuckovic, N.	1079
Thalheim, B.	635, 660	Vujisic, G.	1557, 1562
Tijan, E.	1454	Vukovic, M.	1251, 1298
Tomas, B.	1285	Weidinger, V.	767
Tomic, D.	209	Weiser, M.	57
Tomic, M.	466	Weisshaupt, D.	43
Tomic, S.	238	Wendav, T.	57
Tomicic, I.	1149	Werth, W.	789
Tretinjak, M.	746	Wesiak, G.	619
Trobec, R.	264, 297	Xie, Y.-H.	31
Tucakovic, M.	932	Yakovleva, G.E.	48
Tuncer, T.	171	Yankov, P.V.	136
Turajlic, E.	486	Yrjönkoski, K.	1448
Turán, J.	538	Yudov, D.D.	130
Türk, M.	314	Zagar, I.	979
Tütüncü, K.	613	Zaharija, G.	800
Uglesic, I.	598	Zaikin, O.	1166, 1172
Ul'yanov, S.	1032	Zailskaitė-Jakstė, L.	1373
Ungermanns, C.	789	Zajgar, T.	961
Uroda, I.	949, 1481	Zanlungo, F.	564
Vaccari, A.	21	Zhabinski, A.	1126
Vaitkeviciene, A.	588	Zhabinskii, S.	1126
Vaitkevicius, L.	588	Zidar, M.	1465
Valchev, V.C.	98, 125, 130, 136	Zilak, J.	77, 83
Valligatla, S.	21	Zisko, A.	666
Van den Bossche, A.	136	Zitta, S.	644
Varas, S.	21	Zoroja, J.	1367
Vasilchenko, I.	21	Zouach, F.	53
Veispahic, A.	428	Zufic, J.	961
Vejacka, M.	783	Zupan, A.	598
Velki, T.	1280	Zur, L.	21
Vidacek-Hains, V.	734	Zymbler, M.	1343

IEEE
445 Hoes Lane
Piscataway, NJ 08854-4141

ISBN 978-1-5090-4969-1

2010 IEEE International Reliability Physics Symposium (IRPS 2010)

Garden Grove (Anaheim), California, USA
2-6 May 2010

IEEE Catalog Number: CFP10RPS-POD
ISBN: 978-1-42445-430-3